THE CIRCUITS and FILTERS HANDBOOK

Editor-in-Chief

WAI-KAI CHEN
University of Illinois
Chicago, Illinois

CRC PRESS   IEEE PRESS

A CRC Handbook Published in Cooperation with IEEE Press

Library of Congress Cataloging-in-Publication Data

The circuits and filters handbook / editor-in-chief, Wai-Kai Chen.
p. cm. -- (The electrical engineering handbook series)
Includes bibliographical references and index.
ISBN 0-8493-8341-2 (acid-free paper)
1. Electronic circuits. 2. Electric filters. I. Chen, Wai-Kai.
1936– II. Series.
TK7867.C4977 1995
621.3815 -- dc20
95-7037
CIP

Direct all inquiries to CRC Press, Inc., 2000 Corporate Blvd., N.W., Boca Raton, Florida 33431.

International Standard Book Number 0-8493-8341-2
Library of Congress Card Number 95-7037
Printed in the United States of America 1 2 3 4 5 6 7 8 9 0
Printed on acid-free paper

Preface

Purpose

The purpose of *The Circuits and Filters Handbook* is to provide in a single volume a comprehensive reference work covering the broad spectrum of electrical circuits and filters. It is written and developed for practicing electrical engineers in industry, government, and academia. The goal is to provide the most up-to-date information in classical fields of circuit theory, circuit components, feedback circuits, nonlinear circuits, distributed circuits, active and passive filters, general circuit analysis techniques, and stability analysis, while covering the emerging fields of digital filters, analog integrated circuits, digital and analog VLSI, and computer-aided design and optimization techniques. In addition, the necessary background in mathematics is reviewed in the first section. The handbook is not an all-encompassing digest of everything taught within an electrical engineering curriculum on circuits and filters. Rather it is the engineer's first choice in looking for a solution. Therefore, full references to other sources of contributions are provided. The ideal reader is a B.S. level engineer with a need for a one-source reference to keep abreast of new techniques and procedures as well as review standard practices.

Background

The handbook stresses fundamental theory behind professional applications. In order to do so, it is reinforced with frequent examples. Extensive development of theory and details of proofs have been omitted. The reader is assumed to have a certain degree of sophistication and experience. However, brief reviews of theories, principles, and mathematics of some subject areas are given. These reviews have been done concisely with perception. The handbook is not a textbook replacement, but rather a reinforcement and reminder of material learned as a student. Therefore, important advancements and traditional as well as innovative practices are included.

Since the majority of professional electrical engineers graduated before powerful personal computers were widely available, many computational and design methods may be new to them. Therefore, computers and software use is thoroughly covered. Not only does the handbook use traditional references to cite sources for the contributions, it also contains all *relevant* sources of information and tools that would assist the engineer in performing his/her job. This may include sources of software, databases, standards, seminars, conferences, etc.

Organization

Over the years, the fundamentals of electrical circuits and filters have evolved to include a wide range of topics and a broad range of practice. To encompass such a wide range of knowledge, the handbook focuses on the key concepts, models, and equations that enable the electrical engineer to analyze, design, and predict the behavior of large-scale circuits, devices, filters, and systems. While design formulas and tables are listed, emphasis is placed on the key concepts and theories underlying the applications.

The information is organized into sixteen major sections which encompass the field of electrical circuits and filters. Each section is divided into chapters. In all there are 88 chapters, each of which was written by leading experts in the field to enlighten and refresh knowledge of the mature engineer, and to educate the novice. The first section summarizes the applicable mathematics and symbols underlying other applications. Each section contains introductory material, leading to the appropriate applications. To help the reader, each article includes three important and useful categories: defining terms, references, and further information. *Defining terms* are key definitions and the first occurrence of each term defined is indicated in boldface in the text. The definitions of these terms are mentioned in text. The *references* provide a list of useful books and articles for the reader. Finally, *further information* provides some general and useful sources of additional information on the topic.

Locating Your Topic

Numerous avenues of access to information contained in the handbook are provided. A complete table of contents is presented at the front of the book. In addition, an individual table of contents precedes each of the sixteen sections. Finally, each chapter begins with its own table of contents. The reader is urged to look over these tables of contents to become familiar with the structure, organization, and content of the book. For example, see Section III: General Circuit Theory, then Chapter 16: Theory of Two-Dimensional Hurwitz Polynomials, and then Chapter 16.3: Various Analog Hurwitz Polynomials. This tree-like structure enables the reader to move up the tree to locate information on the topic of interest.

Four indexes have been compiled to provide multiple means of accessing information: a subject index, an index of contributing authors, an index of key tables of data or information, and an index of key illustrations. The subject and author indexes are in alphabetical order; the table and illustration indexes are in order of appearance. The subject index can also be used to locate definitions; the page on which the definition appears for each key defining term is given in this index.

The Circuits and Filters Handbook is designed to provide answers to most inquiries and direct the inquirer to further sources and references. We trust that it will meet your need.

Acknowledgments

The compilation of this book would not have been possible without the dedication and efforts of the editorial board of advisors, the section editors, the publishers, and most of all the contributing authors. I particularly wish to acknowledge Joel Claypool, CRC Press, Publisher, Richard C. Dorf, Handbook Series Editor, Barbara Wehner, who served as my editorial assistant, and my wife, Shiao-Ling, for her patience and understanding.

Wai-Kai Chen
Editor-in-Chief

Editor-in-Chief

Wai-Kai Chen, Professor and Head of the Department of Electrical Engineering and Computer Science at the University of Illinois at Chicago, teaches graduate and undergraduate courses in electrical engineering in the fields of circuits and systems. He received his B.S. and M.S. in electrical engineering at Ohio University where he was later recognized as distinguished Professor. He earned his Ph.D. in electrical engineering at the University of Illinois at Urbana/Champaign.

Professor Chen has extensive experience in education and industry and is very active professionally in the fields of circuits and systems. He has served as a visiting professor at Purdue University and the University of Hawaii at Manoa. He was Editor of the *IEEE Transactions on Circuits and Systems, Series I and II*, President of the IEEE Circuits and Systems Society and is Editor-in-Chief of the *Journal of Circuits, Systems and Computers*. He received the *Lester R. Ford Award* from the Mathematical Association of America, the *Alexander von Humboldt Award* from Germany, the *Ohio University Alumni Medal of Merit for Distinguished Achievement in Engineering Education, the Senior University Scholar Award* from the University of Illinois at Chicago, and the *Distinguished Alumnus Award* from the University of Illinois at Urbana/Champaign. He has also received more than a dozen honorary professor awards from major institutions in China.

A Fellow of the Institute of Electrical and Electronics Engineers and the American Association for the Advancement of Science, Professor Chen is widely known in the profession for his *Applied Graph Theory* (North-Holland), *Theory and Design of Broadband Matching Networks* (Pergamon Press), *Active Network and Feedback Amplifier Theory* (McGraw-Hill), *Linear Networks and Systems* (Brooks/Cole), *Passive and Active Filters: Theory and Implementations* (John Wiley), and *Theory of Nets* (John Wiley).

Advisory Board

Leon O. Chua
University of California
Berkeley, California

John Choma, Jr.
University of Southern California
Los Angeles, California

Lawrence P. Huelsman
University of Arizona
Tucson, Arizona

Contributors

R. Akbari-Dilmaghani
University College of London
London, United Kingdom

Philip E. Allen
Georgia Institute of Technology
Atlanta, Georgia

R. Ansari
Bell Communications Research
Morristown, New Jersey

Andreas Antoniou
University of Victoria
Victoria, Canada

Peter Aronhime
University of Louisville
Louisville, Kentucky

Peter Bauer
University of Notre Dame
Notre Dame, Indiana

Peter Bendix
Technology Modeling Associates, Inc.
Palo Alto, California

Benjamin J. Blalock
Georgia Institute of Technology
Atlanta, Georgia

Bruce W. Bomar
University of Tennessee Space Institute
Tullahoma, Tennessee

Martin A. Brooke
Georgia Institute of Technology
Atlanta, Georgia

Gordon E. Carlson
University of Missouri, Rolla
Rolla, Missouri

A. E. Cetin
Koc University
Istanbul, Turkey

Josephine C. Chang
University of Southern California
Los Angeles, California

Robert C. Chang
University of Southern California
Los Angeles, California

K. S. Chao
Texas Tech University
Lubbock, Texas

Guanrong Chen
University of Houston
Houston, Texas

Ray R. Chen
San Jose State University
San Jose, California

Wai-Kai Chen
University of Illinois
Chicago, Illinois

John Choma, Jr.
University of Southern California
Los Angeles, California

Leon O. Chua
University of California
Berkeley, California

A. G. Constantinides
Imperial College of Science Technology, and Medicine
London, United Kingdom

Artice M. Davis
San Jose State University
San Jose, California

Daniël De Zutter
Universiteit Gent
Gent, Belgium

James F. Delansky
Pennsylvania State University
University Park, Pennsylvania

M. Delgado-Restituto
Universidad de Sevilla
Sevilla, Spain

John R. Deller, Jr.
Michigan State University
East Lansing, Michigan

Allen M. Dewey
IBM Microelectronics
Poughkeepsie, New York

A. Dharchoudhury
University of Illinois
Urbana, Illinois

Igor Djokovic
California Institute of Technology
Pasadena, California

M. H. Er
Nanyang Technological University
Singapore

Joseph B. Evans
University of Kansas
Lawrence, Kansas

Igor M. Filanovsky
University of Alberta
Edmonton, Canada

Norbert J. Fliege
Technische Universitat Hamburg
Hamburg, Germany

Sergio B. Franco
San Francisco State University
San Francisco, California

F. Gail Gray
Virginia Polytechnic Institute and State University
Blacksburg, Virginia

Edwin W. Greeneich
Arizona State University
Tempe, Arizona

Wayne D. Grover
TRLabs, University of Alberta
Edmonton, Canada

D. G. Haigh
University College of London
London, United Kingdom

Ramesh Harjani
University of Minnesota
Minneapolis, Minnesota

Martin Hasler
Swiss Federal Institute of Technology
Lausanne, Switzerland

Marwan M. Hassoun
Iowa State University
Ames, Iowa

Y. F. Huang
University of Notre Dame
Notre Dame, Indiana

S.-C. Huang
Tatung Institute of Technology
Taipei, Taiwan

Lawrence P. Huelsman
University of Arizona
Tucson, Arizona

J. L. Huertas
Universidad de Sevilla
Sevilla, Spain

C.-C. Hung
Ohio State University
Columbus, Ohio

T. K. Ishii
Marquette University
Milwaukee, Wisconsin

M. Ismail
Ohio State University
Columbus, Ohio

W. Kenneth Jenkins
University of Illinois
Urbana, Illinois

S. M. Kang
University of Illinois
Urbana, Illinois

Michael Peter Kennedy
University College
Dublin, Ireland

Jelena Kovačević
AT&T Bell Laboratories
Murray Hill, New Jersey

Stuart S. Lawson
University of Warwick
Coventry, United Kingdom

John Lidgey
Imperial College
London, United Kingdom

Y. C. Lim
National University of Singapore
Singapore

Pen-Min Lin
Purdue University
West Lafayette, Indiana

Erik Lindberg
Technical University of Denmark
Lyngby, Denmark

Ruey-wen Liu
University of Notre Dame
Notre Dame, Indiana

Stephen I. Long
University of California
Santa Barbara, CA

Flavio Lorenzelli
University of California
Los Angeles, California

Wu-Sheng Lu
University of Victoria
Victoria, Canada

Luc Martens
Universiteit Gent
Gent, Belgium

Wolfgang Mathis
University of Wuppertal
Wuppertal, Germany

Wasfy B. Mikhael
University of Central Florida
Orlando, Florida

Stephen W. Milam
Georgia Institute of Technology
Atlanta, Georgia

David G. Nairn
Queen's University
Kingston, Canada

Robert W. Newcomb
University of Maryland
College Park, Maryland

Nhat M. Nguyen
Hewlett Packard Company
Newark, California

Truong Q. Nguyen
University of Wisconsin
Madison, Wisconsin

K. V. Noren
University of Idaho
Moscow, Idaho

Josef A. Nossek
Technical University of Munich
Munich, Germany

Stanisław Nowak
University of Mining and Metallurgy
Krakow, Poland

Alison Payne
Imperial College of Science, Technology, and Medicine
London, United Kingdom

Tomasz W. Postupolski
POLFER Magnetic Materials Company
Warsaw, Poland

Roland Priemer
University of Illinois
Chicago, Illinois

Li Qiu
Hong Kong University of Science and Technology
Kowloon, Hong Kong

Jaime Ramirez-Angulo
New Mexico State University
Las Cruces, New Mexico

Hari C. Reddy
California State University
Long Beach, California
University of California
Irvine, California

B. Redman-White
University of Southampton
Southampton, United Kingdom

A. Rodríguez-Vázquez
Universidad de Sevilla
Sevilla, Spain

Charles E. Rohrs
Tellabs Research Center
University of Notre Dame
Mishawaka, Indiana

J. Gregory Rollins
Technology Modeling Associates, Inc.
Palo Alto, California

Tamás Roska
Hungarian Academy of Sciences
Budapest, Hungary

T. Saether
The Norwegian Institute of Technology
Flatasen, Norway

Michael K. Sain
University of Notre Dame
Notre Dame, Indiana

Edgar Sánchez-Sinencio
Texas A&M University
College Station, Texas

Sachin S. Sapatnekar
Iowa State University
Ames, Iowa

Tapio Saramäki
Tampere University of Technology
Tampere, Finland

Rolf Schaumann
Portland State University
Portland, Oregon

Cheryl B. Schrader
University of Texas
San Antonio, Texas

Naveed Sherwani
Western Michigan University
Kalamazoo, Michigan

Bing J. Sheu
University of Southern California
Los Angeles, California

José Silva-Martínez
National Institute for Research in Astrophysics, Optics and Electronics
Puebla, Mexico

Marwan A. Simaan
University of Pittsburgh
Pittsburgh, Pennsylvania

L. Montgomery Smith
University of Tennessee Space Institute
Tullahoma, Tennessee

Bang-Sup Song
University of Illinois
Urbana, Illinois

F. William Stephenson
Virginia Polytechnic Institute and State University
Blacksburg, Virginia

Maciej A. Styblinski
Texas A&M University
College Station, Texas

James A. Svoboda
Clarkson University
Potsdam, New York

Vladimir Székely
Technical University of Budapest
Budapest, Hungary

Sawasd Tantaratana
University of Massachusetts
Amherst, Massachusetts

Krishnaiyan Thulasiraman
University of Oklahoma
Norman, Oklahoma

Chris Toumazou
Imperial College of Science, Technology, and Medicine
London, United Kingdom

J. Trujillo
University of Southern California
Los Angeles, California

John P. Uyemura
Georgia Institute of Technology
Atlanta, Georgia

P. P. Vaidyanathan
California Institute of Technology
Pasadena, California

L. Vandenberghe
Katholieke Universiteit Leuven
Leuven, Belgium

J. Vandewalle
Katholieke Universiteit Leuven
Leuven, Belgium

F. Vidal
Universidad de Malaga
Malaga, Spain

Jiri Vlach
University of Waterloo
Waterloo, Canada

B. M. Wilamowski
University of Wyoming
Laramie, Wyoming

X. J. Xu
Imperial College of Science, Technology, and Medicine
London, United Kingdom

Andrew T. Yang
University of Washington
Seattle, Washington

Kung Yao
University of California
Los Angeles, California

Contents

ECTION I Mathematics

ECTION II Circuit Components and Their Characterization

SECTION III General Circuit Theory

SECTION IV Linear Circuit Analysis

SECTION V Feedback Circuits

SECTION VI Nonlinear Circuits

SECTION VII Nonlinear Circuits II

SECTION VIII Distributed Circuits

SECTION IX Stability Analysis

SECTION X Computer-Aided Design and Optimization

SECTION XI Analog Integrated Circuits

SECTION XII Digital and Analog VLSI

SECTION XIII Filter Characteristics

SECTION XIV Passive Filters

SECTION XV Active Filters

SECTION XVI Digital Filters

Indexes

THE CIRCUITS and FILTERS HANDBOOK

I

Mathematics

. F. Huang
niversity of Notre Dame

1

Linear Operators and Matrices

ıeryl B. Schrader
ɜ University of Texas

ichael K. Sain
iversity of Notre Dame

1.1 Introduction

It is only after the engineer masters linear concepts—linear models, and circuit and filter theory—that the possibility of tackling nonlinear ideas becomes achievable. Students frequently encounter linear methodologies, and bits and pieces of mathematics that aid in problem solution are stored away. Unfortunately, in memorizing the process of finding the inverse of a matrix or of solving a system of equations, the essence of the problem or associated knowledge may be lost. For example, most engineers are fairly comfortable with the concept of a vector space, but have difficulty in generalizing these ideas to the module level. Therefore, it is the intention of this section to provide a unified view of key concepts in the theory of linear circuits and filters, to emphasize interrelated concepts, to provide a mathematical reference to the handbook itself, and to illustrate methodologies through the use of many and varied examples.

This chapter begins with a basic examination of vector spaces over fields. In relating vector spaces the key ideas of linear operators and matrix representations come to the fore. Standard matrix operations are examined as are the pivotal notions of determinant, inverse, and rank. Next, transformations are shown to determine similar representations, and matrix characteristics such as singular values and eigenvalues are defined. Finally, solutions to algebraic equations are presented in the context of matrices and are related to this introductory chapter on mathematics as a whole.

Standard algebraic notation is introduced first. To denote an element s in a set S, use $s \in S$. Consider two sets S and T. The set of all ordered pairs (s, t) where $s \in S$ and $t \in T$ is defined as the Cartesian product set $S \times T$. A function f from S into T, denoted by $f : S \to T$, is a subset U of ordered pairs $(s, t) \in S \times T$ such that for every $s \in S$ one and only one $t \in T$ exists such that $(s, t) \in U$. The function evaluated at the element s gives t as a solution $(f(s) = t)$,

93-8341-2/95/$0.00+$.50

and each $s \in S$ as a first element in U appears exactly once. A binary operation is a function acting on a Cartesian product set $S \times T$. When $T = S$, one speaks of a binary operation on S.

1.2 Vector Spaces Over Fields

A **field** F is a nonempty set F and two binary operations, sum (+) and product, such that the following properties are satisfied for all $a, b, c \in F$:

1. Associativity: $(a + b) + c = a + (b + c)$; $(ab)c = a(bc)$
2. Commutativity: $a + b = b + a$; $ab = ba$
3. Distributivity: $a(b + c) = (ab) + (ac)$
4. Identities: (Additive) $0 \in F$ exists such that $a + 0 = a$
   (Multiplicative) $1 \in F$ exists such that $a1 = a$
5. Inverses: (Additive) For every $a \in F$, $b \in F$ exists such that $a + b = 0$
   (Multiplicative) For every nonzero $a \in F$, $b \in F$ exists such that $ab = 1$

Examples

- Field of real numbers R
- Field of complex numbers C
- Field of rational functions with real coefficients $R(s)$
- Field of binary numbers

The set of integers Z with the standard notions of addition and multiplication does not form a field because a multiplicative inverse in Z exists only for ± 1. The integers form a **commutative ring.** Likewise, polynomials in the indeterminate s with coefficients from F form a commutative ring $F[s]$. If field property 2 also is not available, then one speaks simply of a **ring.** An **additive group** is a nonempty set G and one binary operation + satisfying field properties 1, 4, and 5 for addition; i.e., associativity and the existence of additive identity and inverse. Moreover, if the binary operation + is commutative (field property 2), then the additive group is said to be **abelian.** Common notation regarding inverses is that the additive inverse for $a \in F$ is $b = -a \in F$. In the multiplicative case $b = a^{-1} \in F$.

An F-**vector space** V is a nonempty set V and a field F together with binary operations $+ : V \times V \to V$ and $* : F \times V \to V$ subject to the following axioms for all elements $v, w \in V$ and $a, b \in F$:

1. V and + form an additive abelian group
2. $a * (v + w) = (a * v) + (a * w)$
3. $(a + b) * v = (a * v) + (b * v)$
4. $(ab) * v = a * (b * v)$
5. $1 * v = v$

Examples

- The set of all n-tuples $(v_1, v_2, \ldots, v_n)$ for $n > 0$ and $v_i \in F$
- The set of polynomials of degree less than n with real coefficients $(F = R)$

Elements of V are referred to as **vectors**, whereas elements of F are **scalars.** Note that the terminology **vector space** V **over the field** F is used often. A **module** differs from a vector space in only one aspect; the underlying field in a vector space is replaced by a ring. Thus, a module is a direct generalization of a vector space.

When considering vector spaces of n-tuples, + is vector addition defined by element using the scalar addition associated with F. Multiplication $(*)$, which is termed scalar multiplication, also is defined by element using multiplication in F. The additive identity in this case is the zero vector (n-tuple of zeros) or null vector, and F^n denotes the set of n-tuples with elements in F, a vector space over F. A nonempty subset $\tilde{V} \subset V$ is called a **subspace** of V if for each $v, w \in \tilde{V}$ and every $a \in F$, $v + w \in \tilde{V}$ and $a * v \in \tilde{V}$. When the context makes things clear, it is customary to suppress the $*$, and write av in place of $a * v$.

A set of vectors $\{v_1, v_2, \ldots, v_m\}$ belonging to an F-vector space V is said to **span** the vector space if any element $v \in V$ can be represented by a linear combination of the vectors v_i. That is, scalars $a_1, a_2, \ldots, a_m \in F$ are such that

$$v = a_1 v_1 + a_2 v_2 + \cdots + a_m v_m \tag{1.1}$$

A set of vectors $\{v_1, v_2, \ldots, v_p\}$ belonging to an F-vector space V is said to be **linearly dependent** over F if scalars $a_1, a_2, \ldots, a_p \in F$, not all zero, exist such that

$$a_1 v_1 + a_2 v_2 + \cdots + a_p v_p = 0 \tag{1.2}$$

If the only solution for (1.2) is that all $a_i = 0 \in F$, then the set of vectors is said to be **linearly independent**.

Examples

- $(1, 0)$ and $(0, 1)$ are linearly independent.
- $(1, 0, 0), (0, 1, 0)$, and $(1, 1, 0)$ are linearly dependent over R. To see this, simply choose $a_1 = a_2 = 1$ and $a_3 = -1$.
- $s^2 + 2s$ and $2s + 4$ are linearly independent over R, but are linearly dependent over $R(s)$ by choosing $a_1 = -2$ and $a_2 = s$.

A set of vectors $\{v_1, v_2, \ldots, v_n\}$ belonging to an F-vector space V is said to form a **basis** for V if it both spans V and is linearly independent over F. The number of vectors in a basis is called the **dimension** of the vector space, and is denoted $\dim(V)$. If this number is not finite, then the vector space is said to be **infinite dimensional**.

Examples

- In an n-dimensional vector space, any n linearly independent vectors form a basis.
- The **natural (standard) basis**

$$e_1 = \begin{bmatrix} 1 \\ 0 \\ 0 \\ \vdots \\ 0 \\ 0 \end{bmatrix}, \quad e_2 = \begin{bmatrix} 0 \\ 1 \\ 0 \\ \vdots \\ 0 \\ 0 \end{bmatrix}, \quad e_3 = \begin{bmatrix} 0 \\ 0 \\ 1 \\ \vdots \\ 0 \\ 0 \end{bmatrix}, \cdots, e_{n-1} = \begin{bmatrix} 0 \\ 0 \\ 0 \\ \vdots \\ 1 \\ 0 \end{bmatrix}, \quad e_n = \begin{bmatrix} 0 \\ 0 \\ 0 \\ \vdots \\ 0 \\ 1 \end{bmatrix}$$

  both spans F^n and is linearly independent over F.

Consider any basis $\{v_1, v_2, \ldots, v_n\}$ in an n-dimensional vector space. Every $v \in V$ can be represented uniquely by scalars $a_1, a_2, \ldots, a_n \in F$ as

$$v = a_1 v_1 + a_2 v_2 + \cdots + a_n v_n \tag{1.3}$$

$$= \begin{bmatrix} v_1 & v_2 & \cdots & v_n \end{bmatrix} \begin{bmatrix} a_1 \\ a_2 \\ \vdots \\ a_n \end{bmatrix} \tag{1.4}$$

$$= \begin{bmatrix} v_1 & v_2 & \cdots & v_n \end{bmatrix} a \tag{1.5}$$

Here, $a \in F^n$ is a coordinate representation of $v \in V$ with respect to the chosen basis. The reader will be able to discern that each choice of basis will result in another representation of the vector under consideration. Of course, in the applications some representations are more popular and useful than others.

1.3 Linear Operators and Matrix Representations

First, recall the definition of a function $f : S \to T$. Alternate terminology for a function is mapping, operator, or transformation. The set S is called the **domain** of f, denoted $D(f)$. The **range** of $f, R(f)$, is the set of all $t \in T$ such that $(s, t) \in U$ $(f(s) = t)$ for some $s \in D(f)$.

Examples

Use $S = \{1, 2, 3, 4\}$ and $T = \{5, 6, 7, 8\}$.

- $\tilde{U} = \{(1, 5), (2, 5), (3, 7), (4, 8)\}$ is a function. The domain is $\{1, 2, 3, 4\}$ and the range is $\{5, 7, 8\}$.
- $\hat{U} = \{(1, 5), (1, 6), (2, 5), (3, 7), (4, 8)\}$ is not a function.
- $\overline{U} = \{(1, 5), (2, 6), (3, 7), (4, 8)\}$ is a function. The domain is $\{1, 2, 3, 4\}$ and the range is $\{5, 6, 7, 8\}$.

If $R(f) = T$, then f is said to be **surjective** (**onto**). Loosely speaking, all elements in T are used up. If $f : S \to T$ has the property that $f(s_1) = f(s_2)$ implies $s_1 = s_2$, then f is said to be **injective** (**one-to-one**). This means that any element in $R(f)$ comes from a unique element in $D(f)$ under the action of f. If a function is both injective and surjective, then it is said to be **bijective** (**one-to-one and onto**).

Examples

- $\tilde{U}$ is not onto because $6 \in T$ is not in $R(f)$. Also $\tilde{U}$ is not one-to-one because $f(1) = 5 = f(2)$, but $1 \neq 2$.
- $\overline{U}$ is bijective.

Now consider an operator $L : V \to W$, where V and W are vector spaces over the same field F. L is said to be a **linear operator** if the following two properties are satisfied for all $v, w \in V$ and for all $a \in F$:

$$L(av) = aL(v) \tag{1.6}$$

$$L(v + w) = L(v) + L(w) \tag{1.7}$$

Equation (1.6) is the property of homogeneity and (1.7) is the property of additivity. Together they imply the principle of **superposition**, which may be written as

$$L(a_1 v_1 + a_2 v_2) = a_1 L(v_1) + a_2 L(v_2) \tag{1.8}$$

for all $v_1, v_2 \in V$ and $a_1, a_2 \in F$. If (1.8) is not satisfied, then L is called a **nonlinear operator**.

Examples

- Consider $V = C$ and $F = C$. Let $L : V \to V$ be the operator that takes the complex conjugate: $L(v) = \overline{v}$ for $v, \overline{v} \in V$. Certainly

$$L(v_1 + v_2) = \overline{v_1 + v_2} = \overline{v_1} + \overline{v_2} = L(v_1) + L(v_2)$$

However,

$$L(a_1 v_1) = \overline{a_1 v_1} = \overline{a_1}\,\overline{v_1} = \overline{a_1} L(v_1) \neq a_1 L(v_1)$$

Then L is a nonlinear operator because homogeneity fails.

- For F-vector spaces V and W, let V be F^n and W be F^{n-1}. Examine $L : V \to W$, the operator that truncates the last element of the n-tuple in V; that is,

$$L((v_1, v_2, \ldots, v_{n-1}, v_n)) = (v_1, v_2, \ldots, v_{n-1})$$

Such an operator is linear.

The **null space (kernel)** of a linear operator $L : V \to W$ is the set

$$\ker L = \{v \in V \text{ such that } L(v) = 0\} \tag{1.9}$$

Equation (1.9) defines a vector space. In fact, $\ker L$ is a subspace of V. The mapping L is injective if and only if $\ker L = 0$; that is, the only solution in the right member of (1.9) is the trivial solution. In this case L is also called **monic.**

The **image** of a linear operator $L : V \to W$ is the set

$$\operatorname{im} L = \{w \in W \text{ such that } L(v) = w \text{ for some } v \in V\} \tag{1.10}$$

Clearly, $\operatorname{im} L$ is a subspace of W, and L is surjective if and only if $\operatorname{im} L$ is all of W. In this case L is also called **epic.**

A method of relating specific properties of linear mappings is the **exact sequence.** Consider a sequence of linear mappings

$$\cdots V \xrightarrow{L} W \xrightarrow{\tilde{L}} U \to \cdots \tag{1.11}$$

This sequence is said to be exact at W if $\operatorname{im} L = \ker \tilde{L}$. A sequence is called exact if it is exact at each vector space in the sequence. Examine the following special cases:

$$0 \to V \xrightarrow{L} W \tag{1.12}$$

$$W \xrightarrow{\tilde{L}} U \to 0 \tag{1.13}$$

Sequence (1.12) is exact if and only if L is monic, whereas (1.13) is exact if and only if $\tilde{L}$ is epic.

Further, let $L : V \to W$ be a linear mapping between finite-dimensional vector spaces. The **rank** of L, $\rho(L)$, is the dimension of the image of L. In such a case

$$\rho(L) + \dim(\ker L) = \dim V \tag{1.14}$$

Linear operators commonly are represented by **matrices.** It is quite natural to interchange these two ideas, because a matrix with respect to the standard bases is indistinguishable from the linear operator it represents. However, insight may be gained by examining these ideas separately. For V and W n- and m-dimensional vector spaces over F, respectively, consider a linear operator $L : V \to W$. Moreover, let $\{v_1, v_2, \ldots, v_n\}$ and $\{w_1, w_2, \ldots, w_m\}$ be respective bases for V and W. Then $L : V \to W$ can be represented uniquely by the matrix $M \in F^{m \times n}$ where

$$M = \begin{bmatrix} m_{11} & m_{12} & \cdots & m_{1n} \\ m_{21} & m_{22} & \cdots & m_{2n} \\ \vdots & \vdots & \ddots & \vdots \\ m_{m1} & m_{m2} & \cdots & m_{mn} \end{bmatrix} \tag{1.15}$$

The ith column of M is the representation of $L(v_i)$ with respect to $\{w_1, w_2, \ldots, w_m\}$. Element $m_{ij} \in F$ of (1.15) occurs in row i and column j.

Matrices have a number of properties. A matrix is said to be **square** if $m = n$. The **main diagonal** of a square matrix consists of the elements m_{ii}. If $m_{ij} = 0$ for all $i > j (i < j)$ a square matrix is said to be **upper (lower) triangular.** A square matrix with $m_{ij} = 0$ for all $i \neq j$ is **diagonal.** Additionally, if all $m_{ii} = 1$, a diagonal M is an **identity matrix.** A **row vector (column vector)** is a special case in which $m = 1$ $(n = 1)$. Also, $m = n = 1$ results essentially in a scalar.

Matrices arise naturally as a means to represent sets of simultaneous linear equations. For example, in the case of Kirchhoff equations, a later section on graph theory shows how incidence, circuit, and cut matrices arise. Or consider a π network having node voltages v_i, $i = 1, 2$ and current sources i_i, $i = 1, 2$ connected across the resistors R_i, $i = 1, 2$ in the two legs of the π. The bridge resistor is R_3. Thus, the unknown node voltages can be expressed in terms of the known source currents in the manner

$$\frac{(R_1 + R_3)}{R_1 R_3} v_1 - \frac{1}{R_3} v_2 = i_1 \tag{1.16}$$

$$\frac{(R_2 + R_3)}{R_2 R_3} v_2 - \frac{1}{R_3} v_1 = i_2 \tag{1.17}$$

If the voltages, v_i, and the currents, i_i, are placed into a voltage vector $v \in R^2$ and current vector $i \in R^2$, respectively, then (1.16) and (1.17) may be rewritten in matrix form as

$$\begin{bmatrix} i_1 \\ i_2 \end{bmatrix} = \begin{bmatrix} \frac{(R_1 + R_3)}{R_1 R_3} & -\frac{1}{R_3} \\ -\frac{1}{R_3} & \frac{(R_2 + R_3)}{R_2 R_3} \end{bmatrix} \begin{bmatrix} v_1 \\ v_2 \end{bmatrix} \tag{1.18}$$

A conductance matrix G may then be defined so that $i = Gv$, a concise representation of the original pair of circuit equations.

1.4 Matrix Operations

Vector addition in F^n was defined previously as element-wise scalar addition. Similarly, two matrices M and N, both in $F^{m \times n}$, can be added (subtracted) to form the resultant matrix $P \in F^{m \times n}$ by

$$m_{ij} \pm n_{ij} = p_{ij} \qquad i = 1, 2, \ldots, m \quad j = 1, 2, \ldots, n \tag{1.19}$$

Matrix addition, thus, is defined using addition in the field over which the matrix lies. Accordingly, the matrix, each of whose entries is $0 \in F$, is an additive identity for the family. One can set up additive inverses along similar lines, which, of course, turn out to be the matrices each of whose elements is the negative of that of the original matrix.

Recall how scalar multiplication was defined in the example of the vector space of n-tuples. Scalar multiplication can also be defined between a field element $a \in F$ and a matrix $M \in F^{m\times n}$ in such a way that the product aM is calculated element-wise:

$$aM = P \iff am_{ij} = p_{ij} \qquad i = 1,2,\ldots,m \quad j = 1,2,\ldots,n \tag{1.20}$$

Examples

$$(F = R): \qquad M = \begin{bmatrix} 4 & 3 \\ 2 & 1 \end{bmatrix} \qquad N = \begin{bmatrix} 2 & -3 \\ 1 & 6 \end{bmatrix} \qquad a = -0.5$$

- $M + N = P = \begin{bmatrix} 4+2 & 3-3 \\ 2+1 & 1+6 \end{bmatrix} = \begin{bmatrix} 6 & 0 \\ 3 & 7 \end{bmatrix}$
- $M - N = \tilde{P} = \begin{bmatrix} 4-2 & 3+3 \\ 2-1 & 1-6 \end{bmatrix} = \begin{bmatrix} 2 & 6 \\ 1 & -5 \end{bmatrix}$
- $aM = \hat{P} = \begin{bmatrix} (-0.5)4 & (-0.5)3 \\ (-0.5)2 & (-0.5)1 \end{bmatrix} = \begin{bmatrix} -2 & -1.5 \\ -1 & -0.5 \end{bmatrix}$

To multiply two matrices M and N to form the product MN requires that the number of columns of M equal the number of rows of N. In this case the matrices are said to be **conformable**. Although vector multiplication cannot be defined here because of this constraint, Chapter 2 examines this operation in detail using the tensor product. The focus here is on matrix multiplication. The resulting matrix will have its number of rows equal to the number of rows in M and its number of columns equal to the number of columns of N. Thus, for $M \in F^{m\times n}$ and $N \in F^{n\times p}$, $MN = P \in F^{m\times p}$. Elements in the resulting matrix P may be determined by

$$p_{ij} = \sum_{k=1}^{n} m_{ik} n_{kj} \tag{1.21}$$

Matrix multiplication involves one row and one column at a time. To compute the p_{ij} term in P, choose the ith row of M and the jth column of N. Multiply each element in the row vector by the corresponding element in the column vector and sum the result. Notice that in general, matrix multiplication is not commutative, and the matrices in the reverse order may not even be conformable. Matrix multiplication is, however, associative and distributive with respect to matrix addition. Under certain conditions, the field F of scalars, the set of matrices over F, and these three operations combine to form an **algebra**. Chapter 2 examines algebras in greater detail.

Examples

$$(F = R): \qquad M = \begin{bmatrix} 4 & 3 \\ 2 & 1 \end{bmatrix} \qquad N = \begin{bmatrix} 1 & 3 & 5 \\ 2 & 4 & 6 \end{bmatrix}$$

- $MN = P = \begin{bmatrix} 10 & 24 & 38 \\ 4 & 10 & 16 \end{bmatrix}$

  To find p_{11}, take the first row of M, $[4 \;\; 3]$, and the first column of N, $\begin{bmatrix} 1 \\ 2 \end{bmatrix}$, and evaluate (1.21): $4(1) + 3(2) = 10$. Continue for all i and j.
- NM does not exist because that product is not conformable.
- Any matrix $M \in F^{m \times n}$ multiplied by an identity matrix $I \in F^{n \times n}$ such that $MI \in F^{m \times n}$ results in the original matrix M. Similarly, $IM = M$ for I an $m \times m$ identity matrix over F. It is common to interpret I as an identity matrix of appropriate size, without explicitly denoting the number of its rows and columns.

The **transpose** $M^T \in F^{n \times m}$ of a matrix $M \in F^{m \times n}$ is found by interchanging the rows and columns. The first column of M becomes the first row of M^T, the second column of M becomes the second row of M^T, and so on. The notations M^t and M' are used also. If $M = M^T$ the matrix is called **symmetric**. Note that two matrices $M, N \in F^{m \times n}$ are equal if and only if all respective elements are equal: $m_{ij} = n_{ij}$ for all i, j. The **Hermitian transpose** $M^* \in C^{n \times m}$ of $M \in C^{m \times n}$ is also termed the **complex conjugate transpose**. To compute M^*, form M^T and take the complex conjugate of every element in M^T. The following properties also hold for matrix transposition for all $M, N \in F^{m \times n}$, $P \in F^{n \times p}$, and $a \in F$: $(M^T)^T = M$, $(M + N)^T = M^T + N^T$, $(aM)^T = aM^T$, and $(MP)^T = P^T M^T$.

Examples

$$(F = C): \qquad M = \begin{bmatrix} j & 1 - j \\ 4 & 2 + j3 \end{bmatrix}$$

- $M^T = \begin{bmatrix} j & 4 \\ 1 - j & 2 + j3 \end{bmatrix}$
- $M^* = \begin{bmatrix} -j & 4 \\ 1 + j & 2 - j3 \end{bmatrix}$

1.5 Determinant, Inverse, and Rank

Consider square matrices of the form $[m_{11}] \in F^{1 \times 1}$. For these matrices, define the **determinant** as m_{11} and establish the notation $\det([m_{11}])$ for this construction. This definition can be used to establish the meaning of $\det(M)$, often denoted $|M|$, for $M \in F^{2 \times 2}$. Consider

$$M = \begin{bmatrix} m_{11} & m_{12} \\ m_{21} & m_{22} \end{bmatrix} \tag{1.22}$$

The **minor** of m_{ij} is defined to be the determinant of the submatrix which results from the removal of row i and column j. Thus, the minors of m_{11}, m_{12}, m_{21}, and m_{22} are m_{22}, m_{21}, m_{12}, and m_{11}, respectively. To calculate the determinant of this M, (1) choose any row i (or column j), (2) multiply each element m_{ik} (or m_{kj}) in that row (or column) by its minor and by $(-1)^{i+k}$ (or $(-1)^{k+j}$), and (3) add these results. Note that the product of the minor with the sign $(-1)^{i+k}$ (or $(-1)^{k+j}$) is called the **cofactor** of the element in question. If row 1 is chosen, the determinant of M is found to be $m_{11}(+m_{22}) + m_{12}(-m_{21})$, a well-known result. The determinant of 2×2 matrices is relatively easy to remember: multiply the two elements along the main diagonal and subtract the product of the other two elements. Note that it makes no difference which row or column is chosen in step 1.

A similar procedure is followed for larger matrices. Consider

$$\det(M) = \begin{vmatrix} m_{11} & m_{12} & m_{13} \\ m_{21} & m_{22} & m_{23} \\ m_{31} & m_{32} & m_{33} \end{vmatrix} \tag{1.23}$$

Expanding about column 1 produces

$$\det(M) = m_{11}\begin{vmatrix} m_{22} & m_{23} \\ m_{32} & m_{33} \end{vmatrix} - m_{21}\begin{vmatrix} m_{12} & m_{13} \\ m_{32} & m_{33} \end{vmatrix} + m_{31}\begin{vmatrix} m_{12} & m_{13} \\ m_{22} & m_{23} \end{vmatrix} \tag{1.24}$$

$$= m_{11}(m_{22}m_{33} - m_{23}m_{32}) - m_{21}(m_{12}m_{33} - m_{13}m_{32}) + m_{31}(m_{12}m_{23} - m_{13}m_{22}) \tag{1.25}$$

$$= m_{11}m_{22}m_{33} + m_{12}m_{23}m_{31} + m_{13}m_{21}m_{32} - m_{13}m_{22}m_{31} - m_{11}m_{23}m_{32} - m_{12}m_{21}m_{33} \tag{1.26}$$

An identical result may be achieved by repeating the first two columns next to the original matrix:

$$\begin{matrix} m_{11} & m_{12} & m_{13} & m_{11} & m_{12} \\ m_{21} & m_{22} & m_{23} & m_{21} & m_{22} \\ m_{31} & m_{32} & m_{33} & m_{31} & m_{32} \end{matrix} \tag{1.27}$$

Then, form the first three products of (1.26) by starting at the upper left corner of (1.27) with m_{11}, forming a diagonal to the right, and then repeating with m_{12} and m_{13}. The last three products are subtracted in (1.26) and are formed by starting in the upper right corner of (1.27) with m_{12} and taking a diagonal to the left, repeating for m_{11} and m_{13}. Note the similarity to the 2×2 case. Unfortunately, such simple schemes fail above the 3×3 case.

Determinants of $n \times n$ matrices for $n > 3$ are computed in a similar vein. As in the earlier cases the determinant of an $n \times n$ matrix may be expressed in terms of the determinants of $(n-1) \times (n-1)$ submatrices; this is termed **Laplace's expansion.** To expand along row i or column j in $M \in F^{n\times n}$, write

$$\det(M) = \sum_{k=1}^{n} m_{ik}\tilde{m}_{ik} = \sum_{k=1}^{n} m_{kj}\tilde{m}_{kj} \tag{1.28}$$

where the m_{ik} (m_{kj}) are elements of M. The $\tilde{m}_{ik}$ ($\tilde{m}_{kj}$) are cofactors formed by deleting the ith (kth) row and the kth (jth) column of M, forming the determinant of the $(n-1) \times (n-1)$ resulting submatrix, and multiplying by $(-1)^{i+k}$ ($(-1)^{k+j}$). Notice that minors and their corresponding cofactors are related by ± 1.

Examples

$$(F = R): \qquad M = \begin{bmatrix} 0 & 1 & 2 \\ 3 & 4 & 5 \\ 2 & 3 & 6 \end{bmatrix}$$

- Expanding about row 1 produces

$$\det(M) = 0\begin{vmatrix} 4 & 5 \\ 3 & 6 \end{vmatrix} - 1\begin{vmatrix} 3 & 5 \\ 2 & 6 \end{vmatrix} + 2\begin{vmatrix} 3 & 4 \\ 2 & 3 \end{vmatrix}$$
$$= -(18 - 10) + 2(9 - 8) = -6$$

- Expanding about column 2 yields

$$\det(M) = -1\begin{vmatrix} 3 & 5 \\ 2 & 6 \end{vmatrix} + 4\begin{vmatrix} 0 & 2 \\ 2 & 6 \end{vmatrix} - 3\begin{vmatrix} 0 & 2 \\ 3 & 5 \end{vmatrix}$$
$$= -(18 - 10) + 4(0 - 4) - 3(0 - 6) = -6$$

- Repeating the first two columns to form (1.27) gives

$$\begin{matrix} 0\,1\,2\,0\,1 \\ 3\,4\,5\,3\,4 \\ 2\,3\,6\,2\,3 \end{matrix}$$

  Taking the appropriate products,

$$0 \cdot 4 \cdot 6 + 1 \cdot 5 \cdot 2 + 2 \cdot 3 \cdot 3 - 1 \cdot 3 \cdot 6 - 0 \cdot 5 \cdot 3 - 2 \cdot 4 \cdot 2$$

  results in -6 as the determinant of M.
- Any square matrix with a zero row and/or zero column will have zero determinant. Likewise, any square matrix with two or more identical rows and/or two or more identical columns will have determinant equal to zero.

Determinants satisfy many interesting relationships. For any $n \times n$ matrix, the determinant may be expressed in terms of determinants of $(n-1) \times (n-1)$ matrices or first-order minors. In turn, determinants of $(n-1) \times (n-1)$ matrices may be expressed in terms of determinants of $(n-2) \times (n-2)$ matrices or second-order minors, etc. Also, the determinant of the product of two square matrices is equal to the product of the determinants:

$$\det(MN) = \det(M)\det(N) \tag{1.29}$$

For any $M \in F^{n\times n}$ such that $|M| \neq 0$, a unique inverse $M^{-1} \in F^{n\times n}$ satisfies

$$MM^{-1} = M^{-1}M = I \tag{1.30}$$

For (1.29) one may observe the special case in which $N = M^{-1}$, then $(\det(M))^{-1} = \det(M^{-1})$. The **inverse** M^{-1} may be expressed using determinants and cofactors in the following manner. Form the matrix of cofactors

$$\tilde{M} = \begin{bmatrix} \tilde{m}_{11} & \tilde{m}_{12} & \cdots & \tilde{m}_{1n} \\ \tilde{m}_{21} & \tilde{m}_{22} & \cdots & \tilde{m}_{2n} \\ \vdots & \vdots & \ddots & \vdots \\ \tilde{m}_{n1} & \tilde{m}_{n2} & \cdots & \tilde{m}_{nn} \end{bmatrix} \tag{1.31}$$

The transpose of (1.31) is referred to as the **adjoint** matrix or adj(M). Then,

$$M^{-1} = \frac{\tilde{M}^T}{|M|} = \frac{\text{adj}(M)}{|M|} \qquad (1.32)$$

Examples

- Choose M of the previous set of examples. The cofactor matrix is

$$\begin{bmatrix} 9 & -8 & 1 \\ 0 & -4 & 2 \\ -3 & 6 & -3 \end{bmatrix}$$

  Because $|M| = -6$, M^{-1} is

$$\begin{bmatrix} -\frac{3}{2} & 0 & \frac{1}{2} \\ \frac{4}{3} & \frac{2}{3} & -1 \\ -\frac{1}{6} & -\frac{1}{3} & \frac{1}{2} \end{bmatrix}$$

  Note that (1.32) is satisfied.
- $M = \begin{bmatrix} 2 & 1 \\ 4 & 3 \end{bmatrix} \quad \text{adj}(M) = \begin{bmatrix} 3 & -1 \\ -4 & 2 \end{bmatrix} \quad M^{-1} = \begin{bmatrix} \frac{3}{2} & -\frac{1}{2} \\ -2 & 1 \end{bmatrix}$

  In the 2×2 case this method reduces to interchanging the elements on the main diagonal, changing the sign on the remaining elements, and dividing by the determinant.
- Consider the matrix equation in (1.18). Because $\det(G) \neq 0$, whenever the resistances are nonzero, with R_1 and R_2 having the same sign, the node voltages may be determined in terms of the current sources by multiplying on the left of both members of the equation using G^{-1}. Then, $G^{-1}i = v$.

The **rank** of a matrix M, $\rho(M)$, is the number of linearly independent columns of M over F, or using other terminology, the dimension of the image of M. For $M \in F^{m \times n}$ the number of linearly independent rows and columns is the same, and is less than or equal to the minimum of m and n. If $\rho(M) = n$, M is of **full column rank**; similarly, if $\rho(M) = m$, M is of **full row rank**. A square matrix with all rows (and all columns) linearly independent is said to be **nonsingular**. In this case $\det(M) \neq 0$. The rank of M also may be found from the size of the largest square submatrix with a nonzero determinant. A full-rank matrix has a full-size minor with a nonzero determinant.

The **null space (kernel)** of a matrix $M \in F^{m \times n}$ is the set

$$\ker M = \{v \in F^n \text{ such that } Mv = 0\} \qquad (1.33)$$

Over F, $\ker M$ is a vector space with dimension defined as the **nullity** of M, $\nu(M)$. The fundamental theorem of linear equations relates the rank and nullity of a matrix $M \in F^{m \times n}$ by

$$\rho(M) + \nu(M) = n \qquad (1.34)$$

If $\rho(M) < n$, then M has a nontrivial null space.

Examples

- $M = \begin{bmatrix} 0 & 1 \\ 0 & 0 \end{bmatrix}$

  The rank of M is 1 because only one linearly independent column of M is found. To examine the null space of M, solve $Mv = 0$. Any element in $\ker M$ is of the form $\begin{bmatrix} f_1 \\ 0 \end{bmatrix}$ for $f_1 \in F$. Therefore, $\nu(M) = 1$.

- $M = \begin{bmatrix} 1 & 4 & 5 & 2 \\ 2 & 5 & 7 & 1 \\ 3 & 6 & 9 & 0 \end{bmatrix}$

  The rank of M is 2 and the nullity is 2.

1.6 Basis Transformations

This section describes a change of basis as a linear operator. Because the choice of basis affects the matrix of a linear operator, it would be most useful if such a basis change could be understood within the context of matrix operations. Thus, the new matrix could be determined from the old matrix by matrix operations. This is indeed possible. This question is examined in two phases. In the first phase the linear operator maps from a vector space to itself. Then a basis change will be called a **similarity transformation**. In the second phase the linear operator maps from one vector space to another, which is not necessarily the same as the first. Then a basis change will be called an **equivalence transformation**. Of course, the first situation is a special case of the second, but it is customary to make the distinction and to recognize the different terminologies. Philosophically, a fascinating special situation exists in which the second vector space, which receives the result of the operation, is an identical copy of the first vector space, from which the operation proceeds. However, in order to avoid confusion, this section does not delve into such issues.

For the first phase of the discussion, consider a linear operator that maps a vector space into itself, such as $L : V \to V$, where V is n-dimensional. Once a basis is chosen in V, L will have a unique matrix representation. Choose $\{v_1, v_2, \ldots, v_n\}$ and $\{\bar{v}_1, \bar{v}_2, \ldots, \bar{v}_n\}$ as two such bases. A matrix $M \in F^{n\times n}$ may be determined using the first basis, whereas another matrix $\bar{M} \in F^{n\times n}$ will result in the latter choice. According to the discussion following (1.15), the ith column of M is the representation of $L(v_i)$ with respect to $\{v_1, v_2, \ldots, v_n\}$, and the ith column of $\bar{M}$ is the representation of $L(\bar{v}_i)$ with respect to $\{\bar{v}_1, \bar{v}_2, \ldots, \bar{v}_n\}$. As in (1.4), any basis element v_i has a unique representation in terms of the basis $\{\bar{v}_1, \bar{v}_2, \ldots, \bar{v}_n\}$. Define a matrix $P \in F^{n\times n}$ using the ith column as this representation. Likewise, $Q \in F^{n\times n}$ may have as its ith column the unique representation of $\bar{v}_i$ with respect to $\{v_1, v_2, \ldots, v_n\}$. Either represents a basis change which is a linear operator. By construction, both P and Q are nonsingular. Such matrices and linear operators are sometimes called **basis transformations**. Notice that $P = Q^{-1}$.

If two matrices M and $\bar{M}$ represent the same linear operator L they must somehow carry essentially the same information. Indeed, a relationship between M and $\bar{M}$ may be established. Consider $a_v, a_w, \bar{a}_v, \bar{a}_w \in F^n$ such that $Ma_v = a_w$ and $\bar{M}\bar{a}_v = \bar{a}_w$. Here, a_v denotes the representation of v with respect to the basis v_i, $\bar{a}_v$ denotes the representation of v with respect to the basis $\bar{v}_i$, and so forth. In order to involve P and Q in these equations it is possible to make use of a sketch:

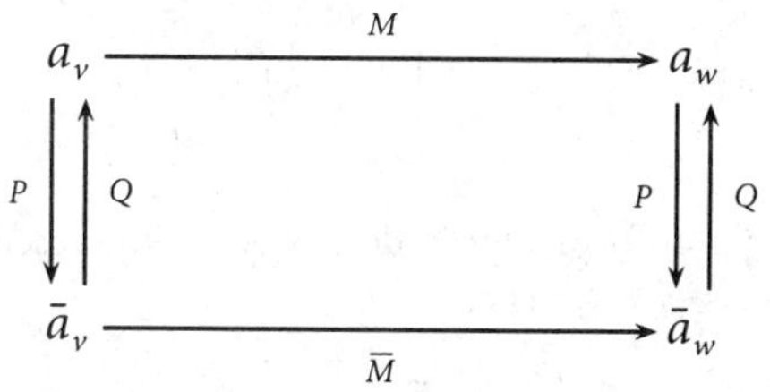

In this sketch a vector at a given corner can be multiplied by a matrix on an arrow leaving the corner and set equal to the vector that appears at the corner at which that arrow arrives. Thus, for example, $a_w = Ma_v$ may be deduced from the top edge of the sketch. It is interesting to perform "chases" around such sketches. By way of illustration, consider the lower right corner. Progress around the sketch counterclockwise so as to reach the lower left corner and set the result equal to that obtained by progressing clockwise to the lower left corner. In equations this is carried out as follows:

$$\bar{a}_w = Pa_w = PMa_v = PMQ\bar{a}_v = \bar{M}\bar{a}_v \tag{1.35}$$

Inasmuch as $\bar{a}_v \in F^n$ is arbitrary it follows that

$$\bar{M} = PMP^{-1} \tag{1.36}$$

Sketches that have this type of property, namely the same result when the sketch is traversed from a starting corner to an finishing corner by two paths, are said to be **commutative**. It is perhaps more traditional to show the vector space F^n instead of the vectors at the corners. Thus, the sketch would be called a **commutative diagram of vector spaces and linear operators.** M and $\bar{M}$ are said to be **similar** because a nonsingular matrix $P \in F^{n\times n}$ is such that (1.36) is true. The matrix P is then called a **similarity transformation.** Note that all matrix representations associated with the same linear operator, from a vector space to itself, are similar. Certain choices of bases lead to special forms for the matrices of the operator, as are apparent in the following examples.

Examples

- Choose $L : R^2 \to R^2$ as the linear operator that rotates a vector by 90°. Pick $\{v_1, v_2\}$ as the natural basis $\{[1\ \ 0]^T, [0\ \ 1]^T\}$ and $\{\bar{v}_1, \bar{v}_2\}$ as $\{[1\ \ 1]^T, [1\ \ 0]^T\}$. Then,

$$L(v_1) = \begin{bmatrix} v_1 & v_2 \end{bmatrix} \begin{bmatrix} 0 \\ 1 \end{bmatrix}$$

$$L(v_2) = \begin{bmatrix} v_1 & v_2 \end{bmatrix} \begin{bmatrix} -1 \\ 0 \end{bmatrix}$$

$$L(\bar{v}_1) = \begin{bmatrix} \bar{v}_1 & \bar{v}_2 \end{bmatrix} \begin{bmatrix} 1 \\ -2 \end{bmatrix}$$

$$L(\bar{v}_2) = \begin{bmatrix} \bar{v}_1 & \bar{v}_2 \end{bmatrix} \begin{bmatrix} 1 \\ -1 \end{bmatrix}$$

so that

$$M = \begin{bmatrix} 0 & -1 \\ 1 & 0 \end{bmatrix} \qquad \bar{M} = \begin{bmatrix} 1 & 1 \\ -2 & -1 \end{bmatrix}$$

To find the similarity transformation P, determine the representations of the basis vectors $\mathbf{v}_i$ in terms of the basis $\{\bar{v}_1, \bar{v}_2\}$. Then

$$P = \begin{bmatrix} 0 & 1 \\ 1 & -1 \end{bmatrix} \qquad P^{-1} = \begin{bmatrix} 1 & 1 \\ 1 & 0 \end{bmatrix}$$

so that $PMP^{-1} = \bar{M}$.

- Suppose that $M \in F^{n \times n}$ is a representation of $L : V \to V$. Assume a $v \in F^n$ exists such that the vectors $\{v, Mv, \ldots, M^{n-1}v\}$ are linearly independent. Thus, these n vectors can be chosen as an alternate basis. (Section 1.7 discusses the characteristic equation of a matrix M.) Using the Cayley-Hamilton theorem, which states that every matrix satisfies its own characteristic equation, it is always possible to write $M^n v$ as a linear combination of these alternate basis vectors

$$-\alpha_n v - \alpha_{n-1} Mv - \cdots - \alpha_1 M^{n-1} v$$

for $\alpha_i \in F$. The matrix representation of L with respect to the alternate basis given by this set of linearly independent vectors is

$$\bar{M} = \begin{bmatrix} 0 & 0 & \cdots & 0 & -\alpha_n \\ 1 & 0 & \cdots & 0 & -\alpha_{n-1} \\ 0 & 1 & \cdots & 0 & -\alpha_{n-2} \\ \vdots & \vdots & \ddots & \vdots & \vdots \\ 0 & 0 & \cdots & 0 & -\alpha_2 \\ 0 & 0 & \cdots & 1 & -\alpha_1 \end{bmatrix}$$

which is the transpose of what is known as the **companion form.**

For the second phase of the discussion, select a pair of bases, one for the vector space V and one for the vector space W, and construct the resulting matrix representation M of $L : V \to W$. Another choice of bases exists for V and W, with the property that the resulting matrix $\bar{M}$ representing L is of the form

$$\begin{bmatrix} I & 0 \\ 0 & 0 \end{bmatrix} \tag{1.37}$$

where I is $\rho(M) \times \rho(M)$. Such a matrix is said to be in **normal form.** It is possible to transform M into $\bar{M}$ with the assistance of three types of transformations which are called **elementary:** (1) interchange any two rows or any two columns, (2) scale any row or column by a nonzero element in F, and (3) add any F-multiple of any row (column) to any other row (column). It is apparent that each of the three transformations involving rows may be accomplished by multiplying M on the left by a nonsingular matrix. Column operations require a corresponding multiplication on the right. The secret to understanding elementary transformations is to recognize that each of them will carry one basis into another. Not as easy to see, but equally true, is that any transformation that carries one basis into another basis must be a product of such elementary transformations. The elementary column transformations are interpreted as changing the basis in the vector space from which the operator takes vectors, whereas the elementary row transformations correspond to changing the basis in the vector space into which the operator places its result. It stands to reason that a simultaneous adjustment of both sets of basis vectors could lead to some quite special forms for the matrices of an operator. Of course, a great deal of linear algebra and its applications is concerned with just such constructions and forms. Space does not permit a complete treatment here.

If a matrix M has been transformed into normal form, certain types of key information become available. For example, one knows the rank of M because $\rho(M)$ is the number of rows and columns of the identity in (1.37). Perhaps more importantly, the normal form is easily factored in a fundamental way, and so such a construction is a natural means to construct two factors of minimal rank for a given matrix. The reader is cautioned, however, to be aware that computational linear algebra is quite a different subject than theoretical linear algrebra. One common saying is that "if an algorithm is straightforward, then it is not numerically desirable". This may be an exaggeration, but it is well to recognize the implications of finite precision on the computer. Space limitations prevent addressing numerical issues.

Many other thoughts can be expressed in terms of elementary basis transformations. By way of illustration, elementary basis transformations offer an alternative in finding the inverse of a matrix. For a nonsingular matrix $M \in F^{n\times n}$, append to M an $n \times n$ identity I to form the $n \times 2n$ matrix

$$\hat{M} = \begin{bmatrix} M & I \end{bmatrix} \tag{1.38}$$

Perform elementary row transformations on (1.38) to transform M into normal form. Then M^{-1} will appear in the last n columns of the transformed matrix.

Examples

$$(F = R): \qquad M = \begin{bmatrix} 2 & 1 \\ 4 & 3 \end{bmatrix}$$

- Transform M into normal form. The process can be carried out in many ways. For instance, begin by scaling row 1 by $\frac{1}{2}$,

$$L_1M = \begin{bmatrix} \frac{1}{2} & 0 \\ 0 & 1 \end{bmatrix}\begin{bmatrix} 2 & 1 \\ 4 & 3 \end{bmatrix} = \begin{bmatrix} 1 & \frac{1}{2} \\ 4 & 3 \end{bmatrix}$$

  Clear the first element in row 2 by

$$L_2L_1M = \begin{bmatrix} 1 & 0 \\ -4 & 1 \end{bmatrix}\begin{bmatrix} 1 & \frac{1}{2} \\ 4 & 3 \end{bmatrix} = \begin{bmatrix} 1 & \frac{1}{2} \\ 0 & 1 \end{bmatrix}$$

  Finally, perform a column operation to produce $\bar{M}$:

$$L_2L_1MR_1 = \begin{bmatrix} 1 & \frac{1}{2} \\ 0 & 1 \end{bmatrix}\begin{bmatrix} 1 & -\frac{1}{2} \\ 0 & 1 \end{bmatrix} = \begin{bmatrix} 1 & 0 \\ 0 & 1 \end{bmatrix}$$

The rank of M is 2.

- Recall M^{-1} from previous examples. Form $\hat{M}$ and transform M into normal form by the following row operations:

$$L_1\hat{M} = \begin{bmatrix} \frac{1}{2} & 0 \\ 0 & 1 \end{bmatrix} \begin{bmatrix} 2 & 1 & 1 & 0 \\ 4 & 3 & 0 & 1 \end{bmatrix} = \begin{bmatrix} 1 & \frac{1}{2} & \frac{1}{2} & 0 \\ 4 & 3 & 0 & 1 \end{bmatrix}$$

$$L_2L_1\hat{M} = \begin{bmatrix} 1 & 0 \\ -4 & 1 \end{bmatrix} \begin{bmatrix} 1 & \frac{1}{2} & \frac{1}{2} & 0 \\ 4 & 3 & 0 & 1 \end{bmatrix} = \begin{bmatrix} 1 & \frac{1}{2} & \frac{1}{2} & 0 \\ 0 & 1 & -2 & 1 \end{bmatrix}$$

$$L_3L_2L_1\hat{M} = \begin{bmatrix} 1 & -\frac{1}{2} \\ 0 & 1 \end{bmatrix} \begin{bmatrix} 1 & \frac{1}{2} & \frac{1}{2} & 0 \\ 0 & 1 & -2 & 1 \end{bmatrix} = \begin{bmatrix} 1 & 0 & \frac{3}{2} & -\frac{1}{2} \\ 0 & 1 & -2 & 1 \end{bmatrix}$$

1.7 Characteristics: Eigenvalues, Eigenvectors, and Singular Values

A matrix has certain characteristics associated with it. Of these, characteristic values or eigenvalues may be determined through the use of matrix pencils. In general a matrix pencil may be formed from two matrices M and $N \in F^{m \times n}$ and an indeterminate λ in the manner

$$\left[\ \lambda N - M\ \right] \in F[\lambda]^{m \times n} \tag{1.39}$$

In determining eigenvalues of a square matrix $M \in F^{n \times n}$ one assumes the special case in which $N = I \in F^{n \times n}$.

Assume that M is a square matrix over the complex numbers. Then, $\lambda \in C$ is called an **eigenvalue** of M if some nonzero vector $v \in C^n$ exists such that

$$Mv = \lambda v \tag{1.40}$$

Any such $v \neq 0$ satisfying (1.40) is said to be an **eigenvector** of M associated with λ. It is easy to see that (1.40) can be rewritten as

$$(\lambda I - M)v = 0 \tag{1.41}$$

Because (1.41) is a set of n linear homogeneous equations, a nontrivial solution ($v \neq 0$) exists if and only if

$$\Delta(\lambda) = \det(\lambda I - M) = 0 \tag{1.42}$$

In other words, $(\lambda I - M)$ is singular. Therefore, λ is an eigenvalue of M if and only if it is a solution of (1.42). The polynomial $\Delta(\lambda)$ is the **characteristic polynomial** and $\Delta(\lambda) = 0$ is the **characteristic equation**. Moreover, every $n \times n$ matrix has n eigenvalues that may be real, complex or both, where complex eigenvalues occur in complex-conjugate pairs. If two or more eigenvalues are equal they are said to be repeated (not distinct). It is interesting to observe that although eigenvalues are unique, eigenvectors are not. Indeed, an eigenvector can be multiplied by any nonzero element of C and still maintain its essential features. Sometimes this lack of uniqueness is resolved by selecting unit length for the eigenvectors with the aid of a suitable norm.

Recall that matrices representing the same operator are similar. One may question if these matrices indeed contain the same characteristic information. To answer this question, examine

$$\det(\lambda I - \bar{M}) = \det(\lambda PP^{-1} - PMP^{-1}) = \det(P(\lambda I - M)P^{-1}) \tag{1.43}$$

$$= \det(P)\det(\lambda I - M)\det(P^{-1}) = \det(\lambda I - M) \tag{1.44}$$

From (1.44) one may deduce that similar matrices have the same eigenvalues because their characteristic polynomials are equal.

For every square matrix M with distinct eigenvalues, a similar matrix $\bar{M}$ is diagonal. In particular, the eigenvalues of M, and hence $\bar{M}$, appear along the main diagonal. Let $\lambda_1, \lambda_2, \ldots, \lambda_n$ be the eigenvalues (all distinct) of M and let $v_1, v_2, \ldots, v_n$ be corresponding eigenvectors. Then, the vectors $\{v_1, v_2, \ldots, v_n\}$ are linearly independent over C. Choose $P^{-1} = Q = [v_1\ v_2 \cdots v_n]$ as the **modal matrix.** Because $Mv_i = \lambda_i v_i$, $\bar{M} = PMP^{-1}$ as before.

For matrices with repeated eigenvalues, a similar approach may be followed wherein $\bar{M}$ is block diagonal, which means that matrices occur along the diagonal with zeros everywhere else. Each matrix along the diagonal is associated with an eigenvalue and takes a specific form depending upon the characteristics of the matrix itself. The modal matrix consists of generalized eigenvectors, of which the aforementioned eigenvector is a special case; thus the modal matrix is nonsingular. The matrix $\bar{M}$ is the **Jordan canonical form**. Space limitations preclude a detailed analysis of such topics here; the reader is directed to Chen (1984) for further development.

Examples

$$(F = C): \qquad M = \begin{bmatrix} 1 & 4 \\ 2 & 3 \end{bmatrix}$$

- The characteristic polynomial is $\Delta(\lambda) = (\lambda - 1)(\lambda - 3) - 8 = (\lambda - 5)(\lambda + 1)$. The eigenvalues are $\lambda_1 = 5, \lambda_2 = -1$. To find the associated eigenvectors recall that for each λ_i, $(\lambda_i I - M)$ is singular, and write (1.41)

$$(\lambda_1 I - M)v_1 = \begin{bmatrix} 4 & -4 \\ -2 & 2 \end{bmatrix}\begin{bmatrix} v_{11} \\ v_{12} \end{bmatrix} = \begin{bmatrix} 0 \\ 0 \end{bmatrix}$$

$$(\lambda_2 I - M)v_2 = \begin{bmatrix} -2 & -4 \\ -2 & -4 \end{bmatrix}\begin{bmatrix} v_{21} \\ v_{22} \end{bmatrix} = \begin{bmatrix} 0 \\ 0 \end{bmatrix}$$

Then, $v_{11} = v_{12}$ and $v_{21} = -2v_{22}$ so that $v_1 = [1\ \ 1]^T$ and $v_2 = [-2\ \ 1]^T$ are eigenvectors associated with λ_1 and λ_2, respectively.

- Because the eigenvalues of M are distinct, M may be diagonalized. For verification, choose $P^{-1} = [v_1\ \ v_2]$. Then

$$\bar{M} = PMP^{-1} = \frac{1}{3}\begin{bmatrix} 1 & 2 \\ -1 & 1 \end{bmatrix}\begin{bmatrix} 1 & 4 \\ 2 & 3 \end{bmatrix}\begin{bmatrix} 1 & -2 \\ 1 & 1 \end{bmatrix}$$

$$= \frac{1}{3}\begin{bmatrix} 15 & 0 \\ 0 & -3 \end{bmatrix} = \begin{bmatrix} 5 & 0 \\ 0 & -1 \end{bmatrix} = \begin{bmatrix} \lambda_1 & 0 \\ 0 & \lambda_2 \end{bmatrix}$$

In general a matrix $M \in F^{m\times n}$ of rank r can be written in terms of its **singular-value decomposition** (SVD),

$$M = U\Sigma V^* \tag{1.45}$$

For any M, **unitary** matrices U and V of dimension $m \times m$ and $n \times n$, respectively, form the decomposition; that is, $UU^* = U^*U = I$ and $VV^* = V^*V = I$. The matrix $\Sigma \in F^{m\times n}$ is of the form

$$\begin{bmatrix} \Sigma_r & 0 \\ 0 & 0 \end{bmatrix} \tag{1.46}$$

for $\Sigma_r \in F^{r\times r}$, a diagonal matrix represented by

$$\begin{bmatrix} \sigma_1 & 0 & \cdots & 0 \\ 0 & \sigma_2 & \cdots & 0 \\ \vdots & \vdots & \ddots & \vdots \\ 0 & 0 & \cdots & \sigma_r \end{bmatrix} \tag{1.47}$$

The elements σ_i, called **singular values**, are related by $\sigma_1 \geq \sigma_2 \geq \cdots \geq \sigma_r > 0$, and the columns of U (V) are referred to as **left (right) singular vectors**. Although the unitary matrices U and V are not unique for a given M, the singular values are unique.

Singular-value decomposition is useful in the numerical calculation of rank. After performing a SVD, the size of the matrix Σ_r may be decided. Additionally, the **generalized inverse** of a matrix M may be found by

$$M^\dagger = V\begin{bmatrix} \Sigma_r^{-1} & 0 \\ 0 & 0 \end{bmatrix}U^* \tag{1.48}$$

It can be verified easily that $MM^\dagger M = M$ and $M^\dagger MM^\dagger = M^\dagger$. In the special case in which M is square and nonsingular,

$$M^\dagger = M^{-1} = V\Sigma^{-1}U^* \tag{1.49}$$

1.8 On Linear Systems

Consider a set of n simultaneous algebraic equations, in m unknowns, written in the customary matrix form $w = Mv$ where

$$\begin{bmatrix} w_1 \\ w_2 \\ \vdots \\ w_n \end{bmatrix} = \begin{bmatrix} m_{11} & m_{12} & \cdots & m_{1m} \\ m_{21} & m_{22} & \cdots & m_{2m} \\ \vdots & \vdots & \ddots & \vdots \\ m_{n1} & m_{n2} & \cdots & m_{nm} \end{bmatrix}\begin{bmatrix} v_1 \\ v_2 \\ \vdots \\ v_m \end{bmatrix} \tag{1.50}$$

In the context of the foregoing discussion (1.50) represents the action of a linear operator. If the left member is a given vector, in the usual manner, then a first basic issue concerns whether the vector represented by the left member is in the image of the operator or not. If it is in the image, the equation has at least one solution; otherwise the equation has no solution. A second basic issue concerns the kernel of the operator. If the kernel contains only the zero vector, then the equation has at most one solution; otherwise more than one solution can occur, provided that at least one solution exists.

When one thinks of a set of simultaneous equations as a "system" of equations, the intuitive transition to the idea of a **linear system** is quite natural. In this case the vector in the left member becomes the **input** to the system, and the solution to (1.50), when it exists and is unique, is the **output** of the system.

Other than being a description in terms of inputs and outputs, as above, linear systems may also be described in terms of sets of other types of equations, such as differential equations or difference equations. When that is the situation, the familiar notion of initial condition becomes an instance of the idea of **state**, and one must examine the intertwining of states and inputs to give outputs. Then, the idea of (1.50), when each input yields a unique output, is said to define a **system function.**

If the differential (difference) equations are linear and have constant coefficients, the possibility exists of describing the system in terms of transforms, for example, in the s-domain or z-domain. This leads to fascinating new interpretations of the ideas of the foregoing sections, this time, for example, over fields of rational functions. Colloquially, such functions are best known as **transfer functions.**

Associated with systems described in the time domain, s-domain, or z-domain some characteristics of the system also aid in analysis techniques. Among the most basic of these are the entities termed **poles** and **zeros**, which have been linked to the various concepts of system **stability**. Both poles and zeros may be associated with matrices of transfer functions, and with the original differential or difference equations themselves. A complete and in-depth treatment of the myriad meanings of poles and zeros is a challenging undertaking, particularly in matrix cases. For a recent survey of the ideas, see Schrader and Sain (1989). However, a great many of the definitions involve such concepts as rank, pencils, eigenvalues, eigenvectors, special matrix forms, vector spaces, and modules—the very ideas sketched out in the sections preceding.

One very commonly known idea for representing solutions to (1.50) is **Cramer's rule**. When $m = n$, and when M has an inverse, the use of Cramer's rule expresses each unknown variable individually by using a ratio of determinants. Choose the ith unknown v_i. Define the determinant M_i as the determinant of a matrix formed by replacing column i in M with w. Then,

$$v_i = \frac{M_i}{\det(M)} \tag{1.51}$$

It turns out that this very interesting idea makes fundamental use of the notion of *multiplying* vectors, which is not part of the axiomatic framework of the vector space. The reader may want to reflect further on this observation, in respect to the foregoing treatment of determinants. When the framework of vector spaces is expanded to include vector multiplication, as in the case of determinants, one gets to the technical subject of **algebras**. The next chapter returns to this concept.

The concepts presented above allow for more detailed considerations in the solution of circuit and filter problems, using various approaches outlined in the remainder of this text. The following chapter provides for the multiplication of vectors by means of the foundational idea of bilinear operators and matrices. The next chapters on transforms—Fourier, z, and Laplace—provide the tools for analysis by allowing a set of differential or difference equations describing a circuit to be written as a system of linear algebraic equations. Moreover, each transform itself can be viewed as a linear operator, and thus becomes a prime example of the ideas of this chapter. The remaining chapters focus on graph-theoretical approaches to the solution of systems of algebraic equations. From this vantage point, then, one can see the entire Section I in the context of linear operators, their addition and multiplication.

A brief treatment cannot deal with all the interesting questions and answers associated with linear operators and matrices. For a more detailed treatment of these standard concepts, see any basic algebra text, for example, Greub (1967).

References

C.-T. Chen, *Linear System Theory and Design*, New York: CBS College Publishing, 1984.

W. H. Greub, *Linear Algebra*, New York: Springer-Verlag, 1967.

C. B. Schrader and M. K. Sain, "Research on system zeros: a survey," *Int. J. Control*, Vol. 50, No. 4, pp. 1407–1433, Oct. 1989.

2

Bilinear Operators and Matrices

chael K. Sain
versity of Notre Dame

eryl B. Schrader
University of Texas

2.1 Introduction

The key player in Chapter 1 was the F-vector space V, together with its associated notion of bases, when they exist, and linear operators taking one vector to another. The idea of basis is, of course, quite central to the applications because it permits a vector v in V to be represented by a list of scalars from F. Such lists of scalars are the quantities with which one computes. No doubt the idea of an F-vector space V is the most common and widely encountered notion in applied linear algebra. It is typically visualized, on the one hand, by long lists of axioms, most of which seem quite reasonable, but none of which is particularly exciting, and on the other hand by images of classical addition of force vectors, velocity vectors, and so forth. The notion seems to do no harm, and helps one to keep his or her assumptions straight. As such it is accepted by most engineers as a plausible background for their work, even if the ideas of matrix algebra are more immediately useful. Perhaps some of the least appreciated but most crucial of the vector space axioms are the four governing the scalar multiplication of vectors. These link the abelian group of vectors to the field of scalars. Along with the familiar distributive covenants, these four agreements intertwine the vectors with the scalars in much the same way that the marriage vows bring about the union of man and woman. This section brings forth a new addition to the marriage.

As useful as it is, the notion of an F-vector space V fails to provide for one of the most important ideas in the applications—the concept of multiplication of vectors. In a vector space one can add vectors and multiply vectors by scalars, but one cannot multiply vectors by vectors. Yet there are numerous situations in which one faces exactly these operations. Consider, for instance, the cross and dot products from field theory. Even in the case of matrices, the

3-8341-2/95/$0.00 + $.50

ubiquitous and crucial matrix multiplication is available, when it is defined. The key to the missing element in the discussion lies in the terminology for matrix operations, which will be familiar to the reader as the **matrix algebra**. What must occur in order for vector-to-vector multiplication to be available is for the vector space to be extended into an algebra.

Unfortunately, the word "algebra" carries a rather imprecise meaning from the most elementary and early exposures from which it came to signify the collection of operations done in arithmetic, at the time when the operations are generalized to include symbols or literals such as a, b, and c or x, y, and z. Such a notion generally corresponds closely with the idea of a field, F, as defined in Chapter 1, and is not much off the target for an environment of scalars. It may, however, come as a bit of a surprise to the reader that algebra is a technical term, in the same spirit as fields, vector spaces, rings, etc. Therefore, if one is to have available a notion of multiplication of vectors, then it is appropriate to introduce the precise notion of an **algebra**, which captures the desired idea in an axiomatic sense.

2.2 Algebras

Chapter 1 mentioned that the integers I and the polynomials $F[s]$ in s with coefficients in a field F were instances of a ring. In this section it is necessary to carry the concept of ring beyond the example stage. Of course, a long list of axioms could be provided, but it may be more direct just to cite the changes necessary to the field axioms already provided in Section 1.2. To be precise, the axioms of a **commutative ring** differ from the axioms of a field only by the removal of the multiplicative inverse. Intuitively, this means that one cannot always divide, even if the element in question is nonzero. Many important commutative rings are found in the applications; however, this chapter is centered on **rings**, wherein one more axiom is removed—the commutativity of multiplication. The ring of $n \times n$ matrices with elements from a field is a classic and familiar example of such a definition. It may be remarked that in some references a distinction is made between **rings** and **rings with identity**, the latter having a multiplicative identity and the former not being so equipped. This treatment has no need for such a distinction, and hereafter the term "ring" is understood to mean ring with identity, or, as described above, a field with the specified two axioms removed.

It is probably true that the field is the most comfortable of axiomatic systems for most persons because it corresponds to the earliest and most persistent of calculation notions. However, it is also true that the ring has an intuitive and immediate understanding as well, which can be expressed in terms of the well-known phrase "playing with one arm behind one's back". Indeed, each time an axiom is removed, it is similar to removing one of the options in a game. This adds to the challenge of a game, and leads to all sorts of new strategies. Such is the case for **algebras**, as is clear from the next definition. What follows is not the most general of possible definitions, but probably that which is most common.

An **algebra** A is an F-vector space A which is equipped with a multiplication a_1a_2 of vectors a_1 and a_2 in such a manner that it is also a ring. First, addition in the ring is simply addition of vectors in the vector space. Second, a special relationship exists between multiplication of vectors and scalar multiplication in the vector space. If a_1 and a_2 are vectors in A, and if f is a scalar in F, then the following identity holds:

$$f(a_1a_2) = (fa_1)a_2 = a_1(fa_2) \tag{2.1}$$

Note that the order of a_1 and a_2 does not change in the above equalities. This must be true because no axiom of commutativity exists for multiplication. The urge to define a symbol for vector multiplication is resisted here so as to keep things as simple as possible. In the same way the notation for scalar multiplication, as introduced in Chapter 1, is suppressed here in the interest of simplicity. Thus, the scalar multiplication can be associated either with the vector product, which lies in A, or with one or other of the vector factors. This is exactly the familiar situation with the matrix algebra.

Hidden in the definition of the **algebra** A above is the precise detail arising from the statement that A is a ring. Associated with that detail is the nature of the vector multiplication represented above with the juxtaposition $a_1 a_2$. Because all readers are familiar with several notions of vector multiplication, the question arises as to just what constitutes such a multiplication. It turns out that a precise notion for multiplication can be found in the idea of a bilinear operator. Thus, an alternative description of Section 2.3 is that of vector spaces equipped with vector multiplication. Moreover, one is tempted to inquire whether a vector multiplication exists that is so general in nature that all other vector multiplications can be derived from it. In fact, this is the case, and the following section sets the stage for introducing such a multiplication.

2.3 Bilinear Operators

Suppose that there are three F-vector spaces: U, V, and W. Recall that $U \times V$ is the Cartesian product of U with V, and denotes the set of all ordered pairs, the first from U and the second from V. Now, consider a mapping b from $U \times V$ into W. For brevity of notation, this can be written b: $U \times V \rightarrow W$. The mapping b is a **bilinear operator** if it satisfies the pair of conditions

$$b(f_1 u_1 + f_2 u_2, v) = f_1 b(u_1, v) + f_2 b(u_2, v) \tag{2.2}$$

$$b(u, f_1 v_1 + f_2 v_2) = f_1 b(u, v_1) + f_2 b(u, v_2) \tag{2.3}$$

for all f_1 and f_2 in F, for all u, u_1, and u_2 in U, and for all v, v_1, and v_2 in V. The basic idea of the bilinear operator is apparent from this definition. It is an operator with two arguments, having the property that if either of the two arguments is fixed, the operator becomes linear in the remaining argument. A moment's reflection will show that the intuitive operation of multiplication is of this type.

One of the important features of a bilinear operator is that its image need not be a subspace of W. This is in marked contrast with the image of a linear operator, whose image is always a subspace. This property leads to great interest in the manipulations associated with vector products. At the same time, it brings about a great deal of nontriviality. The best way to illustrate the point is with an example.

Example. Suppose that U, V, and W have bases $\{u_1, u_2\}$, $\{v_1, v_2\}$, and $\{w_1, w_2, w_3, w_4\}$, respectively. Then, vectors u in U and v in V can be represented in the manner

$$u = f_1 u_1 + f_2 u_2 \tag{2.4}$$

$$v = g_1 v_1 + g_2 v_2 \tag{2.5}$$

where f_i and g_i are elements of F for $i = 1, 2$. Define a bilinear map by the action

$$b(u, v) = 2 f_1 g_1 w_1 + 3 f_1 g_2 w_2 + 3 f_2 g_1 w_3 + 2 f_2 g_2 w_4 \tag{2.6}$$

It is clear that every vector

$$h_1 w_1 + h_2 w_2 + h_3 w_3 + h_4 w_4 \tag{2.7}$$

in the image of b has the property that $9h_1h_4 = 4h_2h_3$. If the $\{h_i, i = 1, 2, 3, 4\}$ are given so as to satisfy the latter condition, consider the task of showing that this vector in W is a vector in the image of b. Suppose that $h_1 = 0$. Then either h_2 or h_3 is zero, or both are zero. If h_2 is zero, one may choose $f_1 = 0$, $f_2 = 1$, $g_1 = h_3/3$, and $g_2 = h_4/2$. If $h_3 = 0$, one may choose $g_1 = 0$, $g_2 = 1$, $f_1 = h_2/3$, and $f_2 = h_4/2$. An analogous set of constructions is available when $h_4 = 0$. For the remainder of the argument, it is assumed that neither h_1 nor h_2 is zero. Accordingly, none of the coordinates $\{h_i, i = 1, 2, 3, 4\}$ is zero. Without loss, assume that $f_1 = 1$. Then, g_1 is given by $h_1/2$, g_2 is found from $h_2/3$, and f_2 is constructed from $h_3/3g_1$, which is then $2h_3/3h_1$. It is easy to check that these choices produce the correct first three coordinates; the last coordinate is $4h_3h_2/9h_1$, which by virtue of the property $9h_1h_4 = 4h_2h_3$ is equal to h_4 as desired. Thus, a vector in W is in the image of b if and only if the relation $9h_1h_4 = 4h_2h_3$ is satisfied. Next, it is shown that the vectors in this class are not closed under addition. For this purpose, simply select a pair of vectors represented by $(1, 1, 9, 4)$ and $(4, 9, 1, 1)$. The sum, $(5, 10, 10, 5)$, does not satisfy the condition.

It is perhaps not so surprising that the image of b in this example is not a subspace of W. After all, the operator b is nonlinear, when both of its arguments are considered. What may be surprising is that a natural and classical way can be used to circumvent this difficulty, at least to a remarkable degree. The mechanism that is introduced in order to address such a question is the **tensor**. The reader should bear in mind that many technical personnel have prior notions and insights on this subject emanating from areas such as the theory of mechanics and related bodies of knowledge. For these persons, the authors wish to emphasize that the following treatment is **algebraic** in character and may exhibit, at least initially, a flavor different from that to which they may be accustomed. This difference is quite typical of the distinctive points of view that often can be found between the mathematical areas of algebra and analysis. Such differences are fortunate insofar as they promote progress in understanding.

2.4 Tensor Product

The notions of tensors and tensor product, as presented in this treatment, have the intuitive meaning of a very general sort of bilinear operator, in fact, the *most* general such operator. Once again, F-vector spaces U, V, and W are assumed. Suppose that b: $U \times V \to W$ is a bilinear operator. Then the pair (b, W) is said to be a **tensor product** of U and V if two conditions are met. The first condition is that W is the smallest F-vector space that contains the image of b. Using alternative terminology this could be expressed as W being the **vector space generated by the image of** b. The term **generated** in this expression refers to the formation of all possible linear combinations of elements in the image of b. The second condition relates b to an arbitrary bilinear operator $\breve{b}$: $U \times V \to X$, in which X is another F-vector space. To be precise, the second condition states that for every such $\breve{b}$, a **linear** operator $\breve{B}$: $W \to X$ exists with the property that

$$\breve{b}(u, v) = \breve{B}(b(u, v)) \tag{2.8}$$

for all pairs (u, v) in $U \times V$. Intuitively, this means that the arbitrary bilinear operator $\breve{b}$ can be factored in terms of the given bilinear operator b, which does not depend upon $\breve{b}$, and a linear operator $\breve{B}$ which does depend upon $\breve{b}$.

The idea of the tensor product is truly remarkable. Moreover, for any bilinear operator $\breve{b}$, the induced linear operator $\breve{B}$ is unique. The latter result is easy to see. Suppose that there are two such induced linear operators, e.g., $\breve{B}_1$ and $\breve{B}_2$. It follows immediately that

$$(\breve{B}_1 - \breve{B}_2)(b(u, v)) = 0 \tag{2.9}$$

for all pairs (u, v). However, the first condition of the tensor product assures that the image of b contains a set of generators for W, and thus that $(\breve{B}_1 - \breve{B}_2)$ must in fact be the zero operator. Therefore, once the tensor product of U and V is put into place, bilinear operations are in a one-to-one correspondence with linear operations. This is the essence of the tensor idea, and a very significant way to parameterize product operations in terms of matrices. In a certain sense, then, the idea of Chapter 2 is to relate the fundamentally nonlinear product operation to the linear ideas of Chapter 1. That this is possible is, of course, classical; nonetheless, it remains a relatively novel idea for numerous workers in the applications. Intuitively what happens here is that the idea of *product* is abstracted in the bilinear operator b, with all the remaining details placed in the realm of the induced linear operator $\breve{B}$.

When a pair (b, W) satisfies the two conditions above, and is therefore a tensor product for U and V, it is customary to replace the symbol b with the more traditional symbol $\otimes$. However, in keeping with the notion that $\otimes$ represents a product and not just a general mapping it is common to write $u \otimes v$ in place of the more correct, but also more cumbersome, $\otimes(u, v)$. Along the same lines, the space W is generally denoted $U \otimes V$. Thus, a tensor product is a pair $(U \otimes V, \otimes)$. The former is called the **tensor product of** U **with** V, and $\otimes$ is loosely termed the **tensor product**. Clearly, $\otimes$ is the most general sort of product possible in the present situation because all other products can be expressed in terms of it by means of linear operators $\breve{B}$. Once again, the colloquial use of the word "product" is to be identified with the more precise algebraic notion of bilinear operation. In this way the tensor product becomes a sort of "grandfather" for all vector products.

Tensor products can be constructed for arbitrary vector spaces. They are not, however, unique. For instance, if $U \otimes V$ has finite dimension, then W obviously can be replaced by any other F-vector space of the same dimension, and $\otimes$ can be adjusted by a vector space isomorphism. Here, the term **isomorphism** denotes an invertible linear operator between the two spaces in question. It can also be said that the two tensor product spaces $U \otimes V$ and W are isomorphic to each other. Whatever the terminology chosen, the basic idea is that the two spaces are essentially the same within the axiomatic framework in use.

2.5 Basis Tensors

Attention is now focused on the case in which U and V are finite-dimensional vector spaces over the field F. Suppose that $\{u_1, u_2, \ldots, u_m\}$ is a basis for U and $\{v_1, v_2, \ldots, v_n\}$ is a basis for V. Consider the vectors

$$u_1 \otimes v_1 \quad u_1 \otimes v_2, \ldots, u_1 \otimes v_n \quad u_2 \otimes v_1, \ldots, u_m \otimes v_n \tag{2.10}$$

which can be represented in the manner $\{u_i \otimes v_j, \; i = 1, 2, \ldots, m; \; j = 1, 2, \ldots, n\}$. These vectors form a basis for the vector space $U \otimes V$. To understand the motivation for this, note that vectors in U and V, respectively, can be written uniquely in the forms

$$u = \sum_{i=1}^{m} f_i u_i \tag{2.11}$$

$$v = \sum_{j=1}^{n} g_j v_j \tag{2.12}$$

Recall that $\otimes$, which is an alternate notation for b, is a bilinear operator. It follows then that

$$u \otimes v = \sum_{i=1}^{m}\sum_{j=1}^{n} f_i g_j u_i \otimes v_j \tag{2.13}$$

which establishes that the proposed basis vectors certainly span the image of $\otimes$, and thus that they span the tensor product space $U \otimes V$. It also can be shown that the proposed set of basis vectors is linearly independent. However, in the interest of brevity for this summary exposition, the details are omitted.

From this point onward, inasmuch as the symbol $\otimes$ has replaced b, it will be convenient to use b in place of $\breve{b}$ and B in place of $\breve{B}$. It is hoped that this leads to negligible confusion. Thus, in the sequel b refers simply to a bilinear operator and B to its induced linear counterpart.

Example. Consider the bilinear form b: $R^2 \times R^3 \to R$ with action defined by

$$b(f_1, f_2, g_1, g_2, g_3) = 2f_2g_3 \tag{2.14}$$

Observe that this can be put into the more transparent form

$$\begin{bmatrix} f_1 & f_2 \end{bmatrix}\begin{bmatrix} 0 & 0 & 0 \\ 0 & 0 & 2 \end{bmatrix}\begin{bmatrix} g_1 \\ g_2 \\ g_3 \end{bmatrix} \tag{2.15}$$

which, in turn, can be written in the compact notation $u^{\mathrm{T}}Mv$. Clearly, U has dimension two, and V has dimension three. Thus, $U \otimes V$ has a basis with six elements. The operator b maps into R, which has a basis with one element. All bases are chosen to be standard. Thus, an ith basis vector contains the multiplicative field element 1 in its ith row, and the additive field element 0 in its other rows. Therefore, the matrix of B has one row and six columns. To compute the entries, it is necessary to agree upon the order of the basis elements in $R^2 \otimes R^3$. It is customary to choose the natural ordering as introduced above:

$$u_1 \otimes v_1 \quad u_1 \otimes v_2 \quad u_1 \otimes v_3 \quad u_2 \otimes v_1 \quad u_2 \otimes v_2 \quad u_2 \otimes v_3 \tag{2.16}$$

The coordinate h_1 associated with the basis vector [1] in R, considered to be a vector space, is given by

$$h_1 = b(1, 0, 1, 0, 0) = 0 \tag{2.17}$$

when u and v are given by the respective first basis vectors in R^2 and R^3, respectively:

$$u = \begin{bmatrix} 1 \\ 0 \end{bmatrix} \tag{2.18}$$

$$v = \begin{bmatrix} 1 \\ 0 \\ 0 \end{bmatrix} \tag{2.19}$$

Similarly, for the other five pairings in order, one obtains

$$h_1 = b(1,0,0,1,0) = 0 \tag{2.20}$$

$$h_1 = b(1,0,0,0,1) = 0 \tag{2.21}$$

$$h_1 = b(0,1,1,0,0) = 0 \tag{2.22}$$

$$h_1 = b(0,1,0,1,0) = 0 \tag{2.23}$$

$$h_1 = b(0,1,0,0,1) = 2 \tag{2.24}$$

in order. In view of these calculations, together with the definitions of matrices in Chapter 1, it follows that the matrix description of B: $R^2 \otimes R^3 \to R$ is given by

$$[B] = \begin{bmatrix} 0 & 0 & 0 & 0 & 0 & 2 \end{bmatrix} \tag{2.25}$$

Observe that all the numerical information concerning B has been arrayed in $[B]$. It becomes increasingly clear then that such numerical entries define all possible bilinear forms of this type.

Example. In order to generalize the preceding example, one has only to be more general in describing the matrix of M. Suppose that

$$[M] = \begin{bmatrix} m_{11} & m_{12} & m_{13} \\ m_{21} & m_{22} & m_{23} \end{bmatrix} \tag{2.26}$$

so that the bilinear operator b has action

$$b(f_1, f_2, g_1, g_2, g_3) = \begin{bmatrix} f_1 & f_2 \end{bmatrix} \begin{bmatrix} m_{11} & m_{12} & m_{13} \\ m_{21} & m_{22} & m_{23} \end{bmatrix} \begin{bmatrix} g_1 \\ g_2 \\ g_3 \end{bmatrix} \tag{2.27}$$

Thus, it is easy to determine that

$$[B] = \begin{bmatrix} m_{11} & m_{12} & m_{13} & m_{21} & m_{22} & m_{23} \end{bmatrix} \tag{2.28}$$

The two examples preceding help in visualizing the linear operator B by means of its matrix. They do not, however, contribute to the understanding of the nature of the tensor product of two vectors. For that purpose, it is appropriate to carry the examples a bit further.

Example. The foregoing example presents the representations

$$\begin{bmatrix} f_1 \\ f_2 \end{bmatrix} \tag{2.29}$$

for **u** and

$$\begin{bmatrix} g_1 \\ g_2 \\ g_3 \end{bmatrix} \tag{2.30}$$

for v. From the development of the ideas of the tensor product, it was established that $b(u,v) = B(u \otimes v)$. The construction of $u \otimes v$ proceeds according to definition in the manner

$$u \otimes v = \left(\sum_{i=1}^{2} f_i u_i\right) \otimes \left(\sum_{j=1}^{3} g_j v_j\right) \tag{2.31}$$

$$= \sum_{i=1}^{2} \sum_{j=1}^{3} f_i g_j u_i v_j \tag{2.32}$$

From this and the basis ordering chosen above, it is clear that the representation of $u \otimes v$ is given by

$$[u \otimes v] = \begin{bmatrix} f_1 g_1 & f_1 g_2 & f_1 g_3 & f_2 g_1 & f_2 g_2 & f_2 g_3 \end{bmatrix}^{\mathrm{T}} \tag{2.33}$$

The total picture for the tensor representation of $b(u,v)$, then, is

$$b(f_1, f_2, g_1, g_2, g_3)$$

$$= \begin{bmatrix} f_1 \\ f_2 \end{bmatrix}^{\mathrm{T}} \begin{bmatrix} m_{11} & m_{12} & m_{13} \\ m_{21} & m_{22} & m_{23} \end{bmatrix} \begin{bmatrix} g_1 \\ g_2 \\ g_3 \end{bmatrix} \tag{2.34}$$

$$= \begin{bmatrix} m_{11} & m_{12} & m_{13} & m_{21} & m_{22} & m_{23} \end{bmatrix} \left(\begin{bmatrix} f_1 \\ f_2 \end{bmatrix} \otimes \begin{bmatrix} g_1 \\ g_2 \\ g_3 \end{bmatrix} \right) \tag{2.35}$$

$$= \begin{bmatrix} m_{11} & m_{12} & m_{13} & m_{21} & m_{22} & m_{23} \end{bmatrix} \begin{bmatrix} f_1 g_1 \\ f_1 g_2 \\ f_1 g_3 \\ f_2 g_1 \\ f_2 g_2 \\ f_2 g_3 \end{bmatrix} \tag{2.36}$$

The reader should have no difficulty extending the notions of these examples to cases in which the dimensions of U and V differ from those used here. The extension to an X with dimension larger than 1 is similar in nature, and can be carried out row by row.

Example. Another sort of example, which is likely to be familiar to most readers, is the formation of the ordinary matrix product $m(P,Q) = PQ$ for compatible matrices P and Q over the field F. Clearly, the matrix product m is a bilinear operator. Thus, a linear operator M exists that has the property

$$m(P,Q) = M(P \otimes Q) \tag{2.37}$$

The matrix $P \otimes Q$ is known in the applications as the **Kronecker product**. If the basis vectors are chosen in the usual way, then its computation has the classical form. Thus, the Kronecker product of two matrices is seen to be the most general of all such products. Indeed, any other product, including the usual matrix product, can be found from the Kronecker product by multiplication with a matrix.

2.6 Multiple Products

It may happen that more than two vectors are multiplied together. Thus, certain famous and well-known field formulas include both crosses and dots. While the notion of multiple product is part and parcel of the concept of ring, so that no further adjustments need be made there, one must undertake the question of how these multiple products are reflected back into the tensor concept. The purpose of this section, therefore, is to sketch the major ideas concerning such questions. A basic and natural step is the introduction of a generalization of bilinear operators.

For obvious reasons, not the least of which is the finite number of characters in the alphabet, it is now necessary to modify notation so as to avoid the proliferation of symbols. With regard to the foregoing discussion, the modification, which is straightforward, is to regard a bilinear operator in the manner b: $U_1 \times U_2 \to V$ in place of the previous $U \times V \to W$. Generalizing, consider p F-vector spaces $U_i, i = 1, 2, \ldots, p$. Let m: $U_1 \times U_2 \times \cdots \times U_p \to V$ be an operator which satisfies the condition

$$\begin{aligned} m(u_1, u_2, \ldots, u_{i-1}, f_i u_i + \check{f}_i \check{u}_i, u_{i+1}, \ldots, u_p) \\ = f_i m(u_1, u_2, \ldots, u_{i-1}, u_i, u_{i+1}, \ldots, u_p) \\ + \check{f}_i m(u_1, u_2, \ldots, u_{i-1}, \check{u}_i, u_{i+1}, \ldots, u_p) \end{aligned} \tag{2.38}$$

for $i = 1, 2, \ldots, p$, for all f_i and $\check{f}_i$ in F, and for all u_i and $\check{u}_i$ in U_i. Thus, m is said to be a p-**linear** operator. Observe in this definition that when $p-1$ of the arguments of m are fixed, m becomes a linear operator in the remaining argument. Clearly, the bilinear operator is a special case of this definition, when $p = 2$. Moreover, the definition captures the intuitive concept of multiplication in a precise algebraic sense.

Next, the notion of tensor product is extended in a corresponding way. To do this, suppose that m and V satisfy two conditions. The first condition is that V is the smallest F-vector space that contains the image of m. Equivalently, V is the F-vector space generated by the image of m. Recall that the image of m is not equal to V, even in the bilinear case $p = 2$. The second condition is that

$$\check{m}(u_1, u_2, \ldots, u_p) = \check{M}(m(u_1, u_2, \ldots, u_p)), \tag{2.39}$$

where $\check{M}$: $V \to W$ is a linear operator, W is an F-vector space, and $\check{m}$: $U_1 \times U_2 \times \cdots \times U_p \to W$ is an arbitrary p-linear operator. If m satisfies these two conditions, the action of m is more traditionally written in the manner

$$m(u_1, u_2, \ldots, u_p) = u_1 \otimes u_2 \otimes \cdots \otimes u_p \tag{2.40}$$

and the space V is given the notation

$$V = U_1 \otimes U_2 \otimes \cdots \otimes U_p \tag{2.41}$$

Once again, existence of the tensor product pair (m, V) is not a problem, and the same sort of uniqueness holds, that is, up to isomorphism.

It is now possible to give a major example of the multiple product idea. The general import of this example far exceeds the interest attached to more elementary illustrations. Therefore, it is accorded its own section. The reason for this will shortly become obvious.

2.7 Determinants

The body of knowledge associated with the theory of determinants tends to occupy a separate and special part of the memory which one reserves for mathematical knowledge. This theory is encountered somewhat indirectly during matrix inversion, and thus is felt to be related to the matrix algebra. However, this association can be somewhat misleading. Multiplication in the matrix algebra is really a multiplication of linear operators, but determinants are more naturally seen in terms of multiplication of vectors. The purpose of this section is to make this idea apparent, and to suggest that a natural way to correlate this body of knowledge is with the concept of an algebra constructed upon a given F-vector space. As such, it becomes a special case of the ideas previously introduced. Fitting determinants into the larger picture is then much less of a challenge than is usually the case, which can save precious human memory.

Consider at the outset a square array of field elements from F, denoted customarily by

$$D = \begin{pmatrix} d_{11} & d_{12} & \cdots & d_{1p} \\ d_{21} & d_{22} & \cdots & d_{2p} \\ \vdots & \vdots & \ddots & \vdots \\ d_{p1} & d_{p2} & \cdots & d_{pp} \end{pmatrix} \tag{2.42}$$

The **determinant** of D will be denoted by $\det(D)$. It is assumed that all readers are comfortable with at least one of the algorithms for computing $\det(D)$. The key idea about $\det(D)$ is that it is a p-linear operator on its columns or upon its rows. In fact, two of the three classical properties of determinants are tantamount precisely to this statement. The third property, which concerns interchanging columns or rows, is also of great interest here (see below).

Without loss of generality, suppose that $\det(D)$ is regarded as a p-linear function of its columns. If the columns, in order, are denoted by $d_1, d_2, \ldots, d_p$, then it is possible to set up a p-linear operator

$$m(d_1, d_2, \ldots, d_p) = \det(D) \tag{2.43}$$

Accordingly, tensor theory indicates that this operator can be expressed in the manner

$$m(d_1, d_2, \ldots, d_p) = M\left(d_1 \otimes d_2 \otimes \cdots \otimes d_p\right) \tag{2.44}$$

It is interesting to inquire about the nature of the matrix $[M]$.

In order to calculate $[M]$, it is necessary to select bases for $U_i, i = 1, 2, \ldots, p$. In this case it is possible to identify U_i for each i with a fixed space U of dimension p. Let $\{u_1, u_2, \ldots, u_p\}$ be a basis for U and represent this basis by the standard basis vectors $\{e_1, e_2, \ldots, e_p\}$ in F^p. Moreover, select a basis for F and represent it by the multiplicative unit 1 in F. Then the elements of $[M]$ are found by calculating

$$\det\left(e_{i_1} \quad e_{i_2} \quad \cdots \quad e_{i_p} \right) \tag{2.45}$$

for all sequences $i_1 i_2 \cdots i_p$ in the increasing numerical order introduced earlier. Thus, if $p = 3$, this set of sequences is 111, 112, 113, 121, 122, 123, 131, 132, 133, 211, 212, 213, 221, 222, 223, 231, 232, 233, 311, 312, 313, 321, 322, 323, 331, 332, 333.

Example. For the case $p = 3$ described above, it is desired to calculate $[M]$. The first few calculations are given by

$$\det\begin{pmatrix} e_1 & e_1 & e_1 \end{pmatrix} = 0 \tag{2.46}$$
$$\det\begin{pmatrix} e_1 & e_1 & e_2 \end{pmatrix} = 0 \tag{2.47}$$
$$\det\begin{pmatrix} e_1 & e_1 & e_3 \end{pmatrix} = 0 \tag{2.48}$$
$$\det\begin{pmatrix} e_1 & e_2 & e_1 \end{pmatrix} = 0 \tag{2.49}$$
$$\det\begin{pmatrix} e_1 & e_2 & e_2 \end{pmatrix} = 0 \tag{2.50}$$
$$\det\begin{pmatrix} e_1 & e_2 & e_3 \end{pmatrix} = +1 \tag{2.51}$$
$$\det\begin{pmatrix} e_1 & e_3 & e_1 \end{pmatrix} = 0 \tag{2.52}$$
$$\det\begin{pmatrix} e_1 & e_3 & e_2 \end{pmatrix} = -1 \tag{2.53}$$
$$\det\begin{pmatrix} e_1 & e_3 & e_3 \end{pmatrix} = 0 \tag{2.54}$$
$$\det\begin{pmatrix} e_2 & e_1 & e_1 \end{pmatrix} = 0 \tag{2.55}$$
$$\det\begin{pmatrix} e_2 & e_1 & e_2 \end{pmatrix} = 0 \tag{2.56}$$
$$\det\begin{pmatrix} e_2 & e_1 & e_3 \end{pmatrix} = -1 \tag{2.57}$$
$$\vdots$$

Rather than provide the entire list in this form, it is easier to give the elements in the right members of the equations. Employing determinant theory, it follows that those sequences with repeated subscripts correspond to 0. Moreover, interchanging two columns changes the sign of the determinant, the third property mentioned previously. Thus, the desired results are

$$0,0,0,0,0,+1,0,-1,0,0,0,-1,0,0,0,+1,0,0,0,+1,0,-1,0,0,0,0,0 \tag{2.58}$$

Then $[M]$ is a row matrix having these numerical entries. It is 1×27.

Example. The preceding example indicates that the formation of the determinant in tensor notation results in the appearance of numerous multiplications by zero. This is inefficient. Moreover, if all the zero entries in $[M]$ are dropped, the result is a product of the form

$$\begin{bmatrix} +1 & -1 & -1 & +1 & +1 & -1 \end{bmatrix} \begin{bmatrix} d_{11}d_{22}d_{33} \\ d_{11}d_{32}d_{23} \\ d_{21}d_{12}d_{33} \\ d_{21}d_{32}d_{13} \\ d_{31}d_{12}d_{23} \\ d_{31}d_{22}d_{13} \end{bmatrix} \tag{2.59}$$

easily seen to be the standard formula for classical calculation of the determinant. In view of this result, one immediately wonders what to do about all the dropped zeros. The following section shows how to do away with all the zeros. In the process, however, more things happen than might have been anticipated; as a result, an entirely new concept appears.

2.8 Skew Symmetric Products

The determinant is an instance of skew symmetry in products. Consider a p-linear operator m: $U_1 \times U_2 \times \cdots \times U_p \rightarrow V$ with the property that each interchange of two arguments changes the sign of the result produced by m. Thus, for example,

$$\begin{aligned} m(u_1, \ldots, u_{i-1}, u_i, \ldots, u_{j-1}, u_j, \ldots, u_p) \\ = -m(u_1, \ldots, u_{i-1}, u_j, \ldots, u_{j-1}, u_i, \ldots, u_p) \end{aligned} \tag{2.60}$$

If a list of k interchanges is performed, the sign is changed k times. Such an operator is described as **skew symmetric.**

Provided that only skew symmetric multiplications are of interest, the tensor construction can be streamlined. Let (m_{skewsym}, V) be a pair consisting of a skew symmetric p-linear operator and an F-vector space V. This pair is said to constitute a **skew symmetric tensor product** for the F-vector spaces $U_1, U_2, \ldots, U_p$, if two conditions hold. The reader can probably guess what these two conditions are. Condition one is that V is the F-vector space generated by the image of m_{skewsym}. Condition two is the property that for every skew symmetric p-linear operator $\breve{m}_{\text{skewsym}}$: $U_1 \times U_2 \times \cdots \times U_p \rightarrow W$, a linear operator $\breve{M}_{\text{skewsym}}$: $V \rightarrow W$ exists having the feature

$$\breve{m}_{\text{skewsym}}(u_1, u_2, \ldots, u_p) = \breve{M}_{\text{skewsym}}(m_{\text{skewsym}}(u_1, u_2, \ldots, u_p)) \tag{2.61}$$

If these two conditions hold for the pair (m_{skewsym}, V), then the custom is to write

$$m_{\text{skewsym}}(u_1, u_2, \ldots, u_p) = u_1 \wedge u_2 \wedge \cdots \wedge u_p \tag{2.62}$$

for the action of m_{skewsym} and

$$V = U_1 \wedge U_2 \wedge \cdots \wedge U_p \tag{2.63}$$

for the product of the vector spaces involved. Once again, this skew symmetric tensor product exists, and is unique in the usual way.

Now suppose that $U_i = U, i = 1, 2, \ldots, p$, and that $\{u_1, u_2, \ldots, u_p\}$ is a basis for U. It is straightforward to show that $u_{i_1} \wedge u_{i_2} \wedge \cdots \wedge u_{i_p}$ vanishes whenever two of its arguments are equal. Without loss, assume that $u_{i_j} = u_{i_k} = u$. If u_{i_j} and u_{i_k} are switched, the sign of the product must change. However, after the switch, the argument list is identical to the previous list. Because the only number whose value is unchanged after negation is zero, the conclusion follows. Accordingly, the basis for $U_1 \wedge U_2 \wedge \cdots \wedge U_p$ is the family

$$\{u_{i_1} \wedge u_{i_2} \wedge \cdots \wedge u_{i_p}\} \tag{2.64}$$

where each $i_1 i_2 \cdots i_p$ consists of p distinct nonzero natural numbers, and where the ordinary convention is to arrange the $\{i_k\}$ so as to increase from left to right. A moment's reflection shows that only one such basis element can exist, which is $u_1 \wedge u_2 \wedge \cdots \wedge u_p$. Thus, if $p = 4$, the basis element in question is $u_1 \wedge u_2 \wedge u_3 \wedge u_4$. If we return to the example of the determinant, and regard it as a skew symmetric p-linear operator, then the representation

$$m_{\text{skewsym}}(d_1, d_2, \ldots, d_p) = M_{\text{skewsym}}(d_1 \wedge d_2 \wedge \cdots \wedge d_p) \tag{2.65}$$

is obtained. Next observe that each of the p columns of the array D can be written as a unique linear combination of the basis vectors $\{u_1, u_2, \ldots, u_p\}$ in the manner

$$d_j = \sum_{i=1}^{p} d_{ij} u_i \tag{2.66}$$

for $j = 1, 2, \ldots, p$. Then it follows that $d_1 \wedge d_2 \wedge \cdots \wedge d_p$ is given by

$$\sum_{i_1=1}^{p} \sum_{i_2=1}^{p} \cdots \sum_{i_p=1}^{p} d_{i_1 1} d_{i_2 2} \cdots d_{i_p p} u_{i_1} \wedge u_{i_2} \wedge \cdots \wedge u_{i_p} \tag{2.67}$$

which is a consequence of the fact that $\wedge$ is a p-linear operator. The only nonzero terms in this p-fold summation are those for which the indices $\{i_k, k = 1, 2, \ldots, p\}$ are distinct. The reader will correctly surmise that these terms are the building blocks of $\det(D)$. Indeed,

$$d_1 \wedge d_2 \wedge \cdots \wedge d_p = \det(D) u_1 \wedge u_2 \wedge \cdots \wedge u_p \tag{2.68}$$

and, if U is R^p with the $\{u_i\}$ chosen as standard basis elements, then

$$d_1 \wedge d_2 \wedge \cdots \wedge d_p = \det(D) \tag{2.69}$$

because $u_1 \wedge u_2 \wedge \cdots \wedge u_p$ becomes 1 in F. Moreover, it is seen from this example how the usual formula for $\det(D)$ is altered if the columns of D are representations with respect to a basis other than the standard basis. In turn, this shows how the determinant changes when the basis of a space is changed. The main idea is that it changes by a constant which is constructed from the determinant whose columns are the corresponding vectors of the alternate basis. Finally, because this new basis is given by an invertible linear transformation from the old basis, it follows that the determinant of the transformation is the relating factor.

It can now be observed that the change from a tensor product based upon $\otimes$ to a tensor product based upon $\wedge$ has indeed eliminated the zero multiplications associated with skew symmetry. However, and this could possibly be a surprise, it has reduced everything to one term, which is the coordinate relative to the singleton basis element in a tensor product space of dimension one. This may be considered almost a tautology, except for the fact that it produces a natural generalization of the determinant to arrays in which the number of rows is not equal to the number of columns. Without loss, assume that the number of columns is less than the number of rows.

Example. Consider, then, an array of field elements from F, with less columns than rows, denoted by

$$D = \begin{pmatrix} d_{11} & d_{12} & \cdots & d_{1p} \\ d_{21} & d_{22} & \cdots & d_{2p} \\ \vdots & \vdots & \ddots & \vdots \\ d_{q1} & d_{q2} & \cdots & d_{qp} \end{pmatrix} \tag{2.70}$$

where $p < q$. The apparatus introduced in this section still permits the formation of a skew symmetric p-linear operation in the manner $d_1 \wedge d_2 \wedge \cdots \wedge d_p$. This is a natural generalization in the sense that the ordinary determinant is recovered when $p = q$. Moreover, the procedure of calculation is along the same lines as before, with the representations

$$d_j = \sum_{i=1}^{q} d_{ij} u_i \tag{2.71}$$

for $j = 1, 2, \ldots, p$. Note that the upper limit on the summation has changed from p to q. The reader will observe, then, that $d_1 \wedge d_2 \wedge \cdots \wedge d_p$ can be found once again by the familiar step

$$\sum_{i_1=1}^{q} \sum_{i_2=1}^{q} \cdots \sum_{i_p=1}^{q} d_{i_1 1} d_{i_2 2} \cdots d_{i_p p} u_{i_1} \wedge u_{i_2} \wedge \cdots \wedge u_{i_p} \tag{2.72}$$

which is a consequence once again of the fact that $\wedge$ is a p-linear operator. In this case, however, there is more than one way to form nonzero products in the family

$$\{u_{i_1} \wedge u_{i_2} \wedge \cdots \wedge u_{i_p}\} \tag{2.73}$$

in which $i_1 i_2 \cdots i_p$ contains p distinct numbers, and where the traditional convention is to arrange the $\{i_k\}$ so that the numbers in each list are increasing, while the numbers which these sequences represent are also increasing. This verbiage is best illustrated quickly.

Example. Suppose that $p = 3$ and $q = 4$. Thus, the sequences $i_1 i_2 i_3$ of interest are 123, 124, 134, 234. It can be observed that each of these sequences describes a 3×3 subarray of D in which the indices are associated with rows. In a sense these subarrays lead to all the interesting 3×3 minors of D, inasmuch as all the others are either zero or negatives of these four. Some of these designated four minors could also be zero, but that would be an accident of the particular problem instead of a general feature.

Example. This example investigates in greater detail the idea of $q > p$. Suppose that $p = 2$ and $q = 3$. Further, let the given array be

$$D = \begin{pmatrix} d_{11} & d_{12} \\ d_{21} & d_{22} \\ d_{31} & d_{32} \end{pmatrix} \tag{2.74}$$

Choose the standard basis $\{e_1, e_2, e_3\}$ for $U = F^3$. Then

$$d_1 = d_{11}e_1 + d_{21}e_2 + d_{31}e_3 \tag{2.75}$$

$$d_2 = d_{12}e_1 + d_{22}e_2 + d_{32}e_3 \tag{2.76}$$

from which one computes that

$$d_1 \wedge d_2 = (d_{11}e_1 + d_{21}e_2 + d_{31}e_3) \wedge (d_{12}e_1 + d_{22}e_2 + d_{32}e_3) \tag{2.77}$$

$$\begin{aligned} &= (d_{11}d_{22} - d_{21}d_{12}) e_1 \wedge e_2 \\ &\quad + (d_{11}d_{32} - d_{31}d_{12}) e_1 \wedge e_3 \\ &\quad + (d_{21}d_{32} - d_{31}d_{22}) e_2 \wedge e_3 \end{aligned} \tag{2.78}$$

that evidently can be rewritten in the form

$$d_1 \wedge d_2 = \det\begin{pmatrix} d_{11} & d_{12} \\ d_{21} & d_{22} \end{pmatrix} e_1 \wedge e_2 + \det\begin{pmatrix} d_{11} & d_{12} \\ d_{31} & d_{32} \end{pmatrix} e_1 \wedge e_3 + \det\begin{pmatrix} d_{21} & d_{22} \\ d_{31} & d_{32} \end{pmatrix} e_2 \wedge e_3 \tag{2.79}$$

making clear the idea that the 2×2 minors of D become the coordinates of the expansion in terms of the basis $\{e_1 \wedge e_2, e_1 \wedge e_3, e_2 \wedge e_3\}$ for $R^3 \wedge R^3$.

2.9 Solving Linear Equations

An important application of the previous section is to relate the skew symmetric tensor algebra to one's intuitive idea of matrix inversion. Consider the linear equation

$$\begin{bmatrix} d_{11} & d_{12} & \cdots & d_{1p} \\ d_{21} & d_{22} & \cdots & d_{2p} \\ \vdots & \vdots & \ddots & \vdots \\ d_{p1} & d_{p2} & \cdots & d_{pp} \end{bmatrix} \begin{bmatrix} x_1 \\ x_2 \\ \vdots \\ x_p \end{bmatrix} = \begin{bmatrix} c_1 \\ c_2 \\ \vdots \\ c_p \end{bmatrix} \tag{2.80}$$

With the aid of the usual notation for columns of D, rewrite this equation in the manner

$$\sum_{i=1}^{p} x_i d_i = c \tag{2.81}$$

where c is a vector whose ith element is c_i. To solve for x_k, multiply both members of this equation by the quantity

$$d_1 \wedge d_2 \wedge \cdots \wedge d_{k-1} \wedge d_{k+1} \wedge \cdots \wedge d_p \tag{2.82}$$

which will be denoted by d_{k-}. This multiplication can be done either on the left or the right, and the vector product which is used is $\wedge$. Multiplying on the left provides the result

$$x_k d_{k-} \wedge d_k = d_{k-} \wedge c \tag{2.83}$$

Now if $\det(D)$ is not zero, then this equation solves for

$$x_k = (d_{k-} \wedge c)/(d_{k-} \wedge d_k) \tag{2.84}$$

and this is essentially Cramer's rule. Using this rule conventionally one performs enough interchanges so as to move c to the kth column of the array. If these interchanges are performed in an analogous way with regard to d_k in the denominator, the traditional form of the rule results. This approach to the result shows that the solution proceeds by selecting multiplication by a factor which annihilates all but one of the terms in the equation, where each term is concerned with one column of D.

This is simply a new way of viewing an already known result. However, the treatment of Section 2.8 suggests the possibility of extending the construction to the case in which are found more unknowns than equations. The latter procedure follows via a minor adjustment of the foregoing discussion, and thus it seems instructive to illustrate the steps by means of an example.

Example. Consider the problem corresponding to $q = 3$ and $p = 2$ and given by three equations in two unknowns as follows:

$$\begin{bmatrix} 1 & 2 \\ 3 & 4 \\ 5 & 6 \end{bmatrix} \begin{bmatrix} x_1 \\ x_2 \end{bmatrix} = \begin{bmatrix} 7 \\ 8 \\ 9 \end{bmatrix} \tag{2.85}$$

Begin by rewriting the equation in the form

$$x_1 \begin{bmatrix} 1 \\ 3 \\ 5 \end{bmatrix} + x_2 \begin{bmatrix} 2 \\ 4 \\ 6 \end{bmatrix} = \begin{bmatrix} 7 \\ 8 \\ 9 \end{bmatrix} \tag{2.86}$$

To solve for x_2, multiply from the left with $\begin{bmatrix} 1 & 3 & 5 \end{bmatrix}^{\mathrm{T}}$. This gives

$$x_2 \begin{bmatrix} 1 \\ 3 \\ 5 \end{bmatrix} \wedge \begin{bmatrix} 2 \\ 4 \\ 6 \end{bmatrix} = \begin{bmatrix} 1 \\ 3 \\ 5 \end{bmatrix} \wedge \begin{bmatrix} 7 \\ 8 \\ 9 \end{bmatrix} \tag{2.87}$$

which then implies

$$x_2 \begin{bmatrix} -2 \\ -4 \\ -2 \end{bmatrix} = \begin{bmatrix} -13 \\ -26 \\ -13 \end{bmatrix} \tag{2.88}$$

which implies that $x_2 = 13/2$. Next, consider a left multiplication by $[2\ \ 4\ \ 6]^{\mathrm{T}}$. Then

$$x_1 \begin{bmatrix} 2 \\ 4 \\ 6 \end{bmatrix} \wedge \begin{bmatrix} 1 \\ 3 \\ 5 \end{bmatrix} = \begin{bmatrix} 2 \\ 4 \\ 6 \end{bmatrix} \wedge \begin{bmatrix} 7 \\ 8 \\ 9 \end{bmatrix} \tag{2.89}$$

which then implies

$$x_1 \begin{bmatrix} 2 \\ 4 \\ 2 \end{bmatrix} = \begin{bmatrix} -12 \\ -24 \\ -12 \end{bmatrix} \tag{2.90}$$

which implies that $x_1 = -6$. It is easy to check that these values of x_1 and x_2 are the unique solution of the equation under study.

The reader is cautioned that the construction of the example above produces necessary conditions on the solution. If any of these conditions cannot be satisfied, no solution can be found. On the other hand, if solutions to the necessary conditions are found, we must check that these solutions satisfy the original equation. Space limitations prevent any further discussion of this quite fascinating point.

..10 Symmetric Products

The treatment presented in Section 2.8 has a corresponding version for this section. Consider a p-linear operator $m: U_1 \times U_2 \times \cdots \times U_p \to V$ with the property that each interchange of two arguments leaves the result produced by m unchanged. Symbolically, this is expressed by

$$\begin{aligned} &m(u_1, \ldots, u_{i-1}, u_i, \ldots, u_{j-1}, u_j, \ldots, u_p) \\ &\quad = m(u_1, \ldots, u_{i-1}, u_j, \ldots, u_{j-1}, u_i, \ldots, u_p) \end{aligned} \tag{2.91}$$

Such an operator is said to be **symmetric**.

If only symmetric multiplications are of interest, the tensor construction can once again be trimmed to fit. Let (m_{sym}, V) be a pair consisting of a symmetric p-linear operator and an F-vector space V. This pair is said to constitute a **symmetric tensor product** for the F-vector spaces $U_1, U_2, \ldots, U_p$ if two conditions hold. First, V is the F-vector space generated by the image of m_{sym}; and, second, for every symmetric p-linear operator $\breve{m}_{\text{sym}}: U_1 \times U_2 \times \cdots \times U_p \to W$, there is a linear operator $\breve{M}_{\text{sym}}: V \to W$ such that

$$\breve{m}_{\text{sym}}(u_1, u_2, \ldots, u_p) = \breve{M}_{\text{sym}}(m_{\text{sym}}(u_1, u_2, \ldots, u_p)) \tag{2.92}$$

In such a case, one writes

$$m_{\text{sym}}(u_1, u_2, \ldots, u_p) = u_1 \vee u_2 \vee \cdots \vee u_p \tag{2.93}$$

to describe the action of m_{sym} and

$$V = U_1 \vee U_2 \vee \cdots \vee U_p \tag{2.94}$$

for the symmetric tensor product of the vector spaces involved. As before, this symmetric tensor product exists and is essentially unique.

Next, let $U_i = U, i = 1, 2, \ldots, p$, and $\{u_1, u_2, \ldots, u_p\}$ be a basis for U. Because the interchange of two arguments does not change the symmetric p-linear operator, the basis elements are characterized by the family

$$\{u_{i_1} \vee u_{i_2} \vee \cdots \vee u_{i_p}\}, \tag{2.95}$$

where each $i_1 i_2 \cdots i_p$ consists of all combinations of p nonzero natural numbers, written in increasing order, and where the ordinary convention is to arrange the basis vectors so that the numbers $i_1 i_2 \cdots i_p$ increase. Unlike the skew symmetric situation, quite a few such basis vectors can, in general, exist. For instance, the first basis element is $u_1 \vee u_1 \vee \cdots \vee u_1$, with p factors, and the last one is $u_p \vee u_p \vee \cdots \vee u_p$, again with p factors.

Example. Suppose that $p = 3$ and that the dimension of U is 4. The sequences $i_1 i_2 i_3$ of interest in the representation of symmetric p-linear products are 111, 112, 113, 114, 122, 123, 124, 133, 134, 144, 222, 223, 224, 233, 234, 244, 333, 334, 344, 444. Section 2.7 showed a related example which produced 27 basis elements for the tensor product space built upon $\otimes$. In this case, it would be 64. The same situation in Section 2.8 produced four basis elements for a tensor product space constructed with $\wedge$. Twenty basis elements are found for the tensor product space produced with $\vee$. Notice that $20 + 4 \neq 64$. This means that the most general space based on $\otimes$ is not just a direct sum of those based on $\wedge$ and $\vee$.

Example. For an illustration of the symmetric product idea, choose $U = R^2$ and form a symmetric bilinear form in the arrangement of a quadratic form $u^T M u$:

$$m_{sym}(f_1, f_2) = \begin{bmatrix} f_1 \\ f_2 \end{bmatrix}^T \begin{bmatrix} m_{11} & m_{12} \\ m_{21} & m_{22} \end{bmatrix} \begin{bmatrix} f_1 \\ f_2 \end{bmatrix} \tag{2.96}$$

A word about the matrix M is in order. Because of the relationship

$$M = \tfrac{1}{2}(M + M^T) + \tfrac{1}{2}(M - M^T) \tag{2.97}$$

it is easy to see that M may be assumed to be symmetric without loss of generality, as the remaining term in this representation of $m_{sym}(f_1, f_2)$ leads to a zero contribution in the result. Thus, one is concerned with a natural candidate for the symmetric tensor mechanism. The tensor construction begins by considering

$$\begin{bmatrix} f_1 \\ f_2 \end{bmatrix} \vee \begin{bmatrix} f_1 \\ f_2 \end{bmatrix} \tag{2.98}$$

Choose a standard basis $\{e_1, e_2\}$ for R^2. Then the expression introduced above becomes

$$\sum_{i=1}^{2} \sum_{j=1}^{2} f_i f_j e_i \vee e_j \tag{2.99}$$

which becomes

$$f_1^2 e_1 \vee e_1 + 2 f_1 f_2 e_1 \vee e_2 + f_2^2 e_2 \vee e_2 \tag{2.100}$$

The result may be represented with the matrix

$$\begin{bmatrix} f_1^2 & 2f_1 f_2 & f_2^2 \end{bmatrix}^T \tag{2.101}$$

Because $p = 2$, the basis vectors of interest are seen to be $\{e_1 \vee e_1, e_1 \vee e_2, e_2 \vee e_2\}$. Inserted into the expression for m_{sym}, these produce the results $m_{11}, m_{12} = m_{21}, m_{22}$, respectively. Thus,

$$[M_{sym}] = \begin{bmatrix} m_{11} & m_{12} & m_{22} \end{bmatrix} \tag{2.102}$$

Finally, the symmetric tensor expression for m_{sym} is

$$m_{sym}(f_1, f_2) = \begin{bmatrix} m_{11} & m_{12} & m_{22} \end{bmatrix} \begin{bmatrix} f_1^2 \\ 2f_1 f_2 \\ f_2^2 \end{bmatrix} \tag{2.103}$$

Example. If M, as in the previous example, is real and symmetric, then it is known to satisfy the equation

$$ME = E\Lambda \tag{2.104}$$

where E is a matrix of eigenvectors of M, satisfying $E^{\mathrm{T}}E = I$, and Λ is a diagonal matrix of eigenvalues $\{\lambda_i\}$, which are real. Then

$$M = E\Lambda E^{\mathrm{T}} \tag{2.105}$$

and the quadratic form $u^{\mathrm{T}}Mu$ becomes

$$u^{\mathrm{T}}Mu = [E^{\mathrm{T}}u]^{\mathrm{T}}\Lambda[E^{\mathrm{T}}u] \tag{2.106}$$

$$= \sum_{i=1}^{p} \lambda_i [E^{\mathrm{T}}u]_i \tag{2.107}$$

When one considers u to be an abitrary vector in R^p, this quadratic form is non-negative for all u if and only if the $\{\lambda_i\}$ are non-negative and is positive for all nonzero u if and only if the $\{\lambda_i\}$ are positive. M is then **non-negative definite** and **positive definite**, respectively.

Example. With reference to the preceding example, it is sometimes of interest to choose $U = C^p$, where the reader is reminded that C denotes the complex numbers. In this case a similar discussion can be carried out, with M as a complex matrix. The quadratic form must be set up a bit differently, with the structure u^*Mu, in which superscript * denotes a combined transposition and conjugation. Also, without loss one can assume that $M^* = M$. A common instance of this sort of situation occurs when M is a function $M(s)$ of the Laplace variable s, which is under consideration on the axis $s = j\omega$.

.11 Summary

The basic idea of this chapter was to examine the axiomatic framework for equipping an F-vector space V with a vector multiplication, and thus develop it into an algebra. Despite the fact that vector multiplication is manifestly a nonlinear operation, it was shown that a very useful matrix theory can be developed for such multiplications. The treatment is based upon notions of algebraic tensor products from which all other multiplications can be derived. The authors showed that the determinant is nothing but a product of vectors, with a special skew symmetric character attached to it. Specializations of the tensor product to such cases, and to the analogous case of symmetric products, were discussed.

As a final remark, it should be mentioned that tensor products themselves develop into a complete algebra of their own. Although space does not permit treatment here, note that the key idea is the definition

$$(v_1 \otimes \cdots \otimes v_p) \otimes (v_{p+1} \otimes \cdots \otimes v_q) = v_1 \otimes \cdots \otimes v_q \tag{2.108}$$

eference

W. H. Greub, *Multilinear Algebra*, New York: Springer-Verlag, 1967.

3

The Laplace Transform

John R. Deller, Jr.
Michigan State University

3.1 Introduction

The Laplace transform (LT) is the cornerstone of classical circuits, systems, and control theory. Developed as a means of rendering cumbersome differential equation solutions simple algebraic problems, the engineer has transcended this original motivation and has developed an extensive toolbag of analysis and design methods based on the "s-plane." After a motivating differential equation (circuit) problem is presented, this chapter introduces the formal principles of the LT, including its properties and methods for forward and inverse computation. The transform is then applied to the analysis of circuits and systems, exploring such topics as the system function and stability analysis. Two appendices conclude the chapter, one of which relates the LT to other signal transforms to be covered in this volume.

3.2 Motivational Example

Series RLC Circuit

Let us motivate our study of the LT with a simple circuit example. Consider the series RLC circuit shown in Fig. 3.1, in which we leave the component values unspecified for the moment.

0-8493-8341-2/95/$0.00+$.50

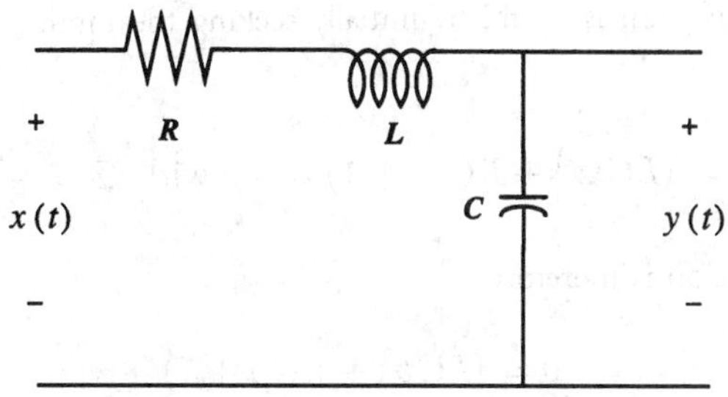

FIGURE 3.1 Series RLC circuit example.

With the input and output to the circuit taken to be the voltages x and y, respectively, the input–output dynamics of this circuit are found to be governed by a linear, constant-coefficient differential equation

$$x(t) = LC\frac{d^2y}{dt^2} + RC\frac{dy}{dt} + y(t) \tag{3.1}$$

This equation arises from a circuit example, but is typical of many second-order systems arising in mechanical, fluid, acoustic, biomedical, chemical, and other engineering models. We can, therefore, view the circuit as a "system," and the theory explored here as having broad applicability in modeling and analysis of systems.

Suppose we are asked to find the complete solution for the output voltage y, given input

$$x(t) = M_x e^{\sigma_x t} \cos(\omega_x t + \theta_x) u(t) \tag{3.2}$$

in which u denotes the **unit step function,**

$$u(t) \stackrel{\text{def}}{=} \left[0,\ t < 0 \quad 1/2,\ t = 0 \quad 1,\ t > 0\right] \tag{3.3}$$

For convenience and without loss of generality we assume that $M_x > 0$. The initial conditions [at time $t(0^-)$] on the circuit are given to be[1] $i(0^-) = i_0$ and $y(0^-) = y_0$, where y_0 and i_0 are known quantities.

Homogeneous Solution and the Natural Response

The **homogeneous** or **complementary solution** of the differential equation, say $\tilde{y}$, represents the **natural**, or **unforced**, **response** of the system. The natural response occurs because of the inequity between the initial conditions and the conditions imposed upon the system by the input at the instant it is applied. The system must adjust to these new conditions and will do so in accordance with its own physical properties (e.g., circuit values). For stable systems (see "Poles and Zeroes—Part II"), the natural response will consist of **transients**, signals which decay exponentially with time. The unforced response will always be present unless the system is stable, and either the input was applied at time $t = -\infty$ (transient will have diminished in the infinite time interval prior to time zero) or the initial conditions on the system exactly nullify the "shock" of the input so that a transient adjustment is not necessary. The form of the natural response (homogeneous solution) is not dependent on the input (except for its use in determining changes around time $t = 0$), but rather on the inherent properties of the system.

[1]The notations 0^+ and 0^- indicate limits at $t = 0$ from the right and left, respectively. That $y(0^+) = y(0^-)$, for example, indicates continuity of $y(t)$ at $t = 0$.

The homogeneous solution is found by initially seeking the input-free response which may be written as

$$0 = (LC\mathscr{D}^2 + RC\mathscr{D} + 1)y(t) \quad \text{with } \mathscr{D}^i \stackrel{\text{def}}{=} \frac{d^i}{dt^i} \tag{3.4}$$

The characteristic equation is therefore

$$0 = (LCp^2 + RCp + 1) \tag{3.5}$$

Solving for the roots, we find

$$p_1, p_2 = \frac{-RC \pm \sqrt{R^2C^2 - 4LC}}{2LC} = -\frac{R}{2L} \pm \sqrt{\frac{R^2}{4L^2} - \frac{1}{LC}} \tag{3.6}$$

In general, p_1 and p_2 can be real and unequal (overdamped case), equal and real (critically damped case), or complex conjugates (underdamped case). Except in the critically damped case (in which $R^2C^2 = 4LC$ so that two identical real roots are found), the homogeneous solution takes the form

$$\tilde{y}(t) = Ae^{p_1 t} + Be^{p_2 t} \tag{3.7}$$

with A and B to be specified at the end of the solution.

For the sake of discussion, assume the underdamped case in which the natural response will be oscillatory, corresponding to a complex-conjugate pair of roots of (3.5). In this case we have $R^2C^2 < 4LC$ and $p_2 = p_1^*$. Let us define $p_h \stackrel{\text{def}}{=} p_1$, so the two roots are p_h and p_h^*. The meaning of the subscript "h" will become clear later. Some manipulation of (3.7) will simplify our future work. We have

$$\tilde{y}(t) = Ae^{p_h t} + Be^{p_h^* t} \tag{3.8}$$

We observe that A and B must be complex conjugates if $\tilde{y}$ is to be a *real* signal. Thus, we write

$$\tilde{y}(t) = Ae^{p_h t} + A^*e^{p_h^* t} \tag{3.9}$$

After some manipulation using Euler's relation, $e^{j\alpha} = \cos(\alpha) + j\sin(\alpha)$, we have

$$\begin{aligned} \tilde{y}(t) &= 2A_{\text{re}}e^{\sigma_h t}\cos(\omega_h t) - 2A_{\text{im}}e^{\sigma_h t}\sin(\omega_h t) \\ &= 2|A|e^{\sigma_h t}\cos(\omega_h t + \theta_A) \end{aligned} \tag{3.10}$$

where $A = A_{\text{re}} + jA_{\text{im}} = |A|e^{j\theta_A}$ and $p_h = \sigma_h + j\omega_h$. The numbers A_{re} and A_{im}, or, equivalently, $|A|$ and θ_A, are to be determined later.[2] Note that the number p_h is often called the **complex frequency** associated with the damped[3] sinusoid. The complex frequency is simply a convenient mathematical way to hold the damping and frequency information in one quantity. Later in the chapter we see that p_h is also a "pole" of the system being analyzed.

[2]In fact, because A_{re} and A_{im} are unknowns, we could replace $2A_{\text{re}}$, $2A_{\text{im}}$, and $2|A|$ with some simpler notations if desired.

[3]We use the term "damping" to refer to the real part of any complex frequency, "σ", with the understanding that two other cases are actually possible: If $\sigma = 0$ the signal is undamped, and if $\sigma > 0$, the signal is exponentially increasing.

Nonhomogeneous Solution and the Forced Response

The **nonhomogeneous**, or **particular, solution**, say $\tilde{\tilde{y}}$, represents the system's **forced**, or **driven, response.** For certain prevalent types of inputs, the forced response represents an attempt to "track" the forcing function in some sense. If the natural response in a particular problem is transient and the forced response is not, then a "long time" (theoretically, $t \to \infty$) after the input is applied, only this "tracking" response will remain. In the present circuit example the forced response is uniquely present as $t \to \infty$ if $\sigma_h < 0$ (natural response exponentially decays) and $\sigma_x \geq 0$ (forcing function persists in driving the circuit for all time). Further, in the special case in which the forcing function is a periodic or constant signal [an **undamped sinusoid, undamped complex exponential**, or a **constant** (in each case $\sigma_x = 0$)] this "tracking" response, in the absence of any transients, is called the **steady-state response** of the system. Note that the forced response will be present for all $t > 0$, but may become uniquely evident only after the transient natural response dies out. Also, note that the forced response might never become evident if it is itself a transient ($\sigma_x < 0$), even though in this case the forced response will still represent an attempt to track the input.

For mathematical convenience in finding $\tilde{\tilde{y}}$, we let x be replaced by

$$\begin{aligned} x(t) &= M_x e^{\sigma_x t} e^{j(\omega_x t + \theta_x)} u(t) \\ &= \left[M_x e^{\sigma_x t} \cos(\omega_x t + \theta_x) + j M_x e^{\sigma_x t} \sin(\omega_x t + \theta_x) \right] u(t) \end{aligned} \tag{3.11}$$

Because of the linearity of the system, the real and imaginary parts of the solution $\tilde{\tilde{y}}$ will correspond to the real and imaginary parts of the complex x. Because we want the response to the real part (the cosine), we simply take the real part of the solution at the end.[4] It is extremely useful to rewrite (3.11) as

$$x(t) = M_x e^{j\theta_x} e^{(\sigma_x + j\omega_x)t} u(t) = M_x e^{j\theta_x} e^{p_x t} u(t) \tag{3.12}$$

We shall call the complex number $\bar{X} = M_x e^{j\theta_x}$ a **generalized phasor** for the sinusoid, noting that the quantity is a conventional phasor (see, e.g., [5]) when $\sigma_x = 0$ (i.e., when x represents an undamped sinusoid). The complex frequency associated with the signal x is $p_x \stackrel{\text{def}}{=} \sigma_x + j\omega_x$.

Any signal that can be written as a sum of exponentials will be an **eigensignal** of a linear, time-invariant (LTI) system such as the present circuit. The forced response of a system to an eigensignal is a scaled, time-shifted version of the eigensignal. This means that an eigensignal generally has its amplitude and phase altered by the system, but *never its frequency*! Many signals used in engineering analysis are eigensignals. This is the case with the present input x.

Because x is an eigensignal, the nonhomogeneous solution will be of the form

$$\tilde{\tilde{y}}(t) = |H(p_x)| M_x e^{\sigma_x t} e^{j(\omega_x t + \theta_x + \arg\{H(p_x)\})} = M_y e^{j\theta_y} e^{p_x t} \tag{3.13}$$

where $|H(p_x)|$ represents the amplitude scaling imposed upon a signal of complex frequency p_x, and $\arg\{H(p_x)\}$ is the phase shift. For the moment, do not be concerned about the seemingly excessive notation $|H(p_x)|$ and $\arg\{H(p_x)\}$. The number $H(p_x) = |H(p_x)| e^{j\arg\{H(p_x)\}}$, called the **eigenvalue** of the system for complex frequency p_x, is a package containing the scaling and phase change induced upon a sinusoid or exponential input of complex frequency p_x. In (3.13) we have implicitly defined $M_y \stackrel{\text{def}}{=} |H(p_x)| M_x$ and $\theta_y = \theta_x + \arg\{H(p_x)\}$, noting that

$$\bar{Y} = M_y e^{j\theta_y} = H(p_x) M_x e^{j\theta_x} = H(p_x)\bar{X} \tag{3.14}$$

is the generalized phasor for the forced response $\tilde{\tilde{y}}$.

[4] Alternatively, we could also find the response to $x^*(t)$, and average the two responses at each t.

Let us now put expressions (3.12) and (3.13) into the differential equation (3.1) [ignoring the $u(t)$ because we are seeking a solution for $t > 0$],

$$M_x e^{j\theta_x} e^{p_x t} = p_x^2 LCM_y e^{j\theta_y} e^{p_x t} + p_x RCM_y e^{j\theta_y} e^{p_x t} + M_y e^{j\theta_y} e^{p_x t} \tag{3.15}$$

Dividing through by $e^{p_x t}$ (note this critical step),

$$M_x e^{j\theta_x} = p_x^2 LCM_y e^{j\theta_y} + p_x RCM_y e^{j\theta_y} + M_y e^{j\theta_y} \tag{3.16}$$

Isolating $M_y e^{j\theta_y}$ on the left side of (3.16), we have

$$M_y e^{j\theta_y} = \frac{M_x e^{j\theta_x}}{LCp_x^2 + RCp_x + 1} \tag{3.17}$$

Because all quantities on the right are known, we can now solve for M_y and θ_y. For example (so that we can compare this result with future work), suppose we have system parameters

$$L = 2\ \text{H}, \quad C = 1\ \text{F}, \quad R = 1\ \Omega, \quad y_0 = \tfrac{1}{2}\ \text{V}, \quad i_0 = 0\ \text{A} \tag{3.18}$$

and signal parameters

$$M_x = 3\ \text{V}, \quad \sigma_x = -0.1/\text{s}, \quad \omega_x = 1\ \text{rd/s}, \quad \theta_x = \pi/4\ \text{rd} \tag{3.19}$$

Then we find $M_y e^{j\theta_y} = 2.076 e^{-j(0.519\pi)}$

Whatever the specific numbers, *let us now assume that* M_y *and* θ_y *are known.* We have

$$\tilde{\tilde{y}}(t) = M_y e^{j\theta_y} e^{(\sigma_x + j\omega_x)t} \tag{3.20}$$

Taking the real part,

$$\tilde{\tilde{y}}(t) = M_y e^{\sigma_x t} \cos(\omega_x t + \theta_y) \tag{3.21}$$

This nonhomogeneous solution is valid for $t > 0$.

Total Solution

Combining the above results, we obtain the complete solution for $t > 0$,

$$y(t) = \tilde{y}(t) + \tilde{\tilde{y}}(t) = 2|A| e^{\sigma_h t} \cos(\omega_h t + \theta_A) + M_y e^{\sigma_x t} \cos(\omega_x t + \theta_y) \tag{3.22}$$

We must apply the initial conditions to find the unknown numbers $|A|$ and θ_A. By physical considerations we know that $y(0^+) = y(0^-) = y_0$ and $i(0^+) = i(0^-) = i_0$, so

$$y(0) = y_0 = 2|A| \cos(\theta_A) + M_y \cos(\theta_y) \tag{3.23}$$

and

$$\begin{aligned} i(0) = i_0 = C \left.\frac{dy}{dt}\right|_{t=0} &= 2|A|C[\sigma_h \cos(\theta_A) - \omega_h \sin(\theta_A)] \\ &\quad + M_y C[\sigma_x \cos(\theta_y) - \omega_x \sin(\theta_y)] \end{aligned} \tag{3.24}$$

These two equations can be solved for $|A|$ and θ_A. For example, for the numbers given in (3.18) and (3.19), using (3.6) we find that $p_h = -1/4 + j\sqrt{7}/4$, and $A = 0.416\, e^{j(0.230\pi)}$. Whatever the specific numbers, *let us assume that* $|A|$ *and* θ_A *are now known.* Then, putting all the known numbers back into (3.22) gives a complete solution for $t > 0$.

Scrutinizing the Solution

The first term in the final solution, (3.22), comprises the unforced response and corresponds to the homogeneous solution of the differential equation in conjunction with the information provided by the initial conditions. Notice that this response involves only parameters dependent upon the circuit components, e.g., σ_h and ω_h, and information provided by the initial conditions, $A = |A|e^{j\theta_A}$. The latter term in (3.22) is the forced response. We reemphasize that this part of the response, which corresponds to the nonhomogeneous solution to the differential equation, is of the same form as the forcing function x and that the system has only scaled and time-shifted (as reflected in the phase angle) the input.

It is important to understand that the natural or unforced response is not altogether unrelated to the forcing function. The adjustment that the circuit must make (using its own natural modes) depends on the discrepancy at time zero between the actual initial conditions on the circuit components and the conditions the input would "like" to impose on the components as the forcing begins. Accordingly, we can identify two parts of the natural solution, one due to the initial energy storage, the other to the "shock" of the input at time zero. We can see this in the example above by reconsidering (3.23) and (3.24) and rewriting them as

$$y_0 = 2A_{\text{re}} + M_y \cos(\theta_y) \tag{3.25}$$

and

$$\frac{i_0}{C} = 2[A_{\text{re}}\sigma_h - A_{\text{im}}\omega_h] + M_y[\sigma_x \cos(\theta_y) - \omega_x \sin(\theta_y)] \tag{3.26}$$

Solving

$$A_{\text{re}} = \frac{y_0}{2} - \frac{M_y \cos(\theta_y)}{2} \tag{3.27}$$

$$A_{\text{im}} = \frac{\sigma_h y_0 - (i_0/C)}{2\omega_h} + \frac{M_y \cos(\theta_y)(\sigma_x - \sigma_h) - M_y \sin(\theta_y)\omega_x}{2\omega_h} \tag{3.28}$$

We see that each of the real and imaginary parts of the complex number A can be decomposed into a part depending on initial circuit conditions, y_0 and i_0, and a part depending on the system's interaction with the input at the initial instant. Accordingly, we may write

$$A = A_{\text{ic}} + A_{\text{input}} \tag{3.29}$$

where $A_{\text{ic}} = A_{\text{re,ic}} + jA_{\text{im,ic}}$ and $A_{\text{input}} = A_{\text{re,input}} + jA_{\text{im,input}}$. In polar form $A_{\text{ic}} = |A_{\text{ic}}|e^{j\theta_{A_{\text{ic}}}}$ and $A_{\text{input}} = |A_{\text{input}}|e^{j\theta_{A_{\text{input}}}}$. Therefore, the homogeneous solution can be decomposed into two parts

$$\begin{aligned} \tilde{y}(t) = \tilde{y}_{\text{ic}}(t) + \tilde{y}_{\text{input}}(t) &= 2|A_{\text{ic}}|e^{\sigma_h t}\cos(\omega_h t + \theta_{A_{\text{ic}}}) \\ &+ 2|A_{\text{input}}|e^{\sigma_h t}\cos(\omega_h t + \theta_{A_{\text{input}}}) \end{aligned} \tag{3.30}$$

Hence, we observe that a natural response may occur even if the initial conditions on the circuit are zero. The combined natural and forced response in this case, $\tilde{\tilde{y}}(t) + \tilde{y}_{\text{input}}(t)$, is called the **zero-state response** to indicate the state of zero energy storage in the system at time $t = 0$. On the other hand, the response to initial energy storage *only*, $\tilde{y}_{\text{ic}}$, is called the **zero-input response** for the obvious reason.

Generalizing the Phasor Concept: Onward to the Laplace Transform

To begin to understand the meaning of the LT, we reflect on the process of solving the circuit problem above. Although we could examine this solution deeply to understand the LT connections to both the natural and forced responses, it is sufficient for current purposes to examine only the forced solution.

Because the input to the system is an eigensignal in the above, we could assume that the form of $\tilde{y}$ would be identical to that of x, with modifications only to the magnitude and phase. In noting that both x and $\tilde{y}$ would be of the form $Me^{j\theta}$ it seems reasonable that the somewhat tedious nonhomogeneous solution would eventually reduce to an algebraic solution to find M_y and θ_y from M_x and θ_x. All information needed and sought is found in the phasor quantities $\bar{X}$ and $\bar{Y}$ in conjunction with the system information. The critical step which converted the differential equation solution to an algebraic one comes in (3.16) in which the superfluous term $e^{p_x t}$ is divided out of the equation.

Also observe that the ratio $H(p_x) = M_y e^{j\theta_y}/M_x e^{j\theta_x}$ depends only on system parameters and the complex frequency of the input, $p_x = \sigma_x + j\omega_x$. In fact, this ratio, when considered a function, e.g., H, of a general complex frequency, say, $s = \sigma + j\omega$, is called the **system function** for the circuit. In the present example, we see that

$$H(s) = \frac{1}{LCs^2 + RCs + 1} \tag{3.31}$$

The complex number $H(s)$, $s = \sigma + j\omega$, contains the scaling and delay (phase) information induced by the system on a signal with damping σ and frequency ω.

Let us now consider a slightly more general class of driving signals. Suppose we had begun the analysis above with a more complicated input of the form[5]

$$x(t) = \sum_{i=1}^{N} M_i e^{\sigma_i t} \cos(\omega_i t + \theta_i) \tag{3.32}$$

which, for convenience, would have been replaced by

$$x(t) = \sum_{i=1}^{N} M_i e^{j\theta_i} e^{(\sigma_i + j\omega_i)t} = \sum_{i=1}^{N} M_i e^{j\theta_i} e^{p_i t} \tag{3.33}$$

in the nonhomogeneous solution. It follows immediately from linearity that the solution could be obtained by entering each of the components in the input individually, and then combining the N solutions at the output. In each case we would clearly need to rid the analysis of the superfluous term of the form $e^{p_i t}$ by division. This information is equivalent to the form $e^{\sigma_i t}\cos(\omega_i t)$ which is known to automatically carry through to the output.

Now, recalling that $M_i e^{j\theta_i}$ is the generalized phasor for the ith component in (3.33), let us rewrite this expression as

$$x(t) = \sum_{i=1}^{N} \bar{X}(p_i) e^{p_i t} \tag{3.34}$$

where $\bar{X}(p_i) \stackrel{\text{def}}{=} M_i e^{j\theta_i}$. Expression (3.34) is similar to a Fourier series (see Chapter 4), except that here (unless all $\sigma_i = 0$) the signal is only "pseudo-periodic" in that all of its sinusoidal

[5]We omit the unit step u which appears in the input above because we are concerned only with the forced response for $t > 0$.

components may be decaying or expanding in amplitude. The generalized phasors $\bar{X}(p_i)$ are similar to Fourier series coefficients and contain all the information (amplitude and phase) necessary to obtain steady-state solutions. These phasors comprise **frequency domain** information as they contain packets of amplitude and phase information for particular complex frequencies.

With the concepts gleaned from this example, we are now prepared to introduce the LT in earnest.

3.3 Formal Developments

Definitions of the Unilateral and Bilateral Laplace Transforms

Most signals used in engineering analysis and design of circuits and filters can be modeled as a sort of limiting case of (3.34). Such a representation includes not just several complex frequency exponentials as in (3.34), but an **uncountably infinite number** of such exponentials, one for every possible value of frequency ω. Each of these exponentials is weighted by a "generalized phasor" of **infinitesimal magnitude.** The exponential at complex frequency $s = \sigma + j\omega$, for example, is weighted by phasor $\bar{X}(\sigma + j\omega)d\omega/2\pi$, where the differential $d\omega$ assures the infinitesimal magnitude and the scale factor 2π is included by convention. The uncountably infinite number of terms is "summed" by integration as follows

$$x(t) = \int_{-\infty}^{\infty} \bar{X}(\sigma + j\omega)\frac{d\omega}{2\pi}e^{(\sigma+j\omega)t} \tag{3.35}$$

The number σ in this representation is arbitrary as long as the integral exists. In fact, if the integral converges for any σ, then the integral exists for an uncountably infinite number of σ.

The complex function $\bar{X}(\sigma + j\omega)$ in (3.35) is the **Laplace transform** (LT) for the signal $x(t)$. Based on the foregoing discussion, we can interpret the LT as a complex-frequency-dependent, uncountably infinite set of "phasor densities" containing all the magnitude and phase information necessary to find forced solutions for LTI systems. We use the word "density" here to indicate that the LT at complex frequency $\sigma + j\omega$ must be multiplied by the differential $d\omega/2\pi$ to become properly analogous to a phasor. The LT, therefore has, for example, units volts per Hertz. However, we find that the LT is much more than just a phasor-like representation, providing a rich set of analysis tools with which to design and analyze systems, including unforced responses, transients, and stability.

As in the simpler examples above, the solution of differential equations will be made easier by ridding the signals of superfluous complex exponentials of form $e^{(\sigma+j\omega)t}$, that is, by working directly with the LTs. Before doing so we change variables, to put (3.35) into a more conventional form. Let s denote the general complex frequency $s \stackrel{\text{def}}{=} \sigma + j\omega$. Then

$$\boxed{x(t) = \frac{1}{j2\pi}\int_{\sigma-j\infty}^{\sigma+j\infty} X(s)e^{st}\,ds} \tag{3.36}$$

where we have dropped the bar over the LT, X. This integral, which we have interpreted as an "expansion" of the signal x in terms of an uncountably infinite set of infinitesimal generalized phasors and complex exponentials, offers a means for obtaining the signal x from the LT, X. Accordingly, (3.36) is known as the **inverse Laplace transform** (*inverse LT*). The inverse LT operation is often denoted

$$x(t) = \mathscr{L}^{-1}\{X(s)\} \tag{3.37}$$

How one would evaluate such an integral and for what values of s it would exist are issues we shall address later.

In order to rid the signal x of the superfluous factors e^{st}, we can simply compute the LT. Without any rigorous attempt to derive the transform from (3.36), it is believable that

$$\boxed{X(s) = \int_{-\infty}^{\infty} \frac{x(t)}{e^{st}}\, ds = \int_{-\infty}^{\infty} x(t)e^{-st}\, ds} \tag{3.38}$$

This is the **bilateral**, or **two-sided, Laplace transform** (BLT). The descriptor "bilateral" or "two-sided" is a reference to the fact that the signal may be nonzero in both positive and negative time. In contrast, the **unilateral**, or **one-sided, Laplace transform** (ULT) is defined as

$$\boxed{X(s) = \int_{0^-}^{\infty} \frac{x(t)}{e^{st}}\, ds = \int_{0^-}^{\infty} x(t)e^{-st}\, ds} \tag{3.39}$$

When a signal is zero for all $t < 0$, the ULT and BLT are identical. The same inverse LT, (3.36), is applied in either case, with the understanding that the resulting time signal is zero by assumption in the ULT case. While the BLT can be used to treat a more general class of signals, we find that the ULT has the advantage of allowing us to find the component of the natural response due to nonzero initial conditions. In other words, ULT is used to analyze signals that "start" somewhere, a time we conventionally call[6] $t = 0$.

These transformations are reminiscent of the process of dividing through by the complex exponential which was first encountered in the forced solution in the motivating circuit problem [see (3.16)].

Existence of the Laplace Transform

The variable $s = \sigma + j\omega$ is a complex variable over which the LT is calculated. The complex plane with σ along the abscissa and $j\omega$ on the ordinate, is called the **s-plane**. We find some powerful tools centered on the s-plane below. Note that the s-plane is *not* the LT, nor can the LT be "plotted" in the s-plane. The LT is a complex function of the complex variable s, and a plot of the LT would require another two dimensions "over" the s-plane. For this reason, we need to place some constraints on either s or $X(s)$ or both to create a plot. For example, we could use the LT to plot $|X(j\omega)|$ as a function of ω, by evaluating the magnitude of $X(s)$ along the $j\omega$ axis in the s-plane.[7] An illustration of these points is found in Fig. 3.2.

We now address the question: For what values of s (i.e., "where" in the s-plane) does the LT exist? Consider first a two-sided (in time) signal, x. x is assumed piecewise continuous in every finite interval of the real line. We assert that the BLT X ordinarily exists in the s-plane in a strip of the form

$$\sigma_+ < \text{Re}\{s\} = \sigma < \sigma_- \tag{3.40}$$

as illustrated in Fig. 3.3. In special cases the BLT may converge in the half-plane $\sigma_+ < \sigma$, or the half-plane $\sigma < \sigma_-$, or even in the entire s-plane.

[6]Note that if we apply the ULT to a signal $x(t)$ that "starts" in negative time, the result will be identical to that for the signal $x(t)u(t)$.

[7]This particular plot is equivalent to the magnitude spectrum of the signal that could be obtained using Fourier techniques discussed in Chapter 4.

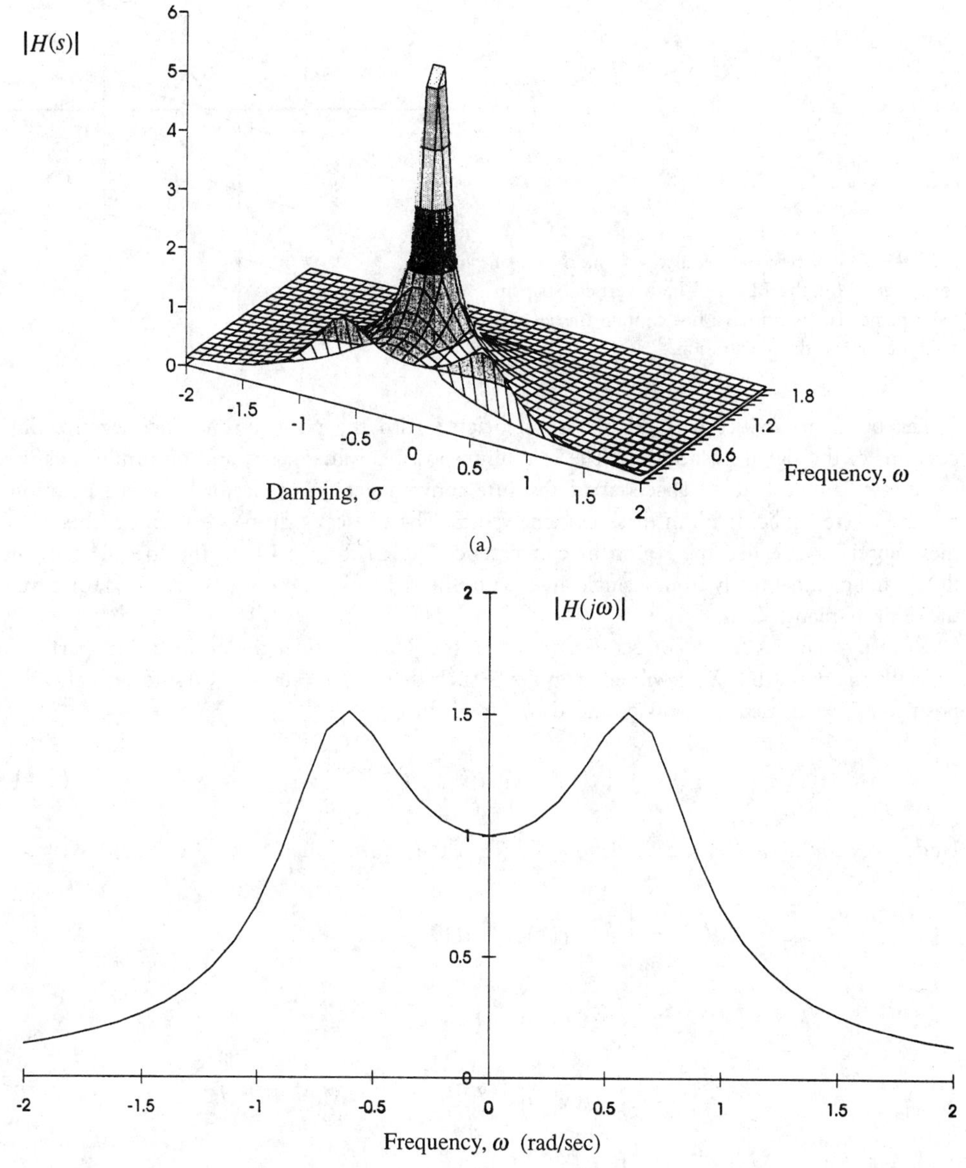

FIGURE 3.2 The LT is called "H" rather than "X" in this figure for a reason to be discovered later. (a) A plot of $|H(s)|$ vs. s for the LT $H(s) = (0.5)/(s^2 + 0.5s + 0.5)$. Only the upper-half s-plane is shown ($\omega \geq 0$). Note that the peak occurs near the value $p_h = -1/4 + j\sqrt{7}/4$, a root of the denominator of $H(s)$ which we shall later call a **pole** of the LT. The LT is theoretically infinite at $s = p_h$. (b) Evaluation of $|H(s)|$ along the $j\omega$ axis (corresponding to $\sigma = 0$) with the magnitude plotted as a function of ω.

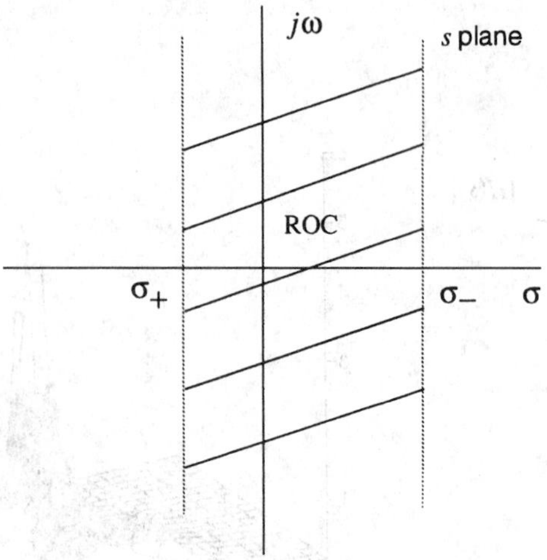

FIGURE 3.3 Except in special cases, the region of convergence for the BLT will be a vertical strip in the s-plane. This strip need **not** contain the $j\omega$ axis as is the case in this illustration.

The boundary values σ_+ and σ_- are associated with the positive-time and negative-time portions of the signal, respectively. The minimum possible value of σ_+, and maximum possible value of σ_-, are called the **abscissas of absolute convergence**. We henceforth use the notations σ_+ and σ_- to explicitly mean these extreme values. The vertical strip between, but exclusive of, these abscissas is called the **region of convergence** (ROC) for the LT. In the special cases the ROC extends indefinitely from a single abscissa to the right ($\sigma_+ < \sigma$) or left ($\sigma < \sigma_-$), or covers the entire s-plane.

To verify that (3.40) is a correct description of the ROC, consider the positive-time part of x first. We maintain that $X(s)$ will exist on any s such that $\text{Re}\{s\} = \sigma > \sigma_+$, if and only if (iff) a positive M exists such that x is bounded by $Me^{\sigma_+ t}$ on $t \geq 0$,

$$|x(t)| < Me^{\sigma_+ t} \quad t \geq 0 \tag{3.41}$$

Under this condition (letting X_+ denote the LT of the non-negative-time part of x),

$$\begin{aligned} |X_+(s)| &= \left| \int_0^\infty x(t)e^{-st}\,dt \right| \\ &\leq \int_0^\infty \left| x(t)e^{-st} \right| dt \\ &= \int_0^\infty |x(t)|\, e^{-\sigma t}\,dt < \int_0^\infty Me^{(\sigma_+ - \sigma)t}\,dt \\ &= \begin{cases} \dfrac{M}{\sigma - \sigma_+}, & \sigma > \sigma_+ \\ \infty, & \text{otherwise} \end{cases} \end{aligned} \tag{3.42}$$

If there is no finite, positive M such that (3.39) holds, then the signal grows faster than $e^{-\sigma_+ t}$ and the LT integral (area under $x(t)e^{-\sigma_+ t}e^{j\omega t}$) will not converge for at least some values of s in the neighborhood of the vertical line $s = \sigma_+$. In this case σ_+ is not a proper abscissa.

By similar means, we can argue that the negative-time part of x must be bounded as

$$|x(t)| < Me^{\sigma_- t} \qquad t < 0 \tag{3.43}$$

for some positive M, if the corresponding part of the LT, say, X_-, is to converge at s such that $\mathrm{Re}\{s\} = \sigma < \sigma_-$. Similarly to (3.42) it is shown that

$$|X_-(s)| = \left| \int_{-\infty}^{0} x(t)e^{-st}\, dt \right| < \int_{-\infty}^{0} Me^{(\sigma_- - \sigma)t}\, dt$$
$$= \begin{cases} \dfrac{M}{\sigma_- - \sigma}, & \sigma < \sigma_- \\ \infty, & \text{otherwise} \end{cases} \tag{3.44}$$

and X_- will not converge in some left neighborhood of $s = \sigma_-$ if the condition (3.43) is not met.

It should be clear from the discussion of the BLT that a ULT will ordinarily converge in the half-plane $\mathrm{Re}\{s\} = \sigma > \sigma_+$ iff a positive M exists such (3.41) is met. This follows immediately from the fact that the ULT of $x(t)$ is equivalent to the BLT of $x(t)u(t)$. The "negative-time" part of $x(t)$ yields $X_-(s) = 0$ which converges everywhere in the s-plane. The ULT may also converge everywhere in the s-plane in special cases.

Example of Laplace Transform Computations and Table of Unilateral Laplace Transforms

A listing of commonly used ULTs with ROCs is given in Table 3.1. Each entry in the table can be verified by direct computation of the appropriate LT integral, or, in many cases, properties of the LT can be exploited to make the computation easier. These properties are developed in the section "Properties of the Laplace Transform."

It is rare to find a table of BLTs in engineering material because most of our work is done with the ULT. However, BLTs can often be found by summing the results of two ULTs in the following way. Suppose x is written as

$$x(t) = x_+(t) + x_-(t) \tag{3.45}$$

where x_+ and x_- are the causal and anticausal parts of the signal, respectively. To obtain X_+, we can use a ULT table. To obtain X_-, note the following easily demonstrated property of the BLT: If $\mathscr{L}\{y(t)\} = Y(s)$ with ROC $\{s\colon \mathrm{Re}\{s\} > a\}$, then $\mathscr{L}\{y(-t)\} = Y(-s)$ with ROC $\{s : \mathrm{Re}\{s\} < -a\}$. Therefore, we can find $\mathscr{L}\{x_-(-t)\}$ in a ULT table, and replace the argument s by $-s$ to obtain $X_-(s)$. Then $X(s) = X_+(s) + X_-(s)$. This strategy is illustrated in Example 1. The ROC of the sum will ordinarily be the intersection of the individual LTs X_+ and X_-, but the total ROC may be larger than this if a pole-zero cancellation occurs (see section following).

Let us consider some examples which illustrate the direct forward computation of the LT and the process discussed in the preceding paragraph.

Example 1. Find the BLT and ULT for the signal

$$x(t) = Ae^{at}u(-t) + Be^{bt}u(t) \qquad |A|, |B| < \infty \tag{3.46}$$

Note that $A, B, a,$ and b may be complex.

TABLE 3.1 Table of Unilateral Laplace Transform Pairs

Signal, $x(t)$	ULT, $X(s)$	ROC
$\delta(t)$	1	Entire s-plane
$\frac{d^k\delta}{dt^k}$	s^k	Entire s-plane
$u(t)$	$\frac{1}{s}$	$\{s : \sigma > 0\}$
$t^{n-1}u(t)$	$\frac{(n-1)!}{s^n}$	$\{s : \sigma > 0\}$
$e^{p_x t}u(t)$	$\frac{1}{(s-p_x)}$	$\{s : \sigma > \sigma_x\}$
$t^n e^{p_x t}u(t)$	$\frac{n!}{(s-p_x)^{n+1}}$	$\{s : \sigma > \sigma_x\}$
$\cos(\omega_x t + \theta_x)u(t)$	$\frac{s\cos(\theta_x) - \omega_x \sin(\theta_x)}{s^2 + \omega_x^2}$	$\{s : \sigma > 0\}$
$e^{\sigma_x t}\cos(\omega_x t + \theta_x)u(t)$	$\frac{(s-\sigma_x)\cos(\theta_x) - \omega_x \sin(\theta_x)}{(s-\sigma_x)^2 + \omega_x^2}$	$\{s : \sigma > \sigma_x\}$
$M_x t^{n-1} e^{\sigma_x t}\cos(\omega_x t + \theta_x)u(t)$	$\frac{E}{(s-p_x)^n} + \frac{E^*}{(s-p_x^*)^n}$; M_x real and positive, $E = \frac{M_x}{2}(n-1)!\,e^{j\theta_x}$	$\{s : \sigma > \sigma_x\}$

Note: In each entry, s is the general complex number $\sigma + j\omega$, and, where relevant, p_x is the specific complex number $\sigma_x + j\omega_x$.

Solution. In the bilateral case, we have

$$X(s) = \int_{-\infty}^{0^-} Ae^{at}e^{-st}\,dt + \int_{0^-}^{\infty} Be^{bt}e^{-st}\,dt$$

$$= \left.\frac{Ae^{(a-s)t}}{a-s}\right|_{-\infty}^{0^-} + \left.\frac{Be^{(b-s)t}}{a-s}\right|_{0^-}^{\infty}$$

$$= \begin{bmatrix} -\dfrac{A}{s-a}, & \mathrm{Re}\{s\} < \mathrm{Re}\{a\} \\ \infty, & \text{otherwise} \end{bmatrix} \tag{3.47}$$

$$+ \begin{bmatrix} \dfrac{B}{s-b}, & \mathrm{Re}\{s\} > \mathrm{Re}\{b\} \\ \infty, & \text{otherwise} \end{bmatrix}$$

The ROC for this LT is $\{s : \mathrm{Re}\{b\} < \mathrm{Re}\{s\} < \mathrm{Re}\{a\}\}$. The LT does not exist for any s for which $\mathrm{Re}\{s\} \geq \mathrm{Re}\{a\}$ or $\mathrm{Re}\{s\} \leq \mathrm{Re}\{b\}$. Note also that when $\mathrm{Re}\{b\} \geq \mathrm{Re}\{a\}$, then no ROC can be found, meaning that the BLT does not exist anywhere for the signal.

The ULT follows immediately from the work above. We have

$$X(s) = \int_{0^-}^{\infty} Be^{bt}e^{-st}dt = \begin{cases} \dfrac{B}{s-b}, & \mathrm{Re}\{s\} > \mathrm{Re}\{b\} \\ \infty, & \text{otherwise} \end{cases} \tag{3.48}$$

The ROC in this case is $\{s : \text{Re}\{s\} > \text{Re}\{b\}\}$. We need not be concerned about the negative-time part of the signal (and the associated ROC) because the LT effectively zeroes the signal on $t < 0$.

Note. The result of this example is worth committing to memory as it will reappear frequently.

Note that if we had found the ULT, (3.48), for the causal part of x in a table [call it $X_+(s)$], then we could employ the trick suggested above to find the LT for the anticausal part. Let x_- denote the negative-time part: $x_-(t) = Ae^{at}u(-t)$. We know that the LT of $x_-(-t) = Ae^{-at}u(t)$ (a causal signal) is

$$X_-(-s) = \frac{A}{s+a}, \quad \text{with ROC } \{s : \text{Re}\{s\} > -a\} \tag{3.49}$$

Therefore,

$$X_-(s) = \frac{-A}{s-a}, \quad \text{with ROC } \{s : \text{Re}\{s\} < a\} \tag{3.50}$$

The overall BLT result is then $X(s) = X_+(s) + X_-(s)$ with ROC equal to the intersection of the individual results. This is consistent with the BLT found by direct integration. □

The simple example above suggests that the BLT can treat a broader class of signals at the expense of greater required care in locating its ROC. A further and related complication of the BLT is the nonuniqueness of the transform with respect to the time signals. Consider the following example.

Example 2. Find the BLT for the following signals:

$$x_1(t) = e^{bt}u(t) \quad \text{and} \quad x_2(t) = -e^{bt}u(-t) \tag{3.51}$$

Solution. From our work in Example 1, we find immediately that

$$\begin{aligned} X_1(s) &= \frac{1}{s-b}, \quad \text{Re}\{s\} > \text{Re}\{b\} \quad \text{and} \\ X_2(s) &= \frac{1}{s-b}, \quad \text{Re}\{s\} < \text{Re}\{b\} \end{aligned} \tag{3.52}$$

Neither X_1 nor X_2 can be unambiguously associated with a time signal without knowledge of its ROC. □

Another drawback of the BLT is its inability to handle initial conditions in problems like the one that motivated our discussion. For this reason, and also because signals tend to be **causal** (occurring only in positive time) in engineering problems, the ULT is more widely used and we shall focus on it exclusively after treating one more important topic in the following section. Before moving to the next section, let us tackle a few more example computations.

Example 3. Find the ULT of the impulse function, $\delta(t)$ (see Section 3.6, Appendix A).

Solution.

$$\Delta(s) = \int_{0^-}^{\infty} \delta(t)e^{-st}\,dt = 1 \quad \text{for all } s \tag{3.53}$$

The LT converges everywhere in the s-plane. We note that the lower limit 0^- is important here to yield the answer 1 (which will provide consistency of the theory) instead of $1/2$. □

Example 4. Find the ULT of the unit step function, $u(t)$ [see (3.3)].

Solution.

$$U(s) = \int_{0^-}^{\infty} 1e^{-st}\,dt = \left.\frac{-e^{-st}}{s}\right|_{0^-}^{\infty} = \frac{1}{s} \quad \text{for } \mathrm{Re}\{s\} > 0 \tag{3.54}$$

The ROC for this transform consists of the entire right-half s-plane exclusive of the $j\omega$ axis. □

Example 5. Find the ULT of the damped ($\sigma_x < 0$), undamped ($\sigma_x = 0$), or expanding ($\sigma_x > 0$) sinusoid, $x(t) = M_x e^{\sigma_x t}\cos(\omega_x t + \theta_x)\,u(t)$.

Solution. Using Euler's relation, write x as

$$x(t) = \frac{M_x}{2} e^{\sigma_x t}\left[e^{j(\omega_x t+\theta_x)} + e^{-j(\omega_x t+\theta_x)}\right] = \frac{M_x}{2}\left[e^{j\theta_x}e^{p_x t} + e^{-j\theta_x}e^{p_x^* t}\right] \tag{3.55}$$

with $p_x \stackrel{\text{def}}{=} \sigma_x + j\omega_x$. Taking the LT,

$$\begin{aligned} X(s) &= \frac{M_x}{2}\int_{0^-}^{\infty}\left[e^{j\theta_x}e^{p_x t} + e^{-j\theta_x}e^{p_x^* t}\right]e^{-st}\,dt \\ &= \int_{0^-}^{\infty}\frac{M_x}{2}e^{j\theta_x}e^{p_x t}e^{-st}\,dt + \int_{0^-}^{\infty}\frac{M_x}{2}e^{-j\theta_x}e^{p_x^* t}e^{-st}\,dt \end{aligned} \tag{3.56}$$

Now using (3.48) on each of the integrals

$$X(s) = \frac{(M_x/2)e^{j\theta_x}}{(s-p_x)} + \frac{(M_x/2)e^{-j\theta_x}}{(s-p_x^*)} \tag{3.57}$$

with the ROC associated with each of the terms being $\mathrm{Re}\{s\} > \mathrm{Re}\{p_x\} = \sigma_x$. Putting the fractions over a common denominator yields

$$\begin{aligned} X(s) &= \frac{M_x}{2}\left[\frac{(s-p_x^*)e^{j\theta_x} + (s-p_x)e^{-j\theta_x}}{(s-p_x)(s-p_x^*)}\right] \\ &= \frac{M_x}{2}\left[\frac{se^{j\theta_x} + se^{-j\theta_x} - p_x^* e^{j\theta_x} - p_x e^{-j\theta_x}}{s^2 - 2\,\mathrm{Re}\{p_x\}s + |p_x|^2}\right] \\ &= M_x\left[\frac{(s-\sigma_x)\cos(\theta_x) - \omega_x\sin(\theta_x)}{(s-\sigma_x)^2 + \omega_x^2}\right] \end{aligned} \tag{3.58}$$

The ROC of X is $\{s : \mathrm{Re}\{s\} > \sigma_x\}$.

Note. The chain of denominators in (3.58) is worth noting because these relations occur frequently in LT work:

$$\boxed{(s-p_x)(s-p_x^*) = s^2 - 2\,\mathrm{Re}\{p_x\}s + |p_x|^2 = s^2 - 2\sigma_x s + |p_x|^2 = (s-\sigma_x)^2 + \omega_x^2} \tag{3.59}$$

□

Poles and Zeroes—Part I

"Pole-zero" analysis is among the most important uses of the LT in circuit and system design and analysis. We need to take a brief look at some elementary theory of functions of complex variables in order to carefully describe the meaning of a pole or zero. When we study methods of inverting the LT in a future section, this side trip will prove to have been especially useful.

Let us begin with a general function, F, of a complex variable s. We stress that $F(s)$ may or may not be a LT. F is said to be **analytic** at $s = a$ if it is differentiable at a and in a neighborhood of a. For example, $F(s) = s - 1$ is analytic everywhere (or **entire**), but $G(s) = |s|$ is nowhere analytic because its derivative exists only at $s = 0$. On the other hand, a point p is an **isolated singular point** of F if the derivative of F does not exist at p, but F is analytic in a neighborhood of p. The function $F(s) = e^{-s}/(s-1)$ has a singular point at $s = 1$. There is a circular analytic domain around any singular point, p, of F, say $\{s : |s - p| < \rho\}$, in which the function F can be represented by a **Laurent series** [3],

$$F(s) = \sum_{i=0}^{\infty} q_{i,p}(s-p)^i + \sum_{i=1}^{\infty} \frac{r_{i,p}}{(s-p)^i} \tag{3.60}$$

The second sum in (3.60) is called the **principle part** of the function F at p. When the principle part of F at p contains terms up to order n, the isolated singular point p is called an ***n*****th-order pole** of F. Evidently from (3.60), F tends to infinity at a pole and the order of infinity is n. For future reference, we note that the complex number $r_p \stackrel{\text{def}}{=} r_{1,p}$ is called the **residue** of F at $s = p$.

A **zero** of F is more simply defined as a value of s, say z, at which F is analytic and for which $F(z) = 0$. If all the derivatives up to the $(m-1)$st are also zero at z, but the mth is nonzero, then z is called an mth-order zero of F. It can be shown that the zeroes of an analytic function F are **isolated**, except in the trivial case $F(s) = 0$ for all s [3].

Most LTs encountered in signal and system problems are quotients of polynomials in s, say

$$X(s) = \frac{N(s)}{D(s)} \tag{3.61}$$

because of the signals employed in engineering work, and because (as we shall see later) rational LTs are naturally associated with LTI systems. N and D connote "numerator" and "denominator." In this case both N and D are analytic everywhere in the s-plane, and the poles and zeroes of X are easily found by factoring the polynomials N and D, to express X in the form

$$X(s) = C\left(\prod_{i=1}^{n_N}(s-z_i) \Big/ \prod_{i=1}^{n_D}(s-p_i)\right) \tag{3.62}$$

where n_N is the number of simple factors in $N(s)$ (order of N in s), n_D the number of simple factors in $D(s)$ (order of D in s), and C is a constant. X is called a **proper rational LT** if $n_D > n_N$. After canceling all factors common to the numerator and denominator, if m terms $(s - z)$ are left in the numerator, then X has an mth order zero at $s = z$. Similarly, if n terms $(s - p)$ are left in the denominator, then X has an nth order pole at $s = p$.

Although the LT does not exist outside the ROC, all of the poles will occur at values of s outside the ROC. None, some, or all of the zeroes may also occur outside the ROC. This does not mean that the LT is valid outside the ROC, but that its poles and zeroes may occur there. A pole is ordinarily indicated in the s-plane by the symbol $\times$; whereas a zero is marked with a small circle ○.

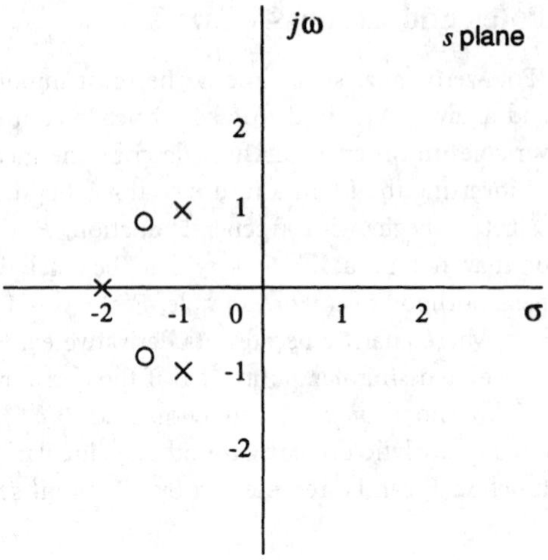

FIGURE 3.4 Pole-zero diagram for Example 6.

Example 6. Find the poles and zeroes of the LT

$$X(s) = \frac{3s^2 + 9s + 9}{(s+2)(s^2 + 2s + 2)} \tag{3.63}$$

Solution. Factoring the top and bottom polynomials to put X in form (3.62), we have

$$X(s) = 3\frac{\left(s + (3 - j\sqrt{3})/2\right)\left(s + (3 + j\sqrt{3})/2\right)}{(s+2)(s+1+j)(s+1-j)} \tag{3.64}$$

There are first-order zeroes at $s = (-3 + j\sqrt{3})/2$ and $s = (-3 - j\sqrt{3})/2$, and first-order poles at $s = -2$, $s = -1 + j$, and $s = -1 - j$. The pole-zero diagram appears in Fig. 3.4.

Two points are worth noting. First, complex poles and zeroes will always occur in conjugate pairs, as they have here, if the LT corresponds to a *real signal.* Second, the denominator of (3.64) can also be expressed as $(s+2)[(s+1)^2 + 1]$ [recall (3.59)]. Comparing this form with the LT obtained in Example 5 suggests that the latter form might prove useful. □

The purpose of introducing poles and zeroes at this point in our discussion is to note the relationship of these singularities to the ROC. The preceding examples illustrate the following facts:

1. For a "right-sided" (non-negative-time only) signal, x, the ROC of LT X (either ULT or BLT) is $\{s : \text{Re}\{s\} > \text{Re}\{p_+\} = \sigma_+\}$, where p_+ is the pole of X with maximum real part, namely, σ_+. If X has no poles, then the ROC is the entire s-plane.
2. For a "left-sided" (negative-time only) signal, x, the ROC of BLT X is $\{s : \text{Re}\{s\} < \text{Re}\{p_-\} = \sigma_-\}$, where p_- is the pole of X with minimum real part, namely, σ_-. If X has no poles, then the ROC is the entire s-plane.
3. For a "two-sided" signal x, the ROC of the BLT X is $\{s : \text{Re}\{p_+\} = \sigma_+ < \text{Re}\{s\} < \text{Re}\{p_-\} = \sigma_-\}$ where p_+ is the pole of maximum real part associated with the right-sided part of x, and p_- is the pole of minimum real part associated with the left-sided part of x. If the right-sided signal has no pole, then the ROC extends indefinitely to the right in the s-plane. If the left-sided signal has no pole, then the ROC extends indefinitely to the

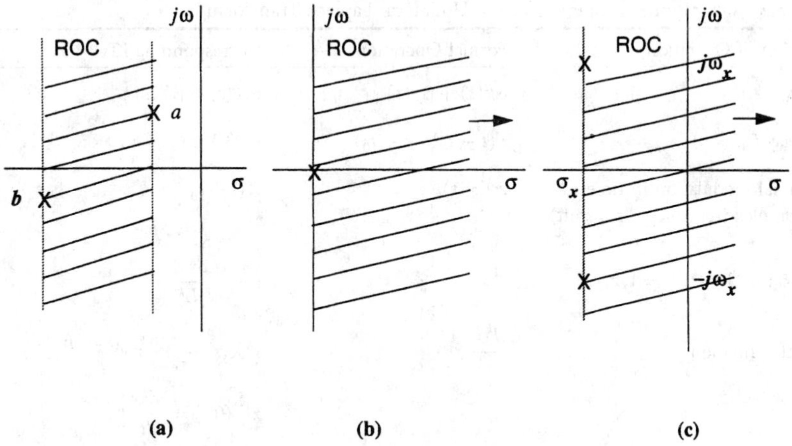

FIGURE 3.5 Pole-zero plots and ROCs for the LTs of (a) Example 1, (b) Example 4, (c) Example 5 above.

left in the s-plane. Therefore, if neither part of the signal has a pole, then the ROC is the entire s-plane.

Let us revisit three of the examples above to verify these claims. In Example 1 we found the ROC for the BLT to be $\{s : \mathrm{Re}\{b\} < \mathrm{Re}\{s\} < \mathrm{Re}\{a\}\}$. The only pole associated with the left-sided sequence is at $s = a$. The only pole associated with the right-sided signal occurs at $s = b$. Following rule 3 in the list above yields exactly the ROC determined by analytical means. The poles of X as well as the ROC are shown in Fig. 3.5(a).

In Example 4 we found the ROC to be the entire right-half s-plane, exclusive of the $j\omega$ axis. The single pole of $U(s) = 1/s$ occurs at $s = 0$. Figure 3.5(b) is consistent with rule 1 above.

The ULT of Example 5 has poles at $s = \sigma_x \pm j\omega_x$ and a zero at $s = \sigma_x + \omega_x \tan(\theta_x)$. Rule 1 therefore specifies that the ROC should be $\{s : \mathrm{Re}\{s\} > \mathrm{Re}\{\sigma_x \pm j\omega_x\} = \sigma_x\}$, which is consistent with the solution to Example 5. The pole-zero plot and ROC are illustrated in Fig. 3.5(c).

Properties of the Laplace Transform[8]

This section considers some properties of the LT which are useful in computing forward and inverse LTs, and in other manipulations occurring in signal and system design and analysis. A list of properties appears in Table 3.2.

In most cases the verification of these properties follows in a straightforward manner from the definition of the transform. Consider the following examples. For convenience, we define the notation

$$x(t) \longleftrightarrow X(s) \tag{3.65}$$

to mean that x and X are a LT pair, $X(s) = \mathscr{L}\{x(t)\}$.

Example 7. Verify the *modulation* property of the LT which states that if $x(t) \longleftrightarrow X(s)$, then $e^{s_0 t}x(t) \longleftrightarrow X(s - s_0)$.

[8]Henceforth this study restricts attention to the ULT and uses the acronym "LT" only.

TABLE 3.2 Operational Properties of the Unilateral Laplace Transform

Description of Operation	Formal Operation	Corresponding LT
Linearity	$\alpha x(t) + \beta y(t)$	$\alpha X(s) + \beta Y(s)$
Time delay ($t_0 > 0$)	$x(t - t_0)u(t - t_0)$	$e^{-st_0}X(s)$
Exponential modulation in time (or complex frequency ("s") shift)	$e^{s_0 t}x(t)$	$X(s - s_0)$
Multiplication by t^k, $k = 1, 2, \ldots$	$t^k x(t)$	$(-1)^k \dfrac{d^k X}{ds^k}$
Time differentiation	$\dfrac{d^k x}{dt^k}$	$s^k X(s) - \sum_{i=0}^{k-1} s^i x^{(k-1-i)}(0^-)$ $x^{(i)}(0^-) \stackrel{\text{def}}{=} \left.\dfrac{d^i x}{dt^i}\right\|_{t=0^-}$
Time integration	$\int_{-\infty}^{t} x(\lambda)\,d\lambda$	$\dfrac{X(s)}{s} + \dfrac{x^{(-1)}(0^-)}{s}$ $x^{(-1)}(0^-) \stackrel{\text{def}}{=} \left.\int_{-\infty}^{t} x(\lambda)\,d\lambda\right\|_{t=0^-}$
Convolution	$\int_0^\infty x(\lambda)y(t-\lambda)\,d\lambda$	$X(s)Y(s)$
Correlation	$\int_0^\infty x(t)y(t-\tau)\,dt$	$X(s)Y(-s)$
Product (s-domain convolution)	$x(t)y(t)$	$\dfrac{1}{j2\pi}\int_{\sigma-j\infty}^{\sigma+j\infty} X(\lambda)Y(s-\lambda)\,d\lambda$
Initial signal value (if time limit exists)	$\lim_{t\to 0^+} x(t)$	$\lim_{s\to\infty} sX(s)$
Final signal value (if time limit exists)	$\lim_{t\to\infty} x(t)$	$\lim_{s\to 0} sX(s)$
Time scaling	$x(\alpha t), \quad \alpha > 0$	$\dfrac{1}{\alpha}X\left(\dfrac{s}{\alpha}\right)$
Periodicity (period T)	$\sum_{i=0}^{\infty} x(t - iT)$ $x(t) = 0, \quad t \notin [0, T)$	$\dfrac{X(s)}{(1 - e^{-sT})}$

Note: Throughout, X and Y are LTs of signals x and y, respectively. x and y are causal signals.

Solution. By definition,

$$\begin{aligned} \mathscr{L}\{e^{s_0 t}x(t)\} &= \int_{0^-}^{\infty} e^{s_0 t}x(t)e^{-st}\,dt \\ &= \int_{0^-}^{\infty} x(t)e^{-(s-s_0)t}\,dt = X(s - s_0) \end{aligned} \tag{3.66}$$

□

Example 8. Verify the periodicity property and find the LT for a square wave of period $T = 2$ and duty cycle $1/2$.

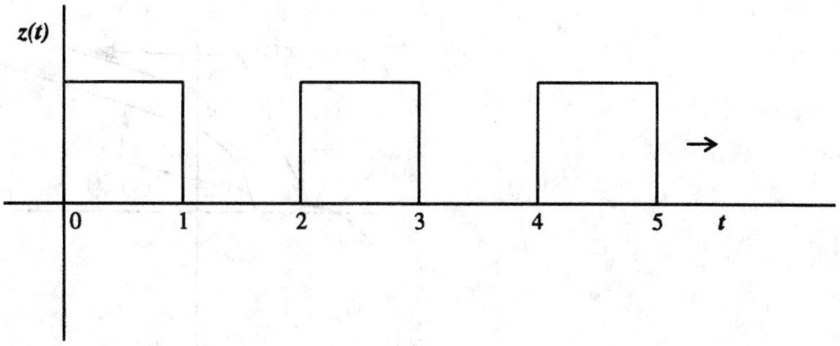

FIGURE 3.6 Square wave of Example 8.

Solution. Using the linearity and time-delay properties of the LT

$$\mathscr{L}\left\{\sum_{i=0}^{\infty} x(t-iT)\right\} = \sum_{i=0}^{\infty} X(s)e^{-siT} = X(s)\sum_{i=0}^{\infty} e^{-siT} = \frac{X(s)}{(1-e^{-sT})} \tag{3.67}$$

Let us call the square wave $z(t)$ and its LT $Z(s)$. Now one period of z can be written as $x(t) = u(t) - u(t-1), 0 \le t < 2$ (see Fig. 3.6). Using the delay property, therefore, $X(s) = (1/s) - (e^{-s}/s)$. Using (3.67) with $T = 2$, we have

$$Z(s) = \frac{(1/s) - (e^{-s}/s)}{(1-e^{-2s})} = \frac{(1-e^{-s})}{s(1-e^{-2s})} \tag{3.68}$$

□

Example 9. Verify the time-differentiation property of the LT which states that if $x(t) \longleftrightarrow X(s)$, then $dx/dt \longleftrightarrow sX(s) - x(0^-)$.

Solution. By definition, $\mathscr{L}\{dx/dt\} = \int_{0^-}^{\infty} (dx/dt)e^{-st}\, dt$. Integrating by parts yields

$$\mathscr{L}\left\{\frac{dx}{dt}\right\} = x(t)e^{-st}\Big|_{0^-}^{\infty} + s\int_{0^-}^{\infty} x(t)e^{-st}dt = sX(s) - x(0^-) \tag{3.69}$$

□

Example 10. Verify the initial value theorem of the LT which states that if $x(0^+) = \lim_{t\downarrow 0} x(t) < \infty$, then $\lim_{s\to\infty} sX(s) = x(0^+)$.

Solution. In case a discontinuity exists in x at $t = 0$, define the signal

$$y(t) = x(t) - cu(t) \tag{3.70}$$

where c is the amplitude shift at the discontinuity, $c = x(0^+) - x(0^-)$. Then y will be continuous at $t = 0$ (see Fig. 3.7). Further,

$$\frac{dx}{dt} = \frac{dy}{dt} + c\delta(t) \tag{3.71}$$

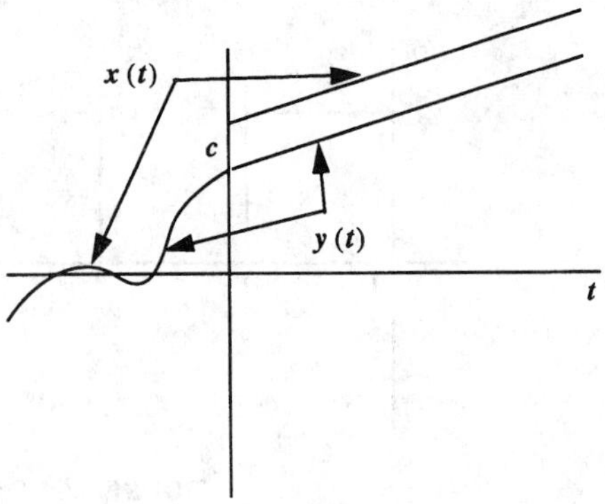

FIGURE 3.7 The signals x and y used in Example 10.

so that using the time-differentiation property and the fact that $\mathscr{L}\{c\delta(t)\} = c$, we have

$$sX(s) - x(0^-) = \int_{0^-}^{\infty} \frac{dy}{dt} e^{-st} dt + c \tag{3.72}$$

Because $c = x(0^+) - x(0^-)$,

$$sX(s) = \int_{0^-}^{\infty} \frac{dy}{dt} e^{-st} dt + x(0^+) \tag{3.73}$$

Assuming that the LT of the signal y has a ROC (y is of exponential order), the integral in (3.73) vanishes as $s \to \infty$. Finally, therefore, we obtain that $\lim_{s\to\infty} sX(s) = x(0^+)$. □

Example 11. Verify the convolution property of the LT which states that if x and h are causal signals with $x(t) \longleftrightarrow X(s)$, and $h(t) \longleftrightarrow H(s)$, then

$$x(t) * h(t) = \int_{-\infty}^{\infty} x(\xi)h(t-\xi)d\xi \longleftrightarrow X(s)H(s) \tag{3.74}$$

Solution. Because $x(t) = 0$ for $t < 0$, we can write

$$\int_{-\infty}^{\infty} x(\xi)h(t-\xi)d\xi = \int_{0^-}^{\infty} x(\xi)h(t-\xi)d\xi \tag{3.75}$$

Now

$$\begin{aligned}
\mathscr{L}\left\{\int_{0^-}^{\infty} x(\xi)h(t-\xi)d\xi\right\} &= \int_{0^-}^{\infty}\int_{0^-}^{\infty} x(\xi)h(t-\xi)e^{-st}d\xi\, dt \\
&= \int_{0^-}^{\infty}\int_{0^-}^{\infty} x(\xi)h(\beta)e^{-s\beta}e^{-s\xi}\, d\beta\, d\xi \qquad (3.76) \\
&= \int_{0^-}^{\infty} x(\xi)e^{-s\xi}d\xi \int_{0^-}^{\infty} h(\beta)e^{-s\beta}d\beta \\
&= X(s)H(s)
\end{aligned}$$

The causality of h is used in line (3.76) in setting the lower limit of integration over β to 0^-. □

The operational properties are used to simplify forward and inverse transform computations and other manipulations involving transforms. To briefly illustrate, three examples follow.

Example 12. Using operational properties, rederive the LT for $x(t) = M_x \cos(\omega_x t + \theta_x)u(t)$ which was first considered in Example 5.

Solution. Write x as

$$x(t) = \frac{M_x}{2} e^{j\theta_x} e^{(\sigma_x + j\omega_x)t} u(t) + \frac{M_x}{2} e^{-j\theta_x} e^{(\sigma_x - j\omega_x)t} u(t) \tag{3.77}$$

The linearity property allows us to ignore the factors $M_x e^{\pm j\theta_x}$ in the process of taking the transform, and returning them afterward. Recalling the previous result, (3.48), we can write immediately

$$\mathcal{L}\{e^{(\sigma_x + j\omega_x)t} u(t)\} = \frac{1}{s - (\sigma_x + j\omega_x)}, \quad \text{Re}\{s\} > \sigma_x \tag{3.78}$$

$$\mathcal{L}\{e^{(\sigma_x - j\omega_x)t} u(t)\} = \frac{1}{s - (\sigma_x - j\omega_x)}, \quad \text{Re}\{s\} > \sigma_x \tag{3.79}$$

so

$$X(s) = \frac{M_x}{2} e^{j\theta_x} \frac{1}{s - (\sigma_x + j\omega_x)} + \frac{M_x}{2} e^{-j\theta_x} \frac{1}{s - (\sigma_x - j\omega_x)} \quad \text{Re}\{s\} > \sigma_x \tag{3.80}$$

Placing the fractions over a common denominator yields the same result as that found using direct integration in Example 5. □

Example 13. Find the time signals corresponding to the following LTs:

$$X(s) = e^{-\pi s} \quad \text{for all } s$$
$$Y(s) = \log(7)\frac{e^{-32s}}{s}, \quad \text{Re}\{s\} > 0 \tag{3.81}$$
$$Z(s) = \frac{e^{\sqrt{2}s}}{s+5} + \frac{\sqrt{3}}{s-5}, \quad \text{Re}\{s\} > 5$$

Solution. Recognize that $X(s) = e^{-\pi s}\Delta(s)$, where $\Delta(s) = 1$ is the LT for the impulse function $\delta(t)$. Using the time-shift property, therefore, we have $x(t) = \delta(t - \pi)$.

Recognize that $Y(s) = \log(7)e^{-32s}U(s)$, where $U(s) = 1/s$ is the LT for the step function $u(t)$. Using linearity and the time-shift property, therefore, $y(t) = \log(7)u(t - 32)$.

In finding z, linearity allows us to treat the two terms separately. Further, from (3.48), we know that $\mathcal{L}^{-1}\{1/(s+5)\} = e^{-5t}u(t)$ and $\mathcal{L}^{-1}\{1/(s-5)\} = e^{5t}u(t)$. Therefore, $z(t) = e^{-5(t+\sqrt{2})}u(t+\sqrt{2}) + \sqrt{3}e^{5t}u(t)$. Note that the first term has a ROC $\{s: \text{Re}\{s\} > -5\}$, while the second has ROC $\{s : \text{Re}\{s\} < 5\}$. The overall ROC is therefore consistent with these two components. □

Inverse Laplace Transform

In principle, finding a time function corresponding to a given LT requires that we compute the integral in (3.36):

$$x(t) = \frac{1}{j2\pi} \int_{\sigma - j\infty}^{\sigma + j\infty} X(s) e^{st} \, ds \tag{3.82}$$

Recall that σ is constant and taken to be in the ROC of X. Direct computation of this line integral requires a knowledge of the theory of complex variables. However, several convenient computational procedures are available that circumvent the need for a detailed understanding of the complex calculus. These measures are the focus of this section. The reader interested in more detailed information on complex variable theory is referred to [3]. "Engineering" treatments of this subject are also found in [10].

We first study the most challenging of the inversion methods, and the one which most directly solves the inversion integral above. The reader interested in quick working knowledge of LT inversion might wish to proceed immediately to the section on partial fraction expansion.

Residue Theory

It is important to be able to compute residues of a function of a complex variable. Recall the Laurent series expansion of a complex function, say, F, which was introduced in "Poles and Zeroes—Part I," equation (3.60). Also, recall that the coefficient $r_{1,p}$ is called the **residue** of F at p, and that we defined the simplified notation $r_p \stackrel{\text{def}}{=} r_{1,p}$ to indicate the residue because the subscript "1" is not useful outside the Laurent series. In the analytic neighborhood of singular point $s = p$ (an nth-order pole) we define the function

$$\begin{aligned} \varphi_p(s) = (s-p)^n F(s) = r_{1,p}(s-p)^{n-1} + r_{2,p}(s-p)^{n-2} \\ + \cdots + r_{n,p} + \sum_{i=0}^{\infty} q_{i,p}(s-p)^{n+i} \end{aligned} \tag{3.83}$$

in which it is important to note that $r_{n,p} \neq 0$. Because F is not analytic at $s = p$, φ_p is not defined, and is therefore not analytic, at $s = p$. We can, however, make φ_p analytic at p by simply *defining* $\varphi_p(p) \stackrel{\text{def}}{=} r_{n,p}$. In this case φ_p is said to have a **removable singular point** (at p). Note that (3.83) can be interpreted as the Taylor series expansion of φ_p about the point $s = p$. Therefore, the residue is apparently given by

$$r_p \stackrel{\text{def}}{=} r_{1,p} = \frac{\varphi_p^{(n-1)}(p)}{(n-1)!} \tag{3.84}$$

where $\varphi_p^{(i)}$ indicates the ith derivative of φ_p. When $n = 1$ (first-order pole), which is frequently the case in practice, this expression reduces to

$$r_p = \varphi_p(p) = \lim_{s \to p} (s-p) F(s) \tag{3.85}$$

The significance of the residues appears in the following key result (e.g., see [3]):

Theorem 1 (Residue Theorem): *Let C be a simple closed contour within and on which a function F is analytic except for a finite number of singularity points, $p_1, p_2, \ldots, p_k$ interior to C. If the respective residues at the singularities are $r_{p_1}, r_{p_2}, \ldots, r_{p_k}$, then*

$$\oint_C F(s) ds = j2\pi \left(r_{p_1} + r_{p_2} + \cdots + r_{p_k} \right) \tag{3.86}$$

where the contour C is traversed in the counterclockwise direction.

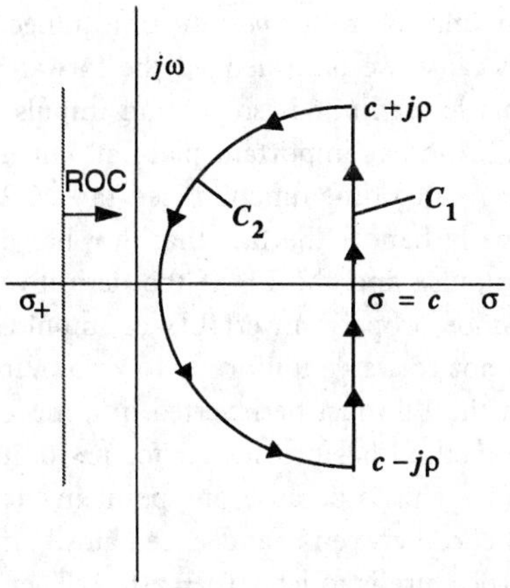

FIGURE 3.8 Contour in the s-plane for evaluating the inverse LT.

The relevance of this theorem in our work is as follows: In principle, according to (3.82), we want to integrate the complex function $F(s) = X(s)e^{st}$ on some vertical line in the ROC, for example, $c - j\infty$ to $c + j\infty$, where $c > \sigma_+$. Instead, suppose we integrate over the contour shown in Fig. 3.8. By the residue theorem, we have

$$\oint_C X(e)e^{st}\,ds = j2\pi\left(r_{p_1} + r_{p_2} + \cdots + r_{p_k}\right) \tag{3.87}$$

where $r_{p_1}, r_{p_2}, \ldots, r_{p_k}$ are the k residues of the function $X(s)e^{st}$. The integral can be decomposed as

$$\oint_C X(s)e^{st}\,ds = \oint_{C_1} X(s)e^{st}\,ds + \oint_{C_2} X(s)e^{st}\,ds \tag{3.88}$$

where, as $\rho \to \infty$, C_1 approaches the line over which we wish to integrate according to (3.82). It can be shown[8] that the integral over C_2 contributes nothing to the answer for $t > 0$, provided that X approaches zero uniformly[9] on C_2. Therefore,

$$\begin{aligned}\lim_{\rho\to\infty} \oint_C X(s)e^{st}\,ds &= \lim_{\rho\to\infty} \oint_{C_1} X(s)e^{st}\,ds \\ &= \lim_{\omega\to\infty} \frac{1}{j2\pi}\int_{c-j\omega}^{c+j\omega} X(s)e^{st}\,ds\end{aligned} \tag{3.89}$$

From (3.87), we have

$$\frac{1}{j2\pi}\int_{\sigma-j\infty}^{\sigma+j\infty} X(s)e^{st}\,ds = r_{p_1} + r_{p_2} + \cdots + r_{p_k}, \quad t > 0 \tag{3.90}$$

Thus, recalling that the left side is the original inversion integral, we have

$$x(t) = \sum_{i=1}^{k} r_{p_i}, \quad t > 0 \tag{3.91}$$

where the r_{p_i} are the residues of $X(s)e^{st}$ at its k singular points.

[8]A rigorous mathematical discussion appears in [3], while a particularly clear "engineering" discussion appears in Appendix B of [6].

[9]At the same rate regardless of the angle considered along C_2.

Note that the residue method returns a time function only over the *positive* time range. We might expect a result beginning at $t = 0$ or $t = 0^-$ because we have defined the forward LT as an integral beginning at $t = 0^-$. The reason for this lower limit is so that an impulse function at the time origin will be transformed "properly." Another important place at which the initial condition "$x(0^-)$" appears is in the LT of a differentiated time function (see Table 3.2 and Example 9). Again, the condition is included to properly handle the fact that if x has a discontinuity at $t = 0$, its derivative should include an impulse and the LT of the derivative should include a corresponding constant. The residue cannot properly invert LTs of impulse functions (constants over the s-plane) because such LTs do not converge uniformly to zero over the semicircular part of the contour C_2. Such constants in the LT must be inverted in a more *ad hoc* way. If no apparent impulses occur and the residue method has provided x for $t > 0$, it is, in fact, possible to assign an arbitrary finite value to $x(0)$. This is because one point in the time signal will not affect the LT, so the proper correspondence between x and X remains.[10] If it is necessary to assign a value to $x(0)$, the most natural procedure is to let $x(0) \stackrel{\text{def}}{=} x(0^+)$. Then when we write the final answer as $x(t) =$ [time signal determined by residues]$u(t)$, the signal takes an implied value $x(0^+)/2$ at $t = 0$.

The discussion above emphasizes that in general, successful application of the residue method depends on the uniform convergence of the LT to zero on the contour segment C_2. In principle each LT to be inverted must be checked against this criterion. However, this seemingly foreboding process is usually not necessary in linear signal and system analysis. The reason is that a practical LT ordinarily will take the form of a ratio of polynomials in s like (3.61). The check for proper convergence is a simple matter of assuring that the order of the denominator in s exceeds that of the numerator. When this is not the case, a simple remedy exists which we illustrate in Example 15. Another frequent problem is the occurrence of a LT of form

$$X(s) = \frac{N(s)e^{st_0}}{D(s)}, \quad t_0 < 0 \tag{3.92}$$

where N and D are polynomials. The trick here is to recognize the factor e^{st_0} as corresponding to a time shift which can be taken care of at the end of the problem, once the rational part of the transform is inverted.

Finally, we remark that similar results apply to the bilateral LT. When the signal is two sided in time, the causal part is obtained using the procedure above, but including only residues of poles known to be associated with the non-negative-time part of the signal. The noncausal part of the signal is found by summing residues belonging to "noncausal" poles. Note that the association of poles with the causal and noncausal parts of the signal follows from a specification of the ROC. Poles belonging to the causal signal are to the left of the ROC, while noncausal poles are to the right.

Let us now illustrate the procedure with two examples.

Example 14. In Example 5 we showed that

$$\mathscr{L}\{M_x e^{\sigma_x t} \cos(\omega_x t + \theta_x)\, u(t)\} = M_x \left[\frac{(s - \sigma_x)\cos(\theta_x) - \omega_x \sin(\theta_x)}{(s - \sigma_x)^2 + \omega_x^2} \right] \tag{3.93}$$

with ROC $\{s : \text{Re}\{s\} > \sigma_x\}$. Verify that this is correct by finding the inverse LT using residues. Call the signal x and the LT X.

[10]In fact, any two signals that differ only on a set of measure zero (e.g., see [7]) will have the same LT.

Solution. We can ignore the scalar M_x until the end due to linearity. Two poles are in the transform: $p_x = \sigma_x + j\omega_x$ and $p_x^* = \sigma_x - j\omega_x$ ($k = 2$ in the discussion above). These can be obtained by expanding the denominator and using the quadratic equation, but it is useful to remember the relationship between a quadratic polynomial written in this form and the conjugate roots [recall (3.59)]. The residue for the pole at p_x is given by

$$\begin{aligned}\varphi_{p_x}(p_x) = X(s)e^{st}(s-p_x)\big|_{s=p_x} &= \left.\frac{[(s-\sigma_x)\cos(\theta_x) - \omega_x\sin(\theta_x)]\, e^{st}}{(s-p_x^*)}\right|_{s=p_x} \\ &= \frac{\left[(p_x-\sigma_x)\cos(\theta_x) - \omega_x\sin(\theta_x)\right] e^{p_x t}}{(p_x-p_x^*)} \\ &= \frac{\left[(j\omega_x)\cos(\theta_x) - \omega_x\sin(\theta_x)\right] e^{(\sigma_x+j\omega_x)t}}{(2j\omega_x)}\end{aligned} \tag{3.94}$$

Similarly, we have for the pole at p_x^*, $\varphi_{p_x^*}(p_x^*) = \left[(-j\omega_x)\cos(\theta_x) - \omega_x\sin(\theta_x)\right] e^{(\sigma_x-j\omega_x)t}/(-2j\omega_x)$. For $t > 0$, therefore,

$$\begin{aligned}x(t) &= \varphi_{p_x}(p_x) + \varphi_{p_x^*}(p_x^*) \\ &= \frac{e^{\sigma_x t}\left[j\omega_x\cos(\theta_x)\left(e^{j\omega_x t} + e^{-j\omega_x t}\right) - \omega_x t\sin(\theta_x)\left(e^{j\omega_x t} - e^{-j\omega_x t}\right)\right]}{2j\omega_x}\end{aligned} \tag{3.95}$$

$$= e^{\sigma_x t}\left[\cos(\theta_x)\cos(\omega_x t) - \sin(\theta_x)\sin(\omega_x t)\right] = e^{\sigma_x t}\cos(\omega_x t + \theta_x) \tag{3.96}$$

and the transform is verified. □

The following example illustrates the technique for handling polynomial quotients in which the order of the denominator does not exceed the numerator (X is not a proper rational LT).

Example 15. Find the causal time signal corresponding to LT

$$X(s) = \frac{N(s)}{D(s)} = \frac{Gs^2}{(s-p)^2}, \quad G, p \text{ are real} \tag{3.97}$$

Solution. To use the residue method, we must reduce the order of the numerator to at most unity. By dividing polynomial D into N using long division, we can express X as

$$X(s) = G + \frac{2Gps - Gp^2}{(s-p)^2} \stackrel{\text{def}}{=} X_1(s) + X_2(s) \tag{3.98}$$

First note that we can use linearity and invert the two terms separately. The first is simple because (see Table 3.1, $\mathcal{L}^{-1}\{X_1(s)\} = \mathcal{L}^{-1}\{G\} = G\delta(t)$. In the second term we find a pole of order $n = 2$ at $s = p$. To use (3.84) to compute the residue, we require $\varphi_p(s) = X_2(s)e^{st}(s-p)^2 = \left[2Gps - Gp^2\right] e^{st}$. Then, for $t > 0$, the residue is

$$r_p = \left.\frac{1}{1!}\frac{d\varphi_p}{ds}\right|_{s=p} = G\left[2pe^{pt} + 2p^2te^{pt} - p^2te^{pt}\right] = Gpe^{pt}\left[2 + pt\right] \tag{3.99}$$

Because r_p is the only residue, we have $x(t) = G\delta(t) + Gpe^{pt}\left[2 + pt\right]u(t)$.

Remark. If the order of the numerator *exceeds* the denominator by $k > 0$, the long division will result in a polynomial of the form $A_k s^k + A_{k-1} s^{k-1} + \cdots + A_0$. Consequently, the time domain signal will contain derivatives of the impulse function. In particular, if $k = 1$, a "doublet" will be present in the time signal. (See Table 3.1 and Appendix A for more infomation.) □

We can usually find several ways to solve inverse LT problems, residues often being among the most challenging. In this example, for instance, we could use (3.98) to write

$$X(s) = G + \frac{2Gps}{(s-p)^2} - \frac{Gp^2}{(s-p)^2} = G + X_3(s) - X_4(s) \tag{3.100}$$

then use Table 3.1 to find $\mathscr{L}^{-1}\{G\} = G\delta(t)$ and $\mathscr{L}^{-1}\{X_4\} = x_4(t)$ [or use residues to find $x_4(t)$]. Noting that $X_3(s) = 2psX_4(s)$, we could then use the s-differentiation property (Table 3.2) to find x_3 from x_4. This alternative solution illustrates a general method that ordinarily is used regardless of the fundamental inversion technique. Linearity allows us to divide the problem into a sum of smaller, easier problems to invert, and then combine solutions at the end. The method below is probably the most popular, and clearly follows this paradigm.

Partial Fraction Expansion

The **partial fraction expansion (PFE)** method can be used to invert only rational LTs, with the exception that factors of the form e^{st_0} can be handled as discussed near (3.92). As noted above, this is not practically restricting for most engineering analyses. Partial fraction expansion is closely related to the residue method, a relationship evident via our examples.

As in the case of residues, a rational LT to be inverted must be proper, having a numerator polynomial whose order is strictly less than that of the denominator. If this is not the case, long division should be employed in the same manner as in the residue method (see Example 15).

Consider the LT $X(s) = N(s)/D(s)$. Suppose X has k poles, $p_1, p_2, \ldots, p_k$ and D is factored as

$$D(s) = (s - p_1)^{n_1}(s - p_2)^{n_2} \cdots (s - p_k)^{n_k} \tag{3.101}$$

where n_i is the order of the ith pole. (Note that $\sum_{i=1}^{k} n_i = n_D$.) Now if x is a real signal, the complex poles will appear in conjugate pairs. It will *sometimes* be convenient to combine the corresponding factors into a quadratic following (3.59),

$$(s - p)(s - p^*) = s^2 + \beta s + \gamma = s^2 - 2\,\mathrm{Re}\{p\}s + |p|^2 = s^2 - 2\sigma_p s + |p|^2 \tag{3.102}$$

for each conjugate pair. Assuming that the poles are ordered in (3.101) so that the first k' are real and the last $2k''$ are complex ($k = k' + 2k''$), D can be written

$$\begin{aligned} D(s) = {} & (s - p_1)^{n_{\mathrm{re},1}}(s - p_2)^{n_{\mathrm{re},2}} \cdots (s - p_{k'})^{n_{\mathrm{re},k'}}(s^2 + \beta_1 s + \gamma_1)^{n_{\mathrm{c},1}} \\ & \times (s^2 + \beta_2 s + \gamma_2)^{n_{\mathrm{c},2}} \cdots (s^2 + \beta_{k''} s + \gamma_{k''})^{n_{\mathrm{c},k''}} \end{aligned} \tag{3.103}$$

Terms of the form $(s - p)$ are called **simple linear factors**, while those of form $(s^2 + \beta s + \gamma)$ are **simple quadratic factors**. When a factor is raised to a power greater than one, we say that there are **repeated linear (or quadratic) factors.** We emphasize that linear factors need not always be combined into quadratics when they represent complex poles. In other words, in the term $(s - p_i)^{n_i}$, the pole p_i may be complex. Whether quadratics are used depends on the approach taken to solution.

The idea behind PFE is to decompose the larger problem into the sum of smaller ones. We expand the LT as

$$X(s)=\frac{N(s)}{D(s)}=\frac{N(s)}{\text{factor}_1\text{factor}_2\cdots\text{factor}_{k'}\,\text{factor}_{k'+1}\cdots\text{factor}_{k'+k''}} \tag{3.104}$$

$$=\frac{N_1(s)}{\text{factor}_1}+\frac{N_2(s)}{\text{factor}_2}+\cdots+\frac{N_{k'}(s)}{\text{factor}_{k'}}+\frac{N_{k'+1}(s)}{\text{factor}_{k'+1}}+\frac{N_{k'+k''}(s)}{\text{factor}_{k'+k''}} \tag{3.105}$$

Now because of linearity, we can invert each of the **partial fractions** in the sum individually, then add the results. Each of these "small" problems is easy and ordinarily can be looked up in a table or solved by memory.

We now consider a series of cases and examples.

Case 1: Simple Linear Factors. Let $X(s)=N(s)/D(s)$. Assume that the order of N is strictly less than the order of D Without loss of generality (for the case under consideration), assume that the first factor in D is a simple linear factor so that D can be written $D(s)=(s-p_1)D_{\text{other}}(s)$ where D_{other} is the product of all remaining factors. Then the PFE will take the form

$$X(s)=\frac{N(s)}{D(s)}=\frac{N(s)}{(s-p_1)D_{\text{other}}(s)}=\frac{N_1(s)}{(s-p_1)}+\left[\text{other PFs corresponding to factors in } D_{\text{other}}(s)\right] \tag{3.106}$$

Now note that

$$(s-p_1)X(s)=\frac{N(s)}{D_{\text{other}}(s)}=N_1(s)+(s-p_1)\left[\text{other PFs corresponding to factors in } D_{\text{other}}(s)\right] \tag{3.107}$$

Letting $s=p_1$ reveals that

$$N_1(p_1)=A_1=\frac{N(p_1)}{D_{\text{other}}(p_1)}=\left[(s-p_1)X(s)\right]\Big|_{s=p_1} \tag{3.108}$$

Note that the number A_1 is the residue of the pole at $s=p_1$. In terms of our residue notation

$$\varphi_{p_1}(s)=(s-p_1)X(s)=\frac{N(s)}{D_{\text{other}}(s)}\quad\text{and}\quad r_{p_1}=A_1=\varphi_{p_1}(p_1) \tag{3.109}$$

Carefully note that we are computing residues of poles of $X(s)$, not $X(s)e^{st}$, in this case.

Example 16. Given LT

$$X(s)=\frac{s^2+3s}{(s+1)(s+2)(s+4)} \tag{3.110}$$

with ROC $\{s\colon \text{Re}\{s\}>-1\}$, find the corresponding time signal, x.

Solution. Check that the order of the denominator exceeds that of the numerator. Because it does, we can proceed by writing

$$X(s) = \frac{A_1}{(s+1)} + \frac{A_2}{(s+2)} + \frac{A_3}{(s+4)} \tag{3.111}$$

Using the method above, we find that

$$A_1 = \left.\frac{s^2+3s}{(s+2)(s+4)}\right|_{s=-1} = -\frac{2}{3} \tag{3.112}$$

In a similar manner we find that $A_2 = 1$ and $A_3 = 2/3$. Therefore

$$X(s) = -\frac{2/3}{(s+1)} + \frac{1}{(s+2)} + \frac{2/3}{(s+4)} \tag{3.113}$$

Now using linearity and Table 3.1 [or recalling (3.48)], we can immediately write

$$x(t) = \left[-\frac{2}{3}e^{-t} + e^{-2t} + \frac{2}{3}e^{-4t}\right] u(t) \tag{3.114}$$

□

Case 2: Simple Quadratic Factors. When $D(s)$ contains a simple quadratic factor, the LT can be expanded as

$$X(s) = \frac{N(s)}{D(s)} = \frac{N(s)}{(s^2 - \beta s + \gamma)D_{\text{other}}(s)} \tag{3.115}$$

$$= \frac{Bs + C}{(s^2 - \beta s + \gamma)} + \left[\text{other PFs corresponding to factors in } D_{\text{other}}(s)\right]$$

The usefulness of this form is illustrated below.

Example 17. Find the time signal x corresponding to LT

$$X(s) = \frac{(s+4)}{(s+2)(s^2+6s+34)}, \quad \text{ROC}\{s\colon \text{Re}\{s\} > -3\} \tag{3.116}$$

Solution. The order of D exceeds the order of N, so we may proceed. The roots of the quadratic term are $p, p^* = (-6 \pm \sqrt{36-136})/2 = -3 \pm j5$, so we leave it as a quadratic. (If the roots were real, we would use the simple linear factor approach.) Expand the LT into PFs

$$X(s) = \frac{A}{(s+2)} + \frac{Bs + C}{(s^2+6s+34)} \tag{3.117}$$

Using the method for simple linear factors, we find that $A = 1/13$. Now multiply both sides of (3.117) by $D(s) = (s+2)(s^2+6s+34)$ to obtain

$$(s+4) = \frac{1}{13}(s^2+6s+34) + (Bs+C)(s+2) \tag{3.118}$$

Equating like powers of s on the two sides of the equation yields $B = -1/13$ and $C = 9/13$, so

$$X(s) = \frac{1/13}{(s+2)} + \frac{[(-1/13)s + (9/13)]}{(s^2 + 6s + 34)} \tag{3.119}$$

The first fraction has become familiar by now and corresponds to time function $(1/13)e^{-2t}u(t)$. Let us focus on the second fraction. Note that [recall (3.59)]

$$s^2 + 6s + 34 = (s - p)(s - p^*) = (s + 3 - j5)(s + 3 + j5) = (s + 3)^2 + 5^2 \tag{3.120}$$

The second fraction, therefore, can be written as

$$\begin{aligned} &-\frac{1}{13}\left[\frac{(s+3)}{(s+3)^2 + 5^2} - \frac{12}{(s+3)^2 + 5^2}\right] \\ &= -\frac{1}{13}\left[\frac{(s+3)}{(s+3)^2 + 5^2} - \frac{(12/5)5}{(s+3)^2 + 5^2}\right] \end{aligned} \tag{3.121}$$

The terms in brackets correspond to a cosine and sine, respectively, according to Table 3.1. Therefore,

$$x(t) = \left[\frac{1}{13}e^{-2t} - \frac{1}{13}e^{-3t}\cos(5t) + \frac{12}{5(13)}e^{-3t}\sin(5t)\right]u(t) \tag{3.122}$$

□

Simple quadratic factors need not be handled by the above procedure, but can be treated as simple linear factors, as illustrated by example.

Example 18. Repeat the previous problem using simple linear factors.

Solution. Expand X into simple linear PFs

$$X(s) = \frac{A}{(s+2)} + \frac{E}{(s-p)} + \frac{E^*}{(s-p^*)}, \quad p = -3 + j5 \tag{3.123}$$

Note that the PFE coefficients corresponding to complex-conjugate pole pairs will themselves be complex conjugates. In order to derive a useful general relationship, we ignore the specific numbers for the moment. Using the familar inverse transform, we can write

$$x(t) = \left[Ae^{-2t} + Ee^{pt} + E^*e^{p^*t}\right]u(t) \tag{3.124}$$

Letting $E = |E|e^{j\theta_E}$ and $p = \sigma_p + j\omega_p$, we have

$$\boxed{Ee^{pt} + E^*e^{p^*t} = |E|e^{j\theta_E}e^{(\sigma_p + j\omega_p)t} + |E|e^{-j\theta_E}e^{(\sigma_p - j\omega_p)t} = 2|E|e^{\sigma_p t}\cos(\omega_p t + \theta_E)} \tag{3.125}$$

This form should be noted for use with complex-conjugate pairs.

Using the method for finding simple linear factor coefficients, we find that $A = 1/13$. Also,

$$E = X(s)(s-p)\big|_{s=p} = \left.\frac{(s+4)}{(s+2)(s-p^*)}\right|_{s=p} = \frac{(p+4)}{(p+2)(j2\,\mathrm{Im}\{p\})}$$
$$= \frac{(1+j5)}{(-1+j5)(j10)} = 0.1\,e^{-j0.626\pi} \tag{3.126}$$

Therefore, the answer can be written:

$$x(t) = \left[\frac{1}{13}e^{-2t} + 0.2e^{-3t}\cos(5t - 0.626\pi)\right]u(t) \tag{3.127}$$

This solution is shown to be consistent with that of the previous example using the trigonometric identity $\cos(\alpha+\beta) = \cos(\alpha)\cos(\beta) - \sin(\alpha)\sin(\beta)$. □

Case 3: Repeated Linear Factors. When $D(s)$ contains a repeated linear factor, for example, $(s-p)^n$, the LT must be expanded as

$$X(s) = \frac{N(s)}{D(s)} = \frac{N(s)}{(s-p)^n D_{\mathrm{other}}(s)} = \frac{A_n}{(s-p)^n} + \frac{A_{n-1}}{(s-p)^{n-1}} + \cdots + \frac{A_1}{(s-p)}$$
$$+ \left[\text{other PFs corresponding to factors in } D_{\mathrm{other}}(s)\right] \tag{3.128}$$

The PFE coefficients of the n fractions are found as follows: Define

$$\varphi_p(s) = (s-p)^n X(s) \tag{3.129}$$

Then

$$A_{n-i} = \left.\frac{1}{i!}\frac{d^i}{ds^i}\varphi_p\right|_{s=p} \tag{3.130}$$

The similarity of these computations to residues is apparent, but only A_1 can be interpreted as the residue of pole p. We illustrate this method by example.

Example 19. Find the time signal x corresponding to LT $X(s) = (s+4)/(s+1)^3$.

Solution. The order of D exceeds the order of N, so we may proceed. X has a third-order pole at $s = -1$, so we expand X as

$$X(s) = \frac{(s+4)}{(s+1)^3} = \frac{A_3}{(s+1)^3} + \frac{A_2}{(s+1)^2} + \frac{A_1}{(s+1)} \tag{3.131}$$

Let $\varphi_{-1}(s) = (s+1)^3 X(s) = (s+4)$. Then

$$A_3 = \varphi_{-1}(-1) = 3, \quad A_2 = \left.\frac{d\varphi_{-1}}{ds}\right|_{s=-1} = 1, \quad \text{and} \quad A_1 = \left.\frac{1}{2}\frac{d^2\varphi_{-1}}{ds^2}\right|_{s=-1} = 0 \tag{3.132}$$

So

$$X(s) = \frac{3}{(s+1)^3} + \frac{1}{(s+1)^2} \tag{3.133}$$

Using Table 3.1, we have

$$x(t) = \left[\frac{3}{2}t^2e^{-t} + te^{-t}\right]u(t) \tag{3.134}$$

□

Case 4: Repeated Quadratic Factors. When $D(s)$ contains a repeated quadratic factor, e.g., $(s^2 + \beta s + \gamma)^n$, the LT may be either inverted by separating the quadratic into repeated linear factors [one nth-order factor for each of the complex roots; see (3.155)], then treated using the method of Case 3, or it can be expanded as

$$X(s) = \frac{N(s)}{D(s)} = \frac{N(s)}{(s^2 + \beta s + \gamma)^n D_{\text{other}}(s)} \tag{3.135}$$

$$= \sum_{i=0}^{n-1} \frac{(B_{n-i}s + C_{n-i})}{(s^2 + \beta s + \gamma)^n} + [\text{other PFs corresponding to factors in } D_{\text{other}}(s)]$$

The PFE coefficients of the n fractions are found algebraically as we illustrate by example.

Example 20. Find the time signal x corresponding to LT $X(s) = 4s^2/(s^2 + 1)^2(s + 1)$ [10].

Solution. Recognize the factor $(s^2 + 1)^2$ as an $n = 2$-order quadratic factor with $\beta = 0$ and $\gamma = 1$. We know from our previous work that $\beta = -2\,\text{Re}\{p\}$, where p and p^* are the poles associated with the quadratic factor, in this case $\pm j$. In turn, the real part of p provides the damping term in front of the sinusoid represented by the quadratic factor. In this case, therefore, we should expect one of the terms in the time signal to be a pure sinusoid.

Write X as

$$X(s) = \frac{(B_2 s + C_2)}{(s^2 + 1)^2} + \frac{(B_1 s + C_1)}{(s^2 + 1)} + \frac{A}{(s + 1)} \tag{3.136}$$

Using the familiar technique for simple linear factors, we find that $A = 1$. Now multiplying both sides of (3.136) by $(s^2+1)^2(s+1)$, we obtain $4s^2 = (B_2 s + C_2)(s + 1) + (B_1 s + C_1)(s^2 + 1)(s + 1) + (s^2 + 1)^2$. Equating like powers of s, we obtain $B_2 = -1$, $C_2 = 1$, $B_1 = 2$, and $C_1 = -2$. Hence,

$$X(s) = \frac{1}{(s + 1)} + \frac{(2s - 2)}{(s^2 + 1)^2} - \frac{(s - 1)}{(s^2 + 1)} \tag{3.137}$$

We can now use Tables 3.1 and 3.2 to invert the three fractions. Note that the third term will yield an undamped sinusoid as predicted. Also note that the middle term is related to the third by differentiation. This fact can be used in obtaining the inverse. □

3.4 Laplace Transform Analysis of Linear Systems

Let us return to the example circuit problem which originally motivated our discussion and discover ways in which the LT can be used in system analysis. Three fundamental means are available for using the LT in such problems. The most basic is the use of LT theory to solve the differential equation governing the circuit dynamics. The differential equation solution methods are quite general and apply to linear, constant coefficient differential equations arising in any context. The second method involves the use of the "system function," a LT-based representation of a LTI system, which embodies all the relevant information about the system dynamics. Finally, we preview "LT equivalent circuits," a LT method which is primarily used for circuit analysis, and which is treated more completely in Chapter 19.

Solution of the System Differential Equation

Consider a system is governed by a linear, constant coefficient differential equation of the form

$$\sum_{\ell=0}^{n_D} a_\ell \frac{d^\ell}{dt^\ell} y(t) = \sum_{\ell=0}^{n_N} b_\ell \frac{d^\ell}{dt^\ell} x(t) \qquad (3.138)$$

with appropriate initial conditions given. (The numbers n_N and n_D should be considered fixed integers for now, but are later seen to be consistent with similar notation used in the discussion of rational LTs.) Using the linearity and time-differentiation properties of the LT, we can transform both sides of the equation to obtain

$$\begin{aligned} \sum_{\ell=0}^{n_D} a_\ell \left[s^\ell Y(s) - \sum_{i=0}^{\ell-1} y^{(i)}(0^-) s^{\ell-i-1} \right] \\ = \sum_{\ell=0}^{n_N} b_\ell \left[s^\ell X(s) - \sum_{i=0}^{\ell-1} x^{(i)}(0^-) s^{\ell-i-1} \right] \end{aligned} \qquad (3.139)$$

where $y^{(i)}$ and $x^{(i)}$ are the ith derivatives of y and x. Rearranging, we have

$$Y(s) = \frac{X(s) \sum_{\ell=0}^{n_N} b_\ell s^\ell - \sum_{\ell=0}^{n_N} b_\ell \sum_{i=0}^{\ell-1} x^{(i)}(0^-) s^{\ell-i-1} + \sum_{\ell=0}^{n_D} a_\ell \sum_{i=0}^{\ell-1} y^{(i)}(0^-) s^{\ell-i-1}}{\sum_{\ell=0}^{n_D} a_\ell s^\ell} \qquad (3.140)$$

Given the input signal x and all necessary initial conditions on x and y, all quantities on the right side of (3.140) are known and can be combined to yield Y. Our knowledge of LT inversion will then, in principle, allow us to deduce y. This process often turns an unwieldy differential equation solution into simple algebraic operations. The price paid for this simplification, however, is that the process of inverting Y to obtain y is sometimes challenging.

Recall that in the motivating example, a similar conversion of the (nonhomogeneous) differential equation solution to algebraic operations occurred [recall (3.16) and surrounding discussion], as the superfluous term $e^{p_x t}$ was divided out of the equation. Except for the lack of attention paid to initial conditions, what remains in (3.16) is tantamount to a LT equation of form (3.139), as shown below. The homogeneous solution and related initial conditions were not included in this discussion to avoid obfuscating the main issue. The reader was encouraged to think of the LT as a process of "dividing" the "e^{st}" term out of the signals before starting the solution. We can now clearly see this fundamental connection between the differential equation and LT solutions.

Example 21. Return to the motivating example in Section 3.2 and solve the problem using LT analysis.

Solution. Recall the differential equation governing the circuit, (3.1), $x(t) = LC(d^2y/dt^2) + RC(dy/dt) + y(t)$. The initial conditions are $y(0^-) = y_0$ and $i(0^-) = i_0$. Recall also that for convenience we seek the solution for $x(t) = M_x e^{j\theta_x} e^{(\sigma_x + j\omega_x)t} u(t) = M_x e^{j\theta_x} e^{p_x t} u(t)$, recognizing that the "correct" solution will be the real part of that obtained.

Taking the LT of each side of the differential equation, we have

$$X(s) = LC\left[s^2 Y(s) - s y(0^-) - y^{(1)}(0^-)\right] + RC\left[sY(s) - y(0^-)\right] + Y(s) \qquad (3.141)$$

or

$$Y(s) = \frac{X(s)}{(LCs^2 + RCs + 1)} + \frac{(sLC - RC)y(0^-) + LCy^{(1)}(0^-)}{(LCs^2 + RCs + 1)} \tag{3.142}$$

Dividing both numerator and denominator of each fraction by LC, and inserting the LT $X(s) = M_x e^{j\theta_x}/(s - p_x)$ and the initial conditions [recall that $Cy^{(1)}(t) = i(t) \Rightarrow Cy^{(1)}(0^-) = i_0$], we have

$$Y(s) = \frac{M_x e^{j\theta_x}/LC}{(s - p_x)[s^2 + (R/L)s + 1]} + \frac{[s - (R/L)]y_0 + i_0/C}{[s^2 + (R/L)s + 1]} \tag{3.143}$$

Using PFE this can be written

$$\begin{aligned} Y(s) = {} & \frac{\dfrac{M_x e^{j\theta_x}/LC}{[p_x^2 + (R/L)p_x + 1]}}{(s - p_x)} + \frac{\dfrac{M_x e^{j\theta_x}/LC}{(p_h - p_x)}}{(s - p_h)} \\ & + \frac{\left(\dfrac{M_x e^{j\theta_x}/LC}{(p_h - p_x)}\right)^*}{(s - p_h^*)} + \frac{[s - (R/L)]y_0 + i_0/C}{(s - p_h)(s - p_h^*)} \end{aligned} \tag{3.144}$$

where p_h and p_h^* are the system poles. We have expanded the first fraction in (3.143) using simple linear factors involving the poles p_x, p_h, and p_h^*. The latter two poles correspond to the system [roots of $(s^2 + (R/L)s + 1)$], and the resulting terms in (3.144) are part of the natural response. The first pole, p_x, is attributable to the forcing function, and the resulting term in (3.144) will yield the forced response in the time domain. Finally, the last term in (3.144), which arises from the last term in (3.143), is also part of the natural response. The separation of the natural response into these three LT terms harkens back to the time-domain discussion about the distinct contributions of the input and initial conditions to the natural response. The last term is clearly related to initial conditions and the circuit's natural means of dissipating that energy. The former terms, which will also yield a damped sinusoid of identical complex frequency to that of the third term, are clearly "caused" by the input.

Remark. The curious reader may wonder why the first fraction in (3.144) and (3.17), both of which represent the forced response, are not identical. After all, we have been encouraged to view the LT as a kind of generalized phasor representation. Recall, however, that the LT at a particular s must be thought of as a "phasor density." If an eigensignal of the system represents one complex frequency, e.g., $s = p_x$, the LT is infinitely dense at that point in the s-plane, corresponding to the existence of a pole there. If the signal does have a conventional phasor representation such as $\bar{Y} = M_y e^{j\theta_y}$, then this phasor will be related to the LT as

$$\bar{Y} = Y(s)\frac{ds}{2\pi}\bigg|_{s=p_x} = \lim_{s \to p_x} Y(s)(s - p_x) \tag{3.145}$$

which we recognize as the residue of the LT Y at the pole p_x. The reader can easily verify this assertion using (3.143). This discussion is closely related to the interconnection between the Fourier series coefficients (which are similar to conventional sinusoidal phasors) and the Fourier transform. These signal representations are discussed in Chapter 4.

The System Function

Definition 1. In our motivating example an input sinusoid with generalized phasor $M_x e^{j\theta_x}$ produced an output sinusoid with generalized phasor $M_y e^{j\theta_y}$. We discovered that the ratio of phasors was dependent only upon system parameters and the complex frequency, p_x. We noted that when considered as a function of the general complex frequency, s, this ratio is called the **system function.** That is, the system function is a complex function, H, of complex frequency, s, such that if a damped sinusoid of complex frequency $s = p_x$ and with generalized phasor $\bar{X} = M_x e^{j\theta_x}$ is used as input to the system, the forced response will be a sinusoid of complex frequency p_x and with generalized phasor

$$M_y e^{j\theta_y} = H(p_x) M_x e^{j\theta_x} \tag{3.146}$$

Preview of Magnitude and Phase Responses

It is often important to know how a system will respond to a pure sinusoid of radian frequency, e.g., ω_x. In particular, we would like to know the amplitude and phase changes imposed on the sinusoid by the system. In terms of the definition just given, we see that this information is contained in the system function evaluated at frequency $p_x = j\omega_x$. In particular, $|H(j\omega_x)|$ represents the magnitude factor and $\arg\{H(j\omega_x)\}$ the phase change at this frequency. When plotted as functions of general frequency ω, the real functions $|H(j\omega)|$ and $\arg\{H(j\omega)\}$ are called the **magnitude** (or sometimes **frequency**) **response**, and **phase response** of the system, respectively. The complex function $H(j\omega)$ will be seen to be the Fourier transform of the impulse response (see Appendix B and Chapter 4) and is sometimes called the **transfer function** for the system. Returning to Fig. 3.2(b), the reader will discover that we have plotted the magnitude response for the series RLC circuit of Section 3.2 with numerical values given in (3.18).

The magnitude and phase responses of the system can be obtained graphically from the pole-zero diagram. Writing $H(s)$ similarly to (3.62), $H(s) = C\prod_{i=1}^{n_N}(s - z_i)/\prod_{i=1}^{n_D}(s - p_i)$. Therefore,

$$|H(j\omega)| = C\frac{\prod_{i=1}^{n_N}|j\omega - z_i|}{\prod_{i=1}^{n_D}|j\omega - p_i|} \quad \text{and}$$
$$\arg\{H(j\omega)\} = \sum_{i=1}^{n_N}\arg\{(j\omega - z_i)\} - \sum_{i=1}^{n_D}\arg\{(j\omega - p_i)\} \tag{3.147}$$

where we have assumed $C > 0$ (if not, add π radians to $\arg\{H(j\omega)\}$). By varying ω the desired plots are obtained as illustrated in Fig. 3.9.

Definition 2. More generally, the **system function,** $H(s)$, for a LTI system can be defined as the ratio of the LT, for example, $Y(s)$, of the output, resulting from any input with LT, for example, $X(s)$, when the system is initially at rest (zero initial conditions). In other words, H is the ratio of the LT of the **zero-state response** to the LT of the input,

$$H(s) \stackrel{\text{def}}{=} \frac{Y(s)}{X(s)} \quad \text{when all initial conditions are zero} \tag{3.148}$$

While the latter definition is more general, the two definitions are consistent when the input is an eigensignal, as we show by example.

Example 22. Show that the two definitions of the system function above are consistent for the RLC circuit of Section 3.2.

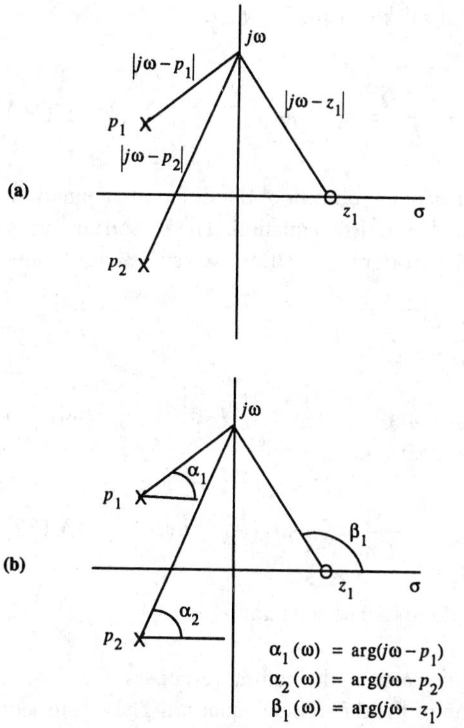

FIGURE 3.9 Pole-zero plot for an example system function, $H(s) = C(s - z_1)/(s - p_1)(s - p_2)$. (a) To obtain the magnitude response $|H(j\omega)|$ at frequency ω, the product of lengths from all zeroes to $s = j\omega$ is divided by the product of lengths from all poles to $s = j\omega$. The result must be multiplied by the gain term $|C|$. (b) To obtain the phase response $\arg\{H(j\omega)\}$ at frequency ω, the sum of angles from all poles to $s = j\omega$ is subtracted from the sum of angles from all zeroes to $s = j\omega$. An additional π radians is added if $C < 0$.

Solution. Replacing the specific complex frequency p_x by a general complex frequency s in the initial example using phasor analysis, we found that [recall (3.31)] $H(s) = 1/(LCs^2 + RCs + 1)$. On the other hand, using more formal LT analysis on the same problem, we derived (3.142). Forming the ratio $Y(s)/X(s)$ where both initial conditions are set to zero, yields an identical result. □

Finally, another very useful definition of the system function is as follows.

Definition 3. The output of a LTI system to an impulse excitation, $x(t) = \delta(t)$ (see Section 3.6, Appendix A), when all initial conditions are zero, is called the **impulse response** of the system. The impulse response is usually denoted $h(t)$. The system function can be defined as the LT of the impulse response $H(s) = \mathcal{L}\{h(t)\}$.

The consistency of this definition with the first two is easy to demonstrate. Let H denote the system function for the system, regardless of how it might be related to $h(t)$. From Definition 2, we have $H(s) = Y(s)/X(s)$ for any valid LTs X and Y. Let $x(t) = \delta(t)$, in which case $X(s) = 1$. By definition, $y(t) = h(t)$, so $\mathcal{L}\{h(t)\} = \mathcal{L}\{y(t)\} = H(s)X(s) = H(s)$.

This interpretation of H enables us to find the impulse response of the system, a task which is not always easy in the time domain because of the pitfalls of working with impulse functions.

Example 23. Find the impulse response, h, for the circuit of Section 3.2.

Solution. Using various means we showed that the system function is $H(s) = 1/(LCs^2+RCs+1)$. Let us find h by computing $h(t) = \mathcal{L}^{-1}\{H(s)\}$. Using the quadratic equation, we find the roots of the denominator (poles of the system) to be [c.f., (3.6)]

$$p_1, p_2 = \frac{-RC \pm \sqrt{R^2C^2 - 4LC}}{2LC} = -\frac{R}{2L} \pm \sqrt{\frac{R^2}{4L^2} - \frac{1}{LC}} \tag{3.149}$$

Assume that these poles are complex conjugates and call them p_h, p_h^*, where

$$p_h = -\frac{R}{2L} + j\sqrt{\frac{1}{LC} - \frac{R^2}{4L^2}} \stackrel{\text{def}}{=} \sigma_h + j\omega_h \tag{3.150}$$

Comparing to our initial work on finding the homogenous solution of the differential equation, we see that these system poles are the roots of the charateristic equation. The reason for using subscript "h" in our early work should now be clear. In terms of (3.150) we can rewrite H as

$$H(s) = \frac{1/LC}{(s - \sigma_h)^2 + \omega_h^2} \tag{3.151}$$

[recall (3.59)]. Now using Table 3.1: $\mathscr{L}\{e^{\sigma_h t}\cos(\omega_h t + \theta_h)\} = [(s - \sigma_h)\cos(\theta_h) - \omega_h \sin(\theta_h)]/[(s - \sigma_h)^2 + \omega_h^2]$. Letting $\theta_h = \pi/2$ and using linearity, we have

$$h(t) = -\frac{1}{LC\omega_h} e^{\sigma_h t} \cos\left(\omega_h t + \frac{\pi}{2}\right) u(t) = \frac{1}{LC\omega_h} e^{\sigma_h t} \sin(\omega_h t)\, u(t) \tag{3.152}$$

We could also note that $|p_h| = \sqrt{1/LC}$ and write the initial scale factor as $|p_h|^2/\omega_h$. □

We see clearly that the impulse response is closely related to the natural responses of a system, which in turn are tied to the homogeneous solution of the differential equation. These transient responses depend only on properties of the system, and not on properties of the input signal (beyond the initial instant of excitation). The form of the homogeneous differential equation solution is specified by the number and values of the roots of the characteristic equation. These roots are, thus, the poles of the system function H. The system function offers an extremely valuable tool for the design and analysis of systems. We now turn to this important topic.

Poles and Zeroes—Part II: Stability Analysis of Systems

We return to the issue of poles and zeroes, this time with attention restricted to rational LTs. In particular, we focus on the poles and zeroes of a system function and their effects on the performance of the system.

Natural Modes

The individual time responses corresponding to the poles of H are often called **natural modes** of the system. These modes are indicators of the physical properties of the system (e.g., circuit values), which in turn determine the natural way in which the system will dissipate, store, amplify, or respond to energy of various frequencies. Consider two general cases. Suppose H has a real pole of order n at $s = p$, so that it can be written

$$H(s) = \sum_{i=1}^{n} \frac{A_i}{(s - p)^i} + [\text{other terms}] \tag{3.153}$$

We know from previous work, therefore, that (see Table 3.1) h has corresponding modal components,

$$h(t) = \sum_{i=1}^{n} \frac{A_i}{(i-1)!} t^{i-1} e^{pt} u(t) + [\text{other terms}] \tag{3.154}$$

When $|p| < 0$, the modal components due to pole p will decay exponentially with time (modulated by the terms t^{i-1}). When $|p| > 0$, these modal components will increase exponentially with time (modulated by the terms t^{i-1}). When $p = 0$, the term will either remain bounded if $n = 1$, or the terms will increase with power n as t increases. These cases are illustrated in Fig. 3.10(a).

Next, let H have a complex pole pair of order n at $s = p, p^*$, so that it can be written

$$H(s) = \sum_{i=1}^{n} \left[\frac{E_i}{(s-p)^i} + \frac{E_i^*}{(s-p^*)^i} \right] + [\text{other terms}] \tag{3.155}$$

Using Table 3.1 we see that h has a corresponding modal component,

$$h(t) = \sum_{i=1}^{n} \frac{2|E_i|}{(i-1)!} t^{i-1} e^{\sigma_p t} \cos(\omega_p t + \theta_{E_i}) u(t) + [\text{other terms}] \tag{3.156}$$

where $p = \sigma_p + j\omega_p$, and $\theta_{E_i} = \arg\{E_i\}$. In this case, when $|p| < 0$, the sinusoidal modal components will decay exponentially with time (modulated by the terms t^{i-1}); when $|p| > 0$, these terms will increase exponentially with time (modulated by the terms t^{i-1}); and when $p = 0$, the terms due to p will either represent a constant or increasing sinusoid, depending on the value of n. These cases are illustrated in Fig. 3.10(b).

BIBO stability

We digress momentarily to discuss the concept of stability. To avoid some unnecessary complications, we restrict the discussion to *causal* systems, and, as usual, to the unilateral LT. There are various ways to define stability, but the most frequent and useful definition for a LTI system is that of BIBO stability. A system is said to be **bounded-input–bounded-output** (BIBO) **stable** iff every bounded input produces a bounded output. Formally, any bounded input, for example, x such that $|x(t)| \le B_x < \infty$ for all t, must result in an output, y, for which $B_y < \infty$ exists so that $|y(t)| \le B_y$ for all t.

A necessary and sufficient condition for BIBO stability of a LTI system is that its impulse response be absolutely integrable,

$$\int_0^{\infty} |h(t)|\, dt < \infty \tag{3.157}$$

This is easy to show. First, assume $|x(t)| < B_x$ for all t. Then, from the convolution integral

$$|y(t)| = \left| \int_0^{\infty} x(t-\lambda) h(\lambda)\, d\lambda \right| \tag{3.158}$$

Now, using the Schwarz inequality (see [7]), we have

$$|y(t)| \le \int_0^{\infty} |x(t-\lambda)|\, |h(\lambda)|\, d\lambda \le B_x \int_0^{\infty} |h(\lambda)|\, d\lambda \tag{3.159}$$

Consequently, it is sufficient that (3.157) be true for $B_y < \infty$ to exist. On the other hand, suppose that condition (3.157) were not true, but the system were BIBO stable. For a fixed t, consider the input

$$x(t-\lambda) = \begin{cases} 1, & h(\lambda) > 0 \\ -1, & h(\lambda) < 0 \\ 0, & h(\lambda) = 0 \end{cases} \tag{3.160}$$

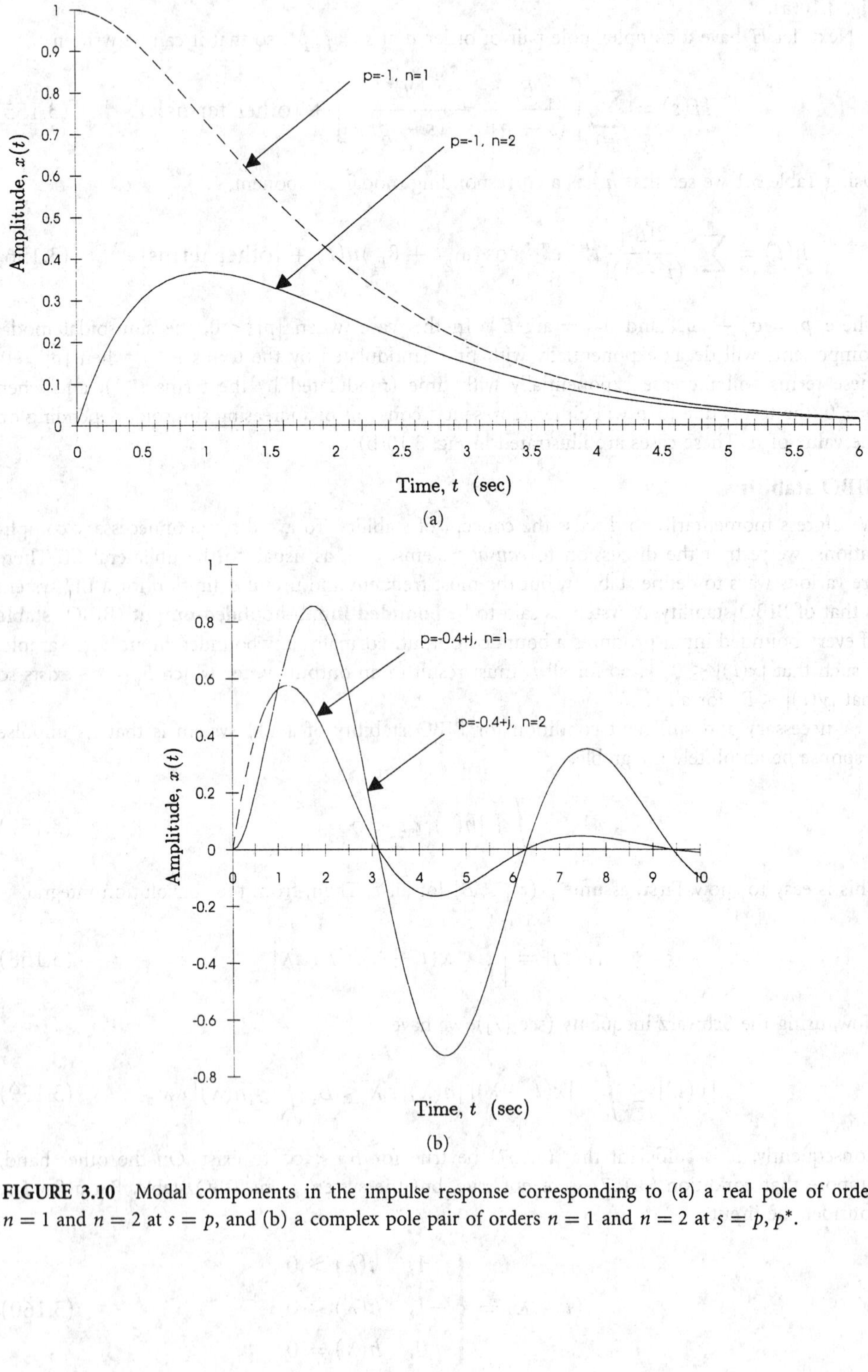

FIGURE 3.10 Modal components in the impulse response corresponding to (a) a real pole of orders $n = 1$ and $n = 2$ at $s = p$, and (b) a complex pole pair of orders $n = 1$ and $n = 2$ at $s = p, p^*$.

For this input, the output at time t is

$$y(t) = \int_0^\infty |h(\lambda)|\, d\lambda \tag{3.161}$$

which is not bounded according to assumption. Therefore, we encounter a contradiction showing that the condition (3.157) is also necessary for BIBO stability.

Stability and Natural Modes

Now we tie the stability analysis to the discussion of modal components above. We assert that *a causal LTI system with proper rational system function H will be BIBO stable iff all of its poles are in the left-half s-plane.* That this is true is easily seen. If any pole is in the right-half s-plane, we know that h will contain at least one mode that will increase without bound. Therefore, (3.157) cannot hold and the system is not BIBO stable. Conversely, if all poles are in the left-half s-plane, (3.157) will hold. The case in which one or more *simple* poles fall exactly on the $s = j\omega$ axis (and none in the right-half s-plane) is called **marginal BIBO stability**. In this case (3.157) does not hold, so the system is not strictly BIBO stable. However, the system does theoretically produce bounded outputs for some inputs. Finally, note that we show by example below Example 24(a) that an improper rational system function cannot represent a BIBO stable system.

Based on our earlier discussion of LT ROCs, an equivalent way to state the BIBO stability condition is as follows: *A causal, LTI system is BIBO stable iff the ROC of its system function H includes the $s = j\omega$ axis in the s-plane.* We recall that the failure to include the $j\omega$ axis would imply at least one pole of H in the right-half s-plane.

Let us conclude this discussion by considering some examples.

Example 24. Comment on the BIBO stability of the following systems:

(a) $H(s) = \dfrac{N(s)}{D(s)} = \dfrac{(1/2)s^3}{s^2 + 2s + 1} = \dfrac{(1/2)s^3}{(s+1)^2}$

(b) $H(s) = \dfrac{(s-3)}{s^2 + 5s + 4}$

(c) $H(s) = \dfrac{(s-3)}{(s^2 + 5s + 4)(s^2 - 2s - 3)}$

(d) $H(s) = \dfrac{6}{[(s+3)^2 + 25](s+8)}$

(e) $H(s) = \dfrac{s}{s^2 + 4}$

Solution. (a) H is not a proper rational fraction because the order of $N(s)$ > order of $D(s)$. We illustrate that such a system is not BIBO stable. Dividing D into N, we can write

$$H(s) = \frac{1}{2}s - 1 + \frac{3/2}{s(s^2 + 2s + 1)} \tag{3.162}$$

Suppose we enter $x(t) = u(t)$ as a (bounded) input. Then

$$Y(s) = H(s)X(s) = \frac{H(s)}{s} = \frac{1}{2} - \frac{1}{s} + \frac{3/2}{s(s^2 + 2s + 1)} \tag{3.163}$$

Therefore,

$$y(t) = \frac{1}{2}\delta(t) - u(t) + \left[\text{terms resulting from } \frac{3/2}{s(s^2 + 2s + 1)}\right] \tag{3.164}$$

The output is unbounded in response to a bounded input and is therefore not BIBO stable. *A similar result will occur whenever H is not proper.*

(b) H has poles at $s = -1, -4$, and a zero at $s = 3$. The system is BIBO stable. Note that the right-half plane zero has no adverse effect on stability.

(c) $H(s) = N(s)/D(s)$ has a second-order pole at $s = -1$, and a simple pole at $s = -4$. Both N and D have roots at $s = 3$, thus, neither a pole nor zero is found there (they cancel). The system is therefore BIBO stable.

Remark. The response to *nonzero initial conditions* of a system which has one or more "canceled poles" in the right-half s-plane will increase without bound, and therefore could be considered "unstable" in some sense, even though it is BIBO stable. The reader is invited to show that the present system will respond in this undesirable manner to nonzero initial conditions. A system is said to be **asymptotically stable** if it is BIBO stable and its response to initial conditions approaches zero as $t \to \infty$. We see that asymptotic stability implies BIBO stability but the converse is not true.

(d) Recalling (3.59) we find that H has poles at $s = -3 \pm j5$ and $s = -8$. No finite zeroes are included. The system is BIBO stable and will have both an oscillatory mode and a damped exponential mode.

(e) H has poles at $s = \pm j2$ which are on the $j\omega$ axis. The system is *marginally* BIBO stable.□

Example 25. Return to the series RLC circuit of Section 3.2 and discuss the system's stability as R varies for a fixed L and C. Assume that $R < 2\sqrt{L/C}$.

Solution. In light of (3.149) the upper bound on R means that the poles of the system will always be complex conjugates, e.g., p_h, p_h^* [h is oscillatory]. From (3.150) we see that the poles of the system can be written in polar form as

$$p_h, p_h^* = \sqrt{\sigma_h^2 + \omega_h^2}\, e^{\pm j \tan^{-1}(\omega_h/\sigma_h)} = \frac{1}{\sqrt{LC}} e^{\pm j \tan^{-1}(\omega_h/\sigma_h)} \tag{3.165}$$

For fixed L and C, the poles remain on a circle of radius $1/\sqrt{LC}$ in the s-plane. Having established this fact, recall the Cartesian form of p_h given in (3.149) and note that as $R \to 0$, $p_h \to j\sqrt{1/LC}$. On the other hand, as $R \to 2\sqrt{L/C}$, $p_h \to -\sqrt{1/LC}$ (see Fig. 3.11). Therefore, the poles remain in the left-half s-plane over the specified range of R, except when $R = 0$, in which case the poles are exactly on the $j\omega$-axis at $s = \pm j(1/\sqrt{LC})$. Therefore, the system is BIBO stable, except when $R = 0$, in which case it is marginally stable. Only if R were to take negative values would the circuit go unstable. This is not realistic unless active circuit elements are present which effectively present negative resistance. The reader is encouraged to explore what happens as R continues to increase beyond the given upper bound. □

Laplace-Domain Phasor Circuits

After completing a sufficient number of circuit problems involving differential equations like Example 21, certain patterns would become apparent in a Laplace-transformed differential equation like (3.141). These patterns occur precisely because of the invariant relationships between currents and voltages across lumped parameter components. For example, because the current and voltage through and inductor are related as $v_L(t) = L(di_L/dt)$, whenever this relationship is encountered and transformed in a circuit problem, it becomes $V_L(s) = sLI_L(s) - Li_L(0^-)$. This relationship may be written without recourse to the time domain by

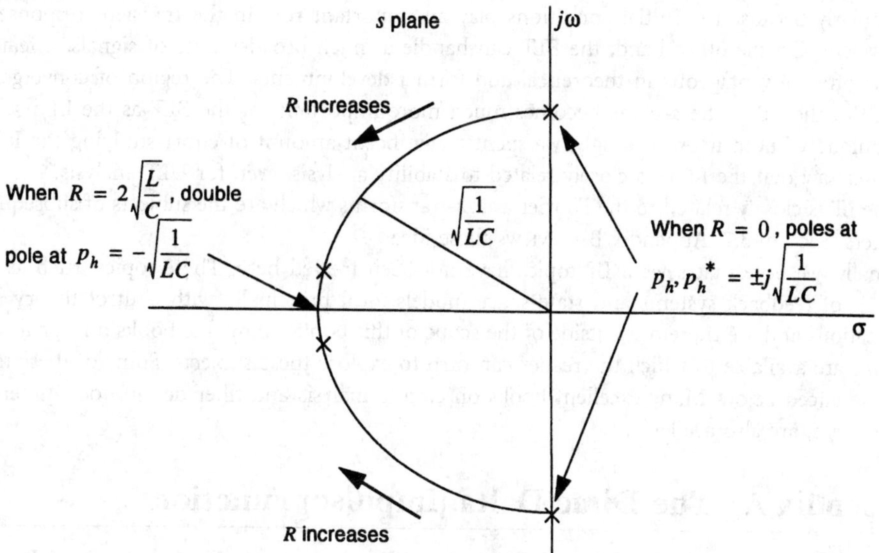

FIGURE 3.11 Locus of the poles as R varies in Example 25.

treating $V_L(s)$ and $I_L(s)$ as a "DC voltage" and a "DC current," by replacing the inductor with a "resistor" with value sL, by adding a voltage source in series with value $Li_L(0^-)$ having proper polarity, a small DC circuit problem.

In general, we can make "Laplace" substitutions for each component in a circuit, then use "DC" analysis to write the Laplace-transformed differential equation directly. The appropriate substitutions and variations for both mesh and nodal analysis are discussed in detail in Chapter 36, along with more advanced frequency-domain uses of such LT replacements. After studying that material, the reader may wish to provide closure to the present discussion by solving the problem below.

Example 26. Derive the Laplace-transformed differential equation, (3.141), using Laplace circuit equivalents.

3.5 Conclusions and Further Reading

The Laplace transform (LT) is a powerful tool in the analysis and design of linear, time-invariant circuits and systems. In this exposition we have developed the LT by appealing to the manner in which it turns differential equation solutions into simple algebraic problems. We have focused not only on the technique for exploiting this advantage, but the reasons that this happens. Along the way we have discovered some useful properties of the LT, as well as the meaning of poles and zeroes. We also discussed numerous techniques for recovering the time signal from the LT, noting that this sometimes difficult task is the price paid for alleviating the difficulties of time-domain solution. In the latter part of the chapter we turned our attention to the analysis of systems, in particular the meaning and derivation of the system function. This study led to the understanding of the relationship between the system function and impulse response, and between the pole locations of the system function and stability of the system.

Most of our work has focused on the unilateral LT (ULT), although the bilateral LT (BLT) was carefully discussed at the outset. The advantage of the ULT is that it provides a way to incorporate initial condition information, a very important property in many design and analysis problems, particularly because the initial conditions play an important role in the transient response of the system. On the other hand, the BLT can handle a much broader class of signals, a feature that is often advantageous in theoretical and formal developments. The region of convergence (ROC) of the LT in the s-plane becomes much more important for the BLT as the LT itself is not unique without it. Accordingly, we spent a significant amount of effort studying the ROC. We later saw that the ROC is closely related to stability analysis, even for ULT analysis.

The LT is closely related to the Fourier and z-transforms which are the subjects of subsequent chapters. Section 3.7, Appendix B previews these ideas.

Finally, we note that several LT topics have not been treated here. These topics, such as the analysis of feedback systems and state-space models, deal principally with control theory and applications and are therefore outside of the scope of this book. Many fine books on signals and systems are available to which the reader can turn to explore these subjects. Some of these texts are referenced below. Many excellent books on circuit analysis and filter design, too numerous to cite here, are also available.

3.6 Appendix A: The Dirac Delta (Impulse) Function

The **Dirac delta, or impulse, function,** $\delta(t)$, is defined as the function with the following properties: If signal x is continuous at t_0 ("where the impulse is located in time"), then,

$$\int_a^b x(t)\delta(t-t_0)\,dt = \begin{cases} 0, & t_0 \notin [a,b] \\ x(t_0), & t_0 \in (a,b) \end{cases} \tag{3.166}$$

Two special cases are noted:

1. Note that (3.166) does not cover the case in which the impulse is located *exactly* at one of the limits of integration. In such cases whether x is continuous at t_0,

   The integral takes the value $\frac{1}{2}x(t_0^+)$ if t_0 is the lower limit, $t_0 = a$.
   The integral takes the value $\frac{1}{2}x(t_0^-)$ if t_0 is the upper limit, $t_0 = b$.

2. The only case not explicitly covered is one in which x is *discontinuous* at t_0 and t_0 is not a limit of integration. In this case the integral takes the value $\frac{1}{2}[x(t_0^-)+x(t_0^+)]$ if $t_0 \in (a,b)$, and 0 otherwise. Note that this answer is also valid if x is continuous at t_0, but it is unnecessarily complicated.

Note what happens in the special case in which $x(t)=1$, $a=-\infty$, $b=t$, and $t_0=0$. From the definition, we can write

$$\int_{-\infty}^{t} \delta(\lambda)\,d\lambda = \begin{cases} 0, & t<0 \\ 1, & t>0 \\ \frac{1}{2}, & t=0 \end{cases} \tag{3.167}$$

We see that

$$\int_{-\infty}^{t} \delta(\lambda)\,d\lambda = u(t) \tag{3.168}$$

Therefore, except at $t=0$,

$$\frac{du}{dt} = \delta(t) \tag{3.169}$$

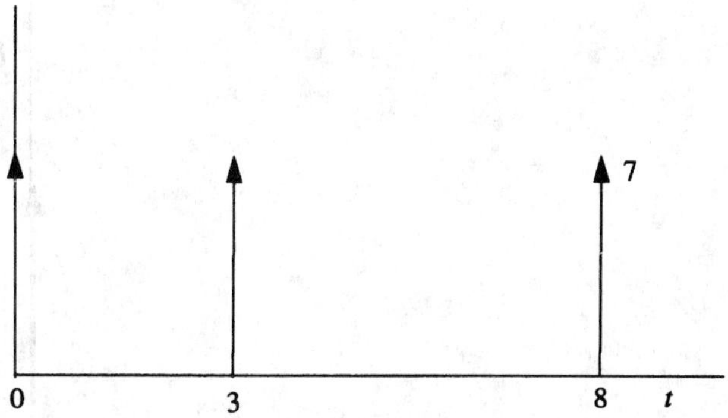

FIGURE 3.12 The impulse functions $\delta(t)$, $\delta(t-3)$, and $7\delta(t-8)$.

What emerges here is a very strange function. We see from (3.168) that δ must be zero *everywhere except at* $t = 0$ because, apparently we accumulate area at only that point. The area under that one point is unity because the integral takes a jump from 0 to 1 as t crosses zero. Because δ has zero width and unity total area, it must have *infinite amplitude* (at that one point!). To indicate the delta function, therefore, we draw "arrows" as shown in Fig. 3.12. It is sometimes mathematically useful to indicate a delta function with area other than unity. In this case we simply label the arrow with a number called the "weight" of the impulse. Note that it makes no sense to draw taller and shorter arrows for different weights, although this is sometimes done in textbooks, because the weight does not indicate the "height" of the function (∞!), rather, its *area*!

Finally, note that computing integrals with delta functions in them is very easy because one need only follow the rules of the definition. For example,

$$\int_{-18}^{\infty} e^{\pi(\lambda-4)}\delta(\lambda-6)\,d\lambda = e^{2\pi} \tag{3.170}$$

and

$$\int_{0}^{2} 3u(t)\delta(t)\,dt = 3\frac{u(0^+)}{2} = \frac{3}{2} \tag{3.171}$$

Remark. The kth derivative of the impulse (usually $k \leq 2$) occasionally appears in LT work. The notation

$$\delta^{(k)}(t) \stackrel{\text{def}}{=} \frac{d^k\delta}{dt^k} \tag{3.172}$$

is often used to denote this signal. The signal $\delta^{(1)}(t)$ is called a **doublet** and is plotted as shown in Fig. 3.13.

3.7 Appendix B: Relationships among the Laplace, Fourier, and z-Transforms

The transforms previewed here are most frequently defined and discussed for two-sided signals. Therefore, it is most natural to base this discussion on the BLT, as defined in (3.38).

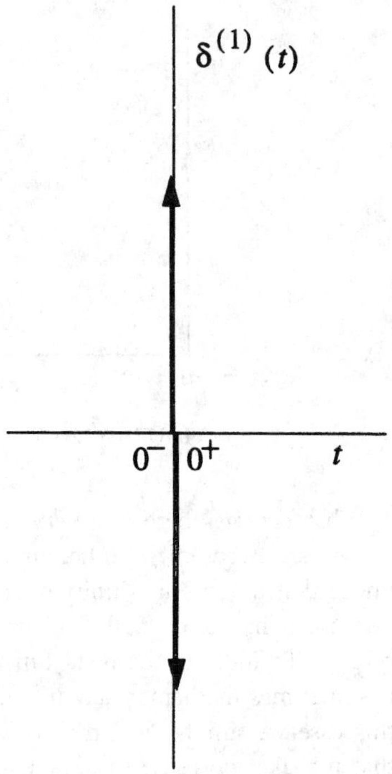

FIGURE 3.13 The doublet.

The **Fourier transform** (FT), X_F, of a signal x is defined by the integral

$$X_F(\omega) = \int_{-\infty}^{\infty} x(t)e^{j\omega t}\, dt \tag{3.173}$$

Upon comparison to (3.38), it is apparent that the FT evaluated at radian frequency ω is equivalent to the BLT of x evaluated at $s = j\omega$ in the s-plane. The FT can, therefore, be obtained over all ω by evaluating the BLT, e.g., X_L, along the $j\omega$ axis:

$$X_F(\omega) = X_L(s)\big|_{s=j\omega} \tag{3.174}$$

Evidently, the FT will only exist for a signal x if its BLT has a ROC that includes the $j\omega$ axis.

One very important class of signals whose BLT ROCs include the entire left-half s-plane, but not the $j\omega$ axis, is the periodic signals. For this purpose, the **Fourier series** (FS) can be used to expand the signal on a set of discrete, harmonically related, basis functions (either complex exponentials or sinusoids). The complex exponential version is

$$x(t) = \sum_{\ell=-\infty}^{\infty} c_\ell e^{j\ell\omega_0 t} \tag{3.175}$$

where $\omega_0 = 2\pi/T_0$ is the fundamental radian frequency, with T_0 the period of the waveform. The complex numbers $c_\ell, \ell = \ldots, -1, 0, 1, 2, \ldots$ are the **FS coefficients** computed as

$$c_\ell = \frac{1}{T_0}\int_{T_0} x(t)e^{-j\ell\omega_0 t}\, dt \tag{3.176}$$

where the integral is taken over any period of the waveform.

Comparing (3.176) and (3.173), we see that the FS coefficients are equivalent to (scaled) samples of the FT *of one period* of x, where the samples are taken at frequencies $\ell\omega_0$. If we have a periodic waveform, therefore, we can always represent it by samples of the FT of one period. Similarly, if we have a "short" signal and want to represent it using only frequency samples, we can let it be periodic and represent it using the FS coefficients. In this case we simply need to recall that the signal is not truly periodic and work with only one period.

Conversely, the FT may be represented using only **samples** of the time waveform by artificially letting the FT become periodic, then letting the time samples play the role of the FS coefficients. In this case we, in effect, let the BLT become periodic along the $j\omega$ axis. This "backward FS" is what is known as the **discrete-time Fourier transform** (DTFT). The DTFT is discussed in Chapter 4 along with the **discrete Fourier transform**, a Fourier-type transform which is discrete and periodic in both time and frequency. For the latter, the connections to the BLT are too obtuse to describe in brief terms here.

Finally, if we let the BLT become periodic in ω with some fixed period *along each* σ *line*, the BLT can also be represented by discrete-time samples. This is similar to writing a "FS" which changes for each σ. This **discrete-time Laplace transform** (DTLT) could, in principle, be used in the design and analysis of discrete-time systems in much the same way the BLT is used in continuous-time work. For historical reasons and for mathematical convenience, however, the **z-transform** (ZT) is almost universally used. The ZT is obtained from the DTLT using the mapping $e^{sT} \to z$, where T is the sample period on the time signal. As a consequence of this mapping, "strips" in the s-plane map into annuli in the z-plane. Therefore, the ROC of a ZT takes the form of an annulus and the unit circle in the z-plane plays the role of the $j\omega$ axis in the s-plane.

These ideas will become clearer and more precise through the study of successive chapters. The reader is also encouraged to see [2] for an elementary approach to discrete FTs.

ferences

[1] G. E. Carlson, *Signal and Linear System Analysis*, Boston: Houghton-Mifflin, 1992.

[2] J. R. Deller, "Tom, Dick, and Mary discover the DFT," *IEEE Signal Processing Mag.*, vol. 11, pp. 36–50, Apr. 1994.

[3] R. V. Churchill, J. W. Brown, and R. F. Verhey, *Complex Variables and Applications*, 3rd ed., New York: McGraw-Hill, 1976.

[4] G. Doetsch, *Guide to the Applications of Laplace Transforms*, New York: Van Nostrand, 1961.

[5] W. H. Hayt and J. E. Kemmerly, *Engineering Circuit Analysis*, New York: McGraw-Hill, 1971.

[6] R. C. Houts, *Signal Analysis in Linear Systems*, Philadelphia: Saunders, 1989.

[7] A. N. Kolmogorov and S. V. Fomin, *Introductory Real Analysis*, New York: Dover, 1975. Translated and edited by R. A. Silverman.

[8] P. Kraniauskas, *Transforms in Signals and Systems*, Reading, MA: Addison-Wesley, 1992.

[9] W. LePage, *Complex Variables and the Laplace Transform for Engineers*, New York: McGraw-Hill, 1961.

[10] C. D. MacGillem and G. R. Cooper, *Continuous and Discrete Signal and System Analysis*, 3rd ed., Philadelphia: Saunders, 1991.

4

Fourier Series, Fourier Transforms and the DFT

W. Kenneth Jenkins
University of Illinois, Urbana-Champaign

4.1 Introduction

Fourier methods are commonly used for signal analysis and system design in modern telecommunications, radar, and image processing systems. Classical Fourier methods such as the Fourier series and the Fourier integral are used for continuous time (CT) signals and systems, i.e., systems in which a characteristic signal, $s(t)$, is defined at all values of t on the continuum $-\infty < t < \infty$. A more recently developed set of Fourier methods, including the discrete time Fourier transform (DTFT) and the discrete Fourier transform (DFT), are extensions of basic Fourier concepts that apply to discrete time (DT) signals. A characteristic DT signal, $s[n]$, is defined only for values of n where n is an integer in the range $-\infty < n < \infty$. The following discussion presents basic concepts and outlines important properties for both the CT and DT classes of Fourier methods, with a particular emphasis on the relationships between these two classes. The class of DT Fourier methods is particularly useful as a basis for digital signal processing (DSP) because it extends the theory of classical Fourier analysis to DT signals and

0-8493-8341-2/95/$0.00+$.50

leads to many effective algorithms that can be directly implemented on general computers or special purpose DSP devices.

The relationship between the CT and the DT domains is characterized by the operations of sampling and reconstruction. If $s_a(t)$ denotes a signal $s(t)$ that has been uniformly sampled every T seconds, then the mathematical representation of $s_a(t)$ is given by

$$s_a(t) = \sum_{n=-\infty}^{\infty} s(t)\delta(t - nT) \tag{4.1}$$

where $\delta(t)$ is a CT impulse function defined to be zero for all $t \neq 0$, undefined at $t = 0$, and has unit area when integrated from $t = -\infty$ to $t = +\infty$. Because the only places at which the product $s(t)\delta(t - nT)$ is not identically equal to zero are at the sampling instances, $s(t)$ in (4.1) can be replaced with $s(nT)$ without changing the overall meaning of the expression. Hence, an alternate expression for $s_a(t)$ that is often useful in Fourier analysis is given by

$$s_a(t) = \sum_{n=-\infty}^{\infty} s(nT)\delta(t - nT) \tag{4.2}$$

The CT sampling model $s_a(t)$ consists of a sequence of CT impulse functions uniformly spaced at intervals of T seconds and weighted by the values of the signal $s(t)$ at the sampling instants, as depicted in Fig. 4.1. Note that $s_a(t)$ is not defined at the sampling instants because the CT impulse function itself is not defined at $t = 0$. However, the values of $s(t)$ at the sampling instants are imbedded as "area under the curve" of $s_a(t)$, and as such represent a useful mathematical model of the sampling process. In the DT domain the sampling model is simply the sequence defined by taking the values of $s(t)$ at the sampling instants, i.e.,

$$s[n] = s(t)|_{t=nT} \tag{4.3}$$

In contrast to $s_a(t)$, which is not defined at the sampling instants, $s[n]$ is well defined at the sampling instants, as illustrated in Fig. 4.2. Thus, it is now clear that $s_a(t)$ and $s[n]$ are different but equivalent models of the sampling process in the CT and DT domains, respectively. They are both useful for signal analysis in their corresponding domains. Their equivalence is established by the fact that they have equal spectra in the Fourier domain, and that the underlying CT signal from which $s_a(t)$ and $s[n]$ are derived can be recovered from either sampling representation, provided a sufficiently large sampling rate is used in the sampling operation (see below).

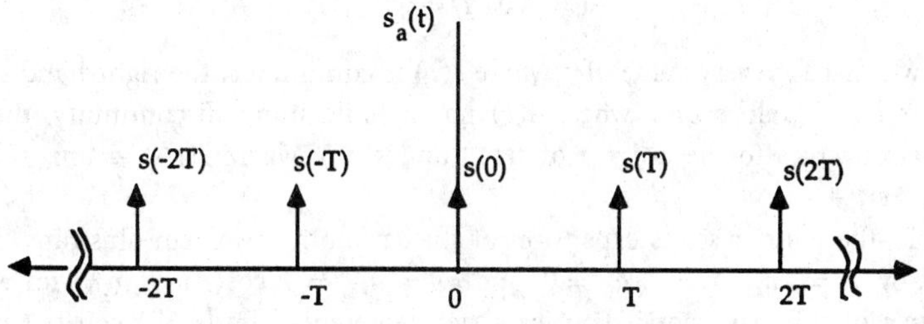

FIGURE 4.1 CT model of a sampled CT signal.

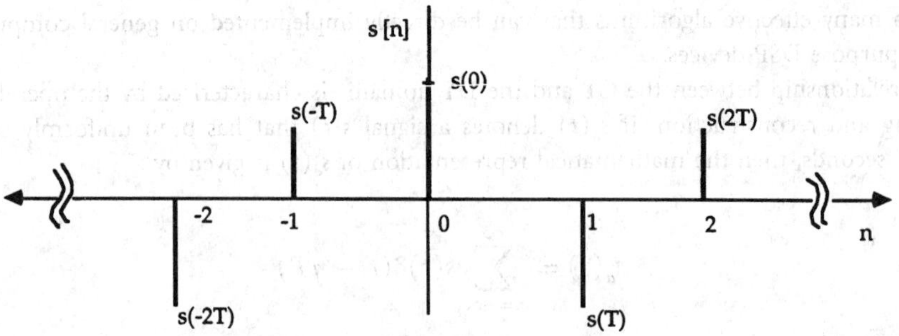

FIGURE 4.2 DT model of a sampled CT signal.

4.2 Fourier Series Representation of Continuous Time Periodic Signals

It is convenient to begin this discussion with the classical Fourier series representation of a periodic time domain signal, and then derive the Fourier integral from this representation by finding the limit of the Fourier coefficient representation as the period goes to infinity. The conditions under which a periodic signal $s(t)$ can be expanded in a Fourier series are known as the Dirichet conditions. They require that in each period $s(t)$ has a finite number of discontinuities, a finite number of maxima and minima, and that $s(t)$ satisfies the following absolute convergence criterion [1]:

$$\int_{-T/2}^{T/2} |s(t)|\, dt < \infty \tag{4.4}$$

It is assumed in the following discussion that these basic conditions are satisfied by all functions that will be represented by a Fourier series.

Exponential Fourier Series

If a CT signal $s(t)$ is periodic with a period T, then the classical complex Fourier series representation of $s(t)$ is given by

$$s(t) = \sum_{n=-\infty}^{\infty} a_n e^{jn\omega_0 t} \tag{4.5a}$$

where $\omega_0 = 2\pi/T$, and where the a_n are the complex Fourier coefficients given by

$$a_n = (1/T) \int_{-T/2}^{T/2} s(t) e^{-jn\omega_0 t}\, dt \tag{4.5b}$$

It is well known that for every value of t where $s(t)$ is continuous, the right-hand side of (4.5a) converges to $s(t)$. At values of t where $s(t)$ has a finite jump discontinuity, the right-hand side of (4.5a) converges to the average of $s(t^-)$ and $s(t^+)$, where $s(t^-) \equiv \lim_{\epsilon \to 0} s(t - \epsilon)$ and $s(t^+) \equiv \lim_{\epsilon \to 0} s(t + \epsilon)$.

For example, the Fourier series expansion of the sawtooth waveform illustrated in Fig. 4.3 is characterized by $T = 2\pi$, $\omega_0 = 1$, $a_0 = 0$, and $a_n = a_{-n} = A\cos(n\pi)/(jn\pi)$ for $n = 1, 2, \ldots,$. The coefficients of the exponential Fourier series represented by (4.5b) can be interpreted as the spectral representation of $s(t)$, because the a_n-th coefficient represents the contribution of

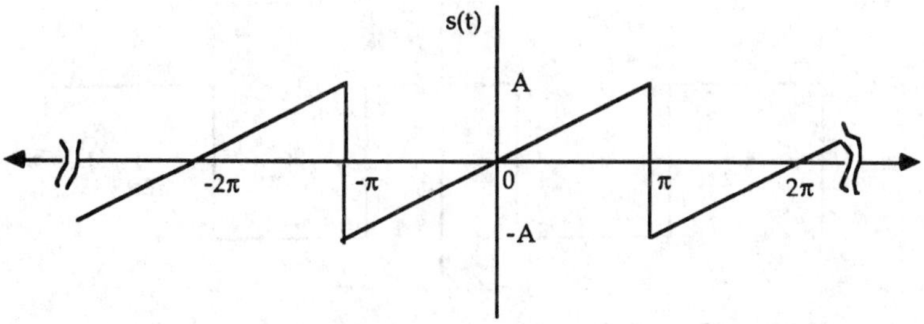

FIGURE 4.3 Periodic CT signal used in Fourier series example.

the $(n\omega_0)$-th frequency to the total signal $s(t)$. Because the a_n are complex valued, the Fourier domain representation has both a magnitude and a phase spectrum. For example, the magnitude of the a_n is plotted in Fig. 4.4 for the sawtooth waveform of Fig. 4.3. The fact that the a_n constitute a discrete set is consistent with the fact that a periodic signal has a "line spectrum", i.e., the spectrum contains only integer multiples of the fundamental frequency ω_0. Therefore, the equation pair given by (4.5a) and (4.5b) can be interpreted as a transform pair that is similar to the CT Fourier transform for periodic signals. This leads to the observation that the classical Fourier series can be interpreted as a special transform that provides a one-to-one invertible mapping between the discrete-spectral domain and the CT domain. The next section shows how the periodicity constraint can be removed to produce the more general classical CT Fourier transform which applies equally well to periodic and aperiodic time domain waveforms.

The Trigonometric Fourier Series

Although Fourier series expansions exist for complex periodic signals, and Fourier theory can be generalized to the case of complex signals, the theory and results are more easily expressed for real-valued signals. The following discussion assumes that the signal $s(t)$ is real-valued for the sake of simplifying the discussion. However, all results are valid for complex signals, although the details of the theory will become somewhat more complicated.

For real-valued signals $s(t)$, it is possible to manipulate the complex exponential form of the Fourier series into a trigonometric form that contains $\sin(\omega_0 t)$ and $\cos(\omega_0 t)$ terms with corresponding real-valued coefficients [1]. The trigonometric form of the Fourier series for a real-valued signal $s(t)$ is given by

$$s(t) = \sum_{n=0}^{\infty} b_n \cos(n\omega_0 t) + \sum_{n=1}^{\infty} c_n \sin(n\omega_0 t) \tag{4.6a}$$

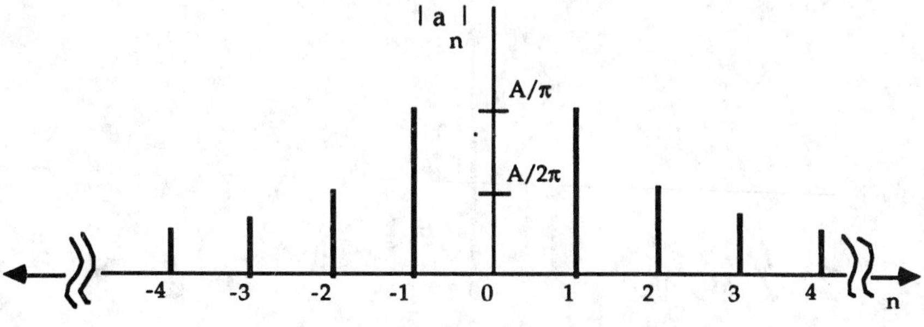

FIGURE 4.4 Magnitude of the Fourier coefficients for example of Figure 4.3.

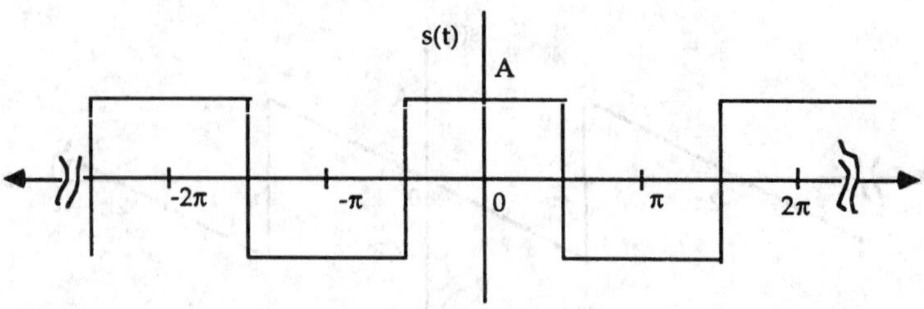

FIGURE 4.5 Periodic CT signal used in Fourier series example 2.

where $\omega_0 = 2\pi/T$. The b_n and c_n are real-valued Fourier coefficients determined by

$$b_0 = (1/T)\int_{-T/2}^{T/2} s(t)\,dt$$

$$b_n = (2/T)\int_{-T/2}^{T/2} s(t)\cos(n\omega_0 t)\,dt, \qquad n = 1, 2, \ldots, \tag{4.6b}$$

$$c_n = (2/T)\int_{-T/2}^{T/2} s(t)\sin(n\omega_0 t)\,dt, \qquad n = 1, 2, \ldots,$$

An arbitrary real-valued signal $s(t)$ can be expressed as a sum of even and odd components, $s(t) = s_{\text{even}}(t) + s_{\text{odd}}(t)$, where $s_{\text{even}}(t) = s_{\text{even}}(-t)$ and $s_{\text{odd}}(t) = -s_{\text{odd}}(-t)$, and where $s_{\text{even}}(t) = [s(t) + s(-t)]/2$ and $s_{\text{odd}}(t) = [s(t) - s(-t)]/2$. For the trigonometric Fourier series, it can be shown that $s_{\text{even}}(t)$ is represented by the (even) cosine terms in the infinite series, $s_{\text{odd}}(t)$ is represented by the (odd) sine terms, and b_0 is the DC level of the signal. Therefore, if it can be determined by inspection that a signal has a DC level, or if it is even or odd, then the correct form of the trigonometric series can be chosen to simplify the analysis. For example, it is easily seen that the signal shown in Fig. 4.5 is an even signal with a zero DC level. Therefore it can be accurately represented by the cosine series with $b_n = 2A\sin(\pi n/2)/(\pi n/2), n = 1, 2, \ldots,$ as illustrated in Fig. 4.6. In contrast, note that the sawtooth waveform used in the previous example is an odd signal with zero DC level; thus, it can be completely specified by the sine terms of the trigonometric series. This result can be demonstrated by pairing each positive frequency component from the exponential series with its conjugate partner, i.e., $c_n = \sin(n\omega_0 t) = a_n e^{jn\omega_0 t} + a_{-n}e^{-jn\omega_0 t}$, whereby it is found that $c_n = 2\text{A}\,\cos(n\pi)/(n\pi)$ for this example. In general it is found that $a_n = (b_n - jc_n)/2$ for $n = 1, 2, \ldots, a_0 = b_0$, and $a_{-n} = a_n^*$. The trigonometric Fourier series is common in the signal processing literature because it replaces complex coefficients with real ones and often results in a simpler and more intuitive interpretation of the results.

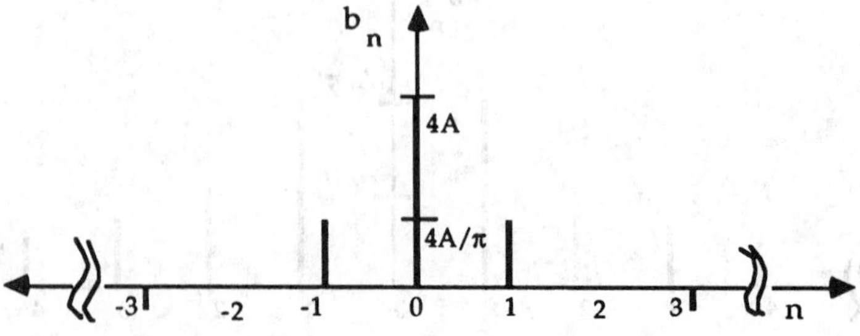

FIGURE 4.6 Fourier coefficients for example of Figure 4.5.

Convergence of the Fourier Series

The Fourier series representation of a periodic signal is an approximation that exhibits mean squared convergence to the true signal. If $s(t)$ is a periodic signal of period T, and $s'(t)$ denotes the Fourier series approximation of $s(t)$, then $s(t)$ and $s'(t)$ are equal in the mean square sense if

$$\text{MSE} = \int_{-T/2}^{T/2} |s(t) - s(t)'|^2 \, dt = 0 \tag{4.7}$$

Even with (4.7) satisfied, mean square error (MSE) convergence does not mean that $s(t) = s'(t)$ at every value of t. In particular, it is known that at values of t, where $s(t)$ is discontinuous, the Fourier series converges to the average of the limiting values to the left and right of the discontinuity. For example, if t_0 is a point of discontinuity, then $s'(t_0) = [s(t_0^-)+s(t_0^+)]/2$, where $s(t_0^-)$ and $s(t_0^+)$ were defined previously. (Note that at points of continuity, this condition is also satisfied by the definition of continuity.) Because the Dirichet conditions require that $s(t)$ have at most a finite number of points of discontinuity in one period, the set S_t, defined as all values of t within one period where $s(t) \neq s'(t)$, contains a finite number of points, and S_t is a set of measure zero in the formal mathematical sense. Therefore, $s(t)$ and its Fourier series expansion $s'(t)$ are *equal almost everywhere*, and $s(t)$ can be considered identical to $s'(t)$ for the analysis of most practical engineering problems.

Convergence almost everywhere is satisfied only in the limit as an infinite number of terms are included in the Fourier series expansion. If the infinite series expansion of the Fourier series is truncated to a finite number of terms, as it must be in practical applications, then the approximation will exhibit an oscillatory behavior around the discontinuity, known as the Gibbs phenomenon [1]. Let $s'_N(t)$ denote a truncated Fourier series approximation of $s(t)$, where only the terms in (4.5a) from $n = -N$ to $n = N$ are included if the complex Fourier series representation is used, or where only the terms in (4.6a) from $n = 0$ to $n = N$ are included if the trigonometric form of the Fourier series is used. It is well known that in the vicinity of a discontinuity at t_0 the Gibbs phenomenon causes $s'_N(t)$ to be a poor approximation to $s(t)$. The peak magnitude of the Gibbs oscillation is 13% of the size of the jump discontinuity $s(t_0^-) - s(t_0^+)$ regardless of the number of terms used in the approximation. As N increases, the region that contains the oscillation becomes more concentrated in the neighborhood of the discontinuity, until, in the limit as N approaches infinity, the Gibbs oscillation is squeezed into a single point of mismatch at t_0.

If $s'(t)$ is replaced by $s'_N(t)$ from (4.7), it is important to understand the behavior of the error MSE_N as a function of N, where

$$\text{MSE}_N = \int_{-T/2}^{T/2} |s(t) - s'_N(t)|^2 \, dt \tag{4.8}$$

An important property of the Fourier series is that the exponential basis functions $e^{jn\omega_0 t}$ (or $\sin(n\omega_0 t)$ and $\cos(n\omega_0 t)$ for the trigonometric form) for $n = 0, \pm 1, \pm 2, \ldots$ (or $n = 0, 1, 2, \ldots$ for the trigonometric form) constitute an orthonormal set, i.e., $t_{nk} = 1$ for $n = k$, and $t_{nk} = 0$ for $n \neq k$, where

$$t_{nk} = (1/T) \int_{-T/2}^{T/2} (e^{-jn\omega_0 t})(e^{jk\omega_0 t}) \, dt \tag{4.9}$$

As terms are added to the Fourier series expansion, the orthogonality of the basis functions guarantees that the error decreases in the mean square sense, i.e., that MSE_N monotonically decreases as N is increased. Therefore, a practitioner can proceed with the confidence that when applying Fourier series analysis more terms are always better than fewer in terms of the accuracy of the signal representations.

4.3 The Classical Fourier Transform for Continuous Time Signals

The periodicity constraint imposed on the Fourier series representation can be removed by taking the limits of (4.5a) and (4.5b) as the period T is increased to infinity. Some mathematical preliminaries are required so that the results will be well defined after the limit is taken. It is convenient to remove the $(1/T)$ factor in front of the integral by multiplying (4.5b) through by T, and then replacing Ta_n by a'_n in both (4.5a) and (4.5b). Because $\omega_0 = 2\pi/T$, as T increases to infinity, ω_0 becomes infinitesimally small, a condition that is denoted by replacing ω_0 with $\Delta\omega$. The factor $(1/T)$ in (4.5a) becomes $(\Delta\omega/2\pi)$. With these algebraic manipulations and changes in notation (4.5a) and (4.5b) take on the following form prior to taking the limit:

$$s(t) = (1/2\pi)\sum_{n=-\infty}^{\infty} a'_n e^{jn\Delta\omega t}\Delta\omega \tag{4.10a}$$

$$a'_n = \int_{-T/2}^{T/2} s(t)e^{-jn\Delta\omega t}\,dt \tag{4.10b}$$

The final step in obtaining the CT Fourier transform is to take the limit of both (4.10a) and (4.10b) as $T \to \infty$. In the limit the infinite summation in (4.10a) becomes an integral, $\Delta\omega$ becomes $d\omega$, $n\Delta\omega$ becomes ω, and a'_n becomes the CT Fourier transform of $s(t)$, denoted by $S(j\omega)$. The result is summarized by the following transform pair, which is known throughout most of the engineering literature as the classical CT Fourier transform (CTFT):

$$s(t) = (1/2\pi)\int_{-\infty}^{\infty} S(j\omega)e^{j\omega t}\,d\omega \tag{4.11a}$$

$$S(j\omega) = \int_{-\infty}^{\infty} s(t)e^{-j\omega t}\,dt \tag{4.11b}$$

Often (4.11a) is called the Fourier integral and (4.11b) is simply called the Fourier transform. The relationship $S(j\omega) = \mathcal{F}\{s(t)\}$ denotes the Fourier transformation of $s(t)$, where $\mathcal{F}\{\cdot\}$ is a symbolic notation for the Fourier transform operator, and where ω becomes the continuous frequency variable after the periodicity constraint is removed. A transform pair $s(t) \leftrightarrow S(j\omega)$ represents a one-to-one invertible mapping as long as $s(t)$ satisfies conditions which guarantee that the Fourier integral converges. (More mathematical details of the CTFT are presented in Chapter 6.)

From (4.11a) it is easily seen that $\mathcal{F}\{\delta(t-t_0)\} = e^{-j\omega t_0}$, and from (4.11b) that $\mathcal{F}^{-1}\{2\pi\delta(\omega-\omega_0)\} = e^{j\omega_0 t}$, so that $\delta(t-t_0) \leftrightarrow e^{-j\omega t_0}$ and $e^{j\omega_0 t} \leftrightarrow 2\pi\delta(\omega-\omega_0)$ are valid Fourier transform pairs. Using these relationships it is easy to establish the Fourier transforms of $\cos(\omega_0 t)$ and $\sin(\omega_0 t)$, as well as many other useful waveforms that are encountered in common signal analysis problems. A number of such transforms are shown in Table 4.1.

The CTFT is useful in the analysis and design of CT systems, i.e., systems that process CT signals. Fourier analysis is particularly applicable to the design of CT filters which are characterized by Fourier magnitude and phase spectra, i.e., by $|H(j\omega)|$ and arg $H(j\omega)$, where $H(j\omega)$ is commonly called the frequency response of the filter. For example, an **ideal transmission channel** is one which passes a signal without distorting it. The signal may be scaled by a real constant A and delayed by a fixed time increment t_0, implying that the impulse response of an ideal channel is $A\delta(t-t_0)$, and its corresponding frequency response is $Ae^{-j\omega t_0}$. Hence, the frequency response of an ideal channel is specified by constant amplitude for all frequencies, and a phase characteristic which is a linear function given by ωt_0.

TABLE 4.1 Some Basic CTFT Pairs

Signal	Fourier Transform	Fourier Series Coefficients (if periodic)
$\sum_{k=-\infty}^{+\infty} a_k e^{jk\omega_0 t}$	$2\pi \sum_{k=-\infty}^{+\infty} a_k \delta(\omega_k \omega_0)$	a_k
$e^{j\omega_0 t}$	$2\pi\delta(\omega + \omega_0)$	$a_1 = 1$ $a_k = 0$, otherwise
$\cos \omega_0 t$	$\pi[\delta(\omega - \omega_0) + \delta(\omega + \omega_0)]$	$a_1 = a_{-1} = \frac{1}{2}$ $a_k = 0$, otherwise
$\sin \omega_0 t$	$\frac{\pi}{j}[\delta(\omega - \omega_0) - \delta(\omega + \omega_0)]$	$a_1 = -a_{-1} = \frac{1}{2j}$ $a_k = 0$, otherwise
$x(t) = 1$	$2\pi\delta(\omega)$	$a_0 = 1$, $a_k = 0$, $k \neq 0$ (has this Fourier series representation for any choice of $T_0 > 0$)
Periodic square wave $x(t) = \begin{cases} 1, & \lvert t\rvert < T_1 \\ 0, & T_1 < \lvert t\rvert \leq \frac{T_0}{2} \end{cases}$ and $x(t + T_0) = x(t)$	$\sum_{k=-\infty}^{+\infty} \frac{2 \sin k\omega_0 T_1}{k} \delta(\omega_k \omega_0)$	$\frac{\omega_0 T_1}{\pi} \sin c\left(\frac{k\omega_0 T_1}{\pi}\right) = \frac{\sin k\omega_0 T_1}{k\pi}$
$\sum_{n=-\infty}^{+\infty} \delta(t - nT)$	$\frac{2\pi}{T} \sum^{+\infty} k = -\infty\delta\left(\omega - \frac{2\pi k}{T}\right)$	$a_k = \frac{1}{T}$ for all k
$x(t) = \begin{cases} 1, & \lvert t\rvert < T_1 \\ 0, & \lvert t\rvert > T_1 \end{cases}$	$2T_1 \sin c\left(\frac{\omega T_1}{\pi}\right) = \frac{2 \sin \omega T_1}{\omega}$	—
$\frac{W}{\pi} \sin c\left(\frac{Wt}{\pi}\right) = \frac{\sin Wt}{\pi t}$	$X(\omega) = \begin{cases} 1, & \lvert\omega\rvert < W \\ 0, & \lvert\omega\rvert > W \end{cases}$	—
$\delta(t)$	1	—
$u(t)$	$\frac{1}{j\omega} + \pi\delta(\omega)$	—
$\delta(t - t_0)$	$e_{j\omega t_0}$	—
$e^{-at}u(t), \mathrm{Re}\{a\} > 0$	$\frac{1}{a + j\omega}$	—
$te^{-at}u(t), \mathrm{Re}\{a\} > 0$	$\frac{1}{(a + j\omega)^2}$	—
$\frac{t^{n-1}}{(n-1)!} e^{-at} u(t)$, $\mathrm{Re}\{a\} > 0$	$\frac{1}{(a + j\omega)^n}$	—

Source: A. V. Oppenheim et al., Signals and Systems, Englewood Cliffs, NJ: Prentice-Hall, 1983.

Properties of the Continuous Time Fourier Transform

The CTFT has many properties that make it useful for the analysis and design of linear CT systems. Some of the more useful properties are stated below. A more complete list of the CTFT properties is given in Table 4.2. Proofs of these properties can be found in [2] and [3]. In the following discussion $\mathcal{F}\{\cdot\}$ denotes the Fourier transform operation, $\mathcal{F}^{-1}\{\cdot\}$ denotes the inverse

TABLE 4.2 Properties of the CTFT

Name	If $\mathcal{F} f(t) = F(j\omega)$, then
Definition	$f(j\omega) = \int_{-\infty}^{\infty} f(t) e_{j\omega t}\, dt$
	$f(t) = \frac{1}{2\pi} \int_{-\infty}^{\infty} F(j\omega) e^{j\omega t}\, d\omega$
Superposition	$\mathcal{F}[af_1(t) + bf_2(t)] = aF_1(j\omega) + bF_2(j\omega)$
Simplification if:	
(a) $f(t)$ is even	$F(j\omega) = 2 \int_0^{\infty} f(t) \cos \omega t\, dt$
(b) $f(t)$ is odd	$F(j\omega) = 2j \int_0^{\infty} f(t) \sin \omega t\, dt$
Negative t	$\mathcal{F} f(-t) = F^*(j\omega)$
Scaling:	
(a) Time	$\mathcal{F} f(at) = \frac{1}{\lvert a \rvert} F\left(\frac{j\omega}{a}\right)$
(b) Magnitude	$\mathcal{F} af(t) = aF(j\omega)$
Differentiation	$\mathcal{F}\left[\frac{d^n}{dt^n} f(t)\right] = (j\omega)^n F(j\omega)$
Integration	$\mathcal{F}\left[\int_{-\infty}^{t} f(x)\, dx\right] = \frac{1}{j\omega} F(j\omega) + \pi F(0)\delta(\omega)$
Time shifting	$\mathcal{F} f(t - a) = F(j\omega) e^{j\omega a}$
Modulation	$\mathcal{F} f(t) e^{j\omega_0 t} = F[j(\omega - \omega_0)]$ $\{\mathcal{F} f(t) \cos \omega_0 t = \frac{1}{2} F[j(\omega - \omega_0)] + F[j(\omega + \omega_0)]\}$ $\{\mathcal{F} f(t) \sin \omega_0 t = \frac{1}{2} j[F[j(\omega - \omega_0)] - F[j(\omega + \omega_0)]\}$
Time convolution	$\mathcal{F}^{-1}[F_1(j\omega) F_2(j\omega)] = \int_{-\infty}^{\infty} f_1(\tau) f_2(\tau) f_2(t_\tau)\, d\tau$
Frequency convolution	$\mathcal{F}[f_1(t) f_2(t)] = \frac{1}{2\pi} \int_{-\infty}^{\infty} F_1(j\lambda) F_2[j(\omega_\lambda)]\, d\lambda$

Source: M. E. VanValkenburg, Network Analysis, (3rd edition), Englewood Cliffs, NJ: Prentice-Hall, 1974.

Fourier transform operation, and $*$ denotes the convolution operation defined as

$$f_1(t) * f_2(t) = \int_{-\infty}^{\infty} f_1(t - \tau) f_2(\tau)\, d\tau$$

1. Linearity (superposition): $\mathcal{F}\{af_1(t) + bf_2(t)\} = a\mathcal{F}\{f_1(t)\} + b\mathcal{F}\{f_2(t)\}$ (a and b, complex constants)
2. Time shifting: $\mathcal{F}\{f(t - t_0)\} = e^{-j\omega t_0} \mathcal{F}\{f(t)\}$
3. Frequency shifting: $e^{j\omega_0 t} f(t) = \mathcal{F}^{-1}\{F(j(\omega - \omega_0))\}$
4. Time domain convolution: $\mathcal{F}\{f_1(t) * f_2(t)\} = \mathcal{F}\{f_1(t)\}\mathcal{F}\{f_2(t)\}$
5. Frequency domain convolution: $\mathcal{F}\{f_1(t) f_2(t)\} = (1/2\pi)\mathcal{F}\{f_1(t)\} * \mathcal{F}\{f_2(t)\}$

6. Time differentiation: $-j\omega F(j\omega) = \mathcal{F}\{d(f(t))/dt\}$
7. Time integration: $\mathcal{F}\{\int_{-\infty}^{t} f(\tau)\,d\tau\} = (1/j\omega)F(j\omega) + \pi F(0)\delta(\omega)$

The above properties are particularly useful in CT system analysis and design, especially when the system characteristics are easily specified in the frequency domain, as in linear filtering. Note that properties 1, 6, and 7 are useful for solving differential or integral equations. Property 4 provides the basis for many signal processing algorithms because many systems can be specified directly by their impulse or frequency response. Property 3 is particularly useful in analyzing communication systems in which different modulation formats are commonly used to shift spectral energy to frequency bands that are appropriate for the application.

Fourier Spectrum of the Continuous Time Sampling Model

Because the CT sampling model $s_a(t)$, given in (4.1), is in its own right a CT signal, it is appropriate to apply the CTFT to obtain an expression for the spectrum of the sampled signal:

$$\mathcal{F}\{s_a(t)\} = \mathcal{F}\left\{\sum_{n=-\infty}^{\infty} s(t)\delta(t - nT)\right\} = \sum_{n=-\infty}^{\infty} s(nT)e^{-j\omega Tn} \qquad (4.12)$$

Becauses the expression on the right-hand side of (4.12) is a function of $e^{j\omega T}$ it is customary to denote the transform as $F(e^{j\omega T}) = \mathcal{F}\{s_a(t)\}$. Later in the chapter this result is compared to the result of operating on the DT sampling model, namely $s[n]$, with the DT Fourier transform to illustrate that the two sampling models have the same spectrum.

Fourier Transform of Periodic Continuous Time Signals

Was saw earlier that a periodic CT signal can be expressed in terms of its Fourier series. The CTFT can then be applied to the Fourier series representation of $s(t)$ to produce a mathematical expression for the "line spectrum" characteristic of periodic signals.

$$\mathcal{F}\{s(t)\} = \mathcal{F}\left\{\sum_{n=-\infty}^{\infty} a_n e^{jn\omega_0 t}\right\} = 2\pi \sum_{n=-\infty}^{\infty} a_n \delta(\omega - n\omega_0) \qquad (4.13)$$

The spectrum is shown pictorially in Fig. 4.7. Note the similarity between the spectral representation of Fig. 4.7 and the plot of the Fourier coefficients in Fig. 4.4, which was heuristically interpreted as a "line spectrum". Figures 4.4 and 4.7 are different but equivalent representations of the Fourier spectrum. Note that Fig. 4.4 is a DT representation of the spectrum, while Fig. 4.7 is a CT model of the same spectrum.

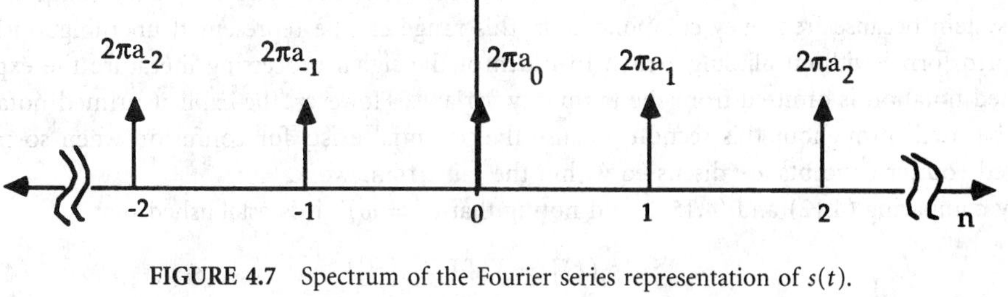

FIGURE 4.7 Spectrum of the Fourier series representation of $s(t)$.

The Generalized Complex Fourier Transform

The CTFT characterized by (4.11a) and (4.11b) can be generalized by considering the variable $j\omega$ to be the special case of $u = \sigma + j\omega$ with $\sigma = 0$, writing (4.11) in terms of u, and interpreting u as a complex frequency variable. The resulting complex Fourier transform pair is given by (4.14a) and (4.14b)

$$s(t) = (1/2\pi j)\int_{\sigma-j\infty}^{\sigma+j\infty} S(u)e^{jut}\,du \tag{4.14a}$$

$$S(u) = \int_{-\infty}^{\infty} s(t)e^{-jut}\,dt \tag{4.14b}$$

The set of all values of u for which the integral of (4.14b) converges is called the region of convergence (ROC). Because the transform $S(u)$ is defined only for values of u within the ROC, the path of integration in (4.14a) must be defined by σ so that the entire path lies within the ROC. In some literature this transform pair is called the **bilateral Laplace transform** because it is the same result obtained by including both the negative and positive portions of the time axis in the classical Laplace transform integral. [Note that in (4.14) the complex frequency variable was denoted by u rather than by the more common s, in order to avoid confusion with earlier uses of $s(\cdot)$ as signal notation.] The complex Fourier transform (bilateral Laplace transform) is not often used in solving practical problems, but its significance lies in the fact that it is the most general form that represents the point at which Fourier and Laplace transform concepts become the same. Identifying this connection reinforces the notion that Fourier and Laplace transform concepts are similar because they are derived by placing different constraints on the same general form. (For further reading on the Laplace transform see Chapter 3.)

4.4 The Discrete Time Fourier Transform

The discrete time Fourier transform (DTFT) can be obtained by using the DT sampling model and considering the relationship obtained in (4.12) to be the definition of the DTFT. Letting $T = 1$ so that the sampling period is removed from the equations and the frequency variable is replaced with a normalized frequency $\omega' = \omega T$, the DTFT pair is defined in (4.15). Note that in order to simplify notation it is not customary to distinguish between ω and ω', but rather to rely on the context of the discussion to determine whether ω refers to the normalized ($T = 1$) or the unnormalized ($T \neq 1$) frequency variable.

$$S(e^{j\omega'}) = \sum_{n=-\infty}^{\infty} s[n]e^{-j\omega' n} \tag{4.15a}$$

$$s[n] = (1/2\pi)\int_{-\pi}^{\pi} S(e^{j\omega'})e^{jn\omega'}\,d\omega' \tag{4.15b}$$

The spectrum $S(e^{j\omega'})$ is periodic in ω' with period 2π. The fundamental period in the range $-\pi < \omega' \leq \pi$, sometimes referred to as the baseband, is the useful frequency range of the DT system because frequency components in this range can be represented unambiguously in sampled form (without aliasing error). In much of the signal processing literature the explicit primed notation is omitted from the frequency variable. However, the explicit primed notation will be used throughout this section because the potential exists for confusion when so many related Fourier concepts are discussed within the same framework.

By comparing (4.12) and (4.15a), and noting that $\omega' = \omega T$, it is established that

$$\mathcal{F}\{s_a(t)\} = \mathrm{DTFT}\{s[n]\} \tag{4.16}$$

where $s[n] = s(t)_{t=nT}$. This demonstrates that the spectrum of $s_a(t)$, as calculated by the CT Fourier transform is identical to the spectrum of $s[n]$ as calculated by the DTFT. Therefore, although $s_a(t)$ and $s[n]$ are quite different sampling models, they are equivalent in the sense that they have the same Fourier domain representation.

A list of common DTFT pairs is presented in Table 4.3. Just as the CT Fourier transform is useful in CT signal system analysis and design, the DTFT is equally useful in the same capacity for DT systems. It is indeed fortuitous that Fourier transform theory can be extended in this way to apply to DT systems.

In the same way that the CT Fourier transform was found to be a special case of the complex Fourier transform (or bilateral Laplace transform), the DTFT is a special case of the bilateral z-transform with $z = e^{j\omega' t}$. The more general bilateral z-transform is given by:

$$S(z) = \sum_{n=-\infty}^{\infty} s[n]z^{-n} \tag{4.17a}$$

$$s[n] = (1/2\pi j)\int_{C} S(z)z^{n-1}\,dz \tag{4.17b}$$

where C is a counterclockwise contour of integration which is a closed path completely contained within the region of convergence of $S(z)$. Recall that the DTFT was obtained by taking the

TABLE 4.3 Some Basic DTFT Pairs

Sequence	Fourier Transform
1. $\delta[n]$	1
2. $\delta[n - n_0]$	$e^{-j\omega n_0}$
3. $1 \quad (-\infty < n < \infty)$	$\sum_{k=-\infty}^{\infty} 2\pi\delta(\omega + 2\pi k)$
4. $a^n u[n] \quad (\lvert a\rvert < 1)$	$\dfrac{1}{1 - ae^{-j\omega}}$
5. $u[n]$	$\dfrac{1}{1 - e^{-j\omega}} + \sum_{k=-\infty}^{\infty} \pi\delta(\omega + 2\pi k)$
6. $(n+1)a^n u[n] \quad (\lvert a\rvert < 1)$	$\dfrac{1}{(1 - ae^{-j\omega})^2}$
7. $\dfrac{r^2 \sin \omega_p(n+1)}{\sin \omega_p} u[n] \quad (\lvert r\rvert < 1)$	$\dfrac{1}{1 - 2r\cos\omega_p e^{-j\omega} + r^2 e^{j2\omega}}$
8. $\dfrac{\sin \omega_c n}{\pi n}$	$Xe^{j\omega} = \begin{cases} 1, & \lvert\omega\rvert < \omega_c \\ 0, & \omega_c < \lvert\omega\rvert \le \pi \end{cases}$
9. $x[n] - \begin{cases} 1, & 0 \le n \le M \\ 0, & \text{otherwise} \end{cases}$	$\dfrac{\sin[\omega(M+1)/2]}{\sin(\omega/2)} e^{-j\omega M/2}$
10. $e^{j\omega_0 n}$	$\sum_{k=-\infty}^{\infty} 2\pi\delta(\omega - \omega_0 + 2\pi k)$
11. $\cos(\omega_0 n + \phi)$	$\pi \sum_{k=-\infty}^{\infty} [e^{j\phi}\delta(\omega - \omega_0 + 2\pi k) + e^{-j\phi}\delta(\omega + \omega_0 + 2\pi k)]$

Source: A. V. Oppenheim and R. W. Schafer, Digital Signal Processing, Englewood Cliffs, NJ: Prentice-Hall, 1975.

CT Fourier transform of the CT sampling model represented by $s_a(t)$. Similarly, the bilateral z-transform results by taking the bilateral Laplace transform of $s_a(t)$. If the lower limit on the summation of (4.17a) is taken to be $n = 0$, then (4.17a) and (4.17b) become the one-sided z-transform, which is the DT equivalent of the one-sided LT for CT signals. The hierarchical relationship among these various concepts for DT systems is discussed later in this chapter, where it will be shown that the family structure of the DT family tree is identical to that of the CT family. For every CT transform in the CT world there is an analogous DT transform in the DT world, and vice versa. (For further reading on the z-transform see Chapter 5.)

Properties of the Discrete Time Fourier Transform

Because the DTFT is a close relative of the classical CT Fourier transform it should come as no surprise that many properties of the DTFT are similar to those presented for the CT Fourier transform in the previous section. In fact, for many of the properties presented earlier an analogous property exists for the DTFT. The following list parallels the list that was presented in the previous section for the CT Fourier transform, to the extent that the same property exists. A more complete list of DTFT pairs is given in Table 4.4. (Note that the primed notation on ω' is dropped in the following to simplify the notation, and to be consistent with standard usage.)

1. Linearity (superposition): $\mathrm{DTFT}\{af_1[n] + bf_2[n]\} = a\mathrm{DTFT}\{f_1[n]\} + b\mathrm{DTFT}\{f_2[n]\}$ (a and b, complex constants)
2. Index shifting: $\mathrm{DTFT}\{f[n - n_0]\} = e^{-j\omega n_0}\mathrm{DTFT}\{f[n]\}$
3. Frequency shifting: $e^{j\omega_0 n}f[n] = \mathrm{DTFT}^{-1}\{F(e^{j(\omega-\omega_0)})\}$
4. Time domain convolution: $\mathrm{DTFT}\{f_1[n] * f_2[n]\} = \mathrm{DTFT}\{f_1[n]\}\mathrm{DTFT}\{f_2[n]\}$
5. Frequency domain convolution: $\mathrm{DTFT}\{f_1[n]f_2[n]\} = (1/2\pi)\mathrm{DTFT}\{f_1[n]\} * \mathrm{DTFT}\{f_2[n]\}$
6. Frequency differentiation: $nf[n] = \mathrm{DTFT}^{-1}\{dF(e^{j\omega})/d\omega\}$

TABLE 4.4 Properties of the DTFT

Sequence		Fourier Transform	
$x[n]$		$X(e^{j\omega})$	
$y[n]$		$Y(e^{j\omega})$	
1. $ax[n] + by[n]$		$aX(e^{j\omega}) + bY(e^{j\omega})$	
2. $x[n - n_d]$	(n_d an integer)	$e^{-j\omega n_d}X(e^{j\omega})$	
3. $e^{j\omega_0 n}x[n]$		$X(e^{j(\omega-\omega_0)})$	
4. $x[-n]$		$X(e^{-j\omega})$	if $x[n]$ is real
		$X^*(e^{j\omega})$	
5. $nx[n]$		$j\dfrac{dX(e^{j\omega})}{d\omega}$	
6. $x[n] * y[n]$		$X(e^{j\omega})Y(e^{j\omega})$	
7. $x[n]y[n]$		$\dfrac{1}{2\pi}\int_{-x}^{x} X(e^{j\theta})Y(e^{j(\omega-\theta)})\,d\theta$	
Parseval's Theorem			
8. $\sum_{n=-\infty}^{\infty} \lvert x[n]\rvert^2 = \frac{1}{2\pi}\int_{-\pi}^{\pi} \lvert X(e^{j\omega})\rvert^2\,d\omega$			
9. $\sum_{n=-\infty}^{\infty} x[n]y^*[n] = \frac{1}{2\pi}\,\mathrm{inf}_{-\pi}^{\pi}\, X(e^{j\omega})Y^*(e^{j\omega})\,d\omega$			

Source: A. V. Oppenheim and R. W. Schafer, Digital Signal Processing, Englewood Cliffs, NJ: Prentice-Hall, 1975.

Note that the time differentiation and time-integration properties of the CTFT do not have analogous counterparts in the DTFT because time domain differentiation and integration are not defined for DT signals. When working with DT systems practitioners must often manipulate difference equations in the frequency domain. For this purpose property 1 and property 2 are very important. As with the CTFT, property 4 is very important for DT systems because it allows engineers to work with the frequency response of the system, in order to achieve proper shaping of the input spectrum or to achieve frequency selective filtering for noise reduction or signal detection. Also, property 3 is useful for the analysis of modulation and filtering operations common in both analog and digital communication systems.

The DTFT is defined so that the time domain is discrete and the frequency domain is continuous. This is in contrast to the CTFT that is defined to have continuous time and continuous frequency domains. The mathematical dual of the DTFT also exists, which is a transform pair that has a continuous time domain and a discrete frequency domain. In fact, the dual concept is really the same as the Fourier series for periodic CT signals presented earlier in the chapter, as represented by (4.5a) and (4.5b). However, the classical Fourier series arises from the assumption that the CT signal is inherently periodic, as opposed to the time domain becoming periodic by virtue of sampling the spectrum of a continuous frequency (aperiodic time) function [8]. The dual of the DTFT, the discrete frequency Fourier transform (DFFT), has been formulated and its properties tabulated as an interesting and useful transform in its own right [5]. Although the DFFT is similar in concept to the classical CT Fourier series, the formal properties of the DFFT [5] serve to clarify the effects of frequency domain sampling and time domain aliasing. These effects are obscured in the classical treatment of the CT Fourier series because the emphasis is on the inherent "line spectrum" that results from time domain periodicity. The DFFT is useful for the analysis and design of digital filters that are produced by frequency sampling techniques.

Relationship between the Continuous and Discrete Time Spectra

Because DT signals often originate by sampling CT signals, it is important to develop the relationship between the original spectrum of the CT signal and the spectrum of the DT signal that results. First, the CTFT is applied to the CT sampling model, and the properties listed above are used to produce the following result:

$$\begin{aligned}\mathcal{F}\{s_a(t)\} &= \mathcal{F}\left\{s(t)\sum_{n=-\infty}^{\infty}\delta(t-nT)\right\}\\ &= (1/2\pi)S(j\omega) * \mathcal{F}\left\{\sum_{n=-\infty}^{\infty}\delta(t-nT)\right\}\end{aligned} \tag{4.18}$$

In this section it is important to distinguish between ω and ω', so the explicit primed notation is used in the following discussion where needed for clarification. Because the sampling function (summation of shifted impulses) on the right-hand side of the above equation is periodic with period T it can be replaced with a CT Fourier series expansion as follows:

$$S(e^{j\omega T}) = \mathcal{F}\{s_a(t)\} = (1/2\pi)S(j\omega) * \mathcal{F}\left\{\sum_{n=-\infty}^{\infty}(1/T)e^{j(2\pi/T)nt}\right\}$$

Applying the frequency domain convolution property of the CTFT yields

$$S(e^{j\omega T}) = (1/2\pi)\sum_{n=-\infty}^{\infty} S(j\omega) * (2\pi/T)\delta(\omega - (2\pi/T)n)$$

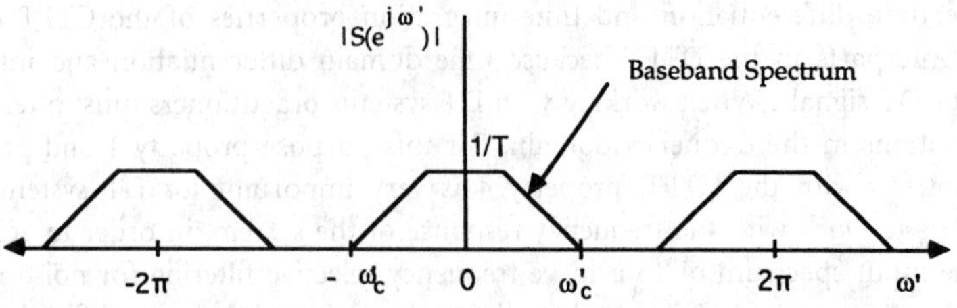

FIGURE 4.8 Illustration of the relationship between the CT and DT spectra.

The result is

$$S(e^{j\omega T}) = (1/T)\sum_{n=-\infty}^{\infty} S(j[\omega - (2\pi/T)n]) = (1/T)\sum_{n=-\infty}^{\infty} S(j[\omega - n\omega_s]) \quad (4.19a)$$

where $\omega_s = (2\pi/T)$ is the sampling frequency expressed in radians per second. An alternate form for the expression of (4.19a) is

$$S(e^{j\omega'}) = (1/T)\sum_{n=-\infty}^{\infty} S(j[(\omega' - n2\pi)/T]) \quad (4.19b)$$

where $\omega' = \omega T$ is the normalized DT frequency axis expressed in radians. Note that $S(e^{j\omega T}) = S(e^{j\omega'})$ consists of an infinite number of replicas of the CT spectrum $S(j\omega)$, positioned at intervals of $(2\pi/T)$ on the ω axis (or at intervals of 2π on the ω' axis), as illustrated in Fig. 4.8. Note that if $S(j\omega)$ is band limited with a bandwidth ω_c, and if T is chosen sufficiently small so that $\omega_s > 2\omega_c$, then the DT spectrum is a copy of $S(j\omega)$ (scaled by $1/T$) in the baseband. The limiting case of $\omega_s = 2\omega_c$ is called the Nyquist sampling frequency. Whenever a CT signal is sampled at or above the Nyquist rate, no aliasing distortion occurs (i.e., the baseband spectrum does not overlap with the higher-order replicas) and the CT signal can be exactly recovered from its samples by extracting the baseband spectrum of $S(e^{j\omega'})$ with an ideal low-pass filter that recovers the original CT spectrum by removing all spectral replicas outside the baseband and scaling the baseband by a factor of T.

4.5 The Discrete Fourier Transform

To obtain the discrete Fourier transform (DFT) the continuous frequency domain of the DTFT is sampled at N points uniformly spaced around the unit circle in the z-plane, i.e., at the points $\omega_k = (2\pi k/N), k = 0, 1, \ldots, N-1$. The result is the DFT pair defined by (4.20a) and (4.20b). The signal $s[n]$ is either a finite length sequence of length N, or it is a periodic sequence with period N.

$$S[k] = \sum_{n=0}^{N-1} s[n]e^{-j2\pi kn/N} \qquad k = 0, 1, \ldots, N-1 \quad (4.20a)$$

$$s[n] = (1/N)\sum_{k=0}^{N-1} S[k]e^{j2\pi kn/N} \qquad n = 0, 1, \ldots, N-1 \quad (4.20b)$$

Regardless of whether $s[n]$ is a finite length or periodic sequence, the DFT treats the N samples of $s[n]$ as though they are one period of a periodic sequence. This is an important feature of the DFT, and one that must be handled properly in signal processing to prevent the introduction of artifacts. Important properties of the DFT are summarized in Table 4.5. The notation $((k))_N$ denotes k modulo N, and $R_N[n]$ is a rectangular window such that $R_N[n] = 1$ for $n = 0, \ldots, N-1$, and $R_N[n] = 0$ for $n < 0$ and $n \geq N$. The transform relationship given by (4.20a) and (4.20b) is also valid when $s[n]$ and $S[k]$ are periodic sequences, each of period N. In this case n and k are permitted to range over the complete set of real integers, and $S[k]$ is referred to as the discrete Fourier series (DFS). The DFS is developed by some authors as a distinct transform pair in its own right [6]. Whether the DFT and the DFS are considered identical or distinct is not very important in this discussion. The important point to be emphasized here is that the DFT treats $s[n]$ as though it were a single period of a periodic sequence, and all signal processing done with the DFT will inherit the consequences of this assumed periodicity.

Properties of the Discrete Fourier Series

Most of the properties listed in Table 4.5 for the DFT are similar to those of the z-transform and the DTFT, although some important differences exist. For example, property 5 (time-shifting property), holds for *circular* shifts of the finite length sequence $s[n]$, which is consistent with the notion that the DFT treats $s[n]$ as one period of a periodic sequence. Also, the multiplication of two DFTs results in the **circular convolution** of the corresponding DT sequences, as specified by property 7. This latter property is quite different from the **linear convolution** property of the DTFT. Circular convolution is the result of the assumed periodicity discussed in the previous paragraph. Circular convolution is simply a linear convolution of the periodic extensions of the

TABLE 4.5 Properties of the DFT

Finite-Length Sequence (Length N)	N-Point DFT (Length N)
1. $x[n]$	$X[k]$
2. $x_1[n], x_2[n]$	$X_1[k], X_2[k]$
3. $ax_1[n] + bx_2[n]$	$aX_1[k] + bX_2[k]$
4. $X[n]$	$Nx[((-k))_N]$
5. $x[((n_m))_N]$	$W_N^{km} X[k]$
6. $W_N^{-ln} x[n]$	$X[((k-l))_N]$
7. $\sum_{m=0}^{N-1} x_1(m) x_2[((n_m))_N]$	$X_1[k]X_2[k]$
8. $x_1[n]x_2[n]$	$\frac{1}{N}\sum_{l=0}^{N-1} X_1(l)X_2[((k-l)_N]$
9. $x^*[n]$	$X^*[((-k))_N]$
10. $x^*[((-n))_N]$	$X^*[k]$
11. $\mathrm{Re}\{x[n]\}$	$X_{ep}[k] = \frac{1}{2}\{X[((k))_N] + K^*[((-k))_N]\}$
12. $j\,\mathrm{Im}\{x[n]\}$	$X_{op}[k] = \frac{1}{2}\{X[((k))_N] - X^*[((-k))_N]\}$
13. $x_{ep}[n] = \frac{1}{2}\{x[n] + x^*[((-n))_N]\}$	$\mathrm{Re}\{X[k]\}$
14. $x_{op}[n] = \frac{1}{2}\{x[n] - x^*[((-n))_N]\}$	$j\,\mathrm{Im}\{X[k]\}$
Properties 15–17 apply only when $x[n]$ is real	
15. Summetry properties	$X[k] = X^*[((-k))_N]$ $\mathrm{Re}\{X[k]\} = \mathrm{Re}\{X[((-k))_N]\}$ $\mathrm{Im}\{X[k]\} = -\mathrm{Im}\{X[((-k))_N]\}$ $\lvert X[k]\rvert = \lvert X[((-k))_N]\rvert$ $\sphericalangle\{X[k]\} = -\sphericalangle\{X[((-k))_N]\}$
16. $x_{ep}[n] = \frac{1}{2}\{x[n] + x[((-n))_N]\}$	$\mathrm{Re}\{X[k]\}$
17. $x_{op}[n] = \frac{1}{2}\{x[n] - x[((-n))_N]\}$	$j\,\mathrm{Im}\{X[k]\}$

Source: A. V. Oppenheim and R. W. Schafer, Discrete-Time Signal Processing, Englewood Cliffs, NJ: Prentice-Hall, 1989.

finite sequences being convolved, in which each of the finite sequences of length N defines the structure of one period of the periodic extensions.

For example, suppose one wishes to implement a digital filter with finite impulse response (FIR) $h[n]$. The output $y(n)$ in response to input $s[n]$ is given by

$$y[n] = \sum_{k=0}^{N-1} h[k]s[n-k] \tag{4.21}$$

where $y(n)$ is obtained by transforming $h[n]$ and $s[n]$ into $H[k]$ and $S[k]$ using the DFT, multiplying the transforms point-wise to obtain $Y[k] = H[k]S[k]$, and then using the inverse DFT to obtain $y[n] = \text{DFT}^{-1}\{Y[k]\}$. If $s[n]$ is a finite sequence of length M, then the results of the circular convolution implemented by the DFT will correspond to the desired linear convolution iff the block length of the DFT, N_{DFT}, is chosen sufficiently large so that $N_{\text{DFT}} \geq N + M - 1$ and both $h[n]$ and $s[n]$ are padded with zeroes to form blocks of length N_{DFT}.

Fourier Block Processing in Real-Time Filtering Applications

In some practical applications either the value of M is too large for the memory available, or $s[n]$ may not actually be finite in length, but rather a continual stream of data samples that must be processed by a filter at real-time rates. Two well-known algorithms are available that partition $s[n]$ into smaller blocks and process the individual blocks with a smaller-length DFT: (1) overlap-save partitioning and (2) overlap-add partitioning. Each of these algorithms is summarized below.

Overlap-Save Processing

In this algorithm N_{DFT} is chosen to be some convenient value with $N_{\text{DFT}} > N$. The signal $s[n]$ is partitioned into blocks which are of length N_{DFT} and which overlap by $N-1$ data points. Hence, the kth block is $s_k[n] = s[n + k(N_{\text{DFT}} - N + 1)], n = 0, \ldots, N_{\text{DFT}} - 1$. The filter is an augmented filter with $N_{\text{DFT}} - N$ zeroes to produce

$$h_{\text{pad}}[n] = \begin{bmatrix} h[n], & n = 0, \ldots, N-1 \\ 0, & n = N, \ldots, N_{\text{DFT}} - 1 \end{bmatrix} \tag{4.22}$$

The DFT is then used to obtain $Y_{\text{pad}}[n] = \text{DFT}\{h_{\text{pad}}[n]\} \cdot \text{DFT}\{s_k[n]\}$, and $y_{\text{pad}}[n] = \text{IDFT}\{Y_{\text{pad}}[n]\}$. From the $y_{\text{pad}}[n]$ array the values that correctly correspond to the linear convolution are saved; values that are erroneous due to wraparound error caused by the circular convolution of the DFT are discarded. The kth block of the filtered output is obtained by

$$y_k[n] = \begin{bmatrix} y_{\text{pad}}[n], & n = N-1, \ldots, N_{\text{DFT}} - 1 \\ 0, & n = 0, \ldots, N-2 \end{bmatrix} \tag{4.23}$$

For the overlap-save algorithm, each time a block is processed there are $N_{\text{DFT}} - N + 1$ points saved and $N - 1$ points discarded. Each block moves forward by $N_{\text{DFT}} - N + 1$ data points and overlaps the previous block by $N - 1$ points.

Overlap-Add Processing

This algorithm is similar to the previous one except that the kth input block is defined as

$$s_k[n] = \begin{bmatrix} s[n + kL], & n = 0, \ldots, L-1 \\ 0, & n = L, \ldots, N_{\text{DFT}} - 1 \end{bmatrix} \tag{4.24}$$

where $L = N_{\text{DFT}} - N + 1$. The filter function $h_{\text{pad}}[n]$ is augmented with zeroes, as before, to create $h_{\text{pad}}[n]$, and the DFT processing is executed as before. In each block $y_{\text{pad}}[n]$ that is obtained at the output the first $N-1$ points are erroneous, the last $N-1$ points are erroneous, and the middle $N_{\text{DFT}} - 2(N-1)$ points correctly correspond to the linear convolution. However, if the last $N-1$ points from block k are overlapped with the first $N-1$ points from block $k+1$ and added pairwise, correct results corresponding to linear convolution are also obtained from these positions. Hence, after this addition the number of correct points produced per block is $N_{\text{DFT}} - N + 1$, which is the same as that for the overlap-save algorithm. The overlap-add algorithm requires approximately the same amount of computation as the overlap-save algorithm, although the addition of the overlapping portions of blocks is extra. This feature, together with the added delay of waiting for the next block to be finished before the previous one is complete, has resulted in more popularity for the overlap-save algorithm in practical applications.

Block filtering algorithms make it possible to efficiently filter continual data streams in real time because the fast Fourier transform (FFT) algorithm can be used to implement the DFT, thereby minimizing the total computation time and permitting reasonably high overall data rates. However, block filtering generates data in bursts, i.e., a delay occurs during which no filtered data appear, and then an entire block is suddenly generated. In real-time systems buffering must be used. The block algorithms are particularly effective for filtering very long sequences of data that are prerecorded on magnetic tape or disk.

Fast Fourier Transform Algorithms

The DFT is typically implemented in practice with one of the common forms of the FFT algorithm. The FFT is not a Fourier transform in its own right, but simply a computationally efficient algorithm that reduces the complexity of the computing DFT from order $\{N^2\}$ to order $\{N \log_2 N\}$. When N is large, the computational savings provided by the FFT algorithm is so great that the FFT makes real-time DFT analysis practical in many situations that would be entirely impractical without it. Fast Fourier transform algorithms abound, including decimation-in-time (D-I-T) algorithms, decimation-in-frequency (D-I-F) algorithms, bit-reversed algorithms, normally ordered algorithms, mixed-radix algorithms (for block lengths that are not powers of 2), prime factor algorithms, and Winograd algorithms [7]. The D-I-T and the D-I-F radix-2 FFT algorithms are the most widely used in practice. Detailed discussions of various FFT algorithms can be found in [3], [6], [7], and [10].

The FFT is easily understood by examining the simple example of $N = 8$. The FFT algorithm can be developed in numerous ways, all of which deal with a nested decomposition of the summation operator of (4.20a). The development presented here is called an **algebraic development** of the FFT because it follows straightforward algebraic manipulation. First, the summation indices (k, n) in (4.20a) are expressed as explicit binary integers, $k = k_2 4 + k_1 2 + k_0$ and $n = n_2 4 + n_1 2 + n_0$, where k_i and n_i are bits that take on the values of either 0 or 1. If these expressions are substituted into (4.20a), all terms in the exponent that contain the factor $N = 8$ can be deleted because $e^{-j2\pi l} = 1$ for any integer l. Upon deleting such terms and regrouping the remaining terms, the product nk can be expressed in either of two ways:

$$nk = (4k_0)n_2 + (4k_1 + 2k_0)n_1 + (4k_2 + 2k_1 + k_0)n_0 \tag{4.25a}$$

$$nk = (4n_0)k_2 + (4n_1 + 2n_0)k_1 + (4n_2 + 2n_1 + n_0)k_0 \tag{4.25b}$$

Substituting (4.25a) into (4.20a) leads to the D-I-T FFT, whereas substituting (4.25b) leads to the D-I-F FFT. Only the D-I-T FFT is discussed further here. The D-I-F and various related forms are treated in detail in [6].

The D-I-T FFT decomposes into $\log_2 N$ stages of computation, plus a stage of bit reversal,

$$x_1[k_0, n_1, n_0] = \sum_{n_2=0}^{1} s[n_2, n_1, n_0] W_8^{4k_0 n_2} \qquad \text{(stage 1)} \tag{4.26a}$$

$$x_2[k_0, k_1, n_0] = \sum_{n_1=0}^{1} x[k_0, n_1, n_0] W_8^{(4k_1+2k_0)n_2} \qquad \text{(stage 2)} \tag{4.26b}$$

$$x_3[k_0, k_1, k_2] = \sum_{n_0=0}^{1} x[k_0, k_1, n_0] W_8^{(4k_2+2k_1+k_0)n_0} \qquad \text{(stage 3)} \tag{4.26c}$$

$$S[k_2, k_1, k_0] = x_3[k_0, k_1, k_2] \qquad \text{(bit reversal)} \tag{4.26d}$$

In each summation above one of the n_i is summed out of the expression, while at the same time a new k_i is introduced. The notation is chosen to reflect this. For example, in stage 3, n_0 is summed out, k_2 is introduced as a new variable, and n_0 is replaced by k_2 in the result. The last operation, called bit reversal, is necessary to correctly locate the frequency samples $X[k]$ in the memory. It is easy to show that if the samples are paired correctly, an **in-place computation** can be done by a sequence of butterfly operations. The term in-place means that each time a butterfly is to be computed, a pair of data samples is read from memory, and the new data pair produced by the butterfly calculation is written back into the memory locations where the original pair was stored, thereby overwriting the original data. An in-place algorithm is designed so that each data pair is needed for only one butterfly, and thus the new results can be immediately stored on top of the old in order to minimize memory requirements.

For example, in stage 3 the $k = 6$ and $k = 7$ samples should be paired, yielding a "butterfly" computation that requires one complex multiply, one complex add, and one subtract:

$$x_3(1,1,0) = x_2(1,1,0) + W_8^3 x_2(1,1,1) \tag{4.27a}$$

$$x_3(1,1,0) = x_2(1,1,0) - W_8^3 x_2(1,1,1) \tag{4.27b}$$

Samples $x_2(6)$ and $x_2(7)$ are read from the memory, the butterfly is executed on the pair, and $x_3(6)$ and $x_3(7)$ are written back to the memory, overwriting the original values of $x_2(6)$ and $x_2(7)$. In general, $N/2$ butterflies are found in each stage and there are $\log_2 N$ stages, so the total number of butterflies is $(N/2)\log_2 N$. Because one complex multiplication per butterfly is the maximum, the total number of multiplications is bounded by $(N/2)\log_2 N$ (some of the multiplies involve factors of unity and should not be counted).

Figure 4.9 shows the signal flow graph of the D-I-T FFT for $N = 8$. This algorithm is referred to as an in-place FFT with normally ordered input samples and bit-reversed outputs. Minor variations that include both bit-reversed inputs and normally ordered outputs and non-in-place algorithms with normally ordered inputs and outputs are possible. Also, when N is not a power of 2, a mixed-radix algorithm can be used to reduce computation. The mixed-radix FFT is most efficient when N is highly composite, i.e., $N = p_1^{r_1} p_2^{r_2} \cdots p_L^{r_L}$, where the p^i are small prime numbers and the r^i are positive integers. It can be shown that the order of complexity of the mixed radix FFT is order$\{N(r_1(p_1 - 1) + r_2(p_2 - 1) + \cdots + r_L(p_L - 1)\}$. Because of the lack of uniformity of structure among stages, this algorithm has not received much attention for hardware implementation. However, the mixed-radix FFT is often used in software applications, especially for processing data recorded in laboratory experiments in which it is not convenient to restrict the block lengths to be powers of 2. Many advanced FFT algorithms, such as higher-radix forms, the mixed-radix form, the prime-factor algorithm, and the Winograd algorithm are

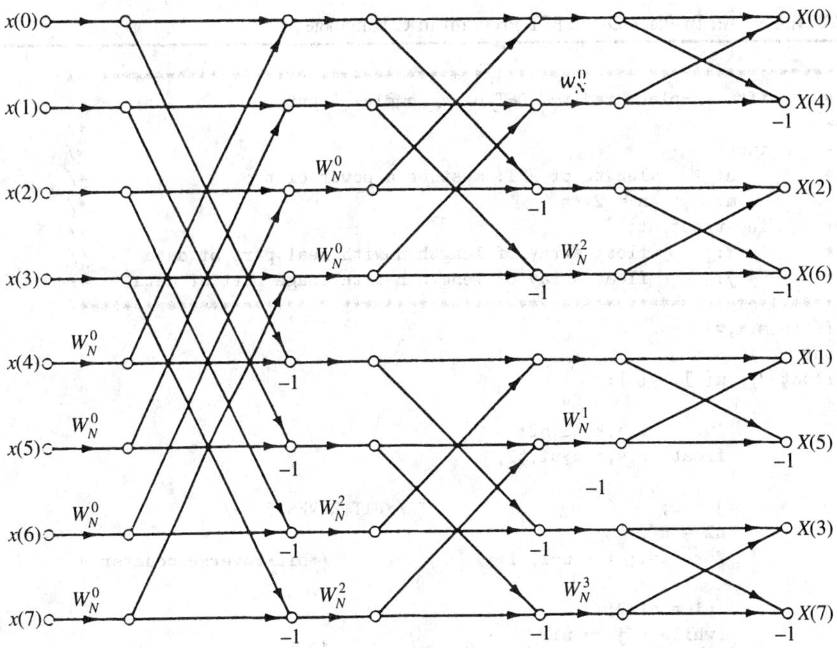

FIGURE 4.9 D-I-T FFT algorithm with normally ordered inputs and bit-reversed outputs.

described in [9]. Algorithms specialized for real-valued data reduce the computational cost by a factor of two. A radix-2 D-I-T FFT program, written in C language, is listed in Table 4.6.

4.6 Family Tree of Fourier Transforms

It is now possible to illustrate the functional relationships among the various forms of Fourier transforms that have been discussed in the previous sections. The family tree of CT Fourier transform is shown in Fig. 4.10, where the most general, and consequently the most powerful, Fourier transform is the classical complex Fourier transform (or equivalently, the bilateral Laplace transform). Note that the complex Fourier transform is identical to the bilateral Laplace transform, and it is at this level that the classical Laplace transform and Fourier transform techniques become identical. Each special member of the CT Fourier family is obtained by impressing certain constraints on the general form, thereby producing special transforms that are simpler and more useful in practical problems where the constraints are met.

The analogous family of DT Fourier techniques is presented in Fig. 4.11, in which the bilateral z-transform is analogous to the complex Fourier transform, the unilateral z-transform is analogous to the classical (one-sided) Laplace transform, the DTFT is analogous to the classical Fourier (CT) transform and the DFT is analogous to the classical (CT) Fourier series.

4.7 Selected Applications of Fourier Methods

Fast Fourier Transform in Spectral Analysis

An FFT program is often used to perform spectral analysis on signals that are sampled and recorded as part of laboratory experiments, or in certain types of data acquisition systems. Several issues must be addressed when spectral analysis is performed on (sampled) analog waveforms that are observed over a finite interval of time.

TABLE 4.6 An In-Place D-I-T FFT Program in C Language

```
***********************************************************************/
*      fft: in-place radix-2 DFT of a complex input                   */
*                                                                      */
*      input:                                                          */
*         n:       length of FFT: must be a power of two               */
*         m:       n = 2**m                                            */
*      input/output:                                                   */
*         x:       float array of length n with real part of data      */
*         y:       float array of lengtn n with image part of data     */
***********************************************************************/
fft(n,m,x,y)
tnt       n,m;
float     x[ ], y[ ]:
{
          int    i,j,k,nl,n2:
          float  c,s,e,a,t1,t2;

          j = 0;                          /*BIT-REVERSE  */
          n2 = n/2;
          for (i=1; 1 < n-1; i++)               /*bit-reverse counter */
          {
           nl = n1/2;
           while ( j >= n1)
            {
             j = j - n1;
             nl = n1/2;
            }
           j = j + nl;
           if (i < j)                           /*swap data */
            {
             t1 = x[i]; x[i] = x[j]; x[j] = t1;
             t1 = y[i]; y[i] = y[j]; y[j] = t1;
            }
          }
          n1 = 0; n2 = 1;                 /* FFT */
          for (i = 0; i < m; i++)               /*state loop */
           {
            n1 = n2;  n2 = n2 + n2;
            e = -6.283185307179586/n2;
            a = 0.0;

            for (j=0; j < n1; j++)              /*flight loop */
             {
              c = cos(a); s=sin (a);
              a = a + e;

              for (k=j; k < n; k=k+n2)          /*butterfly loop */
               {
                t1 = c*x[k+n1] - s*y[k+n1];
                t2 = s*x[k+n1] + c*y[k+n1];
                x[k+n1] = x[k] - t1;
                y[k+n1] = y[k] - t2;
                x[k] = x[k] + t1;
                y[k] = y[k] + t2;
               }
             }
           }
           return;
}
```

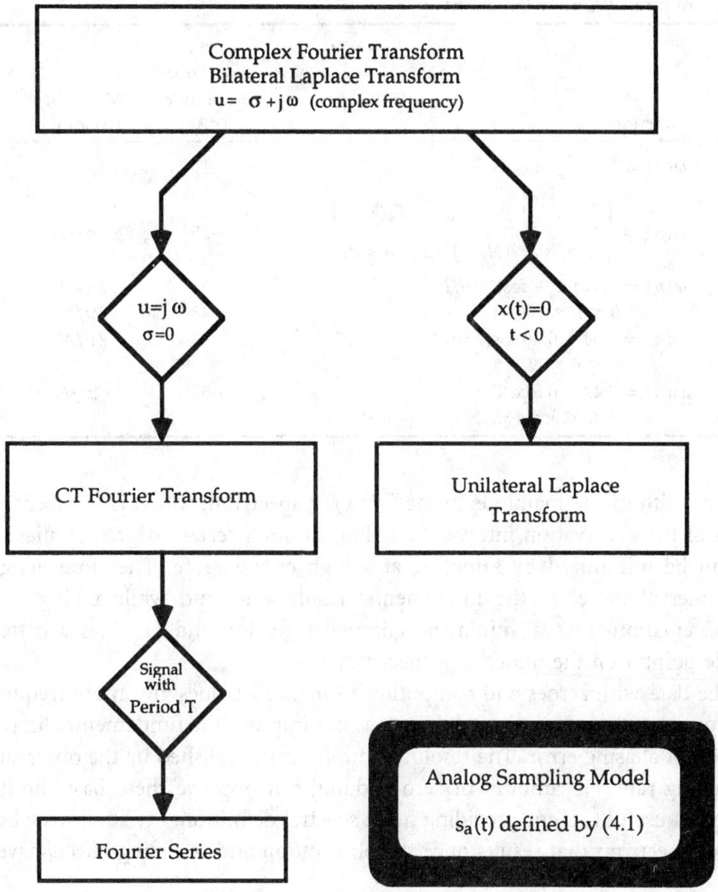

FIGURE 4.10 Relationships among CT Fourier concepts.

Windowing

The FFT treats the block of data as though it were one period of a periodic sequence. If the underlying waveform is not periodic, then harmonic distortion may occur because the periodic waveform created by the FFT may have sharp discontinuities at the boundaries of the blocks. This effect is minimized by removing the mean of the data (it can always be reinserted) and by windowing the data so the ends of the block are smoothly tapered to zero. A good rule of thumb is to taper 10% of the data on each end of the block using either a cosine taper or one of the other common windows shown in Table 4.7. An alternate interpretation of this phenomenon is that the finite length observation has already windowed the true waveform with a rectangular window that has large spectral sidelobes (see Table 4.7). Hence, applying an additional window results in a more desirable window that minimizes frequency domain distortion.

Zero Padding

An improved spectral analysis is achieved if the block length of the FFT is increased. This can be done by taking more samples within the observation interval, increasing the length of the observation interval, or augmenting the original data set with zeroes. First, it must be understood that the finite observation interval results in a fundamental limit on the spectral resolution, even before the signals are sampled. The CT rectangular window has a $(\sin x)/x$ spectrum, which is convolved with the true spectrum of the analog signal. Therefore, the frequency resolution is

TABLE 4.7 Common Window Functions

Name	Function	Peak Side-Lobe Amplitude (dB)	Mainlobe Width	Minimum Stopband Attenuation (dB)
Rectangular	$\omega(n) = 1.\quad 0 \le n \le N-1$	−13	$4\pi/N$	−21
Bartlett	$\omega(n) = \begin{cases} 2/N, & 0 \le n \le (N-1)/2 \\ 22n/N, & (N-1)/2 \le n \le N-1 \end{cases}$	−25	$8\pi/N$	−25
Hanning	$\omega(n) = (1/2)[1 - \cos(2\pi n/N)]$	−31	$8\pi/N$	−44
	$0 \le n \le N-1$	−43	$8\pi/N$	−53
Hamming	$\omega(n) = 0.54 - 0.46\cos(2\pi n/N),$ $0 \le n \le N-1$	−43	$8\pi/N$	−53
Backman	$\omega(n) = 0.42 - 0.5\cos(2\pi n/N)$ $+ 0.08\cos(4\pi n/N),\quad 0 \le n \le N-1$	−57	$12\pi/N$	−74

limited by the width of the mainlobe in the $(\sin x)/x$ spectrum, which is inversely proportional to the length of the observation interval. Sampling causes a certain degree of aliasing, although this effect can be minimized by sampling at a high enough rate. Therefore, lengthening the observation interval increases the fundamental resolution limit, while taking more samples within the observation interval minimizes aliasing distortion and provides a better definition (more sample points) on the underlying spectrum.

Padding the data with zeroes and computing a longer FFT does give more frequency domain points (improved spectral resolution), but it does not improve the fundamental limit, nor does it alter the effects of aliasing error. The resolution limits are established by the observation interval and the sampling rate. No amount of zero padding can improve these basic limits. However, zero padding is a useful tool for providing more spectral definition, i.e., it allows a better view of the (distorted) spectrum that results once the observation and sampling effects have occurred.

Leakage and the Picket Fence Effect

An FFT with block length N can accurately resolve only frequencies $\omega_k = (2\pi/N)k, k = 0, \ldots, N-1$ that are integer multiples of the fundamental $\omega_1 = (2\pi/N)$. An analog waveform that is sampled and subjected to spectral analysis may have frequency components between the harmonics. For example, a component at frequency $\omega_{k+1/2} = (2\pi/N)(k + 1/2)$ will appear scattered throughout the spectrum. The effect is illustrated in Fig. 4.12 for a sinusoid that is observed through a rectangular window and then sampled at N points. The **picket fence effect** means that not all frequencies can be seen by the FFT. Harmonic components are seen accurately, but other components "slip through the picket fence" while their energy is "leaked" into the harmonics. These effects produce artifacts in the spectral domain that must be carefully monitored to assure that an accurate spectrum is obtained from FFT processing.

Finite Impulse Response Digital Filter Design

A common method for designing FIR digital filters is by use of windowing and FFT analysis. In general, window designs can be carried out with the aid of a hand calculator and a table of well-known window functions. Let $h[n]$ be the impulse response that corresponds to some desired frequency response, $H(e^{j\omega})$. If $H(e^{j\omega})$ has sharp discontinuities, such as the low-pass example shown in Fig. 4.13, then $h[n]$ will represent an infinite impulse response (IIR) function. The objective is to time limit $h[n]$ in such a way as to not distort $H(e^{j\omega})$ any more than necessary. If $h[n]$ is simply truncated, a ripple (Gibbs phenomenon) occurs around the discontinuities in the spectrum, resulting in a distorted filter (Fig. 4.13).

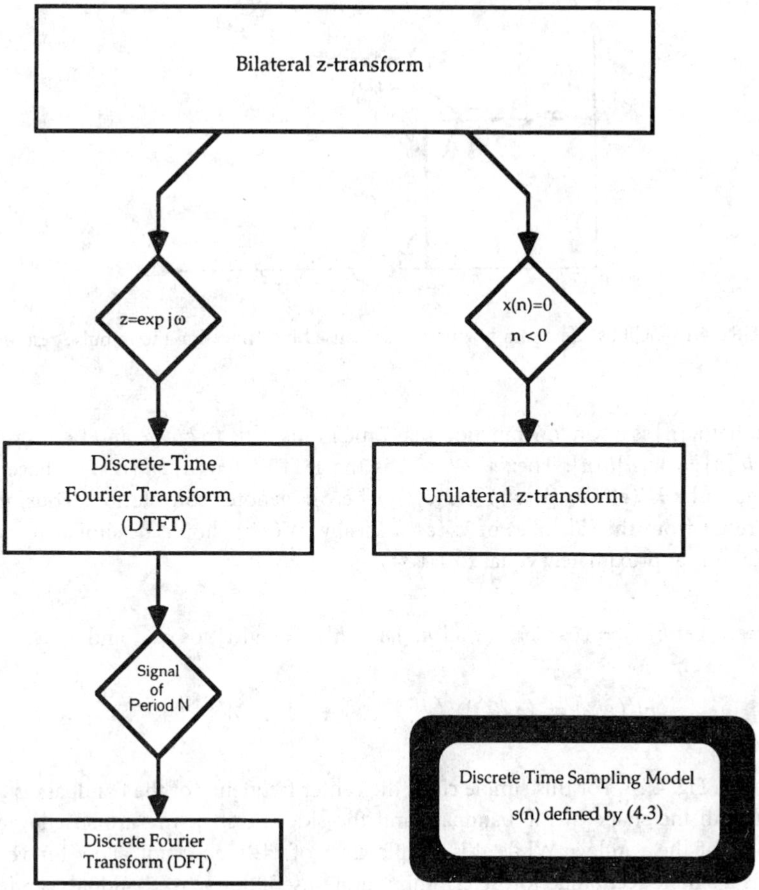

FIGURE 4.11 Relationships among DT concepts.

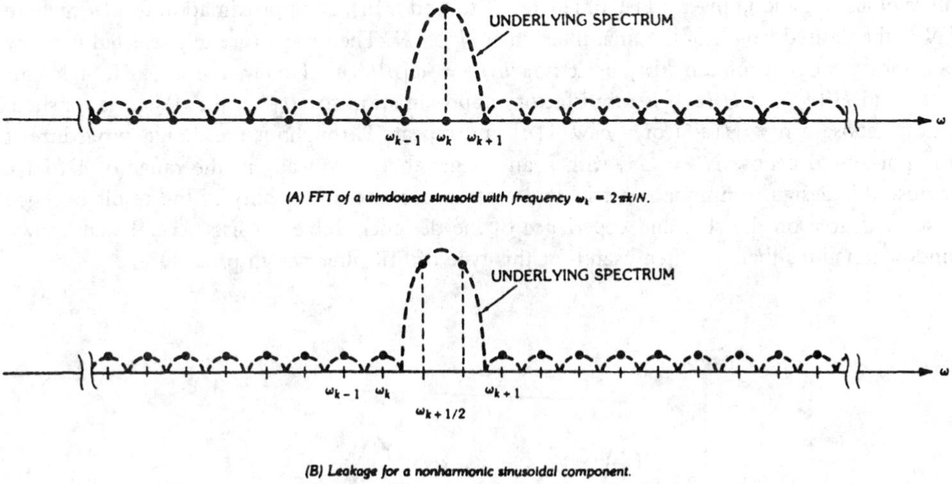

FIGURE 4.12 Illustration of leakage and the picket-fence effects.

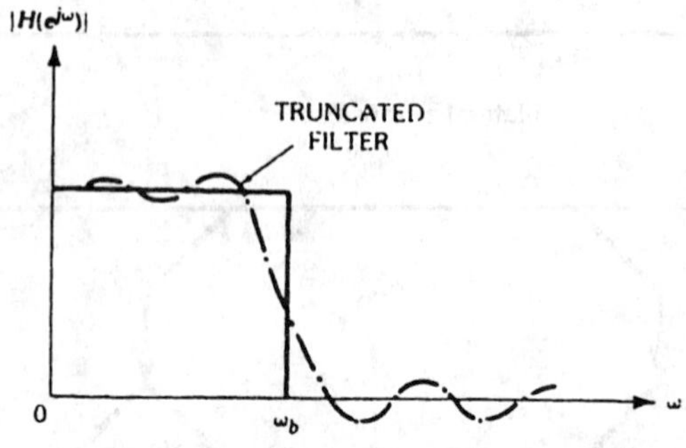

FIGURE 4.13 Gibbs effect in a low-pass filter caused by truncating the impulse response.

Suppose that $w[n]$ is a window function that time limits $h[n]$ to create an FIR approximation, $h'[n]$; i.e., $h'[n] = w[n]h[n]$. Then if $W(e^{j\omega})$ is the DTFT of $w[n]$, $h'[n]$ will have a Fourier transform given by $H'(e^{j\omega}) = W(e^{j\omega}) * H(e^{j\omega})$, where $*$ denotes convolution. Thus, the ripples in $H'(e^{j\omega})$ result from the sidelobes of $W(e^{j\omega})$. Ideally, $W(e^{j\omega})$ should be similar to an impulse so that $H'(e^{j\omega})$ is approximately equal to $H(e^{j\omega})$.

Special Case. Let $h[n] = \cos n\omega_0$, for all n. Then $h[n] = w[n] \cos n\omega_0$, and

$$H'(e^{j\omega}) = (1/2)W(e^{j(\omega+\omega_0)}) + (1/2)W(e^{j(\omega-\omega_0)}) \tag{4.28}$$

as illustrated in Fig. 4.14. For this simple class, the center frequency of the bandpass is controlled by ω_0, and both the shape of the bandpass and the sidelobe structure are strictly determined by the choice of the window. While this simple class of FIRs does not allow for very flexible designs, it is a simple technique for determining quite useful low-pass, bandpass, and high-pass FIRs.

General Case. Specify an ideal frequency response, $H(e^{j\omega})$, and choose samples at selected values of ω. Use a long inverse FFT of length N' to find $h'[n]$, an approximation to $h[n]$, where if N is the desired length of the final filter, then $N' \gg N$. Then use a carefully selected window to truncate $h'[n]$ to obtain $h[n]$ by letting $h[n] = \omega[n]h'[n]$. Finally, use an FFT of length N' to find $H'(e^{j\omega})$. If $H'(e^{j\omega})$ is a satisfactory approximation to $H(e^{j\omega})$, the design is finished. If not, choose a new $H(e^{j\omega})$ or a new $w[n]$ and repeat. Throughout the design procedure it is important to choose $N' = kN$, with k an integer that is typically in the range of 4 to 10. Because this design technique is a trial and error procedure, the quality of the result depends to some degree on the skill and experience of the designer. Table 4.7 lists several well-known window functions that are often useful for this type of FIR filter design procedure.

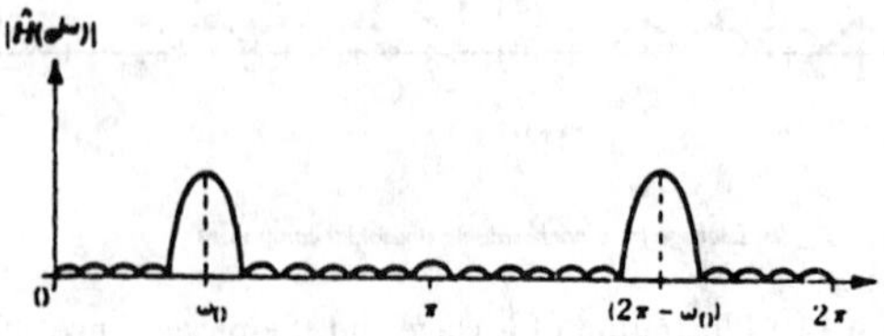

FIGURE 4.14 Design of a simple bandpass FIR filter by windowing.

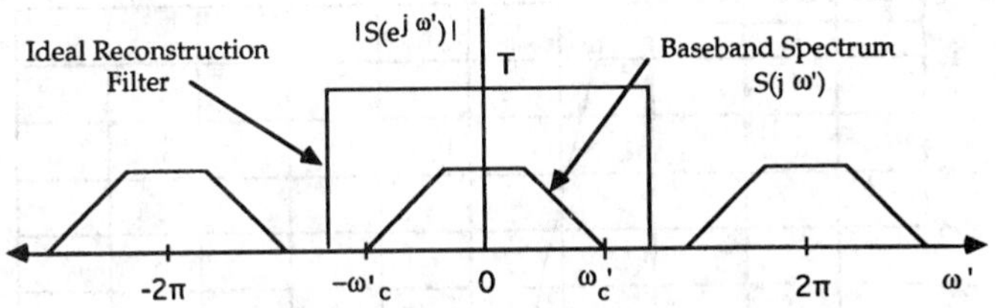

FIGURE 4.15 Illustration of ideal reconstruction.

Fourier Analysis of Ideal and Practical Digital-to-Analog Conversion

From the relationship characterized by (4.19b) and illustrated in Fig. 4.8, a CT signal $s(t)$ can be recovered from its samples by passing $s_a(t)$ through an ideal lowpass filter that extracts only the baseband spectrum. The ideal lowpass filter, shown in Fig. 4.15, is a zero-phase CT filter whose magnitude response is a constant of value T in the range $-\pi < \omega' \leq \pi$, and zero elsewhere. The impulse response of this "reconstruction filter" is given by $h(t) = T\,\mathrm{sinc}((\pi/T)t)$, where $\mathrm{sinc}\,x = (\sin x)/x$. The reconstruction can be expressed as $s(t) = h(t) * s_a(t)$ which, after some mathematical manipulation, yields the following classical reconstruction formula:

$$s(t) = \sum_{n=-\infty}^{\infty} s(nT)\,\mathrm{sinc}((\pi/T)(t - nT)) \tag{4.29}$$

Note that the signal $s(t)$ is exactly recovered from its samples only if an infinite number of terms is included in the summation of (4.29). However, a good approximation of $s(t)$ can be obtained with only a finite number of terms if the lowpass reconstruction filter $h(t)$ is modified to have a finite interval of support, i.e., if $h(t)$ is nonzero only over a finite time interval. The reconstruction formula of (4.29) is an important result in that it represents the inverse of the sampling operation. By this means Fourier transform theory establishes that as long as CT signals are sampled at a sufficiently high rate, the information content contained in $s(t)$ can be represented and processed in either a CT or DT format. Fourier sampling and reconstruction theory provides the theoretical mechanism for translation between one format or the other without loss of information.

A CT signal $s(t)$ can be perfectly recovered from its samples using (4.29) as long as the original sampling rate was high enough to satisfy the Nyquist sampling criterion, because the sampling frequency exceeds the Nyquist rate, i.e., $\omega_s > 2\omega_B$. If the sampling rate does not satisfy the Nyquist criterion the adjacent periods of the analog spectrum will overlap, causing a distorted spectrum. This effect, called **aliasing distortion**, is rather serious because it cannot be corrected easily once it has occurred. In general, an analog signal should always be prefiltered with an CT low-pass filter prior to sampling so that aliasing distortion does not occur.

Figure 4.16 shows the frequency response of a fifth-order elliptic analog low-pass filter that meets industry standards for prefiltering speech signals. These signals are subsequently sampled at an 8-kHz sampling rate and transmitted digitally across telephone channels. The band-pass ripple is less than ± 0.01 dB from DC up to the frequency 3.4 kHz (too small to be seen in Fig. 4.16), and the stopband rejection reaches at least -32.0 dB at 4.6 kHz and remains below this level throughout the stopband.

Most practical systems use digital-to-analog converters for reconstruction, which results in a staircase approximation to the true analog signal, i.e.,

$$\hat{s}(t) = \sum_{n=-\infty}^{\infty} s(nT)\{u(t - nT) - u[t - (n+1)]\}, \tag{4.30}$$

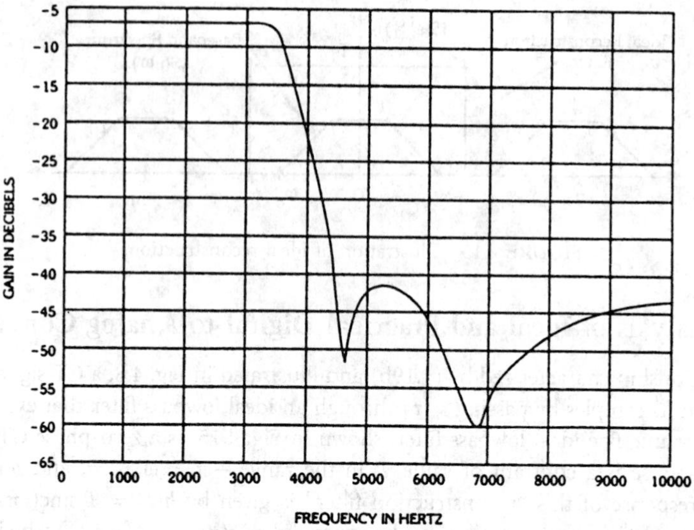

FIGURE 4.16 A fifth-order elliptic analog anti-aliasing filter used in the telecommunications industry with an 8-kHz sampling rate.

where $\hat{s}(t)$ denotes the reconstructed approximation to $s(t)$, and $u(t)$ denotes a CT unit step function. The approximation $\hat{s}(t)$ is equivalent to a result obtained by using an approximate reconstruction filter of the form

$$H_a(j\omega) = 2Te^{-j\omega T/2} \sin c(\omega T/2) \tag{4.31}$$

The approximation $\hat{s}(t)$ is said to contain "sin x/x distortion," which occurs because $H_a(j\omega)$ is not an ideal low-pass filter. $H_a(j\omega)$ distorts the signal by causing a droop near the passband edge, as well as by passing high-frequency distortion terms which "leak" through the sidelobes of $H_a(j\omega)$. Therefore, a practical digital to analog converter is normally followed by an analog postfilter

$$H_p(j\omega) = \begin{bmatrix} H_a^{-1}(j\omega), & 0 \le |\omega| < \pi/T \\ 0, & \omega \text{ otherwise} \end{bmatrix} \tag{4.32}$$

which compensates for the distortion and produces the correct $\hat{s}(t)$, i.e., the correctly constructed CT output. Unfortunately, the postfilter $H_p(j\omega)$ cannot be implemented perfectly, and, therefore, the actual reconstructed signal always contains some distortion in practice that arises from errors in approximating the ideal postfilter. Figure 4.17 shows a digital processor, complete with analog to digital and digital to analog converters, and the accompanying analog pre- and postfilters necessary for proper operation.

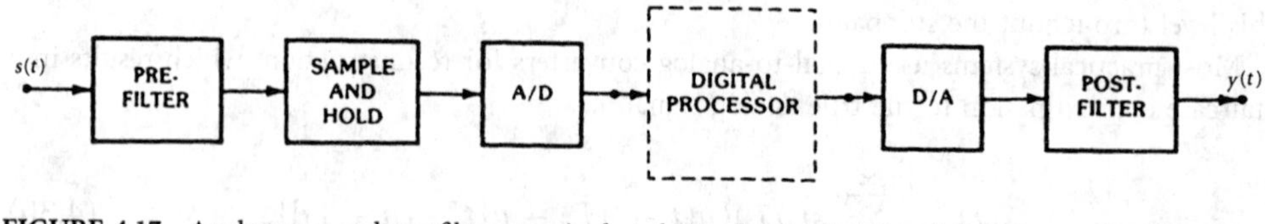

FIGURE 4.17 Analog pre- and postfilters required at the analog to digital and digital to analog interfaces.

4.8 Summary

This chapter presented many different Fourier transform concepts for both continuous time (CT) and discrete time (DT) signals and systems. Emphasis was placed on illustrating how these various forms of the Fourier transform relate to one another, and how they are all derived from more general complex transforms, the complex Fourier (or bilateral Laplace) transform for CT, and the bilateral z-transform for DT. It was shown that many of these transforms have similar properties which are inherited from their parent forms, and that a parallel hierarchy exists among Fourier transform concepts in the CT and the DT worlds. Both CT and DT sampling models were introduced as a means of representing sampled signals in these two different "worlds", and it was shown that the models are equivalent by virtue of having the same Fourier spectra when transformed into the Fourier domain with the appropriate Fourier transform. It was shown how Fourier analysis properly characterizes the relationship between the spectra of a CT signal and its DT counterpart obtained by sampling. The classical reconstruction formula was obtained as an outgrowth of this analysis. Finally, the discrete Fourier transform (DFT), the backbone for much of modern digital signal processing, was obtained from more classical forms of the Fourier transform by simultaneously discretizing the time and frequency domains. The DFT, together with the remarkable computational efficiency provided by the fast Fourier transform (FFT) algorithm, has contributed to the resounding success that engineers and scientists have experienced in applying digital signal processing to many practical scientific problems.

ferences

[1] M. E. VanValkenburg, *Network Analysis*, 3rd ed., Englewood Cliffs, NJ: Prentice-Hall, 1974.

[2] A. V. Oppenheim, A. S. Willsky, and I. T. Young, *Signals and Systems*, Englewood Cliffs, NJ: Prentice-Hall, 1983.

[3] R. N. Bracewell, *The Fourier Transform*, 2nd ed., New York, NY: McGraw-Hill, 1986.

[4] A. V. Oppenheim and R. W. Schafer, *Discrete-Time Signal Processing*, Englewood Cliffs, NJ: Prentice-Hall, 1989.

[5] W. K. Jenkins and M. D. Desai, "The discrete-frequency Fourier transform," *IEEE Trans. Circuits Syst.*, vol. CAS-33, no. 7, pp. 732–734, July 1986.

[6] A. V. Oppenheim and R. W. Schafer, *Digital Signal Processing*, Englewood Cliffs, NJ: Prentice-Hall, 1975.

[7] R. E. Blahut, *Fast Algorithms for Digital Signal Processing*, Reading, MA: Addison-Wesley, 1985.

[8] J. R. Deller, Jr., "Tom, Dick, and Mary discover the DFT," *IEEE Signal Processing Mag.*, vol. 11, no. 2, pp. 36–50, Apr. 1994.

[9] C. S. Burrus and T. W. Parks, *DFT/FFT and Convolution Algorithms*, New York: John Wiley and Sons, 1985.

[10] E. O. Brigham, "The Fast Fourier Transform," Englewood Cliffs, NJ: Prentice-Hall, 1974.

5

z-Transform

Jelena Kovačević
AT&T Bell Laboratories

5.1 Introduction

When analyzing linear systems, one of the problems we often encounter is that of solving linear, constant-coefficient differential equations. A tool used for solving such equations is the Laplace transform. At the same time, to aid the analysis of linear systems, we extensively use Fourier-domain methods. With the advent of digital computers, it has become increasingly necessary to deal with discrete-time signals, or, sequences. These signals can be either obtained by sampling a continuous-time signal, or they could be inherently discrete. To analyze linear discrete-time systems, one needs a discrete-time counterpart of the Laplace transform (LT). Such a counterpart is found in the z transform, which similarly to the LT, can be used to solve linear constant-coefficient difference equations. In other words, instead of solving these equations directly, we transform them into a set of algebraic equations first, and then solve in this transformed domain. On the other hand, the z-transform can be seen as a generalization of the discrete-time Fourier transform (FT)

$$X(e^{j\omega}) = \sum_{n=-\infty}^{+\infty} x[n]e^{-j\omega n} \tag{5.1}$$

The above expression does not always converge, and thus, it is useful to have a representation which will exist for these nonconvergent instances. Furthermore, the use of the z-transform offers considerable notational simplifications. It also allows us to use the extensive body of work on complex variables to aid in analyzing discrete-time systems.

0-8493-8341-2/95/$0.00+$.50

The z-transform, as pointed out by Jury in his classical text [3], is not new. It can be traced back to the early 18th century and the times of DeMoivre, who introduced the notion of the **generating function**, extensively used in probability theory

$$\Gamma(z) = \sum_{n=-\infty}^{+\infty} p[n]z^n \tag{5.2}$$

where $p[n]$ is the probability that the discrete random variable $\mathbf{n}$ will take a value n [8]. By comparing (5.2) and (5.3) below, we can see that the generating function $\Gamma(1/z)$ is the z-transform of the sequence $p[n] = p\{\mathbf{n} = n\}$. After these initial efforts, and due to the fast development of digital computers, a renewed interest in the z-transform occurred in the early 1950s, and the z-transform has been used for analyzing discrete-time systems ever since.

This section is intended as a brief introduction to the theory and application of the z-transform. For a rigorous mathematical treatment of the transform itself, the reader is referred to the book by one of the pioneers in the development of analysis of sampled data systems, Jury [3], and the references therein. For a more succinct account of the z-transform, its properties and use in discrete-time systems, consult, for example, [7]. A few other texts which contain parts on the z-transform include [1, 2, 5, 6, 10].

5.2 Definition of the z-Transform

Suppose we are given a discrete-time sequence $x[n]$, either inherently discrete time, or obtained by sampling a continuous-time signal $x_c(t)$, so that $x[n] = x_c(nT)$, $n \in \mathcal{Z}$, where T is the sampling period. Then the two-sided z-transform of $x[n]$ is defined as

$$X(z) = \sum_{n=-\infty}^{+\infty} x[n]z^{-n} \tag{5.3}$$

Here, z is a complex variable, and depending on its magnitude and the sequence $x[n]$, the above sum may or may not converge. The region in the z-plane where the sum does converge is called the **region of convergence** (ROC), and is discussed in more detail later.

Observe that in (5.3), n ranges from $-\infty$ to $+\infty$. That is why the z-transform defined in this way is called **two-sided.** One could define a **one-sided** z-transform, where n would range from 0 to $+\infty$. Obviously, the two definitions are equivalent only if the signal itself is one-sided, that is, if $x[n] = 0$, for $n < 0$. The advantage of using the one-sided z-transform is that it is useful in solving linear constant-coefficient difference equations with nonzero initial conditions and in the study of sampled-data feedback systems, discussed later. However, from now on, we deal mostly with the two-sided z-transform (see also Chapter 3 where the one-sided LT is used).

A power series given in (5.3) is a Laurent series, and thus for it to converge uniformly, it has to be absolutely summable, that is, the following must hold:

$$\sum_{n=-\infty}^{n=+\infty} |x[n]|\,|z|^{-n} < \infty \tag{5.4}$$

where $|z|$ is the magnitude of the complex variable z, i.e., $z = |z|e^{j\arg z}$. We can now see that if the z-transform converges for $z = z_1$, it will converge for all z such that $|z| = |z_1|$; that is, for all z on the circle $|z| = |z_1|$. In general, therefore, the ROC will be an annular region in the z-plane, and will be of the form

$$0 \le R_- < |z| < R_+ \le \infty \tag{5.5}$$

Here, R_- could be zero, and R_+ could be infinity, so that all of the cases given in Figure 5.1 are possible. Note that the points $z = 0$, or $z = \infty$ could be included in the region of convergence.

A very important notion when talking about the ROC of the z-transform is that of the **unit circle**, because it makes the connection to the discrete-time FT. The unit circle is defined as the set of points in the z-plane for which $|z| = 1$. Let us evaluate the z-transform of a sequence on the unit circle. Because $|z| = 1$, it can be expressed as $z = e^{j\omega}$, where $\omega = \arg z$. Thus,

$$X(z)\Big|_{|z|=1} = X(e^{j\omega}) = \sum_{n=-\infty}^{+\infty} x[n]e^{-j\omega n} \tag{5.6}$$

Note that here we use $X(e^{j\omega})$ rather than $X(\omega)$ to make it explicit that this is a function of $e^{j\omega}$. As can be seen from the equation, the z-transform evaluated on the unit circle is equivalent to the discrete-time FT given in (5.1). This immediately justifies the statement from the beginning of the section that the z-transform can be seen as a generalization of the FT. In other words, instances occur when the FT does not converge, while the z-transform does. One such instance is the unit-step function $u[n]$

$$u[n] = \begin{cases} 1, & n \geq 0, \\ 0, & \text{otherwise} \end{cases}$$

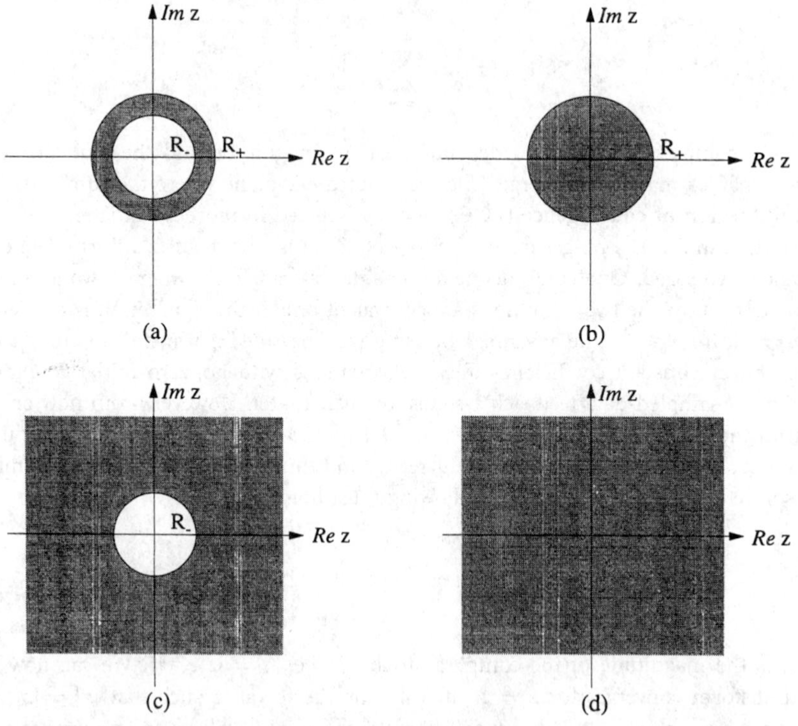

FIGURE 5.1 ROC of the z-transform. (a) General case: $0 \leq R_- < |z| < R_+ \leq \infty$. (b) ROC is the inside of the circle $|z| < R_+$ ($z = 0$ might be excluded). (c) ROC is the outside of the circle $R_- < |z|$ ($z = \infty$ might be excluded). (d) ROC is the whole z-plane (except possibly $z = 0$ or $z = \infty$).

The sum

$$U(e^{j\omega}) = \sum_{n=-\infty}^{+\infty} u[n]e^{-j\omega n} = \sum_{n=0}^{+\infty} e^{-j\omega n}$$

is not absolutely summable and thus its FT does not converge, while

$$U(z) = \sum_{n=-\infty}^{+\infty} u[n]z^{-n} = \sum_{n=0}^{+\infty} z^{-n}$$

converges for $|z| > 1$. As a consequence, if the ROC includes the unit circle, then the discrete-time FT of a given sequence will exist, otherwise it will not.

The unit circle captures the periodicity of the discrete-time FT. If we start evaluating the z-transform on the unit circle at the point $(\text{Re}\, z, \text{Im}\, z) = (1, 0)$ corresponding to $\omega = 0$, going through $(\text{Re}\, z, \text{Im}\, z) = (0, j)$, $(\text{Re}\, z, \text{Im}\, z) = (-1, 0)$ which corresponds to $\omega = \pi$, and $(\text{Re}\, z, \text{Im}\, z) = (0, -j)$, back to $(\text{Re}\, z, \text{Im}\, z) = (1, 0)$ corresponding to $\omega = 2\pi$, we have evaluated one period of the FT and have returned to the same point. Thus, we are effectively warping the linear frequency axis into the unit circle (see Figure 5.2).

We also mentioned that the z-transform is the discrete-time counterpart of the LT. Consider the function

$$x_{cs}(t) = \sum_{n=-\infty}^{+\infty} x_c(nT)\delta(t - nT) \tag{5.7}$$

or, the sampled version of the original continuous time function $x_c(t)$. Here, T is the sampling period, and $\delta(t)$ is the Dirac function. Taking the LT of $x_{cs}(t)$, we obtain

$$X_{cs}(s) = \sum_{n=-\infty}^{+\infty} x_c(nT)e^{-nTs} \tag{5.8}$$

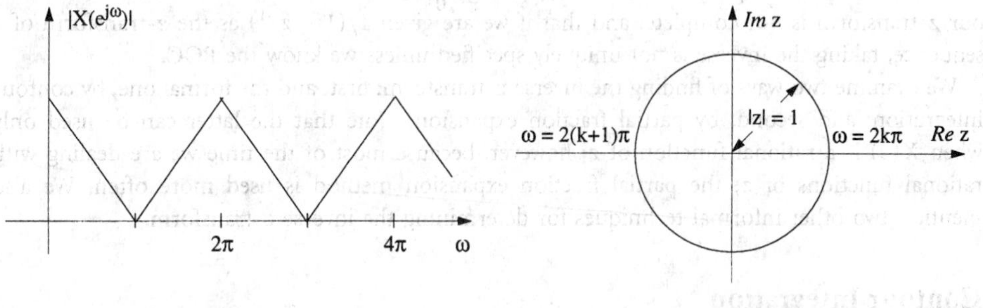

FIGURE 5.2 Warping of the linear frequency axis of the discrete-time FT into the unit circle of the z-transform. For example, the point $(\text{Re}\, z, \text{Im}\, z) = (1, 0)$ on the unit circle corresponds to points $\omega = 2k\pi$ on the ω axis.

If we now replace e^{sT} by z, we obtain the z-transform. Now, observe that $X_{cs}(s)$ in (5.8) is periodic, since

$$X_{cs}\left(s + j\frac{2\pi}{T}\right) = \sum_{n=-\infty}^{+\infty} x_c(nT)e^{-nT(s+j(2\pi)/T)},$$
$$= \sum_{n=-\infty}^{+\infty} x_c(nT)e^{-nTs}e^{-j2\pi n} = X_{cs}(s) \tag{5.9}$$

This means that $X_{cs}(s)$ is periodic along constant lines $\sigma = \sigma_{\text{const}}$ (parallel to the $j\omega$ axis). This further means that any line parallel to the $j\omega$ axis maps into a circle in the z-plane. It is easy to see that the $j\omega$ axis itself would map into the unit circle, while the left (or right) half-planes would map into the inside (or outside) of the unit circle, respectively.

Finally, let us say a few words about a very important class of signals, those whose z-transform is a rational function of z. They arise from systems that can be represented by linear constant-coefficient difference equations and are the signals with which we deal mostly in practice. If we represent such signals by

$$X(z) = \frac{N(z)}{D(z)} \tag{5.10}$$

then the zeroes of the numerator $N(z)$ are called **zeroes** of $X(z)$, while the zeroes of the denominator $D(z)$ are called **poles** of $X(z)$ (more precisely, a pole z_p will be a point at which $\lim_{z \to z_p} X(z)$ does not exist). How the poles can determine the region of convergence is the subject of a discussion later in the chapter.

5.3 Inverse z-Transform

We have seen that specifying the ROC when taking the z-transform of a sequence is an integral part of the process. For example, consider the following sequences: $x[n] = u[n]$ and $y[n] = -u[-n-1]$, where $u[n]$ is the unit-step function. Taking their z-transforms, we obtain

$$X(z) = \frac{1}{1 - z^{-1}}, \qquad |z| > 1 \tag{5.11}$$

$$Y(z) = \frac{1}{1 - z^{-1}}, \qquad |z| < 1 \tag{5.12}$$

They are identical except for their ROCs (see Figure 5.3). This tells us that without the ROC, our z-transform is not complete, and that if we are given $1/(1 - z^{-1})$ as the z-transform of a sequence, taking the inverse is not uniquely specified unless we know the ROC.

We examine two ways of finding the inverse z-transform: first, and the formal one, by contour integration, and second, by partial fraction expansion. Note that the latter can be used only when $X(z)$ is a rational function of z; however, because most of the time we are dealing with rational functions of z, the partial fraction expansion method is used more often. We also mention two other informal techniques for determining the inverse z-transform.

Contour Integration

The formal inversion process for the z-transform is given by contour integration. It is obtained by taking the expression for the z-transform given by (5.3) and multiplying both sides by

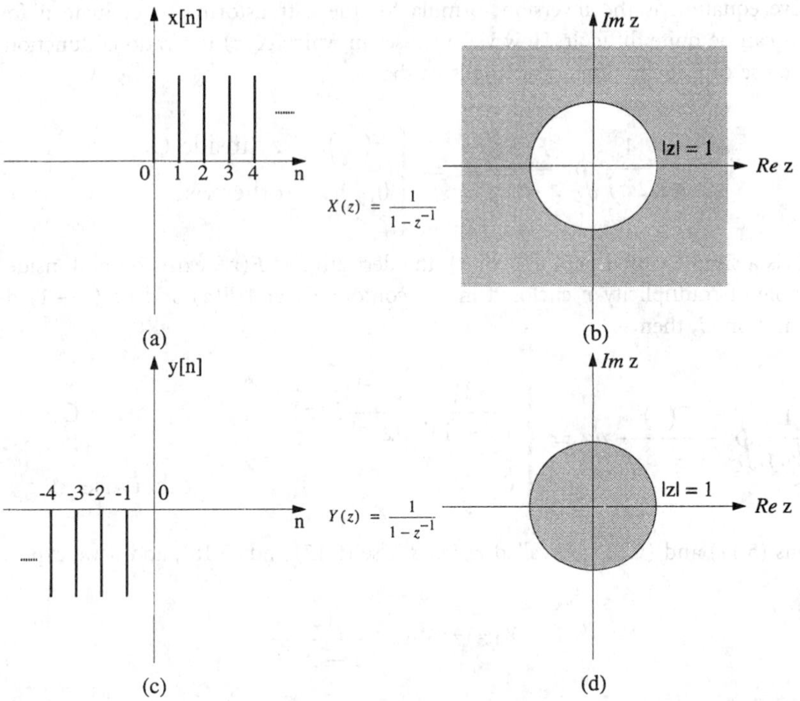

FIGURE 5.3 Two different sequences giving rise to the same z-transforms $1/(1-z^{-1})$ except for their ROCs. (a) Sequence $x[n] = u[n]$ and (b) its ROC $|z| > 1$. (c) Sequence $y[n] = -u[-n-1]$ and (d) its ROC $|z| < 1$.

$(1/2\pi j)z^{k-1}$. Then the result is integrated counterclockwise along a closed contour C in the z-plane containing the origin, leading to

$$\oint_C \frac{1}{2\pi j} X(z) z^{k-1} dz = \oint_C \frac{1}{2\pi j} z^{k-1} \sum_{n=-\infty}^{+\infty} x[n] z^{-n}\, dz \tag{5.13}$$

We choose the contour of integration to lie within the ROC, which will allow us to interchange the order of integration and summation on the right side of (5.13). This leads to

$$\frac{1}{2\pi j} \oint_C X(z) z^{k-1}\, dz = \sum_{n=-\infty}^{+\infty} x[n] \left(\frac{1}{2\pi j} \oint_C z^{-n+k-1}\, dz \right) \tag{5.14}$$

The integral in parentheses on the right side of (5.14) can be evaluated using Cauchy's integral formula, which states that if the contour of integration C contains the origin, and integration is performed counterclockwise, then

$$\frac{1}{2\pi j} \oint_C z^k\, dz = \begin{cases} 1, & k = -1, \\ 0, & \text{otherwise} \end{cases} \tag{5.15}$$

Substituting (5.15) in (5.14), we see that the integral is nonzero only for $n = k$, and thus (5.14) can be rewritten as

$$\frac{1}{2\pi j} \oint_C X(z) z^{k-1}\, dz = x[k] \tag{5.16}$$

The above equation is the inversion formula for the z-transform. To evaluate it for general functions can be quite difficult. However, in cases in which $X(z)$ is a rational function of z, we can make use of Cauchy's formula. It tells us that

$$\frac{1}{2\pi j}\oint_C \frac{F(z)}{z-z_p}\,dz = \begin{cases} F(z_p), & z_p \text{ inside } C, \\ 0, & \text{otherwise} \end{cases} \tag{5.17}$$

where C is a simple closed path and $F'(z)$, the derivative of $F(z)$, exists on and inside C. If we have a pole of multiplicity r enclosed in the contour C, and $F(z)$ and its $(r+1)$ derivatives exist in and on C, then

$$\frac{1}{2\pi j}\oint_C \frac{F(z)}{(z-z_p)^r}\,dz = \begin{cases} \left.\dfrac{1}{(r-1)!}\dfrac{d^{r-1}}{dz^{r-1}}F(z)\right|_{z=z_p}, & z_p \text{ inside } C, \\ 0, & \text{otherwise} \end{cases} \tag{5.18}$$

Equations (5.17) and (5.18) are called *residues.* Use (5.17) and (5.18), and if we express

$$X(z)z^{k-1} = \frac{F(z)}{(z-z_p)^r} \tag{5.19}$$

where $F(z)$ has no poles at $z = z_p$, then Cauchy's residue theorem says that

$$x[k] = \frac{1}{2\pi j}\oint_C X(z)z^{k-1}\,dz = \sum_i R_i \tag{5.20}$$

where R_i are residues of $X(z)z^{k-1}$ at the poles inside the contour C. The poles outside the contour do not contribute to the sum. If no poles are inside the contour of integration for a certain k, then $x[k]$ is zero for that k. Do not ignore the fact that the contour of integration C must lie within the ROC.

In some instances it may be quite cumbersome to evaluate (5.20), for example, when we have a multiple-order pole at $z = 0$, whose order depends on k. In that case we can rewrite (5.20) as

$$x[k] = \sum_i R_i'$$

where R_i' is the residue of $X(1/z)z^{-k-1}$ at the poles inside the contour C', and C' is a circle of radius $1/s$ if C is a circle of radius s. For more details, see [7].

Partial Fraction Expansion

Another method of obtaining the inverse z-transform is by using partial fraction expansion. Note, however, that the partial fraction expansion method can be applied only to rational functions. Thus, suppose that $X(z)$ can be represented as in (5.10). We can then rewrite it as

$$X(z) = \frac{N_0}{D_0}\cdot\frac{\prod_{i=1}^{N}(1-n_i z^{-1})}{\prod_{i=1}^{D}(1-d_i z^{-1})} \tag{5.21}$$

where n_i are nontrivial zeroes of $N(z)$ and d_i are zeroes of $D(z)$. The partial fraction expansion of $X(z)$ can be written as

$$X(z) = \sum_{i=0}^{N-D} A_i z^{-i} + \sum_{i=1}^{D_s} \frac{B_i}{(1 - d_i z^{-1})} + \sum_{i=1}^{D_m} \sum_{m=1}^{p_i} \frac{C_{mi}}{(1 - d_i z^{-1})^m} \tag{5.22}$$

Here, if $N \geq D$, A_i can be obtained by long division of the numerator by the denominator; otherwise, that first sum in (5.22) disappears. In the second sum D_s denotes the number of single poles d_i of $X(z)$, and the coefficients B_i can be obtained as

$$B_i = (1 - d_i z^{-1}) X(z)\big|_{z=d_i} \tag{5.23}$$

The third sum (double sum) represents the part with multiple poles. D_m is the number of multiple poles d_i, and p_i are their respective multiplicities. The coefficients C_{mi} can be obtained as

$$C_{mi} = \frac{1}{(p_i - m)!(-d_i)^{p_i - m}} \frac{d^{p_i - m}}{dz^{p_i - m}} \left[(1 - d_i z)^{p_i} X(z^{-1}) \right] \Bigg|_{z=d_i^{-1}} \tag{5.24}$$

Once we have the expression (5.22), we can recognize each term as the z-transform of a known sequence. For example, $B_i/(1 - d_i z^{-1})$ will be the z-transform of either $B_i d_i^n u[n]$, or $-B_i d_i^n u[-n-1]$, depending on whether $|z| > |d_i|$, or $|z| < |d_i|$.

Other Methods for Obtaining the Inverse z-Transform

While the two methods presented above will work in most cases, sometimes it can be more convenient to use simpler techniques.

One of these is the inspection method [7]. It consists of learning to recognize some often-used z-transform pairs. For example, if we are given $3/(1 - z^{-1})$ with the ROC $|z| < 1$, from (5.12) we can recognize it as the z-transform of $-3u[-n-1]$. In this process the tables of z-transform pairs are an invaluable tool. An extensive list of z-transform pairs can be found in [3].

Another technique can be used if we are given a z-transform in the form of a power series expansion:

$$X(z) = \cdots + x[-1]z + x[0] + x[1]z^{-1} + \cdots$$

Then, we can identify each term with the appropriate power of z. For example, the coefficient with z^{-k} will be $x[k]$.

5.4 Properties of the z-Transform

While we can always obtain a z-transform of a sequence by directly applying its definition as given in (5.3), it is useful to have a list of properties at hand to help calculate a particular z-transform or inverse z-transform more easily. We divide these properties into two categories: properties of the ROC, and properties of the z-transform itself. In what follows R_x will denote the ROC of the signal $x[n]$, while R_{x_-} and R_{x_+} will denote its lower/upper bounds, respectively [as given in (5.5)].

Region of Convergence

The ROC is an integral part of the z-transform of a sequence. Thus, this section goes into more detail on some of the points touched upon earlier. These properties and the order in which they are presented follow those in [7]; therefore, for more details, see [7].

First, we said that the ROC is an annular region in the z-plane, i.e., $0 \le R_- < |z| < R_+ \le \infty$. This follows from the fact that if the z-transform converges for $z = z_1$, it will converge for all z such that $|z| = |z_1|$, that is, for all z on the circle $|z| = |z_1|$. Then, if we put $|z| = 1$ in (5.3), we obtain (5.6), or the discrete-time FT. Therefore, it is obvious that the FT of $x[n]$ converges iff the z-transform of $x[n]$ converges for $|z| = 1$, that is, iff the ROC of the z-transform contains the unit circle. Another useful property is that the ROC cannot contain any poles. This stems from the fact that if it did, the z-transform at the pole would be infinite and would not converge.

Consider now what happens if the sequence is of finite duration—it is zero except in a finite interval $-\infty < N_1 \le n \le N_2 < +\infty$. If all the values are finite, then the sequence is clearly absolutely summable and the z-transform will converge everywhere, except possibly at points $z = 0$ or $z = \infty$. Using the same type of arguments one can conclude that if the sequence is right-sided (it is zero for $n < N_1 < +\infty$), then the ROC will be the annular region outside of the finite pole of $X(z)$ of the largest magnitude. Similarly, if the sequence is left-sided (it is zero for $n > N_2 > -\infty$), then the ROC will be the annular region inside the finite pole of $X(z)$ of the smallest magnitude. As a result, if a sequence is neither left- nor right-sided, the ROC will be an annular region bounded on the interior and the exterior by a pole.

Properties of the Transform

The sequences $x[n], y[n], \cdots$ will have associated z-transforms $X(z), Y(z), \ldots,$ with ROCs $R_x, R_y, \ldots,$ in which each ROC will have its associated lower and upper bounds, as given in (5.5).

Linearity

$$ax[n] + by[n] \longleftrightarrow aX(z) + bY(z) \qquad \text{ROC} \supset R_x \cap R_y \tag{5.25}$$

To prove this, apply the definition given in (5.5). Note that the resulting ROC is at least as large as the intersection of the two starting ROCs. For example, if both $X(z)$ and $Y(z)$ are rational functions of z, and by adding $aX(z)$ to $bY(z)$ we introduce a zero that cancels one of the poles, the resulting ROC is larger than the intersection. If, on the other hand, no pole/zero cancellation exists, the resulting ROC is exactly the intersection.

Shift in Time

$$x[n-i] \longleftrightarrow z^{-i}X(z) \qquad \text{ROC} = R_x \tag{5.26}$$

The proof is straightforward and follows by the change of variables $k = n - i$ in (5.5). Note that the resulting ROC could gain or lose a few poles at $z = 0$ or $z = \infty$.

Time Reversal

$$x[-n] \longleftrightarrow X\left(\frac{1}{z}\right) \qquad \frac{1}{R_{x_+}} < |z| < \frac{1}{R_{x_-}} \tag{5.27}$$

Again, the proof is straightforward and follows by the change of variables $k = -n$ in (5.5).

Multiplication by an Exponential Sequence

One can easily show that the following holds:

$$a^n x[n] \longleftrightarrow X\left(\frac{z}{a}\right) \qquad |a|R_{x_-} < |z| < |a|R_{x_+} \tag{5.28}$$

Because z/a is the variable in the transform domain, we can see that all the poles of the original $X(z)$ have been scaled by a.

Multiplication by a Ramp

This property could also be called "differentiation of $X(z)$".

$$nx[n] \longleftrightarrow -z\frac{dX(z)}{dz}, \qquad \text{ROC} = R_x \tag{5.29}$$

To demonstrate this, differentiate both sides of (5.3) with respect to z and then multiply by $-z$ to obtain (5.29).

Convolution in Time

In transform domain convolution becomes simply the product of the two sequences. If we denote convolution by $*$,

$$x[n] * y[n] = \sum_{k=-\infty}^{+\infty} x[k]y[n-k] = \sum_{k=-\infty}^{+\infty} y[k]x[n-k] \tag{5.30}$$

then

$$x[n] * y[n] \longleftrightarrow X(z) \cdot Y(z), \qquad \text{ROC} \supset R_x \cap R_y \tag{5.31}$$

Although the proof is not difficult, we write it here, because this is one of the most useful properties. Thus, take the convolution in (5.30), multiply it by z^{-n} and sum over n

$$\begin{aligned}
\sum_{n=-\infty}^{+\infty}\sum_{k=-\infty}^{+\infty} x[k]y[n-k]z^{-n} &= \sum_{k=-\infty}^{+\infty} x[k] \sum_{n=-\infty}^{+\infty} y[n-k]z^{-n}, \\
&= \sum_{k=-\infty}^{+\infty} x[k] \sum_{p=-\infty}^{+\infty} y[p]z^{-(p+k)}, \\
&= \sum_{k=-\infty}^{+\infty} x[k]z^{-k} \sum_{p=-\infty}^{+\infty} y[p]z^{-p}, \\
&= X(z)Y(z)
\end{aligned}$$

where we have used change of variables $p = n - k$. As in the case of linearity, if it happens that a pole residing at the border of one of the ROCs is cancelled by a zero from the other transform, then the resulting ROC will be larger than the intersection of the individual ROCs; otherwise, it will be exactly their intersection.

Convolution in z-Domain

Convolution in z-domain is given by (for more details, refer to [3] or [7])

$$x[n]\cdot y[n] \longleftrightarrow \frac{1}{2\pi j}\oint_C X(\lambda)Y\left(\frac{z}{\lambda}\right)\lambda^{-1}d\lambda \qquad R_{x_-}R_{y_-} < |z| < R_{x_+}R_{y_+} \tag{5.32}$$

where C is a closed contour in the intersection of the ROCs of $X(\lambda)$ and $Y(z/\lambda)$, and integration is performed counterclockwise. This property is the generalization of the periodic convolution property of the FT. Suppose that the contour C is the unit circle and $\lambda = e^{j\omega}$, $z = e^{j\theta}$. Also, observe that $d\lambda = je^{j\omega}d\omega$, and that if λ goes around the unit circle, ω ranges from $-\pi$ to π. Then,

$$\frac{1}{2\pi j}\oint_C X(\lambda)Y\left(\frac{z}{\lambda}\right)\lambda^{-1}d\,\lambda = \frac{1}{2\pi}\int_{-\pi}^{\pi} X(e^{j\omega})Y(e^{j(\theta-\omega)})\,d\omega$$

This equation states that the product of two sequences has as its FT the periodic convolution of their FTs.

Conjugation

If we are given a complex sequence $x^*[n]$, then its z-transform pair is

$$x^*[n] \longleftrightarrow X^*(z^*) \qquad \text{ROC} = R_x \tag{5.33}$$

Real Part

This property and the next one use the conjugation property given in (5.33). Thus, if we express

$$\text{Re}\{x[n]\} = \frac{x[n]+x^*[n]}{2}$$

then by (5.33) and the linearity of the z-transform

$$\text{Re}\{x[n]\} \longleftrightarrow \frac{1}{2}\left[X(z)+X^*(z^*)\right] \qquad \text{ROC} \supset R_x \tag{5.34}$$

Imaginary Part

Similarly to the previous property, if we express

$$\text{Im}\{x[n]\} = \frac{x[n]-x^*[n]}{2j}$$

then by (5.33) and the linearity of the z-transform

$$\text{Im}\{x[n]\} \longleftrightarrow \frac{1}{2j}\left[X(z)-X^*(z^*)\right] \qquad \text{ROC} \supset R_x \tag{5.35}$$

Parseval's Relation

Parseval's relation is widely used in various transform domains, usually to find the energy of a signal. We start here with its more general formulation, and then reduce it to its usual form

$$\sum_{n=-\infty}^{+\infty} x[n]y^*[n] = \frac{1}{2\pi j}\oint_C X(\lambda)Y^*\left(\frac{1}{\lambda^*}\right)\lambda^{-1}d\lambda \tag{5.36}$$

where C is a closed contour in the intersection of the ROCs of $X(\lambda)$ and $Y^*(1/\lambda^*)$, and integration is performed counterclockwise. Again, for the proof, we refer the reader to [3] or [7].

Suppose now that $y[n] = x[n]$. Then (5.36) reduces to

$$\sum_{n=-\infty}^{+\infty} |x[n]|^2 = \frac{1}{2\pi j}\oint_C X(\lambda)X^*\left(\frac{1}{\lambda^*}\right)\lambda^{-1}d\lambda \tag{5.37}$$

If both $x[n]$ and $y[n]$ converge on the unit circle, i.e., their FTs exist, then, we can choose $\lambda = e^{j\omega}$ and (5.36) becomes

$$\begin{aligned}\sum_{n=-\infty}^{+\infty} x[n]y^*[n] &= \frac{1}{2\pi j}\oint_C X(e^{j\omega})Y^*(e^{j\omega})e^{-j\omega}d(e^{j\omega}),\\ &= \frac{1}{2\pi}\int_{-\pi}^{\pi} X(e^{j\omega})Y^*(e^{j\omega})d\omega\end{aligned}$$

which is the usual Parseval's relation in Fourier domain.

Initial Value Theorem

If $x[n]$ is causal (it is 0 for $n < 0$), then

$$x[0] = \lim_{z\to\infty} X(z) \tag{5.38}$$

Final Value Theorem

If the poles of $X(z)$ are all inside the unit circle, then

$$\lim_{n\to\infty} x[n] = \lim_{z\to 1}\left[(1 - z^{-1})X(z)\right] \tag{5.39}$$

5.5 Role of the z-Transform in Linear Time-Invariant Systems

The class of systems dealt with mostly are linear time-invariant systems. We discuss here the role of the z-transform in such systems. Recall that if we are given a system with input $x[n]$ and output $y[n]$, described by an operator $H[\cdot]$

$$y[n] = H[x[n]]$$

and $H[\cdot]$ is defined to be linear and time invariant, then

1. If inputs $x_1[n]$ and $x_2[n]$ produce outputs $y_1[n]$ and $y_2[n]$, then the input $ax_1[n]+bx_2[n]$ will produce the output $ay_1[n] + by_2[n]$.
2. If the input $x[n]$ produces output $y[n]$, then the input $x[n-k]$ will produce the output $y[n-k]$.

We recall a few properties of discrete linear time-invariant systems. For more details, refer to [4, 7].

Note first that $x[n]$ can be written as a superposition of unit samples as

$$x[n] = \sum_{k=-\infty}^{+\infty} x[k]\delta[n-k] \tag{5.40}$$

If this is the input and we want to find the corresponding output, then

$$\begin{aligned} y[n] &= H[x[n]], \\ &= H\left[\sum_{k=-\infty}^{+\infty} x[k]\delta[n-k]\right], \\ &= \sum_{k=-\infty}^{+\infty} x[k]H[\delta[n-k]], \\ &= \sum_{k=-\infty}^{+\infty} x[k]h[n-k] \end{aligned} \tag{5.41}$$

Here, $h[n]$ is called the **unit-sample response**, defined as the response of the system to the unit sample $\delta[n]$. Equation (5.41) is a convolution sum and thus it can be written as

$$y[n] = x[n] * h[n] \tag{5.42}$$

One of the most important properties of the z-transform is that the convolution in time has a product as its transform-domain pair. Therefore, using (5.31), we can write (5.42) in z-domain as

$$Y(z) = X(z)H(z) \tag{5.43}$$

Here, $H(z)$ is called the **system transfer function**

$$H(z) = \sum_{n=-\infty}^{+\infty} h[n]z^{-n} \tag{5.44}$$

We may also obtain this transfer function in another manner, if we assume that the systems we are dealing with are those that can be described by linear constant-coefficient difference equations (much in the same way as continuous time systems that can be represented by linear constant-coefficient differential equations). Along the way, the use of z-transform greatly simplifies analysis of such systems and the solutions to these equations reduce to solutions of algebraic equations. Hence, suppose that our system can be described by the following linear constant-coefficient difference equation:

$$\sum_{k=0}^{D} b_k y[n-k] = \sum_{k=0}^{N} a_k x[n-k] \tag{5.45}$$

This equation has many solutions. We assume that the system is causal and moreover, we assume that the initial conditions are satisfied so that the system is linear and time invariant. With these factors in mind, we can write the expression for the output of the system as

$$y[n] = -\sum_{k=1}^{D} \frac{b_k}{b_0} y[n-k] + \sum_{k=0}^{N} \frac{a_k}{b_0} x[n-k] \tag{5.46}$$

This means that the output at time n depends on the outputs from D previous instants as well as from the input at time n and at N previous instants. Taking the z-transform of both sides of (5.45) and using the linearity and shift in time properties of the z-transform, we obtain

$$Y(z) = \sum_{k=0}^{N} a_k z^{-k} \Big/ \sum_{k=0}^{D} b_k z^{-k} X(z) \tag{5.47}$$

Finally, using (5.43) we can identify $H(z)$ here as

$$H(z) = \sum_{k=0}^{N} a_k z^{-k} \Big/ \sum_{k=0}^{D} b_k z^{-k} \tag{5.48}$$

The function $H(z)$ is often referred to as a **filter**. If the denominator is a delay, i.e., $b_i z^{-i}$, such a filter will be a finite impulse response filter (FIR), as opposed to the infinite impulse response filter (IIR). Taking the inverse z-transform of (5.48) we may obtain the unit-sample response of the system $h[n]$.

Finally, let us see how one might obtain the **frequency response** of the system. Evaluate (5.44) on the unit circle (assuming that the ROC of $H(z)$ contains the unit circle)

$$H(z)\big|_{|z|=1} = H(e^{j\omega}) = \sum_{n=-\infty}^{+\infty} h[n] e^{-j\omega n} \tag{5.49}$$

Therefore, the frequency response of the system is the system transfer function evaluated on the unit circle:

$$H(e^{j\omega}) = H(z)\big|_{z=e^{j\omega}} \tag{5.50}$$

A consequence of this is that a linear time-invariant system is bounded-input–bounded-output stable (that is, bounded input produces bounded output), iff the ROC of the transfer function $H(z)$ contains the unit circle.

We mentioned earlier that we are using the double-sided z-transform, while the one-sided z-transform is most useful when solving linear constant-coefficient difference equations with nonzero initial conditions. To transform such an equation into an algebraic equation, we used the linearity and shift in time properties of the double-sided z-transform. We would like to be able to do the same with the one-sided z-transform; however, we must rederive the shift in time property in order to do that. Therefore, assume we are given a sequence $x[n-i]$ as in (5.26).

Taking its one-sided z-transform, we obtain

$$\begin{aligned}\sum_{n=0}^{+\infty} x[n-i]z^{-n} &= \sum_{p=-i}^{+\infty} x[p]z^{-(p+i)}, \\ &= \sum_{p=-i}^{-1} x[p]z^{-(p+i)} + z^{-i}\sum_{p=0}^{+\infty} x[p]z^{-p}, \\ &= x[-i] + \cdots + x[-1]z^{-(i-1)} + z^{-i}X(z)\end{aligned} \tag{5.51}$$

To solve (5.45), we can take the one-sided z-transform of both sides and use the linearity and shift in time properties to obtain

$$\begin{aligned}&\sum_{k=0}^{D} b_k \left(\sum_{p=-k}^{-1} y[p]z^{-(p+k)} + z^{-k}Y(z) \right) \\ &\quad = \sum_{k=0}^{N} a_k \left(\sum_{p=-k}^{-1} x[p]z^{-(p+k)} + z^{-k}X(z) \right)\end{aligned} \tag{5.52}$$

The output $y[n]$ for $n \geq 0$ can be obtained as the inverse z-transform of

$$\begin{aligned}Y(z) = &\left(\sum_{k=0}^{N} a_k z^{-k} \Big/ \sum_{k=0}^{D} b_k z^{-k} \right) X(z) \\ &+ \left(\sum_{k=0}^{N} a_k \sum_{p=-k}^{-1} x[p]z^{-(p+k)} - \sum_{k=0}^{D} b_k \sum_{p=-k}^{-1} y[p]z^{-(p+k)} \Big/ \sum_{k=0}^{D} b_k z^{-k} \right)\end{aligned} \tag{5.53}$$

To solve for $Y(z)$ we need to know initial conditions $y[n]$, for $n = -D, \ldots, -1$, and for $x[n]$ for $n = -N, \ldots, -1$. Note that if $x[n]$ is causal, and the initial conditions are all zero, the above solution is the same as the one when the double-sided z-transform is used; i.e., it reduces to the first term on the right side of (5.53).

5.6 Variations on the z-Transform

Multidimensional z-Transform

Although the two-dimensional and multidimensional z-transforms are related to the one-dimensional z-transform, they are not straightforward generalizations of it. We give here a brief overview of the two-dimensional z-transform. For more details on it and the multidimensional z-transform, see [1, 2].

The two-dimensional z-transform of a sequence $x[n_1, n_2]$ is given by

$$X(z_1, z_2) = \sum_{n_1=-\infty}^{+\infty} \sum_{n_2=-\infty}^{+\infty} x[n_1, n_2] z_1^{-n_1} z_2^{-n_2} \tag{5.54}$$

Here, z_1 and z_2 are complex variables and the region in the four-dimensional (z_1, z_2) space in which the above double sum converges is called the ROC. In one dimension, the ROC is an annular region in the z-plane, while here it is called the **Reinhardt domain**. The analog of the unit circle in one dimension is the **unit bicircle** for $|z_1| = 1$ and $|z_2| = 1$. On the unit bicircle, the two-dimensional z-transform becomes the two-dimensional discrete-time FT.

Properties of the two-dimensional z-transform are similar to those of the one-dimensional z-transform, and they can be found in [1, 2]. Here we list just two. First, for separable signals, the following holds:

$$x[n_1]y[n_2] \longleftrightarrow X(z_1)Y(z_2) \tag{5.55}$$

where (z_1, z_2) is in the ROC, if z_1 is in R_x and z_2 is in R_y. The differentiation property is as follows:

$$n_1 n_2 x[n_1, n_2] \longleftrightarrow z_1 z_2 \frac{\partial^2}{\partial z_1 \partial z_2} X(z_1, z_2), \qquad \text{ROC} = R_x \tag{5.56}$$

In analyzing two-dimensional systems it is useful to identify singularities. Given a two-dimensional linear shift-invariant system with an associated constant-coefficient difference equation

$$\sum_{k_1=0}^{D_1} \sum_{k_2=0}^{D_2} b_{k_1 k_2} y[n_1 - k_1, n_2 - k_2] = \sum_{k_1=0}^{N_1} \sum_{k_2=0}^{N_2} a_{k_1 k_2} x[n_1 - k_1, n_2 - k_2] \tag{5.57}$$

we can find the equivalent transfer function as

$$H(z_1, z_2) = \left(\sum_{k_1=0}^{N_1} \sum_{k_2=0}^{N_2} a_{k_1 k_2} z_1^{-k_1} z_2^{-k_2} \Big/ \sum_{k_1=0}^{D_1} \sum_{k_2=0}^{D_2} b_{k_1 k_2} z_1^{-k_1} z_2^{-k_2} \right) = \frac{N(z_1, z_2)}{D(z_1, z_2)} \tag{5.58}$$

Then, the zero of $H(z_1, z_2)$ is a point at which $A(z_1, z_2) = 0$ and $B(z_1, z_2) \neq 0$, while a pole is a point at which $B(z_1, z_2) = 0$. Note, however, that both zeroes and poles are continuous surfaces rather than a discrete set of points, as in one dimension. Note, also, that unlike in one dimension, no fundamental theorem of algebra tells us how to factorize a multidimensional polynomial into its factors, and thus, it is not easy to isolate poles and zeroes.

Modified z-Transform

The original, one-sided z-transform was developed to deal only with the signal at its sampling instants, and discard the rest. In many systems, particularly in mixed analog-digital systems, it is important to conserve the information about the signal between the sampling instants. Jury, among others, used the **modified z-transform** [3] to take care of this problem. The idea is to delay the continuous-time function $x_c(t)$ by a fictitious delay $(1 - \Delta)T$, where Δ varies from 0 to 1 (see Figure 5.4) in order to get all the values of $x_c(t)$ for $t = (n - 1 + \Delta)T$, $0 \leq \Delta \leq 1$, $n \in \mathscr{Z}$; T is the sampling period. Then, the modified z-transform is defined as follows:

$$X_c(z, \Delta) = \sum_{n=-\infty}^{+\infty} x_c((n - 1 + \Delta)T) z^{-n}, \qquad 0 \leq \Delta \leq 1 \tag{5.59}$$

or, using the change of variable $k = n - 1$,

$$X_c(z, \Delta) = z^{-1} \sum_{k=-\infty}^{+\infty} x_c((k + \Delta)T) z^{-k}, \qquad 0 \leq \Delta \leq 1 \tag{5.60}$$

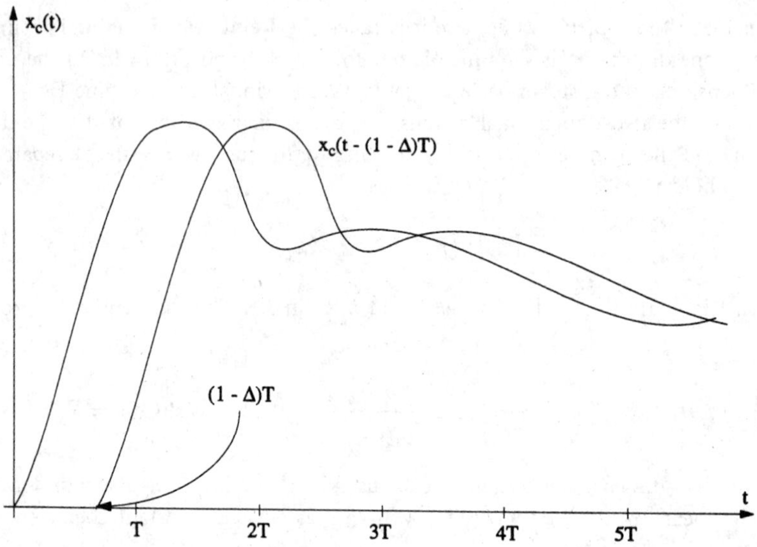

FIGURE 5.4 Modified z-transform takes into account values of the function between sampling instants by creating a fictitious delay $(1 - \Delta)T$.

It is easy to see that the z-transform as defined in (5.3) can be obtained as a special case of the modified z-transform as

$$X_c(z) = X_c(z, \Delta)\big|_{\Delta=1} \tag{5.61}$$

Similarly to the z-transform, the modified z-transform possesses a number of useful properties. For these, the reader is referred to [3].

Chirp z-Transform Algorithm

This is a brief discussion of an algorithm for computing the z-transform of a finite-length sequence, much as was done in the case of the DFT. Suppose we wanted to compute the z-transform on a circle concentric to the unit circle. We could use the DFT algorithm, with some minor modifications. However, the chirp z-transform algorithm used in radar systems [9] can be more efficient and allows one to compute the z-transform of a finite-length sequence on a spiral contour in the z-plane. The samples we compute are equally spaced in angle over some portion of the spiral. The algorithm employs the DFT as well, and has the complexity of

$$(N + M - 1)\log_2(N + M - 1)$$

where N is the number of nonzero values of the sequence $x[n]$, i.e., $x[0], \ldots, x[N-1]$, and M is the number of points at which we evaluate the z-transform, $z_k, k = 1, \ldots, M$.

5.7 Concluding Remarks

This chapter developed the basics of the z-transform. We showed that it is a generalization of the discrete-time FT, and have explored the relationships between the two. We also discussed the connection to the LT. The inverse z-transform was presented, using both contour integration and partial fraction expansion. Although the partial fraction expansion method is easier, it can be used only for rational functions of z. We explored a number of properties of the ROC of the z-transform, as well as important properties of the transform itself. Finally, we showed how the z-transform is used in solving linear constant-coefficient difference equations. To conclude the section, we discussed the modified z-transform and the multidimensional z-transform. We also briefly mentioned the chirp z-transform algorithm, used for computing a few points of the z-transform of a finite-length sequence. For more details on the above topics, refer to [1, 2, 3, 5, 6, 7, 9, 10].

Acknowledgment

I would like to thank Professor Eli Jury for his encouragement and technical insight while writing this manuscript, as well as an anonymous reviewer for his useful remarks.

References

[1] N. K. Bose, *Applied Multidimensional System Theory*, New York: Van Nostrand Reinhold, 1982.

[2] D. E. Dudgeon and R. M. Mersereau, *Multidimensional Digital Signal Processing*, Englewood Cliffs, NJ: Prentice-Hall, 1984.

[3] E. I. Jury, *Theory and Application of the z-Transform Method*, Malabar, FL: Robert E. Krieger, 1986.

[4] T. Kailath, *Linear Systems*, Englewood Cliffs, NJ: Prentice-Hall, 1980.

[5] L. C. Ludeman, *Fundamentals of DSP*, New York: Harper & Row, 1986.

[6] C. D. McGillem and G. R. Cooper, *Continuous and Discrete Signal and System Analysis*, New York: Holt, Rinehart and Winston, 1974.

[7] A. V. Oppenheim and R. W. Shafer, *Discrete-Time Signal Processing*, Englewood Cliffs, NJ: Prentice-Hall, 1989.

[8] A. Papoulis, *Probability, Random Variables and Stochastic Processes*, 2nd ed., New York: McGraw-Hill, 1984.

[9] L. R. Rabiner, R. W. Schafer, and C. M. Rader, "The chirp z-transform algorithm," *IEEE Trans. Audio Electroacoust.*, vol. 17, pp. 86–92, 1969.

[10] R. Vich, *z-Transform Theory and Applications*, Boston: D. Reidel, 1987.

6

Wavelet Transforms

P. P. Vaidyanathan
California Institute of Technology

Igor Djokovic
California Institute of Technology

0-8493-8341-2/95/$0.00+$.50

6.1 Introduction

Transform techniques such as the Fourier and Laplace transforms and the z-transform have long been used in a wide variety of scientific and engineering disciplines [1, 2]. In a number of applications in which we require a joint time-frequency picture, it is necessary to consider other types of transforms or time-frequency representations. Many such methods have evolved. The wavelet transform technique [3–5] in particular has some unique advantages over other kinds of time-frequency representations such as the short-time Fourier transform. For historical developments as well as many technical details and original material see [5]. This chapter describes some of these representations, and explains the advantages of the wavelet transform, and the reason for its recent popularity.

A subclass of wavelet transforms [6] has an intimate connection with the theory of digital filter banks [7–10]. Filter banks have been known to the signal processing community for over 2 decades (see [7] and references therein). It is this relation that makes it possible to construct in a systematic way a wide family of wavelets with several desirable properties such as compact support (i.e., finite duration), smoothness, good time-frequency localization, and basis orthonormality (all these terms will be explained later).

The connection between wavelets and filter banks finds beautiful mathematical expression in the theory of multiresolution [11]. This enables us to compute the wavelet transform coefficients using the so-called **fast wavelet transform (FWT)**, which is essentially a tree-structured filter bank. In addition to the practical value, many deep results from several disciplines find a unified home in the theory and development of the wavelet transform. This includes signal processing, circuit theory, communications, and mathematics. Our emphasis here is on this unification, and the beautiful big picture that it provides. Other tutorials on wavelets with different choices of emphasis can be found in [7, 12–14].

Scope and Outline

The literature on wavelets is enormous, and an attempt to do justice to everything would prove futile. Even a list of references that is fair to all contributors would be too long. We, therefore, restrict discussions to basic, core material. Sections 6.2 to 6.5 provide an overview, with the presentation given at a level that can be comprehended by most engineers. The more advanced results on wavelets, which brought them great attention in recent years, are presented in Sections 6.9 to 6.13. At the heart of these results lie several powerful mathematical tools, which are usually not familiar to engineers, and so we present a fairly extensive math review in three sections (Section 6.6 to 6.8). We suggest that the reader go through this review material once and then use it primarily as a reference.

The advanced sections 6.9 to 6.13 are organized such that the main points, summarized as theorems for convenience of reference, can be appreciated even without the mathematical background material in Sections 6.6 to 6.8. The mathematical sections do, however, facilitate a deeper understanding. It is our hope that these sections will bring most readers to a point where they can pursue wavelet literature without difficulty.

Why Wavelets?

A commonly asked question is "why wavelets?", that is, "what are the advantages offered by wavelets over other types of transform techniques such as, the Fourier transform?" The answer to this question is fairly sophisticated, and also depends on the level at which we address the question. Several discussions addressing this question are scattered throughout this chapter. A convenient listing of the locations of these discussions is given in Section 6.14 under "Why Wavelets?"

General Notations and Acronyms

1. Boldfaced quantities represent matrices and vectors.
2. The notations $\mathbf{A}^T$, $\mathbf{A}^*$ and $\mathbf{A}^\dagger$ represent, respectively, the transpose, conjugate, and transpose-conjugate of the matrix **A**.
3. The accent tilde is defined as follows: $\tilde{\mathbf{H}}(z) = \mathbf{H}^\dagger(1/z^*)$; thus, if $\mathbf{H}(z) = \sum_n \mathbf{h}(n)z^{-n}$, then $\tilde{\mathbf{H}}(z) = \sum_n \mathbf{h}^\dagger(-n)z^{-n}$. On the unit circle $\tilde{\mathbf{H}}(z) = \mathbf{H}^\dagger(z)$.
4. *Acronyms.* BIBO (bounded-input–bounded-output); FIR (finite impulse response); IIR (infinite impulse response); LTI (linear time invariant); PR (perfect reconstruction); STFT (short-time Fourier transform); WT (wavelet transform).
5. For LTI systems, "stability" stands for BIBO stability.
6. $\delta(n)$ denotes the unit pulse or discrete time impulse, defined such that $\delta(0) = 1$ and $\delta(n) = 0$ otherwise. This should be distinguished from the Dirac delta function [2], which is denoted as $\delta_a(t)$.
7. *Figures.* Sampled versions of continuous time signals are indicated with an arrow on the top [e.g., Fig. 6.10(a)]. The sampled versions are impulse trains of the form $\sum_n c(n)\delta_a(t-n)$, and are functions of continuous t.

6.2 Signal Representation Using Basis Functions

The electrical engineer is very familiar with the Fourier transform (FT) and its role in the study of linear time invariant (LTI) systems or **filters**. For example, the frequency response of an LTI system is the FT of its impulse response. The FT is also used routinely in the design and analysis of circuits. As a reminder, the FT of a signal $x(t)$ is given by the familiar integral $X(\omega) = \int_{-\infty}^{\infty} x(t)e^{-j\omega t}\,dt$ and the inverse transform by[1].

$$x(t) = \frac{1}{2\pi}\int_{-\infty}^{\infty} X(\omega)e^{j\omega t}\,d\omega \tag{6.1}$$

From this equation we can say that $x(t)$ has been expressed as a linear superposition (or linear combination) of an infinite number of functions $g_\omega(t) \triangleq e^{j\omega t}$. Because the frequency ω is a continuous variable, *uncountably* many functions $g_\omega(t)$ are superimposed. Electrical engineers, in particular signal processors and communications engineers, are also familiar with two special classes of signals which can be regarded as a superposition of *countably* many functions:

$$x(t) = \sum_{n=-\infty}^{\infty} \alpha_n g_n(t) \tag{6.2}$$

[1]At the moment it is not necessary to worry about the existence, invertibility, and the type (e.g., L^1 or L^2) of the FT. We return to the mathematical subtleties in Section 6.6

where α_n are scalars (possibly complex) uniquely determined by $x(t)$. These two examples are time-limited signals for which we can find a Fourier series (FS), and band-limited signals which can be reconstructed from uniformly spaced samples by weighting them with shifted sinc functions (see below).

First consider a time-limited signal $x(t)$ with duration $0 \le t \le 1$ (Fig. 6.1). Under some mild conditions such a signal can be represented in the form (6.2) with $g_n(t) = e^{j2\pi nt}$. The expression (6.2) is then the FS of $x(t)$, and α_n are the Fourier coefficients. [In contrast we say that (6.1) is the **Fourier integral** of $x(t)$.] The **transform domain** signal $\{\alpha_n\}$ is a sequence, and the transform domain variable is discrete, namely, the frequencies $\omega_n \triangleq 2\pi n$. Because $e^{j2\pi nt}$ is periodic in t with period one, the right-hand side of (6.2) is periodic, and it represents $x(t)$ only in $0 \le t \le 1$. It is sometimes convenient to replace the complex functions $e^{j2\pi nt}$ with the set of real functions $1, \sqrt{2}\cos(2\pi nt), \sqrt{2}\sin(2\pi nt), n > 0$, especially in circuit analysis.

Next, consider a band-limited signal $x(t)$ with FT $X(\omega)$ as demonstrated in Fig. 6.2. If we sample the signal at the Nyquist rate 2β rad/s (i.e., sampling period $T = \pi/\beta$), then multiple copies of the FT are generated [2], and we can recover $x(t)$ from the samples by use of an ideal low-pass filter $F(\omega)$ (Fig. 6.3). The impulse response of the filter is the **sinc function** $f(t) = \sin\beta t/\beta t$ so that the reconstruction formula is

$$x(t) = \sum_{n=-\infty}^{\infty} x(nT)f(t-nT) = \sum_{n=-\infty}^{\infty} x(nT)\frac{\sin\beta(t-nT)}{\beta(t-nT)}, \qquad T = \pi/\beta \quad (6.3)$$

Comparing to (6.2) we see that the **transform domain coefficients** α_n can be regarded as the samples $x(nT)$, whereas the functions $g_n(t)$ are the shifted sinc functions.

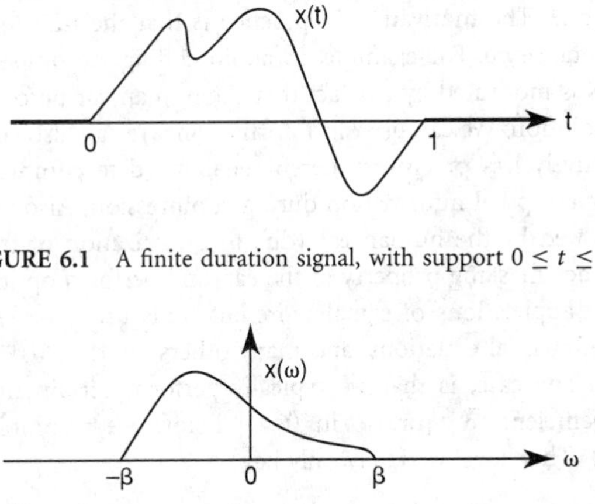

FIGURE 6.1 A finite duration signal, with support $0 \le t \le 1$.

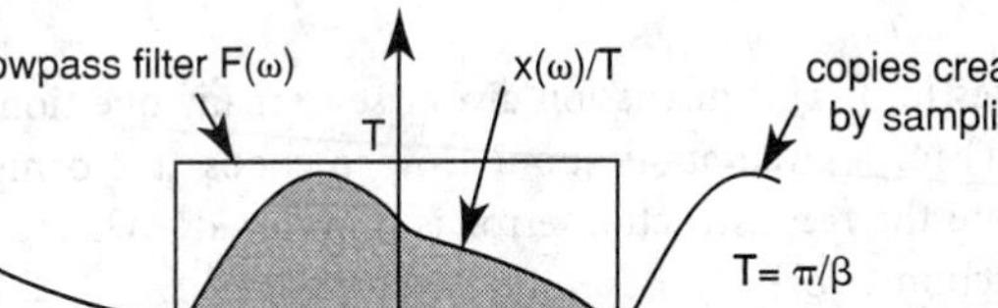

FIGURE 6.2 Fourier transform of a signal band-limited to $|\omega| < \beta$.

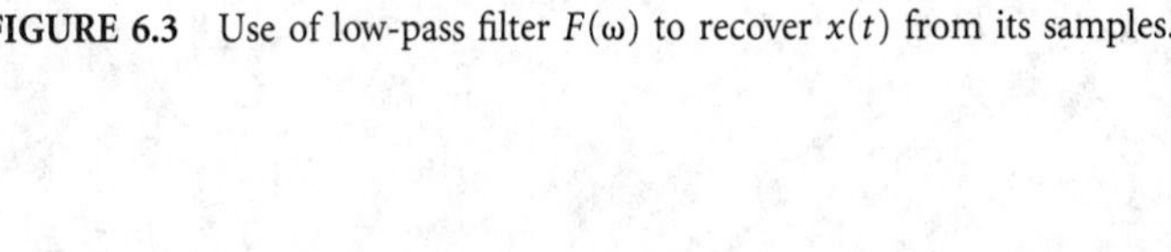

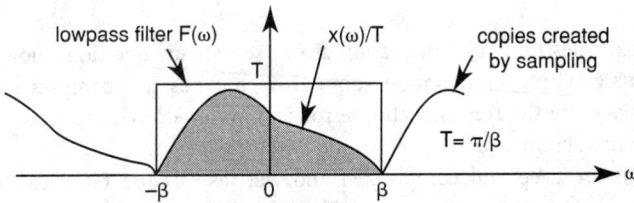

FIGURE 6.3 Use of low-pass filter $F(\omega)$ to recover $x(t)$ from its samples.

If a signal is time-limited or band-limited, we can express it as a countable linear combination of a set of fundamental functions (called **basis** functions, in fact an orthonormal basis; see below). If the signal is more arbitrary, i.e., not limited in time or bandwidth, can we still obtain such a countable linear combination? Suppose we restrict $x(t)$ to be a finite energy signal (i.e., $\int |x(t)|^2 dt < \infty$; also called L^2 signals, see below). Then this is possible. In fact, we can find an unusual kind of basis called the **wavelet basis,** fundamentally different from the Fourier basis. Representation of $x(t)$ using this basis has, in some applications, some advantages over the Fourier representation or the short-time (windowed) Fourier representation. Wavelet bases also exist for many other classes of signals, but this discussion is limited to the L^2 class of signals.

The most common kind of **wavelet representation** takes the form

$$x(t) = \sum_{k=-\infty}^{\infty} \sum_{n=-\infty}^{\infty} c_{kn} \underbrace{2^{k/2}\psi(2^k t - n)}_{\psi_{kn}(t)} \tag{6.4}$$

The functions $\psi_{kn}(t)$ are typically (but not necessarily) linearly independent and form a basis for finite energy signals. The basis is very special in the sense that all the functions $\psi_{kn}(t)$ are derived from a single function $\psi(t)$ called the *wavelet*, by two operations: dilation ($t \rightarrow 2^k t$) and time shift ($t \rightarrow t - 2^{-k}n$). The advantage of such a basis is that it allows us to capture the details of a signal at various scales, while providing time-localization information for these "scales". Examples in future sections clarify this idea.

Why Worry About Signal Representations?

A common feature of all the above discussions is that we have taken a signal $x(t)$ and found an equivalent representation in terms of the transform domain quantity $\{\alpha_n\}$ in (6.2), or $\{c_{kn}\}$ in (6.4). If our only aim is to compute α_n from $x(t)$ and then recompute $x(t)$ from α_n, that would be a futile exercise. The motivation in practice is that the transform domain quantities are better suited in some sense. For example, in audio coding, decomposition of a signal into frequency components is motivated by the fact that the human ear perceives higher frequencies with less frequency resolution. We can use this information. We can also code the high-frequency components with relatively less precision, thereby enabling data compression. In this way we can take into account perceptual information during compression. Also, we can account for the fact that the error allowed by the human ear (due to quantization of frequency components) depends on the frequency masking property of the ear, and perform optimum-bit allocation for a given bit rate. Other applications of signal representations using wavelets include numerical analysis, solution of differential equations, and many others [5, 15, 16].

The main point, in any case, is that we typically perform certain manipulations with the transform domain coefficients α_n [or c_{kn} in (6.4)] before we recombine them to form an approximation of $x(t)$. Therefore, we really only have

$$\hat{x}(t) = \sum_n \hat{\alpha}_n g_n(t) \tag{6.5}$$

where $\{\hat{\alpha}_n\}$ approximates $\{\alpha_n\}$. This discussion gives rise to many questions: how best to choose the basis functions $g_n(t)$ for a given application? How to choose the compressed signal $\{\hat{\alpha}_n\}$ so that for a given data rate the reconstruction error is minimized? What, indeed, is the best way to define the reconstruction error?

These questions are deep and complicated and will take us too far afield. Our goal is to point out the basic advantages (sometimes) offered by the wavelet transform over other kinds of transforms (e.g., the FT).

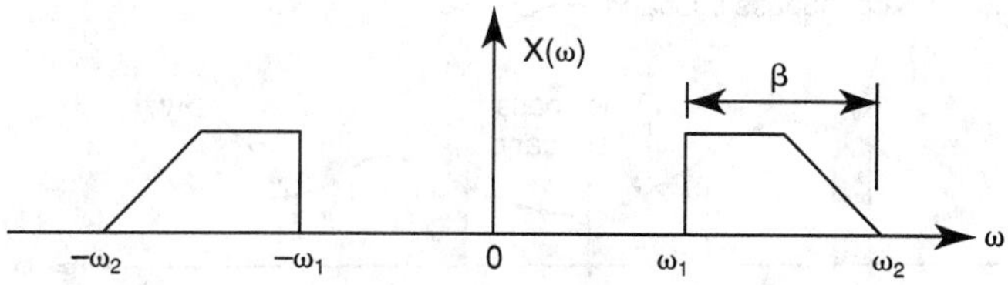

FIGURE 6.4 Fourier transform of a bandpass signal.

Ideal Bandpass Wavelet

Consider a bandpass signal $x(t)$ with FT as shown in Fig. 6.4. Such signals arise in communication applications. The bandedges of the signal are ω_1 and ω_2 (and $-\omega_1$ and $-\omega_2$ on the negative side, which is natural if $x(t)$ is real). Viewed as a low-pass signal, the total bandwidth (counting negative frequencies also) is $2\omega_2$, but viewed as a bandpass signal, the total bandwidth is only 2β where $\beta = \omega_2 - \omega_1$. Does it mean that we can sample it at the rate 2β rad/s (which is the Nyquist rate for the low-pass case)?

In the low-pass case sampling at Nyquist rate was enough to ensure that the copies of the spectrum created by the sampling did not overlap (Fig. 6.3). In the bandpass case we have two sets of such copies; one created by the positive half of the frequency $\omega_1 \leq \omega \leq \omega_2$ and the other by the negative half $-\omega_2 \leq \omega \leq -\omega_1$. This makes the problem somewhat more complicated. It can be shown that for sampling at the rate 2β no overlap of images exists iff one of the edges, ω_1 or ω_2, is a multiple of 2β. This is called the **bandpass sampling theorem.** The reconstruction of $x(t)$ from the samples proceeds exactly as in the low-pass case, except that the reconstruction filter $F(\omega)$ is now a bandpass filter (Fig. 6.5), occupying precisely the signal bandwidth. The first part of the expression (6.3), therefore, is still valid, i.e., $x(t) = \sum_n x(nT)f(t - nT)$, where $T = \pi/\beta$ again, but the sinc function is replaced with the bandpass impulse response $f(t)$.

Given a signal $x(t)$, imagine now that we have split its frequency axis into subbands in some manner (Fig. 6.6). Letting $y_k(t)$ denote the kth subband signal, we can write $x(t) = \sum_k y_k(t)$. This can be visualized as passing $x(t)$ through a bank of filters $\{H_k(\omega)\}$ [Fig. 6.7(a)], with responses as in Fig. 6.7(b). Note that each subband region is symmetric with respect to zero frequency, and therefore supports positive as well as negative frequencies. If the subband region $\omega_k \leq |\omega| < \omega_{k+1}$ satisfies the bandpass sampling condition, then the bandpass signal $y_k(t)$ can be expressed as a linear combination of its samples as before. Thus, $x(t) = \sum_k y_k(t) = \sum_k \sum_n y_k(nT_k) f_k(t - nT_k)$, where $T_k = \pi/\beta_k$. Here, $f_k(t)$ is the impulse response of the reconstruction filter (or synthesis filter) $F_k(\omega)$ shown in Fig. 6.7(c). Fig. 6.7(a) also shows this reconstruction schematic.

Figure 6.8 shows the set of synthesis filters $\{F_k(\omega)\}$ for two examples of frequency splitting arrangement, namely uniform splitting and nonuniform (octave) splitting. We will see later that the uniform splitting arrangement gives an example of the STFT representation (Sections 6.3 and 6.9). In this section we are interested in octave splitting. The bandedges of the filters here are $\omega_k = 2^k\pi$ $(k = \ldots -1, 0, 1, 2, \ldots)$. The bandedges are such that $y_k(t)$ is a signal satisfying

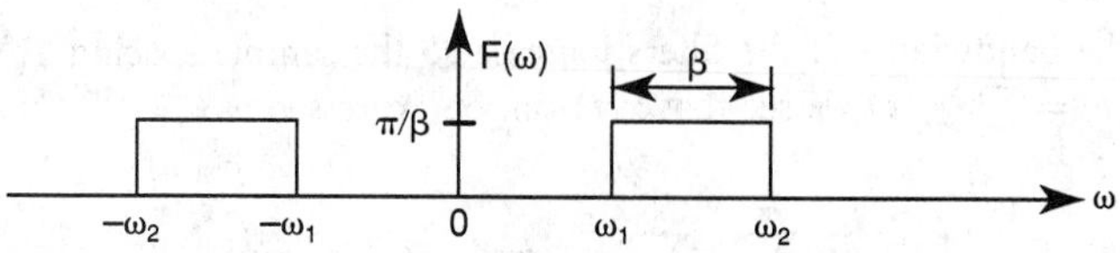

FIGURE 6.5 Bandpass filter to be used in the reconstruction of the bandpass signal from its samples.

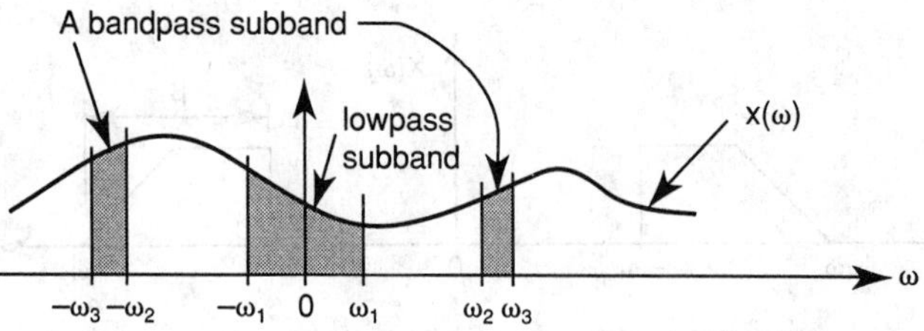

FIGURE 6.6 Splitting a signal into frequency subbands.

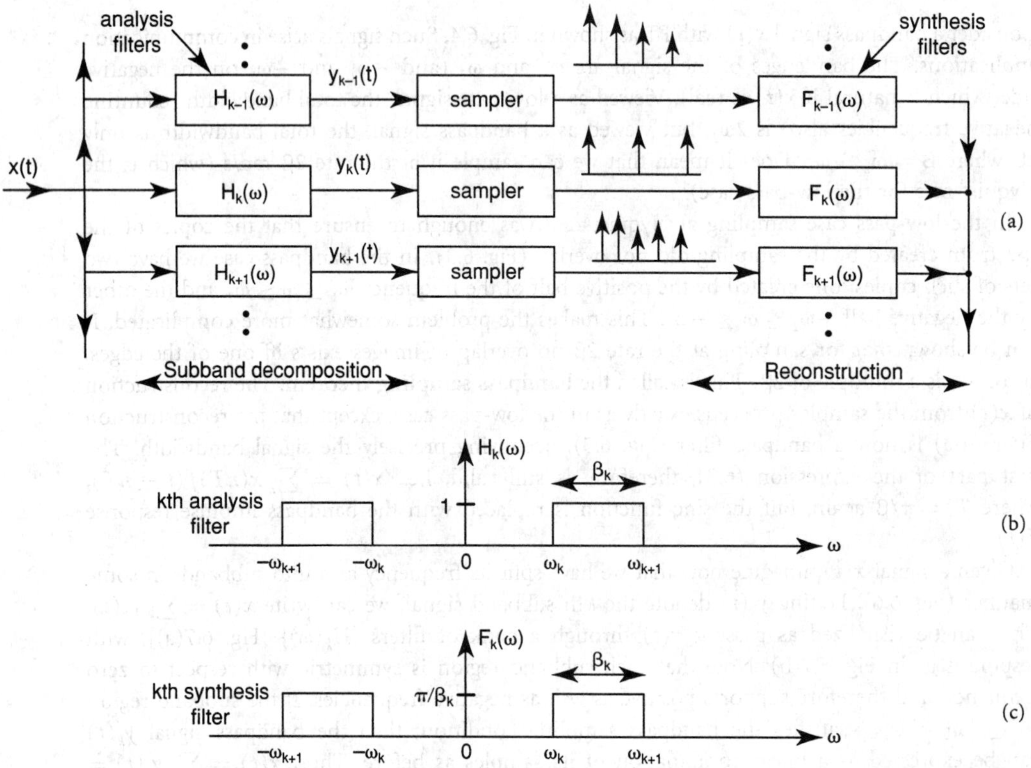

FIGURE 6.7 (a) Splitting a signal into subband signals, sampling, and then recombining; (b) response of the kth analysis filter; and (c) response of kth synthesis filter.

the bandpass sampling theorem. It has $\beta_k = 2^k\pi$, according to the notation of Fig. 6.7. It can be sampled at period $T_k = \pi/\beta_k = 2^{-k}$ without aliasing, and we can reconstruct it from samples as

$$y_k(t) = \sum_{n=-\infty}^{\infty} y_k(2^{-k}n) f_k(t - 2^{-k}n) \tag{6.6}$$

As k increases, the bandwidths of the filters increase so the sample spacing $T_k = 2^{-k}$ becomes finer. Because $x(t) = \sum_k y_k(t)$ we see that $x(t)$ can be expressed as

$$x(t) = \sum_{k=-\infty}^{\infty} \sum_{n=-\infty}^{\infty} y_k(2^{-k}n) f_k(t - 2^{-k}n) \tag{6.7}$$

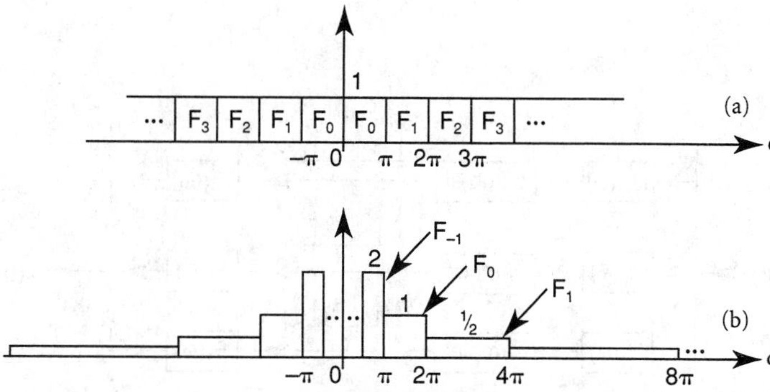

FIGURE 6.8 Two possible schemes to decompose a signal into frequency bands: (a) uniform splitting, and (b) octave-band splitting. The responses shown are those of synthesis filters.

Our definition of the filters shows that the frequency responses are scaled versions of each other, i.e., $F_k(\omega) = 2^{-k}\Psi(2^{-k}\omega)$, with $\Psi(\omega)$ as in Fig. 6.9. The impulse responses are therefore related as $f_k(t) = \psi(2^k t)$, and we can rewrite (6.7) as

$$x(t) = \sum_{k=-\infty}^{\infty} \sum_{n=-\infty}^{\infty} y_k(2^{-k}n)\psi(2^k t - n) \tag{6.8}$$

We will write this as $x(t) = \sum_k \sum_n c_{kn}\psi_{kn}(t)$ by defining $c_{kn} = 2^{-k/2}y_k(2^{-k}n)$ and

$$\psi_{kn}(t) = 2^{k/2}\psi(2^k t - n) = 2^{k/2}\psi\big(2^k(t - 2^{-k}n)\big) \tag{6.9}$$

Then the functions $\psi_{kn}(t)$ will have the same energy $\int |\psi_{kn}(t)|^2 dt$ for all k, n. From the analysis/synthesis filter bank point of view (Fig. 6.7) this is equivalent to making $H_k(\omega) = F_k(\omega)$ and rescaling, as shown in Fig. 6.10. With filters so rescaled, the wavelet coefficients c_{kn} are just samples of the outputs of the analysis filters $H_k(\omega)$.

The function $\psi(2^k t)$ is a dilated version of $\psi(t)$ (squeezed version if $k > 0$ and stretched version if $k < 0$). The dilation factor 2^k is a power of 2, so this is said to be a dyadic dilation. The function $\psi(2^k(t - 2^{-k}n))$ is a shifted version of the dilated version. Thus, we have expressed $x(t)$ as a linear combination of shifted versions of (dyadic) dilated versions of a single function $\psi(t)$. The shifts $2^{-k}n$, are in integer multiples of 2^{-k}, where k governs the dilation. For completeness, note that the impulse response $\psi(t)$ corresponding to the function in Fig. 6.9 is given by

$$\psi(t) = \frac{\sin(\pi t/2)}{\pi t/2}\cos(3\pi t/2) \qquad \text{(ideal bandpass wavelet)} \tag{6.10}$$

This is plotted in Fig. 6.11.

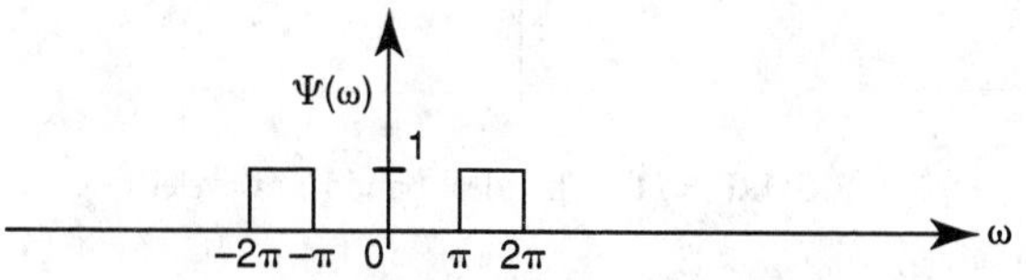

FIGURE 6.9 The fundamental bandpass function that generates a bandpass wavelet.

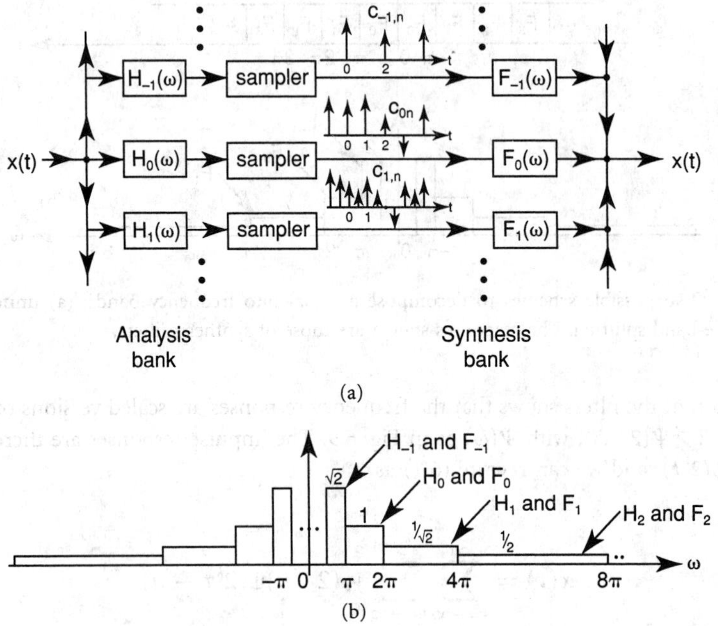

FIGURE 6.10 The octave-band splitting scheme. (a) The analysis bank, samplers, and synthesis bank; and (b) the filter responses.

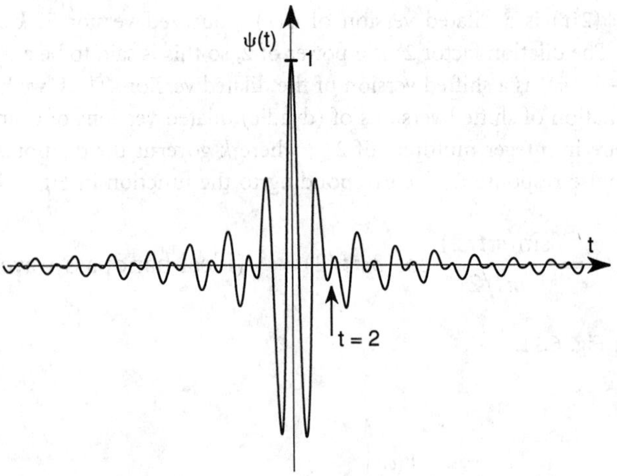

FIGURE 6.11 The ideal bandpass wavelet.

In (6.8) we obtained a wavelet representation for $x(t)$ [compare to (6.4)]. The function $\psi(t)$ is called the **ideal bandpass wavelet**, also known as the Littlewood-Paley wavelet. We now introduce some terminology for convenience and then return to more detailed definitions and discussions of the WT.

L^2 Spaces, Basis Functions, and Orthonormal Bases

Most of our discussions are restricted to the class of L^2 functions or square integrable functions, i.e., functions $x(t)$ for which $\int |x(t)|^2 dt$ exists and has a finite value. The norm, or L^2 norm, of such functions, denoted $\|x(t)\|_2$, is defined as $\|x(t)\|_2 = \left(\int |x(t)|^2 dt\right)^{1/2}$. The notation $L^2[a, b]$ stands for L^2 functions that are zero outside the interval $a \leq t \leq b$. The set $L^2(R)$ is the class of L^2 functions supported on the real line $-\infty < t < \infty$. We often abbreviate $L^2(R)$ as L^2.

The class of L^2 functions forms a (normed) linear vector space, i.e., any linear combination of functions in L^2 is still in L^2. In fact, it forms a special linear space such that a countable basis exists. That is, a sequence of linearly independent functions $\{g_n(t)\}$ exists in L^2 such that any L^2 function $x(t)$ can be expressed as $x(t) = \sum_n \alpha_n g_n(t)$, for a unique set of $\{\alpha_n\}$. We say that $g_n(t)$ are the basis functions. L^2 spaces have orthonormal bases. For such a basis, the basis functions satisfy

$$\langle g_k(t), g_m(t) \rangle = \delta(k - m) \tag{6.11}$$

where the notation $\langle f(t), g(t) \rangle \triangleq \int f(t) g^*(t) dt$ denotes the inner product between $f(t)$ and $g(t)$. For an orthonormal basis, the coefficients α_n in the expansion $x(t) = \sum_{n=-\infty}^{\infty} \alpha_n g_n(t)$ can thus be computed using the exceptionally simple relation

$$\alpha_n = \langle x(t), g_n(t) \rangle \tag{6.12}$$

Two examples of orthonormal basis were shown above. The first is the FS expansion of a time-limited signal ($0 \leq t \leq 1$). Here, the basis functions $\{e^{j2\pi nt}\}$ are clearly orthonormal, with integrals going from 0 to 1. The second example is the expansion (6.3) of a band-limited signal; it can be shown that the shifted versions $f(t - nT)$ of the sinc functions form an orthonormal basis for band-limited signals (integrals going from $-\infty$ to ∞).

Orthogonal Projections

Suppose we consider a subset $\{g_{n_k}(t)\}$ of the orthonormal basis $\{g_n(t)\}$. Let $\mathcal{S}$ denote the subspace generated by $\{g_{n_k}(t)\}$ (an accurate statement would be that $\mathcal{S}$ is the "closure of the span of $\{g_{n_k}(t)\}$"; see Section 6.7). Consider the linear combination $y(t) = \sum_k \alpha_{n_k} g_{n_k}(t)$, where the α_{n_k} are evaluated as above, i.e., $\alpha_{n_k} = \langle x(t), g_{n_k}(t) \rangle$ for some signal $x(t)$. Then, $y(t) \in \mathcal{S}$, and it can be shown that among all functions in $\mathcal{S}$, $y(t)$ is the unique signal closest to $x(t)$ (i.e., $\|x(t) - y(t)\|_2$ is the smallest). We say that $y(t)$ is the orthogonal projection of $x(t)$ onto the subspace $\mathcal{S}$, and write

$$y(t) = P_{\mathcal{S}}[x(t)] \tag{6.13}$$

Wavelet Transforms

If a signal $x(t)$ is in L^2, then its FT $X(\omega)$ exists in the L^2 sense (see Section 6.6). We will see in Section 6.6 that the discussion which resulted in (6.8) is applicable for any signal $x(t)$ in L^2. Equation (6.8) means that the signal can be expressed as a linear combination of the form

$$x(t) = \sum_{k=-\infty}^{\infty} \sum_{n=-\infty}^{\infty} c_{kn} \underbrace{2^{k/2} \psi(2^k t - n)}_{\psi_{kn}(t)} \tag{6.14}$$

where $\psi(t)$ is the impulse response (Fig. 6.11) of the bandpass function $\Psi(\omega)$ in Fig. 6.9.[2] Because the frequency responses for two different values of k do not overlap, the functions $\psi_{kn}(t)$ and $\psi_{mi}(t)$ are orthogonal for $k \neq m$ (use Parseval's relation). For a given k, the functions $\psi_{kn}(t)$ are shifted versions of the impulse responses of the bandpass filter $F_k(\omega)$. From the ideal nature of this bandpass filter we can show that $\psi_{kn}(t)$ and $\psi_{km}(t)$ are also orthonormal for $n \neq m$. Thus, the set of functions $\{\psi_{kn}(t)\}$, with k and n ranging over all integers, forms an orthonormal basis for the class of L^2 functions, i.e., any L^2 function can be expressed as in (6.14) and furthermore,

$$\langle \psi_{kn}(t), \psi_{mi}(t) \rangle = \delta(k-m)\delta(n-i) \tag{6.15}$$

Because of this orthonormality, the coefficients c_{kn} are computed very easily as

$$c_{kn} = \langle x(t), \psi_{kn}(t) \rangle = \int_{-\infty}^{\infty} x(t) 2^{k/2} \psi^*(2^k t - n)\, dt \tag{6.16}$$

Defining

$$\eta(t) = \psi^*(-t) \tag{6.17}$$

this takes the form

$$c_{kn} = \langle x(t), \eta^*_{kn}(-t) \rangle = \int_{-\infty}^{\infty} x(t) 2^{k/2} \eta(n - 2^k t)\, dt \tag{6.18}$$

resembling a convolution.

Wavelet Transform Definitions

A set of basis functions $\psi_{kn}(t)$ derived from a single function $\psi(t)$ by dilations and shifts of the form

$$\psi_{kn}(t) = 2^{k/2} \psi(2^k t - n) \tag{6.19}$$

is said to be a **wavelet basis**, and $\psi(t)$ is called the **wavelet function.** The coefficients c_{kn} are the **wavelet transform coefficients.** The formula (6.16) that performs the transformation from $x(t)$ to c_{kn} is the **wavelet transform** of the signal $x(t)$. Equation (6.14) is the **wavelet representation** or the **inverse wavelet transform.** While this is only a special case of more general wavelet decompositions outlined at the end of this section, it is perhaps the most popular and useful.

Note that the kth dilated version $\psi(2^k t)$ has the shifted versions $\psi(2^k t - n) = \psi(2^k(t - 2^{-k}n))$, so the amount of shift is in integer multiples of 2^{-k}. Thus, the *stretched* versions are shifted by larger amounts and *squeezed* versions by smaller amounts. Even though we developed these ideas based on an example, the above definitions still hold generally for any orthonormal wavelet basis. For the ideal bandpass wavelet, the function $\psi(t)$ is real and symmetric [see (6.10)] so that $\eta(t) = \psi(t)$. For more general orthonormal wavelets we have the relation $\eta(t) = \psi^*(-t)$. We say that $\eta(t)$ is the **analyzing wavelet** [because of (6.18)] and $\psi(t)$ the **synthesis wavelet** [because of (6.14)]. For the nonorthonormal case we still have the transform and inverse transform equations as above, but the relation between $\psi(t)$ and $\eta(t)$ is not as simple as $\eta(t) = \psi^*(-t)$.

Before exploring the properties and usefulness of wavelets let us turn to a distinctly different example. This shows that unlike the Fourier basis functions $\{e^{j2\pi nt}\}$, the wavelet basis functions can be designed by the user. This makes them more flexible, interesting, and useful.

[2]The above equality and the convergence of the summation should be interpreted in the L^2 sense; see Section 6.6.

Haar Wavelet Basis

An orthonormal basis for L^2 functions was found by Haar [5] as early as 1910, which satisfies the definition of a wavelet basis given above. That is, the basis functions $\psi_{kn}(t)$ are derived from a single function $\psi(t)$ using dilations and shifts as in (6.19). To explain this system, first, consider a signal $x(t) \in L^2[0, 1]$. The Haar basis is built from two functions called $\phi(t)$ and $\psi(t)$, as described in Fig. 6.12. The basis function $\phi(t)$ is a constant in $[0, 1]$. The basis function $\psi(t)$ is constant on each half interval, and its integral is zero. After this, the remaining basis functions are obtained from $\psi(t)$ by dilations and shifts as indicated. It is clear from the figure that any two of these functions are mutually orthogonal. We have an orthonormal set, and it can be shown that this set of functions is an orthonormal basis for $L^2[0, 1]$. However, this is not exactly a wavelet basis yet because of the presence of $\phi(t)$.[3]

If we eliminate the requirement that $x(t)$ be supported or defined only on $[0, 1]$ and consider $L^2(R)$ functions then we can still obtain an orthonormal basis of the above form by including the shifted versions $\{\psi(2^k t - n)\}$ for *all* integer values of n, as well as the shifted versions $\{\phi(t - n)\}$.

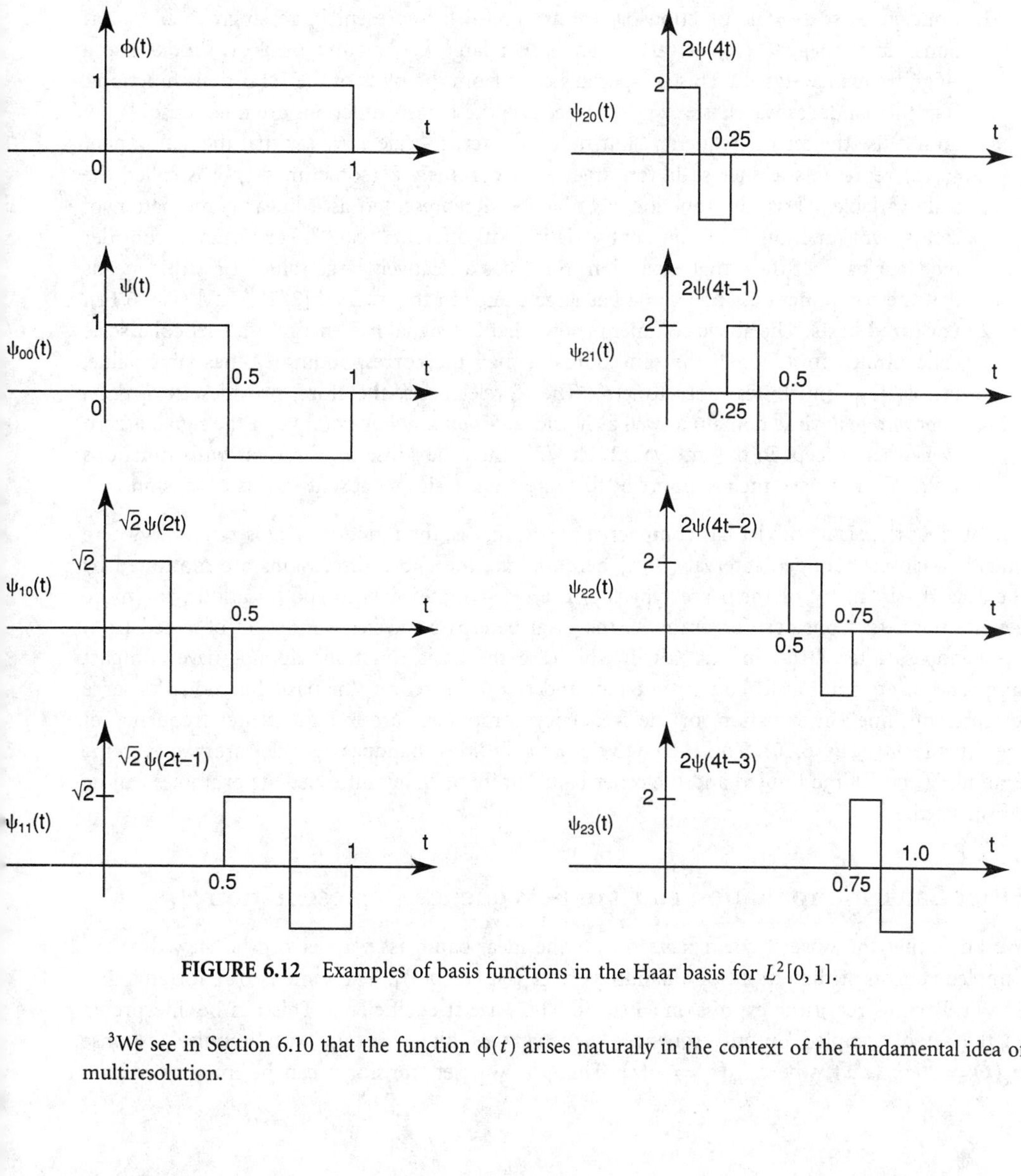

FIGURE 6.12 Examples of basis functions in the Haar basis for $L^2[0, 1]$.

[3]We see in Section 6.10 that the function $\phi(t)$ arises naturally in the context of the fundamental idea of multiresolution.

An alternative to the use of $\{\phi(t-n)\}$ would be to use stretched (i.e., $\psi(2^k t), k < 0$) as well as squeezed (i.e., $\psi(2^k t), k > 0$) versions of $\psi(t)$. The set of functions can thus be written as in (6.19), which has the form of a wavelet basis. It can be shown that this forms an orthonormal basis for $L^2(R)$. The FT of the Haar wavelet $\psi(t)$ is given by

$$\Psi(\omega) = je^{-j\omega/2}\frac{\sin^2(\omega/4)}{\omega/4} \quad \text{(Haar wavelet)} \qquad (6.20)$$

The Haar wavelet has limited duration in time, whereas the ideal bandpass wavelet (6.10), being band-limited, has infinite duration in time.

Basic Properties of Wavelet Transforms

Based on the definitions and examples provided so far we can already draw some very interesting conclusions about wavelet transforms, and obtain a preliminary comparison to the FT.

1. **Concept of scale.** The functions $\psi_{kn}(t)$ are useful to represent increasingly finer "variations" in the signal $x(t)$ at various levels. For large k, the function $\psi_{kn}(t)$ looks like a "high frequency signal". This is especially clear from the plots of the Haar basis functions. (For the bandpass wavelets, see below.) Because these basis functions are not sinusoids, we do not use the term "frequency" but rather the term "scale". We say that the component $\psi_{kn}(t)$ represents a finer scale for larger k. Accordingly k (sometimes $1/k$) is called the scale variable. Thus, the function $x(t)$ has been represented as a linear combination of component functions that represent variations at different "scales". For instance, consider the Haar basis. If the signal expansion (6.14) has a relatively large value of $c_{4,2}$ this means that the component at scale $k = 4$ has large energy in the interval $[2/2^4, 3/2^4]$ (Fig. 6.14).
2. **Localized basis.** The above comment shows that if a signal has energy at a particular scale concentrated in a slot in the time domain, then the corresponding c_{kn} has large value, i.e., $\psi_{kn}(t)$ contributes more to $x(t)$. The wavelet basis, therefore, provides *localization information* in time domain as well as in the *scale domain.* For example, if the signal is zero everywhere except in the interval $[2/2^4, 3/2^4]$ then the subset of the Haar basis functions which do not have their support in this interval are simply absent in this expansion.

Note that the Haar wavelet has **compact support,** that is, the function $\psi(t)$ is zero everywhere outside a closed bounded interval ($[0, 1]$ here). While the above discussions are motivated by the Haar basis, many of them are typically true, with some obvious modifications, for more general wavelets. Consider, for example, the ideal bandpass wavelet (Fig. 6.11) obtained from the bandpass filter $\Psi(\omega)$ in Fig. 6.9. In this case the basis functions do not have compact support, but are still locally concentrated around $t = 0$. Moreover, the basis functions for large k represent "fine" information, or the frequency component around the center frequency of the filter $F_k(\omega)$ (Fig. 6.10). The Haar wavelet and the ideal bandpass wavelet are two extreme examples (one is time limited and the other band-limited). Many intermediate examples can be constructed.

Filter Bank Interpretation and Time-Frequency Representation

We know that the wavelet coefficients c_{kn} for the ideal bandpass wavelet can be viewed as the sampled version of the output of a bandpass filter [Fig. 6.10(a)]. The same is true for any kind of WT. For this recall the expression (6.18) for the wavelet coefficients. This can be interpreted as the set of sampled output sequences of a bank of filters $H_k(\omega)$, with impulse response $h_k(t) = 2^{k/2}h_0(2^k t)$, where $h_0(t) = \eta(t)$. Thus the wavelet transform can be interpreted as a

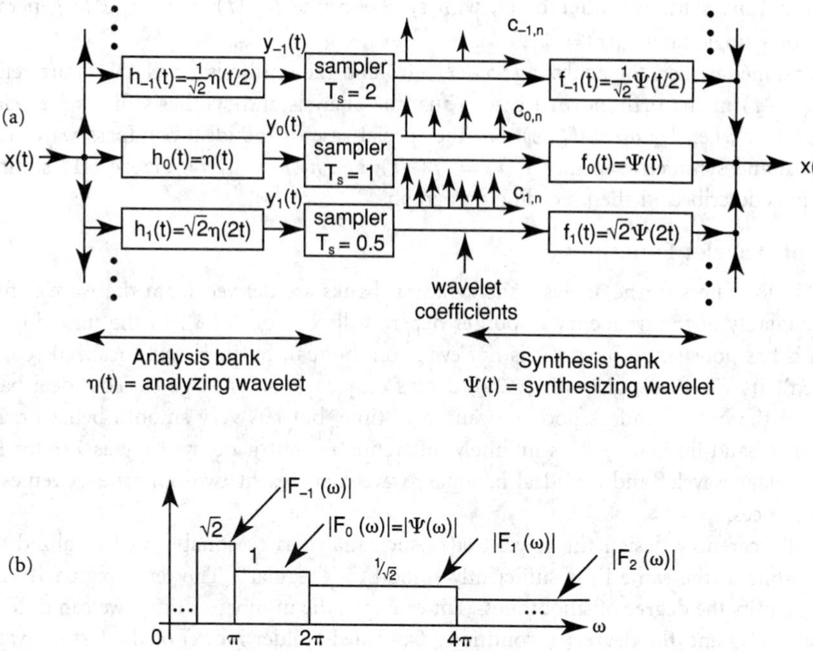

FIGURE 6.13 (a) Representing the dyadic WT as an analysis bank followed by samplers, and the inverse transform as a synthesis bank. For the orthonormal case, $\psi(t) = \eta^*(-t)$, and $f_k(t) = h_k^*(-t)$. (b) Filter responses for the example in which $\psi(t)$ is the ideal bandpass wavelet.

nonuniform continuous time analysis filter bank, followed by samplers. The Haar basis and ideal bandpass wavelet basis are two examples of the choice of these bandpass filters.

The wavelet coefficients c_{kn} for a given scale k are therefore obtained by sampling the output $y_k(t)$ of the bandpass filter $H_k(\omega)$, as indicated in Fig. 6.13(a). The first subscript k (the scale variable) represents the filter number. As k increases by 1, the center frequency ω_k increases by a factor of 2. The wavelet coefficients c_{kn} at scale k are merely the samples $y_k(2^{-k}n)$. As k increases, the filter bandwidth increases, and thus the samples are spaced by a proportionately finer amount 2^{-k}. The quantity $c_{kn} = y_k(2^{-k}n)$ measures the "amount" of the "frequency component" around the center frequency ω_k of the analysis filter $H_k(\omega)$, localized in *time* around $2^{-k}n$.

In wavelet transformation the transform domain is represented by the two integer variables k and n. This means that the transform domain is *two dimensional* (the time-frequency domain), and is *discretized.* We say that c_{kn} is a **time-frequency representation** of $x(t)$. Section 6.3 explains that this is an improvement over another time-frequency representation, the STFT, introduced many years ago in the signal processing literature.

Synthesis Filter Bank and Reconstruction

The inner sum in (6.14) can be interpreted as follows: For each k, convert the sequence c_{kn} into an impulse train[4] $\sum_n c_{kn}\delta_a(t-2^{-k}n)$ and pass it through a bandpass filter $F_k(\omega) = 2^{-k/2}\Psi(2^{-k}\omega)$ with impulse response $f_k(t) = 2^{k/2}\psi(2^k t)$. The outer sum merely adds the outputs of all these filters. Figure 6.13(a) shows this interpretation. Therefore, the reconstruction of the signal $x(t)$ from the wavelet coefficients c_{kn} is equivalent to the implementation of a **nonuniform**

[4] $\delta_a(t)$ is the Dirac delta function [2]. It is used here only as a schematic. The true meaning is that the output of $f_k(t)$ is $\sum_n c_{kn} f_k(t-2^{-k}n)$.

continuous time synthesis filter bank, with synthesis filters $f_k(t) = 2^{k/2} f_0(2^k t)$ generated by dilations of a single filter $f_0(t) \triangleq \psi(t)$.

As mentioned earlier, the analyzing wavelet $\eta(t)$ and the synthesis wavelet $\psi(t)$ are related by $\eta(t) = \psi^*(-t)$ in the orthonormal case. Thus, the analysis and synthesis filters are related as $h_k(t) = f_k^*(-t)$; i.e., $H_k(\omega) = F_k^*(\omega)$. For the special case of the ideal bandpass wavelet (6.10), $\psi(t)$ is real and symmetric so that $f_k(t) = f_k^*(-t)$; i.e., $h_k(t) = f_k(t)$. Figure 6.13 summarizes the relations described in the preceding paragraphs.

Design of Wavelet Functions

Because all the filters in the analysis and synthesis banks are derived from the wavelet function $\psi(t)$, the quality of the frequency responses depends directly on $\Psi(\omega)$. In the time domain the Haar basis has poor smoothness (it is not even continuous), but it is well localized (compactly supported). Its FT $\Psi(\omega)$, given in (6.20), decays only as $1/\omega$ for large ω. The ideal bandpass wavelet, on the other hand, is poorly localized in time, but has very smooth behavior. In fact, because it is band-limited, $\psi(t)$ is infinitely differentiable, but it decays only as $1/t$ for large t. Thus, the Haar wavelet and the ideal bandpass wavelet represent two opposite extremes of the possible choices.

We could carefully design the wavelet $\psi(t)$ such that it is reasonably well localized in time domain, while at the same time sufficiently smooth or "regular". The term **regularity** is often used to quantify the degree of smoothness. For example, the number of times we can differentiate the wavelet $\psi(t)$ and the degree of continuity (so-called Hölder index) of the last derivative are taken as measures of regularity. We return to this in Sections 6.11 to 6.13, where we also present systematic procedures for design of the function $\psi(t)$. This can be designed in such a way that $\{2^{k/2}\psi(2^k t - n)\}$ forms an orthonormal basis with prescribed decay and regularity properties. It is also possible to design $\psi(t)$ such that we obtain other kinds of structures rather than an orthonormal basis, e.g., a Riesz basis or a frame (see Sections 6.7 and 6.8).

Wavelet Basis and Fourier Basis

Returning to the Fourier basis $g_k(t) = \{e^{j2\pi kt}\}$ for functions supported on $[0, 1]$, we see that $g_k(t) = g_1(kt)$, so that all the functions are dilated versions (dilations being integers rather than powers of integers) of $g_1(t)$. However, *these do not have the localization property of wavelets.* To understand this, note that $e^{j2\pi kt}$ has unit magnitude everywhere, and sines and cosines are nonzero almost everywhere. Thus, if we have a function $x(t)$ that is identically zero in a certain time slot (e.g., Fig. 6.14), then in order for the infinite series $\sum_n \alpha_n e^{j2\pi nt}$ to represent $x(t)$, extreme cancellation of terms must occur in that time slot. In contrast, if a compactly supported

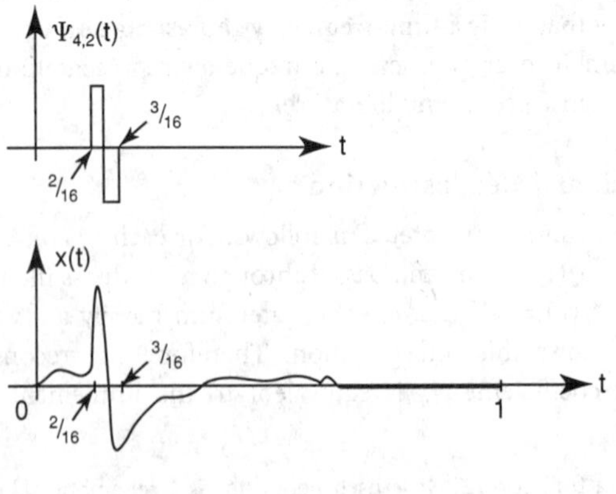

FIGURE 6.14 Example of an $L^2[0, 1]$ signal $x(t)$ for which the Haar component $\psi_{4,2}(t)$ dominates.

wavelet basis is used, it provides localization information as well as information about "frequency contents" in the form of "scales". The "transform domain" in traditional FT is represented by a single continuous variable ω. In the wavelet transform, where the transform coefficients are c_{kn}, the transform domain is represented by two integers k and n.

It is also clear that WT provide a great deal of flexibility because we can choose $\psi(t)$. With FTs, on the other hand, the basis functions (sines and cosines) are pretty much fixed (see, however, Section 6.3 on STFT).

More General Form of Wavelet Transformation

The most general form of the wavelet transform is given by

$$X(a,b) = \frac{1}{\sqrt{|a|}} \int_{-\infty}^{\infty} x(t)\psi\left(\frac{t-b}{a}\right) dt \tag{6.21}$$

where a and b are real. This is called the continuous wavelet transform (CWT) because a and b are continuous variables. The transform domain is a two-dimensional domain (a,b). The restricted version of this, in which a and b take a discrete set of values $a = c^{-k}$ and $b = c^{-k}n$ (where k and n vary over the set of all integers), is called the discrete wavelet transform (DWT). The further special case, in which $c = 2$, i.e., $a = 2^{-k}$ and $b = 2^{-k}n$, is the WT discussed so far [see (6.16)] and is called the **dyadic DWT.** Expansions of the form (6.14) are also called **wavelet series expansions** by analogy with the FS expansion (a summation rather than an integral).

For fixed a, (6.21) is a convolution. Thus, if we apply the input signal $x(t)$ to a filter with impulse response $\psi(-t/a)/\sqrt{|a|}$, its output, evaluated at time b, will be $X(a,b)$. The filter has frequency response $\sqrt{|a|}\Psi(-a\omega)$. If we imagine that $\Psi(\omega)$ has a good bandpass response with center frequency ω_0, then the above filter is bandpass with center frequency $-a^{-1}\omega_0$; i.e., the wavelet transform $X(a,b)$, which is the output of the filter at time b, represents the "frequency content" of $x(t)$ around the frequency $-a^{-1}\omega_0$, "around" time b. Ignoring the minus sign [because $\psi(t)$ and $x(t)$ are typically real anyway], we see that the variable a^{-1} is analogous to frequency. In wavelet literature, the quantity $|a|$ is usually referred to as the "scale" rather than "inverse frequency".

For reasons which cannot be explained with our limited exposure thus far, the wavelet function $\psi(t)$ is restricted to be such that $\int \psi(t)dt = 0$. For the moment notice that this is equivalent to $\Psi(0) = 0$, which is consistent with the bandpass property of $\psi(t)$. In Section 6.10, where we generate wavelets systematically using multiresolution analysis, we see that this condition follows naturally from theoretical considerations.

6.3 The Short-Time Fourier Transform

In many applications we must accommodate the notion of frequency that evolves or changes with time. For example, audio signals are often regarded as signals with a time-varying spectrum, e.g., a sequence of short-lived pitch frequencies. This idea cannot be expressed with the traditional FT because $X(\omega)$ for each ω depends on $x(t)$ for all t.

The STFT was introduced as early as 1946 by Gabor [5] to provide such a time-frequency picture of the signal. Here, the signal $x(t)$ is multiplied with a window $v(t-\tau)$ centered or localized around time τ (Fig. 6.15) and the FT of $x(t)v(t-\tau)$ computed:

$$X(\omega,\tau) = \int_{-\infty}^{\infty} x(t)v(t-\tau)e^{-j\omega t}\, dt \tag{6.22}$$

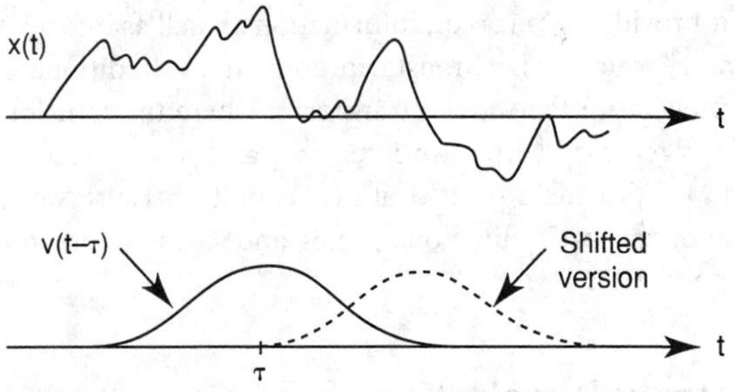

FIGURE 6.15 A signal $x(t)$ and the sliding window $v(t-\tau)$.

This is then repeated for shifted locations of the window, i.e., for various values of τ. That is, we compute not just one FT, but infinitely many. The result is a function of both time τ and frequency ω. If this must be practical we must make two changes: compute the STFT only for discrete values of ω, and use only a discrete number of window positions τ. In the traditional STFT both ω and τ are discretized on uniform grids:

$$\omega = k\omega_s \quad \tau = nT_s. \tag{6.23}$$

The STFT is thus defined as

$$X_{\text{STFT}}(k\omega_s, nT_s) = \int_{-\infty}^{\infty} x(t)v(t - nT_s)e^{-jk\omega_s t}\, dt, \tag{6.24}$$

which we abbreviate as $X_{\text{STFT}}(k, n)$ when there is no confusion. Thus, the time domain is mapped into the **time-frequency domain**. The quantity $X_{\text{STFT}}(k\omega_s, nT_s)$ represents the FT of $x(t)$ "around time nT_s" and "around frequency $k\omega_s$". This, in essence, is similar to the WT: in both cases the transform domain is a two-dimensional discrete domain.

We compare wavelets and STFT on several grounds, giving a filter bank view and comparing time-frequency resolution and localization properties. Section 6.9 provides a comparison on deeper grounds: for example, when can we reconstruct a signal $x(t)$ from the STFT coefficients $X_{\text{STFT}}(k, n)$? Can we construct an orthonormal basis for L^2 signals based on the STFT? The advantage of WTs over the STFT will be clear after these discussions.

Filter Bank Interpretation

The STFT evaluated for some frequency ω_k can be rewritten as

$$X_{\text{STFT}}(\omega_k, \tau) = e^{-j\omega_k\tau} \int_{-\infty}^{\infty} x(t)v(t - \tau)e^{-j\omega_k(t-\tau)}\, dt \tag{6.25}$$

The integral looks like a convolution of $x(t)$ with the filter impulse response

$$h_k(t) \triangleq v(-t)e^{j\omega_k t} \tag{6.26}$$

If $v(-t)$ has a FT looking like a low-pass filter then $h_k(t)$ looks like a bandpass filter with center frequency ω_k (Fig. 6.16). Thus, $X_{\text{STFT}}(\omega_k, \tau)$ is the output of this bandpass filter at time τ, downshifted in frequency by ω_k. The result is a low-pass signal $y_k(t)$ whose output is sampled uniformly at time $\tau = nT_s$. For every frequency ω_k so analyzed, one such filter channel exists.

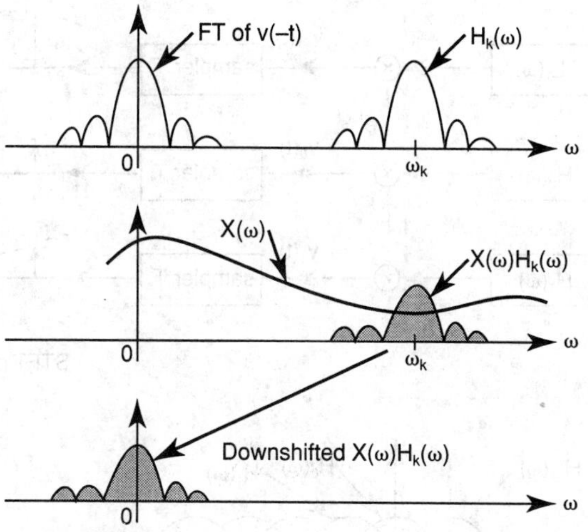

FIGURE 6.16 STFT viewed as a bandpass filter followed by a downshifter.

With the frequencies uniformly located at $\omega_k = k\omega_s$, we get the analysis filter bank followed by downshifters and samplers as shown in Fig. 6.17.

The STFT coefficients $X_{\text{STFT}}(k\omega_s, nT_s)$ therefore can be regarded as the uniformly spaced samples of the outputs of a bank of bandpass filters $H_k(\omega)$, all derived from one filter $h_0(t)$ by modulation: $h_k(t) = e^{jk\omega_s t}h_0(t)$; i.e., $H_k(\omega) = H_0(\omega - k\omega_s)$. (The filters are one sided in frequency so they have complex coefficients in the time domain, but ignore these details for now.) The output of $H_k(\omega)$ represents a portion of the FT $X(\omega)$ around the frequency $k\omega_s$. The downshifted version $y_k(t)$ is therefore a low-pass signal. In other words, it is a slowly varying signal, whose evolution as a function of t represents the evolution of the FT $X(\omega)$ around frequency $k\omega_s$. By sampling this slowly varying signal we can therefore compress the transform domain information.

If the window is narrow in the time domain, then $H_k(\omega)$ has large bandwidth. That is, good time resolution and poor frequency resolution are obtained. If the window is wide the opposite is true. Thus, if we try to capture the local information in time by making a narrow window, we get a fuzzy picture in frequency. Conversely, in the limit, as the filter becomes extremely localized in frequency, the window is very broad and STFT approaches the ordinary FT. That is, the time-frequency information collapses to the all-frequency information of ordinary FT. We see that time-frequency representation is inherently a compromise between time and frequency resolutions (or localizations). This is related to the uncertainty principle: as windows get narrow in time they have to get broad in frequency, and vice versa.

Optimal Time-Frequency Resolution: The Gabor Window

What is the best frequency resolution one can obtain for a given time resolution? For a given duration of the window $v(t)$ how small can the duration of $V(\omega)$ be? If we define duration according to common sense we are already in trouble because if $v(t)$ has finite duration then $V(\omega)$ has infinite duration. A more useful definition of duration is called the **root mean square** (RMS) duration. The RMS time duration D_t and the RMS frequency duration D_f for the window $v(t)$ are defined such that

$$D_t^2 = \frac{\int t^2|v(t)|^2\,dt}{\int |v(t)|^2\,dt} \qquad D_f^2 = \frac{\int \omega^2|V(\omega)|^2 d\omega}{\int |V(\omega)|^2 d\omega} \tag{6.27}$$

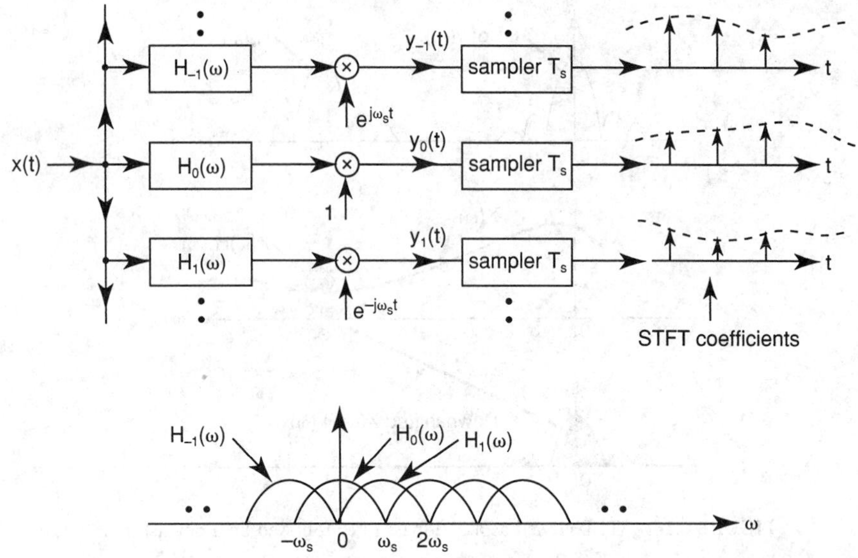

FIGURE 6.17 The STFT viewed as an analysis bank of uniformly shifted filters.

Intuitively, D_t cannot be arbitrarily small for a specified D_f. The uncertainty principle says that $D_t D_f \geq 0.5$. Equality holds iff $v(t)$ has the shape of a Gaussian, i.e., $v(t) = Ae^{-\alpha t^2}, \alpha > 0$. Thus, the best joint time-frequency resolution is obtained by using the Gaussian window. This is also intuitively acceptable for the reason that the Gaussian is its own FT (except for scaling of variables and so forth). Gabor used the Gaussian window as early as 1946. Because it is of infinite duration, a truncated approximation is used in practice. The STFT based on the Gaussian is called the Gabor transform. A limitation of the Gabor transform is that it does not give rise to an orthonormal signal representation; in fact it cannot even provide a "stable basis". (Sections 6.7 and 6.9 explain the meaning of this.)

Wavelet Transform vs. Short-Time Fourier Transform

The STFT works with a fixed window $v(t)$. If a high frequency signal is being analyzed, many cycles are captured by the window, and a good estimate of the FT is obtained. If a signal varies very slowly with respect to the window, however, then the window is not long enough to capture it fully. From a filter bank viewpoint, notice that all the filters have identical bandwidths (Fig. 6.17). This means that the frequency resolution is uniform at all frequencies, i.e., the "percentage resolution" or accuracy is poor for low frequencies and becomes increasingly better at high frequencies. The STFT, therefore, does not provide uniform percentage accuracy for all frequencies; the computational resources are somehow poorly distributed.

Compare this to the WT which is represented by a nonuniform filter bank [Fig. 6.8(b)]. Here the frequency resolution gets poorer as the frequency increases, but the fractional resolution (i.e., the filter bandwidth $\Delta\omega_k$ divided by the center frequency ω_k) is constant for all k (the percentage accuracy is uniformly distributed in frequency). In the time domain this is roughly analogous to having a large library of windows; narrow windows are used to analyze high-frequency components and very broad windows are used to analyze low-frequency components. In electrical engineering language the filter bank representing WT is a **constant Q filter bank**, or an **octave band filter bank**. Consider, for example, the Haar wavelet basis. Here, the narrow basis functions $\psi_{2,n}(t)$ of Fig. 6.12 are useful to represent the highly varying components of the input, and are correspondingly narrower (have shorter support than the functions $\psi_{1,n}(t)$).

A second difference between the STFT and the WTs is the sampling rates at the outputs of the bandpass filters. These are identical for the STFT filters (since all filters have the same bandwidth). For the wavelet filters, these are proportional to the filter bandwidths, hence nonuniform [Fig. 6.10(a)]. This is roughly analogous to the situation that the narrower windows move in smaller steps compared to the wider windows. Compare again to Fig. 6.12 where $\psi_{2,n}(t)$ are moved in smaller steps as compared to $\psi_{1,n}(t)$ in the process of constructing the complete set of basis functions. The nonuniform (constant Q) filter stacking [Fig. 6.8(b)] provided by wavelet filters is also naturally suited for analyzing audio signals and sometimes even as components in the modeling of the human hearing system.

Time-Frequency Tiling

The fact that the STFT performs uniform sampling of time and frequency whereas the WT performs nonuniform sampling is represented by the diagram shown in Fig. 6.18. Here, the vertical lines represent time locations at which the analysis filter bank output is sampled, and the horizontal lines represent the center frequencies of the bandpass filters. The time frequency tiling for the STFT is a simple rectangular grid, whereas for the WT it has a more complicated appearance.

Example 1. Consider the signal $x(t) = \cos(10\pi t) + 0.5\cos(5\pi t) + 1.2\delta_a(t - 0.07) + 1.2\delta_a(t + 0.07)$. It has impulses at $t = \pm 0.07$ in the time domain. Two impulses (or "lines") are found in the frequency domain, at $\omega_1 = 5\pi$ and $\omega_2 = 10\pi$. The function is shown in Fig. 6.19 with impulses replaced by narrow pulses. The aim is to try to compute the STFT or WT such that the impulses in time as well as those in frequency are resolved. Figure 6.20(a) to (c) shows the STFT plot for three widths of the window $v(t)$ and Fig. 6.20(d) shows the wavelet plot. The details of the window $v(t)$ and the wavelet $\psi(t)$ used for this example are described below, but first let us concentrate on the features of these plots.

The STFT plots are time-frequency plots, whereas the wavelet plots are (a^{-1}, b) plots, where a and b are defined by (6.21). As explained in Section 6.2, the quantity a^{-1} is analogous to "frequency" in the STFT, and b is analogous to "time" in the STFT. The brightness of the plots in Fig. 6.20 is proportional to the magnitude of the STFT or WT, so the transform is close to zero in the dark regions. We see that for a narrow window with width equal to 0.1, the STFT resolves the two impulses in time reasonably well, but the impulses in frequency are not resolved. For a wide window with width equal to 1.0, the STFT resolves the "lines" in frequency very well, but not the time domain impulses. For an intermediate window width equal to 0.3, the resolution is poor in both time and frequency. The wavelet transform plot [Fig. 6.20(d)], on the other hand, simultaneously resolves both time and frequency very well. We can clearly see the locations of the two impulses in time, as well as the two lines in frequency.

The STFT for this example was computed using the Hamming window [2] defined as $v(t) = c[0.54 + 0.46\cos(\pi t/D)]$ for $-D \le t \le D$ and zero outside. The "widths" indicated in the figure correspond to $D = 0.1, 1.0$, and 0.3 (although the two-sided width is twice this). The wavelet transform was computed by using an example of the Morlet wavelet [5]. Specifically,

$$\psi(t) = e^{-t^2/16}(e^{j\pi t} - \alpha)$$

First, let us understand what this wavelet function is doing. The quantity $e^{-t^2/16}$ is the Gaussian (except for a constant scalar factor) with Fourier transform $4\sqrt{\pi}e^{-4\omega^2}$, which is again Gaussian, concentrated near $\omega = 0$. Thus, $e^{-t^2/16}e^{j\pi t}$ has a FT concentrated around $\omega = \pi$. Ignoring the second term α in the expression for $\psi(t)$, we see that the wavelet is a narrowband bandpass filter concentrated around π (Fig. 6.21).[5] If we set $a = 1$ in (6.21), then $X(1, b)$ represents the

[5]The quantity α in the expression of $\psi(t)$ ensures that $\int \psi(t)dt = 0$ (Section 6.2). Because α is very small, it does not significantly affect the plots in Fig. 6.20.

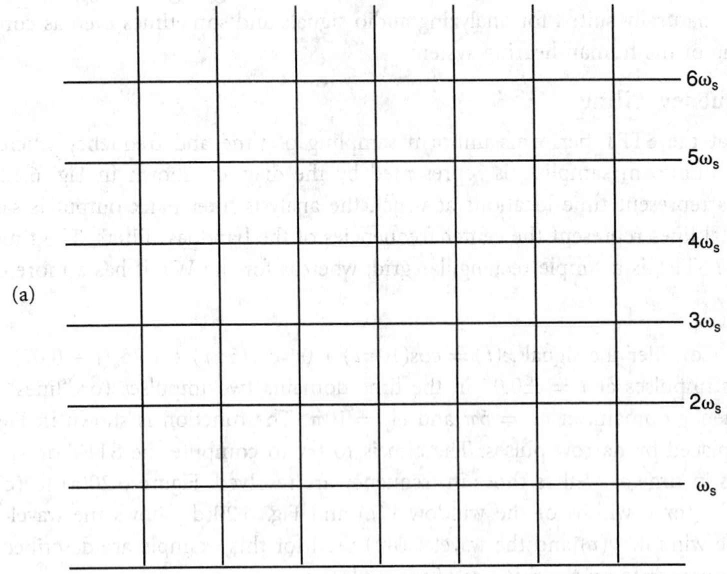

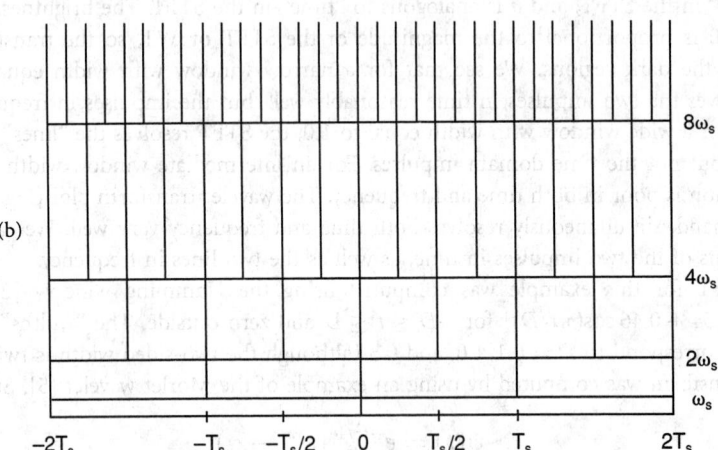

FIGURE 6.18 Time-frequency tiling schemes for (a) the STFT and (b) the WT.

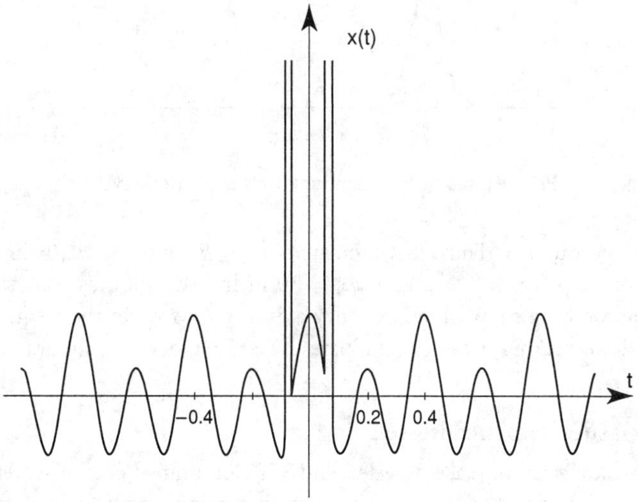

FIGURE 6.19 The signal to be analyzed by STFT and WT.

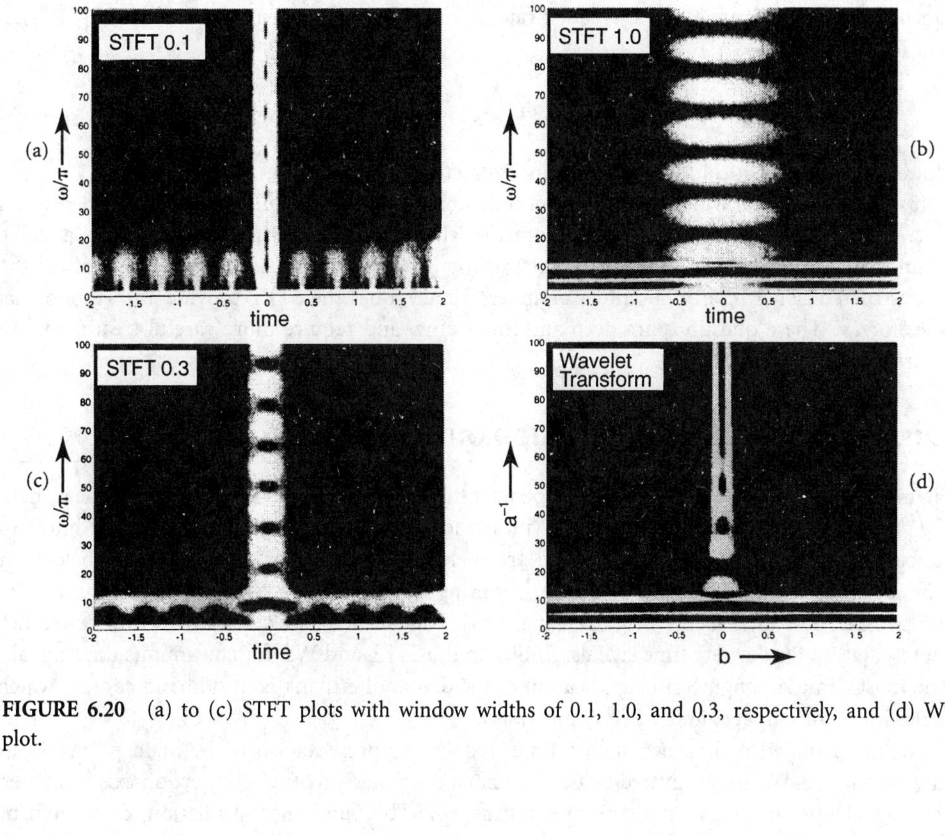

FIGURE 6.20 (a) to (c) STFT plots with window widths of 0.1, 1.0, and 0.3, respectively, and (d) WT plot.

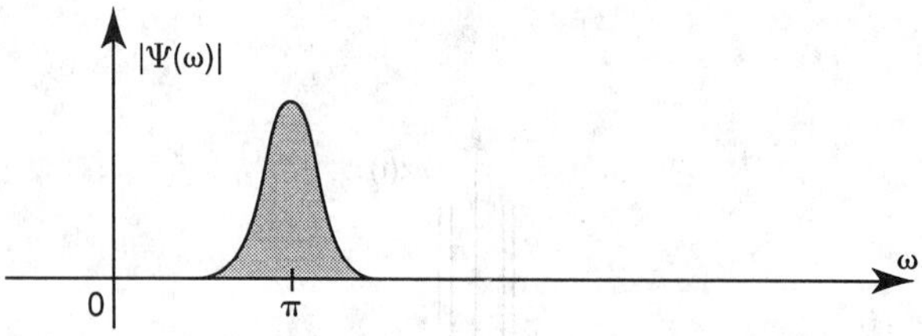

FIGURE 6.21 FT magnitude for the Morlet wavelet.

frequency contents around π. Thus, the frequencies $\omega_1 = 5\pi$ and $\omega_2 = 10\pi$ in the given signal $x(t)$ show up around points $a^{-1} = 5$ and $a^{-1} = 10$ in the WT plot, as seen from Fig. 6.20(d). In the STFT plots we have shown the frequency axis as ω/π so that the frequencies ω_1 and ω_2 show up at 5 and 10, making it easy to compare the STFT plots with the wavelet plot. □

Mathematical Issues to be Addressed

While the filter bank viewpoint places wavelets and STFT on unified ground, several mathematical issues remain unaddressed. It is this deeper study that brings forth further subtle differences, giving wavelets a definite advantage over the STFT.

Suppose we begin from a signal $x(t) \in L^2$ and compute the STFT coefficients $X(k\omega_s, nT_s)$. How should we choose the sampling periods T_s and ω_s of the time and frequency grids so that we can reconstruct $x(t)$ from the STFT coefficients? (Remember that we are not talking about band-limited signals, and no sampling theorem is at work.) If the filters $H_k(\omega)$ are ideal one-sided bandpass filters with bandwidth ω_s, the downshifted low-pass outputs $y_k(t)$ (Fig. 6.16) can be sampled separately at the Nyquist rate ω_s or higher. This then tells us that $T_s \leq 2\pi/\omega_s$, that is,

$$\omega_s T_s \leq 2\pi \tag{6.28}$$

However, the use of ideal filters implies an impractical window $v(n)$.

If we use a practical window (e.g., one of finite duration), how should we choose T_s in relation to ω_s so that we can reconstruct $x(t)$ from the STFT coefficients $X(k\omega_s, nT_s)$? Is this a stable reconstruction? That is, if we make a small error in some STFT coefficient does it affect the reconstructed signal in an unbounded manner? Finally, does the STFT provide an orthonormal basis for L^2? These questions are deep and interesting, and require more careful treatment. We return to these in Section 6.9.

6.4 Digital Filter Banks and Subband Coders

Figure 6.22(a) shows a two-channel filter bank with input sequence $x(n)$ (a discrete-time signal). $G_a(z)$ and $H_a(z)$ are two digital filters, typically low-pass and high-pass. $x(n)$ is split into two subband signals, $x_0(n)$ and $x_1(n)$, which are then downsampled or decimated (see below for definitions). The total subband data rate, counting both subbands, is equal to the number of samples per unit time in the original signal $x(n)$. Digital filter banks provide a time-frequency representation for discrete time signals, similar to the STFT and WT for continuous time signals. The most common engineering application of the digital filter bank is in subband coding, which is used in audio, image, and video compression.

Neither subband coding nor such a time frequency representation is the main point of our discussion here. We are motivated by the fact that a deep mathematical connection exists between this digital filter bank and the continuous time WT. This fundamental relation, discovered by

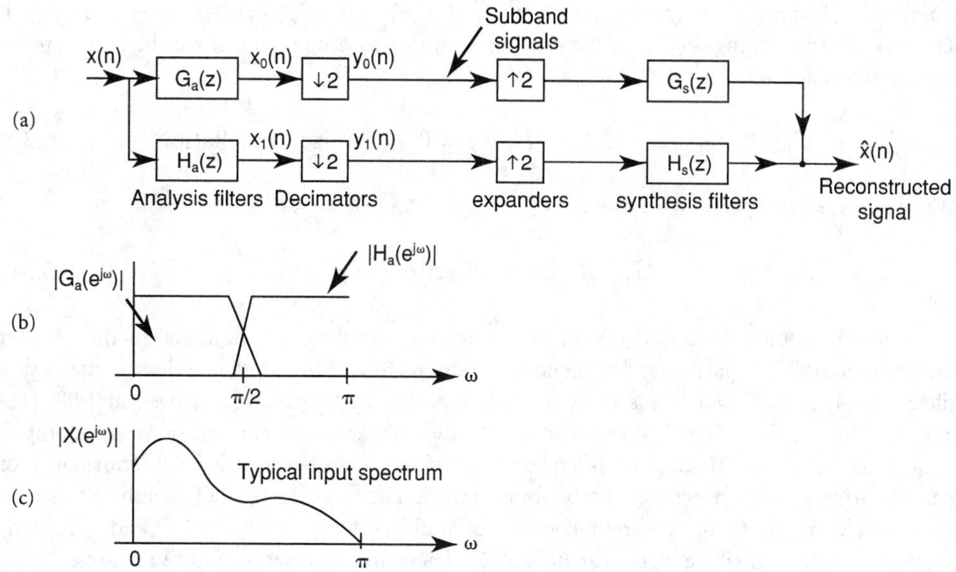

FIGURE 6.22 (a) The two-channel digital filter bank, (b) typical filter responses, and (c) typical input spectrum.

Daubechies [6], is fully elaborated in Section 6.10 to 6.13. This relation is what makes the WT so easy to design and attractive to implement in practice. Several detailed references on the topic of multirate systems and digital filter banks are available [7], and a detailed treatment can be found in Chapter 85 of this handbook, so we will be brief.

The Multirate Signal Processing Building Blocks: The building blocks in the digital filter bank of Fig. 6.22(a) are digital filters, decimators, and expanders. The M-fold **decimator** or **downsampler** (denoted $\downarrow M$) is defined by the input-output relation $y(n) = x(Mn)$. The corresponding z-domain relation is $Y(z) = (1/M)\sum_{k=0}^{M-1} X(z^{1/M}e^{-j2\pi k/M})$. This relation is sometimes abbreviated by the notation $Y(z) = X(z)\big|_{\downarrow M}$ or $Y(e^{j\omega}) = X(e^{j\omega})\big|_{\downarrow M}$. The M-fold **expander** or **upsampler** (denoted $\uparrow M$) is defined by

$$y(n) = \begin{cases} x(n/M), & n = \text{multiple of } M, \\ 0, & \text{otherwise} \end{cases} \tag{6.29}$$

The transform domain relation for the expander is $Y(z) = X(z^M)$, i.e., $Y(e^{j\omega}) = X(e^{jM\omega})$.

Reconstruction from Subbands

In many applications it is desirable to reconstruct $x(n)$ from the decimated subband signals $y_k(n)$ (possibly after quantization). For this, we pass $y_k(n)$ through expanders and combine them with the synthesis filters $G_s(z)$ and $H_s(z)$. The system is said to have the **perfect reconstruction** (PR) property if $\hat{x}(n) = cx(n - n_0)$ for some $c \neq 0$ and some integer n_0. The PR property is not satisfied for several reasons. First, subband quantization and bit allocation are present, which are the keys to data compression using subband techniques. However, because since our interest here lies in the connection between filter banks and wavelets, we will not be concerned with subband quantization. Second, because the filters $G_a(z)$ and $H_a(z)$ are not ideal, aliasing occurs due to decimation. Using the above equations for the decimator and expander building blocks, we can obtain the following expression for the reconstructed signal: $\hat{X}(z) = T(z)X(z) + A(z)X(-z)$, where $T(z) = 0.5[G_a(z)G_s(z) + H_a(z)H_s(z)]$ and $A(z) = 0.5[G_a(-z)G_s(z) + H_a(-z)H_s(z)]$. The second term having $X(-z)$ is the aliasing term due to decimation. It can be eliminated if the filters satisfy

$$G_a(-z)G_s(z) + H_a(-z)H_s(z) = 0 \qquad \text{(alias cancellation)} \tag{6.30}$$

We can then obtain perfect reconstruction $[\hat{X}(z) = 0.5X(z)]$ by setting

$$G_a(z)G_s(z) + H_a(z)H_s(z) = 1 \tag{6.31}$$

A number of authors have developed techniques to satisfy the PR conditions. In this chapter we are interested in a particular technique to satisfy (6.30) and (6.31), the conjugate quadrature filter (CQF) method, which was independently reported by Smith and Barnwell in 1984 [18] and by Mintzer in 1985 [19]. Vaidyanathan [20] showed that these constructions are examples of a general class of M channel filter banks satisfying a property called orthonormality or paraunitariness. More references can be found in [7]. The two-channel CQF solution was later rediscovered in the totally different contexts of multiresolution analysis [11] and compactly supported orthonormal wavelet construction [6]. These are discussed in future sections.

Conjugate Quadrature Filter (CQF) Solution

Suppose the low-pass filter $G_a(z)$ is chosen such that it satisfies the condition

$$\tilde{G}_a(z)G_a(z) + \tilde{G}_a(-z)G_a(-z) = 1 \qquad \text{for all } z \tag{6.32}$$

If we now choose the high-pass filter $H_a(z)$ and the two synthesis filters as

$$H_a(z) = z^{-1}\tilde{G}_a(-z) \qquad G_s(z) = \tilde{G}_a(z) \qquad H_s(z) = \tilde{H}_a(z) \tag{6.33}$$

then (6.30) and (6.31) are satisfied, and $\hat{x}(n) = 0.5x(n)$. In the time domain the above equations become

$$h_a(n) = -(-1)^n g_a^*(-n+1) \qquad g_s(n) = g_a^*(-n) \qquad h_s(n) = h_a^*(-n) \tag{6.34}$$

The synthesis filters are **time-reversed conjugates** of the analysis filters. If we design a filter $G_a(z)$ satisfying the single condition (6.32) and determine the remaining three filters as above, then the system has the PR property. A filter $G_a(z)$ satisfying (6.32) is said to be **power symmetric**. Readers familiar with **half-band filters** will notice that the condition (6.32) says simply that $\tilde{G}_a(z)G_a(z)$ is half-band. To design a perfect reconstruction CQF system, we first design a low-pass half-band filter $G(z)$ with $G(e^{j\omega}) \geq 0$, and then extract a spectral factor $G_a(z)$. That is, find $G_a(z)$ such that $G(z) = \tilde{G}_a(z)G_a(z)$. The other filters can be found from (6.33).

Polyphase Representation

The polyphase representation of a filter bank provides a convenient platform for studying theoretical questions and also helps in the design and implementation of PR filter banks. According to this representation the filter bank of Fig. 6.22(a) can always be redrawn as in Fig. 6.23(a), which in turn can be redrawn as in Fig. 6.23(b) using standard multirate identities. Here, $\mathbf{E}(z)$ and $\mathbf{R}(z)$ are the "polyphase matrices", determined uniquely by the analysis and synthesis filters, respectively.

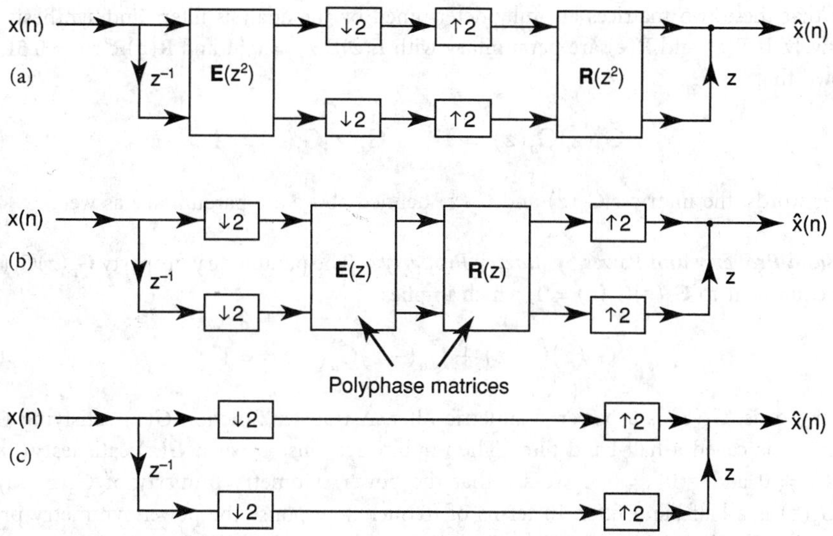

FIGURE 6.23 (a) The polyphase form of the filter bank, (b) further simplification, and (c) equivalent structure when $\mathbf{R}(z) = \mathbf{E}^{-1}(z)$.

If we impose the condition $\mathbf{R}(z)\mathbf{E}(z) = \mathbf{I}$, that is

$$\mathbf{R}(z) = \mathbf{E}^{-1}(z) \tag{6.35}$$

the system reduces to Fig. 6.23(c), which is a perfect reconstruction system with $\hat{x}(n) = x(n)$. Equation (6.35) will be called the PR condition. Notice that insertion of arbitrary scale factors and delays to obtain $\mathbf{R}(z) = cz^{-K}\mathbf{E}^{-1}(z)$ does not affect the PR property.

Paraunitary Perfect Reconstruction System

A transfer matrix[6] $\mathbf{H}(z)$ is said to be paraunitary if $\mathbf{H}(e^{j\omega})$ is unitary; that is, $\mathbf{H}^\dagger(e^{j\omega})\mathbf{H}(e^{j\omega}) = \mathbf{I}$ (more generally $\mathbf{H}^\dagger(e^{j\omega})\mathbf{H}(e^{j\omega}) = c\mathbf{I}, c > 0$), for all ω. In all practical designs the filters are rational transfer functions so that the paraunitary condition implies $\tilde{\mathbf{H}}(z)\mathbf{H}(z) = \mathbf{I}$ for all z, where the notation $\tilde{\mathbf{H}}(z)$ was explained in Section 6.1. Note that $\tilde{\mathbf{H}}(z)$ reduces to transpose conjugation $\mathbf{H}^\dagger(e^{j\omega})$ on the unit circle. A filter bank in which $\mathbf{E}(z)$ is paraunitary and $\mathbf{R}(z) = \tilde{\mathbf{E}}(z)$ enjoys the PR property $\hat{x}(n) = cx(n), c \neq 0$. We often say that the analysis filter pair $\{G_a(z), H_a(z)\}$ is paraunitary instead of saying that the corresponding polyphase matrix is paraunitary.

The paraunitary property has played a fundamental role in electrical network theory [1, 21], and has a rich history (see references in Chapters 6 and 14 of [7]). Essentially, the scattering matrices of lossless (LC) multiports are paraunitary, i.e., unitary on the imaginary axis of the s-plane.

Properties of Paraunitary Filter Banks

Define the matrices $\mathbf{G}_a(z)$ and $\mathbf{H}_a(z)$ as follows:

$$\mathbf{G}_a(z) = \begin{bmatrix} G_a(z) & G_a(-z) \\ H_a(z) & H_a(-z) \end{bmatrix} \qquad \mathbf{G}_s(z) = \begin{bmatrix} G_s(z) & H_s(z) \\ G_s(-z) & H_s(-z) \end{bmatrix} \tag{6.36}$$

[6]Transfer matrices are essentially transfer functions of multi-input multi-output systems. A review is found in Chapter 13 of [7].

Notice that these two matrices are fully determined by the analysis filters and synthesis filters, respectively. If $\mathbf{E}(z)$ and $\mathbf{R}(z)$ are paraunitary with $\tilde{\mathbf{E}}(z)\mathbf{E}(z) = 0.5\mathbf{I}$ and $\tilde{\mathbf{R}}(z)\mathbf{R}(z) = 0.5\mathbf{I}$, it can be shown that

$$\tilde{\mathbf{G}}_a(z)\mathbf{G}_a(z) = \mathbf{I} \qquad \tilde{\mathbf{G}}_s(z)\mathbf{G}_s(z) = \mathbf{I} \tag{6.37}$$

In other words, the matrices $\mathbf{G}_a(z)$ and $\mathbf{G}_s(z)$ defined above are paraunitary as well.

Half-Band Property and Power Symmetry Property. The paraunitary property $\tilde{\mathbf{G}}_a(z)\mathbf{G}_a(z) = \mathbf{I}$ is also equivalent to $\mathbf{G}_a(z)\tilde{\mathbf{G}}_a(z) = \mathbf{I}$, which implies

$$\tilde{G}_a(z)G_a(z) + \tilde{G}_a(-z)G_a(-z) = 1 \tag{6.38}$$

In other words, $G_a(z)$ is a power symmetric filter. A transfer function $G(z)$ satisfying $G(z) + G(-z) = 1$ is called a half-band filter. The impulse response of such $G(z)$ satisfies $g(2n) = 0$ for all $n \neq 0$ and $g(0) = 0.5$. We see that the power symmetry property of $G_a(z)$ says that $\tilde{G}_a(z)G_a(z)$ is a half-band filter. In terms of frequency response the power symmetry property of $G_a(z)$ is equivalent to

$$|G_a(e^{j\omega})|^2 + |G_a(-e^{j\omega})|^2 = 1 \tag{6.39}$$

Imagine that $G_a(z)$ is a real-coefficient low-pass filter so that $|G_a(e^{j\omega})|^2$ has symmetry with respect to zero frequency. Then $|G_a(-e^{j\omega})|^2$ is as demonstrated in Fig. 6.24, and the power symmetry property means that the two plots in the Figure add up to unity. In this figure ω_p and ω_s are the bandedges, and δ_1 and δ_2 are the peak passband ripples of $G_a(e^{j\omega})$ (for definitions of filter specifications see [2] or [7]). Notice in particular that power symmetry of $G_a(z)$ implies that a symmetry relation exists between the passband and stopband specifications of $G_a(e^{j\omega})$. This is given by $\omega_s = \pi - \omega_p$, $\delta_2^2 = 1 - (1 - 2\delta_1)^2$.

Relation Between Analysis Filters. The property $\tilde{\mathbf{G}}_a(z)\mathbf{G}_a(z) = \mathbf{I}$ implies a relation between $G_a(z)$ and $H_a(z)$, namely $H_a(z) = e^{j\theta}z^N\tilde{G}_a(-z)$, where θ is arbitrary and N is an arbitrary odd integer. Let $N = -1$ and $\theta = 0$ for future simplicity. Then

$$H_a(z) = z^{-1}\tilde{G}_a(-z) \tag{6.40}$$

In particular, we have $|H_a(e^{j\omega})| = |G_a(-e^{j\omega})|$. Combining with the power symmetry property (6.39), we see that the two analysis filters are **power complementary**:

$$|G_a(e^{j\omega})|^2 + |H_a(e^{j\omega})|^2 = 1 \tag{6.41}$$

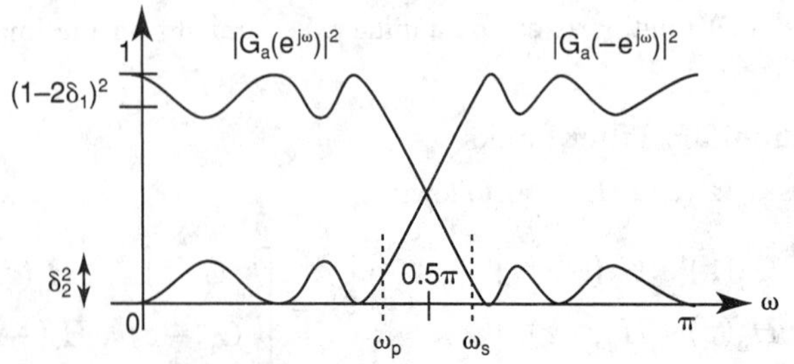

FIGURE 6.24 The magnitude responses $|G_a(e^{j\omega})|^2$ and $|G_a(-e^{j\omega})|^2$ for a real-coefficient power symmetric filter $G_a(z)$.

for all ω. With $G_a(z) = \sum_n g_a(n)z^{-n}$ and $H_a(z) = \sum_n h_a(n)z^{-n}$ we can rewrite (6.40) as

$$h_a(n) = -(-1)^n g_a^*(-n+1) \qquad (6.42)$$

Relation between Analysis and Synthesis Filters. If we use the condition $\mathbf{R}(z) = \tilde{\mathbf{E}}(z)$ in the definitions of $\mathbf{G}_s(z)$ and $\mathbf{G}_a(z)$ we obtain $\mathbf{G}_s(z) = \tilde{\mathbf{G}}_a(z)$, from which we conclude that the synthesis filters are given by $G_s(z) = \tilde{G}_a(z)$ and $H_s(z) = \tilde{H}_a(z)$. We can also rewrite these in the time domain; summarizing all this, we have

$$G_s(z) = \tilde{G}_a(z), \quad H_s(z) = \tilde{H}_a(z) \qquad g_s(n) = g_a^*(-n), \; h_s(n) = h_a^*(-n) \qquad (6.43)$$

The synthesis filter coefficients are time-reversed conjugates of the analysis filters. Their frequency responses are conjugates of the analysis filter responses. In particular, $|G_s(e^{j\omega})| = |G_a(e^{j\omega})|$ and $|H_s(e^{j\omega})| = |H_a(e^{j\omega})|$. In view of the preceding relations the synthesis filters have all the properties of the analysis filters. For example, $G_s(e^{j\omega})$ is power symmetric, and the pair $\{G_s(e^{j\omega}), H_s(e^{j\omega})\}$ is power complementary. Finally, $H_s(z) = z\tilde{G}_s(-z)$, instead of (6.40).

Relation to Conjugate Quadrature Filter Design. The preceding discussions indicate that in a paraunitary filter bank the filter $G_a(z)$ is power symmetric, and the remaining filters are derived from $G_a(z)$ as in (6.40) and (6.43). This is precisely the CQF solution for PR, stated at the beginning of this section.

Summary of Filter Relations in a Paraunitary Filter Bank

If the filter bank of Fig. 6.22(a) is paraunitary, then the polyphase matrices $\mathbf{E}(z)$ and $\mathbf{R}(z)$ (Fig. 6.23) satisfy $\tilde{\mathbf{E}}(z)\mathbf{E}(z) = 0.5\mathbf{I}$ and $\tilde{\mathbf{R}}(z)\mathbf{R}(z) = 0.5\mathbf{I}$. Equivalently the filter matrices $\mathbf{G}_a(z)$ and $\mathbf{G}_s(z)$ satisfy $\tilde{\mathbf{G}}_a(z)\mathbf{G}_a(z) = \mathbf{I}$ and $\tilde{\mathbf{G}}_s(z)\mathbf{G}_s(z) = \mathbf{I}$. A number of properties follow from these:

1. All four filters, $G_a(z), H_a(z), G_s(z)$, and $H_s(z)$, are power symmetric. This property is defined, for example, by the relation (6.38). This means that the filters are spectral factors of half-band filters; for example, $\tilde{G}_s(z)G_s(z)$ is half-band.
2. The two analysis filters are related as in (6.40), so the magnitude responses are related as $|H_a(e^{j\omega})| = |G_a(-e^{j\omega})|$. The synthesis filters are time-reversed conjugates of the analysis filters as shown by (6.43). In particular, $G_s(e^{j\omega}) = G_a^*(e^{j\omega})$ and $H_s(e^{j\omega}) = H_a^*(e^{j\omega})$.
3. The analysis filters form a power complementary pair, i.e., (6.41) holds. The same is true for the synthesis filters.
4. Any two-channel paraunitary system satisfies the CQF equations (6.32, 33) (except for delays, constant scale factors, etc.). Conversely, any CQF design is a paraunitary filter bank.
5. The design procedure for two-channel paraunitary (i.e., CQF) filter banks is as follows: design a zero-phase low-pass half-band filter $G(z)$ with $G(e^{j\omega}) \geq 0$ and then extract a spectral factor $G_a(z)$. That is, find $G_a(z)$ such that $G(z) = \tilde{G}_a(z)G_a(z)$. Then choose the remaining three filters as in (6.33), or equivalently, as in (6.34).

Parametrization of Paraunitary Filter Banks

Factorization theorems exist for matrices which allow the expression of paraunitary matrices as a cascade of elementary paraunitary blocks. For example, let $\mathbf{H}(z) = \sum_{n=0}^{L} \mathbf{h}(n)z^{-n}$ be a 2×2 real causal FIR transfer matrix (thus, $\mathbf{h}(n)$ are 2×2 matrices with real elements). This is paraunitary

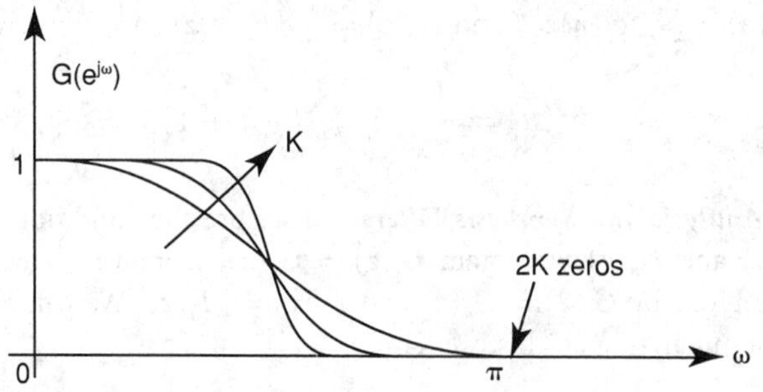

FIGURE 6.25 Maximally flat half-band filter responses with $2K$ zeroes at π.

iff it can be expressed as $\mathbf{H}(z) = \mathbf{R}_N\Lambda(z)\mathbf{R}_{N-1}\ldots\mathbf{R}_1\Lambda(z)\mathbf{R}_0\mathbf{H}_0$ where

$$\mathbf{R}_m = \begin{bmatrix} \cos\theta_m & \sin\theta_m \\ -\sin\theta_m & \cos\theta_m \end{bmatrix} \quad \Lambda(z) = \begin{bmatrix} 1 & 0 \\ 0 & z^{-1} \end{bmatrix} \quad \mathbf{H}_0 = \begin{bmatrix} \alpha & 0 \\ 0 & \pm\alpha \end{bmatrix} \tag{6.44}$$

where α and θ_m are real. For a proof see [7]. The unitary matrix $\mathbf{R}_m$ is called a **rotation operator** or the **Givens rotation.** The factorization gives rise to a cascaded lattice structure that guarantees the paraunitary property structurally. This is useful in the design as well as the implementation of filter banks, as explained in [7]. Thus, if the polyphase matrix is computed using the cascaded structure, $G_a(z)$ is guaranteed to be power symmetric, and the relation $H_a(z) = z^{-1}\tilde{G}_a(-z)$ between the analysis filters automatically holds. Further results on factorizations are described in Chapter 85.

Maximally Flat Solutions

The half-band filter $G(z) \triangleq \tilde{G}_a(z)G_a(z)$ can be designed in many ways. One can choose to have equiripple designs or maximally flat designs [2]. An early technique for designing FIR maximally flat filters was proposed by Herrmann in 1971 [7]. This method gives closed form expressions for the filter coefficients and can be adapted easily for the special case of half-band filters. Moreover, the design automatically guarantees the condition $G(e^{j\omega}) \geq 0$, which in particular implies zero phase.

The family of maximally flat half-band filters designed by Herrmann is demonstrated in Fig. 6.25. The transfer function has the form

$$G(z) = z^K\left(\frac{1+z^{-1}}{2}\right)^{2K}\sum_{n=0}^{K-1}(-z)^n\binom{K+n-1}{n}\left(\frac{1-z^{-1}}{2}\right)^{2n} \tag{6.45}$$

The filter has order $4K-2$. On the unit circle we find $2K$ zeroes, and all of these zeroes are concentrated at the point $z=-1$ (i.e., at $\omega=\pi$). The remaining $2K-2$ zeroes are located in the z-plane such that $G(z)$ has the half-band property described earlier (i.e., $G(z)+G(-z)=1$).

Section 6.13 explains that if the CQF bank is designed by starting from Herrmann's maximally flat half-band filter, then it can be used to design continuous time wavelets with arbitrary regularity (i.e., smoothness) properties.

Tree-Structured Filter Banks

The idea of splitting a signal $x(n)$ into two subbands can be extended by splitting a subband signal further, as demonstrated in Fig. 6.26(a). In this example the low-pass subband is split

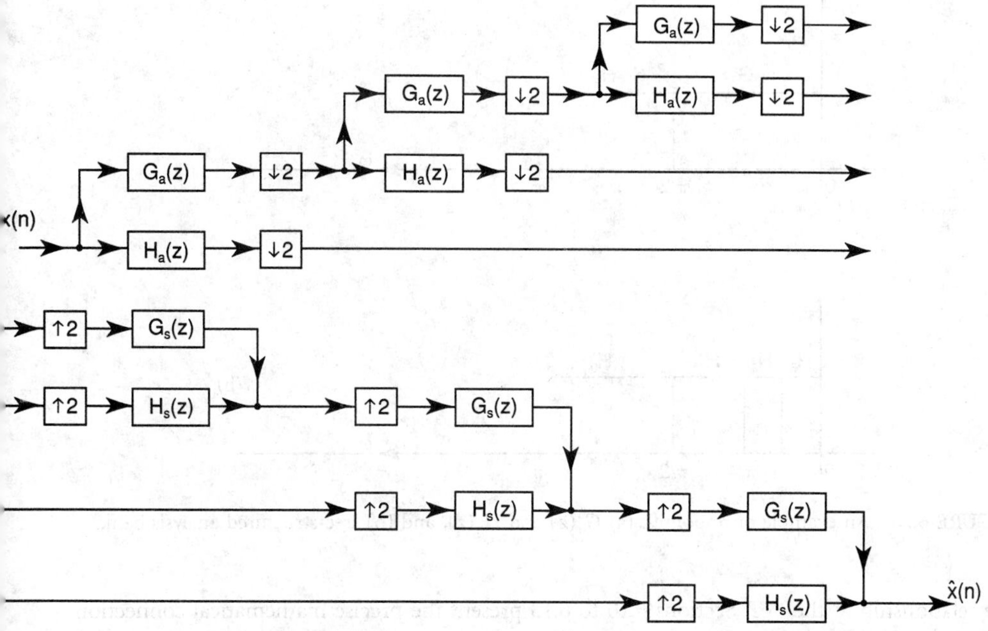

FIGURE 6.26 Tree-structured filter banks: (a) analysis bank and (b) synthesis bank.

repeatedly. This is called a tree-structured filter bank. Each node of the tree is a two-channel analysis filter bank.

The synthesis bank corresponding to Fig. 6.26(a) is shown in Fig. 6.26(b). We combine the signals in pairs in the same manner that we split them. It can be shown that if $\{G_a(z), H_a(z), G_s(z), H_s(z)\}$ is a PR system [i.e., satisfies $\hat{x}(n) = x(n)$ when connected in the form Fig. 6.22(a)] then the tree-structured analysis/synthesis system of Fig. 6.26 has PR $\hat{x}(n) = x(n)$.

The tree-structured system can be redrawn in the form shown in Fig. 6.27. For example, if we have a tree structure similar to Fig. 6.26 with three levels, we have $M = 4$, $n_0 = 2$, $n_1 = 4$, $n_2 = 8$, and $n_3 = 8$. If we assume that the responses of the analysis filters $G_a(e^{j\omega})$ and $H_a(e^{j\omega})$ are as in Fig. 6.28(a), the responses of the analysis filters $H_k(e^{j\omega})$ are as shown in Fig. 6.28(b). Note that this resembles the wavelet transform [Fig. 6.8(b)]. The outputs of different filters are subsampled at different rates exactly as for wavelets. Thus, the tree-structured filter bank bears a

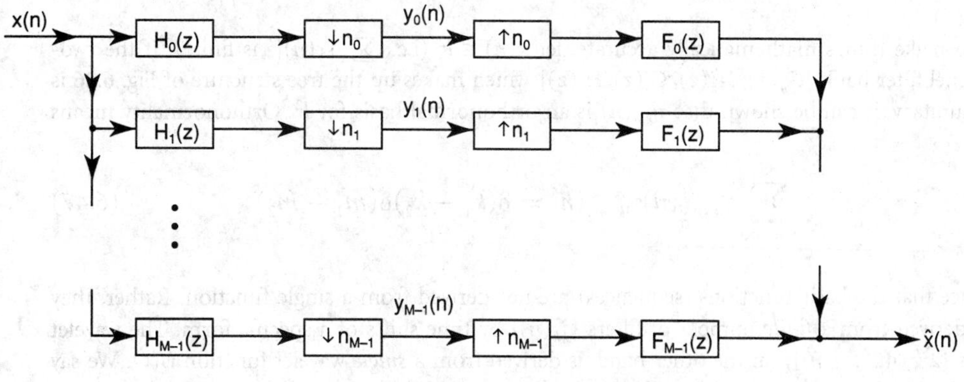

FIGURE 6.27 A general nonuniform digital filter bank.

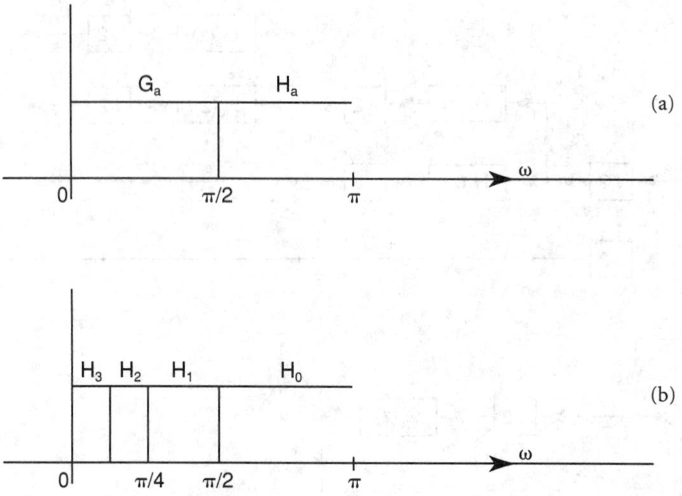

FIGURE 6.28 An example of responses: (a) $G_a(z)$ and $H_a(z)$, and (b) tree-structured analysis bank.

close relationship to the WT. Sections 6.10 to 6.13 present the precise mathematical connection between the two, and the relation to multiresolution analysis.

Filter Banks and Basis Functions

Assuming PR $[\hat{x}(n) = x(n)]$, we can express $x(n)$ as

$$x(n) = \sum_{k=0}^{M-1} \sum_{m=-\infty}^{\infty} y_k(m) \underbrace{f_k(n - n_k m)}_{\eta_{km}(n)} \tag{6.46}$$

where $y_k(n)$ are the decimated subband signals, and $f_k(n)$ are the impulse responses of $F_k(z)$. Thus, the system is analogous to the filter bank systems which represented the continuous time STFT and WT in Sections 6.2 and 6.3. The collection of subband signals $y_k(m)$ can be regarded as a time-frequency representation for the sequence $x(n)$. As before, k denotes the **frequency index** and m the **time index** in the transform domain. If we have a PR filter bank we can recover $x(n)$ from this time-frequency representation using (6.46). The doubly indexed family of discrete time sequences $\{\eta_{km}(n)\}$ can be regarded as "basis functions" for the representation of $x(n)$.

To make things mathematically accurate, let $x(n) \in l^2$ (i.e., $\sum_n |x(n)|^2$ is finite). If the two-channel filter bank $\{G_a(z), H_a(z), G_s(z), H_s(z)\}$ which makes up the tree structure of Fig. 6.26 is paraunitary, it can be shown that $\eta_{km}(n)$ is an orthonormal basis for l^2. Orthonormality means

$$\sum_{n=-\infty}^{\infty} \eta_{k_1 m_1}(n) \eta^*_{k_2 m_2}(n) = \delta(k_1 - k_2)\delta(m_1 - m_2) \tag{6.47}$$

Notice that the basis functions (sequences) are not derived from a single function. Rather, they are derived from a *finite* number of filters $\{f_k(n)\}$ by time shifts of a specific form. The wavelet basis $\{2^{k/2}\psi(2^k t - n)\}$, on the other hand, is derived from a *single* wavelet function $\psi(t)$. We say that $\{\eta_{km}(n)\}$ is a *filter bank* type of basis for the space of l^2 sequences.

6.5 Deeper Study of Wavelets, Filter Banks, and Short-Time Fourier Transforms

We already know what the WT is and how it compares to the STFT, at least qualitatively. We are also familiar with time-frequency representations and digital filter banks. It is now time to fill in several important details, and generally be more quantitative. For example, we would like to mention some major technical limitations of the STFT which are not obvious from its definition, and explain that wavelets do not have this limitation.

For example, if the STFT is used to obtain an orthonormal basis for L^2 signals, the time-frequency RMS durations of the window $v(t)$ should satisfy $D_t D_f = \infty$. That is, either the time or the frequency resolution is very poor (Theorem 6.5). Also, if we have an STFT system in which the time-frequency sampling product $\omega_s T_s$ is small enough to admit redundancy (i.e., the vectors are not linearly independent as they would be in an orthonormal basis), the above difficulty can be eliminated (Section 6.9).

The Gabor transform, while admittedly a tempting candidate because of the optimal time-frequency resolution property ($D_t D_f$ minimized), has a disadvantage. For example, if we want to recover the signal $x(t)$ from the STFT coefficients, the reconstruction is *unstable* in the so-called critically sampled case (Section 6.9). That is, a small error in the STFT coefficients can lead to a large error in reconstruction.

The WT does not suffer from the above limitations of the STFT. Sections 6.11 to 6.13 show how to construct orthonormal wavelet bases with good time and frequency resolutions. We also show that we can start from a paraunitary digital filter bank and construct orthonormal wavelet bases for $L^2(R)$ very systematically (Theorem 6.13). Moreover, this can be done in such a way that many desired properties (e.g., compact support, orthonormality, good time frequency resolution, smoothness, and so forth) can be incorporated during the construction (Section 6.13). Such a construction is placed in evidence by the **theory of multiresolution**, which gives a unified platform for wavelet construction and filter banks (Theorems 6.6. and 6.7).

At this point, the reader may want to preview the above-mentioned theorems in order to get a flavor of things to come. However, to explain these results quantitatively, it is very convenient to review a number of mathematical tools. The need for advanced tools arises because of the intricacies associated with basis functions for infinite dimensional spaces, i.e., spaces in which the set of basis functions is an infinite set. (For finite dimensional spaces an understanding of elementary matrix theory would have been sufficient.) For example, a representation of the form $x(t) = \sum c_n f_n(t)$ in an infinite dimensional space could be *unstable* in the sense that a small error in the transform domain $\{c_n\}$ could be amplified in an unbounded manner during reconstruction. A special type of basis called the *Riesz basis* does not have this problem (orthonormal bases are special cases of these). Also, the so-called **frames** (Section 6.8) share many good properties of the Riesz bases, but may have redundant vectors (i.e., not a linearly independent set of vectors). For example, the concept of frames arises in the comparison of wavelets and the STFT. General STFT frames have an advantage over STFT bases. Frames also come into consideration when the connection between wavelets and paraunitary digital filter banks is explained in Section 6.11. When describing the connection between wavelets and nonunitary filter banks, one again encounters Riesz bases and the idea of **biorthogonality**.

Because it is difficult to find all the mathematical background material in one place, we review a carefully selected set of topics in the next few sections. These are very useful for a deeper understanding of wavelets and STFT. The material in Section 6.6 is fairly standard (Lebesgue integrals, L^p spaces, L^1 and L^2 FTs). The material in Sections 6.7 and 6.8 (Riesz bases and frames) are less commonly known among engineers, but play a significant role in wavelet theory. The reader may want to go through these review sections (admittedly dense), once during first reading and then use them as a reference. Following this review we return to our discussions of wavelets, STFT, and filter banks.

6.6 The Space of L^1 and L^2 Signals

We developed the wavelet representation in Section 6.2 based on the framework of a bank of bandpass filters. To make everything mathematically meaningful it becomes necessary to carefully specify the types of signals, types of FTs, etc. For example, the concept of ideal bandpass filtering is appealing to engineers, but a difficulty arises. An ideal bandpass filter $H(\omega)$ is not stable, that is, $\int |h(t)|\,dt$ does not exist [2]. In other words $h(t)$ does not belong to the space L^1 (see below).

Why should this matter if we are discussing theory? The frequency domain developments based on Fig. 6.2, which finally give rise to the time domain expression (6.8), implicitly rely on the convolution theorem (*convolution in time implies multiplication in frequency*). However, the convolution theorem is typically proved only for L^1 signals and bounded L^2 signals; it is not valid for arbitrary signals. Therefore, care must be exercised when using these familiar engineering notions in a mathematical discussion.

Lebesgue Integrals

In most engineering discussions we think of the integrals as Riemann integrals, but in order to handle several convergence questions in the development of Fourier series, convolution theorems, and wavelet transforms, it is necessary to use Lebesgue integration. Lebesgue integration theory has many beautiful results which are not true for the Riemann integral under comparable assumptions about signals. This includes theorems that allow us to interchange limits, integrals, and infinite sums freely.

All integrals in this chapter are Lebesgue integrals. A review of Lebesgue integration is beyond the scope of this chapter, although many excellent references, for example, [22], are available. A few elementary comparisons between Riemann and Lebesgue integrals are given next.

1. If $x(t)$ is Riemann integrable on a bounded interval $[a, b]$ then it is also Lebesgue integrable on $[a, b]$. The converse is not true, however. For example, if we define $x(t) = -1$ for all rationals and $x(t) = 1$ for all irrationals in [0,1], then $x(t)$ is not Riemann integrable in [0,1], but it is Lebesgue integrable, and $\int_0^1 x(t)\,dt = 1$.
2. A similar statement is not true for the unbounded interval $(-\infty, \infty)$. For the unbounded interval $(-\infty, \infty)$ the Riemann integral is defined only as a limit called the **improper integral.**[7] Consider the sinc function defined as: $s(t) = \sin t/t$ for $t \neq 0$, and $s(0) = 1$. This has improper Riemann integral $= \pi$, but is not Lebesgue integrable.
3. If $x(t)$ is Lebesgue integrable, so is $|x(t)|$. The same is not true for Riemann integrals, as demonstrated by the sinc function $s(t)$ of the preceding paragraph.
4. If $|x(t)|$ is Lebesgue integrable, so is $x(t)$ as long as it is measurable.[8] This, however, is not true for Riemann integrals. If we define $x(t) = -1$ for all rationals and 1 for all irrationals in [0,1], it is not Riemann integrable in [0,1] although $|x(t)|$ is.
5. If $x(t)$ is (measurable and) bounded by a non-negative Lebesgue integrable function $g(t)$ [i.e., $|x(t)| \leq g(t)$] then $x(t)$ is Lebesgue integrable.

[7]Essentially we consider $\int_{-a}^{b} x(t)dt$ and let a and b go to ∞ *separately.* This limit, the improper Riemann integral, should not be confused with the **Cauchy principal value,** which is the limit of $\int_{-a}^{a} x(t)\,dt$ as $a \to \infty$. The function $x(t) = t$ has Cauchy principal value $= 0$, but the improper Riemann integral does not exist.

[8]The notion of a measurable function is very subtle. Any continuous function is measurable, and any Lebesgue integrable function is measurable. In fact, examples of nonmeasureable functions are so rare and so hard to construct that practically no danger exists that we will run into one. We take measurability for granted and never mention it.

Sets of Measure Zero

A subset $\mathcal{S}$ of real numbers is said to have measure zero if, given $\epsilon > 0$, we can find a countable union $\cup_i I_i$ of open intervals I_i [intervals of the form (a_i, b_i), i.e., $a_i < x < b_i$] such that $\mathcal{S} \subset \cup_i I_i$ and the total length of the intervals $< \epsilon$. For example, the set of all integers (in fact, any countable set of real numbers, e.g., rationals) has measure zero. Uncountable sets of real numbers exist which have measure zero, a famous example being the Cantor set [22].

When something is said to be true "almost everywhere" (abbreviated a.e.) or "for almost all t" it means that the statement holds everywhere, except possibly on a set of measure zero. For example, if $x(t) = y(t)$ everywhere except for integer values of t, then $x(t) = y(t)$ a.e. An important fact in Lebesgue integration theory is that if two Lebesgue integrable functions are equal a.e., then their integrals are equal. In particular, if $x(t) = 0$ a.e., the Lebesgue integral $\int x(t)dt$ exists and is equal to zero.

Convergence Theorems

What makes the Lebesgue integral so convenient is the existence of some powerful theorems which allow us to interchange limits with integrals and summations under very mild conditions. These theorems have been at the center of many beautiful results in Fourier and wavelet transform theory.

Let $\{g_k(t)\}, 1 \leq k \leq \infty$ be a sequence of Lebesgue integrable functions. In general, this sequence may not have a limit, and even if it did, the limit may not be integrable. Under some further mild postulates, we can talk about limits and their integrals. In what follows we often say "$g(t)$ is a pointwise limit a.e. of the sequence $\{g_k(t)\}$", or "$g_k(t)$ converges to $g(t)$ a.e." This means that for any chosen value of t (except possibly in a set of measure zero), we have $g_k(t) \to g(t)$ as $k \to \infty$.

Monotone Convergence Theorem: *Suppose* $\{g_k(t)\}$ *is* **nondecreasing** a.e. (*i.e., for almost all values of* t, $g_k(t)$ *is nondecreasing in* k) *and* $\int g_k(t)\,dt$ *is a* **bounded sequence.** *Then* $\{g_k(t)\}$ *converges* a.e. *to a Lebesgue integrable function* $g(t)$ *and* $\lim_k \int g_k(t)\,dt = \int \lim_k g_k(t)\,dt$, i.e., $\lim_k \int g_k(t)\,dt = \int g(t)\,dt$. *That is, we can* **interchange the limit with the integral.**

Dominated Convergence Theorem: *Suppose* $\{g_k(t)\}$ *is dominated by a non-negative Lebesgue integrable function* $f(t)$, *i.e.,* $|g_k(t)| \leq f(t)$ a.e., *and* $\{g_k(t)\}$ *converges to a limit* $g(t)$ a.e. *Then the limit* $g(t)$ *is Lebesgue integrable and* $\lim_k \int g_k(t)dt = \int \lim_k g_k(t)\,dt$, *i.e.,* $\lim_k \int g_k(t)dt = \int g(t)\,dt$. *That is, we can* **interchange the limit with the integral.**

Levi's Theorem: *Suppose* $\int \sum_{k=1}^{m} |g_k(t)|\,dt$ *is a bounded sequence in* m. *Then* $\int \sum_{k=1}^{\infty} g_k(t)\,dt = \sum_{k=1}^{\infty} \int g_k(t)\,dt$. *This means, in particular, that* $\sum_{k=1}^{\infty} g_k(t)$ *converges* a.e. *to a Lebesgue integrable function. This theorem permits us to* **interchange infinite sums with the integrals**.

Fatou's Lemma: *Let* (*a*) $g_k(t) \geq 0$ *a.e.,* (*b*) $g_k(t) \to g(t)$ a.e., *and* (*c*) $\int g_k(t)\,dt \leq A$ *for some* $0 < A < \infty$. *Then the limit* $g(t)$ *is Lebesgue integrable and* $\int g(t) \leq A$. (Stronger versions of this result exist [23], but we shall not require them here.)

L^p Signals

Let p be an integer such that $1 \leq p < \infty$. A signal $x(t)$ is said to be an L^p signal if it is measurable, and if $\int |x(t)|^p dt$ exists. We define the L^p norm of $x(t)$ as $\|x(t)\|_p = [\int |x(t)|^p dt]^{1/p}$. For fixed p the set of L^p signals forms a vector space. It is a normed linear vector space, with norm defined as above. The term "linear" means that if $x(t)$ and $y(t)$ are in L^p, then $\alpha x(t) + \beta y(t)$ is also in L^p for any complex α and β. Because any two signals $x(t)$ and $y(t)$ that are equal a.e. cannot be distinguished (i.e., $\|x(t) - y(t)\|_p = 0$), each element in L^p is in reality "a set of functions that are equal a.e." Each such set becomes an "equivalence class" in mathematical language. For $p = 2$ the quantity $\|x(t)\|_p^2$ is equal to the **energy** of $x(t)$, as defined in signal processing texts. Thus, an L^2 signal is a finite-energy (or square-integrable) signal. For $p = \infty$ the above definitions do not make sense, and we simply define L^∞ to be the space of **essentially bounded signals.** A signal $x(t)$ is said to be essentially bounded if there exists a number $B < \infty$ such that $|x(t)| \leq B$ a.e. We often omit the term "essential" for simplicity; it arises because of the a.e. in the inequality. The norm $\|x(t)\|_\infty$ is taken as essential supremum of $|x(t)|$ over all t. That is, $\|x(t)\|_\infty$ is the smallest number such that $|x(t)| \leq \|x(t)\|_\infty$ a.e.

L^1, L^2, and L^∞ functions are particularly interesting for engineers. Note that neither L^1 nor L^2 contains the other. However, bounded L^1 functions are in L^2, and L^2 functions on bounded intervals are in L^1. That is,

$$L^1 \cap L^\infty \subset L^2 \quad \text{and} \quad L^2[a,b] \subset L^1[a,b] \tag{6.48}$$

Thus, L^2 is already bigger than bounded L^1 functions. Moreover,

$$x(t) \in L^1 \cap L^\infty \quad \Rightarrow \quad x(t) \in L^p \text{ for all } p > 1$$

This follows because $|x(t)|^p \leq |x(t)| \, \|x(t)\|_\infty^{p-1}$. Thus, $|x(t)|^p$ is (measurable and) bounded by a Lebesgue integrable function (because $|x(t)|$ is integrable), and is therefore integrable.

Orthonormal Signals in L^2

The inner product $\langle x(t), y(t)\rangle = \int x(t)y^*(t)\,dt$ always exists for any $x(t)$ and $y(t)$ in L^2. Thus, the product of two L^2 functions is an L^1 function. If $\langle x(t), y(t)\rangle = 0$ we say that $x(t)$ and $y(t)$ are orthogonal. Clearly, $\|x(t)\|_2^2 = \langle x(t), x(t)\rangle$. Consider a sequence $\{g_n(t)\}$ of signals such that any pair of these are orthogonal, and $\|g_n(t)\|_2 = 1$ for all n. This is said to be an orthonormal sequence. The following two results are fundamental.

Theorem 6.1: *Let $\{g_n(t)\}, 1 \leq n \leq \infty$ be an orthonormal sequence in L^2. Define $c_n = \langle x(t), g_n(t)\rangle$ for some $x(t) \in L^2$. Then the sum $\sum_n |c_n|^2$ converges, and $\sum_n |c_n|^2 \leq \|x(t)\|^2$.* □

Theorem 6.2 (Riesz-Fischer Theorem): *Let $\{g_n(t)\}$, $1 \leq n \leq \infty$ be an orthonormal sequence in L^2 and let $\{c_n\}$ be a sequence of complex numbers such that $\sum_n |c_n|^2$ converges. Then there exists $x(t) \in L^2$ such that $c_n = \langle x(t), g_n(t)\rangle$, and $x(t) = \sum_n c_n g_n(t)$ (with equality interpreted in the L^2 sense; see below).* □

The space L^2 is more convenient to work with than L^1. For example, the inner product and the concept of orthonormality are undefined in L^1. Moreover (see following section), the FT in L^2 has more time-frequency symmetry than in L^1. In Section 6.7 we will define **unconditional bases,** which have the property that any rearrangement continues to be a basis. It turns out that any orthonormal basis in L^2 is unconditional, whereas the L^1 space does not even have an unconditional basis.

Equality and Convergence in L^p Sense

Let $x(t)$ and $y(t)$ be L_p functions ($p < \infty$). Then $\|x(t) - y(t)\|_p = 0$ iff $x(t) = y(t)$ a.e. For example, if $x(t)$ and $y(t)$ differ only for every rational t we still have $\|x(t) - y(t)\|_p = 0$. Whenever $\|x(t) - y(t)\|_p = 0$, we say that $x(t) = y(t)$ in L^p sense. Now consider a statement of the form

$$x(t) = \sum_{n=1}^{\infty} c_n g_n(t) \tag{6.49}$$

for $p < \infty$, where $g_n(t)$ and $x(t)$ are in L^p. This means that the sum converges to $x(t)$ in the L^p sense; that is, $\|x(t) - \sum_{n=1}^{N} c_n g_n(t)\|_p$ goes to zero as $N \to \infty$. If we modify the limit $x(t)$ by adding some number to $x(t)$ for all rational t, the result is still a limit of $\sum_{n=1}^{N} c_n g_n(t)$ in the L^p sense. L^p limits are unique only in the a.e. sense. We omit the phrase "in the L^p sense" whenever it is clear from the context.

l^p Spaces

Let p be an integer with $1 \le p \le \infty$. The collection of all sequences $x(n)$ such that $\sum_n |x(n)|^p$ converges to a finite value is denoted l^p. This is a linear space with norm $\|x(n)\|_p$ defined so that $\|x(n)\|_p = \left(\sum_n |x(n)|^p\right)^{1/p}$. Unlike L^p spaces, the l^p spaces satisfy the following inclusion rule:

$$l^1 \subset l^2 \subset l^3 \subset \ldots l^\infty \tag{6.50}$$

The spaces l^1 and l^2 are especially interesting in circuits and signal processing. If $h(n) \in l^1$, $\sum_n |h(n)| < \infty$. This is precisely the condition for the BIBO (bounded-input–bounded-output) stability of a linear time invariant system with impulse response $h(n)$ [2].

Continuity of Inner Products

If $\{x_n(t)\}$ is a sequence in L^2 and has an L^2 limit $x(t)$, then for any $y(t) \in L^2$,

$$\lim_{n\to\infty} \langle x_n(t), y(t)\rangle = \left\langle \lim_{n\to\infty} x_n(t), y(t)\right\rangle = \langle x(t), y(t)\rangle \tag{6.51}$$

with the second limit interpreted in the L^2 sense. Thus, limits can be interchanged with inner product signs. Similarly, infinite summation signs can be interchanged with the inner product sign, that is, $\sum_{n=1}^{\infty}\langle \alpha_n x_n(t), y(t)\rangle = \langle \sum_{n=1}^{\infty} \alpha_n x_n(t),\ y(t)\rangle$, provided the second summation is regarded as an L^2 limit. These follow from the fundamental property that *inner products are continuous* [23].

Next suppose $\{x_n(t)\}$ is a sequence of functions in L^p for some integer $p \ge 1$, and suppose $x_n(t) \to x(t)$ in the L^p sense. Then $\|x_n(t)\|_p \to \|x(t)\|_p$ as well. We can rephrase this as

$$\lim_{n\to\infty} \|x_n(t)\|_p = \left\| \lim_{n\to\infty} x_n(t)\right\|_p = \|x(t)\|_p \tag{6.52}$$

Thus, the limit sign can be interchanged with the norm sign, where the limit in the second expression is in the L^p sense. This follows because

$$\big|\, \|x_n(t)\|_p - \|x(t)\|_p \,\big| \le \|x_n(t) - x(t)\|_p \to 0 \quad \text{as} \quad n \to \infty$$

Fourier Transforms

The Fourier transform is defined for L^1 and L^2 signals in different ways. The properties of these two types of FT are significantly different. In the signal processing literature, in which we ultimately seek engineering solutions (such as filter approximation with rational transfer functions), this distinction often is not necessary. However, when we try to establish that a certain set of signals is a basis for a certain class, we must be careful, especially if we use tools such as the FT, convolution theorem, etc. (as we implicitly did in Section 6.2). Detailed references for this section include [15, 22, and 23].

L^1 Fourier Transform

Given a signal $x(t) \in L^1$, its FT $X(\omega)$ (the L^1 FT) is defined in a manner that is familiar to engineers:

$$X(\omega) = \int_{-\infty}^{\infty} x(t)e^{-j\omega t}\, dt \tag{6.53}$$

The existence of this integral is assured by the fact that $x(t)$ is in L^1.[9] In fact the above integral exists iff $x(t) \in L^1$. The L^1 FT has the following properties:

1. $X(\omega)$ is a continuous function of ω.
2. $X(\omega) \to 0$ as $|\omega| \to \infty$. This is called the **Riemann-Lebesgue lemma.**
3. $X(\omega)$ is bounded, and $|X(\omega)| \leq \|x(t)\|_1$.

In engineering applications we often draw the ideal low-pass filter response ($F(\omega)$ in Fig. 6.3) and consider it to be the FT of the impulse response $f(t)$, but this frequency response is discontinuous and already violates property 1. This is because $f(t)$ is not in L^1 and $F(\omega)$ is not the L^1-FT of $f(t)$. That $f(t)$ is not in L^1 is consistent with the fact that the ideal filter is not BIBO stable (i.e., a bounded input may not produce bounded output because $\int |f(t)|\, dt$ is not finite).

Inverse Fourier Transform. The FT $X(\omega)$ of an L^1 signal generally is not in L^1. For example, if $x(t)$ is the rectangular pulse, then $X(\omega)$ is the sinc function which is not absolutely integrable. Thus, the familiar inverse transform formula

$$x(t) = \frac{1}{2\pi}\int_{-\infty}^{\infty} X(\omega)e^{j\omega t}\, d\omega \tag{6.54}$$

does not make sense in general. However, because $X(\omega)$ is continuous and bounded, it is integrable on any bounded interval, so $\int_{-c}^{c} X(\omega)e^{j\omega t}\, d\omega/2\pi$ exists for any finite c. This quantity may even have a limit as $c \to \infty$, even if the Lebesgue integral or improper Rieman integral, does not exist. Such a limit (the Cauchy principal value) does represent the original function $x(t)$ under some conditions. Two such cases are outlined next.

Case 1. Thus, suppose $x(t) \in L^1$ and suppose that it is of **bounded variation** in an interval $[a, b]$; that is, it can be expressed as the difference of two nondecreasing functions [22]. Then we can show that the above Cauchy principal value exists, and

$$\frac{x(t^+) + x(t^-)}{2} = \lim_{c\to\infty} \frac{1}{2\pi}\int_{-c}^{c} X(\omega)e^{j\omega t}\, d\omega \tag{6.55}$$

for every $t \in (a, b)$. The notations $x(t^-)$ and $x(t^+)$ are the left-hand limit and the right-hand limit, respectively, of $x(\cdot)$ at t; for functions of bounded variation, these limits can be shown to exist. If $x(\cdot)$ is continuous at t, then $x(t^-) = x(t^+) = x(t)$, and the above reduces to the familiar inversion formula.

Case 2. Suppose now that $x(t) \in L^1$ and $X(\omega) \in L^1$ as well. Then the integral $y(t) \triangleq \int_{-\infty}^{\infty} X(\omega)e^{j\omega t}\, d\omega/2\pi$ exists as a Lebesgue integral, and $y(t) = x(t)$ a.e. [23]. In particular, if $x(\cdot)$ is continuous at t, $x(t) = \int_{-\infty}^{\infty} X(\omega)e^{j\omega t}\, d\omega/2\pi$.

If $x(t)$ and $X(\omega)$ are both in L^1 they are both in L^2 as well. This is shown as follows: because $x(t) \in L^1$ implies that $X(\omega)$ is bounded, we see that $X(\omega) \in L^1 \cap L^\infty$. So $X(\omega) \in L^p$ for all integer p (See previous section). In particular, $X(\omega) \in L^2$, so $x(t) \in L^2$ as well (by Parseval's relation; see below).

[9]Because $x(t)$ is Lebesgue integrable (hence, measurable) the product $x(t)e^{-j\omega t}$ is measurable, and it is bounded by the integrable function $|x(t)|$. Thus, $x(t)e^{-j\omega t}$ is integrable.

The L^2 Fourier Transform

The L^1 Fourier transform lacks the convenient property of time-frequency symmetry. For example, even though $x(t)$ is in L^1, $X(\omega)$ may not be in L^1. Also, even though $x(t)$ may not be contiuous, $X(\omega)$ is necessarily continuous. The space L^2 is much easier to work with. Not only can we talk about inner products and orthonormal bases, perfect symmetry also exists between time and frequency domains. We must define the L^2-FT differently because the usual definition (6.53) is meaningful only for L^1 signals. Suppose $x(t) \in L^2$ and we truncate it to the interval $[-n, n]$. This truncated version is in L^1 because of (6.48), and its L^1 FT exists:

$$X_n(\omega) = \int_{-n}^{n} x(t)e^{-j\omega t}\, dt \tag{6.56}$$

It can be shown that $X_n(\omega)$ is in L^2 and that the sequence $\{X_n(\omega)\}$ has a limit in L^2. That is, there exists an L^2 function $X(\omega)$ such that

$$\lim_{n\to\infty} \|X_n(\omega) - X(\omega)\|_2 = 0 \tag{6.57}$$

This limit $X(\omega)$ is defined to be the L^2 FT of $x(t)$. Some of the properties are listed next:

1. $X(\omega)$ is in L^2, and we can compute $x(t)$ from $X(\omega)$ in an entirely analogous manner, namely the L^2 limit of $\int_{-n}^{n} X(\omega)e^{j\omega t}d\omega/2\pi$.
2. If $x(t)$ is in L^1 and L^2, then the above computation gives the same answer as the L^1-FT (6.53) a.e. For example, consider the rectangular pulse $x(t) = 1$ in $[-1, 1]$ and zero otherwise. This is in L^1 and L^2, and the FT using either definition is $X(\omega) = 2\sin\omega/\omega$. This answer is in L^2, but not in L^1. The inverse L^2-FT of $X(\omega)$ is the original $x(t)$.
3. If $x(t) \in L^2$ and $X(\omega) \in L^1$ then the Lebesgue integral $\int_{-\infty}^{\infty} X(\omega)e^{j\omega t}d\omega/2\pi$ exists, and equals $x(t)$ a.e.
4. Parseval's relation holds, i.e., $\sqrt{2\pi}\|x(t)\|_2 = \|X(\omega)\|_2$. Thus, the FT is a linear transformation from L^2 to L^2, which preserves norms except the scale factor $\sqrt{2\pi}$. (Note that this would not make sense if $x(t)$ were only in L^1.) In particular, it is a bounded transformation because the norm $\|X(\omega)\|_2$ in the transform domain is bounded by the norm $\|x(t)\|_2$ in the original domain.
5. Unlike the L^1-FT, the L^2-FT $X(\omega)$ need not be continuous. For example, the impulse response of an ideal low-pass filter (sinc function) is in L^2 and its FT is not continuous.
6. Let $\{f_n(t)\}$ be a sequence in L^2 and let $x(t) = \sum_n c_n f_n(t)$ be a convergent summation (in the L^2 sense). With upper case letters denoting the L^2-FTs, $X(\omega) = \sum_n c_n F_n(\omega)$. This result is obvious for finite summations because of the linearity of the FT. For infinite summations this follows from the property that the L^2-FT is a continuous mapping from L^2 to L^2. (This in turn follows from the result that it is a bounded linear transformation). The continuity allows us to move the FT operation inside the infinite summation.

Thus, complete symmetry exists between the time and frequency domains. The L^2-FT is a one-to-one mapping from L^2 onto L^2. Moreover, because $\sqrt{2\pi}\|x(t)\|_2 = \|X(\omega)\|_2$, it is a norm preserving mapping—one says that the L^2-FT is an isometry from L^2 to L^2.

l^1 Fourier Transform

If a sequence $x(n) \in l^1$ its discrete-time FT $X(e^{j\omega}) = \sum_n x(n)e^{-j\omega n}$ exists, and is the l^1-FT of $x(n)$. It can be shown that $X(e^{j\omega})$ is a continuous function of ω and that $|X(e^{j\omega})|$ is bounded.

Convolutions

Suppose $h(t) \in L^1$ and $x(t) \in L^p$ for some p in $1 \leq p \leq \infty$. The familar convolution integral defined by $(x * h)(t) = \int x(\tau)h(t-\tau)d\tau$ exists for almost all t [23]. If we define a function $y(t)$ to be $x * h$ where it exists and to be zero elsewhere, the result is, in fact, an L^p function. We simply say that the convolution of an L^1 function with an L^p function gives an L^p function. By recalling that an LTI system is stable (i.e., BIBO stable), iff its impulse response is in L^1, we have the following examples:

1. If an L^1 signal is input to a stable LTI system, the output is in L^1. Because the convolution of two L^1 signals is in L^1, the cascade of two stable LTI systems is stable, a readily accepted fact in engineering.
2. If an L^2 signal (finite energy input) is input to a stable LTI system, the output is in L^2.
3. If an L^∞ signal is input to a stable LTI system, the output is in L^∞ (i.e., bounded inputs produce bounded outputs).

If $x(t)$ and $h(t)$ are both in L^1, their convolution $y(t)$ is in L^1, and all three signals have L^1-FT. The **convolution theorem** [23] says that these three are related as $Y(\omega) = H(\omega)X(\omega)$. When signals are not necessarily in L^1 we cannot in general write this, even if convolution might itself be well defined.

Convolution Theorems for L^2 Signals

For all our discussions in the preceding sections, the signals were restricted to be in L^2, but not necessarily in L^1. In fact, even the filters are often only in L^2. For example, ideal bandpass filters (Fig. 6.8) are unstable, and therefore only in L^2. For arbitary L^2 signals $x(t)$ and $h(t)$, the convolution theorem does not hold. We therefore need to better understand L^2 convolution.

Assume that $x(t)$ and $h(t)$ are both in L^2. Their convolution $y(t) = \int x(\tau)h(t-\tau)d\tau$ exists for all t, as the integral is only an inner product in L^2. Using Schwartz inequality [23], we also have $|y(t)| \leq \|x(t)\|_2\|h(t)\|_2$, that is, $y(t) \in L^\infty$. Suppose the filter $h(t)$ has the further property that the frequency response $H(\omega)$ is bounded, i.e., $|H(\omega)| \leq B$ a.e., for some $B < \infty$. Then we can show that $y(t) \in L^2$, and that the convolution theorem holds ($Y(\omega) = H(\omega)X(\omega)$). To prove this, note that

$$y(t) = \int x(\tau)h(t-\tau)d\tau = \frac{1}{2\pi}\int X(\omega)H(\omega)e^{j\omega t}\,d\omega \tag{6.58}$$

from Parseval's relation which holds for L^2 signals [23]. If $|H(\omega)| \leq B$, then $|X(\omega)H(\omega)|^2 \leq B^2|X(\omega)|^2$. Therefore, $|X(\omega)H(\omega)|^2$ is bounded by the integrable function $|X(\omega)|^2$, and is therefore integrable. Thus, $X(\omega)H(\omega) \in L^2$, and the preceding equation establishes that $y(t) \in L^2$. The equation also shows that $y(t)$ and $H(\omega)X(\omega)$ form an L^2-FT pair, so $Y(\omega) = H(\omega)X(\omega)$.

Bounded L^2 Filters

Filters for which $h(t) \in L^2$ and $H(\omega)$ bounded are called **bounded L^2 filters.** The preceding discussion shows that bounded L^2 filters admit the convolution theorem, although arbitrary L^2 filters do not. Another advantage of bounded L^2 filters is that a cascade of two bounded L^2 filters, $h_1(t)$ and $h_2(t)$, is a bounded L^2 filter, just as a cascade of two stable filters would be stable. To see this, note that the cascaded impulse response is the convolution $h(t) = (h_1 * h_2)(t)$. By the preceding discussion, $h(t) \in L^2$, and moreover, $H(\omega) = H_1(\omega)H_2(\omega)$. Clearly, $H(\omega)$ is still bounded. Bounded L^2 filters are therefore very convenient to work with. Fortunately, all filters in the discussion of wavelets and filter banks are bounded L^2 filters, even though they may not be BIBO stable (as are the ideal bandpass filters in Fig. 6.8). We summarize the preceding discussions as follows.

Theorem 6.3 (Convolution of L^2 functions): *We say that $h(t)$ is a bounded L^2 filter if $h(t) \in L^2$ and $|H(\omega)| \leq B < \infty$ a.e.*

1. *Let $x(t) \in L^2$, and let $h(t)$ be a bounded L^2 filter. Then $y(t) = (x * h)(t)$ exists for all t and $y(t) \in L^2$. Moreover, $Y(\omega) = H(\omega)X(\omega)$.*
2. *If $h_1(t)$ and $h_2(t)$ are bounded L^2 filters, then their cascade $h(t) = (h_1 * h_2)(t)$ is a bounded L^2 filter, and $H(\omega) = H_1(\omega)H_2(\omega)$.* □

6.7 Riesz Basis, Biorthogonality, and Other Fine Points

In a finite dimensional space, such as the space of all N-component Euclidean vectors, the ideas of basis and orthonormal basis are easy to appreciate. When we extend these ideas to infinite dimensional spaces (i.e., where the basis $\{g_n(t)\}$ has infinite number of functions), a number of complications and subtleties arise. Our aim is to point these out. References for this section include [5, 15, and 24].

Readers familiar with **Hilbert spaces** will note that the L^2 space is a Hilbert space; all our developments here are valid for any Hilbert space $\mathscr{H}$. Elements in $\mathscr{H}$ (vectors) are typically denoted x, y, etc. When we deal with the Hilbert space L^2, the vectors are functions and are denoted as $x(t)$, $y(t)$, etc. for clarity. Similarly, for the special case of Euclidean vectors we use boldface, e.g., $\mathbf{x}, \mathbf{y}$, etc. The reader not familiar with Hilbert spaces can assume that all discussions are in L^2 and that x is merely a simplification of the notation $x(t)$.

Finite Dimensional Vector Spaces

We first look at the finite dimensional case and then proceed to the infinite dimensional case. Consider an $N \times N$ matrix $\mathbf{F} = [\mathbf{f}_1 \ \mathbf{f}_2 \ \ldots \ \mathbf{f}_N]$. We assume that this is nonsingular, that is, the columns $\mathbf{f}_n$ are linearly independent. These column vectors form a basis for the N-dimensional Euclidean space $\mathscr{C}^N$ of complex N-component vectors. This space is an example of a finite dimensional Hilbert space, with inner product defined as $\langle \mathbf{x}, \mathbf{y} \rangle = \mathbf{y}^\dagger \mathbf{x} = \sum_{n=1}^{N} x_n y_n^*$. The norm $\|\mathbf{x}\|$ induced by this inner product is defined as $\|\mathbf{x}\| = \sqrt{\langle \mathbf{x}, \mathbf{x} \rangle}$. Thus $\|\mathbf{x}\|^2 = \mathbf{x}^\dagger \mathbf{x} = \sum_{n=1}^{N} |x_n|^2$.

Any vector $\mathbf{x} \in \mathscr{C}^N$ can be expressed as $\mathbf{x} = \sum_{n=1}^{N} c_n \mathbf{f}_n$ for some uniquely determined set of scalars c_n. We can abbreviate this as $\mathbf{x} = \mathbf{Fc}$, where $\mathbf{c} = [c_1 \ c_2 \ \ldots \ c_N]^{\mathrm{T}}$. The matrix $\mathbf{F}$ can be regarded as a linear transformation from $\mathscr{C}^N$ to $\mathscr{C}^N$. The nonsingularity of $\mathbf{F}$ means that for every $\mathbf{x} \in \mathscr{C}^N$ we can find a unique $\mathbf{c}$ such that $\mathbf{x} = \mathbf{Fc}$.

Boundedness of *F* and Its Inverse

In practice we have a further requirement that if the norm $\|\mathbf{c}\|$ is "small" then $\|\mathbf{x}\|$ should also be "small", and vice versa. This requirement implies, for example, that if a small error occurs in the transmission or estimate of the vector $\mathbf{c}$, the corresponding error in $\mathbf{x}$ is also small. From the relation $\mathbf{x} = \mathbf{Fc}$ we obtain

$$\|\mathbf{x}\|^2 = \mathbf{x}^\dagger \mathbf{x} = \mathbf{c}^\dagger \mathbf{F}^\dagger \mathbf{F} \mathbf{c} \tag{6.59}$$

Letting λ_M and λ_m denote the maximum and minimum eigenvalues of $\mathbf{F}^\dagger \mathbf{F}$ it then follows that $\|\mathbf{x}\|^2 \geq \lambda_m \|\mathbf{c}\|^2$ and that $\|\mathbf{x}\|^2 \leq \lambda_M \|\mathbf{c}\|^2$. That is,

$$\lambda_m \|\mathbf{c}\|^2 \leq \|\mathbf{x}\|^2 \leq \lambda_M \|\mathbf{c}\|^2 \tag{6.60}$$

with $0 < \lambda_m \leq \lambda_M < \infty$, where $0 < \lambda_m$ follows from the nonsingularity of $\mathbf{F}$. Thus, the transformation $\mathbf{F}$, which converts $\mathbf{c}$ into $\mathbf{x}$, has an amplification factor bounded by λ_M in the sense that $\|\mathbf{x}\|^2 \leq \lambda_M \|\mathbf{c}\|^2$. Similarly, the inverse transformation $\mathbf{G} = \mathbf{F}^{-1}$, which converts $\mathbf{x}$

into $\mathbf{c}$, has amplification bounded by $1/\lambda_m$. Because λ_M is finite, we say that $\mathbf{F}$ is a bounded linear transformation, and because $\lambda_m \neq 0$ we see that the inverse transformation is also bounded.

Using $\mathbf{x} = \sum_n c_n \mathbf{f}_n$ and $\|\mathbf{c}\|^2 = \sum_n |c_n|^2$ we can rewrite the preceding inequality as

$$A \sum_n |c_n|^2 \leq \left\| \sum_n c_n \mathbf{f}_n \right\|^2 \leq B \sum_n |c_n|^2 \tag{6.61}$$

where $A = \lambda_m > 0$ and $B = \lambda_M < \infty$, and all summations are for $1 \leq n \leq N$. Readers familiar with the idea of a Riesz basis in infinite dimensional Hilbert spaces will notice that the above is in the form that agrees with that definition. We will return to this issue later.

Biorthogonality

With $\mathbf{F}^{-1}$ denoted as $\mathbf{G}$, let $\mathbf{g}_n^\dagger$ denote the rows of $\mathbf{G}$:

$$\mathbf{G} = \begin{bmatrix} \mathbf{g}_1^\dagger \\ \mathbf{g}_2^\dagger \\ \vdots \\ \mathbf{g}_N^\dagger \end{bmatrix}, \quad \mathbf{F} = [\, \mathbf{f}_1 \;\; \mathbf{f}_2 \;\; \ldots \;\; \mathbf{f}_N \,] \tag{6.62}$$

The property $\mathbf{GF} = \mathbf{I}$ implies $\mathbf{g}_k^\dagger \mathbf{f}_n = \delta(k-n)$:

$$\langle \mathbf{f}_n, \mathbf{g}_k \rangle = \delta(k-n) \tag{6.63}$$

for $1 \leq k, n \leq N$. Equivalently, $\langle \mathbf{g}_k, \mathbf{f}_n \rangle = \delta(k-n)$.

Two sets of vectors, $\{\mathbf{f}_n\}$ and $\{\mathbf{g}_k\}$, satisfying (6.63) are said to be **biorthogonal.** Because $\mathbf{c} = \mathbf{F}^{-1}\mathbf{x} = \mathbf{Gx}$ we can write the elements of $\mathbf{c}$ as $c_n = \mathbf{g}_n^\dagger \mathbf{x} = \langle \mathbf{x}, \mathbf{g}_n \rangle$. Then $\mathbf{x} = \sum_n c_n \mathbf{f}_n = \sum_n \langle \mathbf{x}, \mathbf{g}_n \rangle \mathbf{f}_n$. Next, $\mathbf{G}^\dagger$ is a nonsingular matrix, therefore, we can use its columns $\mathbf{g}_n$, instead of the columns of $\mathbf{F}$, to obtain a similar development, and express the arbitrary vector $\mathbf{x} \in \mathscr{C}^N$ as $\mathbf{x} = \sum_n \langle \mathbf{x}, \mathbf{f}_n \rangle \mathbf{g}_n$. Summarizing, we have

$$\mathbf{x} = \sum_n \langle \mathbf{x}, \mathbf{g}_n \rangle \mathbf{f}_n = \sum_n \langle \mathbf{x}, \mathbf{f}_n \rangle \mathbf{g}_n \tag{6.64}$$

where the summations are for $1 \leq n \leq N$. By using the expressions $c_n = \langle \mathbf{x}, \mathbf{g}_n \rangle$ and $\mathbf{x} = \sum_n c_n \mathbf{f}_n$ we can rearrange the inequality (6.61) into $B^{-1}\|\mathbf{x}\|^2 \leq \sum_n |\langle \mathbf{x}, \mathbf{g}_n \rangle|^2 \leq A^{-1}\|\mathbf{x}\|^2$. With the columns $\mathbf{g}_n$ of $\mathbf{G}^\dagger$, rather than the columns of $\mathbf{F}$, used as the basis for $\mathscr{C}^N$ we obtain similarly

$$A\|\mathbf{x}\|^2 \leq \sum_n |\langle \mathbf{x}, \mathbf{f}_n \rangle|^2 \leq B\|\mathbf{x}\|^2 \tag{6.65}$$

where $1 \leq n \leq N$, and $A = \lambda_m$, $B = \lambda_M$ again. Readers familiar with the idea of a frame in an infinite dimensional Hilbert space will recognize that the above inequality defines a frame $\{\mathbf{f}_n\}$.

Orthonormality

The basis $\mathbf{f}_n$ is said to be orthonormal if $\langle \mathbf{f}_k, \mathbf{f}_n \rangle = \delta(k-n)$, i.e., $\mathbf{f}_n^\dagger \mathbf{f}_k = \delta(k-n)$. Equivalently, $\mathbf{F}$ is unitary, i.e., $\mathbf{F}^\dagger\mathbf{F} = \mathbf{I}$. In this case the rows of the inverse matrix $\mathbf{G}$ are the quantities $\mathbf{f}_n^\dagger$. But $\mathbf{F}^\dagger\mathbf{F} = \mathbf{I}$, and we have $\lambda_m = \lambda_M = 1$, or $A = B = 1$. With this, (6.60) becomes $\|\mathbf{c}\| = \|\mathbf{x}\|$, or $\sum_n |c_n|^2 = \left\|\sum_n c_n \mathbf{f}_n\right\|^2$. This shows that equation (6.61) is a generalization of the orthonormal situation. Similarly, biorthogonality (6.63) is a generalization of orthonormality.

Basis in Infinite Dimensional Spaces

When the simple idea of a basis in a finite dimensional space (e.g., the Euclidean space $\mathscr{C}^N$) is extended to infinite dimensions, several new issues arise which make the problem nontrivial. Consider the sequence of functions $\{f_n\}, 1 \leq n \leq \infty$ in a Hilbert space $\mathscr{H}$. Because of the infinite range of n, linear combinations of the form $\sum_{n=1}^{\infty} c_n f_n$ must now be considered. The problem that immediately arises is one of convergence. For arbitrary sequences c_n this sum does not converge, so the statement "all linear combinations" must be replaced with something else.[10]

Closure of Span

First define the set of all *finite* linear combinations of the form $\sum_{n=1}^{N} c_n f_n$, where N varies over all integers ≥ 1. This is called the **span of** $\{f_n\}$. Now suppose $x \in \mathscr{H}$ is a vector not necessarily in the span of $\{f_n\}$, but can be approximated as closely as we wish by vectors in the span. In other words, given an $\epsilon > 0$ we can find N and the sequence of constants c_{nN} such that

$$\left\| x - \sum_{n=1}^{N} c_{nN} f_n \right\| < \epsilon \tag{6.66}$$

where $\|x\|$ is the norm defined as $\|x\| = \sqrt{\langle x, x \rangle}$. If we append all such vectors x to the span of $\{f_n\}$ we get the **closure of the span** of $\{f_n\}$.[11] Note that c_{nN} in general depends on ϵ because N depends on ϵ.

Completeness

A sequence of vectors $\{f_n\}$ is said to be **complete** in $\mathscr{H}$ if the closure of the linear span of $\{f_n\}$ equals $\mathscr{H}$. Therefore, any $x \in \mathscr{H}$ can be approximated, as closely as we wish, by finite linear combinations of f_n in the sense of (6.66). This is also expressed by saying that the linear span of $\{f_n\}$ is **dense** in $\mathscr{H}$. Completeness of $\{f_n\}$ in a Hilbert space is equivalent to the statement that the only vector orthogonal to all f_n is the zero vector.

Infinite Summations

When we write $x = \sum_{n=1}^{\infty} c_n f_n$ we mean that the infinite summation converges to x in the norm of $\mathscr{H}$. In other words, given $\epsilon > 0$, there exists n_0 such that

$$\left\| x - \sum_{n=1}^{N} c_n f_n \right\| < \epsilon \qquad \text{for all } N \geq n_0 \tag{6.67}$$

This statement is stronger than saying that x is in the closure of the linear span of $\{f_n\}$. The latter statement only requires (6.66), where N and hence c_{nN} depends on ϵ. In (6.67) $\{c_n\}$ is a fixed sequence.

Linear Independence

Let $\{f_n\}, n = 1, 2, \ldots$ be a sequence of vectors in an infinite dimensional Hilbert space $\mathscr{H}$. Unlike in a finite dimensional space, one must distinguish between several types of linear independence.

[10]Our review uses $1 \leq n \leq \infty$ to be consistent with standard math texts, but all the crucial results hold for doubly infinite sequences and summations, i.e., for the case $-\infty \leq n \leq \infty$. This is what we need in the case of Fourier and wavelet bases; see for example, (6.3) and (6.4).

[11]The term "closure" has its origin from the theory of metric spaces, more generally, topological vector spaces. We do not require the deeper, more general meaning here.

Type 1: $\{f_n\}$ has *finite linear independence* if $\sum_{n=1}^{N} c_n f_n = 0$, for any finite N implies $c_n = 0, 1 \leq n \leq N$.

Type 2: $\{f_n\}$ is *ω-independent* if $\sum_{n=1}^{\infty} c_n f_n = 0$ implies $c_n = 0$ for all n (where the infinite sum is interpreted as explained above).

Type 3: $\{f_n\}$ is *minimal* if none of the f_m is in the closure of the span of the remaining set of f_n.

Type 3 independence implies type 2, which in turn implies type 1. Thus, type 3 is the strongest kind of linear independence. The reason that it is stronger than type 2 is that type 2 implies we cannot have $f_m = \sum_{n \neq m} c_n f_n$. However, for a type 2 independent sequence $\{f_n\}$, it is possible that we can make

$$\left\| f_m - \sum_{\substack{n=1 \\ n \neq m}}^{N} c_{nN} f_n \right\| < \epsilon \tag{6.68}$$

for any given $\epsilon > 0$ by choosing N and c_{nN} properly.[12] Type 3 linear independence prohibits even this. The comments following Example 3 make this distinction more clear. (see p. 178).

Basis or Schauder Basis

A sequence of vectors $\{f_n\}$ in $\mathscr{H}$ is a Schauder basis for $\mathscr{H}$ if any $x \in \mathscr{H}$ can be expressed as $x = \sum_{n=1}^{\infty} c_n f_n$, and the sequence of scalars $\{c_n\}$ is unique for a given x. The second condition can be replaced with the statement that $\{f_n\}$ is ω-independent. A subtle result for Hilbert spaces [24] is that a Schauder basis automatically satisfies minimality (i.e., type 3 independence).

A Schauder basis is ω-independent and complete in the sense defined above. Conversely, ω-independence and completeness do not imply that $\{f_n\}$ is a Schauder basis; completeness only means that we may approximate any vector as closely as we wish in the sense of (6.66), where c_{kN} *depend on* N. In this chapter "independence" (or linear independence) stands for ω-independence. Similarly "basis" stands for Schauder basis unless qualified otherwise.

Riesz Basis

Any basis $\{\mathbf{f}_n\}$ in a finite dimensional space satisfies (6.61), which in turn ensures that the transformation from $\mathbf{x}$ to $\{c_n\}$ and that from $\{c_n\}$ to $\mathbf{x}$ are stable. For a basis in an infinite dimensional space, (6.61) is not automatically guaranteed, as shown by the following example.

Example 2. Let $\{e_n\}$, $1 \leq n \leq \infty$ be an orthonormal basis in a Hilbert space $\mathscr{H}$, and define the sequence $\{f_n\}$ by $f_n = e_n/n$. Then we can show that f_n is still a basis, i.e., it satisfies the definition of a Schauder basis. Suppose we pick $x = \epsilon e_k$ for some k. Then, $x = \sum_n c_n f_n$, with $c_k = \epsilon k$, and $c_n = 0$ for all other n. Thus, $\sum_n |c_n|^2 = \epsilon^2 k^2$ and grows as k increases, although $\|x\| = \epsilon$ for all k. That is, a "small" error in x can become amplified in an unbounded manner. Recall that this could never happen in the finite dimensional case because $A > 0$ in (6.61). For our basis $\{f_n\}$, we can indeed show that no $A > 0$ satisfies (6.61). To see this let $c_n = 0$ for all n, except that $c_k = 1$. Then, $\sum_n c_n f_n = f_k = e_k/k$ and has norm $1/k$. So (6.61) reads $A \leq 1/k^2 \leq B$ for all $k \geq 1$. This is not possible with $A > 0$. □

[12] As we make ϵ increasingly smaller, we may need to change N and *all coefficients* c_{kN}. Therefore, this does not imply $f_m = \sum_{n \neq m} c_n f_n$ for fixed $\{c_n\}$.

If $\{e_n\}$, $1 \leq n \leq \infty$ is an orthonormal basis in an infinite dimensional Hilbert space $\mathscr{H}$, then any vector $x \in \mathscr{H}$ can be expressed uniquely as $x = \sum_{n=1}^{\infty} c_n e_n$ where

$$\|x\|^2 = \sum_{n=1}^{\infty} |c_n|^2.$$

This property automatically ensures the stability of the transformations from x to $\{c_n\}$ and vice versa. The Riesz basis is defined such that this property is made more general.[13]

Definition of a Riesz basis. A sequence $\{f_n\}, 1 \leq n \leq \infty$ in a Hilbert space $\mathscr{H}$ is a Riesz basis if it is *complete* and constants A and B exist such that $0 < A \leq B < \infty$ and

$$A \sum_{n=1}^{\infty} |c_n|^2 \leq \left\| \sum_{n=1}^{\infty} c_n f_n \right\|^2 \leq B \sum_{n=1}^{\infty} |c_n|^2 \tag{6.69}$$

for all choice of c_n satisfying $\sum_n |c_n|^2 < \infty$. □

In a finite dimensional Hilbert space, A and B come from the extreme eigenvalues of a nonsingular matrix $\mathbf{F}^{\dagger}\mathbf{F}$, so automatically $A > 0$ and $B < \infty$. In other words, any basis in a finite dimensional space is a Riesz basis. As Example 2 shows, this may not be the case in infinite dimensions.

Unconditional Basis

It can be shown that the Riesz basis is an **unconditional basis**, that is, any reordering of $\{f_n\}$ is also a basis (and the new c_n are the correspondingly reordered versions). This is a nontrivial statement; an arbitrary (Schauder) basis is not necessarily unconditional. In fact, the space of L^1 functions (which is a Banach space, not a Hilbert space) does not have an unconditional basis.

Role of the Constants *A* and *B*

1. *Strongest linear independence.* The condition $A > 0$ means, in particular, that $\sum_n c_n f_n \neq 0$ unless c_n is zero for all n. This is just ω-independence. Actually the condition $A > 0$ means that the vectors $\{f_n\}$ are independent in the strongest sense (type 3), that is, $\{f_n\}$ is minimal. To see this, assume this is not the case by supposing some vector f_m is in the closure of the span of the others. Then, given arbitrary $\epsilon > 0$ we can find N and c_{nN} satisfying (6.66) with $x = f_m$. Defining $c_n = -c_{nN}$ for $n \neq m$ and $c_m = 1$, (6.69) implies $A\left(1 + \sum_{n \neq m} |c_{nN}|^2\right) \leq \epsilon^2$. Because ϵ is arbitrary, this is not possible for $A > 0$.
2. *Distance between vectors.* The condition $A > 0$ also implies that no two vectors in $\{f_n\}$ can become "arbitrarily close". To see this, choose $c_k = -c_m = 1$ for some k, m and $c_n = 0$ for all other n. Then (6.69) gives $2A \leq \|f_k - f_m\|^2 \leq 2B$. Thus, the distance between any two vectors is at least $\sqrt{2A}$, at most $\sqrt{2B}$.
3. *Bounded basis.* A Riesz basis is a bounded basis in the sense that $\|f_n\|$ cannot get arbitrarily large. In fact, by choosing $c_n = 0$ for all but one value of n, we can see that $0 < A \leq \|f_n\|^2 \leq B < \infty$. That is, the norms of the vectors in the basis cannot become arbitrarily small or large. Note that the basis in Example 2 violates this, because $\|f_n\| = 1/n$. Therefore, Example 2 is only a Schauder basis and not a Riesz basis.

[13]For readers familiar with bounded linear transformations in Hilbert spaces, we state that a basis is a Riesz basis iff it is related to an orthonormal basis via a bounded linear transformation with bounded inverse.

4. *Stability of basis.* The condition $A > 0$ yields $\sum_n |c_n|^2 \leq A^{-1}\|x\|^2$, where $x = \sum_n c_n f_n$. This means that the transformation from the vector x to the sequence $\{c_n\}$ is bounded, so a small error in x is not amplified in an unbounded manner. Similarly, the inequality $\|x\|^2 \leq B\sum_n |c_n|^2$ shows that the role of B is to ensure that the inverse transformation from c_n to x is bounded. Summarizing, the transformation from x to $\{c_n\}$ is numerically stable (i.e., small errors not severely amplified) because $A > 0$, and the reconstruction of x from $\{c_n\}$ is numerically stable because $B < \infty$.
5. *Orthonormality.* For a Riesz basis with $A = B = 1$ the condition (6.69) reduces to $\sum_n |c_n|^2 = \|\sum_n c_n f_n\|^2$. It can be shown that such a Riesz basis is simply an orthonormal basis. The properties listed above show that the Riesz basis is as good as an orthonormal basis in most applications. Any Riesz basis can be obtained from an orthonormal basis by means of a bounded linear transformation with a bounded linear inverse.

Example 3, Mishaps with a System That Is Not a Riesz Basis. Let us modify Example 2 to $f_n = (e_n/n) + e_1$, $n \geq 1$, where $\{e_n\}$ is an orthonormal basis. As $n \to \infty$ the vectors f_n move arbitrarily closer together (although $\|f_n\|$ approaches unity from above). Formally, $f_n - f_m = (e_n/n) - (e_m/m)$, so $\|f_n - f_m\|^2 = (1/n^2) + (1/m^2)$, which goes to zero as $n, m \to \infty$. Thus there does not exist $A > 0$ satisfying (6.69) (because of comment 2 above). This, then, is not a Riesz basis; in fact, this is not even a Schauder basis (see below). This example also has $B = \infty$. To see this let $c_n = 1/n$, then $\sum_n |c_n|^2$ converges, but $\|\sum_{n=1}^N c_n f_n\|^2$ does not converge as $N \to \infty$ (as we can verify), so (6.69) is not satisfied for finite B. *Such mishaps cannot occur with a Riesz basis.* □

In this example $\{f_n\}$ is not minimal (which is type 3 independence). Note that $\|f_1 - f_n\|$ gets arbitrarily small as n increases to infinity, therefore, f_1 is in the closure of the span of $\{f_n\}, n \neq 1$. However, $\{f_n\}$ is ω-independent; no sequence $\{c_n\}$ exists such that $\|\sum_{n=1}^N c_n f_n\| \to 0$ as $N \to \infty$. In any case, the fact that $\{f_n\}$ is not minimal (i.e., not independent in the strongest sense) shows that it is not even a Schauder basis.

Biorthogonal Systems, Riesz Bases, and Inner Products

When discussing finite dimensional Hilbert spaces we found that given a basis $\mathbf{f}_n$ (columns of a nonsingular matrix) we can express any vector $\mathbf{x}$ as a linear combination $\mathbf{x} = \sum_n \langle \mathbf{x}, \mathbf{g}_n \rangle \mathbf{f}_n$, where $\mathbf{g}_n$ is such that the biorthogonality property $\langle \mathbf{f}_m, \mathbf{g}_n \rangle = \delta(m-n)$ holds. A similar result is true for infinite dimensional Hilbert spaces.

Theorem 6.4, Biorthogonality and Riesz Basis: *Let $\{f_n\}$ be a basis in a Hilbert space $\mathcal{H}$. Then there exists a unique sequence $\{g_n\}$ biorthogonal to $\{f_n\}$, that is,*

$$\langle f_m, g_n \rangle = \delta(m-n) \quad \text{(biorthogonality)} \tag{6.70}$$

Moreover, the unique expansion of any $x \in \mathcal{H}$ in terms of the basis $\{f_n\}$ is given by

$$x = \sum_{n=1}^{\infty} \langle x, g_n \rangle f_n \tag{6.71}$$

It is also true that the biorthogonal sequence $\{g_n\}$ is a basis and that $x = \sum_{n=1}^{\infty} \langle x, f_n \rangle g_n$. Moreover, if $\{f_n\}$ is a Riesz basis, then $\sum_n |\langle x, g_n \rangle|^2$ and $\sum_n |\langle x, f_n \rangle|^2$ are finite, and we have

$$A\|x\|^2 \leq \sum_{n=1}^{\infty} |\langle x, f_n \rangle|^2 \leq B\|x\|^2 \tag{6.72}$$

where A and B are the same constants as in the definition (6.69) of a Riesz basis. □

This beautiful result resembles the finite dimensional version, where f_n corresponds to the column of a matrix and g_n corresponds to the rows (conjugated) of the inverse matrix. In this sense we can regard the biorthogonal pair of sequences $\{f_n\}$, $\{g_n\}$ as inverses of each other. Both are bases for $\mathscr{H}$. A proof of the above result can be obtained by combining the ideas on pp. 28 to 32 of [24]. The theorem implies, in particular, that if $\{f_n\}$ is a Riesz basis, then any vector in the space can be written in the form $\sum_{n=1}^{\infty} c_n f_n$, where $c_n \in l^2$.

Summary of Riesz Basis

The Riesz basis $\{f_n\}$ in a Hilbert space is a complete set of vectors, linearly independent in the strongest sense (i.e., type 3 or minimal). It is a bounded basis with bounded inverse. Any two vectors are separated by at least $\sqrt{2A}$, that is, $\|f_n - f_m\|^2 \geq 2A$. The norm of each basis vector is bounded as $\|f_n\| \leq \sqrt{B}$. In the expression $x = \sum_n c_n f_n$ the computation of x from c_n as well as the computation of c_n from x are numerically stable because $B < \infty$ and $A > 0$, respectively. A Riesz basis with $A = B = 1$ is an orthonormal basis. In fact, any Riesz basis can be obtained from an orthonormal basis via a bounded linear transformation with a bounded inverse. Given any basis $\{f_n\}$ in a Hilbert space, a unique biorthogonal sequence $\{g_n\}$ exists such that we can express any $x \in \mathscr{H}$ as $x = \sum_{n=1}^{\infty} \langle x, g_n \rangle f_n$ as well as $x = \sum_{n=1}^{\infty} \langle x, f_n \rangle g_n$; if this basis is also a Riesz basis then $\sum_n |\langle x, f_n \rangle|^2$ and $\sum_n |\langle x, g_n \rangle|^2$ are finite. If $\{f_n\}$ is a Riesz basis, then any vector $x \in \mathscr{H}$ can be written in the form $x = \sum_{n=1}^{\infty} c_n f_n$, where $c_n \in l^2$.

6.8 Frames in Hilbert Spaces

A frame in a Hilbert space $\mathscr{H}$ is a sequence of vectors $\{f_n\}$ with certain special properties. While a frame is not necessarily a basis, it shares some properties of a basis. For example, we can express any vector $x \in \mathscr{H}$ as a linear combination of the frame elements, i.e., $x = \sum_n c_n f_n$. However, frames generally have redundancy—the frame vectors are not necessarily linearly independent, even in the weakest sense defined in Section 6.7. The Riesz basis (hence, any orthonormal basis) is a special case of frames. The concept of a frame is useful when discussing the relation between wavelets, STFTs, and filter banks. Frames were introduced by Duffin and Schaeffer [25], and used in the context of wavelets and STFT by Daubechies [5]. Excellent tutorials can be found in [12] and [24].

Definition of a Frame

A sequence of vectors $\{f_n\}$ in a (possibly infinite dimensional) Hilbert space $\mathscr{H}$ is a frame if there exist constants A and B with $0 < A \leq B < \infty$ such that for any $x \in \mathscr{H}$ we have

$$A\|x\|^2 \leq \sum_{n=1}^{\infty} |\langle x, f_n \rangle|^2 \leq B\|x\|^2 \tag{6.73}$$

The constants A and B are called frame bounds. □

In Section 6.7 we saw that a Riesz basis, which by definition satisfies (6.69), also satisfies (6.72), which is precisely the frame definition. A Riesz basis is, therefore, also a frame, but it is a special case of a frame, where the set of vectors is minimal.

Any frame is complete. That is, if a vector $x \in \mathscr{H}$ is orthogonal to all elements in $\{f_n\}$, then $x = 0$, otherwise $A > 0$ is violated. Thus any $x \in \mathscr{H}$ is in the closure of the span of the frame. In fact, we will see that more is true; for example, we can express $x = \sum c_n f_n$, although $\{c_n\}$ may not be unique. The frame elements are not necessarily linearly independent, as demonstrated by examples below. A frame, then, is not necessarily a basis. Compare (6.73) to the Riesz basis

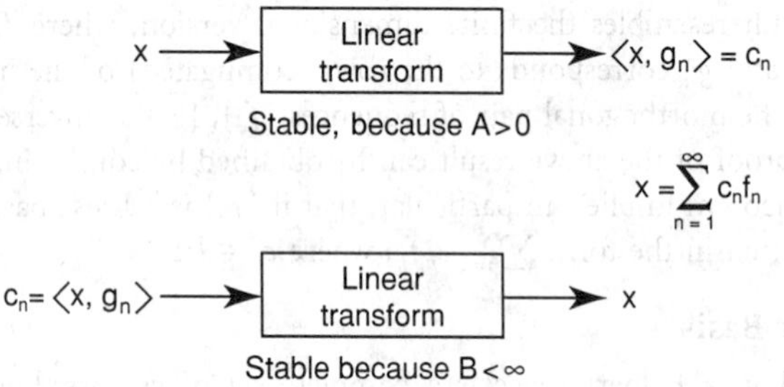

FIGURE 6.29 Representation of x using frame elements $\{f_n\}$. The transformation from x to $\{c_n\}$ and vice versa are stable.

definition (6.69), where the left inequality forced the vectors f_n to be linearly independent (in fact, minimal). The left inequality for a frame only ensures completeness, not linear independence.

Representing Arbitrary Vectors in Terms of Frame Elements

We will see that, given a frame $\{f_n\}$ we can associate with it another sequence $\{g_n\}$ called the dual frame, such that any element $x \in \mathscr{H}$ can be represented as $x = \sum_{n=1}^{\infty} \langle x, g_n \rangle f_n$. We also can write $x = \sum_{n=1}^{\infty} \langle x, f_n \rangle g_n$. This representation in terms of $\{f_n\}$ and $\{g_n\}$ resembles the biorthogonal system discussed in Section 6.7, but some differences are pointed out later.

Stability of Computations

To obtain the representation $x = \sum_{n=1}^{\infty} \langle x, f_n \rangle g_n$ we compute (at least conceptually) the coefficients $\langle x, f_n \rangle$ for all n. This computation is a linear transformation from $\mathscr{H}$ to the space of sequences. The inverse transform computes x from this sequence by using the formula $x = \sum_{n=1}^{\infty} \langle x, f_n \rangle g_n$. The condition $B < \infty$ in the frame definition ensures that the transformation from x to $\langle x, f_n \rangle$ is bounded. Similarly, the condition $A > 0$ ensures that the inverse transformation from $\langle x, f_n \rangle$ to x is bounded. The conditions $A > 0$ and $B < \infty$, therefore, ensure stability; small errors in one domain are not arbitrarily amplified in the other domain. A similar advantage was pointed out earlier for the Riesz basis—for arbitrary bases in infinite dimensional spaces such an advantage cannot be claimed (Example 4).

If we wish to use the dual representation $x = \sum_{n=1}^{\infty} \langle x, g_n \rangle f_n$ instead of $x = \sum_{n=1}^{\infty} \langle x, f_n \rangle g_n$ we must compute $\langle x, g_n \rangle$, etc.; the roles of A and B are taken up by $1/B$ and $1/A$, respectively, and similar discussions hold. This is summarized in Fig. 6.29.

Exact Frames, Tight Frames, Riesz Bases, and Orthonormal Bases

The resemblance between a Riesz basis and a frame is striking. Compare (6.69) to (6.73). One might wonder what the precise relation is. Thus far, we know that a Riesz basis is a frame. To go deeper, we need a definition: a frame $\{f_n\}$ which ceases to be a frame if any element f_k is deleted is said to be an **exact frame**. Such a frame has no redundancy. A frame with $A = B$ is said to be a **tight frame**. The defining property reduces to $\|x\|^2 = A^{-1} \sum_n |\langle x, f_n \rangle|^2$, resembling Parseval's theorem for an orthonormal basis. A frame is *normalized* if $\|f_n\| = 1$ for all n. The following facts concerning exact frames and tight frames are fundamental:

1. A tight frame with $A = B = 1$ and $\|f_n\| = 1$ for all n (i.e., a normalized tight frame with frame bound $= 1$) is an orthonormal basis [5].
2. $\{f_n\}$ is an exact frame iff it is a Riesz basis [24]. Moreover, if a frame is not exact then it cannot be a basis [12]. *Thus, if a frame is a basis it is certainly a Riesz basis.*

3. Because an orthonormal basis is a Riesz basis, a normalized tight frame with frame bound equal to 1 is automatically an exact frame.

Some examples follow which serve to clarify the preceding concepts and definitions. In these examples the sequence $\{e_n\}, n \geq 1$ is an orthonormal basis for $\mathscr{H}$. Thus, $\{e_n\}$ is a tight frame with $A = B = 1$, and $\|e_n\| = 1$.

Example 4. Let $f_n = e_n/n$ as in Example 2. Then $\{f_n\}$ is still a (Schauder) basis for $\mathscr{H}$, but it is not a frame. In fact, this satisfies (6.73) only with $A = 0$; i.e., the inverse transformation (reconstruction) from $\langle x, f_n\rangle$ to x is not bounded. To see why $A = 0$, note that if we let $x = e_k$ for some $k > 0$ then $\|x\| = 1$, whereas $\sum_n |\langle x, f_n\rangle|^2 = 1/k^2$. The first inequality in the frame definition becomes $A \leq 1/k^2$, which cannot be satisfied for all k unless $A = 0$. In this example a finite B works because $|\langle x, f_n\rangle| = |\langle x, e_n\rangle|/n$ for each n. Therefore, $\sum |\langle x, f_n\rangle|^2 \leq \sum |\langle x, e_n\rangle|^2 = \|x\|^2$. □

Example 5. Suppose we modify the above example as follows: define $f_n = (e_n/n) + e_1$. We know that this is no longer a basis (Example 3). We now have $B = \infty$ in the frame definition, so this is not a frame. To verify this, let $x = e_1$ so $\|x\| = 1$. Then $\langle x, f_n\rangle = 1$ for all $n > 1$, so $\sum_n |\langle x, f_n\rangle|^2$ does not converge to a finite value. □

Example 6. Consider the sequence of vectors $\{e_1, e_1, e_2, e_2, \ldots\}$ This is a tight frame with frame bounds $A = B = 2$. Note that even though the vectors are normalized and the frame is tight this is not an orthonormal basis. This has a redundancy of two in the sense that each vector is repeated twice. This frame is not even a basis, therefore, it is not a Riesz basis. □

Example 7. Consider the sequence of vectors $\{e_1, (e_2/\sqrt{2}), (e_2/\sqrt{2}), (e_3/\sqrt{3}), (e_3/\sqrt{3}), (e_3/\sqrt{3}), \ldots\}$. Again, redundancy occurs, so it is not a basis. It is a tight frame with $A = B = 1$, but not an exact frame, and clearly not a basis. It has redundancy (repeated vectors). □

Frame Bounds and Redundancy

For a tight frame with unit norm vectors f_n, the frame bound measures the redundancy. In Example 6 the redundancy is two (every vector repeated twice), and indeed $A = B = 2$. In Example 7, where we still have redundancy, the frame bound $A = B = 1$ does not indicate it. The frame bound of a tight frame measures redundancy only if the vectors f_n have unit norm as in Example 6.

Frame Operator, Dual Frame, and Biorthogonality

The frame operator $\mathscr{F}$ associated with a frame $\{f_n\}$ in a Hilbert space $\mathscr{H}$ is a linear operator defined as

$$\mathscr{F}x = \sum_{n=1}^{\infty} \langle x, f_n\rangle f_n \tag{6.74}$$

The summation can be shown to be convergent by using the definition of the frame. The frame operator $\mathscr{F}$ takes a vector $x \in \mathscr{H}$ and produces another vector in $\mathscr{H}$. The norm of $\mathscr{F}x$ is bounded as follows:

$$A\|x\| \leq \|\mathscr{F}x\| \leq B\|x\| \tag{6.75}$$

The frame operator is a **bounded linear operator** (because $B < \infty$; hence it is a continuous operator [12]. Its inverse is also a bounded linear operator because $A > 0$).

From (6.74) we obtain $\langle \mathscr{F}x, x\rangle = \sum_n |\langle x, f_n\rangle|^2$ by interchanging the inner product with the infinite summation. This is permitted by the continuity of the operator $\mathscr{F}$ and the continuity of inner products; (see Section 6.6). Because $\{f_n\}$ is complete, the right-hand side is positive for $x \neq 0$. Thus, $\langle \mathscr{F}x, x\rangle > 0$ unless $x = 0$, that is, $\mathscr{F}$ is a positive definite operator. The realness of $\langle \mathscr{F}x, x\rangle$ also means that $\mathscr{F}$ is self-adjoint, or $\langle \mathscr{F}x, y\rangle = \langle x, \mathscr{F}y\rangle$ for any $x, y \in \mathscr{H}$.

The importance of the frame operator arises from the fact that if we define $g_n = \mathscr{F}^{-1} f_n$, any $x \in \mathscr{H}$ can be expressed as

$$x = \sum_{n=1}^{\infty} \langle x, g_n\rangle f_n = \sum_{n=1}^{\infty} \langle x, f_n\rangle g_n \tag{6.76}$$

The sequence $\{g_n\}$ is itself a frame in $\mathscr{H}$ called the **dual frame**. It has frame bounds B^{-1} and A^{-1}. Among all representations of the form $x = \sum_n c_n f_n$, the representation $x = \sum_n \langle x, g_n\rangle f_n$ possesses the special property that the energy of the coefficients is minimized, i.e., $\sum_n |\langle x, g_n\rangle|^2 \leq \sum_n |c_n|^2$ with equality iff $c_n = \langle x, g_n\rangle$ for all n [12]. As argued earlier, the computation of $\langle x, f_n\rangle$ from x and the inverse computation of x from $\langle x, f_n\rangle$ are numerically stable operations because $B < \infty$ and $A > 0$, respectively.

For the special case of a tight frame ($A = B$), the frame operator is particularly simple. We have $\mathscr{F}x = Ax$, and so $g_n = \mathscr{F}^{-1} f_n = f_n/A$. Any vector $x \in \mathscr{H}$ can be expressed as

$$x = \frac{1}{A}\sum_{n=1}^{\infty} \langle x, f_n\rangle f_n \quad \text{(tight frames)} \tag{6.77}$$

Notice also that (6.73) gives

$$\sum_{n=1}^{\infty} |\langle x, f_n\rangle|^2 = A\|x\|^2 \quad \text{(tight frames)} \tag{6.78}$$

For a tight frame with $A = 1$, the above equations resemble the representation of x using an orthonormal basis, even though such a tight frame is not necessarily a basis because of possible redundancy (Example 7).

Exact Frames and Biorthogonality

For the special case of an exact frame (i.e., a Riesz basis) the sequence $\{f_n\}$ is minimal, and it is biorthogonal to the dual frame sequence $\{g_n\}$. This is consistent with our observation at the end of Section 6.7.

Summary of Frames

A sequence of vectors $\{f_n\}$ in a Hilbert space $\mathscr{H}$ is a frame if there exist constants $A > 0$ and $B < \infty$ such that (6.73) holds for every vector $x \in \mathscr{H}$. Frames are complete (because $A > 0$), but not necessarily linearly independent. The constants A and B are called the frame bounds. A frame is *tight* if $A = B$. A tight frame with $A = B = 1$ and with normalized vectors ($\|f_n\| = 1$) is an orthonormal basis. For a tight frame with $\|f_n\| = 1$, the frame bound A measures redundancy. Any vector $x \in \mathscr{H}$ can be expressed in either of the two ways shown in (6.76). Here, $g_n = \mathscr{F}^{-1} f_n$, where $\mathscr{F}$ is the *frame operator* defined in (6.74). The frame operator is a bounded linear operator and is self-adjoint (in fact, positive). The sequence $\{g_n\}$ is the *dual frame* and has frame bounds B^{-1} and A^{-1}. For a tight frame the frame representation reduces to (6.77). A frame is *exact* if deletion of any vector f_m destroys the frame property. A sequence $\{f_n\}$ is an exact frame iff it is a Riesz basis. An exact frame $\{f_n\}$ is biorthogonal to the dual frame $\{g_n\}$.

Figure 6.30 is a Venn diagram, which shows the classification of frames and bases and the relationship between these.

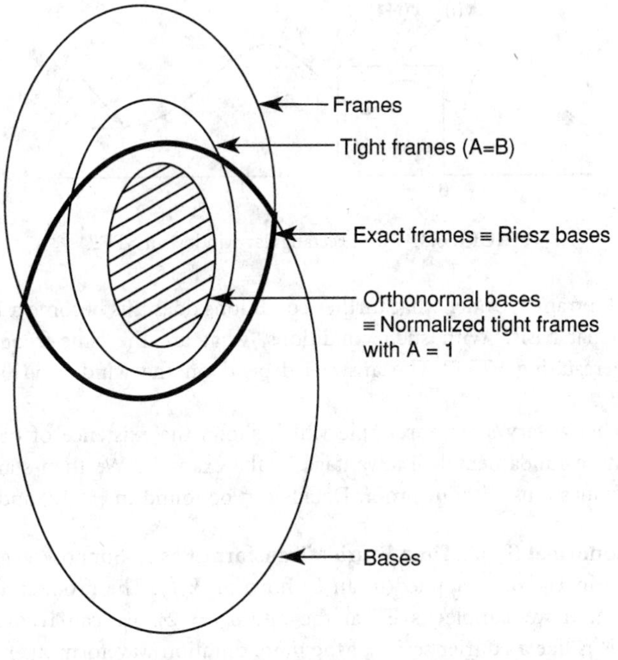

FIGURE 6.30 A Venn diagram showing the relation between frames and bases in a Hilbert space.

6.9 Short-Time Fourier Transform: Invertibility, Orthonormality, and Localization

In Section 6.8 we saw that a vector x in an infinite dimensional Hilbert space (e.g., a function $x(t)$ in L^2) can be expanded in terms of a sequence of vectors $\{f_n\}$ called a frame, that is $x = \sum_{n=1}^{\infty} \langle x, g_n \rangle f_n$. One of the most important features of frames is that the construction of the expansion coefficients $\langle x, g_n \rangle$ from x as well as the reconstruction of x from these coefficients are numerically stable operations because $A > 0$ and $B < \infty$ (see Section 6.8). Riesz and orthonormal bases, which are special cases of a frame (Fig. 6.30), also share this numerical stability.

In Section 6.3 we attempted to represent an L^2 function in terms of the short time Fourier transform (STFT). The STFT coefficients are constructed using the integral (6.24). Denote for simplicity

$$g_{kn}(t) = v^*(t - nT_s)e^{jk\omega_s t} \tag{6.79}$$

The computation of the STFT coefficients can be written as

$$X_{\text{STFT}}(k\omega_s, nT_s) = \left\langle x(t), g_{kn}(t) \right\rangle \tag{6.80}$$

This is a linear transformation which converts $x(t)$ into a two-dimensional sequence because k and n are integers. Our hope is to be able to reconstruct $x(t)$ using an inverse linear transformation (inverse STFT) of the form

$$x(t) = \sum_{k=-\infty}^{\infty} \sum_{n=-\infty}^{\infty} X_{\text{STFT}}(k\omega_s, nT_s) f_{kn}(t) \tag{6.81}$$

We know that this can be done in a numerically stable manner if $\{g_{kn}(t)\}$ is a frame in L^2 and $\{f_{kn}(t)\}$ is the dual frame. The fundamental questions, then, are under what conditions does

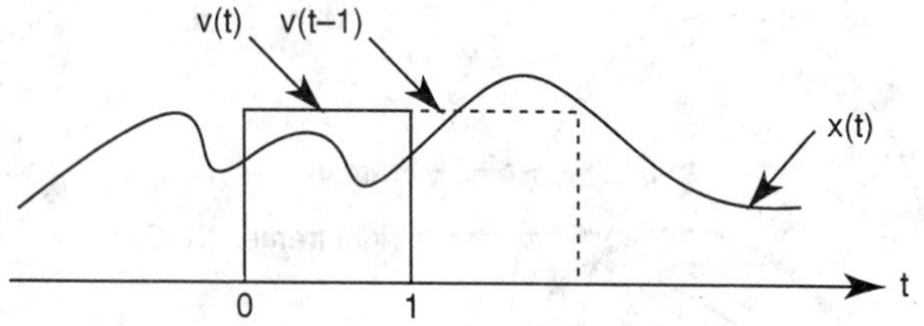

FIGURE 6.31 The rectangular window in STFT.

$\{g_{kn}(t)\}$ constitute a frame? Under what further conditions does this become a Riesz basis, better still, an orthonormal basis? With such conditions, what are the time-frequency localization properties of the resulting STFT? The answers depend on the window $v(t)$ and the sample spacings ω_s and T_s.

We first construct a very simple example which shows the existence of orthonormal STFT bases, and indicate a fundamental disadvantage in the example. We then state the answers to the above general questions without proof. Details can be found in [5, 12, and 16].

Example 8, Orthonormal Short-Time Fourier Transform Basis. Suppose $v(t)$ is the rectangular window shown in Fig. 6.31, applied to an L^2 function $x(t)$. The product $x(t)v(t)$ therefore has finite duration. If we sample its FT at the rate $\omega_s = 2\pi$ we can recover $x(t)v(t)$ from these samples (this is like a Fourier series of the finite duration waveform $x(t)v(t)$). Shifting the window by successive integers (i.e., $T_s = 1$), we can in this way recover successive pieces of $x(t)$ from the STFT, with sample spacing $\omega_s = 2\pi$ in the frequency domain. Thus, the choice $T_s = 1$ and $\omega_s = 2\pi$ (so, $\omega_s T_s = 2\pi$) leads to an STFT $X_{\text{STFT}}(k\omega_s, nT_s)$, from which we can reconstruct $x(t)$ for all t. The quantity $g_{kn}(t)$ becomes

$$g_{kn}(t) = v(t-n)e^{jk\omega_s t} = v(t-n)e^{j2\pi kt} \tag{6.82}$$

Because the successive shifts of the window do not overlap, the functions $g_{kn}(t)$ are orthonormal for different values of n. The functions are also orthonormal for different values of k. Summarizing, the rectangular window of Fig. 6.31, with the time-frequency sampling durations $T_s = 1$ and $\omega_s = 2\pi$, produces an orthonormal STFT basis for L^2 functions. □

This example is reminiscent of the Nyquist sampling theorem in the sense that we can reconstruct $x(t)$ from (time-frequency) samples, but the difference is that $x(t)$ is an L^2 signal, not necessarily band-limited. Note that T_s and ω_s cannot be arbitrarily interchanged (even if $\omega_s T_s = 2\pi$ is preserved). Thus, if we had chosen $T_s = 2$ and $\omega_s = \pi$ (preserving the product $\omega_s T_s$) we would not have obtained a basis because two successive positions of the window would be spaced too far apart and we would miss 50% of the signal $x(t)$.

Time-Frequency Sampling Density for Frames and Orthonormal Bases

Let us assume that $v(t)$ is normalized to have unit energy, i.e., $\int |v(t)|^2\,dt = 1$ so that $\|g_{kn}(t)\| = 1$ for all k, n. If we impose the condition that $g_{kn}(t)$ be a frame, then it can be shown that the frame bounds satisfy the condition

$$A \le \frac{2\pi}{\omega_s T_s} \le B \tag{6.83}$$

regardless of how $v(t)$ is chosen. As an orthonormal basis is a tight frame with $A = B = 1$, *an STFT orthonormal basis must therefore have* $\omega_s T_s = 2\pi$.

It can further be shown that if $\omega_S T_S > 2\pi$, $\{g_{kn}(t)\}$ cannot be a frame. For $\omega_s T_s < 2\pi$ we can find frames (but not orthonormal basis) by appropriate choice of window $v(t)$. The critical time-frequency sampling density is $(\omega_s T_s)^{-1} = (2\pi)^{-1}$. If the density is smaller we cannot have frames, and if it is larger we cannot have orthonormal basis, only frames.

Orthonormal Short-Time Fourier Transform Bases have Poor Time-Frequency Localization

If we wish to have an orthonormal STFT basis, the time-frequency density is constrained so that $\omega_s T_s = 2\pi$. Under this condition suppose we choose $v(t)$ appropriately to design such a basis. The time-frequency localization properties of this system can be judged by computing the mean square durations D_t^2 and D_f^2 defined in (6.27). It has been shown by Balian and Low [5, 16] that one of these is necessarily infinite no matter how $v(t)$ is designed. Thus, an orthonormal STFT basis always satisfies $D_t D_f = \infty$. That is, either the time localization or the frequency resolution is very poor. This is summarized in the following theorem.

Theorem 6.5: *Let the window $v(t)$ be such that $\{g_{kn}(t)\}$ in (6.79) is an orthonormal basis for L^2 (which means, in particular that $\omega_s T_s = 2\pi$). Define the RMS durations D_t and D_f for the window $v(t)$ as usual (6.27). Then, either $D_t = \infty$ or $D_f = \infty$.* □

Return now to Example 8, where we constructed an orthonormal STFT basis using the rectangular window of Fig. 6.31. Here, $T_s = 1$ and $\omega_s = 2\pi$ so that $\omega_s T_s = 2\pi$. The window $v(t)$ has finite mean square duration D_t^2. Its FT $V(\omega)$ has magnitude $|V(\omega)| = |\sin(\omega/2)/(\omega/2)|$ so that $\int \omega^2 |V(\omega)|^2 d\omega$ is not finite. This demonstrates the result of Theorem 6.5. One can try to replace the window $v(t)$ with something for which $D_t D_f$ is finite, but this cannot be done without violating orthonormality.

Instability of the Gabor Transform

Gabor constructed the STFT using the Gaussian window $v(t) = ce^{-t^2/2}$. In this case the sequence of functions $\{g_{kn}(t)\}$ can be shown to be complete in L^2 (in the sense defined in Section 6.7) as long as $\omega_s T_s \leq 2\pi$. However, if $\omega_s T_s = 2\pi$ the system is not a frame because it can be shown that $A = 0$ in (6.73). Thus, the reconstruction of $x(t)$ from $X_{\text{STFT}}(k\omega_s, nT_s)$ is unstable if $\omega_s T_s = 2\pi$ (see Section 6.8), even though $\{g_{kn}(t)\}$ is complete. Although the Gabor transform has the ideal time frequency localization (minimum $D_t D_f$), it cannot provide a *stable basis*, hence, it is certainly not an orthonormal basis, whenever $\omega_s T_s = 2\pi$.

Because orthonormal STFT basis is not possible if $\omega_s T_s \neq 2\pi$, this shows that an orthonormal basis can never be achieved with the Gabor transform (Gaussian windowed STFT), no matter how we choose ω_s and T_s. The Gabor example also demonstrates the fact that even if we successfully construct a complete set of functions (not necessarily a basis) to represent $x(t)$, it may not be useful because of the instabilty of reconstruction. If we construct Riesz bases (e.g., orthonormal bases) or more generally frames, this disadvantage disappears. For example, with the Gabor transform if we let $\omega_s T_s < 2\pi$ then all is well. We obtain a frame (so $A > 0$ and $B < \infty$ in (6.73)); we have stable reconstruction and good time frequency localization, but not orthonormality. Figure 6.32 summarizes these results pertaining to the time-frequency product $\omega_s T_s$ in the STFT.

A major advantage of the WT over the STFT is that it is free from the above difficulties. For example, we can obtain an orthonormal basis for L^2 with excellent time-frequency localization (finite, controllable $D_t D_f$). We will also see how to constrain such a wavelet $\psi(t)$ to have the additional property of **regularity** or smoothness. Regularity is a property which is measured by the continuity and differentiability of $\psi(t)$. More precisely, it is quantified by the Hölder index (defined in Section 6.13). In the next few sections where we construct wavelets based on paraunitary filter banks, we will see how to achieve all this systematically.

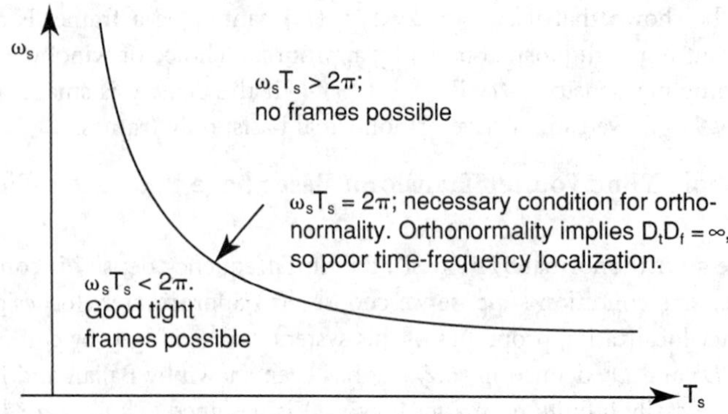

FIGURE 6.32 Behavior of STFT representations for various regions of time-frequency sampling product $\omega_s T_s$. The curve $\omega_s T_s = 2\pi$ is critical; see text.

6.10 Wavelets and Multiresolution

Section 6.11 to 6.13 show how to construct compactly supported wavelets systematically to obtain orthonormal bases for L^2. The construction is such that excellent time-frequency localization is possible. Moreover, the smoothness or regularity of the wavelets can be controlled. The construction is based on the two-channel paraunitary filter bank described in Section 6.4. In that section, we denoted the synthesis filters as $G_s(z)$ and $H_s(z)$, with impulse responses $g_s(n)$ and $h_s(n)$, respectively.

All constructions are based on obtaining the wavelet $\psi(t)$ and an auxiliary function $\phi(t)$, called the scaling function, from the impulse response sequences $g_s(n)$ and $h_s(n)$. We do this by using time domain recursions of the form

$$\phi(t) = 2\sum_{n=-\infty}^{\infty} g_s(n)\phi(2t-n) \qquad \psi(t) = 2\sum_{n=-\infty}^{\infty} h_s(n)\phi(2t-n) \tag{6.84}$$

called **dilation equations.** Equivalently, in the frequency domain

$$\Phi(\omega) = G_s(e^{j\omega/2})\Phi(\omega/2) \qquad \Psi(\omega) = H_s(e^{j\omega/2})\Phi(\omega/2) \tag{6.85}$$

If $\{G_s(z), H_s(z)\}$ is a paraunitary pair with further mild conditions (e.g., that the low-pass filter $G_s(e^{j\omega})$ has a zero at π and no zeroes in $[0, \pi/3]$) the preceding recursions can be solved to obtain $\psi(t)$, which gives rise to an orthonormal wavelet basis $\{2^{k/2}\psi(2^k t - n)\}$ for L^2. By constraining $G_s(e^{j\omega})$ to have a sufficient number of zeroes at π we can further control the Hölder index (or regularity) of $\psi(t)$ (see Section 6.13).

Our immediate aim is to explain the occurrence of the function $\phi(t)$, and the curious recursions (6.84) called the *dilation equations* or **two-scale equations.** These have origin in the beautiful theory of multiresolution for L^2 spaces [4, 11]. Because multiresolution theory lays the foundation for the construction of the most practical wavelets to date, we give a brief description of it here.

The Idea of Multiresolution

Return to Fig. 6.13(a), where we interpreted the wavelet transformation as a bank of continuous time analysis filters followed by samplers, and the inverse transformation as a bank of synthesis filters. Assume for simplicity the filters are ideal bandpass. Figure 6.13(b) is a sketch of the frequency responses. The bandpass filters $F_k(\omega) = 2^{-k/2}\Psi(\omega/2^k)$ become increasingly narrow as k decreases (i.e., as k becomes more and more negative). Instead of letting k be negative, suppose we keep only $k \geq 0$ and include a low-pass filter $\Phi(\omega)$ to cover the low frequency region. Then we get the picture of Fig. 6.33. This is analogous to Fig. 6.12, where we used the pulse function $\phi(t)$ instead of using negative k in $\psi(2^k t - n)$.

Imagine for a moment that $\Phi(\omega)$ is an ideal low-pass filter with cutoff $\pm\pi$. Then we can represent any L^2 function $F(\omega)$ with support restricted to $\pm\pi$ in the form $F(\omega) = \sum_{n=-\infty}^{\infty} a_n \Phi(\omega) e^{-j\omega n}$. This is simply the FS expansion of $F(\omega)$ in $[-\pi, \pi]$, and it follows that $\sum_n |a_n|^2 < \infty$ (Theorem 6.1). In the time domain this means

$$f(t) = \sum_{n=-\infty}^{\infty} a_n \phi(t-n) \tag{6.86}$$

Let us denote by V_0 the closure of the span of $\{\phi(t-n)\}$. Thus, V_0 is the class of L^2 signals that are band-limited to $[-\pi, \pi]$. Because $\phi(t)$ is the sinc function, the shifted functions $\{\phi(t-n)\}$ form an orthonormal basis for V_0.

Consider now the subspace $W_0 \subset L^2$ of bandpass functions band-limited to $\pi < |\omega| \leq 2\pi$. The bandpass sampling theorem (Section 6.2) allows us to reconstruct such a bandpass signal $g(t)$ from its samples $g(n)$ by using the ideal filter $\Psi(\omega)$. Denoting the impulse response of $\Psi(\omega)$ by $\psi(t)$ we see that $\{\psi(t-n)\}$ spans W_0. It can be verified that $\{\psi(t-n)\}$ is an orthonormal basis for W_0. Moreover, as $\Psi(\omega)$ and $\Phi(\omega)$ do not overlap, it follows from Parseval's theorem that W_0 is orthogonal to V_0.

Next, consider the space of all signals of the form $f(t)+g(t)$, where $f(t) \in V_0$ and $g(t) \in W_0$. This space is called the direct sum (in this case, orthogonal sum) of V_0 and W_0, and is denoted as $V_1 = V_0 \oplus W_0$. It is the space of all L^2 signals band-limited to $[-2\pi, 2\pi]$. We can continue in this manner and define the spaces V_k and W_k for all k. Then, V_k is the space of all L^2 signals band-limited to $[-2^k\pi, 2^k\pi]$, and W_k is the space of L^2 functions band-limited to $2^k\pi < |\omega| \leq 2^{k+1}\pi$. The general recursive relation is $V_{k+1} = V_k \oplus W_k$. Figure 6.34 demonstrates this for the case in which the filters are ideal bandpass. Only the positive half of the frequency axis is shown for simplicity.

It is clear that we could imagine V_0 itself to be composed of subspaces V_{-1} and W_{-1}. Thus, $V_0 = V_{-1} \oplus W_{-1}$, $V_{-1} = V_{-2} \oplus W_{-2}$, and so forth. In this way we have defined a sequence of spaces $\{V_k\}$ and $\{W_k\}$ for all integers k such that the following conditions are true:

$$V_{k+1} = V_k \oplus W_k \qquad W_k \perp W_m, \quad k \neq m \tag{6.87}$$

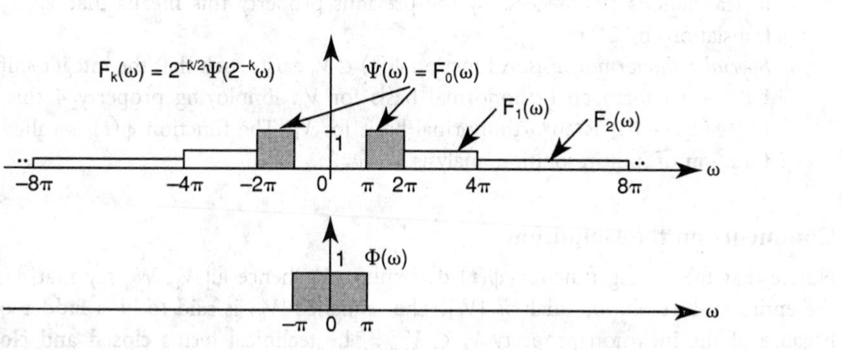

FIGURE 6.33 The low-pass function $\Phi(\omega)$, bandpass function $\Psi(\omega)$, and the stretched bandpass filters $F_k(\omega)$.

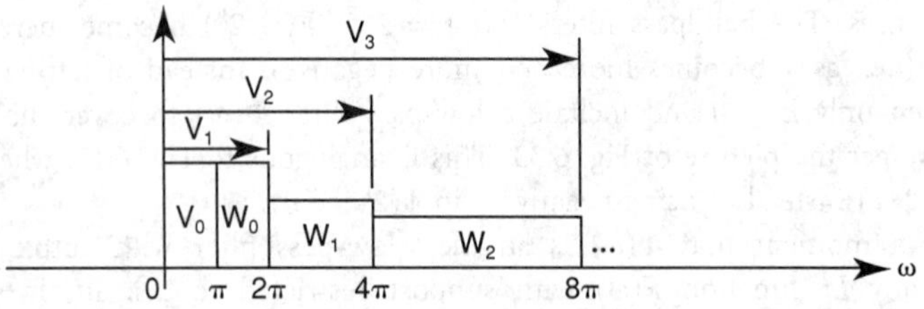

FIGURE 6.34 Toward multiresolution analysis. The spaces $\{V_k\}$ and $\{W_k\}$ spanned by various filter responses.

where $\perp$ means "orthogonal", i.e., the functions in W_k are orthogonal to those in W_m. It is clear that $V_k \subset V_{k+1}$.

We will see later that *even if the ideal filters* $\Phi(\omega)$ *and* $\Psi(\omega)$ *are replaced with nonideal approximations*, we can sometimes define sequences of subspaces V_k and W_k satisfying the above conditions. The importance of this observation is that *whenever* $\Psi(\omega)$ *and* $\Phi(\omega)$ *are such that we can construct such a subspace structure, the impulse response* $\psi(t)$ *of the filter* $\Psi(\omega)$ *can be used to generate an orthonormal wavelet basis.* While this might seem too complicated and convoluted, we will see that the construction of the function $\phi(t)$ is quite simple and elegant, and simplifies the construction of orthonormal wavelet bases. A realization of these ideas based on paraunitary filter banks is presented in Section 6.11. It is now time to be more precise with definitions as well as statements of the results.

Definition of Multiresolution Analysis

Consider a sequence of closed subspaces $\{V_k\}$ in L^2, satisfying the following six properties:

1. *Ladder property.* $\cdots V_{-2} \subset V_{-1} \subset V_0 \subset V_1 \subset V_2 \cdots$.
2. $\bigcap_{k=-\infty}^{\infty} V_k = \{0\}$.
3. Closure of $\bigcup_{k=-\infty}^{\infty} V_k$ is equal to L^2.
4. *Scaling property.* $x(t) \in V_k$ iff $x(2t) \in V_{k+1}$. Because this implies "$x(t) \in V_0$ iff $x(2^k t) \in V_k$", all the spaces V_k are scaled versions of the space V_0. For $k > 0$, V_k is a *finer space* than V_0.
5. *Translation invariance.* If $x(t) \in V_0$, then $x(t-n) \in V_0$; that is, the space V_0 is invariant to translations by integers. By the previous property this means that V_k is invariant to translations by $2^{-k}n$.
6. *Special orthonormal basis.* A function $\phi(t) \in V_0$ exists such that the integer shifted versions $\{\phi(t-n)\}$ form an orthonormal basis for V_0. Employing property 4 this means that $\{2^{k/2}\phi(2^k t - n)\}$ is an orthonormal basis for V_k. The function $\phi(t)$ is called the **scaling function** of multiresolution analysis. □

Comments on the Definition

Notice that the scaling function $\phi(t)$ determines V_0, hence all V_k. We say that $\phi(t)$ generates the entire multiresolution analysis $\{V_k\}$. The sequence $\{V_k\}$ is said to be a **ladder of subspaces** because of the inclusion property $V_k \subset V_{k+1}$. The technical terms **closed** and **closure,** which originate from metric space theory, have simple meanings in our context because L^2 is a Hilbert space. Thus, the subspace V_k is "closed" if the following is true: whenever a sequence of functions

$\{f_n(t)\} \in V_k$ converges to a limit $f(t) \in L^2$ (i.e., $\|f(t) - f_n(t)\| \to 0$ as $n \to \infty$), the limit $f(t)$ is in V_k itself. In general, an infinite union of closed sets is not closed, which is why we need to take "closure" in the third property above. The third property simply means that any element $x(t) \in L^2$ can be approximated arbitrary closely (in the L^2 norm sense) by an element in $\bigcup_{k=-\infty}^{\infty} V_k$.

General meaning of W_k

In the general setting of the above definition, the subspace W_k is defined as the orthogonal complement of V_k with respect to V_{k+1}. Thus, the relation $V_{k+1} = V_k \oplus W_k$, which was valid in the ideal bandpass case (Fig. 6.34), continues to hold.

Haar Multiresolution

A simple example of multiresolution in which $\Phi(\omega)$ is not ideal low-pass is the Haar multiresolution, generated by the function $\phi(t)$ in Fig. 6.35(a). Here, V_0 is the space of all functions that are piecewise constants on intervals of the form $[n, n+1]$. We will see later that the function $\psi(t)$ associated with this example is as in Fig. 6.35(b)—the space W_0 is spanned by $\{\psi(t-n)\}$. The space V_k contains functions which are constants in $[2^{-k}n, 2^{-k}(n+1)]$. Figure 6.35(c) and (d) show examples of functions belonging to V_0 and V_1. For this example, the six properties in the definition of multiresolution are particularly clear (except perhaps property 3, which also can be proved).

The multiresolution analysis generated by the ideal bandpass filters (Fig. 6.33 and 6.34) is another simple example, in which $\phi(t)$ is the sinc function. We see that the two elementary orthonormal wavelet examples (Haar wavelet and the ideal bandpass wavelet) also generate a corresponding multiresolution analysis. The connection between wavelets and multiresolution is deeper than this, and is elaborated in the following section.

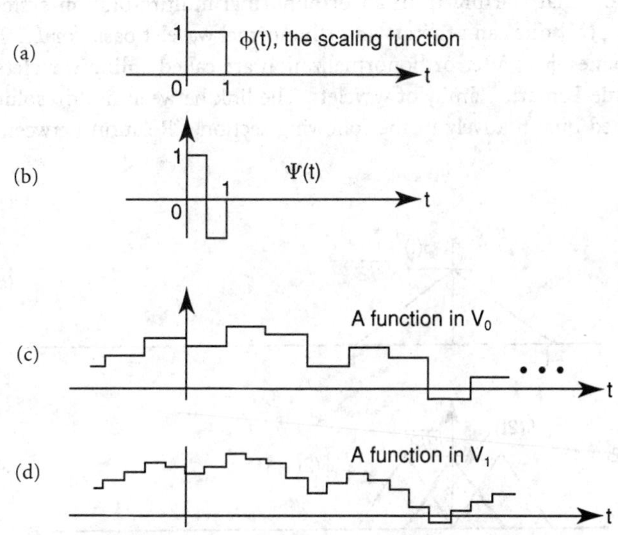

FIGURE 6.35 The Haar multiresolution example. (a) The scaling function $\phi(t)$ that generates multiresolution, (b) the function $\psi(t)$ which generates W_0, (c) example of a member of V_0, and (d) example of a member of V_1.

Derivation of the Dilation Equation

Because $\{\sqrt{2}\phi(2t-n)\}$ is an orthonormal basis for V_1 (see property 6), and because $\phi(t) \in V_0 \subset V_1$, $\phi(t)$ can be expressed as a linear combination of the functions $\{\sqrt{2}\phi(2t-n)\}$:

$$\phi(t) = 2\sum_{n=-\infty}^{\infty} g_s(n)\phi(2t-n) \qquad \text{(dilation equation)} \tag{6.88}$$

Thus, the dilation equation arises naturally out of the multiresolution condition. For example, the Haar scaling function $\phi(t)$ satisfies the dilation equation

$$\phi(t) = \phi(2t) + \phi(2t-1) \tag{6.89}$$

The notation $g_s(n)$ and the factor 2 in the dilation equation might appear arbitrary now, but are convenient for future use. Orthonormality of $\{\phi(t-n)\}$ implies that $\|\phi(t)\| = 1$, and that $\{\sqrt{2}\phi(2t-n)\}$ are orthonormal. Therefore, $\sum_n |g_s(n)|^2 = 0.5$ from (6.88).

Example 9, Nonorthonormal Multiresolution. Consider the triangular function shown in Fig. 6.36(a). This has $\|\phi(t)\| = 1$ and satisfies the dilation equation

$$\phi(t) = \phi(2t) + 0.5\phi(2t-1) + 0.5\phi(2t+1) \tag{6.90}$$

as demonstrated in Fig. 6.36(b). With V_k denoting the closure of the span of $\{2^{k/2}\phi(2^k t - n)\}$ it can be shown that the spaces $\{V_k\}$ satisfy all the conditions in the multiresolution definition, except one. Namely, $\{\phi(t-n)\}$ does not form an orthonormal basis [for example, compare $\phi(t)$ and $\phi(t-1)$]. We will see later (Example 10) that it does form a Riesz basis and that it can be converted into an orthonormal basis by orthonormalization. This example is a special case of a family of scaling functions called **spline functions** [15]. □

We will see below that starting from an orthonormal multiresolution system [in particular from the function $\phi(t)$] one can generate an orthonormal wavelet basis for L^2. The wavelet bases generated from splines $\phi(t)$ after orthonormalization are called **spline wavelets** [15]. These are also called the Battle-Lemarié family of wavelets. The link between multiresolution analysis and wavelets is explained quantitatively in the following section, "Relation between Multiresolution and Wavelets".

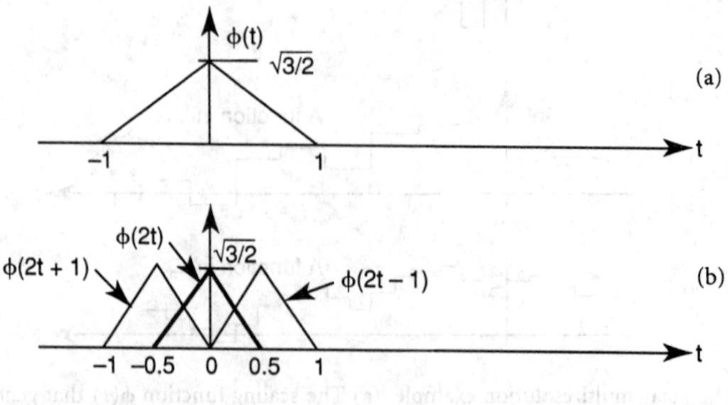

FIGURE 6.36 Example of a scaling function $\phi(t)$ generating nonorthogonal multiresolution. (a) The scaling function, and (b) demonstrating the dilation equation.

Multiresolution Approximation of L^2 Functions

Given a multiresolution analysis, we know that $\bigcap_{k=-\infty}^{\infty} V_k = \{0\}$ and that the closure of $\bigcup_{k=-\infty}^{\infty} V_k = L^2$. From this it can be shown that the W_k make up the entire L^2 space, that is

$$L^2 = \bigoplus_{k=-\infty}^{\infty} W_k \tag{6.91a}$$

We can approximate an arbitrary L^2 function $x(t)$ to a certain degree of accuracy by projecting it onto V_k for appropriate k. Thus, let $x_k(t)$ be this orthogonal projection (see Section 6.2). Suppose we increase k to $k+1$. Because $V_{k+1} = V_k \oplus W_k$ and W_k is orthogonal to V_k, we see that the new approximation $x_{k+1}(t)$ (projection onto the finer space V_{k+1}) is given by $x_{k+1}(t) = x_k(t) + y_k(t)$, where $y_k(t)$ is in W_k.

Thus, when we go from scale k to scale $k+1$ we go to a larger space $V_{k+1} \supset V_k$, which permits a finer approximation. This is nicely demonstrated in the two extreme examples mentioned previously. For the example with ideal filters (Figs. 6.33, 6.34), the process of passing from scale k to $k+1$ is like admitting higher frequency components, which are orthogonal to the existing low-pass components. For the Haar example (Fig. 6.35) where $\psi(t)$ and $\phi(t)$ are square pulses, when we pass from k to $k+1$ we permit finer pulses (i.e., highly localized finer variations in the time domain). For this example, Figs. 6.35(c) and (d) demonstrate the projections $x_k(t)$ and $x_{k+1}(t)$ at two successive resolutions. The projections are piecewise-constant approximations of an L^2 signal $x(t)$.

By repeated application of $V_{k+1} = V_k \oplus W_k$ we can express V_0 as

$$V_0 = \bigoplus_{k=-\infty}^{-1} W_k \tag{6.91b}$$

which, together with (6.91a), yields

$$L^2 = V_0 \oplus W_0 \oplus W_1 \oplus W_2 \oplus \cdots \tag{6.91c}$$

This has a nice interpretation based on Fig. 6.34. The L^2 signal $x(t)$ has been decomposed into orthogonal components belonging to V_0 (low-pass component), W_0 (bandpass component), W_1 (bandpass with higher bandwidth and center frequency), etc.

We can find an infinite number of multiresolution examples by choosing $\phi(t)$ appropriately. It is more important now to obtain systematic techniques for constructing such examples. The quality of the example is governed by the quality of $\psi(t)$ and $\phi(t)$—the time localization and frequency resolution they can provide, the smoothness (regularity) of these functions, and the ease with which we can implement these approximations.

Relation between Multiresolution and Wavelets

Suppose $\phi(t) \in L^2$ generates an orthonormal multiresolution $\{V_k\}$, as defined in the previous section. We know $\phi(t) \in V_0$ and that $\{\phi(t-n)\}$ is an orthonormal basis for V_0. Moreover, $\phi(t)$ satisfies the dilation equation (6.88), and the sequence $\{g_s(n)\} \in \ell^2$ defines the filter $G_s(e^{j\omega})$.

Now consider the finer space $V_1 = V_0 \oplus W_0$, where W_0 is orthonormal to V_0. If $f(t) \in W_0$ then $f(t) \in V_1$, so it is a linear combination of $\sqrt{2}\phi(2t-n)$ (property, 6; see definitions). Using this and the fact that W_0 is orthogonal to V_0 we can show that $F(\omega)$ [the L^2-FT of $f(t)$] has a special form. This is given by

$$F(\omega) = e^{j\omega/2} G_s^*(-e^{j\omega/2}) \Phi(\omega/2) H(e^{j\omega})$$

where $H(e^{j\omega})$ is 2π-periodic. The special case of this with $H(e^{j\omega}) = 1$ is denoted $\Psi(\omega)$; that is,

$$\Psi(\omega) = e^{j\omega/2} G_s^*(-e^{j\omega/2})\Phi(\omega/2). \tag{6.92}$$

The above definition of $\Psi(\omega)$ is equivalent to

$$\psi(t) = 2 \sum_{n=-\infty}^{\infty} (-1)^{n+1} g_s^*(-n-1)\phi(2t-n) \quad \text{[dilation equation for } \psi(t)\text{]} \tag{6.93}$$

The function $\psi(t)$ satisfying this equation has some useful properties. First, it is in L^2. This follows from Theorem 6.2 (Riesz-Fischer Theorem), because $\sum_n |g_s(n)|^2$ is finite. It can be shown that $\psi(t-n) \in W_0$ and that $\{\psi(t-n)\}$ is an orthonormal basis for W_0. This implies that $\{2^{k/2}\psi(2^k t - n)\}$ is an orthonormal basis for W_k because $f(t) \in W_0$ iff $f(2^k t) \in W_k$, which is a property induced by the scaling property (property 4 in the definition of multiresolution). In view of (6.91) we conclude that the sequence $\{2^{k/2}\psi(2^k t - n)\}$, with k and n varying over all integers, forms a basis for L^2. Summarizing we have the following result:

Theorem 6.6 (Multiresolution and Wavelets): *Let $\phi(t) \in L^2$ generate an orthonormal multiresolution, i.e., a ladder of spaces $\{V_k\}$ satisfying the six properties in the definition of multiresolution; $\{\phi(t-n)\}$ is an orthonormal basis for V_0. Then $\phi(t)$ satisfies the dilation equation (6.88) for some $g_s(n)$ with $\sum_n |g_s(n)|^2 = 0.5$. Define the function $\psi(t)$ according to the dilation equation (6.93). Then, $\psi(t) \in W_0 \subset L^2$, and $\{\psi(t-n)\}$ is an orthonormal basis for W_0. Therefore, $\{2^{k/2}\psi(2^k t - n)\}$ is an orthonormal basis for W_k, just as $\{2^{k/2}\phi(2^k t - n)\}$ is an orthonormal basis for V_k (for fixed k). Moreover, with k and n varying over all integers, the doubly indexed sequence $\{2^{k/2}\psi(2^k t - n)\}$ is an orthonormal wavelet basis for L^2.* □

Thus, to construct a wavelet basis for L^2 we have only to construct an orthonormal basis $\{\phi(t-n)\}$ for V_0. Everything else follows from that. All proofs can be found in [5, 11, and 15].

Relation between Multiresolution Analysis and Paraunitary Filter Banks

Denoting

$$h_s(n) = (-1)^{n+1} g_s^*(-1-n), \quad \text{i.e.,} \quad H_s(e^{j\omega}) = e^{j\omega} G_s^*(-e^{j\omega})$$

we see that $\phi(t)$ and $\psi(t)$ satisfy the two dilation equations in (6.84). By construction $\psi(t) \in W_0$ and $\phi(t) \in V_0$. The fact that W_0 and V_0 are mutually orthogonal subspaces can be used to show that $H_s(e^{j\omega})$ and $G_s(e^{j\omega})$ satisfy

$$G_s^*(e^{j\omega})H_s(e^{j\omega}) + G_s^*(-e^{j\omega})H_s(-e^{j\omega}) = 0 \tag{6.94}$$

Also, orthonormality of $\{\phi(t-n)\}$ leads to the power complementary property

$$|G_s(e^{j\omega})|^2 + |G_s(-e^{j\omega})|^2 = 1 \tag{6.95}$$

In other words, $G_s(e^{j\omega})$ is a power symmetric filter. That is, the filter $|G_s(e^{j\omega})|^2$ is a half-band filter. Using $H_s(e^{j\omega}) = e^{j\omega} G_s^*(-e^{j\omega})$, we also have

$$|H_s(e^{j\omega})|^2 + |H_s(-e^{j\omega})|^2 = 1 \tag{6.96}$$

A compact way to express the above three equations is by defining the matrix

$$\mathbf{G}_s(e^{j\omega}) = \begin{bmatrix} G_s(e^{j\omega}) & H_s(e^{j\omega}) \\ G_s(-e^{j\omega}) & H_s(-e^{j\omega}) \end{bmatrix}$$

The three properties (6.94) to (6.96) are equivalent to $\mathbf{G}_s^{\dagger}(e^{j\omega})\mathbf{G}_s(e^{j\omega}) = \mathbf{I}$; i.e., the matrix $\mathbf{G}_s(e^{j\omega})$ is unitary for all ω. This matrix was defined in Section 6.4 in the context of paraunitary digital filter banks. Thus, the filters $G_s(e^{j\omega})$ and $H_s(e^{j\omega})$ constructed from a multiresolution setup constitute a paraunitary (CQF) synthesis bank.

Thus, *orthonormal multiresolution automatically gives rise to paraunitary filter banks.* Starting from a multiresolution analysis we obtained two functions $\phi(t)$ and $\psi(t)$. These functions generate orthonormal bases $\{\phi(t-n)\}$ and $\{\psi(t-n)\}$ for the orthogonal subspaces V_0 and W_0. The functions $\phi(t)$ and $\psi(t)$ generated in this way satisfy the dilation equation (6.84). Defining the filters $G_s(z)$ and $H_s(z)$ from the coefficients $g_s(n)$ and $h_s(n)$ in an obvious way, we find that these filters form a paraunitary pair.

This raises the following fundamental question: If we start from a paraunitary pair $\{G_s(z), H_s(z)\}$ and define the functions $\phi(t)$ and $\psi(t)$ by (successfully) solving the dilation equations, do we obtain an orthonormal basis $\{\phi(t-n)\}$ for multiresolution, and a wavelet basis $\{2^{k/2}\psi(2^k t - n)\}$ for the space of L^2 functions? The answer, fortunately, is in the affirmative, subject to some minor requirements which can be trivially satisfied in practice.

Generating Wavelet and Multiresolution Coefficients from Paraunitary Filter Banks

Recall that the subspaces V_0 and W_0 have the orthonormal bases $\{\phi(t-n)\}$ and $\{\psi(t-n)\}$, respectively. By the scaling property, the subspace V_k has the orthonormal basis $\{\phi_{kn}(t)\}$, and similarly the subspace W_k has the orthonormal basis $\{\psi_{kn}(t)\}$, where, as usual, $\phi_{kn}(t) = 2^{k/2}\phi(2^k t - n)$ and $\psi_{kn}(t) = 2^{k/2}\psi(2^k t - n)$. The orthogonal projections of a signal $x(t) \in L^2$ onto V_k and W_k are given, respectively, by

$$P_k[x(t)] = \sum_{n=-\infty}^{\infty} \langle x(t), \phi_{kn}(t)\rangle \phi_{kn}(t) \quad \text{and}$$
$$Q_k[x(t)] = \sum_{n=-\infty}^{\infty} \langle x(t), \psi_{kn}(t)\rangle \psi_{kn}(t) \tag{6.97}$$

(see Section 6.2). Denote the scale-k projection coefficients as $d_k(n) = \langle x(t), \phi_{kn}(t)\rangle$ and $c_k(n) = \langle x(t), \psi_{kn}(t)\rangle$ for simplicity. (The notation c_{kn} was used in earlier sections, but $c_k(n)$ is convenient for the present discussion.) We say that $d_k(n)$ are the **multiresolution coefficients** at scale k and $c_k(n)$ are the **wavelet coefficients** at scale k.

Assume that the projection coefficients $d_k(n)$ are known for some scale, e.g., $k = 0$. We will show that $d_k(n)$ and $c_k(n)$ for the coarser scales, i.e., $k = -1, -2, \ldots$ can be generated by using a paraunitary analysis filter bank $\{G_a(e^{j\omega}), H_a(e^{j\omega})\}$, corresponding to the synthesis bank $\{G_s(e^{j\omega}), H_s(e^{j\omega})\}$ (Section 6.4). We know $\phi(t)$ and $\psi(t)$ satisfy the dilation equations (6.84). By substituting the dilation equations into the right-hand sides of $\phi_{kn}(t) = 2^{k/2}\phi(2^k t - n)$ and $\psi_{kn}(t) = 2^{k/2}\psi(2^k t - n)$, we obtain

$$\phi_{kn}(t) = \sqrt{2}\sum_{m=-\infty}^{\infty} g_s(m-2n)\phi_{k+1,m}(t) \quad \text{and}$$
$$\psi_{kn}(t) = \sqrt{2}\sum_{m=-\infty}^{\infty} h_s(m-2n)\phi_{k+1,m}(t) \tag{6.98}$$

A computation of the inner products $d_k(n) = \langle x(t), \phi_{kn}(t)\rangle$ and $c_k(n) = \langle x(t), \psi_{kn}(t)\rangle$ yields

$$d_k(n) = \sum_{m=-\infty}^{\infty} \sqrt{2} g_a(2n - m) d_{k+1}(m)$$
$$c_k(n) = \sum_{m=-\infty}^{\infty} \sqrt{2} h_a(2n - m) d_{k+1}(m) \quad (6.99)$$

where $g_a(n) = g_s^*(-n)$ and $h_a(n) = h_s^*(-n)$ are the analysis filters in the paraunitary filter bank.

The beauty of these equations is that they look like **discrete time convolutions.** Thus, if $d_{k+1}(n)$ is convolved with the impulse response $\sqrt{2}g_a(n)$ and the output decimated by 2, the result is the sequence $d_k(n)$. A similar statement follows for $c_k(n)$. The above computation can therefore be interpreted in filter bank form as in Fig. 6.37. Because of the PR property of the two-channel system (Fig. 6.22), it follows that we can reconstruct the projection coefficients $d_{k+1}(n)$ from the projection coefficients $d_k(n)$ and $c_k(n)$.

Fast Wavelet Transform

Repeated application of this idea results in Fig. 6.38, which is a tree-structured paraunitary filter bank (Section 6.4) with analysis filters $\sqrt{2}g_a(n)$ and $\sqrt{2}h_a(n)$ at each stage. Thus, given the projection coefficients $d_0(n)$ for V_0, we can compute the projection coefficients $d_k(n)$ and $c_k(n)$ for the coarser spaces $V_{-1}, W_{-1}, V_{-2}, W_{-2}, \ldots$ This scheme is sometimes referred to as the **fast wavelet transform (FWT)**. Figure 6.39 shows a schematic of the computation. In this figure each node (heavy dot) represents a decimated paraunitary analysis bank $\{\sqrt{2}g_a(n), \sqrt{2}h_a(n)\}$. The subspaces W_m and V_m are indicated in the Figure rather than the projection coefficients.

Computation of the Initial Projection Coefficient. Everything depends on the computation of $d_0(n)$. Note that $d_0(n) = \langle x(t), \phi(t - n)\rangle$, which can be written as the integral $d_0(n) = \int x(t)\phi^*(t - n)\, dt$. An elaborate computation of this integral is avoided in practice. If the

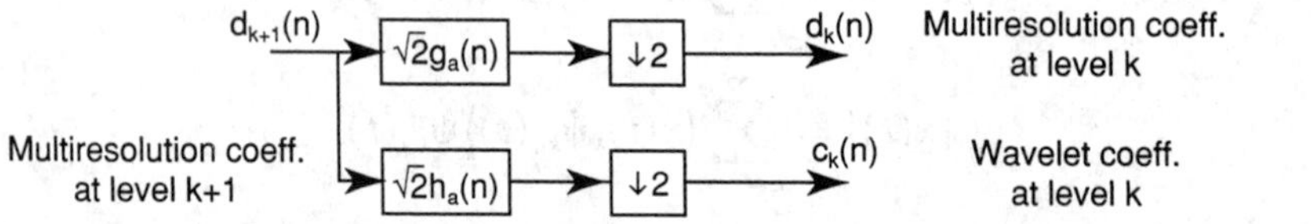

FIGURE 6.37 Generating the wavelet and multiresolution coefficients at level k from level $k + 1$.

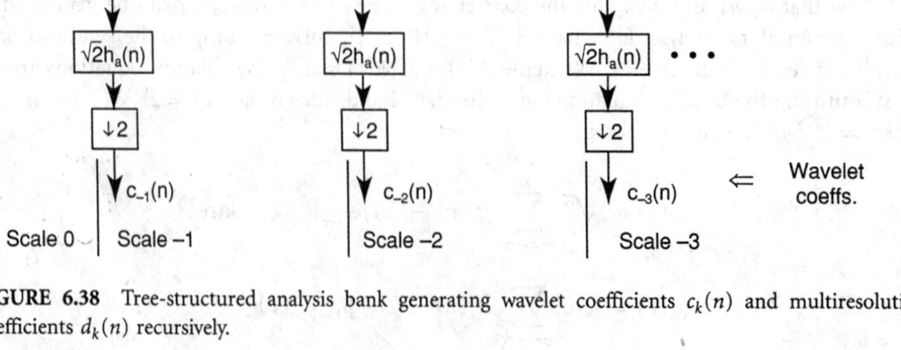

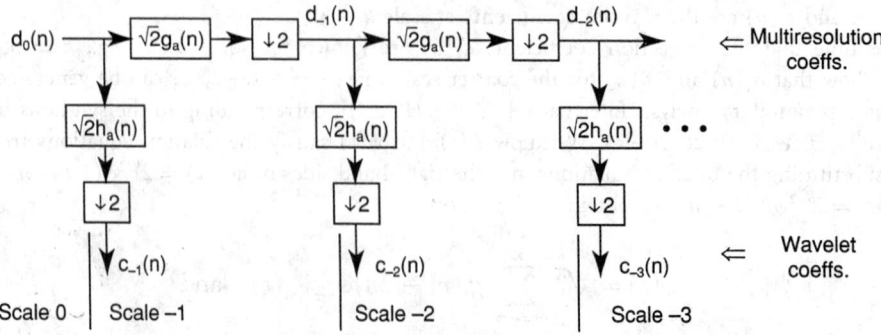

FIGURE 6.38 Tree-structured analysis bank generating wavelet coefficients $c_k(n)$ and multiresolution coefficients $d_k(n)$ recursively.

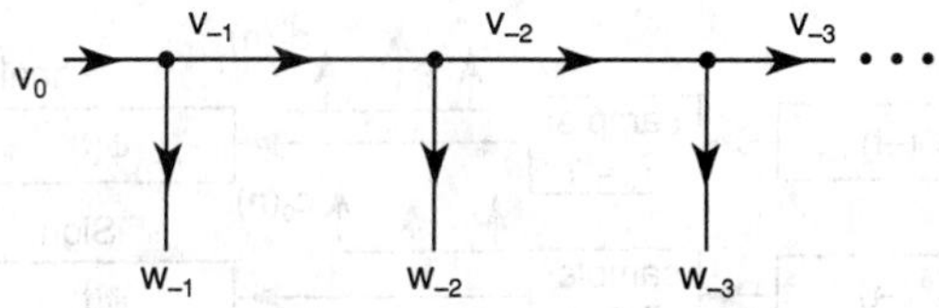

FIGURE 6.39 A schematic of the tree-structured filter bank which generates the coefficients of the projections onto V_k and W_k.

scale $k = 0$ is fine enough—if $x(t)$ does not change much within the duration where $\phi(t)$ is significant—we can approximate this integral with the sample value $x(n)$; i.e., $d_0(n) \approx x(n)$. Improved approximations of $d_0(n)$ have been suggested by other authors, but are not reviewed here.

Continuous Time Filter Banks and Multiresolution

The preceding discussions show the deep connection between orthonormal multiresolution analysis and discrete time paraunitary filter banks. As shown by (6.91c), any L^2 signal $x(t)$ can be written as a sum of its projections onto the mutually orthogonal spaces V_0, W_0, W_1, etc.:

$$x(t) = \sum_n d_0(n)\phi(t-n) + \sum_{k=0}^{\infty}\sum_n c_k(n)2^{k/2}\psi(2^k t - n)$$

This decomposition itself can be given a simple filter bank interpretation, with *continuous time filters and samplers.* For this, first note that the V_0 component $\sum_n d_0(n)\phi(t-n)$ can be regarded as the output of a filter with impulse response $\phi(t)$, with the input chosen as the impulse train $\sum_n d_0(n)\delta_a(t-n)$. Similarly, the W_k component $\sum_n c_k(n)2^{k/2}\psi(2^k t - n)$ is the output of a filter with impulse response $f_k(t) = 2^{k/2}\psi(2^k t)$, in response to the input $\sum_n c_k(n)\delta_a(t - 2^{-k}n)$. This interpretation is shown by the synthesis bank of Fig. 6.40(a).

The projection coefficients $d_0(n)$ and $c_k(n)$ can also be interpreted nicely. For example, we have $d_0(n) = \langle x(t), \phi(t-n)\rangle$ by orthonormality. This inner product can be explicitly written out as

$$d_0(n) = \int x(t)\phi^*(t-n)\,dt$$

The integral can be interpreted as a convolution of $x(t)$ with $\phi^*(-t)$. Consider the output of the filter with impulse response $\phi^*(-t)$, with the input chosen as $x(t)$. This output, sampled at time n, gives $d_0(n)$. Similarly, $c_k(n)$ can be interpreted as the output of the filter $h_k(t) = 2^{k/2}\psi^*(-2^k t)$, sampled at the time $2^{-k}n$. The analysis bank of Fig. 6.40(a) shows this interpretation. Thus, the projection coefficients $d_0(n)$ and $c_k(n)$ are the sampled versions of the outputs of an analysis filter bank.

Notice that all the filters in the filter bank are determined by the scaling function $\phi(t)$ and the wavelet function $\psi(t)$. Every synthesis filter $f_k(t)$ is the time-reversed conjugate of the corresponding analysis filter $h_k(t)$, that is, $f_k(t) = h_k^*(-t)$ (a consequence of orthonormality). In terms of frequency responses this means $F_k(\omega) = H_k^*(\omega)$. For completeness of the picture, Fig. 6.40(b) shows typical frequency response magnitudes of these filters.

Further Manifestations of Orthonormality

The orthonormality of the basis functions $\{\phi(t-n)\}$ and $\{\psi(t-n)\}$ have further consequences, summarized below. A knowledge of these will be useful when we generate the scaling function $\phi(t)$ and the wavelet function $\psi(t)$ systematically in Section 6.11 from paraunitary filter banks.

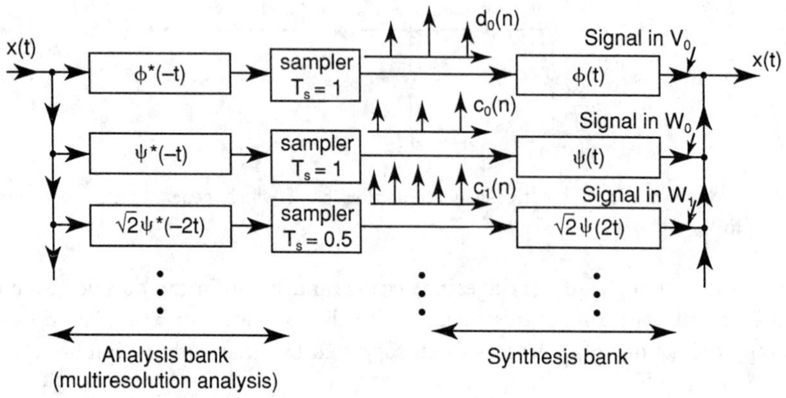

(a)

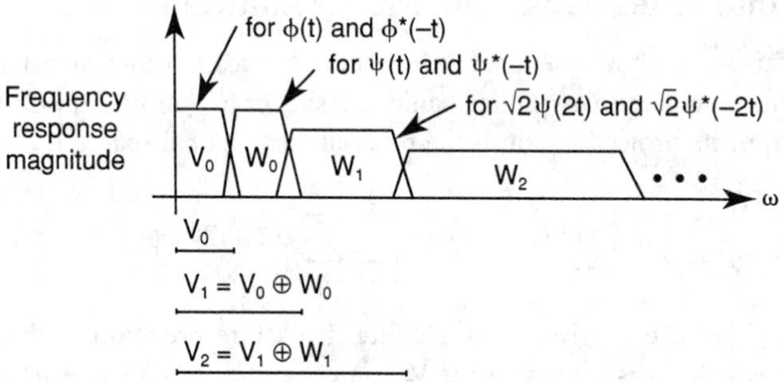

(b)

FIGURE 6.40 (a) The multiresolution analysis and resynthesis in filter bank form, and (b) typical frequency responses.

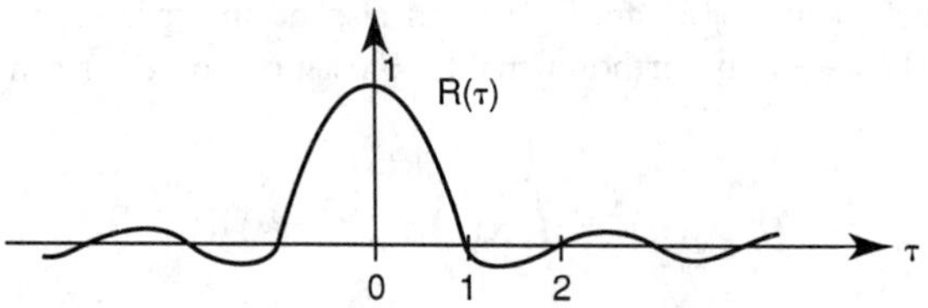

FIGURE 6.41 Example of an autocorrelation of the scaling function $\phi(t)$.

Nyquist Property and Orthonormality

With $\phi(t) \in L^2$, the autocorrelation function $R(\tau) = \int \phi(t)\phi^*(t-\tau)dt$ exists for all τ because this is simply an inner product of two elements in L^2. Clearly, $R(0) = \|\phi(t)\|^2 = 1$. Further, the orthonormality property $\langle \phi(t), \phi(t-n)\rangle = \delta(n)$ can be rewritten as $R(n) = \delta(n)$. Thus, $R(\tau)$ has periodic zero crossings at nonzero integer values of τ (Fig. 6.41). This is precisely the *Nyquist property* familiar to communication engineers. The autocorrelation of the scaling function $\phi(t)$ is a Nyquist function. The same holds for the wavelet function $\psi(t)$.

Next, using Parseval's identity for L^2-FTs, we obtain $\langle \phi(t), \phi(t-n)\rangle = \int \Phi(\omega)\Phi^*(\omega)e^{j\omega n} d\omega/2\pi = \delta(n)$. If we decompose the integral into a sum of integrals over intervals of length 2π and use the 2π-periodicity of $e^{j\omega n}$ we obtain, after some simplification:

$$\sum_{k=-\infty}^{\infty} |\Phi(\omega + 2\pi k)|^2 = 1 \quad \text{a.e.} \tag{6.100}$$

This is the Nyquist condition, now expressed in the frequency domain. The term **a.e., almost everywhere,** arises from the fact that we have drawn a conclusion about an integrand from the value of the integral. Thus, $\{\phi(t-n)\}$ *is orthonormal iff the preceding equation holds.* A similar result follows for $\Psi(\omega)$, so orthonormality of $\{\psi(t-n)\}$ is equivalent to

$$\sum_{k=-\infty}^{\infty} |\Psi(\omega + 2\pi k)|^2 = 1 \quad \text{a.e.} \tag{6.101}$$

Case in Which Equalities Hold Pointwise

If we assume that all FTs are continuous, then equalities in the Fourier domain actually hold pointwise. This is the most common situation; in all examples here, the following are true: the filters $G_s(e^{j\omega})$ and $H_s(e^{j\omega})$ are rational (FIR or IIR), so the frequency responses are continuous functions of ω, and $\phi(t)$ and $\psi(t)$ are not only in L^2, but also in L^1; i.e., $\phi(t), \psi(t) \in L^1 \cap L^2$. Thus, $\Phi(\omega)$ and $\Psi(\omega)$ are continuous functions (Section 6.6).

With the dilation equation $\Phi(\omega) = G_s(e^{j\omega/2})\Phi(\omega/2)$ holding pointwise, we have $\Phi(0) = G_s(e^{j0})\Phi(0)$. In all our applications $\Phi(0) \neq 0$ (it is a low-pass filter), so $G_s(e^{j0}) = 1$. The power symmetry property

$$|G_s(e^{j\omega})|^2 + |G_s(-e^{j\omega})|^2 = 1$$

then implies $G_s(e^{j\pi}) = 0$. Because the high-pass synthesis filter is $H_s(e^{j\omega}) = e^{j\omega} G_s^*(-e^{j\omega})$ we conclude $H_s(e^{j0}) = 0$ and $H_s(e^{j\pi}) = -1$. Thus,

$$G_s(e^{j0}) = 1 \quad G_s(e^{j\pi}) = 0 \quad H_s(e^{j0}) = 0 \quad H_s(e^{j\pi}) = -1 \tag{6.102}$$

In particular, the low-pass impulse response $g_s(n)$ satisfies $\sum_n g_s(n) = 1$. Because we already have $\sum_n |g_s(n)|^2 = 0.5$ (Theorem 6.6), we have both of the following:

$$\sum_{n=-\infty}^{\infty} g_s(n) = 1 \quad \text{and} \quad \sum_{n=-\infty}^{\infty} |g_s(n)|^2 = 0.5 \tag{6.103}$$

From the dilation equation $\Phi(\omega) = G_s(e^{j\omega/2})\Phi(\omega/2)$ we obtain $\Phi(2\pi k) = G_s(e^{j\pi k})\Phi(\pi k)$. By using the fact that $G_s(e^{j\pi}) = 0$, and after elementary manipulations we can show that

$$\Phi(2\pi k) = 0 \qquad k \neq 0 \tag{6.104}$$

In other words, $\Phi(\omega)$ is itself a Nyquist function of ω. If (6.100) is *assumed to hold pointwise*, the above implies that $|\Phi(0)| = 1$. Without loss of generality we will let $\Phi(0) = 1$, i.e., $\int \phi(t)dt = 1$. The dilation equation for the wavelet function $\Psi(\omega)$ in (6.85) shows that $\Psi(0) = 0$ [because $H_s(e^{j0}) = 0$ by (6.102)]. That is, $\int \psi(t)\,dt = 0$. Summarizing, the scaling and wavelet functions satisfy

$$\begin{gathered}\int_{-\infty}^{\infty} \phi(t)\,dt = 1 \quad \int_{-\infty}^{\infty} \psi(t)\,dt = 0 \quad \text{and} \\ \int_{-\infty}^{\infty} |\phi(t)|^2\,dt = \int_{-\infty}^{\infty} |\psi(t)|^2\,dt = 1\end{gathered} \tag{6.105}$$

where property 3 follows from orthonormality. These integrals make sense because of the assumption $\phi(t) \in L^1 \cap L^2$. Another result that follows from $\Phi(2\pi k) = \delta(k)$ is that

$$\sum_{n=-\infty}^{\infty} \phi(t-n) = 1 \quad \text{a.e.} \tag{6.106}$$

Thus, the basis functions of the subspace V_0 themselves add up to unity. Return to the Haar basis and notice how beautifully everything fits together.

Generating Wavelet and Multiresolution Basis by Design of $\phi(t)$

Most of the well-known wavelet basis families of recent times were generated by first finding a scaling function $\phi(t)$ such that it is a valid generator of multiresolution, and then generating $\psi(t)$ from $\phi(t)$. The first step, therefore, is to identify the conditions under which a function $\phi(t)$ will be a valid scaling function (i.e., it will generate a multiresolution). Once this is done and we successfully identify the coefficients $g_s(n)$ in the dilation equation for $\phi(t)$, we can identify the wavelet function $\psi(t)$ using the second dilation equation in (6.84). From Theorem 6.6 we know that if $\psi(t)$ is computed in this way, then $\{2^{k/2}\psi(2^k t - n)\}$ is an orthonormal wavelet basis for L^2. The following results can be deduced from the many detailed results presented in [5].

Theorem 6.7 (Orthonormal Multiresolution): *Let $\phi(t)$ satisfy the following conditions: $\phi(t) \in L^1 \cap L^2$, $\int \phi(t)\,dt \neq 0$ (i.e., $\Phi(0) \neq 0$), $\phi(t) = 2\sum_n g_s(n)\phi(2t-n)$ for some $\{g_s(n)\}$, and $\{\phi(t-n)\}$ is an orthonormal sequence. Then the following are true.*

1. *$\phi(t)$ generates a multiresolution. That is, if we define the space V_k to be the closure of the span of $\{2^{k/2}\phi(2^k t - n)\}$, then the set of spaces $\{V_k\}$ satisfies the six conditions in the definition of multiresolution.*
2. *Define $\psi(t) = 2\sum_n (-1)^{n+1} g_s^*(-n-1)\phi(2t-n)$. Then $\psi(t)$ generates an orthonormal wavelet basis for L^2; that is, $\{2^{k/2}\psi(2^k t - n)\}$, with k and n varying over all integers, is an orthonormal basis for L^2. In fact, for fixed k, the functions $\{2^{k/2}\psi(2^k t - n)\}$ form an orthonormal basis for the subspace W_k (defined following the definition of multiresolution analysis).* □

Comments. In many examples $\phi(t) \in L^2$, and it is compactly supported. Then it is naturally in L^1 as well, so the assumption $\phi(t) \in L^2 \cap L^1$ is not too restrictive. Because $L^1 \cap L^2$ is dense in L^2, the above construction still gives a wavelet basis for L^2. Notice also that the orthonormality of $\{\phi(t-n)\}$ implies orthonormality of $\{\sqrt{2}\phi(2t-n)\}$. The recursion $\phi(t) = 2\sum_n g_s(n)\phi(2t-n)$, therefore, is a Fourier series for $\phi(t)$ in L^2. Thus the condition $\sum_n |g_s(n)|^2 = 0.5$ is automatically implied. This is not explicitly stated as part of the conditions in the theorem.

Orthonormalization

We know that orthonormality of $\{\phi(t-n)\}$ is equivalent to

$$\sum_{k=-\infty}^{\infty} |\Phi(\omega + 2\pi k)|^2 = 1 \tag{6.107}$$

Suppose now that this is not satisfied, but the weaker condition

$$a \leq \sum_{k=-\infty}^{\infty} |\Phi(\omega + 2\pi k)|^2 \leq b \tag{6.108}$$

holds for some $a > 0$ and $b < \infty$. Then, it can be shown that, we can at least obtain a Riesz basis (Section 6.7) of the form $\{\phi(t-n)\}$ for V_0. We can also normalize it to obtain an orthonormal sequence $\{\hat{\phi}(t-n)\}$ from which an orthonormal wavelet basis can be generated in the usual way. The following theorem summarizes the main results.

Theorem 6.8: *Let $\phi(t) \in L^1 \cap L^2$, $\int \phi(t)dt \neq 0$ (i.e., $\Phi(0) \neq 0$), and $\phi(t) = 2\sum_n g_s(n)\phi(2t-n)$ with $\sum_n |g_s(n)|^2 < \infty$. Instead of the orthonormality condition (6.107), let (6.108) hold for some $a > 0$ and $b < \infty$. Then the following are true:*

1. *$\{\phi(t-n)\}$ is a Riesz basis for the closure V_0 of its span.*
2. *$\phi(t)$ generates a multiresolution. That is, if we define the space V_k to be the closure of the span of $\{2^{k/2}\phi(2^k t-n)\}$, the set of spaces $\{V_k\}$ satisfies the six conditions in the definition of multiresolution.* □

If we define a new function $\hat{\phi}(t)$ in terms of its FT as

$$\hat{\Phi}(\omega) = \frac{\Phi(\omega)}{\left(\sum_k |\Phi(\omega + 2\pi k)|^2\right)^{0.5}}, \tag{6.109}$$

then $\hat{\phi}(t)$ generates an orthonormal multiresolution, and satisfies a dilation equation similar to (6.84). Using this we can define a corresponding wavelet function $\hat{\psi}(t)$ in the usual way. That is, if $\hat{\phi}(t) = 2\sum_n g_s(n)\hat{\phi}(2t-n)$, choose $\hat{\psi}(t) = 2\sum_n h_s(n)\hat{\phi}(2t-n)$, where $h_s(n) = (-1)^{n+1}g_s^*(-n-1)$. This wavelet $\hat{\psi}(t)$ generates an orthonormal wavelet basis for L^2. Note that the basis is not necessarily compactly supported if we start with compactly supported $\phi(t)$. An example is seen in Fig. 6.46(b) later.

Example 10, Battle-Lemarié Orthonormal Wavelets from Splines. In Example 9 we considered a triangular $\phi(t)$ (Fig. 6.36), which generates a nonorthonormal multiresolution. In this example we have

$$\Phi(\omega) = \sqrt{\frac{3}{2}}\left(\frac{\sin(\omega/2)}{(\omega/2)}\right)^2 \tag{6.110}$$

and it can be shown that

$$\sum_{k=-\infty}^{\infty} |\Phi(\omega + 2\pi k)|^2 = \frac{2 + \cos\omega}{2} \tag{6.111}$$

The inequality (6.108) is satisfied with $a = 1/2$ and $b = 3/2$. Thus, we have a Riesz basis $\{\phi(t-n)\}$ for V_0. From this scaling function we can obtain the normalized function $\hat{\Phi}(\omega)$ as above and then generate the wavelet function $\hat{\psi}(t)$ as explained earlier. This gives an orthonormal wavelet basis for L^2. $\hat{\phi}(t)$ does not, however, have compact support [unlike $\phi(t)$]. Thus, the wavelet function $\hat{\psi}(t)$ generating the orthonormal wavelet basis is not compactly supported either. □

.11 Orthonormal Wavelet Basis from Paraunitary Filter Banks

The wisdom gained from the multiresolution viewpoint (Section 6.10) tells us a close connection exists between wavelet bases and two-channel digital filter banks. In fact, we obtained the equations of a paraunitary filter bank just by imposing the orthonormality condition on the multiresolution basis functions $\{\phi(t-n)\}$. This section presents the complete story. Suppose we start from a two-channel digital filter bank with the paraunitary property. Can we derive an orthonormal wavelet basis from this? To be more specific, return to the dilation equations (6.84) or equivalently (6.85). Here, $g_s(n)$ and $h_s(n)$ are the impulse response coefficients of the two synthesis filters $G_s(e^{j\omega})$ and $H_s(e^{j\omega})$ in the digital filter bank. Given these two filters, can we "solve" for $\phi(t)$ and $\psi(t)$? If so, does this $\psi(t)$ generate an orthonormal basis for L^2 space? This section answers some of these questions. Unlike any other section, we also indicate a sketch of the proof for each major result, in view of the importance of these in modern signal processing theory.

Recall first that under some mild conditions (Section 6.10) we can prove that the filters must satisfy (6.102) and (6.103), if we need to generate wavelet and multiresolution bases successfully. We impose these at the outset. By repeated application of the dilation equation we obtain $\Phi(\omega) = G_s(e^{j\omega/2})G_s(e^{j\omega/4})\Phi(\omega/4)$. Further indefinite repetition yields an infinite product. Using the condition $\Phi(0) = 1$, which we justified earlier, we obtain the infinite products

$$\Phi(\omega) = \prod_{k=1}^{\infty} G_s(e^{j\omega/2^k}) = G_s(e^{j\omega/2}) \prod_{k=2}^{\infty} G_s(e^{j\omega/2^k}) \tag{6.112a}$$

$$\Psi(\omega) = H_s(e^{j\omega/2}) \prod_{k=2}^{\infty} G_s(e^{j\omega/2^k}) \tag{6.112b}$$

The first issue to be addressed is the convergence of the infinite products above. For this we need to review some preliminaries on infinite products [22, 23].

Ideal Bandpass Wavelet Rederived from the Digital Filter Bank. Before we address the mathematical details, let us consider a simple example. Suppose the pair of filters $G_s(e^{j\omega})$ and $H_s(e^{j\omega})$ are ideal brickwall low-pass and high-pass filters as in Fig. 6.28(a). Then we can verify, by making simple sketches of a few terms in (6.112), that the above infinite products yield the functions $\Phi(\omega)$ and $\Psi(\omega)$ shown in Fig. 6.33. That is, the ideal bandpass wavelet is indeed related to the ideal paraunitary filter bank by means of the above infinite product.

Convergence of Infinite Products

To define convergence of a product of the form $\prod_{k=1}^{\infty} a_k$, consider the sequence $\{p_n\}$ of partial products $p_n = \prod_{k=1}^{n} a_k$. If this converges to a (complex) number A with $0 < |A| < \infty$ we say that the infinite product converges to A. Convergence to zero should be defined more carefully to avoid degenerate situations (e.g., if $a_1 = 0$, then $p_n = 0$ for all n regardless of the remaining terms $a_k, k > 1$). We use the definition in [22]. The infinite product is said to converge to zero iff $a_k = 0$ for a finite nonzero number of values of k, and if the product with these a_k deleted converges to a nonzero value.

Useful Facts About Infinite Products

1. Whenever $\prod_{k=1}^{\infty} a_k$ converges, it can be shown that $a_k \to 1$ as $k \to \infty$. For this reason it is convenient to write $a_k = 1 + b_k$.
2. We say that $\prod_{k=1}^{\infty}(1+b_k)$ converges absolutely if $\prod_{k=1}^{\infty}(1+|b_k|)$ converges. Absolute convergence of $\prod_{k=1}^{\infty}(1+b_k)$ implies its convergence.
3. It can be shown that the product $\prod_{k=1}^{\infty}(1+|b_k|)$ converges iff the sum $\sum_{k=1}^{\infty}|b_k|$ converges. That is, $\prod_{k=1}^{\infty}(1+b_k)$ converges absolutely iff $\sum_{k=1}^{\infty} b_k$ converges absolutely.

Example 11. The product $\prod_{k=1}^{\infty}(1+k^{-2})$ converges because $\sum_{k=1}^{\infty} 1/k^2$ converges. Similarly, $\prod_{k=1}^{\infty}(1-k^{-2})$ converges because it converges absolutely, by the preceding example. The product $\prod_{k=1}^{\infty}(1+k^{-1})$ does not converge because $\sum_{k=1}^{\infty} 1/k$ diverges. Products such as $\prod_{k=1}^{\infty}(1/k^2)$ do not converge because the terms do not approach unity as $k \to \infty$. □

Uniform Convergence

A sequence $\{p_n(z)\}$ of functions of the complex variable z converges *uniformly* to a function $p(z)$ on a set $\mathcal{S}$ in the complex plane if the convergence rate is the same everywhere in $\mathcal{S}$. More precisely, if we are given $\epsilon > 0$, we can find N such that $|p_n(z) - p(z)| < \epsilon$ for every $z \in \mathcal{S}$, as long as $n \geq N$. The crucial thing is that N depends only on ϵ and not on z, as long as $z \in \mathcal{S}$. A similar definition applies for functions of real variables.

We say that an infinite product of functions $\prod_{k=1}^{\infty} a_k(z)$ converges at a point z if the sequence of partial products $p_n(z) = \prod_{k=1}^{n} a_k(z)$ converges as described previously. If this convergence of $p_n(z)$ is uniform in a set $\mathcal{S}$, we say that the infinite product converges *uniformly* on $\mathcal{S}$. Uniform convergence has similar advantages, as in the case of infinite summations. For example, if each of the functions $a_k(\omega)$ is continuous on the real interval $[\omega_1, \omega_2]$, then uniform convergence of the infinite product $A(\omega) = \prod_{k=1}^{\infty} a_k(\omega)$ on $[\omega_1, \omega_2]$ implies that the limit $A(\omega)$ is continuous on $[\omega_1, \omega_2]$. We saw above that convergence of infinite products can be related to that of infinite summations. The following theorem [23] makes the connection between uniform convergence of summations and uniform convergence of products.

Theorem 6.9: *Let $b_k(z), k \geq 1$ be a sequence of bounded functions of the complex variable z, such that $\sum_{k=1}^{\infty} |b_k(z)|$ converges uniformly on a compact set*[14] *$\mathcal{S}$ in the complex z plane. Then the infinite product $\prod_{k=1}^{\infty}(1 + b_k(z))$ converges uniformly on $\mathcal{S}$. This product is zero for some z_0 iff $1 + b_k(z_0) = 0$ for some k.* □

Uniform convergence and analyticity. We know that if a sequence of continuous functions converges uniformly to a function, then the limit is also continuous. A similar result is true for analytic functions. If a sequence $\{f_n(s)\}$ of analytic functions converges uniformly to a function $f(s)$, then $f(s)$ is analytic as well. For a more precise statement of this result see Theorem 10.28 in [23].

Infinite Product Defining the Scaling Function

Return now to the infinite product (6.112a). As justified in Section 6.10, we assume $G_s(e^{j\omega})$ to be continuous, $G_s(e^{j0}) = 1$, and $\Phi(0) \neq 0$. Note that $G_s(e^{j0}) = 1$ is necessary for the infinite product to converge (because convergence of $\prod_k a_k$ implies that $a_k \to 1$; apply this for $\omega = 0$). The following convergence result is fundamental.

Theorem 6.10, Convergence of the Infinite Product: *Let $G_s(e^{j\omega}) = \sum_{n=-\infty}^{\infty} g_s(n)e^{-j\omega n}$. Assume that $G_s(e^{j0}) = 1$, and $\sum_n |ng_s(n)| < \infty$. Then,*

1. *The infinite product (6.112a) converges pointwise for all ω. In fact, it converges absolutely for all ω, and uniformly on compact sets (i.e., closed bounded sets, such as sets of the form $[\omega_1, \omega_2]$).*
2. *The quantity $G_s(e^{j\omega})$ as well as the limit $\Phi(\omega)$ of the infinite product* (6.112a) *are continuous functions of ω.*
3. *$G_s(e^{j\omega})$ is in L^2.* □

[14]For us, a compact set means any closed bounded set in the complex plane or on the real line. Examples are all points on and inside a circle in the complex plane, and the closed interval $[a, b]$ on the real line.

Because the condition $\sum_n |ng_s(n)| < \infty$ implies $\sum_n |g_s(n)| < \infty$, the filter $G_s(e^{j\omega})$ is restricted to be stable, but the above result holds whether $g_s(n)$ is FIR or IIR.

Sketch of Proof. Theorem 6.9 allows us to reduce the convergence of the product to the convergence of an infinite sum. For this we must write $G_s(e^{j\omega})$ in the form $1 - F(e^{j\omega})$ and then consider the summation $\sum_{k=1}^{\infty} |F(e^{j\omega/2^k})|$. Because $G_s(e^{j0}) = 1 = \sum_n g_s(n)$, we can write $G_s(e^{j\omega}) = 1 - (1 - G_s(e^{j\omega})) = 1 - \sum_n g_s(n)(1 - e^{-j\omega n})$. However, $|\sum_n g_s(n)(1 - e^{-j\omega n})| \leq 2\sum_n |g_s(n)\sin(\omega n/2)| \leq |\omega| \sum_n |ng_s(n)|$ (use $|\sin x/x| \leq 1$). $\sum_n |ng_s(n)|$ is assumed to converge, thus we have $|\sum_n g_s(n)(1-e^{-j\omega n})| \leq c|\omega|$. Using this and the fact that $\sum_{k=1}^{\infty} 2^{-k}$ converges, we can complete the proof of part 1 (by applying Theorem 6.9). $\sum_n |ng_s(n)| < \infty$ implies in particular that $g_s(n) \in \ell^1$, therefore, its ℓ^1-FT $G_s(e^{j\omega})$ is continuous (see Section 6.6). The continuity of $G_s(e^{j\omega})$, together with uniform convergence of the infinite product, implies that the pointwise limit $\Phi(\omega)$ is also continuous. Finally, because $\ell^1 \subset \ell^2$ (Section 6.6), we have $g_s(n) \in \ell^2$, that is, $G_s(e^{j\omega}) \in L^2[0, 2\pi]$ as well. □

Orthonormal Wavelet Basis from Paraunitary Filter Bank

We now consider the behavior of the infinite product $\prod_{k=1}^{\infty} G_s(e^{j\omega/2^k})$, when $G_s(e^{j\omega})$ comes from a paraunitary filter bank. The paraunitary property implies that $G_s(e^{j\omega})$ is power symmetric. If we impose some further mild conditions on $G_s(e^{j\omega})$, the scaling function $\phi(t)$ generates an orthonormal multiresolution basis $\{\phi(t-n)\}$. We can then obtain an orthonormal wavelet basis $\{\psi_{kn}(t)\}$ (Theorems 6.6 and 6.7). The main results are given in Theorems 6.11 to 6.15.

First, we define the truncated partial products $P_n(\omega)$. Because $G_s(e^{j\omega})$ has period 2π, the term $G_s(e^{j\omega/2^k})$ has period $2^{k+1}\pi$. For this reason the partial product $\prod_{k=1}^{n} G_s(e^{j\omega/2^k})$ has period $2^{n+1}\pi$, and we can regard the region $[-2^n\pi, 2^n\pi]$ to be the fundamental period. Let us truncate the partial product to this region, and define

$$P_n(\omega) = \begin{cases} \prod_{k=1}^{n} G_s(e^{j\omega/2^k}), & \text{for } -2^n\pi \leq \omega \leq 2^n\pi, \\ 0, & \text{otherwise} \end{cases} \tag{6.113}$$

This quantity will be useful later. We will see that this is in $L^2(R)$, and we can discuss $p_n(t)$, its inverse L^2-FT.

Theorem 6.11: *Let $G_s(e^{j\omega})$ be as in Theorem 6.10. In addition let it be power symmetric; in other words, $|G_s(e^{j\omega})|^2 + |G_s(-e^{j\omega})|^2 = 1$. [Notice in particular that this implies $G_s(e^{j\pi}) = 0$, because $G_s(e^{j0}) = 1$]. Then the following are true.*

1. $\int_0^{2\pi} |G_s(e^{j\omega})|^2 d\omega/2\pi = 0.5$.
2. *The truncated partial product $P_n(\omega)$ is in L^2, and $\int_{-\infty}^{\infty} |P_n(\omega)|^2 d\omega/2\pi = 1$ for all n. Further, the inverse L^2-FT, denoted as $p_n(t)$, gives rise to an orthonormal sequence $\{p_n(t-k)\}$, i.e., $\langle p_n(t-k), p_n(t-i)\rangle = \delta(k-i)$ for any $n \geq 1$.*
3. *The limit $\Phi(\omega)$ of the infinite product (6.112a) is in L^2, hence it has an inverse L^2-FT, $\phi(t) \in L^2$. Moreover, $\|\phi(t)\|_2 \leq 1$.* □

Sketch of Proof. Part 1 follows by integrating both sides of $|G_s(e^{j\omega})|^2 + |G_s(-e^{j\omega})|^2 = 1$. The integral in part 2 is $\int_0^{2^{n+1}\pi} \prod_{k=1}^{n} |G_s(e^{j\omega/2^k})|^2 d\omega/2\pi$, which we can split into two terms such as $\int_0^{2^n\pi} + \int_{2^n\pi}^{2^{n+1}\pi}$. Using the 2π-periodicity and the power symmetric property of $G_s(e^{j\omega})$, we obtain $\int |P_n|^2 d\omega = \int |P_{n-1}|^2 d\omega$. Repeated application of this, together with part 1, yields $\int_{-\infty}^{\infty} |P_n(\omega)|^2 d\omega/2\pi = 1$. The proof of orthonormality of $\{p_n(t-k)\}$ follows essentially similarly by working with the modified integral $\int_{-\infty}^{\infty} |P_n(\omega)|^2 e^{j\omega(k-i)} d\omega/2\pi$, and using the half-band property of $|G_s(e^{j\omega})|^2$.

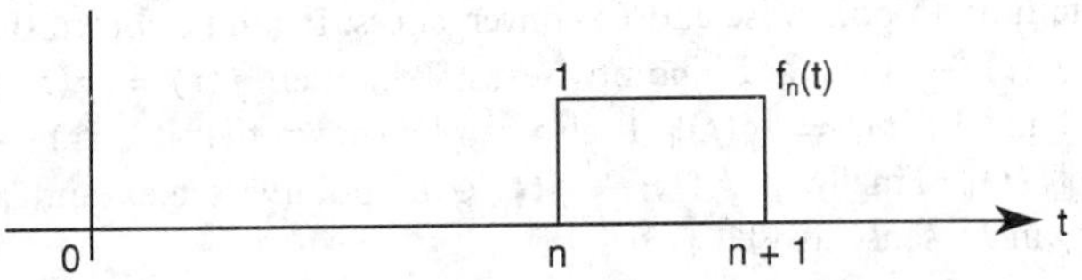

FIGURE 6.42 A sequence $\{f_n(t)\}$ whose pointwise limit is not a limit in the L^2 sense.

The third part is the most subtle, and uses Fatou's lemma for Lebesgue integrals (Section 6.6). For this, define $g_n(\omega) = |P_n(\omega)|^2$. Then $\{g_n(\omega)\}$ is a sequence of non-negative integrable functions such that $g_n(\omega) \to |\Phi(\omega)|^2$ pointwise for each ω. Because $\int g_n(\omega)d\omega = 2\pi$ (from part 2), Fatou's lemma assures us that $|\Phi(\omega)|^2$ is integrable with integral $\leq 2\pi$. This proves part 3. □

It is most interesting that the truncated partial products $P_n(\omega)$ give rise to orthonormal sequences $\{p_n(t-k)\}$. This orthonormality is induced by the paraunitary property, more precisely the power symmetry property of $G_s(e^{j\omega})$. This is consistent with the fact that the *filter bank type of basis* introduced in Section 6.4 is an orthonormal basis for ℓ^2 whenever the filter bank is paraunitary.

As the scaling function $\Phi(\omega)$ is the pointwise limit of $\{P_n(\omega)\}$ as $n \to \infty$, this leads to the hope that $\{\phi(t-k)\}$ is also an orthonormal sequence, so that we can generate a multiresolution and then a wavelet basis as in Theorems 6.6 and 6.7). This, however, is not always true. The crux of the reason is that $\Phi(\omega)$ is only the pointwise limit of $\{P_n(\omega)\}$, and not necessarily the L^2 limit. The distinction is subtle (see below). The pointwise limit property means that for any fixed ω, the function $P_n(\omega)$ approaches $\Phi(\omega)$. The L^2 limit property means that $\int |P_n(\omega)-\Phi(\omega)|^2 d\omega \to 0$. Neither of these limit properties implies the other; neither is stronger than the other. It can be shown that it is the L^2 limit which propagates the orthonormality property, and this is what we want.

Theorem 6.12: *Let $\{p_n(t-k)\}$ be an orthonormal sequence for each n. That is, $\langle p_n(t-k), p_n(t-i)\rangle = \delta(k-i)$. Suppose $p_n(t) \to \phi(t)$ in the L^2 sense. Then $\{\phi(t-k)\}$ is an orthonormal sequence.* □

Proof. If we take limits as $n \to \infty$, we can write

$$\lim_{n\to\infty} \langle p_n(t-k), p_n(t-i)\rangle = \left\langle \lim_{n\to\infty} p_n(t-k), \lim_{n\to\infty} p_n(t-i)\right\rangle \tag{6.114}$$

This movement of the "limit" sign past the inner product sign is allowed (by continuity of inner products, Section 6.6), provided the limits in the second expression are L^2 limits. By the conditions of the theorem, the left side of the above equation is $\delta(k-i)$, whereas the right side is $\langle \phi(t-k), \phi(t-i)\rangle$. So the result follows. □

L^2 Convergence vs. Pointwise Convergence

The fact that L^2 limits are not necessarily pointwise limits is obvious from the fact that differences at a countable set of points do not affect integrals. The fact that pointwise limits are not necessarily L^2 limits is demonstrated by the sequence of L^2 functions $\{f_n(t)\}$, with $f_n(t)$ as in Fig. 6.42. Note that $f_n(t) \to 0$ pointwise for each t, that is, the pointwise limit is $f(t) \equiv 0$. Hence, $\|f_n(t) - f(t)\| = \|f_n(t)\| = 1$ for all n, so $\|f_n(t) - f(t)\|$ does not go to zero as $n \to \infty$, and thus, $f(t)$ is not the L^2 limit of $f_n(t)$. Notice in this example that $1 = \lim_{n\to\infty} \int |f_n(t)|^2 dt \neq \int \lim_{n\to\infty} |f_n(t)|^2 dt = 0$. This is consistent with the fact that the Lebesgue dominated convergence theorem cannot be applied here—no integrable function dominates $|f_n(t)|^2$ for all n. In this example, the sequence $\{f_n(t)\}$ does not converge in the L^2 sense. In fact, $\|f_n(t) - f_m(t)\|^2 = 2$ for $n \neq m$. Thus, $\{f_n\}$ is not a Cauchy sequence [22] in L^2.

Some facts pertaining to pointwise and L^2 convergences: It can be shown that if $f_n(t) \rightarrow f(t)$ in L^2 sense and $f_n(t) \rightarrow g(t) \in L^2$ pointwise as well, then $f(t) = g(t)$ a.e. In particular, $\|f(t) - g(t)\| = 0$ and $\|f(t)\| = \|g(t)\|$. It also can be shown that if $f_n(t) \rightarrow f(t)$ in L^2 sense, then $\|f_n(t)\| \rightarrow \|f(t)\|$. Finally, if $f_n(t) \rightarrow f(t) \in L^2$ pointwise a.e., and $\|f_n(t)\| \rightarrow \|f(t)\|$ then $f_n(t) \rightarrow f(t)$ in L^2 sense as well [23].

Theorem 6.13, Orthonormal Wavelet Basis: *Let the filter* $G_s(e^{j\omega}) = \sum_{n=-\infty}^{\infty} g_s(n)e^{-j\omega n}$ *satisfy the following properties:*

1. $G_s(e^{j0}) = 1$
2. $\sum_n |ng_s(n)| < \infty$
3. $|G_s(e^{j\omega})|^2 + |G_s(-e^{j\omega})|^2 = 1$ *(power symmetry)*
4. $G_s(e^{j\omega}) \neq 0$ *for* $\omega \in [-0.5\pi, 0.5\pi]$

Then the infinite product (6.112a) *converges to a limit* $\Phi(\omega) \in L^2$, *and its inverse FT* $\phi(t)$ *is such that* $\{\phi(t-n)\}$ *is an orthonormal sequence. Defining the wavelet function* $\psi(t)$ *as usual, i.e., as in* (6.93), *the sequence* $\{2^{k/2}\psi(2^k t - n)\}$ *(with k and n varying over all integers) forms an orthonormal wavelet basis for* L^2. □

Sketch of Proof. We will show that the sequence $\{P_n(\omega)\}$ of partial products converges to $\Phi(\omega)$ in the L^2 sense, i.e., $\int |P_n(\omega) - \Phi(\omega)|^2 d\omega \rightarrow 0$, so that $p_n(t) \rightarrow \phi(t)$ in L^2 sense. The desired result then follows in view of Theorems 6.11 and 6.12. The key tool in the proof is the dominated convergence theorem for Lebesgue integrals (Section 6.6). First, the condition $G(e^{j\omega}) \neq 0$ in $[-0.5\pi, 0.5\pi]$ implies that $\Phi(\omega) \neq 0$ in $[-\pi, \pi]$. Because $|\Phi(\omega)|^2$ is continuous (Theorem 6.10) it has a minimum value $c^2 > 0$ in $[-\pi, \pi]$. Now the truncated partial product $P_n(\omega)$ can always be written as $P_n(\omega) = \Phi(\omega)/\Phi(\omega/2^n)$ in its region of support. Because $|\Phi(\omega/2^n)|^2 \geq c^2$ in $[-2^n\pi, 2^n\pi]$, we have $|P_n(\omega)|^2 \leq |\Phi(\omega)|^2/c^2$ for all ω. Define $Q_n(\omega) = |P_n(\omega) - \Phi(\omega)|^2$. Then using $|P_n(\omega)|^2 \leq |\Phi(\omega)|^2/c^2$ we can show that $Q_n(\omega) \leq \alpha|\Phi(\omega)|^2$ for some constant α. Because the right-hand side is integrable, and because $Q_n(\omega) \rightarrow 0$ pointwise (Theorem 6.10) we can use the dominated convergence theorem (Section 6.6) to conclude that $\lim_n \int Q_n(\omega)\, d\omega = \int \lim_n Q_n(\omega)\, d\omega = 0$. This completes the proof. □

Computing the Scaling and Wavelet Functions

Given the coefficients $g_s(n)$ of the filter $G_s(e^{j\omega})$, how do we compute the scaling function $\phi(t)$ and the wavelet function $\psi(t)$? Because we can compute $\psi(t)$ using $\psi(t) = 2\sum_{n=-\infty}^{\infty}(-1)^{n+1}g_s^*(-n-1)\phi(2t-n)$, the key issue is the computation of $\phi(t)$. In the preceding theorems $\phi(t)$ was defined only as an inverse L^2-FT of the infinite product $\Phi(\omega)$ given in (6.112a). Because an L^2 function is determined only in the a.e. sense, this way of defining $\phi(t)$ itself does not fully determine $\phi(t)$. Recall, however, that the infinite product for $\Phi(\omega)$ was only a consequence of the more fundamental equation, the dilation equation $\phi(t) = 2\sum_{n=-\infty}^{\infty} g_s(n)\phi(2t-n)$. In practice $\phi(t)$ is computed using this equation, which is often a finite sum (see Section 6.12). The procedure is recursive; we assume an initial solution for the function $\phi(t)$, substitute it into the right-hand side of the dilation equation, thereby recomputing $\phi(t)$, and then repeat the process. Details of this and discussions on convergence of this procedure can be found in [5, 15, and 26]

Lawton's Eigenfunction Condition for Orthonormality [5]

Equation (6.100) is equivalent to the orthonormality of $\{\phi(t-n)\}$. Let $S(e^{j\omega})$ denote the left-hand side of (6.100), which evidently has period 2π in ω. Using the frequency domain version of the dilation equation (6.85), it can be shown that the scaling function $\phi(t)$ generated from

$G_s(e^{j\omega})$ is such that

$$|G_s(e^{j\omega})|^2 S(e^{j\omega})\Big|_{\downarrow 2} = 0.5S(e^{j\omega}) \qquad (6.115)$$

where the notation $\downarrow 2$ indicates decimation (Section 6.4). Thus, the function $S(e^{j\omega})$ can be regarded as an eigenfunction (with eigenvalue = 0.5) of the operator $\mathcal{F}$, which performs filtering by $|G_s(e^{j\omega})|^2$ followed by decimation.

Now consider the case in which the digital filter bank is paraunitary, so that $G_s(e^{j\omega})$ is power symmetric [i.e., satisfies (6.95)]. The power symmetric condition can be rewritten in the form $|G_s(e^{j\omega})|^2|_{\downarrow 2} = 0.5$. Thus, in the power symmetric case the identity function is an eigenfunction of the operator $\mathcal{F}$. If the only eigenfunction of the operator $\mathcal{F}$ is the identity function, it then follows that $S(e^{j\omega}) = 1$; i.e., (6.100) holds and $\{\phi(t-n)\}$ is orthonormal.

The FIR Case. Section 6.12 shows that restricting $G_s(z)$ to be FIR ensures that $\phi(t)$ has finite duration. For the FIR case, Lawton and Cohen independently showed that the above eigenfunction condition also works in the other direction. That is, if $\{\phi(t-n)\}$ has to be orthonormal, then the trignometric polynomial $S(e^{j\omega})$ satisfying (6.115) must be unique up to a scale factor.[15] Details can be found in [5].

Examples and Counter-Examples

We already indicated after the introduction of (6.112) that the example of the ideal bandpass wavelet can be generated formally by starting from the ideal brickwall paraunitary filter bank. We now discuss some other examples.

Example 12, Haar Basis from Filter Banks. A filter bank of the form Fig. 6.22(a) with filters

$$G_a(z) = \frac{1+z^{-1}}{2} \quad H_a(z) = \frac{z^{-1}-1}{2} \quad G_s(z) = \frac{1+z^{-1}}{2} \quad H_s(z) = \frac{1-z^{-1}}{2}$$

is paraunitary. The magnitude responses of the synthesis filters, $|G_s(e^{j\omega})| = |\cos(\omega/2)|$ and $|H_s(e^{j\omega})| = |\sin(\omega/2)|$, are shown in Fig. 6.43(a). $G_s(z)$ satisfies all the conditions of Theorem 6.13. In this case we can evaluate the infinite products for $\Phi(\omega)$ and $\Psi(\omega)$ explicitly by using the identity $\prod_{m=1}^{\infty} \cos(2^{-m}\omega) = \sin\omega/\omega$. The resulting $\phi(t)$ and $\psi(t)$ are as shown in Fig. 6.43(b) and (c). These are precisely the functions that generate the Haar orthonormal basis. □

Example 13, Paraunitary Filter Bank which Does Not Give Orthonormal Wavelets. Consider the filter bank with analysis filters $G_a(z) = (1+z^{-3})/2$, $H_a(z) = -(1-z^{-3})/2$, and synthesis filters $G_s(z) = (1+z^{-3})/2$, $H_s(z) = (1-z^{-3})/2$. As this is obtained from the preceding example by the substitution $z \to z^3$, it remains paraunitary and satisfies the PR property. $G_s(z)$ satisfies all the properties of Theorem 6.13, except the fourth condition. With $\phi(t)$ and $\psi(t)$ obtained from $G_s(e^{j\omega})$ using the usual dilation equations, the functions $\{\phi(t-n)\}$ are *not orthonormal.* In addition, the wavelet functions $\{2^{k/2}\psi(2^k t-n)\}$ do not form an orthonormal basis. These statements can be verified from the sketches of the functions $\phi(t)$ and $\psi(t)$ shown in Fig. 6.44. Clearly, $\phi(t)$ and $\phi(t-1)$ are not orthogonal, and $\psi(t)$ and $\psi(t-2)$ are not orthogonal. In

[15]A finite sum of the form $\sum_{n=N_1}^{N_2} p_n e^{j\omega n}$ is said to be a trignometric polynomial. If $G_s(e^{j\omega})$ is FIR, it can be shown that the left-hand side of (6.100) is not only periodic in ω, but is in fact a trignometric polynomial.

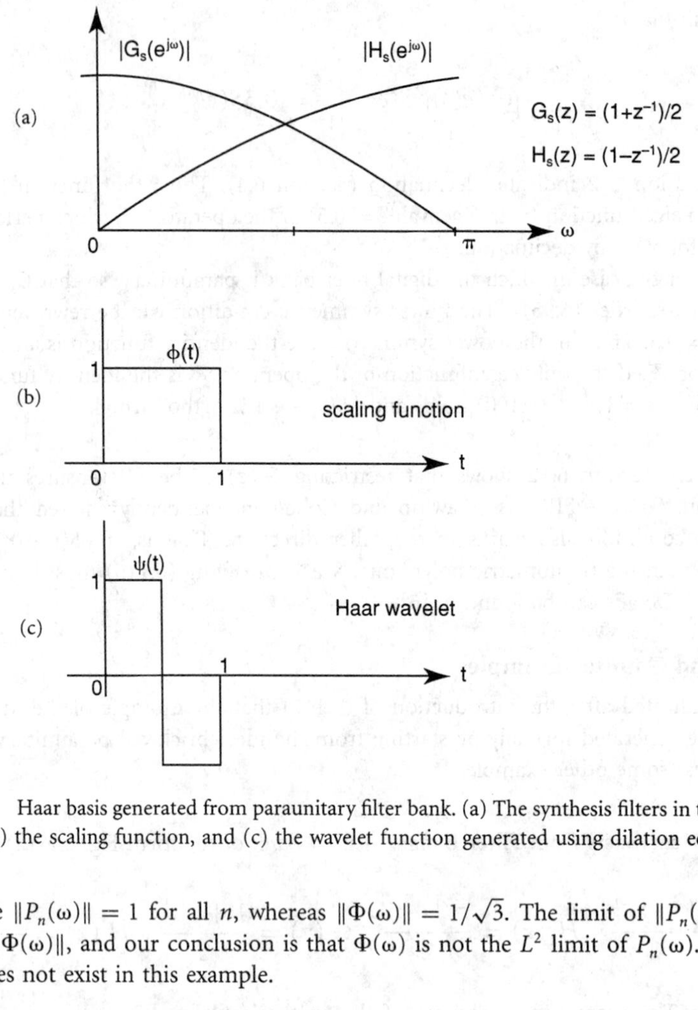

FIGURE 6.43 Haar basis generated from paraunitary filter bank. (a) The synthesis filters in the paraunitary filter bank, (b) the scaling function, and (c) the wavelet function generated using dilation equations.

this example $\|P_n(\omega)\| = 1$ for all n, whereas $\|\Phi(\omega)\| = 1/\sqrt{3}$. The limit of $\|P_n(\omega)\|$ does not agree with $\|\Phi(\omega)\|$, and our conclusion is that $\Phi(\omega)$ is not the L^2 limit of $P_n(\omega)$. The L_2 limit of $P_n(\omega)$ does not exist in this example. □

Thus, a paraunitary filter bank may not generate an orthonormal wavelet basis if the fourth condition in Theorem 6.13 is violated. However, this is hardly of concern in practice, because any reasonable low-pass filter designed for a two-channel filter bank will be free from zeroes in the region $[-0.5\pi, 0.5\pi]$. In fact, a stronger result was proved by Cohen, who derived necessary and sufficient conditions for an FIR paraunitary filter bank to generate an orthonormal wavelet basis. One outcome of Cohen's analysis is that the fourth condition in Theorem 6.13 can be replaced by the even milder condition that $G_s(e^{j\omega})$ not be zero in $[-\pi/3, \pi/3]$. In this sense the condition for obtaining an orthonormal wavelet basis is trivially satisfied in practice. The case in which the fourth condition fails is primarily of theoretical interest; an attractive result in this context is Lawton's tight frame theorem.

Wavelet Tight Frames

Although the wavelet functions $\{2^{k/2}\psi(2^k t - n)\}$ generated from a paraunitary filter bank may not form an orthonormal basis when the fourth condition of Theorem 6.13 is violated, the functions always form a tight frame for L^2. Thus, any L^2 function can be expressed as an infinite linear combination of the functions $\{2^{k/2}\psi(2^k t - n)\}$. More precisely, we have the following result due to Lawton [5].

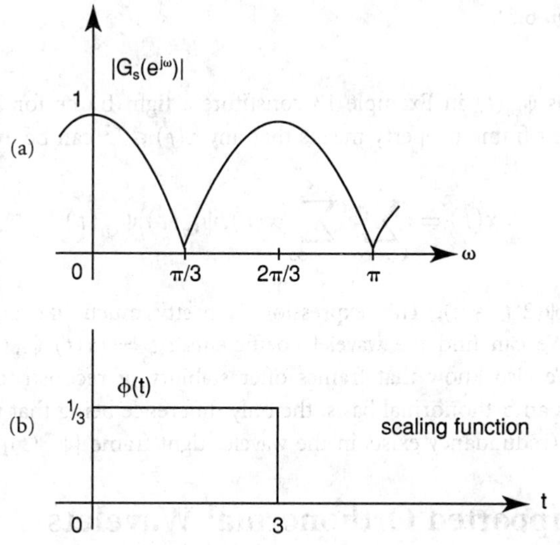

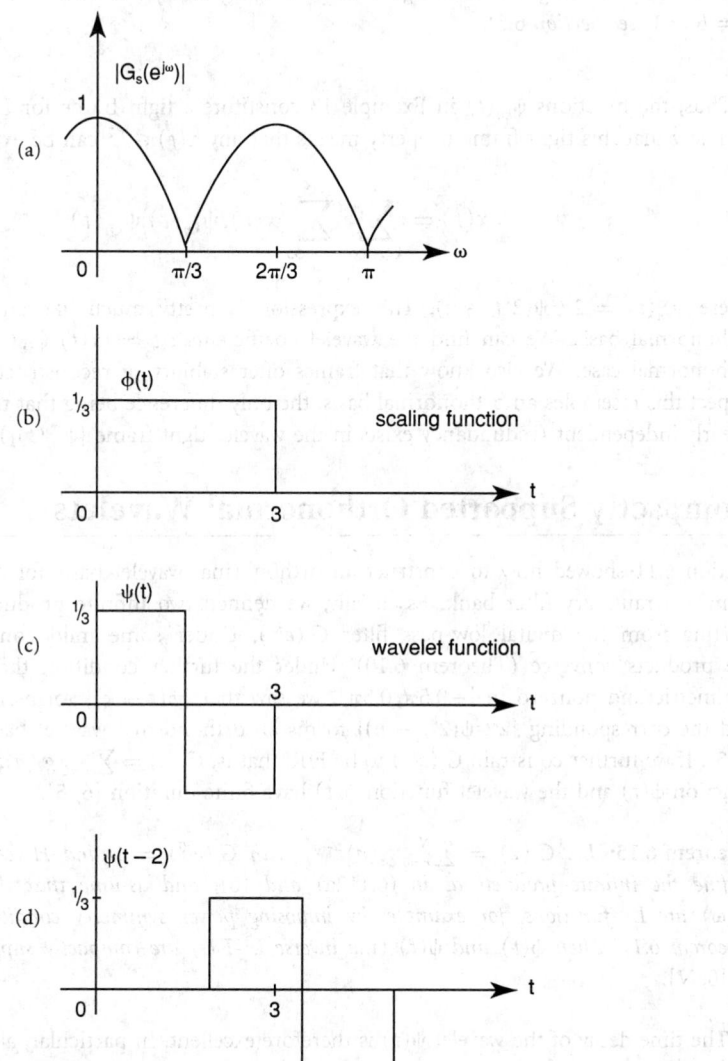

FIGURE 6.44 A paraunitary filter bank generating nonorthonormal $\{\phi(t-n)\}$. (a) The synthesis filter response, (b) the scaling function, (c) the wavelet function, and (d) a shifted version.

Theorem 6.14, Tight Frames from Paraunitary Filter Banks: *Let $G_s(e^{j\omega}) = \sum_{n=0}^{N} g_s(n)e^{-j\omega n}$ be a filter satisfying the following properties:*

1. $G_s(e^{j0}) = 1$.
2. $|G_s(e^{j\omega})|^2 + |G_s(-e^{j\omega})|^2 = 1$ *(power symmetry).*

Then, $\phi(t) \in L^2$. Defining the wavelet function $\psi(t)$ as in (6.93), the sequence $\{2^{k/2}\psi(2^k t - n)\}$ (with k and n varying over all integers) forms a tight frame for L^2, with frame bound unity (i.e., $A = B = 1$; see Section 6.8). □

Thus, the functions $\psi_{kn}(t)$ in Example 13 constitute a tight frame for L^2. From Section 6.8 we know that this tight frame property means that any $x(t) \in L^2$ can be expressed as

$$x(t) = \sum_{k=-\infty}^{\infty} \sum_{n=-\infty}^{\infty} \langle x(t), \psi_{kn}(t)\rangle \psi_{kn}(t) \tag{6.116}$$

where $\psi_{kn}(t) = 2^{k/2}\psi(2^k t - n)$. This expression is pretty much like an expansion into an orthonormal basis. We can find the wavelet coefficients $c_{kn} = \langle x(t), \psi_{kn}(t)\rangle$ exactly as in the orthonormal case. We also know that frames offer stability of reconstruction. Thus, in every respect this resembles an orthonormal basis, the only difference being that the functions are not linearly independent (redundancy exists in the wavelet tight frame $\{\psi_{kn}(t)\}$).

6.12 Compactly Supported Orthonormal Wavelets

Section 6.11 showed how to construct an orthonormal wavelet basis for L^2 space by starting from a paraunitary filter bank. Essentially, we defined two infinite products $\Phi(\omega)$ and $\Psi(\omega)$ starting from the digital low-pass filter $G_s(e^{j\omega})$. Under some mild conditions on $G_s(e^{j\omega})$, the products converge (Theorem 6.10). Under the further condition that $G_s(e^{j\omega})$ be power symmetric and nonzero in $[-0.5\pi, 0.5\pi]$, we saw that $\{\phi(t-k)\}$ forms an orthonormal set, and the corresponding $\{2^{k/2}\psi(2^k t - n)\}$ forms an orthonormal wavelet basis for L^2 (Theorem 6.13). If we further constrain $G_s(e^{j\omega})$ to be FIR, that is, $G_s(z) = \sum_{n=0}^{N} g_s(n)z^{-n}$, then the scaling function $\phi(t)$ and the wavelet function $\psi(t)$ have finite duration [6, 5].

Theorem 6.15: *Let $G_s(z) = \sum_{n=0}^{N} g_s(n)z^{-n}$, with $G_s(e^{j0}) = 1$ and $H_s(e^{j\omega}) = e^{j\omega} G_s^*(-e^{j\omega})$. Define the infinite products as in (6.112a) and (b), and assume that the limits $\Phi(\omega)$ and $\Psi(\omega)$ are L^2 functions, for example, by imposing power symmetry condition on $G_s(z)$ as in Theorem 6.11. Then $\phi(t)$ and $\psi(t)$ (the inverse L^2-FTs) are compactly supported, with support in $[0, N]$.* □

The time decay of the wavelet $\psi(t)$ is therefore excellent. In particular, all the basis functions $2^{k/2}\psi(2^k t - n)$ are compactly supported. By further restricting the low-pass filter $G_s(z)$ to have a sufficient number of zeroes at $\omega = \pi$, we also ensure (Section 6.13) that the FT $\Psi(\omega)$ has excellent decay (equivalently $\psi(t)$ is regular or smooth in the sense to be quantified in Section 6.13).

The rest of this section is devoted to the technical details of the above result. The reader not interested in these details can move to Section 6.13 without loss of continuity. The theorem might seem "obvious" at first sight, and indeed a simple engineering argument based on Dirac delta functions can be given (p. 521 of [7]). However, the correct mathematical justification relies on a number of deep results in function theory. One of these is the celebrated Paley-Wiener theorem for band-limited functions.

Paley-Wiener Theorem. A beautiful result in the theory of signals is that if an L^2 function $f(t)$ is band-limited, that is, $F(\omega) = 0$, $|\omega| \geq \sigma$, then $f(t)$ is the "real-axis restriction of an entire function". We say that a function $f(s)$ of the complex variable s is entire if it is analytic for all s. Examples are polynomials in s, exponentials such as e^s, and simple combinations of these. The function $f(t)$ obtained from $f(s)$ for real values of s ($s = t$) is the real-axis restriction of $f(s)$.

Thus, if $f(t)$ is a band-limited signal then an entire function $f(s)$ exists such that its real-axis restriction is $f(t)$. In particular, therefore, a band-limited function $f(t)$ is continuous and infinitely differentiable with respect to the time variable t. The entire function $f(s)$ associated with the band-limited function has the further property $|f(s)| \leq ce^{\sigma|s|}$ for some $c > 0$. We express this by saying that $f(s)$ is **exponentially bounded** or of the exponential type. What is even more interesting is that the converse of this result is true: if $f(s)$ is an entire function of the exponential type, and the real-axis restriction $f(t)$ is in L^2, then $f(t)$ is band-limited. By interchanging the time and frequency variables we can obtain similar conclusions for time-limited signals; this is what we need in the discussion of time-limited (compactly supported) wavelets.

Theorem 6.16 (Paley-Wiener): *Let $W(s)$ be an entire function such that for all s, we have $|W(s)| \leq c\exp(A|s|)$ for some $c, A > 0$, and the real-axis restriction $W(\omega)$ is in L^2. Then, there exists a function $w(t)$ in L^2 such that $W(s) = \int_{-A}^{A} w(t)e^{-jts}\,dt$.* □

A proof can be found in [23]. Thus, $w(t)$ can be regarded as a compactly supported function with support in $[-A, A]$. Recall (6.48) that $L^2[-A, A] \subset L^1[-A, A]$, so $w(t)$ is in $L^1[-A, A]$ and $L^2[-A, A]$. Therefore, $W(\omega)$ is the L^1-FT of $w(t)$, and agrees with the L^2-FT a.e.

Our aim is to show that the infinite product for $\Phi(\omega)$ satisfies the conditions of the Paley-Wiener theorem, and therefore that $\phi(t)$ is compactly supported. A modified version of the above result is more convenient for this. The modification allows the support to be more general, namely $[-A_1, A_2]$, and permits us to work with the imaginary part of s rather than the absolute value.

Theorem 6.17 (Paley-Wiener, Modified): *Let $W(s)$ be an entire function such that*

$$|W(s)| \leq \begin{cases} c_1 \exp(A_1|\operatorname{Im} s|), & \operatorname{Im} s \geq 0 \\ c_2 \exp(A_2|\operatorname{Im} s|), & \operatorname{Im} s \leq 0 \end{cases} \tag{6.117}$$

for some $c_1, c_2, A_1, A_2 > 0$, and such that the real-axis restriction $W(\omega)$ is in L^2. Then a function $w(t)$ exists in L^2 such that $W(s) = \int_{-A_2}^{A_1} w(t)e^{-jts}\,dt$. We can regard $W(\omega)$ as the FT of the function $w(t)$ supported in $[-A_2, A_1]$. □

This result can be made more general; the condition (6.117) can be replaced with one in which the right-hand sides have the form $P_i(s)\exp(A_i|\operatorname{Im} s|)$, where $P_i(s)$ are polynomials. We are now ready to sketch the proof that $\phi(t)$ and $\psi(t)$ have the compact support $[0, N]$.

1. Using the fact that $G_s(z)$ is FIR and that $G_s(e^{j0}) = 1$, show that the product $\prod_{k=1}^{\infty} G_s(e^{js/2^k})$ converges uniformly on any compact set of the complex s-plane. (For real s, namely $s = \omega$, this holds even for the IIR case as long as $\sum_n |ng_s(n)|$ converges. This was shown in Theorem 6.10.)
2. Uniformity of convergence of the product guarantees that its limit $\Phi(s)$ is an entire function of the complex variable s (Theorem 10.28, [23]).
3. The FIR nature of $G_s(z)$ allows us to establish the exponential bound (6.117) for $\Phi(s)$ with $A_2 = 0$ and $A_1 = N$. This shows that $\phi(t)$ is compactly supported in $[0, N]$. Because $\psi(t)$ is obtained from the dilation equation (6.93), the same result follows for $\psi(t)$ as well.

6.13 Wavelet Regularity

From the preceding section we know that if we construct the power symmetric FIR filter $G_s(z)$ properly, we can get an orthonormal multiresolution basis $\{\phi(t-n)\}$, and an orthonormal wavelet basis $\{2^{k/2}\psi(2^k t-n)\}$ for L^2. Both of these bases are compactly supported. These are solutions to the two-scale dilation equations

$$\phi(t) = 2\sum_{n=0}^{N} g_s(n)\phi(2t-n) \tag{6.118}$$

$$\psi(t) = 2\sum_{n=0}^{N} h_s(n)\phi(2t-n) \tag{6.119}$$

where $h_s(n) = (-1)^{n+1}g_s^*(-n-1)$. In the frequency domain we have the explicit infinite product expressions (6.112) connecting the filters $G_s(z)$ and $H_s(z)$ to the L^2-FTs $\Phi(\omega)$ and $\Psi(\omega)$.

Figure 6.45(a) shows two cases of a ninth-order FIR filter $G_s(e^{j\omega})$ used to generate the compactly supported wavelet. The resulting wavelets are shown in Figs. 6.45(b) and (c). In both cases all conditions of Theorem 6.13 are satisfied so we obtain orthonormal wavelet bases for L^2. The filter $G_s(e^{j\omega})$ has more zeroes at π for case 2 than for case 1. The corresponding wavelet looks much smoother or "regular"; this is an example of a Daubechies wavelet. By designing $G_s(z)$ to have a sufficient number of zeroes at π we can make the wavelet "as regular as we please". A quantitative discussion of the connection between the number of zeroes at π and the smoothness of $\psi(t)$ is given in the following discussions.

Qualitatively, the idea is that if $G_s(e^{j\omega})$ has a large number of zeroes at π, the function $\Phi(\omega)$ given by the infinite product (6.112a) decays "fast", as $\omega \to \infty$. This fast asymptotic decay in the frequency domain implies that the time function $\phi(t)$ is "smooth". Because $\psi(t)$ is derived from $\phi(t)$ using a finite sum (6.119), the smoothness of $\phi(t)$ is transmitted to $\psi(t)$. We will make the ideas more quantitative in the next few sections.

Why Regularity?

The point made above was that if we design an FIR paraunitary filter bank with the additional constraint that the low-pass filter $G_s(e^{j\omega})$ have a sufficient number of zeroes at π, the wavelet basis functions $\psi_{kn}(t)$ are sufficiently smooth. The smoothness requirement is perhaps the main new component brought into the filter bank theory from the wavelet theory. Its importance can be understood in a number of ways.

Consider the expansion $x(t) = \sum_{k,n} c_{k,n}\psi_{kn}(t)$. Suppose we truncate this to a finite number of terms, as is often done in practice. If the basis functions are not smooth, the error can produce perceptually annoying effects in applications such as audio and image coding, even though the L^2 norm of the error may be small.

Next, consider a tree-structured filter bank. An example is shown in Fig. 6.26. In the synthesis bank, the first path can be regarded as an effective interpolation filter, or an expander [e.g., $\uparrow 8$ in Fig. 6.26(b)] followed by a filter of the form $G_s(e^{j\omega})G_s(e^{2j\omega})G_s(e^{4j\omega})\cdots G_s(e^{2^L j\omega})$. The same finite product can be obtained by truncating to $L+1$ terms the infinite product defining $\Phi(\omega)$ (6.112), and making a change of variables. Similarly, the remaining paths can be related to interpolation filters which are various truncated versions of the infinite product defining $\Psi(\omega)$ in (6.112). Imagine that we use the tree-structured system in subband coding. The quantization error in each subband is filtered through an interpolation filter. If the impulse response of the interpolation filter is not smooth enough [e.g., if it resembles Fig. 6.45(b)], the filtered noise tends to show severe perceptual effects, for example, in image reconstruction. This explains, qualitatively, the importance of having "smooth impulse responses" for the synthesis filters.

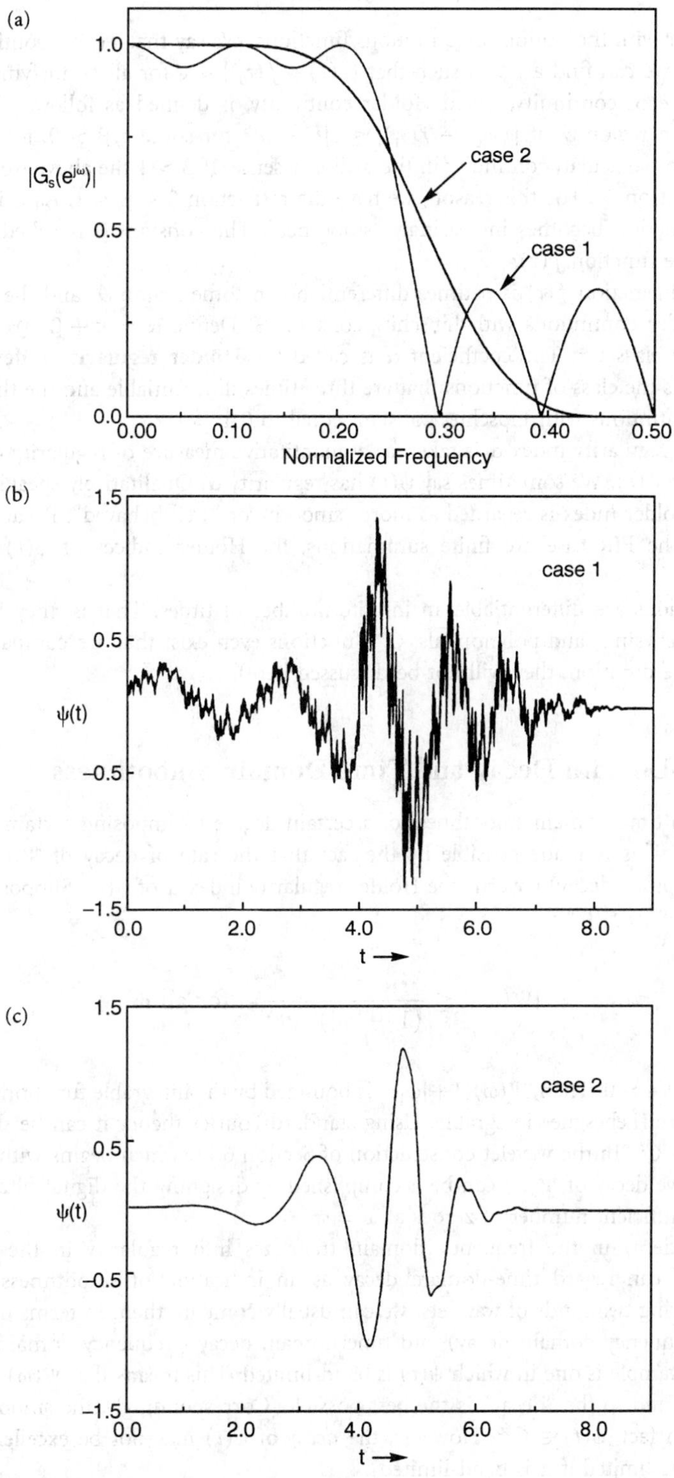

FIGURE 6.45 Demonstrating the importance of zeroes at π. (a) The response of the FIR filter $G_s(z)$ for two cases, and (b) and (c) the corresponding wavelet functions.

Smoothness and Hölder Regularity Index

We are familiar with the notion of continuous functions. We say that $f(t)$ is continuous at t_0 if, for any $\epsilon > 0$, we can find a $\delta > 0$ such that $|f(t) - f(t_0)| < \epsilon$ for all t satisfying $|t - t_0| < \delta$. A stronger type of continuity, called Hölder continuity, is defined as follows: $f(t)$ is Hölder continuous in a region $\mathcal{S}$ if $|f(t_0) - f(t_1)| \leq c|t_0 - t_1|^{\beta}$ for some $c, \beta > 0$, for all $t_0, t_1 \in \mathcal{S}$. This implies, in particular, continuity in the ordinary sense. If $\beta > 1$ the above would imply that $f(t)$ is constant on $\mathcal{S}$. For this reason, we have the restriction $0 < \beta \leq 1$. As β increases from 0 to 1, the function becomes increasingly "smoother". The constant β is called the **Lipschitz constant** of the function $f(t)$.

Suppose the function $f(t)$ is n times differentiable in some region $\mathcal{S}$ and the nth derivative $f^{(n)}(t)$ is Hölder continuous with Lipschitz constant β. Define $\alpha = n + \beta$. We say that $f(t)$ belongs to the class C^{α}. The coefficient α is called the **Hölder regularity index** of $f(t)$. For example, $C^{3.4}$ is the class of functions that are three times differentiable and the third derivatives are Hölder continuous with Lipschitz constant equal to 0.4.

The Hölder regularity index α is taken as a quantitative measure of regularity or smoothness of the function $\psi(t)$. We sometimes say $\psi(t)$ has regularity α. Qualitatively speaking, a function with a large Hölder index is regarded as more "smooth" or "well-behaved". Because the dilation equations in the FIR case are finite summations, the Hölder indices of $\phi(t)$ and $\psi(t)$ are identical.

Some functions are differentiable an infinite number of times. That is, they belong to C^{∞}. Examples are e^t, $\sin t$, and polynomials. C^{∞} functions even exist that are compactly supported (i.e., have finite duration; they will not be discussed here).

Frequency-Domain Decay and Time-Domain Smoothness

We can obtain time-domain smoothness of a certain degree by imposing certain conditions on the FT $\Psi(\omega)$. This is made possible by the fact that the rate of decay of $\Psi(\omega)$ as $\omega \to \infty$ (i.e., the asymptotic decay) governs the Hölder regularity index α of $\psi(t)$. Suppose $\Psi(\omega)$ decays faster than $(1 + |\omega|)^{-(1+\alpha)}$:

$$|\Psi(\omega)| \leq \frac{c}{(1 + |\omega|)^{1+\alpha+\epsilon}} \quad \text{for all } \omega \tag{6.120}$$

for some $c > 0, \epsilon > 0$. Then, $\Psi(\omega)(1+|\omega|)^{\alpha}$ is bounded by the integrable function $c/(1+|\omega|)^{1+\epsilon}$, and is therefore (Lebesgue) integrable. Using standard Fourier theory it can be shown that this implies $\psi(t) \in C^{\alpha}$. In the wavelet construction of Section 6.11, which begins with a digital filter bank, the above decay of $\Psi(\omega)$ can be accomplished by designing the digital filter $G_s(e^{j\omega})$ such that it has a sufficient number of zeroes at $\omega = \pi$.

Thus, the decay in the frequency domain translates into regularity in the time domain. Similarly, one can regard time-domain decay as an indication of smoothness in frequency. When comparing two kinds of wavelets, we can usually compare them in terms of time domain regularity (frequency domain decay) and time domain decay (frequency domain smoothness). An extreme example is one in which $\psi(t)$ is band-limited. This means that $\Psi(\omega)$ is zero outside the passband, and so the "decay" is the best possible. Correspondingly, the smoothness of $\psi(t)$ is excellent; in fact, $\psi(t) \in C^{\infty}$. However, the *decay* of $\psi(t)$ may not be excellent (certainly it cannot be time-limited if it is band-limited).

Return to the two familiar wavelet examples, the Haar wavelet (Fig. 6.12) and the bandpass wavelet (Figs. 6.9 and 6.11). We see that the Haar wavelet has poor decay in the frequency domain because $\Psi(\omega)$ decays only as ω^{-1}. Correspondingly, the time-domain signal $\psi(t)$ is not even continuous, hence, not differentiable.[16] The bandpass wavelet, on the other hand, is band-limited, so the decay in frequency is excellent. Thus, $\psi(t) \in C^\infty$, but it decays slowly, behaving similarly to t^{-1} for large t. These two examples represent two extremes of orthonormal wavelet bases for L^2.

The game, therefore, is to construct wavelets that have good **decay in time** as well as good **regularity in time.** An extreme hope is where $\psi(t) \in C^\infty$, and has compact support as well. It can be shown that such $\psi(t)$ can never give rise to an orthonormal basis, so we must strike a compromise between *regularity in time* and *decay in time.*

Regularity and Decay in Early Wavelet Constructions

In 1982 Stromberg showed how to construct wavelets in such a way that $\psi(t)$ has exponential decay, and at the same time has arbitrary regularity (i.e., $\psi(t) \in C^k$ for any chosen integer k). In 1985 Meyer constructed wavelets with band-limited $\psi(t)$ [so $\psi(t) \in C^\infty$ as for the bandpass wavelet], but he also showed how to design this $\psi(t)$ to decay faster than any chosen inverse polynomial, as $t \to \infty$. Figure 6.46(a) shows an example of a Meyer wavelet; a detailed description of this wavelet can be found in [5]. In both of the above constructions the wavelets gave rise to orthonormal bases for L^2.

In 1987 and 1988 Battle and Lemarié independently constructed wavelets with similar properties as Stromberg's wavelets, namely $\psi(t) \in C^k$ for arbitrary k, and $\psi(t)$ decays exponentially. Their construction is based on spline functions and an orthonormalization step, as described in Section 6.10. The resulting wavelets, while not compactly supported, decay exponentially and generate orthonormal bases. Figure 6.46(b) shows an example of the Battle-Lemarié wavelet.

Table 6.1 gives a summary of the main features of these early wavelet constructions (first three entries). When these examples were constructed, the relation between wavelets and digital filter banks was not known. The constructions were not systematic or unified by a central theory. Moreover, it was not clear whether one could get a compactly supported (i.e., finite duration) wavelet $\psi(t)$ which at the same time had arbitrary regularity (i.e., $\psi(t) \in C^k$ for any chosen k), *and generated an orthonormal wavelet basis.* This was made possible for the first time when the relation between wavelets and digital filter banks was observed by Daubechies in [6]. Simultaneously and independently Mallat invented the multiresolution framework and observed the relation between his framework, wavelets, and paraunitary digital filter banks (the CQF

TABLE 6.1 Summary of Several Types of Wavelet Bases for $L^2(R)$

Type of Wavelet	Decay of $\psi(t)$ in Time	Regularity of $\psi(t)$ in Time	Type of Wavelet Basis
Stromberg, 1982	Exponential	$\psi(t) \in C^k$; k can be chosen arbitrarily large	Orthonormal
Meyer, 1985	Faster than any chosen inverse polnomial	$\psi(t) \in C^\infty$ (band-limited)	Orthonormal
Battle-Lemarié, 1987, 1988 (Splines)	Expotential	$\psi(t) \in C^k$; k can be chosen arbitrarily large	Orthonormal
Daubechies, 1988	Compactly supported	$\psi(t) \in C^\alpha$; α can be chosen as large as we please	Orthonormal

[16]It is true that $\psi(t)$ is differentiable almost everywhere, but the discontinuities at the points $t = 0, 0.5, 1.0$ will be very noticeable if we take linear combinations such as $\sum_{k,n} c_{kn}\psi_{kn}(t)$.

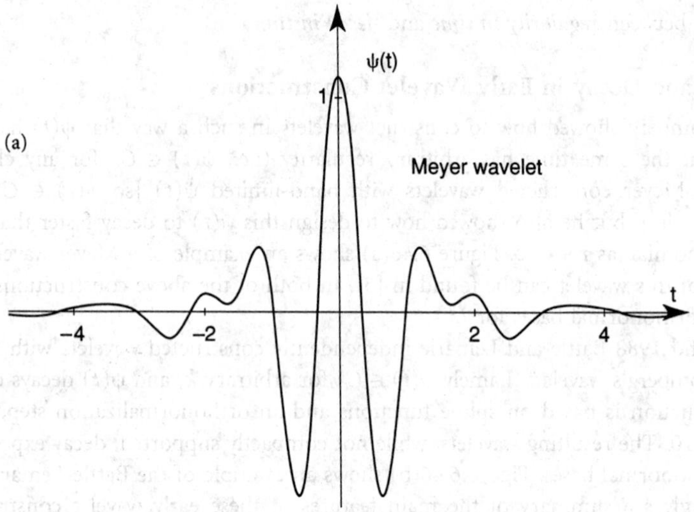

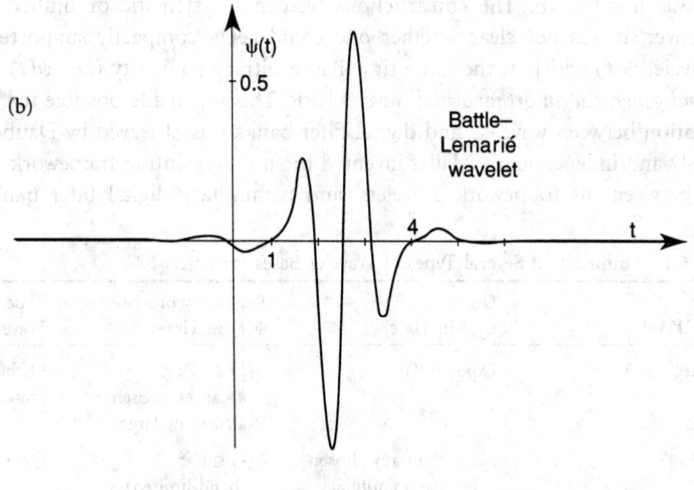

FIGURE 6.46 (a) An example of the Meyer wavelet, and (b) an example of the Battle-Lemarié wavelet.

bank, Section 6.4). These discoveries have made the wavelet construction easy and systematic, as described in Sections 6.11 and 6.12. The way to obtain arbitrary wavelet regularity with this scheme is described next.

Time-Domain Decay and Time-Domain Regularity

We now state a fundamental limitation which arises when trying to impose regularity and decay simultaneously [5].

Theorem 6.18: Vanishing Moments. *Let* $\{2^{k/2}\psi(2^k t - n)\}, -\infty \le k, n \le \infty$ *be an orthonormal set in* L^2. *Suppose the wavelet* $\psi(t)$ *satisfies the following properties:*

1. $|\psi(t)| \le c(1+|t|)^{-(m+1+\epsilon)}$ *for some integer* m *and some* $\epsilon > 0$; *that is, the wavelet decays faster than* $(1+|t|)^{-(m+1)}$.
2. $\psi(t) \in C^m$ *[i.e.,* $\psi(t)$ *differentiable* m *times], and the* m *derivatives are bounded.*

Then the first m *moments of* $\psi(t)$ *are zero, that is,* $\int t^i\psi(t)dt = 0$ for $0 \le i \le m$. □

Impossibility of Compact Support, Infinite Differentiability, and Orthonormality. Suppose we have an orthonormal wavelet basis such that $\psi(t)$ is compactly supported, and infinitely differentiable [i.e., $\psi(t) \in C^\infty$]. Then all the conditions of Theorem 6.18 are satisfied. So the moments of $\psi(t)$ are zero, and therefore $\psi(t) = 0$ for all t violating the unit-norm property of $\psi(t)$. Thus, we cannot design compactly supported orthonormal wavelets which are infinitely differentiable; only a finite Hölder index can be accomplished. A similar observation can be made even when $\psi(t)$ is not compactly supported as long as it decays faster than any inverse polynomial (e.g., exponential decay).

The vanishing moment condition $\int t^i\psi(t)dt = 0, 0 \le i \le m$ implies that the L^2-FT $\Psi(\omega)$ has $m+1$ zeroes at $\omega = 0$. This follows by using standard theorems on the L^1-FT [23].[17] Thus, the first m derivatives of $\Psi(\omega)$ vanish at $\omega = 0$. This implies a certain degree of flatness at $\omega = 0$. Summarizing, we have the following result.

Theorem 6.19, Flatness in Frequency and Regularity in Time: *Suppose we have a compactly supported* $\psi(t)$ *generating an orthonormal wavelet basis* $\{2^{k/2}\psi(2^k t - n)\}$, *and let* $\psi(t) \in C^m$, *with* m *derivatives bounded. Then* $\Psi(\omega)$ *has* $m+1$ *zeroes at* $\omega = 0$. □

Return now to the wavelet construction technique described in Section 6.11. We started from a paraunitary FIR filter bank (Fig. 6.22(a)) and obtained the scaling function $\phi(t)$ and wavelet function $\psi(t)$ as in (6.118) and (6.119). The FIR nature implies that $\psi(t)$ has compact support (Section 6.12). With the mild conditions of Theorem 6.13 satisfied, we have an orthonormal wavelet basis for L^2. We see that if the wavelet $\psi(t)$ has Hölder index α, it satisfies all the conditions of Theorem 6.19, where m is the integer part of α. Thus, $\Psi(\omega)$ has $m+1$ zeroes at $\omega = 0$, but because $\Phi(0) \ne 0$ (Section 6.10), we conclude from the dilation equation $\Psi(\omega) = H_s(e^{j\omega/2})\Phi(\omega/2)$ that the high-pass FIR filter $H_s(z)$ has $m+1$ zeroes at $\omega = 0$ (i.e., at $z = 1$). Using the relation $H_s(e^{j\omega}) = e^{j\omega}G_s^*(-e^{j\omega})$ we conclude that $G_s(e^{j\omega})$ has $m+1$ zeroes at $\omega = \pi$; that is, the low-pass FIR filter $G_s(z)$ has the form $G_s(z) = (1+z^{-1})^{m+1}F(z)$, where $F(z)$ is FIR. Summarizing, we have the theorem below.

Theorem 6.20, Zeroes at π and Regularity: *Suppose we wish to design a compactly supported orthonormal wavelet basis for* L^2 *by designing an FIR filter* $G_s(z)$ *satisfying the conditions of Theorem* 6.13. *If* $\psi(t)$ *must have the Hölder regularity index* α *then it is necessary that* $G_s(z)$ *have the form* $G_s(z) = (1+z^{-1})^{m+1}F(z)$, *where* $F(z)$ *is FIR, and* m *is the integer part of* α. □

[17] Because $\psi(t) \in L^2$ and has compact support, $\psi(t) \in L^1$ as well.

One zero at π *is essential.* From Theorem 6.10 we know that we must have $G_s(e^{j0}) = 1$ for the infinite product (6.112a) to converge. Theorem 6.13 imposes further conditions that enable us to obtain an orthonormal wavelet basis for L^2. One of these conditions is the power symmetric property $|G_s(e^{j\omega})|^2 + |G_s(-e^{j\omega})|^2 = 1$. Together with $G_s(e^{j0}) = 1$, this implies $G_s(e^{j\pi}) = 0$. Thus, it is necessary to have at least one zero of $G_s(e^{j\omega})$ at π. The filter that generates the Haar basis (Example 12) has exactly one zero at π, but the Haar wavelet $\psi(t)$ is not even continuous. If we desire increased regularity (continuity, differentiability, etc.), we need to put additional zeroes at π, as the above theorem shows.

Design techniques for paraunitary filter banks do not automatically yield filters which have zeroes at π. This condition must be incorporated separately. The maximally flat filter bank solution (Section 6.4) does satisfy this property, and in fact even allows us to specify the number of zeroes at π.

Wavelets with Specified Regularity

The fundamental connection between digital filter banks and continuous time wavelets, elaborated in the preceding sections, allows us to construct the scaling function $\phi(t)$ and the wavelet function $\psi(t)$ with specified regularity index α. If $G_s(z)$ has a certain number of zeroes at π, this translates into the Hölder regularity index α. What really matters is not only the number of zeroes at π, but also the order of the FIR filter $G_s(z)$.

For a given order N of the filter $G_s(z)$, suppose we wish to put as many of its zeroes as possible at π. Let this number be K. What is the largest possible K? Not all N zeroes can be at π because we have imposed the power symmetric condition on $G_s(z)$. The best we can do is to put all the unit circle zeroes at π. The power symmetric condition says that $G(z) \triangleq \widetilde{G}_s(z)G_s(z)$ is a half-band filter. This filter has order $2N$, with $2K$ zeroes at π. Because we wish to maximize K for fixed N, the solution for $G(z)$ is the maximally flat FIR filter (Fig. 6.25), given in (6.45). As the filter in (6.45) has $2K$ zeroes at π and order $2N = 4K - 2$, we conclude that $K = (N+1)/2$. For example, if $G_s(z)$ is a fifth-order power symmetric filter it can have at most three zeroes at π.

The 20% Regularity Rule

Suppose $G_s(z)$ has been designed to be FIR power symmetric of order N, with the number K of zeroes at π adjusted to be maximum (i.e., $K = (N+1)/2$). It can be shown that the corresponding scaling and wavelet functions have a Hölder regularity index $\alpha \approx 0.2\,K$. This estimate is poor for small K, but improves as K grows. Thus, every additional zero at π contributes to $\approx 20\%$ improvement in regularity.

For $K = 4$ (i.e., seventh-order $G_s(z)$) we have $\alpha = 1.275$, which means that the wavelet $\psi(t)$ is once differentiable and the derivative is Hölder continuous with Lipschitz constant 0.275. For $K = 10$ [19th-order $G_s(z)$] we have $\alpha = 2.9$, so the wavelet $\psi(t)$ is twice differentiable and the second derivative has Hölder regularity index 0.9.

Design Procedure. The design procedure is therefore very simple. For a specified regularity index α, we can estimate K and hence $N = 2K - 1$. For this K, we compute the coefficients of the FIR half-band maximally flat filter $G(z)$ using (6.45). From this we compute a spectral factor $G_s(z)$ of the filter $G(z)$. Tables of the filter coefficients $g_s(n)$ for various values of N can be found in [5]. From the coefficients $g_s(n)$ of the FIR filter $G_s(z)$, the compactly supported scaling and wavelet functions are fully determined via the dilation equations. These wavelets are called Daubechies wavelets and were first generated in [6]. Figure 6.45(c) is an example, generated with a ninth-order FIR filter $G_s(z)$, whose response is shown as case 2 in Fig. 6.45(a).

The above regularity estimates, based on frequency domain behavior, give a single number α, which represents the regularity of $\psi(t)$ for all t. It is also possible to define **pointwise** or

local regularity of the function $\psi(t)$ so that its smoothness can be estimated as a function of time t. These estimation methods, based on time domain iterations, are more sophisticated, but give a detailed view of the behavior of $\psi(t)$. Detailed discussions on obtaining various kinds of estimates for regularity can be found in [5] and [26].

.14 Concluding Remarks

We introduced the WT, and studied its connection to filter banks and STFTs. A number of mathematical concepts such as frames and Riesz bases were reviewed and used later for a more careful study of wavelets. We introduced the idea of multiresolution analysis, and explained the connections both to filter banks and wavelets. This connection was then used to generate orthonormal wavelet bases from paraunitary filter banks. Such wavelets have compact support when the filter bank is FIR. The regularity or smoothness of the wavelet was quantified in terms of the Hölder exponent. We showed that we can achieve any specified Hölder exponent for compactly supported wavelets by restricting the low-pass filter of the FIR paraunitary filter bank to be a maximally flat power symmetric filter, with a sufficient number of zeroes at π.

Why Wavelets?

Discussions comparing wavelets with other types of time-frequency transforms appear at several places in this chapter. Here is a list of these discussions:

1. Section 6.2 discusses basic properties of wavelets, and gives an elementary comparison of wavelet basis with the Fourier basis.
2. Section 6.3 compares the WT to the STFT, and shows the time-frequency tilings for both cases (e.g., see Figs. 6.18 and 6.20).
3. Section 6.9 gives a deeper comparison with the STFT in terms of stability properties of the inverse, existence of frames, etc.
4. Section 6.13 shows a comparison to the traditional filter bank design approach. In traditional designs the appearance of zero(es) at π is not considered important. At the beginning of Section 6.13 (under "Why Regularity?"), we discuss the importance of these zeroes in wavelets as well as in tree-structured filter banks.

rther Reading

The literature on wavelet theory and applications is enormous. This chapter is only a brief introduction, concentrating on one-dimensional orthonormal wavelets. Many results can be found on the topics of multidimensional wavelets, biorthogonal wavelets, and wavelets based on IIR filter banks. Two special issues of the *IEEE Transactions* have appeared on the topic thus far [27, 28]. Multidimensional wavelets are treated by several authors in the edited volume of [15], and the filter bank perspective can be found in the work by Kovačević and Vetterli [27]. Advanced results on multidimensional wavelets can be found in [29]. Advanced results on wavelets constructed from M-channel filter banks can be found in the chapter by Gopinath and Burrus in the edited volume of [15], and in the work by Steffen et al. [28]. The reader can also refer to the collections of chapters in [15] and [16], and the many references therein.

knowledgment

The authors are grateful to Dr. Ingrid Daubechies, Princeton University, Princeton, NJ, for many useful e-mail discussions on wavelets.

This work was supported in parts by Office of Naval Research grant N00014-93-1-0231, Rockwell International, and Tektronix, Inc.

References

[1] M. E. Van Valkenburg, *Introduction to Modern Network Synthesis,* New York: John Wiley & Sons, 1960.

[2] A. V. Oppenheim and R. W. Schafer, *Discrete-Time Signal Processing,* Englewood Cliffs, NJ: Prentice-Hall, 1989.

[3] A. Grossman and J. Morlet, "Decomposition of Hardy functions into square integrable wavelets of constant shape," *SIAM J. Math. Anal.*, vol. 15, pp. 723–736, 1984.

[4] Y. Meyer, *Wavelets and Operators,* Cambridge: Cambridge University Press, 1992.

[5] I. Daubechies, *Ten Lectures on Wavelets,* SIAM, CBMS Series, Apr. 1992.

[6] I. Daubechies, "Orthonormal bases of compactly supported wavelets," *Commun. Pure Appl. Math.*, vol. 41, pp. 909–996, Nov. 1988.

[7] P. P. Vaidyanathan, *Multirate Systems and Filter Banks,* Englewood Cliffs, NJ: Prentice-Hall, 1993.

[8] M. Vetterli, "A theory of multirate filter banks," *IEEE Trans. Acoust. Speech Signal Process.*, vol. ASSP-35, pp. 356–372, Mar. 1987.

[9] A. N. Akansu and R. A. Haddad, *Multiresolution Signal Decomposition: Transforms, Subbands, and Wavelets,* Orlando, FL: Academic Press, 1992.

[10] H. S. Malvar, *Signal Processing with Lapped Transforms,* Norwood, MA: Artech House, 1992.

[11] S. Mallat, "Multiresolution approximations and wavelet orthonormal bases of $L^2(R)$," *Trans. Am. Math. Soc.*, vol. 315, pp. 69–87, Sept. 1989.

[12] C. E. Heil and D. F. Walnut, "Continuous and discrete wavelet transforms," *SIAM rev.*, vol. 31, pp 628–666, Dec. 1989.

[13] M. Vetterli and C. Herley, "Wavelets and filter banks," *IEEE Trans. Signal Process.*, vol. SP-40, 1992.

[14] R. A. Gopinath and C. S. Burrus, "A tutorial overview of filter banks, wavelets, and interrelations," *Proc. IEEE Int. Symp. Circuits Syst.*, pp. 104–107, May 1993.

[15] C. K. Chui, Vol. 1, *An Introduction to Wavelets,* and Vol. 2 (edited), *Wavelets: A Tutorial in Theory and Applications,* Orlando, FL: Academic Press, 1992.

[16] J. J. Benedetto and M. W. Frazier, *Wavelets: Mathematics and Applications,* Boca Raton, FL: CRC Press, 1994.

[17] J. B. Allen and L. R. Rabiner, "A unified theory of short-time spectrum analysis and synthesis," *Proc. IEEE,* vol. 65, pp. 1558–1564, Nov. 1977.

[18] M. J. T. Smith and T. P. Barnwell, III, "A procedure for designing exact reconstruction filter banks for tree structured subband coders," *Proc. IEEE Int. Conf. Acoust. Speech, Signal Proc.*, pp. 27.1.1–27.1.4, San Diego, CA, Mar. 1984.

[19] F. Mintzer, "Filters for distortion-free two-band multirate filter banks," *IEEE Trans. Acoust., Speech Signal Process.*, vol. ASSP-33, pp. 626–630, June 1985.

[20] P. P. Vaidyanathan, "Theory and design of M-channel maximally decimated quadrature mirror filters with arbitrary M, having perfect reconstruction property," *IEEE Trans. Acoustics, Speech Signal Process.*, vol. ASSP-35, pp. 476–492, Apr. 1987.

[21] V. Belevitch, *Classical Network Theory,* San Francisco: Holden Day, 1968.

[22] T. M. Apostol, *Mathematical Analysis,* Reading, MA: Addison-Wesley, 1974.

[23] W. Rudin, *Real and Complex Analysis,* New York: McGraw-Hill, 1966.

[24] R. M. Young, *An Introduction to Nonharmonic Fourier Series,* New York: Academic Press, 1980.

[25] R. J. Duffin and A. C. Schaeffer, "A class of nonharmonic Fourier series," *Trans. Am. Math. Soc.*, vol. 72, pp. 341–366, 1952.

[26] O. Rioul, "Simple regularity criteria for subdivision schemes," *SIAM J. Math. Anal.*, vol. 23, pp. 1544–1576, Nov. 1992.

[27] Special issue on wavelet transforms and multiresolution signal analysis, *IEEE Trans. Info. Theory*, vol. 38, Mar. 1992.

[28] Special issue on wavelets and signal processing, *IEEE Trans. Signal Process*, vol. 41, Dec. 1993.

[29] A. Cohen and I. Daubechies, "Non-separable bidimensional wavelet bases," *Rev. Mat. IberoAm.*, vol. 9, pp. 51–137, 1993.

7

Graph Theory

Krishnaiyan Thulasiraman
University of Oklahoma

7.1 Introduction

Graph theory had its beginning in Euler's solution of what is known as the Konigsberg Bridge problem. Kirchhoff developed the theory of trees in 1847 as a tool in the study of electrical networks. This was the first application of graph theory to a problem in physical science. Electrical network theorists have since played a major role in the phenomenal advances of graph theory that have taken place in this century. A comprehensive treatment of these developments may be found in [1]. In this chapter we develop most of those results which form the foundation of graph theoretic study of electrical networks.

Our development of graph theory is self-contained, except for the definitions of standard set-theoretic operations and elementary results from matrix theory. We wish to note that the **ring sum** of two sets S_1 and S_2 refers to the set consisting of all those elements which are in S_1 or in S_2 but not in S_1 and S_2.

7.2 Basic Concepts

A graph $G = (V, E)$ consists of two sets: a finite set $V = (v_1, v_2, ..., v_n)$ of elements called **vertices** and a finite set $E = (e_1, e_2, ..., e_m)$ of elements called **edges.** Each edge is identified with a pair of vertices. If the edges of G are identified with ordered pairs of vertices, then G is called a **directed** or an **oriented graph**. Otherwise G is called an **undirected** or a **nonoriented graph.** Graphs are amenable for pictorial representations. In a pictorial representation each vertex is represented by a dot and each edge is represented by a line segment joining the dots associated with the edge. In directed graphs we assign an orientation or direction to each edge. If the edge is associated with the ordered pair (v_i, v_j), then this edge is oriented from v_i to v_j. If an edge e connects vertices v_i and v_j then it is denoted by $e = (v_i, v_j)$. In a directed graph (v_i, v_j) refers

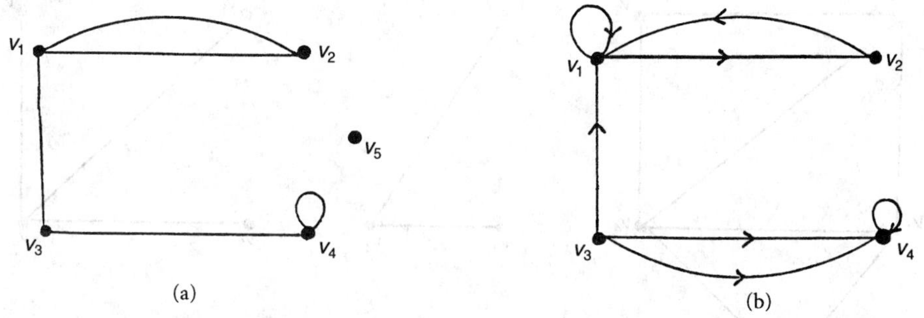

FIGURE 7.1 (a) An undirected graph; (b) a directed graph.

to an edge directed from v_i to v_j. A graph and a directed graph are shown in Fig. 7.1. Unless explicitly stated, the term "graph" may refer to an undirected graph or to a directed graph.

The vertices v_i and v_j associated with an edge are called the **end vertices** of the edge. All edges having the same pair of end vertices are called **parallel edges.** In a directed graph parallel edges refer to edges connecting the same pair of vertices v_i and v_j and oriented in the same way from v_i to v_j or from v_j to v_i. For instance, in the graph of Fig. 7.1(a), the edges connecting v_1 and v_2 are parallel edges. In the directed graph of Fig. 7.1(b) the edges connecting v_3 and v_4 are parallel edges. However, the edges connecting v_1 and v_2 are not parallel edges because they are not oriented the same way. If the end vertices of an edge are not distinct, then the edge is called a **self-loop.** The graph of Fig. 7.1(a) has one self-loop and the graph of Fig. 7.1(b) has two self-loops.

An edge is said to be **incident on** its end vertices. In a directed graph the edge (v_i,v_j) is said to be **incident out** of v_i and is said to be **incident into** v_j. Vertices v_i and v_j are adjacent if an edge connects v_i and v_j.

The number of edges incident on a vertex v_i is called the **degree** of v_i and is denoted by $d(v_i)$. In a directed graph $d_{in}(v_i)$ refers to the number of edges incident into vertex v_i, and it is called the **in-degree** of v_i. $d_{out}(v_i)$ refers to the number of edges incident out of vertex v_i and it is called the **out-degree** of v_i. If $d(v_i) = 0$, then v_i is called an **isolated vertex.** If $d(v_i) = 1$, then v_i is called a **pendant vertex.** A self-loop at a vertex v_i is counted twice while computing $d(v_i)$. As an example, in the graph of Fig. 7.1(a), $d(v_1) = 3$, $d(v_4) = 3$, and v_5 is an isolated vertex. In the directed graph of Fig. 7.1(b) $d_{in}(v_1) = 3$, $d_{out}(v_1) = 2$.

Note that in a directed graph, for every vertex v_i,

$$d(v_i) = d_{in}(v_i) + d_{out}(v_i)$$

Theorem 1:

1. *The sum of the degrees of the vertices of a graph G is equal to 2m, where m is the number of edges of G.*
2. *In a directed graph with m edges, the sum of the in-degrees and the sum of the out-degrees are both equal to m.*

PROOF.

1. Because each edge is incident on two vertices, it contributes 2 to the sum of the degrees of G. Hence, all edges together contribute $2m$ to the sum of the degrees.
2. Proof follows if we note that each edge is incident out of exactly one vertex and incident into exactly one vertex. □

Theorem 2: *The number of vertices of odd degree in any graph is even.*

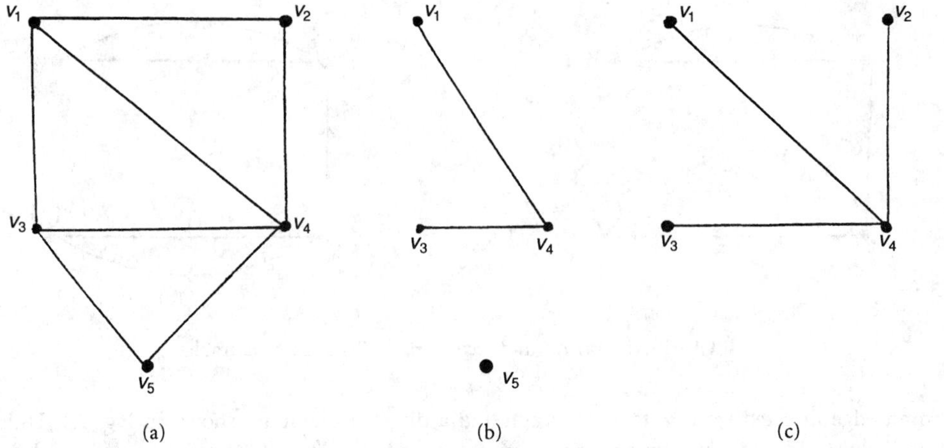

FIGURE 7.2 (a) Graph G; (b) subgraph of G; (c) an edge-induced subgraph of G.

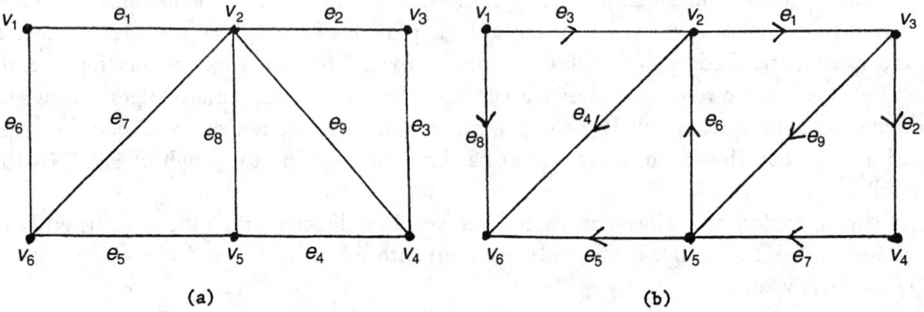

FIGURE 7.3 (a) An undirected graph; (b) a directed graph.

PROOF. By Theorem 1, the sum of the degrees of the vertices is even. Thus, the sum of the odd degrees must be even. This is possible only if the number of vertices of odd degree is even. □

Consider a graph $G = (V, E)$. The graph $G' = (V', E')$ is a subgraph of G if $V' \subseteq V$ and $E' \subseteq E$. If every vertex in V' is an end vertex of an edge in E', then G' is called the **induced subgraph** of G on E'. As an example, a graph G and two subgraphs of G are shown in Fig. 7.2.

In a graph G a **path** P connecting vertices v_i and v_j is an alternating sequence of vertices and edges starting at v_i and ending at v_j, with all vertices except v_i and v_j being distinct. In a directed graph a *path* P connecting vertices v_i and v_j is called a **directed path** from v_i to v_j if all the edges in P are oriented in the same direction as we traverse P from v_i toward v_j. If a path starts and ends at the same vertex, it is called a **circuit**.[1] In a directed graph, a circuit in which all the edges are oriented in the same direction is called a **directed circuit**. It is often convenient to represent paths and circuits by the sequences of edges representing them.

For example, in the undirected graph of Fig. 7.3(a) P: e_1, e_2, e_3, e_4 is a path connecting v_1 and v_5 and C: $e_1, e_2, e_3, e_4, e_5, e_6$ is a circuit. In the directed graph of Fig. 7.3(b) $P : e_1, e_2, e_7, e_5$ is a directed path and $C : e_1, e_2, e_7, e_6$ is a directed circuit. Note that e_7, e_5, e_4, e_1, e_2 is a circuit in this directed graph, although it is not a directed circuit.

[1]In electrical network theory literature the term **loop** is also used to refer to a circuit.

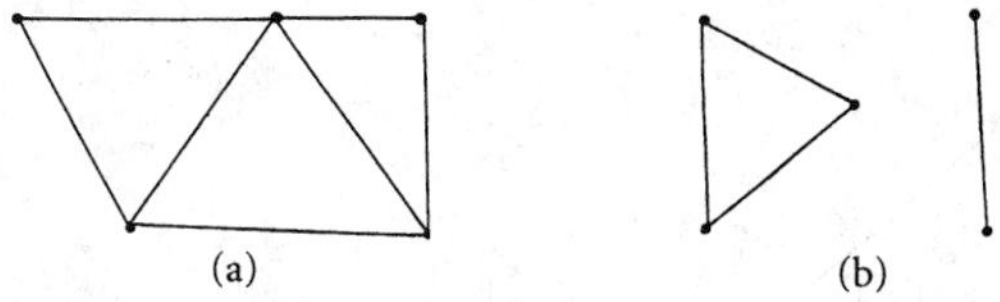

FIGURE 7.4 (a) A connected graph; (b) a disconnected graph.

Two vertices v_i and v_j are said to be **connected** in a graph G if a path in G connects v_i and v_j. A graph G is **connected** if every pair of vertices in G is connected; otherwise it is a **disconnected graph**. For example, the graph G in Fig. 7.4(a) is connected, but the graph in Fig. 7.4(b) is not connected.

A connected subgraph $G' = (V', E')$ of a graph $G = (V, E)$ is a **component** of G if adding to G' an edge $e \in E - E'$ results in a disconnected graph. Thus, a connected graph has exactly one component. For example, the graph in Fig. 7.4(b) is not connected and has two components.

A **tree** is a graph that is connected and has no circuits. Consider a connected graph G. A subgraph of G is a **spanning tree**[2] of G if the subgraph is a tree and contains all the vertices of G. A tree and a spanning tree of the graph of Fig. 7.5(a) are shown in Fig. 7.5(b) and (c), respectively.

The edges of a spanning tree T are called the **branches** of T. Given a spanning tree of a connected graph G, the **cospanning tree**[2] relative to T is the subgraph of G induced by the edges that are not present in T. For example, the cospanning tree relative to the spanning tree T of Fig. 7.5(c) consists of the edges e_3, e_6, and e_7. The edges of a cospanning tree are called **chords.**

A subgraph of a graph G is a **k-tree** of G if the subgraph has exactly k components and has no circuits. For example, a 2-tree of the graph of Fig. 7.5(a) is shown in Fig. 7.5(d). If a graph has k components, then a **forest** of G is a spanning subgraph that has k components and no circuits. Thus, each component of the forest is a spanning tree of a component of G. A graph G and a forest of G are shown in Fig. 7.6.

Consider a directed graph G. A spanning tree T of G is called a **directed spanning tree** with root v_i if T is a spanning tree of G, and $d_{\text{in}}(v_i) = 0$ and $d_{\text{in}}(v_j) = 1$ for all $v_j \neq v_i$. A directed graph G and a directed spanning tree with root v_1 are shown in Fig. 7.7.

It can easily be verified that in a tree exactly one path connects any two vertices.

Theorem 3: *A tree on n vertices has $n - 1$ edges.*

PROOF. Proof is by induction on the number of vertices of the tree. Clearly, the result is true if a tree has one or two vertices. Assume that the result is true for trees on $n \geq 2$ or fewer vertices. Consider now a tree T on $n + 1$ vertices. Pick an edge $e = (v_i, v_j)$ in T. Removing e from T would disconnect it into exactly two components T_1 and T_2. Both T_1 and T_2 are themselves trees. Let n_1 and m_1 be the number of vertices and the number of edges in T_1, respectively. Similarly n_2 and m_2 are defined. Then, by the induction hypothesis

$$m_1 = n_1 - 1$$

and

$$m_2 = n_2 - 1$$

[2]In electrical network theory literature the terms **tree** and **cotree** are usually used to mean spanning tree and cospanning tree, respectively.

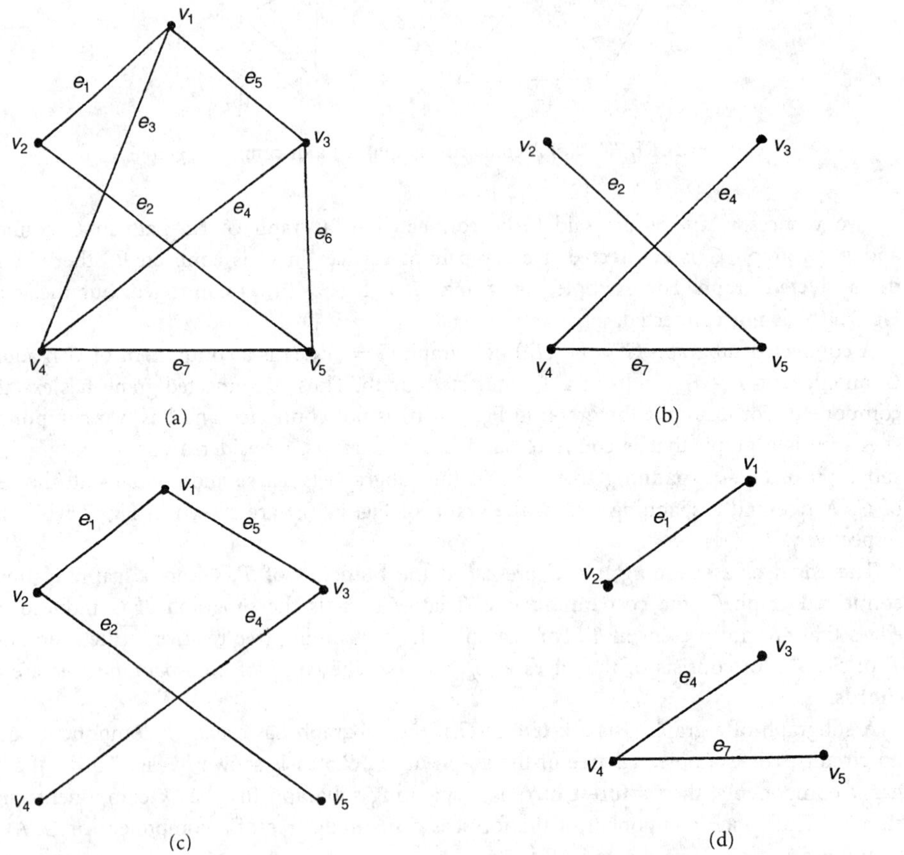

FIGURE 7.5 (a) Graph G; (b) a tree of graph G; (c) a spanning tree of G; (d) a 2-tree of G.

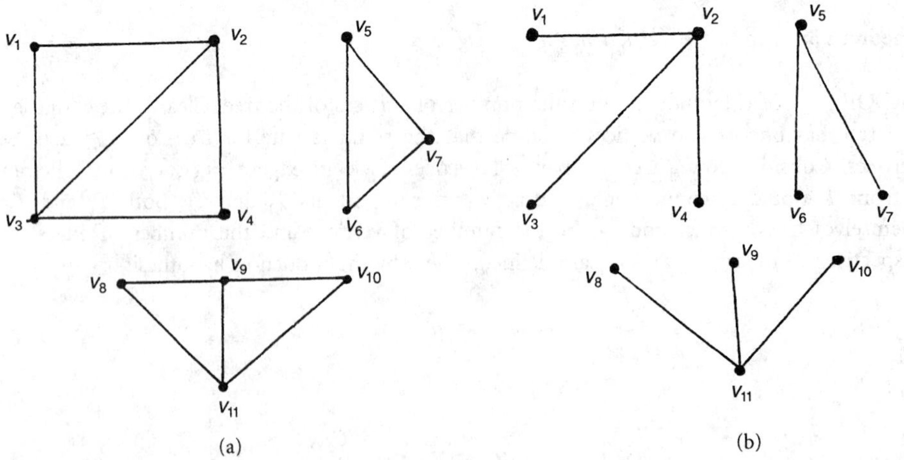

FIGURE 7.6 (a) Graph G; (b) a forest of G.

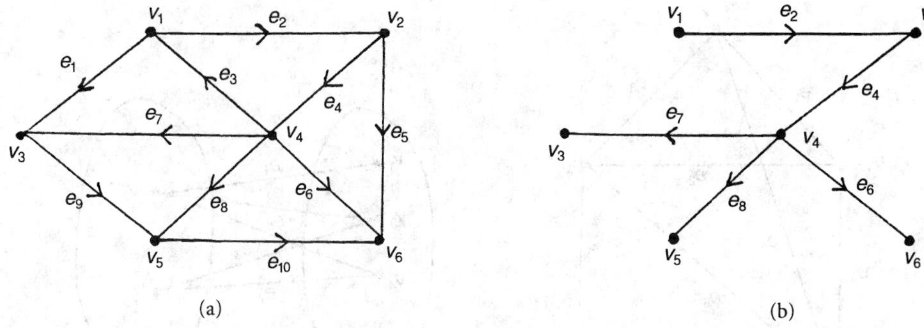

FIGURE 7.7 (a) Directed graph G; (b) a directed spanning tree of G with root v_1.

Thus, the number m of edges in T is given by

$$\begin{aligned} m &= m_1 + m_2 + 1 \\ &= (n_1 - 1) + (n_2 - 1) + 1 \\ &= n_1 + n_2 - 1 \\ &= n - 1 \end{aligned}$$

This completes the proof of the theorem. □

If a connected graph G has n vertices, m edges, and k components, then the **rank** ρ and **nullity** μ of G are defined as follows:

$$\rho(G) = n - k \tag{7.1}$$

$$\mu(G) = m - n + k \tag{7.2}$$

Clearly, if G is connected, then any spanning tree of G has $\rho = n-1$ branches and $\mu = m-n+1$ chords.

We conclude this subsection with the following theorems. Proofs of these theorems may be found in [2].

Theorem 4: *A tree on $n \geq 2$ vertices has at least two pendant vertices.* □

Theorem 5: *A subgraph of an n-vertex connected graph G is a spanning tree of G if and only if the subgraph has no circuits and has $n - 1$ edges.* □

Theorem 6: *If a subgraph G' of a connected graph G has no circuits then there exists a spanning tree of G that contains G'.* □

7.3 Cuts, Circuits, and Orthogonality

We introduce here the notions of a cut and a cutset and develop certain results which bring out the dual nature of circuits and cutsets.

Consider a connected graph $G = (V, E)$ with n vertices and m edges. Let V_1 and V_2 be two mutually disjoint nonempty subsets of V such that $V = V_1 \cup V_2$. Thus, $V_2 = \overline{V_1}$, the

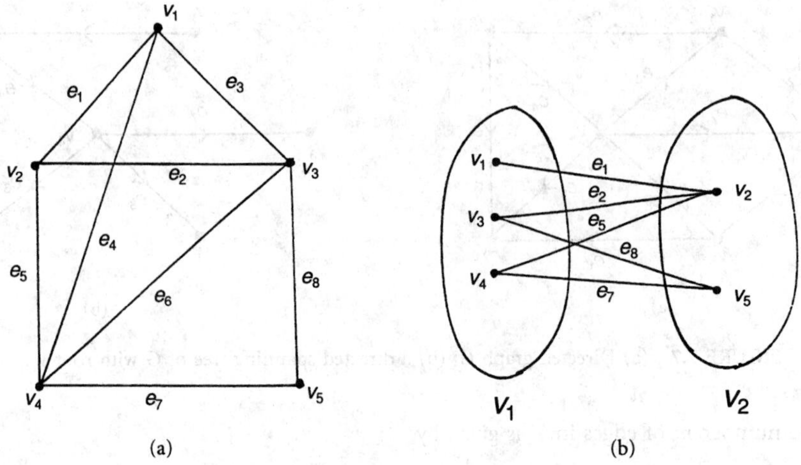

FIGURE 7.8 (a) Graph G; (b) cut $\langle v_1, v_2 \rangle$ of G.

complement of V_1 in V. V_1 and V_2 are also said to form a partition of V. Then the set of all those edges which have one end vertex in V_1 and the other in V_2 is called a **cut** of G and is denoted by $< V_1, V_2 >$. As an example, a graph G and a cut $< V_1, V_2 >$ of G are shown in Fig. 7.8.

The graph G' which results after removing the edges in a cut will have at least two components and so will not be connected. G' may have more than two components. A **cutset** S of a connected graph G is a minimal set of edges of G such that removal of S disconnects G into exactly two components. Thus, a cutset is also a cut. Note that the minimality property of a cutset implies that no proper subset of a cutset is a cutset.

Consider a spanning tree T of a connected graph G. Let b be a branch of T. Removal of the branch b disconnects T into exactly two components, T_1 and T_2. Let V_1 and V_2 denote the vertex sets of T_1 and T_2, respectively. Note that V_1 and V_2 together contain all the vertices of G. We can verify that the cut $< V_1, V_2 >$ is a cutset of G and is called the **fundamental cutset** of G with respect to branch b of T. Thus, for a given connected graph G and a spanning tree T of G, we can construct $n - 1$ fundamental cutsets, one for each branch of T. As an example, for the graph shown in Fig. 7.8, the fundamental cutsets with respect to the spanning tree $T = [e_1, e_2, e_6, e_8]$ are

$$\begin{aligned} \text{Branch } e_1 &: \quad (e_1, e_3, e_4) \\ \text{Branch } e_2 &: \quad (e_2, e_3, e_4, e_5) \\ \text{Branch } e_6 &: \quad (e_6, e_4, e_5, e_7) \\ \text{Branch } e_8 &: \quad (e_8, e_7) \end{aligned}$$

Note that the fundamental cutset with respect to branch b contains b. Furthermore, the branch b is not present in any other fundamental cutset with respect to T.

Next we identify a special class of circuits of a connected graph G. Again, let T be a spanning tree of G. Because exactly one path exists between any two vertices of T, adding a chord c to T produces a unique circuit. This circuit is called the **fundamental circuit** of G with respect to chord c of T. Note again that the fundamental circuit with respect to chord c contains c, and the chord c is not present in any other fundamental circuit with respect to T. As an example, the set of fundamental circuits with respect to the spanning tree $T = (e_1, e_2, e_6, e_8)$ of the graph shown in Fig. 7.8 is

$$\begin{array}{ll} \text{Chord } e_3: & (e_3, e_1, e_2) \\ \text{Chord } e_4: & (e_4, e_1, e_2, e_6) \\ \text{Chord } e_5: & (e_5, e_2, e_6) \\ \text{Chord } e_7: & (e_7, e_8, e_6) \end{array}$$

We now present a result which is the basis of what is known as the orthogonality relationship.

Theorem 7: *A circuit and a cutset of a connected graph have an even number of common edges.*

PROOF. Consider a circuit C and a cutset $S = \langle V_1, V_2 \rangle$ of G. The result is true if C and S have no common edges. Suppose that C and S possess some common edges. Let us traverse the circuit C starting from a vertex, e.g., v_1 in V_1. Because the traversing should end at v_1, it is necessary that every time we encounter an edge of S leading us from V_1 to V_2 an edge of S must lead from V_2 back to V_1. This is possible only if S and C have an even number of common edges. □

The above result is the foundation of the theory of duality in graphs. Several applications of this simple result are explored in different parts of this chapter.

A comprehensive treatment of the duality theory and its relationship to planarity may be found in [2]. The following theorem establishes a close relationship between fundamental circuits and fundamental cutsets.

Theorem 8:

1. *The fundamental circuit with respect to a chord of a spanning tree T of a connected graph consists of exactly those branches of T whose fundamental cutsets contain the chord.*
2. *The fundamental cutset with respect to a branch of a spanning tree T of a connected graph consists of exactly those chords of T whose fundamental circuits contain the branch.*

PROOF. Let C be the fundamental circuit of a connected graph G with respect to a chord c of a spanning tree T of G. Let C contain, in addition to the chord c, the branches $b_1, b_2, \ldots, b_k$ of T. Let S_i be the fundamental cutset with respect to branch b_i.

We first show that each S_i, $1 \le i \le k$ contains c. Note that b_i is the only branch common to S_i and C, and c is the only chord in C. Because by Theorem 7, S_i and C must have an even number of common edges, it is necessary that S_i contains c.

Next we show that no other fundamental cutset of T contains c. Suppose the fundamental cutset S_{k+1} with respect to some branch b_{k+1} of T contains c. Then c will be the only edge common to S_{k+1} and C, contradicting Theorem 7. Thus the chord c is present only in those cutsets defined by the branches $b_1, b_2, \ldots, b_k$.

The proof for item 2 of the theorem is similar to that of item 1. □

7.4 Incidence, Circuit, and Cut Matrices of a Graph

The incidence, circuit, and cut matrices are coefficient matrices of Kirchhoff's equations which describe an electrical network. We develop several properties of these matrices which have proved useful in the study of electrical networks. Our discussions are mainly in the context of directed graphs. The results become valid in the case of undirected graphs if addition and multiplication are in $GF(2)$, the field of integers modulo 2. (Note that $1 + 1 = 0$ in this field.)

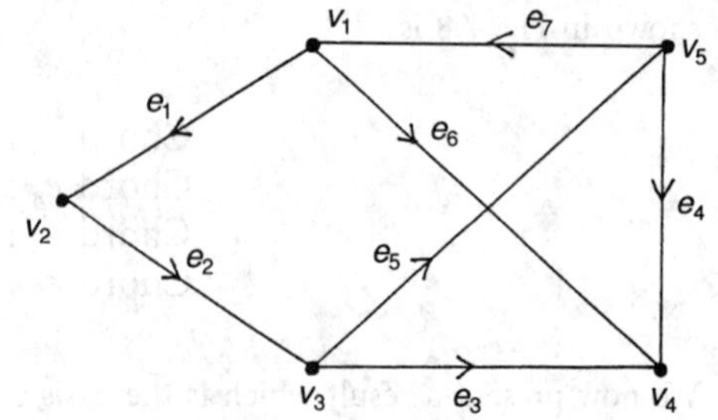

FIGURE 7.9 A directed graph.

Incidence Matrix

Consider a connected directed graph G with n vertices and m edges and having no self-loops. The **all-vertex incidence matrix** $A_c = [a_{ij}]$ of G has n rows, one for each vertex, and m columns, one for each edge. The element a_{ij} of A_c is defined as follows:

$$a_{ij} = \begin{cases} 1, & \text{if the } j\text{th edge is incident out of the } i\text{th vertex,} \\ -1, & \text{if the } j\text{th edge is incident into the } i\text{th vertex,} \\ 0, & \text{if the } j\text{th edge is not incident on the } i\text{th vertex} \end{cases}$$

A row of A_c will be referred to as an **incidence vector**. As an example, for the directed graph shown in Fig. 7.9, the matrix A_c is given below.

$$A_c = \begin{array}{c} \\ v_1 \\ v_2 \\ v_3 \\ v_4 \\ v_5 \end{array} \begin{array}{c} \begin{array}{ccccccc} e_1 & e_2 & e_3 & e_4 & e_5 & e_6 & e_7 \end{array} \\ \begin{bmatrix} 1 & 0 & 0 & 0 & 0 & 1 & -1 \\ -1 & 1 & 0 & 0 & 0 & 0 & 0 \\ 0 & -1 & 1 & 0 & 1 & 0 & 0 \\ 0 & 0 & -1 & -1 & 0 & -1 & 0 \\ 0 & 0 & 0 & 1 & -1 & 0 & 1 \end{bmatrix} \end{array}$$

From the definition of A_c it should be clear that each column of this matrix has exactly two nonzero entries, one $+1$ and one -1, and therefore, we can obtain any row of A_c from the remaining rows. Thus,

$$\text{rank}(A_c) \leq n - 1 \tag{7.3}$$

An $(n-1)$ rowed submatrix of A_c is referred to as an **incidence matrix** of G. The vertex which corresponds to the row of A_c that is not in A is called the **reference vertex** of A.

Theorem 9: *The determinant of an incidence matrix of a tree is* ± 1.

PROOF. Proof is by induction on the number m of edges in the tree. We can easily verify that the result is true for any tree with $m \leq 2$ edges. Assume that the result is true for all trees having $m \geq 2$ or fewer edges. Consider a tree T with $m+1$ edges. Let A be the incidence matrix of T with reference vertex v_r. Because, by Theorem 4, T has at least two pendant vertices, we can find a pendant vertex $v_i \neq v_r$. Let (v_i, v_j) be the only edge incident on v_i. Then the remaining edges form a tree T_1. Let A_1 be the incidence matrix of T_1 with vertex v_r as reference. Now let us rearrange the rows and columns of A so that the first $n-1$ rows correspond to the vertices in T_1, (except v_r) and the first $n-1$ columns correspond to the edges of T_1. Then we have

$$A = \begin{bmatrix} A_1 & A_3 \\ 0 & \pm 1 \end{bmatrix}$$

So

$$\det A = \pm(\det A_1) \tag{7.4}$$

A_1 is the incidence matrix of T_1 and T_1 has m edges, it follows from the induction hypothesis that $\det A_1 = \pm 1$. Hence the theorem. □

Because a connected graph has at least one spanning tree, it follows from the above theorem that any incidence matrix A of a connected graph has a nonsingular submatrix of order $n-1$. Therefore,

$$\text{rank}(A_c) \geq n-1 \tag{7.5}$$

Combining (7.3) and (7.5) yields the following theorem.

Theorem 10: *The rank of any incidence matrix of a connected directed graph G is equal to n − 1, the rank of G.* □

Cut Matrix

Consider a cut $\langle V_a, \overline{V_a}\rangle$ in a connected directed graph G with n vertices and m edges. Recall that $\langle V_a, \overline{V_a}\rangle$ consists of all those edges connecting vertices in V_a to those in $\overline{V_a}$. This cut may be assigned an orientation from V_a to $\overline{V_a}$ or from $\overline{V_a}$ to V_a. Suppose the orientation of $\langle V_a, \overline{V_a}\rangle$ is from V_a to $\overline{V_a}$. Then the orientation of an edge (v_i, v_j) is said to agree with the cut orientation if $v_i \in V_a$, and $v_j \in \overline{V_a}$.

The **cut matrix** $Q_c = [q_{ij}]$ of G has m columns, one for each edge, and has one row for each cut. The element q_{ij} is defined as follows:

$$q_{ij} = \begin{cases} 1, & \text{if the } j\text{th edge is in the } i\text{th cut and its orientation agrees with the cut orientation,} \\ -1, & \text{if the } j\text{th edge is in the } i\text{th cut and its orientation does not agree with the cut orientation,} \\ 0, & \text{if the } j\text{th edge is not in the } i\text{th cut} \end{cases}$$

Each row of Q_c is called a **cut vector**.

The edges incident on a vertex form a cut. Thus it follows that the matrix A_c is a submatrix of Q_c. Next we identify another important submatrix of Q_c.

Recall that each branch of a spanning tree T of a connected graph G defines a fundamental cutset. The submatrix of Q_c corresponding to the $n-1$ fundamental cutsets defined by T is called the **fundamental cutset matrix** Q_f of G with respect to T.

Let $b_1, b_2, \ldots, b_{n-1}$ denote the branches of T. Let us assume that the orientation of a fundamental cutset is chosen so as to agree with that of the defining branch. Suppose we arrange the rows and the columns of Q_f so that the ith column corresponds to branch b_i, and the ith row corresponds to the fundamental cutset defined by b_i. Then the matrix Q_f can be displayed in a convenient form as follows:

$$Q_f = [U \mid Q_{fc}] \tag{7.6}$$

where U is the unit matrix of order $n-1$ and its columns correspond to the branches of T.

As an example, the fundamental cutset matrix of the graph in Fig. 7.9 with respect to the spanning tree $T = (e_1, e_2, e_5, e_6)$ is given below:

$$Q_f = \begin{array}{c} \\ e_1 \\ e_2 \\ e_5 \\ e_6 \end{array} \begin{array}{c} \begin{array}{ccccccc} e_1 & e_2 & e_5 & e_6 & e_3 & e_4 & e_7 \end{array} \\ \left[\begin{array}{ccccccc} 1 & 0 & 0 & 0 & -1 & -1 & -1 \\ 0 & 1 & 0 & 0 & -1 & -1 & -1 \\ 0 & 0 & 1 & 0 & 0 & -1 & -1 \\ 0 & 0 & 0 & 1 & 1 & 1 & 0 \end{array}\right] \end{array}$$

It is clear from (7.6) that the rank of Q_f is $n-1$. Hence,

$$\text{rank}(Q_c) \geq n-1 \tag{7.7}$$

Circuit Matrix

Consider a circuit C in a connected directed graph G with n vertices and m edges. This circuit can be traversed in one of two directions, clockwise or counterclockwise. The direction we choose for traversing C is called the orientation of C. If an edge $e = (v_i, v_j)$ directed from v_i to v_j is in C, and if v_i appears before v_j as we traverse C in the direction specified by its orientation, then we say that the orientation of e agrees with the orientation of C.

The **circuit matrix** $B_c = [b_{ij}]$ of G has m columns, one for each edge, and has one row for each circuit in G. The element b_{ij} is defined as follows:

$$b_{ij} = \begin{cases} 1, & \text{if the } j\text{th edge is in the } i\text{th circuit, and its orientation agrees with the circuit orientation,} \\ -1, & \text{if the } j\text{th edge is in the } i\text{th circuit, and its orientation does not agree with the circuit orientation,} \\ 0, & \text{if the } j\text{th edge is not in the } i\text{th circuit} \end{cases}$$

Each row of B_c is called a **circuit vector.**

The submatrix of B_c corresponding to the fundamental circuits defined by the chords of a spanning tree T is called the **fundamental circuit matrix** B_f of G with respect to the spanning tree T.

Let $c_1, c_2, \ldots, c_{m-n+1}$ denote the chords of T. Suppose we arrange the columns and the rows of B_f so that the ith row corresponds to the fundamental circuit defined by the chord c_i, and the ith column corresponds to the chord c_i.

If, in addition, we choose the orientation of a fundamental circuit to agree with the orientation of the defining chord, we can write B_f as

$$B_f = [U \mid B_{ft}] \tag{7.8}$$

where U is the unit matrix of order $m-n+1$, and its columns correspond to the chords of T.

As an example, the fundamental circuit matrix of the graph shown in Fig. 7.9 with respect to the tree $T = (e_1, e_2, e_5, e_6)$ is given below:

$$B_f = \begin{array}{c} e_3 \\ e_4 \\ e_7 \end{array} \overset{\begin{array}{ccccccc} e_3 & e_4 & e_7 & e_1 & e_2 & e_5 & e_6 \end{array}}{\begin{bmatrix} 1 & 0 & 0 & 1 & 1 & 0 & -1 \\ 0 & 1 & 0 & 1 & 1 & 1 & -1 \\ 0 & 0 & 1 & 1 & 1 & 1 & 0 \end{bmatrix}}$$

It is clear from (7.8) that the rank of B_f is $m - n + 1$. Hence,

$$\text{rank}(B_c) \geq m - n + 1 \tag{7.9}$$

7.5 Orthogonality Relation and Ranks of Circuit and Cut Matrices

Theorem 11: *If a cut and a circuit in a directed graph have 2k edges in common, then k of these edges have the same relative orientation in the cut and in the circuit, and the remaining k edges have one orientation in the cut and the opposite orientation in the circuit.*

PROOF. Consider a cut $\langle V_a, \overline{V_a} \rangle$ and a circuit C in a directed graph. Suppose we traverse C starting from a vertex in V_a. Then, for every edge e_1 that leads from V_a to $\overline{V_a}$, an edge e_2 leads from $\overline{V_a}$ to V_a. Suppose the orientation of e_1 agrees with the orientation of the cut and that of the circuit. Then we can easily verify that e_2 has one orientation in the cut and the opposite orientation in the circuit (see Fig. 7.10). On the other hand, we can also verify that if e_1 has one orientation in the cut and the opposite orientation in the circuit, the e_2 will have the same relative orientation in the circuit and in the cut. This proves the theorem. □

Next we prove the orthogonality relation.

Theorem 12: *If the columns of the circuit matrix B_c and the columns of the cut matrix Q_c are arranged in the same edge order, then*

$$B_c Q_c^t = 0 \tag{7.10}$$

PROOF. Each entry of the matrix $B_c Q_c^t$ is the inner product of a circuit vector and a cut vector. Suppose a circuit and a cut have $2k$ edges in common. The inner product of the corresponding vectors is zero, because by Theorem 11, this product is the sum of k 1's and k $-$ 1's. □

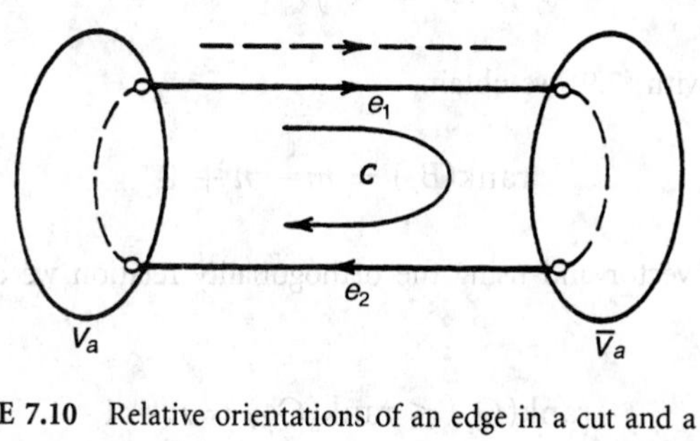

FIGURE 7.10 Relative orientations of an edge in a cut and a circuit.

The orthogonality relation is a profound result with interesting applications in electrical network theory. Consider a connected graph G with m edges and n vertices. Let Q_f be the fundamental cutset matrix and B_f be the fundamental circuit matrix of G with respect to a spanning tree T. If we write Q_f and B_f as in (7.6) and (7.8), then using the orthogonality relation we get

$$B_f Q_f^t = 0$$

that is,

$$[B_{ft} U] \begin{bmatrix} U \\ Q_{fc}^t \end{bmatrix} = 0$$

that is,

$$B_{ft} = -Q_{fc}^t \tag{7.11}$$

Using (7.11) each circuit vector can now be expressed as a linear combination of the fundamental circuit vectors. Consider a circuit vector $\beta = [\beta_1, \beta_2, \ldots, \beta_\rho | \beta_{\rho+1} \cdots \beta_m]$ of G where $\rho = n - 1$, is the rank of G. Then, again by the orthogonality relation we have

$$\beta Q_f^t = [\beta_1, \beta_2, \ldots, \beta_\rho \mid \beta_{\rho+1} \cdots \beta_m] \begin{bmatrix} U \\ Q_{fc}^t \end{bmatrix} = 0 \tag{7.12}$$

Therefore,

$$\begin{aligned} [\beta_1, \beta_2, \ldots, \beta_\rho] &= -[\beta_{\rho+1}, \beta_{\rho+2} \cdots \beta_m] Q_{fc}^t \\ &= [\beta_{\rho+1}, \beta_{\rho+2} \cdots \beta_m] B_{ft} \end{aligned}$$

So,

$$\begin{aligned} [\beta_1, \beta_2, \ldots, \beta_m] &= [\beta_{\rho+1}, \beta_{\rho+2} \cdots \beta_m][B_{ft} \quad U] \\ &= [\beta_{\rho+1}, \beta_{\rho+2} \cdots \beta_m] B_f \end{aligned} \tag{7.13}$$

Thus, any circuit vector can be expressed as a linear combination of the fundamental circuit vectors. So

$$\text{rank}(B_c) \leq \text{rank}(B_f) = m - n + 1$$

Combining the above with (7.9) we obtain

$$\text{rank}(B_c) = m - n + 1 \tag{7.14}$$

Starting from a cut vector and using the orthogonality relation we can prove in an exactly similar manner that

$$\text{rank}(Q_c) \leq \text{rank}(Q_f) = n - 1$$

Combining the above with (7.7) we get

$$\text{rank}(Q_c) = n - 1$$

Summarizing, we have the following theorem.

Theorem 13: *For a connected graph G with m edges and n vertices*

$$\text{rank}(B_c) = m - n + 1$$
$$\text{rank}(Q_c) = n - 1 \qquad \square$$

We wish to note from (7.12) that the vector corresponding to a circuit C can be expressed as an appropriate linear combination of the circuit vectors corresponding to the chords present in C. Similarly, the vector corresponding to a cut can be expressed as an appropriate linear combination of the cut vectors corresponding to the branches present in the cut. Because modulo 2 addition of two vectors corresponds to the ring sum of the corresponding subgraphs we have the following results for undirected graphs.

Theorem 14: *Let G be a connected undirected graph.*

1. *Every circuit can be expressed as a ring sum of the fundamental circuits with respect to a spanning tree.*
2. *Every cut can be expressed as a ring sum of the fundamental cutsets with respect to a spanning tree.* □

We can easily verify the following consequences of the orthogonality relation:

1. A linear relationship exists among the columns of the cut matrix (also of the incidence matrix) which correspond to the edges of a circuit.
2. A linear relationship exists among the columns of the circuit matrix which correspond to the edges of a cut.

The following theorem characterizes the submatrices of A_c, Q_c and B_c which correspond to spanning trees and cospanning trees. Proof follows from the above results and may be found in [2].

Theorem 15: *Let G be a connected graph G with n vertices, and m edges.*

1. *A square submatrix of order $n-1$ of Q_c (also of A_c) is nonsingular iff the edges corresponding to the columns of this submatrix form a spanning tree of G.*
2. *A square submatrix of order $m-n+1$ of B_c is nonsingular iff the edges corresponding to the columns of this submatrix form a cospanning tree of G.* □

7.6 Spanning Tree Enumeration

Here we first establish a formula for counting the number of spanning tress of an undirected graph. We then state a generalization of this result for the case of a directed graph. These formulas have played key roles in the development of topological formulas for electrical network functions. A detailed development of topological formulas for network functions may be found in Swamy and Thulasiraman [1].

The formula for counting the number of spanning trees of a graph is based on Theorem 9 and a result in matrix theory, known as the Binet-Cauchy theorem.

A **major** of a matrix is a determinant of maximum order. Consider a matrix P of order $p \times q$ and a matrix Q of order $q \times p$, with $p \leq q$. The majors of P and Q are of order p. If a major of

P consists of columns $i_1, i_2, \ldots, i_p$ the corresponding major of Q is formed by rows $i_1, i_2, \ldots, i_p$ of Q. For example, if

$$P = \begin{bmatrix} 1 & -2 & -2 & 4 \\ 2 & 3 & -1 & 2 \end{bmatrix} \quad \text{and} \quad Q = \begin{bmatrix} -5 & 0 \\ 2 & 1 \\ -2 & 2 \\ 3 & 1 \end{bmatrix}$$

then for the major

$$\begin{vmatrix} -2 & 4 \\ 3 & 2 \end{vmatrix}$$

of P

$$\begin{vmatrix} 2 & 1 \\ 3 & 1 \end{vmatrix}$$

is the corresponding major of Q.

The **Binet-Cauchy theorem** is stated next. Proof of this theorem may be found in Hohn [3].

Theorem 16: *If P is a $p \times q$ matrix and Q is a $q \times p$ matrix, with $p \le q$, then*

$$\det(PQ) = \sum(\text{product of the corresponding majors of } P \text{ and } Q). \qquad \square$$

Theorem 17: *Let G be a connected undirected graph and A an incidence matrix of a directed graph obtained by assigning orientations to the edges of G. Then*

$$\tau(G) = \det(AA^t) \tag{7.15}$$

where $\tau(G)$ is the number of spanning trees of G.

PROOF. By the Binet-Cauchy theorem we have

$$\det(AA^t) = \sum(\text{product of the corresponding majors of } A \text{ and } A^t) \tag{7.16}$$

Recall from Theorem 15 that a major of A is nonzero iff the edges corresponding to the columns of the major form a spanning tree of G. Also, the corresponding majors of A and A^t have the same value equal to 0, 1, or -1 (Theorem 9). Thus, each nonzero term in the sum on the right-hand side of (7.16) has the value 1, and it corresponds to a spanning tree and vice versa. Hence the theorem. $\square$

For example, consider the undirected graph G shown in Fig. 7.11(a). Assigning arbitrary orientations to the edges of G, we obtain the directed graph in Fig. 7.11(b). If A is the incidence matrix of this directed graph with vertex v_4 as reference vertex then it can be verified that

$$AA^t = \begin{bmatrix} 3 & -1 & -1 \\ -1 & 2 & 0 \\ -1 & 0 & 2 \end{bmatrix}$$

and $\det(AA^t) = 8$. Thus, G has eight spanning trees.

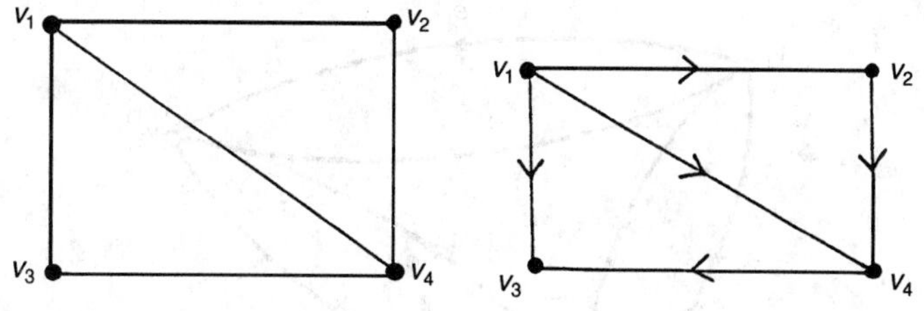

(a) (b)

FIGURE 7.11 (a) An undirected graph G; (b) directed graph obtained after assigning arbitrary orientations to the edges of G.

An interesting and useful interpretation of the matrix AA^t now follows. Let $v_1, v_2, \ldots, v_n$ be the vertices of an undirected graph G. The **degree matrix** $K = [k_{ij}]$ of G is an $n \times n$ matrix defined as follows.

$$k_{ij} = \begin{cases} -p, & \text{if } i \neq j \text{ and } p \text{ parallel edges connect } v_i \text{ and } v_j \\ d(v_i), & \text{if } i = j \end{cases}$$

We may easily verify that $K = A_c A_c^t$, and that it is independent of the choice of orientations for the edges of G. Also, if v_i is the reference vertex, then AA^t is obtained by removing row i and column i of K. In other words, $\det(AA^t)$ is the (i,i) cofactor of K. It then follows from Theorem 17 that all the cofactors of K are equal to the number of spanning trees of G. Thus, Theorem 17 may be stated in the following form originally presented by Kirchhoff [4].

Theorem 18: *All the cofactors of the degree matrix of an undirected graph G have the same value equal to the number of spanning trees of G.* □

Consider a connected undirected graph G. Let A be the incidence matrix of G with reference vertex v_n. Let $\tau_{i,n}$ denote the number of spanning 2-trees of G such that the vertices v_i and v_n are in different components of these spanning 2-trees. Also, let $\tau_{ij,n}$ denote the number of spanning 2-trees such that vertices v_i and v_j are in the same component, and vertex v_n is in a different component of these spanning 2-trees. If Δ_{ij} denotes the (i,j) cofactor of (AA^t), then we have the following result, proof of which may be found in [2].

Theorem 19: *For a connected graph G,*

$$\tau_{i,n} = \Delta_{ii} \tag{7.17}$$

$$\tau_{ij,n} = \Delta_{ij} \tag{7.18}$$

□

Consider next a directed graph $G = (V, E)$ without self-loops and with $V = (v_1, v_2, \ldots, v_n)$. The **in-degree matrix** $K = [k_{ij}]$ of G is an $(n \times n)$ matrix defined as follows:

$$k_{ij} = \begin{cases} -p, & \text{if } i \neq j \text{ and } p \text{ parallel edges are directed from } v_i \text{ to } v_j \\ d_{\text{in}}(v_i), & \text{if } i = j \end{cases}$$

The following result is due to Tutte [5]. Proof of this result may also be found in [2].

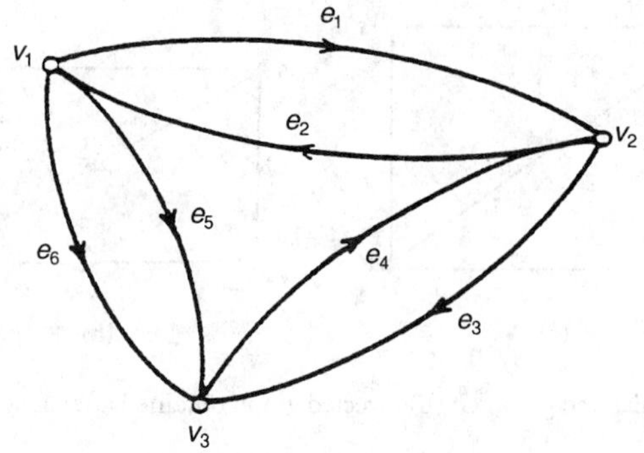

FIGURE 7.12 A directed graph G.

Theorem 20: *Let K be the in-degree matrix of a directed graph G without self-loops. Let the ith row of K correspond to vertex v_i. Then the number τ_d of directed spanning trees of G having v_r as root is given by*

$$\tau_d = \Delta_{rr} \tag{7.19}$$

where Δ_{rr} is the (r, r) co-factor of K. □

Note the similarity between Theorem 18 and Theorem 20.

To illustrate Theorem 20, consider the directed graph G shown in Fig. 7.12. The in-degree matrix K of G is

$$K = \begin{bmatrix} 1 & -1 & -2 \\ -1 & 2 & -1 \\ 0 & -1 & 3 \end{bmatrix}$$

Then

$$\Delta_{11} = \begin{bmatrix} 2 & -1 \\ -1 & 3 \end{bmatrix} = 5$$

The five directed spanning trees of G with vertex v_1 as root are (e_1, e_5), $(e_1, e_6$, (e_1, e_3), (e_4, e_5), and (e_4, e_6)

7.7 Graphs and Electrical Networks

An electrical network is an interconnection of electrical network elements such as resistances, capacitances, inductances, voltage and current sources, etc. Each network element is associated with two variables, the voltage variable $v(t)$ and the current variable $i(t)$. We also assign reference directions to the network elements (see Fig. 7.13) so that $i(t)$ is positive whenever the current is in the direction of the arrow, and $v(t)$ is positive whenever the voltage drop in the network element is in the direction of the arrow. Replacing each element and its associated reference direction by a directed edge results in the directed graph representing the network. For example, a simple electrical network and the corresponding directed graph are shown in Fig. 7.14.

The physical relationship between the current and voltage variables of a network element is specified by Ohm's law. For voltage and current sources, the voltage and current variables are

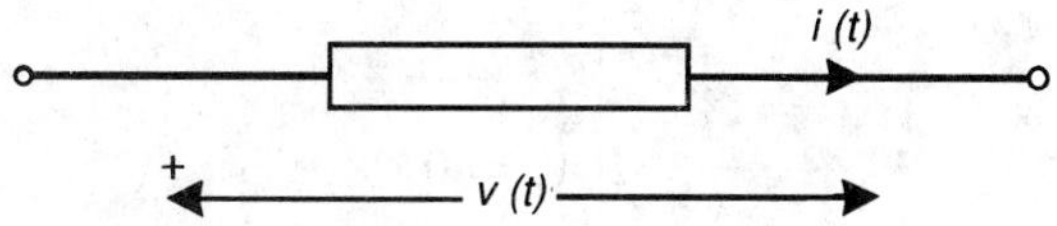

FIGURE 7.13 A network element with reference convention.

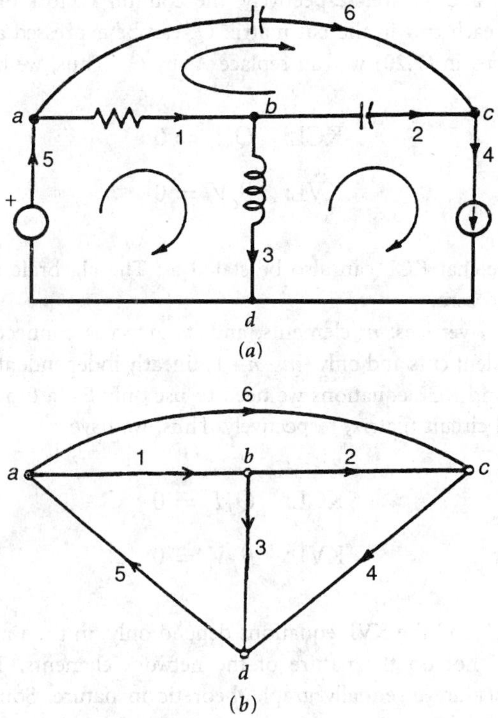

FIGURE 7.14 (a) An electrical network N; (b) directed graph representation of N.

required to have specified values. The linear dependence among the voltage variables in the network and the linear dependence among the current variables are governed by Kirchhoff's voltage and current laws.

Kirchhoff's Voltage Law (KVL): The algebraic sum of the voltages around any circuit is equal to zero.

Kirchhoff's Current Law (KCL): The algebraic sum of the currents flowing out of a node is equal to zero.

As an example, the KVL equation for the circuit 1, 3, 5 and the KCL equation for the vertex b in the graph of Fig. 7.14 are

$$\text{Circuit: } 1, 3, 5 \qquad v_1 + v_3 + v_5 = 0$$
$$\text{Vertex } b: \qquad -i_1 + i_2 + i_3 = 0$$

It can easily be seen that KVL and KCL equations for an electrical network N can be conveniently written as

$$A_c I_e = 0 \tag{7.20}$$

and

$$B_c V_e = 0 \tag{7.21}$$

where A_c and B_c are, respectively, the incidence and circuit matrices of the directed graph representing N, and I_e and V_e are, respectively, the column vectors of element currents and voltages in N. Because each row in the cut matrix Q_c can be expressed as a linear combination of the rows of the matrix, in (7.20) we can replace A_c by Q_c. Thus, we have:

$$\text{KCL:} \quad Q_c I_e = 0 \tag{7.22}$$

$$\text{KVL:} \quad B_c V_e = 0 \tag{7.23}$$

From (7.22) we can see that KCL can also be stated as: The algebraic sum of the currents in any cut of N is equal to zero.

If a network N has n vertices, m elements, and its graph is connected then there are only $(n-1)$ linearly independent cuts and only $(m-n+1)$ linearly independent circuits (Theorem 13). Thus, in writing KVL and KCL equations we need to use only B_f, a fundamental circuit matrix and Q_f, a fundamental circuit matrix, respectively. Thus, we have

$$\text{KCL:} \quad Q_f I_e = 0 \tag{7.24}$$

$$\text{KVL:} \quad B_f V_e = 0 \tag{7.25}$$

We note that the KCL and the KVL equations depend only on the way the network elements are interconnected and not on the nature of the network elements. Thus, several results in electrical network theory are essentially graph theoretic in nature. Some of these results and their usefulness in electrical network analysis are presented in the remainder of this chapter. In the following a network N and its directed graph are both denoted by N.

Theorem 21: *Consider an electrical network N. Let T be a spanning tree of N, and let B_f and Q_f denote the fundamental circuit and the fundamental cutset matrices of N with respect to T. If I_e and V_e are the column vectors of element currents and voltages and I_c and V_t are, respectively, the column vector of currents associated with the chords of T and the column vector of voltages associated with the branches of T, then*

$$\text{Loop Transformation:} \quad I_e = B_f^t I_c \tag{7.26}$$

$$\text{Cutset Transformation:} \quad V_e = Q_f^t V_t \tag{7.27}$$

PROOF. From Kirchhoff's laws we have

$$Q_f I_e = 0 \tag{7.28}$$

and

$$B_f V_e = 0 \tag{7.29}$$

Let us partition I_e and V_e as

$$I_e = \begin{bmatrix} I_c \\ I_t \end{bmatrix}$$

and

$$V_e = \begin{bmatrix} V_c \\ V_t \end{bmatrix}$$

where the vectors which correspond to the chords and branches of T are distinguished by the subscripts c and t, respectively. Then (7.28) and (7.29) can be written as

$$[\,Q_{fc} \quad U\,]\begin{bmatrix} I_c \\ I_t \end{bmatrix} = 0 \qquad (7.30)$$

and

$$[\,U \quad B_{ft}\,]\begin{bmatrix} V_c \\ V_t \end{bmatrix} = 0 \qquad (7.31)$$

Recall (7.11) that

$$B_{ft} = -Q_{fc}^t$$

Then we get from (7.30)

$$\begin{aligned} I_t &= -Q_{fc} I_c \\ &= B_{ft}^t I_c \end{aligned}$$

Thus

$$I_e = \begin{bmatrix} U \\ B_{ft}^t \end{bmatrix} I_c = B_f^t I_c$$

This establishes the loop transformation.

Starting from (7.31) we can show in a similar manner that

$$V_e = Q_f^t V_t$$

thereby establishing the cutset transformation. □

In the special case in which the incidence matrix A is used in place of the fundamental cutset matrix, the cutset transformation (7.27) is called the **node transformation.** The loop, cutset, and node transformations have been extensively employed to develop different methods of network analysis. The loop method of analysis develops a system of network equations which involve only the chord currents as variables. The cutset (node) method of analysis develops a system of equations involving only the branch (node) voltages as variables. Thus, the loop and cutset (node) methods result in systems of equations involving $m - n + 1$ and $n - 1$ variables, respectively. In the mixed-variable method of analysis, which is essentially a combination of both the loop and cutset methods, some of the independent variables are currents and the others are voltages. The minimum number of variables required in the mixed-variable method of analysis is determined by what is known as the principal partition of a graph introduced by Kishi and Kajitani in a classic paper [6]. Ohtsuki, Ishizaki, and Watanabe [7] discuss several issues relating to the mixed-variable method of analysis. A detailed discussion of the principal partition of a graph and the different methods of network analysis including the state-variable method may be found in [1].

7.8 Tellegen's Theorem and Network Sensitivity Computation

Here we first present a simple and elegant theorem due to Tellegen [8]. The proof of this theorem is essentially graph theoretic in nature and is based on the loop and the cutset transformations, (7.26) and (7.27), and the orthogonality relation (Theorem 12). Using this theorem we develop the concept of the adjoint of a network and its application in network sensitivity computations.

Theorem 22: *Consider two electrical networks N and $\hat{N}$ such that the graphs associated with them are identical. Let V_e and ψ_e denote the element voltage vectors of N and $\hat{N}$, respectively, and let I_e and Λ_e be the corresponding element current vectors. Then*

$$V_e^t \Lambda_e = 0$$

$$I_e^t \psi_e = 0$$

PROOF. If B_f and Q_f are the fundamental circuit and cutset matrices of N (and hence also of $\hat{N}$), then from the loop and cutset transformations we obtain

$$V_e = Q_f^t V_t$$

and

$$\Lambda_e = B_f^t \Lambda_c$$

So

$$\begin{aligned} V_e^t \Lambda_e &= V_t^t (Q_f B_f^t) \Lambda_c \\ &= 0, \quad \text{by Theorem 12} \end{aligned}$$

Proof follows in a similar manner. □

The adjoint network was introduced by Director and Rohrer [9], and our discussion is based on their work. A more detailed discussion may be found in [1].

Consider a lumped, linear time-invariant network N. We assume, without loss of generality, that N is a 2-port network. Let $\hat{N}$ be a 2-port network which is topologically equivalent to N. In other words, the graph of $\hat{N}$ is identical to that of N. The corresponding elements of N and $\hat{N}$ are denoted by the same symbol. Our goal now is to define the elements of $\hat{N}$ so that $\hat{N}$ in conjunction with N can be used in computing the sensitivities of network functions of N.

Let V_e and I_e denote, respectively, the voltage and the current associated with the element e in N, and ψ_e and λ_e denote, respectively, the voltage and the current associated with the corresponding element e in $\hat{N}$. Also, V_i and I_i, $i = 1, 2$, denote the voltage and current variables

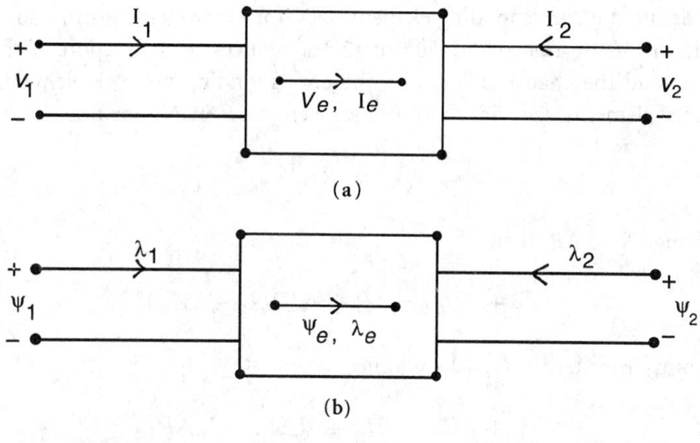

FIGURE 7.15 (a) A 2-port network N; (b) adjoint network $\hat{N}$ of N.

associated with the ports of N, and ψ_i and λ_i, $i = 1, 2$, denote the corresponding variables for the ports of $\hat{N}$ (see Fig. 7.15).

Applying Tellegen's theorem to N and $\hat{N}$ we get

$$V_1\lambda_1 + V_2\lambda_2 = \sum_e V_e\lambda_e \tag{7.32}$$

and

$$I_1\psi_1 + I_2\psi_2 = \sum_e I_e\psi_e \tag{7.33}$$

Suppose we now perturb the values of elements of N and apply Tellegen's theorem to $\hat{N}$ and the perturbed network N:

$$(V_1 + \Delta V_1)\lambda_1 + (V_2 + \Delta V_2)\lambda_2 = \sum_e (V_e + \Delta V_e)\lambda_e \tag{7.34}$$

and

$$(I_1 + \Delta I_1)\psi_1 + (I_2 + \Delta I_2)\psi_2 = \sum_e (I_e + \Delta I_e)\psi_e \tag{7.35}$$

where ΔV and ΔI represent the changes in the voltage and current which result as a consequence of the perturbation of the element values in N. Subtracting (7.32) from (7.34) and subtracting (7.33) from (7.35)

$$\Delta V_1\lambda_1 + \Delta V_2\lambda_2 = \sum_e \Delta V_e\lambda_e \tag{7.36}$$

and

$$\Delta I_1\psi_1 + \Delta I_2\psi_2 = \sum_e \Delta I_e\psi_e \tag{7.37}$$

Subtracting (7.37) from (7.36) yields

$$(\Delta V_1\lambda_1 - \Delta I_1\psi_1) + (\Delta V_2\lambda_2 - \Delta I_2\psi_2) = \sum_e (\Delta V_e\lambda_e - \Delta I_e\psi_e) \tag{7.38}$$

We wish to define the corresponding element of $\hat{N}$ for every element in N so that each term in the summation on the right-hand side of (7.38) reduces to a function of the voltage and current variables and the change in value of the corresponding network element. We illustrate this for resistance elements. Consider a resistance element R in N. For this element we have

$$V_R = RI_R \tag{7.39}$$

Suppose we change R to ΔR, then

$$(V_R + \Delta V_R) = (R + \Delta R)(I_R + \Delta I_R) \tag{7.40}$$

Neglecting second-order terms, (7.40) simplifies to

$$V_R + \Delta V_R = RI_R + R\Delta I_R + I_R \Delta R \tag{7.41}$$

Subtracting (7.39) from (7.41),

$$\Delta V_R = R\Delta I_R + I_R \Delta R \tag{7.42}$$

Now using (7.42) the terms in (7.38) corresponding to the resistance elements of N can be written as

$$\sum_R [R\lambda_R - \psi_R]\,\Delta I_R + I_R \lambda_R \Delta R \tag{7.43}$$

If we now choose

$$\psi_R = R\lambda_R \tag{7.44}$$

then (7.43) reduces to

$$\sum_R I_R \lambda_R \Delta R \tag{7.45}$$

which involves only the network variables in N (before perturbation) and $\hat{N}$ and the changes in resistance values. Equation (7.44) is the relation for a resistance. Therefore, the element in $\hat{N}$ corresponding to a resistance element of value R in N is also a resistance of value R.

Proceeding in a similar manner we can determine the elements of $\hat{N}$ corresponding to other types of network elements (inductance, capacitance, controlled sources, etc.) The network $\hat{N}$ so obtained is called the **adjoint** of N. A table defining adjoint elements corresponding to different types of network elements may be found in [1].

We now illustrate the application of the adjoint network in the computation of the sensitivity of a network function. Note that the sensitivity of a network function F with respect to a parameter x is a measure of the effect on F of an incremental change in x. Computing this sensitivity essentially involves determining $\partial F/\partial x$.

For the sake of simplicity consider the resistance network shown in Fig. 7.16(a). Let us assume that resistance R is perturbed from its nominal value of $3\,\Omega$. Assume that no changes occur in the values of the other resistance elements. We wish to compute $\partial F/\partial R$ where F is the open-circuit voltage ratio, that is,

$$F = \left.\frac{V_2}{V_1}\right|_{I_2=0}$$

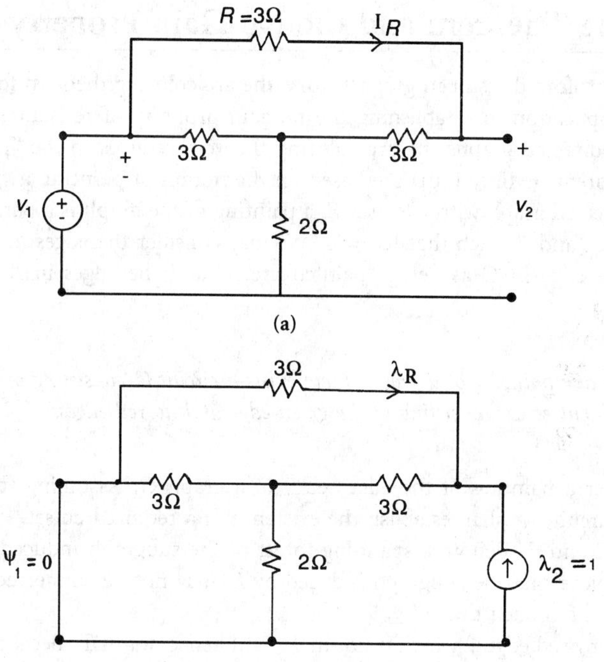

FIGURE 7.16 (a) A 2-port network N; (b) adjoint network $\hat{N}$.

In other words, to compute F, we connect a voltage source of value $V_1 = 1$ across port 1 of N and open-circuit port 2 of N (so that $I_2 = 0$). So, $\Delta V_1 = 0$ and $\Delta I_2 = 0$ and (7.38) reduces to

$$-\Delta I_1 \psi_1 + \Delta V_2 \lambda_2 = I_R \lambda_R \Delta R \tag{7.46}$$

Now we need to determine ΔV_2 as a function of ΔR. This could be achieved if we set $\psi_1 = 0$ and $\lambda_2 = 1$ for the adjoint network $\hat{N}$. Connect a current source of value $\lambda_2 = 1$ across port 2 and short circuit port 1 of $\hat{N}$. The resulting adjoint network is shown in Fig. 7.16(b). With port variables of $\hat{N}$ defined as above, (7.46) reduces to

$$\Delta V_2 = I_R \lambda_R \Delta_R$$

Thus,

$$\partial F/\partial R = \partial V_2/\partial R = I_R \lambda_R$$

where I_R and λ_R are the currents in the networks N and $\hat{N}$ shown in Fig. 7.16. Thus, in general, computing the sensitivity of a network function essentially reduces to the analysis of N and $\hat{N}$ under appropriate excitations at their ports. Note that we do not need to express the network function explicitly in terms of the network elements nor do we need to calculate partial derivatives.

For the example under consideration, we calculate $I_R = 1/12$ A and $\lambda_R = -7/12$ A with the result that $\partial F/\partial R = -7/144$. A further discussion of the adjoint network and related results may be found in Section 3.

7.9 Arc Coloring Theorem and the No-Gain Property

We now derive a profound result in graph theory, the arc coloring theorem for directed graphs, and discuss its application in establishing the no-gain property of resistance networks. In the special case of undirected graphs the arc coloring theorem reduces to the "painting" theorem. Both of these theorems (Minty [10]) are based on the notion of **painting a graph.**

Given an undirected graph with edge set E, a painting of the graph is a partitioning of E into three subsets, R, G, and B, such that $|G| = 1$. We may consider the edges in the set R as being "painted red", the edge in G as being "painted green" and the edges in B as being "painted blue".

Theorem 23: *For any painting of a graph, there exists a circuit C consisting of the green edge and no blue edges, or a cutset C^* consisting of the green edge and no red edges.*

PROOF. Consider a painting of the edge set E of a graph G. Assuming that there does not exist a required circuit, we shall establish the existence of a required cutset.

Let $E' = R \cup G$ and T' denote a spanning forest of the subgraph induced by E', containing the green edge. (Note that the subgraph induced by E' may not be connected). Then construct a spanning tree T of G such that $T' \subseteq T$.

Now consider any red edge y which is not in T', and hence not in T. Because the fundamental circuit of y with respect to T is the same as the fundamental circuit of y with respect to T', this circuit consists of no blue edges. Furthermore, this circuit will not contain the green edge, for otherwise a circuit consisting of the green edge and no blue edges would exist contrary to our assumption. Thus, the fundamental circuit of a red edge with respect to T does not contain the green edge. Then it follows from Theorem 8 that the fundamental cutset of the green edge with respect to T contains no red edges. Thus, this cutset satisfies the requirements of the theorem. □

A painting of a directed graph with edge set E is a partitioning of E into three sets R, G, and B, and the distinguishing of one element of the set G. Again, we may regard the edges of the graph as being colored red, green, or blue with exactly one edge of G being colored dark green. Note that the dark green edge is also to be treated as a green edge.

Next we state and prove Minty's arc coloring theorem.

Theorem 24: *For any painting of a directed graph exactly one of the following is true:*

1. *A circuit exists containing the dark green edge, but no blue edges, in which all the green edges are similarly oriented.*
2. *A cutset exists containing the dark green edge, but no red edges, in which all the green edges are similarly oriented.*

PROOF. Proof is by induction on the number of green edges. If only one green edge exists, then the result will follow from Theorem 23. Assume then that the result is true when the number of green edges is $m \geq 1$. Consider a painting in which $m + 1$ edges are colored green. Pick a green edge x other than the dark green edge (see Fig. 7.17). Color the edge x red. In the resulting painting we find m green edges. If a cutset of type 2 is now found, then the theorem is proved. On the other hand if we color the edge x blue and in the resulting painting a circuit of type 1 exists, then the theorem is proved.

Suppose neither occurs. Then, using the induction hypothesis we have the following:

1. A cutset of type 2 exists when x is colored blue.
2. A circuit of type 1 exists when x is colored red.

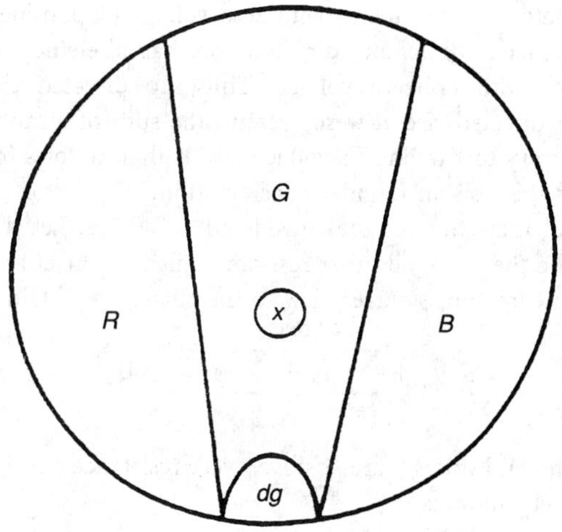

FIGURE 7.17 Painting of a directed graph.

Now let the corresponding rows of the circuit and cutset matrices be

	dg	*R*	*B*	*G*	*x*
Cutset	$+1$	$00\ldots0\ 0$	$1-1\ldots01$	$111\ldots0$	?
Circuit	$+1$	$-11\ldots0-1$	$0\ 0\ldots00$	$011\ldots0$	?

Here we have assumed, without loss of generality, that +1 appears in the dark green position of both rows.

By the orthogonality relation (Theorem 7.12) the inner product of these two row vectors is zero. No contribution is made to this inner product from the red edges or from the blue edges. The contribution from the green edges is a non-negative integer p. The dark green edge contributes 1 and the edge x contributes an unknown integer q which is 0, 1, or -1. Thus, we have $1 + p + q = 0$. This equation is satisfied only for $p = 0$ and $q = -1$. Therefore, in one of the rows, the question mark is +1 and in the other it is -1. The row in which the question mark is 1 corresponds to the required circuit or cutset. Thus, either statement 1 or 2 of the theorem occurs. Both cannot occur simultaneously because the inner product of the corresponding circuit and cutset vectors will then be nonzero. □

Theorem 25: *Each edge of a directed graph belongs to either a directed circuit or to a directed cutset but no edge belongs to both. (Note: A cutset is a* **directed cutset** *if all its edges are similarly oriented.)*

PROOF. Proof will follow if we apply the arc coloring theorem to a painting in which all the edges are colored green and the given edge is colored dark green. □

We next present an application of the arc coloring theorem in the study of electrical networks. We prove what is known as the **no-gain property** of resistance networks. Our proof is the result of the work of Wolaver [11] and is purely graph theoretic in nature.

Theorem 26: *In a network of sources and (linear/nonlinear) positive resistances the magnitude of the current through any resistance with nonzero voltage is not greater than the sum of the magnitudes of the currents through the sources.*

PROOF. Let us eliminate all the elements with zero voltage by considering them to be short-circuits and then assign element reference directions so that all element voltages are positive.

Consider a resistance with nonzero voltage. Thus, no directed circuit can contain this resistance, for if such a directed circuit were present, the sum of all the voltages in the circuit would be nonzero, contrary to Kirchhoff's voltage law. It then follows from Theorem 25 that a directed cutset contains the resistance under consideration.

Pick a directed cutset that contains the considered resistance. Let the current through this resistance be i_o. Let R be the set of all other resistances in this cutset and let S be the set of all sources. Then, applying Kirchhoff's current law to the cutset, we obtain

$$i_o + \sum_{k \in R} i_k + \sum_{s \in S} \pm i_s = 0 \tag{7.47}$$

Because all the resistances and voltages are positive, every resistance current is positive. Therefore, we can write the above equation as

$$|i_o| + \sum_{k \in R} |i_k| + \sum_{s \in S} \pm i_s = 0 \tag{7.48}$$

and so

$$|i_o| \leq \sum_{s \in S} \mp i_s \leq \sum_{s \in S} |i_s| \tag{7.49}$$

Thus follows the theorem. □

The following result is the dual of the above theorem. Proof of this theorem follows in an exactly dual manner, if we replace current with voltage, voltage with current, and circuit with cutset in the proof of the above theorem.

Theorem 27: *In a network of sources and (linear/nonlinear) positive resistances, the magnitude of the voltage across any resistance is not greater than the sum of the voltages across all the sources.* □

Chua and Green [12] used the arc-coloring theorem to establish several properties of nonlinear networks and nonlinear multiport resistive networks.

References

[1] M. N. S. Swamy and K. Thulasiraman, *Graphs, Networks and Algorithms*, New York: Wiley-Interscience, 1981.

[2] K. Thulasiraman and M. N. S. Swamy, *Graphs: Theory and Algorithms*, New York: Wiley-Interscience, 1992.

[3] F. E. Hohn, *Elementary Matrix Algebra*, New York: Macmillan, 1958.

[4] G. Kirchhoff, "Uber die Auflosung der Gleichungen auf welche mon bei der untersuchung der linearen Verteilung galvanischer strome gefuhrt wind," *Ann. Phys. Chem.*, vol. 72, pp. 497–508, 1847.

[5] W. T. Tutte, "The dissection of equilateral triangles into equilateral triangles," *Proc. Cambr. Philos. Soc.*, vol. 44, pp. 203–217, 1948.

[6] G. Kishi and Y. Kajitani, "Maximally distant trees and principal partition of a linear graph," *IEEE Trans. Circuit Theory*, vol. CT-15, pp. 247–276, 1968.

[7] T. Ohtsuki, Y. Ishizaki, and H. Watanabe, "Topological degrees of freedom and mixed analysis of electrical networks," *IEEE Trans. Circuit Theory*, vol. CT-17, pp. 491–499, 1970.

[8] B. D. H. Tellegen, "A general network theorem with applications," Philips Res. Rep., vol. 7, pp. 259–269, 1952.

[9] S. W. Director and R. A. Rohrer, "Automated network design—the frequency domain case," *IEEE Trans. Circuit Theory*, vol. CT-16, pp. 330–337, 1969.

[10] G. J. Minty, "On the axiomatic foundations of the theories of directed linear graphs, electrical networks and network programming," *J. Math. Mech.*, vol. 15, pp. 485–520, 1966.

[11] D. H. Wolaver, "Proof in graph of the 'no-gain' property of resistor networks," *IEEE Trans. Circuit Theory*, vol. CT-17, pp. 436–437, 1970.

[12] L. O. Chua and D. N. Green, "Graph-theoretic properties of dynamic nonlinear networks," *IEEE Trans. Circuit Theory*, vol. CAS-23, pp. 292–312, 1976.

8

Signal Flow Graphs

Krishnaiyan Thulasiraman
University of Oklahoma

8.1 Introduction

Signal flow graph theory is concerned with the development of a graph theoretic approach to solving a system of linear algebraic equations. Two closely related methods proposed by Coates [1] and Mason [2, 3] have appeared in the literature and have served as elegant aids in gaining insight into the structure and nature of solutions of systems of equations. In this chapter we develop these two methods. Our development follows closely [4].

An extensive discussion of signal flow theory may be found in [5]. Applications of signal flow theory in the analysis and synthesis electrical networks may be found in Sections 4 and 5. Coates' and Mason's methods may be viewed as generalizations of a basic theorem in graph theory due to Harary [6], which provides a formula for finding the determinant of the adjacency matrix of a directed graph. Thus, our discussion begins with the development of this theorem. For graph theoretic terminology the reader may refer to Chapter 7.

8.2 Adjacency Matrix of a Directed Graph

Consider a directed graph $G = (V, E)$ with no parallel edges. Let $V = \{v_1, \ldots, v_n\}$. The **adjacency matrix** $M = [m_{ij}]$ of G is an $n \times n$ matrix defined as follows:

$$m_{ij} = \begin{cases} 1, & \text{if } (v_i, v_j) \in E \\ 0, & \text{otherwise} \end{cases}$$

The graph shown in Fig. 8.1 has the following adjacency matrix:

$$M = \begin{array}{c} \\ v_1 \\ v_2 \\ v_3 \\ v_4 \end{array} \begin{array}{c} \begin{array}{cccc} v_1 & v_2 & v_3 & v_4 \end{array} \\ \begin{bmatrix} 1 & 1 & 1 & 0 \\ 0 & 1 & 0 & 0 \\ 1 & 0 & 0 & 1 \\ 1 & 1 & 1 & 1 \end{bmatrix} \end{array}$$

In the following we shall develop a topological formula for $\det M$. Toward this end we introduce some basic terminology. A **1-factor** of a directed graph G is a spanning subgraph of

G in which the in-degree and the out-degree of every vertex are both equal to 1. It is easy to see that a 1-factor is a collection of vertex-disjoint directed circuits. Because a self-loop at a vertex contributes 1 to the in-degree and 1 to the out-degree of the vertex, a 1-factor may have some self-loops. As an example, the three 1-factors of the graph of Fig. 8.1 are shown in Fig. 8.2.

A **permutation** $(j_1, j_2, \ldots, j_n)$ of integers $1, 2, \ldots, n$ **is even (odd)** if an even (odd) number of interchanges are required to rearrange it as $(1, 2, \ldots, n)$. The notation

$$\begin{pmatrix} 1, 2, \ldots, n \\ j_1, j_2, \ldots, j_n \end{pmatrix}$$

is also used to represent the permutation $(j_1, j_2, \ldots, j_n)$. As an example, the permutation $(4, 3, 1, 2)$ is odd because it can be rearranged as $(1, 2, 3, 4)$ using the following sequence of interchanges:

1. Interchange 2 and 4.
2. Interchange 1 and 2.
3. Interchange 2 and 3.

For a permutation $(j) = (j_1, j_2, \ldots, j_n)$, $\varepsilon_{j_1,j_2,\ldots,j_n}$, is defined as equal to 1, if (j) is an even permutation; otherwise, $\varepsilon_{j_1,j_2,\ldots,j_n}$, is equal to -1.

Given an $n \times n$ square matrix $X = [x_{ij}]$, we note that $\det X$ is given by

$$\det X = \sum_{(j)} \varepsilon_{j_1,j_2,j_3,\ldots,j_n} \quad x_{1j_1}, x_{2j_2} \cdots x_{nj_n}$$

where the summation $\sum_{(j)}$ is over all permutations of $1, 2, \ldots, n$ [7].

The following theorem is due to Harary [6].

Theorem 1: *Let H_i, $i = 1, 2, \ldots, p$ be the 1-factors of an n-vertex directed graph G. Let L_i denote the number of directed circuits in H_i, and let M denote the adjacency matrix of G. Then*

$$\det M = (-1)^n \sum_{i=1}^{p} (-1)^{L_i}$$

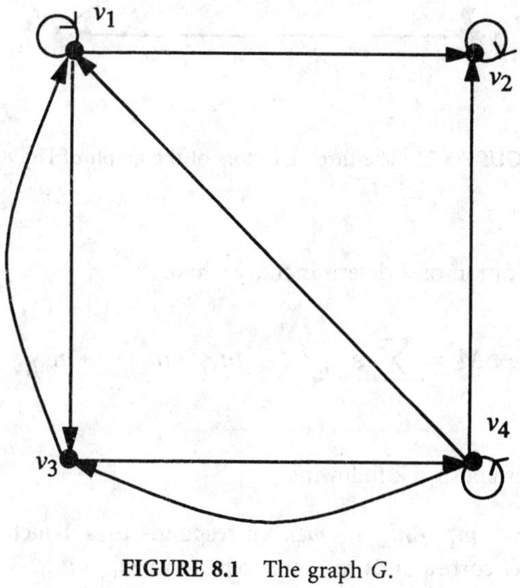

FIGURE 8.1 The graph G.

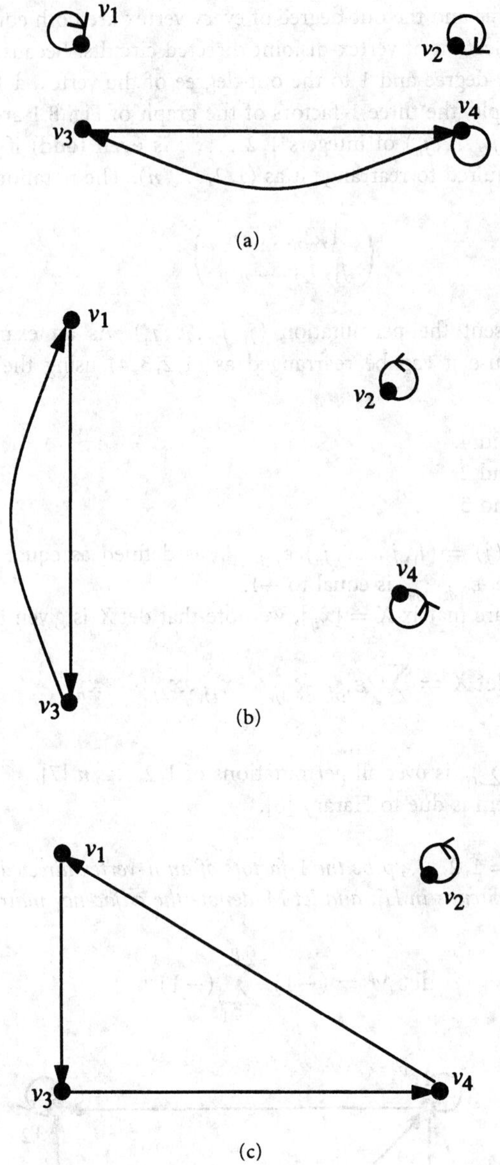

FIGURE 8.2 The three 1-factors of the graph of Fig. 8.1.

PROOF. From the definition of a determinant, we have

$$\det M = \sum_{(j)} \varepsilon_{j_1, j_2, \ldots, j_n} \quad m_{1j_1} \cdot m_{2j_2} \cdots m_{nj_n} \tag{8.1}$$

Proof will follow if we establish the following:

1. Each nonzero term $m_{1j_1} \cdot m_{2j_2} \cdots m_{nj_n}$ corresponds to a 1-factor of G, and conversely, each 1-factor of G corresponds to a non-zero term $m_{1j_1} \cdot m_{2j_2} \cdots m_{nj_n}$.

2. $\varepsilon_{j_1,j_2,\ldots,j_n} = (-1)^{n+L}$ if the 1-factor corresponding to a nonzero $m_{1j_1} \cdot m_{2j_2} \cdots m_{nj_n}$ has L directed circuits.

A nonzero term $m_{1j_1} m_{1j_2} \cdots m_{nj_n}$ corresponds to the set of edges $(v_1, v_{j1}), (v_2, v_{j2}) \cdots (v_n, v_{jn})$. Each vertex appears exactly twice in this set, once as an initial vertex and once as a terminal vertex of a pair of edges. Therefore, in the subgraph induced by these edges, for each vertex its in-degree and its out-degree are both equal to 1, and this subgraph is a 1-factor of G. In other words, each non-zero term in the sum in (8.1) corresponds to a 1-factor of G. The fact that each 1-factor of G corresponds to a non-zero term $m_{1j_1} \cdot m_{2j_2} \cdots m_{nj_n}$ is obvious.

As regards $\varepsilon_{j_1,j_2,\ldots,j_n}$, consider a directed circuit C in the 1-factor corresponding to $m_{1j_1} \cdot m_{2j_2} \cdots m_{nj_n}$. Without loss of generality, assume that C consists of the w edges

$$(v_1, v_2), \quad (v_2, v_3), \quad \ldots, \quad (v_w, v_1)$$

It is easy to see that the corresponding permutation $(2, 3, \ldots, w, 1)$ can be rearranged as $(1, 2, \ldots, w)$ using $w - 1$ interchanges. If the 1-factor has L directed circuits with lengths $w_1, \ldots, w_L$, the permutation $(j_1, \ldots, j_n)$ can be rearranged as $(1, 2, \ldots, n)$ using

$$(w_1 - 1) + (w_2 - 1) + \cdots + (w_L - 1) = n - L$$

interchanges. So,

$$\varepsilon_{j_1,j_2,j_n} = (-1)^{n+L} \qquad \square$$

As an example, for the 1-factors (shown in Fig. 8.2) of the graph of Fig. 8.1, the corresponding L_i are $L_1 = 3$, $L_2 = 3$, and $L_3 = 2$. So, the determinant of the adjacency matrix of the graph of Fig. 8.1 is

$$(-1)^4[(-1)^3 + (-1)^3 + (-1)^2] = -1$$

Consider next a weighted directed graph G in which each edge (v_i, v_j) as associated with a weight w_{ij}. Then we may define the **adjacency matrix** $M = [m_{ij}]$ of G as follows:

$$m_{ij} = \begin{cases} w_{ij} & \text{if } (v_i, v_j) \in E \\ 0, & \text{otherwise} \end{cases}$$

Given a subgraph H of G, let us define weight $w(H)$ of H as the product of the weights of all edges in H. If H has no edges, then we define $w(H) = 1$. The following result is an easy generalization of Theorem 1.

Theorem 2: *The determinant of the adjacency matrix of an n-vertex directed graph G is given by*

$$\det M = (-1)^n \sum_H (-1)^{L_H} w(H),$$

where H is a 1-factor, w(H) is the weight of H, and L_H is the number of directed circuits in H. □

8.3 Coates' Gain Formula

Consider a linear system described by the equation

$$AX = Bx_{n+1} \tag{8.2}$$

where A is a nonsingular $n \times n$ matrix, X is a column vector of unknown variables $x_1, x_2, \ldots, x_n$, B is a column vector of elements $b_1, b_2, \ldots, b_n$ and x_{n+1} is the input variable. It is well known that

$$\frac{x_k}{x_{n+1}} = \frac{\sum_{i=1}^{n} b_i \Delta_{ik}}{\det A} \tag{8.3}$$

where Δ_{ik} is the (i, k) cofactor of A.

To develop Coates' topological formulas for the numerator and the denominator of (8.3), let us first augment the matrix A by adding $-B$ to the right of A and adding a row of zeroes at the bottom of the resulting matrix. Let this matrix be denoted by A'. **The Coates flow graph**[1] $G_c(A')$, or simply the **Coates graph** associated with matrix A', is a weighted directed graph whose adjacency matrix is the transpose of the matrix A'. Thus, $G_c(A')$ has $n+1$ vertices $x_1, x_2, \ldots, x_{n+1}$, and if $a_{ji} \cdots \neq 0$, then $G_c(A')$ has an edge directed from x_i to x_j with weight a_{ji}. Clearly, the Coates graph $G_c(A)$ associated with matrix A can be obtained from $G_c(A')$ by removing the vertex x_{n+1}.

As an example, for the following system of equations

$$\begin{bmatrix} 3 & -2 & 1 \\ -1 & 2 & 0 \\ 3 & -2 & 2 \end{bmatrix} \begin{bmatrix} x_1 \\ x_2 \\ x_3 \end{bmatrix} = \begin{bmatrix} 3 \\ 1 \\ -2 \end{bmatrix} x_4 \tag{8.4}$$

the matrix A' is

$$A' = \begin{bmatrix} 3 & -2 & 1 & -3 \\ -1 & 2 & 0 & -1 \\ 3 & -2 & 2 & 2 \\ 0 & 0 & 0 & 0 \end{bmatrix}$$

The Coates graphs $G_c(A')$ and $G_c(A)$ are shown in Fig. 8.3.

Because a matrix and its transpose have the same determinant value and because A is the transpose of the adjacency matrix of $G_c(A')$, we obtain the following result from Theorem 2.

[1]In network and systems theory literature, the Coates graph is referred to as a **flow graph**.

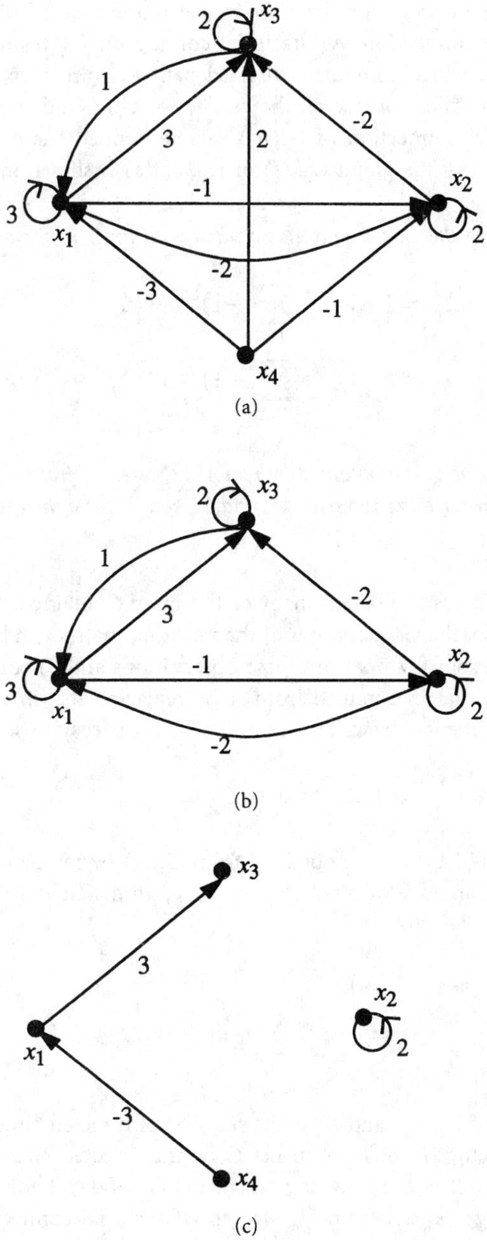

FIGURE 8.3 (a) The Coates graph $G_c(A')$; (b) the graph $G_c(A)$; (c) A-factorial connection $H_{4,3}$ of the graph $G_c(A')$

Theorem 3: *If a matrix A is nonsingular, then*

$$\det A = (-1)^n \sum_H (-1)^{L_H} w(H) \tag{8.5}$$

where H is a 1-factor of $G_c(A)$, $w(H)$ is the weight of H and L_H is the number of directed circuits in H. □

To derive a similar expression for the sum in the numerator of (8.3), we first define the concept of a **1-factorial connection.** A 1-factorial connection H_{ij} from x_i to x_j in $G_c(A)$ is a spanning subgraph of G which contains a directed path P from x_i to x_j and a set of vertex-disjoint directed circuits which include all the vertices of $G_c(A)$ other than those which lie on P. Similarly, a 1-factorial connection of $G_c(A')$ can be defined. As an example, a 1-factorial connection from x_4 to x_3 of the graph $G_c(A')$ of Fig. 8.3(a) is shown in Fig. 8.3(c).

Theorem 4: *Let $G_c(A)$ be the Coates graph associated with an $n \times n$ matrix A. Then*

$$1.\quad \Delta_{ii} = (-1)^{n-1}\sum_{H}(-1)^{L_H}w(H)$$

$$2.\quad \Delta_{ij} = (-1)^{n-1}\sum_{H_{ij}}(-1)^{L'_H}w(H_{ij}) \qquad i \neq j$$

where H is a 1-factor in the graph obtained by removing vertex x_i from $G_c(A)$, H_{ij} is a 1-factorial connection in $G_c(A)$ from vertex x_i to vertex x_j, and L_H and L'_H are the numbers of directed circuits in H and H_{ij}, respectively.

PROOF. 1. Note that Δ_{ii} is the determinant of the matrix obtained from A by removing its row i and column i. Also, the Coates graph of the resulting matrix can be obtained from $G_c(A)$ by removing vertex x_i. Proof follows from these observations and Theorem 3.

2. Let A_α denote the matrix obtained from A by replacing its jth column by a column of zeroes, except for the element in row i, which is 1. Then it is easy to see that

$$\Delta_{ij} = \det A_\alpha$$

Now, the Coates graph $G_c(A_\alpha)$ can be obtained from $G_c(A)$ by removing all edges incident out of vertex x_j and adding an edge directed from x_j to x_i with weight 1. Then from Theorem 3, we get

$$\begin{aligned}\Delta_{ij} &= \det A_\alpha \\ &= (-1)^n \sum_{H_\alpha}(-1)^{L_\alpha}w(H_\alpha)\end{aligned} \tag{8.6}$$

where H_α is a 1-factor of $G_c(A_\alpha)$ and L_α is the number of directed circuits in H_α.

Consider now a 1-factor H_α in $G_c(A_\alpha)$. Let C be the directed circuit of H_α containing x_i. Because in $G_c(A_\alpha)$, (x_j, x_i) is the only edge incident out of x_j, it follows that x_j also lies in C. If we remove the edge (x_j, x_i) from H_α we get a 1-factorial connection, H_{ij}. Furthermore, $L'_H = L_\alpha - 1$ and $w(H_{ij}) = w(H_\alpha)$ because (x_j, x_i) has weight equal to 1. Thus, each H_α corresponds to a 1-factorial connection H_{ij} of $G_c(A_\alpha)$ with $w(H_\alpha) = w(H_{ij})$ and $L'_H = L_\alpha - 1$. The converse of this is also easy to see. Thus, in (8.6) we can replace H_α by H_{ij} and L_α by $(L'_H + 1)$. Then we obtain

$$\Delta_{ij} = (-1)^{n-1}\sum_{H_{ij}}(-1)^{L'_H}w(H_{ij}) \qquad \square$$

Having shown that each Δ_{ij} can be expressed in terms of the weights of the 1-factorial connections H_{ij} in $G_c(A)$, we now show that $\sum b_i\Delta_{ik}$ can be expressed in terms of the weights of the 1-factorial connections $H_{n+1,k}$ in $G_c(A')$.

First, note that adding the edge (x_{n+1}, x_i) to H_{ik} results in a 1-factorial connection $H_{n+1,k}$, with $w(H_{n+1,k}) = -b_i\ w(H_{ik})$. Also, $H_{n+1,k}$ has the same number of directed circuits as H_{ik}. Conversely, from each $H_{n+1,k}$ that contains the edge (x_{n+1}, x_i) we can construct a 1-factorial connection H_{ik} satisfying $w(H_{n+1,k}) = -b_i\ w(H_{ik})$. Also, $H_{n+1,k}$ and the corresponding H_{ik} will have the same number of directed circuits. Thus, a one-to-one correspondence exists between the set of all 1-factorial connections $H_{n+1,k}$ in $G_c(A')$ and the set of all 1-factorial connections in $G_c(A)$ of the form H_{ik} such that each $H_{n+1,k}$ and the corresponding H_{ik} have the same number of directed circuits and satisfy the relation $w(H_{n+1,k}) = -b_i\ w(H_{ik})$. Combining this result with Theorem 4, we get

$$\sum_{i=1}^{n} b_i \Delta_{ik} = (-1)^n \sum_{H_{n+1,k}} (-1)^{L'_H} w(H_{n+1,k}) \tag{8.7}$$

where the summation is over all 1-factorial connections, $H_{n+1,k}$ in $G_c(A')$, and L'_H is the number of directed circuits in $H_{n+1,k}$. From (8.5) and (8.7) we get the following theorem.

Theorem 5: *If the coefficient matrix A is nonsingular, then the solution of* (8.2) *is given by*

$$\frac{x_k}{x_{n+1}} = \frac{\sum_{H_{n+1,k}} (-1)^{L'_H} w(H_{n+1,k})}{\sum_H (-1)^{L_H} w(H)} \tag{8.8}$$

for $k = 1, 2, \ldots, n$, *where* $H_{n+1,k}$ *is a 1-factorial connection of* $G_c(A')$ *from vertex* x_{n+1} *to vertex* x_k, H *is a 1-factor of* $G_c(A)$, *and* L'_H *and* L_H *are the numbers of directed circuits in* $H_{n+1,k}$ *and* H, *respectively.* □

Equation (8.8) is the called **Coates' gain formula.** We now illustrate Coates' method by solving the system (8.4) for x_2/x_4. First, we determine the 1-factors of the Coates' graph $G_c(A)$ shown in Fig. 8.3(b). These 1-factors, along with their weights, are listed below. The vertices enclosed within parentheses represent a directed circuit.

1-Factor H	Weight $w(H)$	L_H
$(x_1)(x_2)(x_3)$	12	3
$(x_2)(x_1, x_3)$	6	2
$(x_3)(x_1, x_2)$	4	2
(x_1, x_2, x_3)	2	1

From the above we get the denominator in (8.8) as

$$\sum_H (-1)^{L_H} w(H) = (-1)^3 \cdot 12 + (-1)^2 \cdot 6 + (-1)^2 \cdot 4 + (-1)^1 \cdot 2 = -4$$

To compute the numerator in (8.8) we need to determine the 1-factorial connections $H_{4,2}$ in the Coates graph $G_c(A')$ shown in Fig. 8.3(a). They are listed below along with their weights. The vertices in a directed path from x_4 to x_2 are given within parentheses.

1-Factorial connection $H_{4,2}$	$w(H_{4,2})$	L'_H
$(x_4, x_1, x_2)(x_3)$	6	1
$(x_4, x_2)(x_1)(x_3)$	−6	2
$(x_4, x_2)(x_1, x_3)$	−3	1
(x_4, x_3, x_1, x_2)	−2	0

From the above we get the numerator in (8.8) as

$$\sum_{H_{4,2}}(-1)^{L'_H}w(H_{4,2}) = (-1)^1\cdot 6 + (-1)^2(-6) + (-1)^1(-3) + (-1)^0(-2) = -11$$

Thus, we get

$$\frac{x_2}{x_4} = \frac{11}{4}$$

8.4 Mason's Gain Formula

Consider again the system of equations

$$AX = Bx_{n+1}$$

We can rewrite the above as

$$x_j = (a_{jj}+1)x_j + \sum_{\substack{k=1\\k\neq j}}^{n} a_{jk}x_k - b_j x_{n+1}, \quad j = 1, 2, \ldots, n, \qquad x_{n+1} = x_{n+1} \tag{8.9}$$

Letting X' denote the column vector of the variables $x_1, x_2, \ldots x_{n+1}$, and U_{n+1} denote the unit matrix of order n, we can write (8.9) in matrix form as follows:

$$(A' + U_{n+1})X' = X' \tag{8.10}$$

where A' is the matrix defined earlier in Section 8.3.

The Coates graph $G_c(A'+U_{n+1})$ is called the **Mason's signal flow graph** or simply the **Mason graph**[2] associated with A', and it is denoted by $G_m(A')$. The Mason graph $G_m(A)$ is defined in a similar manner. The Mason graphs $G_m(A')$ and $G_m(A)$ associated with the system (8.4) are shown in Fig. 8.4. Mason's graph elegantly represents the flow of variables in a system. If we associate each vertex with a variable and if an edge is directed from x_i to x_j, then we may consider the variable x_i as contributing $(a_{ji}x_i)$ to the variable x_j. Thus, x_j is equal to the sum of the products of the weights of the edges incident into vertex x_j and the variables corresponding to the vertices from which these edges emanate.

Note that to obtain the Coates graph $G_c(A)$ from the Mason graph $G_m(A)$ we simply subtract one from the weight of each self-loop. Equivalently, we may add at each vertex of the Mason graph a self-loop of weight -1. Let S denote the set of all such loops of weight -1 added to construct the Coates graph G_c from the Mason graph $G_m(A)$.

Consider now the Coates graph G_c constructed as above and a 1-factor H in G_c having j self-loops from the set S. If H has a total of $L_Q + j$ directed circuits, then removing the j self-loops from H will result in a subgraph Q of $G_m(A)$ which is a collection of L_Q vertex disjoint directed circuits. Also,

$$w(H) = (-1)^j w(Q)$$

[2]In network and systems theory literature Mason graphs are usually referred to as **signal flow graphs**.

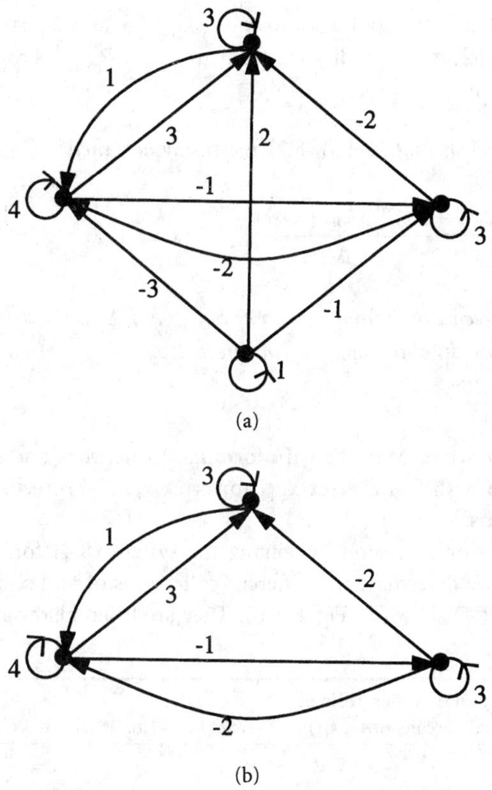

FIGURE 8.4 (a) The Mason graph $G_m(A')$; (b) the Mason graph $G_m(A)$.

Then, from Theorem 3 we get

$$\det A = (-1)^n \sum_H (-1)^{L_Q + j} w(H)$$
$$= (-1)^n \sum_Q (-1)^{L_Q} w(Q) \tag{8.11}$$
$$= (-1)^n \left[1 + \sum_Q (-1)^{L_Q} w(Q) \right]$$

We can rewrite the above as:

$$\det A = (-1)^n \left[1 - \sum_j Q_{j1} + \sum_j Q_{j2} - \sum_j Q_{j3} \cdots \right] \tag{8.12}$$

where each term in $\sum_j Q_{ji}$ is the weight of a collection of i vertex-disjoint directed circuits in $G_m(A)$.

Suppose we refer to $(-1)^n \det A$ as the determinant of the graph $G_m(A)$. Then, starting from $H_{n+1,k}$ and reasoning exactly as above we can express the numerator of (8.3) as

$$\sum_{i=1}^{n} b_i \Delta_{ik} = (-1)^n \sum_j w(P^j_{n+1,k}) \Delta_j \tag{8.13}$$

where $P^j_{n+1,k}$ is a directed path from x_{n+1} to x_k of $G_m(A')$ and Δ_j is the determinant of the subgraph of $G_m(A')$ which is vertex-disjoint from the path $P^j_{n+1,k}$. From (8.12) and (8.13) we get the following theorem.

Theorem 6: *If the coefficient matrix A is (8.2) is nonsingular, then*

$$\frac{x_k}{x_{n+1}} = \frac{\sum_j w(P^j_{n+1,k})\Delta_j}{\Delta}, \qquad k = 1, 2, \ldots, n \tag{8.14}$$

where $P^j_{n+1,k}$ is the jth directed path from x_{n+1} to x_k of $G_m(A')$, Δ_j is the determinant of the subgraph of $G_m(A')$ which is vertex-disjoint from the jth directed path $P^j_{n+1,k}$, and Δ is the determinant of the graph $G_m(A)$. □

Equation (8.14) is known as **Mason's gain formula.** In network and systems theory $P^j_{n+1,k}$ is referred to as a **forward path** from vertex x_{n+1} to vertex x_k. The directed circuits of $G_m(A')$ are called the **feedback loops.**

We now illustrate Mason's method by solving the system (8.4) for x_2/x_4. To compute the denominator in (8.14) we determine the different collections of vertex-disjoint directed circuits of the Mason graph $G_m(A)$ shown in Fig. 8.4(b). They are listed below along with their weights.

Collection of Vertex-Disjoint Directed Circuits of $G_m(A)$	Weight	No. of Directed Circuits
(x_1)	4	1
(x_2)	3	1
(x_3)	3	1
(x_1, x_2)	2	1
(x_1, x_3)	3	1
(x_1, x_2, x_3)	2	1
$(x_1)(x_2)$	12	2
$(x_1)(x_3)$	12	2
$(x_2)(x_3)$	9	2
$(x_2)(x_1, x_3)$	9	2
$(x_3)(x_1, x_2)$	6	2
$(x_1)(x_2)(x_3)$	36	3

From the above we obtain the denominator in (8.14)

$$\begin{aligned}\Delta = 1 &+ (-1)^1[4 + 3 + 3 + 2 + 3 + 2] \\ &+ (-1)^2[12 + 12 + 9 + 9 + 6] + (-1)^3\, 36 = -4\end{aligned}$$

To compute the numerator in (8.14) we need the forward paths in $G_m(A')$ from x_4 to x_2. They are listed below with their weights.

j	$P^j_{4,2}$	Weight
1	(x_4, x_2)	-1
2	(x_4, x_1, x_2)	3
3	(x_4, x_3, x_1, x_2)	-2

The directed circuits which are vertex-disjoint from $P^1_{4,2}$ are $(x_1), (x_3), (x_1, x_3)$. Thus

$$\Delta_1 = 1 - (4 + 3 + 3) + 12 = 1 - 10 + 12 = 3.$$

(x_3) is the only directed circuit which is vertex-disjoint from $P^2_{4,2}$. So,

$$\Delta_2 = 1 - 3 = -2.$$

No directed circuit is vertex-disjoint from $P^3_{4,2}$, so $\Delta_3 = 1$. Thus, the numerator in (8.14) is

$$P^1_{4,2}\Delta_1 + P^1_{4,2}\Delta_2 + P^1_{4,3}\Delta_3 = -3 - 6 - 2 = -11$$

and

$$\frac{x_2}{x_4} = \frac{11}{4}$$

·ferences

[1] C. L. Coates, "Flow graph solutions of linear algebraic equations," *IRE Trans. Circuit Theory*, vol. CT-6, pp. 170–187, 1959.

[2] S. J. Mason, "Feedback theory: some properties of signal flow graphs," *Proc. IRE*, vol. 41, pp. 1144–1156, 1953.

[3] S. J. Mason, "Feedback theory: further properties of signal flow graphs," *Proc. IRE*, vol. 44, pp. 920–926, 1956.

[4] K. Thulasiraman and M. N. S. Swamy, *Graphs: Theory and Algorithms*, New York: Wiley Interscience, 1992.

[5] W. K. Chen, *Applied Graph Theory*, Amsterdam: North Holland, 1971.

[6] F. Harary, "The determinant of the adjacency matrix of a graph," *SIAM Rev.*, vol. 4, pp. 202–210, 1962.

[7] F. E. Hohn, *Elementary Matrix Algebra*, New York: Macmillan, 1958.

II

Circuit Components and Their Characterization

John Choma, Jr.
University of Southern California

II

ircuit Components and heir Characterization

hn Choma, Jr.
iversity of Southern California

9

Passive Circuit Elements

tanisław Nowak
niversity of Mining and
etallurgy

omasz W. Postupolski
OLFER Magnetic Materials
ompany

ordon E. Carlson
niversity of Missouri, Rolla

. M. Wilamowski
niversity of Wyoming

9.1 Resistor

Stanisław Nowak

Linear Resistor

Introduction

An ideal resistor is an electronic component, the fundamental feature of which is resistance R according to Ohm's law expressed by the equation

$$V = RI \tag{9.1}$$

where V represents voltage in volts, I is the current in amperes, and R the resistance in ohms. The main parameters of a resistor are nominal resistance value, nominal power dissipation, and limited voltage value. According to their construction and technology we can divide resistors into five groups: wirewound resistors, foil resistors, thin film resistors, thick film resistors, and bulk resistors. Each group has some advantages and disadvantages; until now it has been impossible to manufacture all of the needed resistors within one technology. It is more interesting to divide resistors with respect to their application into two groups as follows:

1) fixed resistors, including low-power resistors of 0.05 to 2 W, high-power resistors of 2 to 100 W, high-voltage resistors, high-ohmic resistors, chip resistors, resistive networks, and

3493-8341-2/95/$0.00 + $.50

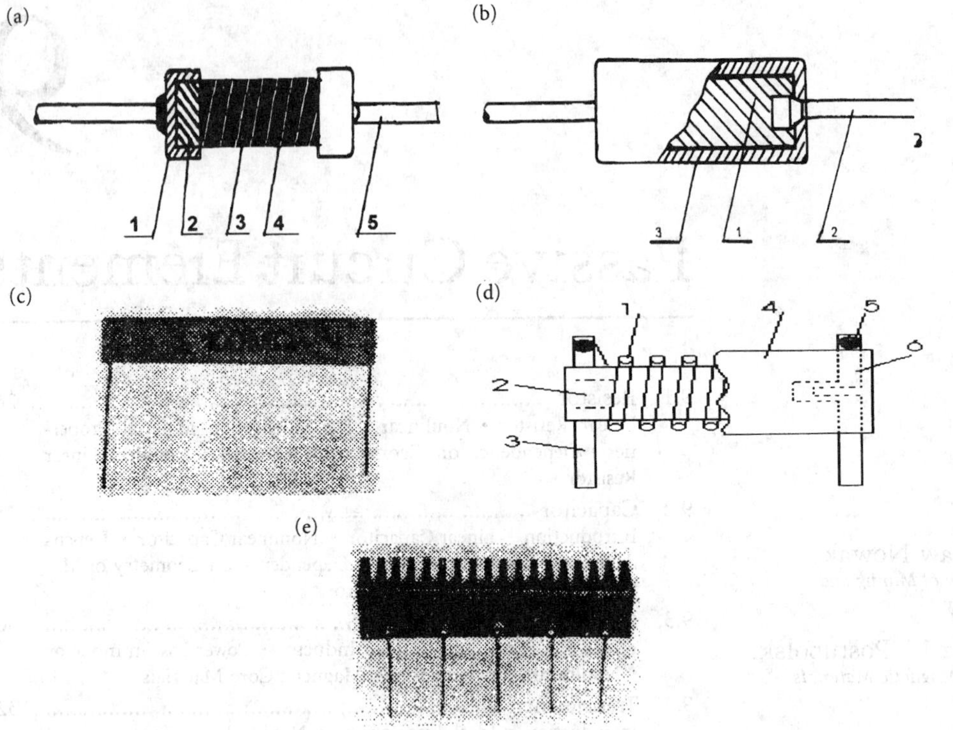

FIGURE 9.1 Typical fixed resistors: (a) film resistor: 1—metal cap, 2—electrode, 3—resistive layer, 4—groove, cut along screw line, 5—termination; (b) bulk composition resistor: 1—resistive composition, 2—termination, 3—pressed encapsulation; (c) high-voltage 100 MΩ thick film resistor; (d) wirewound power resistor, 1—resistive wire, 2—ceramic tube (substrate), 3—termination, 4—cement overcoating, 5—welded point, 6—cut for mechanical fixing of termination; (e) thick film resistor with radiator.

2) variable resistors (potentiometers), including rotary control potentiometers, slide control potentiometers, preset potentiometers, and special potentiometers.

Fixed Resistor

An ideal fixed resistor is an electronic component, the resistance value of which is constant with time and different environmental conditions. In practice we can observe some changes of resistance in time and under high temperature, high humidity, frequency, and electrical load conditions, and so on. Those changes of a resistance, called the instability of resistor, are the basis for classification of resistors according to the requirements of the International Electrical Commission (IEC) and the International Organization for Standardization (ISO 9000–9004) in order to build in a reliability system.

Figure 9.1 shows different kinds of fixed resistors. Each resistor is marked mainly by resistance value R and production tolerance $\delta p(\pm)$. Nominal resistance is rated according to the E6, E12, E24, E48, and E96 series. It is very important for the user to know not only the production deviation δp but also dynamic tolerance Δ.

Dynamic Tolerance Δ and Resistor Class. The author's proposal for calculation of the dynamic tolerance Δ is given by (9.2) and (9.3):

$$\Delta_+ = \delta_{p(+)} + \bar{\delta} + c * s \tag{9.2}$$

$$\Delta_- = \delta_{p(-)} - \bar{\delta} - c * s \tag{9.3}$$

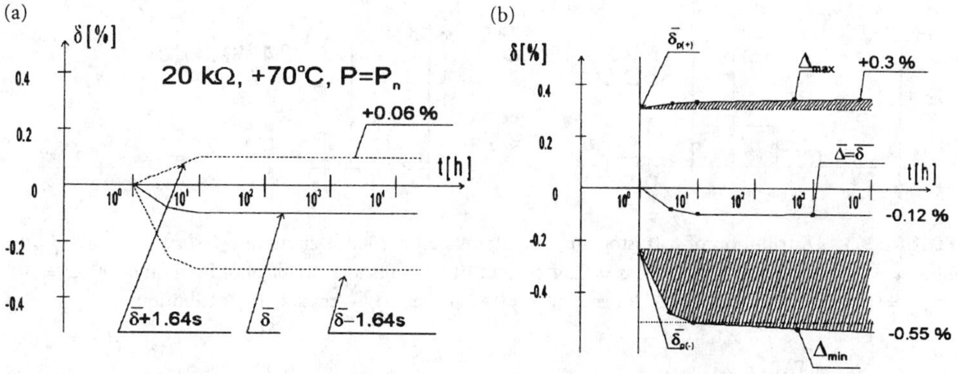

FIGURE 9.2 Resistor tolerance in exploitation. (a) The changes of resistance δ in time t $\bar{\delta} = \varphi_1(t)$, $\bar{\delta} + 1.64 * s = \varphi_2(t)$ $\bar{\delta} - 1.64 * s = \varphi_3(t)$. (b) Dynamic tolerance Δ as a function of time. Test time 10 000 h, $\delta_p = \pm 0.25$ percent.

where

$$\bar{\delta} = \frac{1}{n}\sum_{i=1}^{n} \delta_i \tag{9.4}$$

$$\delta_i = \frac{R_i(t) - R_i(0)}{R_i(0)} * 100\% \tag{9.5}$$

n is the quantity of samples in test, t is the test time, and R_i is the resistance of i, the resistor.

$$s = \sqrt{\frac{1}{n}\sum_{i=1}^{n}(\delta_i - \bar{\delta})^2} \tag{9.6}$$

$c = 1.28$ for probability level 90 percent cases inside the range $\bar{\delta} \pm 1.28 * s$, $c = 1.64$ for probability level 95 percent cases inside the range $\bar{\delta} \pm 1.64 * s$,

Figure 9.2 shows dynamic tolerance Δ as an example of a thick film resistor endurance test prolonged up to 10 000 h. Resistors of 20 kΩ were manufactured by the Telpod factory (Poland) from Birox 1441 Du Pont paste. In (9.2) and (9.3), for the long-life stability test, the coefficient c is chosen for normal distribution of changes δ_i for $t_j = 1000$ h because many experiments performed by author improved it with a high level of confidence. The hot humidity test causes distribution of changes asymmetrical, however.

The results of tests obtained for 400 resistors are shown in Fig. 9.3(a) and (b). For asymmetrical distribution the following values for c parameters are suggested: $c_1 = -1$ for $\delta_i < \bar{\delta}$ and $c_2 = +3$ for $\delta_i > \bar{\delta}$. Dynamic tolerance Δ is recommended by IEc Publ. 115, 1982 because quality classes are connected directly, with instability δ_{max}, which is presented in Table 9.1. In practice, $\delta_{max} = \bar{\delta} + cs$ according to (9.2).

In accordance with IEC Publ. 115-5, 1982, the following classes of resistors are ranked according to precision group (see Table 9.2).

Temperature Coefficient of Resistance—TCR. The influence of temperature on resistance might be observed:

1) When the resistor is exposed to high temperature for a long time; this results in irreversible changes in resistance.

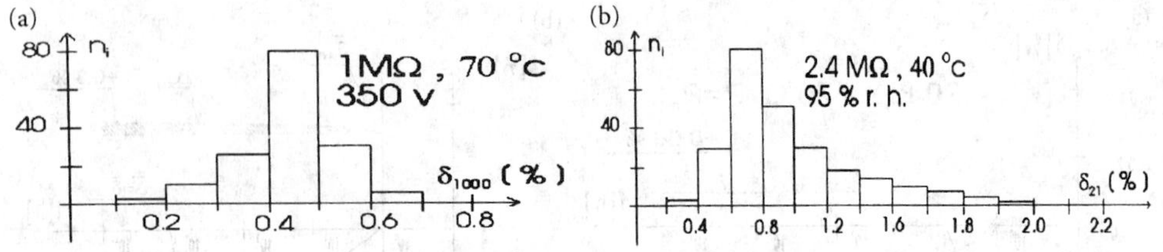

FIGURE 9.3 Distribution of δ (histogram). (a) Results after 1000 h endurance test: $\chi^2 = 5.01$, $\chi^2_{0.05,5} = 11.7$, $n = 150$ pcs, $\bar{\delta} = 0.45$ percent, $s = 0.24$ percent. (b) Results after 21-day hot humidity test: $\chi^2 = 100$, $\chi^2_{0.05,8} = 15.5$, $n = 246$ pcs, $\bar{\delta} = 0.79$ percent, $s = 0.47$ percent, χ^2—chi square distribution.

TABLE 9.1

Classes	δ_{max} after tests: Endurance Test Hot Humidity Test Climatic Cycles Test	Recommended δ_p	Remarks
15	±(15% + 0.5Ω)	±20%, ±10%, ±5%	
10	±(10% + 0.5Ω)	±2%, ±1%	Common use
5	±(5% + 0.1Ω)		resistor
3	±(3% + 0.1Ω)		
2	±(2% + 0.1Ω)	±5%, ±2%, ±1%	
1	±(1% + 0.05Ω)	±0.5%, ±0.25%	Stable resistor
0.5	±(0.5% + 0.05Ω)	±0.1%	
0.25	±(0.25% + 0.05Ω)		

TABLE 9.2

Classes	δ_{max} After Tests	Recommended δ_p
0.5	±(0.5% + 0.05Ω)	±1%, ±0.5%, ±0.25%
0.25	±(0.25% + 0.05Ω)	±0.1%, ±0.05%
0.1	±(0.1% + 0.01Ω)	±0.025%
0.05	±(0.05% + 0.01Ω)	±0.01%

2) When the resistor is exposed to thermal condition of short duration (0.5 h); this results in reversible changes in resistance that are measured as temperature coefficient of resistance (TCR).

TCR may be calculated from (9.7) and in practice from (9.8)

$$\mathrm{TCR} = \frac{1}{R}\frac{\partial R}{\partial T} \tag{9.7}$$

$$\mathrm{TCR} = \frac{1}{R_0}\frac{R_T - R_0}{T - T_0} \tag{9.8}$$

where R_0 is the resistance measured at room temperature T_0, R_T is the resistance measured at temperature T, and T_0 is room temperature.

In Fig. 9.4 some curves $R(T)$ are presented versus temperature for four types of resistors. It can be seen that in the tested range of temperature, TCR is positive and constant for curves

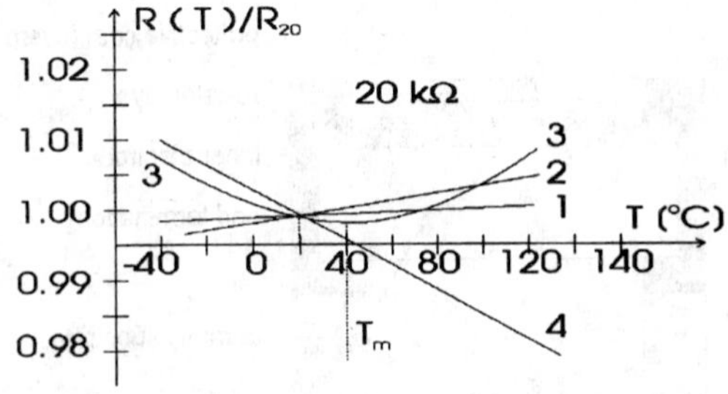

FIGURE 9.4 Dependence of resistance on ambient temperature. 1—wirewound resistor; 2—thin film resistor; 3—thick film ruthenium based resistor; 4—pyrolitic carbon resistor.

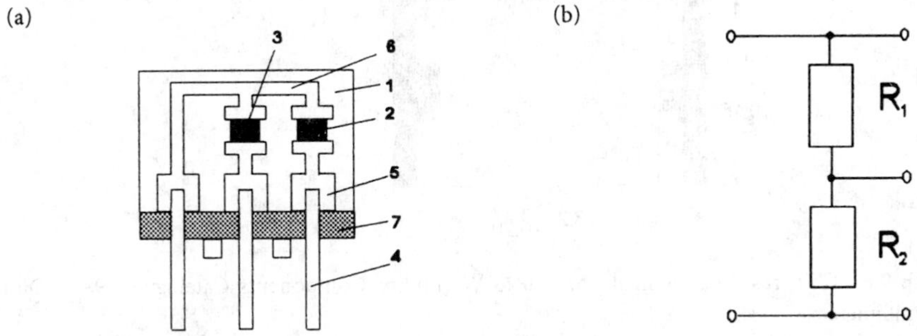

FIGURE 9.5 Thick film resistive network. (a) Topography of divider: 1—substrate; 2—resistor R_1; 3—resistor R_2; 4—wire termination; 5—soldering point; 6—conductive path; 7—insulating saddle; (b) electrical circuit.

1 and 2 but negative and constant for curve 4. A different result is obtained for curve 3; at temperatures lower than T_m, TCR is negative, at higher temperatures than T_m, TCR is positive, and at T_m, TCR = 0. When $T_m = 40°$ C, that type of resistor is the most interesting for users, because in the operating temperature range of 20° C–60° C, TCR is very small, almost zero.

As recommended by IEC Publ. 115, 1982, TCR limit values for different quality classes of resistors are shown in Table 9.3. Data in positions 1–5 refer to common use resistors, in positions 6–10, data refer to stable resistors, and in positions 11–15 to precision resistors.

Resistive Network. In electronic circuits, resistors are often used as elements of dividers. In that case it is more convenient to apply resistive networks (see Fig. 9.5).

Since the resistive network is deposited on a substrate in one technological cycle, both the TCR and time instability δ are almost the same for different resistors in the network. It seems to be more important for a user of a circuit to know differences Δ TCR and Δδ rather than absolute values of TCR or δ. It is estimated that Δ TCR can exhibit values of 1–3 ppm/K for TCR ≤ 50 ppm/K and Δδ can get value of 0.02 percent for δ = 0.5 percent.

TABLE 9.3

Number	TCR [ppm/K]	Number	TCR [ppm/K]	Number	TCR [ppm/K]
1	±2500	6	±250	11	±25
2	−800–+2500	7	±100	12	±20
3	±1000	8	±50	13	±15
4	−400–−1000	9	±25	14	±10
5	±500	10	±15	15	±5

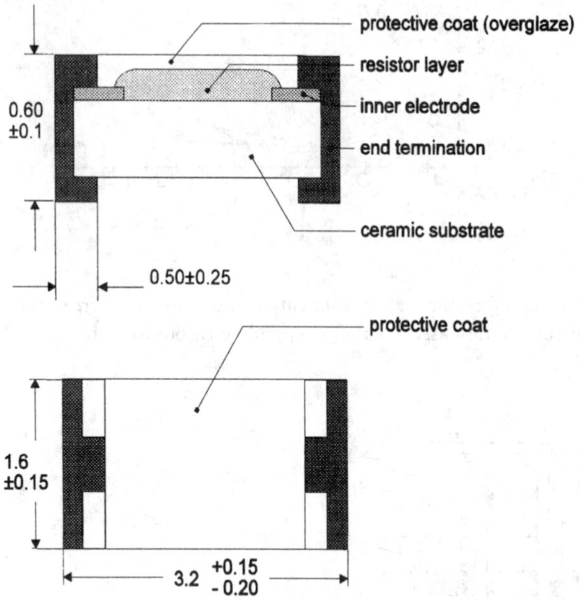

FIGURE 9.6 Chip resistor, nominal power 0.25 W [Philips Components Catalogue, 1989]. Dimensions are in millimeters.

Chip Resistor. The development of electronic circuit mounting technology is going toward reliability and flexibility improvement and this results in a new assembly technique, known as Surface Mounting Technology (SMT). Since SMT components have no terminations, their dimensions can be reduced. The smallest chip resistor is 2.5 mm long. Figure 9.6 shows a 0.25 W chip resistor. Requirement parameters for chip resistors are the same as for fixed resistors with terminations. Thick film technology is often used in manufacturing chip resistors.

High-Ohmic, High-Voltage Resistor. Usually resistors have resistances below 10 MΩ but sometimes resistors up to 10 GΩ are needed (for example in pH measurements, radiation particle detection, and so on). Only thick film technology enables production resistors of such high resistance values. Since the range of sheet resistance of the thick film layer changes from 1 Ω/square to 100 MΩ/square we can easily get a resistance range from 1 to 10 GΩ. Laser trimming and shaping of the layer allows easily to get from 100 to 1000 squares of resistive layer.

A very high value of sheet resistance decreases the thermal and long life stability of resistors, so it is advisable to design stable resistors with inks of 1 MΩ/square and to obtain the required resistance value by multiplying the number of squares.

High-ohmic resistors can be used as high-voltage resistors if their resistive paths are long enough.

The required voltage strength is a maximum of 2 kV/cm of resistive path. These types of resistors are used up to 10 kV in TV focusing systems.

High-Power Resistor. Very often, in electrical systems as well as in some electronic circuits (for example, power suppliers, power amplifiers, R-TV transmitters, and radar equipment), resistors with dissipation power above 5 W are necessary. For dc and low-frequency applications up to 100 W, power resistors are realized by a cement layer but high parasitic inductance makes them useless for higher frequency performances. Film and band resistors have very good high-frequency characteristics and they are suggested for high-frequency applications. Resistive bands are made

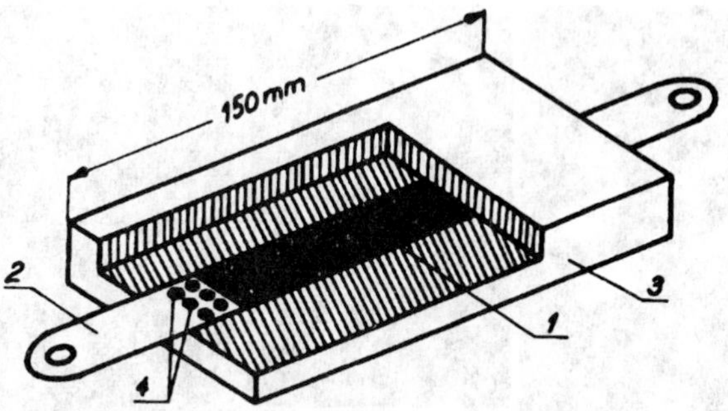

FIGURE 9.7 High-frequency resistor, nominal power 60 W. 1—resistive band; 2—band termination; 3—hot pressed overcoat; 4—holes for decreasing contact resistance.

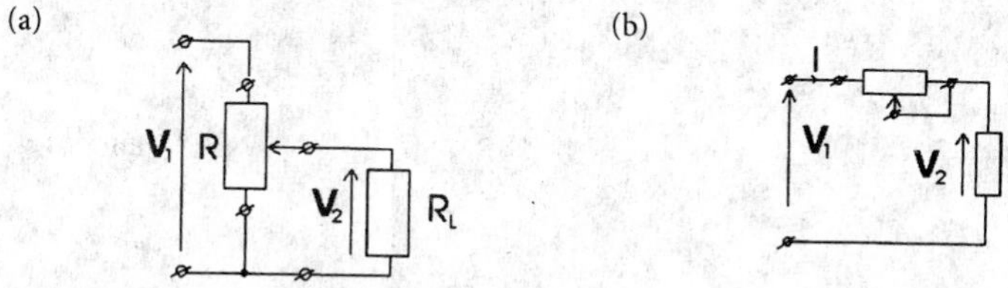

FIGURE 9.8 Potentiometer in an electric circuit: (a) as four-pole, voltage divider; (b) as twin-pole, rheostat.

of boron–carbon or other compositions. Nominal resistance ranges are as follows: 50, 75, 300, 600 Ω. In Fig. 9.7, a band resistor of low inductance is shown.

Variable Resistor

The variable resistor, very often called a potentiometer, is an electronic component in which by movement of mechanical wiper the resistance is regulated. A variable resistor can regulate voltage when it is joined to the circuit as a fourth-pole element [Fig. 9.8(a)]. It can regulate current when it is joined to the circuit as a twin-pole element in series with load R_L [Fig. 9.8(b)]. Requirements for variable resistors are similar as for fixed resistors but several additional problems must be considered: mechanical endurance, rotational noise, contact resistance variation (CRV), the type of regulation curve (taper), and parallel curves in tandem potentiometers (stereo potentiometers).

Variable potentiometers can be divided into three groups: rotary control potentiometers, slide control potentiometers, and preset potentiometers.

In Fig. 9.9, photos of different types of potentiometers are shown. With respect to their application we can divide potentiometers into several categories: standard type (common use) potentiometers with a carbon polymer resistive layer, high-stability potentiometers with a cermet layer, precision potentiometers formed as wirewound, or thin film ones.

To increase sensitivity of regulation, a lead screw actuated potentiometer is used, in which the screw moves the nut connected with the wiper. Slow displacement of the wiper causes fluent resistance regulation.

Specially built potentiometers (helipots) are used for precise regulation (see Fig. 9.10). In that case the wiper is moving along a screw line. This means that for 10 rotations of shaft, the total angle is 3600° and the way of wiper is 10 times longer than in a simple rotary potentiometer. In helipots, precision of adjustment depends on the diameter of potentiometer. This type of a potentiometer is manufactured by Beckman, Bourns, and others.

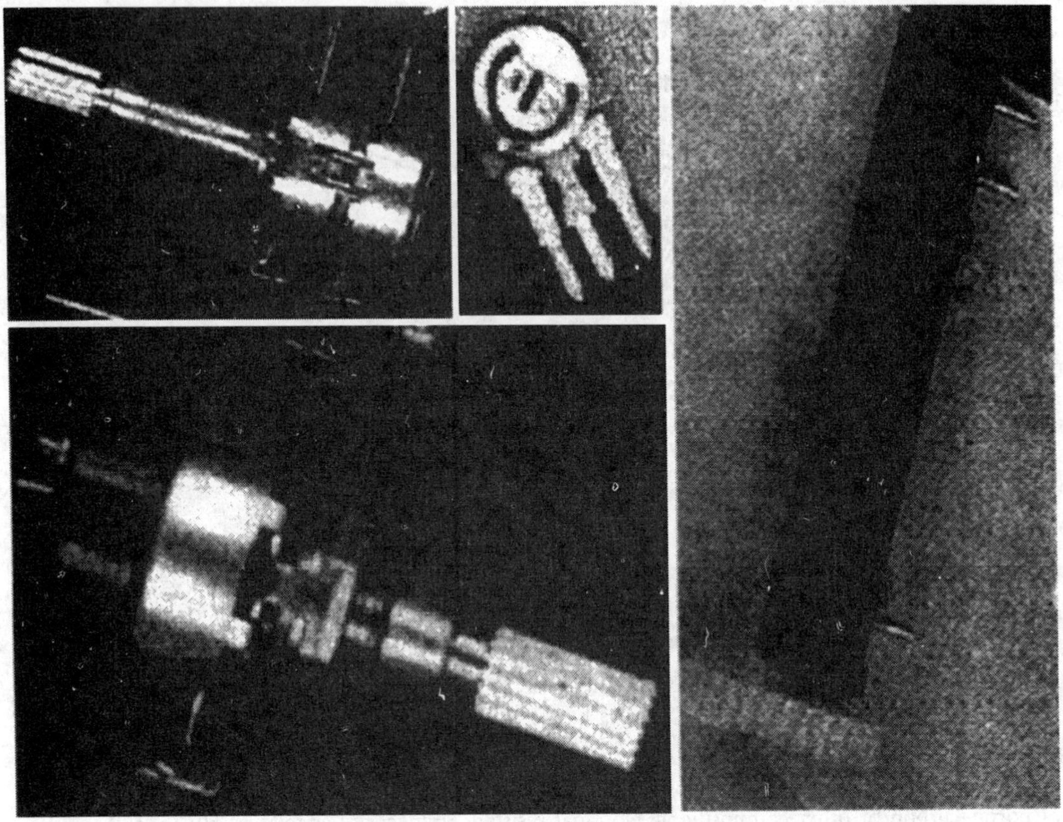

FIGURE 9.9 Photos of different potentiometers.

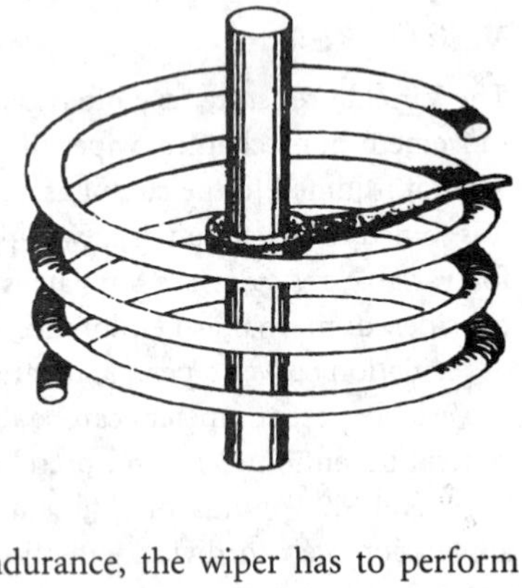

FIGURE 9.10 Helipot—a principle of work *Source:* Bourns, Inc., *The Potentiometer Handbook*, New York: McGraw-Hill, 1975.

Mechanical Endurance. During the test of mechanical endurance, the wiper has to perform many total cycles. For preset potentiometers, the number of cycles is 100–500; for control potentiometers it is 10^5–10^6 cycles.

Regulation Curves (Taper). The most popular regulation curve is a straight line; but as it is well known, our ear has a logarithmic characteristic. Therefore, for volume regulation in radios, audio amplifiers, and television sets, potentiometers with an exponential curve must be used. Figure 9.11 shows typical curves (tapers) of potentiometers.

In practice, a nonlinear curve is exactly realized by a few linear segments of resistive layer. Because each segment has another resistivity, it is necessary to use several segments to obtain a better exponential or antiexponential approximation. The minimum number of segments to

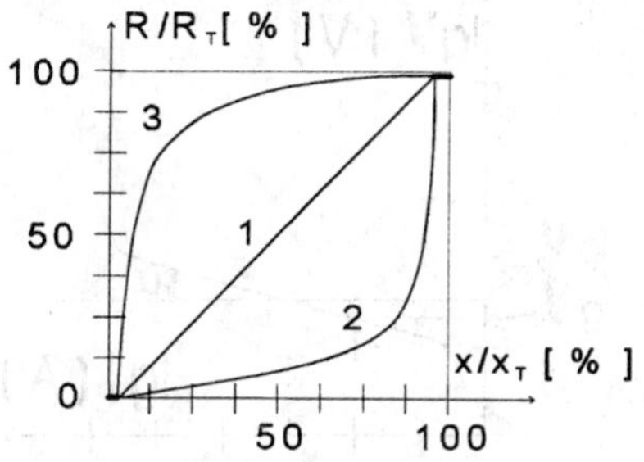

FIGURE 9.11 The main tapers of potentiometers. 1—linear; 2—exponential; 3—antiexponential. X_T—total wiper direction, R_T—total resistance. At the beginning and at the end of wiper movements there are very low resistive paths.

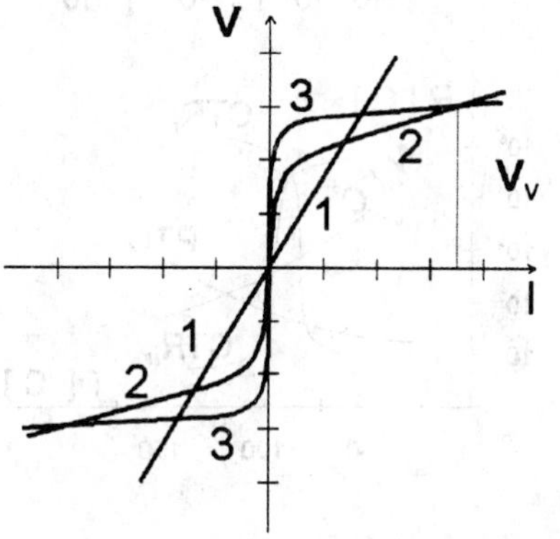

FIGURE 9.12 Voltage versus current characteristics: 1—linear resistor; 2—SiC varistor; 3—ZnO varistor; V_V—varistor voltage.

perform an exponential curve is two, but then some steps in regulation are unavoidable. In production we can achieve potentiometers performing sinus, cosinus, and other curves.

Curve Parallelism of Stereo Tandem Potentiometers. A stereo potentiometer with an exponential curve has to fulfill the additional requirement for parallel curves; both potentiometers are controlled by one roller (for example, at attenuation of 40 dB the difference between both curves must be smaller than 2 dB).

Nonlinear

Varistor

A voltage dependent resistor (VDR), called a varistor, is a resistor whose characteristic V versus I is not a straight line and a small change of voltage causes a significant change of current according to the equation

$$V = CI^{\beta} \tag{9.9}$$

where β is nonlinearity coefficient = 0.03–0.4 (it depends on the material and manufacturing technology), C is varistor dependent coefficient.

The main parameters of varistor are nonlinearity coefficient β and varistor voltage V_v measured at constant current for example 1 mA. Comparisons of the characteristics of a linear resistor and two types of varistors are shown in Fig. 9.12.

A varistor can be used for dc voltage stabilization and especially for electronic circuit protection against overvoltage pulses caused by industrial distortions and atmospheric discharges. Coefficient β is calculated from (9.10):

$$\beta = \frac{\lg(V_2)/(V_1)}{\lg(I_2)/(I_1)} \tag{9.10}$$

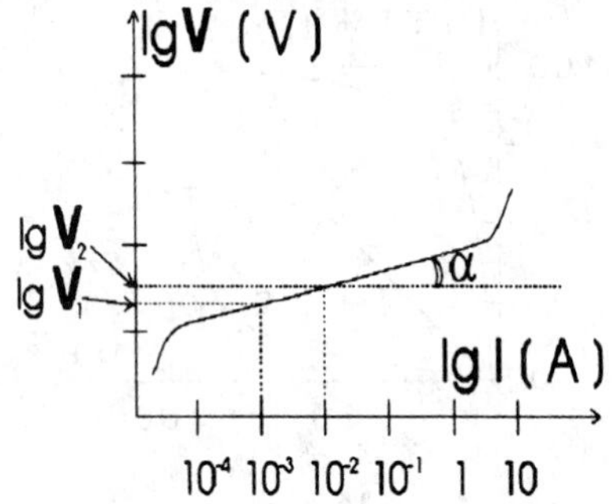

FIGURE 9.13 Log(V)/log(I) characteristic of a varistor for nonlinearity coefficient β description. $\beta = tg\alpha = \log V_2 - \log V_1; I_2/I = 10$.

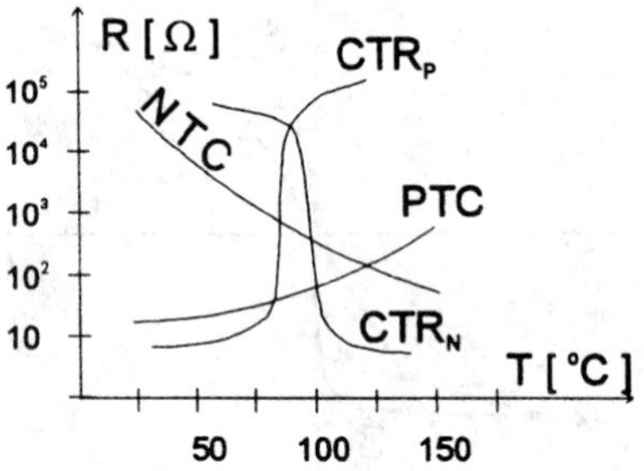

FIGURE 9.14 Characteristic $R = \varphi(T)$ of different thermistors. NTC—negative temperature coefficient thermistor; PTC—positive temperature coefficient thermistor; CTR—thermistor with critical temperature of resistance; index N-curve falling with temperature; index P-curve rising with temperature.

when $I_2/I_1 = 10$, denominator is equal 1 and

$$\beta = \lg V_2 - \lg V_1 \tag{9.11}$$

To explain the above relation, Fig. 9.13 is helpful, where both V and I are in logarithmic scale. The slope of the straight line segment of this curve equals β. For SiC varistors (curve 2 in Fig. 9.12), $\beta = 0.12$–0.4; for ZnO varistors however $\beta = 0.03$–0.1.

Varistor voltage is in the range of 4 V up to 2 kV; it depends on varistor thickness (length). To get a higher operating voltage, disk varistors should be connected in pile. Maximum pulse current is in the range of 0.2 A up to 2kA; it depends on the diameter of the varistor body. For pulse work, the following additional parameters are important: the capacity of the varistor (it is in the range of 100 pF up to 1 μF) and absorption energy (it is in the range of 1 J up to 2200 J) (see [10]).

Thermistor

A temperature dependent resistor, called a thermistor, is a resistor with significant TCR, which can be positive (PTC) or negative (NTC). Some groups of thermistors are characterized by a very rapid change of resistance in temperature. Those thermistors are called critical temperature resistors (CTR). They can be positive CTR_P or negative CTR_N. Figure 9.14 presents typical characteristics R versus T for different types of thermistors.

NTC and PTC thermistors are used for stabilization of the working point in temperature for different electric circuits and as well as for temperature measurement. CTR thermistors are applied as protective elements against overheating in electronic circuits. CTR_P's are used in degaussing circuits in color TV tubes. In some catalogs, CTR_P's are called PTC's (see [7]). The electrical, climatic, and mechanical requirements of thermistors are almost the same as for fixed resistors but some additional parameters were introduced as well, such as thermal time constant (in seconds), heat capacity (in J/K), dissipation factor (in mW/K), switch temperature or critical temperature in °C (for CTR's only). The first three are related to the thermistor dimensions, the remaining to the row material and technological process.

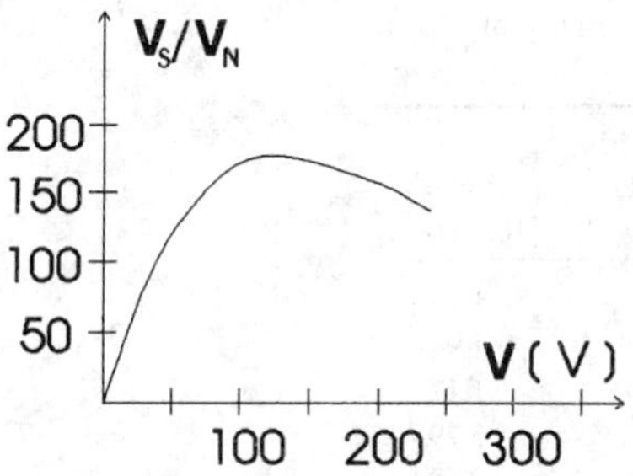

FIGURE 9.15 Signal-to-noise ratio versus polarization voltage V_{DC} of PbS photoresistor, Mullard type, 615 V [3]. V_S—signal voltage, V_N–noise voltage, photo source temperature 473 K, $f = 800$ Hz, $\Delta f = 50$ Hz.

Photoresistor

A photoresistor is a film resistor whose resistance is sensitive to the light, that is, it depends on light intensity and light wavelength λ. The latter, in turn, depends on the kind of material used as follows.

For Cd S: $\lambda = 0.4$–0.7 μm (visible light), for Pb S: $\lambda = 0.9$–3.5 μm (infrared), for Ge Si-doped Zn: $\lambda = 4$–15 μm, for Ge doped Sb: $\lambda = 30$–100 μm. This means that for a different photoresistor there exists an optimal wavelength at which maximum sensitivity (maximum change of resistance between lightness and darkness) occurs. During the design of a circuit with a photoresistor it is necessary to know at which polarization voltage the smallest noise exists. Figure 9.15 presents V_S/V_N versus polarization voltage for a PbS photoresistor made by Mullard.

Magnetoresistor

Some thin film multilayer ferromagnetic structures cause changes in magnetic field H. This phenomenon is called the magnetoresistive effect. An electronic component in which the magnetoresistive effect occurs is called a magnetoresistor and is usually used as a sensor. Special preparation of a ferromagnetic multilayer allows achievement of magnetosensitivity up to 100 MHz. The change of resistance is 1–5 percent at the change of magnetic field H of about 10 Oe. Very often two magnetoresistors are joined in a Wheatstone bridge and then sensitivity of the sensor is doubled.

Dependence on Material Properties

Materials selected for resistors have played a fundamental role in resistor production. Resistive elements are composed of a metal alloy, carbon, metal oxide, and mixtures of insulating and conducting particles such as polymer and black carbon, glass and bismuth ruthenate, as well as glass and metal oxide. Semiconductors are also good materials for resistors, especially for nonlinear resistors such as varistors, thermistors, and photoresistors.

Influence of Resistive Material on TCR

Nonmagnetic Metals. According to Grüneisenn's principle [2], [3] temperature influences resistivity as follows:

$$\rho_T = \rho_\Theta \frac{T - 0.15\Theta}{0.85\Theta} \quad \text{for } T \geq 0.15\Theta \tag{9.12}$$

and

$$\mathrm{TCR}_{20} = \frac{1}{293 - 0.15\Theta} \tag{9.13}$$

$$\Theta = \frac{h\nu_{\max}}{k} \tag{9.14}$$

TABLE 9.4 Debye Temperature and Resistivity of Nonmagnetic Metals [3]

Metal	ρ_{20} at $T = 293$ K $[10^{-8}\Omega*\text{m}]$	Θ [K]	$0.15\ \Theta$ [K]	ρ at Θ $[10^{-8}\Omega*\text{m}]$
Ag	1.62	214	32	1.16
Cu	1.68	320	48	1.94
Au	2.22	160	24	1.17
Al	2.73	374	56	3.79
Zn	6.12	180	27	3.65
Pt	10.6	220	33	7.91
Pb	20.8	84.5	12.7	5.5
W	5.39	346	52	6.76

TABLE 9.5

Metal	T_C [K]
Fe	1043
Ni	635
Co	1400

where $h = 6.625$, Js = Planck constant, $k = 1.38\ J/k$ Boltzmann constant, Θ = the Debye temperature (for several nonmagnetic metals is given in Table 9.4), ν_{max} = the maximal elastic frequency of the atom in a lattice of metal, and ρ_Θ = the resistivity of the metal at the Debye temperature.

From Table 9.4 it is seen that 0.15 Θ is in the range 10–60K, which means that

$$\text{TCR}_{20} = \frac{1}{283} \text{ to } \frac{1}{233} \tag{9.15}$$

that is, TCR = +3500 ppm/K up to +4300 ppm/K.

For nonmagnetic metal, TCR_{20} is constant, and temperature dependency of resistivity can be written as follows:

$$\rho_T = A\rho_\Theta T \tag{9.16}$$

where A = constant.

It appears that for nonmagnetic metal in the range of temperature $T \geq 015\Theta$, resistivity is proportional to the ambient temperature. On this basis the resistive platinum thermometer is built.

Magnetic Metal. For magnetic metal, for example, Fe, Ni, Co, the relation $\rho_T = f(T)$ is nonlinear and given by (9.17):

$$\rho_T = \text{C}T^{1.7} \text{ for } T \leq T_c \tag{9.17}$$

where T_c is the Curie temperature (see Table 9.5).

TCR's measured for iron and nickel are about +4500 ppm/K. Pure metal, both magnetic and nonmagnetic, is not useful for resistor design due to their large TCR and low resistivity.

Metal Alloy. Increasing of resistivity with temperature can be explained by atomic vibrations in the crystalline lattice. Those atoms cause obstructions to free electrons and the higher temperature causes an increase of resistance (obstruction).

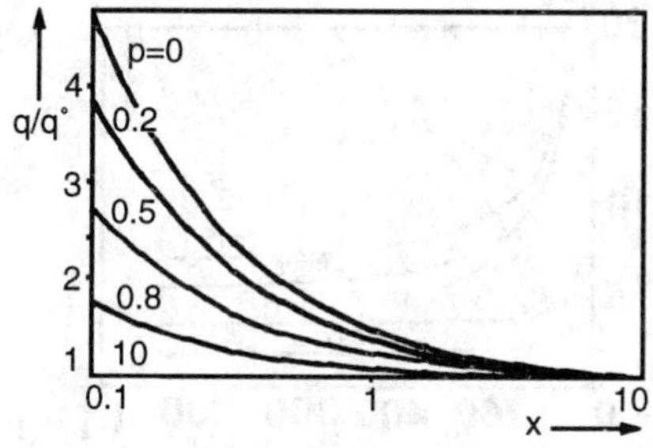

FIGURE 9.16 Influence of layer thickness on resistivity of metal thin film [2]. ρ_0—resistivity of bulk metal; *p*—quantity of reflected electrons; 1–*p*—quantity of absorbed electrons.

Such a resistivity is the first component ρ_s of an alloy resistivity ρ. According to Mathiessen's rule [2] we can add the second component of resistivity ρ_i, that presents obstruction of free electrons by atoms of impurities in the metal lattice.

$$\rho = \rho_s + \rho_i \tag{9.18}$$

$$\rho_i = \text{const},\ \rho_i \neq \varphi(T)$$

where ρ_s is given by (9.16) and (9.17).

When $\rho_i \gg \rho_s$, TCR is very small. In this case $\rho \gg \rho_s$ as well. This means that a resistor made of a specially prepared metal alloy can have very small TCR and large resistivity ρ. That result is very remarkable for a metal alloy resistor design.

Example. Constantin 60 percent Cu+40 percent Ni:TCR_{20} = 1–5 ppm/K, $\rho_{20} = 0.49 \times 10^{-6}\ \Omega$m in contrast with pure copper: TCR = 4000 ppm/K and $\rho_{20} = 0.0168 \times 10^{-6}\ \Omega$m; Canthal 70 percent Fe + 23 percent Cr + 4.5 percent Al + 1 percent Co + 1.5 percent other metals: TCR_{20} = 50 ppm/K, $\rho_{20} = 1.4 \times 10^{-6}\ \Omega$m.

The example above shows that for alloys TCR is from 100 up to 1000 times smaller and ρ_{20} is from 20 up to 100 times greater in comparison to pure metals (see also Table 9.4) This information is useful for wirewound and foil resistor design.

Thin Film Resistor

Over 90 percent of thin film resistors are made of metal alloy. A thin layer of metal alloy is deposited on ceramic or glass substrate. Sheet resistance $R_\square$ (Ω/square) of a metal alloy thin film resistor varies from 10 Ω/square up to 200 Ω/square; the nominal resistance range is from 10 Ω to 10 MΩ; TCR $\approx$ 15–150 ppm/K; classes 5, 3, 2, 1; precision resistors are available.

For a high nominal resistance value, the resistive layer is specially shaped by laser trimming to get enough squares. Some features of the thin film layer are a bit different from bulk metal alloy.

Fuchs's and Sondheimer's Effect [2]. When the thickness of the resistive layer is smaller than a free path of the free electron, the resistivity increases.

Figure 9.16 shows the influence of normalized thickness κ on resistivity ρ:

$$\kappa = \frac{t}{\lambda} \tag{9.19}$$

where t is the thickness of a layer, λ is the free path of the electron, which equals 20–30 nm [2], and κ is expressed by (9.19).

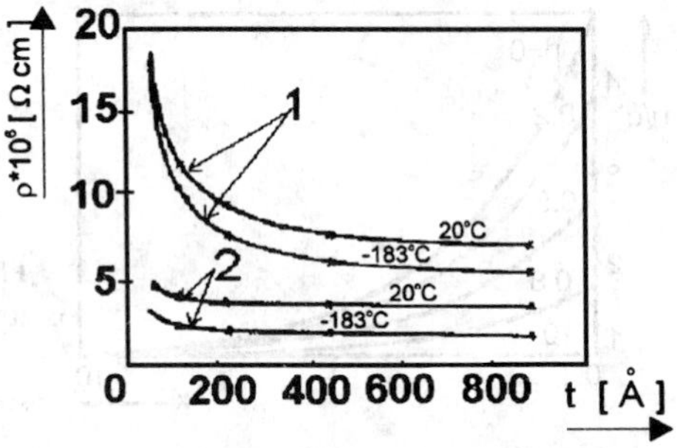

FIGURE 9.17 Influence of annealing process on resistance of thin film [2]. 1—before annealing; 2—after annealing.

Grain Effect. Because free electrons are reflected and absorbed by edges of grains, the resistivity increases when the quantity of grains becomes greater. Annealing processes of the layer limit that effect.

Figure 9.17 shows the result of the annealing process [2]. Small grains of the resistive layer are also the reason for the large absorption of humidity and gases that causes instability of the resistor.

Pyrolytic Carbon Resistor

Carbon deposited in vacuum on a ceramic substrate performs very inexpensively and produces a quite good quality resistive layer. Pyrolytic carbon has graphite structure sp^3; three electrons create bonds and the fourth is a free electron. Surface resistance is 10 Ω/square up to 2 kΩ/square; the nominal resistance value is in the range 10 Ω to 10 MΩ; encountered classes are 15, 10, 5, 3 (see Table 9.1); TCR ≡ −200 to −1000 ppm/K. About 50 percent of the produced resistors are carbon pyrolytic ones.

Thick Film Resistor

A thick film resistive layer is prepared from a composition called paste or ink that is a mixture of conducting particles and glass. Conducting particles have a metallic conduction mechanism. By mixing different quantities of both particles we can obtain a high-range surface resistance of 1Ω/square to 10 MΩ/square. This means that the nominal resistance range might be 0.5 Ω up to 1GΩ; classes are 5, 3, 2, 1, 0.5; TCR = ±100 ppm/K, and ±50 ppm/K; in a resistive network ΔTCR = 3 ppm/K and 1 ppm/K.

The theory of conduction mechanisms is not well known. There are several models of conduction mechanisms but one of them proposed by Pike and Seager is used [9]. Each conductive particle is surrounded by glass. These particles form the chains of MIM (metal–insulator–metal). Electrons are passed through the insulator by tunneling. In the layer of very high resistivity, electrons travel according to the hopping effect [6]. Investigations show that for the very low resistivity layer, chains without glass [6] are formed. The formula for the total resistance of such a chain is as follows:

$$R(T) = R_{\mathrm{M_IM}} + R_M + R_c \tag{9.20}$$

where

$$R_{\mathrm{M_IM}} = \frac{1}{2} R_{bo} \frac{\sin aT}{aT}\left(1 + \exp\frac{\Delta E}{T}\right) \tag{9.21}$$

$$R_m = R_{mo}(1 + bT) \tag{9.22}$$

$$R_c = \kappa_3 p_c (T_s - T)^{-\frac{1}{3}} \tag{9.23}$$

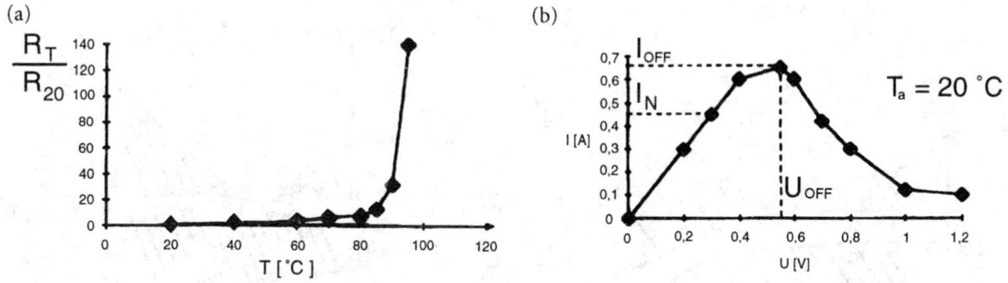

FIGURE 9.18 Characteristics of multifuse resistor. (a) Normalized resistance versus ambient temperature. (b) Current versus voltage; V_{off}—voltage above which current decreases.

where R_{M_IM} is the resistance connected by tunneling [4], R_{bo} is the resistance in $T = 0$, a is the distance between particles, ΔE is the activation energy, k is the Boltzmann constant, R_m is the resistance of the conducting particle (the resistivity value is much higher than in bulk material because its diameter is very small (0.5 μm) and grain effect occurs), R_{mo} is the resistance in $T = 0$ K (obtained by line extrapolation), b is the TCR of the conducting particle, R_c is the resistance of contact between two particles without glass), T_s is the temperature of glass melting, and k_3 is the constant coefficient.

For high resistance, the most important parameter is R_{M_IM}. In that range of resistance TCR is negative and the surrounding glass layers give large voltage coefficients of resistance, which is shown in Fig. 9.30. For low resistance, VCR is negligible, and TCR is positive, which means that R_m and R_c are important while an influence of R_{M_IM} may be neglected.

Equations (9.20)–(9.23) also explain the physical meaning of T_m for curve 3 in Fig. 9.4. For the very high resistivity layer, both tunneling and hopping conduction occur. Parameters for this sort of resistor are not particularly good, but only in that technology can we get high-ohmic resistor in the range of 10 MΩ up to 1GΩ.

Polymer Resistor

About 70 percent of all potentiometers are manufactured as polymer resistors. The resistive layer consists of conductive particles (black carbon, graphite, as well as metal powder) and thermosetting or thermoplastic resin. Also, a polymer layer is used for printing resistors on PC boards or for manufacturing chips in surface mounting technology. The layer is deposited by screen printing, painting, or by some other methods. Classes are 20, 15, 10, 5; TCR $= -1000$ ppm/K or ± 400 ppm/K; surface resistance 50–1 MΩ/square; the nominal resistance range is from 100 Ω to 10 MΩ. There has been great interest in thermoplastic polymer resistors [7] recently. At the softening temperature of a polymer a strong increase of resistance is observed. TCR $\sim$ 100 percent/K; and after cooling resistance retains a previous value. This phenomenon is applied to the multifuse resistor, whose characteristics are shown in Fig. 9.18(a) and (b).

A multifuse is needed for protection of the electronic power circuit against fire. Polymer layers are also used in keyboards as very hard and uncorrodible contact material. For example, that layer contains carbon and copper powder.

Comparison of Parameters of Different Resistors—Suggestion for Application

Figures 9.20 and 9.21 present average instability $\bar{\delta}$ of resistors made in four technologies: metal thin film, thick film, pyrolytic carbon, and polymer carbon. An analysis of these figures as well as Fig. 9.19 gives us a clue as to how to choose a resistor for a specific application. For example, we would like to find a resistor of a nominal resistance value of 1 MΩ to operate under high humidity conditions. A thick film resistor would be the best. Taking into consideration the price, a pyrolytic carbon resistor, class 10 or 5 should be chosen. In an operational amplifier

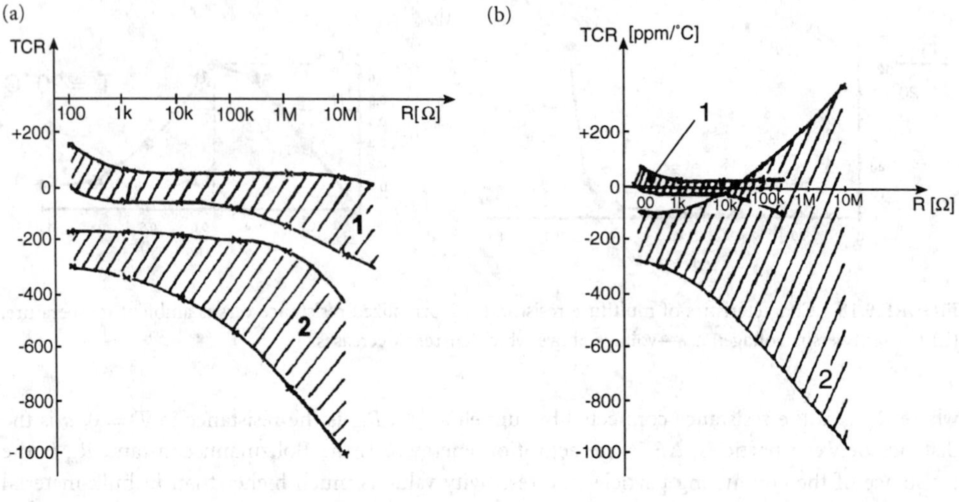

FIGURE 9.19 TCR dependence on nominal resistance value. (a) 1—thick film ruthenium-based resistors; 2—pyrolytic carbon resistors. (b) 1—thin film metal alloy resistors; 2—polymer carbon resistors. The envelope of the TCR range is calculated statistically at ±1.64∗ s.

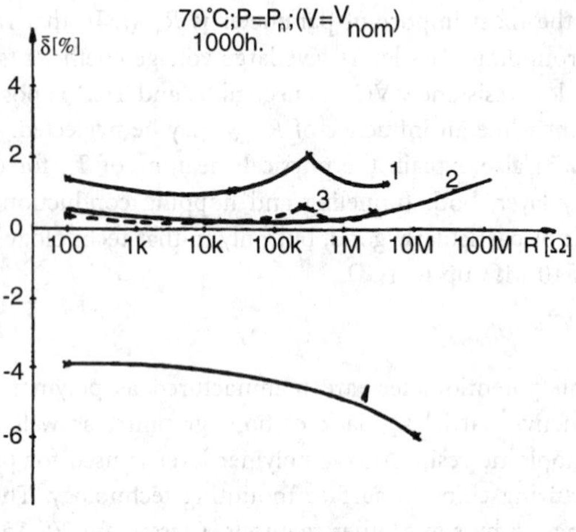

FIGURE 9.20 The average change $\bar{\delta}$ after 1000 h endurance test in relation to a nominal resistance value for different types of resistors. 1—pyrolytic carbon, 2—thick film ruthenium-based resistor, 3—thin film metal alloy, 4—polymer carbon. For the range 100 Ω–240 kΩ $P = P_n$, above 240 kΩ $P < P_n$ and $V = V_{max}$.

application, a thin film resistor or thick film network would be the best choice, although for a small-signal preamplifier considering the small noise level, a thin film resistor would be preferred. Considering our decision we should pay attention to the results of the following tests: TCR $= \varphi(R)$, endurance test $\varphi(R)$ for 1000 h, humidity test, 21 days: $\delta = \varphi(R)$.

Influence of Ceramic Substrate on Parameters of Resistor

All resistive films are deposited on the ceramic substrate. Only polymer film can be put on the fenolic paper or on the epoxy resin substrate. Thick film, pyrolytic carbon, and metal alloy film are also deposited on the ceramic. Some of thin films may be deposited on the glass. It is observed that substrate affects the resistive layer in two ways, as follows.

1) When ion current related to alkali metals in substrate flows. Its destructive effect is shown in Fig. 9.22, where some pyrolytic carbon layers manufactured by the same technology were

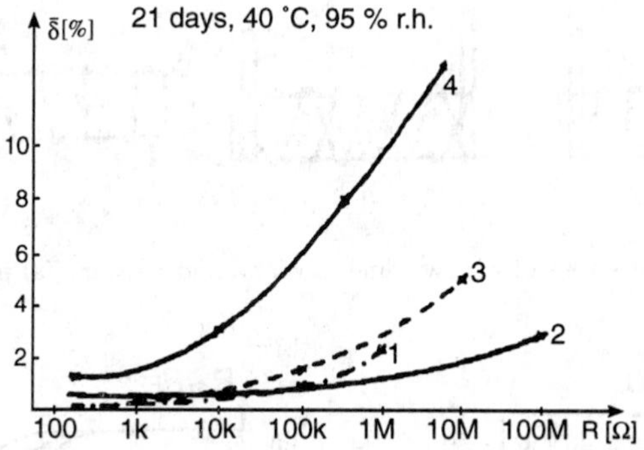

FIGURE 9.21 The average change δ after 21-day humidity test dependence on the nominal resistance value of different types resistors. 1—thin film metal alloy, 2-thick film ruthenium-based resistor, 3—pyrolytic carbon, 4—polymer carbon.

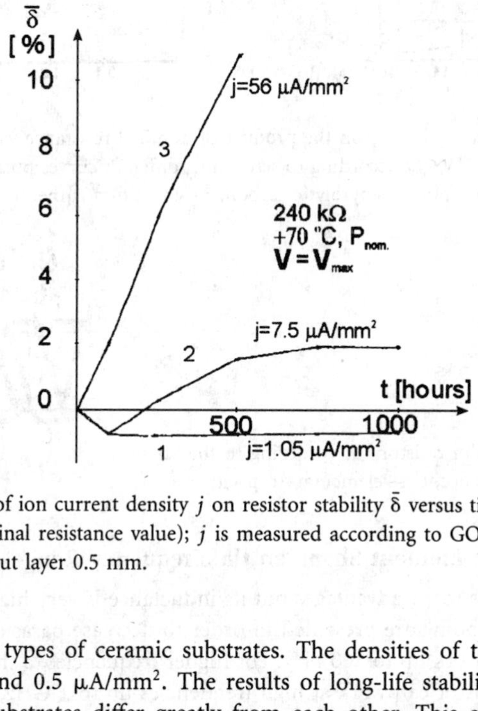

FIGURE 9.22 Influence of ion current density j on resistor stability $\bar{\delta}$ versus time during endurance test. $R = 240$ kΩ (critical nominal resistance value); j is measured according to GOST 10449-63, that is, 400° C and 400 V; distance of cut layer 0.5 mm.

deposited on the three types of ceramic substrates. The densities of the ion current were 56 μA/mm^2, 7 μA/mm^2 and 0.5 μA/mm^2. The results of long-life stability changes for resistors deposited on various substrates differ greatly from each other. This effect is seen only at dc voltages and at high temperatures. The smaller thickness of a layer and the destructive effect of ion current is more remarkable. During long-life tests at high temperatures, this phenomenon can also be observed for high-ohmic wirewound resistors [1].

2) When the thermal expansion coefficient of a layer and substrate are much different, thermal dilatability becomes a large problem and results in plus or minus changes of TCR [1]. Encapsulation of resistors with lacquer, cement, and transfer molding cover causes similar problems.

Dependence on Geometry of Material

The geometry of a resistive element affects either its high-frequency characteristic or its maximum temperature.

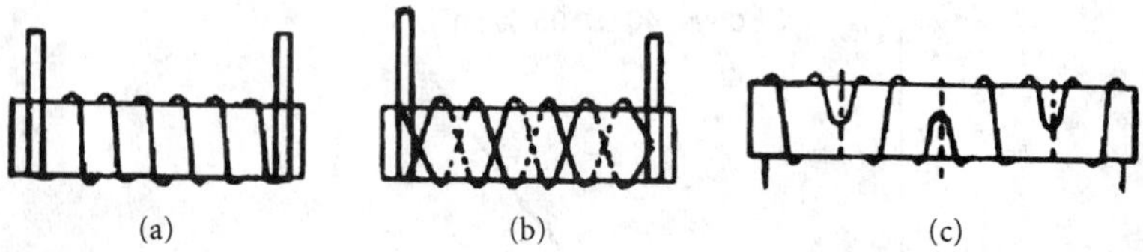

FIGURE 9.23 Different means of wire winding in wirewound resistors: (a) flat; (b) cross; (c) bifilar.

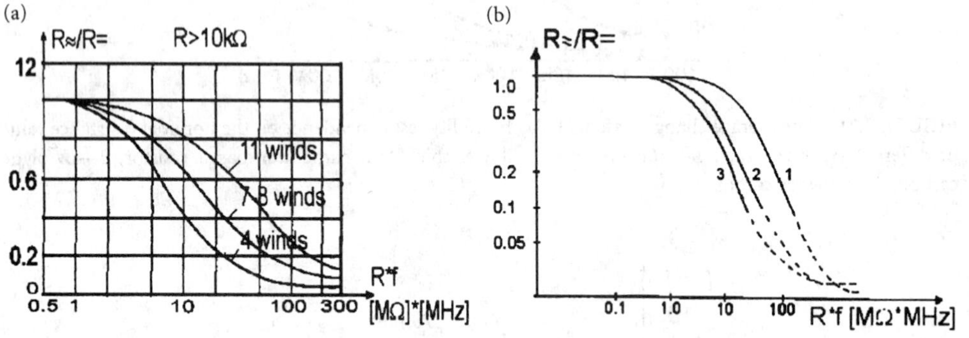

FIGURE 9.24 Dependence of $R_{\sim}/R_{=}$ on the product of nominal resistance value and frequency. (a) Pyrolytic carbon resistors, 0.25 W, cut according to screw line; different curves present resistors with different winds of cutting; (b) 1—thin film, 2—pyrolytic carbon, 3—thick film ruthenium-based resistor [3].

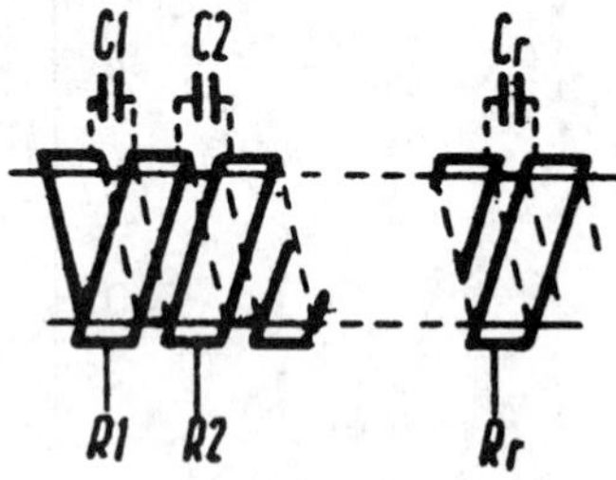

FIGURE 9.25 Part of a film resistor cut according to the screw line: R_r—elementary resistance, C_r—elementary capacitance.

Influence of Resistive Element Shape on the Frequency Range

A wirebound resistor has many advantages but its inductance is very high. In Fig. 9.23, different ways of resistive wire winding are presented in order to decrease parasitic inductance. A special wirewound resistor can work up to 200 kHz. For higher frequencies, a thin film resistor must be used. Though a film resistor can work at high frequencies up to 1 GHz, some limitations occur in this area (see Fig. 9.24). For resistors of a low resistance value, that limitation is inductance of resistive elements and terminations. For resistors of a resistance above 10 kΩ, distributed capacitance is the main problem.

Figure 9.25 presents the part of thin film layer deposited on a cylindrical ceramic substrate. To get many more squares, cutting along the screw line is performed. Each step of the screw wind [Fig. 9.27(a)] gives some elementary resistance R_r and elementary parasitic capacitance C_r. The maximum operating frequency f_m is given by (9.24) as

$$2\pi f_m = \frac{1}{R_r C_r} \tag{9.24}$$

where C_r is bigger when the thickness of the layer is greater and the groove becomes narrower, but also when ϵ_r of the insulating cover is high; R_r is smaller when Ω/square is lower.

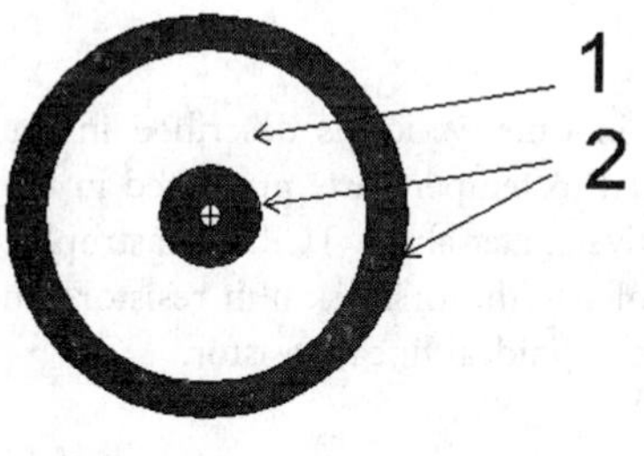

FIGURE 9.26 Coaxial resistor: 1—resistive layer; 2—electrode layer.

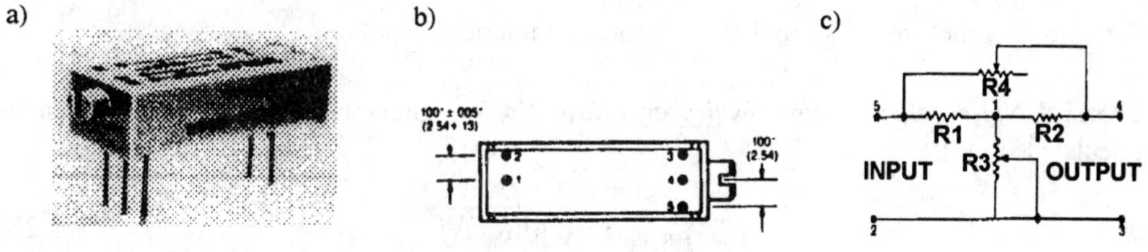

FIGURE 9.27 High-frequency attenuator [7]. (a) Overview; (b) pin localization; (c) electrical circuit.

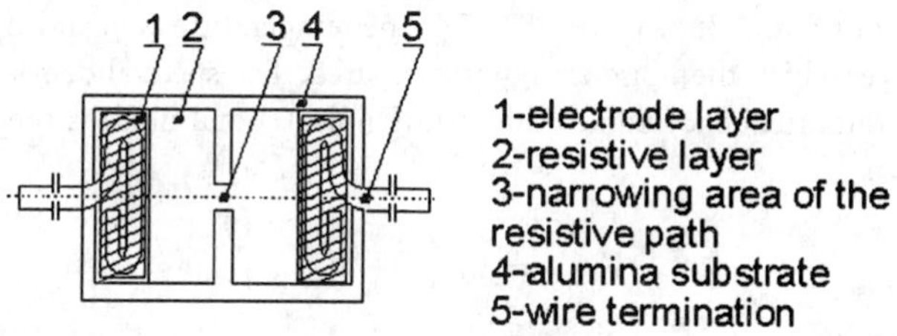

FIGURE 9.28 Fail-safe resistor.

This information suggests how to choose a resistor for high frequency. It is seen that a thin film with a low Ω/square is recommended. To reduce the termination capacitance for coaxial lines or cables, coaxial resistors are used (see Fig. 9.26), which can work up to 20 GHz. For high-frequency applications, special shapes of resistors have been designed (Fig. 9.7). Strip resistors are connected to a microwave strip line. A special potentiometer of 75 Ω input and 75 Ω output works up to 1 GHz. That potentiometer has fluent regulation of the attenuation obtained by specific shapes of three resistive elements inside the potentiometer, that in turn, are regulated by one appropriately shaped slider (a similar attenuator for 300 MHz is shown in Fig. 9.27).

Sensitivity Improvement of Precision Potentiometer by Complication of Shape

In Fig. 9.10, how a helipot works is explained. The wiper makes a rotation of 3600° and the precision is about 0.01 percent. The range of resistance is 100 Ω to 100 kΩ. The producers are Beckman, Bourns, and others.

Influence of Resistor Shape on its Application

The typical requirement for resistor construction is to establish a uniform temperature on the entire surface of the resistive element. This problem does not appear in the designing of a fail-safe resistor. The fail-safe resistor is shown in Fig. 9.28.

At the center of the resistive layer a narrow resistive path is placed [6]. Under normal working conditions, the resistance of the circuit is stable. When a defect occurs in the circuit, the current doubles or triples in the path and breaks the fail-safe resistor. This is important in protection against fire and avalanche devastation of the electronic set. The resistors utilizing the piezoresistive effect for pressure detection are also specially shaped.

Nonideal Linear Resistor

The main feature of the nonideal resistor is its instability in time, which is described in the Introduction to the section on resistors and instability related to temperature, presented in the Introduction and in the section on the influence of resistive material on TCR. Catastrophic failure rate is estimated: for thin film resistors on the level of 10^{-9}/h, for thick film resistors on the level of 10^{-10}/h. Noise is also an important feature of the nonideal linear resistor.

Noise

Total noise is the sum of a number of factors, summarized as follows.

Thermal Noise called Johnson Noise or White Noise. The value of thermal noise can be calculated from (9.25):

$$V_{t(\mathrm{RMS})} = \sqrt{4kRT\Delta f} \tag{9.25}$$

where $V_{t\mathrm{RMS}}$ is the root-mean-square value of the noise voltage $[V]$, R is resistance value $[\Omega]$, k is the Boltzmann constant (1.38×10^{-23} J/K), T is the temperature [K], and Δf is the frequency bandwidth [Hz] over which the noise energy is measured. The spectral density of thermal noise is constant in the total frequency bandwidth (white noise). Total noise is the sum of $V_{t\mathrm{RMS}}$ and current noise $V_{i\mathrm{RMS}}$:

$$V_{\mathrm{RMS}} = V_{t\mathrm{RMS}} + V_{i\mathrm{RMS}} \tag{9.26}$$

Current Noise or Structural Noise. When dc voltage is applied to resistor, dc current causes ac voltage fluctuation. That ac voltage fluctuation depends on the structure of the resistive element and an applied voltage. Density of that noise can be described by (9.27):

$$\frac{V_{i\mathrm{RMS}}}{\Delta f} = cf^{-\gamma}V_{=}^{\alpha} \tag{9.27}$$

where c, γ, and α are constants, V is the applied dc voltage, and Δf is the frequency bandwidth, $\gamma = 0.98$–1.2; very often $\gamma = 1$ and this kind of noise is called "$1/f$ noise"; $\alpha = 1$–2 and depends on the structure of the resistive element.

$V_{i\mathrm{RMS}}$ is measured in [μV/V] or in dB, where 0 dB = 1 [μV/V]. For foil and wirewound resistors, $V_{i\mathrm{RMS}} = 0$ and this means that only thermal noise occurs there. Fig. 9.29 shows simplified characteristics of current noise for different types of film resistors.

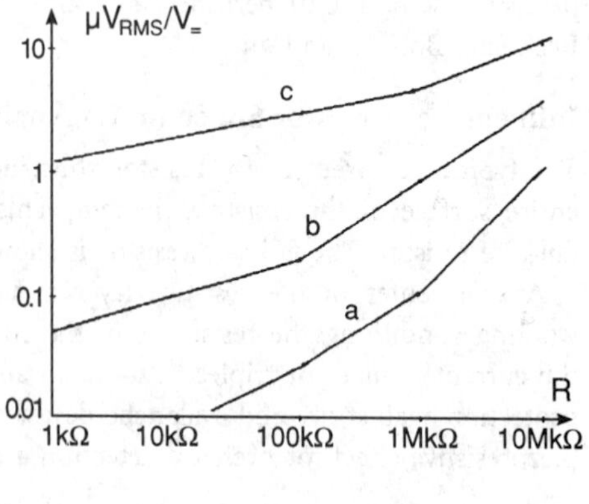

FIGURE 9.29 Simplified current noise characteristics versus nominal resistance value for different types of resistors: (a) Thin film, (b) pyrolytic carbon, (c) ruthenium-based thick film.

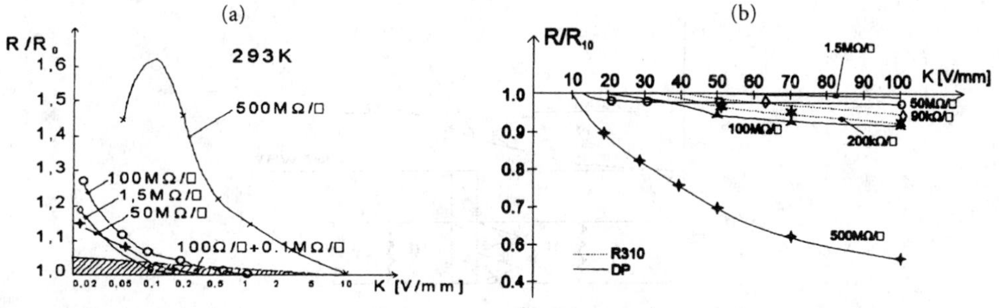

FIGURE 9.30 The change of resistance versus stress voltage for thick film resistors: (a) for low stress; (b) for high stress [6].

Voltage Coefficient of Resistance (VCR).

A linear resistor can exhibit some deviations from Ohm's law (9.1). This nonlinearity, called the voltage coefficient of resistance (VCR), is measured in [%/V] and is calculated from (9.28):

$$\mathrm{VCR} = \frac{R_1 - R_2}{R_2(V_1 - V_2)} \cdot 100[\%/V] \tag{9.28}$$

where R_1 is the resistance at the rated voltage V_1 and R_2 is the resistance at 10 percent of the rated voltage V_2. Metal alloy film resistors and pyrolytic resistors show a negligible small voltage coefficient but polymer resistors and thick film resistors have a remarkable one. In Fig. 9.30 normalized R/R_{10} versus voltage stress for the ruthenium-based thick film resistors are presented, where R_{10} is resistance measured at stress 10 [V/mm]. Results are collected for low stress voltage in Fig. 9.30(a) and for high voltage stress in Fig. 9.30(b).

It is seen that for a resistor made of a low-resistivity ink up to 100 KΩ/square, VCR is small but for inks of high resistance/square (for example, 500 MΩ/square), VCR is large [6]. VCR depends on the ink producer, as well [see R310 in Fig. 9.30(b)].

Rotational Noise and Contact Resistance Variation (CRV)

When a potentiometer is supplied by dc voltage and its wiper is moving from the beginning to the end of resistive layer, some ac voltage appears on the output. It is noise, which is measured in $[mV/V_=]$. The IEC standard requires that for a quality potentiometer this noise has to be smaller than $2[mV/V_=]$. The contact resistance variation (CRV) is important when the resistor works in series with the load or in a very sensitive instrument. This parameter is measured in percentage of total resistance; (1 percent is a typical value). CRV and rotational noise decrease if a multipoint wiper is used in potentiometers, for example, a wiper made from 20 wires. That parameter is also important in the construction of a precision wirewound potentiometer where CRV must be smaller than 1 percent. The proper choice of materials for the slider and resistive wire is the best way to solve this problem.

Smoothness of Regulation Curve of Potentiometer

Exponential Curve. In practice an exponential resistive element consists of two or three linear segments of resistance. As a result, the line is not smooth but has some steps. Rotary noise also increases in that area. More experienced producers use several segments and then the junction is not very sharp but are saw-tooth shaped, as shown in Fig. 9.31.

Linear Curve in a Helipot Precision Potentiometer. In a precision potentiometer, the resistance of the resistive element should be proportional to the distance traveled by wiper.

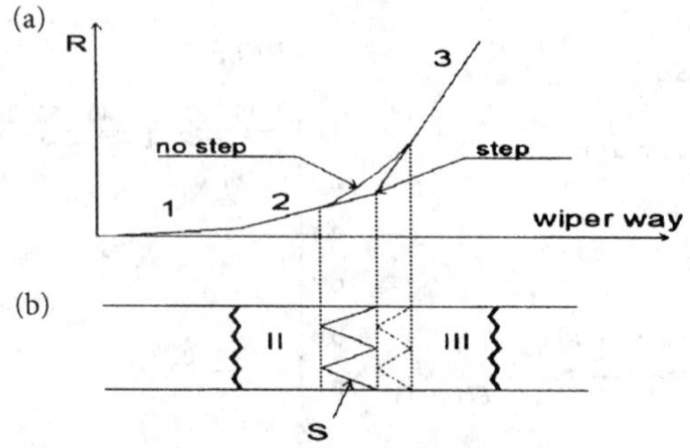

FIGURE 9.31 Shaping of exponential curve: (a) resistance versus wiper movement, the slopes 1, 2, 3 are proportional to resistivity of layer and $\rho_1 < \rho_2 < \rho_3$; (b) junction of resistive layer area marked by *s* is the saw-tooth-shaped junction of segment II and segment III.

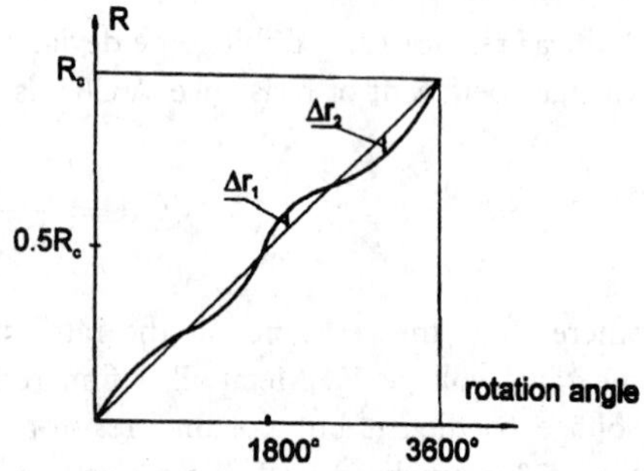

FIGURE 9.32 Resistance versus rotation angle (wiper movement) in precision linear potentiometer.

In Fig. 9.32, it is shown that Δr_1 and Δr_2 are the maximum deviation from a straight line. The nonproportionality *NP* of a precision potentiometer is described by (9.29):

$$NP = \frac{\Delta r_1 + \Delta r_2}{R_c} \cdot 100\% \tag{9.29}$$

The value of *NP* is 0.5–0.01 percent. Such a good result is obtained by fluent control of the proportionality over winding of the wire.

References

[1] G. W. A. Dummer, *Fixed Resistors*, London: Pitman, 1967.

[2] *Handbook of Thin Film Technology*, L. I. Maissel and R. Glang, Eds., New York: McGraw-Hill, 1970.

[3] *Handbuch der Elektronik*, S. Nowak, A. Wenta, and E. Kuzma, vol. 2 and A. Ambroziak et al., vol. 7, Munich: Franzis-Verlag, 1979.

[4] C. A. Neugebauer and M. B. Webb, "Electrical conduction mechanism in ultrathin evaporated metal films," *J. Appl. Phys.*, vol. 33, p. 74, 1962.

[5] S. Nowak and D. Lusniak-Wojcicka, "Thick film fail-safe resistors," *Electrocompon. Sci. Technol.*, vol. 10, no. 4, p. 255, 1983.

[6] S. Nowak, "Nonlinearity of thick film resistors." *Rozprawy Elektrotechniczne*, vol. 4, 1989.

[7] *Philips Components Catalogue*, 1989, Eindhoven.

[8] Bourns, Inc., *The Potentiometer Handbook*, New York: McGraw-Hill, 1975.

[9] Pike, G. E. and C. H. Seager "Electrical properties and conduction mechanism of Ru-based thick film (cermet) resistors," *J. Appl. Phys.*, vol. 48, no. 12, pp. 5152–5168, Dec. 1977.

[10] *Siemens Matsushita Components Catalogue*, Munich, 1993.

[11] R. W. Vest, "Conduction mechanism in thick film microcircuits," Final Technical Report, Purdue Univ. Res. Foundat., Grant DAHC-15-70-67, DAHC-15-73-68, ARPA Order:1642, December 1975.

9.2 Capacitor

Stanisław Nowak

Introduction

A capacitor is a container of the electric energy W. This is expressed by (9.30)

$$W = \frac{CV^2}{2} \tag{9.30}$$

where C is the capacitance expressed in farads and V is the voltage on the capacitor plates expressed in volts.

Capacitance C of the multiplate capacitor can be described by (9.31):

$$C = \frac{x\epsilon_r(N-1)A}{d}10^{-13} \tag{9.31}$$

where ϵ_r is the relative dielectric constant of the insulator, d is the distance between the plates, N is the number of plates, and A is the plate area, where $x = 0.0885$ for A and d expressed in centimeters or $x = 0.225$ for A and d in inches.

The relative dielectric constant value ϵ_r is equal to the ratio of capacity of a capacitor with plates separated by a dielectric to one separated only by vacuum. Dielectric constant values of various materials are presented in Table 9.6.

TABLE 9.6 Comparison of Capacitor Dielectric Constants

Dielectric	ϵ_r (Dielectric Constant)
Air or vacuum	1.0
Paper	2.0–6.0
Plastic	2.1–6.0
Mineral oil	2.2–2.3
Silicone oil	2.7–2.8
Quartz	3.8–4.4
Glass	4.8–8.0
Porcelain	5.1–5.9
Mica	5.4–8.7
Aluminium oxide	8.4
Tantalum pentoxide	26
Ceramic	12–400 000

Source: The Electrical Engineering Handbook, R. Dorf, Ed., 1993. Boca Raton: CRC Press.

Fundamental parameters of capacitor are as follows: capacitance C, nominal voltage V_N, testing voltage V_t (note that V_t =(2–3) V_N), temperature coefficient of capacitance TCC (for class 1 only), insulation resistance R_i (for dc voltage), and power factor PF (for ac voltage).

Power factor PF is described by the (9.32)

$$PF = r\,C\omega = 2\pi f r\,C \tag{9.32}$$

where r is an equivalent series resistance.

Its inversion is a quality factor Q, given by (9.33). Since the power factor expresses total losses in capacitor, it is a sum of the dissipation factor and losses in electrodes as well as in terminations.

$$Q = \frac{1}{PF} = \frac{1}{2\pi f r\,C} \tag{9.33}$$

According to IEC Publ 384/1988, capacitors are divided into two salient groups: class 1 and class 2 but also more detailed classifications are commonly used because of the wide range of capacitances and very different applications. The main applications of capacitors include filtering, coupling, tuning, dc blocking, ac passing, bypassing, phase shifting, compensation, through feeding, isolation, energy storage, noise suppressing, motor starting, and so on. Contemporary capacitors cover the 0.1 pF up to 10 F capacity range and the 2.5 V to 100 kV voltage range. Connecting a dc voltage source to the capacitor plates we can observe the given capacitor is gradually charged and current flowing through the capacitor, large at the beginning, decreases in time to negligible small value. On the other hand, ac source causes the current I, given by the (9.34), flowing permanently through the capacitor.

$$I = \frac{V}{X_c} = \frac{V}{(1)/(2\pi fC)} = V2\pi fC \tag{9.34}$$

That current increases when capacitance, frequency, or applied voltage increase. The ac current can heat the capacitor, whose temperature would depend on its power factor, capacitor size, and cooling conditions. This phenomenon has to be taken into consideration in energetic 50 Hz equipment or in high-frequency power applications. Miniaturization and integration of electronic sets influences miniaturization of capacitors as well. The index υ', called "own volume" (volume per capacitance) and expressed in [$cm^3/\mu F$], might be of use during selection of a capacitor for a given circuit. Nominal voltage V_N strongly affects the index υ' value. Table 9.7 presents the index υ' for capacitors with various dielectrics.

From the user's point of view, capacitors can be divided as follows: linear and nonlinear, fixed capacitors, adjusting capacitors, power energetic capacitors, start motor capacitors, and interference suppression capacitors.

Linear Capacitor

The linearity of a capacitor depends on the polarity of the dielectric used for its manufacture. There are several different polarization mechanisms that can contribute to the total polarization. The most important are the following:

1) Electron polarity that exists in the insulator with covalent bonds between atoms; electrical stress causes deformation of the orbital shape but electrons cannot go out of orbit (relaxation time τ is smaller than 10^{-15} s).
2) Ion polarity that occurs in glass and high-quality ceramic; under electrical stress ion centers are displaced ($\tau < 10^{-13}$ s).

TABLE 9.7 υ' Index of Various Capacitors

Capacitor Definition	Main Parameters	υ' [cm^3/μF]
Variable air	500 pF/250 V	200 000
Mica	10 nF/500 V	250
Ceramic (rutile)	1000 pF/500 V	600
Ferroelectronic	40 nF/250 V	50
Ferroelectric multilayer	0.68 μF/50 V	1.5
Polystyrene	2 μF/160 V	300
Polyester (mylar)	0.1 μF/160 V	12.4
Polycarbonate—metalized	0.15 μF/160 V	5.6
Electrolytic Al(HV)[a]	40 μF/350 V	1.3
Electrolytic Al (LV)[a]	120 μF/7 V	0.008
"Golden" capacitor	1 F/5.5 V	0.00001
Electrolytic Ta (wet)	10 μF/100 V	0.038
Electrolytic Ta (dry)	5.6 μF/10 V	0.0026

[a]HV: High voltage, LV: low voltage.

Source: L. Badian, *Handbuch der Electronik.* Munich: Franzis-Verlag, Vol. 3, 1979.

3) Dipole polarity that occurs in a polar polymer dielectric. Electric field causes rotation of dipoles in the dielectric. Generally dielectric constants ϵ_r depends on frequency, temperature, and voltage, but in limited ranges of these factors ϵ_r is stable.
4) Domain polarity that appears in some insulators, for example, ferrodielectrics. They contain domains that rotate with electric field. This effect is called ferroelectricity because of the analogy to ferromagnetism and gives rise to very high dielectric constant ϵ_r up to 400 000 (see Table 9.6). It strongly depends on voltage, frequency, and temperature. Ceramic capacitors with a domain polarity mechanism and high dielectric constant are very popular in electronic equipment.

Only capacitors with electron polarity and ion polarity are classified in linear capacitor group 1.

The linear capacitor group consists of fixed capacitors in class 1, adjusting capacitors, energetic capacitors, high-voltage capacitors, and interference suppression capacitors.

Fixed Capacitor—Class 1

The main feature of a fixed linear capacitor with class 1 dielectric materials is its stability in time and under temperature. By analogy to resistors we can introduce dynamic tolerance Δ (see the section on the fixed resistor in Section 9.1) for capacitance. Production tolerances $\delta p(\pm)$ for class 1 capacitors are ± 0.25, ± 0.5, ± 1, ± 2, ± 5, ± 10, and ± 20 percent. The instability δ_{max} expressed by (9.35) after the endurance test is up to 3 percent (IEC Publ. 384-8).

$$\delta_{max} = \bar{\delta} + 1.64 * s \tag{9.35}$$

where $\bar{\delta}$ and s are in accordance with the formula (9.4) and (9.6), respectively and

$$\delta_i = \frac{C_i(t) - C_i(0)}{C_i(0)} * 100\% \tag{9.36}$$

δ_{max} for capacitor is larger than for a stable resistor. This also means that dynamic tolerance Δ for capacitors is larger than for resistors, which should be considered during active rc filter design.

TABLE 9.8 TCC and max capacitance of monolithic ceramic capacitor.

Dimension $a \times b$ [mm]		Maximum Capacitance [pF]				
		NP 0	N 75	N 150	N 750	Ferrodielectric
	TCC ppm/K	0 ± 30	-75 ± 30	-150 ± 30	-750 ± 30	Large
4×4		47	47	47	150	10 000
8×8		680	680	680	1600	470 000
10×10		4700	4700	4700	6800	1 000 000[a]

[a]Some producers offer maximum capacitance $C = 4.7$ μF at dimensions of 10×10 mm.

The temperature coefficient of capacitance (TCC), given in Table 9.8 describes temperature stability of capacitors. Negative TCC capacitor may be used to compensate positive temperature coefficient of inductance in resonant LC circuit.

The next important parameter of capacitors in class 1 is a power factor that is required to be smaller than $30 * 10^{-4}$. Ceramic capacitors with low ϵ_r as well as styroflex and mica capacitors meet these conditions, which renders them very suitable for resonant circuits, stable analog filters, integrator circuits, and other circuits where stable capacitance and small losses are necessary.

Ceramic capacitors are produced as tubular, disk, and multilayer (monolithic) capacitors. Disk and tubular ones are inexpensive. Multilayer capacitors are rather expensive but they have small dimensions and low index υ'. Figure 9.33 presents some types of fixed capacitors.

Adjustable (Variable) Capacitor

An adjustable capacitor is an electronic component whose capacitance can be mechanically regulated by the user. For example, an AM radio set tuner is adjusted using a variable capacitor of 10 pF up to 500 pF. The dielectric used in this type of capacitor is either air or plastic foil. The majority of variable capacitors are trimmer capacitors for precision adjustment of reactance in electronic circuits. Their insulator layer is made of air, ceramic of class 1, mica, polystyrene, or teflon. Typical trimmer capacitors are shown on Fig. 9.34.

They have the following capacitance ranges: air trimmer, 1–15 pF, tubular ceramic trimmer, 0.1–7 pF, disk trimmer, 10–50 pF, special disk trimmer 100 pF maximum. The power factor is small $(1\text{–}20)*10^{-4}$. Insulating resistance is above $10^{10}\Omega$. Nominal voltage is 100–500 V.

Energetic Power Capacitor

To compensate inductance in electrical engines and other equipment, energetic power capacitors are used. Besides capacitance it is important to know their reactance power expressed in [VAr] (between vectors of voltage and of current there is an angle 90°). They have rather large size and weight of 10–50 kg per unit. The smaller ones are capacitors used to discharge lamps. They have a capacitance range of 2.5–20 μF and ac working voltage of 150 V effective voltage up to 550 V. Starting motor capacitors also belong to this group of power capacitor. The majority of power capacitors have paper, polypropylene, or mixed dielectrics, and they are impregnated in vacuum against flashover. An impregnant substance, such as mineral or synthetic oil is used. It is very important to match the dielectric constant of the oil with that of the fixed dielectric because it will guarantee uniform distribution of the electric field inside the unit. In some capacitors of smaller size polypropylene without an impregnant is used as an insulator. Electrodes are made of aluminium foil or deposited with thin metal film on the dielectric. The power factor of these capacitors is required to be low with respect to heat dissipation. According to IEC Publ. 384-17 the maximum power factor is $< 50.10^{-4}$ at working frequency and for 1.25 V_n of nominal voltage.

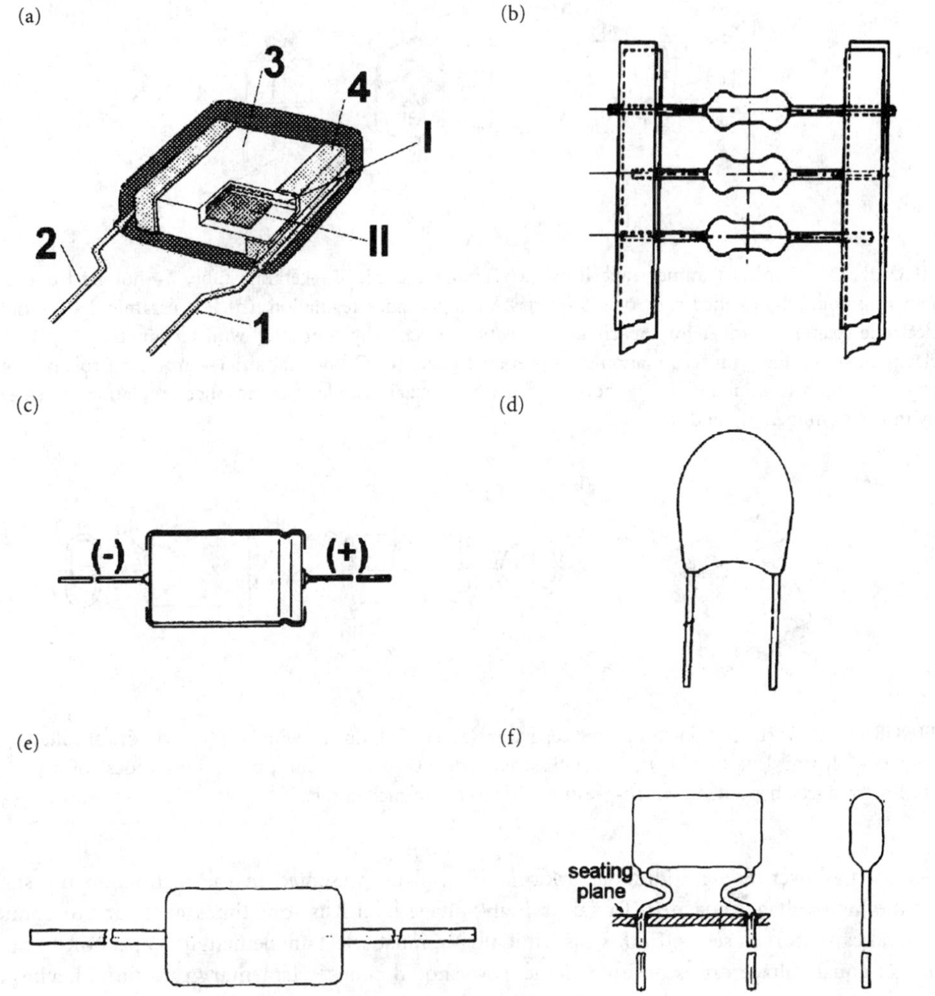

FIGURE 9.33 Typical fixed capacitors. (a) Multilayer ceramic (monolithic) [6]: 1,2—termination, 3—dielectric layer, 4—inner termination; I,II—plates (capacitor electrodes). (b) Tubular taped [8] (connected in tapes for automatic assembling). (c) Aluminium electrolytic (wet) with axial terminations. (d) Tantalum electrolytic (dry) with radial terminations. (e) Polystyrene foil with axial terminations. (f) Metalized film [6].

Under overheating the capacitor reliability falls and lifetime shortens. Polypropylene is chosen as the insulator material because of its good dielectric properties. For example, its PF $=(6–7)*10^{-4}$, $\epsilon_r = 2.1$, TCC $= -200$ ppm/K and over the temperature range up to 110° C it is stable. The elementary cylindrical section of the power capacitor is presented in Fig. 9.35. (a) shows two polypropylene foils on which electrodes (Ia and IIa) were evaporated in vacuum. It is important to provide a fixed, multipoint junction between electrodes (Ia and Ib) and (IIa and IIb). It shortens the path of the current, which at ac voltage is high. That construction reduces the equivalent series inductance of the capacitor at high frequencies.

High-Voltage Capacitor

A high-voltage capacitor for dc voltage and low frequencies is made of polymer foil, but for high-frequencies ceramic capacitors are preferred. They can be disk shaped up to 6 kV (Fig. 9.36) or tube shaped for very high voltage applications. It is easier to protect a tube-shaped capacitor

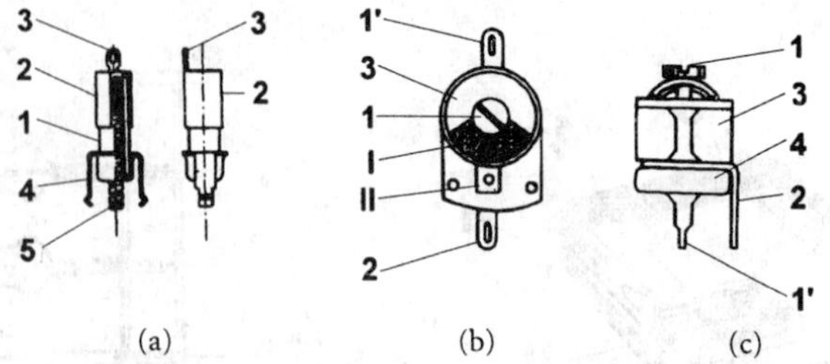

FIGURE 9.34 Typical trimmer capacitors. (a) Tubular ceramic: 1—ceramic tube, 2—hot electrode, 3—soldering point, 4—earthed electrode, 5—screw for capacitance regulation. (b) Flat ceramic: I—grounded electrode (plate), 1—roller for capacitance regulation (electrically connected with termination 1′); II—hot electrode connected with termination 2, 3—ceramic plate. (c) Cylindrical air: 1—grounded roller (screw) connected with termination 1′, 2—hot electrode, 3—coaxial cylinders; capacitance regulation is achieved by moving cylinders up and down.

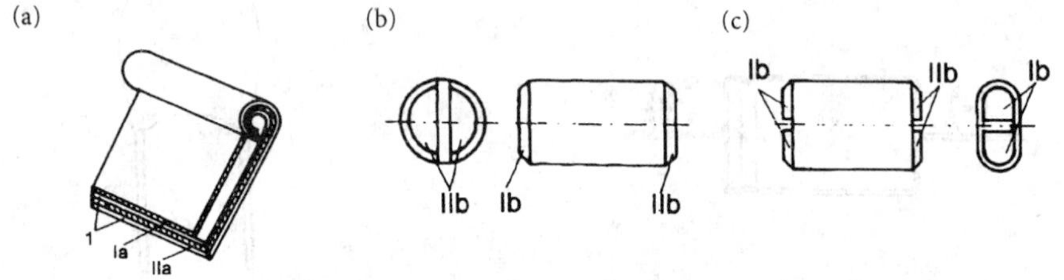

FIGURE 9.35 Metalized polypropylene capacitor. (a) The roll during winding; (b) cylindrical roll; (c) flat roll: 1—polypropylene foils, Ia,IIa—metalized layer deposited in vacuum, plates (electrodes) of capacitor, Ib,IIb—contacts, deposited by high-pressure airbrush with melted metal.

against flashover. Some additional information will be presented in the section on the shape of the high-voltage capacitor. To gain a high-voltage level it is sometimes necessary to connect several capacitors in series. In this case, initially we must select an elementary capacitor. In order to get equal voltage drops on each element we should choose elementary capacitors having the same capacitance for ac voltage and the same insulating resistance for dc voltage.

Dielectric Absorption. According to (9.30), high electrical energy can be stored in power and high-voltage capacitors. Danger to human life may exist even several hours after switching off the circuits that contain such capacitors. Also, capacitors discharging by short circuiting are not nearly safe enough. Because of dielectric absorption some dangerous charge remains on the capacitor electrodes. To prevent accidents, high-energy capacitors should be shunted by high-ohmic resistors.

Interference Suppression Capacitor

This type of capacitor is characterized by low inductance. To accomplish this, a specific construction has been developed, where two or three capacitors are joined in a special way and put into one encapsulation. Figure 9.37 presents an example of this construction. Interference suppression capacitors are divided into two classes: X or Y. Capacitors of class X fulfill lower safety requirements because their breakdown is not dangerous to human life. Capacitors of class Y are used in extremely dangerous environments, where, for example, breakdown causes a short circuit between the body of the equipment and an energetic phase of 50 Hz. Properties of these two types of capacitors vary. For example, the 4 μF/250 V capacitor that belongs to the X class, has a test voltage on the level of 1100 V; however, the one of the Y class has to withstand 2250

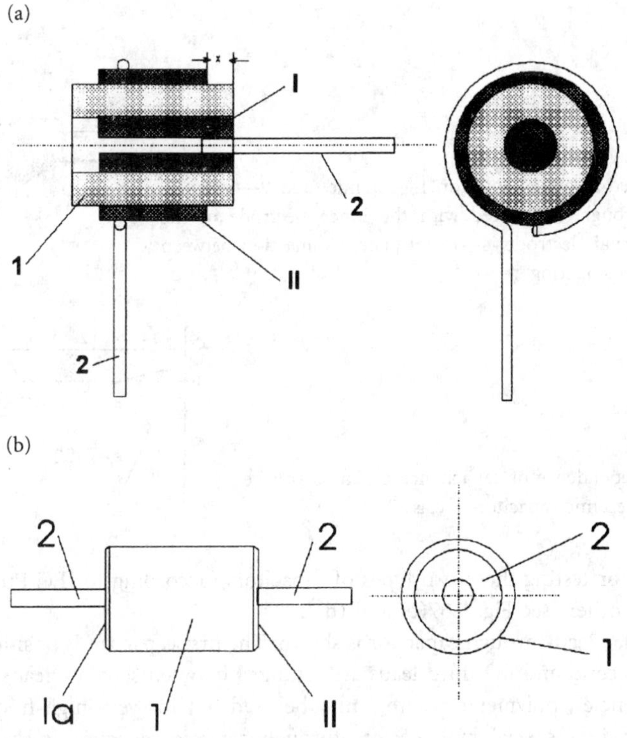

FIGURE 9.36 High-voltage ceramic capacitor. (a) Tubular: I—inner, high-voltage electrode, II—grounded electrode, 1—ceramic tube, 2—wire terminations. (b) Disk (bar) ceramic: 1—ceramic bar, 2—wire termination, Ia,IIa—plates of capacitor, made by screen printing; x—distance for high voltage.

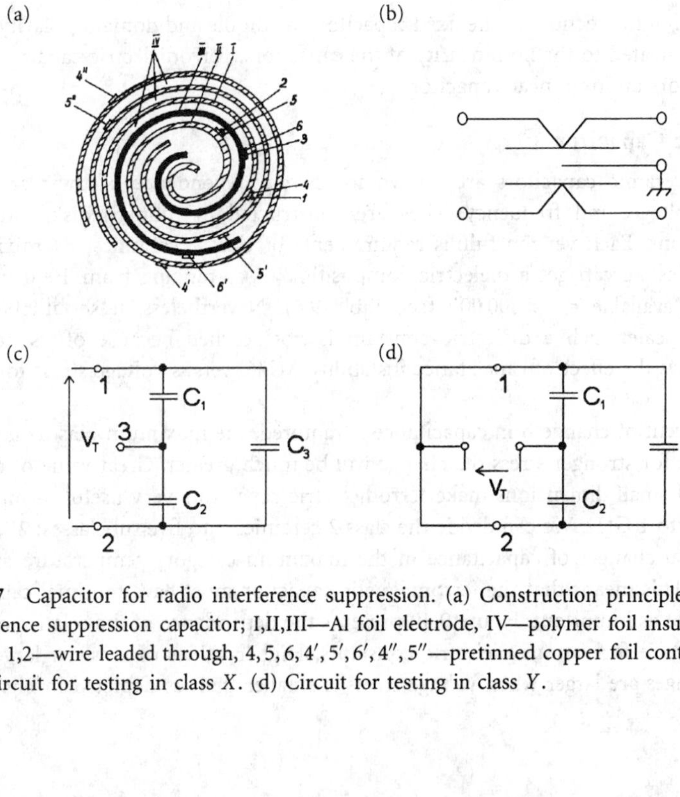

FIGURE 9.37 Capacitor for radio interference suppression. (a) Construction principle of three capacitance interference suppression capacitor; I,II,III—Al foil electrode, IV—polymer foil insulator, 1,2,3—wire terminations, 1,2—wire leaded through, 4, 5, 6, 4′, 5′, 6′, 4″, 5″—pretinned copper foil contact. (b) Electrical circuit. (c) Circuit for testing in class *X*. (d) Circuit for testing in class *Y*.

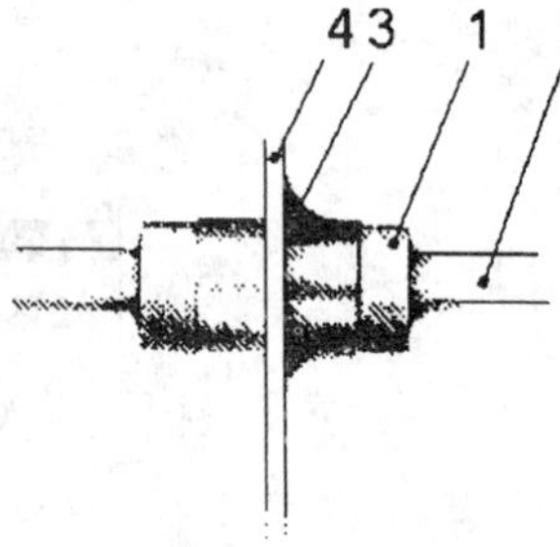

FIGURE 9.38 Feed through capacitor. 1—ceramic tube, 2—wire termination lead through, connected with the inner electrode of the capacitor, 3—external electrode, 4—metal plate. Connection between 3 and 4 is made by soldering.

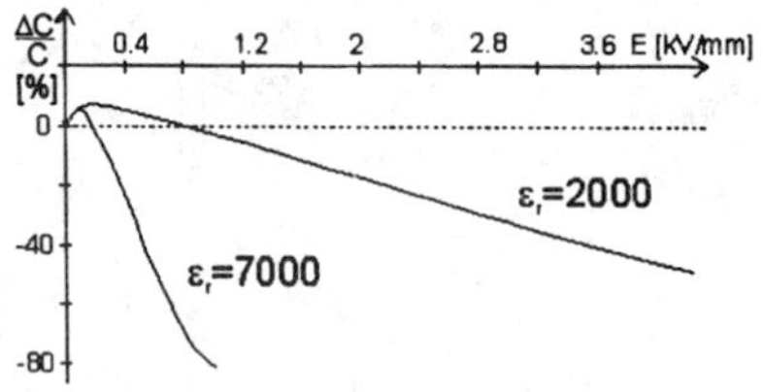

FIGURE 9.39 Dependence of capacitance on stress voltage at different ϵ_r of ceramic capacitor of class 2.

V. Also, circuits for testing these two types of capacitors (according to IEC Publ. 384-14, 1993) differ from each other [see Fig. 9.37(c) and (d)].

In Fig. 9.37(b), a lead-through capacitor is shown. The first capacitor is positioned between the two leads. The second and the third leads are arranged between separate leads and the ground. For radio frequencies, polymer capacitors may be used but for very-high-frequencies, ceramic ones are preferred. A special kind of feed-through ceramic capacitor is shown in Fig. 9.38. It should be noted that automotive capacitors have to fulfill some additional requirements to withstand ignition system pulses.

Nonlinear Capacitor

As was described in the section on the fixed capacitor, the dipole and domain polarity mechanism in a dielectric is related to the nonlinearity of the capacitors. Ferrodielectric capacitors and polar polymer capacitors are nonlinear capacitors.

Ferrodielectric Capacitor

Ferrodielectric ceramic capacitors are known for strong dependence of their capacitance on temperature, voltage, and frequency. They are constructed in tubular, disk and monolithic multilayer versions. Each version fulfills requirements of class 2 ceramics. By mixing different kinds of ceramics we can get a dielectric composition of ϵ_r ranging from 1000 up to 30 000. The maximum available ϵ_r is 400 000 (see Table 9.6). Nevertheless, it should be noted that for production scale, such a dielectric constant is not reached because of its poor stability. Figure 9.39 shows the effect of capacitance instability $\Delta C/C$ versus voltage stress for different ϵ_r values.

If the 25 percent of change δ in capacitance is required, the maximum bias dc is 2000 V/mm (for $\epsilon_r = 2000$); for stronger stress the change will be much greater. Great value of ϵ_r, low series inductance, and small dimensions make ferrodielectric capacitors very useful in high-frequency applications up to 1 GHz. We can divide the class 2 ceramics into five subclasses: 2B, 2C, 2D, 2E, and 2F. Maximal changes of capacitance in the maximum category temperature are presented (in Table 9.9). This means that the ferromagnetic capacitor must have minimal capacitance but its stability is not as important. Figure 9.40 presents the dependence of the relative change $\Delta\epsilon/\epsilon_r$ [3] on temperature for various dielectric constants. Higher values of ϵ_r show larger inequality of curves. Changes are larger when voltage stress is smaller than as depicted in Fig. 9.40(c).

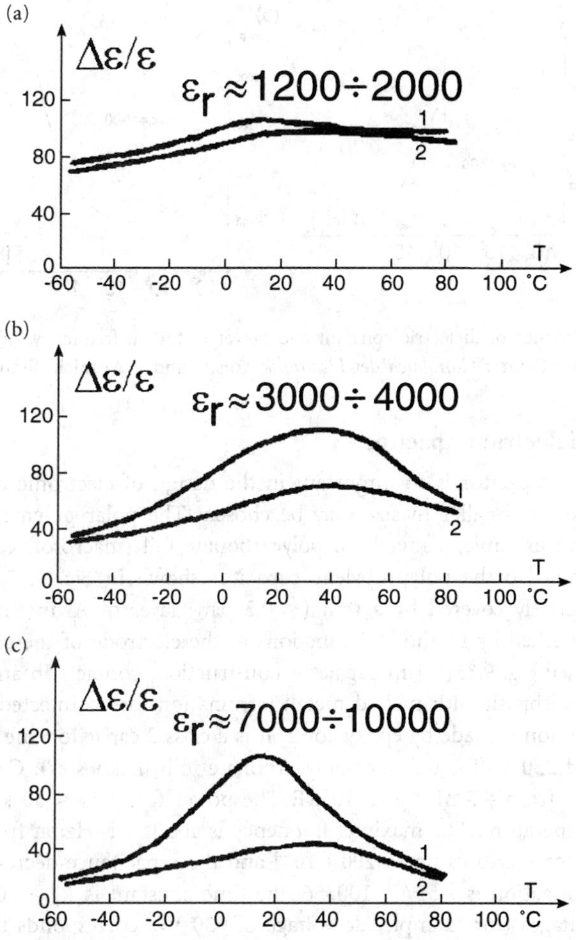

FIGURE 9.40 Normalized change of dielectric constant $\Delta\epsilon/\epsilon_r$ versus temperature for class 2 ceramic *Source: Handbuch der Electronik*, vols. 2 and 3. Munich: Franzis-Verlag, 1979. (a) Dielectric constant $\epsilon_r = 1200$–2000, (b) dielectric constant $\epsilon_r = 3000$–4500, (c) dielectric constant $\epsilon_r = 7000$–10th 000; curve 1 at very low voltage; curve 2 at bias 8 kV/mm.

TABLE 9.9

Subclass	Maximal Changes of C_n [%]
2B	+10, −15
2C	+20, −30
2D	+20, −40
2E	+22, −70
2F	+30, −90

Figure 9.41 shows the influence of frequency on capacitance and power factor. It can be noted that up to 1 GHz the changes are acceptable.

Production tolerance δ_p also differs from values in class 1, for example, −40, +80 percent; −20, +80 percent; −20, +50 percent; ±20, ±10 percent.

Series of nominal voltage are 25, 40, 63, 100, 160, 250 (200), 400(500), 630, 1000, and 1600 V. The maximal power factor is $350 * 10^{-4}$; after an endurance test it can increase up to $500 * 10^{-4}$ or $700 * 10^{-4}$.

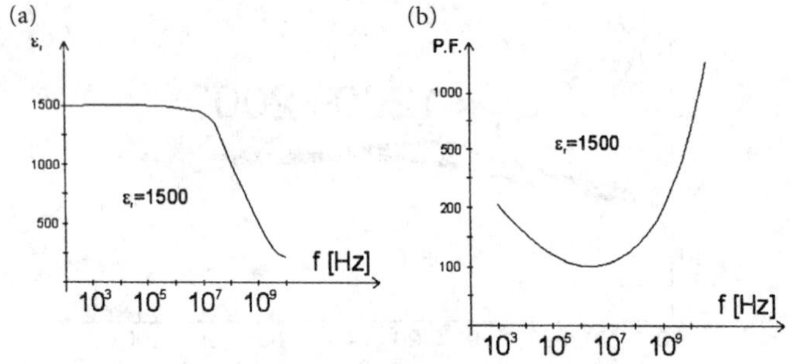

FIGURE 9.41 Dependence of dielectric constant and power factor on frequency. (a) ϵ_r versus frequency; (b) PF versus frequency *Source: Handbuch der Electronik*, vols. 2 and 3. Munich: Franzis-Verlag, 1979.

Polar Polymer Dielectric Capacitor

Miniaturization of a capacitor is so important in the design of electronic circuits that often a capacitor less stable but smaller in size may be chosen. The polar polymer capacitor exhibits such a feature; as an example, a metalized polycarbonate foil (macrofol) capacitor is used. Its construction is similar to the polypropylene capacitor shown in Fig. 9.35. Polycarbonate foil 3-μm thick is selectively covered by a thin ($\leq$ 0.5 μm) layer of Al in the PVD process. An aluminium layer (marked by Ia and IIa) functions as the electrode of the capacitor. Very often a flat coil is used [see Fig. 9.35(c)] in capacitor construction; contacts Ib and IIb are deposited by a high pressure airbrush with melted metal; terminations are connected to the contacts by soldering; encapsulation is made by epoxy cover. It is a class 2 capacitor; the voltage series range is 63, 100, 160, and 250 V (for different polycarbonate foil thicknesses). Capacitances in series E-6 and E-12 range from 4.7 nF up to 10 μF. The power factor is $\leq 30 * 10^{-4}$ and decreases with increasing temperature. The maximal frequency is about 1 MHz; at frequencies above 100 kHz, the power factor increases up to $200 * 10^{-4}$ and the capacitance decreases several percent. The climatic categorization is $-55/+100/56$; the time constant is $R_i \geq 5000$ s. It operates at both dc and ac voltages (for example, dc voltage of 630 V= corresponds to ac voltage of 220 V_{rms}).

Dependence on Material Properties

Dielectrics have played a fundamental role in capacitor performance (see Tables 9.6 and 9.7). Air capacitors, mineral oil capacitors and mica capacitors, as well as polystyrene and teflon capacitors mentioned above, are ranked as class 1 capacitors. Capacitors with natural dielectrics (air, oil, mica) are primarily older types. Perspective dielectrics for capacitors are polymer, ceramics type I and type II, thin film oxide (electrolytic) and electric double layer.

Nonorganic Capacitor

Nonorganic insulating materials are often used as dielectrics in capacitors. Class 1 and class 2 ceramics were described earlier. A large group of capacitors is based on oxide thin film. Commonly used electrolytic capacitors are aluminium oxide and tantalum oxide capacitors. Although the ϵ_r of an oxide is not high, oxide film thickness can be very small, so capacitance per 1 cm^3 is large. Oxide film is made by an electrochemical process. Popular oxides Al_2O_3 and Ta_2O_5 are utilized in aluminium electrolyte and tantalum electrolyte capacitor fabrication. The positive electrode is a metal (Al or Ta) and the negative electrode is a conductive electrolyte. Figure 9.42(a) shows a segment of a wet aluminium foil electrolytic capacitor.

Foil is etched to get a large active surface; the growth coefficient of the surface is above 10 and the theoretically for low voltage is 100. On the active surface, a thin layer of Al_2O_3 is

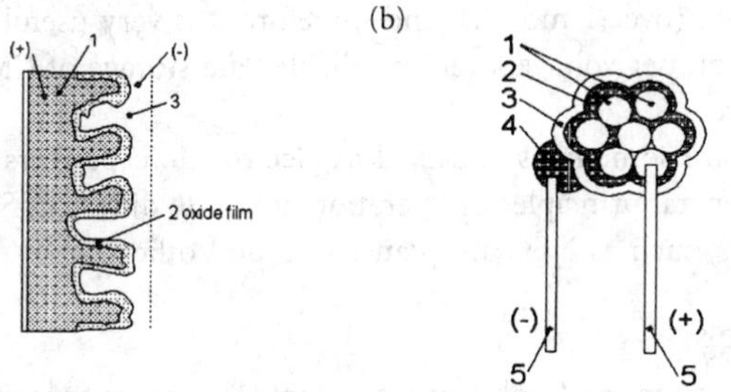

FIGURE 9.42 Segment of electrolytic capacitor. (a) Al_2O_3 wet capacitor: 1—Al foil, 2—oxide layer, 3—fluid (negative electrode). (b) Ta_2O_5 dry capacitor: 1—tantalum balls, 2—Ta_2O_5 layer, 3—conductive layer (negative electrode), 4—contact for negative termination.

produced electrochemically. It functions as a dielectric layer of the capacitor and its ϵ_r equals 8.4. The negative electrode is made as a conductive flux that usually is connected to the Al cover. The dc voltage range is from 6 up to 600 V. The capacitance range is from 4.7 μF to 10 mF in series E-3. In the reverse direction the dielectric does not withstand high voltages and the maximal admitted voltage is only 10 percent of nominal voltage. During operation the capacitors show better parameters than during storage because of the smaller value of the leakage current. A tantalum capacitor is also produced as a foil capacitor but more popular is the dry tantalum capacitor. Fig. 9.42(b) shows a segment of that capacitor where tantalum balls (0.3 mm diameter) are sintered and function as the positive electrode. A thin film of Ta_2O_5 is produced by an electrochemical process on the surface of each ball. In the next step a conductive layer (for example, a colloidal graphite layer) is deposited on Ta_2O_5 film in a chemical process. At the end, a contact for a negative electrode is made. An anode termination is welded to sintered balls and a cathode termination can be soldered to the negative electrode contact. Encapsulation is made using a tixotropic lacquer. Tantalum oxide electrolytic dry capacitors are produced for low dc voltages from 4 to 50V. The tantalum capacitor shows a smaller leakage current and higher work temperature, up to 125°C. The aluminium capacitor can usually work nominally up to 70°C and up to 85°C maximum. Also, the tantalum capacitor exhibits better high-frequency characteristics. It is really a miniature capacitor; its υ' index is very small (see Table 9.7). Al and Ta electrolytic capacitors are applied to dc circuits but can also work at small ac voltages (ac amplitude must be smaller than 10–15 percent of dc nominal values). In a wet electrolytic capacitor self regeneration exists, that is, after a short breakdown, the capacitor can work further because the layer of oxide is regenerated around the breakdown point. Several firms manufacture bipolar electrolytic capacitors for ac voltages.

Super Capacitor (with electric double layer)

The super capacitor, also called the golden capacitor or golden series capacitor, has capacitances up to 10 F and its υ' index is about 0.00001 $cm^3/\mu F$. The origin of that capacitor comes from H.L.F. Helmholtz (1879), who discovered the electric double layer that exists on the interface of two different materials. The layer can store an electric charge. The charge value of the electric double layer increases as the effective contact surface and/or electric field go higher. The right choice of both materials and preparation technology can lead to a very thin (several angstroms) electric double layer for 1.2 V in one cell. Cells are connected in series for higher nominal voltage values, which typically achieve 2.5 V up to 11 V for the whole super capacitor. In known practical constructions of super capacitors, small particles of activated carbon (with a large effective surface) are used in contact with diluted sulfuric acid; carbon in contact with acid yields the electric double layer. Carbon is the positive pole and acid connected with the metal case is the negative pole. The leakage current of supercap is very small, the time constant of the supercap is very high (over 1 month), and therefore it is very useful for backup purposes, such as maintaining the proper voltage level for reliable data storage of CMOS memories during host power supply failure.

Supercapacitors are sometimes classified as electrolytic capacitors because of their inner wet solution, but the general principles of operation are quite different. Supercaps are manufactured by many companies, such as NEC and Panasonic, and others.

Organic Capacitor

This group includes paper and polymer capacitors. The paper capacitor is an obsolete construction and today is not very popular because paper drying and impregnation are rather expensive. Also, the parameters of paper capacitors are not sufficient. The family of polymer capacitors is very large and some of them were described in the section on nonlinear capacitors.

Table 9.10 presents the features of various polymer capacitors. Special attention should be paid to the teflon capacitor, for instance, that the maximum operating temperature is 280°C and the power factor is ca 6×10^{-4}; also, instability δ is < 0.5 percent/1000 h.

For dc and low frequency operation, polyethylentereftalate (mylar) foil capacitor is very popular. It withstands different fluxes and high temperatures up to 150°C. Its stability and power factor are quite good, but only for low frequencies.

Dependence on Geometry of Material

Influence of Capacitor Construction on Equivalent Series Inductance (ESI) and Equivalent Series Resistance (ESR)

Equivalent series inductance with capacitance C causes self resonance at resonant frequency f_0. Above frequency f_0 it has an inductance feature so it cannot fulfill its main function. So reduction of ESL is a very important matter. Figure 9.43 shows the simplest rolling polymer capacitor with two terminations marked 3 and 4.

During charging and discharging, current is flowing through capacitor electrodes that are marked 2 and 5. The electrodes are insulated by polymer foils marked by 6 in Fig. 9.43. Such a construction creates a roll which, in turn, adds significant ESL to the capacitor. The simplest way of decreasing ESL is to supply some additional terminations and connect them in parallel. The maximum number of contacts is shown in Fig. 9.35. Contacts are deposited with melted

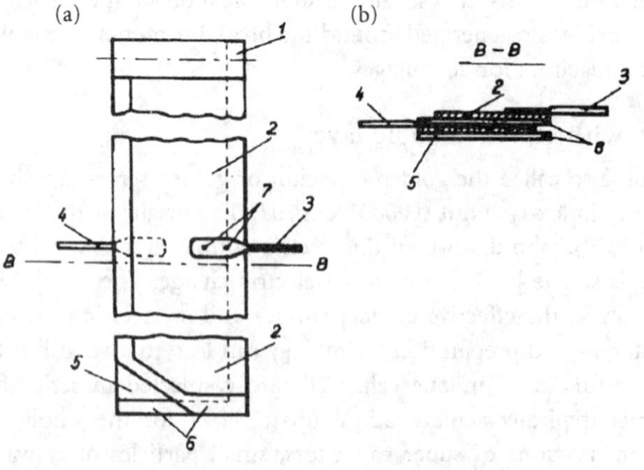

FIGURE 9.43 Simple polymer capacitor construction. (a) 1—start part of roll, 2—Al foil (electrode I), 3,4—wire terminations. 5—Al foil (electrode II), 6—polymer dielectric foil, 7—welding points; (b) cross section B-B.

TABLE 9.10

Capacitor Name	υ' $cm^3/\mu F$	Class	$delta_{max}$ after 1000 h [%]	Smallest $t\ \delta_p$ [%]	Power Factor × 10^{-4}	TCC ppm/K	Maximum Work Temperature [°C]	Remarks	
Polystyrene	300	1	0.5	±0.5	2–5	−100	70°	for telecommunications filter	neutral polymer
Teflon	300	1	0.5	±0.5	6	−150	280	special applications	
Polyethylene	200	1	1	±1	5	−500	100		
Polypropylene	50	2	5	±5	6–8	−200	110		
Metalized polypropylene	10	2	5	±0.5	6–8	−200	85		
Metalized polyester	5.6	2	10	±10	50 (200 at 1 MHz)			for ac pulse	polar polymer
Polyester (polyetylene tereftalate)	12	2	5	±10	50 (200 at 1 MHz)	large	150		
Polycarbonate	12	2	10	±10	20	large	100		
Metalized polycarbonate	5.6	2	10	±10	20	large	100		

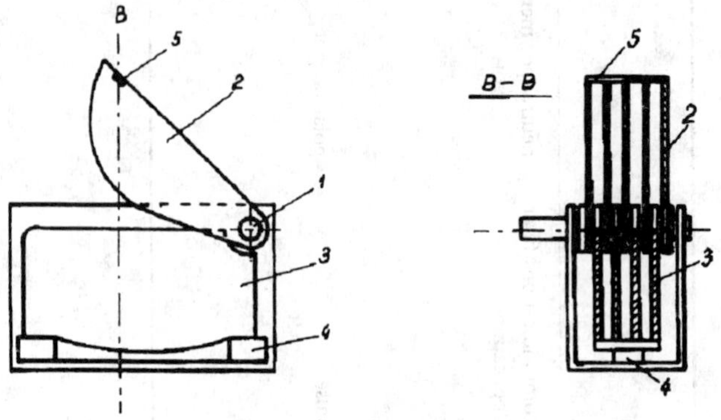

FIGURE 9.44 Rotary capacitor construction. Shape of plates is modified in relation to the circle. 1—shaft, 2—moving plates, 3—standing plates, 4—insulator, 5—insulating traverse.

metal on the electrode by using a high-pressure airbrush. In that case minimal ESL occurs. In this way ESR also decreases, because the distance through which current is flowing becomes much shorter. Very small ESL occurs with the feed-through capacitor shown in Fig. 9.38.

Influence of Plate Shape on Tuning Frequency in Air Variable Capacitors

A variable capacitor consists of standing plates that are insulated usually from the ground and rotary plates that are connected to the ground.

Variable capacitors are used for tuning an LC circuit, for example, to select the proper radio station. As we see from (9.37), resonant frequency f_0 is not proportional to C. Proportionality is achieved by special shaping of the plates as a function of the rotation angle. The modified shape of plate is seen in Fig. 9.44.

$$f_0 = \frac{1}{2\pi\sqrt{LC}} = \frac{b}{\sqrt{C}} \qquad (9.37)$$

Equation (9.37) shows the resonant frequency of the LC circuit.

Shape of the High-Voltage Capacitor

The operation of a capacitor at high voltages is always accompanied by many problems, for example, how to achieve the proper distance between capacitor electrodes.

Two drawings in Fig. 9.36 help us to overcome this obstacle by using either of the following:

1) A disk capacitor with a high dielectric thickness (actually, it is a bar rather than a disk). When the thickness of the dielectric increases, capacitance decreases.
2) A tubular capacitor when the thickness of the dielectric is satisfactory for high voltage but the electrodes are far enough from each other with respect to flashover protection.

Chip Capacitor

Surface mounted technology (SMT) requires electronic components with easily solderable contacts without wire terminations.

Usually those contacts are pretinned during chip production. Figure 9.45(a) and (b) show us a tantalum electrolytic capacitor chip and a ceramic monolythic capacitor chip, respectively. The capacitance of each of these two devices can range from 10 pF up to 10 mF and from 10 pF to 10 nF, respectively, in classes 2 and 1.

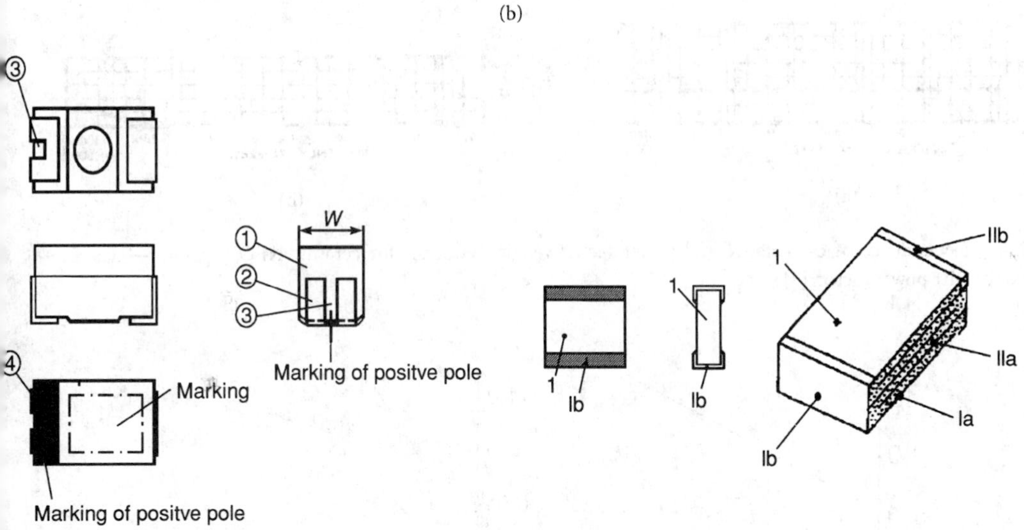

URE 9.45 Chip capacitor. (a) Tantalum pentoxide capacitor. Positive pole is marked by a black bar and double rnal electrode; 1—plastic encapsulation, 2—external electrode for soldering, 3—cutting in positive electrode, lack bar. (b) Ceramic multilayer monolithic capacitor. 1—ceramic layers, Ia,IIa—capacitor electrodes (plates), b—external electrodes.

Nonideal Linear Capacitor

The main source of capacitance instability of the capacitor is the operating frequency. Its influence on dielectric polarity was described in the section on linear capacitors. The change of capacitance with frequency increase in capacitors belonging to the class 1 might be also observed. Furthermore, an equivalent series resistance (ESR) decreases the effective capacitance C_{ef} according to (9.38):

$$C_{ef1} = \frac{C}{1 + \omega^2 r^2 C^2} \tag{9.38}$$

where $r = \text{ESR}$ is calculated from the power factor (PF), which is measured at the operating frequency, and C is the capacitance at low frequencies.

$$r = \frac{\text{PF}}{\omega C} \tag{9.39}$$

On the other hand, an equivalent series inductance (ESL) increases the effective capacitance according to (9.40):

$$C_{ef2} = \frac{C}{1 - \omega^2 LC} \tag{9.40}$$

where $\text{ESL} = L$ can be measured at the self-resonant frequency [see (9.37)].

The ESR and ESL Data Cards for a given capacitor can be obtained from its manufacturer. Figure 9.46 shows the dependence of capacitance and power factor on frequency for class 1 ceramic capacitors.

Figure 9.41 presents data for class 2 ceramic capacitors. In a ceramic multilayer capacitor, migration of metal particles through the dielectric is observed, which causes an increase of capacitance and sometimes breakdown. The problem can be solved by roll pressing of two wet ceramic layers into one dielectric during a production process.

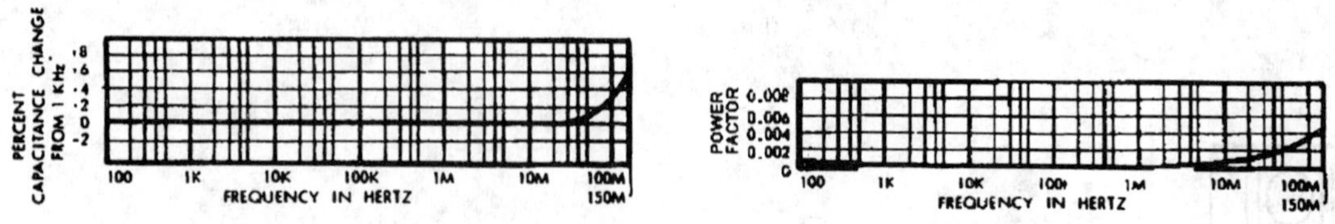

(a) (b)

FIGURE 9.46 Dependence of capacitance and power factor versus frequency for ceramic NPO capacitor—class 1. (a) Capacitance, (b) power factor [7].

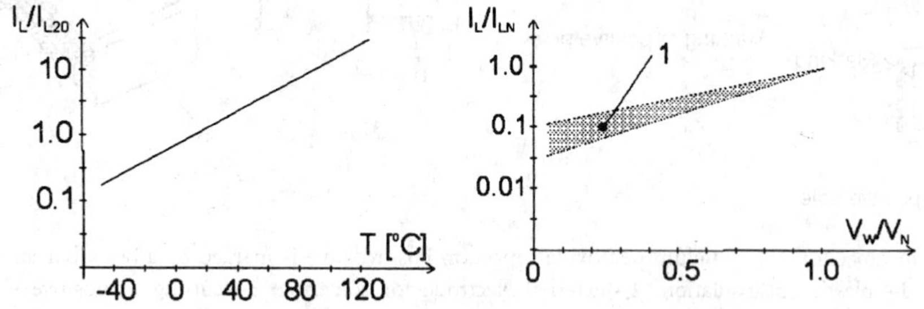

(a) (b)

FIGURE 9.47 Leakage current of tantalum electrolytic capacitor [1]. (a) Dependence of leakage current versus temperature; (b) dependence of leakage current versus working voltage; 1—typical range of I_L.

Self-regeneration of Capacitor. The phenomenon of self-regeneration is exhibited in metalized dielectric capacitors as well as electrolyte capacitors. It is useful in some situations (for example, in long-life tests) but sometimes it is harmful (for example, in pulse counting equipment). To solve that problem we should use a derated capacitor (for 25 V work voltage, 40 V nominal voltage is needed) or change the type of capacitor.

Leakage Current of Electrolytic Capacitor. The insulating resistance R_i in an electrolytic capacitor is not as high as in other capacitors. The result of low R_i is dc current called leakage current I_L. For aluminium electrolytic capacitors, Siemens Matsushita Components Company in their catalog [6] propose to join I_L in μA with capacitance C in μF and working voltage V_w in [V]. According empirical rule (9.41)

$$I_L \leq 0.03CV_W + 20\mu\text{A} \tag{9.41}$$

for normal types and

$$I_L \leq 0.006CV_W + 4\mu\text{A} \tag{9.42}$$

for special long-life capacitors, where $V_w \leq V_N$.

Tantalum electrolytic capacitors show smaller leakage currents and at a temperature 20°C their value is 0.5 up to 20 μA. That value also depends on C, V_w, and temperature. Dependence I_L on ambient temperature T is shown in Fig. 9.47(a), and influence of working voltage value V_w is shown in Fig. 9.47(b). Those data are taken from the 1990 catalogue of the AVX Corporation Ltd. (Great Britain).

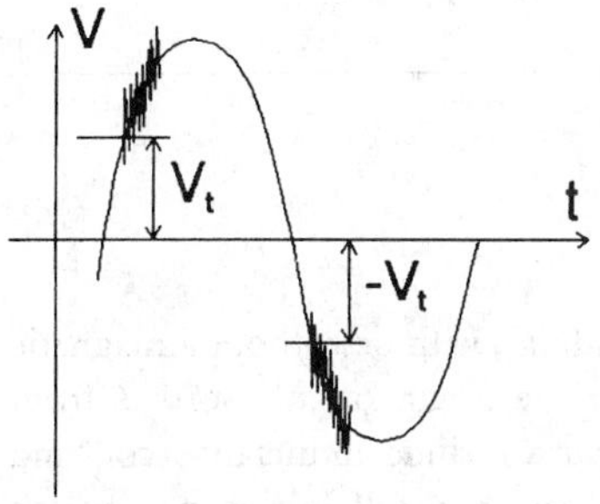

FIGURE 9.48 Ionization threshold in a 50 Hz capacitor. V_t—threshold voltage on which ionization can start.

It is necessary to know that after long-time storage the leakage current is going up. To prevent this one should switch on the testing device containing the capacitors several hours before using.

Noise in Capacitor. Foil capacitors can generate noise when contact between termination and electrode plate is not sufficiently fixed. In Fig. 9.43, the welding points of contact are introduced (marked 7) to improve the quality of a contact. Some producers do not use welding because at nominal voltages the noise generated by contact is negligible. You should pay attention to the fact that for very low signals poor contacts affect noise in a significant way. Welding removes that effect.

Ionization Threshold. In a power capacitor, an ionization of the dielectric is possible. Fig. 9.48 shows the effect of this process.

Although breakdown does not occur, the capacitor life shortens when operating voltage approaches maximum.

Threshold is a minimal value V_T of voltage at which ionization starts. This causes heterogeneity of the dielectric, for example, polypropylene and air between plates. Flat rolling of the dielectric layer is strongly suggested for use in a power capacitor, as well as vacuum impregnation of the capacitor.

Insulating Resistance R_i. When a capacitor is used as a component of an RC time constant, insulation resistance should be taken into consideration. In that case, foil polymers (particularly polystyrene or teflon capacitors, where R_i is greater 100 GΩ) are preferred. Also, protection against dirt and humidity by hermetization improves R_i.

ferences

[1] AVX Corporation Ltd., *AVX Tantalum Capacitors Catalogue*, Great Britain, 1990.
[2] *The Electrical Engineering Handbook* R. Dorf, Ed. Boca Raton, FL: CRC Press, 1993.
[3] L. Badian, *Handbuch der Electronik*, Vol. 3, Munich: Franzis-Verlag, 1979.
[4] *IEC Standard Publ. 384*, pt. 1–17, 1988.
[5] "Capacitors," *Philips Components Catalogue*, 1989, Eindhoven.
[6] "Aluminium electrolytic capacitors, ceramic capacitors, tantalum capacitors," *Siemens Matsushita Components Catalogue*, Munich, 1993.
[7] Sprague Electric Co., "Monolithic ceramic chip capacitors," *Sprague Eng. Bull.* Brussel.
[8] Taiyo Yuden Co. Ltd., *Tubular Ceramic Capacitors Catalogue*, Tokyo, 1988.

9.3 Inductor

Tomasz W. Postupolski

Basics

An inductor is a device consisting of one or more associate windings, with or without a magnetic core, for introducing inductance into an electric circuit [1]. The origin of the word is from Latin: inducere—excite, induce, incite. Other appelations in use include inductance coil and coil, which are synonyms, and inductance, winding, bobbin, self, and self inductance, which are jargon, and reactor and choke. Reactor and choke are inductors but not every inductor is a reactor or choke. The alphabetical symbol of inductor is L. Units of inductance include 1 henry = 1 H = 1 Vs/A and its submultiples nH (10^{-9}), μH (10^{-6}), and mH (10^{-3}).

Inductor graphical symbols:

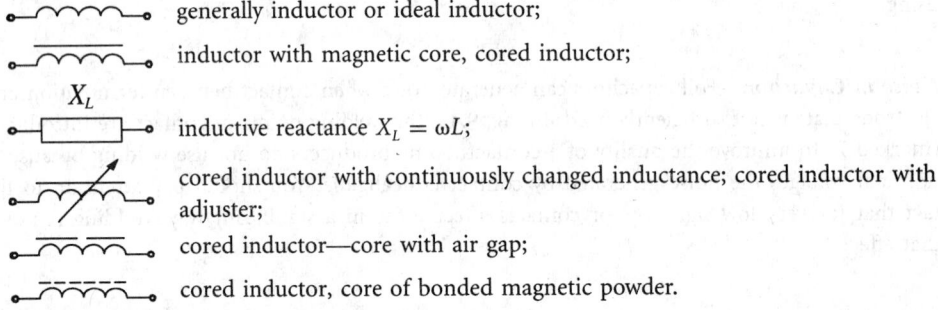

 generally inductor or ideal inductor;

 inductor with magnetic core, cored inductor;

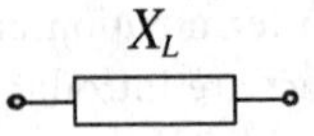

inductive reactance $X_L = \omega L$;

 cored inductor with continuously changed inductance; cored inductor with adjuster;

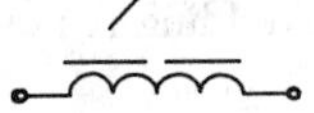 cored inductor—core with air gap;

 cored inductor, core of bonded magnetic powder.

Basic Relationships and Influencing Factors

The main electrical attribute of the inductor is the inductance L. The inductance is the property of the electric circuit, whereby an electromotive force is induced in that circuit by a change of current in the circuit [1].

The electromotive force (voltage) e induced in the winding of the inductor is given by Faraday's law, which can be written in various equivalent forms:

$$e = -n\frac{d\Phi}{dt} = -\frac{d}{dt}(Li) = -\left(L\frac{di}{dt} + i\frac{dL}{dt}\right) \tag{9.43}$$

where $n\Phi$ = total magnetic flux linked with the current i flowing through all the n turns forming the winding of inductor having an inductance L.

Equation (9.43) shows that the circuit "senses" the inductance only when the current in the circuit is changing and/or the inductance itself is changing. Generally, the inductance L can depend on the magnitude of current i and inversely a change in the magnitude of L can cause a change in amplitude of i, so e is an implicit function $e[i(t), L(i, t), t]$. The well-known expression $e = -L\,di/dt$ is fulfilled only if L is a constant independent of current, i, and time t (this restriction is often forgotten).

In terms of the electromagnetic field theory, the inductance L is given by

$$L = \frac{n\Phi}{i} = n\oint_A BdA \Big/ \oint_l Hdl \tag{9.44}$$

B = the normal component of the magnetic induction (flux density) through the area A and H = the magnetic field strength along its path l. For more about various forms of the electromagnetic field equations see courses on the theory of electromagnetic field, for example, [2].

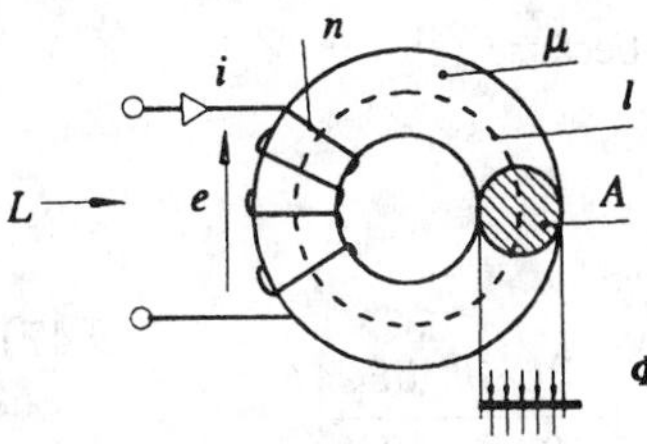

FIGURE 9.49 Inductor of a toroidal shape.

In case of a radially thin toroid (that is, in which $A/l \to 0$) and $\mu = \text{const}$, the magnetic induction B, and field strength H, can be assumed as uniform, that is, $\Phi = BA$; $B = \mu_0\mu H$; $H = ni/l$, the inductance L for a shape shown in Fig. 9.49 is

$$L = \mu_0\mu n^2 A/l \tag{9.45}$$

where $\mu_0 = 4\pi 10^{-7}$ [H/m] is the magnetic constant and μ = magnetic permeability of the medium filling up the shape of the toroid.

Equation (9.45) can be expressed in different equivalent forms: their use is optional.

$$L = A_L n^2 \qquad A_L = \mu_0\mu A/l \qquad A_L = L/n^2 \tag{9.46a}$$
$$L = \mu A_{L0} n^2 \qquad A_{L0} = \mu_0 A/l \qquad A_{L0} = A_L/\mu \tag{9.46b}$$
$$L = \mu_0 G_m n^2 \qquad G_m = \mu A/l \qquad G_m = \mu G_{m0} \tag{9.46c}$$
$$L = \mu_0 n^2 R_m \qquad R_m = l/\mu A \qquad R_m = 1/G_m \tag{9.46d}$$
$$L = \mu_0\mu G_{m0} n^2 \qquad G_{m0} = A/l \qquad G_{m0} = G_m/\mu \tag{9.46e}$$
$$L = \mu_0\mu n^2/R_{m0} \qquad R_{m0} = l/A \qquad R_{m0} = 1/G_{m0} \tag{9.46f}$$

where A_L = inductance factor of inductor, A_{L0} = inductance factor of the winding or core shape (at $\mu = 1$), G_m and R_m = core permeance and reluctance, respectively, G_{m0} and R_{m0} = permeance and reluctance of core shape (at $\mu = 1$); inductance L is in [H] if l is in [m] and A in [m^2].

If μ, B, and H cannot be assumed to be uniform, (9.45) and (9.46) take more complicated forms issuing from (9.44).

The following fundamental observations can be inferred from (9.43) to (9.45):

1) The inductance L of the inductor depends on three factors:
   - square of number of turns n^2 of the winding;
   - geometrical configuration of the magnetic flux Φ produced by the current i; this configuration is represented by the permeance or reluctance of the shape G_{m0} or R_{m0}, containing the magnetic flux;
   - magnetic properties of the medium or media filling up the shape associated with the carrying current winding, producing a magnetic flux Φ; these properties are represented by the magnetic permeability μ. *Note:* magnetic permeability, depending on the approach, can be considered as a scalar (number), complex quantity, or a tensor.

2) The voltage induced across the inductor of inductance L depends on the following:
   - rate of change of the current flowing in the inductor winding di/dt;
   - changes in time of the inductance of inductor dL/dt; these changes result from

the influence of external factors on inductance L, because

$$\begin{aligned}\frac{dL}{dt} &= \sum_{\nu} \frac{\partial L}{\partial \xi_\nu}\frac{\partial \xi_\nu}{\partial t} \\ &= \sum_{\nu}\left[\frac{\partial L}{\partial \mu}\frac{\partial \mu}{\partial \xi_\nu}\frac{\partial \xi_\nu}{\partial t} + \frac{\partial L}{\partial A}\frac{\partial A}{\partial \xi_\nu}\frac{\partial \xi_\nu}{\partial t} + \frac{\partial L}{\partial l}\frac{\partial l}{\partial \xi_\nu}\frac{\partial \xi_\nu}{\partial t}\right]\end{aligned} \tag{9.47}$$

$\nu = 1, 2, 3, \cdots$, ξ = external stressing factors such as time, temperature, mechanical stresses, static and time-varying voltage or magnetic induction, current or magnetic field strength, frequency, moisture, radiation (nuclear, cosmic), and so on.

The term dL/dt, often forgotten in considerations, is however, of importance because it is responsible for any instability or variability introduced into an electric circuit by the inductors. Examples of various dL/dt contributions are listed in Table 9.11.

TABLE 9.11 Exemplification of Various Stressing Factors ξ Causing Changes in Inductance L

	Influencing	
Stressing factor ξ	Permeability μ, $d\mu/d\xi$, such as, for example:	Physical dimensions A and l of winding and/or core; $dA/d\xi$ and $dl/d\xi$, such as, for example:
Temperature	Temperature changes of permeability	Thermal changes of dimensions
Mechanical tensile, contracting or torsion stress	Reversible changes of permeability under mechanical stresses, piezomagnetic effects, mechanomagnetic after-effects (relaxations)	Changes of dimensions under mechanical stresses; mechanical relaxations of dimensions of core or winding
Magnetic induction B or magnetic field strength H (changes of static, time-varying or combined magnitude)	Strong and, generally, nonlinear dependence of μ on B and H; disturbance of the magnetic state of core material	Ponderometric effects in winding at, e.g., high current intensities or pulses; mechanical relaxation of winding or coil formers
Frequency	Strong dependence of μ on frequency: eddy current effects (magnetic skin and proximity effects), magnetic resonances (wall resonance, dimensional resonance, spin resonance), relaxation effects	Skin and proximity effects in wires cause change of current density distribution in conductor, thus, the change of effective dimensions of coil
Humidity, moisture, aggressive agents (salts, acids, etc.)	Corrosion effects in metallic cores, therefore structural changes of magnetic material; changes of dielectric properties in ferrites	Various agents' effects as changes of dimensions and electrical and dielectric properties of coil formers, winding and interwinding insulation layers, substrates, etc.
Irradiation	Structural changes in magnetic material	Structural changes in materials of coils, conductors, insulation, dielectric properties of coil formers, substrates, etc.
Time	Changes of μ: reversible effects (dissaccommodation and accommodation), irreversible effects (so-called aging), thermal fluctuation effects in nonneutral states	For example, aging and rheological effects affecting materials of coil formers, insulation; diffusion effects between metallic conductor deposited e.g., on ceramic substrate

Inductor: Qualifiers and Attributes

Different qualifiers are, in practice, applied to inductors, depending on approach and related attributes. The most frequently used are listed in Table 9.12. These qualifiers are used separately or in combination, e.g., air inductor or linear cored inductor.

TABLE 9.12 Inductor Qualifiers and Attributes

Inductor qualifier	Inductor: Attribute or Quality
Ideal, perfect	Linear inductor having only a "pure" inductance, i.e., there is no power loss related to the flow of time-varying current through the inductor winding. In the ideal inductor, the current of sine wave lags the induced voltage by angle $\varphi = 90°$ ($\pi/2$ rad). The concept of the ideal inductor is used only in idealized or simplified circuit analysis.
Nonideal	Usually a linear inductor in which the power loss in the winding and core is taken into account. The current of sine wave lags the induced voltage by angle $0° \leq \varphi < 90°$ (90° for ideal- power lossfree inductor; 0° for pure resistor). The concept of nonideal inductor is used as a first order approximation of a real inductor.
Linear	Inductor, ideal or nonideal, for which the induced voltage drop across it is proportional to the flowing time-varying current in its steady state. Linear inductor can be described or be used to describe the circuit in terms of transfer function. An air inductor is an example of linear inductor.
Nonlinear	Inductor for which the induced voltage drop is not proportional to the time-varying current flowing by it. As a rule, cored inductors (specifically if a core forms a closed magnetic circuit) are nonlinear. This is a consequence of the strong nonlinear dependence of magnetic induction B, proportional to voltage $u = dL/dt$, on magnetic field strength H, proportional to current i.
Real	Inductor whose electrically behavioral aspects and characteristics are all taken into account, e.g., magnetic power loss, magnetic flux leakage, selfwinding and interwinding capacitances and related dielectric power loss, radiation power loss, parasitic couplings, and so on, and dependences of these factors on frequency, induction, temperature, time, etc.
Air	Inductor not containing magnetic materials as constituents or in its magnetically perceptible vicinity.
Cored	Inductor in which a magnetic material in the form of a core serves intentionally as a path, complete or partial, for guidance of magnetic flux generated by current flowing through inductor winding.
Lumped or discrete	Inductor assumed to be concentrated at a single point.
Distributed	Inductor whose inductance and other properties are distributed over a physical distance(s) that is comparable to a wavelength.

Basic Functions of Inductor

There are four basic functions the inductor has to perform:

- To impede (oppose) any change in the existing current; as the change is more rapid the opposition is stronger.
- To lag (to shift) the current in respect to induced voltage up to 90° for a sine wave.
- To differentiate the current waveform.
- To store the magnetic energy; this when in combination with electrical energy stored in capacitor, results in electrical resonance (LC—resonance).

The more the real inductor approaches the ideal inductor the higher degree in performing these functions.

Parameters Characterizing the Inductor

Primary Parameters.

- inductance L (for all excitation levels),
- power loss at low and middle excitation levels expressed facultatively in terms of

  resistance $R =$ power loss/(square or rms value of current);
  tangent of loss angle δ, $\text{tg}\delta = R/2\pi fL$;
  quality factor $Q = 2\pi fL/R = 1/\text{tg}\delta$;

- power loss for high excitation levels, expressed directly in watts.

Secondary Parameters (Selected).

- dc winding resistance
- self capacitance
- self-resonance frequency
- temperature coefficient or factor of inductance
- time instability
- hysteresis loss
- pulse inductance
- adjustment range
- harmonic distortion
- magnetic radiation
- influence of the static magnetic field
- inductance rise factor
- maximum winding temperature
- temperature rise
- immunity from mechanical stresses and environmental exposures
- acoustic noise

To determine these parameters the following should be specified: frequency, excitation level (voltage or current), temperature, parameters of measuring coil (winding), type of magnetic conditioning to be used, and other details of measuring procedure, as well as conditions needed to ensure the required accuracy and repeatibility of measurement.

The relevant measuring methods are described, e.g., [3], [4] and [5].

Circuit Representation of Inductor

Circuit representation of the ideal inductor is that as shown by the first inductor graphical symbol. For a nonideal linear inductor, a series or parallel representation is used, Fig. 9.50.

The transformation of a nonideal inductor shown in Fig. 9.50 is a purely formal operation serving only for the circuit analysis. Transformation results calculated for one specified frequency and specific operating conditions (e.g., B, temperature) can be extended over other frequencies and conditions only when L and R appear to be independent of these factors: $L =$ const and $R =$ const.

Series and Parallel Connections of Inductors. When individual inductors are connected in series (with no mutual inductive couplings), the total inductance $L_t = L_{ts}$ will be a sum of individual inductances L_i: $L_{ts} = \sum_i L_i$; similarly the equivalent loss resistances: $R_{ts} = \sum_i R_i$.

When individual inductors are connected in parallel, the total inductance $L_t = L_{tp}$ and resistance R_{tp} will be $L_{tp} = 1/\sum_i 1/L_i$ and $R_{tp} = 1/\sum_i 1/R_i$, respectively.

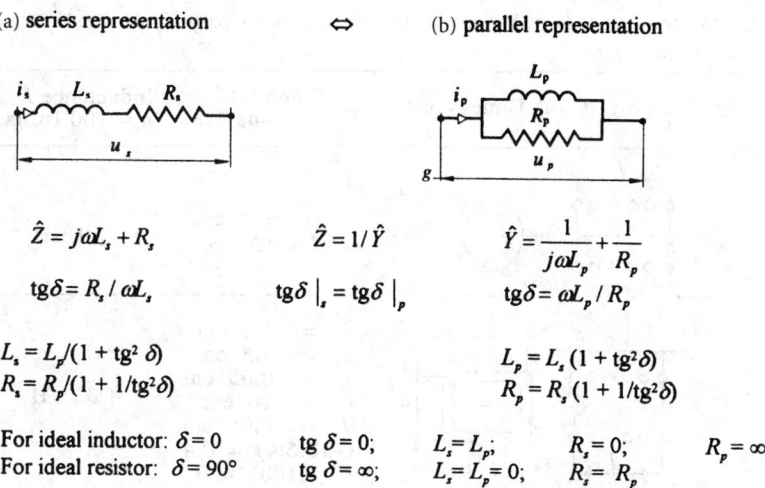

(a) **series representation** ⇔ (b) **parallel representation**

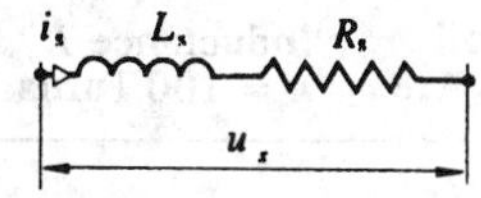

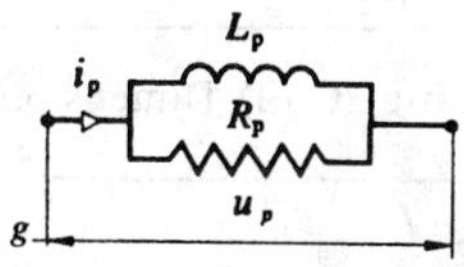

$\hat{Z} = j\omega L_s + R_s$	$\hat{Z} = 1/\hat{Y}$	$\hat{Y} = \dfrac{1}{j\omega L_p} + \dfrac{1}{R_p}$
$\mathrm{tg}\delta = R_s / \omega L_s$	$\mathrm{tg}\delta \mid_s = \mathrm{tg}\delta \mid_p$	$\mathrm{tg}\delta = \omega L_p / R_p$
$L_s = L_p/(1 + \mathrm{tg}^2\,\delta)$		$L_p = L_s\,(1 + \mathrm{tg}^2\delta)$
$R_s = R_p/(1 + 1/\mathrm{tg}^2\delta)$		$R_p = R_s\,(1 + 1/\mathrm{tg}^2\delta)$

For ideal inductor: $\delta = 0$	$\mathrm{tg}\,\delta = 0$;	$L_s = L_p$;	$R_s = 0$;	$R_p = \infty$.
For ideal resistor: $\delta = 90°$	$\mathrm{tg}\,\delta = \infty$;	$L_s = L_p = 0$;	$R_s = R_p$	

FIGURE 9.50 Equivalent circuit representation of the inductor and the respective expressions. *Note:* when subscript *s* is omitted, symbols *L* and *R* always refer to the series representation.

Thus, the series connection of inductors always increases a resultant inductance, parallel connection—decreases. *Note:* these connections rules (laws) apply only for sine waveforms in the steady state and these rules are applicable neither to the nonlinear inductors nor to transient states.

Air Inductor

This type of inductor is assumed to be linear. This means that at stressing factors that are fixed the inductance L and resistance R are characteristic constants, independent of the flowing current or applied voltage.

Selected formulas for the inductance of some air inductors are given below. In Table 9.13 corresponding draftings are depicted. All dimensions of coils shown in Table 9.13 are in [cm]; n = number of turns. For more formulas for calculation of the inductance of air inductors see, e.g., [8] and [9].

- Single-layer cylindrical coil with a circular conductor, Table 9.13(a):

$$L_0 = KDn^2 \cdot 10^{-9}[\mathrm{H}] \tag{9.48a}$$

  where K = coefficient whose values are listed in Appendix 1.

- Single-layer rectangular coil with a circular conductor, Table 9.13(b):

$$L_0 = pn^2(G + H) \cdot 10^{-9}[\mathrm{H}] \tag{9.48b}$$

  where $G = f_g(S_1/S_2; l/S_2)$ and $H = f_h(d/\lambda, n)$ = tabulated coefficients [8], [9].

- Single-layer toroidal coil, Table 9.13(c):

$$L_0 = 2\pi n^2\left(D - \sqrt{D^2 - D_1^2}\right) \cdot 10^{-9}[\mathrm{mH}] \tag{9.48c}$$

TABLE 9.13 Inductance L_0 of Various Air Inductors Dimensionally Similar but Having the Same Number of Turns

	Winding (Coil) Dimensions	When Coil Dimensions Are:	Inductance L_0 for n = 100 Turns is
(a)	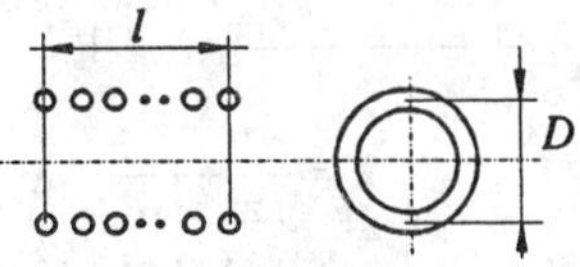	D_1 = 2 cm l = 10 cm	19 μH
(b)	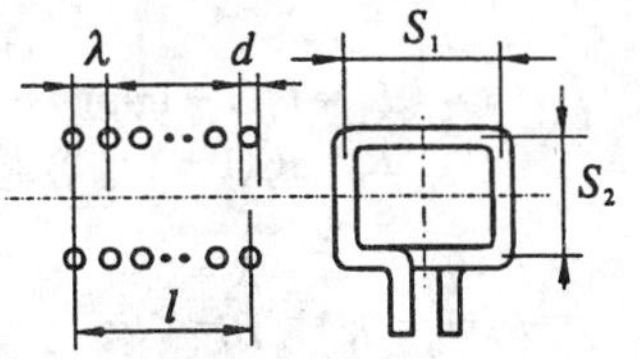	S_1 = 1.5 cm S_2 = 2.5 cm λ = 0.05 cm l = 10 cm d = 0.05 cm G (0.6;4) = 0.4 H (1;100) = 0	32 μH
(c)	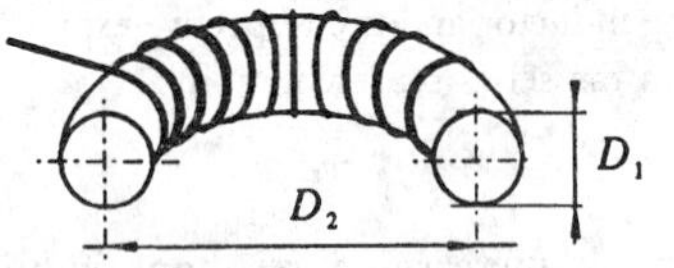	D_1 = 1 cm D_2 = 3 cm	10.3 μH
		D_1 = 2 cm D_2 = 3 cm	41 μH
(d)	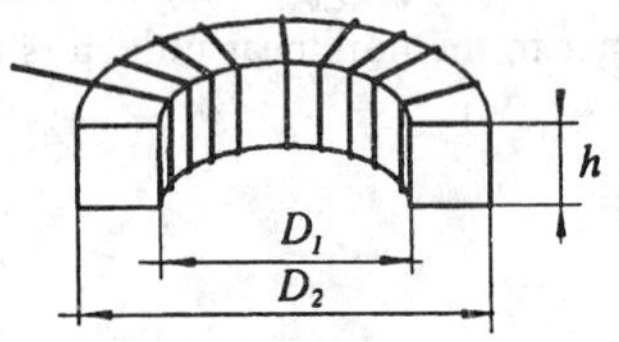	D_1 = 2 cm D_2 = 4 cm h = 1 cm	13.9 μH
		D_1 = 1 cm D_2 = 5 cm h = 2 cm	64.4 μH
(e)	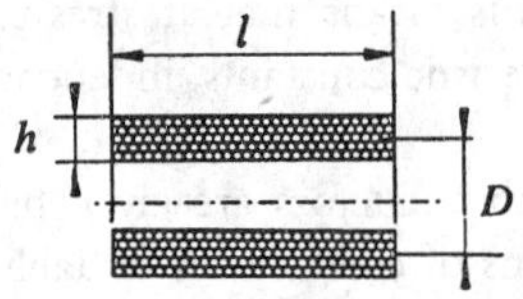	D = 2 cm h = 0.5 cm l = 4 cm	74 μH
(f)	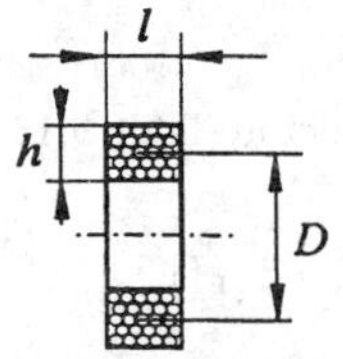	D = 2 cm h = 0.5 cm l = 0.2 cm	245 μH

- Single-layer ring coil, Table 9.13(d):

$$L_0 = 2hn^2 \ln(D_z/D_w) \cdot 10^{-9}[\mathrm{H}] \tag{9.48d}$$

- Multilayer long cylindrical coil, Table 9.13(e):

$$L_0 = Dn^2[K - 2\pi h(0,693 + \gamma)/l] \cdot 10^{-9}[\mathrm{H}] \tag{9.48e}$$

K = coefficient as for the single-layer cylindrical coil; γ = coefficient whose values are listed in Appendix 2.

- Multilayer short cylindrical coil, Table 9.13(f):

$$L_0 \cong \frac{25\pi D^2 n^2}{3D + 9l + 10h} \cdot 10^{-9}\,[\mathrm{H}] \tag{9.48f}$$

Design Guidance. As can be seen, $L = \Psi D n^2$, where Ψ is a function of two factors: l/D and h/d; and the inductance of air inductor increases when:

- outer diameter of winding increases,
- ratio of outer to inner winding diameter decreases,
- width of winding decreases,
- diameter of wire decreases.

This can also be inferred from the last column of Table 9.13 comparing the values of L_0 for dimensionally similar air inductors.

Inductance L_0 can be readily calculated for a specified shape, dimensions, and number of turns of winding. However, this is not a design approach, because the design problem to be solved is inverse: to determine for a given value of L_0 the dimensions, shape, and number of turns of winding. However, such calculations are impossible to perform. This is a consequence of the given geometry of winding of n turns, which results in one and only one geometry of the magnetic flux Φ, therefore in only one value of the inductance L_0. Inversely, to a given value of L_0 an infinite number of winding geometries may be attributed at any n value.

Thus, the theoretical problem: to obtain the dimensions of the winding in combination with the suitable number of turns n from the specified value of L_0 is mathematically insolvable.

Therefore, the solutions are determined by experimental or iterative numerical methods.

Quality Factor of Air Inductors. Usually it is required that the inductor would have the maximum Q-factor or time constant L/R.

For the following cases in Table 9.14: (a) fixed copper (wire) volume; (b) fixed length of wire; (c) fixed coil volume, the L/R reaches maximum at following proportions of the coil winding.

Quality factor of air inductor extends from several units to several hundred and depends on

- shape and dimensional proportions of winding (Table 9.14),
- wire pitch and the wire arrangement,
- operating frequency,
- coil self capacitance,
- material properties of coil former or substrate, wire insulation, proximity of conducting, and/or magnetic parts.

TABLE 9.14 Dimensional Proportions of Coil Providing its Maximum Time Constant L/R

		Case	
Winding		(a) + (b)	(c)
Cylindrical, single layer		$D \approx 3l$	$D = (1.3–2)l$
Cylindrical	long	$D = 5h + 3l$	$D = (1.3–1.6)l$
Multilayer	short		$D = (4–8)l$ $D = (5–7)h$
Single layer	toroidal	$D_2 \approx 3D_1$	—
	ring	$D_2 \approx 1.6D_1$	—

Typical Frequency Ranges of Air Inductors When in Use

In resonant circuits:

- cylindrical mulitlayer 50–1000 kHz
- cylindrical single layer 1 MHz–2 GHz

As chokes:

- multilayer (multisectional) 10–1000 kHz
- single-layer 500 kHz–800 MHz

Conductors (wires) used:

- solid insulated: dc–200 MHz
- solid uninsulated: 100 MHz–1 GHz
- stranded (litz wire): 50 kHz–10 MHz
- tubular: 3 MHz–3 GHz
- Cu-coated metal: 30 MHz–3 GHz
- Sn- or Ag-coated metal: 0.2–3 GHz
- Sn, Ag, Au conducting path: 0.2–3 GHz

Advantages of Air Inductors

- High stability of properties versus temperature, time, magnetic excitation levels (when compared to cored inductors).
- Very low dependence of inductance on frequency (related mainly to changes of effective dimensions of winding due to skin and proximity effects in wires), therefore, a practically linear dependence of $X_L = \omega L$ on frequency.
- Insaturable dependence of B on i (as opposed to cored inductors) allows the air inductors to produce very high pulses of induction B.

Disadvantages of Air Inductors

- Practical inability to obtain dimensionally small inductors of high inductance L, therefore inherent inability to miniaturization.
- Air inductor acts both as an emitting and receiving antenna, therefore it is a) a source of unwanted electromagnetic radiation to surrounding circuitry and environment; b) a receiver of electromagnetic signals introducing them into circuit as disturbance (usually)—electromagnetic interference (EMI). These strong coupling features make the air inductor a highly incompatible component in electronic circuitry, especially miniaturized, and in electronic and electrical environments.
- Air inductors are made, as a rule, as inductors of fixed value of inductance L. If a change of L is required, connections to some wires (taps) positioned between winding extremities, or a sliding contact to wires for single-layer windings enable the inductance to be changed.

Cored Inductor

Reason for Using the Magnetic Material in Inductors. If a winding of inductance L_0 produces in vacuum (in air) a magnetic flux Φ_0 distributed over some volume and that volume is filled up with a magnetic material of permeability μ, then the inductance L of the so combined winding with magnetic material increases μ times: $L = \mu L_0$.

TABLE 9.15 Inductor Related Parameters: Voltage drop u, Current i, Induction B, Magnetic Flux Φ, Field Strength H, Inductance L at the Absence and Presence of Magnetic Material Fully Coupled with Inductor Winding are Shown for Two Limit Cases: (a) Inductor Operates in Circuit of Constant emf ($R_i = 0$) and (b) in Circuit of Constant Current ($R_i = \infty$).

	Magnetic Material of Permeability μ		
	Absent	Present, Fully Coupled with Winding	
Parameter	Any Source	Case (a): Constant emf Source ($R_t = 0$)	Case (b): Constant Current Source ($R_t = \infty$)
$u =$	u_0	u_0	μu_0
$i =$	i_0	i_0/μ	i_0
$B =$	B_0	B_0	μB_0
$\Phi =$	Φ_0	Φ_0	$\mu\Phi_0$
$H =$	H_0	H_0/μ	H_0
$L =$[a]	L_0	μL_0	μL_0
[a]$L = \Phi/i =$	$\Phi_0/i_0 = L_0$	$\Phi_0/(i_0/\mu) = \mu L_0$	$\mu\Phi_0/i_0 = \mu L_0$

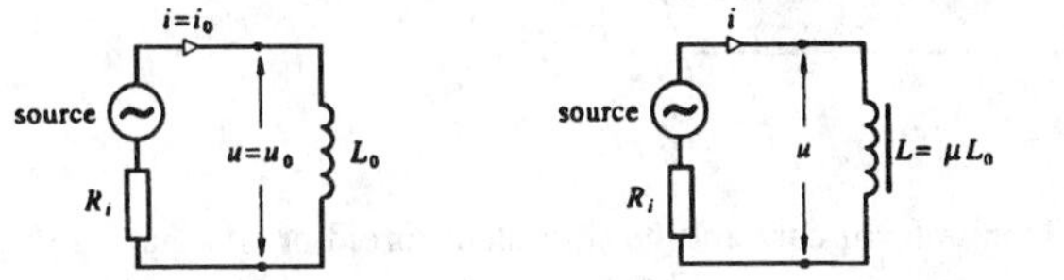

Cored Inductor and its Parameters in Electric Circuit. In Table 9.15 the parameters of the inductor are compared for cases when it is immersed in air ($\mu = 1$) and in magnetic material of permeability μ. It is supposed that the magnetic flux Φ_0 produced by the inductor winding runs completely through the magnetic material (the case of the complete coupling of winding with the magnetic material).

In case (a) the voltage drop across inductor, magnetic induction, and flux remain unchanged while the current through inductor and magnetic field decreases μ times; in case (b) the current and magnetic field remain unchanged, whereas the voltage drop, magnetic induction, and flux increases μ times. The only parameter that increases μ times in all the cases, limit and intermediate ones, $0 < R_i < \infty$, is the inductance $L = \mu L_0$.

The main reason for applying the magnetic material as a core in inductor is to increase its inductance.

Inductor, Magnetic Core, Closed and Open Magnetic Circuit. If the magnetic flux Φ generated by a winding runs for the most of its part through the core being in the possibly strongest coupling with that winding, one says that this core forms a closed or nearly closed magnetic circuit.

If at the possibly strongest coupling of a winding with a core, the flux Φ runs for its nonnegligible part also through the outside of the core, one says that core forms an open magnetic circuit.

The ring core and EE-core illustrate the closed magnetic circuits [Fig. 9.51(a)], whereas the rod core illustrates the open magnetic circuit [Fig. 9.51(b)].

Closed Magnetic Circuit

Equivalent Dimensions and Parameters. The permeance G_m of a toroidal core having a very small radial thickness, magnetic path length l, a uniform cross-section area A, and permeability μ, is given by (9.46c).

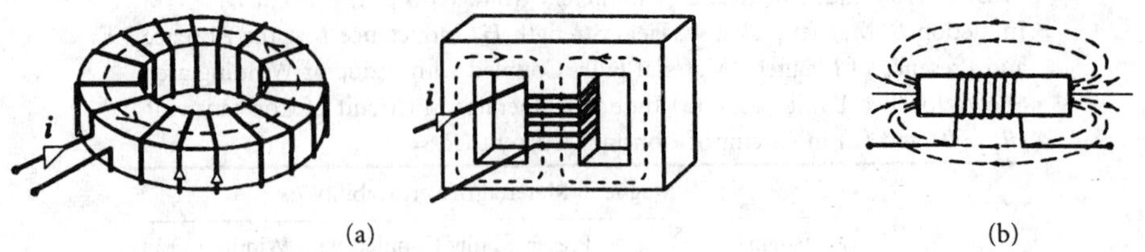

FIGURE 9.51 Examples of (a) closed magnetic circuits; (b) open magnetic circuit.

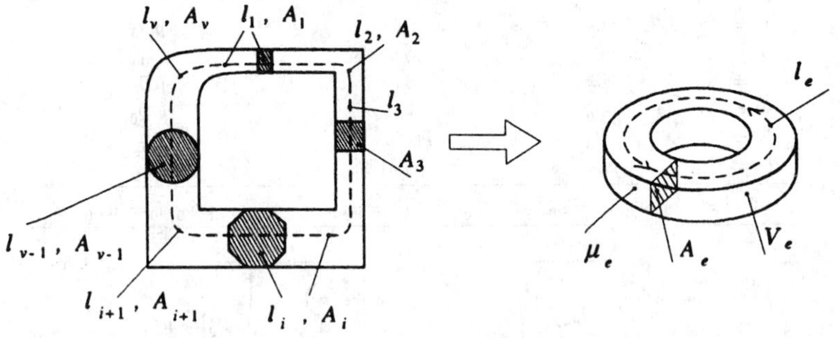

FIGURE 9.52 Nonuniform core and its equivalent toroid or ring having the same permeance.

If the shape of a core differs from the above radially thin toroid, the calculation of permeance becomes very complex, including the problem formulation and mathematics. To overcome these difficulties a simplified standard procedure is recommended [6]: a core having the shape different from radially thin toroid, nonuniform cross-sectional area along magnetic path and often, nonuniform permeability, is replaced by a hypothetic equivalent toroid or ring having the same permeance as the nonuniform core, Fig. 9.52.

The parameters of equivalent toroid: the equivalent magnetic path length l_e, equivalent area A_e, equivalent volume V_e, and equivalent permeability μ_e are calculated from the following formulas:

$$l_e = C_1^2/C_2; \quad A_e = C_1/C_2; \quad V_e = l_e A_e = C_1^3/C_2^2 \tag{9.49}$$

and equivalent magnetic parameters:

$$\mu_e = C_1 \Big/ \sum_{i=1}^{\nu} \frac{l_i}{\mu_i A_i}; \qquad B_e = \frac{u}{\omega n} \cdot \frac{C_2}{C_1};$$
$$H_e = ni \frac{C_2}{C_1^2}; \qquad \Phi_e = \frac{u}{\omega n} \cdot \frac{C_1^2}{C_2^2} \tag{9.50}$$

The core coefficients C_1 and C_2 are calculated as

$$C_1 = \sum_{i=1}^{\nu} \frac{l_i}{A_i} \text{ and } C_2 = \sum_{i=1}^{n} \frac{l_i}{A_i^2} \tag{9.51}$$

Therefore, the permeances G_m and G_{m0} and reluctances R_m and R_{m0} are

$$G_m = \mu/C_1; \qquad G_{m0} = 1/C_1; \qquad R_m = C_1/\mu; \qquad R_{m0} = C_1 \tag{9.52}$$

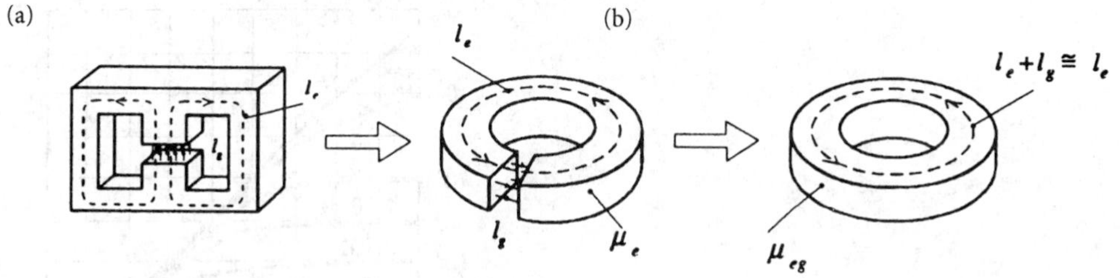

FIGURE 9.53 Ring cores (b) equivalent to the EE gapped core (a).

Inductance L is

$$L = A_L n^2 = \mu_e A_{L0} n^2 = \mu_0 G_m n^2 = \mu_0 \mu_e G_{m0} n^2 = \mu_0 \mu_e n^2 / C_1 \tag{9.53}$$

The core coefficients C_1, C_2, as well as equivalent dimensions l_e, A_e, and V_e for standardized cores are specified in relevant IEC standards and in core manufacturers' catalogs.

The above-described procedure is a standard averaging procedure (line and surface integrals are replaced by summation) of internationally accepted applicability. This procedure is commonly used in engineering, design, and measuring practice.

Air Gap: Reason for Introduction. An intentional air gap is introduced into a magnetic circuit of inductors to improve its performance, namely:

- to decrease the magnetic material nonlinearity and/or to lessen the effect of static magnetic field (bias field) on inductance,
- to lessen the magnetic power loss,
- to lessen excessive instabilities of the core magnetic material (e.g., thermal),
- to make possible a smooth adjustment of inductance to a very precise value by a special adjusting device or a specially shaped air gap.

A closed magnetic circuit (in the form of EE core, made of a material of magnetic permeability in which an air-gap of length l_g, is shown in Fig. 9.53. A succession of equivalent circuits and related permeabilities μ_e and μ_{eg} is also shown there.

The length of intentional air-gap l_g is usually small compared with magnetic path length l. A ratio l_g/l usually does not exceed a fraction of a percent.

Effect of an Air Gap on Core Permeability. The inductance of an inductor having a core with air-gap (at complete coupling between winding and core) is given by

$$L = \frac{\mu_0 \mu_e n^2}{C_1 \left[1 + (\mu_e - 1) \dfrac{l_g}{l_e} \right]} = \frac{\mu_0 \mu_{eg} n^2}{C_1} \tag{9.54}$$

where

$$\mu_{eg} = \mu_e \left[1 + (\mu_e - 1) \frac{l_g}{l_e} \right]^{-1} \tag{9.55}$$

μ_{eg} = equivalent permeability of the core with air-gap. The dependence (9.55) is shown in Fig. 9.54.

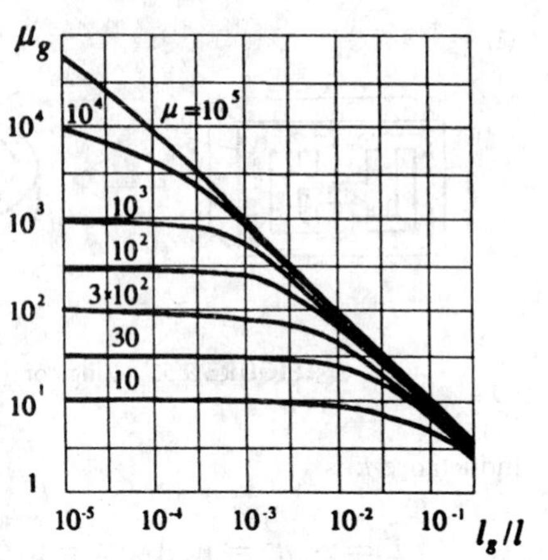

FIGURE 9.54 Equivalent permeability of the core with air-gap versus l_g/l.

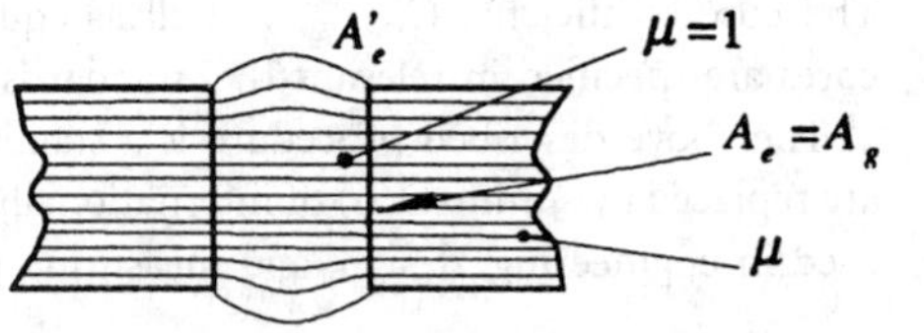

FIGURE 9.55 Bulging of the magnetic flux traversing the air-gap.

Equation (9.55) and Fig. 9.52 ignore a bulging of the magnetic flux traversing the air gap, Fig. 9.55.

Roughly, the bulging is greater as the term $(\mu_e - 1)l_g/l_e = p$ is larger. The bulging, expressed as an apparent increase of the air-gap area $A'_g/A_e = \kappa_g$, becomes perceptible at $p > 1$. The ratio κ_g, known as the bulging factor, can attain an amount of 2 for $p = 30$–60.

A change in flux distribution within the core, resulting from the occurrence of a large reluctance of the air gap in the magnetic circuit, is also ignored in (9.55). In effect, the value of inductance L for larger air gaps is usually higher than that resulting from (9.54).

As the air gap increases, the inductance L becomes less and less dependent on any variation of the magnitude of permeability μ_e. This effect is known as the linearization of cored inductor by the air-gap.

The air-gap effect can be summarized as follows.

The cored inductor winding perceives not the core with material permeability μ, but a core with reduced permeability μ_g lower than μ (or μ_{eg} when dealing with μ_e). Reduction of permeability of gapped core follows (9.55) depicted in Fig. 9.54.

The higher the permeability, the greater the reduction of μ, e.g., the air gap of the same (relative) length $l_g/l = 0.01$ reduces material permeability $\mu = 10\,000$ by a factor of 100 and that of $\mu = 10$ only by a factor of about 1.1. The use of high permeability materials as cores with larger air gaps is not effective.

Inductor with Incomplete Coupling. In real cases, the magnetic material forming the core of the inductor does not occupy the whole space where the magnetic flux Φ_0, emanating from the winding, is distributed. The following simplified quantitative approach is in use.

The flux Φ_0 is a sum of a leakage flux Φ_l, which runs completely outside the core and a hypothetical flux Φ_h, which runs completely inside the hypothetical equivalent toroid of permeability μ_e (or μ_{eg}): $\Phi_0 = \Phi_h + \Phi_l$. These three fluxes are related with three respective inductances: L_0 = inductance of winding without core, L_h = inductance of the shape of the hypothetical equivalent toroid, and L_l = leakage inductance whose flux Φ_l does not penetrate into the core:

$$L_0 = L_h + L_l \tag{9.56}$$

Therefore, the inductance of the cored inductor with leakage is

$$L = \mu_e L_h + L_l = (\mu_e - 1)L_h + L_0 \tag{9.57}$$

This formula is very convenient to determine the equivalent permeability from an actual measurement. Because $L_h = \mu_0 n^2/C_1$, L and L_0 are taken from measurement, the equivalent permeability μ_e (or μ_{eg} for the gapped core) can be determined:

$$\mu_e = \frac{L - L_0}{\mu_0 n^2} C_1 + 1 \text{ or } \mu_e = \frac{A_L - A_{L0}}{\mu_0} C_1 + 1 \tag{9.58}$$

For phenomenal considerations a so-called winding coupling coefficient $L_h/L_0 = k_w$ is introduced. Then

$$L = L_0[(\mu_e - 1)k_w + 1] \tag{9.59}$$

Apparent Permeability. A commonly used parameter quickly characterizing the cored inductor is the apparent permeability μ_{app}:

$$\mu_{app} = L/L_0 = (\mu_e - 1)k_w + 1 \tag{9.60}$$

Equation (9.60) shows whether the winding is completely coupled with the core; then $k_w = 1$, $\mu_{app} = \mu_e$; if $k_w = 0$, $\mu_{app} = 1$: the inductor is coreless; for $0 < k_w < 1$ always $\mu_{app} < \mu_e$ as in the most of real cases. The coupling coefficient is readily calculated from the formula

$$k_w = (\mu_{app} - 1)/(\mu_e - 1) \tag{9.61}$$

μ_{app} shows directly how many times the presence of the core, being in a given coupling with the winding, increases its inductance L_0.

Low winding coupling strongly lowers the effect of high permeability on the inductance of cored inductor. The high permeability materials are effective only for windings strongly coupled with the core. Such are monolayer windings of the thinnest possible wire, tightly fitting the core along its whole length, with no spacing between neighboring turns. This is in distinct opposition to other practical requirements, such as current load, low self capacitance, interwire insulation and electrical breakdown, and ease of manufacturing.

Leakage Coefficient. Also for the phenomenal considerations, the inductor leakage coefficient $\sigma = L_l/L$ is used. It shows what part of the total inductance L is not related with the core and its properties. The leakage coefficient is calculated as

$$\sigma = (\mu_e - \mu_{app})/\mu_{app}(\mu_e - 1) \text{ or } \sigma = (1 - k_w)/\mu_{app} \tag{9.62}$$

If $k_w = 1$ or equivalently $\mu_e = \mu_{app}$, $\sigma = 0$, the inductor is leakage-free; if $k = 0$, then $\mu_{app} = 1$ and $\sigma = 1$, which results in $L = L_0 = L_l$—the leakage inductance is the inductance L_0 of the coreless winding.

Summarizing:

- The coupling coefficient k_w shows the part of the winding flux that is instrumental in carrying the core properties into the inductor properties.
- The leakage coefficient σ shows what part of total inductor properties has no relation with the core properties.
- The core, when not completely coupled with the winding, increases its inductance by amount of only μ_{app}.
- The apparent permeability μ_{app} of the real cored inductor is always lower than the core permeability μ, μ_e, or μ_{eg}.

Joint Effect of the Air Gap and Coupling on the Cored Inductor Inductance. Combining the effect of air gap, (9.55), and that of coupling, (9.59), the formula for the inductance of cored inductor is

$$L = \frac{\mu_e L_0 k_w}{1 + (\mu_e - 1) l_g / \kappa_g l_e} - \left(L_0 - \frac{\mu_0 n^2}{C_1} \right) \tag{9.63}$$

In this formula two terms are to be distinguished: the first term, being roughly proportional to μ_e, is generally of nonlinear character and carries in itself all magnetic material properties including its instabilities. The second term, between parentheses, being effectively an air coreless inductance, is of purely linear character. Magnetic material properties "enter" inductor properties only by the first term. As the air-gap and leakage become larger, the share of the first term decreases for the benefit of the second term: the inductor linearizes or stabilizes. For the sake of example: for an inductor with an EE42/20 ferrite core of material permeability $\mu \cong 1800$, the ratio of the first to the second term for ungapped core is about 450. If an air-gap of length $l_g = 1$ mm is cut in that core, this ratio drops to about 30. Thus, the linearization here reaches an amount of about 15. That effect is called the linearization or stabilization by the air-gap and/or by the leakage (or coupling).

A cored inductor with coupling $k_w = 1$ and with no air gap behaves as a pure magnetic material itself, whereas for $k_w = 0$ it becomes a coreless air inductor having the inductance L_0.

Open Magnetic Circuit.

The most commonly used open magnetic circuit is a cylindrical rod core. The inductor is built by axially placing the rod core inside a cylindrical coil as it is shown in Fig. 9.56.

Inductors of this type are built as variable and fixed inductance inductors. To vary the inductance the rod core is moved axially inside the coil. The central axial position of the rod in relation to the coil corresponds to the maximum inductance. In fixed inductance inductors the movement between the core and coil is not provided.

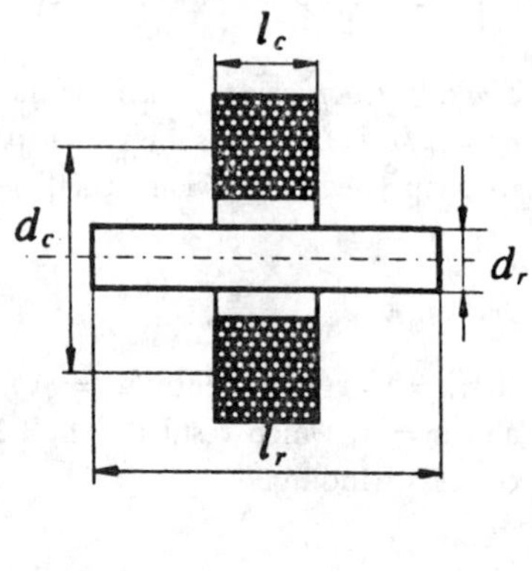

FIGURE 9.56 Open magnetic circuit having a rod core.

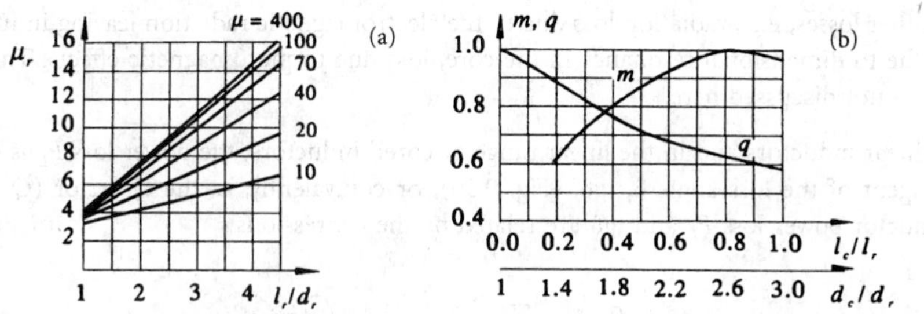

FIGURE 9.57 (a) Equivalent permeability of the rod μ_r versus ratio l_r/d_r for different material permeabilities μ; (b) experimental coefficients m versus l_c/l_r and q versus d_c/d_r.

If the rod core is enclosed by a centrally placed winding, as shown in Fig. 9.56, the inductance of such an inductor can be calculated from the following empirical formula:

$$L = L_0[(\mu_r mq - 1)k_{wr} + 1] = L_0\mu_{\text{app}} \tag{9.64}$$

where L_0 = the inductance of the winding without core; μ_r = equivalent (effective) magnetic permeability of the rod, of similar significance as μ_e for the closed magnetic circuit; m and q = experimentally determined coefficients; k_{wr} = coupling coefficient. The permeability of the rod μ_r is a complicated function of the material permeability μ and of the ratio of the rod length to the rod diameter l_r/d_r [7]. This function is shown in Fig. 9.57(a).

The experimental coefficients $m = f(l_c/l_r)$ and $q = f(d_c/d_r)$ are given in Fig. 9.57(b). The coupling coefficient k_{wr} is

$$k_{wr} = A_{Lh}/A_{L0} \text{ [nH]/[nH]} \tag{9.65}$$

where $A_{Lh} = Kd_r$, K is the coefficient listed in Appendix 1; $A_{L0} = L_0/n^2$, and L_0 = winding inductance.

The inaccuracy of the above calculations of L and μ_{app} does not exceed several percent.

As can be inferred from Fig. 9.57(b), the maximum inductance L is obtained for $mq = 1$, i.e., for $l_c \cong 0.83l_r$ and $d_c = d_r$: i.e., for the winding of width nearly equal to the rod length and closely fitting the rod diameter. As the winding diameter d_c is larger and winding width is narrower, the core contributes less to the inductor inductance L.

It is often of great importance to estimate the material permeability of the rod core. The calculation inverse to that of (9.64) can allow it: first one calculates $\mu_r = (\mu_{\text{app}} - 1 + k_{wr})/mqk_{wr}$, next for the so-determined μ_r and given l_r/d_r, the material permeability μ can be found from Fig. 9.57(a). This method, however, can practically distinguish $\mu < 400$ only and needs to gather very accurate input data and may provide with only approximate μ-values providing that $\mu < 400$.

Power Loss in Inductor

Power loss (in an inductor) is a power that is converted in an inductor into heat and radiation. The total power loss in the inductor is admitted to be a sum of the following contributory losses:

- resistive loss in winding R_{res},
- magnetic loss in core magnetic material R_μ,
- dielectric loss related to self capacitance and other parasitic capacitances of the inductor R_d,

- other losses, e.g., radiation loss due to the electromagnetic radiation leaving inductor, loss due to dimensional resonance in the core, loss due to piezomagnetic effects. These losses are not discussed here.

For linear inductors and in the linear range of cored inductors, the power loss P_L is expressed as a tangent of the loss angle δ, tgδ, (Fig. 9.50), or equivalently, as the Q-factor ($Q = 1/\text{tg}\delta$). The inductor power loss P_L and tgδ are related by the expressions

$$P_L = \frac{u_{\text{rms}}^2}{\omega L_p}\text{tg}\delta = i_{\text{rms}}^2 \omega L_s \text{tg}\delta \tag{9.66}$$

where u_{rms} = the rms value of voltage drop across the inductor and i_{rms} = rms value of current flowing through the inductor.

At high excitation levels, especially in a nonlinear range, the power loss of inductors is expressed directly in watts.

Resistive Loss in Winding. A loss due to a dc resistance (ohmic resistance) of a winding conductor increased by increments due to skin and proximity effects R_{res} referred to the inductive reactance of inductor ωL is

$$\text{tg}\delta_{\text{res}} = R_{\text{res}}/\omega L \tag{9.67}$$

The precise determination of R_{res} and especially, a determination of skin and proximity effect contributions is rather difficult. If an optimization of such winding parameters as a type of conductor (solid, stranded), its diameter, etc., is sought, then a necessary calculation needs a theoretical background [7]. R_{res} has a linear voltage–current character.

Magnetic loss. Depending on whether the excitation level is low 1), middle 2), or high 3), three representations of magnetic loss are accustomed. They are presented below.

1) Low B (or H) Amplitude Excitation. There is a range of B or H amplitudes at which $\mu(B \text{ or } H) = \text{const.}$

This corresponds to the case of cored inductors operating at small-signal levels. For this range the notions of complex inductance $\hat{L}$ and complex permeability $\hat{\mu}$ are used.

In series representation—$\hat{Z} = j\omega L + R_s = j\omega\hat{L}_s$:

$$\hat{L}_s = \hat{\mu}_s L_0; \quad \hat{\mu}_s = \mu_s' - j\mu_s''; \quad L_s = \mu_s' L_0 \text{ and } R_s = \omega L_0 \mu_s'' \tag{9.68}$$

In parallel representation—$\hat{Y} = 1/j\omega L_p + 1/R_p = 1/\omega\hat{L}_p$:

$$\hat{L}_p = \hat{\mu}_p L_0; \quad 1/\hat{\mu}_p = 1/\mu_p' - 1/j\mu_p''; \quad L_p = \mu_p' L_0 \text{ and } R_p = \omega L_0 \mu_p'' \tag{9.69}$$

Magnetic loss tangent tgδ_μ:

$$\text{tg}\delta_\mu = \mu_s''/\mu_s' = \mu_p'/\mu_p'' \tag{9.70}$$

$$\mu_p' = \mu_s'(1 + \text{tg}^2\delta_\mu); \quad \mu_p'' = \mu_s''(1 + 1/\text{tg}^2\delta_\mu) \tag{9.71}$$

In these formulas $\mu'(= \mu_s' \text{ or } \mu_p')$ and $\mu''(= \mu_s'' \text{ or } \mu_p'')$ are real and imaginary components of the complex magnetic permeability $\hat{\mu}$ (= $\hat{\mu}_s$ or $\hat{\mu}_p$) in series or parallel representation, respectively. *Note:* when the subscript at μ is deleted, the symbol always refers to the series representation.

The complex permeability $\hat{\mu}$ and related parameters are restricted to linear voltage–current (or *B*–*H*) dependences of permeability and loss. It means that μ' represents the dynamic initial permeability and μ'' the power loss, which are related solely to only reversible magnetization processes in magnetic material. In that convention the magnetic loss at low excitations is presented usually in form of tg $\delta_\mu(f)$ or relative (magnetic) loss factor tg $\delta_\mu/\mu_{i(st)}$ ($\mu_{i(st)}$ = static initial permeability).

All previous considerations concerning the contribution of magnetic permeability to the cored inductor properties are extendable to deal with the magnetic loss. With that intent the permeability μ or μ_e appearing in the formulas of interest should be replaced by the complex permeability $\hat{\mu}$. This simple formal operation results, however, in a serious complication of calculation because of the dealing with complex numbers.

One of the most important results is that the air-gap reduces the loss resistance R_s and tgδ_μ by the ratio μ'_{eg}/μ':

$$R_{sg} = R_s(\mu'_{eg}/\mu'); \quad \text{tg}\delta_{\mu g} = \text{tg}\delta_\mu(\mu'_{eg}/\mu') \tag{9.72}$$

On the other hand, the power loss in magnetic material P_μ remains unchanged for the same core, gapped or ungapped, if it operates at the same induction *B*; this is because

$$P_\mu = (\omega B_p^2 V_e/2\mu_0) \times (\text{tg}\delta_\mu/\mu') = (\omega B_p^2 V_e/2\mu_0) \times (\text{tg}\delta_{\mu eg}/\mu'_{eg}) \tag{9.73}$$

where B_p = peak value of *B* and V_e = core equivalent volume.

2) Middle B (or H) Amplitude Excitations. There is a range of *B* or *H* amplitudes where magnetic permeability increases linearly with *H*: $\mu = \mu_i + \nu H$, magnetic induction *B* is a parabolic function of *H* and the hysteresis effects occur (Rayleigh range). The magnetic loss tgδ_μ is here expressed as a sum of eddy current loss tg δ_F, residual loss tg δ_n, and hysteresis loss tg δ_h:

$$\text{tg}\delta_\mu = \text{tg}\delta_h + \text{tg}\delta_F + \text{tg}\delta_n \tag{9.74}$$

tgδ_h is proportional to *H* or *B*, tgδ_F to frequency, tgδ_n is assumed to be a constant. This is visible when other notation (given by Legg) is used:

$$R_s/\mu f L = aB_p + ef + c = 2\pi \text{tg}\delta_\mu/\mu' \tag{9.75}$$

where *a*, *e*, and *c* are the hysteresis, eddy current, and residual loss coefficients, respectively; B_p = peak value of induction *B*.

The particular loss tangents can be determined as it is shown in Fig. 9.58.

3) High B (or H) Amplitude Excitations. There is a range of the strongest dependence of the permeability μ on induction *B* (or field strength *H*), therefore, of the strongest nonlinearity of μ versus excitation. As a rule, the power inductors and transformers operate in that range. In this range the magnetic power loss P_μ is being expressed by the heuristic Steinmetz expression:

$$P_\mu = \eta f^m B_p^n \tag{9.76}$$

where η = numerical coefficient, *m* and *n* are experimentally determined Steinmetz exponents, which are approximately constant for a given magnetic material, for example, for power ferrites $1.2 < m < 1.4$ and $2 < n < 3$ at frequencies between 10 and 100 kHz. For Fe–Si alloys: $1.3 < n < 1.4$ and $1.6 < m < 1.8$ for $f < 500$ Hz. The plots log P_μ versus log *f* or versus log B_p are therefore straight lines. If these plots deviate from the straight lines, the constancy of exponents *m* and *n* deteriorates and the applicability of (9.76) becomes less useful.

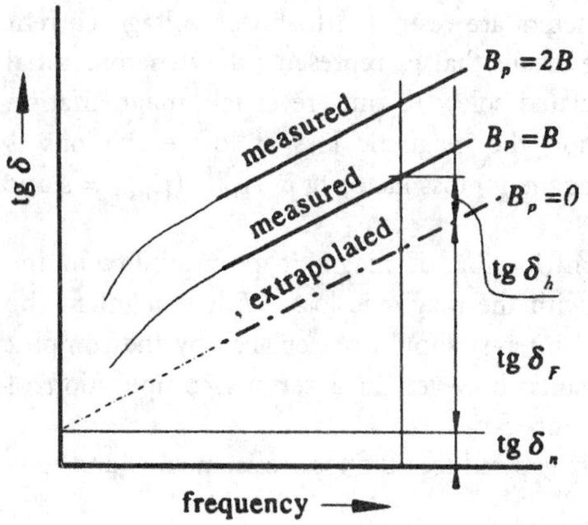

FIGURE 9.58 Determination of the magnetic loss tangents according to (9.74).

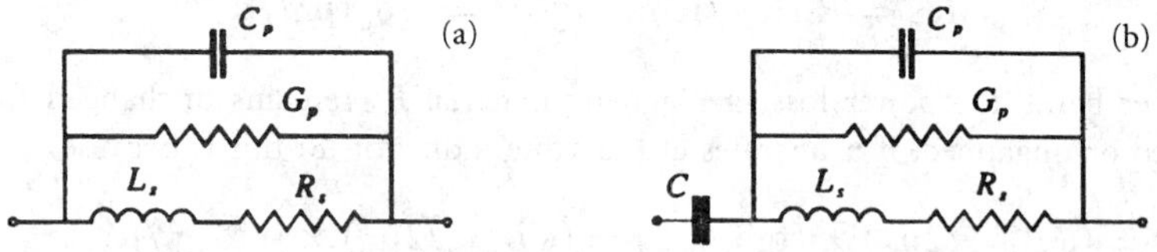

FIGURE 9.59 Self capacitance C_p and conductance G_p of an inductor L_s, R_s in (a) intentionally nonresonant circuit and (b) in an intentionally series resonant circuit.

Dielectric Loss. Every real inductor is inevitably associated with a shunting parasitic capacitance, called self capacitance C_p and accompanied conductance G_p. C_p and G_p result from various capacitive and conductive couplings distributed over a real geometrical assembly of various materials (conducting, insulating) forming the physical body of the inductor. In practice, C_p and G_p are assumed to be as lumped, connected in parallel inductor elements of loss tangent $\text{tg}\delta_d = G_p/\omega C_p$. It is advisable to consider the contribution of $\text{tg}\delta_d$ to overall inductor loss for two cases 1) and 2), [7].

1) Inductor (L_s of $\text{tg}\delta = R_s/\omega L_s$) operating in an intentionally nonresonant circuit, Fig. 9.59(a),

$$\text{tg}\delta_p = \omega^2 L_s C_p \text{tg}\delta_d = \text{tg}\delta_d (f/f_0)^2 \tag{9.77}$$

where $f_0 = (1/2\pi\sqrt{L_s C_p})$ = inductor self-resonant frequency.

2) The same inductor operating in a series LC intentional resonant circuit, Fig. 9.57(b) and, due to the resonant circumstances, $\text{tg}\delta_p$ contributes in that case as $\text{tg}\delta_{pr}$:

$$\text{tg}\delta_{pr} = \omega^2 L_s C_p (2\text{tg}\delta + \text{tg}\delta_d) = (2\text{tg}\delta + \text{tg}\delta_d)(f/f_0)^2 \tag{9.78}$$

Because the value of tg δ_d may come up even to 0.1, the product $L_s C_p$ should be as low as possible to make the self-resonant frequency $f_0 \gg f$. As an example, for an inductor having tg $\delta = 0.004$ (Q-factor = 250) and tg $\delta_d = 0.02$, $(f/f_0)^2 = 0.1$, the degraded Q-factor attains 170 in case 1) and 150 and in case 2). Additionally, tg δ_d is very sensible to any climatic factors and different agents; the case is known that a low value tg δ_d increased by nearly 200 times in a hot humidity environment.

Inductor Total Power Loss. Different representations of total power loss in inductor are summarized in Table 9.16.

TABLE 9.16 Summary of Power-Loss Expressions for the Sinusoidally Excited Inductor

Loss Representation	Low Excitation ($\mu' = \text{const}$)	Middle Excitation ($\mu = \mu_i + \nu H$)	High Excitation (μ—Strongly nonlinear)
$\text{tg}\,\delta_t =$	$\text{tg}\,\delta_{tl} = \underbrace{\text{tg}\,\delta_{\text{res}} + \overbrace{\mu''/\mu'}^{\text{tg}\,\delta_\mu\ \text{b}}}_{\text{tg}\,\delta} + \text{tg}\,\delta_c{}^{\text{a}}$	$\text{tg}\,\delta_{tm} = \underbrace{\text{tg}\,\delta_{\text{res}} + \overbrace{\text{tg}\,\delta_h + \text{tg}\,\delta_F + \text{tg}\,\delta_n}^{\text{tg}\,\delta_\mu\ \text{b}}}_{\text{tg}\,\delta} + \text{tg}\,\delta_c{}^{\text{a}}$	not applicable
$R_s =$	$\omega L_s \cdot \text{tg}\,\delta_{tl}$	$\omega L_s \cdot \text{tg}\,\delta_{tm}$	not applicable
$R_p =$	$R_s(1 + 1/\text{tg}^2\,\delta_{tl})$	$R_s(1 + 1/\text{tg}^2\,\delta_{tm})$	not applicable
$P_L =$	$i^2_{\text{rms}} \cdot \omega L_s\,\text{tg}\,\delta_{tl}$ or $u^2_{\text{rms}} \cdot \text{tg}\,\delta_{tl}/\omega L_p$	$i^2_{\text{rms}} \cdot \omega L_s\,\text{tg}\,\delta_{tm}$ or $u^2_{\text{rms}} \cdot \text{tg}\,\delta_{tm}/\omega L_p$	$u^2_{\text{rms}}\left(\frac{R_{\text{res}}}{R^2_{\text{res}} + \omega^2 L_0^2} + \omega C_p\,\text{tg}\,\delta_d\right)^{\text{c}}$
$P_\mu =$	$(\omega B_p^2 V_e/2\mu_0) \cdot (\mu''/\mu'^2)$	$(\omega B_p^2 V_e/2\mu_0) \cdot (\text{tg}\,\delta_\mu/\mu')$	$\eta f^m B_p^n$

Distinction: low, middle, and high excitation apply only to the cored inductors.

$\text{tg}\,\delta_t$ = total loss tangent; $\text{tg}\,\delta_{tl}$ = total loss tangent at low excitation level; $\text{tg}\,\delta_{tm}$ = total loss tangent at middle excitation level.

[a]
$$\text{tg}\,\delta_c = \begin{cases} \text{tg}\,\delta_p = \text{tg}\,\delta_d \cdot (f/f_0)^2 & \text{for inductor operating in a nonresonant circuit, Fig. 9.59(a)} \\ \text{tg}\,\delta_{pr} = (2\,\text{tg}\,\delta + \text{tg}\,\delta_d) \cdot (f/f_0)^2 & \text{for inductor operating in a series resonant circuit, Fig. 9.59(b).} \end{cases}$$

[b]For air inductors: $\text{tg}\delta_\mu = 0$ because $\mu' = 1$ and μ'', $\text{tg}\delta_h$, $\text{tg}\delta_F$, $\text{tg}\delta_n$, P_μ equal zero.

[c]For air inductor. For cored inductor P_L is practically not calculable because of the distorted signal.

Nonlinearity in Inductors

The cored inductors are nonlinear except when the excitation does not exceed the level where $\mu(B \text{ or } H) = \text{const}$. The most current three characteristics of inductor expose the extent of nonlinearity:

- inductance versus ac current or applied voltage,
- inductance versus dc bias current or voltage with superimposed ac current or voltage,
- harmonic content versus specified excitation.

The nonlinearity of a cored inductor is due to hysteretic properties of the core material, which result in coercive and remanence effects. That peculiarity of magnetic material causes both multivalent and multivalued characters of nonlinearity in cored inductors. In addition, in many circumstances the nonlinearity can also be time dependent, e.g., the harmonic content can vary with time.

Generally, the nonlinearity occurs when the core permeability becomes excitation dependent. Such excitation-dependent permeability is termed as amplitude permeability μ_a.

There are numerous phenomenal expressions for dependence of amplitude permeability μ_a on excitation. The most well known are expressions for the Rayleigh range of excitations given by

$$\text{Rayleigh:} \quad \mu_a = \mu_i + \nu H; \tag{9.79}$$

$$\text{Peterson:} \quad \mu_a = \alpha_0 + \alpha_1 H + \alpha_2 H^2 + \cdots \tag{9.80}$$

where μ_i = initial permeability; ν = Rayleigh constant; α_0, α_1, α_2, and so on = Peterson coefficients.

Today, these expressions are of lesser importance because any experimentally obtained dependence of an inductor on excitation can be numerically processed and used for overall nonlinear analysis (e.g., of signal distortion) in a given circuit. All previously given relationships between the permeability and inductance are to be observed in that analysis.

A practical manner of lowering the nonlinearity of an inductor (for a given core material) is to increase the air-gap length and/or to decrease the coupling coefficient k_w, or to lower the induction B in the core.

For the sake of an example, the third harmonic content, $\text{THC} = \text{emf}_{3f}/\text{emf}_f$ of the ungapped core is related to tangent of material hysteresis loss $\text{tg}\delta_h$ in the Rayleigh range by the formula

$$\text{THC} = 0.6\text{tg}\delta_h \tag{9.81}$$

For the gapped core this formula transforms into

$$\text{THC}_{\text{gapped}} = \text{THC}_{\text{ungapped}} \times (\mu_{eg}/\mu) \tag{9.82}$$

Generally, for a given core material, the most nonlinear inductor is that equipped with a core having an ungapped magnetic circuit (e.g., a toroid); the least nonlinear is the inductor having an open short core (e.g., a rod core of low length to diameter ratio).

The nonlinear core related relationships are strongly frequency dependent and, nearly as a rule, decrease as the frequency increases.

Magnetic Core Materials

For inductor core applications three families of soft magnetic materials are concerned:

- ferrites, mainly Mn–Zn and Ni–Zn, being solid polycrystalline ceramics in whatever form they may be;
- iron and/or nickel based alloys (Fe, Fe–Si, Fe–Ni, Fe–Co, Ni–Mn, Ni–Mn–Mo, etc.) in the form of laminations, strips, and ribbons;
- pure iron and Fe–Ni or Ni–Mn, etc., alloys in the form of fine powders (2–150 μm) whose individual particles are bonded and electrically insulated by polymers.

The most distinctive features of these three families are listed in Table 9.17.

TABLE 9.17 Three Main Families of Magnetic Materials Used as Cores in Inductors

Family	$B_s[T]$	μ_i	$\rho[\Omega m]$	f_w
Ferrites	low: < 0.6	8–20 000	high: 10^0–10^9	1 kHz–2 GHz
Fe, Ni alloys	high: 0.8–2.4	200–100 000	low: $(1–50)10^{-9}$	dc–400 Hz (up to 50 kHz)[a]
Bonded powders	medium: < 1.2	4–150	low: $\sim 10^{-7}$	dc–100 kHz (up to 10 MHz)[a]

B_s = saturation induction; μ_i = initial permeability; ρ = dc resistivity.
f_w = typical working frequency range.
[a]For very small material thickness.

Recently developed amorphous and nanocrystalline magnetic materials have not gained as of yet wider use in inductor applications. Some ring power chokes used as core-wound amorphous ribbons have advantageous properties at the kHz range.

The basic characteristics of magnetic materials for inductor applications are depicted in Fig. 9.60.

efining Terms

Amplitude permeability: The permeability obtained from the peak value of the induction and the peak value of the applied field strength when the field is varying periodically with time, the average of the field over the time is zero and the material is initially in a neutral state.

Complex permeability: The complex quotient of the induction in the material and the product of magnetic constant and field strength, at the excitation range, where both the induction and field vary sinusoidally with time.

Inductor: A device consisting of one or more associate windings, with or without magnetic core, for introducing inductance into an electric circuit.

Initial permeability: The limiting value of the amplitude permeability when the field strength is vanishingly small.

Loss angle of inductor: The angle by which the fundamental component of passing current lags the fundamental component of the fundamental voltage.

Magnetic core: A configuration of magnetic material, which is placed in a specific geometrical relationship to current-carrying conductors and whose magnetic properties are essential to its use.

Magnetic power loss: The power loss in a magnetic material.

Power loss: The power expended without useful work and converted into heat or nonintended radiation.

Quality factor: The magnitude of the ratio of the inductor effective reactance to its effective resistance at a given frequency (reciprocal of the tangent of the loss angle of the inductor).

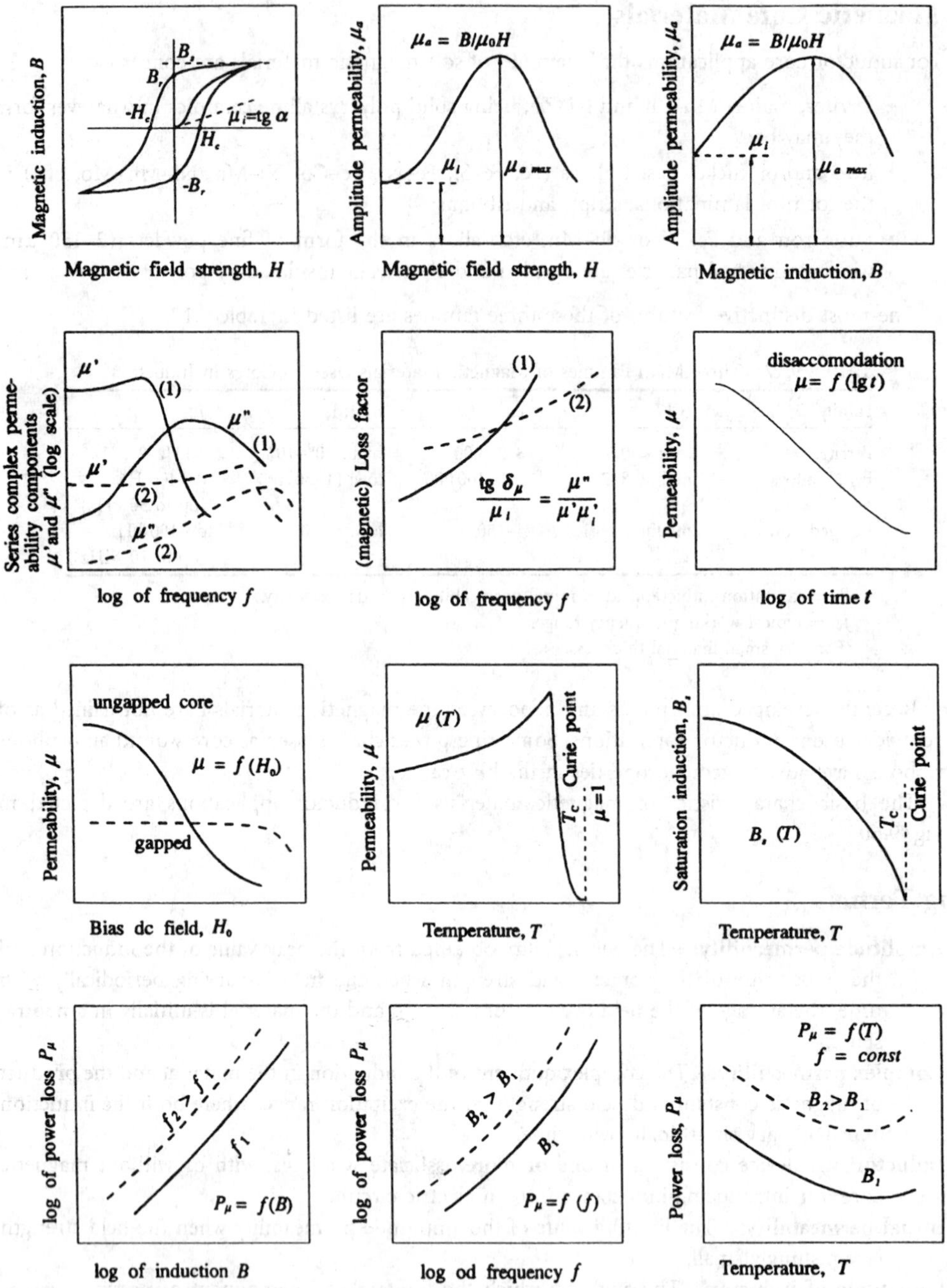

FIGURE 9.60 Basic characteristics of magnetic materials essential for inductor applications.

Tangent of the loss angle (of inductor): The magnitude of the ratio of the inductor effective resistance to its effective reactance at a given frequency (reciprocal of the quality factor of the inductor).

Tangent of the magnetic loss angle: The ratio of the imaginary component of the complex permeability of a material to its real component.

Winding: A conductive path, usually of wire, consisting of one (single-loop winding) or more turns, usually forming a spiral.

eferences

[1] ANSI-IEEE Standard 100-1988. *IEEE Standard Dictionary of Electrical and Electronic Terms,* 4th ed., New York: IEEE, 1988.

[2] M. Zahn, *Electromagnetic Field Theory: A Problem Solving Approach,* New York: Wiley, 1979.

[3] IEC 1007: 1990, "International standard: Transformers and inductors for use in electronic and telecommunication equipment—Measuring methods and procedures," and Amendment 1, Internat. Electrotech. Comm., Geneva, Switzerland, 1993.

[4] IEC 1248—1 to 7: 1995, "International standard: Generic specification: Transformers and inductors for use in electronic telecommunication equipment," Internat. Electrotech. Comm., Geneva, Switzerland, 1995.

[5] CECC 25000 Standard, "Generic specification: Inductor and transformer cores for telecommunications," 1st ed., CECC General Secretary, Frankfurt/M, Germany, 1976.

[6] IEC 205: 1966, "International standard: Calculation of the effective parameters of magnetic piece parts" with Amendments: 1976, no. 1; 1981, no. 2 and supplements: 1968—first; 1974—second, Internat. Electrotech. Comm., Geneva, Switzerland.

[7] E. C. Snelling, *Soft Ferrites, Properties and Applications,* 2nd ed., London: Butterworths, 1988.

[8] H. Hertwig, *Induktivitäten,* Berlin: Verlag für Radio-Foto-Kinotechnik, GmbH, 1954.

[9] F. E. Terman, *Radio Engineering Handbook,* 3rd ed., New York: McGraw-Hill, 1947.

[10] R. C. Dorf, Ed., *Electrical Engineering Handbook,* Boca Raton, FL: CRC Press, 1993.

ırther Information

There are several **books** that provide comprehensive and in-depth information on inductors, choice of suitable magnetic material for core, design methods, and various applications: E. C. Snelling, 1988, *Soft Ferrites. Properties and Applications,* London: Butterworths, 1988; W. Kampczyk and E. Roess, *Ferritkerne,* Berlin, München: Siemens AG, (English version in preparation) 1978; R. Boll, *Weichmagnetische Werkstoffe,* Hanau, Berlin, München: Vacuumschmelze GmbH—Siemens AG, (English version in preparation) 1990; G. E. Fish, 1989, *Soft magnetic materials, Proc. IEEE,* vol. 78, no. 6, p. 947, 1989, review paper.

The following journals and conference proceedings publish articles on inductors and related issues. **Journals:** *IEEE Transactions on Magnetism; Journal of Magnetism and Magnetic Materials; Coil and Winding International.* **Conferences:** *INTERMAG;* (American) *Conference on Magnetism and Magnetic Materials* (MMM); *International Conference on Magnetism* (ICM); *International Conference on Ferrites* (ICF); *Soft Magnetic Materials Conference* (SMM); *European Magnetic Materials and Application Conference* (EMMA); *Intertech Business Conferences on Magnetic Materials,* (American) Annual Applied Power Electronics Conference (APEC).

Because of the enormous diversity of inductor applications, properties of various magnetic materials, inductors, and core shapes, it is suggested that one consult the relevant data published by inductor manufacturers in their catalogs and application notes.

Chip inductors are described in Chapter 56 of this *Handbook.*

The measuring methods concerning the inductors and magnetic core properties are given in **Standards** prepared by Technical Committee No. 51 of the International Electrotechnical Commission (IEC). Copies of these Standards may be obtained from the IEC Central Office: 1, rue de Varembé, Geneva, Switzerland or from the IEC National Committees. Standards on inductors and cores are also issued in the European Union by the CENELEC Electronic Components Committee (CECC), General Secretariat, Gartnerstrasse 179, D-6000 Frankfurt/Main, Germany.

Appendices

APPENDIX 1 Coefficient $K = f(d/l)$ figuring in (9.48a) [8]

D/l	K	D/l	K	D/l	K	D/l	K
0.02	0.1957	0.32	2.769	0.80	5.803	2.20	10.93
0.04	0.3882	0.34	2.919	0.85	6.063	2.40	11.41
0.06	0.5776	0.36	3.067	0.90	6.171	2.60	12.01
0.08	0.7643	0.38	3.212	0.95	6.559	2.80	12.30
0.10	0.9465	0.40	3.355	1.00	6.795	3.00	12.71
0.12	1.126	0.42	3.497	1.10	7.244	3.50	13.63
0.14	1.303	0.44	3.635	1.20	7.670	4.00	14.43
0.16	1.477	0.46	3.771	1.30	8.060	4.50	15.14
0.18	1.648	0.48	3.905	1.40	8.453	5.00	15.78
0.20	1.817	0.50	4.039	1.50	8.811	6.00	16.90
0.22	1.982	0.55	4.358	1.60	9.154	7.00	17.85
0.24	2.144	0.60	4.668	1.70	9.480	8.00	18.68
0.26	2.305	0.65	4.969	1.80	9.569	9.00	19.41
0.28	2.406	0.70	5.256	1.90	10.09	10.00	20.07
0.30	2.616	0.75	5.535	2.00	10.37	12.00	21.21

APPENDIX 2 Coefficient $\gamma = f(l/h)$ figuring in (9.48e) [8]

l/h	γ	l/h	γ	l/h	γ	l/h	γ
1	0.0000	9	0.2730	17	0.3041	25	0.3169
2	0.1202	10	0.2792	18	0.3062	26	0.3180
3	0.1753	11	0.2844	19	0.3082	27	0.3190
4	0.2076	12	0.2888	20	0.3099	28	0.3200
5	0.2292	13	0.2927	21	0.3116	29	0.3209
6	0.2446	14	0.2961	22	0.3131	30	0.3218
7	0.2563	15	0.2991	23	0.3145		
8	0.2656	16	0.3017	24	0.3157		

9.4 Transformer

Gordon E. Carlson

Introduction

The transformer is a two-port passive circuit element that consists of two coils that are coupled magnetically, but have no conductive coupling. It is shown diagrammatically in Fig. 9.61, where the dots by one end of each coil indicate that the magnetic fluxes, ϕ_{m1} and ϕ_{m2} are in the same direction when both currents either enter or leave by the dot marked terminal. Coil 1 is connected to the transformer input terminals and is called the primary winding. Coil 2 is called the secondary winding and is connected to the transformer output terminals.

A transformer can be used to connect a source to a load and comes in a wide range of sizes. The sizes include very large power-distribution transformers and very small transformers used in electronic equipment. The coils of some transformers used in electronic equipment are wound on a nonmagnetic core such as plastic. These transformers are called air-core transformers. All transformers used in power-distribution systems and some transformers used in electronic

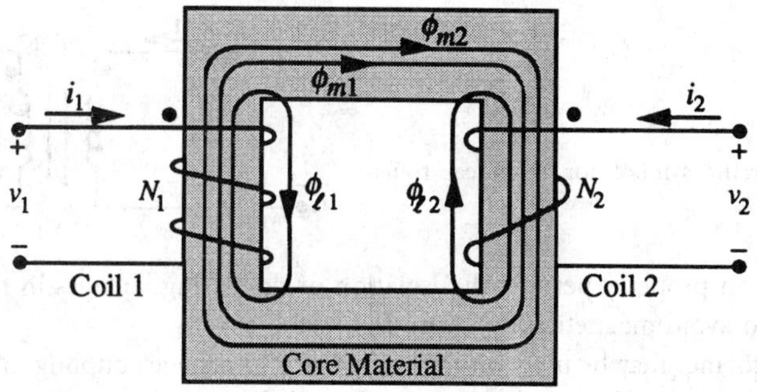

N_1 = number of turns in coil 1 $\qquad$ N_2 = number of turns in coil 2

$\phi_{\ell 1}$ = leakage flux due to current i_1 $\qquad$ $\phi_{\ell 2}$ = leakage flux due to current i_2

ϕ_{m1} = magnetizing flux due to current i_1 $\qquad$ ϕ_{m2} = magnetizing flux due to current i_2

$\phi_1 = \phi_{\ell 1} + \phi_{m1}$ = total flux due to current i_1 $\qquad$ $\phi_2 = \phi_{\ell 1} + \phi_{m2}$ = total flux due to current i_2

$k_1 = \dfrac{\phi_{m1}}{\phi_1}$ = coupling factor for coil 1 $\qquad$ $k_2 = \dfrac{\phi_{m2}}{\phi_2}$ = coupling factor for coil 2

$k = \sqrt{k_1 k_2}$ = transformer coupling coefficient <1

FIGURE 9.61 Transformer diagram and definitions.

equipment use an iron core, which produces a coupling coefficient of nearly unity. The coupling coefficient for an air-core transformer seldom exceeds 0.5.

Four major characteristics of transformers are as follows:

1) accept energy at one voltage and deliver it at a different voltage,
2) change the load impedance as seen by the source,
3) provide conductive isolation between two portions of a circuit, and
4) produce bandpass signal filters when combined with capacitors.

The first characteristic indicated is commonly used in electric power-distribution systems since higher voltages are desired for electric energy transmission than can be safely used by the customer. The higher transmission voltage produces lower transmission current, which requires smaller transmission line conductors.

The output stage of an audio amplifier may include a transformer to provide the second and third characteristics listed. In this way, the low impedance of the speaker is matched to the higher output impedance of the power amplifier to yield maximum power transfer for the

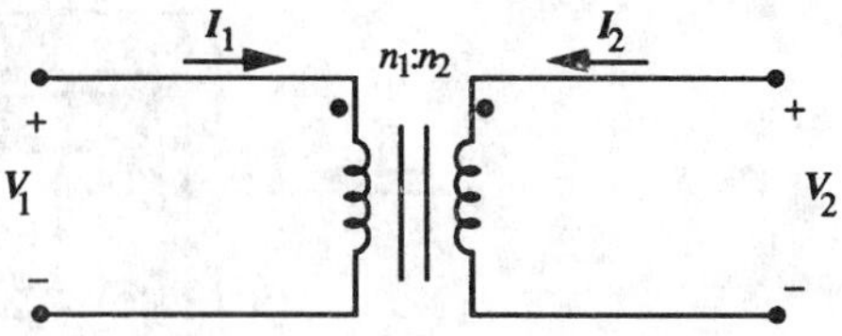

FIGURE 9.62 Circuit symbol for the ideal transformer.

signal. The isolation property permits the isolation of dc biasing voltages in the amplifier from the speaker coil to avoid magnetic flux saturation in the coil.

Finally, a transformer may be used with capacitors for interstage coupling in a radio frequency amplifier. The inductance of the transformer coils and the capacitance of the capacitors can be adjusted to produce bandpass filtering for signals.

This section presents linear mathematical models for transformers. The models are in terms of phasors and impedances; that is, they are frequency-domain models. The section begins with a first-order model known as the ideal transformer. A nonideal linear transformer model that includes transformer inductive and resistive effects is then shown.

Ideal Transformer

The ideal transformer models the first three transformer characteristics listed above with an ideal, lossless, circuit element. It is a reasonably good model for a transformer with a nearly unity coefficient of coupling and primary and secondary coil inductive impedances that are very large with respect to source and load impedances. Well-designed iron-core transformers have approximately these characteristics over a reasonable range of frequencies and terminating impedances. A circuit symbol that is often used for an ideal transformer is shown in Fig. 9.62, where $a = n_1/n_2 \cong N_1/N_2$ is the effective turns ratio. In place of the lines between the coils, which are intended to indicate the similarity to an iron-core transformer, the word ideal or the equation $k = 1$ may be written below the coils to identify an ideal transformer.

The equations that represent the ideal transformer model are $V_2 = V_1/a$ and $I_2 = -aI_1$. Thus, the voltages are in the same ratio as the effective turns ratio. Also, the ideal transformer is lossless since the complex power supplied by the source $V_1 I_1^*$ equals the complex power absorbed by the load $-V_2 I_2^*$. The hybrid-h-parameter two-port equation that represents the ideal transformer model is

$$\begin{bmatrix} V_1 \\ I_2 \end{bmatrix} = \begin{bmatrix} 0 & a \\ -a & 0 \end{bmatrix} \begin{bmatrix} I_1 \\ V_2 \end{bmatrix} \tag{9.83}$$

If the load impedance Z_L is connected to the output terminals of an ideal transformer, then $V_2 = -I_2 Z_L$. In this case, the impedance seen by a source connected to the input terminals is

$$Z_{\text{eq}} = \frac{V_1}{I_1} = \frac{(aV_2)}{\left(-\dfrac{1}{a} I_2\right)} = a^2 Z_L \tag{9.84}$$

Thus the impedance seen by the source is the square of the effective turns ratio times the load impedance.

Nonideal Transformer

The ideal transformer is not an adequate transformer model when the coefficient of coupling is not near unity and/or the load and source impedances are not negligible with respect to the

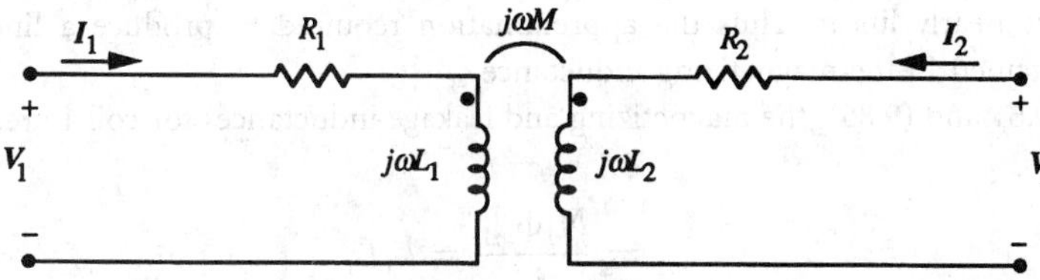

FIGURE 9.63 Linear transformer model.

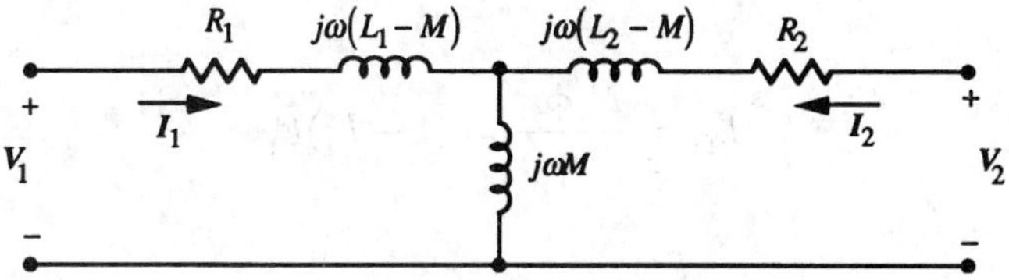

FIGURE 9.64 Equivalent T-network representation of a linear transformer.

transformer-coil inductive impedances. Also it cannot be used to investigate the signal filtering that can be performed with a transformer and capacitors. In these cases a more detailed model is required.

Linear Transformer

The linear transformer model is shown in Fig. 9.63. R_1 and R_2 are the resistances and L_1 and L_2 are the self inductances of the two transformer coils. M is the mutual inductance corresponding to the magnetic coupling of the coils.

The linear transformer is a reasonable model for an air-core transformer since the magnetic flux is proportional to the current. The self inductances are

$$L_1 = N_1 \frac{\phi_1}{i_1} \quad \text{and} \quad L_2 = N_2 \frac{\phi_2}{i_2} \tag{9.85}$$

and the mutual inductance is

$$M = k\sqrt{L_1 L_2} \tag{9.86}$$

From Fig. 9.63, the impedance-parameter two-port equations for the linear-transformer model are

$$\begin{bmatrix} V_1 \\ V_2 \end{bmatrix} = \begin{bmatrix} R_1 + j\omega L_1 & j\omega M \\ j\omega M & R + j\omega L_2 \end{bmatrix} \begin{bmatrix} I_1 \\ I_2 \end{bmatrix} \tag{9.87}$$

Except for its isolation characteristics, the linear transformer in Fig. 9.63 can be represented by the equivalent T-network shown in Fig. 9.64, since this network produces (9.87). The T-network is completely equivalent to the nonisolating linear transformer that is produced when the lower ends of the two coils are conductively connected.

Leakage and Magnetizing Inductances

The nonideal linear-transformer circuit model can be changed so it is expressed in terms of leakage inductance and magnetizing inductance rather than self inductance and mutual inductance. This is convenient since the effects produced by the coil resistance and leakage inductance are nearly linear. Thus the approximation required to produce a linear model is primarily contained in the magnetizing inductance.

From Fig. 9.61 and (9.85), the magnetizing and leakage inductances for coil 1 are, respectively,

$$L_{m1} = \frac{N_1 \phi_{m1}}{i_1} = k_1 L_1 \tag{9.88}$$

and

$$L_{\ell 1} = \frac{N_1 \phi_{\ell 1}}{i_1} = (1 - k_1) L_1 \tag{9.89}$$

Therefore,

$$L_1 = L_{\ell 1} + L_{m1} \tag{9.90}$$

Similarly, for coil 2,

$$L_{m2} = \frac{N_2 \phi_{m2}}{i_2} = k_2 L_2 \tag{9.91}$$

$$L_{\ell 2} = \frac{N_2 \phi_{\ell 2}}{i_2} = (1 - k_2) L_2 \tag{9.92}$$

and

$$L_2 = L_{\ell 2} + L_{m2} \tag{9.93}$$

are the magnetizing, leakage, and self inductances.

Using Fig. 9.61 and (9.86), we can write the mutual inductance as

$$\begin{aligned} M &= \sqrt{k_1 k_2} \sqrt{L_1 L_2} = \sqrt{k_1 L_1 k_2 L_2} \\ &= \sqrt{L_{m1} L_{m2}} \\ &= a L_{m2} = \left(\frac{1}{a}\right) L_{m1} \end{aligned} \tag{9.94}$$

where

$$\sqrt{\frac{L_{m1}}{L_{m2}}} \equiv \frac{n_1}{n_2} = a \cong \frac{N_1}{N_2} \tag{9.95}$$

is the effective turns ratio. How closely the effective turns ratio n_1/n_2 approximates the actual turns ratio N_1/N_2 depends on how completely all magnetic flux links all coil turns. This is a function of the coil and core geometry.

Circuit Model

Substitution of the leakage inductances, magnetizing inductances, and effective turns ratio into (9.87) produces an alternate form for the impedance-parameter two-port equations for the linear transformer model. These equations are

$$\begin{aligned} V_1 &= [R_1 + j\omega(L_{\ell 1} + L_{m1})]I_1 + j\omega L_{m1}\left(\frac{1}{a}\right)I_2 \\ &= (R_1 + j\omega L_{\ell 1})I_1 + j\omega L_{m1}(I_1 - I_a) \end{aligned} \tag{9.96}$$

and

$$\begin{aligned} V_2 &= j\omega L_{m1}\left(\frac{1}{a}\right)I_1 + [R_2 + j\omega(L_{\ell 2} + L_{m2})]I_2 \\ &= \left(\frac{1}{a}\right)V_a + (R_2 + j\omega L_{\ell 2})I_2 \end{aligned} \tag{9.97}$$

where

$$I_a = -\left(\frac{1}{a}\right)I_2 \tag{9.98}$$

and

$$\begin{aligned} V_a &= j\omega L_{m1}I_1 + j\omega a L_{m2}I_2 \\ &= j\omega L_{m1}(I_1 - I_a) \end{aligned} \tag{9.99}$$

The transformer circuit model that produces (9.96)–(9.99) is shown in Fig. 9.65.

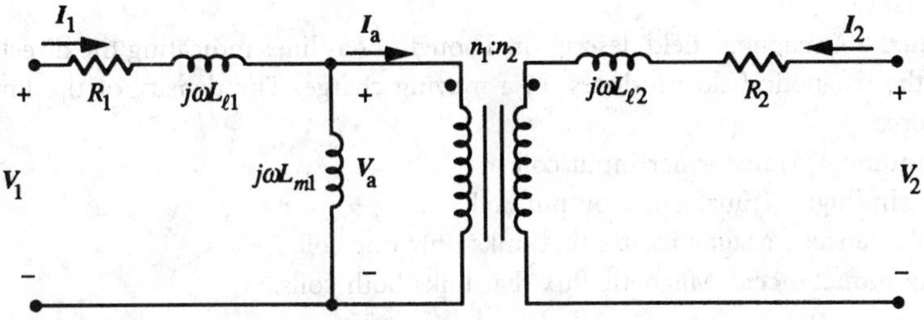

FIGURE 9.65 Linear circuit model for a nonideal transformer.

The only energy losses in the transformer that are accounted for by the linear model shown in Fig. 9.65 are the heating losses in the coils (called copper losses). An iron-core transformer also has heating losses in the core material (called core losses). There are two components to the core losses. The first component is hysteresis losses. Hysteresis is the nonlinear phenomenon that causes the magnetic flux response to increasing current to be different from the response to decreasing current. The plot of flux as a function of current traces out a closed curve that is called a hysteresis loop. The area inside this loop is proportional to the energy that produces core heat.

The second component of transformer core losses is eddy-current losses. This is a heating loss caused by currents (called eddy currents) that flow in the transformer core due to the voltage induced in the core material by the changing flux. The eddy-current losses can be decreased by laminating the core material to reduce the voltage induced in an eddy-current path and thus the eddy-current value.

Core losses are caused by nonlinear effects in the magnetic circuit formed by the transformer core. However, they can be approximately included in a linear transformer circuit model by introducing a resistance in parallel with the magnetizing inductance. This improves the linear model accuracy in modeling terminal-characteristics of the transformer. The resulting circuit model is shown in Fig. 9.66, where I_{e1} is called the excitation current, I_{m1} is called the magnetization current, and R_{c1} is the resistance used to model the core losses.

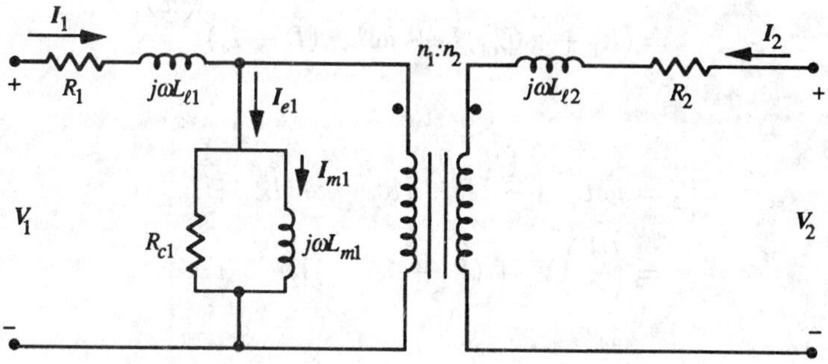

FIGURE 9.66 Linear circuit model for a nonideal transformer—including core losses.

Defining Terms

Transformer: A two-port passive circuit element consisting of two magnetically coupled coils that are not conductively connected.
Air-core transformer: A transformer with a nonmagnetic core.
Source: Signal generator that supplies energy to a network.
Load: Device that converts electrical energy supplied to a useful output. Modeled as an impedance.
Magnetic flux: A magnetic field descriptor. Thought of as lines indicating the direction of force that the magnetic field produces on a moving charge. The density of the lines indicates the force.
Primary winding: Transformer input coil.
Secondary winding: Transformer output coil.
Leakage inductance: Magnetic flux that links only one coil.
Magnetizing inductance: Magnetic flux that links both coils.

References

[1] D. R. Cunningham and J. A. Stuller, *Basic Circuit Analysis*, Boston: Houghton Mifflin, 1991.

[2] A. E. Fitzgerald, C. Kingsley, and S. D. Umans, *Electric Machinery*, 4th ed., New York: McGraw-Hill, 1983.

[3] W. H. Hayt, Jr., and J. E. Kemmerly, *Engineering Circuit Analysis*, 4th ed., New York: McGraw-Hill, 1986.

[4] P. Horowitz and W. Hill, *The Art of Electronics*, 2nd ed., Cambridge: Cambridge, Univ., 1989.

[5] J. G. Kassakian, M. F. Schlecht, and G. C. Verghese, *Principles of Power Electronics*, Reading, MA: Addison-Wesley, 1991.

[6] G. McPherson and R. D. Laramore, *An Introduction to Electrical Machines and Transformers*, 2nd ed., New York: Wiley, 1990.

[7] J. W. Nillson, *Electric Circuits*, 3rd ed., Reading, MA: Addison-Wesley, 1990.

[8] D. L. Schilling and C. Belove, *Electronic Circuits*, 3rd ed., New York: McGraw-Hill, 1989.

urther Information

More detailed developments of linear transformer circuit models can be found in most circuit analysis texts. Two example texts are [1] and [3].

Power transformers and their models are presented in considerable detail in electrical machinery texts such as [6]. These presentations also extend to multiple-coil, three-phase transformers, and tapped transformers. Included in most electrical machinery texts are in-depth discussions of hysteresis and eddy currents and of methods for reducing losses due to these effects. Also included are procedures for measuring the parameters for transformer circuit models.

The use of transformers in electronic circuits is considered in electronic circuit design texts and reference books such as [8] and [4]. Amplifier circuit models that include transformer circuit models are developed in some texts for tuned transformer coupled amplifier stages and transformer coupled load impedances. It is shown how these models can be used to determine amplifier frequency response and power transfer to the load. Other uses indicated for transformers in electronic circuits occur in power supplies and isolation amplifiers.

9.5 Semiconductor Diode

B. M. Wilamowski

Semiconductor diodes are made out of p–n semiconductor junctions. Nonlinear current–voltage characteristics of such junctions are used to rectify and shape electrical signals. Exponential current–voltage characteristics are sometimes used to build logarithmic amplifiers. The variations of junction capacitances with applied voltages are used to tune high-frequency electronic circuits. The semiconductor p–n junction illuminated by light will generate a voltage on its terminals. Such a diode is known as a solar battery. Also, the reverse diode current is proportional to the light intensity at the junction. This phenomenon is used in photodiodes. If a diode is biased in the forward direction, it can generate a light. In order to obtain high emission efficiency the light emitting diode (LED) should be made out of a semiconductor material with a direct energy band structure. This way electrons and holes can recombine directly between valence and conduction bands. Typically, LED's are fabricated using various compositions of $Ga_yAl_{1-y}As_xP_{1-x}$. The wavelength of generated light is inversely proportional to the potential gap of a junction material. When a light intensity is enhanced by additional micromirrors, then laser action occurs. The silicon diodes are not emitting light because the silicon has an *indirect* band structure and the probability of direct band-to-band recombination is very small.

When both sides of the junction are very heavily doped, then for small forward-biasing voltages (0.1–0.3 V) a large tunneling current may occur. For larger forward voltages (0.4–0.5 V) this tunneling current vanishes. This way the current–voltage characteristic has a negative resistance region somewhere between from 0.2 to 0.4 V [Fig. 9.68(d)]. Germanium and other than silicon semiconductors are used to fabricate tunnel diodes. The backward diode has slightly lower impurity concentrations than the tunnel diode and the tunneling current in the forward direction does not occur [Fig. 9.68(e)]. The backward diode is characterized by very sharp knee near zero voltage, and it is used for detection (rectifications) of signals with very small magnitude.

Diodes with high breakdown voltage have a p-i-n structure with an impurity profile shown in Fig. 9.67(d). Similar p-i-n structure is also used in microwave circuits as a switch or as an attenuating resistor. For reverse biasing such a microwave p-i-n diode represents an open circuit with a small parasitic junction capacitance. In the forward direction this diode operates as a resistor whose conductance is proportional to the biasing current. At very high frequencies electrons and holes will oscillate rather than flow. Therefore, the microwave p-i-n diode exhibits linear characteristics even for large modulating voltages. Another interesting "diode" structure has the impurity profile shown in Fig. 9.67(f). When reverse biasing exceeds the breakdown voltage, this element generates a microwave signal with a frequency related to the electron transient time through structure. Such a diode is known as an IMPATT (IMPact Avalanche Transit Time) diode.

The switching time of a p–n junction from the forward to the reverse direction is limited by the storage time of minority carriers injected into the vicinity of the junction. Much faster operation is possible in the Schottky diode, where minority carrier injection does not exist. Another advantage of the Schottky diode is that the forward voltage drop is smaller than in the silicon p–n junction. This diode uses the metal–semiconductor contact for its operation. Schottky diodes are characterized by relatively small reverse breakdown voltage, rarely exceeding 30 V.

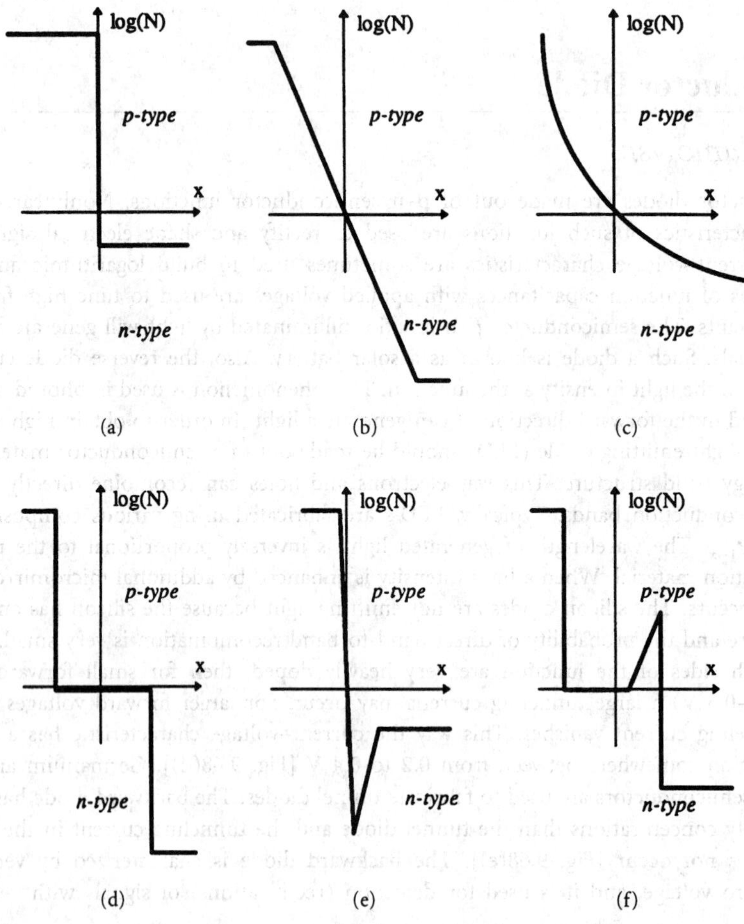

FIGURE 9.67 Impurity profiles for various diodes: (a) step junction, (b) linear junction, (c) diffusion junction, (d) p-i-n junction, (e) p–n$^+$n junction, and (*f*) p-i-p-n junction.

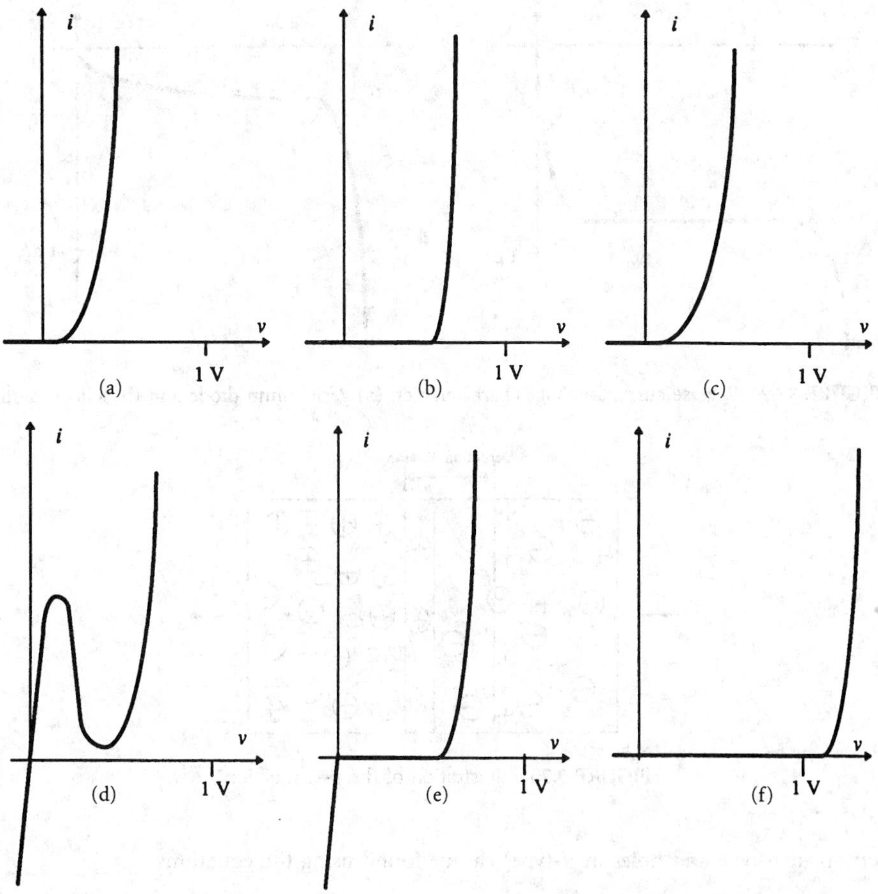

FIGURE 9.68 Forward current–voltage characteristics of various types of diodes: (a) germanium diode, (b) silicon diode, (c) Schottky diode, (d) tunnel diode, (e) backward diode, and (f) LED diode.

Nonlinear Static *I–V* Characteristics

Semiconductor diodes are characterized by nonlinear current–voltage characteristics. Typical *I–V* diode characteristics are shown in Fig. 9.68. In the case of a common silicon diode, the forward-direction current increases exponentially at first, and then is limited by an ohmic resistance of the structure. A very small reverse current at first increases slightly with applied voltage and then starts to multiply near the breakdown voltage (Fig. 9.69). The current at the breakdown is limited by ohmic resistances of the structure.

P–N Junction Equation

The n-type semiconductor material has a positive impurity charge attached to the crystal lattice structure. This fixed positive charge is compensated by free moving electrons with negative charges. Similarly, the p-type semiconductor material has a lattice with a negative charge that is compensated by free moving holes, as Fig. 9.70 shows. The number of majority carriers (electrons in p-type and holes in n-type materials) are approximately equal to the donor or acceptor impurity concentrations, i.e., $n_n = N_D$ and $p_p = N_A$. The number of minority carriers

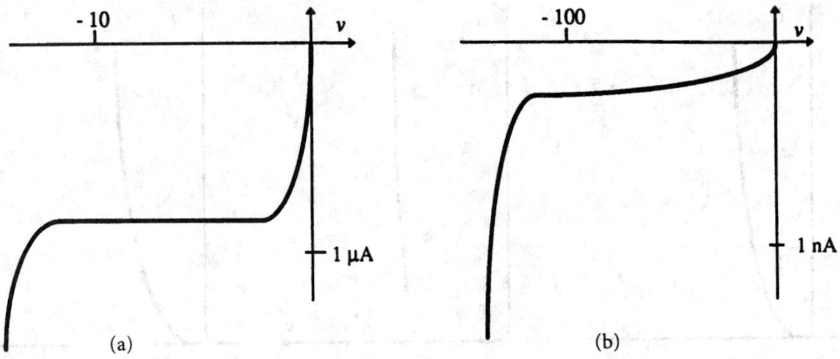

FIGURE 9.69 Reverse current–voltage characteristics: (a) germanium diode and (b) silicon diode.

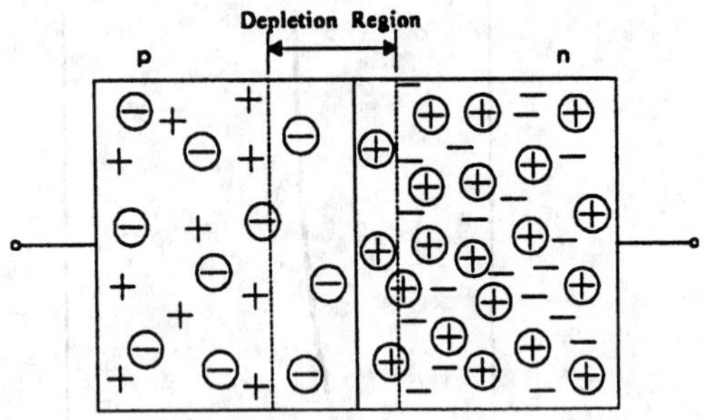

FIGURE 9.70 Illustration of the p–n junction.

(electrons in p-type and holes in n-type) can be found using the equations

$$n_p = \frac{n_i^2}{p_p} \approx \frac{n_i^2}{N_A} \qquad p_n = \frac{n_i^2}{n_n} \approx \frac{n_i^2}{N_D} \tag{9.100}$$

The intrinsic carrier concentration n_i is given by

$$n_i^2 = \xi T^3 \exp\left(-\frac{V_g}{V_T}\right); \qquad V_T = \frac{kT}{q} \tag{9.101}$$

where $V_T = kT/q$ is the thermal potential ($V_T = 25.9$ mV at 300 K), T is absolute temperature in $K, q = 1.6\ 10^{-16}$ C is the electron charge, $k = 8.62\ 10^{-5}$ eV/K is the Boltzmann constant, V_g is potential gap ($V_g = 1.124$ V for silicon), and ξ is a material constant. For silicon, intrinsic concentration n_i is given by

$$n_i = 7.98 \cdot 10^{15} T^{\frac{3}{2}} \exp\left(-\frac{6522}{T}\right) \tag{9.102}$$

For silicon at 300 K, $n_i = 1.5 \cdot 10^{10}$ cm^{-2}.

When a p–n junction is formed, the fixed electrostatic lattice charges form an electrical field at the junction. Electrons are pushed by electrostatic forces deeper into the n-type region and holes into the p-type region, as illustrated in Fig. 9.70. Between n-type and p-type regions there is a depletion layer with a built-in potential which is a function of impurity doping level and intrinsic concentration n_i:

$$V_{pn} = V_T \ln\left(\frac{N_A N_D}{n_i^2}\right) = V_T \ln\left(\frac{n_n p_p}{n_i^2}\right) = V_T \ln\left(\frac{n_n}{n_p}\right) = V_t \ln\left(\frac{p_p}{p_n}\right) \tag{9.103}$$

The junction current as a function of biasing voltage is described by the diode equation:

$$i = I_s\left[\exp\left(\frac{v}{V_T}\right) - 1\right] \tag{9.104}$$

where

$$I_s = Aqn_i^2 V_T\left(\frac{\mu_p}{\int_0^{L_p} n_n\,dx} + \frac{\mu_n}{\int_0^{L_n} p_p\,dx}\right) \tag{9.105}$$

where $n_n \approx N_D$, $p_p \approx N_A$, μ_n and μ_p are mobility of electrons and holes, L_n and L_p are diffusion length for electrons and holes, and A is the device area.

In the case of diodes made of silicon or other semiconductor materials with a high energy gap, the reverse-biasing current cannot be calculated from the diode Eq. (9.104). This is due to the carrier generation-recombination phenomenon. Lattice imperfection and most impurities are acting as generation-recombination centers. Therefore, the more imperfections there are in the structure, the larger the deviation from ideal characteristics.

Forward *I–V* Diode Characteristics

The diode Eq. (9.104) was derived with an assumption that injected carriers are recombining on the other side of the junction. The recombination within the depletion layer was neglected. In real forward-biased diodes, electrons and holes are injected through the depletion region and they may recombine there. The recombination component of the forward-biased diode is given by

$$i_{\text{rec}} = qwA\frac{n_i}{2\tau_0}\exp\left(\frac{v}{2V_T}\right) = I_{r0}\exp\left(\frac{v}{2V_T}\right) \tag{9.106}$$

where w is the depletion layer thickness and τ_0 is the carrier lifetime in the depletion region. The total diode current $i_T = i + i_{\text{rec}}$ where i and i_{rec} are defined by (9.104) and (9.106). The recombination component dominates at low current levels, as Fig. 9.71 illustrates.

Also in very high current levels, the diode Eq (9.104) is not valid. Two phenomena cause this deviation. First, there is always an ohmic resistance that plays an important role for large current values. The second deviation is due to high concentration of injected minority carriers. For very high current levels, the injected minority carrier concentrations may approach, or even become larger than the impurity concentration. An assumption of the quasi-charge neutrality leads to an increase of the majority carrier concentration. Therefore, the effective diode current is lower, as can be seen from (9.105). The high current level in the diode follows the relation

$$i_h = I_{h0}\exp\left(\frac{v}{2V_T}\right) \tag{9.107}$$

Figure 9.71 shows the diode *I–V* characteristics, which include generation-recombination, diffusion, and high current phenomena. For modeling purposes, the forward diode current can be approximated by

$$i_D = I_0\exp\left(\frac{v}{\eta V_T}\right) \tag{9.108}$$

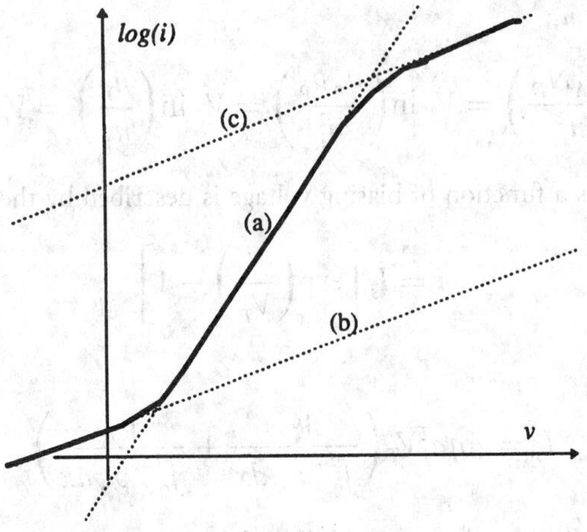

FIGURE 9.71 Current–voltage characteristics of the p–n junction in the forward direction: (a) diffusion current, (b) recombination current, and (c) high-level injection current.

where η has a value between 1.0 and 2.0. Note, that the η coefficient is a function of current, as can be seen in Fig. 9.71. It has a larger value for small and large current regions and it is close to unity in the medium current region.

Reverse *I–V* Characteristics

The reverse leakage current in silicon diodes is mainly caused by the electron–hole generation in the depletion layer. This current is proportional to the number of generation-recombination centers. These centers are formed either by a crystal imperfection or deep impurities, which create energy states near the center of the energy gap. Once the reverse voltage is applied, the size of the depletion region and the number of generation-recombination centers increase. Thus, the leakage current is proportional to the thickness of the depletion layer $w(v)$. For a step-abrupt junction

$$w = \sqrt{\frac{2\epsilon\epsilon_0(V_{pn} - v)}{qN_{\text{eff}}}} \tag{9.109}$$

For other impurity profiles, w can be approximated by

$$w = K(V_{pn} - v)^{\frac{1}{m}} \tag{9.110}$$

The reverse-diode current for small and medium voltages can therefore be approximated by

$$i_{\text{rev}} = Aw(v)\frac{qn_i}{2\tau_0} \tag{9.111}$$

where n_i is given by (9.101) and w by (9.109) or (9.110). The reverse current increases rapidly near the breakdown voltage. This is due to the avalanche multiplication phenomenon. The multiplication factor is often approximated by

$$m = \frac{1}{1 - \left(\dfrac{v}{BV}\right)^m} \tag{9.112}$$

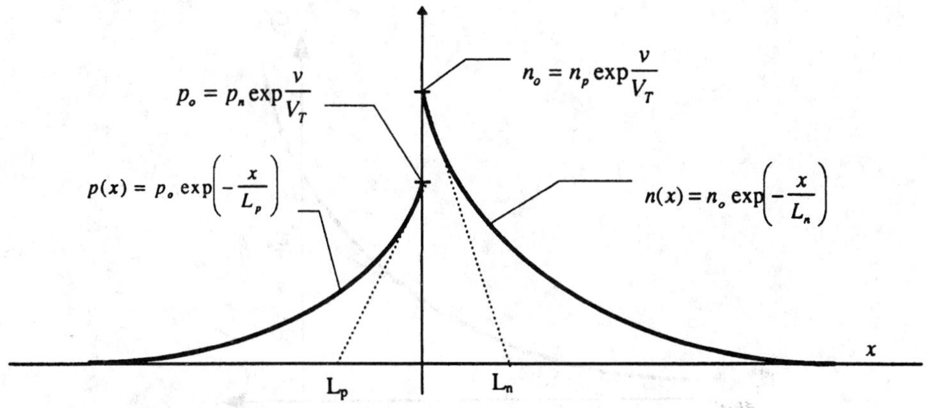

FIGURE 9.72 Minority carrier distribution in the vicinity of the p–n junction biased in the forward direction.

where BV stands for the breakdown voltage and m is an exponent chosen experimentally. Note, that for the reverse biasing both v and BV have negative values and the multiplication factor M reaches an infinite value for $v = BV$.

Diode Capacitances

Two types of capacitances are associated with a diode junction. One capacitance, known as diffusion capacitance, is proportional to the diode current. This capacitance exists only for the forward-biased condition and has the dominant effect there. The second capacitance, known as the depletion capacitance, is a weak function of the applied voltage.

Diffusion Capacitance

In a forward-biased diode, minority carriers are injected into opposite sides of the junction. Those minority carriers diffuse from the junction and recombine with the majority carriers. Fig. 9.72 shows the distribution of minority carriers in the vicinity of the junction of uniformly doped n-type and p-type regions. The electron charge stored in the p-region corresponds to the area under the curve, and it is equal to $Q_n = qn_0L_n$. Similarly, the charge of stored holes $Q_p = qp_0L_p$. The storage charge can also be expressed as $Q_n = I_n\tau_n$ and $Q_p = I_p\tau_p$, where I_n and I_p are electron and hole currents at the junction and τ_n and τ_p are the lifetimes for minority carriers. Assuming $\tau = \tau_n = \tau_p$ and knowing that $I = I_p + I_n$ the total storage charge at the junction is $Q = I\tau$. The diffusion capacitance can be then computed as

$$C_{\text{dif}} = \frac{dQ}{dv} = \frac{d}{dv}\left[\tau I_0 \exp\left(\frac{v}{\eta V_T}\right)\right] = \frac{\tau I_B}{\eta V_T} \tag{9.113}$$

As one can see, the diffusion capacitance C_{dif} is proportional to the storage time τ and to the diode biasing current I_B. Note that the diffusion capacitance does not depend on the junction area, only on the diode current. The diffusion capacitances may have very large values. For example, for 100 mA current and $\tau = 1$ μs the junction diffusion capacitance is about 4 μF. Fortunately, this diffusion capacitance is connected in parallel to the small-signal junction resistance $r = \eta V_T/I_B$, and the time constant $r\,C_{\text{dif}}$ is equal to the storage time τ.

Depletion Capacitance

The reversed-biased diode looks like a capacitor with two "plates" formed of p-type and n-type regions and a dielectric layer (depletion region) between them. The capacitance of a

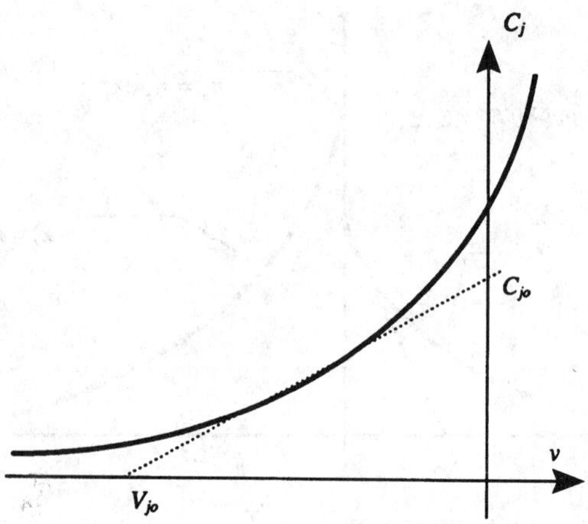

FIGURE 9.73 Capacitance–voltage characteristics for reverse-biased junction.

reversed-biased junction can then be written as

$$C_{dep} = A\frac{\epsilon}{w} \tag{9.114}$$

where A is a junction area, ϵ is the dielectric permittivity of semiconductor material, and w is the thickness of the depletion layer. The depletion layer thickness w is a weak function of the applied reverse-biasing voltage. In the simplest case, with a step-abrupt junction, the depletion capacitance is

$$C_j = \sqrt{\frac{qN_{eff}\epsilon\epsilon_0}{2(V_{pn} - v)}}; \quad \frac{1}{N_{eff}} = \frac{1}{N_D} + \frac{1}{N_A} \tag{9.115}$$

The steepest capacitance–voltage characteristics are in $p^+ - i - p - n^+$ diodes with the impurity profiles shown in Fig. 9.67(f). In general, for various impurity profiles at the junction, the depletion capacitance C_j can be approximated by

$$C_j = C_{j0}\left(1 - \frac{v}{V_{pn}}\right)^{\frac{1}{m}} \tag{9.116}$$

or using linear approximation as shown in Fig. 9.73:

$$C_j = C_{j0}\left(1 - \frac{v}{V_{j0}}\right) \tag{9.117}$$

Diode as a Switch

The switching time of the p–n junction is limited mainly by the storage charge of injected minority carriers into the vicinity of the junction (electrons injected in p-type region and holes injected in n-type region). When a diode is switched from the forward to the reverse direction these carriers may move freely through the junction. Some of the minority carriers recombine with time. Others are moved away to the other side of the junction. The diode cannot recover

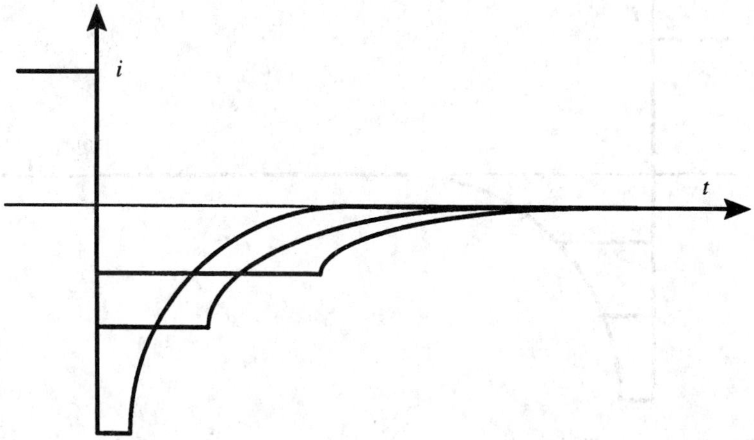

FIGURE 9.74 Currents in diode with large minority carrier lifetimes after switching from the forward to the reverse direction.

its blocking capability as long as a large number of the minority carriers exist and can flow through the junction. An example of the current–time characteristics of a diode switching from the forward to the reverse direction is shown in Fig. 9.74. A few characteristics that are shown in the figure are for the same forward current and different reverse currents. Just after switching, these reverse currents are limited only by external circuitry. In this example, shown in Fig. 9.74, most of the minority carriers are moved to the other side of the junction by the reverse current, and the recombination mechanism is negligible. Note that the larger the reverse current flowing after switching, the shorter the time required to recover the blocking capability. This type of behavior is typical for commonly used high-voltage diodes.

In order to shorten the switching time, diodes sometimes are doped with gold or other deep-level impurities to create more generation centers and to increase the carrier recombination. This way, the minority carrier lifetimes of such switching diodes are significantly reduced. The switching time is significantly shorter, but it is almost independent of the reverse-diode current after switching, as Fig. 9.75 shows. This method of artificially increasing recombination rates has some severe disadvantages. Such switching diodes are characterized by very large reverse leakage currents and small breakdown voltages.

The best switching diodes utilize metal–semiconductor contacts. They are known as Schottky diodes. In such diodes there is no minority carrier injection phenomenon; therefore, these diodes recover the blocking capability instantaneously. Schottky diodes are also characterized by a relatively small (0.2–0.3 V) voltage drop in the forward direction. However, their reverse leakage current is larger, and the breakdown voltage rarely exceeds 20–30 V. Lowering the impurity concentration in the semiconductor material leads to slightly larger breakdown voltages, but at the same time, the series diode resistances increase significantly.

Temperature Properties

Both forward and reverse diode characteristics are temperature dependent. These temperature properties are very important for correct circuit design. The temperature properties of the diode can be used to compensate for the thermal effects of electronic circuits. Diodes can be used also as accurate temperature sensors. The major temperature effect in a diode is caused by the strong temperature dependence of the intrinsic concentration n_i [(9.101) and (9.102)] and by the exponential temperature relationship of the diode Eq (9.104). By combining (9.101) and (9.104) and assuming the temperature dependence of carrier mobilities, the voltage drop on the

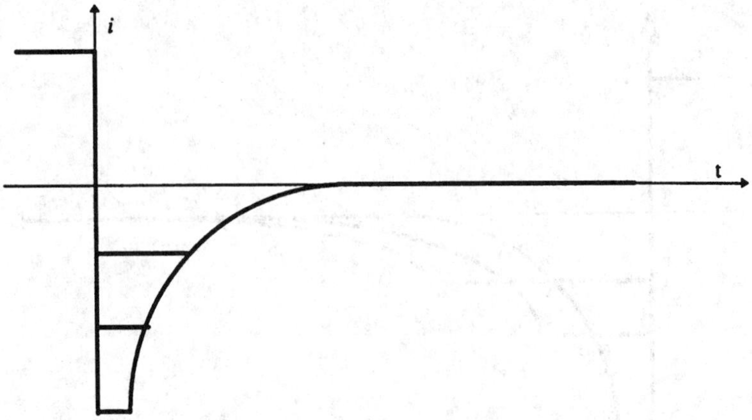

FIGURE 9.75 Currents in diode with small minority carrier lifetimes after switching from the forward to the reverse direction.

forward-biased diode can be written as

$$v = \eta\left[V_T \ln\left(\frac{i}{\xi T^a}\right) + V_g\right] \tag{9.118}$$

or diode current

$$i = I_0\left(\frac{T}{T_0}\right)^\alpha \exp\left(\frac{(v/\eta) - V_g}{V_{T_0}}\frac{T_0}{T}\right) \tag{9.119}$$

where V_g is the potential gap in semiconductor material, V_g = 1.124 V for silicon and $V_g = 1.424$ V for GaAs, and α is a material coefficient ranging between 2.5 and 4.0.

The temperature dependence of the diode voltage drop dv/dT can be obtained by calculating the derivative of (9.118)

$$\frac{dv}{dT} = \frac{v - \eta(V_g + \alpha V_T)}{T} \tag{9.120}$$

For example, in the case of the silicon diode with a 0.6 voltage drop and assuming $\eta = 1.1$, $\alpha = 3.0$, and $T = 300$ K, the $dV/dT = 1.87$ mV/°C.

The reverse-diode current is a very strong function of the temperature. For diodes made of semiconductor material with a small potential gap, such as germanium, the diffusion component dominates. In this case, the reverse current is proportional to

$$i_{rev} \propto T^\alpha \exp\left(-\frac{qV_g}{kT}\right) \tag{9.121}$$

For diodes made of silicon and semiconductors with a higher energy gap, the recombination is the dominant mechanism. In this case, reverse leakage current is proportional to

$$i_{rev} \propto T^\alpha \exp\left(-\frac{qV_g}{2kT}\right) \tag{9.122}$$

Using (9.122) one may calculate that for silicon diodes at room temperatures, the reverse leakage current doubles for about every 10°C.

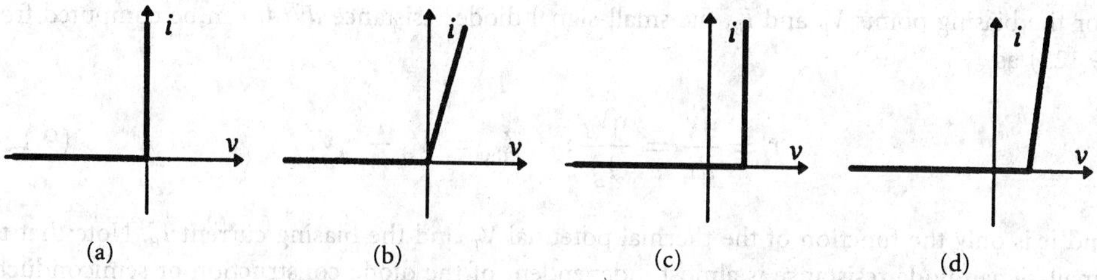

FIGURE 9.76 Various ways to linearize diode characteristics.

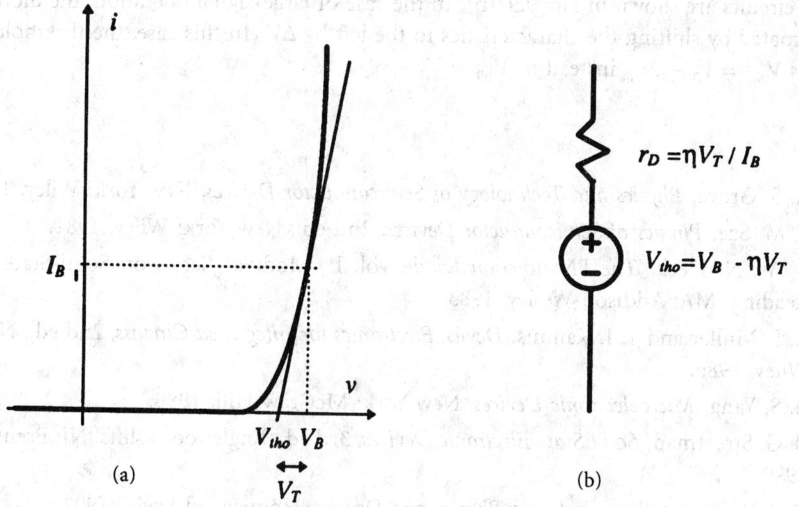

FIGURE 9.77 Linearization of the diode: (a) diode characteristics and (b) equivalent diagram.

The breakdown voltage is also temperature dependent. The tunneling effect dominates in diodes with small breakdown voltages. This effect is often known in literature as the Zener breakdown. In such diodes the breakdown voltage decreases with the temperature. The avalanche breakdown dominates in diodes with large breakdown voltages. When the avalanche mechanism prevails, then the breakdown voltage increases 0.06 percent to 0.1 percent /°C. For medium-range breakdown voltages, one phenomenon compensates the other, and temperature-independent breakdown voltage can be observed. This zero temperature coefficient exists for diodes with breakdown voltages equal to about 5 V_g. In the case of the silicon diode, this breakdown voltage, with a zero temperature coefficient, is equal to about 5.6 V.

Piecewise Linear Model

Nonlinear diode characteristics are often approximated by a piecewise linear model. There are a few possible approaches to linearize the diode characteristics, as Fig. 9.76 shows. The parameters of the most accurate linearized diode model are shown in Fig. 9.77(a), and the linearized diode equivalent circuit is shown in Fig. 9.77(b).

The modified diode Eq. (9.108) also can be written as

$$v = \eta V_T \ln\left(\frac{i}{I_0}\right) \tag{9.123}$$

For the biasing points V_B and I_B, the small-signal diode resistance dv/di can be computed from (9.123) as

$$r = \frac{dv}{di} = \frac{\eta V_T}{I_B}; \quad V_{\text{th0}} = V_B - \eta V_T \tag{9.124}$$

and it is only the function of the thermal potential V_T and the biasing current I_B. Note that the small-signal diode resistance is almost independent of the diode construction or semiconductor material used. If one requires that this linearized diode have I_B current for V_B voltage, then the piecewise diode characteristics should be as in Fig. 9.97(a). The equivalent Thevenin and Norton circuits are shown in Fig. 9.97(b). In the case of large-signal operation, the diode can be approximated by shifting the characteristics to the left by ΔV. In this case, the threshold voltage becomes $V_{\text{th0}} = V_B - 2V_T$ instead of $V_{\text{th0}} = V_B - \eta^{V_T}$.

References

[1] A. S. Grove, *Physics and Technology of Semiconductor Devices*, New York: Wiley, 1967.

[2] S. M. Sze, *Physics of Semiconductor Devices*, 2nd ed., New York: Wiley, 1981.

[3] G. W. Neudeck, *The PN Junction Diode*, vol. II, Modular Series on Solid-State Devices, Reading, MA: Addison-Wesley, 1983.

[4] R. S. Muller and T. I. Kamins, *Device Electronics for Integrated Circuits*, 2nd ed., New York: Wiley, 1986.

[5] E. S. Yang, *Microelectronic Devices*, New York: McGraw-Hill, 1988.

[6] B. G. Streetman, *Solid State Electronic Devices*, 3rd ed., Englewood Cliffs, NJ: Prentice-Hall, 1990.

[7] D. A. Neamen, *Semiconductor Physics and Devices*, Homewood Irwin, 1992.

10

Controlled Circuit Elements

vin W. Greeneich
na State University

ıes F. Delansky
sylvania State University

).1 Controlled Sources

Edwin W. Greeneich

Introduction

Controlled sources generate a voltage or current whose value depends on, or is controlled by, a voltage or current that exists at some other point in the circuit. Four such sources exist: 1) Voltage-Controlled Current Source (VCCS), 2) Voltage-Controlled Voltage Source (VCVS), 3) Current-Controlled Current Source (CCCS), and 4) Current-Controlled Voltage Source (CCVS). In an ideal controlled source the generated voltage or current does not vary with the load to which it is connected; this implies a zero output impedance for a voltage source and an infinite output impedance for a current source. In practice, actual controlled sources have a finite output impedance that causes the generated source to vary somewhat with the load. Circuit representations of the four ideal controlled sources are given in Fig. 10.1. The input terminals on the left represent the controlling voltage or current, and the output terminals on the right represent the controlled voltage or current; the value of the controlled source is proportional to the controlling input through the constants g, μ, β, and r.

Voltage-Controlled Current Source

A voltage-controlled current source produces an output current that is proportional to an input control voltage. The idealized small-signal low-frequency behavior of a field-effect transistor (FET) can be characterized by a VCCS, as illustrated in the equivalent circuit for an n-channel Metal–Oxide–Semiconductor Field-Effect Transistor (MOSFET) shown in Fig. 10.2. In the circuit model, the small-signal drain current i_d of the transistor is proportional to the small-signal gate-to-source voltage v_{gs} through the transconductance parameter g_m.

3-8341-2/95/$0.00 + $.50

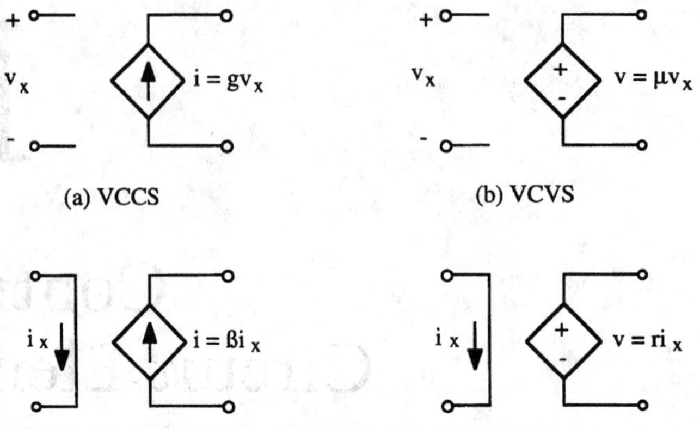

FIGURE 10.1 Circuit representations of ideal controlled sources.

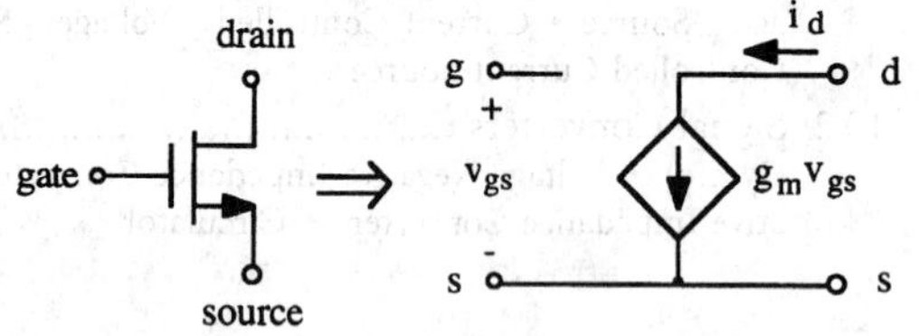

FIGURE 10.2 Small-signal equivalent circuit of the MOSFET.

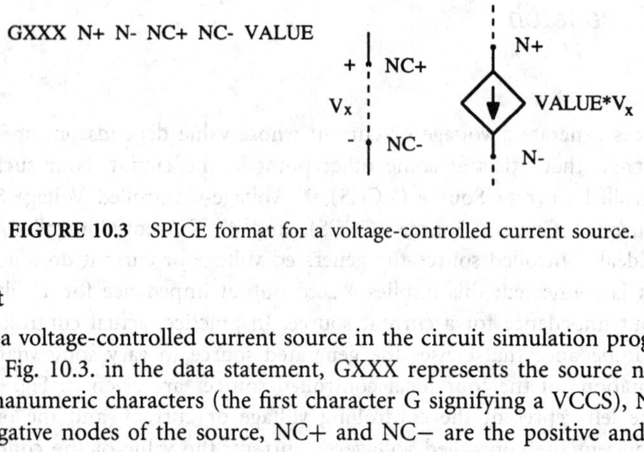

FIGURE 10.3 SPICE format for a voltage-controlled current source.

SPICE Format

The format for a voltage-controlled current source in the circuit simulation program SPICE [1] is illustrated in Fig. 10.3. in the data statement, GXXX represents the source name containing up to eight alphanumeric characters (the first character G signifying a VCCS), N+ and N− are positive and negative nodes of the source, NC+ and NC− are the positive and negative nodes between which the controlling voltage is measured and VALUE is the multiplicative constant giving the value of the current source. The convention used is that *positive* current flows from the N+ node through the source to the N− node.

An example of a circuit using a voltage-controlled current source is shown in Fig. 10.4. The SPICE data specification for the source is GS1 8 6 3 6 3.4.

Circuit Implementation

A circuit implementing a voltage-controlled current source is shown in Fig. 10.5. In the circuit, Q_1 and Q_2 form a current mirror which, due to their equal base-emitter voltages, have the same collector currents. These currents are reflected in the collector current of Q_3 and Q_4, causing them to have the same base-emitter voltages; the voltage at the emitter of Q_4 is thus equal to the input control voltage V_x. The emitter current of Q_4 is thus equal to V_x/R_1, which is then

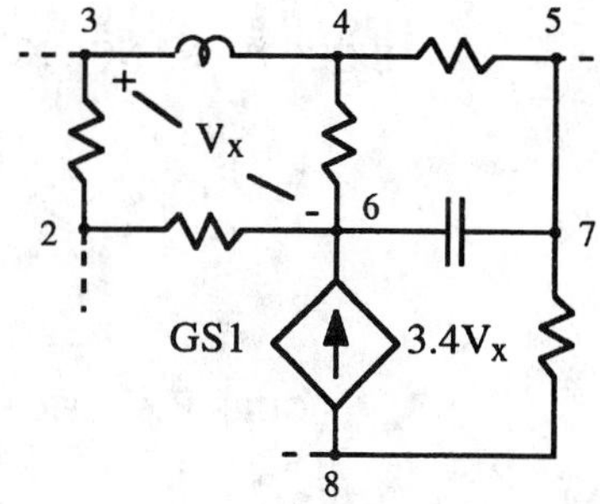

FIGURE 10.4 Example showing a portion of the circuit using a VCCS.

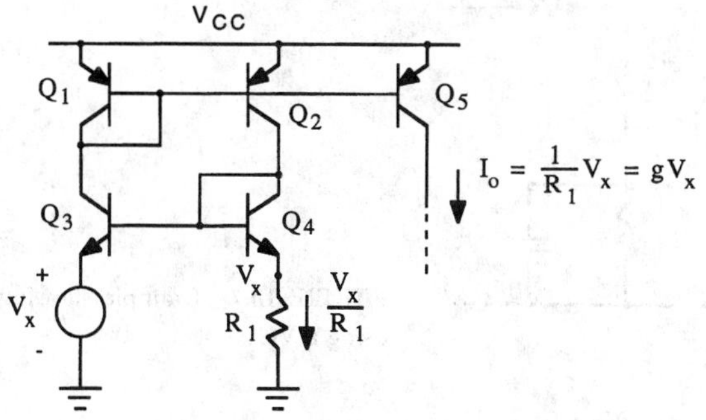

FIGURE 10.5 Circuit that implements a voltage-controlled current source.

(neglecting base currents) equal to the collector current of Q_2. This current is mirrored by Q_5 giving the voltage-controlled output current I_0. This circuit derives from a general form of this configuration called a *current conveyor* [2]. With this circuit (as with all practical current sources) the output current is not totally independent of the output voltage across the source, but rather, I_0 shows a slight increase with increasing voltage. This is due to the finite output resistance of the source; in the circuit of Fig. 10.5, this resistance is equal to the collector-to-emitter resistance of transistor Q_5. For an integrated circuit transistor this resistance may be of the order of 50 kΩ or so.

Voltage-Controlled Voltage Source

A voltage-controlled voltage source produces an output voltage that is proportional to an input control voltage. A voltage amplifier can be thought of as a VCVS; the output voltage is equal to the input voltage multiplied by the voltage gain μ of the amplifier.

SPICE Format

The SPICE format for a voltage-controlled voltage source is given in Fig. 10.6. The first character *E* in the source name signifies a VCVS. The output voltage of the source is given by the product of the VALUE constant and the control voltage V_x.

An example of the circuit using a voltage-controlled voltage source is shown in Fig. 10.7. The SPICE data specification for the source is E41 7 2 5 3 0.8.

Circuit Implementation

The VCCS circuit of Fig. 10.5 can be modified to produce a voltage-controlled voltage source. In Fig. 10.8 the current sensing resistor R_2 develops a voltage drop equal to $V_x R_2/R_1$, which is buffered by a unity-gain stage to reduce loading effects on R_2. The output voltage V_0 is thus proportional to the input control voltage V_x. The buffer should have a low output impedance to minimize variations in the output voltage, V_0, with load current drawn by the source.

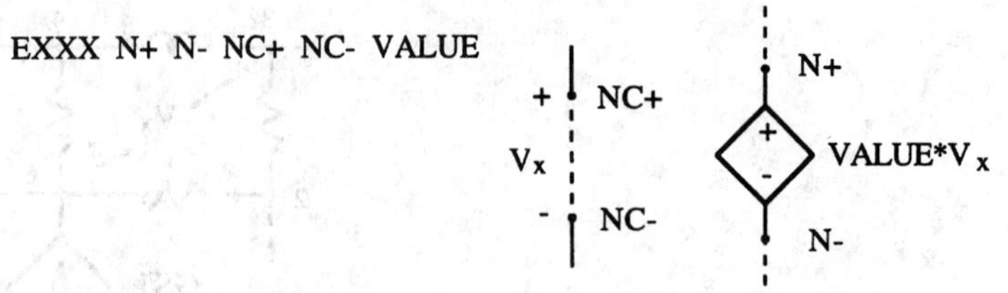

FIGURE 10.6 SPICE format for a voltage-controlled voltage source.

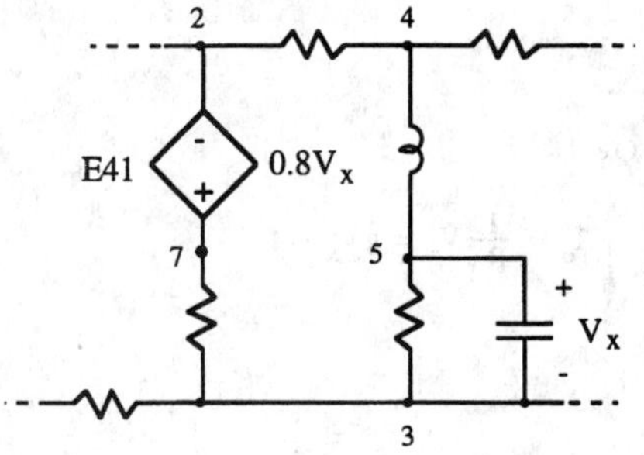

FIGURE 10.7 Example showing a portion of a circuit using a VCVS.

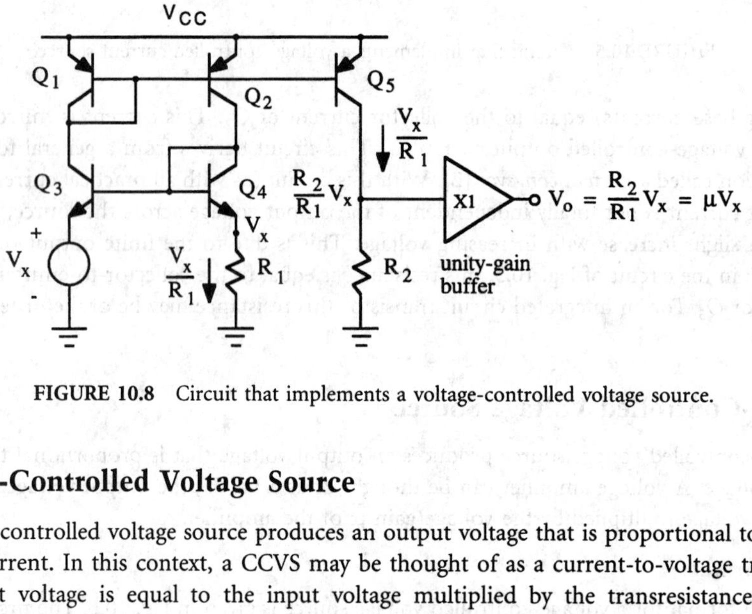

FIGURE 10.8 Circuit that implements a voltage-controlled voltage source.

Current-Controlled Voltage Source

A current-controlled voltage source produces an output voltage that is proportional to an input control current. In this context, a CCVS may be thought of as a current-to-voltage transducer; the output voltage is equal to the input voltage multiplied by the transresistance r of the transducer.

SPICE Format

The SPICE format for a current-controlled voltage source is given in Fig. 10.9. The first character H in the source name signifies a CCVS. There are no ammeters in SPICE, so currents are measured though voltage sources. VNAME is the voltage source through which the control current I_x is measured. The output voltage of the source is given by the product of VALUE and I_x. If the point in the circuit at which the control current is to be measured does not contain a voltage source, a test voltage source of zero value can be inserted in the circuit.

An example of the circuit using a current-controlled voltage source is shown in Fig. 10.10. The data statement for the source is HVS2 3 2 VTEST 0.2. VTEST is a zero-valued voltage

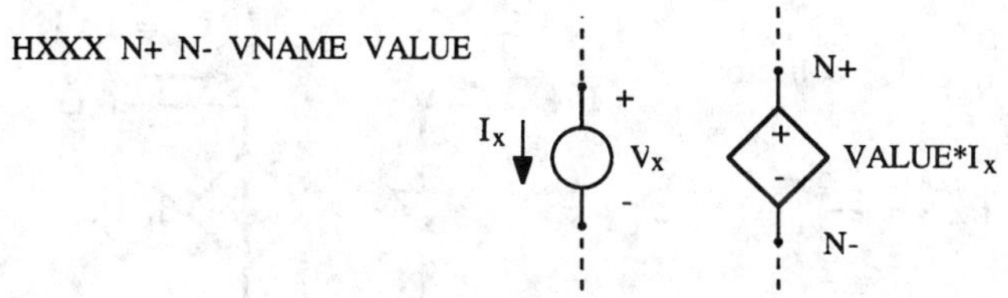

FIGURE 10.9 SPICE format for a current-controlled voltage source.

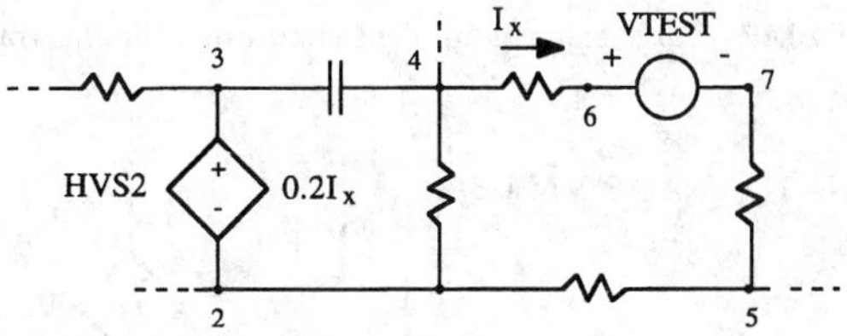

FIGURE 10.10 Example showing a portion of a circuit using a CCVS.

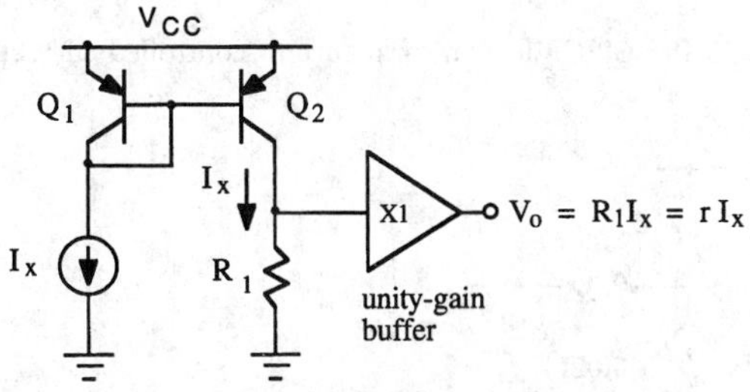

FIGURE 10.11 Circuit that implements a current-controlled voltage source.

source inserted into the circuit to measure the control current I_x. Its data statement is VTEST 6 7 0.

Circuit Implementation

A simple circuit implementing a current-controlled voltage source is shown in Fig. 10.11. In the current mirror comprising transistors Q_1 and Q_2, the collector current of Q_2 is equal (neglecting base currents) to the control current I_x. The voltage across R_1 is then equal to $R_1 I_x$, which after the buffer, is the current-controlled output voltage.

Current-Controlled Current Source

A current-controlled current source produces an output current that is proportional to the input control current. The idealized large-signal (and small-signal as well) low-frequency behavior of a bipolar transistor can be characterized by a CCCS, as illustrated in the equivalent circuit for an NPN transistor in Fig. 10.12. In the circuit model, the collector current I_C is proportional to the base current I_B through the current gain parameter β_F.

SPICE Format

The SPICE format for a current-controlled current source is given in Fig. 10.13. The first character *F* in the source name signifies a CCCS. The output current is given by the product of the VALUE constant and the control current I_x. As with the CCVS, the controlling current is measured through an independent voltage source.

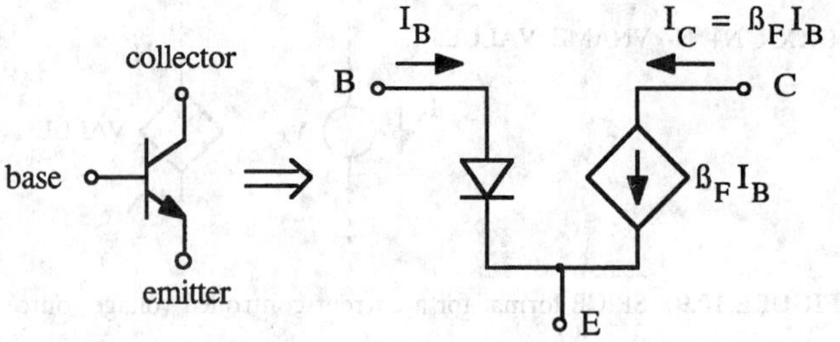

FIGURE 10.12 Large-signal equivalent circuit of a bipolar transistor.

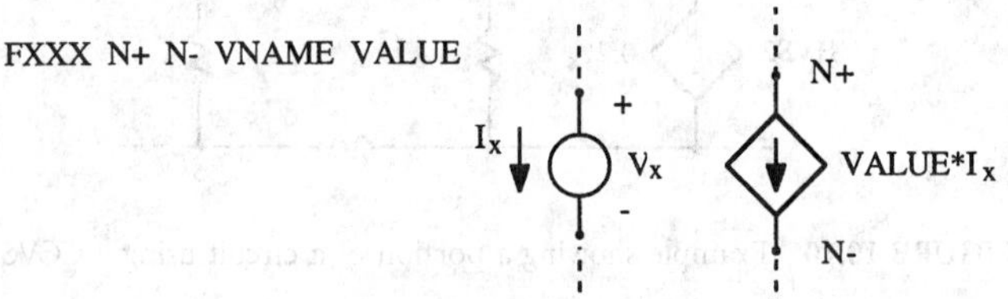

FIGURE 10.13 SPICE format for a current-controlled current source.

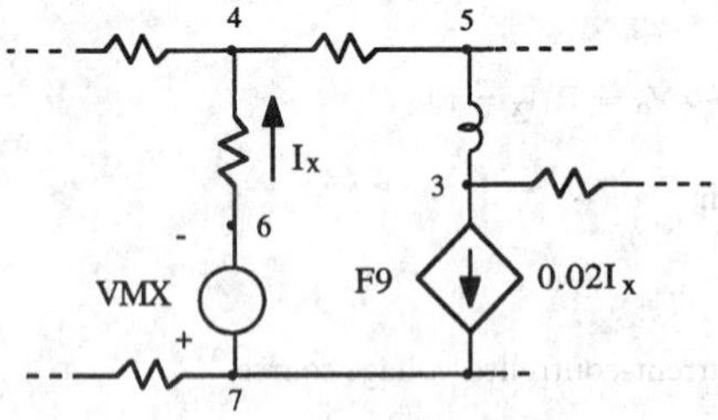

FIGURE 10.14 Example showing a portion of a circuit using a CCCS.

An example of a circuit using a current-controlled current source is shown in Fig. 10.14. The data statement for the source is F9 3 7 VMX 20E-3. VMX is a zero-valued voltage source inserted into the circuit to measure the control current I_x. Its data statement is VMX 7 6 0.

Circuit Implementation

The CCVS circuit in Fig. 10.11 can be combined with the VCCS circuit in Fig. 10.5 to produce a current-controlled current source, as illustrated in Fig. 10.15. The output voltage of the CCVS

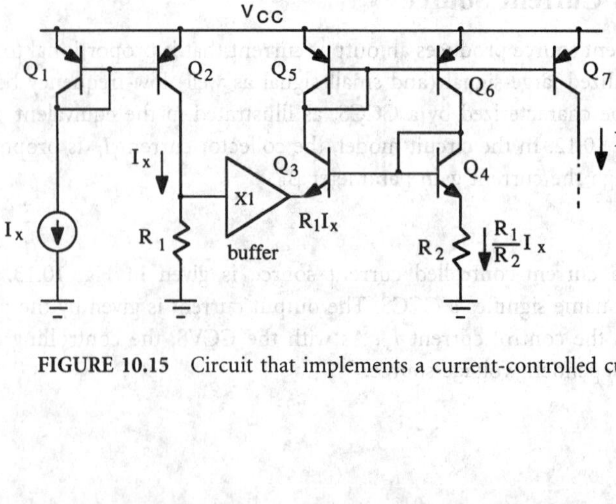

FIGURE 10.15 Circuit that implements a current-controlled current source.

stage is equal to $R_1 I_x$, which, applied at the emitter of Q_3, is the input voltage of the VCCS stage; the output current is equal to $(R_1/R_2)I_x$, and thus proportional to the input control current.

eferences

L. W. Nagel, "SPICE2: A computer program to simulate semiconductor circuits," Electron. Res. Lab., Rep. ERL-M520, Univ. California, Berkeley, 1975.

C. Ioumazou, F. J. Lidgey, and D. G. Haigh, Eds, *Analogue IC Design: the Current-Mode Approach,* London: Peter Peregrinus Ltd., 1990.

0.2 Signal Converters

James F. Delansky

An accessible terminal pair of a network, regarded as a single entity to which an independent 2-terminal signal generator (source or input) is to be connected, is called a *port* of the network. An equivalent view of a port of a network is an accessible terminal pair such that the current entering one of the terminals of the pair is the same current leaving the other terminal of the pair. Thus, an accessible terminal pair of a network is a port when this terminal pair is terminated by (connected to) a 1-port network. For a given port of a network, the port voltage (the voltage drop between the two terminals) and the port current (the current entering one of the terminals) are to be associated as follows: the positive sign for the voltage (drop) is always assumed at the terminal at which the current (positive charges) enters the network. A 1-port, 2-port and general n-port are shown in Fig. 10.16.

In linear network theory, a class of n-ports (particularly $n = 2$), known as signal converters, has become salient as crucial building blocks. These 2-port signal converters include the various versions of the "transformer," "controlled source," and "operational amplifier" found elsewhere in Section II. This part of Chapter 10 will introduce the 2-port signal converters known as the "gyrator" and "negative impedance converter." As extensions of these, the $(n \geq 2)$-port "circulator" is developed and, as special cases, the 1-ports known as "nullator" and "norator" (properly viewed as degenerate cases) and the 2-port known as "nullor" are given.

One of the most general external descriptions of a linear n-port network is the "scattering" matrix. In many cases, a linear n-port network may be externally described by one or more of the "open-circuit impedance," "short-circuit admittance," "hybrid," "inverse hybrid," "chain (transmission)," or "inverse chain (inverse transmission)" matrix. For details of these n-port descriptions, see [1]. As necessary or convenient, any of the above n-port descriptions will be utilized in this part of Chapter 10.

Gyrator

The concept of reciprocity satisfied by a linear n-port network ($n = 2$, in particular) will be useful in the present context.

Definition: A linear n-port network is said to be a reciprocal linear n-port network if the port voltages and currents satisfy

$$\sum_{k=1}^{n} [v_k^{(1)} i_k^{(2)} - v_k^{(2)} i_k^{(1)}] = 0 \tag{10.1}$$

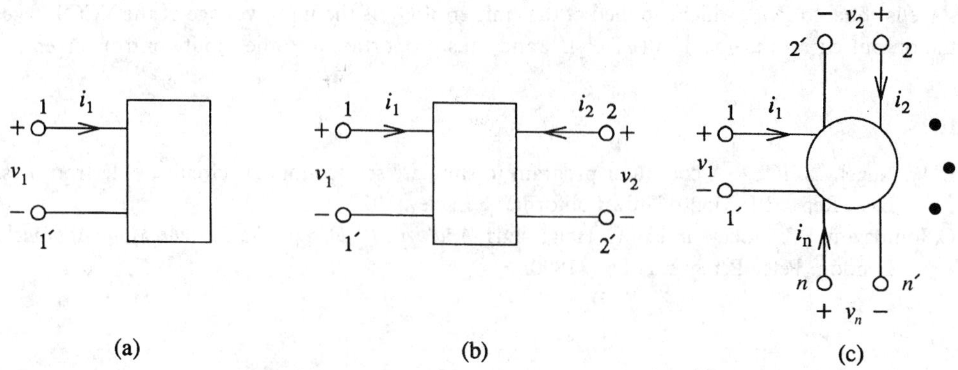

FIGURE 10.16 (a) A 1-port. (b) A 2-port. (c) An n-port.

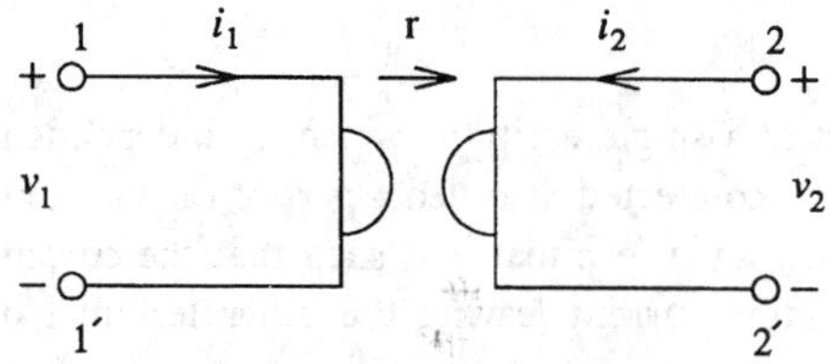

FIGURE 10.17 The ideal 2-port gyrator.

where $v_k^{(1)}, i_k^{(1)}$ and $v_k^{(2)}, i_k^{(2)}$ are any two distinct sets of port voltages and currents that satisfy Kirchhoff's laws for the linear n-port. If (10.1) is not satisfied, the linear n-port is said to be nonreciprocal. [Note: for many useful linear n-port networks, (10.1) can be derived from Tellegen's Theorem in Chapter 7].

The idea of isolating the nonreciprocity of a linear *passive* n-port in a single network building block was first advanced by Tellegen [2]. The linear passive 2-port developed there, christened the *gyrator*, was shown to be necessary and sufficient for this purpose. This (ideal) 2-port gyrator is described by the skew-symmetric open-circuit impedance matrix

$$Z = \begin{bmatrix} z_{11} & z_{12} \\ z_{21} & z_{22} \end{bmatrix} = \begin{bmatrix} 0 & -r \\ r & 0 \end{bmatrix} \tag{10.2}$$

where r (the gyrator transfer impedance parameter) is a real positive number, and is shown in Fig. 10.17.

It is seen that $v_1 = -ri_2$ and $v_2 = ri_1$ (so the signal conversion is clear), and by multiplying these together to obtain $rv_1i_1 = -rv_2i_2$ or $v_1i_1 + v_2i_2 = 0$, then no energy is generated, dissipated, or stored. Thus, the ideal gyrator is a lossless passive nonreciprocal 2-port. Also easily shown is that if port 2 is terminated with the 1-port (driving-point) impedance Z, then the driving-point impedance seen at port 1 is $Z_{11} = -z_{12}z_{21}/Z = r^2/Z$ so that the ideal 2-port gyrator is also an ideal impedance invertor. This leads to another attribute of the ideal 2-port gyrator as a fundamental building block for the synthesis of linear networks. If Z in the above is the impedance of an ideal capacitor C, i.e., $Z = 1/sC$, then

$$Z_{11} = \frac{r^2}{Z} = r^2sC = L_{11}s \tag{10.3}$$

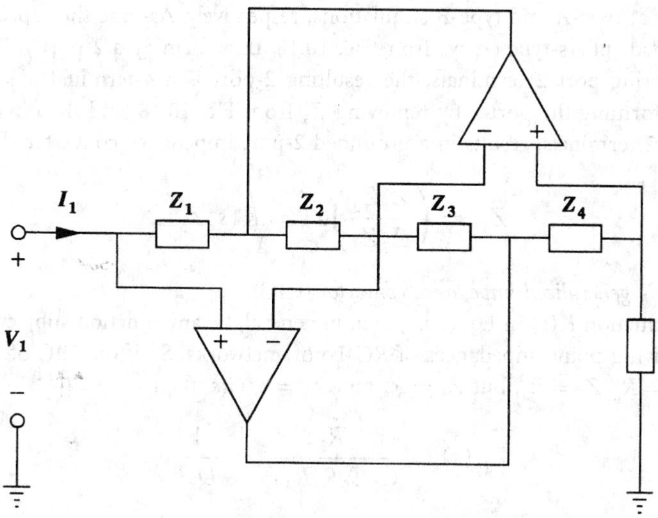

FIGURE 10.18 Antoniou's circuit.

which means that the driving-point impedance Z_{11} is exactly equivalent to the impedance of an ideal inductor $L_{11} = r^2C$. This opens a viable avenue to the inductorless synthesis of linear passive *RLC* networks.

The ideal 2-port gyrator is a 4-terminal passive network (device or element). For network (and physical realization) purposes, however, it is usually implemented with active elements (and thus considered as an "active" building block) and results in a 3-terminal or "common ground" 2-port. For this reason, the simulation of a nongrounded ("floating") inductor in a network is not straightforward. The active realization of the ideal gyrator also results in a nonideal gyrator (i.e., in general $z_{11} \neq 0, z_{22} \neq 0, z_{12} \neq -z_{21}$). These concerns are treated elsewhere [3].

A circuit using two operational amplifiers and *R*'s and a single capacitor for inductor simulation (and thus a gyrator simulation) was proposed by Antoniou [4]. This circuit has subsequently become very widely used. Assuming ideal operational amplifiers and general *RC* 1-ports, this circuit is shown in Fig. 10.18.

Analysis yields

$$Z_{11} = \frac{V_1}{I_1} = \left(\frac{Z_1Z_3}{Z_2Z_4}\right)Z_5 \tag{10.4}$$

For (10.4) to appear as (10.3), there are two choices since either Z_2 and Z_4 must be the impedance of the capacitor. The simplest choices then are

A. $Z_1 = R_1, \quad Z_2 = \dfrac{1}{C_2s}, \quad Z_3 = R_3, \quad Z_4 = R_4, \quad Z_5 = R_5$ results in

$$Z_{11} = \left(\frac{R_1R_3R_5}{R_4}\right)C_2s = L_{11}s$$

B. $Z_1 = R_1, \quad Z_2 = R_2, \quad Z_3 = R_3, \quad Z_4 = \dfrac{1}{C_4s}, \quad Z_5 = R_5$ results in

$$Z_{11} = \left(\frac{R_1R_3R_5}{R_2}\right)C_4s = L_{11}s$$

These are known at type-A and type-B simulations, respectively. Assume the capacitor in A and B above is removed (pliers-type entry) from Fig. 10.18, thus forming a 2-port. With the proper choice of numbering port 2 terminals, the resulting 2-port is a 4-terminal ideal gyrator. On the other hand, forming the port 2 by removing Z_5 from Fig. 10.18, and the obvious choice of numbering port 2 terminals, results in a grounded 2-port impedance converter with

$$Z_{11} = \left(\frac{Z_1 Z_3}{Z_2 Z_4}\right) Z_5 = K(s) Z_5 \tag{10.5}$$

This is Antoniou's *generalized impedance convertor* (GIC).

The rational function $K(s)$ in Eq. (10.5) can, in general, be any function subject to Z_1, Z_2, Z_3, and Z_4 being driving-point impedances of RC 1-port networks. So if the GIC of Eq. (10.5) has $Z_1 = 1/C_1 s$, $Z_2 = R_2$, $Z_3 = R_3$, but $Z_4 = R_4$ and $Z_5 = 1/C_5 s$, then

$$Z_{11}(s) = \frac{R_3}{C_1 C_5 R_2 R_4 s^2} = \frac{1}{Ds^2} \tag{10.6}$$

so that for sinusoidal excitations ($s = j\omega$)

$$Z_{11}(j\omega) = \frac{-1}{D\omega^2} = R_{11}(\omega^2) \tag{10.7}$$

Thus Z_{11} of (10.7) is a frequency-dependent negative resistance (FDNR). Alternate realizations of FDNR are possible. The FDNR element plays an important role in the design of active RC networks.

Voltage Negative Impedance Converter

Suppose a 2-port signal converter such that, with one of its ports terminated with an impedance, the driving-point impedance at the other port is proportional to the negative of the terminating impedance. Such a 2-port is known as a negative impedance converter (NIC) [5]. For a general 2-port terminated at port 2 with the impedance $Z = 1/Y$, the driving-point impedance seen at port 1 is, in terms of the hybrid parameters of the 2-port,

$$Z_{11} = \frac{h_{11} - h_{12} h_{21}}{h_{22} + Y} \tag{10.8}$$

Hence, necessary and sufficient conditions for a 2-port to be an NIC is that $h_{11} = 0$, $h_{22} = 0$, and $h_{12} h_{21} > 0$, so (10.8) becomes

$$Z_{11} = -h_{12} h_{21} Z = -kZ \tag{10.9}$$

where $k > 0$ is called the negative impedance parameter. Now $h_{12} h_{21} > 0$ holds in two cases for h_{12} and h_{21} real:

A. $$h_{12} < 0 \quad \text{and} \quad h_{21} < 0 \tag{10.10}$$

B. $$h_{12} > 0 \quad \text{and} \quad h_{21} > 0 \tag{10.11}$$

Consider an NIC satisfying (10.10). From the hybrid description, this 2-port signal converter has $v_1 = h_{12} v_2$ and since $h_{12} < 0$ this implies a voltage reversal between the ports, while $i_2 = h_{21} i_1$ and since $h_{21} < 0$ this implies the current direction remains the same. For this reason, (10.10) defines a *voltage* inversion *negative impedance converter* (VNIC).

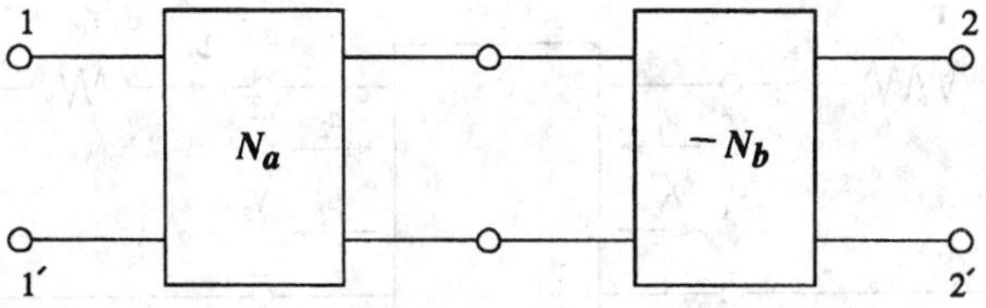

FIGURE 10.19 Cascade of 2-ports designated N_a and $-N_b$.

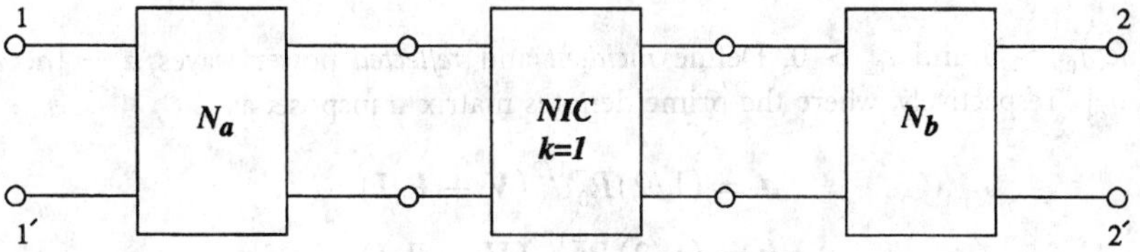

FIGURE 10.20 Equivalent to Fig. 10.19.

Current Negative Impedance Converter

Consider an NIC satisfying (10.11). From the hybrid description, this 2-port signal converter has $v_1 = h_{12}v_2$ and since $h_{12} > 0$, this implies no voltage reversal between the ports, while $i_2 = h_{21}i_1$ and since $h_{21} > 0$ this implies the current directions have reversed. For this reason, (10.11) defines the *current* inversion *negative impedance converter* (INIC).

As with the gyrator, for network (and physical realization) purposes, the NIC is implemented with active devices but for a more fundamental reason. Consider k in (10.9) to be 1 with $h_{12} = h_{21}$ to be either both 1 or -1. Then for VNIC or INIC this implies $v_1i_1 - v_2i_2 = 0$ so the NIC is inherently *active.* Any physical realization of the NIC will result in a nonideal NIC (i.e., in general $h_{11} \neq 0$, $h_{22} \neq 0$, $h_{12}h_{21} \not> 1$). These concerns are treated elsewhere, e.g., [3].

An obvious use of an NIC in active network synthesis is to obtain negative elements, i.e., negative resistor or inductor or capacitor from positive resistor or inductor or capacitor, respectively. Another way an NIC is used in active network synthesis is the partitioning of a network using an NIC. Consider two cascaded 2-ports as shown in Fig. 10.19 to form an overall 2-port.

Suppose in Fig. 10.19, N_a is a 2-port with all positive elements and $-N_b$ is a 2-port with all negative elements. Then it can be shown that the overall 2-port shown in Fig. 10.20 with an NIC of $k = 1$, and with N_b being the same as $-N_b$ in Fig. 10.19, except it is now composed of all positive elements, is equivalent to the overall 2-port shown in Fig. 10.19.

These and other methods of using NIC's in active *RC* synthesis may be found elsewhere, e.g., [3].

Circulator

As discussed in the section on the gyrator, the gyrator is the basic representation of nonreciprocity in linear networks. It is quite natural to believe this property can also be exploited to derive a means of controlling the power flow in n-port ($n \geq 2$) linear networks from the input port to the remaining ports in a prescribed manner. This is indeed the situation and the most important of such n-ports are known as *circulators.*

Circulators are best described in terms of the scattering matrix, since this makes their function very clear. A brief discussion of the scattering description for a 2-port network (extended in a natural way for n-port ($n > 2$) networks) follows. See [6]. Consider the terminated 2-port shown in Fig. 10.21.

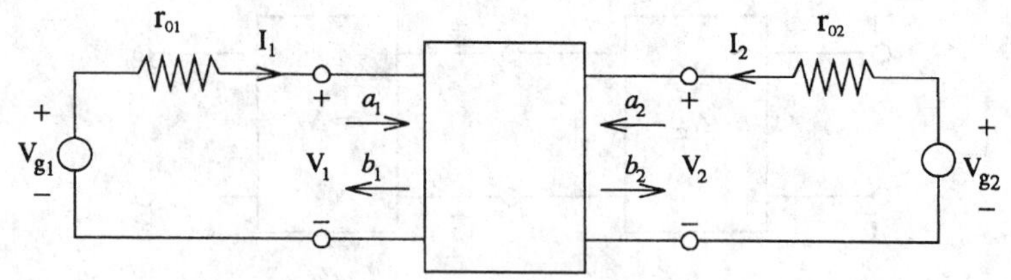

FIGURE 10.21 General terminated 2-port network.

Assume $r_{01} > 0$ and $r_{02} > 0$. Define *incident* and *reflected* power waves $\boldsymbol{a} = [a_1 \; a_2]'$ and $\boldsymbol{b} = [b_1 \; b_2]'$, respectively, where the prime denotes matrix transpose, as

$$\boldsymbol{a} = (1/2)\boldsymbol{R}_0^{-1/2}(\boldsymbol{V} + \boldsymbol{R}_0\boldsymbol{I}) \tag{10.12}$$

$$\boldsymbol{b} = (1/2)\boldsymbol{R}_0^{-1/2}(\boldsymbol{V} - \boldsymbol{R}_0\boldsymbol{I}) \tag{10.13}$$

where $\boldsymbol{R}_0 = \text{diag}(r_{01}, r_{02})$, $\boldsymbol{V} = [V_1 \; V_2]'$ and $\boldsymbol{I} = [I_1 \; I_2]'$. Assume that $\boldsymbol{V} = \boldsymbol{ZI}$ and since Eqs. (10.12) and (10.13) are also linear relations between $\boldsymbol{V}$ and $\boldsymbol{I}$, then $\boldsymbol{a}$ and $\boldsymbol{b}$ are also linearly related as

$$\boldsymbol{b} = \boldsymbol{Sa} \tag{10.14}$$

where $\boldsymbol{S}$ is the scattering matrix with respect to $\boldsymbol{R}_0$. For the 2-port network in Fig. 10.21, (10.14) is explicitly

$$\begin{aligned} b_1 &= S_{11}a_1 + S_{12}a_2 \\ b_2 &= S_{21}a_1 + S_{22}a_2 \end{aligned} \tag{10.15}$$

Hence, the scattering parameters are determined as

$$S_{jj} = \frac{b_j}{a_j}, \quad \text{for } a_k = 0 \text{ and } k \neq j \tag{10.16}$$

and

$$S_{kj} = \frac{b_k}{a_j}, \quad \text{for } a_k = 0 \text{ and } k \neq j \tag{10.17}$$

The parameter S_{jj} is called the reflection coefficient at the jth port and S_{kj} is called the transmission coefficient from port j to port k. The conditions in (10.16) and (10.17) together with (10.12) and (10.13) imply

$$S_{jj} = \frac{(Z_{jj} - r_{0j})}{(Z_{jj} + r_{0j})} \tag{10.18}$$

and

$$S_{kj} = -2\sqrt{r_{0j}r_{0k}}\frac{I_k}{V_{gj}} \tag{10.19}$$

The network is said to be matched at port j if $S_{jj} = 0$, or from (10.18) $r_{0j} = Z_{jj}$. Also from (10.19) and $|S_{kj}|^2 = |b_k|^2/|a_j|^2$, it is clear that $|S_{kj}|^2$ is the power gain of the terminated n-port from the source at port j to the load resistor r_{0k} at port k. For an example, consider the ideal

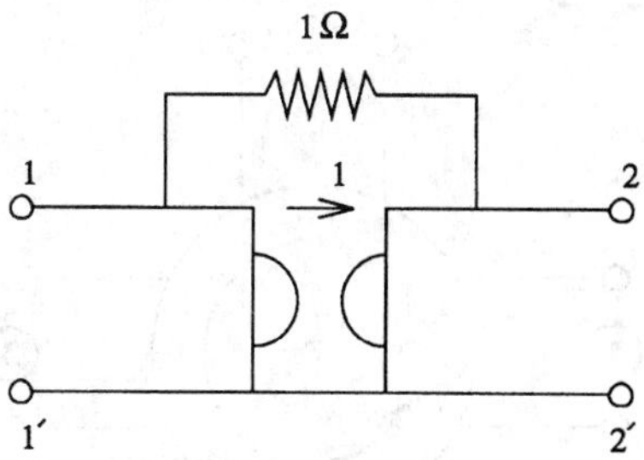

FIGURE 10.22 A 2-port circulator.

2-port gyrator with gyration transfer parameter $r > 0$ and $r_{01} = r_{02} = r$. Here

$$\mathbf{S} = \begin{bmatrix} S_{11} & S_{12} \\ S_{21} & S_{22} \end{bmatrix} = \begin{bmatrix} 0 & -1 \\ 1 & 0 \end{bmatrix} \tag{10.20}$$

The simplest type of circulator is the 2-port circulator. Consider the 2-port shown in Fig. 10.22, which has (for $r_{01} = r_{02} = 1$).

$$\mathbf{S} = \begin{bmatrix} S_{11} & S_{12} \\ S_{21} & S_{22} \end{bmatrix} = \begin{bmatrix} 0 & 0 \\ 1 & 0 \end{bmatrix} \tag{10.21}$$

This 2-port circulator has unity power transmission from port 1 to port 2, but zero power transmission from port 2 to port 1, and is also called an *isolator* or *one-way line.*

Now a 3-port circulator may be defined by the scattering matrix

$$\mathbf{S} = \begin{bmatrix} S_{11} & S_{12} & S_{13} \\ S_{21} & S_{22} & S_{23} \\ S_{31} & S_{32} & S_{33} \end{bmatrix} = \begin{bmatrix} 0 & 0 & S_{13} \\ S_{21} & 0 & 0 \\ 0 & S_{32} & 0 \end{bmatrix} \tag{10.22}$$

with each nonzero $S_{kj} = \pm 1$, so it has unity power transfer from port 1 to 2, 2 to 3, and 3 to 1. This 3-port circulator is usually depicted as shown in Fig. 10.23, where the + and − terminals of each port depend on the selection of each nonzero S_{kj} in (10.22). Likewise, a 4-port circulator may be defined by the scattering matrix

$$\mathbf{S} = \begin{bmatrix} S_{11} & S_{12} & S_{13} & S_{14} \\ S_{21} & S_{22} & S_{23} & S_{24} \\ S_{31} & S_{32} & S_{33} & S_{34} \\ S_{41} & S_{42} & S_{43} & S_{44} \end{bmatrix} = \begin{bmatrix} 0 & 0 & 0 & S_{14} \\ S_{21} & 0 & 0 & 0 \\ 0 & S_{32} & 0 & 0 \\ 0 & 0 & S_{43} & 0 \end{bmatrix} \tag{10.23}$$

with each nonzero $S_{kj} = \pm 1$, and depicted as suggested in Fig. 10.23. Similarly, the n-port ($n > 4$) circulator may be defined and depicted as suggested in Fig. 10.23. See [6] for details.

Nullator

It is a fact that every linear passive n-port network has a scattering description. However, to complete the description of general linear n-port networks (which may not have a scattering description), a few "pathological" or "degenerate" 1- and 2-ports must be included [7]. The first of these is the 1-port linear network denoted as the "nullator." For the *exact* configuration

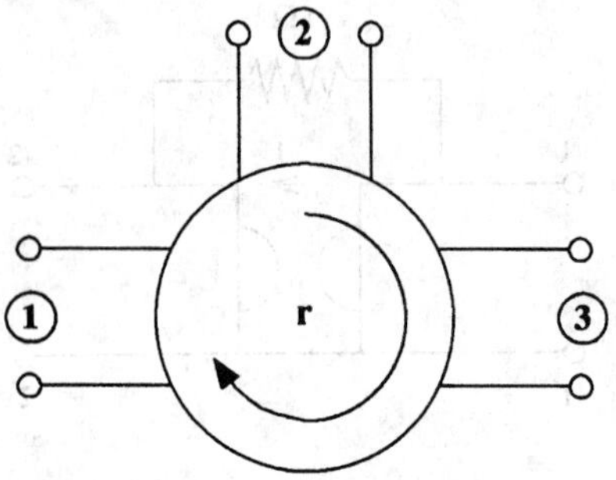

FIGURE 10.23 Matched 3-port circulator.

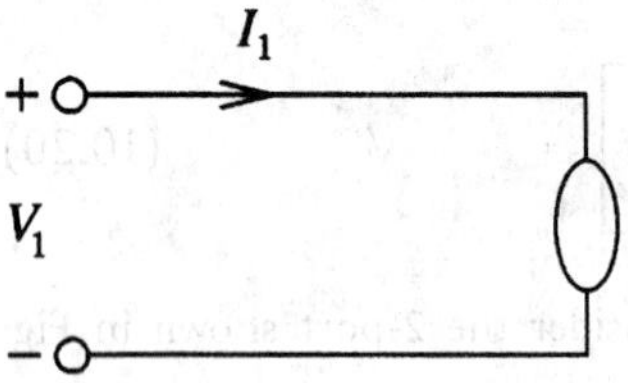

FIGURE 10.24 The nullator ($V_1 = I_1 = 0$).

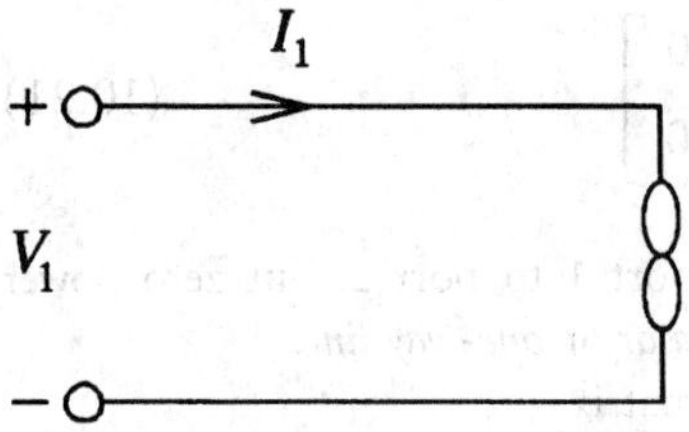

FIGURE 10.25 The norator (V_1 and I_1 arbitrary).

obtained by terminating Fig. 10.23 (with normalization parameters equal to unity) at port 2 with -1Ω and at port 3 with $+1\Omega$, the driving-point relations seen at port 1 are

$$V_1 = I_1 = 0 \tag{10.24}$$

so the resulting 1-port is at once a short and open circuit! This linear 1-port has been designated as the *nullator*. The circuit symbol for the nullator is shown in Fig. 10.24.

Norator

Another degenerate 1-port can be obtained from the *exact* configuration obtained by terminating Fig. 10.23 (with unity normalization parameters) at port 2 with $+1\Omega$ and at port 3 with -1Ω. The resulting driving-point relations seen at port 1 are that

$$V_1 \text{ and } I_1 \text{ are } \textit{completely independent} \tag{10.25}$$

This (nonreciprocal) linear 1-port has been designated as the *norator*. The circuit symbol for the norator is shown in Fig. 10.25.

Nullor

The final building block for the most general n-port linear network is the 2-port designated as the *nullor*. It is defined as a 2-port which, at port 1, demands $V_1 = I_1 = 0$ while simultaneously at port 2, V_2 and I_2 are arbitrary! These relations can be obtained from a 4-port circulator with

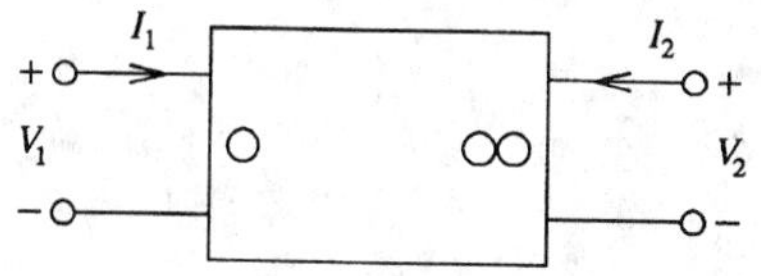

FIGURE 10.26 The nullor ($V_1 = I_1 = 0$, V_2, and I_2 arbitrary).

two of its ports appropriately terminated with negative and positive resistors. The symbol for the nullor is shown in Fig. 10.26. Both 4 and 3 terminal equivalent circuits exist for the nullor.

These singular network elements, the nullator, the norator, and the nullor, have been used to generate realizable circuits for the various signal converters discussed in this section. They have also been used to obtain realizable circuits for other widely used circuit elements such as the family of controlled sources, op amps, transistors, etc. See [3] and [7] for the applications mentioned above.

There are, of course, many references possible for the use of the network elements discussed in this section. Among others, [8] and [9] contain extensive bibliographies.

efining Terms

Reciprocal linear *n*-port: The port voltages and currents satisfy (10.1) and its restrictions.
Ideal gyrator: The basic nonreciprical 2-port satisfying (10.2).
Generalized impedance converter (GIC): A 2-port that satisfies (10.5) and its conditions.
Frequency dependent negative resistor (FDNR): A 1-port that has the driving-point impedance of (10.7).
Voltage negative impedance converter (VNIC): A 2-port satisfying (10.10) and foregoing conditions.
Current negative impedance converter (INIC): A 2-port satisfying (10.11) and foregoing conditions.
The *n*-port ($n \geq 3$) circulator: Described by its *n*-port scattering matrix with main-diagonal elements zero and off-diagonal elements restricted to only one per row or column of unity magnitude.
Nullator: A 1-port such that $V_1 = I_1 = 0$.
Norator: A 1-port such that V_1 and I_1 are arbitrary.
Nullor: A 2-port such that $V_1 = I_1 = 0$, while V_2 and I_2 are arbitrary.

ferences

[1] R. W. Newcomb, *Linear Multiport Synthesis*, New York: McGraw-Hill, 1966.

[2] B. D. H. Tellegen, "The gyrator, a new electric network element," *Phillips Res. Rep.*, vol. 3, pp. 81–101, 1948.

[3] S. K. Mitra, *Analysis and Synthesis of Linear Active Networks*, New York: Wiley, 1969.

[4] A. Antoniou, "Realization of gyrators using operational amplifiers and their use in RC-active network synthesis," *Proc. IEE*, vol. 116 , pp. 1838–1850, 1969.

[5] J. L. Merill Jr., "Theory of the negative impedance converter," *Bell System Tech. J.*, vol. 30, pp. 88–109, 1951.

[6] H. J. Carlin and A. B. Giordano, *Network Theory*, Englewood Cliffs, NJ: Prentice-Hall, 1964.

[7] H. J. Carlin, "Singular network elements," *IEEE Trans. Circuit Theory*, vol. CT-11, pp. 67–72, 1964.

[8] S. K. Mitra, *Active Inductorless Filters*, New York: IEEE, 1971.

[9] M. Herpy and J-C. Berka, *Active RC Filter Design*, New York: Elsevier Science, 1986.

11

Operational Amplifiers

David G. Nairn
Queen's University, Canada

Sergio Franco
San Francisco State University, California

11.1 The Ideal Operational Amplifier

David G. Nairn

The operational amplifier, or op amp, is a fundamental building block for many electronic circuits. Although the op amp itself is composed of numerous transistors, it is usually treated as a single circuit element known as the ideal operational amplifier. The ability to treat the op amp as an ideal circuit element simplifies its use in circuits such as amplifiers, buffers, filters, and data converters. With such varied uses, the op amp has been implemented in many different forms. Nevertheless, the behavior of each of these different forms can still be characterized as an ideal op amp.

Open-Loop Equivalent Circuit

The op amp is primarily a high gain amplifier. Although the op amp can be used on its own, most op amps are part of larger circuits in which feedback is used to determine the circuit's overall transfer function. Consequently, the op amp's precise behavior is only a secondary interest. To simplify preliminary circuit analysis and design, an abstraction of the practical op amp, known as the ideal op amp is often used. The ideal op amp is characterized by the following four parameters:

- infinite gain
- infinite bandwidth
- draws no signal power at its inputs
- is unaffected by loading of its output.

0-8493-8341-2/95/$0.00+$.50

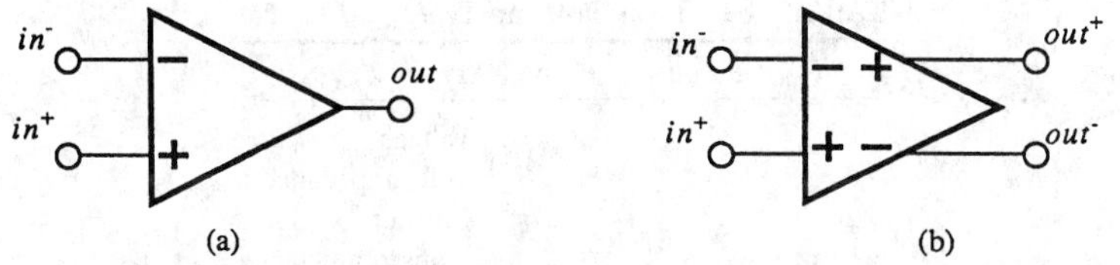

FIGURE 11.1 Circuit symbols for (a) the single-ended op amp and (b) the fully differential op amp.

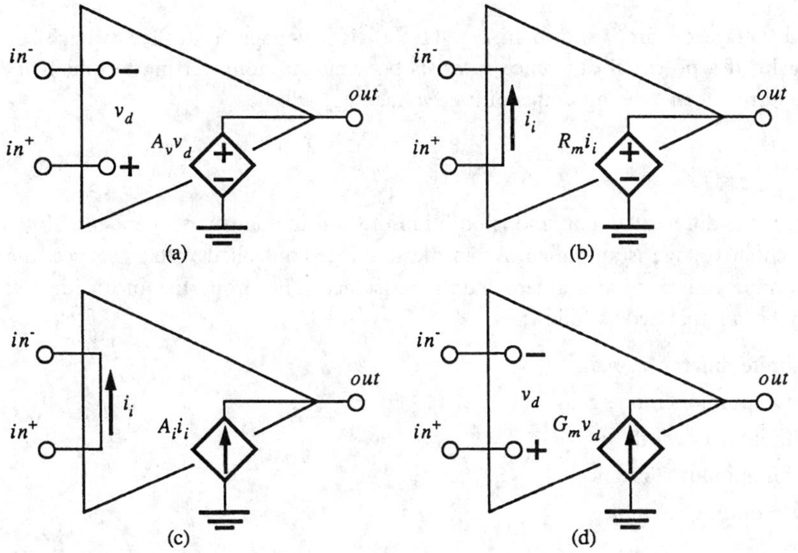

FIGURE 11.2 The four possible op amp configurations: (a) the voltage op amp, (b) the transimpedance op amp, (c) the current op amp, and (d) the transconductance op amp.

Although such specifications are not achieved in practice, the nonidealities of practical op amps (see Section 11.2, On The Nonideal Operational Amplifier) can be neglected in most applications.

The ideal op amp is represented schematically as a triangle with two inputs and either one or two outputs, as shown in Fig. 11.1. For the single output case, the output is referred to ground and the op amp is known as a single-ended op amp. For the two output case, the output is the difference between the *out*+ and *out*− outputs and the op amp is known as a fully differential op amp. Since the op amp provides gain, it requires an external power source. For the ideal op amp, the power supply has no effect on the amplifier's performance and is therefore not indicated in the circuit symbol.

Based on the above description, the op amp's input can be stimulated with either a voltage or current. Also, the controlled output can be either a voltage or a current. Consequently, there are four possible implementations of the ideal op amp: the voltage op amp, the transimpedance op amp, the current op amp, and the transconductance op amp, as shown in Table 11.1 and in Fig. 11.2.

For most applications, the op amp is used in a closed-loop configuration with negative feedback. Due to the negative feedback, all four ideal op amp types perform the same function. When the limitations of practical op amps are considered, it will be found that different op amps are preferred for different applications. Of the four types, the voltage op amp is the most widely known. Therefore, the use of op amps will be considered from the perspective of the voltage op amp. Then the other three types will be considered.

TABLE 11.1 Ideal Op Amp Types

Input	Output	Gain	Type
V	V	A_v	Voltage
I	V	R_m	Transimpedance
I	I	A_i	Current
V	I	G_m	Transconductance

Voltage Op Amps

The ideal voltage op amp, shown in Fig. 11.2(a), is a voltage-controlled voltage source with infinite gain. If a potential difference v_d exists between the noninverting terminal in^+ and the inverting terminal in^- the op amp's output voltage v_{out} will be

$$v_{\text{out}} = A_v v_d \tag{11.1}$$

where A_v is the differential gain and is both infinite and frequency independent. Note that only the differential voltage is amplified. As an ideal voltage-controlled voltage source, the op amp has an infinite resistance and a zero output resistance. The properties of an ideal voltage op amp may be summarized as follow:

- infinite differential gain
- zero common-mode gain
- infinite bandwidth
- infinite input resistance
- zero output resistance.

With these ideal properties, the op amp is relatively easy to use in many circuit applications.

Op Amp Circuit Applications

The op amp can be used in both open-loop configurations and closed-loop configurations. If the op amp is used open-loop, small voltages between the input terminals produced either a positive or negative infinite voltage due to the amplifier's infinite gain. Consequently, the op amp can be used as a comparator. This application is discussed further in the section on comparators. In a closed-loop circuit, feedback allows the op amp's output voltage to influence its input voltage (see Section V). The op amp can then be made to perform many complex operations.

Unity Gain Buffer

The simplest feedback that can be applied to the op amp is shown in Fig. 11.3. The op amp's output is connected to the inverting input and an input signal V_{in} is applied to the noninverting input. The feedback forces the voltage at in^- to equal V_{out}. By multiplying the differential voltage at the op amp's input (i.e., $V_{\text{in}} - V_{\text{out}}$) by the op amp's gain, V_{out} is found to be

$$V_{\text{out}} = A_v(V_{\text{in}} - V_{\text{out}}) \tag{11.2}$$

which can be rewritten as

$$\frac{V_{\text{out}}}{V_{\text{in}}} = \frac{1}{1 + 1/A_v} \tag{11.3}$$

since A_v is infinite, V_{out} equals V_{in}. Therefore, the circuit is a unity gain buffer. It is important to note that the op amp's high gain and the use of negative feedback forces the voltage at the

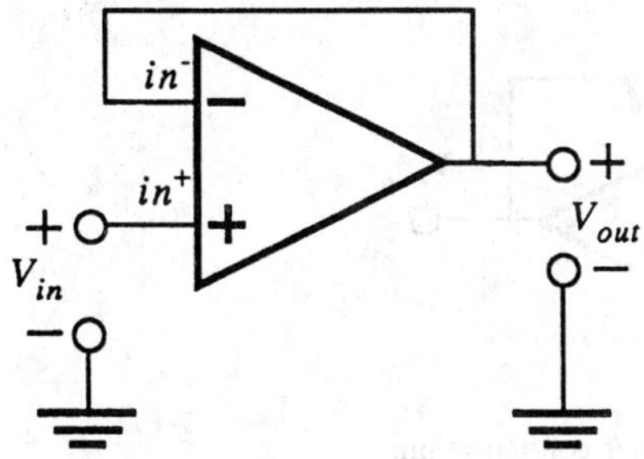

FIGURE 11.3 A unity gain buffer.

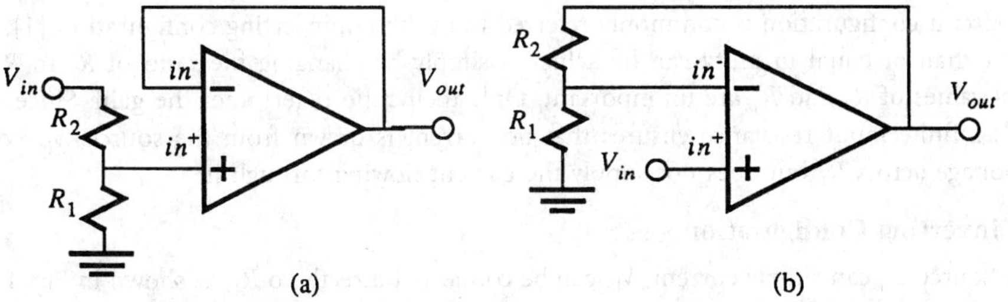

(a) (b)

FIGURE 11.4 (a) A simple buffer attenuator; (b) the noninverting amplifier configuration.

op amp's two input terminals to be equal. Hence if V_{in} is varied, V_{out} will follow or track it. The op amp's two input terminals have the same potential, but no current flows between them. Therefore, a virtual short is said to exists between the inputs. The unity gain buffer draws no current from the signal source due to the op amp's infinite input resistance and the op amp's zero output resistance ensures that loading does not affect the voltage at V_{out}.

Simple Attenuator

If an output equal to a fraction of V_{in} is required, the circuit shown in Fig. 11.4(a). The noninverting terminal is now a fraction of V_{in}. Since V_{out} tracks the voltage at the noninverting terminal, the circuit's output voltage will be

$$V_{out} = V_{in}\left[\frac{R_1}{R_1 + R_2}\right] \tag{11.4}$$

Due to the op amp's infinite input resistance, the voltage divider formed by R_1 and R_2 is not loaded by the op amp. Therefore, large values of R_1 and R_2 can be used to avoid loading the source voltage V_{in}.

The Noninverting Amplifier Configuration

Usually it is more desirable to amplify a signal than to attenuate it. Therefore, instead of matching an attenuated V_{in} to V_{out}, at the op amp's input an attenuated V_{out} can be matched to V_{in}, as shown in Fig. 11.4(b). Due to the negative feedback and the op amp's finite gain, the op amp's two inputs have the same potential. Therefore,

$$V_{in} = V_{out}\left[\frac{R_1}{R_1 + R_2}\right] \tag{11.5}$$

which is more commonly written as

$$\frac{V_{out}}{V_{in}} = 1 + \frac{R_2}{R_1} \tag{11.6}$$

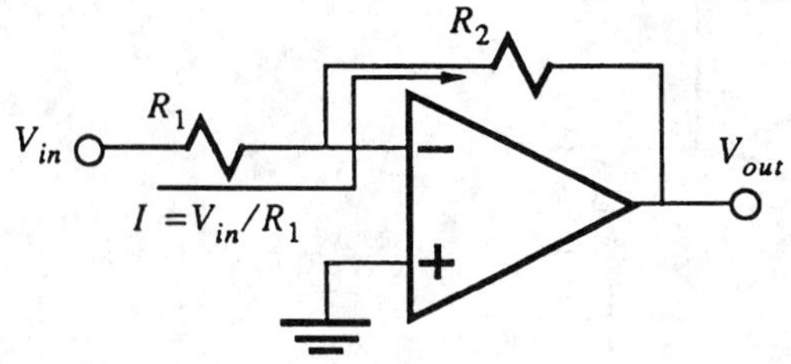

FIGURE 11.5 The inverting amplifier configuration.

This circuit configuration is commonly referred to as the noninverting configuration [1]. Gains greater than or equal to unity can be achieved simply by changing the ratio of R_2 to R_1. The actual values of R_2 and R_1 are unimportant. Only their ratio determines the gain. Since the op amp's infinite input resistance ensures that no current is drawn from the source, V_{in} controls the voltage across R_1 but does not supply the current flowing through it.

The Inverting Configuration

If the source V_{in} can supply current, V_{in} can be connected directly to R_1, as shown in Fig. 11.5. In this case the voltage across R_1 is still equal to V_{in} but of opposite polarity. The voltage at the op amp's inverting input is now at ground potential, but no current flows to ground. Consequently, a virtual ground exists at the inverting input terminal. Since the current flowing through R_1 cannot flow to ground or into the op amp, it flows through R_2, causing an output voltage of

$$V_{out} = 0 - R_2\left[\frac{V_{in}}{R_1}\right] \tag{11.7}$$

or

$$\frac{V_{out}}{V_{in}} = -\frac{R_2}{R_1} \tag{11.8}$$

This circuit configuration is commonly referred to as the inverting configuration [1]. Both amplification and attenuation can be achieved by changing the ratio of R_2 to R_1. A very important difference between the inverting and noninverting configurations is that the inverting configuration draws a current equal to V_{in}/R_1 from the source. Consequently, even though the op amp itself has an infinite input resistance, the inverting configuration only has an input resistance equal to R_1. Fortunately, the gain only depends on the ratio of R_2 to R_1, thereby allowing both resistors to be increased, thus limiting the current drawn from V_{in}.

Frequency Dependent and Nonlinear Transfer Functions

Elements other than resistors can be used in both the inverting and noninverting configuration. By using the frequency dependent elements Z_1 and Z_2 in place of R_1 and R_2, circuits with arbitrary frequency responses can be generated. Two examples of this are illustrated in Fig. 11.6(a) and (c). In Fig. 11.6(a), a capacitor has been added in series with R_1. The inverting amplifier now has the transfer function

$$\frac{V_{out}}{V_{in}}(s) = \frac{-Z_2}{Z_1} \tag{11.9}$$

where Z_1 equals $(1 + R_1Cs)/(Cs)$ and Z_2 equals R_2. Therefore the circuit's transfer function is

$$\frac{V_{out}}{V_{in}}(s) = \frac{-R_2Cs}{1 + R_1Cs} \tag{11.10}$$

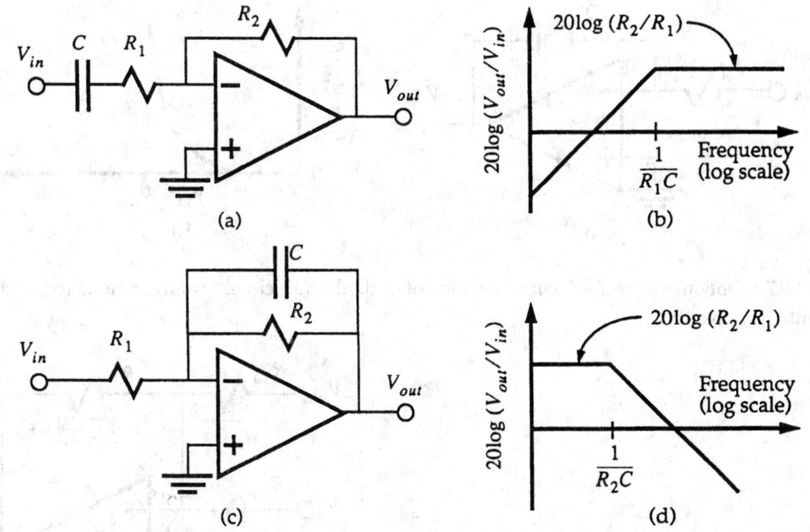

FIGURE 11.6 Frequency dependent circuits using the inverting configuration: (a) a simple high-pass filter and (b) its frequency response; (c) a simple low-pass filter and (d) its frequency response.

which is a simple high-pass filter, as illustrated in Fig. 11.6. (b) Alternatively, a capacitor can be added in parallel with R_2 as shown in Fig. 11.6(c). Now Z_1 equals R_1 and Z_2 equals $R_2/(1 + R_2Cs)$, which results in the transfer function

$$\frac{V_{\text{out}}}{V_{\text{in}}}(s) = \frac{-R_2}{R_1(1 + R_2Cs)} \tag{11.11}$$

This circuit performs as a low-pass filter, as illustrated in Fig. 11.6(d). By selecting Z_1 and Z_2, arbitrary transfer functions can be generated, thereby making op amp circuits useful for implementing active filters (see Section XV). When designing arbitrary transfer functions, the resulting circuits must be stable if they are to perform correctly [2].

If nonlinear elements such as diodes are used in place of R_1 and R_2, nonlinear transfer functions can be obtained [2]. For example, the I–V characteristics of a component can be obtained by replacing R_2 in the inverting configuration with the desired nonlinear element, as shown in Fig. 11.7(a). Then, by applying V_{in} and V_{out} to an oscilloscope, the element's I–V characteristic can be displayed directly as shown in Fig. 11.7(b). Due to the diode's exponential characteristic, the circuit's output is the logarithm of its input

$$V_{\text{out}} = nV_T \ln\left(\frac{V_{\text{in}}}{R_1 I_S}\right) \tag{11.12}$$

The constants n and I_S are determined by the diode while V_T is the thermal voltage (see Chapter 9, Section 9.5). Circuits of this type are made possible by the presence of the virtual ground at the op amp's input.

Multiple Input Circuits

For the inverting configuration, the presence of the virtual ground at the op amp's inverting input allow signals from many sources to be combined. As shown in Fig. 11.8, the currents I_a and I_b are determined independently by V_a and V_b, respectively. These two currents are then

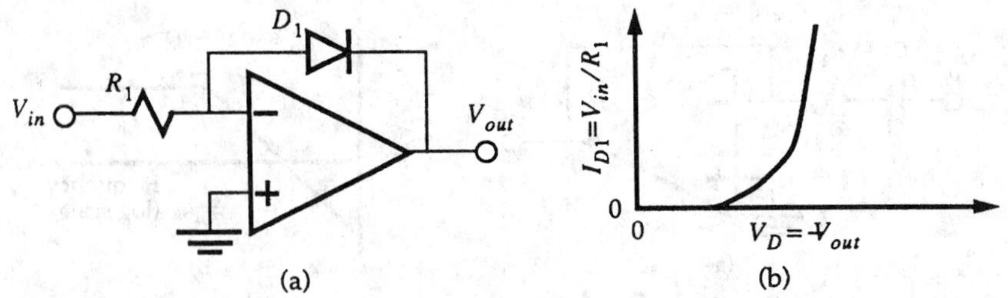

FIGURE 11.7 Obtaining the I–V characteristics of a diode: (a) circuit configuration; (b) oscilloscope display with V_{out} inverted.

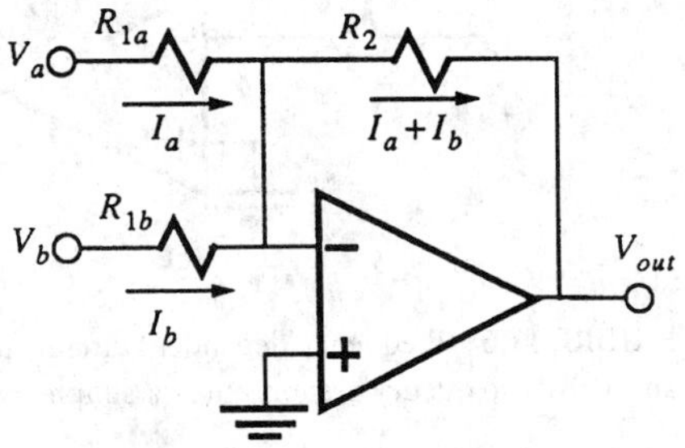

FIGURE 11.8 A weighted summer.

summed at the virtual ground and forced through R_2. The resulting output is a weighted sum of V_a and V_b:

$$V_{out} = \frac{R_2}{R_{1a}} V_a - \frac{R_2}{R_{1b}} V_b \tag{11.13}$$

Any number of additional inputs can be added. The virtual ground prevents the different signals from interacting with each other.

The noninverting input can also be used in the multiinput circuit. In this case though, the op amp's input is no longer at virtual ground and its output ceases to be a weighted sum of the inputs. For example, if V_1 and V_2 are applied to the circuit of Fig. 11.9(a), the output depends on both V_2 and the difference between V_2 and V_1:

$$V_{out} = V_2 + \frac{R_2}{R_1}(V_2 - V_1) \tag{11.14}$$

This output can also be written as a weighted difference between V_2 and V_1

$$V_{out} = \left(1 + \frac{R_2}{R_1}\right) V_2 - \frac{R_2}{R_1} V_1 \tag{11.15}$$

If, as shown in Fig. 11.9(b), V_2 is first attenuated by $R_2/(R_1 + R_2)$, a voltage only proportional to the difference between V_1 and V_2 can be obtained:

$$V_{out} = \frac{R_2}{R_1}(V_2 - V_1) \tag{11.16}$$

where it is evident that the circuit amplifies the difference between its inputs and rejects the common mode component. Consequently, the configuration is referred to as the differential configuration.

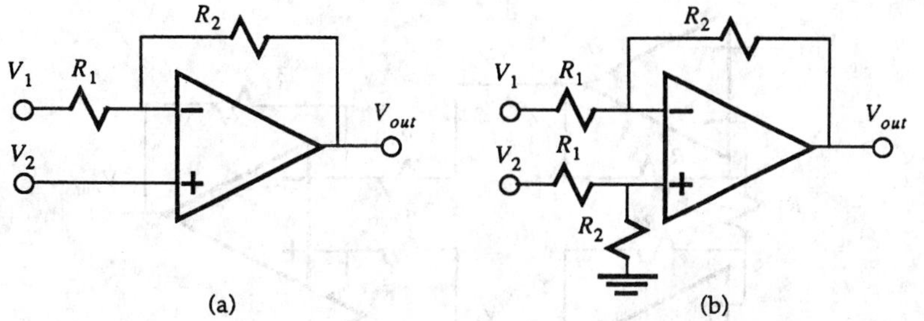

FIGURE 11.9 Circuits for finding weighted differences: (a) a circuit based in the inverting and noninverting configurations, (b) the differential configuration.

Instrumentation Amplifiers

The differential amplifier shown in Fig. 11.9(b) is useful for detecting weak signals in a noisy environment. Unfortunately, its input resistance is only $2R_1$. To circumvent this problem, V_1 and V_2 can be buffered using two unity gain buffers. A better solution is to use the instrumentation amplifier shown in Fig. 11.10. This circuit combines two circuits of the type shown previously in Fig. 11.9(a) and the differential amplifier shown in Fig. 11.9(b). Based on (11.15) the voltage at V_a, will be

$$V_a = \left(1 + \frac{R_2}{R_1}\right)V_1 - \frac{R_2}{R_1}V_2 \tag{11.17}$$

and the voltage at V_b will be

$$V_b = \left(1 + \frac{R_2}{R_1}\right)V_2 - \frac{R_2}{R_1}V_1 \tag{11.18}$$

With V_a and V_b applied to the differential amplifier, the instrumentation amplifier's output voltage will be

$$V_{\text{out}} = \frac{R_2}{R_1}(V_b - V_a) \tag{11.19}$$

Due to the op amp's ideally zero output resistance, the differential amplifier's low input resistance does not load the other two circuits. Hence, by substituting (11.17) and (11.18) for V_a and V_b, respectively, the output voltage can be expressed as

$$V_{\text{out}} = \left[1 + \frac{2R_2}{R_1}\right]\frac{R_4}{R_3}(V_2 - V_1) \tag{11.20}$$

which allows the difference between V_1 and V_2 to be either amplified or attenuated without loading the signal sources. Since there is only one R_1, it can be made variable to allow for an easily adjustable gain. Due to their usefulness, instrumentation amplifiers are available in a single package from many manufacturers.

This last circuit illustrates the two primary characteristics of op amps when used in a closed-loop negative feedback configuration. First, a virtual short exists between the op amp's input terminals allowing a high impedance source to set the potential of a circuit node. Second, the op amp's low output resistance allows op amp circuits to be connected together without altering their individual transfer functions. These two characteristics greatly simplify the analysis and design of circuits containing many op amps.

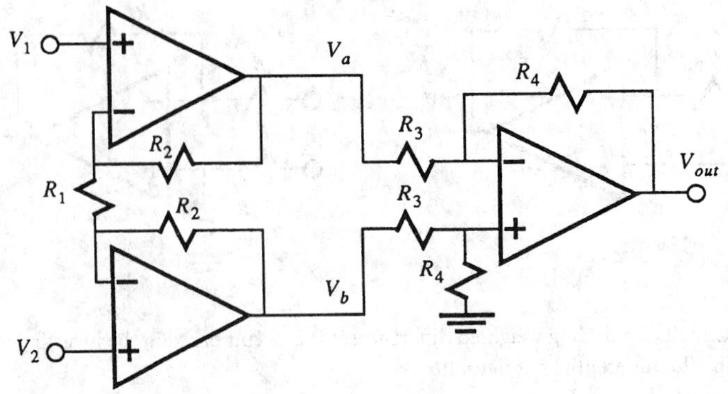

FIGURE 11.10 An instrumentation amplifier.

Comparators

If the op amp is not used in a closed-loop, its output will be either high or low, depending on the polarity of the voltage between its inputs. This appears to make op amps suitable for comparing two closely spaced signal levels. In practice, it is usually better to use a circuit called a comparator for this application. Comparators are similar to op amps but have been specifically designed to operate in an open-loop circuit.

The basic comparator compares the voltage levels at its two inputs. If the voltage at the positive input, in^+, exceeds that at the negative input, in^-, the comparator will generate a logic high. If the voltage at in^- exceeds that at in^+ a logic low will be produced. Often, complementary outputs are also provided. The logic high and low levels are set either by the manufacturer or by the user. Typically, the logic levels are compatible with common logic families such as TTL, CMOS, or ECL.

The comparator's primary objective is to provide the correct output level as fast as possible while op amps are usually used in a closed-loop configuration. To ensure closed-loop stability, most op amps require some form of compensation (see section 11.2 and [3]), which reduces their speed and bandwidth. On the other hand, comparators are specifically designed for open-loop operation thereby, making them better suited for high-speed comparisons.

The comparator is usually used as a threshold comparator and often has added hysteresis. When used as a threshold detector, a reference level is applied to one input and the signal is applied to the other input. The choice of inputs determines the output's polarity. Circuits of this type are commonly employed as level detectors and in analog-to-digital converters (see Chapter 12). In many cases, the input signal contains noise that causes the comparator's output to oscillate as the signal passes the threshold level. To avoid this problem, the use of positive feedback, as shown in Fig. 11.11(a), is used to generate hysteresis, as shown in Fig. 11.11(b). The resistor R_3 is included to reduce the effects of the comparator's bias currents (see the section on input bias currents). For this circuit, the level that causes the output to go low, V_{INL}, is

$$V_{INL} = V_{REF} + (V_{REF} - V_{OH})/K \tag{11.21}$$

where V_{OH} is the output-high level and K is the ratio of the resistors depicted in Fig. 11.11(b). The level that causes the output to go high, V_{INH}, is

$$V_{INH} = V_{REF} + (V_{REF} - V_{OL})/K \tag{11.22}$$

where V_{OL} is the output-low level. By adjusting the ratio K, the amount of noise immunity can be adjusted.

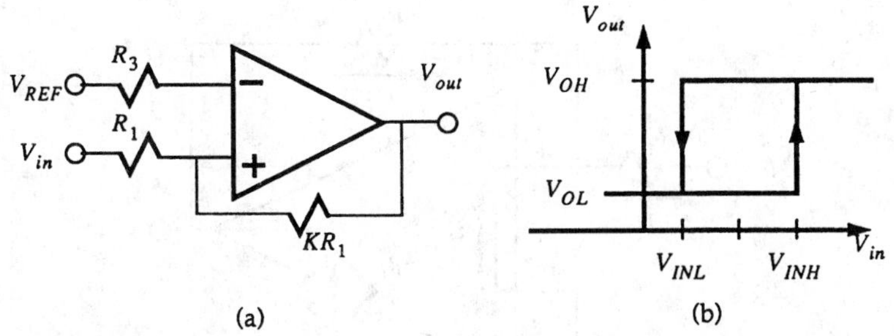

FIGURE 11.11 (a) A comparator with hysteresis and (b) its input–output relationship.

Other Op amp Configurations

As mentioned in the first section, the ideal op amp can be implemented in any of four possible configurations: a voltage amplifier, a current amplifier, a transimpedance amplifier, and a transconductance amplifier. When used in a closed-loop configuration with negative-feedback, all four of the ideal op amps behave the same. In particular, the virtual short circuit between the two inputs remains and the output is unaffected by loading. Consequently, op amps of all four types exist. Due to practical limitations, some configurations are better suited than others for particular applications.

To illustrate that the ideal op amp's configuration does not affect the performance of an op amp circuit, the inverting op amp circuit shown in Fig. 11.12 has been implemented with a current op amp. The ideal current op amp displays zero resistance between its input terminals. The output, which is a current source, has an infinite output resistance. Hence, the current op amp is the dual of the voltage op amp. The circuit's output voltage can be found by summing the currents at the op amp's input. Since there is a physical short between the two input terminals, the current through R_1 will be

$$I_1 = \frac{V_{\text{in}}}{R_1} \tag{11.23}$$

while the current through R_2 will be

$$I_2 = \frac{V_{\text{out}}}{R_2} \tag{11.24}$$

and due to the op amp's current gain, A_i, the op amp's input current will be

$$i_i = \frac{V_{\text{out}}}{A_i R_2} \tag{11.25}$$

The sum of Eqs. (11.23), (11.24), and (11.25) must equal zero. Therefore

$$\frac{V_{\text{out}}}{V_{\text{in}}} = -\frac{R_2}{R_1}\left[\frac{1}{1 + 1/A_i}\right] \tag{11.26}$$

which for $A_i = \infty$ results in the same gain as that produced by the inverting configuration implemented with an ideal voltage op amp. More importantly, the current through R_1 equals the current through R_2 and the op amp's input current goes to zero. Hence, even with the physical short between the op amp's inputs, only a virtual short exists in the closed-loop circuit because

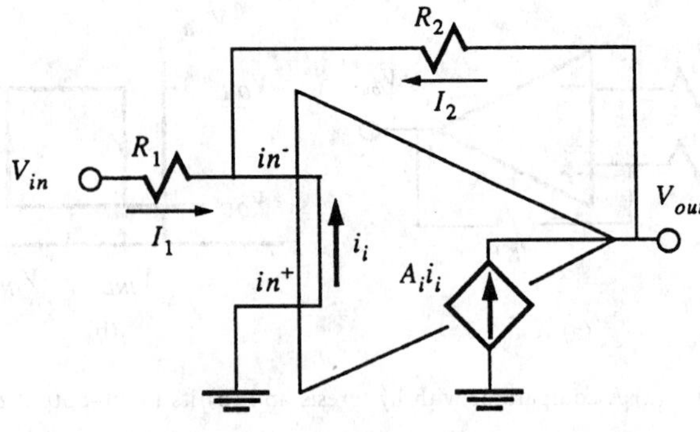

FIGURE 11.12 The inverting amplifier configuration implementated with a current op amp.

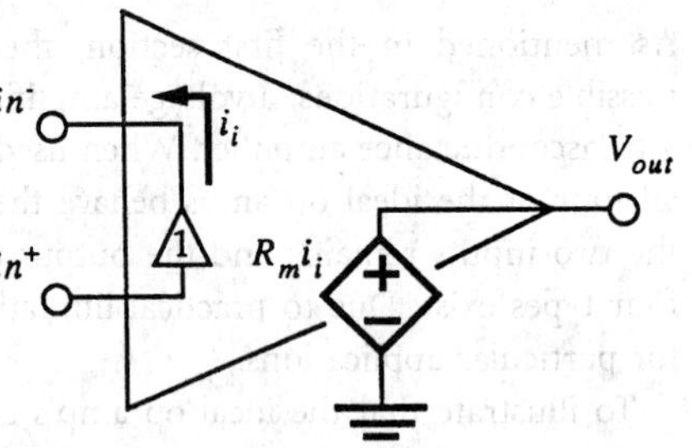

FIGURE 11.13 A practical transimpedance op amp.

no current flows between the two terminals. At the output, the infinite gain of the ideal current op amp ensures that the output stays at the value indicated by (11.26) even if a load is added to the output. Consequently, the ideal op amp's closed-loop behavior is preserved independent of the op amp's configuration.

Current Op Amps

Ideally the current op amp acts as a current-controlled current source with an infinite current gain. In practical current op amps, the gain is relatively low. Therefore, the assumption of an infinite gain is unrealistic for most applications. The primary purpose of current op amps is to boost the output current of a voltage op amp (see Section 11.2).

Transimpedance Op Amps

The ideal transimpedance op amp is a current-controlled voltage source with an infinite transimpedance gain. Op amps of this type are commonly referred to as current feedback op amps. Practical implementations of transimpedance op amps typically display a much higher speed than most voltage op amps (see Section 11.3 and [4]). Due to practical considerations, transimpedance op amps typically have a unity gain buffer between the noninverting and inverting input terminals as shown in Fig. 11.13.

The added buffer has no effect on the ideal closed-loop performance, but it does increase the input resistance of the noninverting input for practical circuits.

Transconductance Op Amps

The ideal transconductance op amp is a voltage-controlled current source with an infinite transconductance gain. Practical transconductance op amps are usually implemented in MOS technologies [5]. MOSFET's themselves are voltage-controlled current sources. Since a practical transimpedance op amp has a less that infinite output resistance, it is not suited for driving resistive loads. This does not pose a problem since most transimpedance op amps are used in switched-capacitor circuits, where they are used to drive capacitive loads.

Summary

The ideal op amp is a high gain circuit element. When used in an open-loop configuration, the op amp can be used to compare closely spaced signal levels. It is generally much more useful when negative feedback is applied to control its output. With negative feedback, the differential voltage at its input approaches zero and the current between its inputs approaches zero. This makes it particularly useful for controlling the voltage or current in a circuit without drawing power from the controlling source. The ideal op amp model, although only approximated in practical op amps, is very useful for quickly analyzing and understanding the operation of larger circuits. Once the circuit's behavior is understood, the effects of the op amp's nonidealities can be considered.

rences

[1] A. S. Sedra and K. C. Smith, *Microelectronic Circuits*, 3rd ed., New York: Holt, Rinehart and Winston, 1991.

[2] D. Sheingold, "Op amps and their characteristics," in *Analog Circuit Design*, J. Williams, Ed., New York: Reed, 1991, ch. 30.

[3] J. K. Roberge, *Operational Amplifiers: Theory and Practice*, New York: Wiley, 1975.

[4] E. Bruun, "Feedback analysis of transimpedance operational amplifier circuits," *IEEE Trans. Circuits Syst.*, vol. 40, pp. 275–278, pt. 1, Apr. 1993.

[5] R. Gregorian and G. C. Temes, *Analog MOS Integrated Circuits*, New York: Wiley, 1986.

.2 The Nonideal Operational Amplifier

David G. Nairn

Practical op amps differ significantly from the ideal op amp. These differences limit the signal levels and range of impedances that can be used in op amp circuits. Fortunately, the nonidealities are only significant in certain applications. For these applications, circuit design precautions often reduce the effects of the nonidealities to acceptable levels. Alternatively, higher performance op amps can be used. The op amp's dominant nonidealities, how they affect various applications, and techniques to compensate for their detrimental effects are discussed in the following sections.

Finite Differential Gain

The op amp's most critical nonideality is its finite gain. Unlike the infinite gain of the ideal op amp, the gain of a practical op amp is typically large at dc and decreases at high frequencies. Most op amps are internally compensated for a frequency dependent gain of the form

$$\frac{v_o}{v_i} = \frac{A_O}{1 + j\omega(A_O/\omega_t)} \tag{11.27}$$

where A_O, is the dc differential open-loop gain, ω is frequency, and ω_t is the op amp's unity gain frequency. Unity gain frequencies are typically in the MHz range. At low frequencies, the op amp's gain simply becomes A_O. At high frequencies, the op amp's gain can be approximated as

$$\frac{v_o}{v_i} = \frac{\omega_t}{j\omega} \tag{11.28}$$

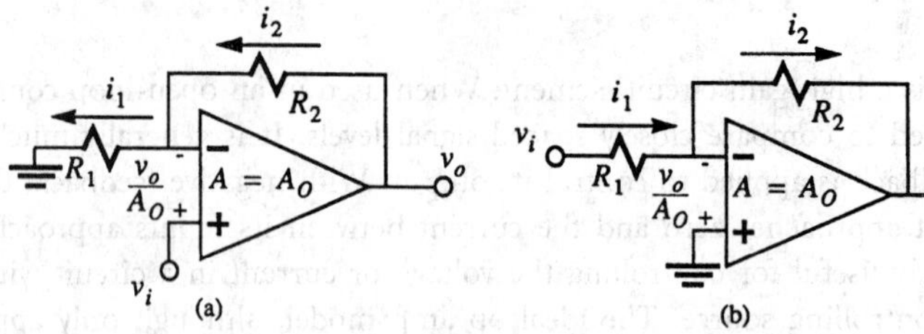

FIGURE 11.14 Analysis of the noninverting (a) and inverting (b) configurations for an op amp with finite gain A_O.

Further details on the amp's high-frequency behavior will be discussed in Section 11.3. The dc gain, A_O, is typically quite large, hence it is usually expressed in decibels (dB). A_O ranges from 40 dB for high-speed op amps to 120 dB for high precision op amps. General-purpose op amps usually have differential gains in the 100 dB region. Since A_O is subject to wide variations, manufacturers usually specify a minimum and typical value.

The op amp's finite A_O reduces the closed-loop gain of most op amp circuits. To illustrate the problem, the noninverting and inverting amplifiers shown in Fig. 11.14 can be analyzed assuming a finite op amp gain. The finite gain results in a nonzero differential voltage at the op amp's input:

$$v_d = \frac{v_o}{A_O} \tag{11.29}$$

Consequently, the voltage at the inverting input is not equal to that at the noninverting input. Hence, the voltage across R_1 and the current through it is changed. By equating the current in R_1 and R_2, it is seen that the non-inverting amplifier's gain is reduced to

$$\frac{v_o}{v_i} = \frac{1 + R_2/R_1}{1 + (1 + R_2/R_1)/A_O} \tag{11.30}$$

while the inverting amplifier's gain is reduced to

$$\frac{v_o}{v_i} = \frac{-R_2/R_1}{1 + (1 + R_2/R_1)/A_O} \tag{11.31}$$

For large values of A_O, (11.30) and (11.31) reduce to the gains that would be obtained with an ideal op amp (i.e., $v_o/v_i = 1 + R_2/R_1$ and $v_o/v_i = -R_2/R_1$, respectively).

It is only when A_O and the desired closed-loop gain become comparable that the op amp's finite gain leads to a significant reduction in the closed-loop gain. To illustrate this problem, the gain deviation for the noninverting amplifier versus $(1 + R_2/R_1)/A_O$ is plotted in Fig. 11.15. Since A_O is subject to wide variations, only gains that are at least 100× lower than A_O should be used to ensure a well-controlled gain. Consequently, the op amp's finite open-loop differential gain places an upper limit on the closed-loop gain that can be provided accurately.

Output Saturation

Although op amps can provide high gains, the op amp's maximum output voltage and current are limited. The maximum output voltage is limited by the op amp's supply voltages while the maximum output current is usually limited by the op amp's allowable power dissipation.

The op amp, like any electronic amplifier, requires a dc power supply. Most op amps require a positive, V^+, and a negative, V^-, power supply. Since V^+ and V^- are typically the same size

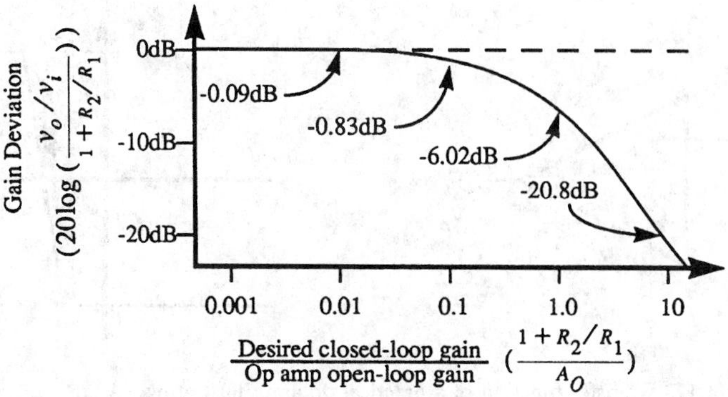

FIGURE 11.15 Gain deviation for the noninverting configuration caused by a finite A_O.

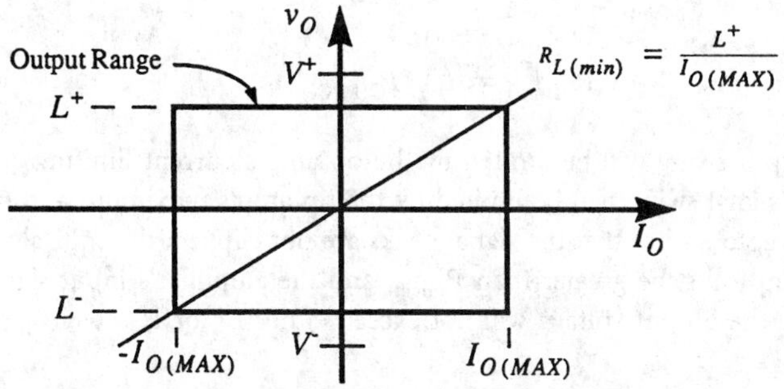

FIGURE 11.16 Voltage and current limitations on the op amp's output signal swing.

and of opposite polarity, they are referred to as dual or split supplies. Usually the op amp has no connection to ground. The supply voltage typically ranges from ±5 V to ±18 V, with ±15 V being the most common. Special-purpose op amps include low-voltage/low-power op amps for use with lower supply voltages, high-voltage op amps for use with supply voltages beyond ±18 V and single supply op amps for use with a single supply.

Regardless of the op amp type, the maximum L^+ and minimum L^- output voltages cannot exceed the supply voltages. Typically, the output saturates within 1 to 3 volts of the supplies as shown in Fig. 11.16. Low-voltage op amps often feature a "rail-to-rail" output swing that allows the output signal to extend to both V^+ and V^-. Due to the op amp's limited output swing, the input signal must be kept small enough to avoid distortion caused by clipping the output signal.

The second limitation on the op amp's output signal is the op amp's maximum output current specification. This limitation is determined by the maximum allowed power dissipation of the op amp. If the power dissipation limit is exceeded, the resulting temperature rise can damage the device. The worst case power dissipation usually occurs when the op amp has a load resistance of zero (i.e., the output is shorted to ground). In this situation the full supply voltage appears across the op amp's output stage and the power dissipation is

$$P_{\text{Disp}} = V^+ I_O + P_Q \tag{11.32}$$

P_Q is the op amp's quiescent power dissipation and usually much smaller than V^+I_O. Hence, to avoid an excessive temperature rise, I_O must be limited to a safe value, $I_{O(\text{MAX})}$. Many op amps are designed with short-circuit protection that limits $I_{O(\text{MAX})}$ to a safe level. For general-purpose op amps $I_{O(\text{MAX})}$ is in the 20-mA range.

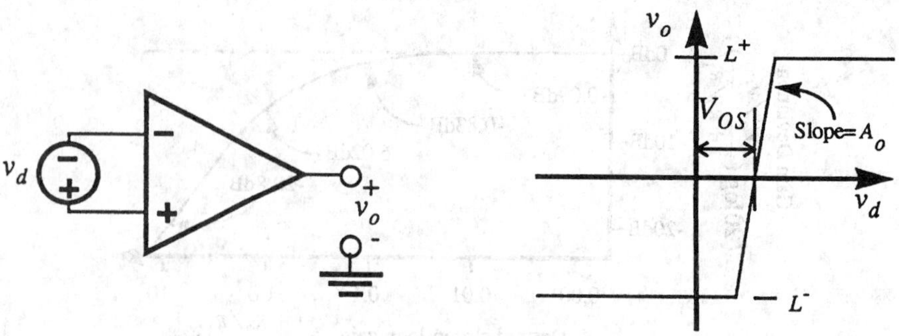

FIGURE 11.17 Transfer function of a practical op amp illustrating A_O, L^+, L^-, and V_{OS}.

The limitations imposed by the combination of L^+, L^-, and $I_{O(\text{MAX})}$ are illustrated in Fig. 11.16. For loads below $R_{L(\text{min})}$:

$$R_{L(\text{min})} = L^+/I_{O(\text{MAX})} \tag{11.33}$$

the op amp's output swing will be limited by the op amp's current limiting circuitry. For large values of R_L, the signal swing will be limited by the op amp's maximum and minimum output voltages. Therefore, to ensure that the signal peaks are not clipped, the equivalent load resistance seen by the op amp must be greater than $R_{L(\text{min})}$ and the amplifier's input signal must be small enough to ensure the output voltage will not exceed either L^+ or L^-.

Offset Voltage

For an ideal op amp, a zero differential input voltage produces a zero output voltage. For a practical op amp, as shown in Fig. 11.17, a zero differential input voltage will, in general, produce a nonzero output. Due to the op amp's high gain, the output will usually saturate at either L^+ or L^- if no feedback is applied. To obtain a zero output voltage, a nonzero input voltage, defined as the input offset voltage, V_{OS}, must be applied between the input terminals.

V_{OS} is generally quite small. It arises from small mismatches in the devices used in the op amp's input stage and circuit asymmetries. General-purpose op amps have offset voltages in the 0.1–10 mV range. Typically op amps with FET input devices will have higher offset voltages than op amps with bipolar input devices. If a very low V_{OS} is required, precision and low offset voltage op amps are available with offset voltages in the μV range. The offset voltage is not constant, it is subject to drift with time and changes in temperature. Consequently its effects will be evident in most circuits.

To analyze the effect of V_{OS}, a voltage source of unknown polarity, equal to V_{OS} is connected to one of the inputs of a ideal op amp. Then using superposition, the circuit's output voltage due to V_{OS} and the input voltage can be determined. Since superposition applies, the effect of V_{OS} on both the inverting and noninverting configurations is identical. As shown in Fig. 11.18(a), in which a practical op amp has been replaced with an ideal op amp in series with V_{OS}, and v_i set to zero, the output v_o, due to V_{OS} alone (i.e., the output offset voltage) is

$$v_o = V_{OS}(1 + R_2/R_1) \tag{11.34}$$

where it is evident that large dc gains result in a large output offset voltage. If dc gain is not required, a capacitor can be placed in series with R_1 to reduce the dc gain to unity. The resulting output offset voltage then becomes V_{OS}. Another group of circuits affected by V_{OS} are integrating circuits such as the one shown in Fig. 11.18(b). If R_f is not included, V_{OS} causes a dc current

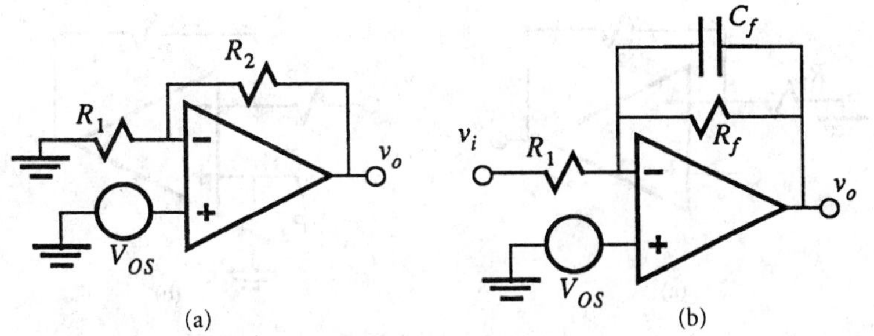

FIGURE 11.18 Some circuits affected by the op amps offset voltage. (a) The inverting and noninverting configurations. (b) Integrator circuits.

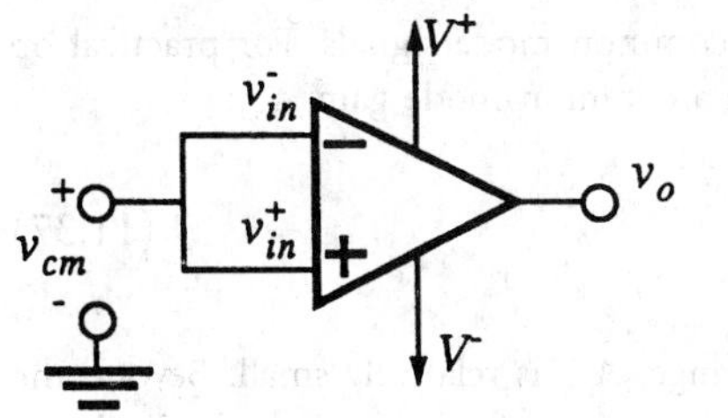

FIGURE 11.19 The op amp with its various input voltages.

V_{OS}/R_I to flow through R_1 and be integrated on C_f thereby causing the output voltage to saturate. Adding R_f limits the dc gain and hence, limits the output offset voltage.

In situations where the smallest possible V_{OS} is required, low V_{OS} op amps can be used. Alternatively, many op amps are provided with one or two terminals to which an offset nulling circuit can be attached. The op amp's V_{OS} can then be trimmed to zero. Since the trimming can only be done for one temperature, V_{OS} will still drift due to temperature and time and hence will limit the circuit's dc accuracy.

Finite CMRR and PSRR

If an op amp's inputs are shorted together, as shown in Fig. 11.19, variations in any one of the three voltages; v_{cm}, the common-mode input voltage, V^+, the positive supply voltage, or V^-, the negative supply voltage, should not affect the output voltage. Nevertheless, if all three voltages are changed by the same amount, it is evident that the output voltage must also change by the same amount [3]. Hence, the op amp's output voltage will be affected by changes in v_{cm}, V^+ and V^-. The relationship between changes in v_{cm} and v_o is usually characterized by the common-mode rejection ratio, CMRR. The effects of changes in the positive and negative supplies on v_o are usually referred to as the power supply rejection ratios, PSRR$^+$ and PSRR$^-$, respectively.

Common-Mode Rejection Ratio

Since the op amp has two input terminals, two signal types exist: differential signals and common-mode signals. Referring to Fig. 11.19, the differential signal v_d is the difference between the two input voltages, v_{in}^+ and v_{in}^-:

$$v_d = v_{\text{in}}^+ - v_{\text{in}}^- \tag{11.35}$$

while the common-mode signal v_{cm} is their average:

$$v_{cm} = (v_{\text{in}}^+ + v_{\text{in}}^-)/2 \tag{11.36}$$

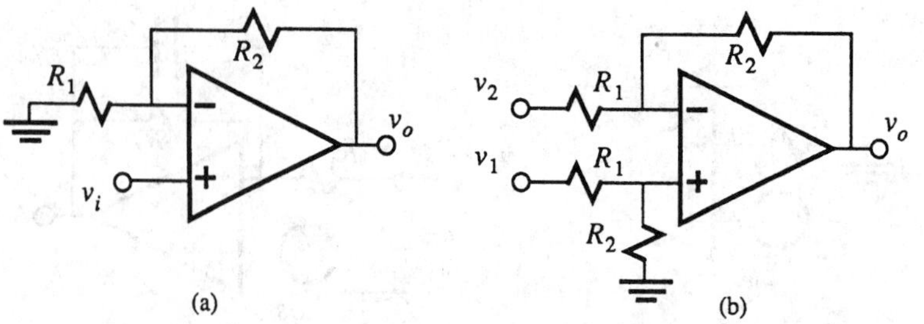

FIGURE 11.20 Some op amp circuits affected by a finite CMRR. (a) The noninverting configuration. (b) The differential configuration.

Ideally, the op amp rejects (i.e., does not respond to) common-mode signals. For practical op amps, changes in v_{cm} lead to changes in v_o, resulting in a common-mode gain A_{cm}:

$$A_{cm} = \frac{v_o}{v_{cm}} \tag{11.37}$$

Over a specified range, known as the common-mode range, A_{cm} is relatively small. Beyond the common-mode range, A_{cm} rises rapidly and the op amp ceases to function properly. Typically, the common-mode range does not extend to either the positive or negative supply. Single-supply op amps though, are usually designed to have a common-mode range that extends down to and often slightly below the lower supply. Within the common-mode range, A_{cm} is usually specified by the common-mode rejection ratio

$$\text{CMRR} = \frac{A_o}{A_{cm}} \tag{11.38}$$

The CMRR is usually expressed in decibels (dB)

$$\text{CMRR} = 20 \log \left| \frac{A_o}{A_{cm}} \right| \tag{11.39}$$

and ranges from 60 dB to over 120 dB. An alternate interpretation of the CMRR is the ratio of a change in v_{cm} to the resulting change in the op amp's V_{OS}.

$$\text{CMRR} = \frac{v_{cm}}{V_{OS}} \tag{11.40}$$

The two interpretations of CMRR are equivalent.

A finite CMRR affects those circuits for which a sizable v_{cm} is applied to the op amp. Hence, the inverting configuration, with a virtual ground at its input, is unaffected by the common-mode gain. On the other hand, circuits such as the noninverting configuration and the differential configuration shown in Fig. 11.20, have a nonzero v_{cm} and hence display common-mode problems.

The effects of a common-mode signal can be determined as follows. Since a finite CMRR can be interpreted as a change in the op amp's V_{OS}, due to the presence of a v_{cm}, the V_{OS} due to the nonzero v_{cm} and the finite CMRR can be found as

$$V_{OS} = v_{cm}/\text{CMRR} \tag{11.41}$$

Then the effect of the resulting V_{OS} can be found by analyzing the op amp assuming the common-mode gain is zero. For example, the noninverting circuit in Fig. 11.20(a), has a v_{cm} approximately equal to v_i, which leads to an equivalent V_{OS} of

$$V_{OS} = v_i/\text{CMRR} \tag{11.42}$$

and a total output voltage of

$$v_o = v_i(1 + 1/\text{CMRR})(1 + R_2/R_1) \tag{11.43}$$

Consequently, the CMRR leads to a gain error. For the differential configuration in Fig. 11.20(b), a finite CMRR leads to an output voltage of the form

$$v_o = \frac{R_2}{R_1}(v_1 - v_2) + \frac{R_2}{R_1}\frac{v_1}{\text{CMRR}} \tag{11.44}$$

where if $(v_1 + v_2)/2 \gg v_1 - v_2$, v_1 will be approximately equal to the signal's common-mode voltage and it can be seen that the differential amplifier responds to both the differential and common-mode components of the signal. Consequently, the op amp's CMRR can lead to problems if both of the op amp's terminal potentials vary with the input signal.

Power Supply Rejection Ratio

Ideally, changing either or both of an op amp's power supplies should not effect the op amp's performance. Practical op amps though, display a power-supply dependent gain and at higher frequencies, power-supply fluctuations are coupled into the op amp's signal path, leading to variations in the output voltage. These problems can be characterized as an equivalent gain, A_{V+} and A_{V-} from V^+ and V^- terminals, respectively, to the output. Alternatively, the variation can be characterized as a power supply dependent variation in the op amp's equivalent input offset voltage. Since the op amp is only supposed to amplify differential signals applied to its input, and reject signals applied to the power supplies, it is desirable to have $A_O \gg A_{V+}$ and $A_O \gg A_{V-}$. To measure this performance, the power-supply rejection ratios are used:

$$\text{PSRR}^+ = \frac{A_O}{A_{V+}} \tag{11.45}$$

$$\text{PSRR}^- = \frac{A_O}{A_{V-}} \tag{11.46}$$

Usually, the PSRRs are expressed in decibels (dB)

$$\text{PSRR}^+ = 20\log\left|\frac{A_0}{A_{V+}}\right| \tag{11.47}$$

$$\text{PSRR}^- = 20\log\left|\frac{A_0}{A_{V-}}\right| \tag{11.48}$$

PSRR's of 60 to 100 dB are common at dc. At higher frequencies, the PSRR decreases.

A noninfinite PSRR may pose a problem if there are variations in the power-supply voltages. Such variations can arise from either the ripple voltage of the supply itself or from large variations in the current being drawn from the supply. To reduce unwanted variations in the op amp's output voltage, either an op amp with a better PSRR can be selected or power supply decoupling capacitors can be used. If decoupling capacitors are used, they should be placed as close as possible to the op amp's power supply terminals.

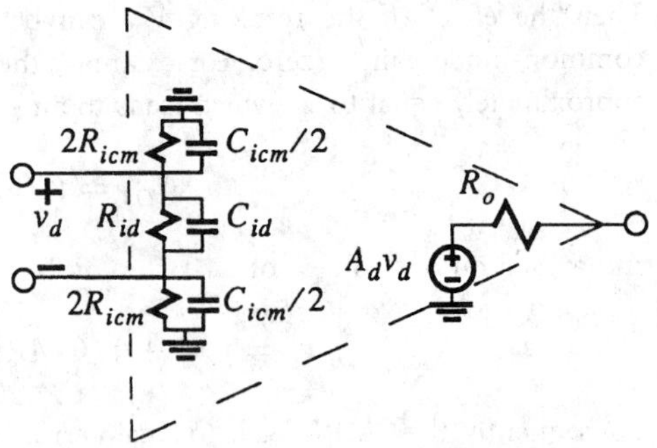

FIGURE 11.21 Input and output impedance of practical op amp.

Finite Input Impedance and Nonzero Output Impedance

Unlike the ideal op amp, practical op amps exhibit a finite input impedance and a nonzero output impedance. The input impedance is composed of a differential and a common-mode component as shown in Fig. 11.21 R_{id} and C_{id} represent the equivalent resistance and capacitance seen between the op amp's two input terminals. R_{icm} and C_{icm} represent the total resistance and capacitance to ground that would be seen by a common-mode signal applied to both input terminals. R_{id} ranges from 100 kΩ to over 100 MΩ. The higher differential input resistances are found in op amps employing FET input stages. R_{icm} is typically two orders of magnitude higher than R_{id}. The input capacitances are generally in the picofarad range. The output resistance R_o is usually in the 50–100 Ω range. Generally, the use of negative feedback reduces the effects of these impedances to levels where they can be neglected. Nevertheless, problems can arise in some applications.

The noninverting configuration is often used as a buffering amplifier due to its high input resistance. At low frequencies, the op amp's input capacitances can be neglected and the negative feedback provided by the op amp's high gain keeps the voltage across and the current through R_{id} negligible. Hence the effective input resistance is approximately $2R_{icm}$. At high frequencies, C_{id} shorts the input and the op amp's decreasing gain causes the voltage across and hence, the current through R_{id} to increase resulting in a significantly decreased input impedance.

Since most op amps employ shunt sampling negative feedback, the op amp's effective output resistance is reduced. Hence, even relatively high values of R_o can be tolerated in most circuits. Nevertheless if the circuit is used to drive a capacitive load, problems can arise. Since the op amp's gain decreases at higher frequencies (see Finite Differential Gain in this section), the amount of negative feedback also decreases leading to an output impedance that appears inductive (i.e., increases with frequency). This can be found by analyzing the output impedance of the circuit shown in Fig. 11.22(a). Approximating the op amp's gain as (see the section on finite differential gain)

$$A = \frac{w_t}{jw} \tag{11.49}$$

and assuming that $R_o \ll R_1 + R_2$, the output impedance becomes

$$Z_{\text{out}} \approx R_o \parallel jwL_{\text{eff}} \tag{11.50}$$

where

$$L_{\text{eff}} = \frac{R_o}{w_t}\left(\frac{R_1 + R_2}{R_1}\right) \tag{11.51}$$

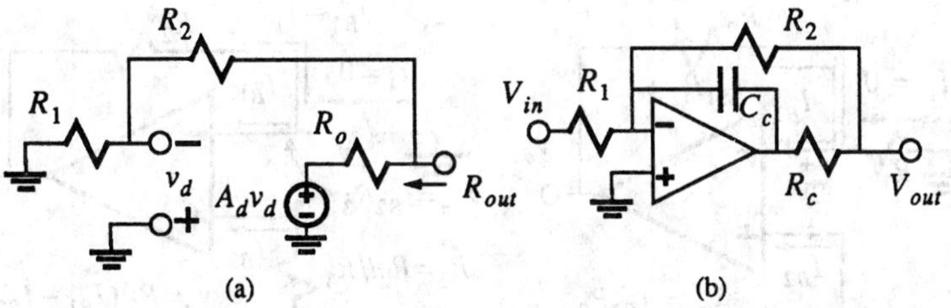

FIGURE 11.22 (a) Determining R_{out} for the inverting and noninverting configurations; (b) Compensating for large capacitive loads.

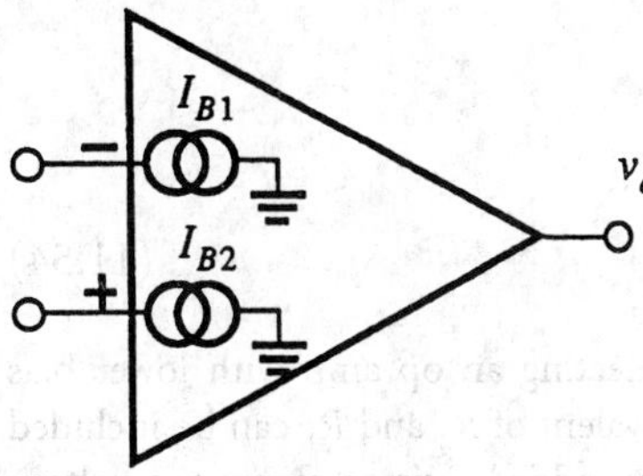

FIGURE 11.23 An op amp showing its input bias currents.

Consequently, if the circuit drives a large capacitor, it may become unstable due to the presence of L_{eff}. To reduce this problem a compensation network as shown in Fig. 11.22(b) is commonly used.

Input Bias Currents

When operating, a current flows in each of the op amp's input leads as shown in Fig. 11.23. These input currents, which are due to the internal structure of the op amp, give rise to errors in many circuits and prevent the practical realization of some circuit configurations. Since these currents cannot be avoided, they should be considered when designing op amp circuits.

The input currents are determined by the devices used to implement the amplifier's input stage. If BJT's are used, their base currents determine the input currents. If FET's are used, the input currents are due to the gate leakage current. In either case, the average of the two input currents, I_{B1} and I_{B2} is referred to as the input bias current, I_B:

$$I_B = (I_{B1} + I_{B2})/2 \tag{11.52}$$

where I_{B1} and I_{B2} are the input currents that cause the op amp's output to go to zero with a zero common-mode input voltage. I_B can range from 0.1 pA to 1 μA, which is much higher than would be expected based on the amplifier's finite input resistance alone. Typically, op amps with FET inputs display a much lower I_B than their bipolar counterparts. Due to mismatches, I_{B1} and I_{B2} are rarely equal. Their difference, referred to as the input offset current, I_{OS} is defined as

$$I_{OS} = |I_{B1} - I_{B2}| \tag{11.53}$$

I_{OS} is typically an order of magnitude lower than I_B. Therefore, in all but the most critical applications, I_{OS} can be neglected.

The effects of nonzero bias currents on both the inverting and noninverting configurations are shown in Fig. 11.24(a). If I_B is not compensated for, both op amp configurations will display

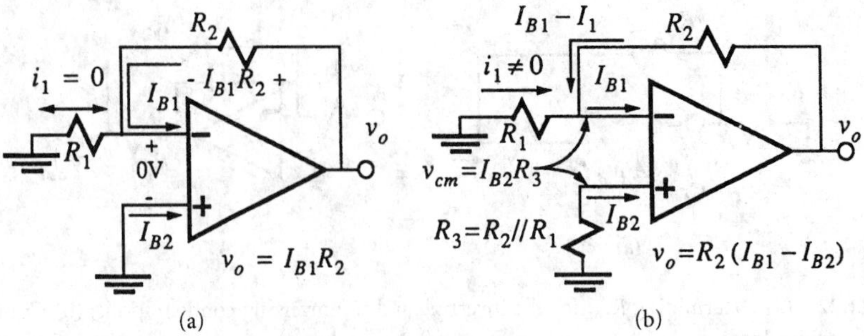

FIGURE 11.24 Analysis of the inverting and noninverting configurations with nonzero I_B. (a) Without bias current compensation. (b) With bias current compensation.

an output voltage of

$$v_O = I_{B1}R_2 \tag{11.54}$$

This voltage can be reduced either by reducing R_2 or by selecting an op amp with lower bias currents. Alternatively, a resistor R_3 equal to the parallel equivalent of R_2 and R_1 can be included in the positive terminal's lead, as shown in Fig. 11.24(b). This added resistor causes the voltage at the op amp's input to be equal to

$$v_{\text{cm}} = -I_{B2}R_1 \parallel R_2 \tag{11.55}$$

which, if two bias currents are equal ($I_{OS} = 0$), causes v_O to be zero. For the practical case of a nonzero offset current, v_O becomes

$$v_O = I_{OS}R_2 \tag{11.56}$$

Since I_{OS} is usually much lower than I_B, the error is greatly reduced. It is important to note that I_B is a dc current. Hence, R_3 should be equal to the equivalent dc resistance seen by the op amp's negative terminal.

Since all op amps require an I_B for proper operation, a dc path between each input terminal and ground must be provided. For example, the ac coupled buffer of Fig. 11.25(a) requires the addition of R_i to provide a path for I_{B2}. Unfortunately, R_i decreases the buffer's input resistance. In Fig. 11.25(b) the difference between I_{B1} and $I_{B2}R_3/R_1$ will flow to C_f, quickly leading to saturation of op amp's output at either L^+ or L^-. By adding R_f, a dc path for the difference current is provided. Unfortunately, R_f makes the integrator nonideal at low frequencies. Consequently the op amp's bias currents restrict the dc accuracy and the frequency range of applications for op amp circuits.

Electrical Noise

Like any electronic component, op amps generate noise that can degrade the system's signal-to-noise ratio (SNR). The amplifier's noise is characterized by the equivalent noise sources shown in Fig. 11.26(a). $\overline{e_n}$ is the equivalent input noise voltage density and is expressed in nV/$\sqrt{\text{Hz}}$. $\overline{i_{n1}}$ and $\overline{i_{n2}}$ are the equivalent input noise current density and are expressed in pA/$\sqrt{\text{Hz}}$. Usually, $\overline{i_{n1}}$ and $\overline{i_{n2}}$ have the same magnitude and are both referred to as $\overline{i_n}$. The typical behavior of $\overline{e_n}$ and $\overline{i_n}$ are illustrated in Fig. 11.26(b). At higher frequencies e_n and i_n are independent of frequency (i.e., white noise). In this range, values of $\overline{e_n}$ and $\overline{i_n}$ range from 50 nV/$\sqrt{\text{Hz}}$ and 0.6 pA/$\sqrt{\text{Hz}}$, respectively, for general-purpose op amps to 2 nV/$\sqrt{\text{Hz}}$ and 10 fA/$\sqrt{\text{Hz}}$ for ultra low-noise op

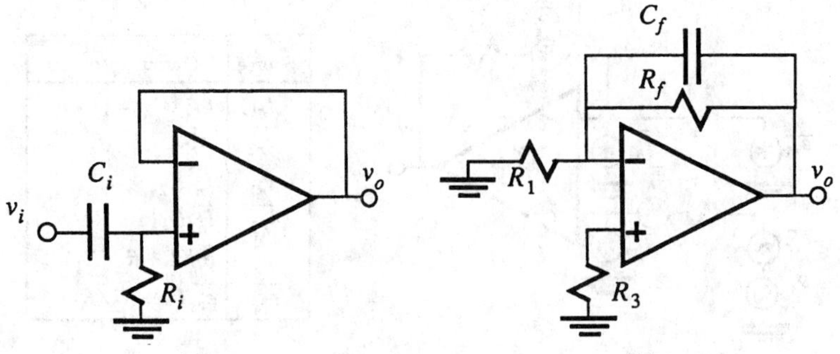

FIGURE 11.25 Some op amp circuits affected by the op amp's bias current. (a) The ac coupled buffer. (b) The inverting integrator.

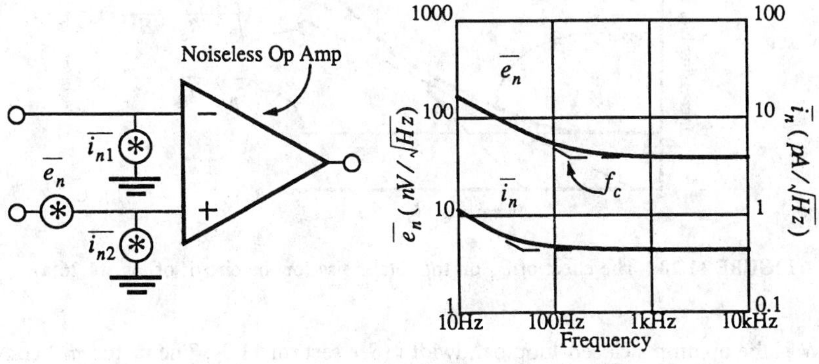

FIGURE 11.26 (a) Noise model of the op amp. (b) Input noise voltage and current densities versus frequency.

amps. At low frequencies, op amps display noise that increases with decreasing frequency (i.e., $1/f$ noise). To specify the low frequency noise, a plot such as that shown in Fig. 11.26(b) may be provided. In some cases only the noise corner frequency f_c as shown in Fig. 11.26(b) may be specified.

To determine the effects of the op amp's noise, the circuit shown in Fig. 11.26 can be used. Since the noise sources are generally uncorrelated, the total noise is the square root of the sum of the square of each noise source acting independently. Therefore the first step is to identify all the noise sources, as shown in Fig. 11.27. Each resistor may be modeled as a noiseless resistor in series with a noise voltage density of

$$\overline{e_r} = \sqrt{4kTR}/\sqrt{\text{Hz}} \qquad (11.57)$$

where k is Boltzmann's constant and T is absolute temperature. Each noise source then gives rise to the output values shown in the table. Since the designer is free to choose R_1 and R_2, the noise sources $\overline{e_n}$, $\overline{e_{rs}}$, and $\overline{i_{n2}}$ typically determine the circuit's total output noise voltage. The total rms white noise voltage at the output is

$$\overline{v_{\text{output}}} = \sqrt{\frac{\pi}{2}\text{BW}}\sqrt{\overline{e_N^2}\left(1+\frac{R_2}{R_1}\right)^2 + \overline{e_{rs}^2}\left(1+\frac{R_2}{R_1}\right)^2 + \overline{i_{n2}^2}R_s\left(1+\frac{R_2}{R_1}\right)^2} \qquad (11.58)$$

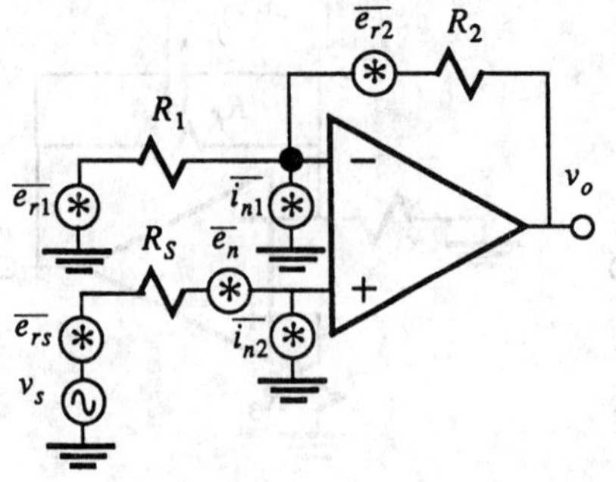

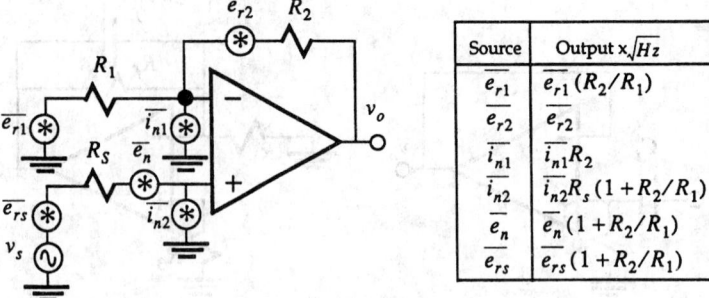

Source	Output x$\sqrt{Hz}$
$\overline{e_{r1}}$	$\overline{e_{r1}}\,(R_2/R_1)$
$\overline{e_{r2}}$	$\overline{e_{r2}}$
$\overline{i_{n1}}$	$\overline{i_{n1}}R_2$
$\overline{i_{n2}}$	$\overline{i_{n2}}R_s\,(1+R_2/R_1)$
$\overline{e_n}$	$\overline{e_n}\,(1+R_2/R_1)$
$\overline{e_{rs}}$	$\overline{e_{rs}}\,(1+R_2/R_1)$

FIGURE 11.27 An op amp circuit showing its noise sources.

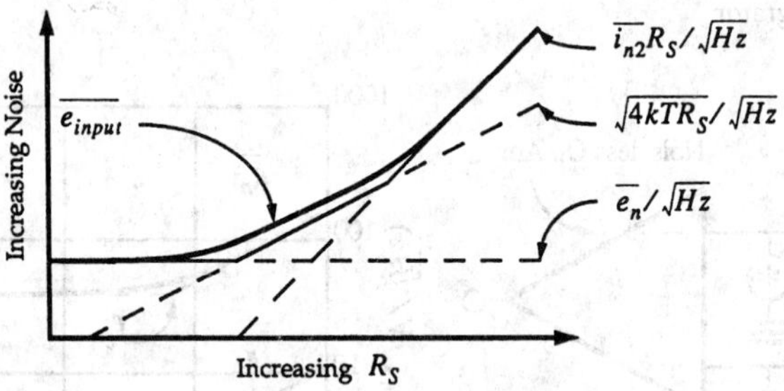

FIGURE 11.28 The effect of R_S on the total noise for the circuit of Fig. 11.26(a).

where BW is the op amp's closed-loop bandwidth (See section 11.3). The factor $\pi/2$ converts the closed-loop bandwidth to the noise equivalent bandwidth, assuming the op amp is characterized by a single pole. The equivalent voltage white-noise density at the input is

$$\overline{e_{\text{input}}} = \sqrt{\overline{e_n^2} + 4kTR_S + \overline{i_{n2}^2}R_S^2}\Big/\sqrt{\text{Hz}} \tag{11.59}$$

The relative importance of each of these three factors depends on the value of R_S as shown in Fig. 11.28. For low values of R_S, $\overline{e_n}$ dominates. At high values of R_S, $\overline{i_{n2}}$ dominates. In the middle region, R_S dominates if $\sqrt{4kT} > \sqrt{\overline{i_n}\,\overline{e_n}}$.

Two measures are used to specify the circuit's noise performance: the SNR and the noise figure (NF). The SNR is the ratio of the signal power to the total noise power, which assuming only white noise is present, can be expressed as

$$\text{SNR} = 20\log\left|\frac{v_s}{\overline{e_{\text{input}}}\sqrt{(\pi\,\text{BW})/2}}\right| \tag{11.60}$$

The second measure, NF, expresses the increase in noise due to the amplifier over that due to the source resistance alone:

$$\text{NF} = 10\log\left|\frac{\overline{e_n^2} + \overline{e_{rs}^2} + \overline{i_{n1}^2}R_S^2}{\overline{e_{rs}^2}}\right| \tag{11.61}$$

While a low NF is desirable, it is usually more important to minimize the total noise to achieve the highest possible SNR.

Summary

Practical op amps suffer from a wide range of nonidealities. The dominant effect of these nonidealities is to limit the range of applications for which an op amp can be used. Problems such as the offset voltage, the CMRR, the bias currents and the electrical noise will limit the accuracy of op amp circuits. The op amp's finite gain and saturation levels will limit the maximum controllable gain of an op amp circuit. Factors such as the op amp's saturation limits and its input and output impedances will limit the range of impedances that can be buffered by or driven by an op amp. In many cases, circuit techniques or special-purpose op amps can be used to reduce the detrimental effects of the op amp's nonidealities.

eferences

[1] E. J. Kennedy, *Operational Amplifier Circuits: Theory and Applications*, New York: Holt, Rinehart and Winston, 1988.

[2] J. K. Roberge, *Operational Amplifier: Theory and Practice*, New York: Wiley, 1975.

[3] E. Säckinger, J. Goette, and W. Guggenbühl, "A general relationship between amplifier parameters, and its application to PSRR improvement," *IEEE Trans. Circuits Syst.*, vol. 38, pp. 1173–1181, Oct. 1991.

[4] A. S. Sedra and K. C. Smith, *Microelectronic Circuits*, 3rd Ed., New York: Holt, Rinehart and Winston, 1991.

1.3 Frequency- and Time-Domain Considerations

Sergio Franco

One of the most important limitations of practical op amps is gain rolloff with frequency. This limitation affects both the frequency-domain and the time-domain behavior of circuits built around op amps. We have linear effects, such as finite small-signal bandwidth and nonzero rise time, and nonlinear effects, such as slew-rate limiting and finite full-power bandwidth. Additional effects are the settling time and intermodulation distortion. We discuss these limitations both for voltage-mode and current-mode op amps [1].

Voltage-Mode Op Amps

Conventional op amps, the most popular representative of which is without doubt the 741 type, are voltage-mode amplifiers because in order to produce an output they require an input imbalance of the voltage type. Consequently, when a negative feedback loop is created around the op amp, the signal returned to the input must be in the form of a voltage.

Block Diagram

Shown in Fig. 11.29 is a simplified circuit diagram [2] that can be used to describe a wide variety of practical voltage-mode op amps, including the popular 741. As illustrated in block-diagram form in Fig. 11.30, the circuit is made up of three stages.

1) The *input stage*, consists of transistors Q_1 through Q_4, whose function is to sense any imbalance between the inverting and noninverting input voltages V_n and V_p, and convert it to a single-ended output current I_{o1}. This stage is also designed to provide high input impedance and draw negligible input currents.

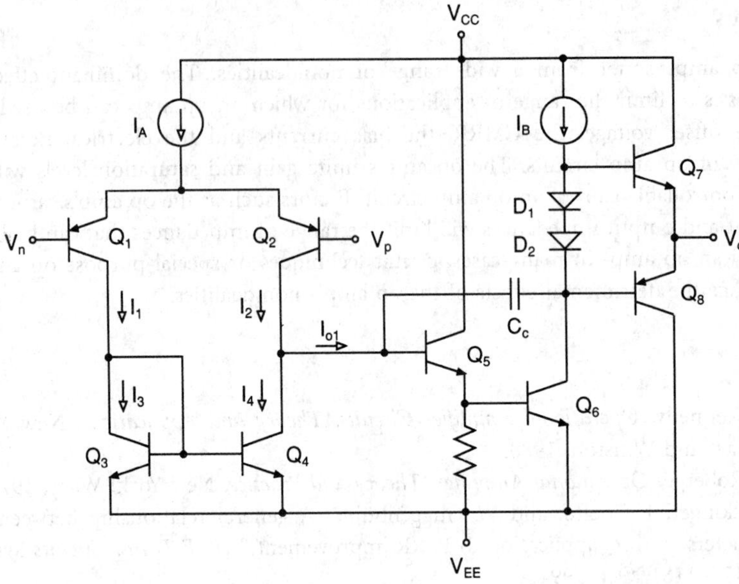

FIGURE 11.29 Simplified circuit diagram of a voltage-mode op amp.

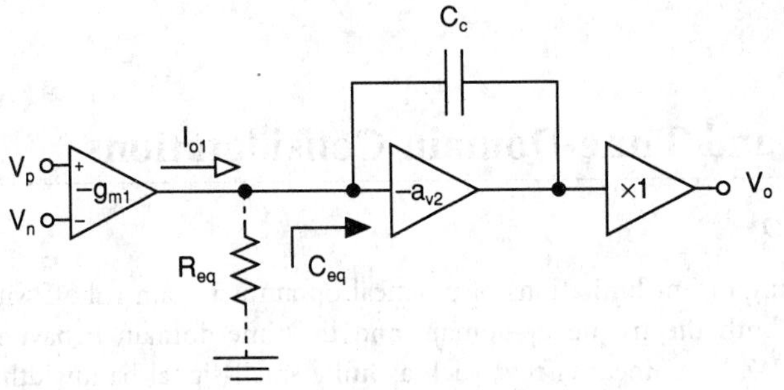

FIGURE 11.30 Voltage-mode op amp block diagram.

Q_1 and Q_2 form a differential pair whose task is to split the bias current I_A into two currents I_1 and I_2 in amounts controlled by the imbalance between V_n and V_p. If this imbalance is sufficiently small, we can write $I_1 - I_2 = g_{m1}(V_p - V_n)$, where g_{m1} is the transconductance of Q_1 and Q_2.

Ignoring transistor base currents, we have $I_3 = I_1$. In response to current I_3, Q_3 develops a base-emitter voltage drop that is then applied to Q_4, forcing the latter to draw the same amount of current as the former, or $I_4 = I_3$. For obvious reasons, Q_3 and Q_4 are said to form a current mirror. Summing currents, we obtain $I_{o1} = I_2 - I_4 = I_2 - I_3 = I_2 - I_1$, or

$$I_{o1} = -g_{m1}(V_p - V_n) = -g_{m1}V_d \tag{11.62}$$

where $V_d = V_p - V_n$ is called the differential input voltage.

2) The **intermediate stage**, consists of Darlington pair Q_5–Q_6 and frequency-compensation capacitance C_c. Its function is to provide additional gain as well as to introduce a dominant pole in the open-loop response of the amplifier. Denoting the net equivalent resistance and

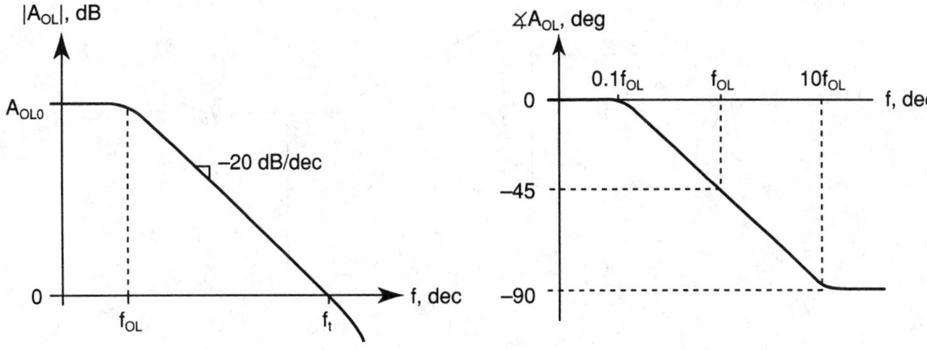

FIGURE 11.31 Frequency plots of the open-loop gain.

capacitance between the input node of this stage and ground as R_{eq} and C_{eq}, the pole frequency is

$$f_{\mathrm{OL}} = \frac{1}{2\pi R_{\mathrm{eq}} C_{\mathrm{eq}}} \tag{11.63}$$

By the Miller effect we have $C_{eq} = (1 + a_{v2})C_c$, where $-a_{v2}$ is the voltage gain of the Darlington pair.

3) The **output stage**, consists of emitter–followers Q_7 and Q_8, and biasing diodes D_1–D_2. Though the voltage gain of this stage is only unity, its current gain can be fairly high, indicating that this stage acts as a power amplifier. Its function is also to provide a low output impedance. Q_7 and Q_8 are referred to as a push-pull pair because in the presence of an output load, Q_7 sources (or pushes) current to the load, and Q_8 sinks (or pulls) current from the load.

The small-signal transfer characteristic of the op amp is

$$V_o = A_{\mathrm{OL}}(jf)V_d \tag{11.64}$$

where $A_{\mathrm{OL}}(jf)$, called the *open-loop voltage gain*, is a complex function of frequency f, and $j = \sqrt{-1}$ is the imaginary unit. With dominant pole compensation, this function can be approximated as

$$A_{\mathrm{OL}}(jf) = \frac{A_{\mathrm{OL}\,0}}{1 + jf/f_{\mathrm{OL}}} \tag{11.65}$$

where $A_{\mathrm{OL}\,0}$ and f_{OL} are, respectively, the *dc value* and *bandwidth* of $A_{\mathrm{OL}}(jf)$. For the circuit shown, $A_{\mathrm{OL}\,0} = g_{m1}R_{\mathrm{eq}}a_{v2}$. As an example, the popular 741 op amp has $A_{\mathrm{OL}\,0} \simeq 2 \times 10^5$ V/V and $f_{\mathrm{OL}} \simeq 5$ Hz. Figure 11.31 shows the Bode plots of $A_{\mathrm{OL}}(jf)$. We make the following observations.

1) For $f \ll f_{\mathrm{OL}}$ we have $|A_{\mathrm{OL}}(jf)| \simeq A_{\mathrm{OL}\,0}$ and $\measuredangle A_{\mathrm{OL}}(jf) \simeq 0°$, indicating an approximately constant gain and negligible delay.

2) For $f = f_{\mathrm{OL}}$ we have $|A_{\mathrm{OL}}(jf)| = A_{\mathrm{OL}\,0}/\sqrt{2}$ and $\measuredangle A_{\mathrm{OL}}(jf) = -45°$. Rewriting as $|A_{\mathrm{OL}}(jf)|_{\mathrm{dB}} = 20\log_{10}|A_{\mathrm{OL}}(jf)| = |A_{\mathrm{OL}\,0}|_{\mathrm{dB}} - 3$ dB explains why f_{OL} is also referred to as the -3 dB frequency or the half-power frequency of the open-loop response.

3) For $f \geq f_{\mathrm{OL}}$ gain rolls off with frequency at a constant rate of -20 dB/dec, and it can be approximated as $|A_{\mathrm{OL}}(jf)| \simeq A_{\mathrm{OL}\,0}/(f/f_{\mathrm{OL}})$ and $\measuredangle A_{\mathrm{OL}}(jf) \simeq -90°$. Rewriting as

$$|A_{\mathrm{OL}}(jf)| \times f \simeq A_{\mathrm{OL}\,0} \times f_{\mathrm{OL}} = f_t \tag{11.66}$$

indicates that in the rolloff region the op amp exhibits a constant gain-bandwidth product (constant GBP). Increasing frequency by a given amount causes gain to decrease by the same

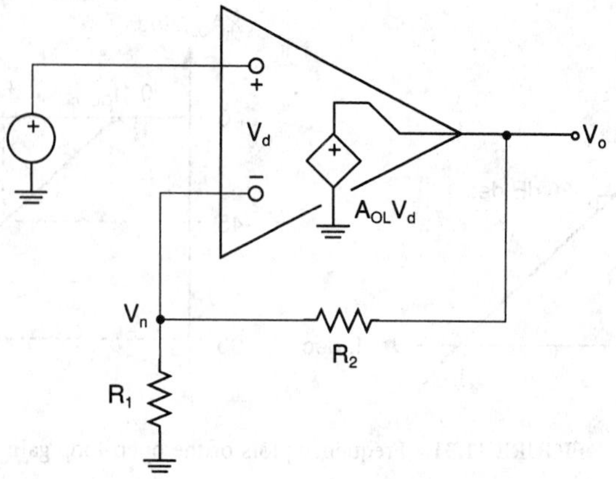

FIGURE 11.32 The noninverting configuration.

amount. The frequency $f_t = A_{OL0} f_{OL}$ at which gain drops to 0 dB is aptly called the transition frequency. For the 741 op amp, $f_t = 2 \times 10^5 \times 5 = 1$ MHz.

Closed-Loop Frequency Response

Figure 11.32 shows a simplified model of the voltage-mode op amp, along with external circuitry to create the popular noninverting configuration. The resistors sample V_o and feed the voltage

$$V_n = \frac{R_1}{R_1 + R_2} V_o = \beta V_o \tag{11.67}$$

back to the inverting input. The parameter

$$\beta = \frac{R_1}{R_1 + R_2} = \frac{1}{1 + R_2/R_1} \tag{11.68}$$

representing the fraction of the output being fed back to the input is called the feedback factor. By inspection,

$$V_o = A_{OL}(jf) V_d = A_{OL}(jf)(V_i - \beta V_o) \tag{11.69}$$

In negative-feedback parlance $V_d = V_i - \beta V_o$ is referred to as the error signal. Collecting and solving for the ratio V_o/V_i yields, after minor algebraic manipulations,

$$A_{CL}(jf) = \frac{V_o}{V_i} = \left(1 + \frac{R_2}{R_1}\right) \frac{1}{1 + 1/T(jf)} \tag{11.70}$$

where $A_{CL}(jf)$ is called the **closed-loop gain**, and

$$T(jf) = O_{OL}(jf)\beta = \frac{A_{OL}(jf)}{1 + R_2/R_1} \tag{11.71}$$

is called the **loop gain**. This designation stems from the fact that a voltage propagating clockwise around the loop is first magnified by $A_{OL}(jf)$, and then attenuated by β, thus experiencing an overall gain of $T(jf) = A_{OL}(jf)\beta$.

By (11.70) we have

$$\lim_{T\to\infty} A_{\text{CL}} = 1 + \frac{R_2}{R_1} \tag{11.72}$$

a value aptly called the ideal closed-loop gain. Clearly, T provides a measure of how close A_{CL} is to ideal: the larger T, the better. To ensure a substantial loop gain for a range of closed-loop gains, op amp manufacturers strive to make A_{OL} as large as possible. Consequently, V_d will assume extremely small values since $V_d = V_o/A_{\text{OL}}$. In the limit $A_{\text{OL}} \to \infty$ we obtain $V_d \to 0$, that is, $V_n \to V_p$. This provides the basis for the familiar ideal voltage-mode op amp rule: When operated with negative feedback, an op amp will provide whatever output is needed to drive its error signal V_d to zero or, equivalently, to force V_n to track V_p.

Substituting (11.65) into (11.71) and then into (11.70), and exploiting the fact that $\beta A_{\text{OL}\,0} \gg 1$, we obtain, after minor algebra,

$$A_{\text{CL}}(jf) = \frac{A_{\text{CL0}}}{1 + jf/f_{\text{CL}}} \tag{11.73}$$

where

$$A_{\text{CL0}} = 1 + \frac{R_2}{R_1} = \frac{1}{\beta} \tag{11.74}$$

is the closed-loop dc gain, and

$$f_{\text{CL}} = \beta f_t = \frac{f_t}{A_{\text{CL0}}} \tag{11.75}$$

is the closed-loop small-signal bandwidth. The quantity $1/\beta$ is also called the **dc noise gain** because this is the gain with which the amplifier will magnify any dc noise present right at its input pins, such as the input offset voltage V_{OS}.

Equation (11.75) indicates a gain-bandwidth trade-off. As we raise the R_2/R_1 ratio to increase A_{CL0}, we also decrease f_{CL} in the process. Moreover, by Eq. (11.71), $T(jf)$ is also decreased, thus leading to a greater departure of $A_{\text{CL}}(jf)$ from the ideal.

The above concepts can be visualized graphically as follows. By (11.70) we can write $|T|_{dB} = 20\log_{10}|T| = 20\log_{10}|A_{\text{OL}}| - 20\log_{10}(1/\beta)$, or

$$|T| = |A_{\text{OL}}|_{\text{dB}} - |1/\beta|_{\text{dB}} \tag{11.76}$$

indicating that the loop gain can be found graphically as the difference between the decibel plot of the open-loop gain and that of the noise gain. This is illustrated in Fig. 11.33. The frequency at which the two curves meet is aptly called the **crossover frequency**. It is readily seen that at this frequency we have $T = 1\underline{/-90^\circ} = -j$, so $|A_{\text{CL}}| = A_{\text{CL0}}/|1+j| = A_{\text{CL0}}/\sqrt{2}$, by (11.70) and (11.74). Consequently, the crossover frequency represents the -3 dB frequency of $A_{\text{CL}}(jf)$, that is, f_{CL}.

We now see that increasing A_{CL0} reduces T and causes the cross-point to move up the $|A_{\text{OL}}|$ curve, thus decreasing f_{CL}. The circuit with the widest bandwidth and the highest loop gain is also the one with the lowest closed-loop gain. This is the familiar voltage follower, obtained by letting $R_1 = \infty$ and $R_2 = 0$. Then, by (11.74) and (11.75) we have $A_{\text{CL0}} = 1$ and $f_{\text{CL}} = f_t$.

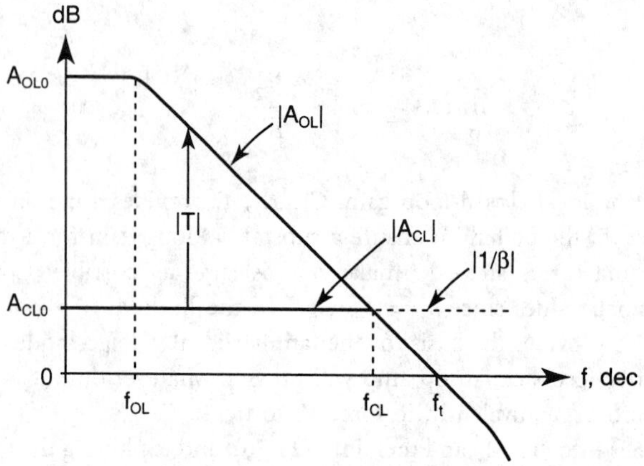

FIGURE 11.33 Graphical interpretation of the loop gain.

Let us now turn to another important configuration, namely, the popular inverting amplifier of Fig. 11.34. Since $V_p = 0$, it follows that $V_d = -V_n$. Applying the superposition principle we have

$$V_d = -\frac{R_2}{R_1 + R_2}V_i - \frac{R_1}{R_1 + R_2}V_o = \frac{-R_2}{R_1 + R_2}V_i - \beta V_o \tag{11.77}$$

indicating that the feedback factor $\beta = R_1/(R_1 + R_2)$ is the same as for the noninverting configuration. Substituting into (11.63), we find the closed-loop gain as

$$A_{CL}(jf) = \frac{V_o}{V_i} = \left(-\frac{R_2}{R_1}\right)\frac{1}{1 + 1/T(jf)} \tag{11.78}$$

Moreover, proceeding as for the noninverting configuration, we get

$$A_{CL}(jf) = \frac{A_{CL0}}{1 + jf/f_{CL}} \tag{11.79}$$

where

$$A_{CL0} = -\frac{R_2}{R_1} \tag{11.80}$$

is the closed-loop dc gain, and

$$f_{CL} = \beta f_t \tag{11.81}$$

is the closed-loop small-signal bandwidth. We can again find this bandwidth as the intercept of the $|A_{OL}|_{dB}$ and $|1/\beta|_{dB}$ curves. However, since we now have $|A_{CL0}| < |1/\beta|$, it follows that the $|A_{CL}|_{dB}$ curve will be shifted downward, as explicitly shown in Fig. 11.34.

Before concluding, we wish to point out that open-loop gain rolloff affects not only the closed-loop gain, but also the closed-loop input and output impedances. The interested reader can find additional information in the literature [3].

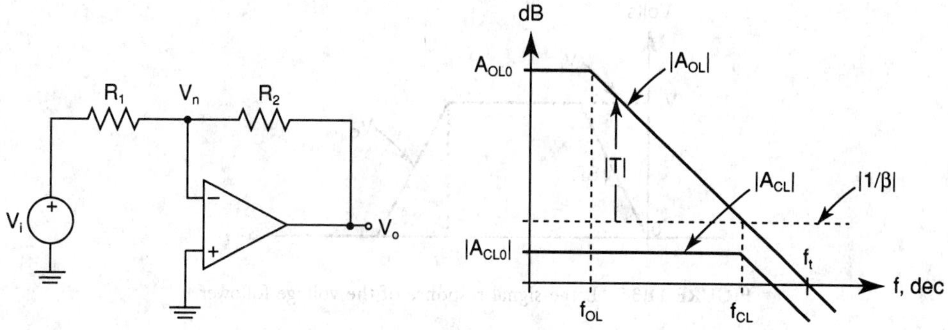

FIGURE 11.34 The inverting configuration,

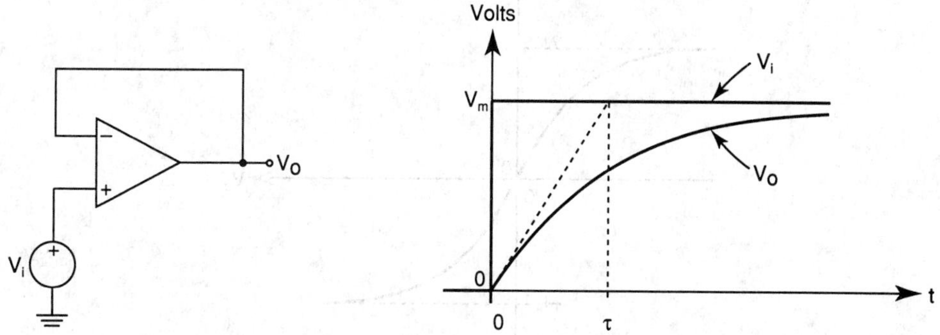

FIGURE 11.35 Voltage follower and its small-signal step response.

Closed-Loop Transient Response

To fully characterize the dynamic behavior of an op amp circuit we also need to know its transient response. This response is usually specified for the case of the op amp operating as a unity-gain voltage follower. As we know, its small-signal transfer characteristic is $V_o = V_i/(1 + jf/f_t)$. This is formally similar to that of an ordinary *RC* circuit. Subjecting a voltage follower to a step of suitably small amplitude V_m, as shown in Fig. 11.35, will cause an exponential output transition with the time constant

$$\tau = \frac{1}{2\pi f_t} \tag{11.82}$$

The rise time t_r, defined as the time it takes for V_o to swing from 10 to 90 percent of V_m, provides a measure of how rapidly the transition takes place. One can readily see that $t_r = \tau \ln 9 \simeq 2.2\tau$. For the 741 op amp we have $\tau = 1/(2\pi \times 10^6) \simeq 159$ ns, and $t_r \simeq 350$ ns.

The rate at which V_o changes with time is highest at the beginning of the exponential transition, when its value is V_m/τ. Increasing the step magnitude V_m increases this initial rate, until a point is reached beyond which the rate saturates at a constant value called the **slew rate** (SR). The transition is now a ramp, rather than an exponential. Figure 11.36 shows the slew-rate limited response to a pulse.

Slew-rate limiting stems from the limited ability of the internal circuitry to charge or discharge the compensation capacitance C_c. To understand this effect, refer to Fig. 11.37. As long as the input imbalance V_d is sufficiently small, (11.62) still holds and the step response is exponential. However, for large values of V_d, I_{o1} is no longer linearly proportional to V_d, but saturates at $\pm I_A$, where I_A is the input-stage bias current depicted in Fig. 11.29. Turning now to Fig. 11.30 and

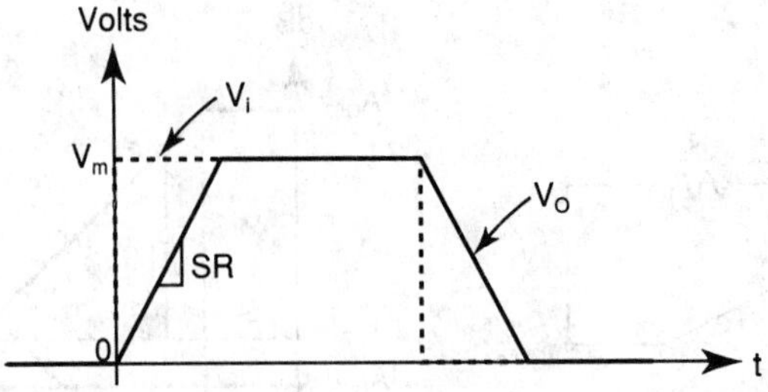

FIGURE 11.36 Large-signal response of the voltage follower.

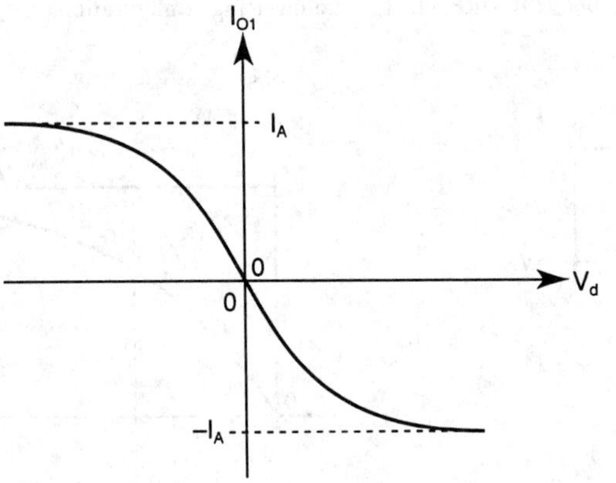

FIGURE 11.37 Actual transfer characteristic of the first stage.

observing that the second stage acts as an integrator, we can state that the maximum rate at which C_c can be charged or discharged is $(dV_o/dt)_{max} = I_A/C_c$. This is precisely the slew rate,

$$SR = \frac{I_A}{C_c} \tag{11.83}$$

The 741 op amp has typically $I_A = 20$ μA and $C = 30$ pF, so SR $= 20 \times 10^{-6}/(30 \times 10^{-12}) = 0.67$ V/μs. To respond to a 10-V input step a 741 follower will take approximately $10/0.67 = 15$ μs.

The step magnitude corresponding to the onset of slew-rate limiting is such that $V_m/\tau = \text{SR}$, or $V_m = \text{SR} \times \tau = \text{SR}/(2\pi f_t)$. For the 741 op amp, $V_m = 0.67 \times 10^6/(2\pi \times 10^6) = 106$ mV. This means that as long as the input step is less than 106 mV, a 741 follower will respond with an exponential transition governed by $\tau \simeq 159$ ns. For a greater input step, however, the output will slew at a constant rate of 0.67 V/μs, and it will do so until it comes within 106 mV of the final value, after which it will complete the transition in exponential fashion.

In certain applications it is important to know the **settling time** t_s, defined as the time it takes for the output to settle within a specified band around its final value, usually for a full-scale output transition. It is apparent that slew-rate limiting plays an important role in the settling-time characteristic of a circuit.

Slew-rate limiting affects also the **full-power bandwidth** (FPBW), defined as the maximum frequency at which the circuit still yields an undistorted full-power output. Letting $V_o = V_m \sin 2\pi ft$, we have $(dV_o/dt)_{max} = (2\pi f V_m \cos 2\pi ft)_{max} = 2\pi f V_m$. Equating it to the slew rate

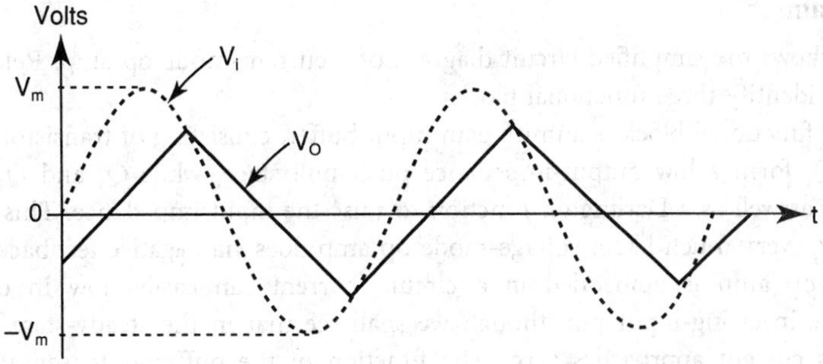

FIGURE 11.38 Distortion when the FPBW is exceeded.

SR and solving for f, whose value is the FPBW, we get

$$\text{FPBW} = \frac{\text{SR}}{2\pi V_m} \tag{11.84}$$

For instance, for $V_m = 10\ V$, the 741 op amp has FPBW $= \text{SR}/(20\pi) = 10.6$ kHz. Figure 11.38 shows the distorted response of a voltage follower to a full-power input with a frequency higher than the FPBW.

From (11.63) and (11.83) it is apparent that the primary cause of frequency and slew-rate limitations is the capacitance C_c. Why not eliminate C_c altogether? Without C_c the open-loop response would exhibit a much wider bandwidth, but also a much greater phase lag because of the various poles introduced by the transistors making up the op amp. We are interested in the situation at the crossover frequency, where $|T| = 1$. Should the phase shift at this frequency reach $-180°$, we would have $T = 1\underline{/-180°} = -1$ which, after substitution into (11.70), would yield $|A_{\text{CL}}| \to (1 + R_2/R_1)/(1-1) \to \infty$! The physical meaning of an infinite closed-loop gain is that the circuit would be capable of sustaining a nonzero output with a vanishingly small external input. But, this is the recipe for sustained oscillation! It is precisely to avoid this possibility that the manufacturer incorporates the frequency-compensation capacitance C_c.

As mentioned, C_c causes gain to roll off, so that by the time the frequency of $-180°$ phase lag is reached, $|A_{\text{OL}}|$ has already dropped well below 0 dB, making it impossible for the circuits of Figs. 11.32, 11.34, and 11.35 to achieve $T = -1$ at the crossover frequency, regardless of the values of R_1 and R_2. This requires that the dominant-pole frequency be suitably low, and thus, by (11.63), that C_{eq} be suitably large. To avoid the need to manufacture impractically large on-chip capacitances, it is customary to start out with a realistic value, such as 30 pF for the 741 op amp, and then exploit the multiplicative action of the Miller effect to raise it to the desired equivalent value.

The hardest configuration to compensate is the unity-gain voltage follower because its crossover frequency is the closest to the frequency region of additional phase lag stemming from the higher-order poles of the op amp. This is why this particular configuration is usually chosen for the specification of the transient response.

Current-Mode Op Amps

Current-mode op amps exploit a special circuit topology, along with high-speed complementary bipolar processes, to achieve much faster dynamics than their voltage-mode amplifier counterparts. The name stems from the fact that these amplifiers respond to an input imbalance of the current type, and the signal propagating around the feedback loop is thus in the form of a current rather than a voltage.

Block Diagram

Figure 11.38 shows the simplified circuit diagram of a current-mode op amp. Referring also to Fig. 11.39, we identify three functional blocks.

1) The first functional block is a unity-gain input buffer, consisting of transistors Q_1 through Q_4. Q_1 and Q_2 form a low output-impedance push–pull stage, while Q_3 and Q_4 provide V_{BE} compensation as well as a Darlington function to raise the input impedance. This buffer forces V_n to follow V_p, very much like a voltage-mode op amp does via negative feedback.

When the op amp is embedded in a circuit, current can easily flow in or out of its low-impedance inverting-input pin, though we shall see that in the steady-state (nonslewing) condition this current approaches zero. The function of the buffer is to sense this current, denoted as I_n, and produce an imbalance

$$I_1 - I_2 = I_n \tag{11.85}$$

between the push–pull transistor currents I_1 and I_2.

2) The second block is a pair of current mirrors Q_5–Q_6 and Q_7–Q_8, which reflect currents I_1 and I_2 and sum them at a common junction node. The current into this node thus equals I_n, as shown.

3) Finally, a unity-gain output buffer, consisting of transistors Q_9 through Q_{12}, buffers the summing node voltage to the outside and provides a low output impedance for the overall op amp.

Denoting the net equivalent resistance and capacitance of the summing node towards ground as R_{eq} and C_{eq}, we can write

$$V_o = Z_{OL}(jf)I_n \tag{11.86}$$

where $Z_{OL}(jf)$, called the *open-loop transimpedance gain*, is the impedance due to the parallel combination of R_{eq} and C_{eq}. This impedance can be expressed as

$$Z_{OL}(jf) = \frac{Z_{OL\,0}}{1 + jf/f_{OL}} \tag{11.87}$$

where $Z_{OL\,0} = R_{eq}$, and

$$f_{OL} = \frac{1}{2\pi R_{eq} C_{eq}} \tag{11.88}$$

As an example in Fig. 11.40, the CLC401 current-mode op amp (Comlinear Co.) has $Z_{OL\,0} \simeq 710$ kΩ, $f_{CL} \simeq 350$ kHz, and $C_{eq} = 1/(2\pi R_{eq} f_{OL}) \simeq 0.64$ pF.

We observe a formal similarity with voltage-mode op amps, except that now the error signal I_n is a current rather than a voltage, and the gain $Z_{OL}(jf)$ is in V/A rather than in V/V. For this reason curent-mode op amps are also referred to as transimpedance op amps. The gain $Z_{OL}(jf)$ is approximately constant from dc to f_{OL}, after which it rolls off with frequency at a constant rate of -1 dec/dec.

Closed-Loop Characteristics

Figure 11.41 shows a simplified model of the current-mode op amp, along with an external feedback network to configure it as a noninverting amplifier. Any attempt to unbalance the inputs will cause the input buffer to source (or sink) an imbalance current I_n to the external network. By (11.86), this imbalance causes V_o to swing in the positive (or negative) direction until the original imbalance current is neutralized via the negative feedback loop.

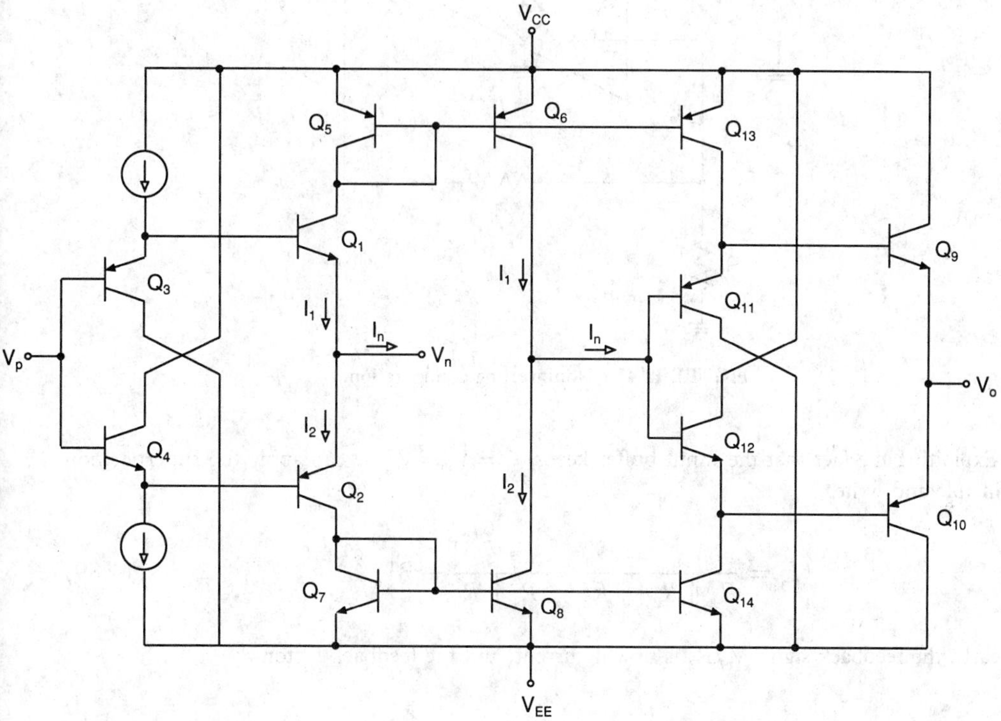

FIGURE 11.39 Simplified circuit diagram of a current-mode op amp.

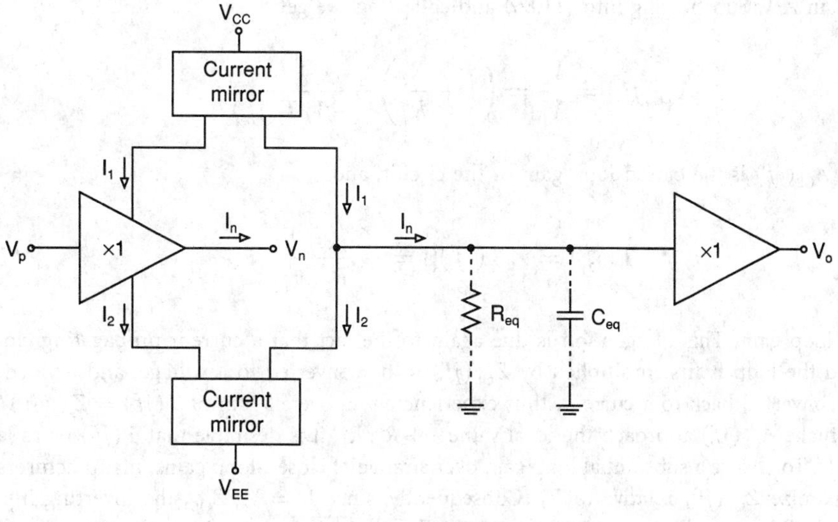

FIGURE 11.40 Current-mode op amp block diagram.

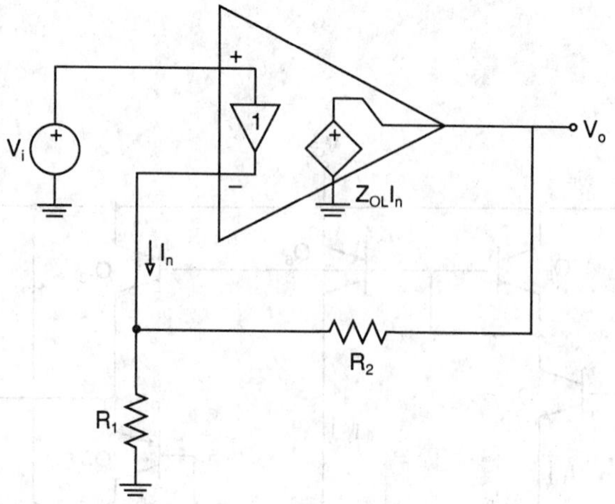

FIGURE 11.41 Noninverting configuration.

Exploiting the fact that the input buffer keeps $V_n = V_p = V_i$, we can apply the superposition principle and write

$$I_n = \frac{V_i}{R_1 \parallel R_2} - \frac{V_o}{R_2} = \frac{V_i}{R_1 \parallel R_2} - \beta V_o \tag{11.89}$$

Clearly, the feedback signal V_o/R_2 is now a current, and the feedback factor

$$\beta = \frac{1}{R_2} \tag{11.90}$$

is now in A/V. Substituting into (11.86) and collecting, we get

$$A_{\text{CL}}(jf) = \frac{V_o}{V_i} = \left(1 + \frac{R_2}{R_1}\right)\frac{1}{1 + 1/T(jf)} \tag{11.91}$$

where $A_{\text{CL}}(jf)$ is the closed-loop gain of the circuit, and

$$T(jf) = Z_{\text{OL}}(jf)\beta = \frac{Z_{\text{OL}}(jf)}{R_2} \tag{11.92}$$

is the loop gain. This designation is due again to the fact that a current propagating clockwise around the loop is first multiplied by $Z_{\text{OL}}(jf)$ to be converted to a voltage, and divided by R_2 to be converted back to a current, thus experiencing an overall gain of $T(jf) = Z_{\text{OL}}(jf)/R_2$.

To make $A_{\text{CL}}(jf)$ approach the ideal value $1 + R_2/R_1$, it is desirable that $T(jf)$ be as large as possible. To ensure a substantial loop gain over a range of closed-loop gains, manufacturers strive to maximize $Z_{\text{OL}}(jf)$ relative to R_2. Consequently, since $I_n = V_o/Z_{\text{OL}}$, the inverting input-pin current will be very small, even though this is a low-impedance node because of the input buffer. In the limit $Z_{\text{OL}} \to \infty$ we obtain $I_n \to 0$, indicating that a current-mode op amp will ideally provide whatever output is needed to drive I_n to zero. Thus, the familiar op amp conditions $V_n \to V_p$, $I_n \to 0$, and $I_p \to 0$ hold also for current-mode op amps, though for different reasons than their voltage-mode counterparts.

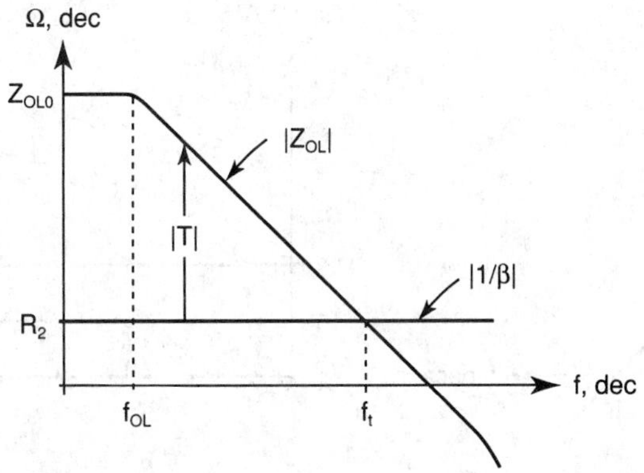

FIGURE 11.42 Graphical interpretation of the loop gain.

Current-Mode Op Amp Dynamics

Substituting (11.87) into (11.92) and then into (11.91), and exploiting the fact that $Z_{OL0}/R_2 \gg 1$, we obtain

$$A_{CL}(jf) = \frac{A_{CL0}}{1 + jf/f_t} \tag{11.93}$$

where

$$A_{CL0} = 1 + \frac{R_2}{R_1} \tag{11.94}$$

is the closed-loop dc gain, and

$$f_t = \frac{Z_{OL0} f_{OL}}{R_2} = \frac{1}{2\pi R_2 C_{eq}} \tag{11.95}$$

is the closed-loop bandwidth. With R_2 in the kΩ range and C_{eq} in the pF range, f_t is typically in range of 10^8 Hz.

We can again visualize $|T|$ and f_t graphically by noting that if we define $|T|_{dec} = \log_{10}|T|$, then we have, by (11.92), $|T|_{dec} = \log_{10}|Z_{OL}(jf)| - \log_{10}|R_2|$, or

$$|T|_{dec} = |Z_{OL}(jf)|_{dec} - |R_2|_{dec} \tag{11.96}$$

As shown in Fig. 11.42, we can visualize the loop gain as the difference between the decade plot of $|Z_{OL}(jf)|$ and that of $|R_2|$, with the latter now acting as the noise gain. Since at the crossover frequency we have $T = 1\angle{-90^\circ} = -j$, (11.91) and (11.94) yield $|A_{CL}| = A_{CL0}/\sqrt{2}$. Consequently, the crossover frequency represents the -3 dB frequency of $A_{CL}(jf)$, that is, f_t.

We are now ready to make two important observations.

1) Equation (11.95) shows that for a given amplifier the closed-loop bandwidth depends only on R_2. We can thus use R_2 to select the bandwidth f_t via (11.95), and R_1 to select the dc gain A_{OL0} via (11.94). The ability to set gain independently of bandwidth, along with the absence of gain-bandwidth trade-off, constitutes the first major advantage of current-mode over voltage-mode op amps, see Fig. 11.43.

2) The other major advantage of current-mode op amps is the absence of slew-rate limiting. To justify, suppose we apply an input step $V_i = V_m$ to the circuit of Fig. 11.41. Referring also

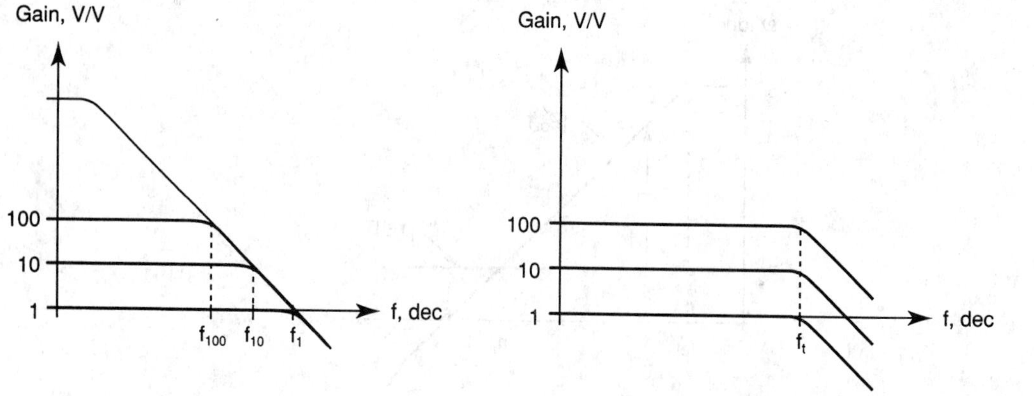

FIGURE 11.43 Comparing the gain-bandwidth characteristics of voltage-mode (left) and current-mode (right) op amps.

to Fig. 11.40, we note that the resulting current imbalance I_n yields an output V_o such that $I_n = C_{eq} dV_o/dt + V_o/R_{eq}$. Substituting into Eq. (11.89), rearranging, and exploiting the fact that $R_2/Z_{OL0} \ll 1$, we get

$$R_2 C_{eq} \frac{dV_o}{dt} + V_o = A_{OL0} V_m \tag{11.97}$$

indicating an *exponential* output transient regardless of V_m. The time constant governing the transient is

$$\tau = R_2 C_{eq} \tag{11.98}$$

and is set by R_2, regardless of A_{CL0}. For instance, a CLC401 op amp with $R_2 = 1.5\ \text{k}\Omega$ has $\tau = R_2 C_{eq} = 1.5 \times 10^3 \times 0.64 \times 10^{-12} \simeq 1$ ns. The rise time is $t_r = 2.2\tau \simeq 2.2$ ns, and the settling time within 0.1 percent of the final value is $t_s \simeq 7\tau \simeq 7$ ns, in reasonable agreement with the data-sheet values $t_r = 2.5$ ns and $t_s = 10$ ns.

Higher-Order Effects

The above analysis indicates that once R_2 has been set, the dynamics are unaffected by the closed-loop gain setting. In practice it is found that bandwidth and rise time do vary with gain somewhat, though not as drastically as for voltage-mode op amps. The main cause is the nonzero output resistance R_n of the input buffer, whose effect is to alter the loop gain and, hence, the closed-loop dynamics. Referring to Fig. 11.44, we again use the superposition principle and write

$$I_n = \frac{V_i}{R_n + R_1 \parallel R_2} - \beta V_o \tag{11.99}$$

where the feedback factor is found using the current divider formula and Ohm's law,

$$\beta = \frac{R_1}{R_1 + R_n} \times \frac{1}{R_2 + (R_n \parallel R_1)} = \frac{1}{R_2 + A_{CL0} R_n} \tag{11.100}$$

Comparing with (11.90), we observe that the effect of R_n is to replace R_2 with $R_2 + A_{CL0} R_n$. The $|1/\beta|$ curve of Fig. 11.42 will thus be shifted upward, leading to a *decrease* in the

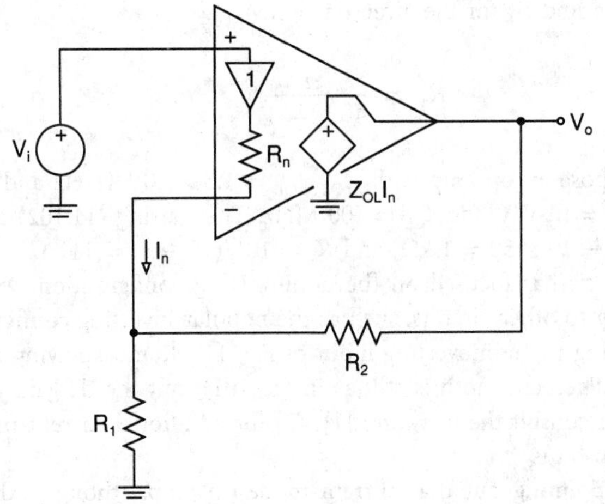

FIGURE 11.44 Investigating the effect of R_n.

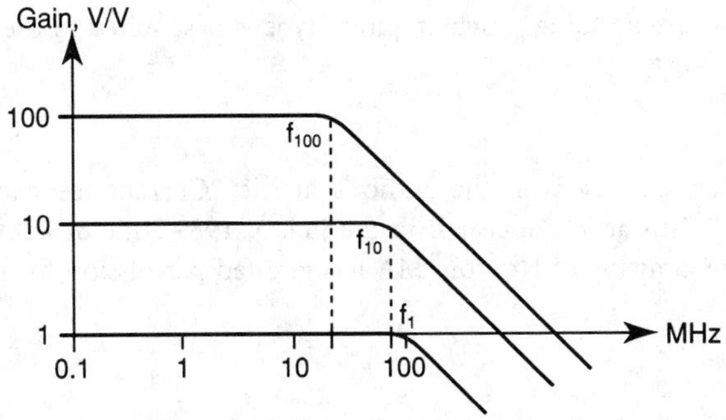

FIGURE 11.45 Bandwidth reduction for different gain settings.

crossover frequency, which we shall now denote as f_{CL}. This frequency is obtained by letting $R_2 \to (R_2 + A_{CL0}R_n)$ in (11.95),

$$f_{CL} = \frac{Z_{OL0}f_{OL}}{R_2 + A_{CL0}R_n} = \frac{1}{2\pi(R_2 + A_{CL0}R_n)C_{eq}} \tag{11.101}$$

As an example, suppose an op amp has $R_n = 50\ \Omega, R_2 = 1.5\ \text{k}\Omega$, and $f_t = 100$ MHz. Then, (11.101) yields $f_{CL} = f_t/(1 + A_{CL0}R_n/R_2) = 10^8/(1 + A_{CL0}/30)$. The bandwidths corresponding to $A_{CL0} = 1$ V/V, 10 V/V, and 100 V/V are, respectively, $f_1 = 96.8$ MHz, $f_{10} = 75.0$ MHz, and $f_{10} = 23.1$ MHz, and are shown in Fig. 11.45. The corresponding rise times are, respectively, $t_1 = 3.6$ ns, $t_{10} = 4.7$ ns, and $t_{100} = 15.2$ ns. We note that the above bandwidth reductions still compare favorably with voltage-mode op amps, where the reduction factors would be, respectively, 1, 10, and 100.

The values of R_1 and R_2 can be predistorted to compensate for the bandwidth reduction. Using (11.101) we find R_2 for a given bandwidth f_{CL} and dc gain A_{CL0},

$$R_2 = \frac{Z_{OL0}f_{OL}}{f_{CL}} - A_{CL0}R_n \tag{11.102}$$

and using (11.94) we find R_1 for the given dc gain A_{CL0},

$$R_1 = \frac{R_2}{A_{CL0} - 1} \tag{11.103}$$

As an example, suppose an op amp with $Z_{OL0}f_{OL} = 1.5 \times 10^{11}$ Ω-Hz and $R_n = 50$ Ω is to be configured for $A_{CL0} = 10$ V/V and $f_{CL} = 100$ MHz. Then, using (11.102) and (11.103) we find $R_2 = 1.5 \times 10^{11}/10^8 - 10 \times 50 = 1$ kΩ, and $R_1 = 10^3/(10 - 1) = 111$ Ω.

Though our analysis has focused on the noninverting configuration, we can readily extend our line of reasoning to other circuits, such as the popular inverting configuration. The latter is obtained by grounding the noninverting input of Fig. 11.44, and applying the source V_i via the bottom lead of R_1. The bandwidth is still as in (11.101), but the dc gain is now $-R_2/R_1$. The interested reader can consult the literature [1], [5] for additional current-mode op amp circuits as well as application hints.

We conclude by pointing out that current-mode op amps, though exhibiting much faster dynamics than their voltage-mode counterparts, in general suffer from poorer input offset voltage and input bias current characteristics. Moreover, having much wider bandwidths, they tend to be noisier. There is no question that the circuit designer must carefully weigh both advantages and disadvantages before deciding which amplifier type is best suited to the application at hand.

Acknowledgment

Parts of this chapter are based on the author's article "Current feedback amplifiers benefit high-speed designs." This article appeared in the Jan. 5, 1989 issue of *EDN* on pages 161–172. Cahners Publishing Company of Newton, MA has granted permission for its appearance.

References

[1] S. Franco, "Current feedback amplifiers benefit high-speed designs," *EDN*, Jan. 1989.

[2] J. E. Solomon, "The monolithic operational amplifier: A tutorial study," *IEEE Journal of Solid-State Circuits*, vol. SC-9, Dec. 1974.

[3] S. Franco, *Design with Operational Amplifiers and Analog ICs*, New York: McGraw-Hill, 1988.

[4] D. Nelson and S. Evans, "A new approach to op amp design," Comlinear Corp. Applicat. Note 300-1, Mar. 1985.

[5] D. Potson, "Current-feedback op amp applications circuit guide," Comlinear Corp. Applicat. Note OA-07, May 1988.

III

;eneral Circuit Theory

hn Choma, Jr.

ıiversity of Southern California

12

Fundamental Circuit Concepts

ın Choma, Jr.
versity of Southern
ifornia

2.1 The Electrical Circuit

An **electrical circuit** or **electrical network** is an array of interconnected elements wired so as to be capable of conducting current. As discussed earlier, the fundamental **two-terminal elements** of an electrical circuit are the **resistor**, the **capacitor**, the **inductor**, the **voltage source**, and the **current source**. The circuit schematic symbols of these elements, together with the algebraic symbols used to denote their respective general values, appear in Fig. 12.1.

As suggested in Fig. 12.1, the value of a resistor is known as its *resistance*, R, and its dimensional units are *ohms*. The case of a wire used to interconnect the terminals of two electrical elements corresponds to the special case of a resistor whose resistance is ideally zero ohms; that is, $R = 0$. For the capacitor in Fig. 12.1(b), the *capacitance*, C, has units of *farads*, and from Fig. 12.1(c), the value of an inductor is its *inductance*, L, the dimensions of which are *henries*. In the case of the voltage sources depicted in Fig. 12.1(d), a constant, time invariant source of voltage, or *battery*, is distinguished from a voltage source that varies with time. The latter type of voltage source is often referred to as a **time varying signal** or simply, a **signal**. In either case the value of the battery voltage, E, and the time varying signal, $v(t)$, is in units of *volts*. Finally, the current source of Fig. 12.1(e) has a value, I, in units of *amperes*, which is typically abbreviated as amps.

Elements having three, four, or more than four terminals can also appear in practical electrical networks. The discrete component **bipolar junction transistor** (BJT), which is schematically portrayed in Fig. 12.2(a), is an example of a three-terminal element, in which the three terminals are the collector, the base, and the emitter. On the other hand, the monolithic *metal-oxide-semiconductor field-effect transistor* (MOSFET) depicted in Fig. 12.2(b) has four terminals: the drain, the gate, the source, and the bulk substrate.

Multiterminal elements appearing in circuits identified for systematic mathematical analyses are routinely represented, or *modeled*, by equivalent subcircuits formed of only interconnected two-terminal elements. Such a respresentation is always possible, provided that the list of two-terminal elements itemized in Fig. 12.1 is appended by an additional type of two-terminal element known as the **controlled source**, or **dependent generator**. Two of the four types of

3-8341-2/95/$0.00+$.50

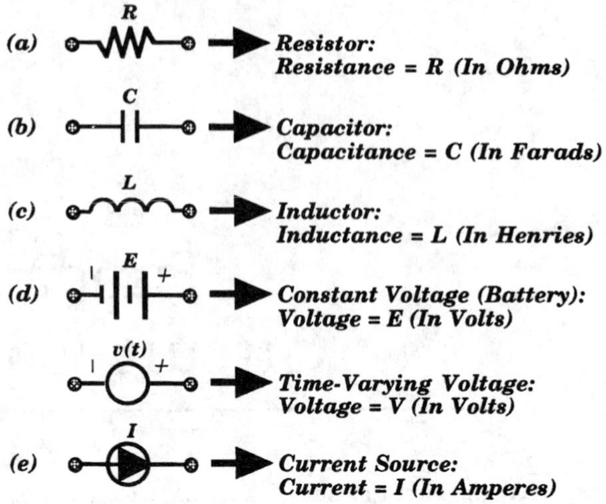

FIGURE 12.1 Circuit schematic symbol and corresponding value notation for (a) resistor, (b) capacitor, (c) inductor, (d) voltage source, and (e) current source. Note that a constant voltage source, or battery, is distinguished from a voltage source that varies with time.

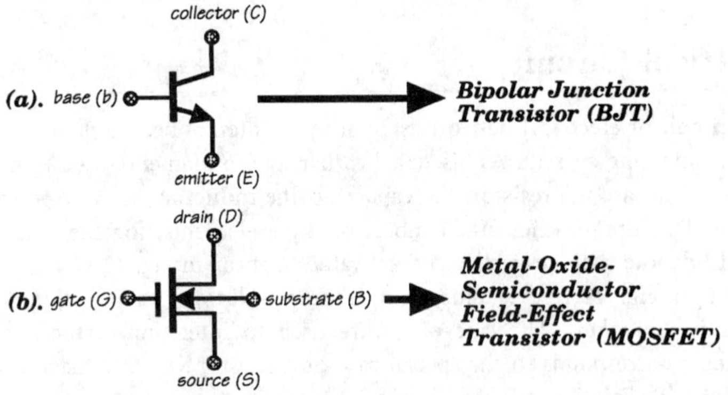

FIGURE 12.2 Circuit schematic symbol for (a) discrete component bipolar junction transistor (BJT) and (b) monolithic metal-oxide-semiconductor field-effect transistor (MOSFET).

controlled sources are voltage sources and two are current sources. In Fig. 12.3(a), the dependent generator is a *voltage-controlled voltage source* (VCVS) in that the voltage, $v_0(t)$, developed from terminal 3 to terminal 4 is a function of, and is therefore dependent on, the voltage, $v_i(t)$, established elsewhere in the considered network from terminal 1 to terminal 2. The *controlled voltage*, $v_0(t)$, as well as the *controlling voltage*, $v_i(t)$, can be constant or time varying. Regardless of the time-domain nature of these two voltages, the value of $v_0(t)$ is not an independent number. Instead, its value is determined by $v_i(t)$ in accordance with a prescribed functional relationship, e.g.,

$$v_0(t) = f[v_i(t)] \tag{12.1}$$

If the function, $f(\cdot)$, is linearly related to its argument, (12.1) collapses to the form

$$v_0(t) = f_{\mu} v_i(t) \tag{12.2}$$

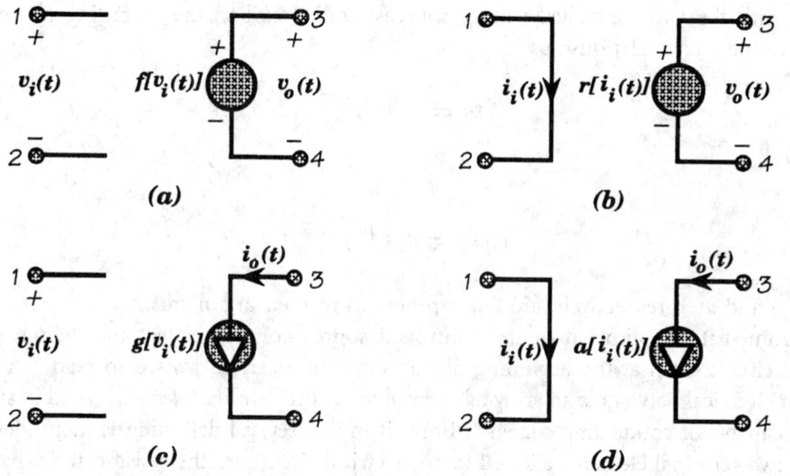

FIGURE 12.3 Circuit schematic symbol for (a) voltage-controlled voltage source (VCVS), (b) current-controlled voltage source (CCVS), (c) voltage-controlled current source (VCCS), and (d) current-controlled current source (CCCS).

where $f\mu$ is a constant, independent of either $v_0(t)$ or $v_i(t)$. When the function on the right-hand side of (12.1) is linear, the subject VCVS becomes known as a *linear voltage-controlled voltage source.*

The second type of controlled voltage source is the *current-controlled voltage source* (CCVS) depicted in Fig. 12.3(b). In this dependent generator, the controlled voltage, $v_0(t)$, developed from terminal 3 to terminal 4 is a function of the *controlling current,* $i_i(t)$, flowing elsewhere in the network between terminals 1 and 2, as indicated. In this case the generalized functional dependence of $v_0(t)$ on $i_i(t)$ is expressible as

$$v_0(t) = r[i_i(t)] \tag{12.3}$$

which reduces to

$$v_0(t) = r_m i_i(t) \tag{12.4}$$

when $r(\cdot)$ is a linear function of its argument.

The two types of dependent current sources are diagrammed symbolically in Figs. 12.3(c) and (d). Figure 12.3(c) depicts a *voltage-controlled current source* (VCCS), for which the controlled current, $i_0(t)$, flowing in the electrical path from terminal 3 to terminal 4, is determined by the controlling voltage, $v_i(t)$, established across terminals 1 and 2. Therefore, the controlled current can be written as

$$i_0(t) = g[v_i(t)] \tag{12.5}$$

In the *current-controlled current source* (CCCS) of Fig. 12.3(d),

$$i_0(t) = a[i_i(t)] \tag{12.6}$$

where the controlled current, $i_0(t)$, flowing from terminal 3 to terminal 4 is a function of the controlling current, $i_i(t)$, flowing elsewhere in the circuit from terminal 1 to terminal 2. As is the case with the two controlled voltage sources studied earlier, the preceding two equations collapse to the linear relationships

$$i_0(t) = g_m v_i(t) \tag{12.7}$$

and

$$i_0(t) = a_\alpha i_i(t) \tag{12.8}$$

when $g(\cdot)$ and $a(\cdot)$, respectively, are linear functions of their arguments.

The immediate implication of the controlled source concept is that the definition for an electrical circuit given at the beginning of this subsection can be revised to read "an electrical circuit or electrical network is an array of *interconnected two-terminal elements* wired in such a way as to be capable of conducting current". Implicit in this revised definition is the understanding that the two-terminal elements allowed in an electrical circuit are the resistor, the capacitor, the inductor, the voltage source, the current source, and any of the four possible types of dependent generators.

In an attempt to reinforce the engineering utility of the foregoing definition, consider the voltage mode **operational amplifier**, or **op-amp**, whose circuit schematic symbol is submitted in Fig. 12.4(a). Observe that the op-amp is a five-terminal element. Two terminals, labeled 1 and 2, are provided to receive input signals that derive either from external signal sources or from the output terminals of subcircuits that feed back a designable fraction of the output signal established between terminal 3 and the system ground. Battery voltages, identified as E_{CC} and E_{BB} in the figure, are applied to the remaining two op-amp terminals (terminals 4 and 5) with respect to ground to *bias* or *activate* the op-amp for its intended application. When E_{CC} and E_{BB} are selected to ensure that the subject op-amp behaves as a linear circuit element, the voltages, E_{CC} and E_{BB}, along with the corresponding terminals at which they are incident, are inconsequential. In this event the op-amp of Fig. 12.4(a) can be modeled by the electrical circuit appearing in Fig. 12.4(b), which exploits a linear VCVS. Thus, the voltage amplifier of Fig. 12.4(c), which interconnects two batteries, a signal source voltage, three resistors, a capacitor, and an op-amp, can be represented by the network given in Fig. 12.4(d). Note that the latter configuration uses only two terminal elements, one of which is a VCVS.

Current and Current Polarity

The concept of an electrical *current* is implicit to the definition of an electrical circuit in that a circuit is said to be an array of two-terminal elements that are connected in such a way as to permit the *conduction of current.* Current flow through an element that is capable of current conduction requires that the net *charge* observed at any elemental cross-section change with time. Equivalently, a net nonzero charge, $q(t)$, must be transferred over finite time across any cross-sectional area of the element. The current, $i(t)$, that actually flows is the time rate of change of this transferred charge;

$$i(t) = \frac{dq(t)}{dt} \tag{12.9}$$

where the MKS unit of charge is the *coulomb*, time t is measured in seconds, and the resultant current is measured in units of amperes. Note that zero current does not necessarily imply a lack of charge at a given cross-section of a conductive element. Rather, zero current implies only that the subject charge is not changing with time; that is, the charge is not moving through the elemental cross-section.

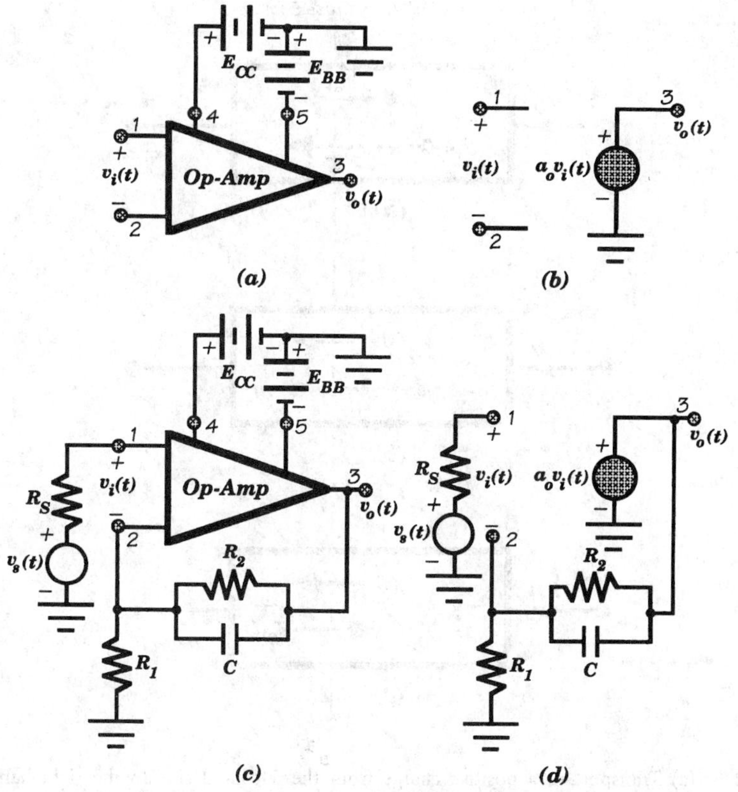

FIGURE 12.4 (a) Circuit schematic symbol for a voltage mode operational amplifier. (b) First-order linear model of the op-amp. (c) A voltage amplifier realized with the op-amp functioning as the gain element. (d) Equivalent circuit of the voltage amplifier in (c).

Electrical charge can be negative, as in the case of *electrons* transported through a cross-section of a conductive element such as aluminum or copper. A single electron has a charge of $-(1.602 \times 10^{-19})$ coulomb. Thus, (12.9) implies a need to transport an average of (6.242×10^{18}) electrons in 1 second through a cross-section of aluminum if the aluminum element is to conduct a constant current of 1 amp. Charge can also be positive, as in the case of *holes* transported through a cross-section of a semiconductor such as germanium or silicon. Hole transport in a semiconductor is actually electron transport at an energy level that is smaller than the energy required to effect electron transport in that semiconductor. To first order, therefore, the electrical charge of a hole is the negative of the charge of an electron, which implies that the charge of a hole is $+(1.602 \times 10^{-19})$ coulomb.

A positive charge, $q(t)$, transported from the left of the cross-section to the right of the cross-section in the element abstracted in Fig. 12.5(a) gives rise to a positive current, $i(t)$, which also flows from left to right across the indicated cross-section. Assume that prior to the transport of such charge, the volumes to the left and to the right of the cross-section are electrically neutral; that is, these volumes have zero initial net charge. Then, the transport of a positive charge, q_0, from the left side to the right side of the element charges the right side to $+1q_0$ and the left side to $-1q_0$.

Alternatively, suppose a negative charge in the amount of $-q_0$ is transported from the right side of the element to its left side, as suggested in Fig. 12.5(b). Then, the left side charges to $-q_0$, and the right side charges to $+q_0$, which is identical to the electrostatic condition incurred by the transport of a positive charge in the amount of q_0 from left- to right-hand sides. As

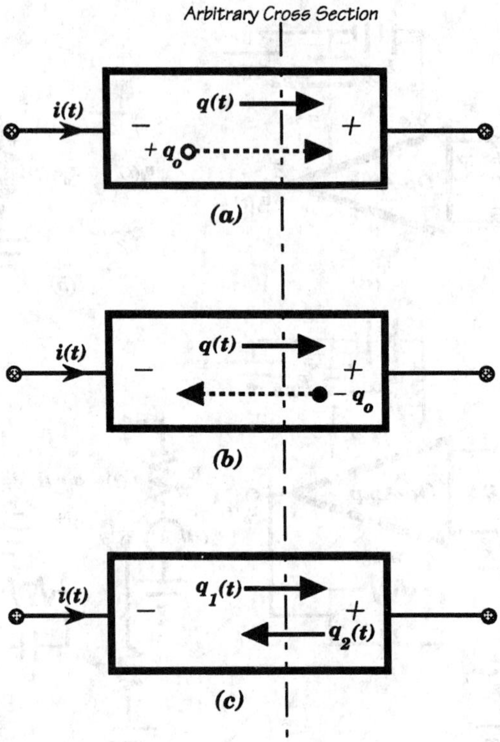

FIGURE 12.5 (a) Transport of a positive charge from the left-hand side to the right-hand side of an arbitrary cross-section of a conductive element. (b) Transport of a negative charge from the right-hand side to the left-hand side of an arbitrary cross-section of a conductive element. (c) Transport of positive or negative charges from either side to the other side of an arbitrary cross-section of a conductive element.

a result, the transport of a net negative charge from right to left produces a positive current, $i(t)$, flowing from left to right, just as positive charge transported from left- to right-hand sides induces a current flow from left to right.

Assume, as portrayed in Fig. 12.5(c), that a positive or a negative charge, say, $q_1(t)$, is transported from the left side of the indicated cross-section to the right side. Simultaneously, a positive or a negative charge in the amount of $q_2(t)$ is directed through the cross-section from right to left. If $i_1(t)$ is the current arising from the transport of the charge $q_1(t)$, and if $i_2(t)$ denotes the current corresponding to the transport of the charge, $q_2(t)$, the net effective current $i_e(t)$, flowing from the left side of the cross-section to the right side of the cross-section is

$$i_e(t) = \frac{d}{dt}[q_1(t) - q_2(t)] = i_1(t) - i_2(t) \tag{12.10}$$

where the charge difference, $[q_1(t) - q_2(t)]$, represents the net charge transported from left to right. Observe that if $q_1(t) \equiv q_2(t)$, the net effective current is zero, even though conceivably large numbers of charges are transported back and forth across the junction.

Energy and Voltage

The preceding section highlights the fundamental physical fact that the flow of current through a conductive electrical element mandates that a net charge be transported over finite time across any arbitrary cross-section of that element. The electrical effect of this charge transport is a net positive charge induced on one side of the element in question and a net negative charge (equal in magnitude to the aforementioned positive charge) mirrored on the other side of the element. This ramification conflicts with the observable electrical properties of an element in equilibrium. In particular, an element sitting in free space, without any electrical connection to a source of *energy*, is necessarily in equilibrium in the sense that the net positive charge in any volume of the element is precisely counteracted by an equal amount of charge of opposite sign in said volume. Thus, if none of the elements abstracted in Fig. 12.5 is connected to an external source of energy, it is physically impossible to achieve the indicated electrical charge differential that materializes across an arbitrary cross-section of the element when charge is transferred from one side of the cross-section to the other.

The energy commensurate with sustaining current flow through an electrical element derives from the application of a *voltage*, $v(t)$, across the element in question. Equivalently, the application of electrical energy to an element manifests itself as a voltage developed across the terminals of an element to which energy is supplied. The amount of applied voltage, $v(t)$, required to sustain the flow of current, $i(t)$, as diagrammed in Fig. 12.6(a), is presciselv the voltage required to offset the electrostatic implications of the differential charge induced across the element through which $i(t)$ flows. This is to say that without the connection of the voltage, $v(t)$, to the element in Fig. 12.6(a), the element cannot be in equilibrium. With $v(t)$ connected, equilibrium for the entire system comprised of element and voltage source is reestablished by allowing for the conduction of the current, $i(t)$.

Instead of viewing the delivery of energy to an electrical element as the ramification of a voltage source applied to the element, the energy delivery may be interpreted as the upshot of a current source used to excite the element, as depicted in Fig. 12.6(b). This interpretation follows from the fact that energy must be applied in an amount that effects charge transport at a desired time rate of change. It follows that the application of a current source in the amount of the desired current is necessarily in one-to-one correspondence with the voltage required to offset

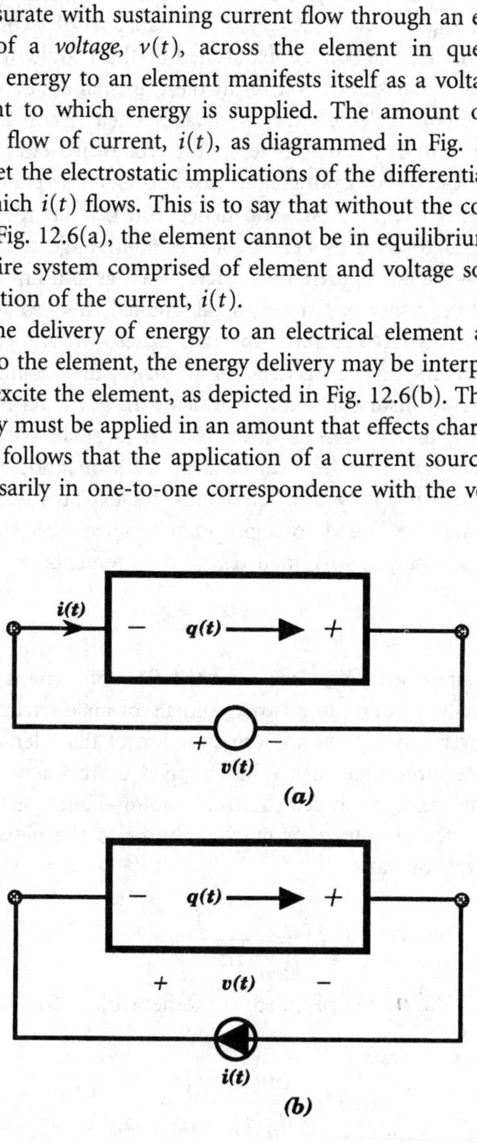

FIGURE 12.6 (a) The application of energy in the form of a voltage applied to an element that is made to conduct a specified current. The applied voltage, $v(t)$, causes the current, $i(t)$, to flow. (b) The application of energy in the form of a current applied to an element that is made to establish a specified terminal voltage. The applied current, $i(t)$, causes the voltage, $v(t)$, to be developed across the terminals of the electrical element.

the charge differential manifested by the charge transport that yields the subject current. To be sure, a voltage source is a physical entity, while a current source is not. But the mathematical modeling of energy delivery to an electrical element can nonetheless be accomplished through either a voltage source or a current source.

In Fig. 12.6 the terminal voltage, $v(t)$, corresponding to the energy, $w(t)$, required to transfer an amount of charge, $q(t)$, across an arbitary cross-section of the element is

$$v(t) = \frac{dw(t)}{dq(t)} \tag{12.11}$$

where $v(t)$ is in units of volts when $q(t)$ is expressed in coulombs, and $w(t)$ is specified in joules. Thus, if 1 joule of applied energy results in the transport of 1 coulomb of charge through an element, the elemental terminal voltage manifested by the 1 joule of applied energy is 1 volt.

It should be understood that the derivative on the right-hand side of (12.11), and thus the terminal voltage demanded of an element that is transporting a certain amount of charge through its cross-section, is a function of the properties of the type of material from which the element undergoing study is fabricated. For example, an insulator such as paper, air, or silicon dioxide is ideally incapable of current conduction and hence, intrinsic charge transport. Thus, $q(t)$ is essentially zero in an insulator and by (12.11), an infinitely large terminal voltage is required for even the smallest possible current. In a conductor such as aluminum, iron, or copper, large amounts of charge can be transported for very small applied energies. Accordingly, the requisite terminal voltage for even very large currents approaches zero in ideal conductors. The electrical properties of semiconductors such as germanium, silicon, and gallium arsenide lie between the extremes of those for an insulator and a conductor. In particular, semiconductor elements behave as insulators when their terminals are subjected to small voltages, while progressively larger terminal voltages render the electrical behavior of semiconductors akin to conductors. This conditional conductive property of a semiconductor explains why semiconductor devices and circuits generally must be biased to appropriate voltage levels before these devices and circuits can function in accordance with their design requirements.

Power

The foregoing material underscores the fact that the flow of current through a two-terminal element, or more generally, through any two terminals of an electrical network, requires that charge be transported over time across any cross-section of that element or network. In turn, such charge transport requires that energy be supplied to the network, usually through the application of an external voltage source. The time rate of change of this applied energy is the *power* delivered by the external voltage or current source to the network in question. If $p(t)$ denotes this power in units of watts

$$p(t) = \frac{dw(t)}{dt} \tag{12.12}$$

where, of course, $w(t)$ is the energy supplied to the network in joules. By rewriting (12.12) in the form

$$p(t) = \frac{dw(t)}{dq(t)}\frac{dq(t)}{dt} \tag{12.13}$$

and applying (12.9) and (12.11), the power supplied to the two terminals of an element or a network becomes the more expedient relationship

$$p(t) = v(t)i(t) \tag{12.14}$$

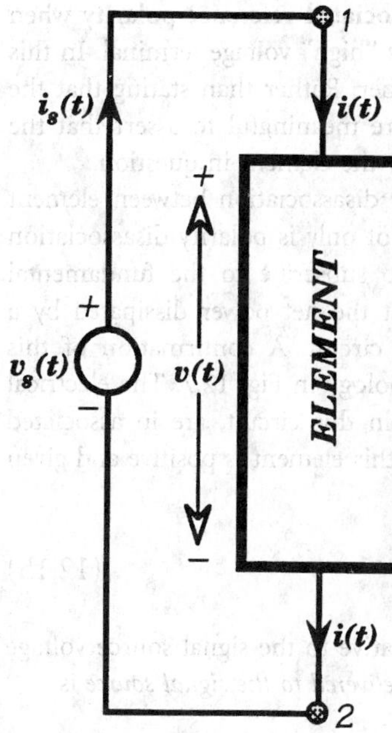

FIGURE 12.7 Circuit used to illustrate power calculations and the associated reference polarity convention.

Equation (12.14) expresses the power delivered to an element as a simple product of the voltage applied across the terminals of the element and the resultant current conducted by that element. However, care must be exercised with respect to relative voltage and current polarity, when applying (12.14) to practical circuits.

To the foregoing end, it is useful to revisit the simple abstraction of Fig. 12.6(a), which is redrawn as the slightly modified form in Fig. 12.7. In this circuit a signal source voltage, $v_s(t)$, is applied across the two terminals, 1 and 2, of an element, which responds by conducting a current $i(t)$, from terminal 1 to terminal 2 and developing a corresponding terminal voltage $v(t)$, as shown. If the wires (zero resistance conductors, as might be approximated by either aluminum or copper interconnects) that connect the signal source to the element are ideal, the voltage, $v(t)$, is identical to $v_s(t)$. Moreover, because the current is manifested by the application of the signal source, which thereby establishes a closed electrical path for current conduction, the element current, $i(t)$, is necessarily the same as the current, $i_s(t)$, that flows through $v_s(t)$.

If attention is focused on only the element in Fig. 12.7, it is natural to presume that the current conducted by the element actually flows from terminal 1 to terminal 2 when (as shown) the voltage developed across the element is positive at terminal 1 with respect to terminal 2. This assertion may be rationalized qualitatively by noting that the positive voltage nature at terminal 1 acts to repel positive charges from terminal 1 to terminal 2, where the negative nature of the developed voltage, $v(t)$, tends to attract the repulsed positive charges. Similarly, the positive nature of the voltage at terminal 1 serves to attract negative charges from terminal 2, where the negative nature of $v(t)$ tends to repel such negative charges. Because current flows in the direction of transported positive charge and opposite to the direction of transported negative charge, either interpretation gives rise to an elemental current, $i(t)$, which flows from terminal 1 to terminal 2. In general, if current is indicated as flowing from the "high" (+) voltage terminal to the "low" (−) voltage terminal of an element, the current conducted by the element and the voltage developed across the element to cause this flow of current are said to be in **associated reference polarity**. When the element current, $i(t)$, and the corresponding element voltage, $v(t)$, as exploited in the defining power relationship of (12.14), are in associated reference polarity, the resulting computed power is a positive number and is said to represent the *power delivered to* the element. In contrast, $v(t)$ and $i(t)$ are said to be in **disassociated reference polarity** when $i(t)$ flows from the "low" voltage terminal of the element to its "high" voltage terminal. In this case the voltage-current product in (12.14) is a negative number. Rather than stating that the resulting negative power is delivered to the element, it is more meaningful to assert that the computed negative power is a *positive power that is generated by* the element in question.

At first glance, it may appear as though the latter polarity disassociation between element voltage and current variables is an impossible circumstance. Not only is polarity disassociation possible, it is absolutely necessary if electrical circuits are to subscribe to the fundamental principle of **conversation of power.** This principle states that the net power dissipated by a circuit must be identical to the net power supplied to that circuit. A confirmation of this basic principle derives from a further consideration of the topology in Fig. 12.7. The electrical variables, $v(t)$ and $i(t)$, pertinent to the element delineated in this circuit, are in associated reference polarity. Accordingly, the power, $p_e(t)$, dissipated by this element is positive and given by (12.14):

$$p_e(t) = v(t)i(t) \tag{12.15}$$

However, the voltage and current variables, $v_s(t)$ and $i_s(t)$, relative to the signal source voltage are in disassociated polarity. It follows that the power, $p_s(t)$, *delivered to the signal source* is

$$p_s(t) = -v_s(t)i_s(t) \tag{12.16}$$

Because, as stated previously, $v_s(t) = v(t)$ and $i_s(t) = i(t)$ for the circuit at hand, (12.16) can be written as

$$p_s(t) = -v(t)i(t) \tag{12.17}$$

The last result implies that the

$$\text{power } \textbf{delivered by} \text{ the signal source} = +v(t)i(t) \equiv p_e(t) \tag{12.18}$$

that is, the power delivered to the element by the signal source is equal to the power dissipated by the element.

An alternative statement to conservation of power, as applied to the circuit in Fig. 12.7 derives from combining (12.15) and (12.17) to arrive at

$$p_s(t) + p_e(t) = 0 \tag{12.19}$$

The foregoing result may be generalized to the case of a more complex circuit comprised of an electrical interconnection of N elements, some of which may be voltage and current sources. Let the voltage across the kth element by $v_k(t)$, and let the current flowing through this kth element, in associated reference polarity with $v_k(t)$, be $i_k(t)$. Then, the power, $p_k(t)$, delivered to the kth electrical element is $v_k(t)i_k(t)$. By conservation of power,

$$\sum_{k=1}^{N} p_k(t) = \sum_{k=1}^{N} v_k(t)i_k(t) = 0 \tag{12.20}$$

The satisfaction of this expression requires that at least one of the $p_k(t)$ be negative, or equivalently, at least one of the N elements embedded in the circuit at hand must be a source of energy.

2.2 Circuit Classifications

It was pointed out earlier that the relationship between the current that is made to flow through an electrical element and the applied energy, and thus voltage, that is required to sustain such current flow is dictated by the material from which the subject element is fabricated. The elemental material and the associated manufacturing methods exploited to realize a particular type of circuit element determine the mathematical nature between the voltage applied across the terminals of the element and the resultant current flowing through the element. To this end, electrical elements and circuits in which they are embedded are generally codified as linear or nonlinear, active or passive, time varying or time invariant, and lumped or distributed.

Linear vs. Nonlinear

A **linear two-terminal circuit element** is one for which the voltage developed across, and the current flowing through, are related to one another by a linear algebraic or a linear integro-differential equation. If the relationship between terminal voltage and corresponding current is nonlinear, the element is said to be *nonlinear.* A linear circuit contains only linear circuit elements, while a circuit is said to be nonlinear if at least one of its embedded electrical elements is nonlinear.

All practical circuit elements, and thus all practical electrical networks, are inherently nonlinear. However, over suitably restricted ranges of applied voltages and corresponding currents, the volt-ampere characteristics of these elements and networks emulate idealized linear relationships. In the design of an electronic linear signal processor, such as an amplifier, an implicit engineering task is the implementation of biasing subcircuits that constrain the voltages and currents of internal semiconductor elements to ranges that ensure linear elemental behavior over all possible operating conditions.

The voltage-current relationship for the linear resistor offered in Fig. 12.8(a) is

$$v(t) = Ri(t) \tag{12.21}$$

where the voltage, $v(t)$, appearing across the terminals of the resistor and the resultant current, $i(t)$, conducted by the resistor are in associated reference polarity. The resistance, R, is independent of either $v(t)$ or $i(t)$. From (12.14), the dissipated resistor power, which is manifested in the form of heat, is

$$p_r(t) = i^2(t)R = \frac{v^2(t)}{R} \tag{12.22}$$

The linear capacitor and the linear inductor, whose schematic symbols appear, respectively, in Figs. 12.8(b) and (c), store energy as opposed to dissipating power. Their volt-ampere equations

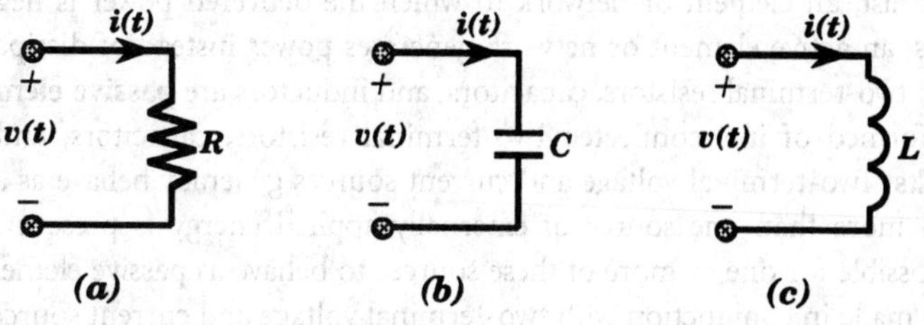

FIGURE 12.8 Circuit schematic symbol and corresponding voltage and current notation for (a) a linear resistor, (b) a linear capacitor, and (c) a linear inductor.

are the linear relationships

$$i(t) = C\frac{dv(t)}{dt} \tag{12.23}$$

for the capacitor, whereas for the inductor in Fig. 12.8(c),

$$v(t) = L\frac{di(t)}{dt} \tag{12.24}$$

Observe from (12.23) and (12.14) that the power, $p_c(t)$, delivered to the linear capacitor is

$$p_c(t) = v(t)i(t) = Cv(t)\frac{dv(t)}{dt} \tag{12.25}$$

From (12.12), this power is related to the energy, e.g., $w_c(t)$, stored in the form of charge deposited on the plates of the capacitor by

$$Cv(t)dv(t) = dw_c(t) \tag{12.26}$$

It follows that the energy delivered to, and hence stored in, the capacitor from time $t = 0$ to time t is

$$w_c(t) = \tfrac{1}{2}Cv^2(t) \tag{12.27}$$

It should be noted that this stored energy, like the energy associated with a signal source or a battery voltage, is available to supply power to other elements in the network in which the capacitor is embedded. For example, if very little current is conducted by the capacitor in question, (12.23) implies that the voltage across the capacitor is essentially constant. However, an element whose terminal voltage is a constant and in which energy is stored and therefore available for use behaves as a battery.

If the preceding analysis is repeated for the inductor of Fig. 12.8(c), it can be shown that the energy, $w_l(t)$, stored in the inductive element from time $t = 0$ to time t is

$$w_l(t) = \tfrac{1}{2}Li^2(t) \tag{12.28}$$

While an energized capacitor conducting almost zero current functions as a voltage source, an energized inductor supporting almost zero terminal voltage emulates a constant current source.

Active vs. Passive

An electrical element or network is said to be *passive* if the power delivered to it, defined in accordance with (12.14), is positive. This definition exploits the requirement that the terminal voltage, $v(t)$, and the element current $i(t)$, appearing in (12.14) be in associated reference polarity. In contrast, an element or network to which the delivered power is negative is said to be *active;* that is, an active element or network generates power instead of dissipating it.

Conventional two-terminal resistors, capacitors, and inductors are **passive elements.** It follows that networks formed of interconnected two-terminal resistors, capacitors, and inductors are **passive networks.** Two-terminal voltage and current sources generally behave as active elements. However, when more than one source of externally applied energy is present in an electrical network, it is possible for one or more of these sources to behave as passive elements. Comments similar to those made in conjunction with two-terminal voltage and current sources apply equally well to each of the four possible dependent generators. Accordingly, multiterminal configurations, whose models exploit dependent sources, can behave as either passive or active networks.

Time Varying vs. Time Invariant

The elements of a circuit are defined electrically by an identifying parameter, such as resistance, capacitance, inductance, and the gain factors associated with dependent voltage or current sources. An element whose indentifying parameter changes as a function of time is said to be a **time varying** element. If said parameter is a constant over time, the element in question is **time invariant.** A network containing at least one time varying electrical element it is said to be a time varying network. Otherwise, the network is time invariant. Excluded from the list of elements whose electrical character establishes the time variance or time invariance of a considered network are externally applied voltage and current sources. Thus, for example, a network whose internal elements are exclusively time-invariant resistors, capacitors, inductors, and dependent sources, but which is excited by a sinusoidal signal source, is nonetheless a time-invariant network.

Although some circuits, and particularly electromechanical networks, are purposely designed to exhibit time varying volt-ampere characteristics, **parametric time variance** is generally viewed as a parasitic phenomena in the majority of practical circuits. Unfortunately, a degree of parametric time variance is unavoidable in even those circuits that are specifically designed to achieve input-output response properties that closely approximate time-invariant characteristics. For example, the best of network elements exhibit a slow aging phenomenon that shifts the values of its intrinsic physical parameters. The upshot of these shifts is electrical circuits whose overall performance deteriorates with time.

Lumped vs. Distributed

Electrons in conventional conductive elements are not transported instantaneously across elemental cross-sections, but their transport velocities are very high. In fact, these velocities approach the speed of light, say c, which is (3×10^8) m/s or about 982 ft/μsec. Electrons and holes in semiconductors are transported at somewhat slower speeds, but generally no less than an order of magnitude or so smaller than the speed of light. The time required to transport charge from one terminal of a two-terminal electrical element to its other terminal, compared to the time required to propagate energy uniformly through the element, determines whether an element is lumped or distributed. In particular, if the time required to transport charge through an element is significantly smaller than the time required to propagate the energy through the element that is required to incur such charge transport, the element in question is said to be **lumped.** On the other hand, if the charge transport time is comparable to the energy propagation time, the element is said to be **distributed.**

The concept of a lumped, as opposed to a distributed, circuit element can be qualitatively understood through a reconsideration of the circuit provided in Fig. 12.7. As argued, the indicated element current, $i(t)$, is identical to the indicated source current, $i_s(t)$. This equality implies that $i(t)$, is effectively circulating around the loop that is electrically formed by the interconnection of the signal source voltage, $v_s(t)$, to the element. Equivalently, the subject equality implies that $i(t)$ is entering terminal 1 of the element and simultaneously is exiting at terminal 2, as shown. Assuming that the element at hand is not a semiconductor, the current, $i(t)$, arises from the transport of electrons through the element in a direction opposite to that of the indicated polarity of $i(t)$. Specifically, electrons must be transported from terminal 2, at the bottom of the element, to terminal 1, at the top of the element, and in turn the requisite amount of energy must be applied in the immediate neighborhoods of both terminals. The implication of presuming that the element at hand is lumped is that $i(t)$ is entering terminal 1 at precisely the same time that it is leaving terminal 2. Such a situation is clearly impossible, for it mandates that electrons be transported through the entire length of the element in zero time. However, given that electrons are transported at a nominal velocity of 982 ft/μsec, a very small physical elemental length renders the approximation of zero electron transport time reasonable. For example, if the element is 1/2 inch long (a typical size for an off-the-shelf resistor), the average transport time for electrons in this unit is only about 42.4 psec. As long as the period of the applied excitation, $v_s(t)$, is significantly larger than 42.4 psec, the electron transport time is significantly smaller than the time commensurate with the propagation of this energy through the entire element. A period of 42.4 psec corresponds to a signal frequency of approximately 23.6 GHz. Thus, a 1/2-in resistive element excited by a signal whose frequency is significantly smaller than 23.6 GHz can be viewed as a lumped circuit element.

In the vast majority of electrical and electronic networks it is difficult not to satisfy the lumped circuit approximation. Nevertheless, several practical electrical systems cannot be viewed as lumped entities. For example, consider the lead-in wire that connects the antenna input terminals of a frequency modulated (FM) radio receiver to an antenna, as diagrammed in Fig. 12.9. Let the signal voltage, $v_a(t)$, across the lead-in wires at point "a" be the sinusoid,

$$v_a(t) = V_M \cos(\omega t) \tag{12.29}$$

where V_M represents the amplitude of the signal, and ω is its frequency in units of radians per second. Consider the case in which $\omega = 2\pi(103.5 \times 10^6)$ rad/s, which is a carrier frequency lying within the commercial FM broadcast band. This high signal frequency makes the length of antenna lead-in wire critically important for proper signal reception.

In an attempt to verify the preceding contention let the voltage developed across the lead-in lines at point "b" in Fig. (3.1-9) be denoted as $v_b(t)$, and let point "b" be 1 foot displaced from point "a"; that is $L_{ab} = 1$ foot. The time, τ_{ab}, required to transport electrons over the indicated length, L_{ab}, is

$$\tau_{ab} = \frac{L_{ab}}{c} = 1.018 \text{ ns} \tag{12.30}$$

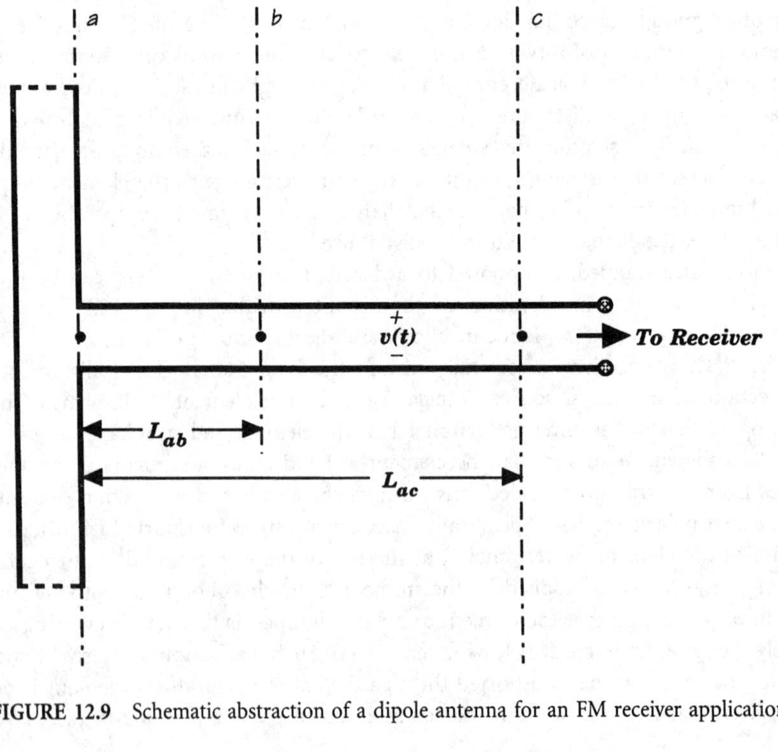

FIGURE 12.9 Schematic abstraction of a dipole antenna for an FM receiver application.

Thus, assuming an idealized line in the sense of zero effective resistance, capacitance, and inductance, the signal, $v_b(t)$, at point "b" is seen as the signal appearing at "a", delayed by approximately 1.02 ns. It follows that

$$v_b(t) = V_M \cos[\omega(t - \tau_{ab})] = V_M \cos(\omega t - 0.662) \qquad (12.31)$$

where the phase angle associated with $v_b(t)$ is 0.662 radian, or almost 38°. Obviously, the signal established at point "b" is a significantly phase-shifted version of the signal presumed at point "a".

An FM receiver can effectively retrieve the signal voltage, $v_a(t)$, by detecting a phase-inverted version of $v_a(t)$ at its input terminals. To this end, it is of interest to determine the length, L_{ac}, such that the signal, $v_c(t)$, established at point "c" in Fig. 12.9 is

$$v_c(t) = V_M \cos(\omega t - \pi) \qquad (12.32)$$

The required phase shift of 180°, or π radians, corresponds to a time delay, τ_{ac}, of

$$\tau_{ac} = \frac{\pi}{\omega} = 4.831 \text{ ns} \qquad (12.33)$$

In turn, a time delay of τ_{ac} implies a required line length, L_{ac}, of

$$L_{ac} = c\tau_{ac} = 4.744 \text{ ft} \qquad (12.34)$$

A parenthetically important point is the observation that the carrier frequency of 103.5 MHz corresponds to a wavelength, λ, of

$$\lambda = \frac{2\pi c}{\omega} = 9.489 \text{ ft} \qquad (12.35)$$

Accordingly, the lead-in length computed in (12.34) is $\lambda/2$; that is a half-wavelength.

13

Network Laws and Theorems

Ray R. Chen
San Jose State University

Artice M. Davis
San Jose State University

Marwan A. Simaan
University of Pittsburgh

13.1 Kirchhoff's Voltage and Current Laws

Ray R. Chen and Artice M. Davis

Circuit analysis, like Euclidean geometry, can be treated as a mathematical system; that is, the entire theory can be constructed upon a foundation consisting of a few fundamental concepts and several axioms relating these concepts. As it happens, important advantages accrue from this approach—it is not simply a desire for mathematical rigor, but a pragmatic need for simplification that prompts us to adopt such a mathematical attitude.

The basic concepts are conductor, element, time, voltage, and current. Conductor and element are axiomatic; thus, they cannot be defined, only explained. A conductor is the idealization of a piece of copper wire; an element is a region of space penetrated by two conductors of finite length termed **leads** and pronounced "leeds". The ends of these leads are called **terminals** and are often drawn with small circles as in Fig. 13.1.

Conductors and elements are the basic objects of circuit theory; we will take time, voltage, and current as the basic variables. The time variable is measured with a clock (or, in more picturesque language, a chronometer). Its unit is the second, s. Thus, we will say that time, like voltage and current, is defined **operationally**, that is, by means of a measuring instrument and a procedure for measurement. Our view of reality in this context is consonant with that branch of philosophy termed **operationalism** [1].

Voltage is measured with an instrument called a *voltmeter*, as shown in Fig. 13.2. As shown there, a voltmeter consists of a readout device and two long, flexible conductors terminated in points called **probes** that can be held against other conductors, thereby making electrical contact with them. These conductors are usually covered with an insulating material. One is often colored red and the other black. The one colored red defines the positive polarity of voltage, and the other the negative polarity. Thus, voltage is always measured between two conductors. If these two conductors are element leads, the voltage is that across the corresponding element. Figure 13.3 shows the symbolic description of such a measurement; the variable v, along with

0-8493-8341-2/95/$0.00 + $.50

FIGURE 13.1 Conductors and elements.

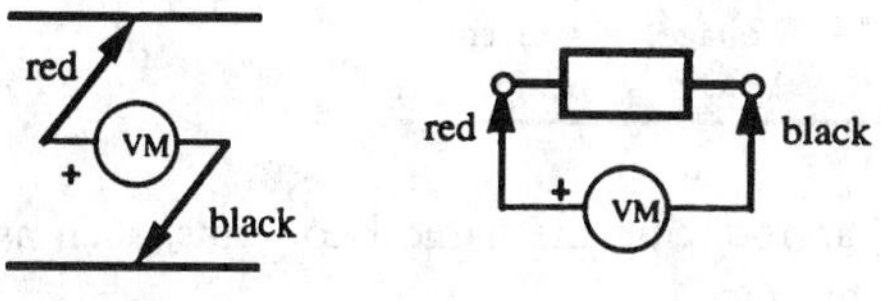

FIGURE 13.2 The operational definition of voltage.

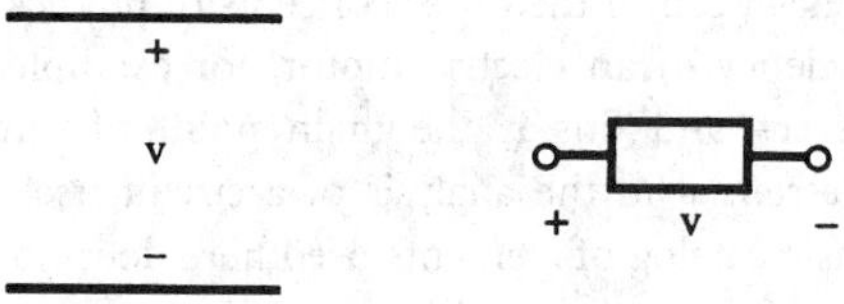

FIGURE 13.3 The symbolic description of the voltage measurement.

the corresponding plus and minus signs, means exactly the experimental procedure shown in Fig. 13.2, neither more nor less. The outcome of the measurement, incidentally, can be either positive or negative. Thus, a reading of $v = -12$ V, for example, has meaning only when viewed within the context of the measurement. If the meter leads are simply reversed after the measurement just described, a reading of $v' = +12$ V will result. The latter, however, is a different variable; hence, we have changed the symbol to v'. The V after the numerical value is the unit of voltage, the volt, V.

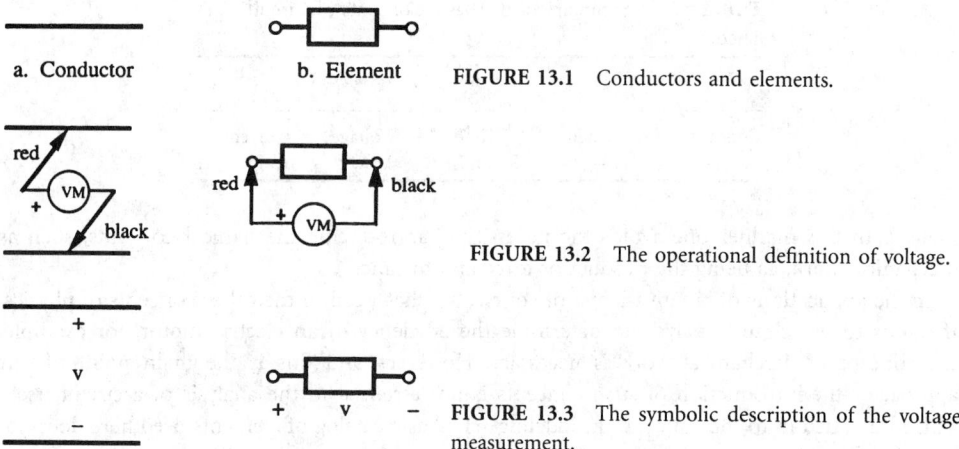

Although voltage is measured across an element (or between conductors), current is measured through a conductor or element. Figure 13.4 provides an operational definition of current. One cuts the conductor or element lead and touches one meter lead against one terminal thus formed and the other against the second. A shorthand symbol for the meter connection is an arrow close to one lead of the ammeter. This arrow, along with the meter reading, defines the current. We show the shorthand symbol for a current in Fig. 13.5. The reference arrow and the symbol i are shorthand for the complete measurement shown in Fig. 13.4, merely this and nothing more. The variable i can be either positive or negative; for example, one possible outcome of the measurement might be $i = -5$ A. The A signifies the unit of current, the ampere. If the red and black leads in Fig. 13.4 were reversed, the reading sign would change.

Table 13.1 provides a summary of the basic concepts of circuit theory: the two basic objects and the three fundamental variables. Notice that we are a bit at variance with the SI system here because although time and current are considered fundamental in that system, voltage is not. Our approach simplifies things, however, for one does not require any of the other SI units or dimensions. All other quantities are derived. For instance, charge is the integral of current and its unit is the ampere·second, or the coulomb, C. Power is the product of voltage and current. Its unit is the watt, W. Energy is the integral of power, and has the unit of the watt-second, or

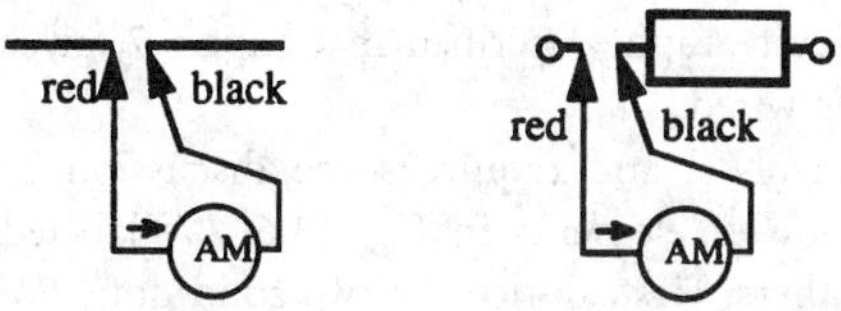

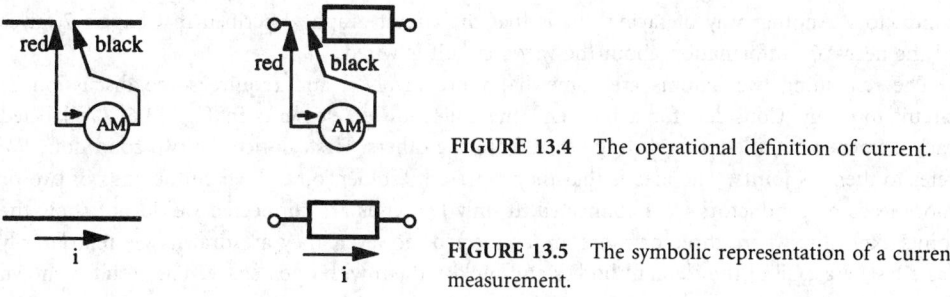

FIGURE 13.4 The operational definition of current.

FIGURE 13.5 The symbolic representation of a current measurement.

TABLE 13.1 Summary of the Basic Concepts of Circuit Theory

Objects		Variables		
Conductor	Element	Time	Voltage	Current
—	—	Seconds, s	Volt, V	Ampere, A

joule, J. In this manner one avoids the necessity of introducing mechanical concepts, such as mechanical work, as being the product of force and distance.

In the applications of circuit theory, of course, one has need of the other concepts of physics. If one is to use circuit analysis to determine the efficiency of an electric motor, for example, the concept of mechanical work is necessary. However—and this is the main point of our approach—the introduction of such concepts is not essential in the analysis of a circuit itself. This idea is tied in to the concept of modeling. The basic catalog of elements used here does not include such things as temperature effects or radiation of electromagnetic energy. Furthermore, a "real" element such as a resistor is not "pure". A real resistor is more accurately modeled, for many purposes, as a resistor plus series inductance and shunt capacitance. The point is this: in order to adequately model the "real world" one must often use complicated combinations of the basic elements. Additionally, to incorporate the influence of variables such as temperature, one must assume that certain parameters (such as resistance or capacitance) are functions of that variable. It is the determination of the more complicated model or the functional relationship of a given parameter to, e.g., temperature that fall within the realm of the practitioner. Such ideas were discussed more fully in the preceding chapter. Circuit analysis merely provides the tools for analyzing the end result.

The radiation of electromagnetic energy is, on the other hand, a quite different aspect of circuit theory. As will be seen, circuit analysis falls within a regime in which such behavior can be neglected. Thus, the theory of circuit analysis we will expound has a limited range of application: to low frequencies or, what is the same in the light of Fourier analysis, to waveforms that do not vary too rapidly.

We are now in a position to state two basic axioms which we will assume all circuits obey:

Axiom 1: The behavior of an element is completely determined by its v–i characteristic, which can be determined by tests made on the element in isolation from the other elements in the circuit in which it is connected.

Axiom 2: The behavior of a circuit is independent of the size or the shape or the orientation of its elements, the conductors which interconnect them, and the element leads.

At this point we loosely consider a circuit to be any collection of elements and conductors, although we will sharpen our definition a bit later. Axiom 1 means that we can run tests on an element in the laboratory, then wire it into a circuit and have the assurance that it will not exhibit any new and different behavior. Axiom 2 means that it is only the *topology* of a circuit that matters, not the way the circuit is stretched or bent or rearranged, so long as we do not change the listing of which element leads are connected to which others or to which conductors. Another way of saying this is that the circuit graph, described in Chapter 7, gives all the necessary information about the way a circuit is wired.

The remaining two axioms are somewhat more involved and require some discussion of circuit topology. Consider, for a moment, the collection of elements in Fig. 13.6. We labeled each element with a letter to distinguish it from the others. First, notice the two solid dots. We refer to them as **joints.** The idea is that they represent "solder joints," where the ends of two or more leads or conductors were connected. If only two ends are connected we do not show the joints explicitly; where three or more are connected, however, they are drawn. We temporarily (as a test) erase all of the element bodies and replace them with open space. The result is shown

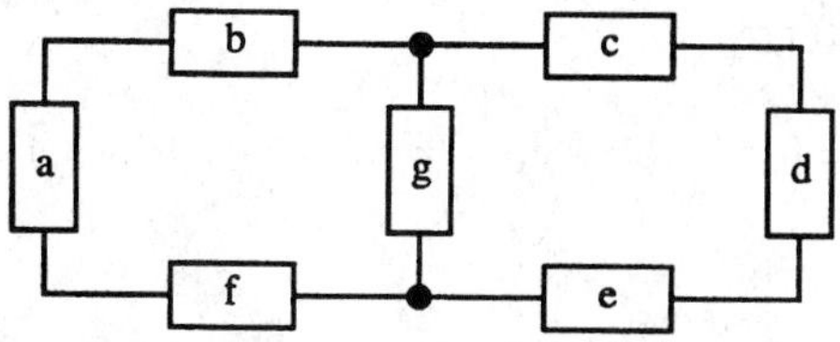

FIGURE 13.6 An example circuit.

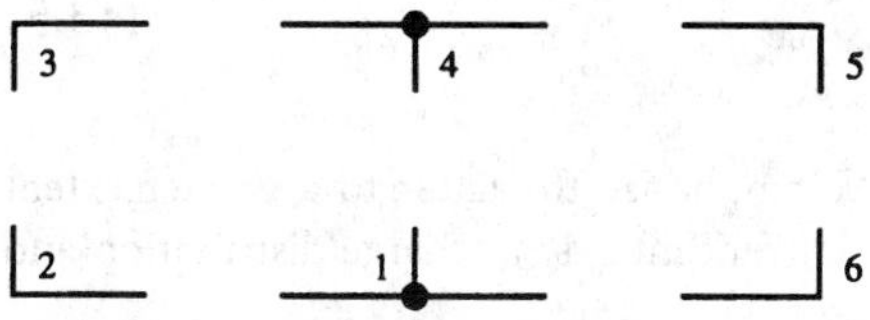

FIGURE 13.7 The nodes of the example circuit.

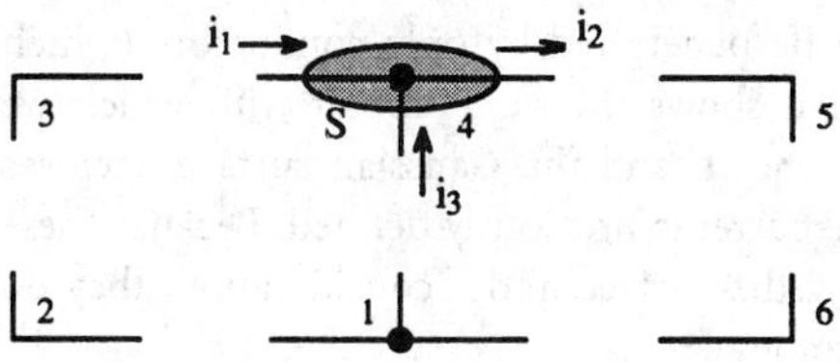

FIGURE 13.8 Illustration of Kirchhoff's current law.

in Fig. 13.7. We refer to each of the interconnected "islands" of a conductor as a **node**. There are six nodes in this example circuit, and we labeled them with the numbers one through six for identification purposes.

> **Axiom 3 (Kirchhoff's Current Law)**: The charge on a node or in an element is identically zero at all instants of time.

Kirchhoff's current law, which we will abbreviate KCL, is not usually phrased in quite this manner. Thus, let us consider a closed (or "Gaussian") surface S as shown in Fig. 13.8. We assume that it is penetrated only by conductors. The elements, of course, are there; we simply do not show them so that we can concentrate on the conductors. We have arbitrarily defined the currents in the conductors penetrating S. Now, recalling that charge is the time integral of the current and thus has the same direction as the current from which it is derived, one can phrase Axiom 3 as follows:

$$\sum_S q_{\text{in}} = q_{\text{enclosed}} = 0 \tag{13.1}$$

at each instant of time. This equation is simply one form of conservation of charge. Because current is the time derivative of voltage, one can also state that

$$\sum_S i_{\text{in}} = 0 \tag{13.2}$$

at each and every time instant. This last equation is the usual phrasing of KCL. The subscript "in" means that a current reference pointed inward is to be considered positive; by default, therefore, a current with its reference pointed outward is to have a negative sign affixed. This sign is in addition to any negative sign that might be present in the value of each variable. For node 4 in Fig. 13.8, KCL in its current form, therefore, reads

$$i_1 - i_2 + i_3 = 0 \tag{13.3}$$

Two other ways of expressing KCL (in current form) are

$$\sum_S i_{\text{out}} = 0 \tag{13.4}$$

and

$$\sum_S i_{\text{in}} = \sum_S i_{\text{out}} \tag{13.5}$$

The equivalent charge forms are clearly also valid. We emphasize the latter to a greater extent than is usual in the classical treatment because of the current interest in charge distribution and transfer circuits.

The Gaussian surface used to express KCL is not constrained to enclose only conductors. It can enclose elements as well, although it still can be penetrated by only conductors (which can be element leads). Thus, consider Fig. 13.9 which shows the same circuit with which we have been working. Now, however, the elements are shown and the Gaussian surface encloses three elements, as well as the conductors carrying the currents previously defined. Because these currents are not carried in the conductors penetrating the surface under consideration, they do not enter into KCL for that surface. Rather, KCl becomes

$$i_x + i_y + i_z = 0 \tag{13.6}$$

As a special case let us look once more at the preceding figure, but use a different surface, one enclosing only the element b. This is shown in Fig. 13.10. If we refer to Axiom 3, which notes that charge cannot accumulate inside an element, and apply charge conservation, we find that

$$i_x = i_1 \tag{13.7}$$

This states that the current into any element in one of its leads is the same as the current leaving in the other lead. In addition, we see that KCL for nodes and KCL for elements (both of which are implied by Axiom 3) imply that KCL holds for any general closed surface penetrated only by conductors such as the one used in connection with Fig. 13.9.

In order to phrase our last axiom, we must discuss circuit topology a bit more, and we will continue to use the circuit just considered above. We define a **path** to be an ordered sequence

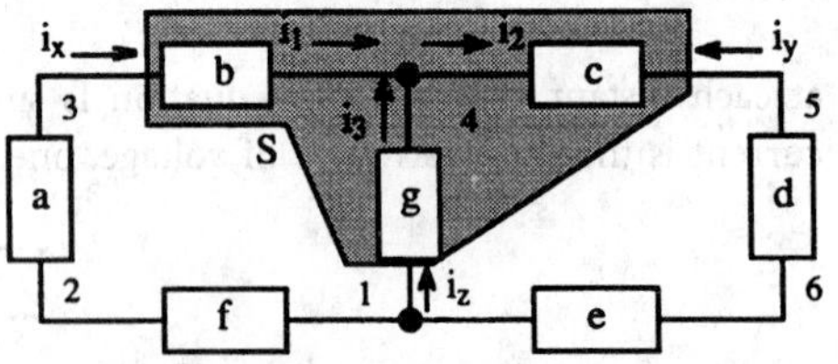

FIGURE 13.9 KCL for a more general surface.

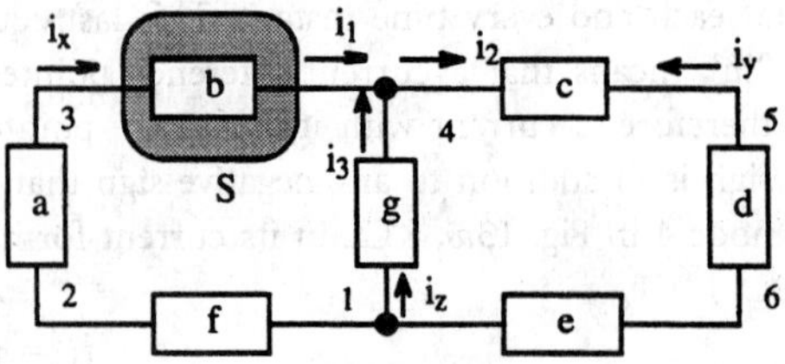

FIGURE 13.10 KCL for a single element.

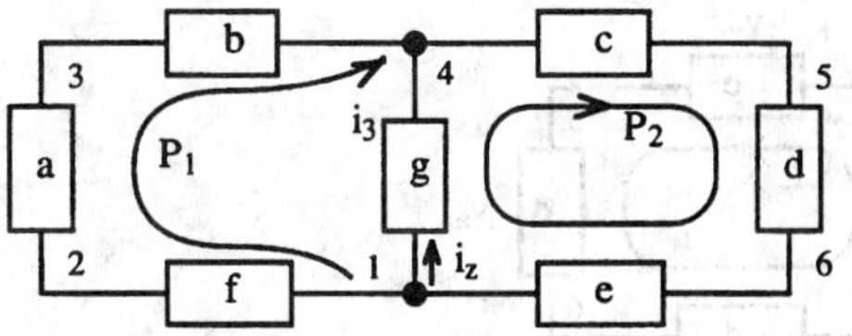

FIGURE 13.11 Circuit paths.

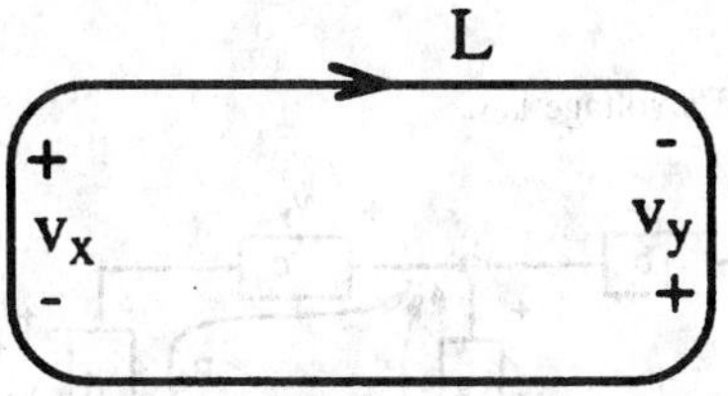

FIGURE 13.12 Voltage rise and drop.

of elements having the property that any two consecutive elements in the sequence share a common node. Thus, referring for convenience back to Fig. 13.10, we see that $\{f, a, b\}$ is a path. The elements f and a share node 2 and a and b share node 3. One lead of the last element in a path is connected to a node that is not shared with the preceding element. Such a node is called the **terminal** node of the path. Similarly, one lead of the first element in the sequence is connected to a node that is not shared with the preceding element[1]. It is called the *initial* node of the path. Thus, in the example just cited, node 1 is the initial node and node 4 is the final node. Thus, a direction is associated with a path, and we can indicate it diagrammatically by means of an arrow on the circuit. This is illustrated in Fig. 13.11 for the path $P_1 = \{f, a, b\}$ and $P_2 = \{g, c, d, e\}$.

If the initial node is identical to the terminal node, then the corresponding path is called a *loop*. An example is $\{f, a, b, g\}$. The path P_2 is a loop. An alternate definition of a loop is as a collection of branches having the property that each node connected to a path branch is connected to precisely two path branches; that is, it has **degree two** relative to the path branches.

We can define the voltage across each element in our circuit in exactly two ways, corresponding to the choices of which lead is designated plus and which is designated minus. Figure 13.12 shows two voltages and a loop L in a highly stylized manner. We have purposely not drawn the circuit itself so that we can concentrate on the essentials in our discussion. If the path enters the given element on the lead carrying the minus and exits on the one carrying the positive, its voltage will be called a **voltage rise**; however, if it enters on the positive and exits on the minus, the voltage will be called a **voltage drop**. If the signs of a voltage are reversed and a negative sign is affixed to the voltage variable, the value of that variable remains unchanged; thus, note that a negative rise is a drop, and vice versa.

We are now in a position to state our fourth and final axiom:

Axiom 4 (Kirchhoff's Voltage Law): The sum of the voltage rises around any loop is identically zero at all instants of time.

We refer to this law as KVL for the sake of economy of space. Just as KCL was phrased in terms of charge, KVL could just as well be phrased in terms of **flux linkage**. Flux linkage is the time integral of voltage, so it can be said that the sum of the flux linkages around a loop is zero. In voltage form we write

$$\sum_{\text{loop}} v_{\text{rises}} = 0 \tag{13.8}$$

[1]We assume that no element has its two leads connected together and that there are more than two elements in the path in this definition.

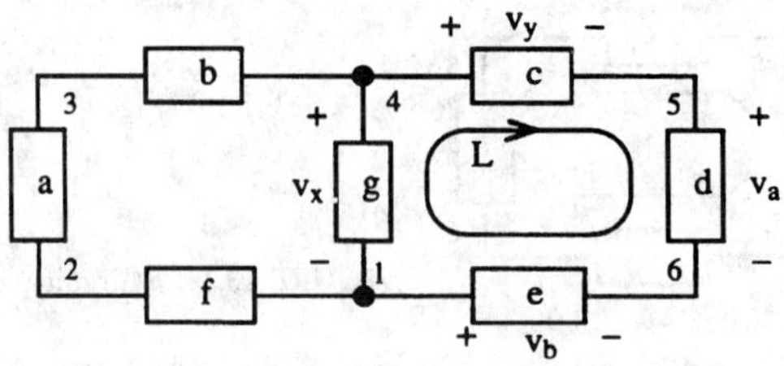

FIGURE 13.13 Illustration of Kirchhoff's voltage law.

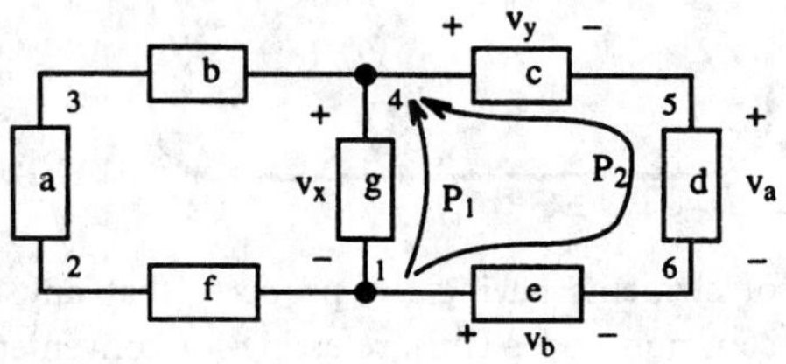

FIGURE 13.14 Path form of KVL.

We observed that a negative rise is a drop, so

$$\sum_{\text{loop}} v_{\text{drops}} = 0 \tag{13.9}$$

or

$$\sum_{\text{loop}} v_{\text{rises}} = \sum_{\text{loop}} v_{\text{drops}} \tag{13.10}$$

Thus, in Fig. 13.13, we could write [should we choose to use the form of (13.8)]

$$v_x - v_y - v_a + v_b = 0 \tag{13.11}$$

Clearly, one can rearrange KVL into many different algebraic forms that are equivalent to those just stated; one form, however, is more useful in circuit computations than many others. It is known as the **path form** of KVL. To better appreciate this form, review Fig. 13.13. This time, however, the paths are defined a bit differently. As shown in Fig. 13.14, we consider two paths, P_1 and P_2, having the same initial and terminal nodes, 1 and 4, respectively.[2] We can rearrange (13.11) into the form

$$v_x = -v_b + v_a + v_y \tag{13.12}$$

This form is often useful for finding one unknown voltage in terms of known voltages along some given path. In general, if P_1 and P_2 are two paths having the same initial and final nodes,

$$\sum_{P_1} v_{\text{rises}} = \sum_{P_2} v_{\text{rises}} \tag{13.13}$$

Be careful to distinguish this equation from (13.10). In the present case two paths are involved; in the former we find only a single loop, and drops are located on one side of the equation and rises on the other. One might call the path form the "all roads lead to Rome" form.

[2]If one defines the **negative** of a path as a listing of the same elements as the original path in the reverse order and summation of two paths as a concatenation of the two listings, one sees that $P_1 - P_2 = L$, the loop in Fig. 13.13.

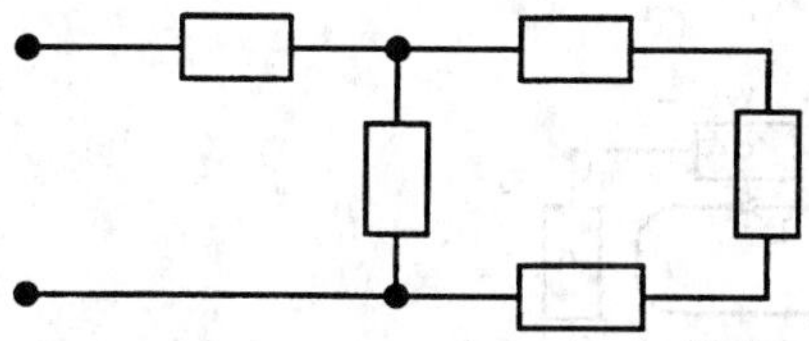

FIGURE 13.15 A "noncircuit."

We covered four basic axioms, and these are all that are needed to construct a mathematical theory of circuit analysis. The first axiom is often referred to by means of the phrase "lumped circuit analysis", for we assume that all the physics of a given element are internal to that element and are of no concern to us; we are only interested in the v–i characteristic. That is, we are treating all the elements as lumps of matter that interact with the other elements in a circuit by means of the voltage and current at their leads. The second axiom says that the physical construction is irrelevant and that the interconnections are completely described by means of the circuit graph. Kirchhoff's current law is an expression of conservation of charge, plus the assumption that neither conductors nor elements can maintain a net charge. In this connection observe that a capacitor maintains a charge separation internally, but it is a separation of two charges of opposite sign; thus, the total algebraic charge within it is zero. Finally, KVL is an expression of conservation of flux linkage. If $\lambda(t) = \int_{-\infty}^{t} v(\alpha)\,d\alpha$ is the flux linkage, then one can write[3] (using one form of KVL)

$$\sum_{\text{loop}} \lambda_{\text{rises}}^{(t)} = 0 \tag{13.14}$$

In the theory of electromagnetics one finds that this equation does not hold exactly; in fact, the right-hand side is equal to the negative of the derivative of the magnetic flux contained within the loop (this is the Faraday-Lenz law). If, however, the time variation of all signals in the circuit are slow, then the right-hand side is approximately zero and KVL can be assumed to hold. A similar result holds also for KCL. For extremely short instants of time, a conductor can support an unbalanced charge. One finds, however, that the "relaxation" time of such unbalanced charge is quite short in comparison with the time variations of interest in the circuits considered in this text.

Finally, we tie up a loose end left hanging at the beginning of this subsection. We consider a circuit to be, not just any collection of elements that are interconnected, but a collection having the property that each element is contained in at least one loop. Thus, the circuit in Fig. 13.15 is not a circuit; instead, it must be treated as a *subcircuit*, that is, as part of a larger circuit in which it is to be inbedded.

The remainder of this section develops the application of the axioms presented here to the analysis of circuits. The reader is referred to [2, 3, 4] for a more detailed treatment.

Nodal Analysis

Nodal analysis of electric circuits, although using all four of the fundamental axioms presented in the introduction, concentrates upon KCL explicitly. Kirchhoff's voltage law is also satisfied automatically in view of the way the basic equations are formulated. This effective method uses the concept of a **node voltage**. Figure 13.16 illustrates the concept. Observe a voltmeter, with

[3]One might anticipate a constant on the right side of (13.14); however, a closer investigation reveals that it is more realistic and pragmatic to assume that all signals are one-sided and that all elements are causal. This implies that the constant is zero. Two-sided signals only arise legitimately within the context of steady-state behavior of stable circuits and systems.

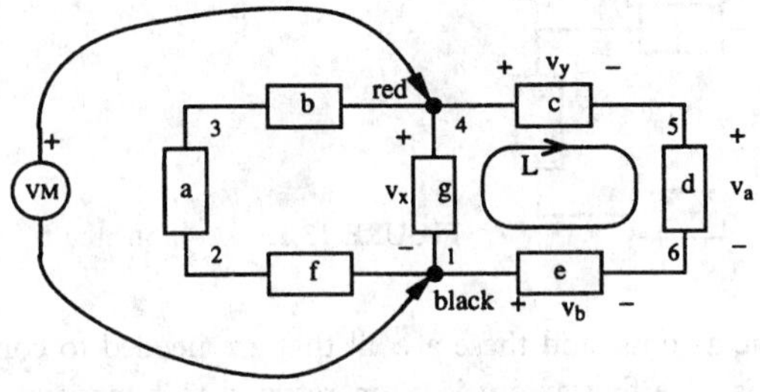

FIGURE 13.16 Node voltages.

its black probe attached to a single node, called the **reference node**, which remains fixed during the course of the investigation. In the case shown node 1 is the reference node. The red probe is shown being touched to node 4; therefore, we call the resulting voltage v_4. The subscript denotes the node and the result is always assumed to have its positive reference on the given node. In the present instance v_4 is identical to the element voltage because element g (across which v_g is defined) is connected between node 4 and the reference node. Note that the voltage of such an element is always either the node voltage or its negative, depending upon the reference polarities of its associated element voltage. If we were to touch the red probe to node 5, however, no element voltage would have this relationship to the resulting node voltage v_5 because no elements are connected directly between nodes 5 and 1.

The concept of reference node is used so often that a special symbol is used for it [see Fig. 13.17(a)]; alternate symbols often seen on circuit diagrams are shown in the figure as well. Often one hears the terms "ground" or "ground reference" used. This is commonly accepted argot for the reference node; however, one should be aware that a safety issue is involved in the *process* of grounding a circuit or appliance. In such cases, sometimes one symbol specifically means "earth ground" and one or more other symbols are used for such things as "signal ground" or "floating ground", although the last term is something of an oxymoron. Here, we use the terms "reference" or "reference node". The reference symbol is quite often used to simplify the drawing of a circuit. The circuit in Fig. 13.16, for instance, can be redrawn as in Fig. 13.18; circuit operation will be unaffected. Note that all four of the reference symbols refer to a single node, node 1, although they are shown separated from one another. In fact, the circuit is not changed electrically if one bends the elements around and thereby separates the ground symbols even more, as we have done in Fig. 13.19. Notice that the loop L shown in the original figure, Fig. 13.16, remains a loop, as in Figs. 13.18 and 13.19. Redrawing a circuit using ground reference symbols does not alter the circuit topology, the circuit graph.

Suppose the red probe were moved to node 5. As above, no element is directly connected between nodes 5 and 1; hence, node voltage v_5 is not an element voltage. However, the element voltages and the node voltages are directly related in a one-to-one fashion. To see how, look at Fig. 13.20. This figure shows a "floating element" e, which is connected between two nodes, k and j, neither of which is the reference node. It is vital here to remember that all node voltages are assumed to have their positive reference polarities on the nodes themselves and their negative reference on the reference node. Now we can define the element voltage in either of two possible

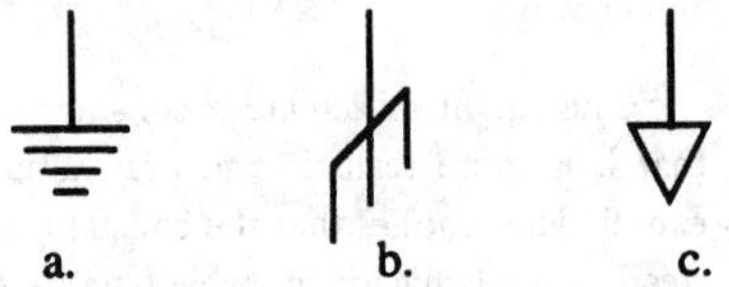

FIGURE 13.17 Reference node symbols.

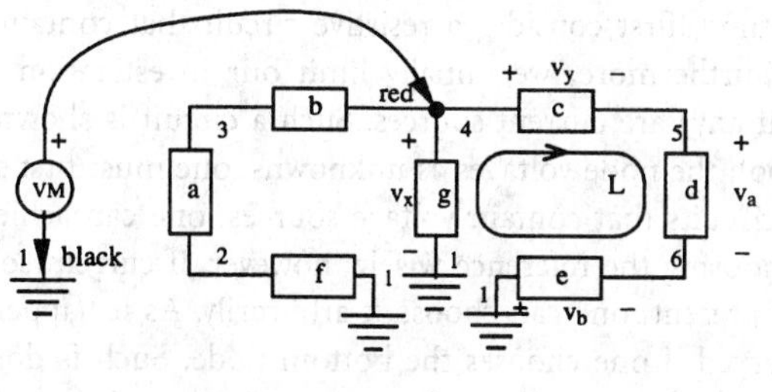

FIGURE 13.18 An alternate drawing.

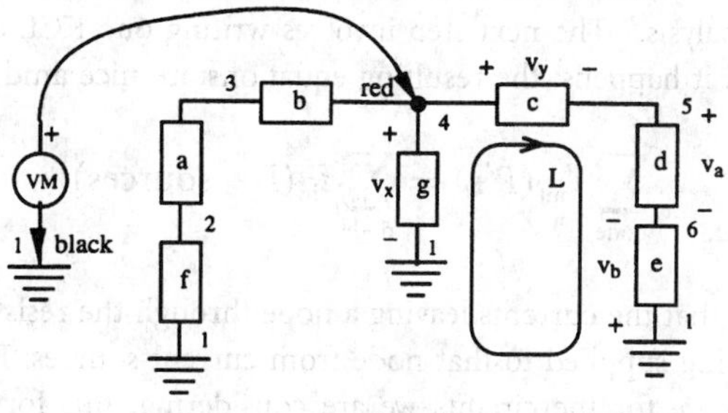

FIGURE 13.19 An equivalent drawing.

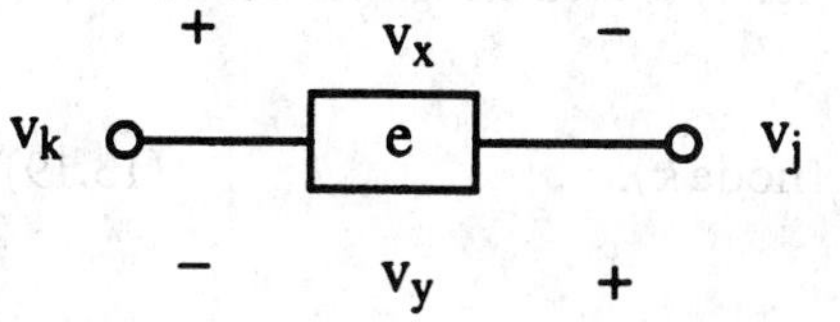

FIGURE 13.20 A floating element and its voltage.

ways, as shown in the figure. Kirchhoff's voltage law (the simplest form perhaps being the path form) shows at once that

$$v_x = v_k - v_j \tag{13.15}$$

and

$$v_y = v_j - v_k \tag{13.16}$$

An easy mnemonic for this result is the following:

$$v_{\text{floating element}} = v_+ - v_- \tag{13.17}$$

where v_+ is the node voltage of the node to which the element lead associated with the positive reference for the element voltage is connected, and v_- is the node voltage of the node to which the lead carrying the negative reference for the element voltage is connected. We refer to an element that is not floating, by the way, as being "grounded".

It is easy to see that a circuit having N nodes has $N - 1$ node voltages; further, if one uses (13.17), any element voltage can be expressed in terms of these $N - 1$ node voltages. Then, for any invertible element,[4] one can determine the element current. The nodal analysis method uses this fact and considers the node voltages to be the unknown variables.

[4]For instance, resistors, capacitors, and inductors are invertible in the sense that one can determine their element currents if their element voltages are known.

To illustrate the method, first consider a resistive circuit that contains only resistors and/or independent sources. Furthermore, we initially limit our investigation to circuits whose only independent sources (if any) are current sources. Such a circuit is shown in Fig. 13.21. Because nodal analysis relies upon the node voltages as unknowns, one must first select an arbitrary node for the reference. For circuits that contain voltage sources, one can achieve some simplification for hand analysis by choosing the reference wisely; however, if current sources are the only type of independent source present, one can choose it arbitrarily. As it happens, physical intuition is almost always better served if one chooses the bottom node. Such is done here and the circuit is redrawn using reference symbols as in Fig. 13.22. Here, we have arbitrarily assigned node voltages to the $N-1=2$ nonreference nodes. In performing these two steps, we have "prepared the circuit for nodal analysis." The next step involves writing one KCL equation at each of the nonreference nodes. As it happens, the resulting equations are nice amd compact if the form

$$\sum_{\text{node}} i_{\text{out}}(R's) = \sum_{\text{node}} i_{\text{in}}(I - \text{sources}) \tag{13.18}$$

is used. Here, we mean that the currents leaving a node through the resistors must sum up to be equal to the current being supplied to that node from current sources. Because these two types of elements are exhaustive for the circuits we are considering, this form is exactly equivalent to the other forms presented in the introduction. Furthermore, for a current leaving a node through a resistor, the floating element KVL result in (13.17) is used along with Ohm's law:

$$\sum_{j=1}^{N-1} \frac{v_k - v_j}{R_{kj}} = \sum_{q=1}^{M_k} i_{sq}(\text{node } k). \tag{13.19}$$

In this equation for node k, R_{kj} is the resistance connected between nodes k and j (or the equivalent resistance of the parallel combination if more than one are found), i_{sq} is the value

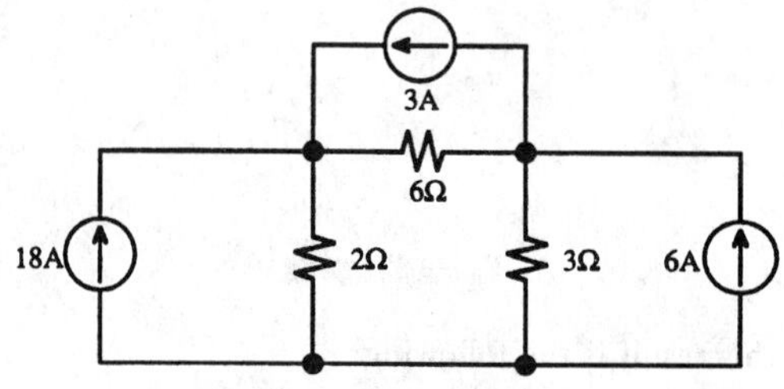

FIGURE 13.21 An example circuit.

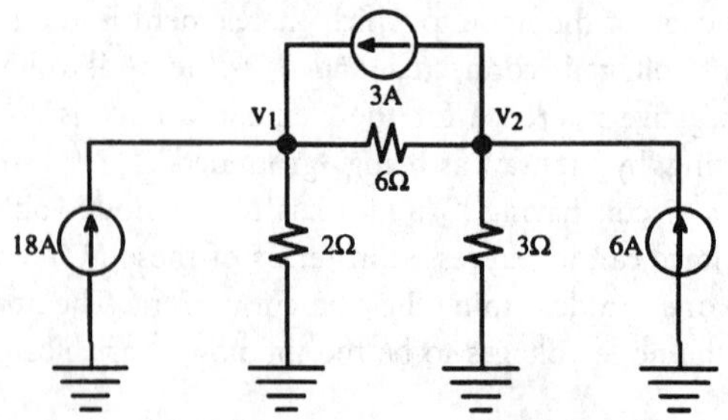

FIGURE 13.22 The example circuit prepared for nodal analysis.

of the qth current source connected to node k (positive if its reference is toward node k), and M_k is the number of such sources. Clearly, one can simply omit the $j = k$ term on the left side because $v_k - v_k = 0$.

The nodal equations for our example circuit are

$$\frac{v_1}{2} + \frac{v_1 - v_2}{6} = 18 + 3 \tag{13.20}$$

and

$$\frac{v_2 - v_1}{6} + \frac{v_2}{3} = 6 - 3 \tag{13.21}$$

Notice, by the way, that we are using units of A, Ω, and V. It is a simple matter to show that KVL, KCL, and Ohm's law remain invariant if we use the consistent units of mA, kΩ, and V. The latter is often a more practical system of units for filter design work. In the present case the matrix form of these equations is

$$\begin{bmatrix} \frac{2}{3} & -\frac{1}{6} \\ -\frac{1}{6} & \frac{1}{2} \end{bmatrix} \begin{bmatrix} v_1 \\ v_2 \end{bmatrix} = \begin{bmatrix} 21 \\ 3 \end{bmatrix} \tag{13.22}$$

It can be verified easily that the solution is $v_1 = 36$ V and $v_2 = 18$ V. To see that one can compute the value of any desired variable from these two voltages, consider the problem of determining the current i_6 (let us call it) through the horizontal 6 Ω resistor from right to left. One can simply use the equation

$$i_6 = \frac{v_2 - v_1}{6} = \frac{18 - 36}{6} = -3\,\text{A} \tag{13.23}$$

The above procedure works for essentially all circuits encountered in practice. If the coefficient matrix on the left in (13.22) (which will always be symmetric for circuits of the type we are considering) is nonsingular, a solution is always possible. It is surprisingly difficult, however, to determine conditions on the circuit under which solutions do not exist, although this is discussed at greater length in a later subsection.

Suppose, now, that our circuit to be solved contains one or more independent voltage sources in addition to resistors and/or current sources. This constrains the node voltages because a given voltage source value must be equal to the difference between two node voltages if it is floating and to a node voltage or its negative if it is grounded. One might expect that this complicates matters, but fortunately the converse is true.

To explore this more fully, examine the example circuit in Fig. 13.23. The algorithm just presented will not work as is because it relies upon balancing the currents between resistors and current sources. Thus, it seems that we must account in some fashion for the currents in the voltage sources. In fact, we do not, as the following analysis shows. The key step in our reasoning is this: the analysis *procedure* should not depend upon the values of the independent circuit variables, that is, on the values of the currents in the current sources and voltages across the voltage sources. This is almost inherent in the definition of an independent source, for it can be adjusted to any value whatsoever. What we are assuming in addition to this is simply that we would not write one given set of equations for a specific set of source values, then change to another set of equations when these values are altered. Thus, let us test the circuit by temporarily deactivating all the independent sources; i.e., by making their values zero. Recalling

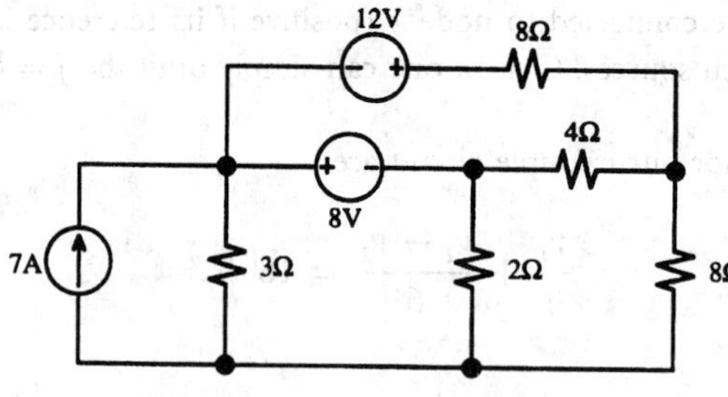

FIGURE 13.23 An example circuit.

that a deactivated voltage source is equivalent to a short circuit and a deactivated current source to an open circuit, we have the resulting configuration of Fig. 13.24. The resulting nodes are shaded for convenience. Note carefully, however, that the nodes in the circuit under test are not the same as those in the original circuit, although they are related. Notice that, for the circuit under test, all of the resistor voltages would be determined by the node voltages as expected; however, *the number of nodes has been reduced by one for each voltage source.* Hence, we suspect that the required number of KCL equations N_{ne} (and the number of independent node voltages) is

$$N_{ne} = N - 1 - N_v \tag{13.24}$$

where N_V is the number of voltage sources. In the example circuit one can easily compute this required number to be $5 - 1 - 2 = 2$. This is compatible with the fact that clearly three nodes ($3 - 1 = 2$ nonreference nodes) are clearly found in Fig. 13.24.

It should also be rather clear that there is only one independent voltage within each of the shaded regions shown in Fig. 13.24. We can use KVL to express any other in terms of that one. For example, in Fig. 13.25 we have redrawn our example circuit with the bottom node arbitrarily chosen as the reference. We have also arbitrarily chosen a node voltage within the top left surface as the unknown v_1. Notice how we have used KVL (the path form, again, is perhaps the most effective) to determine the node voltages of all the other nodes within the top left surface. Any set of connected conductors, leads, and voltage sources to which only one independent voltage can be assigned is called a **generalized node.** If that generalized node does not include the reference node, it is termed a **supernode.** The node within the shaded surface at the top left in Fig. 13.25, however, has no voltage sources; hence, it is called an **essential node.**

As pointed out earlier, the equations that one writes should not depend upon the values of the independent sources. If one were to reduce all the independent sources to zero, each generalized node would reduce to a single node; hence, only one equation should be written for each supernode. One equation should be written for each essential node also; it is unaffected by deactivation of the independent sources. Observe that deactivation of the current sources does not reduce the number of nodes in a circuit.

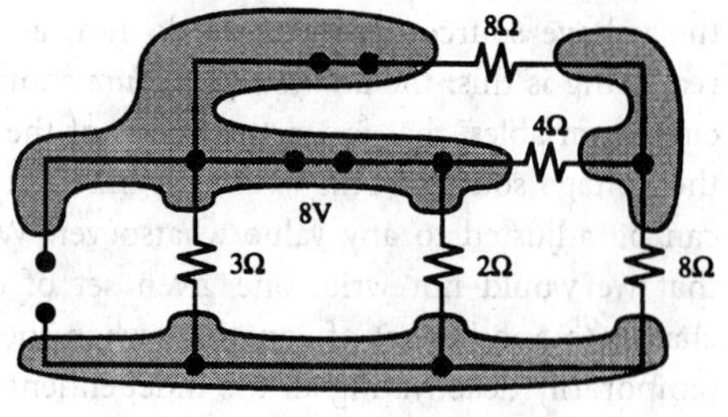

FIGURE 13.24 The example circuit deactivated.

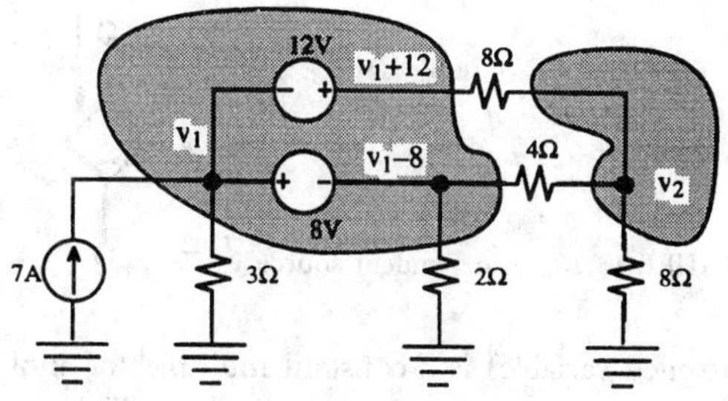

FIGURE 13.25 The example circuit prepared for nodal analysis.

Writing one KCL equation for the supernode and one for the essential node in Fig. 13.25 results in

$$\frac{v_1}{3} + \frac{v_1 - 8}{2} + \frac{v_1 - 8 - v_2}{4} + \frac{v_1 + 12 - v_2}{8} = 7 \tag{13.25}$$

and

$$\frac{v_2}{8} + \frac{v_2 - (v_1 - 8)}{4} + \frac{v_2 - (v_1 + 12)}{8} = 0 \tag{13.26}$$

In matrix form, one has

$$\begin{bmatrix} \frac{29}{24} & -\frac{3}{8} \\ -\frac{3}{8} & \frac{1}{2} \end{bmatrix} \begin{bmatrix} v_1 \\ v_2 \end{bmatrix} = \begin{bmatrix} \frac{23}{2} \\ -\frac{1}{2} \end{bmatrix} \tag{13.27}$$

The solution is $v_1 = 12$ V and $v_2 = 8$ V. Notice once again that the coefficient matrix on the left-hand side is symmetric. This actually follows from our earlier observation about this property for circuits containing only current sources and resistors because the voltage sources only introduce knowns into the nodal equations, thus modifying the right-hand side of (13.27).

The general form for nodal equations in any circuit containing only independent sources and resistors, based upon our foregoing development, is

$$A\bar{v}_n = F_V \bar{v}_s + F_I \bar{i}_s \tag{13.28}$$

where A is a symmetric square matrix of constant coefficients, F_V and F_I are rectangular matrices of constants, and $\bar{v}_n$ is the column matrix of independent node voltages. The vectors $\bar{v}_s$ and $\bar{i}_s$ are column matrices of independent source values. Clearly, if A is a nonsingular matrix, (13.28) can be solved for the node voltages. Then, using KVL and/or Ohm's law, one can solve for any element current or voltage desired. Equally clearly, if a solution exists, it is a multilinear function of the independent source values.[5]

Now suppose that the circuit under consideration contains one or more dependent sources. Recall that the two-terminal characteristics of such elements are indistinguishable from those of the corresponding independent sources except for the fact that their value depends upon some other circuit variable. For instance, in Fig. 13.26 a voltage-controlled voltage source (VCVS) is shown. Its v–i characteristic is identical to that of an independent source except for the fact

[5]That is, it is a linear function of the vector consisting of all of the independent source values.

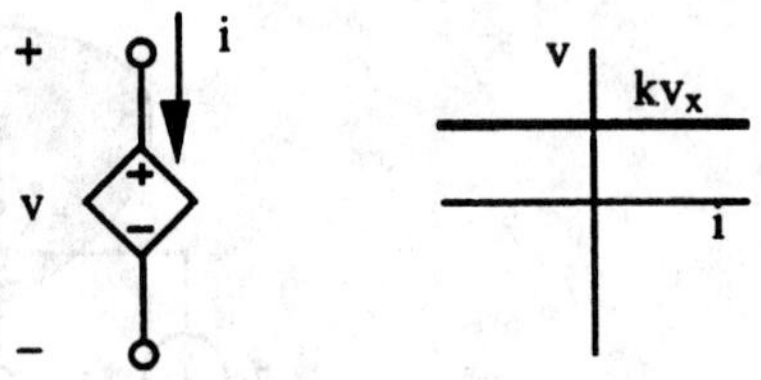

FIGURE 13.26 A dependent source.

that its voltage (the controlled variable) is a constant multiple[6] of another circuit variable (the controlling variable), in this case another voltage. This fact will be relied upon to develop a modification to the nodal analysis procedure.

We will adopt the following attitude: we will imagine that the dependency relationship, kv_x in Fig. 13.26, is a label pasted to the surface of the source in much the same way that a battery is labeled with its voltage. We will imagine ourselves to take a small piece of opaque masking tape and apply it over this label; we will call this process **taping the dependent source**. This means that we are—temporarily—treating it as an independent source. The usual nodal analysis procedure is then followed, which results in (13.28). Then we imagine ourselves to remove the tape from the dependent source(s) and note that the relationship is linear, with the controlling variables as the independent ones and the controlled variables the dependent ones. We next express the controlling variables—and thereby the controlled ones as well—in terms of the node voltages using KVL, KCL, and Ohm's law. The resulting relationships have the forms

$$\bar{v}_c = B'\bar{v}_n + C'\bar{v}_{si} + D'\bar{i}_{si} \tag{13.29}$$

and

$$\bar{i}_c = B''\bar{v}_n + C''\bar{v}_{si} + D''\bar{i}_{si} \tag{13.30}$$

Here, the subscript i refers to the fact that the corresponding sources are the independent ones. Noting that $\bar{v}_c$ and $\bar{i}_c$ appear on the right-hand side of (13.28) because they are source values, one can use the last two results to express the vectors of all source voltages and all source currents in that equation in the form

$$\bar{v}_c = B'''\bar{v}_{si} + C'''\bar{i}_{si} + D'''\bar{v}_n \tag{13.31}$$

and

$$\bar{i}_c = B''''\bar{v}_{si} + C''''\bar{i}_{si} + D''''\bar{v}_n \tag{13.32}$$

Finally, using the last two equations in (13.28), one has

$$A\bar{v}_n = F'_V\bar{v}_s + F'_I\bar{i}_s + B\bar{v}_n \tag{13.33}$$

Now,

$$(A - B)\bar{v}_n = F'_V\bar{v}_s + F'_I\bar{i}_s \tag{13.34}$$

This equation can be solved for the node voltages, provided that $A - B$ is nonsingular. This is even more problematic than for the case without dependent sources because the matrix B is a function of the gain coefficients of the dependent sources; for some set of such values the

[6]Thus, one should actually refer to such a device as a *linear* dependent source.

solution might exist and for others it might not. In any event if $A - B$ is nonsingular, one obtains once more a response that is linear with respect to the vector of independent source values.

Figure 13.27 shows a rather complex example circuit with dependent sources. As pointed out earlier, there are often reasons for preferring one reference node to another. Here, notice that if we choose one of the nodes to which a voltage source is attached it is not necessary to write a nodal equation for the nonreference node because, when the circuit is tested by deactivation of *all* the sources, the node disappears into the ground reference; thus, it is part of a generalized node including the reference called a **nonessential node**. For this circuit, choose the node at the bottom of the $2V$ independent source. The resulting circuit, prepared for nodal analysis, is shown in Fig. 13.28. Surfaces have been drawn around both generalized nodes and the one essential node and they have been shaded for emphasis. Note that we have chosen one node voltage within the one supernode arbitrarily and have expressed the other node voltage within that supernode in terms of the first and the voltage source value; furthermore, we have taped both dependent sources and written in the known value at the one nonessential node.

The nodal equations for the supernode and for the essential node are

$$\frac{v_1 - 2}{2} + \frac{v_1 - (v_2 - v_c)}{1} = -9 \qquad \text{(essential node)} \tag{13.35}$$

and

$$\frac{v_2}{2} + \frac{v_2 - 2}{1} + \frac{v_2 - v_c - v_1}{1} = i_c \qquad \text{(supernode)} \tag{13.36}$$

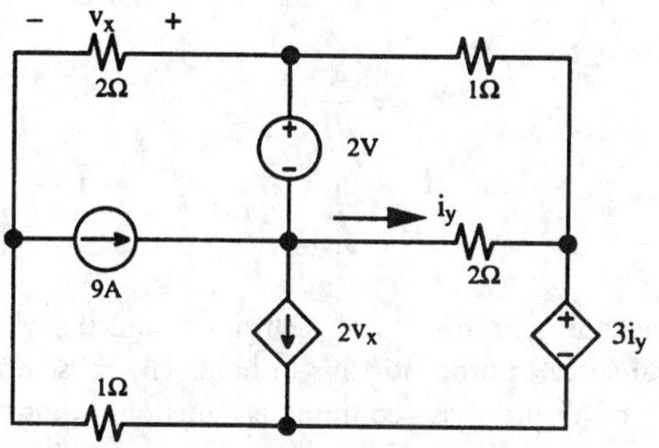

FIGURE 13.27 An example circuit.

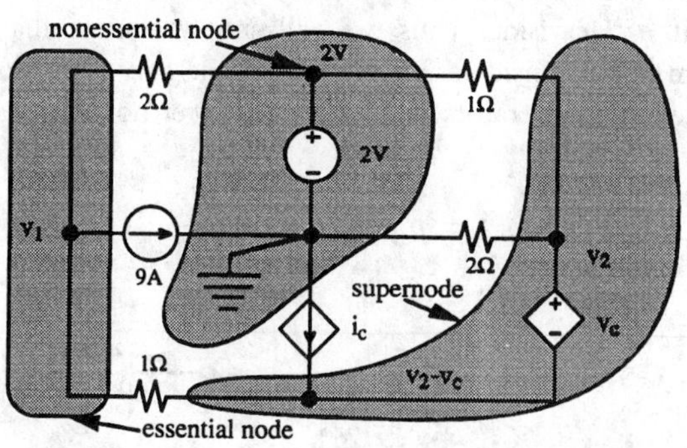

FIGURE 13.28 The example circuit prepared for nodal analysis.

Now the two dependent sources are untaped and their values expressed in terms of the unknown node voltages and known values using KVL, KCL, and Ohm's law. This results in (referring to the original circuit for the definitions)

$$v_c = -\frac{3}{2}v_2 \tag{13.37}$$

and

$$i_c = 4 - 2v_2. \tag{13.38}$$

Solving these four equations simultaneously gives $v_1 = -2$ V and $v_2 = 2$ V.

If the circuit under consideration contains op amps, one can first replace each op amp by a VCVS, using the above procedure, and then allow the voltage gain to go to infinity. This is a bit unwieldy, so one often models the op amp in a different way as a circuit element called a **nullor**. This is explored in more detail elsewhere in the book and is not discussed here.

Thus far, this chapter has consider only nondynamic circuits whose independent sources were all constants (DC). If these independent sources are assumed to possess time-varying waveforms, no essential modification ensues. The only difference is that each node voltage, and hence each circuit variable, becomes a time-varying function. If the circuit considered contains capacitors and/or inductors, however, the nodal equations are no longer algebraic; they become differential equations. The method developed above remains applicable, however. We will now show why.

Capacitors and inductors have the v–i relationships given in Fig. 13.29. The symbols p and $1/p$ are referred to as **operators, differential operators**, or **Heaviside operators.** The last term is in honor of Oliver Heaviside, who first used them in circuit analysis. They are defined by

$$p = \frac{d}{dt} \tag{13.39}$$

$$\frac{1}{p} = \int_{-\infty}^{t} (\,)\,da \tag{13.40}$$

The notation suggests that they are inverses of each other, and this is true; however, one must suitably restrict the signal space in order for this to hold. The most realistic assumption is that the signal space consists of all piecewise continuous functions whose derivatives of all orders exist except on a countable set of points that does not have any finite points of accumulation—*plus all generalized derivatives of such functions.* In fact, Laurent Schwartz, on the first page of the preface of his important work on the theory of distributions, acknowledges that his work was motivated by that of Heaviside. Thus, we will simply assume that all derivatives of all orders of any waveform under consideration exists in a generalized function sense. Higher order differentiation and integration operators are defined in power notation, as expected:

$$p^n = p \cdot p \cdots p = \frac{d^n}{dt^n} \tag{13.41}$$

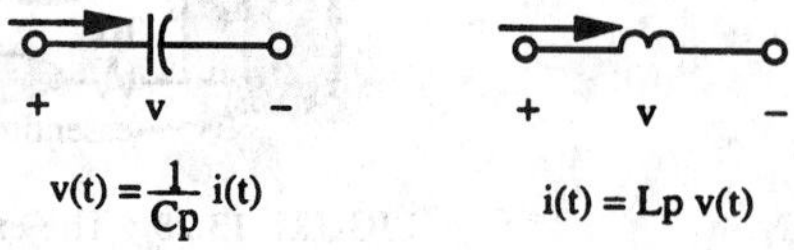

FIGURE 13.29 The dynamic element relationships.

and

$$\frac{1}{p^n} = \frac{1}{p} \cdot \frac{1}{p} \cdots \frac{1}{p} = \int_{-\infty}^{t} \int_{-\infty}^{b} \cdots \int_{-\infty}^{g} (\,)\, da \tag{13.42}$$

Another facet of the above issue often escapes notice, however. Look at any arbitrary function in the above-mentioned signal set, compute its running integral, and differentiate it. This action results in:

$$p\left[\frac{1}{p}x(t)\right] = \frac{d}{dt}\int_{-\infty}^{t} x(\alpha)\, d\alpha = x(t) \tag{13.43}$$

In fact, it is precisely this property that characterizes the set of all generalized functions. It is closed under differentiation. However, suppose the computation is done in the reverse order:

$$\frac{1}{p}[px(t)] = \int_{-\infty}^{t} x'(a)\, da = x(t) = x(t) - x(\infty) \tag{13.44}$$

We have assumed here that the Fundamental Theorem of Calculus holds. This is permissible within the framework of generalized functions, provided that the waveform $x(t)$ has a value in the conventional sense at time t. The problem with the above result is that one does not regain $x(t)$. If it is assumed, however, that $x(t)$ is one sided (that is, $x(t)$ is identically zero for sufficiently large negative values of t), $x(t)$ will be regained and p and $1/p$ will be inverses of one another. Thus, in the following, we will assume that all independent waveforms are one sided. We will, in fact, interpret this as meaning that they are all zero for $t < 0$. We will also assume that all circuit elements possess one property in addition to their defining v–i relationship, namely, that they are causal. Thus, all waveforms in any circuit under consideration will be zero for $t \le 0$ and the above two operators are inverses of one another. The only physically reasonable situation in which two-sided waveforms can occur is that of a stable circuit operating in the steady state, which we recognize as being an approximate mode of behavior derived from the above considerations in the limit as time becomes large.

Referring to Fig. 13.29 once more, we define

$$Z_C(p) = \frac{1}{Cp} \tag{13.45}$$

$$Z_L(p) = Lp \tag{13.46}$$

to be the **impedance operators** (or **operator impedances**) for the capacitor and the inductor, respectively. With our one-sidedness causality assumptions, we can manipulate these quantities just as we would manipulate algebraic functions of a real or complex variable.

The analysis of a dynamic circuit is illustrated by Fig. 13.30. The circuit is shown prepared for nodal analysis, with the reference node at the bottom of the circuit and the dynamic elements

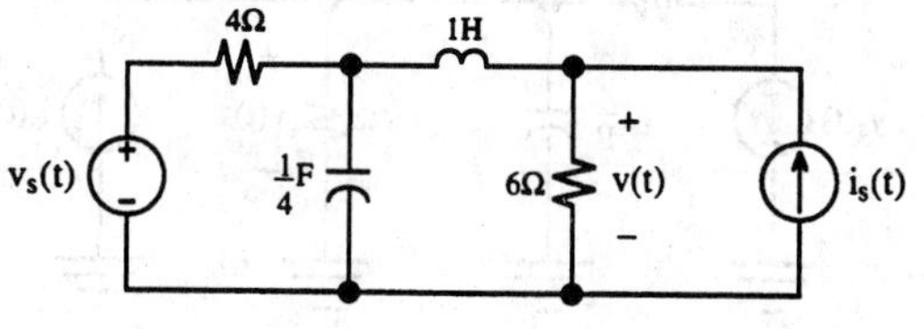

FIGURE 13.30 An example circuit.

expressed in terms of their impedance operators, in Fig. 13.31. Note that if the circuit were to contain dependent sources, we would have taped them at this step. The nodal equations at the two essential nodes are

$$\frac{v_1 - v_2}{4} + \frac{v_1}{4/p} + \frac{v_1 - v_2}{p} = 0 \tag{13.47}$$

and

$$\frac{v_2}{6} + \frac{v_2 - v_1}{p} = i_s \tag{13.48}$$

In matrix form, merely rationalizing and collecting terms,

$$\begin{bmatrix} \frac{1}{4} + \frac{p}{4} + \frac{1}{p} & -\frac{1}{p} \\ -\frac{1}{p} & \frac{1}{6} + \frac{1}{p} \end{bmatrix} \begin{bmatrix} v_1(t) \\ v_2(t) \end{bmatrix} = \begin{bmatrix} \frac{1}{4} v_s(t) \\ i_s(t) \end{bmatrix} \tag{13.49}$$

Notice that the coefficient matrix is once again symmetric because there are no dependent sources. Multiplying the first row of each side by $4p$ and the second by $6p$, thus clearing fractions, one obtains

$$\begin{bmatrix} p^2 + p + 4 & -4 \\ -6 & p + 6 \end{bmatrix} \begin{bmatrix} v_1(t) \\ v_2(t) \end{bmatrix} = \begin{bmatrix} pv_s(t) \\ 6pi_s(t) \end{bmatrix} \tag{13.50}$$

Now multiply both sides by the inverse of the 2×2 coefficient matrix to get

$$\begin{bmatrix} v_1(t) \\ v_2(t) \end{bmatrix} = \frac{1}{p(p^2 + 7p + 10)} \begin{bmatrix} p + 6 & 4 \\ 6 & p^2 + p + 4 \end{bmatrix} \begin{bmatrix} pv_s(t) \\ 6pi_s(t) \end{bmatrix} \tag{13.51}$$

Multiplying the two matrices on the right and cancelling the common p factor (legitimate under our assumptions), we finally have

$$v(t) = v_2(t) = \frac{6v_s(t) + 6(p^2 + p + 4)i_s(t)}{p^2 + 7p + 10} \tag{13.52}$$

We can, on the one hand, consider the result of our nodal analysis process to be a differential equation which we obtain by cross-multiplication:

$$[p^2 + 7p + 10]v(t) = 6v_s(t) + 6(p^2 + p + 4)i_s(t) \tag{13.53}$$

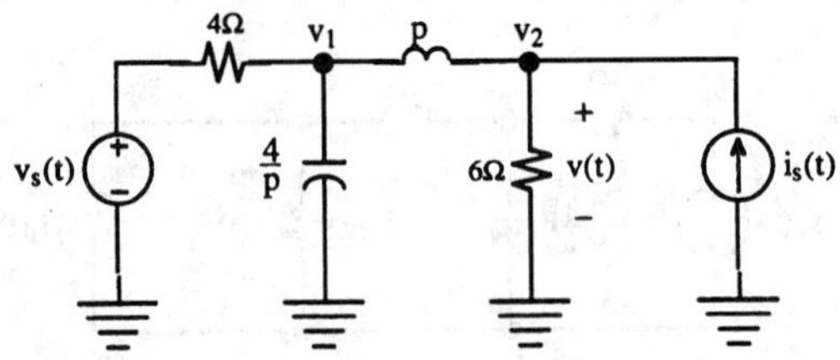

FIGURE 13.31 The example circuit prepared for nodal analysis.

In conventional notation, using the distributive properties of the p operators, one has

$$\frac{d^2v(t)}{dt^2} + 7\frac{dv(t)}{dt} + 10v(t) = 6v_s(t) + 6\frac{d^2i_s(t)}{dt^2} + 6\frac{di_s(t)}{dt} + 24i_s(t). \tag{13.54}$$

On the other hand, it is possible to interpret (13.52) directly as a solution operator equation. We simply note that the denominator factors, then do a partial fraction expansion to get

$$\begin{aligned} v(t) &= \frac{6}{(p+2)(p+5)}v_s(t) + \frac{6(p^2+p+4)}{(p+2)(p+5)}i_s(t) \\ &= \frac{2}{p+2}v_s(t) - \frac{2}{p+2}v_s(t) + 6i_s(t) + \frac{2}{p+2}i_s(t) - \frac{8}{p+2}i_s(t). \end{aligned} \tag{13.55}$$

Thus, we have expressed the two second-order operators in terms of operators of order one. It is quite easy to show that the first-order operator has the following simple form:

$$\frac{1}{p+a}x(t) = e^{-at}\frac{1}{p}[e^{at}x(t)] \tag{13.56}$$

Using this result, one can quickly show that the impulse and step responses of the first-order operator are

$$h(t) = \frac{1}{p+a}\delta(t) = e^{-at}u(t) \tag{13.57}$$

and

$$s(t) = \frac{1}{p+a}u(t) = [1 - e^{-at}]u(t) \tag{13.58}$$

respectively. Thus, if $i_s(t) = \delta(t)$ and $v_s(t) = u(t)$, one has

$$v(t) = 6\delta(t) + \frac{1}{5}[3 + 5e^{-2t} - 38e^{-5t}]u(t) \tag{13.59}$$

References [5, 6] show that all of the usual algebraic results valid for the Laplace transform are also valid for Heaviside operators.

Mesh Analysis

The central concept in nodal analysis is, of course, the node. The central idea in the method we will discuss here is the loop. Just as KCL formed the primary set of equations for nodal analysis, KVL will serve a similar function here. We will begin with the idea of a **mesh**. A mesh is a special kind of loop in a planar circuit (one that can be drawn on a plane) a loop that does not contain any other loop inside it. If one reflects on this definition a bit, one will see that it depends upon how the circuit is drawn. Figure 13.32 illustrates the idea of a mesh. The nodes have been numbered and the elements labeled with letters for clarity. The circuit graph in Fig. 13.32 abstracts all of the information about how the elements are connected, but does not show them explicitly. The lines represent the elements and the solid dots represent the nodes. If we apply the definition given in the introduction to this section, we can quickly verify that $\{h, a, i, k\}$ is a loop. It is a simple loop because each of its elements share only one node with any of the other path elements. It is a mesh because no other loops are inside it.

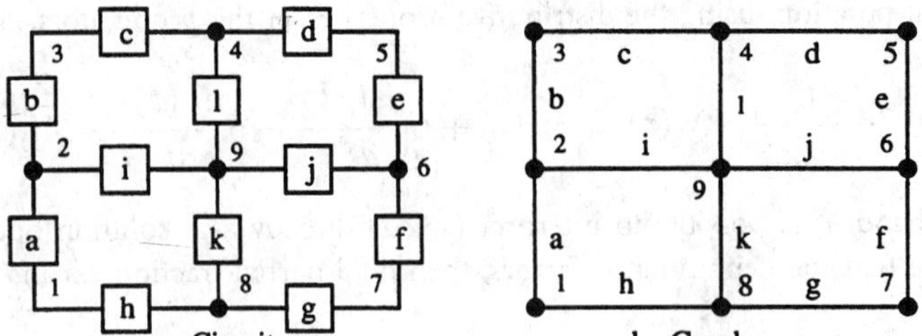

FIGURE 13.32 A circuit and its graph.

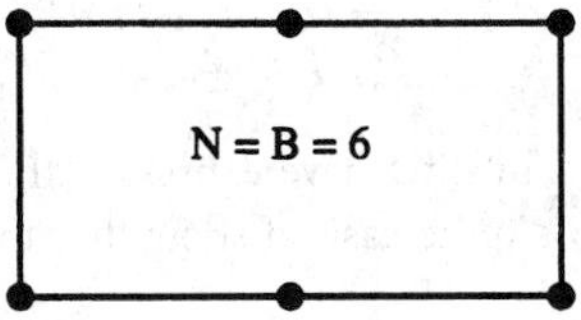

FIGURE 13.33 A one-mesh (series) circuit.

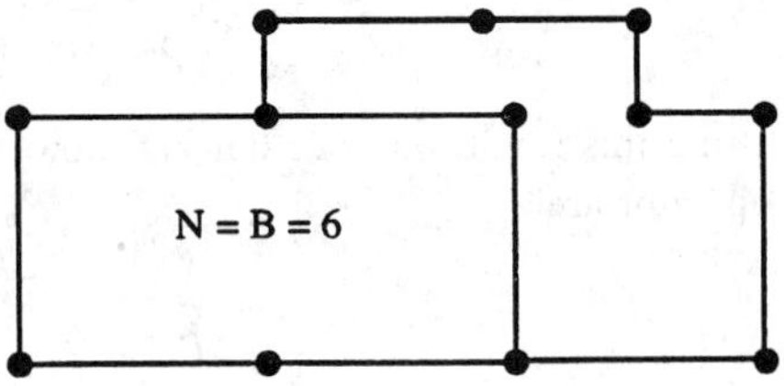

FIGURE 13.34 A two-mesh (series) Circuit

It is an important fact that the number of meshes in a circuit is given by

$$N_m = B - N + 1 \tag{13.60}$$

where B is the number of branches (elements) and, as usual, N is the number of nodes. To see this, just look at the simple one-mesh graph in Fig. 13.33. The number of branches is the same as the number of nodes for such a graph (or circuit). Imagine constructing the graph by placing an element on a planar surface, thereby forming two nodes with the one element. $B - N + 1 = 1 - 2 + 1 = 0$ in this case, and no meshes exist. Now add another element by connecting one of its leads to one of the leads of the first element. Now, $B-N+1 = 2-3+1 = 0$. This can be done indefinitely (or until you tire). At this point, connect one lead of the last element to the free lead of the one immediately preceding and the other lead of the last element to a node already placed. N nodes and $N - 1$ branches will have been put down, and exactly one mesh will have been formed. Thus, it is true that $B - N + 1 = N - (N - 1) + 1 = 1$ mesh and the formula is verified. Now connect a new element to one of the old nodes; the result is that one new element and one new node have been added. A glance at the formula verifies that it remains valid. Again, continue indefinitely, and then connect one new element and no new nodes by connecting the free lead of the last element with one of the nodes in the original one-loop circuit. Clearly, the number of added branches exceeds the number of added nodes by one; once again, the formula is verified. Figure 13.34 shows the new circuit. For the graph shown in the figure, $B = 13$ and $N = 12$, so $B - N + 1 = 2$, as expected. Induction generalizes the result, and (13.60) has been proved.

We now define a fictitious set of currents circulating around the meshes of a circuit. Figure 13.35 illustrates this idea with a circuit graph. All mesh currents are assumed to be circulating in a the clockwise direction, although this is not necessary. We see that i_1 is the only current flowing in the branch in which the element current i_a is defined, therefore, $i_a = i_1$; similarly, i_3

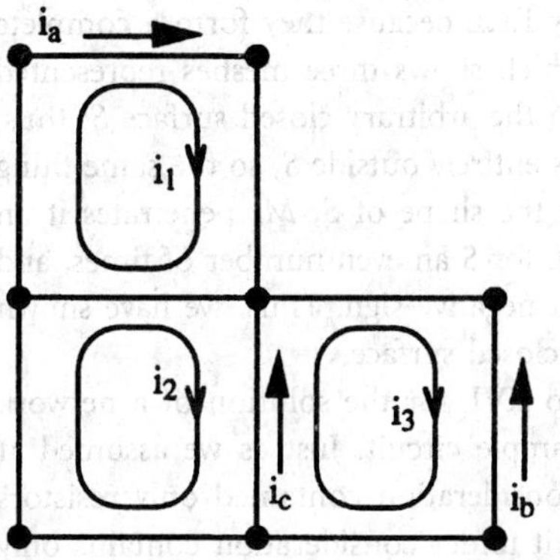

FIGURE 13.35 Mesh currents.

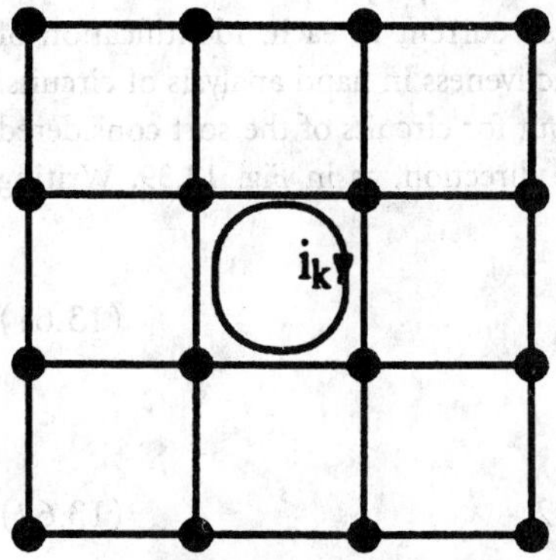

FIGURE 13.36 A fictitious mesh current.

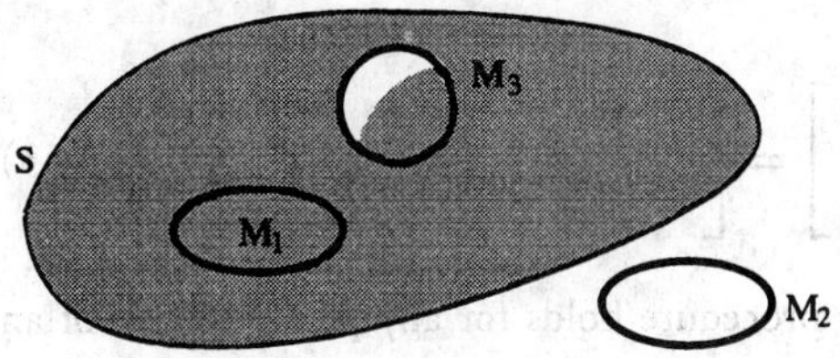

FIGURE 13.37 Illustration of KCL for mesh currents.

is the only mesh current flowing in the element carrying element current i_b, but the two are defined in opposite directions. Thus, one sees that $i_b = -i_3$. The third element whose current is indicated, however, is seen to carry two mesh currents in opposite directions. Hence, its element current is $i_c = i_3 - i_2$. In general, an element that is shared between two meshes has an element current which is the algebraic sum or difference[7] of the two adjacent mesh currents.

We used the term "fictitious" in our definition of mesh current. In the last example, however, we see that it is possible to make a physical measurement of each mesh current because each flows in an element that is not shared with any other mesh. Thus, one need only insert an ammeter in that element to measure the associated mesh current. Circuits exist, however, in which one or more mesh currents are impossible to measure. Figure 13.36 shows the graph of such a circuit. Each of the meshes is assumed to be carrying a mesh current, although only one has been drawn explicitly, i_k. As readily seen, each of the other mesh currents appears in a nonshared branch. For the mesh whose current is shown, however, it is impossible to find an element or a conductor carrying only that current. For this reason, i_k is merely a fiction, though a useful one.

[7]Always the difference if all mesh currents are defined in the same direction: clockwise or counterclockwise.

It is easy to see that mesh currents automatically satisfy KCL because they form a complete loop. Observe the stylized picture shown in Fig. 13.37, which shows three meshes represented for simplicity as small, closed ovals. M_1 lies entirely within the arbitrary closed surface S; thus, its current does not appear in KCL for that surface. M_2 lies entirely outside S, so the same thing is true for its current. Finally, we note that, regardless of the shape of S, M_3 penetrates it an even number of times. Thus, its current will appear in KCL for S an even number of times, and half of its appearances will carry a positive sign and half a negative sign. Thus, we have shown that any mesh current automatically satisfies KCL for any closed surface.

Because KCL is automatically satisfied, we must turn to KVL for the solution of a network in terms of its mesh currents. Figure 13.38 shows an example circuit. Just as we assumed at the outset of the last subsection that any circuit under consideration contained only resistors and current sources, we will assume at first that any circuit under consideration contains only resistors and voltage sources. The one shown in Fig. 13.38 has this property.

The first step is to identify the meshes and assign a mesh current to each. Identification of the meshes is easy, and this is the primary reason for its effectiveness in hand analysis of circuits. The mesh currents can be assigned in arbitrary directions, but for circuits of the sort considered here, it is more convenient to assign them all in the same direction, as in Fig. 13.39. Writing one KVL equation for each mesh results in

$$3i_1 + 2(i_1 - i_2) = 4 \tag{13.61}$$

and

$$2(i_2 - i_1) + 4i_2 = -12 \tag{13.62}$$

In matrix form,

$$\begin{bmatrix} 5 & -2 \\ -2 & 6 \end{bmatrix} \begin{bmatrix} i_1 \\ i_2 \end{bmatrix} = \begin{bmatrix} 4 \\ -12 \end{bmatrix} \tag{13.63}$$

The solution is $i_1 = 0$ A and $i_2 = 2$ A. The same procedure holds for any planar circuit of an arbitrary number of meshes.

Suppose, now, that the circuit being considered has one or more current sources, such as the one in Fig. 13.40. The meshes are readily determined; one need only to look for the "window panes", as meshes have been called. The only problem is this: when we write our mesh equations, what values do we use for the voltages across the current sources? These voltages are not known.

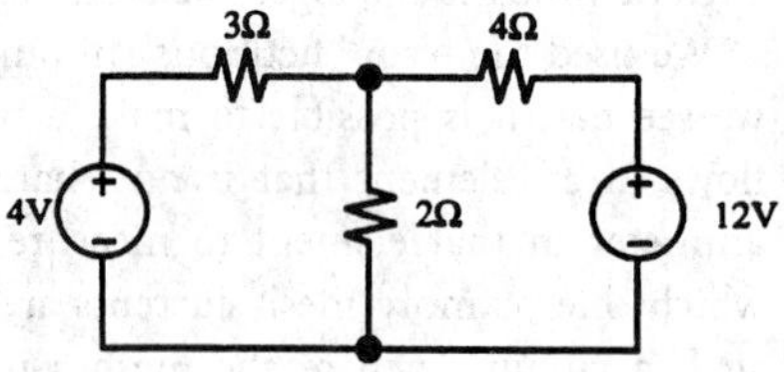

FIGURE 13.38 An example circuit.

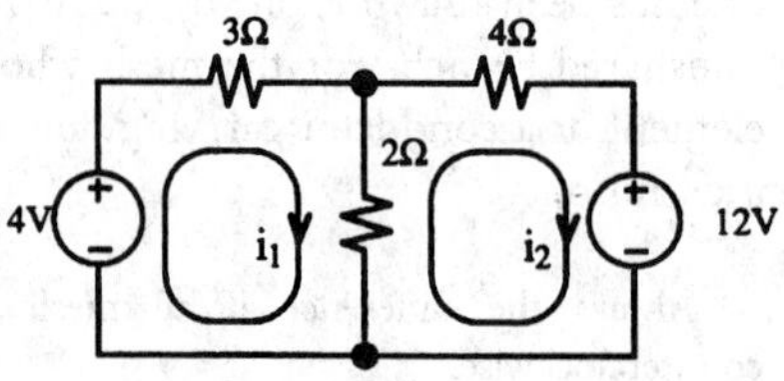

FIGURE 13.39 The example circuit prepared for mesh analysis.

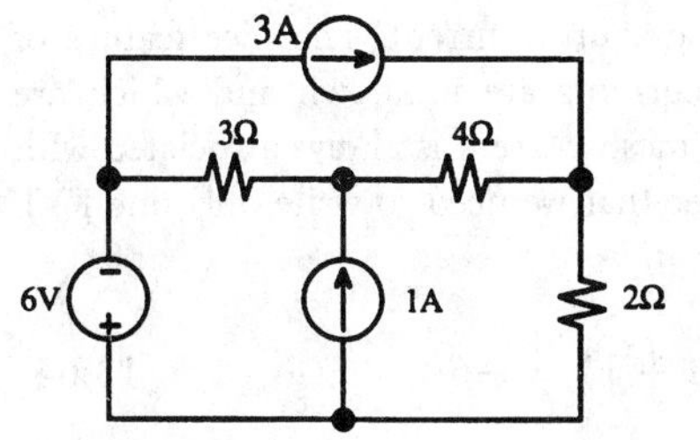

FIGURE 13.40 An example with current sources.

Thus, we could ascribe their voltages as unknowns, but this would lead to a hybrid form of analysis in which the unknowns are both element voltages and mesh currents. There is, however, a more straightforward way. Consider this question: should the variables we use or the loops around which we decide to write KVL change if we alter the *values* of any of the independent sources? The answer, of course, is no. Thus, let us test the circuit by deactivating it-that is, by reducing all sources to zero. Recalling that a zero-valued voltage source is a short circuit and a zero-valued current source is an open circuit, we obtain the test circuit shown in Fig. 13.41.

Notice what has happened. The two bottom meshes merge, thus forming one larger mesh in the deactivated circuit. The top mesh disappears (as a mesh or loop). For this reason, we refer to the former as a **supermesh** and the latter as a **nonessential mesh**. Observe also that it was the deactivation of the current sources that altered the topology; in fact, deactivation of the voltage sources has no effect on the mesh structure at all. Thus, we see that only one KVL equation is required to solve the deactivated circuit. (Reactivation of the source(s) is necessary, otherwise all voltage and currents will have zero values.) The conclusion relative to our example circuit is this: to solve the original circuit in terms of mesh currents, only one equation (KVL around the supermesh) is necessary.

The original circuit, with its three mesh currents arbitrarily defined, is redrawn in Fig. 13.42. Notice that the isolated (nonshared) current source in the top (nonessential) mesh defines the associated mesh current as having the same value as the source itself. On the other hand, the 1 A current source shared by the bottom two meshes introduces a more general constraint: the difference between the two mesh currents must be the same as the source current. This constraint has been used to label the mesh current in the right-hand mesh with a value such

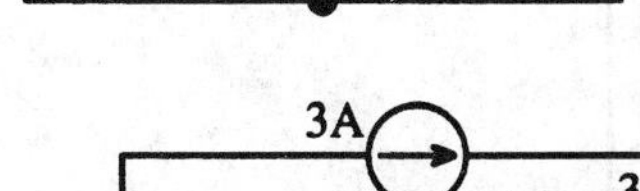

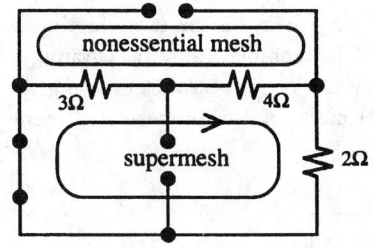

FIGURE 13.41 The deactivated current.

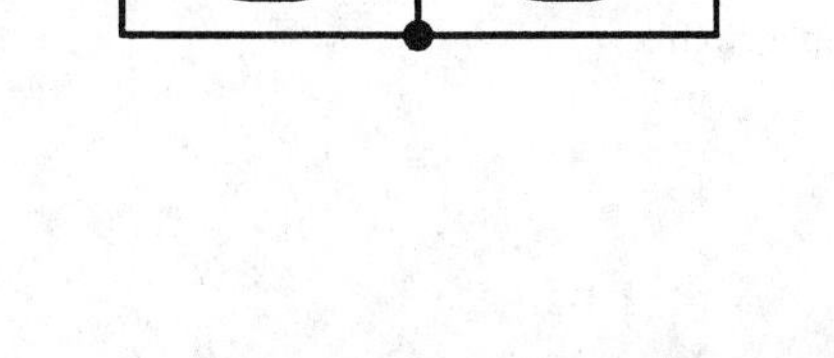

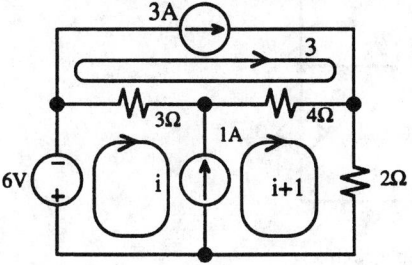

FIGURE 13.42 Assigning the mesh currents.

that it, minus the left-hand mesh current, is equal to the source current. The nice feature of this approach is that one can clearly see which mesh currents are unknown and which are dependent upon the unknowns. Exactly one independent mesh current is always associated with a supermesh. Recalling our test circuit in Fig. 13.41, we see that we need to write only one KVL equation around the supermesh. It is

$$3(i - 3) + 4(i + 1 - 3) + 2(i + 1) = -6 \tag{13.64}$$

or

$$9i - 9 - 8 + 2 = -6 \tag{13.65}$$

The solution is $i = 1$ A. From this, one can compute the mesh current on the bottom right to be $i + 1 = 2$ A and the one in the top loop is already known to be 3 A. With these known mesh currents, we can solve for any circuit variable desired.

The development of mesh analysis seems at first glance to be the complete analog of nodal. This is not quite the case, however, because nodal will work for nonplanar circuits, while mesh works only for planar circuits; furthermore, no global reference exists for mesh currents as it does for node voltages.

Analyzing the problem, we observe that each current source, when deactivated, reduces the number of meshes by one. (A given element can be shared only by two meshes.) Combining this fact with (13.60), we see that the required number of mesh equations is

$$N_{\text{me}} = B - N + 1 - N_I, \tag{13.66}$$

where, as usual, B is the number of branches, N is the number of nodes, and (in this equation) N_I is the number of current sources.

Note that mesh analysis is undertaken for circuits containing dependent sources in exactly the same manner as in nodal analysis—that is, by first taping the dependent sources, writing the mesh equations as above, and then untaping the dependent sources and expressing their controlled variables in terms of the unknown mesh currents. Figure 13.43 shows an example of such a circuit; in fact, it is the same figure investigated with nodal analysis in the preceding subsection.

The first step is to tape the dependent sources, thus placing them on the same footing as their more independent relatives. Then the circuit is tested for complexity by deactivating it as shown in Fig. 13.44. All element labels have been removed merely to avoid obscuring the ideas being discussed. We see one supermesh, one essential mesh, and no nonessential mesh.

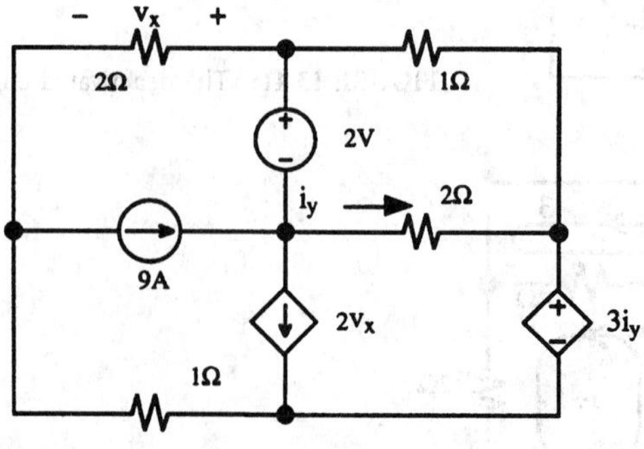

FIGURE 13.43 An example circuit.

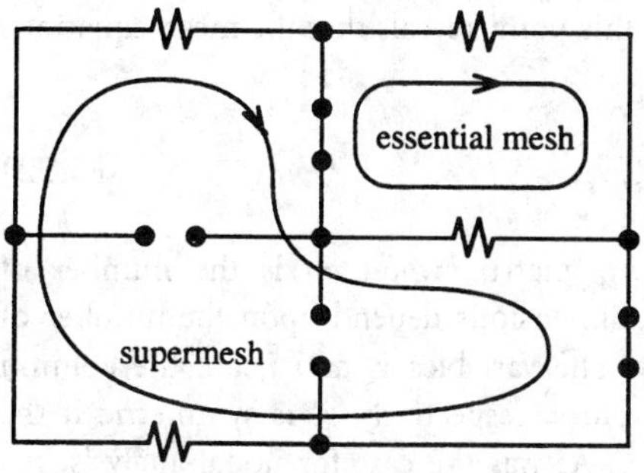

FIGURE 13.44 Testing the example circuit.

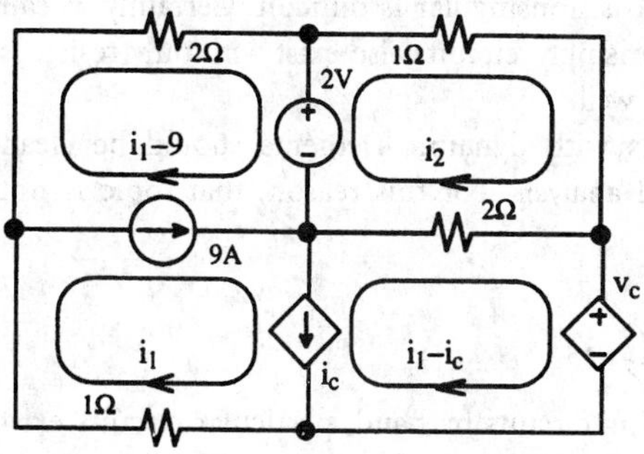

FIGURE 13.45 The example circuit prepared for mesh analysis.

Therefore, two KVL equations must be written in the original circuit, which is shown with the dependent sources taped in Fig. 13.45. Notice that there are only two unknowns, and also that the dependent sources have been taped. For the moment we have turned them into independent sources (albeit with unknown values).

We are now in a position to write KVL equations:

$$2(i_1 - 9) + 2(i_1 - i_c - i_2) + 1i_1 = -2 - v_c \quad \text{(supermesh)} \tag{13.67}$$

and

$$1i_2 + 2(i_2 - i_1 + i_c) = 2 \quad \text{(essential mesh)} \tag{13.68}$$

Observe that ic and v_c are not known quantities, as would be the case were they the values of independent sources. Thus, at this point, we must untape the dependent sources and express their values in terms of the mesh currents. We find that

$$i_c = 2v_x = 2x2x(-i_1 + 9) = -4i_1 + 36 \tag{13.69}$$

and

$$v_c = 3i_y = 3(i_1 - i_c - i_2) = 15i_1 - 3i_2 + 36 \tag{13.70}$$

Inserting the last two results in (13.67) and (13.68) results in the matrix equation

$$\begin{bmatrix} 28 & -5 \\ -10 & 3 \end{bmatrix} \begin{bmatrix} i_1 \\ i_2 \end{bmatrix} = \begin{bmatrix} 196 \\ -70 \end{bmatrix} \tag{13.71}$$

The coefficient matrix is no longer symmetric now that dependent sources have been introduced. (This is also the case with nodal analysis. The example treating this same circuit is found in the last subsection and should be checked to verify this point.) However, the solution is found, as usual, to be $i_1 = 7$ A and $i_2 = 0$ A.

A careful consideration of what we have done up to this point reveals that the mesh equations can be written in the form

$$A\bar{i}_M = B\bar{i}_s + C\bar{v}_s \tag{13.72}$$

In this general formulation A is a square $n_M \times n_M$ matrix, where m is the number of meshes, and B and C are rectangular matrices whose dimensions depend upon the number of independent voltage and current sources, respectively. The variables $\bar{v}_s$ and $\bar{i}_s$ are the column matrices of independent voltage and current source values, respectively. A is symmetric if the circuit contains only resistors and independent sources. As was the case for nodal analysis, the elucidation of conditions under which the A matrix is nonsingular is difficult. Certainly, it can be singular for circuits with dependent sources; surprisingly, circuits also exist with only resistors and independent sources for which A is singular as well.

Finally, the mesh analysis procedure for circuits with dynamic elements should be clear. The algebraic process closely follows that for nodal analysis. For this reason, that topic is not discussed here.

Fundamental Cutset-Loop Circuit Analysis

As effective as nodal and mesh analysis are in treating circuits by hand, particular circuits exist for which they fail. Consider, for example, the circuit in Fig. 13.46. If we were to blindly perform nodal analysis on this circuit, we would perhaps prepare it for analysis as shown in Fig. 13.47. We have three nonreference nodes, hence, we have three nodal equations:

$$\frac{v_1}{2} = 2 \tag{13.73}$$

$$\frac{v_2 - v_3}{2} = 2 \tag{13.74}$$

$$\frac{v_3 - v_2}{2} = -2 \tag{13.75}$$

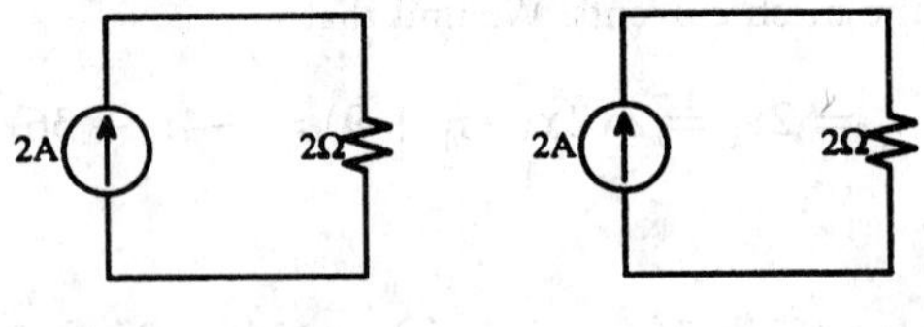

FIGURE 13.46 An example circuit.

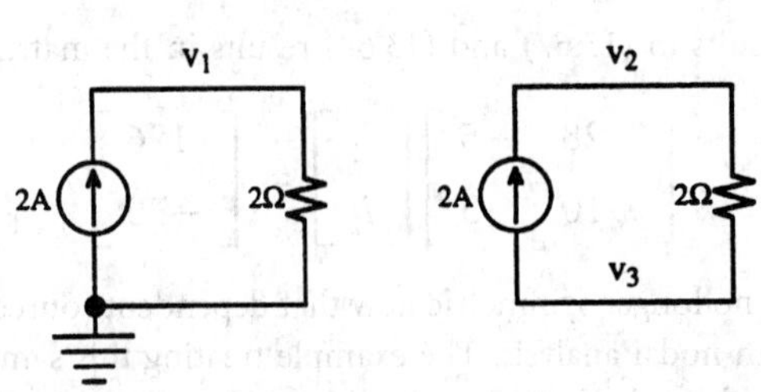

FIGURE 13.47 The example circuit prepared for nodal analysis.

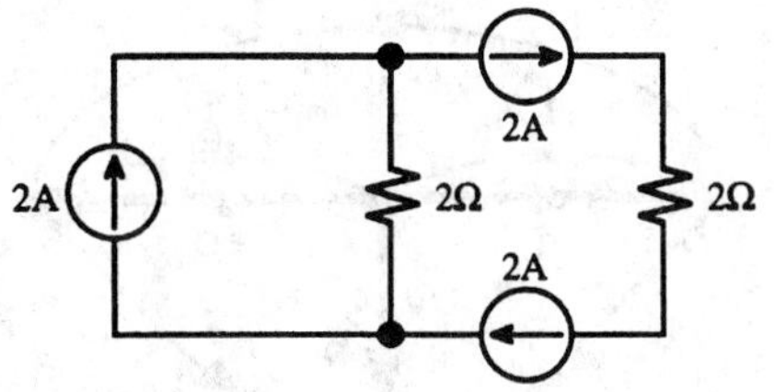

FIGURE 13.48 Another example circuit.

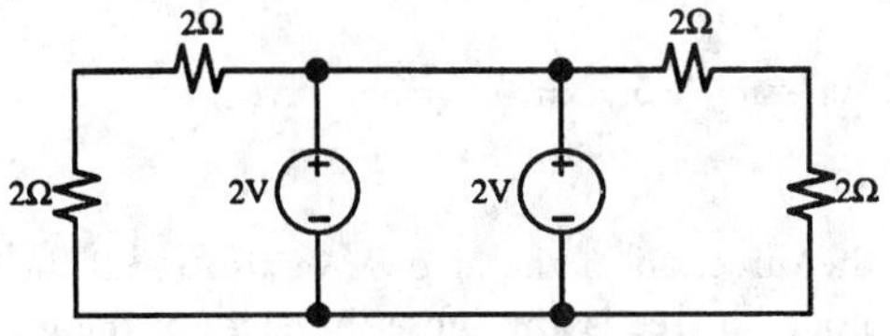

FIGURE 13.49 Another example of a singular circuit.

The third equation is simply the negative of the second; hence, the set of equations is linearly dependent and does not have a unique solution. The reason is quite obvious: the circuit is not connected.[8] It actually consists of two circuits considered as one. Therefore, one should actually select two reference nodes rather than one. A bit of reasoning along this line indicates that the number of nodal equations should be

$$N_{ne} = N - 1 - P \tag{13.76}$$

where P is the number of separate parts, and hence the number of reference nodes required.

Another way nodal analysis can fail is not quite as obvious. Figure 13.48 illustrates the situation. In this case we have a cutset of current sources. Therefore, in reality, at least one of the current sources cannot be independent for it must have the same value as the other in the cutset. We will leave the writing of the nodal equations as an exercise for the reader. They are, however, linearly dependent. The problem here clearly becomes evident if one deactivates the circuit, because the resulting test circuit is not connected.

Analogous problems can occur with mesh analysis, as the circuit in Fig. 13.49 shows. We find a loop of voltage sources and, when the circuit is deactivated, one mesh disappears. Again, it is left as an exercise for the reader to write the mesh equations and show that they are linearly independent (the coefficients of all currents in the KVL equation for the central mesh are zero).

One might question the practicality of such circuits because clearly no one would design such networks to perform a useful function. In the computer automation of circuit analysis, however, dynamic elements are often modeled over a small time increment in terms of independent sources, and singular behavior can result. Furthermore, one would like to be able to include more general elements than R, L, C, and voltage and current sources. For such reasons, a general method that does not fail is desirable. We develop this method below. It is related to the modified nodal analysis technique that is described elsewhere in the book.

To develop this technique, we will examine the graph of a specific circuit: the one shown in Fig. 13.50. Graph theory is covered elsewhere in the book, but salient points will be reviewed here [7, 8]. The graph, of course, is not concerned at all with the v–i characteristics of the elements themselves—only with how they are interconnected. The lines (or edges or branches) represent the elements and the dots represent the nodes. The arrows represent the assumed references for voltage and current, the positive voltage at the "upstream" end of the arrow and

[8]A circuit is connected if there is at least one path between each pair of nodes.

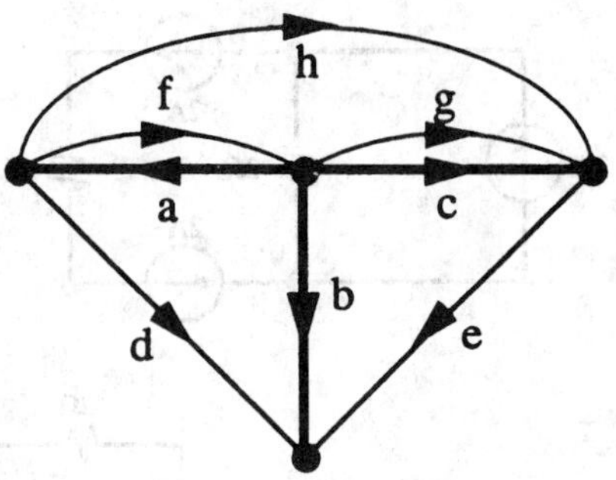

FIGURE 13.50 An example of a circuit graph.

the current reference in the direction of the arrow. We also recall the definition of a tree: for a connected graph of N nodes, a **tree** is any subset of edges of the graph that connects all the nodes, but which contains no loops. Such a tree is shown by means of the darkened edges in the figure: a, b, and c. The complement of a tree is called a **cotree**. Thus, edges d, e, f, g, and h form a cotree in Fig. 13.50. If a graph consists of separate parts (that is, it is not connected), then one calls a subset of edges that connects all N nodes, but forms no loops, a **forest**. The complement of a forest is a **coforest**. The analysis method presented here is applicable to either connected or nonconnected circuits. However, we will use the terms for a graph that is connected for ease of comprehension; one should go through a parallel development for nonconnected circuits to assure oneself that the generalization holds.

Each edge contained in a tree is called a **twig**, and each edge contained in a cotree is called a **link**. (Remember that the set of all nodes in a connected graph is split into exactly two sets of nodes, each of which is individually connected by twigs, when a tree edge is removed). The set of links having one of its nodes in one set and another in the second, together with the associated twig defining the two sets of nodes, is called a **fundamental cutset**, or **f-cutset**. If all links associated with a given tree are removed, then the links replaced one at a time, it can be seen that each link defines a loop called a **fundamental loop** or **f-loop**. Figure 13.51 shows a fundamental cutset and a fundamental loop for the graph shown in Fig. 13.50. The closed surface S is placed around one of the two sets of nodes so defined (in this case consisting of a single node) and is penetrated by the edges in the cutset b, d, and e. A natural orientation of the cutset is provided by the direction of the defining twig, in this case edge b. Thus, a positive sign is assigned to b; then, any link in the fundamental cutset whose direction relative to S agrees with that of the twig receives a positive sign, and each whose direction is opposite receives a negative sign. Similarly, the fundamental loop is given a positive sense by the direction of the defining link, in this case edge h. It is assigned a positive sign; then, each twig in the f-loop is given a positive sign if its direction coincides in the loop with the defining link, and a negative sign if it does not.

The positive and negative signs just defined can be used to write one KCL equation for each f-cutset and one KVL equation for each f-loop, as follows. Consider the example graph with which we are working. The f-cutsets are $\{d, b, e\}$, $\{d, a, f, h\}$, and $\{e, c, g, h\}$. In general, $N-1$ f-cutsets are associated with each tree—exactly the same as the number of twigs in the tree. Using surfaces similar to S in Fig. 13.51 for each of the f-cutsets, we have the following set of KCL equations:

$$i_a - i_d - i_f - i_h = 0 \tag{13.77}$$

$$i_b + i_d + i_e = 0 \tag{13.78}$$

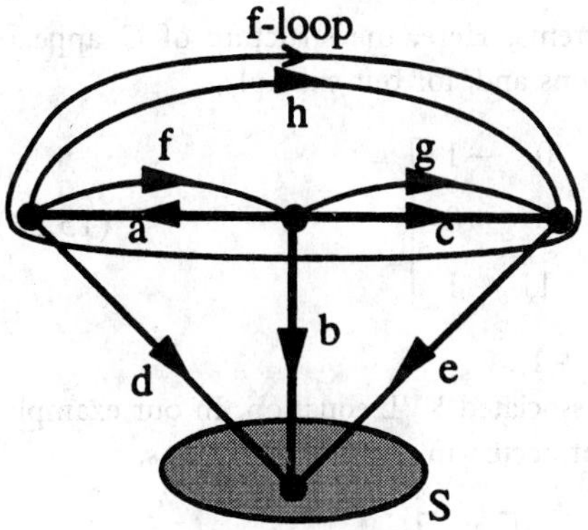

FIGURE 13.51 Fundamental cutsets and loops.

$$i_c - i_e + i_g + i_h = 0 \tag{13.79}$$

In matrix form, these equations become

$$\begin{bmatrix} 1 & 0 & 0 & -1 & 0 & -1 & 0 & -1 \\ 0 & 1 & 0 & 1 & 1 & 0 & 0 & 0 \\ 0 & 0 & 1 & 0 & -1 & 0 & 1 & 1 \end{bmatrix} \begin{bmatrix} i_a \\ i_b \\ i_c \\ i_d \\ i_e \\ i_f \\ i_g \\ i_h \end{bmatrix} = \begin{bmatrix} 0 \\ 0 \\ 0 \end{bmatrix}. \tag{13.80}$$

The coefficient matrix consists of zeroes, and positive and negative ones. It is called the ***f*-cutset matrix**. Each row corresponds to a KCL equation for one of the f-cutsets, and has a zero entry for each edge not in that cutset, $a+1$ for any edge in the cutset with the same orientation as the defining twig, and $a-1$ for each edge in the cutset whose orientation is opposite to the defining twig, and $a-1$ for each edge in the cutset whose orientation is opposite to the defining twig. One often labels the rows and columns,

$$Q = \begin{matrix} & \begin{matrix} a & b & c & d & e & f & g & h \end{matrix} \\ \begin{matrix} a \\ b \\ c \end{matrix} & \begin{bmatrix} 1 & 0 & 0 & -1 & 0 & -1 & 0 & -1 \\ 0 & 1 & 0 & 1 & 1 & 0 & 0 & 0 \\ 0 & 0 & 1 & 0 & -1 & 0 & 1 & 1 \end{bmatrix} \end{matrix}, \tag{13.81}$$

to emphasize the relation between the matrix and the graph. Thus, the first row corresponds to KCL for the f-cutset defined by twig a, the second to that defined by twig b, and the last to the cutset defined by twig c. The columns correspond to each of the edges in the graph, with the twigs occupying the first $N-1$ columns in the same order as that in which they appear in the rows. Notice that a unit matrix of order $N-1 \times N-1$ is located in the leftmost $N-1$ columns. Furthermore, Q has dimensions $(N-1) \times B$, where B is the number of branches (edges). Clearly, Q has maximum rank because of the leading unit matrix. More succinctly, one writes KCL in terms of the f-cutset matrix as

$$Q\bar{i} = [U \vdots H]\bar{i} = 0, \tag{13.82}$$

where $\bar{i}$ is the column matrix of all the branch currents. Here, the structure of Q appears explicitly with the unit matrix in the first $N-1$ columns and, for our example,

$$H = \begin{bmatrix} -1 & 0 & -1 & 0 & -1 \\ 1 & 1 & 0 & 0 & 0 \\ 0 & -1 & 0 & 1 & 1 \end{bmatrix} \tag{13.83}$$

In general, H will have dimensions $(N-1) \times (b-N+1)$.

Each of the links, all $B-N+1$ of them, have an associated KVL equation. In our example, using the same order for these links and equations that occurs in the KCL equations,

$$\begin{bmatrix} 1 & -1 & 0 & 1 & 0 & 0 & 0 & 0 \\ 0 & -1 & 1 & 0 & 1 & 0 & 0 & 0 \\ 1 & 0 & 0 & 0 & 0 & 1 & 0 & 0 \\ 0 & 0 & -1 & 0 & 0 & 0 & 1 & 0 \\ 1 & 0 & -1 & 0 & 0 & 0 & 0 & 1 \end{bmatrix} \begin{bmatrix} v_a \\ v_b \\ v_c \\ v_d \\ v_e \\ v_f \\ v_g \\ v_h \end{bmatrix} = \begin{bmatrix} 0 \\ 0 \\ 0 \\ 0 \\ 0 \end{bmatrix} \tag{13.84}$$

We have one row for each link and, therefore, one for each f-loop. If a given edge is in the given loop, $a+1$ is in the corresponding column if its direction agrees with that of the defining link, and $a-1$ if it disagrees. Notice that a unit matrix of dimensions $(B-N+1) \times (B-N+1)$ is located in the last $B-N+1$ columns. Even more importantly, observe that the matrix in the first $N-1$ columns has a familiar form; in fact, it is $-H'$, the negative transpose of the matrix in (13.83).

This is no accident. In fact, the entries in this matrix are strictly due to twigs in the tree. Focus on a given twig and a given link. The twig defines two twig-connected sets of nodes, as mentioned above. If the given link has both its terminal nodes in only one of these sets, the given twig voltage does not appear in the KVL equation for that f-loop. If, on the other hand, one of the link nodes is in one of those sets and the other in the alternate set, the given twig voltage will appear in the KVL equation for the given link, with $a+1$ multiplier if the directions of the twig agree relative to the f-loop and $a-1$ if they do not. However, a little thought shows that the same result holds for the f-cutset equation defined by the twig, except that the signs are reversed. If the link and twig directions agree for the f-loop, they disagree for the f-cutset, and vice versa. Thus, we can write KVL for the f-loops, in general, as

$$B_f \bar{v} = [-H' \vdots U]\bar{v} = 0. \tag{13.85}$$

B_f is called the fundamental loop matrix, and $\bar{v}$ is the column matrix of all branch voltages.

Suppose that we partition the branch voltages and branch currents according to whether they are associated with twigs or links. Thus, we write

$$\bar{v} = [\bar{v}_T' \bar{v}_C']' \tag{13.86}$$

and

$$\bar{i} = [\bar{i}_T' \bar{i}_C']'. \tag{13.87}$$

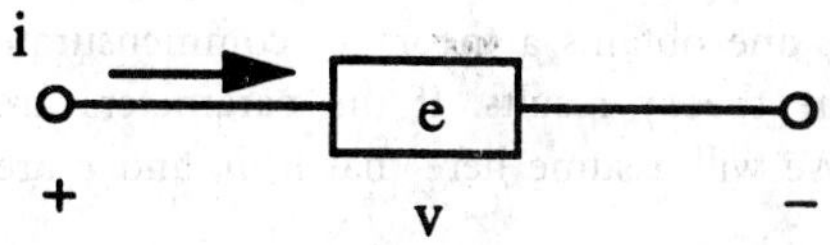

FIGURE 13.52 A two-terminal element.

We use transpose notation to conserve space, and the subscripts T and C represent tree and cotree voltages and currents, respectively. We cannot use (13.82) and (13.85) to write the composite circuit variable vector as

$$\bar{u} = \begin{bmatrix} \bar{v} \\ \bar{i} \end{bmatrix} = \begin{bmatrix} U & 0 \\ H' & 0 \\ 0 & -H \\ 0 & U \end{bmatrix} \begin{bmatrix} \bar{v}_T \\ \bar{i}_C \end{bmatrix} \tag{13.88}$$

The coefficient matrix is of dimensions $2B \times B$ and has rank B because of the two unit matrices. What we have accomplished is a direct sum decomposition of the $2B$-dimensional vector space consisting of all circuit variables in terms of the $N-1$ dimensional vector space of tree voltages and the $B-N+1$ dimensional vector space of link currents. Furthermore, the tree voltages and link currents form a basis for the vector space of all circuit variables.

We have discussed topology enough for our purposes. We now treat the elements. We are looking for a generalized method of circuit analysis that will succeed, not only for circuits containing R, L, C, and source elements, but ideal transformers, gyrators, nullators, and norators, (among others) as well. Thus, we turn to a discussion of elements [9].

Consider elements having two terminals only, as shown in Fig. 13.52. The most general assumption we can make, assuming that we are ruling out "nonlinear" elements, is that the v–i characteristic of each such element be *affine;* that is, it is defined by an operator equation of the form

$$\begin{bmatrix} a & b \\ 0 & c \end{bmatrix} \begin{bmatrix} v \\ \bar{i} \end{bmatrix} = \begin{bmatrix} f(t) \\ g(t) \end{bmatrix} \tag{13.89}$$

where the parameters a, b, and c are operators. It is more classical to assume a scalar form of this equation; that is, with c and $g(t)$ both zero. In a series of papers in the 1960s, however, Carlin, Belevitch, and Tellegen [10–12] proposed that the v–i characteristic be interpreted as a multidimensional relationship. Among other things to come out of the approach was the definition of the nullator and the norator. Now, assuming that this defining characteristic is indeed multidimensional, we see at once that it is not necessary to consider operator matrices of a dimension larger than 2×2. There must be two columns because there are only two scalar terminal variables. If more than two rows were found, the additional equations would be either redundant or inconsistent, depending upon whether row reductions resulted in additional rows of all zeroes or in an inconsistent equation. Finally, the (2, 1) element in the operator matrix clearly can be chosen to be zero as shown, because otherwise it could be reduced to zero with elementary row operations. That is, one could, unless $a = 0$; but here, an exchange of rows produces the desired result shown. Note that any or all of a, b, and c can be the zero operator.

We pause here to remark that a and b can be rather general operators. If they are differential, or Heaviside, operators, (that is, they are real, rational functions of p), a theory of lumped circuits (differential systems) is obtained. On the other hand, they could be rational functions of the delay operator E.[9] In this case, one would obtain the theory of distributed (transmission line)

[9] $Ex(t) = x(t-T)$ for all t and all waveforms $x(t)$.

circuits. Then, if a common delay parameter is used, one obtains a theory of commensurate transmission line circuits; if not, an incommensurate theory results. If the parameters are functions of both p and d, a mixed theory results. We will assume here that a, b, and c are rational functions of p.

Let us suppose that c is the zero operator and that $g(t) = 0$ is the second equality resulting from the stipulation of existence (consistency). This gives the affine scalar relationship

$$av + bi = f(t) \tag{13.90}$$

Special cases are now examined. For instance, if $b = 0$ and $a \neq 0$, one has

$$v(t) = f(t)/a = v_s(t) \tag{13.91}$$

This, of course, is the $v - i$ characteristic for an independent voltage source. If, on the other hand, $a = 0$ and $b \neq 0$, one has

$$i(t) = f(t)/b = i_s(t) \tag{13.92}$$

This is an ideal current source. If, in addition, $f(t)$ is identically zero, one obtains a short circuit and an open circuit, respectively. These results are shown in Fig. 13.53. Now suppose that a and b are both zero. Then, $f(t)$ must be identically zero as well; otherwise, the element does not exist. In this case any arbitrary voltage and current are possible. The resulting element, a "singular one" to be sure, is called a **norator**. Its symbol is shown in Fig. 13.54.

Remaining with the same general case, that is, with $c = 0$ and $g(t) = 0$, we ask what element results if we also assume that neither a nor b are zero, but that $f(t)$ is identically zero. We can solve for either the voltage or the current. In either case one obtains a passive element, as shown in Fig. 13.55. If $-b/a$ is constant, a resistor will result; if $-b/a$ is a constant times the differential operator p, an inductor will result; and if $-b/a$ is reciprocal in p, a capacitor will result. In case the ratio is a more complicated function of p, one would consider the two-terminal object to be a subcircuit, that is, a two-terminal object decomposable into other elements, with $-b/a$ being the driving point impedance operator.

One can, in fact, derive the Thévenin and Norton equivalents from these considerations. Staying with the general case of c and $g(t)$ both zero, but allowing $f(t)$ to be nonzero, we first assume that $a \neq 0$. Then we obtain

$$v(t) = \frac{f(t)}{a} - \frac{b}{a}i(t) = v_{oc}(t) + Z_{eq}(p)i(t) \tag{13.93}$$

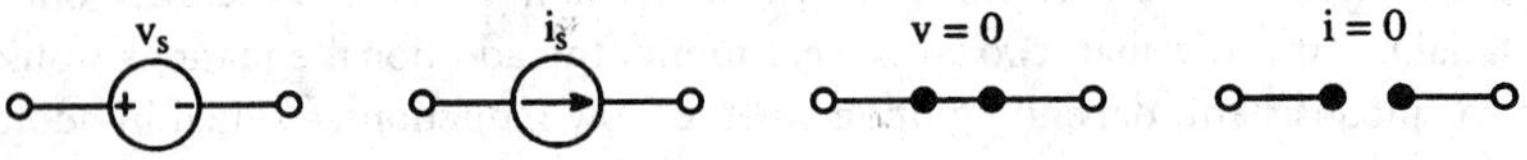

FIGURE 13.53 Conventional two-terminal elements.

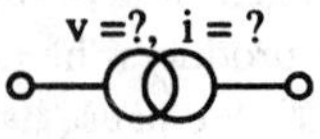

FIGURE 13.54 The norator.

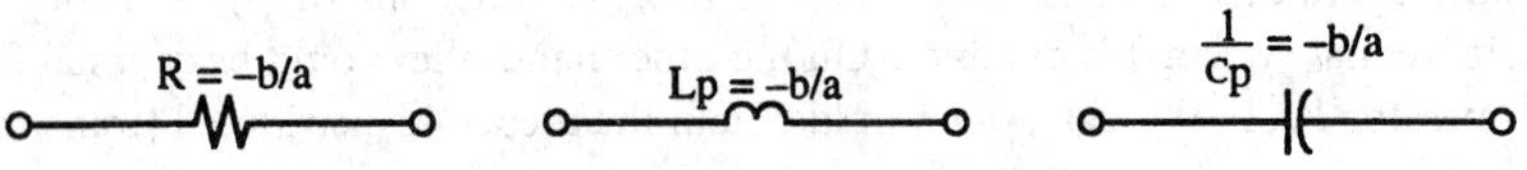

FIGURE 13.55 Passive elements.

which represents the Thévenin equivalent subcircuit shown in Fig. 13.56a. Alternately, if $b \neq 0$ we can write

$$i(t) = \frac{f(t)}{b} - \frac{a}{b}i(t) = i_{sc}(t) + Y_{eq}(p)v(t) \tag{13.94}$$

The latter equation is descriptive of the Norton equivalent shown in Fig. 13.56b. The basic assumption is that the two-terminal object has a v–i characteristic, i.e., an affine relationship; if this object contains only elements characterized by affine relationships having a rank property to be given later, one can use the analysis method being presented here to prove that this assumption is true. At this point, however, we are merely assembling a catalog of elements, so we assume that the two-terminal object is, indeed, a single element (it cannot be decomposed farther).

We have only one other case to consider: that in which c is a nonzero operator. If this is the situation and if, in addition, $a \neq 0$ as well, one can solve (13.90) by inverting the coefficient matrix to obtain

$$\begin{bmatrix} v \\ i \end{bmatrix} = \begin{bmatrix} 1/c & -b/ac \\ 0 & 1/a \end{bmatrix} \begin{bmatrix} f(t) \\ g(t) \end{bmatrix} = \begin{bmatrix} v_s(t) \\ i_s(t) \end{bmatrix} \tag{13.95}$$

Therefore, the voltage and current are independently specified. First, suppose that both $v_s(t)$ and $i_s(t)$ are identically zero. Then one has $v(t) = 0$ and $i(t) = 0$ for t. The associated element is called a **nullator**, and has the symbol shown in Fig. 13.57(a). Finally, if $v_s(t)$ and $i_s(t)$ are nonzero, one can sketch the equivalent subcircuit as in Fig. 13.57(b).

At this point, we have an exhaustive catalog of two-terminal circuit elements: the independent voltage and current sources, the resistor, the inductor, the capacitor, the norator, and the nullator. We would like to include more complex elements with more than two terminals as well. Figure 13.58(a) shows a three-terminal element and Fig. 13.58(b) shows a two-port element. For the former, we see at once that only two voltages and two currents can be independently specified because KVL gives the voltage between the left and right terminals in terms of the two shown, while KCL gives the current in the third lead. As for the latter, it is a basic assumption that only the two-port voltages and the two-port currents are required to specify its operation. In fact, one assumes that the currents coming out of the bottom leads are identical to those going into the top leads. We also assume that the v–i characteristic is independent of the voltages

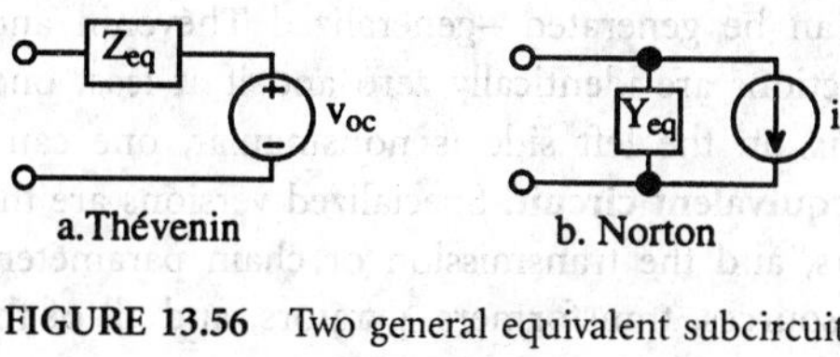

FIGURE 13.56 Two general equivalent subcircuits.

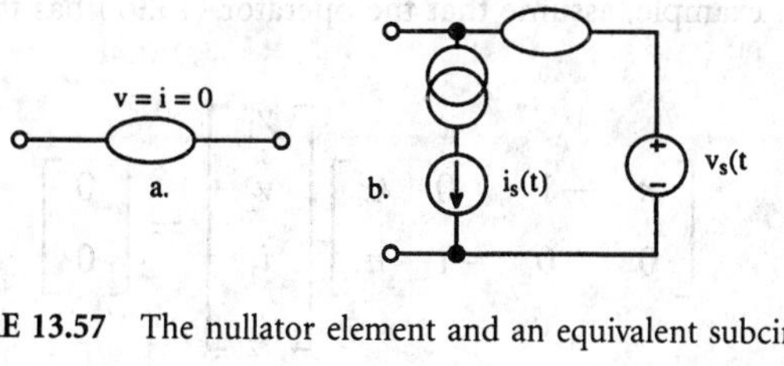

FIGURE 13.57 The nullator element and an equivalent subcircuit.

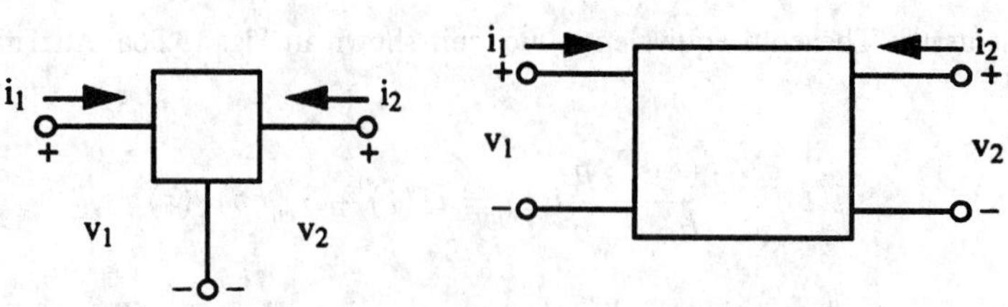

a. Three-terminal b. Two-port

FIGURE 13.58 Three-terminal and two-port elements.

between terminals in different ports. Each of the ports will be an edge in the circuit graph which results when such elements are interconnected.

Because four variables are associated with a three-terminal or two-port element, the dimensionality of the resulting vector space is four; thus, we assume that the describing v–i characteristic is

$$\begin{bmatrix} a_{11} & a_{12} & a_{13} & a_{14} \\ 0 & a_{22} & a_{23} & a_{24} \\ 0 & 0 & a_{33} & a_{34} \\ 0 & 0 & 0 & a_{44} \end{bmatrix} \begin{bmatrix} v_1 \\ v_2 \\ i_1 \\ i_2 \end{bmatrix} = \begin{bmatrix} f_1(f) \\ f_2(f) \\ f_3(f) \\ f_4(f) \end{bmatrix} \tag{13.96}$$

We justify this form exactly as for the case of two-terminal elements. We will not exhaustively catalog all of the possible three-terminal/two-port elements for reasons of space; however, note that the usual case is that in which $a_{ij} = 0$ for $i \geq 3$. In this case one must insist that $f_3(t) = f_4(t) = 0$; then one has the 2×2 system of equations

$$\begin{bmatrix} a_{11} & a_{12} & a_{13} & a_{14} \\ 0 & a_{22} & a_{23} & a_{24} \end{bmatrix} \begin{bmatrix} v_1 \\ v_2 \\ i_1 \\ i_2 \end{bmatrix} = \begin{bmatrix} f_1(t) \\ f_2(t) \end{bmatrix} \tag{13.97}$$

If the two forcing functions on the right are not identically zero, a number of different two-port equivalent circuits can be generated—generalized Thévenin and Norton equivalents. If both of these forcing functions are identically zero and if at least one 2×2 submatrix of the coefficient operator matrix on the left side is nonsingular, one can derive a **hybrid matrix** and a **hybrid parameter equivalent circuit**. Specialized versions are the impedance parameters, the admittance parameters, and the transmission or chain parameters. Furthermore, one can accommodate controlled sources, transformers, gyrators, and all of the other known two-port elements.

To present just one example, assume that the operator (13.97) has the form

$$\begin{bmatrix} n & -1 & 0 & 0 \\ 0 & 0 & -1 & n \end{bmatrix} \begin{bmatrix} v_1 \\ v_2 \\ i_1 \\ i_2 \end{bmatrix} = \begin{bmatrix} 0 \\ 0 \end{bmatrix} \tag{13.98}$$

The parameter n, assumed to be a real scalar multiplier, is called the **turns ratio**, and the element is the ideal transformer. The VCVS (voltage controlled voltage source) obeys

$$\begin{bmatrix} \mu & -1 & 0 & 0 \\ 0 & 0 & 1 & 0 \end{bmatrix} \begin{bmatrix} v_1 \\ v_2 \\ i_1 \\ i_2 \end{bmatrix} = \begin{bmatrix} 0 \\ 0 \end{bmatrix} \tag{13.99}$$

Thus, i_1 is identically zero and $v_2 = \mu v_1$. The quantity μ is the voltage gain.

Similarly, for each element with any number of ports,[10] we can write

$$A_0 \bar{v} + B_0 \bar{i} = \bar{C}_0 \tag{13.100}$$

where the voltage and current vectors are the terminal variables of the element. We can then represent the element equations for any circuit in the same form by forming A_0 and B_0 as quasidiagonal matrices, each of whose diagonal terms is the corresponding A_0 or B_0 for a given element, and stacking up the individual $\bar{C}_0$ column matrices to form the overall matrix. We then rewrite (13.100) in the form

$$[A_0 \;\; B_0] \begin{bmatrix} \bar{v} \\ \bar{i} \end{bmatrix} = \bar{C} \tag{13.101}$$

where the voltage and current vectors are each $B \times 1$ column matrices of the individual element voltages and currents. *We make the assumption that the matrix* $[A \;\; B]$, *which is of dimension* $B \times 2B$ *is of maximum rank* b. This is the only assumption required for the procedure to be outlined to succeed, as will later be demonstrated.

An example will clarify things. Figure 13.59 shows an example circuit. The correspondence between the edge labels and the circuit elements is obvious; that is, for instance, a is the 4 V voltage source and its voltage is -4 V (minus, because of the definition of positive voltage on edge a in the graph). We have shown a tree on the graph. The f-cutset matrix is

$$Q = \begin{matrix} \\ b \\ d \\ e \\ f \end{matrix} \begin{matrix} \begin{matrix} b & d & e & f & a & c & g \end{matrix} \\ \begin{bmatrix} 1 & 0 & 0 & 0 & -1 & 1 & 0 \\ 0 & 1 & 0 & 0 & 0 & -1 & -1 \\ 0 & 0 & 1 & 0 & 1 & 0 & 1 \\ 0 & 0 & 0 & 1 & 0 & 0 & 1 \end{bmatrix} \end{matrix} = [U \vdots H] \tag{13.102}$$

Thus,

$$H = \begin{bmatrix} -1 & 1 & 0 \\ 0 & -1 & -1 \\ 1 & 0 & 1 \\ 0 & 0 & 1 \end{bmatrix} \tag{13.103}$$

[10]A two-terminal element is a one-port device.

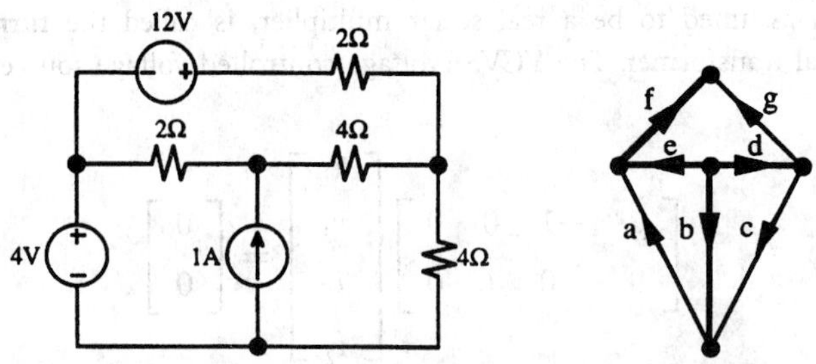

a. Circuit b. Graph

FIGURE 13.59 An example circuit and its graph.

Although we could construct it from the Q matrix, we can just as easily read off the f-loop matrix from the graph:

$$B = \begin{matrix} a \\ c \\ g \end{matrix} \overset{\begin{matrix} b & d & e & f & a & c & g \end{matrix}}{\begin{bmatrix} 1 & 0 & -1 & 0 & 1 & 0 & 0 \\ -1 & 1 & 0 & 0 & 0 & 1 & 0 \\ 0 & 1 & -1 & -1 & 0 & 0 & 1 \end{bmatrix}} = [-H' \vdots U] \tag{13.104}$$

The element constraint equations are

$$\begin{bmatrix} 0 & 0 & 0 & 0 & 0 & 0 & 0 & 1 & 0 & 0 & 0 & 0 & 0 & 0 \\ 0 & 1 & 0 & 0 & 0 & 0 & 0 & 0 & -4 & 0 & 0 & 0 & 0 & 0 \\ 0 & 0 & 1 & 0 & 0 & 0 & 0 & 0 & 0 & -2 & 0 & 0 & 0 & 0 \\ 0 & 0 & 0 & 1 & 0 & 0 & 0 & 0 & 0 & 0 & 0 & 0 & 0 & 0 \\ 0 & 0 & 0 & 0 & 1 & 0 & 0 & 0 & 0 & 0 & 0 & 0 & 0 & 0 \\ 0 & 0 & 0 & 0 & 0 & 1 & 0 & 0 & 0 & 0 & 0 & 0 & -4 & 0 \\ 0 & 0 & 0 & 0 & 0 & 0 & 1 & 0 & 0 & 0 & 0 & 0 & 0 & -2 \end{bmatrix} \begin{bmatrix} v_b \\ v_d \\ v_e \\ v_f \\ v_a \\ v_c \\ v_g \\ i_b \\ i_d \\ i_e \\ i_f \\ i_a \\ i_c \\ i_g \end{bmatrix} = \begin{bmatrix} -1 \\ 0 \\ 0 \\ -12 \\ 4 \\ 0 \\ 0 \end{bmatrix} \tag{13.105}$$

In this case both A_0 and B_0 are diagonal because all the elements are of the two-terminal variety.

The vector of all circuit variables is now expressed in terms of the basis in (13.88), the tree voltages and link currents. We then have

$$
[A\ \ B]\begin{bmatrix} \bar{v} \\ \bar{i} \end{bmatrix} = [A\ \ B]\begin{bmatrix} U & 0 \\ H' & 0 \\ 0 & -H \\ 0 & U \end{bmatrix}\begin{bmatrix} \bar{v}_T \\ \bar{i}_C \end{bmatrix}
$$

$$
= \begin{bmatrix}
0 & 0 & 0 & 0 & 0 & 0 & 0 & 1 & 0 & 0 & 0 & 0 & 0 & 0 \\
0 & 1 & 0 & 0 & 0 & 0 & 0 & 0 & -4 & 0 & 0 & 0 & 0 & 0 \\
0 & 0 & 1 & 0 & 0 & 0 & 0 & 0 & 0 & -2 & 0 & 0 & 0 & 0 \\
0 & 0 & 0 & 1 & 0 & 0 & 0 & 0 & 0 & 0 & 0 & 0 & 0 & 0 \\
0 & 0 & 0 & 0 & 1 & 0 & 0 & 0 & 0 & 0 & 0 & 0 & 0 & 0 \\
0 & 0 & 0 & 0 & 0 & 1 & 0 & 0 & 0 & 0 & 0 & 0 & -4 & 0 \\
0 & 0 & 0 & 0 & 0 & 0 & 1 & 0 & 0 & 0 & 0 & 0 & 0 & -2
\end{bmatrix}
$$

$$
\times \begin{bmatrix}
1 & 0 & 0 & 0 & 0 & 0 & 0 \\
0 & 1 & 0 & 0 & 0 & 0 & 0 \\
0 & 0 & 1 & 0 & 0 & 0 & 0 \\
0 & 0 & 0 & 1 & 0 & 0 & 0 \\
-1 & 0 & 1 & 0 & 0 & 0 & 0 \\
1 & -1 & 0 & 0 & 0 & 0 & 0 \\
0 & -1 & 1 & 1 & 0 & 0 & 0 \\
0 & 0 & 0 & 0 & 1 & -1 & 0 \\
0 & 0 & 0 & 0 & 0 & 1 & 1 \\
0 & 0 & 0 & 0 & -1 & 0 & -1 \\
0 & 0 & 0 & 0 & 0 & 0 & 1 \\
0 & 0 & 0 & 0 & 1 & 0 & 0 \\
0 & 0 & 0 & 0 & 0 & 1 & 0 \\
0 & 0 & 0 & 0 & 0 & 0 & 1
\end{bmatrix}
\begin{bmatrix} v_b \\ v_d \\ v_e \\ v_f \\ v_a \\ i_c \\ i_g \end{bmatrix}
= \begin{bmatrix} -1 \\ 0 \\ 0 \\ -12 \\ 4 \\ 0 \\ 0 \end{bmatrix}
\tag{13.106}
$$

More compactly, after multiplying the two matrices, we have

$$
\begin{bmatrix}
0 & 0 & 0 & 0 & 1 & -1 & 0 \\
0 & 1 & 0 & 0 & 0 & -4 & -4 \\
0 & 0 & 1 & 0 & 2 & 0 & 2 \\
0 & 0 & 0 & 1 & 0 & 0 & 0 \\
-1 & 0 & 1 & 0 & 0 & 0 & 0 \\
1 & -1 & 0 & 0 & 0 & -4 & 0 \\
0 & -1 & 1 & 1 & 0 & 0 & -2
\end{bmatrix}
\begin{bmatrix} v_b \\ v_d \\ v_e \\ v_f \\ i_a \\ i_c \\ i_g \end{bmatrix}
= \begin{bmatrix} -1 \\ 0 \\ 0 \\ -12 \\ 4 \\ 0 \\ 0 \end{bmatrix}
\tag{13.107}
$$

We leave it to the reader to show that the solution is given by (in transpose notation):

$$\begin{bmatrix} v_b & v_d & v_e & v_f & i_a & i_c & i_g \end{bmatrix}' = \begin{bmatrix} 0 & -4 & 4 & -12 & 0 & 1 & -2 \end{bmatrix} \quad (13.108)$$

In general, one must solve the matrix equation

$$[A \;\; B] = \begin{bmatrix} U & 0 \\ H' & 0 \\ 0 & -H \\ 0 & U \end{bmatrix} \begin{bmatrix} \bar{v}_T \\ \bar{i}_C \end{bmatrix} = C \quad (13.109)$$

where H is the nonunit submatrix in the f-cutset matrix and C_0 is the $B \times 1$ column matrix of constants (or independent functions of time). Here is the crucial result: $[A_0 \;\; B_0]$ has dimensions $B \times 2B$; if it has rank B, then the product of it with the next matrix, which also has rank B, will be square (of dimensions $B \times B$) and of rank B by Sylvester's inequality [13]. In this case the resulting coefficient matrix will be invertible, and a solution is possible.

The procedure just described, although involving more computation than, for example, nodal analysis, is general. If one solves for all element currents and voltages for a general circuit, however, one must anticipate additional complexity. Furthermore, the method outlined is algorithmic and can be computer automated. The element constraint matrices A_0 and B_0 consist of stylized submatrices corresponding to each type of element. These are referred to as **stamps** in the modified nodal technique described elsewhere in this volume, and the above method is quite similar. The major difference is that it uses node voltages as a basis for the branch voltage space of a circuit rather than the tree voltages as above.

References

[1] P. W. Bridgman, *The Logic of Modern Physics*, New York: Macmillan, 1927.

[2] W.-K. Chen, *Linear Networks and Systems*, Monterey: Brooks-Cole, 1983.

[3] J. Choma, *Electrical Networks: Theory and Analysis*, New York: Wiley, 1985.

[4] L. P. Huelsman, *Basic Circuit Theory*, Englewood Cliffs, NJ: Prentice-Hall, 1927.

[5] A. M. Davis, "A unified theory of lumped circuits and differential systems based on Heaviside operators and causality," *IEEE Trans. Circuits Syst.*, volume 41, number 11, pp. 712–727, November, 1990.

[6] A. M. Davis, *Linear Circuit Analysis*, text in preparation.

[7] W.-K. Chen, *Applied Graph Theory: Graphs and Electrical Networks*, New York: North-Holland, 1976.

[8] Chan, Shu-Park, *Introductory Topological Analysis of Electrical Networks*, New York: Holt, Rinehart, and Winston, 1969.

[9] A. M. Davis, unpublished notes.

[10] H. J. Carlin and D. C. Youla, "Network synthesis with negative resistors," *Proc. IEEE*, vol. 49, pp. 907–920, May 1961.

[11] V. Belevitch, "Four dimensional transformations of 4-pole matrices with applications to the synthesis of reactance 4-poles," *IRE Trans. Circuit Theory*, vol. CT-3, pp. 105–111, June 1956.

[12] B. D. H. Tellegen, "La Recherche pour une Serie Complete d'Elements de Circuit Ideaux Non-Lineaires," *Rendiconti Del Seminario Mathematico e Fisico di Milano*, vol. 25, pp. 134–144, April 1954.

[13] F. R. Gantmacher, *Theory of Matrices*, New York: Chelsea, 1959.

.2 Network Theorems

Marwan A. Simaan

In Chapter 13.1 we learned how to determine currents and voltages in a resistive circuit. Methods have been developed, which are based on applying Kirchhoff's voltage law (KVL) and current law (KCL), to derive a set of mesh or node equations which, when solved, will yield mesh currents or node voltages, respectively. Frequently, and especially if the circuit is complex with many elements, the application of these methods may be considerably simplified if the circuit itself is simplified. For example, we may wish to replace a portion of the circuit consisting of resistors and sources by an equivalent circuit that has fewer elements in order to write fewer mesh or node equations.

In this context we introduce three important and related theorems known as the **superposition**, the **Thévenin** and the **Norton theorems.** The superposition theorem shows how to solve for a variable in a circuit that has many independent sources, by solving simpler circuits, each excited by only one source. The Thévenin and Norton theorems can be used to replace a portion of a circuit at any two terminals by an equivalent circuit which consists of a voltage source in series with a resistor (i.e., a nonideal voltage source) or a current source in parallel with a resistor (i.e., a nonideal current source). Another important result derived in this section concerns the calculation of power dissipated in a load resistor connected to a circuit. This result is known as the **maximum power transfer theorem**, and is frequently used in circuit design problems. Finally, a result known as the **reciprocity theorem** is also discussed.

An important property of linear resistive circuits is the type of relationship that exists between any variable in the circuit and the independent sources. For linear resistive circuits the solution for a voltage or current variable can always be expressed as a *linear combination of the independent sources.* Let us elaborate on what we mean by this statement through an example.

Consider the circuit shown in Fig. 13.60 and assume that we are interested in the voltage v across R_2. We can solve for v by first applying KCL at node a to get the equation

$$\frac{v - v_1}{R_1} - \beta i_x - i_1 + \frac{v}{R_2} = 0 \tag{13.110}$$

and then by making use of the fact that

$$i_x = \frac{v_1 - v}{R_1} \tag{13.111}$$

This gives

$$v = \frac{(1+\beta)R_2}{R_1 + (1+\beta)R_2} v_1 + \frac{R_1 R_2}{R_1 + (1+\beta)R_2} i_1 \tag{13.112}$$

Here, the voltage v is a linear combination of the independent sources v_1 and i_1.

The above observation indeed applies to every **linear circuit.** In general, if we let y denote a voltage across, or a current in, any element in a linear circuit and if we let $\{x_1, x_2, \ldots, x_N\}$ denote the independent voltage and current sources in that circuit (assuming there is a total of N such sources), then we can write

$$y = \sum_{k=1}^{N} a_k x_k \tag{13.113}$$

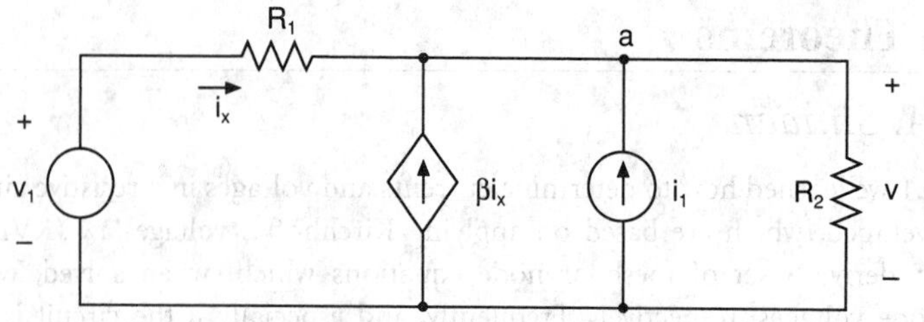

FIGURE 13.60 An example of a linear circuit.

where $a_1, a_2, \ldots, a_N$ are constants which depend on the circuit parameters. Thus, in the circuit of Fig. 13.60, every current or voltage variable can be expressed as a linear combination of the form

$$y = a_1 v_1 + a_2 i_1$$

where a_1 and a_2 are constants that depend on R_1, R_2, and β. For the voltage v across R_2 this relationship is given by expression (13.112).

Mathematically, the relationship between y and $\{x_1, x_2, \ldots, x_N\}$ expressed in (13.113) is said to be *linear* because it satisfies the following two conditions:

1. The *superposition condition*, which requires that:

$$\text{if} \quad \hat{y} = \sum_{k=1}^{N} a_k \hat{x}_k$$

$$\text{and} \quad \tilde{y} = \sum_{k=1}^{N} a_k \tilde{x}_k$$

$$\text{then} \quad \hat{y} + \tilde{y} = \sum_{k=1}^{N} a_k (\hat{x}_k + \tilde{x}_k)$$

2. The *homogeneity condition*, which requires that:

$$\text{if} \quad \hat{y} = \sum_{k=1}^{N} a_k \hat{x}_k$$

$$\text{then} \quad c\hat{y}(t) = \sum_{k=1}^{N} a_k \{c\hat{x}_k\}$$

for any constant c.

The following example illustrates how these two conditions are satisfied for the circuit of Fig. 13.60.

Example 13.1. For the circuit of Fig. 13.60, let $R_1 = 2\ \Omega$, $R_2 = 1\ \Omega$, and $\beta = 2$. Show that the expression for v in terms of v_1 and i_1 satisfies the superposition and homogeneity conditions.

Substituting the values of R_1, R_2, and β in (13.112), the expression for v becomes

$$v = \frac{3}{5} v_1 + \frac{2}{5} i_1 \tag{13.114}$$

To check the superposition property, let $v_1 = \hat{v}_1$ and $i_1 = \hat{i}_1$. Then, the voltage $\hat{v}$ across R_2 is

$$\hat{v} = \frac{3}{5}\hat{v}_1 + \frac{2}{5}\hat{i}_1$$

Similarly, let $v_1 = \tilde{v}_1$ and $i_1 = \tilde{i}_1$. Then, the voltage $\tilde{v}$ across R_2 is

$$\tilde{v} = \frac{3}{5}\tilde{v}_1 + \frac{2}{5}\tilde{i}_1$$

Now, assume that $v_1 = \hat{v}_1 + \tilde{v}_1$ and that $i_1 = \hat{i}_1 + \tilde{i}_1$. Then, according to (13.114) the corresponding voltage v across R_2 is

$$\begin{aligned} v &= \frac{3}{5}(\hat{v}_1 + \tilde{v}_1) + \frac{2}{5}(\hat{i}_1 + \tilde{i}_1) \\ &= \left(\frac{3}{5}\hat{v}_1 + \frac{2}{5}\hat{i}_1\right) + \left(\frac{3}{5}\tilde{v}_1 + \frac{2}{5}\tilde{i}_1\right) \\ &= \hat{v} + \tilde{v} \end{aligned}$$

Hence, the superposition condition is satisfied.

To check the homogeneity condition, let $v_1 = c\hat{v}_1$ and $i_1 = c\hat{i}_1$, where c is an arbitrary constant. Then, according to (13.114) the corresponding voltage v across R_2 is

$$\begin{aligned} v &= \frac{3}{5}(c\hat{v}_1) + \frac{2}{5}(c\hat{i}_1) \\ &= c\left(\frac{3}{5}\hat{v}_1 + \frac{3}{5}\hat{i}_1\right) \\ &= c\hat{v} \end{aligned}$$

The homogeneity condition is also satisfied.

The Superposition Theorem

Let us reexamine expression (13.112) for the voltage v in the circuit of Fig. 13.60. More specifically, let us use this expression to calculate the voltage across R_2 for the two circuits shown in Figs. 13.61(a) and 13.61(b), respectively. Observe that the first circuit is obtained from the original circuit by *deactivating* the current source (i.e. setting $i_1 = 0$) and leaving the voltage source to act alone. The second is obtained by *deactivating* the voltage source (i.e. setting $v_1 = 0$) and leaving the current source to act alone. If we label the voltages across R_2 in these two circuits as v_a and v_b, respectively, then

$$v_a = v\Big|_{\substack{\text{when}\\ i_1=0}} = \frac{(1+\beta)R_2}{R_1 + (1+\beta)R_2}v_1$$

and

$$v_b = v\Big|_{\substack{\text{when}\\ v_1=0}} = \frac{R_1R_2}{R_1 + (1+\beta)R_2}i_1$$

In other words, expression (13.112), which was used to derive the above two expressions, can itself be written as:

$$v = v\Big|_{\substack{\text{when}\\ i_1=0}} + v\Big|_{\substack{\text{when}\\ v_1=0}}$$

or

$$v = v_a + v_b$$

Thus, we conclude that the voltage across R_2 in Fig. 13.60 is actually equal to the sum of two voltages across R_2 due to two independent sources in the circuit acting individually.

The above result is in fact a direct consequence of the linearity property

$$y = a_1x_1 + a_2x_2 + \ldots + a_Nx_N$$

expressed in (13.113). Note that from this expression we can write

$$a_1x_1 = y|_{\text{when } x_1\neq 0, x_2=0, x_3=0,\ldots, x_N=0},$$
$$a_2x_2 = y|_{\text{when } x_1=0, x_2\neq 0, x_3=0,\ldots, x_N=0},$$
$$\vdots \qquad \vdots$$
$$a_Nx_N = y|_{\text{when } x_1=0, x_2=0, x_3=0,\ldots, x_N\neq 0}$$

This means (13.113) can be rewritten as

$$\begin{aligned} y = {} & y|_{\text{when } x_1\neq 0, x_2=0, x_3=0,\ldots, x_N=0} \\ & + y|_{\text{when } x_1=0, x_2\neq 0, x_3=0,\ldots, x_N=0} \\ & \vdots \\ & + y|_{\text{when } x_1=0, x_2=0, x_3=0,\ldots, x_N\neq 0} \end{aligned}$$

The following theorem, known as the **superposition theorem**, is therefore directly implied from the above expression:

The voltage across any element (or current through any element) in a linear circuit may be calculated by adding algebraically the individual voltages across that element (or currents through that element) due to each independent source acting alone with all other independent sources deactivated.

In this statement the word *deactivated* is used to imply that the source is set to zero. In this context we refer to a **deactivated current source** as one that is replaced by an open circuit and a **deactivated voltage source** as one that is replaced by a short circuit. Note that the action of deactivating a source refers only to independent sources. The following example illustrates how the superposition theorem can be used to solve for a variable in a circuit with more than one independent source.

Example 13.2. For the circuit shown in Fig. 13.62, apply superposition to calculate the current I in the 4 Ω resistor.

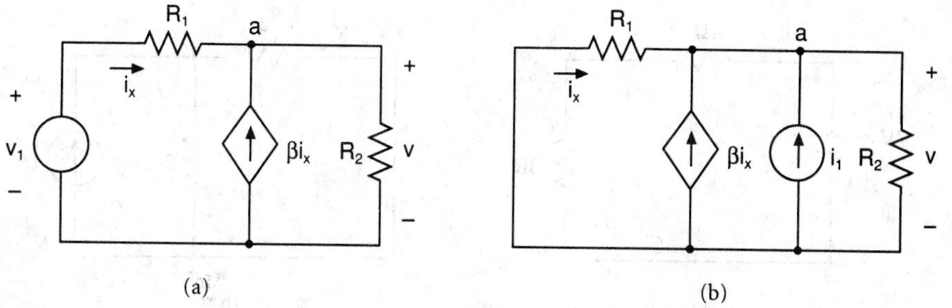

FIGURE 13.61 Circuit of Figure 13.60 with (a) the current source deactivated and with (b) the voltage source deactivated.

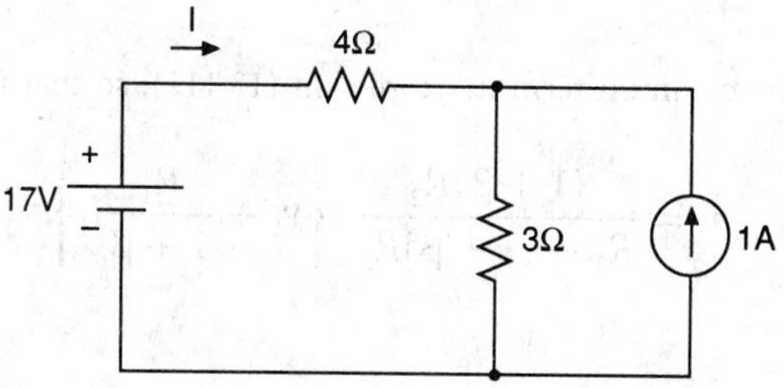

FIGURE 13.62 Circuit for Example 13.2.

Because we are interested in calculating I using the superposition theorem, we need to consider the two circuits shown in Figs. 13.63(a) and (b). The first is obtained by deactivating the current source and the second is obtained by deactivating the voltage source. Let I_a be the current in the 4 Ω resistor in the first circuit and I_b be the current in the same resistor in the second. Then, by superposition

$$I = I_a + I_b$$

We can solve for I_a and I_b independently as follows. From Fig. 13.63(a):

$$I_a = \frac{17}{7}\ A$$

and from Fig. 13.63(b), applying the current divider rule,

$$I_b = -1 \cdot \frac{3}{7}\ A$$

Thus,

$$\begin{aligned} I &= \frac{17}{7} - \frac{3}{7} \\ &= 2\ A \end{aligned}$$

The Thévenin Theorem

In the discussion on the superposition theorem, we interpreted (13.112) for the voltage v in the circuit of Fig. 13.60 as a superposition of two terms. Let us now examine a different interpretation of this expression.

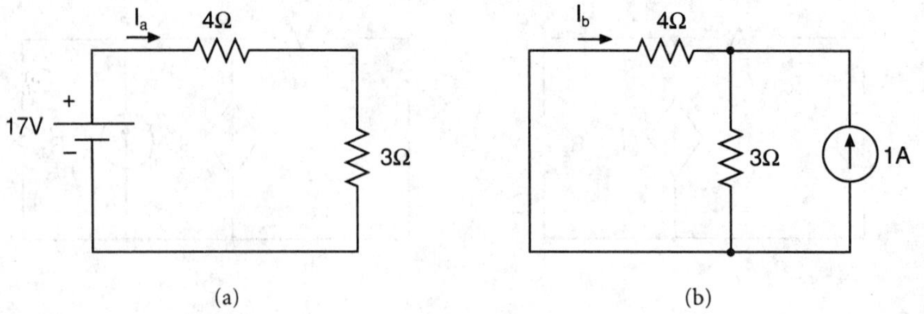

FIGURE 13.63 Circuit Example 2 with (a) the current source deactivated and with (b) the voltage source deactivated.

Suppose we factor the common term in expression (13.112), so that it can be written as

$$v = \frac{(1+\beta)R_2}{R_1 + (1+\beta)R_2}\left\{v_1 + \frac{R_1}{1+\beta}i_1\right\} \tag{13.115a}$$

or

$$v = \frac{R_2}{(R_1/(1+\beta)) + R_2}\left\{v_1 + \left(\frac{R_1}{1+\beta}\right)i_1\right\} \tag{13.115b}$$

Now, suppose we define

$$v_0 = v_1 + \frac{R_1}{1+\beta}i_1 \tag{13.116}$$

and

$$R_0 = \frac{R_1}{1+\beta} \tag{13.117}$$

Then we can write (13.112) in the simple form

$$v = \frac{R_2}{R_0 + R_2}v_0 \tag{13.118}$$

This expression can be interpreted as a voltage divider equation for a two-resistor circuit as shown in Fig. 13.64. This circuit has a voltage source v_0 in series with two resistors R_0 and R_2. When this circuit is compared to Fig. 13.60, the combination of voltage source v_0 in series with R_0 can be interpreted as an equivalent replacement of all the elements in the circuit connected to R_2. That is, we could remove that portion of the circuit of Fig. 13.60 consisting of v_1, R_1, βi_x, and i_1 and replace it with the voltage source v_0 in series with the resistor R_0.

The fact that a portion of a circuit can be replaced by an equivalent circuit consisting of a voltage source in series with a resistor is actually a direct result of the linearity property, and hence is true for linear circuits in general. It is known as **Thévenin's theorem**[11] and is stated as follows:

[11]For an interesting brief discussion on the history of Thévenin's theorem, see an article by James E. Brittain, in *IEEE Spectrum*, p. 42, Mar. 1990.

Any portion of a linear circuit between two terminals a and b can be replaced by an equivalent circuit consisting of a voltage source v_{th} in series with a resistor R_{th}. The voltage v_{th} is determined as the open circuit voltage at terminals a-b. The resistor R_{th} is equal to the input resistance at terminals a-b with all the independent sources deactivated.

The various steps involved in the derivation of the Thévenin equivalent circuit are illustrated in Fig. 13.65. The Thévenin voltage v_{th} is determined by solving for the voltage at terminals a-b when open circuited, and the Thévenin resistance R_{th} is determined by calculating the input resistance of the circuit at terminals a-b when all the independent sources have been deactivated. The following two examples illustrate the application of this important theorem.

Example 13.3. For the circuit shown in Fig. 13.66 determine the Thévenin equivalent of the portion of the circuit to the left of terminals a-b; use it to calculate the current I in the 2 Ω resistor.

First we determine V_{th} from the circuit of Fig. 13.67(a). Note that because terminals a-b are open circuited the current in the branch containing the 2 Ω resistor and 9 V source is equal to 3 A in the direction shown. Applying KCL at the upper node of the 1 Ω resistor, we can calculate the current in this resistor to be $3+2=5$ A as shown. Writing a KVL equation around the inner loop (counterclockwise at terminal b) we have

$$V_{th} + 9 - (2 \times 3) - (1 \times 5) = 0$$

which yields

$$V_{th} = 2\,\text{V}$$

Now, for R_{th} the three sources are deactivated to obtain the circuit shown in Fig. 13.67(b). From this circuit, it is clear that

$$R_{th} = 3\,\Omega$$

The circuit obtained by replacing the portion to the left of terminals a-b with its Thévenin equivalent is shown in Fig. 13.67(c). The current I is now easily computed as

$$I = \frac{2}{2+3} = 0.4\,A$$

Example 13.4. For the circuit shown in Fig. 13.60, determine the Thévenin equivalent circuit for the portion of the circuit to the left of resistor R_2.

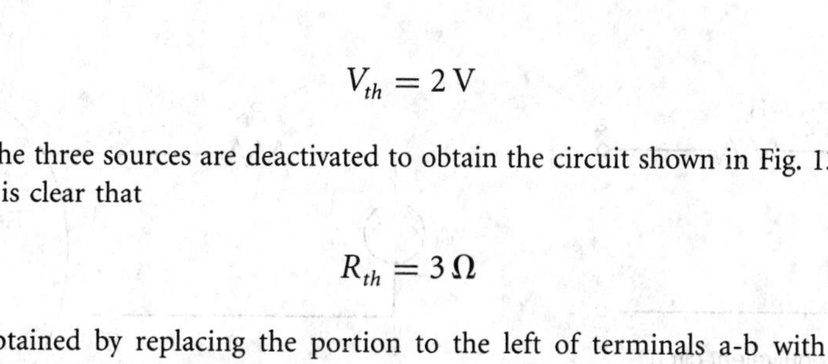

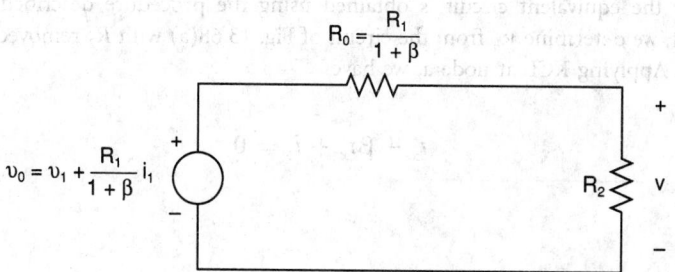

FIGURE 13.64 Voltage divider circuit representing (13.118).

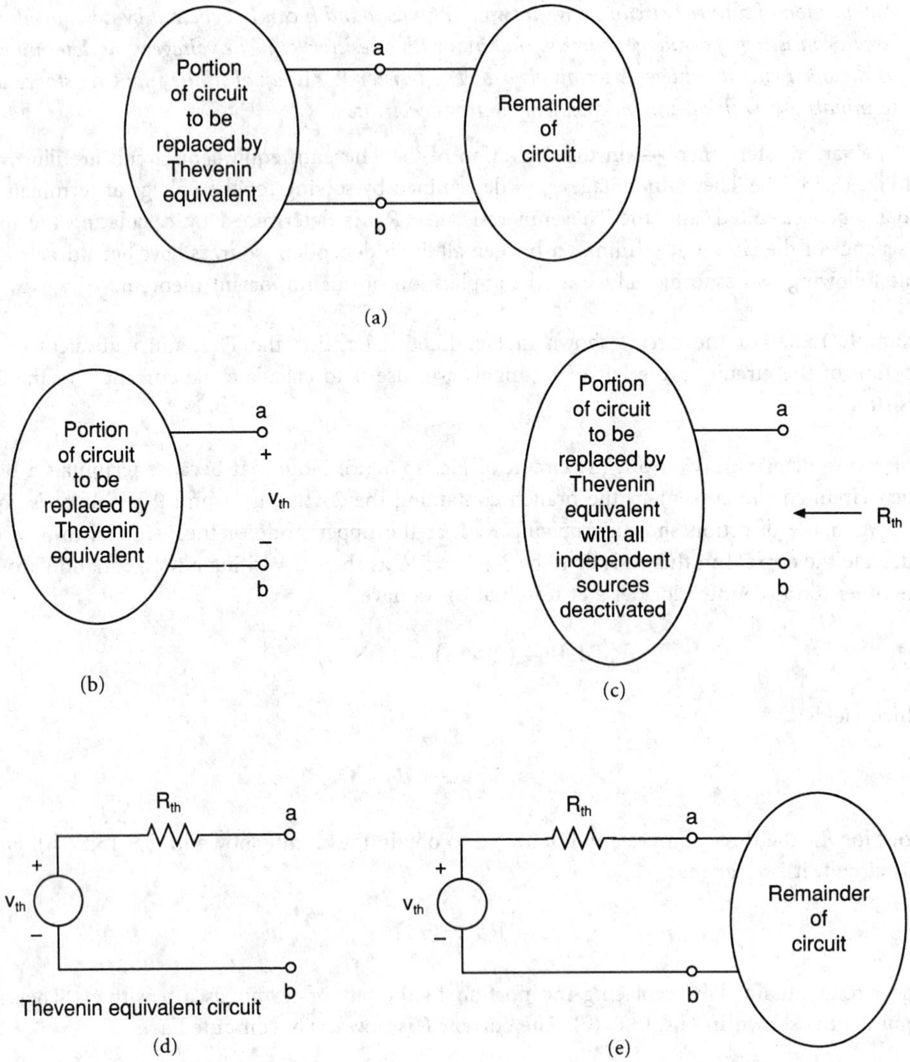

FIGURE 13.65 Steps in determining the Thévenin equivalent circuit.

In deriving (13.118) from (13.110) we actually already determined the Thévenin equivalent for the portion of the circuit to the left of R_2. This was shown in Fig. 13.64. Of course, this procedure is *not* the most efficient way to determine the Thévenin equivalent. Let us now illustrate how the equivalent circuit is obtained using the procedure described in Thévenin's theorem. First, we determine v_{th} from the circuit of Fig. 13.68(a) with R_2 removed and terminals a-b left open. Applying KCL at node a, we have

$$i_x + \beta i_x + i_1 = 0$$

or

$$i_x = -\frac{i_1}{1+\beta}$$

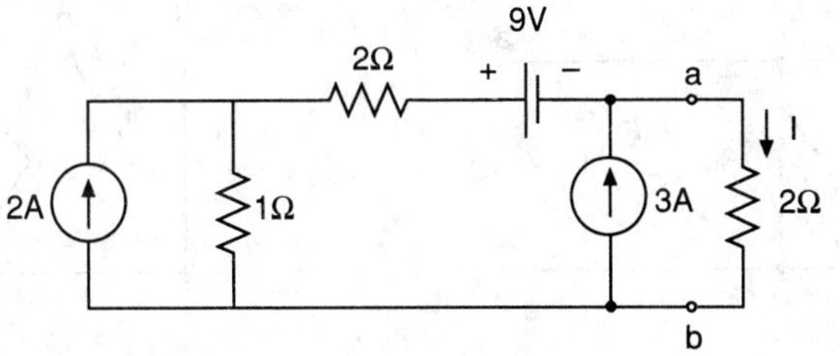

FIGURE 13.66 Circuit for Example 13.3.

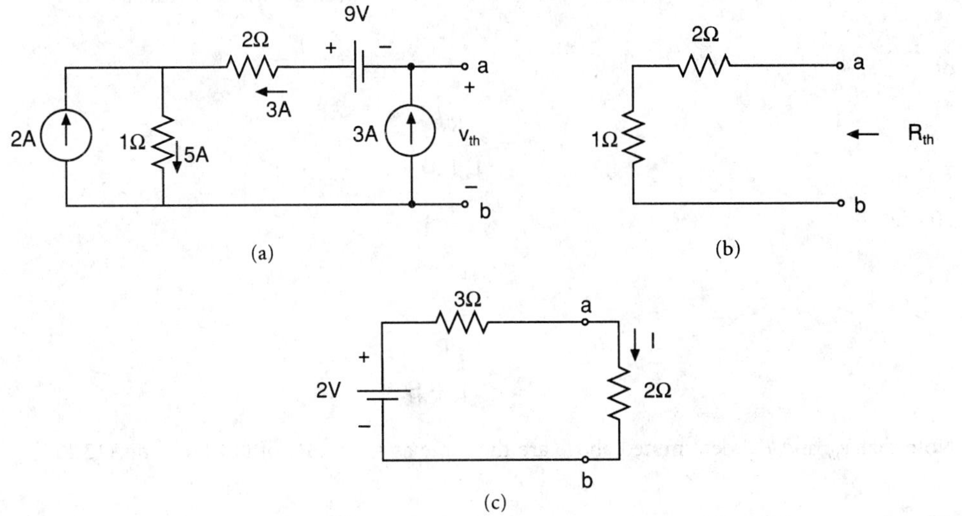

FIGURE 13.67 (a) Calculation of V_{th}, (b) calculation of R_{th}, and (c) the equivalent circuit for Example 13.3.

Hence,

$$\begin{aligned} v_{th} &= v_1 - R_1 i_x \\ &= v_1 + \frac{R_1}{1+\beta} i_1 \end{aligned}$$

As for R_{th}, we need to consider the circuit shown in Fig. 13.68(b), in which the two independent sources were deactivated. Because of the presence of the dependent source βi_x, we determine R_{th} by exciting the circuit with an external source. Let us use a current source i for this purpose and determine the voltage v across it as shown in Fig. 13.68(b). We stress that i is an arbitrary and completely independent source and is in no way related to i_1 which was deactivated. Applying KCL at node a, we have

$$i_x + \beta i_x + i = 0$$

or

$$i_x = -\frac{1}{1+\beta} i$$

Also, applying Ohm's law to R_1,

$$v = -R_1 i_x$$

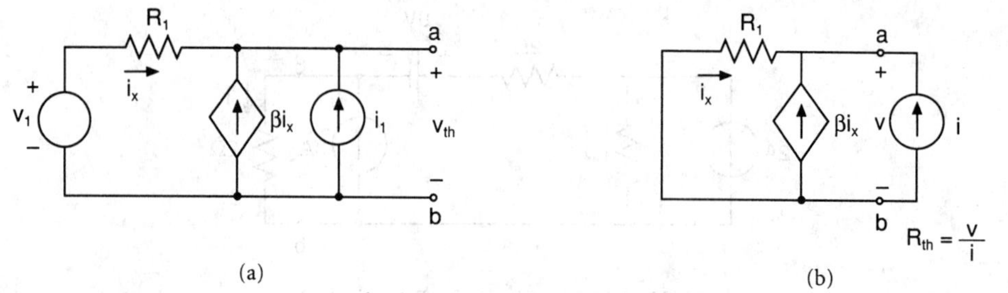

FIGURE 13.68 Calculation of (a) v_{th} and (b) R_{th} for the circuit of Fig. 13.60 (Example 13.4).

or

$$v = \frac{R_1}{1+\beta} i$$

Hence,

$$R_{th} = \frac{v}{i} = \frac{R_1}{1+\beta}$$

Note that v_{th} and R_{th} determined above are the same as v_0 and R_0 of (13.116) and (13.117).

The Norton Theorem

Instead of a voltage source in series with a resistor, it is possible to replace a portion of a circuit by an equivalent current source in parallel with a resistor. This result is formally known as *Norton's theorem* and is stated as follows:

> *Any portion of a linear circuit between two terminals a and b can be replaced by an equivalent circuit consisting of a current source i_n in parallel with a resistor R_n. The current i_n is determined as the current that flows from a to b in a short circuit at terminals a-b. The resistor R_n is equal to the input resistance at terminals a-b with all the independent sources deactivated.*

As in the case of Thévenin's, the above theorem provides a procedure for determining the "Norton" current source and "Norton" resistance in the **Norton equivalent circuit**. The various steps in this procedure are illustrated in Fig. 13.69. The Norton current is determined by solving for the current in a short circuit at terminals a-b and the Norton resistance is determined by calculating the input resistance to the circuit at terminals a-b when all independent sources have been deactivated. The Norton's equivalent of a portion of a circuit is, in effect, a nonideal current source representation of that portion. It should be noted that the procedure to determine R_n is exactly the same as that for R_{th}. That is,

$$R_n = R_{th}$$

Also, if we compare Thévenin's and Norton's equivalent circuits, we see that these are indeed related by the voltage-current source transformation rule discussed in Chapter 13.1. Each circuit is a source transformation of the other. For this reason, the Thévenin and Norton equivalent

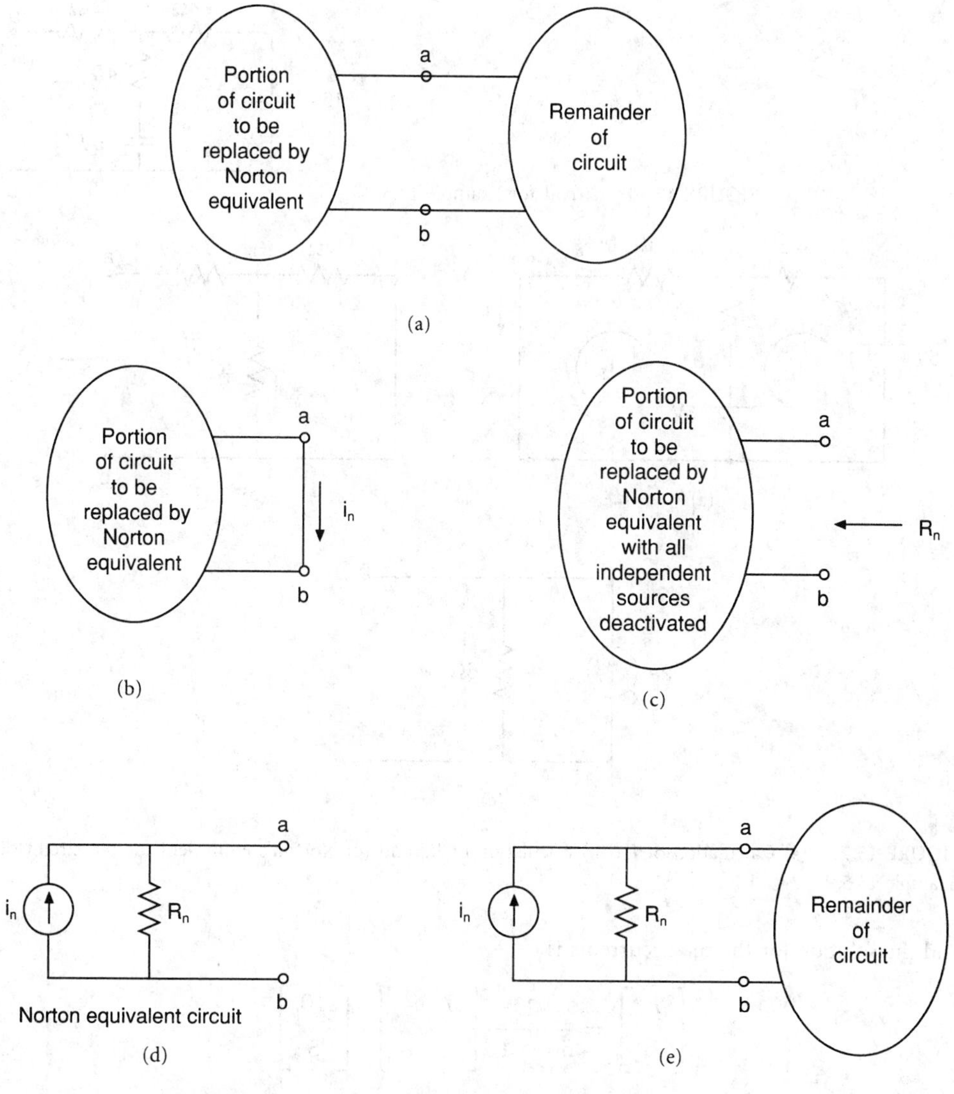

FIGURE 13.69 Steps in determining the Norton equivalent circuit.

circuits are often referred to as **dual circuits**, and the two resistances R_{th} and R_n are frequently referred to as the source resistance and denoted by R_s. Clearly, v_{th} and i_n are related by

$$v_{th} = R_s i_n$$

Example 13.5. For the circuit shown in Fig. 13.70, determine the Norton equivalent circuit at terminals a-b.

We determine Norton's current I_n by placing a short circuit between a and b, as shown in Fig. 13.71(a), and solving for the current in it with a reference direction going from a to b. For this circuit, we could use the mesh equation method. In matrix form the mesh equations are

$$\begin{bmatrix} 8 & -4 \\ -4 & 7 \end{bmatrix} \cdot \begin{bmatrix} I_1 \\ I_2 \end{bmatrix} = \begin{bmatrix} 10 \\ 5 \end{bmatrix}$$

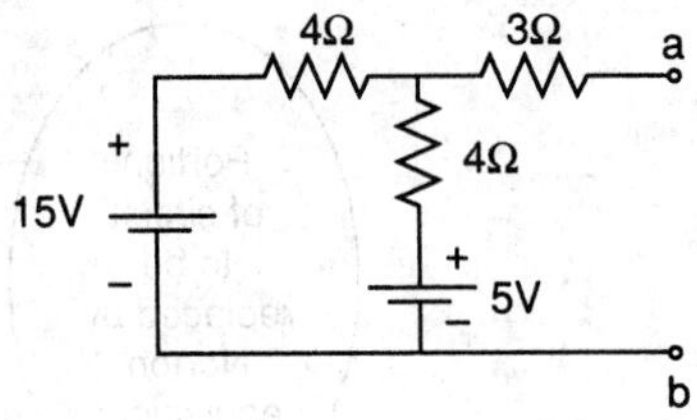

FIGURE 13.70 Circuit for Example 13.5.

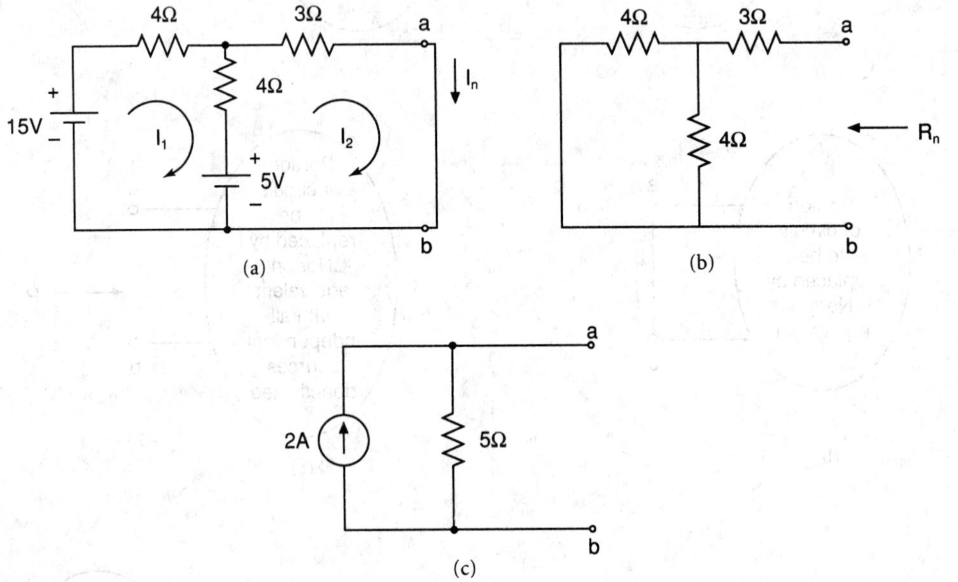

FIGURE 13.71 (a) Calculation of I_n, (b) calculation of R_n and (c) Norton's equivalent for the circuit of Example 13.5.

and the solution for the mesh currents is

$$\begin{bmatrix} I_1 \\ I_2 \end{bmatrix} = \frac{1}{56-16}\begin{bmatrix} 7 & 4 \\ 4 & 8 \end{bmatrix} \cdot \begin{bmatrix} 10 \\ 5 \end{bmatrix}$$

From this, we extract I_n as

$$I_n = I_2 = \frac{40+40}{40} = 2\,A$$

Norton's resistance R_n is determined by deactivating the two voltage sources, as shown in Fig. 13.71(b), and calculating the input resistance at terminals a-b. Clearly,

$$R_n = (4\|4) + 3 = 5\,\Omega$$

Thus, the Norton equivalent for the circuit of Fig. 13.70 is shown in Fig. 13.71(c).

Example 13.6. For the circuit shown in Fig. 13.72 determine the Norton equivalent circuit at terminals a-b and use it to calculate the current and power dissipated in *R*.

With a short circuit placed at terminals a-b, as shown in Fig. 13.73(a), the voltage $V_x = 0$. Hence, the dependent source in this circuit is equal to zero. This means that the 8 A current

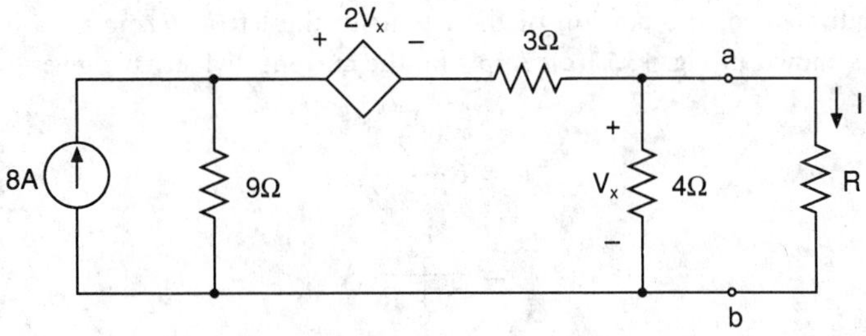

FIGURE 13.72 Circuit for Example 13.6.

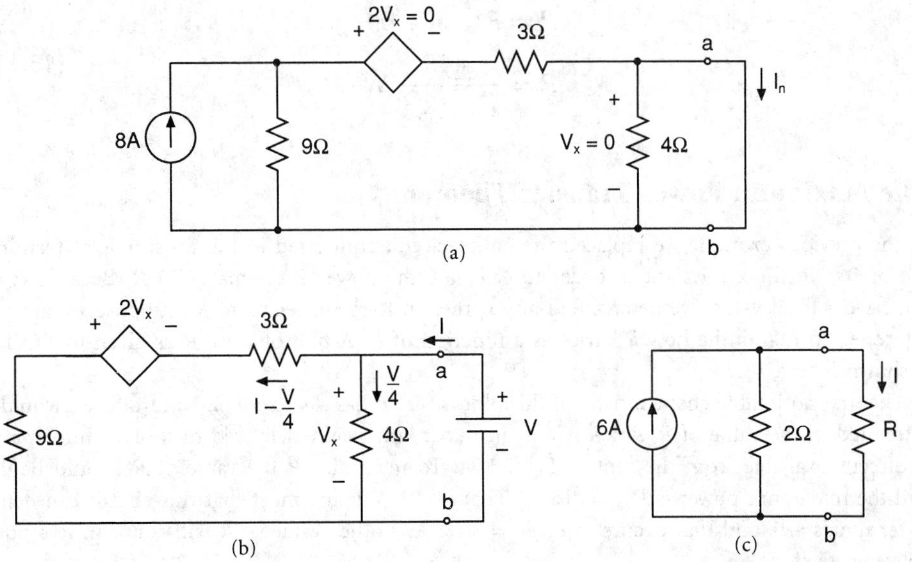

FIGURE 13.73 (a) Calculation of I_n, (b) calculation of R_n and (c) Norton's equivalent for the circuit of Example 13.6.

source has the 9 Ω and 3 Ω resistors in parallel across it, and I_n is the current in the 3 Ω resistor. Using the current divider rule we have

$$I_n = 8\frac{9}{12} = 6\,A$$

Now, deactivating the independent source to determine R_n, we excite the circuit with a voltage source V at terminals a-b. Let I be the current in this source as shown in Fig. 13.73(b). Applying KCL at node a yields the current in the 3 Ω resistor to be $I - (V/4)$ from right to left. Applying KVL around the outer loop and making use of the fact that in this circuit $V_x = V$, we get

$$V - 3\left(I - \frac{V}{4}\right) + 2\,\mathrm{V} - 9\left(I - \frac{V}{4}\right) = 0$$

Solution of this equation yields

$$R_n = \frac{V}{I} = 2\ \Omega.$$

The Norton equivalent of the portion of the circuit to the left of terminals a-b, connected to the resistor R is shown in Fig. 13.73(c). Applying the current divider rule gives

$$I = 6\frac{2}{2+R} = \frac{12}{2+R}\ \text{A}$$

and the power dissipated in R is

$$P = RI^2$$

$$P = \frac{144R}{(2+R)^2}\ \text{W} \tag{13.119}$$

The Maximum Power Transfer Theorem

In the previous example we replaced the entire circuit connected to the resistor R at terminals a-b by its Norton equivalent in order to calculate the power P dissipated in R. Because R did not have a fixed value, we determined an expression for P in terms of R. Suppose we are now interested in examining how P varies as a function of R. A plot of P vs. R as given by (13.119) is shown in Fig. 13.74.

The first noticeable characteristic of this plot is that it has a *maximum*. Naturally, we would be interested in the value of R which results in maximum power delivered to it. This information is directly available from the plot in Fig. 13.74. To maximize P the value of R should be 2 Ω and the maximum power is $P_{max} = 18$ W. That is, 18 W is the most that this circuit can deliver at terminals a-b, and that occurs when $R = 2$ Ω. Any other value of R will result in less power delivered to it.

The problem of finding the value of a load resistor R_L such that maximum power is delivered to it is obviously an important circuit design problem. Because it is possible to reduce any linear circuit connected to R_L into either its Thévenin or Norton equivalent, as illustrated in Fig. 13.75, the problem becomes quite simple. We need to consider only either the circuit of Fig. 13.75(b) or that of 13.75(c). Let us first consider the circuit that uses the Thévenin equivalent. In this case the power P delivered to R_L is given by

$$P = \left(\frac{v_{th}}{R_{th}+R_L}\right)^2 R_L \tag{13.120}$$

In general, we may not always be able to plot P vs. R_L, as we did earlier, therefore, we need to maximize P mathematically. We do this by solving the necessary condition

$$\frac{dP}{dR_L} = 0 \tag{13.121}$$

for R_L. To guarantee that R_L maximizes P, it must also satisfy the sufficiency condition

$$\left.\frac{d^2P}{dR_L^2}\right|_{R_L} < 0 \tag{13.122}$$

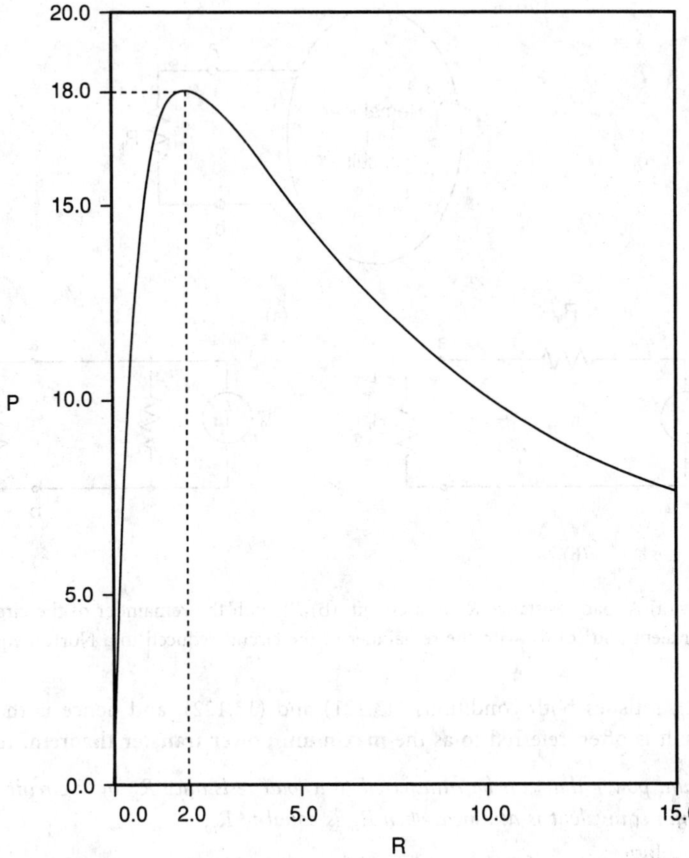

FIGURE 13.74 Plot of P vs. R for the circuit of Example 13.6.

Thus, applying these conditions to (13.120), we have

$$\begin{aligned}\frac{dP}{dR_L} &= v_{th}^2\left[\frac{(R_{th}+R_L)^2 - 2R_L(R_{th}+R_L)}{(R_{th}+R_L)^4}\right], \\ &= v_{th}^2\frac{(R_{th}-R_L)}{(R_{th}+R_L)^3}\end{aligned} \tag{13.123}$$

Equating the right-hand side of (13.123) to zero and solving for R_L yields

$$R_L = R_{th}$$

The sufficiency condition (13.122) yields

$$\frac{d^2P}{dR_L^2} = v_{th}^2\frac{2R_{th} - 4R_L}{(R_{th}+R_L)^4}$$

When R_L is replaced with R_{th}, we get

$$\left.\frac{d^2P}{dR_L^2}\right|_{R_L=R_{th}} = -\frac{v_{th}^2}{8R_{th}^3} < 0$$

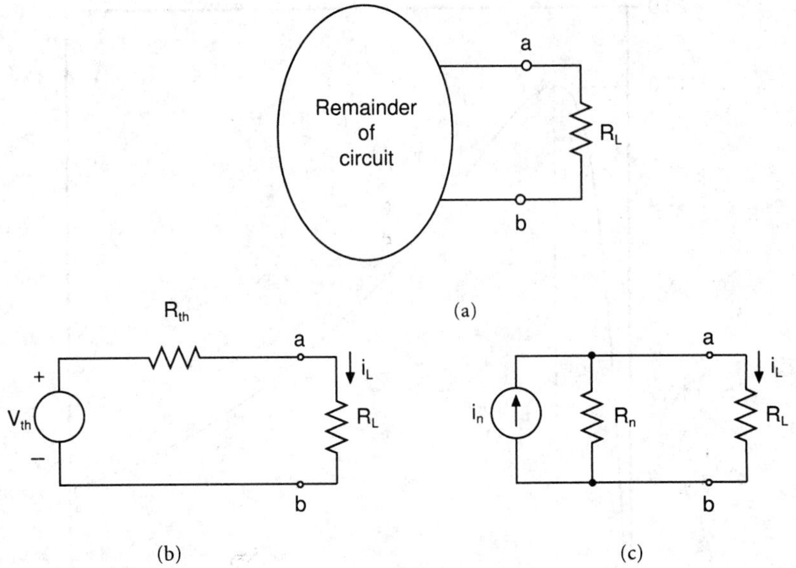

FIGURE 13.75 (a) A load resistance R_L in a circuit. (b) R_L with the remainder of the circuit reduced to a Thévenin equivalent and (c) R_L with the remainder of the circuit reduced to a Norton equivalent.

Thus, $R_L = R_{th}$ satisfies both conditions (13.121) and (13.122), and hence is the maximizing value. This result is often referred to as the **maximum power transfer theorem.** It says

The maximum power that can be transferred to a load resistance R_L by a circuit represented by its Thévenin equivalent is attained when R_L is equal to R_{th}.

The corresponding value of P_{max} is obtained from (13.120) as

$$P_{max} = \frac{v_{th}^2}{4R_{th}} \tag{13.124}$$

In the case of Norton's equivalent circuit of Fig. 13.75(b), a similar derivation can be carried out. The power P delivered to R_L is given by

$$P = \left(\frac{R_n i_n}{R_n + R_L}\right)^2 R_L \tag{13.125}$$

This expression has exactly the same form as (13.120). Consequently, its maximum is achieved when

$$R_L = R_n$$

and the corresponding maximum power is

$$P_{max} = \frac{R_n i_n^2}{4} \tag{13.126}$$

This leads to the following alternate statement of the **maximum power transfer theorem:**

The maximum power that can be transferred to a load resistance R_L by a circuit represented by its Norton equivalent is attained when R_L is equal to R_n.

Example 13.7. Consider the circuit of Example 13.3, shown in Fig. 13.66. Determine the value of a load resistor R_L connected in place of the 2 Ω resistor at terminals a-b in order to achieve maximum power transfer to the load.

Solution. The Thévenin equivalent for the circuit of Fig. 13.66 already was determined and is shown in Fig. 13.67(c). Using the results of the maximum power transfer theorem we should have

$$R_L = 3\ \Omega$$

The corresponding value of maximum power is

$$P_{max} = \frac{2^2}{4 \times 3} = \frac{1}{3}\ \text{W}$$

The Reciprocity Theorem

The reciprocity theorem is an important result that applies to circuits consisting of linear resistors and one independent source (either a current source or a voltage source). It does not apply to nonlinear circuits, and, in general, it does not apply to circuits containing dependent sources. The reciprocity theorem is stated as follows:

The ratio of a voltage (or current) response in one part of the circuit to the current (or voltage) source is the same if the locations of the response and the source are interchanged.

It is important to note that the reciprocity theorem applies only to circuits in which the source and response are voltage and current or current and voltage, respectively. It does not apply to circuits in which the source and response are of the same type (i.e., voltage and voltage or current and current).

Example 13.8. Verify the reciprocity theorem for the circuit shown in Fig. 13.76(a).

Solution. If we interchange the location of the 40 V voltage source and current response I, we obtain the circuit shown in Fig. 13.76(b). The reciprocity theorem implies that I should be the same in both circuits.

For the circuit in Fig. 13.76(a), the current I_1 in the 2 Ω resistor is equal to:

$$I_1 = \frac{40}{2 + 4\|12} = \frac{40}{2+3} = 8\ \text{A}$$

and the response I can be easily determined, using the current divider rule, as:

$$I = 8 \times \frac{4}{4+12} = 2\ \text{A}$$

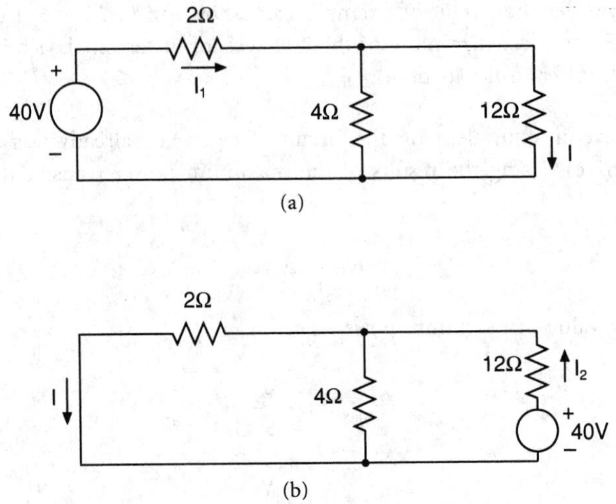

FIGURE 13.76 (a) Circuit for Example 13.8. (b) Circuit with location of voltage source and current response interchanged.

For the circuit in Fig. 13.76(b), the current I_2 in the 12 Ω resistor is equal to:

$$I_2 = \frac{40}{12 + 2\|4}$$
$$= \frac{40}{12 + \frac{4}{3}}$$
$$= 3\ \text{A}$$

and the response I can be determined easily, using the current divider rule, as

$$I = 3 \times \frac{4}{2 + 4}$$
$$= 2\ \text{A}$$

Thus, the reciprocity theorem is satisfied.

14

Analysis in the Time Domain

:rt W. Newcomb
·sity of Maryland

.1 Signal Types

Introduction

Because information into and out of a circuit is carried via time domain signals we look first at some of the basic signals used in continuous time circuits. All signals are taken to depend on continuous time t over the full range $-\infty < t < \infty$. It is important to realize that not all signals of interest are functions in the strict mathematical sense; we must go beyond them to generalized functions (e.g., the impulse), which play a very important part in the signal processing theory of circuits.

Step, Impulse, and Ramp

The unit **step** function, denoted $1(\cdot)$, characterizes sudden jumps, such as when a signal is turned on or a switch is thrown; it can be used to form pulses, to select portions of other functions, and to define the ramp and impulse as its integral and derivative. The unit step function is discontinuous and jumps between two values, 0 and 1, with the time of jump between the two taken as $t = 0$. Precisely

$$1(t) = \begin{cases} 1 & \text{if } t > 0 \\ 0 & \text{if } t < 0 \end{cases} \tag{14.1}$$

which is illustrated in Fig. 14.1 along with some of the functions to follow.

Here, the value at the jump point, $t = 0$, purposely has been left free because normally it is immaterial and specifying it can lead to paradoxical results. Physical step functions used in the laboratory are actually continuous functions which have a continuous rise between 0 and 1,

341-2/95/$0.00 + $.50

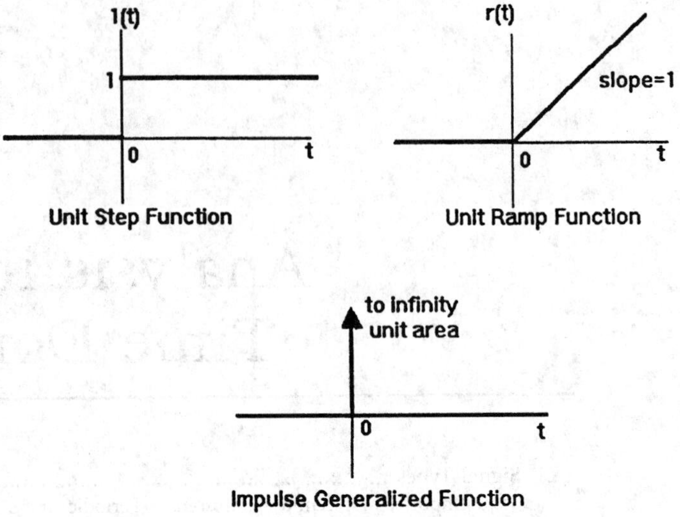

FIGURE 14.1 Step, ramp, and impulse functions.

which occurs over a very short time. Nevertheless, instances occur in which one may wish to set $1(0)$ equal to 0 or to 1 or to 1/2 (the latter, for example, when calculating the values of a Fourier series at a discontinuity). By shifting the time argument the jump can be made to occur at any time, and by multiplying by a factor the height can be changed. For example, $1(t - t_0)$ has a jump at time t_0 and $a[1(t) - 1(t - t_0)]$ is a pulse of width t_0 and height a going up to a at $t = 0$ and down to 0 at time t_0. If $a = a(t)$ is a function of time, then that portion of $a(t)$ between 0 and t_0 is selected. The unit **ramp**, $r(\cdot)$, is the continuous function which ramps up linearly (with unit slope) from zero starting at $t = 0$; the ramp results from the unit step by integration

$$r(t) = \int_{-\infty}^{t} 1(\tau)d\tau = t1(t) = \begin{cases} t & \text{if } t > 0 \\ 0 & \text{if } t < 0 \end{cases} \tag{14.2}$$

As a consequence the unit step is the derivative of the unit ramp, while differentiating the unit step yields the unit **impulse** generalized function, $\delta(\cdot)$, that is

$$\delta(t) = \frac{d1(t)}{dt} = \frac{d^2 r(t)}{dt^2} \tag{14.3}$$

In other words, the unit impulse is such that its integral is the unit step; that is, its area at the origin, $t = 0$, is 1. The impulse acts to sample continuous functions which multiply it, i.e.,

$$a(t)\delta(t - t_0) = a(t_0)\delta(t - t_0) \tag{14.4}$$

This sampling property yields an important integral representation of a signal $x(\cdot)$

$$\begin{aligned} x(t) &= \int_{-\infty}^{\infty} x(\tau)\delta(t - \tau)d\tau \\ &= \int_{-\infty}^{\infty} x(t)\delta(t - \tau)d\tau = x(t)\int_{-\infty}^{\infty} \delta(t - \tau)d\tau \end{aligned} \tag{14.5}$$

where the validity of the first line is seen from the second line, and the fact that the integral of the impulse through its jump point is unity. Equation (14.5) is actually valid even when $x(\cdot)$ is discontinuous and, consequently, is a fundamental equation for linear circuit theory. Differentiating $\delta(t)$ yields an even more discontinuous object, the doublet $\delta'(\cdot)$. Strictly speaking, the impulse, all its derivatives, and signals of that class are not functions in the classical sense, but rather they are operators [1] or functionals [2], called generalized functions or, often, distributions. Their evaluations take place via test functions, just as voltages are evaluated on test meters.

The importance of the impulse lies in the fact that if a linear time-invariant system is excited by the unit impulse, then the response, naturally called the impulse response, is the inverse Laplace transform of the network function. In fact, if $h(t)$ is the impulse response of a linear time-invariant (continuous and continuous time) circuit, the forced response $y(t)$ to any input $u(t)$ can be obtained without leaving the time domain by use of the convolution integral, with the operation of convolution denoted by $*$,

$$y(t) = h * u = \int_{-\infty}^{\infty} h(t-\tau)u(\tau)d\tau \tag{14.6}$$

Equation (14.6) is mathematically rigorous, but justified on physical grounds through (14.5) as follows. If we let $h(t)$ be the output when $\delta(t)$ is the input, then, by time invariance, $h(t-\tau)$ is the output when the input is shifted to $\delta(t-\tau)$. Scaling the latter by $u(\tau)$ and summing via the integral, as designated in (14.5), we obtain a representation of the input $u(t)$. This must result in the output representation being in the form of (14.6) by linearity of the system through similar scaling and summing of $h(t-\tau)$, as was performed on the input.

Sinusoids

Sinusoidal signals are important because they are self-reproducing functions (i.e., eigenfunctions) of linear time-invariant circuits. This is true basically because the derivatives of sinusoids are sinusoidal. As such, sinusoids are also the natural outputs of oscillators and are delivered in power sources, including laboratory signal generators and electricity for the home derived from the power company.

Eternal

Eternal signals are defined as being of the same nature for all time, $-\infty < t < \infty$, in which case an eternal cosine repeats itself eternally in both directions of time, with an origin of time, $t = 0$, being arbitrarily fixed. Because eternal sinusoids have been turned on forever, they are useful in describing the steady operation of circuits. In particular, the signal $A\cos(\omega t + \theta)$ over $-\infty < t < \infty$ defines an eternal cosine of amplitude A, radian frequency $\omega = 2\pi f$ (with f being real frequency, in Hertz, which are cycles per second), at phase angle θ (in radians and with respect to the origin of time), with A, ω, and θ real numbers. When $\theta = \pi/2$ this cosine also represents a sine, so that all eternal sinusoidal signals are contained in the expression $A\cos(\omega t + \theta)$.

At times it is important to work with sinusoids which have an exponential envelope, with the possibility that the envelope increases or decreases with time, that is, with positively or negatively damped sinusoids. These are described by $Ae^{\sigma t}\cos(\omega t + \theta)$, where the real number $-\sigma$ is the damping factor, giving signals that damp out in time when the damping factor is positive and signals that increase with time when the damping factor is negative. Of most importance when working with this class of signals is the identity

$$e^{\sigma t + j\omega t} = e^{st} = e^{\sigma t}[\cos(\omega t) + j\sin(\omega t)] \tag{14.7}$$

where $s = \sigma + j\omega$ with $j = \sqrt{-1}$. Here, s is called the **complex frequency**, with its imaginary part being the real (radian) frequency, ω. When no damping is present, $s = j\omega$, in which case the exponential form of (14.7) represents pure sinusoids. In fact, we see in this expression that the cosine is the real part of an exponential and the sine is its imaginary part. Because exponentials are usually easier than sinusoids to treat analytically, the consequence for real linear networks is that we can do most of the calculations with exponentials and convert back to sinusoids at the end. In other words, if a real linear system has a cosine or a damped cosine as a true input, it can be analyzed by using instead the exponential of which it is the real part as its (fictitious) input, finding the resulting (fictitious) exponential output, and then taking the real part at the end of the calculations to obtain the true output for the true input. Because exponentials are probably the easiest signals to work with in theory, the use of exponentials rather than sinusoids usually greatly simplifies the theory and calculations for circuits operating under steady-state conditions.

Causal

Because practical circuits have not existed since $t = -\infty$ they usually begin to be considered at a suitable starting time, taken to be $t = 0$, in which case the associated signals can be considered to be zero for $t < 0$. Mathematically, these functions are said to have support bounded on the left. The support of a signal is (the closure of) that set of times for which the signal is zero, therefore, the support of these signals is bounded on the left by zero. When signals are discontinuous functions they have the important property that they can be represented by multiplying with unit step functions signals which are differentiable and have nonbounded support. For example, $g(t) = e^{st} \cdot 1(t)$ has a jump at $t = 0$ with support at the half line 0 to ∞ but has e^{st} infinitely differential of "eternal" support.

A **causal** circuit is one for which the response is only nonzero after the input becomes nonzero. Thus, if the inputs are zero for $t < 0$, the outputs of causal circuits are also zero for $t < 0$. In such cases the impulse response, $h(t)$, or the response to an input impulse of "infinite jump" at $t = 0$, satisfies $h(t) = 0$ for $t < 0$ and the convolution form of the output, (14.4), takes the form

$$y(t) = \left[\int_0^t h(t-\tau)u(\tau)d\tau\right]1(t) \tag{14.8}$$

Periodic and Aperiodic Waveforms

The pure sinusoids, although not the sinusoids with nonzero damping, are special cases of periodic signals. In other words, ones which repeat themselves in time every T seconds, where T is the period. Precisely, a time-domain signal $g(\cdot)$ is **periodic** of period T if $g(t) = g(t + T)$, where normally T is taken to be the smallest nonzero T for which this is true. In the case of the sinusoids, $A\cos(\omega t + \theta)$ with $\omega = 2\pi f$, the period is given by $T = 1/f$ because $\{2\pi[f(t + T)] + \theta\} = \{2\pi ft + 2\pi(fT) + \theta\} = \{2\pi ft + (2\pi + \theta)\}$, and sinusoids are unchanged by a change of 2π in the phase angle. Periodic signals need to be specified over only one period of time, e.g., $0 \le t < T$, and then can be extended periodically for all time by using $t = t\text{mod}(T)$ where mod$(\cdot)$ is the modulus function; in other words, periodic signals can be looked upon as being defined on a circle, if we imagine the circle as being a clock face.

Periodic signals represent rhythms of a system and, as such, contain recurring information. As many physical systems, especially biomedical systems, either possess directly or to a very good approximation such rhythms, the periodic signals are of considerable importance. Even though countless periodic signals are available besides the sinusoids, it is important to note that almost all can be represented by a Fourier series. Exponentials are eigenfunctions for linear circuits, thus, the Fourier series is most conveniently expressed for circuit considerations in terms of the exponential form. If $g(t) = g(t+T)$, then

$$g(t) \cong \sum_{n=-\infty}^{\infty} c_n e^{j(2\pi nt/T)} \tag{14.9}$$

where the coefficients are complex and are given by

$$c_n = \frac{1}{T}\int_0^T g(t)e^{-j(2\pi nt/T)}dt = a_n + jb_n \tag{14.10}$$

Strictly speaking, the integral is over the half-open interval $[0,T)$ as seen by considering $g(\cdot)$ defined on the circle. In (14.9) the symbol $\cong$ is used to designate the expression on the right as a representation that may not exactly agree numerically with the left side at every point when $g(\cdot)$ is a function; for example, at discontinuities the average is obtained on the right side. If $g(\cdot)$ is real, that is, $g(t) = g(t)^*$, where the superscript $*$ denotes complex conjugate, then the complex coefficients c_n satisfy $c_n = c_{-n}^*$. In this case the real coefficients a_n and b_n in (14.10) are even and odd in the indices; n and the a_n combine to give a series in terms of cosines, and the b_n gives a series in terms of sines.

As an example the square wave, sqw(t), can be defined by

$$\text{sqw}(t) = 1(t) - 1(t - [T/2]) \quad 0 \le t < T \tag{14.11}$$

and then extended periodically to $-\infty < t < \infty$ by taking $t = t\text{mod}(T)$. The exponential Fourier series coefficients are readily found from (14.10) to be

$$c_n = \begin{cases} 1/2 & \text{if } n = 0 \\ \dfrac{1}{j\pi n}\begin{cases} 0 \\ 1 \end{cases} & \begin{matrix} \text{if } n = 2k \neq 0 \text{ (even} \neq 0) \\ \text{if } n = 2k+1 \text{ (odd)} \end{matrix} \end{cases} \tag{14.12}$$

for which the Fourier series is

$$\text{sqw}(t) \cong \frac{1}{2} + \sum_{k=-\infty}^{\infty} \frac{1}{j\pi[2k+1]} e^{j2\pi[2k+1]t/T} \tag{14.13}$$

The derivative of sqw(t) is a periodic set of impulses

$$\frac{d[\text{sqw}(t)]}{dt} = \delta(t) - \delta(t - [T/2]) \quad 0 \le t < T \tag{14.14}$$

for which the exponential Fourier series is easily found by differentiating (14.13), or by direct calculation from (14.10), to be

$$\sum_{i=-\infty}^{\infty} (\delta(t-i) - \delta(t - i - [T/2]) \cong \sum_{k=-\infty}^{\infty} \frac{2}{T} e^{j(2\pi[2k+1]t/T)} \tag{14.15}$$

Combining the exponentials allows for a sine representation of the periodic generalized function signal. Further differentiation can take place, while by integrating (14.15) we get the Fourier series for the square wave if the appropriate constant of integration is added to give the DC value of the signal. Likewise, a further integration will yield the Fourier series for the sawtooth periodic signal, and so on.

The importance of these Fourier series representations is that a circuit having periodic signals can always be considered to be processing these signals as exponential signals, which are usually self-reproducing signals for the system, making the design or analysis easy. The Fourier series also allows visualization of which radian frequencies, $2\pi n/T$, may be important to filter out or emphasize. In many common cases, especially for periodically pulsed circuits, the series may be expressed in terms of impulses. Thus, the impulse response of the circuit can be used in conjunction with the Fourier series.

References

[1] J. Mikusinski, *Operational Calculus*, 2nd ed., New York: Pergamon Press, 1983.

[2] A. Zemanian, *Distribution Theory and Transform Analysis*, New York: McGraw-Hill, 1965.

14.2 First-Order Circuits

Introduction

First-order circuits are fundamental to the design of circuits because higher order circuits can be considered to be constructed of them. Here, we limit ourselves to single-input–single-output linear time-invariant circuits for which we take the definition of a first-order circuit to be one described by the differential equation

$$d_1 \cdot \frac{dy}{dt} + d_0 \cdot y = n_1 \cdot \frac{du}{dt} + n_0 \cdot u \tag{14.16}$$

where d_0 and d_1 are "denominator" constants and n_0 and n_1 are "numerator" constants, $y = y(\cdot)$ is the output and $u = u(\cdot)$ is the input, and both u and y are generalized functions of time t. So that the circuit truly will be first order we require that $d_1 \cdot n_0 - d_0 \cdot n_1 \neq 0$, which guarantees that at least one of the derivatives is actually present, but if both derivatives occur, the expressions in y and in u are not proportional, which would lead to cancellation, forcing y and u to be constant multiples of each other. Because a factorization of real higher order systems may lead to complex first-order systems, we will allow the numerator and denominator constants to be complex numbers; thus, y and u may be complex-valued functions.

If the derivative is treated as an operator, $p = d[\cdot]/dt$, then (14.16) can be conveniently written as

$$y = \frac{n_1 p + n_0}{d_1 p + d_0} u = \begin{cases} \left[\dfrac{n_1}{d_0} p + \dfrac{n_0}{d_0}\right] u & \text{if } d_1 = 0 \\ \left[\dfrac{n_1}{d_1} + \dfrac{d_1 n_0 - d_0 n_1}{p + (d_0/d_1)}\right] u & \text{if } d_1 \neq 0 \end{cases} \tag{14.17}$$

where the two cases in terms of d_1 are of interest because they provide different forms of responses, each of which frequently occurs in first-order circuits. As indicated by (14.17) the transfer function

$$H(p) = \frac{n_1 p + n_0}{d_1 p + d_0} \tag{14.18}$$

is an operator (as a function of the derivative operator p), which characterizes the circuit. Table 14.1 lists some of the more important types of different first-order circuits along with their transfer functions and causal impulse responses.

TABLE 14.1 Typical Transfer Functions of First-Order Circuits

Transfer Function	Description	Impulse Response
$\frac{n_1}{d_0}p$	Differentiator	$\frac{n_1}{d_0}\delta'(t)$
$\frac{n_0}{d_1 p}$	Integrator	$\frac{n_0}{d_1}1(t)$
$\frac{n_1 p + n_0}{d_1}$	Leaky differentiator	$\frac{n_0}{d_1}\delta(t) + \frac{n_1}{d_1}\delta'(t)$
$\frac{n_0}{d_1 p + d_0}$	Low-pass filter; lossy integrator	$\frac{n_0}{d_1}e^{-\frac{d_0}{d_1}t} \cdot 1(t)$
$\frac{n_1 p}{d_1 p + d_0}$	High-pass filter	$\frac{n_1}{d_1}\delta(t) - \frac{n_1 d_0}{d_1^2}e^{-\frac{d_0}{d_1}t} \cdot 1(t)$
$\frac{n_1}{d_1}\frac{p - (d_0/d_1)}{p + (d_0/d_1)}$	All-pass filter	$\frac{n_1}{d_1}\left[\delta(t) - 2\frac{d_0}{d_1}e^{-\frac{d_0}{d_1}t} \cdot 1(t)\right]$

The following treatment somewhat follows that given in [1], although with a slightly different orientation in order to handle all linear time-invariant continuous time continuous circuits.

Zero Input and Zero State Response

The response of a linear circuit is via the linearity the sum of two responses, one due to the input when the circuit is initially in the zero state, called the **zero state response**, and the other due to the initial state when no input is present, the **zero input response**. By the linearity the total response is the sum of the two separate responses, and thus we may proceed to find each separately. In order to investigate these two types of responses we introduce the state vector $x(\cdot)$ and the state-space representation (as above $p = d[\cdot]/dt$)

$$\begin{aligned} px &= Ax + Bu \\ y &= Cx + Du + Epu \end{aligned} \tag{14.19}$$

where A, B, C, D, E are constant matrices. For our first-order circuit two cases are exhibited, depending upon d_1 being zero or not. In the case of $d_1 = 0$,

$$y = (n_1/d_0)u + (n_1/d_0)pu \qquad d_1 = 0 \tag{14.20a}$$

Here, $C = 0$ and A and B can be chosen anything, including empty. When $d_1 \neq 0$ our first-order circuit has the following set of (minimal size) state-variable equations

$$\begin{aligned} px &= \left[-\frac{d_0}{d_1}\right] \cdot x + [d_1 n_0 - d_0 n_1] \cdot u \\ y &= [1] \cdot x + \left[\frac{n_1}{d_1}\right] \cdot u \end{aligned} \qquad d_1 \neq 0 \tag{14.20b}$$

By choosing $u = 0$ in (14.2) we obtain the equations that yield the zero input response. Specifically, the zero input response is

$$y(t) = \begin{cases} 0 & \text{if } d_1 = 0 \\ e^{-\frac{d_0}{d_1}t} \cdot y(0) & \text{if } d_1 \neq 0 \end{cases} \tag{14.21}$$

which also is seen to be true by direct substitution into (14.16). Here, we have set, in the $d_1 \neq 0$ case, the initial value of the state, $x(0)$, equal to the initial value of the output, $y(0)$, which is valid by our choice of state-space equations. Note that (14.21) is valid for all time and y at $t = 0$ assumes the assigned initial value $y(0)$, which must be zero when the input is zero and no derivative occurs on the output.

The zero state response is explained as the solution of (14.21) when $x(0) = 0$. In the case that $d_1 = 0$ the zero state response is

$$y = \frac{n_0}{d_0}u + \frac{n_1}{d_0}pu = \left\{\frac{n_0}{d_0}\delta(t) + \frac{n_1}{d_0}\delta'(t)\right\} * u \qquad d_1 = 0 \tag{14.22a}$$

where $*$ denotes convolution, $\delta(\cdot)$ is the unit impulse, and $1(\cdot)$ is the unit step function. While in the case that $d_1 \neq 0$

$$y = \left\{\frac{n_1}{d_1}\delta(t) + \left[\frac{d_1 n_0 - d_0 n_1}{d_1}\right] e^{-\frac{d_0}{d_1}t} 1(t)\right\} * u \qquad d_1 \neq 0 \tag{14.22b}$$

which is found by eliminating x from (14.20b) and can be checked by direct substitution into (14.16). The terms in the braces are the causal impulse responses, $h(t)$, which are checked by letting $u = \delta$ with otherwise zero initial conditions, that is, with the circuit initially in the zero state. Actually, infinitely many noncausal impulse responses could be used in (14.22b). One such response is found by replacing $1(t)$ by $-1(-t)]$. However, physically the causal responses are of most interest.

If $d_1 \neq 0$, the form of the responses is determined by the constant d_0/d_1, the reciprocal of which (when $d_0 \neq 0$) is called the **time constant**, t_c, of the circuit because the circuit impulse response decays to $1/e$ at time $t_c = d_1/d_0$. If the time constant is positive, the zero input and the impulse responses asymptotically decay to zero as time approaches positive infinity, and the circuit is said to be **asymptotically stable**. On the other hand, if the time constant is negative, then these two responses grow without bounds as time approaches plus infinity, and the circuit is called unstable. It should be noted that as time goes in the reverse direction to minus infinity, the zero input response decays to zero. If $d_0/d_1 = 0$ the zero input and impulse responses are still stable, but neither decay nor grow as time increases beyond zero.

By linearity of the circuit and its state-space equations, the total response is the sum of the zero state response and the zero input response; thus, even when $d_0 = 0$ or $d_1 = 0$

$$y(t) = e^{-\frac{d_0}{d_1}t} y_0 + h(t) * u(t) \tag{14.23}$$

Assuming that u and h are zero for $t < 0$ their convolution is also zero for $t < 0$, although not necessarily at $t = 0$, where it may even take on impulsive behavior. In such a case we see that y_0 is the value of the output instantaneously before $t = 0$. If we are interested only in the circuit for $t > 0$, surprisingly, an input will yield the zero input response. That is, there is an equivalent input u_0 which will yield the zero input response for $t > 0$, this being $u_0(t) = d_1 y_0 \exp(-td_0/d_1)1(t)$. Thus, $y = h * (u + u_0)$ gives the same result as (14.23).

When $d_1 = 0$ the circuit acts as a differentiator and within the state-space framework it is treated as a special case. However, in practice it is not a special case because the current, i, vs. voltage, v, for a capacitor of capacitance C, in parallel with a resistor of conductance G

is described by $i = Cpv + Gv$. Consequently, it is worth noting that all cases can be handled identically in the semistate description

$$\begin{bmatrix} d_1 & d_1 - 1 \\ 0 & 0 \end{bmatrix} px = \begin{bmatrix} -d_0 & -d_0 \\ 0 & 1 \end{bmatrix} x + \begin{bmatrix} n_0 \\ n_1 \end{bmatrix} u \qquad (14.24)$$
$$y = [1 \quad 1]x$$

where $x(\cdot)$ is the semistate rather than the state, although the first components of the two vectors agree in many cases. In other words, the semistate description is more general than the state description, and handles all circuits in a more convenient fashion [2].

Transient and Steady State Responses

This section considers stable circuits, although the techniques are developed so that they apply to other situations. In the asymptotically stable case the zero input response decays eventually to zero; that is, transient responses due to initial conditions eventually will not be felt and concentration can be placed upon the zero state response. Considering first eternal exponential inputs, $u(t) = U$ exp(st) for $-\infty < t < \infty$ at the complex frequency $s = \sigma + j\omega$, where s is chosen as different from the natural frequency $s_n = -d_0/d_1 = -1/t_c$ and U is a constant, we note that the response is $y(t) = Y(s)$ exp(st), as is seen by direct substitution into (14.16); this substitution yields directly

$$Y(s) = \frac{n_1 s + n_0}{d_1 s + d_0} \cdot U \qquad (14.25)$$

where $y(t) = Y(s)\exp(st)$ for $u(t) = U\exp(st)$ over $-\infty < t < \infty$. That is, an exponential excitation yields an exponential response at the same (complex) frequency $s = \sigma + j\omega$ as that for the input. When $\sigma = 0$ the excitation and response are both sinusoidal and the resulting response is called the **sinusoidal steady state,** (SSS). Equation (14.25) shows that the SSS response is found by substituting the complex frequency $s = j\omega$ into the **transfer function,** now evaluated on complex numbers rather than differential operators as above,

$$H(s) = \frac{n_1 s + n_0}{d_1 s + d_0} \qquad (14.26)$$

This transfer function represents the impulse response, $h(t)$, of which it is actually the Laplace transform, and as we found above, the causal impulse response is

$$h(t) = \begin{cases} \dfrac{n_0}{d_0}\delta(t) + \dfrac{n_1}{d_0}\delta'(t), & \text{if } d_1 = 0 \\ \dfrac{n_1}{d_1}\delta(t) + \left[\dfrac{d_1 n_0 - d_0 n_1}{d_1}\right] e^{-\frac{d_0}{d_1}t} 1(t), & \text{if } d_1 \neq 0 \end{cases} \qquad (14.27)$$

However, practical signals are started at some finite time, normalized here to $t = 0$, rather than at $t = -\infty$, as used for the above exponentials. Thus, consider an input of the same type but applied only for $t > 0$; i.e., let $u(t) = U \exp(st)1(t)$. The output is found by using the convolution $y = h * u$; after a slight amount of calculation this convolution is evaluated to

$$y(t) = h(t) * Ue^{st}1(t)$$
$$= \begin{cases} H(s)Ue^{st}1(t) + \dfrac{n_1}{d_0}U\delta(t) & \text{for } d_1 = 0 \\ H(s)Ue^{st}1(t) - \dfrac{[d_1 n_0 - d_0 n_1]}{d_1 s + d_0} Ue^{-\frac{d_0}{d_1}t}1(t) & \text{for } d_1 \neq 0 \end{cases} \quad (14.28)$$

For $t > 0$ the SSS remains present, while there is another term of importance is when $d_1 \neq 0$. This is a transient term, which disappears after a sufficient waiting time in the case of an asymptotically stable circuit. That is, the SSS is truly a steady state, although one may have to wait for it to dominate. If a nonzero zero input response exists, it must be added to the right side of (14.28), but for $t > 0$ this is of the same form as the transient already present, therefore, the conclusion is identical (the SSS eventually predominates over the transient terms for an asymptotically stable circuit).

Because a cosine is the real part of a complex exponential and the real part is obtained as the sum of two terms, we can use linearity of the circuit to quickly obtain the output to a cosine input when we know the output due to an exponential. We merely write the input as the sum of two complex conjugate exponentials and then take the complex conjugates of the outputs that are summed. In the case of real coefficients in the transfer function this is equivalent to taking the real part of the output when we take the real part of the input; that is, $y = \Re(h * u_e) = h * u$, when $u = \Re(u_e)$, if y is real for all real u.

Network Time Constant

The time constant, t_c, was defined earlier as the time for which a transient decays to $1/e$ of the initial value. As such, the time constant shows up in signals throughout the circuit and is a very useful parameter when identifying a circuit from its responses. In a RC circuit the time constant physically results from the interaction of the equivalent capacitor (of which only one exists in a first-order circuit), of capacitance C_{eq}, and the Thevenin's equivalent resistor, of resistance R_{eq}, that it sees. Thus, $t_c = R_{eq}C_{eq}$.

Closely related to the time constant is the **rise time**. Considering the low-pass case the rise time, t_r, is defined as the time for the unit step response to go between 10% and 90% of its final value from its initial value. This is easily calculated because the unit step response is given by

$$y_{1(\cdot)}(t) = h(t) * 1(t) = \frac{n_0}{d_0}\left[1 - e^{-\frac{d_0}{d_1}t}\right] \cdot 1(t) \quad (14.29)$$

Assuming a stable circuit and setting this equal to 0.1 and 0.9 times the final value, n_0/d_0, it is readily found that

$$t_r = \frac{d_1}{d_0} \cdot \ln(9) = [\ln(9)] \cdot t_c \approx 2.2t_c \quad (14.30)$$

At this point it is worth noting that for theoretical studies the time constant can be normalized to 1 by normalizing the time scale. Thus, assuming d_1 and $d_0 \neq 0$ the differential equation can be written as

$$d_0 \cdot \left[\frac{d_1}{d_0} \cdot \frac{dy}{d(d_1/d_0)(t(d_0/d_1))} + y\right] = d_0\left[\frac{dy}{dt_n} + y\right] \tag{14.31}$$

where $t_n = (d_0/d_1)t$ is the normalized time.

ferences

[1] L. P. Huelsman, *Basic Circuit Theory with Digital Computations*, Englewood Cliffs, NJ: Prentice-Hall, 1972.

[2] R. W. Newcomb and B. Dziurla, "Some circuits and systems applications of semistate theory," *Circuits, Systems, and Signal Processing*, vol. 8, no. 3, pp. 235–260, 1989.

.3 Second-Order Circuits

Introduction

Because real transfer functions can be factored into real second-order transfer functions, second-order circuits are probably the most important circuits available; most designs are based upon them. As with first-order circuits, this chapter is limited to single-input–single-output linear time-invariant circuits, and unless otherwise stated, here real-valued quantities are assumed. By definition a **second-order circuit** is described by the differential equation

$$d_2 \cdot \frac{d^2y}{dt^2} + d_1 \cdot \frac{dy}{dt} + d_0 \cdot y = n_2 \cdot \frac{d^2u}{dt^2} + n_1 \cdot \frac{du}{dt} + n_0 \cdot u \tag{14.32}$$

where d_i and n_i are "denominator" and "numerator" constants, $i = 0, 1, 2$, which, unless mentioned to the contrary, are taken to be real. Continuing the notation used for first-order circuits, $y = y(\cdot)$ is the output and $u = u(\cdot)$ is the input; both u and y are generalized functions of time t. Assume that $d_2 \neq 0$, which is the normal case because any of the other special cases can be considered as cascades of real degree one circuits.

Again, treating the derivative as an operator, $p = d[\cdot]/dt$, (14.32) is written as

$$y = \frac{n_2p^2 + n_1p + n_0}{d_2p^2 + d_1p + d_0}u \tag{14.33}$$

with the **transfer function**

$$\begin{aligned} H(p) &= \frac{1}{d_2}\left[\frac{n_2p^2 + n_1p + n_0}{p^2 + (d_1/d_2)p + (d_0/d_2)}\right] \\ &= \frac{1}{d_2}\left[n_2 + \frac{(n_1 - (d_1/d_2)n_2)p + (n_0 - (d_0/d_2)n_2}{p^2 + (d_1/d_2)p + (d_0/d_2)}\right] \end{aligned} \tag{14.34}$$

where the second form results by long division of the denominator into the numerator. Because they occur most frequently when second-order circuits are discussed, we rewrite the denominator in two equivalent customarily used forms:

$$p^2 + \frac{d_1}{d_2}p + \frac{d_0}{d_2} = p^2 + \frac{\omega_n}{Q}p + \omega_n^2 = p^2 + 2\zeta\omega_np + \omega_n^2 \tag{14.35}$$

where ω_n is the undamped natural frequency ≥ 0, Q is the quality factor, and ζ is the damping factor $= 1/(2Q)$. The transfer function is accordingly

$$H(p) = \frac{1}{d_2}\left[\frac{n_2 p^2 + n_1 p + n_0}{p^2 + (\omega_n/Q)p + \omega_n^2}\right] = \frac{1}{d_2}\left[\frac{n_2 p^2 + n_1 p + n_0}{p^2 + 2\zeta\omega_n p + \omega_n^2}\right] \tag{14.36}$$

Table 14.2 lists several of the more important transfer functions, which, as in the first-order case, are operators as functions of the derivative operator p.

TABLE 14.2 Typical Second-Order Circuit Transfer Functions

Transfer Function	Description	Impulse Response
$\frac{n_0}{d_2}\frac{1}{p^2 + 2\zeta\omega_n p + \omega_n^2}$	Low-pass	$h_{\text{lp}}(t) = \frac{n_0}{d_2}\frac{e^{-\zeta\omega_n t}}{\sqrt{1-\zeta^2}\omega_n}\sin\left(\sqrt{1-\zeta^2}\omega_n t\right)1(t)$
$\frac{n_2}{d_2}\frac{p^2}{p^2 + 2\zeta\omega_n p + \omega_n^2}$	High-pass $\theta = \text{arctan2}\left(\frac{\zeta}{\sqrt{1-\zeta^2}}\right)$	$h_{\text{hp}}(t) = \frac{n_2}{d_2}\left[\delta(t) - \frac{\omega_n e^{-\zeta\omega_n t}}{\sqrt{1-\zeta^2}}\sin\left(\sqrt{1-\zeta^2}\omega_n t + 2\theta\right)1(t)\right]$
$\frac{n_1}{d_2}\frac{p}{p^2 + 2\zeta\omega_n p + \omega_n^2}$	Bandpass $\theta = \text{arctan2}\left(\frac{\zeta}{\sqrt{1-\zeta^2}}\right)$	$h_{\text{bp}}(t) = \frac{n_1}{d_2}\frac{e^{-\zeta\omega_n t}}{\sqrt{1-\zeta^2}}\cos\left(\sqrt{1-\zeta^2}\omega_n t + \theta\right)1(t)$
$\frac{n_2}{d_2}\frac{p^2 + \omega_0^2}{p^2 + 2\zeta\omega_n p + \omega_n^2}$	Band-stop	$h_{\text{bs}}(t) = h_{\text{hp}}(t) + \frac{n_2\omega_0^2}{n_0}h_{\text{lp}}(t)$
$\frac{n_2}{d_2}\frac{p^2 - 2\zeta\omega_n p + \omega_n^2}{p^2 + 2\zeta\omega_n p + \omega_n^2}$	All-pass	$h_{\text{ap}}(t) = \frac{n_2}{d_2}\left[\delta(t) - \frac{4\zeta\omega_n e^{-\zeta\omega_n t}}{\sqrt{1-\zeta^2}}\cos\left(\sqrt{1-\zeta^2}\omega_n t + \theta\right)1(t)\right]$
$\frac{n_0}{d_2}\frac{1}{p^2 + \omega_n^2}$	Oscillator,	$h_{\text{osc}}(t) = \frac{n_0}{d_2}\sin(\omega_n t)\cdot 1(t)$
	when $u = 0$	$y(t)\vert_{u=0} = y(0)\cdot\cos(\omega_n t) + \frac{y'(0)}{\omega_n}\cdot\sin(\omega_n t)$

Zero Input and Zero State Response

Again, as in the first-order case, a convenient tool for investigating the time-domain behavior of a second-order circuit is the **state variable description**. Letting the state vector be $x(\cdot)$, the state-space representation is

$$\begin{aligned} px &= Ax + Bu \\ y &= Cx + Du \end{aligned} \tag{14.37}$$

where, as above, $p = d[\cdot]/dt$, and A, B, C, D are constant matrices. In the present case these matrices are real and one convenient choice, among many, is

$$px = \begin{bmatrix} 0 & 1 \\ -\frac{d_0}{d_2} & -\frac{d_1}{d_2} \end{bmatrix} x + \begin{bmatrix} n_1 - \frac{d_1}{d_2}n_2 \\ \left(n_0 - \frac{d_0}{d_2}n_2\right) - \left(n_1 - \frac{d_1}{d_2}n_2\right) \end{bmatrix} u \tag{14.38}$$

$$y = \begin{bmatrix} \frac{1}{d_2} & 0 \end{bmatrix} x + \begin{bmatrix} \frac{n_1}{d_2} \end{bmatrix} u$$

Here, the state is the 2-vector $x = [x_1\ x_2]^{\text{T}}$, with the superscript T denoting transpose. Normally the state would consist of capacitor voltages and/or inductor currents, although at times one

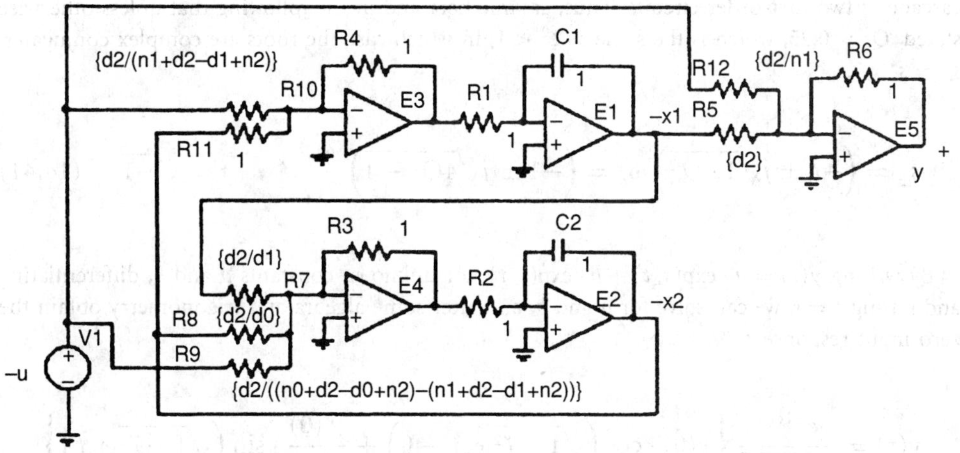

FIGURE 14.2 Generic second-order op-amp RC circuit.

may wish to use linear combinations of these. From these state variable equations a generic operational-amplifier (op-amp) RC circuit to realize any of this class of second-order circuits is readily designed and given in Fig. 14.2. In the figure all voltages are referenced to ground and normalized capacitor and resistor values are listed. Alternate designs in terms of only CMOS differential pairs and capacitors can also be given [3], while a number of alternate circuits exist in the catalog of Sallen and Key [4].

Because (14.38) represents a set of linear constant coefficient differential equations, superposition applies and its solution can again be broken into two parts, the part due to initial conditions, $x(0)$, called the zero input response, and the part due solely to the input u, the zero state response.

The **zero input response** is readily found by solving the state equations with $u = 0$ and initial conditions $x(0)$. The result is $y(t) = C\exp(At)x(0)$, which can be evaluated by several means, including the following. Using a prime to designate the time derivative, first note that when $u = 0$, $x_1(t) = d_2 y(t)$ and $x_1(t)' = x_2(t) = d_2 y(t)'$ (from the first row of A). Thus, $x_1(0) = d_2 y(0)$ and $x_2(0) = d_2 y'(0)$, which allow the initial conditions to be expressed in terms of the measurable output quantities. To evaluate $\exp(At)$, note that its terms are linear combinations of terms whose complex frequencies are zeroes of the characteristic polynomial

$$\begin{aligned}\det(s1_2 - A) &= \det\begin{pmatrix} s & -1 \\ \omega_n^2 & s + 2\zeta\omega_n \end{pmatrix} = s^2 + 2\zeta\omega_n s + \omega_n^2 \\ &= (s - s_-)(s - s_+)\end{aligned} \tag{14.39}$$

For which the roots, called **natural frequencies**, are

$$s_\pm = \left(-\zeta \pm \sqrt{\zeta^2 - 1}\right)\omega_n = \left(-1 \pm \sqrt{1 - 4Q^2}\right)\frac{\omega_n}{2Q} \tag{14.40}$$

The case of equal roots will only occur when $\zeta^2 = 1$, which is the same as $Q^2 = 1/4$, for which the roots are real. Indeed, if the damping factor, ζ, is > 1 in magnitude, or equivalently, if the quality factor, Q, is $< 1/2$ in magnitude, the roots are real and the circuit can be considered a cascade of two first-order circuits. Thus, assume here and in the following that unless otherwise stated, $Q^2 > 0.25$, which is the same as $\zeta^2 < 1$, in which case the roots are complex conjugates, $s_- = s_+^*$

$$s_\pm = \left(-\zeta \pm j\sqrt{1-\zeta^2}\right)\omega_n = \left(-1 \pm j\sqrt{4Q^2-1}\right)\frac{\omega_n}{2Q}, \qquad j = \sqrt{-1} \tag{14.41}$$

By writing $y(t) = a \cdot \exp(s_+ t) + b \cdot \exp(s_- t)$, for unknown constants a and b, differentiating and setting $t = 0$ we can solve for a and b, and after some algebra and trigonometry obtain the zero input response

$$y(t) = \frac{e^{-\zeta\omega_n t}}{\sqrt{1-\zeta^2}}\left\{y(0)\cdot\cos\left(\sqrt{1-\zeta^2}\omega_n t - \theta\right) + \frac{y'(0)}{\omega_n}\cdot\sin\left(\sqrt{1-\zeta^2}\omega_n t\right)\right\} \tag{14.42}$$

where $\theta = \arctan2(\zeta/\sqrt{1-\zeta^2})$ with $\arctan2(\cdot)$ being the arc tangent function which incorporates the sign of its argument.

The form given in (14.42) allows for some useful observations. Remembering that this assumes $\zeta^2 < 1$, first note that if no damping occurs, that is, $\zeta = 0$, then the natural frequencies are purely imaginary, $s_+ = j\omega_n$ and $s_- = -s_+$, and the response is purely oscillatory, taking the form shown in the last line of Table 14.2. If the damping is positive, as it would be for a passive circuit having some loss, usually via positive resistors, then the natural frequencies lie in the left half s-plane, and y decays to zero at positive infinite time so that any transients in the circuit die out after a sufficient wait. The circuit is then called **asymptotically stable.** However, if the damping is negative, as it could be for some positive feedback circuits or those with negative resistance, then the response to nonzero initial conditions increases in amplitude without bound, although in an oscillatory manner, as time increases, and the circuit is said to be **unstable**. In the unstable case, as time decreases through negative time the amplitude also damps out to zero, but usually the responses backwards in time are not of as much interest as those forward in time.

For the **zero state response**, the impulse response, $h(t)$, is convoluted with the input, that is, $y = h * u$, for which we can use the fact that $h(t)$ is the inverse Laplace transform of $H(s) = C[s1_2 - A]^{-1}B$. The denominator of $H(s)$ is $\det(s1_2 - A) = s^2 + 2\zeta\omega_n s + \omega_n^2$, for which the causal inverse Laplace transform is

$$e^{s_+ t}1(t) * e^{s_- t}1(t) = \begin{cases} \dfrac{e^{s_+ t} - e^{s_- t}}{s_+ - s_-}1(t) & \text{if } s_- \neq s_+ \\ te^{s_+ t}1(t) & \text{if } s_- = s_+ \end{cases} \tag{14.43}$$

Here, the bottom case is ruled out when only complex natural frequencies are considered, as with the assumption of handling real natural frequencies in first-order circuits, made above. Consequently,

$$e^{s_+ t}1(t) * e^{s_- t}1(t) = \frac{e^{s_+ t} - e^{s_- t}}{s_+ - s_-}1(t) = \frac{e^{-\zeta\omega_n t}}{\sqrt{1-\zeta^2}\omega_n}\sin\left(\sqrt{1-\zeta^2}\omega_n t\right)\cdot 1(t) \tag{14.44}$$

Again, assuming $\zeta^2 < 1$ using the above calculations give the zero state response as

$$\begin{aligned} y(t) &= \frac{1}{d_2}\left\{\frac{e^{-\zeta\omega_n t}}{\sqrt{1-\zeta^2}\omega_n}\sin\left(\sqrt{1-\zeta^2}\omega_n t\right)1(t)* \right. \\ &\left. \left[\left(n_1 - \frac{d_1}{d_2}n_2\right)\delta'(t) + \left(n_0 - \frac{d_0}{d_2}n_2\right)\delta(t)\right] + n_2\delta(t)\right\} * u(t) \\ &= \frac{1}{d_2}\left\{\frac{e^{-\zeta\omega_n t}}{\sqrt{1-\zeta^2}\omega_n}\sin\left(\sqrt{1-\zeta^2}\omega_n t\right)1(t)* \right. \\ &\left. \left[n_2\delta''(t) + n_1\delta'(t) + n_0\delta(t)\right]\right\} * u(t) \end{aligned} \tag{14.45}$$

The bottom equivalent form is easily seen to result from writing the transfer function $H(p)$ as the product of two terms $1/[d_2(p^2 + 2\zeta\omega_n p + \omega_n^2)]$ and $[n_2p^2 + n_1p + n_0]$ and convoluting the causal impulse response (the inverse of the left half-plane converging Laplace transform), of each term. From (14.45) we directly read the **impulse response** to be

$$\begin{aligned} h(t) = \frac{1}{d_2}\left\{\frac{e^{-\zeta\omega_n t}}{\sqrt{1-\zeta^2}\omega_n}\sin\left(\sqrt{1-\zeta^2}\omega_n t\right)1(t) \right. \\ \left. *[n_2\delta''(t) + n_1\delta'(t) + n_0\delta(t)]\right\} \end{aligned} \tag{14.46}$$

Equations (14.45) and (14.46) are readily evaluated further by noting that the convolution of a function with the second derivative of the impulse, the first derivative of the impulse, and the impulse itself is the second derivative of the function, the first derivative of the function, and the function itself, respectively. For example, in the low-pass case we find the impulse response to be, using (14.46),

$$h_{lp}(t) = \frac{n_0}{d_2}\frac{e^{-\zeta\omega_n t}}{\sqrt{1-\zeta^2}\omega_n}\sin\left(\sqrt{1-\zeta^2}\omega_n t\right)1(t) \tag{14.47}$$

by differentiating once. We then find the bandpass and high-pass impulse responses to be, respectively,

$$h_{bp}(t) = \frac{n_1}{d_2}\frac{e^{-\zeta\omega_n t}}{\sqrt{1-\zeta^2}}\cos\left(\sqrt{1-\zeta^2}\omega_n t + \theta\right)1(t) \tag{14.48}$$

$$h_{hp}(t) = \frac{n_2}{d_2}\left[\delta(t) - \frac{\omega_n e^{-\zeta\omega_n t}}{\sqrt{1-\zeta^2}}\sin\left(\sqrt{1-\zeta^2}\omega_n t + 2\theta\right)1(t)\right] \tag{14.49}$$

In both cases the added phase angle is given, as in the zero input response, via $\theta = \arctan2(\zeta/\sqrt{1-\zeta^2})$. By adding these last three impulse responses suitably scaled the impulse response of the more general second-order circuits are obtained.

Some comments on **normalizations** are worth mentioning in passing. Because $d_2 \neq 0$, one could assume d_2 to be 1 by absorbing its actual value in the transfer function numerator coefficients. If $\omega_n \neq 0$, time could also be scaled so that $\omega_n = 1$ could be taken, in which case a normalized time, t_n, is introduced. Thus, $t = \omega_n t_n$ and, along with normalized time, comes a normalized differential operator $p_n = d[\cdot]/dt_n = d[\cdot]/d(t/\omega_n) = \omega_n p$. This, in turn, leads to a normalized transfer function by substituting $p = p_n/\omega_n$ into $H(p)$. Thus, much of the treatment could be carried out on the normalized transfer function

$$H_n(p_n) = H(p) = \frac{n_{2n}p_n^2 + n_{1n}p_n + n_{0n}}{p_n^2 + 2\zeta p_n + 1} \qquad p_n = \omega_n p \tag{14.50}$$

In this normalized form it appears that the most important parameter in fixing the form of the response is the damping factor $\zeta = 1/(2Q)$.

Transient and Steady State Responses

Let us now excite the circuit with an eternal exponential input, $u(t) = U\exp(st)$ for $-\infty < t < \infty$ at the complex frequency $s = \sigma + j\omega$, where s is chosen as different from either of the natural frequencies, $s_{\pm}$, and U is a constant. As with the first-order and, indeed any higher order case, the response is $y(t) = Y(s)\exp(st)$, as is seen by direct substitution into (14.32). This substitution yields directly

$$Y(s) = \frac{1}{d_2}\left[\frac{n_2 s^2 + n_1 s + n_0}{s^2 + 2\zeta\omega_n s + \omega_n^2}\right]\cdot U \tag{14.51}$$

where $y(t) = Y(s)\exp(st)$ for $u(t) = U\exp(st)$ over $-\infty < t < \infty$. That is, an exponential excitation yields an exponential response at the same (complex) frequency $s = \sigma + j\omega$ as that for the input, as long as s is not one of the two natural frequencies. (s may have positive as well as negative real parts and is best considered as a frequency and not as the Laplace transform variable because the latter is limited to regions of convergence.) Because the denominator polynomial of $Y(s)$ has roots which are the natural frequencies, the magnitude of Y becomes infinite as the frequency of the excitation approaches s_+ or s_-. Thus, the natural frequencies s_+ and s_- are also called **poles** of the transfer function.

When $\sigma = 0$ the excitation and response are both sinusoidal and the resulting response is called the **sinusoidal steady state** (SSS). From (14.51), the SSS response is found by substituting the complex frequency $s = j\omega$ into the transfer function, now evaluated on complex numbers rather than differential operators as above,

$$H(s) = \frac{1}{d_2}\left[\frac{n_2 s^2 + n_1 s + n_0}{s^2 + 2\zeta\omega_n s + \omega_n^2}\right] \tag{14.52}$$

Next, an exponential input is applied, which starts at $t = 0$ rather than at $t = -\infty$; i.e., $u(t) = U\exp(st)1(t)$. Then, the output is found by using the convolution $y = h * u$, which, from the discussion at (14.45) is expressed as

$$y(t) = h * u = \frac{1}{d_2}e^{s_+t}1(t) * e^{s_-t}1(t) * \left[n_2\delta''(t) + n_1\delta'(t) + n_0\delta(t)\right] * e^{st}1(t)$$

$$= H(s)Ue^{st}1(t) + \left\{\frac{1}{d_2(s_+ - s_-)}\left[\left(\frac{N(s)}{s_+ - s} + n_2(s + s_+) + n_1\right)e^{s_+t}\right.\right.$$

$$\left.\left. - \left(\frac{N(s)}{s_- - s} + n_2(s + s_-) + n_1\right)e^{s_-t}\right]1(t)\right\} \tag{14.53}$$

in which $N(s)$ is the numerator of the transfer function and we have assumed that s is not equal to a natural frequency. The second term on the right within the braces varies at the natural frequencies and as such is called the **transient response,** while the first term is the term resulting directly from an eternal exponential, but now with the negative time portion of the response removed. If the system is stable, the transient response decays to zero as time increases and, thus, if we wait long enough the transient response of a stable system can be ignored if the complex frequency of the input exponential has a real part that is greater than that of the natural frequencies. Such is the case for exponentials that yield sinusoids; in that case $\sigma = 0$, or $s = j\omega$. In other words, for an asymptotically stable circuit the output approaches that of the SSS when the input frequency is purely imaginary. If we were to excite at a natural frequency then the first part of (14.53) still could be evaluated using the time-multiplied exponential of (14.43); however, the transient and the steady state are now mixed, both being at the same "frequency."

Because actual sinusoidal signals are real, we use superposition and the fact that the real part of a complex signal is given by adding complex conjugate terms:

$$\cos(\omega t) = \Re[e^{j\omega t}] = \frac{e^{j\omega t} + e^{-j\omega t}}{2} \tag{14.54}$$

This leads to the SSS response for an asymptotically stable circuit excited by $u(t) = U\cos(\omega t)1(t)$ to be

$$\begin{aligned} y(t) &= \frac{H(j\omega)Ue^{j\omega t} + H(-j\omega)U^*e^{-j\omega t}}{2} \\ &= |H(j\omega||U|\cos(\omega t + \angle H(j\omega) + \angle U) \end{aligned} \tag{14.55}$$

Here, we assumed that the circuit has real-valued components such that $H(-j\omega)$ is the complex conjugate of $H(j\omega)$. In which case, the second term in the middle expression is the complex conjugate of the first.

Network Characterization

Although the impulse response is useful for theoretical studies, it is difficult to observe it experimentally due to the impossibility of creating an impulse. However, the unit step response is readily measured, and from it the impulse response actually can be obtained by numerical differentiation if needed. However, it is more convenient to work directly with the unit step response and, consequently, practical characterizations can be based upon it. The treatment most conveniently proceeds from the normalized low-pass transfer function

$$H(p) = \frac{1}{p^2 + 2\zeta p + 1}, \quad 0 < \zeta < 1 \tag{14.56}$$

The **unit step response** follows by applying the input $u(t) = 1(t)$ and noting that the unit step is the special case of an exponential multiplied unit step, where the frequency of the exponential is zero. Conveniently, (14.43) can be used to obtain

$$y_{us}(t) = 1(t) - \frac{e^{-\zeta t}}{\sqrt{1-\zeta^2}}\cos\left(\sqrt{1-\zeta^2}t - \theta\right)\cdot 1(t), \quad \theta = \arctan\left(\frac{\zeta}{\sqrt{1-\zeta^2}}\right) \tag{14.57}$$

Typical unit step responses are shown in Fig. 14.3, where for a small damping factor overshoot can be considerable, with oscillations around the final value, and in addition, a long settling time before reaching the final value. In contrast, with a large damping factor, although no overshoot

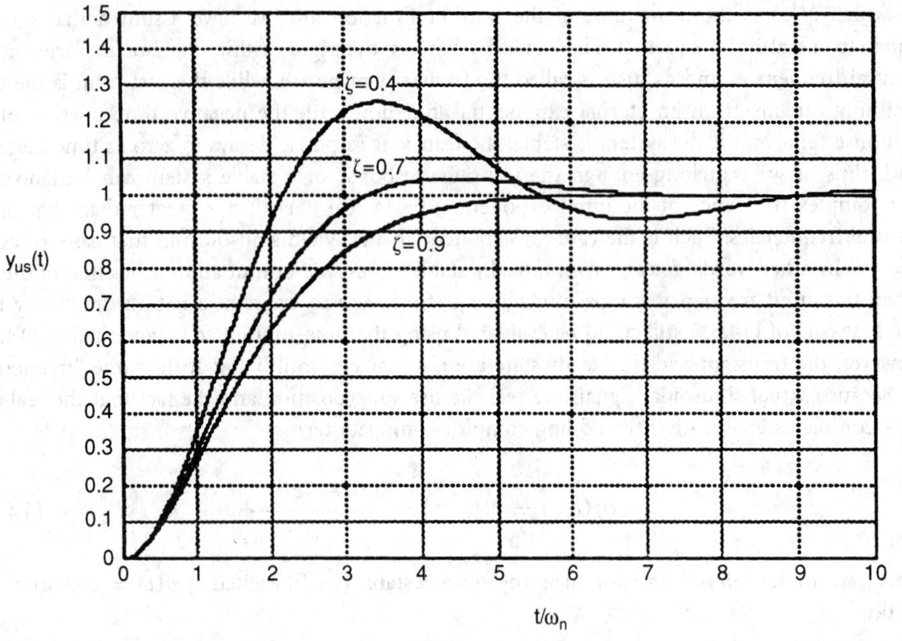

FIGURE 14.3 Unit step response for different damping factors.

or oscillation occurs, the rise to the final value is long. A compromise for obtaining quick rise to the final value with no oscillations is given by choosing a damping factor of 0.7, this being called the **critical value**; i.e., **critical damping** is $\zeta_{crit} = 0.7$.

References

[1] L. P. Huelsman, *Basic Circuit Theory with Digital Computations,* Englewood Cliffs, NJ: Prentice-Hall, 1972.

[2] V. I. Arnold, *Ordinary Differential Equations,* Cambridge, MA: MIT Press, 1983.

[3] J. E. Kardontchik, *Introduction to the Design of Transconductor-Capacitor Filters,* Boston: Kluwer Academic Publishers, 1992.

[4] R. P. Sallen and E. L. Key, "A practical method of designing RC active filters," *IRE Trans. Circuit Theory,* vol. CT-2, no. 1, pp. 74–85, Mar. 1955.

15

Analysis in the Frequency Domain

Vlach
ersity of Waterloo,
da

n Choma, Jr.
esity of Southern
ornia

-Kai Chen
ersity of Illinois, Chicago

.1 Network Functions

Jiri Vlach

Definition of Network Functions

Network functions can be defined if the following constraints are satisfied:

1. The network is linear
2. It is analyzed in the frequency domain using the Laplace transform
3. All initial voltages and currents are zero (zero state conditions)

This chapter will demonstrate how the various functions can be derived, but first we introduce some explanations and definitions. If we analyze any linear network, we can take as output any nodal voltage, or a difference of any two nodal voltages; denote such an output voltage by V_{out}. We can also take as the output a current through any element of the network; we call it output current, I_{out}. If the network is excited by a voltage source, E, then we can also calculate the current delivered into the network by this source; this is the input current, I_{in}. If the network is excited by a current source, J, then the voltage across the current source is the input voltage, V_{in}.

-8341-2/95/$0.00+$.50

Suppose that we analyze the network and keep the letter E or J in our derivations. Then we can define the following network functions:

$$\begin{aligned}
\text{Voltage transfer function,} \quad T_v &= \frac{V_{\text{out}}}{E} \\
\text{Input admittance,} \quad Y_{\text{in}} &= \frac{I_{\text{in}}}{E} \\
\text{Transfer admittance,} \quad Y_{\text{tr}} &= \frac{I_{\text{out}}}{E} \\
\text{Current transfer function,} \quad T_i &= \frac{I_{\text{out}}}{J} \\
\text{Input impedance,} \quad Z_{\text{in}} &= \frac{V_{\text{in}}}{J} \\
\text{Transfer impedance,} \quad Z_{\text{tr}} &= \frac{V_{\text{out}}}{J}
\end{aligned} \tag{15.1}$$

Output impedance or output admittance are also used, but the concept is equivalent to the input impedance or admittance. The only difference is that for calculations the source is placed temporarily at a point from which the output normally will be taken. In the Laplace transform it is common to use capital letters, V for voltages and I for currents. We also deal with impedances, Z, and admittances, Y. Their relationships are

$$V = ZI \qquad I = YV$$

The impedance of a capacitor is $Z_C = 1/sC$, the impedance of an inductor is $Z_L = sL$, and the impedance of a resistor is R. The inverse of these values are admittances: $Y_C = sC$, $Y_L = 1/sL$, and the admittance of a resistor is $G = 1/R$.

To demonstrate the derivations of the above functions two examples are used. Consider the network in Fig. 15.1, with input delivered by the voltage source, E. By Kirchhoff's current law (KCL), the sum of currents flowing *away* from node 1 must be zero:

$$(G_1 + sC_1 + G_2)V_1 - G_2V_2 - EG_1 = 0$$

Similarly, the sum of currents flowing away from node 2 is

$$-V_1G_2 + (G_2 + sC_2 + G_3)V_2 = 0$$

The independent source is denoted by the letter E, and is assumed to be known. In mathematics we transfer known quantities to the right. Doing so and collecting the equations into one matrix equation results in

$$\begin{bmatrix} G_1 + G_2 + sC_1 & -G_2 \\ -G_2 & G_2 + G_3 + sC_2 \end{bmatrix} \begin{bmatrix} V_1 \\ V_2 \end{bmatrix} = \begin{bmatrix} EG_1 \\ 0 \end{bmatrix}$$

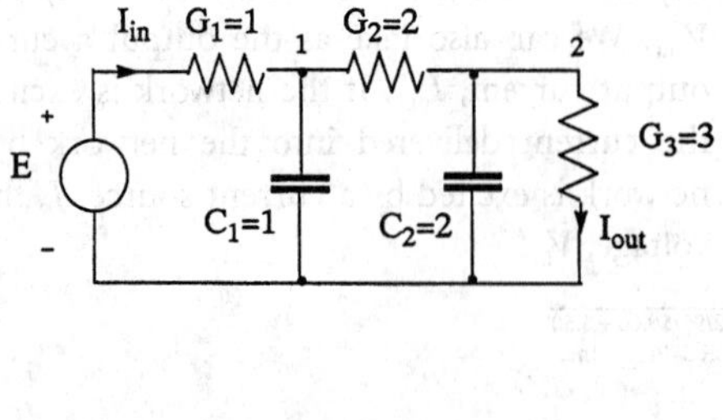

FIGURE 15.1

If numerical values from the figure are used, this system simplifies to

$$\begin{bmatrix} s+3 & -2 \\ -2 & 2s+5 \end{bmatrix} \begin{bmatrix} V_1 \\ V_2 \end{bmatrix} = \begin{bmatrix} E \\ 0 \end{bmatrix}$$

or

$$\boldsymbol{YV} = \boldsymbol{E}$$

Any method can be used to solve this system, but for the sake of explanation it is advantageous to use Cramer's rule. First, find the determinant of the matrix,

$$D = 2s^2 + 11s + 11$$

To obtain the solution for the variable V_1 (V_2), replace the first (second) column of $\boldsymbol{Y}$ by the right-hand side and calculate the determinant of such a modified matrix. Denoting such a determinant by the letter N with an appropriate subscript, evaluate

$$N_1 = \begin{vmatrix} E & -2 \\ 0 & 2s+5 \end{vmatrix} = (2s+5)E$$

Then

$$V_1 = \frac{N_1}{D} = \frac{2s+5}{2s^2+11s+11}E$$

Now divide the equation by E, which results in the voltage transfer function

$$T_v = \frac{V_1}{E} = \frac{2s+5}{2s^2+11s+11}$$

To find the nodal voltage V_2, replace the second column by the elements of the vector on the right-hand side:

$$N_2 = \begin{vmatrix} s+3 & E \\ -2 & 0 \end{vmatrix} = 2E$$

The voltage is

$$V_2 = \frac{N_2}{D} = \frac{2}{2s^2+11s+11}E$$

and another voltage transfer function of the same network is

$$T_v = \frac{V_2}{E} = \frac{2}{2s^2+11s+11}$$

Note that many network functions can be defined for any network. For instance, we may wish to calculate the currents I_{in} or I_{out}, marked in Fig. 15.1. Because the voltages are already known, they are used: For the output current $I_{out} = G_3 V_2$ and dividing by E

$$Y_{tr} = \frac{I_{out}}{E} = \frac{3V_2}{E} = \frac{6}{2s^2+11s+11}$$

The input current $I_{in} = E - G_1V_1 = E - V_1 = E(2s^2 + 9s + 6)/(2s^2 + 11s + 11)$ and dividing by E

$$Y_{in} = \frac{I_{in}}{E} = \frac{2s^2 + 9s + 6}{2s^2 + 11s + 11}$$

In order to define the other possible network functions, we must use a current source, J, as in Fig. 15.2, where we also take the current through the inductor as an output variable. This method of formulating the network equations is called **modified nodal.** The sum of currents flowing away from node 1 is

$$(G_1 + sC_1)V_1 + I_L - J = 0$$

from node 2 it is

$$G_2V_2 - I_L = 0$$

and the properties of the inductor are expressed by the additional equation

$$V_1 - V_2 - sLI_L = 0$$

Inserting numerical values and collecting in matrix form:

$$\begin{bmatrix} s+1 & 0 & 1 \\ 0 & 2 & -1 \\ 1 & -1 & -s \end{bmatrix} \begin{bmatrix} V_1 \\ V_2 \\ I_L \end{bmatrix} = \begin{bmatrix} J \\ 0 \\ 0 \end{bmatrix}$$

The determinant of the system is

$$D = -(2s^2 + 3s + 3)$$

To solve for V_1, we replace the first column by the right-hand side and evaluate the determinant

$$N_1 = \begin{vmatrix} J & 0 & 1 \\ 0 & 2 & -1 \\ 0 & -1 & -s \end{vmatrix} = -(2s + 1)J$$

Then $V_1 = N_1/D$, and dividing by J we obtain the network function

$$Z_{tr} = \frac{V_1}{J} = \frac{2s + 1}{2s^2 + 3s + 3}$$

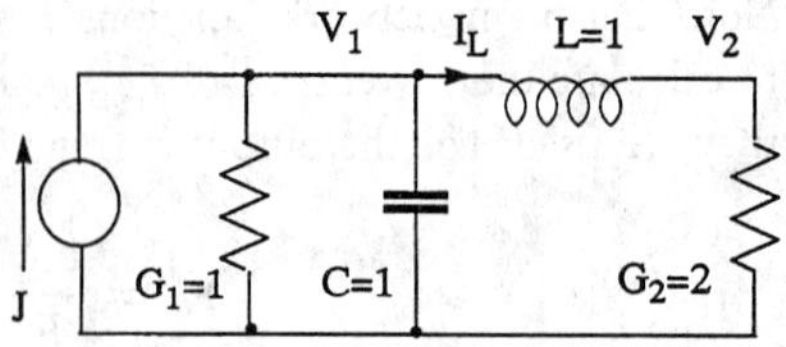

FIGURE 15.2

To obtain the inductor current, evaluate the determinant of a matrix in which the third column is replaced by the right-hand side: $N_3 = -2J$. Then $I_L = N_3/D$ and

$$T_i = \frac{I_L}{J} = \frac{2}{s^2 + 3s + 3}$$

In general,

$$F = \frac{\text{Output variable}}{E \text{ or } J} = \frac{\text{Numerator polynomial}}{\text{Denominator polynomial}} \tag{15.2}$$

Any method that may be used to formulate the equations will lead to the same result. One example shows this is true. Reconsider the network in Fig. 15.2, but use the admittance of the inductor, $Y_L = 1/sL$, and do not consider the current through the inductor. In such a case the nodal equations are

$$\left\{1 + s + \frac{1}{s}\right\} V_1 - \frac{1}{s} V_2 = J$$

$$-\frac{1}{s} V_1 + \left\{\frac{1}{s} + 2\right\} V_2 = 0$$

We can proceed in two ways:

1. We can multiply each equation by s and thus remove the fractions. This provides the system equation

$$\begin{bmatrix} s^2 + s + 1 & -1 \\ -1 & 2s + 1 \end{bmatrix} \begin{bmatrix} V_1 \\ V_2 \end{bmatrix} = \begin{bmatrix} sJ \\ 0 \end{bmatrix}$$

   The determinant of this matrix is $D = 2s^3 + 3s^2 + 3s$. To calculate V_1 find $N_1 = s(2s+1)J$. Their ratio is the same as before because one s in the numerator can be canceled against the denominator.
2. If we do not remove the fractions and go ahead with the solution, we have the matrix equation

$$\begin{bmatrix} s + 1 + 1/s & -1/s \\ -1/s & 1/s + 2 \end{bmatrix} \begin{bmatrix} V_1 \\ V_2 \end{bmatrix} = \begin{bmatrix} J \\ 0 \end{bmatrix}$$

   The determinant is $D = 2s + 3 + 3/s$ and the numerator for V_1 is $N_1 = (1/s + 2)J$. Taking their ratio

$$V_1 = \frac{(1/s + 2)J}{2s + 3 + 3/s} = \frac{2s + 1}{2s^2 + 3s + 3} J$$

   which is the same result as before.

We conclude that it does not matter which method is used to formulate the equations. The result is always a ratio of two polynomials in the variable s.

Many additional conclusions can be drawn from these examples. The most important result so far is that **all network functions of any given network have the same denominator**. It was easy to discover this property because we used Cramer's rule, with its evaluation by the ratio of two determinants. It should be mentioned at this point that we may have network functions in which some terms of the numerator can cancel against the same terms of the denominator. Such a cancellation represents a mathematical simplification which does not change the validity of the above statement.

Occasionally, the network may have more than one source. In such cases we apply the superposition principle of linear networks. The contribution to the output can be calculated separately for each source and the results added. All that must be done is to correctly remove those sources which are not considered at the moment. All *unused independent voltage sources* must be replaced by *short circuits.* All *unused independent current sources* are replaced by *open circuits* (removed from the network). Although we did not use dependent sources in our examples, it is necessary to stress that such removal *must not* take place for dependent sources.

Network functions can be used to find responses to any given input signal. First, multiply the network function by E or J; this will give the expression for the output. Afterward, the letter E or J is replaced by the Laplace transform of the signal. For instance, if the signal is a unit step, then the source is replaced by $1/s$. If it is $\cos \omega t$, then the source is replaced by the Laplace transform, $s/(s^2 + \omega^2)$, and so on.

In the Laplace transform one special signal exists, the Dirac impulse, commonly denoted by $\delta(t)$. It can be represented as a rectangular pulse having width T and height $1/T$. The area of the pulse is always 1, even if we go to $\lim T \rightarrow 0$, which is the Dirac impulse. Its Laplace transform is 1. Because multiplication by 1 does not change the network function, we conclude that any network function is also the Laplace transform of the network response to the Dirac impulse.

A word of caution: In the network function always divide by the *independent* voltage (current) source. We cannot take two analysis results, for instance V_1 and V_2, derived for Fig. 15.1, and take their ratio. This will not be a network function.

Poles and Zeros

Networks with lumped elements have network functions which are always ratios of two polynomials with real coefficients. For some applications the polynomials may be expressed as functions of some (or all) elements, but the principle is unchanged.

Since we have a ratio of two polynomials, the network function can be written in two forms:

$$F = \frac{\sum_{i=0}^{M} a_i s^i}{\sum_{i=0}^{N} b_i s^i} = K \frac{\prod_{i=1}^{M} (s - z_i)}{\prod_{i=1}^{N} (s - p_i)} \tag{15.3}$$

The middle form is what we obtain from analyses similar to those in the examples. Algebraically, a polynomial of order N has exactly N roots. This leads to the form on the right. The multiplicative constant, K, is the ratio

$$K = \frac{a_M}{b_N}$$

and is obtained by dividing each polynomial by the coefficient of its highest power.

It is easy to find roots of a first- and second-order polynomial because formulas are available, but in all other cases iterative methods and a computer are utilized. However, even without actually finding the roots, we can draw a number of important conclusions.

If the highest power of the polynomial is odd, then at least one real root will exist. The other roots may be either real or complex, but if they are complex, then they always appear in complex conjugate pairs. The roots of the numerator are called *zeros*, and those of the denominator are called *poles*. We denote the zeros by

$$z_i = a_i + jb_i$$

where $j = \sqrt{-1}$. Either a or b may be zero. For the poles we have similarly

$$p_i = c_i + jd_i$$

The polynomial also may have multiple roots. For instance, the polynomial $P(s) = (s+1)^2(s+2)^3$ has a double root at $s = -1$ and a triple root at $s = -2$. The positions of the poles and zeros, with the constant K, completely define the network function and also all network properties. The positions can be plotted in a complex plane, the zeros indicated by small circles and poles by crosses. A multiple pole (zero) is indicated by a number appearing at the cross (circle). Figure 15.3 shows a network function with two complex conjugate zeros on the imaginary axis, two complex conjugate poles, and one double real pole.

As derived above, all network functions of any given network have the same poles. Their positions depend only on the structure of the network and are independent of the signal or where the signal is applied. Because of this fundamental property, the poles are also called *natural frequencies* of the network. The zeros depend on the place at which we attach the source and also on the point where we take the output.

It is possible to have networks in which a pole is in exactly the same position as a zero; mathematically, such terms cancel. Figure 15.4 is an example. Writing the sum of currents at nodes 1, 2, and 3, we obtain

$$(2s+2)V_1 - (s+3)V_3 = sE$$
$$(2s+2)V_2 - V_3 = E$$
$$-sV_1 - V_2 + (s+3)V_3 = 0$$

and in matrix form

$$\begin{bmatrix} 2s+2 & 0 & -(s+3) \\ 0 & 2s+2 & -1 \\ -s & -1 & s+3 \end{bmatrix} \begin{bmatrix} V_1 \\ V_2 \\ V_3 \end{bmatrix} = \begin{bmatrix} sE \\ E \\ 0 \end{bmatrix}$$

By carefully evaluating the determinant we discover that we can keep the term $(2s+2)$ separate and get $D = (2s+2)(s^2+2s+5)$. Replacing the third column by the right-hand side we calculate

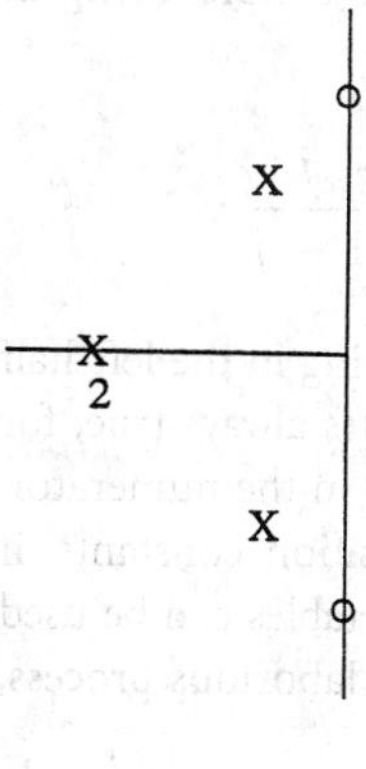

FIGURE 15.3

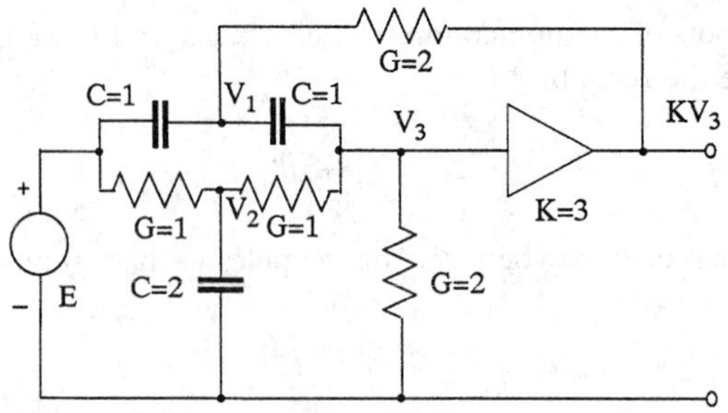

FIGURE 15.4

the numerator $N_3 = (2s + 2)(s^2 + 1)E$. Because the output is KV_3, the voltage transfer function is

$$T_v = \frac{3(2s + 2)(s^2 + 1)}{(2s + 2)(s^2 + 2s + 5)}$$

Mathematically, the term $(2s + 2)$ cancels and the network function is sometimes written as

$$T_v = \frac{3(s^2 + 1)}{s^2 + 2s + 5}$$

Such cancellation makes the denominator different from other network functions that we might derive for the same network, but it is not a correct way to describe the properties of the network. The cancellation gives the impression that we have a second-order network, while it is actually a third-order network.

Network Stability

Stability of the network depends entirely on the positions of its poles. The following is a list of the conditions in order for the network to be stable, with subsequent explanation of the reasons.

1. The network is stable if all its poles are in the left half of the complex plane.
2. The network is unstable if at least one of its poles is in the right-half plane.
3. The network is marginally stable if all its poles are simple and exactly on the imaginary axis.
4. The network is unstable if it has all poles on the imaginary axis, but at least one of them has multiplicity two or more.

Courses on mathematics teach the process of decomposing a rational function into partial fractions. We show an example with one simple real pole and a pair of simple complex conjugate poles,

$$F(s) = \frac{3s^2 + 8s + 6}{(s + 1)(s^2 + 2s + 2)} = \frac{1}{s + 1} + \frac{1 + j}{s + 1 + j} + \frac{1 - j}{s + 1 - j}$$

The poles are $p_1 = -1$ and $p_{2,3} = -1 \pm j$, all with negative real parts and all lying in the left-half plane. Partial fraction decomposition is on the right of the above equation. It is always true, for any lumped network, that the decomposition for a real pole has a real constant in the numerator. Complex poles always appear in complex conjugate pairs and the decomposition constants, if complex, also are complex conjugate. Once such a decomposition is available, tables can be used to invert the functions into time domain. The decomposition may be quite a laborious process, however, only a few types of terms need be considered for lumped networks. All are collected in Table 15.1. Each time domain expression is multiplied by unit step, $u(t)$, which is zero for $t < 0$ and is one for $t \geq 0$. Such multiplication correctly expresses the fact that the time functions start at $t = 0$.

Formula one in Table 15.1 shows that a real, single pole in the left-half plane will lead to a time-domain function which decreases as e^{-ct}. This response is called stable. If $c = 0$, then the response becomes $u(t)$. Should the pole be in the right-half plane, the exponent will be positive and e^{ct} will grow rapidly and without bound. This network is said to be unstable.

Formula two shows what happens if the pole is real, with multiplicity n. If it is in the left-half plane, then t^{n-1} is a growing function, but e^{-ct} decreases faster, and for large t the result tends to zero. The function is still stable.

Formula three considers the case of two simple complex conjugate poles. Their real parts influence the exponent, and the imaginary parts contribute to oscillations. If the real part is negative, the oscillations will be damped, the response will become zero for large t, and the network will be stable. If the real part is zero, then the oscillations continue indefinitely with constant amplitude. For positive real part, the network becomes unstable.

Formula four considers a pair of multiple complex conjugate poles. As long as the real part is negative, the oscillations will decrease with time and the network will be stable. If the real part is zero or positive, the network is unstable because the oscillations will grow.

Initial and Final Value Theorems

Finding the poles and evaluating the time domain responses is a complicated process, which normally requires the use of a computer. It is, therefore, advisable to use all possible steps that may provide information about the network behavior without actually finding the full time-domain response.

Two Laplace transform theorems help in finding how the network behaves at $t = 0$ and at $t \to \infty$. Both theorems are derived from the Laplace transform formula for differentiation,

$$\int_{0^-}^{\infty} f'(t)e^{-st}\,dt = sF(s) - f(0^-) \tag{15.4}$$

where 0^- indicates that we are considering the instant just before the signal is applied. If we let $s \to 0$, then $e^0 = 1$, and the integral of the derivative becomes the function itself. Inserting the integration limits we get

$$f(\infty) - f(0^-) = \lim_{s\to 0}[sF(s) - f(0^-)]$$

Cancelling $f(0^-)$ on both sides we arrive at the *final value theorem*

$$\lim_{t\to\infty} f(t) = \lim_{s\to 0} sF(s) \tag{15.5}$$

TABLE 15.1

Formula	Laplace Domain	Time Domain
1	$\dfrac{K}{s+c}$	$Ke^{-ct}u(t)$
2	$\dfrac{K}{(s+c)^n}$	$K\dfrac{t^{n-1}}{(n-1)!}e^{-ct}u(t)$
3	$\dfrac{A+jB}{s+c+jd} + \dfrac{A-jB}{s+c-jd}$	$2e^{-ct}(A\cos dt + B\sin dt)u(t)$
4	$\dfrac{A+jB}{(s+c+jd)^n} + \dfrac{A-jB}{(s+c-jd)^n}$	$\dfrac{2t^{n-1}}{(n-1)!}e^{-ct}(A\cos dt + B\sin dt)u(t)$

Another possibility is to let $s \to \infty$; then e^{-st} in (15.4) will be zero and the whole left side becomes zero. This can be written as

$$0 = \lim_{s\to\infty} [sF(s) - f(0^-)]$$

and because $f(0^-)$ is nothing but the limit of $f(t)$ for $t \to 0^-$, we obtain the *initial value theorem*

$$\lim_{t\to 0} f(t) = \lim_{s\to\infty} sF(s) \tag{15.6}$$

Note the similarity of the two theorems; we will apply them to the function used in the previous section. Consider

$$sF(s) = \frac{3s^3 + 8s^2 + 6s}{s^3 + 3s^2 + 4s + 2}$$

If we take any large value of s, the highest powers will dominate and in the limit, for $s \to \infty$, we get 3. This is the value of the time-domain response at $t = 0$. The limit for $s = 0$ is zero, and from the final value theorem we know that $f(t)$ will be zero for $t \to \infty$.

To extract still more information, use the example collected in Table 15.2. Scrolling down the table, each Laplace domain function is s times that above it. Each multiplication by s means differentiation in the time domain, as follows from (15.4). Scrolling down the second column of Table 15.2, each function is the derivative of that above it. To apply the limiting theorems, take the Laplace domain formula, which is one level lower, and insert the limits. The limiting is also shown and is confirmed by inserting either $t = 0$ or $t \to \infty$ into the time functions.

While the two theorems are useful, the final value theorem is valid *only if the function is stable.* Consider the unstable function with two poles in the right-half plane

$$F_1(s) = \frac{1}{(s+1)(s-1+j)(s-1-j)} = \frac{1}{s^3 - s^2 + 2}$$

Its time domain response is

$$f_1(t) = \frac{1}{5}[e^{-t} + e^{+t}(2\sin^t - \cos^t)]u(t)$$

and the term e^{+t} will cause the function to grow for large t. If the final value theorem is applied, we consider

$$sF_1(s) = \frac{s}{s^3 - s^2 + 2}$$

Inserting $s = 0$, the theorem predicts that the time function will approach zero for large t. This is disappointing, but some additional simple rules can be applied. The function is unstable if

TABLE 15.2

Laplace Domain $D = s^3 + 3s^2 + 4s + 2$	Time Domain	$s \to \infty$ $t = 0$	$s = 0$ $t \to \infty$
$F(s) = 1/D$	$f(t) = e^{-t}(1 - \cos t)u(t)$	0	0
$G(s) = sF(s) = s/D$	$g(t) = f'(t) = e^{-t}(-1 + \cos t + \sin t)u(t)$	0	0
$H(s) = s^2F(s) = s^2/D$	$h(t) = f''(t) = e^{-t}(1 - 2\sin t)u(t)$	1	0
$K(s) = s^3/D$	$k(t) = f'''(t) = \delta(t) + e^{-t}(2\sin t - 2\cos t - 1)u(t)$	$\delta(t)$	0

some coefficients of the denominator are missing, or if the denominator coefficients do not all have the same sign (all + or all−). Such situations are easily detected, but if all coefficients have the same sign, nothing can be said about stability. Additional theorems exist (Hurwitz theorem), but if in doubt, it is probably simplest to go to the computer and find the poles.

15.2 Advanced Network Analysis Concepts

John Choma, Jr.

Introduction

The systematic analysis of an electrical or electronic network entails formulating and solving the relevant Kirchhoff equations of equilibrium. This analysis is conducted to acquire a theoretically sound understanding of circuit responses. Such an understanding minimally delineates the dynamical effects of topology, controllable circuit branch variables, and observable parameters for active devices embedded in the circuit. It also illuminates circuit node and branch impedances to which the relevant responses of the circuit undergoing investigation are especially sensitive. Unfortunately, the complexity of modern networks, and particularly integrated analog electronic circuits, often inhibits the mathematical tractability that underpins an engineering understanding of circuit behavior. It is therefore not surprising that when mathematical analyses accompany a computer-assisted circuit design venture, the subcircuits identified for manual study are simplified representations of the corresponding subcircuits in the draft design solution. Unless care is exercised, these approximations can mask a satisfying understanding, and they can even lead to erroneous results.

Analytical and modeling approximations notwithstanding, the key to assimilating a satisfying understanding of the electrical characteristics of complex circuits is appropriate studies of simpler partitions of these circuits. To this end, Kron [1, 2] has provided, and others have explained and reinforced [3–5], an elegant theory that allows the circuit response solutions of these network partitions to be coalesced so that the desired response of the interconnected circuit is reconstructed exactly. Aside from formalizing an analytical mechanism for studying complicated circuits in terms of the solutions gleaned for more manageable subcircuits of the composite network [6], Kron's work allows for a computationally efficient study of feedback network responses. The theory also allows for the investigation of the sensitivity of overall network performance with respect to both small and large parametric changes [7]. In view of the exclusive focus on linear circuits in this section, it is worth interjecting that a form of Kron's partitioning theory is also applicable to certain classes of nonlinear circuits [8].

Fundamental Network Analysis Concepts

The derivation of Kron's formula, as well as the development of a general methodology for applying Kron's partitioning mechanism to the analyses of complex circuits, requires a fundamental understanding of the classical techniques exploited in the analysis of linear networks. Such an understanding begins by considering the $(n + 1)$ node, b branch linear network abstracted in Fig. 15.5(a). The input port, which is defined by the node pair, 1-2, is excited by a signal source whose Thévenin voltage is V_S and whose Thévenin impedance is Z_S. In response to this excitation, a load voltage, V_L, is developed across a load impedance, Z_L, which shunts the output port consisting of the node pair, 3-4. Two other nodes, nodes m and p, are explicitly delineated for future reference. In response to the applied signal source, the voltage across the input port is V_I, while the voltage established across the node pair, m-p, is V_k. In addition, the reference, or ground, node is labeled node 0. Either node 2, node 4, or both of these nodes can be incident with the ground node; that is, the signal source and/or the load impedance can be terminated to the network ground. The diagram in Fig. 15.5(b) is identical to that of Fig. 15.5(a) except for the fact that the applied signal source is represented by its Norton equivalent circuit, where the Norton signal current, I_S, is

$$I_S = \frac{V_S}{Z_S} \tag{15.7}$$

Assuming that a nodal admittance matrix exists for the linear $(n + 1)$ node network at hand, the n equilibrium Kirchhoff current law (KCL) equations can be expressed as the matrix relationship

$$\mathbf{J} = \mathbf{YE} \tag{15.8}$$

where $\mathbf{J}$ is an n-vector whose ith entry, J_i, is an independent current flowing into the ith circuit node, $\mathbf{E}$ is an n-vector of node voltages such that its ith entry, E_i, is the ith node voltage referenced to network ground, and $\mathbf{Y}$, a square matrix of order n, is the nodal admittance matrix of the circuit. If $\mathbf{Y}$ is nonsingular, the node voltages follow as

$$\mathbf{E} = \mathbf{Y}^{-1}\mathbf{J} \tag{15.9}$$

Note that (15.9) is useful symbolically, but not necessarily computationally. In particular, (15.9) shows that the n node voltages of the $(n + 1)$ node, b branch network of Fig. 15.5 can be straightforwardly computed in terms of the known independent current source vector and the parameters embedded in the network nodal admittance matrix. In an actual analytical environment, however, the nodal admittance matrix is rarely formulated and inverted. Instead, some or all of the n node voltages of interest are determined merely by algebraically manipulating and solving either the n independent KCL equations or the $(b - n)$ independent Kirchhoff voltage law (KVL) equations that are required to establish the equilibrium conditions of the subject network.

If the n vector, $\mathbf{E}$, is indeed evaluated, all n independent node voltages are known, because

$$\mathbf{E}^{\mathrm{T}} = [E_1, E_2, E_3, \ldots, E_m, \ldots, E_p, \ldots, E_n] \tag{15.10}$$

where the superscript T indicates the operation of *matrix transposition.* In general, E_i, for $i = 1, 2, \ldots, n$, is the voltage developed at node i with respect to ground. It follows that the

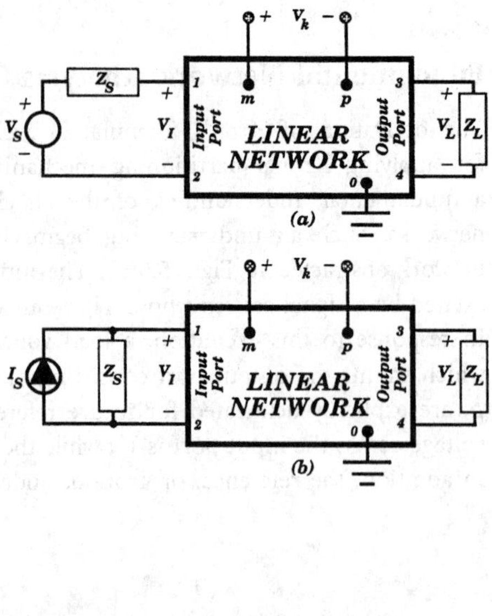

FIGURE 15.5 (a) Generalized linear network driven by a voltage source. (b) The network of (a), but with the signal excitation represented by its Norton equivalent circuit.

voltage between any two nodes derives directly from the network solution inferred by (15.9). For example, the input port voltage, V_I, is $(E_1 - E_2)$, the output port voltage, V_L, is $(E_3 - E_4)$, and the voltage, V_k, from node m to node p is $V_k = (E_m - E_p)$.

The calculation of the voltage appearing between any two circuit nodes can be formalized with the help of the generalized network diagrammed in Fig. 15.6 and through the introduction of the *connection vector* concept. In particular, let $\mathbf{A}_{ij}$ denote the $(n \times 1)$ connection vector for the port defined by the node pair, i-j. Moreover, let the voltage, V, at node i be taken as positive with respect to node j, and allow a current, I (which may be zero), to flow into node i and out of node j, as indicated in the diagram. Then, the elements of the connection vector, $\mathbf{A}_{ij}$, are all zero except for a $+1$ in its ith row and a -1 in its jth row. If node j is the reference node, all elements of $\mathbf{A}_{ij}$, which in this case can be written simply as $\mathbf{A}_i$, are zero except for the ith row element, which remains $+1$. Thus, $\mathbf{A}_{ij}$ has the form

$$\mathbf{A}_{ij}^{\mathrm{T}} = [0 \;\; 0 \;\; \cdots \;\; \underset{\uparrow\, i\text{th column}}{+1} \;\; \cdots \;\; \underset{\uparrow\, j\text{th column}}{-1} \;\; \cdots \;\; \underset{\uparrow\, n\text{th column}}{0}] \tag{15.11}$$

For the special case in which a circuit branch element interconnects every pair of circuit nodes, $\mathbf{A}_{ij}$ is the appropriate column of the *node to branch incidence matrix*, which is a rectangular matrix of order $(n \times b)$, for the $(n + 1)$ node, b branch network at hand [9].

Returning to the calculation of V_I, V_L, and V_k, it follows from (15.9) through (15.11) that

$$V_I = E_1 - E_2 = \mathbf{A}_{12}^{\mathrm{T}}\mathbf{E} = \mathbf{A}_{12}^{\mathrm{T}}\mathbf{Y}^{-1}\mathbf{J} \tag{15.12}$$

$$V_L = E_3 - E_4 = \mathbf{A}_{34}^{\mathrm{T}}\mathbf{E} = \mathbf{A}_{34}^{\mathrm{T}}\mathbf{Y}^{-1}\mathbf{J} \tag{15.13}$$

and

$$V_k = E_m - E_p = \mathbf{A}_{mp}^{\mathrm{T}}\mathbf{E} = \mathbf{A}_{mp}^{\mathrm{T}}\mathbf{Y}^{-1}\mathbf{J} \tag{15.14}$$

Assuming that I_S is the only independent source of excitation in the network of Fig. 15.5

$$\mathbf{J} = \mathbf{A}_{12}I_S \tag{15.15}$$

which is the mathematical equivalent of the observation that the Norton source current, I_S, is entering node 1 of the network and leaving node 2. Accordingly,

$$V_I = (\mathbf{A}_{12}^{\mathrm{T}}\mathbf{Y}^{-1}\mathbf{A}_{12})I_S \tag{15.16}$$

$$V_L = (\mathbf{A}_{34}^{\mathrm{T}}\mathbf{Y}^{-1}\mathbf{A}_{12})I_S \tag{15.17}$$

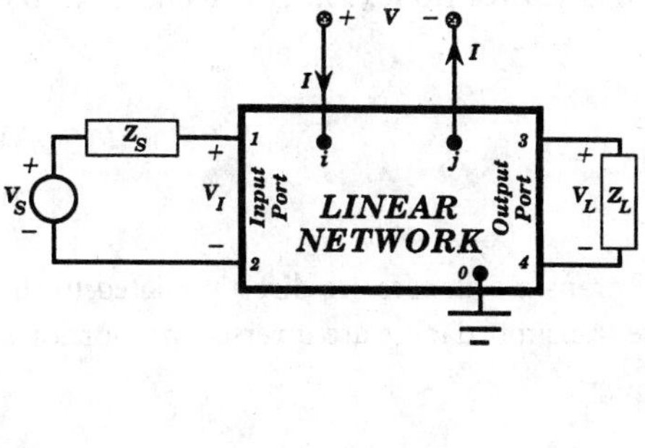

FIGURE 15.6 Generalized network diagram used to define the connection vector concept.

and

$$V_k = (\mathbf{A}_{mp}^{\mathrm{T}}\mathbf{Y}^{-1}\mathbf{A}_{12})I_S \tag{15.18}$$

Several noteworthy features are implicit to the foregoing three relationships. First, each of the three parenthesized matrix products on the right-hand sides of the equations is a scalar. This observation follows from the facts that a transposed connection vector is a row matrix of order $(1 \times n)$, the inverse nodal admittance matrix is an n-square, and a connection vector is an n-vector. Second, these scalar products represent transimpedances from the input port to the port at which the voltage of interest is extracted. In the case of (15.16), the ratio, $V_l\ I_S$, is actually the impedance, Z_{SS}, seen by the Norton current, I_S; that is,

$$Z_{SS} \triangleq \frac{V_S}{I_S} = (\mathbf{A}_{12}^{\mathrm{T}}\mathbf{Y}^{-1}\mathbf{A}_{12}) \tag{15.19}$$

where, as asserted previously, I_S is presumed to be the only source of energy applied to the network undergoing study. Similarly,

$$Z_{LS} \triangleq \frac{V_L}{I_S} = (\mathbf{A}_{34}^{\mathrm{T}}\mathbf{Y}^{-1}\mathbf{A}_{12}) \tag{15.20}$$

is the transimpedance from the input port to the output port, while

$$Z_{kS} \triangleq \frac{V_k}{I_S} = (\mathbf{A}_{mp}^{\mathrm{T}}\mathbf{Y}^{-1}\mathbf{A}_{12}) \tag{15.21}$$

is the transimpedance from the input port to the port defined by the node pair, m-p.

The impedance in (15.19) and the transimpedances given by (15.20) and (15.21) are cast as explicit algebraic functions of the inverse of the network nodal admittance matrix. However, like the node voltages in (15.9) and (15.10), network transimpedances are rarely calculated manually through an actual delineation and inversion of the nodal admittance matrix. Instead, they usually derive from a straightforward analysis of the considered network, subject to the proviso that all excitations applied to the subject network, save for a single test current source, are reduced to zero. For example, in the abstraction shown in Fig. 15.7, the transimpedance, Z_{ij}, from any port j to any port i is

$$Z_{ij} \triangleq \left.\frac{V_{\text{test}}}{I_{\text{test}}}\right|_{\text{all independent sources}=0} \tag{15.22}$$

For the case of $j = i$, this transimpedance becomes the effective impedance seen at port i by the test current source. In view of the preceding discussion, and the node pairs indicated in Fig. 15.7, the transimpedance (or impedance) quantity that derives from (15.22) is identical to the matrix relationship

$$Z_{ij} = (\mathbf{A}_{cd}^{\mathrm{T}}\mathbf{Y}^{-1}\mathbf{A}_{ab}) \tag{15.23}$$

The last result highlights the fact that all network transimpedances are directly related to the inverse of the nodal admittance matrix. Hence, these transimpedances are inversely proportional

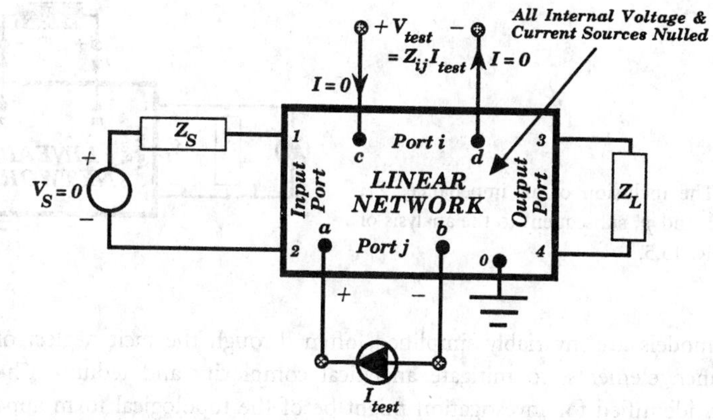

FIGURE 15.7 An illustration of a practical manual technique for computing the transimpedance between any port j to any port i of a linear network.

to the determinant, $\Delta Y(s)$, of the admittance matrix, $\mathbf{Y}$. It follows that the poles of all transimpedances and effective port impedances are the roots of the characteristic polynomial

$$\det(\mathbf{Y}) \triangleq \Delta Y(s) = 0 \tag{15.24}$$

Note from (15.7), (15.17), and (15.20) that the voltage gain of the considered linear network is

$$\frac{V_L}{V_S} = \frac{Z_{LS}}{Z_S} \tag{15.25}$$

Thus, if the source impedance in Fig. 15.5 is a real number, $Z_S = R_S$, the roots of (15.24) also comprise the poles of the voltage transfer function and, indeed, of the linear network.

Kron's Formula

Assume now that the network depicted in Fig. 15.5 has been analyzed in the sense that all network node voltages developed in response to the signal source have been determined. Assume further that subsequent to this analysis, an impedance, Z_k, is appended to nodes m and p, as shown in Fig. 15.8. In addition to causing a current, I, to flow into node m and out of node p, this additional branch element is likely to perturb the values of all of the originally computed circuit node and circuit branch voltages. The matrix, $\mathbf{E}'$, of new node voltages can be evaluated for the modified topology in Fig. 15.8 by determining the new nodal admittance matrix, $\mathbf{Y}'$, and then reapplying (15.9). The tedium associated with a second network analysis, along with the inefficiency of discarding the results of a study performed on a network whose topology differs only modestly from that of the original configuration, can be circumvented through the use of *Kron's theorem*. As illuminated below, this theorem derives from a methodical application of such classical concepts as the theories of superposition, substitution, and Thévenin. In addition to providing a computationally efficient mechanism for determining $\mathbf{E}'$, Kron's technique allows for a direct comparison of $\mathbf{E}'$ to the matrix, $\mathbf{E}$, of original node voltages. It therefore allows for a convenient response sensitivity analysis with respect to the appended branch element.

It is appropriate to interject that the problem postulated above possesses more than mere academic interest. It is, in fact, a problem that is commonly encountered, for example, in the analysis of electronic circuits. In order to linearize these circuits around specified quiescent operating points, it is necessary to supplant the utilized active devices by small signal equivalent

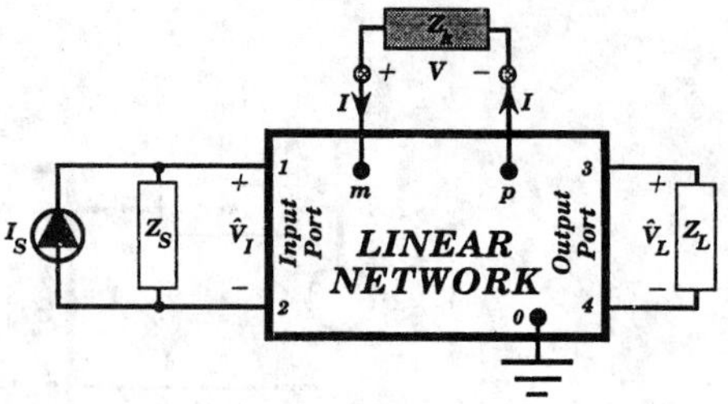

FIGURE 15.8 The inclusion of an impedance, Z_k, between nodes m And p, subsequent to the analysis of the network in Fig. 15.5.

circuits. Such models are invariably simplified, often through the tacit neglect of presumably noncritical branch elements, to mitigate analytical complexity and tedium. Thus, while the circuit properly identified for investigation might be of the topological form appearing in Fig. 15.8, the circuit actually subjected to manual circuit analysis is likely the reduced structure depicted in Fig. 15.5; that is, the ostensibly noncritical impedance, Z_k, is removed in the interest of analytical tractability. Questions naturally arise in regard to the degree of error incurred by the invoked circuit simplification. Kron's method, as developed below, answers these questions in terms of the results already deduced for the approximate network and without requiring explicit analytical results for the "exact" network.

The process of evaluating the perturbation on network node voltages incurred by the action of shunting nodes m and p in the circuit of Fig. 15.5 by the impedance Z_k begins by determining the Thévenin equivalent circuit that drives the appended branch. To this end, Z_k is removed in the diagram of Fig. 15.8, thereby collapsing the network to Fig. 15.5(a). The relevant Thévenin voltage, V_{th}, at the node pair, m-p, is, in fact, V_k, as defined by (15.18). Recalling (15.21), this voltage is

$$V_{th} \equiv V_k = (\mathbf{A}_{mp}^{T}\mathbf{Y}^{-1}\mathbf{A}_{12})I_S = Z_{kS}I_S \tag{15.26}$$

The corresponding Thévenin impedance, Z_{th}, derives from a study of the test configuration of Fig. 15.9, in which the independent source current, I_S, is nulled, the impedance, Z_k, in Fig. 15.8 is replaced by a test current of value I_{test}, and the ratio of the resultant port voltage, V_{test}, to I_{test} is understood to be the desired Thévenin impedance. For this configuration, the network nodal admittance matrix, $\mathbf{Y}$, is unchanged, but the independent network current vector, $\mathbf{J}$, becomes $\mathbf{A}_{mp}I_{test}$. Thus, the resultant n-vector, $\mathbf{E}''$, of nodal voltages is

$$\mathbf{E}'' = \mathbf{Y}^{-1}\mathbf{A}_{mp}I_{test} \tag{15.27}$$

and by (15.8), the voltage, V_{test}, is

$$V_{test} = (\mathbf{A}_{mp}^{T}\mathbf{Y}^{-1}\mathbf{A}_{mp})I_{test} \tag{15.28}$$

It follows that the requisite Thévenin impedance, Z_{th}, is

$$Z_{th} = \frac{V_{test}}{I_{test}} = (\mathbf{A}_{mp}^{T}\mathbf{Y}^{-1}\mathbf{A}_{mp}) \tag{15.29}$$

Insofar as the appended impedance, Z_k, is concerned, the network in Fig. 15.8 behaves in accordance with the circuit abstraction of Fig. 15.9. The current, I, conducted by Z_k is, without

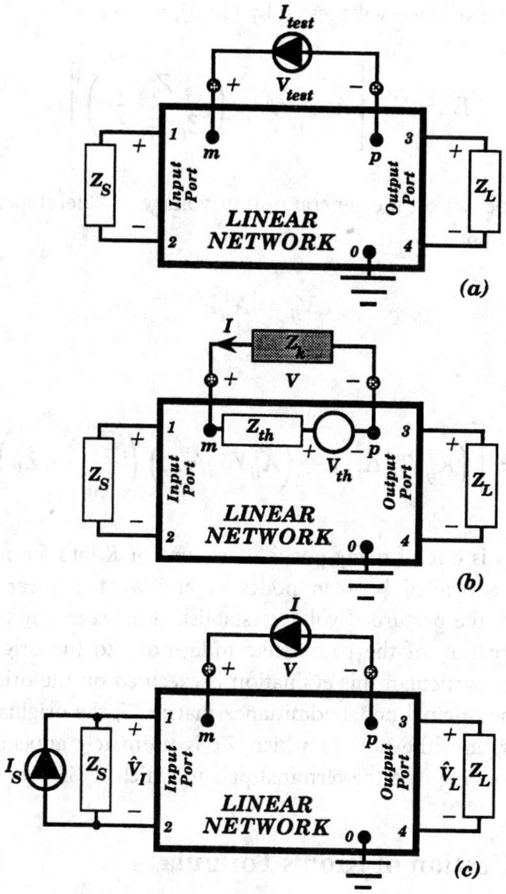

FIGURE 15.9 (a) Circuit diagram for evaluating the Thévenin impedance seen by the appended impedance Z_k. (b) Circuit diagram used to compute the current, I, conducted by Z_k. (c) The application of the substitution theorem with respect to Z_k.

approximation,

$$I = -\frac{V_{\text{th}}}{Z_{\text{th}} + Z_k} = -\left(\frac{Z_{kS}}{Z_{\text{th}} + Z_k}\right) I_S \tag{15.30}$$

where (15.26) has been used, and Z_{th} is understood to be given by (15.29). However, by the substitution theorem, the impedance, Z_k in Fig. 15.8 can be supplanted by an independent current source of value I, as suggested in Fig. 15.9(c). Specifically, this substitution of the impedance of interest with a current source whose value is dictated by (15.30) guarantees that the n-vector of node voltages for the modified circuit in Fig. 15.9(c) is identical to the n-vector, E', of node voltages for the topology given in Fig. 15.8.

The circuit of Fig. 15.9(c) now has two independent excitations: the original current source, I_S, and the current, I, substituted for the appended impedance, Z_k. Accordingly, the current source vector for the subject circuit superimposes two current components and is given by

$$\mathbf{J} = \mathbf{A}_{12} I_S + \mathbf{A}_{mp} I = \mathbf{A}_{12} I_S - \mathbf{A}_{mp}\left(\frac{Z_{kS}}{Z_{\text{th}} + Z_k}\right) I_S \tag{15.31}$$

The corresponding vector of node voltages is, by (15.9),

$$\mathbf{E}' = \mathbf{Y}^{-1}\left[\mathbf{A}_{12} - \mathbf{A}_{mp}\left(\frac{Z_{kS}}{Z_{\text{th}} + Z_k}\right)\right]I_S \tag{15.32}$$

If analytical attention focuses on the general output voltage, $\hat{V}_{ij}$, developed between nodes i and j in the circuit of Fig. 15.8,

$$\hat{V}_{ij} = \mathbf{A}_{ij}^{\text{T}}\mathbf{E}' \tag{15.33}$$

where

$$\hat{V}_{ij} = \left[\left(\mathbf{A}_{ij}^{\text{T}}\mathbf{Y}^{-1}\mathbf{A}_{12}\right) - \left(\mathbf{A}_{ij}^{\text{T}}\mathbf{Y}^{-1}\mathbf{A}_{mp}\right)\left(\frac{Z_{kS}}{Z_{\text{th}}} + Z_k\right)\right]I_S \tag{15.34}$$

The result in (15.34) is one of many possible versions of *Kron's formula.* It states that when an impedance, Z_k, is appended between nodes m and p of a linear network whose nodal admittance matrix is $\mathbf{Y}$, the perturbed voltage established between any two nodes, i and j, can be determined as a function of the parameters indigenous to the original network (prior to the inclusion of Z_k). In particular, this evaluation is executed on the original network (with Z_k absent) and exploits the original nodal admittance matrix, $\mathbf{Y}$, the original transimpedance, Z_{kS}, between the input port and the port to which Z_k is ultimately appended, and the Thévenin impedance, Z_{th}, seen looking into the terminal pair to which Z_k is connected.

Engineering Application of Kron's Formula

The engineering utility of Kron's formula, (15.34), is best demonstrated by examining the voltage transfer function of the network in Fig. 15.8 in terms of the companion gain for the network depicted in Fig. 15.5(a). Using (15.7) and noting that the perturbed output voltage is developed from node 3 to node 4, the perturbed voltage gain, $\hat{A}_v$, is

$$\hat{A}_v \triangleq \frac{\hat{V}_L}{V_S} = \frac{\mathbf{A}_{34}^{\text{T}}\mathbf{Y}^{-1}\mathbf{A}_{12}}{Z_S} - \left(\frac{\mathbf{A}_{34}^{\text{T}}\mathbf{Y}^{-1}\mathbf{A}_{mp}}{Z_S}\right)\left(\frac{Z_{kS}}{Z_{\text{th}} + Z_k}\right) \tag{15.35}$$

The first matrix product on the right-hand side of this relationship represents the transimpedance, Z_{LS}, from the input port to the output port of the original network, as given by (15.20). Moreover, the resultant impedance ratio, Z_{LS}/Z_S, is the voltage gain, A_v, of the original ($Z_k = \infty$) network, as delineated in (15.25). The second matrix product symbolizes the transimpedance, Z_{Lk}, from the port to which the appended impedance, Z_k, is connected to the output port; that is,

$$Z_{Lk} = \mathbf{A}_{34}^{\text{T}}\mathbf{Y}^{-1}\mathbf{A}_{mp} \tag{15.36}$$

Assuming $A_v \neq 0$, (15.35) can then be reduced to

$$\hat{A}_v = A_v\left[1 - \left(\frac{Z_{Lk}}{Z_{LS}}\right)\left(\frac{Z_{kS}}{Z_{\text{th}} + Z_k}\right)\right] \tag{15.37}$$

This result expresses the perturbed voltage gain as a function of the original voltage gain, A_v, the input to output transimpedance, Z_{LS}, the transimpedance, Z_{kS}, from the input port to the port at which Z_k is appended, and Z_{Lk}, the transimpedance from the port to which Z_k is incident to the output port. Observe that when the appended impedance is infinitely large, the perturbed gain reduces to the original voltage gain, as expected.

In an actual analytical situation, however, all of the transimpedances indicated in (15.37) need not be calculated. In order to demonstrate this contention, rewrite (15.37) in the form

$$\hat{A}_v = A_v \left[\frac{1 + Y_k\left(Z_{\text{th}} - (Z_{Lk}Z_{kS}/Z_{LS}) \right)}{1 + Y_k Z_{\text{th}}} \right] \tag{15.38}$$

where

$$Y_k = \frac{1}{Z_k} \tag{15.39}$$

is the admittance of the appended impedance, Z_k. Now consider the test structure of Fig. 15.10(a), which is the modified circuit shown in Fig. 15.8, but with the appended branch supplanted by a test current source, I_{test}. With two sources, I_S and I_{test}, activating the network, superposition yields a resultant output port voltage, V_{LL}, of

$$V_{LL} = Z_{LS}I_S + Z_{Lk}I_{\text{test}} \tag{15.40}$$

and a test port voltage, V_{test}, of

$$V_{\text{test}} = Z_{kS}I_S + Z_{kk}I_{\text{test}} \tag{15.41}$$

For $I_S = 0$, the network in Fig. 15.10(a) reduces to the configuration in Fig. 15.9(a), and (15.41) delivers $V_{\text{test}}/I_{\text{test}} = Z_{kk}$. It follows that the impedance parameter, Z_{kk}, is the Thévenin impedance seen by Z_k, as determined in conjunction with an analytical consideration of Fig. 15.5(a); that is, (15.41) is equivalent to the expression

$$V_{\text{test}} = Z_{kS}I_S + Z_{\text{th}}I_{\text{test}} \tag{15.42}$$

Consider the case, suggested in Fig. 15.10, in which the output port voltage, V_{LL}, is constrained to zero for any and all values of the load impedance, Z_L. From (15.40), this case requires a source excitation that satisfies

$$I_S = -\left(\frac{Z_{Lk}}{Z_{LS}}\right) I_{\text{test}} \tag{15.43}$$

If this result is substituted into (15.42), the ratio, $V_{\text{test}}/I_{\text{test}}$, is found to be

$$\frac{V_{\text{test}}}{I_{\text{test}}} = Z_{\text{th}} - \frac{Z_{Lk}Z_{kS}}{Z_{LS}} \tag{15.44}$$

which mirrors the parenthesized numerator term on the right-hand side of (15.38). The ratio in (15.44) might rightfully be termed a *null Thévenin impedance*, Z_{tho}, seen by Z_k, in the sense that it is indeed the Thévenin impedance witnessed by Z_k, but under the special circumstance of a nonzero source excitation selected to *null* the output response variable of the network undergoing investigation. Thus, in Fig. 15.10,

$$\left.\frac{V_{\text{test}}}{I_{\text{test}}}\right|_{\substack{\text{Source}\neq 0\\ \text{Output}=0}} \triangleq Z_{\text{tho}} = Z_{\text{th}} - \frac{Z_{Lk}Z_{kS}}{Z_{LS}} \tag{15.45}$$

Equation (15.38) now reduces to the simpler result

$$\hat{A}_v = A_v\left(\frac{1 - Y_k Z_{\text{tho}}}{1 + Y_k Z_{\text{th}}}\right) = A_v\left(\frac{1 + (Z_{\text{tho}}/Z_k)}{1 + (Z_{\text{th}}/Z_k)}\right) \tag{15.46}$$

Equation (15.46) is both computationally useful and philosophically important. From a computational viewpoint, it allows for an efficient evaluation of the voltage transfer function of a linear network, perturbed by the addition of an impedance element between two extant nodes of the network, in terms of the voltage gain of the original, unperturbed circuit. As expected, this original voltage gain, A_v, is the voltage gain of the perturbed network for the special case of a perturbing impedance whose admittance is zero (or whose impedance value is infinitely large). Only two other parameters are required to complete the evaluation of the perturbed gain. The first is the Thévenin impedance, Z_{th}, seen by the appended impedance element. This Thévenin impedance is calculated traditionally by nulling all independent sources applied to the subject network. The second parameter is the null Thévenin impedance, Z_{tho}, which is the value of Z_{th} for the special circumstance of a test current source and independent source excitations selected to constrain the output response variable to zero. Once Z_{tho} and Z_{th} are determined, the degree to which the voltage transfer function is dependent on the appended impedance is easily determined. For example, the per-unit change in gain owing to the addition of Z_k between nodes m and p in the network of Fig. 15.8 is

$$\frac{\Delta A_v}{A_v} = \frac{\hat{A}_v - A_v}{A_v} = \frac{Z_{\text{tho}} - Z_{\text{th}}}{Z_k + Z_{\text{th}}} \tag{15.47}$$

It is important to note that the transfer function sensitivity implied by (15.47) imposes no *a priori* restrictions on the value of Z_k. In particular, Z_k can assume any value from that of a short circuit to that of the opposite extreme of an open circuit.

From a philosophical perspective, when analytical attention focuses explicitly on feedback circuits, (15.46) can be derived from signal flow theory [10, 11] and is actually Bode's classical gain equation [12]. In the context of Bode's theory Y_k is referred to as a *reference parameter*, or

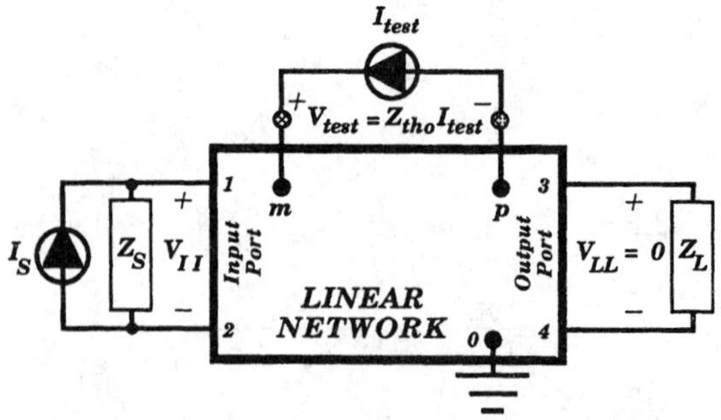

FIGURE 15.10 Network diagram pertinent to the computation of the null Thévenin impedance seen by the impedance appended to the node pair, $m - p$.

a *critical parameter*, of the feedback circuit. The product, $Y_k Z_{th}$, is termed the *return ratio with respect to the critical parameter*, while the product, $Y_k Z_{tho}$, is identified as the *null return ratio with respect to the critical parameter.* Finally, when (15.46) is applied as Bode's equation, the transfer function, A_v, is termed the *null transfer function*, in the sense that it is the actual transfer function of the network at hand, under the special case of a critical parameter constrained to zero.

Example 15.1. In an attempt to demonstrate the engineering utility of the foregoing theoretical disclosures, consider the problem of determining the voltage gain of the common emitter amplifier whose schematic diagram is offered in Fig. 15.11(a). Without detracting from the primary intent of this example, the schematic diagram at hand has been simplified in that requisite biasing subcircuitry is not shown. Assuming that the bipolar junction transistor embedded in the amplifier operates in its linear regime, the pertinent small signal equivalent circuit is the topology depicted in Fig. 15.11(b).

Assume that the amplifier source resistance, R_S, is 600 Ω and that the load resistance, R_L, is 10 kΩ. Assume further that the model parameters for the transistor are as follows: r_b (internal base resistance) = 90 Ω, r_π (base-emitter junction diffusion resistance) = 1.4 kΩ, r_e (internal emitter resistance) = 2.5 Ω, r_0 (forward Early resistance) = 18 kΩ, β (forward short circuit current gain) = 90, and r_c (internal collector resistance) = 70 Ω. Determine the voltage gain of the amplifier and the effect exerted on this gain by neglecting the Early resistance, r_0.

Solution.

1. Analytical simplicity traditionally dictates the tacit neglect of the forward Early resistance, r_0. This commonly invoked approximation reduces the model given in Fig. 15.11(b) to the equivalent circuit in Fig. 15.12(a). By inspection of the latter diagram, the approximate gain of the common emitter voltage is

$$A_v = \frac{V_O}{V_S} = -\frac{\beta R_L}{R_S + r_b + r_\pi + (\beta + 1)r_e} = -388.3 \text{ v/v}$$

2. In order to determine the impact that r_0 has on this voltage gain, r_0 is removed from the equivalent circuit and replaced by a test current source, I_{test}, as depicted in Fig. 15.12(b). With the independent input voltage, V_S, set to zero, the Thévenin resistance seen by r_0,

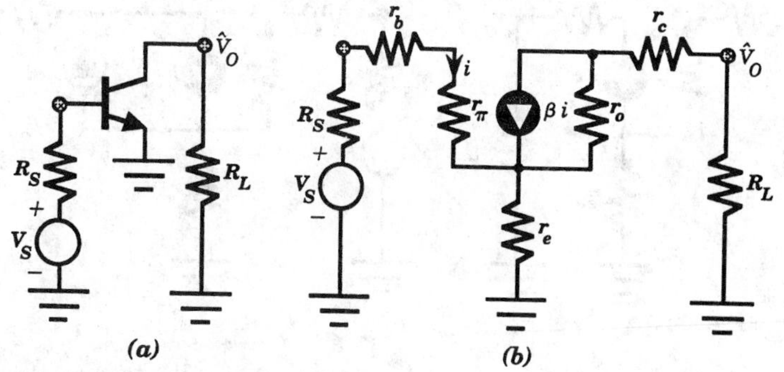

FIGURE 15.11 (a) Simplified schematic diagram of a common emitter amplifier. (b) The small signal equivalent circuit of the common emitter amplifier.

which is the ratio, V_{test}/I_{test}, is easily shown to be

$$R_{th} = r_e \| \left(\frac{R_S + r_b + r_\pi}{\beta + 1} \right) + \frac{r_c + R_L}{1 + \dfrac{\beta r_e}{R_S + r_b + r_\pi + r_e}} = 9.10 \text{ k}\Omega$$

On the other hand, if V_X is constrained to zero, no current flows through the load resistance branch, and therefore, I_{test} is necessarily βi. This condition gives a null Thévenin resistance of

$$R_{tho} = -\frac{r_e}{\beta} = -27.78 \times 10^{-3}\ \Omega$$

3. With $A_v = -388.3$ v/v, $Z_k = r_0 = 18$ kΩ, $Z_{th} = R_{th} = 9.10$ kΩ, and $Z_{tho} = R_{tho} = -27.78 \times 10^{-3}$ Ω, (15.46) produces a corrected voltage gain of

$$\hat{A}_v = A_v \left(\frac{1 + \dfrac{R_{tho}}{r_0}}{1 + \dfrac{R_{th}}{r_0}} \right) = -258.0 \text{ v/v}$$

From (15.47), the presence of r_0, as opposed to its absence, decreases the voltage gain of the subject amplifier by almost 34%.

Example 15.2. As a second example, consider the series-shunt feedback amplifier whose schematic diagram, neglecting requisite biasing circuitry, appears in Fig. 15.13(a). The analysis of this circuit is simplified by the removal of the connection of the feedback resistance, R_F, at the emitter of transistor Q1, as shown in Fig. 15.13(b). If the voltage gain of the simplified topology is denoted as A_v, the voltage gain of the *closed loop* configuration in Fig. 15.13(a) derives from (15.46), provided that the impedance, Z_k, between the indicated node pair, *m-p*, is taken as a short circuit; that is, $Z_k = 0$.

Assume that the amplifier source resistance, R_S, is 300 Ω, the load resistance, R_L, is 3.5 kΩ, the feedback resistance, R_F, is 1.5 kΩ, and the emitter degeneration resistance, R_{EE}, is 100 Ω. The transistor model invoked for small signal analysis is identical to that used in the preceding example, save for the proviso that the Early resistance, r_0, is ignored herewith. Both transistors are presumed to have identical small signal model parameters as follows: $r_b = 90\ \Omega$, $r_\pi = 1.4$ kΩ,

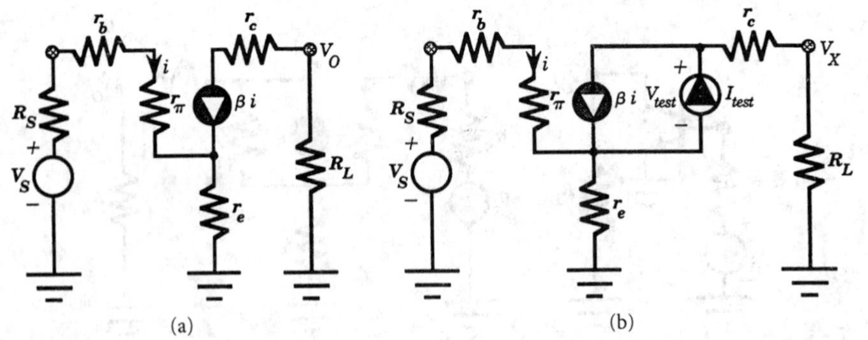

FIGURE 15.12 (a) The approximate small signal equivalent circuit of the common emitter amplifier in Fig. 15.11(a). The approximation entails the tacit neglect of the forward Early resistance, r_0. (b) The test equivalent circuit used to compute the Thévenin and the null Thévenin resistances seen by r_0 in (a).

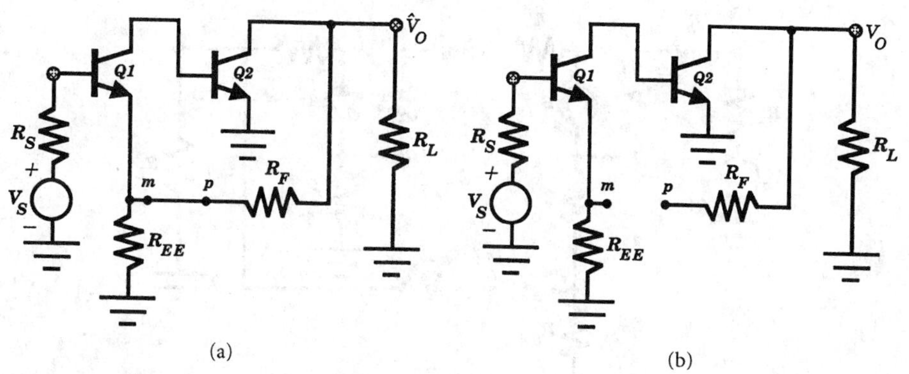

FIGURE 15.13 (a) Simplified schematic diagram of a series-shunt feedback bipolar junction transistor amplifier. The biasing subcircuitry is not shown. (b) The amplifier in (a), but with the feedback resistance connection to the emitter of transistor $Q1$ removed.

$r_e = 2.5\ \Omega$, $\beta = 90$, and $r_c = 70\ \Omega$. Use Kron's theorem to determine the voltage gain of the closed loop series-shunt feedback amplifier.

Solution.

1. The voltage gain of the pertinent equivalent circuit in Fig. 15.14 is straightforwardly derived as

$$A_v = \frac{V_O}{V_S} = \frac{\beta^2 R_L}{R_S + r_b + r_\pi + (\beta + 1)(r_e + R_{EE})} = 2550 \text{ v/v}$$

Observe that the feedback resistance, R_F, does not enter into this calculation because of its disconnection at the emitter of transistor $Q1$. Furthermore, the internal collector resistance, r_c, is inconsequential for both transistor stages because the neglect of the forward Early resistance, r_0, places r_c in series with a controlled current source.

2. The Thévenin resistance, R_{th}, seen by the ultimately appended short circuit between nodes m and p is now calculated through use of the model in Fig. 15.14(b). For this calculation, the signal voltage, V_S, is reduced to zero. With $V_S = 0$,

$$\frac{i_1}{i_{test}} = -\frac{R_{EE}}{R_S + r_b + r_\pi + (\beta + 1)(r_e + R_{EE})}$$

Noting that $i_2 = -\beta i_1$, KVL yields

$$V_{test} = (R_{EE} + R_L + R_F)I_{test} + [(\beta + 1)R_{EE} - \beta^2 R_L]i_1$$

Using the preceding result, introducing the resistance variable, R_X, such that

$$R_X \triangleq r_e + \frac{R_S + r_b + r_\pi}{\beta + 1} = 22.17\ \Omega$$

and letting

$$\alpha \triangleq \frac{\beta}{\beta + 1} = 0.989$$

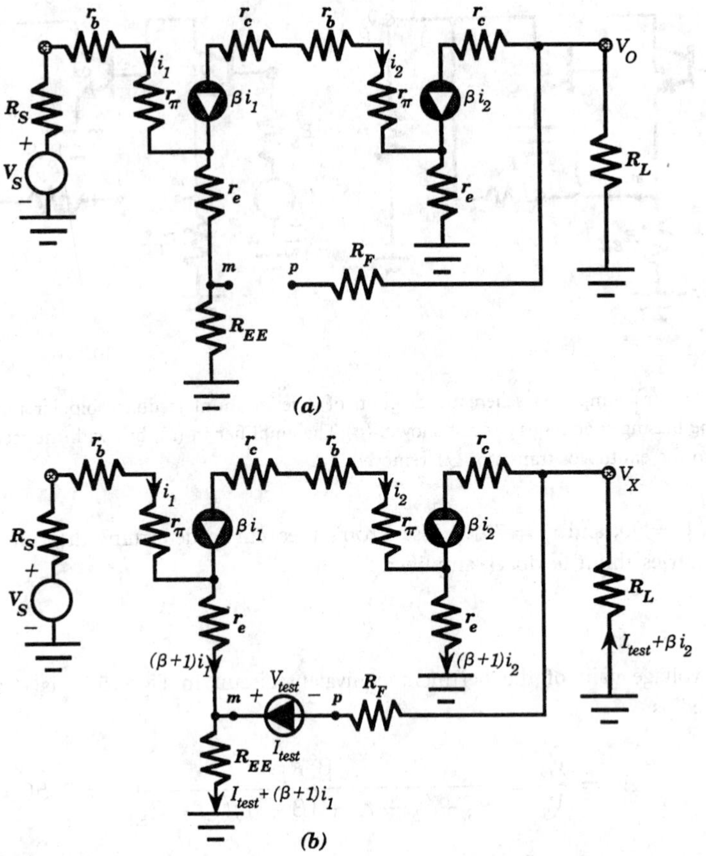

FIGURE 15.14 (a) Small signal equivalent circuit of the feedback amplifier in Fig. 15.13(b). This circuit is used to compute the voltage gain with the feedback resistance disconnected at the emitter of transistor Q1. (b) The small signal model used to compute the Thévenin and the null Thévenin resistances seen by the short circuit that is ultimately appended to the node pair, *m-p*, in Fig. 15.13(b).

R_{th}, which is the ratio V_{test}/I_{test}, is found to be

$$R_{th} = R_F + (R_{EE} \| R_X) + \left(1 + \frac{\alpha\beta R_{EE}}{R_{EE} + R_X}\right) R_L = 260.0\ \text{k}\Omega$$

3. For the evaluation of the null Thévenin resistance, R_{tho}, the output voltage variable, V_X, in Fig. 15.14(b) is nulled, thereby forcing the current relationship, $I_{test} = -\beta i_2 = +\beta^2 i_1$. Accordingly,

$$R_{tho} = R_F + \left(1 + \frac{1}{\alpha\beta}\right) R_{EE} = 1601\ \Omega$$

4. With $Z_k = 0$, $Z_{th} = R_{th} = 260$ kΩ, and $Z_{tho} = R_{tho} = 1601$ Ω, (15.46) provides, after reconnection of the feedback element, an amplifier gain of

$$\hat{A}_v = A_v \left(\frac{R_{tho}}{R_{th}}\right) = 15.70\ \text{V/V}$$

It is interesting to note that if the transistors utilized in the feedback amplifier have very large β, which is tantamount to very small R_X and $a \approx 1$, the voltage gain with the feedback element disconnected is

$$a_v \approx \frac{\beta R_L}{R_{EE}}$$

Moreover,

$$R_{\text{th}} \approx \beta R_L$$

and

$$R_{\text{tho}} \approx R_F + R_{EE}$$

It follows from (15.46) that the approximate voltage gain, subsequent to the reconnection of the feedback resistance, R_F, to the emitter of transistor $Q1$ is

$$\hat{A}_v = A_v\left(\frac{R_{\text{tho}}}{R_{\text{th}}}\right) \approx \left(\frac{\beta R_L}{R_{EE}}\right)\left(\frac{R_F + R_{EE}}{\beta R_L}\right) = 1 + \frac{R_F}{R_{EE}} = 16 \text{ V/V}$$

which is within 2% of the accurately estimated voltage gain.

Generalization of Kron's Formula

The Kron/Bode equation in (15.46) was derived expressly for investigating the voltage transfer function of a linear network to which an impedance element is appended between two network nodes. This equation also can be adapted to the problem of determining the explicit dependence of any type of transfer relationship on any parameter within any linear network.

To this end, consider any linear network, such as the generalization shown in Fig. 15.15, whose load impedance is Z_L and whose source impedance is Z_S. Identify a *critical network parameter*, say P, to which the dependence on, and sensitivity to, the overall transfer performance of the network undergoing study is of particular interest. This parameter can be, for example, a circuit branch impedance or an active element gain factor whose numerical values cannot be determined accurately and/or controlled adequately in view of potentially unacceptable manufacturing tolerances or device fabrication uncertainties. Let the transfer function of interest be

$$H(P, Z_S, Z_L) = \frac{X_R(s)}{X_S(s)} \tag{15.48}$$

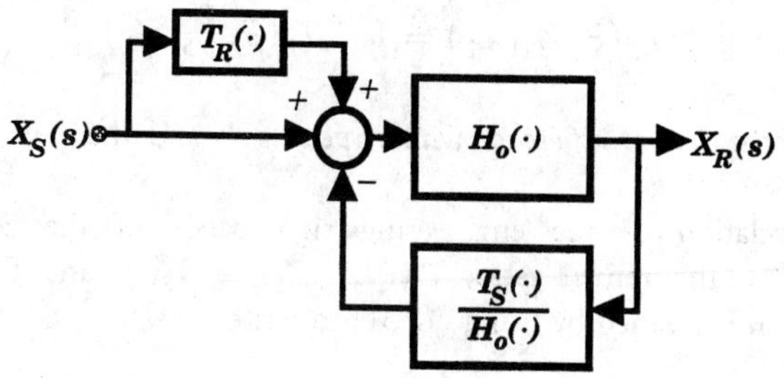

FIGURE 15.15 Generalized block diagram nodel of the I-O transfer characteristics of a linear network.

where $X_R(s)$ denotes the transform of the voltage or current response variable, and $X_S(s)$ is the transform of the voltage or current input variable. The functional notation, $H(P, Z_S, Z_L)$, underscores the observation that the transfer function of the linear network is likely to be dependent on the critical parameter, P, the source impedance, Z_S, and the load impedance, Z_L. The corresponding extension of the Kron/Bode relationship is

$$H(P, Z_S, Z_L) = \frac{X_R(s)}{X_S(s)} = H(0, Z_S, Z_L)\left[\frac{1 + PQ_R(Z_S, Z_L)}{1 + PQ_S(Z_S, Z_L)}\right] \tag{15.49}$$

where $H(0, Z_S, Z_L)$, termed the *null gain or zero parameter gain*, signifies the value of the network transfer function, $H(P, Z_S, Z_L)$, when P is set to zero. This null gain must be finite and nonzero. With reference to the appended impedance formulation in (15.46), observe that the critical parameter, P, is Y_k, the admittance of the appended impedance element, while $H(0, Z_S, Z_L)$ is the gain, A_v, of the network, under the condition of an absent impedance element ($Y_k = 0$).

The product, $PQ_S(Z_S, Z_L)$, is termed the *return ratio with respect to parameter P*, $T_S(P, Z_S, Z_L)$, and the product, $PQ_R(Z_S, Z_L)$, is referred to as the *null return ratio with respect to P*, $T_R(P, Z_S, Z_L)$; that is,

$$T_R(P, Z_S, Z_L) \triangleq PQ_R(Z_S, Z_L) \tag{15.50a}$$

$$T_S(P, Z_S, Z_L) \triangleq PQ_S(Z_S, Z_L) \tag{15.50b}$$

It is to be understood that both $Q_S(Z_S, Z_L)$ and $Q_R(Z_S, Z_L)$ are independent of the critical parameter, P. With reference once again to (15.46), note that $Q_S(Z_S, Z_L)$ is the Thévenin impedance seen by the appended admittance, Y_k, while $Q_R(Z_S, Z_L)$ is the null Thévenin impedance facing Y_k.

Equation (15.49) can now be rewritten as

$$H(P, Z_S, Z_L) = \frac{X_R(s)}{X_S(s)} = H(0, Z_S, Z_L)\left[\frac{1 + T_R(P, Z_S, Z_L)}{1 + T_S(P, Z_S, Z_L)}\right] \tag{15.51}$$

Alternatively,

$$H(P, Z_S, Z_L) = \frac{X_R(s)}{X_S(s)} = H(0, Z_S, Z_L)\left[\frac{F_R(P, Z_S, Z_L)}{F_S(P, Z_S, Z_L)}\right] \tag{15.52}$$

where

$$F_S(P, Z_S, Z_L) \triangleq 1 + T_S(P, Z_S, Z_L) \tag{15.53a}$$

$$F_R(P, Z_S, Z_L) \triangleq 1 + T_R(P, Z_S, Z_L) \tag{15.53b}$$

respectively, denote the *return difference with respect to P* and the *null return difference with respect to P*.

An initial appreciation of the engineering significance of the zero parameter gain, $H(0, Z_S, Z_L) \equiv H_0(\cdot)$, the return ratio, $T_S(P, Z_S, Z_L) \equiv T_S(\cdot)$, and the null return ratio, $T_R(P, Z_S, Z_L) \equiv T_R(\cdot)$, is gleaned by using (15.51) to write

$$X_R(s) = H_0(\cdot)[1 + T_R(\cdot)]X_S(s) - T_S(\cdot)X_R(s) \tag{15.54}$$

In view of the generality of the Kron/Bode formula, this algebraic manipulation of (15.51) implies that the dynamical input/output transfer relationship of all linear networks can be symbolically represented by the block diagram offered in Fig. 15.15. This block diagram makes clear that because $T_S(\cdot)$ and $T_R(\cdot)$ are zero for $P = 0$, $H_0(\cdot)$ is the gain afforded by the network as a result of input/output electrical paths that exclude the parameter, P. The diagram also suggests that $T_S(\cdot)$ is a measure of the amount of feedback incurred by parameter P around that part of the circuit that excludes parameter P. Finally, the diagram at hand implies that $T_R(\cdot)$ is a measure of the amount of *feedforward* incurred by parameter P. In particular, if feedback is removed, two paths remain for the transmission of a signal from the input port to the output port of a linear network. One of these paths, the transmittance of which is measured by $H_0(\cdot)$, is direct and entails the processing of the input signal by that part of the circuit that excludes P. The other nonfeedback path has an effective transmittance of $T_R(\cdot)H_0(\cdot)$. The latter path is the feedforward path in the sense that signal is processed through a signal path that is divorced from feedback and is not a result of direct source coupling through the topological part of the network that excludes parameter P.

Return Ratio Calculations

In the generalized transfer relationship of (15.49), the critical parameter, P, can assume one of only six possible forms: circuit branch admittance, circuit branch impedance, transimpedance, transadmittance, gain associated with a current-controlled current source (CCCS), and gain associated with a voltage-controlled voltage source (VCVS) [13]. The methodology underlying the computation of the return ratio and the null return ratio with respect to each of these critical parameter possibilities is given below. The case of $P = Y_k$, a circuit branch admittance, was investigated in the context of Kron's partitioning theorem. Nevertheless, it is reinvestigated below for the purpose of establishing an analytical common denominator for return ratio calculations with respect to the five other reference parameter possibilities.

Circuit Branch Admittance

Consider the network abstraction of Fig. 15.16(a), which identifies a branch admittance, Y_k, as a critical parameter for analysis; that is, $P = Y_k$ in (15.49). The input excitation can be either a voltage source, or a current source and is therefore indicated as a general transformed input variable, $X_S(s)$. Similarly, the output or response variable can be either a voltage or a current, thereby encouraging the generalized transformed response notation, $X_R(s)$. The source and load impedances (or admittances) are absorbed into the network. In order to evaluate the zero parameter gain, $H(0, Z_S, Z_L)$, Y_k is set to zero by removing it from the network. An analysis is then conducted to determine the ratio, $X_R(s)/X_S(s)$, of output to input variables, as suggested in Fig. (15.16(b)).

As demonstrated for the case of $P = Y_k$, a circuit branch admittance, the function, $Q_S(Z_S, Z_L)$, in (15.49) is the Thévenin impedance, Z_{th}, facing Y_k. This impedance is computed by determining the ratio, V_x/I_x, with the signal source, $X_S(s)$, nulled, as indicated in Fig. 15.16(c). Note in Fig. 15.16(a), that the volt-ampere relationship of the branch housing Y_k is $I_k = Y_k V_k$, where the direction of the branch current, I_k, coincides with the direction of the test current source, I_x, used in the determination of Z_{th}. A comparison of Figs. 15.16(c) and 15.16(a) alludes to the methodology of replacing the admittance branch by a source of excitation (a current source) whose electrical nature is identical to the *dependent* electrical variable (a current, I_k) of the branch volt-ampere characteristic. Note that the polarity of the voltage, V_x, used in the determination of the test ratio, V_x/I_x, is opposite to that of the original branch voltage, V_k. This is to say that while I_k and V_k are in associated reference polarity in Fig. 15.16(a), I_x and V_x in the test cell of Fig. 15.16(c) are in disassociated polarity.

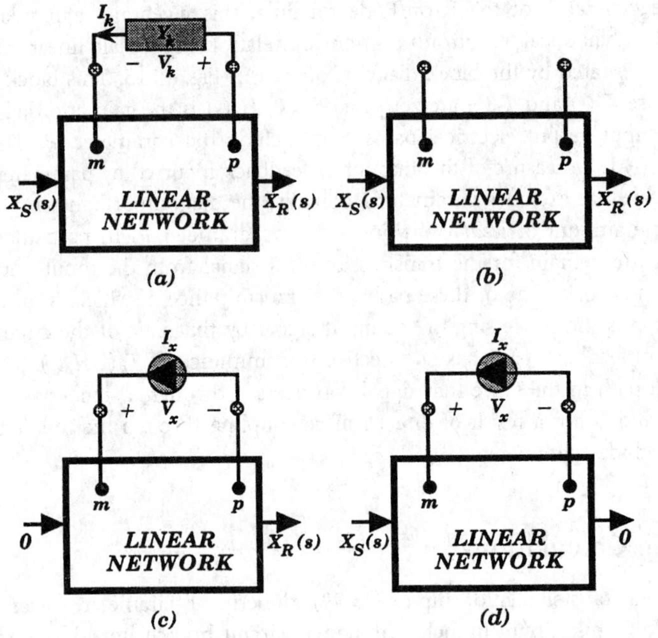

FIGURE 15.16 (a) Linear circuit for which the identified critical parameter is a branch admittance, Y_k. (b) The ratio, $X_R(s)/X_S(s)$, is the zero parameter gain, $H(0, Z_S, Z_L)$. (c) The ratio, V_x/I_x, is the function $Q_S(Z_S, Z_L)$, in (15.49). (d) The ratio, V_x/I_x, is the function, $Q_R(Z_S, Z_L)$, in (15.49).

The computation of the function, $Q_R(Z_S, Z_L)$, in (15.49) mirrors the computation of $Q_S(Z_S, Z_L)$, except for the fact that instead of setting the source excitation to zero, the output response, $X_R(s)$, is nulled. The source excitation, $X_S(s)$, remains at some computationally unimportant nonzero value, such that its effects, when superimposed over those of the test current, I_x, forces $X_R(s)$ to zero. The situation at hand is diagrammed in Fig. 15.16(d).

Example 15.3. Return to the series-shunt feedback amplifier of Fig. 15.29(a). Evaluate the voltage gain of the circuit, but, take the conductance, G_F, of the feedback resistance, R_F, as the critical parameter. The circuit and device model parameters remain the same as in Example 15.2: $R_S = 300\ \Omega$, $R_L = 3.5\ \text{k}\Omega$, $R_F = 1.5\ \text{k}\Omega$, $R_{EE} = 100\ \Omega$, $r_b = 90\ \Omega$, $r_\pi = 1.4\ \text{k}\Omega$, $r_e = 2.5\ \Omega$, $\beta = 90$, and $r_c = 70\ \Omega$.

Solution.

1. The zero parameter voltage gain, A_{vo}, of the subject amplifier is the voltage gain of the circuit with $G_F = 0$. But $G_F = 0$ amounts to a removal of the feedback resistance, R_F. Such removal is electrically equivalent to open circuiting the indicated node pair, *m-p*, as diagrammed in the small signal model of Fig. 15.14(a). Thus, A_{vo} is identical to the gain, computed in Step (1) of Example 15.2. In particular,

$$A_{vo} = \frac{\beta^2 R_L}{R_S + r_b + r_\pi + (\beta + 1)(r_e + R_{EE})} = 2,550\ \text{V/V}$$

2. The model pertinent to computing the functions, $Q_S(Z_S, Z_L)$ and $Q_R(Z_S, Z_L)$, is offered in Fig. 15.17. Note that the test current source, I_x, which replaces the critical conductance element, G_F, and the resultant test response voltage, V_x, are in disassociated reference

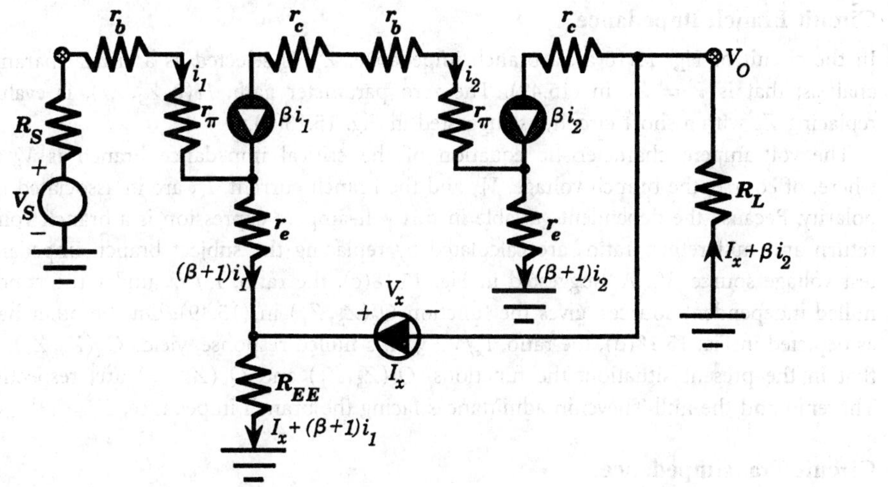

FIGURE 15.17 Circuit used to compute the return ratio and the null return ratio with respect to the conductance, G_F, in the series-shunt feedback amplifier of Fig. 15.13(a).

polarity. As in Example 15.2, let

$$R_X \triangleq r_e + \frac{R_S + r_b + r_\pi}{\beta + 1} = 22.17\ \Omega$$

and

$$\alpha \triangleq \frac{\beta}{\beta + 1} = 0.989$$

Then, with $V_S = 0$, and writing $Q_S(Z_S, Z_L)$ as $Q_S(R_S, R_L)$ because of the lack of energy storage elements in the circuit undergoing study,

$$Q_S(R_S, R_L) = \left.\frac{V_x}{I_x}\right|_{V_S=0} = R_{EE} \| R_X + \left(1 + \frac{\alpha\beta R_{EE}}{R_{EE} + R_X}\right) R_L = 258.5\ \text{k}\Omega$$

On the other hand,

$$Q_R(R_S, R_L) = \left.\frac{V_x}{I_x}\right|_{V_O=0} = \left(1 + \frac{1}{\alpha\beta}\right) R_{EE} = 101.1\ \Omega$$

3. Substituting the preceding results into (15.49), and recalling that $G_F = 1/R_F$, the voltage gain of the series-shunt feedback amplifier is found to be

$$A_v = \frac{V_O}{V_S} = A_{vo}\left[\frac{1 + \dfrac{Q_R(R_S, R_L)}{R_F}}{1 + \dfrac{Q_S(R_S, R_L)}{R_F}}\right] = 15.7\ \text{V/V}$$

which is the gain result deduced previously.

Circuit Branch Impedance

In the circuit of Fig. 15.18(a), a branch impedance, Z_k, is selected as a critical parameter for analysis; that is $P = Z_k$ in (15.49). The zero parameter gain, $H(0, Z_S, Z_L)$, is evaluated by replacing Z_k with a short circuit, as suggested in Fig. 15.18(b).

The volt-ampere characteristic equation of the critical impedance branch is $V_k = Z_k I_k$, where, of course, the branch voltage, V_k, and the branch current, I_k, are in associated reference polarity. Because the dependent variable in this volt-ampere expression is a branch voltage, the return and null return ratios are calculated by replacing the subject branch impedance by a test voltage source, V_x. As suggested in Fig. 15.18(c), the ratio, I_x/V_x, under the condition of nulled independent sources, gives the function $Q_S(Z_S, Z_L)$ in (15.49). On the other hand, and as depicted in Fig. 15.18(d), the ratio, I_x/V_x, with a nulled response, yields $Q_R(Z_S, Z_L)$. Observe that in the present situation, the functions, $Q_S(Z_S, Z_L)$ and $Q_R(Z_S, Z_L)$ are, respectively, the Thévenin and the null Thévenin admittances facing the branch impedance, Z_k.

Circuit Transimpedance

In the circuit of Fig. 15.19(a), a circuit transimpedance, Z_t, is selected as a critical parameter for analysis; that is, $P = Z_t$ in (15.49). The zero parameter gain, $H(0, Z_S, Z_L)$, is evaluated by replacing the current-controlled voltage source (CCVS) by a short circuit, as shown in Fig. 15.19(b).

The volt-ampere characteristic equation of the critical transimpedance branch is $V_k = Z_t I_k$, where I_k is the controlling current for the controlled source branch. Because the dependent variable in this volt-ampere expression is a branch voltage, the return and null return ratios are calculated by replacing the CCVS with a test voltage source, V_x. However, as indicated in Figs. 15.19(c) and (d), the polarity of V_x mirrors that of the voltage, V_k, developed across the controlled branch. With I_x taken as a current flowing in the controlling branch in a direction

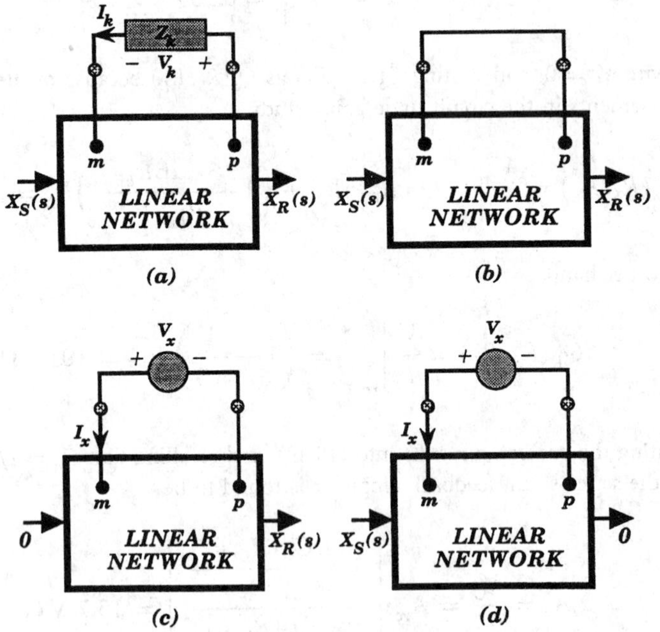

FIGURE 15.18 (a) Linear circuit for which the identified critical parameter is a branch impedance, Z_k. (b) The ratio, $X_R(s)/X_S(s)$, is the zero parameter gain, $H(0, Z_S, Z_L)$. (c) The ratio, I_x/V_x, is the function, $Q_S(Z_S, Z_L)$, in (15.49). (d) The ratio, I_x/V_x, is the function, $Q_R(Z_S, Z_L)$, in (15.49).

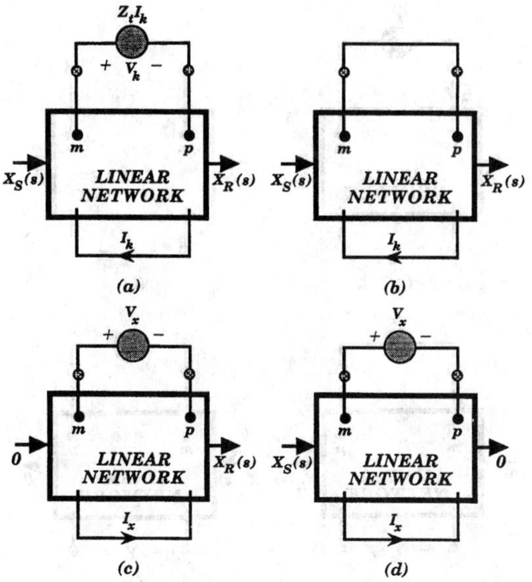

FIGURE 15.19 (a) Linear circuit for which the identified critical parameter is a circuit transimpedance, Z_t. (b) The ratio, $X_R(s)/X_S(s)$, is the zero parameter gain, $H(0, Z_S, Z_L)$. (c) The ratio, I_x/V_x, is the function, $Q_S(Z_S, Z_L)$, in (15.49). (d) The ratio, I_x/V_x, is the function, $Q_R(Z_S, Z_L)$, in (15.49).

opposite to the polarity of the original controlling current, I_k, the ratio, I_x/V_x, under the condition of nulled independent sources, gives the function, $Q_S(Z_S, Z_L)$ in (15.49). On the other hand, and as depicted in Fig. 15.19(d), the ratio, I_x/V_x, with a nulled response, yields $Q_R(Z_S, Z_L)$.

Circuit Transadmittance

In the network of Fig. 15.20(a), a circuit transadmittance, Y_t, is selected as the critical parameter. The zero parameter gain, $H(0, Z_S, Z_L)$, is evaluated by replacing the voltage-controlled current source (VCCS) with an open circuit, as shown in Fig. 15.20(b).

The volt-ampere characteristic equation of the critical transadmittance branch is $I_k = Y_t V_k$, where V_k is the controlling voltage for the VCCS. Because the dependent variable in this volt-ampere expression is a branch current, the return and null return ratios are calculated by replacing the VCCS with a test current source, I_x, where, as indicated in Figs. 15.20(c) and (d), the polarity of I_x mirrors that of the current, I_k, flowing through the controlled branch. With V_x taken as a voltage developed across the controlling branch in a direction opposite to the polarity of the original controlling voltage, V_k, the ratio, V_x/I_x, under the condition of nulled independent sources, gives the function, $Q_S(Z_S, Z_L)$ in (15.43). On the other hand, and as offered in Fig. 15.20(d), the ratio, V_x/I_x, under the condition of a nulled response, yields $Q_R(Z_S, Z_L)$.

Example 15.4. The circuit in Fig. 15.21(a) is a low-frequency, small-signal model of a voltage feedback amplifier. With the transconductance, g_m, selected as the reference parameter of interest, derive a general expression for the voltage gain, $A_v = V_O/V_S$. Approximate the final result for the special case of very large g_m.

Solution.

1. The zero parameter voltage gain, A_{vo}, derives from an analysis of the circuit structure given in Fig. 15.21(b). The diagram differs from Fig. 15.21(a) in that the current conducted

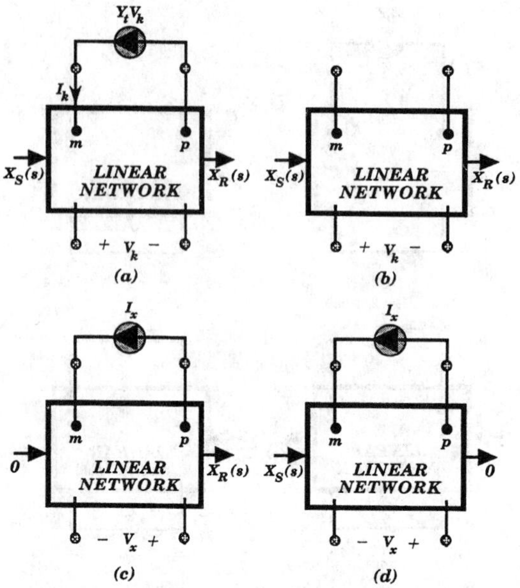

FIGURE 15.20 (a) Linear circuit for which the identified critical parameter is a circuit transadmittance, Y_t. (b) The ratio, $X_R(s)/X_S(s)$, is the zero parameter gain, $H(0, Z_S, Z_L)$. (c) The ratio, V_x/I_x, is the function, $Q_S(Z_S, Z_L)$, in (15.49). (d) The ratio, V_x/I_x, is the function, $Q_R(Z_S, Z_L)$, in (15.49).

by the controlled source branch has been nulled by open circuiting said branch. By inspection of the subject model,

$$A_{vo} = \frac{R_L}{R_L + R_F + R_S}$$

2. The diagram given in Fig. 15.21(c) is appropriate to the computation of the return ratio, $T_S(g_m, Z_S, Z_L) = g_m Q_S(R_S, R_L)$, with respect to the critical transconductance, g_m. A comparison of the model at hand with the diagram in Fig. 15.21(a) confirms that the controlled source, $g_m V$, is replaced by an independent current source, I_x, which flows in a direction identical to that of the controlled source it supplants. The ratio, V_x/I_x, is to be computed, where V_x is developed, antiphase to V, across the branch that supports the original controlling voltage for the VCCS. A straightforward analysis produces

$$Q_S(R_S, R_L) = \frac{V_x}{I_x}\left(\frac{R_L}{R_L + R_F + R_S}\right) R_S \equiv A_{vo} R_S$$

3. The null return ratio, $T_R(g_m, Z_S, Z_L) = g_m Q_R(R_S, R_L)$, with respect to g_m is obtained from an analysis of the circuit in Fig. 15.21(d). Observe a nulled output voltage, with zero current flow through the load resistance, R_L. Observe further that the signal source voltage is nonzero. The specific value of this source voltage is not crucial, and is therefore not delineated. An analysis reveals

$$Q_R(R_S, R_L) = \frac{V_x}{I_x} = -R_F$$

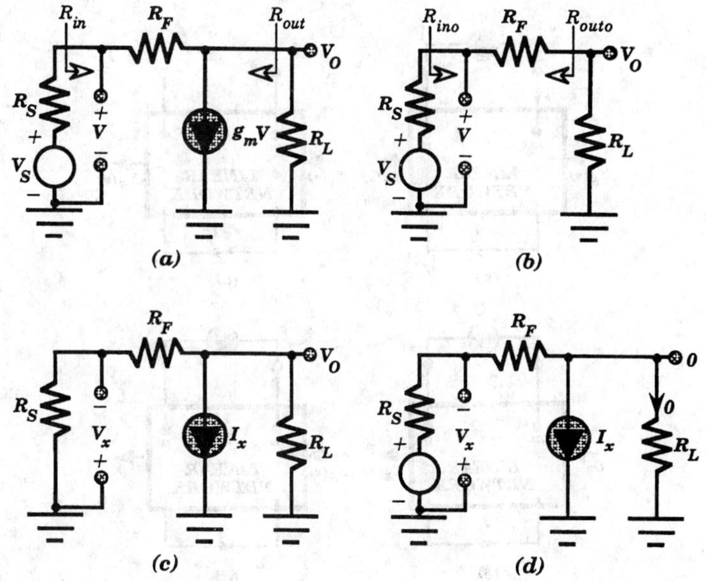

FIGURE 15.21 (a) The low-frequency, small-signal model of a voltage feedback amplifier. (b) The circuit used to evaluate the zero parameter ($g_m = 0$) gain. (c) The circuit used to evaluate the return ratio with respect to g_m. (d) The circuit used to compute the null return ratio with respect to g_m.

4. Using (15.49), the voltage gain of the circuit undergoing study is found to be

$$A_v = \frac{V_O}{V_S} = A_{vo}\left[\frac{1 + g_m(-R_F)}{1 + g_m R_S A_{vo}}\right]$$

which, for large g_m, reduces to

$$A_v \approx -\frac{R_F}{R_S}$$

Gain of a Current-Controlled Current Source

For the network in Fig. 15.22(a), the reference parameter is α_k, the gain associated with a CCCS. The zero parameter gain, $H(0, Z_S, Z_L)$, is evaluated by replacing this CCCS with an open circuit, as depicted in Fig. 15.22(b).

The volt-ampere characteristic equation of the branch in which the reference parameter is embedded is $I_k = \alpha_k I_j$, where I_j is the controlling current for the CCCS. Because the dependent variable in this volt-ampere characteristic is a branch current, the return and null return ratios are calculated by replacing the CCCS with a test current source, I_x. As indicated in Figs. 15.22(c) and (d), the polarity of I_x mirrors that of the current, I_k, flowing through the controlled branch. Let I_y be the resultant current conducted by the controlling branch, and let this current flow in a direction opposite to the polarity of the original controlling current. Then, the current ratio, I_y/I_x, computed under the condition of nulled independent sources, is the function, $Q_S(Z_S, Z_L)$

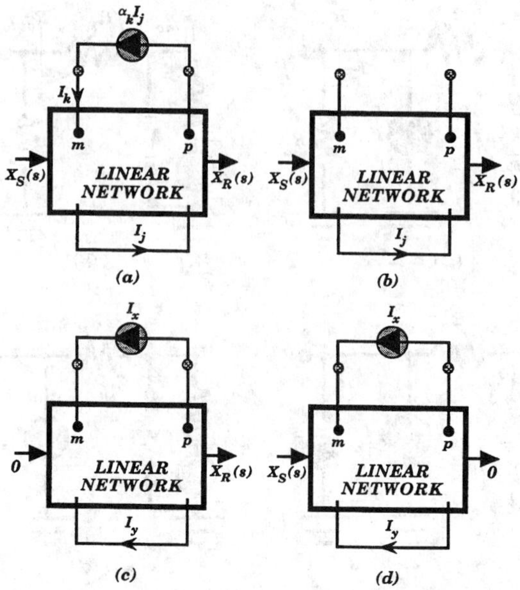

FIGURE 15.22 (a) Linear circuit for which the identified critical parameter is the current gain, α_k, associated with a CCCS. (b) The ratio, $X_R(s)/X_S(s)$, is the zero parameter gain, $H(0, Z_S, Z_L)$. (c) The ratio, I_y/I_x, is the function, $Q_S(Z_S, Z_L)$, in (15.49). (d) The ratio, I_y/I_x, is the function, $Q_R(Z_S, Z_L)$, in (15.49).

in (15.49). Similarly, and as suggested in Fig. 15.22(d), the ratio, I_y/I_x, under the condition of a nulled response, yields $Q_R(Z_S, Z_L)$.

Gain of a Voltage-Controlled Voltage Source

In the network of Fig. 15.23(a), the selected reference parameter is μ_k, the gain corresponding to a VCVS. The zero parameter gain, $H(0, Z_S, Z_L)$, is evaluated by replacing this VCVS with a short circuit, as per Fig. 15.23(b).

The volt-ampere characteristic equation of the dependent generator branch is $V_k = \mu_k V_j$, where V_j is the controlling voltage for the VCVS. Because the dependent variable in this volt-ampere expression is a branch voltage, the return and null return ratios are calculated by replacing the VCVS with a test voltage source, V_x, where as indicated in Figs. 15.23(c) and (d), the polarity of V_x is identical to that of the voltage, V_k, developed across the controlled branch. Let V_y be the resultant voltage established across the controlling branch, and let the polarity of this voltage be in a direction opposite to that of the original controlling voltage. Then, the voltage ratio, V_y/V_x, computed under the condition of nulled independent sources, is the function, $Q_S(Z_S, Z_L)$, in (15.49). As suggested in Fig. 15.23(d), the voltage ratio, V_y/V_x, under the condition of a nulled response, yields $Q_R(Z_S, Z_L)$.

Evaluation of Driving Point Impedances

Having formulated generalized techniques for computing the return ratio and the null return ratio with respect to any of the six possible types of critical circuit parameters, the application of (15.49) is established as a powerful and computationally expedient vehicle for evaluating any transfer function of any linear network. The only restriction limiting the utility of (15.49) is that parameter P must be selected in such a way as to ensure that the zero parameter transfer function is finite and nonzero.

Equation (15.49) is commonly used to evaluate the voltage gain, current gain, transimpedance gain, or transadmittance gain of feedback and other types of complex circuitry. However, the

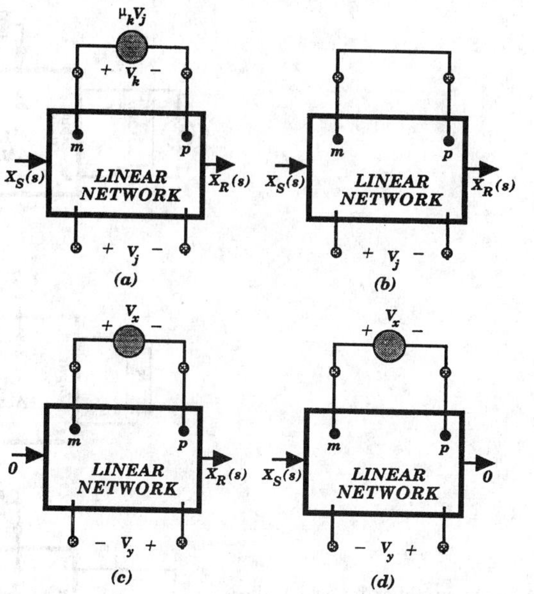

FIGURE 15.23 (a) Linear circuit for which the identified critical parameter is the voltage gain, μ_k, associated with a VCVS. (b) The ratio, $X_R(s)/X_S(s)$, is the zero parameter gain, $H(0, Z_S, Z_L)$. (c) The ratio, V_y/V_x, is the function, $Q_S(Z_S, Z_L)$, in (15.49). (d) The ratio, V_y/V_x, is the function, $Q_R(Z_S, Z_L)$, in (15.49).

expression is equally well suited to determining the driving point input impedance seen by the source impedance, as well as the driving point output impedance seen by the terminating load impedance. In fact, once the return ratios relevant to the gain of interest are found, these I-O impedances can be determined with minimal additional analysis.

Without loss of generality, the foregoing contention is explicitly demonstrated in conjunction with a transimpedance amplifier whose reference parameter is selected to be a branch impedance, Z_k. To this end, consider the circuit abstracted in Fig. 15.24, for which the driving point input impedance, Z_{in}, is to be determined. The input excitation is a current, I_S, and in response to this input, a signal voltage, V_L, is developed across the load impedance, Z_L. Using (15.49), the I-O transimpedance, $Z_T(Z_k, Z_S, Z_L)$, is

$$Z_T(Z_K, Z_S, Z_L) = \frac{V_L(s)}{I_S(s)} = Z_T(0, Z_S, Z_L)\left[\frac{1 + Z_k Q_R(Z_S, Z_L)}{1 + Z_k Q_S(Z_S, Z_L)}\right] \qquad (15.55)$$

where $Z_T(0, Z_S, Z_L)$ is the circuit transimpedance for $Z_k = 0$, $Z_k Q_R(Z_S, Z_L)$ is the null return ratio with respect to Z_k, and $Z_k Q_S(Z_S, Z_L)$ is the return ratio with respect to Z_k. For future reference, the circuit appropriate to the calculation of the function, $Q_S(Z_S, Z_L)$, is drawn in Fig. 15.24(b).

The input impedance derives from an analytical consideration of the cell depicted in Fig. 15.24(c), in which the Norton representation of the signal source has been supplanted by a test current source of value I_z. Note that the load impedance remains as the terminating element for the output port. The transfer relationship of interest is the ratio, V_z/I_z, which is the desired driving point input impedance, Z_{in}. Taking care to choose Z_k, the reference parameter for the gain enumeration, as the reference parameter for the input impedance determination, (15.49) gives

$$Z_{in} = \frac{V_z}{I_z} = Z_{ino}\left[\frac{1 + Z_k Q_{RR}(Z_S, Z_L)}{1 + Z_k Q_{SS}(Z_S, Z_L)}\right] \qquad (15.56)$$

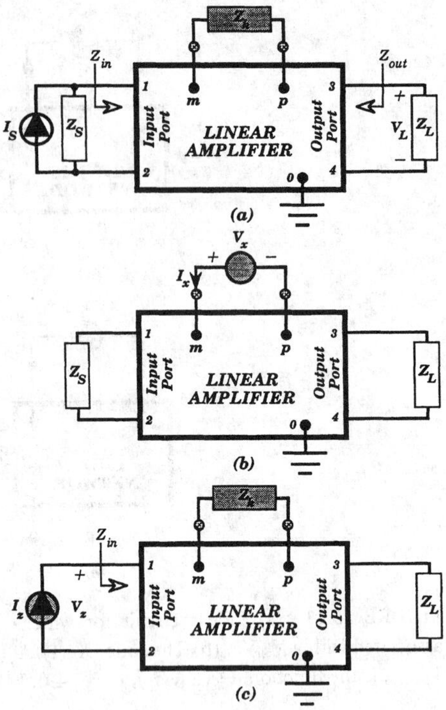

FIGURE 15.24 (a) A linear amplifier for which the input impedance is to be determined. (b) The circuit used for calculating the return ratio with respect to Z_k. (c) The circuit used for calculating the driving point input impedance.

In this expression, Z_{ino} generally derives straightforwardly because it is the $Z_k = 0$ value of Z_{in}; that is, Z_{in} is evaluated for the special case of a nulled reference parameter. Such a null in the present situation is equivalent to short circuiting Z_k, as indicated in Fig. 15.25(a). The function, $Q_{SS}(Z_S, Z_L)$, is the delineated I_x/V_x ratio, for the case of a source excitation (I_z in the present case) set to zero. The pertinent circuit diagram is the structure in Fig. 15.25(b). This last circuit differs from the circuit, shown in Fig. 15.24(b), exploited to find $Q_S(Z_S, Z_L)$ in the gain relationship of (15.50) in only one way: Z_S has been removed, and thus, effectively, Z_S has been set to an infinitely large value. It follows that

$$Q_{SS}(Z_S, Z_L) \equiv Q_S(\infty, Z_L) \tag{15.57}$$

In other words, a circuit analysis aimed toward determining $Q_{SS}(Z_S, Z_L)$ is unnecessary. Instead, $Q_{SS}(Z_S, Z_L)$ is found by evaluating $Q_S(Z_S, Z_L)$, which is already known from the gain analysis, at $Z_S = \infty$.

To evaluate $Q_{RR}(Z_S, Z_L)$, the foregoing I_x/V_x ratio is calculated for the case of zero response. In the present situation the response is the voltage, V_z, and accordingly, the appropriate circuit is as depicted in Fig. 15.25(c). However, a comparison of the circuit at hand with the structure in Fig. 15.24, which is exploited to evaluate the return ratio in the gain equation, indicates that it differs only in that Z_S is now constrained to zero to ensure $V_z = 0$. It is therefore apparent that in (15.56)

$$Q_{RR}(Z_S, Z_L) \equiv Q_S(0, Z_L) \tag{15.58}$$

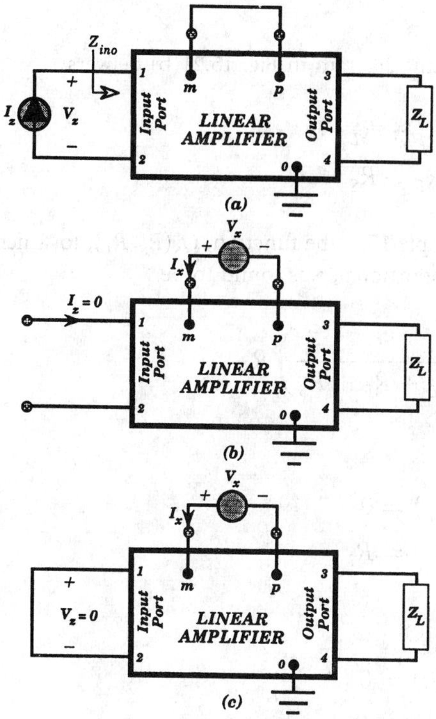

FIGURE 15.25 (a) The circuit used to evaluate the zero parameter driving point input impedance. (b) The computation, relative to input impedance, of the return ratio with respect to Z_k. (c) The computation, relative to input impedance, of the null return ratio with respect to Z_k.

Equation (15.56) is now expressible as

$$Z_{\text{in}} = \frac{V_z}{I_z} = Z_{\text{ino}}\left[\frac{1 + Z_k Q_S(0, Z_L)}{1 + Z_k Q_S(\infty, Z_L)}\right] \tag{15.59}$$

which is occasionally referred to as Blackman's formula [14].

Analogous considerations at the output port in the circuit of Fig. 15.24(a) dictate a driving point output impedance, Z_{out}, of

$$Z_{\text{out}} = Z_{\text{outo}}\left[\frac{1 + Z_k Q_S(Z_S, 0)}{1 + Z_k Q_S(Z_S, \infty)}\right] \tag{15.60}$$

where, like Z_{ino}, Z_{outo}, the $Z_k = 0$ value of Z_{out}, must be finite and nonzero. Although the preceding two relationships were derived for the case in which the selected reference parameter is a branch impedance, both expressions are applicable for any reference parameter, P. In general,

$$Z_{\text{in}} = Z_{\text{ino}}\left[\frac{1 + PQ_S(0, Z_L)}{1 + PQ_S(\infty, Z_L)}\right] \tag{15.61a}$$

$$Z_{\text{out}} = Z_{\text{outo}}\left[\frac{1 + PQ_S(Z_S, 0)}{1 + PQ_S(Z_S, \infty)}\right] \tag{15.61b}$$

Example 15.5. Use the pertinent results of Example 15.4 to derive expressions for the driving point input resistance, R_{in}, and the driving point output resistance, R_{out}, of the feedback amplifier in Fig. 15.21(a).

Solution.

1. With g_m set to zero, an inspection of the circuit diagram in Fig. 15.21(b) delivers

$$R_{\text{ino}} = R_F + R_L$$
$$R_{\text{outo}} = R_F + R_S$$

2. From the second step in the solution to Example 15.4, the function, $Q_S(R_S, R_L)$, to which the return ratio, $T_S(g_m, R_S, R_L)$ is directly proportional, was found to be

$$Q_S(R_S, R_L) = \left(\frac{R_L}{R_L + R_F + R_S}\right) R_S$$

It follows that

$$Q_S(0, R_L) = 0$$
$$Q_S(\infty, R_L) = R_L$$

Moreover,

$$Q_S(R_S, 0) = 0$$
$$Q_S(R_S, \infty) = R_S$$

3. Equations (15.61a) and (b) resultantly yield

$$R_{\text{in}} = R_{\text{ino}} \left[\frac{1 + g_m Q_S(0, Z_L)}{1 + g_m Q_S(\infty, Z_L)}\right] = \frac{R_R + R_L}{1 + g_m R_L}$$

for the driving point input resistance and

$$R_{\text{out}} = R_{\text{outo}} \left[\frac{1 + g_m Q_S(R_S, 0)}{1 + g_m Q_S(R_S, \infty)}\right] = \frac{R_F + R_S}{1 + g_m R_S}$$

for the driving point output resistance.

Sensitivity Analysis

Yet another advantage of the Kron/Bode formula is its amenability to evaluating the impact exerted on a circuit transfer relationship by potentially large fluctuations in the reference parameter P. This convenience stems from the fact that parameter P is isolated in (15.49); that is, $H(0, Z_S, Z_L)$, $Q_R(Z_S, Z_L)$, and $Q_S(Z_S, Z_L)$ are each independent of P. A quantification of this impact is achieved by exploiting the *sensitivity function*,

$$S_P^H \triangleq \frac{\Delta H/H}{\Delta P/P} \tag{15.62}$$

which compares the per unit change in transfer function, $\Delta H/H$, resulting from a specified per unit change, $\Delta P/P$, in a critical parameter. In particular, the notation in this definition is such that if H designates the transfer characteristic, $H(P_0, Z_S, Z_L)$, at the nominal parameter setting, $P = P_0$, $(H + \Delta H)$ signifies the perturbed characteristic, $H(P_0 + \Delta P, Z_S, Z_L)$, when P_0 is altered by an amount, ΔP_0. Using (15.49) and dropping the functional notation in (15.53a) and (15.53b), it can be shown that

$$S_P^H \triangleq \frac{F_R - F_S}{F_R\left[F_S + (F_S - 1)\left(\frac{\Delta P}{P_0}\right)\right]} \tag{15.63}$$

where F_S and F_R are understood to be evaluated at the nominal parameter setting, $P = P_0$. It should be emphasized that unlike a more traditional sensitivity analysis, such as that predicated on the adjoint network [15], (15.63) is easy to use manually and does not rely on an *a priori* assumption of small parametric changes.

References

[1] G. Kron, *Tensor Analysis of Networks*, New York: John Wiley & Sons, 1939.

[2] G. Kron, "A set of principals to interconnect the solutions of physical systems," *J. Appl. Phys.*, vol. 24, pp. 965–980, Aug. 1953.

[3] F. H. Branin, Jr., "The relation between Kron's method and the classical methods of circuit analysis," *IRE Conv. Rec.*, part 2, pp. 3–28, 1959.

[4] F. H. Branin, Jr., "A sparse matrix modification of Kron's method of piecewise analysis," *Proc. IEEE Int. Symp. Circuits Systems*, pp. 383–386, 1975.

[5] R. A. Rohrer, "Circuit partitioning simplified," *IEEE Trans. Circuits Syst.*, vol. 35, pp. 2–5, Jan. 1988.

[6] N. B. Rabbat and H. Y. Hsieh, "A latent macromodular approach to large-scale sparse networks," *IEEE Trans. Circuits Syst.*, vol. CAS-23, pp. 745-752, Dec. 1976.

[7] J. Choma, Jr., "Signal flow analysis of feedback networks," *IEEE Trans. Circuits Syst.*, vol. 37, pp. 455–463, Apr. 1990.

[8] N. B. Rabbat, A. L. Sangiovanni-Vincentelli, and H. Y. Hsieh, "A multilevel Newton algorithm with macromodeling and latency for the analysis of large-scale nonlinear circuits in the time domain," *IEEE Trans. Circuits Syst.*, vol. CAS-26, pp. 733–741, Sept. 1979.

[9] N. Balabanian and T. A. Bickart, *Electrical Network Theory*, New York: John Wiley & Sons, 1969, pp. 73–77.

[10] S. J. Mason, "Feedback theory—some properties of signal flow graphs," *Proc. IRE*, vol. 41, pp. 1144–1156, Sept. 1953.

[11] S. J. Mason, "Feedback theory—further properties of signal flow graphs," *Proc. IRE*, vol. 44, pp. 920–926, July 1956.

[12] H. W. Bode, *Network Analysis and Feedback Amplifier Design*, New York: D. Van Nostrand Company, 1945 (reprinted, 1953), chap. 4.

[13] J. Choma, Jr., *Electrical Networks: Theory and Analysis*, New York: Wiley Interscience, 1985, pp. 590–598.

[14] S. K. Mitra, *Analysis and Synthesis of Linear Active Networks*, New York: John Wiley & Sons, 1969, p. 206.

[15] S. W. Director and R. A. Rohrer, "The generalized adjoint network and network sensitivities," *IEEE Trans. Circuit Theory*, vol. CT-16, pp. 318–328, 1969.

15.3 Linear Two-Port Networks

John Choma, Jr. and Wai-Kai Chen

Two-Port Analysis Concept

Most linear electrical and electronic systems can be viewed as **two-port networks**. A network **port** is a pair of electrical terminals. A two-port network is a circuit that has an input port to which a signal voltage or current is applied and an output port at which the response to the input signal is extracted. In principle, the output port response can be mathematically related to the input port signal by applying Kirchhoff's voltage law (KVL) and Kirchhoff's current law (KCL) to the internal meshes and nodes of the two-port network. Unfortunately, at least three engineering problems limit the utility of this straightforward analytical procedure.

One problem stems from the complexity of practical circuits. A useful linear circuit is likely to contain tens or even hundreds of meshes and nodes. The analytical determination of the response of such a circuit therefore requires a simultaneous solution to the multiorder system of equilibrium equations that derive from the application of KVL and KCL. When the number of requisite equations exceeds two or three, the resultant solution is cumbersome. Complicated solutions are counterproductive, for they dilute the analytical and design insights whose assimulation is, in fact, the fundamental goal of circuit analysis.

A second shortcoming implicit to analyses predicated on the application of KVL and KCL is the necessity that the volt-ampere relationships of all internal network branches be well defined. This constraint is nontrivial when active elements appear in one or more branches of the considered network. It is also nontrivial when an account must be made of presumably second-order circuit effects, such as stray capacitance, parasitic lead inductance, and undesirable electromagnetic coupling from proximate circuits. To be sure, mathematical models of active elements and most second-order phenomena can be constructed, but in the interest of analytical tractability, these models are invariably simplified subcircuits that can produce unacceptable errors in the derived network response. When these models are not simplified, they are often cast in terms of physical parameters that are either nebulously defined or defy reliable measurement. In this case KVL and KCL analyses are an exercise in futility. In particular, the response expressions produced by KVL and KCL may be academically satisfying, but response errors accrue because of numerical uncertainties in the poorly defined parameters associated with certain branch elements intrinsic to the network undergoing study.

Yet another commonly encountered problem is superfluous technical information. The application of KVL and KCL to a linear network produces, in addition to the output voltage or current, all node and branch voltages and all mesh and loop currents for the entire configuration. In many applications, such as those that involve the design of electronic systems in terms of available standard circuit cells, the utility of these internal circuit voltages and currents is questionable. For example, in the design of an active RC filter that exploits commercially available operational amplifiers (op-amps), the only engineering concern is the electrical characteristics registered between the I-O ports of the filter. These characteristics can be deduced without an explicit awareness of the voltages and currents internal to the utilized op-amps. They can, in fact, be unambigously determined as a function of parameters gleaned from measurements conducted at only the I-O ports of each op-amp. These so-called **two-port parameters** are admittedly nonphysical entities in that they generally cannot be cast in terms of the phenomenological processes that underlie the electrical properties observed at the I-O ports of the network they characterize. Nonetheless, these two-port parameters are useful because they do derive from reproducible I-O port measurements, and they do allow for an unambiguous determination of network response.

This chapter defines the commonly used two-port parameters of generalized linear networks and discusses the theory that underlies **two-port parameter equivalent circuits**, or **models.** It develops two-port circuit analysis methods applicable to evaluating the driving-point input, the driving-point output, the transfer, and the stability characteristics of both simple two-ports and two-port systems formed of interconnections of simple two-ports. Basic measurement strategies for the various two-port parameters of linear networks are also addressed.

Generalized Two-Port Network

Figure 15.26 abstracts a generalized linear two-port network. The input port is the terminal pair, 1-2, while the output port, which is terminated in an arbitrary load impedance, Z_L, is the terminal pair, 3-4. No sources of energy exist within the two-port. Because the subject two-port may contain energy storage elements, this presumption implies zero state conditions; that is, all initial voltages associated with internal capacitances and all initial currents of internal inductances are zero. Energy is therefore applied to the two-port network at only its input port. In Fig. 15.26(a), this energy is represented as a Thévenin equivalent circuit consisting of the signal source voltage, V_S, and its internal source impedance, Z_S. Alternatively, the applied energy can be modeled as the Norton circuit depicted in Fig. 15.26(b), where the Norton equivalent current, I_S, is

$$I_S = \frac{V_S}{Z_S} \tag{15.64}$$

As a result of input excitation, a voltage, V_i, is established across the input port, a current, I_i, flows into the input port, a current, I_o, flows into the output port, and a voltage, V_o, is developed across the output port. Note that the output port voltage and current are constrained by the Ohm's law relationship

$$I_o = -\frac{V_o}{Z_L} \tag{15.65}$$

The I-O transfer properties of two-port networks are traditionally defined in terms of one or more of several possible forward transfer characteristics and two driving-point impedance specifications. Among the more commonly used forward transfer specifications is the **voltage gain**, e.g., A_v, which is the ratio of the output voltage developed across the terminating load impedance to the Thévenin equivalent input voltage; that is,

$$A_v \triangleq \frac{V_o}{V_S} \tag{15.66}$$

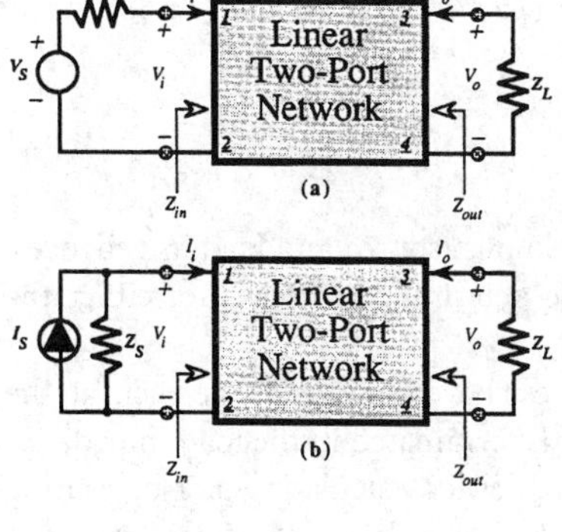

FIGURE 15.26 (a) A linear two-port network, the source circuit of which is modeled by a Thévenin equivalent circuit. (b) The linear two-port network of (a), but with the source circuit represented by its Norton equivalent circuit.

An **ideal voltage amplifier** produces a voltage gain that is independent of source and load impedances. It therefore functions an an ideal voltage controlled voltage source (VCVS). Although ideal voltage amplifiers can never be realized, electronic feedback circuits can be designed so that their terminal electrical properties emulate those of idealized voltage amplifiers. These circuits deliver a voltage gain that, subject to certain parametric restrictions, is approximately independent of source and load impedances.

A second forward transfer characteristic is the **current gain,** A_i, which is the ratio of the output current flowing through the terminating load impedance to the Norton equivalent source current. In particular,

$$A_i \triangleq \frac{I_o}{I_S} \tag{15.67}$$

By (15.64) through (15.65), the current gain of a linear two-port network can be related to its corresponding voltage gain by

$$A_i = -\left(\frac{Z_S}{Z_L}\right)A_v \tag{15.68}$$

An **ideal current amplifier** behaves as an ideal current controlled current source (CCCS) in that it provides a current gain that is independent of source and load impedances. Like ideal voltage amplifiers, idealized current amplification can only be approximated. For example, the low-frequency forward transfer characteristics of a bipolar junction transistor (BJT) can be made to emulate those of a CCCS.

The **forward transadmittance** (or **forward transconductance**, if only low signal frequencies are of interest), Y_f, quantifies the ability of a two-port network to convert input voltages to output currents. It is defined as

$$Y_f \triangleq \frac{I_o}{V_S} \tag{15.69}$$

which, recalling (15.65) and (15.68), is expressible as

$$Y_f = \frac{A_v}{Z_L} = \frac{A_i}{Z_S} \tag{15.70}$$

The metal-oxide-semiconductor field effect transistor (MOSFET), as well as the metal-semiconductor field effect transistor (MESFET), can be operated to approximate transconductor transfer characteristics. Finally, the **forward transimpedance** (or **forward transresistance**, if attention focuses on only low signal frequencies), Z_f, quantifies the ability of a two-port network to convert input currents to output voltages. From (15.64), (15.66), and (15.68),

$$Z_f \triangleq \frac{V_o}{I_S} = Z_S A_v = -Z_L A_i \tag{15.71}$$

The preceding definitions and relationships indicate that for known source and load impedances, any one of the four basic forward transfer functions can be determined in terms of either the current gain or the voltage gain.

The **driving-point input impedance,** Z_{in}, is the effective Thévenin impedance seen at the input port by the signal source circuit, when the output port is terminated in the load impedance used under actual operating conditions. The system level model for calculating and measuring

Z_{in} is offered in Fig. 15.27(a), where

$$Z_{\text{in}} = \frac{V_i}{I_i} \tag{15.72}$$

The equations used to formulate a forward transfer function can be exploited to determine the driving-point input impedance. For example, Fig. 15.26 confirms that

$$V_i = V_S - Z_S I_i \tag{15.73a}$$

and

$$I_i = I_S - \frac{V_i}{Z_S} \tag{15.73b}$$

It follows that

$$Z_{\text{in}} = \lim_{Z_S \to 0} \left(\frac{V_s}{I_i} \right) \tag{15.74a}$$

or

$$Z_{\text{in}} = \lim_{Z_S \to \infty} \left(\frac{V_i}{I_S} \right) \tag{15.74b}$$

The **driving-point output impedance,** Z_{out}, is the Thévenin impedance seen at the output port by the terminating load impedance, when the Thévenin signal source voltage, V_S is set to zero. The system level model for calculating and measuring Z_{out} appears in Fig. 15.27, where

$$Z_{\text{out}} = \frac{V_x}{I_x} \tag{15.75}$$

As in the case of the driving-point input impedance, Z_{out} can be evaluated directly in terms of the equations used to deduce a forward transfer function. To this end, recall that the Thévenin impedance seen looking into a terminal pair of a linear circuit is the ratio of the Thévenin voltage, V_{to}, developed across the terminal pair of interest to the Norton current, I_{no}, that

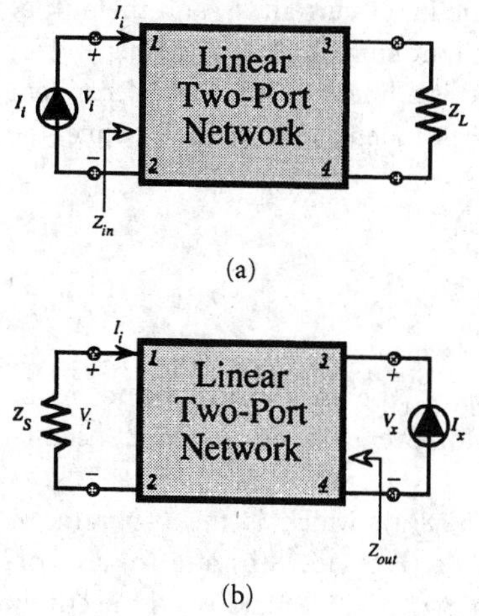

FIGURE 15.27 (a) System level model used to compute the driving-point input impedance of the linear two-port network in Fig. 15.26. (b) System level model used to compute the driving-point output impedance of the linear two-port network in Fig. 15.26.

flows through the short-circuited terminal pair, in associated reference polarity with the original output voltage, V_o. For the network in Fig. 15.26.

$$V_{to} = \lim_{Z_L \to \infty} V_o = \lim_{Z_L \to \infty} (A_v V_S) = \lim_{Z_L \to \infty} (Z_f I_S) \tag{15.76}$$

$$I_{no} = \lim_{Z_L \to 0} (-I_o) = -\lim_{Z_L \to 0} (A_i I_S) = -\lim_{Z_L \to 0} (Y_f V_S) \tag{15.77}$$

Thus, the driving-point output impedance, Z_{out}, is expressible as

$$Z_{out} = \frac{\lim_{Z_L \to \infty} A_V}{\lim_{Z_L \to 0} Y_f} = -\frac{\lim_{Z_L \to \infty} Z_f}{\lim_{Z_L \to 0} A_i} \tag{15.78}$$

Two-Port Parameters

In order to relate the four transfer and two driving-point specifications introduced in the preceding section to the measurable properties of a linear two-port network, it is necessary to interrelate the four two-port network variables, V_i, I_i, V_o, and I_o. Consider the situation in which the internal topology of the linear two-port network in Fig. 15.26 is unknown, inaccessible for measurement and characterization, or too cumbersome for traditional KVL and KCL analyses. Then, only two independent equations of equilibrium can be written: a mesh or a nodal equation for the input loop, comprised of the source termination and the input port, and a mesh or a nodal equation for the output mesh consisting of the load impedance and the output port. A unique solution for the four two-port variables therefore requires that two of these variables be made independent. The two nonindependent variables are necessarily dependent quantities. In principle, any two of the four two-port electrical variables can be selected as independent, but depending on the electrical nature of the considered network, some independent two-port variable pairs may more readily lend themselves to measurement, modeling, and circuit analysis than others. Indeed, some independent variable pairs may be inappropriate for specific types of linear networks, in the sense that they embody two-port parameters that are difficult or impossible to measure. The specific selection determines the type of parameters used to model and characterize a linear two-port network.

Hybrid *h*-Parameters

When a hybrid h-parameter equivalent circuit is selected to model a linear two-port network, the two-port network variables selected as independent are the input current, I_i, and the output voltage, V_o. This means that the input voltage, V_i, and the output current, I_o, are the dependent variables of the model. Because the two-port network at hand is a linear circuit, each dependent two-port variable is a linear superposition of the effects of each independent variable. Thus, the dual volt-ampere relationships,

$$\begin{aligned} V_i &= h_{11} I_i + h_{12} V_o \\ I_o &= h_{21} I_i + h_{22} V_o \end{aligned} \tag{15.79}$$

can be hypothesized. Dimensional consistency mandates that h_{11} be an impedance and h_{22} be an admittance, while h_{12} and h_{21} are both dimensionless. Moreover, the four h_{ij} are constants, independent of all four two-port electrical variables.

The equations in (15.79) produce the model of Fig. 15.28(a), which is the ***h*-parameter equivalent circuit** of a linear two-port network. The models of Figs. 15.28(b) and (c) are corresponding equivalent circuits for the terminated systems in Figs. 15.26(a) and (b), respectively.

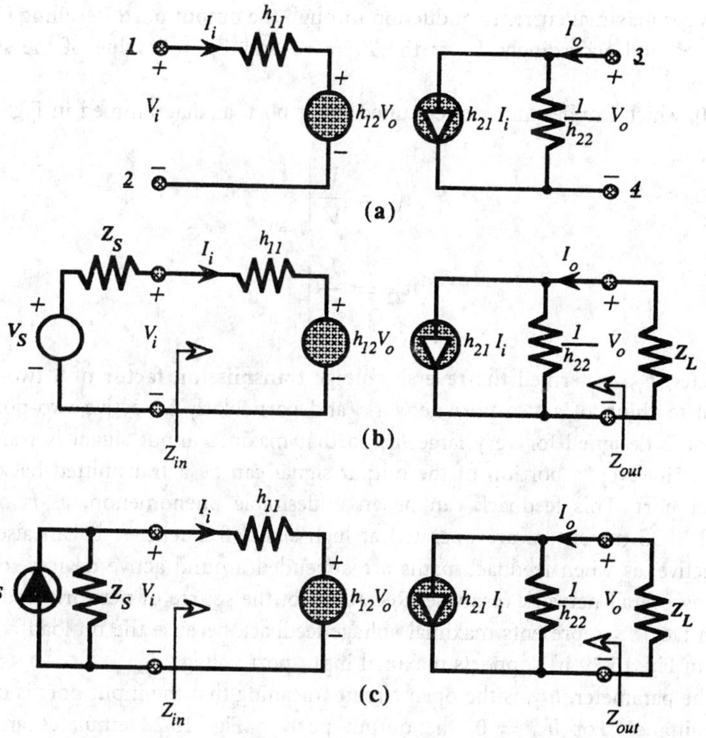

FIGURE 15.28 (a) The h-parameter equivalent circuit of the two-port network in Fig. 15.26. (b) The h-parameter model of the terminated circuit in Fig. 15.26(a). (c) The h-parameter model of the terminated circuit in Fig. 15.26(b).

It is interesting to note in Fig. 15.28(a) that the h-parameter input port model is a Thévenin equivalent circuit, which the h-parameter output port model is a Norton representation. This state of affairs mirrors fundamental circuit theory, which stipulates that any pair of terminals (i.e., any port) of a linear circuit can be supplanted by either a Thévenin or a Norton equivalent circuit. Just as Thévenin and Norton models for simple one-port networks are cast in terms of the independent electrical variables that excite that port, the Thévenin equivalent input port voltage and the Norton equivalent output port current in Fig. 15.28(a) are proportional to the electrical port quantities selected as independent variables in the h-parameter representation of the two-port network.

Measurement procedures for h-parameters derive directly from (15.79). For example, if V_o is clamped to zero, which corresponds to the short-circuited output port drawn in Fig. 15.29(a),

$$h_{11} = \left.\frac{V_i}{I_i}\right|_{V_o=0} \tag{15.80a}$$

$$h_{21} = \left.\frac{I_o}{I_i}\right|_{V_o=0} \tag{15.80b}$$

It follows that h_{11} is the *short-circuit input impedance* (meaning the output port is short circuited) of the subject two-port network. It is, in fact, a particular value of Z_{in}, the previously introduced driving-point input impedance, for the special case of zero load impedance. On the other hand, h_{21} represents the **forward short circuit current gain.** It is an optimistic measure of the ability of the considered two-port network to provide forward current gain because a short-circuited load encourages maximal current conduction through the output port. Recalling (15.73), $I_i = I_S$ when $Z_S = \infty$ and accordingly, h_{21} is the $Z_S = \infty$ and $Z_L = 0$ value of the system current gain, A_i.

For $I_i = 0$, which implies an open circuited input port, as diagrammed in Fig. 15.29(b),

$$h_{12} = \left.\frac{V_i}{V_o}\right|_{I_i=0} \tag{15.81a}$$

$$h_{22} = \left.\frac{I_o}{V_o}\right|_{I_i=0} \tag{15.81b}$$

The parameter, h_{12}, is termed the **reverse voltage transmission factor** of a two-port network. It is natural to think of a two-port network, and particularly an active two-port network, as a circuit that is designed for very large h_{21} so that maximal input signal is transmitted to its output port. However, a portion of the output signal can be a transmitted back, or fed back, to the input port. This feedback can be an undesirable phenomenon, as is observed when bipolar and MOS transistors are operated at high signal frequencies. It can also be a specific design objective, as when feedback paths are appended around active devices for the purpose of optimizing overall network response. Regardless of the source of network feedback, h_{12} is its measure. In fact, h_{12} represents maximal voltage feedback because the no load condition at the input port in Fig. 15.29(b) supports maximal input port voltage, V_i.

Finally, the parameter, h_{22}, is the open circuit (meaning that the input port is open circuited) output admittance. For $h_{22} = 0$, the output port in Fig. 15.28 emulates an ideal current controlled current source. If $Z_S = \infty$, h_{22} is exactly the inverse of the previously introduced driving-point output impedance, Z_{out}.

A parameter measurement complication arises when the two-port network undergoing investigation is an active circuit. This complication stems from the fact that the transistors and other active derives embedded within such a network are inherently nonlinear. These devices behave as approximately linear circuit elements only when static voltages and currents, separate and apart from the input signal source, are applied to bias them in the linear regime of their volt-ampere characteristic curves. Unfortunately, a short-circuited output port and an open-circuited input port are likely to upset requisite biasing. Assuming that the test sources delineated in Fig. 15.29 are sinusoids having zero average value, the biasing dilemma can be circumvented by shunting the original load impedance with a sufficiently large capacitance, as suggested in Fig. 15.30(a). Similarly, a sufficiently large inductance in series with the source impedance at the input port approximates an open circuit for signal test conditions, without disturbing the biasing current flowing into the input port under actual signal input conditions. The latter connection is illustrated in Fig. 15.30(b).

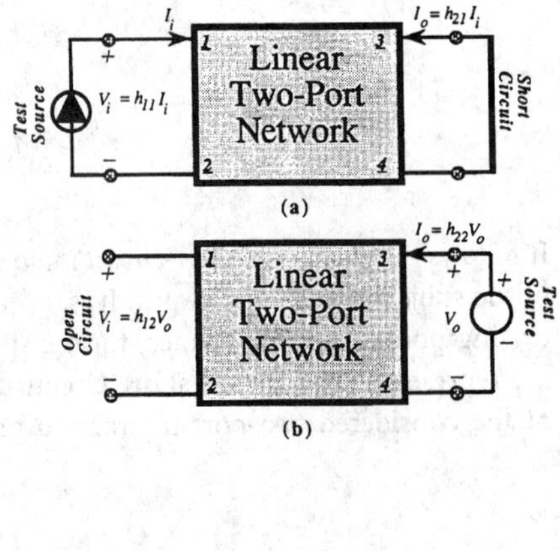

FIGURE 15.29 (a) Measurement of the short circuit h-parameters. (b) Measurement of the open circuit h-parameters.

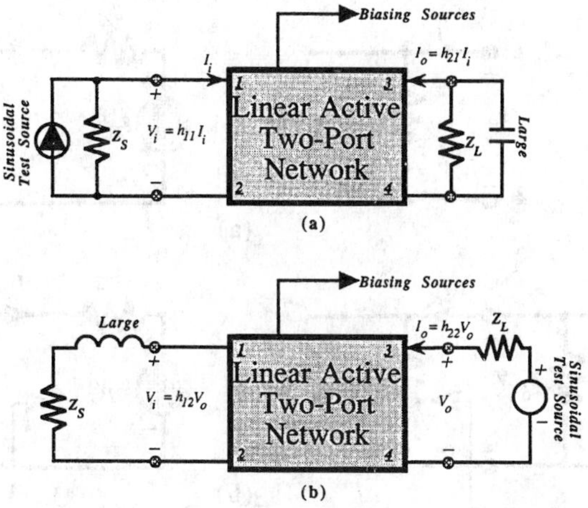

FIGURE 15.30 (a) Measurement of the short circuit h-parameters for a linear active two-port network. (b) Measurement of the open circuit h-parameters for a linear active two-port network.

Although the test fixturing in Fig. 15.30 is conceptually correct, it is impractical when applied to most linear active circuits. This impracticality stems from the fact that almost all active networks are potentially unstable in the sense that a range of passive load or source terminations exists that can support sustained network oscillations. Unfortunately, short-circuited, and especially open-circuited, load and source impedances lie within this range of potentially unstable behavior. For this reason, the h-parameters, as well as the other two-port parameters discussed below, are rarely measured directly for active two-port networks. Instead, they are measured indirectly by calculating them as a function of the measured **scattering** (S-) **parameters** [1, 2]. Like the h-parameters, the S-parameters also measure the I-O immittances forward gain, and feedback properties of a two-port network. Unlike the h-parameters, however, S-parameters are deduced under the condition of finite, nonzero, and equal source and load impedances that are selected to ensure that the network undergoing test operates as a stable linear structure. Invariably, this equal source and load test termination, which is called the **measurement reference impedance**, is a 50-Ω resistance.

Hybrid g-Parameters

The independent and dependent variables used to define the **hybrid g-parameters,** g_{ij}, of a linear two-port network are the converse of those used in conjunction with h-parameters. Specifically, the independent variables for g-parameter modeling are the input voltage, V_i, and the output current, I_o, thereby rendering the input current, I_i, and the output voltage, V_o, dependent electrical quantities. From superposition theory, it follows that

$$\begin{aligned} I_i &= g_{11} V_i + g_{12} I_o \\ V_o &= g_{21} V_i + g_{22} I_o \end{aligned} \tag{15.82}$$

which gives rise to the *g-parameter equivalent circuit* depicted in Fig. 15.31(a). This equivalent circuit represents the input port of a two-port network as a Norton topology, while the output port is modeled by a Thévenin equivalent circuit. The g-parameter equivalent circuit for the system of Fig. 15.26(b) is illustrated in Fig. 15.31(b). A similar structure can be drawn for the system shown in Fig. 15.26(a). Given the Norton topology for the input port model of a g-parameter equivalent circuit, the system model is in Fig. 15.31(b), which supplants the signal source by its Norton equivalent circuit, is more computationally expedient than a system model which represents the signal source as a Thévenin equivalent circuit.

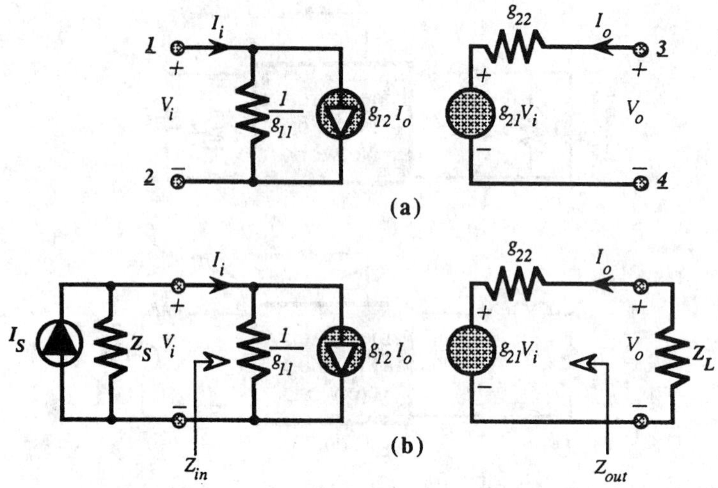

FIGURE 15.31 (a) The g-parameter equivalent circuit of the two-port network in Fig. 15.26. (b) The g-parameter model of the terminated circuit in Fig. 15.26(a).

As is the case with h-parameters, the measurement strategy for g-parameters derives directly from the defining volt-ampere modeling equations. To this end, (15.82) suggests that

$$g_{11} = \left.\frac{I_i}{V_i}\right|_{I_o=0} \tag{15.83a}$$

$$g_{21} = \left.\frac{I_o}{V_i}\right|_{I_o=0} \tag{15.83b}$$

Thus, g_{11} is the **open circuit input admittance** (meaning the output port is open circuited) of a two-port network. It is the value of the driving-point input admittance when the load impedance is an open circuit. The parameter, g_{21}, represents the **forward open circuit voltage gain.** It is an optimistic measure of the ability of a two-port network to provide forward voltage gain because an open circuited load conduces maximal output port voltage. Equation (15.82) also confirms that

$$g_{12} = \left.\frac{I_i}{I_o}\right|_{V_i=0} \tag{15.84a}$$

$$g_{22} = \left.\frac{V_o}{V_o}\right|_{V_i=0} \tag{15.84b}$$

Like h_{12}, g_{12}, which is the **reverse current transmission factor** of a two-port network, is a measure of feedback internal to a two-port network. On the other hand, g_{22} is the **short circuit output impedance** (input port is short circuited). Figure 15.32 outlines the basic measurement strategy inferred by (15.83) and (15.84). The comments offered earlier in regard to the h-parameter characterization of linear active two-port networks apply as well to g-parameter measurements.

At this juncture, two alternatives for the modeling and analysis of linear two-port networks have been formulated. If h-parameters are selected as the modeling vehicle for a given two-port network, the equivalent circuit shown in Fig. 15.28(a) results. On the other hand, g-parameters give rise to the two-port equivalent circuit offered in Fig. 15.31(a). If the two-port network modeled by h-parameters is identical to the two-port network represented by g-parameters,

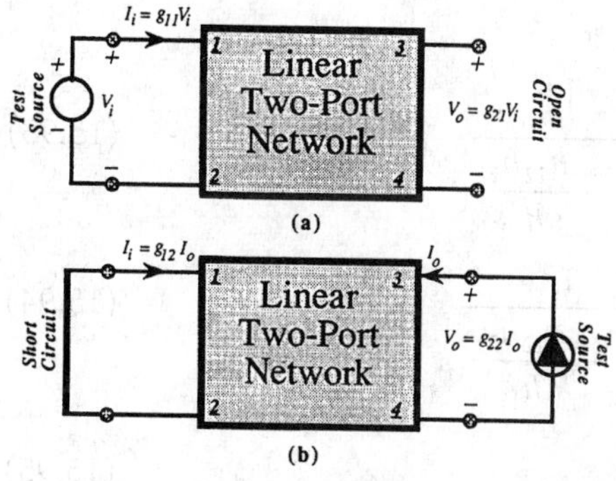

FIGURE 15.32 (a) Measurement of the open circuit g-parameters. (b) Measurement of the short circuit g-parameters.

both equivalent circuits deliver the same system I-O, forward transfer, and reverse transfer characteristics. This homogeneity requirement implies a relationship between the h- and the g-parameters; that is, the h_{ij} and the g_{ij} are not mutually independent.

In order to establish the relationship between h_{ij} and g_{ij}, it is convenient to begin by rewriting (15.79) as the matrix equation

$$\mathbf{Y} = \mathbf{HX} \tag{15.85}$$

where the matrix of h-parameters is

$$\mathbf{H} = \begin{bmatrix} h_{11} & h_{12} \\ h_{21} & h_{22} \end{bmatrix} \tag{15.86}$$

$$\mathbf{X} = \begin{bmatrix} I_i \\ V_o \end{bmatrix} \tag{15.87}$$

is the vector of h-parameter independent electrical variables, and

$$\mathbf{Y} = \begin{bmatrix} V_i \\ I_o \end{bmatrix} \tag{15.88}$$

is the vector of h-parameter dependent variables. Assuming that the h-parameter matrix is not singular, which means that the determinant,

$$\Delta_h = h_{11}h_{22} - h_{12}h_{21} \tag{15.89}$$

of the matrix of h-parameters is nonzero, (15.85) implies

$$\mathbf{X} = \mathbf{H}^{-1}\mathbf{Y} \tag{15.90}$$

The last result is a system of linear algebraic equations that cast I_i and V_o as dependent variables and V_i and I_o as independent variables; that is, (15.90) is identical to (15.82). It follows that if the matrix of g-parameters is

$$\mathbf{G} = \begin{bmatrix} g_{11} & g_{12} \\ g_{21} & g_{22} \end{bmatrix} \tag{15.91}$$

$$\mathbf{G} = \mathbf{H}^{-1} \tag{15.92}$$

Specifically,

$$g_{11} = \frac{h_{22}}{\Delta_h} = \frac{1}{h_{11} - \dfrac{h_{12}h_{21}}{h_{22}}} \tag{15.93}$$

$$g_{22} = \frac{h_{11}}{\Delta_h} = \frac{1}{h_{22} - \dfrac{h_{12}h_{21}}{h_{11}}} \tag{15.94}$$

$$g_{12} = -\frac{h_{12}}{\Delta_h} \tag{15.95}$$

and

$$g_{21} = -\frac{h_{21}}{\Delta_h} \tag{15.96}$$

Note that if $\Delta_h = 0$, the g-parameters of a two-port network cannot be calculated or measured. In other words, the g-parameters of a network do not exist when the matrix of h-parameters is singular. Conversely, the h-parameters of the network cannot be calculated or measured when the matrix of g-parameters is singular. This statement follows from (15.92), which can be rewritten as

$$\mathbf{H} = \mathbf{G}^{-1} \tag{15.97}$$

provided that the determinant,

$$\Delta_g = g_{11}g_{22} - g_{12}g_{21} \tag{15.98}$$

of the matrix of g-parameters in (15.86) is not zero.

Short-Circuit Admittance (y-) Parameters

When **short-circuit admittance parameters**, and **y-parameters**, are used to characterize a linear two-port network, the electrical variables selected as independent quantities are the input port voltage, V_i, and the output port voltage, V_o. The resultant volt-ampere relationships can be cast as the matrix equation

$$\begin{bmatrix} I_i \\ I_o \end{bmatrix} = \begin{bmatrix} y_{11} & y_{12} \\ y_{21} & y_{22} \end{bmatrix} \begin{bmatrix} V_i \\ V_o \end{bmatrix} \tag{15.99}$$

where all of the y-parameters, y_{ij}, are in units of admittance. The corresponding **y-parameter equivalent circuit** of a two-port network is the Norton input port and Norton output structure drawn in Fig. 15.33.

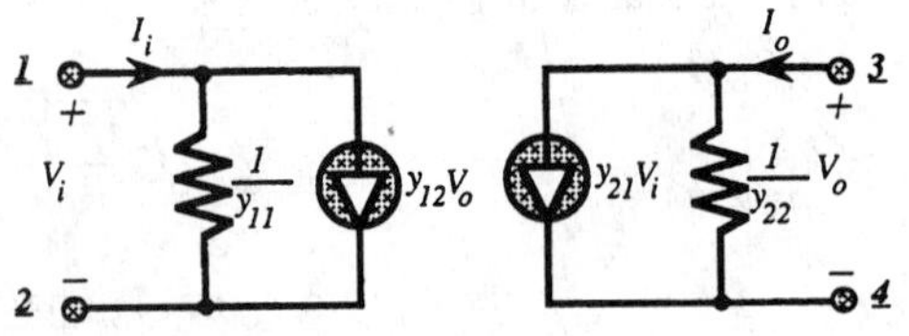

FIGURE 15.33 The y-parameter equivalent circuit of a linear two-port network.

With the output port short circuited ($V_o = 0$), (15.99) yields

$$y_{11} = \left.\frac{I_i}{V_i}\right|_{V_o=0} \tag{15.100a}$$

$$y_{21} = \left.\frac{I_o}{V_i}\right|_{V_o=0} \tag{15.100b}$$

Equation (15.100) suggests that y_{11} is the **short-circuit input admittance** (meaning the output port is short circuited) of a two-port network. It is the value of the inverse of the driving-point input impedance when the load impedance is zero. The parameter, y_{21}, is the **forward short-circuit transadmittance.** Like h_{21} and g_{21}, y_{21} is a measure of the forward signal transmission capability of a linear two-port network. Whereas h_{21} measures this capability through a short circuit current gain, and g_{21} reflects forward gain characteristics through an open circuit voltage gain, y_{21} measures forward gain in terms of a short circuit transadmittance from input to output port. Equation (15.99) additionally shows that

$$y_{12} = \left.\frac{I_i}{V_o}\right|_{V_i=0} \tag{15.101a}$$

$$y_{22} = \left.\frac{I_o}{V_o}\right|_{V_i=0} \tag{15.101b}$$

Thus, y_{22} is the **short-circuit output admittance** (meaning the input port is short circuited). Finally, y_{12} is the **reverse short-circuit transadmittance.** Like h_{12} and g_{12}, y_{12} is a measure of the feedback intrinsic to a linear two-port network.

Because a two-port network study predicated on y-parameters must produce analytical results that mirror a study based on either h-parameter or g-parameter modeling, the y-parameters of a two-port network can be related to either the network h- or g-parameters. For example, from (15.100) and (15.79)

$$y_{11} = \left.\frac{I_i}{V_i}\right|_{V_o=0} = \frac{1}{h_{11}} \tag{15.102}$$

$$y_{21} = \left.\frac{I_o}{V_i}\right|_{V_o=0} = \frac{h_{21}}{h_{11}} \tag{15.103}$$

The result in (15.102) is reasonable in view of the fact that h_{11} symbolizes a short circuit input impedance, while y_{11} is an input admittance that is also evaluated for a short-circuited output port. Because both h_{21} and y_{21} comprise a measure of forward signal transmission, y_{21} is, as expected, directly proportional to h_{21}.

For $V_i = 0$, (15.79) forces the constraint, $h_{11}I_i = -h_{12}V_o$. Accordingly,

$$y_{12} = \left.\frac{I_i}{V_o}\right|_{V_i=0} = -\frac{h_{12}}{h_{11}} \tag{15.104}$$

and

$$i_o = h_{21}\left(-\frac{h_{12}V_o}{h_{11}}\right) + h_{22}V_o$$

where

$$y_{22} = \left.\frac{I_o}{V_o}\right|_{V_i=0} = h_{22} - \frac{h_{12}h_{21}}{h_{11}} \tag{15.105}$$

As expected, y_{12} is proportional to h_{12}. Note, however, that y_{22} is not in general equal to h_{22}, despite the fact that both y_{22} and h_{22} are output port admittances. The engineering explanation of the mathematical observation, $y_{22} \neq h_{22}$, is that y_{22} is a *short circuit* output admittance, whereas h_{22} represents an output admittance with the input port *open circuited.* Similarly, the y_{ij} can be expressed in terms of the g_{ij} by applying the parametric definitions of (15.100) and (15.101) to the defining g-parameter equations of (15.82).

An interesting special case is that of a **reciprocal two-port network.** A linear network is said to be *reciprocal* if its feedback and feedforward y-parameters are identical; that is, $y_{12} \equiv y_{21}$. In view of the defining y-parameter relationships, two-port reciprocity means that a voltage, say V, applied across the input port of a linear two-port network produces a short-circuited output current that is identical to the short-circuited input port current that would result if V were to be removed from the input port and impressed instead across the output port. From (15.103) and (15.104), the h-parameter implication of network reciprocity is $h_{12} = -h_{21}$. Recalling (15.94) and (15.95), the corresponding g-parameter implication is $g_{12} = -g_{21}$.

Open Circuit Impedance (z-) Parameters

The use of **open-circuit impedance parameters,** or **z-parameters,** is premised on selecting the input port current, I_i, and the output port current, I_o, as independent electrical variables. The upshot of this selection is the volt-ampere expression,

$$\begin{bmatrix} V_i \\ V_o \end{bmatrix} = \begin{bmatrix} z_{11} & z_{12} \\ z_{21} & z_{22} \end{bmatrix} \begin{bmatrix} I_i \\ I_o \end{bmatrix} \tag{15.106}$$

where the z_{ij} are called the **open-circuit impedance parameters,** or **z-parameters,** of a two-port network. Equation (15.106) produces the **z-parameter equivalent circuit** of Fig. 15.34, which is seen as exploiting Thévenin's theorem at both the input and the output network ports. A comparison of (15.106) with (15.99) confirms that the matrix of z-parameters is the inverse of the matrix of y-parameters, provided, of course, that the y-parameter matrix is not singular. In particular,

$$z_{11} = \frac{y_{22}}{\Delta_y} = \frac{1}{y_{11} - \dfrac{y_{12}y_{21}}{y_{22}}} \tag{15.107}$$

$$z_{22} = \frac{y_{11}}{\Delta_y} = \frac{1}{y_{22} - \dfrac{y_{12}y_{21}}{y_{11}}} \tag{15.108}$$

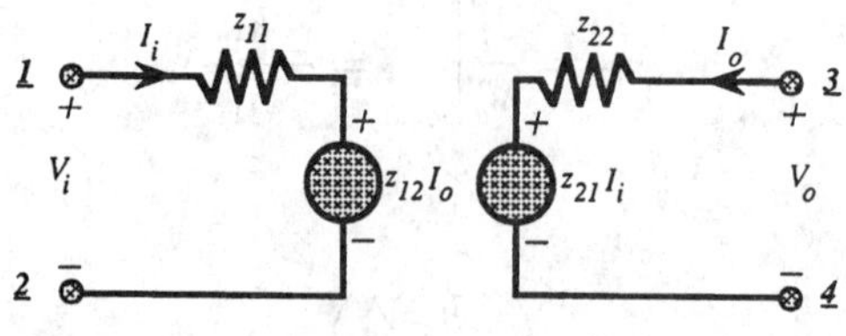

FIGURE 15.34 The z-parameter equivalent circuit of a linear two-port network.

$$z_{12} = \frac{y_{12}}{\Delta_y} - \frac{y_{12}}{\Delta y} \tag{15.109}$$

and

$$z_{21} = -\frac{y_{21}}{\Delta_y} \tag{15.110}$$

where the determinant of y-parameters,

$$\Delta_y = y_{11}y_{22} - y_{12}y_{21} \tag{15.111}$$

is presumed to be nonzero. When the output port is open circuited, the resultant input impedance is z_{11} and the corresponding forward transimpedance is z_{21}. The parameter, z_{12}, is the **open-circuit reverse** transimpedance (a measure of internal feedback), and z_{22} symbolizes the **open-circuit output impedance** (meaning an open circuited input port).

The matrix of y-parameters is, of course, the inverse of the matrix of z-parameters when the z-parameter matrix is nonsingular. When the z-parameter matrix is singular, the y-parameters of a linear two-port network can be neither measured nor computed. Observe from (15.109) and (15.110) that network reciprocity, which implies $y_{12} = y_{21}$, forces $z_{12} = z_{21}$.

Transmission (c-) Parameters

The **transmission parameters**, which are often referred to as the **chain parameters** or **c-parameters**, of a linear two-port network are implicitly defined by the volt-ampere description

$$\begin{bmatrix} V_i \\ I_i \end{bmatrix} = \begin{bmatrix} c_{11} & c_{12} \\ c_{21} & c_{22} \end{bmatrix} \begin{bmatrix} V_o \\ -I_o \end{bmatrix} \tag{15.112}$$

Note that the output voltage, V_o, and the negative of the output current, I_o, are the selected independent variables, thereby constraining the input voltage, V_i, and the input current, I_i, as dependent electrical variables. In contrast to the four other two-port parameter representations, the c-parameter volt-ampere relationships are rarely used to construct an equivalent circuit. Instead, the mathematical description of (15.112) is directly exploited in both analysis and design projects. This description is particularly useful in the design of passive filters [3]. It also proves expedient in the analysis of passive circuits such as those established by parasitic impedances associated with the metallization of a monolithic circuit.

As usual, the strategy that underlies the measurement and calculation of the c-parameters derives from the defining volt-ampere relationships. To this end, let the output port of the system in Fig. 15.26(a) be open circuited so that the output current is nulled. Simultaneously, let the input port be driven by a voltage, V_i, which establishes an input port current of I_i, as abstracted in Fig. 15.35(a). By (15.112) this test fixturing produces

$$\frac{1}{c_{11}} = \left.\frac{V_o}{V_i}\right|_{I_o=0} \tag{15.113a}$$

$$\frac{1}{c_{21}} = \left.\frac{V_o}{I_i}\right|_{I_o=0} \tag{15.113b}$$

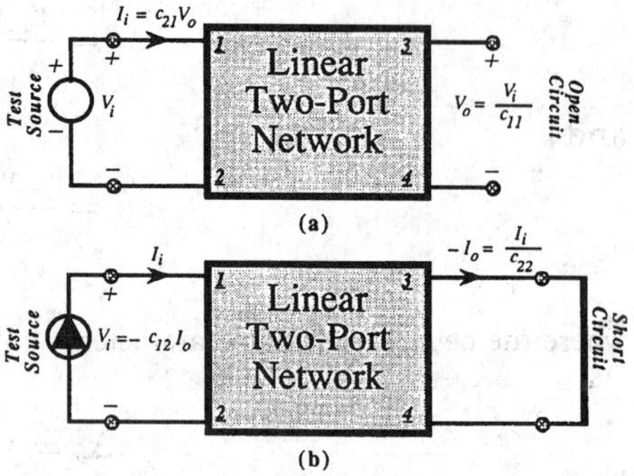

FIGURE 15.35 (a) Test fixturing for the measurement of the open circuit c-parameters. (b) Test fixturing for the measurement of the short circuit c-parameters.

The foregoing result confirms that c_{11} is the **inverse of the open-circuit forward voltage gain** of a two-port network. Recalling (15.82) and (15.106), c_{11} is seen as being equal to the inverse of the g-parameter; g_{21}; that is,

$$c_{11} = \frac{1}{g_{21}} = \frac{z_{11}}{z_{21}} \tag{15.114}$$

Moreover, c_{21} is the **inverse of the open-circuit forward transimpedance**. From (15.106),

$$c_{21} = \frac{1}{z_{21}} \tag{15.115}$$

With the input port of the subject two-port excited by a current source, as depicted in Fig. 15.35(b), let the output port be short circuited instead of open circuited. By (15.112)

$$\frac{1}{c_{12}} = -\frac{I_o}{V_i}\bigg|_{V_o=0} \tag{15.116a}$$

$$\frac{1}{c_{22}} = -\frac{I_o}{I_i}\bigg|_{V_o=0} \tag{15.116b}$$

Because the quantity, $(-I_o)$, is simply an antiphase version of the output current, I_o, delineated in Fig. 15.26(a), c_{12} is the **inverse of the antiphase forward short circuit transadmittance** of a two-port network. From (15.103) and (15.106), this parameter relates to the y- and z-parameters in accordance with

$$c_{12} = -\frac{1}{y_{21}} = \frac{\Delta_z}{z_{21}} \tag{15.117}$$

where

$$\Delta_z = z_{11}z_{22} - z_{12}z_{21} \tag{15.118}$$

is the determinant of the z-parameter matrix. Finally, c_{22} is the **inverse antiphase forward short circuit current gain**. Equations (15.102), (15.103), and (15.106) provide

$$c_{22} = -\frac{y_{11}}{y_{21}} = \frac{z_{22}}{z_{21}} \tag{15.119}$$

Observe that c_{22} and c_{11} are dimensionless, c_{12} has units of impedance, and c_{21} has admittance units.

Two interesting features of the c-parameter model are revealed by investigating the determinant, say Δ_c, of the c-parameter matrix. Using (15.114), (15.115), (15.117), and (15.119),

$$\Delta_c = c_{11}c_{22} - c_{12}c_{21} = \frac{z_{12}}{z_{21}} = \frac{y_{12}}{y_{21}} \tag{15.120}$$

Network reciprocity, which implies $z_{21} = z_{21}$, and also $y_{21} = y_{12}$, is therefore seen to be in one-to-one correspondence with a c-parameter matrix whose determinant is 1. On the other hand, if z_{12} (and y_{12}) is zero, which corresponds to a so-called **unilateral two-port network**, $\Delta_c = 0$. Equivalently, the c-parameter matrix is singular when the two-port network undergoing investigation has no internal feedback.

Unilateralization is often a desirable property of active networks because it implies that input signals are processed only in the forward direction (from input port to output port); that is, no response signal is fed beck to the input port from the output port. As discussed later, the lack of internal feedback renders unilateral networks unconditionally stable in that they cannot support sustained oscillations for any and all passive source and load terminations. Unfortunately, unilateralization is an idealized operating condition. Although active networks can be designed to achieve vanishingly small z_{12} at low signal frequencies, z_{12}, and hence y_{12}, increases with progressively larger signal frequencies.

Another attribute of c-parameter modeling is the ease with which it affords general expressions for the driving-point input and output impedances and the forward voltage gain of a linear two-port network. For example, from (15.65), (15.72), and (15.112)

$$Z_{\text{in}} = \frac{V_i}{I_i} = \frac{c_{11}V_o - c_{12}I_o}{c_{21}V_o - c_{22}I_o} = \frac{c_{11}Z_L + c_{12}}{c_{21}Z_L + c_{22}} \tag{15.121}$$

With the help of Fig. 15.27(b), (15.112) also provides

$$Z_{\text{out}} = \frac{V_x}{I_x} = \frac{c_{22}Z_S + c_{12}}{c_{21}Z_S + c_{11}} \tag{15.122}$$

Because

$$V_i = c_{11}V_o - c_{12}I_o = c_{11}V_o + c_{12}\left(\frac{V_o}{Z_L}\right)$$

the forward voltage gain, say A_f, from the input terminals to the output terminals of a linear two-port network is

$$A_f = \frac{V_o}{V_i} = \frac{Z_L}{c_{11}Z_L + c_{12}} \tag{15.123}$$

Note that for infinitely large Z_L, corresponding to an open-circuited output port, this gain reduces to $1/c_{11}$, which corroborates with (15.113).

An especially laudable aspect of c-parameters is its amenability to the analysis of cascaded two-port networks. To illustrate, consider the two-stage cascade of Fig. 15.36, in which the c-parameter matrix of network 1 is $\mathbf{C}_1$ and that of network 2 is $\mathbf{C}_2$. Then

$$\begin{bmatrix} V_i \\ I_i \end{bmatrix} = \mathbf{C}_1 \begin{bmatrix} V_x \\ -I_{o1} \end{bmatrix} \quad \text{and} \quad \begin{bmatrix} V_x \\ I_{i2} \end{bmatrix} = \mathbf{C}_2 \begin{bmatrix} V_o \\ -I_o \end{bmatrix} \tag{15.124a}$$

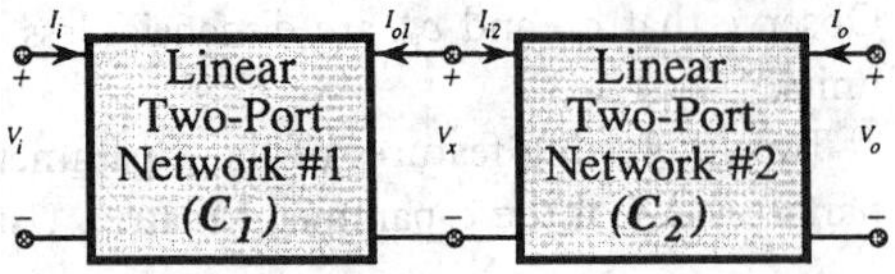

FIGURE 15.36 A two-stage cascade of linear two-port networks.

where the interstage voltage is denoted as V_x. Because the indicated interstage currents are such that $I_{i2} = -I_{o1}$, the preceding result yields

$$\begin{bmatrix} V_i \\ I_i \end{bmatrix} = \mathbf{C}_1 \begin{bmatrix} V_x \\ I_{i2} \end{bmatrix} = \mathbf{C}_1\mathbf{C}_2 \begin{bmatrix} V_o \\ -I_o \end{bmatrix} \stackrel{\Delta}{=} \mathbf{C}_{\text{eff}} \begin{bmatrix} V_o \\ -I_o \end{bmatrix} \tag{15.124b}$$

Equation (15.124) suggests that the effective c-parameter matrix of the two stage cascade is the simple matrix product

$$\mathbf{C}_{\text{eff}} = \mathbf{C}_1\mathbf{C}_2 \tag{15.125}$$

Because matrix multiplication is not commutative, care must be exercised that the requisite multiplication in (15.125) is executed in the proper order. This note of caution highlights the effects of interactive loading between two linear networks. In particular, the terminal electrical properties of network 1 in cascade with network 2 are not necessarily the same—and usually are not the same—as those associated with network 2 placed in cascade with network 1. Equation (15.125) extends readily to the case of an n-stage cascade. In particular, the effective c-parameter matrix for a cascade of n linear two-port networks is the product of this c-parameter matrices pertinent to the individual n stages that comprise the cascade.

The utility of (15.125) can be demonstrated by considering the tee network of Fig. 15.37(a). Each of the three impedances, Z_1, Z_2, and Z_3, can be viewed as elementary two-port networks. As suggested by Fig. 15.37(b), the cascade of these networks gives rise to the topology of Fig. 15.37(a). Network 1 is the series connection of Z_1 between its input and output ports. An application of (15.113) and (15.116) to this structure gives a first network c-parameter matrix of

$$\mathbf{C}_1 = \begin{bmatrix} 1 & Z_1 \\ 0 & 1 \end{bmatrix}$$

Network 3 is topologically identical to network 1 and therefore,

$$\mathbf{C}_3 = \begin{bmatrix} 1 & Z_3 \\ 0 & 1 \end{bmatrix}$$

Network 2 is the impedance, Z_2, connected in shunt with both its input and output ports. Equations (15.113) and (15.116) provide

$$\mathbf{C}_2 = \begin{bmatrix} 1 & 0 \\ Y_2 & 1 \end{bmatrix}$$

where Y_2 is the admittance of the impedance, Z_2. It follows that the effective c-parameter matrix for the circuit of Fig. 15.37(a) is

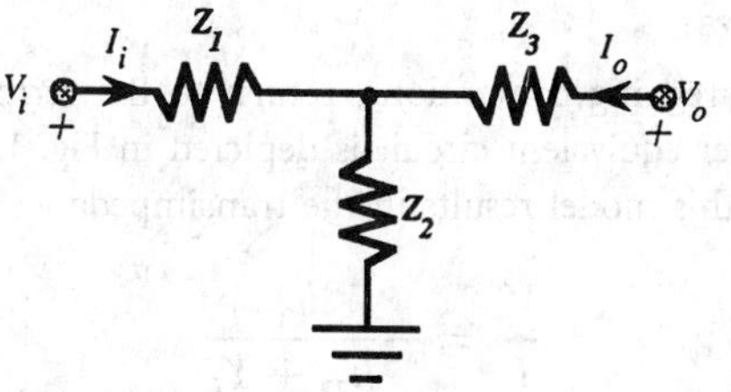

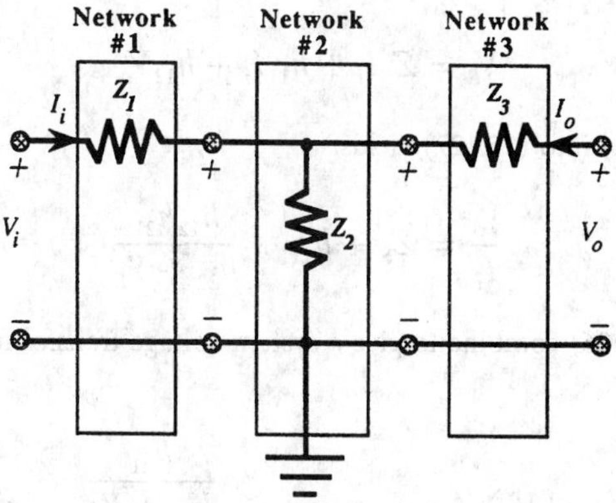

FIGURE 15.37 (a) A two-port tee network formed of three linear impedance elements. (b) The network of (a) viewed as a cascade of three two-port networks.

$$C_{eff} = C_1C_2C_3 = \begin{bmatrix} 1 & Z_1 \\ 0 & 1 \end{bmatrix}\begin{bmatrix} 1 & 0 \\ Y_2 & 1 \end{bmatrix}\begin{bmatrix} 1 & Z_3 \\ 0 & 1 \end{bmatrix}$$

The indicated matrix multiplications produce the effective values of the c_{ij}. For given source and load terminations, (15.121) to (15.123) can be exploited to determine the driving-point input impedance, the driving-point output impedance, and the I-O port voltage gain, respectively, of the tee configuration.

Circuit Analysis with Two-Port Parameters

The two-port parameter sets introduced in the preceding sections allow for an unambiguous volt-ampere definition of a linear two-port network exclusively in terms of electrical characteristics that are observable and measurable at the I-O ports of the network. The respective equivalent circuits corresponding to these parameter sets exploit superposition theory and comprise a modest extension of the classic one-port version of Thévenin's and Norton's theorems. These equivalent circuits comprise a convenient means of analyzing the driving-point and transfer properties of terminated electrical and electronic systems, the intrinsic circuit topologies of which are either unknown or too complicated for conventional analysis predicated on KVL and KCL.

Analysis via *h*-Parameters

In order to illustrate the foregoing contentions, return to the terminated two-port system of Fig. 15.26(a). Its h-parameter equivalent circuit is depicted in Fig. 15.28(b). The application of KCL to the output port of this model results in the transimpedance function

$$\frac{V_o}{I_i} = -\frac{h_{21}}{h_{22} + Y_L} \tag{15.126}$$

where Y_L is the admittance of the load impedance, Z_L. A KVL equation around the input port loop gives

$$V_S = Z_S I_i + h_{11} I_i + h_{12} V_o \tag{15.127}$$

The insertion of (15.126) into (15.127) yields

$$\frac{V_S}{I_i} = Z_S + h_{11} - \frac{h_{12}h_{21}}{h_{22} + Y_L} \tag{15.128}$$

The last result and (15.126) combine to give a system voltage transfer function, or voltage gain, A_v, of

$$A_v = \frac{V_o}{V_S} = \left(\frac{V_o}{I_i}\right)\left(\frac{I_i}{V_S}\right) = -\frac{\dfrac{h_{21}}{h_{22} + Y_L}}{Z_S + h_{11} - \dfrac{h_{12}h_{21}}{h_{22} + Y_L}} \tag{15.129}$$

The driving-point input impedance, Z_{in}, derives directly from an application of (15.74) to (15.128). In particular,

$$Z_{\text{in}} = h_{11} - \frac{h_{12}h_{21}}{h_{22} + Y_L} \tag{15.130}$$

Because the output port in an h-parameter model is a Norton equivalent circuit, the driving-point output admittance, say Y_{out}, is a more convenient measure of output port characteristics than is the driving-point output impedance, Z_{out}. The latter figure of merit can be found by inverting the expression in (15.78), which requires *a priori* knowledge of the forward transadmittance, Y_f. From (15.70) and (15.129),

$$Y_f = -Y_L A_v = \frac{\dfrac{h_{21}Y_L}{h_{22} + Y_L}}{Z_s + h_{11} - \dfrac{h_{12}h_{21}}{h_{22} + Y_L}} \tag{15.131}$$

Then, by (15.78)

$$Y_{\text{out}} = \frac{1}{Z_{\text{out}}} = -\frac{\lim\limits_{Y_L \to \infty} Y_f}{\lim\limits_{Y_L \to 0} A_v} = h_{22} - \frac{h_{12}h_{21}}{h_{11} + Z_S} \tag{15.132}$$

Note the similarity in form between (15.130) and (15.132). Indeed, one expression mirrors the other if the subscripts 1 and 2 are interchanged and the symbol for the load admittance is interchanged with that of the source impedance.

The analysis leading to the expressions for the basic indices of network performance is straightforward, but the fundamental purpose of analysis is not the disclosure of mathematically elegant results. Rather, it is the insightful understanding that is commensurate either with network optimization or with the development of engineering strategies that produce reliable and manufacturable designs. A prerequisite for gaining such understanding is an accurate interpretation of formulated results.

To this end, return to (15.129) and consider the special case of zero internal feedback; that is, the case of $h_{12} = 0$. If feedback from the output port to the input port of a linear network is viewed as a signal path that establishes an electrical loop around the I-O ports of the zero feedback subcircuit, it is logical to refer to the zero feedback value of the gain as the **open loop gain** of the system. From (15.129), the open loop voltage gain, say A_{vo}, is

$$A_{vo} = -\frac{h_{21}}{(h_{22} + Y_L)(h_{11} + Z_S)} \tag{15.133}$$

Therefore, the voltage gain in (15.129) can be written as

$$A_v = \frac{V_o}{V_S} = \frac{A_{vo}}{1 + h_{12}A_{vo}} \tag{15.134}$$

This overall system voltage gain is termed the **closed loop gain** because it incorporates the effects of feedback on the open loop cell, whose gain is given by (15.133). The quantity, $h_{12}A_{vo}$, in the denominator on the right-hand side of (15.134) is the *loop gain*, T_h. It is effectively the gain of the signal path formed by a cascade interconnection of the open loop signal path from the input port to the output port and the feedback path from the output port back to the input port. Accordingly, with

$$T_h \triangleq h_{12}A_{vo} = -\frac{h_{12}h_{21}}{(h_{22} + Y_L)(h_{11} + Z_S)} \tag{15.135}$$

(15.134) becomes

$$A_v = \frac{A_{vo}}{1 + T_h} \tag{15.136}$$

The concepts of open loop gain, closed loop gain, and loop gain are clarified by the block diagram of Fig. 15.38. This abstraction of the two-port system in Fig. 15.26(a) underscores the fact that A_{vo} is the forward gain without feedback (or simply, the open loop gain). Note that for $h_{12} = 0$, the feedback signal, V_f, is nulled, thereby forcing $V_o = A_{vo}V_e = A_{vo}V_S$. Note further that the gain of the loop formed by the forward gain and feedback gain blocks is $h_{12}A_{vo}$, because with $V_S = 0$, $V_f = h_{12}V_o = h_{12}A_{vo}V_e$. The closed loop gain predicted by (15.134) is likewise confirmed in view of the observation

$$V_o = A_{vo}V_e = A_{vo}(V_S - V_f) = A_{vo}(V_S - h_{12}V_o) \tag{15.137}$$

which produces (15.134) directly.

Equations (15.134) and (15.136) loom especially significant in regard to the design of active two-port networks. If over a range of relevant signal frequencies the magnitude, $|T_h|$, of the loop gain is much smaller than 1, the closed loop gain, A_v, collapses to its open loop value, A_{vo}. From (15.133), A_{vo} is seen to be functionally dependent on three h-parameters, the source impedance, and the load admittance. The dependence of the effective gain on Z_S and Y_L means that the

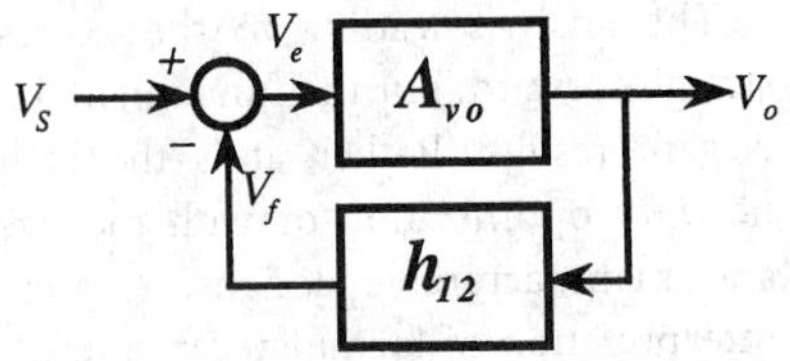

FIGURE 15.38 Block diagram representation of the terminated linear two-port network in Fig. 15.26(a).

observable gain of the active two-port is vulnerable to changes in source and load terminations and uncertainties in their respective values. A particularly strong dependence on source and load terminating immittances limits the utility of the active network in general purpose voltage amplification applications.

Similar statements can be made in regard to the noted functional dependence of A_{vo} on the three h-parameters, h_{11}, h_{12}, and h_{22}, except that this gain sensitivity problem is even more insidious. The h-parameters (and any other type of two-port network parameters) of active cells are determined by the device biasing currents and voltages implemented to achieve reasonable network linearity. This biasing is a function of the degree to which the externally applied bias sources, or **power supplies**, are regulated. They are also dependent on the operating temperature of the utilized devices. Unless special design care is exercised to stabilize the active device biasing levels against power supply uncertainties and changes in operating temperature, the overall system gain is likely to display undesirable variations. However, even if biasing levels are appropriately stabilized, problems can nonetheless arise from the fact that the h-parameters are additionally dependent on phenomenological processes indigenous to the volt-ampere characteristics of the active devices embedded within the two-port network. Many of these processes are related to physical parameters that are difficult or even impossible to measure accurately. Other physical parameters may be relatively easy to measure, but their particular values are subject to routinely nonzero manufacturing tolerances. Some of these physically based parameters are even influenced by electrical coupling among devices and circuit elements laid out in close proximity to one another. In short, significant tolerances can plaque the measured two-port parameters of a linear active network, thereby giving rise to the possibility of relatively unpredictable and ill-controlled system voltage gains.

Many of the foregoing gain sensitivity problems can be circumvented by a design that implements a very large magnitude of loop gain over the signal frequency range of interest. For $|T_h| \gg 1$, (15.134) and (15.136) show that the closed loop gain, A_v, is approximately equal to $1/h_{12}$. Not only is this resultant system gain independent of source and load immittances, it is dependent on only a single h-parameter. To be sure, h_{12} is potentially subject to the vagaries that affect the other three h-parameters, but it is possible to design an active network for which the amount of feedback, and hence, the effective value of h_{12}, is principally determined by ratios of passive element values. For such a design, the system gain is accurately predictable, tightly controllable, and virtually independent of reasonable loading immittances at both the input and the output ports.

The loop gain concept is also expedient from the perspective of studying the relevance of the input and output impedances to the dynamical stability of a two-port network. From (15.130), (15.132), and (15.135),

$$Z_{\text{in}} + Z_S = (h_{11} + Z_S)(1 + T_h) \tag{15.138}$$

and

$$Y_{\text{out}} + Y_L = (h_{22} + Y_L)(1 + T_h) \tag{15.139}$$

Equation (15.138) defines the net series impedance in the loop defined by the input port and the Thévenin source circuit in Fig. 15.26(a). On the other hand, (15.139) gives the net admittance found in the parallel branches of the output port an the load impedance. Now, T_h is in general a complex function of the signal frequency, ω; that is, $T_h \equiv T_h(j\omega)$. If a frequency, say ω_o, exists such that $T_h(j\omega_o) = -1$, $Z_{in}(j\omega_o) = -Z_S(j\omega_o)$, and $Y_{out}(j\omega_o) = -Y_L(j\omega_o)$. The mathematical implication of $Z_{in}(j\omega_o) = -Z_S(j\omega_o)$ is that the current, I_i, flowing into the input port of the subject two-port network is infinitely large. Similarly, the mathematical implication of $Y_{out}(j\omega_o) = -Y_L(j\omega_o)$ is an infinitely large output port voltage, V_o, for all values of I_i. Infinitely large currents and voltages in a presumably linear circuit are assuredly indicative of response instability. In truth, however, infinitely large currents and voltages are never attained. Instead, at the frequency at which such responses are predicted, sinusoidal oscillations are produced throughout the entire two-port system [4]. Because the amplitude of these oscillations is independent of the externally applied signal, the resultant network responses are not controlled, and the network is therefore said to be unstable. Clearly, the design of a linear circuit must embody strategies that ensure that no frequency, ω, exists to render $T_h(j\omega) = -1$.

It should be noted that a unilateral network is incapable of sinusoidal oscillation because unilateralization implies $h_{12} = 0$, and hence, $T_h = 0$. With $T_h = 0$, Z_{in} reduces to h_{11}, while Y_{out} becomes h_{22}. On the assumption that the real parts of h_{11} and of h_{22} are positive functions of frequency, Z_{in} and Y_{out} are necessarily positive real functions. Because Z_S and Y_L are passive terminations and are therefore positive real functions [5], it follows that no frequency can be found to render $Z_{in} = -Z_S$ and $Y_{out} = -Y_L$ in a unilateralized two-port network.

Analysis Via Other Two-Port Parameters

The h-parameter analysis of the two-port system in Fig. 15.26 focuses on the problem of formalizing expressions for the voltage gain, the driving-point input impedance and the driving-point output admittance. When h-parameters comprise the modeling vehicle, two reasons underlie the selection of voltage gain as an analytically expedient measure of the forward transfer characteristics of a linear two-port network. The first reason is that the independent output, or response, variable in an h-parameter equivalent circuit is the output port voltage, V_o. The second reason is that the input port model in an h-parameter equivalent circuit is a Thévenin topology, thereby encouraging the representation of the input signal as a voltage source. Thus, although any of the four basic gain measures of a linear two-port network can be evaluated in terms of h-parameters, the most appropriate measure of forward signal processing properties is the voltage gain. Because the input port is a Thévenin topology, the input impedance is more conveniently evaluated than is the corresponding input admittance. Finally, the Norton nature of the output port model in an h-parameter equivalent circuit suggests the propriety of the output admittance as a convenient measure of driving-point output characteristics.

If the foregoing guidelines are adopted as a basis for the analysis of a two-port network in terms of its hybrid g-, y-, or z-parameters, gain and immittance expressions that mirror the mathematical forms of (15.130), (15.132), and (15.133) through (15.136) are obtained. As a first illustration of this contention, consider the g-parameter equivalent circuit of Fig. 15.31. The independent output variable in a g-parameter model is the output port current, I_o. Moreover, the input port is represented by a Norton equivalent circuit, and the output port is a Thévenin circuit. Hence, the appropriate transfer function measure is the current gain, $A_i = I_o/I_S$, while the I-O port driving-point immittances are, respectively, admittance Y_{in} and impedance Z_{out}.

A conventional analysis of the structure in Fig. 15.31 delivers a **closed-loop current gain**, A_i, of

$$A_i = \frac{I_i}{I_S} = \frac{A_{io}}{1 + T_g} \tag{15.140}$$

where the **open loop current gain**, A_{io}, is

$$A_{io} = -\frac{g_{21}}{(g_{22} + Z_L)(g_{11} + Y_s)} \tag{15.141}$$

and the g-parameter **loop gain**, T_g, is

$$T_g \triangleq g_{12}A_{io} = \frac{g_{12}g_{21}}{(g_{22} + Z_L)(g_{11} + Y_S)} \tag{15.142}$$

In addition, the driving-point input admittance derives from

$$Y_{\text{in}} + Y_S = (g_{11} + Y_S)(1 + T_g) \tag{15.143}$$

and the driving-point output impedance is defined implicitly by

$$Z_{\text{out}} + Z_L = (g_{22} + Z_L)(1 + T_g) \tag{15.144}$$

If the y-parameter equivalent circuit of Fig. 15.33 is used to model the two-port system in Fig. 15.26, the independent response variable is the output port voltage, V_o. Both the input and the output ports are represented by Norton equivalent circuits. It follows that the **closed-loop forward transimpedence**, Z_f, is expressible as

$$Z_f = \frac{V_o}{I_S} = \frac{Z_{fo}}{1 + T_y} \tag{15.145}$$

where the **open-loop transimpedance**, Z_{fo}, is given by

$$Z_{fo} = -\frac{y_{21}}{(y_{22} + Y_L)(y_{11} + Y_S)} \tag{15.146}$$

and the **loop gain**, T_y, is

$$T_y \triangleq y_{12}Z_{fo} = -\frac{y_{12}y_{21}}{(y_{22} + Y_L)(y_{11} + Y_S)} \tag{15.147}$$

The driving-point input and output admittances derive, respectively, from

$$Y_{\text{in}} + Y_S = (y_{11} + Y_S)(1 + T_y) \tag{15.148}$$

and

$$Y_{\text{out}} + Y_L = (y_{22} + Y_L)(1 + T_y) \tag{15.149}$$

When z-parameters are used to evaluate the transfer and driving-point characteristics of a linear two-port network, the independent response variable is the output current, I_o. Both the input and the output ports are respresented by Thévenin equivalent circuits, as depicted in Fig. 15.34. It follows that the **closed-loop forward transadmittance**, Y_f, is

$$Y_f = \frac{I_o}{V_S} = \frac{Y_{fo}}{1 + T_z} \tag{15.150}$$

In (15.150), the **open-loop transadmittance**, Y_{fo}, is given by

$$Y_{fo} = -\frac{z_{21}}{(z_{22} + Z_L)(z_{11} + Z_S)} \tag{15.151}$$

and the **loop gain**, T_z, is

$$T_z \triangleq z_{12} Y_{fo} = -\frac{z_{12} z_{21}}{(z_{22} + Z_L)(z_{11} + Z_S)} \tag{15.152}$$

The driving-point input and output impedances follow immediately from the expressions

$$Z_{\text{in}} + Z_S = (z_{11} + Z_S)(1 + T_z) \tag{15.153}$$

$$Z_{\text{out}} + Z_L = (z_{22} + Z_L)(1 + T_z) \tag{15.154}$$

Example 15.6. The utility of two-port analytical methods can best be demonstrated by a simple, yet reasonably practical, circuit example. Consider the bipolar junction transistor (BJT) amplifier of Fig. 15.39, where, with reference to the generalized system of Fig. 15.26, the "linear two-port network" is the topology formed by the transistor and the circuit resistance, R_{EE}. Note that the ground terminal is common to both the input and the output ports so that terminals 2 an 4 are, in fact, the same electrical node. The indicated BJT requires biasing sources to establish reasonable circuit linearity. For simplicity, these sources are not shown, but their effect is to allow the BJT to be modeled by the linear equivalent circuit offered in Fig. 15.39(b). When this equivalent circuit is exploited as a replacement for the BJT in Fig. 15.39(a), the resultant equivalent circuit for the entire subject system is the topology shown in Fig. 15.39(c).

Let the source and load terminations, R_S and R_L, be purely resistive and equal, respectively, to 300 and 1000 Ω. Additionally, assume that the model in Fig. 15.39(b) has $r_i = 1.1$ kΩ, $r_o = 50$ kΩ, and $\beta = 100$. For $R_{EE} = 150$ Ω, *determine the forward voltage gain, the driving-point input resistance, the driving-point output resistance, and the forward transconductance.*

Solution.

1. The first step toward arriving at the solution is the determination of a set of two-port parameters for the linear network at hand. Any type of parameters can be used. In this example h-parameters are selected. To this end, consider the evaluation of the parameters, h_{11} and h_{21}, which requires a short-circuited output port. The equivalent circuit for evaluating these two parameters is depicted in Fig. 15.40(a). KVL applied around the loop containing the short-circuited output port delivers

$$0 = r_o(I_o - \beta I_i) + R_{EE}(I_o + I_i)$$

where

$$h_{21} = \frac{I_o}{I_i} = \frac{\beta r_o - R_{EE}}{r_o + R_{EE}} = 99.7$$

This value of h_{21} suggests that for each 1 μA of input port current, 99.7 μA are delivered to the short-circuited output port. A KVL equation around the input port loop provides

$$V_i = r_i I_i + R_{EE}(I_o + I_i)$$

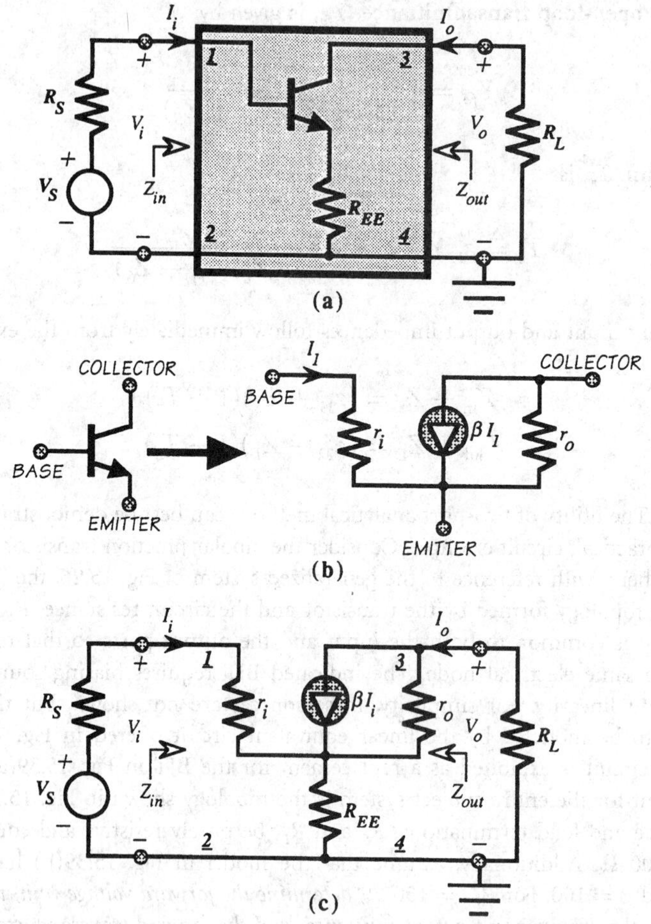

FIGURE 15.39 (a) The simple BJT amplifier addressed in Example 15.26. The biasing required to operate the BJT in its linear regime is not shown. (b) A simplified low-frequency equivalent circuit for the BJT. (c) The equivalent circuit for the two-port system in (a).

Using the preceding disclosure to eliminate the variable, I_o, the short circuit input resistance, h_{11}, is found to be

$$h_{11} = \frac{V_i}{I_i} = r_i + (\beta + 1)(r_o \| R_{EE}) = 16.2 \text{ k}\Omega$$

In order to calculate the h-parameters, h_{12} and h_{22}, the output port of the network is excited under the condition of an open-circuited input port. The pertinent equivalent circuit is offered in Fig. 15.40(b). With the input current, I_i, equal to zero, the CCCS, βI_i, is nulled. It is therefore a simple matter to show that

$$h_{12} = \frac{V_i}{V_o} = \frac{R_{EE}}{R_{EE} + r_o} = 2.99 \text{ mV/V}$$

and

$$h_{22} = \frac{I_o}{V_o} = \frac{1}{R_{EE} + r_o} = 19.94 \text{ } \mu\text{mho}$$

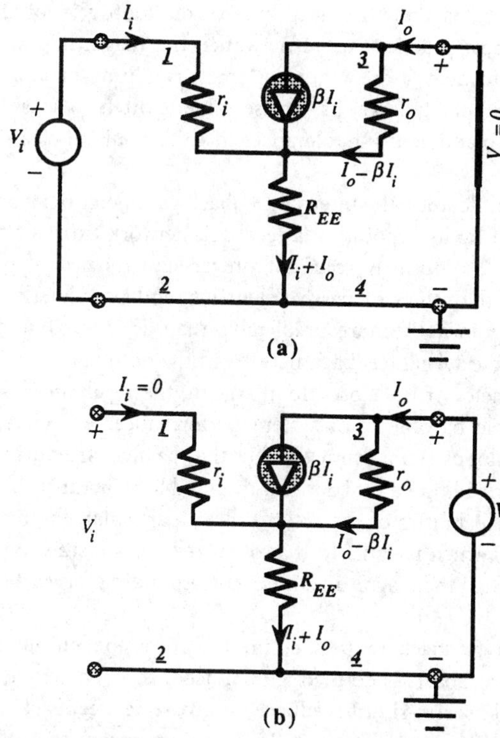

FIGURE 15.40 (a) Short circuit output test fixturing for the computation of the test fixturing for the measurement of the h-parameters, h_{11} and h_{21}. (b) Open circuit input test fixturing for the computation of the h-parameters, h_{12} and h_{22}.

The computed value of h_{12} implies the feedback of almost 3 mV of signal to the open-circuited input port for each volt of signal established across the output port.

2. The voltage gain of the network now follows straightforwardly from (15.133) and (15.134). In particular, with $Y_L = 1/R_L = 1$ mmho and $Z_S = R_S = 300\ \Omega$, (15.133) gives an open loop voltage gain of $A_{vo} = -5.92$, whereas, by (15.134), the closed loop voltage gain is $A_v = -6.03$. Because the loop gain, $h_{12}A_{vo}$, is -17.71 mV/V, (15.138) gives a driving-point input resistance of $R_{in} = 15.91$ kΩ, and a driving-point output resistance of $R_{out} = 534.0$ kΩ. The relatively large input resistance suggests that the input port of the amplifier is amendable to being driven by a voltage source (as long as the Thévenin resistance of this source is significantly smaller than 15.91 kΩ). However, the huge output resistance indicates that the amplifier is best suited for a current mode output port (provided that the load resistance is much smaller than 534 kΩ.
3. From (15.131), the forward transconductance of the amplifier in Fig. 15.39(a) is $G_f = 603$ mmho. This transconductance also can be computed directly by way of (15.150) through (15.152). Such a computation requires that the amplifier be characterized in terms of the open circuit impedance (z-) parameters, which can be found from the available h-parameters. Alternatively, these z-parameters can be determined by applying their defining relationships, (15.106), to the equivalent circuit in Fig. 15.39(c).

Global Feedback

As pointed out earlier, the closed loop gain of a linear two-port network is independent of source and load terminations and is normally dependent on only the two-port feedback parameter

(h_{12}, g_{12}, z_{12}, or y_{12}), provided that the magnitude of the loop gain of the closed loop system is large. However, for most practical active networks, the magnitude of the internal feedback parameter is too small to deliver the requisite large magnitude of loop gain. Even when the feedback parameter is acceptably large, its precise value is often so sensitive to biasing levels or poorly controlled device manufacturing tolerances that the resultant closed loop gain is relatively unpredictable.

To circumvent these difficulties, the internal feedback parameter of an active two-port network is commonly augmented by incorporating a feedback network extrinsic to the active cell. This appended feedback loop is routinely (but not universally) formed of passive elements for at least two reasons. First, the values of passive elements, and particularly the values of ratios of passive elements, are significantly more predictable than are the parameters of active devices. Thus, if the incorporated external feedback network is the dominant vehicle for determining the effective feedback parameter of the closed loop system, the resultant closed loop gain becomes a predictable and reliable barometer of system performance. A second reason justifying the use of passive feedback loops stems from the fact that passive structures are reciprocal circuits that are incapable of providing gain. They are thus stable structures that do not substantively alter the forward gain capabilities of the active cell, In particular, the magnitude of its forward transmission parameter, which is equal to that of its reverse transmission parameter, is likely to be significantly smaller than the magnitude of the corresponding forward transmission parameter of the active cell.

When the appended feedback loop is connected from the output port of an active cell to its input port, the feedback is termed **global feedback**. There are four commonly used forms of global feedback: **series-shunt feedback, shunt-series feedback, shunt-shunt feedback**, and **series-series feedback**. Assuming that the two-port parameters of both the active two-port network and the appended feedback two-port are unaltered by their interconnection, the analysis of the resultant global feedback structure can be formulated conveniently in terms of a mere superposition of appropriate two-port parameters. The preconnection parameter equality to postconnection parameters for both the active and feedback subcircuits is known as **Brune's condition**. Analytical and laboratory procedures that test if Brune's condition is satisfied are available in the literature [6]. These procedures confirm the satisfaction of Brune's condition whenever the active and the feedback two-port are three-terminal networks. In electronic feedback configurations, which may not be formed of interconnected three-terminal networks, Brune's condition rarely comprises an engineering issue if feedback within the active two-port network is negligible and feedforward through the feedback subcircuit is likewise negligible.

Series-Shunt Feedback

In a series-shunt feedback configuration the input port of the two-port network used to implement global feedback is connected in series with the input port of a presumably active two-port cell. As depicted in Fig. 15.41, the output port of the feedback subcircuit shunts the output port of the active network. The analysis of the complete feedback system can proceed subsequent to a selection of any type of two-port parameter model for both of the cells diagrammed in the subject figure. Indeed, the two-port parameters selected for the active network need not be of the same type as those invoked on the feedback circuit, but a prudent selection of two-port parameters can simplify requisite circuit analysis. To this end, the fact that the two respective input ports are in series with one another suggests that these ports are represented by a Thévenin configuration. On the other hand, the shunt-shunt nature of the two output ports suggests that a Norton model for each output port is expedient. In short, compelling reasons exist to choose h-parameters as the modeling vehicle for both the active network and the feedback subcircuit.

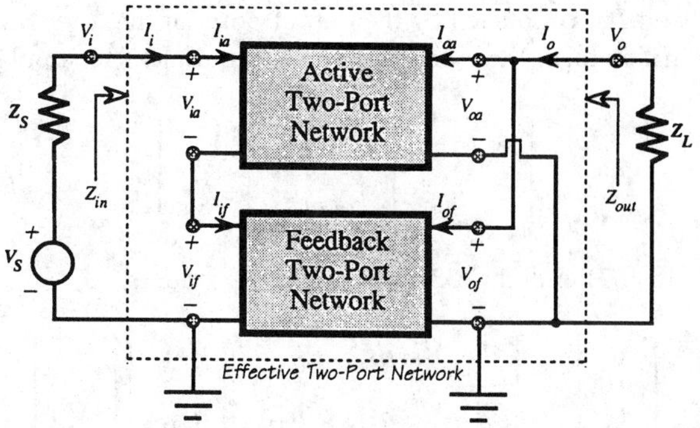

FIGURE 15.41 A linear circuit composed of a feedback network that is connected in series-shunt with an active two-port cell. The h-parameter matrix of the effective two-port network formed by the indicated interconnection is the sum of the respective h-parameter matrices of the active and feedback networks.

Let the terminal volt-ampere characteristic equations of the active two-port network be identified as the h-parameter expression,

$$\begin{bmatrix} V_{ia} \\ I_{oa} \end{bmatrix} = \begin{bmatrix} h_{11a} & h_{12a} \\ h_{21a} & h_{22a} \end{bmatrix} \begin{bmatrix} I_{ia} \\ V_{oa} \end{bmatrix} \tag{15.155}$$

where V_{ia} is the voltage developed across the input port of the active network, and I_{ia} is the current flowing into the input port of this network. Similarly, V_{oa} and I_{oa} are the output port voltage and current, respectively, of the active unit, while h_{ija} is an h-parameter of the active cell. The companion volt-ampere relationship for the feedback structure is

$$\begin{bmatrix} V_{if} \\ I_{of} \end{bmatrix} = \begin{bmatrix} h_{11f} & h_{12f} \\ h_{21f} & h_{22f} \end{bmatrix} \begin{bmatrix} I_{if} \\ V_{of} \end{bmatrix} \tag{15.156}$$

An inspection of Fig. 15.41 reveals that the system output voltage, V_o, the output voltage, V_{oa}, of the active two-port network, and the output voltage, V_{of}, of the feedback cell are identical; that is, $V_o \equiv V_{oa} \equiv V_{of}$. Additionally, assuming that Brune's condition is satisfied, the series connection of the input ports of the active and feedback networks force the input current, I_{ia}, to the active two-port, the input current, I_{if}, to the feedback two-port, and the system input current, I_i, to be identical; namely, $I_i \equiv I_{ia} \equiv I_{if}$. Thus, (15.155) and (15.156) can be rewritten as

$$\begin{bmatrix} V_{ia} \\ I_{oa} \end{bmatrix} = \begin{bmatrix} h_{11a} & h_{12a} \\ h_{21a} & h_{22a} \end{bmatrix} \begin{bmatrix} I_i \\ V_o \end{bmatrix} \tag{15.157}$$

$$\begin{bmatrix} V_{if} \\ I_{of} \end{bmatrix} = \begin{bmatrix} h_{11f} & h_{12f} \\ h_{21f} & h_{22f} \end{bmatrix} \begin{bmatrix} I_i \\ V_o \end{bmatrix} \tag{15.158}$$

A further investigation of the topology in Fig. 15.41 verifies that the input voltage, V_i, of the effective two-port network formed by the series-shunt feedback connection is the sum of the input voltage, V_{ia}, of active network and the voltage, V_{if}, across the input port of the feedback cell. Moreover, the net output current, I_o, conducted by the load termination is the sum of the output currents, I_{oa} and I_{of}, conducted by the output ports of the active network and feedback cell, respectively. In matrix form these observations are equivalent to stipulating

$$\begin{bmatrix} V_i \\ I_o \end{bmatrix} = \begin{bmatrix} V_{ia} \\ I_{oa} \end{bmatrix} + \begin{bmatrix} V_{if} \\ I_{of} \end{bmatrix} \tag{15.159}$$

The substitution of (15.157) and (15.158) into (15.159) produces

$$\begin{bmatrix} V_i \\ I_o \end{bmatrix} = \begin{bmatrix} h_{11a} + h_{11f} & h_{12a} + h_{12f} \\ h_{21a} + h_{21f} & h_{22a} + h_{22f} \end{bmatrix} \begin{bmatrix} I_i \\ V_o \end{bmatrix} \tag{15.160}$$

which defines the terminal volt-ampere characteristics of the effective two-port network formed by the series-shunt interconnection abstracted in Fig. 15.41. Obviously, the interconnected system has an effective h-parameter, h_{ij}, which is simply the sum of the corresponding h-parameters of the active and feedback cells:

$$h_{ij} = h_{ija} + h_{ijf} \tag{15.161}$$

Subsequent to evaluating h_{ija} and h_{ijf}, the voltage gain, A_v, the driving-point input impedance, Z_{in}, and the driving-point output impedance, Z_{out} (or admittance, Y_{out}), derive from the substitution of (15.161) into (15.133), (15.138), and (15.139). Two simplifications are generally possible during the course of such substitution. First, the invariably passive nature of the feedback two-port network precludes its ability to provide a greater than unity forward short circuit current gain. Because the express purpose of the active cell is large forward gain, the magnitude of h_{21a} is generally much larger than that of h_{21f}. Second, the sole purpose of the feedback subcircuit is to augment the ostensibly anemic feedback factor of the active cell. Thus, the magnitude of h_{12f} is likely to be significantly larger than that of h_{12a}. Armed with these approximations, the open loop voltage gain in (15.133) can be approximated as

$$A_{vo} \approx -\frac{h_{21a}}{(h_{22a} + h_{22f} + Y_L)(h_{11a} + h_{11f} + Z_S)} \tag{15.162}$$

and the voltage gain in (15.134) reduces to

$$A_v = \frac{V_o}{V_S} \approx \frac{A_{vo}}{1 + h_{12f} A_{vo}} \tag{15.163}$$

Note that for a large loop gain magnitude, $|h_{12f} A_{vo}| \gg 1$, the voltage gain collapses to $A_v = 1/h_{12f}$, which is independent of source and load terminations and the h-parameters of the active two-port network.

Example 15.7. In simple feedback amplifiers, such as the series-shunt stage whose basic schematic diagram is shown in Fig. 15.42, an explicit evaluation of the two-port parameters of the active cell is often unnecessary. Instead, only the concept underlying (15.161) is exploited in the problem solution methodology.

In the subject amplifier the active two-port network is composed of the two transistors and the resistor, R. As usual, requisite biasing subcircuits are not delineated, and as such, the indicated schematic diagram is known as an **AC schematic diagram**. The feedback two-port network is the voltage divider formed by the two resistances, R_1 and R_2.

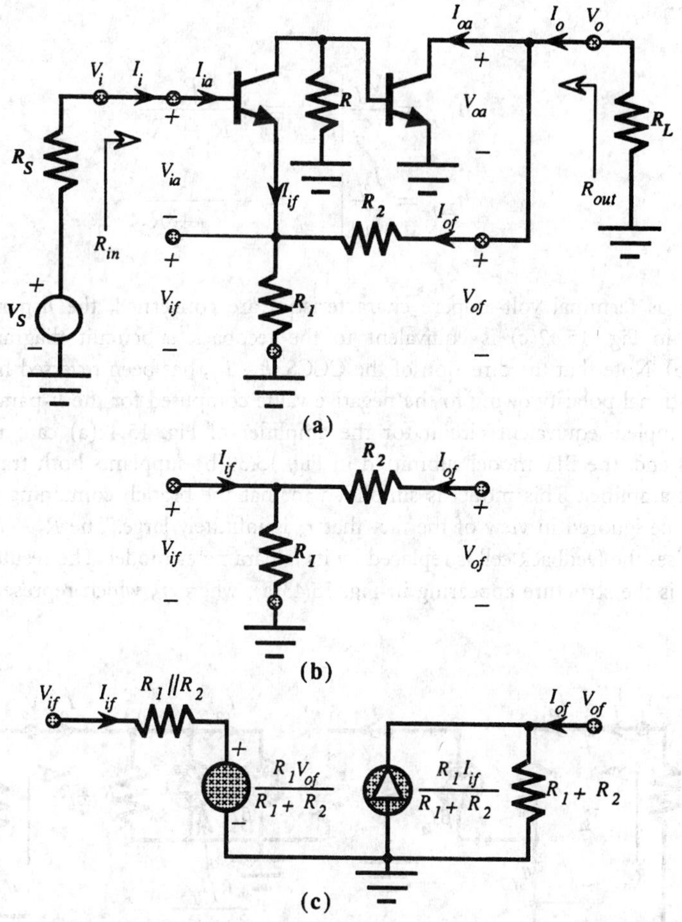

FIGURE 15.42 (a) AC schematic diagram of a practical series-shunt feedback amplifier realized in bipolar technology. (b) The feedback subcircuit for the amplifier in (a). (c) The h-parameter equivalent circuit of the feedback subcircuit in (b).

Let the source and load terminations, R_S and R_L, be purely resistive and equal, respectively, to 300 and 1000 Ω. Additionally, use the BJT equivalent circuit of Fig. 15.39(b) and assume that for both transistors, $r_i = 1.1$ kΩ and $\beta = 100$. For simplicity, the model resistance, r_o, is taken to be infinitely large. *Given that $R = 5.6$ kΩ, $R_1 = 500$ Ω, and $R_2 = 4.5$ kΩ, what is the forward voltage gain, the driving-point input resistance, and the driving-point output resistance?*

Solution.

1. To begin, the h-parameters of the feedback subcircuit must be determined in terms of the circuit resistances, R_1 and R_2. This subcircuit is delineated in Fig. 15.42. Following the definitions of the h-parameters,

$$h_{11f} = \left.\frac{V_{if}}{I_{if}}\right|_{V_{of}=0} = R_1 \| R_2$$

$$h_{21f} = \left.\frac{I_{of}}{I_{if}}\right|_{V_{of}=0} = -\frac{R_1}{R_1 + R_2}$$

Additionally,

$$h_{12f} = \left.\frac{V_{if}}{V_{of}}\right|_{I_{if}=0} = +\frac{R_1}{R_1 + R_2}$$

$$h_{22f} = \left.\frac{I_{of}}{V_{of}}\right|_{I_{if}=0} = -\frac{1}{R_1 + R_2}$$

Insofar as terminal volt-ampere characteristics are concerned, the h-parameter model shown in Fig. 15.42(c) is equivalent to the feedback subcircuit diagrammed in Fig. 15.42(b). Note that the direction of the CCCS, $h_{21f}I_{if}$, has been reversed from that of its conventional polarity owing to the negative value computed for the h-parameters, h_{21f}.

2. The complete equivalent circuit for the amplifier of Fig. 15.42(a) can now be drawn. To this end, the BJT model submitted in Fig. 15.39(b) supplants both transistors in the subject amplifier. This model is simplified in that the branch containing the resistance, r_o, can be ignored in view of the fact that r_o is infinitely large. The $R_1 - R_2$ divider that comprises the feedback cell is replaced by its h-parameter model. The resultant equivalent circuit is the structure appearing in Fig. 15.43(a), where f, which represents a **feedback**

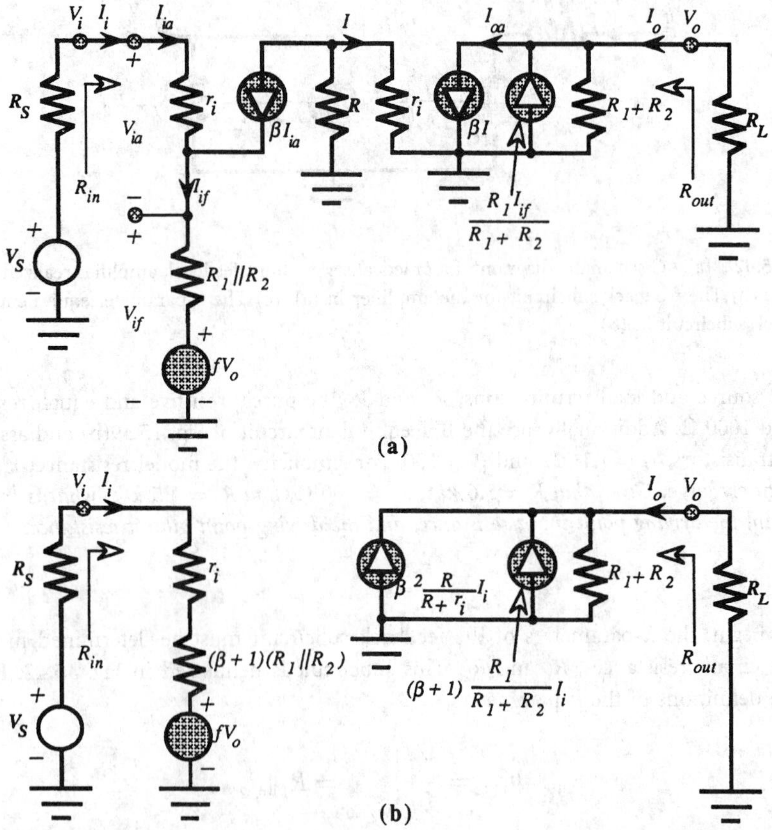

FIGURE 15.43 (a) Equivalent circuit of the feedback amplifier in Fig. 15.42(a). (b) Simplified version of the equivalent circuit in (a).

factor for the feedback subcircuit, is

$$f \equiv h_{12f} = \frac{R_1}{R_1 + R_2}$$

The indicated equivalent circuit can be simplified to render it more mathematically tractable. For example, note that the interstage current, I, relates to the input amplifier current, I_{ia}, by

$$I = -\beta\left(\frac{R}{R + r_i}\right)I_{ia}$$

Because I_{ia} is identical to the system input port current, I_i, the CCCS, βI, at the output port of the system becomes

$$\beta I = -\beta^2\left(\frac{R}{R + r_i}\right)I_i$$

The negative sign in this result means that the direction of the current generator, βI, in Fig. 15.43(a) can be reversed when it is expressed as the indicated proportionality of the system input current, I_i. Note further that the input current, I_{if}, to the feedback network is

$$I_{if} = (\beta + 1)I_{ia} = (\beta + 1)I_i$$

which suggests that the output port generator that models feedforward through the feedback subcircuit can be re-expressed as

$$\left(\frac{R_1}{R_1 + R_2}\right)I_{if} = (\beta + 1)\left(\frac{R_1}{R_1 + R_2}\right)I_i$$

This calculation also confirms that the system input voltage, V_i, is

$$V_i = r_i I_i + (\beta + 1)(R_1 \| R_2)I_i + fV_o$$

These calculations collapse the model of Fig. 15.43(a) to the more compact topology offered in Fig. 15.43(b).

3. An inspection of the model in Fig. 15.43(b) shows that

$$h_{11} = \left.\frac{V_i}{I_i}\right|_{V_o=0} = r_i + (\beta + 1)(R_1 \| R_2) = 46.55\,\text{k}\Omega$$

$$h_{21} = \left.\frac{I_o}{I_i}\right|_{V_o=0} = \frac{\beta^2 R}{R + r_i} - \frac{(\beta + 1)R_1}{R_1 + R_2} = 8.35\,\text{kA/A}$$

It should be noted that in the result for the short circuit system current gain, h_{21}, the feedback subcircuit feedforward term is negligible in comparison to its counterpart amplifier term; that is,

$$\frac{(\beta + 1)R_1}{R_1 + R_2} \ll \frac{\beta^2 R}{R + r_i}$$

or, equivalently,

$$f \ll \frac{\alpha\beta R}{R + r_i}$$

where

$$\alpha = \frac{\beta}{\beta + 1} = 0.990$$

Continuing the h-parameter calculations,

$$h_{12} = \frac{V_i}{V_o}\bigg|_{I_i=0} = f = \frac{R_1}{R_1 + R_2} = 0.10\,\text{V/V}$$

$$h_{22} = \frac{I_o}{V_o}\bigg|_{I_i=0} = \frac{1}{R_1 + R_2} = 200\,\mu\text{mho}$$

Note that to the extent that the utilized BJT model, which inherently neglects internal feedback within the transistor, is valid, the system value of the feedback factor is determined exclusively by the feedback subcircuit value of h_{12f}.

4. From Fig. 15.43(b), the open loop voltage gain, A_{vo}, is (ignoring feedforward transmission through the feedback subcircuit)

$$A_{vo} = \beta^2\left(\frac{R}{R + r_i}\right)\left[\frac{(R_1 + R_2)\|R_L}{R_S + r_i + (\beta + 1)(R_1\|R_2)}\right] = 148.7\,\text{V/V}$$

Using (15.135), the system loop gain is

$$T_h = h_{12}A_{vo} = 14.87$$

whence a closed loop gain, by (15.134), of $A_v = 9.37$ V/V. Note that this closed loop gain is within about 6% of the value of $1/f$. From (15.138) and (15.139), the driving-point input resistance of the series-shunt feedback amplifier is $R_{in} = 743.1$ kΩ, while the driving-point output resistance is $R_{out} = 55.43$ Ω. The last two calculations suggest the propriety of the subject circuit for voltage amplification applications.

Shunt-Series Feedback

Figure 15.44 abstracts the system level diagram of a shunt-series feedback amplifier. As depicted in this diagram, the input ports of the active cell and the feedback subcircuit are in parallel with one another, while the respective output ports are connected in series. The shunt-series interconnection suggests the use of two-port parameter equivalent circuits whose input ports are modeled by a Norton topology and whose output ports are a Thévenin representation. Thus, g-parameters are expedient. With Brune's condition satisfied, the g-parameters of the entire feedback system are sums of the respective g-parameters of the active and feedback subcircuits; that is, the terminal volt-ampere characteristics of the shunt-series feedback amplifier are given by

$$\begin{bmatrix} I_i \\ V_o \end{bmatrix} = \begin{bmatrix} g_{11a} + g_{11f} & g_{12a} + g_{12f} \\ g_{21a} + g_{21f} & g_{22a} + g_{22f} \end{bmatrix}\begin{bmatrix} V_i \\ I_o \end{bmatrix} \tag{15.164}$$

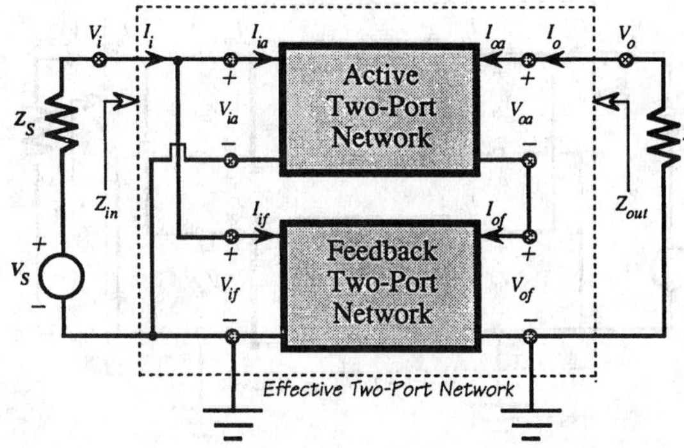

FIGURE 15.44 The generalized shunt-series feedback amplifier. The g-parameter matrix of the effective two-port network formed by the interconnection of the active and feedback two-port networks is the sum of the respective g-parameter matrices of these two circuit cells.

For an amplifier designed well in the sense that $|g_{12f}| \gg |g_{12a}|, |g_{21a}| \gg |g_{21f}|$, and $|T_g|$ (the magnitude of the loop gain) $\gg 1$, the closed loop current gain of a shunt-series feedback amplifier is given approximately by

$$A_i \approx \frac{1}{g_{12f}} \tag{15.165}$$

For a large magnitude of loop gain, the magnitude of the driving-point input impedance, Z_{in}, is small, and the magnitude of the driving-point output impedance, Z_{out}, is large. These impedances are given by, respectively,

$$Z_{\text{in}} \approx \frac{1}{T_g(g_{11a} + g_{11f} + Y_S)} \tag{15.166}$$

$$Z_{\text{out}} \approx T_g(g_{22a} + g_{22f} + Z_L) \tag{15.167}$$

Shunt-Shunt Feedback

In the shunt-shunt feedback configuration of Fig. 15.45 Norton equivalent I-O port models, and thus y-parameter modeling, is appropriate. Assuming Brune's condition is satisfied, it is a simple matter to verify that the terminal volt-ampere characteristics of the shunt-shunt feedback amplifier are given by

$$\begin{bmatrix} I_i \\ I_o \end{bmatrix} = \begin{bmatrix} y_{11a} + y_{11f} & y_{12a} + y_{12f} \\ y_{21a} + y_{21f} & y_{22a} + y_{22f} \end{bmatrix} \begin{bmatrix} V_i \\ V_o \end{bmatrix} \tag{15.168}$$

For $|y_{12f}| \gg |y_{12a}|, |y_{21a}| \gg |y_{21f}|$, and $|T_y|$ (the magnitude of the loop gain) $\gg 1$, the closed-loop forward transimpedance of a shunt-shunt feedback amplifier is

$$Z_f \approx \frac{1}{y_{12f}} \tag{15.169}$$

while the magnitude of the driving-point input and output impedances, Z_{in} and Z_{out}, are very small and given approximately by

$$Z_{\text{in}} \approx \frac{1}{T_y(y_{11a} + y_{11f} + Y_S)} \tag{15.170}$$

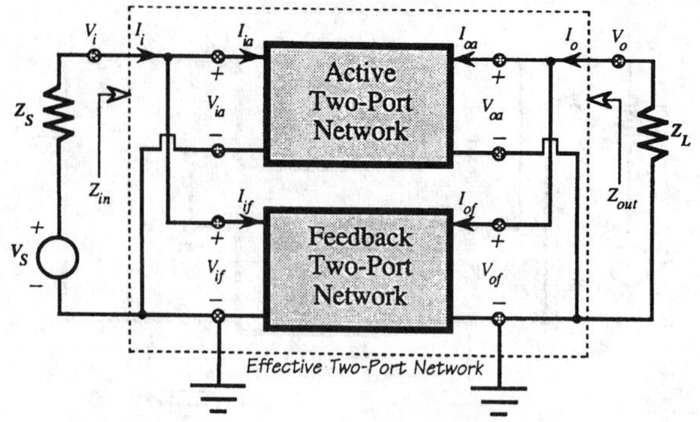

FIGURE 15.45 The generalized shunt-shunt feedback amplifier. The y-parameter matrix of the effective two-port network formed by the interconnection of the active and feedback two-port networks is the sum of the respective y-parameter matrices of these two circuit cells.

$$Z_{out} \approx \frac{1}{T_y(y_{22a} + y_{22f} + Y_L)} \tag{15.171}$$

Series-Series Feedback

In the series-series architecture of Fig. 15.46 Thévenin equivalent I-O port models, and thus z-parameter modeling, is appropriate. If Brune's condition is satisfied, the resultant terminal volt-ampere characteristics of the shunt-shunt feedback amplifier mirror the matrix relationship

$$\begin{bmatrix} V_i \\ V_o \end{bmatrix} = \begin{bmatrix} z_{11a} + z_{11f} & z_{12a} + z_{12f} \\ z_{21a} + z_{21f} & z_{22a} + z_{22f} \end{bmatrix} \begin{bmatrix} I_i \\ I_o \end{bmatrix} \tag{15.172}$$

For $|z_{12f}| \gg |z_{12a}|, |z_{21a}| \gg |y_{21f}|$, and $|T_z| \gg 1$, the closed-loop forward transadmittance of a series-series feedback amplifier is

$$Y_f \approx \frac{1}{z_{12f}} \tag{15.173}$$

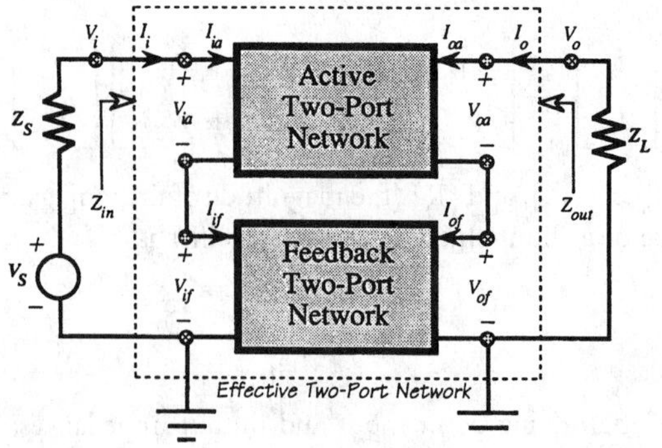

FIGURE 15.46 The generalized series-series feedback amplifier. The z-parameter matrix of the effective two-port network formed by the interconnection of the active and feedback two-port networks is the sum of the respective z-parameter matrices of these two circuit cells.

With a large loop gain magnitude, the magnitude of the driving-point input and output impedances are large. These impedances are given by the approximate expressions

$$Z_{\text{in}} \approx T_z(z_{11a} + z_{11f} + Z_s) \tag{15.174}$$

and

$$Z_{\text{out}} \approx T_z(z_{22a} + z_{22f} + Z_L) \tag{15.175}$$

Power Gains

Refer to the general representation of a two-port network N of Fig. 15.26. The simplest measure of power flow in N is the power gain. **Power gain**, G_p, is defined as the ratio of average power P_2 delivered to the load Z_L to the average power P_1 entering the input port

$$G_p = \frac{P_2}{P_1} \tag{15.176}$$

Clearly, G_p is a function of the two-port parameters and the load impedance Z_L, and does not depend on the source impedance Z_S. For a passive lossless two-port network, $G_p = 1$.

The second measure of power flow is the **available power gain** G_{ava}, which is defined as the ratio of maximum available power $P_{2\text{ava}}$, at the load Z_L to the maximum available average power $P_{1\text{ava}}$ at the source

$$G_{\text{ava}} = \frac{P_{2\text{ava}}}{P_{1\text{ava}}} \tag{15.177}$$

The quantity G_{ava} is a function of the two-port parameters and the source impedance, and is independent of the load impedance.

The third and most useful measure of power flow is the **transducer power gain** G, which is defined as the ratio of average power delivered to the load P_2 to the maximum available average power $P_{1\text{ava}}$ to the source

$$G = \frac{P_2}{P_{1\text{ava}}} \tag{15.178}$$

Observe that the transducer power gain is a function of the two-port parameters and the source and load impedances. Its importance arises from the fact that it compares the power delivered to the load with the power that the source is capable of supplying under optimum conditions. Therefore, it measures the efficacy of using the active device and provides the most meaningful description of the power transfer capabilities of the two-port network.

To illustrate we derive expressions for these power gains in terms of the impedance parameters of Fig. 15.26(a). Substituting $V_o = -I_o Z_L$ in (15.106) and solving for I_o and I_i yields the current gain

$$A_i = \frac{I_o}{I_i} = -\frac{z_{21}}{z_{22} + Z_L} \tag{15.179}$$

The average power P_1 entering the input port and the average power P_2 delivered to the load Z_L are given by

$$P_1 = |I_i|^2 \operatorname{Re} Z_{\text{in}} \tag{15.180a}$$

$$P_2 = |I_o|^2 \operatorname{Re} Z_L \tag{15.180b}$$

where Z_{in} is the input impedance of the two-port network when the output port is terminated in Z_L. Substituting $V_o = -I_o Z_L$ in (15.106) and solving for I_i in terms of V_i, the input impedance Z_{in} is found to be

$$Z_{\text{in}} = z_{11} - \frac{z_{12}z_{21}}{z_{22} + Z_L} \tag{15.181}$$

The maximum available average power P_{lava} from the source is

$$P_{\text{lava}} = \frac{|V_s|^2}{4\,\text{Re}\,Z_S} \tag{15.182}$$

which represents the average power delivered by a source conjugately matched to the network input impedance. Combining (15.179) to (15.182) yields

$$G_p = \frac{P_2}{P_1} = \frac{|z_{21}|^2\,\text{Re}\,Z_L}{|z_{22} + Z_L|^2\,\text{Re}\,Z_{\text{in}}} \tag{15.183}$$

$$G = \frac{P_2}{P_{\text{lava}}} = \frac{4|z_{21}|^2\,\text{Re}\,Z_S\,\text{Re}\,Z_L}{|(z_{11} + Z_S)(z_{22} + Z_L) - z_{12}z_{21}|^2} \tag{15.184}$$

The maximum available average power P_{2ava} at the output port can be appreciated easily by means of the Thévenin equivalent network of Fig. 15.47 when the input port of Fig. 15.26(a) is terminated in V_S in series with Z_S, obtaining

$$Z_{\text{eq}} = z_{22} - \frac{z_{12}z_{21}}{z_{11} + Z_S} \tag{15.185}$$

$$V_{\text{eq}} = \frac{z_{12}V_S}{z_{11} + Z_S} \tag{15.186}$$

The maximum available average power at the output port is obtained when $Z_L = \bar{Z}_{\text{eq}}$, the complex conjugate of Z_{eq}, giving

$$G_{\text{2ava}} = \frac{|z_{21}|^2|Z_S|^2}{4|z_{11} + Z_S|^2\,\text{Re}\,Z_{\text{eq}}} \tag{15.187}$$

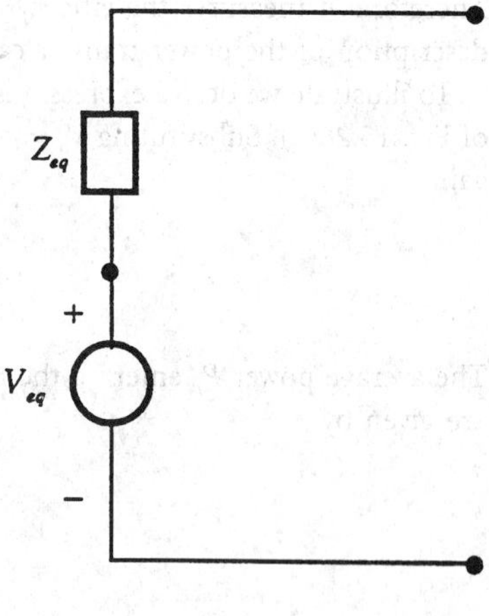

FIGURE 15.47 The Thévenin equivalent network looking into the output port of the network of Fig. 15.26(a).

Thus, the available power gain is given by

$$G_{\text{ava}} = \frac{P_{2\text{ava}}}{P_{1\text{ava}}} = \frac{|z_{21}|^2 \operatorname{Re} Z_S}{|z_{11} + Z_S|^2 \operatorname{Re} Z_{\text{eq}}} \tag{15.188}$$

Let k_{11}, k_{12}, k_{21}, and k_{22} represent any one of the four sets of two-port parameters (z-, y-, h-, and g-parameters), with the corresponding quantities shown in Table 15.3, in which $Y_S = 1/Z_S$ and $Y_L = 1/Z_L$.

Using the general two-port parameters k_{ij}, the general gain formulas are as follows:

$$G_p = \frac{|k_{21}|^2 \operatorname{Re} M_L}{|(k_{22} + M_L)|^2 \left[k_{11} - \dfrac{k_{12}k_{21}}{k_{22} + M_L}\right]} \tag{15.189a}$$

$$G = \frac{4|k_{21}|^2 \operatorname{Re} M_S \operatorname{Re} M_L}{|(k_{11} + M_S)(k_{22} + M_L) - k_{12}k_{21}|^2} \tag{15.189b}$$

$$G_{\text{ava}} = \frac{|k_{21}|^2 \operatorname{Re} M_S}{|(k_{11} + M_S)|^2 \left[k_{22} - \dfrac{k_{12}k_{21}}{k_{11} + M_S}\right]} \tag{15.189c}$$

With appropriate substitutions made from Table 15.3, the various gain functions can be expressed in terms of the chosen two-port parameters and source and load emittances. These power gains are defined at a single real frequency. The variable $s = j\omega$ was dropped in all the expressions for simplicity.

U-Function

We introduce here a useful parameter associated with a two-port network N. We show that this parameter is inherent to the two-port device and provides a unique measure of the degree of inherent transfer capability exhibited by the device. It has the physical meaning of the maximum unilateral power gain of N under a lossless reciprocal imbedding. In fact, it is invariant under all lossless reciprocal imbedding. Before we proceed, we remark on the difference between the words **imbedding** and **terminating.** The general connection of imbedding allows external connections to be made between the ports, whereas the specialized connection of terminating does not. For example, for a two-port network, imbedding permits connection of elements between the I-O ports, whereas termination does not and permits only connection between the two terminals of a port.

TABLE 15.3 Corresponding Quantities in Four Parameter Representations

k	z	y	h	g
k_{11}	z_{11}	y_{11}	h_{11}	g_{11}
k_{22}	z_{22}	y_{22}	h_{22}	g_{22}
k_{12}	z_{12}	y_{12}	h_{12}	g_{12}
k_{21}	z_{21}	y_{21}	h_{21}	g_{21}
M_S	Z_S	Y_S	Z_S	Y_S
M_L	Z_L	Y_L	Y_L	Z_L

Depending on the choice of the two-port parameters k_{ij}, the U-function is defined by the expression

$$U = \frac{|k_{21} - k_{12}|^2}{4(\mathrm{Re}\,k_{11}\,\mathrm{Re}\,k_{22} - \mathrm{Re}\,k_{12}\,\mathrm{Re}\,k_{21})} \tag{15.190a}$$

for z- or y-parameters, and

$$U = \frac{|k_{21} + k_{12}|^2}{4(\mathrm{Re}\,k_{11}\,\mathrm{Re}\,k_{22} + \mathrm{Im}\,k_{12}\,\mathrm{Im}\,k_{21})} \tag{15.190b}$$

for h- or g-parameters. The reason that the U-function, $U(j\omega)$, has different expressions for the immittance and hybrid parameters is that it depends on reciprocity, and reciprocity is expressed differently for the immittance and hybrid parameters.

The U-function is invariant under lossless reciprocal imbedding. The physical significance of this result is that if $U(j\omega)$ is used to characterize the power-amplifying capability of a device, it is unique especially because it is independent of the measuring circuit, provided only that the circuit uses lossless reciprocal elements. Consequently, it measures an inherent characteristic of the device, not the device used in a particular way. This is very similar to the term **gain-bandwidth product** used to characterize a device.

Consider a two-port device N that is imbedded in a lossless reciprocal four-port network N_0, as shown in Fig. 15.48. Let $\mathbf{Y}(s)$ and $\mathbf{Y}_0(s)$ be the admittance matrices characterizing the two-port network N and four-port network N_0, respectively. For the time being, assume that $\mathbf{Y}_0(s)$ exists. Then, from Fig. 15.48, the matrix $\mathbf{Y}_0(s)$ is defined by the equation

$$\begin{bmatrix} \mathbf{I}_a \\ -\mathbf{I} \end{bmatrix} = \begin{bmatrix} \mathbf{Y}_{11}^0 & \mathbf{Y}_{12}^0 \\ \mathbf{Y}_{21}^0 & \mathbf{Y}_{22}^0 \end{bmatrix} \begin{bmatrix} \mathbf{V}_a \\ \mathbf{V} \end{bmatrix} \tag{15.191}$$

where the coefficient matrix is the matrix $\mathbf{Y}_0(s)$ in partitioned form. Substituting

$$\mathbf{I} = \mathbf{Y}(s)\mathbf{V} \tag{15.192}$$

in the second equation, (15.191), and solving for $\mathbf{V}$ yield

$$\mathbf{V} = -(\mathbf{Y} + \mathbf{Y}_{22}^0)^{-1}\mathbf{Y}_{21}^0\mathbf{V}_a \tag{15.193}$$

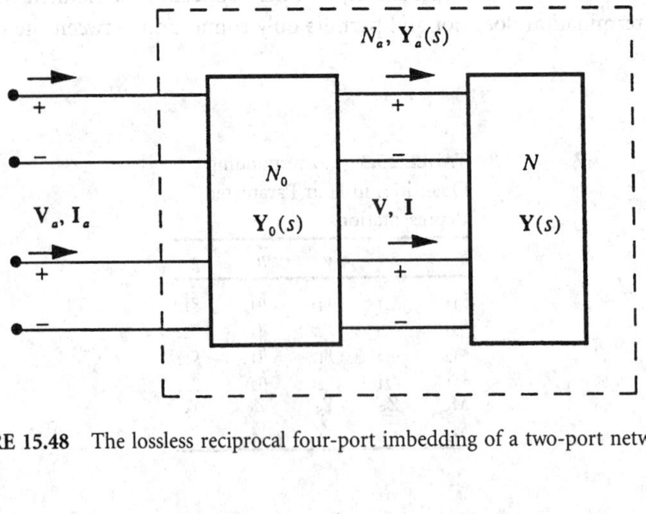

FIGURE 15.48 The lossless reciprocal four-port imbedding of a two-port network N.

where $(\mathbf{Y}+\mathbf{Y}^0_{22})^{-1}$ is assumed to be nonsingular, to be justified shortly. Combining this with the first equation of (15.191) gives

$$\mathbf{I}_a = [\mathbf{Y}^0_{11} - \mathbf{Y}^0_{12}(\mathbf{Y}+\mathbf{Y}^0_{22})^{-1}\mathbf{Y}^0_{21}]\mathbf{V}_a \tag{15.194}$$

showing that the admittance matrix $\mathbf{Y}_a(s)$ of the composite two-port network N_a of Fig. 15.48 is given by

$$\mathbf{Y}_a(s) = \mathbf{Y}^0_{11} - \mathbf{Y}^0_{12}(\mathbf{Y}+\mathbf{Y}^0_{22})^{-1}\mathbf{Y}^0_{21} \tag{15.195}$$

Our objective is to show that on the real-frequency axis, the U-function defined by the elements of $\mathbf{Y}(s)$ and those of $\mathbf{Y}_a(s)$ are the same. To this end, let U and U_a be the U-function of the two-port networks N and N_a, respectively. Then, from the definition of the U-function, U and U_a can be manipulated into the forms

$$U = \frac{|\det[\mathbf{Y}(s) - \mathbf{Y}'(s)]|}{\det[\mathbf{Y}(s) + \bar{\mathbf{Y}}(s)]} \tag{15.196a}$$

$$U_a = \frac{|\det[\mathbf{Y}_a(s) - \mathbf{Y}'_a(s)]|}{\det[\mathbf{Y}_a(s) + \bar{\mathbf{Y}}_a(s)]} \tag{15.196b}$$

where the prime denotes the matrix transpose. Because N_0 is assumed to be lossless and reciprocal, on the $j\omega$-axis

$$\mathbf{Y}^0_{ik}(j\omega) = j\mathbf{B}_{ik}, \qquad i,k = 1,2 \tag{15.197a}$$

$$\mathbf{B}_{ik} = \mathbf{B}'_{ki}, \qquad i,k = 1,2 \tag{15.197b}$$

where $\mathbf{B}_{ik}$ are real matrices. Using these in (15.195) gives

$$\begin{aligned}\mathbf{Y}_a(j\omega) + \bar{\mathbf{Y}}_a(j\omega) &= \mathbf{B}_{12}(\mathbf{W}^{-1} + \bar{\mathbf{W}}^{-1})\mathbf{B}'_{12} \\ &= \mathbf{B}_{12}\mathbf{W}^{-1}[\mathbf{Y}(j\omega) + \bar{\mathbf{Y}}(j\omega)]\bar{\mathbf{W}}^{-1}\mathbf{B}'_{12}\end{aligned} \tag{15.198}$$

whose determinant is

$$\det[\mathbf{Y}_a(j\omega) + \bar{\mathbf{Y}}_a(j\omega)] = \frac{(\det \mathbf{B}_{12})^2 \det[\mathbf{Y}(j\omega) + \bar{\mathbf{Y}}(j\omega)]}{|\det W|^2} \tag{15.199}$$

where

$$\mathbf{W} = \mathbf{Y}(j\omega) + \mathbf{Y}^0_{22}(j\omega) \tag{15.200}$$

Similarly, we obtain

$$\mathbf{Y}_a(j\omega) - \mathbf{Y}'_a(j\omega) = \mathbf{B}_{12}\mathbf{W}^{-1}[\mathbf{Y}'(j\omega) - \mathbf{Y}(j\omega)]\mathbf{W}'^{-1}\mathbf{B}'_{12} \tag{15.201}$$

the determinant of which is given by

$$\det[\mathbf{Y}_a(j\omega) - \mathbf{Y}'_a(j\omega)] = \frac{(\det \mathbf{B}_{12})^2 \det[\mathbf{Y}'(j\omega) - \mathbf{Y}(j\omega)]}{|\det \mathbf{W}|^2} \tag{15.202}$$

Finally, substituting (15.199) and (15.202) in (15.196b) yields

$$U_a(j\omega) = U(j\omega) \tag{15.203}$$

In the case in which the admittance matrix $\mathbf{Y}_0(s)$ does not exist or $\mathbf{W}$ is identically singular, we first insert some lossless reciprocal elements of finite nonzero values in N_0 so that in the resulting four-port network, $\mathbf{Y}_0(s)$ exists and $\mathbf{W}$ is nonsingular. We then take the limit in the resulting $U_a(j\omega)$ of (15.196b) as the values of these added elements approach zero or infinity so that the resulting network becomes N_0. Because $U_a(j\omega)$ is independent of the added elements, we arrive at the same conclusion as in (15.203). We summarize this as a theorem.

Theorem 15.1: *On the real-frequency axis, the U-function of a linear, time-invariant two-port network is invariant under lossless reciprocal imbedding.*

Example 15.8. Figure 15.49 is a shunt-shunt feedback example of the generalized configuration in Fig. 15.45. The active two-port network N_a can be viewed as the lossless reciprocal four-port imbedding of the active two-port device N, with a feedback admittance y_f in shunt from output to input. To compute the admittance matrix $\mathbf{Y}_a(s)$ of N_a, it is convenient to consider N_a as the parallel combination of two two-port networks, so that $\mathbf{Y}_a(s)$ can be expressed as the sum of the admittance matrices of the component two-port networks, as follows:

$$\mathbf{Y}_a = \begin{bmatrix} y_{11} & y_{12} \\ y_{21} & y_{22} \end{bmatrix} + \begin{bmatrix} y_a + y_f & -y_f \\ -y_f & y_b + y_f \end{bmatrix} \tag{15.204}$$

where y_{ij} are the y-parameters of N. Because y_a, y_f, and y_b are assumed to be lossless and reciprocal, on the real-frequency axis their real parts are zero, and the U-function of N_a is found to be

$$U_a(j\omega) = \frac{|y_{a21} - y_{a12}|^2}{4(\mathrm{Re}\, y_{a11}\, \mathrm{Re}\, y_{a22} - \mathrm{Re}\, y_{a12}\, \mathrm{Re}\, y_{a21})} \tag{15.205a}$$

$$= \frac{|y_{21} - y_{12}|^2}{4(\mathrm{Re}\, y_{11}\, \mathrm{Re}\, y_{22} - \mathrm{Re}\, y_{12}\, \mathrm{Re}\, y_{21})} = U(j\omega) \tag{15.205b}$$

which is the U-function of N, where y_{aij} are the y-parameters of N_a, confirming that the U-function is invariant under lossless reciprocal imbedding.

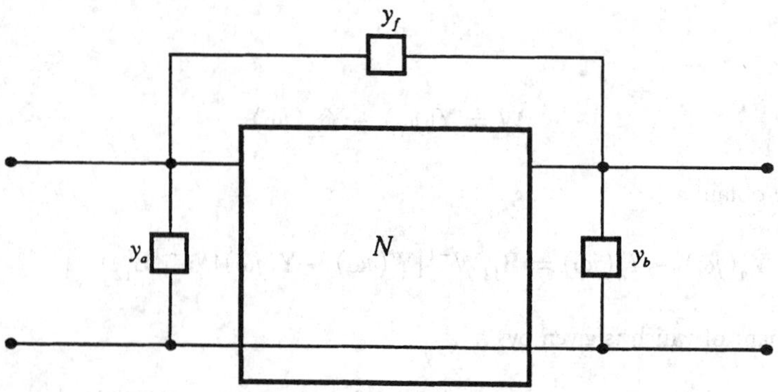

FIGURE 15.49 An active two-port device with a shunt feedback admittance y_f.

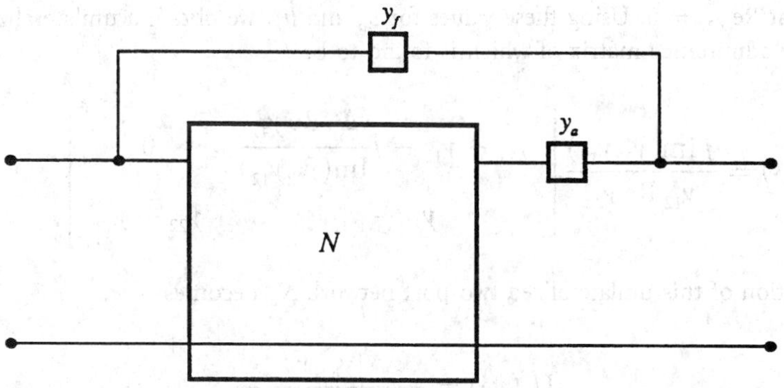

FIGURE 15.50 A three-terminal device N with two lossless reciprocal admittances.

For a given active two-port device, its U-function is identifiable as the maximum unilateral power gain under a lossless reciprocal imbedding, the resulting structure being unilateral. Consider a three-terminal device N such as a transistor. This device is connected with two lossless reciprocal admittances, y_a and y_f, one in series with the output and the other in shunt from output to input, as shown in Fig. 15.50. The network can be viewed as the lossless reciprocal four-port imbedding of the active device N. If y_{ij} are the y-parameters of N and $\Delta_y = y_{11}y_{22} - y_{12}y_{21}$, the admittance matrix $\mathbf{Y}_a(s)$ of the overall network N_a of Fig. 15.50 is found to be

$$\mathbf{Y}_a = \frac{y_a}{y_a + y_{22}} \begin{bmatrix} y_{11} + \dfrac{\Delta_y}{y_a} & y_{12} \\ y_{21} & y_{22} \end{bmatrix} + \begin{bmatrix} y_f & -y_f \\ -y_f & y_f \end{bmatrix} \tag{15.206}$$

Our objective is to determine the admittances, y_a and y_f, so that the resulting network N_a is unilateralized at a specified frequency on the real-frequency axis. Thus, we set the (1,2)-element of $\mathbf{Y}_a(s)$ to zero, and obtain

$$y_f = \frac{y_{12}y_a}{y_{22} + y_a} \tag{15.207}$$

Because y_a and y_f are lossless, we can write

$$y_a(j\omega) = jb_a \qquad y_f(j\omega) = jb_f \tag{15.208}$$

where b_a and b_f are real. Substituting these in (15.207) results in

$$b_a = b_f \frac{\operatorname{Re} y_{22}}{\operatorname{Re} y_{12}} \tag{15.209a}$$

$$b_f = \operatorname{Im} y_{12} - \frac{\operatorname{Re} y_{12}}{\operatorname{Re} y_{22}} \operatorname{Im} y_{22} \tag{15.209b}$$

provided that $\operatorname{Re} y_{22} \neq 0$. Using these values for b_a and b_f, we obtain a unilateralized two-port network, the admittance matrix of which is found to be

$$\mathbf{Y}_a(s) = \frac{j\operatorname{Im}(\bar{y}_{22}y_{12})}{y_{12}\operatorname{Re} y_{22}} \begin{bmatrix} y_{11} + y_{12} - j\dfrac{\Delta_y \operatorname{Re} y_{12}}{\operatorname{Im}(\bar{y}_{22}y_{12})} & 0 \\ y_{21} - y_{12} & y_{22} + y_{12} \end{bmatrix} \tag{15.210}$$

The U-function of this unilateralized two-port network N_a becomes

$$U_a(j\omega) = \frac{|y_{a21}|^2}{4\operatorname{Re} y_{a11} \operatorname{Re} y_{a22}} \tag{15.211}$$

where y_{aij} are the elements of $\mathbf{Y}_a(s)$. An equivalent network for the unilateralized two-port network is shown in Fig. 15.51.

The maximum transducer power gain or the available power gain that can be achieved for N_a is precisely $U(j\omega)$. To this end, we conjugate-match the I-O ports, as shown in Fig. 15.52. Under this condition, the maximum transducer power gain becomes equal to the available power gain and is given by

$$G_{\text{ava}} = \frac{|V_o(j\omega)|^2 \operatorname{Re} y_{a22}(j\omega)}{|V_i(j\omega)|^2 \operatorname{Re} y_{a11}(j\omega)} = \frac{|y_{a21}(j\omega)|^2}{4\operatorname{Re} y_{a11}(j\omega)\operatorname{Re} y_{a22}(j\omega)} = U_a(j\omega) \tag{15.212}$$

Because the U-function is invariant under all lossless reciprocal imbedding, the U-function of N must be the same as $U_a(j\omega)$ or

$$G_{\text{ava}} = U_a(j\omega) = U(j\omega) \tag{15.213}$$

In fact, any lossless reciprocal imbedding that unilaterlizes a three-terminal device will yield the same maximum transducer power gain of (15.213). Thus, the U-function of an active three-terminal device has the physical meaning of the maximum unilateral transducer power gain.

The power-amplifying ability of a common three-terminal device such as the transistor is usually measured in terms of the maximum power gain available in a specified network. Such a specification depends critically on the details of the particular measuring network used. Thus, it is difficult to relate the results to basic transistor properties. Also, the orientation of the transistor terminals such as the common-base, common-emitter, or common-collector configuration results in a critical difference in the measurements. The U-function introduced above provides a unique measure of the degree of inherent power-transfer capability exhibited by a device. The U-function is invariant under all lossless reciprocal imbedding, it is invariant

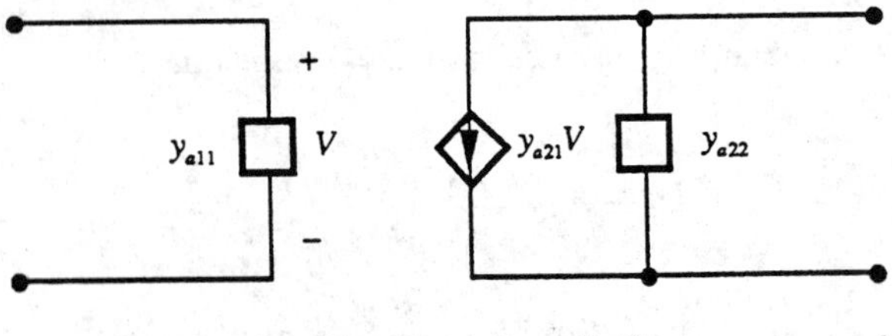

FIGURE 15.51 An equivalent network of the unilateralized two-port network of Fig. 15.50.

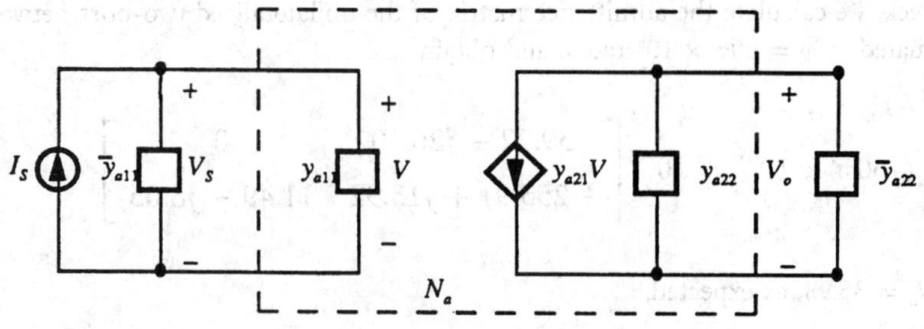

FIGURE 15.52 A conjugately matched unilateral two-port network of Fig. 15.51.

under all orientations of the device. Thus, it measures a general characteristic of the device, not merely the performance of a particular way in which the device is used.

Example 15.9. At 30 MHz, $V_{CE} = 20$ V, and $I_C = 20$ mA, the following values of the y-parameters apply to a sample type 2N697 transistor:

$$y_{11} = y_{ie} = (22.5 + j14.7) \text{ mmho} \tag{15.214a}$$

$$y_{12} = y_{re} = (-0.8 - j0.38) \text{ mmho} \tag{15.214b}$$

$$y_{21} = y_{fe} = (36.6 - j91.6) \text{ mmho} \tag{15.214c}$$

$$y_{22} = y_{oe} = (1.7 - j5.7) \text{ mmho} \tag{15.214d}$$

We wish to design a unilateral two-port network to measure the U-value of this transistor at the frequency of 30 MHz. Referring to the network of Fig. 15.50, let

$$y_f = j\omega C = jb_f, \qquad y_a = \frac{1}{j\omega L} = jb_a \tag{15.215}$$

where $\omega = 60\pi \times 10^6$ rad/s. The required susceptances, b_a and b_f, are computed from (15.209) as

$$b_f = 2.30 \text{ mmho}, \qquad b_a = -4.89 \text{ mmho} \tag{15.216}$$

obtaining from (15.215)

$$C = 12.2 \text{ pF}, \qquad L = 1.08\ \mu H \tag{15.217}$$

The desired unilateral network is shown in Fig. 15.53. The U-value associated with this transistor operating at 30 MHz is found to be

$$U = \frac{|37.4 - j91.22|^2}{4[22.5 \times 1.7 - (-0.8) \times 36.6]} = 35.98 \tag{15.218}$$

This is the maximum transducer power gain or the available power gain of the transistor operating at 30 MHz under all lossless reciprocal imbedding.

As a check, we calculate the admittance matrix of the unilateralized two-port network of Fig. 15.53 evaluated at $\omega = 60\pi \times 10^6$ rad/s, and obtain

$$\mathbf{Y}_a(j60\pi \times 10^6) = 10^{-3}\begin{bmatrix} 39.72 - j26.10 & 0 \\ -255.91 + j13.92 & 11.49 - j8.05 \end{bmatrix} \quad (15.219)$$

yielding $U_a = 35.98$, as expected.

Potential Instability and Absolute Stability

A network is **stable** if all of its natural frequencies are resticted to the left-half of the complex frequency s-plane, not including the $j\omega$-axis; this is referred to simply as the open LHS. The $j\omega$ poles of the transfer function are excluded for stable networks because such poles yield an unbounded time response. For example, if a transfer function has a simple pole, say, at $j\omega_0$, an excitation of the type $\exp(j\omega_0 t)$ results in an unbounded time response.

In the case of a two-port network, its stability cannot be determined by the two-port itself. It also depends on the terminations. Therefore, a two-port network is said to be **potentially unstable** at $j\omega_0$ if two passive one-port immittances exist that when terminated the ports, will produce a natural frequency at $j\omega_0$ in the overall network. A two-port network is said to be **absolutely stable** at $j\omega_0$ if it is not potentially unstable at $j\omega_0$. Using these definitions to test the stability of a two-port network is clearly difficult and tedious because all passive terminations must be examined, and they are not intended for this purpose. In the following, we develop equivalent criteria for these concepts.

To keep the discussion general, we again use the general parameters k_{ij} of Table 15.3. We first introduce a dimensionless quantity

$$\eta = \frac{2\,\mathrm{Re}\,k_{11}\,\mathrm{Re}\,k_{22} - \mathrm{Re}(k_{12}k_{21})}{|k_{12}k_{21}|} \quad (15.220)$$

associated with a two-port network called the **stability parameter**. An important property of this stability parameter is that it is invariant when any one set of the impedance, admittance, hybrid, or inverse hybrid parameters is replaced by another set. This operation is termed the

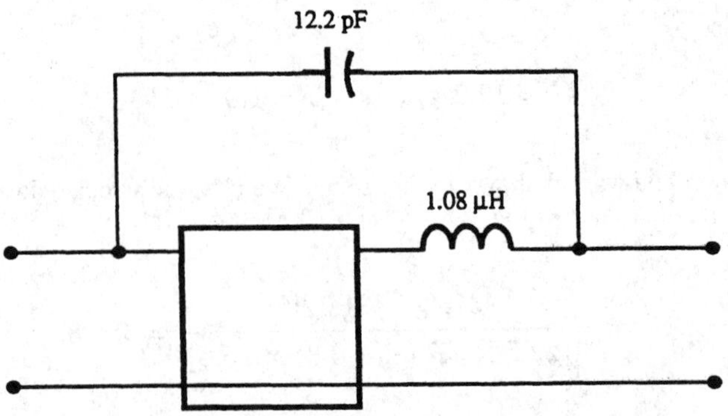

FIGURE 15.53 A two-port network used to measure the U-function of a transistor.

immittance substitution. For example, if we use the y-parameters in (15.220) and then replace them by their equivalent h-parameters, we obtain the same stability parameter of (15.220) by using the h-parameters

$$\begin{aligned}\eta &= \frac{2\,\mathrm{Re}\,y_{11}\,\mathrm{Re}\,y_{22} - \mathrm{Re}(y_{12}y_{21})}{|y_{12}y_{21}|} \\ &= \frac{2\,\mathrm{Re}(1/h_{11})\,\mathrm{Re}(h_{11}h_{22} - h_{12}h_{21}/h_{11}) - \mathrm{Re}((-h_{12}/h_{11}) \times (h_{21}/h_{11}))}{|(-h_{12}/h_{11}) \times (h_{21}/h_{11})|} \\ &= \frac{2\,\mathrm{Re}\,h_{11}\,\mathrm{Re}\,h_{22} - \mathrm{Re}(h_{12}h_{21})}{|h_{12}h_{21}|}\end{aligned} \tag{15.221}$$

In light of these preliminaries, Llewellyn's criteria for absolute stability follows.

Theorem 15.2: *A linear time-invariant two-port network is absolutely stable at $j\omega_0$ if and only if*

$$\mathrm{Re}\,k_{11}(j\omega_0) > 0 \tag{15.222a}$$

$$\mathrm{Re}\,k_{22}(j\omega_0) > 0 \tag{15.222b}$$

$$\eta(j\omega_0) > 1 \tag{15.222c}$$

Observe that the inequalities (15.222) are invariant under the interchange of subscripts 1 and 2, and that if the stability test is performed for one set of parameters, the same conclusions are reached for other sets of parameters. However, the real part quantities may be different for different sets of parameters. Also, for positive $\mathrm{Re}\,k_{11}$ and $\mathrm{Re}\,k_{22}$, the value η lies between -1 and ∞. When η is near unity, the two-port network is close to the boundary between absolute stability and potential instability, which is defined by $\eta = 1$. When $1 \le \eta \le -1$, the two-port network is in the region of potential instability, meaning that we can always choose passive one-port terminations that will result in oscillations.

Example 15.10. Consider a transistor, the high-frequency, small-signal equivalent network of which is shown in Fig. 15.54. We wish to determine the frequency range for which the device is potentially unstable.

The admittance matrix of the two-port network is

$$\mathbf{Y}(j\omega) = \begin{bmatrix} 0.004 + j105 \times 10^{-12}\omega & -j5 \times 10^{-12}\omega \\ 0.2 - j5 \times 10^{-12}\omega & 0.01 + j5 \times 10^{-12}\omega \end{bmatrix} \tag{15.223}$$

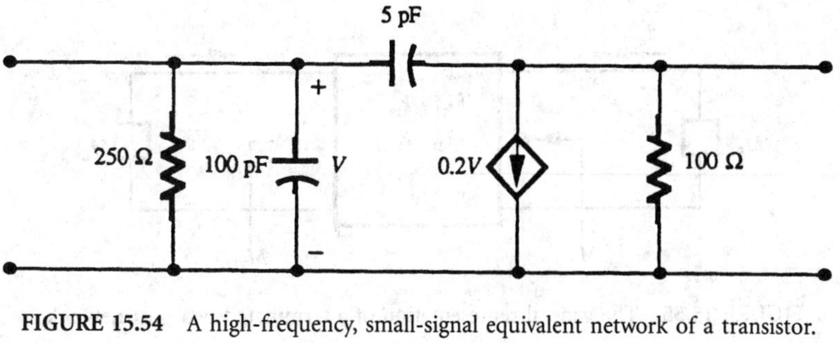

FIGURE 15.54 A high-frequency, small-signal equivalent network of a transistor.

Because

$$\text{Re}\, y_{11} = 0.004 > 0 \tag{15.224a}$$

$$\text{Re}\, y_{22} = 0.01 > 0 \tag{15.224b}$$

for the device to be potentially unstable, it is necessary and sufficient that $\eta \leq 1$ or

$$-25 \times 10^{-18}\omega^2 + \sqrt{625 \times 10^{-36}\omega^4 + 10^{-12}\omega^2} \geq 80 \tag{15.225}$$

yielding

$$\omega \geq 80.16 \times 10^6 \text{ rad/s} \tag{15.226}$$

which corresponds to 12.76 MHz. Thus, the transistor is potentially unstable for all real frequencies not less than 12.76 MHz.

Optimum Terminations for Absolutely Stable Two-Port Networks

In a potentially unstable two-port network, oscillation can occur for certain passive terminations. In such situations, the maximum power gain is infinite and the concept of optimum power gain has no significance. Therefore, we limit our discussion to absolutely stable two-port networks, and find *optimum* source and load immittances in relation to obtaining maximum power gain. The optimum terminations presented below are optimum at a single frequency, which may be at any point on the real-frequency axis. Thus, the results are useful only for narrow-band applications.

One way to determine the optimum terminations is simultaneously to conjugate-match at the I-O ports. For generality, again we use the general two-port parameters k_{ij} of Table 15.3. Refer to Fig. 15.55. The maximum transducer power gain results if the input immittance M_{11} is the conjugate of the source immittance M_s, whereas the output immittance M_{22} is the conjugate of the load immittance M_L. Under these conditions, we set from (15.181).

$$M_{11} = k_{11} - \frac{k_{12}k_{21}}{k_{22} + M_L} = \overline{M}_S \tag{15.227a}$$

$$M_{22} = k_{22} - \frac{k_{12}k_{21}}{k_{11} + M_s} = \overline{M}_L \tag{15.227b}$$

Solving these for M_S and M_L yields the optimum terminations of the two-port network, as

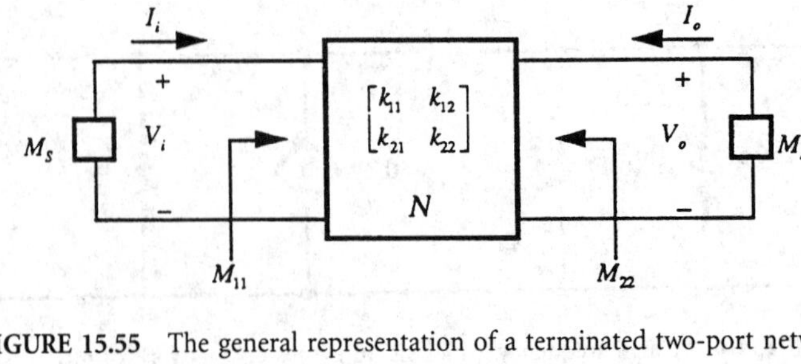

FIGURE 15.55 The general representation of a terminated two-port network.

follows:

$$M_{S,\text{opt}} = \frac{\sqrt{[2\,\text{Re}\,k_{11}\,\text{Re}\,k_{22} - \text{Re}(k_{12}k_{21})]^2 - |k_{12}k_{21}|^2}}{2\,\text{Re}\,k_{22}} + j\left[\frac{\text{Im}(k_{12}k_{21})}{2\,\text{Re}\,k_{22}} - \text{Im}\,k_{11}\right] \tag{15.228a}$$

$$M_{L,\text{opt}} = \frac{\sqrt{[2\,\text{Re}\,k_{11}\,\text{Re}\,k_{22} - \text{Re}(k_{12}k_{21})]^2 - |k_{12}k_{21}|^2}}{2\,\text{Re}\,k_{11}} + j\left[\frac{\text{Im}(k_{12}k_{21})}{2\,\text{Re}\,k_{11}} - \text{Im}\,k_{22}\right] \tag{15.228b}$$

Under the optimum terminations, the power gain $\mathscr{G}_p$, the transducer power gain $\mathscr{G}$, and the available power gain $\mathscr{G}_{ava}$ attain a common maximum value with

$$\mathscr{G}_{\text{max}} = \mathscr{G}_{p,\,\text{max}} = \mathscr{G}_{\text{ava,max}} \tag{15.229}$$

$$\mathscr{G}_{\text{max}} = \frac{|k_{21}|^2}{2\,\text{Re}\,k_{11}\,\text{Re}\,k_{22} - \text{Re}(k_{12}k_{21})^2 + \sqrt{[2\,\text{Re}\,k_{11}\,\text{Re}\,k_{22} - \text{Re}(k_{12}k_{21})]^2 - |k_{12}k_{21}|^2}} \tag{15.230}$$

Recall that for the optimum terminations to be meaningful we assume that the two-port network is absolutely stable at a single frequency $j\omega_0$. This requires that

$$\eta = \frac{2\,\text{Re}\,k_{11}\,\text{Re}\,k_{22} - \text{Re}(k_{12}k_{21})}{|k_{12}k_{21}|} > 1 \tag{15.231}$$

or

$$2\,\text{Re}\,k_{11}\,\text{Re}\,k_{22} - \text{Re}(k_{12}k_{21}) > |k_{12}k_{21}| \tag{15.232}$$

showing that the quantity inside the radical in (15.228) and (15.230) is positive, and therefore resulting in physical terminations. In fact, both the optimum terminations and the maximum power gain can be expressed in terms of the stability parameter η, as follows:

$$M_{S,\text{opt}} = \frac{k_{12}k_{21} + |k_{12}k_{21}|(\eta + \sqrt{\eta^2 - 1})}{2\,\text{Re}\,k_{22}} - k_{11} \tag{15.233a}$$

$$M_{L,\text{opt}} = \frac{k_{12}k_{21} + |k_{12}k_{21}|(\eta + \sqrt{\eta^2 - 1})}{2\,\text{Re}\,k_{11}} - k_{22} \tag{15.233b}$$

$$\mathscr{G}_{\text{max}} = \mathscr{G}_{p,\text{max}} = \mathscr{G}_{\text{ava,max}} = \left|\frac{k_{21}}{k_{21}}\right| \frac{1}{\eta + \sqrt{\eta^2 - 1}} \tag{15.233c}$$

Again, because the two-port network is assumed to be absolutely stable, $\eta > 1$ and the expressions in (15.233) are therefore meaningful.

Example 15.11. Consider the equivalent network of a transistor shown in Fig. 15.56, the admittance matrix of which is

$$\mathbf{Y}(s) = \begin{bmatrix} g_1 + sC & -sC \\ g_m - sC & g_2 + sC \end{bmatrix} \tag{15.234}$$

It is straightforward to show that the device is both active and absolutely stable for all real frequencies ω, satisfying the inequalities

$$\frac{g_m^2}{2} > 2g_1g_2 > \omega C(\sqrt{g_m^2 + \omega^2C^2} - \omega C) \tag{15.235}$$

To calculate the optimum terminating admittances at the frequencies at which the two-port network is absolutely stable, we apply (15.228) and obtain

$$\begin{aligned} Y_{S,\text{opt}} &= \frac{\sqrt{[2\,\text{Re}\,y_{11}\,\text{Re}\,y_{22} - \text{Re}(y_{12}y_{21})]^2 - |y_{12}y_{21}|^2}}{2\,\text{Re}\,y_{22}} + j\left[\frac{\text{Im}(y_{12}y_{21})}{2\,\text{Re}\,y_{22}} - \text{Im}\,y_{11}\right] \\ &= \frac{\sqrt{4g_1^2g_2^2 + \omega^2C^2(4g_1g_2 - g_m^2)}}{2g_2} - j\left[\omega C\left(1 + \frac{g_m}{2g_2}\right)\right] \end{aligned} \tag{15.236a}$$

$$\begin{aligned} Y_{L,\text{opt}} &= \frac{\sqrt{[2\,\text{Re}\,y_{11}\,\text{Re}\,y_{22} - \text{Re}(y_{12}y_{21})]^2}}{2\,\text{Re}\,y_{11}} + j\left[\frac{\text{Im}(y_{12}y_{21})}{2\,\text{Re}\,y_{22}} - \text{Im}\,y_{22}\right] \\ &= \frac{\sqrt{4g_1^2g_2^2 + \omega^2C^2(4g_1g_2 - g_m^2)}}{2g_1} - j\left[\omega C\left(1 + \frac{g_m}{2g_1}\right)\right] \end{aligned} \tag{15.236b}$$

Under these optimum terminations, the three power gains assume a common maximum value determined by (15.230):

$$\begin{aligned} G_{\max} &= \frac{|y_{21}^2|}{2\,\text{Re}\,y_{11}\,\text{Re}\,y_{22} - \text{Re}(y_{12}, y_{21}) + \sqrt{[2\,\text{Re}\,y_{11}\,\text{Re}\,y_{22} - \text{Re}(y_{12}y_{21})]^2 - |y_{12}y_{21}|^2}} \\ &= \frac{2g_1g_2 + \omega^2C^2 - \sqrt{4g_1^2g_2^2 + \omega^2C^2(4g_1g_2 - g_m^2)}}{\omega^2C^2} \end{aligned} \tag{15.237}$$

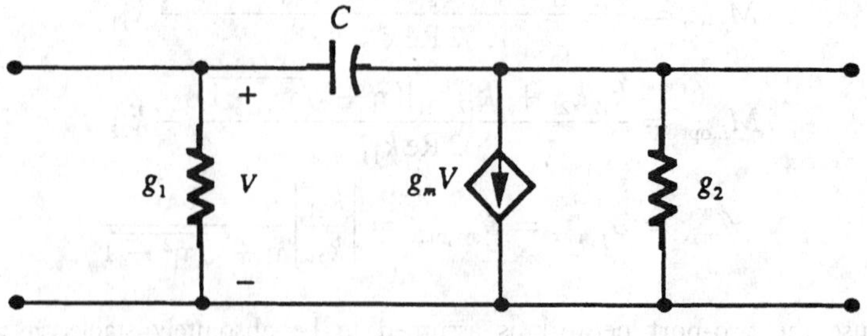

FIGURE 15.56 The equivalent network of a transistor.

ferences

[1] *S-Parameter Techniques for Faster, More Accurate Network Design*, Application Note #95-1, Hewlett-Packard Company, Feb. 1967.

[2] *S-Parameter Design*, Application Note #154, Hewlett-Packard Company, Apr. 1972.

[3] W. C. Yengst, *Procedures of Modern Network Synthesis*, New York: Macmillan, 1964, pp. 171–179.

[4] R. L. Geiger, P. E. Allen, and N. R. Strader, *VLSI Design Techniques for Analog and Digital Circuits*, New York: McGraw-Hill, 1990, pp. 747–750.

[5] G. C. Temes and J. W. LaPatra, *Introduction to Circuit Synthesis and Design*, New York: McGraw-Hill, 1977, pp. 89–92.

[6] A. J. Cote, Jr. and J. B. Oakes, *Linear Vacuum-Tube and Transistor Circuits*, New York: McGraw-Hill, 1961, pp. 40–46.

[7] W. K. Chen, *Active Network and Feedback Amplifier Theory*, New York: McGraw-Hill, 1980, Chapter 3.

[8] W. K. Chen, *Active Network Analysis*, Teaneck, NJ: World Scientific, 1991.

[9] E. F. Bolinder, "Survey of some properties of linear networks," *IRE Trans. Circuit Theory*, vol. **CT-4**, pp. 70–78, 1957 (correction in vol. **CT-5**, p. 139, 1958).

[10] E. S. Kuh and R. A. Rohrer, *Theory of Linear Active Networks*, San Francisco: Holden-Day, 1967.

[11] S. S. Haykin, *Active Network Theory*, Reading, MA: Addison-Wesley, 1970.

16

Theory of Two-Dimensional Hurwitz Polynomials

Hari C. Reddy
California State University, Long Beach and University of California, Irvine

16.1 Introduction

The advances in two-dimensional (2-D) signal and image processing activities have stimulated active research in 2-D circuits and systems area. Two-variable (2-V) or 2-D Hurwitz polynomial study finds application in areas such as generation and testing of (2-V) reactance functions, bounded/positive real functions, and matrices; testing the stability of 2-D digital filters; and the generation of stable 2-D digital transfer functions. Stability analysis is an important aspect of the design of dynamic systems. This analysis is often carried out by examining for the absence of zeroes of the denominator polynomial of a system transfer function in some specified regions of the complex plane. One dimensional (1-D) systems are studied through the characterization whether or not the denominator polynomial is Hurwitz. By expanding this idea we can define and study 2-D (also called bivariate, 2-V) Hurwitz polynomials. In view of the diverse needs of several different applications a number of 2-D Hurwitz polynomials have been defined and their test procedures established. In this chapter, a detailed presentation of various 2-D

0-8493-8341-2/95/$0.00+$.50

Hurwitz polynomials and their relationships to one another is given. We also study their relevant applications.

To highlight the relationships among the various Hurwitz polynomials, the definitions of all the Hurwitz polynomials are presented. This is done in terms of the absence of or the nature of their zerosets in specified regions such as the open or closed right half of the (S_1, S_2)-biplane. The goal is to make a tutorial exposition on 2-D Hurwitz polynomials.

The second section gives some preliminaries and notations. We next present the definitions of a number of (S_1, S_2)-domain Hurwitz polynomials. Based on the continuity property of the zeroes of 2-V polynomials, testsets for the various Hurwitz polynomials are given in the fifth section. Following that, a 2-D, very strict Hurwitz polynomial is discussed in detail because this is the counterpart of strict Hurwitz in the 1-D case. Some of the applications of the various Hurwitz polynomials are described in the final section.

6.2 Preliminaries and Notations

Infinite Distant Points

The following discussion is crucial to the understanding of certain classes of stable 2-D polynomials. The points at infinite distances in the (S_1, S_2)-biplane play an important role in the definition of certain 2-D Hurwitz polynomials. Some of the confusion that resulted in the application of these Hurwitz polynomials can be attributed to the neglect or omission of these infinite distant points. This chapter considers the extended (S_1, S_2)-biplane which includes the infinite distance points. For the sake of clarity, we also explicitly indicate whether the infinite distant points are included in or excluded from the regions considered. The behavior of 2-V polynomials at infinite distant points is well described in the literature [2, 3]. Seemingly many infinite distant points in the 1-D plane, such as the S_1-plane or the S_2-plane, may be assumed to merge to a single point. Thus, infinity is treated as a single point, and any shift from this infinite distant point, however small, leads to a finite distant point.

Analog Biplane

$\mathrm{Re}(s)$ = Real part of the variable s
For $i = 1, 2$

$S_{i+} = \{s_i \,|\, \mathrm{Re}(s_i) > 0, |s_i| < \infty\}$, open right half of the S_i-plane

$S_{i0} = \{s_i \,|\, \mathrm{Re}(s_i) = 0, |s_i| \le \infty\}$, imaginary axis of the the S_i-plane

$S_{i\oplus} = \{s_i \,|\, \mathrm{Re}(s_i) \ge 0, |s_i| \le \infty\}$, closed right half of the S_i-plane

$S^2_{+0} = \{(s_1, s_2) \,|\, \mathrm{Re}(s_1) > 0, \mathrm{Re}(s_2) = 0, |s_1| < \infty, |s_2| \le \infty\}$, open right half of the S_1-plane and the imaginary axis of the S_2-plane

$S^2_{0+} = \{(s_1, s_2) \,|\, \mathrm{Re}(s_1) = 0, \mathrm{Re}(s_2) > 0, |s_1| \le \infty, |s_2| < \infty\}$, open right half of the S_2-plane and the imaginary axis of the S_1-plane

$S^2_{++} = \{(s_1, s_2) \,|\, \mathrm{Re}(s_1) > 0, \mathrm{Re}(s_2) > 0, |s_1| < \infty, |s_2| < \infty\}$, open right half of the (S_1, S_2)-biplane

$S^2_{00} = \{(s_1, s_2) \,|\, \mathrm{Re}(s_1) = 0, \mathrm{Re}(s_2) = 0, |s_1| \le \infty, |s_2| \le \infty\}$, distinguished boundary of the (S_1, S_2)-biplane

$$S^2_{\oplus\oplus} = \{s_1, s_2)| \operatorname{Re}(s_1) \geq 0, \operatorname{Re}(s_2) \geq 0, |s_1| \leq \infty, |s_2| \leq \infty\}, \quad \text{closed right half of the } (S_1, S_2)\text{-biplane}$$

PRF: Positive real function
RF: Reactance function
TPRF: 2-Variable positive real function
TRF: 2-Variable reactance function
TBRF: 2-Variable bounded real function
TLBRF: 2-variable lossless bounded real function
$B_*(s_1, s_2) =$ Paraconjugate of $B(s_1, s_2) = [B(-s_1^*, -s_2^*)]$, where s^* represents complex conjugate of s
$B_e(s_1, s_2) =$ Para-even part of $B(s_1, s_2) = [B(s_1, s_2) + B_*(s_1, s_2)]/2$
$B_0(s_1, s_2) =$ Para-odd part of $B(s_1, s_2) = [B(s_1, s_2) - B_*(s_1, s_2)]/2$

Definition A: A rational function $F(s)$ with real coefficients such that $\operatorname{Re}[F(s)] > 0$ for $\operatorname{Re}(s) > 0$ is called a positive real function (PRF).

Definition B: A PRF $F(s)$ is said to be a strict PRF if $\operatorname{Re}[F(s)] > 0$ for $\operatorname{Re}(s) \geq 0$.

Definition C: A PRF $F(s)$ is said to be a minimum reactive and susceptive if it has neither poles nor zeroes on the imaginary axis of the S-plane.

Definition D: A PRF $F(s)$ is called a reactance function (RF) if $\operatorname{Re}[F(s)] = 0$ for $\operatorname{Re}(s) = 0$.

Definition E: A 2-V rational function $F(s_1, s_2)$ with real coefficients such that $\operatorname{Re}[F(s_1, s_2)] > 0$ for S^2_{++} is called a TPRF. A TPRF $F(s_1, s_2) = -F(-s_1, -s_2)$ is called a TRF.

Definition F: A 2-V rational function $H(s_1, s_2)$ with real coefficients such that $|H(s_1, s_2)| < 1$ for S^2_{++} is called a TBRF. A TBRF $H(s_1, s_2)$ satisfying the condition $|H(s_1, s_2)| = 1$ or 0/0 for S^2_{00} is called TLBRF.

Isolated and Continuum of Zeroes

Some types of Hurwitz polynomials are distinguished on the basis of whether they have isolated zeroes or a continuum of zeroes on S^2_{00}. As a point on S^2_{00} is characterized by $s_1 = jw_1$ and $s_2 = jw_2$, where w_1 and w_2 are real quantities, the region S^2_{00} can be graphically represented by the (W_1, W_2)-plane. The isolated zeroes on S^2_{00} are, thus, points on this plane and the continuum of zeroes is represented by continuous curves. Isolated and continuum of zeroes for the 2-D case are illustrated on the (W_1, W_2)-plane in Fig. 16.1, where zeroes of some simple polynomials are shown.

16.3 Value of a Two-Variable Polynomial at Infinity

Because later in the chapter it is necessary to determine the value of a 2-V polynomial at infinite distance points, the following explanation is in order. In 1-D complex plane S the infinite distant points can be represented by a single point, and the value of any function at this point is found by applying some transformation which transforms the point at infinity to some finite point s', and the value of the transformed function at s' is determined. Often $s = 1/u$ is the transformation used and infinity is mapped onto the origin. Using this transformation, the value of $B(s)$ at infinity can be defined as $B(\infty) = B_T(0)$, where $B_T(u) = B(1/u)$.

In the 2-D biplane (S_1, S_2) consisting of two complex planes, S_1 and S_2, an infinite distant point can have infinite coordinates in either one or both of these planes, and thus an infinite number of infinite distant points exists. They can be classified into three categories [3]:

$$1.\ s_1 = \infty \quad \text{and} \quad s_2 = \text{finite} \tag{16.1a}$$

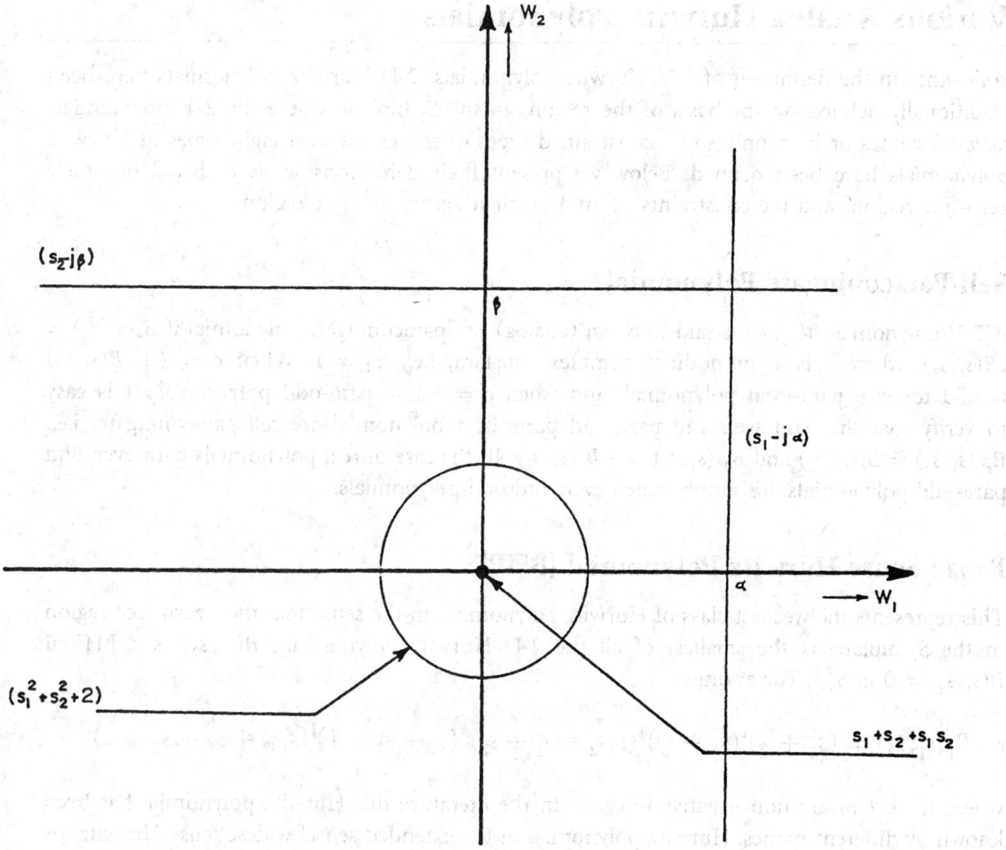

URE 16.1 Zero distribution of some simple polynomials in (W_1, W_2) plane (s_{00}^2) (Polynomial $(s_1 + s_2 + s_1 s_2)$ has ated zero, whereas polynomials $(s_1 - j\alpha)$, $(s_2 - j\beta)$, and $(s_1^2 + s_2^2 + 2)$ have a continuum of zeroes.)

2. $s_1 = \text{finite}$ and $s_2 = \infty$ (16.1b)
3. $s_1 = \infty$ and $s_2 = \infty$ (16.1c)

Applying the transformation method to each variable, the value of the function at each of the above points is defined as:

$$B(\infty, s_2') = B_1(0, s_2') \quad \text{where} \quad B_1(u, s_2) = B(1/u, s_2), |s_2'| < \infty \tag{16.2a}$$

$$B(s_1', \infty) = B_2(s_1', 0) \quad \text{where} \quad B_2(s_1, v) = B(s_1, 1/v), |s_1'| < \infty \tag{16.2b}$$

$$B(\infty, \infty) = B_3(0, 0) \quad \text{where} \quad B_3(u, v) = B(1/u, 1/v) \tag{16.2c}$$

Nonessential Singularities of the Second Kind (NSSK)

It is well known [3] that 2-V polynomials may have NSSK at some infinite distant points where the value of the polynomial is indeterminate. For the sake of notational convenience and to indicate the type of indeterminacy involved we write $B(s_{10}, s_{20}) = 0/0$ to say that $B(s_1, s_2)$ has a NSSK at (s_{10}, s_{20}). Of course, for a polynomial $B(s_1, s_2)$, this can occur only if s_{10}, s_{20}, or both, have infinite value.

16.4 Various Analog Hurwitz Polynomials

Following up the definition of 1-V Hurwitz polynomials, 2-D Hurwitz polynomials have been traditionally defined on the basis of the regions in the S_2-biplane where the 2-V polynomials have no zeroes or have only some constrained types of zeroes. At least eight types of Hurwitz polynomials have been defined. Below we present their definitions so as to bring out their zero-free regions and the constraints (if any) on their zeroes in some regions.

Self-Paraconjugate Polynomial

A 2-V polynomial $B(s_1, s_2)$ is said to be an (analog) self-paraconjugate polynomial if $B_*(s_1, s_2) = cB(s_1, s_2)$, where c is a unimodular complex constant, i.e., $|c| = 1$. When $c = +1$, $B(s_1, s_2)$ is said to be a para-even polynomial, and when $c = -1$, a para-odd polynomial. It is easy to verify that the para-even and para-odd parts of a polynomial are self-paraconjugate, i.e., $B_{e*}(s_1, s_2) = B_e(s_1, s_2)$ and $B_{0*}(s_1, s_2) = -B_0(s_1, s_2)$. In the case of real polynomials para-even and para-odd polynomials are simply called even and odd polynomials.

Broad-Sense Hurwitz Polynomial (BHP)

This represents the weakest class of Hurwitz polynomials in the sense that their zero-free region in the S_2-biplane is the smallest of all the 2-D Hurwitz polynomials. $B(s_1, s_2)$ is a BHP if $B(s_1, s_2) \neq 0$ in S^2_{++}. For example,

$$B_1(s_1, s_2) = (s_2^2 + 4)^i (s_1 + s_2)^j (s_1 s_2 + s_1 + s_2)^k (s_1 + s_2 + 1)^l (s_1 s_2 + s_1 + s_2 + 1)^m$$

where i, j, k, l, m are non-negative integers. In the literature this Hurwitz polynomial has been known by different names: Hurwitz polynomial in the extended sense, widest sense Hurwitz, or simply Hurwitz [4–7, 10].

Narrow-Sense Hurwitz Polynomial (NHP)

A subclass of BHP is NHP, which was introduced by Ansell [1] in the study of two-variable reactance functions (TRFs). They may be characterized by the following equivalent definitions:

1. $B(s_1, s_2)$ is a NHP if $B(s_1, s_2) \neq 0$ in $S^2_{\oplus\oplus} - S^2_{00} = S^2_{++} + S^2_{+0} + S^2_{0+}$, where the minus sign is used to denote set theoretic subtraction and the plus sign, set theoretic union.
2. $B(s_1, s_2)$ is a NHP if $B(s_1, s_2)$ is a BHP [i.e., $B(s_1, s_2) \neq 0$ in S^2_{++}] and $B(s_1, s_2)$ has no 1-V factor having zeroes on the imaginary axis of the S_1- or S_2-plane. For example,

$$B_1(s_1, s_2) = (s_1 + s_2)^i (s_1 s_2 + s_1 + s_2)^j (s_1 + s_2 + 1)^m (s_1 s_2 + s_1 + s_2 + 1)^n$$

   where i, j, m, n are non-negative constants, is a NHP, whereas $(s_1^2 + 1)(s_1 + s_2 + 1)$ is not a NHP because of the factor $(s_1^2 + 1)$.

We see that a NHP may have zeroes on S^2_{00}. The irreducible 2-V factors that give rise to zeroes on S^2_{00} with no zeroes on S^2_{++} can be shown to belong to one of two types: (1) those that become zero only at isolated points on S^2_{00}; (2) those that become zero on a continuum of points on S^2_{00}. The fact that no irreducible factor can have isolated zeroes as well as continuum of zeroes without having zeroes in S^2_{++} can be established based on the continuity property of the zeroes of 2-V polynomials. For example, the factor $(s_1 s_2 + s_1 + s_2)$ with a zero at (0,0) on S^2_{00} corresponds to the first type, and the factor $s_1 + s_2$ with a continuum of zeroes given by $s_1 = -s_2$ on S^2_{00} corresponds to the second type.

Scattering Hurwitz Polynomial (SHP)

Based on the above observation, the next class is identified as a subclass of NHP which can have only discrete (isolated) zeroes on S^2_{00}. Scattering Hurwitz polynomial (SHP) or principal Hurwitz polynomial corresponds to this class [7, 13]. The following equivalent definitions characterize a SHP:

1. A 2-V polynomial $B(s_1, s_2)$ is a SHP if $B(s_1, s_2)$ is a NHP and if $B(s_1, s_2)$ has no (1-D) continuum of zeroes on S^2_{00}.
2. A 2-V polynomial $B(s_1, s_2)$ is a SHP if $B(s_1, s_2)$ is a BHP and if $B(s_1, s_2)$ has no (1-D) continuum of zeroes on S^2_{00}.
3. A 2-V polynomial $B(s_1, s_2)$ is a SHP if $B(s_1, s_2)$ is a BHP and if $B(s_1, s_2)$ has no 1- or 2-V self-paraconjugate factors.
4. A 2-V polynomial $B(s_1, s_2)$ is a SHP if $B(s_1, s_2)$ is a BHP, i.e., $B(s_1, s_2) \neq 0$ in S^2_{++}, and if $B(s_1, s_2)$ and $B_*(s_1, s_2)$ are relatively prime. For example,

$$B(s_1, s_2) = (s_1 s_2 + s_1 + s_2)^i (s_1 + s_2 + 4)^j (3s_1 s_2 + s_1 + s_2 + 1)^k$$

where i, j, k are non-negative integers.

Fettweis [14] identified and popularized this class and suggested the names "principal Hurwitz polynomials" and "scattering Hurwitz polynomials." He studied the properties of SHP in depth and his pioneering efforts established SHP as a very important class of Hurwitz polynomials in stability and passivity studies.

Hurwitz Polynomial in the Strict Sense (HPSS)

Increasing the zero-free regions of a Hurwitz polynomial further by the addition of the whole finite imaginary axes of the S_1- and S_2-planes, we get the definition of a Hurwitz polynomial in the strict sense: $B(s_1, s_2)$ is a HPSS if $B(s_1, s_2) \neq 0$ in $\{S^2_{\oplus\oplus}$ − infinite distant points$\}$. For example,

$$B(s_1, s_2) = (s_1 + s_2 + 7)^m (4s_1 s_2 + s_1 + s_2 + 1)^n$$

where integers m and n are non-negative.

This class seems to have been first defined and used by Saito [4] and Youla [5]. This strict Hurwitz polynomial definition has been used by a number of authors such as Huang [8], in the derivation of stability tests for 2-D digital filters, Strintzis [15], in the extension of digital filter stability tests to continuous domain functions, and Goodman [16, 23], in the study of double bilinear transformation. In order to get a 2-D Hurwitz polynomial that is a counterpart of 1-D strict Hurwitz we need to include the infinite distant points in the 2-D plane. (Note that in certain cases, 2-D scattering Hurwitz appears to be the counterpart of the 1-D strict Hurwitz polynomial.)

Very Strict Hurwitz Polynomial (VSHP)

By including the infinite distant points we get the most strict Hurwitz polynomial, the very strict Hurwitz polynomial, named by Reddy et al. [2, 24]. Similar conclusions were reached by Delsarte et al. [9]. The following equivalent definitions characterize this type of Hurwitz polynomial:

1. $B(s_1, s_2)$ is a VSHP if $B(s_1, s_2) \neq 0$ or if $B(s_1, s_2) \neq 0/0$ in $S^2_{\oplus\oplus}$.

2. $B(s_1, s_2)$ is a VSHP if $1/B(s_1, s_2)$ has no (first or second kind) singularities in $S^2_{\oplus\oplus}$. For example,

$$B(s_1, s_2) = (5s_1s_2 + s_1 + s_2 + 8)$$

We discuss 2-D VSHP in detail in a later section of this chapter.

Self-Paraconjugate Hurwitz Polynomial (SPHP)

$B(s_1, s_2)$ is said to be a SPHP if $B(s_1, s_2)$ is a self-paraconjugate polynomial and if $B(s_1, s_2) \neq 0$ in S^2_{++}. In other words, $B(s_1, s_2)$ is a BHP. For example,

$$B(s_1, s_2) = (s_1^2 + 1)^j (s_1s_2 + 1)^k$$

where j, k are non-negative integers.

Reactance Hurwitz Polynomial (RHP)

A reactance Hurwitz polynomial (RHP) is defined as the para-even or para-odd part of a SHP. $B(s_1, s_2)$ is said to be a RHP if $B(s_1, s_2)$ is a SPHP and if each zero locus of $B(s_1, s_2)$ on S^2_{00} is of multiplicity unity; i.e., $B(s_1, s_2)$ has no repeated factors. For example,

$$B(s_1, s_2) = (s_1^2 + 1)(s_1s_2 + 1)$$

is a RHP whereas

$$B(s_1, s_2) = (s_1^2 + 1)^2(3s_1s_2 + 1)^2$$

is not a RHP.

Immittance Hurwitz Polynomial (IHP)

An immittance Hurwitz polynomial (IHP) is defined as the product of a RHP and a SHP [10]. An alternate definition is given in terms of the zero-free regions.

$B(s_1, s_2)$ is said to be an IHP if $B(s_1, s_2)$ is a BHP and if each continuum zero locus of $B(s_1, s_2)$ on S^2_{00} is of multiplicity unity; i.e., $B(s_1, s_2)$ has no repeated self-paraconjugate factors. For example, $B(s_1, s_2) = (s_1^2 + 1)(s_1s_2 + 1)(2s_1s_2 + 7s_1 + s_2)(s_1 + s_2 + 1)$

Summary

The definitions of BHP, NHP, SHP, HPSS, and VSHP are such that in the above sequence each Hurwitz polynomial satisfies the conditions required for each of the preceding polynomials, and hence form a subset of each. In other words,

$$\{VSHP\} \subset \{HPSS\} \subset \{SHP\} \subset \{NHP\} \subset \{BHP\}$$

$$\{IHP\} \subset \{BHP\}$$

$$\{SPHP\} \subset \{BHP\}$$

$$\{RHP\} = \{IHP\} \cap \{SPHP\}$$

6.5 Testsets for Analog Hurwitz Polynomials

Continuity Property of the Zeroes of 2-V Polynomials

Let $B(s_1, s_2)$ be a 2-V polynomial of degree m in s_1 and n in s_2. Let x_1 be a point in the S_1-plane. Then, $B(x_1, s_2)$ is a 1-V polynomial in s_2 and has n zeroes (possible some at infinity) in the S_2-plane. When x_1 is such that the coefficients of r consecutive terms $s_2^n, s_2^{n-1}, \ldots, s_2^{n-r+1}$ become zero, $B(x_1, s_2)$ becomes a polynomial of degree $n - r$ in s_2 and has only $(n - r)$ zeroes in the finite part of the S_2-plane. If this polynomial is considered an $n - r$ degree polynomial in s_2, then it has n poles at $s_2 = \infty$. Further, we also know that for a polynomial (in general for a rational function), the number of poles and zeroes are equal when those that are at infinity are also counted. Hence, in the present case, where r leading coefficients become zero, the r missing zeroes are accounted to be at infinity.

As discussed earlier, the infinite distant points of the S_2-plane are assumed to converge to a single point at infinity. We further note that at a NSSK the function possesses zeroes as well as poles. Consider a zero s_2' of $B(x_1, s_2)$. When x_1 is moved on a continuous line in the S_1-plane, s_2' either moves continuously or remains stationary. Further, when $K (K \le n)$, zeroes are present at a point, among which K' zeroes are stationary, $(K - K')$ incoming and $(K - K')$ outgoing loci exist at that point. We may associate an incoming locus with an outgoing locus arbitrarily on a one-to-one basis. With these remarks in mind, the continuity property of the zeroes of 2-V polynomials can be stated in the form of a theorem.

Theorem 16.1: *Let $B(x_1, s_2)$ be not identically zero. [If $B(x_1, s_2) \equiv 0, (s_1 - x_1)^K$, $K \le M$, is a factor of $B(s_1, s_2)$. Divide this factor out and consider the resulting polynomial]. Then the locus of a zero $\hat{s}_2$ of $B(x_1, s_2)$ generated by the movement of x_1 on a continuous line in the S_1-plane is a continuous line or a fixed point in the S_2-plane. In the latter case $(s_2 - \hat{s}_2)$ is a factor of $B(s_1, s_2)$.*

PROOF. The proof of Theorem 16.1 can be given based on the properties of algebraic functions discussed by Bliss [27]. Also note that the locus may pass through the point at infinity; this is treated just like any other point in the S_2-plane. Similar property also holds for the zeroes s_1 of $B(s_1, x_2)$.

As a first step to identifying smaller testsets, we show that because of the continuity property of the zeroes of 2-V polynomials, tests at some isolated points can be omitted. This we do for the tests in the region S_{0+}^2 in the following lemma [24].

Lemma 16.1. *A 2-V polynomial $B(s_1, s_2) \ne 0$ or $0/0$ in the region S_{0+}^2 if $B(s_1, s_2) \ne 0$ or $0/0$ in $\{S_{10}$ − some isolated points$\} \times S_{2+}$ (i.e., the Cartesian product of the whole imaginary axis of the S_1-plane, except some isolated points, and the open right half of the S_2-plane), and if $B(\hat{s}_1, s_2) \ne 0$, where $\hat{s}_1$ is any such isolated point in S_{10} [i.e., no 1-V factor of the form $(s_1 - \hat{s}_1)$ is present in $B(s_1, s_2)$].*

Corollary 16.1. *A 2-V polynomial $B(s_1, s_2) \ne 0$ or $0/0$ in the region S_{0+}^2 if $B(jw_1, s_2) \ne 0$ for $-\infty < w_1 < \infty$ and $\operatorname{Re} s_2 > 0$, $|s_2| < \infty$.*

Note that to test for the absence of the zeroes of $B(s_1, s_2)$ in $S_{0\oplus}^2$, we must test all the boundary points and we cannot leave any isolated point untested, as in the case of S_{0+}^2.

Next we discuss test procedures for the various Hurwitz polynomials.

Theorem 16.2: *A 2-V polynomial $B(s_1, s_2)$ is a BHP iff $B(jw_1, s_2) \ne 0$ in S_{2+} for all real finite w_1, except possibly some isolated w_1, and if the polynomial $B(s_1, b)$ has no zeroes in S_{1+} for some $b \in S_{2+}$.*

Theorem 16.3: *A 2-V polynomial $B(s_1, s_2)$ is a NHP iff for all real finite w_1, $B(jw_1, s_2) \neq 0$ in S_{2+}; the polynomial in s_1, $B(s_1, b) \neq 0$ in S_{1+} for some $b \in S_{2+}$, and $B(s_1, s_2)$ has no factor of the type $(s_2 - j\alpha)$, where α is a real constant.*

Theorem 16.4: *A 2-V polynomial $B(s_1, s_2)$ is a SHP iff for all real finite w_1, $B(jw_1, s_2)$ has no zeroes in S_{2+}; the polynomial $B(s_1, b)$ has no zeroes in S_{1+} for some $b \in S_{2+}$, and $B(s_1, s_2)$ and $B_*(s_1, s_2)$ are relatively prime.*

Theorem 16.5: *A 2-V polynomial $B(s_1, s_2)$ is a HPSS iff for all real finite w_1, $B(jw_1, s_2)$ has no zeroes in $\{s_2 : \mathrm{Re}\, s_2 \geq 0, |s_2| < \infty\}$ and the polynomial, $B(s_1, b)$ has no zeroes in $\mathrm{Re}\, s_1 \geq 0$, for some $b \in S_{2+}$. It is easy to verify that only infinite distant points on S^2_{00} are omitted from S^2_{00} in testing for the zero locations of $B(s_1, s_2)$. Hence, $B(s_1, s_2)$ is a HPSS.*

Theorem 16.6: *The necessary and sufficient conditions for a 2-V polynomial $B(s_1, s_2)$ to be a VSHP are* [19]

1. $B(s_1, s_2)$ *is a HPSS*

2a. $B(\infty, s_2) \neq 0/0$ *for* $\mathrm{Re}(s_2) = 0$ *and* $|s_2| < \infty$

2b. $B(s_1, \infty) \neq 0/0$ *for* $\mathrm{Re}(s_1) = 0$ *and* $|s_1| < \infty$

2c. $B(\infty, \infty) \neq 0/0$

The infinite point testing method shown in (16.2a) to (16.2c) could be followed for the testing of the three conditions under number 2 above. Let

$$B(s_1, s_2) = \sum_{i=0}^{M} \sum_{j=0}^{N} b_{ij} s_1^i s_2^j$$

Then conditions 2a to 2c above are equivalent to:

$$A_M(s_2) = \sum_{j=0}^{N} b_{Mj} s_2^j \neq 0 \qquad \text{for } \mathrm{Re}(s_2) = 0$$

$$B_N(s_1) = \sum_{i=0}^{M} b_{iN} s_1^i \neq 0 \qquad \text{for } \mathrm{Re}(s_1) = 0$$

$$b_{MN} \neq 0$$

Theorem 16.7: *Let*

$$B(s_1, s_2) = \sum_{j=0}^{N} a_j(s_1) s_2^j = \sum_{i=0}^{M} b_i(s_2) s_1^i = \sum_{i=0}^{M} \sum_{j=0}^{N} c_{ij} s_1^i s_2^j$$

Then $B(s_1, s_2)$ is a VSHP iff $B(jw_1, jw_2) \neq 0$, $-\infty < w_i < \infty$ $i = 1, 2$, and $a_N(s_1) \neq 0$ in $S_{1\oplus}$, and $b_M(s_2) \neq 0$ in $S_{2\oplus}$, and $c_{MN} \neq 0$.

Testsets for SPHP, RHP, and IHP can be formulated easily based on their definitions and the testset of a BHP.

5.6 Two-Variable Very Strict Hurwitz Polynomials

A brief, additional discussion of VSHPs is given in this section because this class of 2-V Hurwitz polynomials is a counterpart to the 1-V SHP, at least from the domain description of a closed right-half-plane. We now state some of the properties of 2-V VSHPs. Let a two-variable transfer function $T(s_1, s_2)$ be expressed as

$$T(s_1, s_2) = P(s_1, s_2)/Q(s_1, s_2) \tag{16.3}$$

where

$$P(s_1, s_2) = \sum_i \sum_j p_{ij} s_1^i s_2^j \qquad (i = 0, 1, \dots, k; j = 0, 1, \dots, l)$$

$$Q(s_1, s_2) = \sum_i \sum_j q_{ij} s_1^i s_2^j \qquad (i = 0, 1, \dots, m; j = 0, 1, \dots, n)$$

By applying the transformation method of (16.2a) to (16.2c) it can be shown that unless $m \geq k$ and $n \geq l$, polar singularities exist at a set of infinite distant points in the closed right-half of the $\{S_1, S_2\}$-biplane. Therefore, assume that $m \geq k$ and $n \geq 1$. Then the following theorem regarding the singularity in the closed right-half biplane can be stated.

Theorem 16.8: *$T(s_1, s_2)$ does not possess any singularity on the closed right-half of $\{S_1, S_2\}$-biplane defined by $S^2_{\oplus\oplus}$ iff $Q(s_1, s_2)$ is a VSHP.*

PROOF. The proof of the theorem is straightforward if the infinite distant points are also taken into account.

Some other useful properties of VSHPs are [2]

- $B(s_1, s_2) = [B_1(s_1, s_2). B_2(s_1, s_2)]$ is a VSHP iff $B_1(s_1, s_2)$ and $B_2(s_1, s_2)$ are VSHPs.
- If $B(s_1, s_2)$ is a VSHP, then $\partial/\partial s_i[B(s_1, s_2)]$, $i = 1, 2$ are also VSHP's. This property is not true for other 2-D Hurwitz polynomials. Let

$$B(s_1, s_2) = A_M(s_2)s_1^M + A_{M-1}(s_2)s_1^{M-1} + \cdots + A_1(s_2)s_1 + A_0(s_2) \tag{16.4}$$

$$= C_N(s_1)s_2^N + C_{N-1}(s_1)s_2^{N-1} + \cdots + C_1(s_1)s_2 + C_0(s_1) \tag{16.5}$$

- Let $B(s_1, s_2)$ be expressed as in (16.4) and (16.5). Then, $A_i(s_2)$, $i = 0, 1, \dots, M$ and $C_j(s_1)$, $j = 0, 1, \dots, N$ are one-variable strict Hurwitz polynomials. This property readily follows from the partial derivative property above.
- Let $B(s_1, s_2) = \sum_i \sum_j b_{ij} s_1^i s_2^j$ be a real 2-D VSHP. Then $b_{MN} b_{ij} > 0$ for all i and j $(i = 0, 1, \dots, M;\ j = 0, 1, \dots, N)$.
- Let $B(s_1, s_2)$ be expressed as in (16.4) and (16.5). Then, $A_i(s_2)/A_{i-1}(s_2)$ for $i = 1, \dots, M$ and $C_j(s_1)/C_{j-1}(s_1)$ for $j = 1, \dots, N$ are minimum reactive, susceptive, strict PRFs.

The above property gives the following the necessary and sufficient condition for $B(s_1, s_2)$ which has a first degree in s_1 and any degree in s_2.

The necessary and sufficient condition that allows a 2-V polynomial $B(s_1, s_2) = B_1(s_2)s_1 + B_0(s_2)$ to be a VSHP is that the 1-V function $F(s_2) = B_1(s_2)/B_0(s_2)$ be a minimum reactive, susceptive, strict PRF.

Finally, we give a transformation theorem that transforms a 1-D strict Hurwitz polynomial into a 2-D VSHP. This is called a reactance transformation.

Theorem 16.9: *Let $D(s)$ be any strict Hurwitz polynomial of order n. Generate a 2-D polynomial in the following way:*

$$B(s_1, s_2) = [N(s_1, s_2)]^n \cdot \{D(s)\}|s = M(s_1, s_2)/N(s_1, s_2)$$

where M and N are, respectively, the even and odd 2-D polynomials. The necessary and sufficient condition for $B(s_1, s_2)$ to be a VSHP is that $M(s_1, s_2) + N(s_1, s_2)$ be a VSHP [2].

The odd TPRF $Z(s_1, s_2) = M(s_1, s_2)/N(s_1, s_2)$ does not possess NSSK on the distinguished boundary S^2_{00} and is called a proper or strict 2-D reactance function [2].

16.7 Application of Two-Dimensional Hurwitz Polynomials for Two-Variable Passive Networks and Stability

This section enumerates some properties of 2-V passive network functions, with particular reference to the Hurwitz nature of the polynomials [24] . (The following assumes $\text{Re}\, F(1, 1) > 0$)

Let $F(s_1, s_2) = N(s_1, s_2)/D(s_1, s_2)$ be the driving-point immittance of a passive network. Then, $N(s_1, s_2)$ and $D(s_1, s_2)$ are BHPs. Let the common factors of $N(s_1, s_2)$ and $D(s_1, s_2)$ be cancelled out and the resulting polynomials be called $N_1(s_1, s_2)$ and $D_1(s_1, s_2)$. Then, $N(s_1, s_2)$ and $D_1(s_1, s_2)$ are immittance Hurwitz polynomials.

Let $F(s_1, s_2) = A(s_1, s_2)/B(s_1, s_2)$ be a relatively prime 2-V odd rational function. Then, $F(s_1, s_2)$ is a 2-V reactance function if $A(s_1, s_2) + B(s_1, s_2)$ is a VSHP. $F(s_1, s_2)$ is a 2-V reactance function iff $A(s_1, s_2) + B(s_1, s_2)$ is a scattering Hurwitz polynomial. The self-paraconjugate polynomials $A(s_1, s_2)$ and $B(s_1, s_2)$ satisfy the reactance Hurwitz properties.

A relatively prime 2-V odd function $F(s_1, s_2) = A(s_1, s_2)/B(s_1, s_2)$ having no second-kind singularities is a reactance function iff $A(s_1, s_2) + B(s_1, s_2)$ is a VSHP. Such functions are called proper or strict reactance functions [2], and are useful as transformation functions to generate a (structurally stable) 2-D network from a stable 1-D network. This is one of the main applications of VSHP.

Let us now consider a relatively prime function $F(s_1, s_2) = N(s_1, s_2)/D(s_1, s_2)$. $F(s_1, s_2)$ is a TPRF iff $N(s_1, s_2) + D(s_1, s_2)$ is a scattering Hurwitz polynomial. Further, if no second-kind singularities exist for $F(s_1, s_2)$ on S^2_{00}, $N(s_1, s_2) + D(s_1, s_2)$ will be a VSHP. From the above discussion we can conclude that the Hurwitz nature determines important necessary conditions (and in some cases necessary and sufficient conditions) of 2-V positive lossless functions. Hurwitz polynomials can be used to generate 2-V positive and lossless functions as in 1-V case through partial derivative operations [6].

The following property relates to sum separability and Hurwitz nature [20]. Let $F(s_1, s_2) = N(s_1, s_2)/D(s_1, s_2)$ be a 2-V positive function. Assume $D(s_1, s_2)$ is an immittance Hurwitz polynomial having self-paraconjugate factors. In other words, $D(s_1, s_2)$ is written as $D(s_1, s_2) = D_1(s_1, s_2)D(s_1, s_2)$, where $D_1(s_1, s_2)$ is a reactance Hurwitz and $D_2(s_1, s_2)$ is a scattering Hurwitz. Then, $F(s_1, s_2)$ is sum separable as

$$F(s_1, s_2) = \frac{N_1(s_1, s_2)}{D_1(s_1, s_2)} + \frac{N_2(s_1, s_2)}{D_2(s_1, s_2)}$$

where N_1/D_1 is a reactance function and N_2/D_2 is a positive function.

Now we turn our attention to some applications concerning transfer functions. Let $T(s_1, s_2) = A(s_1, s_2)/B(s_1, s_2)$ be the transfer function of a singly terminated or doubly terminated 2-V lossless network. Then, $B(s_1, s_2)$ is a scattering Hurwitz polynomial. References [21] and [22] provide a detailed discussion of networks with transfer functions having scattering and VSHP denominators. It is not necessary that the denominator of all RLC 2-V network transfer functions be scattering Hurwitz. In the most general case it could be a broad-sense Hurwitz.

Another interesting observation is that in the 1-V case the voltage transfer function cannot have a pole at origin and infinity. Extending this to the 2-V situation, we find that the 2-V voltage transfer function, $T(s_1, s_2)$, cannot have first-kind (polar) singularities at $s_i = 0$ or $\infty (i = 1, 2)$, but $T(s_1, s_2)$ can be 0/0 at $s_i = 0$ or $\infty (i = 1, 2)$.

Let $H(s_1, s_2) = P(s_1, s_2)/Q(s_1, s_2)$ be a 2-V bounded real or lossless bounded real function. Then, $Q(s_1, s_2)$ is a scattering Hurwitz polynomial. If $H(s_1, s_2)$ has no NSSK on S^2_{00} then $Q(s_1, s_2)$ must be a VSHP.

Application to Two-Dimensional Analog System Stability

We consider the following important theorem [12].

Theorem 16.10: *The 2-D analog transfer function* $T(s_1, s_2) = A(s_1, s_2)/B(s_1, s_2)$ *is bounded-input–bounded-output (BIBO) stable only if* $B(s_1, s_2)$ *is a scattering Hurwitz polynomial. The* **sufficient** *condition for stability is that* $B(s_1, s_2)$ *be a VSHP (assume that* $T(s_1, s_2)$ *has no polar singularities at infinite distant points).*

We conclude this section with the following unresolved problem of BIBO stability of 2-D continuous time systems [12]:

Conjecture. *The 2-D analog transfer function* $T(s_1, s_2)$ *described in Theorem* 16.10 *is BIBO stable with no NSSK on* S^2_{00} *iff* $B(s_1, s_2)$ *is a VSHP.*

The sufficiency part of this statement is proved. The necessity has yet to be established.

6.8 Conclusions

This chapter provided a comprehensive, yet compact treatment of the theory of two-dimensional (analog) Hurwitz polynomials. With the help of double bilinear transformation $s_i = (1 - z_i)/(1 + z_i) i = 1, 2$, most of the theory could easily be translated to the 2-D discrete case and thus to the stability theory and design of two-dimensional digital filters [2, 24]. As in the 1-D case the 2-D Hurwitz polynomials play a critical role in the study of 2-D circuits, systems, and filters. In this chapter a detailed classification and testing (see theorems) of various 2-D Hurwitz polynomials is presented. Discussion of the properties of 2-D very strict Hurwitz polynomials is also given. The various testing procedures (algorithms) are not discussed. The test procedures can be found in [11, 12, 25, 26]. The chapter concluded by discussing how various Hurwitz polynomials arise in passive 2-V circuit theory and 2-D analog stability.

knowledgment

The author would like to thank Dr. Richard Williams, Dean, College of Engineering, California State University at Long Beach; Dr. Allen Stubberud, Chair, Department of Electrical and Computer Engineering, University of California at Irvine; and Dr. George Moschytz, Director, Institute for Signal and Information Processing, Swiss Federal Institute for Technology, Zurich, for their kind support of this project. The author is also a Visiting Researcher at the Center for High Speed Image and Signal Processing (CHIP) at the University of California, Irvine. The support of Dr. Leonard Ferrari, the director of the Center is appreciated. As always, Dr. P. K. Rajan of the Tennessee Technological University and Dr. E. I. Jury of the University of

California at Berkeley and the University of Miami provided constant encouragement. Dr. Rajan made significant contributions to the theory of stable 2-D polynomials discussed in this chapter. The material in this chapter is based mainly on references [2] & [24] (including Fig. 16.1).

References

[1] H. G. Ansell, "On certain two-variable generalization of circuit theory with applications to networks and transmission lines of lumped reactances," *IEEE Trans. Circuit Theory*, vol. CT-11, pp. 214–223, June 1964.

[2] H. C. Reddy et al., "Generation of two-dimensional digital transfer functions without nonessential singularities of the second kind," in *Proc. IEEE Int. Conf. Acoust., Speech, Signal Processing*, pp. 13–19, Apr. 1979. Also see P. K. Rajan et al., *IEEE Trans. Acoust., Speech, Signal Processing*, pp. 216–223, Apr. 1980.

[3] V. S. Valdimirov, *Methods of Theory of Functions of Many Complex Variables*, Cambridge, MA: MIT Press, 1966, pp. 36–38.

[4] M. Saito, "Synthesis of transmission line networks by multivariable techniques," in *Proc. Symp. Generalized Networks*, PIB, 1966, pp. 353–393.

[5] D. C. Youla, "Synthesis of networks containing lumped and distributed elements," in *Proc. Symp. Generalized Networks*, PIB, 1966, pp. 289–343.

[6] V. Ramachandran, "Some similarities and dissimilarities between single variable and two-variable reactance functions," *IEEE Circuits Syst. Newsl.*, pp. 11–14, 1976.

[7] A. Fettweis, "On the scattering matrix and the scattering transfer matrix of multidimensional lossless two-ports," *Arch. Elk. Ubertragung*, vol. 36, pp. 374–381, Sept. 1982.

[8] T. S. Huang, "Stability of two-dimensional recursive digital filters," *IEEE Trans. Audio Electroacoust.*, vol. AU-20, pp. 158–163, June 1972.

[9] Ph. Delsarte, Y. Genin, and Y. Kamp, "Two-variable stability criteria," *Proc. IEEE Int. Symp. Circuits Syst.*, pp. 495–498, Jul. 1979.

[10] A. Fettweis, "On Hurwitz polynomials in several variables," in *Proc. 1983 IEEE Int. Symp. Circuits and Syst.*, Newport Beach, CA, 1983, pp. 382–385.

[11] N. K. Bose, *Applied Multidimensional Systems Theory*, New York: Van Nostrand Reinhold, 1982.

[12] E. I. Jury, "Stability of multidimensional systems and related problems," in *Multidimensional Systems—Techniques and Applications*, S. G. Tzafestas, Ed., New York: Marcel Dekker, 1986.

[13] A. Fettweis and S. Basu, "On discrete scattering Hurwitz polynomials," *Int. J. Circuit Theory Appl.*, vol. 13, Jan. 1985.

[14] A. Fettweis, "Some properties of scattering Hurwitz polynomials," *Arch. Elk. Ubertragung*, vol. 38, pp. 171–176, 1984.

[15] M. G. Strintzis, "Tests of stability of multidimensional filters," *IEEE Trans. Circuits and Syst.*, vol. CAS-24, Aug. 1977.

[16] D. Goodman, "Some difficulties with double bilinear transformation in 2-D filter design," *Proc. IEEE*, vol. 66, pp. 905–914, June 1977.

[17] D. C. Youla, "The analysis and synthesis of lumped passive *n*-dimensional networks—Part I—analysis," Polytechnic Inst. New York, Brooklyn, Rep. MIR-1437-84, Jul. 1984.

[18] A. Fettweis and S. Basu, "New results on multidimensional Hurwitz polynomials," in *Proc. Int. Symp. Circuits and Systems*, Kyoto, Japan, June 1985, pp. 1359–1362.

[19] H. C. Reddy and P. K. Rajan, "A simpler test-set for very strict Hurwitz polynomials," *Proc. IEEE*, pp. 890–891, June 1986.

[20] H. C. Reddy et al., "Separability of multivariable network driving point functions," *IEEE Trans. Circuits Syst.*, vol. CAS-29, pp. 833–840, Dec. 1982.

[21] H. C. Reddy et al., "Realization of resistively terminated two-variable lossless ladder networks," *IEEE Trans. Circuits Syst.*, vol. CAS-29, pp. 827–832, Dec. 1982.

[22] H. C. Reddy et al., "Design of two-dimensional digital filters using analog reference filters without second kind singularities," in *Proc. IEEE Int. Conf. Acoust., Speech, Signal Processing,* Apr. 1981, pp. 692–695.

[23] D. Goodman, "Some stability properties of linear shift invariant digital filters," *IEEE Trans. Circuits SST.*, vol. 26, pp. 201–208, Apr. 1971.

[24] H. C. Reddy and P. K. Rajan, "A comprehensive study of two-variable Hurwitz polynomials," *IEEE Trans. Educ.*, vol. 32, pp. 198–209, Aug. 1989.

[25] H. C. Reddy and P. K. Rajan, "A test procedure for the Hurwitz nature of two-dimensional analog polynomials using complex lossless function theory," *Proc. IEEE Int. Symp. Circuits Systems,* pp. 702–705, May 1987.

[26] P. K. Rajan and H. C. Reddy, "Hermite matrix test for very strict Hurwitz polynomials," *Proc. Midwest Symp. Circuits Syst.*, pp. 670–673, Aug. 1986.

[27] G. A. Bliss, *Algebraic Functions.* New York: American Mathematical Society, 1933.

IV

inear Circuit Analysis

wrence P. Huelsman
versity of Arizona

17

Terminal and Port Representations

nes A. Svoboda
rkson University,
v York

'.1 Introduction

Frequently it is useful to decompose a large circuit into subcircuits and to consider the subcircuits separately. Subcircuits are connected to other subcircuits using terminals or ports. Terminal and port representations of a subcircuit describe how that subcircuit will act when connected to other subcircuits. Terminal and port representations do not provide the details of mesh or node equations since these details are not required to describe how a subcircuit interacts with other subcircuits.

In this chapter, terminal and port representations of circuits will be described. Particular attention will be given to the distinction between terminals and ports. Applications will show the usefulness of terminal and port representations.

'.2 Terminal Representations

Figure 17.1 shows a subcircuit which can be connected to other subcircuits using terminals. The subcircuit is shown symbolically on the left and an example is shown on the right. Nodes a, b, c, and d are terminals and may be used to connect this circuit to other circuits. Node e is internal to the circuit and is not a terminal. The terminal voltages V_a, V_b, V_c, and V_d are node voltages with respect to an arbitrary reference node. The terminal currents I_a, I_b, I_c, and I_d describe currents that will exist when this subcircuit is connected to other subcircuits. Terminal representations show how the terminal voltages and currents are related. Several equivalent representations are possible, depending on which of the terminal currents and voltages are selected as independent variables.

A terminal representation of the example network in Fig. 17.1 can be obtained by writing a node equation for the network. A general procedure for writing node equations is described in Chapter 22.2 of this handbook. A simpler procedure is available for passive networks, i.e., networks consisting entirely of resistors, capacitors, and inductors. Since the circuit in Fig. 17.1 is a passive circuit, the simpler procedure will be used to write terminal equations to represent this circuit.

8341-2/95/$0.00 + $.50
by CRC Press, Inc.

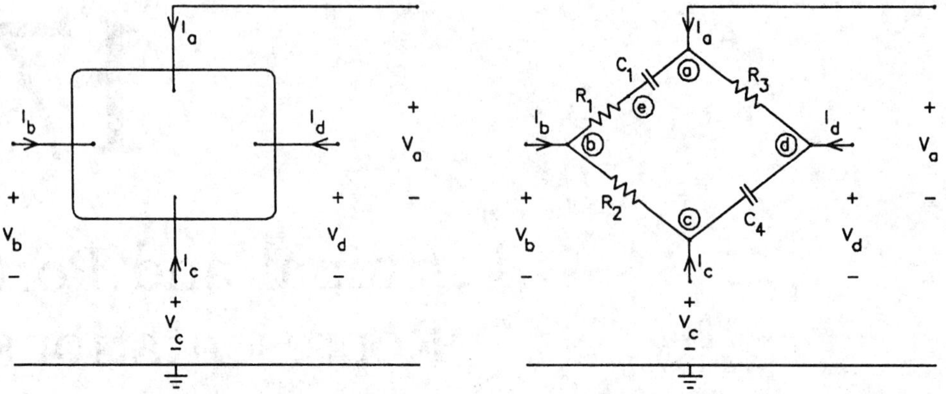

FIGURE 17.1 A four-terminal network.

The nodal admittance matrix of the example circuit will have five rows and five columns. The rows, and also the columns, of this matrix will correspond to the five nodes of the circuit in the order a, b, c, d and e. For example, the fourth row of the nodal admittance matrix corresponds to node d and the third column corresponds to node c. Let y_{ij} denote the admittance of a branch of the network which is incident with nodes i and j and let Y_{ij} denote the element of the nodal admittance matrix that is in row i and column j. Then

$$Y_{ii} = \sum_{\substack{\text{all branches}\\ \text{incident to node } i}} y_{ik} \tag{17.1}$$

i.e., the diagonal element of the nodal admittance matrix in row i and column i is equal to the sum of the admittances of all branches in the circuit that are incident with node i. The off-diagonal elements of the nodal admittance matrix are given by

$$Y_{ij} = -\sum_{\substack{\text{all branches incident}\\ \text{to both nodes } i \text{ and } j}} y_{ij} \tag{17.2}$$

The example circuit is represented by the node equation

$$\begin{pmatrix} I_a \\ I_b \\ I_c \\ I_d \\ 0 \end{pmatrix} = \begin{pmatrix} C_1 s + \frac{1}{R_3} & 0 & 0 & -\frac{1}{R_3} & -C_1 s \\ 0 & \frac{1}{R_1} + \frac{1}{R_2} & -\frac{1}{R_2} & 0 & -\frac{1}{R_1} \\ 0 & -\frac{1}{R_2} & C_4 s + \frac{1}{R_2} & -C_4 s & 0 \\ -\frac{1}{R_3} & 0 & -C_4 s & C_4 s + \frac{1}{R_3} & 0 \\ -C_1 s & -\frac{1}{R_1} & 0 & 0 & C_1 s + \frac{1}{R_1} \end{pmatrix} \begin{pmatrix} V_a \\ V_b \\ V_c \\ V_d \\ V_e \end{pmatrix} \tag{17.3}$$

Suppose that $C_1 = 1$ F, $C_4 = 2$ F, $R_1 = 1/2\ \Omega$, $R_2 = 1/4\ \Omega$, and $R_3 = 1\ \Omega$. Then,

$$\begin{pmatrix} I_a \\ I_b \\ I_c \\ I_d \\ 0 \end{pmatrix} = \begin{pmatrix} s+1 & 0 & 0 & -1 & -1s \\ 0 & 6 & -4 & 0 & -2 \\ 0 & -4 & 2s+4 & -2s & 0 \\ -1 & 0 & -2s & 2s+1 & 0 \\ -s & -2 & 0 & 0 & s+2 \end{pmatrix} \begin{pmatrix} V_a \\ V_b \\ V_c \\ V_d \\ V_e \end{pmatrix}$$

There is no external current I_e since node e is not a terminal. Row 5 of this equation, corresponding to node e, can be solved for V_e and then V_e can be eliminated. Doing so results in

$$\begin{pmatrix} I_a \\ I_b \\ I_c \\ I_d \end{pmatrix} = \begin{pmatrix} \frac{3s+2}{s+2} & -\frac{2s}{s+2} & 0 & -1 \\ -\frac{2s}{s+2} & \frac{6s+8}{s+2} & -4 & 0 \\ 0 & -4 & 2s+4 & -2s \\ -1 & 0 & -2s & 2s+1 \end{pmatrix} \begin{pmatrix} V_a \\ V_b \\ V_c \\ V_d \end{pmatrix} \tag{17.4}$$

The terminal voltages were selected to be the independent variables and the entries in the matrix are admittances. Because none of the nodes of the subcircuit was chosen to be the reference node, the rows and columns of the matrix both sum to zero. This matrix is called an indefinite admittance matrix. Since it is singular, Eq. (17.4) cannot be solved directly. To see the utility of Eq. (17.4), consider Fig. 17.2. Here the four-terminal network has been connected to a voltage source and an amplifier and node c has been grounded. This external circuitry restricts the terminal currents and voltages of the four terminal network. In particular,

$$V_b = V_s, \quad I_a = 0, \quad V_c = 0, \quad I_d = 0 \quad \text{and} \quad V_o = KV_d \tag{17.5}$$

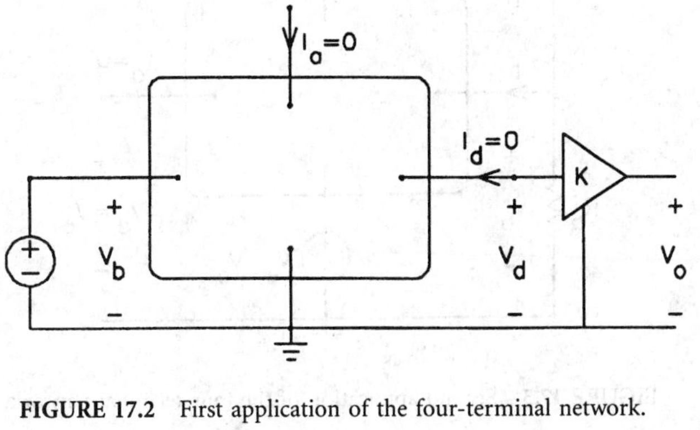

FIGURE 17.2 First application of the four-terminal network.

Under these conditions Eq. (17.4) becomes

$$\begin{pmatrix} 0 \\ I_b \\ I_c \\ 0 \end{pmatrix} = \begin{pmatrix} \dfrac{3s+2}{s+2} & -\dfrac{2s}{s+2} & 0 & -1 \\ -\dfrac{2s}{s+2} & \dfrac{6s+8}{s+2} & -4 & 0 \\ 0 & -4 & 2s+4 & -2s \\ -1 & 0 & -2s & 2s+1 \end{pmatrix} \begin{pmatrix} V_a \\ V_s \\ 0 \\ \dfrac{V_o}{K} \end{pmatrix} \tag{17.6}$$

Since I_b and I_c are not of interest, the second and third rows of this equation can be ignored. The first and fourth rows can be solved to obtain the transfer function of this circuit

$$\frac{V_o(s)}{V_s(s)} = \frac{K}{3(s+1)} \tag{17.7}$$

Next consider Fig. 17.3. The four terminal network is used in a second, different application. In this case

$$I_a + I_b = -\frac{V_b}{R_b}, \quad V_c = V_s, \quad V_a = V_b \quad I_d = 0 \quad \text{and} \quad V_d = V_o \tag{17.8}$$

Since $V_a = V_b$, the first two columns of Eq. (17.4) can be added together, replacing V_a by V_b. Also, it is convenient to add the first two rows together to obtain $I_a + I_b$. Then Eq. (17.4) reduces to

$$\begin{pmatrix} -\dfrac{V_b}{R_b} \\ I_c \\ 0 \end{pmatrix} = \begin{pmatrix} 5 & -4 & -1 \\ -4 & 2s+4 & -2s \\ -1 & -2s & 2s+1 \end{pmatrix} \begin{pmatrix} V_b \\ V_s \\ V_o \end{pmatrix} \tag{17.9}$$

Since I_c is not of interest, the second row of this equation can be ignored. The first and third rows of this equation can be solved to obtain the transfer function

$$\frac{V_o(s)}{V_s(s)} = \frac{(10R_b + 2)s + 4R_b}{(10R_b + 2)s + 4R_b + 1} \tag{17.10}$$

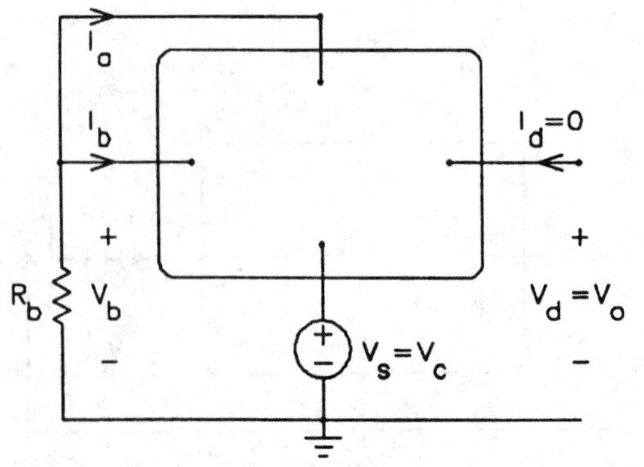

FIGURE 17.3 Second application of the four-terminal network.

These examples illustrate the utility of the terminal equations. In these examples, the problem of analyzing a network was divided into two parts. First the terminal equation was obtained from the node equation by eliminating the rows and columns corresponding to nodes that are not terminals. Second, the terminal equation is combined with equations describing the external circuitry connected to the four terminal network. When the external circuit is changed, only the second step must be redone. The advantage of representing a subnetwork by terminal equations is greater when

1. The subnetwork has many nodes that are not terminals.
2. The subnetwork is expected to be a component of many different networks.

7.3 Port Representations

A port consists of two terminals with the restriction that the terminal currents have the same magnitude but opposite sign. Figure 17.4 shows a two-port network. In this case, the two-port network was constructed from the four-terminal network by pairing terminals a and b to form port 1 and pairing terminals d and c to form port 2. The restrictions

$$I_1 = I_a = -I_b \quad \text{and} \quad I_2 = I_d = -I_c \tag{17.11}$$

must be satisfied in order for these two pairs of terminals to be ports.

The behavior of the two-port network is described using four variables. These variables are the port voltages, V_1 and V_2 and the port currents I_1, and I_2. Several equivalent representations are possible, depending on which two of the port currents and voltages are selected as independent variables. The left column of Table 17.1 shows the two-port representations corresponding to four of the possible choices of independent variables. These representations are equivalent, and the right column of Table 17.1 shows how one representation can be obtained from another.

In row 1 of Table 17.1, the port voltages are selected to be the independent variables. In this case, the two-port circuit is represented by an equation of the form

$$\begin{pmatrix} I_1 \\ I_2 \end{pmatrix} = \begin{pmatrix} y_{11} & y_{12} \\ y_{21} & y_{22} \end{pmatrix} \begin{pmatrix} V_1 \\ V_2 \end{pmatrix} \tag{17.12}$$

The elements of the matrix in this equation have units of admittance. They are denoted using the letter y to suggest admittance and are called the "y parameters" or the "admittance parameters" of the two-port network.

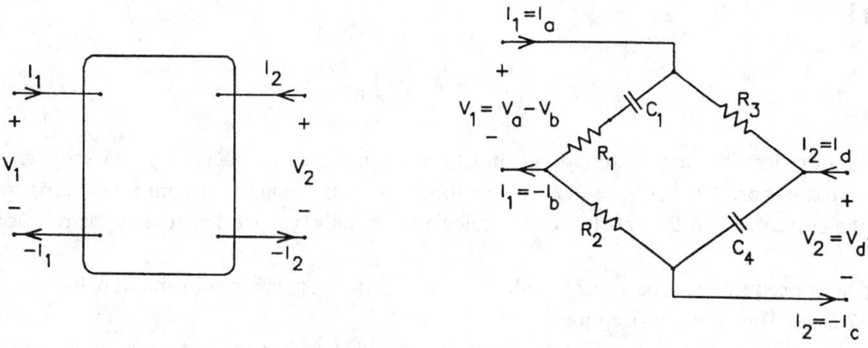

FIGURE 17.4 A two-port network.

TABLE 17.1 Relationships Between Two-Port Representations

$\begin{pmatrix} I_1 \\ I_2 \end{pmatrix} = \begin{pmatrix} y_{11} & y_{12} \\ y_{21} & y_{22} \end{pmatrix} \begin{pmatrix} V_1 \\ V_2 \end{pmatrix}$ $\lvert Y\rvert = y_{11}y_{22} - y_{12}y_{21}$	$\begin{pmatrix} y_{11} & y_{12} \\ y_{21} & y_{22} \end{pmatrix} = \begin{pmatrix} \frac{z_{22}}{\lvert Z\rvert} & -\frac{z_{12}}{\lvert Z\rvert} \\ -\frac{z_{21}}{\lvert Z\rvert} & \frac{z_{11}}{\lvert Z\rvert} \end{pmatrix} = \begin{pmatrix} \frac{1}{h_{11}} & -\frac{h_{12}}{h_{11}} \\ \frac{h_{21}}{h_{11}} & \frac{\lvert H\rvert}{h_{11}} \end{pmatrix} = \begin{pmatrix} \frac{D}{B} & -\frac{\lvert T\rvert}{B} \\ -\frac{1}{B} & \frac{A}{B} \end{pmatrix}$
$\begin{pmatrix} V_1 \\ V_2 \end{pmatrix} = \begin{pmatrix} z_{11} & z_{12} \\ z_{21} & z_{22} \end{pmatrix} \begin{pmatrix} I_1 \\ I_2 \end{pmatrix}$ $\lvert Z\rvert = z_{11}z_{22} - z_{12}z_{21}$	$\begin{pmatrix} z_{11} & z_{12} \\ z_{21} & z_{22} \end{pmatrix} = \begin{pmatrix} \frac{y_{22}}{\lvert Y\rvert} & -\frac{y_{12}}{\lvert Y\rvert} \\ -\frac{y_{21}}{\lvert Y\rvert} & \frac{y_{11}}{\lvert Y\rvert} \end{pmatrix} = \begin{pmatrix} \frac{\lvert H\rvert}{h_{22}} & \frac{h_{12}}{h_{22}} \\ -\frac{h_{21}}{h_{22}} & \frac{1}{h_{22}} \end{pmatrix} = \begin{pmatrix} \frac{A}{C} & -\frac{\lvert T\rvert}{C} \\ \frac{1}{C} & \frac{D}{C} \end{pmatrix}$
$\begin{pmatrix} V_1 \\ I_2 \end{pmatrix} = \begin{pmatrix} h_{11} & h_{12} \\ h_{21} & g_{22} \end{pmatrix} \begin{pmatrix} I_1 \\ V_2 \end{pmatrix}$ $\lvert H\rvert = h_{11}h_{22} - h_{12}h_{21}$	$\begin{pmatrix} h_{11} & h_{12} \\ h_{21} & h_{22} \end{pmatrix} = \begin{pmatrix} \frac{1}{y_{11}} & -\frac{y_{12}}{y_{11}} \\ \frac{y_{21}}{y_{11}} & \frac{\lvert Y\rvert}{y_{11}} \end{pmatrix} = \begin{pmatrix} \frac{\lvert Z\rvert}{z_{22}} & \frac{z_{12}}{z_{22}} \\ -\frac{z_{21}}{z_{22}} & \frac{1}{z_{22}} \end{pmatrix} = \begin{pmatrix} \frac{B}{D} & \frac{\lvert T\rvert}{D} \\ -\frac{1}{D} & \frac{C}{D} \end{pmatrix}$
$\begin{pmatrix} V_1 \\ I_1 \end{pmatrix} = \begin{pmatrix} A & B \\ C & D \end{pmatrix} \begin{pmatrix} V_2 \\ -I_2 \end{pmatrix}$ $\lvert T\rvert = AD - BC$	$\begin{pmatrix} A & B \\ C & D \end{pmatrix} = \begin{pmatrix} -\frac{y_{22}}{y_{21}} & -\frac{1}{y_{21}} \\ -\frac{\lvert Y\rvert}{y_{21}} & -\frac{y_{11}}{y_{21}} \end{pmatrix} = \begin{pmatrix} \frac{z_{11}}{z_{21}} & \frac{\lvert Z\rvert}{z_{21}} \\ \frac{1}{z_{21}} & \frac{z_{22}}{z_{21}} \end{pmatrix} = \begin{pmatrix} -\frac{\lvert H\rvert}{h_{21}} & -\frac{h_{11}}{h_{21}} \\ -\frac{h_{22}}{h_{21}} & -\frac{1}{h_{21}} \end{pmatrix}$

In row 2 of Table 17.1, the port currents are selected to be the independent variables. Now the elements of the matrix have units of impedance. They are denoted using the letter z to suggest impedance and are called the "z parameters" or the "impedance parameters" of the two-port network.

In row 3 of Table 17.1, I_1 and V_2 are the independent variables. The elements of the matrix do not have the same units. In this sense, this is a hybrid matrix. They are denoted using the letter h to suggest hybrid and are called the "h parameters" or the "hybrid parameters" of the two-port network. Hybrid parameters are frequently used to represent bipolar transistors.

In the last row of Table 17.1 the independent signals are V_2 and $-I_2$. In this case the two-port parameters are called "transmission parameters" or "*ABCD* parameters". They are convenient when two-port networks are connected in cascade.

Next, consider the problem of calculating or measuring the parameters of a two-port circuit. Equation (17.12) suggests a procedure for calculating the y parameters of a two-port circuit. The first row of Eq. (17.12) is

$$I_1 = y_{11}V_1 + y_{12}V_2 \tag{17.13}$$

Setting $V_1 = 0$ leads to

$$y_{12} = \frac{I_1}{V_2} \quad \text{when} \quad V_1 = 0 \tag{17.14}$$

This equation describes a procedure that can be used to measure or calculate y_{12}. A short circuit is connected to port 1 to set $V_1 = 0$. A voltage source having voltage V_2 is connected across port 2 and the current, I_1, in the short circuit is calculated. Finally, y_{12} is calculated as the ratio of I_1 to V_2.

Similar procedures can be used to calculate any of the $y, z, h,$ or transmission parameters. Table 17.2 tabulates these procedures.

As an example of how Table 17.2 can be used, consider calculating the y parameters of the two port circuit shown in Fig. 17.3. (Recall that $C_1 = 1$ F, $C_4 = 2$ F, $R_1 = 1/2\ \Omega, R_2 = 1/4\ \Omega,$

TABLE 17.2 Calculation Two-Port Parameters

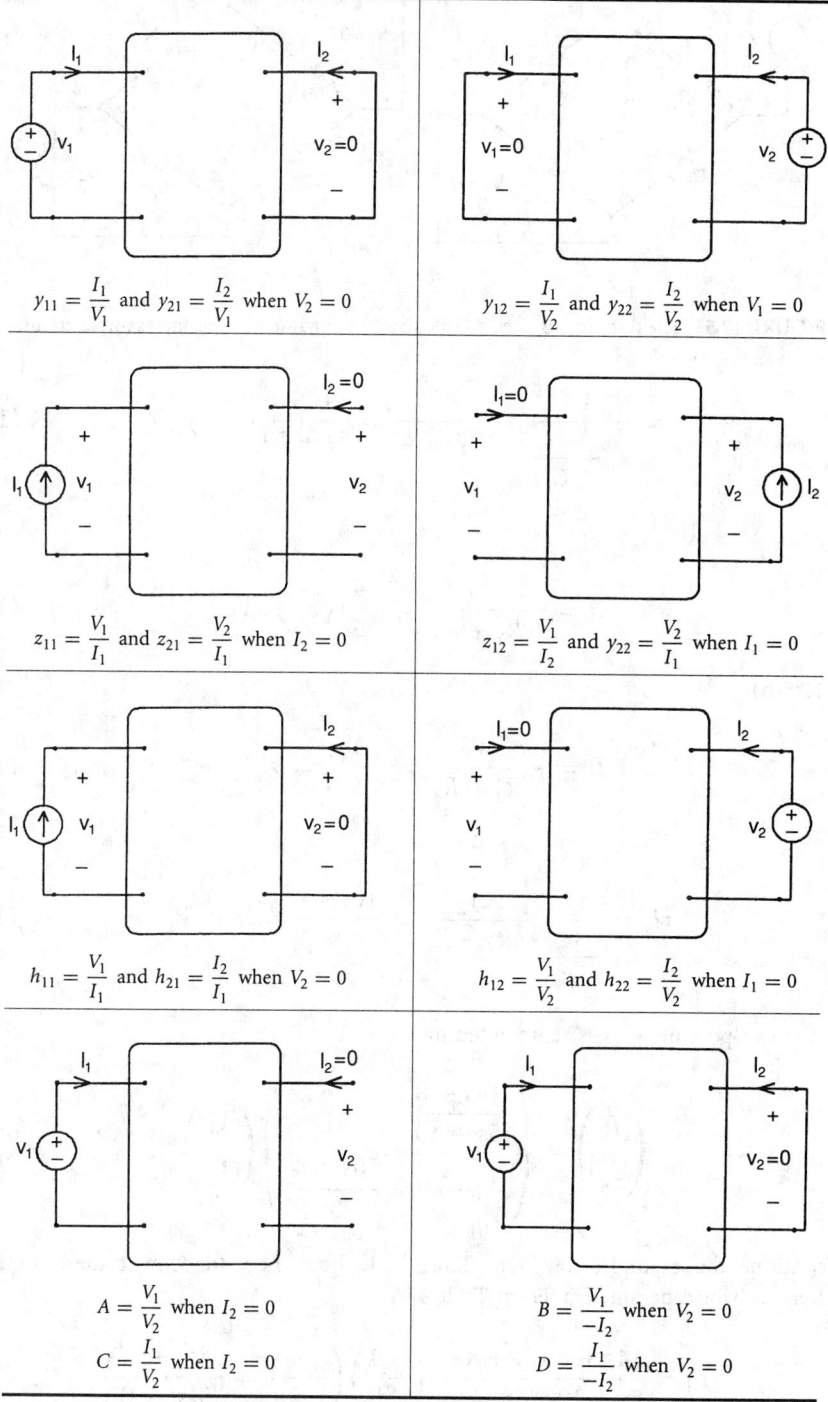

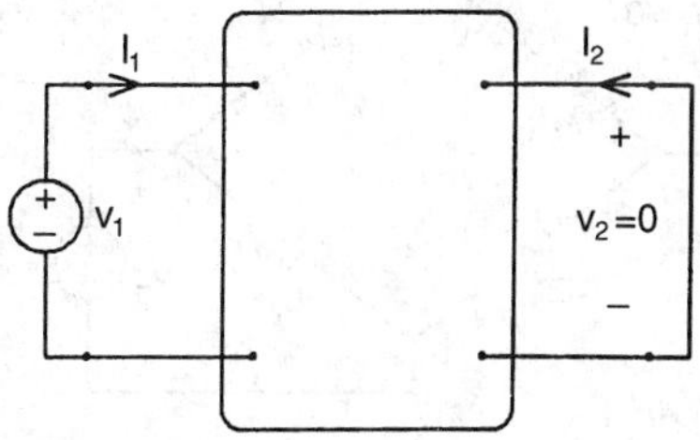

$y_{11} = \frac{I_1}{V_1}$ and $y_{21} = \frac{I_2}{V_1}$ when $V_2 = 0$

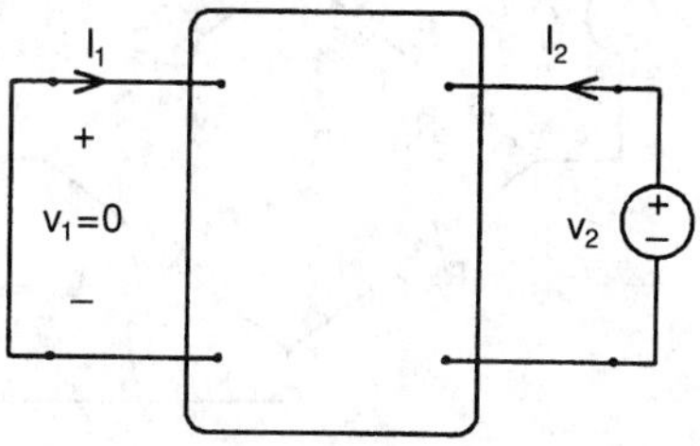

$y_{12} = \frac{I_1}{V_2}$ and $y_{22} = \frac{I_2}{V_2}$ when $V_1 = 0$

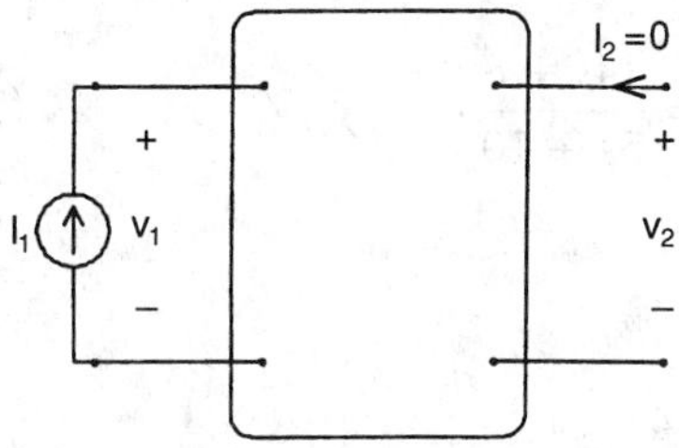

$z_{11} = \frac{V_1}{I_1}$ and $z_{21} = \frac{V_2}{I_1}$ when $I_2 = 0$

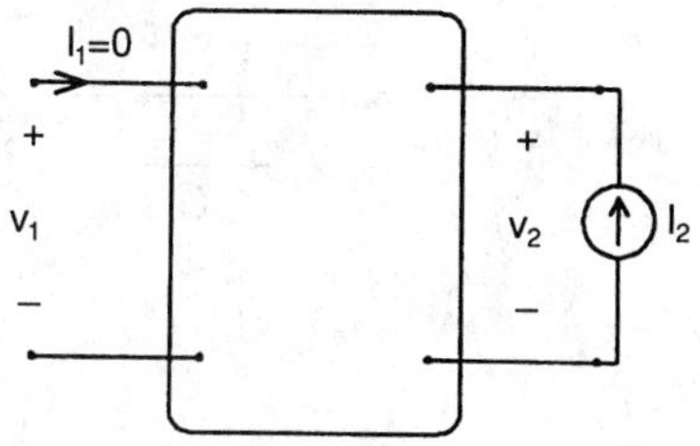

$z_{12} = \frac{V_1}{I_2}$ and $y_{22} = \frac{V_2}{I_1}$ when $I_1 = 0$

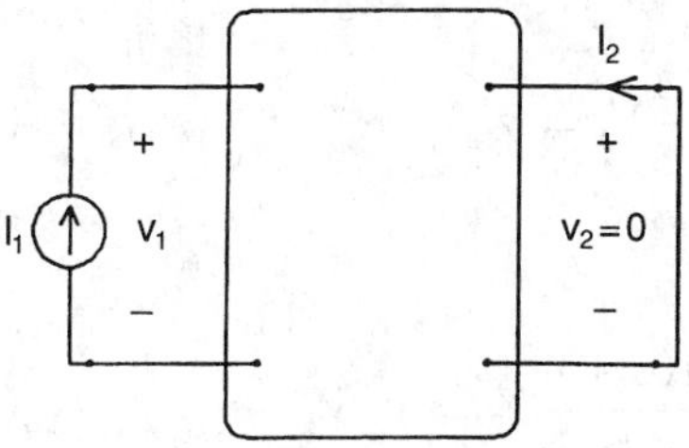

$h_{11} = \frac{V_1}{I_1}$ and $h_{21} = \frac{I_2}{I_1}$ when $V_2 = 0$

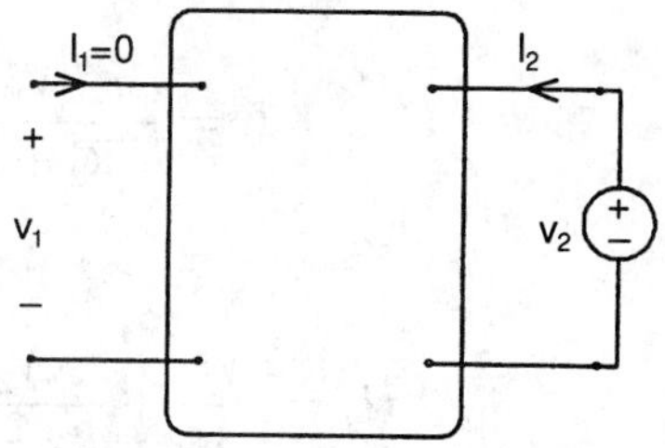

$h_{12} = \frac{V_1}{V_2}$ and $h_{22} = \frac{I_2}{V_2}$ when $I_1 = 0$

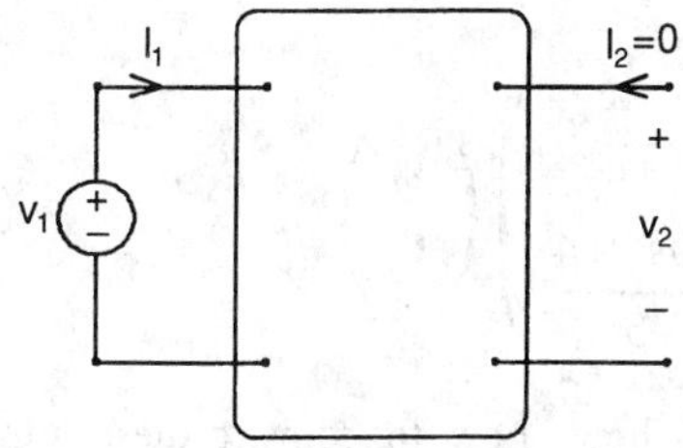

$A = \frac{V_1}{V_2}$ when $I_2 = 0$

$C = \frac{I_1}{V_2}$ when $I_2 = 0$

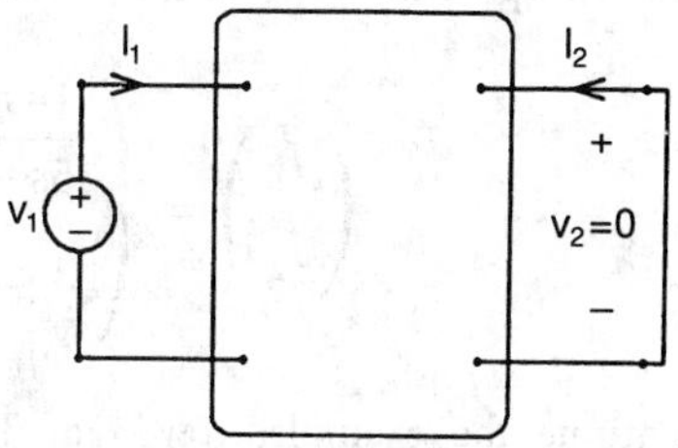

$B = \frac{V_1}{-I_2}$ when $V_2 = 0$

$D = \frac{I_1}{-I_2}$ when $V_2 = 0$

and $R_3 = 1\ \Omega$.) According to Table 17.2, two cases will have to be considered. In the first, a voltage source is connected to port 1 and a short circuit is connected to port 2. In the second, a short circuit is connected across port 1 and a voltage source is connected to port 2. The resulting circuits are shown in Fig. 17.5. In Fig. 17.5(a)

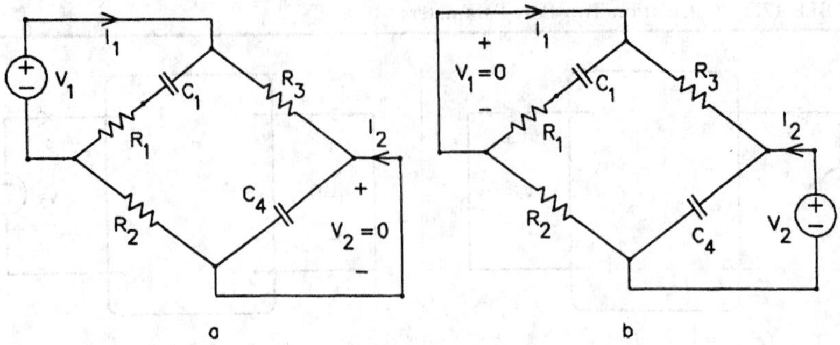

FIGURE 17.5 The test circuits used to calculate the y parameters of the example circuit.

$$I_1 = \frac{V_1}{R_1 + \dfrac{1}{C_1 s}} + \frac{V_1}{R_2 + R_3} = \frac{14s + 8}{5s + 10} V_1 = y_{11} V_1 \tag{17.15}$$

and

$$I_2 = -\frac{V_1}{R_2 + R_3} = -\frac{4}{5} V_1 = y_{21} V_1 \tag{17.16}$$

In Fig. 17.5(b)

$$I_1 = -\frac{V_2}{R_2 + R_3} = -\frac{4}{5} V_2 = y_{12} V_2 \tag{17.17}$$

and

$$I_2 = \frac{V_2}{\dfrac{1}{C_4 s}} + \frac{V_2}{R_2 + R_3} = \frac{10s + 4}{5} V_2 = y_{22} V_2 \tag{17.18}$$

Finally, the two-port network is represented by

$$\begin{pmatrix} I_1 \\ I_2 \end{pmatrix} = \begin{pmatrix} \dfrac{14s + 8}{5s + 10} & -\dfrac{4}{5} \\ -\dfrac{4}{5} & \dfrac{10s + 4}{5} \end{pmatrix} \begin{pmatrix} V_1 \\ V_2 \end{pmatrix} \tag{17.19}$$

To continue the example, row 3 of Table 17.1 shows how to convert these admittance parameters to hybrid parameters. From Table 17.1

$$|Y| = \frac{14s + 8}{5s + 10} \frac{10s + 4}{5} - \left(-\frac{4}{5}\right)\left(-\frac{4}{5}\right) = \frac{28s^2 + 24s}{5(s + 2)} \tag{17.20}$$

$$h_{11} = \frac{1}{y_{11}} = \frac{5s + 10}{14s + 8} \tag{17.21}$$

$$h_{12} = -\frac{y_{12}}{y_{11}} = \frac{2s + 4}{7s + 4} \tag{17.22}$$

$$h_{21} = \frac{y_{21}}{y_{11}} = -\frac{2s+4}{7s+4} \tag{17.23}$$

$$h_{22} = \frac{|Y|}{y_{11}} = \frac{(14s+12)s}{7s+4} \tag{17.24}$$

To complete the example, notice that Table 17.2 shows how to calculate the hybrid parameters directly from the circuit. According to Table 17.2, h_{22} is calculated by connecting an open circuit across port 1 and a voltage source having voltage V_2 across port 2. The resulting circuit is shown in Fig. 17.6. Now h_{22} is determined from Fig. 17.6(b) by calculating the port current I_2

$$I_2 = \frac{V_2}{R_1 + R_2 + R_3 + \dfrac{1}{C_1 s}} + \frac{V_2}{\dfrac{1}{C_4 s}} = \frac{14s^2 + 12s}{7s+4} V_2 = h_{22} V_2 \tag{17.25}$$

Of course, this is the same expression as was calculated earlier from the y parameters of the circuit.

Next, consider the problem of analyzing a circuit consisting of a two-port network and some external circuitry. Figure 17.7 shows such a circuit. The currents in the resistors R_s and R_L are given by

$$I_1 = \frac{V_s - V_1}{R_s} \quad \text{and} \quad I_2 = -\frac{V_2}{R_L} \tag{17.26}$$

Suppose the two-port network used in Fig. 17.7 is the circuit shown in Fig. 17.4 and represented by y parameters in Eq. (17.19). Combining the above expressions for I_1 and I_2 with Eq. (17.19)

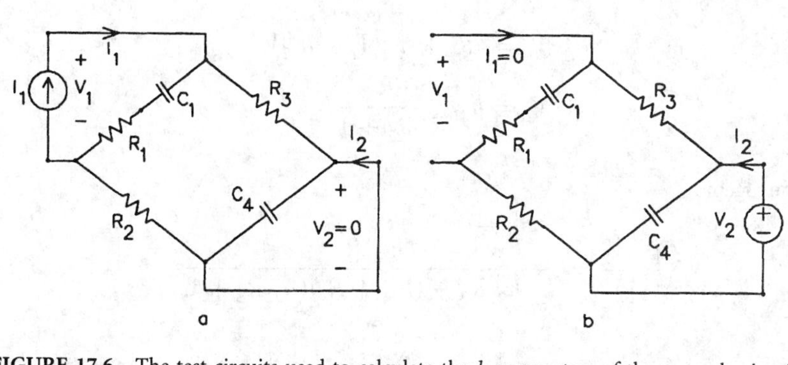

FIGURE 17.6 The test circuits used to calculate the h parameters of the example circuit.

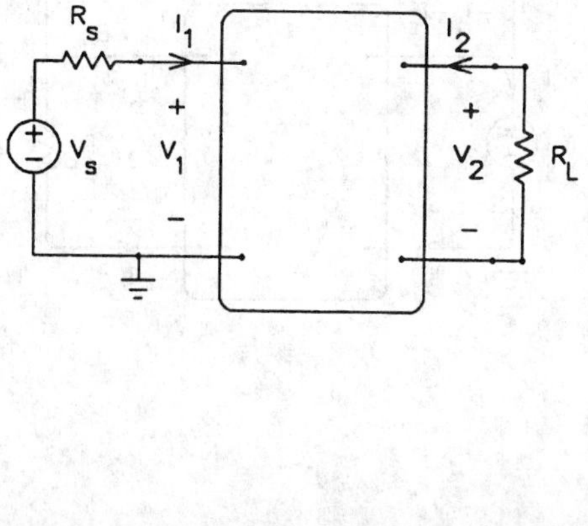

FIGURE 17.7 An application of the two-port network.

yields

$$\begin{pmatrix} \dfrac{V_s}{R_s} \\ 0 \end{pmatrix} = \begin{pmatrix} \dfrac{14s+8}{5s+10} + \dfrac{1}{R_s} & -\dfrac{4}{5} \\ -\dfrac{4}{5} & \dfrac{10s+4}{5} + \dfrac{1}{R_L} \end{pmatrix} \begin{pmatrix} V_1 \\ V_2 \end{pmatrix} \tag{17.27}$$

This equation can then be solved, e.g., using Cramer's Rule, for the transfer function

$$\begin{aligned} \frac{V_2}{V_s}(s) &= \frac{-\left(-\dfrac{4}{5}\right)\left(\dfrac{1}{R_s}\right)}{\left(\dfrac{14s+8}{5s+10} + \dfrac{1}{R_s}\right)\left(\dfrac{10s+4}{5} + \dfrac{1}{R_L}\right) - \left(-\dfrac{4}{5}\right)^2} \\ &= \frac{4(s+2)R_L}{(28s^2+24s)R_LR_s + (10s+4)(s+2)R_L + (14s+8)R_s + 5(s+2)} \end{aligned} \tag{17.28}$$

Next consider Fig. 17.8. This circuit illustrates a caution regarding use of the port convention. The use of ports assumes that the currents in the terminals comprising a port are equal in magnitude and opposite in sign. This assumption is not satisfied in Fig. 17.8 so port equations cannot be used to represent the four-terminal network.

Table 17.3 presents three circuit models for two port networks. These three models are based on y, z and h parameters, respectively. Such models are useful when analyzing circuits that contain subcircuits which are represented by port parameters. As an example, consider the circuit shown in Fig. 17.9(a). This circuit contains a two port network represented by h parameters. In Fig. 17.9(b) the two-port network has been replaced by the model corresponding to h parameters. Analysis of Fig. 17.9(b) yields

$$\begin{aligned} V_s &= (R_s + h_{11})I_1 + h_{12}V_2 \\ V_2 &= h_{21}\frac{-R_2}{R_2h_{22}+1}I_1 \end{aligned} \tag{17.29}$$

After some algebra

$$\frac{V_2}{V_s} = \frac{h_{21}R_2}{h_{12}h_{21}R_2 - (h_{11}+R_s)(h_{22}R_2+1)} \tag{17.30}$$

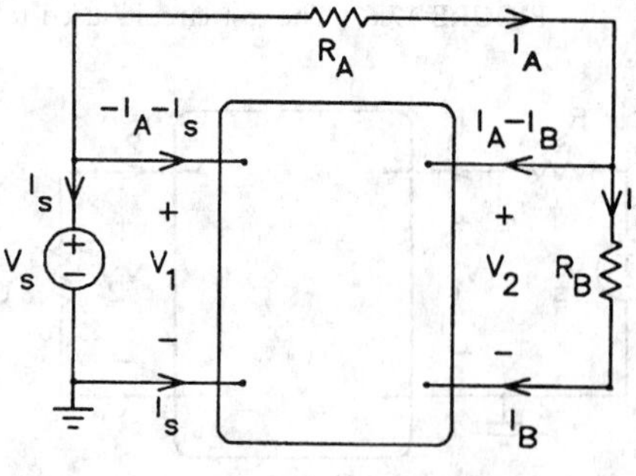

FIGURE 17.8 An incorrect application of the two-port representation.

TABLE 17.3 Models of Two-Port Networks

$\begin{pmatrix} I_1 \\ I_2 \end{pmatrix} = \begin{pmatrix} y_{11} & y_{12} \\ y_{21} & y_{22} \end{pmatrix} \begin{pmatrix} V_1 \\ V_2 \end{pmatrix}$	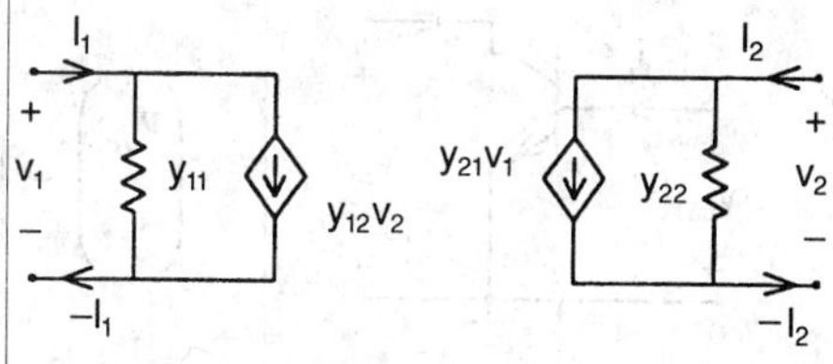
$\begin{pmatrix} V_1 \\ V_2 \end{pmatrix} = \begin{pmatrix} z_{11} & z_{12} \\ z_{21} & z_{22} \end{pmatrix} \begin{pmatrix} I_1 \\ I_2 \end{pmatrix}$	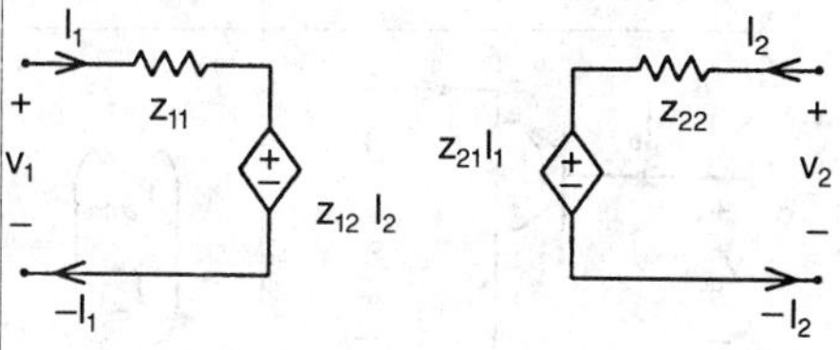
$\begin{pmatrix} V_1 \\ I_2 \end{pmatrix} = \begin{pmatrix} h_{11} & h_{12} \\ h_{21} & h_{22} \end{pmatrix} \begin{pmatrix} I_1 \\ V_2 \end{pmatrix}$	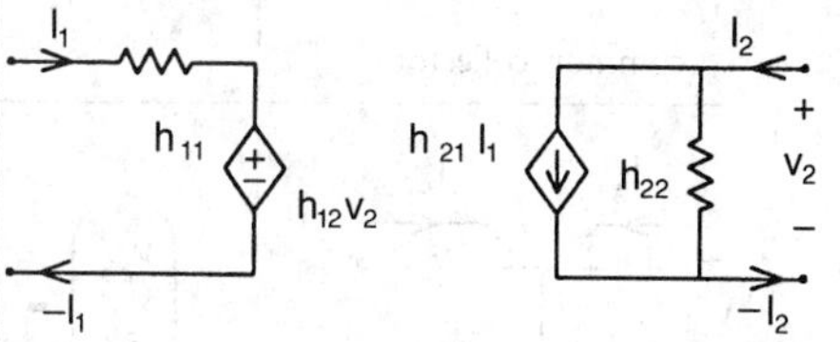

Since the circuits in Fig. 17.9(a) and (b) are equivalent, this is the voltage gain of the circuit in Fig. 17.9(a).

H parameters are frequently used to describe bipolar transistors. Table 17.4 shows the popular methods for converting this three terminal device into a two-port network. The three configurations shown in Table 17.4 are called "common emitter", "common collector", and "common base" to indicate which terminal is common to both ports. Table 17.4 also presents the notation that is commonly used to name the h parameters when they are used to represent a transistor.

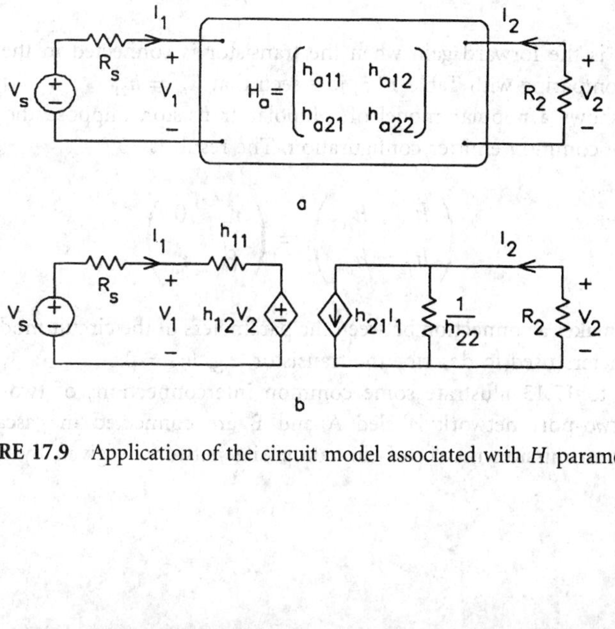

FIGURE 17.9 Application of the circuit model associated with H parameters.

TABLE 17.4 Using H Parameters to Specify Bipolar Transistors

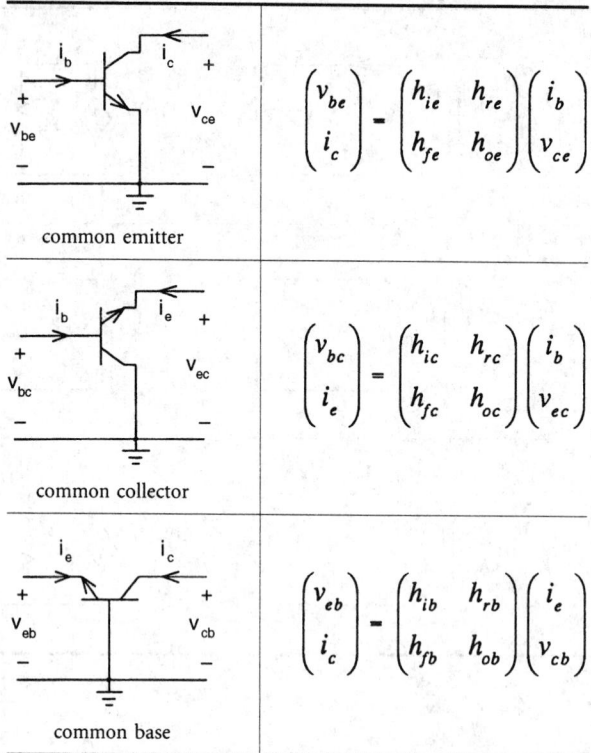

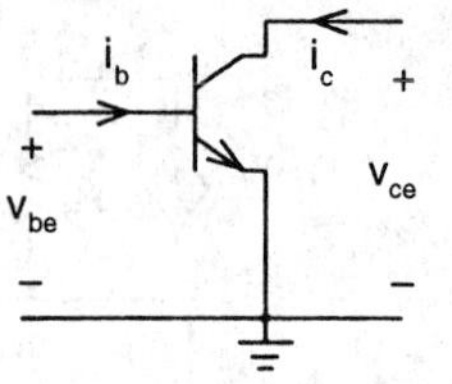 common emitter	$\begin{pmatrix} v_{be} \\ i_c \end{pmatrix} = \begin{pmatrix} h_{ie} & h_{re} \\ h_{fe} & h_{oe} \end{pmatrix} \begin{pmatrix} i_b \\ v_{ce} \end{pmatrix}$
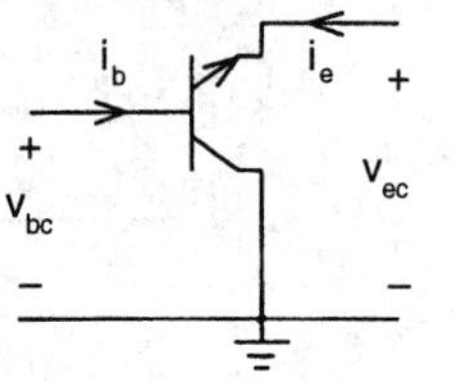 common collector	$\begin{pmatrix} v_{bc} \\ i_e \end{pmatrix} = \begin{pmatrix} h_{ic} & h_{rc} \\ h_{fc} & h_{oc} \end{pmatrix} \begin{pmatrix} i_b \\ v_{ec} \end{pmatrix}$
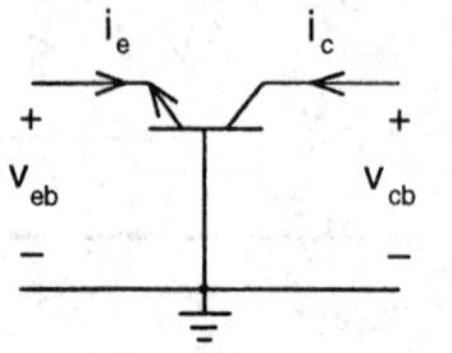 common base	$\begin{pmatrix} v_{eb} \\ i_c \end{pmatrix} = \begin{pmatrix} h_{ib} & h_{rb} \\ h_{fb} & h_{ob} \end{pmatrix} \begin{pmatrix} i_e \\ v_{cb} \end{pmatrix}$

i	input impedance or admittance
o	output impedance or admittance
f	forward gain
r	reverse gain
e	common emitter
b	common base
c	common collector

For example, h_{fe} is the forward gain when the transistor is connected in the common emitter configuration. Comparing with Table 17.3, it is seen that $h_{fe} = h_{21}$.

Figure 17.10 shows a popular model of a bipolar transistor. Suppose the h parameters are calculated for the common emitter configuration. The result is

$$\begin{pmatrix} h_{ie} & h_{re} \\ h_{fe} & h_{oe} \end{pmatrix} = \begin{pmatrix} r_\pi & 0 \\ \beta & r_o \end{pmatrix} \tag{17.31}$$

This calculation makes a connection between the parameters of the circuit model of the transistor and the h parameters used to describe the transistor, e.g. $h_{fe} = \beta$.

Figures 17.11 to 17.13 illustrate some common interconnections of two-port networks. In Fig. 17.11 the two-port network labeled A and B are connected in cascade. As shown in Fig. 17.11, the transmission matrix of the composite network is given by the product of the

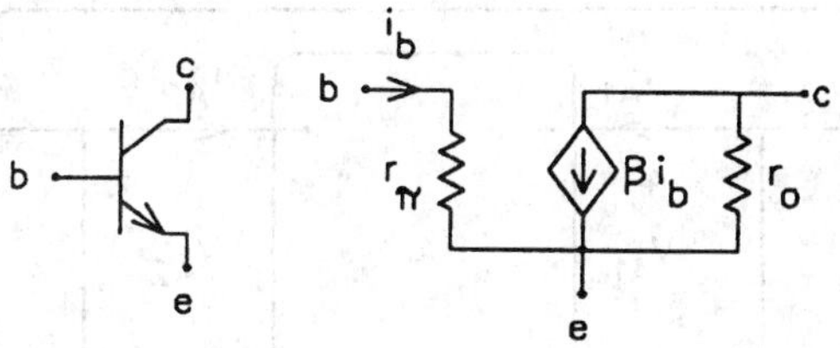

FIGURE 17.10 The hybrid pi model of a transistor.

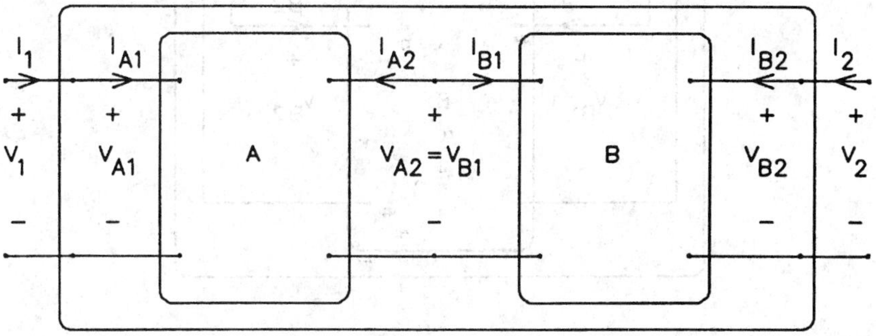

$$\begin{pmatrix} A & B \\ C & D \end{pmatrix} = \begin{pmatrix} A_a & B_a \\ C_a & D_a \end{pmatrix} \begin{pmatrix} A_b & B_b \\ C_b & D_b \end{pmatrix}$$

FIGURE 17.11 Cascade connection of two-port networks.

transmission matrices of the subnetworks. This simple relationship makes it convenient to use transmission parameters to represent a composite network which consists of the cascade of two subnetworks.

In Figure 17.12, the two-port networks labeled A and B are connected in parallel. As shown in Fig. 17.12, the admittance matrix of the composite network is given by the sum of the admittance matrices of the subnetworks. This simple relationship makes it convenient to use admittance parameters to represent a composite network which consists of parallel subnetworks.

In Figure 17.13, the two-port network labeled A and B are connected in series. As shown in Fig. 17.13, the impedance matrix of the composite network is given by the sum of the impedance matrices of the subnetworks. This simple relationship makes it convenient to use impedance parameters to represent a composite network which consists of series subnetworks.

Figure 17.14 shows a circuit that consists of three two-port networks. Two-port parameters representing the entire network can be calculated from the two-port parameters of the subnetworks. Let c denote the two port network consisting of network b connected in parallel with network a. Represent network c with y parameters by converting the h parameters representing network a to y parameters and adding these y parameters to the y parameters representing network b.

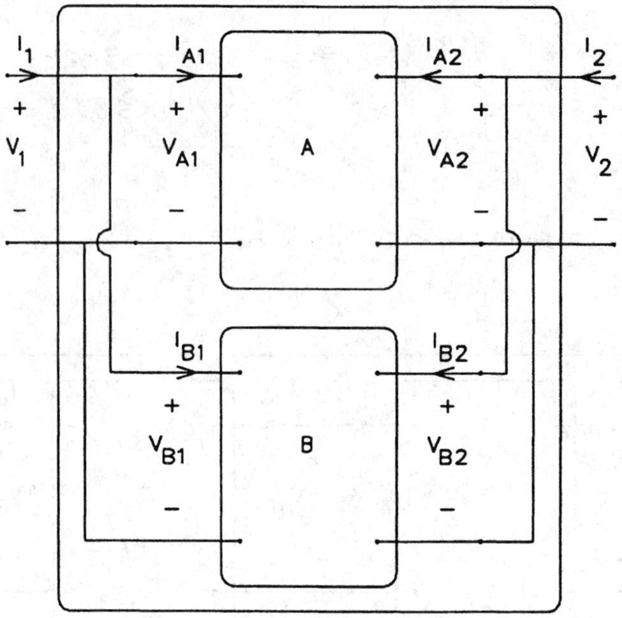

$$\begin{pmatrix} y_{11} & y_{12} \\ y_{21} & y_{22} \end{pmatrix} = \begin{pmatrix} y_{A11} & y_{A12} \\ y_{A21} & y_{A22} \end{pmatrix} + \begin{pmatrix} y_{B11} & y_{B12} \\ y_{B21} & y_{B22} \end{pmatrix}$$

FIGURE 17.12 Parallel connection of two-port networks.

$$Y_c = \begin{pmatrix} \dfrac{1}{h_{a11}} + y_{b11} & -\dfrac{h_{a12}}{h_{a11}} + y_{b12} \\ \dfrac{h_{a21}}{h_{a11}} + y_{b21} & \dfrac{|H_a|}{h_{a11}} + y_{b22} \end{pmatrix} \triangleq \begin{pmatrix} y_{c11} & y_{c12} \\ y_{c21} & y_{c22} \end{pmatrix} \tag{17.32}$$

Next, network c is connected in cascade with network d. Represent network d with transmission parameters by converting the y parameters representing network c to transmission parameters and multiplying these transmissions parameters by the transmission parameters representing network d.

$$T = \begin{pmatrix} -\dfrac{y_{c22}}{y_{c21}} & -\dfrac{1}{y_{c21}} \\ -\dfrac{|Y_c|}{y_{c21}} & -\dfrac{y_{c11}}{y_{c21}} \end{pmatrix} \begin{pmatrix} A_d & B_d \\ C_d & D_d \end{pmatrix} \triangleq \begin{pmatrix} A & B \\ C & D \end{pmatrix} \tag{17.33}$$

Finally, the circuit in Fig. 17.14 is represented by

$$\begin{pmatrix} V_1 \\ I_1 \end{pmatrix} = \begin{pmatrix} A & B \\ C & D \end{pmatrix} \begin{pmatrix} V_2 \\ -I_2 \end{pmatrix} \tag{17.34}$$

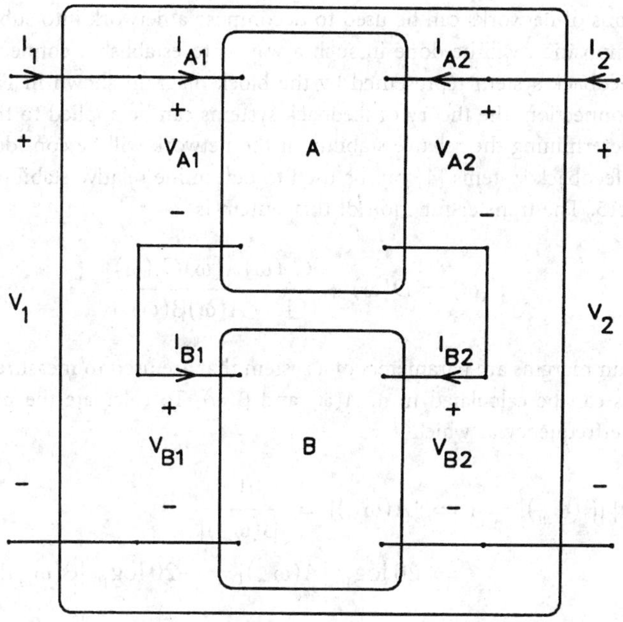

$$\begin{pmatrix} z_{11} & z_{12} \\ z_{21} & z_{22} \end{pmatrix} = \begin{pmatrix} z_{A11} & z_{A12} \\ z_{A21} & z_{A22} \end{pmatrix} + \begin{pmatrix} z_{B11} & z_{B12} \\ z_{B21} & z_{B22} \end{pmatrix}$$

FIGURE 17.13 Series connection of two-port networks.

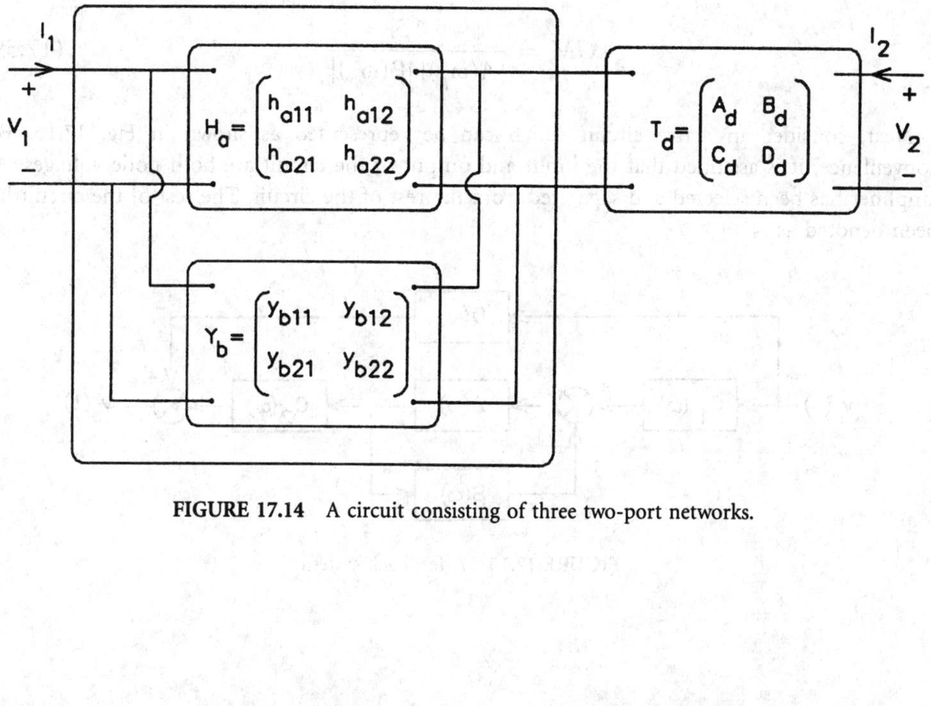

FIGURE 17.14 A circuit consisting of three two-port networks.

17.4 Port Representations and Feedback Systems

Port representations of networks can be used to decompose a network into subnetworks. In this section the decomposition will be done in such a way as to establish a connection between the network and a feedback system represented by the block diagram shown in Fig. 17.15. Having established this connection, the theory of feedback systems can be applied to the network. Here the problem of determining the relative stability of the network will be considered.

The theory of feedback systems [4] can be used to determine relative stability such as the one shown in Fig. 17.15. The transfer function of this system is

$$T(\omega) = D(\omega) + \frac{C_1(\omega)A(\omega)C_2(\omega)}{1 + A(\omega)\beta(\omega)} \tag{17.35}$$

The phase and gain margins are parameters of a system that are used to measure relative stability. These parameters can be calculated from $A(\omega)$ and $\beta(\omega)$. To calculate the phase margin first identify ω_m as the frequency at which

$$\begin{aligned} |A(\omega_m)|\,|\beta(\omega_m)| = 1 \Rightarrow |A(\omega_m)| &= \frac{1}{|\beta(\omega_m)|} \\ \Rightarrow 20\log_{10}|A(\omega_m)| &= -20\log_{10}|\beta(\omega_m)| \end{aligned} \tag{17.36}$$

The phase margin is then given by

$$\begin{aligned} \phi_m &= 180^\circ + \angle(A(\omega_m)\beta(\omega_m)) \\ &= 180^\circ + (\angle A(\omega_m) + \angle\beta(\omega_m)) \end{aligned} \tag{17.37}$$

The gain margin is calculated similarly. First identify ω_p as the frequency at which

$$180^\circ = \angle(A(\omega_p)\beta(\omega_p)) \tag{17.38}$$

The gain margin is then given by

$$GM = \frac{1}{|A(\omega_p)|\,|\beta(\omega_p)|} \tag{17.39}$$

Next, consider an active circuit which can be represented as shown in Fig. 17.16. For convenience, it is assumed that the input and output of the circuit are both node voltages. An amplifier has been selected and separated from the rest of the circuit. The rest of the circuit has been denoted as N.

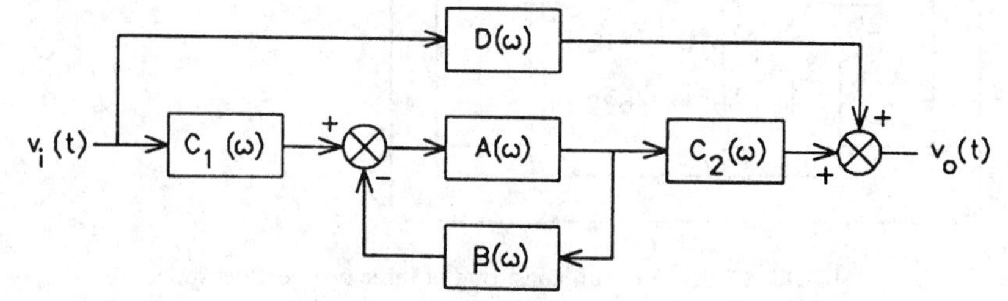

FIGURE 17.15 A feedback system.

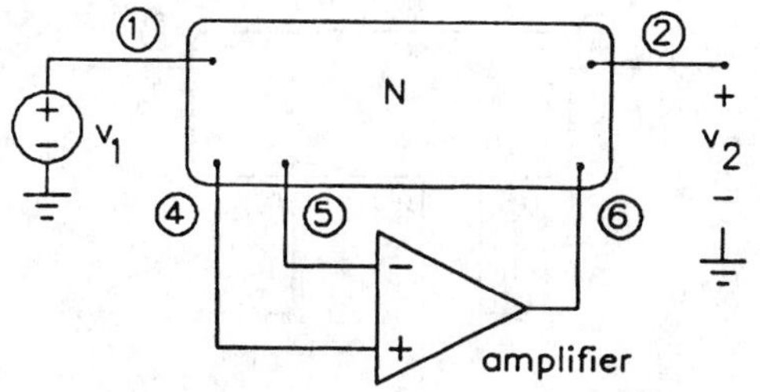

FIGURE 17.16 Identifying the network N.

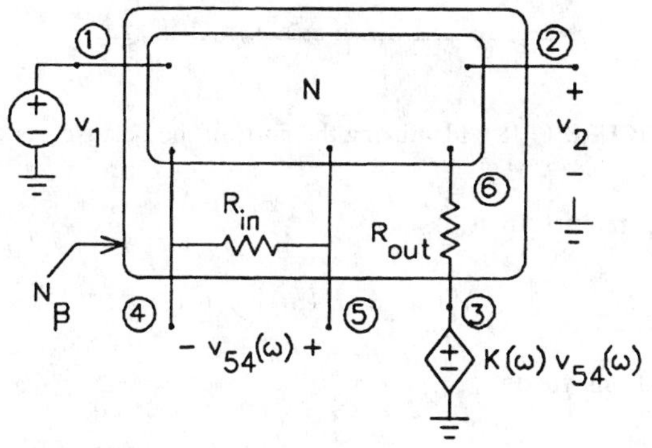

FIGURE 17.17 Identifying the Beta Network, N_β.

Suppose that the amplifier is a voltage controlled voltage source (VCVS), e.g., an inverting or a noninverting amplifier or an op amp. Replacing the VCVS by a simple model yields the circuit shown in Fig. 17.17. (The VCVS model accounts for input and output resistance and frequency dependent gain.) Figure 17.17 shows how to identify the network N_β which consists of the network N from Fig. 17.16 together with the input and output resistances of the VCVS. The network N_β has been called the "Beta Network" [5].

Figure 17.18 shows a way of grouping the terminals of the five-terminal network N_β to obtain a four-port network. This four-port network can be represented by the hybrid equation

$$\begin{pmatrix} I_1 \\ V_2 \\ I_3 \\ V_{54} \end{pmatrix} = \begin{pmatrix} H_{11} & H_{12} & H_{13} & H_{14} \\ H_{21} & H_{22} & H_{23} & H_{24} \\ H_{31} & H_{32} & H_{33} & H_{34} \\ H_{41} & H_{42} & H_{43} & H_{44} \end{pmatrix} \begin{pmatrix} V_1 \\ I_2 \\ V_3 \\ I_5 \end{pmatrix} \tag{17.40}$$

Since I_1 and I_3 are not of interest, the first and third rows of this equation can be set aside. Then setting I_2 and I_5 equal to zero yields

$$\begin{pmatrix} V_2 \\ V_{54} \end{pmatrix} = \begin{pmatrix} H_{21} & H_{23} \\ H_{41} & H_{43} \end{pmatrix} \begin{pmatrix} V_1 \\ V_3 \end{pmatrix} \tag{17.41}$$

where, for example

$$H_{43}(\omega) = \frac{V_{54}(\omega)}{V_3(\omega)} \quad \text{when} \quad V_1(\omega) = 0 \tag{17.42}$$

and $H_{21}(\omega)$, $H_{23}(\omega)$ and $H_{41}(\omega)$ are defined similarly.

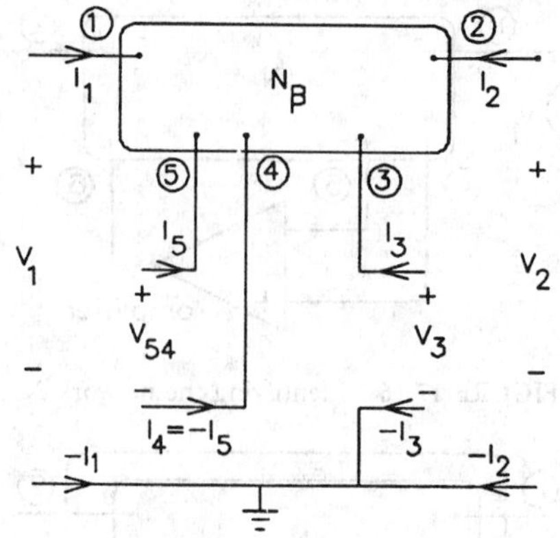

FIGURE 17.18 Identifying the ports of the Beta Network.

The amplifier model requires that

$$V_3(\omega) = K(\omega)V_{54}(\omega) \tag{17.43}$$

Combining these equations yields

$$T(\omega) = \frac{V_2(\omega)}{V_1(\omega)} = H_{21}(\omega) + \frac{K(\omega)H_{23}(\omega)H_{41}(\omega)}{1 - K(\omega)H_{43}(\omega)} \tag{17.44}$$

Comparing this equation to the transfer function of the feedback system yields

$$\begin{aligned} A(\omega) &= -K(\omega) \\ \beta(\omega) &= H_{43}(\omega) \\ C_1(\omega) &= H_{23}(\omega) \\ C_2(\omega) &= H_{41}(\omega) \\ D(\omega) &= H_{21}(\omega) \end{aligned} \tag{17.45}$$

These equations establish a correspondence between the feedback system in Fig. 17.15 and the active circuit in Fig. 17.16. This correspondence can be used to identify $A(s)$ and $\beta(s)$ corresponding to a particular circuit. Once $A(s)$ and $\beta(s)$ are known, the phase or gain margin can be calculated.

Figure 17.19 shows a Tow-Thomas bandpass biquad [3]. When the op amps are ideal devices the transfer function of this circuit is

$$T(s) = \frac{V_{in}(s)}{V_{out}(s)} = \frac{-\dfrac{1}{KRC}s}{s^2 + \dfrac{s}{QRC} + \dfrac{1}{R^2C^2}} \tag{17.46}$$

Figure 17.20 shows circuits that can be used to identify $A(s)$ and $\beta(s)$.

$$\begin{aligned} A(s) &= -1 \\ \beta(s) &= \frac{Q}{QR^2C^2s^2 + RCs} \end{aligned} \tag{17.47}$$

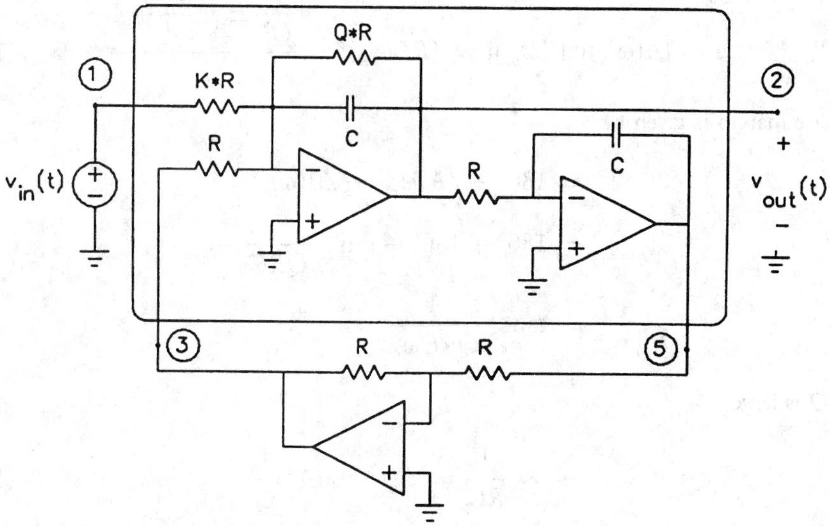

FIGURE 17.19 The Tow-Thomas biquad.

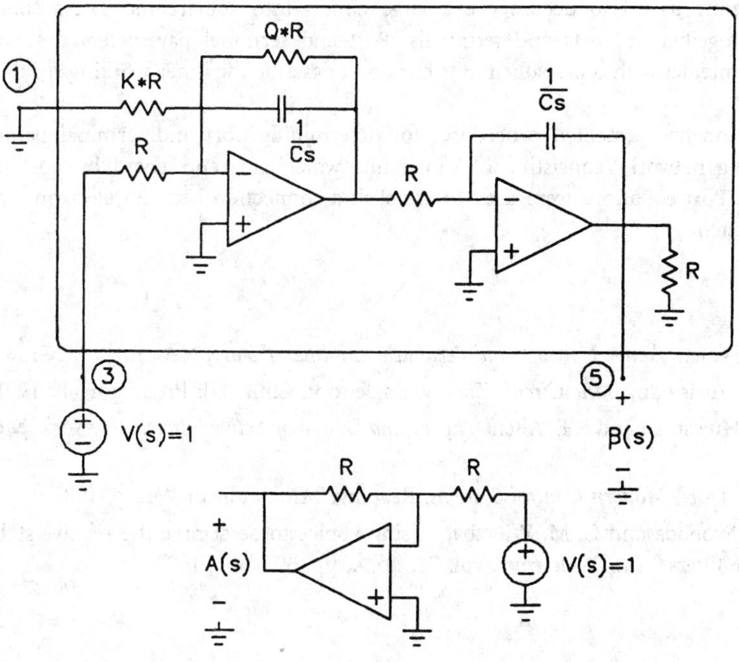

FIGURE 17.20 Calculating $A(s)$ and $\beta(s)$ for the Tow-Thomas biquad.

Next

$$1 = |A(\omega_m)|\,|\beta(\omega_m)| \Rightarrow (RC\omega_m)^2 = \frac{\dfrac{1}{Q^2} \pm \sqrt{\dfrac{1}{Q^4} + 4}}{2} \tag{17.48}$$

The phase margin is given by

$$\begin{aligned} \phi_m &= 180^\circ + \angle A(\omega_m) + \angle \beta(\omega_m) \\ &= 180^\circ + 180^\circ - \tan^{-1} \frac{-1}{QRC\omega_m} \\ &= \tan^{-1} \frac{1}{QRC\omega_m} \end{aligned} \tag{17.49}$$

When Q is large

$$\omega_m \approx \frac{1}{RC} \Rightarrow \phi_m \approx \tan^{-1} \frac{1}{Q} \tag{17.50}$$

When the op amps are not ideal it is not practical to calculate the phase margin by hand. With computer-aided analysis, accurate amplifier models, such as macromodels, can easily be incorporated into this analysis [5].

17.5 Conclusions

It is frequently useful to decompose a large circuit into subcircuits. These subcircuits are connected together at ports and terminals. Port and terminal parameters describe how the subcircuits interact with other subcircuits but don't describe the inner workings of the subcircuit itself.

This section has presented procedures for determining port and terminal parameters and for analyzing networks consisting of subcircuits which are represented by port or terminal parameters. Port equations were used to establish a connection between electronic circuits and feedback systems.

References

[1] W-K. Chen, *Active Network and Feedback Amplifier Theory*, New York: McGraw-Hill, 1980.

[2] L. P. Huelsman, *Basic Circuit Theory*, Englewood Cliffs, NJ: Prentice-Hall, 1991.

[3] L. P. Huelsman and P. E. Allen, *Theory and Design of Active Filters*, New York: McGraw-Hill, 1980.

[4] R. C. Dorf, *Modern Control Systems*, Reading, MA: Addison-Wesley, 1989.

[5] J. A. Svoboda and G. M. Wierzba, "Using PSpice to determine the relative stability of RC active filters," *Int. J. Electron.*, vol. 74, no. 4, pp. 593–604, 1993.

18

Signal Flow Graphs in Filter Analysis and Synthesis

ı-Min Lin
lue University,
ana

8.1 Formulation of Signal Flow Graphs for Linear Networks

Any lumped network obeys three basic laws: the Kirchhoff voltage law (KVL), the Kirchhoff current law (KCL), and the elements' laws (branch characteristics). For filter applications, we write the frequency-domain instead of the time-domain network equations. Three general methods for writing network equations are described in Chapter 13.1. They are the node equations, the loop equations, and the hybrid equations. This section outlines another method, the signal flow graph (abbreviated SFG) method of characterizing a linear network. The basic definitions of terms and theorems related to signal flow graphs are presented in Chapter 8.4. The concepts of tree, co-tree, loop, and cutset, required for the present discussion, are introduced in Chapter 7. Note that the terms loop, tree, and co-tree refer to directed circuit, spanning tree, and co-spanning tree defined in Chapter 7.

Consider first the construction of signal flow graphs for linear networks without controlled sources. For all practical networks, the independent voltage sources (E) contain no loops, and the independent current sources (J) contain no cutsets. Under these conditions, it is always possible to select a tree T, such that all voltage sources are included in the tree and all current sources are included in the co-tree. The network branches are divided into four sets (each set may be empty) indicated by subscripts as follows:

E: independent voltage sources
J: independent current sources
Z: passive branches in the tree, characterized by impedances
Y: passive branches in the co-tree, characterized by admittances

A step-by-step procedure for constructing an SFG is given below.

»3-8341-2/95/$0.00+$.50

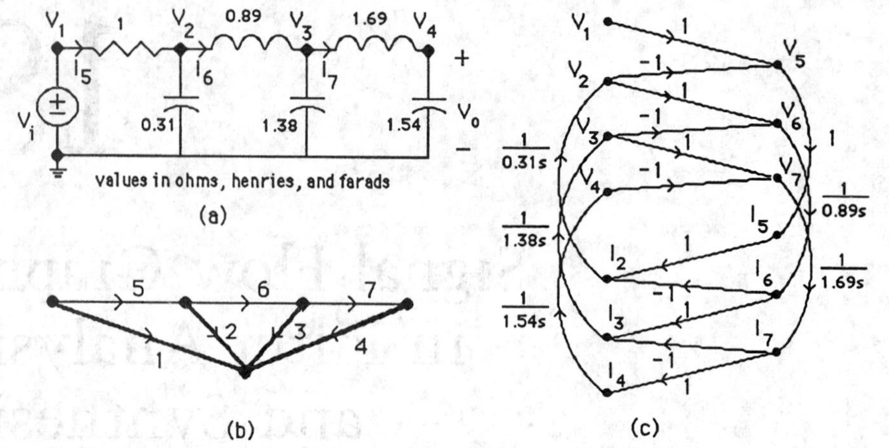

FIGURE 18.1 (a) A low-pass filter network. (b) Directed graph for the network and a chosen tree. (c) SFG based on the chosen tree.

Procedure 1 (for linear networks without controlled sources)

Step 1. Apply KVL to express each V_Y (voltage of a passive branch in the co-tree) in terms of V_E and V_Z.

Step 2. Apply KCL to express each I_Z (current of a passive branch in the tree) in terms of I_J and I_Y.

Step 3. For each passive tree branch, consider its voltage as the product of impedance and current, i.e., $V_Z = Z_Z I_Z$.

Step 4. For each passive co-tree branch, consider its current as the product of admittance and voltage, i.e., $I_Y = Y_Y V_Y$.

Example 1. Construct a signal flow graph for the low-pass filter network shown in Fig. 18.1(a), and use Mason's formula to find the voltage gain function $H(s) = V_o(s)/V_i(s)$.

Solution. The graph associated with the network is shown in Fig. 18.1(b) in which the branch numbers and reference directions (passive sign convention) have been assigned. The complexity of the SFG depends on the choice of the tree. In the case of a ladder network, a good tree to use is a star tree which has all tree branches connected to a common node. For the present network, we choose the tree to be $T = \{1, 2, 3, 4\}$, shown in heavy lines in Fig. 18.1(b).

Step 1 yields: $V_5 = V_i - V_2,\ V_6 = V_2 - V_3,\ V_7 = V_3 - V_4$

Step 2 yields: $I_2 = I_5 - I_6,\ I_3 = I_6 - I_7,\ I_4 = I_7$

Step 3 yields:

$$V_2 = \frac{1}{0.31s} I_2$$

$$V_3 = \frac{1}{1.38s} I_3$$

$$V_4 = \frac{1}{1.54s} I_4$$

Step 4 yields:

$$I_5 = V_5$$

$$I_6 = \frac{1}{0.89s} V_6$$

$$I_7 = \frac{1}{1.69s} V_7$$

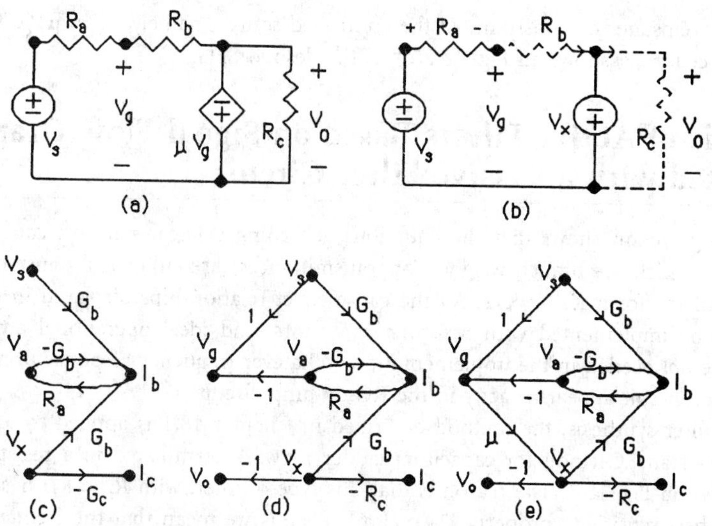

FIGURE 18.2 (a) A linear active network. (b) Result of step 1, procedure 2. (c) Result of step 2, procedure 2. (d) Result of step 3, procedure 2. (e) The desired SFG.

The signal flow graph of Fig. 18.1(c) displays all of the above relationships.

Applying Mason's gain formula to the SFG of Fig. 18.1(c), we find

$$H(s) = \frac{V_o}{V_i} = \frac{V_4}{V_1} = \frac{1}{s^5 + 3.24s^4 + 5.24s^3 + 5.24s^2 + 3.24s + 1}$$

Next consider linear networks containing controlled sources. All four types of controlled sources may be present. Our strategy is to utilize procedure 1 described above with some pre-analysis manipulations. The following is a step-by-step procedure.

Procedure 2 (for linear networks containing controlled sources)

Step 1. *Temporarily* replace each controlled voltage source by an independent voltage source, and each controlled current source by an independent current source, while retaining their original reference directions. The resultant network has no controlled sources.

Step 2. Construct the SFG for the network obtained in step 1 using procedure 1.

Step 3. Express the desired outputs and all controlling variables, if they are not present in the SFG, in terms of the quantities already present in the SFG.

Step 4. Reinstate the constraints of all controlled sources.

Example 2. Construct an SFG for the amplifier circuit shown in Fig. 18.2(a).

Solution. We first replace the controlled voltage source μV_g by an independent voltage source V_x. A tree is chosen to be (V_s, R_a, V_x). The result of step 1 of procedure 2 is shown in Fig. 18.2(b) where dashed lines indicate co-tree branches.

For the links R_b and R_c, we have $I_b = G_b V_b = G_b(V_s - V_a + V_x)$, and $I_c = G_c V_c = -G_c V_x$. For the tree branch R_a we have $V_a = R_a I_a = R_a I_b$. The result of step 2 of procedure 2 is shown in Fig. 18.2(c). Note that the simple relationships $V_b = (V_s - V_a + V_x)$, $V_c = -V_x$ and $I_a = I_b$ have been used to eliminate the variables V_b, V_c and I_a. As a result, these variables do not appear in Fig. 18.2(c).

The desired output is $V_o = -V_x$ and the controlling voltage is $V_g = V_s - V_a$. After expressing these relationships in the SFG, step 3 of procedure 2 results in Fig. 18.2(d).

Finally, we reinstate the constraint of the controlled source, namely, $V_x = \mu V_g$. The result of step 4 of procedure 2, shown in Fig. 18.2(e), is the desired SFG.

18.2 Synthesis of Active Filters Based on Signal Flow Graph Associated with a Passive Filter Circuit

The preceding section shows that the equations governing a linear network can be described by an SFG in which the branch weights (or transmittances) are either real constants or simple expressions of the form Ks or K/s. All the cause-effect relationships displayed in such an SFG can in turn be implemented with resistors, capacitors, and ideal operational amplifiers. The inductors are *not* needed in the implementation. Whatever frequency response prevailing in the original linear circuit appears exactly in the RC-op-amp circuit.

In active filter synthesis, the method described in Chapter 18.1 is applied to a passive filter in the form of an LC (inductor-capacitor) ladder network terminated in a resistance at both ends as shown in Fig. 18.3. The reason is that this type of filter, with $R_s = R_L$, has been proved to have the best sensitivity property [1, p. 196]. By this we mean that the frequency response is least sensitive with respect to the changes in element values, when compared to other types of filter circuits. Since magnitude scaling (i.e., multiplying all impedances in the network by a factor K_m) does not affect the voltage gain function, we always normalize the prototype passive filter network so that the source resistance becomes 1 Ω. The advantage of this normalization will become evident in several examples given in this section.

The SFG illustrated in Fig. 18.1(c) has many branches crossing each other. For a ladder network, with a proper choice of the tree and a rearrangement of the SFG nodes, all crossings can be eliminated. To achieve this, we first label a general ladder network as shown in Fig. 18.4. The following conventions are used in the labels of Fig. 18.4:

1. All series branches are numbered odd and characterized by their admittances.
2. All shunt branches are numbered even and characterized by their impedances.
3. A single arrow is used to indicate the reference directions of both the voltage and the current of each network branch. Passive sign convention is used.
4. If the LC ladder in Fig. 18.4 has a series element at the source end, then Y_1 represents that element in series with R_s.
5. If the LC ladder in Fig. 18.4 has a shunt element at the load end, then Z_{2n} represents that element in parallel with R_L.

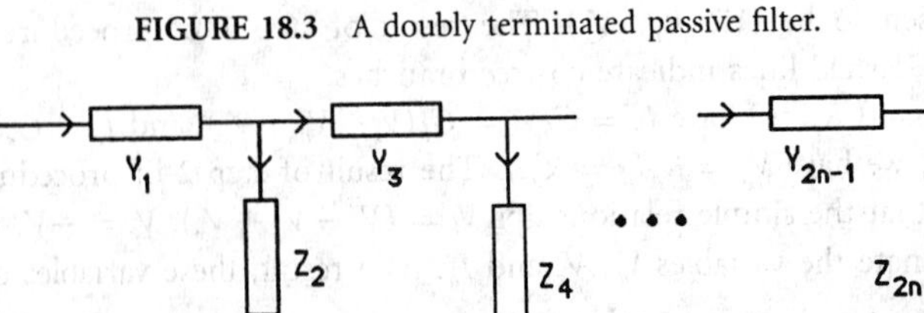

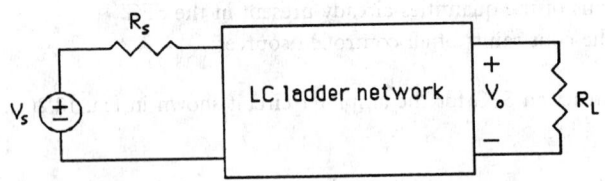

FIGURE 18.3 A doubly terminated passive filter.

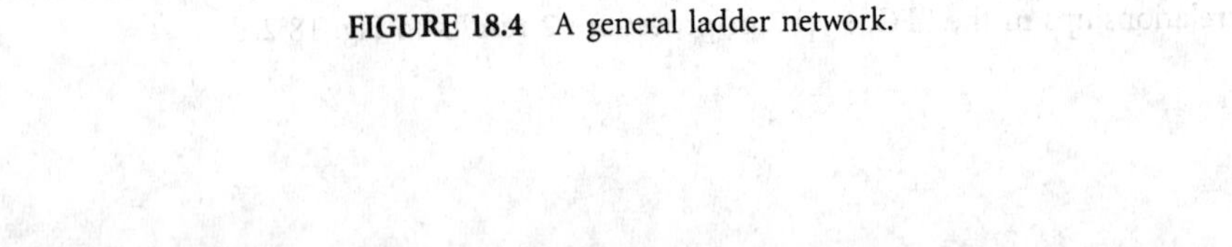

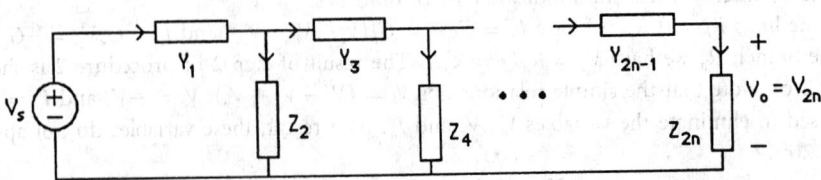

FIGURE 18.4 A general ladder network.

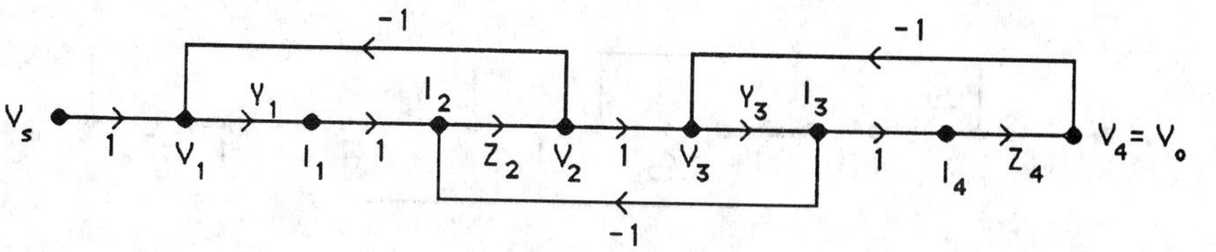

FIGURE 18.5 SFG for a 4-element ladder network.

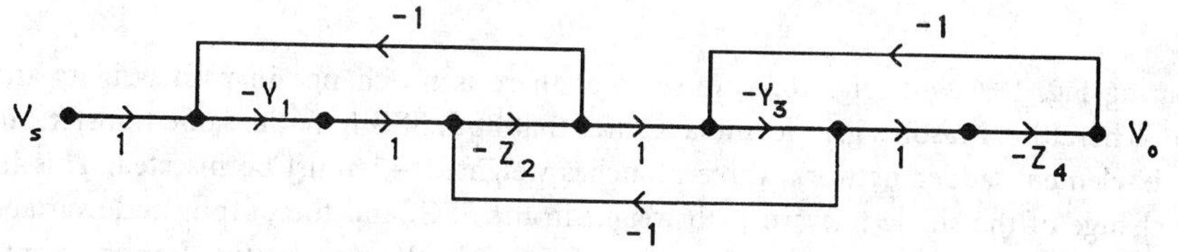

FIGURE 18.6 Inverting integrators are used in this modified SFG.

For constructing the SFG, choose a tree to consist of the voltage source and all shunt branches. The SFG for the circuit may be constructed using procedure 1 of Chapter 18.1. First, list the equations obtained in each step.

Step 1. $V_1 = V_s - V_2,\ V_3 = V_2 - V_4, \ldots, V_{2n-1} = V_{2n-2} - V_{2n}$

Step 2. $I_2 = I_1 - I_3,\ I_4 = I_3 - I_5, \ldots, I_{2n} = I_{2n-1}$

Step 3. $V_2 = Z_2 I_2,\ V_4 = Z_4 I_4, \ldots, V_{2n} = Z_{2n} I_{2n}$

Step 4. $I_1 = Y_1 V_1,\ I_3 = Y_3 V_3, \ldots, I_{2n-1} = Y_{2n-1} V_{2n-1}$

These relationships are represented by the SFG shown in Fig. 18.5 for the case of a four-element ladder network. Note that the SFG graph nodes have been arranged in such a way that there are no branch crossings. The pattern displayed in this SFG suggested the children's game of *leapfrog*. Consequently, an active filter synthesis based on the SFG of Fig. 18.5 is called a leapfrog realization. The transmittance of each SFG branch indicates the type of mathematical operation performed. For example, $1/s$ means integration and is implemented by an op amp integrator. Likewise, $1/(s+a)$ is implemented by a lossy op amp integrator. It is well known that inverting integrators and inverting summers can be designed with singled-ended op amps (i.e., the noninverting input terminal of each op amp is grounded), [2–5]. Noninverting integrators and noninverting summers can also be designed, but require differential-input op amps and more complex circuitry. Therefore, there is an advantage in using the inverting types. To this end, we multiply all Z's and Y's in Fig. 18.5 by -1, with the result shown in Fig. 18.6. Note that in Fig. 18.6 we have removed the labels of internal SFG nodes because they are of no consequence in determining the transfer function. The transfer function V_o/V_s is the same for both Fig. 18.5 and Fig. 18.6. This is quite obvious from Mason's gain formula, as all path weights and loop weights are not affected by the modification. A branch transmittance of -1 indicates an inverting amplifier. In the interest of reducing the total number of op amps used, we want to reduce the number of branches in the SFG that have weight -1. This can be achieved by inserting branches weighted -1 in some strategic places. Each insertion will lead to the change of the signs of one or two feedback branches. The rules are (a) inserting a branch weighted -1 in a forward path segment shared by two feedback loops changes the signs of the two feedback branch weights; (b) inserting a branch weighted -1 in a forward path segment belonging to one feedback loop only changes the signs of that feedback branch weight. Figure 18.6 is modified this way and the result is shown in Fig. 18.7. The inserted branches are shown in heavy lines.

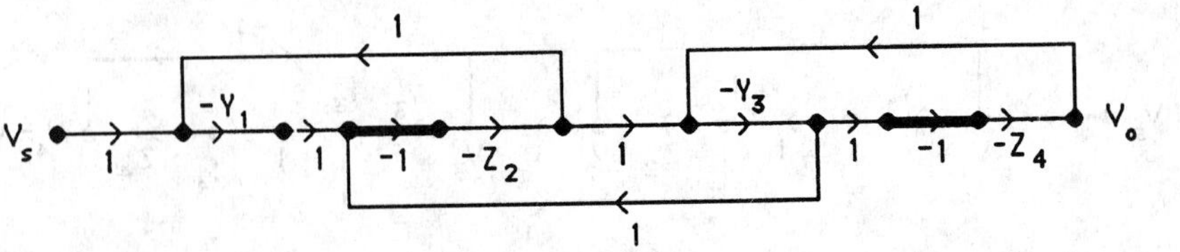

FIGURE 18.7 Modification to reduce the number of inverting amplifiers.

Comparing Fig. 18.6 with Fig. 18.7, we see that there is no change in path weights and loop weights. Therefore, Mason's gain formula assures that both SFG have the same transfer function. For a six-element ladder network, three branches weighted -1 must be inserted. This leads to a sign change of the single forward path weight in the SFG, and the output node variable now becomes $-V_o$. For filter applications this change of sign in the transfer function is acceptable as we are concerned mainly with the magnitude response.

An implementation of the SFG of Fig. 18.7 may be accomplished easily by referring to Table 18.1 and picking the component op amp circuits for realizing the SFG transmittances -1,

TABLE 18.1 Component op amp Circuits for Synthesizing Active low-pass Filters by the Leapfrog Technique

	signal flow graph	RC op amp circuit
(1)	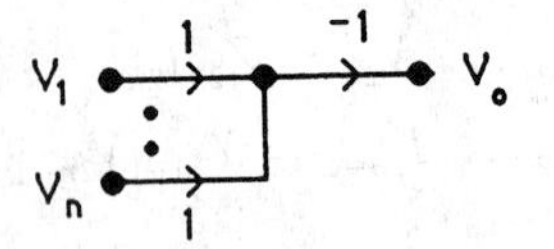 $V_o = -(V_1 + \cdots + V_n)$	Inverting summer 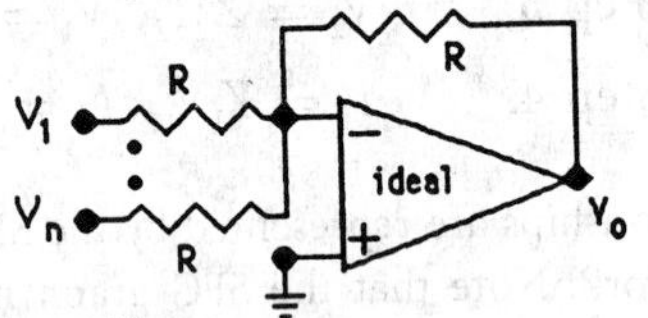 R: arbitrary
(2)	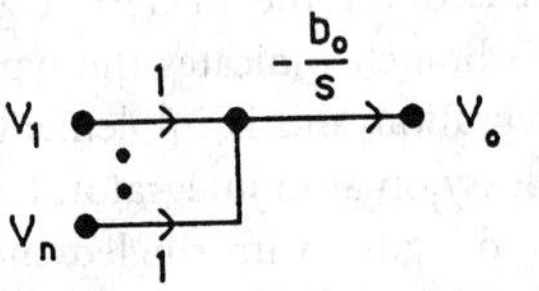 $V_o = -\frac{b_o}{s}(V_1 + \cdots + V_n)$	Inverting summing integrator 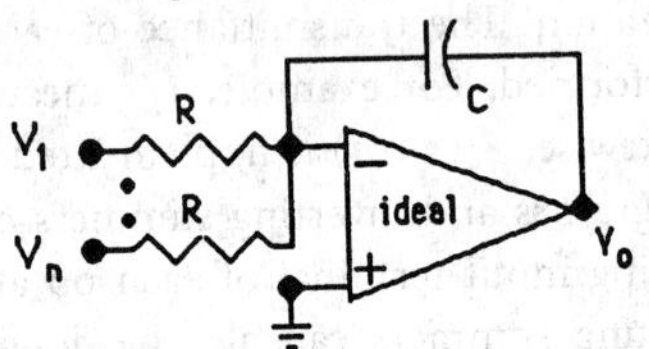 C: arbitrary, $R = \frac{1}{b_o C}$
(3)	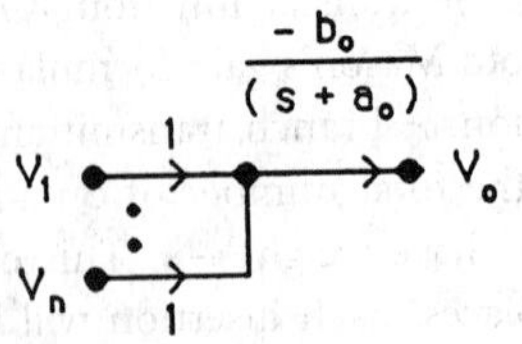 $V_o = \frac{-b_o}{(s + a_o)}(V_1 + \cdots + V_n)$	Inverting summing lossy integrator 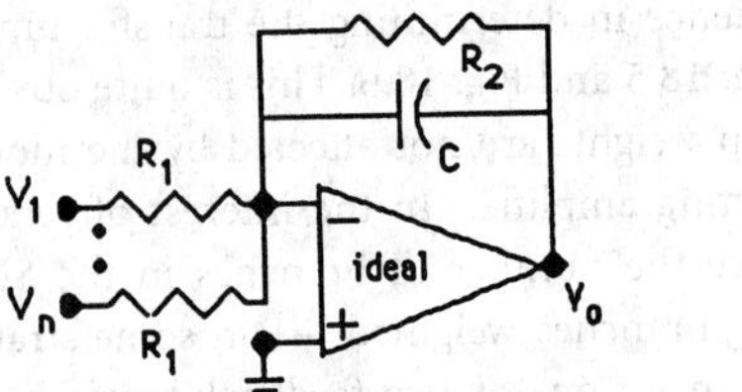 C: arbitrary, $R_1 = \frac{1}{b_o C}$, $R_2 = \frac{1}{a_o C}$

Note: Each component RC op amp circuit in the right column may be magnitude-scaled by an arbitrary factor.

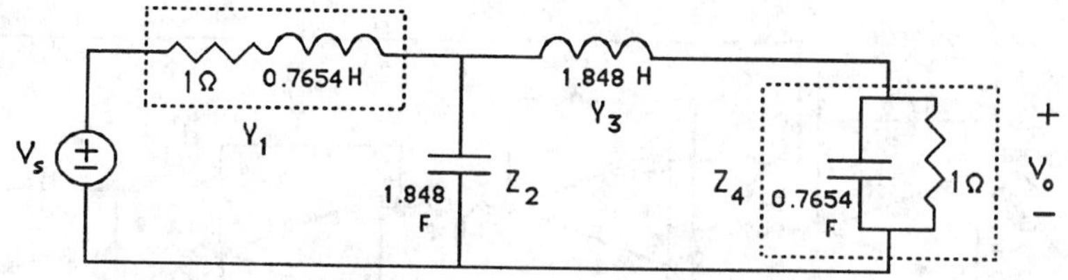

FIGURE 18.8 A fourth-order Butterworth low-pass filter.

$-Y_1$, $-Z_2$, etc. Figure 18.7 dictates how these component op amp circuits are interconnected to produce the desired voltage gain function. An example will illustrate the procedure.

Example 3. Figure 18.8 shows a normalized Butterworth fourth-order low-pass filter where the 1-Ω source resistance has been included in Y_1, and the 1-Ω load resistance included in Z_4.

The leapfrog-type SFG for this circuit, after suitable modifications, is shown in Fig. 18.7, where

$$-Y_1 = -\frac{1}{0.7654s + 1}$$

$$-Z_2 = -\frac{1}{1.848s}$$

$$-Y_3 = -\frac{1}{1.848s}$$

$$-Z_4 = -\frac{1}{0.7654s + 1}$$

The SFG branch transmittances $-Z_2$ and $-Y_3$ are realized using item (2) of Table 18.1 while $-Y_1$ and $-Z_4$ use item (3). The two SFG branches with weight −1 in Fig. 18.7 require item (1). The SFG branches with weight 1 merely indicate how to feed the inputs to each component network. No additional op amps are needed for such SFG branches. Thus, a total of six op amps are required. The interconnection of these component circuits is described by Fig. 18.7. The complete circuit is shown in Fig. 18.9. One-farad capacitances have been used in the circuit. Recall that the original passive low-pass filter has a 3-dB frequency of 1 rad/s. By suitable magnitude scaling and frequency scaling, all element values in the active filter of Fig. 18.9 can be made practical. For example, if the 3-dB frequency is changed to 10^6 rad/s, then the capacitances in Fig. 18.9 are divided by 10^6. We may arbitrarily magnitude scale the resultant circuit by a factor of 10^3. Then, all resistances are multiplied by 10^3 and all capacitances are further divided by 10^3. The final circuit is still the one shown in Fig. 18.9, but with resistances in kΩ and capacitance in nF. The parenthetical quantity beside each op amp indicates the type of transfer function it produces.

If a doubly terminated passive filter has a shunt reactance at the source end and a series reactance the load end, then its dual network has a series reactance at the source end and a shunt reactance at the load end. The voltage gain functions of the original network and its dual differ at most by a multiplying constant. We can apply the method to the dual network.

For doubly terminated Butterworth and Chebyshev low-pass filters of odd orders, the passive filter either has series reactances or shunt reactances at both ends. The next example shows the additional SFG manipulations needed to construct the RC-op-amp circuit.

Example 4. Obtain a leapfrog realization of the third-order Butterworth low-pass filter shown in Fig. 18.10(a).

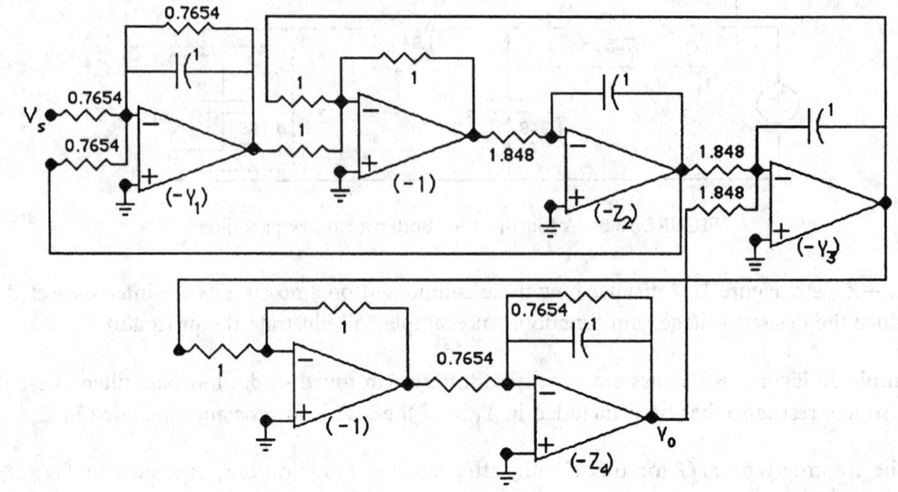

FIGURE 18.9 Leapfrog realization of passive filter of Fig. 18.8. For the normalized case of $\omega_{3dB} = 1$ r/sec, values are in Ω and F. For a practical case of $\omega_{3dB} = 10^6$ rad/s, values are in kΩ and nF.

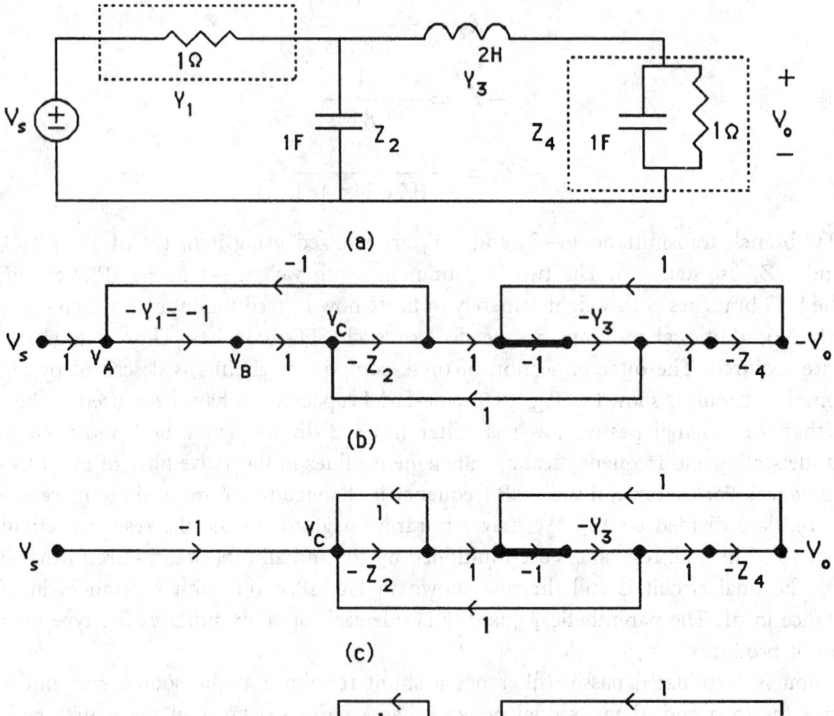

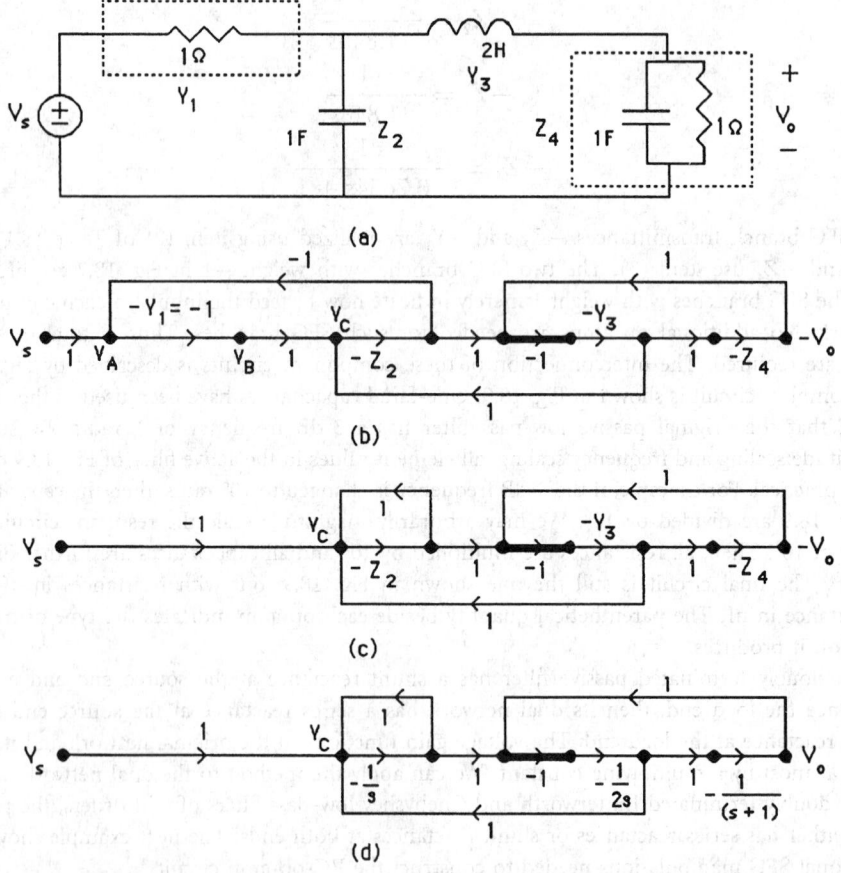

FIGURE 18.10 Leapfrog realization of a third-order Butterworth low-pass filter. (a) The passive prototype. (b) Leapfrog SFG simulation. (c) Absorption of SFG nodes. (d) Final SFG for active filter realization.

Solution. The network is again a four-element ladder network whose modified SFG in terms of the series admittances and shunt impedances is shown in Fig. 18.6. Note that the 1-Ω source resistance alone constitutes the element Y_1. Inserting a branch weighted -1 in front of $-Y_3$ changes the weights of two feedback branches from -1 to 1, and the output from V_o to $-V_o$. Figure 18.10(b) shows the result. Next, apply the node absorption rule to remove nodes V_A and V_B in Fig. 18.10b. The result is Fig. 18.10(c).

Finally, we recognize that the left-most branch weight -1 is not contained in any loop weights, and appears in the single forward path weight. Therefore, if this branch weight is changed from -1 to 1, the output will be changed from $-V_o$ to V_o. When this change is made, and all specific branch weights are used, the final SFG is shown in Fig. 18.10(d). The circuit implementation is now a simple matter of picking component networks from Table 18.1 and connecting them as shown by Fig. 18.10(d). A total of four op amps are required, one each for the branch transmittance $-1/s$, $-1/(2s)$, $-1/(s+1)$, and -1.

Passive bandpass filters may be derived from low-pass filters using the frequency transformation technique described in Chapter 67. The configuration of a bandpass filter derived from the third order Butterworth filter of Fig. 18.10(a) is shown in Fig. 18.11.

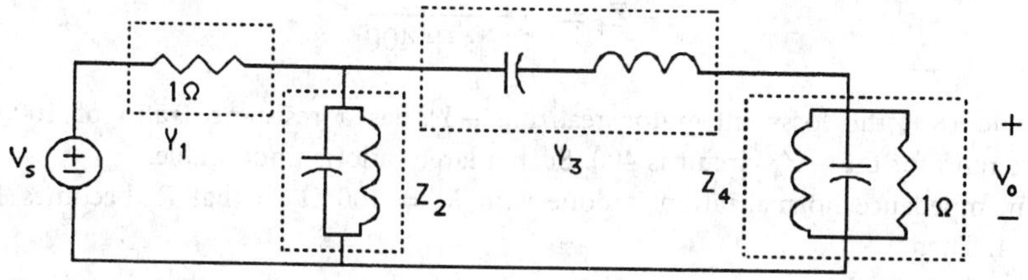

FIGURE 18.11 A bandpass passive filter derived from the circuit of Fig. 18.10(a).

The impedance and admittance functions Z_2, Y_3, and Z_4 are of the form

$$\frac{s}{a_2 s^2 + a_1 s + a_0}$$

The SFG thus contains quadratic branch transmittances. Several single op amp realizations of the quadratic transmittances are discussed in Chapter 77, while some multiple op amp realizations are presented in the next subsection. The interconnection of the component networks, however, is completely specified by an SFG similar to Fig. 18.7 or Fig. 18.10(d). Complete design examples of this type of bandpass active filter may be found in many books [2–5].

The previous example shows the application of the leapfrog technique to low-pass and bandpass filters of the Butterworth or Chebyshev types. The technique, when applied to a low-pass filter having an elliptic response or an inverse Chebyshev response will require the use of some differentiators. The configuration of a third order low-pass elliptic filter or inverse Chebyshev filter is shown in Fig. 18.12. Notice that Y_3 has the expression

$$Y_3 = a_2 s + \frac{a_o}{s} = \frac{a_2 s^2 + a_o}{s}$$

The term $a_2 s$ in the voltage gain function of the component network clearly indicates the need of a differentiator. An example of such a design may be found in [1, pp. 382–385].

As a final point in the leapfrog technique, consider the problem of impedance normalization. In all the previous examples, the passive prototype filter has equal terminations and has been magnitude-scaled so that $R_s = R_L = 1$. There are situations where the passive filter has unequal

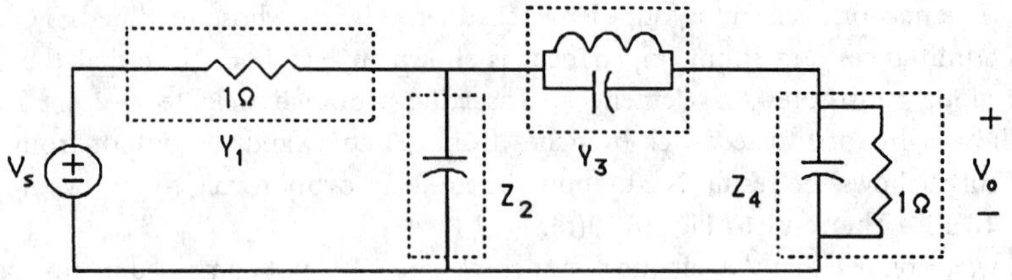

FIGURE 18.12 Network configuration of a doubly terminated filter having a third-order elliptic or inverse Chebyshev low-pass response.

terminations. For example, the passive filter may have $R_s = 100\ \Omega$ and $R_L = 400\ \Omega$ in a four-element ladder network shown in Fig. 18.8. Three possibilities will be considered.

(1) No impedance normalization is done on the passive filter. Then,

$$-Y_1 = -\frac{1}{L_1 s + 100}$$

$$-Z_4 = -\frac{1}{C_4 s + 400}$$

From Table 18.1, the lossy integrator realizing $-Y_1$ has a resistance ratio of 100, and the resistance ratio for the $-Z_4$ circuit is 400. Such a large ratio is undesirable.

(2) An impedance normalization is done with $R_o = 100\ \Omega$ so that R_s becomes 1 and R_L becomes 4. Then

$$-Y_1 = -\frac{1}{L_1 s + 1}$$

$$-Z_4 = -\frac{1}{C_4 s + 4}$$

The resistance ratio in the lossy integrator now becomes 1 for the $-Y_1$ circuit, and 4 for the $-Z_4$ circuit—an obvious improvement over the non-normalized case.

(3) An impedance normalization is done with $R_o = \sqrt{R_s R_L} = 200$. Then $R_s = 0.5$, $R_L = 2$, and

$$-Y_1 = -\frac{1}{L_1 s + 0.5}$$

$$-Z_4 = -\frac{1}{C_4 s + 0.5}$$

The resistance ratio in the lossy integrator is now 2 for both the $-Y_1$ circuit and the $-Z_4$ circuit—a further improvement over case (2) using $R_o = R_s$.

The conclusion is that, in the interest of reducing the spread of resistance values, the best choice of R_o for normalizing the passive filter is $R_o = \sqrt{R_s R_L}$. For the case of equal terminations, this choice leads to $R_s = R_L = 1$.

Instead of starting with a normalized passive filter, one can also construct a leapfrog-type SFG based on the unnormalized passive filter. For a four-element ladder network, the result is shown in Fig. 18.7. We now perform the following SFG manipulation which has the same effect as the impedance normalization of the passive filter: Select a normalization resistance, R_o and divide all Z's in the SFG by R_o, and multiply all Y's by R_o. The resultant SFG is shown in Fig. 18.13.

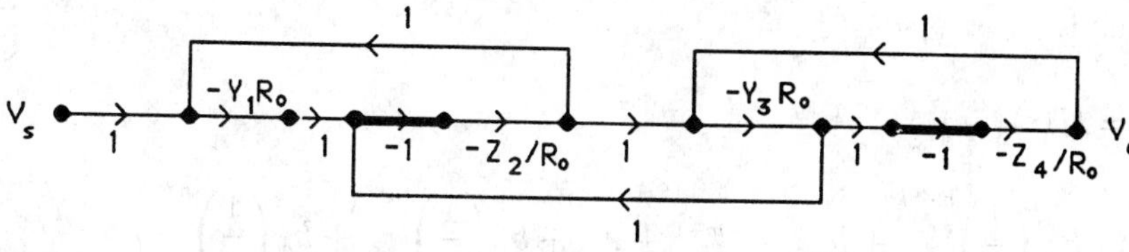

FIGURE 18.13 Result of normalization of the SFG of Fig. 18.7.

It is easy to see that the SFG in both Figs. 18.7 and 18.13 have the same loop weights and single forward path weight. Therefore, the voltage gain function remains unchanged with the normalization process. One advantage of using the normalized SFG is that the branch transmittances $Y_k R_o$ and Z_k/R_o are dimensionless, and truly represent voltage gain function of component op amp circuits [2, p. 288].

18.3 Synthesis of Active Filters Based on Signal Flow Graph Associated with a Filter Transfer Function

The preceding section describes one application of the SFG in the synthesis of active filters. The starting point is a passive filter in the form of a doubly terminated LC ladder network. In this section, we describe another way of using the SFG technique to synthesize an active filter. The starting point in this case is a filter transfer function instead of a passive network.

Let the transfer voltage ratio function of a filter be

$$\frac{V_o}{V_i} = H(s) = \frac{b_m s^m + b_{m-1}s^{m-1} + \ldots + b_1 s + b_o}{s^n + a_{n-1}s^{n-1} + \ldots + a_1 s + a_o}, \quad m \le n \tag{18.1}$$

In Chapters 65 and 66, it is shown that by properly selecting the coefficient a's and b's, all types of filter characteristics can be obtained: low-pass, high-pass, bandpass, band elimination, and all-pass. We assume that these coefficients have been determined. Our problem is how to realize the transfer function using SFG theory and RC-op-amp circuits.

Recall Mason's gain formula of an SFG described in Chapter 8.4. For the present application, we impose two constraints on the signal flow graph:

1. There are no second order or higher order loops. In other words, all loops in the SFG touch each other.
2. Every forward path from the source node to the output node touches all loops.

For such a special kind of SFG, Mason's gain formula reduces to

$$\frac{V_o}{V_i} = H(s) = \frac{\Sigma P_k}{1 - (L_1 + L_2 + \ldots + L_n)} \tag{18.2}$$

where L_n is the nth loop weight, P_k is the kth forward path weight, and summations are over all forward paths and all loops. Our strategy is to manipulate Eq. (18.1) into the form of Eq. (18.2), and then construct an SFG to have the desired loops and paths, meeting constraints (1) and (2). Since integrators are preferred over differentiators in actual circuit implementation, we want $1/s$ instead of s to appear as the SFG branch transmittances. This suggests the division of both the numerator and denominator of Eq. (18.1) by s^n, the highest degree term in the denominator.

The result is

$$\frac{V_o}{V_i} = H(s)$$

$$= \frac{b_m\left(\frac{1}{s}\right)^{n-m} + b_{m-1}s^{n-m+1} + \ldots + b_1\left(\frac{1}{s}\right)^{n-1} + b_o\left(\frac{1}{s}\right)^{n}}{1 + a_{n-1}\left(\frac{1}{s}\right) + \ldots + a_1\left(\frac{1}{s}\right)^{n-1} + a_o\left(\frac{1}{s}\right)^{n}}, \quad m \leq n \tag{18.3}$$

Comparing Eq. (18.3) with Eq. (18.2), we can identify the loop weights

$$\begin{aligned} L_1 &= -a_{n-1}(1/s) \\ L_2 &= -a_{n-2}(1/s)^2 \\ &\ldots \\ L_n &= -a_o(1/s)^n \end{aligned} \tag{18.4}$$

and the forward path weights

$$b_m\left(\frac{1}{s}\right)^{n-m}, \quad b_{m-1}\left(\frac{1}{s}\right)^{n-m+1}, \quad \ldots b_1\left(\frac{1}{s}\right)^{n-1}, \quad b_o\left(\frac{1}{s}\right)^{n} \tag{18.5}$$

Many SFGs may be constructed to have such loop and path weights, and the touching properties stated in (1) and (2) above. Two simple ones are given in Fig. 18.14(a) and (b) for the case $n = m = 3$. The extension to higher order transfer functions is obvious. In control theory, the system represented by Fig. 18.14(a) is said to be of the controllable canonical form, and Fig. 18.14(b) the observable canonical form. In filter application, we need not be concerned about the controllability and observability of the system. The terms are used here merely for the purpose of circuit identification. Our major concern here is how to implement the SFG by an RC-op-amp circuit.

An SFG branch having transmittance $1/s$ indicates an integrator. If the terminating node of the $1/s$ branch has no other incoming branches [as in Fig. 18.14(a)], then that node variable represents the output of the integrator. On the other hand, if $1/s$ is the transmittance of only one of several incoming branches incident at the node V_k [as in Fig. 18.14(b)], then V_k is *not* the output of an integrator. In order to identify the integrator outputs clearly for the purpose of circuit interconnection, we insert some dummy branches with weight 1 in series with the branches weighted $1/s$. When this is done to Fig. 18.14(b), the result is Fig. 18.15 with the inserted dummy branches shown in heavy lines. An SFG branch with weight $-1/s$ represents an inverting integrator. As pointed out in Section 18.2, the circuitry of an inverting integrator is simpler than that of a noninverting integrator. To have an implementation utilizing inverting integrators, we replace all SFG branch weights $1/s$ in Fig. 18.14 by $-1/s$. In order to maintain the original path and loop weights, the signs of some feedback branches and forward path branches must be changed accordingly. When this is done, Fig 18.14(a) and Fig. 18.15 become those shown in Fig. 18.16(a) and (b), respectively. Our next goal is to implement these SFGs by RC-op-amp circuits. Because SFGs of the kind described in this section are widely used in the study of linear systems by the state variable approach, the active filters based such SFGs are called *state variable filters* [6].

Example 5. Synthesize a state variable active filter to have a third order Butterworth low-pass response having 3 dB frequency $\omega_o = 10^6$ rad/s. All op amps used are single-ended.

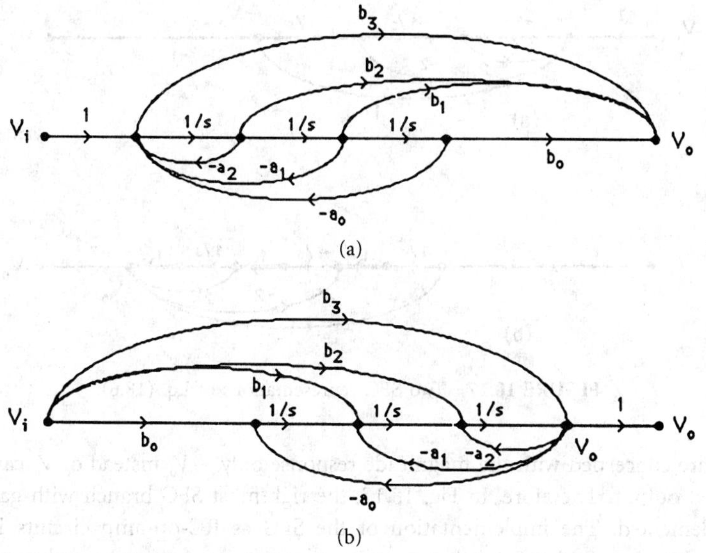

FIGURE 18.14 Two simple SFGs having gain function given by Eq. (18.3). (a) Controllable canonical form. (b) Observable canonical form.

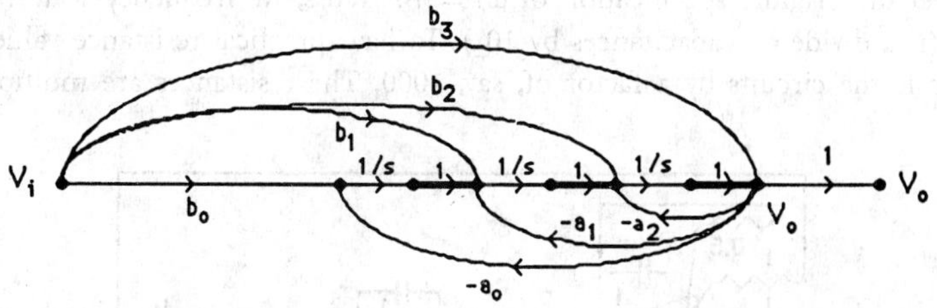

FIGURE 18.15 Insertion of dummy branches to identify integrator outputs.

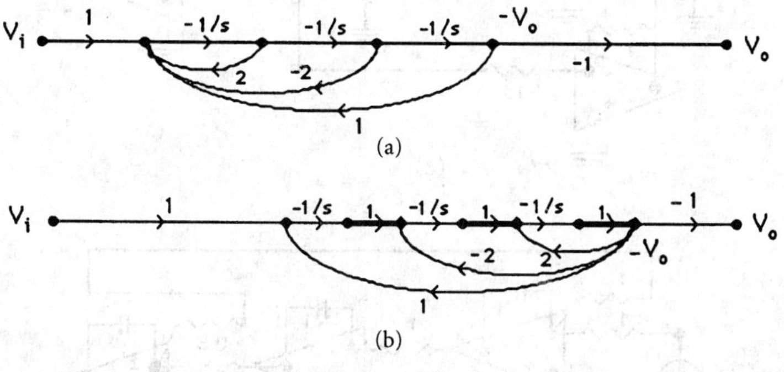

FIGURE 18.16 Simulation of $H(s)$ by an SFG containing inverting integrators.

Solution. As usual in filter synthesis, we first construct the filter for the normalized case, i.e., $\omega_o = 1$ rad/s, and then perform frequency scaling to obtain the required filter. From Chapter 65, the normalized voltage gain function of the filter is

$$H(s) = \frac{V_o}{V_i} = \frac{1}{s^3 + 2s^2 + 2s + 1} \tag{18.6}$$

and the two SFGs in Fig. 18.16 become those shown in Fig. 18.17.

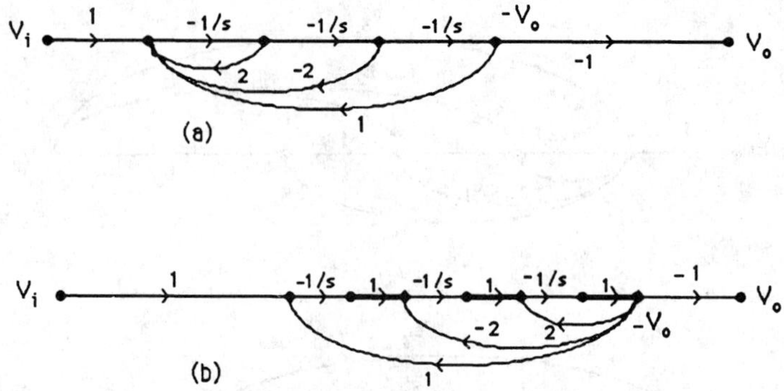

FIGURE 18.17 Two SFG representations of Eq. (18.6).

Since we are concerned with the magnitude response only, $-V_o$ instead of V_o can be accepted as the desired output. Therefore, in Fig. 18.17, the rightmost SFG branch with gain (-1) need not be implemented. The implementation of the SFG as RC-op-amp circuits is now just a matter of looking up Table 18.2, selecting proper component networks and connecting them as specified by Fig. 18.17. The results are given in Fig. 18.18(a) and (b). These circuits, with element values in ohms and farads, realize the normalized transfer function having $\omega_c = 1$ rad/s. To meet the original specification of $\omega_c = 10^6$ rad/s, we frequency-scale the circuits by a factor 10^6 (i.e., divide all capacitances by 10^6). To have practical resistance values, we further magnitude-scale the circuits by a factor of, say, 1000. The resistances are multiplied by 1000,

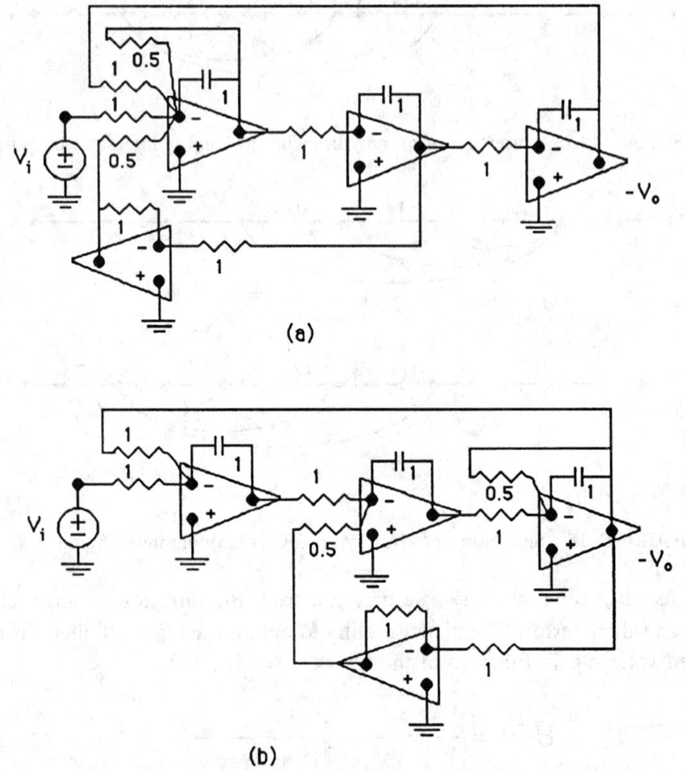

FIGURE 18.18 Two op amp circuit realizations of $H(s)$ given by Eq. (18.6).

and the capacitances are further divided by 1000. The final circuits are still given by Fig. 18.18, but now with element values in kΩ and nF.

In example 5, both realizations require 4 op amps. In general, for an nth order transfer function given by Eq. (18.1) with all coefficients nonzero, a synthesis based on Fig. 18.16(a) (controllable canonical form) requires $n+3$ single-ended op amps. The breakdown is as follows [refer to Fig. 18.16(a)]:

n inverting scaled integrators (item 2, Table 18.2) for the n SFG branches with weight $-1/s$

2 op amps for the bipolarity summer (item 3, Table 18.2) to obtain V_o

1 inverting scaled summer (item 1, Table 18.2) to invert and add up signals from branches with weights $-a_1$, $-a_3$, etc., before applying to the left-most integrator

On the other hand, a synthesis based on Fig. 18.16(b) (observable canonical form) requires only $n+2$ single-ended op amps. To see this we redraw Fig. 18.16(b) as Fig. 18.19 by inserting

TABLE 18.2 Single-Ended op amp Circuits for Implementing State Variable Active Filters.

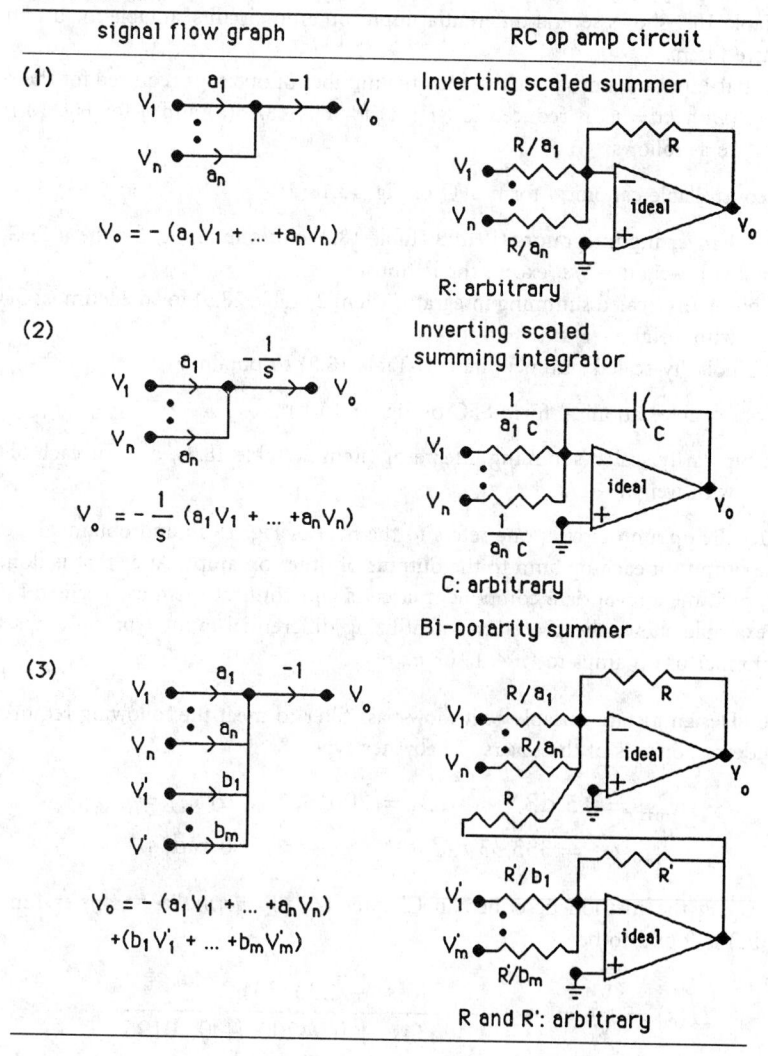

	signal flow graph	RC op amp circuit
(1)	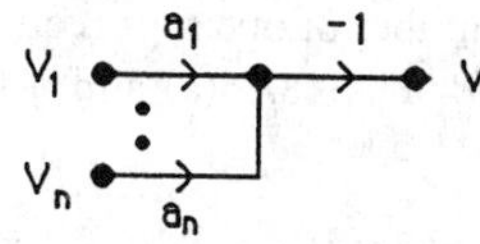 $V_o = -(a_1V_1 + \ldots + a_nV_n)$	Inverting scaled summer 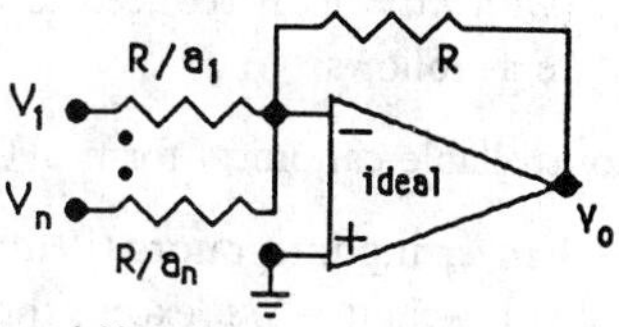 R: arbitrary
(2)	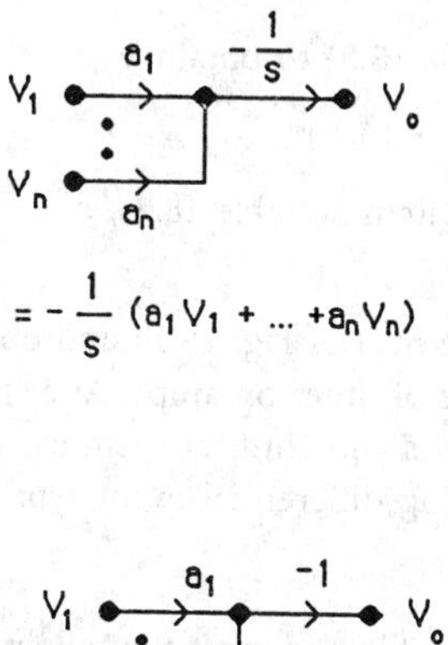 $V_o = -\frac{1}{s}(a_1V_1 + \ldots + a_nV_n)$	Inverting scaled summing integrator 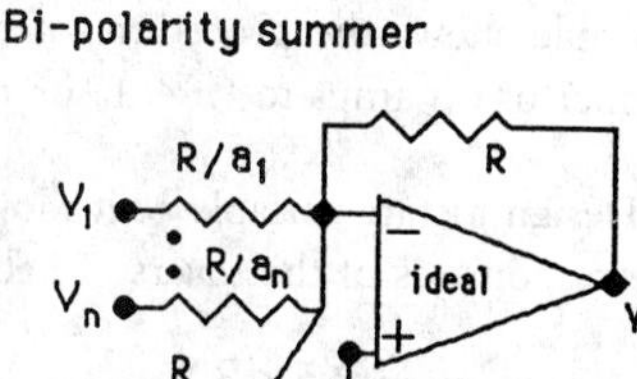 C: arbitrary
(3)	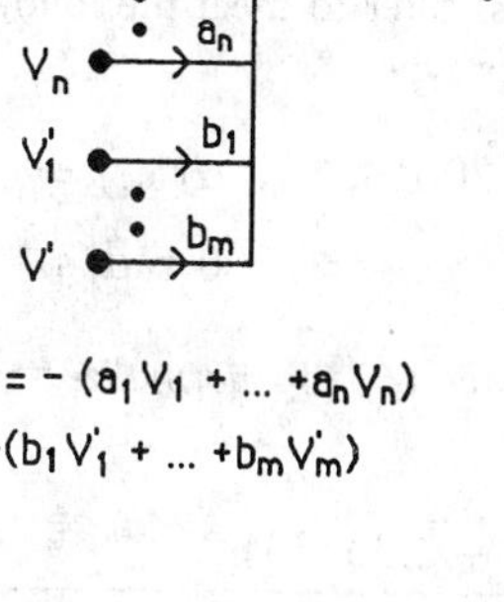 $V_o = -(a_1V_1 + \ldots + a_nV_n) + (b_1V_1' + \ldots + b_mV_m')$	Bi-polarity summer 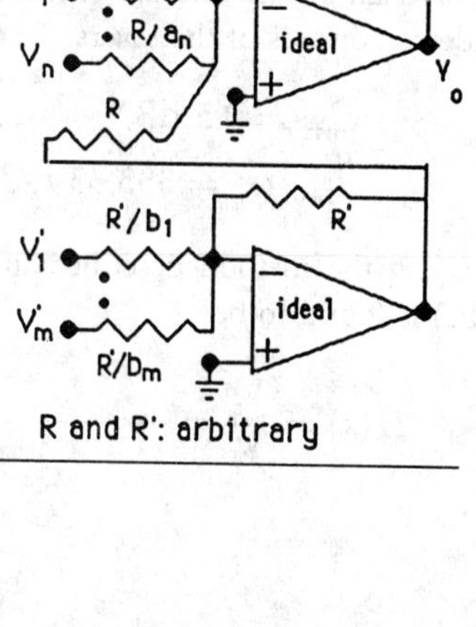 R and R': arbitrary

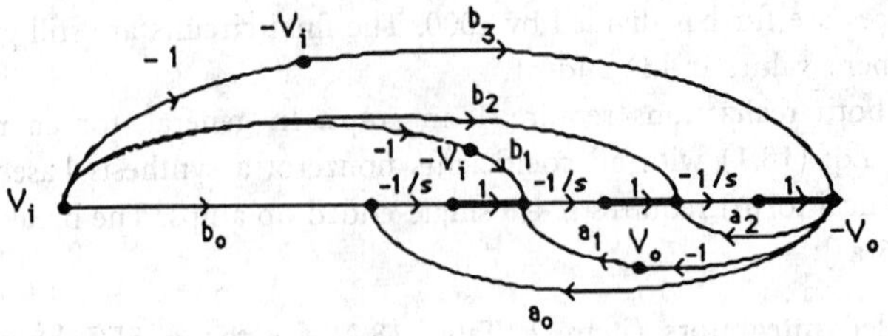

FIGURE 18.19 A modification of Fig. 18.16(b) to use all positive a's and b's.

branches with weight -1, and making all literal coefficients positive. The breakdown is as follows (referring to Fig. 18.19, extended to nth order $H(s)$):

- n inverting scaled integrators (item 2, Table 18.2) for the n SFG branches with weight $-1/s$
- 1 inverting amplifier at the input end to provide $-V_i$
- 1 inverting amplifier at the output end to make available both V_o and $-V_o$

The number of op amps can be reduced if the restriction of using single-ended op amp is removed. Table 18.3 shows several differential-input op amp circuits suitable for use in the state variable active filters.

If differential-input op amps are used, then the number of op amps required for the realization of Eq. (18.1) (with $m = n$) is reduced to $(n + 1)$ for Fig. 18.16(a) and n for Fig. 18.16(b). The breakdowns are as follows:

For the controllable canonical form SFG of Fig. 18.16(a)

- $n - 1$ inverting integrators (item 2, Table 18.2 with one input) for the n SFG branches with weight $-1/s$, except the leftmost
- 1 bipolarity-scaled summing integrator (item 2, Table 18.3) for the leftmost SFG branch with weight $-1/s$
- 1 bipolarity-scaled summer (item 1, Table 18.3) to obtain V_o

For the observable canonical form SFG of Fig. 18.16(b)

- n bipolarity scaled summing integrator (item 2, Table 18.3), one for each SFG branch with weight $-1/s$

To construct the op amp circuit, one refers to the SFG of Fig. 18.14 and obtains the expression relating the output of each op amp to the outputs of other op amps. After that is done, refer to Table 18.3, pick the appropriate component circuits and connect them as specified by the SFG. The next example shows the procedure of utilizing differential-input type op amps to reduce the total number of op amps to $(n + 1)$ or n.

Example 6. Design a state-variable active low-pass filter to meet the following requirements: magnitude response is of the inverse Chebyshev type

$$\alpha_{\max} = 0.5 \text{ dB}, \qquad \alpha_{\min} = 20 \text{ dB}, \qquad \alpha(\omega_s) = \alpha_{\min}$$

$$\omega_p = 333.33 \text{ rad/s}, \qquad \omega_s = 1000 \text{ rad/s}$$

Solution. Using the method described in Chapter 66, the *normalized* transfer function (i.e., $\omega_s = 1$ rad/s) is found to be

$$H(s) = \frac{V_o}{V_i} = \frac{K(s^2 + 1.33333)}{s^3 + 1.40534s^2 + 0.94200s + 0.40196} \tag{18.7}$$

TABLE 18.3 Differential-Input op-amp Circuits

(1) Bi-polarity scaled summer $V_o = -(aV_a + bV_b) + (AV_A + BV_B)$

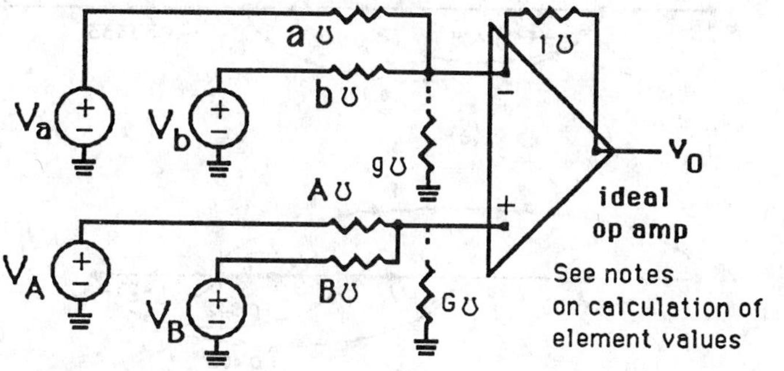

(2) Bi-polarity scaled summing integrator.

$$V_o = \frac{1}{s}[-(aV_a + bV_b) + (AV_A + BV_B)]$$

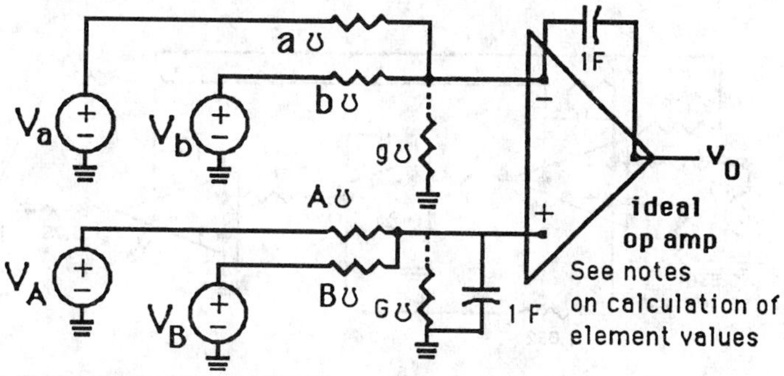

Note: Calculation of element values in Table 18.3 [7].

(i) The initial design uses 1 Ω resistance or 1 F capacitance as the feedback element.

(ii) Either the g mho conductance or the G mho conductance (not both) is connected. Choose the values of g or G such that the sum of all conductances connected to the inverting input terminal equals the sum of all conductances connected to the noninverting input terminal.

(iii) Starting with the initial design, one may magnitude-scale all elements connected to the inverting input terminal by one factor, and all elements connected to the noninverting input terminal by the same or a different factor.

The SFGs for this $H(s)$ are simply obtained from Fig. 18.16 by removing the two branches having weights b_3 and b_1. The results are shown in Fig. 18.20(a) for the case $K = 1$, and in Fig. 18.20(b) for the case $K = -1$. A four-op-amp circuit for the normalized $H(s)$ may be constructed in accordance with the SFG of Fig. 18.20(a). The component op amp circuits are selected from Tables 18.2 and 3 in the following manner:

Relationship from SFG	Component op amp circuit
$V_1 = -\frac{1}{s}(V_i + 1.405V_1 + 0.402V_3 - 0.942V_2)$	item 2, Table 18.3
$V_2 = -\frac{1}{s}V_1$	item 2, Table 18.2
$V_3 = -\frac{1}{s}V_2$	item 2, Table 18.2
$V_o = -V_1 - 1.33333V_3$	item 1, Table 18.2

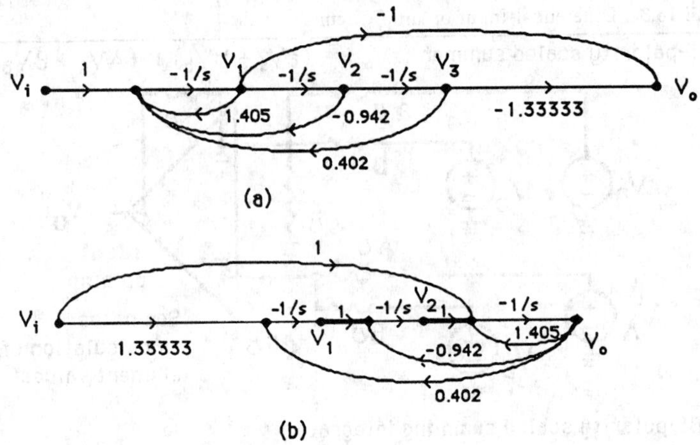

FIGURE 18.20 Two SFGs realizing the transfer function of Eq. (18.7).

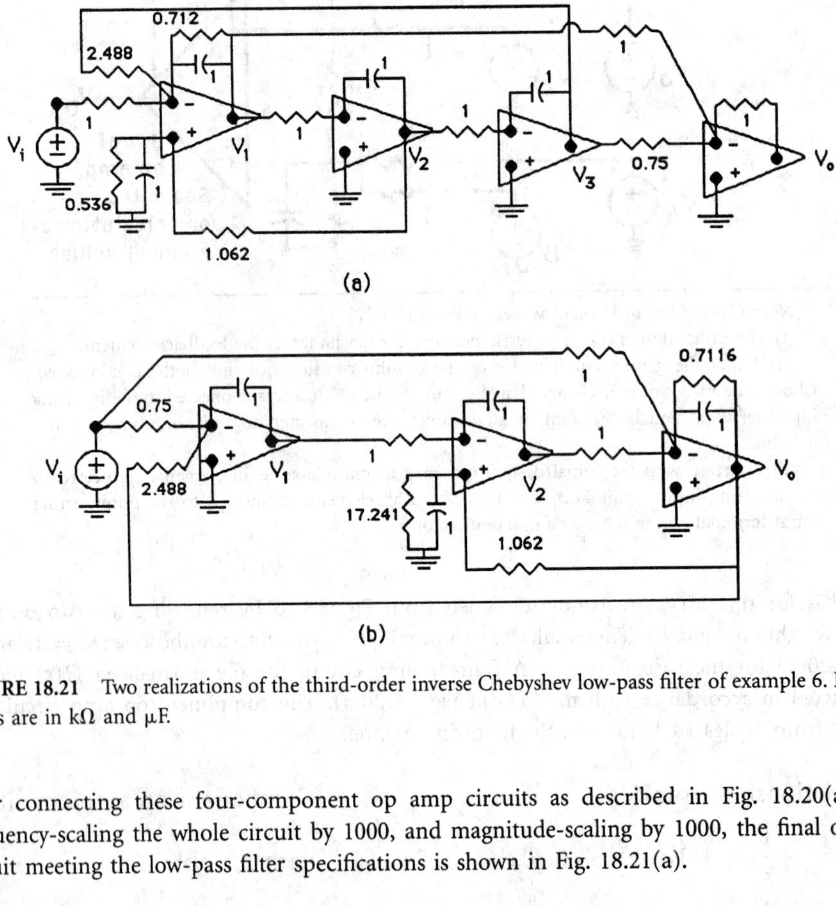

FIGURE 18.21 Two realizations of the third-order inverse Chebyshev low-pass filter of example 6. Element values are in kΩ and μF.

After connecting these four-component op amp circuits as described in Fig. 18.20(a), and frequency-scaling the whole circuit by 1000, and magnitude-scaling by 1000, the final op amp circuit meeting the low-pass filter specifications is shown in Fig. 18.21(a).

In like manner, a three-op-amp circuit for the normalized $H(s)$ may be constructed in accordance with the SFG of Fig. 18.20(b). The final op amp circuit meeting the lowpass filter specifications is shown in Fig. 18.21(b). Both circuits in Fig. 18.21 achieve a gain constant $|K| = 1$ in Eq. (18.7). Should a different value of $|K| = 1/\alpha$ be desired, it is only necessary to multiply the values of all resistors connected to the input V_i by α.

When the method of this subsection is applied to a second order transfer function, the resultant op amp circuit is called a *state variable biquad.* Biquads and first order op amp circuits are used as the basic building blocks in the synthesis of a general nth order transfer function by the "cascade" approach. Single op amp biquads are discussed in Section 15. Depending on the SFGs chosen and the types of op amps allowed (single-ended or differential-input), a state variable biquad may require from 2 to 5 op amps. Some special but useful state variable biquads are listed in Table 18.4 for reference purposes.

TABLE 18.4 Some Special State-Variable Biquads

Normalized transfer functions

(1) Lowpass

$$\frac{V_{LP}}{V_i} = \frac{1}{s^2 + \frac{1}{Q}s + 1}$$

(2) Bandpass

$$\frac{V_{BP}}{V_i} = \frac{s}{s^2 + \frac{1}{Q}s + 1}$$

(3) Highpass

$$\frac{V_{HP}}{V_i} = \frac{s^2}{s^2 + \frac{1}{Q}s + 1}$$

Signal flow graph for transfer functions (1) - (3)

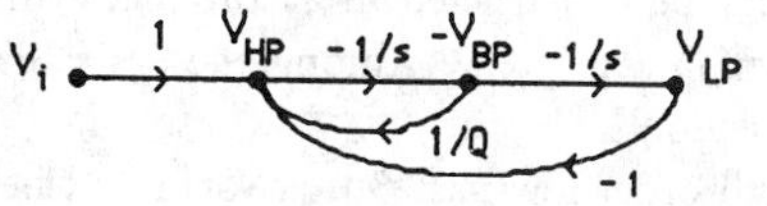

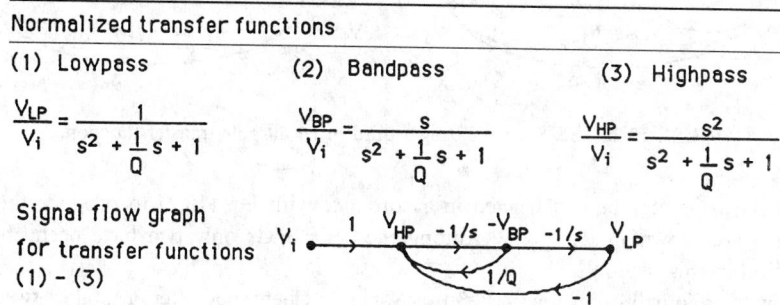

Op amp circuits for (1) -(2). Available outputs: LP and BP

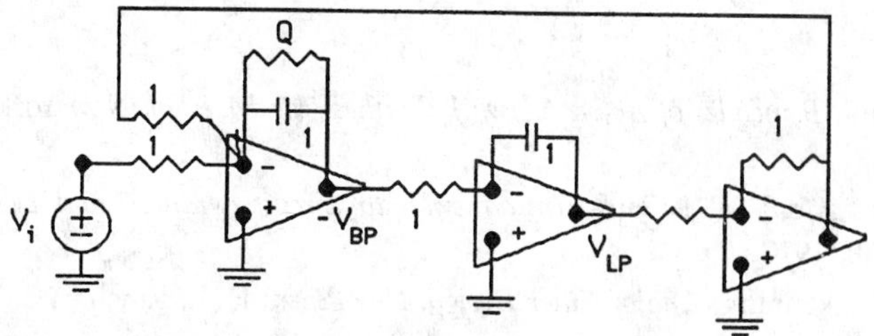

Op amp circuits for (1) - (3). Available outputs: LP, BP and HP.

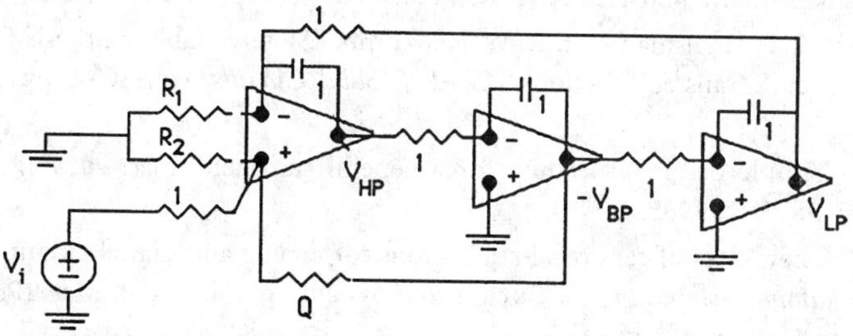

Either R1 or R2 is connected. See notes in Table 18.3 for the determination of their values.

All of the SFGs used in the previous examples are of the two types (controllable and observable canonical forms) illustrated in Fig. 18.16. There are, however, many other possible SFGs that produce the same transfer function. For example, a third order low-pass Butterworth or Chebyshev filter has an all-pole transfer function.

$$H(s) = \frac{V_o}{V_i} = \frac{K}{s^3 + a_2 s^2 + a_1 s + a_o} \tag{18.8}$$

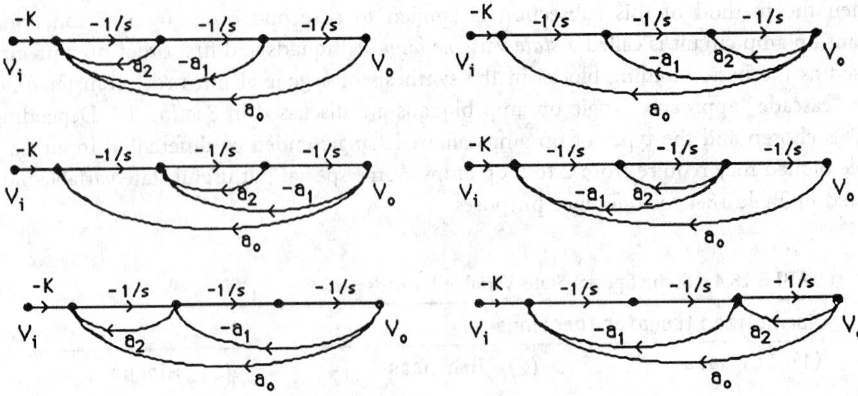

FIGURE 18.22 Six SFGs realizing a third order all-pole transfer function.

A total of six SFGs may be constructed in accordance with Eq. (18.2) to produce the desired $H(s)$. These are shown in Fig. 18.22. Among these six SFGs only two have been chosen for consideration in this section.

Similarly, for a fourth order low-pass Butterworth or Chebyshev filter, a total of twenty SFGs may be constructed. The reader should consult References [8–9] for details.

References

[1] G. Daryanani, *Principles of Active Network Synthesis and Design*, New York: John Wiley & Sons, 1976.

[2] G. C. Temes and J. W. LaPatra, *Introduction to Circuit Synthesis and Design*, New York: McGraw-Hill, 1977.

[3] M. E. Van Valkenburg, *Analog Filter Design*, New York: Holt, Rinehart & Winston, 1982.

[4] W. K. Chen, *Passive and Active Filters*, New York: John Wiley & Sons, 1986.

[5] L. P. Huelsman, *Active and Passive Analog Filter Design*, New York: McGraw-Hill, 1993.

[6] W. J. Kerwin, L. P. Huelsman, and R. W. Newcomb, "State variable synthesis for insensitive integrated circuit transfer functions," *IEEE J. Solid Circuits*, vol. SC-2, pp. 87–92, Sept. 1967.

[7] P. M. Lin, "Simple design procedure for a general summer," *Electron. Eng.*, vol. 57, no. 708, pp. 37–38, Dec. 1985.

[8] N. Fliege, "A new class of canonical realizations for analog and digital circuits," *IEEE Proc. 1984 International Symposium on Circuits and Systems*, pp. 405–408, May 1984.

[9] P. M. Lin, "Topological realization of transfer functions in canonical forms," *IEEE Trans. Automatic Control*, vol. AC-30, pp. 1104–1106, Nov. 1985.

19

Tableau and Modified Nodal Formulations

[J]iri Vlach
[U]niversity of Waterloo,
[C]anada

19.1 Introduction

Network analysis is based on formulation of the relevant equations and on their solutions. Various approaches are possible. If we wish to get as much theoretical information as possible, we may resort to hand analysis and keep the elements as variables (literal parameters). In such cases it is an absolute necessity to use a method which leads to the smallest possible number of equations. If we plan to use a computer, then we can accept methods which lead to larger systems, but the methods must be relatively easy to program. The purpose of this chapter is to give an overview of the various possibilities and point out advantages and disadvantages. Many more details are available in [1, 2].

Section 19.2 is a summary of the nodal and mesh formulations. We review them because they are the best ones for analysis of small networks. Tableau formulation, given in Section 19.3, is very general, but requires special solution routines, probably not available to the reader. Section 19.4 describes the best method for computerized solutions; it is used in many commercial simulators. If nonlinear elements are involved, then iterative solution methods must be used; an introduction on how to deal with nonlinear elements is given in Section 19.5. Finally, Section 19.6 presents a method which is suitable for hand solutions of active networks and which automatically leads to the smallest system of equations.

19.2 Nodal and Mesh Formulations

Classical methods use two types of network equation formulation: the nodal and the mesh. The first one is based on the Kirchhoff current law (KCL): the sum of currents flowing away from a node is equal to zero. The mesh method is based on the Kirchhoff voltage law (KVL): the sum of voltages around any loop is equal to zero.

[0-8]493-8341-2/95/$0.00 + $.50

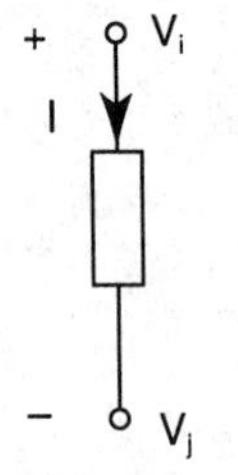

FIGURE 19.1 Definition of positive current direction with respect to the voltage across the element.

For simple problems, both methods are about equivalent, but nodal formulation is more general. The mesh formulation is suitable only for planar networks: it must be possible to draw the network without any element crossing over any other element. Many practical networks are planar, but it is not always easy to see that it is actually the case.

We first introduce some definitions. A positive current flows from a terminal with a higher potential to a terminal with a lower potential. This is sketched on the two-terminal element in Fig. 19.1. If we use this definition, then the product of the current and of the voltage across the element, $V_i - V_j$, expresses the power consumed by the element. If the current flows in opposite direction, then the element is delivering power.

We will use the above definition of a positive current for *all* elements of the network, irrespective of what their role eventually may be. Thus, a positive current through an independent voltage source will flow from the more positive terminal to the less positive terminal, as sketched in Fig. 19.2(a). The current flowing through an independent current source is indicated on its symbol, Fig. 19.2(b), but the voltage across it is not defined; it depends on the network.

The nodal formulation uses the principle that the sum of currents at any node must be equal to zero at any instant of time. To apply this rule in an efficient way, we realize that before we solve the equations, we do not know which way the currents will actually flow. All we know is that *if* a node is more positive than all the other nodes, then all currents must flow *away* from this node.

In nodal formulation the unknowns are nodal voltages and the equations express the sum of currents flowing away from the node. To write the equations we use element admittances: in Laplace transform $Y_C = sC$ for a capacitor, $Y_L = 1/sL$ for an inductor, and $Y_G = G = 1/R$ for a resistor. It is advantageous to use G, since we thus avoid fractions in the equations.

We demonstrate how to set up the equations by considering the network in Fig. 19.3. The nodal voltages are denoted V_1 and V_2 and ground (the lower line) is considered to be at zero potential. We do not know which of these nodes is more positive, but we can *assume* that any node we consider at a given moment is the most positive one. This assumption has the consequence that all currents must flow *away* from this node. For the given network, the current through G_1 will flow down and its value will be $I_{G1} = G_1V_1$, current through G_2 will flow from left to right and will be $I_{G2} = G_2(V_1 - V_2)$. Current from the independent current source flows

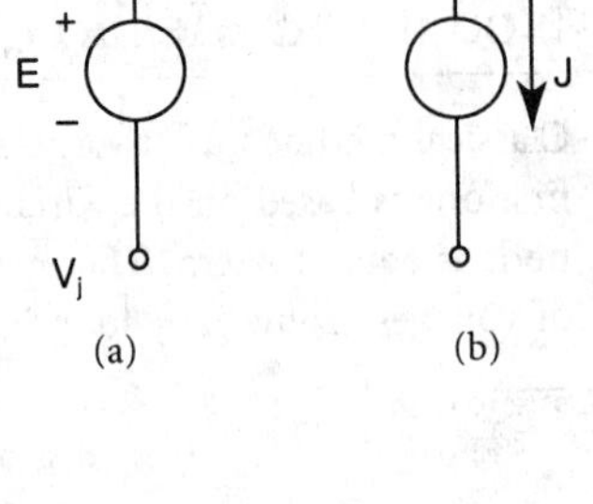

FIGURE 19.2 (a) Direction of positive current through an independent voltage source. (b) Direction of the current through an independent current source.

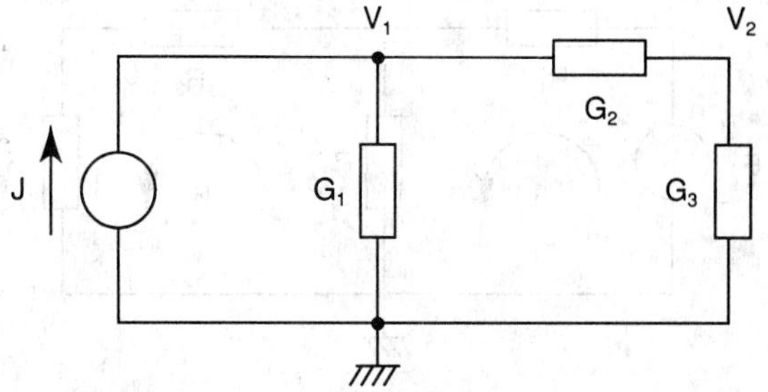

FIGURE 19.3 Example for nodal formulation.

into the node and thus must be subtracted. Together, the sum of the currents at node 1 will be zero:

$$G_1V_1 + G_2(V_1 - V_2) - J = 0$$

Moving to the second node we still do not know which node is more positive, but we can still make the assumption that *now* it is *this* node which is the most positive one. If we make such an assumption, then all currents must flow away from this node: The current through G_3 will be $I_{G3} = G_3V_2$ and will flow down, the current through G_2 will flow from right to left and will be $I_{G2} = G_2(V_2 - V_1)$. In this expression the first voltage within the parentheses must be the assumed higher potential. This is expressed by the equation

$$G_3V_2 + G_2(V_2 - V_1) = 0$$

It is advantageous to put the equations into matrix form, with the known independent source transferred to the right side:

$$\begin{bmatrix} G_1 + G_2 & -G_2 \\ -G_2 & G_2 + G_3 \end{bmatrix} \begin{bmatrix} V_1 \\ V_2 \end{bmatrix} \begin{bmatrix} J \\ 0 \end{bmatrix}$$

The above steps were simple because we selected elements which can be handled by this formulation. Unfortunately, many practical elements are not expressed in terms of currents. For instance, a voltage source connected between nodes i and j, with its positive reference on node i, is described by the equation

$$V_i - V_j = E$$

A positive current does flow through such an element from i to j, but is not available in its defining equation. In fact, all voltage sources, independent or dependent, will create this problem. Another element which cannot be handled directly is a short circuit. It is described by the equation

$$V_i - V_j = 0$$

and current is not a part of its definition.

We can always use transformations by applying various theorems like the Thévenin and Norton transformations or source splitting, and eventually arrive at a network in which all elements have voltage as the independent variable. Such transformations are practical for hand

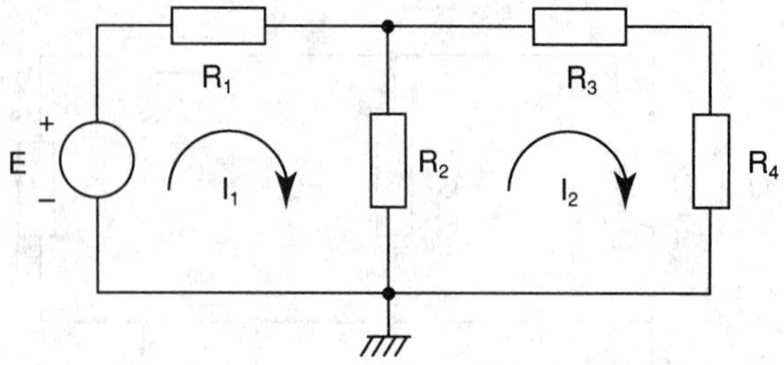

FIGURE 19.4 Example for mesh formulation.

analysis, but are not advantageous for computerized solutions. This is the reason why other formulations have been invented.

Consider next the mesh equations where we use the KVL and impedances of the elements: $Z_L = sL$ for the inductor, $Z_C = 1/sC$ for the capacitor, and R for the resistor. In this formulation we sum the voltages across the elements in a given closed loop. Since this method is suitable for planar networks only, we usually use the concept of circulating mesh currents, indicated on the network in Fig. 19.4. The currents I_1 and I_2 create voltage drops across the resistors. When considering the first mesh, we take the current I_1 as a positive one. The voltage across R_1 is $V_{R1} = R_1 I_1$. The voltage across R_2 is $V_{R2} = R_2(I_1 - I_2)$ and the voltage source contributes a value E to the equation. According to our earlier definition, a positive current flows from plus to minus, but I_1 actually goes in the opposite direction through the voltage source. Thus, the voltage across E must be taken with a negative sign and the sum of voltages around the first mesh is

$$R_1 I_1 + R_2(I_1 - I_2) - E = 0$$

When we move to the second mesh, we consider the current I_2 as positive and the sum of voltage drops around the second mesh is

$$R_2(I_2 - I_1) + R_3 I_2 + R_4 I_4 = 0$$

The equations can be collected into one matrix equation

$$\begin{bmatrix} R_1 + R_2 & -R_2 \\ -R_2 & R_2 + R_3 + R_4 \end{bmatrix} \begin{bmatrix} I_1 \\ I_2 \end{bmatrix} = \begin{bmatrix} E \\ 0 \end{bmatrix}$$

Each of these fundamental formulations has its problems.

In nodal formulation we can deal directly with the following elements:

Current source, J
Conductance, $G = 1/R$
Capacitor admittance, sC
Voltage controlled current source, VC
Inductor admittance, $1/sL$

In mesh formulation we can deal directly with the elements:

Voltage source, E
Resistor, R

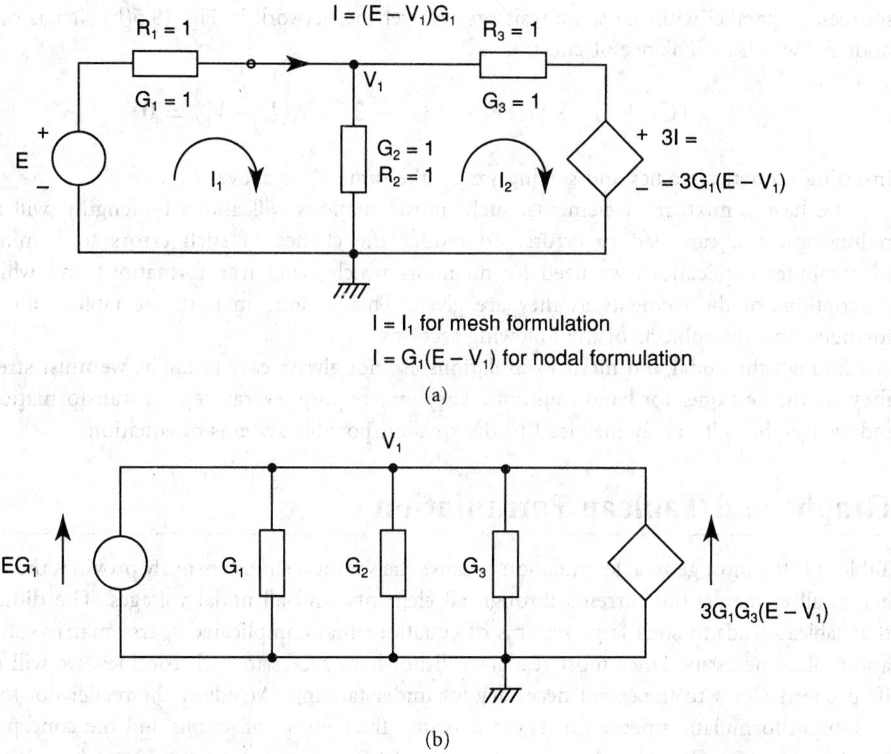

FIGURE 19.5 (a) Example of a network suitable for mesh formulation. (b) Modification of the network in Fig. 5(a) to be suitable for nodal formulation.

Inductor impedance, sL
Current controlled voltage source, CV
Capacitor impedance, $1/sC$.

All other elements create problems and must be dealt with by the Thévenin-Norton theorems and/or source splitting.

As an example we take the network in Fig. 19.5(a). It is directly suitable for mesh formulation, but we demonstrate both. For simplicity, all resistors have unit values.

The mesh formulation, with the indicated circulating currents I_1 and I_2, leads to the equations

$$(R_1 + R_2)I_1 - R_2 I_2 = E$$
$$-R_2 I_1 + (R_2 + R_3)I_2 + 3I_1 = 0$$

Inserting numerical values

$$2I_1 - I_2 = E$$
$$2I_1 + 2I_2 = 0$$

The solution is $I_1 = E/3$, $I_2 = -E/3$ and $V_1 = R_2(I_1 - I_2) = 2E/3$.

To use nodal formulation, we must apply several transformation steps. First, we must express the controlling current as $I = G_1(E - V_1) = E - V_1$ and replace I in the definition of the current controlled voltage source. This has been done in the figure. Afterwards, applying Thévenin-Norton transformations we change the voltage sources, in series with resistors, into current sources, in parallel with the same resistors. We get the network in Fig. 19.5(b). It has only one node for which the balance of currents is

$$(G_1 + G_2 + G_3)V_1 - EG_1 - 3G_1G_3(E - V_1) = 0$$

Inserting numerical values and solving we get the same V_1 as above.

If we have a mixture of elements, such transformations will always be lengthy, will require redrawings, and can lead to errors. To reduce the chance of such errors to a minimum, in computer applications we need formulations which avoid transformations and which use descriptions of the elements as they are given. This is done in both the tableau and nodal formulations, the subjects of the following sections.

Although the nodal and mesh formulations are not always easy to apply, we must stress that they are the best ones for hand solutions. They may require several steps of transformations and redrawings, but ultimately they lead to the smallest possible systems of equations.

19.3 Graphs and Tableau Formulation

Tableau is the most general formulation because the solution simultaneously provides the voltages across all elements, the currents through all elements, and all nodal voltages. The difficulty is that tableau leads to such large systems of equations that complicated sparse matrix solvers are an absolute necessity. Since most readers will not have access to such routines, we will explain its properties only to the extent necessary for understanding. We advise the reader not to use it.

Tableau formulation needs for its construction the concept of graphs and the concept of the incidence matrix. Consider the network in Fig. 19.6(a). A graph of the network replaces each element by a line. We will use oriented graphs, with arrows, because they can be identified with the flow of currents. In all passive elements the current can flow in any direction and the orientation of the graph is entirely our choice. We do not have such freedom when we consider sources. The direction of the current through the current source is given by the arrow marked at its symbol and we use the same direction in the graph. For the voltage source, the direction of the graph will be from plus to minus, in agreement with our previous explanations. Each node is marked by the node voltage and the line representing the element is given the name of the element. Following these rules, we have constructed the graph in Fig. 19.6(b). Since the directions of the arrows are the assumed directions of currents, we can write KCL for the two nodes: positive direction is away from the node, negative is into the node. The sums of currents for the nodes are

$$-I_J + I_C + I_R = 0$$
$$-I_R + I_L = 0$$

This can also be summarized in one matrix equation

$$\begin{bmatrix} -1 & 1 & 1 & 0 \\ 0 & 0 & -1 & 1 \end{bmatrix} \begin{bmatrix} I_J \\ I_C \\ I_R \\ I_L \end{bmatrix} = \begin{bmatrix} 0 \\ 0 \end{bmatrix}$$

or

$$\mathbf{AI} = \mathbf{0} \tag{19.1}$$

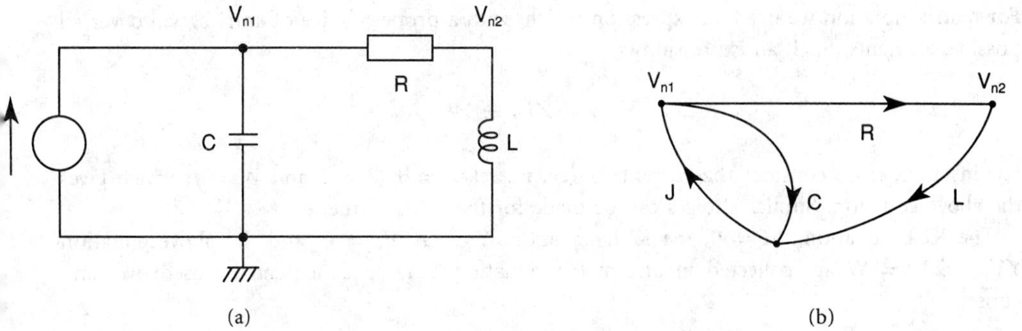

FIGURE 19.6 (a) A simple network, (b) its graph.

The matrix **A** is called the *incidence* matrix. It has as many rows as there are ungrounded nodes, and as many columns as the number of elements. Note that +1 in any given row indicates that we expect the current to flow away from the node, −1 means the opposite.

Still more information can be extracted from this matrix. Denote the nodal voltages by subscripts n, V_{n1}, V_{n2}, as done in Fig. 19.6. The voltages across the elements will have as subscripts the names of the elements. We can write the following set of equations which couple the voltages across the elements with the nodal voltages:

$$
\begin{aligned}
V_J &= -V_{n1}\\
V_C &= V_{n1}\\
V_R &= V_{n1} - V_{n2}\\
V_L &= V_{n2}
\end{aligned}
$$

In matrix form this is equivalent to

$$
\begin{bmatrix} V_J \\ V_C \\ V_R \\ V_L \end{bmatrix} = \begin{bmatrix} -1 & 0 \\ 1 & 0 \\ 1 & -1 \\ 0 & 1 \end{bmatrix} \begin{bmatrix} V_{n1} \\ V_{n2} \end{bmatrix}
$$

The matrix is the transpose of the incidence matrix and we can generalize

$$
\mathbf{V}_{el} - \mathbf{A}^T\mathbf{V}_n = \mathbf{0} \tag{19.2}
$$

Complete formulation needs expressions which couple the element currents and the element voltages. Writing them in the same sequence as for the graph, and using Laplace transformation, we have

$$
\begin{aligned}
I_J &= J\\
I_C &= sCV_C\\
V_R &= RI_R\\
V_L &= sLI_L
\end{aligned}
$$

For matrix notation we need an expression which, with a proper choice of entries, will cover all possible elements. Such an expression is

$$YV_{el} + ZI_{el} = W$$

For instance, if we consider the current source, we set $Y = 0$, $Z = 1$ and $W = J$, which gives the above equation. Similar choices can be made for the other elements.

The KCL equation, $\mathbf{AI} = \mathbf{0}$, the KVL equation, $\mathbf{V}_{el} - \mathbf{A}^T\mathbf{V}_n = \mathbf{0}$, and the above equation $\mathbf{YV}_{el} + \mathbf{ZI}_{el} = \mathbf{W}$ are collected in one matrix equation. Any sequence can be used; we have chosen

$$\begin{aligned} \mathbf{V}_{el} - \mathbf{A}^T\mathbf{V}_n &= \mathbf{0} \\ \mathbf{YV}_{el} + \mathbf{ZI}_{el} &= \mathbf{W} \\ \mathbf{AI}_{el} &= \mathbf{0} \end{aligned} \tag{19.3}$$

and in matrix form

$$\begin{bmatrix} \mathbf{1} & \mathbf{0} & -\mathbf{A}^\mathrm{T} \\ \mathbf{Y} & \mathbf{Z} & \mathbf{0} \\ \mathbf{0} & \mathbf{A} & \mathbf{0} \end{bmatrix} \begin{bmatrix} \mathbf{V}_{el} \\ \mathbf{I}_{el} \\ \mathbf{V}_n \end{bmatrix} = \begin{bmatrix} \mathbf{0} \\ \mathbf{W} \\ \mathbf{0} \end{bmatrix} \tag{19.4}$$

Once the incidence matrix is available, writing this matrix equation is actually quite simple. First determine its size: it will be twice the number of elements plus the number of nodes. For our example it will be 10. The system equation is in Fig. 19.7 where all zero entries were omitted to clearly show the structure. In the top partition is a unit matrix and the negative of the transpose of the **A** matrix. In the bottom partition is the incidence matrix, **A**. The middle portion is filled, element by element, using the above element equations. For better understanding, it is a good idea to write the variables above the matrix, as shown, because each column of the matrix is multiplied by the variable which appears above it.

We have used this simple example to point out the main difficulty of the tableau formulation: the system becomes very large. In nodal formulation this problem would lead to only two

$$\begin{array}{cccc|cccc|cc} V_j & V_C & V_R & V_L & I_J & I_C & I_R & I_L & V_{n1} & V_{n2} \end{array}$$

$$\left[\begin{array}{cccc|cccc|cc} 1 & & & & & & & & 1 & \\ & 1 & & & & & & & -1 & \\ & & 1 & & & & & & -1 & 1 \\ & & & 1 & & & & & & -1 \\ \hline & & & & 1 & & & & & \\ & sC & & & & -1 & & & & \\ & & 1 & & & & R & & & \\ & & & 1 & & & & sL & & \\ & & & & -1 & 1 & 1 & & & \\ & & & & & & -1 & 1 & & \end{array}\right] \begin{bmatrix} V_J \\ V_C \\ V_R \\ V_L \\ I_J \\ I_C \\ I_R \\ I_L \\ V_{n1} \\ V_{n2} \end{bmatrix} = \begin{bmatrix} \\ \\ \\ \\ \\ J \\ \\ \\ \\ \\ \end{bmatrix}$$

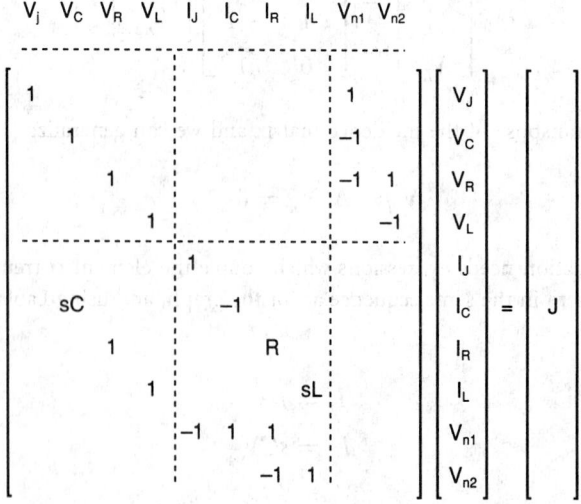

FIGURE 19.7 Tableau formulation for the network in Fig. 6.

equations. However, the tableau system matrix has many zeros and is said to be sparse. Sparse systems are always solved by special routines which, roughly speaking, do not store the zeros and do not operate on them. Such codes are quite difficult to write and in tableau we have the additional difficulty that the matrix has a complicated structure. We discussed this formulation more as a warning rather than a recommendation. Unless a suitable sparse matrix solver is already available, this formulation should be avoided.

9.4 Modified Nodal Formulation

Modified nodal formulation is an extension of the nodal formulation and is the method of choice for computerized analysis. It is used in most commercial simulators and we will explain it in considerable detail.

When nodal formulation is taught in schools, inductors are usually taken as admittances, $Y_L = \frac{1}{sL}$. This is fine, as long as we work by hand and derive the network function. For instance, nodal equations for the network in Fig. 19.8 would be written in the form

$$\left(G_1 + sC_1 + \frac{1}{sL}\right)V_1 - \frac{1}{sL}V_2 = J$$

$$-\frac{1}{sL}V_1 + \left(G_2 + sC_2 + \frac{1}{sL}\right)V_2 = 0$$

However, this creates a problem for computerized solutions. Multiplication by s represents differentiation in the Laplace transform, and $\frac{1}{s}$ represents integration. As a result, the two equations are actually a set of two integro-differential equations and we do not normally have methods to solve them directly in such a form. In all computerized methods we use integration of systems of first order differential equations and the above equations cannot be arranged into such a form. What we need is a method which will keep all frequency-dependent elements in the form sC or sL, with the variable s in the numerator. Such possibility exists if we take into account a new variable, the current through the inductor. Writing KCL for the two nodes of Fig. 19.8

$$(G_1 + sC_1)V_1 + I_L = J$$

$$(G_2 + sC_2)V_2 - I_L = 0$$

We now have two equations but three variables. What we have not used yet is an expression which couples the voltages across the inductor with the current through it. The relationship is $V_1 - V_2 = sLI_L$, but since we do not know any of these three variables, we transfer everything to the left and write the last equation

$$V_1 - V_2 - sLI_L = 0 \qquad (19.5)$$

All three can be put into a matrix form

$$\begin{bmatrix} G_1 + sC_1 & 0 & 1 \\ 0 & G_2 + sC_2 & -1 \\ 1 & -1 & -sL \end{bmatrix} \begin{bmatrix} V_1 \\ V_2 \\ I_L \end{bmatrix} = \begin{bmatrix} J \\ 0 \\ 0 \end{bmatrix}$$

In this equation the nodal portion is the 2×2 matrix in the upper left corner and information about the inductor is collected in the right-most column and the lowest row. A larger network

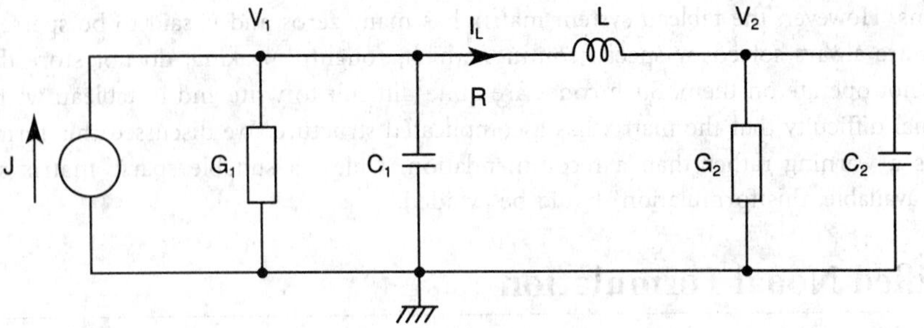

FIGURE 19.8 Example for modified nodal formulation.

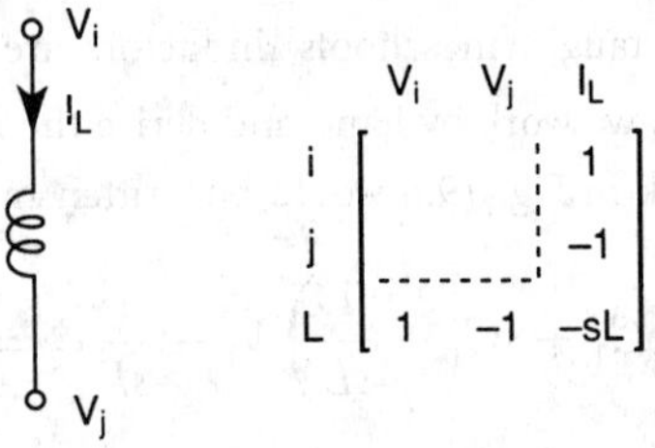

FIGURE 19.9 Stamp of an inductor.

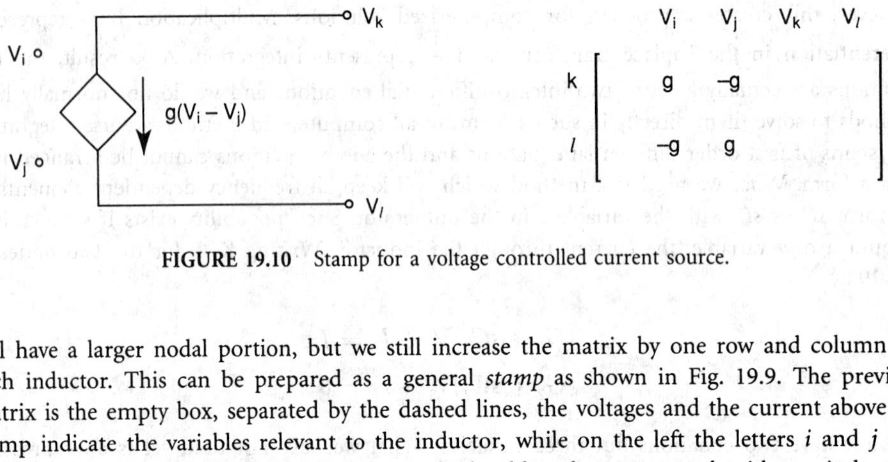

FIGURE 19.10 Stamp for a voltage controlled current source.

will have a larger nodal portion, but we still increase the matrix by one row and column for each inductor. This can be prepared as a general *stamp* as shown in Fig. 19.9. The previous matrix is the empty box, separated by the dashed lines, the voltages and the current above the stamp indicate the variables relevant to the inductor, while on the left the letters i and j give the rows (node numbers) where L is connected. Should we have a network with two inductors, we add one row and one column for each.

The next element we take is the voltage-controlled current source; it was already mentioned in Section 2 as an element which can be taken into the nodal formulation. The *VC* is shown in Fig. 19.10. It adds the current $g(V_i - V_j)$ to node k and subtracts the same current at node l. It permits us to write the stamp, Fig. 19.10. The element influences the balance of currents at node i and j, and the variables which multiply its transconductance are V_i and V_j.

Network theory defines two independent sources: the current and the voltage source. The current source can be taken into consideration in the right hand side (r.h.s.) of the nodal portion; its stamp is in Fig. 19.11. The independent voltage source cannot be taken directly and we must add its current as a new variable. Consider the source with a resistor in series, shown in Fig. 19.12. The voltage relationships are

$$V_i - V_j - rI_E = E \tag{19.6}$$

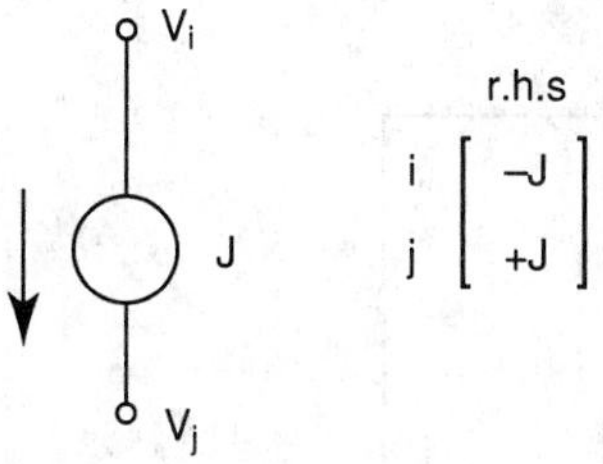

FIGURE 19.11 Stamp for an independent current source.

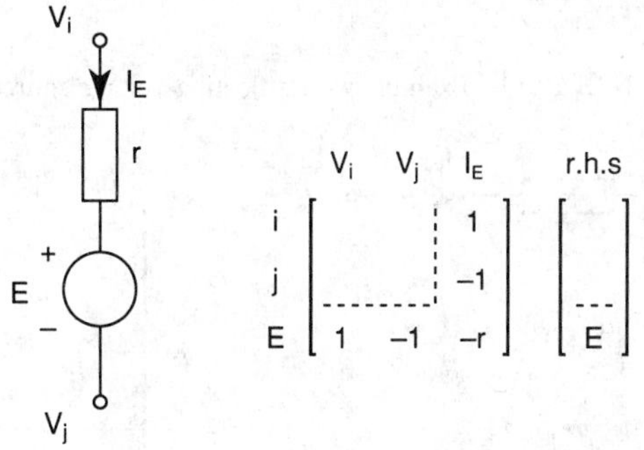

FIGURE 19.12 Stamp for an independent voltage source. The resistor value can be zero.

The current I_E adds to the balance of currents at node i, because it flows away from it. It is subtracted from the balance of currents at node j. The current is taken as a new variable and the equation is attached to previous equations. The stamp is as shown. It is, in fact, a combined stamp for several elements. If we set $r = 0$, we have an ideal voltage source. If we set both r and E equal to zero, we have a stamp for a short circuit. For better understanding consider the example in Fig. 19.13. The network has two ungrounded nodes for which we can write the nodal equations:

$$(G_1 + sC_1)V_1 + I_E = 0$$
$$(G_2 + sC_2)V_2 - I_E = 0$$

To these equations we must add the equation describing the properties of the ideal voltage source,

$$V_i - V_j = E \tag{19.7}$$

Collecting into matrix form

$$\begin{bmatrix} G_1 + sC_1 & 0 & 1 \\ 0 & G_2 + sC_2 & -1 \\ 1 & -1 & 0 \end{bmatrix} \begin{bmatrix} V_1 \\ V_2 \\ I_E \end{bmatrix} \begin{bmatrix} 0 \\ 0 \\ E \end{bmatrix}$$

We still need stamps for the remaining three dependent sources. The voltage controlled voltage source, VV, is shown in Fig. 19.14. For generality, we added the internal resistor. The output is described by the equation

$$\mu(V_i - V_j) + rI_{VV} = V_k - V_l \tag{19.8}$$

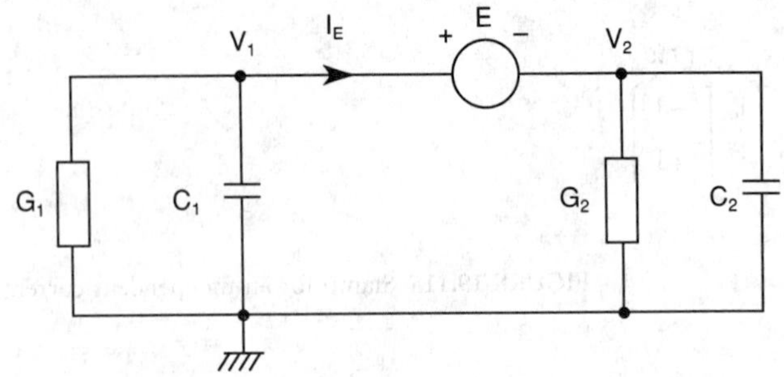

FIGURE 19.13 Example with a floating voltage source.

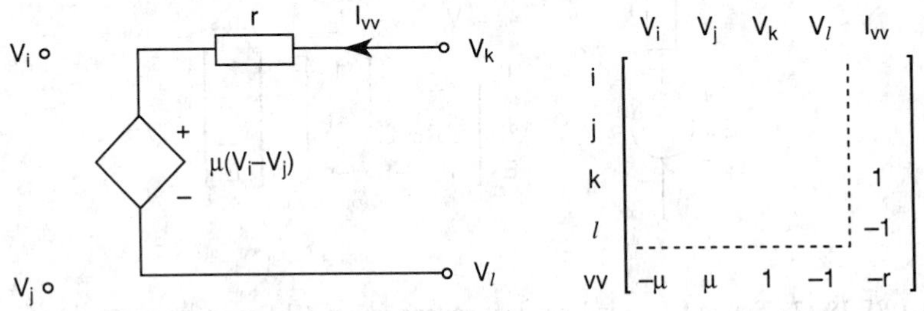

FIGURE 19.14 Stamp for a voltage controlled voltage source. The resistor value can be zero.

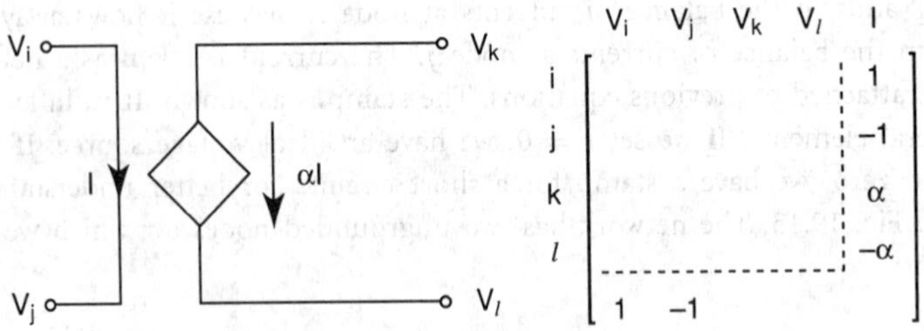

FIGURE 19.15 Stamp for a current controlled current source.

Since none of the voltages is known, we transfer everything to the right side. Input terminals do not influence the balance of currents, but the output terminals do. At node k we must add I_{VV}, and subtract the same at node l. The stamp is in Fig. 19.14.

The current controlled current source, CC, is shown in Fig. 19.15. The input terminals are short circuited and

$$V_i - V_j = 0 \tag{19.9}$$

No information is available about the output voltages, but we know that the current is α times the input current, and thus only one additional variable is needed. Balances of currents are influenced at all four nodes: positive I at node i, negative at node j, positive current αI at node k, and negative αI at node l. The stamp is in Fig. 19.15, where the added column takes care of the currents and the additional row describes the properties of the short circuit.

The most complicated dependent source is the current-controlled voltage source, CV, shown in Fig. 19.16. We can consider it as a combination of a short circuit and a voltage source. The equation for the short circuit is the same as for the CC. The output is defined by the equation

$$V_k - V_l = \rho I_1 + rI_{CV} \tag{19.10}$$

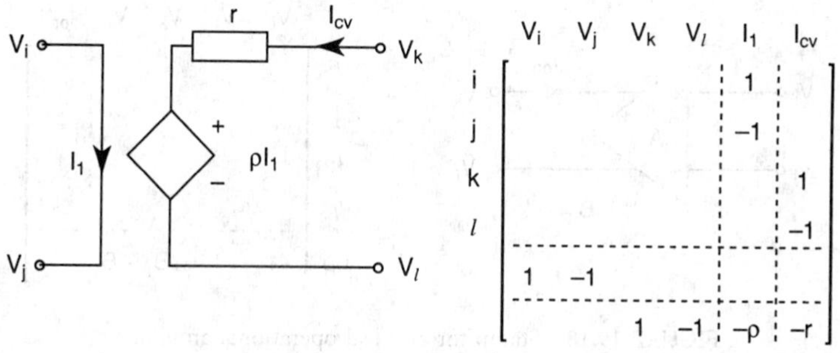

FIGURE 19.16 Stamp for a current controlled voltage source. The resistor value can be zero.

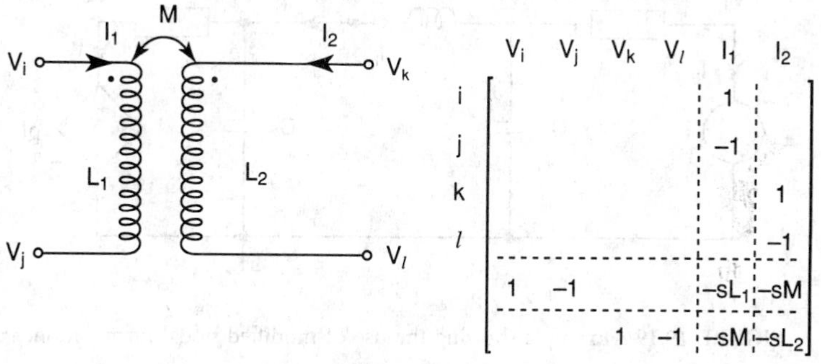

FIGURE 19.17 Stamp for a transformer.

and since none of the variables is known, we transfer everything to the left. This element adds two rows and two columns to the previously defined matrix. Its stamp is in Fig. 19.16. As before, the internal resistor r can be set equal to zero.

Modified nodal formulation easily takes into account a transformer, see Fig. 19.17. It is described by the equations

$$\begin{aligned} V_i - V_j &= sL_1 I_1 + sMI_2 \\ V_k - V_l &= sMI_1 + sL_2 I_2 \end{aligned} \tag{19.11}$$

None of the variables is known and we transfer everything to the left. The currents influence the balance of currents at all nodes. The stamp of the transformer is in Fig. 19.17.

The last element we consider is an ideal operational amplifier. It is sometimes taken as a voltage-controlled voltage source with very high gain, but it is preferable to have a stamp which can take into account ideal properties as well. The element is shown in Fig. 19.18. The terminal l is usually grounded but we will keep it floating, to make the stamp more general. No current flows into the device at the input terminals. The output equation is

$$V_k - V_l = A(V_i - V_j) \tag{19.12}$$

Since a computer cannot handle infinity, it is advantageous to introduce the inverted gain

$$B = -1/A \tag{19.13}$$

and modify the above equation to

$$V_i - V_j + BV_k - BV_l = 0 \tag{19.14}$$

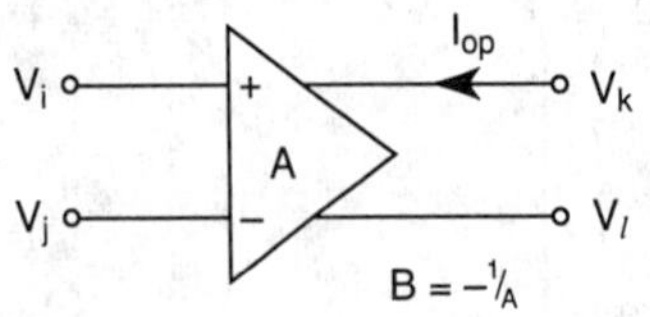

$$
\begin{array}{c|cccc:c|}
 & V_i & V_j & V_k & V_l & I_{op} \\
i & & & & & \\
j & & & & & \\
k & & & & & 1 \\
l & & & & & -1 \\
\hdashline
op & +1 & -1 & +B & -B & \\
\end{array}
$$

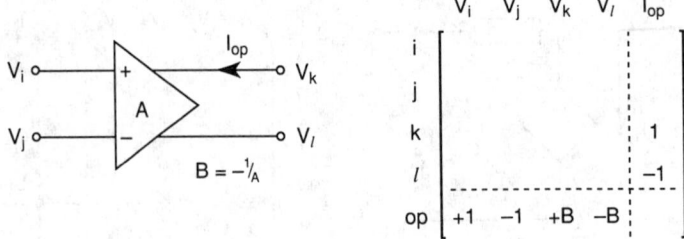

FIGURE 19.18 Stamp for an ideal operational amplifier.

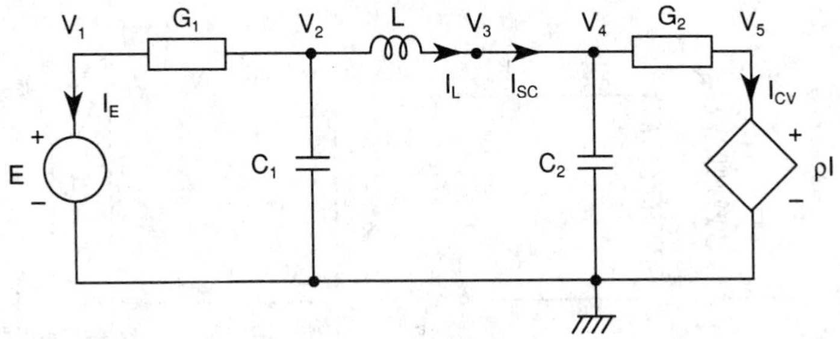

FIGURE 19.19 Example showing the use of modified nodal formulation.

This equation is attached to the set of equations and the balance of currents is influenced at nodes k and l. This leads to the stamp in Fig. 19.18. If we set $B = 0$, the operational amplifier becomes ideal, with no approximation.

An example will show how the stamps are used. Consider the network in Fig. 19.19. It has no practical application, but serves well for the demonstration of how to set up the modified nodal matrix. A short circuit, indicating the controlling current of the current-controlled voltage source is taken into account by increasing the number of nodes. The network has five nongrounded nodes and thus the dimension of its nodal portion will be 5. The voltage source will increase the system matrix by one row and one column, the inductor also, and the CV will need two more rows and columns; altogether the matrix will be 9×9. Write first the nodal portion by disregarding entirely the other elements. This creates the upper left partition. Using the stamps we add first the voltage source, then the inductor, next the short circuit and finally the current-controlled voltage source. The system matrix is

$$
\left[\begin{array}{ccccc:c:c:cc}
G_1 & -G_1 & 0 & 0 & 0 & 1 & 0 & 0 & 0 \\
-G_1 & G_1 + sC_1 & 0 & 0 & 0 & 0 & 1 & 0 & 0 \\
0 & 0 & 0 & 0 & 0 & 0 & -1 & 1 & 0 \\
0 & 0 & 0 & sC_2 + G_2 & -G_2 & 0 & 0 & -1 & 0 \\
0 & 0 & 0 & -G_2 & G_2 & 0 & 0 & 0 & 1 \\
\hdashline
1 & 0 & 0 & 0 & 0 & 0 & 0 & 0 & 0 \\
\hdashline
0 & 1 & -1 & 0 & 0 & 0 & -sL & 0 & 0 \\
\hdashline
0 & 0 & 1 & -1 & 0 & 0 & 0 & 0 & 0 \\
0 & 0 & 0 & 0 & 1 & 0 & 0 & -\rho & 0
\end{array}\right]
\left[\begin{array}{c} V_1 \\ V_2 \\ V_3 \\ V_4 \\ V_5 \\ I_E \\ I_L \\ I_{SC} \\ I_{CV} \end{array}\right]
=
\left[\begin{array}{c} 0 \\ 0 \\ 0 \\ 0 \\ 0 \\ E \\ 0 \\ 0 \\ 0 \end{array}\right]
$$

Modified nodal formulation is the most important method for computer applications. The reader can find additional information in the books [1, 2].

19.5 Nonlinear Elements

In previous sections we used the Laplace transform to explain the various methods of formulation. Since we dealt with linear elements, the systems of equations were linear and it was possible to cast them into matrix forms.

If we must consider nonlinear elements, we face many restrictions. The Laplace transform cannot be used. Various concepts based on it, like the network functions, the poles and the zeros, cannot be applied. Only two types of analysis are available:

The dc solution (operating point)
Time domain solution for a given input signal

Once we have nonlinear elements, we cannot write the equations in matrix form; all we can do is write KCL equations. We must also find another method for the solution of such nonlinear equations.

Consider two differentiable equations in two unknowns

$$\begin{aligned} f_1(v_1, v_2) &= 0 \\ f_2(v_1, v_2) &= 0 \end{aligned} \tag{19.15}$$

the functions can be expanded into Taylor series and the series truncated after the linear terms:

$$f_1(v_1 + \Delta v_1, v_2 + \Delta v_2) = f_1(v_1, v_2) + \frac{\partial f_1}{\partial v_1}\Delta v_1 + \frac{\partial f_1}{\partial v_2}\Delta v_2 + \cdots = 0$$

$$f_2(v_1 + \Delta v_1, v_2 + \Delta v_2) = f_2(v_1, v_2) + \frac{\partial f_2}{\partial v_1}\Delta v_1 + \frac{\partial f_2}{\partial v_2}\Delta v_2 + \cdots = 0$$

Since we are not considering higher order terms, the equation will not be exactly zero, but we can still try to find Δv_1 and Δv_2. Transferring the known function values to the right

$$\frac{\partial f_1}{\partial v_1}\Delta v_1 + \frac{\partial f_1}{\partial v_2}\Delta v_2 = -f_1(v_1, v_2)$$

$$\frac{\partial f_2}{\partial v_1}\Delta v_1 + \frac{\partial f_2}{\partial v_2}\Delta v_2 = -f_2(v_1, v_2)$$

This is a system of *linear* equations and we can rewrite it in matrix form

$$\begin{bmatrix} \dfrac{\partial f_1}{\partial v_1} & \dfrac{\partial f_1}{\partial v_2} \\ \dfrac{\partial f_2}{\partial v_1} & \dfrac{\partial f_2}{\partial v_2} \end{bmatrix} \begin{bmatrix} \Delta v_1 \\ \Delta v_2 \end{bmatrix} = \begin{bmatrix} -f_1(v_1, v_2) \\ -f_2(v_1, v_2) \end{bmatrix} \tag{19.16}$$

The matrix on the left is called the *Jacobian*, on the right is the negative of the functions. Once this *linear* system is solved, we can get new values of the variables by writing

$$\begin{aligned} v_1^{(i+1)} &= v_1^{(i)} + \Delta v_1^{(i)} \\ v_2^{(i+1)} &= v_2^{(i)} + \Delta v_2^{(i)} \end{aligned} \tag{19.17}$$

In this equation we added the superscript to indicate iteration. The process is repeated until all Δv_i become sufficiently small. This iterative method is usually referred to as the Newton-Raphson iteration and is written in the form

$$\begin{aligned} \mathbf{J}(\mathbf{v}^{(i)})\Delta\mathbf{v}^{(i)} &= -\mathbf{f}(\mathbf{v}^{(i)}) \\ \mathbf{v}^{(i+1)} &= \mathbf{v}^{(i)} + \Delta\mathbf{v}^{(i)} \end{aligned} \tag{19.18}$$

Suppose that we now take the network in Fig. 19.3, consider the conductances G_1 and G_3 as linear and replace G_2 by a nonlinear function

$$i = g(v_{el})$$

where v_{el} is the voltage across this element,

$$v_{el} = v_1 - v_2$$

and g represents a nonlinear function. The two KCL equations are

$$\begin{aligned} G_1 v_1 + g(v_{el}) - J &= 0 \\ -g(v_{el}) + G_3 v_2 &= 0 \end{aligned}$$

For the Newton-Raphson equation we need the derivatives with respect to v_1 and v_2. Consider now only the nonlinear element. Using the chain rule of differentiation we can write

$$\begin{aligned} \frac{\partial g(v_{el})}{\partial v_1} &= \frac{\partial g(v_{el})}{\partial v_{el}}\frac{\partial v_{el}}{\partial v_1} = +\frac{\partial g(v_{el})}{\partial v_{el}} \\ \frac{\partial g(v_{el})}{\partial v_2} &= \frac{\partial g(v_{el})}{\partial v_{el}}\frac{\partial v_{el}}{\partial v_2} = -\frac{\partial g(v_{el})}{\partial v_{el}} \end{aligned}$$

With these preliminary steps we can now write the Newton-Raphson equation:

$$\begin{bmatrix} G_1 + \dfrac{\partial g(v_{el})}{\partial v_{el}} & -\dfrac{\partial g(v_{el})}{\partial v_{el}} \\ -\dfrac{\partial g(v_{el})}{\partial v_{el}} & G_3 + \dfrac{\partial g(v_{el})}{\partial v_{el}} \end{bmatrix} \begin{bmatrix} \Delta v_1 \\ \Delta v_2 \end{bmatrix} = -\begin{bmatrix} G_1 v_1 + g(v_{el}) - J \\ -g(v_{el}) + G_3 v_3 \end{bmatrix}$$

Comparing the Jacobian of the Newton-Raphson equation with the nodal formulation we reach the important conclusion, valid for networks of any size:

1. Linear elements will be in the Jacobian in the same position as they were in the linear system matrix.
2. Nonlinear elements will have entries in the same positions as if they were linear, only their numerical values will be equal to the derivative $\dfrac{\partial g}{\partial v_{el}}$, evaluated with already available variables.

This conclusion will also be true for the other formulations, like the tableau or the modified nodal. So far we considered only nonlinear resistive elements and the operating point.

Nonlinear storage elements (capacitors and inductors) contribute to the equations with their fluxes and charges. The current through the nonlinear capacitor is defined by

$$i_C = \frac{dq(v_C)}{dt} \tag{19.19}$$

where $q(v_C)$ is the charge and v_C is the voltage across the capacitor. The voltage across the nonlinear inductor is given by

$$v_L = \frac{d\phi(i_L)}{dt} \tag{19.20}$$

with ϕ denoting the flux.

Integration of systems with storage elements is always done by first replacing the derivative by a suitable algebraic expression and then solving the resulting nonlinear algebraic system by the Newton-Raphson method derived above.

There are many ways to replace the time domain derivatives by algebraic expressions. Books on numerical analysis usually describe the Runge-Kutta method. We mention it here because it is *not* suitable for solution of networks. There are several reasons for this, the main one being that the preferred modified nodal formulation does not lead to systems of differential, but rather to systems of algebraic-differential equations. There are only two widely used two methods, the trapezoidal formula and a family of backward differentiation formulas (BDF). Among the BDFs the simplest is the backward Euler and we will base our explanations on this formula. It replaces the derivative by the difference of the previous and new value, divided by the step size, h,

$$\frac{dq(v_c)}{dt} \sim \frac{q_{new}(v_c) - q_{old}}{h}$$

$$\frac{d\phi(i_L)}{dt} \sim \frac{\phi_{new}(i_L) - \phi_{old}}{h}$$

Consider the network in Fig. 19.20 with nonlinear storage elements and with a linear conductance, G. It can be described by three equations:

$$f_1 = \frac{dq(v_C)}{dt} + G(v_1 - v_2) - j(t) = 0$$

$$f_2 = -G(v_1 - v_2) + i_L = 0$$

$$f_3 = v_2 - \frac{d\phi(I_L)}{dt} = 0$$

The time derivatives are replaced by the backward Euler formula

$$\frac{q_{new}(v_1) - q_{old}}{h} + G(v_1 - v_2) - j(t) = 0$$

$$-G(v_1 - v_2) + i_L = 0$$

$$v_2 - \frac{\phi_{new}(i_L) - \phi_{old}}{h} = 0$$

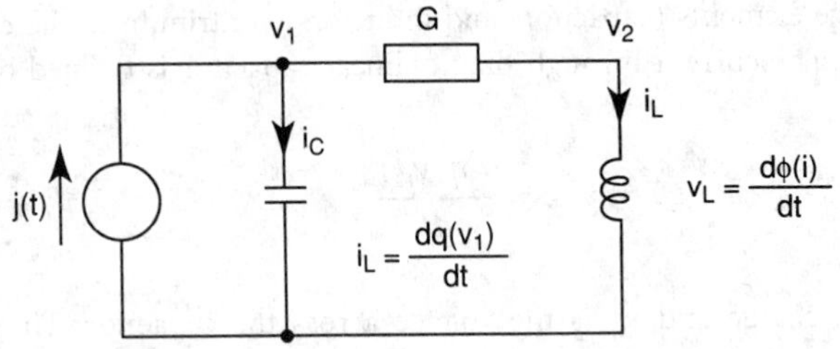

FIGURE 19.20 Network with a nonlinear capacitor and inductor.

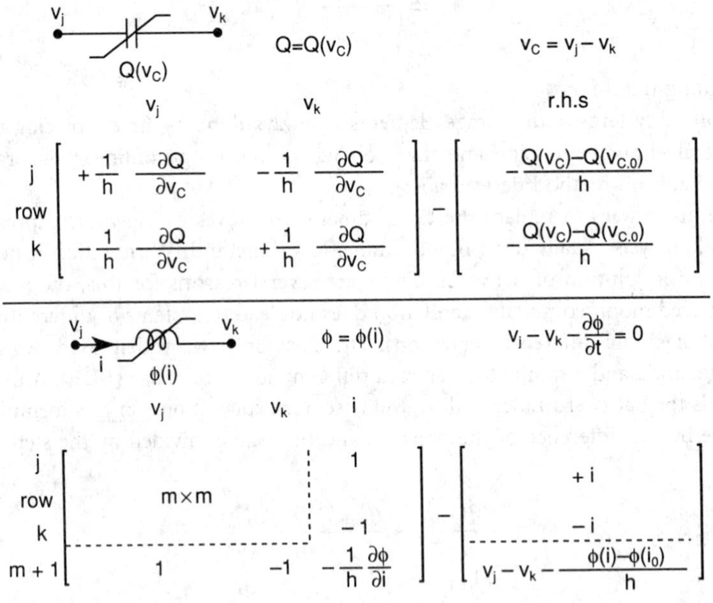

FIGURE 19.21 Stamp for a nonlinear capacitor and inductor.

thus changing the system into an algebraic one. If we now differentiate with respect to the variables v_1, v_2, and i_L, we obtain the Jacobian

$$\begin{bmatrix} \frac{1}{h}\frac{\partial q_{new}(v_1)}{\partial v_1} + G & -G & 0 \\ -G & +G & 1 \\ 0 & 1 & -\frac{1}{h}\frac{\partial \phi_{new}(i_L)}{\partial i} \end{bmatrix}$$

It can be seen that values of the derivatives are in the same places as would be the values of $C(L)$ of linear capacitors (inductors). In addition, the variable s from the Laplace domain is replaced by $\frac{1}{h}$.

The example used a grounded capacitor and a grounded inductor. Figure 19.21 shows the stamps for floating nonlinear elements and for the Newton-Raphson iteration, based on the backward Euler formula.

).6 Nodal Analysis of Active Networks

Low frequency analog filters are often built with active RC networks and the active elements are almost always operational amplifiers. We have seen in Section 19.4 that each such element adds one row and one column to the modified nodal system matrix, thus making the system too large for hand solutions. We need a method which can reduce the size of the matrix to the minimum. Such reduction is possible [1, 2], and becomes extremely simple if the voltage sources (dependent or independent) have one of their terminals grounded. Almost all practical networks meet this condition.

To introduce the method, consider the network in Fig. 19.22. If we are not interested in the current through the voltage source, we can write only one nodal equation for the node on the right:

$$(G_1 + G_2)V_1 - EG_1 = 0$$

The source is known, the term in which it appears is transferred to the right side and instead of three equations of the modified nodal formulation we must solve only one. Consider next the network in Fig. 19.23 with a voltage source and an ideal operational amplifier. One of the output terminals of the operational amplifier is grounded. Such amplifier is described by the equation

$$(V_+ - V_-)A = V_{out} \tag{19.21}$$

where $A \to \infty$. Divide first by A and then substitute

$$B = -1/A \tag{19.22}$$

This changes the equation into

$$V_+ - V_- + BV_{out} = 0 \tag{19.23}$$

If the operational amplifier is ideal, set $B = 0$ and in such case $V_+ = V_-$; the operational amplifier will have the same voltages at its input terminals. We can take this into consideration by simply writing the voltage with the same subscript to both input terminals of the operational amplifier, as was done in Fig. 19.23. We are not interested in the current of the voltage source, nor in the current flowing into the operational amplifier. We mark our lack of interest by crossing out the nodes which have grounded voltage source; this was also done in Fig. 19.23. Each node is given a voltage, but we write the nodal equations only at nodes which were not crossed out. For our example:

$$(G_1 + G_2)V_1 - G_2V_{out} - EG_1 = 0$$
$$(G_3 + G_4 + G_5)V_1 - G_5V_{out} - EG_3 = 0$$

Terms with the known source voltage are transferred to the right and we have the system

$$\begin{bmatrix} G_1 + G_2 & -G_2 \\ G_3 + G_4 + G_5 & -G_5 \end{bmatrix} \begin{bmatrix} V_1 \\ V_{out} \end{bmatrix} = \begin{bmatrix} G_1E \\ G_3E \end{bmatrix}$$

A modified nodal formulation would have required six equations.

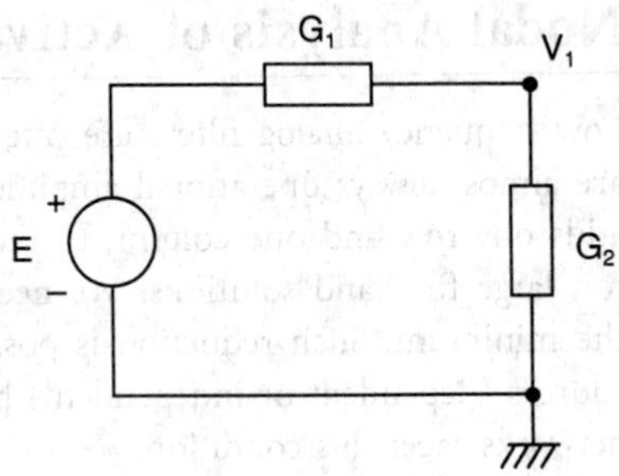

FIGURE 19.22 Example showing how to reduce the number of nodal equations.

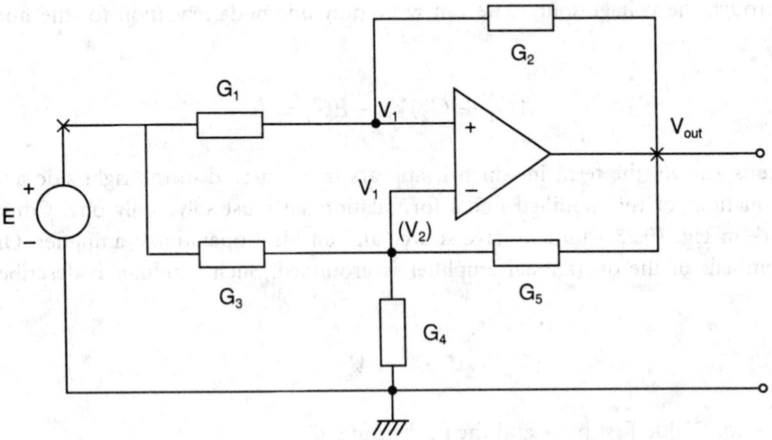

FIGURE 19.23 Nodal analysis of a network with one ideal operational amplifier.

This method can be used for any network if one node of each voltage source, dependent or independent, is grounded. All we have to do is assign every node a voltage, cross out nodes with voltage sources, and write nodal equations for the rest. It is advantageous to use conductances for resistors, because this way we avoid the fractions.

The method remains valid if the operational amplifier is not ideal and has the inverted gain *B*. The only difference is that for a nonideal amplifier we cannot make any assumptions on the voltages at its input terminals and the subscripts of such voltages must be different. This second case is also shown in Fig. 19.23 by the voltage V_2 (in brackets) at the lower node. We still write nodal equations for the two input terminals, but to complete the system we must also attach the equation of the operational amplifier. The result is

$$V_1(G_1 + G_2) - G_2V_{out} = G_1E$$
$$V_1(G_3 + G_4 + G_4)V_2 - G_5V_{out} = G_3E$$
$$V_1 - V_2 + BV_{out} = 0$$

and in matrix form

$$\begin{bmatrix} G_1 + G_2 & 0 & -G_2 \\ 0 & G_3 + G_4 + G_5 & -G_5 \\ 1 & -1 & B \end{bmatrix} \begin{bmatrix} V_1 \\ V_2 \\ V_{out} \end{bmatrix} = \begin{bmatrix} G_1E \\ G_3E \\ 0 \end{bmatrix}$$

where *B* can be set zero for an ideal operational amplifier.

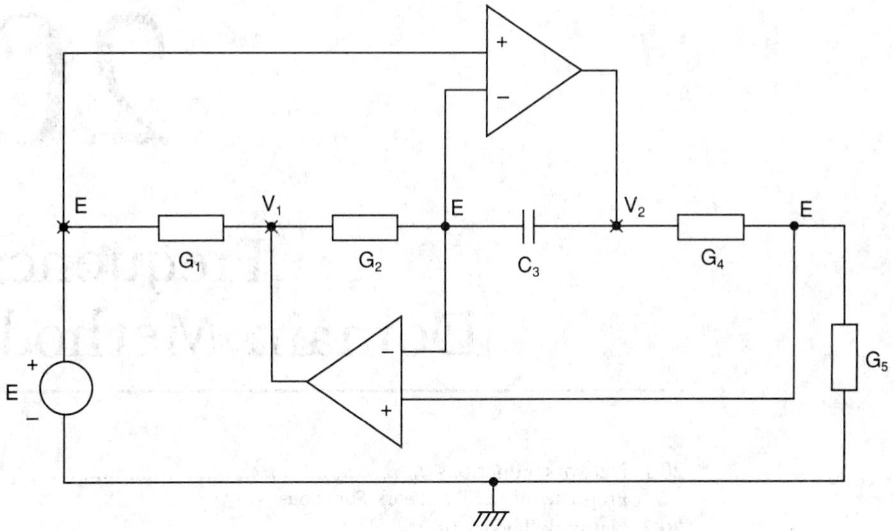

FIGURE 19.24 Nodal analysis of a network with two ideal operational amplifiers.

We will give one example of a practical network, Fig. 19.24. The operational amplifiers are ideal and thus the input voltage, E, appears at three terminals of the network. The other terminal voltages are marked by V_1 and V_2. The terminals with voltage sources are marked by crosses and only nodes 3 and 5, counting from left, remain for writing the KCL. They are

$$(G_2 + sC_3)E - G_2V_1 - sC_3V_2 = 0$$
$$(G_4 + G_5)E - G_4V_2 = 0$$

Transferring terms containing the independent voltage source, E, to the other side of the equation we arrive at the system

$$\begin{bmatrix} G_2 & sC_3 \\ 0 & G_4 \end{bmatrix}\begin{bmatrix} V_1 \\ V_2 \end{bmatrix} = \begin{bmatrix} (G_2 + sC_3)E \\ (G_4 + G_5)E \end{bmatrix}$$

nclusion

It was shown that hand calculations should use nodal or mesh formulations. Computer applications should be based on modified nodal formulation. For active networks, it is advantageous to use the method of Section 19.6.

erences

[1] J. Vlach, *Basic Network Theory with Computer Applications*, New York: Van Nostrand-Reinhold, 1992.

[2] J. Vlach and K. Singhal, *Computer Methods for Circuit Analysis and Design*, 2nd ed., New York: Van Nostrand-Reinhold, 1994.

20

Frequency Domain Methods

Peter Aronhime
University of Louisville

20.1 Network Functions

Network functions are employed to characterize linear, time-invariant networks in the zero state for a single excitation. As described in Chapter 1.4 and other sections of this handbook, network functions contain information concerning a network's stability and natural modes. They allow a designer to focus on obtaining a desired output signal for a given input signal.

In this section, it is shown that the concept of network functions is obtained as an extension of the (transformed) element defining equations for resistors, capacitors, and inductors. The relationships of network functions to transformed loop and node equations are also described. As a result of these relationships, a list of properties of network functions can be generated which is useful in the analysis of linear networks. Much is known about a network function for a given network even before an analysis is performed and the function itself is obtained.

Ohm's law, $v_R(t) = Ri_R(t)$ where R is in ohms, describes the relationship between the voltage across the resistor and the current through the resistor. These variables and their reference polarity and direction are depicted in Fig. 20.1. If elements of the equation for Ohm's law are transformed, we obtain $V(s) = RI(s)$ because R is a constant. Thus, we obtain an Ohm's law-like expression in the frequency domain.

However, the capacitor's voltage and current are related by the integral

$$
\begin{aligned}
v_C(t) &= \frac{1}{C}\int_{-\infty}^{t} i_C(t)\,dt = \frac{1}{C}\int_{-\infty}^{0} i_C(t)\,dt + \frac{1}{C}\int_{0}^{t} i_C(t)\,dt \\
&= V_0 + \frac{1}{C}\int_{0}^{t} i_C(t)\,dt
\end{aligned}
\qquad (20.1)
$$

0-8493-8341-2/95/$0.00+$.50

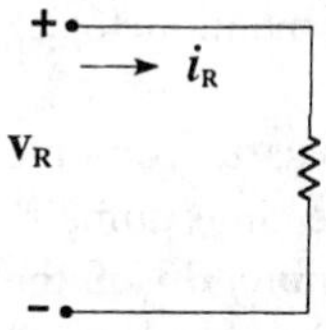

FIGURE 20.1 Reference polarity and direction for Ohm's law.

where the voltage reference polarity and the current reference direction are shown in Fig. 20.2 and the capacitance C is given in farads (F). Unlike the resistor, the relation between the capacitor voltage and the capacitor current is not a simple Ohm's law-like expression in the time domain. In addition, the voltage across the capacitor at any time t, is dependent on the entire history of the current through the capacitor.

The integral expression for the voltage across the capacitor can be split into two terms where the first term is the initial voltage across the capacitor, $V_o = v_C(0)$. If the elements of the equation are transformed, we obtain

$$V_C(s) = \frac{V_o}{s} + \frac{1}{C}\frac{I_C(s)}{s} \tag{20.2}$$

Equation (20.2) shows that if $V_o = 0$, then the expression for the transform of the capacitor voltage becomes more nearly Ohm's law-like in its form. Furthermore, if we associate the s that arises because of the integral of the current with the capacitor C, and define the *impedance* of the capacitor as $Z(s) = V_C(s)/I_C(s) = 1/(sC)$, then the equation becomes Ohm's law-like in the frequency domain.

A similar process can be applied to the inductor. The current through the inductor is expressed as

$$i_L(t) = \frac{1}{L}\int_{-\infty}^{t} v_L(t)\,dt = I_0 + \frac{1}{L}\int_{0}^{t} v_L(t)\,dt \tag{20.3}$$

where L is expressed in henries (H) and $I_o = i_L(0)$ is the initial current through the inductor. Figure 20.3 shows the reference polarity and direction for the inductor voltage and current. If the expression for the current through the inductor is transformed, the result is:

$$I_L(s) = \frac{I_o}{s} + \frac{1}{L}\frac{V_L(s)}{s} \tag{20.4}$$

Again, as with the capacitor, if $I_o = 0$ and if the s that is included because of the integral of $v_L(t)$ is considered as associated with L, then the expression for the transform of the current

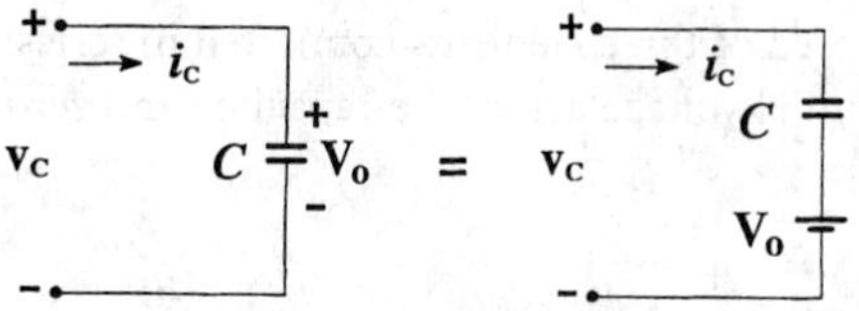

FIGURE 20.2 Capacitor representation showing reference polarities for voltages and reference direction for current.

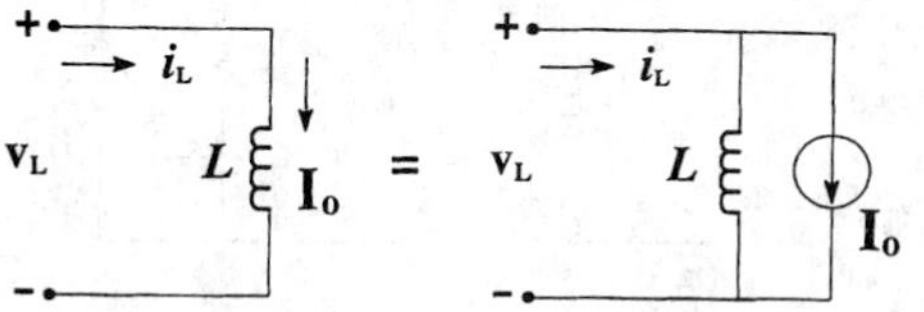

FIGURE 20.3 Inductor representation showing reference directions for currents and reference polarity for voltage.

through the inductor has an Ohm's law-like form if we define the *impedance* of the inductor as $Z(s) = V_L(s)/I_L(s) = sL$.

The impedance concept is an important one in network analysis. It allows us to combine dissimilar elements in the frequency domain, something we cannot do in the time domain. In fact, impedance is a frequency domain concept. It is the ratio of the *transform* of the voltage across the port of the network to the *transform* of the current through the port with all independent sources within the network properly removed and with all initial voltages across capacitors and initial currents through inductors set to zero. Thus, when we indicate that independent sources are to be removed, we mean that initial conditions are to be set to zero as well.

The concept of impedance can be extended to linear, lumped, finite, time-invariant, one-port networks in general. We denote these networks as LLFT networks. These networks are linear. That is, they are composed of elements including resistors, capacitors, inductors, transformers, and dependent sources whose parameters are not functions of the voltage across the element or the current through the element. Thus, the differential equations describing these networks are linear.

These networks are lumped and not distributed. That is, LLFT networks do not contain transmission lines as network elements, and the differential equations describing these networks are ordinary and not partial differential equations.

LLFT networks are finite, meaning that they do not contain infinite networks and require only a finite number of network elements in their representation. Infinite networks are sometimes useful in modeling such things as ground connections in the surface of the earth, but we exclude the discussion of them here.

LLFT networks are time-invariant or constant rather than time-varying. Thus, the ordinary, linear differential equations describing LLFT networks have constant coefficients.

The steps for finding the impedance of an LLFT one-port network are:

1. Properly remove all independent sources in the network. By "properly" removing independent sources we mean that voltage sources are replaced by short circuits and current sources are replaced by open circuits. Dependent sources are not removed.
2. Excite the network with a voltage source or a current source at the port and find an equation or equations to solve for the other port variable.
3. Form $Z(s) = V(s)/I(s)$.

Simple networks do not need to be excited in order to determine their impedance, but in the general case an excitation is required. The next example illustrates these concepts.

Example 1. Find the impedances of the one-port networks shown in Fig. 20.4.

Solution. The network shown in Fig. 20.4(a) is composed of three elements connected in series. There are no independent sources, and there is zero initial voltage across the capacitor and zero

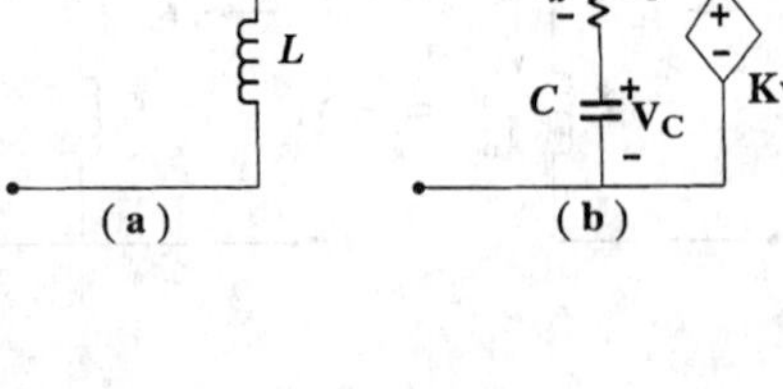

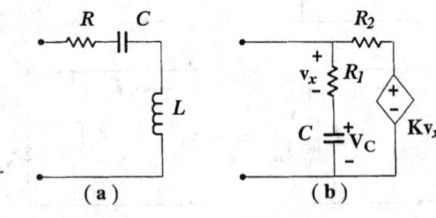

FIGURE 20.4 (a) A simple network. (b) A network containing a dependent source.

initial current through the inductor. The impedance is determined as:

$$Z(s) = R + sL + \frac{1}{sC} = L\left(\frac{s^2 + s\frac{R}{L} + \frac{1}{LC}}{s}\right)$$

The network in Fig. 20.4(b) includes a dependent source that depends on the voltage across R_1. The impedance of this network is not obvious, and so we should excite the port. Also, the capacitor has an initial voltage V_C across it. This voltage is set to zero to find the impedance. Figure 20.5 shows the network in Fig. 20.4(b) prepared for finding the impedance. Using the impedance concept and two-loop equations or two-node equations, we obtain:

$$Z(s) = \frac{V(s)}{I(s)} = \frac{R_1R_2\left(s + \frac{1}{CR_1}\right)}{[R_1(1-K)+R_2]\left(s + \frac{1}{C[R_1(1-K)+R_2]}\right)}$$

The expressions for impedance found in the previous example are rational functions of s, and the coefficients of s are functions of the elements of the network including the coefficient K of the dependent source in the network shown in Fig. 20.4(b). We will show that these observations are general for LLFT networks. But first we will extend the impedance concept in another direction.

We have defined the impedance of a one-port LLFT network. We can also define another network function—the admittance $Y(s)$. The admittance of a one-port LLFT network is the quotient of the *transform* of the current through the port to the *transform* of the voltage across the port with all independent sources within the network properly removed. One-port networks have only two linear network functions, impedance and admittance. Furthermore, $Z(s) = 1/Y(s)$ since both network functions concern the same port of the network, and the impedance or admittance relating the response to the excitation is the same whether a current excitation causes a voltage response or a voltage excitation causes a current response. An additional implication of this observation is that either network function can be determined with either type of excitation, voltage source or current source, applied to the network.

Figure 20.6 depicts a two-port network with the reference polarities and reference directions shown for the port variables. Port one of the two-port network is formed from the two terminals labeled 1 and 1′. The two terminals labeled 2 and 2′ are associated to form port two. A two-port

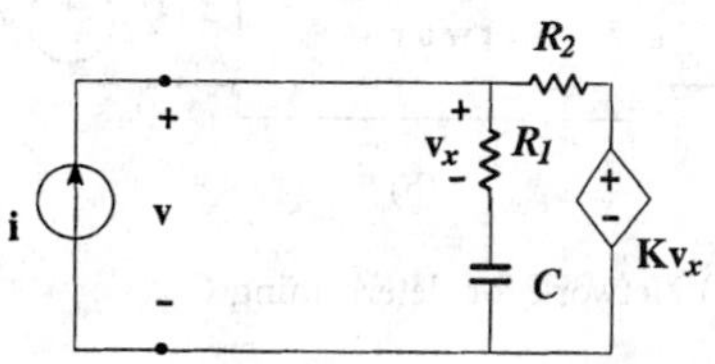

FIGURE 20.5 The network in Fig. 20.4(b) prepared for analysis.

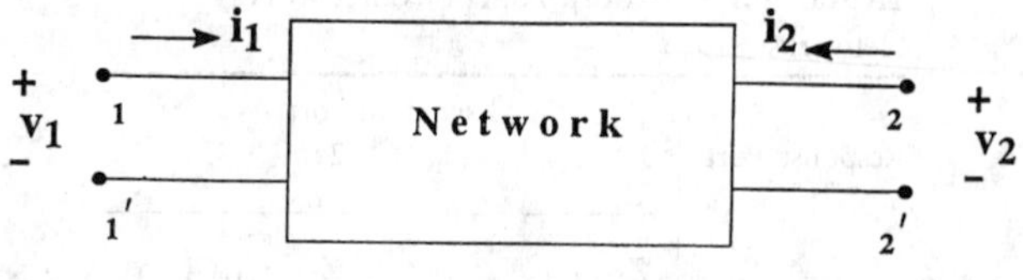

FIGURE 20.6 Reference polarities and reference directions for port variables of a two-port network.

network has 12 network functions associated with it rather than only two, and so we will employ the following notation for these functions:

$$N_{RE}(s) = R(s)/E(s)$$

where $N_{RE}(s)$ is a network function, the subscript "R" is the port at which the response variable exists, the subscript "E" is the port at which the excitation is applied, $R(s)$ is the response variable, and $E(s)$ is the excitation which may be a current source or a voltage source depending on the particular network function. For example, for the two-port networks shown in Fig. 20.7

$$Z_{21}(s) = \frac{V_2(s)}{I_1(s)} \quad \text{and} \quad G_{12}(s) = \frac{V_1(s)}{V_2(s)} \tag{20.5}$$

Note that a load impedance has been placed across port two, the response port for $Z_{21}(s)$, in Fig. 20.7(a). Also, a load has been connected across port one, the response port for $G_{12}(s) = V_1(s)/V_2(s)$, in Fig. 20.7(b). It is assumed that all independent sources have been properly removed in both networks in Fig. 20.7, and this assumption also applies to the loads. Of course, if a load impedance is changed, usually the network function will change. Thus, the load, if any, must be specified.

Table 20.1 lists the network functions of a two-port network. "G" denotes a voltage ratio, and "α" denotes a current ratio. The functions can also be grouped into driving-point and transfer functions. Driving-point functions are ones in which the excitation and response occur at the same port, and transfer network functions are ones in which the excitation and response occur at different ports. For example, $Z_{11}(s) = V_1(s)/I_1(s)$ is a driving-point network function, and $G_{21}(s) = V_2(s)/V_1(s)$ is a transfer network function. Of the twelve network functions for a two-port network, four are driving-point functions and eight are transfer functions. The two network functions for a one-port network are, of necessity, driving-point network functions.

Network functions are related to loop and node equations. Consider the LLFT network shown in Fig. 20.8. Independent sources within the network have been properly removed. Assume the network has n independent nodes plus the ground node, and assume for simplicity that the network has no mutual inductance or dependent sources. Let us determine $Z_{11} = V_1/I_1$ in terms of a quotient of determinants of the nodal admittance matrix. The node equations, all written

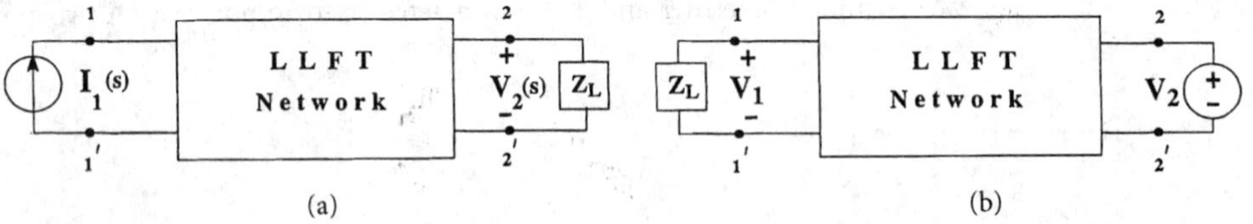

FIGURE 20.7 (a) Network configured for finding $Z_{21}(s)$. (b) Network for determining $G_{12}(s)$.

TABLE 20.1 Network Functions of Two Port Networks

	Excitation Port	
Response Port	1	2
1	Z_{11}, Y_{11}	$Z_{12}, Y_{12}, G_{12}, \alpha_{12}$
2	$Z_{21}, Y_{21}, G_{21}, \alpha_{21}$	Z_{22}, Y_{22}

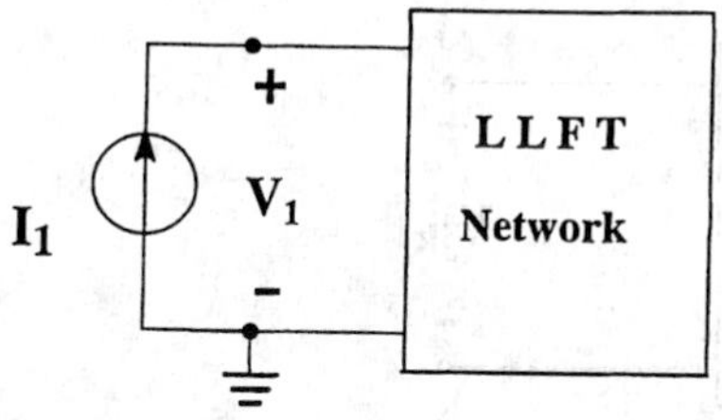

FIGURE 20.8 An LLFT network with n independent nodes plus the ground node.

with currents leaving a node as positive currents, are

$$\begin{bmatrix} y_{11} & y_{12} & \cdots & y_{1n} \\ y_{21} & y_{22} & \cdots & y_{2n} \\ \vdots & \vdots & \ddots & \vdots \\ y_{n1} & y_{n2} & \cdots & y_{nn} \end{bmatrix} \begin{bmatrix} V_1 \\ V_2 \\ \vdots \\ V_n \end{bmatrix} = \begin{bmatrix} I_1 \\ 0 \\ \vdots \\ 0 \end{bmatrix} \tag{20.6}$$

where $V_i, i = 1, 2, \ldots, n$, are the unknown node voltages. The elements $y_{ij}, i, j = 1, 2, \ldots, n$, of the nodal admittance matrix above have the form

$$y_{ij} = \pm\left\{g_{ij} + \frac{\Gamma_{ij}}{s} + sC_{ij}\right\} \tag{20.7}$$

where the plus sign is taken if $i = j$ and the minus sign is taken if i and j are unequal. The quantity g_{ij} is the sum of the conductances connected to node i if $i = j$, and if i does not equal j, it is the sum of the conductances connected between nodes i and j. A similar statement applies to C_{ij}. The quantity Γ_{ij} is the sum of the reciprocal inductances ($\Gamma = 1/L$) connected to node i if $i = j$, and it is the sum of the reciprocal inductances connected between nodes i and j if i does not equal j.

Solving for V_1 using Cramer's rule yields:

$$V_1 = \frac{\begin{vmatrix} I_1 & y_{12} & \cdots & y_{1n} \\ 0 & y_{22} & \cdots & y_{2n} \\ \vdots & \vdots & \ddots & \vdots \\ 0 & y_{n2} & \cdots & y_{nn} \end{vmatrix}}{\Delta'} \tag{20.8}$$

where Δ' is the determinant of the nodal admittance matrix. Thus,

$$V_1 = I_1 \frac{\Delta'_{11}}{\Delta'} \tag{20.9}$$

where Δ'_{11} is the cofactor of element y_{11} of the nodal admittance matrix. Thus, we can write

$$\frac{V_1}{I_1} = Z_{11} = \frac{\Delta'_{11}}{\Delta'} \tag{20.10}$$

If mutual inductance and dependent sources exist in the LLFT network, the nodal admittance matrix elements are modified. Furthermore, there may be more than one entry in the column

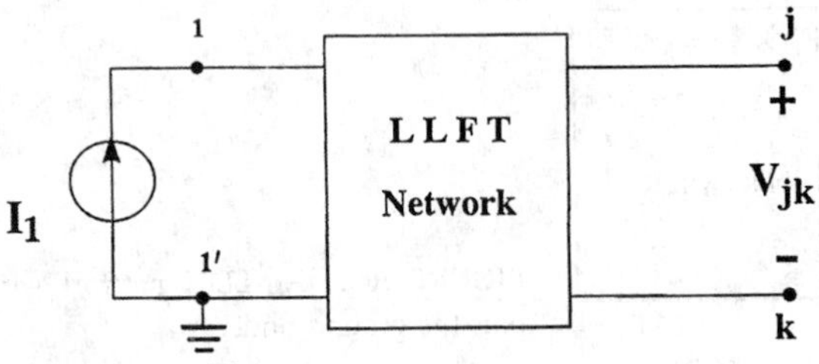

FIGURE 20.9 An LLFT network with two ports indicated.

matrix containing excitations. However, Z_{11} can still be expressed as a quotient of determinants of the nodal admittance matrix.

Next, consider the network, shown in Fig. 20.9, which is assumed to have n independent nodes plus a ground node. The response port exists between terminals j and k. In this network, we are making the pair of terminals j and k serve as the second port. Denote the **transimpedance** V_{jk}/I_1 as Z_{j1}. Let us express this transfer function Z_{j1} as a quotient of determinants. All independent sources within the network have been properly removed. Note that the node voltages are measured with respect to the ground terminal indicated, but the output voltage is the difference of the node voltages V_j and V_k. Thus, we have to solve for these two node voltages. Writing node equations, again taking currents leaving a node as positive currents, and solving for V_j using Cramer's rule we have

$$V_j = \frac{\begin{vmatrix} y_{11} & y_{12} & \cdots & y_{1(j-1)} & I_1 & y_{1k} & \cdots & y_{1n} \\ y_{21} & y_{22} & \cdots & y_{2(j-1)} & 0 & y_{2k} & \cdots & y_{2n} \\ \vdots & \vdots & \ddots & \vdots & \vdots & \vdots & \ddots & \vdots \\ y_{n1} & y_{n2} & \cdots & y_{n(j-1)} & 0 & y_{nk} & \cdots & y_{nn} \end{vmatrix}}{\Delta'} \tag{20.11}$$

which can be written as

$$V_j = I_1 \frac{\Delta'_{1j}}{\Delta'} \tag{20.12}$$

where Δ'_{1j} is the cofactor of element y_{1j} of the nodal admittance matrix. Similarly, we can solve for V_k and obtain

$$V_k = I_1 \frac{\Delta'_{1k}}{\Delta'} \tag{20.13}$$

Then the transimpedance Z_{j1} can be expressed as

$$Z_{j1} = \frac{V_j - V_k}{I_1} = \frac{\Delta'_{1j} - \Delta'_{1k}}{\Delta'} \tag{20.14}$$

If terminal k in Fig. 20.9 is common with the ground node so that the network is a grounded two-port network, then V_k is zero, and Z_{j1} can be expressed as Δ'_{1j}/Δ'. This result can be extended so that if the output voltage is taken between any node h, and ground, then the transimpedance can be expressed as $Z_{h1} = \Delta'_{1h}/\Delta'$.

These results can be used to obtain an expression for G_{21} in terms of the determinants of the nodal admittance matrix. Figure 20.10 shows an LLFT network with a voltage excitation

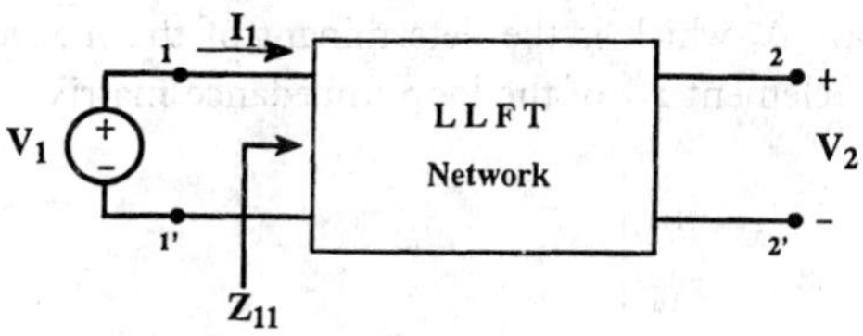

FIGURE 20.10 An LLFT network with port 2 open.

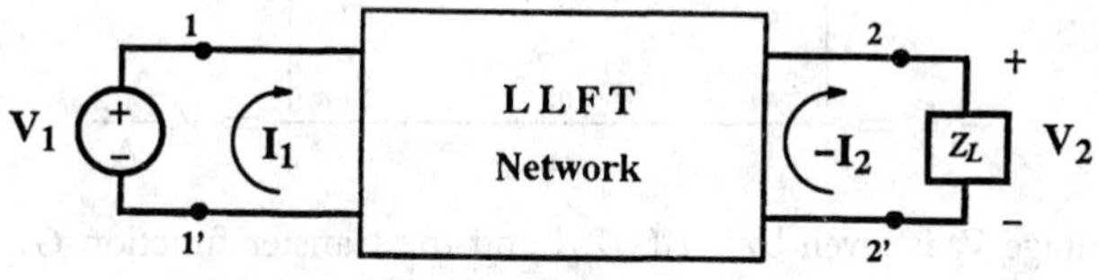

FIGURE 20.11 An LLFT network with load Z_L connected across port 2.

applied at port 1 and with port 2 open. The current $I_1(s)$ is given by V_1/Z_{11}. Then V_2 is given by $V_2(s) = I_1 Z_{21}$. Thus,

$$G_{21} = \frac{V_2}{V_1} = \frac{I_1 Z_{21}}{I_1 Z_{11}} = \frac{\Delta'_{12}}{\Delta'_{11}} \tag{20.15}$$

Note that the determinants in the quotient are of equal order so that G_{21} is dimensionless.

Of course, network functions can also be expressed in terms of determinants of the loop impedance matrix. Consider the two-port network in Fig. 20.11 which is excited with a voltage source applied to port 1 and which has a load Z_L connected across port 2. Let us find the voltage transfer function G_{21} using loop equations. Assume that there are n independent loops of which two are shown explicitly in Fig. 20.11, and assume that there are no independent sources within the network. Also, assume for simplicity that the network contains no dependent sources or mutual inductance and that the loops are chosen so that V_1 is in only one loop. The loop equations are:

$$\begin{bmatrix} z_{11} & z_{12} & \cdots & z_{1n} \\ z_{21} & z_{22} & \cdots & z_{2n} \\ \vdots & \vdots & \ddots & \vdots \\ z_{n1} & z_{n2} & \cdots & z_{nn} \end{bmatrix} \begin{bmatrix} I_1 \\ -I_2 \\ \vdots \\ I_n \end{bmatrix} = \begin{bmatrix} V_1 \\ 0 \\ \vdots \\ 0 \end{bmatrix} \tag{20.16}$$

where $I_j, j = 1, 3, \ldots, n$, and $-I_2$ are the loop currents, and the elements z_{ij} of the loop impedance matrix are given by:

$$z_{ij} = \pm\left(R_{ij} + sL_{ij} + \frac{D_{ij}}{s}\right), \quad i, j = 1, 2, \ldots, n \tag{20.17}$$

where we have assumed that all loop currents are taken in the same direction such as clockwise. The plus sign applies if $i = j$, and the minus sign is used if $i \neq j$. R_{ij} is the sum of the resistances in loop i if $i = j$, and R_{ij} is the sum of the resistances common to loops i and j if $i \neq j$. L_{ij} is the sum of the inductances in loop i if $i = j$, and it is the sum of the inductances common to loops i and j if $i \neq j$. A similar statement applies to the reciprocal capacitances $D_{ij} (D = 1/C)$. However, the element z_{22} includes the extra term Z_L which could be a quotient of determinants itself. Solving for $-I_2$ using Cramer's rule, we have Δ, which is the determinant of the $n \times n$ loop impedance matrix, and Δ_{12} is the cofactor of element z_{12} of the loop impedance matrix.

$$-I_2 = \frac{\begin{vmatrix} z_{11} & V_1 & z_{13} & \cdots & z_{1n} \\ z_{21} & 0 & z_{23} & \cdots & z_{2n} \\ \vdots & \vdots & \vdots & \ddots & \vdots \\ z_{n1} & 0 & z_{n3} & \cdots & z_{nn} \end{vmatrix}}{\Delta} = V_1 \frac{\Delta_{12}}{\Delta} \tag{20.18}$$

The transform voltage V_2 is given by $-I_2(s)Z_L$, and the transfer function G_{21} can be expressed as

$$G_{21} = \frac{V_2}{V_1} = Z_L \frac{\Delta_{12}}{\Delta} \tag{20.19}$$

Thus, G_{21} can be represented as a quotient of determinants of the loop impedance matrix multiplied by Z_L.

In a similar manner, we can write

$$Y_{jk} = \frac{\Delta_{kj}}{\Delta}, \quad j,k = 1,2 \tag{20.20}$$

Then we can use this result to write:

$$V_1 = I_1 \frac{1}{Y_{11}} \quad \text{and} \quad I_2 = V_1 Y_{21} \tag{20.21}$$

Thus,

$$\frac{I_2}{I_1} = \alpha_{21} = \frac{\Delta_{12}}{\Delta_{11}} \tag{20.22}$$

Table 20.2 summarizes some of these results.

TABLE 20.2 Network Functions in Terms of Quotients of Determinants

$Y_{jk} = \Delta_{kj}/\Delta$	$G_{jk} = \Delta'_{kj}/\Delta'_{kk}$
$Z_{jk} = \Delta'_{kj}/\Delta'$	$\alpha_{jk} = \Delta_{kj}/\Delta_{kk}$

Properties of LLFT Network Functions

We now list properties of network functions of LLFT networks. These properties are useful as a check of an analysis of such networks.

1. Network functions of LLFT networks are real, rational functions of s, and therefore have the form

$$N(s) = \frac{P(s)}{Q(s)} = \frac{a_0 s^m + a_1 s^{m-1} + \cdots + a_{m-1}s + a_m}{b_0 s^n + b_1 s^{n-1} + \cdots + b_{n-1}s + b_n}, \tag{20.23}$$

where the coefficients in both the numerator and denominator polynomials are real.

A network function of an LLFT network is a rational function because network functions can be expressed as quotients of determinants of nodal admittance matrices or of loop impedance matrices. The elements of these determinants are at most simple rational functions, and when the determinants are expanded and the fractions cleared, the result is always a rational function. The coefficients $a_i, i = 0, 1, \ldots, m$, and $b_j, j = 0, 1, \ldots, n$, are functions of the real elements of the network R, L, C, M, and coefficients of dependent sources. The constants R, L, C, and M are real. In most networks, the coefficients of dependent sources are real constants. Thus, the coefficients of LLFT network functions are real, and therefore the network function is a real function. It is possible for the "coefficients" of dependent sources in LLFT networks to themselves be real, rational functions of s. But when all the fractions are cleared, the result is a real rational function.

2. An LLFT network function is completely defined by its self poles, self zeros, and the scale factor H.

If the numerator and denominator polynomials are factored and there are no common factors, we have

$$N(s) = \frac{P(s)}{Q(s)} = \frac{a_0}{b_0}\frac{(s - z_1)(s - z_2)\cdots(s - z_m)}{(s - p_1)(s - p_2)\cdots(s - p_n)} \tag{20.24}$$

where $a_0/b_0 = \mathrm{H}$ is the scale factor. The values of $s = z_1, z_2, \ldots, z_m$ are zeros of the polynomial $P(s)$ and self zeros of the network function. Also, $s = p_1, p_2, \ldots, p_n$ are zeros of the polynomial $Q(s)$ and self poles of the network function. In addition, $N(s)$ may have other poles or zeros at infinity. We call these poles and zeros **mutual** poles and **mutual** zeros since they result from the difference in degrees of the numerator and denominator polynomials.

3. Counting both self and mutual poles and zeros, and counting a kth order pole or zero k times, N(s) has the same number of poles as zeros, and this number is equal to the highest power of s in $N(s)$.

If $m = n$, then there are n self zeros and n self poles and no mutual poles or zeros. If $n > m$, then there are m self zeros and n self poles. There are also $n - m$ mutual zeros. Thus, there are n poles and $m + (n - m) = n$ zeros. A similar statement can be constructed for $n < m$.

4. Complex roots of $P(s)$ and $Q(s)$ occur in conjugate pairs.

This property follows from the fact that the coefficients of the numerator and denominator polynomials are real. Thus, complex factors of these polynomials have the form

$$(s + c + jd)(s + c - jd) = [(s + c)^2 + d^2] \tag{20.25}$$

where c and d are real constants.

5. A driving point function of a network having no dependent sources can have neither poles nor zeros in the right-half s-plane (RHP), and poles and zeros on the imaginary axis must be simple. The same restrictions apply to the poles of transfer network functions of such networks but not to the zeros of transfer network functions.

Elsewhere in this handbook it is shown that the denominator polynomials of LLFT networks must not have RHP roots, and roots on the imaginary axis, if any, must be simple. However, the reciprocal of a driving-point network function is also a network function. For example, $1/Y_{22} = Z_{22}$. Thus, restrictions on the locations of poles of driving-point network functions also apply to zeros of driving-point network functions.

However, the reciprocal of a transfer network function is not a network function, see [5]. For example, $1/Y_{21} \neq Z_{21}$. Thus, restrictions on the poles of a transfer function do not apply to its zeros.

We can make a classification of the factors corresponding to the allowed types of poles as follows:

TABLE 20.3 A Classification of Factors of Network Functions of *LLFT* Networks Containing No Dependent Sources

Type	Factors(s)	Conditions
A	$(s + a)$	$a \geq 0$
B	$(s + b + jc)(s + b - jc)$	$b > 0, c > 0$
C	$(s + jd)(s - jd)$	$d > 0$

The Type A factor corresponds to a pole on the $-\sigma$ axis. If $a = 0$, then the factor corresponds to a pole on the imaginary axis, and so only one such factor is allowed. Type B factors correspond to poles in the left-half s-plane (LHP), and Type C factors correspond to poles on the imaginary axis.

6. The coefficients of the numerator and denominator polynomials of a driving-point network function of an LLFT network with no dependent sources are positive. The coefficients of the denominator polynomial of a transfer network function are all one sign. Without loss of generality, we take the sign to be positive. But some or all of the coefficients of the numerator polynomial of a transfer network function may be negative.

A polynomial made up of the factors listed in Table 20.3 would have the form:

$$Q(s) = (s + a_1) \cdots [(s + b_1)^2 + c_1^2] \cdots (s^2 + d_1^2) \cdots$$

Note that all the constants are positive in the expression for $Q(s)$, and so it is impossible for any of the coefficients of $Q(s)$ to be negative.

7. There are no missing powers of s in the numerator and denominator polynomials of a driving-point network function of an LLFT network containing no dependent sources unless all even or all odd powers of s are missing or the constant term is missing. This statement holds for the denominator polynomials of transfer functions of such networks, but there may be missing powers of s in the numerator polynomials of transfer functions.

Property 7 is easily illustrated by combining types of factors from Table 20.3. Thus, a polynomial consisting only of type A factors contains all powers of s between the highest power and the constant term unless one of the "a" constants is zero. Then the constant term is missing. Two a constants cannot be zero because then there would be two roots on the imaginary axis at the same location. The roots on the imaginary axis would not be simple.

A polynomial made up of only type B factors contains all powers of s, and a polynomial containing only type C factors contains only even powers of s. A polynomial constructed from type C factors except for one A type factor with $a = 0$ contains only odd powers of s. If a polynomial is constructed from type B and C factors, then it contains all powers of s.

8. The orders of the numerator and denominator polynomials of a driving-point network function of an LLFT network which contains no dependent sources can differ by no more than one.

The limiting behavior at high frequency must be that of an inductor, a resistor, or a capacitor. That is, if $N_{dp}(s)$ is a driving-point network function, then

$$\lim_{s \to \infty} N_{dp}(s) = \begin{cases} K_1 s \\ K_2 \\ K_3/s \end{cases}$$

where K_i, $i = 1, 2, 3$, are real constants.

9. The terms of lowest order in the numerator and denominator polynomials of a driving-point network function of an LLFT network containing no dependent sources can differ in order by no more than one.

The limiting behavior at low frequency must be that of an inductor, a resistor, or a capacitor. That is,

$$\lim_{s\to 0} N_{dp}(s) = \begin{cases} K_4 s \\ K_5 \\ K_6/s \end{cases}$$

where the constants K_i, $i = 4, 5, 6$, are real constants.

10. The maximum order of the numerator polynomials of the dimensionless transfer functions G_{12}, G_{21}, α_{12}, and α_{21}, of an LLFT network containing no dependent sources is equal to the order of the denominator polynomials. The maximum order of the numerator polynomial of the transfer functions Y_{12}, Y_{21}, Z_{12}, and Z_{21} is equal to the order of the denominator polynomial plus 1. However, the minimum order of the numerator polynomial of any transfer function may be zero.

If dependent sources are included in an LLFT network, then it is *possible* for the network to have poles in the RHP or multiple poles at locations on the imaginary axis. However, an important application of stable networks containing dependent sources is to mimic (simulate) the behavior of LLFT networks that contain no dependent sources. For example, networks that contain resistors, capacitors, and dependent sources can mimic the behavior of networks containing only resistors, capacitors, and inductors. Thus, low frequency filters can be constructed without the need for heavy, expensive inductors that would ordinarily be required in such applications.

0.2 Network Theorems

In this section we provide techniques, strategies, equivalences, and theorems for simplifying the analysis of LLFT networks or for checking the results of an analysis. They can save much work in the analysis of some networks if one remembers to apply them. Thus, it is convenient to have them listed in one place. To begin, we list nine equivalences which are often called source transformations.

Source Transformations

Table 20.4 is a collection of memory aids for the nine source transformations. Source transformations are simple ways the elements and sources externally connected to a network N can be combined or eliminated without changing the voltages and currents within network N thereby simplifying the problem of finding a voltage or current within N.

Source transformation one in Table 20.4 shows the equivalence between two voltage sources connected in series and a single voltage source having a value that is the sum of the voltages of the two sources. A double-headed arrow is shown between the two network representations because it is sometimes advantageous to use this source transformation in reverse. For example, if a voltage source that has both DC and AC components is applied to a linear network N, it may be useful to represent that voltage source as two voltage sources in series—one a DC source and the other an AC source.

Source transformation two shows two voltage sources connected in parallel. Unless V_1 and V_2 are equal, the network would not obey Kirchhoff's law as evidenced by a loop equation written in the loop formed by the two voltage sources. A network that does not obey Kirchhoff's

TABLE 20.4 Source Transformations

1.

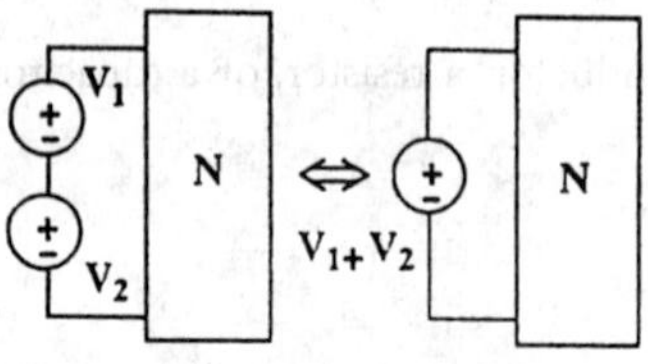

2.

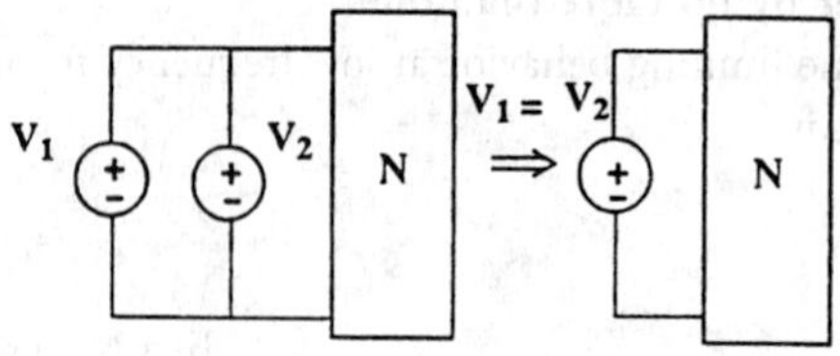

3.

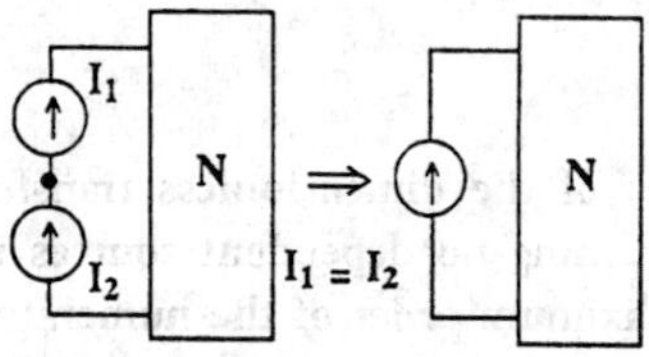

4.

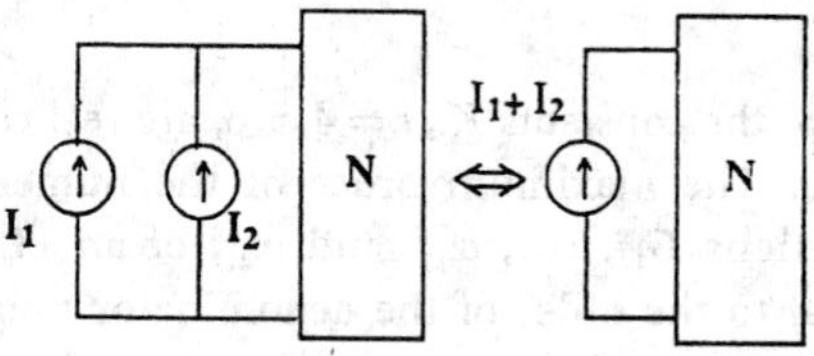

5.

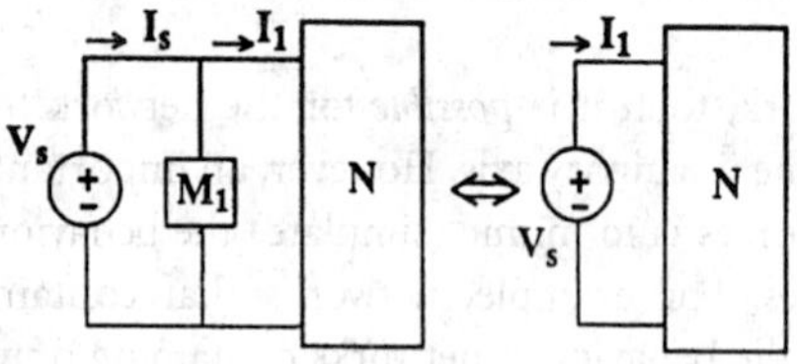

6.

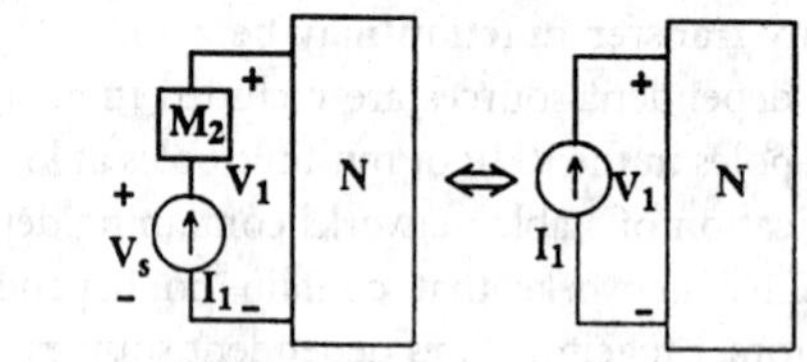

7.

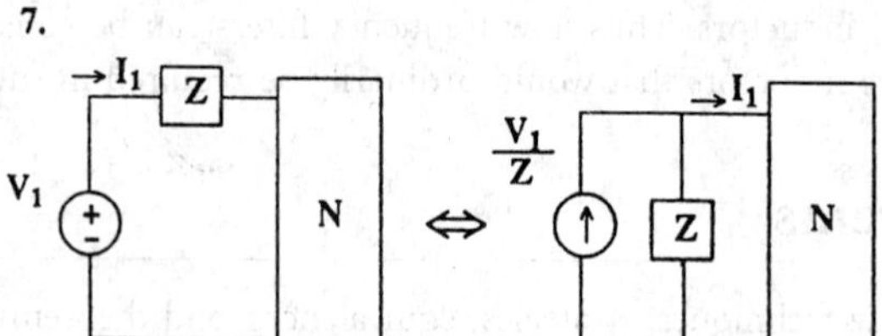

8.

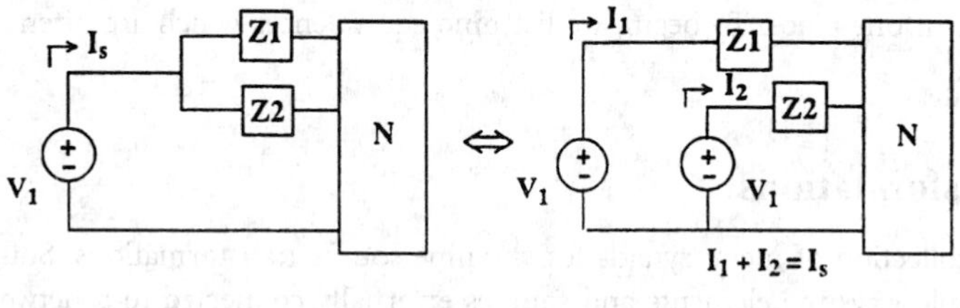

9.

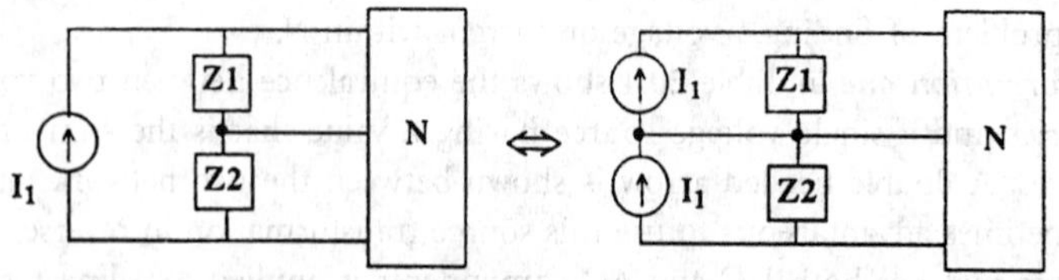

N is an arbitrary network in which analysis for a voltage or current is to be performed. M_1 is an arbitrary one-port network or network element except a voltage source. M_2 is an arbitrary one-port network or network element except a current source. It is assumed there is no magnetic coupling between N and M_1 or M_2. There are no dependent sources in N in Source Transformation 5 that depend on I_s. Furthermore, there are no dependent sources in N in Source Transformation 6 that depend on V_s. However, M_1 and M_2 can have dependent sources that depend on voltages or currents in N. Z, Z_1, and Z_2 are one-port impedances.

laws is termed a contradiction. Thus, a single-headed arrow is shown between the two network representations.

Source transformations three and four are duals, respectively, of source transformations two and one. The current sources must be equal in transformation three or else Kirchhoff's law would not be valid at the node indicated, and the circuit would be a contradiction.

Source transformation five shows that the circuit M_1 can be removed without altering any of the voltages and currents inside N. Whether M_1 is connected as shown or is removed, the voltage applied to N remains V_s. However, the current supplied by the source V_s changes from I_s to I_1.

Source transformation six shows that circuit M_2 can be replaced by a short circuit without affecting voltages and currents in N. Whether M_2 is in series with the current source I_1 as shown or removed, the current applied to N is the same. However, if network M_2 is removed, then the voltage across the current source changes from V_s to V_1.

Source transformation seven is sometimes termed a Thévenin circuit to Norton circuit transformation. This transformation, as shown by the double-headed arrow, can be used in the reverse direction. Thévenin's theorem is discussed thoroughly later in this section.

Source transformation eight is sometimes described as "pushing a voltage source through a node", but we will term it as "splitting a voltage source". Loop equations remain the same with this transformation, and the current leaving network N through the lowest wire continues to be I_s.

Source transformation nine shows that if a current source is not in parallel with one element, then it can be "split" as shown. Now, each one of the current sources I_1 has an impedance in parallel. Thus, analysis of network N may be simplified because source transformation seven can be applied.

Source transformations cannot be applied to all networks, but when they can be employed they usually yield useful simplifications of the network.

Example 2. Use source transformations to find V_o for the network shown. Initial current through the inductor in the network is zero.

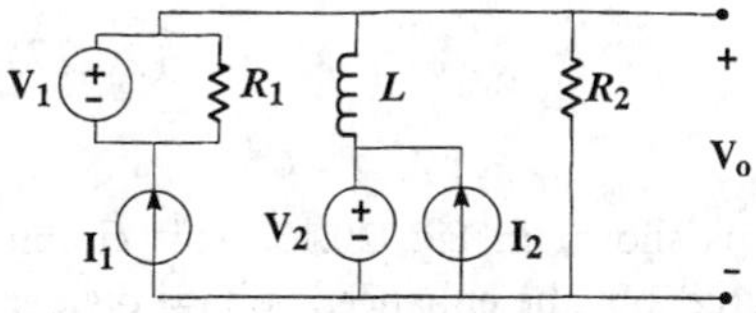

FIGURE 20.12 Network for example 2.

Solution. The network can be readily simplified by employing source transformation five from Table 20.4 to eliminate R_1 and also I_2. Then source transformation six can be used to eliminate V_1 because it is an element in series with a current source. The results to this point are shown in Fig. 20.13(a). If we then apply transformation seven, we obtain the network shown in Fig. 20.13(b). Next, we can apply transformation four to obtain the single loop network shown in Fig. 20.13(c). The output voltage can be written in the frequency domain as

$$V_o = \left(I_1 + \frac{V_2}{sL}\right)\left(\frac{sLR_2}{sL + R_2}\right)$$

Source transformations can often be used advantageously with the following theorems.

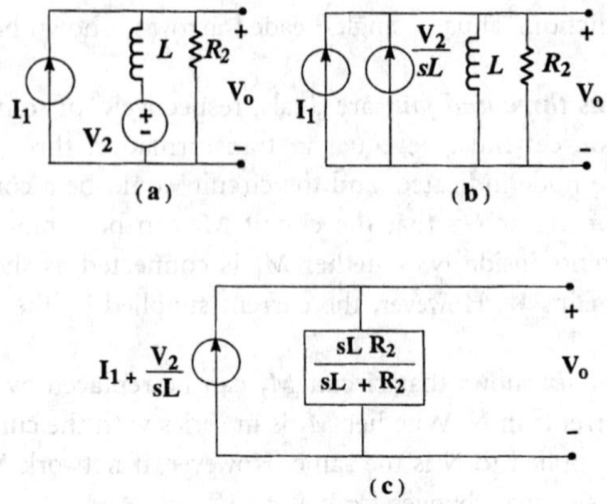

FIGURE 20.13 (a, b, c) Applications of source transformations to the network in Fig. 20.12.

Dividers

Current dividers and voltage dividers are circuits that are employed frequently, especially in the design of electronic circuits. Thus, there is a need to analyze dividers quickly. The relationships derived below satisfy this need.

Figure 20.14 shows a current divider circuit. The source current I_s divides between the two impedances, and we wish to determine the current through Z_2. Writing a loop equation for the loop indicated, we have

$$I_2 Z_2 - (I_s - I_2) Z_1 = 0 \tag{20.26}$$

from which we obtain

$$I_2 = I_s \frac{Z_1}{Z_1 + Z_2} \tag{20.27}$$

A circuit which we term an **enhanced** voltage divider is shown in Fig. 20.15. This circuit contains two voltage sources instead of the usual single source, but the enhanced voltage divider occurs more often in electronic circuits. Writing a node equation at node A and solving for V_o, we obtain

$$V_o = \frac{V_1 Z_2 + V_2 Z_1}{Z_1 + Z_2} \tag{20.28}$$

If V_2, for example, is zero, then the results from the enhanced voltage divider reduce to those of the single source voltage divider.

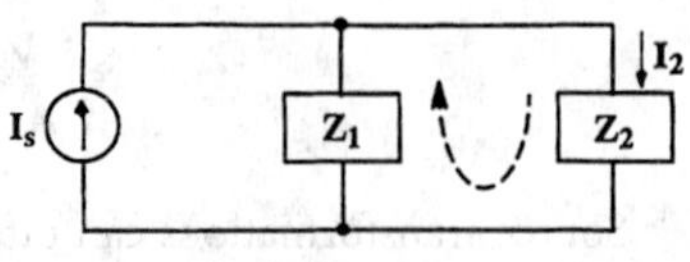

FIGURE 20.14 A current divider.

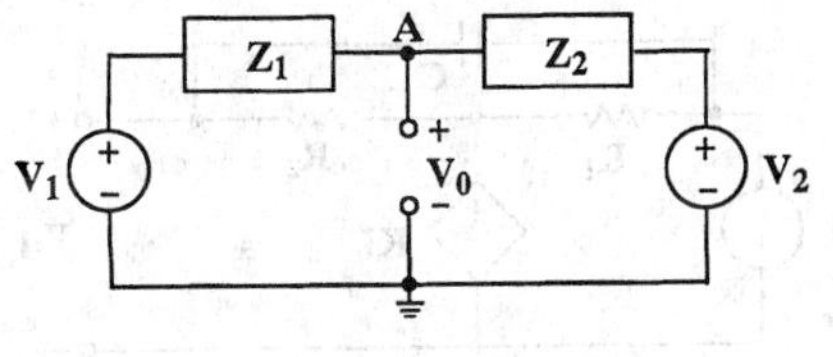

FIGURE 20.15 Enhanced voltage divider.

Example 3. Use (20.28) to find V_o for the network shown in Fig. 20.16.

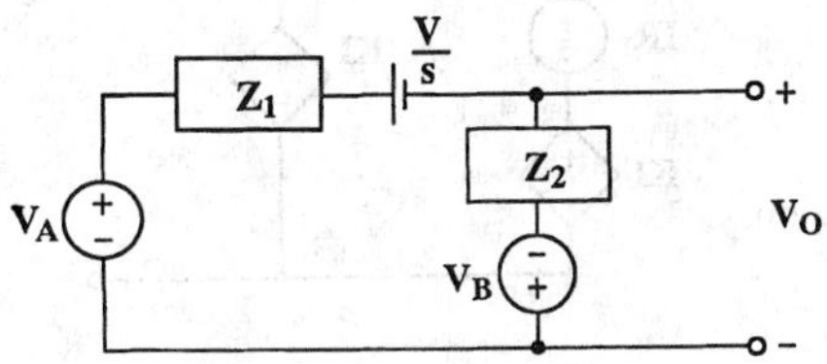

FIGURE 20.16 Circuit for Example 3.

Solution. The network in Fig. 20.16 matches with the network used to derive (20.28) even though it is drawn somewhat differently and has three voltage sources rather than two. However, we can use (20.28) to write the answer for V_o by inspection.

$$V_o = \frac{(V_A - (V/s))Z_2 - V_B Z_1}{Z_1 + Z_2}$$

The following example illustrates the use of source transformations together with the voltage divider.

Example 4. Find V_o for the network shown in Fig. 20.17. The units of K, the coefficient of the dependent source, are ohms, and the capacitor is initially uncharged.

Solution. We note that the dependent voltage source is not in series with any one particular element and that the independent current source is not in parallel with any one particular element. However, we can split both the voltage source and the current source using source transformations eight and nine, respectively, from Table 20.4. Then, employing transformations five and seven, we obtain the network configuration shown in Fig. 20.18 for which we can use the voltage divider to write:

$$V_o = \frac{I(K + R_1)R_2 + KI\left(R_1 + (1/sC)\right)}{R_1 + R_2 + (1/sC)}$$

It should be mentioned that the method used to find V_o in this example is not the most efficient one. For example, loops can be chosen for the network in Fig. 20.17 so there is only one unknown loop current. However, source transformations and dividers become more powerful analysis tools as they are coupled with additional network theorems.

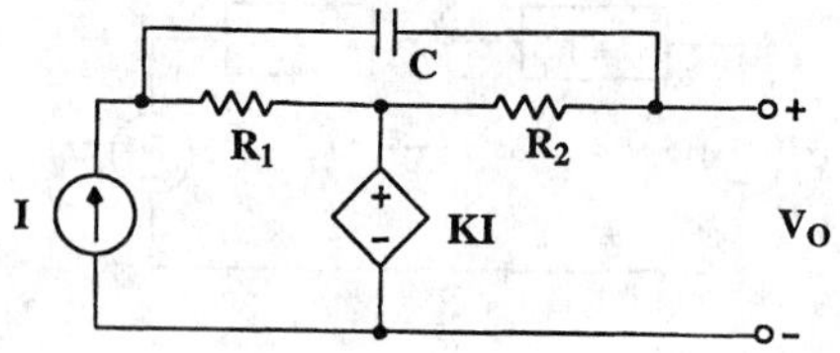

FIGURE 20.17 Network for Example 4.

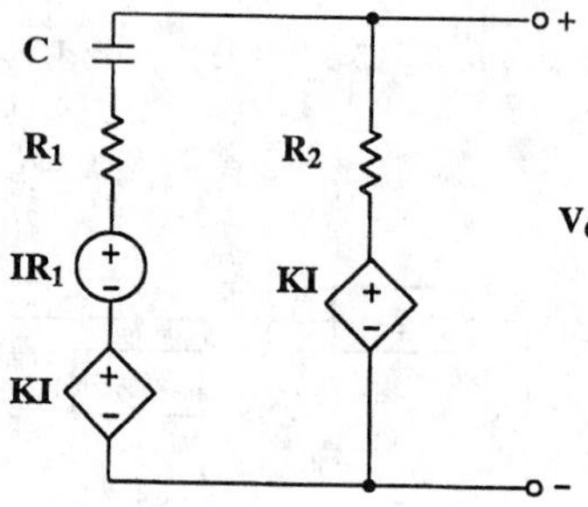

FIGURE 20.18 Results after employing source transformations on the network in Fig. 20.17.

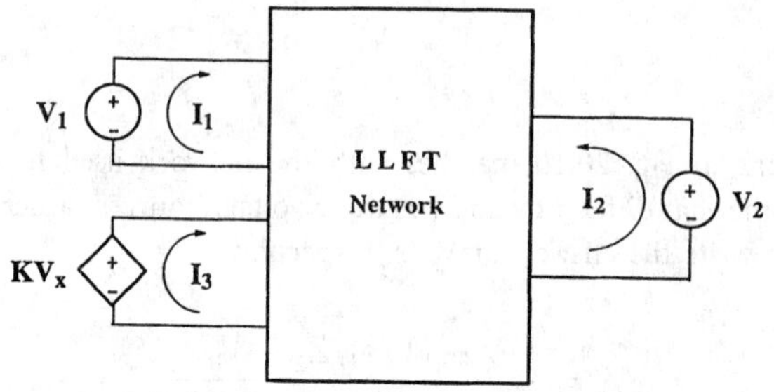

FIGURE 20.19 LLFT network with three voltage sources of which one is dependent.

Superposition

Superposition is a property of all linear networks, and whether it is used directly in the analysis of a network or not, it is a concept that is valuable in thinking about LLFT networks. Consider the LLFT network shown in Fig. 20.19 in which, say, we wish to solve for I_1. Assume the network has n independent loops, and, for simplicity, assume there are no sources within the box in the figure and that initial voltages across capacitors and initial currents through inductors are zero or are represented by independent sources external to the box. Note that one dependent source is shown in Fig. 20.19 that depends on a voltage V_x in the network and that there are two independent sources V_1 and V_2 applied to the network. If loops are chosen so that each source has only one loop current flowing through it as indicated in Fig. 20.19, then the loop equations can be written as

$$\begin{bmatrix} V_1 \\ V_2 \\ KV_x \\ 0 \\ \vdots \\ 0 \end{bmatrix} = \begin{bmatrix} z_{11} & z_{12} & \cdots & z_{1n} \\ z_{21} & z_{22} & \cdots & z_{2n} \\ \cdots & \cdots & \ddots & \cdots \\ z_{n1} & z_{n2} & \cdots & z_{nn} \end{bmatrix} \begin{bmatrix} I_1 \\ I_2 \\ \vdots \\ I_n \end{bmatrix} \tag{20.29}$$

where the elements of the loop impedance matrix are defined in the section describing network functions. Solving for I_1 using Cramer's rule, we have:

$$I_1 = V_1\frac{\Delta_{11}}{\Delta} + V_2\frac{\Delta_{21}}{\Delta} + KV_x\frac{\Delta_{31}}{\Delta} \tag{20.30}$$

where Δ is the determinant of the loop impedance matrix, and $\Delta_{j1}, j = 1, 2, 3$, are cofactors. The expression for I_1 given in (20.30) is an intermediate and not a finished solution. The finished solution would express I_1 in terms of the independent sources and the parameters (Rs, Ls, Cs, Ms, and Ks) of the network and not in terms of an unknown V_x. Thus, one normally has to eliminate V_x from the expression for I_1. But the intermediate expression for I_1 illustrates superposition. There are three components that add up to I_1 in (20.30)—one for each source including one for the dependent source. Furthermore, we see that each source is multiplied by a transadmittance (or a driving-point admittance in the case of V_1). Thus, we can write:

$$I_1 = V_1Y_{11} + V_2Y_{12} + KV_xY_{13} \tag{20.31}$$

where each admittance is found from the port at which a voltage source (whether independent or dependent) is applied. The response variable for each of these admittances is I_1 at port 1.

The simple derivation that led to (20.30) is easily extended to both types of independent excitations (voltage sources and current sources) and to all four types of dependent sources. The generalization of (20.30) leads to the conclusion:

> To apply superposition in the analysis of a network containing at least one independent source and a variety of other sources, dependent or independent, one finds the contribution to the response from each source in turn with all other sources, dependent or independent, properly removed and then adds the individual contributions to obtain the total response. No distinction is made between independent and dependent sources in the application of superposition other than requiring the network to have at least one independent source.

However, if dependent sources are present in the network, the quantities (call them V_x and I_x) on which they depend must often be eliminated from the answer by additional analysis if the answer is to be useful unless V_x or I_x are themselves the variables of independent sources or the quantities sought in the analysis.

Some examples will illustrate the procedure.

Example 5. Find V_o for the circuit shown using superposition. In this circuit only independent sources are present.

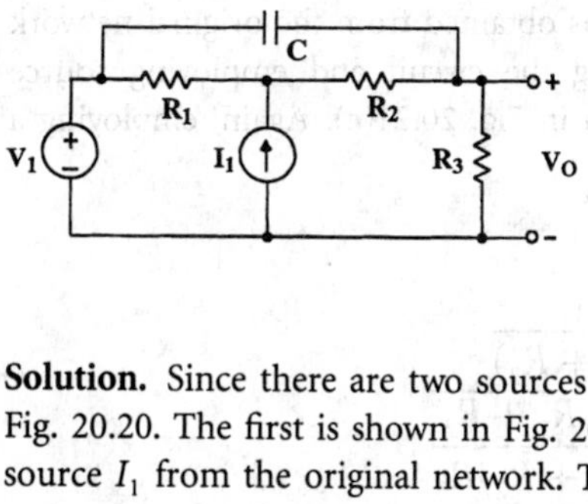

FIGURE 20.20 Network for Example 5.

Solution. Since there are two sources in the network, we abstract two fictitious networks from Fig. 20.20. The first is shown in Fig. 20.21(a) and is obtained by properly removing the current source I_1 from the original network. The impedance of the capacitor can then be combined in

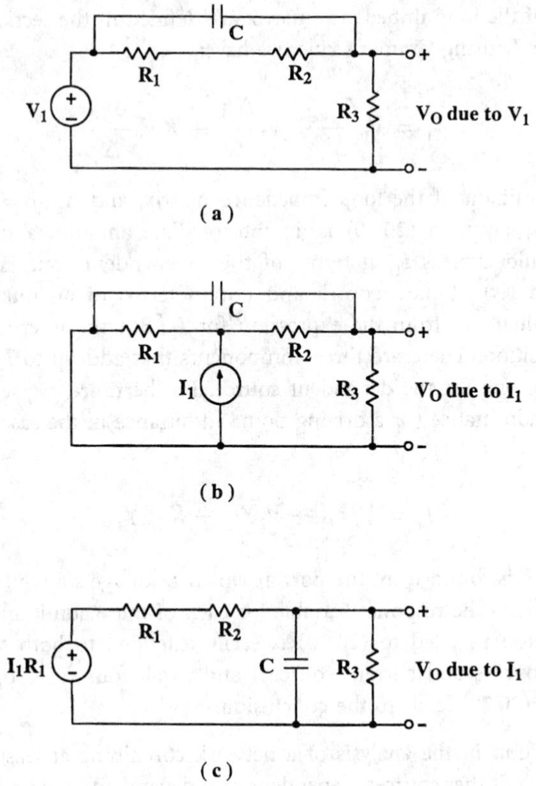

FIGURE 20.21 (a, b, c) Steps in the use of superposition for finding the response to two independent sources.

parallel with $R_1 + R_2$, and the contribution to V_o from V_1 can be found using a voltage divider. The result is

$$V_o \text{ due to } V_1 = V_1 \frac{s + \dfrac{1}{C(R_1 + R_2)}}{s + \dfrac{R_1 + R_2 + R_3}{C(R_1 + R_2)R_3}}$$

The second fictitious network, shown in Fig. 20.21(b), is obtained from the original network by properly removing the voltage source V_1. Redrawing the circuit and employing source transformation seven (in reverse) yields the circuit shown in Fig. 20.21(c). Again, employing a voltage divider, we have

$$V_o \text{ due to } I_1 = I_1 \frac{\dfrac{R_1}{C(R_1 + R_2)}}{s + \dfrac{R_1 + R_2 + R_3}{C(R_1 + R_2)R_3}}$$

Then adding the two contributions, we obtain

$$V_o = \frac{V_1\left[s + \dfrac{1}{C(R_1+R_2)}\right] + I_1\dfrac{R_1}{C(R_1+R_2)}}{s + \dfrac{R_1+R_2+R_3}{C(R_1+R_2)R_3}}$$

The next example includes a dependent source.

Example 6. Find i in the network shown in Fig. 20.22 using superposition.

Solution. Since there are two sources, we abstract two fictitious networks from Fig. 20.22. The first one is shown in Fig. 20.23(a) and is obtained by properly removing the dependent current source. Thus,

$$i \text{ due to } v_1 = \frac{v_1}{R_1+R_2}$$

Next, voltage source v_1 is properly removed yielding the fictitious network in Fig. 20.23(b). An important question immediately arises about this network. Namely, why is not i in this network zero? The reason i is not zero is that the network in Fig. 20.23(b) is merely an abstracted network that concerns a step in the analysis of the original circuit. It is an artifice in the application of superposition, and the dependent source is considered to be independent for this step. Thus,

$$i \text{ due to } \beta i = -\frac{\beta i R_2}{R_1+R_2}$$

Adding the two contributions, we obtain the intermediate result:

$$i = \frac{v_1}{R_1+R_2} - \frac{\beta i R_2}{R_1+R_2}$$

Collecting the terms containing i, we obtain the finished solution for i:

$$i = \frac{v_1}{(\beta+1)R_2+R_1}$$

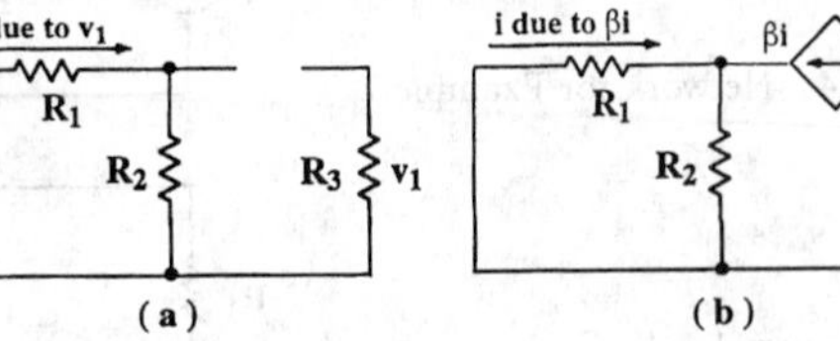

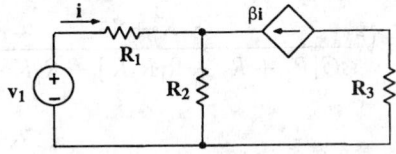

FIGURE 20.22 Network for Example 6.

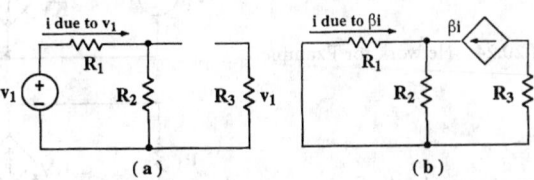

FIGURE 20.23 (a, b) Steps in the application of superposition to the network in Fig. 20.22.

We note that the finished solution depends only on the independent source v_1 and parameters of the network which are R_1, R_2, and β.

The following example involves a network in which a dependent source depends on a voltage that is neither the voltage of an independent source nor the voltage being sought in the analysis.

Example 7. Find V_o using superposition for the network shown in Fig. 20.24. Note that K, the coefficient of the VCCS, has the units of siemens.

Solution. When the dependent current source is properly removed, the network reduces to a simple voltage divider, and the contribution to V_o due to V_1 can be written as:

$$V_o \text{ due to } V_1 = V_1 \frac{1}{sC(R_1 + R_2) + 1}$$

Then reinserting the current source and properly removing the voltage source, we obtain the fictitious network shown in Fig. 20.25. Using the current divider to obtain the current flowing through the capacitor and then multiplying this current by the impedance of the capacitor we have:

$$V_o \text{ due to } KV_x = \frac{KV_x R_1}{sC(R_1 + R_2) + 1}$$

Adding the individual contributions to V_o provides the equation

$$V_o = \frac{V_1 + KV_x R_1}{sC(R_1 + R_2) + 1}$$

This is a valid expression for V_o. But it is not a finished expression because it includes V_x, an unknown voltage. Superposition has taken us to this point in the analysis, but more work must be done to eliminate V_x. However, superposition can be applied again to solve for V_x, or other analysis tools can be used. The results for V_x are:

$$V_x = \frac{V_1 sCR_1}{sC[R_1 + R_2 + R_1R_2K] + R_1K + 1}$$

Then eliminating V_x from the equation for V_o, we obtain the finished solution as:

$$V_o = V_1 \frac{R_1K + 1}{sC[R_1 + R_2 + R_1R_2K] + R_1K + 1}$$

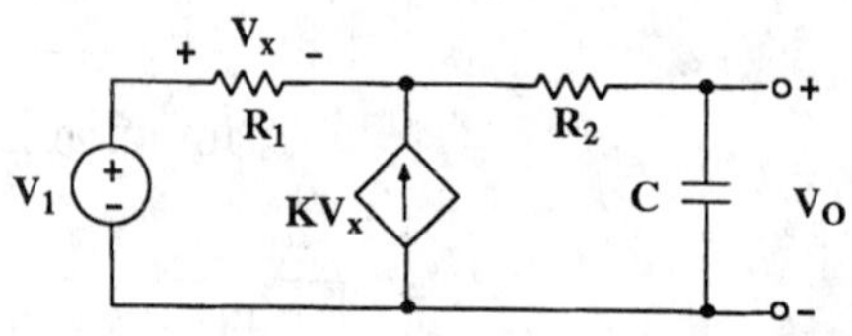

FIGURE 20.24 Network for Example 7.

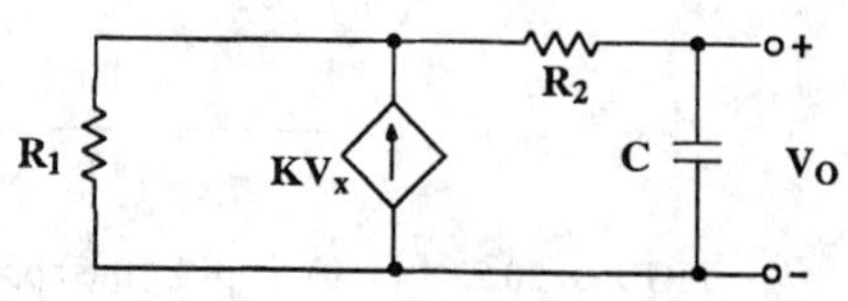

FIGURE 20.25 Fictitious network obtained when the voltage source is properly removed in Fig. 20.24.

Clearly, superposition is not the most efficient technique to use to analyze the network shown in Fig. 20.24. Analysis based on a node equation written at the top end of the current source would yield a finished result for V_o with less algebra. However, this example does illustrate the application of superposition when a dependent source depends on a rather arbitrary voltage in the network.

If the dependent current source in the previous example depended on V_1 rather than on the voltage across R_1, the network would be a different network. This is illustrated by the next example.

Example 8. Use superposition to determine V_o for the circuit shown in Fig. 20.26.

Solution. If the current source is properly removed, the results are the same as for the previous example. Thus,

$$V_o \text{ due to } V_1 = \frac{V_1}{sC(R_1 + R_2) + 1}$$

Then, if the current source is reinserted, and the voltage source is properly removed, we have the circuit depicted in Fig. 20.27. A question that can be asked for this circuit is why include the dependent source KV_1 if the voltage on which it depends, namely V_1, has been set to zero? However, the network shown in Fig. 20.27 is merely a fictitious network that serves as an aid in the application of superposition, and superposition deals with all sources, whether they are dependent or independent, as if they were independent. Thus, we can write:

$$V_o \text{ due to } KV_1 = \frac{KV_1R_1}{sC(R_1 + R_2) + 1}$$

Adding the contributions to V_o we obtain

$$V_o = V_1\frac{KR_1 + 1}{sC(R_1 + R_2) + 1}$$

and this is the finished solution.

In this example, we did not have the task of eliminating an unknown quantity from an intermediate result for V_o because the dependent source depended on an independent source V_1 which is assumed to be known.

Superposition is often useful in the analysis of circuits having only independent sources, but it is especially useful in the analysis of some circuits having both independent and dependent sources because it deals with all sources as if they were independent.

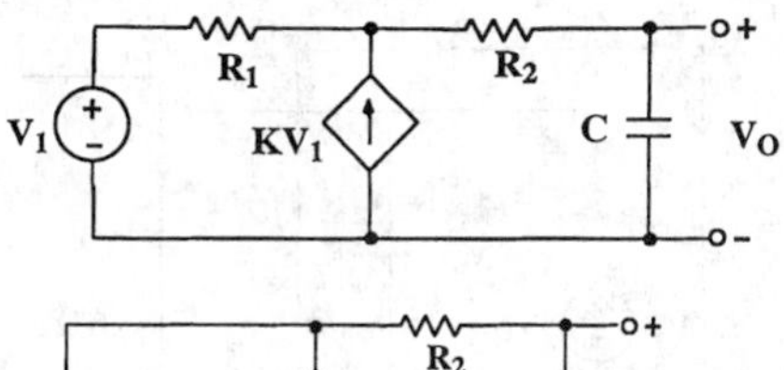

FIGURE 20.26 Network for Example 8.

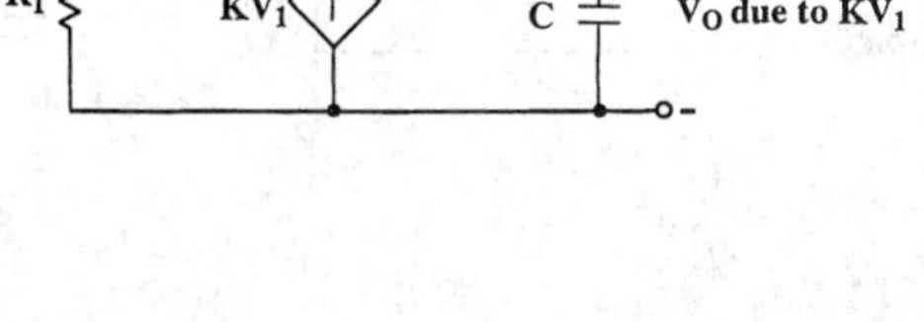

FIGURE 20.27 A step in the application of superposition to the network in Fig. 20.26.

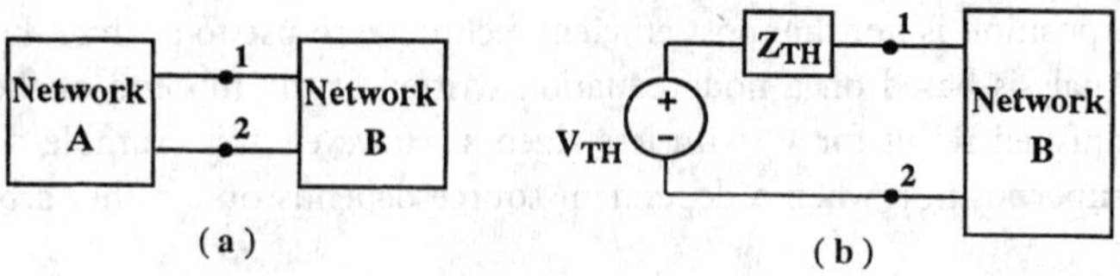

FIGURE 20.28 (a) Two subnetworks having a common pair of terminals. (b) The Thévenin equivalent for subnetwork A.

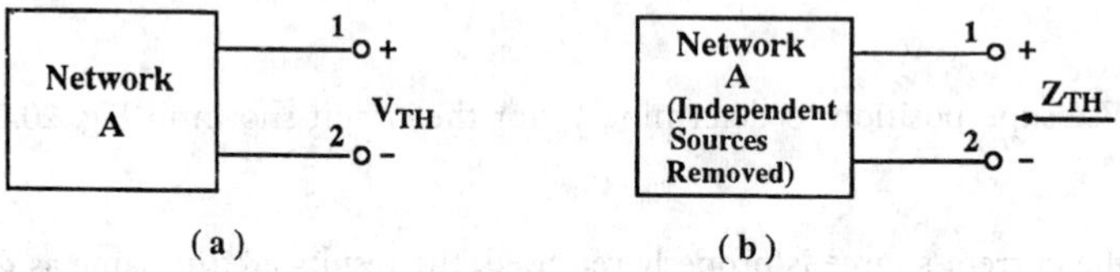

FIGURE 20.29 (a) Network used for finding V_{TH}. (b) Network used for obtaining Z_{TH}.

Thévenin's Theorem

Thévenin's theorem is useful in reducing the complexity of a network so that analysis of the network for a particular voltage or current can be performed more easily. For example, consider Fig. 20.28(a) which is composed of two subnetworks A and B that have only two nodes in common. In order to facilitate analysis in subnetwork B, it is convenient to reduce subnetwork A to the network shown in Fig. 20.28(b) which is termed the Thévenin equivalent of subnetwork A. The requirement on the Thévenin equivalent network is that when it replaces subnetwork A, the voltages and currents in subnetwork B remain unchanged. We assume that there is no inductive coupling between the subnetworks, and that dependent sources in B are not dependent on voltages or currents in A. We also assume that subnetwork A is an LLFT network, but subnetwork B does not have to meet this assumption.

To find the Thévenin equivalent network, we need only determine V_{TH} and Z_{TH}. V_{TH} is found by unhooking B from A and finding the voltage that appears across the terminals of A. In other words, we abstract a fictitious network from the complete network as shown in Fig. 20.29(a), and find the voltage that appears between the terminals that were common to B. This voltage is V_{TH}.

Z_{TH} is also obtained from a fictitious network that is created from the fictitious network used for finding V_{TH} by properly removing all independent sources. The effects that dependent sources have on the procedure are discussed later in this section. The fictitious network used for finding Z_{TH} is shown in Fig. 20.29(b). Oftentimes, the expression for Z_{TH} cannot be found by mere inspection of this network, and, therefore, we must excite the network in Fig. 20.29(b) by a voltage source or a current source and find an expression for the other variable at the port in order to find Z_{TH}.

Example 9. Find the Thévenin equivalent of subnetwork A in Fig. 20.30.

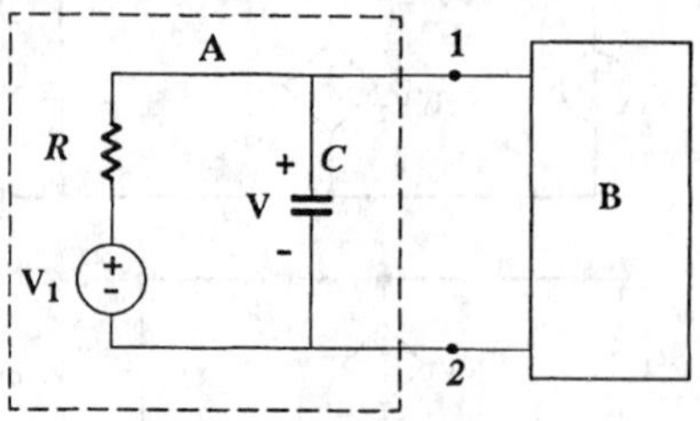

FIGURE 20.30 Network for Example 9.

Solution. There are no dependent sources in subnetwork A, but the capacitor has an initial voltage V across it. However, the charged capacitor can be represented by an uncharged capacitor in series with a transformed voltage source V/s. The fictitious network used for finding V_{TH} is shown in Fig. 20.31(a).

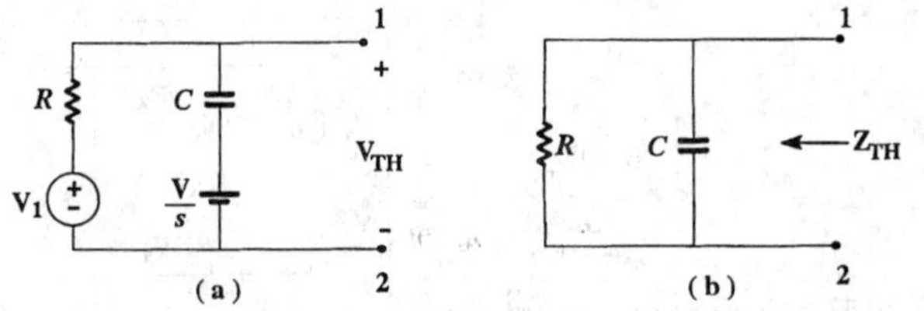

FIGURE 20.31 (a) Network for finding V_{TH}. (b) Network which yields Z_{TH}.

It should be noted that although subnetwork B has been removed and the two terminals that were connected to B are now "open circuited" in Fig. 20.31(a), there is still current flowing in network A. V_{TH} is easily obtained using a voltage divider:

$$V_{TH} = \frac{V_1(s) + VCR}{sCR + 1}$$

Z_{TH} is obtained from the fictitious network shown in Fig. 20.31(b) which is obtained by properly removing the independent source and the voltage representing the initial voltage across the capacitor in Fig. 20.31(a). We see by inspection that $Z_{TH} = R/(sCR+1)$. Thus, if subnetwork A is removed from Fig. 20.30 and replaced by the Thévenin equivalent network, the voltages and currents in subnetwork B remain unchanged.

It is assumed that B in Fig. 20.28 has no dependent sources that depend on voltages or currents in A, although dependent sources in B can depend on voltages and currents in B. However, A can have dependent sources, and these dependent sources create a modification in the procedure for finding the Thévenin equivalent network. There may be dependent sources in A that depend on voltages and currents that also exist in A. We call these dependent sources Case I-dependent sources. There may also be dependent sources in A that depend on voltages and currents in B, and we label these sources as Case II-dependent sources. Then the procedure for finding the Thévenin equivalent network is:

V_{TH} is the voltage across the terminals of Fig. 20.29(a). The voltages and currents that Case I-dependent sources depend on must be eliminated from the expression for V_{TH} unless they happen to be the voltages of independent voltage sources or the currents of independent current sources in A. Otherwise, the expression for V_{TH} would not be a finished solution. However, Case II-dependent sources are handled as if they were independent sources. That is, Case II-dependent sources are included in the results for V_{TH} just as independent sources would be.

Z_{TH} is the impedance looking into the terminals in Fig. 20.29(b). In this fictitious network, independent sources are properly removed and *Case II-dependent sources are properly removed.* Case I-dependent sources remain in the network and influence the result for the Thévenin impedance. The finished solution for Z_{TH} is a function only of the parameters of the network in Fig. 20.29(b) which are Rs, Ls, Cs, Ms (there may be inductive coupling between coils in this network), and the coefficients of the Case I-dependent sources.

Thus, Case II-dependent sources, sources that depend on voltages or currents in subnetwork B, are uniformly treated as if they were independent sources in finding the Thévenin equivalent network. Some examples will clarify the issue.

Example 10. Find the Thévenin equivalent network for subnetwork A in Fig. 20.32. Assume the initial current through the inductor is zero.

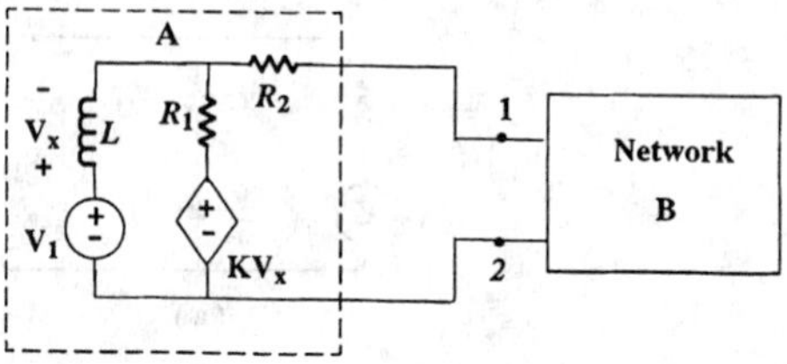

FIGURE 20.32 Network for Example 10.

Solution. There is one independent source and one Case I-dependent source. Figure 20.33(a) depicts the fictitious network to be analyzed to obtain V_{TH}. Since there is no current flowing through R_2 in this figure, we can write $V_{TH} = V_1 - V_x$. To eliminate V_x from our intermediate expression for V_{TH}, we can use the results of the enhanced voltage divider to write:

$$V_{TH} = \frac{V_1 R_1 + sKL(V_1 - V_{TH})}{R_1 + sL}$$

The finished solution for V_{TH} is

$$V_{TH} = V_1 \frac{sKL + R_1}{(K+1)sL + R_1}$$

Z_{TH} is obtained from Fig. 20.33(b) where a current source excitation is shown already applied to the fictitious network. Two node equations, with unknown node voltages V and $-V_x$, enable us to obtain I in terms of V while eliminating V_x. We also note that Z_{TH} consists of resistor R_2 in series with some unknown impedance, so we could remove R_2 (replace it by a short) if we remember to add it back later. The finished result for Z_{TH} is

$$Z_{TH} = R_2 + \frac{sLR_1}{(K+1)sL + R_1}$$

The following example involves a network having a Case II-dependent source.

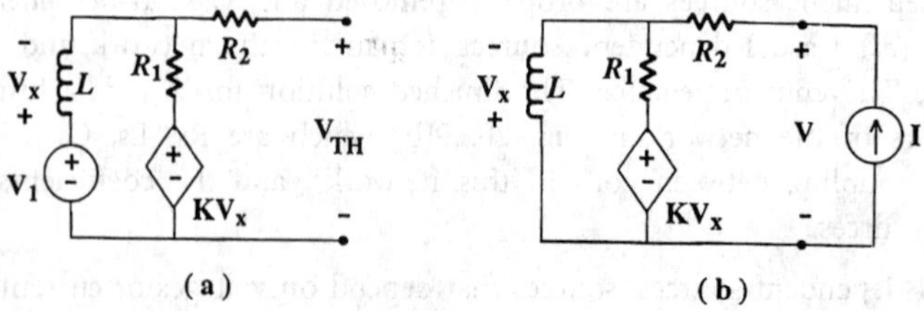

FIGURE 20.33 (a) Abstracted network for finding V_{TH}. (b) Abstracted network for finding Z_{TH}.

Example 11. Find the Thévenin equivalent network for subnetwork A in the network shown in Fig. 20.34. In this instance, subnetwork B is shown explicitly.

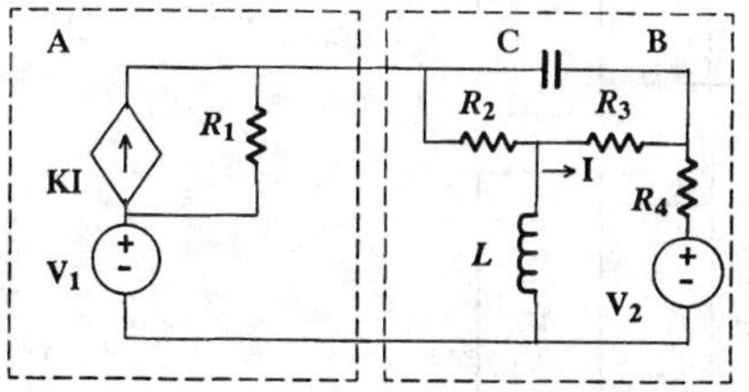

FIGURE 20.34 Network for Example 11.

Solution. Subnetwork A contains one independent source and one Case II-dependent source. Figure 20.35(a) shows the abstracted network for finding V_{TH}. Thus,

$$V_{TH} = V_1 + KIR_1$$

Then, both V_1 and the dependent source KI are deleted from Fig. 20.35(a) to obtain Fig. 20.35(b), the network used for finding Z_{TH}. Thus, $Z_{TH} = R_1$.

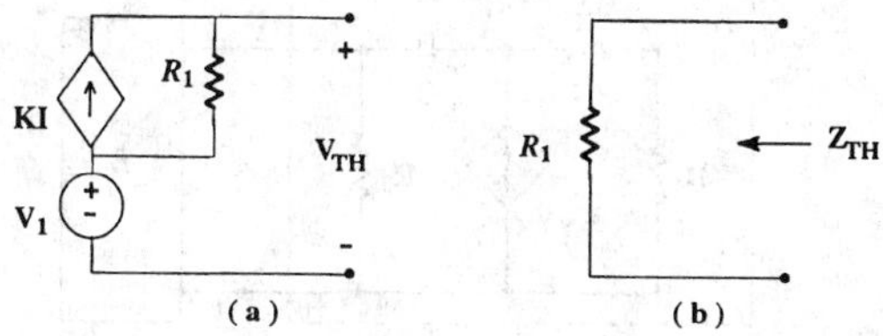

FIGURE 20.35 (a) Network used to find V_{TH}. (b) Network for finding Z_{TH}.

Of course, the subnetwork for which the Thévenin equivalent is being determined may have both Case I- and Case II-dependent sources, but these sources can be handled concurrently using the procedures given above.

There are special conditions that can arise in the application of Thévenin's theorem. One condition is $Z_{TH} = 0$ and the other is $V_{TH} = 0$. The conditions for which Z_{TH} is zero are:

1. If the circuit (subnetwork A) for which the Thévenin equivalent is being determined has an independent voltage source connected between terminals 1 and 2, then $Z_{TH} = 0$. Figure 20.36(a) illustrates this case.
2. If subnetwork A has a *dependent* voltage source connected between terminals 1 and 2, then Z_{TH} is zero *provided* neither of the port variables associated with the port formed by terminals 1 and 2 is coupled back into the network. Figure 20.36(b) shows a subnetwork A for which Z_{TH} is zero. However, Fig. 20.36(c) depicts a subnetwork A in which the port variable I is coupled back into A by the dependent source K_1I. If I is considered to be a variable of subnetwork A so that K_1I is a Case I–dependent source, then Z_{TH} is not zero.

The other special condition, $V_{TH} = 0$, occurs if subnetwork A contains only Case I–dependent sources, no independent sources, and no Case II–dependent sources. An example of such a network is shown in Fig. 20.36(d). With subnetwork B disconnected, subnetwork A is a dead network, and its Thévenin voltage is zero.

The network shown in Fig. 20.36(c) is of interest because the dependent source K_1I can be considered as a Case I– or a Case II–dependent source hinging on whether I is considered a variable of subnetwork A or B.

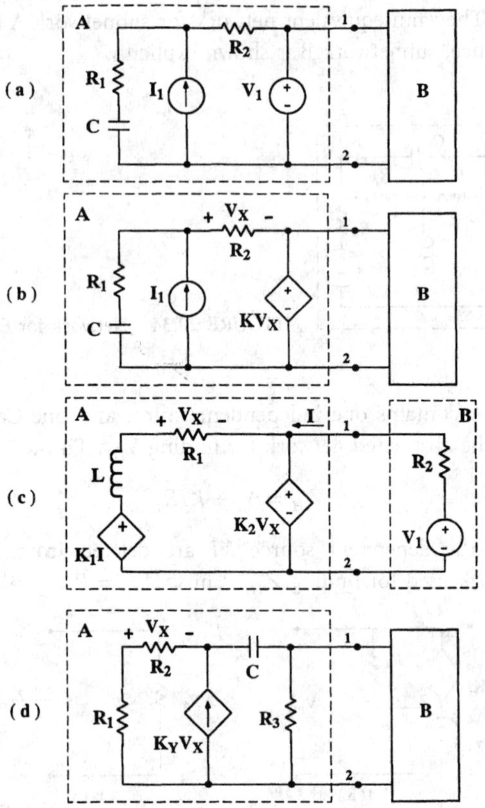

FIGURE 20.36 Special cases of Thévenin's theorem. (a, b) Z_{TH} equals zero. (c) A port variable is coupled back into A. (d) V_{TH} is zero.

Example 12. Solve for I in Fig. 20.36(c) using two versions of the Thévenin equivalent for subnetwork A. For the first version, consider I to be associated with A and therefore both dependent sources are Case I–dependent sources. In the second version, consider I to be associated with B.

Solution. If I is considered as associated with A, then V_{TH} is zero by inspection since A contains only Case I–dependent sources. Figure 20.37(a) shows subnetwork A with a current excitation applied in order to determine Z_{TH}. Clearly, $V = K_2V_x$. Also, writing a loop equation in the loop encompassed by the two dependent sources, we obtain

$$K_1I - \frac{V_x}{R_1}(sL + R_1) = V$$

Eliminating V_x, we have

$$\frac{V}{I} = Z_{TH} = \frac{K_1K_2R_1}{sL + R_1(K_2 + 1)}$$

Once Z_{TH} is obtained, it is an easy matter to write

$$I = \frac{V_1}{R_2 + Z_{TH}} = \frac{V_1[sL + R_1(K_2 + 1)]}{sLR_2 + R_1R_2(K_2 + 1) + K_1K_2R_1}$$

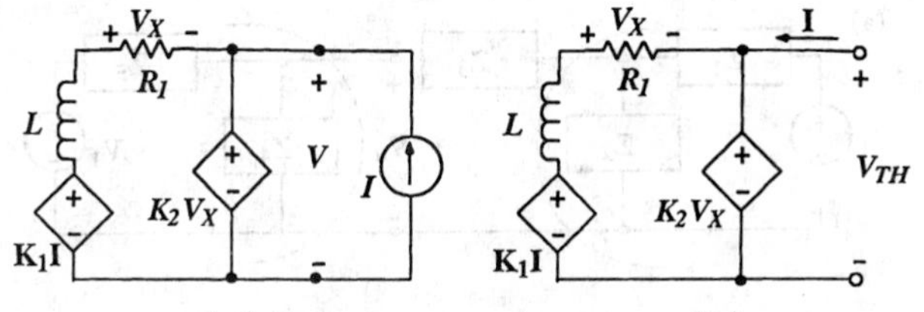

FIGURE 20.37 (a) Network for finding Z_{TH} when both sources are Case I–dependent sources. (b) Network for finding V_{TH} when a Case II–dependent source exists in the network.

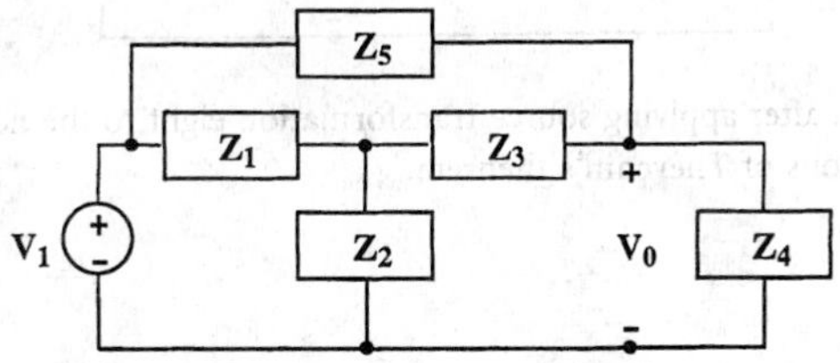

FIGURE 20.38 Network for example 13.

If I is associated with B, then V_{TH} is found from the network shown in Fig. 20.37(b) with the source K_1I treated as if it were independent. The equation for V_{TH} may contain the variable I, but V_x must be eliminated from the finished expression for V_{TH}. We obtain

$$V_{TH} = I\frac{K_1K_2R_1}{sL + R_1(K_2 + 1)}$$

Also, Z_{TH} is zero since if K_1I is removed, subnetwork A reduces to a network with only a Case I-dependent source and a port variable is not coupled back into the network. Finally, I can be written as:

$$I = \frac{V_1 - V_{TH}}{R_2}$$

which yields the same result for I as was found previously.

The following example illustrates the interplay that can be achieved among these theorems and source transformations.

Example 13. Find V_o/V_1 for the bridged T network shown in Fig. 20.38.

Solution. The application of source transformation eight to the network yields the ladder network shown in Fig. 20.39(a). Thévenin's theorem is particularly useful in analyzing ladder networks. If it is applied to the left and right sides of the network, taking care not to obscure the nodes between which V_o exists, we obtain the single loop network shown in Fig. 20.39(b). Then, using a voltage divider, we obtain

$$\frac{V_o}{V_1} = \frac{Z_4[Z_1(Z_2 + Z_3) + Z_2Z_3 + Z_2Z_5]}{(Z_1 + Z_2)(Z_3Z_4 + Z_3Z_5 + Z_4Z_5) + Z_1Z_2(Z_4 + Z_5)}$$

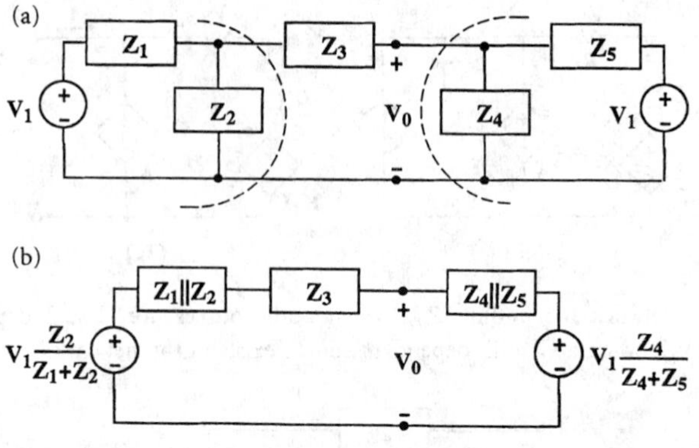

FIGURE 20.39 (a) Results after applying source transformation eight to the network shown in Fig. 20.38. (b) Results of two applications of Thévenin's theorem.

Norton's Theorem

If a source transformation is applied to the Thévenin equivalent network consisting of V_{TH} and Z_{TH} in Fig. 20.28(b), then a Norton equivalent network, shown in Fig. 20.40(a), is obtained. The current source $I_{sc} = V_{TH}/Z_{TH}, Z_{TH} \neq 0$. If Z_{TH} equals zero in Fig. 20.28(b), then the Norton equivalent network does not exist. The subscripts "*sc*" on the current source stand for short circuit and indicate a procedure for finding the value of this current source. To find I_{sc} for subnetwork A in Fig. 20.28(a), we disconnect subnetwork B and place a short circuit between nodes 1 and 2 of subnetwork A. Then I_{sc} is the current flowing through the short circuit in the direction indicated in Fig. 20.40(b). I_{sc} is zero if subnetwork A has only Case I–dependent sources and no other sources. Z_{TH} is found in the same manner as for Thévenin's theorem.

It is sometimes more convenient to find I_{sc} and V_{TH} rather than Z_{TH}.

Example 14. Find the Norton equivalent for the network "seen" by Z_L in Fig. 20.41. That is, Z_L is subnetwork B and the rest of the network is A, and we wish to find the Norton equivalent network for A.

Solution. Figure 20.42(a) shows the network with Z_L replaced by a short circuit. An equation for I_{sc} can be obtained quickly using superposition. This yields

$$I_{sc} = I_1 + \frac{KI_1}{R_2}$$

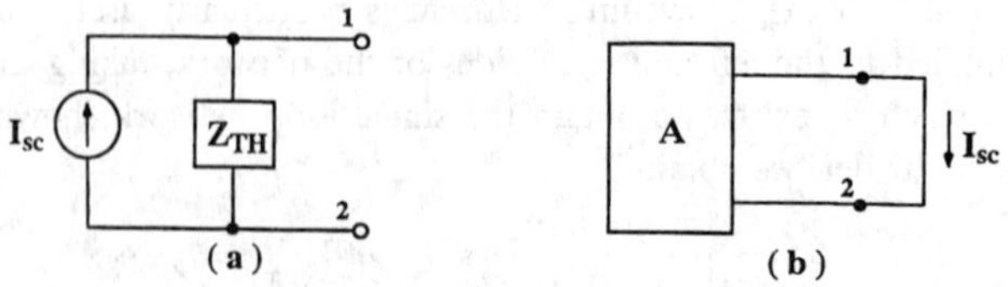

FIGURE 20.40 (a) Norton equivalent network. (b) Reference direction for I_{sc}.

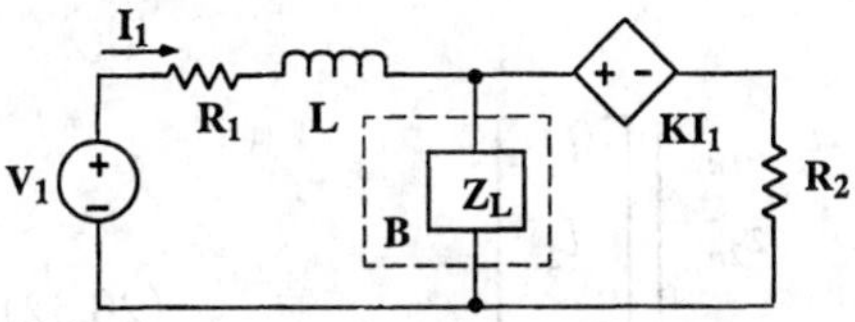

FIGURE 20.41 Network for Example 14.

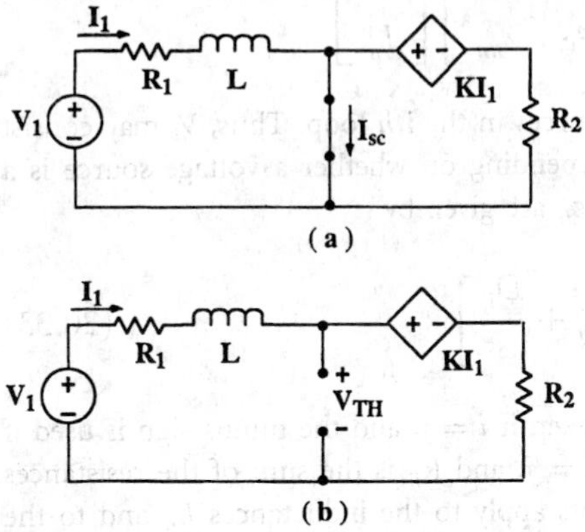

FIGURE 20.42 (a) Network for finding I_{sc}. (b) Network for V_{TH}.

but I_1 must be eliminated from this equation. I_1 is obtained as: $I_1 = V_1/(sL + R_1)$. Thus,

$$I_{sc} = \frac{V_1\left(1 + \dfrac{K}{R_2}\right)}{sL + R_1}$$

V_{TH} is found from the network shown in Fig. 20.42(b). The results are:

$$V_{TH} = \frac{V_1(K + R_2)}{sL + R_1 + R_2 + K}$$

Z_{TH} can be found as V_{TH}/I_{sc}.

Thévenin's and Norton's Theorems and Network Equations

Thévenin's and Norton's theorems can be related to loop and node equations. Here, we examine the relationship to loop equations by means of the LLFT network shown in Fig. 20.43. Assume that the network N in Fig. 20.43 has n independent loops with all the loop currents chosen in the same direction. Without loss of generality, assume that only one loop current, say I_1, flows through Z_L as shown so that Z_L appears in only one loop equation. For simplicity, assume that there are no dependent sources or inductive couplings in N and that all current sources have been source-transformed so that only voltage source excitations remain. Then the loop

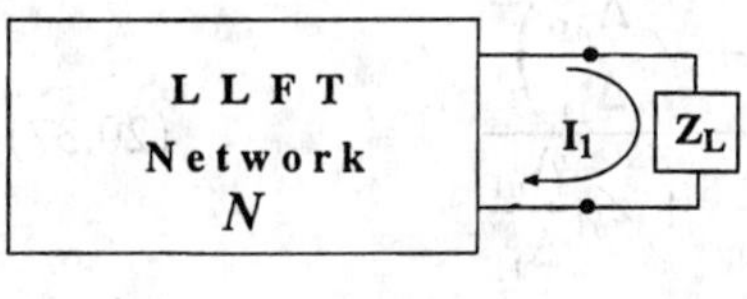

FIGURE 20.43 An LLFT network N with n independent loops.

equations are

$$\begin{bmatrix} V_1 \\ V_2 \\ \vdots \\ V_n \end{bmatrix} = \begin{bmatrix} z_{11} & z_{12} & \cdots & z_{1n} \\ z_{21} & z_{22} & \cdots & z_{2n} \\ \vdots & \vdots & \ddots & \vdots \\ z_{n1} & z_{n2} & \cdots & z_{nn} \end{bmatrix} \begin{bmatrix} I_1 \\ I_2 \\ \vdots \\ I_n \end{bmatrix} \tag{20.32}$$

where $V_i, i = 1, 2, \ldots, n$, is the sum of all voltage sources in the ith loop. Thus, V_i may consist of several terms, some of which may be negative depending on whether a voltage source is a voltage rise or a voltage drop. Also, the impedances z_{ij} are given by

$$z_{ij} = \pm\left[R_{ij} + sL_{ij} + \frac{D_{ij}}{s}\right] \tag{20.33}$$

where $i, j = 1, 2, \ldots, n$, and where the plus sign is taken if $i = j$, and the minus sign is used if $i \neq j$. R_{ij} is the sum of the resistances in loop i if $i = j$, and R_{ij} is the sum of the resistances common to loops i and j if $i \neq j$. Similar statements apply to the inductances L_{ij} and to the reciprocal capacitances D_{ij}. The currents $I_i, i = 1, 2, \ldots, n$, are the unknown loop currents.

Note that Z_L is included only in z_{11} so that z_{11} can be written as $z_{11} = z_{11A} + Z_L$, where z_{11A} is the sum of all the impedances around loop one except Z_L. Solving for I_1 using Cramer's rule, we have

$$I_1 = \frac{\begin{vmatrix} V_1 & z_{12} & \cdots & z_{1n} \\ V_2 & z_{22} & \cdots & z_{2n} \\ \vdots & \vdots & \ddots & \vdots \\ V_n & z_{n2} & \cdots & z_{nn} \end{vmatrix}}{\Delta} \tag{20.34}$$

where Δ is the determinant of the loop impedance matrix. Thus, we can write

$$I_1 = \frac{V_1\Delta_{11} + V_2\Delta_{21} + \cdots + V_n\Delta_{n1}}{z_{11}\Delta_{11} + z_{21}\Delta_{21} + \cdots + z_{n1}\Delta_{n1}} \tag{20.35}$$

or

$$I_1 = \frac{V_1 + V_2\dfrac{\Delta_{21}}{\Delta_{11}} + \cdots + V_n\dfrac{\Delta_{n1}}{\Delta_{11}}}{z_{11} + z_{21}\dfrac{\Delta_{21}}{\Delta_{11}} + \cdots + z_{n1}\dfrac{\Delta_{n1}}{\Delta_{11}}} \tag{20.36}$$

where Δ_{ij} are cofactors of the loop impedance matrix. Then, forming the product of I_1 and Z_L, we have:

$$I_1 Z_L = \frac{Z_L\left(V_1 + V_2\dfrac{\Delta_{21}}{\Delta_{11}} + \cdots + V_n\dfrac{\Delta_{n1}}{\Delta_{11}}\right)}{Z_L + z_{11A} + z_{21}\dfrac{\Delta_{21}}{\Delta_{11}} + \cdots + z_{n1}\dfrac{\Delta_{n1}}{\Delta_{11}}} \tag{20.37}$$

If we take the limit of $I_1 Z_L$ as Z_L approaches infinity, we obtain the "open circuit" voltage V_{TH}. That is,

$$\lim_{Z_L \to \infty} I_1 Z_L = V_{TH} = \left(V_1 + V_2 \frac{\Delta_{21}}{\Delta_{11}} + \cdots + V_n \frac{\Delta_{n1}}{\Delta_{11}} \right) \tag{20.38}$$

and if we take the limit of I_1 as Z_L approaches zero, we obtain the "short circuit" current I_{sc}:

$$\lim_{Z_L \to 0} I_1 = I_{sc} = \frac{V_{TH}}{z_{11A} + z_{21} \frac{\Delta_{21}}{\Delta_{11}} + \cdots + z_{n1} \frac{\Delta_{n1}}{\Delta_{11}}} \tag{20.39}$$

Finally, the quotient of V_{TH} and I_{sc} yields:

$$\frac{V_{TH}}{I_{sc}} = Z_{TH} = z_{11A} + z_{21} \frac{\Delta_{21}}{\Delta_{11}} + \cdots + z_{n1} \frac{\Delta_{n1}}{\Delta_{11}} \tag{20.40}$$

If network N contains coupled inductors (but not coupled to Z_L), then some elements of the loop impedance matrix may be modified in value and sign. If network N contains dependent sources, then auxiliary equations can be written to express the quantities on which the dependent sources depend in terms of the independent excitations and/or the unknown loop currents. Thus, dependent sources may modify the elements of the loop impedance matrix in value and sign, and they may modify the elements of the excitation column matrix $[V_i]$. Nevertheless, we can obtain expressions similar to those above for V_{TH} and I_{sc} Of course, we must exclude from this illustration dependent sources which depend on the voltage across Z_L because they violate the assumption that Z_L appears in only one loop equation and are beyond the scope of this discussion.

The π-T Conversion

The π-T conversion is employed for the simplification of circuits, especially in power systems analysis. The "π" refers to a circuit having the topology shown in Fig. 20.44. In this figure, the left-most and right-most loop currents have been chosen to coincide with the port currents for convenience of notation only.

A circuit having the topology shown in Fig. 20.45 is referred to as a "T" or as a "Y". We wish to determine equations for Z_1, Z_2, and Z_3 in terms of Z_A, Z_B, and Z_C so that the π can be replaced by a T without affecting any of the port variables. In other words, if an overall circuit contains a π subcircuit, we wish to replace the π subcircuit with a T subcircuit without disturbing any of the other voltages and currents within the overall circuit. To determine what Z_1, Z_2, and Z_3 should be, we first write loop equations for the π network. The results are:

$$V_1 = I_1 Z_A - I_3 Z_A \tag{20.41}$$

$$V_2 = I_2 Z_B + I_3 Z_B \tag{20.42}$$

$$0 = I_3 (Z_A + Z_B + Z_C) - I_1 Z_A + I_2 Z_B \tag{20.43}$$

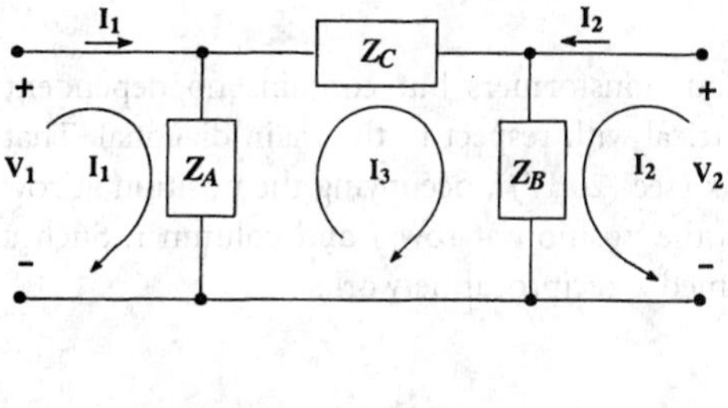

FIGURE 20.44 A π network shown with loop currents.

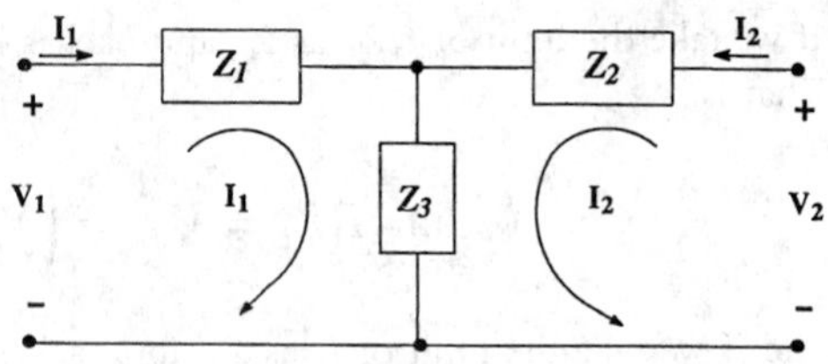

FIGURE 20.45 A T network.

But the T circuit has only two loop equations given by:

$$V_1 = I_1(Z_1 + Z_3) + I_2 Z_3 \tag{20.44}$$

$$V_2 = I_1 Z_3 + I_2(Z_2 + Z_3) \tag{20.45}$$

We must eliminate one of the loop equations for the π circuit, and so we solve for I_3 in (20.43) and substitute the result into (20.41) and (20.42) to obtain:

$$V_1 = I_1\left[\frac{Z_A(Z_B + Z_C)}{Z_A + Z_B + Z_C}\right] + I_2\left[\frac{Z_A Z_B}{Z_A + Z_B + Z_C}\right] \tag{20.46}$$

$$V_2 = I_1\left[\frac{Z_A Z_B}{Z_A + Z_B + Z_C}\right] + I_2\left[\frac{Z_B(Z_A + Z_C)}{Z_A + Z_B + Z_C}\right] \tag{20.47}$$

From a comparison of the coefficients of the currents in (20.46) and (20.47) with those in (20.44) and (20.45), we obtain the following relationships.

Replacing π with T

$$Z_1 = \frac{Z_A Z_C}{S_Z}; \quad Z_2 = \frac{Z_B Z_C}{S_Z}; \quad Z_3 = \frac{Z_A Z_B}{S_Z} \tag{20.48}$$

where

$$S_Z = Z_A + Z_B + Z_C$$

We can also replace a T network by a π network. To do this we need equations for Z_A, Z_B, and Z_C in terms of Z_1, Z_2, and Z_3. The required equations can be obtained algebraically from (20.48).

From T to π

$$Z_A = Z_1 + Z_3 + \frac{Z_1 Z_3}{Z_2}; \quad Z_B = Z_2 + Z_3 + \frac{Z_2 Z_3}{Z_1}; \quad Z_C = Z_1 + Z_2 + \frac{Z_1 Z_2}{Z_3} \tag{20.49}$$

Reciprocity

If an LLFT network contains only Rs, Ls, Cs, and transformers but contains no dependent sources, then its loop impedance matrix is symmetrical with respect to the main diagonal. That is, if z_{ij} is an element of the loop impedance matrix (see (20.17)), occupying the position at row i and column j, then $z_{ji} = z_{ij}$, where z_{ji} occupies the position at row j and column i. Such a network has the property of reciprocity and is termed a reciprocal network.

Assume that a reciprocal network, shown in Fig. 20.46, has m loops and is in the zero state. It has only one excitation—a voltage source in loop j. To solve for the loop current in loop k, we write the loop equations:

$$\begin{bmatrix} z_{11} & z_{12} & \cdots & z_{1m} \\ z_{21} & z_{22} & \cdots & z_{2m} \\ \vdots & \vdots & \ddots & \vdots \\ z_{j1} & z_{j2} & \cdots & z_{jm} \\ z_{k1} & z_{k2} & \cdots & z_{km} \\ \vdots & \vdots & \ddots & \vdots \\ z_{m1} & z_{m2} & \cdots & z_{mm} \end{bmatrix} \begin{bmatrix} I_1 \\ I_2 \\ \vdots \\ I_j \\ I_k \\ \vdots \\ I_m \end{bmatrix} = \begin{bmatrix} 0 \\ 0 \\ \vdots \\ V_j \\ 0 \\ \vdots \\ 0 \end{bmatrix} \tag{20.50}$$

The column excitation matrix has only one nonzero entry. To determine I_k using Cramer's rule, we replace column k by the excitation column and then expand along this column. Since there is only one nonzero term in the column, we obtain a single term for I_k:

$$I_k = V_j \frac{\Delta_{jk}}{\Delta} \tag{20.51}$$

where Δ_{jk} is the cofactor, and Δ is the determinant of the loop impedance matrix.

Next, we replace the voltage source by a short circuit in loop j, cut the wire in loop k, and insert a voltage source V_k. Figure 20.47 shows the modifications to the circuit. Then we solve for I_j obtaining

$$I_j = V_k \frac{\Delta_{kj}}{\Delta} \tag{20.52}$$

Since the network is reciprocal, $\Delta_{jk} = \Delta_{kj}$ so that

$$\frac{I_k}{V_j} = \frac{I_j}{V_k} \tag{20.53}$$

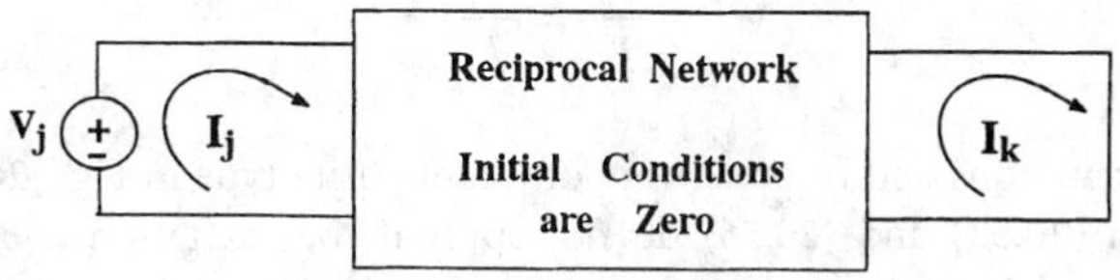

FIGURE 20.46 A reciprocal network with l independent loops.

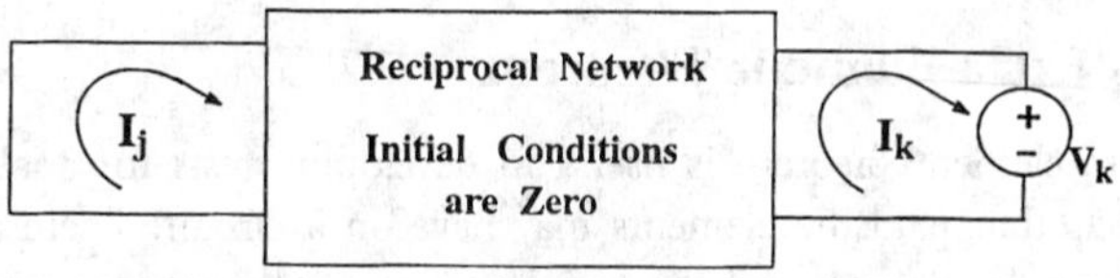

FIGURE 20.47 Interchange of the ports of excitation in the network shown in Fig. 20.46.

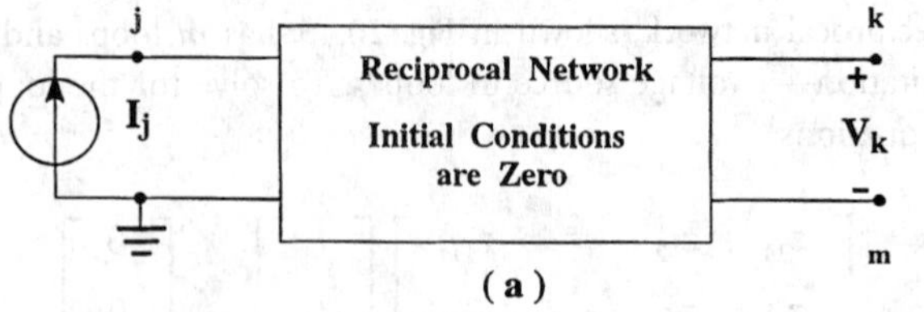

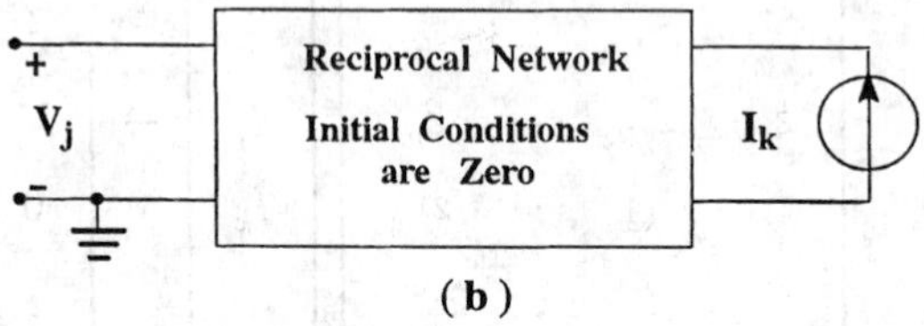

FIGURE 20.48 (a) Reciprocal ungrounded network with a current source excitation. (b) Interchange of the ports of excitation and response.

Eq. (20.53) is the statement of reciprocity for the network in Figs. 20.46 and 20.47 with the excitations shown.

Figure 20.48(a) shows a reciprocal network with a current excitation applied to node j and a voltage response, labeled V_k, taken between nodes k and m. We assume the network has n independent nodes plus the ground node indicated and is not a grounded network (does not have a common connection between the input and output ports shown). If we write node equations to solve for V_k in Fig. 20.48(a) and use Cramer's rule, we have:

$$V_k = I_j \frac{\Delta'_{jk} - \Delta'_{jm}}{\Delta'} \tag{20.54}$$

where the primes indicate node-basis determinants. Then we interchange the ports of excitation and response as shown in Fig. 20.48(b). If we solve for V_j in Fig. 20.48(b), we obtain

$$V_j = I_k \frac{\Delta'_{kj} - \Delta'_{mj}}{\Delta'} \tag{20.55}$$

Since the corresponding determinants in (20.54) and (20.55) are equal because of reciprocity, we have:

$$\frac{V_k}{I_j} = \frac{V_j}{I_k} \tag{20.56}$$

Note that the excitations and responses are of the opposite type in Figs. 20.46 and 20.48. The results obtained in (20.53) and (20.56) do not apply if the excitation and response are both voltages or both currents because when the ports of excitation and response are interchanged, the impedance levels of the network are changed [2].

Middlebrook's Extra Element Theorem

Middlebrook's extra element theorem is useful in developing tests for analog circuits and for predicting the effects that parasitic elements may have on a circuit. There are two versions of this theorem: the parallel version and the series version. Both versions present the results of connecting an extra network element in the circuit as the product of the network function obtained without the extra element times a correction factor. This is a particularly convenient form for the results since it shows exactly the effects of the extra element on the network function.

Parallel Version. Consider an arbitrary LLFT network having a transfer function $A_1(s)$. In the parallel version of the theorem, an impedance is added between any two independent nodes of the network. The modified transfer function is then obtained as $A_1(s)$ multiplied by a correction factor. Figure 20.49 shows an arbitrary network in which the quantities U_i and U_o represent a general input and a general output, respectively, whether they are voltages or currents. The extra element is to be connected between terminals 1 and 1′ in Fig. 20.49, and the port variables for this port are V_2 and I_2.

We can write:

$$\begin{aligned} U_o &= A_1 U_i + A_2 I_2 \\ V_2 &= B_1 U_i + B_2 I_2 \end{aligned} \tag{20.57}$$

where

$$\begin{aligned} A_1 &= \left.\frac{U_o}{U_i}\right|_{I_2=0} \qquad A_2 = \left.\frac{U_o}{I_2}\right|_{U_i=0} \\ B_1 &= \left.\frac{V_2}{U_i}\right|_{I_2=0} \qquad B_2 = \left.\frac{V_2}{I_2}\right|_{U_i=0} \end{aligned} \tag{20.58}$$

Note that A_1 is assumed to be known.

The extra element Z to be added across terminals 1 and 1′ is shown in Fig. 20.50. It can be described as $Z = V_2/(-I_2)$ which yields $I_2 = V_2/(-Z)$. Substituting this expression for I_2 into (20.57), results in:

$$\begin{aligned} U_o &= A_1 U_i + A_2\left(\frac{-V_2}{Z}\right) \\ V_2\left(1 + \frac{B_2}{Z}\right) &= B_1 U_i \end{aligned} \tag{20.59}$$

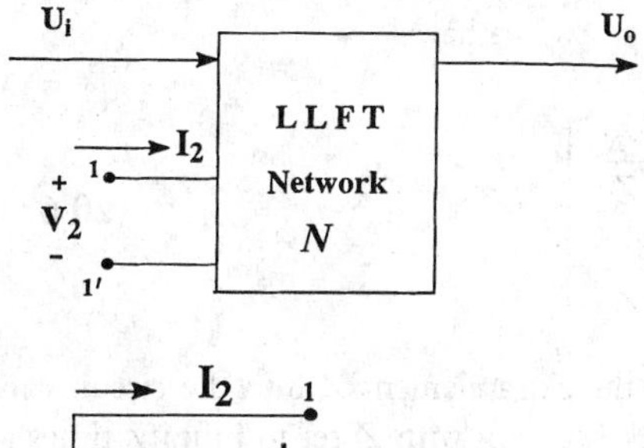

FIGURE 20.49 An arbitrary LLFT network.

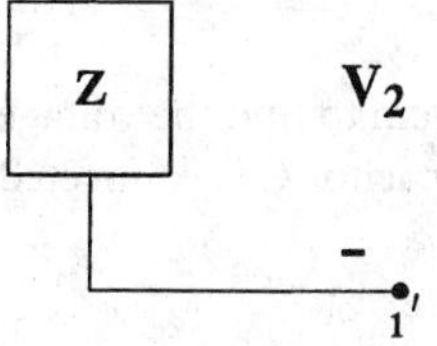

FIGURE 20.50 The extra element Z.

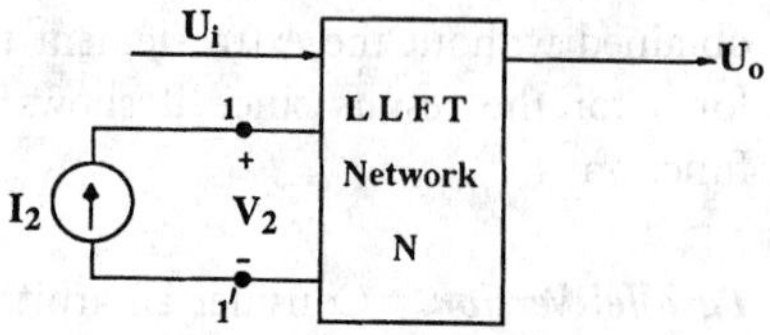

FIGURE 20.51 Network of Fig. 20.49 with two excitations applied.

After eliminating V_2 and solving for U_o/U_i, we obtain:

$$\frac{U_o}{U_i} = A_1\left[\frac{1+\dfrac{1}{Z}\left(\dfrac{A_1B_2 - A_2B_1}{A_1}\right)}{1+\dfrac{B_2}{Z}}\right] \tag{20.60}$$

Next, we provide physical interpretations for the terms in (20.60). Clearly, B_2 is the impedance seen looking into the network between terminals 1 and 1′ with $U_i = 0$. Thus, rename $B_2 = Z_d$ where d stands for "dead network."

To find a physical interpretation of $(A_1B_2 - A_2B_1)/A_1$, examine the network shown in Fig. 20.51. Here, two excitations are applied to the network, namely U_i and I_2. Simultaneously adjust both inputs so as to null output U_o. Thus, with $U_o = 0$, we have from (20.57),

$$U_i = \frac{-A_2}{A_1}I_2 \tag{20.61}$$

Substituting this result into the equation for V_2 in (20.57), we have:

$$V_2 = B_1\left(\frac{-A_2}{A_1}\right)I_2 + B_2I_2$$

or (20.62)

$$\left.\frac{V_2}{I_2}\right|_{U_o=0} = \frac{A_1B_2 - A_2B_1}{A_1}$$

Since the quantity $(A_1B_2 - A_2B_1)/A_1$ is the ratio of V_2 to I_2 with the output *nulled*, we rename this quantity as Z_N. Then rewriting (20.60) with Z_d and Z_N, we have:

$$\frac{U_o}{U_i} = A_1\left[\frac{1+\dfrac{Z_N}{Z}}{1+\dfrac{Z_d}{Z}}\right] \tag{20.63}$$

Equation (20.63) shows that the results of connecting the extra element Z into the circuit can be expressed as the product of A_1, which is the network function with Z set to infinity, times a correction factor given in the brackets in (20.63).

Example 15. Use the parallel version of Middlebrook's extra element theorem to find the voltage transfer function of the ideal op amp circuit shown in Fig. 20.52 when a capacitor C is connected between terminals 1 and 1′.

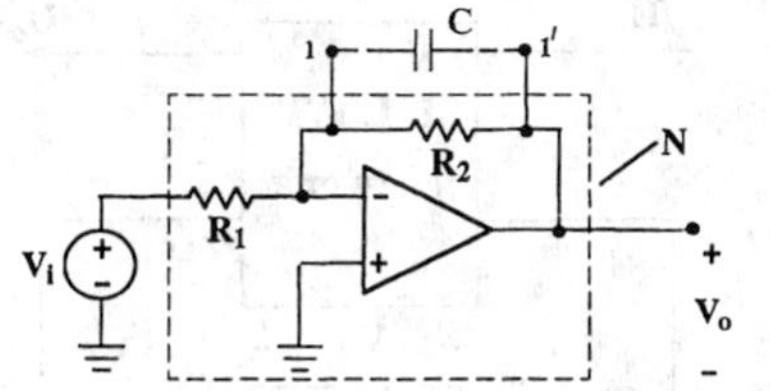

FIGURE 20.52 Ideal op amp circuit for Example 15.

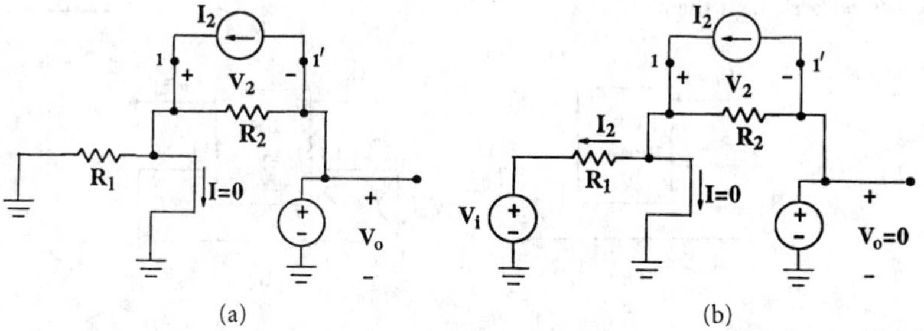

FIGURE 20.53 (a) Network for finding Z_d. (b) Network used to determine Z_N.

Solution. With the capacitor not connected, the voltage transfer function is

$$\left.\frac{V_o}{V_i}\right|_{Z=\infty} = -\frac{R_2}{R_1} = A_1$$

Next, we determine Z_d from the circuit shown in Fig. 20.53(a), where a model has been included for the ideal op amp, the excitation V_i has been properly removed, and a current excitation I_2 has been applied to the port formed by terminals 1 and 1′. Since there is no voltage across R_1 in Fig. 20.53(a), there is no current through it, and all the current I_2 flows through R_2. Thus, $V_2 = I_2 R_2$, and $Z_d = R_2$. We next find Z_N from Fig. 20.53(b). We observe in this figure that the right end of R_2 is zero volts above ground because V_i and I_2 have been adjusted so that V_o is zero. Furthermore, the left end of R_2 is zero volts above ground because of the virtual ground of the op amp. Thus, there is zero current flowing through R_2, and so V_2 is zero. Consequently, $Z_N = V_2/I_2 = 0$. Following the format of (20.63), we have:

$$\frac{V_o}{V_i} = -\frac{R_2}{R_1}\left(\frac{1}{1+sCR_2}\right)$$

Note that for V_o to be zero in Fig. 20.53(b), V_i and I_2 must be adjusted so that $V_i/R_1 = -I_2$, although this information was not needed to work the example.

Series Version. The series version of the theorem allows us to cut a loop of the network, add an impedance Z in series, and obtain the modified network function as $A_1(s)$ multiplied by a correction factor. A_1 is the network function when $Z = 0$. Figure 20.54 shows an LLFT network with part of a loop shown explicitly. The quantities U_i and U_o represent a general input and a general output, respectively, whether they be a voltage or a current. Define

$$A_1 = \left.\frac{U_o}{U_i}\right|_{V_2=0} \qquad A_2 = \left.\frac{U_o}{V_2}\right|_{U_i=0}$$
$$B_1 = \left.\frac{I_2}{U_i}\right|_{V_2=0} \qquad B_2 = \left.\frac{I_2}{V_2}\right|_{U_i=0} \tag{20.64}$$

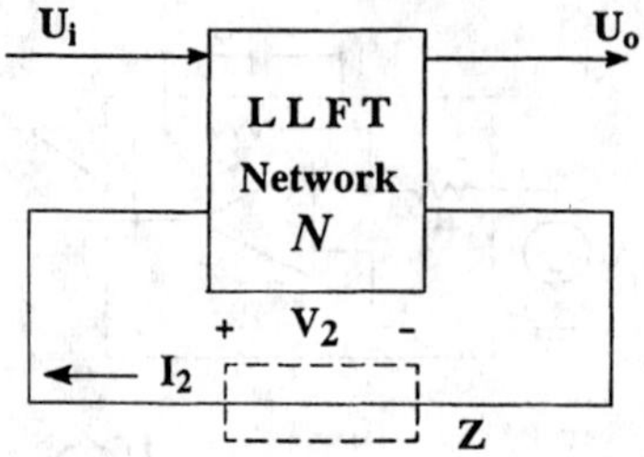

FIGURE 20.54 LLFT network used for the series version of Middlebrook's extra element theorem.

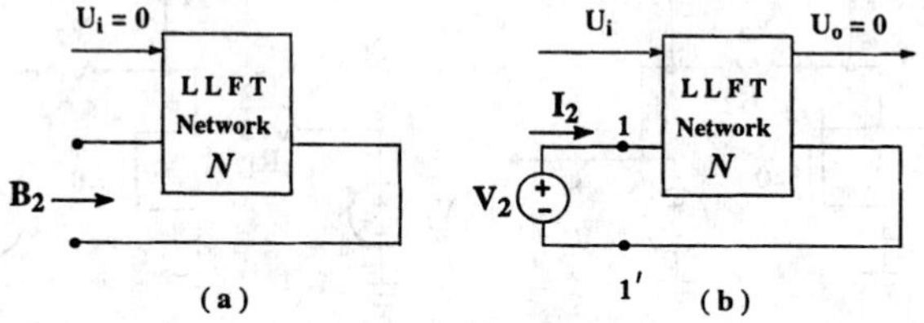

FIGURE 20.55 (a) Looking into the network with U_i equal zero. (b) U_i and V_2 are simultaneously adjusted to null the output U_o.

where V_2 and I_2 are depicted in Fig. 20.54, and A_1 is assumed to be known. Then using superposition, we have:

$$\begin{aligned} U_0 &= A_1 U_i + A_2 V_2 \\ I_2 &= B_1 U_i + B_2 V_2 \end{aligned} \tag{20.65}$$

The impedance of the extra element Z can be described by $Z = V_2/(-I_2)$ so that $V_2 = -I_2 Z$. Substituting this relation for V_2 into (20.65) and eliminating I_2, we have:

$$\frac{U_o}{U_i} = A_1 \left[\frac{1 + Z\left(B_2 - B_1 \dfrac{A_2}{A_1}\right)}{1 + B_2 Z} \right] \tag{20.66}$$

Again, as we did for the parallel version of the theorem, we look for physical interpretations of the quantities in the square bracket in (20.66). From (20.65) we see that B_2 is the admittance looking into the port formed by cutting the loop in Fig. 20.54 with $U_i = 0$. This is depicted in Fig. 20.55(a). Thus, B_2 is the admittance looking into a dead network, and so let $B_2 = 1/Z_d$.

To find a physical interpretation of the quantity $(A_1B_2 - A_2B_1)/A_1$, we examine Fig. 20.55(b) in which both inputs, V_2 and U_i, are adjusted to null the output U_o. From (20.65) with $U_o = 0$, we have:

$$U_i = -\frac{A_2}{A_1} V_2 \tag{20.67}$$

Then eliminating U_i in (20.65) we obtain:

$$\frac{A_1B_2 - A_2B_1}{A_1} = \left.\frac{I_2}{V_2}\right|_{U_o=0} \tag{20.68}$$

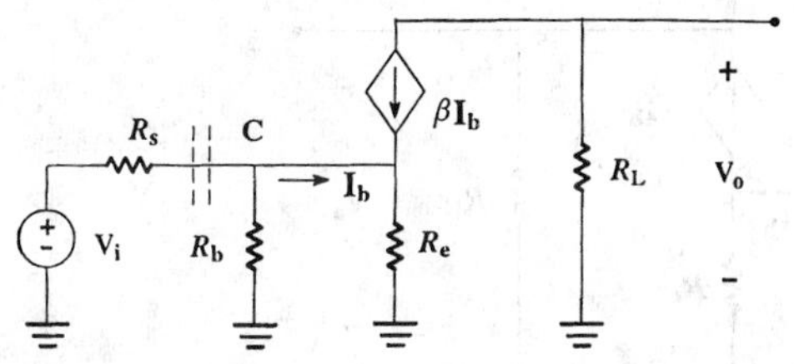

FIGURE 20.56 Network for Example 16.

Since this quantity is the admittance looking into the port formed by terminals 1 and 1′ in Fig. 20.55(b) with U_o nulled, rename it as $1/Z_N$. Thus, from (20.66) we can write

$$\frac{U_o}{U_i} = A_1 \left[\frac{1 + \dfrac{Z}{Z_N}}{1 + \dfrac{Z}{Z_d}} \right] \tag{20.69}$$

Eq. (20.69) is particularly convenient for determining the effects of adding an impedance Z into a loop of a network.

Example 16. Use the series version of Middlebrook's extra element theorem to determine the effects of inserting a capacitor C in the location indicated in Fig. 20.56.

Solution. The voltage transfer function for the network without the capacitor is found to be:

$$A_1 = \left.\frac{V_o}{V_i}\right|_{Z=0} = \frac{-\beta R_L}{R_s + (\beta + 1)R_e\left(1 + \dfrac{R_s}{R_b}\right)}$$

Next we find Z_d from Fig. 20.57(a). This yields:

$$Z_d = \frac{V}{I} = R_s + [R_b \| (\beta + 1)R_e]$$

The impedance Z_N is found from Fig. 20.57(b) where the two input sources, V_i and V, are adjusted so that V_o equals zero. If V_o equals zero, then βI_b equals zero since there is no current through R_L. Thus, I_b equals zero, which implies that V_{Re}, the voltage across R_e as indicated in Fig. 20.57(b), is also zero. We see that the null is propagating through the circuit. Continuing to analyze Fig. 20.57(b), we see that I_{Rb} is zero so that we conclude that I is zero. Since $Z_N = V/I$, we conclude that Z_N is infinite. Using the format given by (20.69) with $Z = 1/(sC)$, we obtain the result as:

$$\frac{V_o}{V_i} = A_1 \left\{ \frac{1}{1 + \dfrac{1/(sC)}{R_s + [R_b \| (\beta + 1)R_e]}} \right\}$$

It is interesting to note that to null the output so that Z_N could be found in Example 16, V_i is set to V, although this fact is not needed in the analysis.

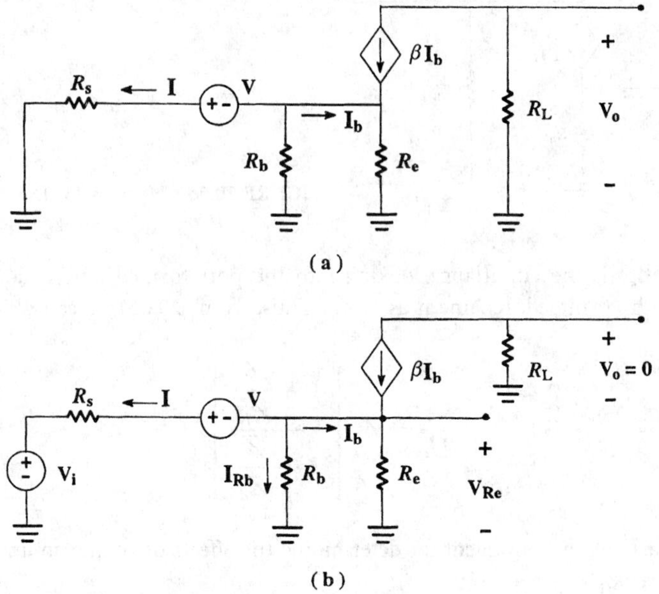

FIGURE 20.57 (a) Network used to obtain Z_d. (b) Network which yields Z_N.

Substitution Theorem

Figure 20.58(a) shows an LLFT network consisting of two subnetworks A and B which are connected to each other by two wires. If the voltage $v(t)$ is known, the voltages and currents in subnetwork A remain unchanged if a voltage source of value $v(t)$ is substituted for subnetwork B as illustrated in Fig. 20.58(b).

Example 17 Determine $i_1(t)$ in the circuit shown in Fig. 20.59. The voltage $v_1(t)$ is known from a previous analysis.

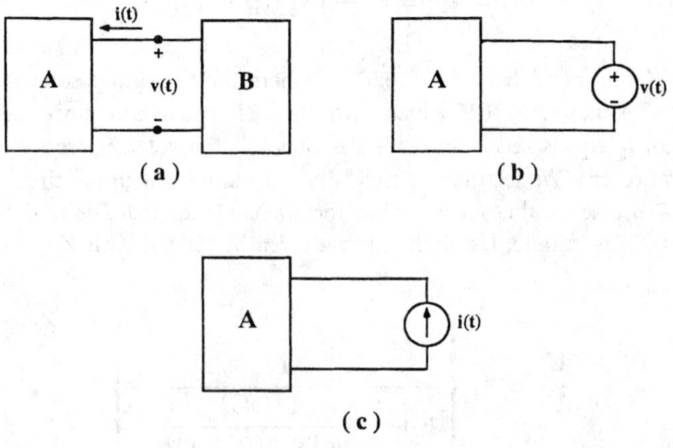

FIGURE 20.58 (a) An LLFT network consisting of two subnetworks A and B connected by two wires. (b) A voltage source can be substituted for subnetwork B if $v(t)$ is known in (a). (c) A current source can be substituted for B if i is a known current.

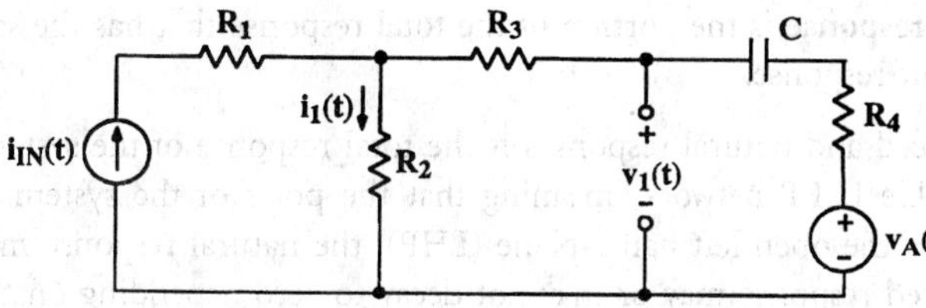

FIGURE 20.59 Circuit for example 17.

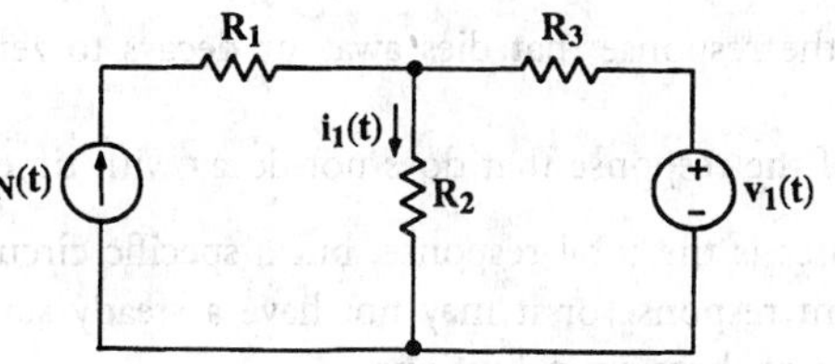

FIGURE 20.60 Circuit that results when the substitution theorem is applied to the circuit in Fig. 20.59.

Solution. Since $v_1(t)$ is known, the substitution theorem can be applied to obtain the circuit in Fig. 20.60. Analysis of this simplified circuit yields:

$$i_1 = i_{in}\frac{R_3}{R_2 + R_3} + v_1\frac{1}{R_2 + R_3}$$

If the current $i(t)$ is known in Fig. 20.58(a), then the substitution shown in Fig. 20.58(c) can be employed.

.3 Sinusoidal Steady State Analysis and Phasor Transforms

Sinusoidal Steady State Analysis

In this section we develop techniques for analyzing lumped, linear, finite, time invariant (LLFT), networks in the sinusoidal steady state. These techniques are important for analyzing and designing networks ranging from AC power generation systems to electronic filters.

To put the development of sinusoidal steady state analysis in its context, we list the following definitions of responses of circuits:

A. The **zero-input response** is the response of a circuit to its initial conditions when the input excitations are set to zero.
B. The **zero-state response** is the response of a circuit to a given input excitation or set of input excitations when the initial conditions are all set to zero.

The sum of the zero-input response and the zero-state response yields the total response of the system being analyzed. However, the total response can also be decomposed into the **forced response** and the **natural response** if the input excitations are DC, real exponentials, sinusoids, and/or sinusoids multiplied by real exponentials and if the exponent(s) in the input excitation differs from the exponents appearing in the zero-input response. These excitations are very common in engineering applications, and the decomposition of the response into forced and natural components corresponds to the particular and complementary (homogeneous) solutions, respectively, of the linear, constant coefficient, ordinary differential equations that characterize LLFT networks in the time domain. Therefore, we define:

C. The **forced response** is the portion of the total response that has the same exponents as the input excitations.

D. The **natural response** is the portion of the total response that has the same exponents as the zero-input response.

The sum of the forced and natural responses is the total response of the system.

For a strictly stable LLFT network, meaning that the poles of the system transfer function $T(s)$ are confined to the open left-half s-plane (LHP), the natural response must decay to zero eventually. The forced response may or may not decay to zero depending on the excitation and the network to which it is applied, and so it is convenient to define the terms **steady state response** and **transient response**:

E. The **transient response** is the portion of the response that dies away or decays to zero with time.

F. The **steady-state response** is the portion of the response that does not decay with time.

The sum of the transient and steady state responses is the total response, but a specific circuit with a specific excitation may not have a transient response or it may not have a steady state response. The following example illustrates aspects of these six definitions.

Example 18. Find the total response of the network shown in Fig. 20.61, and identify the zero-state, zero-input, forced, natural, transient, and steady state portions of the response.

Solution. Note that there is a nonzero initial condition represented by V_c. Using superposition and the simple voltage divider concept, we can write:

$$V_o(s) = \left[\frac{V_1}{s+1} + \frac{V_2}{s^2+1}\right]\frac{2}{s+2} + \frac{V_c}{s}\left(\frac{s}{s+2}\right)$$

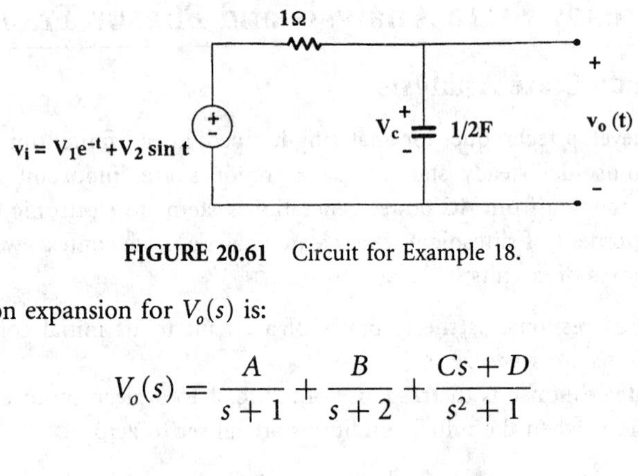

FIGURE 20.61 Circuit for Example 18.

The partial fraction expansion for $V_o(s)$ is:

$$V_o(s) = \frac{A}{s+1} + \frac{B}{s+2} + \frac{Cs+D}{s^2+1}$$

where

$$A = 2V_1$$
$$B = -2V_1 + 0.4V_2 + V_c$$
$$C = -0.4V_2$$
$$D = 0.8V_2$$

Thus, for $t \geq 0$, $v_o(t)$ can be written as:

$$v_o(t) = 2V_1e^{-t} - 2V_1e^{-2t} + 0.4V_2e^{-2t} + V_ce^{-2t} - 0.4V_2\cos t + 0.8V_2\sin t$$

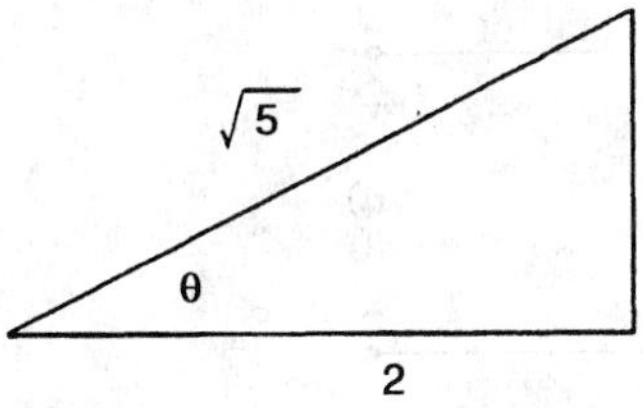

FIGURE 20.62 Sketch for determining the phase angle θ.

With the aid of the angle-sum and difference formula

$$\sin(\alpha \pm \beta) = \sin\alpha \cos\beta \pm \cos\alpha \sin\beta$$

and the sketch in Fig. 20.62. we can combine the last two terms in the expression for $v_o(t)$ to obtain:

$$v_o(t) = 2V_1e^{-t} - 2V_1e^{-2t} + 0.4V_2e^{-2t} + V_ce^{-2t} + 0.4\sqrt{5}V_2 \sin(t + \theta)$$

where

$$\theta = -\tan^{-1}\left(\tfrac{1}{2}\right)$$

The terms of $v_o(t)$ are characterized by our definitions as follows:

$$\text{zero-input response} = V_ce^{-2t}$$

$$\text{zero-state response} = 2V_1e^{-t} - 2V_1e^{-2t} + 0.4V_2e^{-2t} + 0.4\sqrt{5}V_2 \sin(t + \theta)$$

$$\text{natural response} = [-2V_1 + 0.4V_2 + V_c]e^{-2t}$$

$$\text{forced response} = 2V_1e^{-t} + 0.4\sqrt{5}V_2 \sin(t + \theta)$$

$$\text{transient response} = 2V_1e^{-t} + [-2V_1 + 0.4V_2 + V_c]e^{-2t}$$

$$\text{steady state response} = 0.4\sqrt{5}V_2 \sin(t + \theta)$$

As can be seen by comparing the terms above with the total response, part of the forced response is the steady state response, and the rest of the forced response is included in the transient response in this example.

If the generator voltage in the previous example had been $v_i = V_1 + V_2 \sin(t)$, then there would have been two terms in the steady state response—a DC term and a sinusoidal term. On the other hand, if the transfer function from input to output had a pole at the origin and the excitation were purely sinusoidal, there would also have been a DC term and a sinusoidal term in the steady state response. The DC term would have arisen from the pole at the origin in the transfer function and therefore would also be classed as a term in the natural response.

Oftentimes, it is desirable to obtain only the sinusoidal steady state response, without having to solve for other portions of the total response. The ability to solve for just the sinusoidal steady state response is the goal of sinusoidal steady state analysis.

The sinusoidal steady state response can be obtained based on analysis of the network using Laplace transforms. Figure 20.63 shows an LLFT network which is excited by the voltage sine wave $v_i(t) = V \sin(\omega t)$, where V is the peak amplitude of the sine wave and ω is the frequency of the sine wave in radians/second. Assume that the poles of the network transfer function

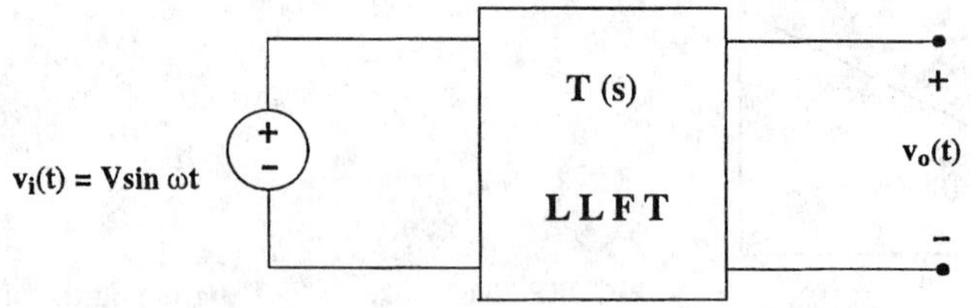

FIGURE 20.63 LLFT network with transfer function $T(s)$.

$V_o(s)/V_i(s) = T(s)$ are confined to the open left-half s-plane (LHP) except possibly for a single pole at the origin. Then the forced response of the network is

$$v_{oss}(t) = V|T(j\omega)| \sin(\omega t + \theta) \tag{20.70}$$

where the extra subscripts "ss" on $v_o(t)$ indicate sinusoidal steady state and where

$$\theta = \tan^{-1}\left(\frac{\mathcal{I}T(j\omega)}{\mathcal{R}T(j\omega)}\right) \tag{20.71}$$

The symbols $\mathcal{I}$ and $\mathcal{R}$ are read as "imaginary part of" and "real part of", respectively. In other words, the LLFT network modifies the sinusoidal input signal in only two ways at steady state. The network multiplies the amplitude of the signal by $|T(j\omega)|$ and shifts the phase by θ. If the transfer function of the network is known beforehand, then the sinusoidal steady state portion of the total response can be easily obtained by means of Eqs. (20.70) and (20.71).

To prove (20.70) and (20.71), we assume that $T(s) = V_o(s)/V_i(s)$ in Fig. 20.63 is real for s real and that the poles of $T(s)$ are confined to the open LHP except possibly for a single pole at the origin. Without loss of generality, assume the order of the numerator of $T(s)$ is at most one greater than the order of the denominator. Then the transform of the output voltage is

$$V_o(s) = V_i(s)T(s) = \frac{V\omega}{s^2 + \omega^2} T(s)$$

If $V_o(s)$ is expanded into partial fractions, we have:

$$V_o(s) = \frac{A}{s - j\omega} + \frac{B}{s + j\omega} + \text{other terms due to the poles of } T(s)$$

The residue A is

$$A = [(s - j\omega)V_o(s)]_{s=j\omega} = \left[\frac{V\omega}{s + j\omega} T(s)\right]_{s=j\omega}$$
$$= \frac{V}{2j} T(j\omega)$$

But

$$T(j\omega) = |T(j\omega)|e^{j\theta} \quad \text{where} \quad \theta = \tan^{-1}\frac{\mathcal{I}T(j\omega)}{\mathcal{R}T(j\omega)}$$

Thus, we can write the residue A as

$$A = \frac{V}{2j}|T(j\omega)|e^{j\theta}$$

Also, $B = A^*$ where "*" denotes "conjugate", and so

$$B = -\frac{V}{2j}T(-j\omega) = -\frac{V}{2j}T((j\omega)^*)$$
$$= -\frac{V}{2j}T^*(j\omega) = -\frac{V}{2j}|T(j\omega)|e^{-j\theta}$$

In the equation for the residue B, we can write $T((j\omega)^*) = T^*(j\omega)$ because of the assumption that $T(s)$ is real for s real (see Property 1 in Section 1 on Properties of LLFT Network Functions).

All other terms in the partial fraction of $V_o(s)$ will yield, when inverse transformed, functions of time that decay to zero except for a term arising from a pole at the origin of $T(s)$. A pole at the origin yields, when its partial fraction is inverse transformed, a DC term which is part of the steady state solution in the time domain. However, only the first two terms in $V_o(s)$ will ultimately yield a sinusoidal function. We can rewrite these two terms as:

$$V_{oss}(s) = \frac{V}{2j}|T(j\omega)|\left[\frac{e^{j\theta}}{s - j\omega} - \frac{e^{-j\theta}}{s + j\omega}\right]$$

The extra subscripts "*ss*" denote sinusoidal steady state. The time domain equation for the sinusoidal steady state output voltage is

$$v_{oss}(t) = \frac{V}{2j}|T(j\omega)|[e^{j\theta}e^{j\omega t} - e^{-j\theta}e^{-j\omega t}]$$
$$= V|T(j\omega)|\sin(\omega t + \theta)$$

where θ is given by (20.71). This completes the proof.

Example 19. Verify the expression for the sinusoidal steady state response found in the previous example.

Solution. The transfer function for the network in Fig. 20.61 is $T(s) = 2/(s+2)$, and the frequency of the sinusoidal portion of $v_i(t)$ is $\omega = 1$ rad/s. Thus,

$$T(j\omega) = \frac{2}{j+2} = \frac{2}{\sqrt{4+1}}e^{j\theta}$$

where

$$\theta = \tan^{-1}\left(\frac{-2}{4}\right) = -\tan^{-1}\left(\frac{1}{2}\right)$$

If the excitation in Fig. 20.63 were $v_i(t) = V\sin(\omega t + \Phi)$, then the sinusoidal steady state response of the network would be:

$$v_{oss}(t) = V|T(j\omega)|\sin(\omega t + \Phi + \theta) \tag{20.72}$$

where θ is given by (20.71). Similarly, if the excitation were $v_i(t) = V[\cos(\omega t + \Phi)]$, then the sinusoidal steady state response would be expressed as:

$$v_{oss}(t) = V|T(j\omega)| \cos(\omega t + \Phi + \theta) \tag{20.73}$$

with θ again given by (20.71).

Phasor Transforms

In the sinusoidal steady state analysis of stable LLFT networks, we find that both the inputs and outputs are sine waves of the same frequency. The network only modifies the amplitudes and the phases of the sinusoidal input signals; it does not change their nature. Thus, we need only keep track of the amplitudes and phases, and we do this by using phasor transforms. Phasor transforms are closely linked to Euler's identity:

$$e^{\pm j\omega t} = \cos(\omega t) \pm j \sin(\omega t) \tag{20.74}$$

If, say, $v_i(t) = V \sin(\omega t + \Phi)$, then we can write $v_i(t)$ as

$$v_i(t) = \mathscr{I}[Ve^{j(\omega t+\Phi)}] = \mathscr{I}[Ve^{j\Phi} e^{j\omega t}] \tag{20.75}$$

Similarly, if $v_i(t) = V \cos(\omega t + \Phi)$, then we can write

$$v_i(t) = \mathscr{R}[Ve^{j(\omega t+\Phi)}] = \mathscr{R}[Ve^{j\Phi} e^{j\omega t}] \tag{20.76}$$

If we confine our analysis to single-frequency sine waves, then we can drop the imaginary sign and the term $e^{j\omega t}$ in (20.75) to obtain the phasor transform. That is,

$$\wp[v_i(t)] = \wp[V \sin(\omega t + \Phi)] = \wp\{\mathscr{I}[Ve^{j\Phi} e^{j\omega t}]\} = Ve^{j\Phi} \tag{20.77}$$

The first and last terms in (20.77) are read as "the phasor transform of $v_i(t)$ equals $Ve^{j\Phi}$". Note that $v_i(t)$ is not equal to $Ve^{j\Phi}$ as can be seen from the fact that $v_i(t)$ is a function of time while $Ve^{j\Phi}$ is not. Phasor transforms will be denoted with **bold** letters that are underlined as in $\wp[v_i(t)] = \underline{\mathbf{V}}_i$.

If our analysis is confined to single-frequency cosine waves, we perform the phasor transform in the following manner:

$$\wp[V \cos(\omega t + \Phi)] = \wp\{\mathscr{R}[Ve^{j\Phi} e^{j\omega t}]\} = Ve^{j\Phi} = \underline{\mathbf{V}} \tag{20.78}$$

In other words, to perform the phasor transform of a cosine function, we drop both the real sign and the term $e^{j\omega t}$. Both sines and cosines are sinusoidal functions, but when we transform them, they lose their identities. Thus, before starting an analysis, we must decide whether to perform the analysis all in sines or all in cosines. The two functions must not be mixed when using phasor transforms. Furthermore, we cannot simultaneously employ the phasor transforms of sinusoids at two different frequencies. However, if a network has two excitations which have different frequencies, we can use superposition in an analysis for a voltage or current, and add the solutions in the time-domain.

There are three equivalent representations for a phasor $\underline{\mathbf{V}}$. These are

$$\mathbf{V} = \begin{cases} Ve^{j\Phi} & \text{exponential form} \\ V(\cos\Phi + j\sin\Phi) & \text{rectangular form} \\ V\angle\Phi & \text{polar form} \end{cases} \tag{20.79}$$

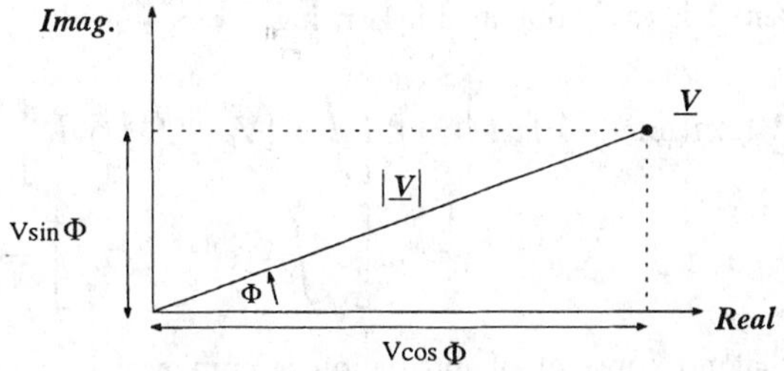

FIGURE 20.64 Relationships among phasor representations.

If phasors are to be multiplied or divided by a complex number, the exponential or polar forms are the most convenient. If phasors are to be added or subtracted, the rectangular form is the most convenient. The relationships among the equivalent representations are illustrated in Fig. 20.64. In this figure, the phasor $\underline{V}$ is denoted by a point in the complex plane. The magnitude of the phasor, $|\underline{V}| = V$, is illustrated by the length of the line drawn from the origin to the point. The phase of the phasor, Φ, is shown measured counterclockwise from the horizontal axis. The real part of $\underline{V}$ is $V\cos\Phi$ and the imaginary part of $\underline{V}$ is $V\sin\Phi$.

Phasors can be developed in a way that parallels, to some extent, the usual development of Laplace transforms. In the following, we assume that the constants V_1, V_2, Φ_1, and Φ_2 are real.

Theorem 1: *For sinusoids of the same type (either sines or cosines) and of the same frequency* ω, $\wp[V_1\sin(\omega t+\Phi_1)+V_2\sin(\omega t+\Phi_2)] = V_1\wp[\sin(\omega t+\Phi_1)]+V_2\wp[\sin(\omega t+\Phi_2)]$. *A similar relation can be written for cosines. This theorem shows that the phasor transform is a linear transform.*

Theorem 2: *If* $\wp[V_1\sin(\omega t+\Phi)] = V_1e^{j\Phi}$, *then*

$$\wp\left[\frac{d}{dt}V_1\sin(\omega t+\Phi)\right] = j\omega V_1e^{j\Phi} \tag{20.80}$$

To prove theorem 2, we can write:

$$\wp\left[\frac{d}{dt}V_1\mathscr{I}(e^{j\Phi}e^{j\omega t})\right] = \wp\left[V_1\mathscr{I}\left(e^{j\Phi}\frac{d}{dt}e^{j\omega t}\right)\right]$$
$$= \wp[V_1\mathscr{I}e^{j\Phi}j\omega e^{j\omega t}] = V_1 j\omega e^{j\Phi}$$

Note the interchange of the derivative and the imaginary sign in the proof of the theorem. Also, theorem 2 can be generalized to:

$$\wp\left[\frac{d^n}{dt^n}V_1\sin(\omega t+\Phi)\right] = (j\omega)^nV_1e^{j\Phi} \tag{20.81}$$

These results are useful for finding the sinusoidal steady state solutions of linear, constant-coefficient, ordinary differential equations assuming the roots of the characteristic polynomials lie in the open LHP with possibly one at the origin.

Theorem 3: *If* $\wp[V_1\sin(\omega t+\Phi)] = V_1e^{j\Phi}$, *then*

$$\wp\left[\int V_1\sin(\omega t+\Phi)\,dt\right] = \frac{1}{j\omega}V_1e^{j\Phi} \tag{20.82}$$

The proof of theorem 3 is easily obtained by writing:

$$\wp\left[\int V_1 \sin(\omega t + \Phi)\, dt\right] = \wp\left[\int \mathscr{I}[V_1 e^{j(\omega t+\Phi)}]\, dt\right]$$
$$= \wp\left[\mathscr{I}\int V_1 e^{j(\omega t+\Phi)}\, dt\right] = \frac{V_1}{j\omega} e^{j\Phi}$$

It should be noted that no constant of integration is employed because a constant is not a sinusoidal function and is therefore not permitted when using phasors. A constant of integration arises in LLFT network analysis because of initial conditions, and we are interested only in the sinusoidal state response and not in a zero-input response. No limits are used with the integral either, because the (constant) lower limit would also yield a constant which would imply that we are not at steady state.

Theorem 3 is easily extended to the case of n integrals:

$$\wp\left[\int \cdots \int V_1 \sin(\omega t + \Phi)(dt)^n\right] = \frac{V_1}{(j\omega)^n} e^{j\Phi} \tag{20.83}$$

This result is useful for finding the sinusoidal steady state solution of integro-differential equations.

Inverse Phasor Transforms

To obtain time domain results, we must be able to inverse transform phasors. The inverse transform operation is denoted by $\wp^{-1}$. This is an easy operation that consists of restoring the term $e^{j\omega t}$, restoring the imaginary sign (the real sign if cosines are used), and dropping the inverse transform sign. That is,

$$\begin{aligned}\wp^{-1}[V_1 e^{j\Phi}] &= \mathscr{I}[V_1 e^{j\Phi} e^{j\omega t}] \\ &= V_1 \sin(\omega t + \Phi)\end{aligned} \tag{20.84}$$

The following example illustrates both the use of theorem 2 and the inverse transform procedure.

Example 20. Determine the sinusoidal steady state solution for the differential equation:

$$\frac{d^2 f(t)}{dt^2} + 4\frac{df(t)}{dt} + 3f(t) = V \sin(\omega t + \Phi)$$

Solution. We note that the characteristic polynomial, $D^2 + 4D + 3$, has all its roots in the open LHP. The next step is to phasor
transform each term of the equation to obtain:

$$-\omega^2\, \underline{F} + 4j\omega\, \underline{F} + 3\, \underline{F} = Ve^{j\Phi}$$

where $\underline{F}(j\omega) = \wp[f(t)]$. Therefore, when we solve for $\underline{F}$ we obtain

$$\underline{F} = \frac{Ve^{j\Phi}}{(3-\omega^2) + j4\omega}$$
$$= \frac{Ve^{j\Phi} e^{j\theta}}{\sqrt{(3-\omega^2)^2 + 16\omega^2}}$$

where

$$\theta = \tan^{-1}\frac{-4\omega}{3-\omega^2} = \tan^{-1}\frac{4\omega}{\omega^2-3}$$

Thus,

$$\underline{F} = \frac{V}{\sqrt{\omega^4+10\omega^2+9}}e^{j(\Phi+\theta)}$$

To obtain a time domain function, we inverse transform $\underline{F}$ to obtain:

$$\wp^{-1}[\underline{F}\,(j\omega)] = f(t) = \frac{V}{\sqrt{\omega^4+10\omega^2+9}}\sin(\omega t+\Phi+\theta)$$

In this example, we see that the sinusoidal steady state solution consists of the sinusoidal forcing term, $V\sin(\omega t+\Phi)$, modified in amplitude and shifted in phase.

Phasors and Networks

Phasors are time independent representations of sinusoids. Thus, we can define impedances in the phasor transform domain and obtain Ohm's law-like expressions relating currents through network elements with the voltages across those elements. In addition, the impedance concept allows us to combine dissimilar elements, such as resistors with inductors, in the transform domain.

The time-domain expressions relating the voltages and currents for Rs, Ls, and Cs, repeated here for convenience, are:

$$v_R(t) = i_R(t)R \qquad v_L(t) = \frac{Ldi_L}{dt} \qquad v_c(t) = \frac{1}{C}\int i_C\,dt$$

Note that initial conditions are set to zero. Then performing the phasor transform of the time domain variables, we have

$$Z_R = R \qquad Z_L = j\omega L \qquad Z_C = \frac{1}{j\omega C}$$

We can also write the admittances of these elements as $Y_R = 1/Z_R, Y_L = 1/Z_L$, and $Y_C = 1/Z_C$. Then we can extend the impedance and admittance concepts for two-terminal elements to multiport networks in the same manner as was done in the development of Laplace transform techniques for network analysis. For example, the transfer function of the circuit shown in Fig. 20.65 can be written as:

$$\frac{\underline{V_o}\,(j\omega)}{\underline{V_i}(j\omega)} = G_{21}(j\omega)$$

where the "$j\omega$" indicates that the analysis is being performed at sinusoidal steady state [1]. It is also assumed that there are not other excitations in N in Fig. 20.65. With impedances and transfer functions defined, then all the theorems developed for Laplace transform analysis, including source transformations, have a phasor transform counterpart.

Example 21. Use phasor analysis to find the transfer function $G_{21}(j\omega)$ and $v_{oss}(t)$ for the circuit shown in Fig. 20.66.

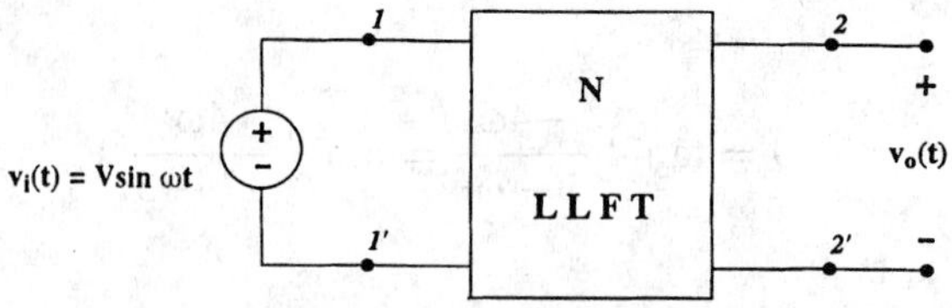

FIGURE 20.65 An LLFT network excited by a sinusoidal voltage source.

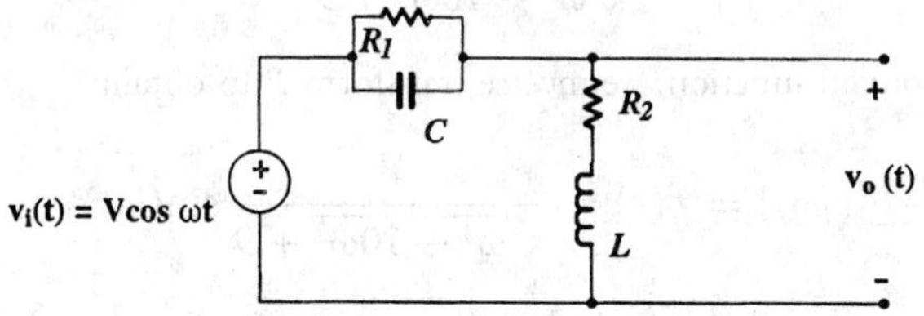

FIGURE 20.66 Circuit for Example 21.

Solution. The phasor transform of the output voltage can be obtained easily by means of the simple voltage divider. Thus,

$$\underline{V_o} = \underline{V_i} \frac{j\omega L + R_2}{j\omega L + R_2 + \dfrac{R_1}{1 + j\omega CR_1}}$$

To obtain $G_{21}(j\omega)$, we form $\underline{V_o} / \underline{V_i}$ which yields

$$\frac{\underline{V_o}}{\underline{V_i}} = G_{21}(j\omega) = \frac{(R_2 + j\omega L)(1 + j\omega CR_1)}{(R_2 + j\omega L)(1 + j\omega CR_1) + R_1}$$

Expressing the numerator and denominator of G_{21} in exponential form produces:

$$G_{21} = \frac{\sqrt{(R_2 - \omega^2 LCR_1)^2 + (\omega L + \omega CR_1 R_2)^2}\, e^{j\alpha}}{\sqrt{(R_1 + R_2 - \omega^2 LCR_1)^2 + (\omega L + \omega CR_1 R_2)^2}\, e^{j\beta}}$$

where

$$\alpha = \tan^{-1} \frac{(\omega L + \omega CR_1 R_2)}{R_2 - \omega^2 LCR_1}$$

$$\beta = \tan^{-1} \frac{(\omega L + \omega CR_1 R_2)}{R_1 + R_2 - \omega^2 LCR_1}$$

Thus,

$$G_{21}(j\omega) = Me^{j\theta}$$

where

$$M = \sqrt{\frac{(R_2 - \omega^2 LCR_1)^2 + (\omega L + \omega CR_1 R_2)^2}{(R_1 + R_2 - \omega^2 LCR_1)^2 + (\omega L + \omega CR_1 R_2)^2}}$$

and

$$\theta = \alpha - \beta$$

The phasor transform of $v_i(t)$ is

$$\underline{V_i} = \wp[V\mathcal{R}e^{j\omega t}] = Ve^{j0} = V$$

and, therefore, the time domain expression for the sinusoidal steady state output voltage is:

$$v_{oss}(t) = VM\cos(\omega t + \theta)$$

Driving point impedances and admittances as well as transfer functions are not phasors because they do not represent sinusoidal waveforms. But an impedance or a transfer function is a complex number at a particular real frequency, and the product of a complex number times a phasor is a new phasor.

The product of two arbitrary phasors is not ordinarily defined because $\sin^2(\omega t)$ or $\cos^2(\omega t)$ are not sinusoidal and have no phasor transforms. However, as we will see later, power relations for AC circuits can be expressed in efficient ways as functions of products of phasors. Since such products have physical interpretations, we permit them in the context of power calculations.

Division of one phasor by another is permitted only if the two phasors are related by a driving point or transfer network function such as $\underline{V_o} / \underline{V_i} = G_{21}(j\omega)$.

Phase Lead and Phase Lag

The terms "phase lead" and "phase lag" are used to describe the phase shift between two or more sinusoids of the same frequency. This phase shift can be expressed as an angle in degrees or radians, or it can be expressed in time as seconds. For example, suppose we have three sinusoids given by:

$$v_1(t) = V_1\sin(\omega t) \qquad v_2(t) = V_2\sin(\omega t + \Phi) \qquad v_3(t) = V_3\sin(\omega t - \Phi)$$

where V_1, V_2, V_3, and Φ are all positive. Then we say that v_2 **leads** v_1 and that v_3 **lags** v_1. To see this more clearly, we rewrite v_2 and v_3 as:

$$v_2 = V_2\sin[\omega(t + t_o)] \qquad v_3 = V_3\sin[\omega(t - t_0)]$$

where the constant $t_o = \Phi/\omega$. Figure 20.67 shows the three sinusoids sketched on the same axis, and from this graph we see that the zero crossings of $v_2(t)$ occur t_0 seconds before the zero crossings of $v_1(t)$. Thus, $v_2(t)$ leads $v_1(t)$ by t_o seconds. Similarly, we see that the zero crossings of $v_3(t)$ occur t_o seconds after the zero crossings of $v_1(t)$. Thus, $v_3(t)$ lags $v_1(t)$. We can also say that $v_3(t)$ lags $v_2(t)$. When comparing the phases of sine waves with $V\sin(\omega t)$, the key thing to look for in the arguments of the sines are the signs of the angles following ωt. A positive sign means lead and a negative sign means lag. If two sines or two cosines have the same phase angle, then they are called "in phase".

If we have $i_1(t) = I_1[\cos(\omega t - \pi/4)]$ and $i_2(t) = I_2[\cos(\omega t - \pi/3)]$, then i_2 lags i_1 by $\pi/12$ rad or 15° because even though the phases of both cosines are negative, the phase of $i_1(t)$ is less negative than the phase of $i_2(t)$. We can also say that i_1 leads i_2 by 15°.

Example 22. Suppose we have five signals with equal peak amplitudes and equal frequencies but with differing phases. The signals are: $i_1 = I[\sin(\omega t)]$, $i_2 = I[\cos(\omega t)]$, $i_3 = I[\cos(\omega t + \theta)]$, $i_4 = -I[\sin(\omega t + \psi)]$, and $i_5 = -I[\cos(\omega t - \Phi)]$. Assume I, θ, ψ, and Φ are positive.

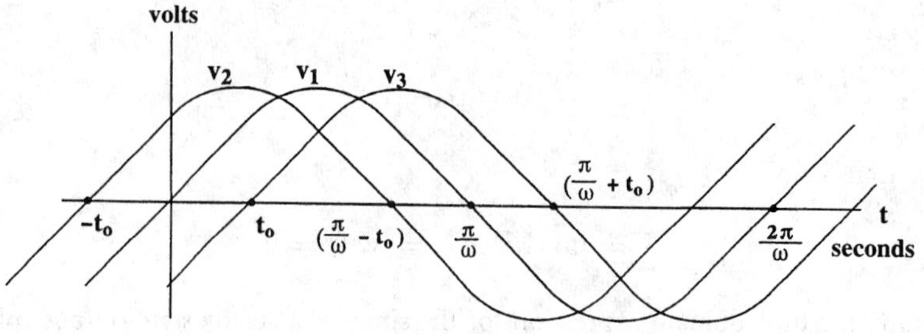

FIGURE 20.67 Three sinusoids sketched on a time axis.

A. How much do the signals i_2 through i_5 lead i_1?
B. How much do the signals i_1 and i_3 through i_5 lead i_2?

Solution. For part (A), we express i_2 through i_5 as sines with lead angles. That is,

$$i_2 = I\cos(\omega t) = I\sin\left(\omega t + \frac{\pi}{2}\right)$$

$$i_3 = I\cos(\omega t + \theta) = I\sin\left(\omega t + \theta + \frac{\pi}{2}\right)$$

$$i_4 = -I\sin(\omega t + \psi) = I\sin(\omega t + \psi \pm \pi)$$

$$i_5 = -I\cos(\omega t - \Phi) = I\cos(\omega t - \Phi \pm \pi)$$

$$= I\sin\left(\omega t - \Phi \pm \pi + \frac{\pi}{2}\right)$$

Thus, i_2 leads i_1 by $\pi/2$ rad, and i_3 leads i_1 by $\theta + \pi/2$. For i_4, we can take the plus sign in the argument of the sign to obtain $\psi + \pi$ or we can take the minus sign to obtain $\psi - \pi$. The current i_5 leads i_1 by $(3\pi/2 - \Phi)$ or by $(-\pi/2 - \Phi)$. An angle of $\pm 2\pi$ can be added to the argument without affecting lead or lag relationships.

For part (B), we express i_1 and i_3 through i_5 as cosines with lead angles yielding:

$$i_1 = I\sin(\omega t) = I\cos\left(\omega t - \frac{\pi}{2}\right)$$

$$i_3 = I\cos(\omega t + \theta)$$

$$i_4 = -I\sin(\omega t + \psi) = I\sin(\omega t + \psi \pm \pi)$$

$$= I\cos\left(\omega t + \psi \pm \pi - \frac{\pi}{2}\right)$$

$$i_5 = -I\cos(\omega t - \Phi) = I\cos(\omega t - \Phi \pm \pi)$$

We conclude that i_1 leads i_2 by $(-\pi/2)$ rad. (We could also say that i_1 lags i_2 by $(\pi/2)$ rad.) Also, i_3 leads i_2 by θ. The current i_4 leads i_2 by $(\psi + \pi/2)$ where we have chosen the plus sign in the argument of the cosine. Finally, i_5 leads i_2 by $(\pi - \Phi)$, where we have chosen the plus sign in the argument.

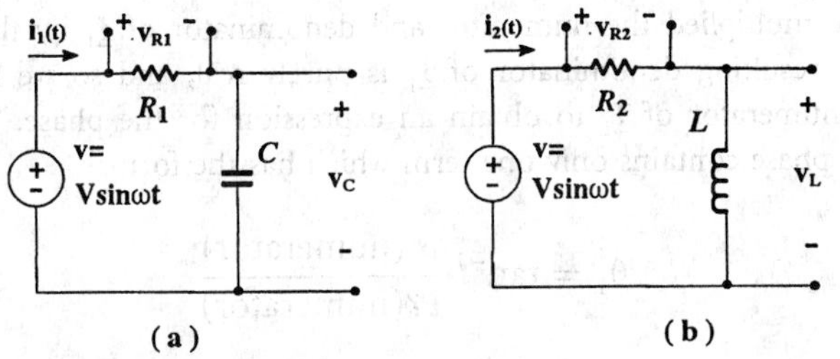

FIGURE 20.68 (a) An RC network. (b) An RL network.

In the example above, we have made use of the identities:

$$\cos(\alpha) = \sin\left(\alpha + \frac{\pi}{2}\right); \quad -\sin(\alpha) = \sin(\alpha \pm \pi)$$

$$-\cos(\alpha) = \cos(\alpha \pm \pi); \quad \sin(\alpha) = \cos\left(\alpha - \frac{\pi}{2}\right)$$

The concepts of phase lead and phase lag are clearly illustrated by means of phasor diagrams which are described in the next section.

Phasor Diagrams

Since phasors are complex numbers that represent sinusoids, phasors can be depicted graphically on a complex plane. Such graphical illustrations are called phasor diagrams. Phasor diagrams are valuable because they present a clear picture of the relationships among the currents and voltages in a network. Furthermore, addition and subtraction of phasors can be performed graphically on a phasor diagram. The construction of phasor diagrams is demonstrated in the next example.

Example 23. For the network shown in Fig. 20.68(a), find $\underline{I}_1$, $\underline{V}_{R1}$, and $\underline{V}_C$. For Fig. 20.68(b), find $\underline{I}_2$, $\underline{V}_{R2}$, and $\underline{V}_L$. Construct phasor diagrams which illustrate the relations of the currents to the voltage excitation and the other voltages of the networks.

Solution. For Fig. 20.68(a), we have

$$\wp[v(t)] = V\angle 0^\circ \quad \text{and} \quad \wp[i_1(t)] = \underline{I}_1 = \frac{V}{R_1 + \dfrac{1}{j\omega C}}$$

Rewriting $\underline{I}_1$ we have:

$$\underline{I}_1 = \frac{Vj\omega C}{1 + j\omega CR_1} = \frac{Vj\omega C}{1 + j\omega CR_1}\left[\frac{1 - j\omega CR_1}{1 - j\omega CR_1}\right]$$

$$= V\left[\frac{\omega^2 C^2 R_1 + j\omega C}{\omega^2 C^2 R_1^2 + 1}\right] = \frac{V\omega C}{\sqrt{\omega^2 C^2 R_1^2 + 1}} e^{j\theta_1}$$

where

$$\theta_1 = \tan^{-1}\frac{\omega C}{\omega^2 C^2 R_1} = \tan^{-1}\frac{1}{\omega C R_1}$$

Note that we have multiplied the numerator and denominator of $\underline{I}_1$ by the conjugate of the denominator. The resulting denominator of $\underline{I}_1$ is purely real, and so we need only consider the terms in the numerator of $\underline{I}_1$ to obtain an expression for the phase. Thus, the resulting expression for the phase contains only one term which has the form:

$$\theta_1 = \tan^{-1} \frac{\mathscr{I}(\text{numerator})}{\mathscr{R}(\text{numerator})}$$

We could have obtained the same results without application of this artifice. In this case, we would have obtained

$$\theta_1 = \frac{\pi}{2} - \tan^{-1} \omega CR_1$$

For $\omega CR_1 \geq 0$, it is easy to show that the two expressions for θ_1 are equivalent.

Since the same current flows through both network elements, we have

$$\underline{V_{R1}} = \frac{V\omega CR_1}{\sqrt{\omega^3 C^2 R_1^2 + 1}} e^{j\theta_1}$$

and

$$\underline{V_C} = \underline{I_1}\left(\frac{1}{j\omega C}\right) = \frac{-jV}{\sqrt{\omega^2 C^2 R_1^2 + 1}} e^{j\theta_1} = \frac{V}{\sqrt{\omega^2 C^2 R_1^2 + 1}} e^{j\psi}$$

where

$$\psi = -\frac{\pi}{2} + \theta_1 = -\tan^{-1} \omega CR_1$$

For $\underline{I_2}$ in Fig. 20.68(b), we obtain

$$\underline{I_2} = \frac{V\angle 0°}{R_2 + j\omega L} = \frac{V}{\sqrt{R_2^2 + \omega^2 L^2}} e^{j\theta_2}$$

where θ_2 is given by

$$\theta_2 = -\tan^{-1} \frac{\omega L}{R_2}$$

The phasor current $\underline{I_2}$ flows through both R_2 and L. So we have:

$$\underline{V_{R2}} = \underline{I_2}\, R_2$$

and

$$\underline{V_L} = j\omega L\, \underline{I_2} = \frac{V\omega L}{\sqrt{\omega^2 L^2 + R_2^2}} e^{j\Phi}$$

where

$$\Phi = \frac{\pi}{2} + \theta_2$$

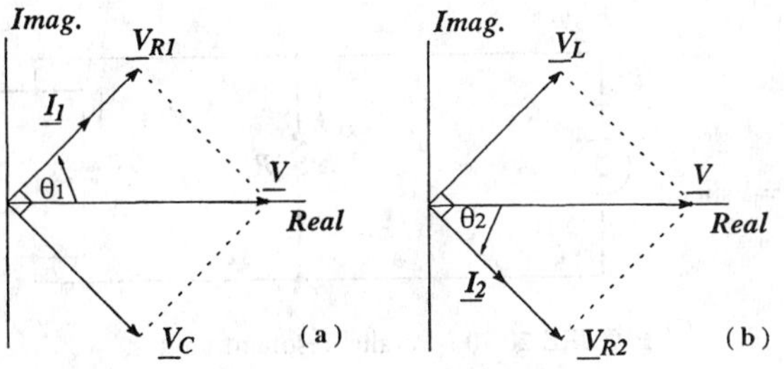

FIGURE 20.69 (a) Phasor diagram for the voltages and currents in Fig. 20.68(a). (b) Phasor diagram for Fig. 20.68(b).

To construct the phasor diagram shown in Fig. 20.69(a) for the RC network in Fig. 20.68(a), we first draw a vector corresponding to the phasor transform $\underline{V} = V\angle 0°$ of the excitation. Since the phase of this phasor is zero, it is represented as a vector along the positive real axis. The length of this vector is $|\underline{V}|$. Then we construct the vector representing $\underline{I}_1 = |\underline{I}_1| e^{j\theta 1}$. Again, the length of the vector is $|\underline{I}_1|$, and it is drawn at the angle θ_1. The vector representing $\underline{V}_{R1}$ lies along $\underline{I}_1$ because the voltage across a resistor is always in phase or 180° out of phase with the current flowing through the resistor. The vector representing the current leads $\underline{V}_C$ by exactly 90°. It should be noted from the phasor diagram that $\underline{V}_{R1}$ and $\underline{V}_C$ add to produce $\underline{V}$ as required by Kirchhoff's law.

Figure 20.69(b) shows the phasor diagram for the RL network in Fig. 20.68(b). For this network, $\underline{I}_2$ lags $\underline{V}_L$ by exactly 90°. Also, the vector sum of the voltages $\underline{V}_L$ and $\underline{V}_{R2}$ must be the excitation voltage $\underline{V}$ as indicated by the dotted lines in Fig. 20.69(b).

If the excitation $V\sin(\omega t)$ had been $V\sin(\omega t + \Phi)$ in Fig. 20.68 in the previous example, then the vectors in the phasor diagrams in Fig. 20.69 would have just been rotated around the origin by Φ. Thus, for example, $\underline{I}_1$ in Fig. 20.69(a) would have an angle equal to $\theta_1 + \Phi$. The lengths of the vectors and the relative phase shifts between the vectors would remain the same.

If R_1 in Fig. 20.68(a) is decreased, then from the expression for $\theta_1 = \tan^{-1}(1/(\omega C R_1))$ we see that the phase of $\underline{I}_1$ is increased. As R_1 is reduced further, θ_1 approaches 90°, and the circuit becomes more nearly like a pure capacitor. However, as long as $\underline{I}_1$ leads $\underline{V}$, we label the circuit as capacitive.

As R_2 in Fig. 20.68(b) is decreased, then θ_2 in Fig. 20.69(b) decreases (becomes more negative) and approaches −90°. Nevertheless, as long as $\underline{I}_2$ lags $\underline{V}$, we refer to the circuit as inductive.

If there are both inductors and capacitors in a circuit, then it is possible for the circuit to appear capacitive at some frequencies and inductive at others. An example of such a circuit is provided in the next section.

Resonance

There are two basic types of resonant networks—the parallel resonant network and the series resonant (sometimes called antiresonant) network. More complicated networks may contain a variety of both types of resonant circuits. To see what happens at resonance, we examine a parallel resonant network at sinusoidal steady state [1]. Figure 20.70 shows a network consisting of a capacitor and inductor connected in parallel, often called a tank circuit or tank, and an additional resistor R_1 connected in parallel with the tank. The phasor transforms of the excitation and the currents through the elements in Fig. 20.70 are:

$$\underline{V} = V\angle 0°; \quad \underline{I}_{R1} = \frac{V}{R_1}; \quad \underline{I}_C = j\omega C V; \quad \underline{I}_L = \frac{V}{j\omega L} \tag{20.85}$$

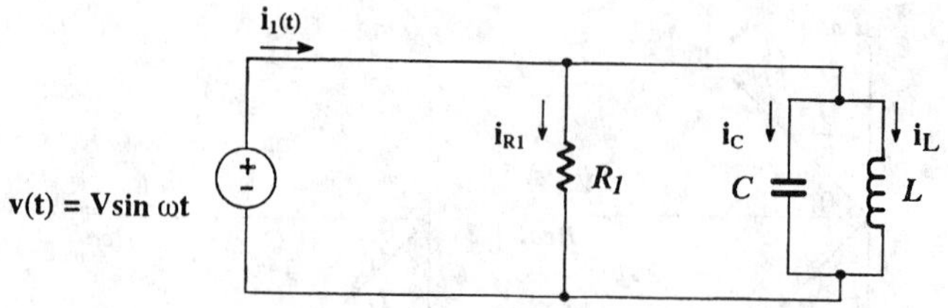

FIGURE 20.70 Parallel resonant circuit.

where V is the peak value of the excitation. The transform of the current supplied by the source is

$$\underline{I_1} = I_1\angle\theta_1 = \underline{I_{R1}} + \underline{I_C} + \underline{I_L} = V\left[\frac{1}{R_1} + j\omega C\left(1 - \frac{1}{\omega^2 LC}\right)\right] \tag{20.86}$$

The peak value of the current $i_1(t)$ at steady state is

$$I_1 = V\sqrt{\left(\frac{1}{R_1}\right)^2 + \omega^2 C^2\left(1 - \frac{1}{\omega^2 LC}\right)^2} \tag{20.87}$$

It is not difficult to determine that the minimum value of I_1 occurs at

$$\omega = \frac{1}{\sqrt{LC}} \tag{20.88}$$

which is the condition for resonance, and $I_{1\,min}$ is given by

$$I_{1min} = \frac{V}{R_1} \tag{20.89}$$

This result is somewhat surprising since it means that at resonance, the source in Fig. 20.70 delivers no current to the tank at steady state. However, this result does not mean that the currents through the capacitor and inductor are zero. In fact, for $\omega^2 = 1/(LC)$, we have:

$$\underline{I_C} = jV\sqrt{\frac{C}{L}} \quad \text{and} \quad \underline{I_L} = -jV\sqrt{\frac{C}{L}}$$

That is, the current through the inductor is 180° out of phase with the current through the capacitor, and, since their magnitudes are equal, their sum is zero. Thus, at steady state and at the frequency given by (20.88), the tank circuit looks like an open circuit to the voltage source. Yet there is a circulating current in the tank, labeled $\underline{I_T}$ in Fig. 20.71, which can be quite large depending on the values of C and L. That is, at resonance,

$$\underline{I_T} = jV\sqrt{\frac{C}{L}} = \underline{I_C} = -\underline{I_L} \tag{20.90}$$

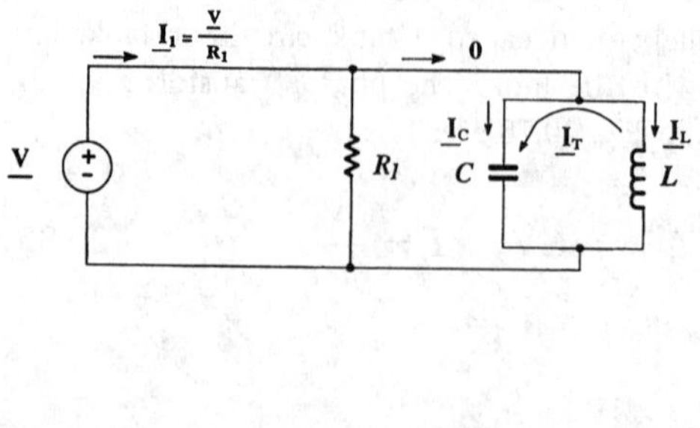

FIGURE 20.71 Circuit of Fig. 20.70 at resonance. No current is supplied to the tank by the source, but there is a circulating current in the tank.

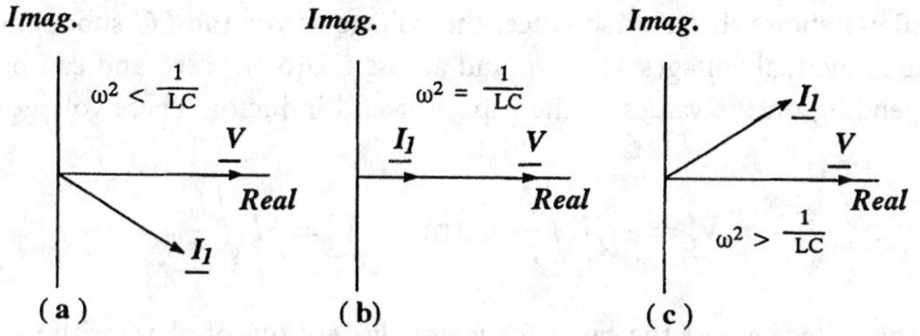

FIGURE 20.72 Phasor diagrams for the circuit in Fig. 20.70. (a) $\omega^2 < 1/(LC)$. (b) Diagram at resonance. (c) $\omega^2 > 1/(LC)$.

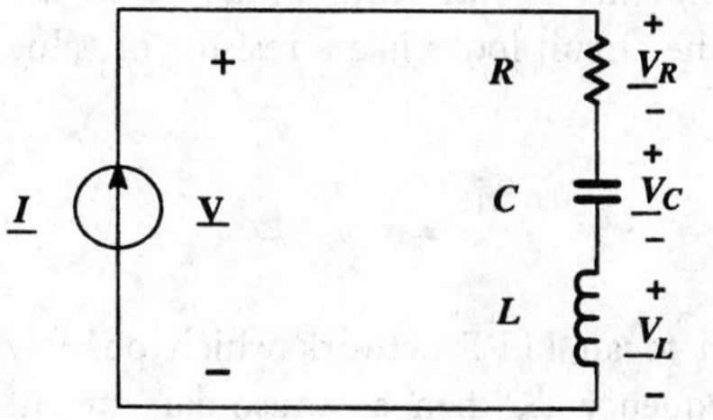

FIGURE 20.73 Series resonant circuit.

Therefore, energy is being transferred back and forth between the inductor and the capacitor. If the inductor and capacitor are ideal, the energy transferred would never decrease. In practice, parasitic resistances, especially in a physical inductor, would eventually dissipate this energy. Of course, parasitic resistances can be modeled as additional elements in the network.

Another interesting aspect of the network in Fig. 20.70 is that at low frequencies ($\omega^2 < 1/(LC)$), $\underline{I}_1$ lags $\underline{V}$, and so the network appears inductive to the voltage source. At high frequencies ($\omega^2 > 1/(LC)$), $\underline{I}_1$ leads $\underline{V}$, and the network looks capacitive to the voltage source. At resonance, the network appears as only a resistor R_1 to the source. Figure 20.72 depicts phasor diagrams of $\underline{V}$ and $\underline{I}_1$ at low frequency, at resonance, and at high frequency.

Figure 20.73 shows the second basic type of resonant circuit—a series resonant circuit which is excited by a sinusoidal current source with phasor transform $\underline{I} = I\angle 0°$. This circuit is dual to the circuit in Fig. 20.70. The voltages across the network elements can be expressed as:

$$\underline{V_R} = IR; \quad \underline{V_C} = -j\left(\frac{1}{\omega C}\right)I; \quad \underline{V_L} = j\omega LI \tag{20.91}$$

Then the voltage $\underline{V}$ is

$$\underline{V} = I\left[R + j\left(\omega L - \frac{1}{\omega C}\right)\right] \tag{20.92}$$

The peak value of $\underline{V}$ is

$$V = I\sqrt{R^2 + \left(\omega L - \frac{1}{\omega C}\right)^2} \tag{20.93}$$

where I is the peak value of $\underline{I}$. The minimum value of V, is

$$V_{\min} = IR \tag{20.94}$$

and this occurs at the frequency $\omega = 1/\sqrt{(LC)}$ which is the same resonance condition as for the circuit in Fig. 20.70.

Equation (20.94) shows that at resonance, the voltage across the LC subcircuit in Fig. 20.73 is zero. But the individual voltages across L and across C are not zero and can be quite large in magnitude depending on the values of the capacitor and inductor. These voltages are given by:

$$\underline{V_C} = -jI\sqrt{\frac{L}{C}} \quad \text{and} \quad \underline{V_L} = jI\sqrt{\frac{L}{C}} \tag{20.95}$$

and therefore the voltage across the capacitor is exactly 180° out of phase with the voltage across the inductor.

At frequencies below resonance, $\underline{V}$ lags $\underline{I}$ in Fig. 20.73, and therefore the circuit looks capacitive to the source. Above resonance, $\underline{V}$ leads $\underline{I}$, and the circuit looks inductive to the source. If the frequency of the source is $\omega = 1/\sqrt{(LC)}$, the circuit looks like a resistor of value R to the source.

Power in AC Circuits

If a sinusoidal voltage $v(t) = V\sin(\omega t + \theta_v)$ is applied to an LLFT network which possibly contains other sinusoidal sources having the same frequency ω, then a sinusoidal current $i(t) = I\sin(\omega t + \theta_I)$ flows at steady state as depicted in Fig. 20.74. The instantaneous power delivered to the circuit by the voltage source is

$$p(t) = v(t)i(t) = VI\sin(\omega t + \theta_V)\sin(\omega t + \theta_I) \tag{20.96}$$

where the units of $p(t)$ are watts (W). With the aid of the trigonometric identity

$$\sin\alpha\sin\beta = \frac{1}{2}[\cos(\alpha - \beta) - \cos(\alpha + \beta)]$$

we rewrite (20.96) as

$$p(t) = \frac{1}{2}VI[\cos(\theta_V - \theta_I) - \cos(2\omega t + \theta_V + \theta_I)] \tag{20.97}$$

The instantaneous power delivered to the network in Fig. 20.74 has a component that is constant and has a component that has a frequency twice that of the excitation. At different instances of time, $p(t)$ can be positive or negative, meaning that the voltage source is delivering power to the network or receiving power from the network, respectively.

In AC circuits, however, it is usually the average power P that is of more interest than the instantaneous power $p(t)$ because average power generates the heat or performs the work.

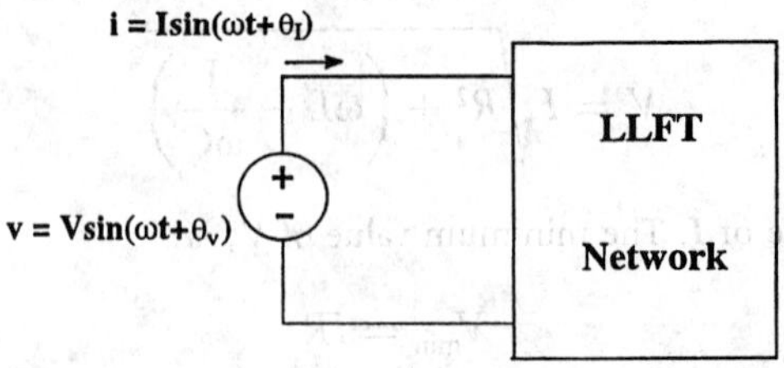

FIGURE 20.74 LLFT network that may contain other sinusoidal sources at the same frequency as the external generator.

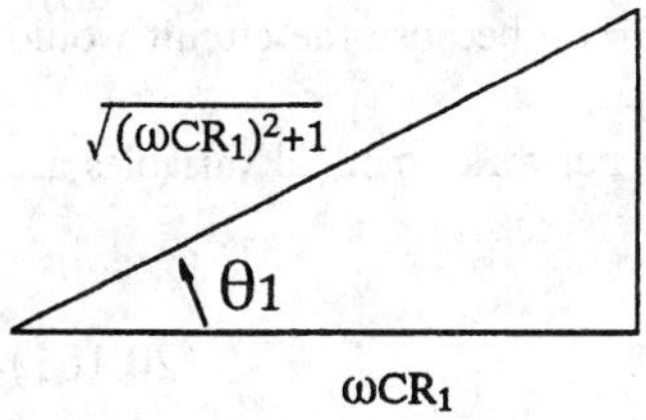

FIGURE 20.75 Triangle for determining *PF*.

The average over a period of a periodic function $f(t)$ with period T is

$$[f(t)]_{avg} = F = \frac{1}{T}\int_0^T f(t)\,dt \tag{20.98}$$

The period of $p(t)$ in (20.97) is $T = \pi/\omega$, and so

$$[p(t)]_{avg} = P = \frac{\omega}{\pi}\int_0^{\frac{\pi}{\omega}} p(t)\,dt = \frac{1}{2}VI\cos(\theta_V - \theta_I) \tag{20.99}$$

The cosine term in (20.99) plays an important role in power calculations and so is designated as the Power Factor (PF). Thus,

$$\text{Power Factor} = PF = \cos(\theta_V - \theta_I) \tag{20.100}$$

If $|\theta_V - \theta_I| = \pi/2$, then $PF = 0$, and the average power delivered to the network in Fig. 20.74 is zero. But if $PF = 1$, then P delivered to the network by the source is $VI/2$. If $0 < |\theta_V - \theta_I| < \pi/2$, then P is positive, and the source is delivering average power to the network. However, the network delivers average power to the source when P is negative, and this occurs if $\pi/2 < |\theta_V - \theta_I| < 3\pi/2$.

If the current leads the voltage in Fig. 20.74, the convention is to consider PF as leading, and if current lags the voltage, then PF is regarded as lagging. However, it is not possible from PF alone to determine whether a current leads or lags a voltage.

Example 24. Determine the average power delivered to the network shown in Fig. 20.68(a).

Solution. The phasor transform of the applied voltage is $\underline{V} = V\angle 0°$, and we determined in Example 23 that the current supplied was

$$\underline{I_1} = \frac{V\omega C e^{j\theta_1}}{\sqrt{\omega^2 C^2 R_1^2 + 1}}, \qquad \theta_1 = \tan^{-1}\frac{1}{\omega C R_1}$$

The power factor is

$$PF = \cos(0 - \theta_1) = \cos(\theta_1)$$

which, with the aid of the triangle in Fig. 20.75, can be rewritten as

$$PF = \frac{\omega C R_1}{\sqrt{(\omega C R_1)^2 + 1}}$$

Thus, the average power delivered to the circuit is

$$P = \frac{1}{2}\frac{V^2\omega C}{\sqrt{(\omega C R_1)^2 + 1}}\left[\frac{\omega C R_1}{\sqrt{(\omega C R_1)^2 + 1}}\right] = \frac{V^2\omega^2 C^2 R_1}{2(\omega^2 C^2 R_1^2 + 1)} = \frac{I_1^2 R_1}{2}$$

We note that if R_1 were zero in the previous example, then $P = 0$ because the circuit would be purely capacitive, and PF would be zero.

If there are no sources in the network in Fig. 20.74, then the network terminal variables are related by:

$$\underline{V} = \underline{I}\, Z(j\omega) \tag{20.101}$$

where $Z(j\omega)$ is the input impedance of the network. Since Z is, in general, complex, we can write it as:

$$Z(j\omega) = R(\omega) + jX(\omega) = |Z|e^{j\theta_Z} \tag{20.102}$$

where

$$R(\omega) = \mathcal{R}Z(j\omega); \qquad X(\omega) = \mathcal{I}Z(j\omega)$$
$$\text{and} \quad \theta_Z = \tan^{-1}\left(\frac{X(\omega)}{R(\omega)}\right) \tag{20.103}$$

In (20.103), the (real) function $X(\omega)$ is termed the reactance. Employing the polar form of the phasors, we can rewrite (20.101) as

$$V\angle\theta_V = I\angle\theta_I|Z|\angle\theta_Z = I|Z|\angle(\theta_I + \theta_Z) \tag{20.104}$$

Equating magnitudes and angles, we obtain

$$V = I|Z| \quad \text{and} \quad \theta_V = \theta_I + \theta_Z \tag{20.105}$$

Thus, we can express P delivered to the network as

$$P = \frac{1}{2}VI\cos(\theta_V - \theta_I) = \frac{1}{2}I^2|Z|\cos\theta_Z \tag{20.106}$$

But $|Z|\cos(\theta_Z) = R(\omega)$ so that

$$P = \frac{1}{2}I^2R(\omega) \tag{20.107}$$

Eq. (20.107) indicates that the real part of the impedance absorbs the power. The imaginary part of the impedance, $X(\omega)$, does not absorb average power. Example 24 in this section provides an illustration of (20.107).

An expression for average power in terms of the input admittance $Y(j\omega) = 1/Z(j\omega)$ can also be obtained. Again, if there are no sources within the network, then the terminal variables in Fig. 20.74 are related by

$$\underline{I} = \underline{V}\, Y(j\omega) \tag{20.108}$$

The admittance $Y(j\omega)$ can be written as

$$Y(j\omega) = |Y(j\omega)|e^{j\theta_Y} = G(\omega) + jB(\omega) \tag{20.109}$$

where $G(\omega)$ is conductance and $B(\omega)$ is susceptance, and where

$$G(\omega) = \mathcal{R}Y(j\omega); \qquad B(\omega) = \mathcal{I}Y(j\omega)$$
$$\text{and} \quad \theta_Y = \tan^{-1}\left[\frac{B(\omega)}{G(\omega)}\right] \tag{20.110}$$

Then average power delivered to the network can be expressed as:

$$P = \frac{1}{2}V^2|Y|\cos\theta_Y = \frac{1}{2}V^2G(\omega) \tag{20.111}$$

If the network contains sinusoidal sources, then (20.99) should be employed to obtain P rather than (20.107) or (20.111).

Consider a resistor R with a voltage $v(t) = V\sin(\omega t)$ across it and therefore a current $i(t) = I\sin(\omega t) = v(t)/R$ through it. The instantaneous power dissipated by the resistor is

$$p(t) = v(t)i(t) = \frac{v^2(t)}{R} = i^2(t)R \tag{20.112}$$

The average power dissipated in R is

$$P = \frac{1}{T}\int_0^T i^2(t)R\,dt = I_{eff}^2R \tag{20.113}$$

where we have introduced the new constant I_{eff}. From (20.113), we can express I_{eff} as

$$I_{eff} = \sqrt{\frac{1}{T}\int_0^T i^2(t)dt} \tag{20.114}$$

This expression for I_{eff} can be read as "the square root of the mean (average) of the square of $i(t)$" or, more simply, as "the root mean square value of $i(t)$", or, even more compactly, as "the *RMS* value of $i(t)$." Another designation for this constant is I_{rms}. Equation (20.114) can be extended to any periodic voltage or current.

The *RMS* value of a pure sine wave such as $i(t) = I\sin(\omega t + \theta_I)$ or $v(t) = V\sin(\omega t + \theta_V)$ is

$$I_{rms} = \frac{I}{\sqrt{2}} \quad \text{or} \quad V_{rms} = \frac{V}{\sqrt{2}} \tag{20.115}$$

where I and V are the peak values of the sine waves. Normally, the voltages and currents listed on the nameplates of power equipment and household appliances are given in terms of *RMS* values rather than peak values. For example, a 120-V, 100-W lightbulb is expected to dissipate 100 W when a voltage $120(\sqrt{2})[\sin(\omega t)]$ is impressed across it. The peak value of this voltage is 170 V.

If we employ *RMS* values, (20.99) can be rewritten as

$$P = V_{rms}I_{rms}PF \tag{20.116}$$

Eq. (20.116) emphasizes the fact that the concept of *RMS* values of voltages and currents was developed in order to simplify the calculation of average power.

Since $PF = \cos(\theta_V - \theta_I)$, we can rewrite (20.116) as

$$P = V_{rms} I_{rms} \cos(\theta_V - \theta_I) = \mathscr{R}[V_{rms} e^{j\theta_V} I_{rms} e^{-j\theta_I}] \\ = \mathscr{R}[\underline{V}\,\underline{I}^*] \tag{20.117}$$

where $\underline{I}^*$ is the conjugate of $\underline{I}$. If $P = \mathscr{R}[\underline{V}\,\underline{I}^*]$, the question arises as to what the imaginary part of $\underline{V}\,\underline{I}^*$ represents. This question leads naturally to the concept of complex power, denoted by the bold letter **S**, which has the units of volt-amperes (VA). If P represents real power, then we can write

$$S = P + jQ \tag{20.118}$$

where

$$S = \underline{V}\,\underline{I}^* \tag{20.119}$$

and where

$$Q = \mathscr{I}[\underline{VI}^*] = V_{rms} I_{rms} \sin(\theta_V - \theta_I) \tag{20.120}$$

Thus, Q represents imaginary or reactive power. The units of Q are VARs which stand for Volt-Amperes Reactive. Reactive power is not available for conversion into useful work. It is needed to establish and maintain the electric and magnetic fields associated with capacitors and inductors [4]. It is an overhead required for delivering P to loads, such as electric motors, that have a reactive part in their input impedances.

The components of complex power can be represented on a power triangle. Figure 20.76 shows a power triangle for a capacitive circuit. Real and imaginary power are added as shown to yield the complex power **S**. Note that $(\theta_V - \theta_I)$ and Q are both negative for capacitive circuits. The following example illustrates the construction of a power triangle for an RL circuit.

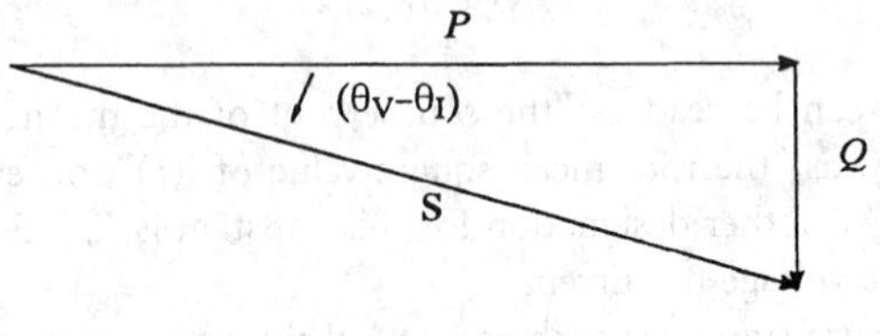

FIGURE 20.76 Power triangle for a capacitive circuit.

Example 25. Determine the components of power delivered to the RL circuit shown in Fig. 20.77. Provide a phasor diagram for the current and the voltages, construct a power triangle for the circuit, and show how the power diagram is related to the impedances of the circuit.

Solution. We have

$$\underline{V} = Ve^{j0} \quad \text{and} \quad \underline{I} = \frac{Ve^{j\theta_I}}{\sqrt{R^2 + (\omega L)^2}}$$

where

$$\theta_I = -\tan^{-1}\frac{\omega L}{R}$$

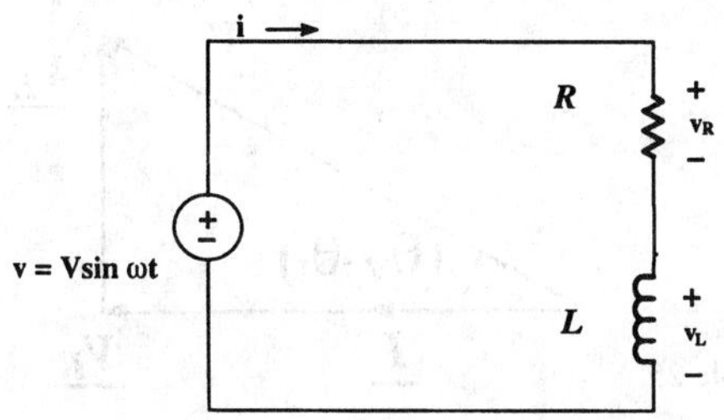

FIGURE 20.77 Network for Example 25.

Since $\theta_V = 0, PF$ is

$$PF = \cos(\theta_V - \theta_I) = \frac{R}{\sqrt{R^2 + (\omega L)^2}}$$

and is lagging. The voltages across R and L are given by:

$$\underline{\mathbf{V_R}} = \underline{\boldsymbol{I}}\, R = \frac{VRe^{j\theta_I}}{\sqrt{R^2 + (\omega L)^2}}$$

$$\underline{\mathbf{V_L}} = j\omega L\, \underline{\boldsymbol{I}} = \frac{V\omega L}{\sqrt{R^2 + (\omega L)^2}} e^{j(\frac{\pi}{2}+\theta_I)}$$

and Z is

$$Z = R + j\omega L = \sqrt{R^2 + (\omega L)^2}\, e^{-j\theta_I}$$

The real and imaginary components of the complex power are simply calculated as:

$$P = I_{rms}^2 R(\omega) = \frac{V_{rms}^2 R}{R^2 + (\omega L)^2}$$

$$Q = \frac{V_{rms}^2 \omega L}{R^2 + (\omega L)^2}$$

Figure 20.78 shows the phasor diagram for this circuit in which we have taken the reference phasor as $\underline{\boldsymbol{I}}$ and therefore have shown $\underline{\boldsymbol{V}}$ leading $\underline{\boldsymbol{I}}$ by $(\theta_V - \theta_I)$. Also, we have moved $\underline{\boldsymbol{V}_L}$ parallel to itself to form a triangle. These operations cause the phasor diagram to be similar to the power triangle. Figure 20.79(a) shows a representation for the impedance in Fig. 20.77. If each side of the triangle in Fig. 20.79(a) is multiplied by I_{rms}, then we obtain the voltage triangle shown in Fig. 20.79(b). Next, we multiply the sides of the voltage triangle by I_{rms} again to obtain the power triangle in Fig. 20.79(c). The horizontal side is the average power P, the vertical side is Q, and the hypotenuse has a length that represents the magnitude of the complex power **S**. All three triangles in Fig. 20.79 are similar. The angles between sides are preserved.

If P remains constant in Fig. 20.76, but the magnitude of the angle becomes larger so that the magnitude of Q increases, then $|\mathbf{S}|$ increases. If the magnitude of the voltage is fixed, then the magnitude of the current supplied must increase. But then, either power would be lost in the form of heat in the wires supplying the load or larger diameter, and more expensive wires would be needed. For this reason power companies that supply power to large manufacturing firms that have many large motors impose unfavorable rates. However, the manufacturing firm can improve its rates if it improves its power factor. The following example illustrates how improving (correcting) PF is done.

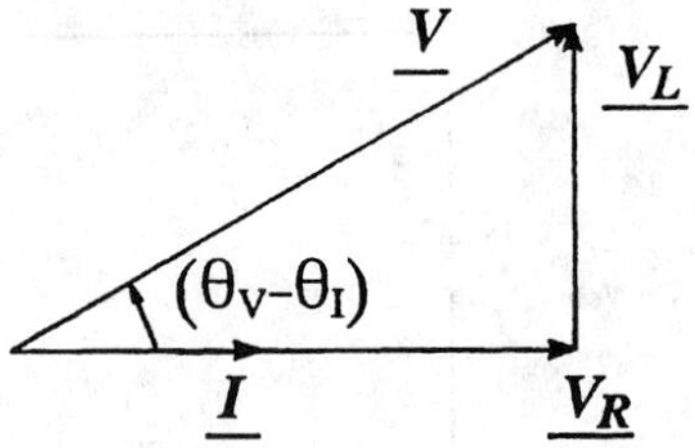

FIGURE 20.78 Phasor diagram for Example 25.

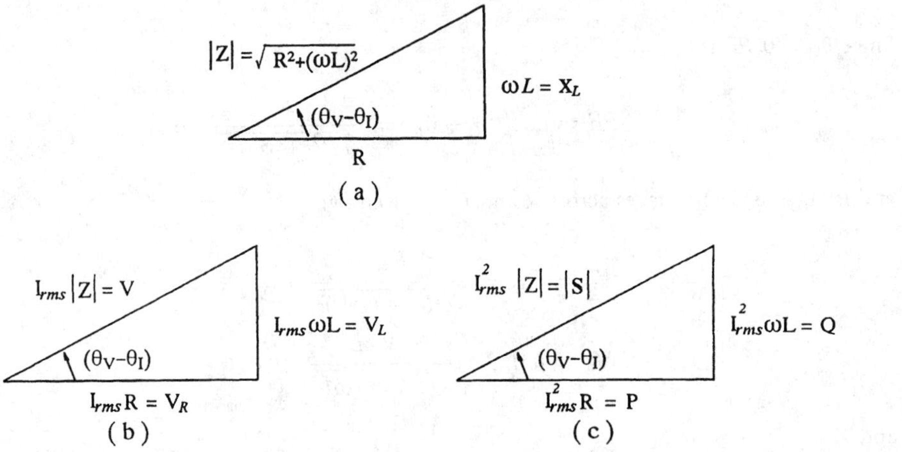

FIGURE 20.79 (a) Impedance triangle for circuit in Example 25. (b) Corresponding voltage triangle. (c) Power triangle.

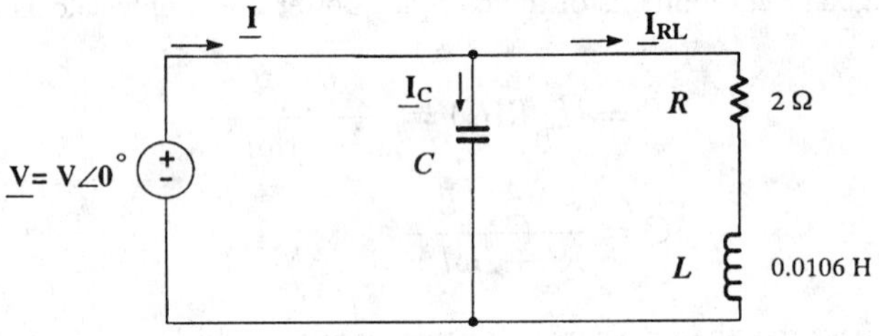

FIGURE 20.80 Circuit for Example 26.

Example 26. Determine the value of the capacitor to be connected in parallel with the RL circuit in Fig. 20.80 to improve the *PF* of the overall circuit to one. The excitation is a voltage source having an amplitude of 120 V *RMS* and frequency $2\pi(60\text{ Hz}) = 377$ rad/s. What are the RMS values of the current supplied by the source at steady state before and after the capacitor is connected?

Solution. The current through the RL branch in Fig. 20.80 is

$$\underline{I_{RL}} = \frac{Ve^{j\theta}}{\sqrt{R^2 + (\omega L)^2}}; \qquad \theta = -\tan^{-1}\frac{\omega L}{R}$$

and the current through the capacitor is

$$\underline{I_C} = j\omega CV = V\omega Ce^{j(\pi/2)}$$

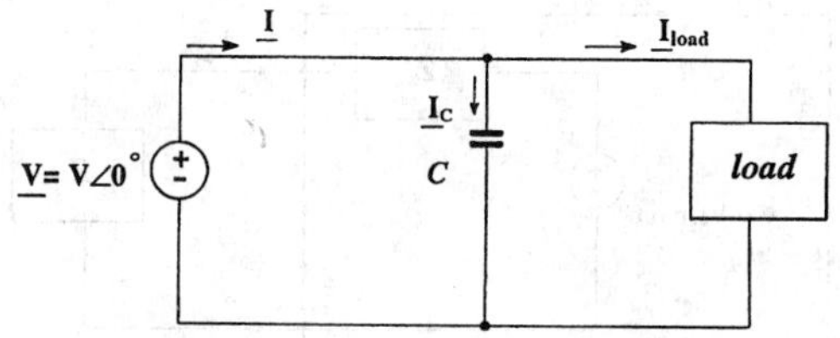

FIGURE 20.81 Circuit for Example 27 showing the load and the capacitor to be connected in parallel with the load to improve the power factor.

Thus, the current supplied by the source to the *RLC* network is

$$\underline{I} = \underline{I_{RL}} + \underline{I_C}$$
$$= \frac{V\cos\theta}{\sqrt{R^2 + (\omega L)^2}} + jV\left[\frac{-\omega L}{R^2 + (\omega L)^2} + \omega C\right]$$

To improve the *PF* to one, the current $\underline{I}$ should be in phase with $\underline{V}$. Thus, we set the imaginary term in the equation for $\underline{I}$ equal to zero, yielding:

$$C = \frac{L}{R^2 + (\omega L)^2} = 530\ \mu F$$

a rather large capacitor. Before this capacitor is connected, the *RMS* value of the current supplied by the voltage source is $I_{rms} = 26.833$ amps. After the capacitor is connected, the source has to supply only 12 amps *RMS*, a considerable reduction. In both cases, *P* delivered to the load is the same.

The following example also illustrates *PF* improvement.

Example 27. A load with $PF = 0.7$ lagging, shown in Fig. 20.81, consumes 12 kW of power. The line voltage supplied is 220 V *RMS* at 60 Hz. Find the size of the capacitor needed to correct the *PF* to 0.9 lagging, and determine the values of the currents supplied by the source both before and after the *PF* is corrected.

Solution. We will take the phase of the line voltage to be 0°. From $P = V_{rms} I_{rms} PF = 12$ kW, we obtain $I_{rms} = 77.922$ amps. Since *PF* is 0.7 lagging, the phase of the current through the load relative to the phase of the line voltage is $-\cos^{-1}(0.7) = -45.57°$. Therefore, $\underline{I_{load}} = 77.922\angle(-45.57°)$ amps *RMS*. When *C* is connected in parallel with the load,

$$\underline{I} = \underline{I_C} + \underline{I_{load}} = 220(377)jC + 77.922e^{-j0.7954}$$
$$= 54.54 - j[55.64 - 82,940C]$$

If the *PF* were to be corrected to unity, we would set the imaginary part of the previous expression for current to zero. But this would require a larger capacitor (671 μF) which may be uneconomical. Instead, to retain a lagging, but improved $PF = 0.9$, corresponding to the current lagging the voltage by 25.84°, we write

$$0.9 = \frac{54.54}{\sqrt{54.54^2 + (55.64 - 82,940C)^2}}$$

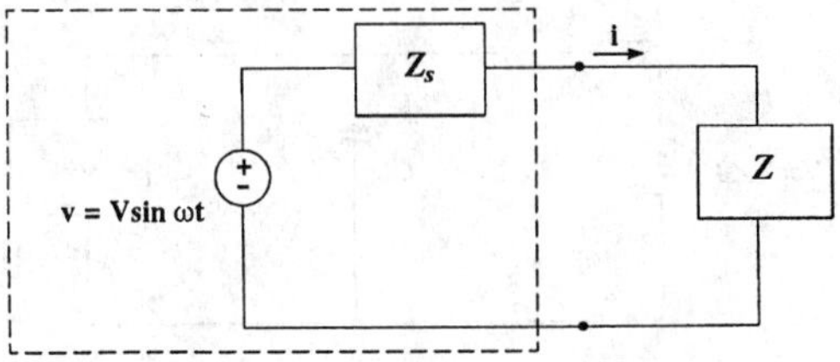

FIGURE 20.82 Z_s is fixed, and Z is to be chosen so that maximum average power is transferred to Z.

Therefore, $C = 352\ \mu F$. The line current is now

$$\underline{I} = \underline{I_C} + \underline{I_{load}} = 60.615\angle(-25.87°) \text{ amps RMS}$$

Previous examples have employed ideal voltage sources to supply power to networks. However, in many electronic applications, the source has a fixed impedance associated with it, and the problem is to obtain the maximum average power transferred to the load [2]. Here, we assume that the resistance and reactance of the load can be independently adjusted. Let the source impedance be:

$$Z_S(j\omega) = R_S(\omega) + jX_S(\omega)$$

The load impedance is denoted as

$$Z(j\omega) = R(\omega) + jX(\omega)$$

Figure 20.82 depicts these impedances. We assume that all the elements, including the voltage source, within the box formed by the dotted lines are fixed. The voltage source is $v(t) = V\sin(\omega t)$, and thus $i(t) = I\sin(\omega t + \theta)$, where V and I are peak values and

$$\theta = -\tan^{-1}\left[\frac{X_S(\omega) + X(\omega)}{R_S(\omega) + R(\omega)}\right]$$

The average power delivered to Z is

$$P = I_{rms}^2 R(\omega)$$

where $I_{rms} = I/\sqrt{2}$ and

$$I = \frac{V}{\sqrt{[R_S(\omega) + R(\omega)]^2 + [X_S(\omega) + X(\omega)]^2}} \tag{20.121}$$

Thus, the average power delivered to Z can be written as

$$P = \frac{V_{rms}^2 R(\omega)}{[R_S(\omega) + R(\omega)]^2 + [X_S(\omega) + X(\omega)]^2} \tag{20.122}$$

To maximize P, we first note that the term $[X_S(\omega) + X(\omega)]^2$ is always positive, and so this term always contributes to a larger denominator unless it is zero. Thus, we set

$$X(\omega) = -X_S(\omega) \tag{20.123}$$

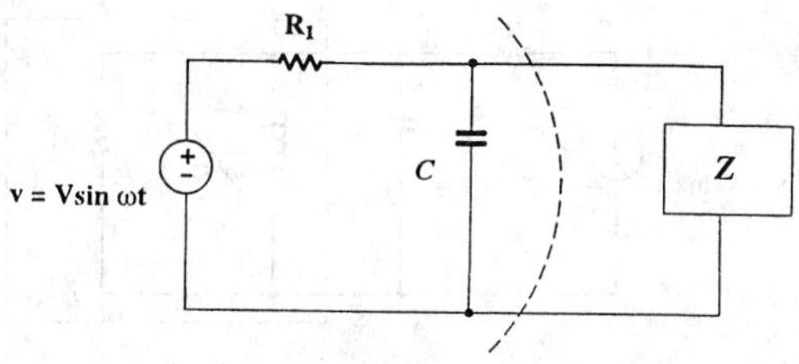

FIGURE 20.83 Circuit for Example 28.

and (20.122) becomes

$$P = \frac{V_{rms}^2 R(\omega)}{[R_S(\omega) + R(\omega)]^2} \tag{20.124}$$

Secondly, we set the partial derivative with respect to $R(\omega)$ of the expression in (20.124) to zero to obtain

$$\frac{\partial P}{\partial R} = V_{rms}^2 \frac{(R_S + R)^2 - 2R(R_S + R)}{(R_S + R)^4} = 0 \tag{20.125}$$

Eq. (20.125) is satisfied for

$$R(\omega) = R_S(\omega) \tag{20.126}$$

and this value of $R(\omega)$, together with $X(\omega) = -X_S(\omega)$, yields maximum average power transferred to Z. Thus, we should adjust Z to:

$$Z(j\omega) = Z_S^*(j\omega) \tag{20.127}$$

and we obtain

$$P_{max} = \frac{V_{rms}^2}{4R(\omega)} \tag{20.128}$$

Example 28. Find Z for the network in Fig. 20.83 so that maximum average power is transferred to Z. Determine the value of P_{max}.

Solution. We first obtain the Thévenin equivalent of the circuit to the left of the dotted arc in Fig. 20.83 in order to reduce the circuit to the form of Fig. 20.82.

$$\underline{V_{TH}} = \frac{\underline{V}}{1 + j\omega R_1 C}$$

$$Z_{TH} = \frac{R_1}{1 + j\omega R_1 C}$$

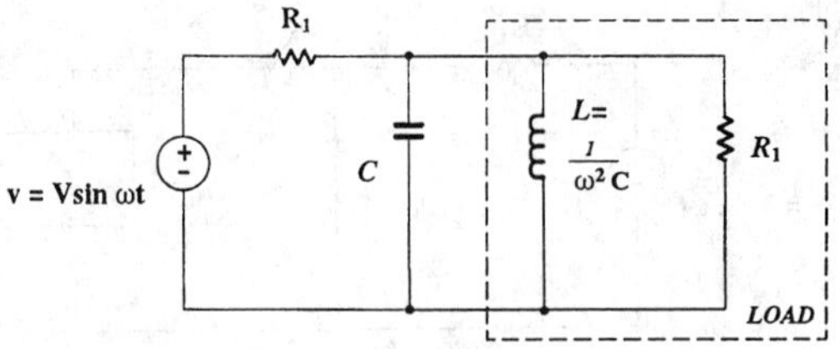

FIGURE 20.84 Circuit with load chosen to obtain maximum average power.

Thus,

$$Z = Z_{TH}^* = \frac{R_1}{1 - j\omega R_1 C} = \frac{R_1}{1 + \dfrac{\omega R_1 C}{j}}$$

$$= \frac{jR_1}{j + \omega R_1 C} = \frac{j\dfrac{1}{\omega C}R_1}{R_1 + j\dfrac{1}{\omega C}}$$

The term $j/(\omega C)$ appears inductive (at a single frequency) and so we equate it to $j\omega L$ to obtain:

$$L = \frac{1}{\omega^2 C}$$

The impedance Z is therefore formed by the parallel connection of a resistor R_1 with the inductor L. Figure 20.84 depicts the resulting circuit. To determine P_{max}, we note that the capacitor and inductor constitute a parallel circuit which is resonant at the frequency of excitation. It therefore appears as an open circuit to the source. Thus, P_{max} is easily obtained as:

$$P_{\max} = I_{rms}^2 R_1 = \frac{V^2}{8R_1}$$

where V is the peak value of $v(t)$.

Suppose Z is fixed and Z_S is adjustable in Fig. 20.82. What should Z_S be so that maximum average power is delivered to Z? This is a problem that is applicable in the design of electronic amplifiers. The average power delivered to Z is given by (20.122), and to maximize P, we set $X_S(\omega) = -X(\omega)$ as before. We therefore obtain (20.124) again. But if R_S is adjustable rather than R, we see from (20.124) that P_{max} is obtained when R_S equals zero.

Acknowledgments

The author wishes to convey his gratitude to Dr. Jacek Zurada and to Mr. K. Wang for their help in proofreading this manuscript and to Mr. Zbigniew J. Lata and Mr. Peichu (Peter) Sheng for producing the drawings.

References

[1] A. Budak, *Circuit Theory Fundamentals and Applications,* 2nd ed., Englewood Cliffs, NJ: Prentice-Hall, 1987.

[2] L. P. Huelsman, *Basic Circuit Theory with Digital Computations,* Englewood Cliffs, NJ: Prentice-Hall, 1972.

[3] L. P. Huelsman, *Basic Circuit Theory,* 3rd ed., Englewood Cliffs, NJ: Prentice-Hall, 1991.

[4] S. Karni, *Applied Circuit Analysis,* New York: John Wiley & Sons, 1988.

[5] L. Weinberg, *Network Analysis and Synthesis,* New York: McGraw-Hill, 1962.

21

Symbolic Analysis

Marwan M. Hassoun
Iowa State University

21.1 Introduction and Definition

Symbolic circuit analysis, simply stated, is a term that describes the process of studying the behavior of electrical circuits using symbols instead of, or in conjunction with, numerical values. As an example to illustrate the concept, consider the input resistance of the simple circuit in Fig. 21.1. Analyzing the circuit using the unique symbols for each resistor without assigning any numerical values to them yields the input resistance of the circuit in the form:

$$\frac{V_{in}}{I_{in}} = \frac{R_1R_2 + R_1R_3 + R_1R_4 + R_2R_3 + R_2R_4}{R_2 + R_3 + R_4} \tag{21.1}$$

Equation (21.1) is the symbolic expression for the input resistance of the circuit in Fig. 21.1. The formal definition of symbolic circuit analysis is given in Definition 1.

Definition 1. Symbolic circuit analysis (simulation) is the process of producing an expression that describes a certain behavioral aspect of the circuit with one, some, or all of the circuit elements represented as symbols.

The idea of symbolic circuit analysis (also referred to as symbolic circuit simulation) is not new; engineers and scientists have been using the process to study circuits since the inception of the concept of circuits. Every engineer has used symbolic circuit analysis during his or her education process. Most still use it in their everyday job functions. As an example, all electrical engineers have symbolically analyzed the circuit in Fig. 21.2. The equivalent resistance between nodes i and j is known to be:

$$\frac{1}{R_{ij}} = \frac{1}{R_1} + \frac{1}{R_2} \tag{21.2}$$

0-8493-8341-2/95/$0.00+$.50

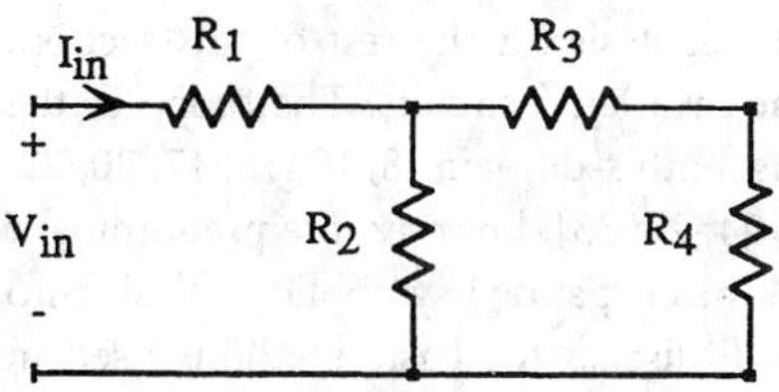

FIGURE 21.1 Symbolic circuit analysis example.

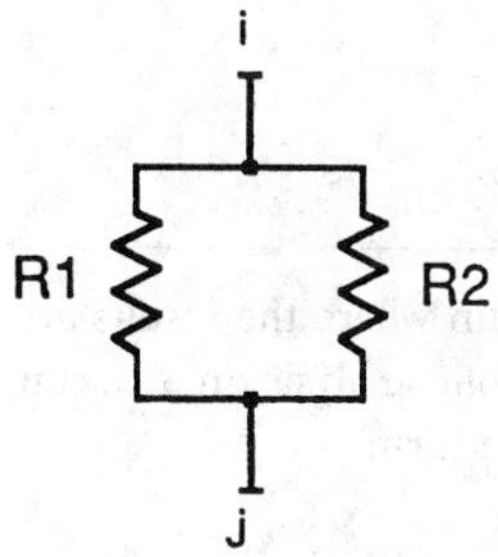

FIGURE 21.2 Common symbolic analysis problem.

or

$$R_{ij} = \frac{R_1 R_2}{R_1 + R_2} \tag{21.3}$$

This is the most primitive form of symbolic circuit analysis.

The basic justification for performing symbolic analysis rather than numerical analysis on a circuit can be illustrated by considering the circuit in Fig. 21.1 again. Assuming that the values of all the resistances R_1 through R_4 are given as 1 Ω and that the input resistance is to be analyzed numerically, the result obtained would be

$$\frac{V_{in}}{I_{in}} = \frac{5}{3} = 1.667\ \Omega \tag{21.4}$$

Now consider the problem of increasing the input resistance of the circuit by adjusting only one of the resistor values. Equation (21.4) provides no insight into which resistor has the greatest impact on the input resistance. However, Eq. (21.1) clearly shows that changing R_2, R_3, or R_4 would have very little impact on the input resistance since the terms appear in both the numerator and the denominator of the symbolic expression. It can also be seen that R_1 should be the resistor to change since it only appears in the numerator of the expression. Symbolic analysis has provided an insight into the problem.

From a circuit design perspective, numerical results from the simulation of a circuit can be obtained by evaluating the results of the symbolic analysis at a specific numerical point for each symbol. So ideally, only one simulation run is needed in order to analyze the circuit, and successive evaluations of the results replace the need for any extra iterations through the simulator. Other applications include sensitivity analysis, circuit stability analysis, device modeling, and circuit optimization [13, 25, 32].

While the above "hand calculations" and somewhat trivial examples are used to illustrate symbolic circuit analysis, the thrust of the methods developed for symbolic analysis are aimed at computer implementations that are capable of symbolically analyzing circuits that cannot be analyzed "by hand". Several such implementations have been developed over the years [4, 8, 10, 12, 17, 20, 22–24, 26, 27, 29, 34, 38, 39, 41–43, 46–48].

Symbolic circuit analysis, referred to as simply symbolic analysis for the rest of this section, in its current form is limited to linear,[1] lumped, and time-invariant[2] circuits. The scope of the analysis is primarily concentrated in the frequency domains, both s-domain [8, 10, 12, 17, 20, 22, 23, 26, 29, 34, 38, 39, 41–43, 46–48, 50] and z-domain [4, 24, 27, 35], however, the predominant development has been in the s-domain. Also, recent work has expanded symbolic analysis into the time domain [3, 19, 28]. The next few subsections will discuss the basic methods used in symbolic analysis for mainly s-domain frequency analysis; however, Section 21.5 highlights the currently known time domain techniques.

21.2 Frequency Domain Analysis

Traditional symbolic circuit analysis is performed in the frequency domain where the results are in terms of the frequency variable s. The main goal of performing symbolic analysis on a circuit in the frequency domain is to obtain a symbolic transfer function of the form

$$H(s, \mathbf{X}) = \frac{N(s, \mathbf{X})}{D(s, \mathbf{X})}, \qquad \mathbf{X} = [x_1, x_2, \ldots, x_n], \quad n \le n_{all} \tag{21.5}$$

The expression is a function of the complex frequency variable s, and the variables x_1 through x_n representing the variable circuit elements, where n is the number of variable circuit elements and n_{all} is the total number of circuit elements. A more recent approach to representing the above circuit function emerged in the 1980s and is based on a decomposed hierarchical form of Eq. (21.5) [17, 20, 48]. This hierarchical representation is referred to as a **sequence of expressions** representation to distinguish it from the **single expression** representation of Eq. (21.5) and is addressed in Section 21.4.

Several methodologies exist to perform symbolic analysis in the frequency domain. The early work was to produce a transfer function $H(s)$ with the frequency variable s being the only symbolic variable. Computer programs with these capabilities include: CORNAP [41] and NASAP [38]. The interest in symbolic analysis today is in the more general case where some or all of the circuit elements are represented by symbolic variables. The methods developed for this type of analysis fall under one of the following categories:

Traditional methods (single expression)

1. Tree-enumeration method
2. Signal flowgraph methods
3. Parameter extraction method
4. Interpolation method

Hierarchical methods (sequence of expressions)

1. Signal flowgraph methods
2. Modified nodal analysis-based methods

The next two sections will discuss the theory behind these methods. Circuit examples are shown for all the methods except for the interpolation method due to its limited current usage and its inability to analyze fully symbolic circuits.

[1] Some references are made to the ability to analyze "weakly nonlinear" circuits [13]; however, the actual symbolic analysis is performed on a linearized model of the weakly nonlinear circuit.

[2] There is one method reported in [28] that does deal with a limited class of time-variant circuits.

21.3 Traditional Methods (Single Expressions)

This class of methods attempts to produce a single transfer function in the form of Eq. (21.5). The major advantage of having a symbolic expression in that form is the insight that can be gained by observing the terms in both the numerator and the denominator. The effects of the different terms can, perhaps, be determined by inspection. This process is valid for the cases where there are relatively few symbolic terms in the expression.

Before indulging in the explanation of the different methods covered by this class, some definition of terms is in order.

Definition 2 *$RLCg_m$ circuit:* Is one that may contain only resistors, inductors, capacitors, and voltage-controlled current sources with the gain designated as g_m.

Definition 3 *Term cancellation:* Is the process in which two equal symbolic terms cancel out each other in the symbolic expression. This can happen in one of two ways: by having two equal terms with opposite signs added together, or by having two equal terms (regardless of their signs) divided by each other. For example, the equation

$$\frac{ab(ab + cd) - ab(cd - ef)}{ab(cd - gh)} \tag{21.6}$$

where a, b, c, d, e, f, g, and h are symbolic terms, can be reduced by observing that the terms ab in the numerator and denominator cancel each other and the terms $+cd$ and $-cd$ cancel each other in the numerator. The result is

$$\frac{ab + ef}{cd - gh} \tag{21.7}$$

Definition 4 *Cancellation-free:* Equation (21.7) is said to be a cancellation-free equation while Eq. (21.6) is not. That is, no possible cancellations exist in the expression.

Definition 5 *Cancellation-free algorithm:* The process of term cancellation can occur during the execution of an algorithm where a cancellation-free equation is generated directly rather than generating an expression with possible term cancellations in it. Cancellation-free algorithms are more desirable because, otherwise, an overhead is needed to generate and keep the terms that are to be canceled later.

The different methods that fall under the traditional class are explained next.

1. The tree enumeration method

Several programs have been produced based on this method [5, 11, 33, 37]. Practical implementations of the method can only handle small circuits in the range of 15 nodes and 30 branches [30]. The main reason is the exponential growth in the number of symbolic terms generated. The method can only handle one type of controlled source, namely voltage-controlled current sources. So only $RLCg_m$ circuits can be analyzed. Also, the method does not produce any symbolic term cancellations for RLC circuits, and produces only a few for $RLCg_m$ circuits.

The basic idea of the tree enumeration method is to construct an augmented circuit (a slightly modified version of the original circuit) its associated directed graph, and then enumerating all the directed trees of the graph. The admittance products of these trees are then used to find the nodal admittance matrix determinant (which itself is never constructed) and cofactors to produce the required symbolic transfer functions. For a circuit with n nodes, where the input is an excitation between nodes 1 and n and the output is taken between nodes 2 and n, the transfer functions of the circuit can be written as:

$$Z_{in} = \frac{V_1}{I_1} = \frac{\Delta_{11}}{\Delta} \tag{21.8}$$

$$\frac{V_o}{I_{in}} = \frac{V_2}{I_1} = \frac{\Delta_{12}}{\Delta} \tag{21.9}$$

$$\frac{V_o}{V_{in}} = \frac{V_2}{V_1} = \frac{\Delta_{12}}{\Delta_{11}} \tag{21.10}$$

where Δ is the determinant of the nodal admittance matrix $\mathbf{Y}_n$ (dimensions $n - 1 \times n - 1$) and Δ_{ij} is the ijth cofactor of $\mathbf{Y}_n$. It can be shown that a simple method for obtaining Δ, Δ_{11}, Δ_{12}, is to construct another circuit comprised of the original circuit with an extra admittance y_s' in parallel with a voltage-controlled current source ($g_m' V_{2,n}$) draped across the input terminals (nodes 1 and n). The determinant of $\mathbf{Y}_n'$ (the nodal admittance matrix for the new, slightly modified, circuit) can be written as:

$$\Delta' = \Delta + y_s' \Delta_{11} + g_m' \Delta_{12} \tag{21.11}$$

This simple trick allows the construction of the determinant expression of the original circuit and its two needed cofactors by simply constructing the expression for the new augmented circuit. Example 1 below illustrates this process.

The basic steps of the tree enumeration algorithm are (condensed from [6])

1. Construct the augmented circuit from the original circuit by adding an admittance y_s' and a transconductance ($g_m' V_o$) in parallel between the input node and the reference node.
2. Construct a directed graph $\boldsymbol{G}_{ind}$ associated with the augmented circuit. The stamps used to generate $\boldsymbol{G}_{ind}$ are shown in Fig. 21.3.
3. Find all directed trees for $\boldsymbol{G}_{ind}$. A directed tree rooted at node i is a subgraph of $\boldsymbol{G}_{ind}$ with node i having no incoming branches and each other node having exactly one incoming branch.
4. Find the admittance product for each directed tree. An admittance product of a directed tree is simply a term that is the product of all the weights of the branches in that tree.
5. Apply the following theorem:

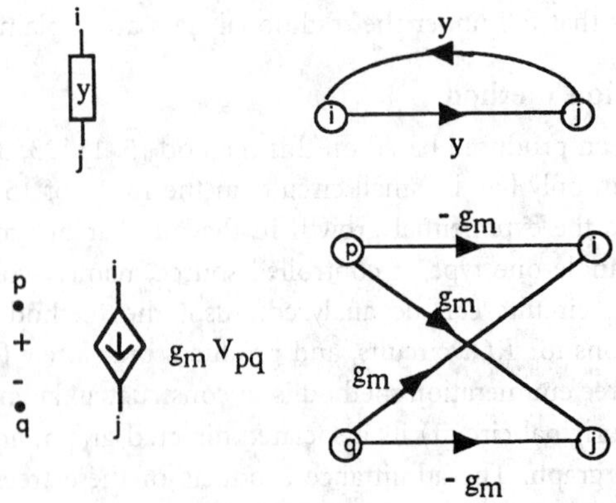

FIGURE 21.3 Element stamps for generating $\boldsymbol{G}_{ind}$.

Theorem 1 [6]: *For any RLCg_m circuit, the determinant of the node admittance matrix (with any node as the reference node) is equal to the sum of all directed tree admittance products of* $\mathbf{G}_{ind}$ *(with any node as the root).*

In other words

$$\Delta' = \sum \textit{tree admittance products} \tag{21.12}$$

Arranging Eq. (21.12) in the form of Eq. (21.11) results in the necessary determinant and cofactors of the original circuit and the needed transfer functions are generated from Eqs. (21.8), (21.9), and (21.10).

Example 1. A circuit and its augmented counterpart are shown in Fig. 21.4. The circuit is the small-signal model of a simple inverting CMOS amplifier shown with the coupling capacitance C_c taken into account. Figure 21.5 shows the directed graph associated with the augmented circuit constructed using the rules shown in Fig. 21.3. The figure also shows all the directed trees rooted at node 3 of the graph. Parallel branches heading in the same direction are combined into one branch with a weight equal to the sum of the weights of the individual parallel branches. Applying Eq. (21.12) and rearranging the terms results in:

$$\Delta' = (sC_c - g'_m)(g_m + y'_s) + (sC_c - g_m)(g'_m + g_o) + (g_m + y'_s)(g'_m + g_o) \tag{21.13}$$

$$= sC_c g_m + sC_c g_o \qquad (\Delta) \tag{21.14}$$

$$+y'_s(sC_c + g_o) \qquad (y'_s\,\Delta_{11}) \tag{21.15}$$

$$+g'_m(sC_c - g_m) \qquad (g'_m\,\Delta_{12}) \tag{21.16}$$

Note the fact that Eq. (21.13), which is the direct result of the algorithm, is not cancellation-free. Some terms cancel out to result in the determinant of the original circuit and its two cofactors of interest. The final transfer functions can be readily obtained by substituting the above results into Eqs. (21.8) through (21.10).

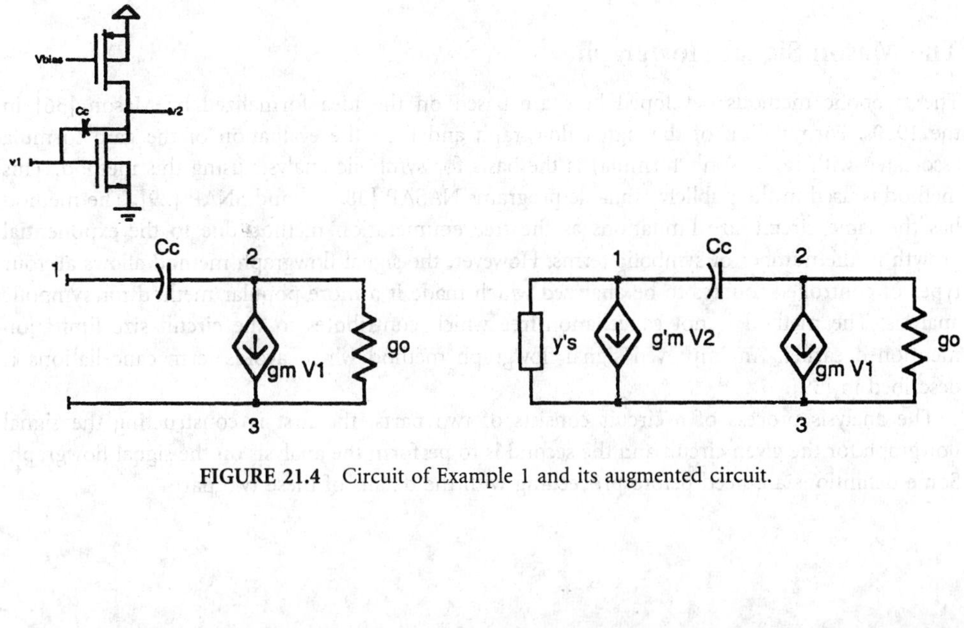

FIGURE 21.4 Circuit of Example 1 and its augmented circuit.

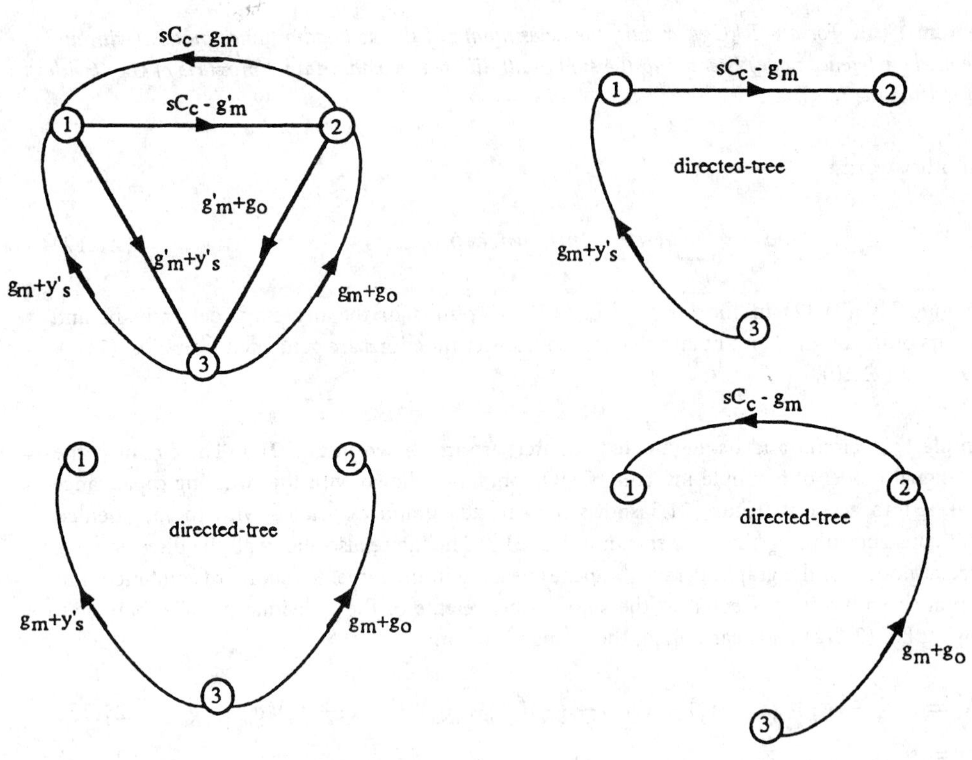

FIGURE 21.5 Graph and directed trees of Example 1.

2. The signal flowgraph method

Two types of flowgraphs are used in symbolic analysis. The first is referred to as a Mason's signal flowgraph and the second as Coates graph. Mason's signal flowgraph is by far a more popular and well known flowgraph which has been used extensively in symbolic analysis in addition to many other applications. Both the Mason signal flowgraph and the Coates graph are used as basis for hierarchical symbolic analysis. However, the Coates graph was introduced to symbolic analysis by Starzyk and Konczykowska [4] solely for the purpose of performing hierarchical symbolic analysis. This section will cover both approaches.

The Mason Signal Flowgraph

The symbolic methods developed here are based on the idea formalized by Mason [36] in the 1950s. Formulation of the signal flowgraph and then the evaluation of the gain formula associated with it (Mason's formula) is the basis for symbolic analysis using this method. This method is used in the publicly available programs NASAP [38, 40] and SNAP [29]. The method has the same circuit size limitations as the tree enumeration method due to the exponential growth in the number of symbolic terms. However, the signal flowgraph method allows all four types of controlled sources to be analyzed which made it a more popular method for symbolic analysis. The method is not cancellation-free which contributes to the circuit size limitation mentioned earlier. An improved signal flowgraph method which avoids term cancellations is described in [39].

The analysis process of a circuit consists of two parts; the first is constructing the signal flowgraph for the given circuit and the second is to perform the analysis on the signal flowgraph. Some definitions are need before proceeding with the details of these two parts.

Definition 6 *Signal Flowgraph:* A signal flowgraph is a weighted directed graph representing a system of simultaneous linear equations. Each node (x_i) in the signal flowgraph represents a circuit variable (node voltage, branch voltage, branch current, capacitor charge or inductor flux) and each branch weight (w_{ij}) represents a coefficient relating x_i to x_j.

Every node in the signal flowgraph can be looked at as a summer. For a node x_k with m incoming branches

$$x_k = \sum_i w_{ik} x_i \tag{21.17}$$

where i is the indices of all incoming branches from x_i to x_k.

Definition 7 *Path Weight:* The weight of a path (P_{ij}) from x_i to x_j is the product of all the branch weights in the path.

Definition 8 *Loop Weight:* The weight of a loop is the product of all the branch weights in that loop. This also holds for a loop with only one branch in it (self-loop).

Definition 9 *nth order loop:* An nth order loop is a set of n loops that have no common nodes between any two of them. The weight of an nth order loop is the product of the weights of all n loops.

Any transfer function x_j/x_i, where x_i is a source node, can be found by the application of Mason's formula:

$$\frac{x_j}{x_i} = \frac{1}{\Delta} \sum_k P_k \Delta_k \tag{21.18}$$

where

$$\begin{aligned} \Delta = 1 &- (\text{sum of all } L_i'\text{s}) \\ &+ (\text{sum of all 2nd order loop weights}) \\ &- (\text{sum of all 3rd order loop weights}) \\ &+ \ldots \end{aligned} \tag{21.19}$$

$$\begin{aligned} P_k &= \text{weight of the } k\text{th path from the source node } x_i \text{ to } x_j \\ \Delta_k &= \Delta \text{ with all loop contributions that are touching } P_k \text{ eliminated} \end{aligned} \tag{21.20}$$

The use of the above equations can be illustrated via an example.

Example 2. Consider the circuit in Fig. 21.6. The formulation of the signal flowgraph for this circuit takes on the following steps:

1. Find a tree and co-tree of the circuit such that all current sources are in the co-tree and all the voltage sources are in the tree.

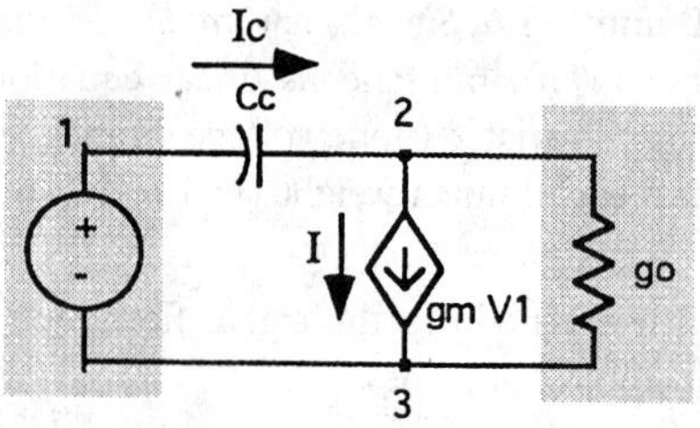

FIGURE 21.6 Circuit for Example 2 with its highlighted tree.

2. Use Kirchhoff's Current Law (KCL), branch admittances, and tree branch voltages to find an expression for every co-tree link current. In the case of a controlled source, simply use the branch relationship. For the example above, this yields:

$$I_c = sC_c(V_1 - V_2) = sC_cV_1 - sC_cV_2 \tag{21.21}$$

$$I = g_m V_1 \tag{21.22}$$

3. Use Kirchhoff's Voltage Law (KVL), branch impedances, and co-tree link currents to find an expression for every tree branch voltage. In the case of a controlled source, simply use the branch relationship. For the example above, this yields

$$V_{go} = V_2 = \frac{1}{g_o}(-I + I_c) \tag{21.23}$$

4. Create the signal flowgraph by drawing a node for each current source, voltage source, tree branch voltage, and co-tree link current.
5. Use Eq. (21.17) to draw the branches between the nodes that realize the linear equations developed in the previous steps.

Figure 21.7 shows the result of executing the above five steps on the example circuit. This formulation is referred to as the compact signal flowgraph. Any other variables that are linear combinations of the variables in the signal flowgraph (e.g., node voltages) can be added to the signal flowgraph by simply adding the extra node and implementing the linear relationship using signal flowgraph branches. A more detailed discussion of signal flowgraphs can be found in [6, 31].

Now applying Eqs. (21.19) and (21.20) yields:

$$P_1 = -\frac{g_m}{g_o} \qquad P_2 = \frac{sC_c}{g_o} \qquad {}_1 = -\frac{sC_c}{g_o}$$

$$\Delta = 1 - \left(-\frac{sC_c}{g_o}\right) \qquad \Delta_1 = 1 \qquad \Delta_2 = 1$$

Equation (21.18) then produces the final transfer function

$$\frac{V_2}{V_1} = \frac{1}{1 + \dfrac{sC_c}{g_o}}\left(-\frac{g_m}{g_o} + \frac{sC_c}{g_o}\right) = \frac{sC_c - g_m}{sC_c + g_o} \tag{21.24}$$

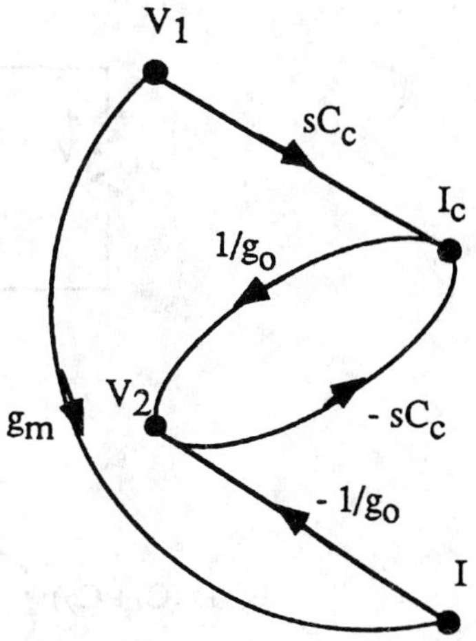

FIGURE 21.7 Signal flowgraph for Example 2.

The Coates Graph

For the Coates graph, the resulting transfer function is expressed as the ratio of the graph's 1-connection, which is dependent on the input variable and the output variable selected, to the graph's 0-connection, which is global to all the circuit's transfer functions (i.e., independent of the I/O variables selected) [7]. A generalization of Coates 1-connection and 0-connection is referred to as a k-connection or a multiconnection [48]. To illustrate the concept of a multiconnection, the proper definition of a directed graph must be presented first.

Definition 10. A directed graph $G = (V, E)$ consists of a finite, nonempty set of vertices V, and a set of edges E, which are ordered pairs $e_i = (v_i, u_i)$ of vertices; v_i is called the tail and u_i the head of edge e_i. Each edge $e_i \in E$ has a weight w_i associated with it.

Definition 11 [48]. A k-connection (multiconnection) of a graph G with n nodes, is a subgraph p_j with n nodes, α node-disjoint directed edges, β isolated nodes and λ node-disjoint directed loops, where $k = \alpha + \beta$. The weight of a k-connection is given by

$$|p_j| = \prod_{e_i \in p_j} w_i \tag{21.25}$$

and the weight of a set of multiconnections P is given by

$$|P| = \sum_{p_j \in P} \mathbf{sign}\, p_j \tag{21.26}$$

where

$$\mathbf{sign}\, p = (-1)^{n+k+l_p}\, \mathbf{ord}(v_1, \ldots, v_k) \cdot \mathbf{ord}(u_1, \ldots, u_k)$$

$$\mathbf{ord}(x_1, \ldots, x_k) = \begin{cases} 1 & \textit{when the number of permutations ordering the set is even} \\ -1 & \textit{otherwise} \end{cases} \tag{21.27}$$

(l_p is the number of loops in multiconnection p).

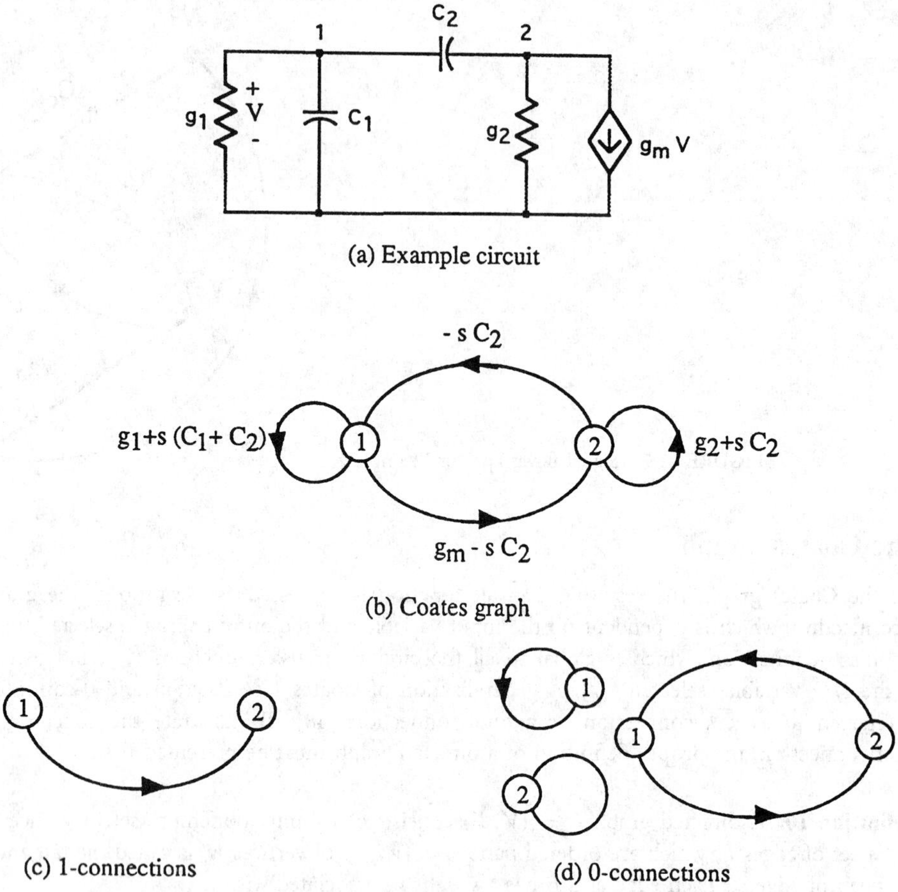

FIGURE 21.8 A circuit, its Coates graph, and its k-connections.

The transfer function of a circuit with input variable v_1 and output variable v_2 is therefore,

$$H(s, v_3, v_4, \ldots) = \frac{|P_{1\text{-}connection_{v_1 \to v_2}}|}{|P_{0\text{-}connection}|} \tag{21.28}$$

The 1-$connection_{v_1 \to v_2}$ is the set of all 1-connections that include a directed path (a set of directed edges) that connect graph nodes v_1 to v_2.

Example 3. As an example, consider the circuit in Fig. 21.8(a). To find V_2/V_1, the 1-$connection_{v_1 \to v_2}$ is needed. Only one such subgraph exists, Fig. 21.8(b). However the 0-$connection$ set has two members illustrated in Fig. 21.8(c), so

$$\begin{aligned} H(s, g_1, g_2, g_m, C_1, C_2) &= \frac{|P_{1\text{-}connection_{v_1 \to v_2}}|}{|P_{0\text{-}connection}|} \\ &= -\frac{(-sC_2 + g_m)}{(g_2 + sC_2)} \end{aligned} \tag{21.29}$$

3. The parameter extraction method

This method is best suited when few parameters in a circuit are symbolic while the rest of the parameters are in numeric form (s being one of the symbolic variables). The method was introduced in 1973 [2]. Other variations on the method were proposed later [42, 46]. The advantage of the method is that it is directly related to the basic determinant properties of widely used equation formulation methods like the modified nodal method [21] and the tableau method [16]. As the name of the method implies, it provides a mechanism for extracting the symbolic parameters out of the matrix formulation breaking the matrix solution problem into a numeric part and a symbolic part. The numeric part can then be solved using any number of standard techniques and then recombined with the extracted symbolic part. The method has the advantage of being able to handle larger circuits than the previously discussed fully symbolic methods if only a few parameters are represented symbolically. If the number of symbolic parameters in a circuit is high, the method will exhibit the same exponential growth in the number of symbolic terms generated and will have the same circuit size limitations as the other algorithms previously discussed.

The method does not limit the type of matrix formulation used to analyze the circuit. However, the extraction rules depend on the pattern of the symbolic parameters in the matrix. Alderson and Lin [1] use the indefinite admittance matrix as the basis of the analysis; rules depend on the appearance of a symbolic parameter in four locations in the matrix: (i, i), (i, j), (j, i) and (j, j). Singhal and Vlach [46] use the tableau equations which can handle a symbolic parameter that appears only once in the matrix. The method used by Sannuti and Puri [42] forces the symbolic parameters to appear only on the matrix diagonal using a two-graph method [6] to write the tableau equations. To illustrate the concept of parameter extraction, this section will concentrate on the indefinite admittance matrix (IAM) formulation and $RLCg_m$ circuits. Details of other formulations can be found in [31, 39, 42, 46].

One of the basic properties of the IAM matrix is the symmetric nature of the entries sometimes referred to as quadrantal entries [6, 31]. A symbolic variable α will always appear in four places in the indefinite admittance matrix, $+\alpha$ in entries (i, k) and (j, m), and $-\alpha$ in entries (i, m) and (j, k) as shown in the following equation:

$$\begin{array}{c} \\ i \\ \\ j \\ \\ \end{array}\begin{array}{c} \begin{array}{ccc} k & & m \end{array} \\ \begin{bmatrix} \vdots & & \vdots \\ \alpha & \cdots & -\alpha \\ \vdots & & \vdots \\ -\alpha & \cdots & \alpha \\ \vdots & & \vdots \end{bmatrix} \end{array} \tag{21.30}$$

where $i \neq j$ and $k \neq m$. For the case of an admittance y between nodes i and j, then $k = i$ and $m = j$. The basic process of extracting the parameter (the symbol) α can be performed by applying the following equation [2, 6].

$$\text{cofactor of } \mathbf{Y}_{ind} = \text{cofactor of } (\mathbf{Y}_{ind,\alpha=0}) + (-1)^{j+m}\alpha(\text{cofactor of } \mathbf{Y}_{\alpha}) \tag{21.31}$$

where $\mathbf{Y}_{\alpha}$ is a matrix that does not contain α and is obtained by:

1. Adding row j to row i.
2. Adding column m to column k.
3. Deleting row j and column m.

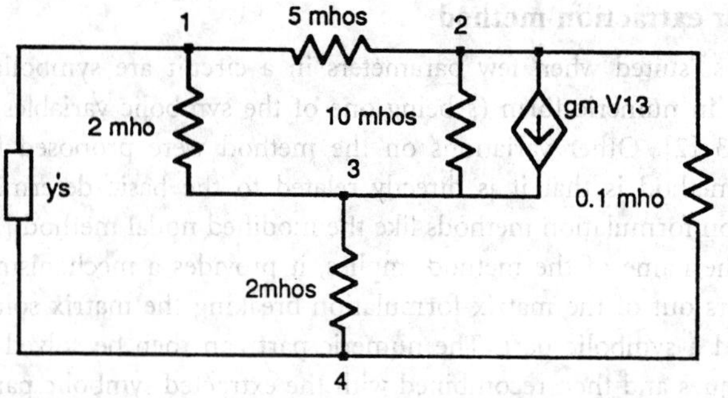

FIGURE 21.9 Circuit for the parameter extraction method.

For the case where several symbols exist, the above extraction process can be repeated and would result in

$$\text{cofactor of } \mathbf{Y}_{ind} = \sum P_j(\text{cofactor of } \mathbf{Y}_j) \tag{21.32}$$

where P_j is some product of symbolic parameters including the sign and $\mathbf{Y}_j$ is a matrix with the frequency variable s, possibly, being the only symbolic variable. The cofactor of $\mathbf{Y}_j$ may be evaluated using any of the typical evaluation methods [6].

Example 4 [6]. Consider the resistive circuit in Fig. 21.9. The goal is to find the input impedance Z_{14} using the parameter extraction method where g_m is the only symbolic variable in the circuit. In order to use Eqs. (21.8) and (21.11), a source admittance y'_s is added across the input terminals of the circuit to create the augmented circuit. (note that all numerical values are admittances rather than resistances). The IAM matrix is then written as

$$\mathbf{Y}'_{ind} = \begin{bmatrix} 6+y'_s & -5 & -1 & -y'_s \\ g_m-5 & 15.1 & -g_m-10 & -0.1 \\ -g_m-1 & -10 & g_m+13 & -2 \\ -y'_s & -0.1 & -2 & y'_s+2.1 \end{bmatrix} \tag{21.33}$$

Applying Eq. (21.31) to extract y'_s results in

$$\begin{aligned}\text{cofactor of } \mathbf{Y}'_{ind} &= \text{cofactor of} \begin{bmatrix} 6 & -5 & -1 & 0 \\ g_m-5 & 15.1 & -g_m-10 & -0.1 \\ -g_m-1 & -10 & g_m+13 & -2 \\ 0 & -0.1 & -2 & 2.1 \end{bmatrix} \\ &+ y'_s * \text{cofactor of} \begin{bmatrix} 8.1 & -5.1 & -3 \\ g_m-5.1 & 15.1 & -g_m-10 \\ -g_m-3 & -10 & g_m+13 \end{bmatrix}\end{aligned} \tag{21.34}$$

Applying Eq. (21.31) again to extract g_m results in

$$\text{cofactor of } \mathbf{Y}'_{ind} = \text{cofactor of} \begin{bmatrix} 6 & -5 & -1 & 0 \\ -5 & 15.1 & -10 & -0.1 \\ 1 & -10 & 13 & -2 \\ 0 & -0.1 & -2 & 2.1 \end{bmatrix}$$
$$+g_m * \text{cofactor of} \begin{bmatrix} 5 & -5 & 0 \\ -3 & 5.1 & -2.1 \\ -2 & -0.1 & 2.1 \end{bmatrix} \tag{21.35}$$
$$+y'_s * \text{cofactor of} \begin{bmatrix} 8.1 & -5.1 & -3 \\ -5.1 & 15.1 & -10 \\ -3 & -10 & 13 \end{bmatrix}$$
$$+y'_s g_m * \text{cofactor of} \begin{bmatrix} 5.1 & -5.1 \\ -5.1 & 5.1 \end{bmatrix}$$

After evaluating the cofactors numerically, the equation reduces to

$$\text{cofactor of } \mathbf{Y}'_{ind} = 137.7 + 10.5g_m + 96.3y'_s + 5.1y_s g_m \tag{21.36}$$

From Eq. (21.11) this results in

$$Z_{14} = \frac{\Delta_{11}}{\Delta} = \frac{96.3 + 5.1g_m}{137.7 + 10.5g_m} \tag{21.37}$$

4. The interpolation method

This method is best suited when s is the only symbolic variable. The method requires the finding of the coefficient of the determinant's polynomial by evaluating it at different values of s. A serious disadvantage of this method is that, for circuits with over 20 nodes, using real values for s leads to ill-conditioned equations and results in inaccurate solutions [45]. Therefore, it is best to use complex values for s. This can be done by the manipulation of the nodal admittance matrix $\mathbf{Y}_n$ to write it in the form:

$$\mathbf{Y}_n = s\mathbf{A} + \mathbf{B} \tag{21.38}$$

where $\mathbf{A}$ and $\mathbf{B}$ are $n \times n$ real matrices. The determinant of $\mathbf{Y}_n$, which is a function of s, can then be written as a polynomial of degree n or lower

$$\Delta(s) = \sum_{i=0}^{n} k_i s^i = k_0 + k_1 s + k_2 s^2 + k_3 s^3 + \ldots + k_n s^n \tag{21.39}$$

The k_i's in the above equations are numeric constants that are calculated by evaluating the polynomial at $n+1$ different values of s and then solving the following set of linear equations

$$\begin{bmatrix} 1 & s_0 & s_0^2 & \cdots & s_0^n \\ 1 & s_1 & s_1^2 & \cdots & s_1^n \\ & & \vdots & & \\ 1 & s_n & s_n^2 & \cdots & s_n^n \end{bmatrix} \begin{bmatrix} k_0 \\ k_1 \\ \vdots \\ k_n \end{bmatrix} = \begin{bmatrix} \Delta(s_0) \\ \Delta(s_1) \\ \vdots \\ \Delta(s_n) \end{bmatrix} \tag{21.40}$$

An advantage of the mostly numeric nature of this method is that the Fast Fourier Transform can be used to find the coefficients which greatly enhances the execution time of the method [31]. Singhal and Vlach [45] extended the method to handle several symbolic variables in addition to s. The program implementation [49] allows a maximum of five symbolic parameters in a circuit.

There are other classifications of symbolic methods that have been reported [13]. These methods can be considered as variations on the above basic four methods. The reported methods include elimination algorithms, recursive determinant-expansion algorithms, and the nonrecursive nested-minors method. All three are based on the use of Cramer's rule to find the determinant and the cofactors of a matrix. Another reported class of algorithms uses Modified Nodal Analysis [21] as the basis of the analysis, sometimes referred to as a direct network approach [17, 28]. This class of methods is covered in the next section.

The first generation of computer programs available for symbolic circuit simulation based on these methods includes NASAP [38] and SNAP [29]. Research in the late 1980s and early 1990s has produced newer symbolic analysis programs. These programs include ISAAC [21], SCAPP [17], ASAP [8], EASY [47], SYNAP [44], SAPEC [34], SCYMBAL [24], GASCAP [23], and SSPICE [50].

21.4 Hierarchical Methods (Sequence of Expressions)

All of the methods presented in the previous section have circuit size limitations. The main problem is the exponential growth of the number of symbolic terms involved in the expression for the transfer function in Eq. (21.1) as the circuit gets larger. The solution to analyzing large-scale circuits lies in a total departure from the traditional procedure of trying to state the transfer function as a single expression and using a **sequence of expressions** procedure instead. The idea is to produce a succession of small expressions with a backward hierarchical dependency on each other. The growth of the number of expressions in this case will be, at worst case, quadratic [17].

The advantage of having the transfer function stated in a single expression lies in the ability to gain insight in the relationship between the transfer function and the circuit elements by inspection [30]. For large expressions though, this is not possible, and the single expression loses that advantage. ISAAC [51], ASAP [9], and SYNAP [44] attempt to handle larger circuits by maintaining the single expression method and using circuit-dependent approximation techniques. The tradeoff is accuracy for insight. Therefore, the sequence of expressions approach is more suitable for accurately handling large-scale circuits. The following example illustrates the features of the **sequence of expressions.**

Example 5. Consider the resistive ladder circuit in Fig. 21.10. The goal is to obtain the input impedance function of the circuit, $Z_{in} = V_{in}/I_{in}$. The single expression transfer function Z_4 is

$$Z_4 = \frac{R_4R_1 + R_4R_2 + R_4R_3 + R_3R_1 + R_3R_2}{R_1 + R_2 + R_3} \tag{21.41}$$

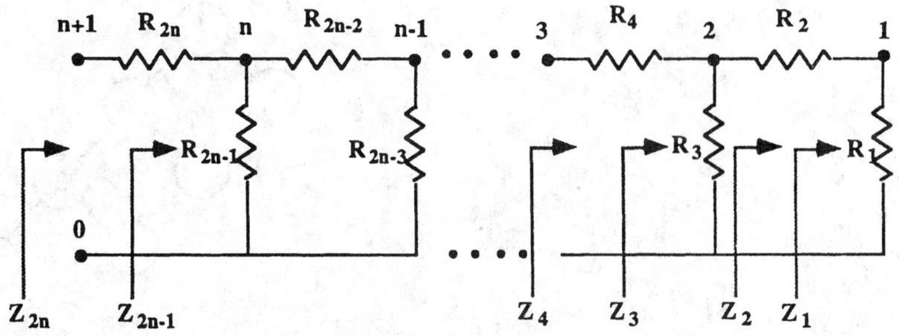

FIGURE 21.10 Resistive ladder circuit.

The number of terms in the numerator and denominator are given by the Fibonacci numbers satisfying the following difference equation:

$$y(k+2) = y(k+1) + y(k) \qquad k = 0, 1, 2, \ldots \qquad y(0) = 0,\ y(1) = 1 \quad (21.42)$$

An explicit solution to the above equation is

$$\begin{aligned} y(n) &= \left[\frac{1}{\sqrt{5}}\left(\frac{1+\sqrt{5}}{2}\right)^n - \left(\frac{1-\sqrt{5}}{2}\right)^n\right] \\ &\approx 0.447 \times 1.618^n \textit{for large } n \end{aligned} \quad (21.43)$$

The solution shows that the number of terms in Z_n increases exponentially with n. Any single expression transfer function has this inherent limitation.

Now using the **sequence of expressions** procedure the input impedance can be obtained from the following expressions:

$$Z_1 = R_1 \qquad Z_2 = R_2 + Z_1 \qquad Z_3 = \frac{Z_2 R_3}{Z_2 + R_3} \qquad Z_4 = R_4 + Z_3 \quad (21.44)$$

It is obvious for each additional resistance added, the sequence of expressions will grow by one expression, either of the form $R_i + Z_{i-1}$ or $R_i Z_{i-1}/(R_i + Z_{i-1})$. The number of terms in the sequence of expressions can be given by

$$y(n) = \begin{cases} 2.5n - 2 & \text{for } n \text{ even} \\ 2.5n - 1.5 & \text{for } n \text{ odd} \end{cases} \quad (21.45)$$

which exhibits a linear growth with respect to n. So, to find the input impedance of a 100-resistor ladder circuit, the single expression method would produce 7.9×10^{20} terms which requires unrealistically huge computer storage capabilities. On the other hand, the **sequence of expressions** method would produce only 248 terms, which is even within the scope of some desk calculators.

Another advantage of the **sequence of expressions** is the number of mathematical operations needed to evaluate the transfer function. To evaluate Z_9, for example, the single expression methods would require 302 multiplications and 87 additions. The **sequence of expressions** method would only require 8 multiplications and 8 additions, a large reduction in computer

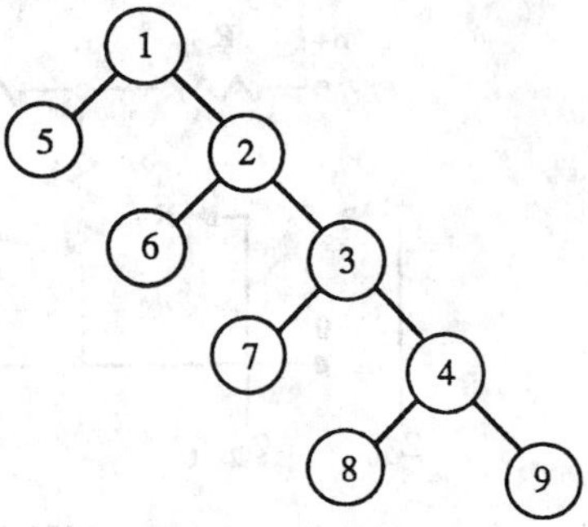

FIGURE 21.11 Hierarchical analysis.

evaluation time. All this makes the concept of symbolic circuit simulation of large-scale circuits very possible.

Two flowgraph methods for symbolic simulation of large-scale circuits have been proposed in [48] and in [20]. The first method utilizes the **sequence of expressions** idea to obtain the transfer functions. The method operates on the Coates graph [7] representing the circuit. A partitioning is proposed onto the flowgraph and not the physical circuit. The second method also utilizes the **sequence of expressions** and a Mason's signal flowgraph [36] representation of the circuit. The method makes use of partitioning on the physical level rather than on the graph level. Therefore, for a hierarchical circuit, the method can operate on the subcircuits in a hierarchical fashion in order to produce a final solution. The fundamentals of both signal flowgraph methods were described in the previous section.

Another hierarchical approach is one that is based on Modified Nodal Analysis [21]. This method [17] exhibits a linear growth (for practical circuits) in the number of terms in the symbolic solutions. The analysis methodology introduces the concept of the RMNA (Reduced Modified Nodal Analysis) matrix. This allows the characterization of symbolic circuits in terms of only a small subset of the circuit variables (external variables) rather than the complete set of variables. The analysis algorithm is most efficient when circuit partitioning is used. Partitioning results in a reduction in the number of terms in the symbolic solutions.

The analysis process is divided into the following parts: (1) binary circuit partitioning, (2) subcircuit analysis, referred to as terminal block analysis, and (3) upward hierarchical analysis, referred to as middle block analysis. The process is modeled using a binary tree, Fig. 21.11. The main circuit is partitioned into subcircuits which are represented by the leaves of the binary tree and each nonleaf vertex in the tree represents a binary partitioning operation. The analysis process is noniterative. The symbolic solutions of the leaf vertices are found only once and then are recombined by tracing the tree upward until the root node is reached. A detailed discussion of the method can be found in [17, 18].

21.5 Time Domain Analysis

The previous subsections have discussed the different frequency domain techniques for symbolic analysis. Symbolic analysis methods in the transient domain did not appear until the beginning of the 1990s [3, 19, 28]. The main limitation to symbolic time domain analysis is the difficulty in handling the symbolic integration and differentiation needed for the energy storage elements (mainly capacitors and inductors). This problem, of course, does not exist in the frequency domain because of the use of Laplace transforms to represent these elements. While there exist symbolic algebra software packages such as MATHEMATICA, MAXIMA, and MAPLE that can be used to perform integration and differentiations, they have not been applied to transient symbolic analysis due to the execution time complexity of these programs. All but one of the approaches in the time domain are actually semisymbolic. The semisymbolic algorithms use a mixture of symbolic and numeric techniques to perform the analysis. The work here is still in its infancy. This subsection will lightly discuss the three contributions published in the literature so far.

The symbolic time domain techniques all deal with linear circuits and can be classified under one of the following two categories.

Fully Symbolic

Only one method has been reported in the literature that is fully symbolic [15]. This method utilizes a direct and hierarchical symbolic transient analysis approach similar to the one reported in [17]. The formulation is based on well-known discrete models for numerical integration of linear differential equations. Three of these integration methods are implemented symbolically; the Backward Euler, the Trapezoidal, and Gear's 2nd Order Backward Differentiation [15]. The inherent accuracy problems due to the approximations in these methods show up when the symbolic expressions are evaluated numerically. A detailed discussion of this method can be found in [15].

Semi-Symbolic

Three such algorithms have been reported in the literature so far. Two of them [19, 28] simply take the symbolic expressions in the frequency domain, evaluate them numerically for a range of frequencies, and then perform a numeric Inverse Laplace Transformation or a Fast Fourier Transformation (FFT) on the results. The approach reported in [28] uses a modified nodal analysis then a state-variable symbolic formulation to get the frequency domain response and can handle time-varying circuits, namely switch power converters. The approach in [19] uses a hierarchical network approach [17] to generate the symbolic frequency domain response. The third algorithm reported in [3] is a hierarchical approach that uses an MNA and a state-variable symbolic formulation and then uses the eigenvalues of the system to find a closed-form numerical transient solution.

eferences

[1] G. E. Alderson and P. M. Lin, "Integrating topological and numerical methods for semi-symbolic network analysis," Proceedings of the Midwest Symposium on Circuit Theory, 1970.

[2] G. E. Alderson and P. M. Lin, "Computer generation of symbolic network functions—a new theory and implementation, *IEEE Trans. Circuit Theory*, vol. CT-20, pp. 48–56, Jan. 1973.

[3] B. Alspaugh and M. Hassoun, "A mixed symbolic and numeric method for closed-form transient analysis," Proc. of *ECCTD*, Davos, Switzerland, pp. 1687–1692, 1993.

[4] Z. Arnautovic and P. M. Lin, "Symbolic analysis of mixed continuous and sampled data systems," *Proc. ISCAS*, pp. 798–801, Singapore, 1991.

[5] D. A. Calahan, "Linear network analysis and realization digital computer programs, and instruction manual," University of Ill. Bull., Champaign-Urbana vol. 62, Feb. 1965.

[6] L. O. Chua and P. M. Lin, *Computer Aided Analysis of Electronic Circuits—Algorithms and Computational Techniques*, Englewood Cliffs, NJ: Prentice-Hall, 1975.

[7] C. L. Coates, "Flow graph solutions of linear algebraic equations," *IRE Trans. Circuit Theory*, vol. CT-6, pp. 170–187, 1959.

[8] F. V. Fernandez, A. Rodriguez-Vazquez, and J. L. Huertas, "An advanced symbolic analyzer for the automatic generation of analog circuit design equations," Proc. IEEE ISCAS, Singapore, pp. 810–813, June 1991.

[9] F. V. Fernandez, J. Martin, A. Rodriguez-Vazquez, and J. L. Huertas, "On simplification techniques for symbolic analysis of analog integrated circuits," Proc. IEEE ISCAS, San Diego, pp. 1149–1152, May 1992.

[10] J. K. Fidler and J. I. Sewell, "Symbolic analysis for computer-aided circuit design—the interpolative approach," *IEEE Trans. Circuit Theory*, vol. CT-20, Nov. 1973.

[11] T. F. Gatts and N. R. Malik, "Topological analysis program for linear active networks (TAPLAN)," Proc. 13th Midwest Symposium on Circuit Theory, 1970.

[12] G. Gielen, H. Walscharts, and W. Sansen, "ISSAC: a symbolic simulator for analog integrated circuits," *IEEE J. Solid-State Circuits*, vol. SC-24, pp. 1587–1597, Dec. 1989.

[13] G. Gielen and W. Sansen, *Symbolic Analysis for Automated Design of Analog Integrated Circuits*, Boston: Kluwer Academic, 1991.

[14] S. Greenfield, Transient Analysis for Symbolic Simulation, M.S. thesis, Iowa State University, Ames, Dec. 1993.

[15] S. Greenfield and M. Hassoun, "Direct hierarchical symbolic transient analysis of linear circuits," Proc. ISCAS, pp. 29–32, May 1994.

[16] G. D. Hachtel et al., "The sparse tableau approach to network and design," *IEEE Trans. Circuit Theory*, vol. CT-18, pp. 101–113, Jan. 1971.

[17] M. M. Hassoun and P. M. Lin, "A new network approach to symbolic simulation of large-scale networks," Proc. 1989 IEEE Int. Symp. on Circuits and Systems, pp. 806–809, May 1989.

[18] M. M. Hassoun, and P. M. Lin, "An efficient partitioning algorithm for large-scale circuits," Proc. IEEE ISCAS, New Orleans, pp. 2405–2408, May 1990.

[19] M. M. Hassoun and J. E. Ackerman, "Symbolic simulation of large-scale circuits in both frequency and time domains," Proc. of the 33rd IEEE Midwest Symp. on CAS, Calgary, pp. 707–710, Aug. 1990.

[20] M. Hassoun and K. McCarville, "Symbolic analysis of large-scale networks using a hierarchical signal flowgraph approach," *J. Analog Integrated Circuits and Signal Processing*, pp. 31–42, Jan. 1993.

[21] C. Ho, A. E. Ruehli, and Brennan, "The modified nodal approach to network analysis," *IEEE Trans. on Circuits and Systems*, vol. CAS-25, pp. 504–509, June 1975.

[22] J. J. Hsu and C. Sechen, "Low-frequency symbolic analysis of large analog integrated circuits," Proc. CICC, pp. 14.7.1–14.7.4, 1993.

[23] L. Huelsman, "Personal computer symbolic analysis programs for undergraduate engineering courses," Proc. ISCAS, pp. 798–801, 1989.

[24] A. Konczykowska and M. Bon, "Automated design software for switched capacitor ICs with symbolic simulator SCYMBAL," Proc. DAC, pp. 363–368, 1988.

[25] A. Konczykowska, M. Bon, H. Wang, and R. Mezui-Mintsa, "Symbolic analysis as a tool for circuit optimization," IEEE Int. Symp. on Circuits and Systems, San Diego, pp. 1161–1164, May 1992.

[26] J. Lee and R. Rohrer, "AWEsymbolic: compiled analysis of linear(ized) circuits using asymptotic waveform evaluation," Proc. DAC, pp. 213–218, 1992.

[27] B. Li and D. Gu, "SSCNAP: a program for symbolic analysis of switched capacitor circuits," *IEEE Trans. Computer-Aided Design*, vol. 11, pp. 334–340, 1992.

[28] A. Liberatore et al., "Simulation of switching power converters using symbolic techniques," *Alt Frequenza,* vol. 5, no. 6, pp. 16–23, Nov. 1993.

[29] P. M. Lin and G. E. Alderson, SNAP—A Computer Program for Generating Symbolic Network Functions, School of EE, Purdue University, West Lafayette, IN, Rep. TR-EE 70-16, Aug. 1970.

[30] P. M. Lin, "A survey of applications of symbolic network functions," *IEEE Trans. Circuit Theory,* vol. CT-20, pp. 732–737, Nov. 1973.

[31] P. M. Lin, *Symbolic Network Analysis,* Elsevier Science, Amsterdam, 1991.

[32] P. M. Lin, "Sensitivity analysis of large linear networks using symbolic programs," IEEE Int. Symp. Circuits and Systems, San Diego, pp. 1145–1148, May 1992.

[33] V. K. Manaktala and G. L. Kelly, "On the symbolic analysis of electrical networks," Proceedings of the 15th Midwest Symp. on Circuit Theory, 1972.

[34] S. Manetti, "New approaches to automatic symbolic analysis of electric circuits," *Proc. Inst. Elec. Eng.,* pp. 22–28, Feb. 1991.

[35] M. Martins et al., "A computer-assisted tool for the analysis of multirate SC networks by symbolic signal flow graphs," *Alt Frequenza,* vol. 5, no. 6, pp. 6–10, Nov. 1993.

[36] S. J. Mason, "Feedback theory—further properties of signal flow graphs," *Proc. IRE,* vol. 44, pp. 920–926, July 1956.

[37] J. O. McClanahan and S. P. Chan, "Computer analysis of general linear networks using digraphs," *Int. J. Electron.,* no. 22, pp. 153–191, 1972.

[38] L. P. McNamee and H. Potash, A User's and Programmer's Manual for NASAP, University of California at Los Angeles, Rep. 63-38, Aug. 1968.

[39] R. R. Mielke, "A new signal flowgraph formulation of symbolic network functions," *IEEE Trans. Circuits and Systems,* vol. CAS-25, pp. 334–340, June 1978.

[40] H. Okrent and L. P. McNamee, NASAP-70 User's and Programmer's Manual, University of California at Los Angeles, Tech. Rep. ENG-7044, 1970.

[41] C. Pottle, CORNAP User's Manual, School of Electrical Engineering, Cornell University, Ithaca, NY, 1968.

[42] P. Sannuti and N. N. Puri, "Symbolic network analysis—an algebraic formulation," *IEEE Trans. Circuits and Systems,* vol. CAS-27, pp. 679–687, Aug. 1980.

[43] G. DiDomenico, S. Seda, and M. Khaifa et al., "BRAINS: a symbolic solver for electronic circuits," Int. Workshop on Symbolic Methods and Applications to Circuit Design, Paris, Oct. 1991.

[44] S. Seda, M. Degrauwe, and W. Fichtner, "Lazy-expansion symbolic expression approximation in SYNAP," 1992 Int. Conf. on Computer-Aided Design, Santa Clara, CA, pp. 310–317, 1992.

[45] K. Singhal and J. Vlach, "Generation of immittance functions in symbolic form for lumped distributed active networks," *IEEE Trans. Circuits and Systems,* vol. CAS-21, pp. 57–67, Jan. 1974.

[46] K. Singhal and J. Vlach, "Symbolic analysis of analog and digital circuits," *IEEE Trans. Circuits and Systems,* vol. CAS-24, pp. 598–609, Nov. 1977.

[47] R. Sommer, "EASY—an experimental analog design system framework," Int. Workshop on Symbolic Methods and Applications to Circuit Design, Paris, Oct. 1991.

[48] J. A. Starzyk and A. Konczykowska, "Flowgraph analysis of large electronic networks," *IEEE Trans. Circuits and Systems,* vol. CAS-33, pp. 302–315, March 1986.

[49] J. Vlach and K. Singhal, *Computer Methods for Circuit Analysis and Design,* New York: Van Nostrand Reinhold, 1983.

[50] G. Wierzba et al., "SSPICE—a symbolic SPICE program for linear active circuits," Proc. Midwest Symp. on Circuits and Systems, 1989.

[51] P. Wombacq, G. Gielen, and W. Sansen, "A cancellation-free algorithm for the symbolic simulation of large analog circuits," IEEE Int. Symp. on Circuits and Systems, San Diego, pp. 1157–1160, May 1992.

22

Using Nullors to Analyze Linear Networks

nes A. Svoboda
rkson University,
v York

2.1 Introduction

This chapter describes a procedure for computer analysis of a linear network. This procedure uses nullors to model the linear network with an equivalent network. The equivalent network will consist of only resistors, capacitors, and nullors and will be excited only by current sources. It is natural to write a node equation to represent the equivalent network. Solving this node equation provides the node voltages of the equivalent network. The branch voltages and currents of the linear network can be calculated from the node voltages of the equivalent network.

Figure 22.1 shows the symbol that will be used for a nullor. A nullor consists of two branches, one called a nullator and the other called a norator. The branch current and voltage of the nullator are both zero. The branch current and voltage of the norator are unspecified. A pair consisting of one nullator and one norator is called a nullor.

Consider the circuit shown in Fig. 22.2(a). When the op amp is ideal, the op amp input current and voltage are both zero, regardless of the values of R_1 and R_2. The op amp output current and voltage are given by

$$i_{oa} = \frac{V_s}{R_1} \quad \text{and} \quad v_{oa} = -\frac{R_2}{R_2} V_s \tag{22.1}$$

The op amp output current and voltage depend on the parameters of the resistors and voltage source. The op amp determines its input current and voltage but the rest of the circuit determines the op amp output current and voltage. A model of an ideal op amp will specify the input current and voltage to be zero but will not specify the output current or voltage. It is in this sense that the op amp output current and voltage are said to be arbitrary. A nullor is well suited to model an ideal op amp (see Fig. 22.5).

93-8341-2/95/$0.00+$.50

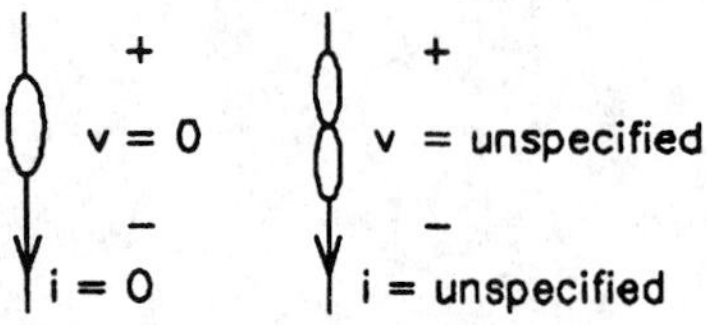

FIGURE 22.1 A nullor.

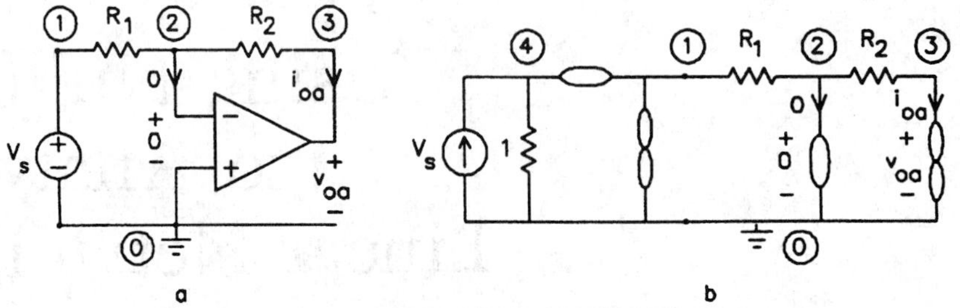

FIGURE 22.2 Inverting amplifier and equivalent R-nullor circuit.

Figure 22.2(b) illustrates the use of nullors for modeling linear circuits. In this case the linear circuit contains an op amp and is excited by a voltage source. The voltage source and op amp are replaced by models involving nullors to obtain an equivalent circuit. This equivalent circuit isn't particularly convenient for hand analysis but it is well suited to computer analysis. This computer analysis consists of several steps. First some bookkeeping is done by considering the nullors. Next, node equations are formed by considering the resistors, capacitors, and current sources. Solving the node equation gives the node voltages of the equivalent network. There is a correspondence between the nodes of the linear network and the equivalent network. This correspondence provides a way to determine the node voltages of the linear network from the node voltages of the equivalent network. Finally, the branch voltages and currents of the linear network are calculated.

Computer analysis of linear networks using nullors consists of the following five steps

1. Model the linear network using nullors.
2. Do some bookkeeping.
3. Form node equations representing the equivalent network.
4. Solve the node equations.
5. Obtain the solution of the linear network from the solution of the equivalent network.

22.2 Modeling Linear Networks Using Nullors

Every linear network can be modeled by an equivalent network consisting of only resistors, capacitors, and nullors and excited by only current sources [1, 2, 4]. The procedure for constructing the equivalent circuit is straightforward. Each device in the original circuit is replaced by an RC-nullor model. Figures 22.3, 22.4, and 22.5 provide models for some devices. Figure 22.6 illustrates the use of these models. Here the original circuit is an RL circuit excited by a voltage source. To construct the equivalent circuit, replace the voltage source and inductor by the models shown in Fig. 22.3. (The model of a resistor is a resistor. Similarly, capacitor and current sources are modeled as capacitors and current sources.)

Notice that the equivalent circuit in Fig. 22.6 does indeed consist of only resistors, capacitors, and nullors and is excited by a current source. Inductors are not needed to model reactive circuits—capacitors are sufficient. The inputs in the equivalent circuit are all current sources. Both of these facts simplify the algorithms used to analyze the circuit.

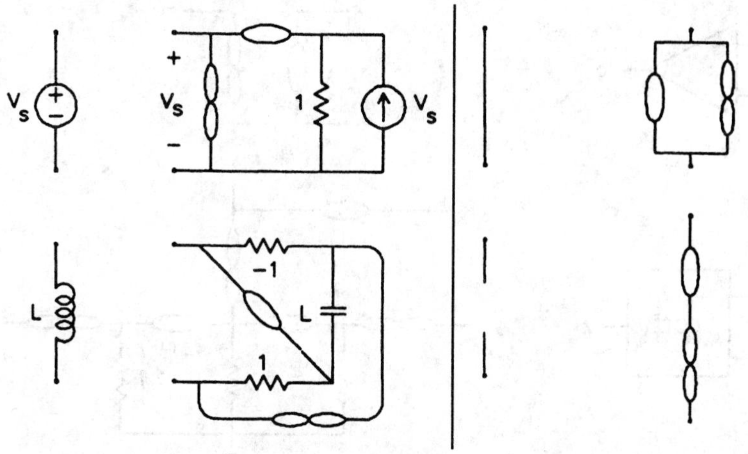

FIGURE 22.3 RC-nullor models for voltage sources, inductors, short circuits, and open circuits.

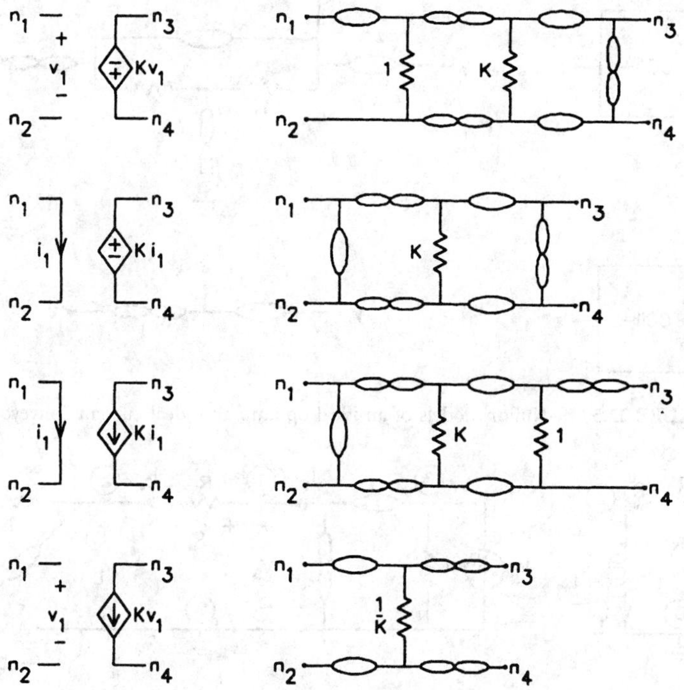

FIGURE 22.4 R-Nullor models of dependent sources.

It is convenient to adopt some node numbering conventions. The reference node will be numbered as node 0 in both the original network and the equivalent network. The remaining nodes in the original network will be numbered consecutively from 1 to n_o. (The algorithms don't require the nodes be numbered consecutively. This assumption is made to simplify the description of the algorithms.) For each node of the original network there is a corresponding node in the equivalent network. This correspondence is used to number nodes 1 to n_o of the equivalent network. The equivalent network can contain additional nodes. These nodes are numbered from $n_o + 1$ to n_e.

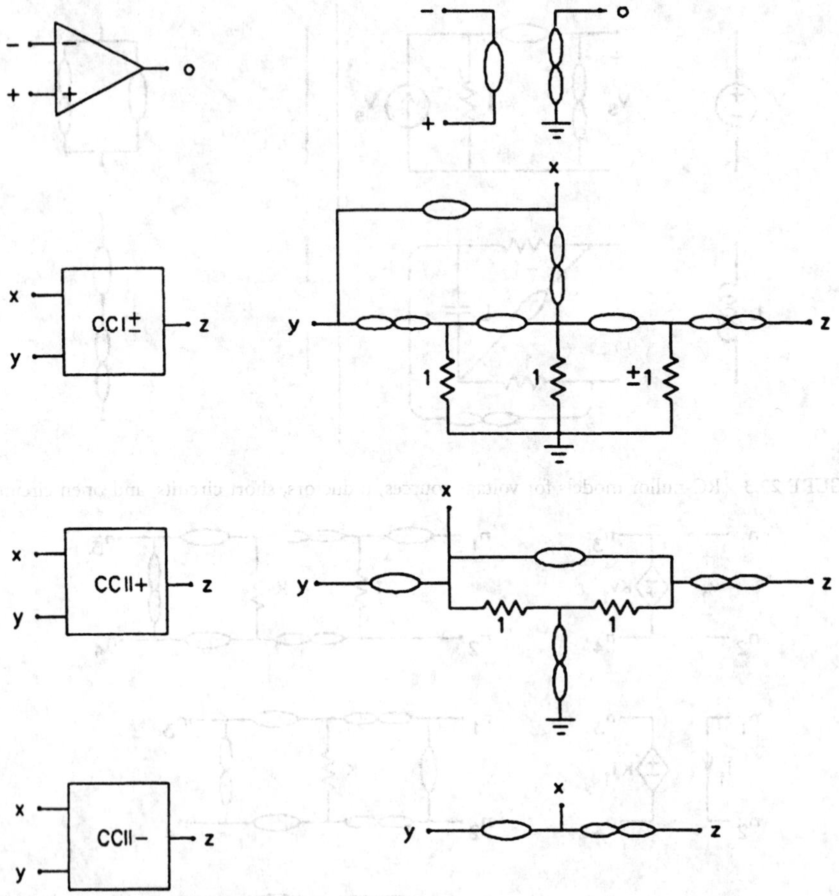

FIGURE 22.5 R-Nullor models of an ideal op amp and ideal current conveyors.

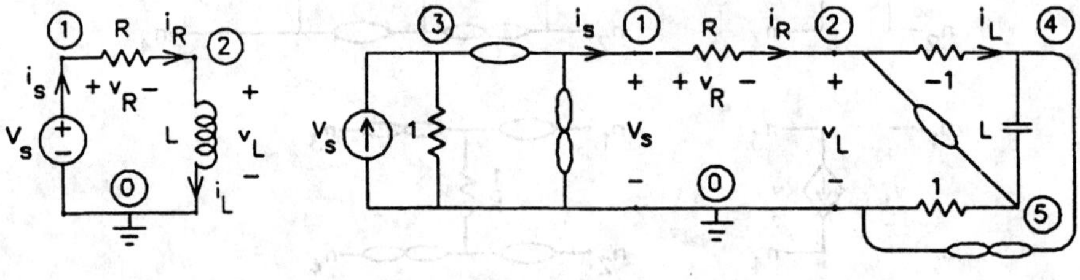

FIGURE 22.6 Correspondence between the original circuit and the RC-nullor equivalent circuit.

This node numbering convention is used in Fig. 22.6. The nodes of the original network and corresponding nodes of the equivalent network are numbered 1, 2. (Notice that the resistor is connected between nodes 1 and 2 in both the original and equivalent networks.) Nodes 3, 4, and 5 of the equivalent network are internal to the models of the voltage source and inductor and do not correspond to nodes of the original network.

There is also a correspondence between the voltages and currents in the equivalent network and the voltages and currents in the original network. Because of this correspondence, it is possible to determine the branch voltages and currents of the original network by analyzing the equivalent network. The node voltages at corresponding nodes of the original and equivalent networks are equal. For each branch current and for each branch voltage in the original network there is a corresponding branch current or voltage in the equivalent network. These corresponding voltages and currents are equal.

Refer again to Fig. 22.6 where corresponding branch currents and voltages have been labeled. Notice in particular that the branch currents and voltages i_s, v_L, i_L, i_R and v_R of the original network can all be calculated by analyzing the equivalent network.

22.3 Analysis of RC-Nullor Networks

The first step in the analysis of an RC-nullor network is bookkeeping. The RC-nullor network will be represented by a node equation. A row of this node equation and a column of the nodel admittance matrix will correspond to each node of the RC-nullor network. The purpose of the bookkeeping is to determine this correspondence. (This is not a 1:1 correspondence. For example, several nodes could correspond to the same row of the node equation.)

Consider the RC-nullor circuit shown in Fig. 22.7. This circuit can be represented by the node equation

$$\begin{matrix} 1: \\ 2,3: \end{matrix}\begin{pmatrix} I \\ 0 \end{pmatrix} = \begin{pmatrix} G_1 + G_4 & -G_4 \\ -G_4 & C_2 s + G_3 + G_4 \end{pmatrix}\begin{pmatrix} V_{1,2} \\ V_3 \end{pmatrix} \tag{22.2}$$

where, for convenience, the resistors have been represented by their conductances.

This node equation illustrates the correspondence between nodes of the circuit and columns of the nodal admittance matrix. Notice that the nullator forces the voltage at nodes 1 and 2 to be equal. Let $V_{1,2}$ denote $V_1 = V_2$. Since $V_{1,2}$ multiplies the first column of the nodal admittance matrix, nodes 1 and 2 are associated with column 1 of this matrix. Similarly, column 2 of the matrix is associated with node 3. No column of the nodal admittance matrix is associated with the reference node, node 0.

The correspondence between the columns of the nodal admittance matrix and the nodes of the network will be represented by a vector called *COLUMN*. An entry in *COLUMN*

$$j = COLUMN[i] \quad i = 0, 1, \ldots n_e \tag{22.3}$$

indicates that column j of the nodal admittance matrix corresponds to node i of the circuit. For the circuit in Fig. 22.7

$$COLUMN = [0 \quad 1 \quad 1 \quad 2] \tag{22.4}$$

The four entries of this vector correspond to nodes 0, 1, 2, and 3. The first entry indicates that column 0 corresponds to the reference node, node 0. There is no column 0 of the matrix. A zero entry in *COLUMN* indicates that no column of the nodal admittance matrix corresponds to that node. The third entry indicates that column 1 of the matrix corresponds to node 2 of the network.

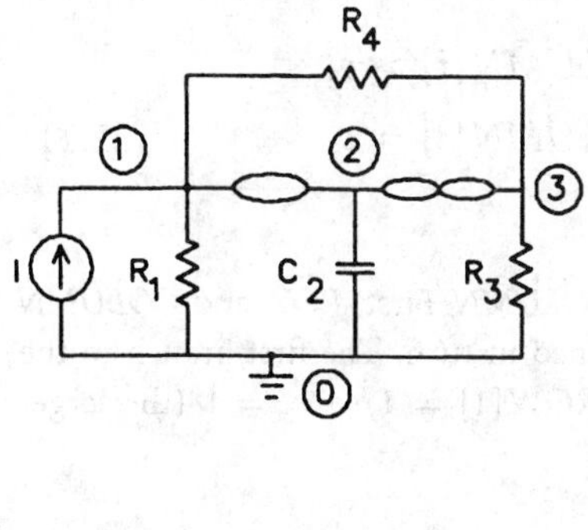

FIGURE 22.7 An R-nullor circuit.

The node equation also illustrates the correspondence between nodes of the circuit and rows of the node equation. The first row of the node equation is obtained by applying Kirchhoff's Current Law (KCL) at node 1 of the network, i.e., to the cutset that separates node 1 from the remaining nodes of the network. The second row of the node equation is obtained by applying KCL to the cutset that separates the nodes of the norator, nodes 2 and 3, from the remaining nodes of the network. These observations associate row 1 of the node equation with node 1 of the network and row 2 of the node equation with nodes 2 and 3 of the network.

The correspondence between the rows of the node equation and the nodes of the network will be represented by a vector called *ROW*. An entry in *ROW*

$$j = ROW[i] \quad i = 0, 1, \ldots n_e \tag{22.5}$$

indicates that row j of the node equation corresponds to node i of the circuit. For the circuit in Fig. 22.7

$$ROW = [0 \quad 1 \quad 2 \quad 2] \tag{22.6}$$

The four entries of this vector correspond to nodes 0, 1, 2, and 3. The first entry in *ROW* indicates that no row of the node equation corresponds to the reference node. The third entry indicates that row 2 of the node equation corresponds to node 2 of the network.

A simple algorithm is available for forming *ROW* and *COLUMN* [4]. First, initialize *ROW* and *COLUMN*

$$ROW[i] = i \quad \text{and} \quad COLUMN[i] = i \quad \text{for} \quad i = 0, 1, 2, \ldots n_e \tag{22.7}$$

Examine each branch of the network in turn. *ROW* will be updated whenever a norator is encountered and *COLUMN* will be updated whenever a nullator is encountered.

Suppose that a norator having nodes n_1 and n_2 is encountered. Suppose further that

$$r_1 = ROW[n_1] < r_2 = ROW[n_2] \tag{22.8}$$

before updating. Then *ROW* is updated as follows

$$ROW[i] = \begin{cases} ROW[i] - 1 & \text{when } ROW[i] > r_2 \\ r_1 & \text{when } ROW[i] = r_2 \\ ROW[i] & \text{otherwise} \end{cases} \tag{22.9}$$

Similarly, suppose a nullator having nodes n_1 and n_2 is encountered. Suppose further that

$$c_1 = COLUMN[n_1] < c_2 = COLUMN[n_2] \tag{22.10}$$

before updating. Then *COLUMN* is updated as follows

$$COLUMN[i] = \begin{cases} COLUMN[i] - 1 & \text{when } COLUMN[i] > c_2 \\ c_1 & \text{when } COLUMN[i] = c_2 \\ COLUMN[i] & \text{otherwise} \end{cases} \tag{22.11}$$

Fig. 22.8 illustrates the algorithm for generating *ROW* and *COLUMN*. First, *ROW* and *COLUMN* are initialized. Next, the branches of the network are examined in turn. The first branch in the list is a norator with nodes 1 and 0. $ROW[0] = 0$ and $ROW[1] = 1$ so $r_2 = 1$ (the larger

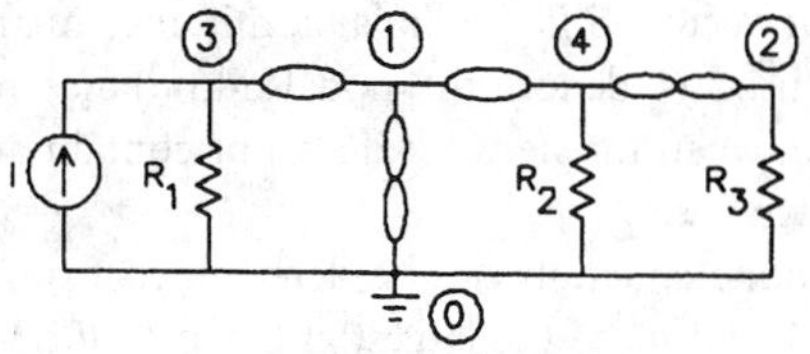

				ROW	COLUMN
				[0 1 2 3 4]	[0 1 2 3 4]
NOR	1	0		[0 0 1 2 3]	[0 1 2 3 4]
NOR	2	4		[0 0 1 2 1]	[0 1 2 3 4]
RES	2	0	R3	[0 0 1 2 1]	[0 1 2 3 4]
NUL	3	1		[0 0 1 2 1]	[0 1 2 1 3]
RES	3	0	R1	[0 0 1 2 1]	[0 1 2 1 3]
CS	0	3	I	[0 0 1 2 1]	[0 1 2 1 3]
NUL	1	4		[0 0 1 2 1]	[0 1 2 1 1]
RES	4	0	R2	[0 0 1 2 1]	[0 1 2 1 1]

FIGURE 22.8 Generating the bookkeeping vectors, *ROW* and *COLUMN*.

of the two row numbers) and $r_1 = 0$. *ROW* is updated by replacing entries having value 1 ($= r_2$) by 0 ($= r_1$) and decrementing all entries having value greater than 1 ($= r_2$). The second branch of the network is also a norator, this time having nodes 2 and 4. Now $ROW[2] = 1$ and $ROW[4] = 3$ so $r_2 = 3$ and $r_1 = 1$. *ROW* is updated by replacing entries having value 3 ($= r_2$) by 1 ($= r_1$). None of the entries of *ROW* have value greater than 3 ($= r_2$) so none are decremented. The third branch in the list is a resistor, so neither *ROW* nor *COLUMN* needs to be updated. The fourth branch in the list is a nullator with nodes 3 and 1. $COLUMN[1] = 1$ and $COLUMN[3] = 3$ so $c_2 = 3$ (the larger of the two column numbers) and $c_1 = 1$. *COLUMN* is updated by replacing entries having value 3 ($= c_2$) by 1 ($= c_1$) and decrementing all entries having value greater than 3 ($= c_2$). The fifth and sixth branches in the list are a resistor and a current source, so neither *ROW* nor *COLUMN* needs to be updated. The seventh branch in the list is a nullator with nodes 4 and 1. $COLUMN[4] = 3$ and $COLUMN[1] = 1$ so $c_2 = 3$ (the larger of the two column numbers) and $c_1 = 1$. *COLUMN* is updated by replacing entries having value 3 ($= c_2$) by 1 ($= c_1$). None of the entries of *COLUMN* have value greater than 3 ($= r_2$) so none are decremented. Finally, the last branch in the list is a resistor, so neither *ROW* nor *COLUMN* needs to be updated.

Once *ROW* and *COLUMN* are available, a node equation representing the network can be formed. First the node equation is initialized. Next, each branch of the network is examined in turn. The node equation is updated whenever the branch is a resistor, capacitor, or current source.

For ease of exposition, suppose that the original does not contain any reactive devices. In this case the equivalent network does not contain any capacitors. The node equation will be of the form

$$Gv_n = J \tag{22.12}$$

where v_n is the node voltage vector, G is the nodal admittance matrix, and J is the nodal current vector. (The symbol G is used to denote the nodal admittance matrix to emphasize that this matrix will be a real matrix when the network does not contain any reactive devices. Entries of G will have units of conductance.)

The dimensions of the node equation can be determined from *ROW* and *COLUMN*. Let n_r be the largest entry in *ROW* and n_c be the largest entry in *COLUMN*. Then n_r is the number of rows of the node equation and n_c is the number of columns of nodal admittance matrix. The nodal admittance matrix must be square so n_r must equal n_c. Let $n = n_r = n_c$. Initialize the nodal admittance matrix and current vector as follows.

$$G(i, j) = 0 \quad \text{and} \quad J(i) = 0 \quad \text{for } i, j = 0, 1, \ldots n \tag{22.13}$$

Next, each branch of the network is examined in turn. Consider the case when the branch is a resistor having resistance R and nodes n_i and n_j. Let r_i, r_j, c_i, and c_j denote the rows and columns of G corresponding to n_i and n_j. (For example, $r_i = ROW[n_i]$.) The nodal admittance matrix is updated as follows

$$\begin{aligned}
G(r_i, c_i) &= G(r_i, c_i) + \frac{1}{R} \quad \text{if } r_i > 0 \text{ and } c_i > 0 \\
G(r_i, c_j) &= G(r_i, c_j) - \frac{1}{R} \quad \text{if } r_i > 0 \text{ and } c_j > 0 \\
G(r_j, c_i) &= G(r_j, c_i) - \frac{1}{R} \quad \text{if } r_j > 0 \text{ and } c_i > 0 \\
G(r_j, c_j) &= G(r_j, c_j) + \frac{1}{R} \quad \text{if } r_j > 0 \text{ and } c_j > 0
\end{aligned} \tag{22.14}$$

Consider the case when the branch is a current source having current I directed from node n_i toward node n_j. Let r_i, r_j, denote the rows of the node equation corresponding to n_i and n_j. (For example, $r_i = ROW[n_i]$.) The nodal current vector is updated as follows

$$\begin{aligned}
J(r_i) &= J(r_i) - I \quad \text{if } r_i > 0 \\
J(r_j) &= J(r_j) + I \quad \text{if } r_j > 0
\end{aligned} \tag{22.15}$$

The procedure for forming the matrices G and J is illustrated in Fig. 22.9. First G and J are initialized. Next each branch of the network is examined in turn. The matrix G is updated whenever the branch is a resistor and the vector J is updated whenever the branch is a current source. In this example the first two branches are norators. Neither G nor J is updated until the third branch of the network is encountered. This branch is a resistor having conductance G_3 and connected between nodes 2 and 0. The vectors *ROW* and *COLUMN* are used to determine the rows and columns of G corresponding to these nodes. Because $r_0 = 0$ and $c_0 = 0$ (no row or column of G corresponds to node 0) the operations $G(r_0, c_0) = G(r_0, c_0) + G_3$, $G(r_0, c_2) = G(r_0, c_2) - G_3$ and $G(r_0, c_0) = G(r_0, c_0) - G_3$ are not performed. Since both $r_2 = 1 > 0$ and $c_2 = 2 > 0$, $G(r_2, c_2)$ is updated: $G(r_2, c_2) = G(r_2, c_2) + G_3$. The other resistors are treated similarly. The sixth branch of the network is the current source directed from node 0 to node 3. The vector J is updated to account for this branch. Since $r_0 = 0$, no entry in J corresponds to node zero and the $J(r_0) = J(r_0) - I$ is not performed. Since $r_3 > 0$ and the current I is directed toward node 3, $J(r_3)$ is updated by adding I to $J(r_3)$.

The node equation representing the equivalent network can be solved using standard techniques for solving linear equations, e.g., LU factorization [9]. Let V_n denote the solution vector

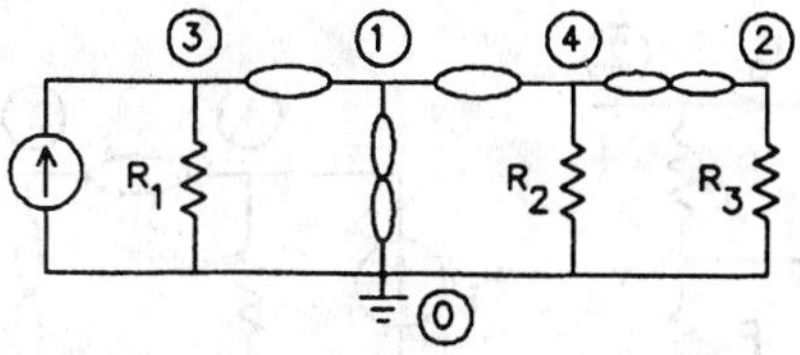

$$G = \begin{pmatrix} 0 & 0 \\ 0 & 0 \end{pmatrix} \quad J = \begin{pmatrix} 0 \\ 0 \end{pmatrix}$$

ROW = [0 0 1 2 1] COLUMN = [0 1 2 1 1]

NOR	1	0			
NOR	2	4			
RES	2	0	R3	$r_2=1,\ r_0=0,\ c_2=2,\ c_0=0$	add G_3 to G(1, 2)
NUL	3	1			
RES	3	0	R1	$r_3=2,\ r_0=0,\ c_3=1,\ c_0=0$	add G_1 to G(2, 1)
CS	0	3	I	$r_0=0,\ r_3=2$	add I to J(2)
NUL	1	4			
RES	4	0	R2	$r_4=1,\ r_0=0,\ c_4=1,\ c_0=0$	add G_2 to G(1, 1)

FIGURE 22.9 Constructing the nodal admittance matrix G and current vector J.

of this node equation. The node voltage at the node i of the equivalent network is given by

$$v_i = V_n(COLUMN[i]) \tag{22.16}$$

Because of the correspondence between the nodes of the original and equivalent networks, the voltage at node i of the original network is equal to the voltage at node i of the equivalent network. Thus Eq. (22.16) can be used to calculate the node voltages of the original network from the solution of the node equation representing the equivalent network.

Notice that the solution of this problem can be given in terms of the original network rather than the equivalent network. A computer program based on this algorithm could accept a description of the original network from the user and return the node voltages of the original network. The user would not need to be aware of nullors.

Figure 22.10 provides another illustration of the complete procedure for calculating the node voltages of a linear network. Suppose that $v_i = 4$ V, $R_i = 20\ K\Omega$, and $R_f = 10\ K\Omega$. Using the units volts, milliamps, and millimhos, the node equation representing the equivalent network is

$$\begin{pmatrix} 0.15 & -0.1 \\ 1 & 0 \end{pmatrix} \begin{pmatrix} v_{1,3,4} \\ v_2 \end{pmatrix} = \begin{pmatrix} 0 \\ 4 \end{pmatrix} \tag{22.17}$$

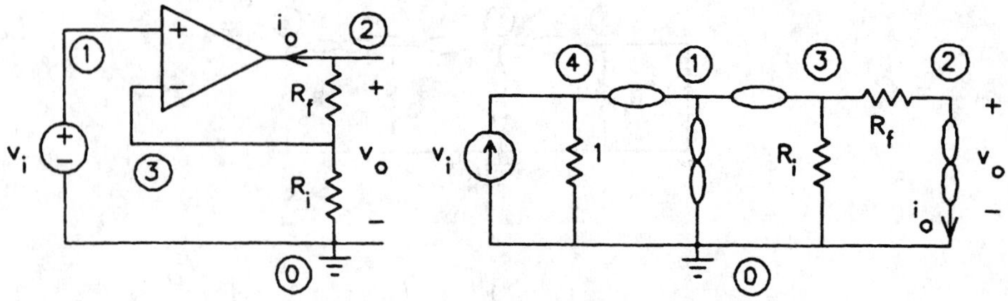

node: 0 1 2 3 4
ROW = [0 0 0 1 2]
COLUMN = [0 1 2 1 1]

$$\begin{matrix} 3: \\ 4: \end{matrix} \begin{bmatrix} G_i + G_f & -G_f \\ 1 & 0 \end{bmatrix} \begin{bmatrix} v_{1,3,4} \\ v_2 \end{bmatrix} = \begin{bmatrix} 0 \\ v_i \end{bmatrix}$$

FIGURE 22.10 Using nullors to represent a noninverting amplifier by a node equation.

The solution of this linear equation is

$$V_n = \begin{pmatrix} 4 \\ 6 \end{pmatrix} \tag{22.18}$$

The voltage at node 4 of the equivalent network is

$$v_4 = V_n(COLUMN[4]) = V_n(1) = 4 \text{ volts} \tag{22.19}$$

The voltage at node 3 of the original network (or the equivalent network) is

$$v_3 = V_n(COLUMN[3]) = V_n(1) = 4 \text{ volts} \tag{22.20}$$

Similarly

$$v_2 = V_n(COLUMN[2]) = V_n(2) = 6 \text{ volts} \tag{22.21}$$

and

$$v_1 = V_n(COLUMN[1]) = V_n(1) = 4 \text{ volts} \tag{22.22}$$

Once the node voltages have been calculated, all branch voltages and resistor currents can be easily calculated. The currents of the nullators and current sources are known. The only variables that still need to be calculated are the norator currents. Notice that it is only necessary to calculate some of the norator currents. In general, not all of the norator currents correspond to a current in the original network. Further, some norator currents may correspond to currents in the original network that are not of interest. The method suggested here is to augment the node voltage vector with some norator currents. This will be accomplished in two steps.

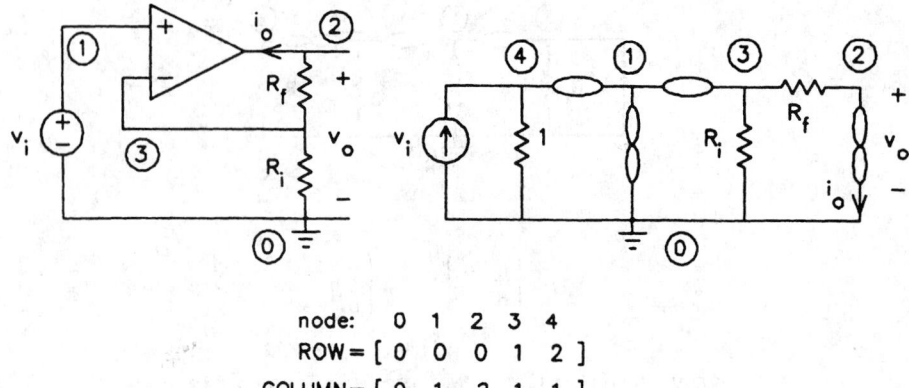

Suppose the norator current directed from node n_1 to n_2 is required. The algorithm will be modified two ways:

1. The vector *ROW* will *NOT* be modified when this norator is encountered.
2. A column will be added to the matrix *G*. The entries in this column will all be 0 except for a 1 in row $ROW[n_1]$ and $a - 1$ in row $ROW[n_2]$.

For example, consider again the circuit shown in Fig. 22.8. Suppose that the norator current, directed from node 2 toward node 4, is to be calculated. *ROW* is not updated when this norator is encountered during the bookkeeping step so

$$\begin{array}{rcccccc} \text{node} & & 0 & 1 & 2 & 3 & 4 \\ ROW & = & [0 & 0 & 1 & 2 & 3] \\ COLUMN & = & [0 & 1 & 2 & 1 & 1] \end{array} \tag{22.23}$$

(Notice that these vectors describe a matrix having three rows and two columns. A column corresponding to the norator current will be added, producing a 3 × 3 matrix.)

The nodal conductance matrix *G* and current source vector *J* in much the same way as before. First the node equation is initialized. Next, each branch of the network is examined in turn. The node equation is updated as before, whenever the branch is a resistor or current source. When the norator with nodes $n_1 = 2$ and $n_2 = 4$ is encountered a column is added to *G*. This column consists of 0s except for a 1 in row $ROW[2] = 1$ and $a - 1$ in row $ROW[4] = 3$. The node equation is

$$\begin{array}{r} 2: \\ 3: \\ 4: \end{array} \begin{pmatrix} 0 \\ I \\ 0 \end{pmatrix} = \begin{pmatrix} 0 & G_3 & -1 \\ G_1 & 0 & 0 \\ G_2 & 0 & 1 \end{pmatrix} \begin{pmatrix} v_1 \\ v_2 \\ i_1 \end{pmatrix} \tag{22.24}$$

Where i_1 is the required norator current. (Notice that rows 1 and 3 of this equation are the KCL equations at nodes 2 and 4. Both equations involve the norator current, i_1. When this example was considered earlier, i_1 did not need to be calculated and these equations were replaced with the KCL equation corresponding to the cutset that separates nodes 2 and 4 from the rest of the network.)

The example provided in Fig. 22.11 shows that the procedures described in this section can be used to obtain the indefinite admittance matrix of a linear active network. The original network in this example is a three-terminal network excited by current sources. The equivalent R-nullor network is generated and represented by a node equation. Since the three-terminal network is ungrounded, the nodal admittance matrix in this node equation will be an indefinite matrix. Notice that this matrix provides a relationship between the voltages at the nodes of the original network and the currents of the current sources. This matrix is indeed the indefinite admittance matrix of the original network.

This section has described an algorithm based on node equations for analyzing linear networks. Nullors have also been used in algorithms based on loop analysis [2, 11]. Algorithms based on nullors have been shown to be effective for analyzing switched capacitor networks [11, 12].

2.4 Computing the Transfer Function of an RC-Nullor Network

Consider the problem of finding the transfer function of an RC-nullor circuit that is excited by a current source. This problem can be solved in three steps:

1. The problem of finding a transfer function is reduced to the problem of calculating the determinant of a nodal admittance matrix.

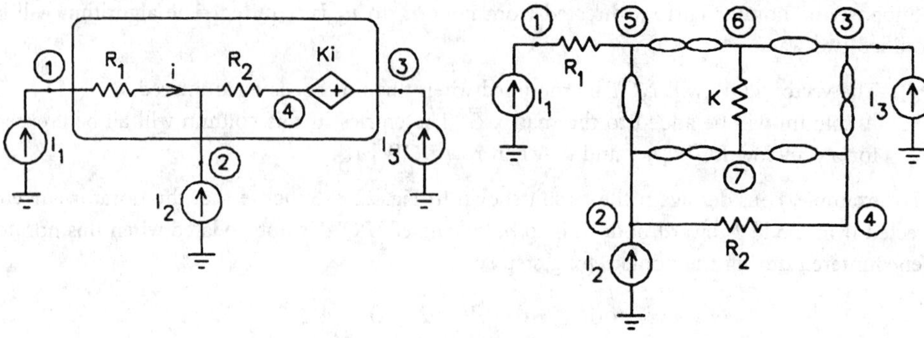

node: 0 1 2 3 4 5 6 7

ROW = [0 1 2 3 3 4 4 2]

COLUMN = [0 1 2 3 4 2 3 4]

$$\begin{array}{r} 1: \\ 2,7: \\ 3,4: \\ 5,6: \end{array} \begin{bmatrix} G_1 & -G_1 & 0 & 0 \\ 0 & G_2 & -\frac{1}{K} & -G_2+\frac{1}{K} \\ 0 & -G_2 & 0 & G_2 \\ -G_1 & G_1 & \frac{1}{K} & -\frac{1}{K} \end{bmatrix} \begin{bmatrix} V_1 \\ V_{2,5} \\ V_{3,6} \\ V_{4,7} \end{bmatrix} = \begin{bmatrix} I_1 \\ I_2 \\ I_3 \\ 0 \end{bmatrix}$$

FIGURE 22.11 Using nullors to obtain the indefinite admittance of a network.

2. Elementary row and column operations are used to manipulate the nodal admittance matrix into a convenient form.
3. The Leverrier algorithm is used to calculate the determinant.

The problem of calculating the transfer function of a single-input single-output circuit is simplified by modeling the original network using nullors. Both the input and output of the original network could be either a voltage or a current, so four cases require consideration. Since a voltage source in the original network is modeled using a current source, the input to the equivalent network is always a current. Only two cases need to be considered.

Figure 22.12(a) shows one of those two cases, the case when the output is a voltage. Suppose the transfer function of this circuit is

$$\frac{V_{out}(s)}{I_{in}(s)} = T(s) = \frac{a_m s^m + a_{m-1} s^{m-1} + \cdots + a_1 s + a_0}{b_n s^n + b_{n-1} s^{n-1} + \cdots + b_1 s + b_0} = \frac{A(s)}{B(s)} \qquad (22.25)$$

In Fig. 22.12(b), the input to the network has been replaced with a dependent source. The characteristic polynomial of this circuit is $B(s) + KA(s)$. (Replace I_{in} by $I_{in} - KV_{out}$ and calculate the transfer function. Then let $I_{in} = 0$.) Let $Y(s)$ denote the nodal admittance of the circuit shown in Fig. 22.12(b) (including the dependent source). Then

$$\det Y(s) = B(s) + KA(s) \qquad (22.26)$$

This equation shows that the transfer function of the network shown in Fig. 22.12(a) can be obtained by calculating the characteristic polynomial of the augmented network shown in Fig. 22.12(b). The gain of the dependent source, K, can be used to separate the part of $\det Y(s)$ that corresponds to $A(s)$ from the part of $\det Y(s)$ that corresponds to $B(s)$.

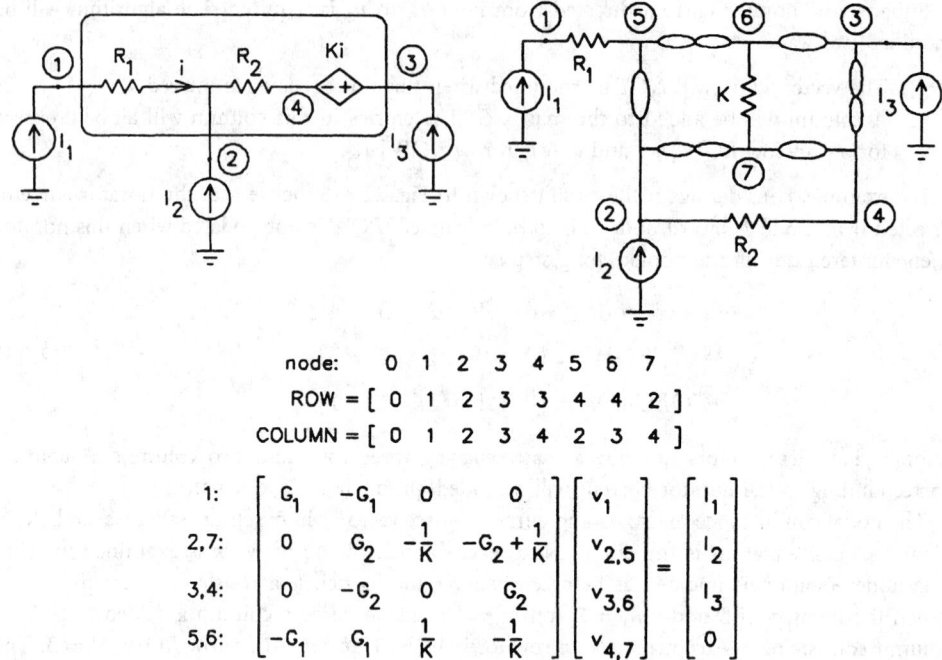

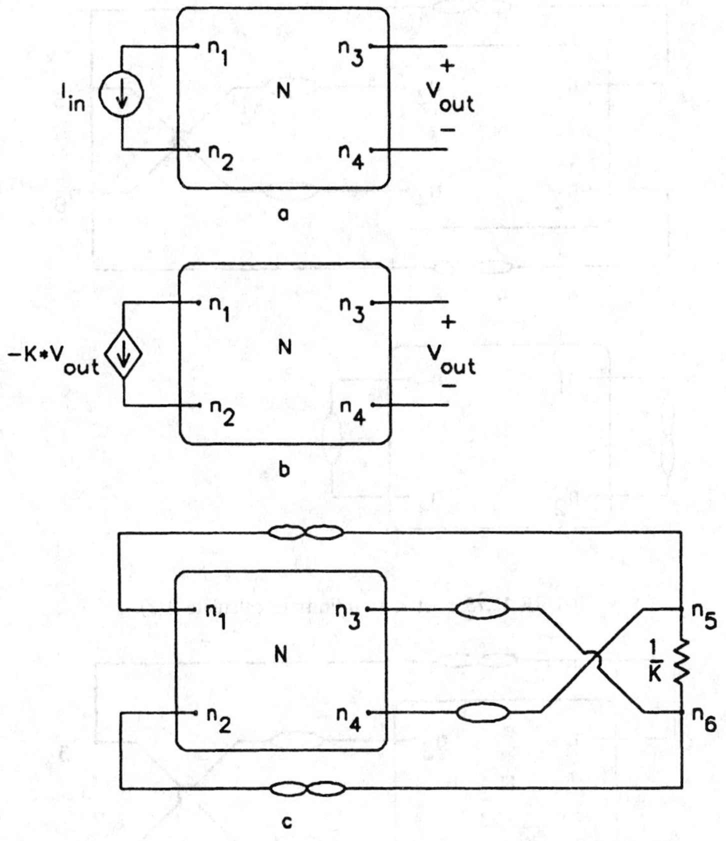

FIGURE 22.12 A dependent source is added to facilitate calculation of the transfer function.

In Fig. 22.12(c), the dependent source has been replaced by an equivalent circuit (see Fig. 22.4). The determinant of the nodal admittance of this network can be expressed as [3]:

$$\det Y(s) = \det\left(Y(s)\Big|_{K=0}\right) + K(-1)^{(r_2+c_3)}\det\left(Y(s)\Big|_{K=\infty}\right) \tag{22.27}$$

where $r_2 = ROW[n_2]$ and $c_3 = COLUMN[n_3]$. Comparing these equations yields

$$B(s) = \det\left(Y(s)\Big|_{K=0}\right) \quad \text{and} \quad A(s) = (-1)^{(r_2+c_3)}\det\left(Y(s)\Big|_{K=\infty}\right) \tag{22.28}$$

Setting $K = \infty$ corresponds to replacing the resistor in Fig. 22.12(c) by a short circuit. This is shown in Fig. 22.13(a). The circuit in Fig. 22.13(a) can be simplified to the circuit shown in Fig. 22.13(b) [2]. Setting $K = 0$ corresponds to replacing the resistor in Fig. 22.12(c) by an open circuit. This is shown in Fig. 22.14(a). A series nullator and norator is equivalent to an open circuit so the circuit in Fig. 22.14(a) can be simplified to the circuit shown in Fig. 22.14(b).

The right-hand column of Fig. 22.15 summarizes these results. An RC-nullor network N is excited by a single current source. The input current is directed from node n_1 to node n_2. The output of the network is a voltage across an open circuit. The nodes of the open circuit are denoted as n_3 and n_4. The output voltage is the voltage at n_3 with respect to n_4. Finally, $c_3 = COLUMN[n_3]$ and $r_2 = ROW[n_2]$.

Two circuit analyses are required to compute the transfer function. The denominator of the transfer function is computed from the nodal admittance matrix of a network formed by

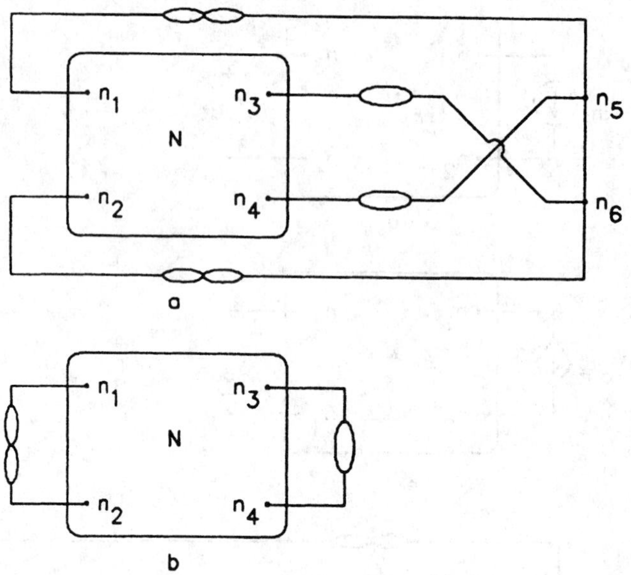

FIGURE 22.13 Set K to infinity to calculate $A(s)$.

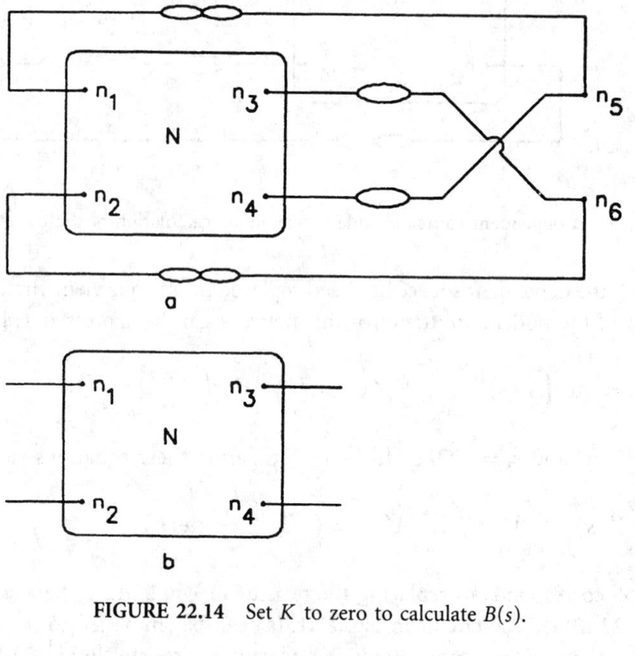

FIGURE 22.14 Set K to zero to calculate $B(s)$.

replacing the current source by a norator and the open circuit by a nullator. The numerator of the transfer function is computed from the nodal admittance matrix of a network formed by replacing the current source by an open circuit.

The case when the output is a current rather than a voltage can be analyzed in the same way. The left-hand column of Fig. 22.15 summarizes these results. In this case, the output of the network is a current in a short circuit. The nodes of the short circuit are denoted as n_3 and n_4. The output current is direct from n_3 to n_4. Finally, $c_2 = COLUMN[n_2]$ and $r_3 = ROW[n_3]$.

Figure 22.16 illustrates this procedure for computing the transfer function. This example starts in Fig. 22.16(a), with a circuit whose input and output are both voltages. Modeling the

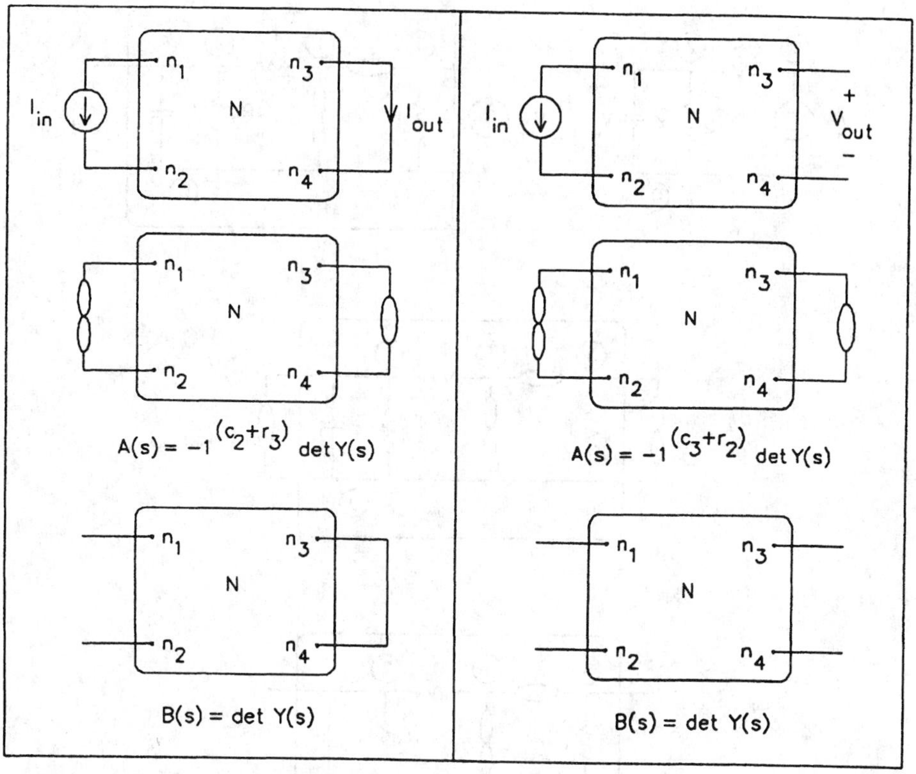

FIGURE 22.15 The transfer function of a network can be determined by calculating the determinants of the nodal admittance matrices of two related networks.

voltage source using nullors produces the circuit shown in Fig. 22.16(b). Here the input is a current and the output is a voltage across an open circuit. This is the case summarized in the right-hand column of Fig. 22.15. Fig. 22.16(c) shows the circuit that results from replacing the current source by a norator and the open circuit by a nullator. The + node of the output voltage is node 2 and the input current is directed toward node 3. In the notation of Fig. 22.15, this means that $n_2 = 3$ and $n_3 = 2$. Then, $r_2 = ROW[3]$ and $c_3 = COLUMN[2]$. For this circuit

$$\begin{aligned} &\text{node} \quad 0 \quad 1 \quad 2 \quad 3 \\ ROW &= [0 \quad 0 \quad 1 \quad 0] \Rightarrow ROW[3] = 0 \\ COLUMN &= [0 \quad 1 \quad 0 \quad 1] \Rightarrow COLUMN[2] = 0 \end{aligned} \tag{22.29}$$

The largest entry in *ROW*, and in *COLUMN* is a 1, so the nodal admittance matrix has only one row and one column. The denominator of the transfer function is

$$A(s) = -1^{(0+0)} \det\left(-\frac{1}{R_1}\right) = -\frac{1}{R_1} \tag{22.30}$$

Figure 22.16(d) shows the circuit that results from replacing the current source by an open circuit. The numerator is calculated from the nodal admittance matrix of this circuit.

$$B(s) = \det\begin{pmatrix} -\frac{1}{R_1} & \frac{1}{R_1} + \frac{1}{R_2} + Cs \\ 1 & 0 \end{pmatrix} = -\left(\frac{1}{R_1} + \frac{1}{R_2} + Cs\right) \tag{22.31}$$

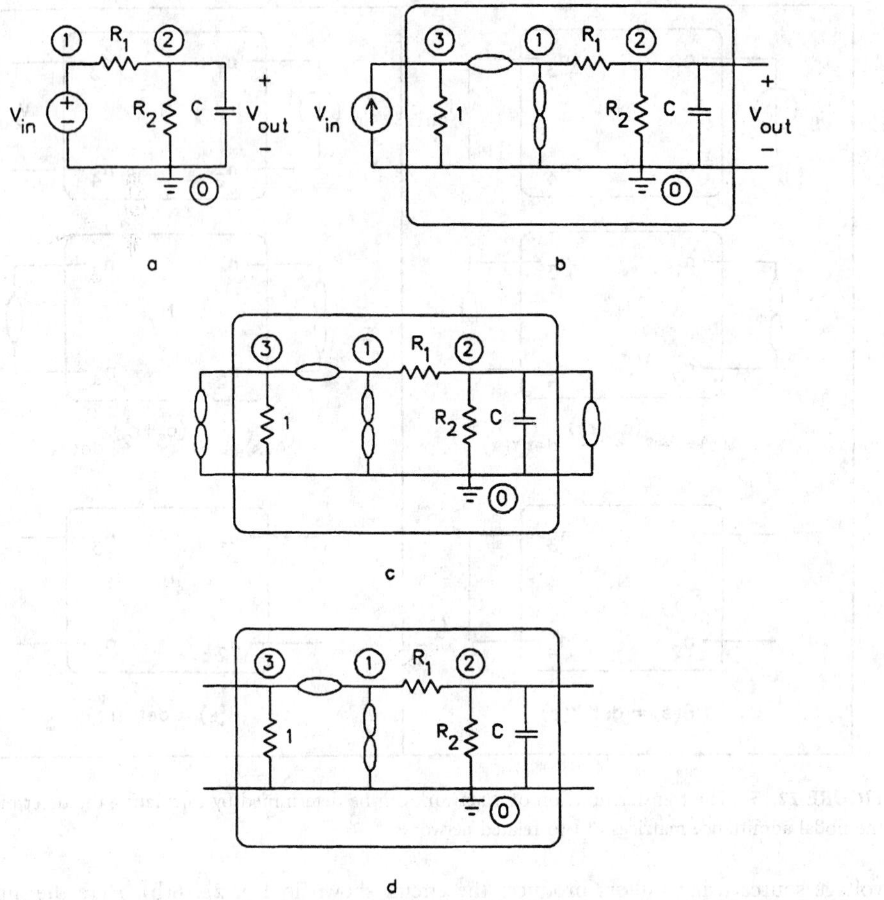

FIGURE 22.16 An example of the procedure for computing the transfer function.

Finally,

$$T(s) = \frac{-\dfrac{1}{R_1}}{-\left(\dfrac{1}{R_1} + \dfrac{1}{R_2} + Cs\right)} = \frac{\dfrac{1}{R_1 C}}{s + \dfrac{R_1 + R_2}{R_1 R_2 C}} \tag{22.32}$$

A second example is shown in Fig. 22.17. The circuit in Fig. 22.17(b) is the RC nullor equivalent of the circuit in Fig. 22.17(a). The input to the equivalent circuit is a current and the output is a voltage across an open circuit. This is the case summarized in the right-hand column of Fig. 22.15. Figure 22.17(c) shows the circuit that results from replacing the current source by a norator and the open circuit by a nullator. The + node of the output voltage is node 4 and the input current is directed toward node 5. In the notation of Fig. 22.15, this means that $n_2 = 5$ and $n_3 = 4$. Then $r_2 = ROW[5]$ and $c_3 = COLUMN[4]$. For this circuit

$$\begin{array}{rllllll} \text{node} & 0 & 1 & 2 & 3 & 4 & 5 \\ ROW = [0 & 0 & 1 & 2 & 0 & 0] & \Rightarrow ROW[5] = 0 \\ COLUMN = [0 & 1 & 0 & 2 & 0 & 1] & \Rightarrow COLUMN[4] = 0 \end{array} \tag{22.33}$$

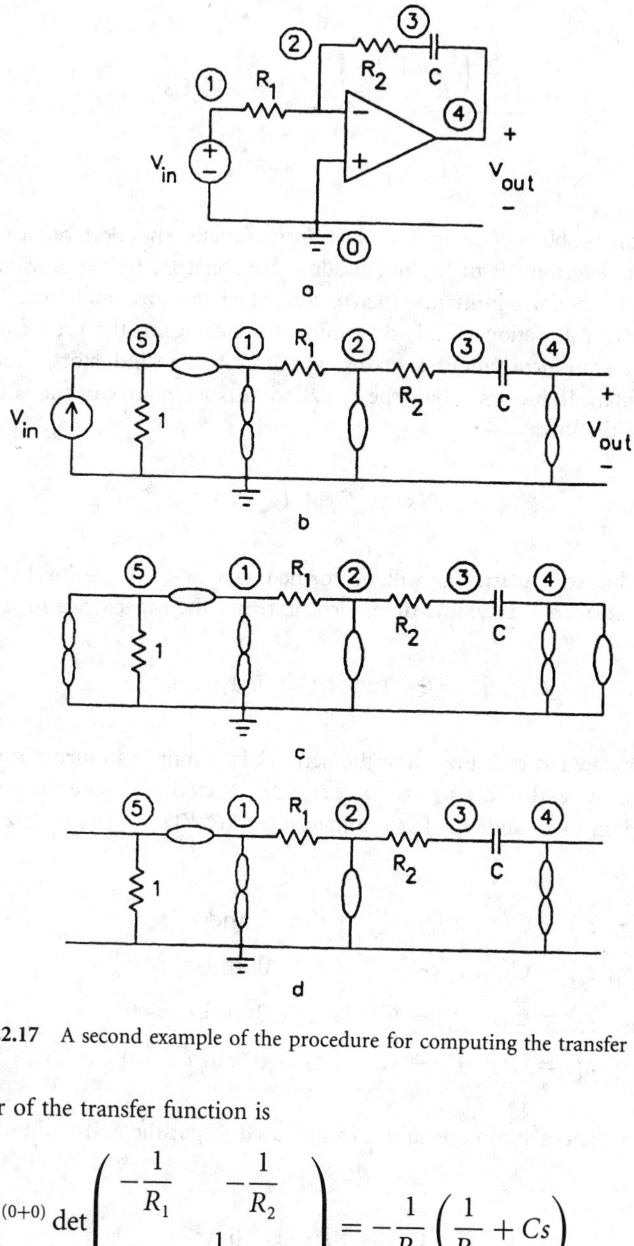

FIGURE 22.17 A second example of the procedure for computing the transfer function.

The denominator of the transfer function is

$$A(s) = -1^{(0+0)} \det \begin{pmatrix} -\dfrac{1}{R_1} & -\dfrac{1}{R_2} \\ 0 & \dfrac{1}{R_2} + Cs \end{pmatrix} = -\frac{1}{R_1}\left(\frac{1}{R_2} + Cs\right) \tag{22.34}$$

Figure 22.16(d) shows the circuit that results from replacing the current source by an open circuit. The numerator is calculated from the nodal admittance matrix of this circuit.

$$B(s) = \det \begin{pmatrix} -\dfrac{1}{R_1} & -\dfrac{1}{R_2} & 0 \\ 0 & \dfrac{1}{R_2} + Cs & -Cs \\ 1 & 0 & 0 \end{pmatrix} = \frac{Cs}{R_2} \tag{22.35}$$

Finally,

$$T(s) = \frac{-\frac{1}{R_1}\left(\frac{1}{R_2} + Cs\right)}{\frac{Cs}{R_2}} = -\frac{1 + R_2 Cs}{R_1 Cs} \tag{22.36}$$

At this point, the problem of computing the transfer function has been reduced to the problem of calculating the determinant of the nodal admittance matrix. In the previous examples, the determinant of the nodal admittance matrix was calculated by hand. Next, a procedure is given for computer calculation of this determinant. Once again, the use of nullors simplifies the problem. The equivalent circuit contains capacitors but no inductors, though the original network may contain inductors. Thus, the nodal admittance matrix of the equivalent network can be written in the form

$$Y(s) = Cs + G \tag{22.37}$$

The matrix conductance matrix G will be formed as described previously. The capacitance matrix C will be formed at the same time. First, initialize the capacitance matrix

$$C(i,\, j) = 0 \quad \text{for } i,\, j = 0,\, 1, \ldots n \tag{22.38}$$

Next C will be updated as each branch of the network is examined in turn. Consider a capacitor having capacitance C and nodes n_i and n_j. Let r_i, r_j, c_i, and c_j denote the rows and columns of C corresponding to n_i and n_j. (For example, $r_i = ROW[n_i]$.) The matrix C is updated as follows

$$\begin{aligned} C(r_i,\, c_i) &= C(r_i,\, c_i) + C \quad \text{if } r_i > 0 \text{ and } c_i > 0 \\ C(r_i,\, c_j) &= C(r_i,\, c_j) - C \quad \text{if } r_i > 0 \text{ and } c_j > 0 \\ C(r_j,\, c_i) &= C(r_j,\, c_i) - C \quad \text{if } r_j > 0 \text{ and } c_i > 0 \\ C(r_j,\, c_j) &= C(r_j,\, c_j) + C \quad \text{if } r_j > 0 \text{ and } c_j > 0 \end{aligned} \tag{22.39}$$

Elementary row and column operations can be used to put the nodal admittance matrix into the form [5]

$$\begin{pmatrix} I & 0 \\ 0 & 0 \end{pmatrix} s + \begin{pmatrix} G_{11} & 0 \\ G_{21} & I \end{pmatrix} \tag{22.40}$$

This is accomplished in two, or possibly three, steps. First, elementary row and column operations are used to make the matrix C zero except for an identity matrix in the upper left hand corner. The size of this identity matrix determines the partitioning of the matrices C and G. Next, elementary row operations are used to reduce the submatrix in the lower right hand corner of G to an identity matrix. These row operations do not disturb the matrix C. When the submatrix in the lower right hand corner of G is singular, an additional elementary operation is required. This row operation involves adding αs times a lower row to an upper row. This operation does disturb the matrix C, and the first two steps must be repeated [5].

Putting the nodal admittance matrix into this standard form facilitates calculation of the determinant of the nodal admittance matrix.

$$\det(Cs+G) = k \cdot \det\left(\begin{pmatrix} I & 0 \\ 0 & 0 \end{pmatrix} s + \begin{pmatrix} G_{11} & 0 \\ G_{21} & I \end{pmatrix}\right) \\ = k \cdot \det(Is + G_{11}) \tag{22.41}$$

where the multiplier k accounts for the effect of elementary operations on the determinant. A standard numerical method, the Leverrier algorithm [10], can be used to calculate $\det(sI + G_{11})$.

The circuit shown in Fig. 22.18 will be used to illustrate the use of elementary operations to simplify the nodal admittance matrix so that the Leverrier algorithm can be used to compute the determinant. The circuit shown in Fig. 22.18(b) is the RC-nullor equivalent of the circuit shown in Fig. 22.18(a). The vectors *ROW* and *COLUMN* for the equivalent network are

$$\begin{array}{rcccccccc} \text{node} & 0 & 1 & 2 & 3 & 4 & 5 & 6 & 7 \\ ROW = & [0 & 1 & 2 & 3 & 4 & 0 & 0 & 5] \\ COLUMN = & [0 & 0 & 1 & 2 & 3 & 4 & 5 & 5] \end{array} \tag{22.42}$$

The nodal admittance matrix is

$$Cs + G = \begin{pmatrix} 0 & 0 & 0 & 0 & 0 \\ 1 & -1 & 0 & 0 & 0 \\ -1 & 2 & 0 & 0 & 0 \\ 0 & 0 & 0 & 0 & 0 \\ 0 & 0 & 0 & 0 & 0 \end{pmatrix} s + \begin{pmatrix} -0.5 & 0 & 0 & -0.5 & -1 \\ 0.5 & 0 & 0 & 0 & 0 \\ 0 & 2 & -1 & 0 & 0 \\ 0 & -1 & 2 & -0.5 & 0 \\ 0 & 0 & 0 & 0 & 1 \end{pmatrix} \tag{22.43}$$

The scalar k is used to keep track of any changes to the determinant of the nodal admittance matrix that result from elementary row and column operations. Initialize k to 1. Elementary row operations can be used to put the nodal admittance matrix into the form

$$\begin{pmatrix} 1 & 0 & 0 & 0 & 0 \\ 0 & 1 & 0 & 0 & 0 \\ 0 & 0 & 0 & 0 & 0 \\ 0 & 0 & 0 & 0 & 0 \\ 0 & 0 & 0 & 0 & 0 \end{pmatrix} s + \begin{pmatrix} 1 & 2 & -1 & 0 & 0 \\ 0.5 & 2 & -1 & 0 & 0 \\ -0.5 & 0 & 0 & -0.5 & -1 \\ 0 & -1 & 2 & -0.5 & 0 \\ 0 & 0 & 0 & 0 & 1 \end{pmatrix} \tag{22.44}$$

These row operations included two row interchanges. The scalar k is multiplied by -1 when a row or column interchange is performed. Now $k = 1 * (-1) * (-1) = 1$. The identity matrix in the upper left hand corner of C determines the way these matrices are partitioned. In this case, the first two rows (columns) are partitioned away from the last three rows (columns).

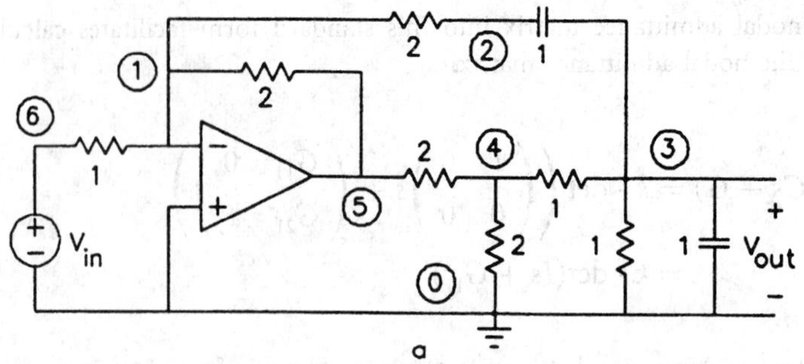

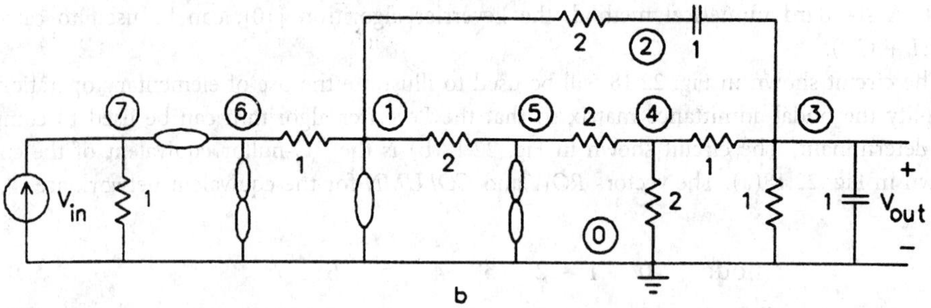

FIGURE 22.18 Example illustrating the use of elementary operations to simplify the nodal admittance matrix.

Next, elementary row operations are used to reduce the lower right-hand submatrix of G to an identity matrix and to reduce the upper right-hand submatrix of G to a zero matrix.

$$\begin{pmatrix} 1 & 0 & 0 & 0 & 0 \\ 0 & 1 & 0 & 0 & 0 \\ 0 & 0 & 0 & 0 & 0 \\ 0 & 0 & 0 & 0 & 0 \\ 0 & 0 & 0 & 0 & 0 \end{pmatrix} s + \begin{pmatrix} 1.25 & 1.5 & 0 & 0 & 0 \\ 0.75 & 1.5 & 0 & 0 & 0 \\ 0.25 & -0.5 & 1 & 0 & 0 \\ 1 & 0 & 0 & 1 & 0 \\ 0 & 0 & 0 & 0 & 1 \end{pmatrix} \tag{22.45}$$

These elementary operations included another row interchange. Now $k = 1 * (-1) = -1$. Three of the elementary row operations consisted of multiplying a row by a constant. (The constants were 2, 0.5, and -1.) When a row or column is multiplied by a constant, k must be multiplied by that same constant. Now, $k = -1 * 2 * 0.5 * (-1) = 1$.

Finally,

$$\det(Cs + G) = 1 \cdot \det\left(\begin{pmatrix} 1 & 0 \\ 0 & 1 \end{pmatrix} s + \begin{pmatrix} 1.25 & 1.5 \\ 0.75 & 1.5 \end{pmatrix}\right) \tag{22.46}$$

This determinant can be computed using the Leverrier algorithm [10].

The circuit shown in Fig. 22.19 provides a second example of the use of elementary operations to simplify the nodal admittance matrix so that the Leverrier algorithm can be used to compute the determinant. In this example an additional complication is encountered. The lower right-hand submatrix of G will be singular. This example illustrates the procedure used to overcome this difficulty.

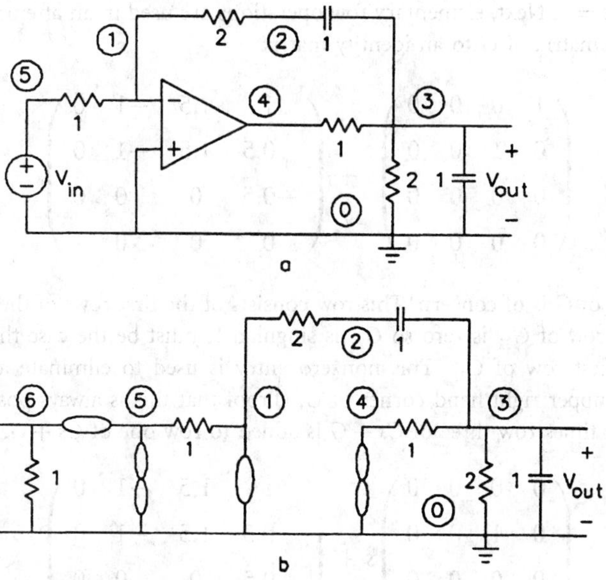

FIGURE 22.19 Second example illustrating the use of elementary operations to simplify the nodal admittance matrix.

The circuit shown in Fig. 22.19(b) is the RC-nullor equivalent of the circuit shown in Fig. 22.19(a). The vectors *ROW* and *COLUMN* for the equivalent network are

$$\begin{aligned} &\text{node} \quad 0 \quad 1 \quad 2 \quad 3 \quad 4 \quad 5 \quad 6 \\ &ROW = [0 \quad 1 \quad 2 \quad 3 \quad 0 \quad 0 \quad 4] \\ &COLUMN = [0 \quad 0 \quad 1 \quad 2 \quad 3 \quad 4 \quad 4] \end{aligned} \tag{22.47}$$

The nodal admittance matrix is

$$Cs + G = \begin{pmatrix} 0 & 0 & 0 & 0 \\ 1 & -1 & 0 & 0 \\ -1 & 2 & 0 & 0 \\ 0 & 0 & 0 & 0 \end{pmatrix} s + \begin{pmatrix} -0.5 & 0 & 0 & -1 \\ 0.5 & 0 & 0 & 0 \\ 0 & 1.5 & -1 & 0 \\ 0 & 0 & 0 & 1 \end{pmatrix} \tag{22.48}$$

Initialize k to 1. Elementary row operations can be used to put the nodal admittance matrix into the form

$$\begin{pmatrix} 1 & 0 & 0 & 0 \\ 0 & 1 & 0 & 0 \\ 0 & 0 & 0 & 0 \\ 0 & 0 & 0 & 0 \end{pmatrix} s + \begin{pmatrix} 1 & 1.5 & -1 & 0 \\ 0.5 & 1.5 & -1 & 0 \\ -0.5 & 0 & 0 & -1 \\ 0 & 0 & 0 & 1 \end{pmatrix} \tag{22.49}$$

At this point $k = 1$. Next, elementary row operations are used in an attempt to reduce the lower right-hand submatrix of G to an identity matrix.

$$\begin{pmatrix} 1 & 0 & 0 & 0 \\ 0 & 1 & 0 & 0 \\ 0 & 0 & 0 & 0 \\ 0 & 0 & 0 & 0 \end{pmatrix} s + \begin{pmatrix} 1 & 1.5 & -1 & 0 \\ 0.5 & 1.5 & -1 & 0 \\ -0.5 & 0 & 0 & 0 \\ 0 & 0 & 0 & 1 \end{pmatrix} \tag{22.50}$$

The third row of G is of concern. This row consists of the first rows of the submatrices G_{21} and G_{22}. The first row of G_{22} is zero so G_{22} is singular. It must be the case that there is a nonzero entry in the first row of G_{21}. This nonzero entry is used to eliminate a 1 from the identity matrix in the upper right hand corner of C. (Proof that this is always possible is given in [5].) In this case $2s$ times row three of $Cs + G$ is added to row one of $Cs + G$. The result is

$$\begin{pmatrix} 0 & 0 & 0 & 0 \\ 0 & 1 & 0 & 0 \\ 0 & 0 & 0 & 0 \\ 0 & 0 & 0 & 0 \end{pmatrix} s + \begin{pmatrix} 1 & 1.5 & -1 & 0 \\ 0.5 & 1.5 & -1 & 0 \\ -0.5 & 0 & 0 & 0 \\ 0 & 0 & 0 & 1 \end{pmatrix} \tag{22.51}$$

The matrix C has been disturbed. Elementary row and column operations are used to put an identity matrix into the upper left hand corner of C. This results in

$$\begin{pmatrix} 1 & 0 & 0 & 0 \\ 0 & 0 & 0 & 0 \\ 0 & 0 & 0 & 0 \\ 0 & 0 & 0 & 0 \end{pmatrix} s + \begin{pmatrix} 1.5 & 0.5 & -1 & 0 \\ 1.5 & 1 & -1 & 0 \\ 0 & -0.5 & 0 & 0 \\ 0 & 0 & 0 & 1 \end{pmatrix} \tag{22.52}$$

At this point $k = 1$. Now, elementary row operations are used to reduce the lower right-hand submatrix of G to an identity matrix and to reduce the upper right-hand submatrix of G to a zero matrix.

$$\begin{pmatrix} 1 & 0 & 0 & 0 \\ 0 & 0 & 0 & 0 \\ 0 & 0 & 0 & 0 \\ 0 & 0 & 0 & 0 \end{pmatrix} s + \begin{pmatrix} 0 & 0 & 0 & 0 \\ 0 & 1 & 0 & 0 \\ -1.5 & 0 & 1 & 0 \\ 0 & 0 & 0 & 1 \end{pmatrix} \tag{22.53}$$

Now $k = -2$. Finally,

$$\det(Cs + G) = -2 \cdot \det[s + 0] = -2s$$

22.5 Equivalent Networks

Nullors can be used to establish the equivalence of linear circuits, even when these circuits are implemented using different active devices [2, 6–8].

The transfer functions of linear networks can be calculated by analyzing the equivalent RC-nullor network. When two linear networks have the same RC-nullor equivalent circuit (after

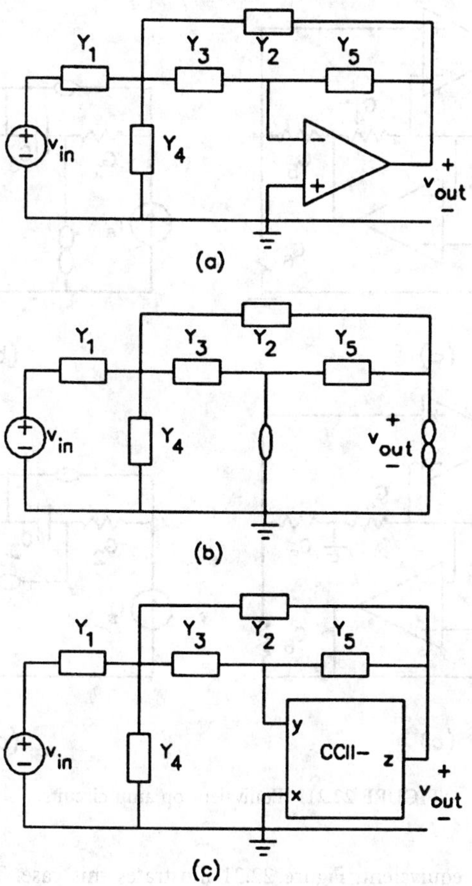

FIGURE 22.20 Equivalent implementations of the Multiple Feedback Filter (MFF).

renumbering the nodes and branches, if necessary) then the linear networks can be represented by the same transfer function. Such linear networks are said to be equivalent networks.

Figure 22.20(a) shows an implementation of the multiple feedback filter (MFF) that uses op amps. Figure 22.20(c) shows a second implementation of the MFF, this time using current conveyors [8]. R-Nullor models of the op amp and current conveyor are given in Fig. 22.5. The circuit shown in Fig. 22.20(b) is the RC-nullor equivalent circuit of both implementations of the MFF. All three circuits shown in Fig. 22.20 can be represented by the same transfer function.

$$T(s) = -\frac{Y_1 Y_3}{(Y_1 + Y_2 + Y_3 + Y_4)Y_5 + Y_2 Y_3} \tag{22.54}$$

In this sense the circuits shown in Fig. 22.20(a) and Fig. 22.20(c) are equivalent circuits. These exhibit identical performance when the op amp and current conveyor are ideal devices. Differences in the performance of these "equivalent circuits" may be detected when nonideal active devices are used.

The nullators and norators of a circuit can sometimes be rearranged without altering the transfer function of the network [2, 6–8]. (Suppose the nullators of a network contain a tree. This tree can be replaced by any other tree of nullators having the same nodes. A similar statement applies to the norators of the network.) Two RC-nullor networks that are related in this way are said to be equivalent. Linear networks that are equivalent to equivalent RC-nullor

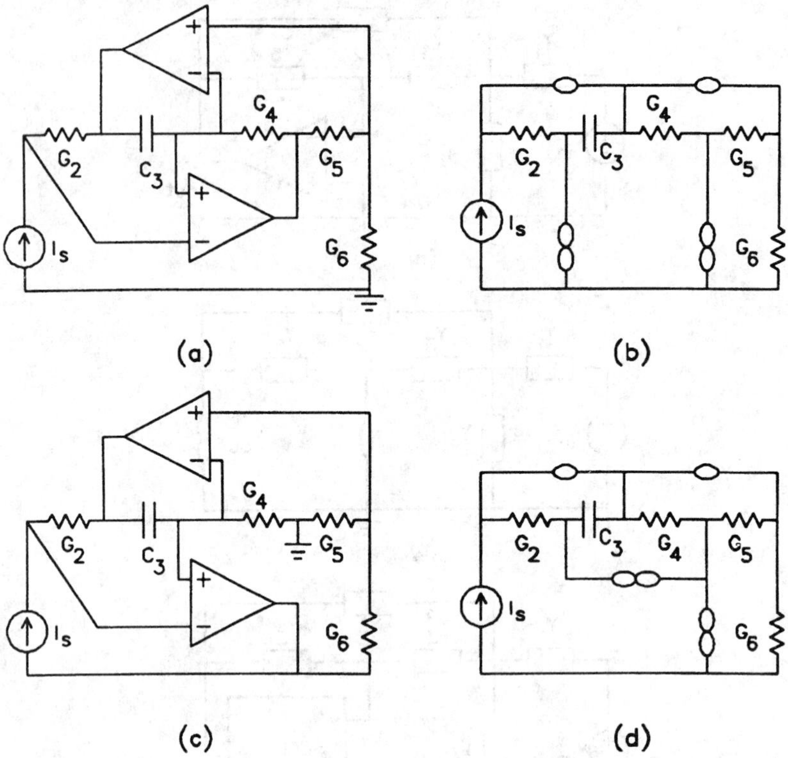

FIGURE 22.21 Equivalent op amp circuits.

networks are themselves equivalent. Figure 22.21 illustrates this case. The RC-nullor networks shown in Figs. 22.21(b) and 22.21(d) are equivalent [7]. The op amp circuits in Figs. 22.21(a) and 22.21(c) are equivalent to equivalent circuits and so are equivalent.

22.6 Conclusions

This chapter described methods for using nullors to analyze linear networks. Two types of analysis were considered: the node voltages of a DC circuit were calculated and the transfer function of a circuit containing reactive elements was calculated.

When linear circuits are modeled using nullors, the resulting network contains only resistors, capacitors, nullors, and current sources. Restricting consideration to so small a set of devices simplifies the algorithms for circuit analysis.

The network solutions can be expressed in terms of the original network, not the RC-nullor equivalent circuit. A computer program implementing these algorithms can be written to accept the original network as input and to provide the solution in terms of the original network. The user of this program would not be required to know anything about nullors.

References

[1] A. C. Davies, "The significance of nullators, norators and nullors in active network theory," *Radio and Electronic Engineer*, vol. 34, pp. 259–267, 1967.

[2] L. T. Bruton, *RC-Active Circuits: Theory and Design*, Englewood Cliffs, NJ: Prentice-Hall, 1980.

[3] G. E. Alderson and P. M. Lin, "Computer generation of symbolic network functions—a new theory and implementation," *IEEE Trans. Circuit Theory*, CT-20, no. 1, pp. 48–56, Jan. 1973.

[4] J. A. Svoboda, "Using nullors to analyze linear networks," *Int. J. Circuit Theory and Appl.*, vol. 14, pp. 169–180, 1986.

[5] J. A. Svoboda and D. P. Brown, "On modifying state equations to represent changes in RLC nullor networks," *J. Franklin Inst.*, vol. 317, no. 4, pp. 213–226, April 1984.

[6] G. M. Wierzba, "Op amp relocation: a topological active network synthesis," *IEEE Trans. Circuits and Systems*, CAS-33, pp. 469–475, 1986.

[7] J. A. Svoboda, "Op amp relocation and the complementary transformation," *Int. J. Circuit Theory and Appl.*, vol. 18, pp. 535–540, 1990.

[8] J. A. Svoboda, "Current conveyors, operational amplifiers and nullors," *IEE Proc.*, vol. 136, pt. G, no. 6, pp. 317–322, 1989.

[9] J. Vlach and K. Singhal, *Computer Methods for Circuit Analysis and Design*, New York: Van Nostrand-Reinhold, 1983.

[10] T. Kailath, *Linear Systems*, Englewood Cliffs, NJ: Prentice-Hall, 1980.

[11] L. T. Bruton, G. R. Bailey, and G. Battacharjee, "Loop equation formulation for switched capacitor networks containing nullors," *Int. J. Circuit Theory and Appl.*, vol. 11, pp. 52–72, 1983.

[12] L. T. Bruton and G. Battacharjee, "Formulation of the nodal charge equations of switched capacitor networks containing nullors," *Proceedings of the European Conference on Circuit Theory and Design*, Delft University Press, pp. 110–117, 1981.

[13] D. Coldham and L. T. Bruton, "Computer analysis of nullor networks," *Electron. Lett.*, vol. 9, pp. 22–23, 1973.

23

State-Variable Techniques

K. S. Chao
Texas Tech University

23.1 The Concept of States

For resistive (or memoryless) circuits, given the circuit structure, the present output depends only on the present input. In order to analyze a dynamic circuit, however, in addition to the present input it is also necessary to know the state of the circuit at some time t_o. The state of the circuit at t_o represents the condition of the circuit at $t = t_0$, and is related to the energy storage of the circuit, or the voltage (or electric charge) across the capacitor and the currents (or magnetic fluxes) through the inductors. These voltages and currents are considered as the state of the circuit at $t = t_o$. For $t > t_o$, the behavior of the circuit is completely characterized by these variables. In view of the above, a definition for the state of a circuit can now be given.

Definition: The state of a circuit at time t_o is the minimum amount of information at t_o that, along with the input to the circuit for $t \geq t_o$, uniquely determines the behavior of the circuit for $t \geq t_o$.

The concept of states is closely related to the order of complexity of the circuit. The order of complexity of a circuit is the minimum number of initial conditions which, along with the input, is sufficient to determine the future behavior of the circuit. Furthermore, if a circuit is described by an nth-order linear differential equation, it is well known that the general solution for $t \geq t_o$ contains n arbitrary constants which are determined by n initial conditions. This set of n initial conditions contains information concerning the circuit prior to $t = t_o$ and constitutes the state of the circuit at $t = t_o$. Thus, the order of complexity or the order of a circuit is the same as the order of the differential equation that describes the circuit, and it is also the same as the number of state variables that can be defined in a circuit. For an nth-order circuit, the state of the circuit at $t = t_o$ consists of a set of n numbers which denotes a vector in an n-dimensional state space spanned by the n corresponding state variables. This key number n can simply be obtained by inspection of the given circuit. Knowing the total number of energy storage elements, n_{LC}, the total number of independent capacitive loops, n_C, and the total number of independent inductive cutsets, n_L, the order of complexity n of a circuit is given by

$$n = n_{LC} - n_L - n_C. \tag{23.1}$$

0-8493-8341-2/95/$0.00 + $.50

A capacitive loop is defined as one that consists of only capacitors and possibly voltage sources while an inductive cutset represents a cutset that contains only inductors and possibly current sources. The following two examples illustrate the concept of states.

Example 1. Consider a simple RC-circuit shown in Fig. 23.1. The circuit equation is

$$RC\frac{dv_c(t)}{dt} + v_c(t) = v_{in} \quad \text{for} \quad t \geq t_o \tag{23.2}$$

and the corresponding capacitor voltage is easily obtained as

$$v_c(t) = [v_c(t_0) - v_{in}]e^{-\frac{1}{RC}(t-t_0)} + v_{in} \quad \text{for} \quad t \geq t_0 \tag{23.3}$$

For this first-order circuit, it is clear from (23.3) the capacitor voltage for $t \geq t_o$ is uniquely determined by the initial condition $v_c(t_o)$ and the input voltage v_{in} for $t \geq t_o$. This is independent of the charging circuit for the capacitor prior to t_o. Hence $v_c(t_o)$ is the state of the circuit at $t = t_o$ and $v_c(t)$ is regarded as the state variable of the circuit.

Example 2. As another illustration, consider the circuit of Fig. 23.2 which is a slight modification of the circuit considered in the previous example. The circuit equation and its corresponding solution are readily obtained as

$$\frac{dv_{C_1}(t)}{dt} = -\frac{1}{R(C_1 + C_2)}v_{C_1}(t) + \frac{1}{R(C_1 + C_2)}v_{in} \tag{23.4}$$

and

$$v_{C_1}(t) = (v_{C_1}(t_o) - v_{in})e^{\frac{-1}{R(C_1+C_2)}(t-t_o)} + v_{in} \quad \text{for} \quad t \geq t_o \tag{23.5}$$

respectively. Even though there are two energy storage elements, one can only arbitrarily specify one independent initial condition. Once the initial condition on C_1, $v_{C_1}(t_o)$, is specified, the initial voltage on C_2 is automatically constrained by the loop equation $v_{C_2}(t) = V_{C_1}(t) - E$ at t_o. The circuit is thus still first order and only one state variable can be assigned for the circuit. It is clear from (23.5) that with the input v_{in}, $v_{C_1}(t_o)$ is the minimum amount of information that is needed to uniquely determine the behavior of this circuit. Hence, $v_{C_1}(t)$ is the state variable of

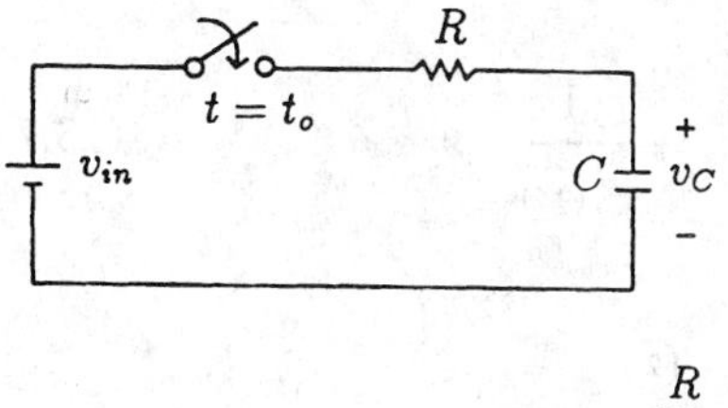

FIGURE 23.1 A simple RC circuit.

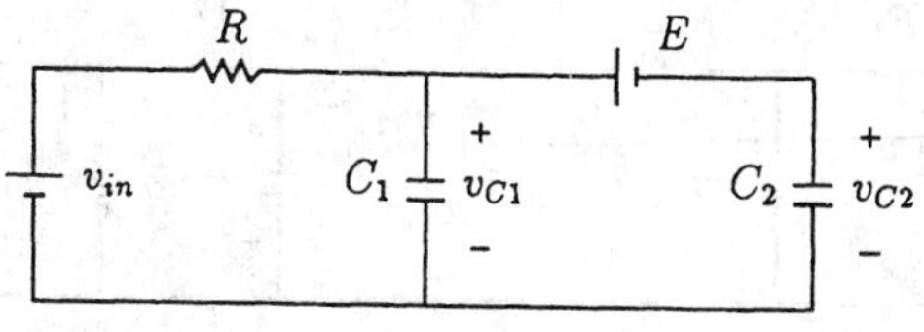

FIGURE 23.2 The circuit for Example 2.

the circuit. One can just as well analyze the circuit by solving a first-order differential equation in terms of $v_{C_2}(t)$ with $v_{C_2}(t_o)$ defined as the state of the circuit at $t = t_o$. The selection of state variables is thus not unique. In this example, either $v_{C_1}(t)$ or $v_{C_2}(t)$ can be defined as the state variable of the circuit. In fact, it is easily shown that any linear combination of $v_{C_1}(t)$ and $v_{C_2}(t)$ can also be regarded as state variables.

23.2 State-Variable Formulation Via Network Topology

Various mathematical descriptions of circuits are available. Depending on the type of analysis used, different formulations of circuit equations may result. In the state variable formulation, a system of n first-order differential equations is written in the form

$$\dot{\mathbf{x}} = \mathbf{f}(\mathbf{x}, t) \tag{23.6}$$

where $\mathbf{x}$ is an $n \times 1$ vector consisting of n state variables for an nth-order circuit and t represents the time variable. This set of equations is usually referred to as the state equation in normal form.

When compared to other circuit descriptions, the state-variable representation is not necessarily the simplest. It does, however, simultaneously provide the solution of all state variables and hence yields the behavior of the entire circuit. The state equation is also particularly suitable for analysis by numerical techniques. Another distinct advantage of the state-variable approach is that it can be easily extended to nonlinear and/or time varying circuits.

Example 3. Consider the linear circuit of Fig. 23.3. By inspection, the order of complexity of this circuit is three. Hence, three state variables are selected as $x_1 = v_{C_1}, x_2 = v_{C_2}$, and $x_3 = i_L$. Since the left-hand side of the normal form equation is the derivative of the state vector, it is necessary to express the voltages across the inductors and the currents through the capacitors in terms of the state variables and the input sources.

The current through C_1 can be obtained by writing a KCL equation at node 1 to yield

$$C_1 \frac{dv_{C_1}}{dt} = i_{R_1} - i_L$$
$$= \frac{1}{R_1}(v_s - v_{C_1}) - i_L$$

or

$$\frac{dv_{C_1}(t)}{dt} = -\frac{1}{R_1 C_1} v_{C_1} - \frac{1}{C_1} i_L + \frac{1}{R_1 C_1} v_s \tag{23.7}$$

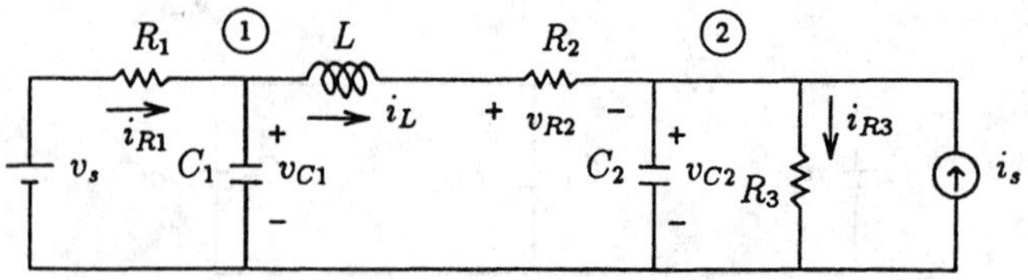

FIGURE 23.3 The circuit for Example 3.

In a similar manner, applying KCL to node 2 gives

$$C_2 \frac{dv_{C_2}}{dt} = i_L - i_{R_3} + i_s$$
$$= i_L - \frac{1}{R_3} v_{C_2} + i_s$$

or

$$\frac{dv_{C_2}(t)}{dt} = -\frac{1}{R_3 C_2} v_{C_2} + \frac{1}{C_2} i_L + \frac{1}{C_2} i_s \tag{23.8}$$

The expression for the inductor voltage is derived by applying KVL to the mesh containing L, R_2, C_2, and C_1 yielding

$$L \frac{di_L}{dt} = v_{C_1} - v_{C_2} - R_2 i_L$$

or

$$\frac{di_L}{dt} = \frac{1}{L} v_{C_1} - \frac{1}{L} v_{C_2} - \frac{R_2}{L} i_L \tag{23.9}$$

Eqs. (23.7), (23.8), and (23.9) are the state equations which can be expressed in matrix form as

$$\begin{bmatrix} \frac{dv_{C_1}}{dt} \\ \frac{dv_{C_2}}{dt} \\ \frac{di_L}{dt} \end{bmatrix} = \begin{bmatrix} -\frac{1}{R_1 C_1} & 0 & -\frac{1}{C_1} \\ 0 & -\frac{1}{R_3 C_2} & \frac{1}{C_2} \\ \frac{1}{L} & -\frac{1}{L} & -\frac{R_2}{L} \end{bmatrix} \begin{bmatrix} v_{C_1} \\ v_{C_2} \\ i_L \end{bmatrix} + \begin{bmatrix} \frac{1}{R_1 C_1} & 0 \\ 0 & \frac{1}{C_2} \\ 0 & 0 \end{bmatrix} \begin{bmatrix} v_s \\ i_s \end{bmatrix} \tag{23.10}$$

Any number of branch voltages and/or currents may be chosen as output variables. If i_{R_1} and v_{R_2} are considered as outputs for this example, then the output equations, written as a linear combination of state variables and input sources become

$$i_{R_1} = \frac{1}{R_1}(v_s - v_{C_1}) \tag{23.11}$$

$$v_{R_2} = R_2 i_L \tag{23.12}$$

or in matrix form

$$\begin{bmatrix} i_{R_1} \\ v_{R_2} \end{bmatrix} = \begin{bmatrix} -\frac{1}{R_1} & 0 & 0 \\ 0 & 0 & R_2 \end{bmatrix} \begin{bmatrix} v_{C_1} \\ v_{C_2} \\ i_L \end{bmatrix} + \begin{bmatrix} \frac{1}{R_1} & 0 \\ 0 & 0 \end{bmatrix} \begin{bmatrix} v_s \\ i_s \end{bmatrix} \tag{23.13}$$

In general, for an nth-order linear circuit with r input sources and m outputs, the state and output equations are represented by

$$\dot{\mathbf{x}} = \mathbf{Ax} + \mathbf{Bu} \tag{23.14}$$

and

$$\mathbf{y} = \mathbf{C}\mathbf{x} + \mathbf{D}\mathbf{u} \tag{23.15}$$

where $\mathbf{x}$ is an $n \times 1$ state vector, $\mathbf{u}$ is an $r \times 1$ vector representing the r input sources, $m \times 1$-vector $\mathbf{y}$ denotes the m output variables, matrices $\mathbf{A}, \mathbf{B}, \mathbf{C}$, and $\mathbf{D}$ are of order $n \times n, n \times r, m \times n$, and $m \times r$, respectively.

In the preceding example, the state equations are obtained by inspection for a simple circuit by writing voltage equations for inductors and current equations for capacitors and properly eliminating the nonstate variables. For more complicated circuits, a systematic procedure for eliminating the nonstate variables is desirable. Such a procedure can be generated with the aid of a proper tree. A proper tree is a tree obtained from the associated network graph that contains all capacitors, independent voltage sources, and possibly some resistive elements, but does not contain inductors and independent current sources. The selection of such a tree is always possible if the circuit contains no capacitive loops and no inductive cutsets. The reason for providing such a tree for writing state equations is obvious. With each tree branch, there is a unique cutset known as the fundamental cutset that contains only one tree branch and some links. Thus, if capacitors are in the tree, a fundamental cutset equation may be written for the corresponding currents through the capacitors. Similarly, every link (together with some tree branches) forms a unique loop called a fundamental loop. If inductors are selected as links, inductor voltages may be obtained by writing the corresponding fundamental loop equations. With the selection of a proper tree, state variables can be defined as the capacitor tree-branch voltages and inductive link currents. In view of the above observation, a systematic procedure for writing state equations can now be stated as follows:

STEP 1: From the associated directed graph, pick a proper tree.

STEP 2: Write fundamental cutset equations for the capacitive tree branches and express the capacitor currents in terms of link currents.

STEP 3: Write fundamental loop equations for the inductive links and express the inductor voltages in terms of tree-branch voltages.

STEP 4: Define the state variables. Capacitive tree-branch voltages and inductive link currents are selected as state variables. Other quantities such as capacitor charges and inductor fluxes may also be used.

STEP 5: Group the branch relations and the remaining fundamental equations according to their element types into three sets: resistor, inductor, and capacitor equations. Solve for the nonstate variables that appeared in the equations obtained in Steps 2 and 3 from the corresponding set of equations in terms of the state variables and independent sources.

STEP 6: Substitute the result of Step 5 into the equations obtained in Steps 2 and 3, and rearrange them in normal form.

Example 4. Consider again the same circuit shown in Fig. 23.3. The various steps outlined above are used to write the state equations.

STEP 1: The associated graph and the proper tree of the circuit are shown in Fig. 23.4. The tree branches include v_s, C_1, C_2, and R_2.

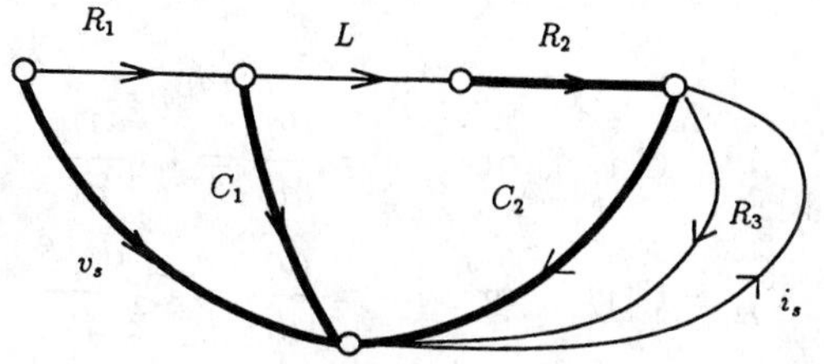

FIGURE 23.4 The directed graph associated with the circuit of Fig. 23.3.

STEP 2: The fundamental cutset associated with C_1 consists of tree branch C_1 and two links R_1 and L. By writing the current equation for this cutset, the capacitor current i_{C_1} is expressed in terms of link currents as

$$i_{C_1} = i_{R_1} - i_L \tag{23.16}$$

Similarly, the fundamental cutset $\{L, C_2, R_3, i_s\}$ associated with C_2 leads to

$$i_{C_2} = i_L - i_{R_3} + i_s \tag{23.17}$$

STEP 3: The fundamental loop associated with link L consists of L and tree branches R_2, C_2, and C_1. By writing the voltage equation around this loop, the inductor voltage can be written in terms of tree-branch voltages as

$$v_L = v_{C_1} - v_{C_2} - v_{R_2} \tag{23.18}$$

STEP 4: The tree-branch capacitor voltages v_{C_1}, v_{C_2}, and inductive link current i_L are defined as the state variables of the circuit.

STEP 5: The branch relation and the remaining two fundamental loops for R_1 and R_2, and the fundamental cutset equation for R_2 are grouped into three sets.

Resistor equations:

$$v_{R_1} + v_{C_1} - v_s = 0 \tag{23.19}$$

$$i_{R_1} = \frac{1}{R_1} v_{R_1} \tag{23.20}$$

$$i_{R_2} - i_L = 0 \tag{23.21}$$

$$v_{R_2} = R_2 i_{R_2} \tag{23.22}$$

$$v_{R_3} - v_{C_2} = 0 \tag{23.23}$$

$$v_{R_3} = R_3 i_{R_3} \tag{23.24}$$

Inductor equations:

$$\phi_L = L i_L \quad \text{or} \quad v_L = \frac{d\phi_L}{dt} = L\frac{di_L}{dt} \tag{23.25}$$

Capacitor equations:

$$q_1 = C_1 v_{C_1} \quad \text{or} \quad i_{C_1} = \frac{dq_1}{dt} = C_1 \frac{dv_{C_1}}{dt} \tag{23.26}$$

$$q_2 = C_2 v_{C_2} \quad \text{or} \quad i_{C_2} = \frac{dq_2}{dt} = C_2 \frac{dv_{C_2}}{dt} \tag{23.27}$$

The resistive link currents i_{R_1}, i_{R_3}, and resistive tree-branch voltage v_{R_2} are solved from (23.19)–(23.24) in terms of the inductive link current i_L, the capacitive tree-branch voltages v_{C_1} and v_{C_2}, and sources as

$$i_{R_1} = \frac{1}{R_1}(v_s - v_{C_1}) \tag{23.28}$$

$$i_{R_3} = \frac{1}{R_3} v_{C_2} \tag{23.29}$$

and

$$v_{R_2} = R_2 i_L \tag{23.30}$$

For this example i_L, v_{C_1}, and v_{C_2} have already been defined as state variables.

STEP 6: Substituting (23.28)–(23.30) into (23.16), (23.17), and (23.18) yields the desired state equation in matrix form:

$$\begin{bmatrix} \frac{dv_{C_1}}{dt} \\ \frac{dv_{C_2}}{dt} \\ \frac{di_L}{dt} \end{bmatrix} = \begin{bmatrix} -\frac{1}{R_1 C_1} & 0 & -\frac{1}{C_1} \\ 0 & -\frac{1}{R_3 C_2} & \frac{1}{C_2} \\ \frac{1}{L} & -\frac{1}{L} & -\frac{R_2}{L} \end{bmatrix} \begin{bmatrix} v_{C_1} \\ v_{C_2} \\ i_L \end{bmatrix} + \begin{bmatrix} \frac{1}{R_1 C_1} & 0 \\ 0 & \frac{1}{C_2} \\ 0 & 0 \end{bmatrix} \begin{bmatrix} v_s \\ i_s \end{bmatrix} \tag{23.31}$$

which, as expected, is the same as (23.10) obtained previously by inspection.

As mentioned earlier, the selection of state variables is not unique. Instead of using capacitor voltages and inductor currents as state variables, basic quantities such as the capacitor charges and inductor fluxes may also be considered. If q_1, q_2, and ϕ_L are defined as state variables in Step 4, the inductive link current i_L and capacitive tree-branch voltages, v_{C_1} and v_{C_2}, can be solved from the inductor and capacitor equations in terms of state variables and possibly sources in Step 5 as

$$i_L = \frac{1}{L}\phi_L \tag{23.32}$$

$$v_{C_1} = \frac{1}{C_1} q_1 \tag{23.33}$$

$$v_{C_2} = \frac{1}{C_2} q_2 \tag{23.34}$$

Finally, state equations are obtained by substituting eqs. (23.28)–(23.30) and (23.32)–(23.34) into (23.16)–(23.18) as

$$\begin{bmatrix} \frac{dq_1}{dt} \\ \frac{dq_2}{dt} \\ \frac{d\phi_L}{dt} \end{bmatrix} = \begin{bmatrix} -\frac{1}{R_1C_1} & 0 & -\frac{1}{L} \\ 0 & -\frac{1}{R_3C_2} & \frac{1}{L} \\ \frac{1}{C_1} & -\frac{1}{C_2} & -\frac{R_2}{L} \end{bmatrix} \begin{bmatrix} q_1 \\ q_2 \\ \phi_L \end{bmatrix} + \begin{bmatrix} \frac{1}{R_1} & 0 \\ 0 & 1 \\ 0 & 0 \end{bmatrix} \begin{bmatrix} v_s \\ i_s \end{bmatrix} \tag{23.35}$$

In the systematic procedure outlined above, it is assumed that the network exists with neither inductive cutsets nor capacitive loops so that the selection of a proper tree is always guaranteed. For networks that do have these constraints, it is not possible to include all the capacitors in a tree without forming a closed path. Also, in order for a tree to contain all the nodes, some inductors will have to be included in a tree. A tree that includes independent voltage sources, some resistors, and a maximum number of capacitors but no independent current sources is called a modified proper tree. In writing a state equation for such networks, the same systematic procedure can be applied with the selection of a modified proper tree. However, if capacitor tree-branch voltages and inductive link currents are defined as the state variables, the standard $(\mathbf{A}, \mathbf{B}, \mathbf{C}, \mathbf{D})$ description (23.14) and (23.15) may not exist. In fact, if inductive cutsets contain independent current sources and/or capacitive loops contain independent voltage sources, the derivative of these sources will appear in the state equation and the general equation is of the form

$$\dot{\mathbf{x}} = \mathbf{A}\mathbf{x} + \mathbf{B}_1\mathbf{u} + \mathbf{B}_2\dot{\mathbf{u}} \tag{23.36}$$

where $\mathbf{B}_1$ and $\mathbf{B}_2$ are $n \times r$ matrices and $\mathbf{A}$, $\mathbf{x}$, and $\mathbf{u}$ are defined as before. To recast (23.36) into the standard form, it is necessary to redefine

$$\mathbf{z} = \mathbf{x} - \mathbf{B}_2\mathbf{u} \tag{23.37}$$

as new state variables. Substituting (23.37) into (23.36), yields

$$\dot{\mathbf{z}} = \mathbf{A}\mathbf{z} + \mathbf{B}\mathbf{u} \tag{23.38}$$

where

$$\mathbf{B} = \mathbf{B}_1 + \mathbf{A}\mathbf{B}_2 \tag{23.39}$$

It is noted from (23.37), the new state variables represent a linear combination of sources and capacitor voltages or inductor currents which, except for the mathematical convenience, may not have sound physical significance. To avoid such state variables and transformation (23.37), Step 4 of the systematic procedure described earlier needs to be modified. By defining state variables as the algebraic sum of capacitor charges in the fundamental cutset associated with each of the capacitor tree branches, and the algebraic sum of inductor fluxes in the fundamental loop associated with each of the inductive links, the resulting state equation will be in the standard form. The above generalizations are illustrated by the following two examples.

Example 5. As a simple illustration, consider the same circuits given in Fig. 23.2, where the constant DC voltage source E is replaced by a time-varying source $e(t)$. It can easily be shown that the equation describing the circuit now becomes

$$\frac{dv_{C_1}(t)}{dt} = -\frac{1}{R(C_1+C_2)}v_{C_1} + \frac{1}{R(C_1+C_2)}v_{in}(t) + \frac{C_2}{R(C_1+C_2)}\frac{de(t)}{dt} \tag{23.40}$$

The above equation is the same as the state eq. (23.4) with the exception of an additional term involving the first-order derivative of source $e(t)$. Equation (23.40) is clearly not the standard state equation described in (23.14) with capacitor voltage v_{C_1} defined as the state variable.

Example 6. As another illustration, consider the circuit shown in Fig. 23.5 which consists of an inductive cutset $\{L_1, L_2, i_s\}$ and a capacitive loop (C_1, v_{s_2}, C_2). The state equations are determined from the systematic procedure by first using the transformation (23.37) and then by defining the algebraic sum of charges and fluxes as state variables.

STEP 1: The directed graph of the circuit is shown in Fig. 23.6 where branches denoted by $v_{s_1}, v_{s_2}, C_1, R_2$, and L_2 are selected to form a modified proper tree.

STEP 2: The fundamental cutset associated with C_1 consists of branches R_1, C_1, L_1, i_s, C_2, and R_3. Applying KCL to this cutset yields

$$i_{C_1} = -i_{R_1} - i_{L_1} - i_s - i_{C_2} - i_{R_3} \tag{23.41}$$

STEP 3: The fundamental loop equation associated with the inductive link L_1 is given by

$$v_{L_1} = v_{C_1} + v_{L_2} - v_{R_2} \tag{23.42}$$

where the link voltage v_{L_1} has been expressed in terms of tree-branch voltages.

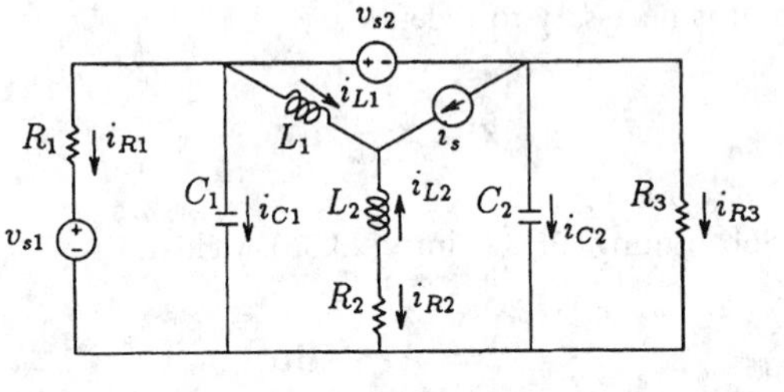

FIGURE 23.5 A circuit with a capacitive loop and an inductive cutset.

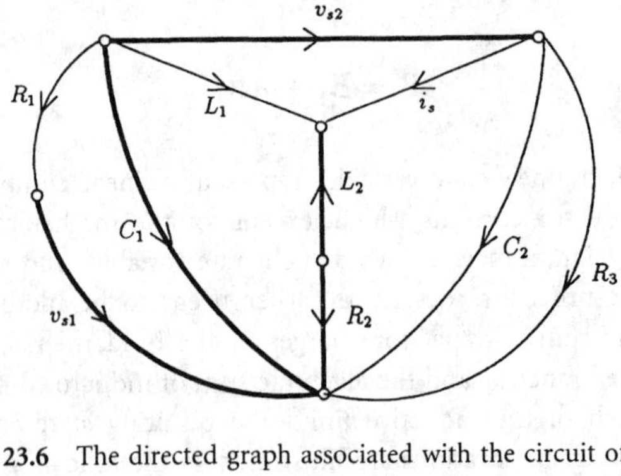

FIGURE 23.6 The directed graph associated with the circuit of Fig. 23.5.

STEP 4: In the first illustration, the tree-branch capacitor voltage v_{C_1} and the inductive link current i_{L_1} are defined as the state variables.

STEP 5: The branch relation and the remaining two fundamental equations are grouped into the following three sets:

Resistor equations:

$$v_{R_1} + v_{s_1} - v_{C_1} = 0 \tag{23.43}$$

$$v_{R_1} = R_1 i_{R_1} \tag{23.44}$$

$$i_{R_2} - i_{L_1} - i_s = 0 \tag{23.45}$$

$$v_{R_2} = R_2 i_{R_2} \tag{23.46}$$

$$v_{R_3} - v_{C_1} + v_{s_2} = 0 \tag{23.47}$$

$$v_{R_3} = R_3 i_{R_3} \tag{23.48}$$

Inductor equations:

$$i_{L_2} + i_{L_1} + i_s = 0 \tag{23.49}$$

$$\phi_{L_1} = L_1 i_{L_1} \quad \text{or} \quad v_{L_1} = L_1 \frac{di_{L_1}}{dt} \tag{23.50}$$

$$\phi_{L_2} = L_2 i_{L_2} \quad \text{or} \quad v_{L_2} = L_2 \frac{di_{L_2}}{dt} \tag{23.51}$$

Capacitor equations:

$$v_{C_2} - v_{C_1} + v_{s_2} = 0 \tag{23.52}$$

$$q_1 = C_1 v_{C_1} \quad \text{or} \quad i_{C_1} = C_1 \frac{dv_{C_1}}{dt} \tag{23.53}$$

$$q_2 = C_2 v_{C_2} \quad \text{or} \quad i_{C_2} = C_2 \frac{dv_{C_2}}{dt} \tag{23.54}$$

For this example, the nonstate variables are identified as $i_{R_1}, v_{R_2}, i_{R_3}, v_{L_2}$, and i_{C_2} from (23.41) and (23.42). These variables are now solved from the corresponding group of equations in terms of state variables and independent sources:

$$i_{R_1} = \frac{1}{R_1}(v_{C_1} - v_{s_1}) \tag{23.55}$$

$$v_{R_2} = R_2(i_{L_1} + i_s) \tag{23.56}$$

$$i_{R_3} = \frac{1}{R_3}(v_{C_1} - v_{s_2}) \tag{23.57}$$

$$v_{L_2} = -L_2 \frac{di_{L_1}}{dt} - L_2 \frac{di_s}{dt} \tag{23.58}$$

$$i_{C_2} = C_2 \frac{dv_{C_1}}{dt} - C_2 \frac{dv_{s_2}}{dt} \tag{23.59}$$

STEP 6: Assuming the existence of the first-order derivatives of sources with respect to time and susbtituting eqs. (23.50), (23.53), and (23.55)–(23.59) into (23.41) and (23.42) yields

$$\begin{bmatrix} \dfrac{dv_{C_1}}{dt} \\ \dfrac{di_{L_1}}{dt} \end{bmatrix} = \begin{bmatrix} -\dfrac{R_1 + R_3}{R_1 R_3 (C_1 + C_2)} & -\dfrac{1}{C_1 + C_2} \\ \dfrac{1}{L_1 + L_2} & -\dfrac{R_2}{L_1 + L_2} \end{bmatrix} \begin{bmatrix} v_{C_1} \\ i_{L_1} \end{bmatrix} + \begin{bmatrix} \dfrac{1}{R_1(C_1 + C_2)} & \dfrac{1}{R_3(C_1 + C_2)} & -\dfrac{1}{C_1 + C_2} \\ 0 & 0 & -\dfrac{R_2}{L_1 + L_2} \end{bmatrix} \begin{bmatrix} v_{s_1} \\ v_{s_2} \\ i_s \end{bmatrix} + \begin{bmatrix} 0 & \dfrac{C_2}{C_1 + C_2} & 0 \\ 0 & 0 & -\dfrac{L_2}{L_1 + L_2} \end{bmatrix} \begin{bmatrix} \dfrac{dv_{s_1}}{dt} \\ \dfrac{dv_{s_2}}{dt} \\ \dfrac{di_s}{dt} \end{bmatrix} \tag{23.60}$$

Clearly, eq. (23.60) is not in the standard form. Applying transformation (23.37) with $x_1 = v_{C_1}$, $x_2 = i_{L_2}$, $u_1 = v_{s_1}$, $u_2 = v_{s_2}$, and $u_3 = i_s$ gives the state equation in normal form

$$\begin{bmatrix} \dfrac{dz_1}{dt} \\ \dfrac{dz_2}{dt} \end{bmatrix} = \begin{bmatrix} -\dfrac{R_1 + R_3}{R_1 R_3 (C_1 + C_2)} & -\dfrac{1}{C_1 + C_2} \\ \dfrac{1}{L_1 + L_2} & -\dfrac{R_2}{L_1 + L_2} \end{bmatrix} \begin{bmatrix} z_1 \\ z_2 \end{bmatrix} + \begin{bmatrix} \dfrac{1}{R_1(C_1 + C_2)} & \dfrac{R_1 C_1 - R_3 C_2}{R_1 R_3 (C_1 + C_2)^2} & -\dfrac{L_1}{(L_1 + L_2)(C_1 + C_2)} \\ 0 & \dfrac{C_2}{(L_1 + L_2)(C_1 + C_2)} & -\dfrac{R_2 L_1}{(L_1 + L_2)^2} \end{bmatrix} \times \begin{bmatrix} v_{s_1} \\ v_{s_2} \\ i_s \end{bmatrix} \tag{23.61}$$

where new state variables are defined as

$$\mathbf{z} = \begin{bmatrix} z_1 \\ z_2 \end{bmatrix} = \begin{bmatrix} v_{C_1} - \dfrac{C_2}{C_1 + C_2} v_{s_2} \\ i_{L_1} + \dfrac{L_2}{L_1 + L_2} i_s \end{bmatrix} \tag{23.62}$$

Alternatively, if the state variables are defined in Step 4 as

$$q_a = q_1 + q_2 \tag{23.63}$$

$$\phi_b = \phi_1 - \phi_2 \tag{23.64}$$

then eqs. (23.41) and (23.42) become

$$\frac{dq_a}{dt} = \frac{dq_1}{dt} + \frac{dq_2}{dt} = -i_{R_1} - i_{L_1} - i_s - i_{R_3} \tag{23.65}$$

$$\frac{d\phi_b}{dt} = \frac{d\phi_1}{dt} - \frac{d\phi_2}{dt} = v_{L_1} - v_{L_2} = v_{C_1} - v_{R_2} \tag{23.66}$$

respectively. In Step 5, the resistive link currents i_{R_1}, i_{R_3}, and the resistive tree-branch voltage v_{R_2} are solved from resistive eqs. (23.43)–(23.48) in terms of inductive link currents, capacitive tree-branch voltages, and independent sources. The results are those given in (23.55)–(23.57). By solving the inductor eqs. (23.49), (23.50), and (23.64), inductive link current i_{L_1} is expressed as a function of state variables and independent sources:

$$i_{L_1} = \frac{1}{L_1 + L_2}(\phi_b - L_2 i_s) \tag{23.67}$$

Similarly, solving v_{C_1} from capacitor eqs. (23.52)–(23.54), and (23.63), yields the capacitor tree-branch voltage

$$v_{C_1} = \frac{1}{C_1 + C_2}(q_a + C_2 v_{s_2}) \tag{23.68}$$

Finally, in Step 6, eqs. (23.55)–(23.57), (23.67), and (23.68) are substituted into (23.65) and (23.66) to form the state equation in normal form:

$$\begin{bmatrix} \dfrac{dq_a}{dt} \\[2mm] \dfrac{d\phi_b}{dt} \end{bmatrix} = \begin{bmatrix} -\dfrac{R_1 + R_3}{R_1 R_3 (C_1 + C_2)} & -\dfrac{1}{L_1 + L_2} \\[2mm] \dfrac{1}{C_1 + C_2} & -\dfrac{R_2}{L_1 + L_2} \end{bmatrix} \begin{bmatrix} q_a \\ \phi_b \end{bmatrix} + \begin{bmatrix} \dfrac{1}{R_1} & \dfrac{R_1 C_1 - R_3 C_2}{R_1 R_3 (C_1 + C_2)} & -\dfrac{L_1}{(L_1 + L_2)} \\[2mm] 0 & \dfrac{C_2}{(C_1 + C_2)} & -\dfrac{R_2 L_1}{(L_1 + L_2)} \end{bmatrix} \begin{bmatrix} v_{s_1} \\ v_{s_2} \\ i_s \end{bmatrix} \tag{23.69}$$

.3 Natural Response and State Transition Matrix

In the preceding section, the state-variable description has been presented for linear time-invariant circuits. The response of the circuit depends on the solution of the state equation. The behavior of the circuit due to any arbitrary input sources can easily be obtained once the zero-input response or the natural response of the circuit is known. In order to find its natural response, the homogeneous state equation of the circuit

$$\dot{\mathbf{x}} = \mathbf{A}\mathbf{x} \tag{23.70}$$

is considered, where independent source term $u(t)$ has been set equal to zero. The above state equation is analogous to the scalar equation

$$\dot{x} = ax \tag{23.71}$$

whose solution is given by

$$x(t) = e^{at}x(0) \tag{23.72}$$

for any arbitrary initial condition $x(0)$ given at $t = 0$, or

$$x(t) = e^{a(t-t_o)}x(t_o) \tag{23.73}$$

if the initial time is specified at $t = t_o$.

It is thus reasonable to assume a solution for (23.70) of the form

$$\mathbf{x}(t) = e^{(t-t_o)\lambda}\mathbf{p} \tag{23.74}$$

where λ is a scalar constant and $\mathbf{p}$ is a constant n-vector. Substituting (23.74) into (23.70) leads to

$$\mathbf{Ap} = \lambda\mathbf{p} \tag{23.75}$$

Therefore, (23.74) is a solution of (23.70) precisely when $\mathbf{p}$ is an eigenvector of $\mathbf{A}$ associated with the eigenvalue λ. For simplicity, it is assumed that $\mathbf{A}$ has n distinct eigenvalues $\lambda_1, \lambda_2, \ldots, \lambda_n$. Since the corresponding eigenvectors denoted by $\mathbf{p}_1, \mathbf{p}_2, \ldots, \mathbf{p}_n$ are linearly independent, the general solution of (23.70) can be uniquely written as a linear combination of n distinct normal modes of the form (23.74):

$$\mathbf{x}(t) = c_1 e^{(t-t_o)\lambda_1}\mathbf{p}_1 + c_2 e^{(t-t_o)\lambda_2}\mathbf{p}_2 + \cdots + c_n e^{(t-t_o)\lambda_n}\mathbf{p}_n \tag{23.76}$$

where $c_1, c_2, \ldots, c_n$ are n arbitrary constants determined by the given initial conditions. Specifically,

$$\mathbf{x}(t_o) = c_1\mathbf{p}_1 + c_2\mathbf{p}_2 + \cdots + c_n\mathbf{p}_n \tag{23.77}$$

The general solution (23.76) can also be written in the form

$$\mathbf{x}(t) = e^{(t-t_o)\mathbf{A}}\mathbf{x}(t_o) \tag{23.78}$$

where the exponential function of a matrix is defined by a power series:

$$\begin{aligned} e^{(t-t_o)\mathbf{A}} &= \mathbf{I} + (t-t_o)\mathbf{A} + \frac{(t-t_o)^2}{2!}\mathbf{A}^2 + \cdots \\ &= \sum_{k=0}^{\infty}\frac{(t-t_o)^k}{k!}\mathbf{A}^k \end{aligned} \tag{23.79}$$

In fact, taking the derivative of (23.78) with respect to t yields

$$\begin{aligned} \frac{d\mathbf{x}}{dt} &= \frac{d}{dt}\left[\mathbf{I} + (t-t_o)\mathbf{A} + \frac{(t-t_o)^2}{2!}\mathbf{A}^2 + \cdots\right]\mathbf{x}(t_o) \\ &= \left[\mathbf{A} + (t-t_o)\mathbf{A}^2 + \frac{(t-t_o)^2}{2!}\mathbf{A}^3 + \cdots\right]\mathbf{x}(t_o) \\ &= \mathbf{A}\left[\mathbf{I} + (t-t_o)\mathbf{A} + \frac{(t-t_o)^2}{2!}\mathbf{A}^2 + \cdots\right]\mathbf{x}(t_o) \\ &= \mathbf{A}e^{(t-t_o)\mathbf{A}}\mathbf{x}(t_o) = \mathbf{A}\mathbf{x}(t) \end{aligned} \tag{23.80}$$

Also, at $t = t_0$, (23.78) gives

$$\mathbf{x}(t_o) = \mathbf{I}\mathbf{x}(t_o) = \mathbf{x}(t_o) \tag{23.81}$$

Thus, expression (23.78) satisfies both eq. (23.70) and the initial conditions and hence is the unique solution. The matrix $e^{(t-t_o)\mathbf{A}}$, usually denoted by $\mathbf{\Phi}(t - t_o)$, is called the state transition matrix or the fundamental matrix of the circuit described by (23.70). The transition of the initial state $\mathbf{x}(t_o)$ to the state $\mathbf{x}(t)$ at any time t is thus governed by

$$\mathbf{x}(t) = \mathbf{\Phi}(t - t_o)\mathbf{x}(t_o) \tag{23.82}$$

where

$$\mathbf{\Phi}(t - t_o) = e^{(t-t_o)\mathbf{A}} \tag{23.83}$$

is an $n \times n$ matrix with the following properties:

$$\mathbf{\Phi}(t_o - t_o) = \mathbf{\Phi}(0) = \mathbf{I} \tag{23.84}$$

$$\mathbf{\Phi}(t + \tau) = \mathbf{\Phi}(t)\mathbf{\Phi}(\tau) \tag{23.85}$$

$$\mathbf{\Phi}(t_2 - t_1)\mathbf{\Phi}(t_1 - t_o) = \mathbf{\Phi}(t_2 - t_o) \tag{23.86}$$

$$\mathbf{\Phi}(t_2 - t_1) = \mathbf{\Phi}^{-1}(t_1 - t_2) \tag{23.87}$$

$$\mathbf{\Phi}^{-1}(t) = \mathbf{\Phi}(-t) \tag{23.88}$$

Once the state transition matrix is known, the solution of the state equation can be obtained from (23.82). In general, it is rather difficult to obtain a closed-form solution from the infinite series representation of the state transition matrix. The formula given by (23.79) is useful only if numerical solution by digital computer is desired. Several methods are available for finding a closed form expression for $\mathbf{\Phi}(t - t_o)$. The relationship between solution (23.76) and the state transition matrix is first established.

For simplicity, let $t_o = 0$. According to (23.82), the first column of $\mathbf{\Phi}(t)$ is the solution of the state equation generated by the initial condition

$$\mathbf{x}(0) = \mathbf{x}^{(1)}(0) = \begin{bmatrix} 1 \\ 0 \\ 0 \\ \vdots \\ 0 \end{bmatrix} \tag{23.89}$$

Indeed, if (23.89) is substituted into (23.82), then

$$\mathbf{x}(t) \triangleq \mathbf{x}^{(1)}(t) = \mathbf{\Phi}(t)\mathbf{x}^{(1)}(0) = \begin{bmatrix} \phi_{11} & \phi_{12} & \cdots & \phi_{1n} \\ \phi_{21} & \phi_{22} & \cdots & \phi_{2n} \\ \vdots & \vdots & \vdots & \vdots \\ \phi_{n1} & \phi_{n2} & \cdots & \phi_{nn} \end{bmatrix} \begin{bmatrix} 1 \\ 0 \\ 0 \\ \vdots \\ 0 \end{bmatrix} = \begin{bmatrix} \phi_{11} \\ \phi_{21} \\ \vdots \\ \phi_{n1} \end{bmatrix} \tag{23.90}$$

which can be computed from (23.76) and the arbitrary constants $c_i \triangleq c_i^{(1)}$ for $i = 1, 2, \ldots, n$ are solved from (23.77). The first column of the state transition matrix is thus given by

$$\begin{bmatrix} \phi_{11} \\ \phi_{21} \\ \vdots \\ \phi_{n1} \end{bmatrix} = c_1^{(1)} e^{\lambda_1 t} + c_2^{(1)} e^{\lambda_2 t} + \cdots + c_n^{(1)} e^{\lambda_n t} \tag{23.91}$$

Instead of (23.89), if

$$\mathbf{x}(0) = \mathbf{x}^{(2)}(0) = \begin{bmatrix} 0 \\ 1 \\ 0 \\ \vdots \\ 0 \end{bmatrix} \tag{23.92}$$

is used, the arbitrary constants $c_1, c_2, \ldots, c_n$ denoted by $c_1^{(2)}, c_2^{(2)}, \ldots, c_n^{(2)}$ are solved. Then the second column of $\mathbf{\Phi}(t)$ is given

$$\begin{bmatrix} \phi_{12} \\ \phi_{22} \\ \vdots \\ \phi_{n2} \end{bmatrix} = c_1^{(2)} e^{\lambda_1 t} + c_2^{(2)} e^{\lambda_2 t} + \cdots + c_n^{(2)} e^{\lambda_n t} \tag{23.93}$$

In a similar manner, the remaining columns of $\mathbf{\Phi}(t)$ are determined.

The closed form expression for state transition matrix can also be obtained by means of a similarity transformation of the form

$$\mathbf{AP} = \mathbf{PJ}$$

or

$$\mathbf{J} = \mathbf{P}^{-1}\mathbf{AP} \tag{23.94}$$

where $\mathbf{P}$ is a nonsingular matrix. If the eigenvalues of $\mathbf{A}$, $\lambda_1, \lambda_2, \ldots, \lambda_n$, are assumed to be distinct, $\mathbf{J}$ is a diagonal matrix with eigenvalues on its main diagonal:

$$\mathbf{J} = \begin{bmatrix} \lambda_1 & 0 & \cdots & 0 \\ 0 & \lambda_2 & \cdots & 0 \\ \vdots & \vdots & \ddots & \vdots \\ 0 & 0 & \cdots & \lambda_n \end{bmatrix} \tag{23.95}$$

and

$$\mathbf{P} = \begin{bmatrix} \mathbf{p}_1 & \mathbf{p}_2 & \cdots & \mathbf{p}_n \end{bmatrix} \tag{23.96}$$

where $\mathbf{p}_i$'s, for $i = 1, 2, \ldots, n$, are the corresponding eigenvectors associated with the eigenvalue λ_i, for $i = 1, 2, \ldots, n$. Substituting (23.94) into (23.83), the state transition matrix can now be written in the closed form

$$\begin{aligned} \mathbf{\Phi}(t - t_o) &= e^{(t-t_o)\mathbf{A}} = e^{(t-t_o)\mathbf{PJP}^{-1}} \\ &= \mathbf{P}e^{(t-t_o)\mathbf{J}}\mathbf{P}^{-1} \end{aligned} \tag{23.97}$$

where

$$e^{(t-t_o)\mathbf{J}} = \begin{bmatrix} e^{(t-t_o)\lambda_1} & 0 & \cdots & 0 \\ 0 & e^{(t-t_o)\lambda_2} & \cdots & 0 \\ \vdots & \vdots & \ddots & \vdots \\ 0 & 0 & \cdots & e^{(t-t_o)\lambda_n} \end{bmatrix} \tag{23.98}$$

is a diagonal matrix.

In the more general case, where the **A** matrix has repeated eigenvalues, a diagonal matrix of the form (23.95) may not exist. However, it can be shown that any square matrix **A** can be transformed by a similarity transformation to the Jordan canonical form

$$\mathbf{J} = \begin{bmatrix} \mathbf{J}_1 & 0 & \cdots & 0 \\ 0 & \mathbf{J}_2 & \cdots & 0 \\ \vdots & \vdots & \ddots & \vdots \\ 0 & 0 & \cdots & \mathbf{J}_l \end{bmatrix} \tag{23.99}$$

where $\mathbf{J}_i$'s, for $i = 1, 2, \ldots, l$ are known as Jordan blocks. Assuming that **A** has m distinct eigenvalues, λ_i, with multiplicity r_i, for $i = 1, 2, \ldots, m$, and $r_1 + r_2 + \cdots + r_m = n$. Associated with each λ_i there may exist several Jordan blocks. A Jordan block is a block diagonal matrix of order $k \times k (k \le r_i)$ with λ_i on its main diagonal, all 1's on the superdiagonal, and zeros elsewhere. In the special case when $k = 1$, the Jordan block reduces to a 1×1 scalar block with only one element λ_i.

In fact, the number of Jordan blocks associated with the eigenvalue λ_i is equal to the dimension of the null space of $(\lambda_i\mathbf{I} - \mathbf{A})$. For each $k \times k$ Jordan block $\mathbf{J}(k)$ associated with the eigenvalue λ_i of the form

$$\mathbf{J}(k) = \begin{bmatrix} \lambda_i & 1 & 0 & 0 & \cdots & 0 \\ 0 & \lambda_i & 1 & 0 & \cdots & 0 \\ \vdots & \vdots & \vdots & \vdots & \ddots & 1 \\ 0 & 0 & 0 & 0 & \cdots & \lambda_i \end{bmatrix} \tag{23.100}$$

the exponential function of $\mathbf{J}(k)$ takes the form

$$e^{(t-t_o)\mathbf{J}(k)} = \begin{bmatrix} 1 & t & \dfrac{t^2}{2!} & \cdots & \dfrac{t^{k-1}}{(k-1)!} \\ 0 & 1 & t & \cdots & \dfrac{t^{k-2}}{(k-2)!} \\ \vdots & \vdots & \vdots & \ddots & \vdots \\ 0 & 0 & 0 & \cdots & 1 \end{bmatrix} e^{(t-t_o)\lambda_i} \tag{23.101}$$

and the corresponding k columns of $\mathbf{P}$, known as the generalized eigenvectors, satisfy the equations

$$\begin{aligned} (\lambda_i\mathbf{I} - \mathbf{A})\mathbf{p}_i^{(1)} &= 0 \\ (\lambda_i\mathbf{I} - \mathbf{A})\mathbf{p}_i^{(2)} &= -\mathbf{p}_i^{(1)} \\ &\vdots \\ (\lambda_i\mathbf{I} - \mathbf{A})\mathbf{p}_i^{(k)} &= -\mathbf{p}_i^{(k-1)} \end{aligned} \tag{23.102}$$

The closed form expression $\mathbf{\Phi}(t - t_o)$ for this general case now becomes

$$\mathbf{\Phi}(t - t_o) = \mathbf{P}e^{(t-t_o)\mathbf{J}}\mathbf{P}^{-1} \tag{23.103}$$

where

$$e^{(t-t_o)\mathbf{J}} = \begin{bmatrix} e^{(t-t_o)\mathbf{J}_1} & 0 & \cdots & 0 \\ 0 & e^{(t-t_o)\mathbf{J}_2} & \cdots & 0 \\ \vdots & \vdots & \ddots & \vdots \\ 0 & 0 & \cdots & e^{(t-t_o)\mathbf{J}_l} \end{bmatrix} \tag{23.104}$$

and each of the $e^{(t-t_o)\mathbf{J}_i}$, for $i = 1, 2, \ldots, l$, is of the form given in (23.101).

The third approach for obtaining closed form expression for the state transition matrix involves the Laplace transform technique. Taking the Laplace transform of (23.70) yields

$$s\mathbf{X}(s) - \mathbf{x}(0) = \mathbf{A}\mathbf{X}(s)$$

or

$$\mathbf{X}(s) = (s\mathbf{I} - \mathbf{A})^{-1}\mathbf{x}(0) \tag{23.105}$$

where $(s\mathbf{I} - \mathbf{A})^{-1}$ is known as the resolvent matrix. The time response

$$\mathbf{x}(t) = \mathscr{L}^{-1}[(s\mathbf{I} - \mathbf{A})^{-1}]\mathbf{x}(0) \tag{23.106}$$

is obtained by taking the inverse Laplace transform of (23.105). It is seen by comparing (23.106) to (23.82) and (23.83) with $t_o = 0$ that

$$\mathbf{\Phi}(t) = e^{t\mathbf{A}} = \mathscr{L}^{-1}[(s\mathbf{I} - \mathbf{A})^{-1}] \tag{23.107}$$

By way of illustration, the following example is considered. The state transition matrix is obtained by using each of the three approaches presented above.

Example 7. Consider the parallel RLC circuit shown in Fig. 23.7. The state equation of the circuit is obtained as

$$\begin{bmatrix} \frac{di_L}{dt} \\ \frac{dv_C}{dt} \end{bmatrix} = \begin{bmatrix} 0 & \frac{1}{L} \\ -\frac{1}{C} & -\frac{1}{RC} \end{bmatrix} \begin{bmatrix} i_L \\ v_C \end{bmatrix} + \begin{bmatrix} 0 \\ \frac{1}{C} \end{bmatrix} i_s \tag{23.108}$$

With $R = \frac{2}{3}\Omega, L = 1H$, and $C = \frac{1}{2}F$, the **A** matrix becomes

$$\mathbf{A} = \begin{bmatrix} 0 & 1 \\ -2 & -3 \end{bmatrix} \tag{23.109}$$

(a) Normal Mode Approach: The eigenvalues and the corresponding eigenvectors of the **A** are found to be

$$\lambda_1 = -1 \qquad \lambda_2 = -2 \tag{23.110}$$

and

$$\mathbf{p}_1 = \begin{bmatrix} 1 \\ -1 \end{bmatrix}, \quad \mathbf{p}_2 = \begin{bmatrix} 1 \\ -2 \end{bmatrix} \tag{23.111}$$

Therefore, the natural response of the circuit is given as a linear combination of the two distinct normal modes as

$$\begin{bmatrix} i_L(t) \\ v_C(t) \end{bmatrix} = c_1 e^{-t} \begin{bmatrix} 1 \\ -1 \end{bmatrix} + c_2 e^{-2t} \begin{bmatrix} 1 \\ -2 \end{bmatrix} \tag{23.112}$$

When evaluated at $t = 0$, (23.112) becomes

$$\begin{bmatrix} i_L(0) \\ v_C(0) \end{bmatrix} = \begin{bmatrix} c_1 + c_2 \\ -c_1 - 2c_2 \end{bmatrix} \tag{23.113}$$

In order to find the first column of $\mathbf{\Phi}(t)$, it is assumed that

$$\begin{bmatrix} i_L(0) \\ v_C(0) \end{bmatrix} = \begin{bmatrix} 1 \\ 0 \end{bmatrix} \tag{23.114}$$

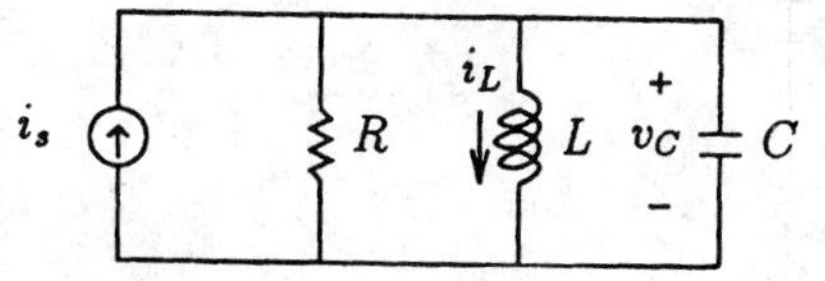

FIGURE 23.7 A parallel RLC circuit.

With this initial condition, the solution of (23.113) becomes

$$c_1 \triangleq c_1^{(1)} = 2 \quad \text{and} \quad c_2 \triangleq c_2^{(1)} = -1 \tag{23.115}$$

Substituting (23.115) into (23.112) results in the first column of $\mathbf{\Phi}(t)$:

$$\begin{bmatrix} \phi_{11} \\ \phi_{21} \end{bmatrix} = \begin{bmatrix} 2e^{-t} - e^{-2t} \\ -2e^{-t} + 2e^{-2t} \end{bmatrix} \tag{23.116}$$

Similarly, for

$$\begin{bmatrix} i_L(0) \\ v_C(0) \end{bmatrix} = \begin{bmatrix} 0 \\ 1 \end{bmatrix} \tag{23.117}$$

constants c_1 and c_2 are solved from (23.113) to give

$$c_1 \triangleq c_1^{(2)} = 1 \quad \text{and} \quad c_2 \triangleq c_2^{(2)} = -1 \tag{23.118}$$

The second column of $\mathbf{\Phi}(t)$:

$$\begin{bmatrix} \phi_{12} \\ \phi_{22} \end{bmatrix} = \begin{bmatrix} e^{-t} - e^{-2t} \\ -e^{-t} + 2e^{-2t} \end{bmatrix} \tag{23.119}$$

is obtained by substituting (23.118) into (23.112). Combining (23.116) and (23.119) yields the state transition matrix in closed form

$$\mathbf{\Phi}(t) = \begin{bmatrix} 2e^{-t} - e^{-2t} & e^{-t} - e^{-2t} \\ -2e^{-t} + 2e^{-2t} & -e^{-t} + 2e^{-2t} \end{bmatrix} \tag{23.120}$$

(b) Similarity Transformation Method: Since the eigenvalues are distinct, the nonsingular transformation $\mathbf{P}$ is constructed from (23.96) by the eigenvectors of $\mathbf{A}$:

$$\mathbf{P} = \begin{bmatrix} \mathbf{p}_1 & \mathbf{p}_2 \end{bmatrix} = \begin{bmatrix} 1 & 1 \\ -1 & -2 \end{bmatrix} \tag{23.121}$$

with

$$\mathbf{P}^{-1} = \begin{bmatrix} 2 & 1 \\ -1 & -1 \end{bmatrix} \tag{23.122}$$

Substituting λ_1, λ_2, and $\mathbf{P}$ into (23.97) and (23.98) yields the desired state transition matrix

$$\begin{aligned} \mathbf{\Phi}(t) = \mathbf{P}e^{t\mathbf{J}}\mathbf{P}^{-1} &= \begin{bmatrix} 1 & 1 \\ -1 & -2 \end{bmatrix} \begin{bmatrix} e^{-t} & 0 \\ 0 & e^{-2t} \end{bmatrix} \begin{bmatrix} 2 & 1 \\ -1 & -1 \end{bmatrix} \\ &= \begin{bmatrix} 2e^{-t} - e^{-2t} & e^{-t} - e^{-2t} \\ -2e^{-t} + 2e^{-2t} & -e^{-t} + 2e^{-2t} \end{bmatrix} \end{aligned} \tag{23.123}$$

which is in agreement with (23.120).

(c) Laplace Transform Technique: The state transition matrix can also be computed in the frequency domain from (23.107). The resolvent matrix is

$$
\begin{aligned}
(s\mathbf{I} - \mathbf{A})^{-1} &= \begin{bmatrix} s & -1 \\ 2 & s+3 \end{bmatrix}^{-1} \\
&= \begin{bmatrix} \dfrac{s+3}{(s+1)(s+2)} & \dfrac{1}{(s+1)(s+2)} \\ \dfrac{-2}{(s+1)(s+2)} & \dfrac{s}{(s+1)(s+2)} \end{bmatrix} \\
&= \begin{bmatrix} \dfrac{2}{s+1} - \dfrac{1}{s+2} & \dfrac{1}{s+1} - \dfrac{1}{s+2} \\ -\dfrac{2}{s+1} + \dfrac{2}{s+2} & -\dfrac{1}{s+1} + \dfrac{2}{s+2} \end{bmatrix}
\end{aligned}
\tag{23.124}
$$

where partial-fraction expansion has been applied. Taking the inverse Laplace transform of (23.124) yields the same closed form expression as given previously in (23.120) for $\mathbf{\Phi}(t)$.

.4 Complete Response

When independent sources are present in the circuit, the complete response depends on the initial states of the circuits as well as the input sources. It is well known that the complete response is the sum of the zero-input (or natural) response and the zero-state (or forced) response and satisfies the nonhomogeneous state equation

$$\dot{\mathbf{x}}(t) = \mathbf{A}\mathbf{x}(t) + \mathbf{B}\mathbf{u}(t) \tag{23.125}$$

subject to the given initial condition $\mathbf{x}(t_o) = \mathbf{x}_o$. Equation (23.125) is again analogous to the scalar equation

$$\dot{x}(t) = ax(t) + bu(t) \tag{23.126}$$

which has the unique solution of the form

$$x(t) = e^{(t-t_o)a}x(t_o) + \int_{t_o}^{t} e^{(t-\tau)a}bu(\tau)\,d\tau \tag{23.127}$$

It is thus assumed that the solution to the state equation is given by

$$
\begin{aligned}
\mathbf{x}(t) &= e^{(t-t_o)\mathbf{A}}\mathbf{x}(t_o) + \int_{t_o}^{t} e^{(t-\tau)\mathbf{A}}\mathbf{B}\mathbf{u}(\tau)\,d\tau \\
&= \mathbf{\Phi}(t-t_o)\mathbf{x}(t_o) + \int_{t_o}^{t} \mathbf{\Phi}(t-\tau)\mathbf{B}\mathbf{u}(\tau)\,d\tau
\end{aligned}
\tag{23.128}
$$

Indeed, one can show by direct substitution that (23.128) satisfies the state eq. (23.125). Differentiating both sides of (23.128) with respect to t yields

$$\begin{aligned}
\dot{\mathbf{x}}(t) &= \frac{d}{dt}\boldsymbol{\Phi}(t-t_o)\mathbf{x}(t_o) + \frac{d}{dt}\int_{t_o}^{t}\boldsymbol{\Phi}(t-\tau)\mathbf{B}\mathbf{u}(\tau)\,d\tau \\
&= \mathbf{A}\boldsymbol{\Phi}(t-t_o)\mathbf{x}(t_o) + \int_{t_o}^{t}\frac{d}{dt}\boldsymbol{\Phi}(t-\tau)\mathbf{B}\mathbf{u}(\tau)\,d\tau + \boldsymbol{\Phi}(t-t)\mathbf{B}\mathbf{u}(t) \\
&= \mathbf{A}\boldsymbol{\Phi}(t-t_o)\mathbf{x}(t_o) + \int_{t_o}^{t}\mathbf{A}\boldsymbol{\Phi}(t-\tau)\mathbf{B}\mathbf{u}(\tau)d\tau + \mathbf{B}\mathbf{u}(t) \\
&= \mathbf{A}\left[\boldsymbol{\Phi}(t-t_o)\mathbf{x}(t_o) + \int_{t_o}^{t}\boldsymbol{\Phi}(t-\tau)\mathbf{B}\mathbf{u}(\tau)d\tau\right] + \mathbf{B}\mathbf{u}(t) \\
&= \mathbf{A}\mathbf{x}(t) + \mathbf{B}\mathbf{u}(t)
\end{aligned} \tag{23.129}$$

Also, at $t = t_o$, (23.128) becomes

$$\begin{aligned}
\mathbf{x}(t_o) &= \boldsymbol{\Phi}(t_o - t_o)\mathbf{x}(t_o) + \int_{t_o}^{t_o}\boldsymbol{\Phi}(t_o-\tau)\mathbf{B}\mathbf{u}(\tau)\,d\tau \\
&= \mathbf{I}\mathbf{x}(t_o) + 0 = \mathbf{x}(t_o)
\end{aligned} \tag{23.130}$$

The assumed solution (23.128) thus satisfies both the state eq. (23.125) and the given initial condition. Hence, $\mathbf{x}(t)$ as given by (23.128) is the unique solution.

It is seen from (23.128) that if $\mathbf{u}(t)$ is set to zero, the solution reduces to the zero-input response or the natural response given in (23.82). On the other hand, if the original circuit is relaxed, i.e., $\mathbf{x}(t_o) = 0$, the solution represented by the convolution integral, the second term on the right-hand side of (23.128), is the forced response or the zero-state response. Thus, eq. (23.128) verifies the fact that the complete response is the sum of the zero-input response and the zero-state response. The above result is illustrated by means of the following example.

Example 8. Consider again the same circuit given in Example 7 where the input current source is assumed to be a unit step function applied to the circuit at $t = 0$.

The state equation of the circuit is found from (23.108) to be

$$\begin{bmatrix} \dfrac{di_L}{dt} \\ \dfrac{dv_C}{dt} \end{bmatrix} = \begin{bmatrix} 0 & 1 \\ -2 & -3 \end{bmatrix}\begin{bmatrix} i_L \\ v_C \end{bmatrix} + \begin{bmatrix} 0 \\ 2 \end{bmatrix} i_s(t) \tag{23.131}$$

where the state transition matrix $\Phi(t)$ is given in (23.120).

The zero-state response for $t > 0$ is obtained by evaluating the convolution integral indicated in (23.128):

$$\begin{aligned}\int_0^t \mathbf{\Phi}(t-\tau)\mathbf{B}\mathbf{u}(\tau)\,d\tau &= \int_0^t \begin{bmatrix} 2e^{-(t-\tau)} - e^{-2(t-\tau)} & e^{-(t-\tau)} - e^{-2(t-\tau)} \\ -2e^{-(t-\tau)} + 2e^{-2(t-\tau)} & -e^{-(t-\tau)} + 2e^{-2(t-\tau)} \end{bmatrix} \begin{bmatrix} 0 \\ 2 \end{bmatrix} d\tau \\ &= 2\int_0^t \begin{bmatrix} e^{-(t-\tau)} - e^{-2(t-\tau)} \\ -e^{-(t-\tau)} + 2e^{-2(t-\tau)} \end{bmatrix} d\tau \\ &= \begin{bmatrix} 1 - 2e^{-t} + e^{-2t} \\ 2e^{-t} - 2e^{-2t} \end{bmatrix}\end{aligned} \tag{23.132}$$

By adding the zero-input response represented by $\mathbf{\Phi}(t)\mathbf{x}(0)$ to (23.132), the complete response for any given initial condition $\mathbf{x}(0)$ becomes

$$\begin{bmatrix} i_L(t) \\ v_C(t) \end{bmatrix} = \begin{bmatrix} 2e^{-t} - e^{-2t} & e^{-t} - e^{-2t} \\ -2e^{-t} + 2e^{-2t} & -e^{-t} + 2e^{-2t} \end{bmatrix} \begin{bmatrix} i_L(0) \\ v_C(0) \end{bmatrix} + \begin{bmatrix} 1 - 2e^{-t} + e^{-2t} \\ 2e^{-t} - 2e^{-2t} \end{bmatrix} \tag{23.133}$$

for $t > 0$.

erences

[1] T. C. Chen, *Linear System Theory and Design*, New York: Holt, Rinehart and Winston, 1970.

[2] W. K. Chen, *Linear Networks and Systems*, Monterey, CA: Brooks/Cole Engineering Division, 1983.

[3] L. O. Chua and P. M. Lin, *Computer-Aided Analysis of Electronics Circuits: Algorithms and Computational Techniques*, Englewood Cliffs, NJ: Prentice-Hall, 1969.

[4] P. M. DeRusso, R. J. Roy, and C. M. Close, *State Variables for Engineers*, New York: Wiley, 1965.

[5] C. A. Desoer and E. S. Kuh, *Basic Circuit Theory*, New York: McGraw-Hill, 1969.

[6] B. C. Kuo, *Linear Networks and Systems*, New York: McGraw-Hill, 1967.

[7] K. Ogata, *State Space Analysis of Control Systems*, Englewood Cliffs, NJ: Prentice-Hall, 1967.

[8] R. A. Rohrer, *Circuit Theory: An Introduction to the State Variable Approach*, New York: McGraw-Hill, 1970.

[9] D. G. Schultz and J. L. Melsa, *State Functions and Linear Control Systems*, New York: McGraw-Hill, 1967.

[10] T. E. Stern, *Theory of Nonlinear Networks and Systems: An Introduction*, Reading, MA: Addison-Wesley, 1965.

[11] L. K. Timothy and B. E. Bona, *State Space Analysis: An Introduction*, New York: McGraw-Hill, 1968.

[12] L. A. Zadeh and C. A. Desoer, *Linear System Theory*, New York: McGraw-Hill, 1963.

V

eedback Circuits

ai-Kai Chen
versity of Illinois

24

Feedback Amplifier Theory

ı Choma, Jr.
ırsity of Southern
ırnia

.1 Introduction

Feedback, whether intentional or parasitic, is pervasive of all electronic circuits and systems. In general, feedback is comprised of a subcircuit that allows a fraction of the output signal of an overall network to modify the effective input signal in such a way as to produce a circuit response that can differ substantially from the response produced in the absence of such feedback. If the magnitude and relative phase angle of the fed back signal decreases the magnitude of the signal applied to the input port of an amplifier, the feedback is said to be *negative* or **degenerative**. On the other hand, *positive* (or **regenerative**) feedback, which gives rise to oscillatory circuit responses, is the upshot of a feedback signal that increases the magnitude of the effective input signal. Because negative feedback produces stable circuit responses, the majority of all intentional feedback architectures is degenerative [1], [2]. But parasitic feedback incurred by the energy storage elements associated with circuit layout, circuit packaging, and second-order high-frequency device phenomena often degrades an otherwise degenerative feedback circuit into either a potentially regenerative or severely underdamped network.

Intentional degenerative feedback applied around an analog network produces four circuit performance benefits. First, negative feedback desensitizes the gain of an **open-loop amplifier** (an amplifier implemented without feedback) with respect to variations in circuit element and active device model parameters. This desensitization property is crucial in view of parametric uncertainties caused by aging phenomena, temperature variations, biasing perturbations, and nonzero fabrication and manufacturing tolerances. Second, and principally because of the foregoing desensitization property, degenerative feedback reduces the dependence of circuit responses on the parameters of inherently nonlinear active devices, thereby reducing the total harmonic distortion evidenced in open loops. Third, negative feedback broadbands the dominant pole of an open-loop amplifier, thereby affording at least the possibility of a closed-loop network

341-2/95/$0.00+$.50

with improved high-frequency performance. Finally, by modifying the driving-point input and output impedances of the open-loop circuit, negative feedback provides a convenient vehicle for implementing voltage buffers, current buffers, and matched interstage impedances.

The disadvantages of negative feedback include gain attenuation, a closed-loop configuration that is disposed to potential instability, and, in the absence of suitable frequency compensation, a reduction in the open-loop gain-bandwidth product. In uncompensated feedback networks, open-loop amplifier gains are reduced in almost direct proportion to the amount by which closed-loop amplifier gains are desensitized with respect to open-loop gains. Although the 3-dB bandwidth of the open-loop circuit is increased by a factor comparable to that by which the open-loop gain is decreased, the closed-loop gain-bandwidth product resulting from uncompensated degenerative feedback is never greater than that of the open-loop configuration [3]. Finally, if feedback is incorporated around an open-loop amplifier that does not have a dominant pole [4], complex conjugate closed-loop poles yielding nonmonotonic frequency responses are likely. Even positive feedback is possible if substantive negative feedback is applied around an open-loop amplifier for which more than two poles significantly influence its frequency response.

Although the foregoing detail is common knowledge deriving from Bode's pathfinding disclosures [5], most circuit designers remain uncomfortable with analytical procedures for estimating the frequency responses, I/O impedances, and other performance indices of practical feedback circuits. The purposes of this section are to formulate systematic feedback circuit analysis procedures and ultimately, to demonstrate their applicability to six specific types of commonly used feedback architectures. Four of these feedback types, the series-shunt, shunt-series, shunt-shunt, and series-series configurations, are single-loop architectures, while the remaining two types are the series-series/shunt-shunt and series-shunt/shunt-series dual-loop configurations.

24.2 Methods of Analysis

There are several standard techniques for analyzing linear feedback circuits [6]. The most straightforward of these entails writing the Kirchhoff equilibrium equations for the small-signal model of the entire feedback system. This analytical tack presumably leads to the idealized feedback circuit block diagram abstracted in Fig. 24.1. In this model, the circuit voltage or current response, X_R, is related to the source current or voltage excitation, X_S, by

$$G_{\text{cl}} \triangleq \frac{X_R}{X_S} = \frac{G_o}{1 + fG_o} \equiv \frac{G_o}{1 + T} \tag{24.1}$$

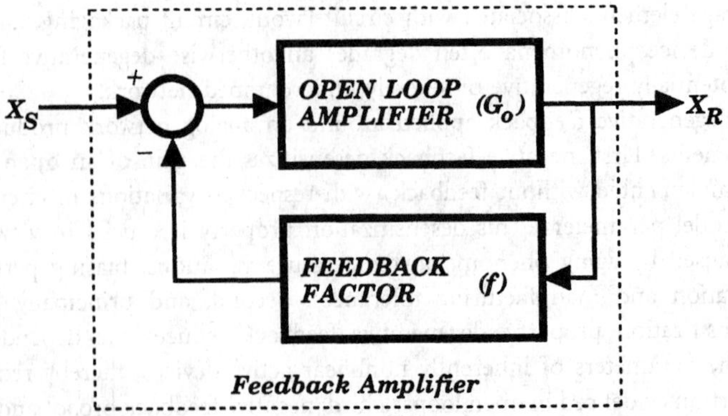

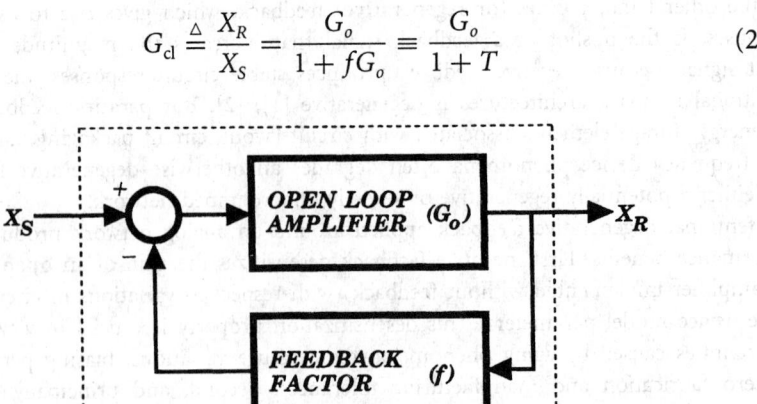

FIGURE 24.1 Block diagram model of a feedback network.

where G_{cl} is the closed-loop gain of the feedback circuit, the feedback factor f is the proportion of circuit response fed back for antiphase superposition with the source signal, and G_o represents the open-loop gain. The product fG_o is termed the loop gain T.

Equation (24.1) shows that for loop gains whose magnitudes are much larger than one, the closed-loop gain collapses to $1/f$, which is independent of the open-loop gain. To the extent that the open-loop amplifier, and not the feedback subcircuit, contains circuit elements and other parameters that are susceptible to modeling uncertainties, variations in the fabrication of active and passive elements, and nonzero manufacturing tolerances, large loop gain achieves a desirable parametric desensitization. Unfortunately, the determination of G_o and f directly from the Kirchhoff relationships is a nontrivial task, especially since G_o is rarely independent of f in practical electronics. Moreover, (24.1) does not illuminate the manner in which the loop gain modifies the driving-point input and output impedances of the open-loop amplifier.

A second approach to feedback network analysis involves modeling the open-loop, feedback, and overall closed-loop networks by a homogeneous set of two-port parameters [7]. When the two-port parameter model is selected judiciously, the two-port parameters for the closed-loop network derive from a superposition of the respective two-port parameters of the open-loop and feedback subcircuits. Given the resultant parameters of the closed-loop circuit, standard formulas can then be exploited to evaluate closed-loop values of the circuit gain and the driving-point input and output impedances.

Unfortunately, several limitations plague the utility of feedback network analysis predicated on two-port parameters. First, the computation of closed-loop two-port parameters is tedious if the open-loop configuration is a multistage amplifier, or if multiloop feedback is utilized. Second, the two-port method of feedback circuit analysis is straightforwardly applicable to only those circuits that implement **global feedback** (feedback applied from output port-to-input port). Many single-ended feedback amplifiers exploit only **local feedback**, wherein a fraction of the signal developed at the output port is fed back to a terminal pair other than that associated with the input port. Finally, the appropriate two-port parameters of the open-loop amplifier can be superimposed with the corresponding parameter set of the feedback subcircuit if and only if the Brune condition is satisfied [8]. This requirement mandates equality between the preconnection and postconnection values of the two-port parameters of open-loop and feedback cells, respectively. The subject condition is often not satisfied when the open-loop amplifier is not a simple three-terminal two-port configuration.

The third method of feedback circuit analysis exploits Mason's signal flow theory [9]–[11]. The circuit level application of this theory suffers few of the shortcomings indigenous to block diagram and two-port methods of feedback circuit analysis [12]. Signal flow analyses applied to feedback networks efficiently express I/O transfer functions, driving-point input impedances, and driving-point output impedances in terms of an arbitrarily selected critical or reference circuit parameters, say P.

An implicit drawback of signal flow methods is the fact that unless P is selected to be the feedback factor f, which is not always transparent in feedback architectures, expressions for the loop gain and the open-loop gain of feedback amplifiers are obscure. But by applying signal flow theory to a feedback circuit model engineered from insights that derive from the results of two-port network analyses, the feedback factor can be isolated. The payoff of this hybrid analytical approach includes a conventional block diagram model of the I/O transfer function, as well as convenient mathematical models for evaluating the closed-loop driving-point input and output impedances. Yet another attribute of hybrid methods of feedback circuit analysis is its ability to delineate the cause, nature, and magnitude of the feedforward transmittance produced by interconnecting a certain feedback subcircuit to a given open-loop amplifier. This information is crucial in feedback network design because feedforward invariably decreases gain and often causes undesirable phase shifts that can lead to significantly underdamped or unstable closed-loop responses.

24.3 Signal Flow Analysis

Guidelines for feedback circuit analysis by hybrid signal flow methods can be established with the aid of Fig. 24.2 [13]. Figure 24.2(a) depicts a linear network whose output port is terminated in a resistance, R_L. The output signal variable is the voltage V_O, which is generated in response to an input port signal whose Thévenin voltage and resistance are respectively, V_S and R_S. Implicit to the linear network is a current controlled voltage source (CCVS) Pi_b, whose value is directly proportional to the indicated network branch current i_b. The problem at hand is the deduction of the voltage gain $G_v(R_S, R_L) = V_O/V_S$, the driving-point input resistance (or impedance) R_{in}, and the driving-point output resistance (or impedance) R_{out}, as explicit functions of the critical transimpedance parameter P. Although the following systematic procedure is developed in conjunction with the diagram in Fig. 24.2, with obvious changes in notation, it is applicable to determining any type of transfer relationship for any linear network in terms of any type of reference parameter [14].

1. Set $P = 0$, as depicted in Fig. 24.2(b), and compute the resultant voltage gain $G_{vo}(R_S, R_L)$, where the indicated notation suggests an anticipated dependence of gain on source and load resistances. Also, compute the corresponding driving-point input and output resistances R_{ino}, and R_{outo}, respectively. In this case, the "critical" parameter P is associated

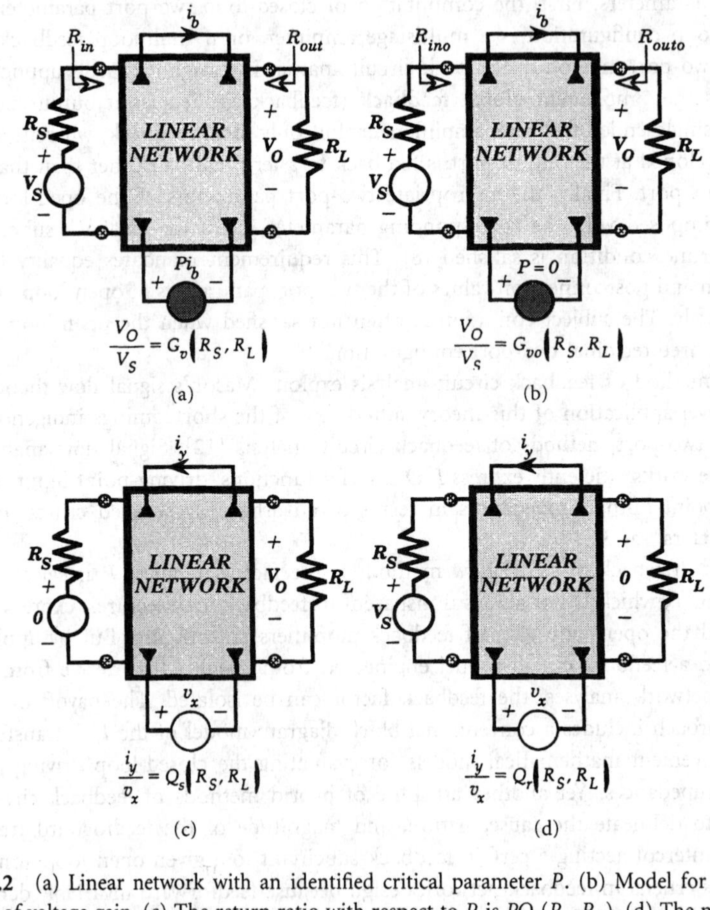

FIGURE 24.2 (a) Linear network with an identified critical parameter P. (b) Model for calculating the $P = 0$ value of voltage gain. (c) The return ratio with respect to P is $PQ_s(R_S, R_L)$. (d) The null return ratio with respect to P is $PQ_r(R_S, R_L)$.

with a controlled voltage source. Accordingly, $P = 0$ requires that the branch containing the controlled source be supplanted by a short circuit. If, for example, P is associated with a controlled current source, $P = 0$ mandates the replacement of the controlled source by an open circuit.

2. Set the Thévenin source voltage V_S to zero, and replace the original controlled voltage source Pi_b by an independent voltage source of symbolic value, v_x. Then, calculate the ratio, i_y/v_x, where, as illustrated in Fig. 24.2(c), i_y flows in the branch that originally conducts the controlling current i_b. Note, however, that the reference polarity of i_y is opposite to that of i_b. The computed transfer function i_y/v_x is denoted by $Q_s(R_S, R_L)$. This transfer relationship, which is a function of the source and load resistances, is used to determine the *return ratio* $T_s(P, R_S, R_L)$, with respect to parameter P of the original network. In particular,

$$T_s(P, R_S, R_L) = PQ_s(R_S, R_L) \tag{24.2}$$

If P is associated with a controlled current source, the controlled generator Pi_b is replaced by a current source of value i_x. If the controlling variable is a voltage, instead of a current, the ratio v_y/v_x, is computed, where v_y, whose polarity is opposite to that of the original controlling voltage, is the voltage developed across the controlling branch.

3. The preceding computational step is repeated, but instead of setting V_S to zero, the output variable, which is the voltage V_O in the present case, is nulled, as indicated in Fig. 24.2(d). Let the computed ratio i_y/v_x, be symbolized as $Q_r(R_S, R_L)$. In turn, the *null return ratio* $T_r(P, R_S, R_L)$, with respect to parameter P is

$$T_r(P, R_S, R_L) = PQ_r(R_S, R_L) \tag{24.3}$$

4. The desired voltage gain $G_v(R_S, R_L)$, of the linear network undergoing study can be shown to be [5, 12]

$$G_v(R_S, R_L) = \frac{V_O}{V_S} = G_{vo}(R_S, R_L)\left[\frac{1 + PQ_r(R_S, R_L)}{1 + PQ_s(R_S, R_L)}\right] \tag{24.4}$$

5. Given the function $Q_s(R_S, R_L)$, the driving-point input and output resistances follow straightforwardly from [12]

$$R_{\text{in}} = R_{\text{ino}}\left[\frac{1 + PQ_s(0, R_L)}{1 + PQ_s(\infty, R_L)}\right] \tag{24.5}$$

$$R_{\text{out}} = R_{\text{outo}}\left[\frac{1 + PQ_s(R_S, 0)}{1 + PQ_s(R_S, \infty)}\right] \tag{24.6}$$

An important special case entails a controlling electrical variable i_b associated with the selected parameter P that is coincidentally the voltage or current output of the circuit under investigation. In this situation, a factor P of the circuit response is fed back to the port (not necessarily the input port) defined by the terminal pair across which the controlled source is incident. When the controlling variable i_b is the output voltage or current of the subject circuit $Q_r(R_S, R_L)$, which is evaluated under the condition of a nulled network response, is necessarily zero. With $Q_r(R_S, R_L) = 0$, the algebraic form of (24.4) is identical to that of (24.1), whence the loop gain T is the return ratio with respect to parameter P; that is,

$$PQ_s(R_S, R_L)\Big|_{Q_r(R_S, R_L)=0} = T \tag{24.7}$$

Moreover, a comparison of (24.4) to (24.1) suggests that $G_v(R_S, R_L)$ symbolizes the closed-loop gain of the circuit, $G_{vo}(R_S, R_L)$ represents the corresponding open-loop gain, and the circuit feedback factor f is

$$f = \frac{PQ_s(R_S, R_L)}{G_{vo}(R_S, R_L)} \tag{24.8}$$

24.4 Global Single-Loop Feedback

Consider the global feedback scenario illustrated in Fig. 24.3(a), in which a fraction P of the output voltage V_O is fed back to the voltage-driven input port. Figure 24.3(b) depicts the model used to calculate the return ratio $Q_s(R_S, R_L)$, where, in terms of the branch variables in the schematic diagram, $Q_s(R_S, R_L) = v_y/v_x$. An inspection of this diagram confirms that the transfer function v_y/v_x, is identical to the $P = 0$ value of the gain V_O/V_S, which derives from an analysis of the structure in Fig. 24.3(a). Thus, for global voltage feedback in which a fraction of the output voltage is fed back to a voltage-driven input port, $Q_s(R_S, R_L)$ is the open-loop voltage gain; that is, $Q_s(R_S, R_L) \equiv G_{vo}(R_S, R_L)$. It follows from (24.8) that the feedback factor f is identical to the selected critical parameter P. Similarly, for the global current feedback architecture of Fig. 24.4(a), in which a fraction P of the output current, I_O, is fed back to the current-driven input port $f = P$. As implied by the model of Fig. 24.4(b), $Q_s(R_S, R_L) \equiv G_{io}(R_S, R_L)$, the open-loop current gain.

Driving-Point *I/O* Resistances

Each of the two foregoing circuit architectures has a closed-loop gain whose algebraic form mirrors (24.1). It follows that for sufficiently large loop gain [equal to either $PG_{vo}(R_S, R_L)$ or $PG_{io}(R_S, R_L)$], the closed-loop gain approaches $(1/P)$ and is therefore desensitized with respect to open-loop gain parameters. However, such a desensitization with respect to the driving-point input and output resistances (or impedances) cannot be achieved. For the voltage feedback circuit in Fig. 24.3(a), $Q_s(\infty, R_L)$ is the $R_S = \infty$ value, $G_{vo}(\infty, R_L)$, of the open-loop voltage gain. This particular open-loop gain is zero, since $R_S = \infty$ decouples the source voltage from

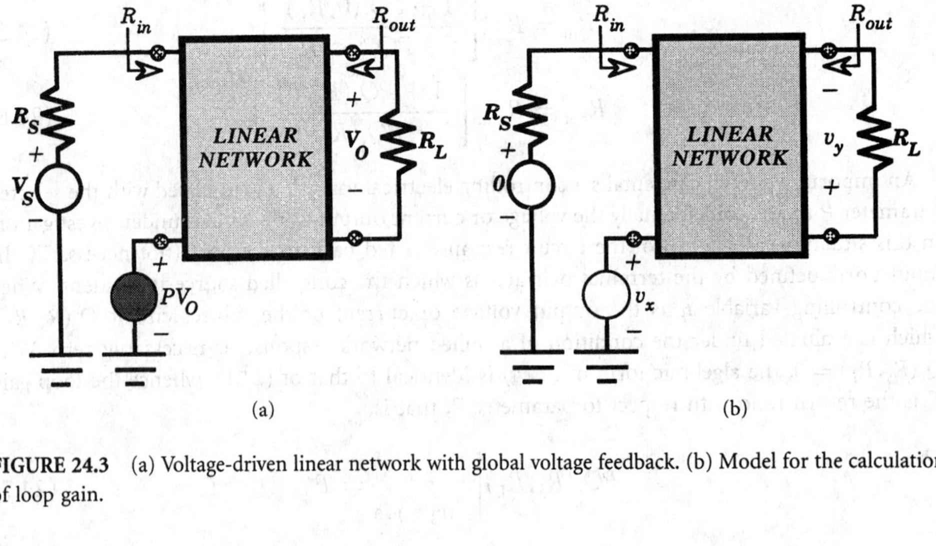

FIGURE 24.3 (a) Voltage-driven linear network with global voltage feedback. (b) Model for the calculation of loop gain.

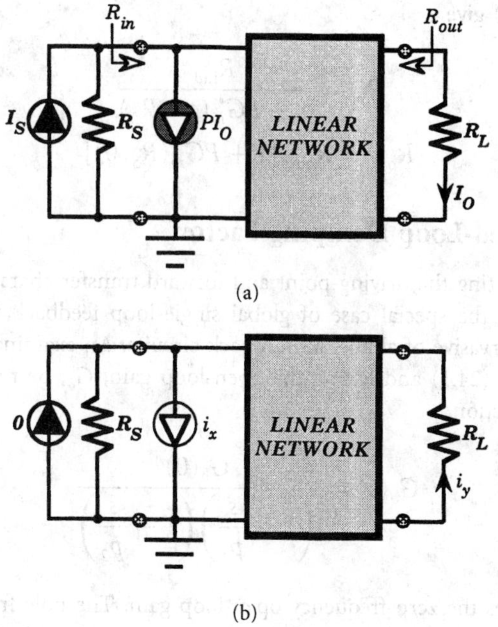

FIGURE 24.4 (a) Current-driven linear network with global current feedback. (b) Model for the calculation of loop gain.

the input port of the amplifier. On the other hand, $Q_s(0, R_L)$ is the $R_S = 0$ value, $G_{vo}(0, R_L)$, of the open-loop voltage gain. This gain is at least as large as $G_{vo}(R_S, R_L)$, since a short circuited Thévenin source resistance implies lossless coupling of the Thévenin signal to the amplifier input port. Recalling (24.5), the resultant driving-point input resistance of the voltage feedback amplifier is

$$R_{\text{in}} = R_{\text{ino}}[1 + PG_{vo}(0, R_L)] \geq R_{\text{ino}}[1 + PG_{vo}(R_S, R_L)] \tag{24.9}$$

which shows that the closed-loop driving-point input resistance is larger than its open-loop counterpart and is dependent on open-loop voltage gain parameters.

Conversely, the corresponding driving-point output resistance in Fig. 24.3(a) is smaller than the open-loop output resistance and approximately inversely proportional to the open-loop voltage gain. These assertions derive from the facts that $Q_s(R_S, 0)$ is the $R_L = 0$ value of the open-loop voltage gain $G_{vo}(R_S, R_L)$. Since $R_L = 0$ corresponds to a short-circuited load resistance, $G_{vo}(R_S, 0) = 0$. In contrast, $Q_s(R_S, \infty)$ is the $R_L = \infty$ value, $G_{vo}(R_S, \infty)$, of the open-loop gain, which is at least as large as $G_{vo}(R_S, R_L)$. By (24.6),

$$R_{\text{out}} = \frac{R_{\text{outo}}}{1 + PG_{vo}(R_S, \infty)} \leq \frac{R_{\text{outo}}}{1 + PG_{vo}(R_S, R_L)} \tag{24.10}$$

Similarly, the driving-point input and output resistances of the global current feedback configuration of Fig. 24.4(a) are sensitive to open-loop gain parameters. In contrast to the voltage amplifier of Fig. 24.3(a), the closed-loop driving-point input resistance of the current amplifier is smaller than its open-loop value, while the driving-point output resistance is larger than its open-loop counterpart. Noting that the open-loop current gain $G_{io}(R_S, R_L)$ is zero for both $R_S = 0$ (which short circuits the input port), and $R_L = \infty$ (which open circuits the load port), (24.5) and (24.6) give

$$R_{\text{in}} = \frac{R_{\text{ino}}}{1 + PG_{io}(\infty, R_L)} \tag{24.11}$$

$$R_{\text{out}} = R_{\text{outo}}[1 + PG_{io}(R_S, 0)] \tag{24.12}$$

Diminished Closed-Loop Damping Factor

In addition to illuminating the driving-point and forward transfer characteristics of single-loop feedback architectures, the special case of global single-loop feedback illustrates the potential instability problems pervasive of almost all feedback circuits. An examination of these problems begins by returning to (24.1) and letting the open-loop gain, G_o, be replaced by the two-pole frequency-domain function,

$$G_o(s) = \frac{G_o(0)}{\left(1 + \dfrac{s}{p_1}\right)\left(1 + \dfrac{s}{p_2}\right)} \tag{24.13}$$

where $G_o(0)$ symbolizes the zero-frequency open-loop gain. The pole frequencies p_1 and p_2 in (24.13) are either real numbers or complex conjugate pairs. Alternatively, (24.13) is expressible as

$$G_o(s) = \frac{G_o(0)}{1 + \dfrac{2\zeta_{\text{ol}}}{\omega_{\text{nol}}}s + \dfrac{s^2}{\omega_{\text{nol}}^2}} \tag{24.14}$$

where

$$\omega_{\text{nol}} = \sqrt{p_1 p_2} \tag{24.15}$$

represents the *undamped natural frequency* of oscillation of the open-loop configuration, and

$$\zeta_{\text{ol}} = \frac{1}{2}\left[\sqrt{\frac{p_2}{p_1}} + \sqrt{\frac{p_1}{p_2}}\right] \tag{24.16}$$

is the *damping factor* of the open-loop circuit.

In (24.1), let the feedback factor f be the single left-half-plane zero function,

$$f(s) = f_o\left(1 + \frac{s}{z}\right) \tag{24.17}$$

where z is the frequency of the real zero introduced by feedback, and f_o is the zero-frequency value of the feedback factor. The resultant loop gain is

$$T(s) = f_o\left(1 + \frac{s}{z}\right)G_o(s) \tag{24.18}$$

the zero-frequency value of the loop gain is

$$T(0) = f_o G_o(0) \tag{24.19}$$

and the zero frequency closed-loop gain $G_{cl}(0)$, is

$$G_{cl}(0) = \frac{G_o(0)}{1 + f_o G_o(0)} = \frac{G_o(0)}{1 + T(0)} \tag{24.20}$$

Upon inserting (24.14) and (24.17) into (24.1), the closed-loop transfer function is determined to be

$$G_{cl}(s) = \frac{G_{cl}(0)}{1 + \frac{2\zeta_{cl}}{\omega_{ncl}}s + \frac{s^2}{\omega_{ncl}^2}} \tag{24.21}$$

where the closed-loop undamped natural frequency of oscillation ω_{ncl} relates to its open-loop counterpart ω_{nol}, in accordance with

$$\omega_{ncl} = \omega_{nol}\sqrt{1 + T(0)} \tag{24.22}$$

Moreover, the closed-loop damping factor ζ_{cl} is

$$\zeta_{cl} = \frac{\zeta_{ol}}{\sqrt{1 + T(0)}} + \left[\frac{T(0)}{1 + T(0)}\right]\frac{\omega_{ncl}}{2z} = \frac{\zeta_{ol}}{\sqrt{1 + T(0)}} + \left[\frac{T(0)}{\sqrt{1 + T(0)}}\right]\frac{\omega_{nol}}{2z} \tag{24.23}$$

A frequency invariant feedback factor $f(s)$ applied to the open-loop configuration whose transfer function is given by (24.13) implies an infinitely large frequency, z, of the feedback zero. For this case, (24.23) confirms a closed-loop damping factor that is always less than the open-loop damping factor. Indeed, for a smaller than unity open-loop damping factor (which corresponds to complex conjugate open-loop poles) and reasonable values of the zero-frequency loop gain $T(0)$, $\zeta_{cl} \ll 1$. Thus, constant feedback applied around an underdamped two-pole open-loop amplifier yields a severely underdamped closed-loop configuration. It follows that the closed-loop circuit has a transient step response plagued by overshoot and a frequency response that displays response peaking within the closed-loop passband. Observe that underdamping is likely even in critically damped (identical real open-loop poles) or overdamped (distinct real poles) open-loop amplifiers, which, respectively, correspond to $\zeta_{ol} = 1$ and $\zeta_{ol} > 1$, when a large zero-frequency loop gain is exploited.

Underdamped closed-loop amplifiers are not unstable systems, but they are nonetheless unacceptable. From a practical design perspective, closed-loop underdamping predicted by relatively simple mathematical models of the loop gain portend undesirable amplifier responses or even closed-loop instability. The problem is that simple transfer function models invoked in a manual circuit analysis are oblivious to presumably second-order parasitic circuit layout and device model energy storage elements whose effects include a deterioration of phase and gain margins.

Frequency Invariant Feedback Factor

Let the open-loop amplifier be overdamped, such that its real poles satisfy the relationship

$$p_2 = \kappa^2 p_1 \tag{24.24}$$

If the open-loop amplifier pole p_1 is dominant, κ^2 is a real number that is greater than the magnitude, $|G_o(0)|$, of the open-loop zero frequency gain, which is presumed to be much larger than one. The open-loop damping factor in (24.16) resultantly reduces to $\zeta_{ol} \approx \kappa/2$. With

$\kappa^2 > |G_o(0)| \gg 1$, which formally reflects the *dominant pole approximation*, the 3-dB bandwidth B_{ol} of the open-loop amplifier is given approximately by [15]

$$B_{ol} \approx \frac{\omega_{nol}}{2\zeta_{ol}} = \frac{1}{\dfrac{1}{p_1} + \dfrac{1}{p_2}} = \left(\frac{\kappa^2}{\kappa^2 + 1}\right) p_1 \tag{24.25}$$

As expected, (24.25) predicts an open-loop 3-dB bandwidth that is only slightly smaller than the frequency of the open-loop dominant pole.

The frequency, z, in (24.23) is infinitely large if frequency invariant degenerative feedback is applied around an open-loop amplifier. For a critically damped or overdamped closed-loop amplifier, $\zeta_{cl} > 1$. Assuming open-loop pole dominance, this constraint imposes the open-loop pole requirement,

$$\frac{p_2}{p_1} \geq 4[1 + T(0)] \tag{24.26}$$

Thus, for large zero-frequency loop gain, $T(0)$, an underdamped closed-loop response is avoided if and only if the frequency of the nondominant open-loop pole is substantially larger than that of the dominant open-loop pole. Unless frequency compensation measures are exploited in the open loop, (24.26) is difficult to satisfy, especially if feedback is implemented expressly to realize a substantive desensitization of response with respect to open-loop parameters. On the chance that (24.26) can be satisfied, and if the closed-loop amplifier emulates a dominant pole response, the closed-loop bandwidth is, using (24.22), (24.23), and (24.25),

$$B_{cl} \approx \frac{\omega_{ncl}}{2\zeta_{cl}} \approx [1 + T(0)]B_{ol} \approx [1 + T(0)]p_1 \tag{24.27}$$

Observe from (24.27) and (24.26) that the maximum possible closed-loop 3-dB bandwidth is 2 octaves below the minimum acceptable frequency of the nondominant open-loop pole.

Although (24.27) theoretically confirms the broadbanding property of negative feedback amplifiers, the attainment of very large closed-loop 3-dB bandwidths is nevertheless a challenging undertaking. The problem is that (24.26) is rarely satisfied. As a result, the open-loop configuration must be suitably compensated, usually by pole splitting methodology [16–18], to force the validity of (24.26). But since the open-loop poles are not mutually independent, any compensation that increases p_2 is accompanied by decreases in p_1. The pragmatic upshot of the matter is that the closed-loop 3-dB bandwidth is not directly proportional to the uncompensated value of p_1 but rather, it is proportional to the smaller, compensated value of p_1.

Frequency Variant Feedback Factor (Compensation)

Consider now the case where the frequency, z, of the compensating feedback zero is finite and positive. Equation (24.23) underscores the stabilizing property of a left-half-plane feedback zero in that a sufficiently small positive z renders a closed-loop damping factor ζ_{cl} that can be made acceptably large, regardless of the value of the open-loop damping factor ζ_{ol}. To this end, $\zeta_{cl} > 1/\sqrt{2}$ is a desirable design objective in that it ensures a monotonically decreasing closed-loop frequency response. If, as is usually a design goal, the open-loop amplifier subscribes to pole dominance, (24.23) translates the objective, $\zeta_{cl} > 1/\sqrt{2}$, into the design constraint

$$z \leq \frac{\left[\dfrac{T(0)}{1 + T(0)}\right]\omega_{ncl}}{\sqrt{2} - \dfrac{\omega_{ncl}}{[1 + T(0)]B_{ol}}} \tag{24.28}$$

where use is made of (24.25) to cast ζ in terms of the open-loop bandwidth B_{ol}. When the closed-loop damping factor is precisely equal to $1/\sqrt{2}$, a maximally flat magnitude closed-loop response results for which the 3-dB bandwidth is ω_{ncl}. Equation (24.28) can then be cast into the more useful form

$$zG_{cl}(0) = \frac{\text{GBP}_{ol}}{\sqrt{2}\left(\frac{\text{GBP}_{ol}}{\text{GBP}_{cl}}\right) - 1} \tag{24.29}$$

where (24.20) is exploited, GBP_{ol} is the gain-bandwidth product of the open-loop circuit, and GBP_{cl} is the gain-bandwidth product of the resultant closed-loop network.

For a given open-loop gain-bandwidth product GPB_{ol}, a desired low-frequency closed-loop gain, $G_{cl}(0)$, and a desired closed-loop gain-bandwidth product, GBP_{cl}, (24.29), provides a first-order estimate of the requisite feedback compensation zero. Additionally, note that (24.29) imposes an upper limit on the achievable high-frequency performance of the closed-loop configuration. In particular, since z must be positive to ensure acceptable closed-loop damping, (24.29) implies

$$\text{GBP}_{ol} > \frac{\text{GBP}_{cl}}{\sqrt{2}} \tag{24.30}$$

In effect, (24.30) imposes a lower limit on the required open-loop gain-bandwidth product commensurate with feedback compensation implemented to achieve a maximally flat closed-loop frequency response.

4.5 Pole Splitting Open-Loop Compensation

Equation (24.26) underscores the desirability of achieving an open-loop dominant pole frequency response in the design of a feedback network. In particular, (24.26) shows that if the ultimate design goal is a closed-loop dominant pole frequency response, the frequency, p_2, of the nondominant open-loop amplifier pole must be substantially larger than its dominant pole counterpart, p_1. Even if closed-loop pole dominance is sacrificed as a trade-off for other performance merits, open-loop pole dominance is nonetheless a laudable design objective. This contention follows from (24.23) and (24.16), which combine to suggest that the larger p_2 is in comparison to p_1, the larger is the open-loop damping factor. In turn, the unacceptably underdamped closed-loop responses that are indicative of small closed-loop damping factors are thereby eliminated. Moreover, (24.23) indicates that larger open-loop damping factors impose progressively less demanding restrictions on the feedback compensation zero that may be required to achieve acceptable closed-loop damping. This observation is important because in an actual circuit design setting, small z in (24.23) generally translates into a requirement of a correspondingly large RC time constant, whose implementation may prove difficult in monolithic circuit applications.

Unfortunately, many amplifiers, and particularly broadbanded amplifiers, earmarked for use as open-loop cells in degenerative feedback networks, are not characterized by dominant pole frequency responses. The frequency response of these amplifiers is therefore optimized in accordance with a standard design practice known as **pole splitting compensation**. Such compensation entails the connection of a small capacitor between two high impedance, phase inverting nodes of the open-loop topology [17, 19–21]. Pole splitting techniques increase the frequency p_2 of the uncompensated nondominant open-loop pole to a compensated value, say p_{2c}. The frequency, p_1, of the uncompensated dominant open-loop pole is simultaneously reduced to a smaller frequency, say p_{1c}. Although these pole frequency translations complement the design requirement implicit to (24.26) and (24.23), they do serve to limit the resultant closed-loop bandwidth, as discussed earlier. And, as highlighted below, they also impose other performance limitations on the open loop.

The Open-Loop Amplifier

The engineering methods, associated mathematics, and engineering trade-offs underlying pole splitting compensation are best revealed in terms of the generalized, phase inverting linear network abstracted in Fig. 24.5. Although this amplifier may comprise the entire open-loop configuration, in the most general case, it is an interstage of the open loop. Accordingly, R_{st} in this diagram is viewed as the Thévenin equivalent resistance of either an input signal source or a preceding amplification stage. The response to the Thévenin driver, V_{st}, is the indicated output voltage, V_l, which is developed across the Thévenin load resistance, R_{lt}, seen by the stage under investigation. Note that the input current conducted by the amplifier is I_s, while the current flowing into the output port of the unit is denoted as I_l. The dashed branch containing the capacitor C_c, which is addressed later, is the pole splitting compensation element.

Since the amplifier under consideration is linear, any convenient set of two-port parameters can be used to model its terminal volt–ampere characteristics. Assuming the existence of the short circuit admittance, or y parameters,

$$\begin{bmatrix} I_s \\ I_l \end{bmatrix} = \begin{bmatrix} y_{11} & y_{12} \\ y_{21} & y_{2} \end{bmatrix} \begin{bmatrix} V_i \\ V_l \end{bmatrix} \tag{24.31}$$

Defining

$$\begin{aligned} y_i &\triangleq y_{11} + y_{12} \\ y_o &\triangleq y_{22} + y_{12} \\ y_f &\triangleq y_{21} - y_{12} \\ y_r &\triangleq -y_{12} \end{aligned} \tag{24.32}$$

(24.31) implies

$$I_s = y_i V_i + y_r (V_i - V_l) \tag{24.33}$$

$$I_l = y_f V_i + y_o V_l + y_r (V_l - V_i) \tag{24.34}$$

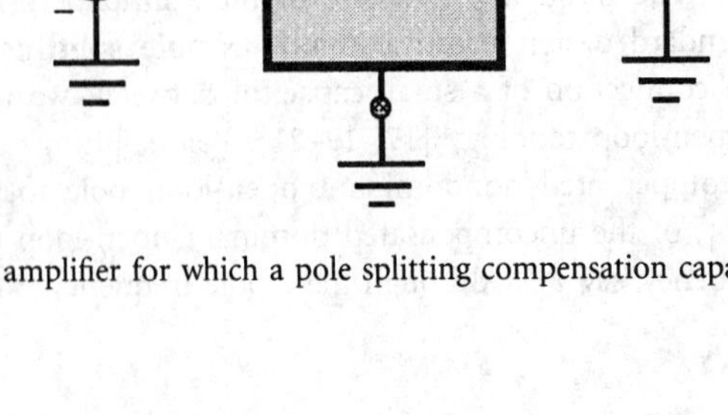

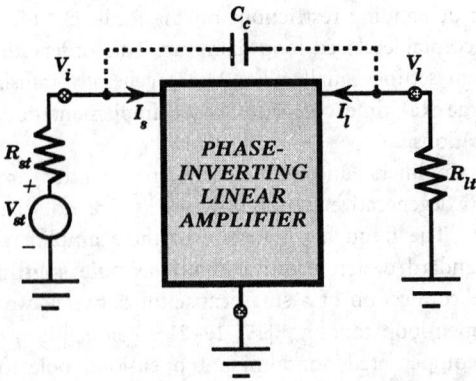

FIGURE 24.5 A linear amplifier for which a pole splitting compensation capacitance C_c is incorporated.

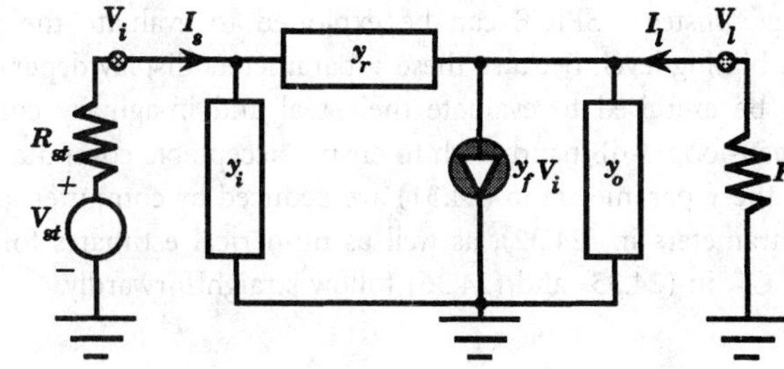

(a)

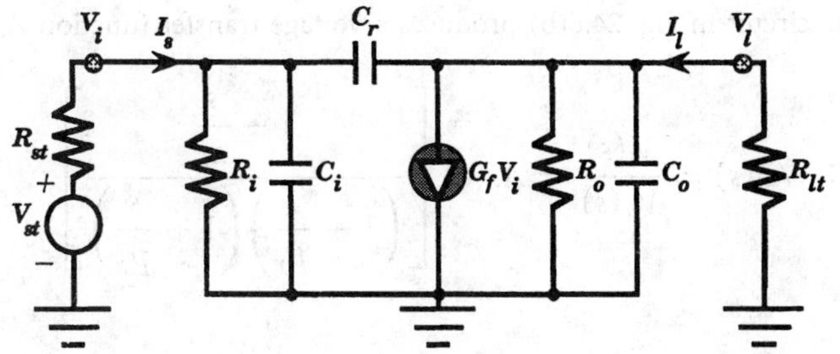

(b)

FIGURE 24.6 (a) The y-parameter equivalent circuit of the phase-inverting linear amplifier in Fig. 24.5. (b) An approximate form of the model in (a).

The last two expressions produce the y-parameter model depicted in Fig. 23.6(a), in which y_i represents an effective shunt input admittance, y_o is a shunt output admittance, y_f is a forward transadmittance, and y_r reflects voltage feedback intrinsic to the amplifier.

Amplifiers amenable to pole splitting compensation have capacitive input and output admittances; that is, y_i and y_o are of the form

$$y_i = \frac{1}{R_i} + sC_i$$
$$y_o = \frac{1}{R_o} + sC_o \tag{24.35}$$

Similarly,

$$y_f = G_f - sC_f$$
$$y_r = \frac{1}{R_r} + sC_r \tag{24.36}$$

In (24.36), the conductance component G_f of the forward transadmittance y_f is positive in a phase-inverting amplifier. Moreover, the reactive component $-sC_f$ of y_f produces an *excess phase angle*, and hence, a *group delay*, in the forward gain function. This component, which deteriorates phase margin, can be ignored to first order if the signal frequencies of interest are not excessive in comparison to the upper-frequency limit of performance of the amplifier. Finally, the feedback internal to many practical amplifiers is predominantly capacitive so that the feedback resistance R_r can be ignored. These approximations allow the model in Fig. 23.6(a) to be drawn in the form offered in Fig. 24.6(b).

It is worthwhile interjecting that the six parameters indigenous to the model in Fig. 24.6(b) need not be deduced analytically from the small-signal models of the active elements embedded in the subject interstage. Instead, SPICE can be exploited to evaluate the y parameters in (24.31) at the pertinent biasing level. Because these y parameters display dependencies on signal frequency, care should be exercised to evaluate their real and imaginary components in the neighborhood of the open-loop 3-dB bandwidth to ensure acceptable computational accuracy at high frequencies. Once the y parameters in (24.31) are deduced by computer-aided analysis, the alternate admittance parameters in (24.32), as well as numerical estimates for the parameters, R_i, C_i, R_o, C_o, C_r, and G_f, in (24.35) and (24.36) follow straightforwardly.

Pole Splitting Analysis

An analysis of the circuit in Fig. 24.6(b) produces a voltage transfer function $A_v(s)$ of the form

$$A_v(s) = \frac{V_l(s)}{V_{st}(s)} = A_v(0)\left[\frac{1 - \dfrac{s}{z_r}}{\left(1 + \dfrac{s}{p_1}\right)\left(1 + \dfrac{s}{p_2}\right)}\right] \tag{24.37}$$

Letting

$$R_{ll} = R_{lt} \| R_o \tag{24.38}$$

an inspection of the circuit in Fig. 24.6(b) confirms that

$$A_v(0) = -G_f R_{ll}\left(\frac{R_i}{R_i + R_{st}}\right) \tag{24.39}$$

is the zero frequency voltage gain. Moreover, the frequency, z_r, of the right-half-plane zero is

$$z_r = \frac{G_f}{C_r} \tag{24.40}$$

The lower pole frequency, p_1, and the higher pole frequency, p_2, derive implicitly from

$$\frac{1}{p_1} + \frac{1}{p_2} = R_{ll}(C_o + C_r) + R_{ss}[C_i + (1 + G_f R_{ll})C_r] \tag{24.41}$$

and

$$\frac{1}{p_1 p_2} = R_{ss} R_{ll} C_o\left[C_i + \left(\frac{C_o + C_i}{C_o}\right)C_r\right] \tag{24.42}$$

where

$$R_{ss} = R_{st} \,\|\, R_i \tag{24.43}$$

Most practical amplifiers, and particularly amplifiers realized in bipolar junction transistor technology, have very large forward transconductance, G_f, and small internal feedback capacitance, C_r. The combination of large G_f and small C_r renders the frequency in (24.40) so large as to be inconsequential to the passband of interest. When utilized in a high-gain application, such as the open-loop signal path of a feedback amplifier, these amplifiers also operate with a large effective load resistance, R_{ll}. Accordingly, (24.41) can be used to approximate the pole frequency p_1 as

$$p_1 \approx \frac{1}{R_{ss}[C_i + (1 + G_f R_{ll})C_r]} \tag{24.44}$$

Substituting this result into (24.42), the approximate frequency p_2 of the high-frequency pole is

$$p_2 \approx \frac{C_i + (1 + G_f R_{ll})C_r}{R_{ll}C_o\left[C_i + \left(\frac{C_o + C_i}{C_o}\right)C_r\right]} \tag{24.45}$$

Figure 24.7 illustrates asymptotic frequency responses corresponding to pole dominance and to a two-pole response. Figure 24.7(a) depicts the frequency response of a dominant pole amplifier, which does not require pole splitting compensation. Observe that its high-frequency response is determined by a single pole (p_1 in this case) through the signal frequency at which the gain ultimately degrades to unity. In this interpretation of a dominant pole amplifier, p_2 is not only much larger than p_1, but is in fact, larger than the unity gain frequency, which is indicated as ω_u in the figure. This unity gain frequency, which can be viewed as an upper limit to the useful passband of the amplifier, is approximately, $|A_v(0)|p_1$. To the extent that p_1 is essentially the 3-dB bandwidth when $p_2 \gg p_1$, the unity gain frequency is also the gain-bandwidth product, GBP, of the subject amplifier. In short, with $|A_v(j\omega_u)| \triangleq 1, p_2 \gg p_1$ in (24.37) implies

$$\omega_u \approx |A_v(0)|p_1 \approx \text{GBP} \tag{24.46}$$

The contrasting situation of a response indigenous to the presence of two significant open loop poles is illustrated in Fig. 24.7(b). In this case, the higher pole frequency p_2 is smaller than ω_u and hence, the amplifier does not emulate a single-pole response throughout its theoretically useful frequency range. The two critical frequencies, p_1 and p_2, remain real numbers, and as long as $p_2 \neq p_1$, the corresponding damping factor, is greater than one. But the damping factor of the two pole amplifier whose response is plotted in Fig. 24.7(b) is nonetheless smaller than that of the dominant pole amplifier. It follows that for reasonable loop gains, unacceptable underdamping is more likely when feedback is invoked around the two-pole amplifier, as opposed to the same amount of feedback applied around a dominant pole amplifier. Pole splitting attempts to circumvent this problem by transforming the pole conglomeration of the two pole amplifier into one that emulates the dominant pole situation inferred by Fig. 24.7(a).

To the foregoing end, append the compensation capacitance C_c between the input and the output ports of the phase-inverting linear amplifier, as suggested in Fig. 24.5. With reference to the equivalent circuit in Fig. 24.6(b), the electrical impact of this additional element is the effective replacement of the internal feedback capacitance C_r by the capacitance sum $(C_r + C_c)$. Letting

$$C_p \triangleq C_r + C_c \tag{24.47}$$

it is apparent that (24.40)–(24.42) remain applicable, provided that C_r in these relationships is supplanted by C_p. But since C_p is conceivably significantly larger than C_c, the approximate expressions for the resultant pole locations differ from those of (24.44) and (24.45). In particular,

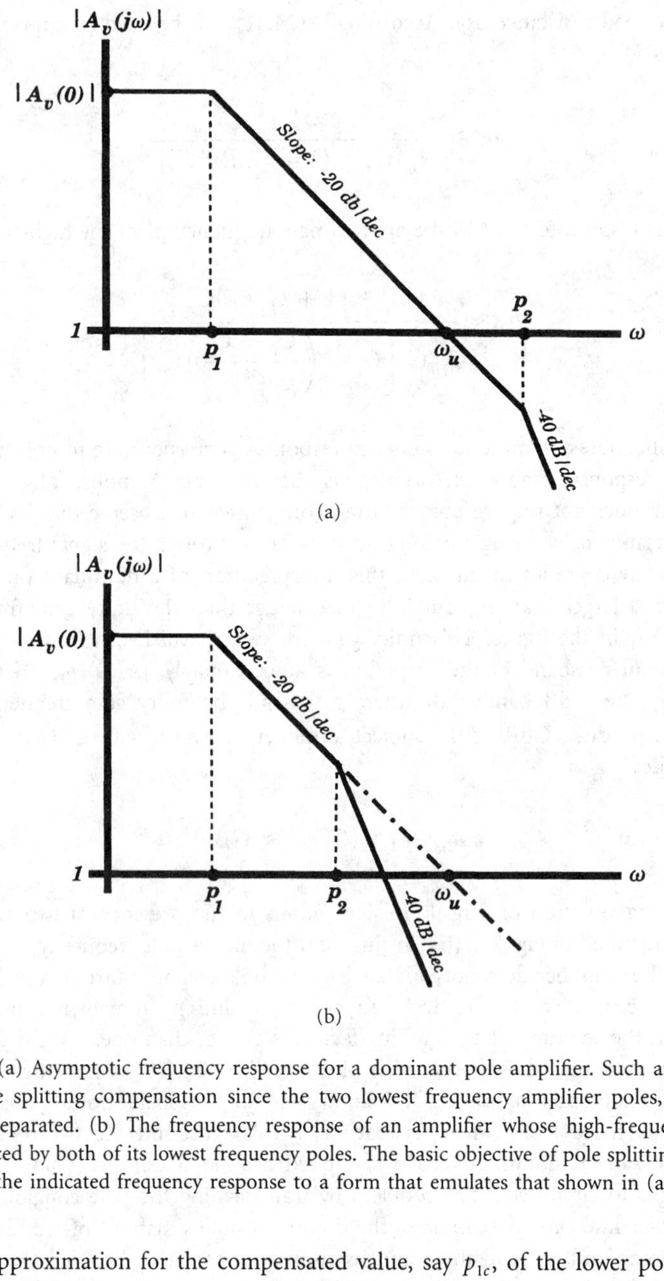

FIGURE 24.7 (a) Asymptotic frequency response for a dominant pole amplifier. Such an amplifier does not require pole splitting compensation since the two lowest frequency amplifier poles, p_1 and p_2, are already widely separated. (b) The frequency response of an amplifier whose high-frequency response is strongly influenced by both of its lowest frequency poles. The basic objective of pole splitting compensation is to transform the indicated frequency response to a form that emulates that shown in (a).

a reasonable approximation for the compensated value, say p_{1c}, of the lower pole frequency is now

$$p_{1c} \approx \frac{1}{[R_{ll} + (1 + G_f R_{ll})R_{ss}]C_p} \tag{24.48}$$

while the higher pole frequency, p_{2c}, becomes

$$p_{2c} \approx \frac{1}{\left(R_{ss} \| R_{ll} \| \dfrac{1}{G_f}\right)(C_o + C_i)} \tag{24.49}$$

Clearly, $p_{1c} < p_1$, and $p_{2c} > p_2$. Moreover, for large G_f, p_{2c} is potentially much larger than p_{1c}. It should also be noted that the compensated value, say, z_{rc}, of the right-half-plane zero is smaller than its uncompensated value, z_r, since (24.40) shows that

$$z_{rc} = \frac{G_f}{C_p} = z_r\left(\frac{C_r}{C_r + C_c}\right) \tag{24.50}$$

Although z_{rc} can conceivably exert a significant influence on the high-frequency response of the compensated amplifier, the following discussion presumes tacitly that $z_{rc} > p_{2c}$ [2].

Assuming a dominant pole frequency response, the compensated unity gain frequency, ω_{uc} is, using (24.39), (24.46), and (24.48),

$$\omega_{uc} \approx |A_v(0)|\, p_{1c} \approx \left(\frac{1}{R_{st}C_p}\right)\left[G_f\left(R_{ss}\|R_{ll}\|\frac{1}{G_f}\right)\right] \tag{24.51}$$

It is interesting to note that

$$\omega_{uc} < \left(\frac{1}{R_{st}C_p}\right) \tag{24.52}$$

that is, the unity gain frequency is limited by the inverse of the RC time constant formed by the Thévenin source resistance R_{st} and the net capacitance C_p appearing between the input port and the phase inverted output port. The subject inequality comprises a significant performance limitation, for if p_{2c} is indeed much larger than p_{ic}, ω_{uc} is approximately the gain-bandwidth product of the compensated cell. Accordingly, for a given source resistance, a required open-loop gain, and a desired open-loop bandwidth, (24.52) imposes an upper limit on the compensation capacitance that can be exploited for pole splitting purposes.

In order for the compensated amplifier to behave as a dominant pole configuration, p_{2c} must exceed ω_{uc}, as defined by (24.51). Recalling (24.49), the requisite constraint is found to be

$$R_{st}C_p > G_f\left(R_{ss}\|R_{ll}\|\frac{1}{G_f}\right)^2(C_o + C_i) \tag{24.53}$$

Assuming $G_f(R_{ss}/R_{ll}) \ll 1$, (24.53) reduces to the useful simple form,

$$G_f R_{st} > \frac{C_o + C_i}{C_p} \tag{24.54}$$

which confirms the need for large forward transconductance G_f if pole splitting is to be an effective compensation technique.

24.6 Summary

The use of negative feedback is fundamental to the design of reliable and reproducible analog electronic networks. Accordingly, this paper documents the salient features of the theory that underlies the efficient analysis and design of commonly used feedback networks. Four especially significant points are postulated in this section.

1. By judiciously exploiting signal flow theory, the classical expression, (24.1), for the I/O transfer relationship of a linear feedback system is rendered applicable to a broad range of electronic feedback circuits. This expression is convenient for design-oriented analysis because it clearly identifies the open-loop gain, G_o, and the loop gain, T. The successful application of signal flow theory is predicated on the requirement that the feedback factor, to which T is proportional and that appears in the signal flow literature as a "critical" or "reference" parameter, can be identified in a given feedback circuit.
2. Signal flow theory, as applied to electronic feedback architectures, proves to be an especially expedient analytical tool because once the loop gain T is identified, the driving-point input and output impedances follow with minimal additional calculations. Moreover, the functional dependence of T on the Thévenin source and terminating load impedances unambiguously brackets the magnitudes of the driving point I/O impedances attainable in particular types of feedback arrangements.
3. The damping factor concept is advanced herewith as a simple way of assessing the relative stability of both the open and closed loops of a feedback circuit. The open-loop damping factor derives directly from the critical frequencies of the open-loop gain, while these frequencies and any zeros appearing in the loop gain unambiguously define the corresponding closed-loop damping factor. Signal flow theory is once again used to confirm the propensity of closed loops toward instability unless the open-loop subcircuit functions as a dominant pole network. Also confirmed is the propriety of the common practice of implementing a feedback zero as a means of stabilizing an otherwise potentially unstable closed loop.
4. Pole splitting as a means to achieve dominant pole open-loop responses is definitively discussed. Generalized design criteria are formulated for this compensation scheme, and limits of performance are established. Of particular interest is the fact that pole splitting limits the gain-bandwidth product of the compensated amplifier to a value that is determined by a source resistance-compensation capacitance time constant.

References

[1] J. A. Mataya, G. W. Haines, and S. B. Marshall, "IF amplifier using C_c-compensated transistors," *IEEE J. Solid-State Circuits*, vol. SC-3, pp. 401–407, Dec. 1968.

[2] W. G. Beall and J. Choma, Jr., "Charge-neutralized differential amplifiers," *J. Analog Integrat. Circuits Signal Process.*, vol. 1, pp. 33–44, Sept. 1991.

[3] J. Choma, Jr., "A generalized bandwidth estimation theory for feedback amplifiers," *IEEE Trans. Circuits Syst.*, vol. CAS-31, pp. 861–865, Oct. 1984.

[4] R. D. Thornton, C. L. Searle, D. O. Pederson, R. B. Adler, and E. J. Angelo, Jr., *Multistage Transistor Circuits*, New York: Wiley, 1965, chs. 1 and 8.

[5] H. W. Bode, *Network Analysis and Feedback Amplifier Design*, New York: Van Nostrand, 1945.

[6] P. J. Hurst, "A comparison of two approaches to feedback circuit analysis," *IEEE Trans. Education*, vol. 35, pp. 253–261, Aug. 1992.

[7] M. S. Ghausi, *Principles and Design of Linear Active Networks*, New York: McGraw-Hill, 1965, pp. 40–56.

[8] A. J. Cote, Jr. and J. B. Oakes, *Linear Vacuum-Tube And Transistor Circuits*, New York: McGraw-Hill, 1961, pp. 40–46.

[9] S. J. Mason, "Feedback theory—Some properties of signal flow graphs," *Proc. IRE*, vol. 41, pp. 1144–1156, Sept. 1953.

[10] S. J. Mason, "Feedback theory—Further properties of signal flow graphs," *Proc. IRE*, vol. 44, pp. 920–926, July 1956.

[11] N. Balabanian and T. A. Bickart, *Electrical Network Theory*, New York: Wiley, 1969, pp. 639–669.

[12] J. Choma, Jr., "Signal flow analysis of feedback networks," *IEEE Trans. Circuits Syst.*, vol. 37, pp. 455–463, Apr. 1990.

[13] J. Choma, Jr., *Electrical Networks: Theory and Analysis*, New York: Wiley-Interscience, 1985, pp. 589–605.

[14] P. J. Hurst, "Exact simulation of feedback circuit parameters, *IEEE Trans. Circuits Syst.*, vol. 38, pp. 1382–1389, Nov. 1991.

[15] J. Choma, Jr. and S. A. Witherspoon, "Computationally efficient estimation of frequency response and driving point impedance in wideband analog amplifiers," *IEEE Trans. Circuits Syst.*, vol. 37, pp. 720–728, June 1990.

[16] R. G. Meyer and R. A. Blauschild, "A wide-band low-noise monolithic transimpedance amplifier," *IEEE J. Solid-State Circuits*, vol. SC-21, pp. 530–533, Aug. 1986.

[17] Y. P. Tsividis, "Design considerations in single-channel MOS analog integrated circuits," *IEEE J. Solid-State Circuits*, vol. SC-13, pp. 383–391, June 1978.

[18] J. J. D'Azzo and C. H. Houpis, *Feedback Control System Analysis and Synthesis*, New York: McGraw-Hill, 1960, pp. 230–234.

[19] P. R. Gray and R. G. Meyer, *Analysis and Design of Analog Integrated Circuits*, New York: Wiley, 1977, pp. 512–521.

[20] P. R. Gray, "Basic MOS operational amplifier design—An overview," in *Analog MOS Integrated Circuits*, P. R. Gray, D. A. Hodges, and R. W. Brodersen, Eds., New York: IEEE, 1980, pp. 28–49.

[21] J. E. Solomon, "The monolithic op-amp: A tutorial study," *IEEE J. Solid-State Circuits*, vol. SC-9, pp. 314–332, Dec. 1974.

25

Feedback Amplifier Configurations

John Choma, Jr.
University of Southern California

25.1 Introduction

There are four basic types of single-loop feedback amplifiers: the **series-shunt, shunt-series, shunt-hunt**, and **series-series** architectures [1]. Each of these cells is capable of a significant reduction of the dependence of forward transfer characteristics on the ill-defined or ill-controlled parameters implicit to the open-loop gain. But none of these architectures can simultaneously offer controlled driving-point input and output impedances. Such additional control is afforded only by dual global loops comprised of series and/or shunt feedback signal paths appended to an open-loop amplifier [2], [3]. There are only two types of global dual-loop feedback architectures: the **series-series/shunt-shunt feedback amplifier** and the **series-shunt/shunt-series feedback amplifier**.

Although only bipolar technology is exploited in the analysis of the aforementioned four single-loop and two dual-loop feedback cells, all disclosures are generally applicable to metal–oxide–silicon (MOS), heterostructure bipolar transistor (HBT), and III–V compound metal–semiconductor field effect transistor (MESFET) technologies. All analytical results derive from an application of a hybrid, signal flow/two-port parameter analytical tack. Since the thought processes underlying this technical approach apply to all feedback circuits, the subject analytical procedure is developed in detail for only the series-shunt feedback amplifier.

25.2 Series-Shunt Feedback Amplifier

Circuit Modeling and Analysis

Figure 25.1(a) depicts the **ac schematic diagram** (a circuit diagram divorced of biasing details) of a series-shunt feedback amplifier. In this circuit, the output voltage V_O, which is established in

0-8493-8341-2/95/$0.00 + $.50

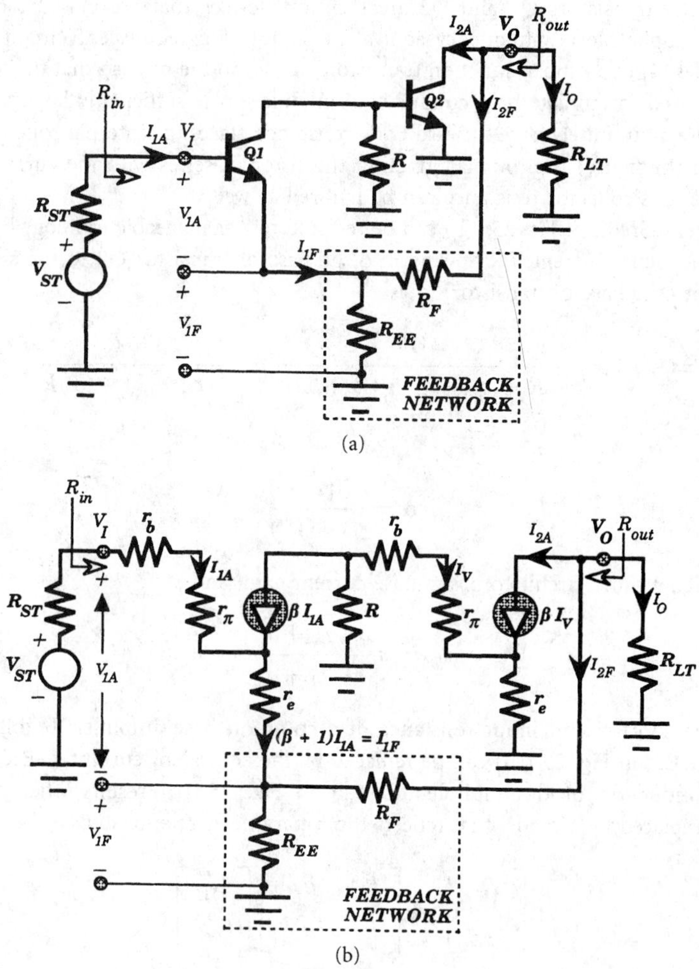

FIGURE 25.1 (a) The ac schematic diagram of a bipolar series-shunt feedback amplifier. (b) Low-frequency small-signal equivalent circuit of the feedback amplifier.

response to a signal source represented by the Thévenin voltage V_{ST}, and the Thévenin resistance, R_{ST}, is sampled by the feedback network composed of the resistances, R_{EE} and R_F. The sampled voltage is fed back in such a way that the closed-loop input voltage, V_I, is the sum of the voltage, V_{1A}, across the input port of the amplifier and the voltage V_{1F}, developed across R_{EE} in the feedback subcircuit. Since $V_I = V_{1A} + V_{1F}$, the output port of the feedback configuration can be viewed as connected in series with the amplifier input port. On the other hand, output voltage sampling constrains the net load current, I_O, to be the algebraic sum of the amplifier output port current, I_{2A}, and the feedback network input current, I_{2F}. Accordingly, the output topology is indicative of a shunt connection between the feedback subcircuit and the amplifier output port. The fact that voltage is fed back to a voltage-driven input port renders the driving point input resistance, R_{in}, of the closed-loop amplifier large, whereas the driving-point output resistance, R_{out}, seen by the terminating load resistance R_{LT}, is small. The resultant closed-loop amplifier is therefore best suited for voltage amplification, in the sense that the closed-loop voltage gain, V_O/V_{ST}, can be made approximately independent of source and load resistances. For large loop gain, this voltage transfer function is also nominally independent of transistor parameters.

Assuming that transistors $Q1$ and $Q2$ are identical devices that are biased identically, Fig. 25.1(b) is the applicable low-frequency equivalent circuit. This equivalent circuit exploits the hybrid-π model [4] of a bipolar junction transistor, subject to the proviso that the forward Early resistance [5] used to emulate base conductivity modulation is sufficiently large to warrant its neglect. Because an infinitely large forward Early resistance places the internal collector resistance (not shown in the figure) of a bipolar junction transistor in series with the current controlled current source, this collector resistance can be ignored as well.

The equivalent circuit of Fig. 25.1(b) can be reduced to a manageable topology by noting that the ratio of the signal current, I_V, flowing into the base of transistor $Q2$ to the signal current, I_{1A}, flowing into the base of transistor $Q1$ is

$$\frac{I_V}{I_{1A}} \triangleq -K_\beta = -\frac{\beta R}{R + r_b + r_\pi + (\beta + 1) r_e} = -\frac{\alpha R}{r_{ib} + (1 - \alpha) R} \tag{25.1}$$

where

$$\alpha = \frac{\beta}{\beta + 1} \tag{25.2}$$

is the small-signal short-circuit common base current gain, and

$$r_{ib} = r_e + \frac{r_\pi + r_b}{\beta + 1} \tag{25.3}$$

symbolizes the short-circuit input resistance of a common base amplifier. It follows that the current source βI_v in Fig. 25.1(b) can be replaced by the equivalent current $(-\beta K_\beta I_{1A})$.

A second reduction of the equivalent circuit in Fig. 25.1(b) results when the feedback subcircuit is replaced by a model that reflects the h-parameter relationships

$$\begin{bmatrix} V_{1F} \\ I_{2F} \end{bmatrix} = \begin{bmatrix} h_{if} & h_{rf} \\ h_{ff} & h_{of} \end{bmatrix} \begin{bmatrix} I_{1F} \\ V_O \end{bmatrix} \tag{25.4}$$

where $V_{1F}(V_O)$ represents the signal voltage developed across the output (input) port of the feedback subcircuit and $I_{1F}(I_{2F})$ symbolizes the corresponding current flowing into the feedback output (input) port. Although any homogeneous set of two-port parameters can be used to model the feedback subcircuit, h parameters are the most convenient selection herewith. In particular, the feedback amplifier undergoing study is a series-shunt configuration. Since the h-parameter equivalent circuit represents its input port as a Thévenin circuit and its output port as a Norton configuration, the h-parameter equivalent circuit is likewise a series-shunt structure.

For the feedback network at hand, which is redrawn for convenience in Fig. 25.2(a), the h-parameter equivalent circuit is as depicted in Fig. 25.2(b). The latter diagram exploits the facts that the short-circuit input resistance h_{if} is the parallel combination of the resistances R_{EE} and R_F, and the open-circuit output conductance h_{of}, is $1/(R_{EE} + R_F)$. The open-circuit reverse voltage gain h_{rf} is

$$h_{rf} = \frac{R_{EE}}{R_{EE} + R_F} \tag{25.5}$$

while the short-circuit forward current gain h_{ff} is

$$h_{ff} = -\frac{R_{EE}}{R_{EE} + R_F} = -h_{rf} \tag{25.6}$$

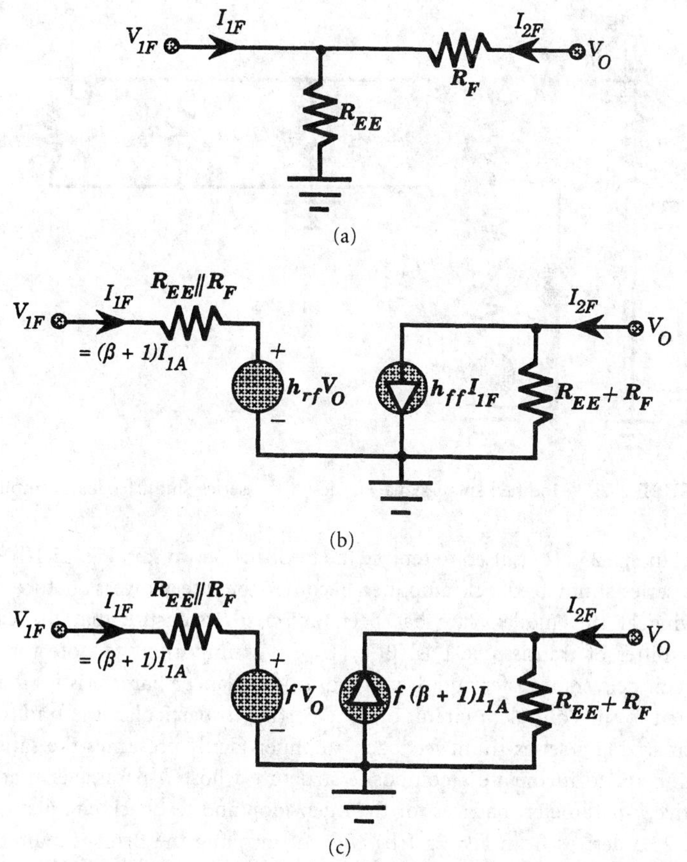

FIGURE 25.2 (a) The feedback subcircuit in the series-shunt feedback amplifier of Fig. 25.1(a). (b) The h-parameter equivalent circuit of the feedback subcircuit. (c) Alternative form of the h-parameter equivalent circuit.

Figure 25.2(c) modifies the equivalent circuit in Fig. 25.2(b) in accordance with the following two arguments. First, h_{rf} in (25.5) is recognized as the fraction of the feedback subcircuit input signal that is fed back as a component of the feedback subcircuit output voltage, V_{1F}. But this subcircuit input voltage is identical to the closed-loop amplifier output signal V_O. Moreover, V_{1F} superimposes with the Thévenin input signal applied to the feedback amplifier to establish the amplifier input port voltage, V_{1A}. It follows that h_{rf} is logically referenced as a feedback factor, say f, of the amplifier under consideration; that is,

$$h_{rf} = \frac{R_{EE}}{R_{EE} + R_F} \stackrel{\Delta}{=} f \tag{25.7}$$

whence by (25.6),

$$h_{ff} = -\frac{R_{EE}}{R_{EE} + R_F} = -f \tag{25.8}$$

Second, the feedback subcircuit output current, I_{1F}, is, as indicated in Fig. 25.1(b), the signal current, $(\beta + 1)I_{1A}$. Thus, in the model of Fig. 25.2(b),

$$h_{ff} I_{1F} = -f(\beta + 1)I_{1A} \tag{25.9}$$

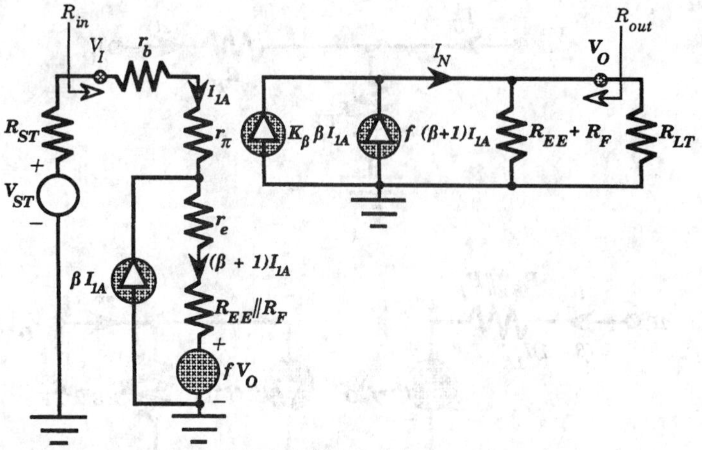

FIGURE 25.3 Modified small-signal model of the series-shunt feedback amplifier.

If the model in Fig. 25.2(c) is used to replace the feedback network in Fig. 25.1(b) the equivalent circuit of the series-shunt feedback amplifier becomes the alternative structure offered in Fig. 25.3. In arriving at this model, care has been exercised to ensure that the current flowing through the emitter of transistor Q1 is $(\beta + 1)I_{1A}$. It is important to note that the modified equivalent circuit delivers transfer and driving point impedance characteristics that are identical to those implicit to the equivalent circuit of Fig. 25.1(b). In particular, the traditional analytical approach to analyzing a series-shunt feedback amplifier tacitly presumes the satisfaction of the Brune condition [6] to formulate a composite structure whose h-parameter matrix is the sum of the respective h-parameter matrices for the open-loop and feedback circuits. In contrast, the model of Fig. 25.3 derives from Fig. 25.1(b) without invoking the Brune requirement, which is often not satisfied. It merely exploits the substitution theorem; that is, the feedback network in Fig. 25.1(b) is substituted by its h-parameter representation.

In addition to modeling accuracy, the equivalent circuit in Fig. 25.3 boasts at least three other advantages. The first is an illumination of the vehicle by which feedback is implemented in the series-shunt configuration. This vehicle is the voltage controlled voltage source, fV_O, which feeds back a fraction of the output signal to produce a branch voltage that algebraically superimposes with, and thus modifies, the applied source voltage effectively seen by the input port of the open-loop amplifier. Thus, with $f = 0$, no feedback is evidenced, and the model at hand emulates an open-loop configuration. But even with $f = 0$, the transfer and driving-point impedance characteristics of the resultant open-loop circuit are functionally dependent on the feedback elements, R_{EE} and R_F, because appending the feedback network to the open-loop amplifier incurs additional impedance loads at both the input and the output ports of the amplifier.

The second advantage of the subject model is its revelation of the magnitude and nature of feed-forward through the closed loop. In particular, note that the signal current, I_N, driven into the effective load resistance comprised of the parallel combination of $(R_{EE} + R_F)$ and R_{LT}, is the sum of two current components. One of these currents, $\beta K_\beta I_{1A}$, materializes from the transfer properties of the two transistors utilized in the amplifier. The other current, $f(\beta + 1)I_{1A}$, is the feed-forward current resulting from the bilateral nature of the passive feedback network. In general, negligible feed-forward through the feedback subcircuit is advantageous, particularly in high-frequency signal-processing applications. To this end, the model in Fig. 25.3 suggests the design requirement,

$$f \ll \alpha K_\beta \tag{25.10}$$

When the resistance, R, in Fig. 25.1(a) is the resistance associated with the output port of a PNP current source used to supply biasing current to the collector of transistor $Q1$ and the base of transistor $Q2$, K_β approaches β, and (25.10) is easily satisfied. But PNP current sources are undesirable in broadband low-noise amplifiers. In these applications, the requisite biasing current must be supplied by a passive resistance, R, connected between the positive supply voltage and the junction of the $Q1$ collector and the $Q2$ base. Unfortunately, the corresponding value of K_β can be considerably smaller than β, with the result that (25.10) may be difficult to satisfy. Circumvention schemes for this situation are addressed later.

A third attribute of the model in Fig. 25.3 is its disposition to an application of signal flow theory. For example, with the feedback factor f selected as the reference parameter for signal flow analysis, the open-loop voltage gain $G_{vo}(R_{ST}, R_{LT})$, of the series-shunt feedback amplifier is computed by setting f to zero. Assuming that (25.10) is satisfied, circuit analysis reveals this gain as

$$G_{vo}(R_{ST}, R_{LT}) = \alpha K_\beta \left[\frac{(R_{EE} + R_F) \| R_{LT}}{r_{ib} + (1-\alpha)R_{ST} + (R_{EE} \| R_F)} \right] \tag{25.11}$$

The corresponding input and output driving point resistances, R_{ino} and R_{outo}, respectively, are

$$R_{ino} = r_B + r_\pi + (\beta + 1)(r_E + R_{EE} \| R_F) \tag{25.12}$$

and

$$R_{outo} = R_{EE} + R_F \tag{25.13}$$

It follows that the closed-loop gain $G_v(R_{ST}, R_{LT})$ of the series-shunt feedback amplifier is

$$G_v(R_{ST}, R_{LT}) = \frac{G_{vo}(R_{ST}, R_{LT})}{1 + T} \tag{25.14}$$

where the loop gain T is

$$\begin{aligned} T = fG_{vo}(R_{ST}, R_{LT}) &= \left(\frac{R_{EE}}{R_{EE} + R_F} \right) G_{vo}(R_{ST}, R_{LT}) \\ &= \alpha K_\beta \left(\frac{R_{EE}}{R_{EE} + R_F + R_{LT}} \right) \left[\frac{R_{LT}}{r_{ib} + (1-\alpha)R_{ST} + (R_{EE} \| R_F)} \right] \end{aligned} \tag{25.15}$$

For $T \gg 1$, which mandates a sufficiently large K_β in (24.11), the closed-loop gain collapses to

$$G_v(R_{ST}, R_{LT}) \approx \frac{1}{f} = 1 + \frac{R_F}{R_{EE}} \tag{25.16}$$

which is independent of active element parameters. Moreover, to the extent that $T \gg 1$, the series-shunt feedback amplifier behaves as an ideal voltage controlled voltage source in the sense that its closed-loop voltage gain is independent of source and load terminations. The fact that the series-shunt feedback network behaves approximately as an ideal voltage amplifier implies that its closed-loop driving point input resistance is very large and its closed-loop driving point output resistance is very small. These facts are confirmed analytically by noting that

$$\begin{aligned} R_{in} &= R_{ino}[1 + fG_{vo}(0, R_L)] \approx fR_{ino}G_{vo}(0, R_L) \\ &= \beta K_\beta \left(\frac{R_{EE}}{R_{EE} + R_F + R_{LT}} \right) R_{LT} \end{aligned} \tag{25.17}$$

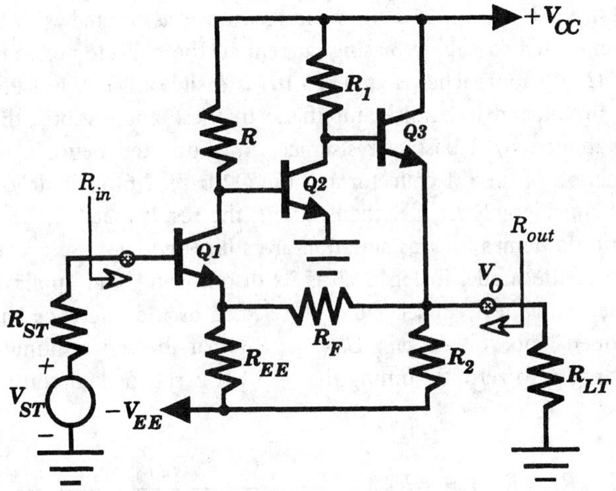

FIGURE 25.4 A series-shunt feedback amplifier that incorporates an emitter follower output stage to reduce the effects of feed-forward through the feedback network.

and

$$R_{out} = \frac{R_{outo}}{1 + fG_{vo}(R_S, \infty)} \approx \frac{R_{outo}}{fG_{vo}(R_S, \infty)}$$
$$= \left(1 + \frac{R_F}{R_{EE}}\right)\left[\frac{r_{ib} + (1-\alpha)R_{ST} + R_{EE}\|R_F}{\alpha K_\beta}\right] \tag{25.18}$$

To the extent that the interstage biasing resistance, R, is sufficiently large to allow K_β to approach β, observe that R_{in} in (25.17) is nominally proportional to β^2, while R_{out} in (25.18) is inversely proportional to β.

Feed-Forward Compensation

When practical design restrictions render the satisfaction of (25.10) difficult, feed-forward problems can be circumvented by inserting an emitter follower between the output port of transistor Q2 in the circuit diagram of Fig. 25.1(a) and the node to which the load termination and the input terminal of the feedback subcircuit are incident [2]. The resultant circuit diagram, inclusive now of simple biasing subcircuits, is shown in Fig. 25.4. The buffer transistor Q3 increases the original short-circuit forward current gain, $K_\beta\beta$, of the open-loop amplifier by a factor approaching $(\beta + 1)$, while not altering the feed-forward factor implied by the feedback network in Fig. 25.1(a). In effect, K_β is increased by a factor of almost $(\beta + 1)$, thereby making (25.10) easy to satisfy. Because of the inherently low output resistance of an emitter follower, the buffer also reduces the driving-point output resistance achievable by the original configuration.

The foregoing contentions can be confirmed through an analysis of the small-signal model for the modified amplifier in Fig. 25.4. Such an analysis is expedited by noting that the circuit to the left of the current controlled current source, $K_\beta\beta I_{1A}$, in Fig. 25.3 remains applicable. For zero feedback, it follows that the small-signal current I_{1A} flowing into the base of transistor Q1 derives from

$$\left.\frac{I_{1A}}{V_{ST}}\right|_{f=0} = \frac{1-\alpha}{r_{ib} + (1-\alpha)R_{ST} + (R_{EE}\|R_F)} \tag{25.19}$$

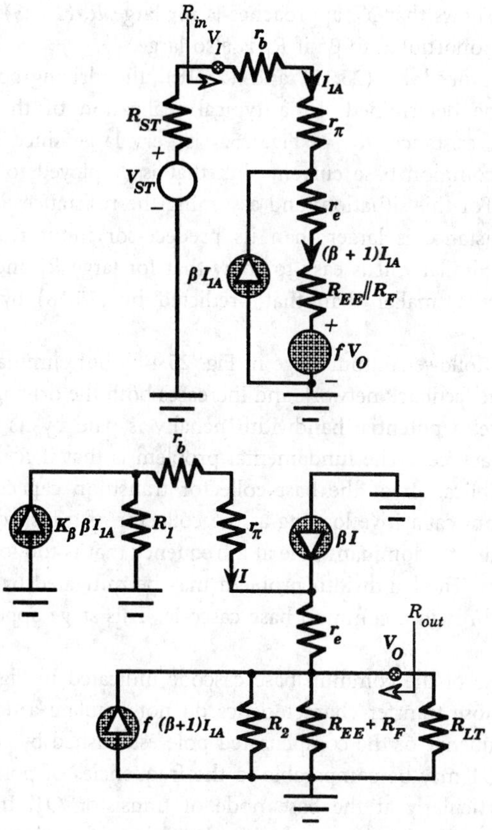

FIGURE 25.5 Small-signal model of the buffered series-shunt feedback amplifier.

The pertinent small-signal model for the buffered series-shunt feedback amplifier is resultantly the configuration offered in Fig. 25.5.

Letting

$$R' = R_2 \| (R_{\text{EE}} + R_F) \| R_{\text{LT}} \tag{25.20}$$

an analysis of the structure in Fig. 25.5 reveals

$$\frac{V_O}{I_{1A}} = (\beta + 1)\left[\frac{R'}{R' + r_{ib} + (1-\alpha)R_1}\right]\{\alpha K_\beta R_1 + f[r_{ib} + (1-\alpha)R_1]\} \tag{25.21}$$

which suggests negligible feed-forward for

$$f \ll \frac{\alpha K_\beta R_1}{r_{ib} + (1-\alpha)R_1} \tag{25.22}$$

Note that for large R_1, (25.22) implies the requirement $f \ll \beta K_\beta$, which is easier to satisfy than is (25.10). Assuming the validity of (25.22), (25.21), and (25.19) deliver an open-loop voltage gain, $G_{vo}(R_{\text{ST}}, R_{\text{LT}})$, of

$$G_{vo}(R_{\text{ST}}, R_{\text{LT}}) = \alpha K_\beta \left[\frac{R'}{r_{ib} + (1-\alpha)R_{\text{ST}} + R_{\text{EE}} \| R_F}\right]\left[\frac{R_1}{R' + r_{ib} + (1-\alpha)R_1}\right] \tag{25.23}$$

Recalling (25.1), which shows that K_β approaches β for large R, (25.23) suggests an open-loop gain that is nominally proportional to β^2 if R_1 is also large.

Using the concepts evoked by (25.17) and (25.18), the driving-point input and output impedances can now be determined. In a typical realization of the buffered series-shunt feedback amplifier, the resistance, R_2, in Fig. 25.4 is very large since it is manifested as the output resistance of a common base current sink that is employed to stabilize the operating point of transistor Q3. For this situation, and assuming the resistance R_1 is large, the resultant driving-point input resistance is larger than its predecessor input resistance by a factor of approximately $(\beta + 1)$. Similarly, it is easy to show that for large R_1 and large R_2, the driving-point output resistance is smaller than that predicted by (25.18) by a factor approaching $(\beta + 1)$.

Although the emitter follower output stage in Fig. 25.4 all but eliminates feed-forward signal transmission through the feedback network and increases both the driving point input resistance and output conductance, a potential bandwidth penalty is paid by its incorporation into the basic series-shunt feedback cell. The fundamental problem is that if R_1 is too large, potentially significant Miller multiplication of the base-collector transition capacitance of transistor Q2 materializes. The resultant capacitive loading at the collector of transistor Q1 is exacerbated by large R, which may produce a dominant pole at a frequency that is too low to satisfy closed-loop bandwidth requirements. The bandwidth problem may be mitigated by coupling resistance R_1 to the collector of Q2 through a common base cascode. This stage appears as transistor Q4 in Fig. 25.6.

Unfortunately, the use of the common base cascode indicated in Fig. 25.6 may produce an open-loop amplifier whose transfer characteristics do not emulate a dominant pole response. In other words, the frequency of the compensated pole established by capacitive loading at the collector of transistor Q1 may be comparable to the frequencies of poles established elsewhere in the circuit, and particularly at the base node of transistor Q1. In this event, frequency compensation aimed toward achieving acceptable closed-loop damping can be implemented by replacing the feedback resistor R_F with the parallel combination of R_F and a feedback capacitance, say C_F, as indicated by the dashed branch in Fig. 25.6. The resultant frequency-domain feedback

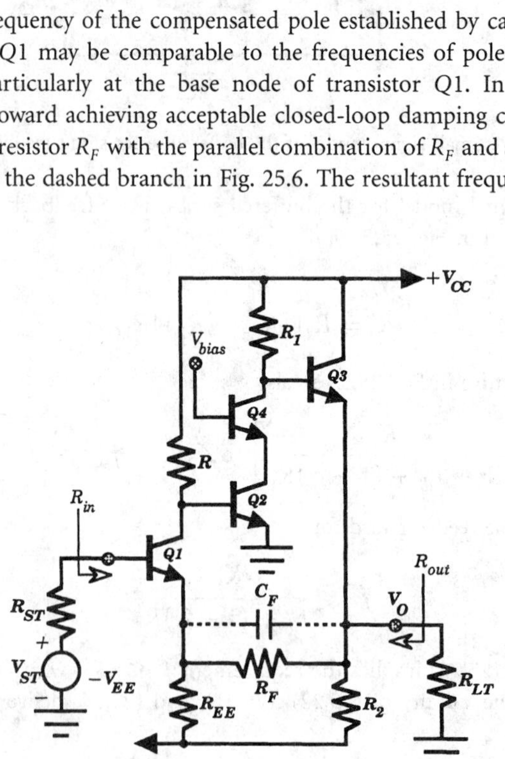

FIGURE 25.6 Buffered series-shunt feedback amplifier with common base cascode compensation of the common emitter amplifier formed by transistor Q2. A feedback zero is introduced by the capacitance C_F to achieve acceptable closed-loop damping.

factor $f(s)$ is

$$f(s) = f\left[\frac{1+\frac{s}{z}}{1+\frac{fs}{z}}\right] \tag{25.24}$$

where f is the feedback factor given by (25.7) and z is the frequency of the introduced compensating zero, is

$$z = \frac{1}{R_F C_F} \tag{25.25}$$

The pole in (25.24) is inconsequential if the closed-loop amplifier bandwidth B_{cl} satisfies the restriction, $fB_{cl}R_F C_F = B_{cl}(R_{EE} \| R_F)C_F \ll 1$.

25.3 Shunt-Series Feedback Amplifier

While the series-shunt circuit functions as a voltage amplifier, the shunt-series configuration, whose ac schematic diagram is depicted in Fig. 25.7(a), is best suited as a current amplifier. In the subject circuit, the Q2 emitter current, which is a factor of $(1/\alpha)$ of the output signal current, I_O, is sampled by the feedback network formed of the resistances, R_{EE} and R_F. The sampled current is fed back as a current in shunt with the amplifier input port. Since output current is fed back as a current to a current-driven input port, the resultant driving point output resistance is large, and the driving-point input resistance is small. These characteristics allow for a closed-loop current gain, $G_I(R_{ST}, R_{LT}) = I_O/I_{ST}$, that is relatively independent of source and load resistances and insensitive to transistor parameters.

In the series-shunt amplifier, h parameters were selected to model the feedback network because the topology of an h-parameter equivalent circuit is, like the amplifier in which the feedback network is embedded, a series shunt, or Thévenin-Norton, topology. An analogous train of thought compels the use of g-parameters to represent the feedback network in Fig. 25.7(a). With reference to the branch variables defined in the schematic diagram,

$$\begin{bmatrix} I_{1F} \\ V_{2F} \end{bmatrix} = \begin{bmatrix} \dfrac{1}{R_{EE}+R_F} & -\dfrac{R_{EE}}{R_{EE}+R_F} \\ \dfrac{R_{EE}}{R_{EE}+R_F} & R_{EE} \| R_F \end{bmatrix} \begin{bmatrix} V_{1F} \\ I_{2F} \end{bmatrix} \tag{25.26}$$

Noting that the feedback network current, I_{2F}, relates to the amplifier output current, I_O, in accordance with

$$I_{2F} = -\frac{I_O}{\alpha} \tag{25.27}$$

and letting the feedback factor, f, be

$$f = \frac{1}{\alpha}\left(\frac{R_{EE}}{R_{EE}+R_F}\right) \tag{25.28}$$

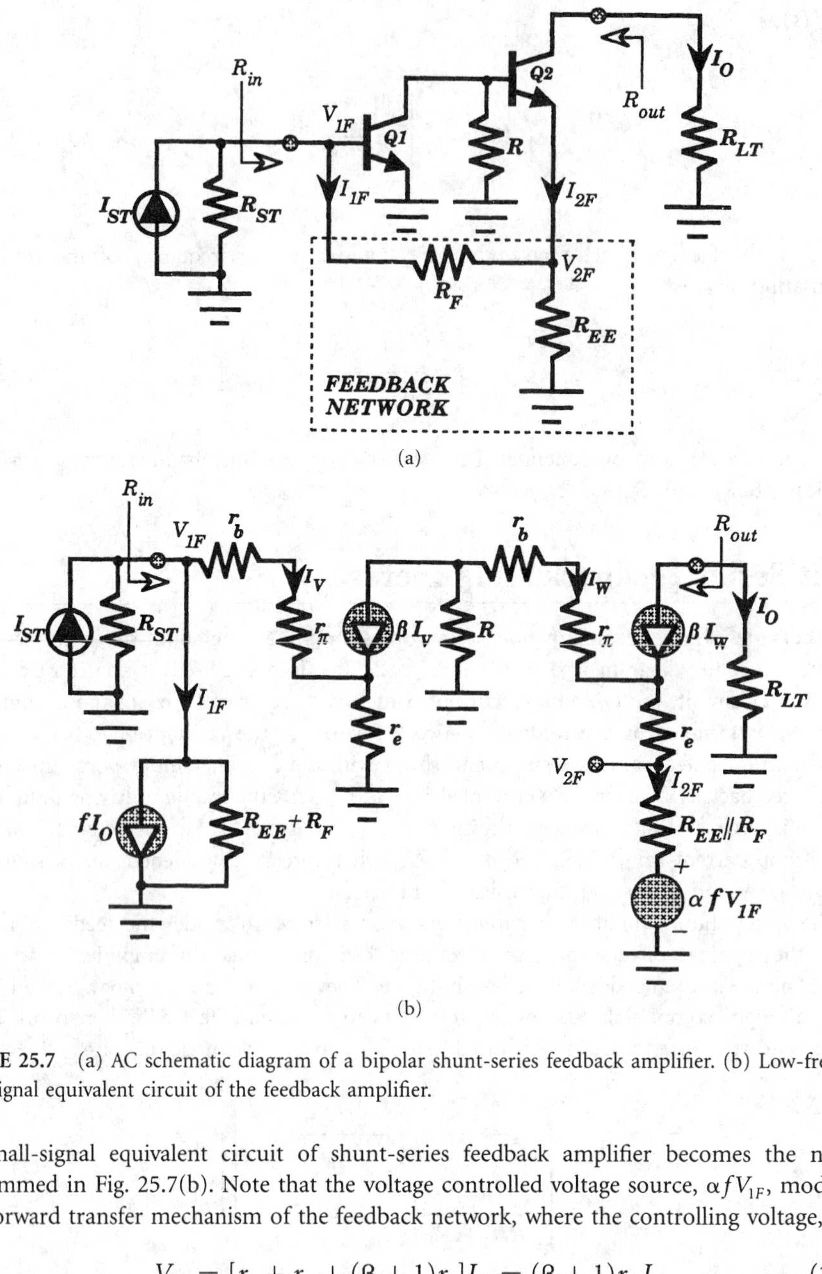

FIGURE 25.7 (a) AC schematic diagram of a bipolar shunt-series feedback amplifier. (b) Low-frequency small-signal equivalent circuit of the feedback amplifier.

the small-signal equivalent circuit of shunt-series feedback amplifier becomes the network diagrammed in Fig. 25.7(b). Note that the voltage controlled voltage source, $\alpha f V_{1F}$, models the feed-forward transfer mechanism of the feedback network, where the controlling voltage, V_{1F}, is

$$V_{1F} = [r_b + r_\pi + (\beta + 1)r_e]I_V = (\beta + 1)r_{ib}I_V \tag{25.29}$$

An analysis of the model in Fig. 25.7(b) confirms that the second-stage signal-base current I_W relates to the first-stage signal-base current I_V as

$$\frac{I_W}{I_V} = -\frac{\alpha(R + fr_{ib})}{r_{ib} + R_{EE} \| R_F + (1 - \alpha)R} \tag{25.30}$$

For

$$f \ll \frac{R}{r_{ib}} \tag{25.31}$$

which offsets feed-forward effects,

$$\frac{I_W}{I_V} \approx -\frac{\alpha R}{r_{ib} + R_{EE} \| R_F + (1-\alpha)R} \triangleq -K_r \tag{25.32}$$

Observe that the constant K_r tends toward β for large R, as can be verified by an inspection of Fig. 25.7(b).

Using (25.32), the open-loop current gain, found by setting f to zero, is

$$G_{IO}(R_{ST}, R_{LT}) = \left.\frac{I_O}{I_{ST}}\right|_{f=0} = \alpha K_r \left\{\frac{R_{ST} \| (R_{EE} + R_F)}{r_{ib} + (1-\alpha)[R_{ST} \| (R_{EE} + R_F)]}\right\} \tag{25.33}$$

whence, recalling (25.28), the loop gain T is

$$\begin{aligned} T = fG_{IO}(R_{ST}, R_{LT}) &= \frac{1}{\alpha}\left(\frac{R_{EE}}{R_{EE} + R_F}\right) G_{IO}(R_{ST}, R_{LT}) \\ &= K_r \left(\frac{R_{EE}}{R_{EE} + R_F + R_{ST}}\right)\left\{\frac{R_{ST}}{r_{ib} + (1-\alpha)[R_{ST} \| (R_{EE} + R_F)]}\right\} \end{aligned} \tag{25.34}$$

By inspection of the model in Fig. 25.7(b), the open-loop input resistance, R_{ino}, is

$$R_{ino} = (R_{EE} + R_F) \| [(\beta + 1) r_{ib}] \tag{25.35}$$

and, within the context of an infinitely large Early resistance, the open-loop output resistance, R_{outo}, is infinitely large.

The closed-loop current gain of the shunt-series feedback amplifier is now found to be

$$G_I(R_{ST}, R_{LT}) = \frac{G_{IO}(R_{ST}, R_{LT})}{1+T} \approx \alpha\left(1 + \frac{R_F}{R_{EE}}\right) \tag{25.36}$$

where the indicated approximation exploits the presumption that the loop gain T is much larger than one. As a result of the large loop-gain assumption, note that the closed-loop gain is independent of the source and load resistances and is invulnerable to uncertainties and perturbations in transistor parameters. The closed-loop output resistance, which exceeds its open-loop counterpart, remains infinitely large. Finally, the closed-loop driving point input resistance of the shunt-series amplifier is

$$R_{in} = \frac{R_{ino}}{1 + fG_{IO}(\infty, R_{LT})} \approx \left(1 + \frac{R_F}{R_{EE}}\right)\frac{r_{ib}}{K_r} \tag{25.37}$$

25.4 Shunt-Shunt Feedback Amplifier

Circuit Modeling and Analysis

The ac schematic diagram of the third type of single-loop feedback amplifier, the shunt-shunt triple, is drawn in Fig. 25.8(a). A cascade interconnection of three transistors $Q1, Q2$, and $Q3$, forms the open loop, while the feedback subcircuit is the single resistance, R_F. This resistance samples the output voltage, V_O, as a current fed back to the input port. Since output voltage is fed back as a current to a current-driven input port, both the driving point input and output resistances are very small. Accordingly, the circuit operates best as a transresistance amplifier in

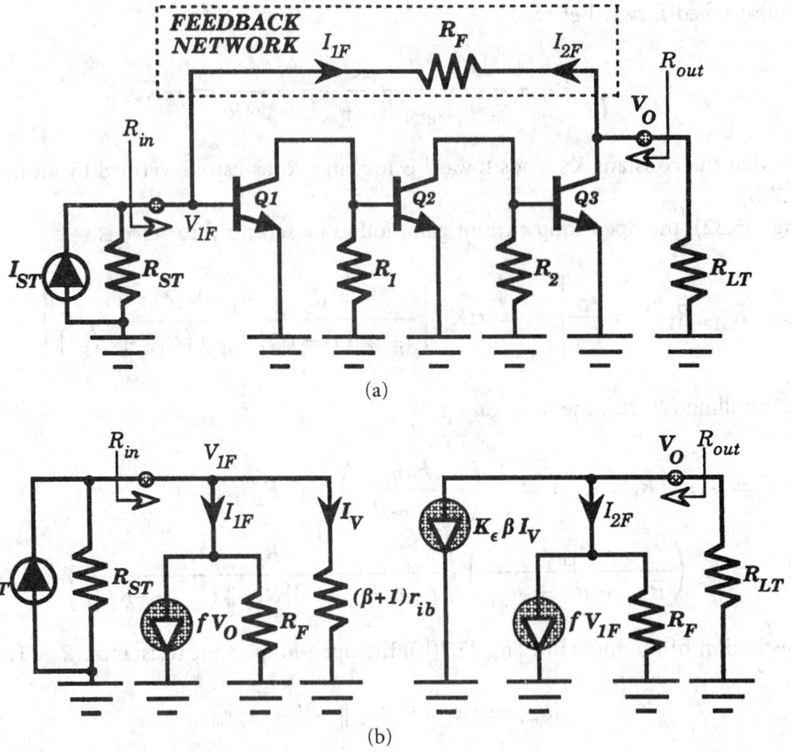

FIGURE 25.8 (a) AC schematic diagram of a bipolar shunt-shunt feedback amplifier. (b) Low-frequency small-signal equivalent circuit of the feedback amplifier.

that its closed-loop transresistance, $R_M(R_{ST}, R_{LT}) = V_O/I_{ST}$, is nominally invariant with source resistance, load resistance, and transistor parameters.

The shunt-shunt nature of the subject amplifier suggests the propriety of y-parameter modeling of the feedback network. For the electrical variables indicated in Fig. 25.8(a),

$$\begin{bmatrix} I_{1F} \\ I_{2F} \end{bmatrix} = \begin{bmatrix} \dfrac{1}{R_F} & -\dfrac{1}{R_F} \\ -\dfrac{1}{R_F} & \dfrac{1}{R_F} \end{bmatrix} \begin{bmatrix} V_{1F} \\ V_O \end{bmatrix} \tag{25.38}$$

which implies that a resistance, R_F, loads both the input and the output ports of the open-loop three-stage cascade. The short-circuit admittance relationship in (25.38) also suggests a feedback factor, f, given by

$$f = \frac{1}{R_F} \tag{25.39}$$

The foregoing observations and the small-signal modeling experience gained with the preceding two feedback amplifiers lead to the equivalent circuit submitted in Fig. 25.8(b). For analytical simplicity, the model reflects the assumption that all three transistors in the open loop have identical small-signal parameters. Moreover, the constant, K_ϵ, which symbolizes the ratio of the signal base current flowing into transistor Q3 to the signal base current conducted by transistor

Q1, is given by

$$K_\epsilon = \left[\frac{\alpha R_1}{r_{ib} + (1-\alpha)R_1}\right]\left[\frac{\alpha R_2}{r_{ib} + (1-\alpha)R_2}\right] \tag{25.40}$$

Finally, the voltage controlled current source, fV_{1F}, accounts for feed-forward signal transmission through the feedback network. If such feed-forward is to be negligible, the magnitude of this controlled current must be significantly smaller than $K_\epsilon \beta I_V$, a current that emulates feed-forward through the open-loop amplifier. Noting that the input port voltage, V_{1F}, in the present case remains the same as that specified by (25.29), negligible feed-forward through the feedback network mandates

$$R_F \gg \frac{r_{ib}}{\alpha K_\epsilon} \tag{25.41}$$

Since the constant K_ϵ in (25.40) tends toward β^2 if R_1 and R_2 are large resistances, (25.41) is relatively easy to satisfy.

With feed-forward through the feedback network ignored, an analysis of the model in Fig. 25.8(b) provides an open-loop transresistance, $R_{MO}(R_{ST}, R_{LT})$, of

$$R_{MO}(R_{ST}, R_{LT}) = -\alpha K_\epsilon \left[\frac{R_F \| R_{ST}}{r_{ib} + (1-\alpha)(R_F \| R_{ST})}\right](R_F \| R_{LT}) \tag{25.42}$$

while the loop gain is

$$\begin{aligned} T = fR_{MO}(R_{ST}, R_{LT}) &= -\frac{R_{MO}(R_{ST}, R_{LT})}{R_F} \\ &= \alpha K_\epsilon \left[\frac{R_{ST}}{R_{ST} + R_F}\right]\left[\frac{R_F \| R_{LT}}{r_{ib} + (1-\alpha)(R_F \| R_{ST})}\right] \end{aligned} \tag{25.43}$$

For $T \gg 1$, the corresponding closed-loop transresistance $R_M(R_{ST}, R_{LT})$ is

$$R_M(R_{ST}, R_{LT}) = \frac{R_{MO}(R_{ST}, R_{LT})}{1 + T} \approx -R_F \tag{25.44}$$

Finally, the approximate driving-point input and output resistances are, respectively,

$$R_{in} \approx \left(\frac{r_{ib}}{\alpha K_\epsilon}\right)\left(1 + \frac{R_F}{R_{LT}}\right) \tag{25.45}$$

$$R_{out} \approx \left[\frac{r_{ib} + (1-\alpha)(R_F \| R_{ST})}{\alpha K_\epsilon}\right]\left(1 + \frac{R_F}{R_{ST}}\right) \tag{25.46}$$

Design Considerations

Because the shunt-shunt triple uses three gain stages in the open-loop amplifier, its loop gain is significantly larger than the loop gains provided by either of the previously considered feedback cells. Accordingly, the feedback triple affords superior desensitization of the closed-loop gain with respect to transistor parameters and source and load resistances. But the presence of a cascade of three common emitter gain stages in the open loop of the amplifier complicates frequency compensation and limits the 3-dB bandwidth. The problem is that although each common

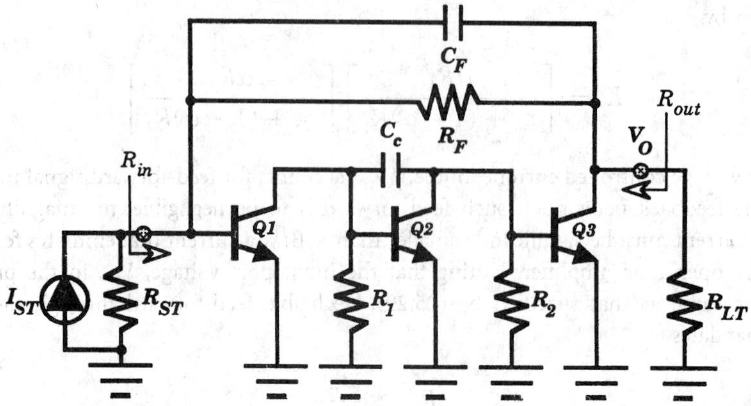

FIGURE 25.9 AC schematic diagram of a frequency compensated shunt-shunt triple. The capacitance, C_c, achieves open-loop pole splitting, while the capacitance, C_F, implements a compensating feedback network zero.

emitter stage approximates a dominant pole amplifier, none of the critical frequencies in the cluster of poles established by the cascade interconnection of these units is likely to be dominant. The uncompensated closed loop is therefore predisposed to unacceptable underdamping, thereby making compensation via an introduced feedback zero difficult.

At least three compensation techniques can be exploited to optimize the performance of the shunt-shunt feedback amplifier [3], [7]–[9]. The first of these techniques entails pole splitting of the open-loop interstage through the introduction of a capacitance, C_c, between the base and the collector terminals of transistor Q2, as depicted in the ac schematic diagram of Fig. 25.9. In principle, pole splitting can be invoked on any one of the three stages of the open loop. But pole splitting of the interstage is most desirable since such compensation of the first stage proves effective only for large source resistance. Moreover, the resultant dominant pole becomes dependent on the source termination. On the other hand, pole splitting of the third stage produces a dominant pole that is sensitive to load termination. In conjunction with pole splitting, a feedback zero can be introduced, if necessary, to increase closed-loop damping by replacing the feedback resistance, R_F, by the parallel combination of R_F and a feedback capacitance, C_F, as shown in Fig. 25.9. This compensation produces left-half-plane zero in the feedback factor at $s = -(1/R_F)$.

A second compensation method broadbands the interstage of the open-loop amplifier through local current feedback introduced by the resistance, R_X, in Fig. 25.10. Simultaneously, the third stage is broadbanded by way of a common base cascode transistor Q4. Because emitter degeneration of the interstage reduces the open-loop gain, an emitter follower (transistor Q5) is embedded between the feedback network and the output port of the open-loop third stage. As in the case of the series-shunt feedback amplifier, the first-order effect of this emitter follower is to increase feed-forward signal transmission through the open-loop amplifier by a factor that approaches $(\beta + 1)$.

A final compensation method is available if shunt-shunt feedback is implemented as the balanced differential architecture whose ac schematic diagram is offered in Fig. 25.11. By exploiting the antiphase nature of opposite collectors in a balanced common emitter topology, a shunt-shunt feedback amplifier can be realized with only two gain stages in the open loop. The resultant closed loop 3-dB bandwidth is invariably larger than that of its three-stage single-ended counterpart, since the open loop is now characterized by only two, as opposed to three, fundamental critical frequencies. Because the forward gain implicit to two amplifier stages is smaller than the gain afforded by three stages of amplification, a balanced emitter follower

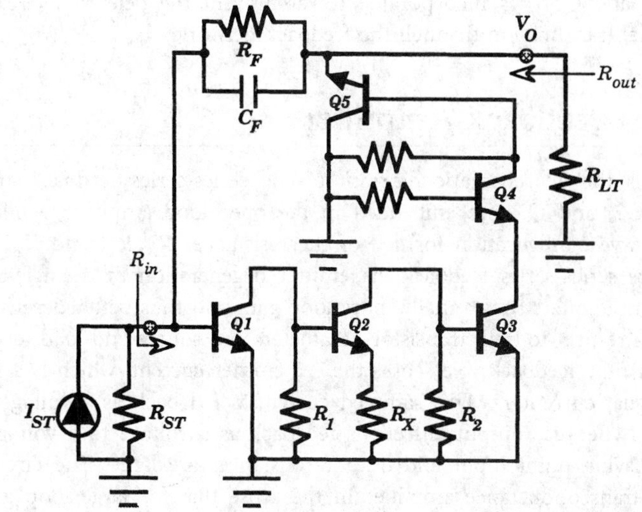

FIGURE 25.10 AC schematic diagram of an alternative compensation scheme for the shunt-shunt triple. Transistor $Q2$ is broadbanded by the emitter degeneration resistance R_X and transistor $Q3$ is broadbanded by the common base cascode transistor $Q4$. The emitter follower transistor, $Q5$, minimizes feed-forward signal transmission through the feedback network.

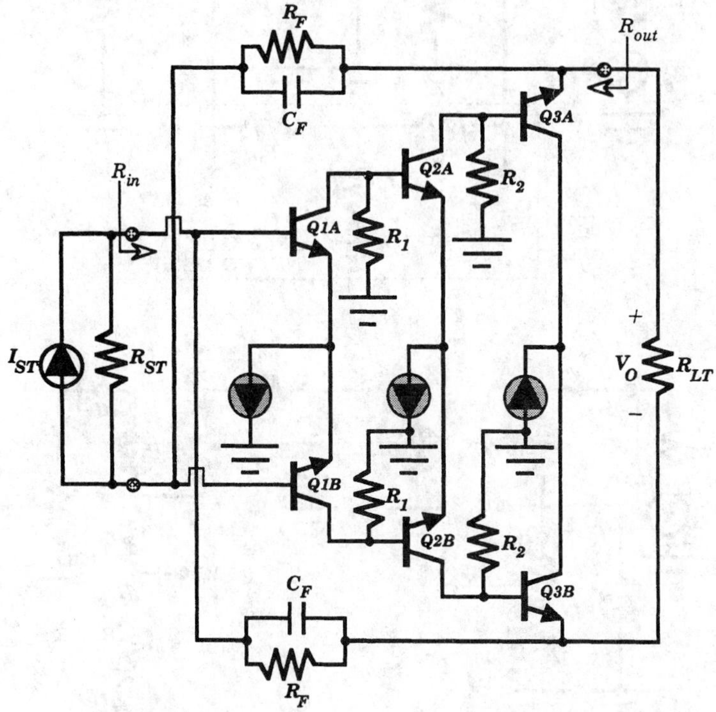

FIGURE 25.11 AC schematic diagram of a differential realization of the compensated shunt-shunt feedback amplifier. The balanced stage boasts improved bandwidth over its single-ended counterpart because of its use of only two high-gain stages in the open loop. The emitter follower pair $Q3A$ and $Q3B$ diminishes feed-forward transmission through the feedback network composed of the shunt interconnection of resistor R_F with capacitor C_F.

(transistors $Q3A$ and $Q3B$) is incorporated to circumvent the deleterious relative effects of feed-forward signal transmission through the feedback network.

25.5 Series-Series Feedback Amplifier

Figure 25.12(a) is the ac schematic diagram of the series-series feedback amplifier. Three transistors, $Q1$, $Q2$, and $Q3$, are embedded in the open-loop amplifier, while the feedback subcircuit is the wye configuration formed of the resistances R_X, R_Y, and R_Z. Although it is possible to realize series-series feedback via emitter degeneration of a single-stage amplifier, the series-series triple offers substantially more loop gain and thus better desensitization of the forward gain with respect to both transistor parameters and source and load terminations.

In Fig. 25.12(a), the feedback wye senses the $Q3$ emitter current, which is a factor of $(1/\alpha)$ of the output signal current I_o. This sampled current is fed back as a voltage in series with the emitter of $Q1$. Because output current is fed back as a voltage to a voltage-driven input port, both the driving point input and output resistances are large. The circuit is therefore best suited as a transconductance amplifier in the sense that for large loop gain, its closed-

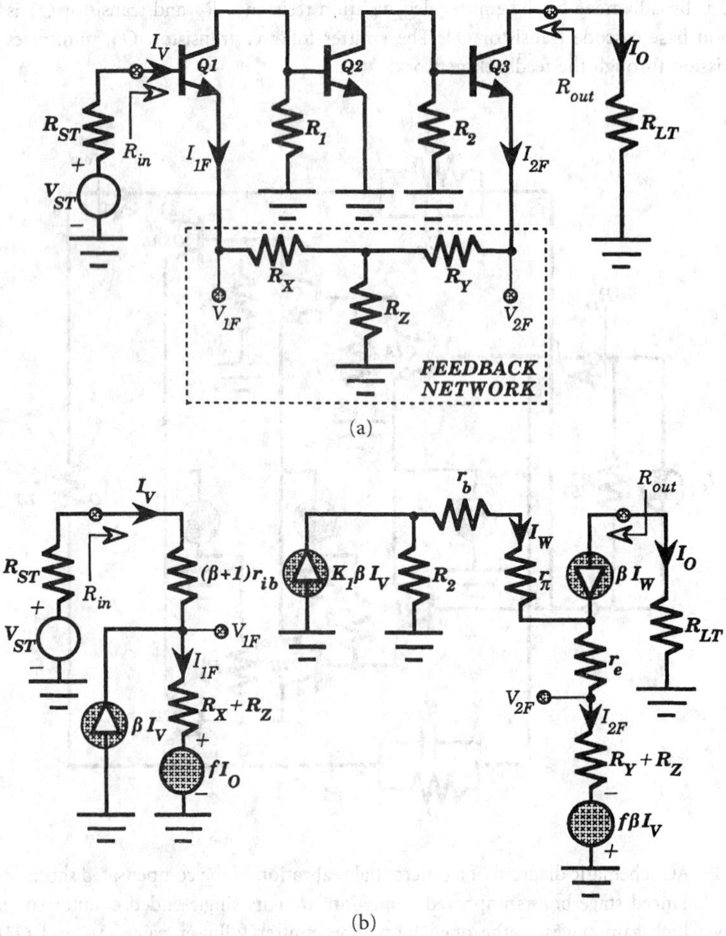

FIGURE 25.12 (a) AC schematic diagram of a bipolar series-series feedback amplifier. (b) Low-frequency, small-signal equivalent circuit of the feedback amplifier.

loop transconductance $G_M(R_{ST}, R_{LT}) = I_O/V_{ST}$, is almost independent of the source and load resistances.

The series-series topology of the subject amplifier conduces z-parameter modeling of the feedback network. Noting the electrical variables delineated in the diagram of Fig. 25.12(a),

$$\begin{bmatrix} V_{1F} \\ V_{2F} \end{bmatrix} = \begin{bmatrix} R_X + R_Z & R_Z \\ R_Z & R_Y + R_Z \end{bmatrix} \begin{bmatrix} I_{1F} \\ I_{2F} \end{bmatrix} \tag{25.47}$$

Equation (25.47) suggests that the open-circuit feedback network resistances loading the emitters of transistors $Q1$ and $Q3$ are $(R_X + R_Z)$ and $(R_Y + R_Z)$, respectively, and the voltage fed back to the emitter of transistor $Q1$ is $R_Z I_{2F}$. Since the indicated feedback network current I_{2F} is $(-I_O/\alpha)$, this fed back voltage is equivalent to $(-R_Z I_O/\alpha)$, which suggests a feedback factor, f, of

$$f = -\frac{R_Z}{\alpha} \tag{25.48}$$

Finally, the feed-forward through the feedback network is $R_Z I_{1F}$. Since I_{1F} relates to the signal base current I_V flowing into transistor $Q1$ by $I_{1F} = (\beta + 1)I_V$, this feed-forward voltage is also expressible as $(-f\beta I_V)$. The foregoing observations and the hybrid-pi model of a bipolar junction transistor produce the small-signal model depicted in Fig. 25.12(b). In this model, all transistors are presumed to have identical corresponding small-signal parameters, and the constant, K_1, is

$$K_1 = \frac{\alpha R_1}{r_{ib} + (1 - \alpha)R_1} \tag{25.49}$$

An analysis of the model of Fig. 25.12(b) confirms that the ratio of the signal current, I_W, flowing into the base of transistor $Q3$ to the signal base current, I_V, of transistor $Q1$ is

$$\frac{I_W}{I_V} = \frac{\alpha K_1 R_2 \left(1 + \dfrac{f}{K_1 R_2}\right)}{r_{ib} + R_Y + R_Z + (1 - \alpha)R_2} \tag{25.50}$$

This result suggests that feed-forward effects through the feedback network are negligible if $|f| \ll K_1 R_2$, which requires

$$R_Z \ll \alpha K_1 R_2 \tag{25.51}$$

In view of the fact that the constant, K_1, approaches β for large values of the resistance, R_1, (25.51) is not a troublesome inequality. Introducing a second constant, K_2, such that

$$K_2 \triangleq \frac{\alpha R_2}{r_{ib} + R_Y + R_Z + (1 - \alpha)R_2} \tag{25.52}$$

the ratio I_W/I_V in (25.50) becomes

$$\frac{I_W}{I_V} \approx K_1 K_2 \tag{25.53}$$

assuming (25.51) is satisfied.

Given the propriety of (25.50) and using (25.53) the open-loop transconductance, $G_{MO}(R_{ST}, R_{LT})$ is found to be

$$G_{MO}(R_{ST}, R_{LT}) = -\left\{\frac{\alpha K_1 K_2}{r_{ib} + R_X + R_Z + (1-\alpha)R_{ST}}\right\} \tag{25.54}$$

and recalling (25.48), the loop gain T is

$$T = -\left(\frac{R_Z}{\alpha}\right) G_{MO}(R_{ST}, R_{LT}) = \frac{K_1 K_2 R_Z}{r_{ib} + R_X + R_Z + (1-\alpha)R_{ST}} \tag{25.55}$$

It follows that for $T \gg 1$, the closed-loop transconductance is

$$G_M(R_{ST}, R_{LT}) = \frac{G_{MO}(R_{ST}, R_{LT})}{1+T} \approx -\frac{\alpha}{R_Z} \tag{25.56}$$

Since the Early resistance is large enough to justify its neglect, the open-loop, and thus the closed-loop, driving-point output resistances are infinitely large. On the other hand, the closed-loop driving point input resistance R_{in} can be shown to be

$$R_{in} = R_{ino}[1 + fG_{MO}(0, R_{LT})] \approx (\beta + 1)K_1 K_2 R_Z \tag{25.57}$$

Like its shunt-shunt counterpart, the series-series feedback amplifier uses three open-loop gain stages to produce large loop gain. But also like the shunt-shunt triple, frequency compensation via an introduced feedback zero is difficult unless design care is exercised to realize a dominant pole open-loop response. To this end, the most commonly used compensation is pole splitting in the open loop, combined, if required, with the introduction of a zero in the feedback factor. The relevant ac schematic diagram appears in Fig. 25.13 where the indicated capacitance, C_c, inserted across the base-collector terminals of transistor Q3 achieves the aforementioned pole splitting compensation. The capacitance, C_F, in Fig. 25.13 delivers a frequency-dependent feedback factor, $f(s)$ of

$$f(s) = f\left[\frac{1 + \dfrac{s}{z}}{1 + \dfrac{s}{z}\left(\dfrac{R_Z}{R_Z + R_X \| R_Y}\right)}\right] \tag{25.58}$$

where the frequency z of the introduced zero derives from

$$\frac{1}{z} = (R_X + R_Y)\left(1 + \frac{R_X \| R_Y}{R_Z}\right) C_F \tag{25.59}$$

The corresponding pole in (25.58) is insignificant if the closed-loop amplifier is designed for a bandwidth, B_{cl} that satisfies the inequality, $B_{cl}(R_X + R_Y)C_F \ll 1$.

As is the case with shunt-shunt feedback, an alternative frequency compensation scheme is available if series-series feedback is implemented as a balanced differential architecture. The pertinent ac schematic diagram, inclusive of feedback compensation, appears in Fig. 25.14. This diagram exploits the fact that the feedback wye consisting of the resistances, R_X, R_Y, and R_Z as utilized in the single-ended configurations of Figs. 25.12(a) and 25.13 can be transformed into the feedback delta of Fig. 25.15. The terminal volt–ampere characteristics of the two networks in

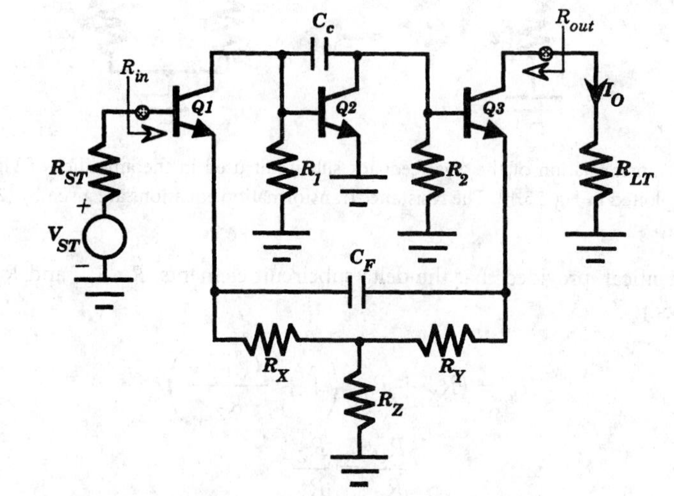

FIGURE 25.13 AC schematic diagram of a frequency compensated series-series feedback triple. The capacitance, C_c, achieves pole splitting in the open-loop configuration, while the capacitance, C_F, introduces a zero in the feedback factor of the closed-loop amplifier.

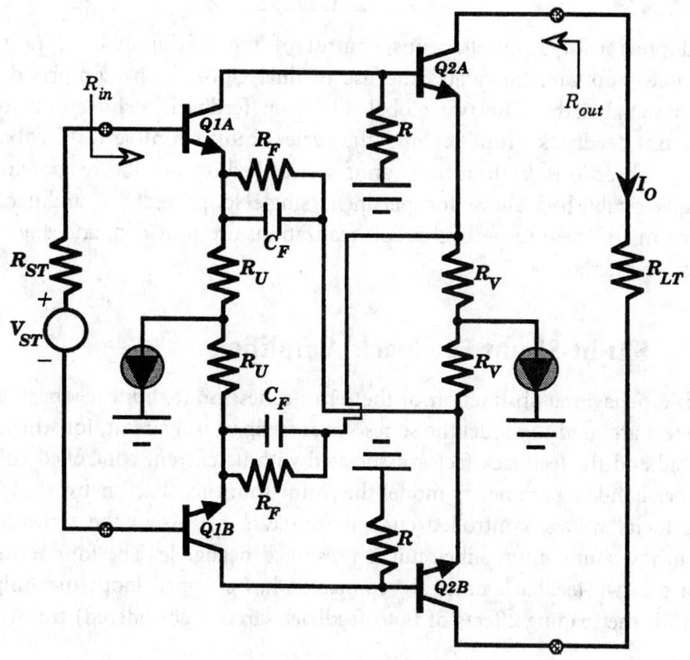

FIGURE 25.14 AC schematic diagram of a balanced differential version of the series-series feedback amplifier. The circuit utilizes only two, as opposed to three, gain stages in the open loop.

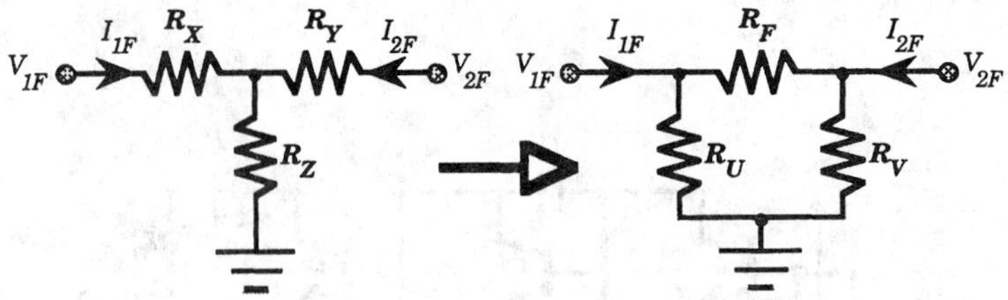

FIGURE 25.15 Transformation of the wye feedback subcircuit used in the amplifier of Fig. 25.13 to the delta subcircuit exploited in Fig. 25.14. The resistance transformation equations are given by (25.60)–(25.62).

Fig. 25.15 are identical, provided that the delta subcircuit elements, R_F, R_U, and R_V, are chosen in accordance with

$$R_F = (R_X + R_Y)\left(1 + \frac{R_X \| R_Y}{R_Z}\right) \tag{25.60}$$

$$\frac{R_U}{R_F} = \frac{R_Z}{R_Y} \tag{25.61}$$

$$\frac{R_V}{R_F} = \frac{R_Z}{R_X} \tag{25.62}$$

25.6 Dual-Loop Feedback

As mentioned previously, a simultaneous control of the driving point I/O resistances, as well as the closed-loop gain, mandates the use of dual global loops comprised of series and shunt feedback signal paths. The two global dual-loop feedback architectures are the **series-series/shunt-shunt feedback amplifier** and the **series-shunt/shunt-series feedback amplifier.** In the following subsections, both of these units are studied by judiciously applying the relevant analytical results established above for pertinent single loop feedback architectures. The ac schematic diagrams of these respective circuit realizations are provided, and engineering design considerations are offered.

Series-Series/Shunt-Shunt Feedback Amplifier

Figure 25.16 is a behavioral abstraction of the series-series/shunt-shunt feedback amplifier. Two port z parameters are used to model the series-series feedback subcircuit, for which feed-forward is tacitly ignored and the feedback factor associated with its current controlled voltage source is f_{ss}. On the other hand, y parameters model the shunt-shunt feedback network, whose feedback factor relative to its voltage controlled current source is f_{pp}. As in the series-series network, feed-forward in the shunt-shunt subcircuit is presumed negligible. The four terminal amplifier around which the two feedback units are connected has an open loop (meaning $f_{ss} = 0$ and $f_{pp} = 0$, but with the loading effects of both feedback circuits considered) transconductance of $G_{MO}(R_{ST}, R_{LT})$.

With f_{pp} set to zero to deactivate shunt-shunt feedback, the resultant series-series feedback network is a transconductance amplifier whose closed loop transconductance, $G_{MS}(R_{ST}, R_{LT})$, is

$$G_{MS}(R_{ST}, R_{LT}) = \frac{I_O}{V_{ST}} = \frac{G_{MO}(R_{ST}, R_{LT})}{1 + f_{ss}G_{MO}(R_{ST}, R_{LT})} \approx \frac{1}{f_{ss}} \tag{25.63}$$

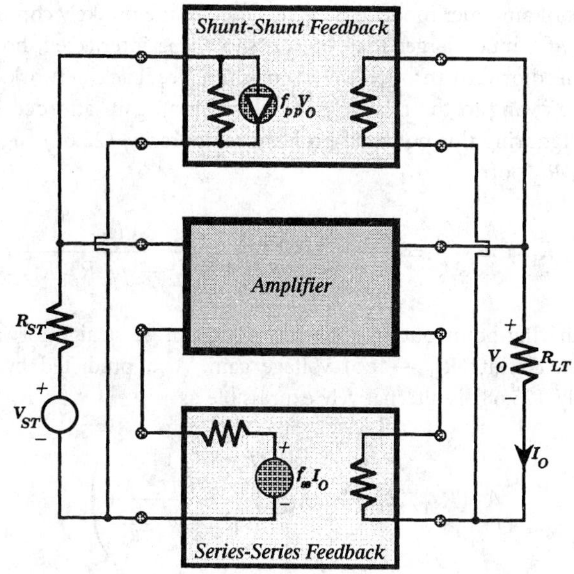

FIGURE 25.16 System level diagram of a series-series/shunt-shunt dual-loop feedback amplifier. Note that feed-forward signal transmission through either feedback network is ignored.

where the loop gain, $f_{ss}G_{MO}(R_{ST}, R_{LT})$, is presumed much larger than one, and the loading effects of both the series-series feedback subcircuit and the deactivated shunt-shunt feedback network are incorporated into $G_{MO}(R_{ST}, R_{LT})$. The transresistance, $R_{MS}(R_{ST}, R_{LT})$, implied by (25.63), which expedites the study of the shunt-shunt component of the feedback configuration, is

$$R_{MS}(R_{ST}, R_{LT}) = \frac{V_O}{I_{ST}} = R_{ST}R_{LT}\frac{I_O}{V_{ST}} \approx \frac{R_{ST}R_{LT}}{f_{ss}} \tag{25.64}$$

The series-series feedback input and output resistances R_{ins} and R_{outs}, respectively, are large and given by

$$R_{ins} = R_{ino}[1 + f_{ss}G_{MO}(0, R_{LT})] \tag{25.65}$$

and

$$R_{outs} = R_{outo}[1 + f_{ss}G_{MO}(R_{ST}, 0)] \tag{25.66}$$

where the zero feedback ($f_{ss} = 0$ and $f_{pp} = 0$) values, R_{ino} and R_{outo}, of these driving point quantities are computed with due consideration given to the loading effects imposed on the amplifier by both feedback subcircuits.

When shunt-shunt feedback is applied around the series-series feedback cell, the configuration becomes a transresistance amplifier. The effective open-loop transresistance is $R_{MS}(R_{ST}, R_{LT})$, as defined by (25.64). Noting a feedback factor of f_{pp}, the corresponding closed loop transresistance is

$$R_M(R_{ST}, R_{LT}) \approx \frac{\dfrac{R_{ST}R_{LT}}{f_{ss}}}{1 + f_{pp}\left(\dfrac{R_{ST}R_{LT}}{f_{ss}}\right)} \tag{25.67}$$

which is independent of amplifier model parameters despite the unlikely condition of an effective loop gain, $f_{pp}R_{ST}R_{LT}/f_{ss}$, much larger than one. It should be interjected, however, that (25.67) presumes negligible feedforward through the shunt-shunt feedback network. This presumption may be inappropriate owing to the relatively low closed loop gain afforded by the series-series feedback subcircuit. Ignoring this potential problem temporarily, (25.67) suggests a closed loop voltage gain $A_V(R_{ST}, R_{LT})$ of

$$A_V(R_{ST}, R_{LT}) = \frac{V_O}{V_S} = \frac{R_M(R_{ST}, R_{LT})}{R_{ST}} \approx \frac{R_{LT}}{f_{ss} + f_{pp}R_{ST}R_{LT}} \tag{25.68}$$

The closed-loop driving-point output resistance R_{out}, can be straightforwardly calculated by noting that the open circuit ($R_{LT} \to \infty$) voltage gain, A_{VO}, predicted by (25.68) is $A_{VO} = 1/f_{pp}R_{ST}$. Accordingly, (25.68) is alternatively expressible as

$$A_V(R_{ST}, R_{LT}) \approx A_{VO}\left(\frac{R_{LT}}{R_{LT} + \frac{f_{ss}}{f_{pp}R_{ST}}}\right) \tag{25.69}$$

Since (25.69) is a voltage divider relationship stemming from a Thévenin model of the output port of the dual loop feedback amplifier, as delineated in Fig. 25.17, it follows that the driving point output resistance is

$$R_{out} \approx \frac{f_{ss}}{f_{pp}R_{ST}} \tag{25.70}$$

Observe that like the forward gain characteristics, the driving-point output resistance is nominally insensitive to changes and other uncertainties in open-loop amplifier parameters. Moreover, this output resistance is directly proportional to the ratio f_{ss}/f_{pp} of feedback factors. As is illustrated in preceding sections, the individual feedback factors, and thus the ratio of feedback factors, is likely to be proportional to a ratio of resistances. In view of the fact that resistance ratios can be tightly controlled in a monolithic fabrication process R_{out} in (25.70) is accurately prescribed for a given source termination.

The driving point input resistance R_{in} can be determined from a consideration of the input port component of the system level equivalent circuit depicted in Fig. 25.17. This resistance is the ratio of V_{ST} to I, under the condition of $R_S = 0$. With $R_S = 0$, (25.68) yields $V_O = R_{LT}V_{ST}/f_{ss}$ and thus, KVL applied around the input port of the model at hand yields

$$R_{in} = \frac{R_{ins}}{1 + \frac{f_{pp}R_{LT}R_{ins}}{f_{ss}}} \approx \frac{f_{ss}}{f_{pp}R_{LT}} \tag{25.71}$$

where the "open-loop" input resistance R_{ins}, defined by (25.65) is presumed large. Like the driving point output resistance of the series-series/shunt-shunt feedback amplifier, the driving point input resistance is nominally independent of open-loop amplifier parameters.

It is interesting to observe that the input resistance in (25.71) is inversely proportional to the load resistance by the same factor (f_{ss}/f_{pp}) that the driving-point output resistance in (25.70) is inversely proportional to the source resistance. As a result,

$$\frac{f_{ss}}{f_{pp}} \approx R_{in}R_{LT} \equiv R_{out}R_{ST} \tag{25.72}$$

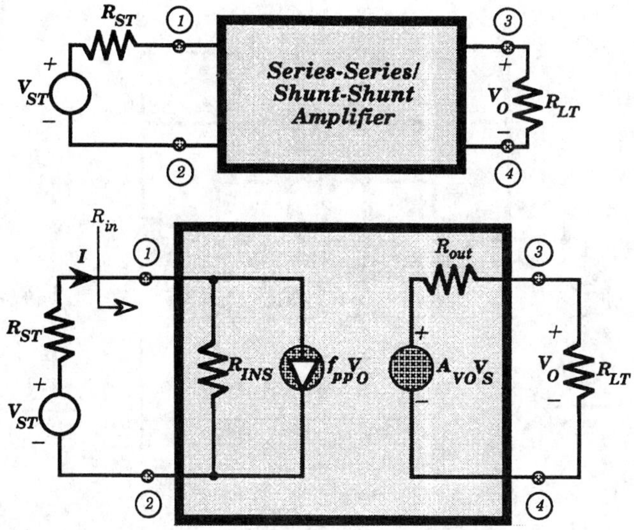

FIGURE 25.17 Norton equivalent input and Thévenin equivalent output circuits for the series-series/shunt-shunt dual-loop feedback amplifier.

Thus, in addition to being stable performance indices for well-defined source and load terminations, the driving-point input and output resistances track one another, despite manufacturing uncertainties and changes in operating temperature that might perturb the individual values of the two feedback factors f_{ss} and f_{pp}.

The circuit property stipulated by (25.72) has immediate utility in the design of wideband communication transceivers and other high-speed signal-processing systems [10]–[14]. In these and related applications, a cascade of several stages is generally required to satisfy frequency response, distortion, and noise specifications. A convenient way of implementing a cascade interconnection is to force each member of the cascade to operate under the match terminated case of $R_{\mathrm{ST}} = R_{\mathrm{in}} = R_{\mathrm{LT}} = R_{\mathrm{out}} \triangleq R$. From (25.72) match terminated operation demands feedback factors selected so that

$$R = \sqrt{\frac{f_{ss}}{f_{pp}}} \tag{25.73}$$

which forces a match terminated closed-loop voltage gain A_V^* of

$$A_V^* \approx \frac{1}{2f_{pp}R} = \frac{1}{2\sqrt{f_{pp}f_{ss}}} \tag{25.74}$$

The ac schematic diagram of a practical single ended series-series/shunt-shunt amplifier is submitted in Fig. 25.18. An inspection of this diagram reveals a topology that coalesces the series-series and shunt-shunt triples studied earlier. In particular, the wye network formed of the three resistances, R_X, R_Y, and R_Z, comprises the series-series component of the dual-loop feedback amplifier. The capacitor, C_c, narrowbands the open-loop amplifier to facilitate frequency compensation of the series-series loop through the capacitance, C_{F1}. Compensated shunt feedback of the network is achieved by the parallel combination of the resistance, R_F, and the capacitance, C_{F2}. If C_{F1} and C_c combine to deliver a dominant pole series-series feedback amplifier, C_{F2} is not necessary. Conversely, C_{F1} is superfluous if C_{F2} and C_c interact to provide a

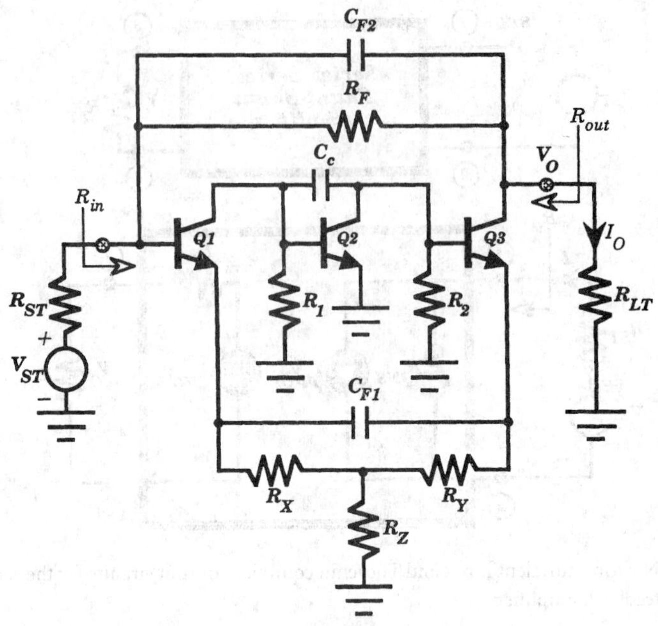

FIGURE 25.18 AC schematic diagram of a frequency-compensated series-series/shunt-shunt dual-loop feedback amplifier. The compensation is affected by the capacitances C_{F1} and C_{F2}, while C_c achieves pole splitting in the open-loop amplifier.

dominant pole shunt-shunt feedback amplifier. As in the single ended series-series configuration, transistor $Q3$ can be broadbanded via a common base cascode. Moreover, if feedback through the feedback networks poses a problem, an emitter follower can be inserted at the port to which the shunt feedback path and the load termination are incident.

A low-frequency analysis of the circuit in Fig. 25.18 is expedited by assuming high beta transistors having identical corresponding small-signal model parameters. This analysis, which, in contrast to the simplified behavioral analysis, does not ignore the electrical effects of the aforementioned feedforward through the shunt-shunt feedback network, yields a voltage gain $A_V(R_{ST}, R_{LT})$, of

$$A_V(R_{ST}, R_{LT}) \approx -\left(\frac{R_{in}}{R_{in}+R_{ST}}\right)\left(\frac{R_{LT}}{R_{LT}+R_F}\right)\left(\frac{\alpha R_F}{R_Z}-1\right) \quad (25.75)$$

where the driving-point input resistance of the amplifier R_{in} is

$$R_{in} \approx \frac{R_F + R_{LT}}{1 + \dfrac{\alpha R_{LT}}{R_Z}} \quad (25.76)$$

The driving-point output resistance R_{out} can be shown to be

$$R_{out} \approx \frac{R_F + R_{ST}}{1 + \dfrac{\alpha R_{ST}}{R_Z}} \quad (25.77)$$

As predicted by the behavioral analysis R_{in}, R_{out}, and $A_V(R_{ST}, R_{LT})$, are nominally independent of transistor parameters. Observe that the functional dependence of R_{in} on the load resistance,

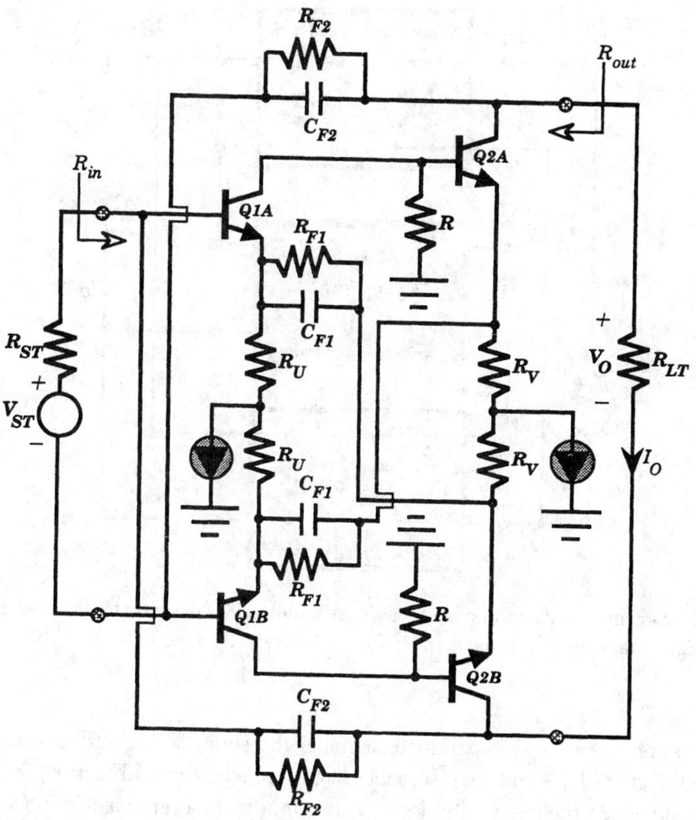

FIGURE 25.19 AC schematic diagram of the differential realization of a compensated series-series/shunt-shunt feedback amplifier.

R_{LT}, is identical to the manner in which R_{out} is related to the source resistance R_{ST}. In particular, $R_{in} \equiv R_{out}$ if $R_{ST} \equiv R_{LT}$. For the match terminated case in which $R_{ST} = R_{in} = R_{LT} = R_{out} \triangleq R$,

$$R \approx \sqrt{\frac{R_F R_Z}{\alpha}} \tag{25.78}$$

The corresponding match terminated voltage gain in (25.75) collapses to

$$A_V^* \approx -\left(\frac{R_F - R}{2R}\right) \tag{25.79}$$

Like the series-series and shunt-shunt triples, many of the frequency compensation problems implicit to the presence of three open-loop stages can be circumvented by realizing the series-series/shunt-shunt amplifier as a two stage differential configuration. Figure 25.19 is the ac schematic diagram of a compensated differential series-series/shunt-shunt feedback dual.

Series-Shunt/Shunt-Series Feedback Amplifier

The only other type of global dual loop architecture is the series-shunt/shunt-series feedback amplifier, whose behavioral diagram appears in Fig. 25.20. The series-shunt component of this system, which is modeled by h-parameters, has a negligibly small feed-forward factor and a

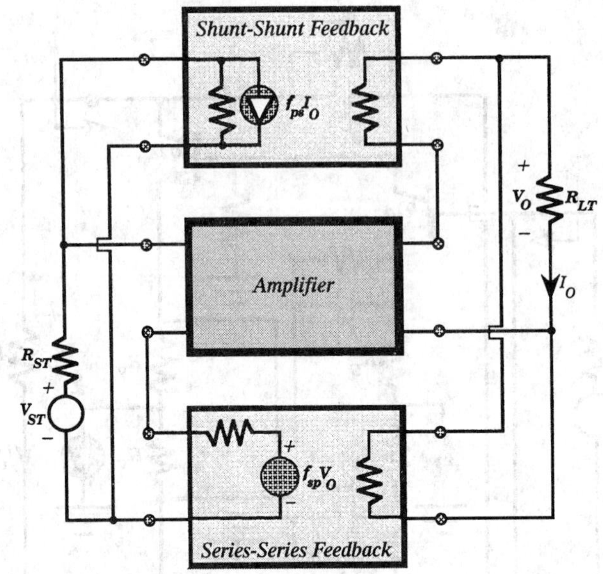

FIGURE 25.20 System level diagram of a series-shunt/shunt-series dual-loop feedback amplifier. Note that feed-forward signal transmission through either feedback network is ignored.

feedback factor of f_{sp}. Hybrid g-parameters model the shunt-series feedback structure, which has a feedback factor of f_{ps} and a presumably negligible feed-forward factor. The four-terminal amplifier around which the two feedback units are connected has an open-loop (meaning $f_{sp} = 0$ and $f_{ps} = 0$, but with the loading effects of both feedback circuits considered) voltage gain of $A_{VO}(R_{ST}, R_{LT})$.

For $f_{ps} = 0$, the series-shunt feedback circuit voltage gain $A_{VS}(R_{ST}, R_{LT})$, is

$$A_{VS}(R_{ST}, R_{LT}) = \frac{V_O}{V_{ST}} = \frac{A_{VO}(R_{ST}, R_{LT})}{1 + f_{sp}A_{VO}(R_{ST}, R_{LT})} \approx \frac{1}{f_{sp}}, \tag{25.80}$$

where the approximation reflects an assumption of a large loop gain. When the shunt-series component of the feedback amplifier is activated, the dual-loop configuration functions as a current amplifier. Its effective open-loop transfer function is the current gain, $A_{IS}(R_{ST}, R_{LT})$, established by the series-shunt amplifier; namely,

$$A_{IS}(R_{ST}, R_{LT}) = \frac{I_O}{I_{ST}} = \left(\frac{R_{ST}}{R_{LT}}\right)\frac{V_O}{V_{ST}} \approx \frac{R_{ST}}{f_{sp}R_{LT}} \tag{25.81}$$

It follows that the current gain, $A_I(R_{ST}, R_{LT})$, of the closed loop is

$$A_I(R_{ST}, R_{LT}) \approx \frac{\dfrac{R_{ST}}{f_{sp}R_{LT}}}{1 + f_{ps}\left(\dfrac{R_{ST}}{f_{sp}R_{LT}}\right)} = \frac{R_{ST}}{f_{sp}R_{LT} + f_{ps}R_{ST}} \tag{25.82}$$

while the corresponding voltage gain, $A_V(R_{ST}, R_{LT})$, assuming negligible feed-forward through the shunt-series feedback network, is

$$A_V(R_{ST}, R_{LT}) = \frac{R_{LT}}{R_{ST}} A_I(R_{ST}, R_{LT}) \approx \frac{R_{LT}}{f_{sp}R_{LT} + f_{ps}R_{ST}} \tag{25.83}$$

Repeating the analytical strategy employed to determine the input and output resistances of the series-series/shunt-shunt configuration, (25.83) delivers a driving-point input resistance of

$$R_{in} \approx \frac{f_{sp}R_{LT}}{f_{ps}} \tag{25.84}$$

and a driving-point output resistance of

$$R_{out} \approx \frac{f_{ps}R_{ST}}{f_{sp}} \tag{25.85}$$

Like the forward voltage gain, the driving-point input and output resistances of the series-shunt/shunt-series feedback amplifier are nominally independent of active element parameters. But note that the input resistance is directly proportional to the load resistance by a factor (f_{sp}/f_{ps}), that is the inverse of the proportionality constant that links the output resistance to the source resistance. Specifically,

$$\frac{f_{sp}}{f_{ps}} = \frac{R_{in}}{R_{LT}} = \frac{R_{ST}}{R_{out}} \tag{25.86}$$

Thus, although R_{in} and R_{out} are reliably determined for well-defined load and source terminations, they do not track one another as well as they do in the series-series/shunt-shunt amplifier. Using (25.86), the voltage gain in (25.83) is expressible as

$$A_V(R_{ST}, R_{LT}) \approx \frac{1}{f_{sp}\left(1 + \sqrt{\dfrac{R_{out}R_{ST}}{R_{in}R_{LT}}}\right)} \tag{25.87}$$

The simplified ac schematic diagram of a practical series-shunt/shunt-series feedback amplifier appears in Fig. 25.21. In this circuit, series-shunt feedback derives from the resistances, R_{EE1} and R_{F1}, and shunt-series feedback is determined by the resistances, R_{EE2} and R_{F2}. Since this circuit topology merges the series-shunt and shunt-series pairs, requisite frequency compensation, which is not shown in the subject figure, mirrors the relevant compensation schemes studied earlier. Note, however, that a cascade of only two open-loop gain stages renders compensation easier to implement and larger 3-dB bandwidths easier to achieve in the series-series/shunt-shunt circuit, which requires three open-loop gain stages for a single-ended application.

For high beta transistors having identical corresponding small-signal model parameters, a low-frequency analysis of the circuit in Fig. 25.21 gives a voltage gain of

$$A_V(R_{ST}, R_{LT}) \approx \left(\frac{\alpha R_{in}}{R_{in} + \alpha R_S}\right)\left(1 + \frac{R_{F1}}{R_{EE1}}\right) \tag{25.88}$$

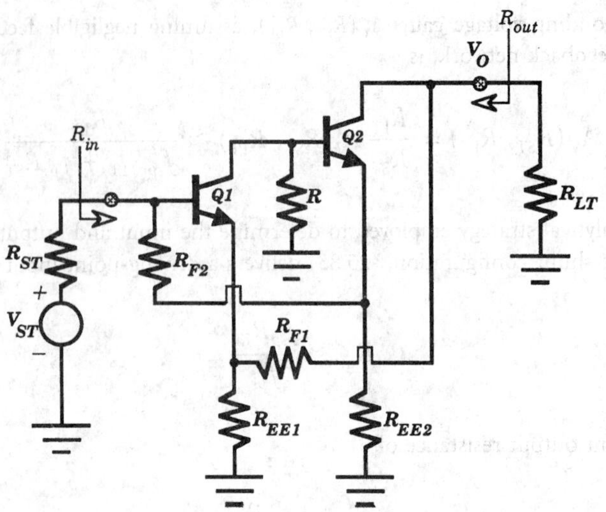

FIGURE 25.21 AC schematic diagram of a series-shunt/shunt-series dual-loop feedback amplifier.

where the driving-point input resistance, R_{in}, of the subject amplifier is

$$R_{\text{in}} \approx \alpha R_{\text{LT}} \left(\frac{1 + \dfrac{R_{F2}}{R_{\text{EE2}}}}{1 + \dfrac{R_{F1}}{R_{\text{EE1}}} + \dfrac{R_{\text{LT}}}{R_{\text{EE1}} \| R_{\text{EE2}}}} \right) \tag{25.89}$$

The driving point output resistance, R_{out}, is

$$R_{\text{out}} \approx R_{\text{ST}} \left(\frac{1 + \dfrac{R_{F1}}{R_{\text{EE1}}}}{1 + \dfrac{R_{F2}}{R_{\text{EE2}}} + \dfrac{R_{\text{ST}}}{R_{\text{EE1}} \| R_{\text{EE2}}}} \right) \tag{25.90}$$

25.7 Summary

This section documents small-signal performance equations, general operating characteristics, and engineering design guidelines for the six most commonly used global feedback circuits. These observations derive from analyses based on the judicious application of signal flow theory to the small-signal model that results when the subject feedback network is supplanted by an appropriate two-port parameter equivalent circuit.

Four of the six fundamental feedback circuits are single-loop architectures.

1. The series-shunt feedback amplifier functions best as a voltage amplifier in that its input resistance is large, and its output resistance is small. Because only two gain stages are required in the open loop, the amplifier is relatively easy to compensate for acceptable closed-loop damping and features potentially large 3-dB bandwidth. A computationally efficient analysis aimed toward determining loop gain, closed-loop gain, *I*/*O* resistances, and the condition that renders feed-forward through the feedback network inconsequential is predicated on replacing the feedback subcircuit with its *h*-parameter model.
2. The shunt-series feedback amplifier is a current amplifier in that its input resistance is small, and its output resistance is large. Like its series-shunt dual, only two gain stages are

required in the open loop. Computationally efficient analyses are conducted by replacing the feedback subcircuit with its g-parameter model.

3. The shunt-shunt feedback amplifier is a transresistance signal processor in that both its input and output resistances are small. Although this amplifier can be realized theoretically with only a single open-loop stage, a sufficiently large loop gain generally requires a cascade of three open-loop stages. As a result, pole splitting is invariably required to ensure an open-loop dominant pole response, thereby limiting the achievable closed-loop bandwidth. In addition, compensation of the feedback loop may be required for acceptable closed-loop damping. The bandwidth and stability problems implicit to the use of three open-loop gain stages can be circumvented by a balanced differential realization, which requires a cascade of only two open-loop gain stages. Computationally efficient analyses are conducted by replacing the feedback subcircuit with its y-parameter model.
4. The series-series feedback amplifier is a transconductance signal processor in that both its input and output resistances are large. Like its shunt-shunt counterpart, its implementation generally requires a cascade of three open-loop gain stages. Computationally efficient analyses are conducted by replacing the feedback subcircuit with its z-parameter model.

The two remaining feedback circuits are dual loop topologies that can stabilize the driving point input and output resistances, as well as the forward gain characteristics, with respect to shifts in active element parameters. One of these latter architectures, the series-series/shunt-shunt feedback amplifier, is particularly well suited to electronic applications that require a multistage cascade.

1. The series-series/shunt-shunt feedback amplifier coalesces the series-series architecture with its shunt-shunt dual. It is particularly well suited to applications, such as wideband communication networks, which require match terminated source and load resistances. Requisite frequency compensation and broadbanding criteria mirror those incorporated in the series-series and shunt-shunt single-loop feedback topologies.
2. The series-shunt/shunt-series feedback amplifier coalesces the series-shunt architecture with its shunt-series dual. Although its input resistance can be designed to match the source resistance seen by the input port of the amplifier, and its output resistance can be matched to the load resistance driven by the amplifier, match terminated operation ($R_{in} = R_{ST} = R_{LT} = R_{out}$) is not feasible. Requisite frequency compensation and broadbanding criteria mirror those incorporated in the series-shunt and shunt-series single-loop feedback topologies.

erences

[1] J. Millman and A. Grabel, *Microelectronics* 2nd ed., New York: McGraw-Hill, 1987, ch. 12.

[2] A. B. Grebene, *Bipolar and MOS Analog Integrated Circuit Design*, New York: Wiley-Interscience, 1984, pp. 424–432.

[3] R. G. Meyer, R. Eschenbach, and R. Chin, "A wideband ultralinear amplifier from DC to 300 MHz," *IEEE J. Solid-State Circuits*, vol. SC-9, pp. 167–175, Aug. 1974.

[4] A. S. Sedra and K. C. Smith, *Microelectronic Circuits*, New York: Holt, Rinehart, and Winston, 1987, pp. 428–441.

[5] J. M. Early, "Effects of space–charge layer widening in junction transistors," *Proc. IRE*, vol. 46, pp. 1141–1152, Nov. 1952.

[6] A. J. Cote Jr. and J. B. Oakes, *Linear Vacuum-Tube And Transistor Circuits*, New York: McGraw-Hill, 1961, pp. 40–46.

[7] R. G. Meyer and R. A. Blauschild, "A four-terminal wideband monolithic amplifier," *IEEE J. Solid-State Circuits*, vol. SC-17, pp. 634–638, Dec. 1981.

[8] M. Ohara, Y. Akazawa, N. Ishihara, and S. Konaka, "Bipolar monolithic amplifiers for a gigabit optical repeater," *IEEE J. Solid-State Circuits*, vol. SC-19, pp. 491–497, Aug. 1985.

[9] M. J. N. Sibley, R. T. Univin, D. R. Smith, B. A. Boxall, and R. J. Hawkins, "A monolithic transimpedance preamplifier for high speed optical receivers," *British Telecommunicat. Tech. J.*, vol. 2, pp. 64–66, July 1984.

[10] J. F. Kukielka and C. P. Snapp, "Wideband monolithic cascadable feedback amplifiers using silicon bipolar technology," *IEEE Microwave Millimeter-Wave Circuits Symp. Dig.*, vol. 2, pp. 330, 331, June 1982.

[11] R. G. Meyer, M. J. Shensa, and R. Eschenbach, "Cross modulation and intermodulation in amplifiers at high frequencies," *IEEE J. Solid-State Circuits*, vol. SC-7, pp. 16–23, Feb. 1972.

[12] K. H. Chan and R. G. Meyer, "A low distortion monolithic wide-band amplifier," *IEEE J. Solid-State Circuits*, vol. SC-12, pp. 685–690, Dec. 1977.

[13] A. Arbel, "Multistage transistorized current modules," *IEEE Trans. Circuits Syst.*, vol. CT-13, pp. 302–310, Sept. 1966.

[14] A. Arbel, *Analog Signal Processing and Instrumentation*, London: Cambridge Univ., 1980, ch. 3.

[15] W. G. Beall, "New feedback techniques for high performance monolithic wideband amplifiers," Electron. Res. Group, Univ. Southern California, Tech. Memo., Jan. 1990.

26

General Feedback Theory*

.i-Kai Chen
'ersity of Illinois,
ago

.1 Introduction

In Chapter 24.2, we used the ideal feedback model to study the properties of feedback amplifiers. The model is useful only if we can separate a feedback amplifier into the basic amplifier $\mu(s)$ and the feedback network $\beta(s)$. The procedure is difficult and sometimes virtually impossible, because the forward path may not be strictly unilateral, the feedback path is usually bilateral, and the input and output coupling networks are often complicated. Thus, the ideal feedback model is not an adequate representation of a practical amplifier. In the remainder of this section, we shall develop Bode's feedback theory, which is applicable to the general network configuration and avoids the necessity of identifying the transfer functions $\mu(s)$ and $\beta(s)$.

Bode's feedback theory [2] is based on the concept of return difference, which is defined in terms of network determinants. We show that the return difference is a generalization of the concept of the feedback factor of the ideal feedback model, and can be measured physically from the amplifier itself. We then introduce the notion of null return difference and discuss its physical significance. Since the feedback theory will be formulated in terms of the first- and second-order cofactors of the elements of the indefinite-admittance matrix of a feedback circuit, we first review briefly the formulation of the indefinite-admittance matrix.

.2 The Indefinite-Admittance Matrix

Figure 26.1 is an n-terminal network N composed of an arbitrary number of active and passive network elements connected in any way whatsoever. Let $V_1, V_2, \cdots, V_n$ be the Laplace-transformed potentials measured between terminals $1, 2, \cdots, n$ and some arbitrary but unspecified reference point, and let $I_1, I_2, \cdots, I_n$ be the Laplace-transformed currents entering the terminals $1, 2, \cdots, n$ from outside the network. Since the network N together with its load is linear, the terminal

*References for this chapter can be found on page 897.

·8341-2/95/$0.00+$.50

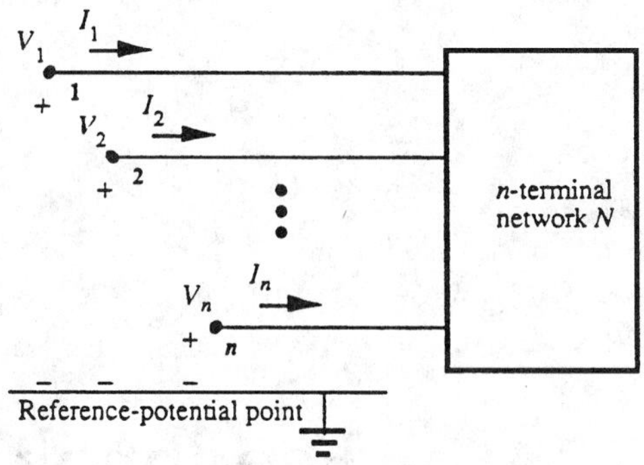

FIGURE 26.1 The general symbolic representation of an n-terminal network.

currents and voltages are related by the equation

$$\begin{bmatrix} I_1 \\ I_2 \\ \vdots \\ I_n \end{bmatrix} = \begin{bmatrix} y_{11} & y_{12} & \cdots & y_{1n} \\ y_{21} & y_{22} & \cdots & y_{2n} \\ \vdots & \vdots & \vdots & \vdots \\ y_{n1} & y_{n2} & \cdots & y_{nn} \end{bmatrix} \begin{bmatrix} V_1 \\ V_2 \\ \vdots \\ V_n \end{bmatrix} + \begin{bmatrix} J_1 \\ J_2 \\ \vdots \\ J_n \end{bmatrix} \tag{26.1}$$

or more compactly as

$$\mathbf{I}(s) = \mathbf{Y}(s)\mathbf{V}(s) + \mathbf{J}(s) \tag{26.2}$$

where J_k $(k = 1, 2, \cdots, n)$ denotes the current flowing into the kth terminal when all terminals of N are grounded to the reference point. The coefficient matrix $\mathbf{Y}(s)$ is called the **indefinite-admittance matrix** because the reference point for the potentials is some arbitrary but unspecified point outside the network. Notice that the symbol $\mathbf{Y}(s)$ is used to denote either the admittance matrix or the indefinite-admittance matrix. This should not create any confusion because the context will tell. In the remainder of this section, we shall deal exclusively with the indefinite-admittance matrix.

We remark that the short-circuit currents J_k result from the independent sources and/or initial conditions in the interior of N. For our purposes, we shall consider all independent sources outside the network and set all initial conditions to zero. Hence, $\mathbf{J}(s)$ is considered to be zero, and (26.2) becomes

$$\mathbf{I}(s) = \mathbf{Y}(s)\mathbf{V}(s) \tag{26.3}$$

where the elements y_{ij} of $\mathbf{Y}(s)$ can be obtained as

$$y_{ij} = \left.\frac{I_i}{V_j}\right|_{V_x=0,\, x\neq j} \tag{26.4}$$

As an illustration, consider a small-signal equivalent model of a transistor shown in Fig. 26.2.

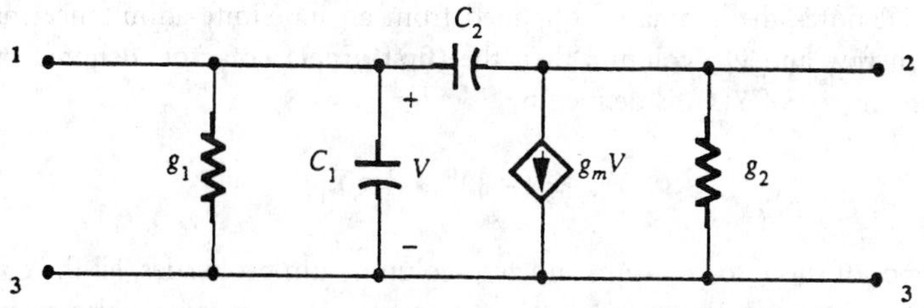

FIGURE 26.2 A small-signal equivalent network of a transistor.

Its indefinite-admittance matrix is found to be

$$\mathbf{Y}(s) = \begin{bmatrix} g_1 + sC_1 + sC_2 & -sC_2 & -g_1 - sC_1 \\ g_m - sC_2 & g_2 + sC_2 & -g_2 - g_m \\ -g_1 - sC_1 - g_m & -g_2 & g_1 + g_2 + g_m + sC_1 \end{bmatrix} \tag{26.5}$$

Observe that the sum of elements of each row or column is equal to zero. That these properties are valid in general for the indefinite-admittance matrix will now be demonstrated.

To see that the sum of the elements in each column of $\mathbf{Y}(s)$ equals zero, we add all n equations of (26.1) to yield

$$\sum_{i=1}^{n}\sum_{j=1}^{n} y_{ji}V_i = \sum_{m=1}^{n} I_m - \sum_{m=1}^{n} J_m = 0 \tag{26.6}$$

The last equation is obtained by appealing to Kirchhoff's current law for the node corresponding to the reference point. Setting all the terminal voltages to zero except the kth one, which is nonzero, gives

$$V_k \sum_{j=1}^{n} y_{jk} = 0 \tag{26.7}$$

Since $V_k \neq 0$, it follows that the sum of the elements of each column of $\mathbf{Y}(s)$ equals zero. Thus, the indefinite-admittance matrix is always singular.

To demonstrate that each row sum of $\mathbf{Y}(s)$ is also zero, we recognize that because the point of zero potential may be chosen arbitrarily, the currents J_k and I_k remain invariant when all the terminal voltages V_k are changed by the same but arbitrary constant amount. Thus, if $\mathbf{V}_0$ is an n-vector, each element of which is $v_0 \neq 0$, then

$$\mathbf{I}(s) - \mathbf{J}(s) = \mathbf{Y}(s)[\mathbf{V}(s) + \mathbf{V}_0] = \mathbf{Y}(s)\mathbf{V}(s) + \mathbf{Y}(s)\mathbf{V}_0 \tag{26.8}$$

which after invoking (26.2) yields that

$$\mathbf{Y}(s)\mathbf{V}_0 = \mathbf{0} \tag{26.9}$$

or

$$\sum_{j=1}^{n} y_{ij} = 0, \qquad i = 1, 2, \cdots, n \tag{26.10}$$

showing that each row sum of $\mathbf{Y}(s)$ equals zero.

Thus, if $\mathbf{Y}_{uv}$ denotes the submatrix obtained from an indefinite-admittance matrix $\mathbf{Y}(s)$ by deleting the uth row and vth column, then the **(first-order) cofactor**, denoted by the symbol Y_{uv}, of the element y_{uv} of $\mathbf{Y}(s)$, is defined by

$$Y_{uv} = (-1)^{u+v} \det \mathbf{Y}_{uv} \tag{26.11}$$

As a consequence of the zero-row-sum and zero-column-sum properties, all the cofactors of the elements of the indefinite-admittance matrix are equal. Such a matrix is also referred to as the **equicofactor matrix**. If Y_{uv} and Y_{ij} are any two cofactors of the elements of $\mathbf{Y}(s)$, then

$$Y_{uv} = Y_{ij} \tag{26.12}$$

for all u, v, i and j. For the indefinite-admittance matrix $\mathbf{Y}(s)$ of (26.5) it is straightforward to verify that all of its nine cofactors are equal to

$$Y_{uv} = s^2 C_1 C_2 + s(C_1 g_2 + C_2 g_1 + C_2 g_2 + g_m C_2) + g_1 g_2 \tag{26.13}$$

for $u, v = 1, 2, 3$.

Denote by $\mathbf{Y}_{rp,sq}$ the submatrix obtained from $\mathbf{Y}(s)$ by striking out rows r and s and columns p and q. Then the **second-order cofactor**, denoted by the symbol $Y_{rp,sq}$ of the elements y_{rp} and y_{sq} of $\mathbf{Y}(s)$ is a scalar quantity defined by the relation

$$Y_{rp,sq} = \operatorname{sgn}(r - s)\operatorname{sgn}(p - q)(-1)^{r+p+s+q} \det \mathbf{Y}_{rp,sq} \tag{26.14}$$

where $r \neq s$ and $p \neq q$, and

$$\operatorname{sgn} u = +1 \quad \text{if } u > 0 \tag{26.15a}$$

$$\operatorname{sgn} u = -1 \quad \text{if } u < 0 \tag{26.15b}$$

The symbols $\mathbf{Y}_{uv}$ and Y_{uv} or $\mathbf{Y}_{rp,sq}$ and $Y_{rp,sq}$ should not create any confusion because one is in boldface whereas the other is italic. Also, for our purposes, it is convenient to define

$$Y_{rp,sq} = 0, \qquad r = s \quad \text{or} \quad p = q \tag{26.16a}$$

or

$$\operatorname{sgn} 0 = 0 \tag{26.16b}$$

This convention will be followed throughout the remainder of this section.

As an example, consider the hybrid-pi equivalent network of a transistor shown in Fig. 26.3. Assume that each node is an accessible terminal of a four-terminal network. Its indefinite-admittance matrix is found to be

$$\mathbf{Y}(s) = \begin{bmatrix} 0.02 & 0 & -0.02 & 0 \\ 0 & 5 \times 10^{-12}s & 0.2 - 5 \times 10^{-12}s & -0.2 \\ -0.02 & -5 \times 10^{-12}s & 0.024 + 105 \times 10^{-12}s & -0.004 - 10^{-10}s \\ 0 & 0 & -0.204 - 10^{-10}s & 0.204 + 10^{-10}s \end{bmatrix} \tag{26.17}$$

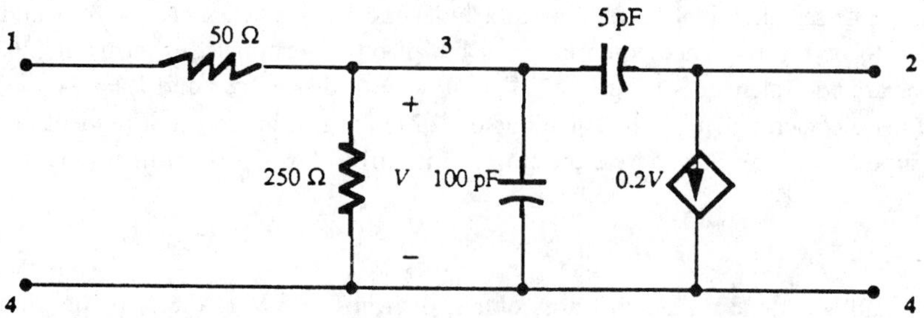

FIGURE 26.3 The hybrid-pi equivalent network of a transistor.

The second-order cofactor $Y_{31,42}$ and $Y_{11,34}$ of the elements of $\mathbf{Y}(s)$ of (26.17) are computed as follows:

$$Y_{31,42} = \text{sgn}(3-4)\text{sgn}(1-2)(-1)^{3+1+4+2}\det\begin{bmatrix} -0.02 & 0 \\ 0.2 - 5\times 10^{-12}s & -0.2 \end{bmatrix}$$

$$= 0.004 \qquad (26.18\text{a})$$

$$Y_{11,34} = \text{sgn}(1-3)\text{sgn}(1-4)(-1)^{1+1+3+4}\det\begin{bmatrix} 5\times 10^{-12}s & 0.2 - 5\times 10^{-12}s \\ 0 & -0.204 - 10^{-10}s \end{bmatrix}$$

$$= 5\times 10^{-12}s(0.204 + 10^{-10}s) \qquad (26.18\text{b})$$

The usefulness of the indefinite-admittance matrix lies in the fact that it facilitates the computation of the driving-point or transfer functions between any pair of nodes or from any pair of nodes to any other pair. In the following, we present elegant, compact, and explicit formulas that express the network functions in terms of the ratios of the first- and/or second-order cofactors of the elements of the indefinite-admittance matrix.

Assume that a current source is connected between any two nodes r and s so that a current I_{sr} is injected into the rth node and at the same time is extracted from the sth node. Suppose that an ideal voltmeter is connected from node p to node q so that it indicates the potential rise from q to p, as depicted symbolically in Fig. 26.4. Then the **transfer impedance**, denoted by the symbol $z_{rp,sq}$, between the node pairs rs and pq of the network of Fig. 26.4 is defined by the relation

$$z_{rp,sq} = \frac{V_{pq}}{I_{sr}} \qquad (26.19)$$

with all initial conditions and independent sources inside N set to zero. The representation is, of course, quite general. When $r = p$ and $s = q$, the transfer impedance $z_{rp,sq}$ becomes the *driving-point impedance* $z_{rr,ss}$ between the terminal pair rs.

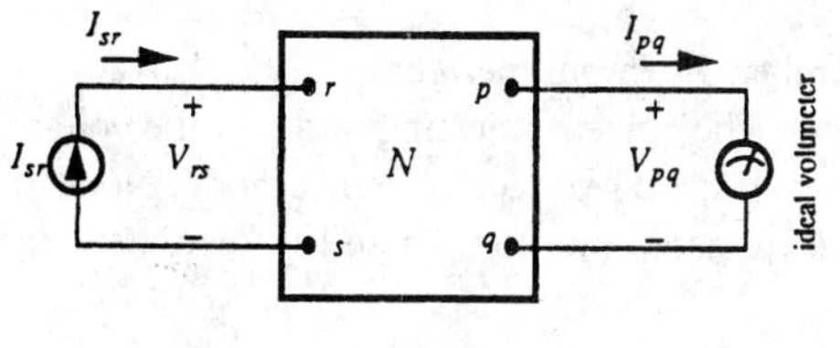

FIGURE 26.4 The symbolic representation for the measurement of the transfer impedance.

In Fig. 26.4, set all initial conditions and independent sources in N to zero and choose terminal q to be the reference-potential point for all other terminals. In terms of (26.1) these operations are equivalent to setting $\mathbf{J} = \mathbf{0}$, $V_q = 0$, $I_x = 0$ for $x \neq r$, s and $I_r = -I_s = I_{sr}$. Since $\mathbf{Y}(s)$ is an equicofactor matrix, the equations of (26.1) are not linearly independent and one of them is superfluous. Let us suppress the sth equation from (26.1), which then reduces to

$$\mathbf{I}_{-s} = \mathbf{Y}_{sq}\mathbf{V}_{-q} \tag{26.20}$$

where $\mathbf{I}_{-s}$ and $\mathbf{V}_{-q}$ denote the subvectors obtained from $\mathbf{I}$ and $\mathbf{V}$ of (26.3) by deleting the sth row and qth row, respectively. Applying Cramer's rule to solve for V_p yields

$$V_p = \frac{\det \tilde{\mathbf{Y}}_{sq}}{\det \mathbf{Y}_{sq}} \tag{26.21}$$

where $\tilde{\mathbf{Y}}_{sq}$ is the matrix derived from $\mathbf{Y}_{sq}$ by replacing the column corresponding to V_p by $\mathbf{I}_{-s}$. We recognize that $\mathbf{I}_{-s}$ is in the pth column if $p < q$ but in the $(p-1)$th column if $p > q$. Furthermore, the row in which I_{sr} appears is the rth row if $r < s$ but is the $(r-1)$th row if $r > s$. Thus, we obtain

$$(-1)^{s+q} \det \tilde{\mathbf{Y}}_{sq} = I_{sr} Y_{rp,sq} \tag{26.22}$$

In addition, we have

$$\det \mathbf{Y}_{sq} = (-1)^{s+q} Y_{sq} \tag{26.23}$$

Substituting these in (26.21) in conjunction with (26.19), we obtain

$$z_{rp,sq} = \frac{Y_{rp,sq}}{Y_{uv}} \tag{26.24}$$

$$z_{rr,ss} = \frac{Y_{rr,ss}}{Y_{uv}} \tag{26.25}$$

in which we have invoked the fact that $Y_{sq} = Y_{uv}$.

The *voltage gain*, denoted by $g_{rp,sq}$, between the node pairs rs and pq of the network of Fig. 26.4 is defined by

$$g_{rp,sq} = \frac{V_{pq}}{V_{rs}} \tag{26.26}$$

again with all initial conditions and independent sources in N being set to zero. Thus, from (26.24) and (26.25) we obtain

$$g_{rp,sq} = \frac{z_{rp,sq}}{z_{rr,ss}} = \frac{Y_{rp,sq}}{Y_{rr,ss}} \tag{26.27}$$

The symbols have been chosen to help us remember. In the numerators of (26.24), (26.25), and (26.27), the order of the subscripts is as follows: r, the current injecting node; p, the voltage measurement node; s, the current extracting node; and q, the voltage reference node. Nodes r and p designate the input and output transfer measurement, and nodes s and q form a sort of double datum.

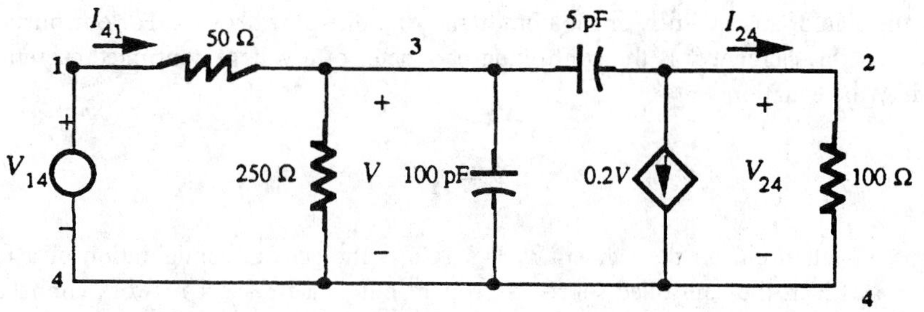

FIGURE 26.5 A transistor amplifier used to illustrate the computation of $g_{rp,sq}$.

As an illustration, we consider the hybrid-pi transistor equivalent network of Fig. 26.3. For this transistor, suppose that we connect a 100-Ω load resistor between nodes 2 and 4, and excite the resulting circuit by a voltage source V_{14}, as depicted in Fig. 26.5. To simplify our notation, let $p = 10^{-9}s$. The indefinite-admittance matrix of the amplifier is found to be

$$\mathbf{Y}(s) = \begin{bmatrix} 0.02 & 0 & -0.02 & 0 \\ 0 & 0.01 + 0.005p & 0.2 - 0.005p & -0.21 \\ -0.02 & -0.005p & 0.024 + 0.105p & -0.004 - 0.1p \\ 0 & -0.01 & -0.204 - 0.1p & 0.214 + 0.1p \end{bmatrix} \tag{26.28}$$

To compute the voltage gain $g_{12,44}$, we appeal to (26.27) and obtain

$$g_{12,44} = \frac{V_{24}}{V_{14}} = \frac{Y_{12,44}}{Y_{11,44}} = \frac{p - 40}{5p^2 + 21.7p + 2.4} \tag{26.29}$$

The input impedance facing the voltage source V_{14} is determined by

$$z_{11,44} = \frac{V_{14}}{I_{41}} = \frac{Y_{11,44}}{Y_{uv}} = \frac{Y_{11,44}}{Y_{44}} = \frac{50p^2 + 217p + 24}{p^2 + 4.14p + 0.08} \tag{26.30}$$

To compute the current gain defined as the ratio of the current I_{24} in the 100-Ω resistor to the input current I_{41}, we apply (26.24) and obtain

$$\frac{I_{24}}{I_{41}} = 0.01\frac{V_{24}}{I_{41}} = 0.01z_{12,44} = 0.01\frac{Y_{12,44}}{Y_{44}} = \frac{0.1p - 4}{p^2 + 4.14p + 0.08} \tag{26.31}$$

Finally, to compute the transfer admittance defined as the ratio of the load current I_{24} to the input voltage V_{14}, we appeal to (26.27) and obtain

$$\frac{I_{24}}{V_{14}} = 0.01\frac{V_{24}}{V_{14}} = 0.01g_{12,44} = 0.01\frac{Y_{12,44}}{Y_{11,44}} = \frac{p - 40}{500p^2 + 2170p + 240} \tag{26.32}$$

5.3 The Return Difference

In the study of feedback amplifier response, we are usually interested in how a particular element of the amplifier affects that response. This element is either crucial in terms of its effect on the entire system or of primary concern to the designer. It may be the transfer function of an active device, the gain of an amplifier, or the immittance of a one-port network. For our purposes, we assume that this element x is the controlling parameter of a voltage-controlled current source defined by the equation

$$I = xV \tag{26.33}$$

To focus our attention on the element x, Fig. 26.6 is the general configuration of a feedback amplifier in which the controlled source is brought out as a two-port network connected to a general four-port network, along with the input source combination of I_s and admittance Y_1 and the load admittance Y_2.

We remark that the two-port representation of a controlled source (26.33) is quite general. It includes the special situation where a one-port element is characterized by its immittance. In this case, the controlling voltage V is the terminal voltage of the controlled current source I, and x becomes the one-port admittance.

The *return difference* $F(x)$ of a feedback amplifier with respect to an element x is defined as the ratio of the two functional values assumed by the first-order cofactor of an element of its indefinite-admittance matrix under the condition that the element x assumes its nominal value and the condition that the element x assumes the value zero. To emphasize the importance of the feedback element x, we express the indefinite-admittance matrix $\mathbf{Y}$ of the amplifier as a function of x, even though it is also a function of the complex-frequency variable s, and write $\mathbf{Y} = \mathbf{Y}(x)$. Then we have [3]

$$F(x) \equiv \frac{Y_{uv}(x)}{Y_{uv}(0)} \tag{26.34}$$

where

$$Y_{uv}(0) = Y_{uv}(x)\Big|_{x=0} \tag{26.35}$$

The physical significance of the return difference will now be considered. In the network of Fig. 26.6, the input, the output, the controlling branch, and the controlled source are labeled as indicated. Then the element x enters the indefinite-admittance matrix $\mathbf{Y}(x)$ in a rectangular

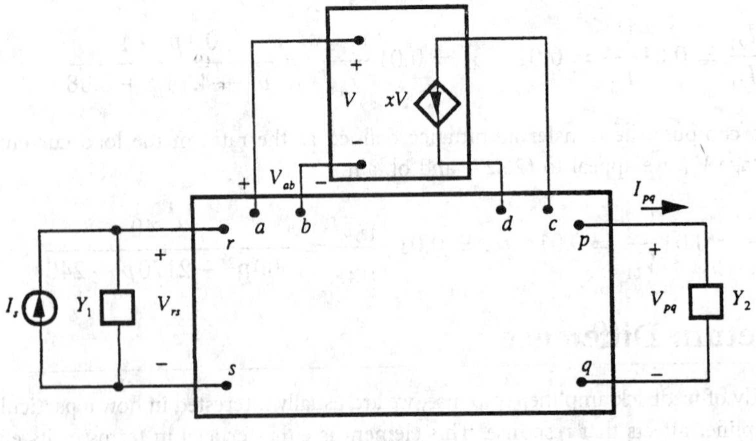

FIGURE 26.6 The general configuration of a feedback amplifier.

pattern as shown below:

$$\mathbf{Y}(x) = \begin{array}{c} \\ a \\ b \\ c \\ d \end{array} \begin{array}{c} \begin{array}{cccc} a & b & c & d \end{array} \\ \left[\begin{array}{cccc} & & & \\ & & & \\ x & -x & & \\ -x & x & & \end{array}\right] \end{array} \tag{26.36}$$

If in Fig. 26.6 we replace the controlled current source xV by an independent current source of x A and set the excitation current source I_s to zero, the indefinite-admittance matrix of the resulting network is simply $\mathbf{Y}(0)$. By appealing to (26.24) the new voltage V'_{ab} appearing at terminals a and b of the controlling branch is found to be

$$V'_{ab} = x\frac{Y_{da,cb}(0)}{Y_{uv}(0)} = -x\frac{Y_{ca,db}(0)}{Y_{uv}(0)} \tag{26.37}$$

Notice that the current injecting point is terminal d, not c.

The above operation of replacing the controlled current source by an independent current source and setting the excitation I_s to zero can be represented symbolically as in Fig. 26.7. Observe that the controlling branch is broken off as marked and a 1-V voltage source is applied to the right of the breaking mark. This 1-V sinusoidal voltage of a fixed angular frequency produces a current of x A at the controlled current source. The voltage appearing at the left of the breaking mark caused by this 1-V excitation is then V'_{ab} as indicated. This returned voltage V'_{ab} has the same physical significance as the loop transmission $\mu\beta$ defined for the ideal feedback model in Chapter 24. To see this, we set the input excitation to the ideal feedback model to zero, break the forward path, and apply a unit input to the right of the break, as shown in Fig. 26.8. The signal appearing at the left of the break is precisely the loop transmission.

For this reason, we introduce the concept of **return ratio** T, which is defined as the negative of the voltage appearing at the controlling branch when the controlled current source is replaced by an independent current source of x A and the input excitation is set to zero. Thus, the return ratio T is simply the negative of the returned voltage V'_{ab}, or $T = -V'_{ab}$. With this in mind, we

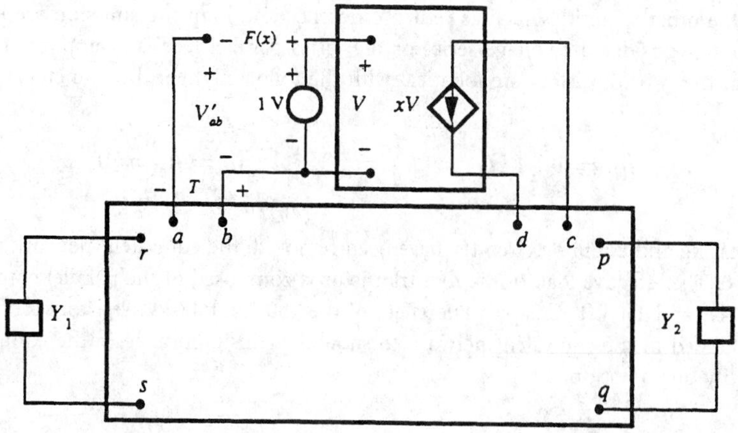

FIGURE 26.7 The physical interpretation of the return difference with respect to the controlling parameter of a voltage-controlled current source.

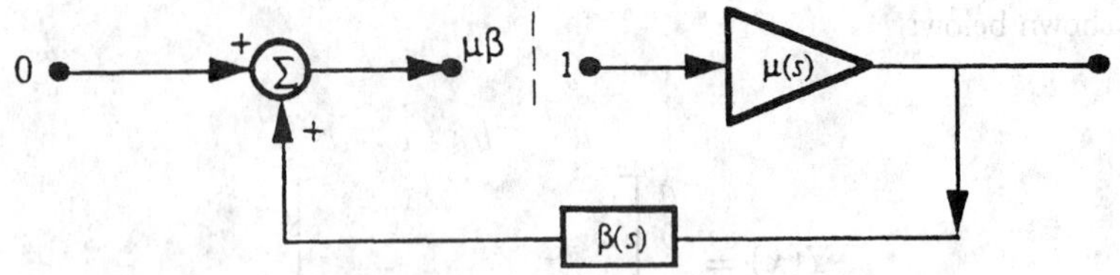

FIGURE 26.8 The physical interpretation of the loop transmission.

next compute the difference between the 1-V excitation and the returned voltage V'_{ab}, obtaining

$$1 - V'_{ab} = 1 + x\frac{Y_{ca,db}}{Y_{uv}(0)} = \frac{Y_{uv}(0) + xY_{ca,db}}{Y_{uv}(0)} = \frac{Y_{db}(0) + xY_{ca,db}}{Y_{db}(0)}$$
$$= \frac{Y_{db}(x)}{Y_{db}(0)} = \frac{Y_{uv}(x)}{Y_{uv}(0)} = F(x) \tag{26.38}$$

in which we have invoked the identities $Y_{uv} = Y_{ij}$ and

$$Y_{db}(x) = Y_{db}(0) + xY_{ca,db} \tag{26.39}$$

We remark that we write $Y_{ca,db}(x)$ as $Y_{ca,db}$ because it is independent of x. In other words, the return difference $F(x)$ is simply the difference of the 1-V excitation and the returned voltage V'_{ab} as illustrated in Fig. 26.7, and hence its name. Since

$$F(x) = 1 + T = 1 - \mu\beta \tag{26.40}$$

we conclude that the return difference has the same physical significance as the feedback factor of the ideal feedback model. The significance of the above physical interpretations is that it permits us to determine the return ratio T or $-\mu\beta$ by measurement. Once the return ratio is measured, the other quantities such as return difference and loop transmission are known.

To illustrate, consider the voltage-series or the series-parallel feedback amplifier of Fig. 26.9. Assume that the two transistors are identical with the following hybrid parameters:

$$h_{ie} = 1.1\ \text{k}\Omega, \qquad h_{fe} = 50, \qquad h_{re} = h_{oe} = 0 \tag{26.41}$$

After the biasing and coupling circuitry have been removed, the equivalent network is presented in Fig. 26.10. The effective load of the first transistor is composed of the parallel combination of the 10, 33, 47, and 1.1-kΩ resistors. The effect of the 150- and 47-kΩ resistors can be ignored; they are included in the equivalent network to show their insignificance in the computation.

To simplify our notation, let

$$\tilde{\alpha}_k = \alpha_k \times 10^{-4} = \frac{h_{fe}}{h_{ie}} = 455 \times 10^{-4}, \qquad k = 1, 2 \tag{26.42}$$

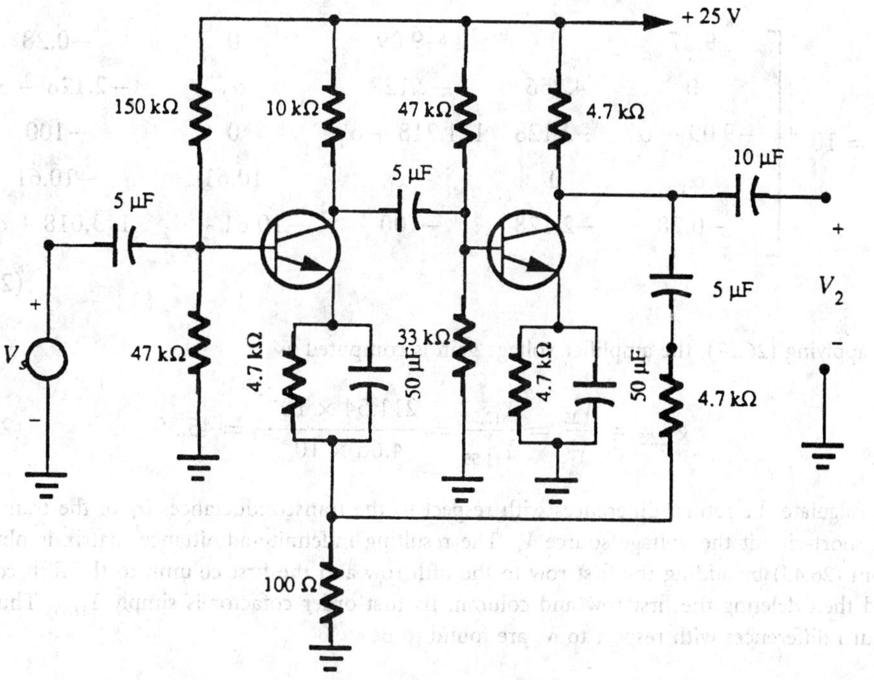

FIGURE 26.9 A voltage-series feedback amplifier together with its biasing and coupling circuitry.

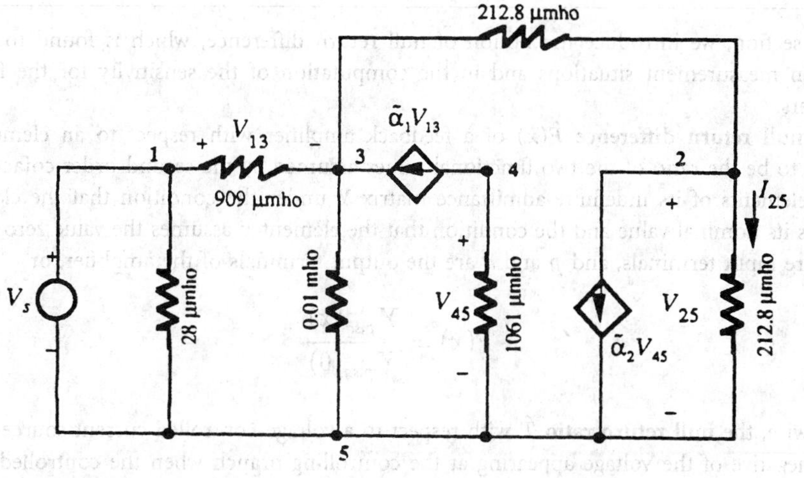

FIGURE 26.10 An equivalent network of the feedback amplifier of Fig. 26.9.

The subscript k is used to distinguish the transconductances of the first and the second transistors. The indefinite-admittance matrix of the feedback amplifier of Fig. 26.9 is found to be

$$\mathbf{Y} = 10^{-4}\begin{bmatrix} 9.37 & 0 & -9.09 & 0 & -0.28 \\ 0 & 4.256 & -2.128 & \alpha_2 & -2.128-\alpha_2 \\ -9.09-\alpha_1 & -2.128 & 111.218+\alpha_1 & 0 & -100 \\ \alpha_1 & 0 & -\alpha_1 & 10.61 & -10.61 \\ -0.28 & -2.128 & -100 & -10.61-\alpha_2 & 113.018+\alpha_2 \end{bmatrix} \tag{26.43}$$

By applying (26.27), the amplifier voltage gain is computed as

$$g_{12,55} = \frac{V_{25}}{V_s} = \frac{Y_{12,55}}{Y_{11,55}} = \frac{211.54 \times 10^{-7}}{4.66 \times 10^{-7}} = 45.39 \tag{26.44}$$

To calculate the return differences with respect to the transconductances $\tilde{\alpha}_k$ of the transistors, we short-circuit the voltage source V_s. The resulting indefinite-admittance matrix is obtained from (26.43) by adding the first row to the fifth row and the first column to the fifth column and then deleting the first row and column. Its first-order cofactor is simply $Y_{11,55}$. Thus, the return differences with respect to $\tilde{\alpha}_k$ are found to be

$$F(\tilde{\alpha}_1) = \frac{Y_{11,55}(\tilde{\alpha}_1)}{Y_{11,55}(0)} = \frac{466.1 \times 10^{-9}}{4.97 \times 10^{-9}} = 93.70 \tag{26.45a}$$

$$F(\tilde{\alpha}_2) = \frac{Y_{11,55}(\tilde{\alpha}_2)}{Y_{11,55}(0)} = \frac{466.1 \times 10^{-9}}{25.52 \times 10^{-9}} = 18.26 \tag{26.45b}$$

26.4 The Null Return Difference

In this section, we introduce the notion of null return difference, which is found to be very useful in measurement situations and in the computation of the sensitivity for the feedback amplifiers.

The **null return difference** $\hat{F}(x)$ of a feedback amplifier with respect to an element x is defined to be the ratio of the two functional values assumed by the second-order cofactor $Y_{rp,sq}$ of the elements of its indefinite-admittance matrix **Y** under the condition that the element x assumes its nominal value and the condition that the element x assumes the value zero where r and s are input terminals, and p and q are the output terminals of the amplifier, or

$$\hat{F}(x) = \frac{Y_{rp,sq}(x)}{Y_{rp,sq}(0)} \tag{26.46}$$

Likewise, the **null return ratio** $\hat{T}$ with respect to a voltage-controlled current source $I = xV$ is the negative of the voltage appearing at the controlling branch when the controlled current source is replaced by an independent current source of x A and when the input excitation is adjusted so that the output of the amplifier is identically zero.

Now we demonstrate that the null return difference is simply the return difference in the network under the situation that the input excitation I_s has been adjusted so that the output is identically zero. In the network of Fig. 26.6 suppose that we replace the controlled current source by an independent current source of x A. Then by applying formula (26.24) and the superposition principle, the output current I_{pq} at the load is found to be

$$I_{pq} = Y_2\left[I_s\frac{Y_{rp,sq}(0)}{Y_{uv}(0)} + x\frac{Y_{dp,cq}(0)}{Y_{uv}(0)}\right] \tag{26.47}$$

Setting $I_{pq} = 0$ or $V_{pq} = 0$ yields

$$I_s \equiv I_0 = -x\frac{Y_{dp,cq}(0)}{Y_{rp,sq}(0)} \tag{26.48}$$

in which $Y_{dp,cq}$ is independent of x. This adjustment is possible only if there is a direct transmission from the input to the output when x is set to zero. Thus, in the network of Fig. 26.7, if we connect an independent current source of strength I_0 at its input port, the voltage V'_{ab} is the negative of the null return ratio $\hat{T}$. Using (26.24) we obtain [4]

$$\begin{aligned}\hat{T} = -V'_{ab} &= -x\frac{Y_{da,cb}(0)}{Y_{uv}(0)} - I_0\frac{Y_{ra,sb}(0)}{Y_{uv}(0)}\\ &= -\frac{x[Y_{da,cb}(0)Y_{rp,sq}(0) - Y_{ra,sb}(0)Y_{dp,cq}(0)]}{Y_{uv}(0)Y_{rp,sq}(0)}\\ &= \frac{x\dot{Y}_{rp,sq}}{Y_{rp,sq}(0)} = \frac{Y_{rp,sq}(x)}{Y_{rp,sq}(0)} - 1\end{aligned} \tag{26.49}$$

where

$$\dot{Y}_{rp,sq} \equiv \frac{dY_{rp,sq}(x)}{dx} \tag{26.50}$$

This leads to

$$\hat{F}(x) = 1 + \hat{T} = 1 - V'_{ab} \tag{26.51}$$

showing that the null return difference $\hat{F}(x)$ is simply the difference of the 1-V excitation applied to the right of the breaking mark of the broken controlling branch of the controlled source and the returned voltage V'_{ab} appearing at the left of the breaking mark under the situation that the input signal I_s is adjusted so that the output is identically zero.

As an illustration, consider the voltage-series feedback amplifier of Fig. 26.9, an equivalent network of which is presented in Fig. 26.10. Using the indefinite-admittance matrix of (26.43) in conjunction with (26.42), the null return differences with respect to $\tilde{\alpha}_k$ are found to be

$$\hat{F}(\tilde{\alpha}_1) = \frac{Y_{12,55}(\tilde{\alpha}_1)}{Y_{12,55}(0)} = \frac{211.54 \times 10^{-7}}{205.24 \times 10^{-12}} = 103.07 \times 10^3 \tag{26.52a}$$

$$\hat{F}(\tilde{\alpha}_2) = \frac{Y_{12,55}(\tilde{\alpha}_2)}{Y_{12,55}(0)} = \frac{211.54 \times 10^{-7}}{104.79 \times 10^{-10}} = 2018.70 \tag{26.52b}$$

Alternatively, $\hat{F}(\tilde{\alpha}_1)$ can be computed by using its physical interpretation as follows. Replace the controlled source $\tilde{\alpha}_1 V_{13}$ in Fig. 26.10 by an independent current source of $\tilde{\alpha}_1$ A. We then

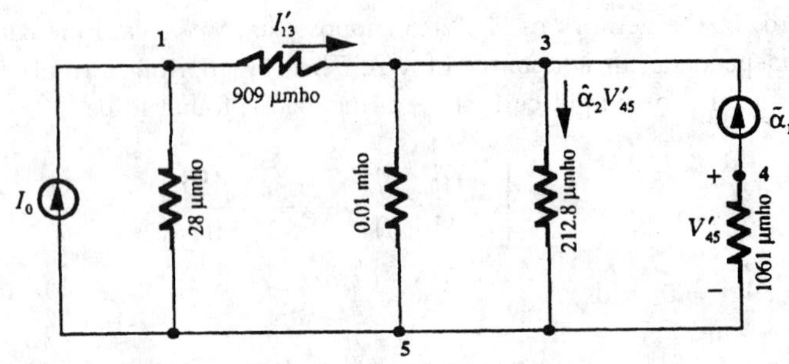

FIGURE 26.11 The network used to compute the null return difference $\hat{F}(\tilde{\alpha}_1)$ by its physical interpretation.

adjust the voltage source V_s so that the output current I_{25} is identically zero. Let I_0 be the input current resulting from this source. The corresponding network is presented in Fig. 26.11. From this network we obtain

$$\hat{F}(\tilde{\alpha}_1) = 1 + \hat{T} = 1 - V'_{13} = 1 - \frac{100V'_{35} + \alpha_2 V'_{45} - \alpha_1}{9.09} = 103.07 \times 10^3 \qquad (26.53)$$

Likewise, we can use the same procedure to compute the return difference $\hat{F}(\tilde{\alpha}_2)$.

27

The Network Functions and Feedback*

-Kai Chen
ersity of Illinois,
ago

We now study the effects of feedback on amplifier impedance and gain and obtain some useful relations among the return difference, the null return difference, and impedance functions in general.

Refer to the general feedback configuration of Fig. 26.6. Let w be a transfer function. As before, to emphasize the importance of the feedback element x, we write $w = w(x)$. To be definite, let $w(x)$ for the time being be the current gain between the output and input ports. Then from (26.24) we obtain

$$w(x) = \frac{I_{pq}}{I_s} = \frac{Y_2 V_{pq}}{I_s} = \frac{Y_{rp,sq}(x)}{Y_{uv}(x)} Y_2 \tag{27.1}$$

yielding

$$\frac{w(x)}{w(0)} = \frac{Y_{rp,sq}(x)}{Y_{uv}(x)} \frac{Y_{uv}(0)}{Y_{rp,sq}(0)} = \frac{\hat{F}(x)}{F(x)} \tag{27.2}$$

provided that $w(0) \neq 0$. This gives a very useful formula for computing the current gain:

$$w(x) = w(0)\frac{\hat{F}(x)}{F(x)} \tag{27.3}$$

Equation (27.3) remains valid if $w(x)$ represents the transfer impedance $z_{rp,sq} = V_{pq}/I_s$ instead of the current gain.

.1 Blackman's Formula

In particular, when $r = p$ and $s = q$, $w(x)$ represents the driving-point impedance $z_{rr,ss}(x)$ looking into the terminals r and s, and we have a somewhat different interpretation. In this case, $F(x)$ is the return difference with respect to the element x under the condition $I_s = 0$. Thus, $F(x)$ is the return difference for the situation when the port where the input impedance is defined is left open without a source and we write $F(x) = F(\text{input open circuited})$. Likewise, from Fig. 26.6, $\hat{F}(x)$ is the return difference with respect to x for the input excitation I_s and

*References for this chapter can be found on page 897.

-8341-2/95/$0.00+$.50

output response V_{rs} under the condition I_s is adjusted so that V_{rs} is identically zero. Thus, $\hat{F}(x)$ is the return difference for the situation when the port where the input impedance is defined is short circuited, and we write $\hat{F}(x) = F(\text{input short circuited})$. Consequently, the input impedance $Z(x)$ looking into a terminal pair can be conveniently expressed as

$$Z(x) = Z(0)\frac{F(\text{input short circuited})}{F(\text{input open circuited})} \tag{27.4}$$

This is the well-known **Blackman's formula** for computing an active impedance. The formula is extremely useful because the right-hand side can usually be determined rather easily. If x represents the controlling parameter of a controlled source in a single-loop feedback amplifier, then setting $x = 0$ opens the feedback loop and $Z(0)$ is simply a passive impedance. The return difference for x when the input port is short circuited or open circuited is relatively simple to compute because shorting out or opening a terminal pair frequently breaks the feedback loop. In addition, Blackman's formula can be used to determine the return difference by measurements. Because it involves two return differences, only one of them can be identified and the other must be known in advance. In the case of a single-loop feedback amplifier, it is usually possible to choose a terminal pair so that either the numerator or the denominator on the right-hand side of (27.4) is unity. If $F(\text{input short circuited}) = 1$, $F(\text{input open circuited})$ becomes the return difference under normal operating condition and we have

$$F(x) = \frac{Z(0)}{Z(x)} \tag{27.5}$$

On the other hand, if $F(\text{input open circuited}) = 1$, $F(\text{input short circuited})$ becomes the return difference under normal operating condition and we obtain

$$F(x) = \frac{Z(x)}{Z(0)} \tag{27.6}$$

Example 1. The network of Fig. 27.1 is a general active RC one-port realization of a rational impedance. We use Blackman's formula to verify that its input admittance is given by

$$Y = 1 + \frac{Z_3 - Z_4}{Z_1 - Z_2} \tag{27.7}$$

Appealing to (27.4), the input admittance written as $Y = Y(x)$ can be written as

$$Y(x) = Y(0)\frac{F(\text{input open circuited})}{F(\text{input short circuited})} \tag{27.8}$$

where $x = 2/Z_3$. By setting x to zero, the network used to compute $Y(0)$ is shown in Fig. 27.2. Its input admittance is found to be

$$Y(0) = \frac{Z_1 + Z_2 + Z_3 + Z_4 + 2}{Z_1 + Z_2} \tag{27.9}$$

When the input port is open-circuited, the network of Fig. 27.1 degenerates to that shown in Fig. 27.3. The return difference with respect to x is found to be

$$F(\text{input open circuited}) = 1 - V_3' = \frac{Z_1 + Z_3 - Z_2 - Z_4}{2 + Z_1 + Z_2 + Z_3 + Z_4} \tag{27.10}$$

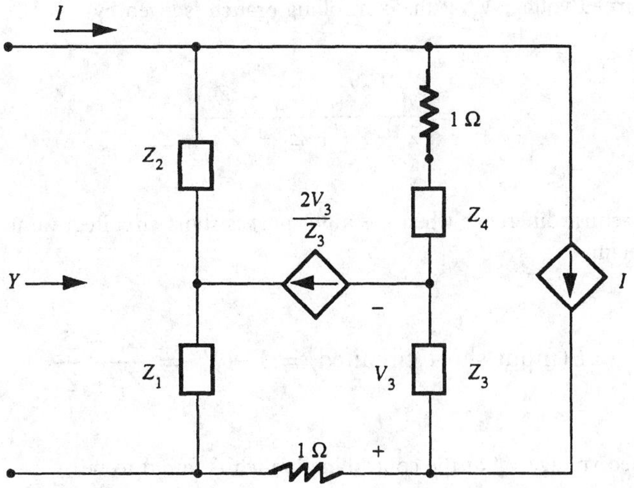

FIGURE 27.1 A general active *RC* one-port realization of a rational function.

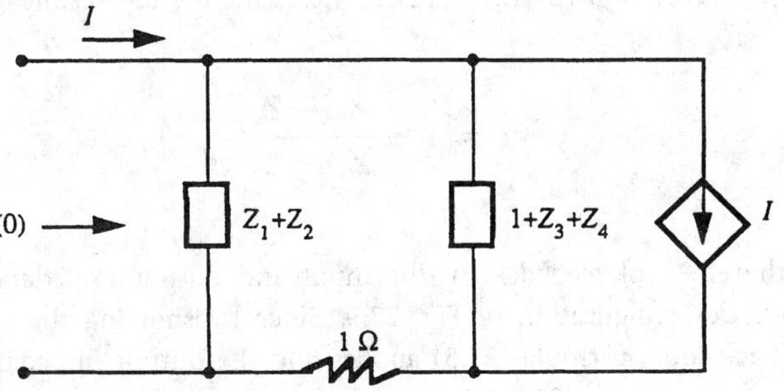

FIGURE 27.2 The network used to compute $Y(0)$.

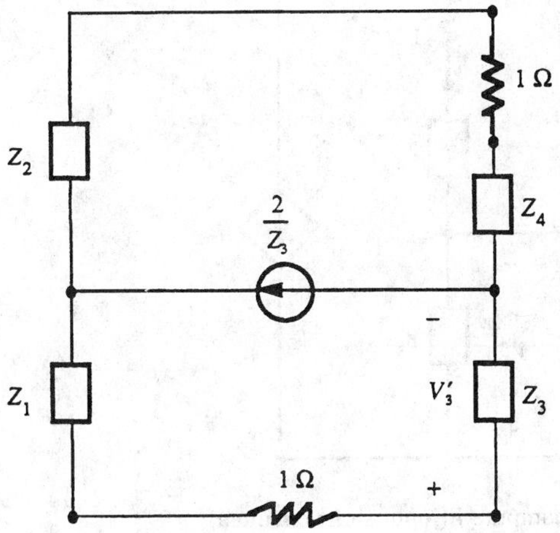

FIGURE 27.3 The network used to compute F(input open circuited).

where the returned voltage V_3' at the controlling branch is given by

$$V_3' = \frac{2(1 + Z_2 + Z_4)}{2 + Z_1 + Z_2 + Z_3 + Z_4} \tag{27.11}$$

To compute the return difference when the input port is short circuited, we use the network of Fig. 27.4, and obtain

$$F(\text{input short circuited}) = 1 - V_3'' = \frac{Z_1 - Z_2}{Z_1 + Z_2} \tag{27.12}$$

where the returned voltage V_3'' at the controlling branch is found to be

$$V_3'' = \frac{2Z_2}{Z_1 + Z_2} \tag{27.13}$$

Substituting (27.9), (27.10), and (27.12) in (27.8) yields the desired result.

$$Y = 1 + \frac{Z_3 - Z_4}{Z_1 - Z_2} \tag{27.14}$$

To determine the effect of feedback on the input and output impedances, we choose the series-parallel feedback configuration of Fig. 27.5. Since by shorting the terminals of Y_2, we interrupt the feedback loop, formula (27.5) applies and the output impedance across the load

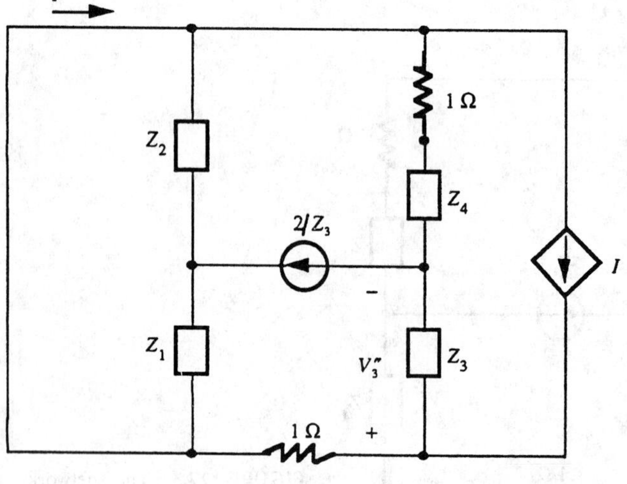

FIGURE 27.4 The network used to compute F(input short circuited).

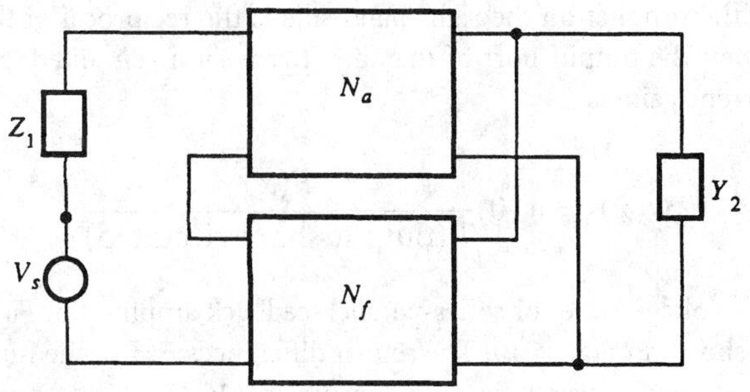

FIGURE 27.5 The series–parallel feedback configuration.

admittance Y_2 becomes

$$Z_{\text{out}}(x) = \frac{Z_{\text{out}}(0)}{F(x)} \tag{27.15}$$

showing that the impedance measured across the path of the feedback is reduced by the factor that is the normal value of the return difference with respect to the element x, where x is an arbitrary element of interest. For the input impedance of the amplifier looking into the voltage source V_s of Fig. 27.5, by open circuiting or removing the voltage source V_s, we break the feedback loop. Thus, formula (27.6) applies and the input impedance becomes

$$Z_{\text{in}}(x) = F(x)Z_{\text{in}}(0) \tag{27.16}$$

meaning that the impedance measured in series lines is increased by the same factor $F(x)$. Similar conclusions can be reached for other types of configurations discussed in Chapter 25 by applying Blackman's formula.

Again refer to the general feedback configuration of Fig. 26.6. If $w(x)$ represents the voltage gain V_{pq}/V_{rs} or the transfer admittance I_{pq}/V_{rs}. Then from (26.27) we can write

$$\frac{w(x)}{w(0)} = \frac{Y_{rp,sq}(x)}{Y_{rp,sq}(0)} \frac{Y_{rr,ss}(0)}{Y_{rr,ss}(x)} \tag{27.17}$$

The first term in the product on the right-hand side is the null return difference $\hat{F}(x)$ with respect to x for the input terminals r and s and output terminals p and q. The second term is the reciprocal of the null return difference with respect to x for the same input and output port at terminals r and s. This reciprocal can then be interpreted as the return difference with respect to x when the input port of the amplifier is short circuited. Thus, the voltage gain or the transfer admittance can be expressed as

$$w(x) = w(0)\frac{\hat{F}(x)}{F(\text{input short circuited})} \tag{27.18}$$

Finally, if $w(x)$ denotes the short circuit current gain I_{pq}/I_s as Y_2 approaches infinity, we obtain

$$\frac{w(x)}{w(0)} = \frac{Y_{rp,sq}(x)}{Y_{rp,sq}(0)} \frac{Y_{pp,qq}(0)}{Y_{pp,qq}(x)} \tag{27.19}$$

The second term in the product on the right-hand side is the reciprocal of the return difference with respect to x when the output port of the amplifier is short circuited, giving a formula for the short circuit current gain as

$$w(x) = w(0)\frac{\hat{F}(x)}{F(\text{output short circuited})} \tag{27.20}$$

Again consider the voltage-series or series-parallel feedback amplifier of Fig. 26.9 an equivalent network of which is shown in Fig. 26.10. The return differences $F(\tilde{\alpha}_k)$, the null return differences $\hat{F}(\tilde{\alpha}_k)$ and the voltage gain w were computed earlier in (26.45), (26.52) and (26.44), and are repeated below:

$$F(\tilde{\alpha}_1) = 93.70, \qquad F(\tilde{\alpha}_2) = 18.26 \tag{27.21a}$$

$$\hat{F}(\tilde{\alpha}_1) = 103.07 \times 10^3, \qquad \hat{F}(\tilde{\alpha}_2) = 2018.70 \tag{27.21b}$$

$$w = \frac{V_{25}}{V_s} = w(\tilde{\alpha}_1) = w(\tilde{\alpha}_2) = 45.39 \tag{27.21c}$$

We apply (27.18) to calculate the voltage gain w, as follows:

$$w(\tilde{\alpha}_1) = w(0)\frac{\hat{F}(\tilde{\alpha}_1)}{F(\text{input short circuited})} = 0.04126\frac{103.07 \times 10^3}{93.699} = 45.39 \tag{27.22}$$

where

$$w(0) = \left.\frac{Y_{12,55}(\tilde{\alpha}_1)}{Y_{11,55}(\tilde{\alpha}_1)}\right|_{\tilde{\alpha}_1=0} = \frac{205.24 \times 10^{-12}}{497.41 \times 10^{-11}} = 0.04126 \tag{27.23a}$$

$$F(\text{input short circuited}) = \frac{Y_{11,55}(\tilde{\alpha}_1)}{Y_{11,55}(0)} = \frac{466.07 \times 10^{-9}}{4.9741 \times 10^{-9}} = 93.699 \tag{27.23b}$$

and

$$w(\tilde{\alpha}_2) = w(0)\frac{\hat{F}(\tilde{\alpha}_2)}{F(\text{input short circuited})} = 0.41058\frac{2018.70}{18.26} = 45.39 \tag{27.24}$$

where

$$w(0) = \left.\frac{Y_{12,55}(\tilde{\alpha}_2)}{Y_{11,55}(\tilde{\alpha}_2)}\right|_{\tilde{\alpha}_2=0} = \frac{104.79 \times 10^{-10}}{255.22 \times 10^{-10}} = 0.41058 \tag{27.25a}$$

$$F(\text{input short circuited}) = \frac{Y_{11,55}(\tilde{\alpha}_2)}{Y_{11,55}(0)} = \frac{466.07 \times 10^{-9}}{25.52 \times 10^{-9}} = 18.26 \tag{27.25b}$$

27.2 The Sensitivity Function

One of the most important effects of negative feedback is its ability to make an amplifier less sensitive to the variations of its parameters because of aging, temperature variations, or other environment changes. A useful quantitative measure for the degree of dependence of an amplifier on a particular parameter is known as the sensitivity. The **sensitivity function**, written as $\mathcal{S}(x)$, for a given transfer function with respect to an element x is defined as the ratio of the fractional change in a transfer function to the fractional change in x for the situation when all changes concerned are differentially small. Thus, if $w(x)$ is the transfer function, the sensitivity function can be written as

$$\mathcal{S}(x) = \lim_{\Delta x \to 0} \frac{\Delta w/w}{\Delta x/x} = \frac{x}{w}\frac{\partial w}{\partial x} = x\frac{\partial \ln w}{\partial x} \tag{27.26}$$

Refer to the general feedback configuration of Fig. 26.6, and let $w(x)$ represent either the current gain I_{pq}/I_s or the transfer impedance V_{pq}/I_s for the time being. Then we obtain from (26.24)

$$w(x) = Y_2\frac{Y_{rp,sq}(x)}{Y_{uv}(x)} \quad \text{or} \quad \frac{Y_{rp,sq}(x)}{Y_{uv}(x)} \tag{27.27}$$

As before, we write

$$\dot{Y}_{uv}(x) = \frac{\partial Y_{uv}(x)}{\partial x} \tag{27.28a}$$

$$\dot{Y}_{rp,sq}(x) = \frac{\partial Y_{rp,sq}(x)}{\partial x} \tag{27.28b}$$

obtaining

$$Y_{uv}(x) = Y_{uv}(0) + x\dot{Y}_{uv}(x) \tag{27.29a}$$

$$Y_{rp,sq}(x) = Y_{rp,sq}(0) + x\dot{Y}_{rp,sq}(x) \tag{27.29b}$$

Substituting (27.27) in (27.26) in conjunction with (27.29) yields

$$\begin{aligned}\mathcal{S}(x) &= x\frac{\dot{Y}_{rp,sq}(x)}{Y_{rp,sq}(x)} - x\frac{\dot{Y}_{uv}(x)}{Y_{uv}(x)} = \frac{Y_{rp,sq}(x) - Y_{rp,sq}(0)}{Y_{rp,sq}(x)} - \frac{Y_{uv}(x) - Y_{uv}(0)}{Y_{uv}(x)} \\ &= \frac{Y_{uv}(0)}{Y_{uv}(x)} - \frac{Y_{rp,sq}(0)}{Y_{rp,sq}(x)} = \frac{1}{F(x)} - \frac{1}{\hat{F}(x)}\end{aligned} \tag{27.30}$$

Combining this with (27.3), we obtain

$$\mathcal{S}(x) = \frac{1}{F(x)}\left[1 - \frac{w(0)}{w(x)}\right] \tag{27.31}$$

Observe that if $w(0) = 0$, (27.31) becomes

$$\mathcal{S}(x) = \frac{1}{F(x)} \tag{27.32}$$

meaning that sensitivity is equal to the reciprocal of the return difference. For the ideal feedback model, the feedback path is unilateral. Hence, $w(0) = 0$ and

$$\mathcal{S} = \frac{1}{F} = \frac{1}{1+T} = \frac{1}{1-\mu\beta} \tag{27.33}$$

For a practical amplifier, $w(0)$ is usually very much smaller than $w(x)$ in the passband, and $F \approx 1/\mathcal{S}$ may be used as a good estimate of the reciprocal of the sensitivity in the same frequency band. A single-loop feedback amplifier composed of a cascade of common-emitter stages with a passive network providing the desired feedback fulfills this requirements. If in such a structure any one of the transistors fails, the forward transmission is nearly zero and $w(0)$ is practically zero. Our conclusion is that if the failure of any element will interrupt the transmission through the amplifier as a whole to nearly zero, the sensitivity is approximately equal to the reciprocal of the return difference with respect to that element. In the case of driving-point impedance, $w(0)$ is not usually smaller than $w(x)$, and the reciprocity relation is not generally valid.

Now assume that $w(x)$ represents the voltage gain. Substituting (26.27) in (27.26) results in

$$\begin{aligned}\mathcal{S}(x) &= x\frac{\dot{Y}_{rp,sq}(x)}{Y_{rp,sq}(x)} - x\frac{\dot{Y}_{rr,ss}(x)}{Y_{rr,ss}(x)} = \frac{Y_{rp,sq}(x) - Y_{rp,sq}(0)}{Y_{rp,sq}(x)} - \frac{Y_{rr,ss}(x) - Y_{rr,ss}(0)}{Y_{rr,ss}(x)} \\ &= \frac{Y_{rr,ss}(0)}{Y_{rr,ss}(x)} - \frac{Y_{rp,sq}(0)}{Y_{rp,sq}(x)} = \frac{1}{F(\text{input short circuited})} - \frac{1}{\hat{F}(x)}\end{aligned} \tag{27.34}$$

Combining this with (27.18) gives

$$\mathcal{S}(x) = \frac{1}{F(\text{input short circuited})}\left[1 - \frac{w(0)}{w(x)}\right] \tag{27.35}$$

Finally, if $w(x)$ denotes the short circuit current gain I_{pq}/I_s as Y_2 approaches infinity, the sensitivity function can be written as

$$\mathcal{S}(x) = \frac{Y_{pp,qq}(0)}{Y_{pp,qq}(x)} - \frac{Y_{rp,sq}(0)}{Y_{rp,sq}(x)} = \frac{1}{F(\text{output short circuited})} - \frac{1}{\hat{F}(x)} \tag{27.36}$$

which when combined with (27.20) yields

$$\mathcal{S}(x) = \frac{1}{F(\text{output short circuited})}\left[1 - \frac{w(0)}{w(x)}\right] \tag{27.37}$$

We remark that formulas (26.31), (26.35), and (26.37) are quite similar. If the return difference $F(x)$ is interpreted properly, they can all be represented by the single relation (26.31). As before, if $w(0) = 0$, the sensitivity for the voltage gain function is equal to the reciprocal of the return difference under the situation that the input port of the amplifier is short circuited, whereas the sensitivity for the short circuit current gain is the reciprocal of the return difference when the output port is short circuited.

Example 2. The network of Fig. 27.6 is a common-emitter transistor amplifier. After removing the biasing circuit and using the common-emitter hybrid model for the transistor at low frequencies, an equivalent network of the amplifier is presented in Fig. 27.7 with

$$I_s' = \frac{V_s}{R_1 + r_x} \tag{27.38a}$$

$$G_1' = \frac{1}{R_1'} = \frac{1}{R_1 + r_x} + \frac{1}{r_\pi} \tag{27.38b}$$

$$G_2' = \frac{1}{R_2'} = \frac{1}{R_2} + \frac{1}{R_c} \tag{27.38c}$$

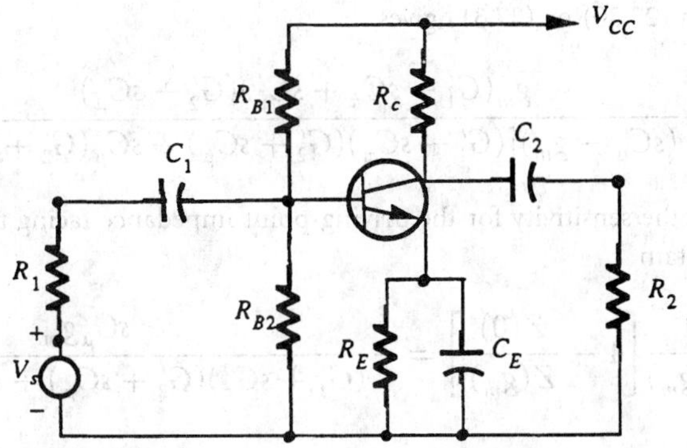

FIGURE 27.6 A common-emitter transistor feedback amplifier.

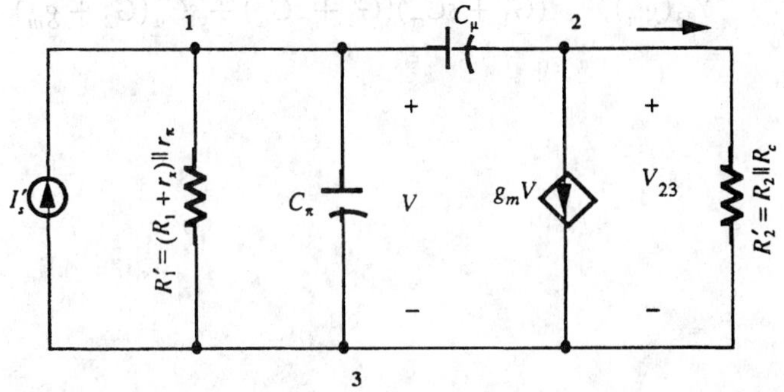

FIGURE 27.7 An equivalent network of the feedback amplifier of Fig. 27.6.

The indefinite-admittance matrix of the amplifier is found to be

$$\mathbf{Y} = \begin{bmatrix} G_1' + sC_\pi + sC_\mu & -sC_\mu & -G_1' - sC_\pi \\ g_m - sC_\mu & G_2' + sC_\mu & -G_2' - g_m \\ -G_1' - sC_\pi - g_m & -G_2' & G_1' + G_2' + sC_\pi + g_m \end{bmatrix} \quad (27.39)$$

Assume that the controlling parameter g_m is the element of interest. The return difference and the null return difference with respect to g_m in Fig. 27.7 with I_s' as the input port and R_2' as the output port, are found to be

$$F(g_m) = \frac{Y_{33}(g_m)}{Y_{33}(0)} = \frac{(G_1' + sC_\pi)(G_2' + sC_\mu) + sC_\mu(G_2' + g_m)}{(G_1' + sC_\pi)(G_2' + sC_\mu) + sC_\mu G_2'} \quad (27.40)$$

$$\hat{F}(g_m) = \frac{Y_{12,33}(g_m)}{Y_{12,33}(0)} = \frac{sC_\mu - g_m}{sC_\mu} = 1 - \frac{g_m}{sC_\mu} \quad (27.41)$$

The current gain I_{23}/I_s', as defined in Fig. 27.7, is computed as

$$w(g_m) = \frac{Y_{12,33}(g_m)}{R_2' Y_{33}(g_m)} = \frac{sC_\mu - g_m}{R_2'[(G_1' + sC_\pi)(G_2' + sC_\mu) + sC_\mu(G_2' + g_m)]} \quad (27.42)$$

Substituting these in (27.30) or (27.31) gives

$$\mathscr{S}(g_m) = -\frac{g_m(G_1' + sC_\pi + sC_\mu)(G_2' + sC_\mu)}{(sC_\mu - g_m)[(G_1' + sC_\pi)(G_2' + sC_\mu) + sC_\mu(G_2' + g_m)]} \tag{27.43}$$

Finally, we compute the sensitivity for the driving-point impedance facing the current source I_s'. From (27.31) we obtain

$$\mathscr{S}(g_m) = \frac{1}{F(g_m)}\left[1 - \frac{Z(0)}{Z(g_m)}\right] = -\frac{sC_\mu g_m}{(G_1' + sC_\pi)(G_2' + sC_\mu) + sC_\mu(G_2' + g_m)} \tag{27.44}$$

where

$$Z(g_m) = \frac{Y_{11,33}(g_m)}{Y_{33}(g_m)} = \frac{G_2' + sC_\mu}{(G_1' + sC_\pi)(G_2' + sC_\mu) + sC_\mu(G_2' + g_m)} \tag{27.45}$$

28

Measurement of Return Difference*

-Kai Chen
rsity of Illinois,
go

The zeros of the network determinant are called the **natural frequencies**. Their locations in the complex-frequency plane are extremely important in that they determine the stability of the network. A network is said to be **stable** if all of its natural frequencies are restricted to the open left-half side of the complex-frequency plane (LHS). If a network determinant is known, its roots can readily be computed explicitly with the aid of a computer if necessary, and the stability problem can then be settled directly. However, for a physical network there remains the difficulty of getting an accurate formulation of the network determinant itself, because every equivalent network is, to a greater or lesser extent, an idealization of the physical reality. As frequency is increased, parasitic effects of the physical elements must be taken into account. What is really needed is some kind of experimental verification that the network is stable and will remain so under certain prescribed conditions. The measurement of the return difference provides an elegant solution to this problem.

The return difference with respect to an element x in a feedback amplifier is defined by

$$F(x) = \frac{Y_{uv}(x)}{Y_{uv}(0)} \tag{28.1}$$

Since $Y_{uv}(x)$ denotes the nodal determinant, the zeros of the return difference are exactly the same as the zeros of the nodal determinant provided that there is no cancellation of common factors between $Y_{uv}(x)$ and $Y_{uv}(0)$. Therefore, if $Y_{uv}(0)$ is known to have no zeros in the closed right-half side of the complex-frequency plane (RHS), which is usually the case in a single-loop feedback amplifier when x is set to zero, $F(x)$ gives precisely the same information about the stability of a feedback amplifier as does the nodal determinant itself. The difficulty inherent in the measurement of the return difference with respect to the controlling parameter of a controlled source is that, in a physical system, the controlling branch and the controlled source both form part of a single device such as a transistor, and cannot be physically separated. In the following, we present a scheme that does not require the physical decomposition of a device.

Let a device of interest be brought out as a two-port network connected to a general four-port network as shown in Fig. 28.1. For our purposes, assume that this device is characterized by its y parameters, and represented by its y-parameter equivalent two-port network as indicated in Fig. 28.2, in which the parameter y_{21} controls signal transmission in the forward direction

*References for this chapter can be found on page 897.

-8341-2/95/$0.00 + $.50

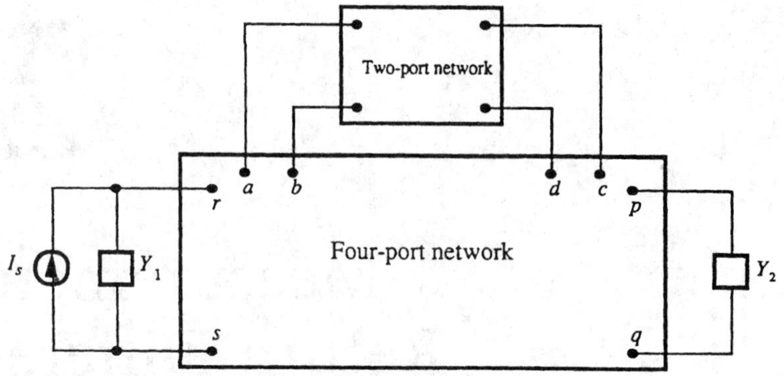

FIGURE 28.1 The general configuration of a feedback amplifier with a two-port device.

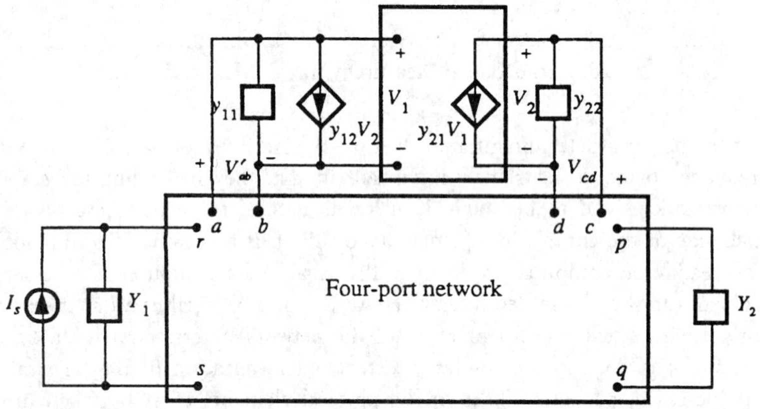

FIGURE 28.2 The representation of a two-port device in Fig. 28.1 by its y parameters.

through the device, whereas y_{12} gives the reverse transmission, accounting for the internal feedback within the device. Our objective is to measure the return difference with respect to the forward short circuit transfer admittance y_{21}.

28.1 Blecher's Procedure [1]

Let the two-port device be a transistor operated in the common-emitter configuration with terminals a, $b = d$, and c representing, respectively, the base, emitter, and collector terminals. To simplify our notation, let $a = 1$, $b = d = 3$ and $c = 2$, as exhibited explicitly in Fig. 28.3.

To measure $F(y_{21})$, we break the base terminal of the transistor and apply a 1-V excitation at its input as exhibited in Fig. 28.3. To ensure that the controlled current source $y_{21}V_{13}$ drives a replica of what it sees during normal operation, we connect an active one-port network composed of a parallel combination of the admittance y_{11} and a controlled current source $y_{12}V_{23}$ at terminals 1 and 3. The returned voltage V_{13} is precisely the negative of the return ratio with respect to the element y_{21}. If, in the frequency band of interest, the externally applied feedback is large compared with the internal feedback of the transistor, the controlled source $y_{12}V_{23}$ can be ignored. If, however, we find that this internal feedback cannot be ignored, we can simulate it by using an additional transistor, connected as shown in Fig. 28.4. This additional transistor must be matched as closely as possible to the one in question. The one-port admittance y_o denotes the admittance presented to the output port of the transistor under consideration as indicated in Figs. 28.3 and 28.4. For a common-emitter state, it is perfectly reasonable to assume that

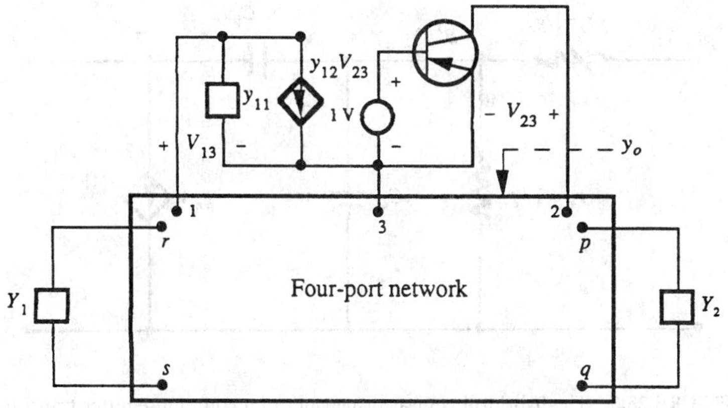

FIGURE 28.3 A physical interpretation of the return difference $F(y_{21})$ for a transistor operated in the common-emitter configuration and represented by its y parameters y_{ij}.

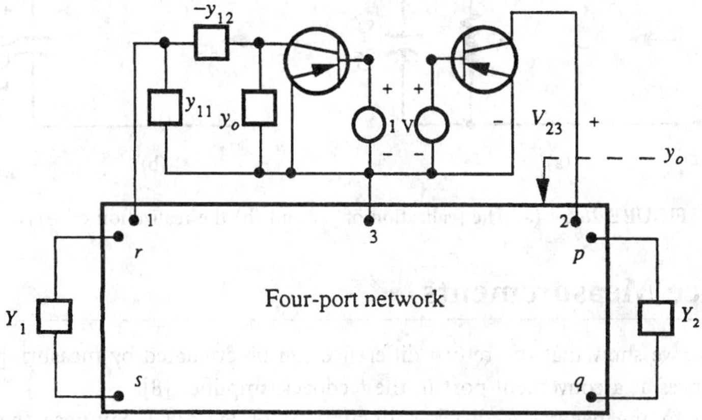

FIGURE 28.4 The measurement of return difference $F(y_{21})$ for a transistor operated in the common-emitter configuration and represented by its y parameters y_{ij}.

$|y_o| \gg |y_{12}|$ and $|y_{11}| \gg |y_{12}|$. Under these assumptions, it is straightforward to show that the Norton equivalent network looking into the two-port network at terminals 1 and 3 of Fig. 28.4 can be approximated by the parallel combination of y_{11} and $y_{12}V_{23}$, as indicated in Fig. 28.3. In Fig. 28.4 if the voltage sources have very low internal impedances, we can join together the two base terminals of the transistors and feed them both from a single voltage source of very low internal impedance. In this way, we avoid the need of using two separate sources. For the procedure to be feasible, we must demonstrate that the admittances y_{11} and $-y_{12}$ can be realized as the input admittances of one-port *RC* networks.

Consider the hybrid-pi equivalent network of a common-emitter transistor of Fig. 28.5, the short circuit admittance matrix of which is found to be

$$\mathbf{Y}_{sc} = \frac{1}{g_x + g_\pi + sC_\pi + sC_\mu} \begin{bmatrix} g_x(g_\pi + sC_\pi + sC_\mu) & -g_x sC_\mu \\ g_x(g_m - sC_\mu) & sC_\mu(g_x + g_\pi + sC_\pi + g_m) \end{bmatrix} \tag{28.2}$$

It is easy to confirm that the admittance y_{11} and $-y_{12}$ can be realized by the one-port networks of Fig. 28.6.

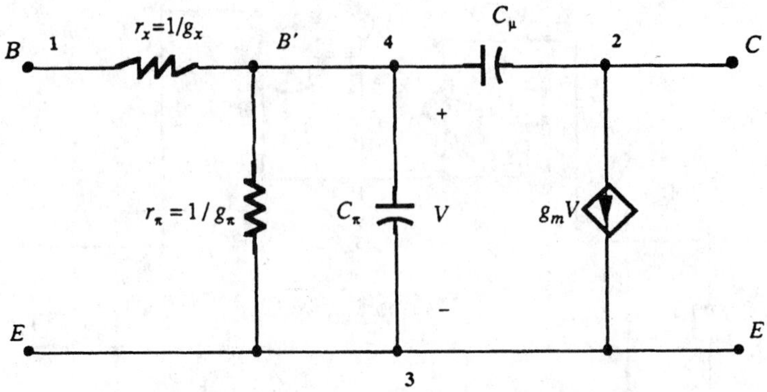

FIGURE 28.5 The hybrid-pi equivalent network of a common-emitter transistor.

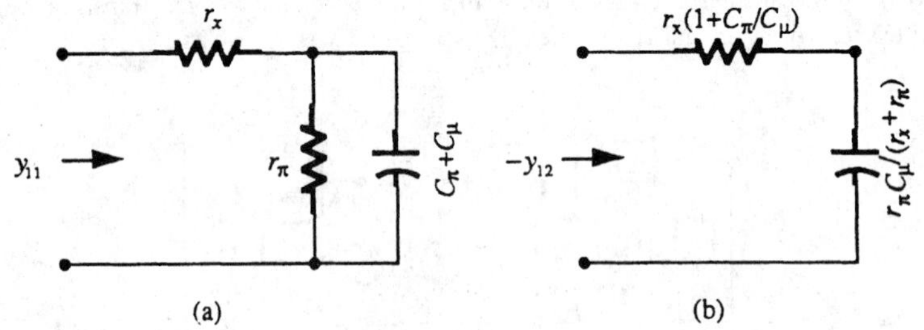

FIGURE 28.6 (a) The realization of y_{11} and (b) the realization of $-y_{12}$.

28.2 Impedance Measurements

In this section, we show that the return difference can be evaluated by measuring two driving-point impedances at a convenient port in the feedback amplifier [8].

Refer again to the general feedback configuration of Fig. 28.2. Suppose that we wish to evaluate the return difference with respect to the forward short circuit transfer admittance y_{21}. The controlling parameters y_{12} and y_{21} enter the indefinite-admittance matrix **Y** in the rectangular patterns as shown below:

$$\mathbf{Y}(x) = \begin{array}{c} \\ a \\ b \\ c \\ d \end{array} \begin{array}{c} \begin{array}{cccc} a & b & c & d \end{array} \\ \begin{bmatrix} & & y_{12} & -y_{12} \\ & & -y_{12} & y_{12} \\ y_{21} & -y_{21} & & \\ -y_{21} & y_{21} & & \end{bmatrix} \end{array} \tag{28.3}$$

To emphasize the importance of y_{12} and y_{21}, we again write $Y_{uv}(x)$ as $Y_{uv}(y_{12}, y_{21})$ and $z_{aa,bb}(x)$ as $z_{aa,bb}(y_{12}, y_{21})$. By appealing to formula (26.25), the impedance looking into terminals a and b of Fig. 28.2 is found to be

$$z_{aa,bb}(y_{12}, y_{21}) = \frac{Y_{aa,bb}(y_{12}, y_{21})}{Y_{dd}(y_{12}, y_{21})} \tag{28.4}$$

The return difference with respect to y_{21} is given by

$$F(y_{21}) = \frac{Y_{dd}(y_{12}, y_{21})}{Y_{dd}(y_{12}, 0)} \tag{28.5}$$

Combining these yields

$$F(y_{21})z_{aa,bb}(y_{12}, y_{21}) = \frac{Y_{aa,bb}(y_{12}, y_{21})}{Y_{dd}(y_{12}, 0)} = \frac{Y_{aa,bb}(0, 0)}{Y_{dd}(y_{12}, 0)}$$
$$= \frac{Y_{aa,bb}(0, 0)}{Y_{dd}(0, 0)} \frac{Y_{dd}(0, 0)}{Y_{dd}(y_{12}, 0)} = \frac{z_{aa,bb}(0, 0)}{F(y_{12})\big|_{y_{21}=0}} \tag{28.6}$$

obtaining a relation

$$F(y_{12})\Big|_{y_{21}=0} F(y_{21}) = \frac{z_{aa,bb}(0, 0)}{z_{aa,bb}(y_{12}, y_{21})} \tag{28.7}$$

among the return differences and the driving-point impedances. $F(y_{12})|_{y_{21}=0}$ is the return difference with respect to y_{12} when y_{21} is set to zero. This quantity can be measured by the arrangement of Fig. 28.7. $z_{aa,bb}(y_{12}, y_{21})$ is the driving-point impedance looking into terminals a and b of the network of Fig. 28.2. Finally, $z_{aa,bb}(0, 0)$ is the impedance to which $z_{aa,bb}(y_{12}, y_{21})$ reduces when the controlling parameters y_{12} and y_{21} are both set to zero. This impedance can be measured by the arrangement of Fig. 28.8. Note that in all three measurements, the independent current source I_s is removed.

Suppose that we wish to measure the return difference $F(y_{21})$ with respect to the forward transfer admittance y_{21} of a common-emitter transistor shown in Fig. 28.2. Then the return

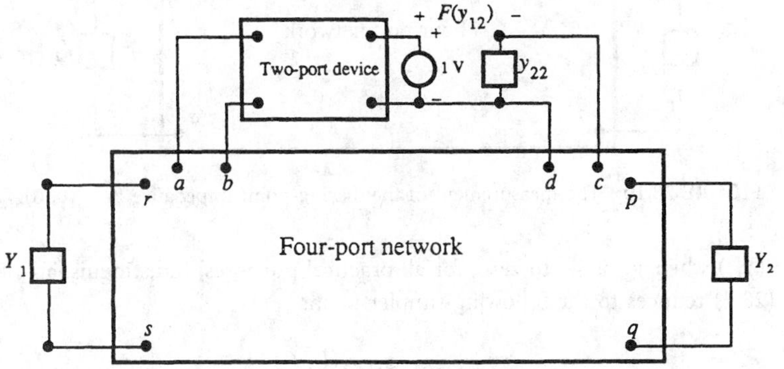

FIGURE 28.7 The measurement of the return difference $F(y_{12})$ with y_{21} being set to zero.

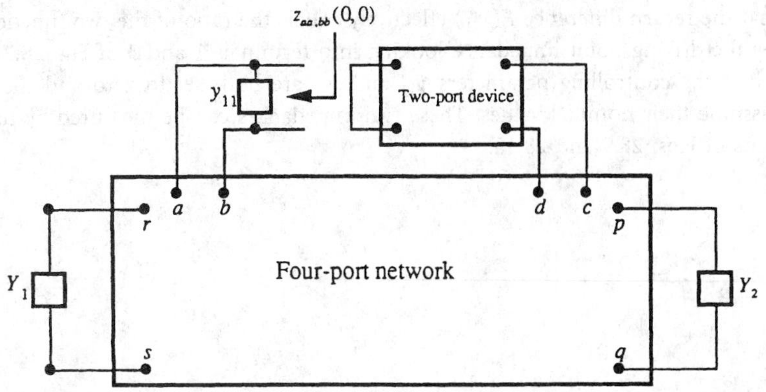

FIGURE 28.8 The measurement of the driving-point impedance $z_{aa,bb}(0, 0)$.

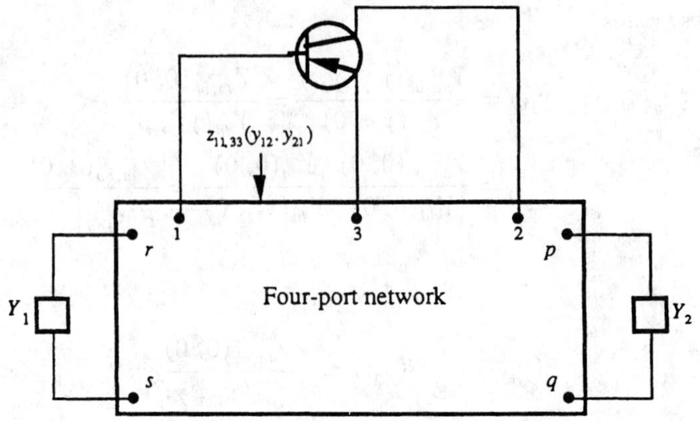

FIGURE 28.9 The measurement of the driving-point impedance $z_{11,33}(y_{12}, y_{21})$.

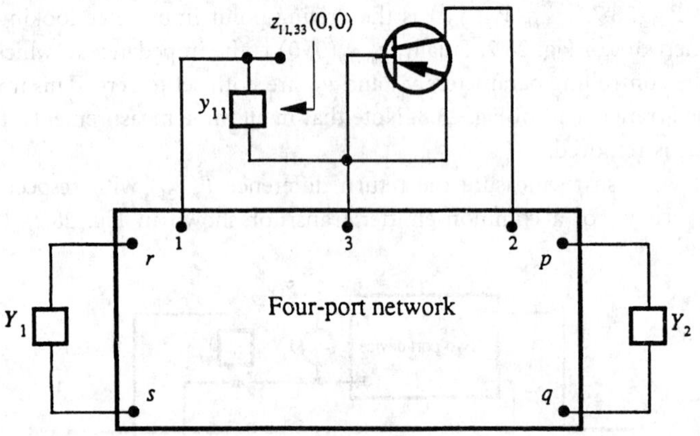

FIGURE 28.10 The measurement of the driving-point impedance $z_{11,33}(0, 0)$.

difference $F(y_{12})$ when y_{21} is set to zero, for all practical purposes, indistinguishable from unity. Therefore, (28.7) reduces to the following simpler form:

$$F(y_{21}) \approx \frac{z_{11,33}(0, 0)}{z_{11,33}(y_{12}, y_{21})} \tag{28.8}$$

showing that the return difference $F(y_{21})$ effectively equals the ratio of the two functional values assumed by the driving-point impedance looking into terminals 1 and 3 of Fig. 28.2 under the condition that the controlling parameters y_{12} and y_{21} are both set to zero and the condition that they assume their nominal values. These two impedances can be measured by the network arrangements of Figs. 28.9 and 28.10.

29

Multiple-Loop Feedback Amplifiers

i-Kai Chen
versity of Illinois,
cago

So far we have studied the single-loop feedback amplifiers. The concept of feedback was introduced in terms of return difference. We found that return difference is the difference between the unit applied signal and the returned signal. The returned signal has the same physical meaning as the loop transmission in the ideal feedback model. It plays an important role in the study of amplifier stability, its sensitivity to the variations of the parameters, and the determination of its transfer and driving-point impedances. The fact that return difference can be measured experimentally for many practical amplifiers indicates that we can include all the parasitic effects in the stability study, and that stability problem can be reduced to a Nyquist plot.

In this section, we study amplifiers that contain a multiplicity of inputs, outputs, and feedback loops. They are referred to as the **multiple-loop feedback amplifiers**. As might be expected, the notion of return difference with respect to an element is no longer applicable, because we are dealing with a group of elements. For this we generalize the concept of return difference for a controlled source to the notion of return difference matrix for a multiplicity of controlled sources. For measurement situations, we introduce the null return difference matrix and discuss its physical significance. We show that the determinant of the overall transfer function matrix can be expressed explicitly in terms of the determinants of the return difference and the null return difference matrices, thereby allowing us to generalize Blackman's formula for the input impedance.

9.1 Multiple-Loop Feedback Amplifier Theory

The general configuration of a multiple-input, multiple-output, and multiple-loop feedback amplifier is presented in Fig. 29.1, in which the input, output, and feedback variables may be either currents or voltages. For the specific arrangement of Fig. 29.1, the input and output

)3-8341-2/95/$0.00+$.50

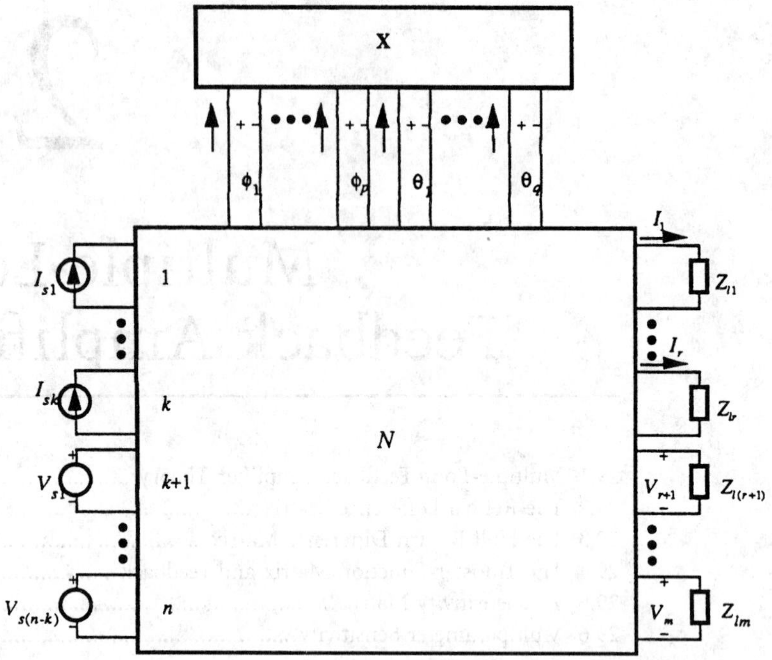

FIGURE 29.1 The general configuration of a multiple-input, multiple-output, and multiple-loop feedback amplifier.

variables are represented by an n-dimensional vector $\mathbf{u}$ and an m-dimensional vector $\mathbf{y}$ as

$$\mathbf{u}(s) = \begin{bmatrix} u_1 \\ u_2 \\ \vdots \\ u_k \\ u_{k+1} \\ u_{k+2} \\ \vdots \\ u_n \end{bmatrix} = \begin{bmatrix} I_{s1} \\ I_{s2} \\ \vdots \\ I_{sk} \\ V_{s1} \\ V_{s2} \\ \vdots \\ V_{s(n-k)} \end{bmatrix}, \qquad \mathbf{y}(s) = \begin{bmatrix} y_1 \\ y_2 \\ \vdots \\ y_r \\ y_{r+1} \\ y_{r+2} \\ \vdots \\ y_m \end{bmatrix} = \begin{bmatrix} I_1 \\ I_2 \\ \vdots \\ I_r \\ V_{r+1} \\ V_{r+2} \\ \vdots \\ V_m \end{bmatrix} \tag{29.1}$$

respectively. The elements of interest can be represented by a rectangular matrix $\mathbf{X}$ of order $q \times p$ relating the controlled and controlling variables by the matrix equation

$$\mathbf{\Theta} = \begin{bmatrix} \theta_1 \\ \theta_2 \\ \vdots \\ \theta_q \end{bmatrix} = \begin{bmatrix} x_{11} & x_{12} & \cdots & x_{1p} \\ x_{21} & x_{22} & \cdots & x_{2p} \\ \vdots & \vdots & \vdots & \vdots \\ x_{q1} & x_{q2} & \cdots & x_{qp} \end{bmatrix} \begin{bmatrix} \phi_1 \\ \phi_2 \\ \vdots \\ \phi_p \end{bmatrix} = \mathbf{X\Phi} \tag{29.2}$$

where the p-dimensional vector $\mathbf{\Phi}$ is called the **controlling vector**, and the q-dimensional vector $\mathbf{\Theta}$ the **controlled vector**. The controlled variables θ_k and the controlling variables ϕ_k can either be currents or voltages. The matrix $\mathbf{X}$ can represent either a transfer-function matrix or a driving-point function matrix. If $\mathbf{X}$ represents a driving-point function matrix, the vectors $\mathbf{\Theta}$ and $\mathbf{\Phi}$ are of the same dimension ($q = p$) and their components are the currents and voltages of a p-port network.

The general configuration of Fig. 29.1 can be represented equivalently by the block diagram of Fig. 29.2 in which N is a $(p + q + m + n)$-port network and the elements of interest are exhibited explicitly by the block $\mathbf{X}$. For the $(p + q + m + n)$-port network N, the vectors $\mathbf{u}$ and $\mathbf{\Theta}$ are its inputs, and the vectors $\mathbf{\Phi}$ and $\mathbf{y}$ its outputs. Since N is linear, the input and output vectors are related by the matrix equations

$$\mathbf{\Phi} = \mathbf{A\Theta} + \mathbf{Bu} \tag{29.3a}$$

$$\mathbf{y} = \mathbf{C\Theta} + \mathbf{Du} \tag{29.3b}$$

where $\mathbf{A}, \mathbf{B}, \mathbf{C}$, and $\mathbf{D}$ are transfer-function matrices of orders $p \times q, p \times n, m \times q$, and $m \times n$, respectively. The vectors $\mathbf{\Theta}$ and $\mathbf{\Phi}$ are not independent and are related by

$$\mathbf{\Theta} = \mathbf{X\Phi} \tag{29.3c}$$

The relationships among the above three linear matrix equations can also be represented by a matrix signal-flow graph as shown in Fig. 29.3 known as the **fundamental matrix feedback-flow graph**. The overall closed-loop **transfer-function matrix** of the multiple-loop feedback amplifier is defined by the equation

$$\mathbf{y} = \mathbf{W(X)u} \tag{29.4}$$

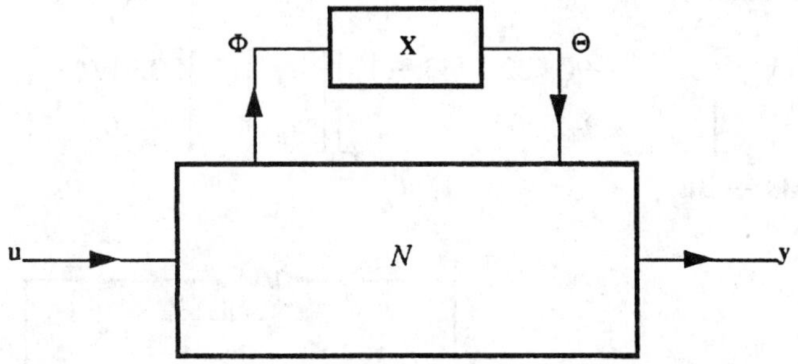

FIGURE 29.2 The block diagram of the general feedback configuration of Fig. 29.1.

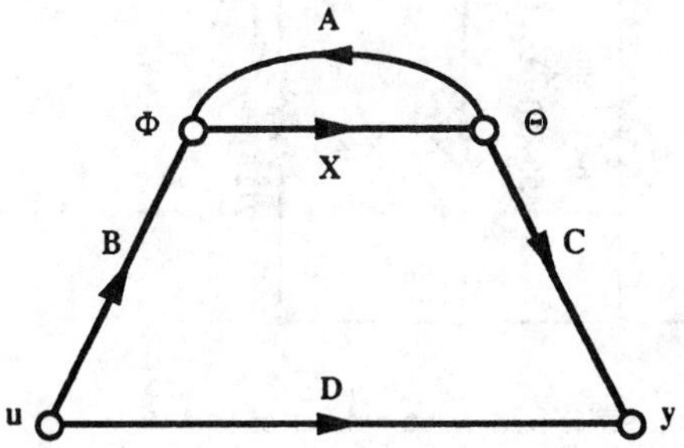

FIGURE 29.3 The fundamental matrix feedback-flow graph.

where $\mathbf{W}(\mathbf{X})$ is of order $m \times n$. As before, to emphasize the importance of $\mathbf{X}$, the matrix $\mathbf{W}$ is written as $\mathbf{W}(\mathbf{X})$ for the present discussion, even though it is also a function of the complex-frequency variable s. Combining the above matrix equations, the transfer-function matrix is found to be

$$\mathbf{W}(\mathbf{X}) = \mathbf{D} + \mathbf{C}\mathbf{X}(\mathbf{1}_p - \mathbf{A}\mathbf{X})^{-1}\mathbf{B} \tag{29.5a}$$

or

$$\mathbf{W}(\mathbf{X}) = \mathbf{D} + \mathbf{C}(\mathbf{1}_q - \mathbf{X}\mathbf{A})^{-1}\mathbf{X}\mathbf{B} \tag{29.5b}$$

where $\mathbf{1}_p$ denotes the identity matrix of order p. Clearly, we have

$$\mathbf{W}(\mathbf{0}) = \mathbf{D} \tag{29.6}$$

In particular, when $\mathbf{X}$ is square and nonsingular, (29.5) can be written as

$$\mathbf{W}(\mathbf{X}) = \mathbf{D} + \mathbf{C}(\mathbf{X}^{-1} - \mathbf{A})^{-1}\mathbf{B} \tag{29.7}$$

Example 3. Consider the voltage-series feedback amplifier of Fig. 26.9. An equivalent network is shown in Fig. 29.4 in which we have assumed that the two transistors are identical with $h_{ie} = 1.1$ kΩ, $h_{fe} = 50$, $h_{re} = h_{oe} = 0$. Let the controlling parameters of the two controlled sources be the elements of interest. Then we have

$$\boldsymbol{\Theta} = \begin{bmatrix} I_a \\ I_b \end{bmatrix} = 10^{-4}\begin{bmatrix} 455 & 0 \\ 0 & 455 \end{bmatrix}\begin{bmatrix} V_{13} \\ V_{45} \end{bmatrix} = \mathbf{X}\boldsymbol{\Phi} \tag{29.8}$$

Assume that the output voltage V_{25} and input current I_{51} are the output variables. Then the seven-port network N defined by the variables $V_{13}, V_{45}, V_{25}, I_{51}, I_a, I_b$, and V_s can be characterized by the matrix equations

$$\boldsymbol{\Phi} = \begin{bmatrix} V_{13} \\ V_{45} \end{bmatrix} = \begin{bmatrix} -90.782 & 45.391 \\ -942.507 & 0 \end{bmatrix}\begin{bmatrix} I_a \\ I_b \end{bmatrix} + \begin{bmatrix} 0.91748 \\ 0 \end{bmatrix}[V_s]$$
$$= \mathbf{A}\boldsymbol{\Theta} + \mathbf{B}\mathbf{u} \tag{29.9a}$$

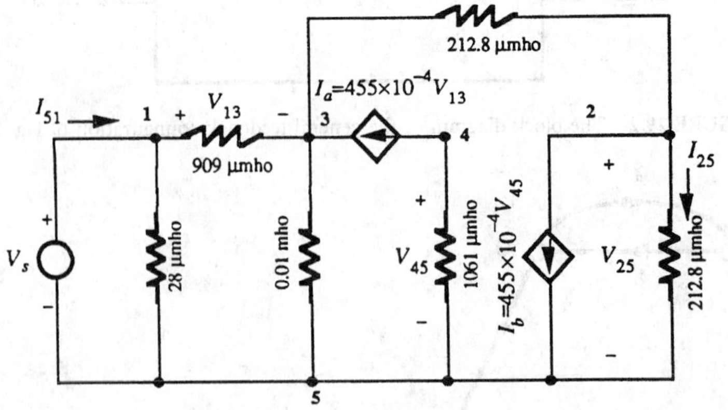

FIGURE 29.4 An equivalent network of the voltage-series feedback amplifier of Fig. 26.9.

$$\mathbf{y} = \begin{bmatrix} V_{25} \\ I_{51} \end{bmatrix} = \begin{bmatrix} 45.391 & -2372.32 \\ -0.08252 & 0.04126 \end{bmatrix} \begin{bmatrix} I_a \\ I_b \end{bmatrix} + \begin{bmatrix} 0.041260 \\ 0.000862 \end{bmatrix} [V_s]$$
$$= \mathbf{C\Theta} + \mathbf{Du} \tag{29.9b}$$

According to (29.4), the transfer-function matrix of the amplifier is defined by the matrix equation

$$\mathbf{y} = \begin{bmatrix} V_{25} \\ I_{51} \end{bmatrix} = \begin{bmatrix} w_{11} \\ w_{21} \end{bmatrix} [V_s] = \mathbf{W(X)u} \tag{29.10}$$

Since $\mathbf{X}$ is square and nonsingular, we can use (29.7) to calculate $\mathbf{W(X)}$:

$$\mathbf{W(X)} = \mathbf{D} + \mathbf{C}(\mathbf{X}^{-1} - \mathbf{A})^{-1}\mathbf{B} = \begin{bmatrix} 45.387 \\ 0.369 \times 10^{-4} \end{bmatrix} = \begin{bmatrix} w_{11} \\ w_{21} \end{bmatrix} \tag{29.11}$$

where

$$(\mathbf{X}^{-1} - \mathbf{A})^{-1} = 10^{-4} \begin{bmatrix} 4.856 & 10.029 \\ -208.245 & 24.914 \end{bmatrix} \tag{29.12}$$

obtaining the closed-loop voltage gain w_{11} and input impedance Z_{in} facing the voltage source V_s as

$$w_{11} = \frac{V_{25}}{V_s} = 45.387, \qquad Z_{in} = \frac{V_s}{I_{51}} = \frac{1}{w_{21}} = 27.1\ \text{k}\Omega \tag{29.13}$$

29.2 The Return Difference Matrix

In this section, we extend the concept of return difference with respect to an element to the notion of return difference matrix with respect to a group of elements.

In the fundamental feedback-flow graph of Fig. 29.3, suppose that we break the input of the branch with transmittance $\mathbf{X}$, set the input excitation vector $\mathbf{u}$ to zero, and apply a signal p-vector $\mathbf{g}$ to the right of the breaking mark, as depicted in Fig. 29.5. Then the returned signal

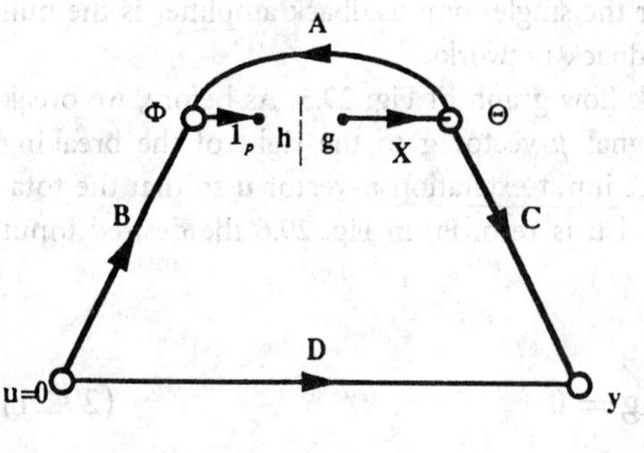

FIGURE 29.5 The physical interpretation of the loop-transmission matrix.

p-vector **h** to the left of the breaking mark is found to be

$$\mathbf{h} = \mathbf{AXg} \tag{29.14}$$

The square matrix **AX** is called the **loop-transmission matrix** and its negative is referred to as the **return ratio matrix** denoted by

$$\mathbf{T(X)} = -\mathbf{AX} \tag{29.15}$$

The difference between the applied signal vector **g** and the returned signal vector **h** is given by

$$\mathbf{g} - \mathbf{h} = (\mathbf{1}_p - \mathbf{AX})\mathbf{g} \tag{29.16}$$

The square matrix $\mathbf{1}_p - \mathbf{AX}$ relating the applied signal vector **g** to the difference of the applied signal vector **g** and the returned signal vector **h** is called the **return difference matrix** with respect to **X** and is denoted by

$$\mathbf{F(X)} = \mathbf{1}_p - \mathbf{AX} \tag{29.17}$$

Combining this with (29.15) gives

$$\mathbf{F(X)} = \mathbf{1}_p + \mathbf{T(X)} \tag{29.18}$$

For the voltage-series feedback amplifier of Fig. 29.4, let the controlling parameters of the two controlled current sources be the elements of interest. Then the return ratio matrix is found from (29.8) and (29.9a)

$$\begin{aligned}\mathbf{T(X)} = -\mathbf{AX} &= -\begin{bmatrix} -90.782 & 45.391 \\ -942.507 & 0 \end{bmatrix}\begin{bmatrix} 455 \times 10^{-4} & 0 \\ 0 & 455 \times 10^{-4} \end{bmatrix} \\ &= \begin{bmatrix} 4.131 & -2.065 \\ 42.884 & 0 \end{bmatrix}\end{aligned} \tag{29.19}$$

obtaining the return difference matrix as

$$\mathbf{F(X)} = \mathbf{1}_2 + \mathbf{T(X)} = \begin{bmatrix} 5.131 & -2.065 \\ 42.884 & 1 \end{bmatrix} \tag{29.20}$$

29.3 The Null Return Difference Matrix

A direct extension of the null return difference for the single-loop feedback amplifier is the null return difference matrix for the multiple-loop feedback networks.

Refer again to the fundamental matrix feedback-flow graph of Fig. 29.3. As before, we break the branch with transmittance **X** and apply a signal p-vector **g** to the right of the breaking mark, as illustrated in Fig. 29.6. We then adjust the input excitation n-vector **u** so that the total output m-vector **y** resulting from the inputs **g** and **u** is zero. From Fig. 29.6 the desired input excitation **u** is found to be

$$\mathbf{Du} + \mathbf{CXg} = \mathbf{0} \tag{29.21}$$

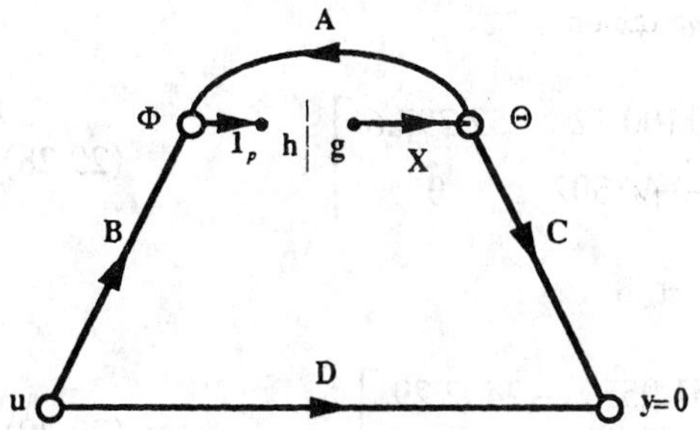

FIGURE 29.6 The physical interpretation of the null return difference matrix.

or

$$\mathbf{u} = -\mathbf{D}^{-1}\mathbf{CXg} \tag{29.22}$$

provided that the matrix **D** is square and nonsingular. This requires that the output **y** be of the same dimension as the input **u** or $m = n$. Physically, this requirement is reasonable because the effects at the output caused by **g** can be neutralized by a unique input excitation **u** only when **u** and **y** are of the same dimension. With these inputs **u** and **g**, the returned signal **h** to the left of the breaking mark in Fig. 29.6 is computed as

$$\mathbf{h} = \mathbf{Bu} + \mathbf{AXg} = (-\mathbf{BD}^{-1}\mathbf{CX} + \mathbf{AX})\mathbf{g} \tag{29.23}$$

obtaining

$$\mathbf{g} - \mathbf{h} = (\mathbf{1}_p - \mathbf{AX} + \mathbf{BD}^{-1}\mathbf{CX})\mathbf{g} \tag{29.24}$$

The square matrix

$$\hat{\mathbf{F}}(\mathbf{X}) = \mathbf{1}_p + \hat{\mathbf{T}}(\mathbf{X}) = \mathbf{1}_p - \mathbf{AX} + \mathbf{BD}^{-1}\mathbf{CX} = \mathbf{1}_p - \hat{\mathbf{A}}\mathbf{X} \tag{29.25}$$

relating the input signal vector **g** to the difference of the input signal vector **g** and the returned signal vector **h** is called the **null return difference matrix** with respect to **X**, where

$$\hat{\mathbf{T}}(\mathbf{X}) = -\mathbf{AX} + \mathbf{BD}^{-1}\mathbf{CX} = -\hat{\mathbf{A}}\mathbf{X} \tag{29.26a}$$

$$\hat{\mathbf{A}} = \mathbf{A} - \mathbf{BD}^{-1}\mathbf{C} \tag{29.26b}$$

The square matrix $\hat{\mathbf{T}}(\mathbf{X})$ is known as the *null return ratio matrix.*

Example 4. Consider again the voltage-series feedback amplifier of Fig. 26.9, an equivalent network of which is shown in Fig. 29.4. Assume that the voltage V_{25} is the output variable. Then from (29.9)

$$\begin{aligned}\mathbf{\Phi} &= \begin{bmatrix} V_{13} \\ V_{45} \end{bmatrix} = \begin{bmatrix} -90.782 & 45.391 \\ -942.507 & 0 \end{bmatrix}\begin{bmatrix} I_a \\ I_b \end{bmatrix} + \begin{bmatrix} 0.91748 \\ 0 \end{bmatrix}[V_s] \\ &= \mathbf{A\Theta} + \mathbf{B}u \end{aligned} \tag{29.27a}$$

$$\begin{aligned} y &= [V_{25}] = \begin{bmatrix} 45.391 & -2372.32 \end{bmatrix}\begin{bmatrix} I_a \\ I_b \end{bmatrix} + [0.04126][V_s] \\ &= \mathbf{C\Theta} + Du \end{aligned} \tag{29.27b}$$

Substituting the coefficient matrices in (29.26b) we obtain

$$\hat{\mathbf{A}} = \mathbf{A} - \mathbf{B}\mathbf{D}^{-1}\mathbf{C} = \begin{bmatrix} -1100.12 & 52{,}797.6 \\ -942.507 & 0 \end{bmatrix} \tag{29.28}$$

giving the null return difference matrix with respect to **X** as

$$\hat{\mathbf{F}}(\mathbf{X}) = \mathbf{1}_2 - \hat{\mathbf{A}}\mathbf{X} = \begin{bmatrix} 51.055 & -2402.29 \\ 42.884 & 1 \end{bmatrix} \tag{29.29}$$

Suppose that the input current I_{51} is chosen as the output variable. Then from (29.9b) we have

$$y = [I_{51}] = \begin{bmatrix} -0.08252 & 0.04126 \end{bmatrix} \begin{bmatrix} I_a \\ I_b \end{bmatrix} + [0.000862][V_s] = \mathbf{C}\boldsymbol{\Theta} + Du \tag{29.30}$$

The corresponding null return difference matrix becomes

$$\hat{\mathbf{F}}(\mathbf{X}) = \mathbf{1}_2 - \hat{\mathbf{A}}\mathbf{X} = \begin{bmatrix} 1.13426 & -0.06713 \\ 42.8841 & 1 \end{bmatrix} \tag{29.31}$$

where

$$\hat{\mathbf{A}} = \begin{bmatrix} -2.95085 & 1.47543 \\ -942.507 & 0 \end{bmatrix} \tag{29.32}$$

29.4 The Transfer-Function Matrix and Feedback

In this section, we show the effect of feedback on the transfer-function matrix $\mathbf{W}(\mathbf{X})$. Specifically, we express $\det \mathbf{W}(\mathbf{X})$ in terms of the $\det \mathbf{X}(\mathbf{0})$ and the determinants of the return difference and null return difference matrices, thereby generalizing Blackman's impedance formula for a single input to a multiplicity of inputs.

Before we proceed to develop the desired relation, we state the following determinant identity for two arbitrary matrices **M** and **N** of orders $m \times n$ and $n \times m$:

$$\det(\mathbf{1}_m + \mathbf{MN}) = \det(\mathbf{1}_n + \mathbf{NM}) \tag{29.33}$$

a proof of which may be found in [5, 6]. Using this we next establish the following generalization of Blackman's formula for input impedance.

Theorem 1. *In a multiple-loop feedback amplifier, if* $\mathbf{W}(\mathbf{0}) = \mathbf{D}$ *is nonsingular, then the determinant of the transfer-function matrix* $\mathbf{W}(\mathbf{X})$ *is related to the determinants of the return difference matrix* $\mathbf{F}(\mathbf{X})$ *and the null return difference matrix* $\hat{\mathbf{F}}(\mathbf{X})$ *by*

$$\det \mathbf{W}(\mathbf{X}) = \det \mathbf{W}(\mathbf{0}) \frac{\det \hat{\mathbf{F}}(\mathbf{X})}{\det \mathbf{F}(\mathbf{X})} \tag{29.34}$$

PROOF: From (29.5a) we obtain

$$\mathbf{W}(\mathbf{X}) = \mathbf{D}[\mathbf{1}_n + \mathbf{D}^{-1}\mathbf{C}\mathbf{X}(\mathbf{1}_p - \mathbf{A}\mathbf{X})^{-1}\mathbf{B}] \tag{29.35}$$

yielding

$$\begin{aligned} \det \mathbf{W}(\mathbf{X}) &= [\det \mathbf{W}(\mathbf{0})] \det[\mathbf{1}_n + \mathbf{D}^{-1}\mathbf{C}\mathbf{X}(\mathbf{1}_p - \mathbf{A}\mathbf{X})^{-1}\mathbf{B}] \\ &= [\det \mathbf{W}(\mathbf{0})] \det[\mathbf{1}_p + \mathbf{B}\mathbf{D}^{-1}\mathbf{C}\mathbf{X}(\mathbf{1}_p - \mathbf{A}\mathbf{X})^{-1}] \\ &= [\det \mathbf{W}(\mathbf{0})] \det[\mathbf{1}_p - \mathbf{A}\mathbf{X} + \mathbf{B}\mathbf{D}^{-1}\mathbf{C}\mathbf{X}] \det(\mathbf{1}_p - \mathbf{A}\mathbf{X})^{-1} \\ &= \frac{\det \mathbf{W}(\mathbf{0}) \det \hat{\mathbf{F}}(\mathbf{X})}{\det \mathbf{F}(\mathbf{X})} \end{aligned} \tag{29.36}$$

The second line follows directly from (29.33). This completes the proof of the theorem.

As indicated in (27.4), the input impedance $Z(x)$ looking into a terminal pair can be conveniently expressed as

$$Z(x) = Z(0)\frac{F(\text{input short circuited})}{F(\text{input open circuited})} \tag{29.37}$$

A similar expression can be derived from (29.34) if $\mathbf{W}(\mathbf{X})$ denotes the impedance matrix of an n-port network of Fig. 29.1. In this case, $\mathbf{F}(\mathbf{X})$ is the return difference matrix with respect to $\mathbf{X}$ for the situation when the n ports where the impedance matrix are defined are left open without any sources, and we write $\mathbf{F}(\mathbf{X}) = \mathbf{F}(\text{input open circuited})$. Likewise, $\hat{\mathbf{F}}(\mathbf{X})$ is the return difference matrix with respect to $\mathbf{X}$ for the input port-current vector $\mathbf{I}_s$ and the output port-voltage vector $\mathbf{V}$ under the condition that $\mathbf{I}_s$ is adjusted so that the port-voltage vector $\mathbf{V}$ is identically zero. In other words, $\hat{\mathbf{F}}(\mathbf{X})$ is the return difference matrix for the situation when the n ports where the impedance matrix is defined are short circuited, and we write $\hat{\mathbf{F}}(\mathbf{X}) = \mathbf{F}(\text{input short circuited})$. Consequently, the determinant of the impedance matrix $\mathbf{Z}(\mathbf{X})$ of an n-port network can be expressed from (29.34) as

$$\det \mathbf{Z}(\mathbf{X}) = \det \mathbf{Z}(\mathbf{0})\frac{\det \mathbf{F}(\text{input short circuited})}{\det \mathbf{F}(\text{input open circuited})} \tag{29.38}$$

Example 5. Refer again to the voltage-series feedback amplifier of Fig. 26.9, an equivalent network of which is shown in Fig. 29.4. As computed in (29.20), the return difference matrix with respect to the two controlling parameters is given by

$$\mathbf{F}(\mathbf{X}) = \mathbf{1}_2 + \mathbf{T}(\mathbf{X}) = \begin{bmatrix} 5.131 & -2.065 \\ 42.884 & 1 \end{bmatrix} \tag{29.39}$$

the determinant of which is found to be

$$\det \mathbf{F}(\mathbf{X}) = 93.68646 \tag{29.40}$$

If V_{25} of Fig. 29.4 is chosen as the output and V_s as the input, the null return difference matrix is, from (29.29),

$$\hat{\mathbf{F}}(\mathbf{X}) = \mathbf{1}_2 - \hat{\mathbf{A}}\mathbf{X} = \begin{bmatrix} 51.055 & -2402.29 \\ 42.884 & 1 \end{bmatrix} \tag{29.41}$$

the determinant of which is found to be

$$\det \hat{\mathbf{F}}(\mathbf{X}) = 103{,}071 \tag{29.42}$$

By appealing to (29.34), the feedback amplifier voltage gain V_{25}/V_s can be written as

$$w(\mathbf{X}) = \frac{V_{25}}{V_s} = w(\mathbf{0})\frac{\det \hat{\mathbf{F}}(\mathbf{X})}{\det \mathbf{F}(\mathbf{X})} = 0.04126\frac{103{,}071}{93.68646} = 45.39 \tag{29.43}$$

confirming (26.44), where $w(\mathbf{0}) = 0.04126$, as given in (29.27b).

Suppose, instead, that the input current I_{51} is chosen as the output and V_s as the input. Then from (29.31), the null return difference matrix becomes

$$\hat{\mathbf{F}}(\mathbf{X}) = \mathbf{1}_2 - \hat{\mathbf{A}}\mathbf{X} = \begin{bmatrix} 1.13426 & -0.06713 \\ 42.8841 & 1 \end{bmatrix} \tag{29.44}$$

the determinant of which is found to be

$$\det \hat{\mathbf{F}}(\mathbf{X}) = 4.01307 \tag{29.45}$$

By applying (29.34), the amplifier input admittance is obtained as

$$\begin{aligned} w(\mathbf{X}) &= \frac{I_{51}}{V_s} = w(\mathbf{0})\frac{\det \hat{\mathbf{F}}(\mathbf{X})}{\det \mathbf{F}(\mathbf{X})} \\ &= 8.62 \times 10^{-4}\frac{4.01307}{93.68646} = 36.92\ \mu\text{mho} \end{aligned} \tag{29.46}$$

or 27.2 kΩ, confirming (29.13), where $w(\mathbf{0}) = 862\ \mu$mho is found from (29.30).

Another useful application of the generalized Blackman's formula (29.38) is that it provides the basis of a procedure for the indirect measurement of return difference. Refer to the general feedback network of Fig. 29.2. Suppose that we wish to measure the return difference $F(y_{21})$ with respect to the forward short circuit transfer admittance y_{21} of a two-port device characterized by its y parameters y_{ij}. Choose the two controlling parameters y_{21} and y_{12} to be the elements of interest. Then from Fig. 28.2 we obtain

$$\boldsymbol{\Theta} = \begin{bmatrix} I_a \\ I_b \end{bmatrix} = \begin{bmatrix} y_{21} & 0 \\ 0 & y_{12} \end{bmatrix} \begin{bmatrix} V_1 \\ V_2 \end{bmatrix} = \mathbf{X}\boldsymbol{\Phi} \tag{29.47}$$

where I_a and I_b are the currents of the voltage-controlled current sources. By appealing to (29.38), the impedance looking into terminals a and b of Fig. 28.2 can be written as

$$z_{aa,bb}(y_{12}, y_{21}) = z_{aa,bb}(0,0)\frac{\det \mathbf{F}(\text{input short circuited})}{\det \mathbf{F}(\text{input open circuited})} \tag{29.48}$$

When the input terminals a and b are open circuited, the resulting return difference matrix is exactly the same as that found under normal operating conditions, and we have

$$\mathbf{F}(\text{input open circuited}) = \mathbf{F}(\mathbf{X}) = \begin{bmatrix} F_{11} & F_{12} \\ F_{21} & F_{22} \end{bmatrix} \tag{29.49}$$

Since

$$\mathbf{F}(\mathbf{X}) = \mathbf{1}_2 - \mathbf{AX} \tag{29.50}$$

the elements F_{11} and F_{21} are calculated with $y_{12} = 0$, whereas F_{12} and F_{22} are evaluated with $y_{21} = 0$. When the input terminals a and b are short circuited, the feedback loop is interrupted and only the second row and first column element of the matrix **A** is nonzero, and we obtain

$$\det \mathbf{F}(\text{input short circuited}) = 1 \tag{29.51}$$

Since **X** is diagonal, the return difference function $F(y_{21})$ can be expressed in terms of $\det \mathbf{F}(\mathbf{X})$ and the cofactor of the first row and first column element of $\mathbf{F}(\mathbf{X})$:

$$F(y_{21}) = \frac{\det \mathbf{F}(\mathbf{X})}{F_{22}} \tag{29.52}$$

Substituting these in (29.48) yields

$$F(y_{12})\Big|_{y_{21}=0} F(y_{21}) = \frac{z_{aa,bb}(0,0)}{z_{aa,bb}(y_{12}, y_{21})} \tag{29.53}$$

where

$$F_{22} = 1 - a_{22}y_{12}\Big|_{y_{21}=0} = F(y_{12})\Big|_{y_{21}=0} \tag{29.54}$$

and a_{22} is the second row and second column element of **A**. Formula (29.53) was derived earlier in (28.7) using the network arrangements of Figs. 28.7 and 28.8 to measure the elements $F(y_{12})|_{y_{21}=0}$ and $z_{aa,bb}(0,0)$, respectively.

29.5 The Sensitivity Matrix

We have studied the sensitivity of a transfer function with respect to the change of a particular element in the network. In a multiple-loop feedback network, we are usually interested in the sensitivity of a transfer function with respect to the variation of a set of elements in the network. This set may include either elements that are inherently sensitive to variation or elements whose effect on the overall amplifier performance is of paramount importance to the designers. For this we introduce a sensitivity matrix and develop formulas for computing multiparameter sensitivity functions for a multiple-loop feedback amplifier [7].

Figure 29.7 is the block diagram of a multivariable open-loop control system with n inputs and m outputs, whereas Fig. 29.8 shows the general feedback structure. If all feedback signals are obtainable from the output and if the controllers are linear, there is no loss of generality by assuming the controller to be of the form shown in Fig. 29.9.

Denote the set of Laplace-transformed input signals by the n-vector **u**, the set of inputs to the network **X** in the open-loop configuration of Fig. 29.7 by the p-vector $\mathbf{\Phi}_o$, and the set of

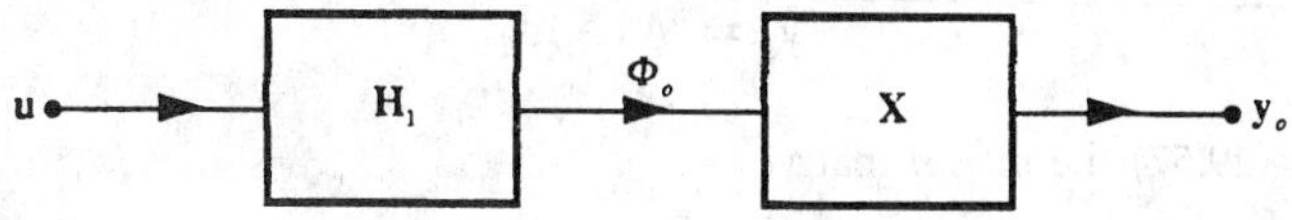

FIGURE 29.7 The block diagram of a multivariable open-loop control system.

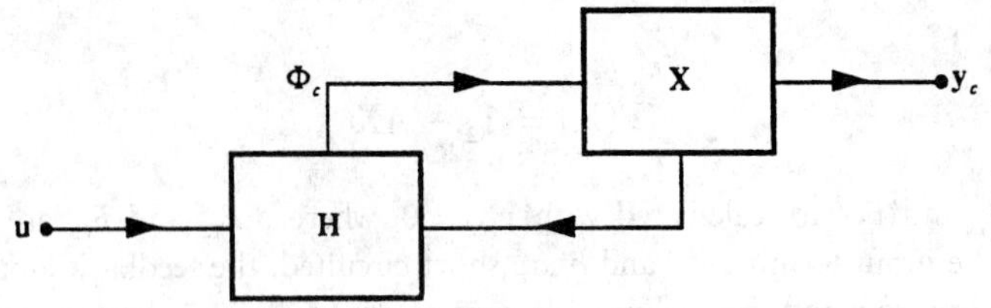

FIGURE 29.8 The general feedback structure.

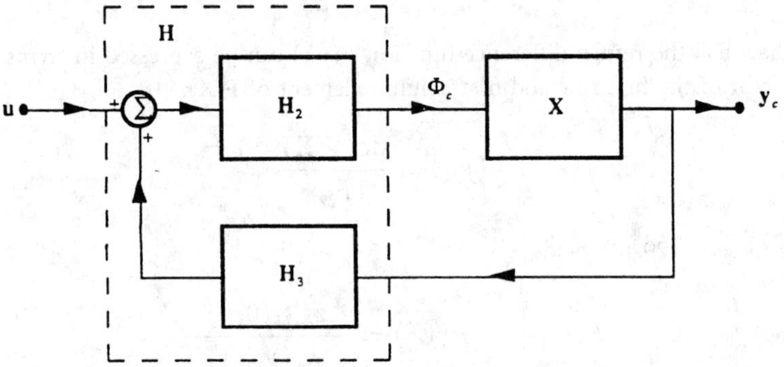

FIGURE 29.9 A general feedback configuration.

outputs of the network **X** of Fig. 29.7 by the m-vector $\mathbf{y}_o$. Let the corresponding signals for the closed-loop configuration of Fig. 29.9 be denoted by the n-vector $\mathbf{u}$, the p-vector $\mathbf{\Phi}_c$, and the m-vector $\mathbf{y}_c$, respectively. Then from Figs. 29.7 and 29.9 we obtain the following relations:

$$\mathbf{y}_o = \mathbf{X}\mathbf{\Phi}_o \tag{29.55a}$$

$$\mathbf{\Phi}_o = \mathbf{H}_1\mathbf{u} \tag{29.55b}$$

$$\mathbf{y}_c = \mathbf{X}\mathbf{\Phi}_c \tag{29.55c}$$

$$\mathbf{\Phi}_c = \mathbf{H}_2(\mathbf{u} + \mathbf{H}_3\mathbf{y}_c) \tag{29.55d}$$

where the transfer-function matrices $\mathbf{X}, \mathbf{H}_1, \mathbf{H}_2$ and $\mathbf{H}_3$ are of orders $m \times p, p \times n, p \times n$, and $n \times m$, respectively. Combining (29.55c) and (29.55d) yields

$$(\mathbf{1}_m - \mathbf{X}\mathbf{H}_2\mathbf{H}_3)\mathbf{y}_c = \mathbf{X}\mathbf{H}_2\mathbf{u} \tag{29.56}$$

or

$$\mathbf{y}_c = (\mathbf{1}_m - \mathbf{X}\mathbf{H}_2\mathbf{H}_3)^{-1}\mathbf{X}\mathbf{H}_2\mathbf{u} \tag{29.57}$$

The closed-loop transfer function matrix **W(X)** that relates the input vector **u** to the output vector $\mathbf{y}_c$ is defined by the equation

$$\mathbf{y}_c = \mathbf{W}(\mathbf{X})\mathbf{u} \tag{29.58}$$

identifying from (29.57) the $m \times n$ matrix

$$\mathbf{W}(\mathbf{X}) = (\mathbf{1}_m - \mathbf{X}\mathbf{H}_2\mathbf{H}_3)^{-1}\mathbf{X}\mathbf{H}_2 \tag{29.59}$$

Now suppose that $\mathbf{X}$ is perturbed from $\mathbf{X}$ to $\mathbf{X} + \boldsymbol{\delta}\mathbf{X}$. The outputs of the open-loop and closed-loop systems of Figs. 29.7 and 29.9 will no longer be the same as before. Distinguishing the new from the old variables by the superscript +, we have

$$\mathbf{y}_o^+ = \mathbf{X}^+\boldsymbol{\Phi}_o \tag{29.60a}$$

$$\mathbf{y}_c^+ = \mathbf{X}^+\boldsymbol{\Phi}_c^+ \tag{29.60b}$$

$$\boldsymbol{\Phi}_c^+ = \mathbf{H}_2(\mathbf{u} + \mathbf{H}_3\mathbf{y}_c^+) \tag{29.60c}$$

where $\boldsymbol{\Phi}_o$ remains the same.

We next proceed to compare the relative effects of the variations of $\mathbf{X}$ on the performance of the open-loop and the closed-loop systems. For a meaningful comparison, we assume that $\mathbf{H}_1, \mathbf{H}_2$, and $\mathbf{H}_3$ are such that when there is no variation of $\mathbf{X}, \mathbf{y}_o = \mathbf{y}_c$. Define the error vectors resulting from perturbation of $\mathbf{X}$ as

$$\mathbf{E}_o = \mathbf{y}_o - \mathbf{y}_o^+ \tag{29.61a}$$

$$\mathbf{E}_c = \mathbf{y}_c - \mathbf{y}_c^+ \tag{29.61b}$$

A square matrix relating $\mathbf{E}_o$ to $\mathbf{E}_c$ is called the **sensitivity matrix** $\mathscr{S}(\mathbf{X})$ for the transfer function matrix $\mathbf{W}(\mathbf{X})$ with respect to the variations of $\mathbf{X}$:

$$\mathbf{E}_c = \mathscr{S}(\mathbf{X})\mathbf{E}_o \tag{29.62}$$

In the following, we express the sensitivity matrix $\mathscr{S}(\mathbf{X})$ in terms of the system matrices $\mathbf{X}, \mathbf{H}_2$, and $\mathbf{H}_3$.

The input and output relation similar to that given in (29.57) for the perturbed system can be written as

$$\mathbf{y}_c^+ = (\mathbf{1}_m - \mathbf{X}^+\mathbf{H}_2\mathbf{H}_3)^{-1}\mathbf{X}^+\mathbf{H}_2\mathbf{u} \tag{29.63}$$

Substituting (29.57) and (29.63) in (29.61b) gives

$$\begin{aligned}\mathbf{E}_c = \mathbf{y}_c - \mathbf{y}_c^+ &= [(\mathbf{1}_m - \mathbf{X}\mathbf{H}_2\mathbf{H}_3)^{-1}\mathbf{X}\mathbf{H}_2 - (\mathbf{1}_m - \mathbf{X}^+\mathbf{H}_2\mathbf{H}_3)^{-1}\mathbf{X}^+\mathbf{H}_2]\mathbf{u} \\ &= (\mathbf{1}_m - \mathbf{X}^+\mathbf{H}_2\mathbf{H}_3)^{-1}\left\{[\mathbf{1}_m - (\mathbf{X} + \boldsymbol{\delta}\mathbf{X})\mathbf{H}_2\mathbf{H}_3](\mathbf{1}_m - \mathbf{X}\mathbf{H}_2\mathbf{H}_3)^{-1}\mathbf{X}\mathbf{H}_2 \right. \\ &\quad \left. - (\mathbf{X} + \boldsymbol{\delta}\mathbf{X})\mathbf{H}_2\right\}\mathbf{u} \\ &= (\mathbf{1}_m - \mathbf{X}^+\mathbf{H}_2\mathbf{H}_3)^{-1}[\mathbf{X}\mathbf{H}_2 - \boldsymbol{\delta}\mathbf{X}\mathbf{H}_2\mathbf{H}_3(\mathbf{1}_m - \mathbf{X}\mathbf{H}_2\mathbf{H}_3)^{-1}\mathbf{X}\mathbf{H}_2 - \mathbf{X}\mathbf{H}_2 - \boldsymbol{\delta}\mathbf{X}\mathbf{H}_2]\mathbf{u} \\ &= -(\mathbf{1}_m - \mathbf{X}^+\mathbf{H}_2\mathbf{H}_3)^{-1}\boldsymbol{\delta}\mathbf{X}\mathbf{H}_2[\mathbf{1}_n + \mathbf{H}_3\mathbf{W}(\mathbf{X})]\mathbf{u}\end{aligned} \tag{29.64}$$

From (29.55d) and (29.58), we obtain

$$\boldsymbol{\Phi}_c = \mathbf{H}_2[\mathbf{1}_n + \mathbf{H}_3\mathbf{W}(\mathbf{X})]\mathbf{u} \tag{29.65}$$

Since by assumption $\mathbf{y}_o = \mathbf{y}_c$, we have

$$\boldsymbol{\Phi}_o = \boldsymbol{\Phi}_c = \mathbf{H}_2[\mathbf{1}_n + \mathbf{H}_3\mathbf{W}(\mathbf{X})]\mathbf{u} \tag{29.66}$$

yielding

$$\mathbf{E}_o = \mathbf{y}_o - \mathbf{y}_o^+ = (\mathbf{X} - \mathbf{X}^+)\boldsymbol{\Phi}_o = -\boldsymbol{\delta}\mathbf{X}\boldsymbol{\Phi}_o = -\boldsymbol{\delta}\mathbf{X}\mathbf{H}_2[\mathbf{1}_n + \mathbf{H}_3\mathbf{W}(\mathbf{X})]\mathbf{u} \tag{29.67}$$

Combining (29.64) and (29.67) yields an expression relating the error vectors $\mathbf{E}_c$ and $\mathbf{E}_o$ of the closed-loop and open-loop systems by

$$\mathbf{E}_c = (\mathbf{1}_m - \mathbf{X}^+\mathbf{H}_2\mathbf{H}_3)^{-1}\mathbf{E}_o \tag{29.68}$$

obtaining the sensitivity matrix as

$$\mathscr{S}(\mathbf{X}) = (\mathbf{1}_m - \mathbf{X}^+\mathbf{H}_2\mathbf{H}_3)^{-1} \tag{29.69}$$

For small variation of $\mathbf{X}$, $\mathbf{X}^+$ is approximately equal to $\mathbf{X}$. Thus, in Fig. 29.9 if the matrix triple product $\mathbf{X}\mathbf{H}_2\mathbf{H}_3$ is regarded as the *loop-transmission matrix* and $-\mathbf{X}\mathbf{H}_2\mathbf{H}_3$ as the *return ratio matrix*, then the difference between the unit matrix and the loop-transmission matrix,

$$\mathbf{1}_m - \mathbf{X}\mathbf{H}_2\mathbf{H}_3 \tag{29.70}$$

can be defined as the *return difference matrix*. Therefore, (29.69) is a direct extension of the sensitivity function defined for a single-input, single-output system and for a single parameter. Recall that in (27.33) we demonstrated that, using the ideal feedback model, the sensitivity function of the closed-loop transfer function with respect to the forward amplifier gain is equal to the reciprocal of its return difference with respect to the same parameter.

In particular, when $\mathbf{W}(\mathbf{X})$, $\boldsymbol{\delta}\mathbf{X}$, and $\mathbf{X}$ are square and nonsingular, from (29.55a), (29.55b), and (29.58), (29.61) can be rewritten as

$$\mathbf{E}_c = \mathbf{y}_c - \mathbf{y}_c^+ = [\mathbf{W}(\mathbf{X}) - \mathbf{W}^+(\mathbf{X})]\mathbf{u} = -\boldsymbol{\delta}\mathbf{W}(\mathbf{X})\mathbf{u} \tag{29.71a}$$

$$\mathbf{E}_o = \mathbf{y}_o - \mathbf{y}_o^+ = (\mathbf{X}\mathbf{H}_1 - \mathbf{X}^+\mathbf{H}_1)\mathbf{u} = -\boldsymbol{\delta}\mathbf{X}\mathbf{H}_1\mathbf{u} \tag{29.71b}$$

If $\mathbf{H}_1$ is nonsingular, $\mathbf{u}$ in (29.71b) can be solved for and substituted in (29.71a) to give

$$\mathbf{E}_c = \boldsymbol{\delta}\mathbf{W}(\mathbf{X})\mathbf{H}_1^{-1}(\boldsymbol{\delta}\mathbf{X})^{-1}\mathbf{E}_o \tag{29.72}$$

As before, for meaningful comparison, we require that $\mathbf{y}_o = \mathbf{y}_c$ or

$$\mathbf{X}\mathbf{H}_1 = \mathbf{W}(\mathbf{X}) \tag{29.73}$$

From (29.72), we obtain

$$\mathbf{E}_c = \boldsymbol{\delta}\mathbf{W}(\mathbf{X})\mathbf{W}^{-1}(\mathbf{X})\mathbf{X}(\boldsymbol{\delta}\mathbf{X})^{-1}\mathbf{E}_o \tag{29.74}$$

identifying that

$$\mathscr{S}(\mathbf{X}) = \boldsymbol{\delta}\mathbf{W}(\mathbf{X})\mathbf{W}^{-1}(\mathbf{X})\mathbf{X}(\boldsymbol{\delta}\mathbf{X})^{-1} \tag{29.75}$$

This result is to be compared with the scalar sensitivity function defined in (27.26), which can be put in the form

$$\mathscr{S}(x) = (\delta w)w^{-1}x(\delta x)^{-1} \tag{29.76}$$

.6 Multiparameter Sensitivity

In this section, we derive formulas for the effect of change of **X** on a scalar transfer function $w(\mathbf{X})$.

Let x_k, $k = 1, 2, \ldots, pq$, be the elements of **X**. The multivariable Taylor series expansion of $w(\mathbf{X})$ with respect to x_k is given by

$$\delta w = \sum_{k=1}^{pq} \frac{\partial w}{\partial x_k} \delta x_k + \sum_{j=1}^{pq} \sum_{k=1}^{pq} \frac{\partial^2 w}{\partial x_j \partial x_k} \frac{\delta x_j \delta x_k}{2!} + \cdots \tag{29.77}$$

The first-order perturbation can then be written as

$$\delta w \approx \sum_{k=1}^{pq} \frac{\partial w}{\partial x_k} \delta x_k \tag{29.78}$$

Using (27.26), we obtain

$$\frac{\delta w}{w} \approx \sum_{k=1}^{pq} \mathscr{S}(x_k) \frac{\delta x_k}{x_k} \tag{29.79}$$

This expression gives the fractional change of the transfer function w in terms of the scalar sensitivity functions $\mathscr{S}(x_k)$.

Refer to the fundamental matrix feedback-flow graph of Fig. 29.3. If the amplifier has a single input and a single output, from (29.35) the overall transfer function $w(\mathbf{X})$ of the multiple-loop feedback amplifier becomes

$$w(\mathbf{X}) = D + \mathbf{CX}(\mathbf{1}_p - \mathbf{AX})^{-1}\mathbf{B} \tag{29.80}$$

When **X** is perturbed to $\mathbf{X}^+ = \mathbf{X} + \boldsymbol{\delta}\mathbf{X}$, the corresponding expression of (29.80) is given by

$$w(\mathbf{X}) + \delta w(\mathbf{X}) = D + \mathbf{C}(\mathbf{X} + \boldsymbol{\delta}\mathbf{X})(\mathbf{1}_p - \mathbf{AX} - \mathbf{A}\boldsymbol{\delta}\mathbf{X})^{-1}\mathbf{B} \tag{29.81}$$

or

$$\delta w(\mathbf{X}) = \mathbf{C}[(\mathbf{X} + \boldsymbol{\delta}\mathbf{X})(\mathbf{1}_p - \mathbf{AX} - \mathbf{A}\boldsymbol{\delta}\mathbf{X})^{-1} - \mathbf{X}(\mathbf{1}_p - \mathbf{AX})^{-1}]\mathbf{B} \tag{29.82}$$

As $\boldsymbol{\delta}\mathbf{X}$ approaches zero, we obtain

$$\begin{aligned}
\delta w(\mathbf{X}) &= \mathbf{C}[(\mathbf{X} + \boldsymbol{\delta}\mathbf{X}) - \mathbf{X}(\mathbf{1}_p - \mathbf{AX})^{-1}(\mathbf{1}_p - \mathbf{AX} - \mathbf{A}\boldsymbol{\delta}\mathbf{X})] \\
&\quad \times (\mathbf{1}_p - \mathbf{AX} - \mathbf{A}\boldsymbol{\delta}\mathbf{X})^{-1}\mathbf{B} \\
&= \mathbf{C}[\boldsymbol{\delta}\mathbf{X} + \mathbf{X}(\mathbf{1}_p - \mathbf{AX})^{-1}\mathbf{A}\boldsymbol{\delta}\mathbf{X}](\mathbf{1}_p - \mathbf{AX} - \mathbf{A}\boldsymbol{\delta}\mathbf{X})^{-1}\mathbf{B} \\
&= \mathbf{C}(\mathbf{1}_q - \mathbf{XA})^{-1}(\boldsymbol{\delta}\mathbf{X})(\mathbf{1}_p - \mathbf{AX} - \mathbf{A}\boldsymbol{\delta}\mathbf{X})^{-1}\mathbf{B} \\
&\approx \mathbf{C}(\mathbf{1}_q - \mathbf{XA})^{-1}(\boldsymbol{\delta}\mathbf{X})(\mathbf{1}_p - \mathbf{AX})^{-1}\mathbf{B}
\end{aligned} \tag{29.83}$$

where **C** is a row q vector and **B** is a column p vector. Write

$$\mathbf{C} = [\, c_1 \;\; c_2 \;\; \cdots \;\; c_q \,] \tag{29.84a}$$

$$\mathbf{B}' = [\, b_1 \;\; b_2 \;\; \cdots \;\; b_p \,] \tag{29.84b}$$

$$\tilde{\mathbf{W}} = \mathbf{X}(\mathbf{1}_p - \mathbf{AX})^{-1} = (\mathbf{1}_q - \mathbf{XA})^{-1}\mathbf{X} = [\tilde{w}_{ij}] \tag{29.84c}$$

The increment $\delta w(\mathbf{X})$ can be expressed in terms of the elements of (29.84) and those of $\mathbf{X}$. In the case where $\mathbf{X}$ is diagonal with

$$\mathbf{X} = \text{diag}\begin{bmatrix} x_1 & x_2 & \cdots & x_p \end{bmatrix} \tag{29.85}$$

where $p = q$, the expression for $\delta w(\mathbf{X})$ can be compactly written as

$$\begin{aligned} \delta w(\mathbf{X}) &= \sum_{i=1}^{p}\sum_{k=1}^{p}\sum_{j=1}^{p} c_i\left(\frac{\tilde{w}_{ik}}{x_k}\right)(\delta x_k)\left(\frac{\tilde{w}_{kj}}{x_k}\right)b_j \\ &= \sum_{i=1}^{p}\sum_{k=1}^{p}\sum_{j=1}^{p} \frac{c_i\tilde{w}_{ik}\tilde{w}_{kj}b_j}{x_k}\frac{\delta x_k}{x_k} \end{aligned} \tag{29.86}$$

Comparing this to (29.79) we obtain an explicit form for the single-parameter sensitivity function as

$$\mathscr{S}(x_k) = \sum_{i=1}^{p}\sum_{j=1}^{p} \frac{c_i\tilde{w}_{ik}\tilde{w}_{kj}b_j}{x_k w(\mathbf{X})} \tag{29.87}$$

Thus, knowing (29.84) and (29.85) we can calculate the multiparameter sensitivity function for the scalar transfer function $w(\mathbf{X})$ immediately.

Example 6. Consider again the voltage-series feedback amplifier of Fig. 26.9, an equivalent network of which is shown in Fig. 29.4. Assume that V_s is the input and V_{25} the output. The transfer function of interest is the amplifier voltage gain V_{25}/V_s. The elements of main concern are the two controlling parameters of the controlled sources. Thus, we let

$$\mathbf{X} = \begin{bmatrix} \tilde{\alpha}_1 & 0 \\ 0 & \tilde{\alpha}_2 \end{bmatrix} = \begin{bmatrix} 0.0455 & 0 \\ 0 & 0.0455 \end{bmatrix} \tag{29.88}$$

From (29.27) we have

$$\mathbf{A} = \begin{bmatrix} -90.782 & 45.391 \\ -942.507 & 0 \end{bmatrix} \tag{29.89a}$$

$$\mathbf{B}' = \begin{bmatrix} 0.91748 & 0 \end{bmatrix} \tag{29.89b}$$

$$\mathbf{C} = \begin{bmatrix} 45.391 & -2372.32 \end{bmatrix} \tag{29.89c}$$

yielding

$$\tilde{\mathbf{W}} = \mathbf{X}(\mathbf{1}_2 - \mathbf{A}\mathbf{X})^{-1} = 10^{-4}\begin{bmatrix} 4.85600 & 10.02904 \\ -208.245 & 24.91407 \end{bmatrix} \tag{29.90}$$

Also, from (29.13) we have

$$w(\mathbf{X}) = \frac{V_{25}}{V_s} = 45.387 \tag{29.91}$$

To compute the sensitivity functions with respect to $\tilde{\alpha}_1$ and $\tilde{\alpha}_2$, we apply (29.87) and obtain

$$\mathcal{S}(\tilde{\alpha}_1) = \sum_{i=1}^{2}\sum_{j=1}^{2}\frac{c_i\tilde{w}_{i1}\tilde{w}_{1j}b_j}{\tilde{\alpha}_1 w(\mathbf{X})} = \frac{c_1\tilde{w}_{11}\tilde{w}_{11}b_1 + c_1\tilde{w}_{11}\tilde{w}_{12}b_2 + c_2\tilde{w}_{21}\tilde{w}_{11}b_1 + c_2\tilde{w}_{21}\tilde{w}_{12}b_2}{\tilde{\alpha}_1 w}$$
$$= 0.01066 \tag{29.92a}$$

$$\mathcal{S}(\tilde{\alpha}_2) = \frac{c_1\tilde{w}_{12}\tilde{w}_{21}b_1 + c_1\tilde{w}_{12}\tilde{w}_{22}b_2 + c_2\tilde{w}_{22}\tilde{w}_{21}b_1 + c_2\tilde{w}_{22}\tilde{w}_{22}b_2}{\tilde{\alpha}_2 w} = 0.05426 \tag{29.92b}$$

As a check, we use (27.30) to compute these sensitivities. From (26.45) and (26.52), we have

$$F(\tilde{\alpha}_1) = 93.70 \tag{29.93a}$$

$$F(\tilde{\alpha}_2) = 18.26 \tag{29.93b}$$

$$\hat{F}(\tilde{\alpha}_1) = 103.07 \times 10^3 \tag{29.93c}$$

$$\hat{F}(\tilde{\alpha}_2) = 2018.70 \tag{29.93d}$$

Substituting these in (27.30) the sensitivity functions are found to be

$$\mathcal{S}(\tilde{\alpha}_1) = \frac{1}{F(\tilde{\alpha}_1)} - \frac{1}{\hat{F}(\tilde{\alpha}_1)} = 0.01066 \tag{29.94a}$$

$$\mathcal{S}(\tilde{\alpha}_2) = \frac{1}{F(\tilde{\alpha}_2)} - \frac{1}{\hat{F}(\tilde{\alpha}_2)} = 0.05427 \tag{29.94b}$$

confirming (29.92).

Suppose that $\tilde{\alpha}_1$ is changed by 4 percent and $\tilde{\alpha}_2$ by 6 percent. The fractional change of the voltage gain $w(\mathbf{X})$ is found from (29.79) as

$$\frac{\delta w}{w} \approx \mathcal{S}(\tilde{\alpha}_1)\frac{\delta\tilde{\alpha}_1}{\tilde{\alpha}_1} + \mathcal{S}(\tilde{\alpha}_2)\frac{\delta\tilde{\alpha}_2}{\tilde{\alpha}_2} = 0.003683 \tag{29.95}$$

or 0.37 percent.

ferences

[1] F. H. Blecher, "Design principles for single loop transistor feedback amplifiers," *IRE Trans. Circuit Theory*, vol. CT-4, pp. 145–156, 1957.

[2] H. W. Bode, *Network Analysis and Feedback Amplifier Design*, Princeton, NJ: Van Nostrand, 1945.

[3] W. K. Chen, "Indefinite-admittance matrix formulation of feedback amplifier theory," *IEEE Trans. Circuits Syst.*, vol. CAS-23, pp. 498–505, 1976.

[4] W. K. Chen, "On second-order cofactors and null return difference in feedback amplifier theory," *Int. J. Circuit Theory Applicat.*, vol. 6, pp. 305–312, 1978.

[5] W. K. Chen, *Active Network and Feedback Amplifier Theory*, New York: McGraw-Hill, 1980, ch. 2, 4, 5, 7.

[6] W. K. Chen, *Active Network Analysis*, Singapore: World Scientific, 1991, chs. 2, 4, 5, 7.

[7] J. B. Cruz, Jr., and W. R. Perkins, "A new approach to the sensitivity problem in multivariable feedback system design," *IEEE Trans. Automat. Control*, vol. AC-9, pp. 216–223, 1964.

[8] S. S. Haykin, *Active Network Theory*, Reading, MA: Addison-Wesley, 1970.

[9] E. S. Kuh and R. A. Rohrer, *Theory of Linear Active Networks*, San Francisco, CA: Holden-Day, 1967.

[10] I. W. Sandberg, "On the theory of linear multi-loop feedback systems," *Bell Syst. Tech. J.*, vol. 42, pp. 355–382, 1963.

VI

Nonlinear Circuits

on O. Chua
iversity of California

30

Circuit Elements, Modeling, and Equation Formulation

:f A. Nossek
nical University of
ich, Germany

).1 Lumped Circuit Approximation

Most texts on circuits, whether they deal with linear or nonlinear circuits, consider only lumped circuits. If this is not the case, such is normally stated explicitly (see, e.g., Section VIII on distributed circuits). A physical circuit is considered to be a lumped circuit, if its size is small enough that for the situation under discussion, electromagnetic waves propagate across the circuit virtually instantaneously. If this is satisfied, voltages across ports and currents through terminals are well defined and, therefore, well suited to describe and analyze the behavior of a circuit.

To check whether an actual circuit is lumped or not, the largest extension d of the circuit in any spatial dimension is compared to the shortest wavelength λ of interest or with the shortest time interval τ of interest. If

$$d \ll \lambda = c/f, \qquad d \ll \tau c \tag{30.1}$$

is fullfilled, the circuit is lumped. In (30.1) c is the propagation velocity of electromagnetic waves in the medium under consideration, and f is the frequency corresponding to the wavelength λ as well as to the period τ.

).2 Circuit Elements and Connecting Multiport

It is obvious that a nonlinear circuit is described by a set of nonlinear equations, which, generally speaking, can be solved only approximately. Moreover, we may not find a unique solution, but a set of different solutions. Because of this complicated situation (compared to the simple solution of a linear circuit), it is even more important here to exploit the structure of the equations that reflect the structural properties of the circuit. The most important step in this direction

3-8341-2/95/$0.00 + $.50

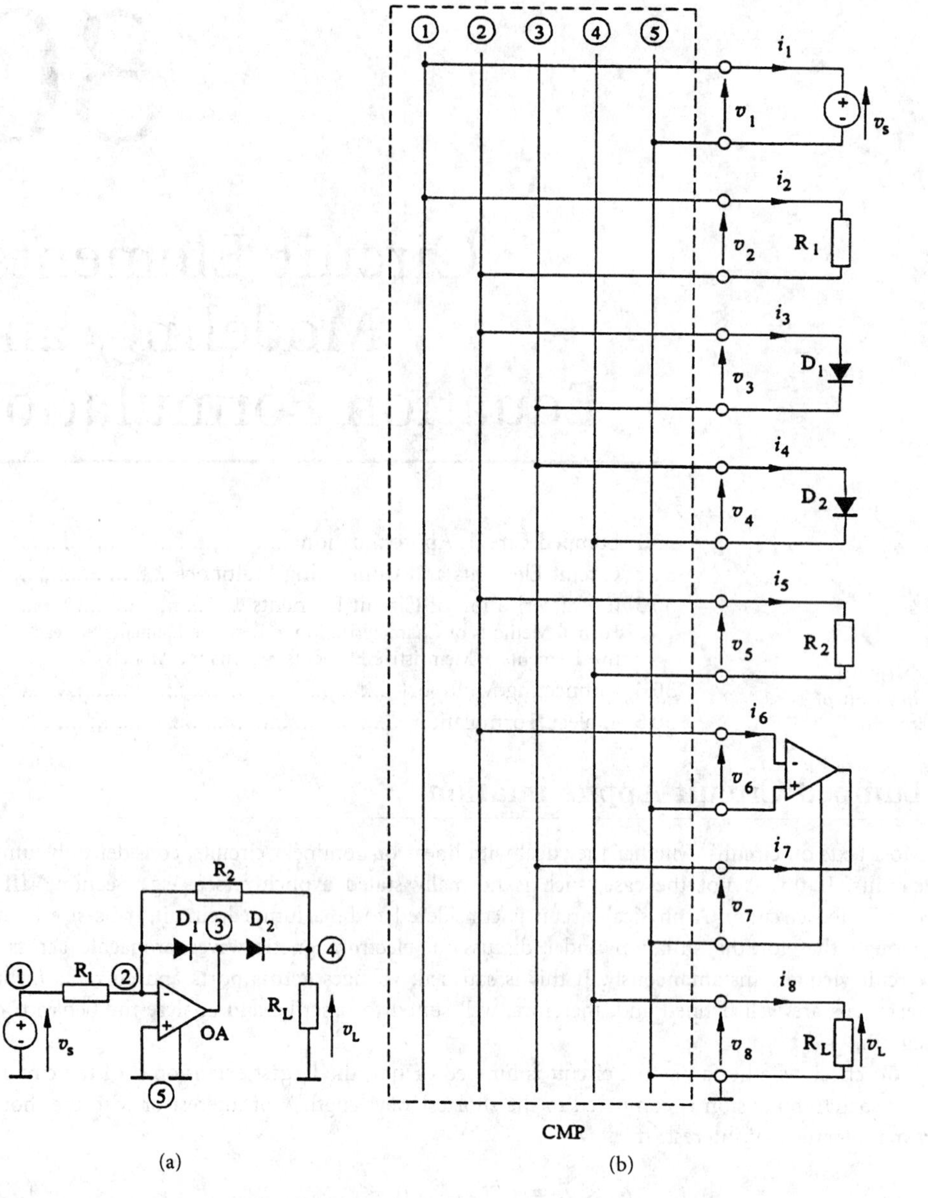

FIGURE 30.1 (a) A nonlinear circuit conventionally drawn. (b) Partitioning into CMP and circuit elements.

is to partition the circuit into two parts: one containing all circuit elements separately and one containing the interconnections thereof only. The latter is called the connecting multiport (CMP). This partitioning is demonstrated with a real circuit in Fig. 30.1; it is obvious that this partitioning is completely independent of the nature of the utilized circuit elements: linear or nonlinear, two-terminal or multiterminal, time-variant or time-invariant, passive or active, and so forth.

The equations that describe the CMP are merely Kirchhoff's circuit and voltage laws, (KCL, KVL) (see "Connecting Multiport"), which are, of course, linear, while nonlinearities show up in the description of the circuit elements. First, the circuit element is described, and then some details of the CMP are discussed.

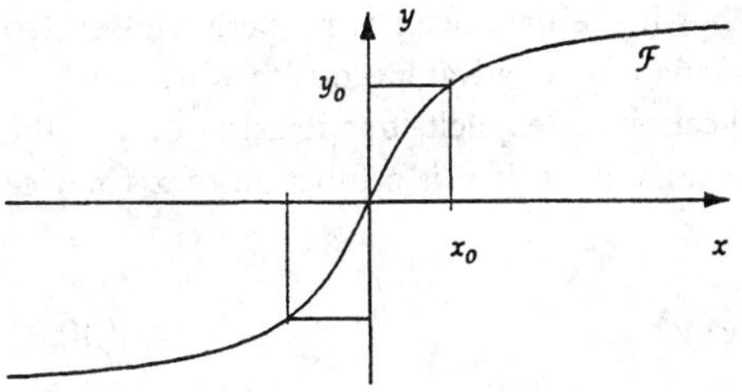

FIGURE 30.2 Plot of the one-port characteristic (Fig. 30.1).

).3 Characterization of Circuit Elements

This section characterizes a circuit element, two-terminal or multiterminal, with algebraic equations. This necessitates a proper choice of variables. Because of the algebraic nature of this discussion, differential or integral operators cannot be used. Let us begin with a formal discussion, using variables x and y without elaborating on their physical meaning. Later, these methods are applied to specific circuit elements of practical relevance, and voltages, currents, charges, and fluxes are utilized instead of x and y.

Formal Methods of Characterization

A relation between variables x and y can be given in an implicit form

$$\mathcal{F} \subset \mathbb{R} \times \mathbb{R}, x \in \mathbb{R}, y \in \mathbb{R} \qquad \mathcal{F} = \{(x, y) | f(x, y) = 0\} \tag{30.2}$$

Here, $\mathcal{F}$ is the characteristic of a two-terminal device which is described by a single implicit equation, $f(x, y) = 0$. Note that this equation is not unique. Various equivalent forms of $f(x, y) = 0$, exist which look quite different, but define the same one-port:

$$g(x, y) = e^{f(x,y)} - 1 = 0 \Leftrightarrow f(x, y) = 0 \tag{30.3}$$

The two functions f and g are quite different, but the tuples (x, y) defined by them constitute the same set $\mathcal{F}$.

A simple example for such an implicit description is

$$f(x, y) = (y/y_0) - \arctan(x/x_0) = 0 \tag{30.4}$$

the plot of which in the (x, y)-plane is shown in Fig. 30.2.

An alternative to the aforementioned implicit form is a parametric description, in which we use an additional parameter $\lambda \in \mathbb{R}$ to express the port variables as functions of this parameter:

$$x = f_x(\lambda) \qquad y = f_y(\lambda) \tag{30.5}$$

with every tuple

$$(x, y) = (f_x(\lambda), f_y(\lambda)) \in \mathcal{F}$$

being an admissible element of $\mathcal{F}$. Using (30.4) with $\lambda = y/y_0$ we have $x/x_0 = \tan \lambda$ and, therefore,

$$x = f_x(\lambda) = x_0 \tan \lambda \quad y = f_y(\lambda) = y_0\lambda \quad \lambda \in (-\pi/2, \pi/2) \tag{30.6}$$

Parameterized descriptions are also not unique, but if the parameter is properly chosen (so that f_x and f_y are continuous and differentiable), they are quite advantageous to work with.

The most favorable description in practical applications is explicit, but it exists only if the relation (30.2) is unique in at least one of the variables x or y. If y is a function of x (or vice versa), we write

$$y = f(x) \quad x = g(y) \tag{30.7}$$

For this example, both explicit versions do exist:

$$y = y_0 \arctan(y/y_0) \quad x = x_0 \tan(y/y_0) \quad y \in (-y_0\pi/2, y_0\pi/2) \tag{30.8}$$

All these descriptive methods can be extended in a straightforward manner to the multiterminal case by simply replacing the scalars x and y by vectors $\mathbf{x}$ and $\mathbf{y}$ and the functions f and g by vectors of functions $\mathbf{f}$ and $\mathbf{g}$.

The formal approach is applied to actual circuit elements in the following section.

Resistive Elements

A resistive element is by definition uniquely characterized by one of the aforementioned algebraic descriptions, where x and y are replaced by voltage, v, and current, i. This relation may depend on time, t (time-variant circuit element), but not on the history of the variables v and i.

Many important circuit elements can be modeled resistively, as far as their main property is concerned. This is true for most semiconductor devices such as diodes, bipolar transistors, field effect transistors, operational amplifiers (op amps), and so forth. This section concentrates on the main effect, leaving a more detailed description, including parasitics, to later sections.

An example of a time-variant (nonautonomous) resistive one-port is given in Fig. 30.3

$$\mathcal{F}(t) = \{(v(t), i(t)) | i(t) = I_s(\exp(v(t)/V_T) - 1) - i_L(t)\} \tag{30.9}$$

Using a reverse saturation current $I_s = 10\ \mu A$ and a thermal voltage $V_T = 25$ mV the individual characteristics in Fig. 30.3(b) are obtained, illustrating the nonautonomous nature of the device with the photocurrent i_L as the controlling parameter, which itself depends on the light intensity, which is assumed to be time dependent.

An ordinary pn-junction diode is nothing more than a special case of (30.9), i.e., $i_L = 0$.

Given the device characteristic in graphical form such as Fig. 30.3(b), which may be the summary of a set of measurements, it is easy to check as to whether explicit descriptions $i = g(u)$ or $u = f(i)$ exist or not. A simple example of a device for which $i = g(u)$ does exist, but g does not have an inverse, is the tunnel diode (Fig. 30.4).

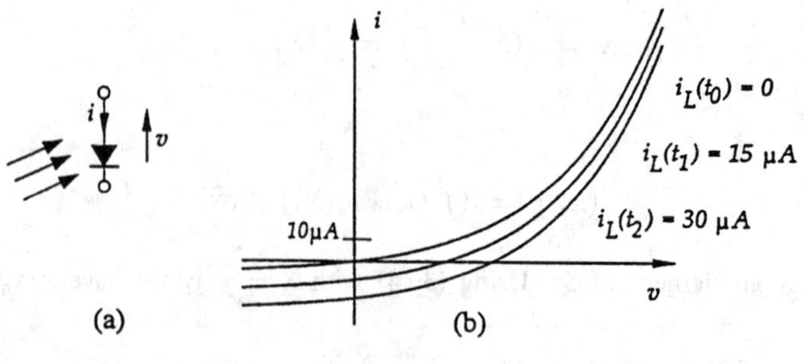

FIGURE 30.3 (a) Device symbol of a photodiode. (b) Device $i - v$ characteristic.

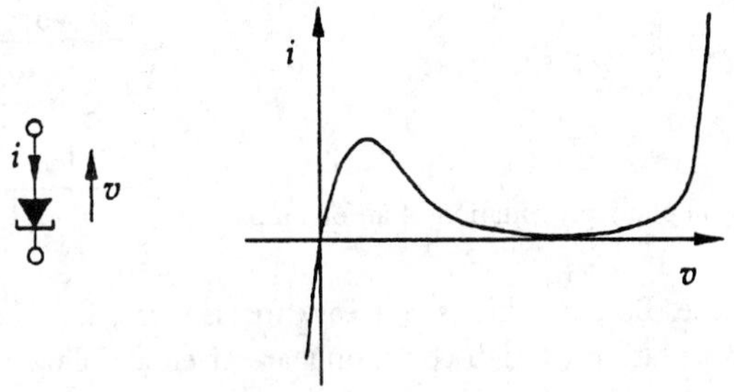

FIGURE 30.4 (a) Symbol and (b) $v - i$ characteristic of a tunnel diode.

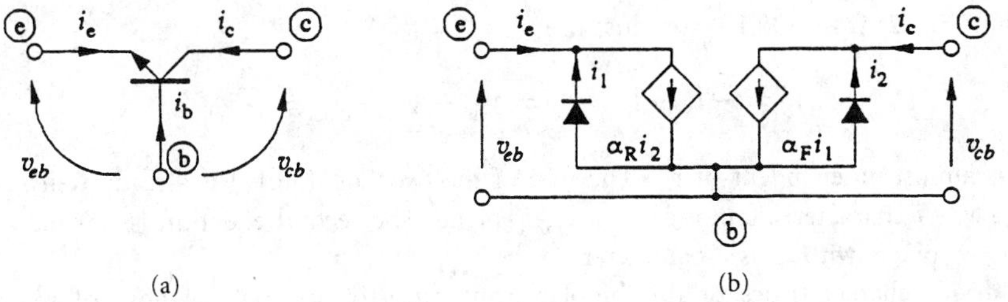

FIGURE 30.5 (a) Symbol and (b) equivalent circuit with two linear CCCSs and two diodes exactly representing Ebers-Moll eq. (30.10).

Many more nonlinear resistive one-ports or models thereof can be found, but the basic concept is always the same. Therefore, we proceed with an important example of a multiterminal device, the transistor. Again, we have a multitude of various transistors [bipolar *npn* and *pnp*, unipolar field-effect insulated gate transistors (MOSFET) and junction type (JFET), *n*- and *p*-channel, enhancement and depletion, etc.]. Here, we demonstrate only the basic idea of a resistive two-port (or three-terminal *e*mitter, *c*ollector, *b*ase) model of a bipolar *npn* transistor. Many more details are given in Section XII.

The so-called Ebers-Moll equations describe a bipolar *npn* transistor:

$$\begin{aligned} i_e &= -I_{es}(\exp(-v_{eb}/V_T) - 1) + \alpha_R I_{cs}(\exp(-v_{cb}/V_T) - 1) = -i_1 + \alpha_R i_2, \\ i_c &= \alpha_F I_{es}(\exp(-v_{eb}/V_T) - 1) - I_{cs}(\exp(-v_{cb}/V_T) - 1) = \alpha_F i_1 - i_2, \end{aligned} \tag{30.10}$$

which is an explicit two-port description

$$\mathbf{i} = \mathbf{f}(\mathbf{v}) \tag{30.11}$$

with

$$\mathbf{i} = \begin{bmatrix} i_e \\ i_c \end{bmatrix} \quad \mathbf{v} = \begin{bmatrix} v_{eb} \\ v_{cb} \end{bmatrix}$$

using the base b as the common terminal. α_F and α_R are the forward and reverse current gain of the transistor in common base configuration as shown in Fig. 30.5.

As with any device having more than two terminals, the device characteristics cannot be represented simply as a curve in a plane. In general it is a hypersurface in a multidimensional space. Especially in the case of the three-terminal transistor, it is a two-dimensional surface

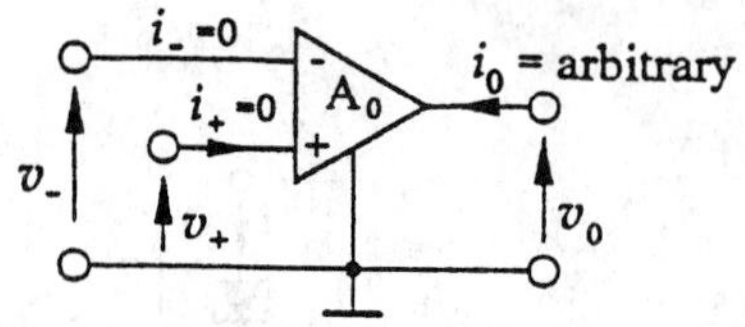

FIGURE 30.6 Symbolic representation of an op amp.

in four-dimensional space. Because this is not easy to visualize, it is normally split into two three-dimensional representations, which commonly are given as follows:

$$\begin{aligned} i_b &= f_1(v_{be}, v_{ce}) \approx f_1(v_{be}) \\ i_c &= f_2(v_{be}, v_{ce}) \approx f_2'(v_{ce}, i_b) \end{aligned} \tag{30.12}$$

To obtain (30.12) from (30.10) we must use

$$i_b = -i_e - i_c \quad v_{ce} = v_{cb} - v_{eb} \quad v_{be} = -v_{eb}$$

and i_b is almost independent of v_{ce}. The first of the two Eqs. (30.12) is already well suited for having a $v - i$ characteristic plotted in a v_{eb}-i_b-plane. The second equation is normally plotted in the v_{ce}-i_c-plane, with i_b as a parameter.

The device characteristics of this bipolar *npn* transistor, as well as those of many other multiterminal semiconductor devices, are today quite standard; they are given in data sheets and used by designers (discussed further in Section XII). These nonlinear models are the basis for deriving linearized small signal models, where they are needed.

Finally, we look at a higher level model of a multiterminal device, and model a complete op amp (containing a multitude of transistors) using a very simple, but nevertheless very powerful resistive model. An op amp, at a rather high level of abstraction, is a four-terminal device, as depicted in Fig. 30.6. In this figure everything dealing with power supply, biasing, offset compensation, etc. has been hidden. In this very simple model we assume the input currents to be zero and the output voltage to depend only on the difference of the two input voltages (common mode gain is zero):

$$v_d = v_+ - v_- \tag{30.13}$$

in the following way

$$v_o = \begin{cases} V_{sat} & v_d \geq \quad V_{sat}/A_0 \\ A_0 v_d & |v_d| \leq \quad V_{sat}/A_0 \\ -V_{sat} & v_d \leq \; -V_{sat}/A_0 \end{cases} \tag{30.14}$$

Therefore, we have a piecewise linear transfer characteristic, consisting of three pieces (Fig. 30.7). According to the three pieces (I, II, and III) we have three equivalent circuits (Fig. 30.8). Even if we increase our idealization to $A_0 \to \infty$, the equivalent circuit in Fig. 30.8(b) reduces from the voltage-controlled voltage source (VCVS) to a nullor. (It is worth emphasizing that this surprisingly simple model of such a complex nonlinear functional unit as an op amp is capable of capturing all of the main effects of such a multitransistor circuit. For many practically important applications it provides an accurate prediction of the behavior of real circuits.)

If we are not satisfied with resistive models because of the bandwidth of signals to be processed, the model must be refined by including elements with memory. First, we need to set up a way to describe memory-possessing elements ("Reactive Elements") and then combine this with the resistive model into a dynamic model (see "Dynamic Models").

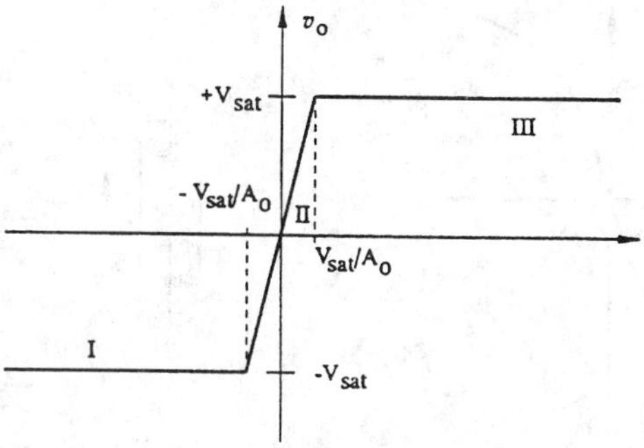

FIGURE 30.7 PWL transfer of an op amp.

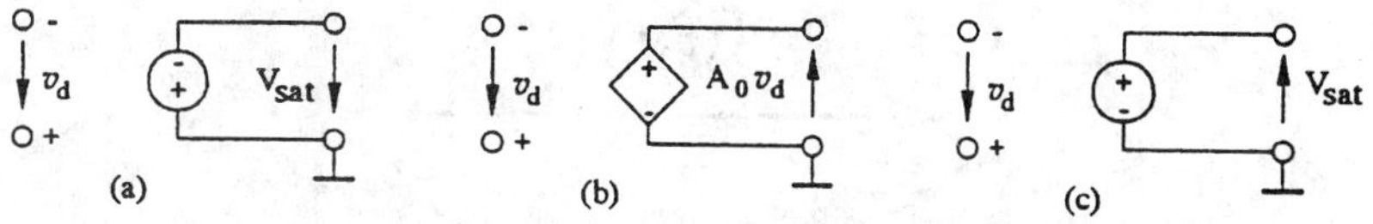

FIGURE 30.8 Equivalent circuit for an op amp (a) in the negative saturation region I ($v_d \leq -V_{sat}/A_0$), (b) in the linear region II ($v_d \leq V_{sat}/A_0$), and (c) in the positive saturation region III ($v_d \geq V_{sat}/A_0$).

Reactive Elements

To use algebraic descriptions and to plot a characteristic such as a curve in an x-y-plane, we have to extend the set of variables from v and i (resistive case) to charge q and flux ϕ:

$$q(t) = q(t_0) + \int_{t_0}^{t} i(\tau)d\tau \quad \phi(t) = \phi(t_0) + \int_{t_0}^{t} v(\tau)d\tau \tag{30.15}$$

If the integrals exist for $t_0 \to -\infty$ and if $q(-\infty) = 0$ and $\phi(-\infty) = 0$, we can write

$$q(t) = \int_{-\infty}^{t} i(\tau)d\tau \quad \phi(t) = \int_{-\infty}^{t} v(\tau)d\tau$$

which simply means to ignore the initial conditions of charge q and flux ϕ, which are unimportant for the electrical behavior of the component and the circuit.

With this in mind we define a capacitive (inductive) one-port in the following way:

$$\mathcal{F}_C = \{(v, q) | f_C(v, q) = 0\} \quad \mathcal{F}_L = \{(i, \phi) | f_L(i, \phi) = 0\} \tag{30.16}$$

Other than this implicit algebraic description, parameterized or even explicit descriptions may exist, similar to the resistive case dealt with previously.

Figure 30.9 gives examples of some nonlinear characteristics of a capacitive and an inductive reactance. It is obvious that this concept can be extended to the multiterminal case by replacing the scalars v and q or i and ϕ by vectors $\mathbf{v}$ and $\mathbf{q}$ or $\mathbf{i}$ and ϕ, respectively. This is useful when creating a first-order model of a multiport transformer.

Memristive Elements

After having dealt with resistive and reactive elements and considering all the variables and their interrelations which we used (Fig. 30.10), an interesting question remains: what about an element with an algebraic characterization in the q-ϕ-plane? Because this missing element is

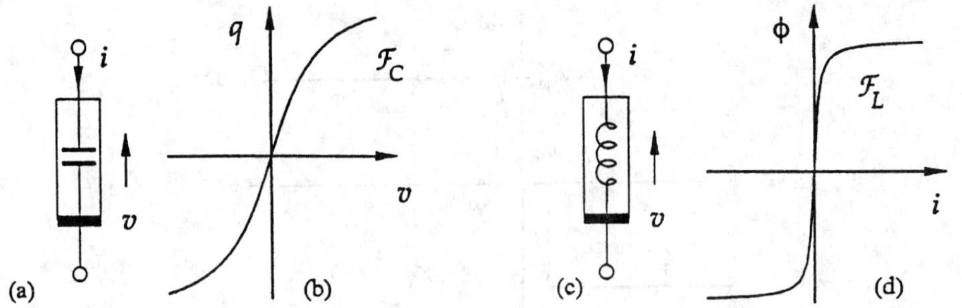

FIGURE 30.9 (a) Symbol and (b) characteristic of capacitor with dielectric material and (c) symbol and (d) characteristic of an inductor with ferromagnetic material.

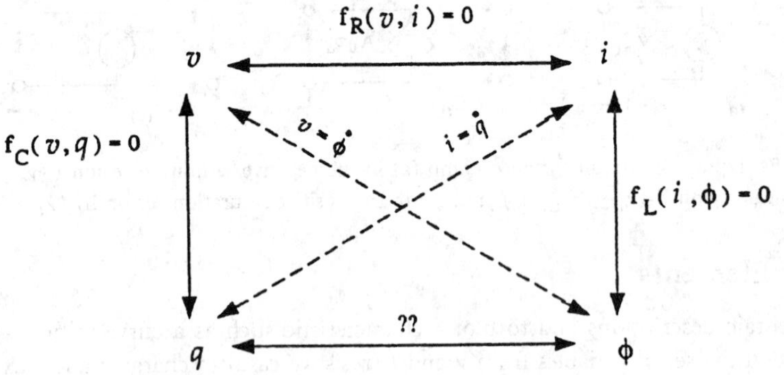

FIGURE 30.10 Interrelation between variables.

characterized by algebraic relation between the integral of current and the integral of voltage, it is a resistive element with memory, and is therefore called a memristor:

$$\mathscr{F}_M = \{(q, \phi) | f_M(q, \phi) = 0\} \tag{30.17}$$

A real-world example of a memristor is the so-called Coulomb cell, which consists of a gold anode immersed in an electrolyte in a silver can cathode. Memristive descriptions, although not widely utilized, are very useful for modeling the behavior of electrochemical elements.

Dynamic Models

Only a few examples are shown in this section, comprising resistive and reactive elements, in order to achieve a realistic description of electronic devices, including dynamical effects.

A *pn*-junction was described resistively earlier in the chapter, with an exponential $v - i$ characteristic, neglecting dynamical effects. To remove this shortcoming, we use the more elaborate dynamic model depicted in Fig. 30.11, which consist of a resistive *pn*-junction in parallel with a nonlinear capacitor. This extension can, of course, be carried on for the (multiterminal) transistor case, where we use dynamic diode models (Fig. 30.12).

The combination of the resistive and reactive models in one circuit leads to a description that makes use of differential and integral operators. It is almost always possible to reduce this to a set of nonlinear ordinary differential and algebraic equations. We return to this point later, but first we conclude this section with a dynamic model of an op amp (Fig. 30.13), making use of nonlinear controlled sources, (VCVS) and voltage-controlled circuit sources (VCCS). This simple

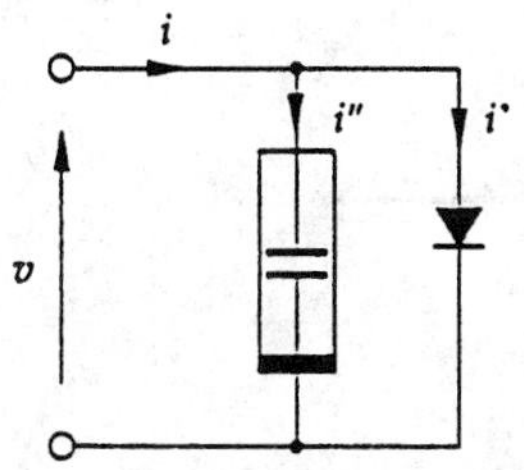

FIGURE 30.11 Dynamic *pn*-junction model.

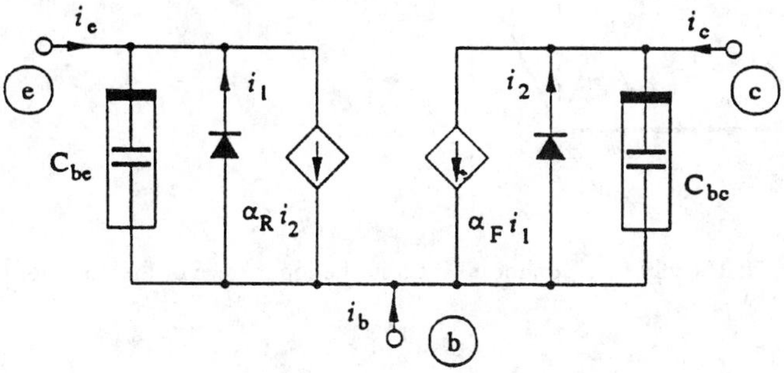

FIGURE 30.12 Dynamic transistor model.

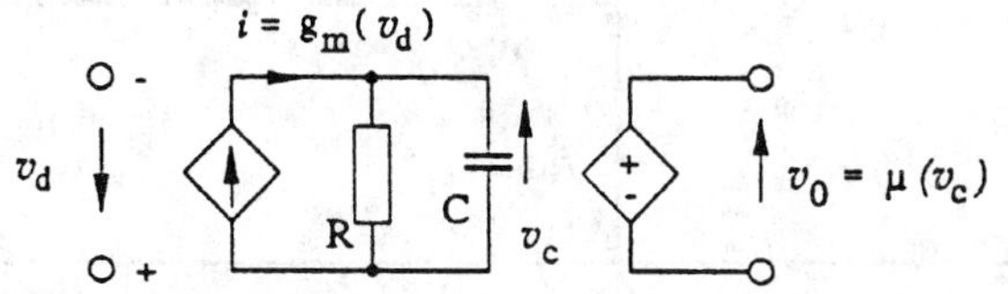

FIGURE 30.13 Dynamic op amp model.

model accounts for the following important practical properties of a real-world op amp:

- First-order low-pass behavior in the linear region of the controlled sources
- Slew rate limitation is incorporated by the nonlinearity of $g_m(v_d)$ of the VCCS (Fig. 30.14)
- Output voltage saturation is modeled with the aid of the nonlinearity of $\mu(v_C)$ (Fig. 30.15)

The 3-dB bandwidth of this op amp is given by $\omega_{3dB} = 1/(RC)$ with a 20 dB/decade roll-off. The slew rate is given by SR=i_o/C, while the DC open-loop gain $A_0 = g_{m0}R\mu_0$.

Using this simple model, not all, but some, of the most important nonlinear dynamical effects are properly described.

30.4 Connecting Multiport

After having dealt with the description of the circuit elements, we return to the interconnection structure, which is summarized in the CMP [see Fig. 30.1(b)]. This multiport is linear, lossless, and reciprocal and its description stated in implicit form is simply KCL and KVL equations

$$\begin{bmatrix} \mathbf{B} & 0 \\ 0 & \mathbf{A} \end{bmatrix}\begin{bmatrix} \mathbf{v} \\ \mathbf{i} \end{bmatrix} = \begin{bmatrix} \mathbf{B} \\ 0 \end{bmatrix}\mathbf{v} + \begin{bmatrix} 0 \\ \mathbf{A} \end{bmatrix}\mathbf{i} = \begin{bmatrix} 0 \\ 0 \end{bmatrix} \Rightarrow \mathbf{Bv} = 0 \text{ and } \mathbf{Ai} = 0 \tag{30.18}$$

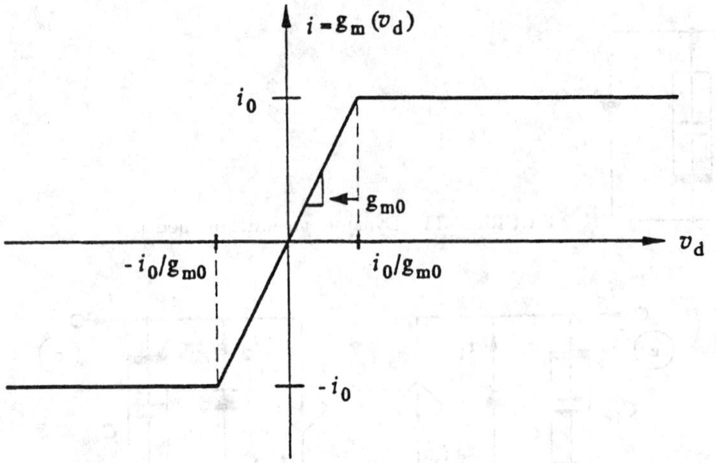

FIGURE 30.14 Nonlinear VCCS in the input stage of the op amp.

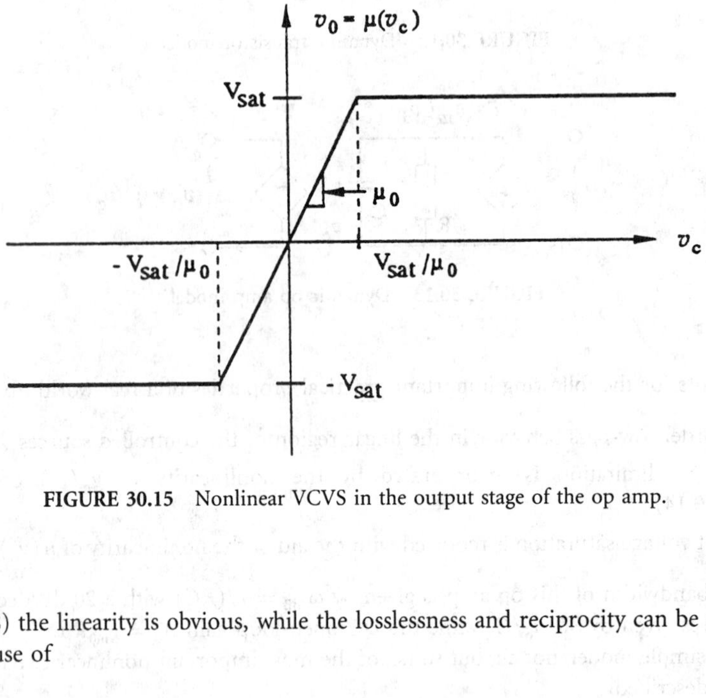

FIGURE 30.15 Nonlinear VCVS in the output stage of the op amp.

From (30.18) the linearity is obvious, while the losslessness and reciprocity can be proven easily by making use of

$$\mathbf{BA}^T = 0 \quad \mathbf{AB}^T = 0 \quad \text{rank}\,\mathbf{A} + \text{rank}\,\mathbf{B} = b \tag{30.19}$$

(b is the number of ports of the CMP).

In addition to the perfect wires, which are the ingredients of CMP, ideal transformers (which are also linear, lossless, and reciprocal) can be accommodated in the CMP without changing the structure of (30.18) and (30.19). **A** is an $(n-1) \times b$ incidence matrix containing the coefficients of any $(n-1)$ linearly independent nodal equations (or supernodal or cutset equations), while **B** is a $(b-(n-1)) \times b$ incidence matrix, the entries of which are the coefficients of any $(b-(n-1))$ linearly independent loop equations (or fundamental-loop equations).

30.5 Tableau Formulation

Combining the description of the CMP and the description of all circuit elements into one tableau, all information about the circuit under consideration is at our fingertips:

$$\begin{bmatrix} \mathbf{B} & 0 \\ 0 & \mathbf{A} \end{bmatrix} \begin{bmatrix} \mathbf{v} \\ \mathbf{i} \end{bmatrix} = \begin{bmatrix} 0 \\ 0 \end{bmatrix}, \quad b \text{ linear algebraic equations}$$

$$\mathbf{f}(\dot{\mathbf{v}}, \mathbf{v}, \dot{\mathbf{i}}, \mathbf{i}, t) = \mathbf{0}, \quad b \text{ nonlinear differential equations} \tag{30.20}$$

This set of equations is not unique, although their solution is unique for properly modeled circuits. Instead of using derivatives $\dot{\mathbf{v}} = d\mathbf{v}/dt, \dot{\mathbf{i}} = d\mathbf{i}/dt$ we could have used integrals, or we could have formulated the equations with $\mathbf{q}, \boldsymbol{\phi}, \dot{\mathbf{q}}, \dot{\boldsymbol{\phi}}$ as variables.

It is important to note that at least half of the equations are linear. To solve the nonlinear equations numerical techniques are commonly used, however, this is beyond the scope of this chapter.

References

[1] L. O. Chua, C. A. Desoer, and E. S. Kuh, *Linear and Nonlinear Circuits*, New York: McGraw-Hill, 1987.

[2] J. E. Solomon, "The monolithic opamp: a tutorial study," *IEEE J. Solid-State Circuits*, vol. SC-9, pp. 314–332, Dec. 1974.

[3] W. Mathis, *Theorie Nichtlinearer Netzwerke*, Berlin: Springer-Verlag, 1987.

[4] R. K. Brayton, L. O. Chua, J. D. Rhodes, and R. Spence, *Modern Network Theory—An Introduction*, Saphorin, Switzerland: Georgi Publishing, 1978.

31

Qualitative Analysis

Martin Hasler
Swiss Federal Institute of Technology

31.1 Introduction

The main goal of circuit analysis is to determine the solution of the circuit, i.e., the voltages and the currents in the circuit, usually as functions of time. The advent of powerful computers and circuit analysis software has greatly simplified this task. Basically, the circuit to be analyzed is fed to the computer through some circuit description language, or it is analyzed graphically, and the software will produce the desired voltage or current waveforms. Progress has rendered the traditional paper-and-pencil methods obsolete, in which the engineer's skill and intuition led the way through a series of clever approximations, until the circuit equations can be solved analytically.

A closer comparison of the numerical and the approximate analytical solution reveals, however, that the two are not quite equivalent. While the former is precise, it only provides the solution of the circuit with given parameters, whereas the latter is an approximation, but the approximate solution most often is given explicitly as a function of some circuit parameters. Therefore, it allows us to assess the influence of these parameters on the solution.

If we rely entirely on the numerical solution of a circuit, we never get a global picture of its behavior, unless we carry out a huge number of analyses. Thus, the numerical analysis should be complemented by a qualitative analysis, one that concentrates on general properties of the circuit, properties that do not depend on the particular set of circuit parameters.

31.2 Resistive Circuits

The term *resistive circuits* is not used, as one would imagine, for circuits that are composed solely of resistors. It admits all circuit elements that are not dynamic, i.e., whose constitutive relations do not involve time derivatives, integrals over time, or time delays, etc. Expressed positively, resistive circuit elements are described by constitutive relations that involve only currents and voltages at the same time instants.

0-8493-8341-2/95/$0.00 + $.50

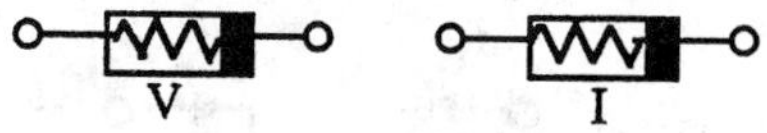

FIGURE 31.1 Symbols of the V- and the I-resistor.

Physical circuits can never be modeled in a satisfactory way by resistive circuits, but resistive circuits appear in many contexts as auxiliary constructs. The most important problem that leads to a resistive circuit is the determination of the equilibrium points, or, as is current use in electronics, the **DC-operating points**, of a dynamic circuit. The DC-operating points of a circuit correspond in a one-to-one fashion to the solutions of the resistive circuit obtained by removing the capacitors and by short circuiting the inductors. The resistive circuit associated with the state equations of a dynamic circuit is discussed in [1].

Among the resistive circuit elements we find, of course, the resistors. For the purposes of this introduction, we distinguish between linear resistors, V-resistors and I-resistors. **V-resistors** are **voltage controlled**, i.e., defined by constitutive relations of the form

$$i = g(v) \tag{31.1}$$

In addition, we require that g is a continuous, increasing function of v, defined for all real v. Dually, an **I-resistor** is **current controlled**, i.e., defined by a constitutive relation of the form

$$v = h(i) \tag{31.2}$$

In addition, we require that h is a continuous, increasing function of i, defined for all real i. We use the symbols of Fig. 31.1 for V- and I-resistors. Linear resistors are examples of both I- and V-resistors. An example of a V-resistor that is not an I-resistor is the junction diode, modeled by its usual exponential constitutive relation

$$i = I_s(e^{v/nV_T} - 1) \tag{31.3}$$

While (31.3) could be solved for v and thus the constitutive relation could be written in the form (31.2), the resulting function h would be defined only for currents between $-I_s$ and $+\infty$, which is not enough to qualify for an I-resistor. For the same reason, the static model for a Zener diode would be an I-resistor, but not a V-resistor. Indeed, the very nature of the Zener diode limits its voltages on the negative side.

A somewhat strange by-product of our definition of V- and I-resistors is that **independent voltage sources** are I-resistors and **independent current sources** are V-resistors. Indeed, a voltage source of value E has the constitutive relation

$$v = E \tag{31.4}$$

which clearly is of the form (31.2), with a constant function h, and a current source of value I has the form

$$i = I \tag{31.5}$$

which is of the form (31.1) with a constant function g. Despite this, we shall treat the independent sources as a different type of element.

Another class of resistive elements is the **controlled sources**. We consider them to be two-ports, e.g., a **voltage-controlled voltage source** (VCVS) is the two-port of Fig. 31.2, whose constitutive relations are

$$v_1 = \alpha v_2 \tag{31.6}$$

$$i_1 = 0 \tag{31.7}$$

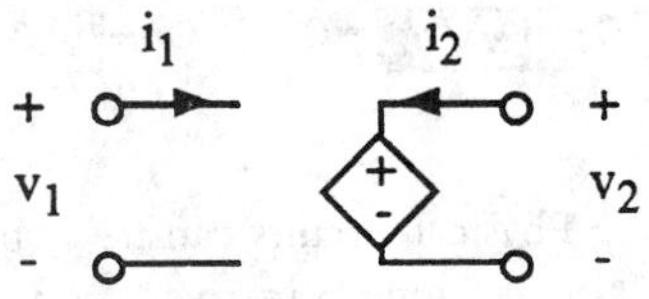

FIGURE 31.2 VCVS as a two-port.

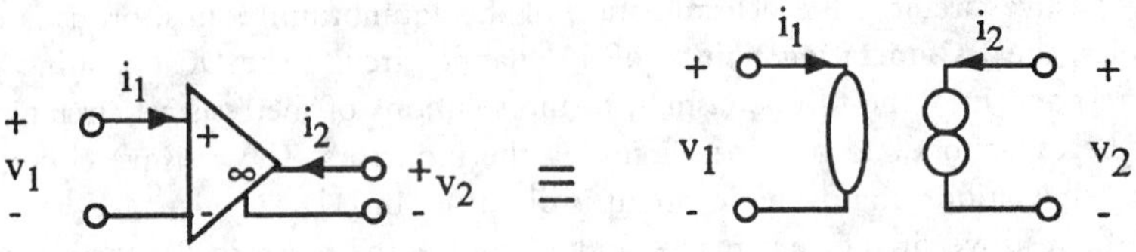

FIGURE 31.3 Operational amplifier as a juxtaposition of a nullator and a norator.

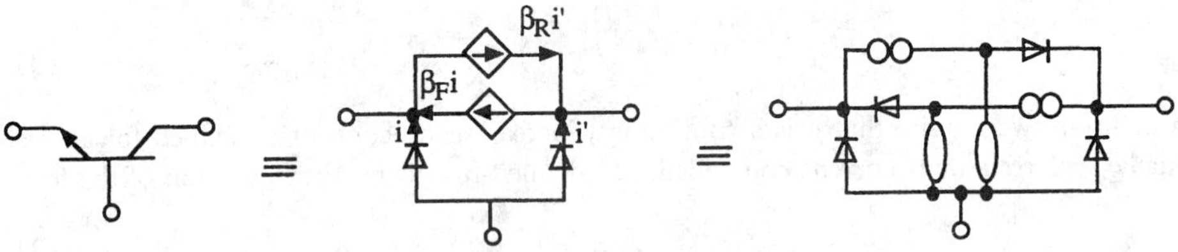

FIGURE 31.4 Equivalent circuit of a bipolar npn transistor.

The other controlled sources have similar forms. Another useful resistive circuit element is the **ideal operational amplifier**. It is a two-port defined by the two constitutive relations

$$v_1 = 0 \tag{31.8}$$

$$i_1 = 0 \tag{31.9}$$

This two-port can be decomposed into the juxtaposition of two singular one-ports, the nullator and the norator, as shown in Fig. 31.3. The nullator has two constitutive relations:

$$v = 0 \quad i = 0 \tag{31.10}$$

whereas the norator has no constitutive relation.

For all practical purposes, the resistive circuit elements mentioned thus far are sufficient. By this we mean that all nonlinear resistive circuits encountered in practice possess an equivalent circuit composed of nonlinear resistors, independent and controlled sources, and nullator-norator pairs. Figure 31.4 illustrates this fact. Here, the equivalent circuit of the bipolar transistor is modeled by the Ebers-Moll equations:

$$\begin{pmatrix} i_1 \\ i_2 \end{pmatrix} = \begin{pmatrix} 1 + \dfrac{1}{\beta_F} & -1 \\ -1 & 1 + \dfrac{1}{\beta_R} \end{pmatrix} \begin{pmatrix} g(v_1) \\ g(v_2) \end{pmatrix} \tag{31.11}$$

The function g is given by the right-hand side of (31.3).

Actually, the list of basic resistive circuit elements given so far is redundant, and the nullator-norator pair renders the controlled sources superfluous. An example of a substitution of controlled sources by nullator-norator pairs is given in Fig. 31.4. Equivalent circuits exist for all four types of controlled sources with nullator-norator pairs. Figure 31.5 gives an equivalent

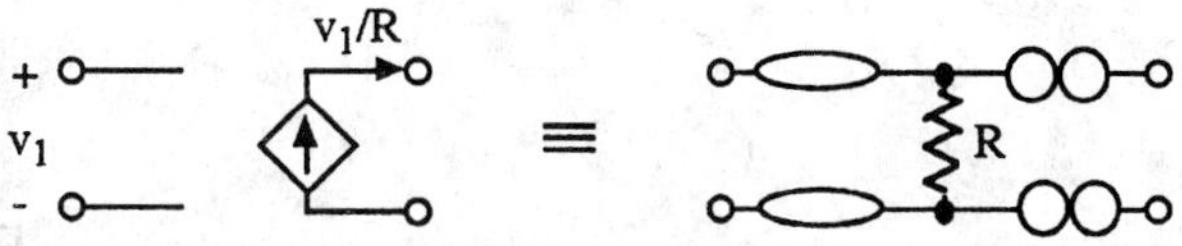

FIGURE 31.5 Equivalent circuit for a floating voltage-controlled current source.

circuit for a voltage-controlled current source (VCCS), where the input port is floating with respect to the output port.

The system of equations that describes a resistive circuit is the collection of Kirchhoff equations and the constitutive relations of the circuit elements. It has the following form (if we limit ourselves to resistors, independent sources, nullators, and norators):

$$\mathbf{Ai} = \mathbf{0} \quad \text{(Kirchoff's current law)} \tag{31.12}$$

$$\mathbf{Bv} = \mathbf{0} \quad \text{(Kirchoff's voltage law)} \tag{31.13}$$

$$i_k = g(v_k) \quad (V - \text{resistor}) \tag{31.14}$$

$$v_k = h(i_k) \quad (I - \text{resistor}) \tag{31.15}$$

$$v_k = E_k \quad \text{(independent voltage source)} \tag{31.16}$$

$$i_k = I_k \quad \text{(independent current source)} \tag{31.17}$$

$$\left.\begin{aligned} v_k &= 0 \\ i_k &= 0 \end{aligned}\right\} \quad \text{(nullators)} \tag{31.18}$$

In this system of equations the unknowns are the branch voltages and the branch currents

$$\mathbf{v} = \begin{pmatrix} v_1 \\ v_2 \\ \vdots \\ v_b \end{pmatrix} \quad \mathbf{i} = \begin{pmatrix} i_1 \\ i_2 \\ \vdots \\ i_b \end{pmatrix} \tag{31.19}$$

where the b is the number of branches. Because we have b linearly independent Kirchhoff equations [2], the system contains $2b$ equations and $2b$ unknowns. A solution $\boldsymbol{\xi} = \begin{pmatrix} \mathbf{v} \\ \mathbf{i} \end{pmatrix}$ of the system is called a **solution of the circuit**. It is a collection of branch voltages and currents that satisfy (31.12) to (31.19).

Number of Solutions of a Resistive Circuit

As we found earlier, the number of equations of a resistive circuit equals the number of unknowns. One may therefore expect a unique solution. While this may be the norm, it is far from being generally true. It is not even true for linear resistive circuits. In fact, the equations for a linear resistive circuit are of the form

$$\mathbf{Hx} = \mathbf{e} \tag{31.20}$$

where the $2b \times 2b$ matrix $\mathbf{H}$ contains the resistances and elements of value 0, ± 1, whereas the vector $\mathbf{e}$ contains the source values and zeroes. The solution of (31.20) is unique iff the

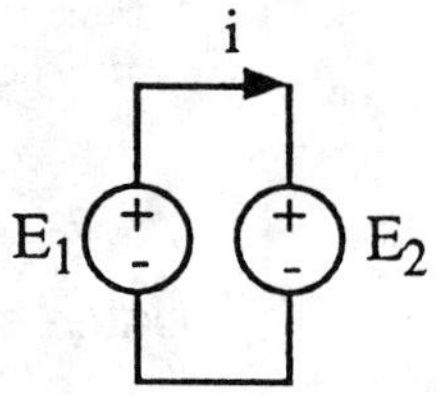

FIGURE 31.6 Circuit with zero or infinite solutions.

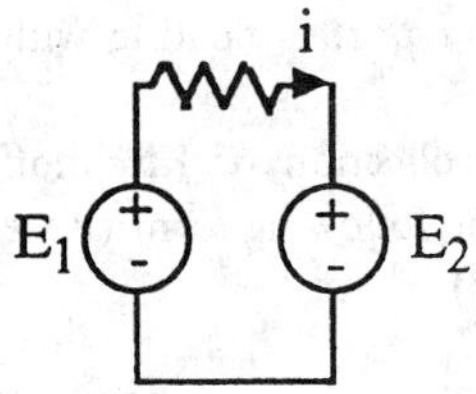

FIGURE 31.7 Circuit with exactly one solution.

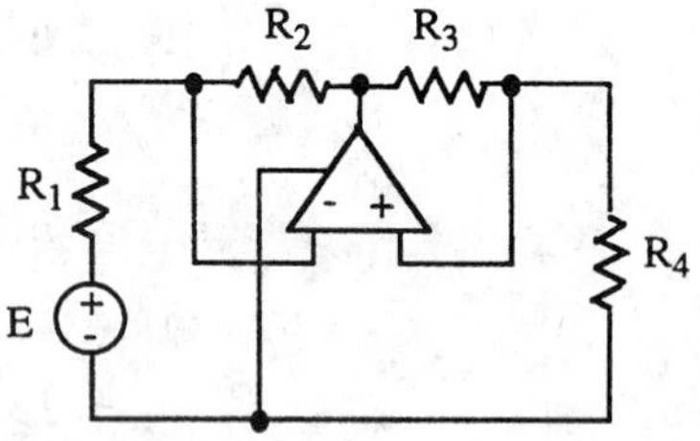

FIGURE 31.8 Circuit with one, zero, or infinite solutions.

determinant of H differs from zero. If it is zero, then the circuit has either infinitely many solutions or no solution at all. Is such a case realistic? The answer is yes and no. Consider two voltage sources connected as shown in Fig. 31.6.

If $E_1 \neq E_2$, the constitutive relations of the sources are in contradiction with Kirchhoff's voltage law (KVL), and thus the circuit has no solution, whereas when $E_1 = E_2$, the current i in Fig. 31.6 is not determined by the circuit equations, and thus the circuit has infinitely many solutions. One may object that the problem is purely academic, because in practice wires as connections have a small, but positive, resistance, and therefore one should instead consider the circuit of Fig. 31.7 which has exactly one solution.

Examples of singular linear resistive circuits exist that are much more complicated. However, the introduction of parasitic elements always permits us to obtain a circuit with a single solution, and thus the special case in which the matrix **H** in (31.9) is singular can be disregarded. Within the framework of linear circuits, this attitude is perfectly justified. When a nonlinear circuit model is chosen, however, the situation changes. An example clarifies this point.

Consider the linear circuit of Fig. 31.8. It is not difficult to see that it has exactly one solution, except when

$$R_1 R_3 = R_2 R_4 \tag{31.21}$$

In this case the matrix **H** in (31.29) is singular and the circuit of Fig. 31.8 has zero or infinitely many solutions, depending on whether E differs from zero. From the point of view of linear circuits, we can disregard this singular case because it arises only when (31.21) is exactly satisfied with infinite precision.

Now replace resistor R_4 by a nonlinear resistor, whose characteristic is represented by the bold line in Fig. 31.9. The resulting circuit is equivalent to the connection of a voltage source, a linear resistor, and the nonlinear resistor, as shown in Fig. 31.10. Its solutions correspond to the intersections of the nonlinear resistor characteristic and the load line (Fig. 31.9). Depending on the value of E, either one, two, or three solutions are available. While we still need infinite

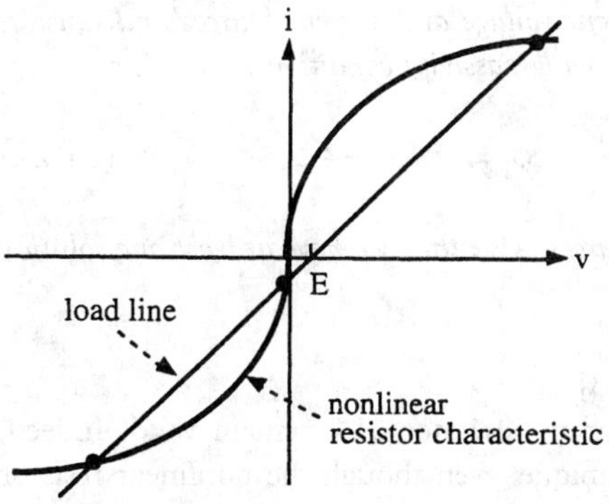

FIGURE 31.9 Characteristic of the nonlinear resistor and solutions of the circuit of Fig. 31.10.

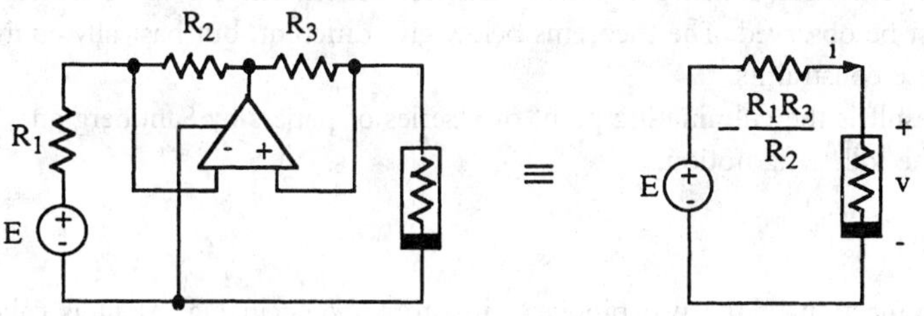

FIGURE 31.10 Circuit with one, two, or three solutions.

precision to obtain two solutions, this is not the case for one or three solutions. Thus, more than one DC-operating point may be observed in electronic circuits. Indeed, for static memories, and multivibrators in general, multiple DC-operating points are an essential feature.

The example of Fig. 31.10 shows an important aspect of the problem. The number of solutions depends on the parameter values of the circuit. In the example the value of E determines whether one, two, or three solutions are available. This is not always the case. An important class of nonlinear resistive circuits always has exactly one solution, irrespective of circuit parameters. In fact, for many applications, e.g., amplification, signal shaping, logic operations, etc., it is necessary that a circuit has exactly one DC-operating point. Circuits that are designed for these functionalities should thus have a unique DC-operating point for any choice of element values.

If a resistive circuit contains only two-terminal resistors with increasing characteristics and sources, but no nonreciprocal element such as controlled sources, operational amplifiers, or transistors, the solution is usually unique. The following theorem gives a precise statement.

Theorem 31.1: *A circuit composed of independent voltage and current sources and strictly increasing resistors without loop of voltage sources and without cutset of current sources has at most one solution.*

The interconnection condition concerning the sources is necessary. The circuit of Fig. 31.6 is an illustration of this statement. Its solution is not unique because of the loop of voltage sources. This loop is no longer present in the circuit of Fig. 31.7, which satisfies the conditions of Theorem 31.1, and which indeed has a unique solution.

If the resistor characteristics are not *strictly* increasing but only increasing (i.e., if the v-i curves have horizontal or vertical portions), the theorem still holds, if we exclude loops of voltage sources and I-resistors, and cutsets of current sources and V-resistors.

While Theorem 31.1 guarantees the uniqueness of the solution, it cannot assure its existence. On the other hand, we do not need increasing resistor characteristics for the existence.

Theorem 31.2: *Let a circuit be composed of independent voltage and current sources and resistors whose characteristics are continuous and satisfy the following passivity condition at infinity:*

$$v \to +\infty \Leftrightarrow i \to +\infty \quad \text{and} \quad v \to -\infty \Leftrightarrow i \to -\infty \tag{31.22}$$

If no loop of voltage sources and no cutset of current sources exist, then we have at least one solution of the circuit.

For refinements of this theorem, refer to [1] and [3].

If we admit nonreciprocal elements, neither Theorem 31.1 nor 31.2 remain valid. Indeed, the solution of the circuit of Fig. 31.10 may be nonunique, even though the nonlinear resistor has a strictly increasing characteristic. In order to ensure the existence and uniqueness of a nonreciprocal nonlinear resistive circuit, nontrivial constraints on the interconnection of the elements must be observed. The theorems below give different, but basically equivalent, ways to formulate these constraints.

The first result is the culminating point of a series of papers by Sandberg and Willson [3]. It is based on the following notion.

Definition 31.1.

- The connection of the two bipolar transistors shown in Fig. 31.11 is called a **feedback structure.** The type of the transistors and the location of the collectors and emitters is arbitrary.
- A circuit composed of bipolar transistors, resistors, and independent sources contains a feedback structure, if it can be reduced to the circuit of Fig. 31.11 by replacing each voltage source by a short circuit, each current source by an open circuit, each resistor and diode by an open or a short circuit, and each transistor by one of the five short-open-circuit combinations represented in Fig. 31.12.

Theorem 31.3: *Let a circuit be composed of bipolar transistors, described by the Ebers-Moll model, positive linear resistors, and independent sources. Suppose we have no loop of voltage sources and no cutset of current sources. If the circuit contains no feedback structure, it has exactly one solution.*

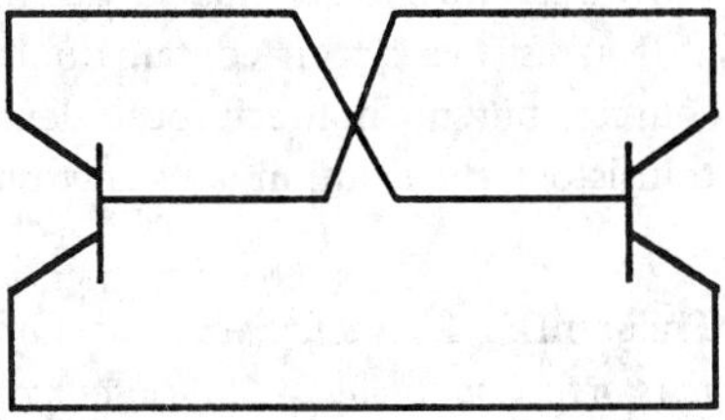

FIGURE 31.11 Feedback structure.

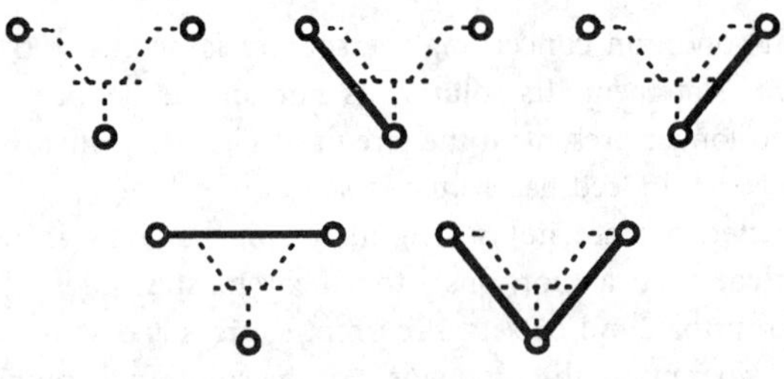

FIGURE 31.12 Short-open-circuit combinations for replacing the transistors.

This theorem [4] is extended in [5] to MOS transistors.

The second approach is due to Nishi and Chua. Instead of transistors, it admits controlled sources. In order to formulate the theorem, two notions must be introduced.

Definition 31.2. A circuit composed of controlled sources, resistors, and independent sources satisfies the **interconnection condition**, if the following conditions are satisfied:

- No loop is composed of voltage sources, output ports of (voltage or current) controlled voltage sources, and input ports of current controlled (voltage or current) sources.
- No cutset is composed of current sources, output ports of (voltage or current) controlled current sources, and input ports of voltage controlled (voltage or current) sources.

Definition 31.3. A circuit composed exclusively of controlled sources has a **complementary tree structure** if both the input and output ports each form a tree. The fundamental loop matrix of the input port tree has the form

$$B = [B_T \mid 1] \tag{31.23}$$

The circuit is said to have a **positive (negative) complementary tree structure**, if the determinant of B_T is positive (negative).

Theorem 31.4: *Suppose a circuit composed of controlled sources, strictly increasing resistors satisfying (31.22), and independent sources satisfies the interconnection condition. If, by replacing each resistor either by a short circuit or an open circuit, all independent and dependent voltage sources by short circuits, all independent and some dependent voltage sources by a short circuit, and all independent and some dependent current sources by an open circuit, one never obtains a negative complementary tree structure, the circuit has exactly one solution [6].*

A similar theorem for circuits with operational amplifiers instead of controlled sources is proved in [7].

The third approach is that of Hasler [1, 8]. The nonreciprocal elements here are nullator-norator pairs. Instead of reducing the circuit by some operations in order to obtain a certain structure, we must orient the resistors in a certain way. Again, we must first introduce a new concept.

Definition 31.4. Let a circuit be composed of nullator-norator pairs, resistors, and independent voltage and current sources. A **partial orientation** of the resistors **is uniform**, if the following two conditions are satisfied:

- Every oriented resistor is part of an evenly directed loop composed only of oriented resistors and voltage sources
- Every oriented resistor is part of an evenly directed cutset composed only of norators, oriented resistors, and voltage sources

Theorem 31.5: *Let a circuit be composed of nullator-norator pairs, V- and I-resistors, and independent voltage and current sources. If the following conditions are satisfied, the circuit has exactly one solution:*

- *The norators, I-resistors, and the voltage sources together form a tree*
- *The nullators, I-resistors, and the voltage sources together form a tree*
- *The resistors have no uniform partial orientation, except for the trivial case, in which no resistor is oriented.*

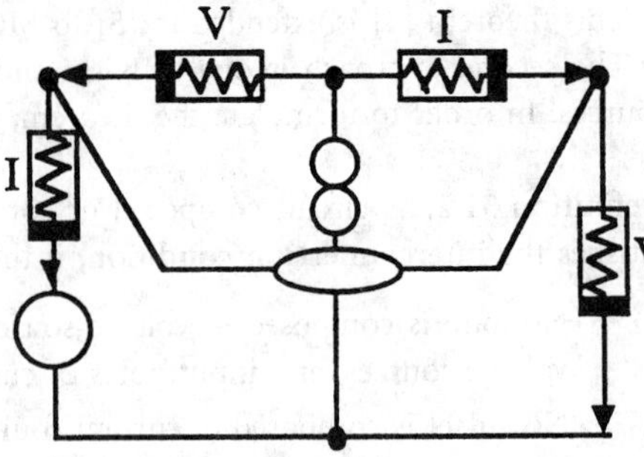

FIGURE 31.13 Circuit of Fig. 31.10 with nullator and norator.

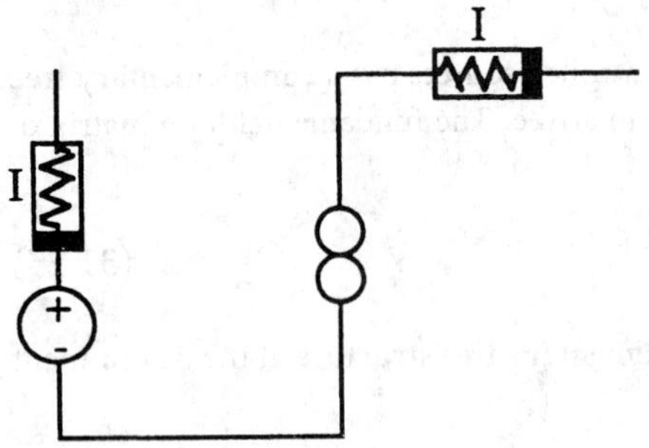

FIGURE 31.14 Norator–I-resistor–voltage source tree.

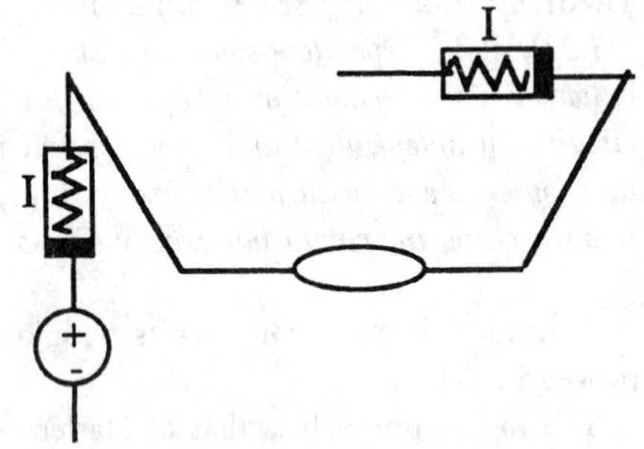

FIGURE 31.15 Nullator–I-resistor–voltage source tree.

We illustrate the conditions of this theorem with the example of Fig. 31.10. In Fig. 31.13 the resistors are specified as V- and I-resistors and a uniform orientation of the resistors is indicated. Note that the nonlinear resistor is a V-resistor, but not an I-resistor, because its current saturates. The linear resistors, however, are both V- and I-resistors. The choice in Fig. 31.13 is made in order to satisfy the first two conditions of Theorem 31.5. Correspondingly, in Figs. 31.14 and 31.15 the norator–I-resistor–voltage source tree and the nullator–I-resistor–voltage source tree are represented. Because the third condition is not satisfied, Theorem 31.5 cannot guarantee a unique solution. Indeed, as explained earlier, this circuit may have three solutions.

Theorem 31.5 has been generalized to controlled sources, to resistors that are increasing but neither voltage nor current controlled (e.g., the ideal diode), and to resistors that are decreasing instead of increasing [9].

Theorems 31.3, 31.4, and 31.5 have common features. Their conditions concern the **circuit structure**—the circuit graph that expresses the interconnection of the elements and the type of elements that occupy the branches of the graph, but not the element values. Therefore, the theorems guarantee the existence and uniqueness of the solution for whole classes of circuits, in which the individual circuits differ by their element values and parameters. In this sense the conditions are not only sufficient, but also necessary. This means, for example, in the case of Theorem 31.5 that if all circuits with the same structure have exactly one solution, then the three conditions must be satisfied. However, by logical contraposition, if one of the three conditions is not satisfied for a given circuit structure, a circuit with this structure exists which has either no solution or more than one solution.

On the other hand, if we consider a specific circuit, the conditions are only sufficient. They permit us to prove that the solution exists and is unique, but some circuits do not satisfy the conditions and still have exactly one solution. However, if the parameters of such a circuit are varied, one eventually falls onto a circuit with no solution or more than one solution.

The main conditions of Theorems 31.3 and 31.4 have an evident intuitive meaning. The orientations to look for in Theorem 31.5 are linked to the sign of the currents and the voltages of the difference of two solutions. Because the resistors are increasing, these signs are the same for the voltage and current differences. If we extend the analysis of the signs of solutions or solution differences to other elements, we must differentiate between voltages and currents. This approach, in which two orientations for all branches are considered, one corresponding to the currents and one corresponding to the voltages, is pursued in [10].

The conditions of Theorems 31.3 to 31.5 can be verified by inspection for small circuits. For larger circuits, one must resort to combinatorial algorithms. Such algorithms are proposed in [11, 12]. As can be expected from the nature of the conditions, the algorithms grow exponentially with the number of resistors. It is not known whether algorithms of polynomial complexity exist.

Some circuits always have either no solution or an infinite number of solutions, irrespective of the element and parameter values. Figure 31.6 gives the simplest example. Such circuits clearly are not very useful in practice. The remaining circuits are those that may have a finite number $n > 1$ of solutions if the circuit parameters are chosen suitably. These are the circuits that are useful for static memories and for multivibrators in general. This class is characterized by the following theorem.

Theorem 31.6: *Let a circuit be composed of nullator-norator pairs, V- and I-resistors, and independent voltage and current sources. If the following three conditions are satisfied, the circuit has more than one, but a finite number of solutions for a suitable choice of circuit parameters:*

- *The norators, I-resistors, and the voltage sources together form a tree*
- *The nullators, I-resistors, and the voltage sources together form a tree*
- *There is a nontrivial, uniform partial orientation of the resistors*

Can we be more precise and formulate conditions on the circuit structure that guarantee four solutions, for example? This is not possible because changing the parameters of the circuit will lead to another number of solutions. Particularly with a circuit structure that satisfies the conditions of Theorem 31.6, a linear circuit always has an infinite number of solutions. If we are more restrictive on the resistor characteristics, e.g., imposing convex or concave characteristics for certain resistors, it is possible to determine the **maximum number of solutions**. A method to determine an upper bound is given in [14], whereas the results of [15] allow us to determine the actual maximum number under certain conditions. Despite these results, however, the maximum number of solutions is still an open problem.

Bounds on Voltages and Currents

It is common sense for electrical engineers that in an electronic circuit all node voltages lie between 0 and the power supply voltage, or between the positive and the negative power supply voltages, if both are present. Actually, this is only true for the DC-operating point, but can we prove it in this case? The following theorems give the answer. They are based on the notion of passivity.

Definition 31.5. A resistor is **passive** if it can only absorb, but never produce power. This means that for any point (v, i) on its characteristic we have

$$v \cdot i \geq 0 \tag{31.24}$$

A resistor is **strictly passive,** if in addition to (31.24) it satisfies the condition

$$v \cdot i = 0 \rightarrow v = i = 0 \tag{31.25}$$

Theorem 31.7: *Let a circuit be composed of strictly passive resistors and independent voltage and current sources. Then, for every branch k of the circuit the following bounds can be given:*

$$|v_k| \leq \sum_{\text{source branches } j} |v_j| \tag{31.26}$$

$$|i_k| \leq \sum_{\text{source branches } j} |i_j| \tag{31.27}$$

If, in addition, the circuit is connected and all sources have a common node, the ground node, then the maximum and the minimum node voltage are at a source terminal.

The theorem implies in particular that in a circuit with a single voltage source, all branch voltages are bounded by the source voltage in magnitude, and all node voltages lie between zero and the source voltage. Similarly, if a circuit has a single current source, all branch currents are bounded by the source current in magnitude. Finally, if several voltage sources are present that are all connected to ground and have positive value, then the node voltages lie between zero and the maximum source voltage. If some sources have positive values and others have negative values, then all node voltages lie between the maximum and the minimum source values.

This theorem and various generalizations can be found in [1]. The main drawback is that it does not admit nonreciprocal elements. A simple counterexample is the voltage amplifier of Fig. 31.16. The voltage of the output node of the operational amplifier is

$$v = \frac{R_1 + R_2}{R_1} E \tag{31.28}$$

Thus, the output node voltage is higher than the source voltage. Of course, the reason is that the operational amplifier is an active element. It is realized by transistors and needs a positive and a negative voltage source as the power supply. The output voltage of the operational amplifier cannot exceed these supply voltages. This fact is not contained in the model of the ideal operational amplifier, but follows from the extension of Theorem 31.7 to bipolar transistors [1, 16].

Theorem 31.8: *Let a circuit be composed of bipolar transistors modeled by the Ebers-Moll equations, of strictly passive resistors, and of independent voltage and current sources. Then the conclusions of Theorem 31.7 hold.*

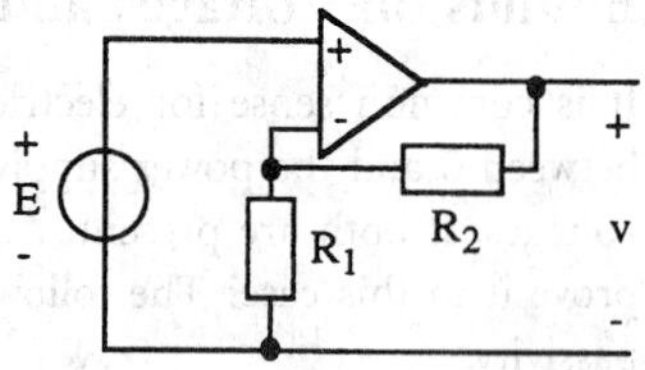

FIGURE 31.16 Voltage amplifier.

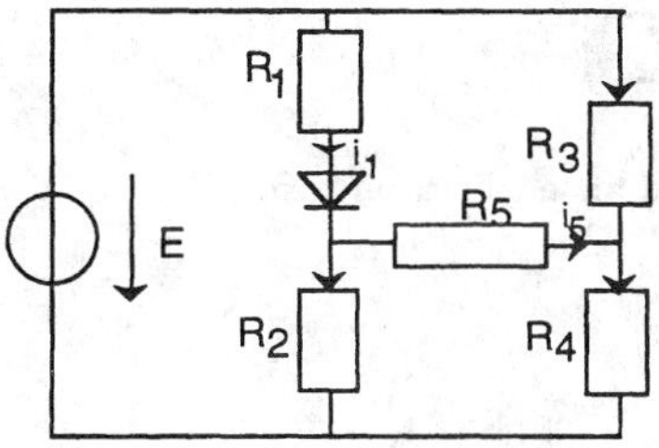

FIGURE 31.17 Circuit example for source dependence.

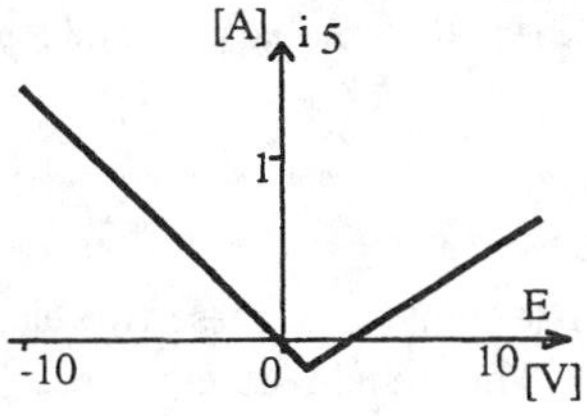

FIGURE 31.18 Nonmonotonic dependence.

At first sight, Theorem 31.8 seems to imply that it is impossible to build an amplifier with bipolar transistors. Indeed, it is impossible to build such an amplifier with a single source, the input signal. We need at least one power supply source that sets the limits of dynamic range of the voltages according to Theorem 31.8. The signal source necessarily has a smaller amplitude and the signal can be amplified roughly up to the limit set by the power supply source.

Theorem 31.8 can be extended to MOS transistors. The difficulty is that the nonlinear characteristic of the simplest model is not strictly increasing, and therefore some interconnection condition must be added to avoid parts with undetermined node voltages.

Monotonic Dependence

Rather than looking at single solutions of resistive circuits, as done earlier in the chapter, we consider here a solution as a function of a parameter. The simplest and at the same time the most important case is the dependence of a solution on the value of a voltage or current source. To have a well-defined situation, we suppose that the circuit satisfies the hypotheses of Theorem 31.5. In this case [1, 8] the solution is a continuous function of the source values.

As an example, let us consider the circuit of Fig. 31.17. We are interested in the dependence of the various currents on the source voltage E. Because the circuit contains only strictly increasing resistors, we expect all currents to be strictly monotonic functions of E. This is not true. In Fig. 31.18 the current $i_5(E)$ is represented for $R_1 = R_2 = R_3 = 2R_4 = R_5 = 1\,\Omega$ and for standard diode model parameters. Clearly, it is nonmonotonic.

1.3 Autonomous Dynamic Circuits

Introduction

This section adds to the resistive elements of the previous section, the capacitors and the inductors. A nonlinear capacitor is defined by the constitutive relation

$$v = h(q) \tag{31.29}$$

where the auxiliary variable q is the charge of the capacitor, which is linked to the current by

$$i = \frac{dq}{dt} \tag{31.30}$$

FIGURE 31.19 Symbols of the nonlinear capacitor and the nonlinear inductor.

The dual element, the nonlinear inductor, is defined by

$$i = g(\varphi) \tag{31.31}$$

where the auxiliary variable φ, the flux, is linked to the voltage by

$$v = \frac{d\varphi}{dt} \tag{31.32}$$

The symbols of these two elements are represented in Fig. 31.19.

The system of equations that describes an autonomous dynamic circuit is composed of (31.12) to (31.17), completed with (31.29) and (31.30) for capacitor branches and (31.31) and (31.32) for inductor branches. Hence, it becomes a mixed differential–nondifferential system of equations. Its solutions are the voltages, current charges, and fluxes as functions of time. Because it contains differential equations, we have infinitely many solutions, each one determined by some set of initial conditions.

If all variables except the charges and fluxes are eliminated from the system of equations, one obtains a reduced, purely differential system of equations

$$\frac{d\mathbf{q}}{dt} = \mathbf{f}(\mathbf{q}, \varphi) \tag{31.33}$$

$$\frac{d\varphi}{dt} = \mathbf{g}(\mathbf{q}, \varphi) \tag{31.34}$$

where $\mathbf{q}$ and φ are the vectors composed of, respectively, the capacitor charges and the inductor fluxes. These are the **state equations** of the circuit. Under mild assumptions on the characteristics of the nonlinear elements (local Lipschitz continuity and eventual passivity) it can be shown that the solutions are uniquely determined by the initial values of the charges and fluxes at some time t_0, $\mathbf{q}(t_0)$, and $\varphi(t_0)$, and that they exist for all times $t_0 \leq t < \infty$ [1, 17].

It cannot be taken for granted, however, that the circuit equations actually can be reduced to the state Eqs. (31.33) and (31.34). On the one hand, the charges and fluxes may be dependent and thus their initial values cannot be chosen freely. However, the state equations may still exist, in terms of a subset of charges and fluxes. This means that only these charges and fluxes can be chosen independently as initial conditions. On the other hand, the reduction, even to some alternative set of state variables, may be simply impossible. This situation is likely to lead to **impasse points**, i.e., nonexistence of the solution at a finite time. We refer the reader to the discussion in [1]. In the sequel we suppose that the solutions exist from the initial time t_0 to $+\infty$ and that they are determined by the charges and fluxes at t_0.

We are interested in the asymptotic behavior, i.e., the behavior of the solutions when the time t goes to infinity. If the dynamic circuit is linear and strictly stable, i.e., if all its natural frequencies are in the open left half of the complex plane, then all solutions converge to 1 and the same DC-operating (equilibrium) point. This property still holds for many nonlinear circuits, but not for all by far. In particular, the solutions may converge to different DC-operating points, depending on the initial conditions (static memories), they may converge to periodic solutions (free running oscillators), or they may even show chaotic behavior (e.g., Chua's circuit). Here, we give conditions that guarantee the solutions converge to a unique solution or one among several DC-operating points.

Convergence to DC-Operating Points

The method to prove convergence to one or more DC-operating points is based on **Lyapunov functions.** A Lyapunov function is a continuously differentiable function $W(\xi)$, where ξ is the vector composed of the circuit variables (the voltages, currents, charges, and fluxes). In the case of autonomous circuits a Lyapunov function must have the following properties:

1. W is bounded below, i.e., there exists a constant W_0 such that

$$W(\xi) \geq W_0 \quad \text{for all } \xi \tag{31.35}$$

2. The set of voltages, currents, charges, and fluxes of the circuit such that $W(\xi) \leq E$ is bounded for any real E.
3. For any solution $\xi(t)$ of the circuit

$$\frac{d}{dt} W(\xi(t)) \leq 0 \tag{31.36}$$

4. If

$$\frac{d}{dt} W(\xi(t)) = 0 \tag{31.37}$$

then $\xi(t)$ is a DC-operating point.

If an autonomous circuit has a Lyapunov function and if it has at least one, but a finite number of DC-operating points, then every solution converges to a DC-operating point. The reason is that the Lyapunov function must decrease along each solution, and thus must result in a local minimum, a stable DC-operating point. If more than one DC-operating point exists, it may, as a mathematical exception that cannot occur in practice, end up in a saddle point, i.e., an unstable DC-operating point.

The problem with the Lyapunov function method is that it gives no indication as to how to find such a function. Basically, three methods are available to deal with this problem:

1. Some standard candidates for Lyapunov functions, e.g., the stored energy.
2. Use a certain kind of function and adjust the parameters in order to satisfy 2 and 3 in the list above. Often, quadratic functions are used.
3. Use an algorithm to generate Lyapunov functions [18–20].

The following theorems were obtained via approach 1, and we indicate which Lyapunov was used to prove them. At first sight, this may seem irrelevant from an engineering point of view. However, if we are interested in designing circuits to solve optimization problems, we are likely to be interested in Lyapunov functions. Indeed, as mentioned above, along any solution of the circuit, the Lyapunov function decreases and approaches a minimum of the function. Thus, the dynamics of the circuit solve a minimization problem. In this case we look for a circuit with a given Lyapunov function, however, usually we look for a Lyapunov function for a given circuit.

Theorem 31.9: *Let a circuit be composed of capacitors and inductors with a strictly increasing characteristic, resistors with a strictly increasing characteristic, and independent voltage and current sources. Suppose the circuit has a DC-operating point* $\bar{\xi}$. *By Theorem 31.1, this DC-operating point is unique. Finally, suppose the circuit has no loop composed of capacitors, inductors, and voltage sources and no cutset composed of capacitors, inductors, and current sources. Then all solutions of the circuit converge to* $\bar{\xi}$.

The Lyapunov function for this circuit is given by a variant of the stored energy in the capacitors and the resistors, the stored energy with respect to $\bar{\xi}$ [1, 17]. If the constitutive relations of the capacitors and the inductors are given by $v_k = h_k(q_k)$ and $i_k = g_k(v_k)$, respectively, then this Lyapunov function becomes

$$W(\xi) = \sum_{\substack{\text{capacitor} \\ \text{branches } k}} \int_{\bar{q}_k}^{q_k} (h_k(q) - h_k(\bar{q}_k))dq + \sum_{\substack{\text{induction} \\ \text{branches } k}} \int_{\bar{\varphi}_k}^{\varphi_k} (g_k(\varphi) - g_k(\bar{\varphi}_k))d\varphi \tag{31.38}$$

The main condition (31.36) for a Lyapunov function follows from the fact that the derivative of the stored energy is the absorbed power, here in incremental form:

$$\frac{d}{dt}W(\xi) = \sum_{\substack{\text{capacitor} \\ \text{and inductor} \\ \text{branches } k}} \Delta v_k \, \Delta i_k = - \sum_{\substack{\text{resistor} \\ \text{branches } k}} \Delta v_k \, \Delta i_k \leq 0 \tag{31.39}$$

Various generalizations of Theorem 31.9 have been given. The condition "strictly increasing resistor characteristic" has been relaxed to a condition that depends on $\bar{\xi}$ in [1, 17] and mutual inductances and capacitances have been admitted in [17].

The next theorem admits resistors with nonmonotonic characteristics. However, it does not allow for both inductors and capacitors.

Theorem 31.10: *Let a circuit be composed of capacitors with a strictly increasing characteristic, voltage-controlled resistors such that*

$$v \to +\infty \Rightarrow i > I_+ > 0 \quad \text{and} \quad v \to -\infty \Rightarrow i < I_- < 0 \tag{31.40}$$

and independent voltage sources. Furthermore, suppose that the circuit has a finite number of DC-operating points. Then every solution of the circuit converges toward a DC-operating point.

This theorem is based on the following Lyapunov function, called **cocontent**:

$$W(\xi(t)) = \sum_{\substack{\text{resistor} \\ \text{branches } k}} \int_0^{v_k} g_k(v)dv \tag{31.41}$$

where $i_k = g_k(v_k)$ is the constitutive relation of the resistor on branch k. The function W is decreasing along a solution of the circuit because

$$\begin{aligned} \frac{d}{dt}W(\xi(t)) &= \sum_{\substack{\text{resistor} \\ \text{branches } k}} \frac{dv_k}{dt} i_k = - \sum_{\substack{\text{capacitor} \\ \text{branches } k}} \frac{dv_k}{dt} i_k \\ &= - \sum_{\substack{\text{capacitor} \\ \text{branches } k}} \frac{dh_k}{dq} i_k^2 \leq 0 \end{aligned} \tag{31.42}$$

where $h_k(q_k)$ is the constitutive relation of the capacitor on branch k.

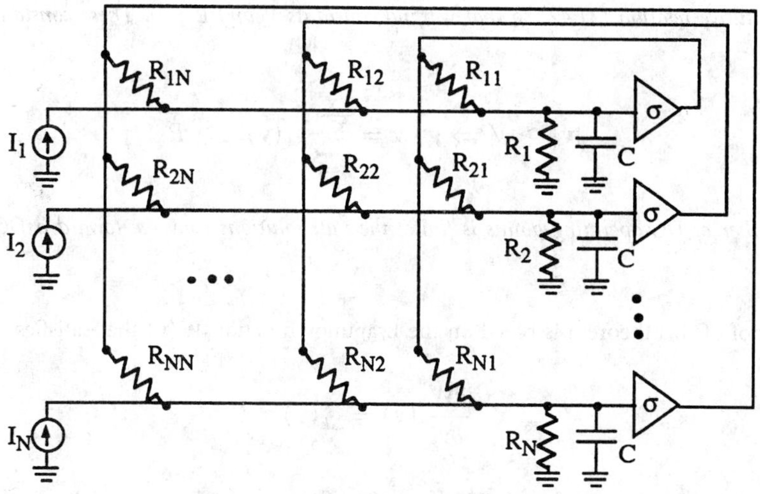

FIGURE 31.20 Analog neural network.

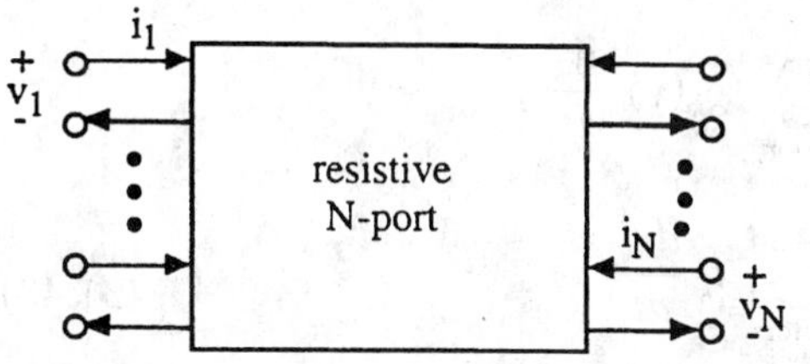

FIGURE 31.21 Resistive N-port.

Theorem 31.10 has a dual version. It admits inductors instead of capacitors, current-controlled resistors, and current sources. The corresponding Lyapunov function is the **content**:

$$W(\xi) = \sum_{\substack{\text{resistor} \\ \text{branches } k}} \int_0^{i_k} h_k(i)di \tag{31.43}$$

where $v_k = h_k(i_k)$ is the constitutive relation of the resistor on branch k.

The main drawback of the two preceding theorems is that they do not admit nonreciprocal elements such as controlled sources, operational amplifiers, etc. In other words, no statement about the analog neural network of Fig. 31.20 can be made. In this network the nonreciprocal element is the VCVS with the nonlinear characteristic $v_2 = \sigma(v_1)$. However, Theorem 31.10 can be generalized to a reciprocal voltage controlled N-port resistor closed on capacitors and voltage sources. Such an N-port (Fig. 31.21) is described by a constitutive relation of the form

$$i_k = g_k(v_1, \ldots, v_N) \tag{31.44}$$

and it is **reciprocal**, if for all **v**, and all k, j we have

$$\frac{\partial g_k}{\partial v_j}(\mathbf{v}) = \frac{\partial g_j}{\partial v_k}(\mathbf{v}) \tag{31.45}$$

Theorem 31.11: *Let a circuit be composed of charge-controlled capacitors with a strictly increasing characteristic and independent voltage sources that terminate a reciprocal voltage-controlled N-port* *with constitutive relation (31.42) so that we find constants V and P > 0. These constants are such that*

$$\|\mathbf{v}\| \geq V \Rightarrow \mathbf{g} \cdot \mathbf{v} = \sum_{k=1}^{N} g_k(\mathbf{v}) v_k \geq P \tag{31.46}$$

If the number of DC-operating points is finite, then all solutions converge toward a DC-operating point.

The proof of this theorem is based on the Lyapunov function $W(\mathbf{v})$ that satisfies

$$\frac{\partial W}{\partial v_k}(\mathbf{v}) = g_k(\mathbf{v}) \tag{31.47}$$

Thanks to condition (31.45), function W exists. The first two conditions for a Lyapunov function are a consequence of (31.46). Finally,

$$\begin{aligned} \frac{d}{dt} W(\xi(t)) &= \sum_{\substack{\text{resistor} \\ \text{branches } k}} g_k(\mathbf{v}) \frac{dv_k}{dt} \\ &= \sum_{\substack{\text{resistor} \\ \text{branches } k}} i_k \frac{dv_k}{dt} \\ &= -\sum_{\substack{\text{capacitor} \\ \text{branches } k}} \frac{dh_k}{dq} i_k^2 \leq 0 \end{aligned} \tag{31.48}$$

where $h_k(q_k)$ is the constitutive relation of the capacitor on branch k.

To illustrate how Theorem 31.11 can be applied when Theorem 31.10 fails, consider the analog neural network of Fig. 31.20. If the capacitor voltages are denoted by u_i and the voltages at the output of the voltage sources by v_i, the state equations for the network of Fig. 31.1 become

$$-C_i \frac{du_i}{dt} = \frac{u_i}{R_i} + \sum_{j=1}^{N} \frac{u_i - v_j}{R_{ij}} + I_i \tag{31.49}$$

Suppose that the nonlinear characteristic $\sigma(u)$ is invertible. The state equations can be written in terms of the voltages v_i:

$$-C \frac{d\sigma^{-1}}{dv}(v_i) \frac{dv_i}{dt} = G_i \sigma^{-1}(v_i) - \sum_{j=1}^{N} \frac{v_j}{R_{ij}} + I_i \tag{31.50}$$

where

$$G_i = \frac{1}{R_i} + \sum_{j=1}^{N} \frac{1}{R_{ij}} \tag{31.51}$$

Equations (31.40) can be reinterpreted as the equations of a resistive N-port with the constitutive relations

$$g_i(\mathbf{v}) = G_i\sigma^{-1}(v_i) - \sum_{j=1}^{N} \frac{v_j}{R_{ij}} + I_i \tag{31.52}$$

closed on nonlinear capacitors with the constitutive relation

$$v = \sigma\left(\frac{q}{C}\right) \tag{31.53}$$

If σ is a sigmoidal function, as is most often supposed in this context (i.e., a strictly increasing function with $s(u) \to \pm 1$ for $u \to \pm\infty$), then the capacitors have a strictly increasing characteristic, as required by Theorem 31.11. Furthermore, the resistive N-port is reciprocal if for $i \neq j$

$$\frac{\partial g_i}{\partial v_j} = -\frac{1}{R_{ij}} = \frac{\partial g_j}{\partial v_i} = -\frac{1}{R_{ji}} \tag{31.54}$$

In other words, if for all i, j

$$R_{ij} = R_{ji} \tag{31.55}$$

On the other hand, inequality (31.46) must be modified because the sigmoids have values only in the interval $[-1, +1]$ and thus (31.50) are defined only on the invariant bounded set $S = \{\mathbf{v} | -1 < v_i < +1\}$. Therefore, inequality (31.50) must be satisfied for vectors $\mathbf{v}$ sufficiently close to the boundary of S. This is indeed the case, because $\sigma^{-1}(v) \to \pm\infty$ as $v \to \pm 1$, whereas the other terms of the right-hand side of (31.52) remain bounded.

It follows that all solutions of the analog neural network of Fig. 31.20 converge to a DC-operating point as $t \to \infty$, provided σ is a sigmoid function and the connection matrix R_{ij} (synaptic matrix) is symmetrical. The Lyapunov function can be given explicitly:

$$W(\mathbf{v}) = \sum_{i=1}^{N} G_i \int_0^{v_i} \sigma^{-1}(v)dv - \frac{1}{2}\sum_{i,j=1}^{N} \frac{v_i v_j}{R_{ij}} + \sum_{i=1}^{N} v_i I_i \tag{31.56}$$

31.4 Nonautonomous Dynamic Circuits

Introduction

This section is a consideration of circuits which contain elements whose constitutive relations depend explicitly on time. However, we limit time dependence to the independent sources. For most practical purposes this is sufficient. A time-dependent voltage source has a constitutive relation

$$v = e(t) \tag{31.57}$$

and a time-dependent current source

$$i = e(t) \tag{31.58}$$

where $e(t)$ is a given function of time which we suppose here to be continuous. In information processing circuits $e(t)$ represents a signal that is injected into the circuit, whereas in energy transmission circuits $e(t)$ usually is a sinusoidal or nearly sinusoidal function related to a generator.

The time-dependent sources may drive the voltages and the currents to infinity, even if they only inject bounded signals into the circuit. Therefore, the discussion begins with the conditions that guarantee the boundedness of the solutions.

Boundedness of the Solutions

In electronic circuits even active elements become passive, when the voltages and currents grow large. This is the reason solutions remain bounded.

Definition 31.6. A resistor is **eventually passive** if for sufficiently large voltages and/or currents, it can only absorb power. More precisely, eventual passivity means that constants V and I exist such that for all points (v, i) on the resistor characteristic with $|v| > V$ or $|i| > I$ we have

$$v \cdot i \geq 0 \tag{31.59}$$

Note that sources are not eventually passive, but as soon as an internal resistance of a source is taken into account, the source becomes eventually passive. The notion of eventual passivity can be extended to time-varying resistors.

Definition 31.7. A time-varying resistor is eventually passive if constants V and I are independent of time and are such that all points (v, i) with $|v| > V$ or $|i| > I$ that at some time lie on the characteristic of the resistor satisfy the passivity condition (31.59). According to this definition, time-dependent sources with internal resistance are eventually passive if the source signal remains bounded.

Eventual passivity allows us to deduce bounds for the solutions. These bounds are uniform in the sense that they do not depend on the particular solution. To be precise, this is true only asymptotically, as $t \to \infty$.

Definition 31.8. The solutions of a circuit are **eventually uniformly bounded** if constants V, I, Q, and Φ are such that for any solution, T is time, for any $t > T$ the voltages $v_k(t)$ are bounded by V, the currents $i_k(t)$ are bounded by I, the charges $q_k(t)$ are bounded by Q, and the fluxes $\varphi_k(t)$ are bounded by Φ.

Another manner of expressing the same property is to say that an attracting domain exists in state space [1].

Theorem 31.12: *A circuit composed of eventually passive resistors with $v \cdot i \to +\infty$ as $|v| \to \infty$ or $|i| \to \infty$, capacitors with $v \to \pm\infty$ as $q \to \pm\infty$, and inductors with $i \to \pm\infty$ as $\varphi \to \infty$ has eventually uniformly bounded solutions if no loop or cutset exists without a resistor [1, 17].*

Again, this theorem is proved by using a Lyapunov function, namely the stored energy

$$W(\xi) = \sum_{\substack{\text{capacitor} \\ \text{branches } k}} \int_0^{q_k} h_k(q)dq + \sum_{\substack{\text{inductor} \\ \text{branches } k}} \int_0^{\varphi_k} g_k(\varphi)d\varphi \tag{31.60}$$

Inequality (31.36) holds only outside of a bounded domain.

Unique Asymptotic Behavior

In the presence of signals with complicated waveforms that are injected into a circuit we cannot expect simple waveforms for the voltages and the currents, not even asymptotically, as $t \to \infty$. However, we can hope that two solutions, starting from different initial conditions, but subject to the same source signals, have the same steady-state behavior. The latter term needs a more formal definition.

Definition 31.9. A circuit has **unique asymptotic behavior** if the following two conditions are satisfied

1. All solutions are bounded
2. For any two solutions $\xi_1(t)$ and $\xi_2(t)$

$$\left\|\xi_1(t) - \xi_2(t)\right\| \to_{t\to\infty} 0 \tag{31.61}$$

In order to prove unique asymptotic behavior, it is necessary to extend the notion of the Lyapunov function [1]. This does not lead very far, but at least it permits us to prove the following theorem.

Theorem 31.13: *Suppose a circuit is composed of resistors with a strictly increasing characteristic such that $v \cdot i \to +\infty$ as $|v| \to \infty$ or $|i| \to \infty$, positive linear capacitors, positive linear inductors, time-dependent voltage (current) sources with bounded voltage (current) and a positive resistor in series (parallel). If no loop or cutset is composed exclusively of capacitors and inductors, the circuit has unique asymptotic behavior [1, 17].*

This theorem is unsatisfactory because linear reactances are required and real devices are never exactly real. It has been shown that slight nonlinearities can be tolerated without losing the unique asymptotic behavior [21]. On the other hand, we cannot expect to get much stronger general results because nonautonomous nonlinear circuits may easily have multiple steady-state regimes and even more complicated dynamics, such as chaos, even if the characteristics of the nonlinear elements are all strictly increasing.

Another variant of Theorem 31.13 considers linear resistors and nonlinear reactances [17].

ferences

[1] M. Hasler and J. Neirynck, *Nonlinear Circuits*, Boston: Artech House, 1986.

[2] L. O. Chua, C. A. Desoer, and E. S. Kuh, *Linear and Nonlinear Circuits*, Electrical & Electronic Engineering Series, Singapore: McGraw-Hill International Editions, 1987.

[3] A. N. Willson (Ed.), *Nonlinear Networks: Theory and Analysis*, New York: IEEE Press, 1974.

[4] R. O. Nielsen and A. N. Willson, "A fundamental result concerning the topology of transistor circuits with multiple equilibria," *Proc. IEEE*, vol. 68, pp. 196–208, 1980.

[5] A. N. Willson, "On the topology of FET circuits and the uniqueness of their dc operating points," *IEEE Trans. Circuits Syst.*, vol. 27, pp. 1045–1051, 1980.

[6] T. Nishi and L. O. Chua, "Topological criteria for nonlinear resistive circuits containing controlled sources to have a unique solution," *IEEE Trans. Circuits Syst.*, vol. 31, pp. 722–741, Aug. 1984.

[7] T. Nishi and L. O. Chua, "Nonlinear op-amp circuits: existence and uniqueness of solution by inspection," *Int. J. Circuit Theory Appl.*, vol. 12, pp. 145–173, 1984.

[8] M. Hasler, "Nonlinear nonreciprocal resistive circuits with a unique solution," *Int. J. Circuit Theory Appl.*, vol. 14, pp. 237–262, 1986.

[9] M. Fosséprez, *Topologie et Comportement des Circuits non Linéaires non Réciproques*, Lausanne: Presses Polytechnique Romandes, 1989.

[10] M. Hasler, "On the solution of nonlinear resistive networks," *J. Commun. (Budapest, Hungary)*, special issue on nonlinear circuits, July 1991.

[11] T. Parker, M. P. Kennedy, Y. Liao, and L. O. Chua, "Qualitative analysis of nonlinear circuits using computers," *IEEE Trans. Circuits Syst.*, vol. 33, pp. 794–804, 1986.

[12] M. Fosséprez and M. Hasler, "Algorithms for the qualitative analysis of nonlinear resistive circuits," *IEEE ISCAS Proc.*, pp. 2165–2168, May 1989.

[13] M. Fosséprez and M. Hasler, "Resistive circuit topologies that admit several solutions," *Int. J. Circuit Theory Appl.*, vol. 18, pp. 625–638, Nov. 1990.

[14] M. Fosséprez, M. Hasler, and C. Schnetzler, "On the number of solutions of piecewise linear circuits," *IEEE Trans. Circuits Syst.*, vol. CAS-36, pp. 393–402, Mar. 1989.

[15] T. Nishi and Y. Kawane, "On the number of solutions of nonlinear resistive circuits," *IEICE Trans.*, vol. E74, pp. 479–487, 1991.

[16] A. N. Willson, "The no-gain property for networks containing three-terminal elements," *IEEE Trans. Circuits Syst.*, vol. 22, pp. 678–687, 1975.

[17] L. O. Chua, "Dynamic nonlinear networks: state of the art," *IEEE Trans. Circuits Syst.*, vol. 27, pp. 1059–1087, 1980.

[18] R. K. Brayton and C. H. Tong, "Stability of dynamical systems," *IEEE Trans. Circuits Syst.*, vol. 26, pp. 224–234, 1979.

[19] R. K. Brayton and C. H. Tong, "Constructive stability and asymptotic stability of dynamical systems," *IEEE Trans. Circuits Syst.*, vol. 27, pp. 1121–1130, 1980.

[20] L. Vandenberghe and S. Boyd, "A polynomial-time algorithm for determining quadratic Lyapunov functions for nonlinear systems," *Proc. ECCTD-93*, pp. 1065–1068, 1993.

[21] M. Hasler and Ph. Verburgh, "Uniqueness of the steady state for small source amplitudes in nonlinear nonautonomous circuits," *Int. J. Circuit Theory Appl.*, vol. 13, pp. 3–17, 1985.

32

Synthesis and Design of Nonlinear Circuits

Rodríguez-Vázquez
iversidad de Sevilla,
in

. Delgado-Restituto
iversidad de Sevilla,
in

L. Huertas
iversidad de Sevilla,
in

Vidal
iversidad de Malaga
in

2.1 Introduction

Nonlinear synthesis and design can be informally defined as a constructive procedure to interconnect components from a catalog of available primitives, and to assign values to their constitutive parameters to meet a specific nonlinear relationship among electrical variables. This relationship is represented as an implicit integro-differential operator, although we primarily focus on the synthesis of *explicit algebraic* functions,

$$y = f(\mathbf{x}) \tag{32.1}$$

where y is a voltage or current, $f(\cdot)$ is a nonlinear real-valued function, and $\mathbf{x}$ is a vector whose components include voltages and currents. This synthesis problem is found in two different circuit-related areas: device **modeling** [8, 76] and analog **computation** [26]. The former uses ideal circuit elements as primitives to build computer models of real circuits and devices (see Chapter 31). The latter uses real circuit components, available either off the shelf or integrable in a given fabrication technology, to realize hardware for nonlinear signal processing tasks. We

3-8341-2/95/$0.00+$.50

focus on this second area, and intend to outline systematic approaches to devise electronic function generators. Synthesis relies upon hierarchical decomposition, conceptually shown in Fig. 32.1, which encompasses several subproblems listed from top to bottom:

- Realization of nonlinear **operators** (multiplication, division, squaring, square rooting, logarithms, exponentials, sign, absolute value, etc.) through the interconnection of primitive components (transistors, diodes, operational amplifiers, or op amps, etc.).
- Realization of **elementary functions** (polynomials, truncated polynomials, Gaussian functions, etc.) as the interconnection of the circuit blocks devised to build nonlinear operators.
- **Approximation** of the target as a combination of elementary functions and its realization as the interconnection of the circuit blocks associated with these functions.

Figure 32.1 illustrates this hierarchical decomposition of the synthesis problem through an example in which the function is approximated as a linear combination of truncated polynomials [30], whose realization involves analog multipliers, built by exploiting the nonlinearities of bipolar junction transistors (BJTs) [63]. Also note that the subproblems cited above are closely interrelated and, depending on the availability of primitives and the nature of the nonlinear function, some of these phases can be by-passed. For instance, a logarithmic function can be realized exactly using BJTs [63], but requires approximation if our catalog includes only field-effect transistors whose nonlinearities are polynomic [44].

The technical literature contains excellent contributions to the solution of all these problems. These contributions can hardly be summarized or even quoted in just one section. Many authors follow a block-based approach which relies on the pervasive voltage op amp, the rectification properties of junction diodes, and the availability of voltage multipliers, in the tradition of classical analog computation. Examples are [7], [59], and [80]. Remarkable contributions have been made which focus on qualitative features such as negative resistance or hysteresis, rather than the realization of well-defined approximating functions [9, 20, 67]. Other contributions focus on the realization of nonlinear operators in the form of IC units. **Translinear** circuits,

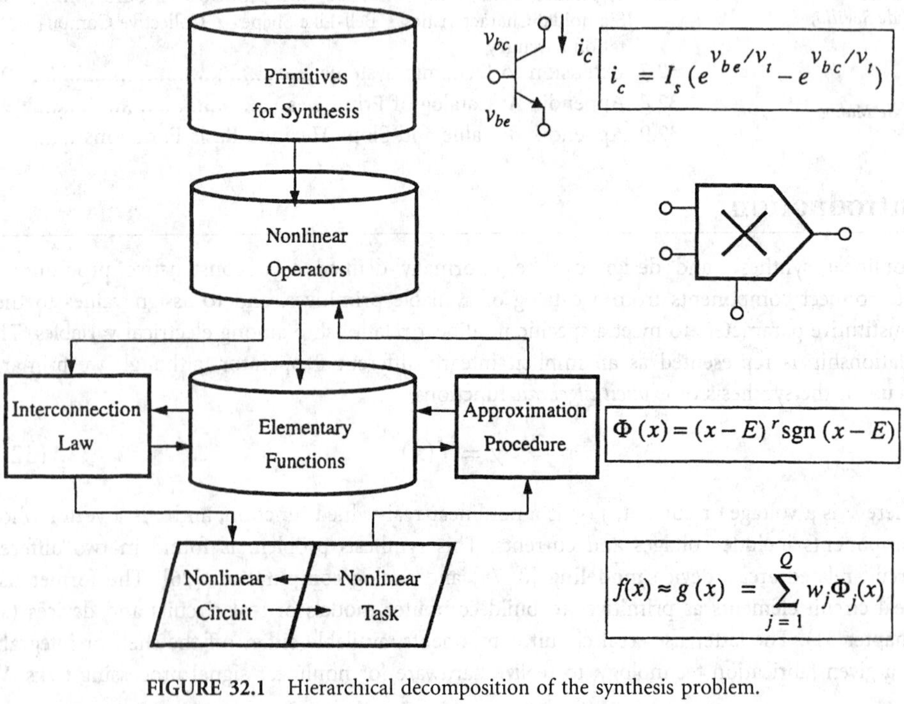

FIGURE 32.1 Hierarchical decomposition of the synthesis problem.

BJTs [23, 62] and MOSFETs [79] are particularly well suited to realize algebraic functions in IC form. This IC orientation is shared by recent developments in analog VLSI computational and signal processing systems for neural networks [75], fuzzy logic [81], and other nonlinear signal processing paradigms [56, 57, 71].

This chapter is organized to fit the hierarchical approach shown in Fig. 32.1. We intend to review a wide range of approximation techniques and circuit design styles, for both discrete and monolithic circuits. It is based on the catalog of primitives shown in Appendix A. In addition to the classical op amp-based continuous-time circuits, we include current-mode circuitry because nonlinear operators are realized simply and accurately by circuits that operate in *current domain* [23, 57, 62, 79]. We also cover discrete-time circuits realized using analog dynamic techniques based on charge transfer, which is very significant for mixed-signal processing and computational microelectronic systems [27, 72]. Section 32.2 is devoted to approximation issues and outlines different techniques for uni- and multidimensional functions, emphasizing hardware-oriented approaches. These techniques involve several nonlinear operators and the linear operations of scaling and aggregation (covered in Section 32.3, which also presents circuits to perform transformations among different kinds of characteristics). Subsections 32.4 and 32.5 present circuits for piecewise-linear (PWL) and piecewise-polynomial (PWP) functions, 32.6 covers neural and fuzzy approximation techniques, and 32.7 outlines an extension to dynamic circuits.

2.2 Approximation Issues

Unidimensional Functions

Consider a target function, $f(x)$, given analytically or as a collection of measured data at discrete values of the independent variable. The approximation problem consists of finding a multiparameter function, $g(x, \mathbf{w})$, which yields proper fitting to the target, and implies solving two different subproblems: (1) which approximating functions to use, and (2) how to adjust the parameter vector, **w**, to render optimum fitting. We only outline some issues related to this first point. Detailed coverage of both problems can be found in mathematics and optimization textbooks [73, 78]. Other interesting views are found in circuit-related works [6, 11, 30], and the literature on neural and fuzzy networks [12, 21, 33, 43, 51].

An extended technique to design nonlinear electronic hardware for both discrete [63, 80] and monolithic [35, 62, 79] design styles uses **polynomial approximating functions,**

$$g(x) = \sum_{j=0}^{Q} \alpha_j x^j \tag{32.2}$$

obtained through expansion by either Taylor series or orthogonal polynomials (Chebyshev, Legendre, or Laguerre) [26]. Other related approaches use **rational functions,**

$$g(x) = \frac{\sum_{j=0,Q} \alpha_j x^j}{\sum_{j=0,R} \beta_j x^j} \tag{32.3}$$

to improve accuracy in the approximation of certain classes of functions [14]. These can be realized by polynomial building blocks connected in feedback configuration [63]. In addition, [39] presents an elegant synthesis technique relying on linearly controlled resistors and conductors to take advantage of linear circuit synthesis methods (further extended in [28]).

From a more general point of view, hardware-oriented approximating functions can be classified into two major groups:

1. Those involving the linear combination of basis functions

$$g(x) = \sum_{j=1}^{Q} w_j \Phi_j(x) \tag{32.4}$$

which include polynomial expansions, PWL and PWP interpolation, and **radial basis** functions (RBF). The hardware for these functions consists of two layers, as shown in Fig. 32.2(a). The first layer contains Q nonlinear processing nodes to evaluate the basis functions; the second layer scales the output of these nodes and aggregates these scaled signals in a summing node.
2. Those involving a multilayer of nested **sigmoids** [51]; for instance, in the case of two layers [82],

$$g(x) = h\left[\left\{\sum_{j=1,Q} w_{2j} h(w_{1j}x - \delta_{1j})\right\} - \delta_2\right] \tag{32.5}$$

with the sigmoid function given by

$$h(x) = \frac{2}{1 + \exp(-\lambda x)} - 1 \tag{32.6}$$

where $\lambda > 0$ determines the steepness of the sigmoid. Figure 32.2(b) shows a hardware concept for this approximating function, also consisting of two layers.

Piecewise-Linear and Piecewise-Polynomial Approximants

A drawback of polynomial and rational approximants is that their behavior in a small region determines their behavior in the whole region of interest [78]. Consequently, they are not appropriate to fit functions which are uniform throughout the whole region [see Fig. 32.3(a)]. Another drawback is their lack of modularity, a consequence of the complicated dependence of each fitting parameter on multiple target data, which complicates the calculation of optimum parameter values. These drawbacks can be overcome by splitting the target definition interval

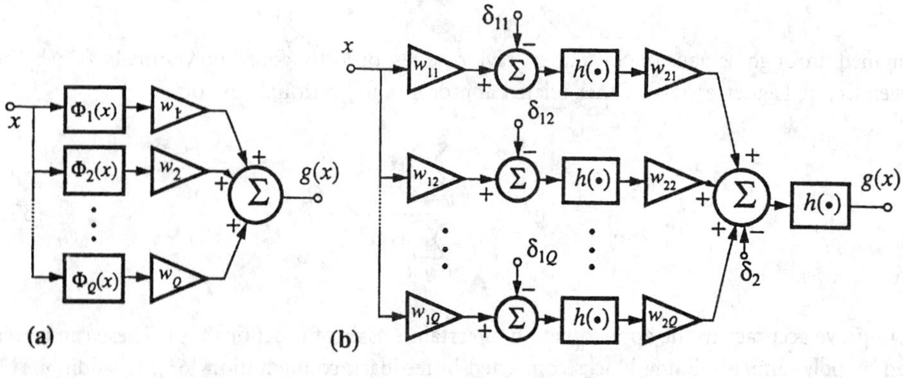

FIGURE 32.2 Block diagram for approximating function hardware. (a) Using linear combination of basis functions; (b) using two layers of nested sigmoids.

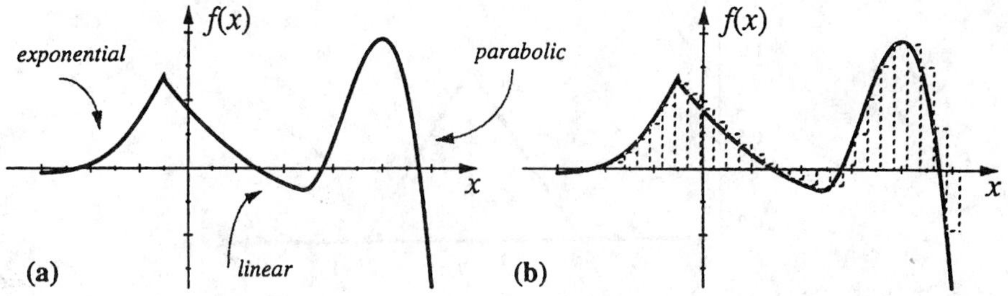

FIGURE 32.3 Example of nonuniform function.

into Q subintervals, and then expressing the approximating function as a linear combination of basis functions, each having **compact** support over only one subinterval, i.e., zero value outside this subinterval. For the limiting case in which $Q \to \infty$, this corresponds to interpolating the function by its samples associated to infinitely small subintervals [Fig. 32.3(b)]. Such action is functionally equivalent to expressing a signal as its convolution with a delta of Dirac [10].

This splitting and subsequent approximation can be performed ad hoc, by using different functional dependences to fit each subregion. However, to support the systematic design of electronic hardware it is more convenient to rely on well-defined classes of approximating functions. In particular, Hermite PWPs provide large modularity by focusing on the interpolation of measured data taken from the target function. Any lack of flexibility as compared to the ad hoc approach may be absorbed in the splitting of the region.

Consider the more general case in which the function, $y = f(x)$, is defined inside a real interval $[\delta_0, \delta_{N+1}]$ and described as a collection of data measured at *knots* of a given interval partition, $\Delta = \{\delta_0, \delta_1, \delta_2, \ldots, \delta_N, \delta_{N+1}\}$. These data may include the function values at these points, as well as their derivatives, up to the $(M - 1)$th order,

$$f^{(k)}(\delta_i) = \left. \frac{d^k}{dx^k} f(x) \right|_{x=\delta_i} \qquad i = 0, 1, 2, \ldots, N, N+1 \tag{32.7}$$

where k denotes the order of the derivative and is zero for the function itself. These data can be interpolated by a linear combination of basis polynomials of degree $2M - 1$,

$$g(x) = \sum_{i=0}^{N+1} \sum_{k=0}^{M-1} f^{(k)}(\delta_i) \Phi_{ik}(x) \tag{32.8}$$

where the expressions for these polynomials are derived from the interpolation data and continuity conditions [78]. Note that for a given basis function set and a given partition of the interval, each coefficient in (32.8) corresponds to a single interpolation datum.

The simplest case uses *linear* basis functions to interpolate only the function values,

$$g(x) = \sum_{i=0}^{N+1} f(\delta_i) l_i(x) \tag{32.9}$$

with no function derivatives interpolated. Figure 32.4 shows the shape of the inner jth linear basis function, which equals 1 at δ_j and decreases to 0 at δ_{j-1} and δ_{j+1}. By increasing the degree of the polynomials, the function derivatives also can be interpolated. In particular, two sets of third-degree basis functions are needed to retain modularity in the interpolation of the function

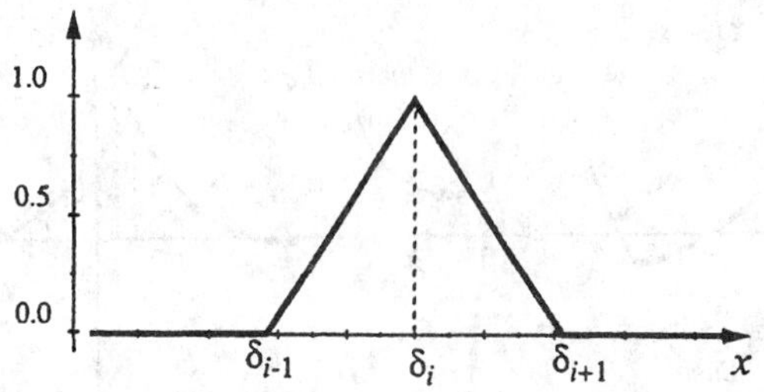

FIGURE 32.4 Hermite linear basis function.

and its first derivative at the knots

$$g(x) = \sum_{i=0}^{N+1} f(\delta_i)v_i(x) + \sum_{i=0}^{N+1} f^{(1)}(\delta_i)s_i(x) \tag{32.10}$$

where Appendix B shows the shapes and expressions of the value, $v_i(x)$, and slope, $s_i(x)$, basis functions.

The modularity of Hermite polynomials is not free; their implementation is not the cheapest in terms of components and, consequently, may not be optimal for applications in which the target function is fixed. These applications are more conveniently handled by the so-called **canonical representations** of PWP functions. A key concept is the **extension** operator introduced in [6]; the basic idea behind this concept is to build the approximating function following an iterative procedure. At each iteration, the procedure starts from a function that fits the data on a subinterval, enclosing several pieces of the partition interval, and then adds new terms to also fit the data associated to the next piece. Generally, some pieces are fit from left to right and others from right to left, to yield

$$g(x) = g^0(x) + \sum_{i=1}^{N_+} \Delta^+ g_i(x) + \sum_{i=-N_-}^{-1} \Delta^- g_i(x) \tag{32.11}$$

It is illustrated in Fig. 32.5. The functions in (32.11) have the following general expressions

$$\begin{aligned} \Delta^+ g(x) &= wu_+(x-\delta) \equiv w(x-\delta)\,\mathrm{sgn}(x-\delta) \\ \Delta^- g(x) &= wu_-(x-\delta) \equiv w(x-\delta)\,\mathrm{sgn}(\delta - x) \\ g^0(x) &= ax + b \end{aligned} \tag{32.12}$$

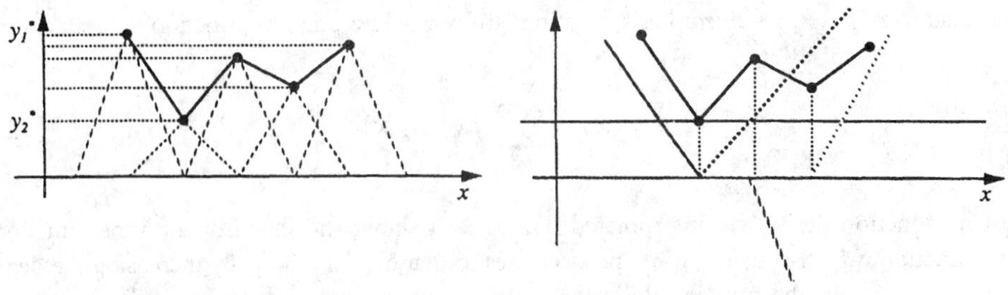

FIGURE 32.5 Decomposition of a PWL function using the extension operator.

where sgn(·) denotes the *sign* function, defined as an application of the real axis onto the discrete set {0, 1}.

This representation, based on the extension operator, is elaborated in [6] to obtain the following canonical representation for unidimensional PWL functions:

$$g(x) = ax + b + \sum_{i=1}^{N} w_i |x - \delta_i| \tag{32.13}$$

which has the remarkable feature of involving only one nonlinearity: the **absolute value** function.

The extension operator concept was applied in [30] to obtain canonical representations for cubic Hermite polynomials and B-splines. Consequently, it demonstrates that a PWP function admits a global expression consisting of a linear combination of powers of the input variable, plus truncated powers of shifted versions of this variable. For instance, the following expression is found for a cubic B-spline:

$$g(x) = \sum_{r=0}^{3} \alpha_r x^r + \sum_{i=1}^{N} \beta_i (x - \delta_i)^3 \, \text{sgn}(x - \delta_i) \tag{32.14}$$

with α_r and β_i obtainable through involved operations using the interpolation data. Other canonical PWP representations devised by these authors use

$$(x - \delta_i)^r \, \text{sgn}(x - \delta_i) = \tfrac{1}{2}\{|x - \delta_i| + (x - \delta_i)\}(x - \delta_i)^{r-1} \tag{32.15}$$

to involve the absolute value, instead of the sign function, in the expression of the function.

Gaussian and Bell-Shaped Basis Functions

The Gaussian basis function belongs to the general class of radial basis functions [51, 52], and has the following expression:

$$\Phi(x) = \exp\left(-\frac{(x - \delta)^2}{2\sigma^2}\right) \tag{32.16}$$

shown in Fig. 32.6. The function value is significant only for a small region of the real axis centered around its *center*, δ, and its shape is controlled by the *variance* parameter, σ^2. Thus, even though the support of Gaussian functions is not exactly compact, they are negligible except for well-defined *local* domains of the input values.

By linear combination of a proper number of Gaussians, and a proper choice of their centers and variances, as well as the weighting coefficients, it is possible to approximate nonlinear functions to any degree of accuracy [51]. Also, the local feature of these functions renders this

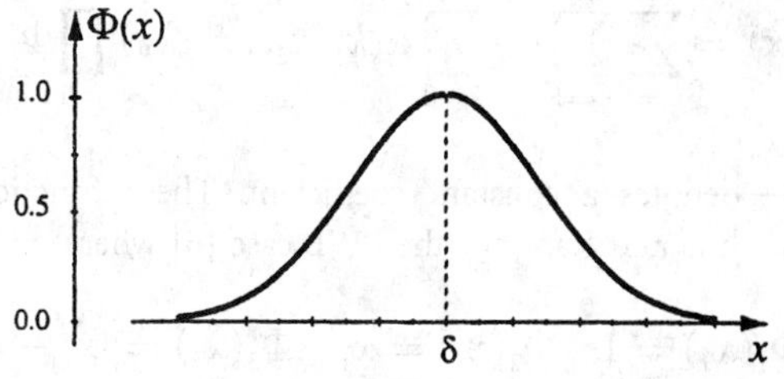

FIGURE 32.6 Gaussian basis function.

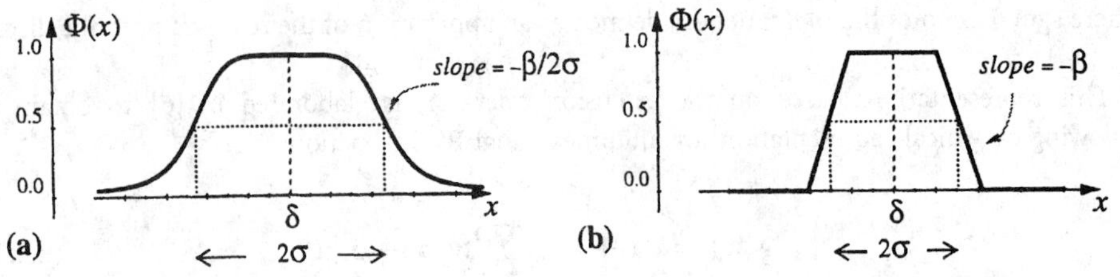

FIGURE 32.7 Fuzzy membership functions: (a) polynomial; (b) piecewise-linear.

adjustment process simpler than for multilayer networks composed of nested sigmoids, whose components are global [43, 50].

A similar interpolation strategy arises in the framework of fuzzy reasoning, which is based on local *membership* functions whose shape resembles a Gaussian. For instance, in the ANFIS system proposed by Jang [33]

$$\Phi(x) = \frac{1}{1 + \left[\left(\frac{x-\delta}{\sigma}\right)^2\right]^{\beta}} \tag{32.17}$$

as shown in Fig. 32.7(a) and whose shape is controlled by β and σ, and whose position is controlled by δ. Other authors, for instance, Yamakawa [81], use the PWL membership function shape of Fig. 32.7(b), which is to the Hermite linear basis function of Fig. 32.4. From a more general point of view, cubic B-splines [78] used to build hardware [59] and for device modeling [76] also can be considered to be members of this class of functions.

Multidimensional Functions

Approximation techniques for multidimensional functions are informally classified into five groups:

1. Sectionwise piecewise polynomial functions [6, 30]
2. Canonical piecewise linear representations [11]
3. Neuro-fuzzy interpolation [33, 81]
4. Radial basis functions [51, 52]
5. Multilayers of nested sigmoids [82]

Sectionwise Piecewise Polynomial Functions

This technique reduces the multidimensional function to a sum of products of functions of only one variable:

$$g(\mathbf{x}) = \sum_{k_1=1}^{M_1} \sum_{k_2=1}^{M_2} \cdots \sum_{k_P=1}^{M_P} \alpha(k_1, k_2, \ldots, k_P) \prod_{j=1}^{P} \Phi_{k_j}(x_j) \tag{32.18}$$

where $\alpha(k_1, k_2, \ldots, k_P)$ denotes a constant coefficient. These function representations were originally proposed by Chua and Kang for the PWL case [6] where

$$\begin{aligned} \Phi_1(x_j) = 1 \quad \Phi_2(x_j) = x_j \quad \Phi_3(x_j) = |x_j - \delta_{j1}| \\ \cdots \Phi_{M_P}(x_j) = |x_j - \delta_{jM_P-2}| \end{aligned} \tag{32.19}$$

Like the unidimensional case, the only nonlinearity involved in these basis functions is the absolute value. However, multidimensional functions not only require weighted summations, but also multiplications. The extension of (32.18) to PWP functions was covered in [30], and involves the same kind of nonlinearities as (32.14) and (32.15).

Canonical Piecewise Linear Representations

The canonical PWL representation of (32.13) can be extended to the multidimensional case, based on the following representation:

$$g(\mathbf{x}) = \mathbf{a}^{\mathrm{T}}\mathbf{x} + b + \sum_{i=1}^{Q} c_i |\mathbf{w}_i^{\mathrm{T}}\mathbf{x} - \delta_i| \tag{32.20}$$

where $\mathbf{a}$ and $\mathbf{w}_i$ are P-vectors; b, c_i, and δ_i are scalars; and Q represents the number of hyperplanes that divide the whole space R^P into a finite number of polyhedral regions where $g(\cdot)$ can be expressed as an affine representation. Note that (32.20) avoids the use of multipliers. Thus, $g(\cdot)$ in (32.20) can be realized through the block diagram of Fig. 32.8, consisting of Q absolute value nonlinearities and weighted summers.

Radial Basis Functions

The idea behind radial basis function expansion is to represent the function at each point of the input space as a linear combination of kernel functions whose arguments are the radial distance of the input point to a selected number of centers

$$g(\mathbf{x}) = \sum_{j=1}^{Q} \mathbf{w}_j \Phi_j(\|\mathbf{x} - \boldsymbol{\delta}_j\|) \tag{32.21}$$

where $\|\cdot\|$ denotes a norm imposed on R^P, usually assumed Euclidean. The most common basis function is a Gaussian kernel similar to (32.16),

$$\Phi(\mathbf{x}) = \exp\left(-\frac{\|\mathbf{x} - \boldsymbol{\delta}\|^2}{2\sigma^2}\right) \tag{32.22}$$

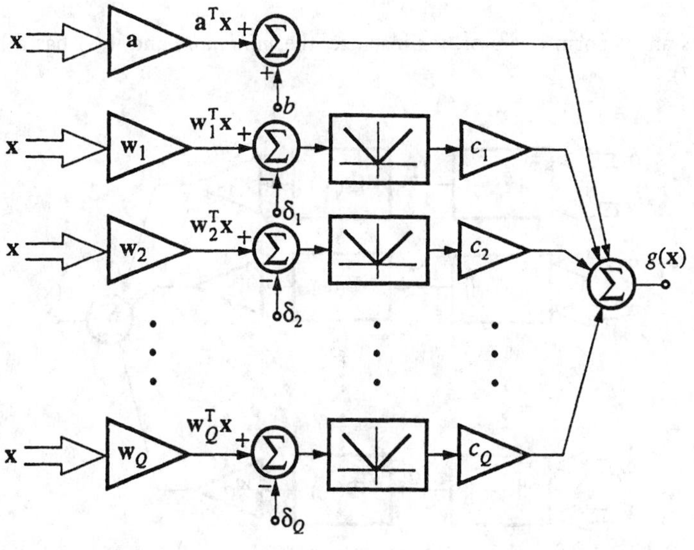

FIGURE 32.8 Canonical block diagram for a canonical PWL function.

although there are many other alternatives [51], for instance,

$$\Phi(r) = (\sigma^2 + r^2)^{-\alpha} \qquad \Phi(r) = r \qquad \alpha \geq -1 \tag{32.23}$$

where r is the radial distance to the center of the basis function, $r \equiv \|\mathbf{x} - \boldsymbol{\delta}\|$. Micchelli [42] showed that any function whose first derivative is monotonic qualifies as a radial basis function. As an example, as (32.23) displays, the identity function $\Phi(r) = r$ falls into this category, which enables connecting the representation by radial basis functions to the canonical PWL representation [40]. Figure 32.9 shows a block diagram for the hardware realization of the radial basis function model.

Neuro-Fuzzy Interpolation

This technique exploits the interpolation capabilities of fuzzy inference, and can be viewed as the multidimensional extension of the use of linear combination of bell-shaped basis functions to approximate nonlinear functions of a single variable [see (32.4) and (32.17)]. Apart from its connection to approximate reasoning and artificial intelligence, this extension exhibits features similar to the sectionwise PWP representation, namely, it relies on a well-defined class of unidimensional functions. However, neuro-fuzzy interpolation may be advantageous for hardware implementation because it requires easy-to-build collective computation operators instead of multiplications.

Figure 32.10 depicts the block diagram of a neuro-fuzzy interpolator for the simplest case in which inference is performed using the singleton algorithm [33] to obtain

$$g(\mathbf{x}) = \sum_{j=1}^{Q} w_j \frac{s_j(\mathbf{x})}{\sum_{i=1,Q} s_i(\mathbf{x})} \tag{32.24}$$

where the functions $s_j(\mathbf{x})$, called activities of the fuzzy *rules*, are given as

$$s_j(\mathbf{x}) = \Gamma\left\{\Phi_{j1}(x_1), \Phi_{j2}(x_2), \ldots, \Phi_{jP}(x_P)\right\} \tag{32.25}$$

where $\Gamma(\cdot)$ is any T-norm operator, for instance, the *minimum*, and $\Phi(\cdot)$ has a bell-like shape (see Fig. 32.7).

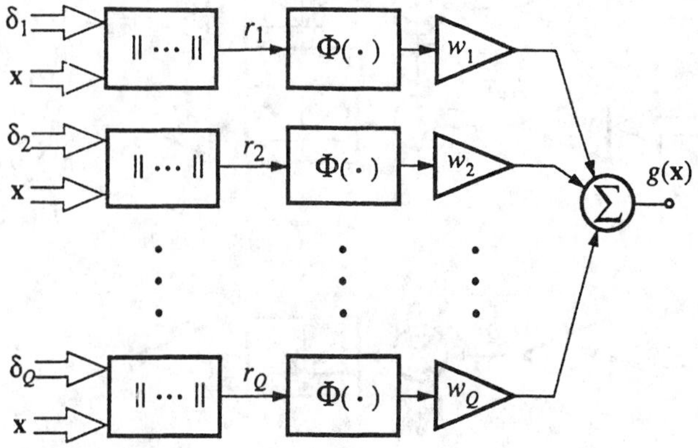

FIGURE 32.9 Concept of radial basis function hardware.

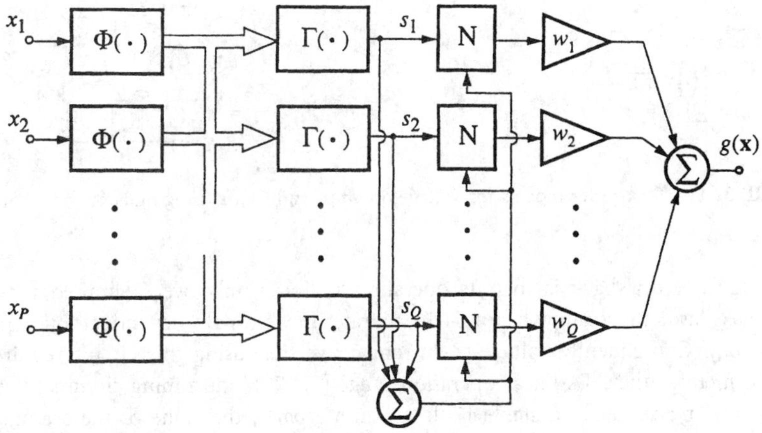

FIGURE 32.10 Conceptual architecture of a neuro-fuzzy interpolator.

Multilayer Perceptron

Similar to (32.5), but consists of the more general case of several layers, with the input to each nonlinear block given as a linear combination of the multidimensional input vector [82].

2.3 Aggregation, Scaling, and Transformation Circuits

The mathematical techniques presented in Section 32.2 require several nonlinear operators and the linear operators of scaling and aggregation (covered for completeness in this section). This section also covers transformation circuits. This is because in many practical situations we aim to exploit some nonlinear mechanism which intrinsically involves a particular kind of characteristics. For instance, a MOS transistor has inherent square-law transconductance, while a diode exhibits an exponential driving-point. Similarly, many nonlinear operators are naturally realized in current-mode domain and involve currents at both the input and the output. Thus, transformation circuits are needed to exploit these mechanisms for other types of characteristics.

Transformation Circuits

Two basic problems encountered in the design of transformation circuits are how to convert a voltage node into a current node and vice versa. We know no unique way to realize these functions. Instead, there are many alternatives which depend on which active component from Appendix A is used. The OTA can be represented to a first-order model as a voltage-controlled current source (VCCS) with linear transconductance parameter g_m. Regarding the op amp and CCII, it is convenient to represent them by the first-order models of Fig. 32.11, which contain **nullators** and **norators.**[1] A common appealing feature of both models is the virtual ground created by the input nullator. It enables us to sense the current drawn by nodes with fixed voltage—fully exploitable to design transformation circuits.

Voltage-to-Current Transformation

A straightforward technique for voltage-to-current conversion exploits the operation of the OTA as a VCCS [see Fig. 32.12(a)] to obtain $i_o = g_m v_i$, where g_m is the OTA transconductance

[1]A nullator simultaneously yields a short circuit and an open circuit, while the voltage and the current at a norator are determined by the external circuitry. The use of a nullator to model the input port of an op amp is valid only if the component is embedded in a negative feedback configuration. With regard to the CCII, the required feedback is created by the internal circuitry.

FIGURE 32.11 First-order models for voltage op amps and CCIIs using nullators and norators.

parameter [22]. A drawback is that its operation is linear only over a limited range of the input voltage. Also, the scaling factor is inaccurate and strongly dependent on temperature and technology. Consequently voltage-to-current conversion using this circuit requires circuit strategies to increase the OTA linear operation range [17, 70], and tuning circuits to render the scaling parameter accurate and stable [70]. As counterparts, the value of the scaling factor is continuously adjustable through a bias voltage or current. Also, because the OTA operates in open loop, its operation speed is not restricted by feedback-induced pole displacements.

The use of feedback attenuates the linearity problem of Fig. 32.12(a) by making the conversion rely on the constitutive equation of a passive resistor. Figure 32.12(b) shows a concept commonly found in op amp-based voltage-mode circuits [29, 59]. The idea is to make the voltage at node ① of the resistor change linearly with v_o, $v_1 = v_o + av_i$, and thus render the output current independent of v_o, to obtain $i_o = G(v_o + av_i - v_o) = aGv_i$. The summing node in Fig. 32.12(b) is customarily realized using op amps and resistors, which is very costly in the more general case in which the summing inputs must have high impedance. The circuits of Fig. 32.12(c) and (d) reduce this cost by direct exploitation of the virtual ground at the input of current conveyors [Fig. 32.12(c)] and op amps [Fig. 32.12(d)]. For both circuits, the virtual ground forces the input voltage v_i across the resistor. The resulting current is then sensed at the virtual ground node and routed to the output node of the conveyor, or made to circulate through the feedback circuitry of the op amp, to obtain $i_o = Gv_i$.

Those implementations of Fig. 32.12(b), (c), and (d) that use off-the-shelf passive resistors overcome the accuracy problems of Fig. 32.12(a). However, the values of monolithic components are poorly controlled. Also, resistors may be problematic for standard VLSI technologies, where high-resistivity layers are not available and consequently, passive resistors occupy a large area. A common IC-oriented alternative uses the ohmic region of the MOS transistor to realize an

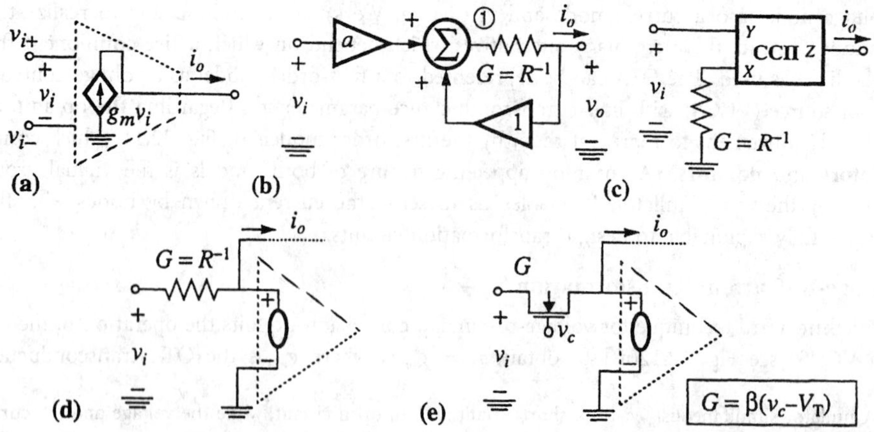

FIGURE 32.12 Voltage-to-current transformation: (a) using an OTA; (b) using voltage feedback; (c) using a current conveyor; (d) using virtual ground of an op amp; (e) same as d, but with active resistors.

active resistor [69] [Fig. 32.12(e)]. Tuning and linearity problems are similar to those for the OTA. Circuit strategies to overcome the latter are ground in [13, 32, 66, 69].

Current-to-Voltage Transformation

The most straightforward strategy consists of a single resistor to draw the input current. It may be passive [Fig. 32.13(a)] or active [Fig. 32.13(b)]. Its drawback is that the node impedance coincides with the resistor value, and thus makes difficult impedance matching to driving and loading stages. These matching problems are overcome by Fig. 32.13(c), which obtains low impedances at both the input and the output ports. On the other hand, Fig. 32.13(d) obtains low impedance at only the input terminal, but maintains the output impedance equal to the resistor value. All circuits in Fig. 32.13 obtain $v_o = Ri_i$, where $R = g_m^{-1}$ for the OTA.

Voltage/Charge Domain Transformations for Sampled-Data Circuits

The linearity and tuning problems of previous IC-related transformation approaches are overcome through the use of *dynamic* circuit design techniques based on **switched-capacitors** [72]. The price is that the operation is no longer asynchronous: relationships among variables are only valid for a discrete set of time instants. Variables involved are voltage and charge, instead of current, and the circuits use capacitors, switches, and op amps.

Figure 32.14(a) is for voltage-to-charge transformation, while Fig. 32.14(b) is for charge-to-voltage transformation. The switches in Fig. 32.14(a) are controlled by nonoverlapping clock signals, so that the structure delivers the following incremental charge to the op amp virtual ground node:

$$\Delta q^e = C(v_{i+} - v_{i-}) = -\Delta q^o \tag{32.26}$$

where the superscript denotes the clock phase during which the charge is delivered. Complementarily, the structure of Fig. 32.14(b) initializes the capacitor during the even clock phase, and senses the incremental charge that circulates through the virtual ground of the op amp during the odd clock phase. Thus, it obtains

$$v_o^o = C(\Delta q^o) \tag{32.27}$$

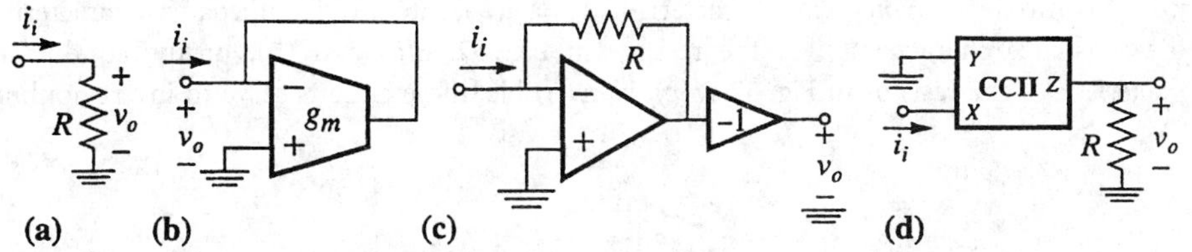

FIGURE 32.13 Current-to-voltage transformation: (a) using a resistor; (b) using a feedback OTA; (c) using op amps; (d) using current conveyors.

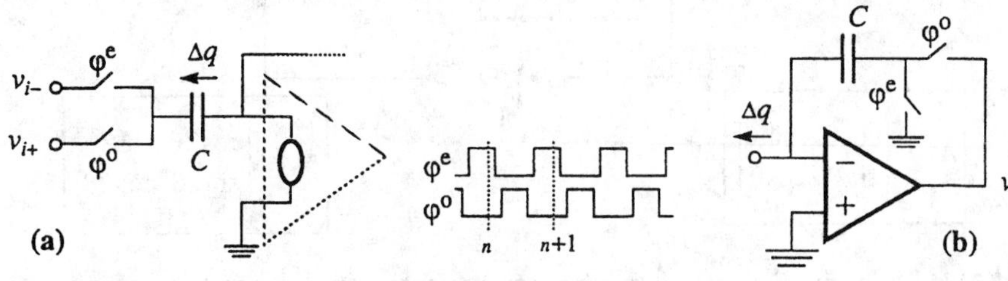

FIGURE 32.14 Transformations for sampled-data circuits: (a) V-to-q; (b) q-to-V.

References [45, 46] and [68] contain alternative circuits for the realization of the scaling function. The circuits have superior performance in the presence of parasitics of actual monolithic op amps and capacitors.

Transformation among Transfer Characteristics

Figure 32.15 shows the general architecture needed to convert one kind of transfer characteristics, e.g., voltage transfer, into another, e.g., current transfer. Variables x' and y' of the original characteristics can be either voltage or current, and the same occurs for x and y of the converted characteristic. The figure shows the more general case, which also involves a linear transformation of the characteristics themselves:

$$\begin{bmatrix} x \\ y \end{bmatrix} = \mathbf{A}\begin{bmatrix} x' \\ y' \end{bmatrix} = \begin{bmatrix} a_{11} & a_{12} \\ a_{21} & a_{22} \end{bmatrix}\begin{bmatrix} x' \\ y' \end{bmatrix} \tag{32.28}$$

For example, the figure encloses the matrices to rotate the characteristics by an angle θ, and to reflect the characteristics with respect to an edge with angle θ. This concept of linear transformation converters and its applications in the synthesis of nonlinear networks was proposed initially by Chua [5] for driving-point characteristics, and further extended by different authors [24, 29].

In the simplest case, in which the nondiagonal entries in (32.28) are zero, the transformation performed over the characteristics is scaling, and the circuits of Figs. 32.12 and 32.13 can be used directly to convert x into x' at the input front end, and y into y' at the output front end. Otherwise, aggregation operation is also required, which can be realized using the circuits described elsewhere.

From Driving-Point to Transfer and Vice Versa

Figure 32.16 shows circuits to transform driving-point characteristics into related transfer characteristics. Figure 32.16(a) and (b) use the same principle as Fig. 32.12(c) and (d) to transform a voltage-controlled *driving-point* characteristic, $i_i = f(v_i)$, into a *transconductance* characteristics. On the other hand, Fig. 32.16(c) operates similarly to Fig. 32.13(c) to transform a current-controlled *driving-point* characteristic, $v_i = f(i_i)$, into a transimpedance characteristic. If the resistance characteristic of the resistor in Fig. 32.16(a) and (b), or the conductance characteristic of the resistor in Fig. 32.16(c), is invertible, these circuits serve to invert nonlinear

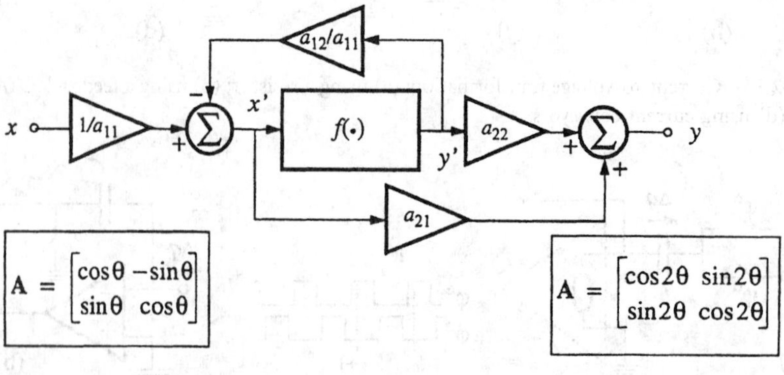

FIGURE 32.15 Concept of linear transformation converter for transfer characteristics: general architecture, and transformation matrices for rotation and reflection.

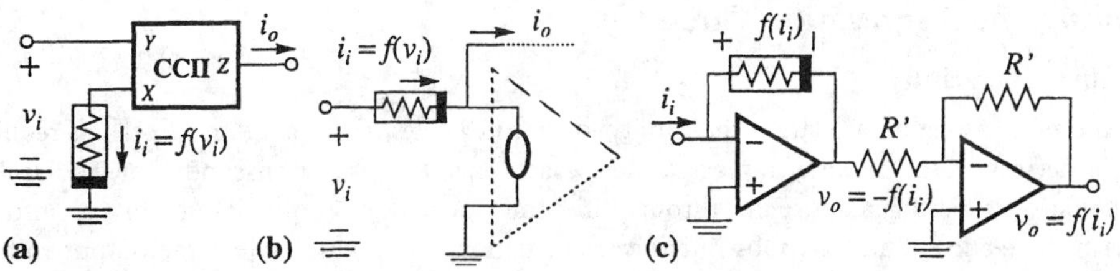

FIGURE 32.16 From driving-point to transfer characteristics: (a) and (b) transconductance from voltage-controlled driving-point; (c) transimpedance from current-controlled driving-point.

functions [63]. For instance, using a common base BJT in Fig. 32.16(c) obtains a logarithmic function from the BJT exponential transconductance. Also, the use of a MOST operating in the ohmic region serves to realize a division operation.

Lastly, let us consider how to obtain driving-point characteristics from related transfer characteristics. Figure 32.17(a) and (b) correspond to the common situation found in op amp-based circuits, where the transfer is between voltages. Figure 32.17(a) is for the voltage-controlled case and Fig. 32.17(b) is for the current-controlled case. They use feedback strategies similar to Fig. 32.12(b) to render either the input voltage or the input current independent of the linear contributions of the other port variable. A general theory for this kind of transformation converter can be found in [29].

Note that these figures rely on a Thevenin representation. Similar concepts based on Norton representations allow us to transform current transfer characteristics into driving-point characteristics. However, careful design is needed to preserve the input current while sensing it.

Other interesting transformation circuits are shown in Fig. 32.17(c) and (d). The block in Fig. 32.17(c) is a transconductor that obtains $i_o = -f(v_i)$ with very large input impedance. Then, the application of feedback around it obtains a voltage-controlled resistor, $i_i = f(v_i)$. Figure 32.17(d) obtains a current-controlled resistor, $v_i = f(i_i)$, using a current conveyor to sense the input current and feedback the output voltage of a transimpedance device with $v_o = f(i_i)$.

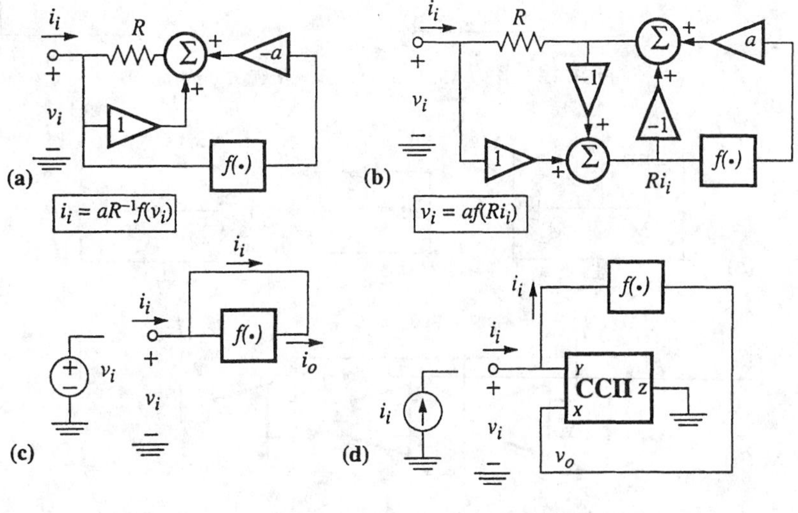

FIGURE 32.17 From transfer to driving-point characteristics.

Scaling and Aggregation Circuitry

Scaling Operation

Whenever the weights are larger than unity, or are negatives, the operation of scaling requires active devices. Also, because any active device acts basically as a transconductor, the scaling of voltages is performed usually through the transformation of the input voltage into an intermediate current and the subsequent transformation of this current into the output voltage. Figure 32.18 illustrates this for an op amp-based amplifier and an OTA-based amplifier. The input voltage is first scaled and transformed into i_o, and then this current is scaled again and transformed into the output voltage. Thus, the scaling factor depends on two design parameters. Extra control is achieved by also scaling the intermediate current.

Let us now consider how to scale currents. The most convenient strategy uses a **current mirror**, whose simplest structure consists of two matched transistors connected as shown in Fig. 32.19(a) [25]. Its operating principle relies on functional cancellation of the transistor nonlinearities to yield a linear relationship

$$i_o = p_2 f(v_i) = p_2 f\left[f^{-1}\left(\frac{i_i}{p_1}\right)\right] = \frac{p_2}{p_1} i_i \tag{32.29}$$

where p_1 and p_2 are parameters whose value can be designer controlled; for instance, β of the MOST or I_S of the BJT (see Appendix A and [44]). The input and output currents in Fig. 32.19(a) must be positive. Driving the input and output nodes with bias currents I_B and $(p_2/p_1)I_B$, respectively, one obtains $i_i = i_i' + I_B$ and $i_o = i_o' + (p_2/p_1)I_B$, and this enables bilateral operation on i_i' and i_o'.

In practical circuits this simple design concept must be combined with circuit strategies to reduce errors due to non-negligible input current of BJTs, DC voltage mismatch between input and output terminals, finite input resistance, and finite output resistance. Examples of these

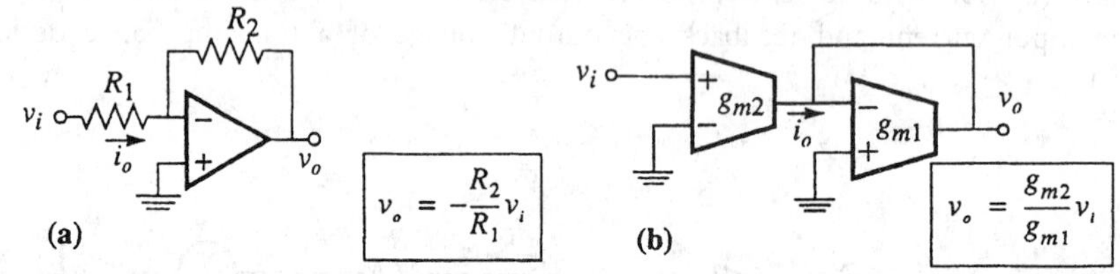

FIGURE 32.18 Mechanisms for voltage scaling.

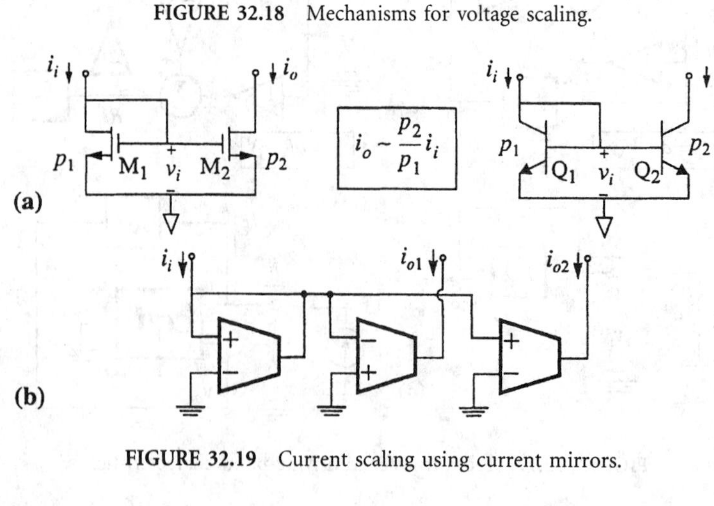

FIGURE 32.19 Current scaling using current mirrors.

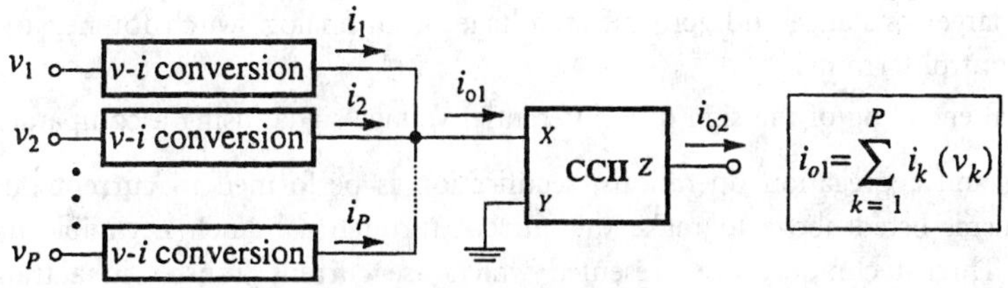

FIGURE 32.20 Aggregation of voltages through intermediate currents and current conveyor.

strategies can be found in [25, 56, 77]. On the other hand, sizing and layout strategies for other problems related to random mismatches between input and output devices are found in [41] and [48], which are applicable to most matching problems in MOS IC design.

The current mirror concept is extensible to any pair of matched transconductors, provided their transconductance characteristics are invertible and parameterized by a designer-controlled scale factor p, and that the dependence of the output current with the output voltage is negligible. In particular, the use of differential transconductors enables us to obtain bilateral operation simply, requiring no current-shifted biasing at the input and output nodes. It also simplifies achieving noninverting amplification (that is, positive scale factors), as Fig. 32.19(b) illustrates. This figure also serves to illustrate the extension of the mirror concept to multiple current outputs. Note that except for loading considerations, no other limitations exist on the number of output transconductors that can share the input voltage. Also, because fan-out of a current source is strictly one, this replication capability is needed to enable several nodes to be excited by a common current. On the other hand, the fact that the different current output replicas can be scaled independently provides additional adjusting capability for circuit design.

Signal Aggregation

As for the scaling operation, aggregation circuitry operates in current domain, based on Kirchhoff's current law (KCL). Thus, the aggregation of voltages requires that first they be transformed into currents (equivalently, charge packets in the switched-capacitor circuitry) and then the intermediate currents are added through KCL, while current and incremental charge are added by routing all the components to a common node. If the number of components is large, the output impedance of the driving nodes is not large enough, and/or the input impedance of the load is not small enough, this operation will encompass significant loading errors due to variations of the voltage at the summing node. This is overcome by clamping the voltage of this node using a virtual ground, which in practical circuits is realized by using either the input port of an op amp, or terminals X and Y of a current conveyor. Figure 32.20 illustrates the current conveyor case.

2.4 Piecewise-Linear Circuitry

Consider the elementary PWL functions that arise in connection with the different methods of representation covered in Section 32.2:

- Two-piece concave and convex characteristics [see (32.12)]
- Hermite linear basis function (see Fig. 32.4 and Appendix B)
- Absolute value [see (32.13)]

where **rectification** is the only nonlinear operator involved. The circuit primitives in Appendix A exhibit several mechanisms which are exploitable in order to realize rectification:

- Cut-off of diodes and transistors. Specifically, current through a diode negligible for negative voltage, and output current of BJTs and MOSTs negligible under proper biasing.

- Very large resistance and zero offset voltage of an analog switch for negative biasing of the control terminal.
- Digital encoding of the sign of a differential voltage signal using a comparator.

Like scaling and aggregation operations, rectification is performed in current domain, using the mechanisms listed above to make the current through a branch negligible under certain conditions. Three techniques are presented which use current transfer in a transistor-based circuit, current-to-voltage transfer using diodes and op amps, and charge transfer using switches and comparators, respectively.

Current Transfer Piecewise-Linear Circuitry

Figure 32.21(a) and (b) shows the simplest technique to rectify the current transferred from node ① to node ②. They exploit the feature of diodes and diode-connected transistors to support only positive currents. Figure 32.21(a) operates by precluding negative currents to circulate from node ① to node ②, while Fig. 32.21(b) also involves the nonlinear transconductance of the output transistor M_o; negative currents driving the node ① force v_i to become smaller than the cut-in voltage and, consequently, the output current becomes negligible. A drawback to both circuits is that they do not provide a path for negative input currents, which accumulates spurious charge at the input node and forces the driving stage to operate outside its linear operating regime. Solutions to these problems can be found in [57] and [61]. Also, Fig. 32.21(a) produces a voltage displacement equal to the cut-in voltage of the rectifying device, which may be problematic for applications in which the voltage at node ① bears information. A common strategy to reduce the voltage displacements uses feedback to create superdiodes (shown in Fig. 32.21(c) for the grounded case and Fig. 32.21(d) for the floating case), and where the reduction of the voltage displacement is proportional to the DC gain of the amplifier.

Figure 32.22(a), called **current switch**, provides paths for positive and negative currents entering node ①, and obtains both kinds of elementary PWL characteristics exploiting cut-off of either BJTs or MOSTs. It consists of two complementary devices: npn (bottom) and pnp BJTs, or n-channel (bottom) and p-channel MOSTs. Its operation is very simple: any positive input current increases the input voltage, turning the bottom device ON. Because both devices share the input voltage, the top device becomes OFF. Similarly, the input voltage decreases for negative input currents, so that the top device becomes ON and the bottom OFF. In sum, positive input currents are drawn to the bottom device, while negative currents are drawn to the top device.

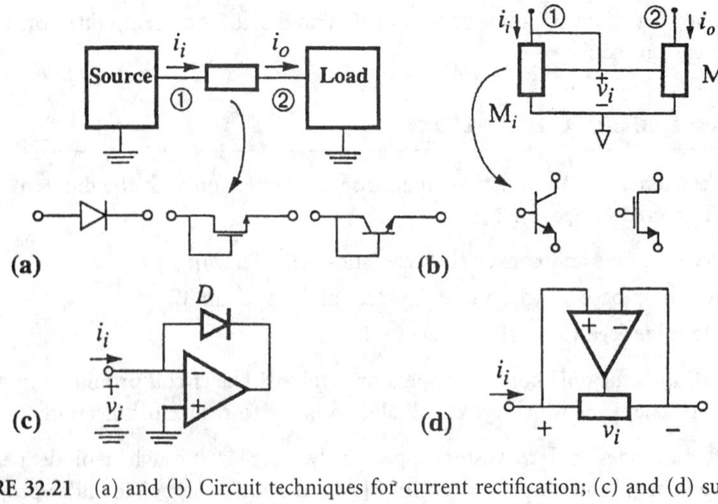

FIGURE 32.21 (a) and (b) Circuit techniques for current rectification; (c) and (d) superdiodes.

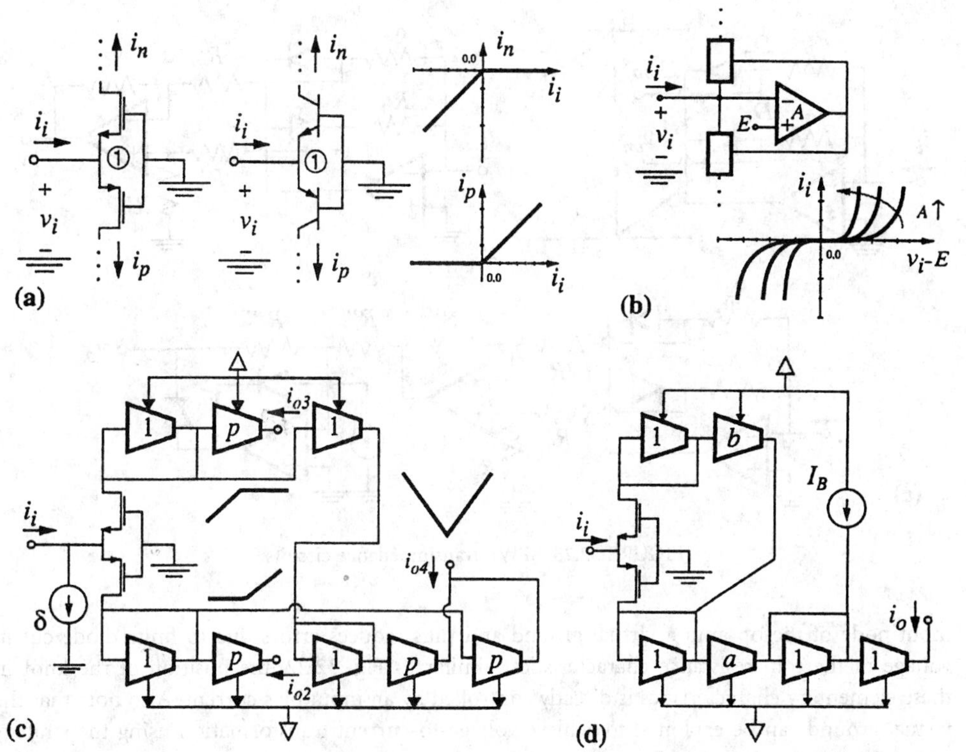

FIGURE 32.22 Current switch and its application for different basic PWL curves.

An inconvenience of Fig. 32.22(a) is the dead zone exhibited by its input driving-point characteristics, which is very wide for MOSTs. It may produce errors due to nonlinear loading of the circuitry that drives the input node. Figure 32.22(b) overcomes this by using a circuit strategy similar to that of the superdiodes. The virtual ground at the op amp input renders the dead-zone centered around the voltage level E, and its amplitude is reduced by a factor proportional to the amplifier DC gain. Some considerations related to the realization of this amplifier are found in [58].

Proper routing and scaling of the currents i_p and i_n in Fig. 32.22(a) gives us the concave and convex basic characteristics with full control of the knot and position and the slope in the conducting region. Figure 32.33(c) shows the associated circuit, in which the input bias current controls the knot position, and the slope in the conducting region is given by the gain of the current mirrors. Note that this circuit also obtains the absolute value characteristics, while Fig. 32.22(d) obtains the Hermite linear basis function. The way to obtain the PWL fuzzy membership function from this latter circuit is straightforward, and can be found in [58].

Transresistance Piecewise-Linear Circuitry

The circuit strategies involved in PWL current transfer can be combined in different ways with the transformation circuits discussed previously to obtain transconductance and voltage-transfer PWL circuits. In many cases design ingenuity enables optimum merging of the components and consequently, simpler circuits. Figure 32.23(a) shows what constitutes the most extended strategy to realize the elementary PWL functions using off-the-shelf components [63, 80]. The input current is split by the feedback circuitry around the op amp to make negative currents circulate across D_n and positive currents circulate across D_p. Consequently, this feedback renders the

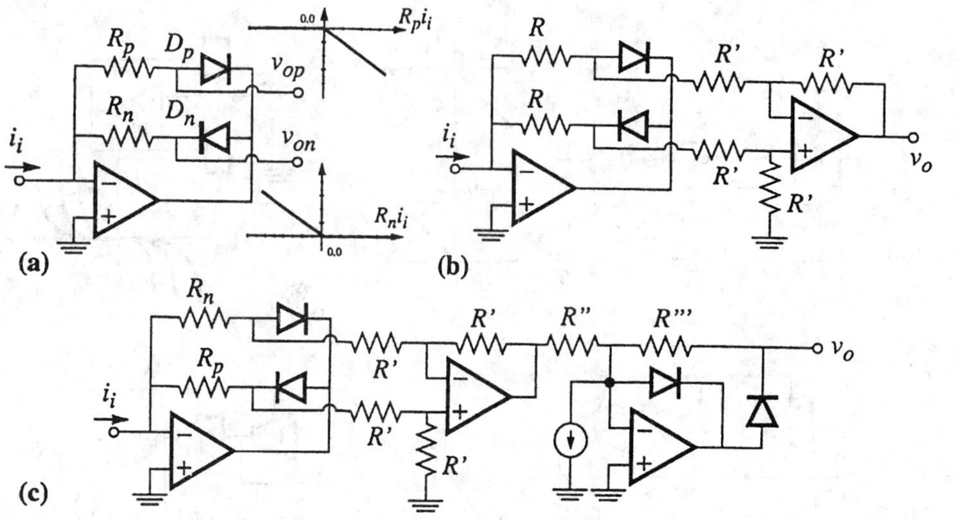

FIGURE 32.23 PWL transimpedance circuits.

input node of the op amp a virtual ground and thus reduces errors due to finite diode cut-in voltage in the transresistance characteristics. Similar to Fig. 32.22, the position of the knot in these elementary characteristics is directly controlled by an input bias current. Also note that the virtual ground can be exploited to achieve voltage-to-current transformation using the strategy of Fig. 32.12(d) and thus, voltage transfer operation.

Algebraic combination of the elementary curves provided by Fig. 32.23(a) requires transforming the voltages v_{on} and v_{op} into currents and then aggregating these currents by KCL. For example, Fig. 32.23(b) shows the circuit for the absolute value; Fig. 32.23(c) shows a possible implementation of the Hermite basis function.

Other related contributions found in the literature focus on the systematic realization of PWL driving-point resistors, and can be found in [7] and [10].

Piecewise-Linear Shaping of Voltage-to-Charge Transfer Characteristics

The realization of PWL relationships among sampled-data signals is based on nonlinear voltage-to-charge transfer and uses analog switches and comparators. Figure 32.24(a) shows a circuit structure, where one of the capacitor terminals is connected to virtual ground and the other to a switching block. Assume that nodes ① and ② are both grounded. Note that for $(v - \delta) > 0$ the switch arrangement set node ④ to δ, while node ⑤ is set to v. For $(v - \delta) < 0$, nodes ④ and ⑤ are both grounded. Consequently, voltage at node ③ in this latter situation does not change from one clock phase to the next, and consequently, the incremental charge becomes null for $(v - \delta) < 0$. On the other hand, for $(v - \delta) > 0$, the voltage at node ③ changes from one clock phase to the next, and generates an incremental charge

$$\Delta q^e = C(v - \delta) = -\Delta q^o \tag{32.30}$$

which enables us to obtain negative and positive slopes using the same circuit, as shown in Fig. 32.24(a). To make the characteristics null for $(v - \delta) > 0$, it suffices to interchange the comparator inputs. Also, the technique is easily extended to the absolute value operation by connecting terminal ① to v, and terminal ② to δ. The realization of the Hermite linear basis function is straightforward and can be found in [55].

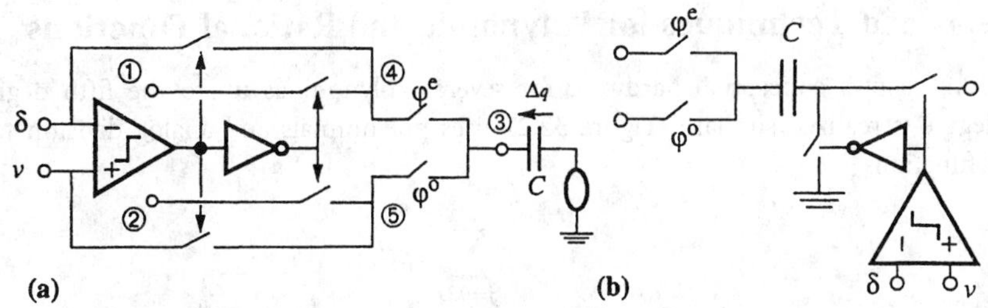

FIGURE 32.24 Circuits for rectification in voltage-to-charge domain.

Other approaches to the realization of PWL switched-capacitor circuitry use series rectification of the circulating charge through a comparator-controlled switch [Fig. 32.24(b)], and can be found in [16] and [31]. The latter also discusses exploitation of these switched-capacitor circuits to realize continuous time driving-point characteristics, the associated transformation circuits, and the dynamic problematics.

32.5 Polynomials, Rational, and Piecewise-Polynomial Functions

These functions use rectification (required for truncation operation in the PWP case) and analog *multiplication*,

$$z = \frac{xy}{\alpha} \tag{32.31}$$

as basic nonlinear operators.[2] Joining the two inputs of the multiplier realizes the *square* function. Analog *division* is realized by applying feedback around a multiplier, illustrated at the conceptual level in Fig. 32.25(a); the multiplier obtains $e = (zy)/\alpha$, and, for $A \to \infty$, the feedback forces $x = e$. Thus, if $y \neq 0$, the circuit obtains $z = \alpha(x/y)$. Joining y and z terminals, the circuit realizes the *square root,* $z = (\alpha x)^{1/2}$. This concept of division is applicable regardless of the physical nature of the variables involved. In the special case in which e and x are current and z is a voltage, the division can be accomplished using KCL to yield $x = e$. Figure 32.25(b) shows a circuit for the case in which the multiplication is in voltage domain, and Fig. 32.25(c) is for the case in which multiplication is performed in transconductance domain. The transconductance gain for input z of the latter case must be negative to guarantee stability.

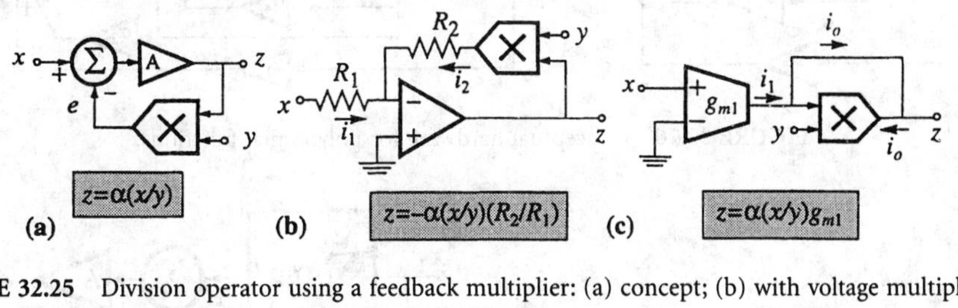

FIGURE 32.25 Division operator using a feedback multiplier: (a) concept; (b) with voltage multiplier and op amp; (c) with transconductance multiplier and OTA.

[2]Scale factor α in (32.31) must be chosen to guarantee linear operation in the full variation range of inputs and outputs.

Concepts and Techniques for Polynomic and Rational Functions

Figure 32.26 shows conceptual hardware for several polynomials up to the fifth degree. Any larger degree is realized similarly. Figure 32.27 uses polynomials and analog division to realize rational functions

$$g(x) = \frac{\sum_{j=0,Q} \alpha_j x^j}{\sum_{j=0,R} \beta_j x^j} \tag{32.32}$$

For simplicity, we have assumed that the internal scaling factors of the multipliers in Fig. 32.26 and Fig. 32.27 equal one.

An alternative technique to realize rational functions is based on **linearly controlled resistors**, described as $v = (Lx)i$, and **linearly controlled conductors**, $i = (Cx)v$, where L and C are real parameters. This technique exploits the similarity between these characteristics and those which describe inductors and capacitors in the frequency domain, to take advantage of the synthesis techniques for rational transfer functions in the s-plane through interconnection of these linear components [28] [39] (Fig. 32.28). As for the previous cases, realization of linearly controlled resistors and conductors require only multipliers and, depending upon the nature of the variables involved in the multiplier, voltage-to-current and current-to-voltage transformation circuits.

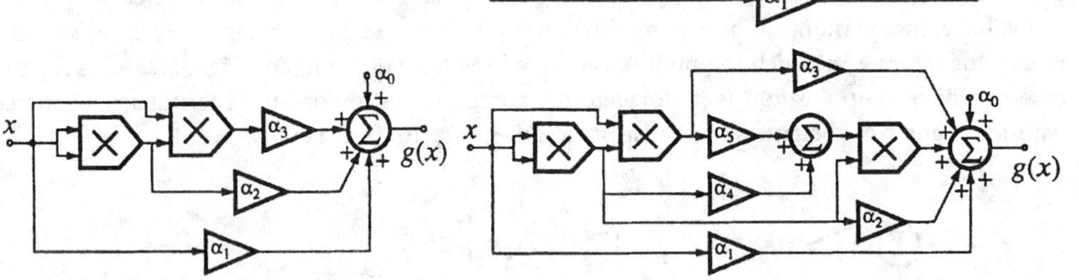

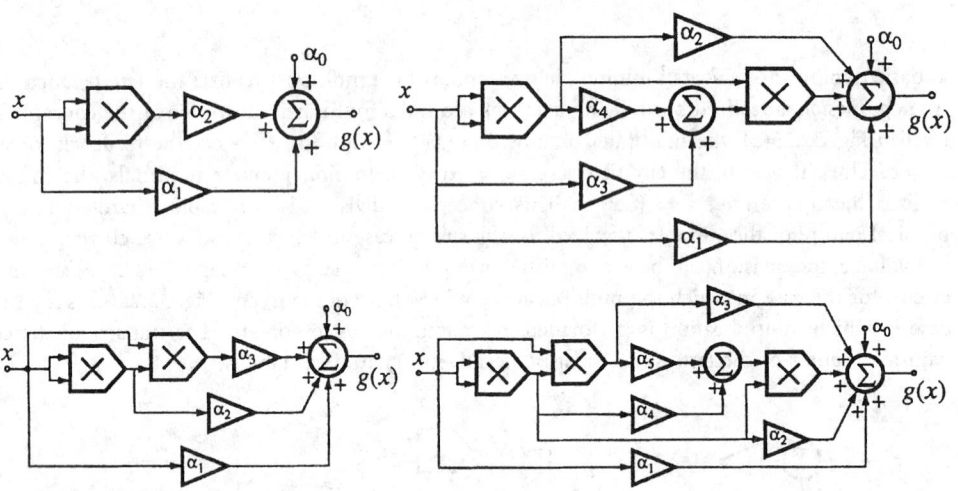

FIGURE 32.26 Conceptual hardware for polynomial functions.

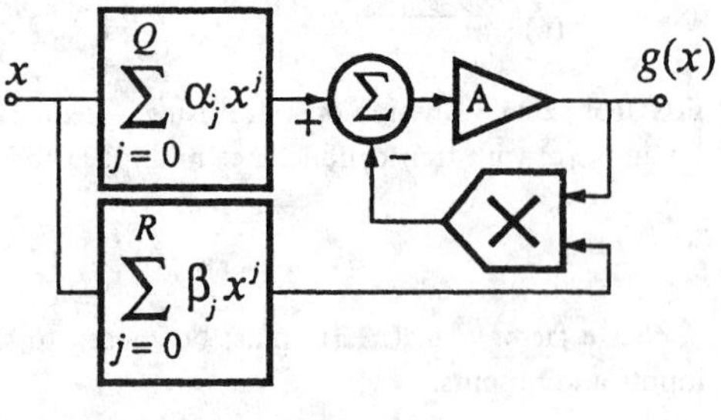

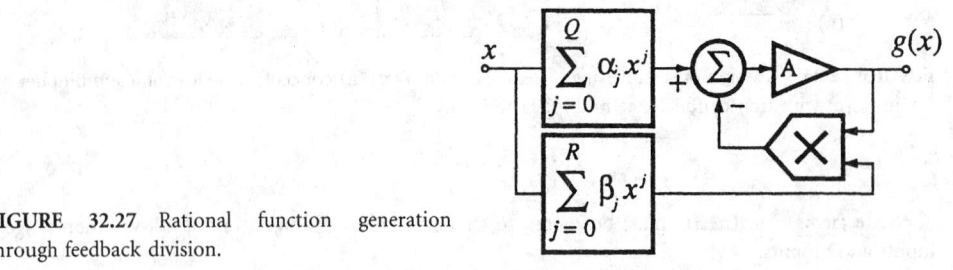

FIGURE 32.27 Rational function generation through feedback division.

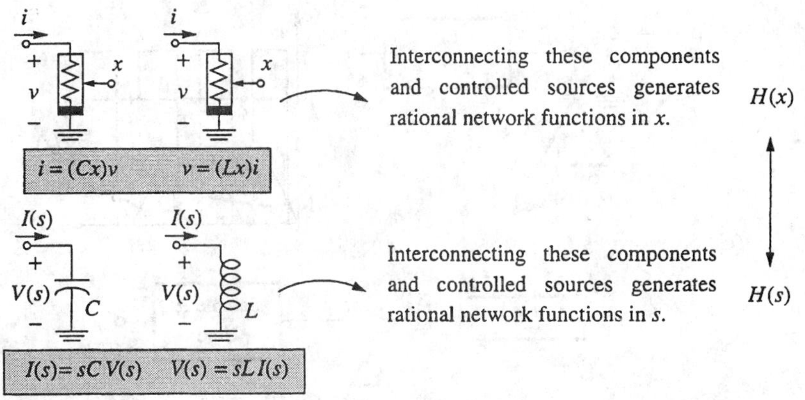

FIGURE 32.28 Usage of linearly controlled resistors to synthesize rational network functions.

Multiplication Circuitry

Two basic strategies realize multiplication circuitry: using signal processing and exploiting some nonlinear mechanism of the primitive components. Signal processing multipliers rely on the generation of a pulsed signal whose amplitude is determined by one of the multiplicands and its duty cycle by the other, so that the area is proportional to the result of the multiplication operation. Figure 32.29(a) shows an implementation concept based on averaging. This is performed by a low-pass filter whose input is a pulse train with amplitude proportional to x and duty cycle proportional to y. The latter proportionality is achieved through nonlinear sampling by comparing y with a time reference sawtooth signal. Thus, the area under each pulse in the train is the product of $x \times y$, extracted by the low-pass filter. This implementation concept is discussed in further detail in classical texts on analog computation [63], and applied more recently to analog VLSI signal processing [72].

Figure 32.29(b) shows an alternative implementation concept based on signal shaping in the time domain. It uses two linear blocks with normalized unit step response given as $h_z(t)$ and $h_y(t)$. The first is driven by level x to obtain

$$z(t) = xh_z(t) \quad 0 \leq t < \tau \tag{32.33}$$

where τ denotes the amplitude of the time interval during which the switch S remains closed. The other is driven by a reference level α, to render τ given by

$$\tau = h_y^{-1}\left(\frac{y}{\alpha}\right) \tag{32.34}$$

Assuming both linear blocks are identical and the time function invertible, one obtains the steady-state value of z, $z(\tau)$, as the product of levels x and y.

The simplest implementation of Fig. 32.29 uses integrators, i.e., $h(t) = t$, as linear blocks [see Fig. 32.41(b)]. Also note that the principle can be extended to the generation of powers of an input signal by higher order shaping in time domain. In this case both linear blocks are driven by reference levels. The block $h_y(t)$ consists of a single integrator, $\tau = y/\alpha$. The other consists of the cascade of P integrators, and obtains $z(t) = \beta t^P$. Thus, $z(t) = \beta(y/\alpha)^P$. Realizations suitable for integrated circuits are found in [34] and [55].

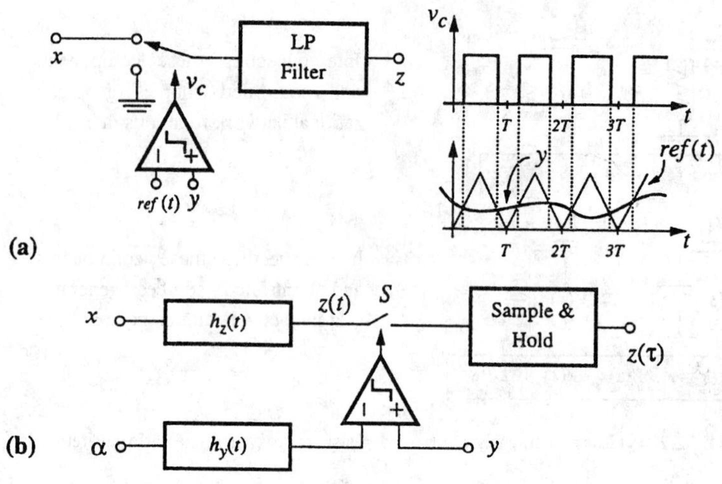

FIGURE 32.29 Signal processing multipliers: (a) by averaging; (b) by shaping in time domain.

Multipliers Based on Nonlinear Devices

The primitives in Appendix A display several mechanisms that are exploitable to realize analog multipliers:

- Exponential functionals associated to the large-signal transconductance of BJTs, and the possibility of obtaining logarithmic dependencies using feedback inversion.
- Square-law functionals associated to the large-signal transconductance of the MOS transistor operating in saturation region.
- Small-signal transconductance of a BJT in active region as a linear function of collector current. Equivalent, small-signal transconductance of a MOST in saturation as a linear function of gate voltage.
- Small-signal self-conductance of a MOS transistor in ohmic region as a linear function of gate voltage.

These and related mechanisms have been exploited in different ways and have resulted in a huge catalog of practical circuits. To quote all the related published material is beyond the scope of this section. The references listed at the end were selected because of their significance, and their cross-references contain a complete view of the state of the art. Also, many of the reported structures can be grouped according to the theory of *translinear circuits*, which provides a unified framework to realize nonlinear algebraic functions through circuits [23, 62, 79].

Log-Antilog Multipliers

Based on the exponential large-signal transconductance of the BJT, and the following relationships,

$$\begin{aligned} z' &= \ln(x) + \ln(y) = \ln(xy) \\ z &= e^{z'} = e^{\ln(xy)} = xy \end{aligned} \tag{32.35}$$

which can be realized as shown in Fig. 32.30(a) [65]. This circuit operates on positive terminal currents to obtain $i_o = (i_1 i_2)/i_3$, which can be understood from translinear circuit principles by noting that the four base-to-emitter voltages define a **translinear loop**,

$$\begin{aligned} 0 &= v_{be1} + v_{be2} - v_{be3} - v_{be4} \\ &= \ln\left(\frac{i_1}{I_s}\right) + \ln\left(\frac{i_2}{I_s}\right) - \ln\left(\frac{i_3}{I_s}\right) - \ln\left(\frac{i_o}{I_s}\right) \end{aligned} \tag{32.36}$$

The circuit can be made to operate in four-quadrant mode, though restricted to currents larger than $-I_B$, by driving each terminal with a bias current source of value I_B. Also, because all input terminals are virtual ground the circuit can be made to operate on voltages by using the voltage-to-current transformation concept of Fig. 32.12(d). Similarly, the output current can be transformed into a voltage by using an extra op amp and the current-to-voltage transformation concept of Fig. 32.13(c). Extension of this circuit structure to generate arbitrary powers is discussed in [23]. Figure 32.30(b) [1] uses similar techniques, based on introducing scaling factors in the translinear loop, to obtain

$$i_y = i_\alpha^{1-k} i_x^k \tag{32.37}$$

Square-Law Multipliers

These are based on the algebraic properties of the square function, most typically

$$z = \tfrac{1}{4}[(x+y)^2 - (x-y)^2] = xy \tag{32.38}$$

shown conceptually in Fig. 32.31(a), and the possibility of obtaining the square of a signal using circuits, typically consisting of a few MOS transistors operating in saturation region. Figure 32.31(b) through (f) depict some squarer circuits reported in the literature.

The completeness of square-law operators for the realization of nonlinear circuits was demonstrated from a more general point of view in [47] and their exploitation has evolved into systematic circuit design methodologies to perform both linear and nonlinear functions [3].

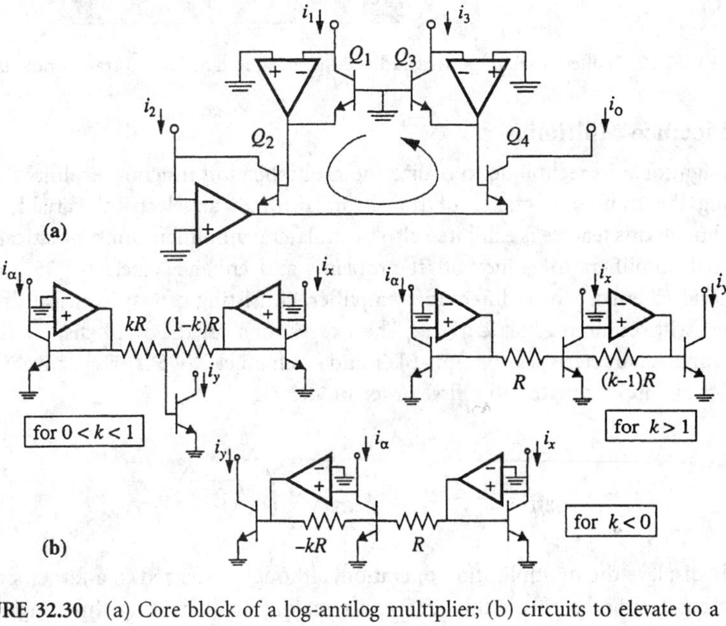

FIGURE 32.30 (a) Core block of a log-antilog multiplier; (b) circuits to elevate to a power.

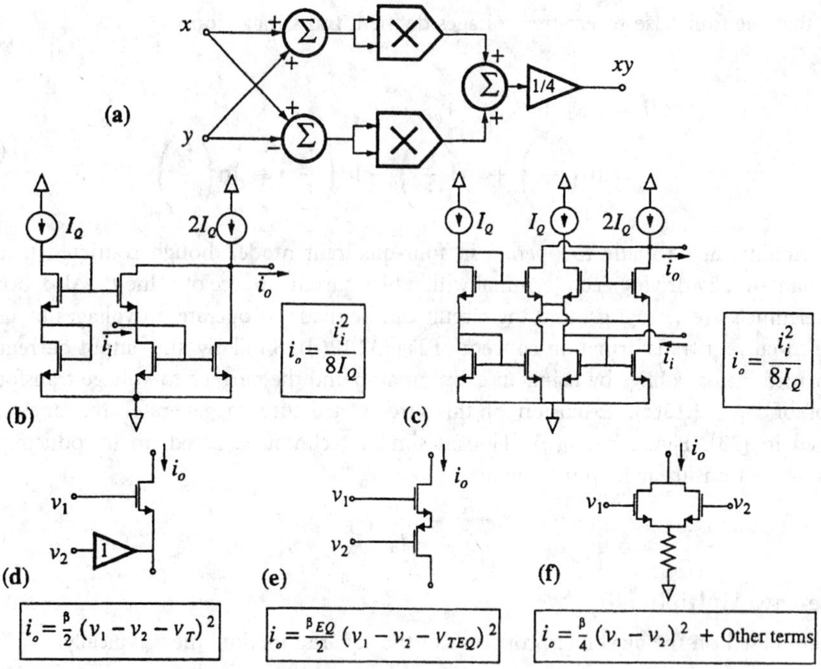

FIGURE 32.31 (a) Block diagram of the quarter-square multiplier; (b) current-mode squarer circuit in [3]; (c) current-mode squarer circuit in [79]; (d) voltage-mode squarer circuit in [36]; (e) voltage-mode squarer circuit in [60]; (f) voltage-mode squarer circuit in [49].

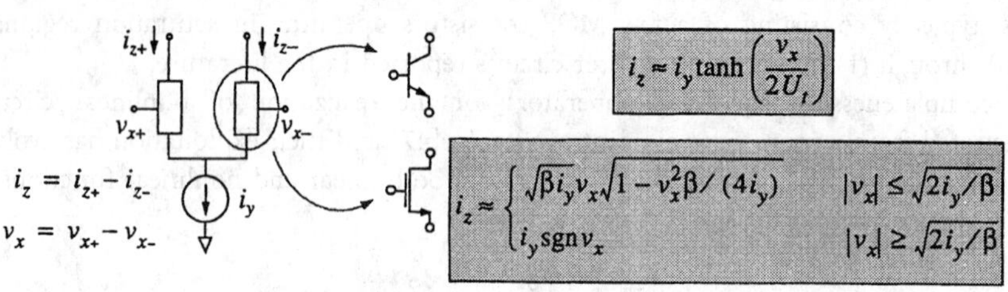

FIGURE 32.32 Differential amplifiers and their associated large-signal transconductances.

Transconductance Multipliers

A direct, straightforward technique to realize the multiplication function exploits the possibility of controlling the transconductance of transistors through an electrical variable (current or voltage). Although this feature is exhibited also by unilateral amplifiers, most practical realizations use differential amplifiers to reduce offset problems and enhance linearity [25]. Figure 32.32 shows a generic schematic for a differential amplifier, consisting of two identical three-terminal active devices with common bias current. The expressions on the right display its associated transconductance characteristics for npn-BJTs and n-channel MOSTs, respectively [25]. These characteristics are approximated to a first-order model as

$$i_{zBJT} \approx \frac{i_y}{4U_t} v_x \quad i_{zMOST} \approx \left(\sqrt{\beta i_y}\right) v_x \tag{32.39}$$

which clearly displays the multiplication operation, although restricted to a rather small linearity range. Practical circuits based on this idea focus mainly on increasing this range of linearity, and follow different design strategies. Figure 32.33 shows an example, known as the Gilbert cell, or Gilbert multiplier [23]. Corresponding realizations using MOS transistors are discussed in [2] and [53]. Sánchez-Sinencio et al. [61] present circuits to realize this multiplication function using OTA blocks. On the other hand, [17] presents a tutorial discussion of different linearization techniques for MOS differential amplifiers.

Multiplier Based in the Ohmic Region of MOS Transistors

The ohmic region of JFETs has been used to realize amplifiers with controllable gain for automatic gain control [54]. It is based on controlling the equivalent resistance of the JFET transistor in its ohmic region through a bias voltage. More recently, MOS transistors operating in the ohmic region were used to realize linear [69, 70] and nonlinear [35] signal processing tasks in VLSI chips. There exist many ingenious circuits to eliminate second- and higher order nonlinearities in the equivalent resistance characteristics. The circuit shown in Fig. 32.34(a) achieves very good nonlinearity cancellation through cross-coupling and fully differential operation, obtaining

$$i_{z+} - i_{z-} = 2\beta(v_{x+} - v_{x-})(v_{y+} - v_{y-}) \tag{32.40}$$

whose use in multiplication circuits is discussed in [35] and [66]. A more general view is presented in Fig. 32.34(b) [35], where the conductance as well as the resistance of the MOS ohmic region are used to obtain a versatile amplifier-divider building block. Enomoto and Yasumoto [18] report another interesting multiplier that combines the ohmic region of the MOS transistor and sampled-data circuits.

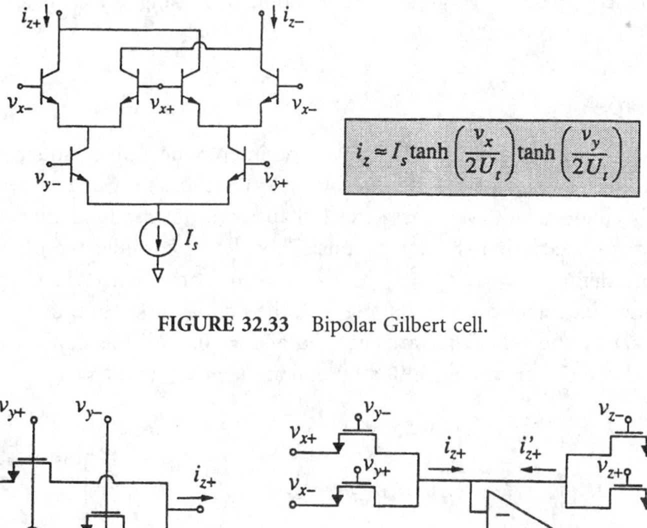

FIGURE 32.33 Bipolar Gilbert cell.

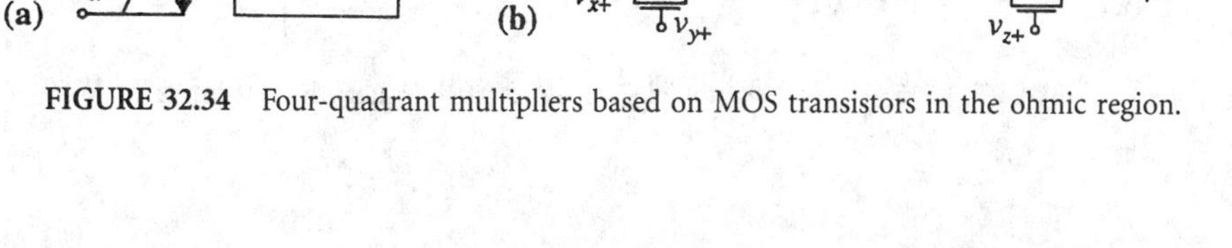

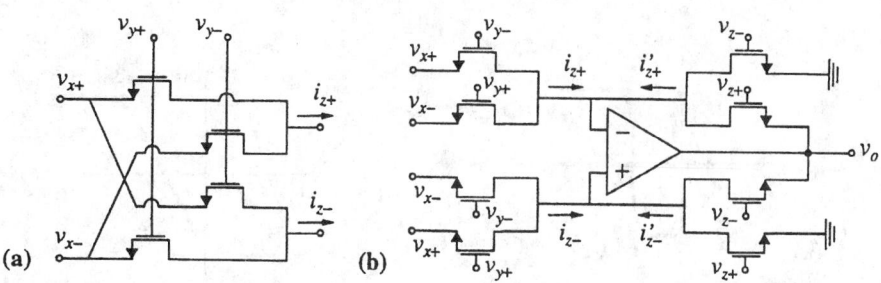

FIGURE 32.34 Four-quadrant multipliers based on MOS transistors in the ohmic region.

32.6 Sigmoids, Bells, and Collective Computation Circuits

Sigmoidal Characteristics

As (32.5) illustrates, approximating a nonlinear function through a multilayer perceptron requires the realization of sigmoidal functions, with arguments given as linear combinations of several variables. The exact shape of the sigmoidal is not critical for the approximation itself, although it may play an important role in fitting [82]. Figure 32.35 depicts two shapes used in practice. Figure 32.35(a), the hard limiter, has an inner piece of large (ideally infinite) slope, while for Fig. 32.35(b), the soft limiter, this slope is smaller and can be used as a fitting parameter.

Most amplifiers have large-signal transfer characteristics whose shape is a sigmoid or an inverted sigmoid. We present only those circuits whose inputs are currents because this simplifies the circuitry needed to obtain these inputs as linear combinations of other variables. The op amp circuit of Fig. 32.36(a) realizes the soft limiter characteristics in transimpedance form. The center is set by the input bias current and the slope through the resistor ($\beta = R$). If the branch composed of the two Zener diodes is eliminated, the saturation levels E_+ and E_- are determined through the internal op amp circuitry, inappropriate for accurate control. (Otherwise, they are determined through the Zener breakdown voltages.) On the other hand, Fig. 32.36(b) also realizes the hard sigmoid in transimpedance domain [58]. The output saturation levels for this structure are $E_+ = V_{Tn}$ and $E_- = |V_{Tp}|$, where V_{Tn} and V_{Tp} are the threshold voltages of the NMOS transistor and the PMOS transistor, respectively. To obtain the output represented by a current, one can use voltage-to-current transformation circuits. References [15, 57, 58] discuss simpler alternatives operating directly in current domain. For instance, Fig. 32.36(c) and (d) show circuits for the soft limiter characteristics and the hard limiter characteristics.

With regard to the calculation of the input to the sigmoid as a linear combination of variables, note that the input node of all circuits in Fig. 32.36 is virtual ground. Consequently, the input current can be obtained as a linear combination of voltages or currents using the techniques for signal scaling and aggregation presented in Section 32.3.

Transconductance circuits for bell-shaped function: (a) using OTAs; (b) using differential amplifiers.

Bell-Like Shapes

The exact shapes of (32.16) and (32.17) involve the interconnection of squarers, together with blocks to elevate to power, and exponential blocks—all realizable using techniques previously discussed in this chapter. However, these exact shapes are not required in many applications, and can be approximated using simpler circuits. Thus, let us consider the differential amplifier of Fig. 32.32, and define $v_i = v_x$, $I_B = i_y$, and $i_o = i_z$ for convenience. The expressions for the large-signal transconductance displayed along with the figures shows that they are sigmoids with saturation levels at I_B and $-I_B$. They are centered at $v_i = 0$, with the slope at this center point given by (32.39). The center can be shifted by making $v_i = v_{x+}$ and $\delta = v_{x-}$.

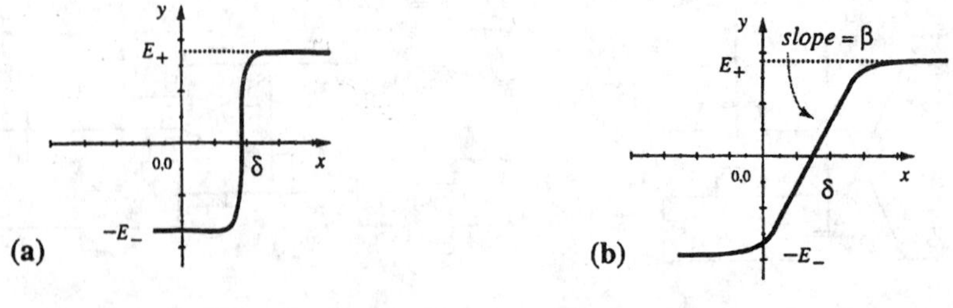

FIGURE 32.35 Typical sigmoidal shapes: (a) hard limiter; (b) soft limiter.

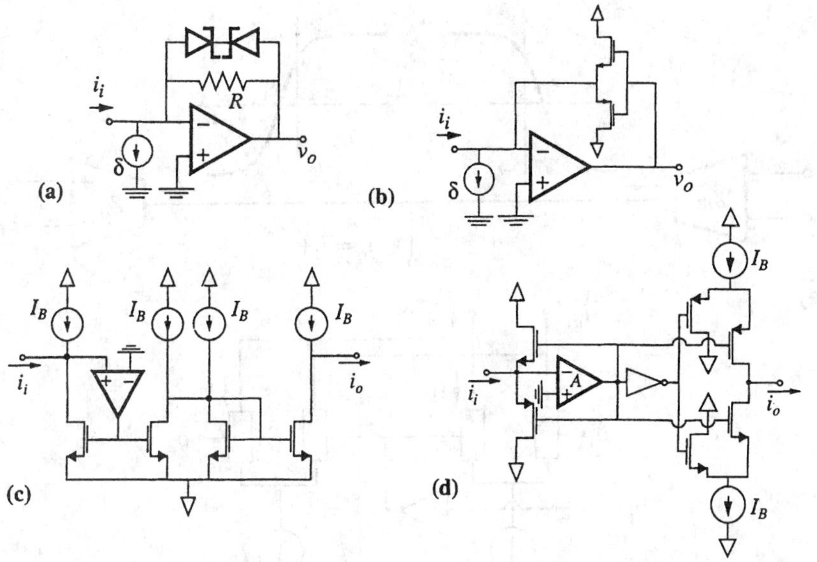

FIGURE 32.36 Realization of sigmoidal characteristics with input current: (a) transimpedance soft limiter; (b) transimpedance hard limiter; (c) and (d) soft and hard limiters in current transfer domain.

Like the differential amplifier, most OTAs exhibit sigmoid-like characteristics under large-signal operation, exploitable to realize nonlinear functions [19, 37, 56, 61, 71]. This may rely on the mathematical techniques behind multilayer perceptrons, or on those behind radial basis functions and fuzzy interpolation.

Figure 32.37(a) obtains a bell-shaped transconductance through a linear, KCL combination of the two sigmoidal characteristics, one of negative slope and the other of positive slope. The width and center of the bell (see Fig. 32.7) are given respectively by

$$2\sigma = \delta_2 - \delta_1 \quad \delta = \frac{\delta_2 + \delta_1}{2} \tag{32.41}$$

controlled by the designer. The slope of the bell at the crossover points is also controlled through the transconductance of the OTAs.

For simpler circuit realizations, this technique can be used directly with differential amplifiers, as shown in Fig. 32.37(b). The differential output current provided by the circuit can be transformed into a unilateral one using a p-channel current mirror. Equation (32.41) also applies for this circuit, and the slope at the crossovers is

$$\text{slope}_{\text{MOST}} = k\sqrt{\beta I_B} \quad \text{slope}_{\text{BJT}} = \frac{kI_B}{4U_t} \tag{32.42}$$

Note that the control of this slope through the bias current changes the height of the bell. It motivates the use of a voltage gain block in Fig. 32.37. Thus, the slope can be changed through its gain parameter k. The slope can also be changed through β for the MOSTs. Practical realizations of this concept are found in [4], [71], and [74]. The voltage amplifier block can be realized using the techniques presented in this chapter. Simpler circuits based on MOS transistors are found in [53].

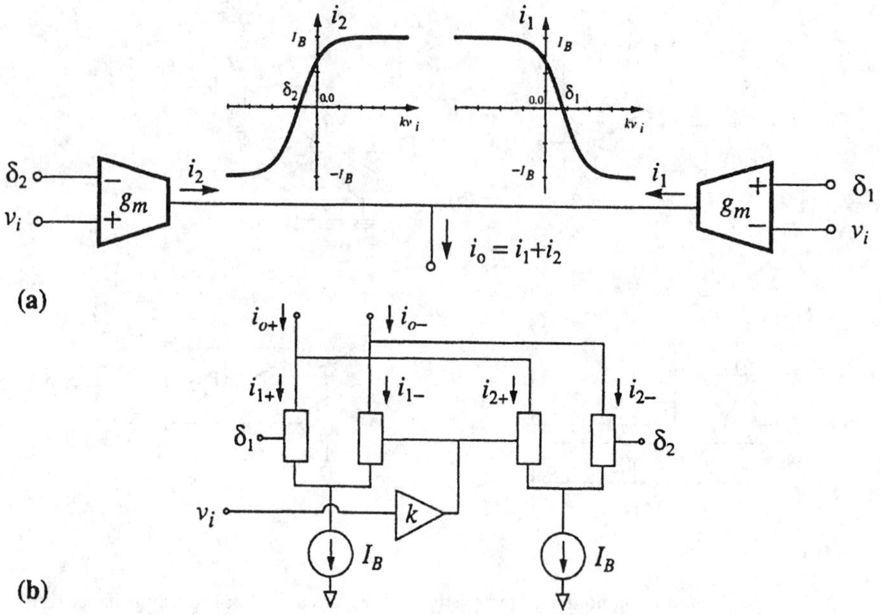

FIGURE 32.37 Transconductance circuits for bell-shaped function: (a) using OTAs; (b) using differential amplifiers.

Collective Computation Circuitry

Radial basis functions and fuzzy inference require multidimensional operators to calculate radial distances in the case of radial basis functions, and to normalize vectors and calculate T-norms in the case of fuzzy inference. These operators can be expressed as the interconnection of the nonlinear blocks discussed previously, or realized in a simpler manner through dedicated collective computation circuitry. Most of these circuits operate intrinsically in current domain and are worth mentioning because of this simplicity and relevance for parallel information processing systems.

Euclidean Distance

Figure 32.38 [38] shows a current-mode circuit to compute

$$i_y = \sqrt{\sum_{k=1.P} i_{xk}^2} \tag{32.43}$$

based on the square-law of MOS transistors in the saturation region. If the current i_k at each terminal is shifted through a bias current of value δ_k, the circuit serves to compute the Euclidean distance between the vector of input currents and the vector δ.

Normalization Operation

Figure 32.39 depicts circuits to normalize an input current vector, for the BJT [23] and the CMOS [74] cases, respectively. Their operation is based on KCL and the current mirror principle. Kirchhoff's circuit law forces the sum of the output currents at node ① to be constant. On the other hand, the current mirror operation forces a functional dependency between each pair of

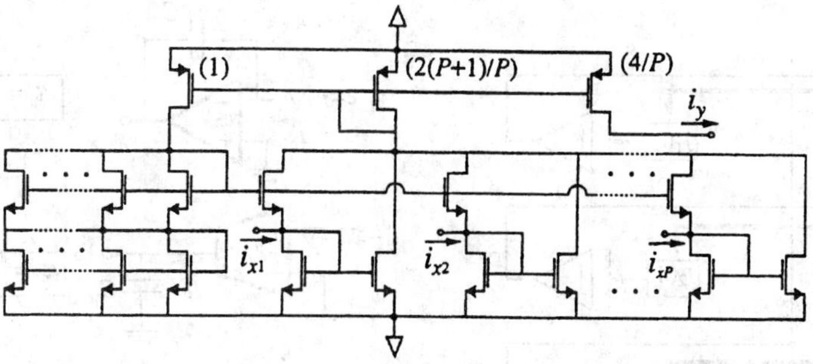

FIGURE 32.38 CMOS self-biased Euclidean distance circuit [38].

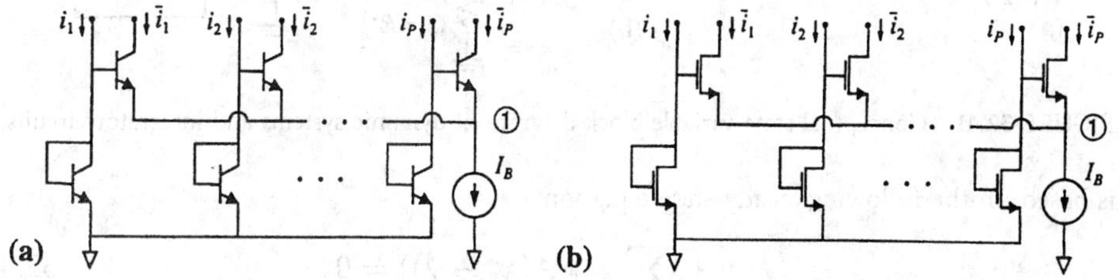

FIGURE 32.39 Current-mode normalization circuits: (a) BJT; (b) CMOS.

input and output currents. Thus, they obtain

$$\bar{i}_k \approx \frac{i_k}{\sum_{j=1,P} i_j} \tag{32.44}$$

for each current component.

T-Norm Operator

The calculation of the minimum of an input vector **x** is functionally equivalent to obtaining the complement of the maximum of the complements of its components. Figure 32.40(a) illustrates a classical approach used in analog computation to calculate the maximum of an input vector **x**.

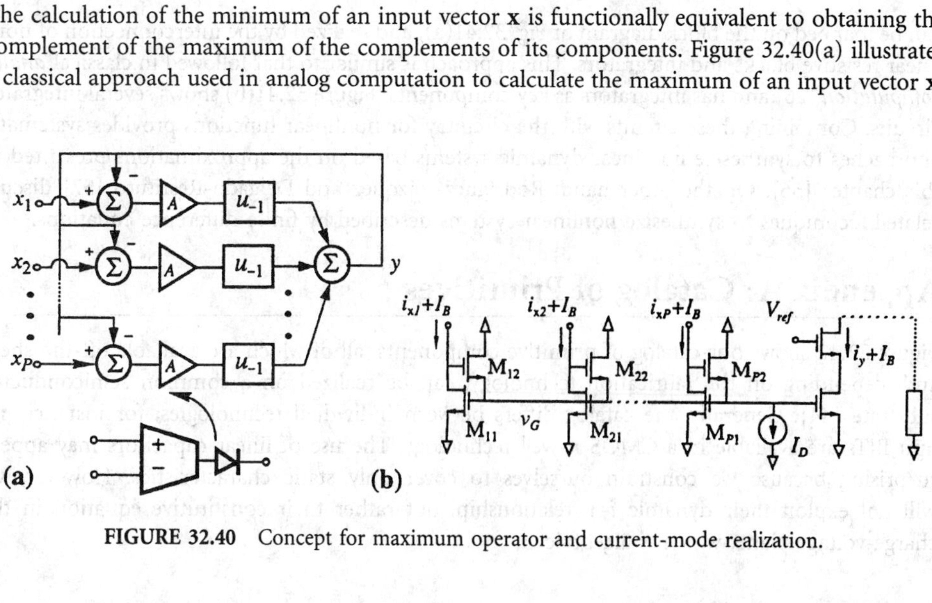

FIGURE 32.40 Concept for maximum operator and current-mode realization.

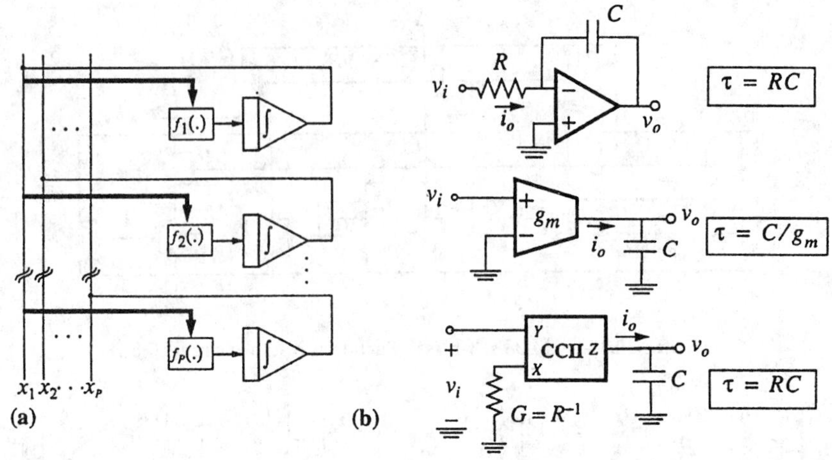

FIGURE 32.41 Conceptual state-variable block diagram of dynamic systems and integrator circuits.

It is based on the following steady-state equation:

$$-y + \sum_{k=1,P} u_{-1}(A(x_k - y)) = 0 \tag{32.45}$$

where A is large. This concept can be realized in practice using OTAs, op amps, or diodes. Both of these have voltage input and output. Alternatively, Fig. 32.40(b) shows a CMOS current-mode realization [74]. In this circuit the maximum current determines the value of the common gate voltage, v_G. The only input transistor operating in the saturation region is that which is driven by maximum input current; the rest operate in the ohmic region.

32.7 Extension to Dynamic Systems

A dynamic system with state vector $\mathbf{x}$ and dynamics represented as

$$\tau_k \frac{dx_k}{dt} = f_k(\mathbf{x}), \quad 1 \le k \le P \tag{32.46}$$

can be mapped on the block diagram of Fig. 32.41(a), and realized by the interconnection of nonlinear resistive blocks and integrators. This approach is similar to that followed in classical *analog computation* [26] and has integrators as key components. Figure 32.41(b) shows several integrator circuits. Combining these circuits with the circuitry for nonlinear functions provides systematic approaches to synthesize nonlinear dynamic systems based on the approximations presented in this chapter [56]. On the other hand, Rodríguez-Vázquez and Delgado-Restituto [57] discuss related techniques to synthesize nonlinear systems described by finite-difference equations.

32.8 Appendix A: Catalog of Primitives

Figure 32.42 shows our catalog of primitive components, all of which are available off-the-shelf, and, depending on the fabrication technology, can be realized on a common semiconductor substrate [44]. Generally, the catalog differs between individual technologies; for instance, no npn BJTs are available in a CMOS n-well technology. The use of linear capacitors may appear surprising because we constrain ourselves to cover only static characteristics. However, we will not exploit their dynamic i–v relationship, but rather their constitutive equation in the charge-voltage plane, which is algebraic.

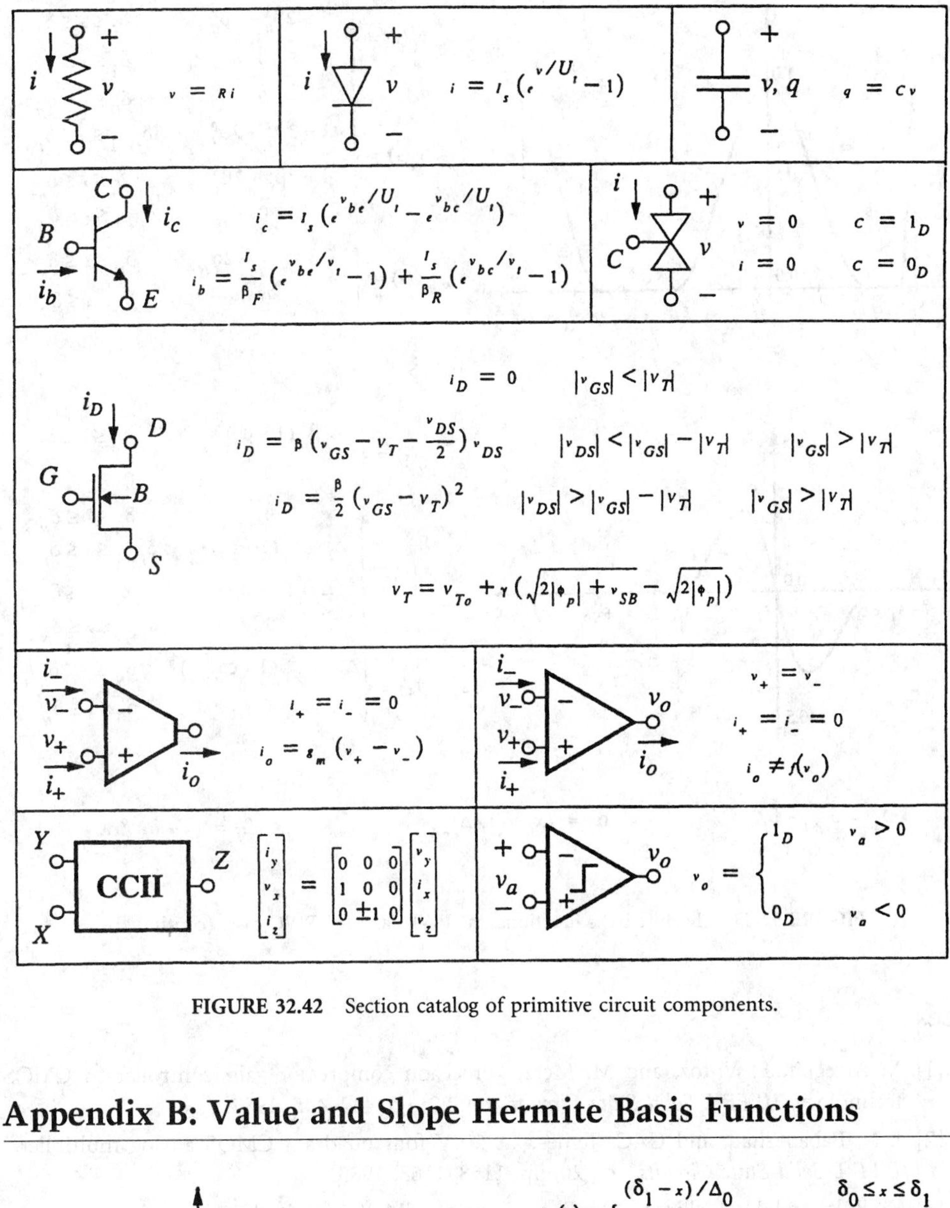

$v = Ri$
$i = I_s(e^{v/U_t} - 1)$
$q = Cv$
$i_c = I_s(e^{v_{be}/U_t} - e^{v_{bc}/U_t})$
$i_b = \frac{I_s}{\beta_F}(e^{v_{be}/v_t} - 1) + \frac{I_s}{\beta_R}(e^{v_{bc}/v_t} - 1)$
$v = 0 \quad c = 1_D$
$i = 0 \quad c = 0_D$
$i_D = 0 \quad |v_{GS}| < |v_T|$
$i_D = \beta(v_{GS} - v_T - \frac{v_{DS}}{2})v_{DS} \quad |v_{DS}| < |v_{GS}| - |v_T| \quad |v_{GS}| > |v_T|$
$i_D = \frac{\beta}{2}(v_{GS} - v_T)^2 \quad |v_{DS}| > |v_{GS}| - |v_T| \quad |v_{GS}| > |v_T|$
$v_T = v_{To} + \gamma(\sqrt{2|\phi_p| + v_{SB}} - \sqrt{2|\phi_p|})$
$i_+ = i_- = 0$
$i_o = g_m(v_+ - v_-)$
$v_+ = v_-$
$i_+ = i_- = 0$
$i_o \neq f(v_o)$
CCII
$v_o = 1_D \quad v_a > 0$; $v_o = 0_D \quad v_a < 0$

FIGURE 32.42 Section catalog of primitive circuit components.

2.9 Appendix B: Value and Slope Hermite Basis Functions

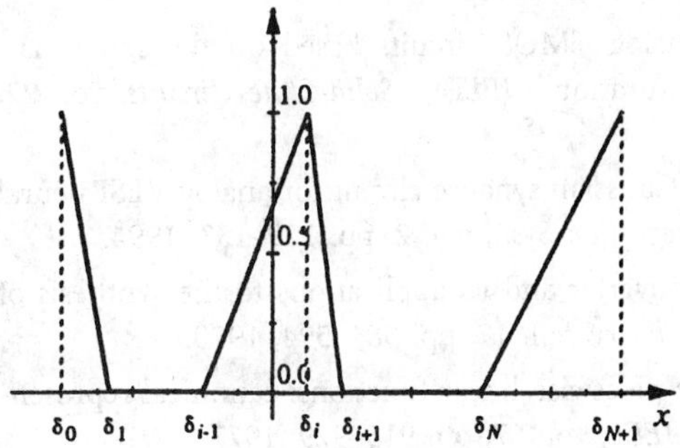

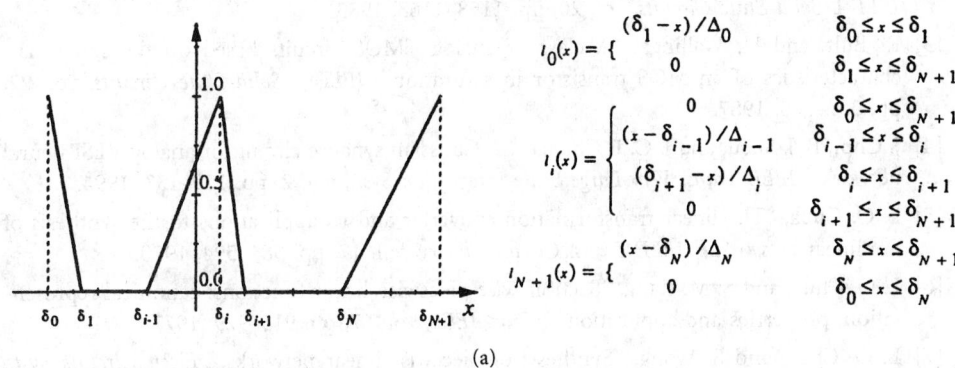

$$l_0(x) = \begin{cases} (\delta_1 - x)/\Delta_0 & \delta_0 \le x \le \delta_1 \\ 0 & \delta_1 \le x \le \delta_{N+1} \end{cases}$$

$$l_i(x) = \begin{cases} 0 & \delta_0 \le x \le \delta_{N+1} \\ (x - \delta_{i-1})/\Delta_{i-1} & \delta_{i-1} \le x \le \delta_i \\ (\delta_{i+1} - x)/\Delta_i & \delta_i \le x \le \delta_{i+1} \\ 0 & \delta_{i+1} \le x \le \delta_{N+1} \end{cases}$$

$$l_{N+1}(x) = \begin{cases} (x - \delta_N)/\Delta_N & \delta_N \le x \le \delta_{N+1} \\ 0 & \delta_0 \le x \le \delta_N \end{cases}$$

(a)

FIGURE 32.43 Hermite basis functions: (a) PWL case; (b) PWC case.

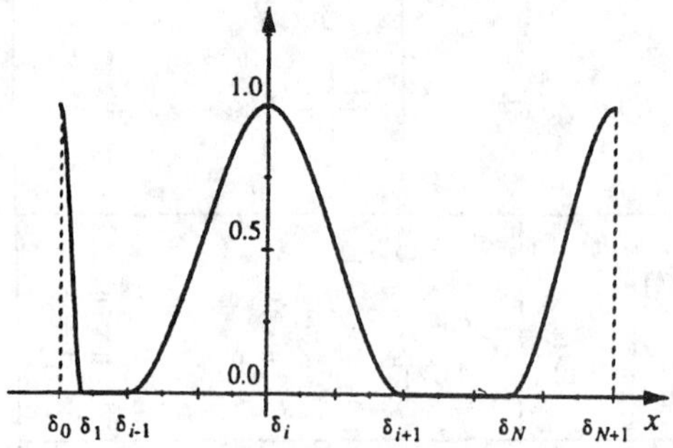

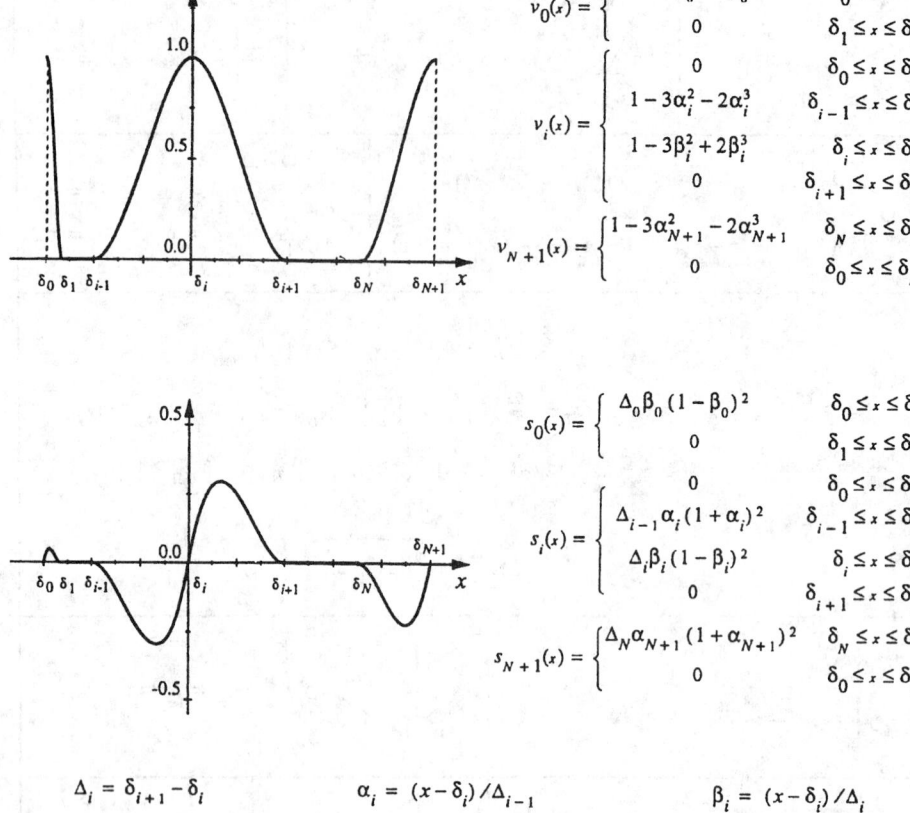

$$v_0(x) = \begin{cases} 1 + 2\beta_0^3 - 3\beta_0^2 & \delta_0 \le x \le \delta_1 \\ 0 & \delta_1 \le x \le \delta_{N+1} \end{cases}$$

$$v_i(x) = \begin{cases} 0 & \delta_0 \le x \le \delta_{N+1} \\ 1 - 3\alpha_i^2 - 2\alpha_i^3 & \delta_{i-1} \le x \le \delta_i \\ 1 - 3\beta_i^2 + 2\beta_i^3 & \delta_i \le x \le \delta_{i+1} \\ 0 & \delta_{i+1} \le x \le \delta_{N+1} \end{cases}$$

$$v_{N+1}(x) = \begin{cases} 1 - 3\alpha_{N+1}^2 - 2\alpha_{N+1}^3 & \delta_N \le x \le \delta_{N+1} \\ 0 & \delta_0 \le x \le \delta_N \end{cases}$$

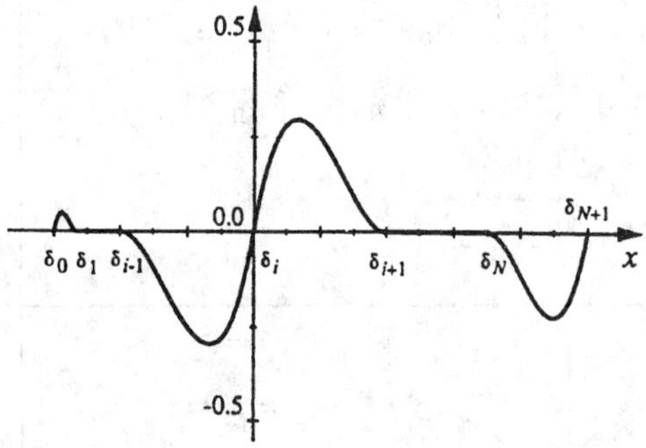

$$s_0(x) = \begin{cases} \Delta_0\beta_0(1 - \beta_0)^2 & \delta_0 \le x \le \delta_1 \\ 0 & \delta_1 \le x \le \delta_{N+1} \end{cases}$$

$$s_i(x) = \begin{cases} 0 & \delta_0 \le x \le \delta_{N+1} \\ \Delta_{i-1}\alpha_i(1 + \alpha_i)^2 & \delta_{i-1} \le x \le \delta_i \\ \Delta_i\beta_i(1 - \beta_i)^2 & \delta_i \le x \le \delta_{i+1} \\ 0 & \delta_{i+1} \le x \le \delta_{N+1} \end{cases}$$

$$s_{N+1}(x) = \begin{cases} \Delta_N\alpha_{N+1}(1 + \alpha_{N+1})^2 & \delta_N \le x \le \delta_{N+1} \\ 0 & \delta_0 \le x \le \delta_N \end{cases}$$

$$\Delta_i = \delta_{i+1} - \delta_i \qquad \alpha_i = (x - \delta_i)/\Delta_{i-1} \qquad \beta_i = (x - \delta_i)/\Delta_i$$

(b)

FIGURE 32.43 Hermite basis functions: (a) PWL case; (b) PWC case. (continued).

References

[1] X. Arreguit, E. Vittoz, and M. Merz, "Precision compressor gain controller in CMOS technology," *IEEE J. Solid-State Circuits*, vol. 22, pp. 442–445, 1987.

[2] J. N. Babanezhad, and G. C. Temes, "A 20-V four quadrant CMOS analog multiplier," *IEEE J. Solid-State Circuits*, vol. 20, pp. 1158–1168, 1985.

[3] K. Bult, and H. Wallinga, "A class of analog CMOS circuits based on the square-law characteristics of an MOS transistor in saturation," *IEEE J. Solid-State Circuits*, vol. 22, pp. 357–365, 1987.

[4] J. Choi, B. J. Sheu, and J. C. F. Change, "A Gaussian synapse circuit for analog VLSI neural network," *IEEE Trans. Very Large Scale Integration Syst.*, vol. 2, pp. 129–133, 1994.

[5] L. O. Chua, "The linear transformation converter and its applications to the synthesis of nonlinear networks," *IEEE Trans. Circuit Theory*, vol. 17, pp. 584–594, 1970.

[6] L. O. Chua, and S. M. Kang, "Section-wise piecewise-linear functions: canonical representation, properties and applications," *Proc. IEEE*, vol. 67, pp. 915–929, 1977.

[7] L. O. Chua, and S. Wong, "Synthesis of piecewise-linear networks," *Elect. Circuits Syst.*, vol. 2, pp. 102–108, 1978.

[8] L. O. Chua, "Nonlinear circuits," *IEEE Trans. Circuits Syst.*, vol. 31, pp. 69–87, 1984.

[9] L. O. Chua, and G. Zhong, "Negative resistance curve tracer," *IEEE Trans. Circuits Syst.*, vol. 32, pp. 569–582, 1985.

[10] L. O. Chua, C. A. Desoer, and E. S. Kuh, *Linear and Nonlinear Circuits*, New York: McGraw-Hill, 1987.

[11] L. O. Chua, and A. C. Deng, "Canonical piecewise linear representation," *IEEE Trans. Circuits Syst.*, vol. 35, pp. 101–111, 1988.

[12] G. Cybenko, "Approximation by superposition of a sigmoidal function," *Math. Control. Syst. Signals*, vol. 2, pp. 303–314, 1989.

[13] Z. Czarnul, "Modification of Banu-Tsividis continuous-time integrator structure," *IEEE Trans. Circuits Syst.*, vol. 33, pp. 714–716, 1986.

[14] R. W. Daniels, "Approximation Methods for Electronic Filter Design," New York: McGraw-Hill, 1974.

[15] M. Delgado-Restituto, and A. Rodríguez-Vázquez, "Switched-current chaotic neutrons," *Electr. Lett.*, vol. 30, pp. 429–430, 1994.

[16] M. Delgado-Restituto, A. Rodríguez-Vázquez, S. Espejo, and J. L. Huertas, "A chaotic switched-capacitor circuit for $1/f$ noise generation," *IEEE Trans. Circuits Syst.*, vol. 39, pp. 325–328, 1992.

[17] S. T. Dupuie, and M. Ismail, "High-frequency CMOS transconductors," in *Analogue IC Design: The Current-Mode Approach*, C. Toumazou, F. J. Lidgey, and D. G. Haigh, (Eds.), London: Peter Peregrinus Ltd., 1990.

[18] T. Enomoto, and M. Yasumoto, "Integrated MOS four-quadrant analog multiplier using switched capacitor technology for analog signal processor IC's," *IEEE J. Solid-State Circuits*, vol. 20, pp. 852–859, 1985.

[19] J. W. Fattaruso, and R. G. Meyer, "MOS analog function synthesis," *IEEE J. Solid-State Circuits*, vol. 22, pp. 1059–1063, 1987.

[20] I. M. Filanovsky, and H. Baltes, "CMOS Schmitt trigger design," *IEEE Trans. Circuits Syst.*, vol. 41, pp. 46–49, 1994.

[21] K. Funahashi, "On the approximate realization of continuous mappings by neural networks," *Neural Networks*, vol. 2, pp. 183–192, 1989.

[22] R. L. Geiger, and E. Sánchez-Sinencio, "Active filter design using operational transconductance amplifiers: a tutorial," *IEEE Circuits Devices Mag.*, vol. 1, pp. 20–32, 1985.

[23] B. Gilbert, "Current-mode circuits from a translinear view point: a tutorial," in *Analogue IC Design: The Current-Mode Approach*, C. Toumazou, F. J. Lidgey, and D. G. Haigh, (Eds.), London: Peter Peregrinus Ltd., 1990.

[24] J. Glover, "Basic T matrix patterns for 2-port linear transformation networks in the real domain," *IEEE Trans. Circuit Theory*, vol. 10, pp. 495–497, 1974.

[25] P. R. Gray, and R. G. Meyer, *Analysis and Design of Analog Integrated Circuits*, 3rd ed., New York: John Wiley, 1993.

[26] A. Hausner, *Analog and Analog/Hybrid Computer Programming*, Englewood Cliffs, NJ: Prentice-Hall, 1971.

[27] B. J. Hosticka, W. Brockherde, U. Kleine, and R. Schweer, "Design of nonlinear analog switched-capacitor circuits using building blocks," *IEEE Trans. Circuits Syst.*, vol. 31, pp. 354–368, 1984.

[28] J. L. Huertas, J. A. Acha, and A. Gago, "Design of general voltage or current controlled resistive elements and their applications to the synthesis of nonlinear networks," *IEEE Trans. Circuits Syst.*, vol. 27, pp. 92–103, 1980.

[29] J. L. Huertas, "DT-adaptor: applications to the design of nonlinear n-ports," *Int. J. Circuit Theory Appl.*, vol. 8, pp. 273–290, 1980.

[30] J. L. Huertas, and A. Rueda, "Sectionwise piecewise polynomial functions: applications to the analysis and synthesis on nonlinear n-port Networks," *IEEE Trans. Circuits Syst.*, vol. 31, pp. 897–905, 1984.

[31] J. L. Huertas, L. O. Chua, A. Rodríguez-Vázquez, and A. Rueda, "Nonlinear switched-capacitor networks: basic principles and piecewise-linear design," *IEEE Trans. Circuits Syst.*, vol. 32, pp. 305–319, 1985.

[32] M. Ismail, "Four-transistor continuous-time MOS transconductors," *Electr. Lett.*, vol. 23, pp. 1099–1100, 1987.

[33] J. S. Jang, Neuro-Fuzzy Modeling: Architectures, Analyses and Applications, Ph.D. dissertation, University of California, Berkeley, 1992.

[34] C. Jansson, K. Chen, and C. Svensson, "Linear, polynomial and exponential ramp generators with automatic slope adjustment," *IEEE Trans. Circuits Syst.*, vol. 41, pp. 181–185, 1994.

[35] N. I. Khachab, and M. Ismail, "Linearization techniques for nth-order sensor models in MOS VLSI technology," *IEEE Trans. Circuits Syst.*, vol. 38, pp. 1439–1449, 1991.

[36] Y. H. Kim, and S. B. Park, "Four-quadrant CMOS analogue multiplier," *Electr. Lett.*, vol. 28, pp. 649–650, 1992.

[37] K. Kimura, "Some circuit design techniques using two cross-coupled, emitter-coupled pairs," *IEEE Trans. Circuits Syst.*, vol. 41, pp. 411–423, 1994.

[38] O. Landolt, E. Vittoz, and P. Heim, "CMOS selfbiased Euclidean distance computing circuit with high dynamic range," *Electr. Lett.*, vol. 28, pp. 352–354, 1992.

[39] N. R. Malik, G. L. Jackson, and Y. S. Kim, "Theory and applications of resistor, linear controlled resistor, linear controlled conductor networks," *IEEE Trans. Circuits Syst.*, vol. 31, pp. 222–228, 1976.

[40] A. I. Mees, M. F. Jackson, and L. O. Chua, "Device modeling by radial basis functions," *IEEE Trans. Circuits Syst.*, vol. 39, pp. 19–27, 1992.

[41] C. Michael, and M. Ismail, *Statistical Modeling for Computer-Aided Design of MOS VLSI Circuits*, Boston: Kluwer Academic, 1993.

[42] C. A. Micchelli, "Interpolation of scattered data: distance matrices and conditionally positive definite functions," *Constr. Approx.*, vol. 2, pp. 11–22, 1986.

[43] J. Moody, and C. Darken, "Fast learning in networks of locally-tuned processing units," *Neural Comput.*, vol. 1, pp. 281–294, 1989.

[44] R. S. Muller, and T. I. Kamins, *Device Electronics for Integrated Circuits*, New York: John Wiley & Sons, 1986.

[45] K. Nagaraj, K. Singhal, T. R. Viswanathan, and J. Vlach, "Reduction of finite-gain effect in switched-capacitor filters," *Electr. Lett.*, vol. 21, pp. 644–645, 1985.

[46] K. Nagaraj, "A parasitic-insensitive area-efficient approach to realizing very large time constants in switched-capacitor circuits," *IEEE Trans. Circuits Syst.*, vol. 36, pp. 1210–1216, 1989.

[47] R. W. Newcomb, "Nonlinear differential systems: a canonic multivariable theory," *Proc. IEEE*, vol. 65, pp. 930–935, 1977.

[48] M. J. M. Pelgrom, A. C. J. Duinmaijer, and A. P. G. Welbers, "Matching properties of MOS transistors," *IEEE J. Solid-State Circuits*, vol. 24, pp. 1433–1440, 1989.

[49] J. S. Peña-Finol, and J. A. Connelly, "A MOS four-quadrant analog multiplier using the quarter-square technique," *IEEE J. Solid-State Circuits*, vol. 22, pp. 1064–1073, 1987.

[50] J. Platt, "A resource allocating neural network for function interpolation," *Neural Comput.*, vol. 3, pp. 213–215, 1991.

[51] T. Poggio, and F. Girosi, "Networks for approximation and learning," *Proc. IEEE*, vol. 78, pp. 1481–1497, 1990.

[52] M. J. D. Powell, "Radial basis functions for multivariate interpolation: a review," Rep. 1985/NA12, Dept. Applied Mathematics and Theoretical Physics, Cambridge University, Cambridge, U.K., 1985.

[53] S. Qin, and R. L. Geiger, "A 5-V CMOS analog multiplier," *IEEE J. Solid-State Circuits*, vol. 22, pp. 1143–1146, 1987.

[54] J. K. Roberge, *Operational Amplifiers: Theory and Practice*, New York: John Wiley, 1975.

[55] A. Rodríguez-Vázquez, J. L. Huertas, and A. Rueda, "Low-order polynomial curve fitting using switched-capacitor circuits," *Proc. IEEE Int. Symp. Circuits Syst.*, May, pp. 1123–1125, 1984.

[56] A. Rodríguez-Vázquez, and M. Delgado-Restituto, "CMOS design of chaotic oscillators using state variables: a monolithic Chua's circuit," *IEEE Trans. Circuits Syst.*, vol. 40, pp. 596–613, 1993.

[57] A. Rodríguez-Vázquez, and M. Delgado-Restituto, "Generation of chaotic signals using current-mode techniques," *J. Intelligent Fuzzy Syst.*, vol. 2, pp. 15–37, 1994.

[58] A. Rodríguez-Vázquez, R. Domínguez-Castro, F. Medeiro, and M. Delgado-Restituto, "High-resolution CMOS current comparators: design and applications to current-mode function generation," *Analog Integrated Circuits Signal Process.*, in press.

[59] A. Rueda, J. L. Huertas, and A. Rodríguez-Vázquez, "Basic circuit structures for the synthesis of piecewise polynomial one-port resistors," *IEEE Proc.*, vol. 132 (Part G), pp. 123–129, 1985.

[60] S. Sakurai, and M. Ismail, "High frequency wide range CMOS analogue multiplier," *Elect. Lett.*, vol. 28, pp. 2228–2229, 1992.

[61] E. Sánchez-Sinencio, J. Ramírez-Angulo, B. Linares-Barranco, and A. Rodríguez-Vázquez, "Operational transconductance amplifier-based nonlinear function synthesis," *IEEE J. Solid-State Circuits*, vol. 24, pp. 1576–1586, 1989.

[62] E. Seevinck, *Analysis and Synthesis of Translinear Integrated Circuits*, Amsterdam: Elsevier, 1988.

[63] D. H. Sheingold, *Nonlinear Circuits Handbook*, Norwood, CA: Analog Devices Inc., 1976.

[64] J. Silva-Martínez, and E. Sánchez-Sinencio, "Analogue OTA multiplier without input voltage swing restrictions, and temperature compensated," *Electr. Lett.*, vol. 22, pp. 559–560, 1986.

[65] S. Soclof, *Analog Integrated Circuits*, Englewood Cliffs, NJ: Prentice-Hall, 1985.

[66] B. Song, "CMOS RF circuits for data communication applications," *IEEE J. Solid-State Circuits*, vol. 21, pp. 310–317, 1986.

[67] L. Strauss, *Wave Generation and Shaping*, New York: McGraw-Hill, 1970.

[68] G. C. Temes, and K. Haug, "Improved offset compensation schemes for switched-capacitor circuits," *Electr. Lett.*, vol. 20, pp. 508–509, 1984.

[69] Y. P. Tsividis, M. Banu, and J. Khoury, "Continuous-time MOSFET-C filters in VLSI," *IEEE J. Solid-State Circuits*, vol. 21, pp. 15–30, 1986.

[70] Y. P. Tsividis, and J. O. Voorman, (Eds.), *Integrated Continuous-Time Filters*, New York: IEEE Press, 1993.

[71] C. Turchetti, and M. Conti, "A general approach to nonlinear synthesis with MOS analog circuits," *IEEE Trans. Circuits Syst.*, vol. 40, pp. 608–612, 1993.

[72] R. Unbehauen, and A. Cichocki, *MOS Switched-Capacitor and Continuous-Time Integrated Circuits and Systems*, Berlin Springer-Verlag, 1989.

[73] G. N. Vanderplaats, *Numerical Optimization Techniques for Engineering Design: With Applications*, New York: McGraw-Hill, 1984.

[74] F. Vidal, and A. Rodríguez-Vázquez, "A basic building block approach to CMOS design of analog neuro/fuzzy systems," *Proc. IEEE Int. Conf. Fuzzy Systems*, vol. 1, pp. 118–123, 1994.

[75] E. Vittoz, "Analog VLSI signal processing: why, where and how?," *Analog Integrated Circuits Signal Process.*, vol. 6, pp. 27–44, 1994.

[76] J. Vlach, and K. Singhal, *Computer Methods for Circuit Analysis and Design*, 2nd ed., New York: Van Nostrand Reinhold, 1994.

[77] Z. Wang, "Analytical determination of output resistance and DC matching errors in MOS current mirrors," *IEEE Proc.*, vol. 137 (Part G), pp. 397–404, 1990.

[78] G. A. Watson, *Approximation Theory and Numerical Methods*, Chichester, U.K.: John Wiley, 1980.

[79] R. Wiegerink, *Analysis and Synthesis of MOS Translinear Circuits*, Boston: Kluwer Academic, 1993.

[80] Y. J. Wong, and W. E. Ott, *Function Circuits; Design and Applications*, New York: McGraw-Hill, 1976.

[81] T. Yamakawa, "A fuzzy inference engineer in nonlinear analog mode and its application to fuzzy control," *IEEE Trans. Neural Networks*, vol. 4, pp. 496–522, 1993.

[82] J. M. Zurada, *Introduction to Artificial Neural Systems*, St. Paul, MN: West Publishing, 1992.

33

Representation, Approximation, and Identification

ıanrong Chen
iversity of Houston

3.1 Introduction

Representation, approximation, and identification of physical systems, linear or nonlinear, deterministic or random, or even chaotic, are three fundamental issues in systems theory and engineering. To describe a physical system, such as a circuit or a microprocessor, we need a mathematical formula or equation that can represent the system both qualitatively and quantitatively. Such a formulation is what we call a mathematical representation of the physical system. If the physical system is so simple that the mathematical formula or equation, or the like, can describe it perfectly without error, then the representation is ideal and ready to use for analysis, computation, and synthesis of the system. An ideal representation of a real system is generally impossible, so that system approximation becomes necessary in practice. Intuitively, approximation is always possible. However, the key questions are what kind of approximation is good, where the sense of "goodness" must first be defined, of course, and how to find such a good approximation. On the other hand, when looking for either an ideal or an approximate mathematical representation for a physical system, one must know the system structure (the form of the linearity or nonlinearity) and parameters (their values). If some of these are unknown, then one must identify them, leading to the problem of system identification.

This chapter is devoted to a brief description of mathematical representation, approximation, and identification of, in most cases, nonlinear systems. As usual, a linear system is considered to be a special case of a nonlinear system, but we do not focus on linear systems in this chapter on nonlinear circuits. It is known that a signal, continuous or discrete, is represented by a

-8341-2/95/$0.00 + $.50

function of time. Hence, a signal can be approximated by other functions and also may be identified using its sampled data. These are within the context of "representation, approximation and identification", but at a lower level—one is dealing with functions. A system, in contrast, transforms input signals to output signals, namely, maps functions to functions, and is therefore at a higher level—it can only be represented by an operator (i.e., a mapping). Hence, while talking about representation, approximation, and identification in this chapter, we essentially refer to operators. However, we notice that two systems are considered to be equivalent over a set of input signals if and only if (iff) they map the same input signal from the set to the same output signal, regardless of the distinct structures of the two systems. From this point of view, one system is a good approximation of the other if the same input produces outputs that are approximately the same under certain measure. For this reason, we also briefly discuss the classical function approximation theory in this chapter.

The issue of system representation is addressed in Section 33.2, while approximation (for both operators and functions) is discussed in Section 33.3, leaving the system identification problem to Section 33.4. Limited by space, we can discuss only deterministic systems. Topics on stochastic systems are hence referred to some standard textbooks (e.g., [13, 17]).

It is impossible to cover all the important subjects and to mention many significant results in the field in this short and sketchy chapter. The selections made only touch upon the very elementary theories, commonly used methods, and basic results related to the central topics of the chapter, reflecting the author's personal preference. In order to simplify the presentation, we elected to cite only those closely related references known to us, which may or may not be the original sources. From our citations, the reader should be able to find more references for further reading.

33.2 Representation

The scientific term "representation" as used here refers to a mathematical description of a physical system. The fundamental issue in representing a physical system by a mathematical formulation, called a **mathematical model**, is its correct symbolization, accurate quantization, and strong ability to illustrate and reproduce important properties of the original system.

A circuit consisting of some capacitor(s), inductor(s), and/or resistor(s), and possibly driven by a voltage source or a current source, is a physical system. In order to describe this system mathematically for the purpose of analysis, design, and/or synthesis, a mathematical model is needed. Any mathematical model, which can correctly describe the physical behavior of the circuit, is considered a **mathematical representation** of the circuit. A lower level mathematical representation of a circuit can, for instance, be a signal flow chart or a circuit diagram like the nonlinear Chua's circuit shown in Fig. 33.1, which is discussed below.

A circuit, such as that shown in Fig. 33.1, can be used to describe a physical system, including its components and its internal as well as external connections. However, it is not convenient for carrying out theoretical analysis or numerical computations. This is because no qualitative or quantitative description exists about the relations among the circuit elements and their dynamic behavior. Hence, a higher level mathematical model is needed to provide a qualitative and quantitative representation of the real physical circuit.

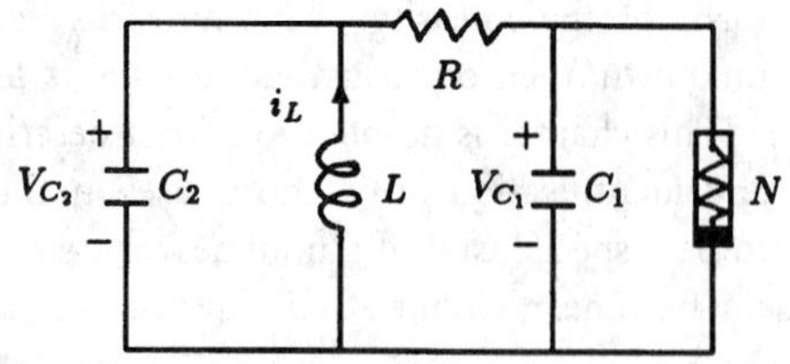

FIGURE 33.1 Chua's circuit.

Among several commonly used mathematical modeling approaches for various physical systems, differential equations, state-space formulations, I-O mappings, and functional series (particularly, the Volterra series) are the most important and useful, which have been very popular in the field of circuits and systems engineering. In the following we introduce these mathematical representation methods, along with some brief discussions of other related issues. Limited by space, detailed derivations are omitted.

Differential Equation and State-Space Representations

Mathematical modeling via differential equations and via state-space descriptions are the most basic mathematical representation methods. We illustrate the concept of mathematical modeling and the two representation methods by a simple, yet representative example: the nonlinear circuit shown in Fig. 33.1. This circuit consists of one inductor L, two capacitors C_1 and C_2, one linear resistor R, and one nonlinear resistor, N, which is a nonlinear function of the voltage across its two terminals: $N = N(V_{C_1}(t))$. Let $i_L(t)$ be the current through the inductor L, and $V_{C_1}(t)$ and $V_{C_2}(t)$ be the voltages across C_1 and C_2, respectively. For the time being, let us remove the nonlinear resistor N from Fig. 33.1 and consider the remaining linear circuit. This nonlinear resistor N is readded to the circuit with detailed discussions in (33.6).

For this linear circuit without the resistor N, it follows from Kirchhoff's laws that

$$C_1 \frac{d}{dt} V_{C_1}(t) = \frac{1}{R}[V_{C_2}(t) - V_{C_1}(t)] \tag{33.1}$$

$$C_2 \frac{d}{dt} V_{C_2}(t) = \frac{1}{R}[V_{C_1}(t) - V_{C_2}(t)] + i_L(t) \tag{33.2}$$

$$L \frac{d}{dt} i_L(t) = -V_{C_2}(t) \tag{33.3}$$

By simple calculation we can eliminate both V_{C_2} and i_L, leaving a single ordinary differential equation on the unknown voltage V_{C_1} as follows:

$$\begin{aligned} &\frac{d^3}{dt^3} V_{C_1}(t) + \frac{1}{R}\left(\frac{1}{C_1} + \frac{1}{C_2}\right) \frac{d^2}{dt^2} V_{C_1}(t) \\ &\quad + \frac{1}{C_2 L} \frac{d}{dt} V_{C_1}(t) + \frac{1}{C_1 C_2 R L} V_{C_1}(t) = 0 \end{aligned} \tag{33.4}$$

Once V_{C_1} is obtained from (33.4), based on certain initial conditions, the other two unknowns, V_{C_2} and i_L, can be obtained by using (33.1) and (33.3), successively. Hence, this third-order ordinary differential equation describes both qualitatively and quantitatively the circuit shown in Fig. 33.1 (without the nonlinear resistor N). For this reason, (33.4) is considered to be a mathematical representation, called a **differential equation representation**, of the physical linear circuit.

Very often, a higher-order single-variable ordinary differential equation similar to (33.4) is not as convenient as a first-order multivariable system of ordinary differential equations as is the original system of (33.1) to (33.3), even when an analytic formulation of the solution is desired. Hence, a more suitable way for modeling a physical system is to introduce the concept of system state variables, which leads to a first-order higher dimensional system of ordinary differential equations.

If we introduce three **state variables** in (33.1) to (33.3):

$$x_1(t) = V_{C_1}(t) \quad x_2(t) = V_{C_2}(t) \quad x_3(t) = i_L(t)$$

then we can rewrite those equations in the following vector form:

$$\begin{cases} \dot{\mathbf{x}}(t) = A\mathbf{x}(t) + B\mathbf{u}(t) & t \geq 0 \\ \mathbf{x}(0) = \mathbf{x}_0 \end{cases} \tag{33.5}$$

with an initial value $\mathbf{x}_0$ (usually given), where

$$\mathbf{x}(t) = \begin{bmatrix} x_1(t) \\ x_2(t) \\ x_3(t) \end{bmatrix} \quad \text{and} \quad A = \begin{bmatrix} -\frac{1}{RC_1} & \frac{1}{RC_1} & 0 \\ \frac{1}{RC_2} & -\frac{1}{RC_2} & \frac{1}{C_2} \\ 0 & -\frac{1}{L} & 0 \end{bmatrix}$$

in which $\mathbf{x}(t)$ is called the **state vector** of the system. Here, to be more general and for convenience in the discussions following, we formally added the term $B\mathbf{u}(t)$ to the system, in which B is a constant matrix and $\mathbf{u}(t)$ is called the **control input** of the system. In the present case, of course, $\mathbf{u} = 0$ and it is not important to specify B. However, note that $\mathbf{u}$ can be a nonzero external input to the circuit [19], which is discussed in more detail below.

This first-order vector-valued linear ordinary differential equation is equivalent to the third-order differential equation representation, (33.4), of the same physical circuit. A special feature of this state vector formulation is that with different initial state vectors and with zero control inputs, all the possible system state vectors together constitute a linear space of the same dimension [31]. Hence, (33.5) is also called a **linear state-space representation** (or, a **linear state-space description**) for the circuit.

A few important remarks are in order.

First, if the circuit is nonlinear, its state vectors do not constitute a linear space in general. Hence, its mathematical model in the state vector form should not be called a "state-space" representation. Note, however, that some of the linear system terminology such as state variables and state vectors usually make physical sense for nonlinear systems. Therefore, we use the term **nonlinear state-variable representation** to describe a first-order vector-valued nonlinear ordinary differential equation of the form $\dot{\mathbf{x}}(t) = \mathbf{f}(\mathbf{x}(t), \mathbf{u}(t), t)$, where $\mathbf{f}(\cdot, \cdot, t)$ is generally a vector-valued nonlinear function. This is illustrated in more detail shortly.

Second, a linear state-space representation for a given physical system is not unique because one can choose different state variables. For example, in (33.1) to (33.3) if we instead define $x_1 = V_{C_2}$ and $x_2 = V_{C_1}$, we arrive at a different linear state-space representation of the same circuit. However, we should note that if a linear nonsingular transformation of state vectors can map one state-space representation to another, then these two seemingly different representations are actually equivalent in the sense that the same initial values and control inputs will generate the same outputs (perhaps in different forms) through these two representations. Also worth noting is that not every circuit element can be used as a state variable, particularly for nonlinear systems. A basic requirement is that all the chosen state variables must be "linearly independent" in that the first-order vector-valued ordinary differential equation has a unique solution (in terms of the control input) for any given initial values of the chosen state variables.

Finally, because A and B in the state-space representation (33.5) are both constant (independent of time), the representation is called a **linear time-invariant system.** If A or B is a matrix-valued function of time, then it will be called a **linear time-varying system.** Clearly, a time-invariant system is a special case of a time-varying system.

Now, let us return to the nonlinear circuit, with the nonlinear resistor N being connected to the circuit, as shown in Fig. 33.1. Similar to (33.1) to (33.3), we have the following circuit equations:

$$C_1 \frac{d}{dt} V_{C_1}(t) = \frac{1}{R}[V_{C_2}(t) - V_{C_1}(t)] - N(V_{C_1}(t)) \tag{33.6}$$

$$C_2 \frac{d}{dt} V_{C_2}(t) = \frac{1}{R}[V_{C_1}(t) - V_{C_2}(t)] + i_L(t) \tag{33.7}$$

$$L \frac{d}{dt} i_L(t) = -V_{C_2}(t) \tag{33.8}$$

Note that if the nonlinear resistor N is given by

$$\begin{aligned} N(V_{C_1}(t)) &= N(V_{C_1}(t); m_0, m_1) \\ &= m_0 V_{C_1}(t) + \tfrac{1}{2}(m_1 - m_0)(|V_{C_1}(t) + 1| - |V_{C_1}(t) - 1|) \end{aligned} \tag{33.9}$$

with $m_0 < 0$ and $m_1 < 0$ being two appropriately chosen constant parameters, then this nonlinear circuit is the well-known Chua's circuit [24].

It is clear that compared to the linear case, it would be rather difficult to eliminate two unknowns, particularly V_{C_1}, in order to obtain a simple third-order nonlinear differential equation that describes the nonlinear circuit. That is, it would often be inconvenient to use a higher-order single-variable differential equation representation for a nonlinear physical system in general. By introducing suitable state variables, however, one can easily obtain a nonlinear state-variable representation in a first-order vector-valued nonlinear differential equation form. For instance, we may choose the following state variables:

$$x(\tau) = V_{C_1}(t) \quad y(\tau) = V_{C_2}(t) \quad \text{and} \quad z(\tau) = Ri_L(t) \quad \text{with} \quad \tau = t/RC_2$$

where the new variable $z(\tau) = Ri_L(t)$ and the rescaled time variable $\tau = t/RC_2$ are introduced to simplify the resulting representation of this particular circuit. Under this nonsingular linear transform, the above circuit equations are converted to the following state-variable representation:

$$\begin{cases} \dot{x}(\tau) = p[-x(\tau) + y(\tau) - \tilde{N}(x(\tau))] \\ \dot{y}(\tau) = x(\tau) - y(\tau) + z(\tau) \\ \dot{z}(\tau) = -qy(\tau) \end{cases} \tag{33.10}$$

where $p = C_2/C_1, q = R^2 C_2/L$, and

$$\begin{aligned} \tilde{N}(x(\tau)) &= N(x(\tau); \tilde{m}_0, \tilde{m}_1) \\ &= \tilde{m}_0 x(\tau) + \tfrac{1}{2}(\tilde{m}_1 - \tilde{m}_0)(|x(\tau) + 1| - |x(\tau) - 1|) \end{aligned} \tag{33.11}$$

with $\tilde{m}_0 = Rm_0$ and $\tilde{m}_1 = Rm_1$.

It is easy to see that this state-variable representation can be written as a special case in the following form, known as a **canonical representation** of Chua's circuit family:

$$\dot{\mathbf{x}}(\tau) = \mathbf{a} + A\mathbf{x}(\tau) + \sum_{i=1}^{k} |\mathbf{h}_i^{\top}\mathbf{x}(\tau) - \beta_i| \mathbf{c}_i + B\mathbf{u}(\tau) \tag{33.12}$$

namely, with $\mathbf{a} = 0$, $k = 2$, $\mathbf{h}_1 = \mathbf{h}_2 = [1\ 0\ 0]^T$, $\beta_1 = -\beta_2 = -1$, $\mathbf{c}_1 = -\mathbf{c}_2 = \frac{1}{2}(\tilde{m}_1 - \tilde{m}_0)$, $B\mathbf{u}(\tau)$ being a possible control input to the circuit [19], and

$$A = \begin{bmatrix} -\tilde{m}_0 - p & p & 0 \\ 1 & -1 & 1 \\ 0 & -q & 0 \end{bmatrix}$$

The canonical (piecewise-linear) representation given by (33.12) describes a large class of circuits that have very rich nonlinear dynamics [25, 55].

Now we return to (33.6) to (33.8) and Fig. 33.1. If we replace the L-C_2 part of Chua's circuit by a lossless transmission line (with the spatial variable ξ) of length l terminated on its left-hand side (at $\xi = 0$) by a short circuit, as shown in Fig. 33.2, then we obtain a time-delayed Chua's circuit [89]. This circuit has a **partial differential equation representation** of the form

$$\begin{cases} \partial v(\xi,t)/\partial\xi = -L\partial i(\xi,t)/\partial t \\ \partial i(\xi,t)/\partial\xi = -C_1\partial v(\xi,t)/\partial t \\ \quad v(0,t) = 0 \\ \quad i(l,t) = N(v(l,t) - e - Ri(l,t)) + C_1\partial[v(l,t) - Ri(l,t)]/\partial t \end{cases} \tag{33.13}$$

where $v(\xi,t)$ and $i(\xi,t)$ are the voltage and current, respectively, at the point $\xi \in [0,l]$ at time t, and $V_{C_1} = e > 0$ is a constant, with the nonlinear resistor N satisfying

$$N(V_{C_1} - e) = \begin{cases} m_0(V_{C_1} - e) & |V_{C_1} - e| < 1 \\ m_1(V_{C_1} - e) - (m_1 - m_0)\mathrm{sgn}(V_{C_1} - e) & |V_{C_1} - e| \geq 1 \end{cases}$$

In general, systems that are described by (linear or nonlinear) partial differential equations, with initial-boundary value conditions, are studied under a unified framework of (linear or nonlinear) operator semigroup theory, and are considered to have an **infinite-dimensional system representation** [7].

Input-Output Representation

A state-variable representation of a nonlinear physical system generally can be written as

$$\begin{cases} \dot{\mathbf{x}}(t) = \mathbf{f}(\mathbf{x}(t), \mathbf{u}(t), t) & t \geq 0 \\ \mathbf{x}(0) = \mathbf{x}_0 \end{cases} \tag{33.14}$$

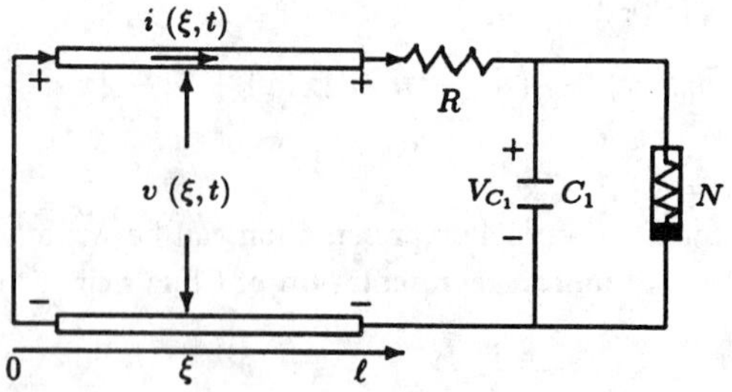

FIGURE 33.2 Time-delayed Chua's circuit.

where $\mathbf{f}(\cdot,\cdot,t)$ is a nonlinear vector-valued function, $\mathbf{x}_0$ a (given) initial value for the state vector $\mathbf{x}$ at $t=0$, and $\mathbf{u}$ a control input to the system.

Because not all state variables in the state vector $\mathbf{x}$ can be measured (observed) in a physical system, let us suppose that what can be measured is only part of $\mathbf{x}$, or a mixture of its components, expressed by a vector-valued function of $\mathbf{x}$ in the form

$$\mathbf{y}(t) = \mathbf{g}(\mathbf{x}(t), t) \qquad t \geq 0 \tag{33.15}$$

where $\mathbf{y}$ is called a (**measurement** or **observation**) **output** of the physical system, and $\mathbf{g}$ is in general a lower dimensional vector-valued nonlinear function. As a particular case, $\mathbf{g}$ can be linear or, even more so, be $\mathbf{g}(\mathbf{x}(t), t) = \mathbf{x}(t)$ when all the components of the state vector are directly measurable.

If both $\mathbf{f} = \mathbf{f}(\mathbf{x}(t), \mathbf{u}(t))$ and $\mathbf{g} = \mathbf{g}(\mathbf{x}(t))$ are not explicit functions of the independent time variable t, the corresponding state-variable representation (33.14) and (33.15) is said to be **autonomous.**

It is clear that with both the system input $\mathbf{u}$ and output $\mathbf{y}$, one can simply represent the overall physical system by its input-output (I-O) relationship, as illustrated in Fig. 33.3.

Now, under certain mild conditions on the nonlinear function $\mathbf{f}$, for a given control input $\mathbf{u}$ and an initial value $\mathbf{x}_0$, the state-variable representation (33.14) has a unique solution, $\mathbf{x}$, which depends on both $\mathbf{u}$ and $\mathbf{x}_0$. If we denote the solution as

$$\mathbf{x}(t) = \mathcal{F}(t; \mathbf{u}(t), \mathbf{x}_0) \tag{33.16}$$

where $\mathcal{F}$ is called an **input-state mapping**, then the overall I-O relationship shown in Fig. 33.3 can be formulated as

$$\mathbf{y}(t) = \mathbf{g}(\mathcal{F}(t; \mathbf{u}(t), \mathbf{x}_0), t) \tag{33.17}$$

This is an **I-O representation** of the physical system having the state-variable representation (33.14) and (33.15).

As a simple example, let us consider the linear state-space representation (33.5), with a special linear measurement equation of the form $\mathbf{y}(t) = C\mathbf{x}(t)$, where C is a constant matrix. It is well known [31] that

$$\mathbf{y}(t) = C\mathcal{F}(t; \mathbf{u}(t), \mathbf{x}_0) = C\left\{e^{tA}\mathbf{x}_0 + \int_0^t e^{(t-\tau)A}B\mathbf{u}(\tau)d\tau\right\} \qquad t \geq 0 \tag{33.18}$$

yielding an explicit representation formula for the I-O relationship of the linear circuit (together with the assumed measurement equation).

Note that because the state-variable representation (33.14) is not unique, as mentioned above, this I-O representation is not unique in general. However, we note that if two state-variable representations are equivalent, then their corresponding I-O relationships also will be equivalent.

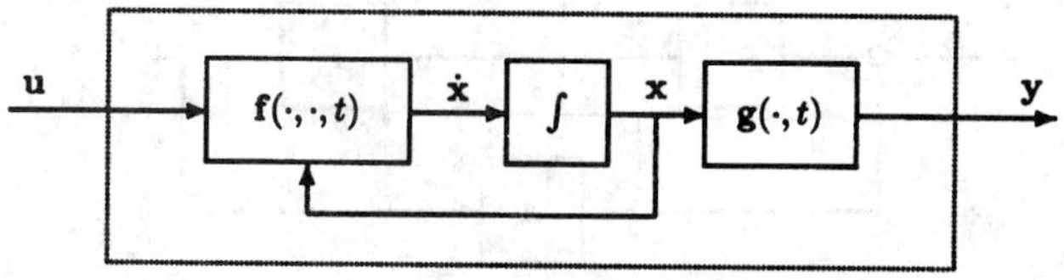

FIGURE 33.3 System I-O relationship.

It is also important to note that although the above I-O relationship is formulated for a finite-dimensional open-loop system, it can also be applied to infinite-dimensional [7] and closed-loop systems [39]. In particular, similar to linear systems, many finite-dimensional closed-loop nonlinear systems possess an elegant **coprime factorization representation**. The (left or right) coprime factorization representation of a nonlinear feedback system is a general I-O relationship which can be used as a fundamental framework, particularly suitable for studies of stabilization, tracking, and disturbance rejection. The problem is briefly described as follows. Let a nonlinear system (mapping) P be given, not necessarily stable, and assume that it has a right-coprime factorization $P = ND^{-1}$, where both N and D are stable (D^{-1} usually has the same stability as P). One is looking for two stable nonlinear subsystems (mappings), A and B^{-1}, representing feedback and feedforward controllers, respectively, satisfying the Bezout identity

$$AN + BD = I$$

which are connected as shown in Fig. 33.4, where B is also stable. If two controllers, A and B, can be found to satisfy such conditions, then even with an unstable P, the resulting closed-loop control system will be I-O, as well as internally, stable. In this sense A and B together stabilize P.

For the left-coprime factorization, one simply uses formulas $P = D^{-1}N$ and $NA + DB = I$ instead and interchanges the two blocks of A and B^{-1} in Fig. 33.4.

Taking into account causality and well-posedness of the overall closed-loop system, it is a technical issue as to how to construct the four subsystems A, B, D, and N, such that the above requirements can be satisfied. Some characterization results and construction methods are available in the literature [38, 45, 51, 95].

Volterra Series Representation

Recall from the fundamental theory of ordinary differential equations that an explicit I-O representation of the overall system still can be found, even if the linear state-space representation (33.5) is time varying, via the state transition matrix $\Phi(t,\tau)$ determined by

$$\begin{cases} \dfrac{d}{dt}\Phi(t,\tau) = A(t)\Phi(t,\tau) & t \geq \tau \\ \Phi(\tau,\tau) = I \end{cases} \tag{33.19}$$

where I is the identity matrix. The formula, for the simple case $\mathbf{y}(t) = C(t)\mathbf{x}(t)$, is

$$\mathbf{y}(t) = C(t)\left\{\Phi(t,0)\mathbf{x}_0 + \int_0^t \Phi(t,\tau)B(\tau)\mathbf{u}(\tau)d\tau\right\} \qquad t \geq 0 \tag{33.20}$$

For linear time-invariant systems, we actually have $\Phi(t,\tau) = e^{(t-\tau)A}$, so that (33.20) reduces to the explicit formula (33.18).

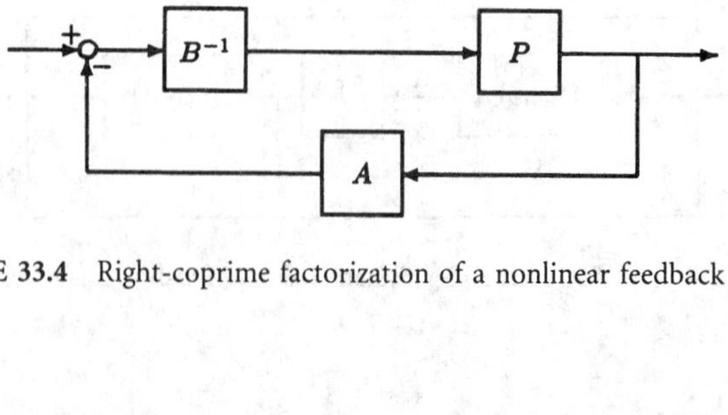

FIGURE 33.4 Right-coprime factorization of a nonlinear feedback system.

For a nonlinear system, a simple explicit I-O representation with a single integral of the form (33.18) or (33.20) is generally impossible. A natural generalization of such an integral formulation is the Volterra series representation. For simplicity, let us consider the one-dimensional case in which $y(t) = g(x(t), t) = x(t)$ below. A Volterra series representation for a nonlinear I-O relationship $\mathcal{F}(\cdot)$, convergent in some measure, is an infinite sum of integrals in the following form:

$$\mathcal{F}(t, u(t)) = \phi_0(t; x_0) + \int_0^t \phi_1(t, \tau_1)u(\tau_1)d\tau_1 + \cdots$$
$$+ \int_0^t \cdots \int_0^{\tau_2} \phi_n(t, \tau_1, \ldots, \tau_n)u(\tau_1)\cdots u(\tau_n)d\tau_1 \cdots d\tau_n + \cdots \tag{33.21}$$

where $\{\phi_n\}_{n=0}^{\infty}$ are called the **Volterra kernels** of the series. Here, we note that this Volterra series representation can be extended easily to higher dimensional systems.

For some representations $\mathcal{F}$, the corresponding Volterra series may have only finitely many nonzero terms in the above infinite sum. In this case it is called a **Volterra polynomial**, which does not have the convergence problem for bounded inputs, provided that all the integrals exist. In particular, when $\mathcal{F}$ is affine (or linear, if initial conditions are zero, so that $\phi_0 = 0$), its Volterra series has at most two nonzero terms, as given by (33.18) and (33.20), and is called a first-order Volterra polynomial. In general, however, the Volterra series (33.21) is an infinite sum. Hence, the convergence of a Volterra series is a crucial issue in formulating such a representation for a given nonlinear I-O relationship [5, 12, 59, 85].

In order to state a fundamental result about the convergence of a Volterra series, we must first recall that a mapping that takes a function to a (real or complex) value is called a **functional** and a mapping that takes a function to another function is called an **operator**. A functional may be considered to be a special operator if one views a value as a constant function in the image of the mapping. Clearly, the I-O relationship (33.17) and the Volterra series (33.21), including Volterra polynomials, are nonlinear operators. Recall also that an operator $\mathcal{T}: X \to Y$, where X and Y are normed linear spaces, is said to be continuous at $x \in X$ if $\|x_n - x\|_X \to 0$ implies $\|\mathcal{T}(x_n) - \mathcal{T}(x)\|_Y \to 0$ as $n \to \infty$. Note that for a linear operator, if it is continuous at a point, then it is also continuous on its entire domain [34], but this is not necessarily true for nonlinear operators.

As usual, we denote by $C[0, T]$ and $L_p[0, T]$, respectively, the space of continuous functions defined on $[0, T]$ and the space of measurable functions f satisfying $\int_0^T |f(t)|^p \, dt < \infty$ for $1 \le p < \infty$ or $\sup_{t\in[0,T]} |f(t)| < \infty$ for $p = \infty$. The following result [5] is an extension of the classical Stone-Weierstrass theorem [22, 36, 40].

Theorem 33.1: *Let X be either $C[0, T]$ or $L_p[0, T]$ with $1 \le p < \infty$, and Ω be a compact subset in X. Then, for any continuous operator $\mathcal{F}: \Omega \to L_q[0, T]$, where $(1/p) + (1/q) = 1$, and for any $\varepsilon > 0$, a Volterra polynomial $P_n(\cdot)$ exists, with n determined by ε, such that*

$$\sup_{x\in\Omega} \left\| \mathcal{F}(x) - P_n(x) \right\|_{L_q} < \varepsilon$$

In other words, $P_n \to \mathcal{F}$ uniformly on the compact subset $\Omega \subset X$ as $n \to \infty$.

In the literature many variants of this fundamental convergence theorem exist under various conditions in different forms, including the $L_\infty[0, T]$ case [45, 59, 84, 85]. We may also find different methods for constructing the Volterra kernels $\{\phi_n\}_{n=0}^{\infty}$ for $\mathcal{F}$ [83]. In addition, specially structured Volterra series representations abound for nonlinear systems, such as the Volterra series with finite memory [5], approximately-finite memory [86], and fading memory [10].

Finally, it should be mentioned that in a more general manner, a few abstract functional series representations exist, including the **generating power series representation** for certain nonlinear systems [48], from which the Volterra series can be derived. Briefly, an important result is the following theorem [6, 54, 71, 91].

Theorem 33.2: *Consider a nonlinear control system of the form*

$$\begin{cases} \dot{x}(t) = g_0(x(t)) + \sum_{k=1}^{m} g_k(x(t))u_k(t) & t \in [0, T] \\ y(t) = h(x(t)) \end{cases}$$

where $h(\cdot)$ *and* $\{g_i(\cdot)\}_{i=0}^{m}$ *are su' ciently smooth functionals, with an initial state* x_0. *If the control inputs satisfy* $\max_{0 \le \tau \le T} |u_k(\tau)| < 1$, *then the corresponding output of this nonlinear system has a convergent functional series of the form*

$$y(t) = h(x_0) + \sum_{i=0}^{\infty} \sum_{k_0,\ldots,k_i=0}^{m} L_{g_{k_0}} \cdots L_{g_{k_i}} h(x_0) \int_0^t d\xi_{k_i} \cdots d\xi_{k_0} \tag{33.22}$$

where $L_g h(x_0) := [\partial h/\partial x] g(x)|_{x=x_0}$ *and* ξ_k *are defined by*

$$\xi_0(t) = t \quad \xi_k(t) = \int_0^t u_k(\tau)d\tau \quad k = 1, \ldots, m$$

with the notation

$$\int_0^t d\xi_{k_i} \cdots d\xi_{k_0} := \int_0^t d\xi_{k_i}(\tau) \int_0^\tau d\xi_{k_{i-1}} \cdots d\xi_{k_0}$$

Note that in order to guarantee the convergence of the functional series (33.22), in many cases it may be necessary for T to be sufficiently small.

Analogous to the classical Taylor series of smooth functions, a fairly general series representation for some nonlinear systems is still possible using **polynomial operators**, or the like [90]. As usual, however, the more general the presentation is, the less concrete the results. Moreover, a very general series expansion is likely to be very local, and its convergence is difficult to analyze.

33.3 Approximation

The mathematical term "approximation" used here refers to the theory and methodology of function (functional or operator) approximation. Mathematical approximation theory and techniques are important in engineering when one seeks to represent a set of discrete data by a continuous function, to replace a complicated signal by a simpler one, or to approximate an infinite-dimensional system by a finite-dimensional model, etc., under certain optimality criteria.

Approximation is widely used in system modeling, reduction, and identification, as well as in many other areas of control systems and signal processing [32]. A Volterra polynomial as a truncation of the infinite Volterra series (discussed earlier) serves as a good example of system (or operator) approximation, where the question "in what sense is this approximation good?" must be addressed further.

Best Approximation of Systems (Operators)

Intuitively, approximation is always possible. However, two key issues are the quality of the approximation and the efficiency of its computation (or implementation). Whenever possible, one would like to have the best (or optimal) approximation, based on the available conditions and subject to all the requirements. A commonly used criterion for best (or optimal) approximations is to achieve a minimum norm of the approximation error using a norm that is meaningful to the problem. Best approximations of systems (operators) include the familiar least-squares technique, and various other uniform approximations.

Least-Squares Approximation and Projections

Let us start with the most popular "best approximation" technique (the least-squares method), which can also be thought of as a projection, and a special min-max approximation discussed in the next section. Discrete data fitting by a continuous function is perhaps the best-known example of least-squares. The special structure of Hilbert space, a complete inner-product space of functions, provides a general and convenient framework for exploring the common feature of various least-squares approximation techniques. Because we are concerned with approximation of nonlinear systems rather than functions, a higher-level framework, the Hilbert space of operators, is needed. We illustrate such least-squares system (or operator) approximations with the following two examples.

First, we consider the linear space, H, of certain nonlinear systems that have a convergent Volterra series representation (33.21) mapping an input space X to an output space Y. Note that although a nontrivial Volterra series is a nonlinear operator, together they constitute a linear space just like nonlinear functions.

To form a Hilbert space, we first need an inner product between any two Volterra series. One way to introduce an inner product structure into this space is as follows. Suppose that all the Volterra series, $\mathscr{F}: X \to Y$, where both X and Y are Hilbert spaces of real-valued functions, have bounded admissible inputs from the set

$$\Omega = \{x \in X \,|\, \|x\|_X \le \gamma < \infty\}$$

For any two convergent Volterra series of the form (33.21), say $\mathscr{F}$ and $\mathscr{G}$, with the corresponding Volterra kernel sequences $\{\phi_n\}$ and $\{\psi_n\}$, respectively, we can define an inner product between them via the convergent series formulation

$$\langle \mathscr{F}, \mathscr{G} \rangle_H := \sum_{n=0}^{\infty} \frac{\rho_n}{n!} |\phi_n \psi_n|$$

with the induced norm $\|\mathscr{F}\|_H = \langle \mathscr{F}, \mathscr{F} \rangle_H^{1/2}$, where the weights $\{\rho_n\}$ satisfy

$$\sum_{n=0}^{\infty} \frac{1}{\rho_n} \frac{\gamma^{2n}}{n!} < \infty$$

Recall also that a **reproducing kernel Hilbert space** $\tilde{H}$ is a Hilbert space (of real-valued functions or operators) defined on a set S, with a **reproducing kernel** $K(x, y)$, which belongs to $\tilde{H}$ for each fixed x or y in S and has the property

$$\langle K(x, y), \mathscr{F}(y) \rangle_{\tilde{H}} = \mathscr{F}(x) \qquad \forall \mathscr{F} \in \tilde{H} \quad \text{and} \quad \forall x, y \in S$$

Using the notation defined above, the following useful result was established [43, 45] and is useful for nonlinear systems identification (see Theorem 33.23).

Theorem 33.3: *The family of all the convergent Volterra series of the form* (33.21) *that maps the bounded input set* Ω *to* Y *constitutes a reproducing kernel Hilbert space with the reproducing kernel*

$$K(x,y) = \sum_{n=0}^{\infty} \frac{1}{n!}\frac{1}{\rho_n}\langle x,y\rangle_X^n \qquad x,y \in \Omega \subset X \tag{33.23}$$

The reproducing kernel Hilbert space H defined above is called a **generalized Fock space** [46]. For the special case in which $\rho_n \equiv 1$, its reproducing kernel has a nice closed-form formula as an exponential operator $K(x,y) = e^{\langle x,y\rangle}$.

Now, suppose that a nonlinear system $\mathcal{F}$ is given, which has a convergent Volterra series representation (33.21) with infinitely many nonzero terms in the series. For a fixed integer $n \geq 0$, if we want to find an nth-order Volterra polynomial, denoted V_n^*, from the Hilbert space H such that

$$\|\mathcal{F} - V_n^*\|_H = \inf_{V_n \in H}\|\mathcal{F} - V_n\|_H \tag{33.24}$$

then we have a best approximation problem in the least-squares sense. To solve this optimization problem is to find the best Volterra kernels $\{\phi_k(t)\}_{k=0}^n$ over all the possible kernels that define the Volterra polynomial V_n, such that the minimization (33.24) is achieved.

Note that if we view the optimal solution V_n^* as the projection of $\mathcal{F}$ onto the $(n+1)$-dimensional subspace of H, then this least-squares minimization is indeed a projection approximation. It is then clear, even from the Hilbert space geometry (see Fig. 33.5), that such an optimal solution, called a **best approximant**, always exists due to the norm-completeness of Hilbert space and is unique by the convexity of inner product space.

As a second example, let H be a Hilbert space consisting of all the linear and nonlinear systems that have an nth-order Taylor series representation of the form

$$P_n(\cdot) = \sum_{k=0}^{n} \alpha_k(t)M_k(\cdot) = \alpha_0(t) + \alpha_1(t)(\cdot)(t) + \cdots + \alpha_n(t)(\cdot)^n(t) \tag{33.25}$$

where $M_k(\cdot) := (\cdot)^k$ is the monomial operator of degree k and $\{\alpha_k\}_{k=0}^n$ are continuous real-valued functions satisfying certain conditions arising from some basic properties of both the domain and the range of the operator. Suppose that the monomial operators $\{M_k\}_{k=0}^{\infty}$ are orthogonal under the inner product of H. Given an lth-order polynomial operator P_l with $\alpha_l(t) \neq 0$ almost everywhere, if for a fixed integer $n < l$ we want to find an nth-order polynomial operator P_n^* of the form (33.25) from H, such that

$$\|P_l - P_n^*\|_H = \inf_{P_n \in H}\|P_l - P_n\|_H \tag{33.26}$$

then we have a best approximation problem in the least-squares sense. To solve this optimization problem is to find the best coefficient functions $\{\alpha_k(t)\}_{k=0}^n$ over all possible functions that define

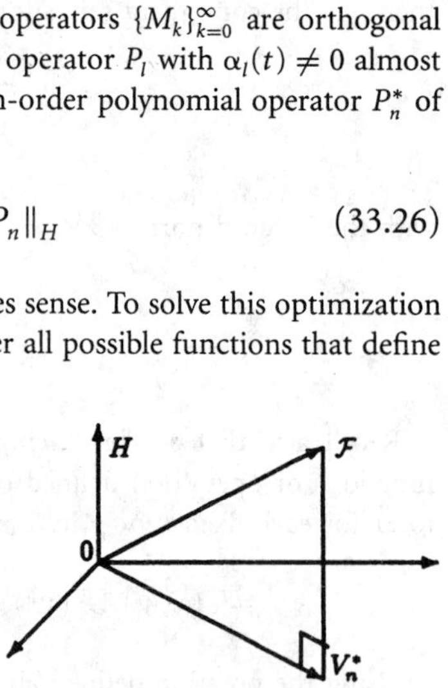

FIGURE 33.5 Projection in a Hilbert space.

the polynomial operator P_n. Again, because the optimal solution is the projection of P_l onto the $(n+1)$-dimensional subspace H of a Hilbert space to which P_l belongs, it always exists and is unique.

We now state a general result of least-squares approximation for systems, which is a straightforward generalization of the classical result of least-squares approximation for functions [22, 36].

Theorem 33.4: *Let H be a Hilbert space of nonlinear operators, and let H_n be its n-dimensional subspace. Then, given an $\mathcal{F} \in H$, the least-squares approximation problem*

$$\|\mathcal{F} - \mathcal{N}_n^*\|_H = \inf_{\mathcal{N}_n \in H_n} \|\mathcal{F} - \mathcal{N}_n\|_H$$

is always uniquely solvable, with the optimal solution given by

$$\mathcal{N}^*(\cdot) = \sum_{k=1}^{n} \langle \mathcal{F}, h_k \rangle_H h_k(\cdot)$$

where $\{h_k\}_{k=1}^{n}$ is an orthonormal basis of H_n.

A more general setting is to replace the Hilbert space H by a Banach space (a complete normed linear space, such as L_1 and L_∞, which may not have an inner product structure). This extension includes the Hilbert space setting as a special case, but generally does not have so many special features. Even the existence and uniqueness of best approximants cannot be taken for granted in general—not even for the simpler case of best approximation of real-valued functions—if a Banach (non-Hilbert) space is considered [73]. Nevertheless, the following result is still convenient to use [22].

Theorem 33.5: *Let B be a uniformly convex Banach space and Ω be a closed convex set in B. Then, for any given $\mathcal{F} \in B$, the optimal approximation problem*

$$\|\mathcal{F} - \omega^*\|_B = \inf_{\omega \in \Omega} \|\mathcal{F} - \omega\|_B$$

has a unique solution.

Here, a space (or subset) B is said to be **uniformly convex** if for any $\varepsilon > 0$ there exists a $\delta = \delta(\varepsilon) > 0$ such that $\|f\|_B = \|g\|_B = 1$ and $\|\frac{1}{2}(f+g)\|_B > 1-\delta$ together imply $\|f-g\|_B < \varepsilon$. Geometrically, a disk is uniformly convex while a triangle is only convex, but not uniformly so. It is then intuitively clear that for a given point outside (or inside) a disk, only a single point exists in the disk that has the shortest distance to the given point. However, this is not always true for a nonuniform case. In fact, a best approximation problem in the general Banach space setting has either a unique solution or has infinitely many solutions (if it is solvable), as can be seen from the next result [32].

Theorem 33.6: *Let Ω be a closed convex set in a Banach space B, and ω_1^* and ω_2^* be two optimal solutions of the best approximation problem*

$$\|\mathcal{F} - \omega^*\|_B = \inf_{\omega \in \Omega} \|\mathcal{F} - \omega\|_B$$

Then any convex combination of ω_1^* and ω_2^* in the form

$$\omega^* = a\omega_1^* + (1-a)\omega_2^* \quad 0 \le a \le 1$$

is also an optimal solution of the problem.

Usually, a best approximant (if it exists) for an optimal approximation problem in a Banach space is also called a **(minimal) projection** of the given operator from a higher-dimensional subspace onto a lower-dimensional subspace. In this extension the projection has no simple geometric meaning of "orthonormality" due to the lack of an inner product structure. However, a projection operator with a unity norm in the Banach space setting is a natural generalization of the orthonormal projection in the Hilbert space framework.

Min-Max (Uniform) Approximation

It is clear from the least-squares approximation formulation that if the given nonlinear representation (operator) $\mathscr{F}$ and the lower-order approximant (used to approximate $\mathscr{F}$) do not have the same structure (the same type of series), then the least-squares approximation cannot be applied directly or efficiently.

To introduce another approach, we first recall that for two given normed linear spaces X and Y and for a given bounded subset Ω of X, with $0 \in \Omega$, the operator norm of a nonlinear operator $\mathscr{N}: \Omega \to Y$ satisfying $\mathscr{N}(0) = 0$, can be defined as

$$|||\mathscr{N}||| = \sup_{\substack{x,y\in\Omega \\ x\neq y}} \frac{\|\mathscr{N}(x) - \mathscr{N}(y)\|_Y}{\|x - y\|_X} \tag{33.27}$$

Thus, given a norm-bounded nonlinear operator $\mathscr{F}$, representing a given physical system, we may consider the problem of finding another norm-bounded nonlinear operator $\mathscr{N}^*$ from a certain class N of desired nonlinear operators (systems), not necessarily having the same structure as $\mathscr{F}$, to best approximate $\mathscr{F}$ in the sense that

$$|||\mathscr{F} - \mathscr{N}^*||| = \inf_{\mathscr{N}\in N} |||\mathscr{F} - \mathscr{N}||| \tag{33.28}$$

For example, N can be the family of nth-order Volterra polynomials or nth-order polynomial operators discussed above. Commonly used function spaces X and Y include the space of all continuous functions, the standard L_p space (or l_p for the discrete case), and the Hardy space H_p (for complex-variable functions [32]), with $1 \leq p \leq \infty$.

Because the nonlinear operator norm defined by (33.27) is a sup (max) norm and this optimization is an inf (min) operation, the best approximation problem (33.28) is called a **min-max approximation.** Note also that because the nonlinear operator norm (33.27) is defined over all the bounded inputs in the set Ω, this approximation is uniform, and thus independent of each individual input function of the set Ω. For this reason, this approximation is also called a **uniform approximation**, indicating that the best approximant is the optimal solution over *all* input functions.

It should be noted that both existence and uniqueness of best approximation solutions to the min-max approximation problem (33.28) must be investigated according to the choice of the operator family N and the I-O spaces X and Y, which generally cannot be taken for granted, as previously discussed.

An important and useful class of nonlinear operators which can be put into a Banach space setting with great potential in systems and control engineering is the family of **generalized Lipschitz operators** [45]. To introduce this concept, we first need some notation. Let X be a Banach space of real-valued functions defined on $[0, \infty)$ and, for any $f \in X$ and any $T \in [0, \infty)$, define

$$[f]_T(t) = \begin{cases} f(t) & t \leq T \\ 0 & t > T \end{cases}$$

Then, form a normed linear space X^e, called the **extended linear space associated with X**, by

$$X^e = \{f \in X \| \|[f]_T\|_X < \infty, \forall T < \infty\}$$

For a subset $D \subseteq X^e$, any (linear or nonlinear) operator $\mathcal{G}: D \to Y^e$ satisfying

$$\left\| [\mathcal{G}(x_1)]_T - [\mathcal{G}(x_2)]_T \right\|_Y \leq L \left\| [x_1]_T - [x_2]_T \right\|_X, \quad \forall x_1, x_2 \in D, \forall T \in [0, \infty)$$

for some constant $L < \infty$, is called a generalized Lipschitz operator defined on D. The least of such constants L is given by the semi-norm of the operator $\mathcal{G}$:

$$\|\mathcal{G}\| := \sup_{T \in [0,\infty)} \sup_{\substack{x_1, x_2 \in D \\ [x_1]_T \neq [x_2]_T}} \frac{\left\| [\mathcal{G}(x_1)]_T - [\mathcal{G}(x_2)]_T \right\|_Y}{\left\| [x_1]_T - [x_2]_T \right\|_X}$$

and the operator norm of $\mathcal{G}$ is defined via this semi-norm by

$$\|\mathcal{G}\|_{\text{Lip}} = \left\| \mathcal{G}(x_0) \right\|_Y + \|\mathcal{G}\|$$

for an arbitrarily chosen and fixed $x_0 \in D$. The following result has been established [45].

Theorem 33.7: *The family of generalized Lipschitz operators*

$$\text{Lip}(D, Y^e) = \{\mathcal{G}: D \subseteq X^e \to Y^e \| \|\mathcal{G}\|_{\text{Lip}} < \infty \text{ on } D\}$$

is a Banach space.

Based on this theorem, a best approximation problem for generalized Lipschitz operators can be similarly formulated, and many fundamental approximation results can be obtained. In addition, generalized Lipschitz operators provide a self-unified framework for both left and right coprime factorization representations of nonlinear feedback systems. Under this framework, the overall closed-loop system shown in Fig. 33.4 can have a causal, stable, and well-posed coprime factorization representation, which can be applied to optimal designs such as tracking and disturbance rejection [45].

We now discuss briefly a different kind of min-max (uniform) approximation: the **best Hankel-norm approximation**, where the norm (33.27) is replaced by the operator norm of a Hankel operator defined as follows [32, 77]. Consider, for instance, the transfer function

$$H(z) = \alpha_0 + \alpha_1 z^{-1} + \alpha_2 z^{-2} + \cdots$$

of a discrete time linear time-invariant system. The **Hankel operator** associated with this series is defined as the infinite matrix

$$\Gamma_\alpha := [\alpha_{|i-j|}] = \begin{bmatrix} \alpha_0 & \alpha_1 & \alpha_2 & \cdots \\ \alpha_1 & \alpha_2 & \cdots & \\ \alpha_2 & \cdots & & \\ \vdots & & & \end{bmatrix}$$

which is a linear operator on a normed linear space of sequences. The operator norm of Γ_α over the l_2-space is called the **Hankel norm** of Γ_α.

One important feature of the Hankel operators is reflected in the following theorem [32, 77].

Theorem 33.8: *An infinite Hankel matrix has a finite rank iff its corresponding functional series is rational (it sums up to a rational function); and this is true iff the rational series corresponds to a finite-dimensional bilinear system.*

Another useful property of Hankel operators in system approximation is represented in the following theorem [28].

Theorem 33.9: *The family of compact Hankel operators is an M-ideal in the space of Hankel operators that are defined on a Hilbert space of real-valued functions.*

Here, a compact operator is one that maps bounded sets to compact closures and an M-ideal is a closed subspace X of a Banach space Z such that $X^{\perp}$, the orthogonal complemental subspace of X in Z, is the range of the projection P from the dual space Z^* to $X^{\perp}$ that has the property

$$\|f\| = \|P(f)\| + \|f - P(f)\| \quad \forall f \in Z^*$$

The importance of the M-ideal is that it is a proximinal subspace with certain useful approximation characteristics, where the proximinal property is defined as follows. Let $L(X)$ and $C(X)$ be the classes of bounded linear operators and compact operators, respectively, both defined on a Banach space X. If every $\mathscr{L} \in L(X)$ has at least one best approximant from $C(X)$, then $C(X)$ is said to be proximinal in $L(X)$. A typical result would be the following: for any $1 < p < \infty$, $C(l_p)$ is proximinal in $L(l_p)$. However, $C(X)$ is not proximinal in $L(X)$ if $X = C[a, b]$, the space of continuous functions defined on $[a, b]$, or $X = L_p[a, b]$ for all $1 < p < \infty$ except $p = 2$.

Best (Uniform) Approximation of Signals (Functions)

Best approximations of signals for circuits and systems are also important. For example, two (different) systems (e.g., circuits) are considered to be equivalent over a set Ω of admissible input signals iff the same input from Ω yields the same outputs through the two systems. Thus, the problem of using a system to best approximate another may be converted, in many cases, to the best approximation problem for their output signals.

A signal is a function of time, usually real valued and one dimensional. The most general formulation for best approximation of functions can be stated as follows. Let X be a normed linear space of real-valued functions and Ω be a subset of X. For a given f in X but not in Ω, find a $g^* \in \Omega$ such that

$$\|f - g^*\|_X = \inf_{g \in \Omega} \|f - g\|_X \tag{33.29}$$

In particular, if $X = L_\infty, l_\infty$, or H_∞, the optimal solution is the best result for the worst case.

If such a g^* exists, then it is called a **best approximant** of f from the subset Ω. In particular, if $\Omega_1 \subset \Omega_2 \subset \cdots$ is a sequence of subspaces in X, such that $\overline{\bigcup \Omega_n} = X$, an important practical problem is to find a sequence of best approximants $g_n^* \in \Omega_n$ satisfying the requirement (33.29) for each $n = 1, 2, \ldots$, such that $\|g_n^* - g^*\|_X \to 0$ as $n \to \infty$. In this way for each n one may be able to construct a simple approximant g_n^* for a complicated (even unknown) function f, which is optimal in the sense of the min-max approximation (33.29).

Existence of a solution is the first question about this best approximation. The fundamental result is the following [22, 36].

Theorem 33.10: *For any $f \in X$, a best approximant g^* of f in Ω always exists, if Ω is a compact subset of X; or Ω is a finite-dimensional subspace of X.*

Uniqueness of a solution is the second question in approximation theory, but it is not as important as the existence issue in engineering applications. Instead, characterization of a best approximant for a specific problem is significant in that it is often useful for constructing a best approximant.

As a special case, the above best approximation reduces to the least-squares approximation if X is a Hilbert space. The basic result is the following (compare it to Theorem 33.4, and see Fig. 33.5).

Theorem 33.11: *Let H be a Hilbert space of real-valued functions, and let H_n be its n-dimensional subspace. Then, given an $f \in H$, the least-squares approximation problem*

$$\|f - h_n^*\|_H = \inf_{h_n \in H_n} \|f - h_n\|_H$$

is always uniquely solvable, with the optimal solution given by

$$h_n^*(t) = \sum_{k=1}^{n} \langle f, h_k \rangle_H h_k(t)$$

where $\{h_k\}_{k=1}^n$ is an orthonormal basis of H_n.

Here, the orthonormal basis of H_n is a **Chebyshev system**, a system of functions which satisfy the **Haar condition** that the determinant of the matrix $[h_i(t_j)]$ is nonzero at n distinct points $t_1 < \cdots < t_n$ in the domain. Chebyshev systems include many commonly used functions, such as algebraic and trigonometric polynomials, splines, and radial functions. Best approximation by these functions is discussed in more detail below.

We remark that the least-squares solution shown in Theorem 33.11 is very general, which includes the familiar truncations of the Fourier series [36] and the wavelet series [29] as best approximation.

Polynomial and Rational Approximations

Let π_n be the space of all algebraic polynomials $p_n(t)$ of degree not greater than n. For any continuous function $f(t)$ defined on $[a,b]$, one is typically looking for a best approximant $p_n^* \in \pi_n$ for a fixed n, such that

$$\|f - p_n^*\|_{L_\infty[a,b]} = \min_{p_n \in \pi_n} \|f - p_n\|_{L_\infty[a,b]} \tag{33.30}$$

This is a best (min-max and uniform) **algebraic polynomial approximation** problem. Replacing the algebraic polynomials by the nth-order trigonometric polynomials of the form $\sum_{k=0}^{n}(a_k \cos(kt) + b_k \sin(kt))$ changes the problem to the best **trigonometric polynomial approximation**, in the same sense as the best algebraic polynomial approximation, for a given function $f \in C[-\pi, \pi]$. This can be much further extended to any Chebyshev system, such as the radial basis functions and polynomial spline functions, which are discussed later. According to the second part of Theorem 33.10, the best uniform polynomial approximation problem (33.30) always has a solution which, in this case, is unique. Moreover, this best approximant is characterized by the following important sign-alternation theorem. This theorem is also valid for the best uniform approximation from any other Chebyshev system [22, 36].

Theorem 33.12: *The algebraic polynomial p_n^* is a best uniform approximant of $f \in C[a,b]$ from π_n iff there exist $n+2$ points $a \le t_0 < \cdots < t_{n+1} \le b$ such that*

$$f(t_k) - p_n^*(t_k) = c(-1)^k \|f - p_n^*\|_{L_\infty[a,b]} \qquad k = 0, 1, \ldots, n+1$$

where $c = 1$ or -1.

An efficient **Remes (exchange) algorithm** is available for constructing such a best approximant [79].

Another type of function is related to algebraic polynomials: the algebraic rational functions of the form $r_{n,m}(t) = p_n(t)/q_m(t)$, which has finite values on $[a,b]$ with coprime $p_n \in \pi_n$ and $q_m \in \pi_m$. We denote by $R_{n,m}$ the family of all such rational functions, or a subset of them, with fixed integers $n \geq 0$ and $m \geq 1$. Although $R_{n,m}$ is not a compact set or a linear space, the following result can be established [22].

Theorem 33.13: *For any given function $f \in C[a,b]$, there exists a unique $r^*_{n,m}(t) \in R_{n,m}$ such that*

$$\|f - r^*_{n,m}\|_{L_\infty[a,b]} = \inf_{r_{n,m} \in R_{n,m}} \|f - r_{n,m}\|_{L_\infty[a,b]} \qquad (33.31)$$

The optimal solution $r^*_{n,m}(t)$ of (33.31) is called the **best uniform rational approximant** of $f(t)$ on $[a,b]$ from $R_{n,m}$.

Note that the unique best rational approximant may have different expressions unless it is coprime, as assumed above. The following theorem [22] characterizes such a best approximant, in which we use $d(p_n)$ to denote the actual degree of $p_n, 0 \leq d(p_n) \leq n$.

Theorem 33.14: *A rational function $r^*_{n,m} = p^*_n/q^*_m$ is a best uniform approximant of $f \in C[a,b]$ from $R_{n,m}$ iff there exist s points $a \leq t_1 < \cdots < t_s \leq b$, with $s = 2 + \min\{n + d(q_m), m + d(p_n)\}$, such that*

$$f(t_k) - r^*_{n,m}(t_k) = c(-1)^k \|f - r^*_{n,m}\|_{L_\infty[a,b]} \qquad k = 1, \ldots, s$$

where $c = 1$ or -1.

The Remes (exchange) algorithm [79] also can be used for constructing a best rational approximant.

An important type of function approximation, which utilizes rational functions, is the **Padé approximation.** Given a formal power series of the form

$$f(t) = c_0 + c_1 t + c_2 t^2 + \cdots \qquad t \in [-1,1]$$

not necessarily convergent, the question is to find a rational function $p_n(t)/q_m(t)$, where n and m are both fixed, to best approximate $f(t)$ on $[-1,1]$, in the sense that

$$\left| f(t) - \frac{p_n(t)}{q_m(t)} \right| \leq c|t|^l \qquad t \in [-1,1] \qquad (33.32)$$

for a "largest possible" integer l. It turns out that normally the largest possible integer is $l = n+m+1$. If such a rational function exists, it is called the **$[n, m]$th-order Padé approximant** of $f(t)$ on $[-1,1]$. The following result is important [22].

Theorem 33.15: *If $f(t)$ is $(n + m + 1)$ times continuously differentiable in a neighborhood of $t = 0$, then the $[n, m]$th-order Padé approximant of $f(t)$ exists, with $l > n$. If $l \leq n + m + 1$, then the coefficients $\{a_k\}_{k=0}^n$ and $\{b_k\}_{k=0}^m$ of $p_n(t)$ and $q_m(t)$ are determined by the following linear system of algebraic equations:*

$$\sum_{j=0}^{i} \frac{f^j(0)}{j!} b_{i-j} = a_i \qquad i = 0, 1, \ldots, l-1$$

with $a_{n+j} = b_{m+j} = 0$ *for all* $j = 1, 2, \ldots$. *Moreover, if* p_n/q_m *is the* $[n, m]$*th-order Padé approximant of* $f(t) = \sum_{k=0}^{\infty} f_k t^k$, *then the approximation error is given by*

$$\left| f(t) - \frac{p_n(t)}{q_m(t)} \right| = \sum_{k=n+1}^{\infty} \left(\sum_{j=0}^{m} f_{k-j} b_j \right) \frac{t^k}{q_m(t)} \qquad t \in [-1, 1]$$

Padé approximation can be extended from algebraic polynomials to any other Chebyshev systems [22].

Approximation via Splines and Radial Functions

Roughly speaking, spline functions, or simply splines, are piecewise smooth functions that are structurally connected and satisfy some special properties. The most elementary and useful splines are polynomial splines, which are piecewise algebraic polynomials, usually continuous, with a certain degree of smoothness at the connections. More precisely, let

$$a = t_0 < t_1 < \cdots < t_n < t_{n+1} = b$$

be a partition of interval $[a, b]$. The **polynomial spline** of degree m with knots $\{t_k\}_{k=1}^{n}$ on $[a, b]$ is defined to be the piecewise polynomial $g_m(t)$ that is a regular algebraic polynomial of degree m on each subinterval $[t_k, t_{k+1}]$, $k = 0, \ldots, n$, and is $(m-1)$ times continuously differentiable at all knots [41, 88]. We denote the family of these algebraic polynomial splines by $S_m(t_1, \ldots, t_n)$, which is an $(n + m + 1)$-dimensional linear space.

Given a continuous function $f(t)$ on $[a, b]$, the best uniform spline approximation problem is to find a $g_m^* \in S_m(t_1, \ldots, t_n)$ such that

$$\|f - g_m^*\|_{L_\infty[a,b]} = \inf_{g_m \in S_m} \|f - g_m\|_{L_\infty[a,b]} \tag{33.33}$$

According to the second part of Theorem 33.10, this best uniform approximation problem always has a solution. A best spline approximant can be characterized by the following sign-alteration theorem [72], which is a generalization of Theorem 33.12, from polynomials to polynomial splines.

Theorem 33.16: *The polynomial spline* $g_m^*(t)$ *is a best uniform approximant of* $f \in C[a, b]$ *from* $S_m(t_1, \ldots, t_n)$ *iff there exists a subinterval* $[t_r, t_{r+s}] \subset [a, b]$, *with integers* r *and* $s \geq 1$, *such that the maximal number* γ *of sign-alteration points on this subinterval* $[t_r, t_{r+s}]$, namely,

$$f(t_k) - g_m(t_k) = c(-1)^k \|f - g_m\|_{L_\infty[a,b]} \qquad t_k \in [t_r, t_{r+s}] \qquad k = 1, \ldots, \gamma$$

satisfies $\gamma \geq m + s + 1$, *where* $c = 1$ *or* -1.

Polynomial splines can be used for least-squares approximation, just like regular polynomials, if the L_∞-norm is replaced by the L_2-norm in (33.33). For example, B-splines, i.e., basic splines with a compact support, are very efficient in least-squares approximation. The spline quasi-interpolant provides another type of efficient approximation, which has the following structure

$$g_m(t) = \sum_k f(t_k) \phi_k^m(t) \tag{33.34}$$

and can achieve the optimal approximation order, where $\{\phi_k^m\}$ is a certain linear combination of B-splines of order m [18].

Spline functions have many variants and generalizations, including natural splines, perfect splines, various multivariate splines, and some generalized splines defined by linear ordinary or partial differential operators with initial-boundary conditions [27, 41, 42, 44, 88].

Splines are essentially local, in the sense of having compact supports, perhaps with the exception perhaps of the **thin-plate splines** [94], whose domains do not have a boundary.

Radial functions are global, with the property $\phi(r) \to \infty$ as $r \to \infty$ and, normally, $\phi(0) = 0$. Well-conditioned radial functions include $|r|^{2m+1}, r^{2m}\log(r), (r^2 + a^2)^{\pm 1/2}, 0 < a \ll 1$, etc. [80]. Many radial functions are good candidates for modeling nonlinear circuits and systems [63, 64]. For example, for l distinct points $\mathbf{t}_1, \ldots, \mathbf{t}_l$ in R^n, the radial functions $\{\phi(|\mathbf{t} - \mathbf{t}_k|)\}_{k=1}^{l}$ are linearly independent, and thus the minimization

$$\min_{\{c_k\}} \left| f(\mathbf{t}) - \sum_{k=1}^{l} c_k \phi(|\mathbf{t} - \mathbf{t}_k|) \right|^2 \tag{33.35}$$

at some scattered points can yield a best least-squares approximant for a given function $f(\mathbf{t})$, with some especially desirable features [81]. In particular, an affine plus radial function in the form

$$\mathbf{a} \cdot \mathbf{t} + b + \sum_{k=1}^{l} c_k \phi(|\mathbf{t} - \mathbf{t}_k|) \qquad \mathbf{t} \in R^n \tag{33.36}$$

where $\mathbf{a}, b, \{c_k\}_{k=1}^{l}$ are constants, provides a good modeling framework for the canonical piecewise linear representation (33.12) of a nonlinear circuit [63].

Approximation by Means of Interpolation

Interpolation plays a central role in function approximation theory. The main theme of interpolation is this: suppose that an unknown function exists for which we are given some measurement data such as its function values, and perhaps some values of its derivatives, at some discrete points in the domain. How can we use this information to construct a new function that interpolates these values at the given points as an approximant of the unknown function, preferably in an optimal sense? Constructing such a function, called an **interpolant**, is usually not a difficult problem, but the technical issue that remains is what kind of functions should be used as the interpolant so that a certain meaningful and optimal objective is attained?

Algebraic polynomial interpolation is the simplest approach for the following **Lagrange interpolation** problem [22, 36].

Theorem 33.17: *For arbitrarily given $n + 1$ distinct points $0 \le t_0 < t_1 < \cdots < t_n \le 1$ and $n + 1$ real values $v_0, v_1, \ldots, v_n$, there exists a unique polynomial $p_n(t)$ of degree n, which satisfies*

$$p_n(t_k) = v_k \qquad k = 0, 1, \ldots, n$$

This polynomial is given by

$$p_n(t) = \sum_{k=0}^{n} v_k L_k(t)$$

with the Lagrange basis polynomials

$$L_k(t) := \frac{(t - t_0) \cdots (t - t_{k-1})(t - t_{k+1})(t - t_n)}{(t_k - t_0) \cdots (t_k - t_{k-1})(t_k - t_{k+1})(t_k - t_n)} \qquad k = 0, \ldots, n$$

Moreover, if $f(t)$ is l $(\leq n+1)$ times continuously differentiable on $[a,b]$, then the interpolation error is bounded by

$$\|f - p_n\|_{L_\infty[0,1]} \leq \frac{1}{n!}\|f^{(l)}\|_{L_\infty[0,1]}\|h\|_{L_\infty[0,1]}$$

where $h(t) = \prod_{k=0}^{n-1}(t - t_k)$, and $\|h\|_{L_\infty[0,1]}$ attains its minimum at the Chebyshev points $t_k = \cos((2k+1)\pi/2(n+1)), k = 0, 1, \ldots, n$.

Note that the set $\{L_k(t)\}_{k=0}^n$ is a Chebyshev system on the interval $[t_0, t_n]$, which guarantees the existence and uniqueness of the solution. This set of basis functions can be replaced by any other Chebyshev system to obtain a unique interpolant.

If not only functional values, but also derivative values, are available and required to be interpolated by the polynomial,

$$p_n^{(i_k)}(t_k) = v_{k,i_k} \qquad i_k = 0, \ldots, m_k \qquad k = 0, 1, \ldots, n$$

then we have a **Hermite interpolation** problem. An algebraic polynomial of degree $d = n + \sum_{k=0}^{n} m_k$ always exists as a Hermite interpolant. An explicit closed-form formula for the Hermite interpolant also can be constructed. For example, if only the functional values $\{v_k\}_{k=0}^n$ and the first derivative values $\{w_k\}_{k=0}^n$ are given and required to be interpolated, then the Hermite interpolant is given by

$$p_{2n}(t) = \sum_{k=0}^{n}\{v_k A_k(t) + w_k B_k(t)\}$$

where, with the notation $L_k'(t_k) := (d/dt)L_k(t)\big|_{t=t_k}$,

$$A_k(t) = [1 - 2(t - t_k)L_k'(t_k)]L_k^2(t) \quad \text{and} \quad B_k(t) = (t - t_k)L_k^2(t)$$

in which $L_k(t)$ are Lagrange basis polynomials, $k = 0, 1, \ldots, n$.

However, if those derivative values are not consecutively given, we have a **Hermite–Birkhoff interpolation** problem, which is not always uniquely solvable [61].

The above discussions did not take into consideration any optimality. The unique algebraic polynomial interpolant obtained above may not be a good result in many cases. A well-known example is provided by Runge, in interpolating the continuous and smooth function $f(t) = 1/(1+25t^2)$ at $n+1$ equally spaced points on the interval $[-1, 1]$. The polynomial interpolant $p_n(t)$ shows extremely high oscillations near the two end-points ($|t| > 0.726, \ldots$). Hence, it is important to impose an additional optimality requirement (e.g., a uniform approximation requirement) on the interpolant. In this concern, the following result is useful [36].

Theorem 33.18: *Given a continuous function $f \in C[-1,1]$, let $\{t_k\}_{k=1}^n$ be the Chebyshev points on $[-1,1]$; namely, $t_k = \cos((2k-1)\pi/(2n)), k = 1, \ldots, n$. Let also $P_{2n-1}(t)$ be the polynomial of degree $2n-1$ that satisfies the following special Hermite interpolation conditions: $P_{2n-1}(t_k) = f(t_k)$ and $P_{2n-1}'(t_k) = 0, k = 1, \ldots, n$. Then the interpolant $P_{2n}(t)$ has the uniform approximation property*

$$\|f - P_{2n-1}\|_{L_\infty[-1,1]} \to 0 \quad \text{as} \quad n \to \infty$$

Because polynomial splines are piecewise algebraic polynomials, similar uniform approximation results for polynomial spline interpolants may be established [41, 72, 88].

Finally, a simultaneous interpolation and uniform approximation for a polynomial of a finite (and fixed) degree may be very desirable in engineering applications. The problem is that given an $f \in C[a,b]$ with $n+1$ points $a \le t_0 < t_1 < \cdots < t_n \le b$ and a given $\varepsilon > 0$, find a polynomial $p(t)$ of finite degree (usually, larger than n) that satisfies both

$$\|f - p\|_{L_\infty[a,b]} < \varepsilon \quad \text{and} \quad p(t_k) = f(t_k) \qquad k = 0, 1, \ldots, n$$

The answer to this question is the Walsh theorem, which states that this is always possible, even for complex polynomials [36]. Note that natural splines can also solve this simultaneous interpolation and uniform-approximation problem.

Best Approximation of Linear Functionals

As already mentioned, a functional is a mapping that maps functions to values. Definite integrals, derivatives evaluated at some points, and interpolation formulas are good examples of linear functionals.

The best approximation problem for a given bounded linear functional, L, by a linear combination of n independent and bounded linear functionals $L_1, \ldots, L_n$, all defined on the same normed linear space X of functions, can be similarly stated as follows: determine n constant coefficients $\{a_k^*\}_{k=1}^n$ such that

$$\left\| L - \sum_{k=1}^{n} a_k^* L_k \right\|_{X^*} = \min_{\{a_k\}} \left\| L - \sum_{k=1}^{n} a_k L_k \right\|_{X^*} \tag{33.37}$$

where X^* is the dual space of X, which is also a normed linear space. A basic result is described by the following theorem [36].

Theorem 33.19: *If X is a Hilbert space, then the best approximation problem* (33.37) *is uniquely solvable. Moreover, if r and $\{r_k\}_{k=1}^n$ are the functional representors of L and $\{L_k\}_{k=1}^n$, respectively, then*

$$\left\| r - \sum_{k=1}^{n} a_k r_k \right\|_{X^*} = \min \Rightarrow \left\| L - \sum_{k=1}^{n} a_k L_k \right\|_{X^*} = \min$$

It is important to note that for linear functionals, we have an interpolation problem: given bounded linear functionals L and $\{L_k\}_{k=1}^n$, all defined on a normed linear space X, where the last n functionals are linearly independent on X, and given also n points $x_k \in X, k = 1, \ldots, n$, determine n constant coefficients $\{a_k\}_{k=1}^n$ such that

$$\sum_{k=1}^{n} a_k L_k(x_i) = L(x_i) \qquad i = 1, \ldots, n$$

Obviously, this problem is uniquely solvable. Depending on the specific formulation of the linear functionals, a bulk of the approximation formulas in the field of numerical analysis can be derived from this general interpolation formulation.

Finally, convergence problems also can be formulated and discussed for bounded linear functionals in a manner similar to interpolation and approximation of functions. The following result is significant [36].

Theorem 33.20: *Let L and $\{L_k\}_{k=1}^{\infty}$ be bounded linear functionals defined on a Banach space X. A necessary and sufficient condition for*

$$\lim_{k\to\infty} \|L_k - L\|_{X^*} = 0$$

is that $\{L_k\}_{k=1}^{\infty}$ are uniformly bounded:

$$\|L_k\|_{X^*} \le M < \infty \quad \forall k = 1, 2, \ldots$$

and there is a convergent sequence $\{x_i\}_{i=1}^{\infty} \in X$, such that

$$\lim_{k\to\infty} L_k(x_i) = L(x_i) \quad \text{for each } i = 1, 2, \ldots$$

Artificial Neural Network for Approximation

Artificial neural networks offer a useful framework for signal and system approximations, including approximation of continuous and smooth functions of multivariables. Due to its usually multilayered structure with many weights, an artificial neural network can be "trained", and hence has a certain "learning" capability in data processing. For this reason artificial neural networks can be very efficient in performing various approximations. The main concern with a large-scale artificial neural network is its demand on computational speed and computer memory.

Both parametrized and nonparametrized approaches to approximations use artificial neural networks. In the parametrized approach the activation function, basic function, and network topology are all predetermined; hence, the entire network structure is fixed, leaving only a set of parameters (weights) to be adjusted to best fit the available data. In this way the network with optimal weights becomes a best approximant, usually in the least-squares sense, to a nonlinear system. Determining the weights from the data is called a **training process.** Back-propagation multilayered artificial neural networks are a typical example of the parametrized framework. The nonparametrized approach requires that the activation and/or basic functions also be determined, which turns out to be difficult in general.

To illustrate how an artificial neural network can be used as a system or signal approximant, we first describe the structure of a network. The term neuron used here refers to an operator or processing unit, which maps R^n to R, with the mathematical expression

$$o_i = f_a(f_b(\mathbf{i}_i, \mathbf{w}_i)) \tag{33.38}$$

where $\mathbf{i}_i = [i_1 \cdots i_n]^\top$ is the input vector, $\mathbf{w}_i = [w_{i1} \cdots w_{in}]^\top$ is the weight vector associated with the ith neuron, o_i the output of the ith neuron, f_a the activation function (usually sigmoidal or Gaussian), and f_b the basic function (which can be linear, affine, or radial). For example, if an affine basic function is used, (33.38) takes on the form

$$o_i = f_a(\mathbf{i}_i \cdot \mathbf{w}_i + b_i) \tag{33.39}$$

where b_i is a constant.

A fully connected feedforward artificial neural network is generally a multi-input/multi-output network, where the output from each neuron of each layer is an input to each neuron of the next layer. Such a network, arranged in one input layer, multiple hidden layers, and one output layer, can be constructed as follows (see Fig. 33.6). Suppose we have n-inputs, n_L-outputs and

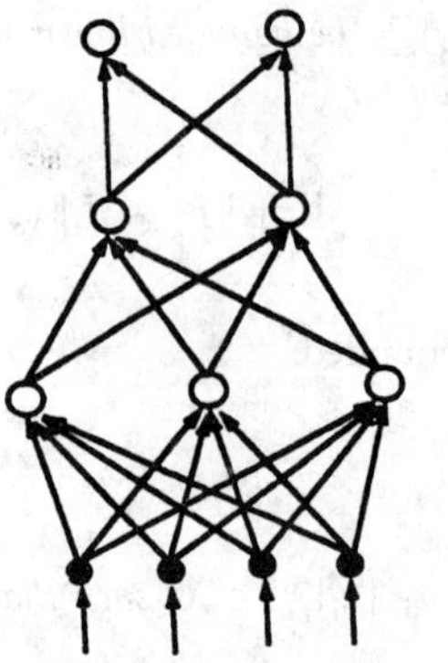

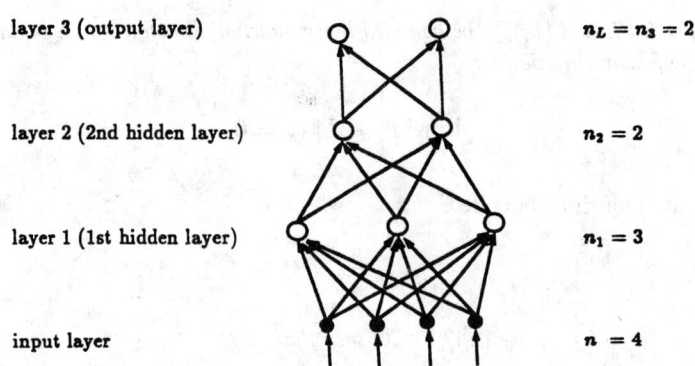

FIGURE 33.6 A two-hidden-layer feedforward artificial neural network.

$L-1$ hidden layers, and a linear basic function is used with a sigmoidal activation function $f_a(t) = \sigma(t)$:

$$\sigma(t) \to \begin{cases} 1 & \text{as } t \to +\infty \\ 0 & \text{as } t \to -\infty \end{cases}$$

Also, let $o_{l,i}$ be the output of the ith neuron at the lth layer and $\mathbf{w}_{l,i} = [w_{l,i,1} \cdots w_{l,i,s}]^\top$ be the weight vector associated with the same neuron connected to the neurons at the $(l-1)$st layer. Then, we have

$$o_{l,i} = \sigma\left(\sum_{j=1}^{n_l} o_{l-1,j} w_{l,i,j} + w_{l,i,0}\right) \tag{33.40}$$

Inductively, the output of the ith neuron in the last (the Lth) layer is given by

$$o_{L,i} = \sigma\left(\sum_{j=1}^{n_{L-1}} w_{L,i,j}\sigma\left(\cdots\sigma\left(\sum_{q=1}^{n_i} w_{1,p,q} i_{n_0} + w_{1,p,0}\right) + w_{2,p,0}\right) + \cdots + w_{L,i,o}\right) \tag{33.41}$$

where $i = 1, \ldots, n_L$.

The following best uniform approximation property of an artificial neural network is a fundamental result in neural-network approximation [35].

Theorem 33.21: *Let $f(\mathbf{t})$ be a continuous function defined on a compact subset $\Omega \subset R^n$. Then for any $\varepsilon > 0$, there exist an integer $m \geq 1$ and real parameters $\{c_k, w_{ki}, b_k\}_{k=1}^m$ such that using any nonconstant, bounded and monotonically increasing continuous function f_a as the activation function, the artificial neural network can uniformly approximate f on Ω, in the sense that*

$$\|f - N\|_{L_\infty(\Omega)} < \varepsilon$$

where the network has the form

$$N(\mathbf{t}) = \sum_{k=1}^{m} c_k f_a\left(\sum_{i=1}^{n} w_{k,i} t_i + b_k\right) \qquad \mathbf{t} = [t_1 \cdots t_n]^\top \in \Omega$$

Neural networks also can provide approximation for a mapping together with its derivatives [52]. On the other hand, neural networks can provide localized approximation, which is advantageous in that if a certain portion of the data is perturbed, only a few weights in the network need to be retrained. It was shown that a single hidden layered network cannot provide localized approximation of continuous functions on any compact set of a Euclidean space with dimension higher than one; however, two hidden layers are sufficient for the purpose [33].

As mentioned above, the basic function f_b in a network need not be linear. An artificial neural network, using a radial function for f_b, can also give very good approximation results [76]. Also, as a system approximation framework, stability of a network is very important [68]. Finally, a major issue that must be addressed in designing a large-scale network is the computer memory, which requires some special realization techniques [67].

3.4 Identification

System identification is a problem of finding a good mathematical model, preferably optimal in some sense, for an unknown physical system, using some available measurement data. These data usually include system outputs and sometimes also inputs. Very often, the available data are discrete, but the system to be identified is continuous [97].

A general formulation of the system identification problem can be described as follows. Let S be the family of systems under consideration (linear or nonlinear, deterministic or stochastic, or even chaotic), with input u and output y, and let $R_{(u,y)}$ be the set of I-O data. Define a mapping $M: S \to R_{(u,y)}$. Then, a system $\mathcal{F} \in S$ is said to be **(exactly) identifiable** if the mapping M is invertible, and the problem is to find the $\mathcal{F} = M^{-1}(\tilde{u}, \tilde{y})$ using the available data $(\tilde{u}, \tilde{y}) \in R_{(u,y)}$. Here, how to define the mapping M, linear or not, is the key to the identification problem. Usually, we also want M^{-1} be causal for the implementation purpose.

The first question about system identification is of course the *identifiability* [82]. Not all systems, not even linear deterministic systems, are exactly identifiable [21]. Because many physical systems are not exactly identifiable, system identification in a weaker sense is more realistic.

Suppose that some inputs and their corresponding outputs of an unknown system, $\mathcal{S}_1$, are given. We want to identify this unknown system by an approximate model, $\mathcal{S}_2$, using the available I-O data, such that the corresponding outputs produced by any input through $\mathcal{S}_1$ and $\mathcal{S}_2$, respectively, are "very close" under certain meaningful measure. If the structure of $\mathcal{S}_1$ (hence, $\mathcal{S}_2$) is known *a priori*, then what we need is to identify some system parameters. If the structure of $\mathcal{S}_1$ is not clear, the task becomes much more difficult because we must determine what kind of model to choose in approximating the unknown system [50]. This includes many crucial issues such as the linearity and dimension (or order) of the model used. In particular, if the system is nonlinear and contains uncertainties, special techniques from set-valued mapping and differential inclusion theories may be needed [58].

Usually the basic requirement is that $\mathcal{S}_2$ should be a best approximant of $\mathcal{S}_1$ from a desired class of simple and realizable models, under a suitably chosen criterion. For example, the least-squares operator approximation discussed previously can be thought of as an identification scheme. For this reason, identification in the weak sense is traditionally considered to be one of the typical best approximation problems in mathematics. If a minimal worst-case model-matching error bound is required, the approximation is known as the **optimal recovery** problem, for either functions or functionals [65, 66], or for operators [15, 45]. In system engineering it usually refers to **system identification** or **reconstruction**, with an emphasis on obtaining an identified model or a reconstruction scheme.

Generally speaking, system identification is a difficult problem, often leading to nonunique solutions when it is solvable. This is typically true for nonlinear circuits and systems. In systems and control engineering an unknown system is identified by a desired model such that they can produce "close enough" outputs from the same input, measured by a norm in the signal space, such as L_p, l_p or $H_p (1 \leq p \leq \infty)$. For dynamic systems, however, this norm-measure is generally not a good choice because one is concerned with nonlinear dynamics of the unknown system, such as limit cycles, attractors, bifurcations, and chaos. Hence, it is preferable to have an identified model that preserves the same dynamic behavior. This is a very challenging research topic; its fundamental theories and methodologies are still open for further exploration.

Linear Systems Identification

Compared to nonlinear systems, linear systems, either ARMA (autoregressive with moving-average) or state-space models, can be relatively easily identified, especially when the system dimension (order) is fixed. The mainstream theory of linear system identification has the following characteristics [37]:

1. The model class consists of linear, causal, stable, finite-dimensional systems with constant parameters.
2. Both system inputs and their corresponding outputs are available as discrete or continuous data.
3. Noise, if any, is stationary and ergodic (usually with rational spectral densities), white and uncorrelated with state vectors in the past.
4. Criteria for measuring the closeness in model-matching are of least-squares type (in the deterministic case) or of maximum likelihood type (in the stochastic case).
5. Large-scale linear systems are decomposed into lower-dimensional subsystems, and nonlinear systems are decomposed into linear and simple (e.g., memoryless) nonlinear subsystems.

Because for linear systems, ARMA models and state-space models are equivalent under a nonsingular linear transformation [17, 32], we discuss only ARMA models here.

An (n, m, l)th-order ARMAX model (an ARMA model with exogenous noisy inputs) has the general form

$$a(z^{-1})\mathbf{y}(t) = b(z^{-1})\mathbf{u}(t) + c(z^{-1})\varepsilon(t) \qquad t = \ldots, -1, 0, 1, \ldots \tag{33.42}$$

in which z^{-1} is the time-delay operator defined by $z^{-l}\mathbf{f}(t) = \mathbf{f}(t-l)$, and

$$a(z^{-1}) = \sum_{i=0}^{n} A_i z^{-i} \quad b(z^{-1}) = \sum_{j=0}^{m} B_j z^{-j} \quad c(z^{-1}) = \sum_{k=1}^{l} C_k z^{-k}$$

with constant coefficient matrices $\{A_i\}, \{B_j\}, \{C_k\}$ of appropriate dimensions, where $A_0 = I$ (or, is nonsingular). In the ARMAX model (33.42) $\mathbf{u}(t), \mathbf{y}(t)$, and $\varepsilon(t)$ are considered to be system input, output, and noise vectors, respectively, where the input can be either deterministic or random. In particular, if $l = 0$ and $n = 0$ (or $m = 0$), then (33.42) reduces to a simple MA (or AR) model. Kolmogorov [56] proved that every linear system can be represented by an infinite-order AR model. It is also true that every nonlinear system with a Volterra series representation can be represented by a nonlinear AR model of infinite order [53].

The system identification problem for the ARMAX model (33.42) can now be described as follows. Given the system I-O data $(\mathbf{u}(t), \mathbf{y}(t))$ and the statistics of $\varepsilon(t)$, determine integers (n, m, l) (system-order determination) and constant coefficient matrices $\{A_i\}, \{B_j\}, \{C_k\}$ (system-parameter identification). While many successful methods exist for system parameter identification [3, 23, 49, 60], system order determination is a difficult problem [47].

As already mentioned, the identifiability of an unknown ARMAX model using the given I-O data is a fundamental issue. We discuss the exact model identification problem here. The ARMAX model (33.42) is said to be **exactly identifiable** if $(\tilde{a}(z^{-1}), \tilde{b}(z^{-1}), \tilde{c}(z^{-1}))$ is an ARMAX model with $\tilde{n} \le n$, $\tilde{m} \le m$, and $\tilde{l} \le l$, such that

$$\begin{cases} [\tilde{a}(z^{-1})]^{-1}\tilde{b}(z^{-1}) = [a(z^{-1})]^{-1}b(z^{-1}) \\ [\tilde{a}(z^{-1})]^{-1}\tilde{c}(z^{-1}) = [a(z^{-1})]^{-1}c(z^{-1}) \end{cases}$$

Note that not all ARMAX models are exactly identifiable in this sense. A basic result about this identifiability is the following [21].

Theorem 33.22: *The ARMAX model (33.42) (with $t \ge 0$) is exactly identifiable iff $a(z^{-1}), b(z^{-1})$, and $c(z^{-1})$ have no common left factor and the rank of the constant matrix $[A_n, B_m, C_l]$, consisting of the highest order coefficient terms in $a(z^{-1}), b(z^{-1}), c(z^{-1})$, respectively, is equal to the dimension of the system output* $\mathbf{y}$.

Even if an unknown system is exactly identifiable and its identification is unique, how to find the identified system is still a very technical issue. For simple AR models, the well-known Levinson-Durbin algorithm is a good scheme for constructing the identified model; for MA models, one can use Trench-Zohar and Berlekamp-Massey algorithms. There exist some generalizations of these algorithms in the literature [23]. For stochastic models with significant exogenous noise inputs, various statistical criteria and estimation techniques, under different conditions, are available [82]. Various recursive least-squares schemes, such as the least-mean-square (LMS) algorithm [96], and various stochastic searching methods, such as the stochastic gradient algorithm [49], are popular. Because of their simplicity and efficiency, the successful (standard and extended) Kalman filtering algorithms [16, 30] have also been widely applied in parameters identification for stochastic systems [13, 17, 62], with many real-world applications [92].

Finally, for linear systems, a new framework, called the **behavioral approach**, is proposed for mathematical system modeling and some other related topics [2, 98].

Nonlinear Systems Identification

Identifying a nonlinear system is much more difficult than identifying a linear system in general, whether it is in the exact or in the weak sense, as is commonly known and can be seen from its information-based complexity analysis [15, 45].

For some nonlinear systems with simple Volterra series representations, the least-squares approximation technique can be employed for the purpose of identification in the weak sense [8]. As a simple illustrative example, consider the cascaded nonlinear system with noise input shown in Fig. 33.7. In this figure $h_1(t)$ and $h_2(t)$ are unit impulse responses of two linear subsystems, respectively, and $V_n(\cdot)$ is a memoryless nonlinear subsystem which is assumed to

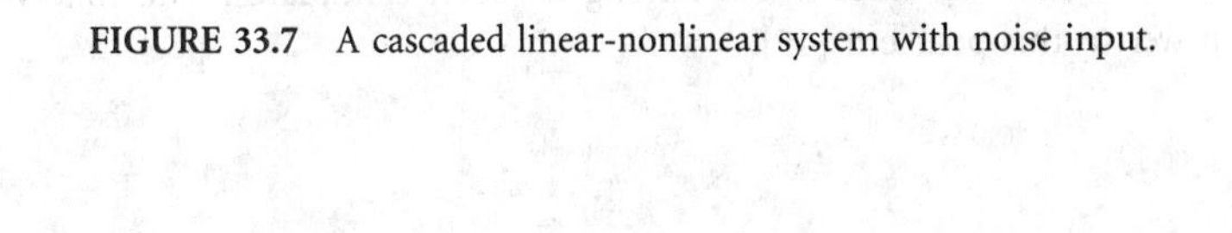

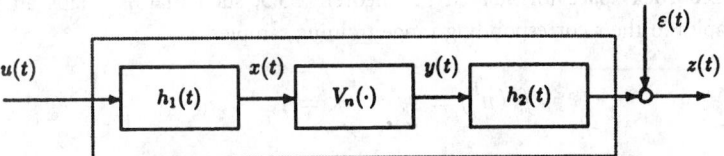

FIGURE 33.7 A cascaded linear-nonlinear system with noise input.

have an nth-order Volterra polynomial in the special form

$$y(t) = \sum_{k=1}^{n} c_k \int_0^t \cdots \int_0^t \phi_k(\tau_1, \ldots, \tau_k) x(\tau_1) \cdots x(\tau_k) d\tau_1 \cdots d\tau_k \tag{33.43}$$

where all the Volterra kernels $\{\phi_k\}_{k=0}^n$ are assumed to be known, but the constant coefficients $\{c_k\}_{k=0}^n$ must be identified.

It is clear from Fig. 33.7 that the output of the cascaded system can be expressed via convolution-type integrals as

$$z(t) = c_1 \left(\int h_2 \phi_1 h_1 u \right)(t) + \cdots + c_n \left(\int \cdots \int h_2 \phi_n h_1 \cdots h_1 u \cdots u \right)(t) + \varepsilon(t) \tag{33.44}$$

Now, because all the integrals can be computed if the input function $u(t)$ is given, the standard least-squares technique can be used to determine the unknown constant coefficients $\{c_k\}_{k=0}^n$, using the measured system output $z(t)$.

A neural network implementation of Volterra series model identification is described in [1]. Neural network for system identification has been used in many different cases, as can also be seen from [70].

Finally, we consider one approach to nonlinear systems identification which combines the special structure of the generalized Fock space of Volterra series (Theorem 33.3) and the "training" idea from neural networks discussed previously (Theorem 33.21). For simplicity, consider the scalar nonlinear system

$$y(t) + f(y(t-1), y(t-2), \ldots, y(t-n)) = u(t) \qquad t = 0, 1, \ldots \tag{33.45}$$

where n is a fixed integer, with the given initial conditions $y(-1) = y_1, \ldots, y(-n) = y_n$. Introducing a simple notation

$$y^{t-1} = (y(t-1), \ldots, y(t-n)) \tag{33.46}$$

we first rewrite this system as

$$y(t) + f(y^{t-1}) = u(t) \tag{33.47}$$

Then, we denote by E^n the n-dimensional Euclidean space of continuous functions and let $u^1, \ldots, u^m \in E^n$, called the **domain training samples**, be given data vectors that are componentwise nonzero and distinct, namely, $u_i^k \neq u_j^k$ if $i \neq j$ for all $k = 1, \ldots, m$, $1 \leq i$, $j \leq n$. Here, (33.46) also has been used for these domain training samples. Also, let $r_1, \ldots, r_m$, be given real numbers, called the corresponding **range training samples**. The identification problem is to find an approximate system, $f^*(\cdot)$, among all Volterra series representations from the generalized Fock space formulated in Theorem 33.3, such that f^* maps all the domain training samples to their corresponding range training samples:

$$f^*(u^k) = r_k \qquad k = 1, \ldots, m \tag{33.48}$$

and f^* has the minimum operator-norm among all such candidates. The following theorem provides an answer to this problem [45].

Theorem 33.23: *There is a unique element f^* of minimum norm in the generalized Fock space defined in Theorem* 33.3, *with the domain $\Omega = E^n$ therein, that satisfies the constraint* (33.48). *Moreover, f^* has the following expression:*

$$f^*(v) = \sum_{k=1}^{m} a_k K(u^k, v) \qquad \forall v \in E^n$$

where $K(\cdot, \cdot)$ is the reproducing kernel defined in Theorem 33.3, and the system parameters are determined by

$$\begin{bmatrix} a_1 \\ \vdots \\ a_m \end{bmatrix} = \begin{bmatrix} K(u^1, u^1) & \cdots & K(u^1, u^m) \\ \vdots & & \vdots \\ K(u^m, u^1) & \cdots & K(u^m, u^m) \end{bmatrix}^{-1} \begin{bmatrix} r_1 \\ \vdots \\ r_m \end{bmatrix}$$

Here, it should be noted that because K is a reproducing kernel, the set of functions $\{K(u^k, \cdot)\}_{k=1}^{m}$ are linearly independent, so that the above inverse matrix exists.

Also note that this system identification method can be applied to higher dimensional systems and the continuous-time setting [45].

Nonlinear Dynamic Systems Identification from Time Series

Measurement (observation) data obtained from an unknown system are often available in the form of time series. There are some successful techniques for identification of linear and nonlinear systems from time series if the time series is generated from Gaussian white noise. For example, for linear systems we have the Box-Jenkins scheme and for nonlinear systems, a statistical method using a nonlinear filter [75].

Concerned with nonlinear dynamic systems, however, statistical methods and the commonly used norm-measure criterion may not be capable of identifying the system dynamics in general. This is because the main issue in this concern is the nonlinear dynamic behavior of the unknown system, such as limit cycles, attractors, bifurcations, and chaos. Hence, it is preferable that the identified model can preserve the nonlinear dynamics of the unknown system. This turns out to be a very challenging task; many fundamental theories and methodologies for this task remain to be developed.

When an unknown nonlinear dynamic system is measured to produce a set of continuous or discrete data (a time series), a natural approach for studying its dynamics from the available time series is to take an integral transform of the series, so as to convert the problem from the time domain to the frequency domain. Then, some well-developed engineering frequency domain methods can be applied to perform analysis and computation of the nonlinear dynamics [69].

One common approach formulated in the time domain is the **(delay-coordinate) embedding method** that can be applied to reconstruct (identify) an unknown nonlinear dynamic model from which only a set of discrete measurement data (a time series) is available.

Let us consider the problem of identifying a periodic trajectory of an unknown nonlinear dynamic system using only an experimental time series measured from the system. Let $\{r_k\}$ be the available data. The embedding theory guarantees this can be done in the space R^m with the embedding dimension $m \geq 2n + 1$, where n is the dimension of the dynamic system [93], or $m \geq 2d_A$, where d_A is the dimension of the attractor [87]. A way to achieve this is to use the delay-coordinate technique, which approximates the unknown nonlinear dynamics in R^m by introducing the embedding vector

$$\mathbf{r}_k = [r_k r_{k-\mu} \cdots r_{k-(m-1)\mu}]^\top \tag{33.49}$$

where μ is the time-delay step. This embedding vector provides enough information to characterize the essence of the system dynamics and can be used to obtain an experimental Poincaré map, which helps in understanding the dynamics. For example, one may let the map be the equation of the first component of the vector being equal to a constant: r_{k_i} = constant. This procedure yields the successive points

$$\xi_i := [r_{k_i-\mu} \cdots r_{k_i-(m-1)\mu}]^\top \tag{33.50}$$

at the ith piercing of the map by the trajectory (or the vector $\mathbf{r}_k$), where k_i is the time index at the ith piercing. Then, one can locate the periodic trajectories of the unknown system using the experimental data [4, 14]. In this approach, however, determining a reasonably good time-delay step size, i.e., the real number μ in (33.49), remains an open technical problem.

Finally, we note that the embedding method discussed above has been applied to the control of chaotic circuits and systems [19, 20, 74].

References

[1] A. Abdulaziz and M. Farsi, "Non-linear system identification and control based on neural and self-tuning control," *Int. J. Adapt. Control Signal Process.*, vol. 7, no. 4, pp. 297–307, 1993.

[2] A. C. Antoulas and J. C. Willems, "A behavioral approach to linear exact modeling," *IEEE Trans. Autom. Control,* vol. 38, no. 12, pp. 1776–1802, 1993.

[3] K. J. Aström and P. Eykhoff, "System identification: a survey," *Automatica,* vol. 7, pp. 123–162, 1971.

[4] H. Atmanspacher, H. Scheingraber, and W. Voges, "Attractor reconstruction and dimensional analysis of chaotic signals," *Data Analysis in Astronomy III,* V. di Gesú, et al., Eds., New York: Plenum Press, 1988, pp. 3–19.

[5] I. Baesler and I. K. Daugavet, "Approximation of nonlinear operators by Volterra polynomials," *Am. Math. Soc. Transl.*, vol. 155, no. 2, pp. 47–57, 1993.

[6] S. P. Banks, *Mathematical Theories of Nonlinear Systems,* New York: Prentice-Hall, 1988.

[7] A. Bensoussan, G. de Prado, M. C. Felfour, and S. K. Mitter, *Representation and Control of Infinite Dimensional Systems,* Boston: Birkhäuser, 1992.

[8] S. A. Billings, "Identification of nonlinear systems—a survey," *IEE Proc., Pt D.*, vol. 127, pp. 272–285, 1980

[9] S. Boyd and L. O. Chua, "Fading memory and the problem of approximating nonlinear operators with Volterra series," *IEEE Trans. Circuits Syst.*, vol. 32, no. 11, pp. 1150–1161, 1985.

[10] S. Boyd, L. O. Chua, and C. A. Desoer, "Analytic foundations of Volterra series," *IMA J. Info. Control,* vol. 1, pp. 243–282, 1984.

[11] D. Braess, *Nonlinear Approximation Theory,* New York: Springer-Verlag, 1986.

[12] R. W. Brockett, "Volterra series and geometric control theory," *Automatica,* vol. 12, pp. 167–176, 1976.

[13] P. E. Caines, *Linear Stochastic Systems,* New York: Wiley, 1988.

[14] M. Casdagli, S. Eubank, J. D. Farmer, and J. Gibson, "State space reconstruction in the presence of noise," *Physica D,* vol. 51, pp. 52–98, 1991.

[15] G. Chen, "Optimal recovery of certain nonlinear analytic mappings," in *Optimal Recovery*, B. Bojanov and H. Wozniakowski, Eds., New York: Nova Science, 1992, pp. 141–144.

[16] G. Chen, (Ed.), *Approximate Kalman Filtering*, New York: World Scientific, 1993.

[17] G. Chen, G. Chen, and H. S. Hsu, *Linear Stochastic Control Systems*, Boca Raton, FL: CRC Press, 1995.

[18] G. Chen, C. K. Chui, and M. J. Lai, "Construction of real-time spline quasi-interpolation schemes," *Approx. Theory Appl.*, vol. 4, no. 4, pp. 61–75, 1988.

[19] G. Chen and X. Dong, "From chaos to order: perspectives and methodologies in controlling chaotic nonlinear dynamical systems," *Int. J. Bifurcation Chaos*, vol. 3, no. 6, pp. 1343–1389, 1993.

[20] G. Chen and J. Moiola, "An overview of bifurcation, chaos and nonlinear dynamics in control systems," *The Franklin Institute Journal*, in press.

[21] H. F. Chen and J. F. Zhang, "On identifiability for multidimensional ARMAX model," *Acta Math. Appl. Sin.*, vol. 9, no. 1, pp. 1–8, 1993.

[22] E. W. Cheney, *Introduction to Approximation Theory*, New York: McGraw-Hill, 1966.

[23] B. Choi, *ARMA Model Identification*, New York: Springer-Verlag, 1992.

[24] L. O. Chua, "The genesis of Chua's circuit," *AEU Int. J. Electr. Commun.*, vol. 46, pp. 187–257, 1992.

[25] L. O. Chua and A. C. Deng, "Canonical piecewise-linear modeling," *IEEE Trans. Circuits Syst.*, vol. 33, no. 5, pp. 511–525, 1986.

[26] L. O. Chua and A. C. Deng, "Canonical piecewise-linear representation," *IEEE Trans. Circuits Syst.*, vol. 35, no. 1, pp. 101–111, 1988.

[27] C. K. Chui, *Multi-Variate Splines*, Philadelphia, MA: SIAM, 1988.

[28] C. K. Chui, "Approximations and expansions," in *Encyclopedia of Physical Science and Technology*, vol. 2, New York: Academic Press, 1992, pp. 1–29.

[29] C. K. Chui, *An Introduction to Wavelets*, New York: Academic Press, 1992.

[30] C. K. Chui and G. Chen, *Kalman Filtering with Real-Time Applications*, New York: Springer-Verlag, 1987.

[31] C. K. Chui and G. Chen, *Linear Systems and Optimal Control*, New York: Springer-Verlag, 1989.

[32] C. K. Chui and G. Chen, *Signal Processing and Systems Theory: Selected Topics*, New York: Springer-Verlag, 1992.

[33] C. K. Chui, X. Li, and H. N. Mhaskar, "Neural networks for localized approximation," *Math. Comput.*, vol. 63, pp. 607–623, 1994.

[34] J. B. Conway, *A Course in Functional Analysis*, New York: Springer-Verlag, 1985.

[35] G. Cybenko, "Approximation by superposition of sigmoidal functions," *Math. Control, Signals Syst.*, vol. 2, pp. 303–314, 1989.

[36] P. J. Davis, *Interpolation and Approximation*, New York: Dover, 1975.

[37] M. Deistler, "Identification of Linear Systems (a survey)," in *Stochastic Theory and Adaptive Control*, T. E. Duncan and B. P. Duncan, Eds., New York: Springer-Verlag, 1991, pp. 127–141.

[38] C. A. Desoer and M. G. Kabuli, "Right factorizations of a class of time-varying nonlinear systems," *IEEE Trans. Autom. Control*, vol. 33, pp. 755–757, 1988.

[39] C. A. Desoer and M. Vidyasagar, *Feedback Systems: Input-Output Properties*, New York: Academic Press, 1975.

[40] R. A. DeVore and G. G. Lorentz, *Constructive Approximation*, New York: Springer-Verlag, 1993.

[41] C. de Boor, *A Practical Guide to Splines*, New York: Springer-Verlag, 1978.

[42] C. de Boor, K. Höllig, and S. Riemenschneider, *Box Splines*, New York: Springer-Verlag, 1993.

[43] R. J. P. de Figueiredo and G. Chen, "Optimal interpolation on a generalized Fock space of analytic functions," in *Approximation Theory VI*, C. K. Chui, L. L. Schumaker, and J. D. Ward, Eds., New York: Academic Press, 1989, pp. 247–250.

[44] R. J. P. de Figueiredo and G. Chen, "PDLG splines defined by partial differential operators with initial and boundary value conditions," *SIAM J. Numerical Anal.*, vol. 27, no. 2, pp. 519–528, 1990.

[45] R. J. P. de Figueiredo and G. Chen, *Nonlinear Feedback Control Systems: An Operator Theory Approach*, New York: Academic Press, 1993.

[46] R. J. P. de Figueiredo and T. A. Dwyer, "A best approximation framework and implementation for simulation of large scale nonlinear systems," *IEEE Trans. Circuits Syst.*, vol. 27, no. 11, pp. 1005–1014, 1980.

[47] J. G. de Gooijer, B. Abraham, A. Gould, and L. Robinson, "Methods for determining the order of an autoregressive-moving average process: a survey," *Int. Statist. Rev.*, vol. 53, pp. 301–329, 1985.

[48] M. Fliess, M. Lamnabhi, and F. L. Lagarrigue, "An algebraic approach to nonlinear functional expansions," *IEEE Trans. Circuits Syst.*, vol. 30, no. 8, pp. 554–570, 1983.

[49] G. C. Goodwin and K. S. Sin, *Adaptive Filtering, Prediction and Control*, Englewood Cliffs, NJ: Prentice-Hall, 1984.

[50] R. Haber and H. Unbehauen, "Structure identification of nonlinear dynamic systems—a survey on input/output approaches," *Automatica*, vol. 26, no. 4, pp. 651–677, 1990.

[51] J. Hammer, "Fraction representations of nonlinear systems: a simplified approach," *Int. J. Control*, vol. 46, pp. 455–472, 1987.

[52] K. Hornik, M. Stinchcombe, and H. White, "Universal approximation of an unknown mapping and its derivatives using multilayer feedforward networks," *Neural Networks*, vol. 3, pp. 551–560, 1990.

[53] L. R. Hunt, R. D. DeGroat, and D. A. Linebarger, "Nonlinear AR modeling," *Circuits, Syst. Signal Process.*, in press.

[54] A. Isidori, *Nonlinear Control Systems*, New York: Springer-Verlag, 1985.

[55] C. Kahlert and L. O. Chua, "A generalized canonical piecewise-linear representation," *IEEE Trans. Circuits Syst.*, vol. 37, no. 3, pp. 373–383, 1990.

[56] A. N. Kolmogorov, "Interpolation and extrapolation von stationaren zufallingen folgen," *Bull. Acad. Sci. USSR Ser. Math.*, vol. 5, pp. 3–14, 1941.

[57] S. Y. Kung, *Digital Neural Networks*, Englewood Cliffs, NJ: Prentice-Hall, 1993.

[58] A. B. Kurzhanski, "Identification—a theory of guaranteed estimates," in *From Data to Model*, J. C. Willems, Ed., New York: Springer-Verlag, 1989, pp. 135–214.

[59] C. Lesiak and A. J. Krener, "The existence and uniqueness of Volterra series for nonlinear systems," *IEEE Trans. Autom. Control*, vol. 23, no. 6, pp. 1090–1095, 1978.

[60] L. Ljung, *System Identification: Theory for the User*, Englewood Cliffs, NJ: Prentice-Hall, 1987.

[61] G. G. Lorentz, K. Jetter, and S. D. Riemenschneider, *Birkhoff Interpolation*, Reading, MA: Addison-Wesley, 1983.

[62] H. Lütkepohl, *Introduction to Multiple Time Series Analysis,* New York: Springer-Verlag, 1991.

[63] A. I. Mees, M. F. Jackson, and L. O. Chua, "Device modeling by radial basis functions," *IEEE Trans. Circuits Syst.,* vol. 39, no. 11, pp. 19–27, 1992.

[64] A. I. Mees, "Parsimonious dynamical reconstruction," *Int. J. Bifurcation Chaos,* vol. 3, no. 3, pp. 669–675, 1993.

[65] C. A. Micchelli and T. J. Rivlin, "A survey of optimal recovery," in *Optimal Estimation in Approximation Theory,* C. A. Micchelli and T. J. Rivlin, Eds., New York: Plenum Press, 1977, pp. 1–54.

[66] C. A. Micchelli and T. J. Rivlin, "Lectures on optimal recovery," in *Numerical Analysis,* P. R. Turner, Ed., New York: Springer-Verlag, 1984, pp. 21–93.

[67] A. N. Michel and J. A. Farrell, "Associative memories via artificial neural networks," *IEEE Control Syst. Mag.,* pp. 6–17, Apr. 1990.

[68] A. N. Michel, J. A. Farrell, and W. Porod, "Qualitative analysis of neural networks," *IEEE Trans. Circuits Syst.,* vol. 36, no. 2, pp. 229–243, 1989.

[69] J. L. Moiola and G. Chen, "Frequency domain approach to computation and analysis of bifurcations and limit cycles: a tutorial," *Int. J. Bifurcation Chaos,* vol. 3, no. 4, pp. 843–867, 1993.

[70] K. S. Narendra and K. Parthasarathy, "Identification and control of dynamic systems using neural networks," *IEEE Trans. Neural Networks,* vol. 1, no. 1, pp. 4–27, 1990.

[71] H. Nijmeijer and A. J. van der Schaft, *Nonlinear Dynamical Control Systems,* New York: Springer-Verlag, 1990.

[72] G. Nürnberger, *Approximation by Spline Functions,* New York: Springer-Verlag, 1989.

[73] W. Odyniec and G. Lewicki, *Minimal Projections in Banach Spaces,* New York: Springer-Verlag, 1990.

[74] E. Ott, C. Grebogi and J. A. Yorke, "Controlling chaos," *Phys. Rev. Lett.,* vol. 64, pp. 1196–1199, 1990.

[75] T. Ozaki, "Identification of nonlinearities and non-Gaussianities in time series," in *New Directions in Time Series Analysis, Part I,* D. Brillinger et al., Eds., New York: Springer-Verlag, 1993, pp. 227–264.

[76] J. Park and I. W. Sandberg, "Approximation and radial-basic-function networks," *Neural Comput.,* vol. 5, pp. 305–316, 1993.

[77] J. R. Partington, *An Introduction to Hankel Operators,* London: Cambridge University Press, 1988.

[78] P. P. Petrushev and V. A. Popov, *Rational Approximation of Real Functions,* London: Cambridge University Press, 1987.

[79] M. J. D. Powell, *Approximation Theory and Methods,* London: Cambridge University Press, 1981.

[80] M. J. D. Powell, "Radial basis functions for multi-variable interpolation: a review," in *IMA Conf. Algorithms for Approximation of Functions and Data,* RMCS, Shrivenham, U.K., 1985.

[81] E. Quak, N. Sivakumar, and J. D. Ward, "Least-squares approximation by radial functions," *SIAM J. Math. Anal.,* vol. 24, no. 4, pp. 1043–1066, 1993.

[82] B. L. S. P. Rao, *Identifiability in Stochastic Models,* New York: Academic Press, 1992.

[83] W. J. Rugh, *Nonlinear Systems Theory: The Volterra/Wiener Approach,* Baltimore: Johns Hopkins University Press, 1981.

[84] I. W. Sandberg, "A perspective on system theory," *IEEE Trans. Circuits Syst.,* vol. 31, no. 1, pp. 88–103, 1984.

[85] I. W. Sandberg, "Criteria for the global existence of functional expansions for input/output maps," *AT&T Tech. J.*, vol. 64, no. 7, pp. 1639–1658, 1985.

[86] I. W. Sandberg, "Approximately-finite memory and input-output maps," *IEEE Trans. Circuits Syst.*, vol. 39, no. 7, pp. 549–556, 1992.

[87] T. Sauer, J. A. Yorke, and M. Casdagli, "Embedology," *J. Statist. Phys.*, vol. 65, pp. 579–616, 1991.

[88] L. L. Schumaker, *Spline Functions: Basic Theory*, New York: Wiley, 1981.

[89] A. N. Sharkovsky, "Chaos from a time-delayed Chua's circuit," *IEEE Trans. Circuits Syst.*, vol. 40, no. 10, pp. 781–783, 1993.

[90] E. D. Sontag, *Polynomial Response Maps*, New York: Springer-Verlag, 1979.

[91] E. D. Sontag, *Mathematical Control Theory*, New York: Springer-Verlag, 1990.

[92] H. W. Sorenson, (Ed.), *Kalman Filtering: Theory and Applications*, New York: IEEE Press, 1985.

[93] F. Takens, "Detecting strange attractors in turbulence," in *Lecture Notes in Mathematics*, vol. 898, D. A. Rand and L. S. Yong, Eds., New York: Springer-Verlag, 1981, pp. 366–381.

[94] F. I. Utreras, "Positive thin plate splines," *J. Approx. Theory Appl.*, vol. 1, pp. 77–108, 1985.

[95] M. S. Verma and L. R. Hunt, "Right coprime factorizations and stabilization for nonlinear systems," *IEEE Trans. Autom. Control*, vol. 38, pp. 222–231, 1993.

[96] B. Widrow and S. D. Stearns, *Adaptive Signal Processing*, Englewood Cliffs, NJ: Prentice-Hall, 1985.

[97] J. C. Willems, (Ed.), *From Data to Model*, New York: Springer-Verlag, 1989.

[98] J. C. Willems, "Paradigms and puzzles in the theory of dynamical systems," *IEEE Trans. Autom. Control*, vol. 36, pp. 259–294, 1991.

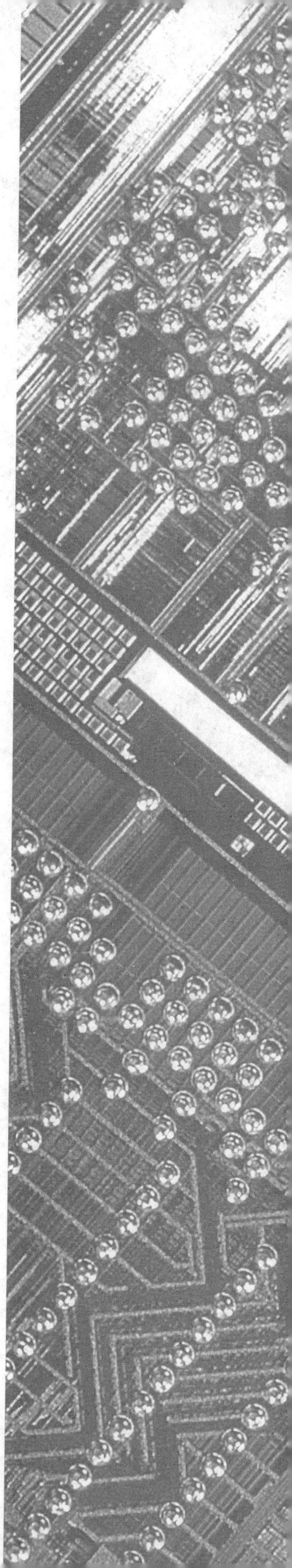

VII

Nonlinear Circuits II

Leon O. Chua
University of California

34

Transformation and Equivalence

Volfgang Mathis
University of Wuppertal,
Germany

34.1 General Equivalence Theorems for Nonlinear Circuits

One of the basic problems in the study of linear and nonlinear dynamical electrical networks is the analysis of the underlying descriptive equations and their solution manifold. These equations can be formulated using methods discussed in Chapter 30. In the case of linear or affine networks, the constitutive relations of network elements are restricted to classes of linear or affine functions and, therefore, possess rather restricted types of solutions. In contrast, the solution manifold of nonlinear networks may consist of many different types. Naturally, it is useful to decompose nonlinear networks into classes that possess certain similarities. One approach, for example, is to consider the solution manifold and to decompose solutions into similar classes. Furthermore, if the descriptive differential equations of dynamic networks are considered to be mathematical sets, their decompositions will be of interest.

The technique of equivalence relations is preferred method used to decompose a set of mathematical objects into certain classes. A well-known approach to define equivalence relations uses transformation groups. For example, real symmetric $n \times n$ matrices $\mathbb{R}_s^{n\times n}$ can be decomposed into equivalence classes by using the general linear transformation group $GL(n; \mathbb{R})$, and by applying the following similarity transformation:

$$\mathbf{M} \mapsto \mathbf{U}^{-1}\mathbf{M}\mathbf{U} \tag{34.1}$$

where $\mathbf{U} \in GL(n; \mathbb{R})$. By applying $GL(n; \mathbb{R})$, the set $\mathbb{R}_s^{n\times n}$ is decomposed into similarity classes that are characterized by certain eigenvalues. Furthermore, each class $\mathbb{R}_s^{n\times n}$ contains a diagonal matrix $\mathbf{D}$ with the eigenvalues on the main diagonal [33]. These eigenvalues are invariants of the group and characterize the classes. These and other related results can be applied to classify linear and affine dynamical networks [20]. Thus, properties of the A-matrix of the state-space equations are used for the classification. Note that each linear or affine network can be described in state-space form. We will discuss the theory of equivalence of linear and affine

3493-8341-2/95/$0.00 + $.50

dynamical networks only as special cases of nonlinear networks. A fine reformulation of the classical material of the decomposition of real matrices by using similarity transformations in the framework of one-parameter groups in $GL(n; \mathbb{R})$ is given by [22].

A classification of the vector fields is needed in order to classify differential equations of the type $\dot{\mathbf{x}} = \mathbf{f}(\mathbf{x})$, where $\mathbf{x} \in \mathbb{R}^n$ and $\mathbf{f}: \mathbb{R}^n \to \mathbb{R}^n$ is a vector field on $\mathbb{R}^n$. A first concept is established by a k-times differentiable change of coordinates that transforms the differential equation $\dot{\mathbf{x}} = \mathbf{f}(\mathbf{x})$ into $\dot{\mathbf{y}} = \mathbf{g}(\mathbf{y})$ by a function $\mathbf{h} \in C^k$. C^k is the set of k-fold continuously differentiable functions $\mathbf{h}: \mathbb{R}^n \to \mathbb{R}^n$. In other words, two vector fields are called C^k conjugate if there exists a C^k diffeomorphism $\mathbf{h}$ ($k \geq 1$) such that $\mathbf{h} \circ \mathbf{f} = \mathbf{g} \circ \mathbf{h}$. An equivalent formulation uses the concept of flows associated with differential equations $\dot{\mathbf{x}} = \mathbf{f}(\mathbf{x})$. A flow is a continuously differentiable function $\varphi: \mathbb{R} \times \mathbb{R}^n \to \mathbb{R}^n$ such that, for each $t \in \mathbb{R}$, the restriction $\varphi(t, \cdot) =: \varphi_t(\cdot)$ satisfies $\varphi_0 = \mathrm{id}_{\mathbb{R}^n}$ and $\varphi_t \circ \varphi_s = \varphi_{t+s}$ for all $t, s \in \mathbb{R}$. The relationship to a differential equation is given by

$$\mathbf{f}(\mathbf{x}) := \left.\frac{d\varphi_t}{dt}(\mathbf{x})\right|_{t=0} = \lim_{\varepsilon \to 0} \left\{ \frac{\varphi(\varepsilon, \mathbf{x}) - \varphi(0, \mathbf{x})}{\varepsilon} \right\}.$$

For more details see, for example, [3]. Two flows φ_t and ψ_t (associated with $\mathbf{f}$ and $\mathbf{g}$, respectively) are called C^k conjugate if there exists a C^k diffeomorphism $\mathbf{h}$ ($k \geq 1$) such that $\mathbf{h} \circ \varphi_t = \psi_t \circ \mathbf{h}$. In case when $k = 0$, the term C^k conjugate needs to be replaced by C^0 or topologically conjugate and $\mathbf{h}$ is a homeomorphism. Clearly, differential equations, vector fields, and flows are only alternative ways of presenting the same dynamics.

By above definitions, equivalence relations can be generated and the set of differential equations $\dot{\mathbf{x}} = \mathbf{f}(\mathbf{x})$ (as well as vector fields and flows) can be decomposed in certain classes. Although C^k conjugacy seems to be a natural concept for classifying differential equations, vector fields, and flows, this approach leads to a very refined classification (up to a diffeomorphism). In other words, too many systems become inequivalent. The following two examples illustrate this statement.

Consider the nonlinear dynamical circuit (see Fig. 34.1) with the descriptive equations (dimensionless form)

$$C\frac{dv_C}{dt} = i(v_C, i_L) - i_L \tag{34.2}$$

$$L\frac{di_L}{dt} = v_C + v(v_C, i_L) \tag{34.3}$$

If the nonlinear controlled sources are defined by

$$\begin{aligned} i = i(v_C, i_L) &:= -\frac{1}{2}\left(\sqrt{v_C^2 + i_L^2} - 1\right) v_C, \\ v = v(v_C, i_L) &:= -\frac{1}{2}\left(\sqrt{v_C^2 + i_L^2} - 1\right) i_L \end{aligned} \tag{34.4}$$

the following equations are derived:

$$C\frac{dv_C}{dt} = -\frac{1}{2}\left(\sqrt{v_C^2 + i_L^2} - 1\right) v_C - i_L \tag{34.5}$$

$$L\frac{di_L}{dt} = v_C - \frac{1}{2}\left(\sqrt{v_C^2 + i_L^2} - 1\right) i_L \tag{34.6}$$

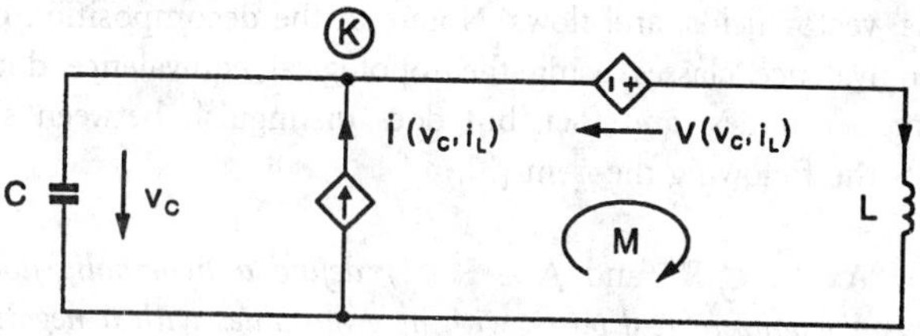

FIGURE 34.1 A nonlinear network.

These equations can be transformed to polar coordinates (r, ϕ) by $v_C := r\cos(\phi t)$ and $i_L := r\sin(\phi t)$ $(C = 1$ and $L = 1)$

$$\frac{dr}{dt} = \frac{1}{2}(1 - r)r \quad \text{and} \quad \dot{\phi} = 1 \tag{34.7}$$

If we consider the same circuit with parameters $C/2$ and $L/2$, the descriptive equations formulated in polar coordinates become

$$\frac{dr}{dt} = (1 - r)r \quad \text{and} \quad \dot{\phi} = 2 \tag{34.8}$$

Note that both differential equations differ only by a time rescaling $(t \mapsto 2t)$.

It can be shown (see [1], [28]) that if a diffeomorphism converts a singular point of a vector field into a singular point of another vector field, then the derivative of the diffeomorphism converts the Jacobian matrix of the first vector field at its singular point into the Jacobian matrix of the second field at its singular point. Consequently, these two Jacobian matrices are in the same similarity class and, therefore, have the same eigenvalues. In other words, the eigenvalues of the Jacobian matrices are invariants with respect to diffeomorphism, and the corresponding decomposition of the set of vector fields (differential equations and flows) is continuous rather than discrete. Obviously, the eigenvalues of (34.7) and (34.8) $(\lambda_1 = 1/2$, $\lambda_2 = 1$ and $\tilde{\lambda}_1 = 1$, $\tilde{\lambda}_2 = 2$, respectively) are different and, in conclusion, the two vector fields are not C^1 conjugate. Moreover, these two vector fields are not topologically or C^0 conjugate. A more "coarse" equivalence relation is needed in order to classify those vector fields, differential equations, and flows. As mentioned above, a time rescaling transforms the differential equations (34.7) and (34.8) into one another. This motivates the following definition.

Definition 1. Two flows φ_t and ψ_t are called C^k equivalent $(k \geq 1)$ if there exists a C^k diffeomorphism $\mathbf{h}$ that takes each orbit of φ_t into an orbit of ψ_t, preserving their orientation. In the case of $k = 0$, the flows are called C^0 or topologically equivalent.

Since C^k equivalence preserves the orientation of orbits, the relation $\mathbf{h}[\varphi_t(\mathbf{x})] = \varphi_{\tau_y(t)}(\mathbf{y})$ with $\mathbf{y} = \mathbf{h}(\mathbf{x})$ between φ_t and ψ_t is allowed, where τ_y is an increasing function of t for every $\mathbf{y}$.

It can be shown (see [28]) that the eigenvalues of the Jacobian matrices of two vector fields must be in the same ratio if a monotonic time rescaling is allowed. Therefore, the two vector fields (34.7) and (34.8) are C^1 equivalent. But, even the two linear vector fields of the equations

$$\begin{pmatrix} \dot{x} \\ \dot{y} \end{pmatrix} = \begin{pmatrix} 1 & 0 \\ 0 & 1 \end{pmatrix} \begin{pmatrix} x \\ y \end{pmatrix} \quad \begin{pmatrix} \dot{x} \\ \dot{y} \end{pmatrix} = \begin{pmatrix} 1 & 0 \\ 0 & 1 + \varepsilon \end{pmatrix} \begin{pmatrix} x \\ y \end{pmatrix} \tag{34.9}$$

are not C^1 equivalent for any $\varepsilon \neq 0$, although the solutions are very close for small ε in a finite time interval. In conclusion, topological equivalence is the appropriate setting for classifying differential equations, vector fields, and flows. Note that the decomposition of the set of linear vector fields into equivalence classes using the topological equivalence does not distinguish between nodes, improper nodes, and foci, but does distinguish between sinks, saddles, and sources. This suggests the following theorem [3].

Theorem 1: *Let* $\dot{\mathbf{x}} = \mathbf{A}\mathbf{x}$ $(\mathbf{x} \in \mathbb{R}^n$ and $\mathbf{A} \in \mathbb{R}^{n\times n})$ *define a hyperbolic flow on* $\mathbb{R}^n$*, i.e., the eigenvalues of* **A** *have only nonzero real parts, with* n_s *eigenvalues with a negative real part. Then,* $\dot{\mathbf{x}} = \mathbf{A}\mathbf{x}$ *is topologically equivalent to the system* $(n_u := n - n_s)$

$$\dot{\mathbf{x}}_s = -\mathbf{x}_s, \qquad \mathbf{x}_s \in \mathbb{R}^{n_s} \tag{34.10}$$

$$\dot{\mathbf{x}}_u = +\mathbf{x}_u, \qquad \mathbf{x}_u \in \mathbb{R}^{n_u} \tag{34.11}$$

Therefore, it follows that hyperbolic linear flows can be classified in a finite number of types using topological equivalence.

A local generalization of this theorem to nonlinear differential equations is known as the theorem of Hartman and Grobman (see [3]).

Theorem 2: *Let* $\mathbf{x}^*$ *be a hyperbolic fixed point of* $\mathbf{x} = \mathbf{f}(\mathbf{x})$ *with the flow* $\varphi_t : U \subseteq \mathbb{R}^n \to \mathbb{R}^n$*, i.e., the eigenvalues of the Jacobian matrix* $\mathbf{J}_f(\mathbf{x}^*)$ *have only nonzero real parts. Then, there is a neighborhood N of* $\mathbf{x}^*$ *on which* $\dot{\mathbf{x}} = \mathbf{f}(\mathbf{x})$ *is topologically equivalent to* $\dot{\mathbf{x}} = \mathbf{J}_f(\mathbf{x}^*)\mathbf{x}$.

The combination of the two theorems implies that a very large set of differential equations can be classified in an isolated hyperbolic fixed point by a finite number of types [namely, n_s(or n_u)]. The reason behind this interesting result is that the theorem of Hartman and Grobman, based on homeomorphisms, leads to a coarse decomposition of the set under consideration.

As a consequence of the above theorems, the behavior of nonlinear differential equations near a hyperbolic fixed point is equivalent up to a homeomorphism to the behavior of a simple system of linear differential equations. In the theory of nonlinear circuits these mathematical results can be interpreted in the following way: the behavior of nonlinear circuits near an operational point where the Jacobian matrix of descriptive equations (in state-space form) has only eigenvalues with nonzero real parts is "similar" to that of a suitable linear circuit (so-called small-signal behavior).

In the following section we discuss more general methods for the analysis of vector fields that have at least one nonhyperbolic fixed point.

34.2 Normal Forms

In Section 34.1 we presented theorems useful for classifying the "local" behavior of nonlinear differential equations near hyperbolic fixed points by using "global" results from the theory of linear differential equations. A main remaining problem is to calculate a homeomorphic transformation **h** in a concrete case. An alternative way to circumvent some of the difficulties is to apply the theory of normal forms that goes back to the beginning of this century and is based on classical ideas of Poincaré and Dulac. Detailed investigations of this subject are beyond the scope of this section and an interested reader should consult the monographs of [1], [3], [10], as well as [28], where further references the theory of normal forms can be found. In this section we present only the main ideas to illustrate its areas of application.

In contrast to theory described in Section 34.1, which is dedicated to hyperbolic cases, the theory of normal forms applies diffeomorphisms instead of homoemorphisms. This is necessary in order to distinguish the dynamical behavior in more detail. To classify the topological types of fixed points of the nonlinear differential equation $\dot{\mathbf{x}} = \mathbf{f}(\mathbf{x})$ one proceeds in two steps:

a) construction of a "normal form" in which the nonlinear terms of the vector field **f** take their "most simple" form, and b) determination of the topological type of the fixed point (under consideration) from the normal form. We present the main aspects of this "algorithm" without a proof.

First, we suppose that the vector field $\mathbf{f}(\mathbf{x})$ of the nonlinear differential equation $\dot{\mathbf{x}} = \mathbf{f}(\mathbf{x})$ satisfies $\mathbf{f}(\mathbf{0}) = \mathbf{0}$ (using a suitable transformation) and that it is represented by a formal Taylor expansion

$$\mathbf{f}(\mathbf{x}) = \mathbf{A}\mathbf{x} + \tilde{\mathbf{f}}(\mathbf{x}) \tag{34.12}$$

where $\mathbf{A} = \mathbf{J}_f(\mathbf{0})$ and $\tilde{\mathbf{f}}(\mathbf{x}) = \mathcal{O}(\|\mathbf{x}\|^2)$ is of class C^r. Power (Taylor) series with no assumptions about convergence are called *formal series.* In practice, we begin with formal series and then we determine the corresponding region of converence. Then, we apply a diffeomorphic C^r change of coordinates $\mathbf{h}: \mathbb{R}^n \to \mathbb{R}^n$ with $\mathbf{y} \mapsto \mathbf{x} = \mathbf{h}(\mathbf{y})$ ($\mathbf{h}(\mathbf{0}) = \mathbf{0}$) in the form of a near identity transformation

$$\mathbf{h}(\mathbf{y}) := \mathbf{y} + \mathbf{h}^k(\mathbf{y}) \tag{34.13}$$

where $\mathbf{h}^k(\mathbf{y})$ is a homogeneous polynomial of order k in $\mathbf{y}$ ($k \geq 2$). The result of the transformation is

$$\dot{\mathbf{y}} = \{\mathrm{id} + \mathbf{h}_y^k(\mathbf{y})\}^{-1}\mathbf{A}\{\mathbf{y} + \mathbf{h}^k(\mathbf{y})\} + \{\mathrm{id} + \mathbf{h}_y^k(\mathbf{y})\}^{-1}\tilde{\mathbf{f}}[\mathbf{y} + \mathbf{h}^k(\mathbf{y})] \tag{34.14}$$

$$= \mathbf{A}\mathbf{y} + \mathbf{g}(\mathbf{y}) \tag{34.15}$$

where $\mathbf{h}_y^k$ is the Jacobian matrix of $\mathbf{h}^k$ with respect to $\mathbf{y}$, and $\mathbf{g}(\mathbf{y}) = \mathcal{O}(\|\mathbf{y}\|^2)$ is of class C^r. It is useful to define $\mathbf{f}^k$ of a function $\mathbf{f}: \mathbb{R}^n \to \mathbb{R}^n$ as the "truncated" Taylor series in $\mathbf{0}$ expansion that have a degree less or equal k; the ith-order terms are denoted by $\mathbf{f}_k$. The set of f^k's forms a real vector space of functions whose components are homogeneous polynomials in n variables of degree less or equal k. The vector space of f_k's, the homogeneous polynomials in n variables of degree k, is denoted by H_k^n. Using the k-jet notation and expanding $\mathbf{g}$ into a formal Taylor series, (34.15) can be reformulated as

$$\dot{\mathbf{y}} = \mathbf{g}^{k-1}(\mathbf{y}) + \overset{k}{\mathbf{g}}(\mathbf{y}) \tag{34.16}$$

where $\overset{k}{\mathbf{g}}(\mathbf{y})$ contains all terms of degree k or higher. Expanding $\tilde{\mathbf{f}}$ into a formal Taylor series

$$\tilde{\mathbf{f}} = \tilde{\mathbf{f}}_2 + \tilde{\mathbf{f}}_3 + \cdots, \tag{34.17}$$

(34.16) and (34.17) can be represented by

$$\dot{\mathbf{y}} = \mathbf{g}^{k-1}(\mathbf{y}) + \{\tilde{\mathbf{f}}_k - [\mathbf{A}\mathbf{y}, \mathbf{h}^k]\} + \mathcal{O}(\|\mathbf{y}\|^{k+1}) \tag{34.18}$$

where $[\mathbf{A}\mathbf{y}, \mathbf{h}^k]$ is the so-called *Lie bracket* of the linear vector field $\mathbf{A}\mathbf{y}$ and $\mathbf{h}^k(\mathbf{y})$ is defined by

$$[\mathbf{A}\mathbf{y}, \mathbf{h}^k] := \mathbf{h}_y^k(\mathbf{y})\mathbf{A}\mathbf{y} - \mathbf{A}\mathbf{h}^k(\mathbf{y}) \tag{34.19}$$

Define a linear operator $L_A^k: H_k^n \to H_k^n$ on H_k^n by

$$L_A^k\mathbf{h}^k: \mathbf{y} \mapsto [\mathbf{A}\mathbf{y}, \mathbf{h}^k] \tag{34.20}$$

with the range $\mathscr{R}_k$, and let $\mathscr{C}_k$ be any complementary subspace to $\mathscr{R}_k$ in H_k^n, that is $H_k^n = \mathscr{R}_k \oplus \mathscr{C}_k$ $(k \geq 2)$. Then, the following theorem implies a simplification of a nonlinear differential equation $\dot{\mathbf{x}} = \mathbf{f}(\mathbf{x})$.

Theorem 3: *Let* $\mathbf{f}: \mathbb{R}^n \rightarrow \mathbb{R}^n$ *be a* C^r *vector field with* $\mathbf{f}(\mathbf{0}) = \mathbf{0}$ *and* $\mathbf{A} \in \mathbb{R}^{n \times n}$, *and let the decomposition* $H_k^n = \mathscr{R}_k \oplus \mathscr{C}_k$ *of* H_k^n *be given. Then, there exists a series of near identity transformations* $\mathbf{x} = \mathbf{y} + \mathbf{h}_k(\mathbf{y})$ $(\mathbf{y} \in \Omega,\ k = 2, 3, \ldots, r)$ *in some region* Ω, *where* $\mathbf{h}_k \in H_k^n$ *and* Ω *is a neighborhood of the origin, such that the equation* $\dot{\mathbf{x}} = \mathbf{f}(\mathbf{x})$ *is transformed to*

$$\dot{\mathbf{y}} = \mathbf{A}\mathbf{y} + g_2(\mathbf{y}) + \cdots + \mathbf{g}_k(\mathbf{y}) + \mathscr{O}(\|\mathbf{y}\|^{r+1}) \qquad \mathbf{y} \in \Omega \tag{34.21}$$

where $\mathbf{g}_k \in \mathscr{C}_k$ *for* $k = 2, 3, \cdots$.

The proof of this theorem and the following definition can be found in [4].

Definition 2. Let $\mathscr{R}_k \oplus \mathscr{C}_k$ be decompositions of H_k^n for $k = 2, 3, \ldots, r$. The truncated equation (34.21)

$$\dot{\mathbf{y}} = \mathbf{A}\mathbf{y} + \mathbf{g}_2(\mathbf{y}) + \cdots + \mathbf{g}_r(\mathbf{y}) \tag{34.22}$$

where $\mathbf{g}_k \in \mathscr{C}_k$ $(k = 2, 3, \ldots, r)$, is called normal form of $\dot{\mathbf{x}} = \mathbf{A}\mathbf{x} + \tilde{\mathbf{f}}(\mathbf{x})$ associated with matrix $\mathbf{A}$ up to order $r \geq 2$ (with respect to the decomposition $\mathscr{R}_k \oplus \mathscr{C}_k$), or an A-normal form of $\dot{\mathbf{x}} = \mathbf{f}(\mathbf{x})$.

Theorem 3 suggests an equivalence relation in the set of vector fields $\mathbf{f}$ that decomposes the set into equivalence classes. Each class can be represented by using definition of normal forms. Because a concrete normal form depends on the choice of complementary subspaces $\mathscr{C}_k$, it is not unique. In practical problems a constructive method for finding these subspaces is needed. An elegant way to find the subspaces is to start with the introduction of a suitable inner product $\langle \cdot | \cdot \rangle_n$ in H_k^n that is needed to define the adjoint operator $(L_A^k)^*$ of L_A^k (in a usual way) by

$$\langle \eta \,|L_A^k(\xi)\rangle_n := \langle (L_A^k)^*(\eta)\,|\xi\rangle_n, \quad \text{for all } \eta,\ \xi \in H_k^n \tag{34.23}$$

It can be shown that $(L_A^k)^* = L_{A^*}^k$, where $\mathbf{A}^* = \mathbf{A}^T$, is the transposed matrix of $\mathbf{A}$. The desired construction is available as an application of the following theorem.

Theorem 4: *Vector space* $\ker\{L_{A^*}^k\}$ *that is the solution space of the equation* $L_{A^*}^k \xi = 0$ *is a complementary subspace of* $\mathscr{R}_k$ *in* H_k^n, *i.e.,*

$$H_k^n = \mathscr{R}_k \oplus \ker\{L_{A^*}^k\} \tag{34.24}$$

Interested reader is referred to [4] for a detailed discussion of this subject. As a consequence, finding a normal form in the above defined sense up to the order r, requires solving the partial differential equation. From the algebraic point of view this means that a base of $\ker\{L_{A^*}^k\}$ has to be chosen, but this can be done, again, with some degree of freedom. For example, the two sets of differential equations are distinct normal forms of $\dot{\mathbf{x}} = \mathbf{A}\mathbf{x} + \tilde{\mathbf{f}}(\mathbf{x})$ associated with the same matrix $\mathbf{A}$ (see [4]):

$$\frac{d}{dt}\begin{pmatrix} x_1 \\ x_2 \end{pmatrix} = \begin{pmatrix} 0 & 1 \\ 0 & 0 \end{pmatrix}\begin{pmatrix} x_1 \\ x_2 \end{pmatrix} + \begin{pmatrix} ax_1^2 \\ bx_2^2 + ax_1x_2 \end{pmatrix} \tag{34.25}$$

$$\frac{d}{dt}\begin{pmatrix} x_1 \\ x_2 \end{pmatrix} = \begin{pmatrix} 0 & 1 \\ 0 & 0 \end{pmatrix}\begin{pmatrix} x_1 \\ x_2 \end{pmatrix} + \begin{pmatrix} 0 \\ ax_1^2 + bx_1x_2 \end{pmatrix} \tag{34.26}$$

To reduce the number of nonlinear monomials in the normal form, a more useful base of $\mathscr{C}_k$ must be determined. If a nonlinear differential equation of the form $\dot{\mathbf{x}} = \mathbf{A}\mathbf{x} + \tilde{\mathbf{f}}(\mathbf{x})$ is given with an arbitrary matrix **A**, several partial differential equations need to be solved. This, in general, is not an easy task. If **A** is diagonal or has an upper triangular form, methods for constructing a base are available. For this purpose we introduce the following definition.

Definition 3. Let $\mathbf{A} \in \mathbb{R}^{n\times n}$ possess the eigenvalues $\lambda_1, \ldots, \lambda_n$ and let $x_1^{\alpha_1} x_2^{\alpha_2} \cdots x_n^{\alpha_n} \mathbf{e}_j$ be a monomial in n variables. It is called a resonant monomial if the so-called resonance condition

$$\alpha^T \lambda - \lambda_j = 0 \tag{34.27}$$

is satisfied $[\alpha^T := (\alpha_1 \cdots \alpha_n),\ \lambda^T := (\lambda_1 \cdots \lambda_n)]$. If the resonance condition holds for some $\alpha_1 + \alpha_2 + \cdots + \alpha_n \geq 2$ and some $j \in \{1, \ldots, n\}$, we say that **A** has a resonant set of eigenvalues.

The next theorem shows that if **A** is diagonal, a minimal normal form exists (in certain sense).

Theorem 5: *Let* $\mathbf{A} = \mathrm{diag}(\lambda_1, \ldots, \lambda_n)$. *Then an A-normal form equation up to order r can be chosen to contain all resonant monomials up to order r.*

If some eigenvalues of **A** are complex, a linear change to complex coordinates is needed to apply this theorem. Furthermore, theorems and definitions need to be modified to complex cases. In the case of differential equations

$$\frac{d}{dt}\begin{pmatrix} x_1 \\ x_2 \end{pmatrix} = \begin{pmatrix} 0 & -1 \\ +1 & 0 \end{pmatrix}\begin{pmatrix} x_2 \\ x_2 \end{pmatrix} + \mathscr{O}(\|\mathbf{x}\|^2) \tag{34.28}$$

that can be used to describe oscillatory circuits, the coordinates are transformed to

$$\dot{z}_1 := x_1 + jx_2 \tag{34.29}$$

$$\dot{z}_2 := x_1 - jx_2 \tag{34.30}$$

with the resonant set of eigenvalues $\{-j, +j\}$ that can be written in a complex normal form (z presents one of the z_i's):

$$\dot{z} = jz + a_1|z|^2 z + \cdots + a_k|z|^{2k} z \tag{34.31}$$

This last normal form equation—Poincaré's normal form—is used intensively in the theory of the Poincaré–Andronov–Hopf bifurcation. The book of Hale and Kocak [31], is worth reading for the illustration of this phenomenon in nonlinear dynamical systems. A proof of the Poincaré–Andronov–Hopf theorem can be found in [32]. These authors emphasize that further preparation of the system of nonlinear differential equations is needed before the normal form theorem is applicable. In general, the linearized part of $\dot{\mathbf{x}} = \mathbf{A}\mathbf{x} + \tilde{\mathbf{f}}(\mathbf{x})$ (in a certain fixed point of the vector field **f**) has two classes of eigenvalues: *central* eigenvalues that lie on the imaginary axis, and *noncentral* eigenvalues. The dynamical behavior that is associated with the noncentral eigenvalues is governed by the theorem of Grobman and Hartman (see Section 34.1). A systematic procedure to eliminate these noncentral eigenvalues from the system is based on the so-called center manifold theorem (see [11]) that is used also in the paper of Hassard and Wan. A numerical algorithm is presented by Hassard, Kazarinoff and Wan [31]. Discussion of

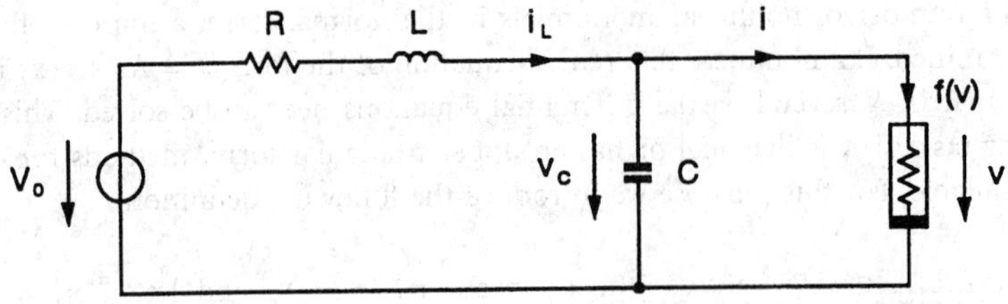

FIGURE 34.2 Nonlinear network with tunnel diode.

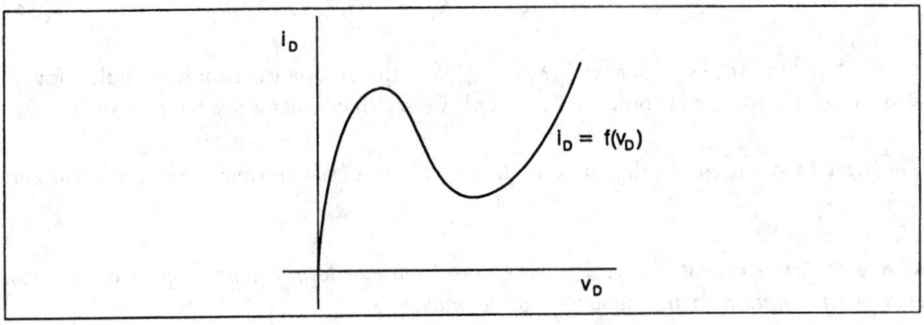

FIGURE 34.3 Characteristic of a tunnel diode.

this essential theorem goes beyond the scope of this section. A detailed presentation of the theory of normal forms of vector fields and its applications can be found in [21], [22].

Until now we have discussed differential equations of the type $\dot{\mathbf{x}} = \mathbf{f}(\mathbf{x})$. Descriptive equations of circuits consist, in general, of linear and nonlinear algebraic equations, as well as differential equations. Therefore, so-called constrained differential equations need to be considered for applications in network theory. A typical example is the well-known circuit shown in Fig. 34.2, containing a model of a tunnel diode (see [9], [41]). Circuit equations can be written as

$$C\frac{dv_C}{dt} = i_L - f(v_C) \tag{34.32}$$

$$L\frac{di_L}{dt} = -v_C - Ri_L + V_0 \tag{34.33}$$

where $i_D = f(v_D)$ is the characteristics of the tunnel diode (see Fig. 34.3).

If the behavior of the circuit is considered with respect to the time scale $\tau := RC$, then the differential equations (34.32) and (34.33) can be reformulated as

$$\frac{dv}{d\theta} = i - F(v) \tag{34.34}$$

$$\varepsilon\frac{di}{d\theta} = -v - i + 1 \tag{34.35}$$

with $\theta := t/\tau$, $i := (R/V_0)i_L$, $v := v_C/V_0$, $\varepsilon := L/(CR^2)$, and $F(v) := (R/V_0)f(V_0 \cdot v)$. The behavior of the *idealized* circuit, where $\varepsilon = 0$ is often of interest. These types of descriptive equations are called **constrained equations.** The theory of normal forms cannot be applied on descriptive equations of this type. Fortunately, a generalized theory is available, as presented (with applications) in [24], [25].

34.3 Dimensionless Form

In Section 34.1 the problem of equivalence is considered in a rather complex manner by adopting the mathematical point of view of dynamical systems. We will use this approach in Section 34.6 to classify nonlinear dynamical networks. In contrast to this approach, another kind of equivalence of descriptive equations for linear time-invariant circuits and for certain nonlinear circuits is well known. It is illustrated by means of a simple example. The *RLC* parallel circuit in Fig. 34.4, analyzed in the frequency domain, can be described by its impedance

$$Z = \frac{1}{\frac{1}{R} + j\left(\omega C - \frac{1}{\omega L}\right)} \tag{34.36}$$

In the case of small damping, this formula can be simplified using the so-called Q-factor that is defined as the ratio of the current flowing through L (or C) to the current through the circuit in the case of resonance. The case of small damping is characterized by the condition $R/(2L) \ll \omega_0$, where $\omega_0 := 1/\sqrt{LC}$ (Thompson's formula for LC circuits with $R = 0$). The Q-factor is given by $Q := 1/(\omega_0 CR) = \omega_0 L/R$. Simple calculations lead to a so-called normalized impedance

$$\tilde{Z} := \frac{Z}{R} = \left[1 - jQ^{-1}\left(\frac{\omega}{\omega_0} - \frac{\omega_0}{\omega}\right)\right]^{-1} \tag{34.37}$$

that contains only two instead of three parameters. By using this method, a whole class of *RLC* circuits may be described by the same normalized impedance. Therefore, an equivalence relation is defined in this manner. Handbooks written for practical electrical engineers contain diagrams of those and of similar curves of normalized impedances and admittances. Note that the formula is exact, although the interpretations of the parameters Q and ω_0 depend on the condition $R/(2L) \ll \omega_0$.

Methods for normalizing descriptive equations of circuits and for reducing the number of parameters are known in linear and nonlinear circuit theory (see [20]). Unfortunately, these methods are stated without a presentation of their mathematical foundations. The main ideas for justification of normalization procedures are based on so-called dimensional analysis. Their first applications in physics and the development of their mathematical foundations can be traced to the end of the last century. In this section, we discuss only a few aspects of this subject. Interested reader may find more details about the theory and the applications of dimensional analysis in the paper by Mathis and Chua [38]. It was shown there that for a complete mathematical discussion of physical quantities, several algebraic introductions are necessary. In this section, a concise introduction is preferred and, therefore, an intuitive introduction based on Lie groups is presented. The following presentation uses ideas from the book of Ovsiannikov [42].

For describing the physical arrangements, we require descriptive quantities that can be measured. To perform a physical measurement, we need at least one measuring instrument that provides a value on its scale. Intuitively, a mathematical model of a physical quantity ϕ consists of two parts: a real number $|\phi|$ that characterizes its value, and symbol E_ϕ that is due to the

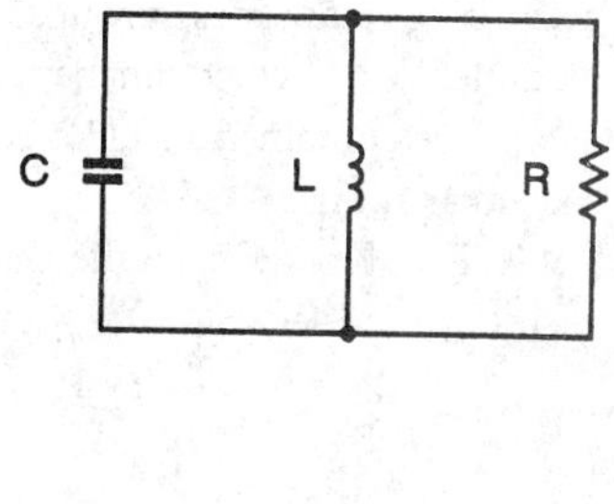

FIGURE 34.4 LCR network.

measuring arrangement of ϕ. In general, a measuring arrangement is composed of elementary measuring instruments to evaluate, for example, time or frequency, length, voltage, and charge. Each elementary measuring instrument will have associated a symbol E_k. Therefore, a physical quantity ϕ is defined by

$$\phi := |\phi| E_1^{\lambda_1} E_2^{\lambda_2} \cdots E_r^{\lambda_r} \tag{34.38}$$

where integers λ_k determine how many times an instrument is applied and whether the value on the scale needs to be multiplied or divided. The *dimensionality* of a physical quantity ϕ is defined by

$$[\phi] := E_1^{\lambda_1} E_2^{\lambda_2} \cdots E_r^{\lambda_r} \tag{34.39}$$

where $r \leq n$. A quantity is called **dimensionless** if its dimensionality is null, that is if $\lambda_1 = 0$, $\lambda_2 = 0, \ldots, \lambda_r = 0$. Moreover, a set of n physical quantities $\phi_1, \phi_2, \ldots, \phi_n$ is called *dependent* (in the sense of their dimensionality) if there exist integers $\chi_1, \chi_2, \ldots, \chi_n$ (not all equal zero) such that the product of these quantities

$$\phi_1^{\chi_1} \phi_2^{\chi_2} \cdots \phi_n^{\chi_n} \tag{34.40}$$

is dimensionless. Otherwise, $\phi_1, \phi_2, \ldots, \phi_n$ are called **independent**.

The main problem of dimension theory is to determine how many independent physical quantities are in a given set $\phi_1, \phi_2, \ldots, \phi_n$, to find them, and then to express the other quantities in terms of these independent quantities. As an application, a systematic procedure to normalize physical descriptive equations can be derived.

In order to solve the main problem, the change of measuring instruments and measuring scales needs to be introduced. Obviously, in terms of modeling physical quantities, the same ϕ can be represented in different ways. Using two sets of measuring instruments, denoted by $E_1, E_2, \ldots, E_r$ and $\tilde{E}_1, \tilde{E}_2, \ldots, \tilde{E}_r$, respectively, ϕ is given by

$$\phi = |\phi| E_1^{\lambda_1} E_2^{\lambda_2} \cdots E_r^{\lambda_r} = |\tilde{\phi}| \tilde{E}_1^{\tilde{\lambda}_1} \tilde{E}_2^{\tilde{\lambda}_2} \cdots \tilde{E}_\rho^{\tilde{\lambda}_\rho} \tag{34.41}$$

where in general $r \neq \rho$. This suggests the so-called **analogy transformation** (see [38]).

$$E_k = a_1^{\alpha_1} \cdot a_k^{\alpha_k} \tilde{E}_1^{\alpha_1} \cdots \tilde{E}_k^{\alpha_k} \qquad k = 1, \ldots, r \tag{34.42}$$

Transformations of the scales of measuring instruments (in the following *scale transformation*) are special analogy transformations

$$E_k = a_k \tilde{E}_k, \qquad k = 1, \ldots, r \tag{34.43}$$

It can be shown that analogy transformations of dimension theory are special cases of so-called *extension groups*. These groups belong to the *r-parameter Lie groups*, and, subsequently, all results of dimension theory can be interpreted by theorems from this mathematical theory. Further details are contained in [42]. To introduce the main ideas, we consider in this section only scale transformations.

Let $Z := \mathbb{R}^n \times \mathbb{R}^m$ be the Cartesian product of the set of n-column vectors $\mathbf{x}$ and m-column vectors $\mathbf{y}$, and let $\{\mathbf{e}_1, \ldots, \mathbf{e}_n\}$ and $\{\mathbf{f}_1, \ldots, \mathbf{f}_m\}$, be, respectively, arbitrary (but fixed) bases of this vector spaces. Endowing Z with the structure of a direct sum $\mathbb{R}^n \oplus \mathbb{R}^m$, each $\mathbf{z} \in Z$ can be represented by (with respect to the bases)

$$\mathbf{z} = \sum_{i=1}^{n} x^i \mathbf{e}_i + \sum_{k=1}^{n} y^k \mathbf{f}_k \tag{34.44}$$

An *extension* of Z (with respect to the bases) is defined by the transformation

$$h: \mathbf{z} \mapsto \sum_{i=1}^{n} c^i x^i \mathbf{e}_i + \sum_{k=1}^{n} d^k y^k \mathbf{f}_k \qquad (c^i > 0,\ d^k > 0) \tag{34.45}$$

Obviously, the set of all extensions generates Abel's group of transformations on Z that is a $(n+m)$-parameter Lie group. This group is denoted by $\mathrm{Diag}\{\mathbf{e}_i, \mathbf{f}_k\}$. Any subgroup $H \subset \mathrm{Diag}\{\mathbf{e}_i, \mathbf{f}_k\}$ is called an *extension group* of Z. We now consider extension groups H^r, with $0 < r \leq n + m$.

Ossiannikov showed that extensions of H^r can be represented, choosing a parametric group, in the form

$$\tilde{x}^i = x^i \prod_{\alpha=1}^{r} (a_\alpha)^{\lambda_\alpha^i} \qquad \tilde{y}^k = y^k \prod_{\alpha=1}^{r} (a_\alpha)^{\mu_\alpha^k} \tag{34.46}$$

where $i = 1, \ldots, n$ and $k = 1, \ldots, m$.

The main property of transformation groups is that they induce equivalence relations decomposing the subjects into equivalence classes on which the group acts. If h_p acts on elements $x \in X$, and $\mathbf{p} \in \mathbb{R}^p$ is the vector of parameters, an *orbit* U *of a point* $x \in X$ is defined by the set $U := \{\xi \in X | \xi = h_p(x, \mathbf{p}), \text{for all } \mathbf{p} \in \mathbb{R}^p\}$. In this sense, the points of an orbit can be identified by a transformation group. A transformation group acts *transitive on* X if there exists an orbit U that is an open subset of X, with $\bar{U} = X$.

To study so-called local Lie groups whose actions are defined near a null neighborhood of the parameter space (includes its vector $\mathbf{0}$), we can discuss the Lie algebra that characterizes the local behavior of the associated local Lie group. In finite dimensional parameter spaces, a Lie algebra is generated by certain partial differential operators. Using the representations (34.46) of H^r, the operators are of the form

$$\sum_{i=1}^{n} \lambda_\alpha^i x^i \frac{\partial}{\partial x^i} + \sum_{k=1}^{m} \mu_\alpha^k y^k \frac{\partial}{\partial y^k}, \tag{34.47}$$

where $i = 1, \ldots, n$ and $k = 1, \ldots, m$. These operators can be represented in a matrix form

$$\mathbf{M}(\mathbf{z}) := \mathbf{M}_1 \circ \mathrm{diag}\{x^1, \ldots, x^n; y^1, \ldots, y^m\} \tag{34.48}$$

where

$$\mathbf{M}_1 := \begin{pmatrix} \lambda_1^1 & \cdots & \lambda_1^n, & \mu_1^1 & \cdots & \mu_1^m \\ \vdots & \cdots & & & \cdots & \vdots \\ \lambda_r^1 & \cdots & \lambda_r^n, & \mu_r^1 & \cdots & \mu_r^m \end{pmatrix} \tag{34.49}$$

Obviously, H^r is intransitive if $r < n + m$.

In order to solve the main problem of dimension theory, we need to introduce invariants of a Lie group. Let $F: X \rightarrow Y$ be a function on X and let transformation h_p of a transformation group act on X, then F is an invariant of the group if $F[h_a(x)] = F(x)$ holds for any $x \in X$ and $\mathbf{p} \in \mathbb{R}^p$. The invariant $J: X \rightarrow Y$ is called a *universal invariant* if there exists, for any invariant $F: X \rightarrow Y$ of the group, a function Φ such that $F = \Phi \circ J$. The following main theorem can be proved for the extension group.

Theorem 6: *For the extension group H^r on Z, there exists a universal invariant $J: Z \rightarrow \mathbb{R}^{n+m-r}$ if the condition $r < n + m$ is satisfied. The independent components of J have the monomial form*

$$J^{\tau}(z) = \prod_{i=1}^{n} (x^i)^{\theta_i^{\tau}} . \prod_{k=1}^{m} (y^k)^{\sigma_k^{\tau}} \tag{34.50}$$

where $\tau = 1, \ldots, n + m - r$.

If dimensional analysis considers only scale transformations (34.43), this theorem contains the essential result of the so-called **Pi-theorem**. For this purpose we present a connection between the dimensionalities and the extension group H^r (see [42]). The group H^r of the space $\mathbb{R}^n$, defined only by the dimensions of the physical quantities ϕ_k with respect to the set of symbols $\{E_\alpha\}$, has a one-to-one correspondence with every finite set $\{\phi_k\}$ of n physical quantities, which can be measured in the system of symbols $\{E_\alpha\}$ consisting of r independent measurement units [see (34.41)]. The transformations belonging to the group H^r give the rule of change, in the form

$$|\tilde{\phi}| = |\phi| \prod_{\alpha=1}^{r} (a^{\alpha})^{\lambda_\alpha} \tag{34.51}$$

of the numerical values $|\phi_k|$ as a result of the transition from the units $\{E_\alpha\}$ to $\{\tilde{E}_\alpha\}$ by means of (34.43).

As a consequence of this relationship, a quantity ϕ is dimensionless if and only if its numerical value is an invariant of the group H^r. Thus, the problem to determine the independent physical quantities of a given set of quantities is solved by the construction of a universal invariant of H^r stated by the Pi-theorem (see also [5]). Normalization, as well as the popular method of *dimension comparison*, are consequences of the invariance of physical equations with respect to the group of analogy transformations. In applications of dimensional theory, a normal form that has certain advantageous properties is desired. For example, it is useful to reduce the number of parameters in physical equations. Normal forms of this type are used very often in practice, but with no clarification of their mathematical foundation.

Network equations, like other physical equations, contain numerous parameters. In applications, it is often desired to suppress some of these properties and they should be replaced by the numerical value 1. For this purpose Desloge [27], chooses a new system of units $\{E_\alpha\}$. A theory of Desloge's method, based on analogy transformations (34.42) instead of scale transformations (34.43), was presented by Mathis and Chua [38]. The main idea behind this method is that, beside the foundational units time $[T]$, voltage $[E]$, and charge $[Q]$ that are useful in network theory, the units of other parameters are considered foundational units. We denote the units by $[A_\alpha]$ instead of E_α. For example, in the case of the tunnel-diode circuit (see Fig. 34.2) $[T], [E]$, and $[Q]$, as well as $[R]$, $[C]$, and $[L]$ need to be discussed. As a consequence of Desloge's method, three of the four parameters can be suppressed and the other variables will be normalized. The method works in the case of linear as well as nonlinear networks.

The method is illustrated using the tunnel-diode circuit (see (34.32) and (34.33). At first, the dimensional matrix is determined by

$$\begin{array}{c} \\ [R] \\ [L] \\ [C] \end{array} \begin{array}{c} \begin{array}{ccc} [T] & [E] & [Q] \end{array} \\ \begin{pmatrix} 1 & 1 & -1 \\ 2 & 1 & -1 \\ 0 & -1 & 1 \end{pmatrix} \end{array} \tag{34.52}$$

that characterizes the relation between the dimensions of t, v, q, R, C, L.

Desloge now considers another set of power independent dimensional scalars A_1, A_2, A_3 with

$$[A_i] = [T]^{a_i^1}[E]^{a_i^2}[Q]^{a_i^3} \qquad (i = 1, 2, 3) \tag{34.53}$$

These relations are interpreted as an analogy transformation (34.42). Applying the map $L(\cdot)$ that has the same properties as the logarithmic function (see [38]) to (34.53), the symbols $L([A_1])$, $L([A_2])$, $L([A_3])$ are represented by linear combinations of $L([T])$, $L([E])$, $L([Q])$. The coefficient matrix in (34.53) is regular and contains the exponents. Solving these linear equations and using "antilog," the $[T]$, $[E]$, $[Q]$ are products of powers of $[A_1]$, $[A_2]$, $[A_3]$. In this manner dimensionless versions of differential equations of the tunnel diode can be derived.

By using the independent units $A_1 := L$, $A_2 := C$, $A_3 := V_0$ to replace $|V_0|$, $|L|$, $|C| \to 1$ (with respect to the new units) the following equation is derived by the approach sketched above:

$$\begin{array}{c} \\ [V_0] \\ [L] \\ [C] \end{array} \begin{array}{c} \begin{array}{ccc} [T] & [E] & [Q] \end{array} \\ \begin{pmatrix} 0 & 1 & 0 \\ 2 & 1 & -1 \\ 0 & -1 & 1 \end{pmatrix} \end{array} \begin{pmatrix} \ln([T]) \\ \ln([E]) \\ \ln([Q]) \end{pmatrix} = \begin{pmatrix} \ln([V_0]) \\ \ln([L]) \\ \ln([C]) \end{pmatrix} \tag{34.54}$$

Multiplying (34.54) with the inverse of the dimensional matrix

$$\begin{array}{c} \\ [T] \\ [E] \\ [Q] \end{array} \begin{array}{c} \begin{array}{ccc} [V_0] & [L] & [C] \end{array} \\ \begin{pmatrix} 0 & 1/2 & 1/2 \\ 1 & 0 & 0 \\ 1 & 0 & 1 \end{pmatrix} \end{array} \tag{34.55}$$

and apply "antilog" to the result we obtain

$$[T] = [L]^{1/2}[C]^{1/2} \qquad [E] = [V_0] \qquad [Q] = [V_0][C] \tag{34.56}$$

From these equations the relations between the old and the new units can be derived (see [38]). T, E, and Q are expressed by the new units L, C, and V_0 and the parameters and variables in (34.34) and (34.35) can be reformulated if the numerical values of V_0, L, and C are added:

$$T = |L|^{-1/2}|C|^{-1/2}L^{1/2}C^{1/2}, \qquad E = |V_0|^{-1}V_0, \qquad Q = |V_0|^{-1}|C|^{-1}V_0C \tag{34.57}$$

These relations represent parameters and variables of the tunnel-diode network with respect to the new units V_0, L and C.

$$R = \frac{|R|\,|C|^{1/2}}{|L|^{1/2}} L^{1/2}C^{-1/2}, \qquad V_0 = 1 \cdot V_0, \qquad L = 1 \cdot L, \qquad C = 1 \cdot C \tag{34.58}$$

$$i_L = \frac{|i_L|\,|L|^{1/2}}{|V_0|\,|C|^{1/2}} V_0 L^{-1/2}C^{1/2}, \quad v_C = \frac{|v_C|}{|V_0|} V_0, \quad t = \frac{|t|}{|L|^{1/2}|C|^{1/2}} L^{1/2}C^{1/2} \tag{34.59}$$

The dimensional exponents for these quantities can be found by finding the inverse dimensional matrix (34.55):

1. T, E, Q: their exponents correspond the associated rows of (34.55).
2. V_0, L, C, R: premultiply (34.55) with the corresponding row (34.52).

For example, taking $[C] \hat{=} (0 \ \ -1 \ \ 1)$ results in

$$[C] \overset{\begin{matrix}[T] & [E] & [Q]\end{matrix}}{\begin{pmatrix} 0 & -1 & 1 \end{pmatrix}} \begin{matrix}[T]\\ [E]\\ [Q]\end{matrix} \overset{\begin{matrix}[V_0] & [L] & [C]\end{matrix}}{\begin{pmatrix} 0 & 1/2 & 1/2 \\ 1 & 0 & 0 \\ 1 & 0 & 1 \end{pmatrix}} = [C] \overset{\begin{matrix}[V_0] & [L] & [C]\end{matrix}}{\begin{pmatrix} 0 & 0 & 1 \end{pmatrix}} \tag{34.60}$$

or with (34.52) $[R] \hat{=} (1 \ \ 1 \ \ -1)$

$$[R] \overset{\begin{matrix}[T] & [E] & [Q]\end{matrix}}{\begin{pmatrix} 1 & 1 & -1 \end{pmatrix}} \begin{matrix}[T]\\ [E]\\ [Q]\end{matrix} \overset{\begin{matrix}[V_0] & [L] & [C]\end{matrix}}{\begin{pmatrix} 0 & 1/2 & 1/2 \\ 1 & 0 & 0 \\ 1 & 0 & 1 \end{pmatrix}} = [R] \overset{\begin{matrix}[V_0] & [L] & [C]\end{matrix}}{\begin{pmatrix} 0 & 1/2 & 1/2 \end{pmatrix}} \tag{34.61}$$

With these representations of the dimensional quantities, we can obtain a dimensionless representation of (34.34) and (34.35)

$$\frac{d\bar{v}_C}{dt} = \bar{i}_L - \bar{f}(\bar{v}_C) \tag{34.62}$$

$$\frac{d\bar{i}_L}{dt} = 1 - \sqrt{\varepsilon}\,\bar{i}_L - \bar{v}_C \tag{34.63}$$

where

$$\bar{v}_C := \frac{|v_C|}{|V_0|}, \qquad \bar{t} := \frac{|t|}{\sqrt{|L|\,|C|}} \qquad \sqrt{\varepsilon} := \frac{|R|\,|C|^{1/2}}{|L|^{1/2}} \tag{34.64}$$

$$\bar{i}_L := \frac{|i_L|\,|L|^{1/2}}{|V_0|\,|C|^{1/2}} \tag{34.65}$$

Furthermore, the dimensionless tunnel-diode current $\bar{f}$ is defined by

$$\bar{f}(\bar{v}_C) := V_0^{-1}L^{1/2}C^{-1/2}f(V_0\bar{v}_C) \tag{34.66}$$

The associated dimensionless form of the (34.34) and (34.35) can be derived by another scaling of the current $\bar{\bar{i}}_L := \sqrt{\varepsilon}\,\bar{i}_L$. Obviously, the dimensionless normal form is not unique.

The classical dimensional analysis shows that R^2C/L is the only dimensionless constant of (34.32) and (34.33). Since the parallel LCR circuit includes the same constants and variables, the results of the above dimensional analysis of the tunnel-diode circuit can be used to normalize (34.37).

Further interesting applications of Desloge's approach of suppressing superfluous parameters can be found in the theory of singular perturbations. The reader is referred to the monograph of Smith [43] for further details. Miranker [41] showed that the differential equations of the tunnel-diode circuit can be studied on three time scales $\tau_1 = L/R$, $\tau_2 = RC$, and $\tau_3 = \sqrt{LC}$ with different phenomena arising. The corresponding dimensionless equations can be derived in a systematic manner by Desloge's method. In this way, normalized differential equations describing Chua's circuit (see [39]) can be obtained but other representations of these differential equations are possible using dimensional analysis.

34.4 Equivalence Between Nonlinear Resistive Circuits

In this section we consider equivalence of nonlinear resistive n-ports. (We do not discuss resistive networks without accessible ports.) Although the explanations that follow are restricted to resistive n-ports, this theory can be extended to capacitive and inductive n-ports (see [23]). In Section 34.5 we give a definition of those n-ports.

At first, we consider linear resistive 1-ports that contain Ohmic resistors described by $v_k = R_k i_k$ or/and $i_k = G_k i_k$, and independent current and voltages sources $v_k = V_0^k$ and $i_k = I_0^k$. We can use Thévenin's or Norton's theorem to compare any two of those 1-ports and reduce a complex 1-port to a simple "normal" form. Therefore, two of those 1-ports are called equivalent if they have the same Thévenin (or Norton) 1-port. Clearly, by this approach, an equivalence relation is defined in the set of linear resistive 1-ports and it is decomposed into "rich" classes of 1-ports. To calculate these normal forms, Δ–Y and/or Y–Δ transformations are needed (see [20]). It is known that this approach is not applicable to nonlinear resistive networks because Δ–Y and Y–Δ transformations generally do not exist for nonlinear networks. (This was observed by Millar [40] for the first time.) Certain networks where these transformations can be performed were presented by Chua [14]. More recently, Boyd and Chua [6], [7] clarified the reasons behind this difficulty from the point of view of a Volterra series. As a conclusion, the set of nonlinear resistive 1-ports can be decomposed into equivalence classes, but, there is no reasonably large class of equivalent 1-ports. More general studies of this subject are based on the well-known substitution theorem, which can be extended to a certain class of nonlinear networks (see [26], [29]). Some results applicable to 1-ports can be generalized to linear resistive n-ports ("extraction of independent sources") but this point of view is not suitable for nonlinear resistive n-ports.

Better understanding of nonlinear resistive n-ports and the problem of equivalence cannot be based on the "operational" approach mentioned above. Rather, a geometric approach that was developed by Brayton and Moser [9] is more useful. These authors (see also [8]) characterize a resistive n-port in a generic manner by n independent relations between the $2n$ port variables, n-port currents $i_1, \ldots, i_n$ and n-port voltages $v_1, \ldots, v_n$. Geometrically, this means that in the $2n$-dimensional space of port variables the external behavior of a resistive n-port can be represented generically by an n-dimensional surface. The classical approach formulates a system of equations

$$\begin{aligned} y_1 - f_1(x_1, \ldots, x_n) &= 0 \\ &\vdots \\ y_n - f_n(x_1, \ldots, x_n) &= 0 \end{aligned} \tag{34.67}$$

where x's and y's are the port variables. The zero set of equations (34.67) corresponds to the n-dimensional surface. Therefore, two n-ports are called equivalent if they are different parameterizations of the same surface. As an application of this point of view, Brayton and Moser [9] showed that a 2-port consisting of a Y-circuit and a circuit consisting of a Δ-circuit cannot be equivalent, in general. For example, they proved by means of Legendre transformations that a Y-circuit with two ohmic resistors and a third resistor can be equivalent to a Δ-circuit if and only if the third resistor is also linear. Therefore, the operational approach is not a very useful concept for nonlinear n-ports.

The subject of synthesizing a prescribed input–output behavior of nonlinear resistive n-ports is closely related to the problem of equivalence. Several results were published in this area using ideal diodes, concave and convex resistors, dc voltage and current sources, ideal op amps, and controlled sources. Therefore, we give a short review of some results. We do not consider here the synthesis of resistive n-ports.

Although the synthesis of nonlinear resistive n-ports was of interest to many circuit designers since the beginning of this century, the first systematic studies of this subject were published by Chua [13], [14]. Chua's synthesis approach is based on the introduction of new linear 2-ports (R-rotator, R-reflector, and scalors) as well as their electronic realizations. Now, curves in the i–v space of port current i and port voltage v that characterize a (nonlinear) resistive 1-port can be reflected and scaled in a certain manner. Chua showed that a prescribed behavior of an active or passive nonlinear resistive 1-port can be reduced essentially to the realization of passive i–v curves. Piecewise-linear approximations of characteristics of different types of diodes, as well as the above mentioned 2-ports, are used to realize a piecewise-linear approximation of any prescribed passive i–v curve. In a succeeding paper, Chua [15] discussed a unified procedure to synthesize a nonlinear dc circuit mode that represents a prescribed family of input and output curves of any strongly passive 3-terminal device (e.g, transistor). It was assumed that the desired curves are piecewise-linear. Since then, this research area has grown very rapidly and piecewise-linear synthesis and modeling has become an essential tool in the simulation of nonlinear circuits. (see [19], [35], [37] for further references.)

34.5 Equivalence of Lumped n-Port Networks

In this section we consider more general n-ports that can be used in for device modeling (see [16]). Although there are many different *lumped* multiterminal and multiport networks, a decomposition into two mutually-exclusive classes is possible: *algebraic* and *dynamic* multiterminal and multiport networks. Adopting the definition of Chua [16], an $(n+1)$-terminal or n-port network is called an *algebraic element* if and only if its constitutive relations can be expressed symbolically by algebraic relationships involving at most two dynamically independent variables for each port. In the case of a 1-port a so-called memristor is described by flux and charge, a resistor by voltage and current, a inductor by flux and current, and a capacitor by voltage and charge. An element is called a *dynamic element* if and only if it is not an algebraic element.

In spite of the fact that the class of all dynamic elements is much larger than that of algebraic ones, the following theorem of Chua [16] shows that resistive multiports are essential for dynamic elements, too.

Theorem 7: *Every lumped $(n+1)$-terminal or n-port element can be synthesized using only a finite number m of linear 2-terminal capacitors (or inductors) and one (generally nonlinear) $(n+m)$-port resistor with n accessible ports and m ports for the capacitors.*

Theorem 7 shows that any n-port made of lumped multiterminal and/or multiport elements is *equivalent* to a multiterminal network where all of its nonlinear elements are *memoryless*.

This fact offers a possibility to classify $(n+1)$-terminal and n-port elements in an operational manner.

The proof of this theorem provides the answer of a fundamental question: what constitutes a *minimal set* if network elements from which *all* lumped elements can be synthesized?

Theorem 8: *The following set $\mathcal{M}$ of network elements constitutes the minimal basic building blocks in the sense that any lumped multiterminal or multiport element described by a continuous constitutive relation on any closed and bounded set can be synthesized using only a finite number of elements of $\mathcal{M}$, and that this statement is false if even one element is deleted from $\mathcal{M}$:*

1. *Linear 2-terminal capacitors (or inductors)*
2. *Nonlinear 2-terminal resistors*
3. *Linear 2-port current-controlled voltage sources (CCVS) defined by $v_1 = 0$ and $v_2 = ki_1$*
4. *Linear 2-port current-controlled current sources (VCCS) defined by $i_1 = 0$ and $i_2 = kv_1$.*

The proof of Theorem 8 (see [16]), is based on a remarkable theorem of Kolmogoroff, which asserts that a continuous function $\mathbf{f}\colon \mathbb{R}^n \to \mathbb{R}$ can always be decomposed over the unit cube of $\mathbb{R}^n$ into a certain sum of functions of a *single* variable. Although the proof of Theorem 8 is constructive, it is mainly of theoretical interest because the number of controlled sources needed in the realization is often excessive.

34.6 Equivalence Between Nonlinear Dynamic Circuits

As already mentioned in Section 34.1, a set of networks can be decomposed into classes of equivalent networks by some type of equivalence relation. Such equivalence relations are introduced in a direct manner with respect to the descriptive equations, using a transformation group or classifying the behavior of the solution of the descriptive equations. In the last three sections, several useful ideas for defining equivalence relations were discussed that can be suitable for circuit theory. In this section, equivalent dynamic circuits are discussed in more details. It should be emphasized again that equivalence has a different meaning depending of the applied equivalence relation.

As the so-called state-space equations in network and system theory arose in the early 1960's, a first type of equivalence was defined because various networks can be described by the same state-space equations that induced an equivalence relation. (For further references, see [46].) Although this approach is interesting, in some cases different choices of variables for describing nonlinear networks exist that need not lead to equivalent state-space equations (see [17]). In other words, the transformations of coordinates are not well conditioned. This approach was applied also to nonlinear input–output systems.

A study of equivalence of a subclass of nonlinear input–output networks was presented by Verma [45] and Varaiya and Verma [44]. These authors discussed nonlinear reciprocal networks that can be formulated by a so-called mixed potential function. This approach was developed by Brayton and Moser [9]. If $\mathbf{x} \in \mathbb{R}^n$ is the state-space vector, $\mathbf{u} \in \mathbb{R}^m$ the input vector, and $\mathbf{e} \in \mathbb{R}^m$ is the output vector, then the state-space equations can be generated by a matrix-valued function $\mathbf{A}(\mathbf{x})\colon \mathbb{R}^n \to \mathbb{R}^{n\times n}$ and a real-valued function $P\colon \mathbb{R}^n \times \mathbb{R}^m \to \mathbb{R}$

$$\mathbf{A}(\mathbf{x})\frac{d\mathbf{x}}{dt} = -\frac{\partial P}{\partial \mathbf{x}}(\mathbf{x}, \mathbf{u}) \tag{34.68}$$

$$\mathbf{e} = \frac{\partial P}{\partial \mathbf{u}}(\mathbf{x}, \mathbf{u}) \tag{34.69}$$

For two such networks $N_1 = \{\mathbf{A}_1, P_1\}$ and $N_2 = \{\mathbf{A}_2, P_2\}$, Varaiya and Verma defined the following equivalence.

Definition 4. Networks N_1 and N_2

$$\mathbf{A}_1(\mathbf{x})\frac{d\mathbf{x}}{dt} = -\frac{\partial P_1}{\partial \mathbf{x}}(\mathbf{x}, \mathbf{u}) \tag{34.70}$$

$$\mathbf{e}_1 = \frac{\partial P_1}{\partial \mathbf{u}}(\mathbf{x}, \mathbf{u}) \tag{34.71}$$

and

$$\mathbf{A}_2(\mathbf{y})\frac{d\mathbf{y}}{dt} = -\frac{\partial P_2}{\partial \mathbf{y}}(\mathbf{y}, \mathbf{u}) \tag{34.72}$$

$$\mathbf{e}_2 = \frac{\partial P_2}{\partial \mathbf{u}}(\mathbf{y}, \mathbf{u}) \tag{34.73}$$

are equivalent if there exists a diffeomorphism $\mathbf{y} = \phi(\mathbf{x})$ such that for all $\mathbf{x}_0 \in \mathbb{R}^n$, all input functions $\mathbf{u}$, and all $t \geq 0$:

1. $\phi[\mathbf{x}(t, \mathbf{x}_0, \mathbf{u})] = \mathbf{y}(t, \phi(\mathbf{x}_0), \mathbf{u})$
2. $\mathbf{e}_1(t, \mathbf{x}_0, \mathbf{u}) = \mathbf{e}_2(t, \phi(\mathbf{x}_0), \mathbf{u})$.

The diffeomorphism ϕ is called the equivalence map.

Thus, two networks are equivalent if their external behavior is identical, i.e., if for the same input and corresponding states they yield the same output. It is clear that this definition yields an equivalence relation on the set of all dynamical networks under consideration. In their paper, Varaiya and Verma showed that, under an additional assumption of controllability, the diffeomorphism ϕ establishes an isometry between the manifold with the (local) pseudo-Riemannian metric $(d\mathbf{x}, d\mathbf{x}) := \langle d\mathbf{x}, \mathbf{A}_1 d\mathbf{x}\rangle$ and the manifold with the (local) pseudo-Riemannian metric $(d\mathbf{y}, d\mathbf{y}) := \langle d\mathbf{y}, \mathbf{A}_2 d\mathbf{y}\rangle$ in many interesting cases of reciprocal nonlinear networks. This statement has an interesting interpretation in the network context. It can be proven that ϕ must relate the reactive parts of the networks N_1 and N_2 in such a way that if N_1 is in the state $\mathbf{x}$ and N_2 is in the state $\mathbf{y} = \phi(\mathbf{x})$ and if the input $\mathbf{u}$ is applied, then

$$\left\langle \frac{d\mathbf{i}}{dt}, \mathbf{L}(\mathbf{i})\frac{d\mathbf{i}}{dt}\right\rangle - \left\langle \frac{d\mathbf{v}}{dt}, \mathbf{C}(\mathbf{v})\frac{d\mathbf{v}}{dt}\right\rangle = \left\langle \frac{d\tilde{\mathbf{i}}}{dt}, \tilde{\mathbf{L}}(\tilde{\mathbf{i}})\frac{d\tilde{\mathbf{i}}}{dt}\right\rangle - \left\langle \frac{d\tilde{\mathbf{v}}}{dt}, \tilde{\mathbf{C}}(\tilde{\mathbf{v}})\frac{d\tilde{\mathbf{v}}}{dt}\right\rangle \tag{34.74}$$

The concept of equivalence defined in a certain subset of nonlinear dynamic networks with input and output terminals given by Varaiya and Verma is based on diffeomorphic coordinate transformations (the transformation group of diffeomorphisms). Unfortunately, the authors presented no ideas about the kind of "coarse graining" produced in the set of networks by their equivalence relation. However, a comparison to C^k conjugacy or C^k equivalence of vector fields in Section 34.1 implies that input–output equivalence leads to a "fine" decomposition in the set of these networks. To classify the main features of the dynamics of networks, the concept of topological equivalence (the transformation group of homeomorphisms) is useful. On the other hand, in the case of networks with nonhyperbolic fixed points, the group of diffeomorphisms is needed to distinguish the interesting features. An interesting application of C^1 equivalence of vector fields is given by Chua [18]. To compare nonlinear networks that generate chaotic signals, Chua applied the concept of equivalence relation and concluded that the class of networks and systems that are C^1 equivalent to Chua's circuit (Fig. 34.5) is relatively small. The nonlinearity

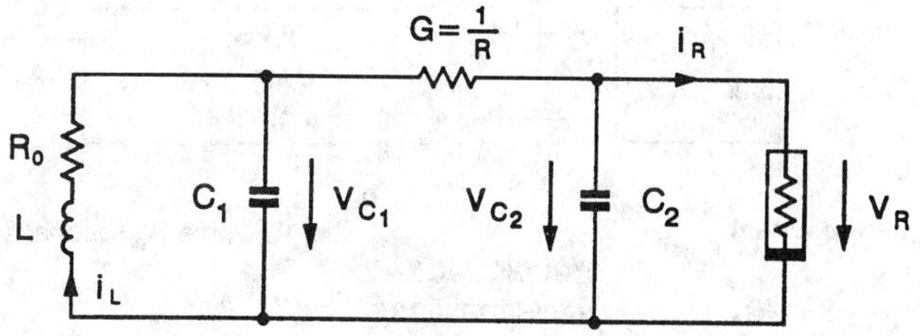

FIGURE 34.5 Modified Chua's circuit.

in this network is described by a piecewise linear i–v characteristic. (See [39] for further details.) The equations describing the circuit are

$$\frac{dv_{C_1}}{dt} = \frac{1}{C_1}[G(v_{C_2} - v_{C_1}) - f(v_{C_1})] \tag{34.75}$$

$$\frac{dv_{C_2}}{dt} = \frac{1}{C_2}(G(v_{C_1} - v_{C_2}) + i_L) \tag{34.76}$$

$$\frac{di_L}{dt} = \frac{1}{L}(v_{C_2} + R_0 i_L) \tag{34.77}$$

where $R_0 = 0$ and the piecewise linear function is defined by

$$f(v_{C_1}) := G_b v_{C_1} + \frac{1}{2}(G_a - G_b)(|v_{C_1} + E| - |v_{C_1} - E|) \tag{34.78}$$

Chua's extended approach to study the set of the piecewise linear networks that include Chua's circuit introduces the concept of global unfoldings. This concept can be considered as an analogy to the theory of "local unfoldings" of nonhyperbolic systems in a small neighborhood of singularities [3], [30]. Heuristically, a minimum number of parameters in a given nonhyperbolic system is introduced, and, as the parameters are varied, "any other system" near the nonhyperbolic system is obtained. Chua showed that Chua's circuit with arbitrary $R_0(\neq 0)$ can be considered as an "unfolding" of the original circuit. Furthermore, he proved that a class of networks that can be described without loss of generality by

$$\dot{\mathbf{x}} = \mathbf{A}\mathbf{x} + \mathbf{b}, \quad x_1 \leq -1 \tag{34.79}$$

$$= \mathbf{A}_0\mathbf{x}, \quad -1 \leq x_1 \leq 1 \tag{34.80}$$

$$= \mathbf{A}\mathbf{x} + \mathbf{b}, \quad x_1 \geq 1 \tag{34.81}$$

is equivalent to the unfolded Chua's circuit if certain conditions are satisfied. In the associated parameter space, these conditions define a set of measure zero. The proof of this theorem as well as some applications are included in [18].

The ideas of normal forms presented in Section 34.2 can be applied to nonlinear networks with hyperbolic and nonhyperbolic fixed points. A similar theory of normal forms of maps can be used to study limit cycles, but, this subject is beyond our scope. (See [3] for further details.) In any case the vector field has to be reduced to lower dimensions and that can be achieved by

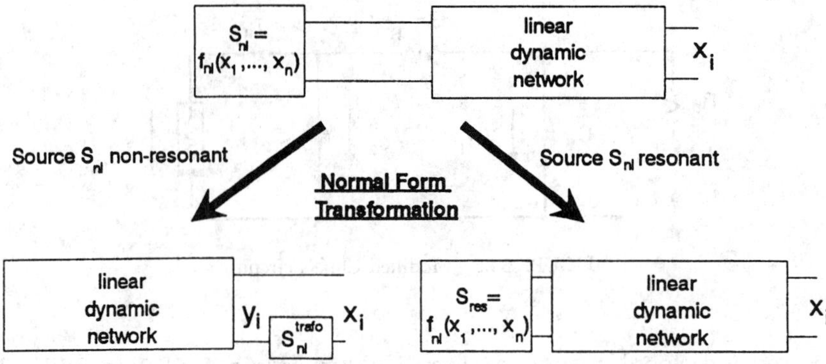

FIGURE 34.6 Decomposition of nonlinear dynamic networks.

the application of the so-called center manifold theorem. Recently, Altman [2] illustrated this approach by calculating the center manifold and a normal form of Chua's circuit in a tutorial style. To perform the analytical computations the piecewise nonlinearity (34.78) is replaced by a cubic function $f(x) = c_0x + c_1x^3$. Based on this normal form, Altman studied bifurcations of Chua's circuit.

Another application of normal forms in nonlinear dynamical networks is discussed by Keidies and Mathis [36]. In this approach, nonlinear dynamical networks with constant sources are considered and are described by nonlinear differential equations in state-space form:

$$\dot{\mathbf{x}} = \mathbf{f}(\mathbf{x}), \quad \mathbf{f}\colon \mathbb{R}^n \to \mathbb{R}^n \tag{34.82}$$

where all nonlinear reactances are replaced by linear reactances, nonlinear resistors, and linear controlled sources. The nonlinearities are interpreted as nonlinear controlled sources. The network is decomposed into a linear part that consists of linear reactances and resistive elements, and the nonlinear sources that are used as input sources (Fig. 34.6). The network is described by the vector of state-space variables $\mathbf{x}$. Now, normal form theorems are used to transform the nonlinear sources to the input. In other words, if the RHS $\mathbf{f}$ of (34.82) is decomposed into a linear and a nonlinear part, $f(\mathbf{x}) = \mathbf{A}\mathbf{x} + \tilde{\mathbf{f}}(\mathbf{x})$, where $\tilde{\mathbf{f}}$ corresponds the nonlinear sources, the system can be decomposed into two equations:

$$\dot{\mathbf{y}} = \mathbf{A}\mathbf{y} \tag{34.83}$$

$$\mathbf{x} = \mathbf{y} + \mathbf{F}(\mathbf{y}) \tag{34.84}$$

We now have to define the nonresonant and resonant terms of vector fields that depend on the eigenvalues of the linear part $\mathbf{A}$ of $\mathbf{f}$ and the degrees of the polynomial nonlinearities. Under certain conditions, a finite recursive process exists such that all nonlinear sources can be transformed to the input of the linear part of a network. In these cases, the networks are described by (34.82) and (34.83). In other cases, a number of new sources are generated during the recursive process that cannot transform sources to the input. This effect is shown in Fig. 34.7(a) and (b). It should be mentioned that this idea is related in certain sense to the so-called exact linearization that is studied in the theory of nonlinear control systems (see [34]). Therefore, this application of normal form theorems can be interpreted as a kind of extraction of nonlinear controlled sources from a nonlinear dynamic network.

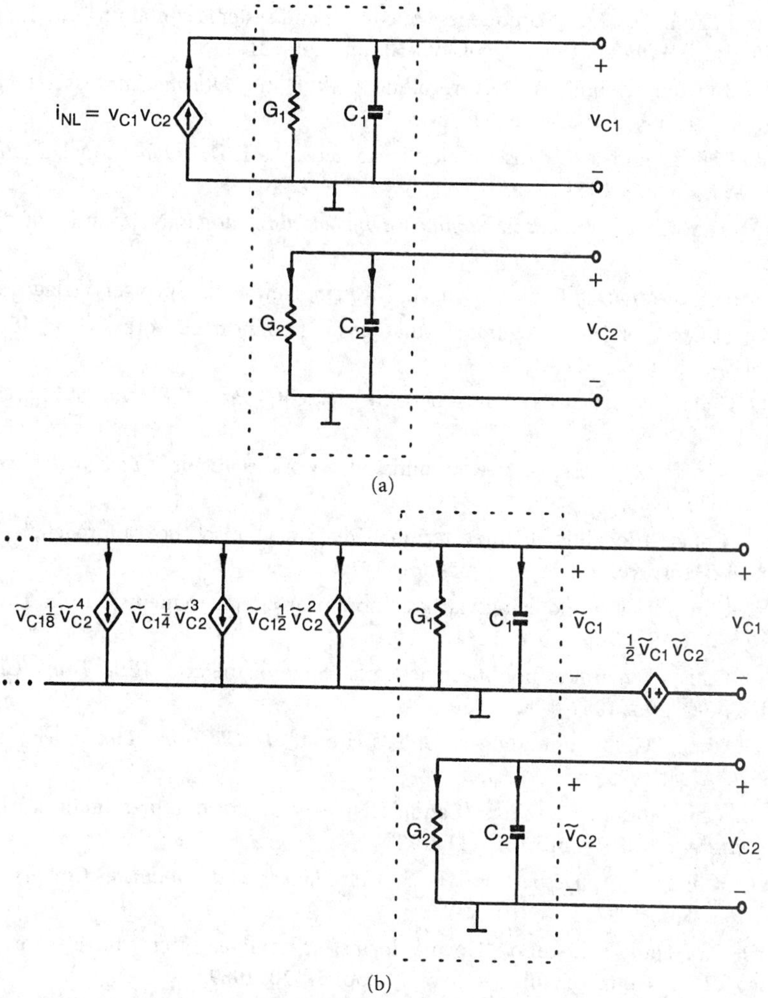

FIGURE 34.7 Decomposition of a simple nonlinear network.

eferences

[1] V. I. Arnol'd, *Geometrical Methods in the Theory of Ordinary Differential Equations*, New York: Springer-Verlag, 1983.

[2] E. J. Altman, "Bifurcation analysis of Chua's circuit with application for low-level visual sensing," in *Chua's Circuit: A Paradigm for Chaos*, R. N. Madan, Ed., Singapore: World Scientific, 1993.

[3] D. K. Arrowsmith and C. M. Place, *An Introduction to Dynamical Systems*, Cambridge: Cambridge Univ., 1990.

[4] M. Ashkenazi and S.-N. Chow, "Normal forms near critical points for differential equations and maps," *IEEE Trans. Circuits Syst.*, vol. 35, pp. 850–862, 1988.

[5] G. W. Bluman and S. Kumei, *Symmetries and Differential Equations*, New York: Springer-Verlag, 1989.

[6] S. Boyd and L. O. Chua, "Uniqueness of a Basic Nonlinear Structure," *IEEE Trans. Circuits Syst.*, vol. CAS-30, pp. 648–651, 1983.

[7] S. Boyd and L. O. Chua, "Uniqueness of circuits and systems containing one nonlinearity," *IEEE Trans. Automat. Control,* vol. AC-30, pp. 674–681, 1985.

[8] R. K. Brayton, "Nonlinear reciprocal networks," in *Mathematical Aspects of Electrical Network Analysis,* Providence, RI: AMS, 1971.

[9] R. K. Brayton and J. K. Moser, "Nonlinear networks I, II," *Quart. Appl. Math.*, vol. 23, pp. 1–33, 81–104, 1964.

[10] A. D. Bruno, *Local Methods in Nonlinear Differential Equations,* New York: Springer-Verlag, 1989.

[11] J. Carr, *Applications of Center Manifold Theorem,* New York: Springer-Verlag, 1981.

[12] L. O. Chua, "Δ–Y and Y–Δ transformation for nonlinear networks," *Proc. IEEE,* vol. 59, pp. 417–419, 1971.

[13] L. O. Chua, "The rotator—A new network element," *Proc. IEEE,* vol. 55, pp. 1566–1577, 1967.

[14] L. O. Chua, "Synthesis of new nonlinear network elements," *Proc. IEEE,* vol. 56, pp. 1325–1340, 1968.

[15] L. O. Chua, "Modeling of three terminal devices: A black box approach," *IEEE Trans. Circuit Theory,* vol. CT-19, pp. 555–562, 1972.

[16] L. O. Chua, "Device modeling via basic nonlinear circuit elements," *IEEE Trans. Circuits Syst.*, vol. CAS-27, pp. 1014–1044, 1980.

[17] L. O. Chua, "Dynamical nonlinear networks: State of the art," *IEEE Trans. Circuits Syst.*, vol. CAS-27, pp. 1059–1087, 1980.

[18] L. O. Chua, "Global unfolding of Chua's circuit," *IEICE Trans. Fundament.*, vol. E76-A, pp. 704–734, 1993.

[19] L. O. Chua and A. C. Deng, "Canonical piecewise linear representation," *IEEE Trans. Circuits Syst.*, vol. 33, pp. 101–111, 1988.

[20] L. O. Chua, C. A. Desoer, and E. S. Kuh, *Linear and Nonlinear Circuits,* New York: McGraw-Hill, 1987.

[21] L. O. Chua and H. Kokubo, "Normal forms for nonlinear vector fields—Part I: Applications," *IEEE Trans. Circuits Syst.*, vol. 36, pp. 51–70, 1989.

[22] L. O. Chua and H. Kokubo, "Normal forms for nonlinear vector fields—Part II: Theory and algorithm," *IEEE Trans. Circuits Syst.*, vol. 35, pp. 863–880, 1988.

[23] L. O. Chua and Y.-F. Lam, "A theory of algebraic n-ports," *IEEE Trans. Circuit Theory,* vol. CT-20, pp. 370–382, 1973.

[24] L. O. Chua and H. Oka, "Normal forms of constrained nonlinear differential equations—Part I: Theory," *IEEE Trans. Circuits Syst.*, vol. 35, pp. 881–901, 1988.

[25] L. O. Chua and H. Oka, "Normal forms of constrained nonlinear differential equations—Part II: Bifurcation," *IEEE Trans. Circuits Syst.*, vol. 36, pp. 71–88, 1989.

[26] C. A. Desoer and E. S. Kuh, *Basic Circuit Theory,* New York: McGraw-Hill, 1969.

[27] E. A. Desloge, "Suppression and restoration of constants in physical equations," *Amer. J. Phys.*, vol. 52, pp. 312–316, 1984.

[28] J. Guckenheimer and P. Holmes, *Nonlinear Oscillations, Dynamical Systems, and Bifurcations of Vector Fields,* New York: Springer-Verlag, 1990.

[29] J. Haase, "On generalizations and applications of the substitution theorem," in *Proc. ECCTD '85,* Praha, Sept. 2–6, 1985, pp. 220–223.

[30] S. Hale and H. Kocak, *Dynamics and Bifurcations,* New York: Springer-Verlag, 1991.

[31] B. D. Hassard, N. D. Kazarinoff, and Y.-H. Wan, *Theory and Applications of the Hopf Bifurcation,* Cambridge: Cambridge Univ., 1980.

[32] B. D. Hassard and Y.-H. Wan, "Bifurcation formulae derived from center manifold theorem," *J. Math. Anal. Applicat.*, vol. 63, pp. 297–312, 1978.

[33] R. A. Horn and C. R. Johnson, *Matrix Analysis*, Cambridge: Cambridge Univ., 1992.

[34] A. Isidori, *Nonlinear Control Systems*, Berlin: Springer-Verlag, 1989.

[35] J. Jess, "Piecewise Linear Models for Nonlinear Dynamic Systems," *Frequenz*, vol. 42, pp. 71–78, 1988.

[36] C. Keidies and W. Mathis, "Applications of normal forms to the analysis of nonlinear circuits," in *Proc. 1993 Int. Symp. Nonlinear Theory, Applicat.*, Hawaii, Dec. 1993.

[37] T. A. M. Kevenaar and D. M. W. Leenaerts, "A comparison of piecewise-linear model descriptions," *IEEE Trans. Circuits Syst. I*, vol. 39, pp. 996–1004, 1992.

[38] W. Mathis and L. O. Chua, "Applications of dimensional analysis to network theory," *Proc. ECCTD'91*, Copenhagen, Sept. 4–6, 1991.

[39] R. N. Madan, Ed., *Chua's Circuit: A Paradigm for Chaos*, Singapore: World Scientific, 1993.

[40] W. Millar, "The nonlinear resistive 3-pole: Some general concepts," in *Proc. Symp. Nonlinear Circuit Anal.*, Polytech. Instit., Brooklyn, NY, New York: Interscience, 1957.

[41] W. L. Miranker, *Numerical Methods for Stiff Equations and Singular Perturbation Problems*, Dordrecht: D. Reidel, 1981.

[42] L. V. Ovsiannikov, *Group Analysis of Differential Equations*, New York: Academic, 1982.

[43] D. R. Smith, *Singular-Perturbation Theory*, Cambridge: Cambridge Univ., 1985.

[44] P. P. Varaiya and J. P. Verma, "Equivalent nonlinear reciprocal networks," *IEEE Trans. Circuit Theory*, vol. CT-18, pp. 214–217, 1971.

[45] J. P. Verma, "Equivalence of nonlinear networks," Ph.D. dissertation, Univ. California, Berkeley, 1969.

[46] A. N. Willson, Jr., *Nonlinear Networks: Theory and Analysis*, New York: IEEE, 1975.

35

Piecewise-Linear Circuits and Piecewise-Linear Analysis

J. Vandewalle
Katholieke Universiteit Leuven, Belgium

L. Vandenberghe
Katholieke Universiteit Leuven, Belgium

35.1 Introduction and Motivation

In this chapter, we present a comprehensive description of the use of piecewise-linear methods in modeling, analysis, and structural properties of nonlinear circuits. The main advantages of piecewise linear circuits are fourfold. 1) Piecewise-linear circuits are the easiest in the class of nonlinear circuits to analyze exactly, because many methods for linear circuits can still be used. 2) The piecewise-linear approximation is an adequate approximation for most applications. Moreover, certain op amp, operational transconductance amplifier, diode and switch circuits are essentially piecewise linear. 3) Quite a number of methods exist to analyze piecewise-linear circuits. 4) Last, but not least, piecewise-linear circuits exhibit most of the phenomena of nonlinear circuits while still being manageable. Hence, PWL circuits provide unique insight in nonlinear circuits.

The section consists of six parts. First, the piecewise-linear models will be presented and interrelated. A complete hierarchy of models and representations of models is presented. Rather than proving many relations, simple examples are given. Second, the piecewise-linear models for several important electronic components are presented. Third, since many PWL properties are preserved by interconnection, a short discussion on the structural properties of piecewise-linear circuits is given in Section 35.4. Fourth, analysis methods of PWL circuits are presented, ranging from the Katzenelson algorithm to the linear complementarity methods and the homotopy methods. Fifth, we discuss PWL dynamic circuits, such as the famous Chua circuit, which

0-8493-8341-2/95/$0.00+$.50

produces chaos. Finally, in Section 35.7, efficient computer-aided analysis of PWL circuits and the hierarchical mixed-mode PWL analysis are described. A comprehensive reference list is included. For the synthesis of PWL circuits, we refer to Chapter 32.

In order to situate these subjects in the general framework of nonlinear circuits it is instructive to interrelate the PWL circuit analysis methods (Fig. 35.1). In the horizontal direction of the diagrams, one does the PWL approximation of the dc analysis from left to right. In the vertical

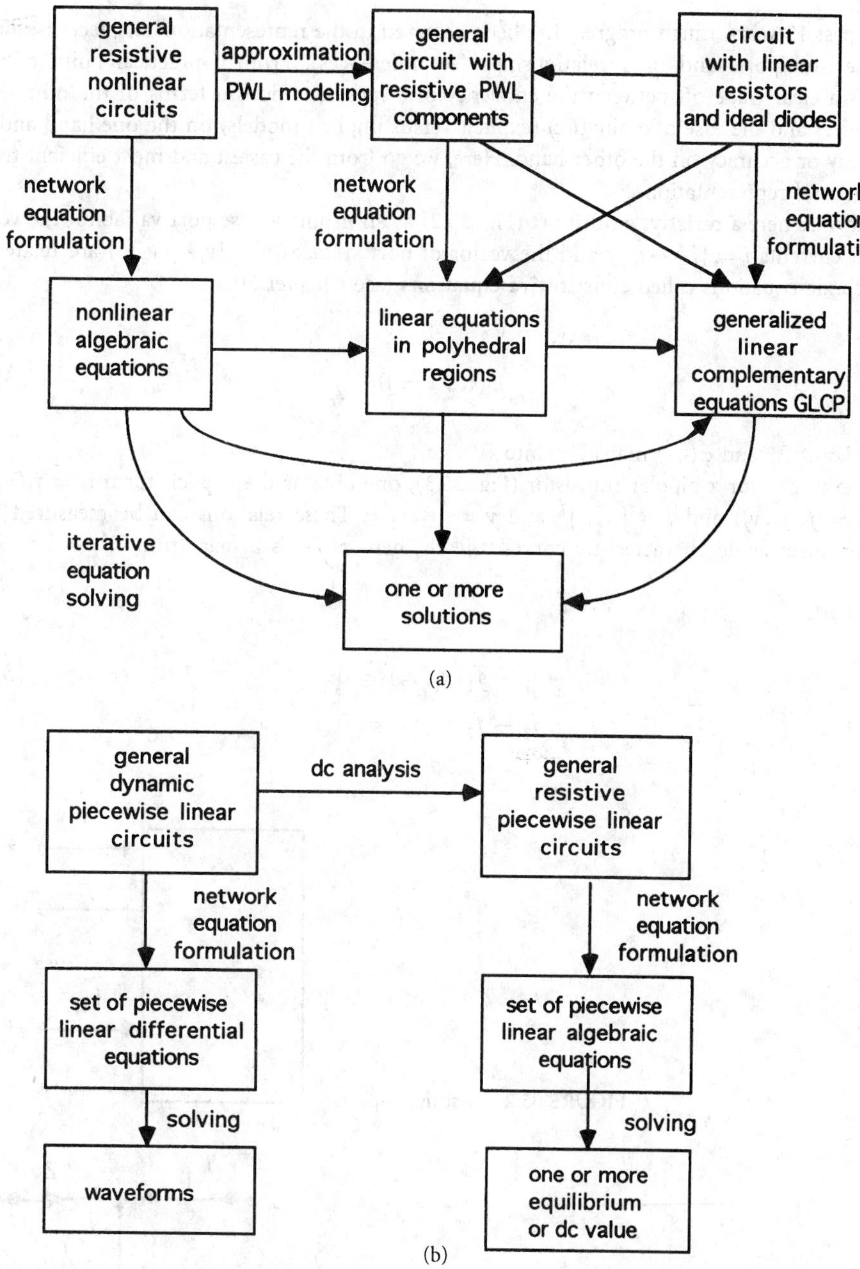

FIGURE 35.1 Interrelation of PWL circuit analysis methods: (a) resistive and (b) dynamic nonlinear circuits.

direction we show the conversion from a circuit to a set of equations by network equation formulation and the conversion from equations to solutions (waveforms or dc values) by solution methods. The specific methods and names used in the figure are described in detail in the different parts.

35.2 Hierarchy of Piecewise-Linear Models and Their Representations

In the past 15 years, much progress has been achieved in the representations of piecewise-linear resistive multiports and their relationships (see references). From a practical point of view there is a clear trade-off between the efficiency of a representation in terms of the number of parameters and the ease of evaluation (explicit versus implicit models) on the one hand and the generality or accuracy on the other hand. Here, we go from the easiest and most efficient to the most general representations.

We define here a **resistive multiport** (Fig. 35.2) as an n-port whose port variables (the vector of port currents $i = [i_1 \cdots i_n]^T$ and the vector of port voltages $v = [v_1 \cdots v_n]^T$) are related by m algebraic equations called **constitutive equations** (see Chapter 30.3)

$$\varphi(i, v) = 0 \tag{35.1}$$

where $i, v \in \mathbb{R}^n$ and $\varphi(\cdot, \cdot)$ maps $\mathbb{R}^{2n}$ into $\mathbb{R}^m$.

For example, for a bipolar transistor (Fig. 35.3), one obtains the explicit form $i_1 = f_1(v_1, v_2)$ and $i_2 = f_2(v_1, v_2)$ and $i = [i_1, i_2]^T$ and $v = [v_1, v_2]^T$. These relations can be measured with a curve tracer as dc characteristic curves. Clearly, here $\varphi(\cdot, \cdot)$ is a map from $\mathbb{R}^4 \to \mathbb{R}^2$ in the form

$$i_1 - f_1(v_1, v_2) = 0 \tag{35.2}$$

$$i_2 - f_2(v_1, v_2) = 0 \tag{35.3}$$

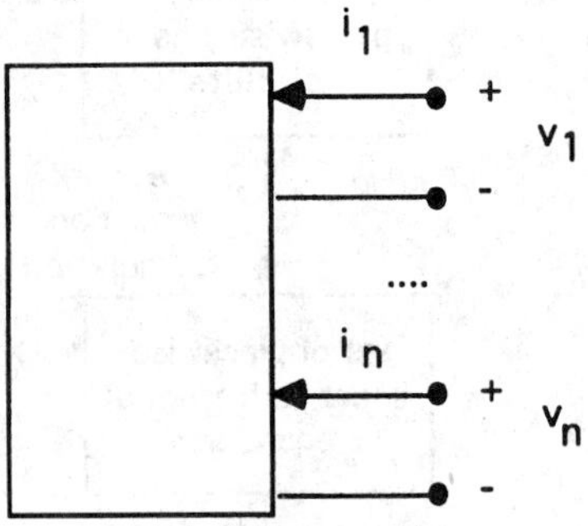

FIGURE 35.2 Resistive n-port.

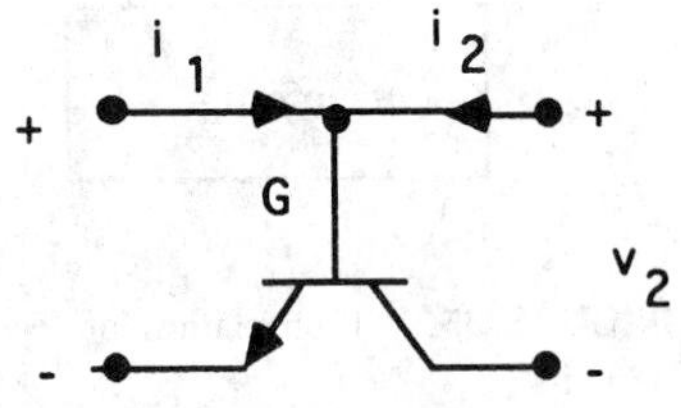

FIGURE 35.3 2-port configuration of a bipolar transistor.

It is easy to see that a complete table of these relationships would require an excessive amount of computer storage already for a transistor. Hence, it is quite natural to describe a resistive n-port with a piecewise-linear map f over polyhedral regions P_k by

$$v = f(i) = a_k + B_k i, \qquad i \in P_k, \quad k \in \{0, 1, \cdots, 2^l - 1\} \tag{35.4}$$

where the Jacobian $B_k \in \mathbb{R}^{n \times n}$ and the offset vector $a_k \in \mathbb{R}^n$ are defined over the polyhedral region P_k, separated by hyperplanes $c_i^T x - d_i = 0$, $i = 1, \cdots, l$ and defined by

$$P_k = \{x \in \mathbb{R}^n | c_j^T x - d_j \geq 0, j \in I_k, c_j^T x - d_j \leq 0, j \notin I_k\} \tag{35.5}$$

where $k = \sum_{j \in I_k} 2^{j-1}$, $I_k \subseteq \{1, 2, \cdots, l\}$ and $c_j \in \mathbb{R}^n$, $d_j \in \mathbb{R}$.

In other words the hyperplanes $c_i^T x - d_i = 0$, $i = 1, \cdots, l$ separate the space $\mathbb{R}^n$ into 2^l polyhedral regions P_k (see Fig. 35.4) where the constitutive equations are linear.

The computer storage requirements for this representation is still quite large, especially for large multiports. A more fundamental problem with this rather intuitive representation is that it is not necessarily continuous at the boundaries between two polyhedral regions. In fact, the continuity of the nonlinear map is usually desirable for physical reasons and also in order to avoid problems in the analysis.

The canonical piecewise-linear representation [6] is a very simple, attractive, and explicit description for a resistive multiport that solves both problems:

$$v = f(i) = a + Bi + \sum_{j=1}^{l} e_j |c_j^T i - d_j| \tag{35.6}$$

One can easily understand this equation by looking at the wedge form of the modulus map (see Fig. 35.5). It has two linear regions: in the first $x \geq 0$ and $y = x$, while in the second $x \leq 0$

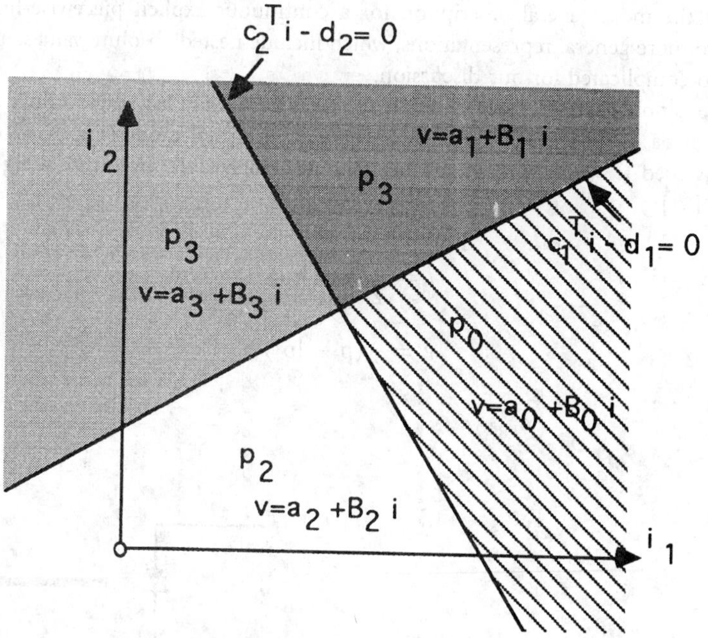

FIGURE 35.4 A PWL function defined in four polyhedral regions in $\mathbb{R}^2$ defined by $c_1^T i - d_1 \lesseqgtr 0$ and $c_2^T i - d_2 \lesseqgtr 0$.

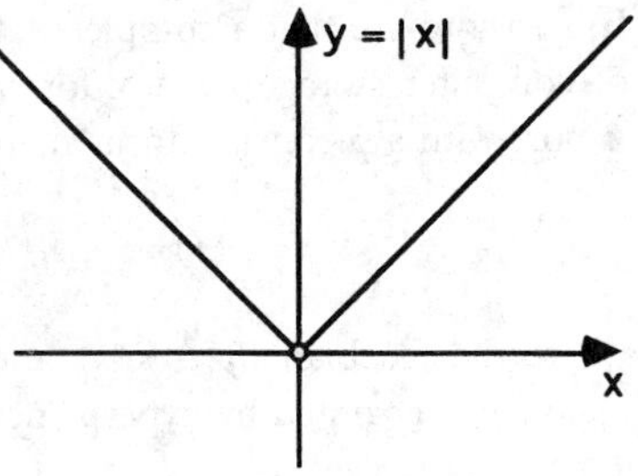

FIGURE 35.5 The absolute function $y = |x|$.

and $y = -x$. At the boundary the function is clearly continuous. Equation (35.6) is hence also continuous and is linear in each of the polyhedral regions P_k described by (35.5). If there are l modulus terms in (35.6), there are 2^l polyhedral regions where the map (35.6) is linear. Since the map is represented canonically with $n + n^2 + l(n+1)$ real parameters, this is a very compact and explicit representation.

Several examples of canonical PWL models for components are given in Section 35.3.

From Fig. 35.5, it should be clear that the right and left derivative of $y = |x|$ at 0 are different, their difference being 2. Hence the Jacobian J_+ and J_- of (35.6) will be different on the boundary between the two neighboring polyhedral regions where $(c_j i - d_j) \geq 0$ and $(c_j i - d_j) \leq 0$

$$J_+ - J_- = 2e_j c_j^T \tag{35.7}$$

Observe that this difference is a rank 1 matrix, which is also called a dyadic or outer vector product of e_j and c_j. Moreover, this difference is independent of the location of the independent variable i on the boundary. This important observation is made in [24], and is called the consistent variation property [10] and essentially says that the variation of the Jacobian of a canonical piecewise-linear representation is independent of the place where the hyperplane $c_j i - d_j = 0$ is crossed. Of course, this implies that the canonical piecewise-linear representation (35.6) is not the most general description for a continuous explicit piecewise-linear map. In [26], [29] two more general representations, which include nested absolute values, are presented. These are too complicated for our discussion.

Clearly the canonical PWL representation (35.6) is valid only for single-valued functions. It can clearly not be used for an important component: the ideal diode (Fig. 35.6) characterized by the multivalued (i, v) relation. It can be presented analytically by introducing a real scalar parameter ρ [31].

$$i = \frac{1}{2}(\rho + |\rho|) \tag{35.8}$$

$$v = \frac{1}{2}(\rho - |\rho|) \tag{35.9}$$

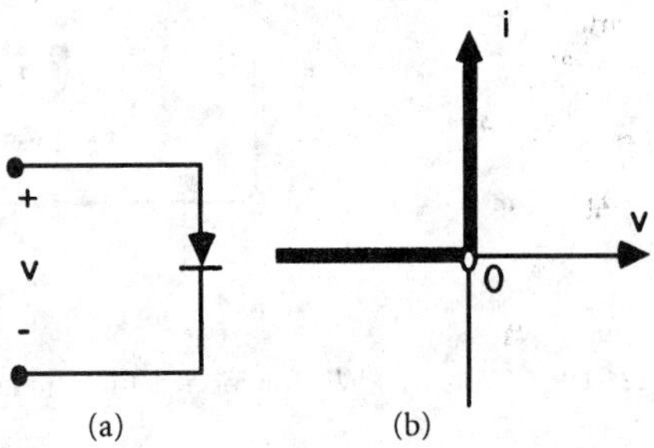

FIGURE 35.6 (a) The ideal diode and (b) the $(i–v)$ relation of an ideal diode.

This parametric description can easily be seen to correspond to Fig. 35.6(b) because $i = \rho$ and $v = 0$ for $\rho \geq 0$, while $i = 0$ and $v = \rho$ when $\rho \leq 0$. Such a parametric description $i = f(\rho)$ and $v = g(\rho)$ with f and g PWL can be obtained for a whole class of unicursal curves (see [6]).

When we allow implicit representations between v and i for a multiport, we obtain an LCP (linear complementarity problem) model (35.10)–(35.12) with an interesting state space like form [55]:

$$v = Ai + Bu + f \tag{35.10}$$

$$s = Ci + Du + g \tag{35.11}$$

$$u \geq 0, \quad s \geq 0, \quad u^T s = 0 \tag{35.12}$$

where $A \in \mathbb{R}^{n\times n}$, $B \in \mathbb{R}^{n\times l}$, $f \in \mathbb{R}^n$, $g \in \mathbb{R}^l$, $c \in \mathbb{R}^{l\times n}$, $D \in \mathbb{R}^{l\times l}$ are the parameters that characterize the relationship between v and i. In this model, u and s are called the state vectors and we say that $u \geq 0$ when all its components are nonnegative. Clearly, (35.12) dictates that all components of u and s should be nonnegative and that whenever a component u_j satisfies $u_j > 0$; then $s_j = 0$ and vice versa when $s_j > 0$ then $u_j = 0$. This is called the linear complementarity property, which we have seen already in the ideal diode (35.8) and (35.9) where $i \geq 0$, $v \geq 0$ and $iv = 0$. Hence, an implicit or LCP model for the ideal diode (35.8) and (35.9) is

$$v = u \tag{35.13}$$

$$s = i \tag{35.14}$$

$$u \geq 0 \quad s \geq 0 \quad us = 0 \tag{35.15}$$

In order to understand that the general equations (35.10)–(35.12) describe a PWL relation like (35.4)–(35.5) between i and v over polyhedral regions, one should observe first that $v = Ai + f$ is linear when $u = 0$ and $s = Ci + g \geq 0$. Hence, the relation is linear in the polyhedral region determined by $Ci + g \geq 0$. In general, one can consider 2^l possibilities for u and s according to

$$(u_j \geq 0 \quad \text{and} \quad s_j = 0) \quad \text{or} \quad (u_j = 0 \quad \text{and} \quad s_j \geq 0), \quad \text{for } j = 1, 2, \cdots, l$$

Denote sets of indexes U and S for certain values of u and s satisfying (35.12)

$$U = \{j \mid u_j \geq 0 \text{ and } s_j = 0\} \tag{35.16}$$

$$S = \{j \mid u_j = 0 \text{ and } s_j \geq 0\} \tag{35.17}$$

then, clearly U and S are complementary subsets of $\{1, 2, \cdots, l\}$ when for any j, u_j, and s_j cannot be both zero. Clearly each of these 2^l possibilities corresponds to a polyhedral region P_U in $\mathbb{R}^n$, which can be determined from

$$u_j \geq 0, \quad (Ci + Du + g)_j = 0 \quad \text{for } j \in U \tag{35.18}$$

$$u_j = 0, \quad (Ci + Du + g)_j \geq 0 \quad \text{for } j \in S \tag{35.19}$$

The PWL map in region P_U is determined by solving the u_j for $j \in U$ from (35.18) and substituting these along with $u_j = 0$ for $j \in S$ into (35.10). This generates, of course, a map that is linear in the region P_U.

When (35.11) is replaced by the implicit equation

$$Es + Ci + Du + g\alpha = 0 \qquad \alpha \geq 0$$

in (35.10)–(35.13) we call the problem a generalized linear complementarity problem (GLCP).

A nontrivial example of an implicit PWL relation (LCP model) is the hysteresis one port resistor (see Fig. 35.7). Its equations are

$$v = -i + [-1 \quad 1]\begin{bmatrix} u_1 \\ u_2 \end{bmatrix} + 1 \tag{35.20}$$

$$\begin{bmatrix} s_1 \\ s_2 \end{bmatrix} = \begin{bmatrix} -1 \\ 1 \end{bmatrix} i + \begin{bmatrix} -1 & 1 \\ 1 & -1 \end{bmatrix}\begin{bmatrix} u_1 \\ u_2 \end{bmatrix} + \begin{bmatrix} 1 \\ 0 \end{bmatrix} \tag{35.21}$$

$$s_1 \geq 0, \qquad s_2 \geq 0, \qquad u_1 \geq 0, \qquad u_2 \geq 0, \qquad u_1 s_1 + u_2 s_2 = 0 \tag{35.22}$$

In the first region P we have

$$s_1 = -i + 1 \geq 0, \qquad s_2 = i \geq 0, \quad \text{and thus} \quad v = -i + 1 \tag{35.23}$$

The region $P_{\{1,2\}}$, on the other hand, is empty since the following set of equations is contradictory:

$$s_1 = s_2 = 0, \qquad -i - u_1 + u_2 + 1 = 0, \qquad i + u_1 - u_2 = 0 \tag{35.24}$$

The region $P_{\{1\}}$ is

$$u_1 \geq 0, \qquad s_1 = -i - u_1 + 1 = 0, \qquad u_2 = 0, \qquad s_2 = i + u_1 \geq 0 \tag{35.25}$$

Hence $u_1 = -i + 1$ and $s_2 = 1$ and $v = -i + i - 1 + 1 = 0$, while $i \leq 1$.

Finally, the region $P_{\{2\}}$ is

$$u_1 = 0, \qquad s_1 = -i + u_2 + 1 \geq 0, \qquad u_2 \geq 0, \qquad s_2 = i - u_2 = 0$$

Hence

$$u_2 = i \quad \text{and} \quad s_1 = 1 \quad \text{and} \quad v = -i + i + 1 = 1 \quad \text{while} \quad i \geq 0 \tag{35.26}$$

It is now easy to show in general that the canonical PWL representation is a special case of the LCP model. Just choose $u_j \geq 0$ and $s_j \geq 0$ for all j as follows:

$$|c_j^T i - d_j| = \frac{1}{2}(u_j + s_j) \tag{35.27}$$

$$c_j^T i - d_j = \frac{1}{2}(u_j - s_j) \tag{35.28}$$

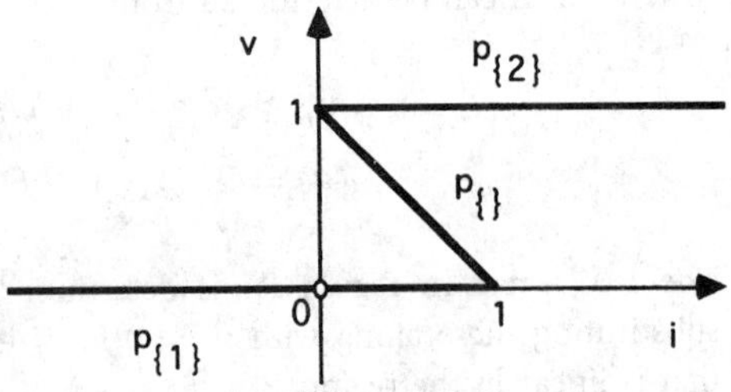

FIGURE 35.7 The hysteresis nonlinear resistor.

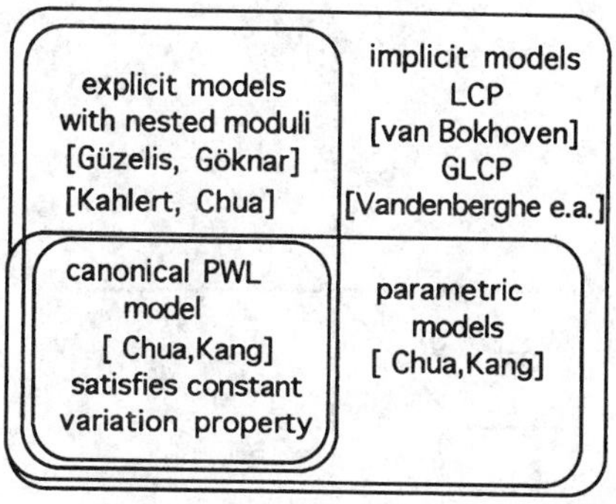

FIGURE 35.8 The interrelation of the PWL models.

then, u and s are complementary vectors, i.e.,

$$u \geq 0 \qquad s \geq 0 \qquad u^T s = 0$$

Observe that the moduli in (35.6) can be eliminated with (35.27) to produce an equation of the form (35.10) and that (35.28) produces equations of the form (35.11).

More generally it has been proven [36] that the implicit model includes all explicit models. Since it also includes the parametric models, one obtains the general hierarchy of models as depicted in Fig. 35.8.

A general remark should be made about all models that have been presented until now. Although the models have been given for resistive multiports where the voltages v at the ports are expressed in terms of the currents i, analogous equations can be given for the currents i in terms of the voltages or hybrid variables. It can even be adapted for piecewise linear capacitors, inductors or memristors, where the variables are respectively q, v for capacitors, φ, i for inductors, and q, φ for memristors.

5.3 Piecewise-Linear Models for Electronic Components

In order to simulate nonlinear networks with a circuit- or network simulator, the nonlinear behavior of the components must be modeled first. During this modeling phase, properties of the component that are not considered important for the behavior of the system may be neglected. Since the nonlinear behavior is often important, nonlinear models have to be used. In typical simulators like SPICE, nonlinear models often involve polynomials and transcendental functions for bipolar and MOS transistors. Since these consume a large part of the simulation time, table lookup methods have been worked out. However the table lookup methods need much storage for an accurate description of multiports and complex components.

The piecewise-linear models constitute an attractive alternative that is both efficient in memory use and in computation time. We discuss here the most important components. The derivation of a model usually requires two steps: first, the PWL approximation of constitutive equations, and second, the algebraic representation.

Two PWL models for an ideal diode (Fig. 35.6) have been derived, that is, a parametric model (35.8) and (35.9) and an implicit model (35.13)–(35.15), while a canonical PWL model does not exist.

The piecewise-linear models for operational amplifiers (op amps) and operational transconductance amplifiers (OTA's) are also simple and frequently used. The piecewise-linear approximation of op amps and OTA's of Fig. 35.9 is quite accurate. It leads to the following representation for the op amp, which is in the linear region for $-E_{sat} \leq v_0 \leq E_{sat}$ with voltage

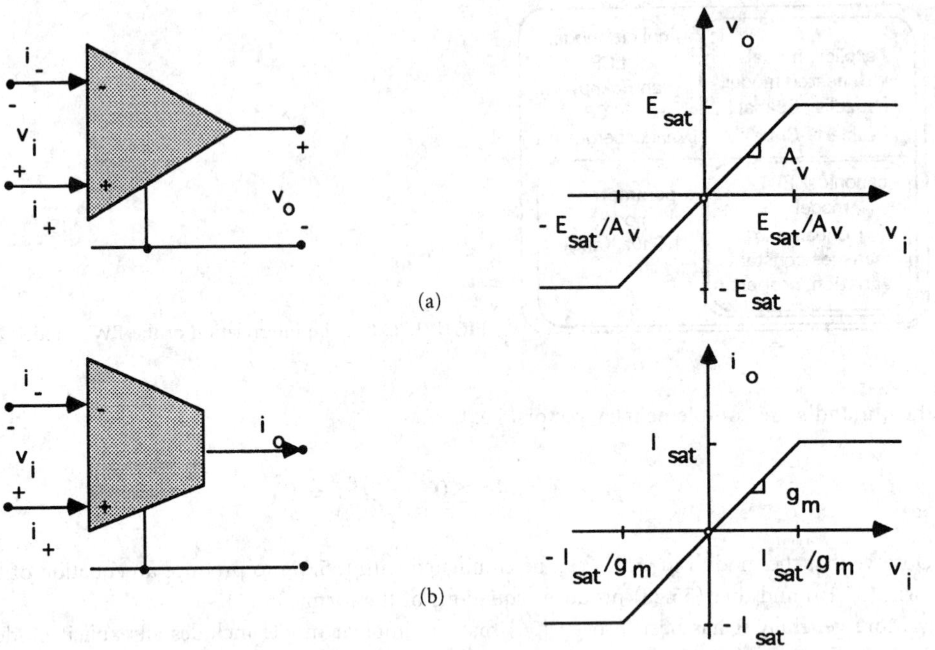

FIGURE 35.9 (a) Op amp and PWL model and (b) OTA and PWL model.

amplification A_v and positive and negative saturation E_{sat} and $-E_{sat}$

$$v_0 = \frac{A_v}{2}\left(\left|v_i + \frac{E_{sat}}{A_v}\right| - \left|v_i - \frac{E_{sat}}{A_v}\right|\right) \tag{35.29}$$

$$i_- = i_+ = 0 \tag{35.30}$$

This is called the op amp finite-gain model. In each of the three regions the op amp can be replaced by a linear circuit.

For the OTA we have similarly in the linear region for $-I_{sat} \leq i_0 \leq I_{sat}$ with transconductance gain g_m and positive and negative saturation I_{sat} and $-I_{sat}$

$$i_0 = \frac{g_m}{2}\left(\left|v_i + \frac{I_{sat}}{g_m}\right| - \left|v_i - \frac{I_{sat}}{g_m}\right|\right) \tag{35.31}$$

$$i_- = i_+ = 0 \tag{35.32}$$

Next for a tunnel diode one can perform a piecewise-linear approximation for the tunnel-diode characteristic as shown in Fig. 35.10. It clearly has three regions with conductances g_1, g_2, and g_3. This PWL characteristic can be realized by three components (Fig. 35.10(b)) with conductances, voltage sources, and diodes. The three parameters G_0, G_1, and G_2 of Fig. 35.10(b) must satisfy

$$\text{in Region 1:} \quad G_0 = g_1 \tag{35.33}$$

$$\text{in Region 2:} \quad G_0 + G_1 = g_2 \tag{35.34}$$

$$\text{in Region 3:} \quad G_0 + G_1 + G_2 = g_3 \tag{35.35}$$

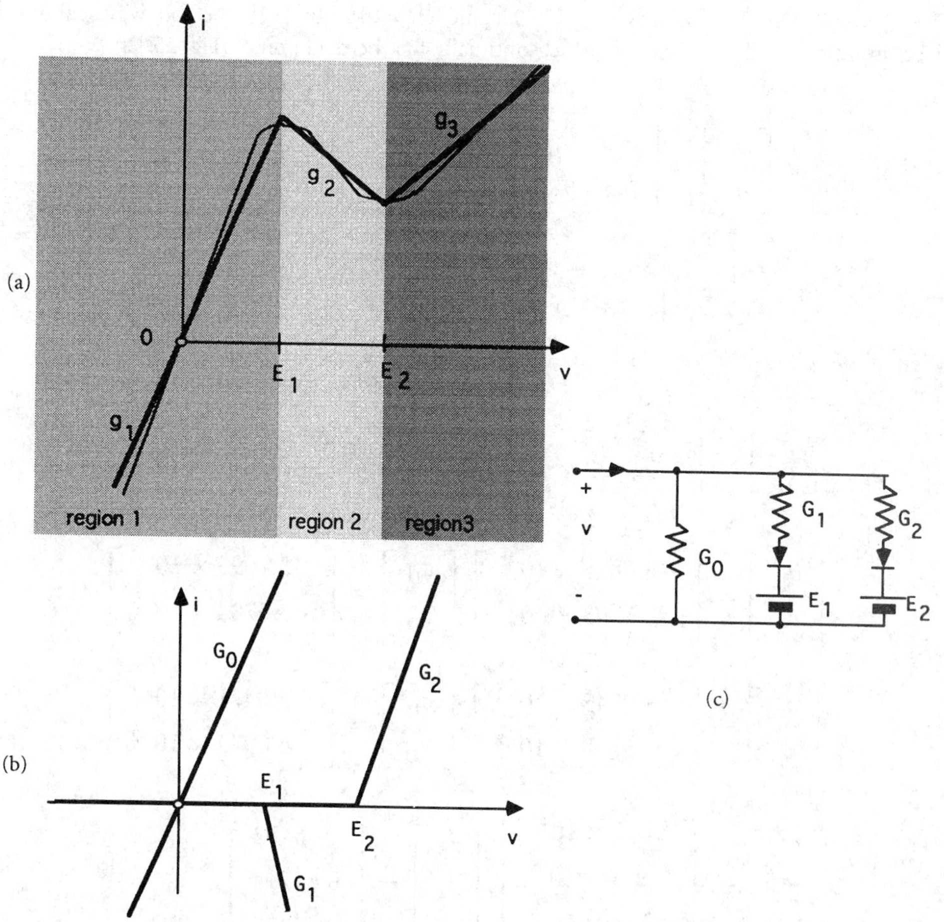

FIGURE 35.10 (a) Piecewise-linear approximation of the tunnel-diode characteristic. The three-segment approximation defines the three regions indicated. (b) Decomposition of the piecewise-linear characteristic (a) into three components, and (c) the corresponding circuit.

Thus, $G_0 = g_1$, $G_1 = -g_1 + g_2$, and $G_2 = -g_2 + g_3$. We can derive the canonical PWL representation as follows:

$$i = -\frac{1}{2}(G_1E_1 + G_2E_2) + \left(G_0 + \frac{1}{2}G_1 + \frac{1}{2}G_2\right)v + \frac{1}{2}G_1|v - E_1| + \frac{1}{2}G_2|v - E_2| \quad (35.36)$$

Next we present a canonical piecewise-linear bipolar transistor model [12]. Assume a *npn* bipolar transistor is connected in the common base configuration with $v_1 = v_{BE}$, $v_2 = v_{BC}$, $i_1 = i_E$, and $i_2 = i_C$, as shown in Fig. 35.3. We consider data points in a square region defined by $0.4 \leq v_1 \leq 0.7$ and $0.4 \leq v_2 \leq 0.7$, and assume the terminal behavior of the transistor follows the Ebers–Moll equation; namely

$$i_1 = \frac{I_s}{\alpha_f}(e^{v_1/V_T} - 1) - I_s(e^{v_2/V_T} - 1) \quad (35.37)$$

$$i_2 = \frac{I_s}{\alpha_r}(e^{v_2/V_T} - 1) - I_s(e^{v_1/V_T} - 1) \quad (35.38)$$

with $I_s = 10^{-14}$ A, V_T =26 mV, $\alpha_f = 0.99$, and $\alpha_r = 0.5$. In [12], the following canonical piecewise-linear model is obtained, which optimally fits the data points (Fig. 35.11)

$$\begin{bmatrix} i_1 \\ i_2 \end{bmatrix} = \begin{bmatrix} a_1 \\ a_2 \end{bmatrix} + \begin{bmatrix} b_{11} & b_{21} \\ b_{12} & b_{22} \end{bmatrix} \begin{bmatrix} v_1 \\ v_2 \end{bmatrix} + \begin{bmatrix} c_{11} \\ c_{21} \end{bmatrix} |m_1 v_1 - v_2 + t_1| + \begin{bmatrix} c_{12} \\ c_{22} \end{bmatrix} |m_2 v_1 - v_2 + t_2| + \begin{bmatrix} c_{13} \\ c_{23} \end{bmatrix} |m_3 v_1 - v_2 + t_3| \quad (35.39)$$

where

$$\begin{bmatrix} a_1 \\ a_2 \end{bmatrix} = \begin{bmatrix} 5.8722 \times 10^{-3} \\ -3.2652 \times 10^{-2} \end{bmatrix} \begin{bmatrix} b_{11} \\ b_{21} \end{bmatrix} = \begin{bmatrix} 3.2392 \times 10^{-2} \\ -3.2067 \times 10^{-2} \end{bmatrix}$$

$$\begin{bmatrix} b_{12} \\ b_{22} \end{bmatrix} = \begin{bmatrix} -4.0897 \times 10^{-2} \\ 8.1793 \times 10^{-2} \end{bmatrix} \begin{bmatrix} c_{11} \\ c_{21} \end{bmatrix} = \begin{bmatrix} 3.1095 \times 10^{-6} \\ -3.0784 \times 10^{-6} \end{bmatrix}$$

$$\begin{bmatrix} c_{12} \\ c_{22} \end{bmatrix} = \begin{bmatrix} -9.9342 \times 10^{-3} \\ 1.9868 \times 10^{-2} \end{bmatrix} \begin{bmatrix} c_{13} \\ c_{23} \end{bmatrix} = \begin{bmatrix} -3.0471 \times 10^{-2} \\ 6.0943 \times 10^{-2} \end{bmatrix}$$

$$\begin{bmatrix} m_1 \\ m_2 \\ m_3 \end{bmatrix} = \begin{bmatrix} 1.002 \times 10^{4} \\ -1.4 \times 10^{-4} \\ 1.574 \times 10^{-6} \end{bmatrix} \begin{bmatrix} t_1 \\ t_2 \\ t_3 \end{bmatrix} = \begin{bmatrix} -6472 \\ 0.61714 \\ 0.66355 \end{bmatrix}$$

Next, a canonical piecewise-linear MOS transistor model is presented. Assume the MOS transistor is connected in the common source configuration with $v_1 = v_{GS}$, $v_2 = v_{DS}$, $i_1 = i_G$, and $i_2 = i_D$, as shown in Fig. 35.12, where both v_1, v_2 are in volts, and i_1, i_2 are in microamperes. The data points are uniformly spaced in a grid within a rectangular region defined by $0 \le v_1 \le 5$, and $0 \le v_2 \le 5$. We assume the data points follow the Shichman–Hodges model, namely:

$$\begin{aligned} i_1 &= 0 \\ i_2 &= k[(v_1 - V_t)v_2 - 0.5v_2^2], \quad \text{if} \quad v_1 - V_t \ge v_2 \end{aligned}$$

or

$$i_2 = 0.5k(v_1 - V_t)^2[1 + \lambda(v_2 - v_1 + V_t)], \quad \text{if} \quad v_1 - V_t < v_2 \quad (35.40)$$

with $k = 50\ \mu\text{A/V}^2$, $V_t = 1$ V, $\lambda = 0.02\ \text{V}^{-1}$. Applying the optimization algorithm of [11] we obtain the following canonical piecewise-linear model (see Fig. 35.13):

$$\begin{aligned} i_2 = {} & a_2 + b_{21}v_1 + b_{22}v_2 + c_{21}|m_1 v_1 - v_2 + t_1| \\ & + c_{22}|m_2 v_1 - v_2 + t_2| + c_{23}|m_3 v_1 - v_2 + t_3| \end{aligned} \quad (35.41)$$

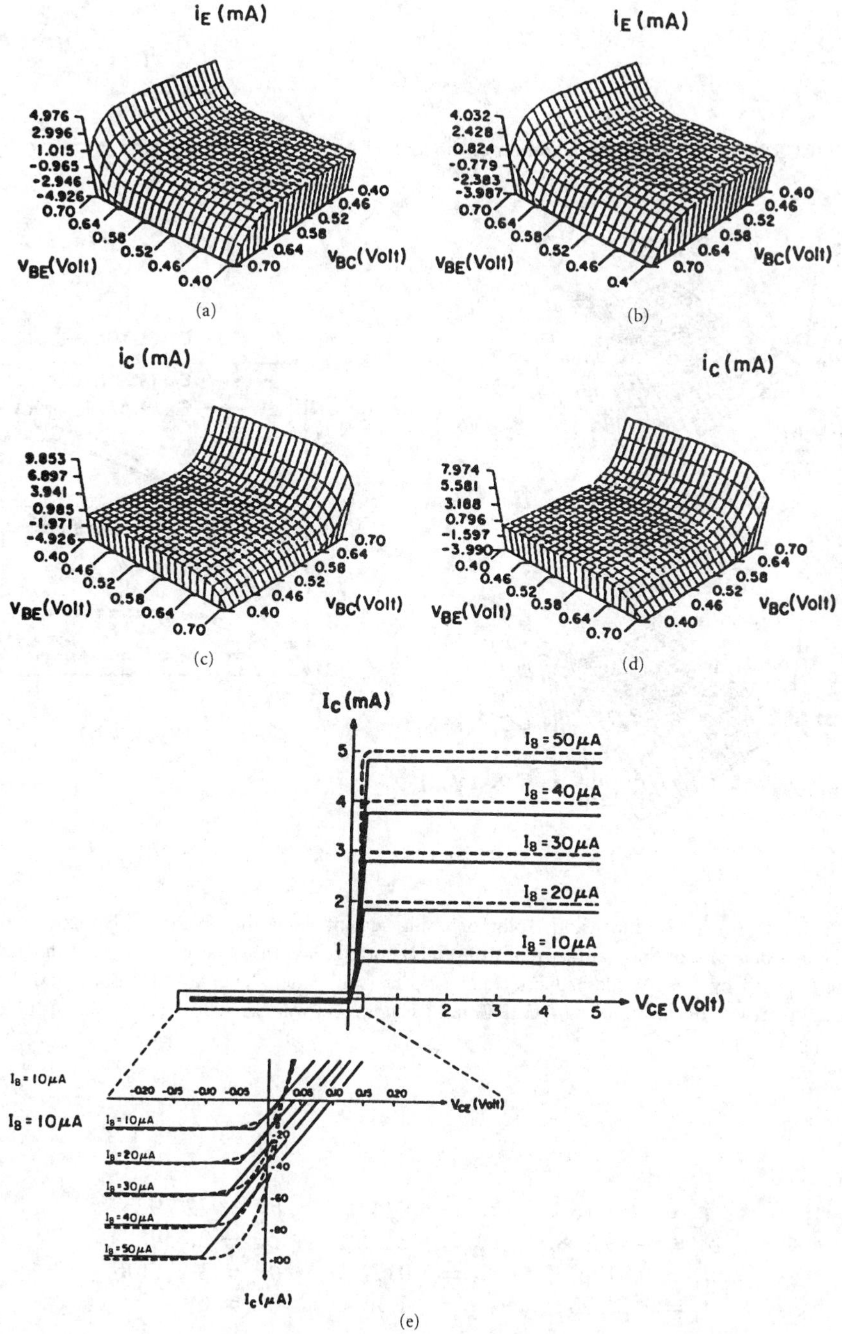

FIGURE 35.11 Three-dimensional plots for the emitter current in the Ebers–Moll model given by (35.37) and (35.38). (b) Three-dimensional plot for the emitter current in the canonical piecewise-linear model given by [10, (B.1)] (low-voltage version). (c) Three-dimensional plot for the collector current in the Ebers–Moll model given by (35.37) and (35.38). (d) Three-dimensional plot for the collector current in the canonical piecewise-linear model given by [10, (B.1)] (low-voltage version). (e) Comparison between the family of collector currents in the Ebers–Moll model (dashed line) and the canonical piecewise-linear model (solid line). *Source:* L. O. Chua and A. Deng, "Canonical piecewise linear modeling." *IEEE Trans. Circuits Syst.*, vol. CAS-33, p. 519, ©1986, IEEE.

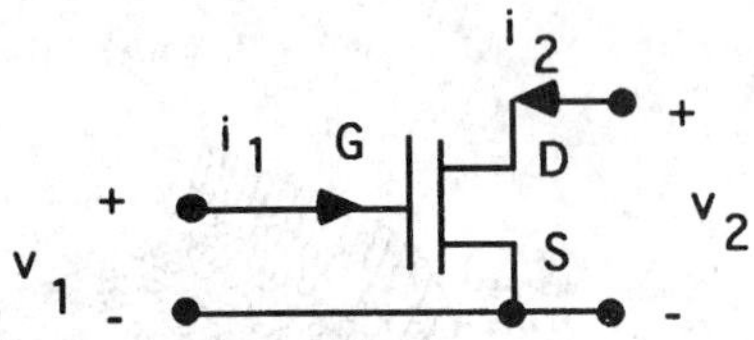

FIGURE 35.12 2-port configuration of the MOSFET.

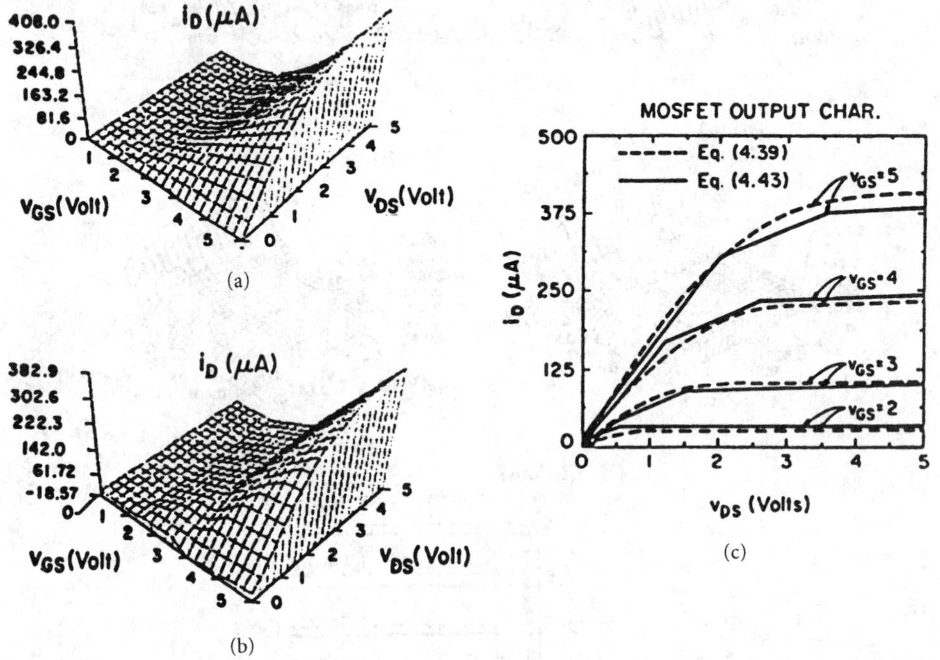

FIGURE 35.13 (a) Three-dimensional plot of drain current from the Shichman–Hodges model. (b) Three-dimensional plot of the drain current from the canonical piecewise-linear model. (c) Family of drain currents modeled by (35.40) (dashed line) and (35.41) (solid line). *Source:* L. O. Chua and A. Deng, "Canonical piecewise-linear modeling," *IEEE Trans. Circuits Syst.*, vol. CAS-33, p. 520, 1986. © 1986 IEEE.

where

$$
\begin{aligned}
a_2 &= -61.167, & b_{21} &= 30.242, & b_{22} &= 72.7925\\
c_{21} &= -49.718, & c_{22} &= -21.027, & c_{23} &= 2.0348\\
m_1 &= 0.8175, & m_2 &= 1.0171, & m_3 &= -23.406\\
t_1 &= -2.1052, & t_2 &= -1.4652, & t_3 &= 69
\end{aligned}
$$

Finally a canonical piecewise-linear model of GaAs FET is presented. The GaAs FET has become increasingly important in the development of microwave circuits and high-speed digital IC's due to its fast switching speed.

$$
\begin{aligned}
i_2 = a_2 + b_{21}v_1 + b_{22}v_2 + c_{21}|m_1v_1 - v_2 + t_2|\\
+ c_{22}|m_2v_1 - v_2 + t_2| + c_{23}|m_3v_1 - v_2 + t_3|
\end{aligned}
\tag{35.42}
$$

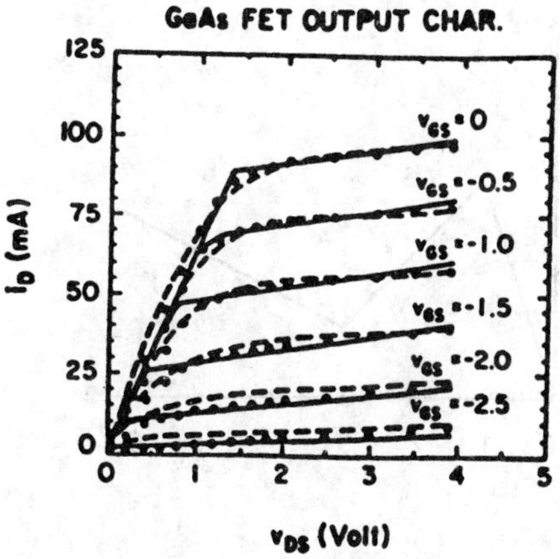

FIGURE 35.14 Comparison of the canonical piecewise-linear model described by (35.42) (solid line) and the analytical model (dashed line) for the ion-implanted GaAs FET. *Source:* L. O. Chua and A. Deng, "Canonical piecewise-linear modeling," *IEEE Trans. Circuits Syst.*, vol. CAS-33, p. 522, 1986. © 1986 IEEE.

where $v_1 = v_{GS}$ (volt), $v_2 = v_{DS}$ (volt), $i_2 - i_D$ (mA), and

$$a_2 = 6.3645, \quad b_{21} = 2.4961, \quad b_{22} = 32.339$$
$$c_{21} = 0.6008, \quad c_{22} = 0.9819, \quad c_{23} = -29.507$$
$$m_1 = -19.594, \quad m_2 = -6.0736, \quad m_3 = 0.6473$$
$$t_1 = -44.551, \quad t_2 = -8.9962, \quad t_3 = 1.3738$$

Observe that this model requires only three absolute-value functions and 12 numerical coefficients and compares rather well to the analytical model (Fig. 35.14).

More piecewise-linear models for timing analysis of logic circuits can be found in [21]. In the context of analog computer design, even PWL models of other nonlinear relationships have been derived in [51].

35.4 Structural Properties of Piecewise-Linear Resistive Circuits

When considering interconnections of PWL resistors (components), it follows from the linearity of KVL and KCL that the resulting multiport is also a piecewise-linear resistor. However, if the components have a canonical PWL representation, the resulting multiport may not have a canonical PWL representation. This can be illustrated by graphically deriving the equivalent one port of the series connection of two tunnel diodes [3] (Fig. 35.15). Since both resistors have the same current, we have to add the corresponding voltages $v = v_1 + v_2$ and obtain an i–v plot with two unconnected parts. There are values of i that correspond to 3 values of v_1 for R_1 and 3 values of v_2 for R_2 and hence to 9 values of the equivalent resistor [Fig. 35.15(d)]. This illustrates once more that nonlinear circuits may have more solutions than expected at first sight. Although the two tunnel diodes R_1 and R_2 have a canonical PWL representation, the equivalent one port of their series connection has neither a canonical PWL voltage description, nor a current one. It however has a GLCP description since KVL, KCL, and the LCP of R_1 and R_2 constitute a GLCP. If the v–i PWL relation is monotonic, the inverse i–v function exists and then some uniqueness properties hold.

These observations are, of course, also valid for the parallel connection of two PWL resistors and for more complicated interconnections.

In Section 35.3 we illustrated with an example how a PWL one-port resistor can be realized with linear resistors and ideal diodes. This can be proven in general. One essentially needs a diode for each breakpoint in the PWL characteristic. Conversely, each one port with diodes and resistors is a PWL one port resistor.

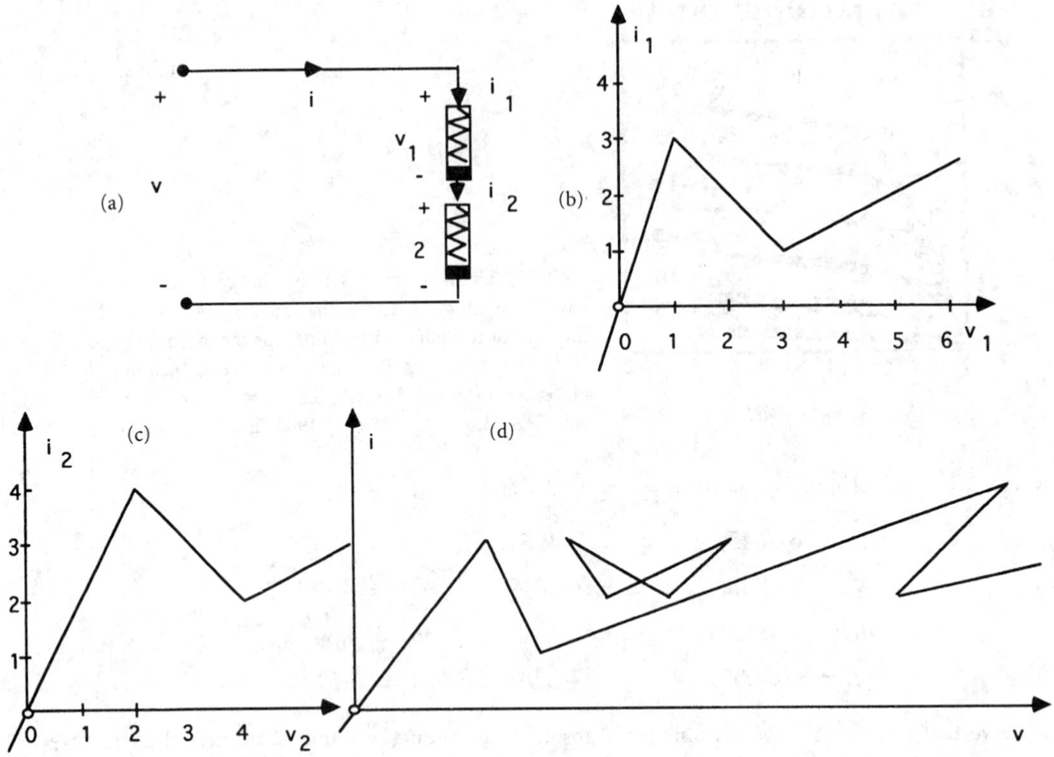

FIGURE 35.15 (a) The series connection of two tunnel diodes, (b) and (c), their i–v characteristics, and (d) the composite i–v plot, which consists of two unconnected parts.

This brings us to an interesting class of circuits composed of linear resistors, independent sources, linear controlled sources, and ideal diodes. These circuits belong to the general class of circuits with PWL components [see Fig. 35.1(a)] and can be described by GLCP equations. Such networks have not only shown their importance in analysis but also in the topologic study of the number of solutions and more general qualitative properties. When there are only short-circuit and open-circuit branches, one independent voltage source with internal resistance and ideal diodes, an interesting loop cut set exclusion property holds that is also called the colored branch theorem or the arc coloring theorem, (see Chapter 7.9). It says that the voltage source either forms a conducting loop with forward-oriented diodes and some short circuits or there is a cut set of the voltage source, some open circuits, and blocking diodes. Such arguments have been used to obtain [23] topologic criteria for upper bounds of the number of solutions of PWL resistive circuits. In fact, diode resistor circuits have been used extensively in PWL function generators for analog computers [51]. These electrical analogs can also be used for mathematical programming problems (like linear programming) and have reappeared in the neural network literature.

35.5 Analysis of Piecewise Linear Resistive Circuits

It is first shown that all conventional network formulation methods (nodal, cut set, hybrid, modified nodal, and tableau) can be used for PWL resistive circuits where the components are described with canonical or with LCP equations. These network equations may have one or more solutions. In order to find solutions, one can either search through all the polyhedral regions P_k by solving the linear equations for that region or by checking whether its solution is located inside that region P_k.

Since there are often many regions, this is a time-consuming method, but there are several ways to reduce the search [28], [61]. If one is interested in only one solution, one can use solution tracing methods, also called continuation methods or homotopy methods, of which the Katzenelson method is best known. If one is interested in all solutions, the problem is more complicated, but some algorithms exist.

Theorem Canonical PWL (Tableau Analysis) [8]

Consider a connected resistive circuit N containing only linear two-terminal resistors, dc independent sources, current-controlled and voltage-controlled piecewise-linear two-terminal resistors, linear- and piecewise-linear-controlled sources (all four types) and any linear multi-terminal resistive elements. A composite branch of this circuit is given in Fig. 35.16. If each piecewise-linear function is represented in the canonical form (35.6), then the *tableau formulation* also has the canonical PWL form

$$f(x) = a + Bx + \sum_{i+1}^{p} c_i |\alpha_i^T x - \beta_i| = 0 \tag{35.43}$$

where $x = [i^T, v^T, v_n^T]^T$ and i, respectively v, is the branch current voltage vector (Fig. 35.16) and v_n is the node-to-datum voltage vector.

PROOF. Let A be the reduced incidence matrix of N relative to some datum node, then KCL, KVL, and element constitutive relations give

$$Ai = AJ \tag{35.44}$$

$$v = A^T v_n + E \tag{35.45}$$

$$f_V(i) + f_V(v) = S \tag{35.46}$$

where we can express $f_I(\cdot)$ and $f_V(\cdot)$ in the canonical form (31.6)

$$f_I(i) = a_I + B_I i + C_I \operatorname{abs}(D_I^T e - e_I) \tag{35.47}$$

$$f_V(v) = a_V + B_V v + C_V \operatorname{abs}(D_V^T v - e_V) \tag{35.48}$$

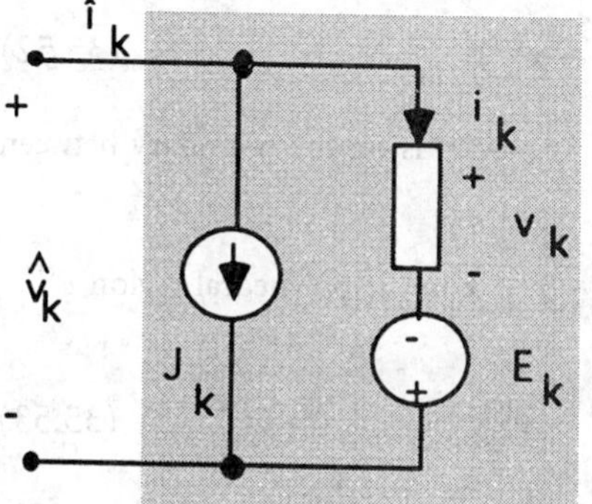

FIGURE 35.16 A composite branch.

Substituting (35.47) and (35.48) into (35.46), we obtain

$$\begin{bmatrix} -AJ \\ -E \\ a_I + a_V - S \end{bmatrix} + \begin{bmatrix} A & 0 & 0 \\ 0 & 1 & A^T \\ B_I & B_V & 0 \end{bmatrix} \begin{bmatrix} i \\ v \\ v_n \end{bmatrix} + \begin{bmatrix} 0 & 0 & 0 \\ 0 & 0 & 0 \\ C_I & C_V & 0 \end{bmatrix} \text{abs}\left[\begin{bmatrix} D_I & 0 & 0 \\ 0 & D_V & 0 \\ 0 & 0 & 0 \end{bmatrix}^T \begin{bmatrix} i \\ v \\ v_n \end{bmatrix} - \begin{bmatrix} e_I \\ e_V \\ 0 \end{bmatrix} \right] = 0 \quad (35.49)$$

Clearly (35.49) is in the canonical form of (35.43). □

Of course, an analogous theorem can be given when the PWL resistors are given in LCP form. Then the tableau equations constitute a GLCP. Moreover, completely in line with the section on circuit analysis (see Chapter 19), one can derive nodal, cut set, loop, hybrid, and modified nodal analysis from the tableau analysis by eliminating certain variables. Alternatively, one can also directly derive these equations.

Whatever the description for the PWL components may be, one can always formulate the network equations as linear equations

$$0 = f(x) = a_k + B_k x \qquad x \in P_k \quad (35.50)$$

in the polyhedral region P_k defined by (35.50). The map f is a continuous PWL map. A solution x of (35.50) can then be computed in a finite number of steps with the Katzenelson algorithm [4], [33], by tracing the map f from an initial point $(x^{(1)}, y^{(1)})$ to a value $(x^*, 0)$ (see Fig. 35.18).

Algorithm

STEP 1. Choose an initial point $x^{(1)}$ and determine its polyhedral region $P^{(1)}$ and compute

$$y^{(1)} = f(x^{(1)}) = a^{(1)} + B^{(1)}x \quad \text{and set} \quad j = 1$$

STEP 2. Compute

$$\hat{x} = x^{(j)} + (B^{(j)})^{-1}(0 - y^{(j)}) \quad (35.51)$$

STEP 3. If $\hat{x} \in P^{(j)}$ we have obtained a solution $\hat{x}$ of $f(\hat{x}) = 0$. Stop.

STEP 4. Otherwise compute

$$x^{(j+1)} = x^{(j)} + \lambda^{(j)}(\hat{x} - x^{(j)}) \quad (35.52)$$

where $\lambda^{(j)}$ is the largest number such that $x^{(j+1)} \in P^{(j)}$, i.e., $x^{(j+1)}$ is on the boundary between $P^{(j)}$ and $P^{(j+1)}$ (see Fig. 35.17).

STEP 5. Identify $P^{(j+1)}$ and the linear map $y = a^{(j+1)} + B^{(j+1)}x$ in the polyhedral region $P^{(j+1)}$ and compute

$$y^{(j+1)} = y^{(j)} + \lambda^{(j)}(y^* - y^{(j)}) \quad (35.53)$$

Set $j = j + 1$. Go to step 2.

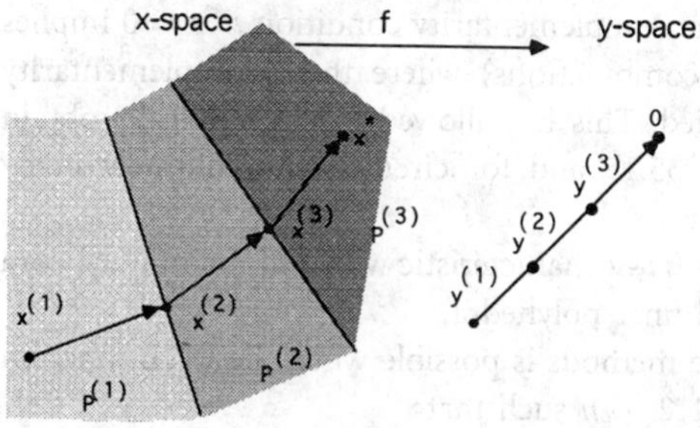

FIGURE 35.17 The iteration in the Katzenelson algorithm for solving $y = f(x) = 0$.

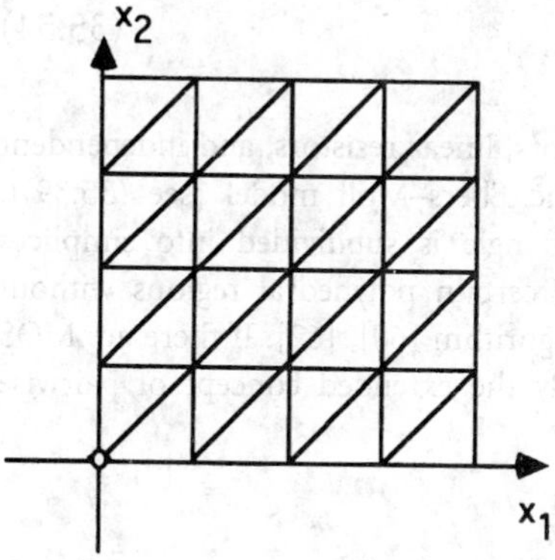

FIGURE 35.18 Simplicial subdivision.

This algorithm has been shown to converge to a solution in a finite number of steps if the determinants of all matrices $B^{(j)}$ have the same sign. This condition is satisfied when the i–v curves for the PWL one port resistors are monotonic. The Katzenelson algorithm has been extended in [45] by taking the sign of the determinants into account in (35.52) and (35.53). This requires the PWL resistors to be globally coercive. If by accident in the iteration the point $x^{(j+1)}$ is not on a single boundary and instead is located on a corner, the region $P^{(j+1)}$ is not uniquely defined. However with a small perturbation [1] one can avoid this corner and still be guaranteed to converge.

This algorithm has been adapted to the canonical PWL equation (35.49) in [8]. It can also be adapted to the GLCP. However, there exist circuits where this algorithm fails to converge. For the LCP problem, one can then use other algorithms [20], [40], [56]. One can also use other homotopy methods [43], [57], [60], which can be shown to converge based on eventual passivity arguments. In fact, this algorithm extends the rather natural method of source stepping, where the PWL circuit is solved by first making all sources zero and then tracing the solution for increasing (stepping up) the sources. It is instructive to observe here that these methods can be used successfully in another sequence of the steps in Fig. 35.1(a). Until now we always first performed the horizontal step of PWL approximation or modeling and then the vertical step of network equation formulation. With these methods, one can first perform the network equation formulation and then the PWL approximation. The advantage is that one can use a coarser grid in the simplicial subdivision far away from the solution, and hence dynamically adapt the accuracy of the PWL approximation.

In any case if all solutions are requested, all these homotopy-based methods are not adequate, because not all solutions can be found even if the homotopy method is started with many different $x^{(1)}$. Hence, special methods have been designed. It is beyond the scope of this text to give a complete algorithm [39], [59], but the solution of the GLCP basically involves two parts. First, calculate the solution set of all nonnegative solutions to (35.10) and (35.11). It is shown that this is a polyhedral cone where extremal rays can be easily determined [44], [54]. Second, this solution set is intersected with a hyperplane and the complementarity condition $u^T s = 0$ implies the elimination of vertices (respectively, convex combinations) where these complementarity (respectively, cross complementarity) is not satisfied. This has allowed to systematically obtain the complete solution set for the circuit of Fig. 35.15 and for circuits with infinitely many solutions.

A recently discovered method [46] covers the PWL i–v characteristic with a union of polyhedra and hierarchically solves the circuit with finer and finer polyhedra.

An important improvement in efficiency for the methods is possible when the PWL function $f(\cdot)$ is separable, i.e., there exist $f^i: \mathbb{R} \to \mathbb{R}^n$ $i = 1, 2 \cdots n$ such that

$$f(x) = \sum_{i=1}^{n} f^i(x_i) \tag{35.54}$$

This happens when there are only two terminal PWL resistors, linear resistors, and independent sources and if the bipolar transistors are modeled by the Ebers–Moll model [see (35.39)]. Then the subdivision for x is rectangular and each rectangle is subdivided into simplices (see Fig. 35.18). This property can be used to eliminate certain polyhedral regions without solutions [62] and also to speed up the Katzenelson-type algorithm [60], [62]. If there are MOS transistors, the map f is not separable but one can apply the extended concept of pairwise separable map [62].

35.6 Piecewise-Linear Dynamic Circuits

As mentioned at the end of Section 35.2, the piecewise linear descriptions of Section 35.2 can be used also for PWL capacitors, respectively, inductors and memristors, by replacing the port voltages v and currents i by a, v, respectively, φ, i and φ, q. Whenever we have a network obtained by interconnecting linear and/or PWL resistors, inductors, capacitors, and memristors, we have a dynamic piecewise-linear circuits. Of course, such networks are often encountered because it includes the networks with linear R, L, C, and linear dependent sources, diodes, switches, op amps, and components like bipolar and MOS transistors, and GaAs FET's with PWL resistive models. This includes several important and famous nonlinear circuits like Chua's circuit [18], [19], and the cellular neural networks (CNN's) [48], which are discussed in Chapter 37 and Chapter 38.2.

Of course, PWL dynamic circuits are much more interesting and much more complicated and can exhibit a much more complex behavior than resistive circuits and hence this subject is much less explored. It is clear from the definition of a PWL dynamic circuit that it can be described by linear differential equations over polyhedral regions. Hence, it can exhibit many different types of behavior. They may have many equilibria, which can essentially be determined by solving the resistive network (see Section 35.5 and Fig. 35.1) obtained by opening the capacitive ports and short circuiting the inductive ports (dc analysis). When there is no input waveform the circuit is said to be autonomous and has transients. Some transients may be periodic and are called limit cycles, but they may also show chaotic behavior. Next, one may be interested in the behavior of the circuit for certain input waveforms (transient analysis). This can be performed by using integration rules in simulations.

For the analysis of limit cycles, chaos, and transients, one can of course use the general methods for nonlinear circuits, but some improvements can be made based on the PWL nature of the nonlinearities. Here we only describe the methods briefly. If one is interested in the periodic behavior of a PWL dynamic circuit (autonomous or with a periodic input) then one can for each PWL nonlinearity make some approximations.

First, consider the case that one is only interested in the dc and fundamental sinusoidal contributions in all signals of the form $i(t) = A_0 + A_1 \cos \omega t$. The widely used describing function method [6] for PWL resistors $v = f(i)$ consists of approximating this resistor by an approximate resistor where $\hat{v}(t) = D_0 + D_1 \cos \omega t$ has only the dc and fundamental contribution of $v(t)$. This is often a good approximation since the remainder of the circuit often filters out all higher harmonics anyway. Using a Fourier series, one can then find D_0 and D_1 as

$$D_0(A_0, A_1) = \frac{1}{2\pi} \int_0^{2\pi} f(A_0 + A_1 \cos \phi)\, d\phi$$

$$D_1(A_0, A_1) = \frac{1}{\pi A_1} \int_0^{2\pi} f(A_0 + A_1 \cos \phi)\, d\phi$$

By replacing all PWL components by their describing functions, one can use linear methods to set up the network equations in the Laplace–Fourier domain. When this approximation is not sufficient, one can include more harmonics. Then one obtains the famous harmonic balance method, because one is balancing more harmonic components.

Alternatively, one can calculate the periodic solution by simulating the circuit with a certain initial condition and considering the map $F: x_0 \to x_1$ from the initial condition x_0 to the state x_1 one period later. Of course, a fixed point $x^* = F(x^*)$ of the map corresponds to a periodic solution. It has been shown [27] that the map F is differentiable for PWL circuits. This is very useful in setting up an efficient iterative search for a fixed point of F. This map is also useful in studying the eventual chaotic behavior and is then called the Poincaré return map.

In transient analysis of PWL circuits, one is often interested in the sensitivity of the solution to certain parameters in order to optimize the behavior. As a natural extension of the adjoint network for linear circuits in [22] the adjoint PWL circuit is defined and used to determine simple sensitivity calculations for transient analysis.

Another important issue is whether the PWL approximation of a nonlinear characteristic in a dynamic circuit has a serious impact on the transient behavior. In [63], error bounds have been obtained on the differences of the waveforms.

5.7 Efficient Computer-Aided Analysis of PWL Circuits

Transient analysis and timing verification is an essential part of the VLSI system design process. The most reliable way of analyzing the timing performance of a design is to use analog circuit analysis methods. Here as well, a set of algebraic–differential equations has to be solved. This can be done by using implicit integration formulas that convert these equations into a set of algebraic equations, which can be solved by iterative techniques like Newton–Raphson (see Chapter 36). The computation time then becomes excessive for large circuits. It mainly consists of linearizations of the nonlinear component models and the solution of the linear equations. In addition, the design process can be facilitated substantially if this simulation tool can be used at many different levels from the top level of specifications over the logic and switch level to the circuit level. Such a hierarchical simulator can support the design from top to bottom and allow for mixtures of these levels. In limited space we describe here the main simulation methods for improving the efficiency and supporting the hierarchy of models with piecewise-linear methods. We refer the reader to Chapter 61 for general simulation of VLSI circuits and to the literature for more details on the methods and for more descriptions on complete simulators.

It is clear from our discussion before that PWL models and circuit descriptions can be used at many different levels. An op amp, for example, can be described by the finite gain model [see Fig. 35.9 and (35.29) and (35.30)], but when it is designed with a transistor circuit it can be described by PWL circuit equations as in Section 35.5. Hence, it is attractive to use a simulator that can support this top-down design process [35]. One can then even incorporate logic gates into the PWL models. One can organize the topological equations of the network hierarchically, so that it is easy to change the network topology. The separation between topological equations and model descriptions allows for an efficient updating of the model when moving from one polyhedral region into another. Several other efficiency issues can be built into a hierarchical PWL simulator.

An important reduction in computation time needed for solving the network equations can be obtained by using the consistent variation property. In fact, there is only a rank one difference between the matrices of two neighboring polyhedral regions, and hence, one inverse can be easily derived from the other [8], [35]. In the same spirit one can at the circuit level take advantage of the PWL transistor models (see [62] and separability discussion in Section 32.5). In [53] the circuit is partitioned dynamically into subcircuits, during the solution process depending on the transistor region of operation. Then, the subcircuits are dynamically ordered and solved with block Gauss–Seidel for minimal or no coupling among them.

Interesting savings can be obtained [34] by solving the linear differential equations in a polyhedral region with Laplace transformations and by partitioning the equations. However, the computation of the intersection between trajectories in neighboring polyhedral regions can be a disadvantage of this method.

References

[1] M. J. Chien, "Piecewise-linear homeomorphic resistive networks," *IEEE Trans. Circuits Syst.*, vol. CAS-24, pp. 118–127, Mar. 1977.

[2] M. J. Chien and E. S. Kuh, "Solving nonlinear resistive networks using piecewise-linear analysis and simplicial subdivision," *IEEE Trans. Circuits Syst.*, vol. CAS-24, pp. 305–317, 1977.

[3] L. O. Chua, "Analysis and synthesis of multivalued memoryless nonlinear networks," *IEEE Trans. Circuit Theory*, vol. CT-14, pp. 192–209., June 1967.

[4] L. O. Chua and P. M. Lin, *Computer-Aided Analysis of Electronic Circuits: Algorithms and Computational Techniques*, Englewood Cliffs, NJ: Prentice-Hall, 1975.

[5] L. O. Chua and P. M. Lin, "A Switching-parameter algorithm for finding multiple solutions of nonlinear resistive circuits," *Int. J. Circuit Theory, Applicat.*, vol. 4, pp. 215–239, 1976.

[6] L. O. Chua and S. M. Kang, "Section-wise piecewise linear functions: Canonical representation properties and applications," *Proc. IEEE*, vol. 65, pp. 915–929, June 1977.

[7] L. O. Chua and D. J. Curtin, "Reciprocal n-port resistor represented by continuous n-dimensional piecewise-linear function realized by circuit with 2-terminal piecewise-linear resistor and $p+g$ port transformer," *IEEE Trans. Circuits Syst.*, vol. CAS-27, pp. 367–380, May 1980.

[8] L. O. Chua and R. L. P. Ying, "Finding all solutions of piecewise-linear circuits," *Int. J. Circuit Theory, Applicat.*, vol. 10, pp. 201–229, 1982.

[9] L. O. Chua and R. L. P. Ying, "Canonical piecewise-linear analysis," *IEEE Trans. Circuits Syst.*, vol. CAS-30, pp. 125–140, 1983.

[10] L. O. Chua and A. C. Deng, "Canonical piecewise-linear analysis—Part II: Tracing driving point and transfer characteristics," *IEEE Trans. Circuits Syst.*, vol. CAS-32, pp. 417–444, May 1985.

[11] L. O. Chua and A. C. Deng, "Canonical piecewise-linear modeling, unified parameter optimization algorithm; application to pn junctions, bipolar transistors, MOSFETs, and GaAs FET's, *IEEE Trans. Circuits Syst.*, vol. CAS-33, pp. 511–525, May 1986.

[12] L. O. Chua and A. C. Deng, "Canonical piecewise linear modeling," *IEEE Trans. Circuits Syst.*, vol. CAS-33, pp. 511–525, May 1986.

[13] L. O. Chua and A. C. Deng, "Canonical piecewise linear analysis: Generalized breakpoint hopping algorithm," *Int. J. Circuit Theory, Applicat.*, vol. 14, pp. 35–52, 1986.

[14] L. O. Chua and A. C. Deng, "Canonical piecewise linear representation," *IEEE Trans. Circuits Syst.*, vol. 35, pp. 101–111, Jan. 1988.

[15] L. O. Chua and A. C. Deng, "Canonical piecewise-linear modeling," ERL Memo. UCB/ERL M85/35, Univ. California, Berkeley, Apr. 26, 1985.

[16] L. O. Chua and G. Lin, "Canonical realization of Chua's circuit family," *IEEE Trans. Circuits Syst.*, vol. 37, pp. 885–902, July 1990.

[17] L. O. Chua and G. Lin, "Intermittency in piecewise-linear circuit," *IEEE Trans. Circuits Syst.*, vol. 38, pp. 510–520, May 1991.

[18] L. O. Chua, "The genesis of Chua's circuit," *Archiv Elektronik Übertragungstechnik*, vol. 46, no. 4, pp. 250–257, 1992.

[19] L. O. Chua, C.-W. Wu, A.-S. Huang, and G.-Q. Zhong, "A Universal circuit for studying and generating chaos," *IEEE Trans. Circuits Syst. I*, vol. 40, no. 10, pp. 732–744, 745–761, Oct. 1993.

[20] R. Cottle, J.-S. Pang, and R. Stone, *The Linear Complementarity Problem.* New York: Academic, 1992.

[21] A. C. Deng, "Piecewise-linear timing model for digital CMOS circuits," *IEEE Trans. Circuits Syst.*, vol. 35, pp. 1330–1334, Oct. 1988.

[22] Y. Elcherif and P. Lin, "Transient analysis and sensitivity computation in piecewise-linear circuits," *IEEE Trans. Circuits Syst.*, vol. 38, pp. 1525–1533, Dec. 1988.

[23] M. Fossepréz, M. J. Hasler, and C. Schnetzler, "On the number of solutions of piecewise-linear resistive circuits," *IEEE Trans. Circuits Syst.*, vol. 36, pp. 393–402, Mar. 1989.

[24] T. Fujisawa and E. S. Kuh, "Piecewise linear theory of nonlinear networks," *SIAM J. Appl. Math.*, vol. 22, no. 2, pp. 307–328, Mar. 1972.

[25] T. Fujisawa, E. S. Kuh, and T. Ohtsuki, "A sparse matrix method for analysis of piecewise-linear resistive networks," *IEEE Trans. Circuit Theory*, vol. 19, pp. 571–584, Nov. 1972.

[26] G. Güzelis and I. Göknar, "A canonical representation for piecewise affine maps and its applications to circuit analysis," *IEEE Trans. Circuits Syst.*, vol. 38, pp. 1342–1354, Nov. 1991.

[27] I. N. Hajj and S. Skelboe, "Dynamic systems: Steady-state analysis," *IEEE Trans. Circuits Syst.*, vol. CAS-28, pp. 234–242, Mar. 1981.

[28] Q. Huang and R. W. Liu, "A simple algorithm for finding all solutions of piecewise-linear networks," *IEEE Trans. Circuits Syst.*, vol. 36, pp. 600–609, Apr. 1989.

[29] C. Kahlert and L. O. Chua, "A generalized canonical piecewise linear representation," *IEEE Trans. Circuits Syst.*, vol. 37, pp. 373–383, Mar. 1990.

[30] C. Kahlert and L. O. Chua, "Completed canonical piecewise-linear representation: Geometry of domain space," *IEEE Trans. Circuits Syst.*, vol. 39, pp. 222–236, Mar. 1992.

[31] S. M. Kang and L. O. Chua, "A global representation of multidimensional piecewise linear functions with linear partitions," *IEEE Trans. Circuits Syst.*, vol. CAS-25, pp. 938–940, Nov. 1978.

[32] S. Karamardian, "The complementarity problem," *Mathemat. Program.*, vol. 2, pp. 107–129, 1972.

[33] S. Karamardian, J. Katzenelson, "An algorithm for solving nonlinear resistive networks," *Bell Syst. Tech. J.*, vol. 44, pp. 1605–1620, 1965.

[34] R. J. Kaye and A. Sangiovanni-Vincentelli, "Solution of piecewise-linear ordinary differential equations using waveform relaxation and Laplace transforms," *IEEE Trans. Circuits Syst.*, vol. CAS-30, pp. 353–357, June 1983.

[35] T. A. M. Kevenaar and D. M. W. Leenaerts, "A flexible hierarchical piecewise linear simulator," *Integrat., VLSI J.*, vol. 12, pp. 211–235, 1991.

[36] T. A. M. Kevenaar and D. M. W. Leenaerts, "A comparison of piecewise linear model descriptions," *IEEE Trans. Circuits Syst.*, vol. 39, pp. 996–1004, Dec. 1992.

[37] M. Kojima and Y. Yamamoto, "Variable dimension algorithms: Basic theory, interpretations and extensions of some existing methods," *Mathemat. Program.*, vol. 24, pp. 177–215, 1982.

[38] S. Lee and K. Chao, "Multiple solutions of piecewise-linear resistive networks," *IEEE Trans. Circuits Syst.*, vol. CAS-30, pp. 84–89, Feb. 1983.

[39] D. M. W. Leenaerts and J. A. Hegt, "Finding all solutions of piecewise linear functions and the application to circuit design," *Int. J. Circuit Theory, Applicat.*, vol. 19, pp. 107–123, 1991.

[40] C. E. Lemke, "On complementary pivot theory," in *Nonlinear Programming*, J. B. Rosen, O. L. Mangasarian, and K. Ritten, Eds., New York: Academic, 1968, pp. 349–384.

[41] J. Lin and R. Unbehauen, "Canonical piecewise-linear approximations," *IEEE Trans. Circuits Syst.*, vol. 39, pp. 697–699, Aug. 1992.

[42] R. Lum and L. O. Chua, "Generic properties of continuous piecewise-linear vector fields in 2-D space," *IEEE Trans. Circuits Syst.*, vol. 38, pp. 1043–1066, Sep. 1991.

[43] R. Melville, L. Trajkovic, S.-C. Fang, and L. Watson, "Artificial homotopy methods for the DC operating point problem," *IEEE Trans. Comput.-Aided Design Integrat. Circuits Syst.*, vol. 12, pp. 861–877, June 1993.

[44] T. S. Motzkin, H. Raiffa, G. L. Thompson, and R. M. Thrall, "The double description method," in *Contributions to the Theory of Games, Ann. Mathemat. Studies*, H. W. Kuhn and A. W. Tucker, Eds., Princeton: Princeton Univ. Press, 1953, pp. 51–73.

[45] T. Ohtsuki, T. Fujisawa, and S. Kumagai, "Existence theorem and a solution algorithm for piecewise-linear resistor circuits," *SIAM J. Math. Anal.*, vol. 8, no. 1, pp. 69–99, 1977.

[46] S. Pastore and A. Premoli, "Polyhedral elements: A new algorithm for capturing all the equilibrium points of piecewise-linear circuits," *IEEE Trans. Circuits Syst. I*, vol. 40, pp. 124–132, Feb. 1993.

[47] V. C. Prasad and V. P. Prakash, "Homeomorphic piecewise-linear resistive networks," *IEEE Trans. Circuits Syst.*, vol. 35, pp. 251–253, Feb. 1988.

[48] T. Roska and J. Vandewalle, *Cellular Neural Networks*, New York: Wiley, 1993.

[49] I. W. Sandberg, "A note on the operating-point equations of semiconductor-device networks," *IEEE Trans. Circuits Syst.*, vol. 37, p. 966, July 1990.

[50] A. S. Solodovnikov, *Systems of Linear Inequalities*, transl. by L. M. Glasser, T. P. Branson, Chicago: Univ. Chicago, 1980.

[51] T. E. Stern, *Theory of Nonlinear Networks and Systems:* An Introduction, Reading, MA: Addison-Wesley, 1965.

[52] S. Stevens and P.-M. Lin, "Analysis of piecewise-linear resistive networks using complementary pivot theory," *IEEE Trans. Circuits Syst.*, vol. CAS-28, pp. 429–441, May 1981.

[53] O. Tejayadi and I. N. Hajj, "Dynamic partitioning method for piecewise-linear VLSI circuit simulation," *Int. J. Circuit Theory, Applicat.*, vol. 16, pp. 457–472, 1988.

[54] S. N. Tschernikow, *Lineare Ungleichungen.* Berlin: VEB Deutscher Verlag der Wissenschaften, 1971; translation from H. Weinert and H. Hollatz, *Lineinye Neravenstva,* 1968, into German.

[55] W. M. G. van Bokhoven, *Piecewise-linear Modelling and Analysis.* Deventer: Kluwer, 1980.

[56] C. van de Panne, "A complementary variant of Lemke's method for the linear complementary problem," *Mathemat. Program.*, vol. 7, pp. 283–310, 1974.

[57] L. Vandenberghe and J. Vandewalle, "Variable dimension algorithms for solving resistive circuits," *Int. J. Circuit Theory Applicat.*, vol. 18, pp. 443–474, 1990.

[58] L. Vandenberghe and J. Vandewalle, "A continuous deformation algorithm for DC-analysis of active nonlinear circuits," *J. Circuits, Syst., Comput.*, vol. 1, pp. 327–351, 1991.

[59] L. Vandenberghe, B. L. De Moor, and J. Vandewalle, "The generalized linear complementarity problem applied to the complete analysis of resistive piecewise-linear circuits," *IEEE Trans. Circuits Syst.*, vol. 36, pp. 1382–1391, 1989.

[60] K. Yamamura and K. Horiuchi, "A globally and quadratically convergent algorithm for solving resistive nonlinear resistive networks," *IEEE Trans. Computer-Aided Design Integrat. Circuits Syst.*, vol. 9, pp. 487–499, May 1990.

[61] K. Yamamura and M. Ochiai, "Efficient algorithm for finding all solutions of piecewise-linear resistive circuits," *IEEE Trans. Circuits Syst.*, vol. 39, pp. 213–221, Mar. 1992.

[62] K. Yamamura, "Piecewise-linear approximation of nonlinear mappings containing Gummel–Poon models or Shichman–Hodges models," *IEEE Trans. Circuits Syst.*, vol. 39, pp. 694–697, Aug. 1992.

[63] M. E. Zaghloul and P. R. Bryant, "Nonlinear network elements; error bounds," *IEEE Trans. Circuits Syst.*, vol. CAS-27, pp. 20–29, Jan. 1980.

36
Simulation

Erik Lindberg
Technical University of Denmark

This chapter deals with the **simulation** or analysis of a nonlinear electrical circuit by means of a computer program. The program creates and solves the differential–algebraic equations of a **model** of the circuit. The basic tools in the solution process are *linearization, difference approximation,* and *the solution of a set of linear equations.* The output of the analysis may consist of 1) all node and branch voltages and all branch currents of a bias point (dc analysis); 2) a linear small-signal model of a bias point, which may be used for analysis in the frequency domain (ac analysis); or 3) all voltages and currents as functions of time in a certain time range for a certain excitation (transient analysis). A model is satisfactory if there is good agreement between measurements and simulation results. In this case, simulation may be used instead of measurement for obtaining a better understanding of the nature and abilities of the circuit. The crucial point is to set up a model that is as simple as possible, in order to obtain a fast and inexpensive simulation, but sufficiently detailed to give the proper answer to the questions concerning the behavior of the circuit under study. **Modeling is the bottleneck of simulation**.

The **model** is an **equivalent scheme** or a **branch table** describing the basic components (n-terminal elements) of the circuit and their connection. It is always possible to model an n-terminal element by means of a number of 2-terminals (branches). These internal 2-terminals may be coupled. By pairing the terminals of an n-terminal element, a port description may be obtained. The branches are either admittance branches or impedance branches. All branches may be interpreted as controlled sources. An **admittance branch** is a current source primarily controlled by its own voltage or primarily controlled by the voltage or current of another branch (transadmittance). An **impedance branch** is a voltage source primarily controlled by its own current or primarily controlled by the current or voltage of another branch. Control by a signal (voltage or current) and control by a time derivative of a signal is allowed. Control by several variables is allowed. Examples of admittance branches: 1) the conductor is a current source controlled by its own voltage; 2) the capacitor is a current source controlled by the time derivative of its own voltage; 3) the open circuit is a zero-valued current source (a conductor with value zero). Examples of impedance branches: 1) the resistor is a voltage source controlled by its own current; 2) the inductor is a voltage source controlled by the time derivative of its own current; 3) the short circuit is a zero-valued voltage source (a resistor with value zero).

A component may often be modeled in different ways. A diode, for example, is normally modeled as a current source controlled by its own voltage such that the model can be linearized

0-8493-8341-2/95/$0.00+$.50

into a **dynamic conductor** in parallel with a current source during the iterative process of finding the bias point of the diode. The diode may also be modeled as 1) a voltage source controlled by its own current (a dynamic resistor in series with a voltage source); 2) a **static conductor** being a function of the voltage across the diode; or 3) a static resistor being a function of the current through the diode. Note that in the case where a small-signal model is wanted, for frequency analysis, only the dynamic model is appropriate.

The **primary variables** of the model are the currents of the impedance branches and the node potentials. The current law of Kirchhoff (the sum of all the currents leaving a node is zero) and the voltage–current relations of the impedance branches are used for the creation of the equations describing the relations between the primary variables of the model. The contributions to the equations from the branches are taken one branch at a time based on the question: will this branch add new primary variables? If yes, then new columns (variables) and new rows (equations) must be created and updated, or else the columns and rows corresponding to the existing primary variables of the branch must be updated. This approach to equation formulation is called the **extended nodal approach** or the **modified nodal approach (MNA).**

In the following, some algorithms for solving a set of nonlinear algebraic equations and nonlinear differential equations are very briefly described. Because we are dealing with physical systems and because we are responsible for the models, we assume that at least one solution is possible. The zero solution is of course always a solution. It might happen that our models become invalid if we, for example, increase the amplitudes of the exciting signals, diminish the risetime of the exciting signals, or by mistake create unstable models. It is important to define the range of validity for our models. What are the consequences of our assumptions? Can we believe in our models?

5.1 Numerical Solution of Nonlinear Algebraic Equations

Let the equation system to be solved be $f(x, u) = 0$, where x is the vector of primary variables and u is the excitation vector. Denote the solution by x_s. Then, if we define a new function $g(x) = \alpha[f(x, u)] + x$, where α may be some function of $f(x, u)$ which is zero for $f(x, u) = 0$, we can define an **iterative scheme** where $g(x)$ converges to the solution x_s by means of the iteration: $x_{k+1} = g(x_k) = \alpha[f(x_k, u)] + x_k$, where k is the iteration counter.

If for all x in the interval $[x_a, x_b]$ the condition $\|g(x_a) - g(x_b)\| \leq L\|x_a - x_b\|$ for some $L < 1$ is satisfied, the iteration is called a **contraction mapping**. The condition is called a **Lipschitz condition**. Note that a function is a contraction if it has a derivative less than 1.

For $\alpha = -1$ the iterative formula becomes $x_{k+1} = g(x_k) = -f(x_k, u) + x_k$. This scheme is called the **Picard method**, the **functional method**, or the **contraction mapping algorithm**. At each step, each nonlinear component is replaced by a linear **static component** corresponding to the solution x_k. A nonlinear conductor for example is replaced by a linear conductor defined by the straight line through the origin and the solution point. Each iterative solution is calculated by solving a set of linear equations. All components are updated and the next iteration is made. When two consecutive solutions are within a prescribed tolerance the solution point is accepted.

For $\alpha = -1/(df/dx)$ the iterative formula becomes $x_{k+1} = g(x_k) = -f(x_k, u)/(df(x_k, u)/dx) + x_k$. This scheme is called the **Newton–Raphson method** or the **derivative method.** At each step each nonlinear component is replaced by a linear **dynamic component** plus an independent source corresponding to the solution x_k. A nonlinear conductor, for example, is replaced by a linear conductor defined by the derivative of the branch current with respect to the branch voltage (the slope of the nonlinearity) in parallel with a current source corresponding to the branch voltage of the previous solution. A new solution is then calculated by solving a set of linear equations. The components are updated and the next iteration is made. When the solutions converge within a prescribed tolerance the solution point is accepted.

It may of course happen that the above mentioned iterative schemes do not **converge** before an iteration limit k_{max} is reached. One reason might be that the nonlinearity $f(x)$ changes very rapidly for a small change in x. Another reason could be that $f(x)$ possess some kind of symmetry that causes cycles in the Newton–Raphson iteration scheme. If **convergence problems** are detected, the iteration scheme can be *modified* by introducing a limiting of the actual step size. Another approach may be to change the modeling of the nonlinear branches from voltage control to current control or vice versa. Often the user of a circuit analysis program may be able to solve convergence problems by means of proper modeling and adjustment of the program options. [1, 2, 3, 11]

36.2 Numerical Integration of Nonlinear Differential Equations

The dynamics of a nonlinear electronic circuit may be described by a set of coupled first-order differential equations–algebraic equations of the form $dx/dt = f(x, y, t)$ and $g(x, y, t) = 0$, where x is the vector of **primary variables** (node potentials and impedance branch currents), y is the vector of variables that cannot be explicitly eliminated, and f and g are nonlinear vector functions. It is always possible to express y as a function of x and t by inverting the function g and inserting it into the differential equations such that the general differential equation form $dx/dt = f(x, t)$ is obtained. The task is then to obtain a solution $x(t)$, when an initial value of x is given. The usual methods for solving differential equations reduce to the solution of **difference equations**, with either the derivatives or the integrals expressed *approximately* in terms of finite differences.

Assume, at a given time t_0, we have a known solution point $x_0 = x(t_0)$. At this point the function f can be expanded in Taylor series: $dx/dt = f(x_0, t) + A(x_0)(x - x_0) + \cdots$, where $A(x_0)$ is the Jacobian of f evaluated at x_0. Truncating the series we obtain a linearization of the equations such that the small-signal behavior of the circuit in the neighborhood of x_0 is described by $dx/dt = Ax + k$, where A is a constant matrix equal to the Jacobian and k is a constant vector.

The most simple scheme for the approximate solution of the differential equation $dx/dt = f(x, t) = Ax + k$ is the **forward Euler formula**: $x(t) = x(t_0) + hA(t_0)$ where $h = t - t_0$ is the integration time step. From the actual solution point at time t_0 the next solution point at time t is found along the tangent of the solution curve. It is obvious that we will rapidly leave the vicinity of the exact solution curve if the integration step is too large. To guarantee stability of the computation, the time step h must be smaller than $2/|\lambda|$ where λ is the largest **eigenvalue** of the Jacobian A. Typically, h must not exceed $0.2/|\lambda|$.

The forward Euler formula is a linear *explicit* formula based on forward Taylor expansion from t_0. If we make a backward Taylor expansion from t we arrive at the **backward Euler** formula: $x(t) = x(t_0) + hA(t)$. Because the unknown appears on both sides of the equation it must in general be found by iteration and the formula is a linear *implicit* formula. From a stability point of view the backward Euler formula has a much larger stability region than the forward Euler formula has. The truncation error for the Euler formulas is of order h^2.

The two Euler formulas can be thought of as polynomials of degree one that approximate $x(t)$ in the interval $[t_0, t]$. If we compute $x(t)$ from a second-order polynomial $p(t)$ that matches the conditions that $p(t_0) = x(t_0)$, $dp/dt(t_0) = dx/dt(t_0)$, and $dp/dt(t) = dx/dt(t)$, we arrive at the **trapezoidal rule**: $x(t) = x(t_0) + 0.5hA(t_0) + 0.5hA(t)$. In this case, the truncation error is of order h^3.

At each integration step the size of the local truncation error can be estimated. If it is too large, the step size must be reduced. An explicit formula like the forward Euler may be used as a **predictor** giving a starting point for an implicit formula like the trapezoidal, which in turn is used as a **corrector**. The use of a predictor–corrector pair provides the base for the estimate of the local truncation error. The trapezoidal formula with varying integration step size is the main formula used in the **SPICE program.**

The two Euler formulas and the trapezoidal formula are special cases of a general linear multistep formula: $\sum(a_i x_{n-i} + b_i h(dx/dt)_{n-i})$, where i goes from -1 to $m-1$ and m is the degree of the polynomial used for the approximation of the solution curve. The trapezoidal rule, for example, is obtained by setting $a_{-1} = -1$, $a_0 = +1$, and $b_{-1} = b_0 = 0.5$, all other coefficients being zero. The formula can be regarded as being derived from a polynomial of degree r, which matches $r+1$ of the solution points x_{n-i} or their derivatives $(dx/dt)_{n-i}$.

Very fast transients often occur together with very slow transients in electronic circuits. We observe widely different time constants. The large spread in component values, for example, from large decoupling capacitors to small parasitic capacitors, imply a large spread in the modules of the eigenvalues. We say that the circuits are **stiff.** A family of **implicit multistep methods** suitable for stiff differential equations has been proposed by C. W. Gear. The methods are stable up to the polynomial of order 6. For example the second-order **Gear formula** for fixed integration step size may be stated as $x_{n+1} = -(1/3)x_{n-1} + (4/3)x_n + (2/3)h(dx/dt)_{n+1}$.

By changing both the *order* of the **approximating polynomial** and the integration *step size,* the methods adapt themselves dynamically to the performance of the solution curve. The family of Gear formulas is modified into a "stiff-stable variable-order variable-step predictor–corrector" method based on implicit approximation by means of **backward difference formulas (BDF's).** The resulting set of nonlinear equations is solved by **modified Newton–Raphson iteration.** Note that numerical integration in a sense is a kind of low-pass filtering defined by means of the minimum integration step. [1, 2, 3, 11]

6.3 Use of Simulation Programs

Since around 1960 a large number of circuit-simulation programs have been developed by universities, industrial companies, and commercial software companies. In particular, the SPICE program has become a standard simulator both in the industry and in academia. Here only a few programs that together cover a very large number of simulation possibilities are presented. Due to competition there is a tendency to develop programs that are supposed to cover any kind of analysis so that only one program should be sufficient ("the Swiss Army Knife Approach"). Unfortunately this implies that the programs become very large and complex to use. Also, it may be difficult to judge the correctness and accuracy of the results of the simulation having only one program at your disposal. If you try to make the same analysis of the same model with different programs you will frequently see that the results from the programs may not agree completely. By comparing the results you may obtain a better feel for the correctness and accuracy of the simulation. The program **SPICE**, supplemented with the programs **NAP2** and **ESACAP**, has proven to be a good choice in the case where a large number of different kinds of circuits and systems are to be modeled and simulated ("the Tool Box Approach"). The programs are available in inexpensive evaluation versions running on IBM compatible personal computers. The input languages are very close, so it is possible to transfer input data easily between the programs. In the following, short descriptions of the programs are given and a small circuit is simulated in order to give an idea of the capabilities of the programs.

SPICE

The first versions of **SPICE** (**Simulation Program with Integrated Circuit Emphasis** version 2), based on the modified nodal approach, were developed in 1975 at the Electronics Research Laboratory, College of Engineering, University of California, Berkeley, CA.

SPICE is a general-purpose circuit analysis program. Circuit models may contain resistors, capacitors, inductors, mutual inductors, independent sources, controlled sources, transmission lines, and the most common semiconductor devices: diodes, bipolar junction transistors, and field-effect transistors. SPICE has very detailed built-in models for the semiconductor devices, some of which may require about 50 parameters. Besides the normal dc, ac, and transient analyses the program can make sensitivity, noise and distortion analysis, and analysis at different temperatures. In the various commercial versions of the program many other possibilities have been added; for example, behavior modeling (poles and zeros) and statistical analysis. SPICE is available on a number of platforms from DOS, UNIX/X-window (SUN, HP, DEC) to Cray. The size of the PSPICE1.EXE file is 729,056 bytes.

In order to give an idea of the **input language** the syntax of the statements describing controlled sources is as follows:

```
Voltage Controlled Current Source:  Gxxx  N+ N-  NC+ NC-  VALUE
Voltage Controlled Voltage Source:  Exxx  N+ N-  NC+ NC-  VALUE
Current Controlled Current Source:  Fxxx  N+ N-  VNAM     VALUE
Current Controlled Voltage Source:  Hxxx  N+ N-  VNAM     VALUE
```

where the initial characters of the branch name: G, E, F, and H indicate the type of the branch, N+ and N- are integers ("node numbers") indicating the placement and orientation of the branch, respectively, NC+, NC-, and VNAM indicate from where the control comes (VNAM is a dummy dc voltage source with value 0 inserted as an ammeter!), and VALUE specifies the numerical value of the control, which may be a constant or a polynomial expression in case of nonlinear dependent sources. Independent sources are specified with Ixxx for current and Vxxx for voltage sources.

The input file below describes an analysis of the **Chua oscillator circuit**. It is a simple harmonic oscillator with losses (C2, L2, and RL2) loaded with a linear resistor (R61) in series with a capacitor (C1) in parallel with a nonlinear resistor. The circuit is influenced by a sinusoidal voltage source VRS through a coil L1.

```
PSpice input file CRC2CHUA.SPI                                          :
*  :                                                                    :
*  *:   The Chua Oscillator, sinusoidal excitation, F = 150mV >         :
*  :                RL2 = 1 ohm, RL1 = 0 ohm f = 1286.336389 Hz         :
*  :                ------------------------------------------          :
*  : ref. K. Murali and M. Lakshmanan,                                  :
*  :      Effect of Sinusoidal Excitation on the Chua's Circuit,        :
*  :      IEEE Transactions on Circuits and Systems - 1:                :
*  :      Fundamental Theory and Applications,                          :
*  :      vol. 39, No. 4, April 1992, pp. 264-270                       :
*  :--------------------------------------------------------------------:
*  : input source;                                                      :
     vrs    7   0    sin(0   150m    1.2863363889332e+3 0 0 )           :
*  : choke                                                              :
     L1      6  17    80m                                               :
     VRL1   17   7    DC  0                                             :
*  :--------------------------------------------------------------------:
*  : harmonic oscillator;                                               :
     L2      6  16    13m                                               :
     RL2    16   0    1                                                 :
     C2      6   0    1.250u                                            :
*  :--------------------------------------------------------------------:
*  : load                                                               :
     r61     6  10    1310                                              :
     vrrC1  10  11    DC   0                                            :
     C1     11   0    0.017u                                            :
*    i(vrr10) = current of nonlinear resistor                           :
     vrr10  10   1    DC 0                                              :
```

```
*   :---------------------------------------------------------------:
*   : nonlinear circuit;                                            :
*   :                                          vt=n*k*T/q           :
      .model n4148 d ( is=0.1p    rs=16       n = 1 )               :
      d13     1    3 n4148                                          :
      d21     2    1 n4148                                          :
      rm9     2   22 47k                                            :
      vrm9   22    0 DC      -9                                     :
      rp9     3   33 47k                                            :
      vrp9   33    0 DC      +9                                     :
      r20     2    0 3.3k                                           :
      r30     3    0 3.3k                                           :
*   :                                                               :
*   : ideal op. amp.                                                :
*   :                                                               :
      evop   4 0   1 5   1e+20                                      :
      r14    1 4         290                                        :
      r54    5 4         290                                        :
      r50    5 0         1.2k                                       :
*   :---------------------------------------------------------------:
      .TRAN      0.05m   200m   0   0.018m   UIC                    :
      .plot   tran   v(11)                                          :
      .probe                                                        :
      .OPTIONS  acct  nopage  opts  gmin=1e-20  reltol=1e-3         :
+     abstol=1e-16      vntol=1e-16  tnom=25  itl5=0                :
+     limpts=15000                                                  :
      .end                                                          :
```

The program is controlled by means of the statements: .TRAN, where the maximum integration step is set to 18 μs, and .OPTIONS, where for example the relative truncation error is set to 1e-3. The result of the analysis is presented in Fig. 36.1. It can be seen that transition from **chaotic behavior** to a period-5 **limit cycle** takes place at about 150 ms. [4, 6, 7]

NAP2

The first versions of **NAP2 (Nonlinear Analysis Program version 2)**, [8] based on the extended nodal equation formulation, were developed in 1973 at the Institute of Circuit Theory and Telecommunication, Technical University of Denmark, Lyngby, Denmark.

NAP2 is a general-purpose circuit analysis program. Circuit models may contain resistors, conductors, capacitors, inductors, mutual inductors, ideal operational amplifiers, independent sources, controlled sources, and the most common semiconductor devices: diodes, bipolar junction transistors, and field-effect transistors. NAP2 has only simple built-in models for the semiconductor devices, which require about 15 parameters. Besides the normal dc, ac, and transient analyses the program can make parameter variation analysis. Any parameter (e.g., component value or temperature) may be varied over a range in an arbitrary way and a dc, ac, or transient analysis may be performed for each value of the parameter. **Optimization** of dc bias point ("given: voltages, find: resistors") is possible. **Event detection** is included such that it is possible to interrupt the analysis when a certain signal, for example, goes from a positive to a negative value. The results may be combined into one output plot. It is also possible to calculate the **poles** and **zeros** of driving point and transfer functions for the linearized model at a certain bias point. **Eigenvalue technique** (based on the **QR algorithm** by J. G. F. Francis) is the method behind the calculation of poles and zeros. **Group delay** (i.e., the derivative of the phase with respect to the angular frequency) is calculated from the poles and zeros. This part of the program is available as an independent program named **ANP3 (Analytical Network Program version 3)**. NAP2 is only available in a version running under DOS. The original versions were

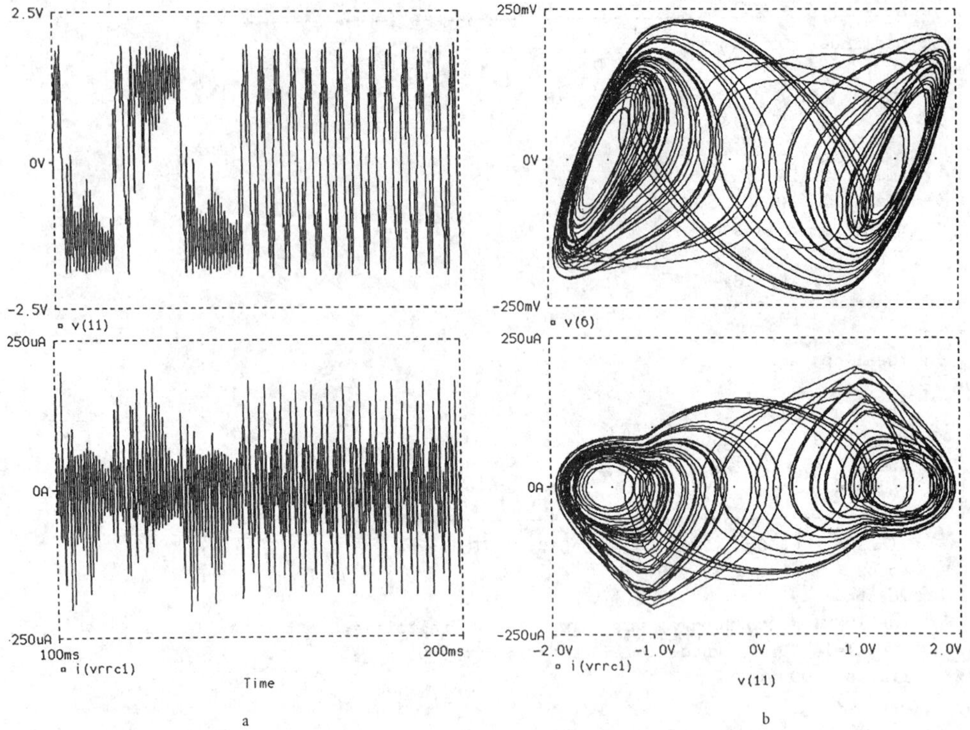

FIGURE 36.1 (a) PSpice analysis. The voltage of C1: v(11) and the current of C1: i(vrrc1) as functions of time in the time interval from 100 to 200 ms. (b) PSpice analysis. The voltage of C2: v(6) and the current of C1: i(vrrc1) as functions of the voltage of C1: v(11) in the time interval from 100 to 200 ms.

implemented on IBM and other mainframes under OS, VM, and other operating systems. The size of the NAP2X.EXE file is 271,535 bytes.

In order to give an idea of the input language, the syntax of the statements describing controlled sources is as follows:

```
Voltage Controlled Current Source:  Ixxx  N+ N-  VALUE  VByyy
Voltage Controlled Voltage Source:  Vxxx  N+ N-  VALUE  VByyy
Current Controlled Current Source:  Ixxx  N+ N-  VALUE  IByyy
Current Controlled Voltage Source:  Vxxx  N+ N-  VALUE  IByyy
```

where the initial characters of the branch name I and V indicate the type of the branch, N+ and N- are integers ("node numbers") indicating the placement and orientation of the branch, respectively, and VALUE specifies the numerical value of the control, which may be a constant or an arbitrary functional expression in case of nonlinear control. IB and VB refer to the current or voltage of the branch, respectively, from where the control comes. If the control is the time derivative of the branch signal, SI or SV may be specified. Independent sources must be connected to a resistor R or a conductor G as follows: Rxxx N+ N- VALUE E=VALUE and Gxxx N+ N- VALUE J=VALUE, where VALUE may be any function of time, temperature, and components.

The input file below describes the same analysis of the **Chua oscillator circuit** as performed above by means of SPICE. The circuit is a simple harmonic oscillator with losses (C2, L2, and RL2) loaded with a linear resistor (R61) in series with a capacitor (C1) in parallel with a nonlinear resistor. The circuit is excited by a sinusoidal voltage source RS.E through a coil L1.

The frequency is specified as angular frequency in rps. It is possible to specify more than one statement on one line.

```
*circuit;    *list    9;     :file: CRC-CHUA.NAP                    :
:                                                                   :
*: The Chua Oscillator, sinusoidal excitation, F = 150mV >          :
:           RL2 = 1 ohm, RL1 = 0 ohm f = 1286.336389 Hz             :
:           ---------------------------------------                 :
: ref. K. Murali and M. Lakshmanan,                                 :
:      Effect of Sinusoidal Excitation on the Chua's Circuit,       :
:      IEEE Transactions on Circuits and Systems - 1:               :
:      Fundamental Theory and Applications,                         :
:      vol.39, No.4, April 1992, pp. 264-270                        :
:-------------------------------------------------------------------:
: input source;                                                     :
:                                                                   :
  sin/sin/ ; rs  7   0  0   e=150m*sin(8.0822898994674e+3*time)     :
: choke    ; l1  6  17  80mH;   RL1  17   7  0                      :
:-------------------------------------------------------------------:
: harmonic oscillator; l2  6  16  13mH   ;   RL2   16   0  1        :
                       c2  6   0  1.250uF                           :
:-------------------------------------------------------------------:
: load; r61  6  10  1310 ;  rrc1  10 11  0;  c1  11  0  0.017uF     :
:      rr10 10   1     0 : irr10 = current of nonlinear resistor    :
:-------------------------------------------------------------------:
: nonlinear circuit;                                                :
:                                                                   :
  n4148   /diode/    is=0.1p gs=62.5m           vt=25mV ;           :
:                                               vt=n*k*T/q          :
  td13   1 3  n4148;        td21  2 1  n4148;                       :
  rm9    2 0  47k  e=-9;  rp9   3 0  47k  e=+9;                     :
  r20    2 0  3.3k;         r30   3 0  3.3k;                        :
:                                                                   :
: ideal op. amp.                                                    :
:                                                                   :
  gop   1 5    0; vop  4 0  vgop : no value means infinite value;   :
  r14   1 4  290; r54  5 4  290;    r50  5 0  1.2k;                 :
:-------------------------------------------------------------------:
  *time  time   0   200m   : variable order, variable step          :
  *tr   vnall  *plot(50 v6) v1   *probe                             :
  *RUN cycle=15000  minstep=1e-20 >                                 :
       trunc=1e-3  step=50n                                         :
*end                                                                :
```

The program options are set by means of the statement *RUN, where, for example, the minimum integration step is set to 1e-20 s and the relative truncation error is set to 1e-3. The result of the analysis is presented in Fig. 36.2. It can be seen that transition from **chaotic behavior** to a period-5 **limit cycle** takes place at about 40 ms. If we compare to the results obtained above by means of SPICE, we see that although the two programs are "modeled and set" the same way (e.g., same relative tolerance), the results are different due to the chaotic nature of the circuit and possibly also due to the different strategies of equation formulation and solution used in the two programs (e.g., SPICE uses the **trapezoidal integration method** with variable step and NAP2 uses the **Gear integration methods** with variable order and variable step).

ESACAP

The first versions of **ESACAP (Engineering System And Circuit Analysis Program)**, based on the extended nodal equation formulation, were developed in 1979 at Elektronik Centralen,

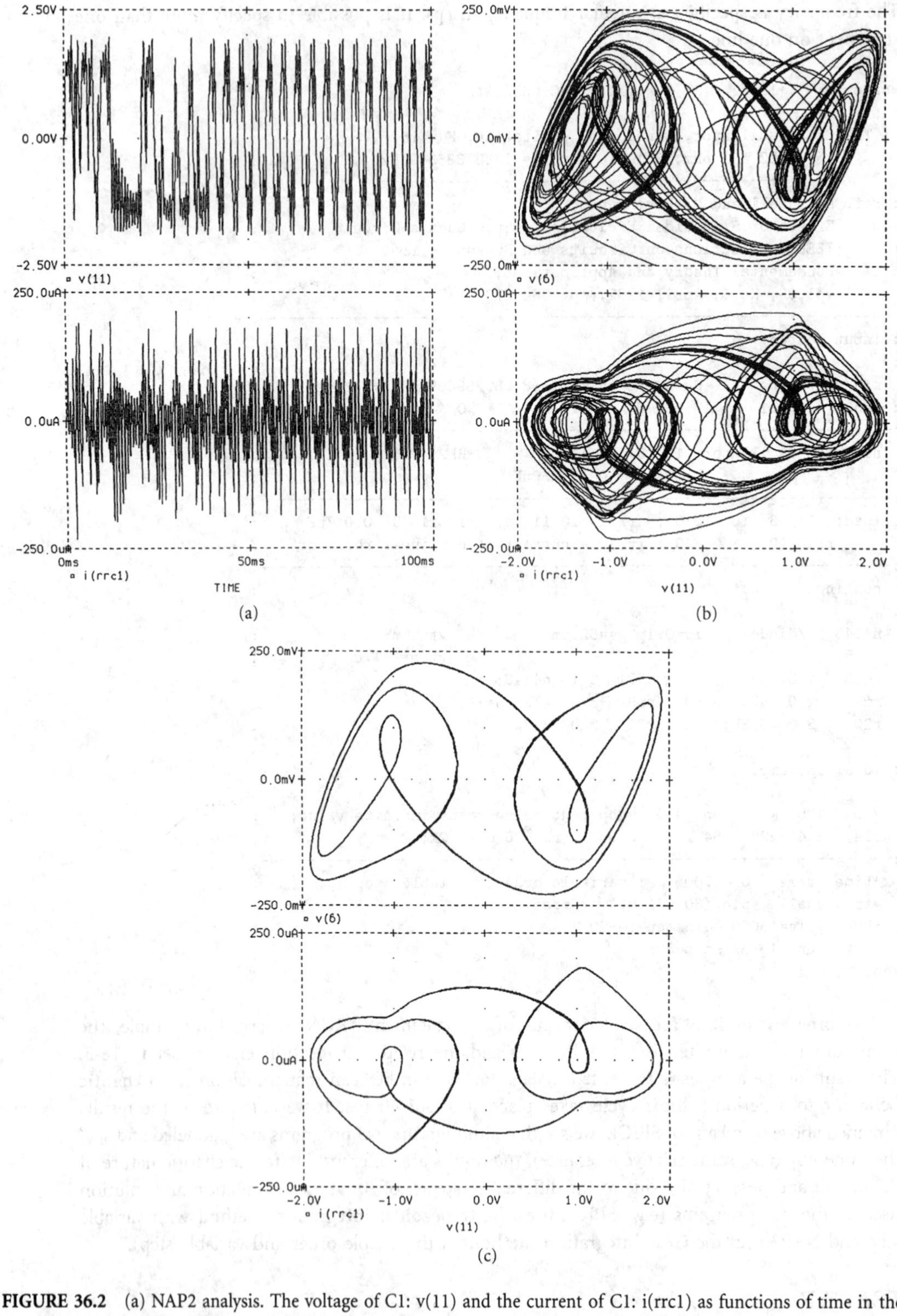

FIGURE 36.2 (a) NAP2 analysis. The voltage of C1: v(11) and the current of C1: i(rrc1) as functions of time in the time interval from 0 to 100 ms. (b) NAP2 analysis. The voltage of C2: v(6) and the current of C1: i(rrc1) as functions of the voltage of C1: v(11) in the time interval from 0 to 100 ms. (c) NAP2 analysis. The voltage of C2: v(6) and the current of C1: i(rrc1) as functions of the voltage of C1: v(11) in the time interval from 500 to 600 ms.

Hoersholm, Denmark, for ESA—the European Space Agency—as a result of the strong need for a **simulation language** capable of handling interdisciplinary problems (e.g., coupled electrical and thermal phenomena). ESACAP was therefore born with facilities that have only recently been implemented in other simulation programs (e.g., facilities referred to as behavioral or functional modeling).

ESACAP carries out analyses on nonlinear systems in dc and in the time domain. The nonlinear equations are solved by a hybrid method combining the robustness of the gradient method with the good convergence properties of the **Newton–Raphson method**. The **derivatives** required by the Jacobian matrix are **symbolically evaluated** from arbitrarily complex arithmetic expressions and are therefore exact. The time-domain solution is found by numerical integration implemented as backward difference formulas, **BDF's**, of variable step and orders 1 through 6 (modified **Gear method**). An efficient **extrapolation method** (the **epsilon algorithm**) accelerates the asymptotic solution in the periodic steady-state case.

Frequency-domain analyses may be carried out on linear or linearized systems (e.g., after a dc analysis). Besides complex transfer functions, special outputs such as **group delay** and **poles/zeros** are available. The group delay is computed as the sum of the frequency sensitivities of all the reactive components in the system. These sensitivities are provided by the adjoint network method. Poles and zeros are found by a numerical interpolation of transfer functions evaluated on a circle in the complex frequency plane. ESACAP also includes a complex number postprocessor by means of which any function of the basic outputs can be generated (e.g., stability factor, *S*-parameters, complex ratios).

The ESACAP language combines procedural facilities, such as if-then-else, assignment statements, and do-loops, with the usual description by structure (nodes/branches). Arbitrary expressions containing system variables and their derivatives are allowed for specifying branch values, thereby establishing any type of nonlinearities. The language also accepts the specification of nonlinear differential equations. Besides all the standard functions known from high-level computer languages, ESACAP provides a number of useful functions. One of the most important of these functions is the delay function. The delay function returns one of its arguments delayed by a specified value, which in turn may depend on system variables. Another important function is the threshold switch—the ZEROREF function—used in if-then-else constructs for triggering discontinuities. The ZEROREF function interacts with the integration algorithm, which may be reinitialized at the exact threshold crossing. The ZEROREF function is an efficient means for separating cause and action in physical models thereby eliminating many types of causality problems. **Causality problems** are typical examples of bad modeling techniques and the most frequent reason for divergence in the simulation of dynamic systems.

Typical ESACAP applications include electronics as well as **thermal and hydraulic systems.** The frequency-domain facilities have been a powerful tool for designing stable control systems, including nonelectronics engineering disciplines. ESACAP is available on a number of platforms from DOS, UNIX/X-window (SUN, HP, DEC) to Cray. The size of the ESACAPX.EXE file is 1,302,576 bytes.

In order to give an idea of the input language, the syntax of the statements describing sources is as follows:

```
Current Source:   Jxxx(N+, N-) = VALUE;
Voltage Source:   Exxx(N+, N-) = VALUE;
```

where the initial characters of the branch name J and E indicate the type of the branch, N+ and N− are integers ("node numbers") indicating the placement and orientation of the branch, respectively, and VALUE specifies the numerical value of the source, which may be an arbitrary function of time, temperature and other variables and parameters. Control with the time derivative of a signal is allowed. Reference to the time derivative of a system variable is done by adding an apostrophe, e.g., V(N1,N2)' is the time derivative of the voltage drop from node N1 to node N2.

The input file below describes an analysis of a tapered transmission line. The input language is a little more complicated than the languages of SPICE and NAP2. Data is specified in a number of blocks ("chapters" and "sections") starting with $$ and $. Note how the line model is specified in a do-loop, where ESACAP creates the nodes and branches of a ladder network. [9]

```
ESA.040  DISCRETIZATION OF TAPERED-LINE BY ESACAP DO-LOOP              #
#                                                                      #
# This example shows the use of the do-loop in connection with         #
# the descretization of a tapered line into a 30-section ladder        #
# network.                                                             #
#                                                                      #
$$DES           # DEScription chapter                                  #
#                                                                      #
$CONSTANTS:   sections=30;   END;         # Number of sections         #
#                                                                      #
# Transmission line section modelled by delay function.                #
#                                                                      #
# Z0: characteristic impedance, LEN: length in m                       #
# JR and JF are controlled current sources.                            #
# G1 and G2 conductances                                               #
#                                                                      #
$MODEL:                                                                #
 LINE(IN,OUT): Z0, LEN;                                                #
  delay=LEN/3E8;                                                       #
  JR(0,IN)=DEL(2*V(OUT)/Z0-I(JF),delay);                               #
  JF(0,OUT)=DEL(2*V(IN)/Z0-I(JR),delay);                               #
  G1(IN,0)=1/Z0; G2(OUT,0)=1/Z0;                                       #
END;                                                                   #
 #                                                                     #
 $NETWORK:                               # Main network. Linear impedance- #
                                         # tapering between Z1 and Z2  #
 length=1;  Z1=50;  Z2=100;                                            #
 #                                                                     #
 #-------------------          Line model called in do-loop            #
 FOR(I=1,sections) DO                                                  #
  X[I]([I],[I+1]) = LINE(  Z1+I*(Z2-Z1)/sections,                      #
                                     length/sections );                #
  ENDDO;                                                               #
  #                                                                    #
 RG(G,1)=Z1;    RL(31,0)=Z2;                                           #
  #--- independent voltage source, applied step at 1 nS                #
  #                              risetime 1pS                          #
  E1(G,0)=TABLE(TIME,(0,0),(1n,0),(1.001n,1),(10n,1));                 #
  #                                                                    #
  END;                                                                 #
  #-------------------- output specification                           #
  $OUT:V(1); V(31);END;                                                #
  #                                                                    #
  $$TRA             # TRAnsient analysis chapter                       #
  $PAR: TIME=0,20n;  HMAX=.05n;  END;                                  #
  $PLO:X=TIME; Y(-.2,1)=V(1)!; Y(-.2,1)=V(31)!; END;                   #
  #                                                                    #
  $$STOP                                                               #
```

The result of the analysis is presented in Fig. 36.3.

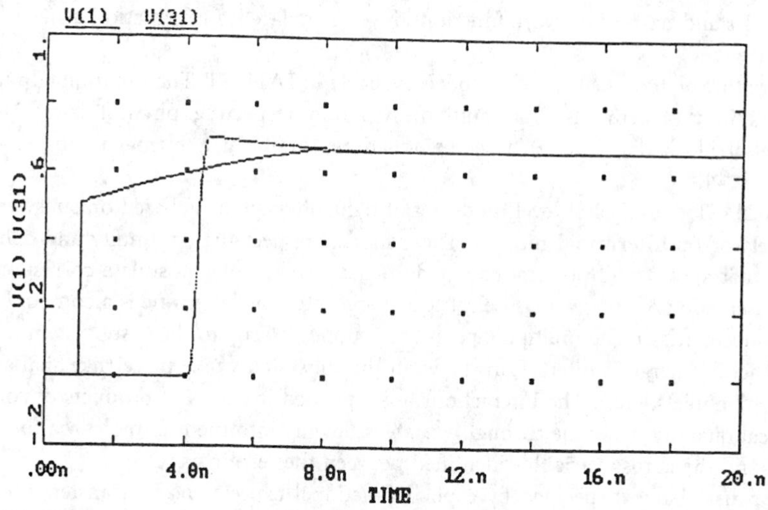

FIGURE 36.3 ESACAP analysis. The input voltage of the tapered-line: V(1) and the output voltage of the tapered-line: V(31) as functions of time in the time interval from 0 to 20 ns.

Other Programs

APLAC (originally **Analysis Program for Linear Active Circuits**) [10] has been under constant development at the Helsinki University of Technology since 1972. Over time it has developed into an object-oriented analog circuits and systems simulation and design tool. Inclusion of a new model into APLAC requires only the labor of introducing the parameters and equations defining the model under the control of "C-macros." The code of APLAC itself remains untouched. The APLAC *Interpreter* immediately understands the syntax of the new model. A separate utility program SPIC2A transforms a SPICE netlist into an APLAC input file.

The program is capable of carrying out dc, ac, transient, noise, oscillator, and multitone harmonic steady-state analyses and measurements using an **IEEE-488 bus**. Transient analysis correctly handles, through convolution, components defined by frequency-dependent characteristics. **Monte Carlo analysis** is available in all basic analysis modes and **sensitivity analysis** in dc and ac modes. N-port z, y, and s parameters as well as two-port h parameters are available in ac analysis. In addition APLAC includes a versatile collection of system level blocks for the simulation and design of analog and digital communication systems. APLAC version 6.1 includes seven different **optimization methods**. Any parameter in the design problem can be used as a variable and any user-defined function may act as an objective. Combined time- and frequency-domain optimization is possible.

Full versions of APLAC are available only for UNIX workstations. The IBM PC compatible version of APLAC operates only under control of WINDOWS. The size of the APLAC directory is 4,480,760 bytes. The size of the IAPLAC.EXE file is 2,896,896 bytes.

DYNAST (DYNAmic Simulation Tool) [5] was developed in 1992 in a cooperation between Czech Technical University, Prague, the Czech Republic, and Katholieke Universiteit Leuven, Afdeling PMA, Produktietechnieken, Machinebouw en Automatisering, Heverlee, Belgium. The program is developed as an interdisciplinary simulation and design tool in the field of "mechatronics" (mixed mechanical/electrical systems).

The main purpose of DYNAST is to simulate dynamic systems decomposed into subsystems defined independently of the system structure. The structure can be hierarchical. DYNAST is a versatile software tool for modeling, simulation, and analysis of general linear as well as nonlinear dynamic systems, both in the time- and frequency-domains. **Semisymbolic analysis** is possible (poles and zeros of network functions, inverse Laplace transformation using closed-form formulas).

There are three types of subsystem models available in DYNAST. The program admits systems descriptions in the form of 1) a multipole diagram respecting physical laws, 2) a causal or an acausal block diagram, 3) a set of equations, or 4) in a form combining the above approaches.

1) In DYNAST the physical-level modeling of dynamic systems is based on subsystem **multipole models** or **multiterminal models.** These models respect the continuity and compatibility postulates that apply to all physical energy domains. (The former postulate corresponds to the laws of conservation of energy, mass, electrical charge, etc.; the latter one is a consequence of the system connectedness.) The multipole poles correspond directly to those subsystem locations in which the actual energetic interactions between the subsystems take place (like shafts, electrical terminals, pipe inlets, etc.). The interactions are expressed in terms of products of complementary physical quantity pairs: the **through variables** flowing into the multipoles via the individual terminals, and the **across variables** identified between the terminals.

2) The **causal blocks**, specified by explicit functional expressions or transfer functions, are typical for any simulation program. But the variety of basic blocks is very poor in DYNAST, as its language permits to define the block behavior in a very flexible way. Besides the built-in basic blocks, also user specified multiinput multioutput macroblocks are available. The causal block interconnections are restricted by the rule, that only one block output may be connected to one or several block inputs. In the DYNAST block variety, however, causal blocks are also available with no restrictions imposed on their interconnections, as they are defined by implicit-form expressions.

3) DYNAST can also be used as an **equation solver** for systems of nonlinear first-order algebro–differential and algebraic equations in the implicit form. The equations can be submitted in a natural way (without converting them into block diagrams) using a rich variety of functions including the Boolean, event-dependent, and tabular ones. The equations, as well as any other input data, are directly interpreted by the program without any compilation.

The equation formulation approach used for both multipoles and block diagrams evolved from the **extended method of nodal voltages (MNA)** developed for electrical systems. Because all the equations of the diagrams are formulated simultaneously, there are no problems with the "algebraic loops." As the formulated equations are in the implicit form, it does not create any problems with the **causality** of the physical models.

The integration method used to solve the nonlinear algebro–differential and algebraic equations is based on a stiff-stable implicit backward-differentiation formula (a modified **Gear method**). During the integration, the step length as well as the order of the method is varied continuously to minimize the computational time, while respecting the admissible computational error. Jacobians necessary for the integration are computed by **symbolic differentiation**. Their evaluation as well as their LU decomposition, however, is not performed at each iteration step if the convergence is fast enough. Considerable savings of computational time and memory are achieved by a consistent **matrix sparsity** exploitation.

To accelerate the computation of periodic responses of weakly damped dynamic systems the iterative **epsilon-algorithm** is utilized. Also, fast Fourier transformation is available for **spectral analysis** of the periodic-steady state responses.

DYNAST runs under DOS control on IBM compatible personal computers. Being coded in FORTRAN 77 and C languages, it is easily implemented on other platforms. It is accompanied by a menu-driven *graphical environment.* The block and multiport diagrams can be submitted in a graphical form by a schematic capture editor. DYNAST can be easily augmented by various pre- and postprocessors because all its input and output data are available in the ASCII code. The size of the DYNAST.EXE file is 210,077 bytes. There are a total of 115 files in the directory, occupying 1,351,680 bytes.

ferences

[1] D. A. Calahan, *Computer-Aided Network Design*, New York: McGraw-Hill, 1972.

[2] L. O. Chua, and P.-M. Lin, *Computer-Aided Analysis of Electronic Circuits*, Englewood Cliffs, NJ: Prentice-Hall, 1975.

[3] M. Dertouzos *et al.*, *Systems, Networks, and Computation: Basic Concepts*, New York: McGraw-Hill, 1972.

[4] Intusoft, *IsSpice3—ICAPS System Packages*, San Pedro, CA: Intusoft, 1994.

[5] H. Mann, "DYNAST—A multipurpose engineering simulation tool," Czech Tech. Univ., Prague, Czech Republic, 1994.

[6] Meta-Software, *HSPICE User's Manual H9001*, Campbell, CA: Meta-Software, 1990.

[7] MicroSim, *PSpice—The Design Center*, Irvine, CA: MicroSim, 1994.

[8] T. Rübner-Petersen, "NAP2—A nonlinear analysis program for electronic circuits, version 2, users manual 16/5-73, Rep. IT-63, Instit. Circuit Theory, Telecommun., Tech. Univ. Denmark, Lyngby, Denmark, 1973.

[9] P. Stangerup, *ESACAP user's manual*, Nivaa, Denmark: StanSim Research Aps, 1990.

[10] M. Valtonen, *et al.*, *APLAC—An Object-Oriented Analog Circuit Simulator and Design Tool*, Espoo, Finland, Circuit Theory Lab., Helsinki Univ. Technol., Nokia Corp. Center, 1992.

[11] J. Vlach, and K. Singhal, *Computer Methods for Circuit Analysis and Design*, New York: Van Nostrand Reinhold, 1983.

[12] D. G. Funk and D. Christiansen (edt.) Electronic Engineers' Handbook, 3rd edition, McGraw-Hill, 1989.

37

Cellular Neural Networks

Tamás Roska
Hungarian Academy of Sciences, Budapest

37.1 Introduction: Definition and Classification

Current VLSI technologies provide for the fabrication of chips with several million transistors. With these technologies a single chip may contain one powerful digital processor, a huge memory containing millions of very simple units placed in a regular structure, and other complex functions. A powerful combination of a simple logic processor placed in a regular structure is the cellular automaton invented by John von Neumann. The cellular automaton is a highly parallel computer architecture. Although many living neural circuits resemble this architecture, the neurons do not function in a simple logical mode: they are analog "devices." The cellular neural network architecture, invented by Leon O. Chua and his graduate student Lin Yang [1] has both properties: the cell units are nonlinear continuous time dynamic elements placed in a cellular array. Of course, the resulting nonlinear dynamics in space could be extremely complex. The inventors, however, showed that these networks can be designed and used for a variety of engineering purposes, while maintaining stability and keeping the dynamic range within well-designed limits. Subsequent developments have uncovered the many inherent capabilities of this architecture (IEEE conferences: CNNA-90, CNNA-92, CNNA-94; Special issues: *Int. J. Circuit Theory and Applications*, September–October 1993 and *IEEE Transactions on Circuits and Systems, I* and *II*, Mar. 1994, etc.). In the circuit implementation, unlike analog computers or general neural networks, the CNN cells are not the ubiquitous high-gain operational amplifiers. In most practical cases, they are either simple unity gain amplifiers or simple second- or third-order simple dynamic circuits with one or two simple nonlinear components. Tractability in the design and the possibility for exploiting the complex nonlinear dynamic phenomena in space, as well as the trillion operations per second computing speed in a single chip are but some of the many attractive properties of cellular neural networks. The trade-off is in the accuracy; however, in

0-8493-8341-2/95/$0.00 + $.50

many cases, the accuracy achieved with current technologies is enough to solve a lot of real-life problems.

The cellular neural/nonlinear network (henceforth called CNN) is a new paradigm for multidimensional, nonlinear, dynamic processor arrays [1], [23]. The mainly uniform *processing elements*, called *cells* or artificial neurons, are placed on a regular *geometric grid* (with a square, hexagonal, or other pattern). This grid may consist of several two-dimensional layers packed upon each other (Fig. 37.1). Each processing element or cell is an analog dynamical system, the state (x), the input (u), and the output (y) signals are analog (real-valued) functions of time (both continuous-time and discrete-time signals are allowed). The *interconnection* and *interaction pattern* assumed at each cell is mainly *local* within a neighborhood N_r, where N_r denotes the first "r" circular layers of surrounding cells. Figure 37.2 shows a two-dimensional layer with a square grid of interconnection radius of 1 (nearest neighborhood). Each vertex contains a cell and the edges represent the interconnections between the cells. The pattern of interaction strengths between each cell and its neighbors is the "program" of the CNN array. It is called a *cloning template* (or just template).

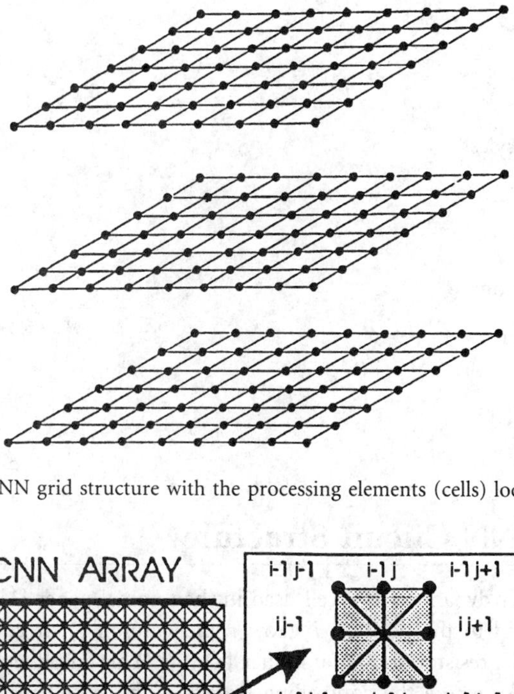

FIGURE 37.1 A CNN grid structure with the processing elements (cells) located at the vertices.

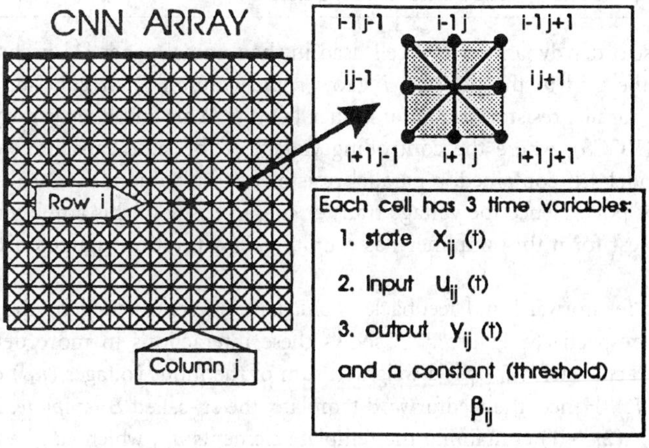

FIGURE 37.2 A single two-dimensional CNN layer and a magnified cell with its neighbor cells with the normal neighborhood radius $r = 1$.

Depending on the types of grids, processors (cells), interactions, and modes of operation, several classes of CNN architectures and models have been introduced. Although the summary below is not complete, it gives an impression of vast diversities.

Typical CNN Models

- Grid type:

  square
  hexagonal
  planar
  circular
  equidistant
  logarithmic

- Processor type:

  linear
  sigmoid
  first order
  second order
  third order

- Interaction type:

  linear memoryless
  nonlinear
  dynamic
  delay-type

- Mode of operation:

  continuous-time
  discrete-time
  equilibrium
  oscillating
  chaotic

37.2 The Simple CNN Circuit Structure

The simplest first-order dynamic CNN cell used in the seminal paper [1] is shown in Fig. 37.3. It is placed on the grid in the position ij (row i and column j). It consists of a single state capacitor with a parallel resistor and an amplifier $[f(x_{ij})]$. This amplifier is a voltage-controlled current source (VCCS), where the controlling voltage is the state capacitor voltage. To make the amplifier model self-contained, a parallel resistor of unit value is assumed to be connected across the output port. Hence the voltage transfer characteristic of this amplifier is also equal to $f(\cdot)$. In its simplest form this amplifier has a unity gain saturation characteristic (see Fig. 37.7 for more details).

The aggregate feedforward and feedback *interactions* are represented by the current sources i_{input} and i_{output}, respectively. Figure 37.4 shows these interactions in more detail. In fact, the feedforward interaction term i_{input} is a weighted sum of the input voltages (u_{kl}) of all cells in the neighborhood (N_r). Hence, the feedforward template, the so-called ***B** template*, is a small matrix of size $(2r+1) \times (2r+1)$ containing the template elements b_{kl}, which can be implemented by an array of linear voltage-controlled current sources. The controlling voltages of these controlled sources are the input voltages of the cells within the neighborhood of radius r. This means, for

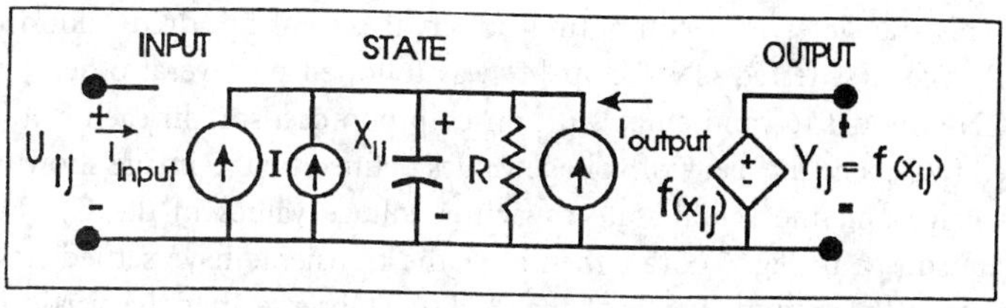

FIGURE 37.3 The simple first-order CNN cell.

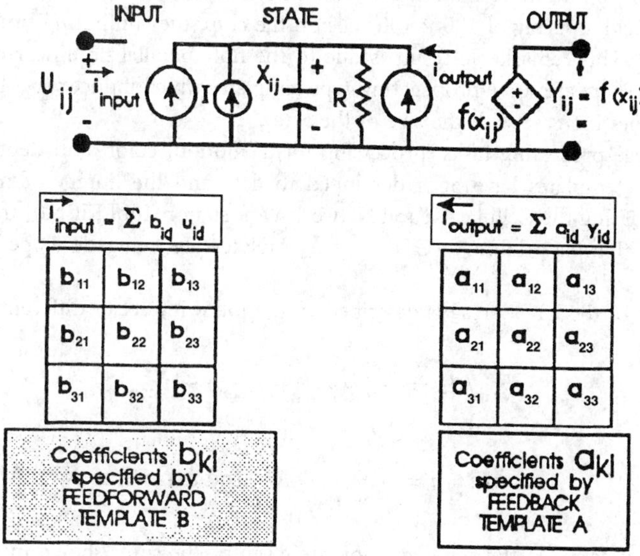

FIGURE 37.4 The 19 numbers (a program) that govern the CNN array (the 19th number is the constant bias term I but is not shown in the figure) define the cloning template (***A***, ***B***, and ***I***).

example, that b_{12} is the VCCS controlled by the input voltage of the cell lying north from the cell ij. In most practical cases the ***B*** template is translation invariant, i.e., the interaction pattern (the ***B*** template) is the same for all cells, hence the chip layout will be very regular (as in memories or PLAs). The feedback interaction term i_{output} is a weighted sum of the output voltages (y_{kl}) of all cells in the neighborhood (N_r). The weights are the elements of a small matrix ***A*** called the ***A*** template (or feedback template). Similar arguments apply for the ***A*** template as for the ***B*** template above. If the constant threshold term is translation invariant as denoted by the constant current source ***I***, then in the case of $r = 1$, the complete cloning template contains only 19 numbers (***A*** and ***B*** and ***I***, i.e., $9+9+1$ terms), irrespective of the size of the CNN array. These 19 numbers define the task which the CNN array can solve.

What kind of tasks are we talking about? The simplest, and perhaps the most important, are image-processing tasks. In the CNN array computer the input and output images are coded as follows. For each picture element (called pixel) in the image a single cell is assigned in the CNN. This means that there is a one-to-one correspondence between the pixels and the CNN cells. Voltages in the CNN cells code the gray scale values of the pixels. Black is coded by +1 V, white is −1 V, and the gray scale values are in between. Two independent input images can be defined pixel by pixel: the input voltages u_{ij} and the initial voltage values of the capacitors $x_{ij}(0)$ (cell by cell). Placing these input images onto the cell array and starting the transient, the steady state outputs y_{ij} will encode the output image. The computing time is equal to the settling time of the CNN array. This time is below one microsecond using a CNN chip made with a 1.0–1.5 μm technology containing thousands of CNN processing elements, i.e., pixels, in an area of about 2 cm². This translates to a computing power of several hundred billion operations per second (GXPS). The first tested CNN chip [4] was followed by several others implementing a discrete-time CNN model [6] and chips with on-chip photosensors in each cell [5].

For example, if we place the array of voltage values defined by the image shown in Fig. 37.5(b) as the input voltage and the initial state capacitor voltage values in the CNN array with the cloning template shown in Fig. 37.5(a), then after the transients have settled down, the output voltages will encode the output image of Fig. 37.5(c). Observe that the vertical line has been deleted. Since the image contains 40 × 40 pixels, the CNN array contains 40 × 40 cells. It is quite interesting that if we had more than one vertical line, the computing time would be the same. Moreover, if we had an array of 100 × 100 cells on the chip, the computing time would remain the same as well. This remarkable result is due to the fully parallel nonlinear dynamics of the CNN computer. There are some propagating-type templates which induce wave-like phenomena. Their settling times increase with the size of the array.

For other image-processing tasks, processing form, motion, color, and depth, more than a hundred cloning templates have been developed to date and the library of new templates is growing rapidly. Using the Cellular Neural Network Workstation Tool Kit [10] they can be called in from a CNN Template Library (CTL). New templates are being developed and published continually.

The dynamics of the CNN array is described by the following set of differential equations:

$$dx_{ij}/dt = -x_{ij} + I + i_{\text{output}} + i_{\text{input}}$$

$$y_{ij} = f(x_{ij})$$

$$i = 1, 2, \cdots, N \quad \text{and} \quad j = 1, 2, \cdots, M \text{ (the array has } N \times M \text{ cells)}$$

where the last two terms in the state equation are given by the sums shown in Fig. 37.4.

We can generalize the domain covered by the original CNN defined via linear and time-invariant templates by introducing the "nonlinear" templates (denoted by ˆ) and the "delay" templates (indicated by τ in the superscript) as well, to obtain the generalized state equation

$$A = \begin{matrix} 0 & 0 & 0 \\ 0 & 2 & 0 \\ 0 & 0 & 0 \end{matrix} \qquad B = \begin{matrix} 0 & -0.25 & 0 \\ 0 & 0 & 0 \\ 0 & -0.25 & 0 \end{matrix} \qquad I = -1.5$$

(a)

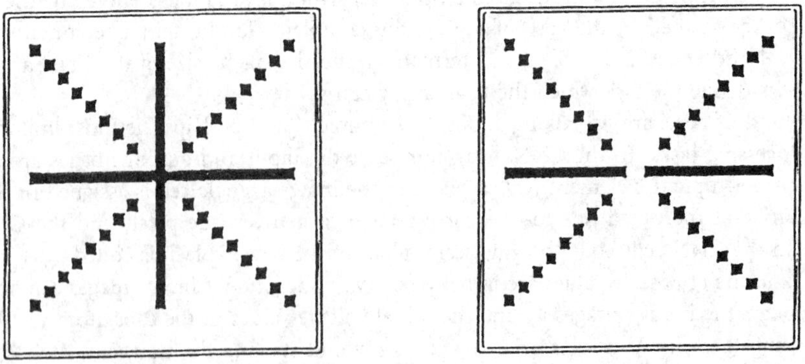

(b) (c)

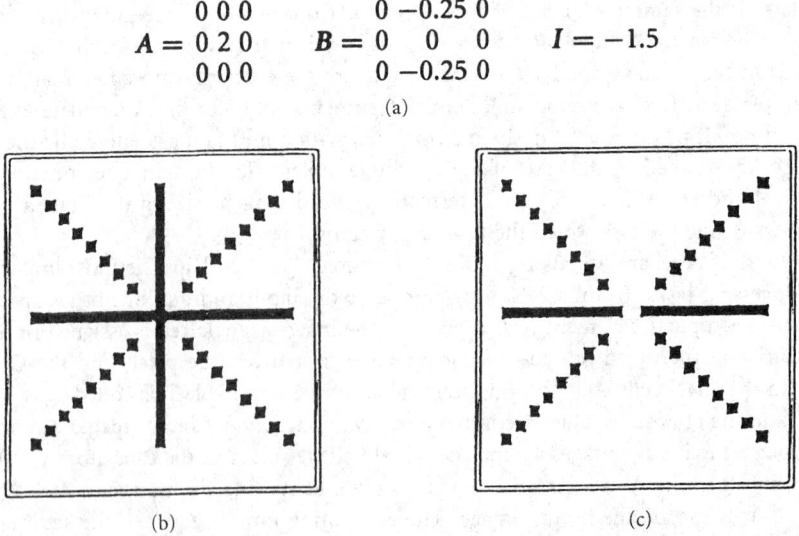

FIGURE 37.5 An input and output image where the vertical line was deleted.

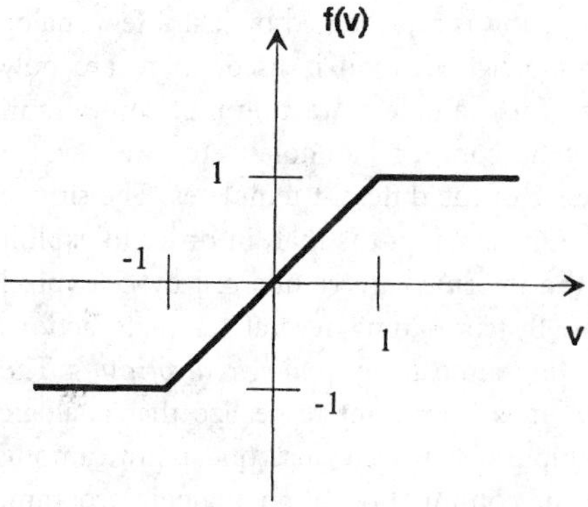

FIGURE 37.6 The simple unity-gain sigmoid characteristics.

shown below. The unity-gain nonlinear sigmoid characteristics f are shown in Fig. 37.6.

$$\begin{aligned}\frac{dv_{xij}}{dt} &= -v_{xij} + I_{ij} + \sum_{kl\in N_r(ij)} \hat{A}_{ij;kl}(v_{ykl}(t), v_{yij}(t)) \\ &+ \sum_{kl\in N_r(ij)} \hat{B}_{ij;kl}(v_{ukl}(t), v_{uij}(t)) + \sum_{kl\in N_r(ij)} A^{\tau}_{ij;kl} v_{ykl}(t-\tau) \\ &+ \sum_{kl\in N_r(ij)} B^{\tau}_{ij;kl} v_{ukl}(t-\tau)\end{aligned}$$

Several strong results have been proved that assure stable and reliable operations. If the ***A*** template is symmetric then the CNN is stable. Several other results have extended this condition [6, 7]. The sum of the absolute values of all the 19 template elements plus one defines the dynamic range within which the state voltage remains bounded during the entire transient, if the input and initial state signals are less than 1 V in absolute value [1].

In a broader sense, the CNN is defined [2] as shown in Fig. 37.7.

- 1-, 2-, 3-, or n-dimensional *array of* mainly identical *dynamical systems,* called cells or processor units, which satisfies two properties:
- most *interactions are local* within a finite radius r, and
- all *state variables are continuous valued* signals

FIGURE 37.7 The CNN definition.

7.3 The Stored Program CNN Universal Machine and the Analogic Supercomputer Chip

For different tasks, say image-processing, we need different cloning templates. If we want to implement them in hardware, we would need different chips. This is inefficient except for dedicated, mass-production applications.

The invention of the CNN universal machine [8] has overcome the problem above. It is the first stored program array computer with analog nonlinear array dynamics. One CNN operation, for example, solving thousands of nonlinear differential equations in a microsecond, is just one single instruction. In addition a single instruction is represented by just a few analog (real) values (numbers). In the case when the nearest neighborhood is used, there are only 19 numbers. When combining several CNN templates, for example extracting first contours in a gray scale image, then detecting those areas where the contour has holes, etc., we have to design a flowchart-logic that satisfies the correct sequence of the different templates. The simple flowchart for the above example is shown in Fig. 37.8. One key point is that, in order to exploit the high speed of the CNN chips, we have to store the intermediate results cell by cell (pixel by pixel). Therefore we need a local analog memory. By combining several template actions we can write more complex flowcharts for implementing almost any *analogic algorithms.* The name analogic is an acronym for "analog and logic." It is important to realize that analogic computation is completely different from hybrid computing. To cite just one point, among others, there are no A/D or D/A conversions during the computation of an analogic program. As with digital microprocessors, to control the execution of an analogic algorithm we need a global programming unit. The global architecture of the CNN universal machine is shown in Fig. 37.9.

As we can see from this figure, the CNN nucleus described in the previous section has been generalized to include several crucial functions depicted in the periphery. We have already discussed the role of the local analog memory (LAM) that provides the local (on-chip) storage of intermediate analog results. Since the results of many detection tasks in applications involve only black-and-white logic values, adding a local logic memory (LLM) in each cell is crucial. After applying several templates in a sequence, it is often necessary to combine their results. For example, to analyze motion, consecutive snapshots processed by CNN templates are compared. The local analog output unit (LAOU) and the local logic unit (LLU) perform these tasks, both on the local analog (gray scale) and the logical (black-and-white) values. The local communication and control unit (LCCU) of each cell decodes the various instructions coming from the global analogic program unit (GAPU).

The global control of each cell is provided by the global analogic program unit (GAPU). It consists of four parts.

- The analog program (instruction) register (APR) stores the CNN template values (19 values for each CNN template instruction in the case of nearest interconnection). The templates stored here will be used during the run of the prescribed analogical algorithm.
- The global logic program register (LPR) stores the code for the local logic units.

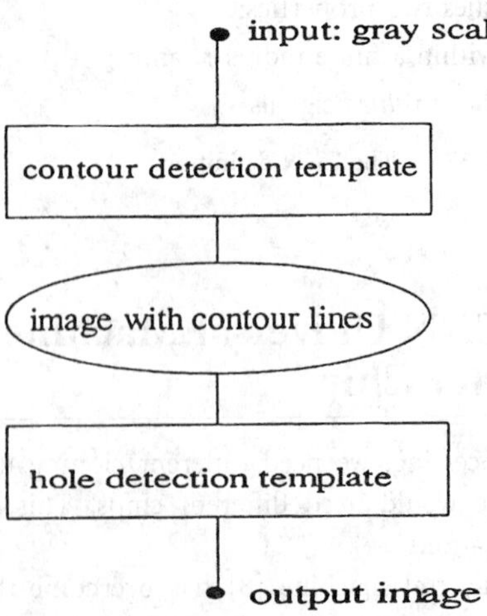

FIGURE 37.8 A flowchart representing the logic sequence of two templates.

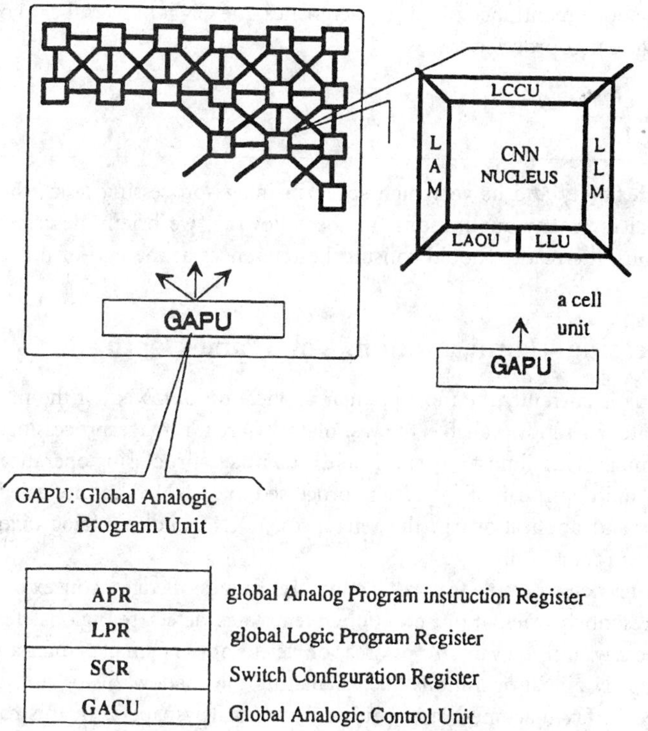

FIGURE 37.9 The global architecture of the CNN universal machine. *Source:* T. Roska and L. O. Chua, "The CNN universal machine: An analogic array computer," *IEEE Trans. Circuits Syst. I,* vol. 40, pp. 163–173, 1993. © 1993 IEEE.

- The flexibility of the extended CNN cells is provided by embedding controllable switches in each cell. By changing the switch configurations of each cell simultaneously, we can execute many tasks using the same cell. For example, the CNN program starts by loading a given template, storing the results of this template action in the local analog memory, placing this intermediate result back on the input to prepare the cell, starting the action with another template, etc. The switch configurations of the cells are coded in the switch configuration register (SCR).
- Finally, the heart of the GAPU is the global analogic control unit (GACU) which contains the physical machine code of the logic sequence of an analogical algorithm. It is important to emphasize that here the control code is *digital*; hence, although its internal operation is analog and logical, a CNN universal chip can be programmed with the same flexibility and ease as a digital microprocessor—except the language is much simpler. Indeed, a high-level language, a compiler, an operating system, an algorithm development system are available for CNN universal chip architectures. Moreover, by fabricating optical sensors cell-by-cell on the chip [5], the image input is directly interfaced.

The CNN universal chip is called a supercomputer chip because the execution speed of an analogic algorithm falls in the same range as the computing power of today's average digital supercomputers (hundred billion operations per second). Another reason for this enormous computing power is that the reprogramming time of a new analog instruction (template) is of the same order, or less, than the analog array execution time (less than a microsecond). This is about a million times faster than some fully-interconnected analog chips.

Based on the above-mentioned novel characteristics, the CNN universal chip can be considered to be an analogic microprocessor.

37.4 Applications

In view of its flexibility and its very high speed in image-processing tasks, the CNN Universal machine is ideal for many applications. In the following, we briefly describe three areas. For more applications, the reader should consult the references at the end of this chapter.

Image Processing—Form, Motion, Color, and Depth

Image processing is currently the most popular application of CNN. Of the more than hundred different templates currently available, the vast majority are for image-processing tasks. Eventually, we will have templates for almost all conceivable local image-processing operations. Form (shape), motion, color, and depth can all be ideally processed via CNN. The interested reader can find many examples and applications in the references. CNN handles analog pixel values, so gray scale images are processed directly.

Many templates detect simple features like different types of edges, convex or concave corners, lines with a prescribed orientation, etc. Other templates detect semiglobal features like holes, groups of objects within a given size of area, or delete objects smaller than a given size. There are also many CNN global operations like calculating the shadow, histogram, etc. Halftoning is commonly used in fax machines, laser printers, and in newspapers. In this case the local gray level is represented by black dots of identical size, whose density varies in accordance with the gray level. CNN templates can do this job as well. A simple example is shown in Fig. 37.10. The original gray scale image is shown on the left-hand side, the halftoned image is shown on the right-hand side. The "smoothing" function of our eye completes the image processing task.

More complex templates detect patterns defined within the neighborhood of interaction. In this case the patterns of the ***A*** and ***B*** templates reflect somehow the pattern of the object to be detected.

Since the simplest templates are translation invariant, the detection or pattern recognition is translation invariant as well. By clever design, however, some rotationally invariant detection procedures have been developed as well.

Combining several templates according to some prescribed logic sequence, more complex pattern detection tasks can be performed, e.g., color halftoning.

FIGURE 37.10 Halftoning: an original gray scale image (LHS) and its halftoned version (RHS). A low resolution is deliberately chosen in (b) in order to reveal the differing dot densities at various regions of the image.

Color-processing CNN arrays represent the three basis colors by single layers via a multilayer CNN. For example, using the red–green–blue (RGB) representation in a three-layer CNN, simple color-processing operations can be performed. Combining them with logic, conversions between various color representations are possible.

One of the most complex tasks that has been undertaken by an analogic CNN algorithm is the recognition of bank notes. Recognition of bank notes in a few milliseconds is becoming more and more important. Recent advances in the copy machine industry have made currency counterfeiting easier. Therefore, automatic bank note detection is a pressing need. Figure 37.11 shows a part of this process (which involves color processing as well). The dollar bill shown in the foreground is analyzed and the circles of a given size are detected (colors are not shown). The "color cube" means that each color intensity is within prescribed lower and upper limit values.

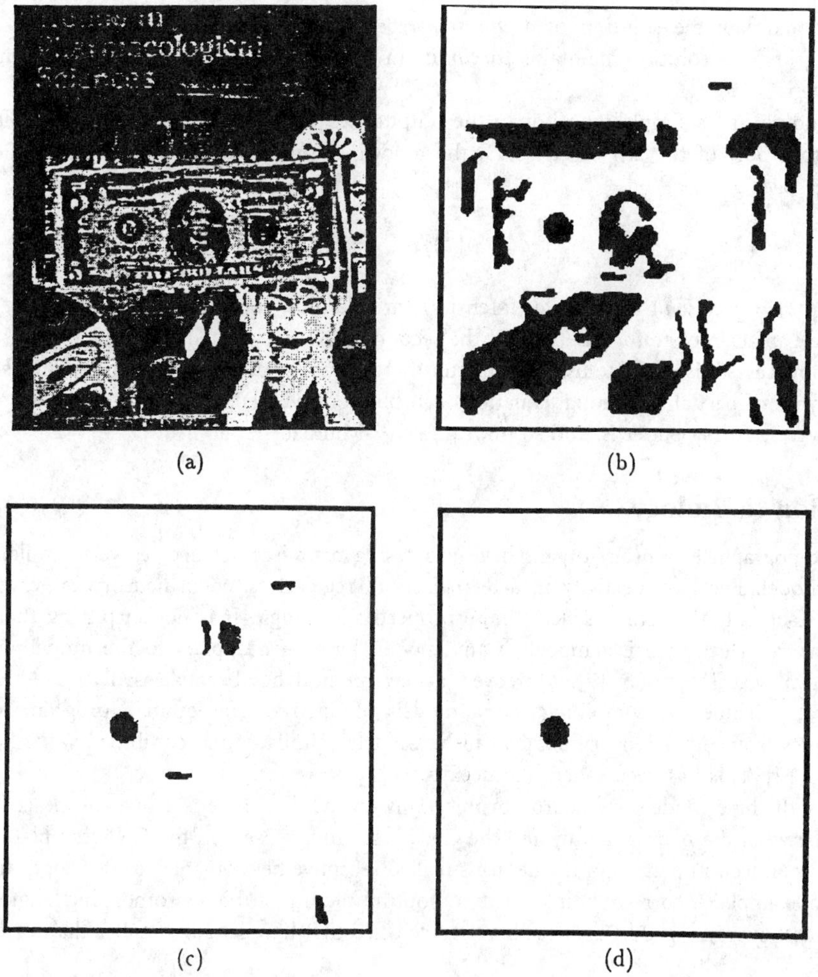

FIGURE 37.11 Some intermediate steps in the dollar bill recognition process. An input image (a) shown here in single color, results in the "color cube" (b), the convex objects (c), and the size classification (d). *Source:* A. Zarándy, F. Werblin, T. Roska, L. O. Chua, and "Novel type of analogical CNN algorithms for recognizing bank notes," Memorandum UCB/ERL, M94/29 1994, Electron. Res. Lab., Univ. California, Berkeley, 1994.

Motion detection can be achieved by CNN in many ways. One approach to process motion is to apply two consecutive snapshots to the input and the initial state of the CNN cell. The CNN array calculates the various combinations between the two snapshots. The simplest case is just taking the difference to detect motion. Detecting direction, shape, etc., of moving objects are only the simplest problems that can be solved via CNN. In fact, even depth detection can be included as well.

Partial Differential Equations

As noted in the original paper [1], even the simple-cell CNN with the linear template

$$A = \begin{matrix} 0 & 1 & 0 \\ 1 & -3 & 1 \\ 0 & 1 & 0 \end{matrix} \qquad B = 0 \qquad I = 0$$

can approximate the solution of a *diffusion-type* partial differential equation on a discrete spatial grid. This solution maintains *continuity* in time, a nice property not possible in digital computers.

By adding just a simple capacitor to the output, i.e., by placing a parallel RC circuit across the output port of the cell of Fig. 37.3, the following *wave equation* will be represented in a discrete space grid:

$$d^2p(t)/dt^2 = \Delta p$$

where $p(t) = p(x, y, t)$ is the state (intensity) variable on a two-dimensional plane (x, y), and Δ is the Laplacian operator (the sum of the second derivatives related to x and y).

In some cases it is useful to use a cell circuit that is chaotic. Using the canonical Chua's circuit, other types of partial differential equations can be modeled, generating effects like auto-waves, spiral waves, Turing patterns, and so on, e.g., Perez-Munuzuri et al. in [7].

Relation to Biology

Many topographical sensory organs have processing neural-cell structures very similar to the CNN model. Local connectivity in a few sheets of regularly situated neurons is very typical. Vision, especially the retina, reflects these properties strikingly. It is not surprising that, based on standard neurobiological models, CNN models have been applied to the modeling of the subcortical visual pathway [9]. Moreover, a new method has been devised to use the CNN universal machine for combining retina models of different species in a programmed way. Modalities from other sensory organs can be modeled similarly and combined with the retina models [12]. This has been called: Bionic Eye.

Many of these models are neuromorphic. This means that there is a one-to-one correspondence between the neuroanatomy and the CNN structure. Moreover, the CNN template reflects the interconnection pattern of the neurons (called receptive field organization). Length tuning is such an example. A corresponding input and output picture of the neuromorphic length tuning model is shown in Fig. 37.12. Those bars are detected that have lengths smaller than or equal to 3 pixels.

37.5 Template Library: Analogical CNN Algorithms

During the last few years, after the invention of the cellular neural network paradigm and the CNN universal machine, many new cloning templates have been discovered. In addition, the

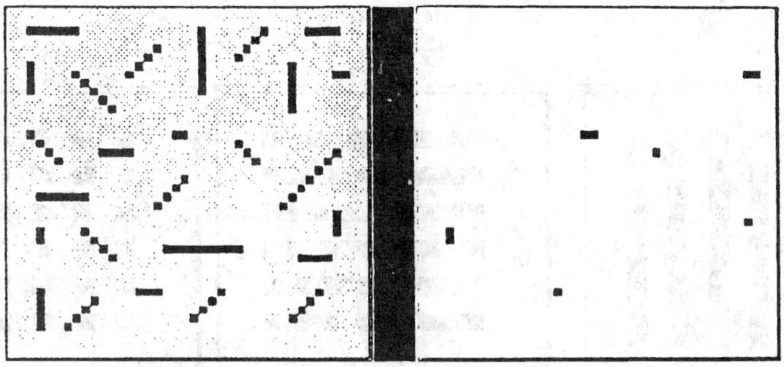

FIGURE 37.12 The length tuning effect. The input image on the LHS contains bars of different lengths. The output image on the RHS contains only those that are smaller than a given length. *Source:* T. Roska, J. Hámori, E. Lábos, K. Lotz, J. Takács, P. Venetianer, Z. Vidnyánszki, and A. Zarándy, "The use of CNN models in the subcortical visual pathway," *IEEE Trans. Circuits Syst. I*, vol. 40, pp. 182–195, 1993. © 1993 IEEE.

number of innovative analogical CNN algorithms, combining both analog cloning templates and local as well as global logic, is presently steadily increasing at a rapid rate.

As an illustration, let us choose a couple of cloning templates from the CNN library [1], [11]. In each case, a name, a short description of the function, the cloning templates, and a representative input–output image pair are shown. With regard to the inputs: the default case means that the input and initial state are the same. If $B = 0$, then the input picture is chosen as the initial state.

Name: AVERAGE

Function. Spatial averaging of pixel intensities over the $r = 1$ convolutional window.

$$A = \begin{bmatrix} 0 & 1 & 0 \\ 1 & 2 & 1 \\ 0 & 1 & 0 \end{bmatrix} \qquad B = \begin{bmatrix} 0 & 0 & 0 \\ 0 & 0 & 0 \\ 0 & 0 & 0 \end{bmatrix} \qquad I = 0$$

Example. Input and output picture.

Name: AND

Function. Logical "AND" function of the input and the initial state pictures.

$$A = \begin{bmatrix} 0 & 0 & 0 \\ 0 & 1.5 & 0 \\ 0 & 0 & 0 \end{bmatrix} \qquad B = \begin{bmatrix} 0 & 0 & 0 \\ 0 & 1.5 & 0 \\ 0 & 0 & 0 \end{bmatrix} \qquad I = -1$$

Example. Input, initial state, and output picture.

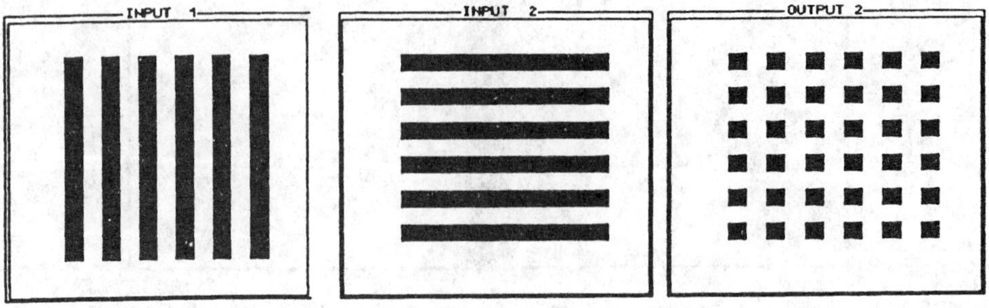

Name: CONTOUR

Function. Gray-scale contour detector.

$$A = \begin{bmatrix} 0 & 0 & 0 \\ 0 & 2 & 0 \\ 0 & 0 & 0 \end{bmatrix} \qquad B = \begin{bmatrix} a & a & a \\ a & a & a \\ a & a & a \end{bmatrix} \qquad I = 0.7$$

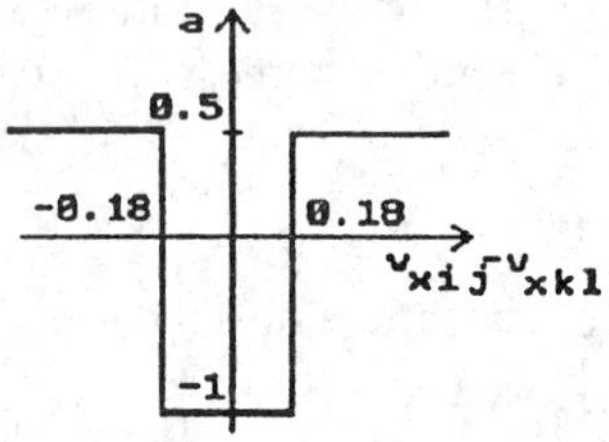

Example. Input and output picture.

Name: CORNER

Function. Convex corner detector.

$$A = \begin{bmatrix} 0 & 0 & 0 \\ 0 & 2 & 0 \\ 0 & 0 & 0 \end{bmatrix} \qquad B = \begin{bmatrix} -0.25 & -0.25 & -0.25 \\ -0.25 & 2 & -0.25 \\ -0.25 & -0.25 & -0.25 \end{bmatrix} \qquad I = -3$$

Example. Input and output picture.

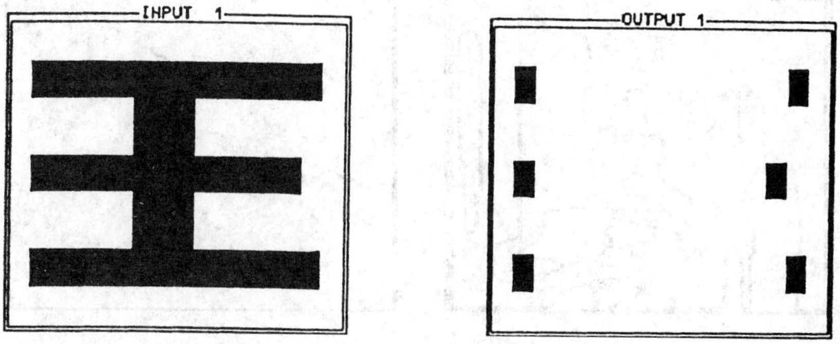

Name: DELDIAG1

Function. Deletes one pixel wide diagonal lines (5).

$$A = \begin{bmatrix} 0 & 0 & 0 \\ 0 & 2 & 0 \\ 0 & 0 & 0 \end{bmatrix} \qquad B = \begin{bmatrix} -0.25 & 0 & -0.25 \\ 0 & 0 & 0 \\ -0.25 & 0 & -0.25 \end{bmatrix} \qquad I = -2$$

Example. Input and output picture.

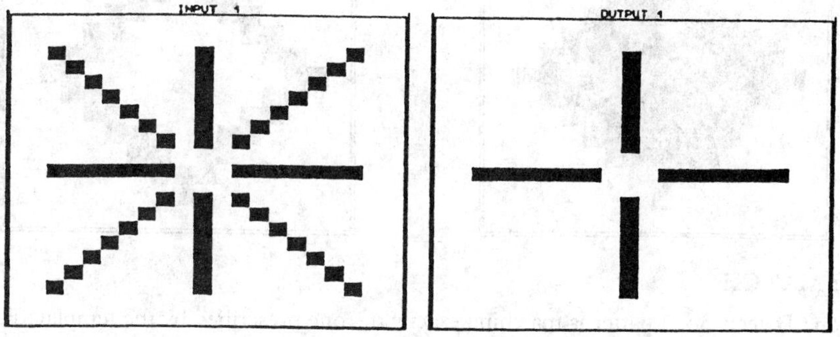

Name: DIAG

Function. Detects approximately diagonal lines being in the SW–NE direction.

$$A = \begin{bmatrix} 0 & 0 & 0 & 0 & 0 \\ 0 & 0 & 0 & 0 & 0 \\ 0 & 0 & 2 & 0 & 0 \\ 0 & 0 & 0 & 0 & 0 \\ 0 & 0 & 0 & 0 & 0 \end{bmatrix} \qquad B = \begin{bmatrix} -1 & -1 & -0.5 & 0.5 & 1 \\ -1 & -0.5 & 1 & 1 & 0.5 \\ -0.5 & 1 & 5 & 1 & -0.5 \\ 0.5 & 1 & 1 & -0.5 & -1 \\ 1 & 0.5 & -0.5 & -1 & -1 \end{bmatrix}$$

$$I = -9$$

Example. Input and output picture.

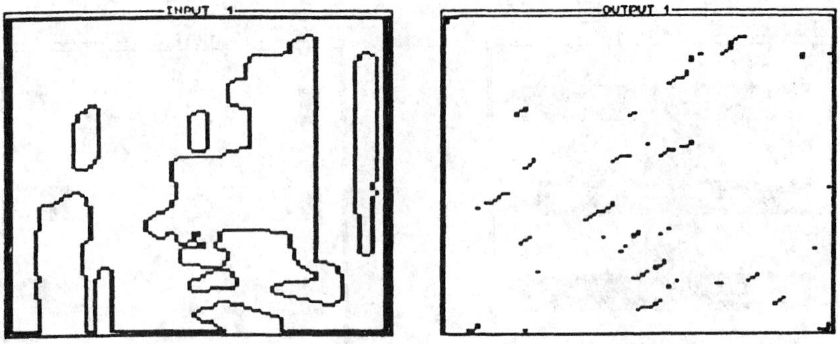

Name: EDGE

Function. Black and white edge detector.

$$A = \begin{bmatrix} 0 & 0 & 0 \\ 0 & 2 & 0 \\ 0 & 0 & 0 \end{bmatrix} \qquad B = \begin{bmatrix} -0.25 & -0.25 & -0.25 \\ -0.25 & 2 & -0.25 \\ -0.25 & -0.25 & -0.25 \end{bmatrix} \qquad I = -1.5$$

Example. Input and output picture.

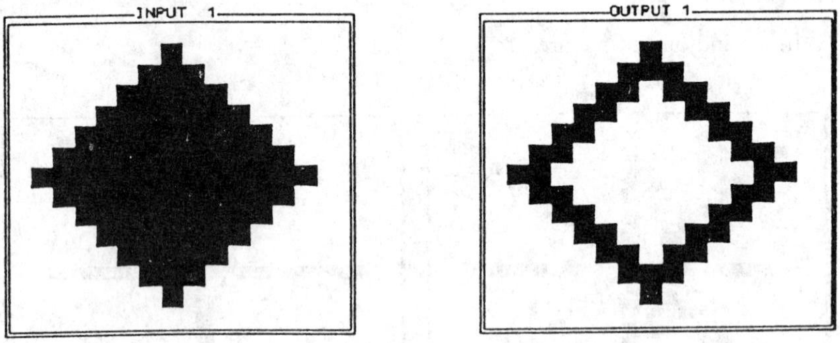

Name: MATCH

Function. Detects 3 × 3 patterns matching exactly the one prescribed by the template B, namely having a black/white pixel where the template value is $+1/-1$, respectively.

$$A = \begin{bmatrix} 0 & 0 & 0 \\ 0 & 1 & 0 \\ 0 & 0 & 0 \end{bmatrix} \qquad B = \begin{bmatrix} v & v & v \\ v & v & v \\ v & v & v \end{bmatrix} \qquad I = -N + 0.5$$

where $v = +1$, if corresponding pixel is required to be black; $v = 0$, if corresponding pixel is don't care; $v = -1$, if corresponding pixel is required to be white; N = number of pixels required to be either black or white, i.e., the number of nonzero values in the $\boldsymbol{B}$ template.

Example. Input and output picture, using the following values:

$$A = \begin{bmatrix} 0 & 0 & 0 \\ 0 & 1 & 0 \\ 0 & 0 & 0 \end{bmatrix} \qquad B = \begin{bmatrix} 1 & -1 & 1 \\ 0 & 1 & 0 \\ 1 & -1 & 1 \end{bmatrix} \qquad I = -6.5$$

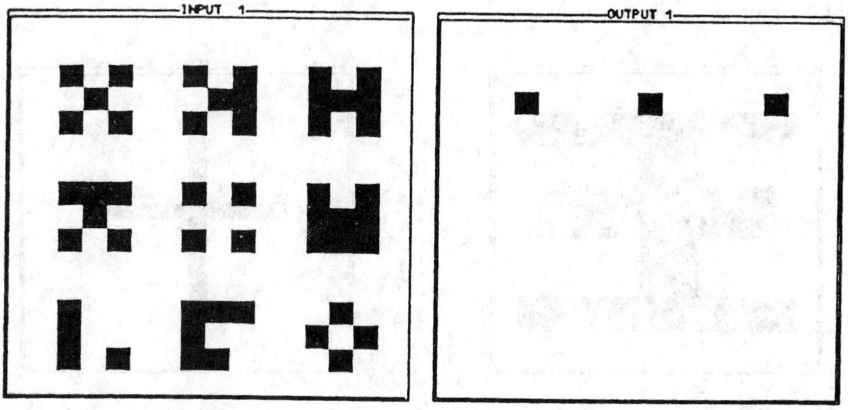

Name: OR

Function. Logical "OR" function of the input and the initial state.

$$A = \begin{bmatrix} 0 & 0 & 0 \\ 0 & 3 & 0 \\ 0 & 0 & 0 \end{bmatrix} \qquad B = \begin{bmatrix} 0 & 0 & 0 \\ 0 & 3 & 0 \\ 0 & 0 & 0 \end{bmatrix} \qquad I = 2$$

Example. Input, initial state, and output picture.

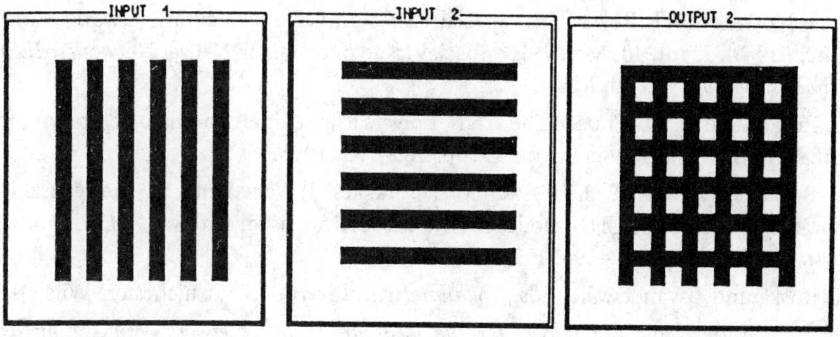

Name: PEEL1PIX

Function. Peels one pixel from all directions.

$$A = \begin{bmatrix} 0 & 0.4 & 0 \\ 0.4 & 1.4 & 0.4 \\ 0 & 0.4 & 0 \end{bmatrix} \qquad B = \begin{bmatrix} 4.6 & -2.8 & 4.6 \\ -2.8 & 1 & -2.8 \\ 4.6 & -2.8 & 4.6 \end{bmatrix} \qquad I = -7.2$$

Example. Input and output picture.

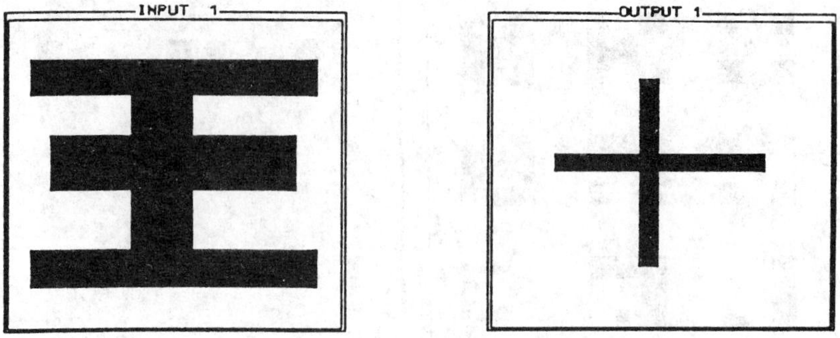

References

[1] L. O. Chua and L. Yang, "Cellular neural networks: Theory," *IEEE Trans. Circuits Syst.*, vol. 35, pp. 1257–1272, 1988;

[2] "Cellular neural networks: Applications," *IEEE Trans. Circuits Syst.*, vol. 35, pp. 1273–1290, 1988.

[3] L. O. Chua and T. Roska, "The CNN paradigm," *IEEE Trans. Circuits Syst. I*, vol. 40, pp. 147–156, 1993.

[4] J. Cruz and L. O. Chua, "A CNN chip for connected component detection," *IEEE Trans. Circuits Syst.*, vol. 38, pp. 812–817, 1991.

[5] R. Domínguez-Castro, S. Espejo, A. Rodríguez-Vázquez, and R. Carmona, "A CNN Universal Chip in CMOS Technology," Proc. IEEE 3rd Int. Workshop on CNN and Applications, (CNNA-94), Rome, pp. 91–96, 1994.

[6] H. Harrer, J. A. Nossek, and R. Stelzl, "An analog implementation of discrete-time cellular neural networks," *IEEE Trans. Neural Networks*, vol. 3, pp. 466–476, 1992.

[7] J. A. Nossek and T. Roska, Eds. Special Issues on Cellular Neural Networks, *IEEE Trans. Circuits Syst. I*, vol. 40, Mar. 1993; Special Issue on Cellular Neural Networks, *IEEE Trans. Circuits Syst. II*, vol. 40, Mar. 1993.

[8] T. Roska, and L. O. Chua, "The CNN universal machine: An analogic array computer," *IEEE Trans. Circuits Syst. II*, vol. 40, pp. 163–173, 1993.

[9] T. Roska, J. Hámori, E. Lábos, K. Lotz, J. Takács, P. Venetianer, Z. Vidnyánszki, and Á. Zarándy. "The use of CNN models in the subcortical visual pathway," *IEEE Trans. Circuits Syst. I*, vol. 40, pp. 182–195, 1993.

[10] T. Roska and J. Vandewalle, Eds., *Cellular Neural Networks.* Chichester: Wiley, 1993.

[11] "User's guide," in *Version 5.2, for the Cellular Neural Network Workstation Took Kit.* Budapest: MTA SzTAKI, 1993.

[12] F. Werblin, T. Roska, and L. O. Chua, "The analogic cellular neural network as a Bionic Eye," Report M94170, ERL, Univ. California, Berkeley, 1993.

38

Bifurcation and Chaos

chael Peter
nnedy
versity College, Dublin
ınd

8.1 Introduction to Chaos

Electrical and Electronic Circuits as Dynamical Systems

A **system** is something having parts that may be perceived as a single entity. A dynamical system is one that changes with time; what changes is the state of the system. Mathematically, a dynamical system consists of a space of states (called the state space or phase space) and a rule, called the **dynamic**, for determining which state corresponds at a given future time to a given present state [8]. A deterministic dynamical system is one whose state at any time is completely determined by its initial state and dynamic. In this section, we consider only deterministic dynamical systems.

A deterministic dynamical system may have a continuous or discrete state space and a continuous-time or discrete-time dynamic.

3-8341-2/95/$0.00 + $.50

A lumped[1] circuit containing resistive elements (resistors, voltage and current sources) and energy-storage elements (capacitors and/or inductors) may be modeled as a continuous-time deterministic dynamical system in $\mathbb{R}^n$. The evolution of the state of the circuit is described by a system of ordinary differential equations called state equations.

Discrete-time deterministic dynamical systems occur in electrical engineering as models of switched-capacitor and digital filters, sampled phase-locked loops, and sigma–delta modulators. Discrete-time dynamical systems also arise when analyzing the stability of steady-state solutions of continuous-time systems. The evolution of a discrete-time dynamical system is described by a system of *difference equations.*

Continuous-Time Dynamical Systems

Theorem 1: (Existence and Uniqueness of Solution for a Differential Equation) *Consider a continuous-time deterministic dynamical system defined by a system of ordinary differential equations of the form*

$$\dot{\mathbf{X}}(t) = \mathbf{F}(\mathbf{X}(t), t) \qquad (38.1)$$

where $\mathbf{X}(t) \in \mathbb{R}^n$ *is called the state,* $\dot{\mathbf{X}}(t)$ *denotes the derivative of* $\mathbf{X}(t)$ *with respect to time,* $\mathbf{X}(t_0) = \mathbf{X}_0$ *is called the initial condition, and the map* $\mathbf{F}(\cdot, \cdot)\colon \mathbb{R}^n \times \mathbb{R}_+ \to \mathbb{R}^n$ *is i) continuous almost everywhere*[2] *on* $\mathbb{R}^n \times \mathbb{R}_+$ *and ii) globally Lipschitz*[3] *in* $\mathbf{X}$. *Then, for each* $(\mathbf{X}_0, t_0) \in \mathbb{R}^n \times \mathbb{R}_+$, *there exists a continuous function* $\phi(\cdot\,; \mathbf{X}_0, t_0)\colon \mathbb{R}_+ \to \mathbb{R}^n$ *such that*

$$\phi(t_0; \mathbf{X}_0, t_0) = \mathbf{X}_0$$

and

$$\dot{\phi}(t; \mathbf{X}_0, t_0) = \mathbf{F}(\phi(t; \mathbf{X}_0, t_0), t) \qquad (38.2)$$

Furthermore, this function is unique.

The function $\phi(\cdot\,; \mathbf{X}_0, t_0)$ *is called the solution or trajectory through* $(\mathbf{X}_0, t_0)$ *of the differential equation (38.1).*

The image $\{\phi(t; \mathbf{X}_0, t_0) \in \mathbb{R}^n \mid t \in \mathbb{R}_+\}$ of the trajectory through $(\mathbf{X}_0, t_0)$ is a continuous curve in $\mathbb{R}^n$ called the orbit through $(\mathbf{X}_0, t_0)$.

$\mathbf{F}(\cdot, \cdot)$ is called the vector field of (38.1) because its image $\mathbf{F}(\mathbf{X}, t)$ is a vector that defines the direction and speed of the trajectory through $\mathbf{X}$ at time t.

The vector field $\mathbf{F}$ generates the *flow* ϕ, where $\phi(\cdot\,; \cdot, \cdot)\colon \mathbb{R}_+ \times \mathbb{R}^n \times \mathbb{R}_+ \to \mathbb{R}^n$ is a collection of continuous maps $\{\phi(t; \cdot, \cdot)\colon \mathbb{R}^n \times \mathbb{R}_+ \to \mathbb{R}^n \mid t \in \mathbb{R}_+\}$.

In particular, a point $\mathbf{X}_0 \in \mathbb{R}^n$ at t_0 is mapped by the flow into $\mathbf{X}(t) = \phi_t(t; \mathbf{X}_0, t_0)$ at time t.

[1] A *lumped* circuit is one whose physical dimensions are small compared to the wavelengths of its voltage and current waveforms [2].

[2] By *continuous almost everywhere* we mean the following: let D be a set in $\mathbb{R}_+$ that contains a countable number of discontinuities and for each $\mathbf{X} \in \mathbb{R}^n$, assume that the function $t \in \mathbb{R}_+ \backslash D \to \mathbf{F}(\mathbf{X}, t) \in \mathbb{R}^n$ is *continuous* and for any $\tau \in D$ the left-hand and right-hand limits $\mathbf{F}(\mathbf{X}, \tau_-)$ and $\mathbf{F}(\mathbf{X}, \tau_+)$ respectively are *finite* in $\mathbb{R}^n$ [1]. This condition includes circuits that contain switches and/or squarewave voltage and current sources.

[3] There is a piecewise continuous function $k(\cdot)\colon \mathbb{R}_+ \to \mathbb{R}_+$ such that $\|\mathbf{F}(\mathbf{X}, t) - \mathbf{F}(\mathbf{X}', t)\| \leq k(t)\|\mathbf{X} - \mathbf{X}'\|, \forall t \in \mathbb{R}_+, \forall \mathbf{X}, \mathbf{X}' \in \mathbb{R}^n$.

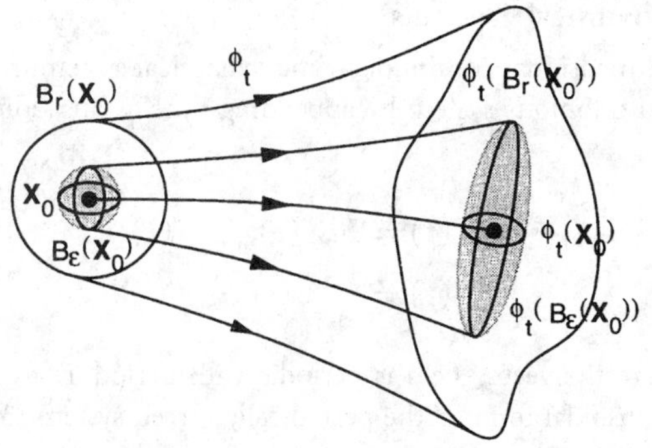

FIGURE 38.1 The vector field **F** of an autonomous continuous-time dynamical system generates a flow ϕ that maps a point $\mathbf{X}_0$ in the state space to its image $\phi_t(\mathbf{X}_0)$ t seconds later. A volume of state space $B_r(\mathbf{X}_0)$ evolves under the flow into a region $\phi_t[B_r(\mathbf{X}_0)]$. Sufficiently close to the trajectory $\phi_t(\mathbf{X}_0)$, the linearized flow maps a sphere of radius ε into an ellipsoid.

Autonomous Continuous-Time Dynamical Systems

If the vector field of a continuous-time deterministic dynamical system depends only on the state and is *independent* of time t, then the system is said to be *autonomous* and may be written as

$$\dot{\mathbf{X}}(t) = \mathbf{F}[\mathbf{X}(t)]$$

or simply

$$\dot{\mathbf{X}} = \mathbf{F}(\mathbf{X}) \tag{38.3}$$

If, in addition, the vector field $\mathbf{F}(\cdot)\colon \mathbb{R}^n \to \mathbb{R}^n$ is *Lipschitz*[4], then there is a *unique* continuous function $\phi(\cdot, \mathbf{X}_0)\colon \mathbb{R}_+ \to \mathbb{R}^n$ (called the trajectory through $\mathbf{X}_0$), which satisfies

$$\dot{\phi}(t, \mathbf{X}_0) = \mathbf{F}[\phi(t, \mathbf{X}_0)], \qquad \phi(t_0, \mathbf{X}_0) = \mathbf{X}_0 \tag{38.4}$$

Since the vector field is independent of time, we choose $t_0 \equiv 0$. For shorthand, we denote the flow by ϕ and the map $\phi(t, \cdot)\colon \mathbb{R}^n \to \mathbb{R}^n$ by ϕ_t.

The t-advance map ϕ_t takes a state $\mathbf{X}_0 \in \mathbb{R}^n$ to state $\mathbf{X}(t) = \phi_t(\mathbf{X}_0)$ t seconds later. In particular, ϕ_0 is the identity mapping. Furthermore, $\phi_{t+s} = \phi_t\phi_s$, since the state $\mathbf{Y} = \phi_s(\mathbf{X})$ to which $\mathbf{X}$ evolves after time s evolves after an additional time t into the same state $\mathbf{Z}$ as that to which $\mathbf{X}$ evolves after time $t + s$:

$$\mathbf{Z} = \phi_t(\mathbf{Y}) = \phi_t[\phi_s(\mathbf{X})] = \phi_{t+s}(\mathbf{X})$$

A bundle of trajectories emanating from a ball $B_r(\mathbf{X}_0)$ of radius r centered at $\mathbf{X}_0$ is mapped by the flow into some region $\phi_t[B_r(\mathbf{X}_0)]$ after t seconds (see Fig. 38.1). Consider a short segment of the trajectory $\phi_t(\mathbf{X}_0)$ along which the flow is differentiable with respect to $\mathbf{X}$: in a sufficiently small neighborhood of this trajectory, the flow is almost linear, so the ball $B_\varepsilon(\mathbf{X}_0)$ of radius ε about $\mathbf{X}_0$ evolves into an ellipsoid $\phi_t[B_\varepsilon(\mathbf{X}_0)]$, as shown.

An important consequence of Lipschitz continuity in an autonomous vector field and the resulting uniqueness of solution of (38.3) is that a trajectory of the dynamical system cannot go through the same point twice in two different directions. In particular, no two trajectories may cross each other; this is called the **noncrossing property** [18].

[4]There exists a finite $k \in \mathbb{R}$ such that $\|\mathbf{F}(\mathbf{X}) - \mathbf{F}(\mathbf{X}')\| \le k\|\mathbf{X} - \mathbf{X}'\|, \forall \mathbf{X}, \mathbf{X}' \in \mathbb{R}^n$.

Nonautonomous Dynamical Systems

A nonautonomous n-dimensional continuous-time dynamical system may be transformed to an $(n+1)$-dimensional autonomous system by appending time as an additional state variable and writing

$$\begin{aligned}\dot{\mathbf{X}}(t) &= \mathbf{F}[\mathbf{X}(t), X_{n+1}(t)] \\ \dot{X}_{n+1}(t) &= 1\end{aligned} \tag{38.5}$$

In the special case where the vector field is periodic with period T, as for example in the case of an oscillator with sinusoidal forcing, the periodically forced system (38.5) is equivalent to the $(n+1)$st order autonomous system

$$\begin{aligned}\dot{\mathbf{X}}(t) &= \mathbf{F}(\mathbf{X}(t), \theta(t)T) \\ \dot{\theta}(t) &= \frac{1}{T}\end{aligned} \tag{38.6}$$

where $\theta(t) = X_{n+1}/T$.

By identifying the n-dimensional hyperplanes corresponding to $\theta = 0$ and $\theta = 1$, the state space may be transformed from $\mathbb{R}^n \times \mathbb{R}_+$ into an equivalent cylindrical state space $\mathbb{R}^n \times S^1$, where S^1 denotes the circle. In the new coordinate system, the solution through $(\mathbf{X}_0, t_0)$ of (38.6) is

$$\begin{pmatrix} \mathbf{X}(t) \\ \theta_{S^1}(t) \end{pmatrix} = \begin{pmatrix} \phi_t(\mathbf{X}_0, t_0) \\ t/T \bmod 1 \end{pmatrix}$$

where $\theta(t) \in \mathbb{R}_+$ is identified with a point on S^1 (which has normalized angular coordinate $\theta_{S^1}(t) \in [0,1)$) via the transformation $\theta_{S^1}(t) = \theta(t) \bmod 1$. Using this technique, periodically-forced nonautonomous systems can be treated like autonomous systems.

Discrete-Time Dynamical Systems

Consider a discrete-time deterministic dynamical system defined by a system of difference equations of the form

$$\mathbf{X}(k+1) = \mathbf{G}(\mathbf{X}(k), k) \tag{38.7}$$

where $\mathbf{X}(k) \in \mathbb{R}^n$ is called the state, $\mathbf{X}(k_0) = \mathbf{X}_0$ is the initial condition, and $\mathbf{G}(\cdot,\cdot)\colon \mathbb{R}^n \times \mathbb{Z}_+ \to \mathbb{R}^n$ maps the current state $\mathbf{X}(k)$ into the next state $\mathbf{X}(k+1)$, where $k_0 \in \mathbb{Z}_+$.

By analogy with the continuous-time case, there exists a function $\phi(\cdot, \mathbf{X}_0, k_0)\colon \mathbb{Z}_+ \to \mathbb{R}^n$ such that

$$\phi(k_0; \mathbf{X}_0, k_0) = \mathbf{X}_0$$

and

$$\phi(k+1; \mathbf{X}_0, k_0) = \mathbf{G}(\phi(k; \mathbf{X}_0, k_0), k)$$

The function $\phi(\cdot\,; \mathbf{X}_0, k_0)\colon \mathbb{Z}_+ \to \mathbb{R}^n$ is called the solution or trajectory through $(\mathbf{X}_0, k_0)$ of the difference equation (38.7).

The image $\{\phi(k; \mathbf{X}_0, k_0) \in \mathbb{R}^n \mid k \in \mathbb{Z}_+\}$ in $\mathbb{R}^n$ of the trajectory through $(\mathbf{X}_0, k_0)$ is called an orbit through $(\mathbf{X}_0, k_0)$.

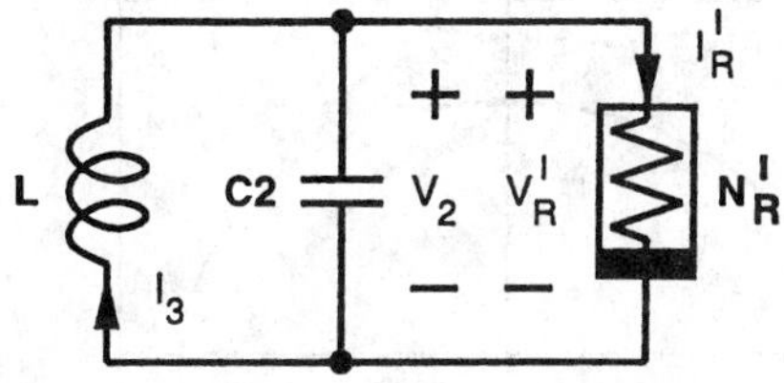

FIGURE 38.2 Parallel *RLC* circuit whose nonlinear resistor N_R' has a DP characteristic as shown in Fig. 38.3. By Kirchhoff's voltage law, $V_R' = V_2$.

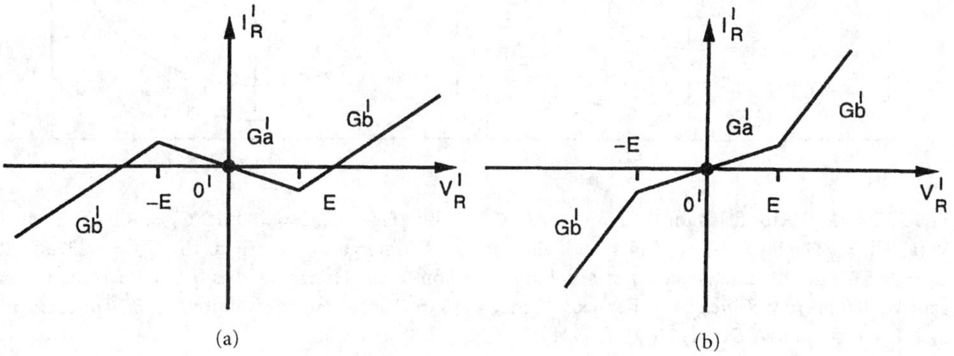

FIGURE 38.3 DP characteristic of N_R' in Fig. 38.2 when (a) $G_a' < 0$, and (b) $G_a' > 0$.

If the map $\mathbf{G}(\cdot,\cdot)$ of a discrete-time dynamical system depends only on the state $\mathbf{X}(k)$ and is independent of k, then the system is said to be autonomous and may be written more simply as

$$\mathbf{X}_{k+1} = \mathbf{G}(\mathbf{X}_k) \tag{38.8}$$

where $\mathbf{X}_k$ is shorthand for $\mathbf{X}(k)$ and the initial iterate k_0 is chosen, without loss of generality, to be zero. Using this notation, $\mathbf{X}_k$ is the image of $\mathbf{X}_0$ after k iterations of the map $\mathbf{G}(\cdot)\colon \mathbb{R}^n \rightarrow \mathbb{R}^n$.

Example: Nonlinear Parallel RLC Circuit. Consider the parallel *RLC* circuit shown in Fig. 38.2. This circuit contains a linear inductor L, a linear capacitor C_2, and a nonlinear resistor N_R', whose continuous piecewise-linear driving-point (DP) characteristic (shown in Fig. 38.3) has slope G_a' for $|V_R'| \le E$ and slope G_b' for $|V_R'| > E$. The DP characteristic of N_R' may be written explicitly

$$I_R'(V_R') = G_b' V_R' + \tfrac{1}{2}(G_a' - G_b')(|V_R' + E| - |V_R' - E|)$$

This circuit may be described by a pair of ordinary differential equations and is therefore a second-order continuous-time dynamical system. Choosing I_3 and V_2 as state variables, we write

$$\frac{dI_3}{dt} = -\frac{1}{L}V_2$$
$$\frac{dV_2}{dt} = \frac{1}{C_2}I_3 - \frac{1}{C_2}I_R'(V_2)$$

with $I_3(0) = I_{3_0}$ and $V_2(0) = V_{2_0}$.

We illustrate the vector field by drawing vectors at uniformly-spaced points in the two-dimensional state space defined by (I_3, V_2). Starting from a given initial condition (I_{3_0}, V_{2_0}), a solution curve in state space is the locus of points plotted out by the state as it moves through

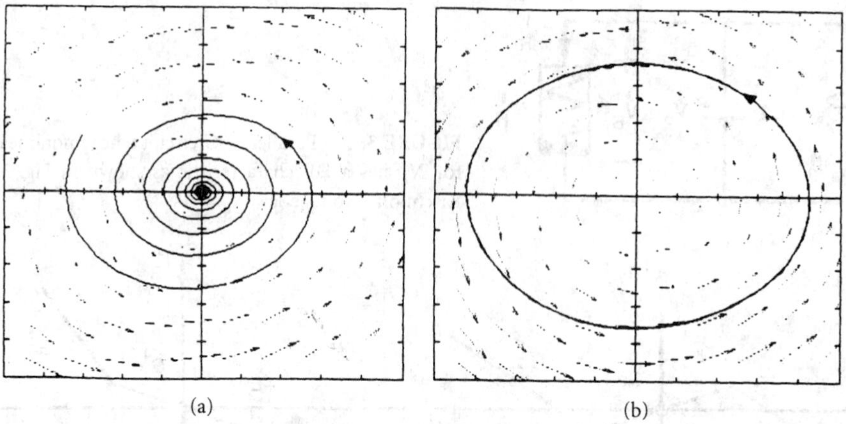

FIGURE 38.4 Vector fields for the nonlinear *RLC* circuit in Fig. 38.2. $L = 18$ mH, $C_2 = 100$ nF, $E = 0.47$ V. (a) $G'_a = 242.424\mu$S, $G'_b = 1045.455\mu$S: all trajectories converge to the origin. (b) $G'_a = -257.576\mu$S, $G'_b = 545.455\mu$S: the unique steady-state solution is a limit cycle. Horizontal axis: I_3, 400 μA/div; vertical axis: V_2, 200 mV/div. *Source:* M. P. Kennedy, "Three steps to chaos—Part I: Evolution," *IEEE Trans. Circuits Syst. I,* vol. 40, p. 647, Oct. 1993. © 1993 IEEE.

the vector field, following the direction of the arrow at every point. Fig. 38.4 shows typical vector fields and trajectories of the circuit.

If L, C_2, and G'_b are positive, the steady-state behavior of the circuit depends on the sign of G'_a. When $G'_a > 0$, the circuit is dissipative everywhere and all trajectories collapse toward the origin. The unique steady-state solution of the circuit is the stable dc equilibrium condition $I_3 = V_2 = 0$.

If $G'_a < 0$, N'_R looks like a negative resistor close to the origin and injects energy into the circuit, pushing trajectories away. Further out, where the characteristic has positive slope, trajectories are pulled in by the dissipative vector field. The resulting balance of forces produces a steady-state orbit called a *limit cycle,* which is approached asymptotically by all initial conditions of this circuit.

This limit cycle is said to be *attracting* because nearby trajectories move toward it and is structurally stable in the sense that, for almost all values of G'_a, a small change in the parameters of the circuit has little effect on it. In the special case when $G'_a \equiv 0$, a perturbation of G'_a causes the steady-state behavior to change from an equilibrium point to a limit cycle; this is called a *bifurcation.*

In the following subsections, we consider in detail steady-state behaviors, stability, structural stability, and bifurcations.

Classification and Uniqueness of Steady-State Behaviors

A trajectory of a dynamical system from an initial state $\mathbf{X}_0$ settles, possibly after some transient, onto a set of points called a *limit set.* The $\boldsymbol{\omega}$-limit set corresponds to the asymptotic behavior of the system as $t \to +\infty$ and is called the steady-state response. We use the idea of recurrent states to determine when the system has reached steady-state.

A state $\mathbf{X}$ of a dynamical system is called recurrent under the flow ϕ if, for every neighborhood $B_\varepsilon(\mathbf{X})$ of $\mathbf{X}$ and for every $T > 0$, there is a time $t > T$ such that $\phi_t(\mathbf{X}) \cap B_\varepsilon(\mathbf{X}) \neq \varnothing$. Thus, a state $\mathbf{X}$ is recurrent if, by waiting long enough, the trajectory through $\mathbf{X}$ repeatly returns arbitrarily close to $\mathbf{X}$ [7].

Wandering points correspond to transient behavior, while steady-state or asymptotic behavior corresponds to orbits of recurrent states.

A point $\mathbf{X}_\omega$ is an ω-limit point of $\mathbf{X}_0$ if and only if $\lim_{k\to+\infty} \phi_{t_k}(\mathbf{X}_0) = \mathbf{X}_\omega$ for some sequence $\{t_k \mid k \in \mathbb{Z}_+\}$ such that $t_k \to +\infty$. The set $L(\mathbf{X}_0)$ of ω-limit points of $\mathbf{X}_0$ is called the ω-limit set of $\mathbf{X}_0$.[5]

A limit set L is called attracting if there exists a neighborhood U of L such that $L(\mathbf{X}_0) = L$ for all $\mathbf{X}_0 \in U$. Thus, nearby trajectories converge toward an attracting limit set as $t \to \infty$.

An attracting set $\mathscr{A}$ that contains at least one orbit that comes arbitrarily close to every point in $\mathscr{A}$ is called an attractor [7].

In an asymptotically stable linear system the limit set is independent of the initial condition and unique so it makes sense to talk of the steady-state behavior. By contrast, a nonlinear system may possess several different limit sets and therefore may exhibit a variety of steady-state behaviors, depending on the initial condition.

The set of all points in the state space that converge to a particular limit set L is called the **basin of attraction** of L.

Since nonattracting limit sets cannot be observed in physical systems, the asymptotic or steady-state behavior of a real electronic circuit corresponds to motion on an attracting limit set.

Equilibrium Point

The simplest steady-state behavior of a dynamical system is an equilibrium point. An equilibrium point or stationary point of (38.3) is a state $\mathbf{X}_Q$ at which the vector field is zero. Thus $\mathbf{F}(\mathbf{X}_Q) = 0$ and $\phi_t(\mathbf{X}_Q) = \mathbf{X}_Q$; a trajectory starting from an equilibrium point remains indefinitely at that point.

In state space, the limit set consists of a single nonwandering point $\mathbf{X}_Q$. A point is a zero-dimensional object. Thus, an equilibrium point is said to have dimension zero.

In the time domain, an equilibrium point of an electronic circuit is simply a dc solution or operating point.

An equilibrium point or fixed point of a discrete-time dynamical system is a point $\mathbf{X}_Q$ that satisfies

$$\mathbf{G}(\mathbf{X}_Q) = \mathbf{X}_Q$$

Example: Nonlinear Parallel *RLC* Circuit. The nonlinear *RLC* circuit shown in Fig. 38.2 has just one equilibrium point $(I_{3_Q}, V_{2_Q}) = (0, 0)$. When G_a' is positive, a trajectory originating at any point in the state space converges to this attracting dc steady-state [as shown in Fig. 38.4(a)]. The basin of attraction of the origin is the entire state space.

Since all trajectories, and not just those that start close to it, converge to the origin, this equilibrium point is said to be a **global** attractor.

When $G_a' < 0$, the circuit possesses two steady-state solutions: the equilibrium point at the origin, and the limit cycle Γ. The equilibrium point is unstable in this case. All trajectories, except that which starts at the origin, are attracted to Γ.

Periodic Steady-State

A state X is called periodic if there exists $T > 0$ such that $\phi_T(\mathbf{X}) = \mathbf{X}$. A periodic orbit which is not a stationary point is called a **cycle**.

[5]The set of points to which trajectories converge from $\mathbf{X}_0$ as $t \to -\infty$ is called the α-limit set of $\mathbf{X}_0$. Since we consider only positive time, by *limit set* we mean the ω-limit set.

A limit cycle Γ is an isolated periodic orbit of a dynamical system [see Fig. 38.5(b)]. The limit cycle trajectory visits every point on the simple closed curve Γ with period T. Indeed, $\phi_t(\mathbf{X}) = \phi_{t+T}(\mathbf{X}) \forall \mathbf{X} \in \Gamma$. Thus, every point on the limit cycle Γ is a nonwandering point.

A limit cycle is said to have dimension one because a small piece of it looks like a one-dimensional object: a line. The n components $X_i(t)$ of a limit cycle trajectory $\mathbf{X}(t) = [X_1(t), X_2(t), \ldots, X_n(t)]^T$ in $\mathbb{R}^n$ are periodic time waveforms with period T.

Every periodic signal $X(t)$ may be decomposed into a Fourier series—a weighted sum of sinusoids at integer multiples of a fundamental frequency. Thus, a periodic signal appears in

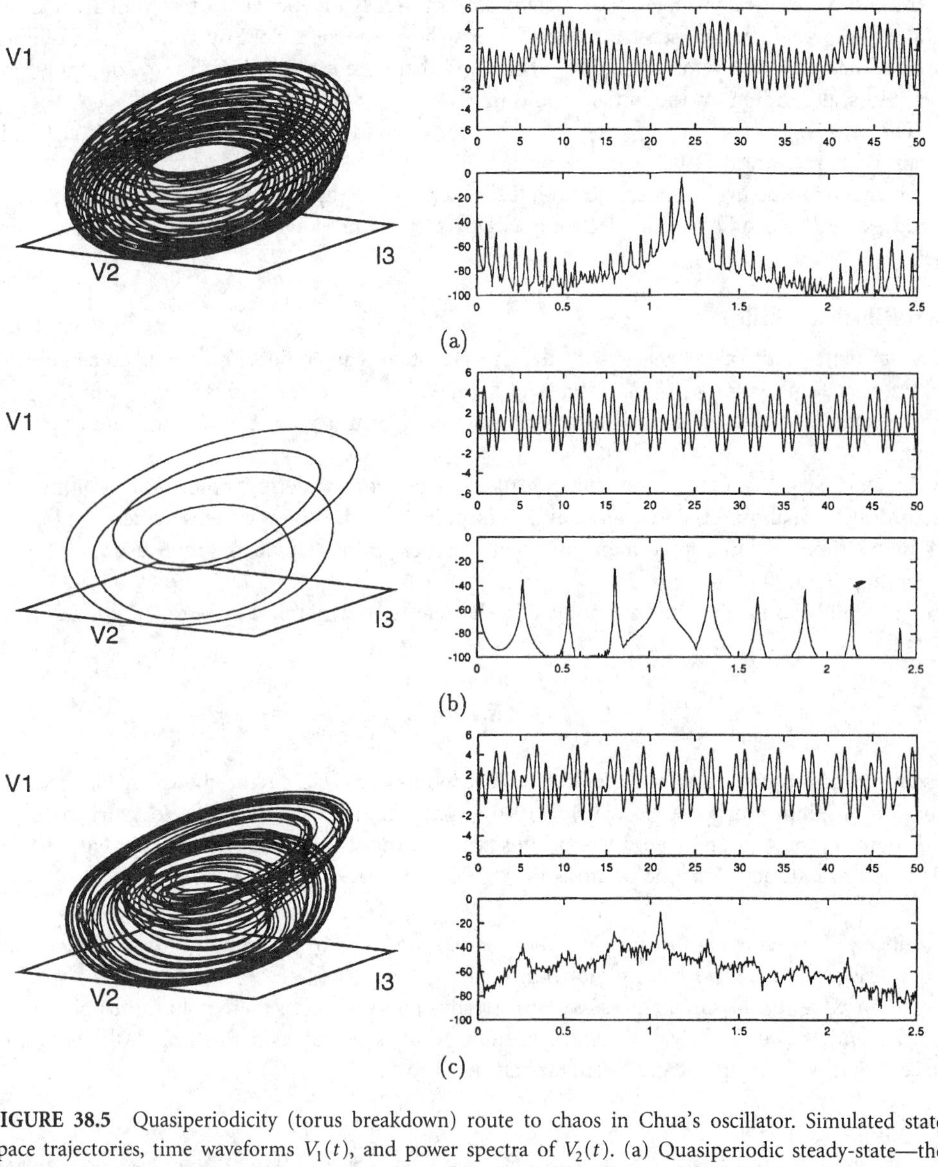

FIGURE 38.5 Quasiperiodicity (torus breakdown) route to chaos in Chua's oscillator. Simulated state space trajectories, time waveforms $V_1(t)$, and power spectra of $V_2(t)$. (a) Quasiperiodic steady-state—the signal is characterized by a discrete power spectrum with incommensurate frequency components; (b) periodic window—all spikes in the power spectrum are harmonically related to the fundamental frequency; (c) chaotic steady-state following breakdown of the torus—the waveform has a broadband power spectrum. Time plots: horizontal axis—t (ms); vertical axis—V_1 (V). Power spectra: horizontal axis—frequency (kHz); vertical axis—power (mean squared amplitude) of $V_2(t)$ (dB).

the frequency domain as a set of spikes at integer multiples (**harmonics**) of the fundamental frequency. The amplitudes of these spikes correspond to the coefficients in the Fourier series expansion of $X(t)$. The Fourier transform is an extension of these ideas to aperiodic signals; one considers the distribution of the signal's power over a continuum of frequencies rather than on a discrete set of harmonics.

The distribution of power in a signal $X(t)$ is most commonly quantified by means of the power density spectrum, often simply called the power spectrum. The simplest estimator of the power spectrum is the periodogram [17], which given N uniformly spaced samples $X(k/f_s)$, $k = 0, 1, \ldots, N-1$ of $X(t)$, yields $N/2+1$ numbers $P(nf_s/N)$, $n = 0, 1, \ldots, N/2$, where f_s is the sampling frequency.

If one considers the signal $X(t)$ as being composed of sinusoidal components at discrete frequencies, then $P(nf_s/N)$ is an estimate of the power in the component at frequency nf_s/N. By Parseval's theorem, the sum of the power in each of these components equals the mean squared amplitude of the N samples of $X(t)$ [17].

If $X(t)$ is periodic with period T, then its power will be concentrated in a dc component, a fundamental frequency component $1/T$, and harmonics. In practice, the discrete nature of the sampling process causes power to "leak" between adjacent frequency components; this leakage may be reduced by "windowing" the measured data before calculating the periodogram [17].

Example: Periodic Steady-State Solution. Figure 38.5(b) shows a state-space orbit, time waveform, and power spectrum of a periodic steady-state solution of a third-order autonomous continuous-time dynamical system.

The orbit in state space is an asymmetric closed curve consisting of four loops. In the time domain, the waveform has four crests per period and a dc offset. In the power spectrum, the dc offset manifests itself as a spike at zero frequency. The period of approximately 270 Hz produces a fundamental component at that frequency. Notice that the fourth harmonic (arising from "four crests per period") has the largest magnitude. This power spectrum is reminiscent of subharmonic mode locking in a forced oscillator.

Subharmonic Periodic Steady-State

A subharmonic periodic solution or period-K orbit of a discrete-time dynamical system is a set of K points $\{\mathbf{X}_1, \mathbf{X}_2, \ldots, \mathbf{X}_K\}$ that satisfy

$$\mathbf{X}_2 = \mathbf{G}(\mathbf{X}_1) \qquad \mathbf{X}_3 = \mathbf{G}(\mathbf{X}_2) \cdots \mathbf{X}_K = \mathbf{G}(\mathbf{X}_{K-1}) \qquad \mathbf{X}_1 = \mathbf{G}(\mathbf{X}_K)$$

More compactly, we may write $\mathbf{X}_i = \mathbf{G}^{(K)}(\mathbf{X}_i)$, where $\mathbf{G}^{(K)} = \mathbf{G}[\mathbf{G}(\cdots[\mathbf{G}(\cdot)]\cdots)]$ denotes $\mathbf{G}$ applied K times to the argument of the map; this is called the Kth iterate of $\mathbf{G}$.

Subharmonic periodic solutions occur in systems that contain two or more competing frequencies, such as forced oscillators or sampled-data circuits. Subharmonic solutions also arise following period-doubling bifurcations (see the section on structural stability and bifurcations).

Quasiperiodic Steady-State

The next most complicated form of steady-state behavior is called **quasiperiodicity**. In state space, this corresponds to a *torus* [see Fig. 38.5(a)]. While a small piece of a limit cycle in $\mathbb{R}^3$ looks like a line, a small section of two-torus looks like a plane; a two-torus has dimension two.

A quasiperiodic function is one that may be expressed as a countable sum of periodic functions with incommensurate frequencies, i.e., frequencies that are not rationally related. For example, $X(t) = \sin(t) + \sin(2\pi t)$ is a quasiperiodic signal. In the time domain, a quasiperiodic signal may look like an amplitude- or phase-modulated waveform.

While the Fourier spectrum of a periodic signal consists of a discrete set of spikes at integer multiples of a fundamental frequency, that of a quasiperiodic solution comprises a discrete set of spikes at incommensurate frequencies, as shown in Fig. 38.5(a).

In principle, a quasiperiodic signal may be distinguished from a periodic one by determining whether the frequency spikes in the Fourier spectrum are harmonically related. In practice, it is impossible to determine whether a measured number is rational or irrational; therefore, any spectrum that appears to be quasiperiodic may simply be periodic with an extremely long period.

A two-torus in a three-dimensional state space looks like a doughnut. Quasiperiodic behavior on a higher dimensional torus is more difficult to visualize in state space but appears in the power spectrum as a set of discrete components at incommensurate frequencies. A K-torus has dimension K.

Quasiperiodic behavior occurs in discrete-time systems where two incommensurate frequencies are present. A periodically-forced or discrete-time dynamical system has a frequency associated with the period of the forcing or sampling interval of the system; if a second frequency is introduced that is not rationally related to the period of the forcing or the sampling interval, then quasiperiodicity may occur.

Example: Discrete Torus. Consider a map from the circle S^1 onto itself. In polar coordinates, a point on the circle is parametrized by an angle θ. Assume that θ has been normalized so that one complete revolution of the circle corresponds to a change in θ of 1. The state of this system is determined by the normalized angle θ and the dynamics by

$$\theta_{k+1} = (\theta_k + \Omega) \bmod 1$$

If Ω is a rational number (of the form J/K where $J, K \in \mathbb{Z}_+$), then the steady-state solution is a period-K (subharmonic) orbit. If Ω is irrational, we obtain quasiperiodic behavior.

Chaotic Steady-State

DC equilibrium, periodic, and quasiperiodic steady-state behaviors have been correctly identified and classified since the pioneering days of electronics in the 1920s. By contrast, the existence of more exotic steady-state behaviors in electronic circuits has been acknowledged only in the past 30 years. While the notion of chaotic behavior in dynamical systems has existed in the mathematics literature since the turn of the century, unusual behaviors in the physical sciences as recently as the 1960s were described as "strange." Today, we classify as *chaos* recurrent[6] motion in a deterministic dynamical system which is characterized by a positive Lyapunov exponent.

From an experimentalist's point of view, chaos may be defined as bounded steady-state behavior in a deterministic dynamical system that is not an equilibrium point, not periodic, and not quasiperiodic [15].

Chaos is characterized by repeated stretching and folding of bundles of trajectories in state space. Two trajectories started from almost identical initial conditions diverge and soon become uncorrelated; this is called sensitive dependence on initial conditions and gives rise to long-term unpredictability.

In the time domain, a chaotic trajectory is neither periodic nor quasiperiodic, but looks "random." This "randomness" manifests itself in the frequency domain as a broad "noise-like" Fourier spectrum, as shown in Fig. 38.5(c).

While an equilibrium point, a limit cycle, and a K-torus each have integer dimension, the repeated stretching and folding of trajectories in a chaotic steady state gives the limit set a more

[6]Because a chaotic steady state does not settle down onto a single well-defined trajectory, the definition of recurrent states must be used to identify posttransient behavior.

complicated structure that, for three-dimensional continuous-time circuits, is something more than a surface but not quite a volume.

Dimension

The structure of a limit set $L \subset \mathbb{R}^n$ of a dynamical system may be quantified using a generalized notion of dimension that considers not just the geometrical structure of the set, but also the time evolution of trajectories on L.

Capacity (D_0 Dimension). The simplest notion of dimension, called capacity (or D_0 dimension) considers a limit set simply as a set of points, without reference to the dynamical system that produced it.

To estimate the capacity of L, cover the set with n-dimensional cubes having side length ε. If L is a D_0-dimensional object, then the minimum number $N(\varepsilon)$ of cubes required to cover L is proportional to ε^{-D_0}. Thus, $N(\varepsilon) \propto \varepsilon^{-D_0}$.

The D_0 dimension is given by

$$D_0 = \lim_{\varepsilon \to 0} -\frac{\ln N(\varepsilon)}{\ln \varepsilon}$$

When this definition is applied to a point, a limit cycle (or line), or a two-torus (or surface) $\mathbb{R}^3$, the calculated dimensions are 0, 1, and 2, respectively, as expected. When applied to the set of nonwandering points that comprise a chaotic steady state, the D_0 dimension is typically noninteger. An object that has noninteger dimension is called a **fractal.**

Example: The Middle-Third Cantor Set. Consider the set of points that is obtained by repeatedly deleting the middle third of an interval, as indicated in Fig. 38.6(a). At the first iteration, the unit interval is divided into 2^1 pieces of length 1/3 each; after k iterations, the set is covered by 2^k pieces of length $1/3^k$. By contrast, the set that is obtained by dividing the intervals into thirds but *not* throwing away the middle third each time [Fig. 38.6(b)] is covered at the kth step by 3^k pieces of length $1/3^k$.

Applying the definition of capacity, the dimension of the unit interval is

$$\lim_{k \to \infty} \frac{k \ln 3}{k \ln 3} = 1.00$$

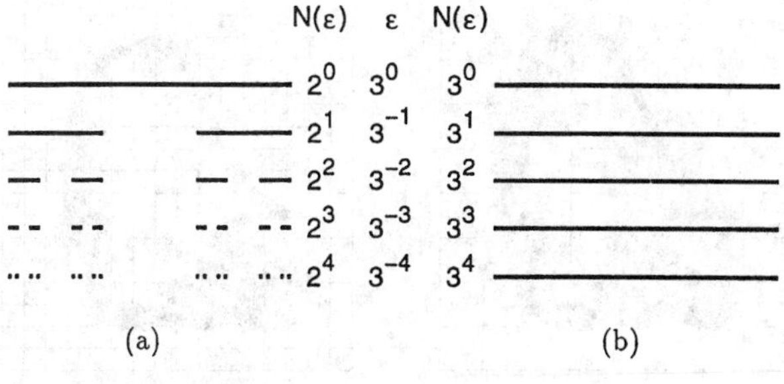

FIGURE 38.6 (a) The middle-third Cantor set is obtained by recursively removing the central portion of an interval. At the kth step, the set consists of $N(\varepsilon) = 2^k$ pieces of length $\varepsilon = 3^{-k}$. The limit set has capacity 0.63. (b) By contrast, the unit interval is covered by 3^k pieces of length 3^{-k}. The unit interval has dimension 1.00.

By contrast, the middle-third Cantor set has dimension

$$\lim_{k\to\infty} \frac{k \ln 2}{k \ln 3} \approx 0.63$$

The set is something more than a zero-dimensional object (a point) but not quite one-dimensional (like a line segment); it is a fractal.

Correlation **(D_2)** ***Dimension.*** The D_2 dimension considers not just the geometry of a limit set, but also the time evolution of trajectories on the set.

Consider the two limit sets L_a and L_b in $\mathbb{R}^2$ shown in Fig. 38.7(a) and (b) respectively. The D_0 dimension of these sets may be determined by iteratively covering them with squares (two-dimensional "cubes") of sidelength $\varepsilon = \varepsilon_0/2^k$, $k = 0, 1, 2, \ldots$, counting the required number of squares $N(\varepsilon)$ for each ε, and evaluating the limit

$$D_0 = \lim_{k\to\infty} -\frac{\ln N(\varepsilon)}{\ln(\varepsilon)}$$

For the smooth curve L_a, the number of squares required to cover the set grows linearly with $1/\varepsilon$; hence $D_0 = 1.0$. By contrast, if the kinks and folds in set L_b are present at all scales, then the growth of $N(\varepsilon)$ versus $1/\varepsilon$ is superlinear and the object has a noninteger D_0 dimension between 1.0 and 2.0.

Imagine now that L_a and L_b are not simply static geometrical objects but are orbits of discrete-time dynamical systems. In this case, a steady-state trajectory corresponds to a sequence of points moving around the limit set.

Cover the limit set with the minimum number $N(\varepsilon)$ of "cubes" with sidelength ε, and label the boxes $1, 2, \ldots, i, \ldots, N(\varepsilon)$. Count the number of times $n_i(N, \varepsilon)$ that a typical steady-state

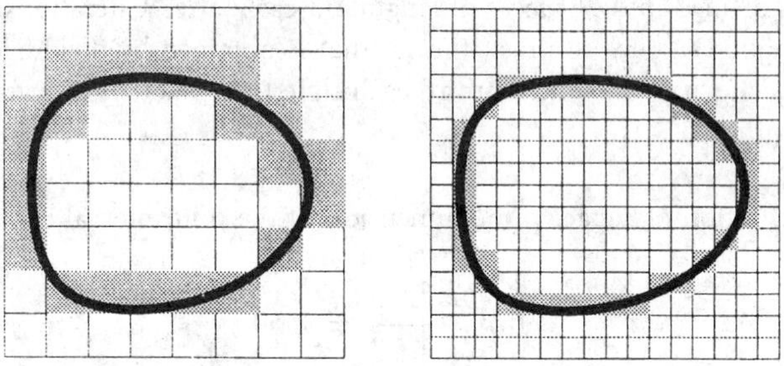

(a)

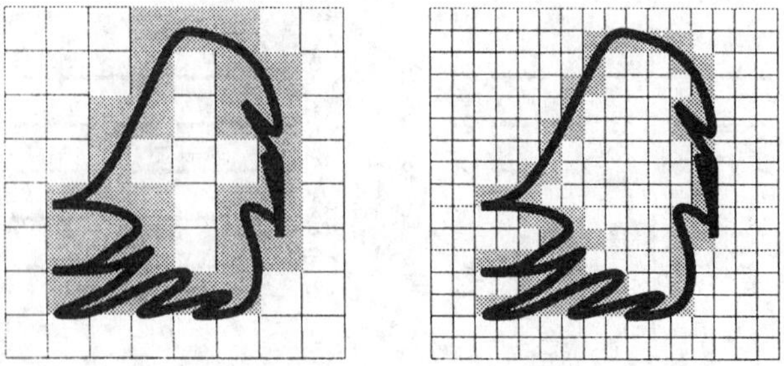

(b)

FIGURE 38.7 Coverings of two limit sets L_a (a) and L_b (b) with squares of sidelength ε_0 and $\varepsilon_0/2$, respectively.

trajectory of length N visits box i and define

$$p_i = \lim_{N\to\infty} \frac{n_i(N,\varepsilon)}{N}$$

p_i is the relative frequency with which a trajectory visits the ith cube. The D_2 dimension is defined as

$$D_2 = \lim_{\varepsilon\to 0} \frac{\ln \sum_{i=1}^{N(\varepsilon)} p_i^2}{\ln \varepsilon}$$

In general, $D_2 \le D_0$, with equality when a typical trajectory visits all $N(\varepsilon)$ cubes with the same relative frequency $p = 1/N(\varepsilon)$. In this special case,

$$\begin{aligned} D_2 &= \lim_{\varepsilon\to 0} \frac{\ln \sum_{i=1}^{N(\varepsilon)} \frac{1}{N(\varepsilon)^2}}{\ln \varepsilon} \\ &= \lim_{\varepsilon\to 0} -\frac{\ln N(\varepsilon)}{\ln \varepsilon} \\ &= D_0 \end{aligned}$$

An efficient algorithm (due to Grassberger and Procaccia) for estimating D_2 is based on the approximation $\sum_{i=1}^{N(\varepsilon)} p_i^2 \approx C(\varepsilon)$ [15], where

$$C(\varepsilon) = \lim_{N\to\infty} \frac{1}{N^2} (\text{the number of pairs of points}(X_i, X_j)\text{such that} \|X_i - X_j\| < \varepsilon)$$

is called the correlation. The D_2 or correlation dimension is given by

$$D_2 = \lim_{\varepsilon\to 0} \frac{\ln C(\varepsilon)}{\ln \varepsilon}$$

Example: Correlation (D_2) Dimension. The correlation dimension of the chaotic attractor shown in Fig. 38.5(c), estimated using INSITE, is 2.1, while D_2 for the uniformly covered torus in Fig. 38.5(a) is 2.0.

Stability of Steady-State Trajectories

Consider once more the nonlinear *RLC* circuit shown in Fig. 38.2. If G_a' is negative, this circuit settles to a periodic steady state from almost every initial condition. However, a trajectory started from the origin will, in principle, remain indefinitely at the origin since this is an equilibrium point. The circuit has two possible steady-state solutions. Experimentally, only the limit cycle will be observed. Why?

If trajectories starting from states close to a limit set converge to that steady-state, the limit set is called an attracting limit set. If, in addition, the attracting limit set contains at least one trajectory that comes arbitrarily close to every point in the set, then it is an attractor. If nearby points diverge from the limit set, it is called a repellor.

In the nonlinear *RLC* circuit with $G_a' < 0$, the equilibrium point is a repellor and the limit cycle is an attractor.

Stability of Equilibrium Points

Qualitatively, an equilibrium point is said to be stable if trajectories starting close to it remain nearby for all future time and unstable otherwise. Stability is a local concept, dealing with trajectories in a small neighborhood of the equilibrium point.

To analyze the behavior of the vector field in the vicinity of an equilibrium point $\mathbf{X_Q}$, we write $\mathbf{X} = \mathbf{X_Q} + \mathbf{x}$ and substitute into (38.3) to obtain

$$\dot{\mathbf{X}}_\mathbf{Q} + \dot{\mathbf{x}} = \mathbf{F}(\mathbf{X_Q} + \mathbf{x})$$
$$\mathbf{F}(\mathbf{X_Q}) + \dot{\mathbf{x}} \approx \mathbf{F}(\mathbf{X_Q}) + \mathbf{D_X F}(\mathbf{X_Q})\mathbf{x}$$

where we have kept just the first two terms of the Taylor series expansion of $\mathbf{F}(\mathbf{X})$ about $\mathbf{X_Q}$. The Jacobian matrix $\mathbf{D_X F}(\mathbf{X})$ is the matrix of partial derivatives of $\mathbf{F}(\mathbf{X})$ with respect to X:

$$\mathbf{D_X F}(\mathbf{X}) = \begin{bmatrix} \frac{\partial F_i(\mathbf{X})}{\partial X_1} & \frac{\partial F_1(\mathbf{X})}{\partial X_2} & \cdots & \frac{\partial F_1(\mathbf{X})}{\partial X_n} \\ \frac{\partial F_2(\mathbf{X})}{\partial X_1} & \frac{\partial F_2(\mathbf{X})}{\partial X_2} & \cdots & \frac{\partial F_2(\mathbf{X})}{\partial X_n} \\ \vdots & \vdots & \ddots & \vdots \\ \frac{\partial F_n(\mathbf{X})}{\partial X_1} & \frac{\partial F_n(\mathbf{X})}{\partial X_2} & \cdots & \frac{\partial F_n(\mathbf{X})}{\partial X_n} \end{bmatrix}$$

Subtracting $\mathbf{F}(\mathbf{X_Q})$ from both sides of (38.9) we obtain the linear system

$$\dot{\mathbf{x}} = \mathbf{D_X F}(\mathbf{X_Q})\mathbf{x} \tag{38.9}$$

where the Jacobian matrix is evaluated at $\mathbf{X_Q}$. This linearization describes the behavior of the circuit in the vicinity of $\mathbf{X_Q}$; we call this the local behavior.

Note that the linearization is simply the small-signal equivalent circuit at the operating point $\mathbf{X_Q}$. In general, the local behavior of a circuit depends explicitly on the operating point $\mathbf{X_Q}$. For example, a *pn*-junction diode exhibits a small incremental resistance under forward bias, but a large small-signal resistance under reverse bias.

Eigenvalues

If $\mathbf{X_Q}$ is an equilibrium point of (38.3), a complete description of its stability is contained in the eigenvalues of the linearization of (38.3) about $\mathbf{X_Q}$. These are defined as the roots λ of the characteristic equation

$$\det[\lambda \mathbf{I} - \mathbf{D_X F}(\mathbf{X_Q})] = 0 \tag{38.10}$$

where $\mathbf{I}$ is the identity matrix.

If the real parts of all of the eigenvalues of $\mathbf{D_X F}(\mathbf{X_Q})$ are strictly negative, then the equilibrium point $\mathbf{X_Q}$ is asymptotically stable and is called a sink because all nearby trajectories converge toward it.

If any of the eigenvalues has a positive real part, the equilibrium point is unstable; if all of the eigenvalues have positive real parts, the equilibrium point is called a source. An equilibrium point that has eigenvalues with both negative and positive real parts is called a saddle. A saddle is unstable.

An equilibrium point is said to be hyperbolic if all of the eigenvalues of $\mathbf{D_X F}(\mathbf{X_Q})$ have nonzero real parts. All hyperbolic equilibrium points are either unstable or asymptotically stable.

Discrete-Time Systems

The stability of a fixed point $\mathbf{X}_Q$ of a discrete-time dynamical system

$$\mathbf{X}_{k+1} = \mathbf{G}(\mathbf{X}_k)$$

is determined by the eigenvalues of the linearization $\mathbf{D_X G}(\mathbf{X}_Q)$ of the vector field $\mathbf{G}$, evaluated at $\mathbf{X}_Q$.

The equilibrium point is classified as stable if all of the eigenvalues of $\mathbf{D_X G}(\mathbf{X}_Q)$ are strictly less than unity in modulus, and unstable if any has modulus greater than unity.

Eigenvectors, Eigenspaces, Stable and Unstable Manifolds

Associated with each distinct eigenvalue λ of the Jacobian matrix $\mathbf{D_X F}(\mathbf{X}_Q)$ is an eigenvector $\vec{\nu}$ defined by

$$\mathbf{D_X F}(\mathbf{X}_Q)\vec{\nu} = \lambda\vec{\nu}$$

A real eigenvalue γ has a real eigenvector $\vec{\eta}$. Complex eigenvalues of a real matrix occur in pairs of the form $\sigma \pm j\omega$. The real and imaginary parts of the associated eigenvectors $\vec{\eta}_r \pm j\vec{\eta}_c$ span a plane called a complex eigenplane.

The n_s-dimensional subspace of $\mathbb{R}^n$ associated with the stable eigenvalues of the Jacobian matrix is called the stable eigenspace, denoted $E^s(\mathbf{X}_Q)$. The n_u-dimensional subspace corresponding to the unstable eigenvalues is called the unstable eigenspace, denoted $E^u(\mathbf{X}_Q)$.

The analogs of the stable and unstable eigenspaces for a general nonlinear system are called the local stable and unstable *manifolds*[7] $W^s(\mathbf{X}_Q)$ and $W^u(\mathbf{X}_Q)$.

The stable manifold $W^s(\mathbf{X}_Q)$ is defined as the set of all states from which trajectories remain in the manifold and converge under the flow to $\mathbf{X}_Q$. The unstable manifold $W^u(\mathbf{X}_Q)$ is defined as the set of all states from which trajectories remain in the manifold and diverge under the flow from $\mathbf{X}_Q$.

By definition, the stable and unstable manifolds are invariant under the flow (if $\mathbf{X} \in W^s$, then $\phi_t(\mathbf{X}) \in W^s$). Furthermore, the n_s- and n_u-dimensional tangent spaces to W^s and W^u at $\mathbf{X}_Q$ are E^s and E^u (as shown in Fig. 38.8). In the special case of a linear or affine vector field $\mathbf{F}$, the stable and unstable manifolds are simply the eigenspaces E^s and E^u themselves.

Chaos is associated with two characteristic connections of the stable and unstable manifolds. A homoclinic orbit [shown in Fig. 38.9(a)] joins an isolated equilibrium point $\mathbf{X}_Q$ to itself along its stable and unstable manifolds. A heteroclinic orbit [Fig. 38.9(b)] joins two distinct equilibrium points, $\mathbf{X}_{Q_1}$ and $\mathbf{X}_{Q_2}$, along the unstable manifold of one and the stable manifold of the other.

Stability of Limit Cycles

While the stability of an equilibrium point may be determined by considering the eigenvalues of the linearization of the vector field near the point, how does one study the stability of a limit cycle, torus, or chaotic steady-state trajectory?

The idea introduced by Poincaré is to convert a continuous-time dynamical system into an equivalent discrete-time dynamical system by taking a transverse slice through the flow. Intersections of trajectories with this so-called Poincaré section define a Poincaré map from

[7]An m-dimensional manifold is a geometrical object every small section of which looks like $\mathbb{R}^m$. More precisely, M is an m-dimensional manifold if, for every $x \in M$, there exists an open neighborhood U of x and a smooth invertible map which takes U to some open neighborhood of $\mathbb{R}^m$. For example, a limit cycle of a continuous-time dynamical system is a one-dimensional manifold.

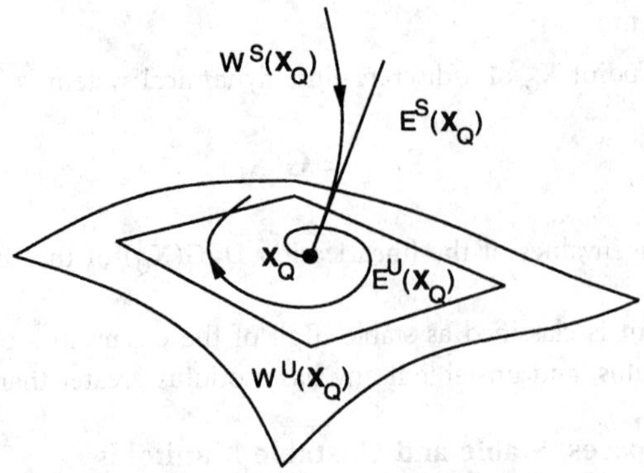

FIGURE 38.8 Stable and unstable manifolds $W^s(\mathbf{X_Q})$ and $W^u(\mathbf{X_Q})$ of an equilibrium point $\mathbf{X_Q}$. The stable and unstable eigenspaces $E^s(\mathbf{X_Q})$ and $E^u(\mathbf{X_Q})$ derived from the linearization of the vector field at $\mathbf{X_Q}$ are tangent to the corresponding manifolds W^s and W^u at $\mathbf{X_Q}$. A trajectory approaching the equilibrium point along the stable manifold is tangential to $E^s(\mathbf{X_Q})$ at $\mathbf{X_Q}$; a trajectory leaving $\mathbf{X_Q}$ along the unstable manifold is tangential to $E^u(\mathbf{X_Q})$ at $\mathbf{X_Q}$.

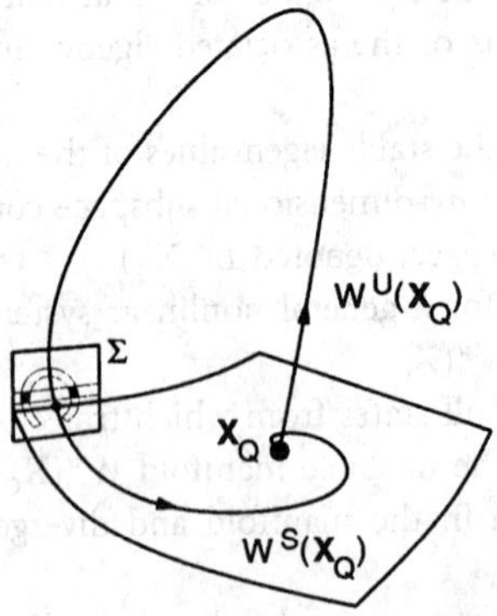

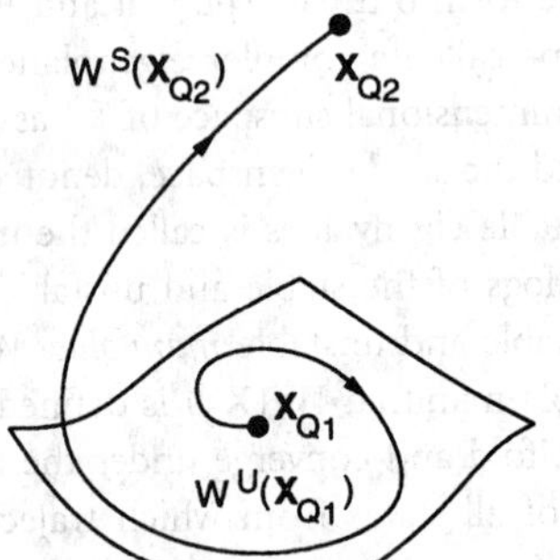

FIGURE 38.9 (a) A homoclinic orbit joins an isolated equilibrium point $\mathbf{X_Q}$ to itself along its stable and unstable manifolds. (b) A heteroclinic orbit joins two distinct equilibrium points, $\mathbf{X_{Q_1}}$ and $\mathbf{X_{Q_2}}$, along the unstable manifold of one and the stable manifold of the other.

the section to themselves. Since the limit cycle is a fixed point $\mathbf{X_Q}$ of the associated discrete-time dynamical system, its stability may be determined by examining the eigenvalues of the linearization of the Poincaré map at $\mathbf{X_Q}$.

Poincaré Sections

A Poincaré section of an n-dimensional autonomous continuous-time dynamical system is an $(n-1)$-dimensional hyperplane Σ in the state space that is intersected transversally[8] by the flow.

Let Γ be a closed orbit of the flow of a smooth vector field $\mathbf{F}$, and let $\mathbf{X_Q}$ be a point of intersection of Γ with Σ. If T is the period of Γ and $\mathbf{X} \in \Sigma$ is sufficiently close to $\mathbf{X_Q}$, then the trajectory $\phi_t(\mathbf{X})$ through $\mathbf{X}$ will return to Σ after a time $\tau(\mathbf{X}) \approx T$ and intersect the hyperplane at a point $\phi_{\tau(\mathbf{X})}(\mathbf{X})$, as shown in Fig. 38.10.

[8]A *transverse* intersection of manifolds in $\mathbb{R}^n$ is an intersection of manifolds such that, from any point in the intersection, all directions in $\mathbb{R}^n$ can be generated by linear combinations of vectors tangent to the manifolds.

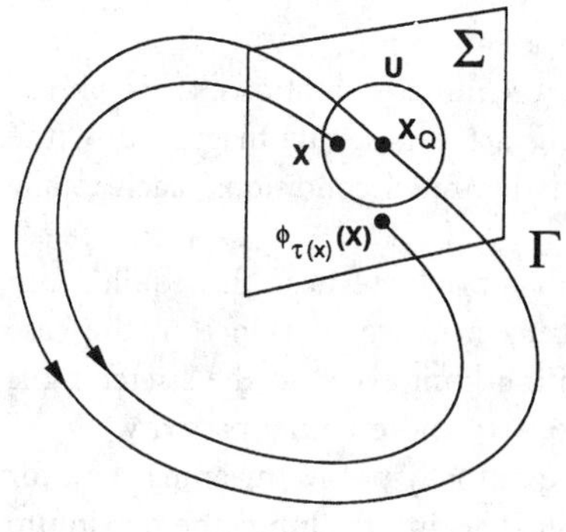

FIGURE 38.10 A transverse Poincaré section Σ through the flow of a dynamical system induces a discrete Poincaré map from a neighborhood U of the point of intersection $\mathbf{X}_Q$ to Σ.

This construction implicitly defines a function (called a Poincaré map or first return map) $\mathbf{G}: U \to \Sigma$

$$\mathbf{G}(\mathbf{X}) = \phi_{\tau(\mathbf{X})}(\mathbf{X})$$

where U is a small region of Σ close to $\mathbf{X}_Q$. The corresponding discrete-time dynamical system

$$\mathbf{X}_{k+1} = \mathbf{G}(\mathbf{X}_k)$$

has a fixed point at $\mathbf{X}_Q$.

The stability of the limit cycle is determined by the eigenvalues of the linearization $D_{\mathbf{X}}\mathbf{G}(\mathbf{X}_Q)$ of $\mathbf{G}$ at $\mathbf{X}_Q$. If all of the eigenvalues of $D_{\mathbf{X}}\mathbf{G}(\mathbf{X}_Q)$ have modulus less than unity, the limit cycle is asymptotically stable; if any has modulus greater than unity, the limit cycle is unstable.

Note that the stability of the limit cycle is independent of the position and orientation of the Poincaré plane, provided that the intersection is chosen *transverse* to the flow. For a nonautonomous system with periodic forcing, a natural choice for the hyperplane is at a fixed phase θ_0 of the forcing.

In the Poincaré section, a limit cycle looks like a fixed point. A period-K subharmonic of a nonautonomous system with periodic forcing appears as a period-K orbit of the corresponding map [see Fig. 38.11(b)].

The Poincaré section of a quasiperiodic attractor consisting of two incommensurate frequencies looks like a closed curve—a transverse cut through a two-torus [Fig. 38.11(a)].

The Poincaré section of a chaotic attractor has fractal structure, as shown in Fig. 38.11(c).

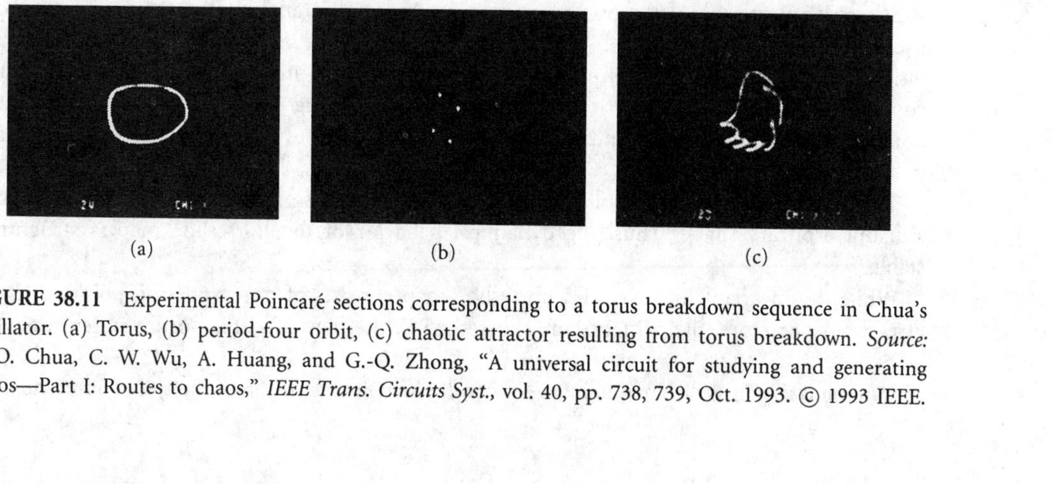

FIGURE 38.11 Experimental Poincaré sections corresponding to a torus breakdown sequence in Chua's oscillator. (a) Torus, (b) period-four orbit, (c) chaotic attractor resulting from torus breakdown. *Source:* L. O. Chua, C. W. Wu, A. Huang, and G.-Q. Zhong, "A universal circuit for studying and generating chaos—Part I: Routes to chaos," *IEEE Trans. Circuits Syst.*, vol. 40, pp. 738, 739, Oct. 1993. © 1993 IEEE.

Horseshoes and Chaos

Chaotic behavior is characterized by sensitive dependence on initial conditions. This phrase emphasizes the fact that small differences in initial conditions are persistently magnified by the dynamics of the system so that trajectories starting from nearby initial conditions reach totally different states in a finite time.

Trajectories of the nonlinear *RLC* circuit in Fig. 38.2 that originate near the equilibrium point are initially stretched apart exponentially by the locally negative resistance in the case $G'_a < 0$. Eventually, however, they are squeezed together onto a limit cycle, so the stretching is not persistent. This is a consequence of the noncrossing property and eventual passivity.

Although perhaps locally active, every physical resistor is eventually passive meaning that, for a large enough voltage across its terminals, it dissipates power. This in turn limits the maximum values of the voltages and currents in the circuit giving a bounded steady-state solution. All physical systems are bounded, so how can small differences be magnified persistently in a real circuit?

Chaos in the Sense of Shil'nikov

Consider a flow ϕ in $\mathbb{R}^3$ that has an equilibrium point at the origin with a real eigenvalue $\gamma > 0$ and a pair of complex conjugate eigenvalues $\sigma \pm j\omega$ with $\sigma < 0$ and $\omega \neq 0$. Assume that the flow has a homoclinic orbit Γ through the origin.

One may define a Poincaré map for this system by taking a transverse section through the homoclinic orbit, as shown in Fig. 38.9(a).

Theorem 2 (Shil'nikov): *If* $|\sigma/\gamma| < 1$, *the flow* ϕ *can be perturbed to* ϕ' *such that* ϕ' *has a homoclinic orbit* Γ' *near* Γ *and the Poincaré map of* ϕ' *defined in a neighborhood of* Γ' *has a countable number of horseshoes in its discrete dynamics.*

The characteristic horseshoe shape in the Poincaré map stretches and folds trajectories repeatedly (see Fig. 38.12). The resulting dynamics exhibit extreme sensitivity to initial conditions [7].

The presence of horseshoes in the flow of a continuous-time system that satisfies the assumptions of Shil'nikov's theorem implies the existence of a countable number of unstable periodic orbits of arbitrarily long period as well as an uncountable number of complicated bounded nonperiodic chaotic solutions [7].

Horseshoes. The action of the Smale horseshoe map is to take the unit square [Fig. 38.12(a)], stretch it, fold it into a horseshoe shape [Fig. 38.12(b)], and lay it down on itself [Fig. 38.12(c)]. Under the action of this map, only four regions of the unit square are returned to the square.

Successive iterations of the horseshoe map return smaller and smaller regions of square to itself, as shown in Fig. 34.12(d)–(f). If the map is iterated *ad infinitum*, the unit square is ultimately mapped onto a set of points. These points form an invariant (fractal) limit set L that contains a countable set of periodic orbits of arbitrarily long periods, an uncountable set of bounded nonperiodic orbits, and at least one orbit that comes arbitrarily close to every point in L.

The properties of the map still hold if the horseshoe is distorted by a perturbation of small size but arbitrary shape. Thus, the dynamical behavior of the horseshoe map is structurally stable.[9]

While the invariant limit set of a horseshoe map consists of nonwandering points, it is not attracting. Therefore, the existence of a horseshoe in the flow of a third-order system does not

[9] *Structural stability* is discussed in more detail in the section on structural stability and bifurcations.

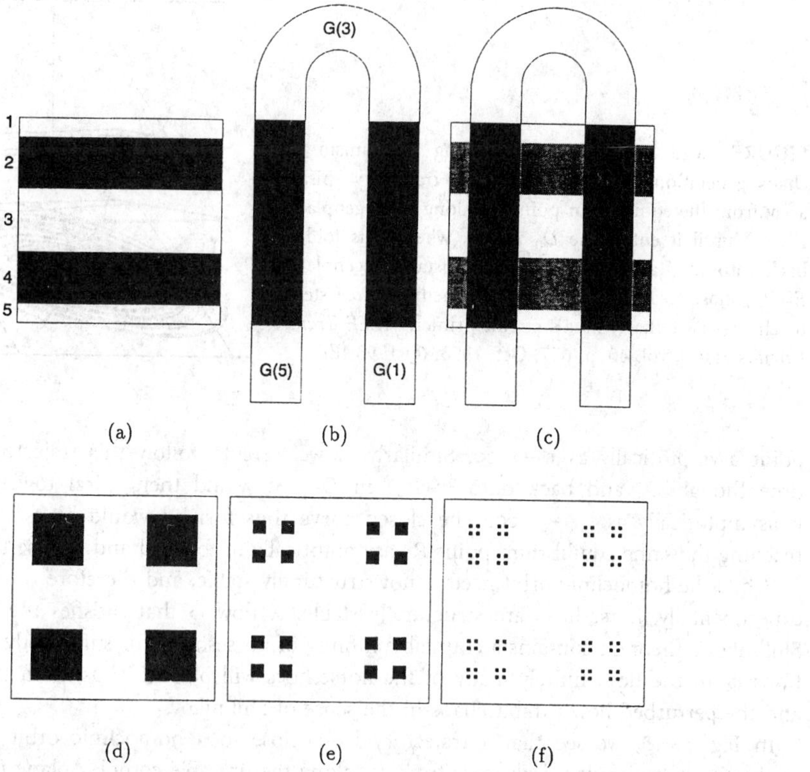

FIGURE 38.12 The Smale horseshoe map stretches the unit square (a), folds it into a horseshoe (b), and lays it back on itself (c), so that only points lying in bands 2 and 4 of (a) are mapped into the square. At the next iteration, only those points in $(\mathbf{G}(2) \cup \mathbf{G}(4)) \cap (2 \cup 4)$ (d) are mapped back to the square. Repeated iterations of the map (d)–(f) remove all points from the square except an invariant (fractal) set of fixed points.

imply that the system will exhibit chaotic steady-state behavior. However if a typical trajectory in the Poincaré map remains in a neighborhood of the invariant set, then the system may exhibit chaos. Thus, while Shil'nikov's theorem is a strong indicator of chaos, it does not provide definitive proof that a system is chaotic.

Example: Chaos in a Piecewise-Linear System. Although we have stated them for the case $\sigma < 0$, $\gamma > 0$, Shil'nikov's theorem also applies when the equilibrium point at the origin has an unstable pair of complex conjugate eigenvalues and a stable real eigenvalue. In that case, it is somewhat easier to visualize the stretching and folding of bundles of trajectories close to a homoclinic orbit.

Consider the trajectory in a three-region piecewise-linear vector field shown in Fig. 38.13. We assume that the equilibrium point P_- has a stable real eigenvalue γ_1 [whose eigenvector is $E^r(P_-)$] and an unstable complex conjugate pair of eigenvalues $\sigma_1 \pm j\omega_1$, the real and imaginary parts of whose eigenvectors span the plane $E^c(P_-)$ [2], as shown. A trajectory originating from a point $\mathbf{X}_0$ on $E^c(P_-)$ spirals away from the equilibrium point along $E^c(P_-)$ until it enters the D_0 region, where it is folded back into D_{-1}. Upon reentering D_{-1}, the trajectory is pulled toward P_- roughly in the direction of the real eigenvector $E^r(P_-)$ as shown.

Now imagine what would happen if the trajectory entering D_{-1} from D_0 were in precisely the direction $E^r(P_-)$. Such a trajectory would follow $E^r(P_-)$ toward P_-, reaching the equilibrium

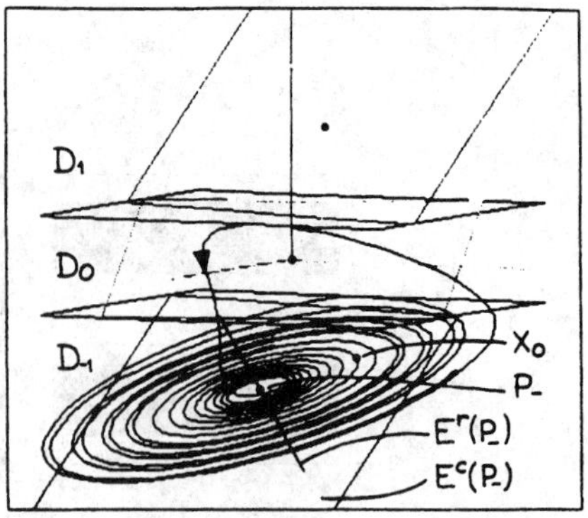

FIGURE 38.13 Stretching and folding mechanism of chaos generation in Chua's circuit. A trajectory spirals away from the equilibrium point P_- along the eigenplane $E^c(P_-)$ until it enters the D_0 region, where it is folded back into D_{-1} and returns to the unstable eigenplane $E^c(P_-)$ close to P_-. *Source:* M. P. Kennedy, "Three steps to chaos—Part II: A Chua's circuit primer," *IEEE Trans. Circuits Syst. I,* vol. 40, p. 657, Oct. 1993. © 1993 IEEE.

point asymptotically as $t \to \infty$. Similarly, if we were to follow this trajectory backward in time though D_0 and back onto $E^c(P_-)$ in D_{-1}, it would then spiral toward P_-, reaching it asymptotically as $t \to -\infty$. The closed curve thus formed would be a homoclinic orbit, reaching the same equilibrium point P_- asymptotically in forward and reverse time.

While the homoclinic orbit itself is not structurally stable, and therefore cannot be observed experimentally, horseshoes are structurally stable. A flow ϕ that satisfies the assumptions of Shil'nikov's theorem contains a countable infinity of horseshoes; for sufficiently small perturbations ϕ' of the flow, finitely many of the horseshoes will persist. Thus, both the original flow and the perturbed flow exhibit chaos in the sense of Shil'nikov.

In Fig. 38.13, we see that a trajectory lying close to a homoclinic orbit exhibits similar qualitative behavior: it spirals away from P_- along the unstable complex plane $E^c(P_-)$, is folded in D_0, reenters D_{-1} above $E^c(P_-)$, and is pulled back toward $E^c(P_-)$ only to be spun away from P_- once more.

Thus, two trajectories starting from distinct initial states close to P_- on $E^c(P_-)$ are stretched apart exponentially along the unstable eigenplane before being folded in D_1 and reinjected close to P_-; this gives rise to sensitive dependence on initial conditions. The recurrent stretching and folding continues *ad infinitum*, producing a chaotic steady-state solution.

Lyapunov Exponents

The notion of sensitive dependence on initial conditions may be made more precise through the introduction of Lyapunov exponents (LEs). Lyapunov exponents quantify the average exponential rates of separation of trajectories along the flow.

The flow in a neighborhood of asymptotically stable trajectory is contracting so the LEs are zero or negative.[10] Sensitive dependence on initial conditions results from a positive LE.

To determine the stability of an equilibrium point, we considered the eigenvalues of the linearization of the vector field in the vicinity of the equilibrium trajectory. This idea can be generalized to any trajectory of the flow.

The local behavior of the vector field along a trajectory $\phi_t(\mathbf{X}_0)$ of an autonomous continuous-time dynamical system (38.3) is governed by the linearized dynamics

$$\begin{aligned}\dot{\mathbf{x}} &= \mathbf{D_X F(X) x}, \qquad \mathbf{x}(0) = \mathbf{x}_0 \\ &= \mathbf{D_X F}[\phi_t(\mathbf{X}_0)]\mathbf{x}\end{aligned}$$

[10]A continuous flow that has a bounded trajectory not tending to an equilibrium point has a zero Lyapunov exponent (in the direction of the flow).

This is a linear time-varying system whose state transition matrix $\Phi_t(\mathbf{X}_0)$ maps a point $\mathbf{x}_0$ into $\mathbf{x}(t)$. Thus,

$$\mathbf{x}(t) = \Phi_t(\mathbf{X}_0)\mathbf{x}_0$$

Note that Φ_t is a linear operator. Therefore, a ball $B_\varepsilon(\mathbf{X}_0)$ of radius ε about $\mathbf{X}_0$ is mapped into an ellipsoid as shown in Fig. 38.1. The principal axes of the ellipsoid are determined by the singular values of Φ_t.

The singular values $\sigma_1(t), \sigma_2(t), \ldots, \sigma_n(t)$ of Φ_t are defined as the square roots of the eigenvalues of $\Phi_t^H \Phi_t$, where Φ_t^H is the complex conjugate transpose of Φ_t. The singular values are ordered so that $\sigma_1(t) > \sigma_2(t) > \cdots > \sigma_n(t)$.

In particular, a ball of radius ε is mapped by the linearized flow into an ellipsoid (see Fig. 38.1), the maximum and minimum radii of which are bounded by $\sigma_1(t)\varepsilon$ and $\sigma_n(t)\varepsilon$ respectively.

The stability of a steady-state orbit is governed by the average local rates of expansion and contraction of volumes of state space close to the orbit. The Lyapunov exponents (LEs) λ_i are defined by

$$\lambda_i = \lim_{t \to \infty} \frac{1}{t} \ln \sigma_i(t)$$

whenever this limit exists. The LEs quantify the average exponential rates of separation of trajectories along the flow.

The LEs are a property of a steady-state trajectory. Any transient effect is averaged out by taking the limit as $t \to \infty$. Furthermore, the LEs are global quantities of an attracting set that depend on the local stability properties of a trajectory within the set.

The set $\{\lambda_i, i = 1, 2, \ldots, n\}$ is called the Lyapunov spectrum. An attractor has the property that the sum of its LEs is negative.

Lyapunov Exponents of Discrete-Time Systems. The local behavior along an orbit of the autonomous discrete-time dynamical system (38.8) is governed by the linearized dynamics

$$\mathbf{x}_{k+1} = \mathbf{D_X G}(\mathbf{X}_k)\mathbf{x}_k, \quad k = 0, 1, 2, \cdots$$

whose state transition matrix $\Phi_k(\mathbf{X}_0)$ maps a point $\mathbf{x}_0$ into $\mathbf{x}_k$. Thus,

$$\mathbf{x}_k = \Phi_k(\mathbf{X}_0)\mathbf{x}_0$$

The Lyapunov exponents λ_i for the discrete-time dynamical system (38.8) are defined by

$$\lambda_i = \lim_{k \to \infty} \frac{1}{k} \ln \sigma_i(k)$$

whenever this limit exists. $\sigma_i(k)$ denotes the ith singular value of $\Phi_k^H \Phi_k$.

Lyapunov Exponents of Steady-State Solutions. Consider once more the continuous-time dynamical system (38.3). If $\mathbf{D_X F}$ were constant along the flow, with n distinct eigenvalues $\tilde{\lambda}_i$,

$i = 1, 2, \ldots, n$, then

$$\Phi_t = \begin{pmatrix} \exp(\tilde{\lambda}_1 t) & 0 & \cdots & 0 \\ 0 & \exp(\tilde{\lambda}_2 t) & \cdots & 0 \\ \vdots & \vdots & \ddots & \vdots \\ 0 & 0 & \cdots & \exp(\tilde{\lambda}_n t) \end{pmatrix}$$

and

$$\Phi_t^H \Phi_t = \begin{pmatrix} \exp(2\,\mathrm{Re}\,(\tilde{\lambda}_1)t) & 0 & \cdots & 0 \\ 0 & \exp(2\,\mathrm{Re}\,(\tilde{\lambda}_2)t) & \cdots & 0 \\ \vdots & \vdots & \ddots & \vdots \\ 0 & 0 & \cdots & \exp(2\,\mathrm{Re}\,(\tilde{\lambda}_n)t) \end{pmatrix}$$

giving $\sigma_i(t) = \exp(\mathrm{Re}\,(\tilde{\lambda}_i)t)$ and

$$\begin{aligned} \lambda_i &= \lim_{t\to\infty} \frac{1}{t} \ln(\exp[\mathrm{Re}\,(\tilde{\lambda}_i)t]) \\ &= \mathrm{Re}\,(\tilde{\lambda}_i) \end{aligned}$$

In this case, the LEs are simply the real parts of the eigenvalues of $\mathbf{D_X F}$.

All of the eigenvalues of a stable equilibrium point have negative real parts and therefore the largest Lyapunov exponent of an attracting equilibrium point is negative.

Trajectories close to a stable limit cycle converge onto the limit cycle. Therefore, the largest LE of a periodic steady-state is zero (corresponding to motion along the limit cycle [15]) and all of its other LEs are negative.

A quasiperiodic K-torus has K zero LEs since the flow is locally neither contracting nor expanding along the surface of the K-torus.

S chaotic trajectory is locally unstable and therefore has a positive LE; this produces sensitive dependence on initial conditions. Nevertheless, in the case of a chaotic attractor, this locally unstable chaotic trajectory belongs to an attracting limit set to which nearby trajectories converge.

The steady-state behavior of a four-dimensional continuous-time dynamical system which has two positive, one zero, and one negative LE is called hyperchaos.

The Lyapunov spectrum may be used to identify attractors, as summarized in Table 38.1.

Structural Stability and Bifurcations

Structural stability refers to the sensitivity of a phenomenon to small changes in the parameter of a system. A structurally stable vector field $\mathbf{F}$ is one for which sufficiently close vector fields $\mathbf{F}'$ have equivalent[11] dynamics [18].

The behavior of a typical circuit depends on a set of parameters one or more of which may be varied in order to optimize some performance criteria. In particular, one may think of a one-parameter family of systems

$$\dot{\mathbf{X}} = \mathbf{F}_\mu(\mathbf{X}) \tag{38.11}$$

[11]**Equivalent** means that there exists a continuous invertible function h that transforms $\mathbf{F}$ into $\mathbf{F}'$.

TABLE 38.1 Classification of Steady-State Behaviors According to their Limit Sets, Power Spectra, LEs, and Dimension

Steady State	Limit Set	Spectrum	LEs	Dimension
DC	Fixed Point	Spike at DC	$0 > \lambda_1 \geq \cdots \geq \lambda_n$	0
Periodic	Closed Curve	Fundamental Plus Integer Harmonics	$\lambda_1 = 0$ $0 > \lambda_2 \geq \cdots \geq \lambda_n$	1
Quasiperiodic	K-Torus	Incommensurate Frequencies	$\lambda_1 = \cdots = \lambda_K = 0$ $0 > \lambda_{K+1} \geq \cdots \geq \lambda_n$	K
Chaotic	Fractal	Broad Spectrum	$\lambda_1 > 0$ $\sum_{i=1}^{n} \lambda_i < 0$	Noninteger

whose vector field is parametrized by a control parameter μ. A value μ_0 of (38.11) for which the flow of (38.11) is not structurally stable is a bifurcation value of μ [7].

The dynamics in the state space may be qualitatively very different from one value of μ to another. In the nonlinear *RLC* circuit example, the steady-state solution is a limit cycle if the control parameter G'_a is negative and an equilibrium point if G'_a is positive. If G'_a is identically equal to zero, trajectories starting from $I_{3_0} = 0$, $V_{2_0} < E$ yield sinusoidal solutions. These sinusoidal solutions are not structurally stable because the slightest perturbation of G'_a will cause the oscillation to decay to zero or converge to the limit cycle, depending on whether G'_a is made slightly larger or smaller than zero.

If we think of this circuit as being parametrized by G'_a, then its vector field is not structurally stable at $G'_a \equiv 0$. We say that the equilibrium point undergoes a bifurcation (from stability to instability) as the value of the bifurcation parameter G'_a is reduced through the bifurcation point $G'_a = 0$.

Bifurcation Types

In this section, we consider three types of local bifurcation: the Hopf bifurcation, the saddle-node bifurcation, and the period-doubling bifurcation [18]. These bifurcations are called local because they may be understood by linearizing the system close to an equilibrium point or limit cycle.

Hopf Bifurcation. A Hopf bifurcation occurs in a continuous-time dynamical system (38.3) when a simple pair of complex conjugate eigenvalues of the linearization $\mathbf{D_X F}(\mathbf{X_Q})$ of the vector field at an equilibrium point $\mathbf{X_Q}$ crosses the imaginary axis.

Typically, the equilibrium point changes stability from stable to unstable and a stable limit cycle is born. The bifurcation at $G'_a \equiv 0$ in the nonlinear *RLC* circuit is Hopf-like.[12]

When an equilibrium point undergoes a Hopf bifurcation, a limit cycle is born. When a limit cycle undergoes a Hopf bifurcation, motion on a two-torus results.

Saddle-Node Bifurcation. A saddle-node bifurcation occurs when a stable and an unstable equilibrium point merge and disappear; this typically manifests itself as the abrupt disappearance of an attractor.

A common example of a saddle-node bifurcation in electronic circuits is switching between equilibrium states in a Schmitt trigger. At the threshold for switching, a stable equilibrium point corresponding to the "high" saturated state merges with the high-gain region's unstable saddle-type equilibrium point and disappears. After a switching transient, the trajectory settles to the other stable equilibrium point, which corresponds to the "low" state.

[12]Note that the Hopf bifurcation theorem is proven for sufficiently smooth systems and does not strictly apply to piecewise-linear systems. However, a physical implementation of a piecewise-linear characteristic, such as that of N_R, is always smooth.

A saddle-node bifurcation may also manifest itself as a switch between periodic attractors of different size, between a periodic attractor and a chaotic attractor, or between a limit cycle at one frequency and a limit cycle at another frequency.

Period-Doubling Bifurcation. A period-doubling bifurcation occurs in a discrete-time dynamical system (38.8) when a real eigenvalue of the linearization $\mathbf{D_X G(X_Q)}$ of the map $\mathbf{G}$ at an equilibrium point crosses the unit circle at -1 [7].

In a continuous-time system, a period-doubling bifurcation occurs only from a periodic solution (an equilibrium point of the Poincaré map). At the bifurcation point, a periodic orbit with period T changes smoothly into one with period $2T$, as shown in Fig. 38.14(a) and (b).

Blue Sky Catastrophe. A blue sky catastrophe is a global bifurcation that occurs when an attractor disappears "into the blue," usually because of a collision with a saddle-type limit set. Hysteresis involving a chaotic attractor is often caused by a blue sky catastrophe [18].

Routes to Chaos

Each of the three local bifurcations may give rise to a distinct route to chaos, and all three have been reported in electronic circuits. These routes are important because it is often difficult to conclude from experimental data alone whether irregular behavior is due to measurement noise or to underlying chaotic dynamics. If, upon adjusting a control parameter, one of the three prototype routes is observed, this indicates that the dynamics might be chaotic.

Periodic-Doubling Route to Chaos. The period-doubling route to chaos is characterized by a cascade of period-doubling bifurcations. Each period-doubling transforms a limit cycle into one at half the frequency, spreading the energy of the system over a wider range of frequencies. An infinite cascade of such doublings results in a chaotic trajectory of infinite period and a broad frequency spectrum that contains energy at all frequencies. Figure 38.14 is a set of snapshots of the period-doubling route to chaos in Chua's oscillator.

An infinite number of period-doubling bifurcations to chaos can occur over a finite range of the bifurcation parameter because of a geometric relationship between the intervals over which the control parameter must be moved to cause successive bifurcations. Period-doubling is governed by a universal scaling law that holds in the vicinity of the bifurcation point to chaos μ_∞.

Define the ratio δ_k of successive interval μ, in each of which there is a constant period of oscillation, as follows:

$$\delta_k = \frac{\mu_{2^k} - \mu_{2^{k-1}}}{\mu_{2^{k+1}} - \mu_{2^k}}$$

where μ_{2^k} is the bifurcation point for the period from $2^k T$ to $2^{k+1}T$. In the limit as $k \to \infty$, a universal constant called the Feigenbaum number δ is obtained:

$$\lim_{k\to\infty} \delta_k = \delta = 4.6692\cdots$$

The period-doubling route to chaos is readily identified from a state-space plot, time series, power spectrum, or a Poincaré map.

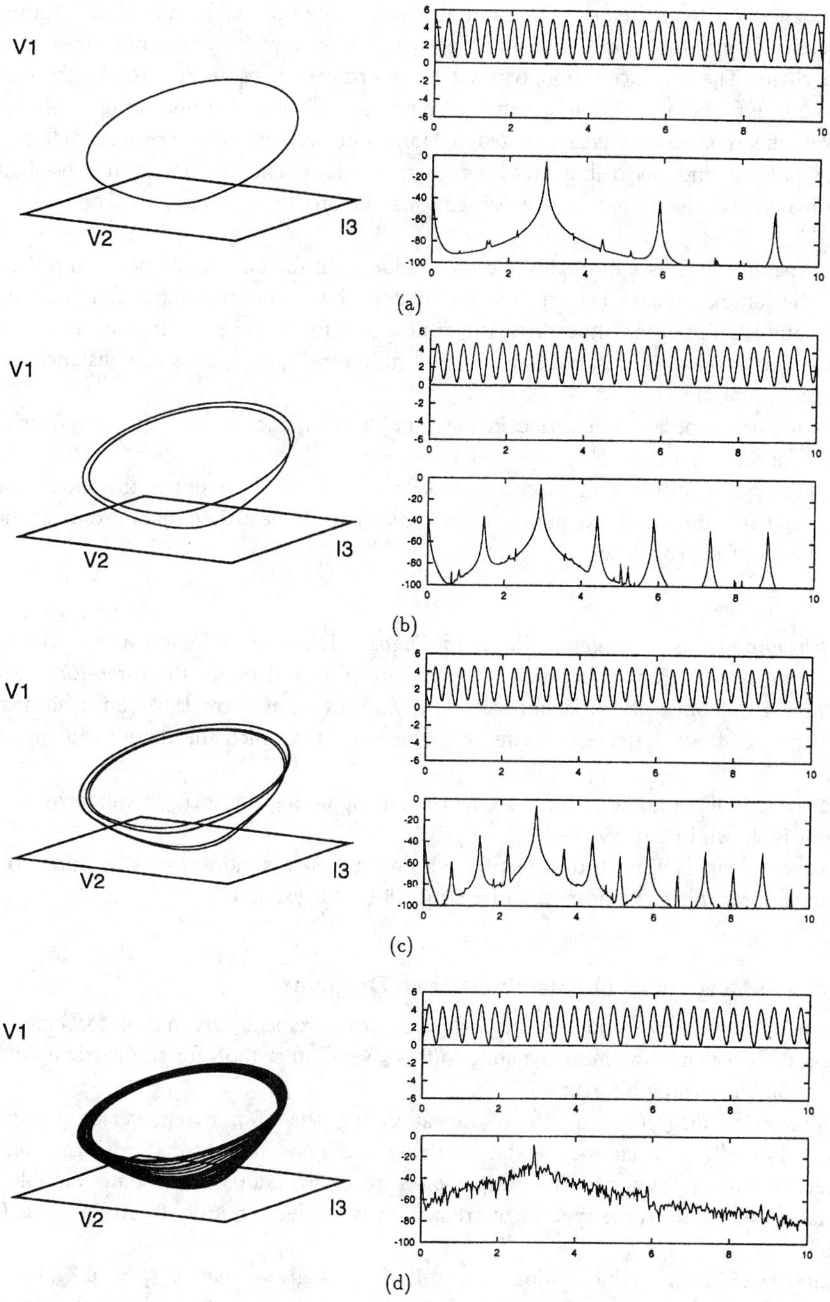

FIGURE 38.14 Period-doubling route to chaos in Chua's oscillator. Simulated state space trajectories, time waveforms $V_1(t)$, and power spectra of $V_2(t)$. (a) $G = 530\mu S$: periodic steady state—the signal is characterized by a discrete power spectrum with energy at integer multiples of the fundamental frequency f_0; (b) $G = 537\mu S$: period-two—after a period-doubling bifurcation, the period of the signal is approximately twice that of (a). In the power spectrum, a spike appears at the new fundamental frequency $\approx f_0/2$. (c) $G = 539\mu S$: period-four—a second period-doubling bifurcation gives rise to a fundamental frequency of $\approx f_0/4$; (d) $G = 541\mu S$: spiral Chua's attractor—a cascade of period doublings results in a chaotic attractor that has a broadband power spectrum. Time plots: horizontal axis—t (ms); vertical axis—V_1 (V). Power spectra: horizontal axis—frequency (kHz); vertical axis—power [mean squared amplitude of $V_2(t)$] (dB).

Intermittency Route to Chaos. The route to chaos caused by saddle-node bifurcations comes in different forms, the common feature of which is a direct transition from regular motion to chaos. The most common type is the intermittency route and results from a single saddle-node bifurcation. This is a route and not just a jump because straight after the bifurcation, the trajectory is characterized by long intervals of almost regular motion (called laminar phases) and short bursts of irregular motion. The period of the oscillations is approximately equal to that of the system just before the bifurcation. This is shown in Fig. 38.15.

As the parameter passes through the critical value μ_c at the bifurcation point into the chaotic region, the laminar phases become shorter and the bursts become more frequent, until the regular intervals disappear altogether. The scaling law for the average interval of the laminar phases depends on $|\mu - \mu_c|$, so chaos is not fully developed until some distance from the bifurcation point [13].

Intermittency is best characterized in the time domain since its scaling law governs on the length of laminar phases.

Another type of bifurcation to chaos associated with saddle-nodes is the direct transition from a regular attractor (fixed point or limit cycle) to a coexisting chaotic one, without the phenomenon of intermittency.

Quasiperiodic (Torus Breakdown) Route to Chaos. The quasiperiodic route to chaos results from a sequence of Hopf bifurcations. Starting from a fixed point, the three-torus generated after three Hopf bifurcations is not stable in the sense that there exists an arbitrarily small perturbation of the a system (in terms of parameters) for which the three-torus gives way to chaos.

A quasiperiodic-periodic-chaotic sequence corresponding to torus breakdown in Chua's oscillator is shown in Fig. 38.5.

Quasiperiodicity is difficult to detect from a time series; it is more readily identified by means of a power spectrum or Poincaré map (see Fig. 38.11, for example).

Bifurcation Diagrams and Parameter Space Diagrams

While state-space, time- and frequency-domain measurements are useful for characterizing steady-state behaviors, nonlinear dynamics offers several other tools for summarizing qualitative information concerning bifurcations.

A bifurcation diagram is a plot of the attracting sets of a system versus a control parameter. Typically, one chooses a state variable and plots this against a single control parameter. In discrete systems, one simply plots successive values of a state variable. In the continuous-time case, some type of discretization is needed, typically by means of a Poincaré section.

Figure 38.16 shows a bifurcation diagram of the logistic map $X_{k+1} = \mu X_k(1 - X_k)$ for $\mu \in [2.5, 4]$ and $X_k \in [0, 1]$. Period doubling from period-one to period-two occurs at μ_2; the next two doublings in the period-doubling cascade occur at μ_2 and μ_4, respectively. A periodic window in the chaotic region is indicated by μ_3. The map becomes chaotic by the period-doubling route if μ is increased from μ_3 and by the intermittency route if μ is reduced out of the window.

When there is more than one control parameter in a system, the steady-state behavior may be summarized in a series of bifurcation diagrams, where one parameter is chosen as the control parameter, with the others held fixed, and only changed from one diagram to the next. This provides a complete but cumbersome representation of the dynamics [13].

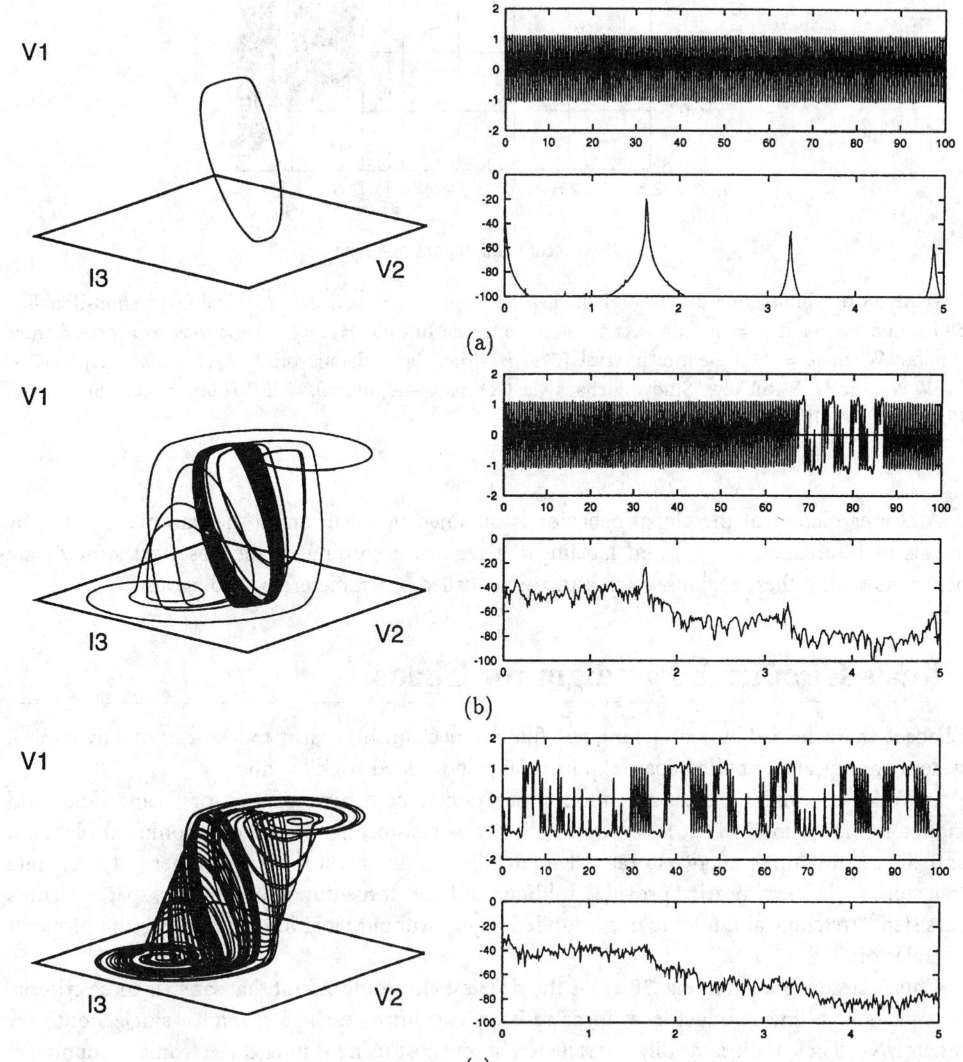

FIGURE 38.15 **Intermittency route to chaos** in **Chua's oscillator.** Simulated state-space trajectories, time waveforms $V_1(t)$, and power spectra of $V_2(t)$. (a) Periodic steady state—the signal is characterized by a discrete power spectrum with energy at integer multiples of the fundamental frequency; (b) onset of intermittency—the time signal contains long regular **"laminar" phases** and occasional **"bursts"** of irregular motion—in the frequency domain, intermittency manifests itself as a raising of the noise floor; (c) fully developed chaos—laminar phases are infrequent and the power spectrum is broad. Time plots: horizontal axis—t (ms); vertical axis—V_1 (V). Power spectra: horizontal axis—frequency (kHz); vertical axis—power [mean squared amplitude of $V_2(t)$] (dB).

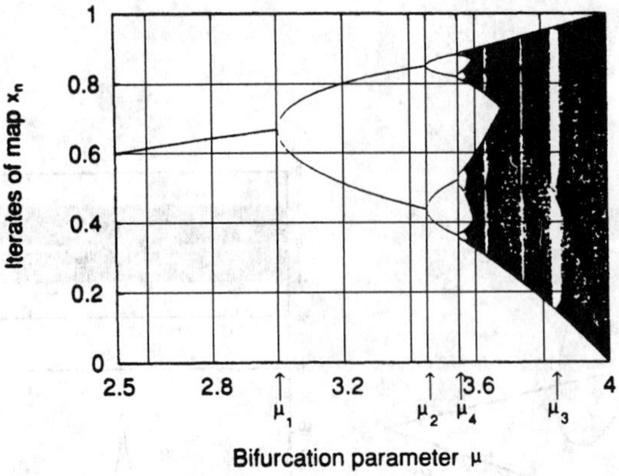

FIGURE 38.16 Bifurcation diagram for the logistic map: $X_{k+1} = \mu X_k(1 - X_k)$. The first period-doubling bifurcation occurs at $\mu = \mu_1$, the second at μ_2, and the third at $\mu_4 \cdot \mu_3$ corresponds to a period-three window. When $\mu = 4$, the entire interval (0, 1) is visited by a chaotic orbit $\{X_k, k = 0, 1, \ldots\}$. *Source:* C. W. Wu and N. F. Rul'kov, "Studying chaos via 1-D maps—A tutorial," *IEEE Trans. Circuits Syst. I,* vol. 40, p. 708, Oct. 1993. © 1993 IEEE.

A clearer picture of the global behavior is obtained by partitioning the parameter space by means of bifurcation curves, and labeling the regions according to the observed steady-state behaviors within these regions. Such a picture is called a parameter space diagram.

38.2 Chua's Circuit: A Paradigm for Chaos

Chaos is characterized by a stretching and folding mechanism; nearby trajectories of a dynamical system are repeatedly pulled apart exponentially and folded back together.

In order to exhibit chaos, an autonomous circuit consisting of resistors, capacitors, and inductors must contain i) at least one locally active resistor, ii) at least one nonlinear element, and iii) at least three energy-storage elements. The active resistor supplies energy to separate trajectories, the nonlinearity provides folding, and the three-dimensional state space permits persistent stretching and folding in a bounded region without violating the noncrossing property of trajectories.

Chua's circuit (shown in Fig. 38.17) is the simplest electronic circuit that satisfies these criteria. It consists of a linear inductor, a linear resistor, two linear capacitors, and a single nonlinear resistor N_R. The circuit is readily constructed at low cost using standard electronic components and exhibits a rich variety of bifurcations and chaos [10].

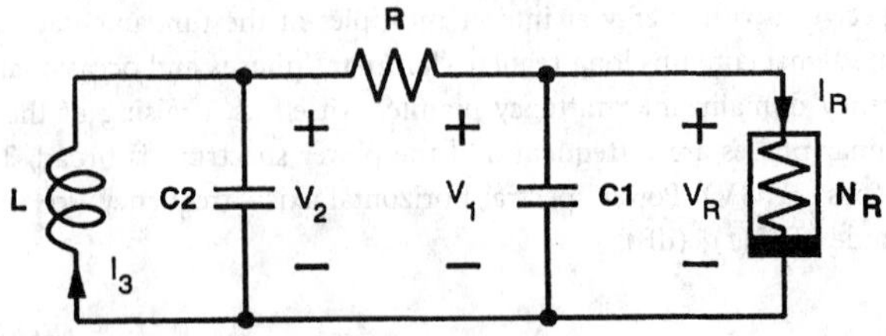

FIGURE 38.17 Chua's circuit consists of a linear inductor L, two linear capacitors (C_2, C_1), a linear resistor R, and a voltage-controlled nonlinear resistor N_R.

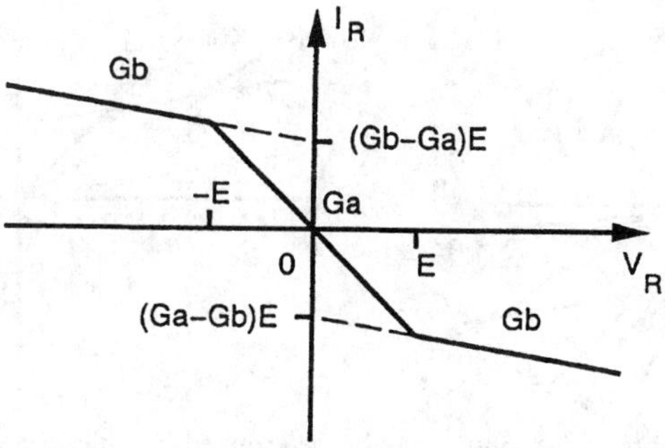

FIGURE 38.18 The driving-point characteristic of the nonlinear resistor N_R in Chua's circuit has breakpoints at $\pm E$ and slopes G_a and G_b in the inner and outer regions, respectively.

Dynamics of Chua's Circuit

State Equations

Chua's circuit may be described by three ordinary differential equations. Choosing V_1, V_2, and I_3 as state variables, we write

$$\begin{aligned}\frac{dV_1}{dt} &= \frac{G}{C_1}(V_2 - V_1) - \frac{1}{C_1} f(V_1)\\ \frac{dV_2}{dt} &= \frac{G}{C_2}(V_1 - V_2) + \frac{1}{C_2} I_3\\ \frac{dI_3}{dt} &= -\frac{1}{L} V_2\end{aligned} \tag{38.12}$$

where $G = 1/R$ and $f(V_R) = G_b V_R + 1/2(G_a - G_b)(|V_R + E| - |V_R - E|)$, as shown in Fig. 38.18.

Because of the piecewise-linear nature of N_R, the vector field of Chua's circuit may be decomposed into three distinct affine regions: $V_1 < -E$, $|V_1| \leq E$, and $V_1 > E$. We call these the D_{-1}, D_0, and D_1 regions, respectively. The global dynamics may be determined by considering separately the behavior in each of the three regions (D_{-1}, D_0, and D_1) and then gluing the pieces together along the boundary planes U_{-1} and U_1.

Piecewise-Linear Dynamics

In each region, the circuit is governed by a three-dimensional autonomous affine dynamical system of the form

$$\dot{\mathbf{X}} = \mathbf{AX} + \mathbf{b} \tag{38.13}$$

where $\mathbf{A}$ is the (constant) system matrix and $\mathbf{b}$ is a constant vector.

The equilibrium points of the circuit may be determined graphically by intersecting the load line $I_R = -GV_R$ with the DP characteristic $I_R = f(V_R)$ of the nonlinear resistor N_R, as shown in Fig. 38.19 [2]. When $G > |G_a|$ or $G < |G_b|$, the circuit has a unique equilibrium point at the origin (and two virtual equilibria P_- and P_+); otherwise, it has three equilibrium points at P_-, 0, and P_+.

The dynamics close to an equilibrium point $\mathbf{X_Q}$ are governed locally by the linear system

$$\dot{\mathbf{x}} = \mathbf{Ax} \tag{38.14}$$

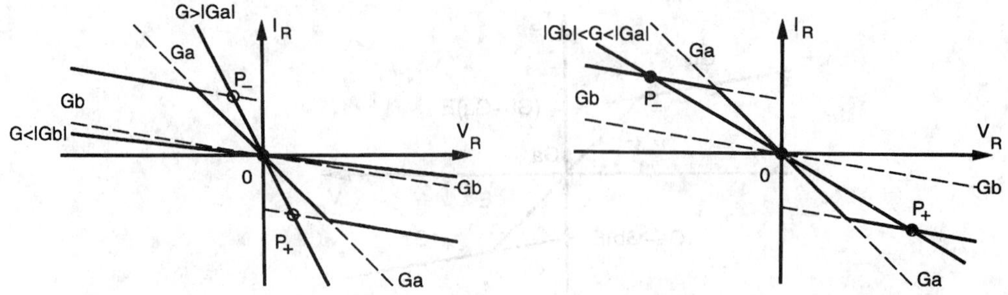

FIGURE 38.19 DC equilibrium points of Fig. 38.17 may be determined graphically by intersecting the load line $I_R = -GV_R$ with the DP characteristic of N_R. (a) If $G > |G_a|$ or $G < |G_b|$, the circuit has a unique equilibrium point at the origin (P_- and P_+ are virtual equilibria in this case). (b) When $|G_b| < G < |G_a|$, the circuit has three equilibrium points at P_-, 0, and P_+.

If the eigenvalues λ_1, λ_2, and λ_3 of $\mathbf{A}$ are distinct, then every solution $\mathbf{x}(t)$ of (38.14) may be expressed in the form

$$\mathbf{x}(t) = c_1 \exp(\lambda_1 t)\vec{\xi}_1 + c_2 \exp(\lambda_2 t)\vec{\xi}_2 + c_3 \exp(\lambda_3 t)\vec{\xi}_3$$

where $\vec{\xi}_1$, $\vec{\xi}_2$, and $\vec{\xi}_3$ are the (possibly complex) eigenvectors associated with the eigenvalues λ_1, λ_2, and λ_3, respectively, and the c_k's are (possibly complex) constants that depend on the initial state $\mathbf{X}_0$.

In the special case when $\mathbf{A}$ has one real eigenvalue γ and a complex conjugate pair of eigenvalues $\sigma \pm j\omega$, the solution of (38.14) has the form

$$\mathbf{x}(t) = c_r \exp(\gamma t)\vec{\xi}_\gamma + 2c_c \exp(\sigma t)[\cos(\omega t + \phi_c)\vec{\eta}_r - \sin(\omega t + \phi_c)\vec{\eta}_i]$$

where $\vec{\eta}_r$ and $\vec{\eta}_i$ are the real and imaginary parts of the eigenvectors associated with the complex conjugate pair of eigenvalues, $\vec{\xi}_\gamma$ is the eigenvector defined by $\mathbf{A}\vec{\xi}_\gamma = \gamma\vec{\xi}_\gamma$, and c_r, c_c, and ϕ_c are real constants that are determined by the initial conditions.

Let us relabel the real eigenvector E^r, and define E^c as the complex eigenplane spanned by $\vec{\eta}_r$ and $\vec{\eta}_i$.

We can think of the solution $\mathbf{x}(t)$ of (38.14) as being the sum of two distinct components $\mathbf{x}_r(t) \in E^r$ and $\mathbf{x}_c(t) \in E^c$:

$$\begin{aligned}\mathbf{x}_r(t) &= c_r \exp(\gamma t)\vec{\xi}_\gamma \\ \mathbf{x}_c(t) &= 2c_c \exp(\sigma t)[\cos(\omega t + \phi_c)\vec{\eta}_r - \sin(\omega t + \phi_c)\vec{\eta}_i]\end{aligned}$$

The complete solution $\mathbf{X}(t)$ of (38.13) may be found by translating the origin of the linearized coordinate system to the equilibrium point $\mathbf{X}_Q$. Thus,

$$\begin{aligned}\mathbf{X}(t) &= \mathbf{X}_Q + \mathbf{x}(t) \\ &= \mathbf{X}_Q + \mathbf{x}_r(t) + \mathbf{x}_c(t)\end{aligned}$$

We can determine the qualitative behavior of the complete solution $\mathbf{X}(t)$ by considering separately the components $\mathbf{x}_r(t)$ and $\mathbf{x}_c(t)$ along E^r and E^c, respectively.

If $\gamma > 0$, $\mathbf{x}_r(t)$ grows exponentially in the direction of E^r; if $\gamma < 0$, the component $\mathbf{x}_r(t)$ tends asymptotically to zero. When $\sigma > 0$ and $\omega \neq 0$, $\mathbf{x}_c(t)$ spirals away from $\mathbf{X}_Q$ along the complex eigenplane E^c, and if $\sigma < 0$, $\mathbf{x}_c(t)$ spirals toward $\mathbf{X}_Q$ along E^c.

We remark that the vector E^r and plane E^c are invariant under the flow of (38.13): if $\mathbf{X}(0) \in E^r$, then $\mathbf{X}(t) \in E^r$ for all t; if $\mathbf{X}(0) \in E^c$, then $\mathbf{X}(t) \in E^c$ for all t. An important consequence of this is that a trajectory $\mathbf{X}(t)$ cannot cross through the complex eigenspace E^c; suppose $\mathbf{X}(t_0) \in E^c$ at some time t_0, then $\mathbf{X}(t) \in E^c$ for all $t > t_0$.

Chaos in Chua's Circuit

In the following discussion, we consider a fixed set of component values: $L = 18$ mH, $C_2 = 100$ nF, $C_1 = 10$ nF, $G_a = -50/66$ mS $= -757.576$ μS, $G_b = -9/22$ $mS = -409.091$ μS, and $E = 1$ V. When $G = 550$ μS, there are three equilibrium points at P_+, 0, and P_-. The equilibrium point at the origin (0) has one unstable real eigenvalue γ_0 and a stable complex pair $\sigma_0 \pm j\omega_0$. The outer equilibria (P_- and P_+) each have a stable real eigenvalue γ_1 and an unstable complex pair $\sigma_1 \pm j\omega_1$.

Dynamics of D_0

A trajectory starting from some initial state $\mathbf{X}_0$ in the D_0 region may be decomposed into its components along the complex eigenplane $E^c(0)$ and along the eigenvector $E^r(0)$. When $\gamma_0 > 0$ and $\sigma_0 < 0$, the component along $E^c(0)$ spirals toward the origin along this plane while the component in the direction $E^r(0)$ grows exponentially. Adding the two components, we see that a trajectory starting slightly above the stable complex eigenplane $E^c(0)$ spirals toward the origin along the $E^c(0)$ direction, all the while being pushed away from $E^c(0)$ along the unstable direction $E^r(0)$. As the (stable) component along $E^c(0)$ shrinks in magnitude, the (unstable) component grows exponentially, and the trajectory follows a helix of exponentially decreasing radius whose axis lies in the direction of $E^r(0)$; this is illustrated in Fig. 38.20.

Dynamics of D_{-1} and D_1

Associated with the stable real eigenvalue γ_1 in the D_1 region is the eigenvector $E^r(P_+)$. The real and imaginary parts of the complex eigenvectors associated with $\sigma_1 \pm j\omega_1$ define a complex eigenplane $E^c(P_+)$.

A trajectory starting from some initial state $\mathbf{X}_0$ in the D_1 region may be decomposed into its components along the complex eigenplane $E^c(P_+)$ and the eigenvector $E^r(P_+)$. When $\gamma_1 < 0$ and $\sigma_1 > 0$, the component on $E^c(P_+)$ spirals away from P_+ along this plane while the component in the direction of $E^r(0)$ tends asymptotically toward P_+. Adding the two components, we see that a trajectory starting close to the stable real eigenvector $E^r(P_+)$ above the complex eigenplane moves toward $E^c(P_+)$ along a helix of exponentially increasing radius. Since the component along $E^r(P_+)$ shrinks exponentially in magnitude and the component on $E^c(P_+)$ grows exponentially, the trajectory is quickly flattened onto $E^c(P_+)$, where it spirals away from P_+ along the complex eigenplane; this is illustrated in Fig. 38.21.

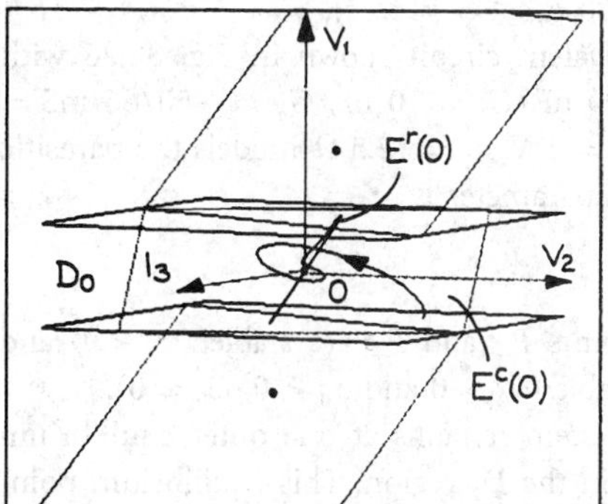

FIGURE 38.20 Dynamics of the D_0 region. A trajectory starting slightly above the stable complex eigenplane $E^c(0)$ spirals toward the origin along this plane and is repelled close to 0 in the direction of the unstable eigenvector $E^r(0)$. *Source:* M. P. Kennedy, "Three steps to chaos—Part II: A Chua's circuit primer," *IEEE Trans. Circuits Syst. I*, vol. 40, p. 660, Oct. 1993. © 1993 IEEE.

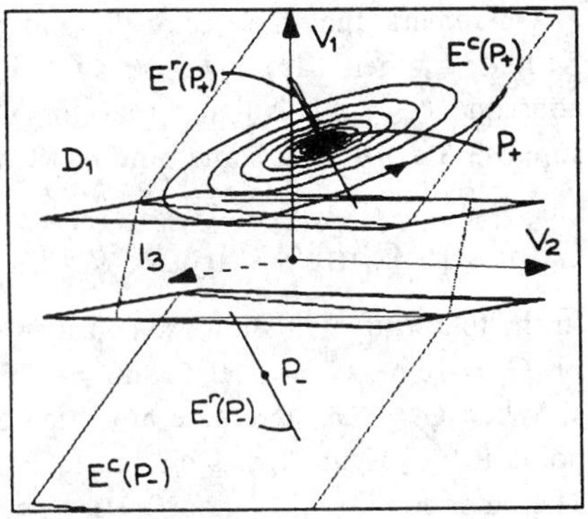

FIGURE 38.21 Dynamics of the D_1 region. A trajectory starting above the unstable complex eigenplane $E^c(P_+)$ close to the eigenvector $E^r(P_+)$ moves toward the plane and spirals away from P_+ along $E^c(P_+)$. By symmetry, the D_{-1} region has equivalent dynamics. *Source:* M. P. Kennedy, "Three steps to chaos—Part II: A Chua's circuit primer," *IEEE Trans. Circuits Syst.*, vol. 40, p. 662, Oct. 1993. © 1993 IEEE.

By symmetry, the equilibrium point P_- in the D_{-1} region has three eigenvalues: γ_1 and $\sigma_1 \pm j\omega_1$. The eigenvector $E^r(P_-)$ is associated with the stable real eigenvalue γ_1; the real and imaginary parts of the eigenvectors associated with the unstable complex pair $\sigma_1 \pm j\omega_1$ define an eigenplane $E^c(P_-)$, along which trajectories spiral away from P_-.

Global Dynamics

With the given set of parameter values, the equilibrium point at the origin has an unstable real eigenvalue and a stable pair of complex conjugate eigenvalues; the outer equilibrium point P_- has a stable real eigenvalue and an unstable complex pair.

In particular, P_- has a pair of unstable complex conjugate eigenvalues $\sigma_1 \pm \omega_1$ $(\sigma_1 > 0, \omega_1 \neq 0)$ and a stable real eigenvalue γ_1, where $|\sigma_1| < |\omega_1|$. In order to prove that the circuit is chaotic in the sense of Shil'nikov, it is necessary to show that it possesses a homoclinic orbit for this set of parameter values.

A trajectory starting on the eigenvector $E^r(0)$ close to 0 moves away from the equilibrium point until it crosses the boundary U_1 and enters D_1, as shown in Fig. 38.20. If this trajectory is folded back into D_0 by the dynamics of the outer region, and reinjected toward 0 along the stable complex eigenplane $E^c(0)$, then a homoclinic orbit is produced.

That Chua's circuit is chaotic in the sense of Shil'nikov was first proven by Chua *et al.* [21] in 1985. Since then, there has been an intensive effort to understand every aspect of the dynamics of this circuit with a view to developing it as a paradigm for learning, understanding, and teaching about nonlinear dynamics and chaos [3].

Steady States and Bifurcations in Chua's Circuit

In the following discussion, we consider the global behavior of the circuit using our chosen set of parameters with R in the range $0 \leq R \leq 2000\ \Omega$ ($500\ \mu$ s$\leq G < \infty$ s).

Figure 38.14 is a series of simulations of the equivalent circuit shown in Fig. 38.26 with the following parameter values: $L = 18$ mH, $C_2 = 100$ nF, $C_1 = 10$ nF, $G_a = -50/66$ mS $= -757.576\ \mu$S, $G_b = -9/22$ mS $= -409.091\ \mu$S, and $E = 1$ V. $R_0 = 12.5\ \Omega$ models the parasitic series resistance of a real inductor. R is the bifurcation parameter.

Equilibrium Point and Hopf Bifurcation

When R is large (2000 Ω), the outer equilibrium points P_- and P_+ are stable ($\gamma_1 < 0$ and $\sigma_1 < 0$, $\omega_1 \neq 0$); the inner equilibrium point 0 is unstable ($\gamma_0 > 0$ and $\sigma_0 < 0$, $\omega_0 \neq 0$).

Depending on the initial state of the circuit, the system remains at one outer equilibrium point or the other. Let us assume that we start at P_+ in the D_1 region. This equilibrium point has one negative real eigenvalue and a complex pair with negative real parts. The action of the negative real eigenvalue γ_1 is to squeeze trajectories down onto the complex eigenplane $E^c(P_+)$, where they spiral toward the equilibrium point P_+.

As the resistance R is decreased, the real part of the complex pair of eigenvalues changes sign and becomes positive. Correspondingly, the outer equilibrium points become unstable as σ_1 passes through 0; this is a Hopf-like bifurcation.[13] The real eigenvalue of P_+ remains negative so trajectories in the D_1 region converge toward the complex eigenplane $E^c(P_+)$. However, they spiral away from the equilibrium point P_+ along $E^c(P_+)$ until they reach the dividing plane U_1 (defined by $V_1 \equiv E$) and enter the D_0 region.

The equilibrium point at the origin in the D_0 region has a stable complex pair of eigenvalues and an unstable real eigenvalue. Trajectories that enter the D_0 region on the complex eigenplane $E^c(0)$ are attracted to the origin along this plane. Trajectories that enter D_0 from D_1 below or above the eigenplane either cross over to D_{-1} or are turned back toward D_1, respectively. For R sufficiently large, trajectories that spiral away from P_+ along $E^c(P_+)$ and enter D_0 above $E^c(0)$ are returned to D_1, producing a stable period-one limit cycle. This is illustrated in Fig. 38.14.

Period-Doubling Cascade

As the resistance R is decreased further, a period-doubling bifurcation occurs. The limit cycle now closes on itself after encircling P_+ twice; this is called a period-two cycle because a trajectory takes approximately twice the time to complete this closed orbit as to complete the preceding period-one orbit [see Fig. 38.14(b)].

Decreasing the resistance R still further produces a cascade of period-doubling bifurcations to period-four [Fig. 38.14(c)], period-eight, period-sixteen, and so on until an orbit of infinite period is reached, beyond which we have chaos [see Fig. 38.14(d)]. This is a spiral Chua's **chaotic attractor.**

The spiral Chua's attractor in Fig. 38.14(d) looks like a ribbon or band that is smoothly folded on itself; this folded band is the simplest type of chaotic attractor [18]. A trajectory from an initial condition $\mathbf{X}_0$ winds around the strip repeatedly, returning close to $\mathbf{X}_0$, but never closing on itself.

Periodic Windows

Between the chaotic regions in the parameter space of Chua's circuit, there exist ranges of the bifurcation parameter R over which stable periodic motion occurs. These regions of periodicity are called periodic windows and are similar to those that exist in the bifurcation diagram of the logistic map (see Fig. 38.16).

Periodic windows of periods three and five are readily found in Chua's circuit. These limit cycles undergo period-doubling bifurcations to chaos as the resistance R is decreased.

For certain sets of parameters, Chua's circuit follows the intermittency route to chaos as R is increased out of the period-three window.

Spiral Chua's Attractor

Figure 38.22 shows three views of another simulated spiral Chua's chaotic attractor. Figure 38.22(b) is a view along the edge of the outer complex eigenplanes $E^c(P_+)$ and $E^c(P_-)$; notice how trajectories in the D_1 region are compressed toward the complex eigenplane $E^c(P_+)$ along the direction of the stable real eigenvector $E^r(P_+)$ and they spiral away from the equilibrium point P_+ along $E^c(P_+)$.

When a trajectory enters the D_0 region through U_1 from D_1, it is twisted around the unstable real eigenvector $E^r(0)$ and returned to D_1.

[13]Recall that the Hopf bifurcation theorem strictly applies only for sufficiently smooth systems but that physical implementations of piecewise-linear characteristics are typically smooth.

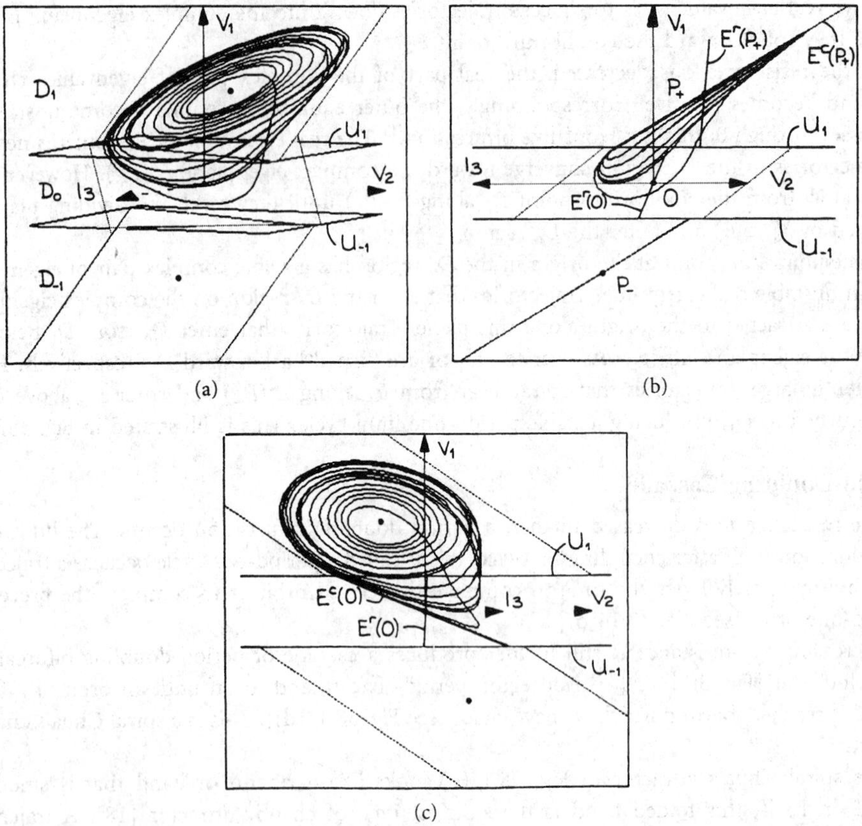

FIGURE 38.22 Three views of a simulated spiral Chua's attractor in Chua's oscillator with $G = 550\mu S$. (a) Reference view [compare with Fig. 38.14(d)]. (b) View along the edge of the outer complex eigenplanes $E^c(P_+)$ and $E^c(P_-)$; note how the trajectory in D_1 is flattened onto $E^c(P_+)$. (c) View along the edge of the complex eigenplane $E^c(0)$; trajectories cannot cross this plane. *Source:* M. P. Kennedy, "Three steps to chaos—Part II: A Chua's circuit primer," *IEEE Trans. Circuits Syst. I*, vol. 40, p. 664, Oct. 1993. © 1993 IEEE.

Figure 38.22(c) shows clearly that when the trajectory enters D_0 from D_1, it crosses U_1 above the eigenplane $E^c(0)$. The trajectory cannot cross through this eigenplane and therefore it must return to the D_1 region.

Double-Scroll Chua's Attractor

Because we chose a nonlinear resistor with a symmetric nonlinearity, every attractor that exists in the D_1 and D_0 regions has a counterpart (mirror image) in the D_{-1} and D_0 regions. As the coupling resistance R is decreased further, the spiral Chua's attractor "collides" with its mirror image and the two merge to form a single compound attractor called a double-scroll Chua's chaotic attractor [10], as shown in Fig. 38.23.

Once more, we show three views of this attractor in order to illustrate its geometrical structure. Figure 38.23(b) is a view of the attractor along the edge of the outer complex eigenplanes $E^c(P_+)$ and $E^c(P_-)$. Upon entering the D_1 region from D_0, the trajectory collapses onto $E^c(P_+)$ and spirals away from P_+ along this plane.

Figure 38.23(c) is a view of the attractor along the edge of the complex eigenplane $E^c(0)$ in the inner region. Notice once more that when the trajectory crosses U_1 into D_0 above $E^c(0)$, it must remain above $E^c(0)$ and so returns to D_1. Similarly, if the trajectory crosses U_1 below

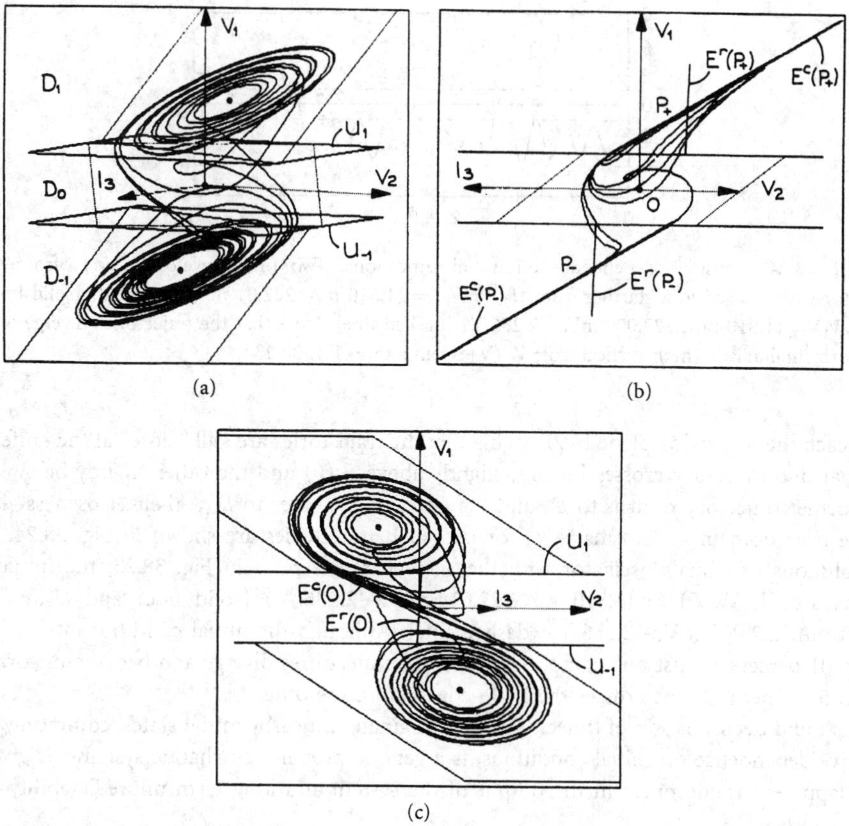

FIGURE 38.23 Three views of a simulated double-scroll Chua's attractor in Chua's oscillator with $G = 565\mu S$. (a) Reference view [compare to Fig. 38.14(d)]. (b) View along the edge of the outer complex eigenplanes $E^c(P_+)$ and $E^c(P_-)$; note how the trajectory in D_1 is flattened onto $E^c(P_+)$ and onto $E^c(P_-)$ in D_{-1}. (c) View along the edge of the complex eigenplane $E^c(0)$; a trajectory entering D_0 from D_1 above this plane returns to D_1 while one entering D_0 below $E^c(0)$ crosses to D_{-1}. *Source:* M. P. Kennedy, "Three steps to chaos—Part II: A Chua's circuit primer," *IEEE Trans. Circuits Syst. I*, vol. 40, p. 665, Oct. 1993. © 1993 IEEE.

$E^c(0)$, it must remain below $E^c(0)$ and therefore crosses over to the D_{-1} region. Thus, $E^c(0)$ presents a knife-edge to the trajectory as it crosses U_1 into the D_0 region, forcing it back toward D_1 or across D_0 to D_{-1}.

Boundary Crisis

Reducing the resistance R still further produces more regions of chaos, interspersed with periodic windows. Eventually, for a sufficiently small value of R, the unstable saddle trajectory that normally resides outside the stable steady-state solution collides with the double-scroll Chua's attractor and a blue sky catastrophe called a boundary crisis [10] occurs. After this, all trajectories become unbounded.

Manifestations of Chaos

Sensitive Dependence on Initial Conditions

Consider once more the double-scroll Chua's attractor shown in Fig. 38.23. Two trajectories starting from distinct but almost identical initial states in D_1 will remain "close together" until

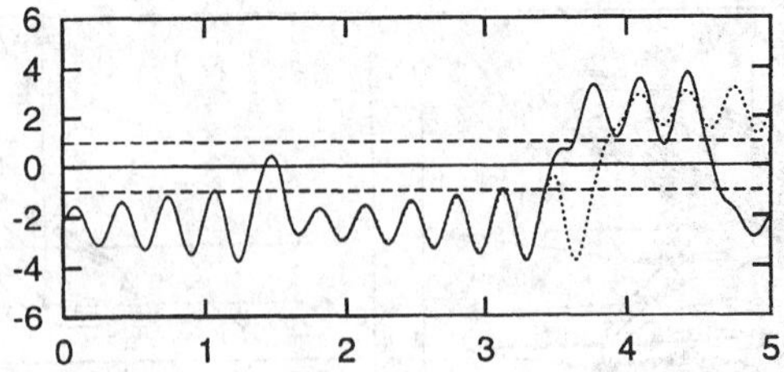

FIGURE 38.24 Sensitive dependence on initial conditions. Two time waveforms $V_1(t)$ from Chua's oscillator with $G = 550\mu S$, starting from $(I_3, V_2, V_1) = (1.810\text{ mA}, 222.01\text{ mV}, -2.286\text{ V})$ [solid line] and $(I_3, V_2, V_1) = (1.810\text{ mA}, 222.000\text{ mV}, -2.286\text{ V})$ [dashed line]. Note that the trajectories diverge within 5 ms. Horizontal axis: t (ms); vertical axis: V_1 (V). Compare to Fig. 38.23.

they reach the separating plane U_1. Imagine that the trajectories are still "close" at the knife-edge, but that one trajectory crosses into D_0 slightly above $E^c(0)$ and the other slightly below $E^c(0)$. The former trajectory returns to D_1 and the latter crosses over to D_{-1}: their "closeness" is lost.

The time-domain waveforms $V_1(t)$ for two such trajectories are shown in Fig. 38.24. These are solutions of Chua's oscillator with the same parameters as in Fig. 38.23; the initial conditions are $(I_3, V_2, V_1) = (1.810\text{ mA}, 222.014\text{ mV}, -2.286\text{ V})$ [solid line] and $(I_3, V_2, V_1) = (1.810\text{ mA}, 222.000\text{ mV}, -2.286\text{ V})$ [dashed line]. Although the initial conditions differ by less than 0.01 percent in just one component (V_2), the trajectories diverge and become uncorrelated within 5 ms because one crosses the knife-edge before the other.

This rapid decorrelation of trajectories that originate in nearby initial states, commonly called sensitive dependence on initial conditions, is a generic property of chaotic systems. It gives rise to an apparent randomness in the output of the system and long-term unpredictability of the state.

"Randomness" in the Time Domain

Figures 38.14(a), (b), (c), and (d) show the state-space trajectories of period-one, period-two, and period-four periodic attractors, a spiral Chua's chaotic attractor, respectively, and the corresponding voltage waveforms $V_1(t)$.

The "period-one" waveform is periodic; it looks like a slightly distorted sinusoid. The "period-two" waveform is also periodic. It differs qualitatively from the "period-one" in that the pattern of a large peak followed by a small peak repeats approximately once every two cycles of the period-one signal; that is why it is called "period-two."

In contrast with these periodic time waveforms, $V_1(t)$ for the spiral Chua's attractor is quite irregular and does not appear to repeat itself in any observation period of finite length. Although it is produced by a third-order deterministic differential equation, the solution looks "random."

Broadband "Noise-Like" Power Spectrum

In the following discussion, we consider 8192 samples of $V_2(t)$ recorded at 200 kHz; leakage in the power spectrum is controlled by applying a Welch window [17] to the data.

We remarked earlier that the period-one time waveform corresponding to the attractor in Fig. 38.14(a), is almost sinusoidal; we expect, therefore, that most of its power should be concentrated at the fundamental frequency. The power spectrum of the period-one waveform $V_2(t)$ shown in Fig. 38.14(a) consists of a sharp spike at approximately 3 kHz and higher harmonic components which are over 30 dB below the fundamental.

Because the period-two waveform repeats roughly once every 0.67 ms, this periodic signal has a fundamental frequency component at approximately 1.5 kHz [see Fig. 38.14(b)]. Notice, however, that most of the power in the signal is concentrated close to 3 kHz.

The period-four waveform repeats roughly once every 1.34 ms, corresponding to a fundamental frequency component at approximately 750 Hz [see Fig. 38.14(c)]. Note once more that most of the power in the signal is still concentrated close to 3 kHz.

The spiral Chua's attractor is qualitatively different from these periodic signals. The aperiodic nature of its time-domain waveforms is reflected in the broadband noise-like power spectrum [Fig. 38.14(d)]. No longer is the power of the signal concentrated in a small number of frequency components; rather, it is distributed over a broad range of frequencies. This broadband structure of the power spectrum persists even if the spectral resolution is increased by sampling at a higher frequency f_s. Notice that the spectrum still contains a peak at approximately 3 kHz that corresponds to the average frequency of rotation of the trajectory about the fixed point.

Practical Realization of Chua's Circuit

Chua's circuit can be realized in a variety of ways using standard or custom-made electronic components. All of the linear elements (capacitor, resistor, and inductor) are readily available as two-terminal devices. A nonlinear resistor N_R with the prescribed DP characteristic (called a Chua diode [10]) may be implemented by connecting two negative resistance converters in parallel as shown in Fig. 38.25. A complete list of components is given in Table 38.2.

The op amp subcircuit consisting of A_1, A_2 and R_1–R_6 functions as **a negative resistance converter** N_R with driving-point characteristic as shown in Fig. 38.28(b). Using two 9 V batteries to power the op amps gives $V^+ = 9$ V and $V^- = -9$ V. From measurements of the saturation levels of the AD712 outputs, $E_{sat} \approx 8.3$ V, giving $E \approx 1$ V. With $R_2 = R_3$ and $R_5 = R_6$, the nonlinear characteristic is defined by $G_a = -1/R_1 - 1/R_4 = -50/66$ mS, $G_b = 1/R_3 - 1/R_4 = -9/22$ mS, and $E = R_1 E_{sat}/(R_1 + R_2) \approx 1$ V [10].

The equivalent circuit of Fig. 38.25 is shown in Fig. 38.26, where the real inductor is modeled as a series connection of an ideal linear inductor L and a linear resistor R_0. When the inductor's resistance is modeled explicitly in this way, the circuit is called **Chua's oscillator** [5].

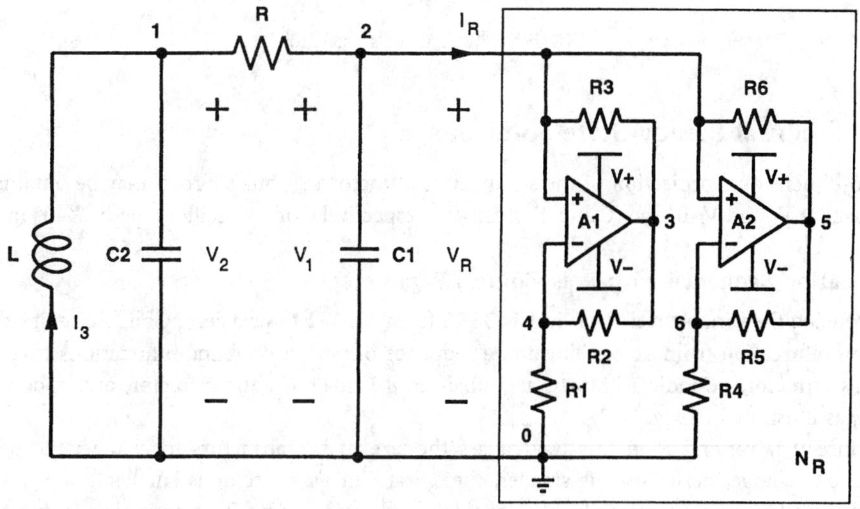

FIGURE 38.25 Practical implementation of Chua's circuit using two op amps and six resistors to realize the Chua diode [10]. Component values are listed in Table 38.2.

TABLE 38.2 Component List for the Practical Implementation of Chua's Circuit, shown in Fig. 38.25

Element	Description	Value	Tolerance
A_1	Op Amp $\left(\frac{1}{2}\text{AD712, TL082, or Equivalent}\right)$		
A_2	Op Amp $\left(\frac{1}{2}\text{AD712, TL082, or Equivalent}\right)$		
C_1	Capacitor	10 nF	±5%
C_2	Capacitor	100 nF	±5%
R	Potentiometer	2 kΩ	
R_1	$\frac{1}{4}$ W Resistor	3.3 kΩ	±5%
R_2	$\frac{1}{4}$ W Resistor	22 kΩ	±5%
R_3	$\frac{1}{4}$ W Resistor	22 kΩ	±5%
R_4	$\frac{1}{4}$ W Resistor	2.2 kΩ	±5%
R_5	$\frac{1}{4}$ W Resistor	220 Ω	±5%
R_6	$\frac{1}{4}$ W Resistor	220 Ω	±5%
L	Inductor (TOKO-Type 10 RB, or Equivalent)	18 mH	±10%

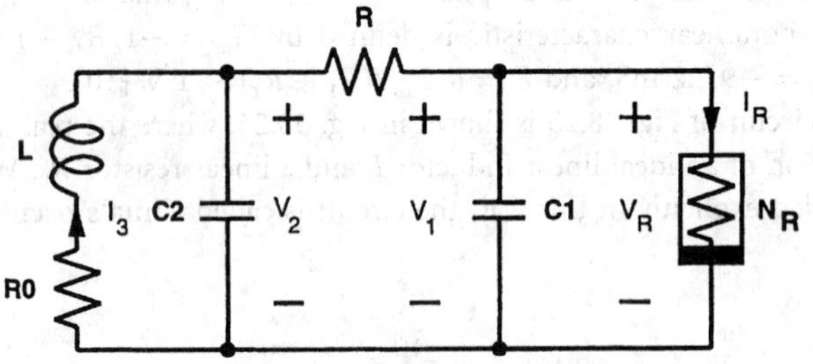

FIGURE 38.26 Chua's oscillator.

Experimental Steady-State Solutions

A two-dimensional projection of the steady-state attractor in Chua's circuit may be obtained by connecting V_2 and V_1 to the X and Y channels, respectively, of an oscilloscope in X–Y mode.

Bifurcation Sequence with R as Control Parameter

By reducing the variable resistor R in Fig. 38.25 from 2000 Ω toward zero, Chua's circuit exhibits a Hopf bifurcation from dc equilibrium, a sequence of period-doubling bifurcations to a spiral Chua's attractor, periodic windows, a double-scroll Chua's chaotic attractor, and a boundary crisis, as illustrated in Fig. 38.27.

Notice that varying R in this way causes the size of the attractors to change: the period-one orbit is large, period-two is smaller, the spiral Chua's attractor is smaller again, and the double-scroll Chua's attractor shrinks considerably before it dies. This shrinking is due to the equilibrium points P_+ and P_- moving closer toward the origin as R is decreased. Consider the load line in Fig. 38.19(b): as R is decreased, the slope G increases, and the equilibrium points

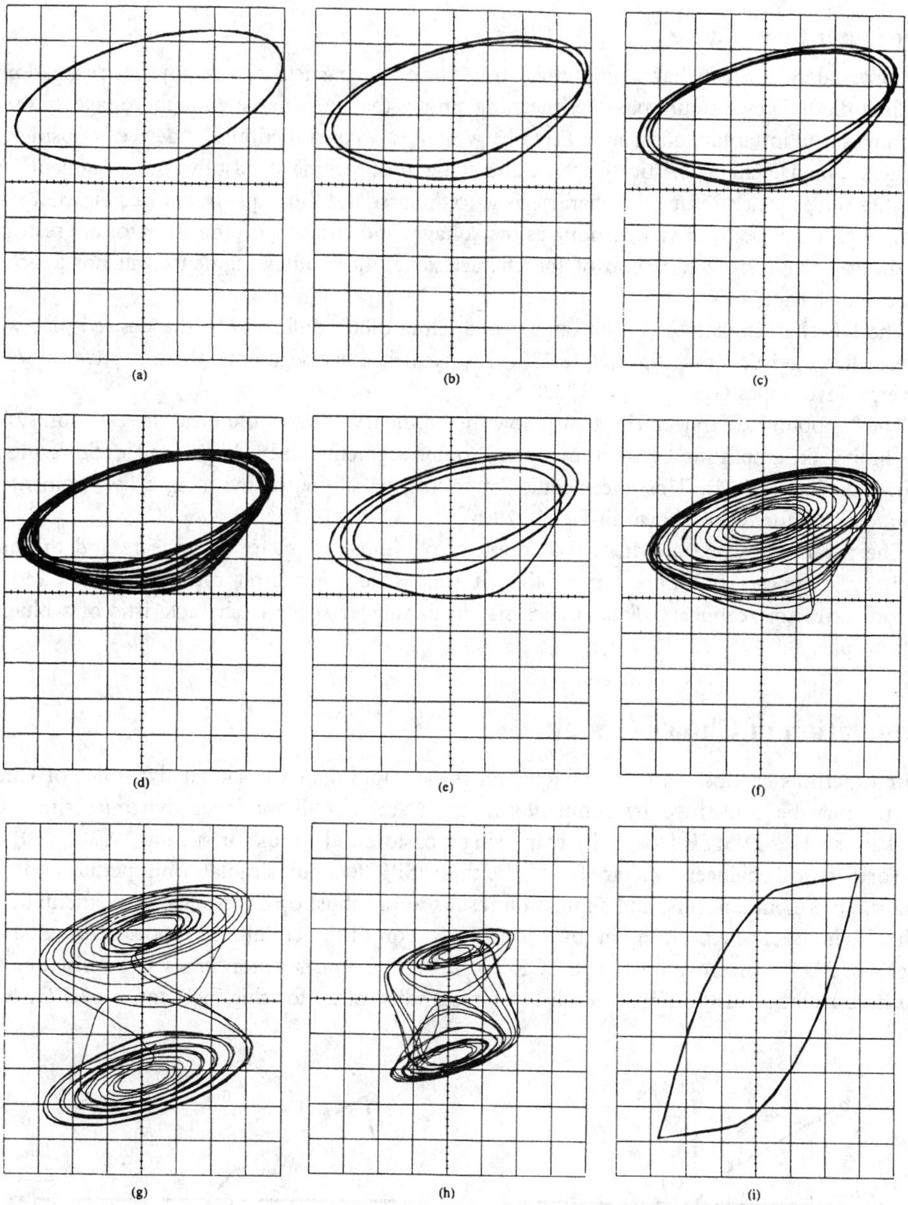

FIGURE 38.27 Typical experimental bifurcation sequence in Chua's circuit (component values as in Table 38.2) recorded using a digital storage oscilloscope. Horizontal axis V_2 (a)–(h) 200 mV/div, (i) 2 V/div; vertical axis V_1 (a)–(h) 1 V/div, (i) 2 V/div. (a) $R = 1.83$ kΩ, period-1; (b) $R = 1.82$ kΩ, period-2; (c) $R = 1.81$ kΩ, period-4; (d) $R = 1.80$ kΩ, spiral Chua's attractor; (e) $R = 1.797$ kΩ, period-3 window; (f) $R = 1.76$ kΩ, spiral Chua's attractor; (g) $R = 1.73$ kΩ, double-scroll Chua's attractor; (h) $R = 1.52$ kΩ, double-scroll Chua's attractor; (i) $R = 1.42$ kΩ, large limit cycle corresponding to the outer segments of the Chua diode's DP characteristic. *Source:* M. P. Kennedy, "Three steps to chaos—Part II: A Chua's circuit primer," *IEEE Trans. Circuit Syst. I,* vol. 40, pp. 669, 670, Oct. 1993. © 1993 IEEE.

P_- and P_+ move toward the origin. Compare also the positions of P_+ in Figs. 38.22(a) and 38.23(a).

The Outer Limit Cycle

No physical system can have unbounded trajectories. In particular, any physical realization of a Chua diode is **eventually passive**, meaning simply that for a large enough voltage across its terminals, the instantaneous power $P_R(t)$ [$= V_R(t)I_R(t)$] consumed by the device is positive.

Hence, the DP characteristic of a real Chua diode must include at least two outer segments with positive slopes which return the characteristic to the first and third quadrants [see Fig. 38.28(b)]. From a practical point of view, as long as the voltages and currents on the attractor are restricted to the negative resistance region of the characteristic, these outer segments will not affect the circuit's behavior.

The DP characteristic of the op amp-based Chua diode differs from the desired piecewise-linear characteristic shown in Fig. 38.28(a) in that it has five segments, the outer two of which have positive slopes $G_c = 1/R_5 = 1/220$ S.

The "unbounded" trajectories that follow the boundary crisis in the ideal three-region system are limited in amplitude by these dissipative outer segments and a large limit cycle results, as shown in Fig. 38.27(i). This effect could, of course, be simulated by using a five-segment DP characteristic for N_R as shown in Fig. 38.28(b).

The parameter value at which the double-scroll Chua's attractor disappears and the outer limit cycle appears is different from that at which the outer limit cycle disappears and the chaotic attractor reappears. This "hysteresis" in parameter space is characteristic of a blue sky catastrophe.

Simulation of Chua's Circuit

Our experimental observations and qualitative description of the global dynamics of Chua's circuit may be confirmed by simulation using a specialized nonlinear dynamics simulation package such as **INSITE** [15] or by employing a customized simulator such as "**ABC**" [10].

For electrical engineers who are familiar with the **SPICE** circuit simulator but perhaps not with chaos, we present a net-list and simulation results for a robust op amp-based implementation of Chua's circuit. The AD712 op amps in this realization of the circuit are modeled using Analog Devices' AD712 macro-model. The TOKO 10RB inductor has a nonzero series resistance that we have included in the SPICE net-list; a typical value of RO for this inductor is 12.5 Ω. Node

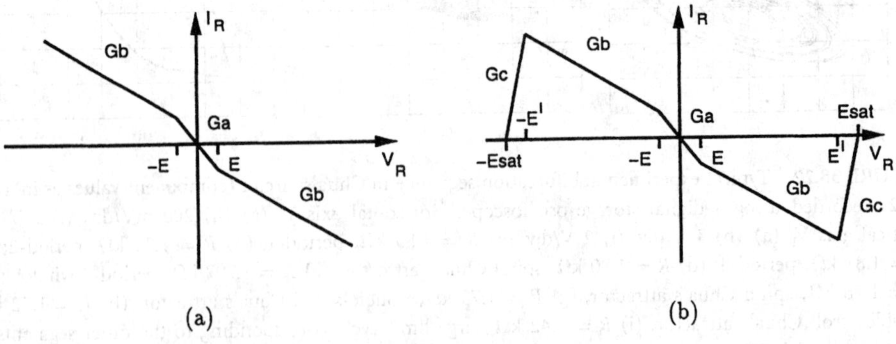

FIGURE 38.28 (a) Required three-segment piecewise-linear DP characteristic for the Chua diode in Fig. 38.17. (b) Every physically realizable nonlinear resistor N_R is eventually passive—the outermost segments (while not necessarily linear as shown here) must lie completely within the first and third quadrants of the V_R–I_R plane for sufficiently large $|V_R|$ and $|I_R|$.

numbers are as in Fig. 38.25: the power rails are 111 and 222; 10 is the "internal" node of the physical inductor, where its series inductance is connected to its series resistance.

A double-scroll Chua's attractor results from a PSpice simulation using the input deck shown in Fig. 38.29; this attractor is plotted in Fig. 38.30.

Dimensionless Coordinates and the $\alpha - \beta$ Parameter-Space Diagram

Thus far, we have discussed Chua's circuit equations in terms of seven parameters: L, C_2, G, C_1, E, G_a, and G_b. We can reduce the number of parameters by normalizing the nonlinear resistor such that its breakpoints are at ± 1 V instead of $\pm E$ V. Furthermore, we may write Chua's circuit equations (38.12) in normalized dimensionless form by making the following change of

```
ROBUST OP AMP REALIZATION OF CHUA'S CIRCUIT

V+  111 0    DC 9
V-  0   222  DC 9
L    1 10   0.018
RO   10 0    12.5
R    1 2    1770
C2   1 0     100.0N
C1   2 0     10.0N
XA1  2 4 111 222 3 AD712
R1   2 3 220
R2   3 4 220
R3   4 0 2200
XA2  2 6 111 222 5 AD712
R4   2 5 22000
R5   5 6 22000
R6   6 0 3300

* AD712 SPICE Macro-model          1/91, Rev. A
* Copyright 1991 by Analog Devices, Inc. (reproduced with permission)
*
.SUBCKT AD712 13 15 12 16 14
*
VOS 15 8 DC 0
EC 9 0 14 0 1
C1 6 7 .5P
RP 16 12 12K
GB 11 0 3 0 1.67K
RD1 6 16 16K
RD2 7 16 16K
ISS 12 1 DC 100U
CCI 3 11 150P
GCM 0 3 0 1  1.76N
GA 3 0 7 6  2.3M
RE 1 0 2.5MEG
RGM 3 0 1.69K
VC 12 2 DC 2.8
VE 10 16 DC 2.8
RO1 11 14 25
CE 1 0 2P
RO2 0 11 30
RS1 1 4 5.77K
RS2 1 5 5.77K
J1 6 13 4 FET
J2 7 8 5 FET
DC 14 2 DIODE
DE 10 14 DIODE
DP 16 12 DIODE
D1 9 11 DIODE
D2 11 9 DIODE
IOS 15 13 5E-12
.MODEL DIODE D
.MODEL FET PJF(VTO=-1 BETA=1M IS=25E-12)
.ENDS

.IC V(2)=0.1 V(1)=0
.TRAN 0.01MS 100MS 50MS
.OPTIONS RELTOL=1.0E-4 ABSTOL=1.0E-4
.PRINT TRAN V(2) V(1)
.END
```

FIGURE 38.29 SPICE deck to simulate the transient response of the dual op amp implementation of Chua's circuit. Node numbers are as in Fig. 38.25. The op amps are modeled using the Analog Devices AD712 macro-model. RO models the series resistance of the real inductor L.

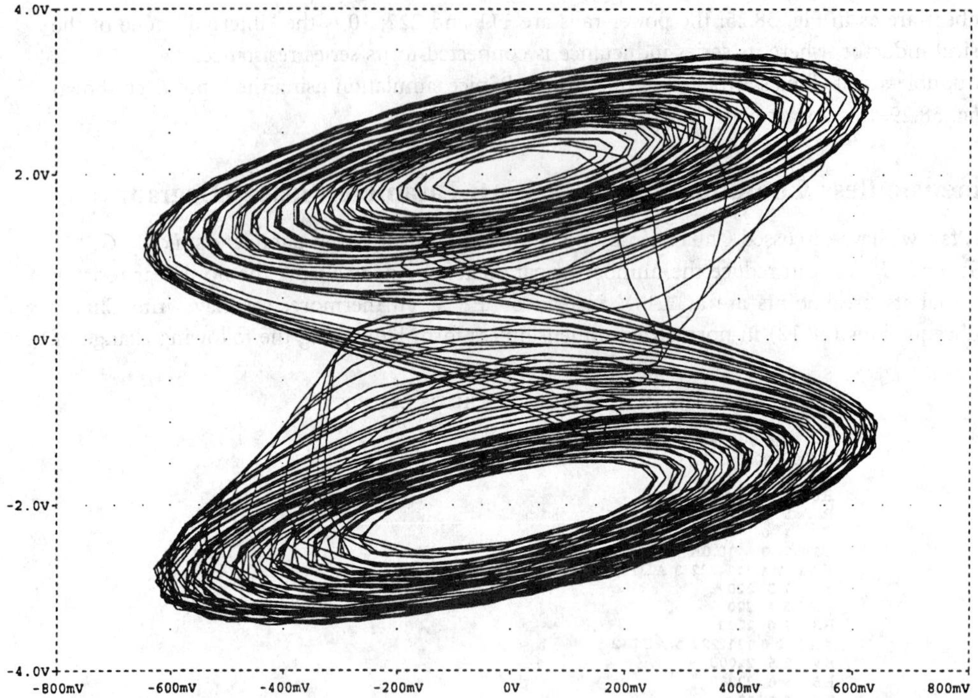

FIGURE 38.30 **PSpice** (evaluation version 5.4, July 1993) simulation of Fig. 38.25 using the input deck shown in Fig. 38.29 yields this double-scroll Chua's attractor. Horizontal axis V_2 (V); vertical axis V_1 (V).

variables: $X_1 = V_1/E$, $X_2 = V_2/E$, $X_3 = I_3/(EG)$, and $\tau = tG/C_2$. The resulting state equations are

$$\begin{aligned}\frac{dX_1}{d\tau} &= \alpha[X_2 - X_1 - f(X_1)]\\ \frac{dX_2}{d\tau} &= X_1 - X_2 + X_3 \\ \frac{dX_3}{d\tau} &= -\beta X_2\end{aligned} \tag{38.15}$$

where $\alpha = C_2/C_1$, $\beta = C_2/(LG^2)$, and $f(X) = bX + 1/2(a-b)(|X+1| - |X-1|)$; $a = G_a/G$ and $b = G_b/G$. Thus, each set of seven circuit parameters has an equivalent set of four normalized dimensionless parameters $\{\alpha, \beta, a, b\}$. If we fix the values of a and b (which correspond to the slopes G_a and G_b of the Chua diode), we can summarize the steady-state dynamical behavior of Chua's circuit by means of a two-dimensional parameter-space diagram.

Figure 38.31 shows the (α, β) parameter-space diagram with $a = -8/7$ and $b = -5/7$. In this diagram, each region denotes a particular type of steady-state behavior: for example, an equilibrium point, period-one orbit, period-two, spiral Chua's attractor, double-scroll Chua's attractor. Typical state-space behaviors are shown in the insets. For clarity, we show chaotic regions in a single shade; it should be noted that these chaotic regions are further partitioned by periodic windows and "islands" of periodic behavior.

To interpret the α–β diagram, imagine fixing the value of $\beta = C_2/(LG^2)$ and increasing $\alpha = C_2/C_1$ from a positive value to the left of the curve labeled "Hopf at $P^{\pm}$"; experimentally, this corresponds to fixing the parameters L, C_2, G, E, G_a, and G_b, and reducing the value of C_1—this is called a "C_1 bifurcation sequence."

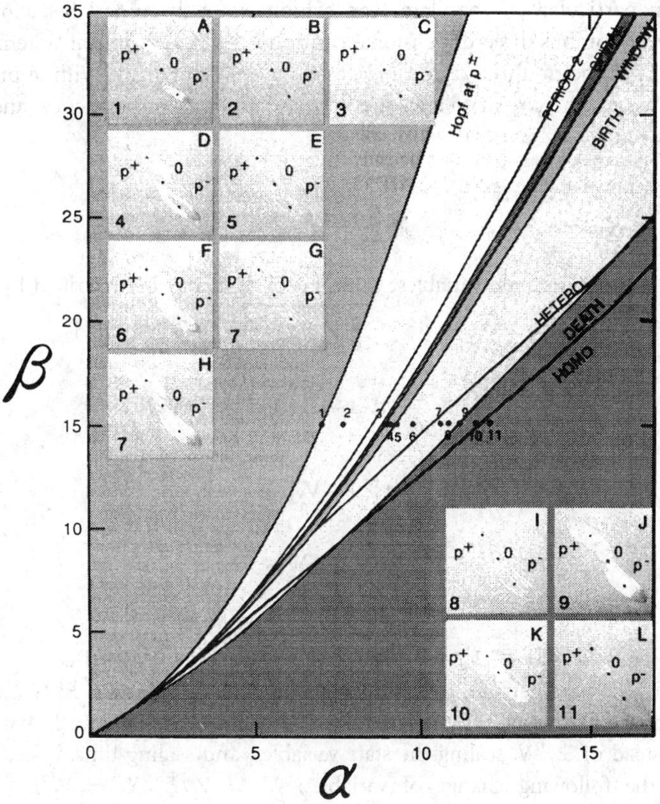

FIGURE 38.31 α–β parameter space diagram for the normalized dimensionless Chua's circuit equations (38.15) with $a = -8/7$ and $b = -5/7$. *Source:* M. P. Kennedy, "Three steps to chaos—Part II: A Chua's circuit primer," *IEEE Trans. Circuits Syst. I*, vol. 40, p. 673, Oct. 1993. © 1993 IEEE.

Initially, the steady-state solution is an equilibrium point. As the value of C_1 is reduced, the circuit undergoes a Hopf bifurcation when α crosses the "Hopf at $P^{\pm}$" curve. Decreasing C_1 still further, the steady-state behavior bifurcates from period-one to period-two to period-four and so on to chaos, periodic windows, and a double-scroll Chua's attractor. The right-hand side edge of the chaotic region is delimited by a curve corresponding to the boundary crisis and "death" of the attractor. Beyond this curve, trajectories diverge toward infinity. Because of eventual passivity in a real circuit, these divergent trajectories will of course converge to a limit cycle in any physical implementation of Chua's circuit.

8.3 Chua's Oscillator

Chua's oscillator [5] (shown in Fig. 38.26) is derived from Chua's circuit by adding a resistor R_0 in series with the inductor L. The oscillator contains a linear inductor, two linear resistors, two linear capacitors, and a single Chua diode N_R. N_R is a voltage-controlled piecewise-linear resistor whose continuous odd-symmetric three-segment driving-point characteristic (shown in Fig. 38.18) is described explicitly by the relationship

$$I_R = G_b V_R + \tfrac{1}{2}(G_a - G_b)(|V_R + E| - |V_R - E|)$$

The primary motivation for studying this circuit is that the vector field of Chua's oscillator is topologically conjugate to the vector field of a large class of three-dimensional piecewise-linear vector fields. In particular, the oscillator can exhibit every dynamical behavior known to be possible in an autonomous three-dimensional continuous-time dynamical system described by a continuous odd-symmetric three-region piecewise-linear vector field. With appropriate choices of component values, the circuit follows the period-doubling, intermittency, and quasiperiodic routes to chaos.

State Equations

Choosing V_1, V_2, and I_3 as state variables, Chua's oscillator may be described by three ordinary differential equations:

$$\frac{dV_1}{dt} = \frac{G}{C_1}(V_2 - V_1) - \frac{1}{C_1} f(V_1)$$
$$\frac{dV_2}{dt} = \frac{G}{C_2}(V_1 - V_2) + \frac{1}{C_2} I_3$$
$$\frac{dI_3}{dt} = -\frac{1}{L} V_2 - \frac{R_0}{L} I_3$$

where $G = 1/R$ and $f(V_R) = G_b V_R + 1/2(G_a - G_b)(|V_R + E| - |V_R - E|)$.

The vector field is parametrized by eight constants: L, C_2, G, C_1, R_0, E, G_a, and G_b. We can reduce the number of parameters by normalizing the nonlinear resistor such that its breakpoints are at ± 1 V instead of $\pm E$ V, scaling the state variables, and scaling time.

By making the following change of variables: $X_1 = V_1/E$, $X_2 = V_2/E$, $X_3 = I_3/(EG)$, $\tau = t|G/C_2|$, and $k = \text{sgn}(G/C_2)$,[14] we can rewrite the state equations (38.16) in normalized dimensionless form:

$$\frac{dX_1}{d\tau} = k\alpha[X_2 - X_1 - f(X_1)]$$
$$\frac{dX_2}{d\tau} = k(X_1 - X_2 + X_3)$$
$$\frac{dX_3}{d\tau} = -k(\beta X_2 + \gamma X_3)$$

where $\alpha = C_2/C_1$, $\beta = C_2/(LG^2)$, $\gamma = R_0 C_2/(LG)$, and $f(X) = bX + 1/2(a-b)(|X+1| - |X-1|)$ with $a = G_a/G$ and $b = G_b/G$. Thus, each set of eight circuit parameters has an equivalent set of six normalized dimensionless parameters $\{\alpha, \beta, \gamma, a, b, k\}$.

Topological Conjugacy

Two vector fields F and **F**′ are **topologically conjugate** if there exists a continuous map h (which has a continuous inverse) such that h maps trajectories of **F** into trajectories of **F**′, preserving time orientation and parametrization of time. If ϕ_t and ϕ'_t are the flows of **F** and **F**′ respectively, then $\phi_t \circ h = h \circ \phi'_t$ for all t. This means that the dynamics of **F** and **F**′ are qualitatively the same. If h is linear, then **F** and **F**′ are said to be linearly conjugate.

[14]The signum function is defined by $\text{sgn}(x) = x$ if $x > 0$, $\text{sgn}(x) = -x$ if $x < 0$, and $\text{sgn}(0) = 0$.

Class $\mathscr{C}$

A three-dimensional autonomous continuous-time dynamical system defined by the state equation

$$\dot{\mathbf{X}} = \mathbf{F}(\mathbf{X}) \qquad \mathbf{X} \in \mathbb{R}^3$$

is said to belong to class $\mathscr{C}$ iff

1. $\mathbf{F}$: $\mathbb{R}^3 \to \mathbb{R}^3$ is continuous,
2. F is odd-symmetric, i.e., $\mathbf{F}(-X) = -\mathbf{F}(X)$,
3. $\mathbb{R}^3$ is partitioned by two parallel boundary planes U_1 and U_{-1} into an inner region D_0, which contains the origin, and two outer regions D_1 and D_{-1}, and $\mathbf{F}$ is affine in each region.

Without loss of generality, the boundary planes and the regions they separate can be chosen as follows:

$$\begin{aligned} D_{-1} &= \{\mathbf{X}: X_1 \le -1\} \\ U_{-1} &= \{\mathbf{X}: X_1 = -1\} \\ D_0 &= \{\mathbf{X}: |X_1| \le 1\} \\ U_{-1} &= \{\mathbf{X}: X_1 = 1\} \\ D_1 &= \{\mathbf{X}: X_1 \ge 1\} \end{aligned}$$

Any vector field in the family $\mathscr{C}$ can then be written in the form

$$\dot{\mathbf{X}} = \begin{cases} \mathbf{A}_{-1}\mathbf{X} - \mathbf{b} & X_1 \le -1 \\ \mathbf{A}_0\mathbf{X} & -1 \le X_1 \le 1 \\ \mathbf{A}_1\mathbf{X} + \mathbf{b} & X_1 \ge 1 \end{cases}$$

where

$$\mathbf{A}_{-1} = \mathbf{A}_1 = \begin{bmatrix} a_{11} & a_{12} & a_{13} \\ a_{21} & a_{22} & a_{23} \\ a_{31} & a_{32} & a_{33} \end{bmatrix} \quad \text{and} \quad \mathbf{b} = \begin{bmatrix} b_1 \\ b_2 \\ b_3 \end{bmatrix}$$

By continuity of the vector field across the boundary planes,

$$\mathbf{A}_0 = \begin{bmatrix} (a_{11} + b_1) & a_{12} & a_{13} \\ (a_{21} + b_2) & a_{22} & a_{23} \\ (a_{31} + b_3) & a_{32} & a_{33} \end{bmatrix}$$

Equivalent Eigenvalue Parameters

Let (μ_1, μ_2, μ_3) denote the eigenvalues associated with the linear vector field in the D_0 region and let (ν_1, ν_2, ν_3) denote the eigenvalues associated with the affine vector fields in the outer regions D_1 and D_{-1}. Define

$$\left.\begin{aligned} p_1 &= \mu_1 + \mu_2 + \mu_3 \\ p_2 &= \mu_1\mu_2 + \mu_2\mu_3 + \mu_3\mu_1 \\ p_3 &= \mu_1\mu_2\mu_3 \\ q_1 &= \nu_1 + \nu_2 + \nu_3 \\ q_2 &= \nu_1\nu_2 + \nu_2\nu_3 + \nu_3\nu_1 \\ q_3 &= \nu_1\nu_2\nu_3 \end{aligned}\right\} \tag{38.16}$$

Since the six parameters $\{p_1, p_2, p_3, q_1, q_2, q_3\}$ are uniquely determined by the eigenvalues $\{\mu_1, \mu_2, \mu_3, \nu_1, \nu_2, \nu_3\}$ and vice versa, the former are called the equivalent eigenvalue parameters. Note that the equivalent eigenvalues are real; they are simply the coefficients of the characteristic polynomials:

$$(s-\mu_1)(s-\mu_2)(s-\mu_3) = s^3 - p_1 s^2 + p_2 s - p_3$$
$$(s-\nu_1)(s-\nu_2)(s-\nu_3) = s^3 - q_1 s^2 + q_2 s - q_3$$

Theorem 3 (Chua *et al.*) [5]: *Let $\{\mu_1, \mu_2, \mu_3; \nu_1, \nu_2, \nu_3\}$ be the eigenvalues associated with a vector field* $\mathbf{F}(\mathbf{X}) \in \mathscr{C}\backslash\mathscr{E}_0$, *where $\mathscr{E}_0$ is the set of measure zero in the space of equivalent eigenvalue parameters where one of (38.17) is satisfied. Then Chua's oscillator with parameters defined by (38.18) and (38.19) is linearly conjugate to this vector field.*

$$\left.\begin{aligned} p_1 - q_1 &= 0 \\ p_2 - \left(\frac{p_3 - q_3}{p_1 - q_1}\right) - \left(\frac{p_2 - q_2}{p_1 - q_1}\right)\left(p_1 - \frac{p_2 - q_2}{p_1 - q_1}\right) &= 0 \\ -\left(\frac{p_2 - q_2}{p_1 - q_1}\right) - \frac{k_1}{k_2} &= 0 \\ -k_1 k_3 + k_2\left(\frac{p_3 - q_3}{p_1 - q_1}\right) &= 0 \\ \det \mathbf{K} = \det\begin{bmatrix} 1 & 0 & 0 \\ a_{11} & a_{12} & a_{13} \\ K_{31} & K_{32} & K_{33} \end{bmatrix} = a_{12}K_{33} - a_{13}K_{32} &= 0 \end{aligned}\right\} \tag{38.17}$$

where

$$K_{3i} = \sum_{j=i}^{3} a_{1j} a_{ji} \qquad i = 1, 2, 3$$

We denote by $\tilde{\mathscr{C}}$ the set of vector fields $\mathscr{C}\backslash\mathscr{E}_0$. Two vector fields in $\tilde{\mathscr{C}}$ are linearly conjugate if they have the same eigenvalues in each region.

Eigenvalues-to-Parameters Mapping Algorithm for Chua's Oscillator

Every continuous third-order odd-symmetric three-region piecewise-linear vector field $\mathbf{F}'$ in $\tilde{\mathcal{C}}$ may be mapped onto a Chua's oscillator (whose vector field $\mathbf{F}$ is topologically conjugate to $\mathbf{F}'$) by means of the following algorithm [5].

1. Calculate the eigenvalues (μ_1', μ_2', μ_3') and (ν_1', ν_2', ν_3') associated with the linear and affine regions, respectively, of the vector field $\mathbf{F}'$ of the circuit or system whose attractor is to be reproduced (up to linear conjugacy) by Chua's oscillator.
2. Find a set of circuit parameters $\{C_1, C_2, L, R, R_0, G_a, G_b, E\}$ (or dimensionless parameters $\{\alpha, \beta, \gamma, a, b, k\}$) so that the resulting eigenvalues μ_j and ν_j for Chua's oscillator satisfy $\mu_j = \mu_j'$ and $\nu_j = \nu_j'$, $j = 1, 2, 3$.

Let $\{p_1, p_2, p_3, q_1, q_2, q_3\}$ be the equivalent eigenvalue parameters defined by (38.16). Furthermore, let

$$\left.\begin{aligned}
k_1 &= -p_3 + \left(\frac{p_3 - q_3}{p_1 - q_1}\right)\left(p_1 - \frac{p_2 - q_2}{p_1 - q_1}\right)\\
k_2 &= p_2 - \left(\frac{p_3 - q_3}{p_1 - q_1}\right) - \left(\frac{p_2 - q_2}{p_1 - q_1}\right)\left(p_1 - \frac{p_2 - q_2}{p_1 - q_1}\right)\\
k_3 &= -\left(\frac{p_2 - q_2}{p_1 - q_1}\right) - \frac{k_1}{k_2}\\
k_4 &= -k_1 k_3 + k_2\left(\frac{p_3 - q_3}{p_1 - q_1}\right)
\end{aligned}\right\} \tag{38.18}$$

The corresponding circuit parameters are given by

$$\left.\begin{aligned}
C_1 &= 1\\
C_2 &= -\frac{k_2}{k_3^2}\\
L &= -\frac{k_3^2}{k_4}\\
R &= -\frac{k_3}{k_2}\\
R_0 &= -\frac{k_1 k_3^2}{k_2 k_4}\\
G_a &= -p_1 + \left(\frac{p_2 - q_2}{p_1 - q_1}\right) + \frac{k_2}{k_3}\\
G_b &= -q_1 + \left(\frac{p_2 - q_2}{p_1 - q_1}\right) + \frac{k_2}{k_3}
\end{aligned}\right\} \tag{38.19}$$

The breakpoint E of the piecewise-linear Chua diode can be chosen arbitrarily since the choice of E does not affect either the eigenvalues or the dynamics; it simply scales the circuit variables. In a practical realization of the circuit, one should scale the voltages and currents so that they lie within the inner three segments of the nonlinear resistor N_R.

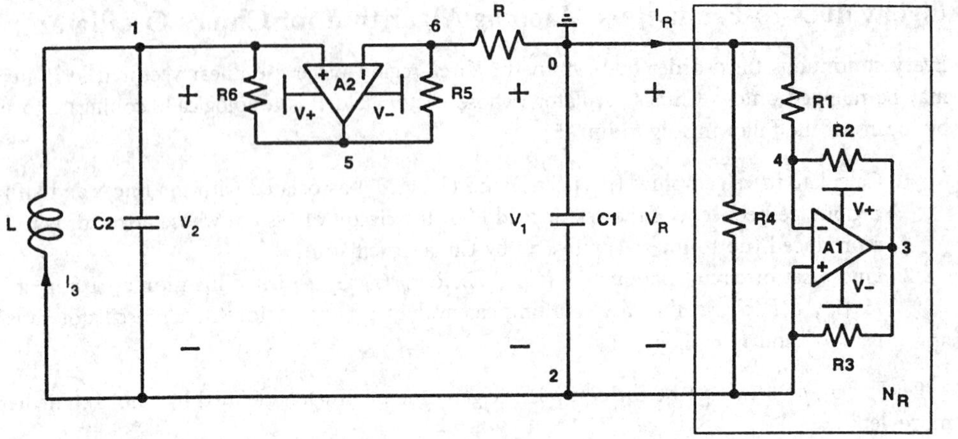

FIGURE 38.32 Practical implementation of Chua's oscillator using an op amp and four resistors to realize the Chua diode [10]. The negative resistor G is realized by means of a negative resistance converter (A_2, R_5, R_6, and positive resistor R). If $R_2 = R_3$ and $R_5 = R_6, G_a = 1/R_4 - 1/R_1, G_b = 1/R_4 + 1/R_2$, and $G = -1/R$. Component values are listed in Table 38.3.

The dimensionless parameters may be calculated as follows:

$$\left.\begin{aligned}
\alpha &= -\frac{k_2}{k_3^2} \\
\beta &= \frac{k_4}{k_2 k_3^2} \\
\gamma &= \frac{k_1}{k_2 k_3} \\
a &= -1 + \frac{k_3}{k_2}\left(p_1 - \frac{p_2 - q_2}{p_1 - q_1}\right) \\
b &= -1 + \frac{k_3}{k_2}\left(q_1 - \frac{p_2 - q_2}{p_1 - q_1}\right) \\
k &= \operatorname{sgn}(k_3)
\end{aligned}\right\} \tag{38.20}$$

Example: Torus

Figure 38.32 shows a practical implementation of Chua's oscillator that exhibits a transition to chaos by torus breakdown. A complete list of components is given in Table 38.3.

A SPICE simulation of this circuit produces a quasiperiodic voltage `V(2)` ($= -V_1$), as expected (see Fig. 38.33). The resistor `R0` is not explicitly added to the circuit, but models the dc resistance of the inductor.

38.4 Van der Pol Neon Bulb Oscillator

In a paper entitled "Frequency demultiplication" the eminent Dutch electrical engineer Balthazar van der Pol described an experiment in which, by tuning the capacitor in a neon bulb *RC*

TABLE 38.3 Component List for the Chua Oscillator Shown in Fig. 38.32

Element	Description	Value	Tolerance
A_1	Op Amp $\left(\frac{1}{2}\text{AD712, TL082, or Equivalent}\right)$		
A_2	Op Amp $\left(\frac{1}{2}\text{AD712, TL082, or Equivalent}\right)$		
C_1	Capacitor	47 nF	±5%
C_2	Capacitor	820 nF	±5%
R_1	$\frac{1}{4}$ W Resistor	6.8 kΩ	±5%
R_2	$\frac{1}{4}$ W Resistor	47 kΩ	±5%
R_3	$\frac{1}{4}$ W Resistor	47 kΩ	±5%
R_4	Potentiometer	2 kΩ	
R_5	$\frac{1}{4}$ W Resistor	220 Ω	±5%
R_6	$\frac{1}{4}$ W Resistor	220 Ω	±5%
R	Potentiometer	2 kΩ	
L	Inductor (TOKO-Type 10 RB, or Equivalent)	18 mH	±10%

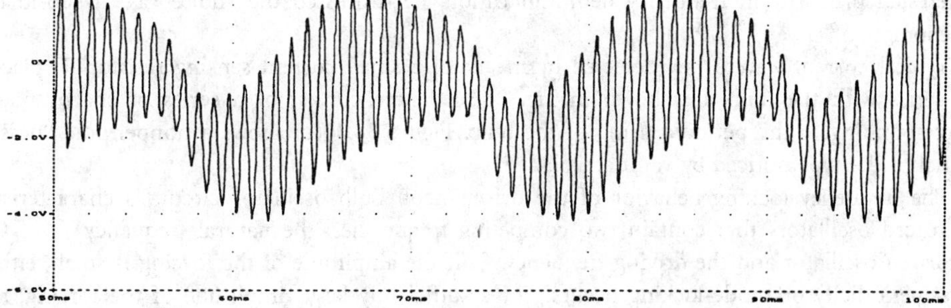

FIGURE 38.33 PSpice simulation (`.TRAN 0.01MS 100MS 50MS`) of Fig. 38.32 with initial conditions `.IC V(2)=-0.1 V(1)=-0.1` and tolerances `.OPTIONS RELTOL=1E-4 ABSTOL=1E-4` yields this quasiperiodic voltage waveform at node 2.

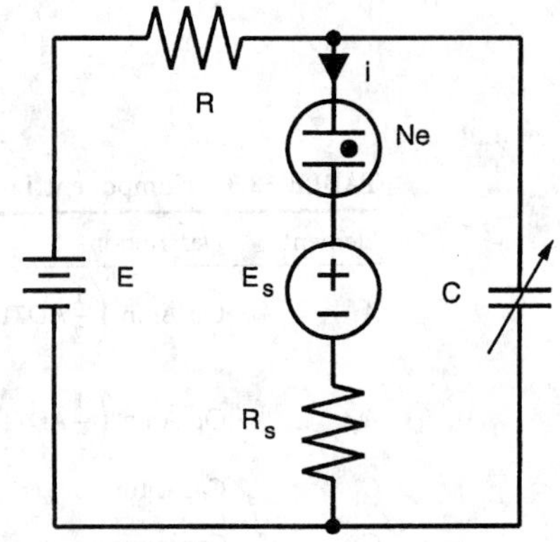

FIGURE 38.34 Sinusoidally-driven **neon bulb relaxation oscillator.** Ne is the neon bulb.

relaxation oscillator driven by a sinusoidal voltage source (shown in Fig 38.34), "currents and voltages appear in the system which are whole submultiples of the driving frequency" [11].

The circuit consists of a high-voltage dc source E attached via a large series resistance R to a neon bulb and capacitor C that are connected in parallel; this forms the basic relaxation oscillator. Initially, the capacitor is discharged and the neon bulb is nonconducting. The dc source charges C with time constant RC until the voltage across the neon bulb is sufficient to turn it on. Once lit, the bulb presents a shunt low resistance path to the capacitor. The voltage across the capacitor falls exponentially until the neon arc is quenched, the bulb is returned to its "off" state, and the cycle repeats.

In series with the neon bulb is inserted a sinusoidal voltage source $E_s = E_0 \sin(2\pi f_s t)$ whose effect is to perturb the "on" and "off" switching thresholds of the capacitor voltage.

Experimental results for this circuit are summarized in Fig. 38.35, where the ratio of the system period (time interval before the pattern of current pulses repeats itself) to the period T of the forcing is plotted versus the capacitance C.

Van der Pol noted that as the capacitance was increased from that value (C_0) for which the natural frequency f_0 of the undriven relaxation oscillator equaled that of the sinusoidal source (system period/$T = 1$), the system frequency made "discrete jumps from one whole submultiple of the driving frequency to the next" (detected by means of "a telephone coupled loosely in some way to the system"). Van der Pol noted that "often an irregular noise is heard in the telephone receiver before the frequency jumps to the next lower value"; van der Pol had observed chaos. Interested primarily in frequency demultiplication, he dismissed the "noise" as "a subsidiary phenomenon."

Typical current waveforms, detected by means of a small current-sensing resistor R_s placed in series with the bulb are shown in Fig. 38.36. These consist of a series of sharp spikes, corresponding to the periodic firing of the bulb. Figure 38.36(c) shows a nonperiodic "noisy" signal of the type noticed by van der Pol.

The frequency locking behavior of the driven neon bulb oscillator circuit is characteristic of forced oscillators that contain two competing frequencies: the natural frequency f_0 of the undriven oscillator and the driving frequency f_s. If the amplitude of the forcing is small, either quasiperiodicity or mode-locking occurs. For a sufficiently large amplitude of the forcing, the system may exhibit chaos.

Winding Numbers

Subharmonic frequency locking in a forced oscillator containing two competing frequencies f_1 and f_2 may be understood in terms of **a winding number**. The concept of a winding number was introduced by Poincaré to describe periodic and quasiperiodic trajectories on a torus.

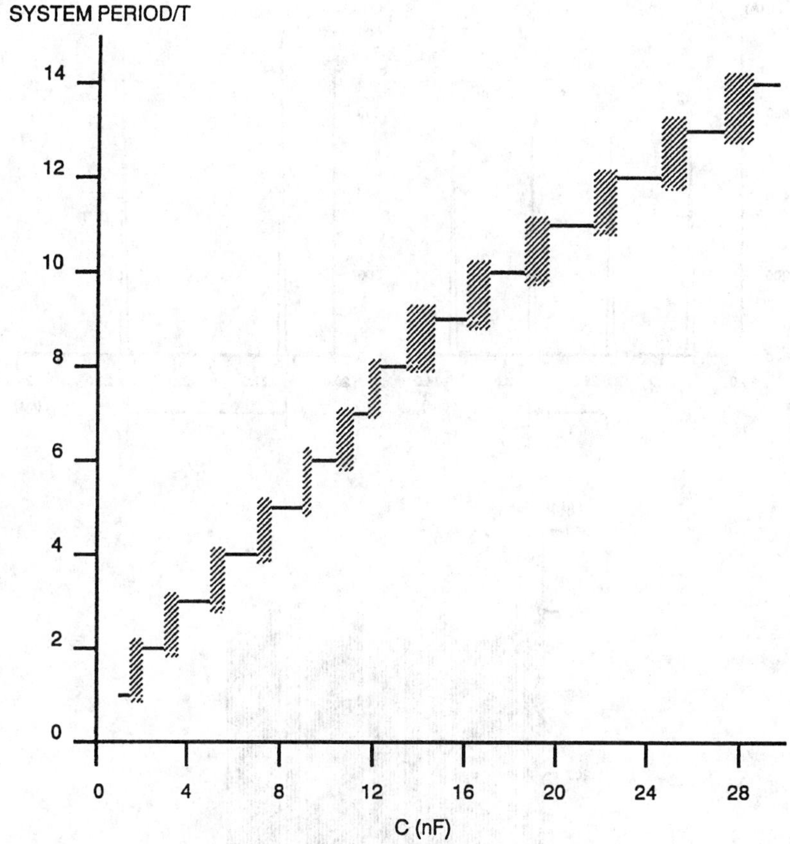

FIGURE 38.35 Normalized current pulse pattern repetition rate versus C for the sinusoidally-driven neon relaxation oscillator in Fig. 38.34, showing a coarse staircase structure of mode-lockings. *Source:* M. P. Kennedy and L. O. Chua, "Van der Pol and Chaos," *IEEE Trans. Circuits Syst.*, vol. CAS-33, p. 975, Oct. 1986 © 1986 IEEE.

A trajectory on a torus that winds around the minor axis of the torus with frequency f_1 revolutions per second, and completes revolutions of the major axis with frequency f_2, may be parametrized by two angular coordinates $\theta_1 \equiv f_1 t$ and $\theta_2 \equiv f_2 t$, as shown in Fig. 38.37. The angles of rotation θ_1 and θ_2 about the major and minor axes of the torus are normalized so that one revolution corresponds to a change in θ of 1.

A Poincaré map for this system can be defined by sampling the state θ_1 with period $\tau = 1/f_2$. Let $\theta_k = \theta_1(k\tau)$. The Poincaré map has the form

$$\theta_{k+1} = G(\theta_k), \qquad k = 0, 1, 2, \cdots$$

If $f_1/f_2 = p/q$ is rational, then the trajectory is periodic, closing on itself after completing q revolutions about the major axis of the torus. In this case, we say that the system is periodic with period q and completes p cycles per period. If the ratio p/q is irrational then the system is quasiperiodic; a trajectory covers the surface of the torus, coming arbitrarily close to every point on it, but does not close on itself.

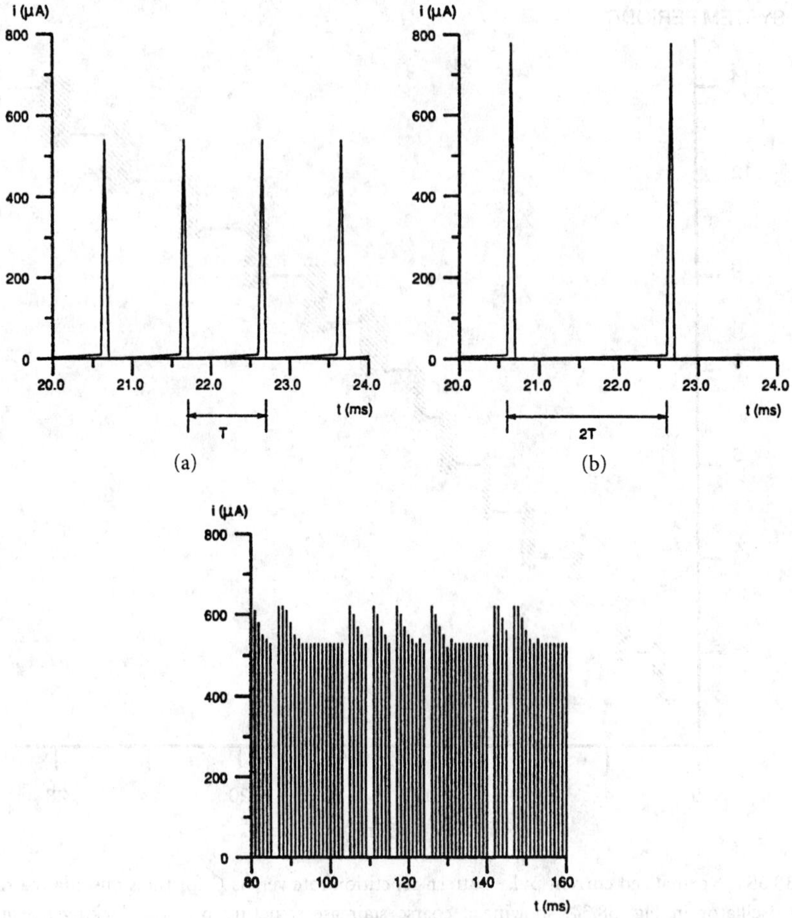

FIGURE 38.36 Periodic and chaotic neon bulb current waveforms. (a) One current pulse per cycle of E_s: $f_s/f_d = 1/1$; (b) one current pulse every two cycles of E_s: $f_s/f_d = 2/1$; (c) "noisy" current waveform.

The winding number w is defined by

$$w = \lim_{k \to \infty} \frac{G^{(k)}(\theta_0)}{k}$$

where $G^{(k)}$ denotes the k-fold iterate of G and θ_0 is the initial state.

The winding number counts the average number of revolutions in the Poincaré section per iteration. Equivalently, w equals the average number of turns about the minor axis per revolution about the major axis of the torus.[15] Periodic orbits possess rational winding numbers and are called resonant; quasiperiodic trajectories have irrational winding numbers.

[15]Since either frequency may be chosen to correspond to the major axis of the torus, the winding number and its reciprocal are equivalent.

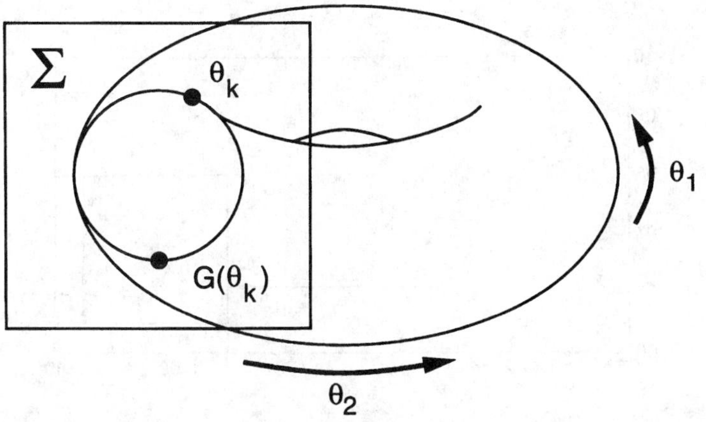

FIGURE 38.37 A trajectory on a torus is characterized by two normalized angular coordinates. $\theta_1 = f_1 t$ is the angle of rotation about the minor axis of the torus, while $\theta_2 = f_2 t$ is the angle of rotation along the major axis, where f_1 and f_2 are the frequencies of rotation about the corresponding axes. A Poincaré map $\theta_{k+1} = G(\theta_k)$ is defined by sampling the trajectory with frequency $1/f_2$. The winding number w counts the average number of revolutions in the Poincaré section per iteration of the map.

The Circle Map

A popular paradigm for explaining the behavior of coupled nonlinear oscillators with two competing frequencies is the circle map:

$$\theta_{k+1} = \left(\theta_k + \frac{K}{2\pi}\sin(2\pi\theta_k) + \Omega\right) \bmod 1, \quad k = 0, 1, 2, \cdots \tag{38.21}$$

so-called because it maps the circle into itself. The sinusoidal term represents the amplitude of the forcing, and Ω is the ratio of the natural frequency of the unperturbed system and the forcing frequency [18].

When $K \equiv 0$, the steady state of the discrete-time dynamical system (38.22) is either periodic or quasiperiodic, depending on whether Ω is rational or irrational.

If the amplitude K of the forcing is nonzero but less than unity, the steady-state is q-periodic when $\Omega = p/q$ is rational. In this case, there is a nonzero mode-locked window $[\Omega_{\min}(w), \Omega_{\max}(w)]$ over which $w = p/q$. A mode-locked region is delimited by saddle-node bifurcations at $\Omega_{\min}(w)$ and $\Omega_{\max}(w)$ [18].

The function $w(\Omega)$, shown in Fig. 38.38, is monotone increasing and forms a **Devil's staircase** with plateaus at every rational value of w—for example, the step with winding number 1/2 is centered at $\Omega = 0.5$. An experimental Devil's staircase for the driven neon bulb circuit, with low-amplitude forcing, is shown in Fig. 38.39.

As the amplitude K is increased, the width of each locked interval in the circle map increases so that mode-locking becomes more common and quasiperiodicity occurs over smaller ranges of driving frequencies. The corresponding (K, Ω) parameter space diagram (shown in Fig. 38.40) consists of a series of distorted triangles, known as **Arnold Tongues**, whose apexes converge to rational values of Ω at $K = 0$.

Within a tongue, the winding number is constant, yielding one step of the Devil's staircase. The winding numbers of adjacent tongues are related by a *Farey tree* structure. Given two periodic windows with winding numbers $w_1 = p/q$ and $w_2 = r/s$, another periodic window with winding number $w = (\alpha p + \beta r)/(\alpha q + \beta s)$ can always be found, where p and q and r and s are relatively prime and α and β are strictly positive integers. Furthermore, the widest

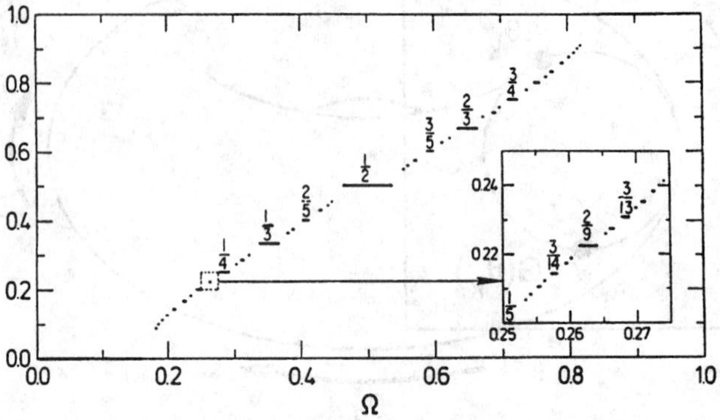

FIGURE 38.38 Devil's staircase for the circle map with $K = 1$. The steps indicate the regions in which w is constant. The staircase is self-similar in the sense that its structure is reproduced qualitatively at smaller scales (see inset). *Source:* J. A. Glazier and A. Libchaber, "Quasi-periodicity and dynamical systems: An experimentalist's view," *IEEE Trans. Circuits Syst.*, vol. 35, p. 793, July 1988. © 1988 IEEE.

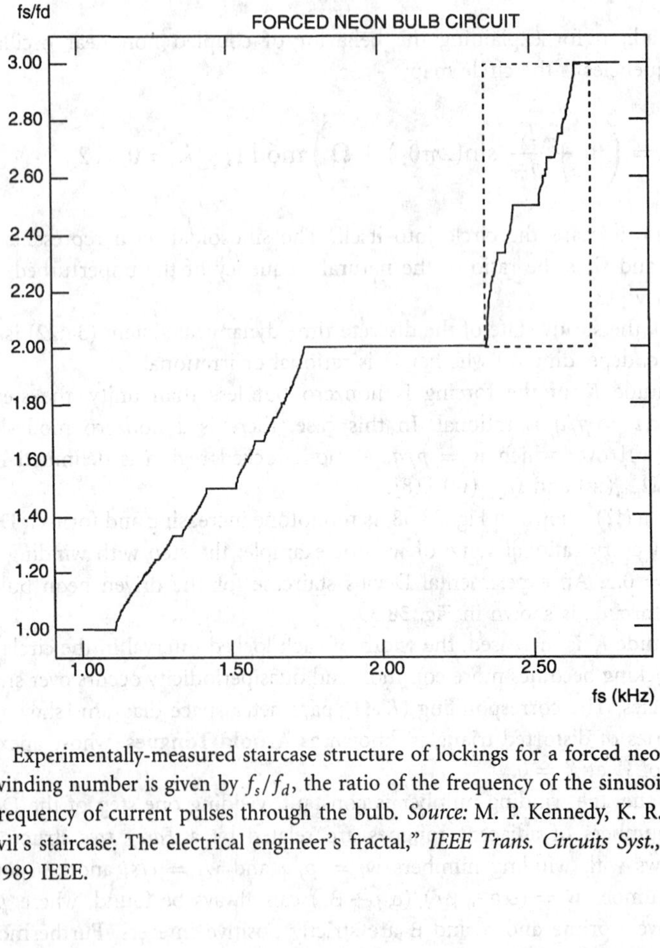

FIGURE 38.39 Experimentally-measured staircase structure of lockings for a forced neon bulb relaxation oscillator. The winding number is given by f_s/f_d, the ratio of the frequency of the sinusoidal driving signal to the average frequency of current pulses through the bulb. *Source:* M. P. Kennedy, K. R. Krieg, and L. O. Chua, "The Devil's staircase: The electrical engineer's fractal," *IEEE Trans. Circuits Syst.*, vol. 36, p. 1137, Aug. 1989. © 1989 IEEE.

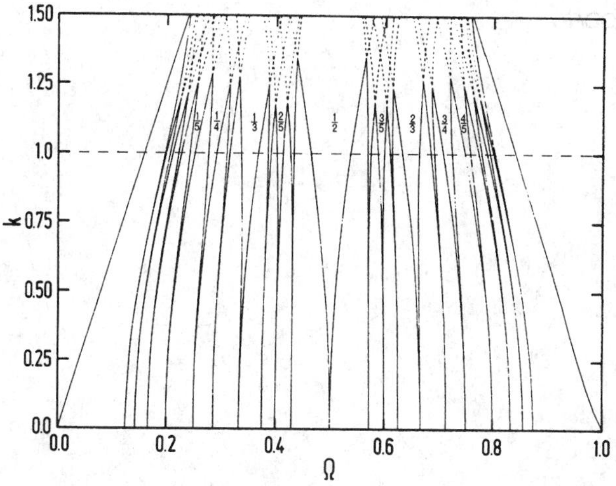

FIGURE 38.40 Parameter space diagram for the circle map showing Arnold tongue structure of lockings in the K–Ω plane. The relative widths of the tongues decrease as the denominator of the winding number increases. Below the critical line $K = 1$, the tongues bend away from each other and do not overlap; for $K > 1$, the Poincaré map develops a fold and chaos can occur. *Source:* J. A. Glazier and A. Libchaber, "Quasi-periodicity and dynamical systems: An experimentalist's view," *IEEE Trans. Circuits Syst.*, vol. 35, p. 793, July 1988. © 1988 IEEE.

mode-locked window between w_1 and w_2 has winding number $(p + r)/(q + s)$. For example, the widest step between those with winding numbers 1/2 and 2/3 in Fig. 38.38 has $w = 3/5$.

The sum of the widths of the mode-locked states increases monotonically from zero at $K = 0$ to unity at $K = 1$. Below the critical line $K = 1$, the tongues bend away from each other and do not overlap. At $K = 1$, tongues begin to overlap, a kink appears in the Poincaré section and the Poincaré map develops a horseshoe; this produces coexisting attractors and chaos.

The transition to chaos as K is increased through $K = 1$ may be by a period-doubling cascade within a tongue, intermittency, or directly from a quasiperiodic trajectory by the abrupt disappearance of that trajectory (a blue sky catastrophe). This qualitative behavior is observed in van der Pol's neon bulb circuit.

Experimental Observations of Mode-Locking and Chaos in van der Pol's Neon Bulb Circuit

With the signal source E_s zeroed, the natural frequency of the undriven relaxation oscillator is set to 1 kHz by tuning capacitance C to C_0. A sinusoidal signal with frequency 1 kHz and amplitude E_0 is applied as shown in Fig. 38.34. The resulting frequency of the current pulses (detected by measuring the voltage across R_s) is recorded with C as the bifurcation parameter.

C Bifurcation Sequence

If a fixed large-amplitude forcing E_s is applied and C is increased slowly, the system at first continues to oscillate at 1 kHz [Fig. 38.36(a)] over a range of C, until the frequency "suddenly" drops to 1000/2 Hz [Fig. 38.36(b)], stays at that value over an additional range of capacitance, drops to 1000/3 Hz, then 1000/4 Hz, 1000/5 Hz, and so on as far as 1000/20 Hz. These results are summarized in Fig. 38.35.

Between each two submultiples of the oscillator driving frequency, a further rich structure of submultiples is found. At the macroscopic level (the coarse structure examined by van der Pol) increasing the value of C causes the system periods to step from T (1 ms) to $2T, 3T, 4T, \ldots,$

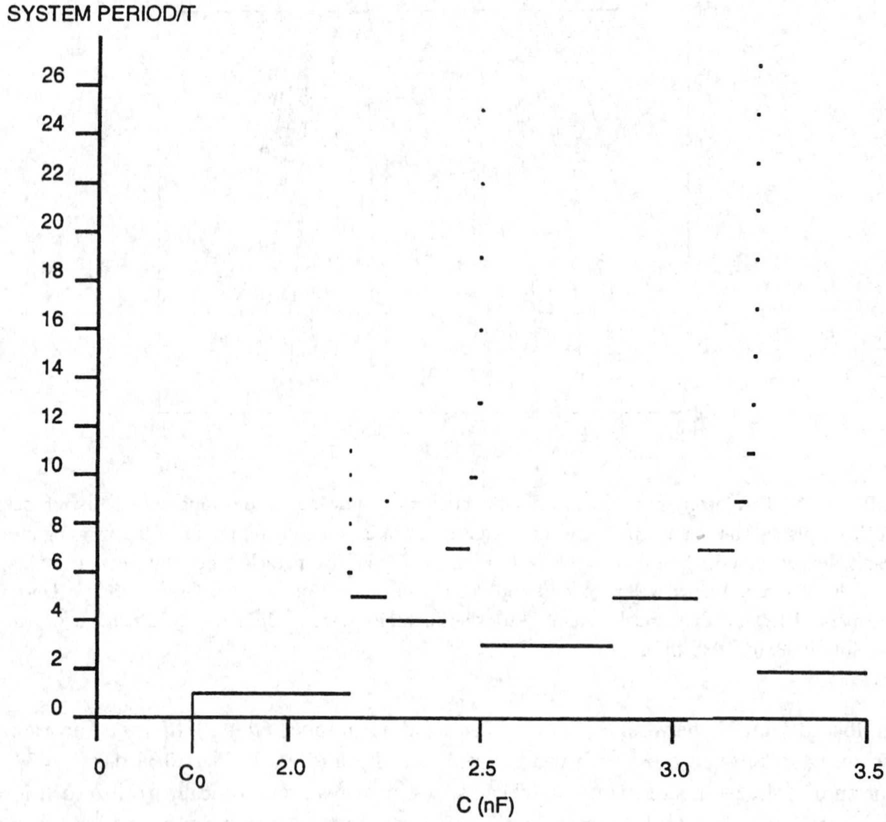

FIGURE 38.41 Experimental pulse pattern repetition rate versus C for van der Pol's forced neon bulb oscillator, showing fine period-adding structure. *Source:* M. P. Kennedy and L. O. Chua, "Van der Pol and chaos," *IEEE Trans. Circuits Syst.*, vol. CAS-33, p. 975, Oct. 1986. © 1986 IEEE.

where the range of C for which the period is fixed is much greater than that over which the transitions occur (Fig. 38.35).

Examining the shaded transition regions more closely, one finds that between any two "macroscopic" regions where the period is fixed at $(n-1)T$ and nT $(n > 1)$, respectively, there lies a narrower region over which the system oscillates with stable period $(2n-1)T$. Further, between $(n-1)T$ and $(2n-1)T$, one finds a region of C for which the period is $(3n-2)T$, and between $(2n-1)T$ and nT, a region with period $(3n-1)T$. Indeed, between any two stable regions with periods $(n-1)T$ and nT, respectively, a region with period $(2n-1)T$ can be found. Figure 38.41 shows an enlargement of the C axis in the region of the T to $2T$ macro-transition, showing the finer period-adding structure.

Between T and $2T$ is a region with stable period $3T$. Between this and $2T$, regions of periods $5T, 7T, 9T, \cdots$ up to $25T$ are detected. Current waveforms corresponding to period-3 and period-5 steps, with winding numbers 3/2 and 5/3, respectively, are shown in Fig. 38.42.

A region of period $4T$ lies between T and $3T$, with steps $7T, 10T, 13T, \cdots$ up to $25T$ between that and $3T$.

In practice, it becomes difficult to observe cycles with longer periods because stochastic noise in the experimental circuit can throw the solution out of the narrow window of existence of a high period orbit.

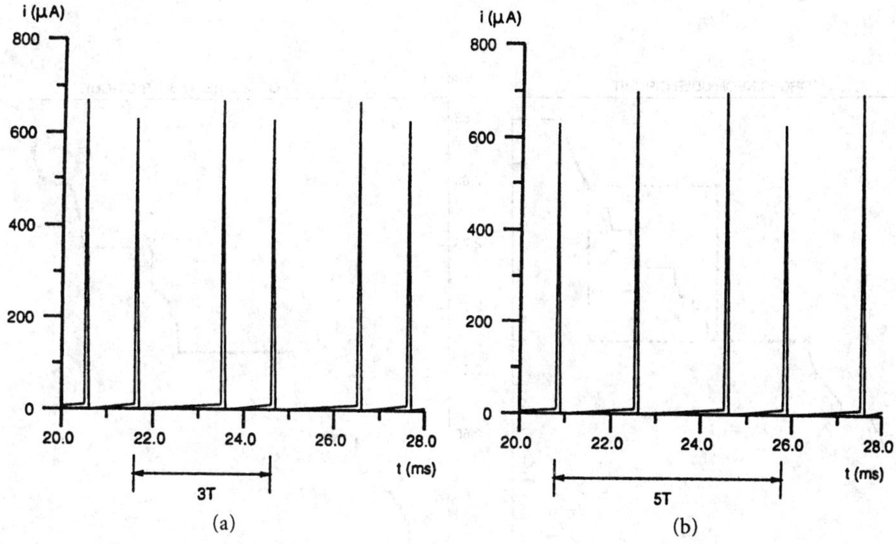

FIGURE 38.42 Neon bulb current waveforms. (a) Two pulses every three cycles of E_s: $f_s/f_d = 3/2$; (b) three pulses every five cycles $f_s/f_d = 5/3$.

Experimental f_s Bifurcation Sequence with Low Amplitude Forcing

An experimental Devil's staircase may be plotted for this circuit by fixing the parameters of the relaxation oscillator and the amplitude of the sinusoidal forcing signal, and choosing the forcing frequency f_s as the bifurcation parameter. The quantity f_s/f_d is the equivalent winding number in this case, where f_d is the average frequency of the current pulses through the neon bulb.

Experimental results for the neon bulb circuit with low-amplitude forcing are shown in Figs. 38.39 and 38.43. The monotone staircase of lockings is consistent with a forcing signal of small amplitude. Note that the staircase is self-similar in the sense that its structure is reproduced qualitatively at smaller scales of the bifurcation parameter.

If the amplitude of the forcing is increased, the onset of chaos is indicated by a nonmonotonicity in the staircase.

Circuit Model

The experimental behavior of van der Pol's sinusoidally driven neon bulb circuit may be reproduced in simulation by an equivalent circuit (shown in Fig. 38.44) in which the only nonlinear element (the neon bulb) is modeled by a nonmonotone current-controlled resistor with a series parasitic inductor L_p.

The corresponding state equations are

$$\frac{dV_C}{dt} = -\frac{1}{RC}V_C - \frac{1}{C}I_L + \frac{E}{RC}$$

$$\frac{dI_L}{dt} = \frac{1}{L_p}V_C - \frac{R_s}{L_p}I_L - \frac{f(I_L)}{L_p} - \frac{E_0 \sin(2\pi f_s t)}{L_p}$$

where $V = f(I)$ is the driving-point characteristic of the current-controlled resistor (shown in Fig. 38.45).

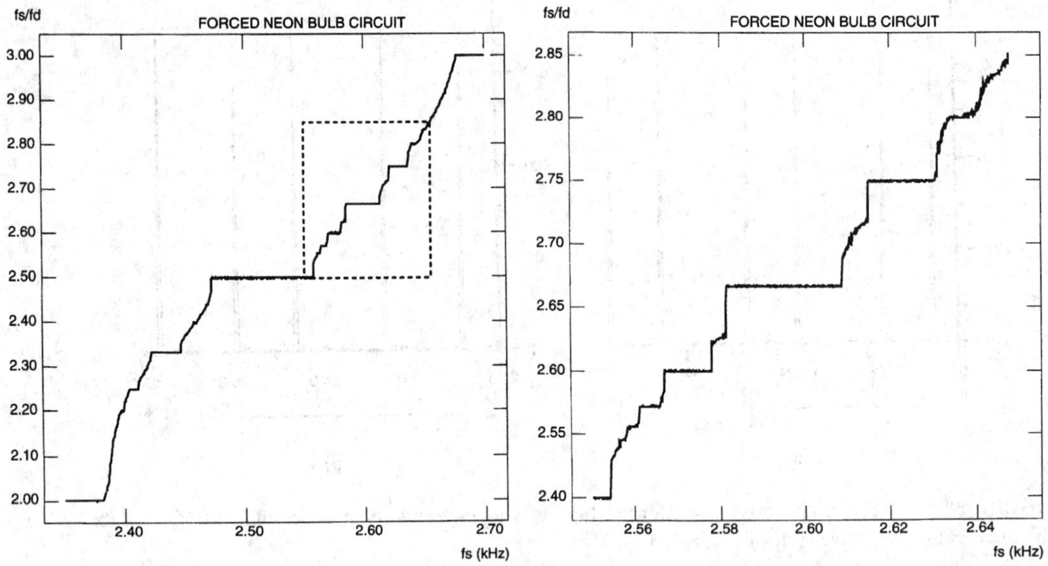

FIGURE 38.43 Magnification of Fig. 38.39 showing self-similarity. *Source:* M. P. Kennedy, K. R. Krieg, and L. O. Chua, "The Devil's staircase: The electrical engineer's fractal," *IEEE Trans. Circuits Syst.*, vol. 36, p. 1137, Aug. 1989. © 1989 IEEE.

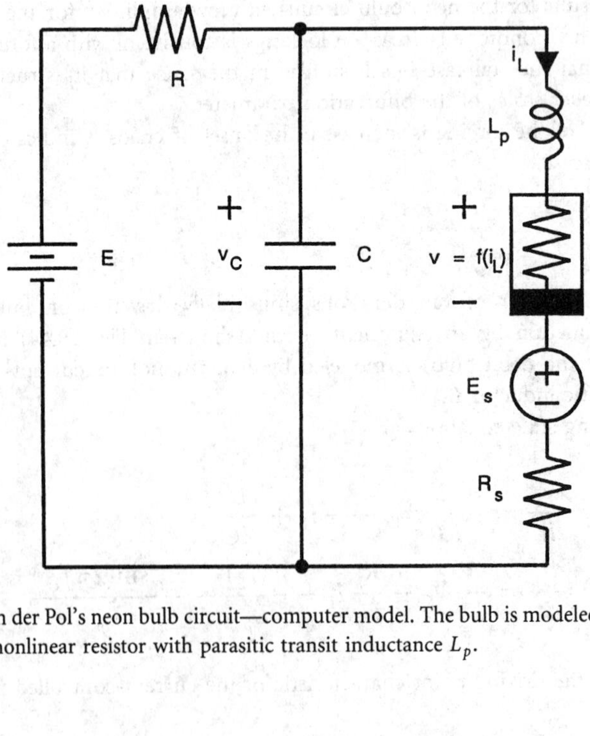

FIGURE 38.44 Van der Pol's neon bulb circuit—computer model. The bulb is modeled by a nonmonotonic current-controlled nonlinear resistor with parasitic transit inductance L_p.

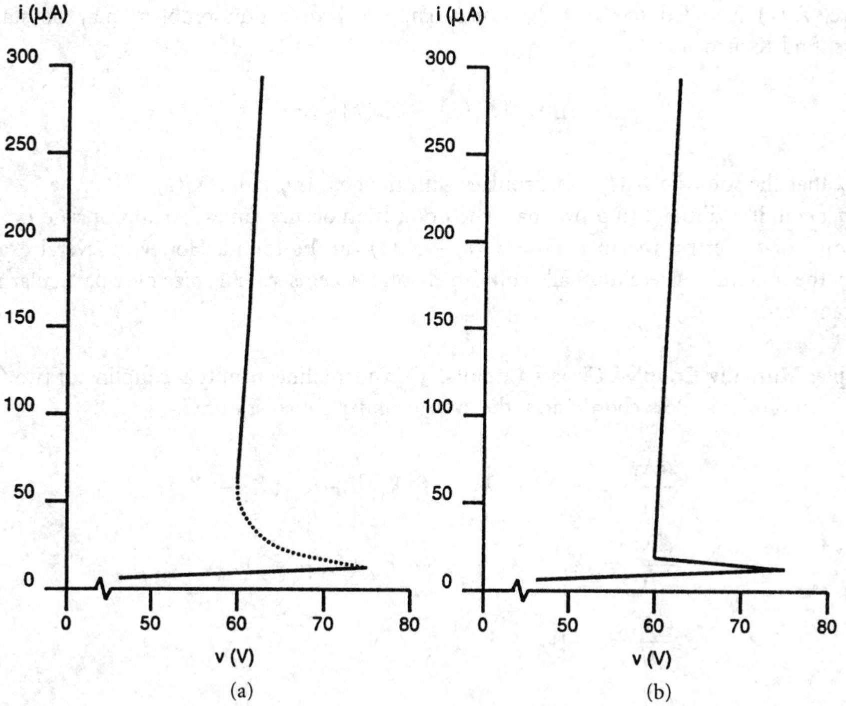

FIGURE 38.45 Neon bulb driving-point characteristics: (a) measured and (b) simulated. *Source:* M. P. Kennedy and L. O. Chua, "Van der Pol and chaos," *IEEE Trans. Circuit Syst.*, vol. CAS-33, p. 976, Oct. 1986. © 1986 IEEE.

8.5 Synchronization of Chaotic Circuits

Chaotic steady-state solutions are characterized by sensitive dependence on initial conditions: trajectories of two identical autonomous continuous-time dynamical systems started from slightly different initial conditions quickly become uncorrelated. Surprisingly perhaps, it is nevertheless possible to synchronize these systems in the sense that a trajectory of one asymptotically approaches that of the other. Two trajectories $\mathbf{X}_1(t)$ and $\mathbf{X}_2(t)$ are said to *synchronize* if

$$\lim_{t\to\infty} \left\|\mathbf{X}_1(t) - \mathbf{X}_2(t)\right\| = 0$$

In this section we describe two techniques for synchronizing chaotic trajectories.

Linear Mutual Coupling

The simplest technique for synchronizing two dynamical systems

$$\dot{\mathbf{X}}_1 = \mathbf{F}_1(\mathbf{X}_1) \qquad \mathbf{X}_1(0) = \mathbf{X}_{1_0}$$

$$\dot{\mathbf{X}}_2 = \mathbf{F}_2(\mathbf{X}_2) \qquad \mathbf{X}_2(0) = \mathbf{X}_{2_0}$$

is by linear mutual coupling of the form

$$\begin{aligned} \dot{\mathbf{X}}_1 &= \mathbf{F}_1(\mathbf{X}_1) + \mathbf{K}(\mathbf{X}_2 - \mathbf{X}_1) \qquad \mathbf{X}_1(0) = \mathbf{X}_{1_0} \\ \dot{\mathbf{X}}_2 &= \mathbf{F}_2(\mathbf{X}_2) + \mathbf{K}(\mathbf{X}_1 - \mathbf{X}_2) \qquad \mathbf{X}_2(0) = \mathbf{X}_{2_0} \end{aligned} \tag{38.22}$$

where $\mathbf{X}_1, \mathbf{X}_2 \in \mathbb{R}^n$ and $\mathbf{K} = \text{diag}(K_{11}, K_{22}, \ldots, K_{nn})^T$.

Here, $\mathbf{X}_1(t)$ is called the goal dynamics. The synchronization problem may be stated as follows: find K such that

$$\lim_{t\to\infty} \|\mathbf{X}_1(t) - \mathbf{X}_2(t)\| = 0$$

that is, that the solution $\mathbf{X}_2(t)$ synchronizes with the goal trajectory $\mathbf{X}_1(t)$.

In general, it is difficult to prove that synchronization occurs, unless an appropriate Lyapunov function[16] of the error system $\mathbf{E}(t) = \mathbf{X}_1(t) - \mathbf{X}_2(t)$ can be found. However, several examples exist in the literature where mutually coupled chaotic systems synchronize over particular ranges of parameters.

Example: Mutually Coupled Chua's Circuits. Consider a linear mutual coupling of two Chua's circuits. In dimensionless coordinates, the system under consideration is

$$\frac{dX_1}{dt} = \alpha[X_2 - X_1 - f(X_1)] + K_{11}(X_4 - X_1)$$
$$\frac{dX_2}{dt} = X_1 - X_2 - X_3 + K_{22}(X_5 - X_2)$$
$$\frac{dX_3}{dt} = -\beta y + K_{33}(X_6 - X_3)$$
$$\frac{dX_4}{dt} = \alpha(X_5 - X_4 - f(X_4)) + K_{11}(X_1 - X_4)$$
$$\frac{dX_5}{dt} = X_4 - X_5 - X_6 + K_{22}(X_2 - X_5)$$
$$\frac{dX_6}{dt} = -\beta X_5 + K_{33}(X_3 - X_6)$$

Two mutually coupled Chua's circuits characterized by $\alpha = 10.0$, $\beta = 14.87$, $a = -1.27$, $b = -0.68$ will synchronize (the solutions of the two systems will approach each other asymptotically) for the following matrices K:

X_1 – **coupling** $K_{11} > 0.5$, $K_{22} = K_{33} = 0$,
X_2 – **coupling** $K_{22} > 5.5$, $K_{11} = K_{33} = 0$, and
X_3 – **coupling** $0.7 < K_{33} < 2$, $K_{11} = K_{22} = 0$.

Coupling between states X_1 and X_4 may be realized experimentally by connecting a resistor between the tops of the nonlinear resistors, as shown in Fig. 38.46. States X_2 and X_5 may be coupled by connecting the tops of capacitors C_2 by means of a resistor.

An INSITE simulation of the system, which confirms synchronization of the two chaotic Chua's circuits in the case of linear mutual coupling between states V_{C_1} and V'_{C_1}, is shown in Fig. 38.47.

Pecora–Carroll Drive-Response Concept

The drive-response synchronization scheme proposed by Pecora and Carroll applies to systems that are drive-decomposable [16]. A dynamical system is called drive-decomposable if it can be partitioned into two subsystems that are coupled so that the behavior of the second (called the

[16]For a comprehensive exposition of Lyapunov stability theory, see [19].

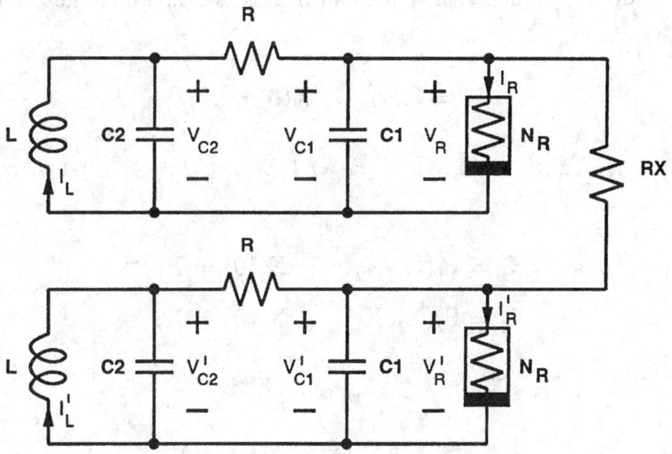

FIGURE 38.46 Synchronization of two Chua's circuits by means of resistive coupling between V_{C_1} and V'_{C_1}.

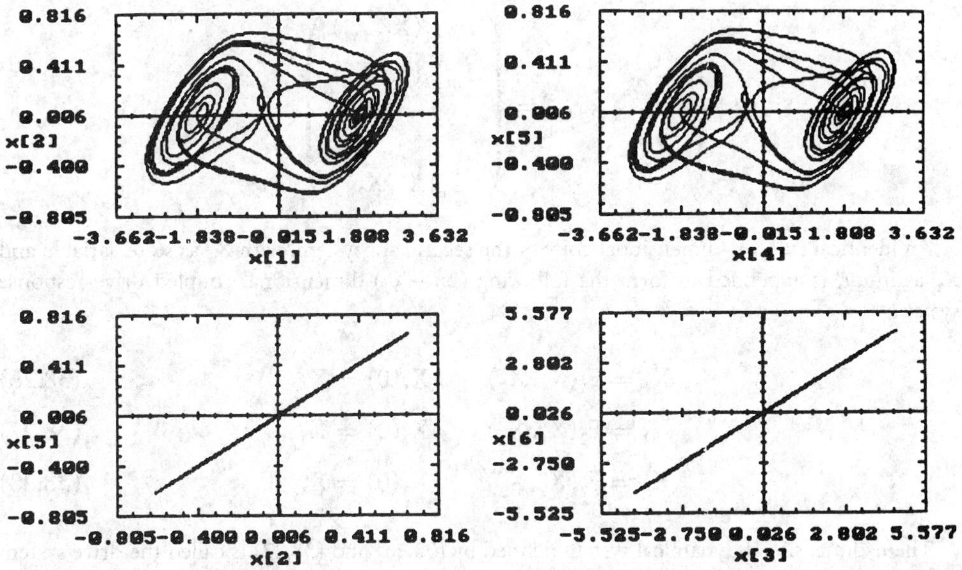

FIGURE 38.47 INSITE simulation of the normalized dimensionless form of Fig. 38.46, showing synchronization by mutual coupling of state variables. Identify $\{x[1], x[2], x[3]\}$ with $\{V_{C_1}, V_{C_2}, I_L\}$ and $\{x[4], x[5], x[6]\}$ with $\{V'_{C_1}, V'_{C_2}, I'_L\}$. $V'_{C_2}(t)$ synchronizes with $V_{C_2}(t)$ and $I'_L(t)$ synchronizes with $I_L(t)$. *Source:* L. O. Chua, M. Itoh, L. Kočarev, and K. Eckert, "Chaos synchronization in Chua's Circuit," *J. Circuits Syst. Comput.*, vol. 3, no. 1, p. 99, Mar. 1993.

response subsystem) depends on that of the first, but the behavior of the first (called the drive subsystem) is independent of that of the second.

To construct a drive-decomposable system, an n-dimensional autonomous continuous-time dynamical system

$$\dot{\mathbf{X}} = \mathbf{F}(\mathbf{X}) \qquad \mathbf{X}(0) = \mathbf{X}_0 \tag{38.23}$$

where $\mathbf{X} = (X_1, X_2, \ldots, X_n)^T$ and $\mathbf{F}(\mathbf{X}) = [F_1(\mathbf{X}), F_2(\mathbf{X}), \ldots, F_n(\mathbf{X})]^T$, is first partitioned into two subsystems

$$\dot{\mathbf{X}}_1 = \mathbf{F}_1(\mathbf{X}_1, \mathbf{X}_2) \qquad \mathbf{X}_1(0) = \mathbf{X}_{1_0} \tag{38.24}$$

$$\dot{\mathbf{X}}_2 = \mathbf{F}_2(\mathbf{X}_1, \mathbf{X}_2) \qquad \mathbf{X}_2(0) = \mathbf{X}_{2_0} \tag{38.25}$$

where $\mathbf{X}_1 = (X_1, X_2, \ldots, X_m)^T$, $\mathbf{X}_2 = (X_{m+1}, X_{m+2}, \ldots, X_n)^T$,

$$\mathbf{F}_1(\mathbf{X}_1, \mathbf{X}_2) = \begin{pmatrix} F_1(\mathbf{X}_1, \mathbf{X}_2) \\ F_2(\mathbf{X}_1, \mathbf{X}_2) \\ \vdots \\ F_m(\mathbf{X}_1, \mathbf{X}_2) \end{pmatrix}$$

and

$$\mathbf{F}_2(\mathbf{X}_1, \mathbf{X}_2) = \begin{pmatrix} F_{m+1}(\mathbf{X}_1, \mathbf{X}_2) \\ F_{m+2}(\mathbf{X}_1, \mathbf{X}_2) \\ \vdots \\ F_n(\mathbf{X}_1, \mathbf{X}_2) \end{pmatrix}$$

An identical $(n-m)$-dimensional copy of the second subsystem, with $\mathbf{X}_3$ as state variable and $\mathbf{X}_1$ as input, is appended to form the following $(2n - m)$-dimensional coupled drive-response system:

$$\dot{\mathbf{X}}_1 = \mathbf{F}_1(\mathbf{X}_1, \mathbf{X}_2) \qquad \mathbf{X}_1(0) = \mathbf{X}_{1_0} \tag{38.26}$$

$$\dot{\mathbf{X}}_2 = \mathbf{F}_2(\mathbf{X}_1, \mathbf{X}_2) \qquad \mathbf{X}_2(0) = \mathbf{X}_{2_0} \tag{38.27}$$

$$\dot{\mathbf{X}}_3 = \mathbf{F}_2(\mathbf{X}_1, \mathbf{X}_3) \qquad \mathbf{X}_3(0) = \mathbf{X}_{3_0} \tag{38.28}$$

The n-dimensional dynamical system defined by (38.26) and (38.27) is called the drive system and (38.28) is called the response subsystem.

Note that the second drive subsystem (38.27) and the response subsystem (38.28) lie in state spaces of dimension $\mathbb{R}^{(n-m)}$ and have identical vector fields $\mathbf{F}_2$ and inputs $\mathbf{X}_1$.

Consider a trajectory $\mathbf{X}_3(t)$ of (38.29) that originates from an initial state $\mathbf{X}_{3_0}$ "close" to $\mathbf{X}_{2_0}$. We may think of $\mathbf{X}_2(t)$ as a perturbation of $\mathbf{X}_3(t)$. In particular, define the error $\mathbf{x}_3(t) = \mathbf{X}_2(t) - \mathbf{X}_3(t)$. The trajectory $\mathbf{X}_2(t)$ approaches $\mathbf{X}_3(t)$ asymptotically (synchronizes) if $\|\mathbf{x}_3\| \to 0$ as $t \to \infty$. Equivalently, the response subsystem (38.29) is asymptotically stable when driven with $\mathbf{X}_1(t)$.

The stability of an orbit of a dynamical system may be determined by examining the linearization of the vector field along the orbit. The linearized response subsystem is governed by

$$\dot{\mathbf{x}}_3 = \mathbf{D}_{\mathbf{X}_3}\mathbf{F}_2(\mathbf{X}_1(t), \mathbf{X}_3)\mathbf{x}_3, \qquad \mathbf{x}_3(0) = \mathbf{x}_{3_0}$$

where $\mathbf{D}_{\mathbf{X}_3}\mathbf{F}_2(\mathbf{X}_1(t), \mathbf{X}_3)$ denotes the partial derivatives of the vector field $\mathbf{F}_2$ of the response subsystem with respect to $\mathbf{X}_3$. This is a linear time-varying system whose state transition matrix $\Phi_t(\mathbf{X}_{1_0}, \mathbf{X}_{3_0})$ maps a point $\mathbf{x}_{3_0}$ into $\mathbf{x}_3(t)$. Thus

$$\mathbf{x}_3(t) = \Phi_t(\mathbf{X}_{1_0}, \mathbf{X}_{3_0})\mathbf{x}_{3_0}$$

Note that Φ_t is a linear operator. Therefore, an $(n-m)$-dimensional ball $B_\varepsilon(\mathbf{X}_{3_0})$ of radius ε about $\mathbf{X}_{3_0}$ is mapped into an ellipsoid whose principal axes are determined by the singular values of Φ_t. In particular, a ball of radius ε is mapped by Φ_t into an ellipsoid, the maximum and minimum radii of which are bounded by the largest and smallest singular values, respectively, of Φ_t.

The conditional Lyapunov exponents $\lambda_i(\mathbf{X}_{1_0}, \mathbf{X}_{2_0})$ (hereafter denoted CLE) are defined by

$$\lambda_i(\mathbf{X}_{1_0}, \mathbf{X}_{2_0}) = \lim_{t\to\infty} \frac{1}{t} \ln \sigma_i[\phi_t(\mathbf{X}_{1_0}, \mathbf{X}_{2_0})], \quad i = 1, 2, \ldots, (n-m)$$

whenever the limit exists.

The term conditional refers to the fact that the exponents depend explicitly on the trajectory $\phi_t(\mathbf{X}_{1_0}, \mathbf{X}_{2_0})$ of the drive system.

Given that ε remains infinitesimally small, one is considering the local linearized dynamics along the flow determined by $\phi_t(\mathbf{X}_{1_0}, \mathbf{X}_{2_0})$ and determining the average local exponential rates of expansion and contraction along the principal axes of an ellipsoid. If all CLEs are negative, the response subsystem is asymptotically stable. A subsystem, all of whose CLEs are negative, is called a stable subsystem.

A stable subsystem does not necessarily exhibit dc steady-state behavior. For example, while an asymptotically stable linear parallel *RLC* circuit has all negative LEs, the system settles to a periodic steady-state solution when driven with a sinusoidal current. Although the *RLC* subcircuit has negative CLEs in this case, the complete forced circuit has one nonnegative LE corresponding to motion along the direction of the flow.

Theorem 4 (Pecora and Carroll): *The trajectories* $\mathbf{X}_2(t)$ *and* $\mathbf{X}_3(t)$ *will synchronize only if the CLEs of the response system (38.28) are all negative.*

Note that this is a necessary but not sufficient condition for synchronization. If the response and second drive subsystems are identical and the initial conditions $\mathbf{X}_{2_0}$ and $\mathbf{X}_{3_0}$ are sufficiently close, and the CLEs of (38.28) are all negative, synchronization will occur. However, if the systems are not identical or the initial conditions are not sufficiently close, synchronization might not occur, even if all of the CLEs are negative.

Although we have described it only for an autonomous continuous-time system, the drive-response technique may also be applied for synchronizing nonautonomous and discrete-time circuits.

Cascaded Drive-Response Systems

The drive-response concept may be extended to the case where a dynamical system can be partitioned into more than two parts. A simple two-level drive-response cascade is constructed as follows. Divide the dynamical system

$$\dot{\mathbf{X}} = \mathbf{F}(\mathbf{X}), \qquad \mathbf{X}(0) = \mathbf{X}_0 \tag{38.29}$$

into three parts:

$$\dot{\mathbf{X}}_1 = \mathbf{F}_1(\mathbf{X}_1, \mathbf{X}_2, \mathbf{X}_3) \qquad \mathbf{X}_1(0) = \mathbf{X}_{1_0} \tag{38.30}$$

$$\dot{\mathbf{X}}_2 = \mathbf{F}_2(\mathbf{X}_1, \mathbf{X}_2, \mathbf{X}_3) \qquad \mathbf{X}_2(0) = \mathbf{X}_{2_0} \tag{38.31}$$

$$\dot{\mathbf{X}}_3 = \mathbf{F}_3(\mathbf{X}_1, \mathbf{X}_2, \mathbf{X}_3) \qquad \mathbf{X}_3(0) = \mathbf{X}_{3_0} \tag{38.32}$$

Now construct an identical copy of the subsystems corresponding to (38.31) and (38.32) with $\mathbf{X}_1(t)$ as input:

$$\dot{\mathbf{X}}_4 = \mathbf{F}_2(\mathbf{X}_1, \mathbf{X}_4, \mathbf{X}_5) \qquad \mathbf{X}_4(0) = \mathbf{X}_{4_0} \tag{38.33}$$

$$\dot{\mathbf{X}}_5 = \mathbf{F}_3(\mathbf{X}_1, \mathbf{X}_4, \mathbf{X}_5) \qquad \mathbf{X}_5(0) = \mathbf{X}_{5_0} \tag{38.34}$$

If all of the CLEs of the driven subsystem composed of (38.33) and (38.34) are negative then, after the transient decays, $\mathbf{X}_4(t) = \mathbf{X}_2(t)$ and $\mathbf{X}_5(t) = \mathbf{X}_3(t)$.

Note that (38.30)–(38.34) together define one large coupled dynamical system. Hence, the response subsystem can exhibit chaos even if all of its CLEs are negative.

Proceeding one step further, we reproduce subsystem (38.30):

$$\dot{\mathbf{X}}_6 = \mathbf{F}_1(\mathbf{X}_6, \mathbf{X}_4, \mathbf{X}_5) \qquad \mathbf{X}_6(0) = \mathbf{X}_{6_0} \tag{38.35}$$

As before, if all of the conditional Lyapunov exponents of (38.35) are negative, then $\|\mathbf{X}_6(t) - \mathbf{X}_1(t)\| \to 0$. If the original system could be partitioned so that (38.30) is one-dimensional, then using (38.33)–(38.35) as a driven response system, *all* of the variables in the drive system could be reproduced by driving with just *one* variable $X_1(t)$. This principle can be exploited for spread-spectrum communication using a chaotic carrier signal.

Example: Synchronization of Chua's Circuits Using the Drive-Response Concept. Chua's circuit may be partitioned in three distinct ways to form five-dimensional drive-decomposable systems:

X_1-drive configuration

$$\frac{dX_1}{dt} = \alpha[X_2 - X_1 - f(X_1)]$$

$$\frac{dX_2}{dt} = X_1 - X_2 - X_3$$

$$\frac{dX_3}{dt} = -\beta X_2$$

$$\frac{dX_4}{dt} = X_1 - X_4 - X_5$$

$$\frac{dX_5}{dt} = -\beta X_5$$

With $\alpha = 10.0$, $\beta = 14.87$, $a = -1.27$, and $b = -0.68$, the CLEs for the (X_2, X_3) subsystem (calculated using INSITE) are $(-0.5, -0.5)$.

X_2-drive configuration

$$\frac{dX_2}{dt} = X_1 - X_2 - X_3$$
$$\frac{dX_1}{dt} = \alpha[X_2 - X_1 - f(X_1)]$$
$$\frac{dX_3}{dt} = -\beta X_2$$
$$\frac{dX_4}{dt} = \alpha[X_2 - X_4 - f(X_4)]$$
$$\frac{dX_5}{dt} = -\beta X_5$$

This case is illustrated in Fig. 38.48. The CLEs of the (X_1, X_3) subsystem are 0 and -2.5 ± 0.05. Because of the zero CLE, states $X_5(t)$ and $X_3(t)$ remain a constant distance $|X_{3_0} - X_{5_0}|$ apart, as shown in Fig. 38.49.

X_3-drive configuration

$$\frac{dX_3}{dt} = -\beta X_2$$
$$\frac{dX_1}{dt} = \alpha[X_2 - X_1 - f(X_1)]$$
$$\frac{dX_2}{dt} = X_1 - X_2 - X_3$$
$$\frac{dX_4}{dt} = \alpha[X_5 - X_4 - f(X_4)]$$

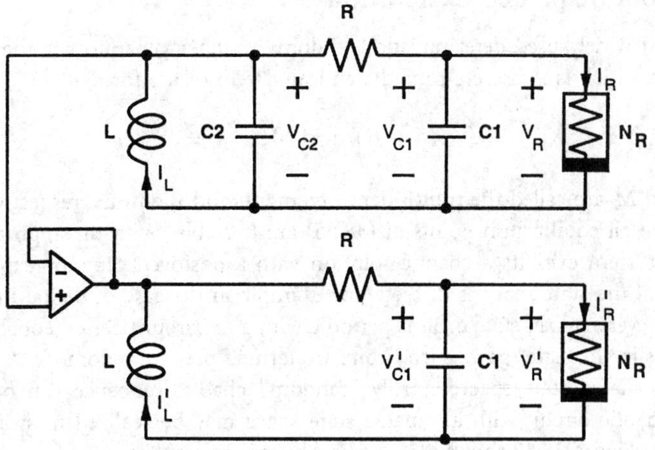

FIGURE 38.48 Synchronization of two Chua's circuits using the Pecora–Carroll drive-response method with V_{C_2} as drive variable.

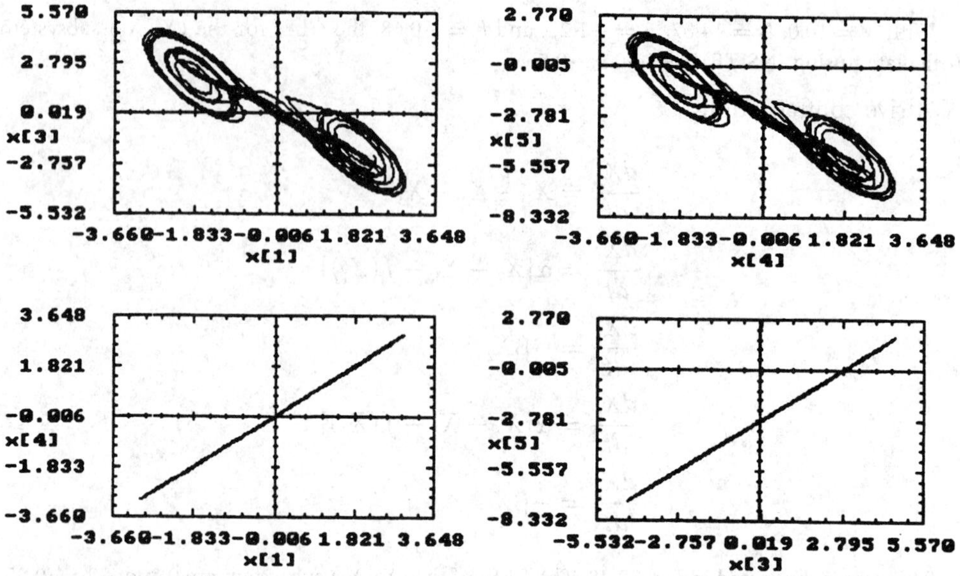

FIGURE 38.49 INSITE simulation of the normalized dimensionless form of Fig. 38.48, showing synchronization of state variables. Identify $\{x[1], x[2], x[3]\}$ with $\{V_{C_1}, V_{C_2}, I_L\}$ and $\{x[4], x[5]\}$ with $\{V'_{C_1}, I'_L\}$. $V'_{C_1}(t)$ synchronizes with $V_{C_1}(t)$ and $I'_L(t)$ synchronizes with $I_L(t)$. Because one of the CLEs of the response subsystem is zero, the difference in the initial conditions $I_{L_0} - I_{L_0}$ does not decay to zero. *Source:* L. O. Chua, M. Itoh, L. Kočarev, and K. Eckert, "Chaos synchronization in Chua's circuit," *J. Circuits Syst. Comput.*, vol. 3, no. 1, p. 106, Mar. 1993.

$$\frac{dX_5}{dt} = X_4 - X_5 - X_3$$

The CLEs in this case are 1.23 ± 0.03 and -5.42 ± 0.02. Because the (X_1, X_2) subsystem has a positive CLE, the response subsystem does not synchronize.

38.6 Applications of Chaos

Pseudorandom Sequence Generation

One of the most widely-used deterministic "random" number generators is the linear congruential generator, which is a discrete-time dynamical system of the form

$$X_{k+1} = (AX_k + B) \bmod M, \quad k = 0, 1, \cdots \tag{38.36}$$

where A, B, and M are called the **multiplier, increment,** and **modulus**, respectively.

If $A > 1$, then all equilibrium points of (38.36) are unstable. With the appropriate choice of constants, this system exhibits a chaotic solution with a positive Lyapunov exponent equal to $\ln A$. However, if the state space is discrete, for example in the case of digital implementations of (38.37), then every steady-state orbit is periodic with a maximum period equal to the number of distinct states in the state space; such orbits are termed pseudorandom.

By using an analog state space, a truly "random" chaotic sequence can be generated. A discrete-time chaotic circuit with an analog state space may be realized in switched-capacitor (SC) technology. Figure 38.50 shows a SC realization of the parabolic map

$$x_{k+1} = V - 0.5x_k^2 \tag{38.37}$$

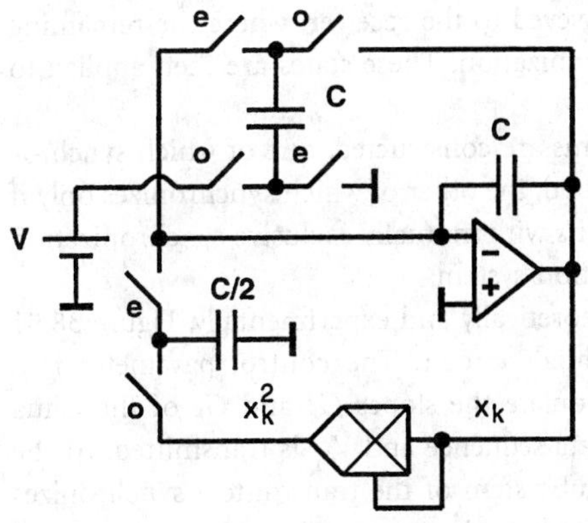

FIGURE 38.50 Switched-capacitor (SC) realization of the parabolic map $x_{k+1} = V - 0.5X_k^2$. The switches labeled *o* and *e* are driven by the odd and even phases, respectively, of a nonoverlapping two-phase clock.

which, by the change of variables

$$X_k = Ax_k + B$$

with $\mu = 1/(2A)$, $B = 0.5$, and $A = (-1 \pm \sqrt{1+2V})/(4V)$, is equivalent to the logistic map

$$X_{k+1} = \mu X_k(1 - X_k) \tag{38.38}$$

The logistic map is chaotic for $\mu = 4$ with Lyapunov exponent ln 2 [18]. Figure 38.16 is a bifurcation diagram of (38.38) with $0 \leq \mu \leq 4$.

For $V < 1.5$, the steady-state solution of the SC parabolic map described by (38.37) is a fixed point. As the bifurcation parameter V is increased from 1.5 to 3 V, the circuit undergoes a series of period-doubling bifurcations to chaos. $V = 4$ corresponds to fully developed chaos on the open interval $(0 < X_k < 1)$ in the logistic map with $\mu = 4$.

Spread-Spectrum and Secure Communications

Modulation and coding techniques for mobile communication systems are driven by two fundamental requirements: that the communication channel should be secure and the modulation scheme should be tolerant of multipath effects. Security is ensured by coding and immunity from multipath degradation may be achieved by using a spread-spectrum transmission.

With appropriate modulation and demodulation techniques, the "random" nature and "noise-like" spectral properties of chaotic circuits can be exploited to provide simultaneous coding and spreading of a transmission.

Chaotic Switching

The simplest idea for data transmission using a chaotic circuit is use the data to modulate some parameter(s) of the transmitter. This technique is called **parameter modulation, chaotic switching**, or **chaos shift keying** (CSK).

In the case of binary data, the information signal is encoded as a pair of circuit parameter sets which produce distinct attractors in a dynamical system (the transmitter). In particular, a single control parameter μ may be switched between two values μ_0, corresponding to attractor $\mathscr{A}_0$, and μ_1, corresponding to $\mathscr{A}_1$. By analogy with FSK and PSK, this technique is known as chaos shift keying.

The binary sequence to be transmitted is mapped into the appropriate control signal $\mu(t)$ and the corresponding trajectory switches, as required, between $\mathscr{A}_1$ and $\mathscr{A}_0$.

One of the state variables of the transmitter is conveyed to the receiver, where the remaining state variables are recovered by drive-response synchronization. These states are then applied to the second stage of a drive-response cascade.

At the second level, two matched receiver subsystems are constructed, one of which synchronizes with the incoming signal if a "zero" was transmitted, the other of which synchronizes only if a "one" was transmitted. The use of two receiver circuits with mutually exclusive synchronization properties improves the reliability of the communication system.

Chaos shift keying has been demonstrated both theoretically and experimentally. Figure 38.51 shows a CSK transmitter and receiver based on Chua's circuit. The control parameter is a resistor with conductance ΔG whose effect is to modulate the slopes G_a and G_b of the Chua diode. Switch S is opened and closed by the binary data sequence and V_{C_1} is transmitted. At the receiver, the first subsystem (a copy of the (V_{C_2}, I_L) subsystem of the transmitter) synchronizes with the incoming signal, recovering $V_{C_2}(t)$. Thus, $V_{C_{21}}(t) \to V_{C_2}(t)$.

The synchronized local copy of $V_{C_2}(t)$ is then used to synchronize two further subsystems corresponding to the V_{C_1} subsystem of the transmitter with and without the resistor ΔG.

If the switch is closed at the transmitter, $V'_{C_{12}}$ (but not $V_{C_{12}}$) synchronizes with V_{C1} and if the switch is open, $V_{C_{12}}$ (but not $V'_{C_{12}}$) synchronizes with V_{C_1}.

Figure 38.52 shows simulated results for a similar system consisting of two Chua's circuits. At the receiver, a decision must be made as to which bit has been transmitted. In this case, b_{out} was derived using the rule

$$b_{\text{out}} = \begin{pmatrix} 0, b_{\text{old}} = 0 & \text{for} & a_0 < \varepsilon, a_1 > \varepsilon \\ 1, b_{\text{old}} = 1 & \text{for} & a_0 > \varepsilon, a_1 < \varepsilon \\ b_{\text{old}} & \text{for} & a_0 < \varepsilon, a_1 < \varepsilon \\ 1 - b_{\text{old}} & \text{for} & a_0 > \varepsilon, a_1 > \varepsilon \end{pmatrix}$$

where b_{old} is the last bit received and b_{out} is the current bit [14].

Chaotic Masking

Chaotic masking is a method of hiding an information signal by adding it to a chaotic carrier at the transmitter. The drive-response synchronization technique is used to recover the carrier at the receiver.

Figure 38.53 shows a block diagram of a communication system using matched Chua's circuits. The receiver has the same two-layer structure as in the previous example. The first subcircuit, which has very negative conditional Lyapunov exponents, synchronizes with the incoming signal, despite the perturbation $s(t)$, and recovers V_{C_2}. The second subcircuit, when driven by V_{C_2}, produces the receiver's copy of V_{C_1}. The information signal $r(t)$ is recovered by subtracting the local copy of V_{C_1} from the incoming signal $V_{C_1} + s(t)$.

Vector Field Modulation

With *vector field modulation,* an information-carrying signal is added to the vector field at the transmitter and recovered at the receiver [9].

A dynamical system is partitioned as follows:

$$\dot{\mathbf{X}}_1 = \mathbf{F}_1(\mathbf{X}_1, \mathbf{X}_2) \qquad \mathbf{X}_1(0) = \mathbf{X}_{1_0}$$
$$\dot{\mathbf{X}}_2 = \mathbf{F}_2(\mathbf{X}_1, \mathbf{X}_2) \qquad \mathbf{X}_2(0) = \mathbf{X}_{2_0}$$

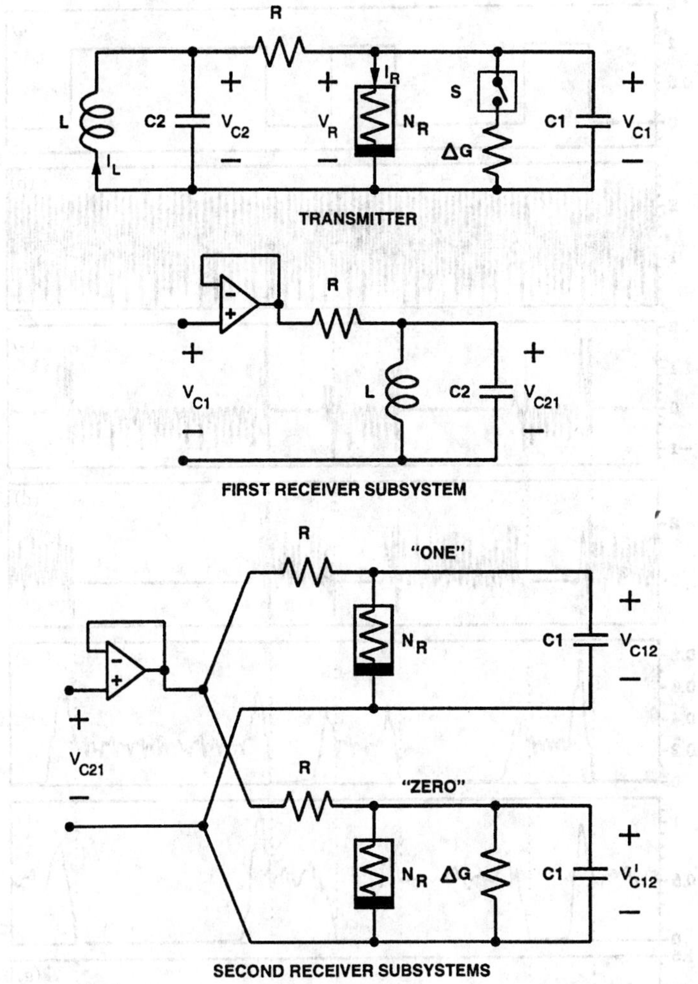

FIGURE 38.51 Chaos shift keying communication system using Chua's circuit. When a "one" is transmitted, switch S remains open, $V_{C_{21}}(t)$ synchronizes with $V_{C_2}(t)$, $V_{C_{12}}(t)$ synchronizes with $V_{C_1}(t)$, and $V'_{C_{12}}(t)$ falls out of synchronization with $V_{C_1}(t)$. When a "zero" is transmitted, switch S is closed, $V_{C_{21}}(t)$ synchronizes with $V_{C_2}(t)$, $V_{C_{12}}(t)$ falls out of synchronization with $V_{C_1}(t)$, and $V'_{C_{12}}(t)$ synchronizes with $V_{C_1}(t)$.

The information signal $\mathbf{S}(t)$ is added into the transmitter

$$\dot{\mathbf{X}}_1 = \mathbf{F}_1(\mathbf{X}_1, \mathbf{X}_2) + \mathbf{S}(t) \qquad \mathbf{X}_1(0) = \mathbf{X}_{1_0} \tag{38.39}$$

$$\dot{\mathbf{X}}_2 = \mathbf{F}_2(\mathbf{X}_1, \mathbf{X}_2) \qquad \mathbf{X}_2(0) = \mathbf{X}_{2_0} \tag{38.40}$$

and the state $\mathbf{X}_1(t)$ transmitted.

The receiver contains a copy of the second drive subsystem (38.41)

$$\dot{\mathbf{X}}_3 = \mathbf{F}_2(\mathbf{X}_1, \mathbf{X}_3) \qquad \mathbf{X}_3(0) = \mathbf{X}_{3_0}$$

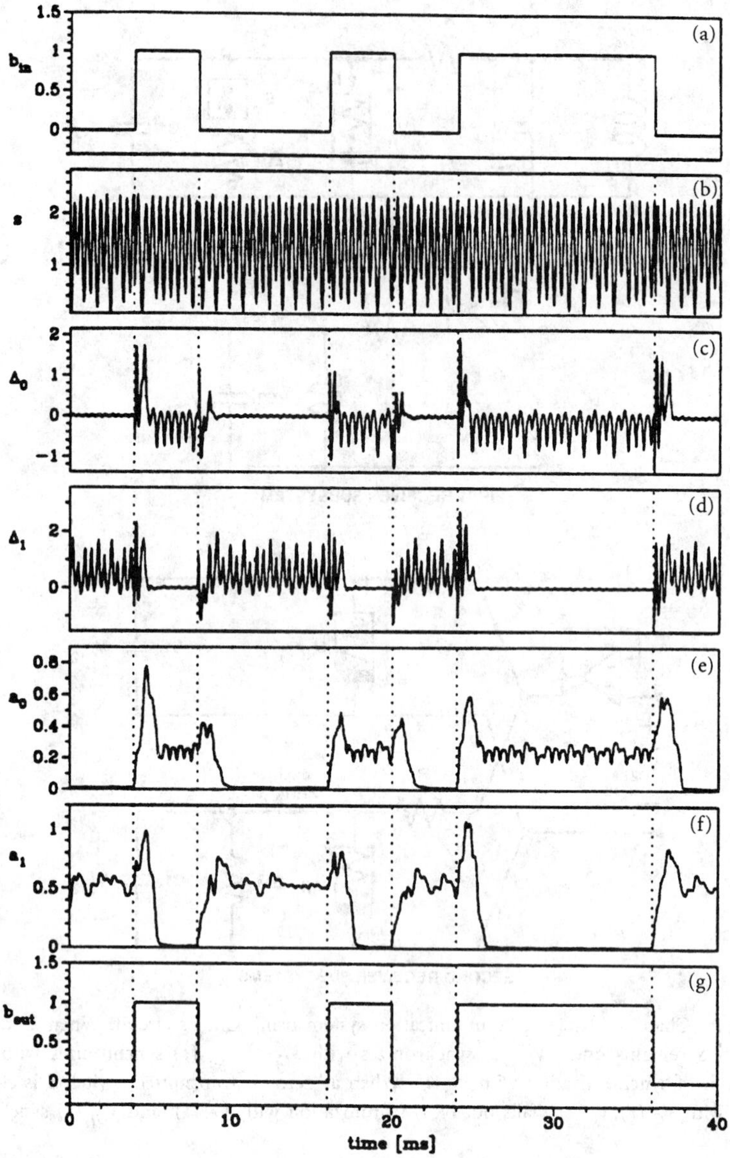

FIGURE 38.52 Chaos shift keying waveforms. (a) Binary input signal b_{in}; (b) transmitted signal $s(t)$; (c) response $\Delta_0 = V_{C_{12}} - V_{C_1}$; (d) response $\Delta_1 = V'_{C_{12}} - V_{C_1}$; (e) 40-point moving average of Δ_0; (f) 40-point moving average of Δ_1; (g) output binary signal b_{out} when $\varepsilon = 0.1$. *Source:* M. Ogorzałek, "Taming chaos—Part I: Synchronization," *IEEE Trans. Circuits Syst. I,* vol. 40, p. 696, Oct. 1993. © 1993 IEEE.

whose state $\mathbf{X}_3(t)$ synchronizes with $\mathbf{X}_2(t)$, and a demodulator,

$$\mathbf{R}(t) = \dot{\mathbf{X}}_1 - \mathbf{F}_1(\mathbf{X}_1, \mathbf{X}_3)$$

which is used to recover $\mathbf{S}(t)$.

If all of the CLEs of the response system are negative, and the initial conditions are sufficiently close, then $\mathbf{X}_3(t) \to \mathbf{X}_2(t)$ and the recovered signal $\mathbf{R}(t)$ equals $\mathbf{S}(t)$.

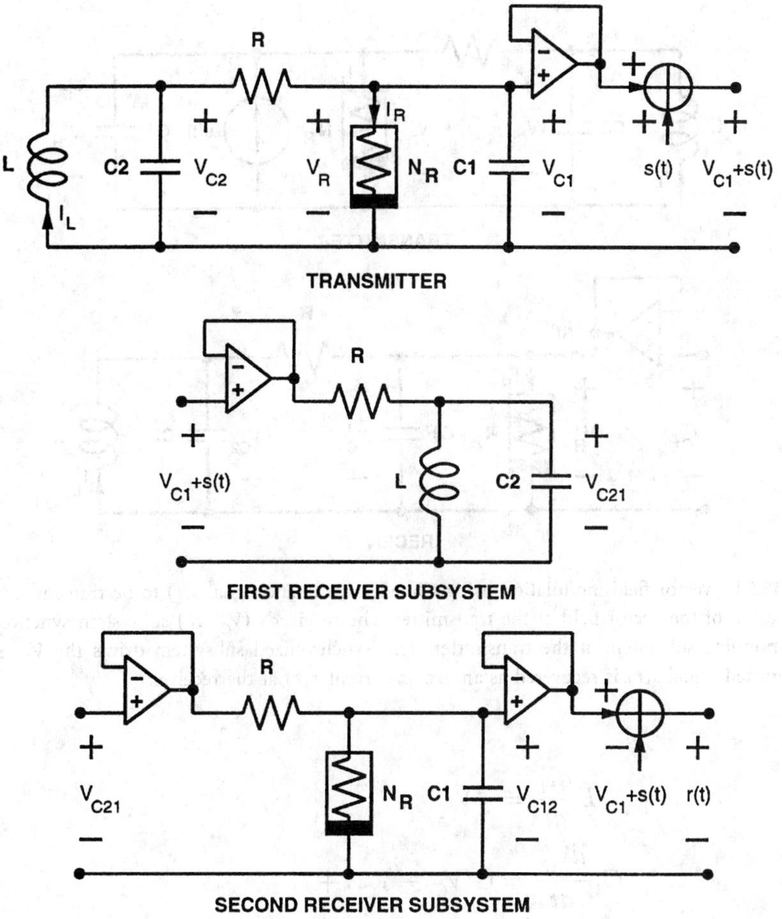

FIGURE 38.53 Chaos masking using Chua's circuits. At the transmitter, the information signal $s(t)$ is added to the chaotic carrier signal $V_{C_1}(t)$. Provided $s(t)$ is sufficiently small and the first receiver subsystem is sufficiently stable, $V_{C_{21}}(t)$ synchronizes with $V_{C_2}(t)$. This signal is applied to a second receiver subsystem, from which an estimate $V_{C_{12}}(t)$ of the unmodulated carrier $V_{C_1}(t)$ is derived. $V_{C_{12}}$ is subtracted from the incoming signal $V_{C_1} + s(t)$ to yield the received signal $r(t)$. The method works well only if $s(t)$ is much smaller than $V_{C_1}(t)$.

Example: Communication via Vector Field Modulation Using Chua's Circuits

A communication system based on Chua's circuit that uses the vector field modulation technique and a chaotic carrier is shown in Fig. 38.54. In this case, the signal to be transmitted is the scalar current $s(t)$, which is recovered as a current $r(t)$ at the receiver.

The state equations of the coupled drive-response system are

$$C_1 \frac{dV_{C_1}}{dt} = G(V_{C_2} - V_{C_1}) - f(V_{C_1}) + s(t)$$

$$C_2 \frac{dV_{C_2}}{dt} = G(V_{C_1} - V_{C_2}) + I_L$$

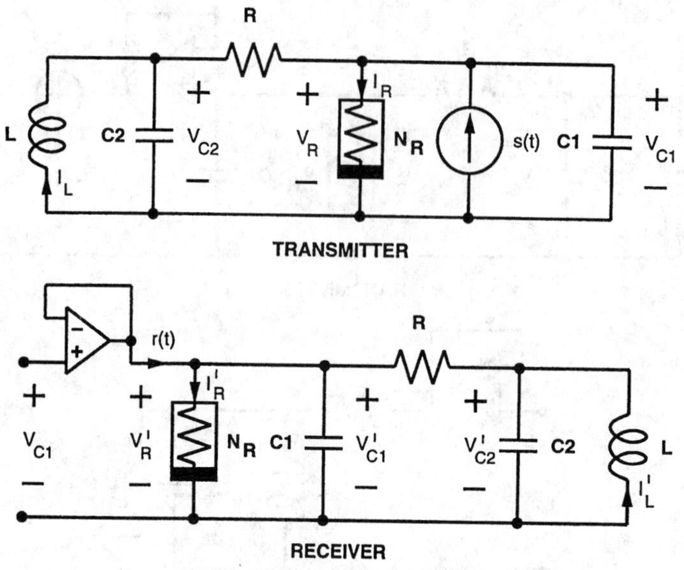

FIGURE 38.54 Vector field modulation using Chua's circuit. The signal $s(t)$ to be transmitted is applied as perturbation of the vector field at the transmitter. The receiver's (V'_{C_2}, I'_L) subsystem synchronizes with the corresponding subsystem at the transmitter. This synchronized subsystem drives the V'_{C_1} subsystem. The transmitted signal $s(t)$ is recovered as an excess current $r(t)$ at the receiver.

$$
\begin{aligned}
L\frac{dI_L}{dt} &= -V_{C_2} \\
C_2\frac{dV'_{C_2}}{dt} &= G(V_{C_1} - V'_{C_2}) + I'_L \\
L\frac{dI'_L}{dt} &= -V'_{C_2}
\end{aligned}
$$

Since the CLEs of the (V_{C_2}, I_L) subsystem are negative, $V'_{C_2}(t) \to V_{C_2}(t)$ and $I'_L(t) \to I_L(t)$. The current $r(t)$ at the receiver is given by

$$
\begin{aligned}
r(t) &= C_1\frac{dV_{C_1}}{dt} - G(V'_{C_2} - V_{C_1}) - f(V_{C_1}) \\
&= C_1\frac{dV_{C_1}}{dt} - G(V_{C_2} - V_{C_1}) - f(V_{C_1}) \\
&= s(t)
\end{aligned}
$$

Miscellaneous

Chaotic circuits may also be used for suppressing spurious tones in ΣΔ modulators, for modeling musical instruments, fractal pattern generation, image-processing, and pattern recognition [3].

A chaotic attractor contains an infinite number of unstable periodic trajectories of different periods. Various control schemes for stabilizing particular orbits in chaotic circuits have been successfully demonstrated [14].

ferences

[1] F. M. Callier and C. A. Desoer, *Linear System Theory*, New York: Springer-Verlag, 1991.

[2] L. O. Chua, C. A. Desoer, and E. S. Kuh, *Linear and Nonlinear Circuits*, New York: McGraw-Hill, 1987.

[3] L. O. Chua and M. Hasler, Eds., *Special Issue on Chaos in Nonlinear Electronic Circuits*, Part A: Tutorials and Reviews, *IEEE Trans. Circuits Syst. I, Fundament. Theory Applicat.*, vol. 40, Oct. 1993; Part B: Bifurcation and Chaos, *IEEE Trans. Circuits Syst. I, Fundament. Theory Applicat.*, vol. 40, Nov. 1993; Part C: Applicat., *IEEE Trans. Circuits Syst. II, Analog and Digital Signal Process.*, vol. 40, Oct. 1993.

[4] L. O. Chua, M. Itoh, L. Kočarev, and K. Eckert, "Chaos synchronization in Chua's circuit," *J. Circuits Syst. Comput.*, vol. 3, no. 1, pp. 93–108, Mar. 1993.

[5] L. O. Chua, C. W. Wu, A. Huang, and G.-Q. Zhong, "A universal circuit for studying and generating chaos—Part I: Routes to chaos," *IEEE Trans. Circuits Syst. I, Fundamental Theory Applicat.*, vol. 40, pp. 732–744, Oct. 1993.

[6] J. A. Glazier and A. Libchaber, "Quasi-periodicity and dynamical systems: An experimentalist's view," *IEEE Trans. Circuits Syst.*, vol. 35, pp. 790–809, July 1988.

[7] J. Guckenheimer and P. Holmes, *Nonlinear Oscillations, Dynamical Systems, and Bifurcations of Vector Fields*, New York: Springer-Verlag, 1983.

[8] M. W. Hirsch, "The dynamical systems approach to differential equations," *Bull. Amer. Math. Soc.*, vol. 11, no. 1, pp. 1–64, July 1984.

[9] M. Itoh, H. Murakami, K. S. Halle, and L. O. Chua, "Transmission of signals by chaos synchronization," IEICE Tech. Rep., CAS93-39, NLP93-27, pp. 89–96, 1993.

[10] M. P. Kennedy, "Three steps to chaos—Part I: Evolution," *IEEE Trans. Circuits and Systems I, Fundament. Theory Applicat.*, vol. 40, pp. 640–656, Oct. 1993; "Three steps to chaos—Part II: A Chua's circuit primer," *IEEE Trans. Circuits Syst. I, Fundament. Theory Applicat.*, vol. 40, pp. 657–674, Oct. 1993.

[11] M. P. Kennedy and L. O. Chua, "Van der Pol and chaos," *IEEE Trans. Circuits Syst.*, vol. CAS-33, pp. 974–980, Oct. 1986.

[12] M. P. Kennedy, K. R. Krieg, and L. O. Chua, "The Devil's staircase: The electrical engineer's fractal," *IEEE Trans. Circuits Syst.*, vol. 36, pp. 1133–1139, 1989.

[13] W. Lauterborn and U. Parlitz, "Methods of chaos physics and their application to acoustics," *J. Acoust. Soc. Amer.*, vol. 84, no. 6, pp. 1975–1993, Dec. 1988.

[14] M. Ogarzałek, "Taming chaos—Part I: Synchronization," *IEEE Trans. Circuits Syst. I, Fundament. Theory Applicat.*, vol. 40, pp. 693–699, Oct. 1993; "Taming chaos—Part II: Control," *IEEE Trans. Circuits Syst. I, Fundament. Theory Applicat.*, vol. 40, pp. 700–706, Oct. 1993.

[15] T. S. Parker and L. O. Chua, *Practical Numerical Algorithms for Chaotic Systems*, New York: Springer-Verlag, 1989.

[16] L. M. Pecora and T. Carroll, "Driving systems with chaotic signals," *Phys. Rev.*, vol. 44, no. 4, pp. 2374–2383, Aug. 15, 1991.

[17] W. H. Press, B. P. Flannery, S. A. Teukolsky, and W. T. Vetterling, *Numerical Recipes in C*, Cambridge: Cambridge Univ., 1988.

[18] J. M. T. Thompson and H. B. Stewart, *Nonlinear Dynamics and Chaos*, New York: Wiley, 1986.

[19] M. Vidyasagar, *Nonlinear Systems Analysis*, Englewood Cliffs, NJ: Prentice-Hall, 1978.

[20] C. W. Wu and N. F. Rul'kov, "Studying chaos via 1-D maps—A tutorial," *IEEE Trans. Circuits Syst. I, Fundament. Theory Applicat.*, vol. 40, pp. 707–721, Oct. 1993.

[21] L. O. Chua, M. Komuro, and T. Matsumoto, "The Double Scroll Family, Parts I and II," *IEEE Trans. Circuits Syst.*, vol. 33, pp. 1073–1118, Nov. 1986.

Further Information

Current Research in Chaotic Circuits. The August 1987 issue of the *Proceedings of the IEEE* is devoted to "Chaotic Systems." The *IEEE Transactions on Circuits and Systems*, July 1988, focuses on "Chaos and Bifurcations of Circuits and Systems."

A three-part special issue of the *IEEE Transactions on Circuits and Systems* on "Chaos in Electronic Circuits" appeared in October (parts I and II) and November 1993 (part I). This Special Issue contains 42 papers on various aspects of bifurcations and chaos.

Two Special Issues (March and June 1993) of the *Journal of Circuits, Systems, and Computers* are devoted to "Chua's Circuit: A Paradigm for Chaos." These works, along with several additional papers, a pictorial guide to forty-five attractors in Chua's oscillator, and the ABC simulator, have been compiled into a book of the same name—*Chua's Circuit: A Paradigm for Chaos,* R. N. Madan, Ed. Singapore: World Scientific, Singapore 1993.

Developments in the field of bifurcations and chaos, with particular emphasis on the applied sciences and engineering, are reported in *International Journal of Bifurcation and Chaos,* which is published quarterly by World Scientific, Singapore 9128.

Research in chaos in electronic circuits appears regularly in the *IEEE Transactions on Circuits and Systems.*

Simulation of Chaotic Circuits A variety of general-purpose and custom software tools has been developed for studying bifurcations and chaos in nonlinear circuits systems.

ABC (Adventures in Bifurcations and Chaos) is a graphical simulator of Chua's oscillator, which runs on IBM-compatible PCs. ABC contains a database of component values for all known attractors in Chua's oscillator, initial conditions, and parameter sets corresponding to homoclinic and heteroclinic trajectories, and bifurcation sequences for the period-doubling, intermittency, and quasiperiodic routes to chaos. The program and database are available from

Dr. Michael Peter Kennedy,
Department of Electronic and Electrical Engineering,
University College Dublin,
Dublin 4, Ireland

E-mail: mpk@midir.ucd.ie

In "Learning About Chaotic Circuits with SPICE," *IEEE Transactions on Education,* vol. 36, pp. 28–35, Jan. 1993. David Hamill describes how to simulate a variety of smooth chaotic circuits using the general-purpose circuit simulator SPICE. A commercial variant of SPICE called PSpice is available from

MicroSim Corporation,
20 Fairbanks,
Irvine, CA 92718

Telephone: (714) 770-3022

The free student evaluation version of this program is sufficiently powerful for studying simple chaotic circuits. PSpice runs on both workstations and PCs.

INSITE is a software toolkit that was developed at the University of California, Berkeley, for studying continuous-time and discrete-time dynamical systems. The suite of nine programs calculates and displays trajectories, power spectra, and state delay maps, draws vector fields of two-dimensional systems, reconstructs attractors from time series, calculates Poincaré sections, dimension, and Lyapunov exponents. The package runs on DEC, HP, IBM, and Sun workstations under UNIX, and on IBM-compatible PCs under DOS. For additional information, contact

INSITE Software,
P.O. Box 9662,
Berkeley, CA 94709

Telephone: (510) 530-9259

VIII

)istributed Circuits

K. Ishii
rquette University

39

Transmission Lines

K. Ishii
rquette University,
sconsin

9.1 Generic Relations

Equivalent Circuit

The equivalent circuit of a generic transmission line in monochromatic single frequency operation is shown in Fig. 39.1 [1–3], where $\dot{Z}$ is a series impedance per unit length of the transmission line (Ω/m) and $\dot{Y}$ is a shunt admittance per unit length of the transmission line (S/m). For a uniform, nonlinear transmission line, either $\dot{Z}$ or $\dot{Y}$ or both $\dot{Z}$ and $\dot{Y}$ are functions of the transmission line voltage and current, but both $\dot{Z}$ and $\dot{Y}$ are not functions of location on the transmission line. For a nonuniform linear transmission line, both $\dot{Z}$ and $\dot{Y}$ are functions of location, but not functions of voltage and current on the transmission line. For a nonuniform and nonlinear transmission line, both $\dot{Z}$ and $\dot{Y}$ or $\dot{Z}$ or $\dot{Y}$ are functions of the voltage, the current, and the location on the transmission line.

Transmission Line Equations

Ohm's law of a transmission line, which is the amount of voltage drop on a transmission line per unit distance of voltage transmission is expressed as

$$\frac{d\dot{V}}{dz} = -\dot{I}\dot{Z} \quad \text{(V/m)} \tag{39.1}$$

93-8341-2/95/$0.00+$.50

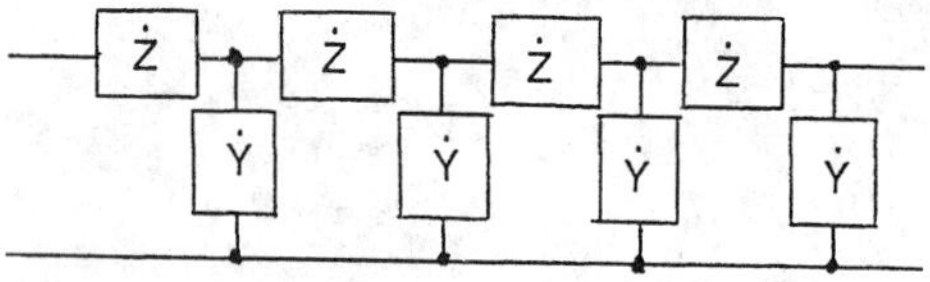

FIGURE 39.1 Equivalent circuit of a generic transmission line.

where $\dot{V}$ is the transmission line voltage (volt), $\dot{I}$ is the transmission line current (ampere), Z is the series impedance per unit length of the transmission line (Ω), and z is a one-dimensional coordinate placed in parallel to the transmission line (meter). The equation of current decrease per unit length is

$$\frac{d\dot{I}}{dz} = -\dot{Y}\dot{V} \quad (\text{A/m}) \tag{39.2}$$

where $\dot{Y}$ is the shunt admittance per unit distance of the transmission line (S/m).

Combining (39.1) and (39.2), the Telegrapher's equation or Helmholtz's wave equation for the transmission line voltage is [1–3].

$$\frac{d^2\dot{V}}{dz^2} - \dot{Z}\dot{Y}\dot{V} = 0 \quad (\text{V/m})^2 \tag{39.3}$$

The Telegrapher's equation or Helmholtz's wave equation for the transmission line current is [1–3]

$$\frac{d^2\dot{I}}{dz^2} - \dot{Z}\dot{Y}\dot{I} = 0 \quad (\text{A/m}^2) \tag{39.4}$$

General Solutions and Propagation Constant

The general solution of the Telegrapher's equation for transmission line voltage is [1–3]

$$\dot{V} = \dot{V}_F \varepsilon^{-\dot{\gamma}z} + \dot{V}_R \varepsilon^{\dot{\gamma}z} \tag{39.5}$$

where $|\dot{V}_F|$ is the amplitude of the voltage waves propagating in $+z$-direction, $|\dot{V}_R|$ is the amplitude of the voltage waves propagating in $-z$-direction, and $\dot{\gamma}$ is the propagation constant of the transmission line [1].

$$\dot{\gamma} = \pm\sqrt{\dot{Z}\dot{Y}} = \alpha + \dot{\gamma}\beta \quad (\text{m}^{-1}) \tag{39.6}$$

where the + sign is for forward propagation or propagation in $+z$-direction, and the − sign is for the backward propagation or propagation in $-z$-direction. α is the attenuation constant and it is the real part of propagation constant $\dot{\gamma}$:

$$\alpha = \Re\dot{\gamma} = \Re\sqrt{\dot{Z}\dot{Y}} \quad (\text{m}^{-1}) \tag{39.7}$$

In (39.6), β is the phase constant and it is the imaginary part of the propagation constant $\dot{\gamma}$:

$$\beta = \Im\dot{\gamma} = \Im\sqrt{\dot{Z}\dot{Y}} \quad (\text{m}^{-1}) \tag{39.8}$$

Characteristic Impedance

The characteristic impedance of the transmission line is [1–3]

$$Z_0 = \sqrt{\frac{\dot{Z}}{\dot{Y}}} \quad (\Omega) \tag{39.9}$$

Wavelength

The wavelength of the transmission line voltage wave and transmission line current wave is

$$\lambda = \frac{2\pi}{\beta} = \frac{2\pi}{\Im\sqrt{\dot{Z}\dot{Y}}} \quad (\text{m}) \tag{39.10}$$

Phase Velocity

The phase velocity of voltage wave propagation and current wave propagation is

$$v_p = f\lambda = \frac{\omega}{\beta} \quad (\text{m/s}) \tag{39.11}$$

where f is the frequency of operation and the phase constant is

$$\beta = \frac{2\pi}{\lambda} \quad (\text{m}^{-1}) \tag{39.12}$$

Voltage Reflection Coefficient at the Load

If a transmission line of characteristic impedance Z_o is terminated by a mismatched load impedance $\dot{Z}_L$, as shown in Fig. 39.2, a voltage wave reflection occurs at the load impedance $\dot{Z}_L$. The voltage reflection coefficient is [1–3]

$$\dot{\rho}(l) = \frac{\dot{V}^r(l)}{\dot{V}^i(l)} = \frac{\dot{Z}_L - Z_0}{\dot{Z}_L + Z_0} = \frac{\dot{\tilde{Z}}_L - 1}{\dot{\tilde{Z}}_L + 1} \tag{39.13}$$

where $\dot{\rho}(l)$ is the voltage reflection coefficient at $z = l$, and $\dot{\tilde{Z}} = \dot{Z}_L/Z_0$ is the normalized load impedance. $\dot{V}_i(l)$ is the incident voltage at the load at $z = l$. When Z_0 is a complex quantity

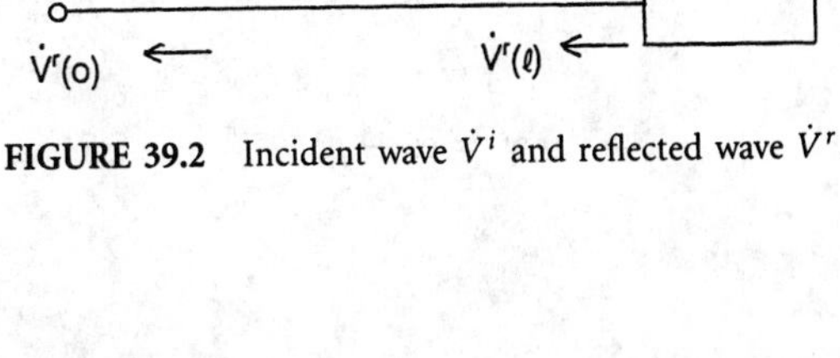

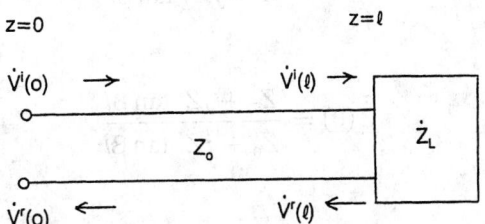

FIGURE 39.2 Incident wave $\dot{V}^i$ and reflected wave $\dot{V}^r$.

$\dot{Z}_0$, then

$$\dot{\rho}(l) = \frac{\dot{Z}_L - Z_0^*}{\dot{Z}_L + Z_0^*} \tag{39.14}$$

where Z_0^* is a conjugate of $\dot{Z}_0$.

Voltage Reflection Coefficient at the Input

In Fig. 39.2 $\dot{V}^i(l)$ is caused by $\dot{V}^i(0)$, which is the incident voltage at the input $z = 0$ of the transmission line. $\dot{V}^r(l)$ produces $\dot{V}^r(0)$, which is the reflected voltage at the input $z = 0$. The voltage reflection coefficient at the input occurs, then, by omitting transmission line loss [1–3]

$$\dot{\rho} = \frac{\dot{V}^r(0)}{\dot{V}^i(0)} = \dot{\rho}(l)\varepsilon^{-2j\beta l} \tag{39.15}$$

Input Impedance

At the load $z = l$, from (39.13),

$$\dot{\tilde{Z}}_L = \frac{1 + \dot{\rho}(l)}{1 - \dot{\rho}(l)} \tag{39.16}$$

or

$$\dot{Z}_L = Z_0 \frac{1 + \dot{\rho}(l)}{1 - \dot{\rho}(l)} \tag{39.17}$$

At the input of the transmission line $z = 0$

$$\dot{\tilde{Z}}(0) = \frac{1 + \dot{\rho}(0)}{1 - \dot{\rho}(0)} \tag{39.18}$$

Using (39.15),

$$\dot{\tilde{Z}}(0) = \frac{1 + \dot{\rho}(l)\varepsilon^{-2j\beta l}}{1 - \dot{\rho}(l)\varepsilon^{-2j\beta l}} \tag{39.19}$$

Inserting (39.13), [1–3]

$$\dot{\tilde{Z}}(0) = \frac{\dot{\tilde{Z}} + j\tan\beta l}{1 + j\dot{\tilde{Z}}_L \tan\beta l} \tag{39.20}$$

or

$$\dot{\tilde{Z}}(0) = \frac{\dot{Z}_L + jZ_0 \tan\beta l}{Z_0 + j\dot{Z}_L \tan\beta l} \tag{39.21}$$

If the line is lossy and

$$\gamma = \alpha + j\beta \tag{39.22}$$

then [1]

$$\dot{Z}(0) = \dot{Z}_0^* \frac{\dot{Z}_L + \dot{Z}_0^* \tanh \dot{\gamma} l}{\dot{Z}_0^* + \dot{Z}_L \tanh \dot{\gamma} l} \tag{39.23}$$

where $\dot{Z}_0^*$ is the complex conjugate of $\dot{Z}_0$.

Input Admittance

The input admittance at $z = 0$ is

$$\dot{Y}(0) = \dot{Y}_0^* \frac{\dot{Y}_L + \dot{Y}_0^* \tanh \gamma l}{\dot{Y}_0^* + \dot{Y}_L \tanh \gamma l} \tag{39.24}$$

where $\dot{Y}(0)$ is the input admittance of the transmission line, $\dot{Y}_0$ is the characteristic admittance of the transmission line, which is

$$\dot{Y}_0 = \frac{1}{\dot{Z}_0} \tag{39.25}$$

and $\dot{Y}_0^*$ is the conjugate of $\dot{Y}_0$. $\dot{Y}_L$ is the load admittance; i.e.,

$$\dot{Y}_L = \frac{1}{\dot{Z}_L} \tag{39.26}$$

When the line is lossless,

$$\gamma = j\beta \tag{39.27}$$

then [1–3]

$$\dot{Z}(0) = \frac{\dot{Z}_L + jZ_0 \tan \beta l}{Z_0 + j\dot{Z}_L \tan \beta l} \tag{39.28}$$

39.2 Two-Wire Lines

Geometric Structure

A structural diagram of the cross-sectional view of a commercial two-wire line is shown in Fig. 39.3. As seen from this figure, two parallel conductors, in most cases made of hard-drawn copper, are positioned by a plastic dielectric cover.

Transmission Line Parameters

In a two-wire line

$$\dot{Z} = R + jX = R + j\omega L \quad (\Omega/\text{m}) \tag{39.29}$$

where $\dot{Z}$ is the impedance per unit length of the two-wire line (Ω/m), R is the resistance per unit length (Ω/m), X is the reactance per unit length (Ω/m), $\omega = 2\pi f$ is the operating angular frequency (s^{-1}), and L is the inductance per unit length (H/m).

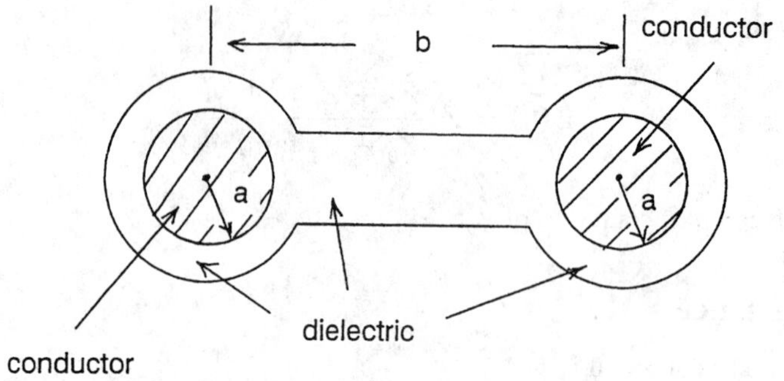

FIGURE 39.3 Cross-sectional view of a two-wire line.

For a two-wire line made of hard-drawn copper [4]

$$R = 8.42\frac{\sqrt{f}}{a} \quad (\mu\Omega/\mathrm{m}) \tag{39.30}$$

where f is the operating frequency and a is the radius of the conductor [4]:

$$L = 0.4\ln\frac{b}{a} \quad (\mu\mathrm{H/m}) \tag{39.31}$$

where b is the wire separation or the center-to-center distance of the two-wire line as shown in Fig. 39.3.

$$\dot{Y} = G + jB = G + j\omega C \quad (\mathrm{S/m}) \tag{39.32}$$

where $\dot{Y}$ is a shunt admittance per unit length of the two-wire line (S/m), G is a shunt conductance per unit length of the two-wire line (S/m), B is a shunt susceptance per unit length (S/m), and C is a shunt capacitance per unit length (F/m)

$$G = \frac{3.14\sigma_d}{\cosh^{-1}\left(\frac{b}{2a}\right)} \quad (\mathrm{pS/m}) \tag{39.33}$$

where σ_d is the insulation conductivity of the plastic dielectric surrounding the two parallel conductors, and

$$C = \frac{27.8\varepsilon_r}{\cosh^{-1}\left(\frac{b}{2a}\right)} \quad (\mathrm{pF/m}) \tag{39.34}$$

where ε_r is the relative permittivity of the plastic insulating material [5].

If R and G are negligibly small, the characteristic impedance is [6]

$$Z_0 = 277\log_{10}\frac{b}{a} \quad (\Omega) \tag{39.35}$$

The attenuation constant of a generic two-wire line is, including both R and G [1],

$$\alpha = \frac{R}{2Z_0} + \frac{GZ_0}{2} \quad (\mathrm{m}^{-1}) \tag{39.36}$$

and the phase constant is [1]

$$\beta = \left\{ \frac{(\omega LB - RG) + \sqrt{(RG - \omega LB)^2 + (\omega LG + BR)^2}}{2} \right\}^{\frac{1}{2}} \quad (\mathrm{m}^{-1}) \tag{39.37}$$

Wavelength and Phase Velocity

The wavelength on a lossless two-wire line ($R = 0, G = 0$) is

$$\lambda_0 = \frac{\omega}{\beta_0} = \frac{1}{\sqrt{LC}} \quad (\mathrm{m}) \tag{39.38}$$

where β_0 is the phase constant of the lossless two-wire line

$$\beta_0 = \omega\sqrt{LC} \quad (\mathrm{m}^{-1}) \tag{39.39}$$

The wavelength on a lossy two-wire line ($R \neq 0, G \neq 0$) is

$$\lambda = \frac{2\pi}{\beta} = \left\{ \frac{8\pi^2}{(\omega LB - RG) + \sqrt{(RG - \omega LB)^2 + (\omega LG + BR)}} \right\}^{\frac{1}{2}} \tag{39.40}$$

The phase velocity of transverse electromagnetic (TEM) waves on a lossless two-wire line is

$$v_0 = f\lambda_0 = \frac{\omega}{\beta_0} = \frac{1}{\sqrt{LC}} \quad (\mathrm{m/s}) \tag{39.41}$$

The phase velocity of TEM waves on a lossy two-wire line is

$$v = f\lambda = \frac{\omega}{\beta} = \left\{ \frac{2\omega^2}{(\omega LB - RG) + \sqrt{(RG - \omega LB)^2 + (\omega LB + BR)^2}} \right\}^{\frac{1}{2}} \tag{39.42}$$

).3 Coaxial Lines

Geometric Structure

A generic configuration of a coaxial line is shown in Fig. 39.4. The center and outer conductors are coaxially situated and separated by a coaxial insulator. Generally, coaxial lines are operated in the TEM mode, in which both the electric and magnetic fields are perpendicular to the direction of propagation. The propagating electric fields are in the radial direction and propagating magnetic fields are circumferential to the cylindrical surfaces.

In Fig. 39.4 a is the radius of the center conductor, b is the inner radius of the outer conductor, and c is the outer radius of the outer conductor.

Transmission Line Parameters

The series resistance per unit length of the line for copper is [7],

$$R = 4.16\sqrt{f}\left(\frac{1}{a} + \frac{1}{b}\right) \quad (\mu\Omega/\mathrm{m}) \tag{39.43}$$

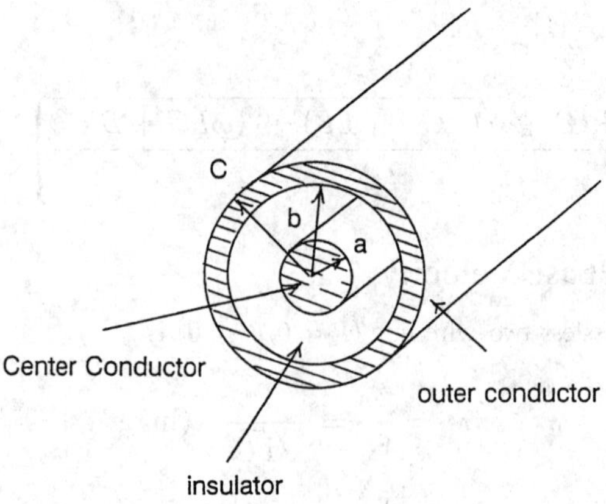

FIGURE 39.4 Generic configuration of a coaxial line.

The series inductance per unit length is

$$L = 0.2 \ln \frac{b}{a} \quad (\mu\text{H/m}) \tag{39.44}$$

The shunt conductance per unit length is

$$G = \frac{6.28\sigma_i}{\ln(b/a)} \quad (\text{S/m}) \tag{39.45}$$

where σ_i is the conductivity of the *insulator* between the conductors. The shunt capacitance per unit length is

$$C = \frac{55.5\varepsilon_r}{\ln(b/a)} \quad (\text{pF/m}) \tag{39.46}$$

where ε_r is the relative permittivity of the insulator. When the loss of the line is small, the characteristic impedance of the coaxial line is

$$Z_0 = \frac{138}{\sqrt{\varepsilon_r}} \log_{10} \frac{b}{a} \quad (\Omega) \tag{39.47}$$

when the line is lossy [1]

$$\dot{Z}_0 = \sqrt{\frac{\dot{Z}}{\dot{Y}}} = \sqrt{\frac{R + j\omega L}{G + j\omega C}} = R_0 + jX_0 \tag{39.48}$$

$$R_0 = \frac{\left\{RG + \omega^2 LC + \sqrt{(RG + \omega^2 LC)^2 + (\omega LG - \omega RC)^2}\right\}^{\frac{1}{2}}}{\sqrt{G^2 + \omega^2 C^2}} \tag{39.49}$$

$$X_0 = \frac{1}{2R_0} \cdot \frac{\omega LG - \omega CR}{G^2 + \omega^2 C^2} \tag{39.50}$$

The propagation constant of the coaxial line is

$$\dot{\gamma} = \alpha + j\beta \tag{39.51}$$

The attenuation constant is [1]

$$\alpha = \frac{\omega LG + \omega CR}{2\beta} \tag{39.52}$$

where the phase constant is

$$\beta = \left\{ \frac{(\omega^2 LC - RG) + \sqrt{(RG - \omega^2 LC)^2 + (\omega LG + \omega RC)^2}}{2} \right\}^{\frac{1}{2}} \tag{39.53}$$

Wavelength and Phase Velocity

The phase velocity on the coaxial line is

$$\upsilon_p = \frac{\omega}{\beta} \tag{39.54}$$

The wavelength on the line is

$$\lambda_l = \frac{2\pi}{\beta} \tag{39.55}$$

9.4 Waveguides

Rectangular Waveguides

Geometric Structure

A rectangular waveguide is a hollow conducting pipe of rectangular cross-section as shown in Fig. 39.5. Electromagnetic microwaves are launched inside the waveguide through a coupling antenna at the transmission site. The launched waves are received at the receiving end of the waveguide by a coupling antenna. In this case a rectangular coordinate system is set up on the rectangular waveguide (Fig. 39.5). The z-axis is parallel to the axis of the waveguide and is set coinciding with the lower left corner of the waveguide. The wider dimension of the cross-section of the waveguide is a and the narrower dimension of the cross-section of the waveguide is b, as shown in the figure.

Modes of Operation

The waveguide can be operated in either H- or E-modes, depending on the excitation configuration. An H-mode is a propagation mode in which the magnetic field, H, has a z-component, H_z, as referred to in Fig. 39.5. In this mode the electric field, E, is perpendicular to the direction of propagation, which is the $+z$-direction. Therefore, an H-mode is also called a transverse electric (TE) mode. An E-mode is a propagation mode in which the electric field, E, has a z-component, E_z, as referred to in Fig. 39.5. In this mode the magnetic field, H, is perpendicular to the direction of propagation, which is the $+z$-direction. Therefore, an E-mode is also called a transverse magnetic (TM) mode.

Solving Maxwell's equations for H-modes [1],

$$\dot{H}_z = \dot{H}_0 \cos\frac{m\pi x}{a} \cos\frac{m\pi y}{b} \varepsilon^{-\dot{\gamma} z + j\omega t} \tag{39.56}$$

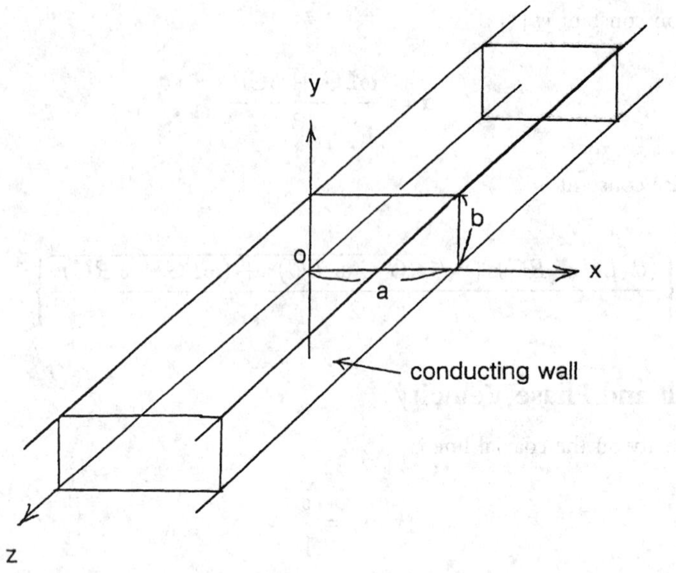

FIGURE 39.5 A rectangular waveguide and rectangular coordinate system.

where $|\dot{H}_0|$ is the amplitude of $\dot{H}_z$. Waveguide loss was neglected. Constants m and n are integral numbers $0, 1, 2, 3, \ldots$ and are called the mode number, and $\dot{\gamma}$ is the propagation constant. Both m and n cannot equal 0 simultaneously. Solving Maxwell's equations for E-modes [1],

$$\dot{E}_z = \dot{E}_0 \sin \frac{m\pi x}{a} \sin \frac{n\pi y}{b} \varepsilon^{-\dot{\gamma}z + j\omega t} \tag{39.57}$$

where $|\dot{E}_0|$ is the amplitude of $\dot{E}_z$. Neither m nor n can equal 0. The waveguide loss was neglected.

An H-mode is expressed as the H_{mn}-mode or TE_{mn}-mode. An E-mode is expressed as the E_{mn}-mode or TM_{mn}-mode.

Waveguide Parameters

Propagation constant $\dot{\gamma}$ of a rectangular waveguide made of a good conductor [1]:

$$\dot{\gamma} = j\beta_0 \sqrt{1 - \left(\frac{\lambda}{\lambda_c}\right)^2} \quad (\text{m}^{-1}) \tag{39.58}$$

where β_0 is the phase constant of free space, which is

$$\beta_0 = \frac{2\pi}{\lambda} \quad (\text{m}^{-1}) \tag{39.59}$$

Here, λ is the wavelength in free space, and λ_c is the cutoff wavelength of the waveguide. Electromagnetic waves with $\lambda > \lambda_c$ cannot propagate inside the waveguide. It is given for both E_{mn}-mode and H_{mm}-mode operations by [1]:

$$\lambda_c = \frac{2}{\sqrt{\left(\frac{m}{a}\right)^2 + \left(\frac{n}{b}\right)^2}} \quad (\text{m}) \tag{39.60}$$

This means that if the waveguide is made of a good conductor, the attenuation constant

$$\alpha \approx 0 \quad (\mathrm{m}^{-1}) \tag{39.61}$$

and the phase constant is

$$\beta_g = \beta_0 \sqrt{1 - \left(\frac{\lambda}{\lambda_c}\right)^2} \tag{39.62}$$

The wavelength in the waveguide, i.e., waveguide wavelength λ_g, is longer than the free space wavelength λ:

$$\lambda_g = \frac{\lambda}{\sqrt{1 - \left(\frac{\lambda}{\lambda_c}\right)^2}} \tag{39.63}$$

Then the speed of propagation v_p is $> c = f\lambda$.

$$v_p = f\lambda_g = \frac{f\lambda}{\sqrt{1 - \left(\frac{\lambda}{\lambda_c}\right)^2}} = \frac{c}{\sqrt{1 - \left(\frac{\lambda}{\lambda_c}\right)^2}} \tag{39.64}$$

For an H_{mn}-mode the wave impedance is

$$\eta_H = \frac{-\dot{E}_y}{\dot{H}_x} = \frac{\dot{E}_x}{\dot{H}_y} = \frac{\sqrt{\frac{\mu_0}{\varepsilon_0}}}{\sqrt{1 - \left(\frac{\lambda}{\lambda_c}\right)^2}} \quad (\Omega) \tag{39.65}$$

For an E_{mn}-mode the wave impedance is

$$\eta_E = \frac{-\dot{E}_y}{\dot{H}_x} = \frac{\dot{E}_x}{\dot{H}_y} = \sqrt{1 - \left(\frac{\lambda}{\lambda_c}\right)^2} \Big/ \sqrt{\frac{\mu_0}{\varepsilon_0}} \quad (\Omega) \tag{39.66}$$

Circular Waveguides

Geometric Structure

A circular waveguide is a hollow conducting pipe of circular cross-section, as shown in Fig. 39.6. Electromagnetic microwaves are launched inside the waveguide through a coupling antenna at the transmission site. The launched waves are received at the receiving end of the waveguide by a coupling antenna. In this case a circular coordinate system (r, ϕ, z) is set up in the circular waveguide, as shown in Fig. 39.6. The z-axis is coincident with the axis of the cylindrical waveguide. The inside radius of the circular waveguide is a.

Modes of Operation

The circular waveguide can be operated in either H- or E-modes, depending on the excitation configuration. An H-mode is a propagation mode in which the magnetic field, H, has a z-component, H_z, as referred to in Fig. 39.6. In this mode the electric field, E, is perpendicular to the direction of propagation, which is the $+z$-direction. Therefore, an H-mode is also called a TE mode. In this mode $E_z = 0$. An E-mode is a propagation mode in which the electric field, E, has a z-component, E_z, as referred to in Fig. 39.6. In this mode the magnetic field, H, is perpendicular to the direction of propagation, which is the $+z$-direction. Therefore, an E-mode is also called a TM mode.

Solving Maxwell's equations [1],

$$\dot{H}_z = \dot{H}_0 J_n(k'_{cm} r) \cos n\phi \varepsilon^{-\dot{\gamma}z + j\omega t} \tag{39.67}$$

Here, $|\dot{H}_0|$ is the amplitude of $\dot{H}_z$, n and m are integral numbers $0, 1, 2, 3, \ldots$ and are called the mode number, $J_n(k'_{cm} r)$ is the Bessel function of nth order, with argument $k'_{cm} r$, k'_{cm} is the mth room of $J'_n(k_{cm} a) = 0$ which is

$$k'_{cm} = \frac{u'_{nm}}{a} \tag{39.68}$$

where u'_{nm} is the mth root of the derivative of the Bessel function of order n, i.e., $J'_n(x) = 0$, where x is a generic real argument. The propagation constant is $\dot{\gamma}$.

Solving Maxwell's equations for E-modes,

$$\dot{E}_z = \dot{E}_0 J_n(k_{cm} r) \cos n\phi \varepsilon^{-\dot{\gamma}z + j\omega t} \tag{39.69}$$

k_{cm} is an mth root of $J_n(k_c a) = 0$, which is

$$k_{cm} = \frac{u_{nm}}{a} \tag{39.70}$$

where u_{nm} is the mth root of the Bessel function of order n, i.e., $J_n(x) = 0$, where x is a generic real argument.

An H-mode in a circular waveguide is expressed as the H_{nm}-mode or the TE_{nm}-mode. An E-mode is expressed as the E_{nm}-mode or the TM_{nm}-mode.

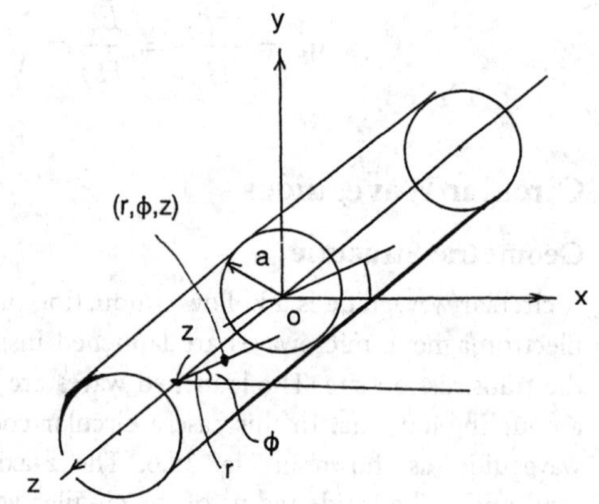

FIGURE 39.6 A circular waveguide and cylindrical coordinate system.

Waveguide Parameters

The propagation constant $\dot{\gamma}$ of a circular waveguide made of a good conductor is [1],

$$\dot{\gamma} = j\beta_0\sqrt{1 - \left(\frac{\lambda}{\lambda_c}\right)} \quad (\mathrm{m}^{-1}) \tag{39.71}$$

where β_0 is the phase constant of free space, which is

$$\beta_0 = \frac{2\pi}{\lambda} \quad (\mathrm{m}^{-1}) \tag{39.72}$$

Here, λ is the wavelength in free space, and λ_c is the cutoff wavelength of the waveguide. Electromagnetic waves with $\lambda > \lambda_c$ cannot propagate inside the waveguide. It is given for an H_{nm}-mode [1],

$$\lambda_{CH} = \frac{2\pi a}{u'_{nm}} \quad (\mathrm{m}) \tag{39.73}$$

For E_{nm}-mode operation,

$$\lambda_{CE} = \frac{2\pi a}{u_{nm}} \quad (\mathrm{m}) \tag{39.74}$$

This means that if the waveguide is made of a good conductor, the attenuation constant is

$$\alpha \approx 0 \quad (\mathrm{m}^{-1}) \tag{39.75}$$

and the phase constant is

$$\beta_g = \beta_0\sqrt{1 - \left(\frac{\lambda}{\lambda_c}\right)^2} \quad (\mathrm{m}^{-1}) \tag{39.76}$$

The waveguide wavelength is

$$\lambda_g = \frac{\lambda}{\sqrt{1 - \left(\frac{\lambda}{\lambda_c}\right)^2}} \quad (\mathrm{m}^{-1}) \tag{39.77}$$

The speed of propagation (phase velocity) is

$$v_p = f\lambda_g = \frac{c}{\sqrt{1 - \left(\frac{\lambda}{\lambda_c}\right)^2}} \quad (\mathrm{m/s}) \tag{39.78}$$

For an H_{nm}-mode the wave impedance is

$$\begin{aligned} \eta_H &= \frac{\dot{E}_r}{\dot{H}_\phi} = \frac{-\dot{E}\phi}{\dot{H}_r} = \frac{j\omega\mu_0}{\dot{\gamma}} \\ &= \frac{\omega\mu_0}{\beta_0\sqrt{1 - \left(\frac{\lambda}{\lambda_c}\right)^2}} \quad (\Omega) \end{aligned} \tag{39.79}$$

For an E_{nm}-mode the wave impedance is

$$\eta_E = \frac{\dot{E}_r}{\dot{H}_\phi} = \frac{-\dot{E}\phi}{\dot{H}_r} = \frac{\dot{\gamma}}{j\omega\varepsilon_0} = \frac{\beta_0\sqrt{1-\left(\frac{\lambda}{\lambda_c}\right)^2}}{\omega\varepsilon_0} \quad (\Omega) \tag{39.80}$$

39.5 Microstrip Lines

Geometric Structure

Figure 39.7 shows a general geometric structure of a microstrip line. A conducting strip of width, w, and thickness, t, is laid on an insulating substrate of thickness, H, and permittivity, $\varepsilon = \varepsilon_0\varepsilon_r$. The dielectric substrate has a groundplate underneath, as seen in Fig. 39.7.

Transmission Line Parameters

The characteristic impedance, Z_0, of a microstrip line, as shown in Fig. 39.7, is given by [8–10]:

$$Z_0 = \frac{42.4}{\sqrt{\varepsilon_r + 1}} \ln\left\{1 + \left(\frac{4H}{w'}\right)\left[\left(\frac{14 + 8/\varepsilon_r}{11}\right)\left(\frac{4H}{w'}\right) + \sqrt{\left(\frac{14 + 8\varepsilon_r}{11}\right)^2\left(\frac{4H}{w'}\right) + \frac{1 + 1/\varepsilon_r}{2}\pi^2}\right]\right\} \tag{39.81}$$

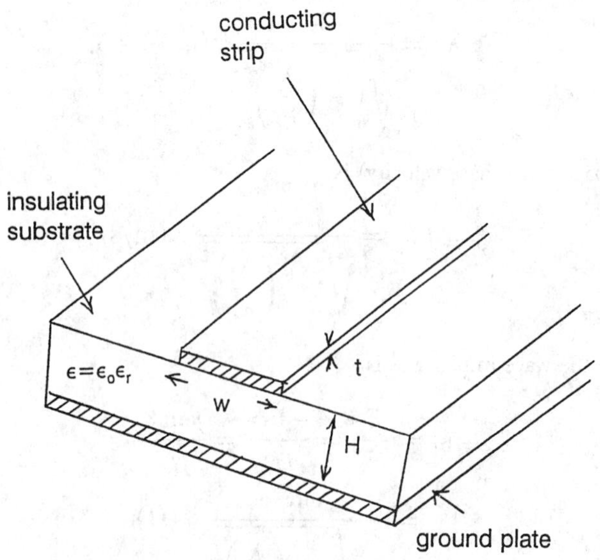

FIGURE 39.7 Geometric structure of a microstrip line.

where w' is an effective width of the microstrip, which is given by

$$w' = w + \frac{1 + 1/\varepsilon_r}{2} \cdot \frac{t}{\pi} \ln \frac{10.87}{\sqrt{\left(\frac{t}{H}\right)^2 + \left(\frac{1/\pi}{w/t + 1.10}\right)^2}} \tag{39.82}$$

The attenuation constant of a microstrip line is

$$\alpha = \frac{p + p'}{2} \frac{Z_{01}}{Z_0} \frac{2\pi}{\lambda_0} \quad \text{(Np/m)} \tag{39.83}$$

where

$$Z_{01} \equiv 30 \ln \left\{ 1 + \frac{1}{2}\left(\frac{8H}{w'}\right)\left[\frac{8H}{w'} + \sqrt{\left(\frac{8H}{w'}\right)^2 + \pi^2}\right] \right\} \tag{39.84}$$

and

$$p \equiv 1 - \frac{Z_{01}}{Z_{0\delta}} \tag{39.85}$$

where

$$Z_{0\delta} = 30 \ln \left\{ 1 + \frac{1}{2}\left(\frac{8(H+\delta)}{w'}\right)\left[\frac{8(H+\delta)}{w'} + \sqrt{\left(\frac{8(H+\delta)}{w'}\right)^2 + \pi^2}\right] \right\} \tag{39.86}$$

and

$$\delta = \frac{1}{\sqrt{\pi f \mu \sigma}} \tag{39.87}$$

is the skin depth of the conducting strip of conductivity σ

$$p' \equiv \frac{P_K}{1 - \dfrac{1/q - 1}{\varepsilon_r}} \tag{39.88}$$

where

$$\varepsilon_r = \varepsilon_r' - j\varepsilon_r'' \tag{39.89}$$

and

$$P_K = \sin\left[\tan^{-1} \frac{\varepsilon_r''}{\varepsilon_r'}\right] \tag{39.90}$$

and

$$q = \frac{1}{\varepsilon_r - 1}\left[\left(\frac{Z_{01}}{Z_0}\right)^2 - 1\right] \tag{39.91}$$

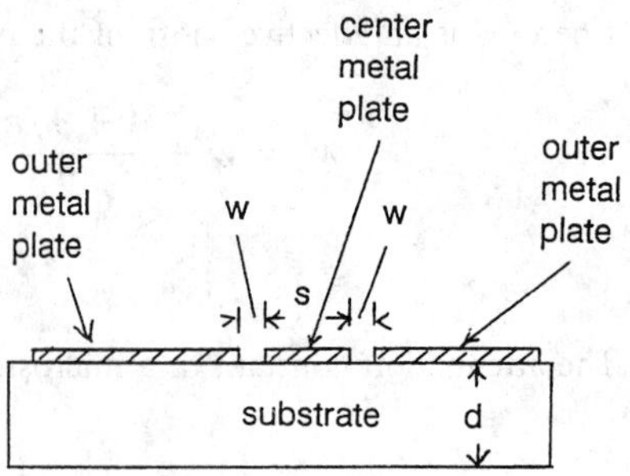

FIGURE 39.8 Cross-sectional view of a coplanar waveguide.

Wavelength and Phase Velocity

The transmission line wavelength is

$$\lambda_l = \frac{Z_0}{Z_{01}} \lambda_0 \tag{39.92}$$

The phase constant of the microstrip line is then

$$\beta = \frac{2\pi}{\lambda_0} \frac{Z_{01}}{Z_0} \tag{39.93}$$

The phase velocity of electromagnetic waves on the microstrip line is

$$v_p = 3 \times 10^8 \frac{Z_0}{Z_{01}} \tag{39.94}$$

39.6 Coplanar Waveguide

Geometric Structure

A cross-sectional view of a coplanar waveguide (CPW) is shown in Fig. 39.8. The waveguide consists of a narrow, central conducting strip of width s(m) and very wide conducting plates on both sides of the central conducting strip, with gap widths w. These conductors are developed on a surface of dielectric substrate of thickness d, as shown in the figure. The electromagnetic waves propagate in the gap between the outer conducting plates and the center conducting strip.

Transmission Line Parameters

The attenuation constant of a coplanar waveguide is given by [11]

$$\alpha = \left(\frac{\pi}{2}\right)^5 \cdot 2 \cdot \frac{1 - \dfrac{\varepsilon_{\text{eff}}(f)}{\varepsilon_r}}{\sqrt{\dfrac{\varepsilon_{\text{eff}}(f)}{\varepsilon_r}}} \cdot \frac{(s+2w)^2 \varepsilon_r^{3/2}}{c^3 K'(k) K(k)} \quad (\text{Np/m}) \tag{39.95}$$

where the effective dielectric constant is given by

$$\sqrt{\varepsilon_{\text{eff}}(f)} = \sqrt{\varepsilon_q} + \frac{\sqrt{\varepsilon_r} - \sqrt{\varepsilon_q}}{1 + a\left(\dfrac{f}{f_{\text{TE}}}\right)^{-b}} \tag{39.96}$$

and

$$\varepsilon_q = \frac{\varepsilon_r + 1}{2} \tag{39.97}$$

ε_r is the relative permittivity of the substrate material,

$$f_{\mathrm{TE}} = \frac{c}{4d\sqrt{\varepsilon_r - 1}} \tag{39.98}$$

is the TE mode cutoff frequency,

$$k \equiv s/(s + 2w) \tag{39.99}$$

$K(k)$ is the complete elliptic integral of the first kind of the argument k, and c is the speed of light in vacuum, which is 3×10^8 m/s. The parameter a is [11]

$$a \approx \log^{-1}\left[u \log \frac{s}{w} + v\right] \tag{39.100}$$

$$u \approx 0.54 - 0.64q + 0.015q^2 \tag{39.101}$$

$$v \approx 0.43 - 0.86q + 0.54q^2 \tag{39.102}$$

$$q \approx \log \frac{s}{d} \tag{39.103}$$

The parameter b is an experimentally determined constant

$$b \approx 1.8 \tag{39.104}$$

$$K'(k) = K(\sqrt{1 - k^2}) \tag{39.105}$$

The phase constant of the coplanar waveguide is

$$\beta(f) = 2\pi \frac{f}{c} \sqrt{\varepsilon_{\mathrm{eff}}(f)} \quad \text{(rad/m)} \tag{39.106}$$

The characteristic impedance of the coplanar waveguide is [11]

$$Z_0 = \frac{120\pi}{\sqrt{\varepsilon_{\mathrm{eff}}(f)}} \frac{K'(k)}{4K(k)} \quad (\Omega) \tag{39.107}$$

Wavelength and Phase Velocity

The wavelength of electromagnetic waves propagating on a coplanar waveguide is obtained from (39.106):

$$\lambda_l = \frac{2\pi}{\beta(f)} = \frac{c}{f} \cdot \frac{1}{\sqrt{\varepsilon_{\mathrm{eff}}(f)}} = \frac{\lambda}{\sqrt{\varepsilon_{\mathrm{eff}}(f)}} \quad \text{(m)} \tag{39.108}$$

The phase velocity of the waves on the coplanar waveguide is, then,

$$v_p = f\lambda_l = \frac{c}{\sqrt{\varepsilon_{\mathrm{eff}}(f)}} \quad \text{(m/s)} \tag{39.109}$$

References

[1] T. K. Ishii, *Microwave Engineering*, San Diego, CA: Harcourt, Brace, Jovanovich, 1989.

[2] J. R. Wait, *Electromagnetic Wave Theory*, New York: Harper & Row, 1985.

[3] V. F. Fusco, *Microwave Circuits*, Englewood Cliffs, NJ: Prentice-Hall, 1987.

[4] L. A. Ware and H. R. Reed, *Communications Circuits*. New York: John Wiley & Sons, 1949.

[5] E. A. Guillemin, *Communications Networks*, New York: John Wiley & Sons, 1935.

[6] F. E. Terman, *Radio Engineering*, New York: McGraw-Hill, 1941.

[7] H. J. Reich, P. F. Ordung, H. L. Krauss, and J. G. Skalnik, *Microwave Theory and Techniques*, Princeton, NJ: D. Van Nostrand, 1953.

[8] H. A. Wheeler, "Transmission-line properties of parallel strips separated by a dielectric sheet," *IEEE Trans. MTT*, vol. MTT-13, pp. 172–185, Mar. 1965.

[9] H. A. Wheeler, "Transmission-line properties of parallel strips by a conformal-mapping approximation," *IEEE Trans. MTT*, vol. MTT-12, pp. 280–289, May 1964.

[10] H. A. Wheeler, "Transmission-line properties of a strip on a dielectric sheet on a plane," *IEEE Trans. MTT*, vol. MTT-25, pp. 631–647, Aug. 1977.

[11] M. Y. Frankel, S. Gupta, J. A. Valdmanis, and G. A. Mourou, "Terahertz attenuation and dispersion characteristics of coplanar transmission lines," *IEEE Trans. MTT*, vol. 39, no. 6, pp. 910–916, June 1991.

40

Multiconductor Transmission Lines

niël De Zutter
versiteit Gent,
ium
: Martens
versiteit Gent,
ium

.1 Introduction: Frequency vs. Time Domain Analysis

Multiconductor transmission lines (MTL), or multiconductor buses as they are also often called, are found in almost every electrical packaging technology and on every technology level from digital chips, over MMICs (monolithic microwave integrated circuits) to MCMs (multichip modules), boards, and backplanes. Multiconductor transmission lines are electrical conducting structures with a constant cross-section (the x, y-plane) which propagates signals in the direction perpendicular to that cross-section (the z-axis) (see also Fig. 40.1). Being restricted to a constant cross-section we are in fact dealing with the so-called uniform MTL. The more general case using a nonuniform cross-section is much more difficult to handle and constitutes a fully three-dimensional problem.

It is not the purpose of this chapter to give a detailed account of the physical properties and the use of the different types of MTL. The literature on this subject is abundant and any particular reference is bound to be both subjective and totally incomplete. Hence, we put forward only [1], [2], and [3] as references here, as they contain a wealth of information and additional references.

In the frequency domain, i.e., for harmonic signals, solution of Maxwell's sourceless equations yields a number of (evanescent and propagating) modes characterized by modal propagation factors $\exp(\pm j\beta z)$ and by a modal field distribution, which depends only upon the (x, y)-coordinates of the cross-section. In the presence of losses and for evanescent modes β can take complex values, and $j\beta$ is then replaced by $\gamma = \alpha + j\beta$ (see Eq. 40.30). In the propagation direction the modal field amplitudes essentially behave as voltage and current along a transmission line. This immediately suggests that MTL should be represented on the circuit level by a set of coupled circuit transmission lines. The relationship between the typical circuit quantities, such as voltages, currents, coupling impedances, and signal velocities, on the one hand, and the original field quantities (modal fields and modal propagation factors) is not straightforward [4].

3-8341-2/95/$0.00 + $.50

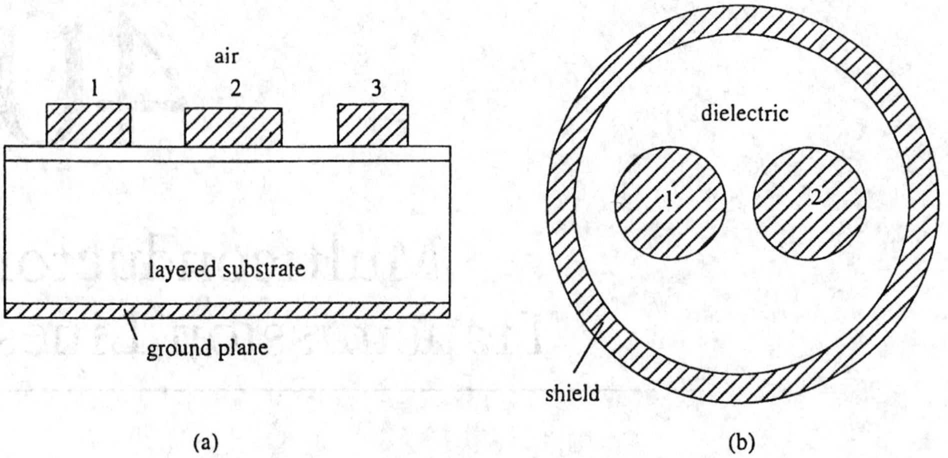

FIGURE 40.1 Two examples of cross-sections of multiconductor lines.

In general, the circuit quantities will be frequency dependent. The frequency domain circuit model parameters can be used as the starting point for time domain analysis of networks, including multiconductor lines. This is, again, a vast research topic with important technical applications.

Section 40.2 describes the circuit modeling in the frequency domain of uniform MTL based on the Telegrapher's equations. The meaning of the voltages and currents in these equations is explained both at lower frequencies in which the quasi-transverse electromagnetic (TEM) approach is valid as well as in the so-called full-wave regime valid for any frequency. The notions TEM, quasi-TEM, and full-wave are elucidated. We introduce the capacitance, inductance, resistance, and conductance matrices together with the characteristic impedance matrix of the coupled transmission line model. Finally, for some simple MTL configurations analytical formulas are presented expressing the above quantities and the propagation factors as a function of the geometric and electrical parameters of these configurations.

It would be a formidable task to give a comprehensive overview of all the methods that are actually used for the time domain analysis of MTL. In the remaining part of this paragraph a very short overview (both for uniform and nonuniform structures) is presented along with some references. In the case of linear loads and drivers frequency domain methods in combination with (fast) Fourier transform techniques are certainly most effective [5–7]. In the presence of nonlinear loads and drivers other approaches must be used. Simulations based on harmonic balance techniques [8, 9] are, again, mainly frequency domain methods. All signals are approximated by a finite sum of harmonics and the nonlinear loads and drivers are taken into account by converting their time domain behavior to the frequency domain. Kirchhoff laws are then imposed for each harmonic in an iterative way. Harmonic balance techniques are not very well suited for transient analysis or in the presence of strong nonlinearities, but are excellent for mixers, amplifiers, filters, etc. Many recent efforts were directed toward the development of time domain simulation methods (for both uniform and nonuniform interconnection structures) based on advanced convolution-type approaches. It is, of course, impossible to picture all the ramifications in this research field. We refer the reader to a recent special issue of *IEEE Circuits and Systems Transactions* [10], to the "Simulation Techniques for Passive Devices and Structures" section of a special issue of *IEEE Microwave Theory and Techniques Transactions* [11], and to a 1994 special issue of the *Analog Integrated Circuits and Signal Processing* journal [12] and to the wealth of references therein.

Both frequency and time domain experimental characterization techniques for uniform and nonuniform multiconductor structures can be found in Chapter 41.

40.2 Telegrapher's Equations for Uniform Multiconductor Transmission Lines

Generalities

Figures 40.1(a) and (b) show the cross-sections of two general coupled lossy MTLs consisting of $N+1$ conductors. These conductors either can be perfectly conducting or exhibit finite conductivity. Their cross-section remains constant along the propagation or longitudinal direction z. The $(N+1)$th conductor is taken as reference conductor. In many practical cases this will be the ground plane at the top or bottom (Fig. 40.1(a)) of the layered dielectric in which the conductors are embedded or is the shielding surrounding the other conductors (Fig. 40.1(b)). We restrict the analysis to the frequency domain, i.e., all field components and all voltages and currents have a common time dependence, $\exp(j\omega t)$, which is suppressed in the sequel. The generalized Telegrapher's equations governing the circuit representation of the MTL of Fig. 40.1 in terms of a set of C-coupled circuit transmission lines is given by [4]:

$$\begin{aligned} \frac{d\mathbf{V}}{dz} + \mathbf{Z}\mathbf{I} &= 0 \\ \frac{d\mathbf{I}}{dz} + \mathbf{Y}\mathbf{V} &= 0 \end{aligned} \tag{40.1}$$

$\mathbf{V}$ and $\mathbf{I}$ are column vectors, the C elements of which are the voltages and currents of the circuit model; $\mathbf{Z}$ and $\mathbf{Y}$ are the $C \times C$ impedance and admittance matrices. Eq. (40.1) is a good circuit description of the wave phenomena along a MTL if only the fundamental modes of the corresponding field problem are of importance. In that case $C = N$ ($C = 3$ in Fig. 40.1(a) and $C = 2$ in Fig. 40.1(b)) if a ground plane is present, and $C = N - 1$ in the absence of a ground plane. For the relationship between the actual electromagnetic field description in terms of modes and the circuit model (40.1), we refer the reader to [4] and [13]. The general solution to (40.1) is given by

$$\begin{aligned} \mathbf{V}(z) &= 2(\mathbf{I}_m^T)^{-1} e^{-j\boldsymbol{\beta} z}\mathbf{K}^+ + 2(\mathbf{I}_m^T)^{-1} e^{j\boldsymbol{\beta} z}\mathbf{K}^- \\ \mathbf{I}(z) &= \mathbf{I}_m e^{-j\boldsymbol{\beta} z}\mathbf{K}^+ - \mathbf{I}_m e^{j\boldsymbol{\beta} z}\mathbf{K}^- \end{aligned} \tag{40.2}$$

$\mathbf{K}^+$ and $\mathbf{K}^-$ are column vectors with C elements. $\boldsymbol{\beta}$ is a diagonal $C \times C$ matrix with the propagation factors β_f ($f = 1, 2, \ldots, C$) of the C fundamental eigenmodes as diagonal elements. This matrix reduces to a single propagation factor for a single transmission line (see Eq. 39.6). For the calculation of the fields and of the propagation constants many different methods can be found in the literature [14]. Solution (40.2) of the differential equations (40.1) is the extension of the corresponding equations (39.5) for a single transmission line to the coupled line case. It also consists of waves respectively traveling in positive and negative z-directions. The $\mathbf{I}_m$ is a $C \times C$ matrix, the columns of which are the current eigenvectors of the circuit model. The following relationships hold:

$$\begin{aligned} \mathbf{Z} &= j\omega\mathbf{L} + \mathbf{R} = 2j(\mathbf{I}_m^T)^{-1}\boldsymbol{\beta}(\mathbf{I}_m)^{-1} \\ \mathbf{Y} &= j\omega\mathbf{C} + \mathbf{G} = \frac{j}{2}\mathbf{I}_m\boldsymbol{\beta}\mathbf{I}_m^T \end{aligned} \tag{40.3}$$

L, **R**, **C** and **G** are the $C \times C$ (frequency dependent) inductance, resistance, capacitance, and conductance matrices. The $C \times C$ characteristic impedance matrix of the transmission line models is given by $\mathbf{Z}_{\text{char}} = 2(\mathbf{I}_m^T)^{-1}(\mathbf{I}_m)^{-1}$. The matrix $\mathbf{Z}_{\text{char}}$ replaces the simple characteristic impedance number of (39.9). In general, the mapping of the wave phenomenon onto the circuit model [(40.1) to (40.3)] depends on the choice of $\mathbf{I}_m$. We refer the reader to the detailed discussions in [4]. For MTLs the most adopted definition for the elements of $\mathbf{I}_m$ is [15]

$$I_{m,jf} = \oint_{c_j} \mathbf{H}_{\text{tr},f} \cdot \mathbf{dl} \qquad j, f = 1, 2, \ldots, C \tag{40.4}$$

where c_j is the circumference of conductor j and where $\mathbf{H}_{\text{tr},f}$ is the transversal component of the magnetic field of eigenmode f. This means that $I_{m,jf}$ is the total current through conductor j due to eigenmode f. This definition is used in the power-current impedance definition for microstrip and stripline problems [9]. For slotline circuits, a formulation that parallels the one given above must be used, but in this case it makes much more sense to introduce the voltage eigenvectors and to define them as line integrals of the electric field (see Appendix B of [16]).

As **Z** and **Y** in (40.3) are frequency dependent, the time domain equivalent of (40.1) involves convolution integrals between **Z** and **I** on the one hand and **Y** and **V** on the other hand. The propagation factors β_f are also frequency dependent, hence the signal propagation will show dispersion, i.e., the signal waveform becomes distorted while propagating.

Low-Frequency or Quasi-Transverse Electromagnetic Description

In the previous section the reader was given a very general picture of the MTL problem, valid for any frequency. This analysis is the so-called full-wave analysis. The present section is restricted to the low-frequency or quasi-static regime. Here, the cross-section of the MTL is small with respect to the relevant wavelengths, the longitudinal field components can be neglected, and the transversal field components can be found from the solution of an electrostatic or magnetostatic problem in the cross-section of the MTL. A detailed discussion of the theoretical background can be found in [17]. In the quasi-TEM limit and in the absence of losses **R** and **G** are zero and **L** and **C** become frequency independent and take their classical meaning. Both skin-effect losses and small dielectric losses can be accounted for by a perturbation approach [18]. In that case a frequency dependent **R** and **G** must be reintroduced. If **R** and **G** are zero and **L** and **C** are frequency independent, the following Telegrapher's equations hold:

$$\begin{aligned} \frac{\partial \mathbf{v}}{\partial z} &= -\mathbf{L} \cdot \frac{\partial \mathbf{i}}{\partial t} \\ \frac{\partial \mathbf{i}}{\partial z} &= -\mathbf{C} \cdot \frac{\partial \mathbf{v}}{\partial t} \end{aligned} \tag{40.5}$$

Equation (40.5) is the time domain counterpart of (40.1). We have replaced the capital letters for voltages and currents with lower case letters to distinguish between time and frequency domain. **L** and **C** are related to the total charge Q_i per unit length carried by each conductor and to the total magnetic flux F_i between each conductor and the reference conductor:

$$\begin{aligned} \mathbf{Q} &= \mathbf{C} \cdot \mathbf{V} \\ \mathbf{F} &= \mathbf{L} \cdot \mathbf{I} \end{aligned} \tag{40.6}$$

where **Q** and **F** are $C \times 1$ column vectors with elements Q_i and F_i, respectively. For a piecewise homogeneous medium, one can prove that the inductance matrix **L** can be derived from an equivalent so-called vacuum capacitance matrix $\mathbf{C}_v$ with $\mathbf{L} = \mathbf{C}_v^{-1}$ and where $\mathbf{C}_v$ is calculated in the same way as **C**, but with the piecewise constant ε everywhere replaced by the corresponding value of $1/\mu$. For nonmagnetic materials this operation corresponds with taking away all dielectrics and working with vacuum, thus explaining the name of the matrix $\mathbf{C}_v$. Other properties of **C** and **L** are

C and **L** are symmetric, i.e., $C_{ij} = C_{ji}$ and $L_{ij} = L_{ji}$
C is real, $C_{ii} > 0$ and $C_{ij} < 0$ $(i \neq j)$
L is real, $L_{ii} > 0$ and $L_{ij} > 0$

The propagation factors β_f which form the elements of $\boldsymbol{\beta}$ in (40.2) are now given by the eigenvalues of $(\mathbf{LC})^{1/2}$ or equivalently of $(\mathbf{CL})^{1/2}$. The current eigenvectors which form the columns of $\mathbf{I}_m$ are now solutions of the following eigenproblem (where ω is the circular frequency):

$$\omega^2(\mathbf{CL})\mathbf{I} = (\beta_f)^2\mathbf{I} \tag{40.7}$$

The corresponding eigenvoltages are solutions of:

$$\omega^2(\mathbf{LC})\mathbf{V} = (\beta_f)^2\mathbf{V} \tag{40.8}$$

Hence, corresponding voltage and current eigenmodes propagate with the same propagation factors and as **L**, **C**, **V**, and **I** are frequency independent, β_f is proportional with ω and can be rewritten as $\beta_f = \omega\beta'_f$, proving that the propagation is nondispersive with velocity $v_f = 1/\beta'_f$. Remember that the subindex f takes the values $1, 2, \ldots, C$, i.e., for a three-conductor problem above a ground plane ($N = C = 3$, see Fig. 40.1(a)), three distinct propagation factors and corresponding eigenmode profiles exist for currents and voltages. Note, however, that for the same β_f the eigenvector for the currents differs from the eigenvector of the voltages.

We conclude this section by remarking that for MTL embedded in a homogeneous medium (such as the simple stripline or the coaxial cable with homogeneous filling) $(\mathbf{LC}) = \varepsilon\mu\mathbf{1}$, where $\mathbf{1}$ is the unity matrix. Thus, the eigenmodes are purely TEM, i.e., electric and magnetic fields have only transversal components and the longitudinal ones are exactly zero. All propagation factors β_f take the same value $[c/(\varepsilon_r\mu_r)^{1/2}]$, where c is the velocity of light in vacuum. Note, however, that even for identical β_f different eigenmodes will be found.

Numerical calculation of **L** and **C** can be performed by many different numerical methods (see the reference section of [18]), and for sufficiently simple configurations analytical formulas are available. For a line with one conductor and a ground plane ($N = C = 1$) the characteristic impedance $Z_0 = (L/C)^{1/2}$ and the signal velocity v_p is $v_p = (LC)^{-1/2}$.

Analytical Expressions for Some Simple Multiconductor Transmission Line Configurations

Symmetric Strip Line

Sections 39.3 through 39.6 presented a number of MTL consisting of a single conductor and a ground plane ($N = C = 1$). Here, we add another important practical example, the symmetric stripline configuration of Fig. 40.2. We restrict ourselves to the lossless case. A perfectly conducting strip of width w and thickness t is symmetrically placed between two perfectly conducting ground planes with spacing b. The insulating substrate has a permittivity $\varepsilon = \varepsilon_0\varepsilon_r$, and is nonmagnetic. This case has a single fundamental mode. The characteristic

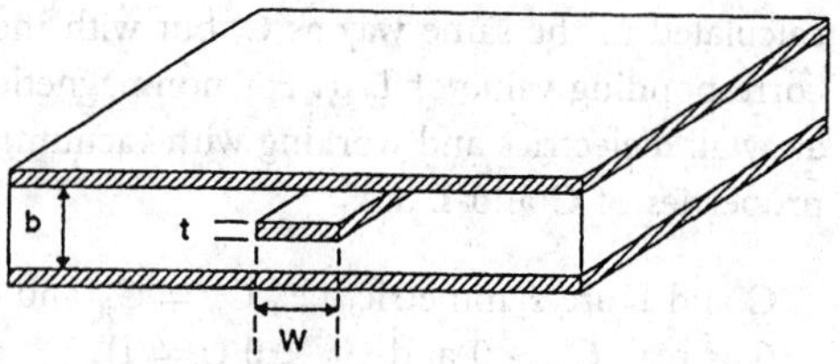

FIGURE 40.2 Strip line configuration.

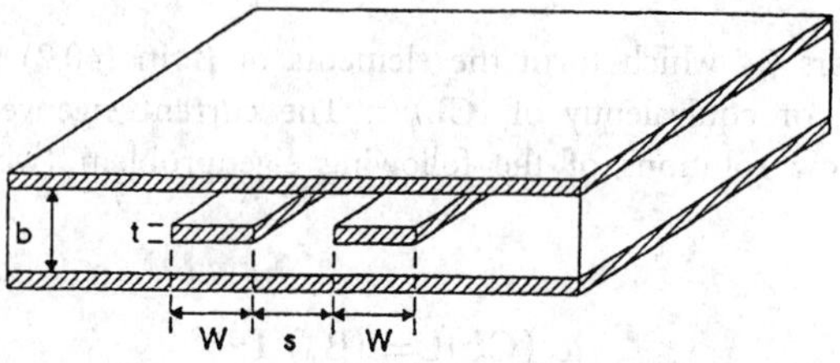

FIGURE 40.3 Coupled symmetric strip line configuration.

impedance Z_0 is given by [19]:

$$Z_0\sqrt{\varepsilon_r} = 30\ln\left\{1+\frac{4}{\pi}\frac{b-t}{W'}\left[\frac{8}{\pi}\frac{b-t}{W'}+\sqrt{\left(\frac{8}{\pi}\frac{b-t}{W'}\right)^2+6.27}\right]\right\} \tag{40.9}$$

with

$$\frac{W'}{b-t} = \frac{W}{b-t}+\frac{\Delta W}{b-t} \tag{40.10}$$

where

$$\frac{\Delta W}{b-t} = \frac{x}{\pi(1-x)}\left\{1-\frac{1}{2}\ln\left[\left(\frac{x}{2-x}\right)^2+\left(\frac{0.0796x}{W/b+1.1x}\right)^m\right]\right\}$$

$$m = 2\left[1+\frac{2}{3}\frac{x}{1-x}\right]^{-1}, \quad x = t/b$$

For $W'/(b-t) < 10$ (40.9) is 0.5% accurate. The signal velocity v_p is given by $c/(\varepsilon_r)^{1/2}$, where c is the velocity of light in vacuum. The corresponding L and C are given by $L = Z_0/v_p$ and $C = 1/(Z_0 v_p)$.

Coupled Strip Lines

The configuration is depicted in Fig. 40.3. It consists of the symmetric combination of the structure of Fig. 40.2. There are now two fundamental TEM modes ($N = C = 2$). The even mode (index e) corresponds to the situation in which both central conductors are placed at the same voltage (speaking in low-frequency terms, of course). The odd mode (index o) corresponds to the situation where the central conductors are placed at opposite voltages. The impedances of the modes (respectively, $Z_{0,e}$ and $Z_{0,o}$) are given by:

$$Z_{0e,o} = \frac{30\pi(b-t)}{\sqrt{\varepsilon_r}\left(W+\frac{bC_f}{2\pi}A_{e,o}\right)} \tag{40.11}$$

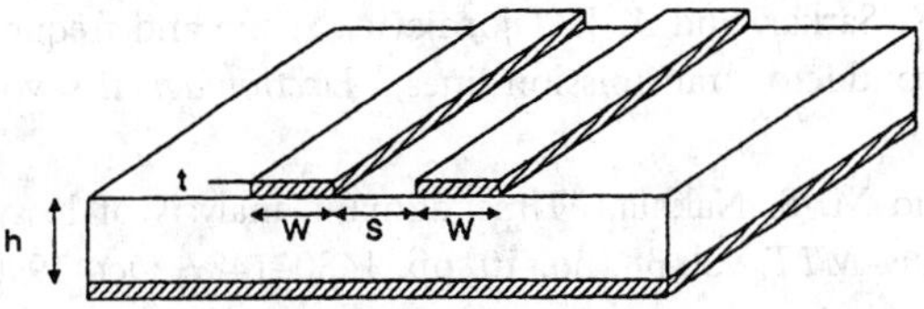

FIGURE 40.4 Coupled symmetric microstrip configuration.

with

$$\begin{aligned} A_e &= 1 + \frac{\ln(1+\tanh\theta)}{\ln 2} \\ A_o &= 1 + \frac{\ln(1+\coth\theta)}{\ln 2} \\ \theta &= \frac{\pi S}{2b} \\ C_f(t/b) &= 2\ln\left(\frac{2b-t}{b-t}\right) - \frac{t}{b}\ln\left[\frac{t(2b-t)}{(b-t)^2}\right] \end{aligned} \tag{40.12}$$

The signal velocity is the same for both modes $(c/(\varepsilon_r)^{1/2})$ and the L and C of both modes can be found by replacing Z_0 in the "Symmetric Strip Line" section by $Z_{0,e}$ and $Z_{0,o}$, respectively.

Coupled Microstrip Lines

The configuration is depicted in Fig. 40.4. It consists of the symmetric combination of the structure of Fig. 39.7. Again, we have two fundamental modes, but the modes are hybrid, i.e., not purely TEM. Much work has been done on this configuration. The formulas proposed in the literature are quite lengthy; and we refer the reader to [20] and [21]. Reference [20] gives a very good overview together with some simple approximations, and [21] gives the most accurate formulas, taking into account the frequency dependence. It is important to remark here that the two impedances, $Z_{0,e}$ and $Z_{0,o}$, can be found. They depend upon frequency. Both modes now have a different velocity. The data found in literature are typically expressed in terms of the effective dielectric constant. The modal field lines are both found in the air above the substrate and in the substrate itself. Hence, the field experiences an effective dielectric constant which is smaller than the dielectric constant of the substrate. The effective dielectric constant for the even mode $(\varepsilon_{r,e})$ will be higher than for the odd mode $(\varepsilon_{r,o})$, and are frequency dependent. The corresponding modal velocities are given by $c/(\varepsilon_{r,e})^{1/2}$ and $c/(\varepsilon_{r,o})^{1/2}$.

Two-Wire Line

See Section 39.2.

ferences

[1] T. Itoh (Ed.), *Numerical Techniques for Microwave and Millimeter-Wave Passive Structures*, New York: Wiley, 1989.

[2] Jin Au Kong (Ed.), *Progress in Electromagnetics Research*, Volumes 1–5, New York: Elsevier, 1989–1991.

[3] C. F. Coombs, (Ed.), *Printed Circuits Handbook*. 3rd ed. New York: McGraw-Hill, 1988.

[4] N. Faché, F. Olyslager, and D. De Zutter, *Electromagnetic and Circuit Modelling of Multiconductor Transmission Lines*. Oxford Engineering Series 35, Oxford: Clarendon Press, 1993.

[5] T. R. Arabi, T. K. Sarkar, and A. R. Djordjevic, "Time and frequency domain characterization of multiconductor transmission lines," *Electromagnetics*, vol. 9, no. 1, pp. 85–112, 1989.

[6] J. R. Griffith and M. S. Nakhla, "Time domain analysis of lossy coupled transmission lines," *IEEE Trans. MTT*, vol. 38, no. 10, pp. 1480–1487, Oct. 1990.

[7] B. J. Cooke, J. L. Prince, and A. C. Cangellaris, "*S*-parameter analysis of multiconductor, integrated circuit interconnect systems," *IEEE Trans. Comput. Aided Design*, vol. CAD-11, no. 3, pp. 353–360, Mar. 1992.

[8] V. Rizzoli et al., "State of the art and present trends in nonlinear microwave CAD techniques," *IEEE Trans. MTT*, vol. 36, no. 2, pp. 343–363, Feb. 1988.

[9] R. Gilmore, "Nonlinear circuit design using the modified harmonic balance algorithm," *IEEE Trans. MTT*, vol. 34, no. 12, pp. 1294–1307, Dec. 1986.

[10] *IEEE Trans. Circuits Syst. Transactions, I: Fundamental Theory and Applications, Special Issue on Simulation, Modelling and Electrical Design of High-Speed and High-Density Interconnects*, vol. 39, no. 11, Nov. 1992.

[11] *IEEE Trans. MTT, Special Issue on Process-Oriented Microwave CAD and Modeling*, vol. 40, no. 7, July 1992.

[12] *Analog Integrated Circuits Signal Process., Special Issue on High-Speed Interconnects*, vol. 5, no. 1, pp. 1–107, Jan. 1994.

[13] F. Olyslager, D. De Zutter, and A. T. de Hoop, "New reciprocal circuit model for lossy waveguide structures based on the orthogonality of the eigenmodes," *IEEE Trans. MTT*, vol. 42, no. 12, pp. 2261–2269, Dec. 1994.

[14] F. Olyslager and D. De Zutter, "Rigorous boundary integral equation solution for general isotropic and uniaxial anisotropic dielectric waveguides in multilayered media including losses, gain and leakage," *IEEE Trans. MTT*, vol. 41, no. 8, pp. 1385–1392, Aug. 1993.

[15] R. H. Jansen and M. Kirschning, "Arguments and an accurate model for the power-current formulation of microstrip characteristic impedance," *Arch. Elek. Übertragung*, vol. 37, no. 3/4, pp. 108–112, 1983.

[16] T. Dhaene and D. De Zutter, "CAD-oriented general circuit description of uniform coupled lossy dispersive waveguide structures," *IEEE Trans. MTT, Special Issue on Process-Oriented Microwave CAD and Modeling*, vol. 40, no. 7, pp. 1445–1554, July 1992.

[17] I. V. Lindell, "On the quasi-TEM modes in inhomogeneous multiconductor transmission lines," *IEEE Trans. MTT*, vol. 29, no. 8, pp. 812–817, 1981.

[18] F. Olyslager, N. Faché, and D. De Zutter, "New fast and accurate line parameter calculation of general multiconductor transmission lines in multilayered media," *IEEE Trans. MTT*, vol. MTT-39, no. 6, pp. 901–909, June 1991.

[19] H. A. Wheeler, "Transmission line properties of a stripline between parallel planes," *IEEE Trans. MTT*, vol. 26, pp. 866–876, Nov. 1978.

[20] T. Edwards, *Foundations for Microstrip Circuit Design*. 2nd ed., Chichester, U.K.: John Wiley & Sons, 1992.

[21] M. Kirschning and R. H. Jansen, "Accurate wide-range design equations for the frequency-dependent characteristics of parallel coupled microstrip lines," *IEEE Trans. MTT*, vol. 32, no. 1, pp. 83–90, Jan. 1984.

41

Time and Frequency Domain Responses

c Martens
iversiteit Gent,
gium
niël De Zutter
iversiteit Gent,
gium

1.1 Time Domain Reflectometry

Principles

Time domain reflectometry is used to characterize interconnections in the time domain. The setup essentially consists of a time domain step generator and a digital sampling oscilloscope (Fig. 41.1) [1]. The generator produces a positive-going step signal with a well-defined rise time. The step is applied to the device under test. The reflected and the transmitted signals are shown on the oscilloscope. Measuring the reflected signal is called time domain reflectometry (TDR); the transmitted signal is measured using the time domain transmission (TDT) option.

The characteristic impedance levels and delay through an interconnection structure can be derived from the TDR measurements. The TDT measurement gives information about the losses (decrease of magnitude) and degradation of the rise time (filtering of high-frequency components). The TDR/TDT measurements also are used to extract an equivalent circuit consisting of transmission lines and lumped elements.

The fundamentals of TDR are discussed in detail in [2] and [3]. Reference [4] describes the applications of TDR in various environments including PCB/backplane, wafer/hybrids, IC packages, connectors, and cables.

One-Port Time Domain Reflectometry

Figure 41.2 shows that the device under test is a simple resistor with impedance Z_L. In this case a mismatch with respect to the reference or system impedance Z_0 exists. A reflected voltage wave will appear on the oscilloscope display algebraically added to the incident wave. The amplitude of

93-8341-2/95/$0.00 + $.50

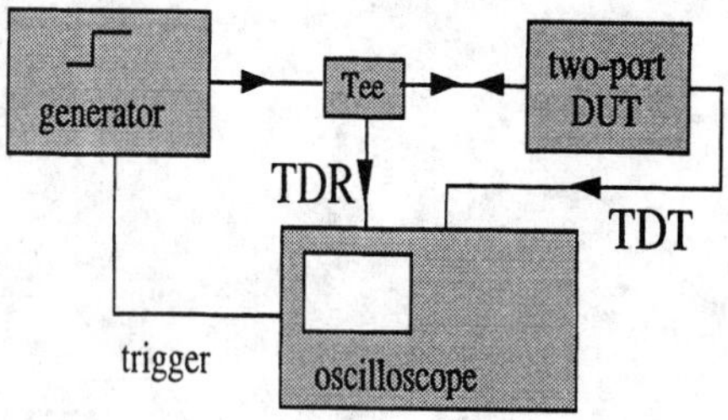

FIGURE 41.1 Setup for TDR/TDT measurements. DUT = Device under test.

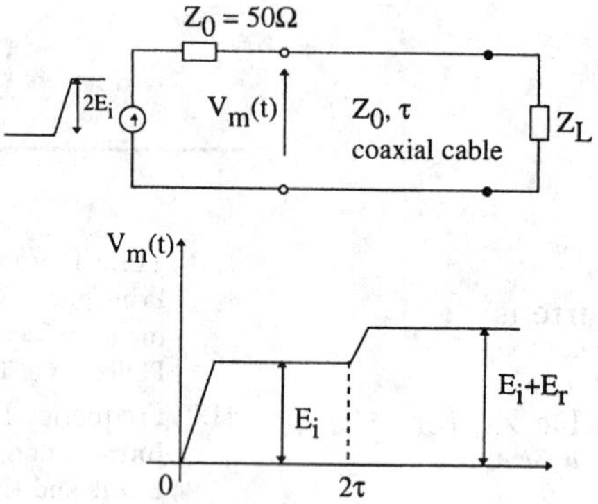

FIGURE 41.2 TDR measurement of an impedance Z_L.

the reflected voltage wave, E_r, is determined by the reflection coefficient of the load impedance, Z_L, with respect to the system impedance Z_0:

$$E_r = \rho E_i = \frac{Z_L - Z_0}{Z_L + Z_0} E_i \tag{41.1}$$

Figure 41.2 also depicts the time domain picture shown on the oscilloscope for a load, the impedance, Z_L, of which is larger than Z_0. From the measurement of the magnitude E_r of the reflected voltage wave, the load impedance Z_L can be derived.

Time Domain Reflectometry Pictures for Typical Loads

The most simple loads to be measured are the open circuit and the short circuit. For ideal open-circuit and short-circuit loads the reflection coefficient is, respectively, 1 and −1. This means that the measured voltage doubles in the first case and goes to zero in the second case, when the reflected voltage wave arrives at the oscilloscope (Fig. 41.3).

For any other real load impedance the reflection coefficient lies between −1 and 1. If the real load impedance is larger than the reference impedance, the reflected voltage wave is a positive-going step signal. In this case the amplitude of the voltage is increased when the reflected wave is added to the input step (Fig. 41.4). The reverse happens when the load impedance is lower than the reference impedance.

For complex load impedances, the step response is more complicated. For example, in the case of a series connection of a resistance and an inductance or a parallel connection of a resistance and a capacitance, a first-order step response. From the two pictures shown in Fig. 41.5 we learn that a series inductance gives a positive dip, while the capacitance produces a negative dip.

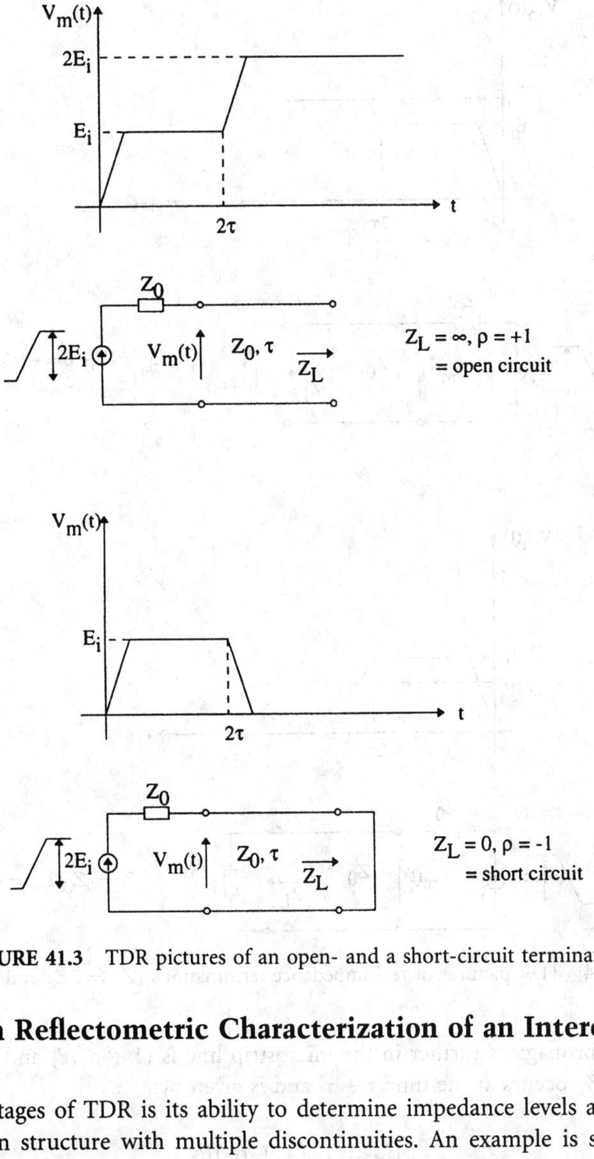

FIGURE 41.3 TDR pictures of an open- and a short-circuit termination.

Time Domain Reflectometric Characterization of an Interconnection Structure

One of the advantages of TDR is its ability to determine impedance levels and delays through an interconnection structure with multiple discontinuities. An example is shown in Fig. 41.6 for a microstrip line connected to the measurement cable. The line is terminated in a load with impedance Z_L. It is seen that two mismatches produce reflections that can be analyzed separately. The mismatch at the junction of the two transmission lines generates a reflected wave $E_{r1} = \rho_1 E_i$. Similarly, the mismatch at the load creates a reflection due to its reflection coefficient ρ_2. Both reflection coefficients are defined as:

$$\rho_1 = \frac{Z_0' - Z_0}{Z_0' + Z_0}$$
$$\rho_2 = \frac{Z_L - Z_0'}{Z_L + Z_0'} \tag{41.2}$$

After a time τ the reflection at the junction of the transmission lines occurs. The voltage wave associated with this reflection adds to the oscilloscope's picture at the time instant 2τ. The

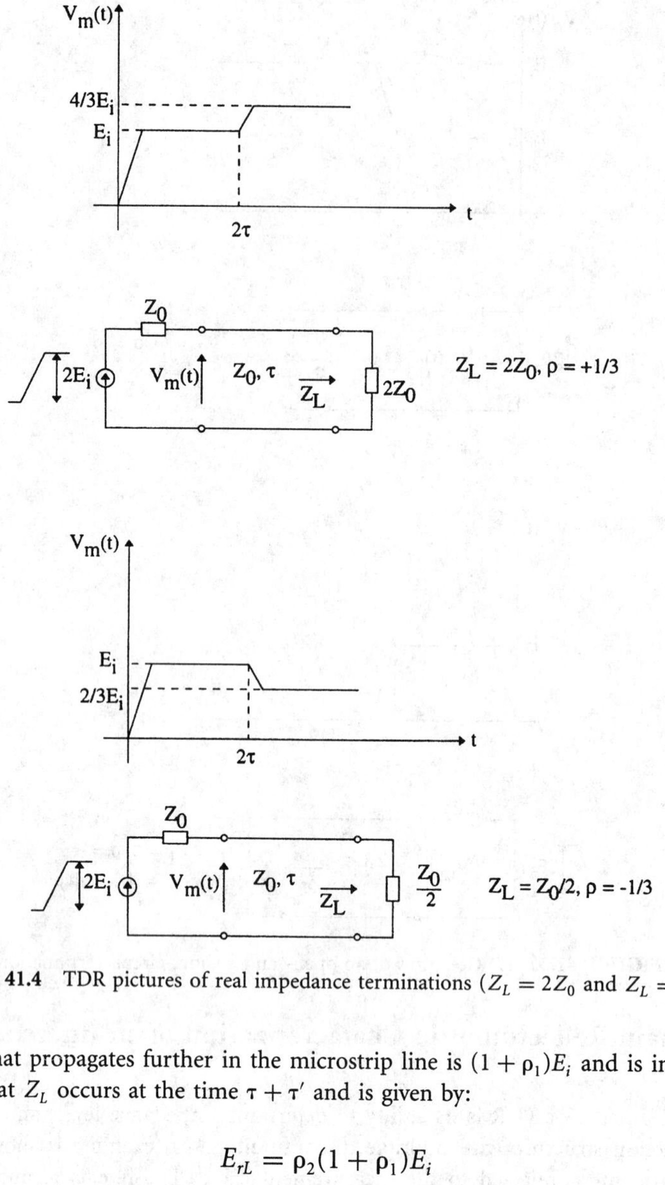

FIGURE 41.4 TDR pictures of real impedance terminations ($Z_L = 2Z_0$ and $Z_L = Z_0/2$).

voltage wave that propagates further in the microstrip line is $(1+\rho_1)E_i$ and is incident on Z_L. The reflection at Z_L occurs at the time $\tau + \tau'$ and is given by:

$$E_{rL} = \rho_2(1+\rho_1)E_i \tag{41.3}$$

After a time $\tau + 2\tau'$ a second reflection is generated at the junction. The magnitude of the reflection is now determined by the reflection coefficient $\rho_1' = -\rho_1$. The voltage wave E_{r2} that is transmitted through the junction and propagates in the direction of the generator adds to the time domain picture at time instant $2\tau + 2\tau'$ and is given by:

$$E_{r2} = (1+\rho_1)E_{rL} = (1-\rho_1)\rho_2(1+\rho_1)E_i = (1-\rho_1^2)\rho_2 E_i \tag{41.4}$$

If ρ_1 is small in comparison to 1, then

$$E_{r2} \approx \rho_2 E_i \tag{41.5}$$

which means that ρ_2 can be determined from the measurement of E_{r2}.

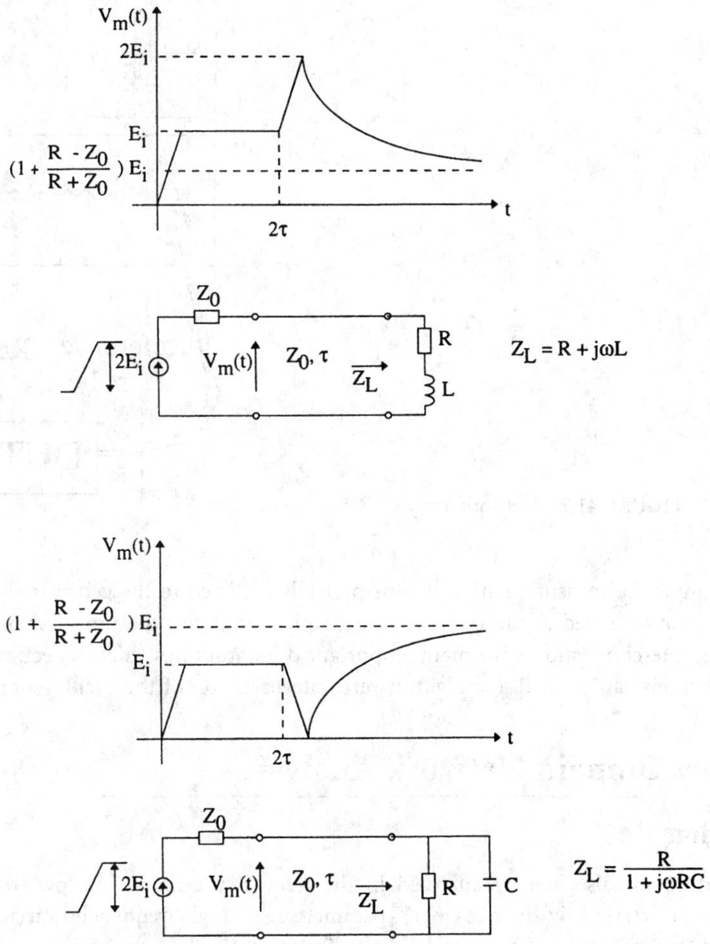

FIGURE 41.5 TDR pictures of two complex impedance terminations.

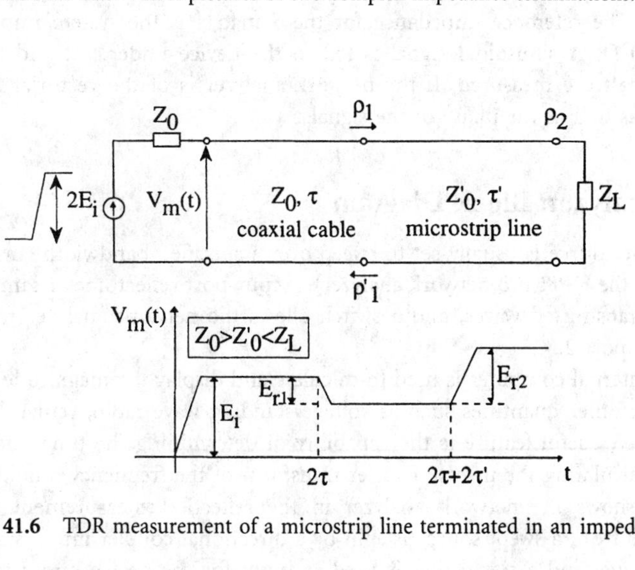

FIGURE 41.6 TDR measurement of a microstrip line terminated in an impedance Z_L.

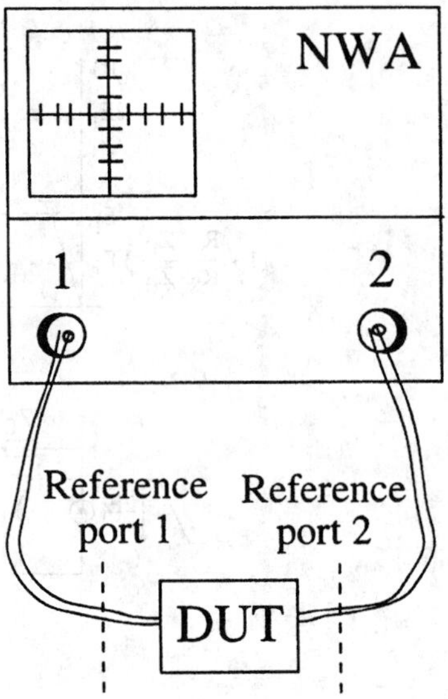

FIGURE 41.7 Network analyzer (NWA) setup.

In this example the measurement cable was perfectly matched to the generator impedance so that no reflection occurred at the generator side, which simplifies the time domain picture. In the case of an interconnection with many important discontinuities (high reflection coefficient) multiple reflections can prevent a straightforward interpretation of the oscilloscope's display.

41.2 Frequency Domain Network Analysis

Introduction

A distributed circuit also can by analyzed in the frequency domain. At low frequencies the circuits are characterized by their Z- or Y-parameters. At high frequencies circuits are better characterized by S-parameters. We focus only on S-parameter characterization.

The network analyzer is an instrument that measures the S-parameters of a two-port device (Fig. 41.7) [5]. The reference impedance for the S-matrix is the system impedance, which is the standard 50 Ω. A sinusoidal signal is fed to the device under test and the reflected and transmitted signals are measured. If the network analyzer is of the vectorial type, it measures the magnitude as well as the phase of the signals.

Network Analyzer: Block Diagram

In operation, the source is usually set to sweep over a specified bandwidth (maximally 45 MHz to 26.5 GHz for the HP8510B network analyzer). A four-port reflectometer samples the incident, reflected, and transmitted waves, and a switch allows the network analyzer to be driven from either port 1 or port 2.

A powerful internal computer is used to calculate and display the magnitude and phase of the S-parameters or other quantities such as voltage standing wave radio, return loss, group delay, impedance, etc. A useful feature is the capability of determining the time domain response of the circuit by calculating the inverse Fourier transform of the frequency domain data.

Figure 41.8 shows the network analyzer in the reflection measurement configuration. A sinusoidal signal from a swept source is split by a directional coupler into a signal that is fed to a reference channel and a signal that is used as input for the device under test. The reflected

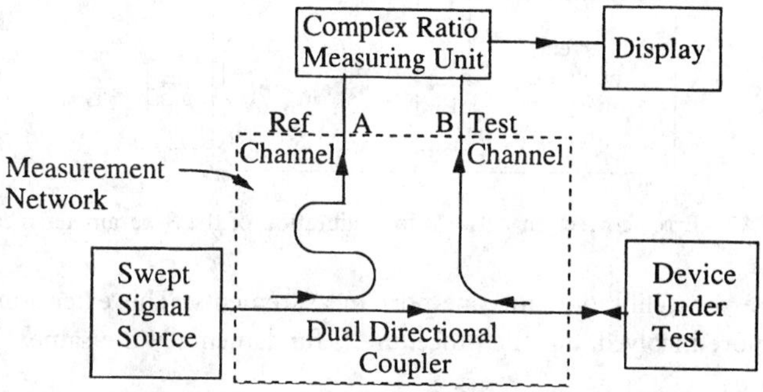

FIGURE 41.8 Block diagram for reflection measurements with the network analyzer.

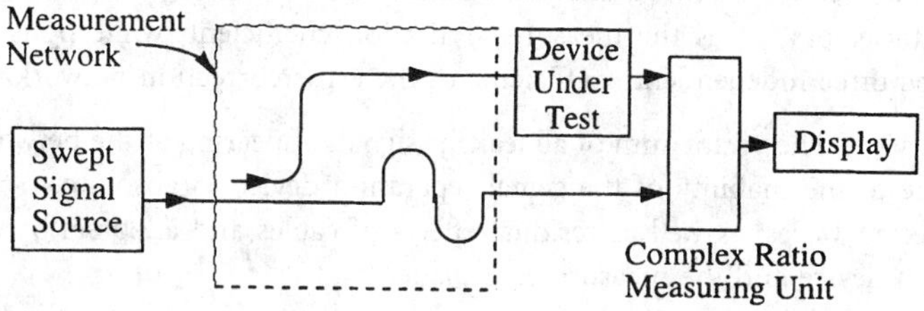

FIGURE 41.9 Block diagram for transmission measurements with the network analyzer.

signal is fed back through the directional coupler to the test channel. In the complex ratio measuring unit the amplitude and phase of the test and reference signals are compared.

The principle of the transmission measurement is the same as for the reflection configuration (Fig. 41.9). In this setup the signal transmitted through the device under test is fed to the test channel. The other part of the signal of the swept source is transmitted through the directional coupler and fed to the reference channel. Again, amplitude and phase of the two signals are compared.

An important issue concerning the measurement of *S*-parameters is the calibration of the results. In order to perform a good calibration, the nature of the measurement errors must be well understood. These items are the subject of the following section.

Measurement Errors and Calibration

Measurement Errors

The network analyzer measurement errors [6] can be classified in two catagories:

1. Random Errors: Nonrepeatable measurement variations that occur due to noise, environmental changes, and other physical changes in the test setup. These are any errors that the system itself may not be able to measure.
2. Systematic Errors: These are repeatable errors. They include the mismatch and leakage terms in the test setup, the isolation characteristic between the reference and test signal paths, and the system frequency response.

In most *S*-parameters measurements, the systematic errors are those that produce the most significant measurement uncertainty. Because each of these errors produces a predictable effect upon the measured data, their effects can be removed to obtain a corrected value for the test device response. The procedure of removing these systematic errors is called calibration.

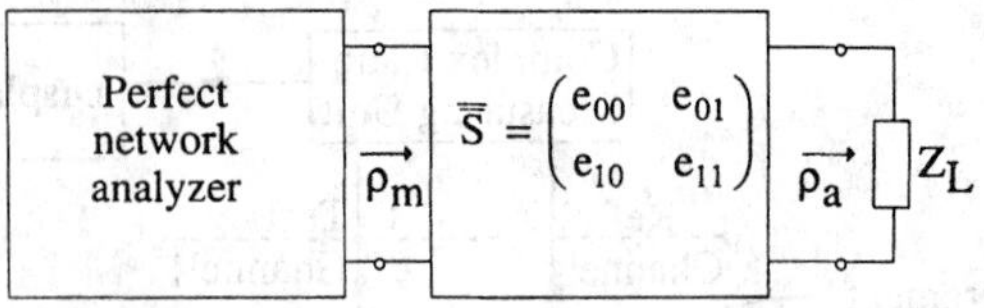

FIGURE 41.10 Error correction network for calibration of the *S*-parameter measurements.

We illustrate here the calibration for one-port measurements. The extension to two-ports is mathematically more involved, but the procedure is fundamentally the same.

One-Port Calibration: Error Correction Network

In the configuration for one-port measurements the real network analyzer can be considered as a connection of a "perfect" network analyzer and a two-port whose parameters are determined by the systematic errors. ρ_m is the measured reflection coefficient, while ρ_a is the actual one (Fig. 41.10). The three independent coefficients of the error correction network [7] are

1. Directivity e_{00}: The vector sum of all leakage signals appearing at the network analyzer test input due to the inability of the signal separation device to absolutely separate incident and reflected waves, as well as residual effects of cables and adapters between the signal separation device and the measurement plane.
2. Source impedance mismatch e_{11}: The vector sum of the signals appearing at the network analyzer test input due to the inability of the source to maintain absolute constant power at the test device input as well as cable and adaptor mismatches and losses outside the source leveling loop.
3. Tracking or frequency response $e_{10}e_{01}$: The vector sum of all test setup variations in the magnitude and phase frequency response between the reference and test signal paths.

The relations between ρ_m and ρ_a are given by:

$$\rho_m = e_{00} + \frac{e_{01}e_{10}\rho_a}{1 - e_{11}\rho_a} \tag{41.6}$$

$$\rho_a = \frac{\rho_m - e_{00}}{e_{11}(\rho_m - e_{00}) + e_{10}e_{01}} \tag{41.7}$$

The error coefficients are determined by measuring standard loads. In the case of the SOL calibration, a short-circuit, open-circuit, and matched load are measured [5]. The standards are characterized over a specified frequency range (e.g., from 45 MHz to 26.5 GHz for the network analyzer HP8510B). The best-specified standards are coaxial. The quality of the coaxial standards is determined by the precision of the mechanical construction. The short circuit and the load are in most cases nearly perfect ($\rho_a^1 = -1$ and $\rho_a^3 = 0$, respectively). The open circuit behaves as a frequency-dependent capacitance:

$$\rho_a^2 = e^{-j2\,\mathrm{arctg}(\omega C Z_0)} \quad \text{with } C = C_0 + f^2 C_2 \tag{41.8}$$

Z_0 is the characteristic impedance of the coaxial part of the standard. C_0 and C_2 are specified by the constructor of the hardware. Once the reflection coefficients of the standards are specified and the standards are measured, a set of linear equations in the error coefficients is derived:

$$a\rho_a^i + b - c\rho_a^i\rho_m^i = \rho_m^i \qquad i = 1, 2, 3 \tag{41.9}$$

where $a = e_{01}e_{10} - e_{00}e_{11}$, $b = e_{00}$, and $c = -e_{11}$. This set is easily solved in the error coefficients which are then used together with the measurement of ρ_m to determine ρ_a.

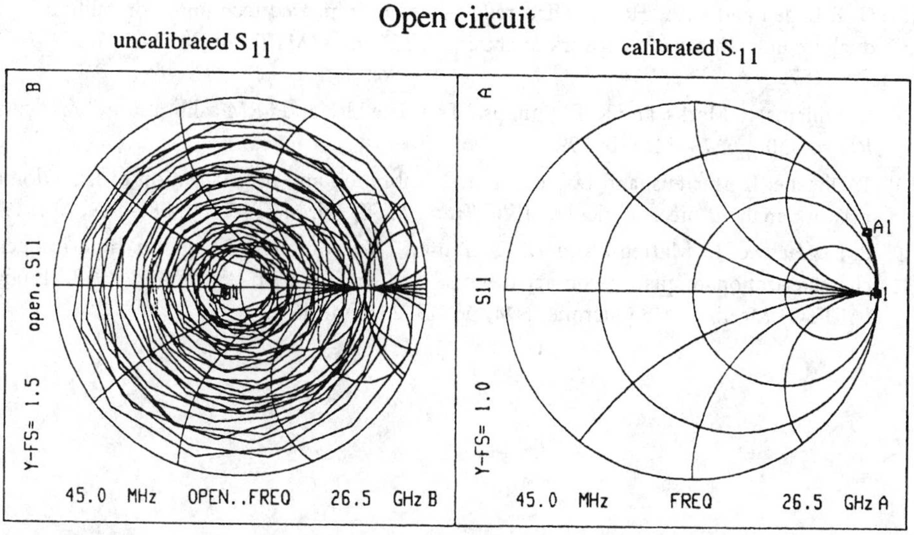

FIGURE 41.11 Smith chart representation of the uncalibrated and calibrated S_{11} data of an open circuit (measurements in the range of 45 MHz to 26.5 GHz).

Other calibration methods (TRL, LRM) are described in [8] and [9]. Analogous calibration techniques also were developed for time domain measurements (TDR/TDT) [10]. Applications of S-parameter measurements on circuits on PCBs are described in [11].

Error Checking and Verification of the Calibration

Checking errors in the calibration is done by applying the calibration to a measured standard. The specified reflection response of the standard must be found. In Fig. 41.11, the uncalibrated measurement of the open-circuit standard is shown. After calibration we obtain the specified response (capacitance model) of the standard (Fig. 41.11). The calibration can be further verified on a standard transmission line.

References

[1] A. M. Nicolson, C. L. Benett, Jr., D. Lammensdorf, and L. Susman, "Applications of time domain metrology to the automation of broad-band microwave measurements," *IEEE Trans. MTT*, vol. MTT-20, no. 1, pp. 3–9, Jan. 1972.

[2] Hewlett Packard, "TDR Fundamentals for Use with HP54120T Digitizing Oscilloscope and TDR," HP-Appl. Note 62, Apr. 1988.

[3] Hewlett Packard, "Improving Time Domain Network Analysis Measurements," HP-Appl. Note 62-1, Apr. 1988.

[4] Hewlett Packard, "Advanced TDR Techniques," HP-Appl. Note 62-3, May 1990.

[5] R. A. Hackborn, "An automatic network analyzer system," *Microwave J.*, vol. 11, pp. 45–52, May 1968.

[6] B. Donecker, "Accuracy predictions for a new generation network analyzer," *Microwave J.*, vol. 27, pp. 127–141, June 1984.

[7] J. Williams, "Accuracy Enhancement Fundamentals for Network Analyzers," *Microwave J.*, vol. 32, pp. 99–114, Mar. 1989.

[8] G. F. Engen and C. A. Hoer, "Thru-reflect-line: an improved technique for calibrating the dual six-port automatic network analyzer," *IEEE Trans. MTT*, vol. MTT-27, pp. 987–993, Dec. 1979.

[9] D. Wiliams, R. Marks, and K. R. Phillips, "Translate LRL and LRM calibrations," *Microwaves RF*, vol. 30, pp. 78–84, Feb. 1991.

[10] T. Dhaene, L. Martens, and D. De Zutter, "Calibration and normalization of time domain network analyzer measurements," *IEEE Trans. MTT*, vol. MTT-42, pp. 580–589, Apr. 1994.

[11] P. Degraeuwe, L. Martens, and D. De Zutter, "Measurement set-up for high-frequency characterization of planar contact devices," in Proc. of 39th Automatic RF Techniques (ARFTG) Meeting, Albuquerque, NM, pp. 19–25, June 1992.

42

Distributed RC Networks

ladimir Székely
chnical University of
ıdapest, Hungary

In everyday practice one often encounters RC networks that are inherently distributed (DRC lines). All the resistors and the interconnection lines of an IC, the channel regions of the FET, and MOS transistors are DRC lines. The electrical behavior of such networks is discussed in the current section.

42.1 Uniform Distributed RC Lines

First, let us consider the so-called lossless RC lines. The structure in Fig. 42.1 has only serial resistance and parallel capacitance. Both the resistance and the capacitance are distributed along the x-axis. r and c denote the resistance and the capacitance pro unit length, respectively. In this section these values are considered constants, which means that the RC line is *uniform.*

Consider a Δx length section of the structure. Using the notations of Fig. 42.1 the following relations can be given between the currents and voltages:

$$\Delta v = -ir\,\Delta x \quad \Delta i = -\frac{\partial v}{\partial t}\,c\,\Delta x \tag{42.1}$$

By taking the limit as $\Delta x \to 0$ we obtain a pair of differential equations

$$\frac{\partial v}{\partial x} = -ri \quad \frac{\partial i}{\partial x} = -c\,\frac{\partial v}{\partial t} \tag{42.2}$$

which describe the behavior of the lossless RC line. Substituting the first equation of (42.2) after

3493-8341-2/95/$0.00+$.50

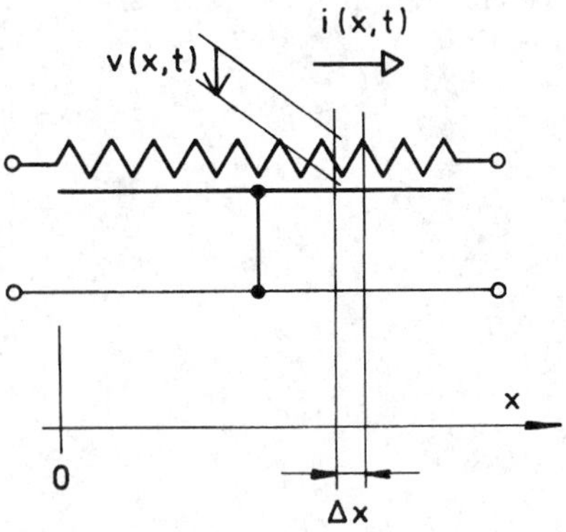

FIGURE 42.1 Lossless DRC line.

derivation into the second equation of (42.2) results in

$$\frac{\partial v}{\partial t} = \frac{1}{rc}\frac{\partial^2 v}{\partial x^2} \tag{42.3}$$

This shows that the time-dependent voltage along the line $v(x, t)$ is determined by a homogeneous, constant-coefficient, second-order partial differential equation, with the first time derivative on one side and the second spatial derivative on the other side. This kind of equation is called a **diffusion equation**, because diffusion processes are described using such equations.

Solution in the Time Domain

Differential Eq. (42.3) is fulfilled by the simple, spatially Gauss-like function

$$v(x,\, t) = \frac{Q/c}{2\sqrt{\pi t/rc}} \exp\left(-\frac{x^2}{4t/rc}\right) \tag{42.4}$$

This can be proved easily by substituting the function into (42.3).

Solution (42.4) describes the physical situation in which the uniform RC structure is of infinite length in both directions, and a short, impulse-like Q charge injection is applied at $x = 0$ in the $t = 0$ instant. This means a Dirac-δ excitation at $x = 0$ in both charge and voltage. As time elapses, the charge spreads equally in both directions. The extension of the charge and voltage wave is every increasing, but their amplitude is decreasing (see Fig. 42.2(a)).

The same result is represented in a different way in Fig. 42.2(b). The time dependence of the voltage is shown at the $x = H$, $x = 2H$, etc. spatial positions. As we move away from the $x = 0$ point, the maximum of the impulse appears increasingly later in time, and the originally sharp impulse is increasingly extended. This means that the RC line delays the input pulse and at the same time strongly spreads it. The RC line is *dispersive.*

Superposition

Equation (42.3) is homogeneous and linear. This means that any sum of the solutions is again a solution. Based on this fact, the problem of Fig. 42.3(a) can be solved. At $t = 0$ the voltage distribution of the RC line is given by an arbitrary $U(x)$ function. For $t > 0$ the distribution can be calculated by dividing the $U(x)$ function into $\Delta\xi \to 0$ elementary slices. These may be considered to be individual Dirac-δ excitations and the responses given to them can be summarized by integration:

$$v(x,\, t) = \frac{1}{2\sqrt{\pi t/rc}} \int_{-\infty}^{\infty} U(\xi) \exp\left(-\frac{(x-\xi)^2}{4t/rc}\right) d\xi \tag{42.5}$$

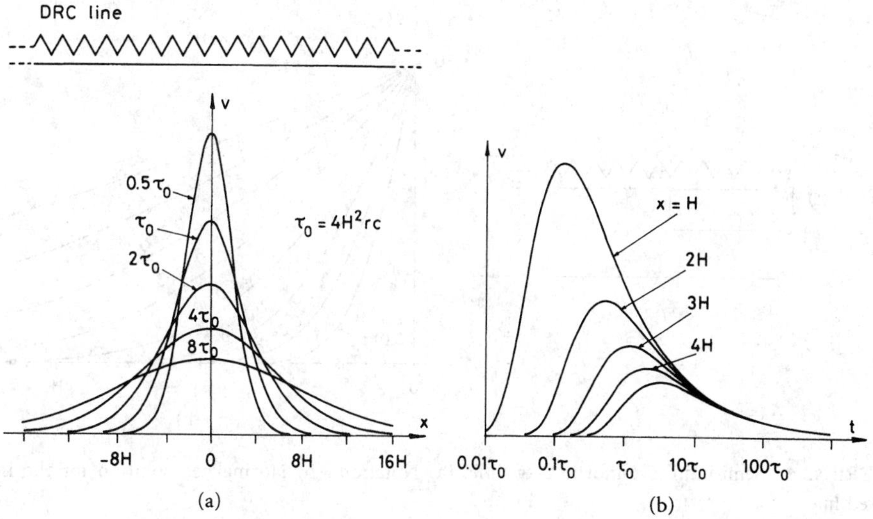

FIGURE 42.2 Effect of an impulse-like charge injection at $x = 0$. (a) Voltage distribution in subsequent time instants. (b) Voltage transients in different distances from the injection point.

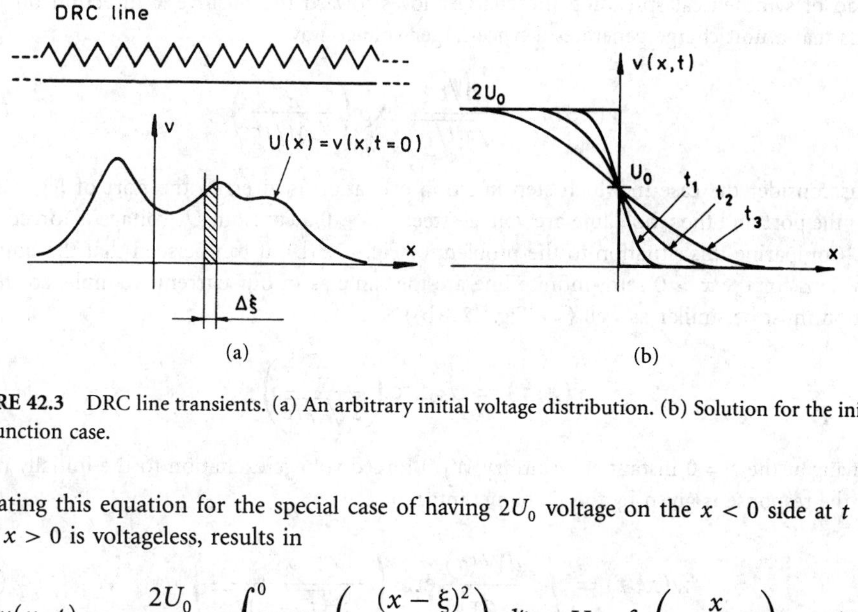

FIGURE 42.3 DRC line transients. (a) An arbitrary initial voltage distribution. (b) Solution for the initially step-function case.

Evaluating this equation for the special case of having $2U_0$ voltage on the $x < 0$ side at $t = 0$, while $x > 0$ is voltageless, results in

$$v(x, t) = \frac{2U_0}{2\sqrt{\pi t/rc}} \int_{-\infty}^{0} \exp\left(-\frac{(x-\xi)^2}{4t/rc}\right) d\xi = U_0 \operatorname{erfc}\left(\frac{x}{2\sqrt{t/rc}}\right) \tag{42.6}$$

where the integral of the Gauss function is notated by erfc(x), the complementary error function.[1] The originally abrupt voltage step is getting increasingly less steep with time (Fig. 42.3(b)). In the middle at $x = 0$ the voltage remains U_0.

Semi-Infinite Line

Our next model is a bit closer to practice; the uniform RC line extends to $x \geq 0$ only. At $x = 0$ a port is characterized by the $V(t)$ voltage and the $I(t)$ current (Fig. 42.4).

[1] $\operatorname{erfc}(x) = 1 - \frac{2}{\sqrt{\pi}} \int_0^x \exp(-y^2)\, dy = \frac{2}{\sqrt{\pi}} \int_x^{\infty} \exp(-y^2)\, dy$

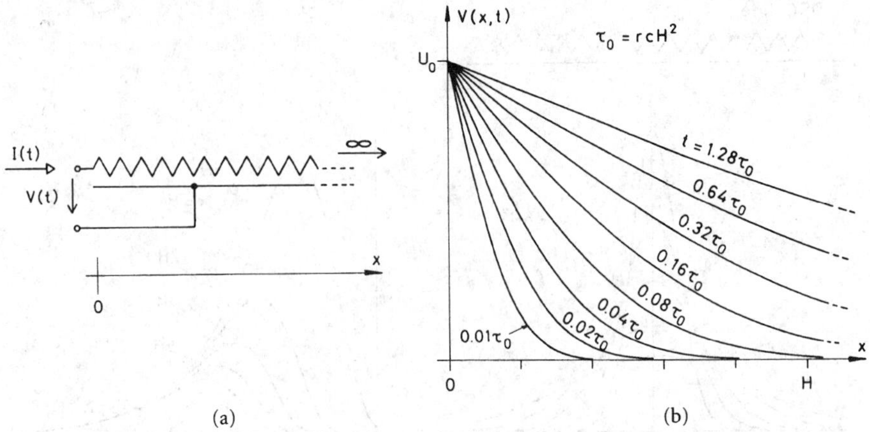

FIGURE 42.4 Semi-infinite uniform DRC line. (a) Notation. (b) Normalized solution for the initially relaxed line.

If the line is relaxed and a $I(t) = \delta(t)$ current (a Dirac-δ current pulse) is forced to the port, a unit charge is introduced at $x = 0$. The result will be similar to that of Fig. 42.2(a), but instead of symmetrical spreading the charge moves toward the positive x direction only. This means that a unit charge generates a twice-larger voltage wave

$$v(x, t) = \frac{1/c}{\sqrt{\pi t/rc}} \exp\left(-\frac{x^2}{4t/rc}\right) \tag{42.7}$$

Let us consider the case in which step function excitation is given to the port of Fig. 42.4. At $t < 0$ the port and the whole line are voltage-free; at $t \geq 0$ a constant U_0 voltage is forced to the port. Comparing this situation to the problem of Fig. 42.3(b), it can be seen that the boundary conditions for the $x > 0$ semi-infinite line are the same as in our current example, so that the solution must be similar as well (see Fig. 42.4(b)):

$$v(x, t) = U_0 \operatorname{erfc}\left(\frac{x}{2\sqrt{t/rc}}\right) \tag{42.8}$$

Applying at the $t = 0$ instant an arbitrary $W(t)$ forced voltage excitation to the initially relaxed line, the response is given by the Duhamel integral

$$w(x, t) = \int_0^t \frac{dW(\tau)}{d\tau} \operatorname{erfc}\left(\frac{x}{2\sqrt{(t-\tau)/rc}}\right) d\tau \tag{42.9}$$

Finite DRC Line

Let the DRC line of length L of Fig. 42.5(a) be closed at $x = L$ with a short circuit. Let the line at $t < 0$ be relaxed and assume a W $(t > 0)$ voltage excitation at the $x = 0$ port. Using the $w(x, t)$ voltage response of the semi-finite line (42.9), for the short-terminated line of length L

$$\begin{aligned} v(x, t) &= \sum_{i=0}^{\infty} (-1)^i \cdot w(2iL + (-1)^i x, t) \\ &= \sum_{k=0}^{\infty} (w(2kL + x, t) - w(2kL + 2L - x, t)) \end{aligned} \tag{42.10}$$

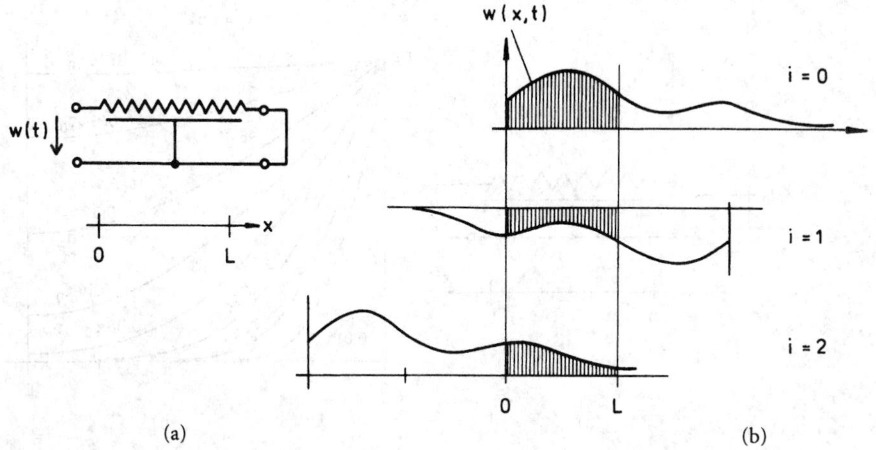

FIGURE 42.5 Finite-length uniform DRC line. (a) DRC line with short circuit at $x = L$. (b) Visualization of the mirroring procedure.

This result is illustrated in Fig. 42.5(b). The $v(x, t)$ function is given as the sum of the shifted, negated, and mirrored replicas of the $w(x, t)$ function, so that it is a valid solution as well. The $x = L$ boundary condition is the short circuit $v(x = L, t) = 0$. The $i = 0$ and $i = 1$ functions are the same size with different signs at $x = L$, so they cancel each other. The same is true for $i = 2$ and 3, and so on. The $x = 0$ boundary condition $v = W(t)$ is fulfilled by the $i = 0$ function, while the further functions cancel each other in pairs (the $i = 1$ and 2, etc.).

The result can be interpreted as the response of the semi-infinite line being mirrored with negative sign on the short termination. In the case of Fig. 42.5(a) the termination on both the $x = 0$ and $x = L$ ends is assured by zero impedance short circuit; the resultant voltage function comes from the successive back and forth mirroring between these two "mirrors".

It is easy to understand that a termination with an open circuit results in mirroring without sign change. (At this termination the current equals zero so that the dv/dx derivative equals zero as well. This requirement is always fulfilled in the mirroring point summarizing the continuous incident function with its mirrored version.) According to this, the voltage on the open-terminated line of Fig. 42.6(a) is

$$v(x, t) = \sum_{k=0}^{\infty} (-1)^k (w(2kL + x, t) + w(2kL + 2L - x, t)) \tag{42.11}$$

which in the case of step function excitation with U_0 amplitude is

$$v(x, t) = U_0 \sum_{k=0}^{\infty} (-1)^k \left[\operatorname{erfc}\left(\frac{2kL + x}{2\sqrt{t/rc}}\right) + \operatorname{erfc}\left(\frac{2kL + 2L - x}{2\sqrt{t/rc}}\right) \right] \tag{42.12}$$

This function is given in Fig. 42.6(b) for some time instants.

Solution in the Frequency Domain

To find the solution of the differential Eq. (42.3), the following trial function can be used

$$v(x, t) = v \cdot e^{j\omega t} \cdot e^{\gamma x} \tag{42.13}$$

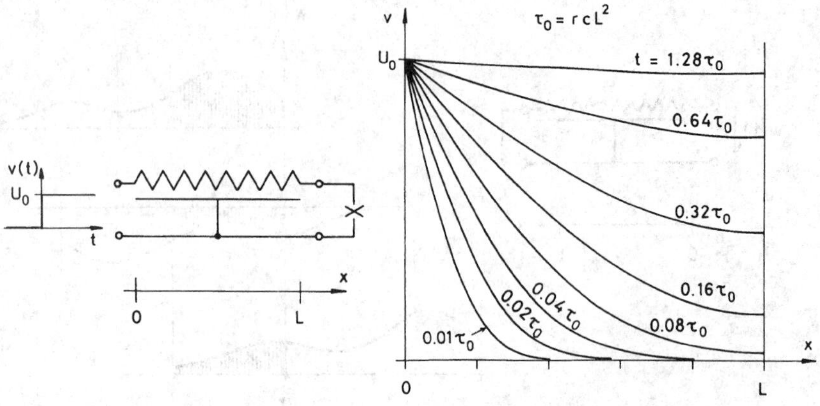

(a) (b)

FIGURE 42.6 Finite-length uniform DRC line. (a) Open circuit at the far end. (b) Normalized solution for the initially relaxed line.

Substituting this function into (42.3) results in the following so-called *dispersion equation:*

$$\gamma = \sqrt{j\omega rc} = (1+j)\frac{1}{\sqrt{2}}\sqrt{\omega rc} \tag{42.14}$$

This means that a wave-like solution exists as well. However, it is strongly collapsing because the real and imaginary parts of γ are always equal, which means that the attenuation on a path of λ wavelength is $\exp(-2\pi) \cong 1/535$.

The lossless DRC line can be considered to be a special telegraph line having neither serial inductance nor shunt conductance. The telegraph line theory can be conveniently used at the calculation of uniform DRC networks. The γ propagation constant and the $\mathbf{Z}_0$ characteristic impedance for the present case are

$$\gamma = \sqrt{src} \quad \mathbf{Z}_0 = \sqrt{\frac{r}{sc}} \tag{42.15}$$

With these the two-port impedance parameters and chain parameters of an RC line of length L can be given as follows (see 39.23)

$$\mathbf{Z}_{ij} = \mathbf{Z}_0 \begin{bmatrix} \operatorname{cth}\gamma L & 1/\operatorname{sh}\gamma L \\ 1/\operatorname{sh}\gamma L & \operatorname{cth}\gamma L \end{bmatrix} \quad \begin{bmatrix} A & B \\ C & D \end{bmatrix} = \begin{bmatrix} \operatorname{ch}\gamma L & \mathbf{Z}_0 \operatorname{sh}\gamma L \\ \dfrac{1}{\mathbf{Z}_0}\operatorname{sh}\gamma L & \operatorname{ch}\gamma L \end{bmatrix} \tag{42.16}$$

If one of the ports is terminated by an impedance of $\mathbf{Z}_t$, the following impedance can be "seen" on the opposite port (see 39.23)

$$\mathbf{Z}_{\text{in}} = \mathbf{Z}_0 \frac{\mathbf{Z}_t \operatorname{ch}\gamma L + \mathbf{Z}_0 \operatorname{sh}\gamma L}{\mathbf{Z}_t \operatorname{sh}\gamma L + \mathbf{Z}_0 \operatorname{ch}\gamma L} \tag{42.17}$$

Uniform, Lossy DRC Lines

In some cases the DRC structure also has shunt conductance, which means that it is lossy. The value of this conductance for the unit length is notated by g. In such a case, without giving the details of the calculation, the $v(x, t)$ line voltage can be determined by the solution of the equation

$$\frac{\partial v}{\partial t} = \frac{1}{rc}\frac{\partial^2 v}{\partial x^2} - \frac{g}{c}v \tag{42.18}$$

The following forms of the characteristic impedance and the propagation constant can be used now in the frequency domain:

$$\gamma = \sqrt{r(g + sc)} \quad \mathbf{Z}_0 = \sqrt{\frac{r}{g + sc}} \tag{42.19}$$

It is an interesting fact that the charge carrier motion in the base region of homogeneously doped bipolar transistors can be described by formally similar equations, so that the intrinsic transients of the bipolar transistors can be exactly modeled by lossy DRC two-ports [4].

Example 42.1 Wiring Delays: Neither the series resistance nor the stray capacitance of the interconnection leads of integrated circuits are negligible. As an example, in the case of a polysilicon line of 1.5 μm width $r \cong 33$ KΩ/mm, $c \cong 0.06$ pF/mm. This means that these wires should be considered to be DRC lines. The input logical levels appear on their output with a finite delay. Let us determine the delay of a wire of length L. From (42.12)

$$v(L, t) = U_0 \sum_{k=0}^{\infty} 2(-1)^k \operatorname{erfc}((2k + 1)\vartheta)$$

where

$$\vartheta = \frac{L}{2\sqrt{t/rc}}$$

The sum on the left will reach the 0.9 value at $\vartheta = 0.5$, so that the voltage at the end of the line will reach the 90% of U_0 after a time delay of

$$t_{\text{delay}} \cong rcL^2 \tag{42.20}$$

Note that the time delay increases with the square of the length of the wire. In the case of $L = 1$ mm the time delay of this polysilicon wire is already 2 ns, which is more than the time delay of a CMOS logical gate. For lengthy wires (> 0.2 ÷ 0.5 mm) metal wiring must be applied with its inherently small resistivity.

Example 42.2 Parasitic Effects of IC Resistors: In an IC amplifier stage the transistor is loaded with, e.g., $R = 10$ kΩ (Fig. 42.7(a)). This resistor has been fabricated by the base diffusion, the sheet resistance is 200 Ω, and the parasitic capacitance is 125 pF/mm². The minimum feature size of the technology is 4 μm. Let us determine the impedance of the resistor in the 10 to 10^4 MHz range.

The resistance can be realized in 4 × 200 μm size. The total parasitic capacitance is $C_p = 0.1$ pF. The impedance of the transistor-side port can be calculated according to (42.17), considering that the opposite port is short-terminated, as

$$\mathbf{Z}_{\text{port}} = \mathbf{Z}_0 \text{th}\ \gamma L = \sqrt{\frac{r}{sc}}\ \text{th}\ (\sqrt{src}\, L) = \sqrt{\frac{R}{sC_p}}\ \text{th}\ \sqrt{sRC_p} \tag{42.21}$$

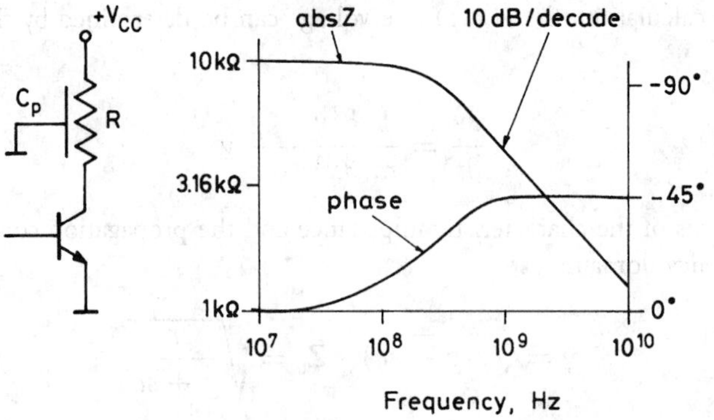

(a) (b)

FIGURE 42.7 A simple IC amplifier stage. (a) The load resistor is in fact a DRC line. (b) The amplitude and phase plot of the load.

Using the $\mathbf{s} = j\omega$ substitution, with the actual data the amplitude and phase functions of Fig. 42.7(b) can be obtained for the impedance. At 10 MHz the phase shift caused by the parasitic capacitance is negligible, but at 100 MHz it is already considerable.

It is important to recognize that in the case of half as wide linewidths the size of the resistor will be only 2 × 100 μm, which results in one fourth of the previous value in C_p. This means that the capacitance becomes disturbing only at four times larger frequencies.

Note in Fig. 42.7(b) that the amplitude function shows a 10 dB/decade decay and the phase keeps to 45°, as if the load would be characterized by a "half pole". This 10 dB/decade frequency dependence often can be experienced at DRC lines.

42.2 Nonuniform Distributed RC Lines

In some cases the capacitance and/or the resistance of the DRC line shows a spatial dependency. This happens if the width of the lead strip is modulated in order to reach some special effects (Fig. 42.8(a), *tapered* RC line). In case of a biased IC resistance the capacitance changes along the length of the structure as well because of the voltage dependency of the junction capacitance. These structures are referred to as *nonuniform* DRC lines.

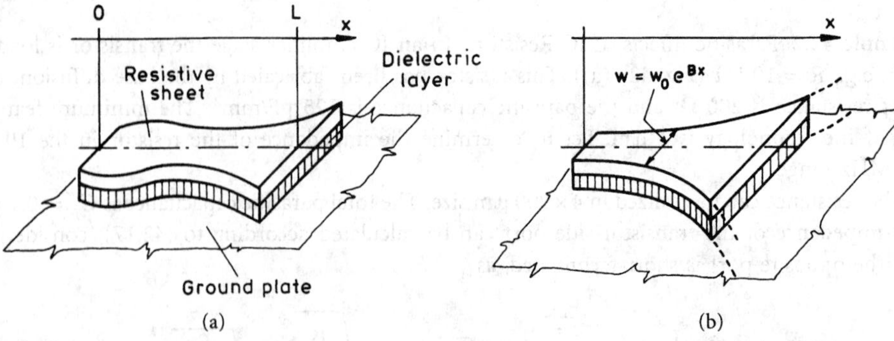

(a) (b)

FIGURE 42.8 Nonuniform DRC lines. (a) Tapered line. (b) Exponentially tapered line.

Let the spatially dependent resistance and capacitance pro unit length be notated by $r(x)$ and $c(x)$, respectively. The following equations can be given for the structure

$$\frac{\partial v}{\partial x} = -r(x)i \quad \frac{\partial i}{\partial x} = -c(x)\frac{\partial v}{\partial t} \tag{42.22}$$

With these, the following differential equation can be written:

$$\frac{\partial v}{\partial t} = \frac{1}{c(x)}\frac{\partial}{\partial x}\left(\frac{1}{r(x)}\frac{\partial v}{\partial x}\right) \tag{42.23}$$

We can obtain a more convenient form if we consider as the independent variable (instead of the x spatial coordinate) the total ρ resistance related to a given reference point (e.g., to the $x = 0$ point), as follows

$$\rho(x) = \int_0^x r(\xi)\,d\xi \quad r(x) = \frac{\partial \rho}{\partial x} \tag{42.24}$$

The variable defined this way can be considered as a kind of arc-length parameter. It has been introduced by [2]. With this new variable

$$\frac{\partial v}{\partial t} = \frac{1}{K(\rho)}\frac{\partial^2 v}{\partial \rho^2} \tag{42.25}$$

where

$$K(\rho) = K(\rho(x)) = \frac{c(x)}{r(x)} \tag{42.26}$$

The $K(\rho)$ function describes well the spatial parameter changes of the RC line, so that the structure of the line. Therefore the $K(\rho)$ function is now called the **structure function.** (Reference [2] uses the $\sigma = \int K d\rho$ function for the structure description.) Those DRC structures for which the $K(\rho)$ functions are the same can be considered to be electrically equivalent.

The differential Eq. (42.25) is homogeneous and linear, therefore, superposition can be used. Because this equation is of variable coefficient type, however, analytic solution can be expected only rarely. Such a case is that of the $K = K_0/\rho^4$ structure function for which

$$v(\rho, t) = \text{const}\,\frac{1}{t^{3/2}}\exp\left(-\frac{K_0}{4\rho^2 t}\right) \tag{42.27}$$

Another form of (42.25) is also known. To obtain this form, we should turn to the **s** domain with

$$\mathbf{s}v = \frac{1}{K(\rho)}\frac{\partial^2 v}{\partial \rho^2} \tag{42.28}$$

Let us introduce the following new variable:

$$\mathbf{Z}(\rho) = \frac{v(\mathbf{s}, \rho)}{i(\mathbf{s}, \rho)} = \frac{v}{-\dfrac{1}{r}\dfrac{\partial v}{\partial x}} = -\frac{v}{\dfrac{\partial v}{\partial \rho}} \tag{42.29}$$

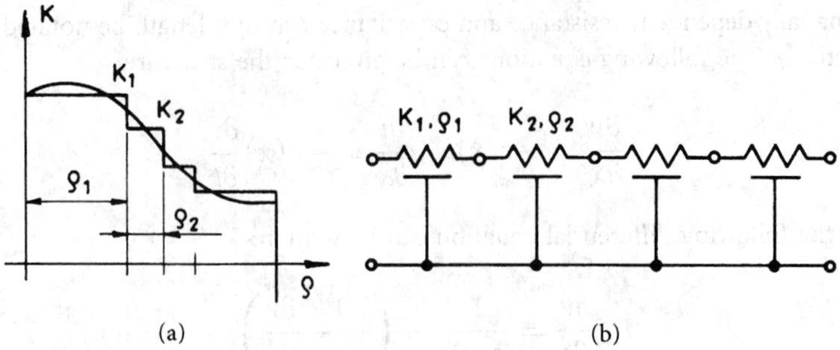

FIGURE 42.9 Approximation with concatenated uniform sections. (a) Stepwise approximation of the structure function. (b) Approximate model.

This variable is in fact the impedance of the line at the location of ρ. After rearrangements the

$$\frac{d\mathbf{Z}}{d\rho} = 1 + \mathbf{s}K(\rho)\mathbf{Z}^2 \tag{42.30}$$

equation can be obtained. This is called the Riccati differential equation. In the case of a known $K(\rho)$ structure function the one-port impedance of the nonuniform line can be determined from it by integration. In some cases even the analytic solution is known. Such a case is the exponentially tapered line of Fig. 42.8(b), for which

$$r(x) = \frac{R_\square}{w_0}e^{-Bx} \quad c(x) = C_p w_0 e^{Bx} \quad K(\rho) = \frac{R_\square C_p}{B^2}\frac{1}{\rho^2} \tag{42.31}$$

where $R_\square$ is the sheet resistance of the structure, C_p is the capacitance per unit area, and ρ is related to the point in the infinity. If the port in the infinity is considered to be shorted, the impedance of the location of ρ is

$$\mathbf{Z(s)} = \frac{\sqrt{1 + 4\mathbf{s}R_\square C_p/B^2} - 1}{2\mathbf{s}R_\square C_p/B^2}\rho \tag{42.32}$$

In other cases numerical integration of (42.30) leads to the solution.

Approximation with Concatenated Uniform Sections

The following model can be used for approximate calculation of nonuniform structures. We split the structure function into sections (see Fig. 42.9(a)) and use stepwise approximation. Inside the sections K = constant, so that they are uniform sections. Concatenating them according to Fig. 42.9(b), an approximate model is obtained.

In the frequency domain the overall parameters of the resultant two-port can be easily calculated. The chain parameter matrices of the concatenated sections have to be multiplied in the appropriate order. The time domain behavior can be calculated by inverse Laplace transformation.

Asymptotic Approximation for Large s

The chain parameters of a nonuniform DRC line can be approximately written as

$$\begin{bmatrix} A & B \\ C & D \end{bmatrix} \cong \frac{1}{2}\exp(\Delta\sqrt{\mathbf{s}})\begin{bmatrix} \lambda & \mu/\sqrt{\mathbf{s}} \\ \sqrt{\mathbf{s}}/\mu & 1/\lambda \end{bmatrix} \tag{42.33}$$

where

$$\Delta = \int_0^{R_0} \sqrt{K(\rho)}\, d\rho \quad R_0 = \rho(L) \quad \lambda = \left(\frac{K(R_0)}{K(0)}\right)^{1/4} \quad \mu = (K(R_0)K(0))^{-1/4}$$

This approximation is valid for large **s** values and for a sufficiently smooth function $K(\rho)$ [2].

Lumped Element Approximation

Distributed RC networks can be approximated by lumped element RC networks as well. The case of a lossless line is shown in Fig. 42.10(a). Two ways to determine the element values follow.

1. In the case of a known structure spatial discretization can be used. The nonuniform line must be split into sections of width h (see Fig. 42.10(b)). A node of the network is associated with the middle of each section. The total capacitance of the section must be calculated and this gives the value of the lumped capacitance connected to the node of the section. The resistance between the middle points of two adjacent sections must be calculated, and this must be connected between the nodes of the appropriate sections. It is obvious that the accuracy can be increased by decreasing h. The price is the increasing number of lumped elements. With $h \to 0$ we obtain the exact model.
2. When we know the impedance function we can build the model using the pole–zero pattern of the network. For example, let us investigate a uniform RC line of finite length L, short circuited at the far end. The corresponding impedance expression, according to (42.21), is

$$\mathbf{Z(s)} = \frac{1}{\sqrt{\mathbf{s}K_0}} \text{ th } R_0\sqrt{\mathbf{s}K_0} \tag{42.34}$$

where $K_0 = c/r$, $R_0 = r \cdot L$. This function has poles and zeroes on the negative real axis in an infinite number. The zero and pole frequencies are

$$\sigma_{zi} = (2i)^2 \frac{\pi^2}{4} \frac{1}{R_0^2 K_0} \qquad \sigma_{pi} = (2i+1)^2 \frac{\pi^2}{4} \frac{1}{R_0^2 K_0} \tag{42.35}$$

where $i = 1, 2, \ldots, \infty$. Neglecting all the poles and zeroes situated well above the frequency range of interest and eliminating successively the remainder poles and zeroes

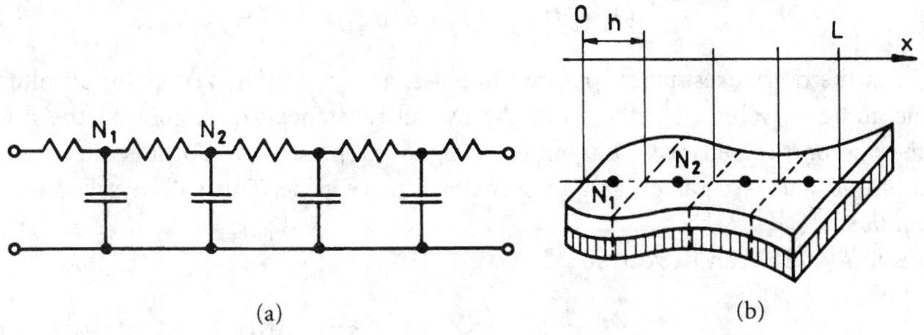

FIGURE 42.10 Lumped element approximation. (a) Network model. (b) The line split into sections.

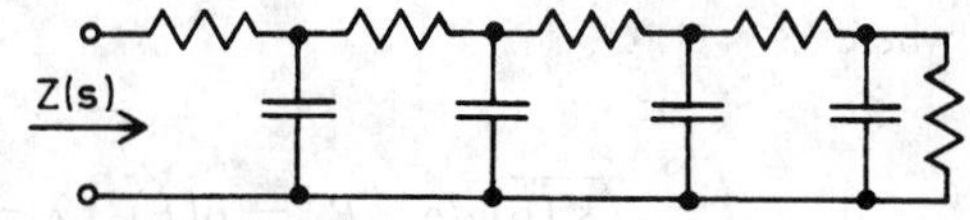

FIGURE 42.11 Cauer equivalent circuit (ladder structure).

from the (42.36) impedance function the element values of the ladder network in Fig. 42.11 (Cauer equivalent) can be obtained.

$$\mathbf{Z(s)} = R_0 \frac{\prod_{i=1}^{z}(1 + \mathbf{s}/\sigma_{zi})}{\prod_{i=1}^{p}(1 + \mathbf{s}/\sigma_{pi})} \tag{42.36}$$

42.3 Infinite-Length RC Lines

It was demonstrated earlier in the chapter that the DRC network can be described with the help of the pole–zero set of their impedance, as in the case of lumped element circuits, however, the number of these poles and zeroes is infinite. The *infinite-length* DRC lines generally do not have this property. For this network category the exact description by discrete poles and zeroes is not possible.

For example, let us consider an infinitely long uniform DRC line. Its input impedance is the characteristic impedance

$$\mathbf{Z(s)} = \sqrt{\frac{r}{\mathbf{s}c}} \tag{42.37}$$

Evidently this impedance functions *does not have poles and zeroes* on the negative σ-axis. This is the general case for a more complex, nonuniform distributed network if the length of the structure is infinite. The characteristic feature of these impedance functions is that $\sqrt{j\omega}$ factors appear in them. This is the reason that in the logarithmic amplitude vs. frequency diagram (Bode plot) regions appear with a 10 dB/decade slope, as pointed out in [1].

This section provides a generalization of the pole and zero notion and the time-constant representation in order to make them suitable to describe infinitely long distributed one-ports as well.

Before developing new ideas let us summarize the normal, well-known descriptions of a lumped element RC one-port. The port impedance of such a circuit is described by a rational function with real coefficients, as

$$\mathbf{Z(s)} = R_0 \frac{(1 + \mathbf{s}/\sigma_{z1})(1 + \mathbf{s}/\sigma_{z2}) \cdots (1 + \mathbf{s}/\sigma_{zn-1})}{(1 + \mathbf{s}/\sigma_{p1})(1 + \mathbf{s}/\sigma_{p2}) \cdots (1 + \mathbf{s}/\sigma_{pn})} \tag{42.38}$$

where R_0 is the overall resistance, σ_p are the poles, and σ_z are the zeroes (as absolute values). The pole and zero values, together with the overall resistance value, hold all the information about the one-port impedance. Thus, an unambiguous representation of this impedance is given by a set of pole and zero values, and an overall resistance value. This will be called the *pole–zero representation.*

Expression (40.38) can be rearranged as

$$\mathbf{Z(s)} = \sum_{i=1}^{n} \frac{R_i}{1 + \mathbf{s}/\sigma_{pi}} = \sum_{i=1}^{n} \frac{R_i}{1 + \mathbf{s}\tau_i} \tag{42.39}$$

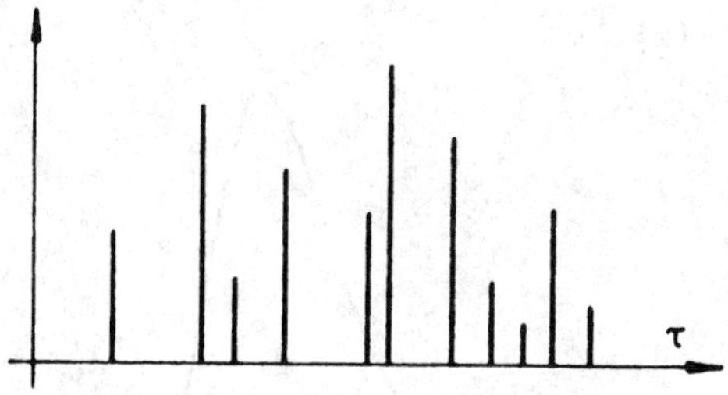

FIGURE 42.12 A lumped element one-port can be represented with a discrete set of time constants.

where

$$\tau_i = 1/\sigma_{pi} \tag{42.40}$$

which corresponds directly to the $v(t)$ voltage response for a step-function current excitation:

$$v(t) = \sum_{i=1}^{n} R_i(1 - \exp(-t/\tau_i)) \tag{42.41}$$

In this case the impedance is described in terms of the τ_i time constants of its response and of the R_i magnitudes related to it. This will be called the **time constant representation.**

Generalization of the Time Constant Representation

A lumped element one-port can be represented by a finite number of τ time constants and R magnitudes. A graphic representation of this is demonstrated in Fig. 42.12. Each line of this plot represents a time constant, and the height of the line is proportional to the magnitude. This figure can be regarded as some kind of a spectrum, the spectrum of the time constants that appeared in the step-function response of the network.

The port-impedance of a lumped element network has discrete "spectrum lines" in finite number. An infinite distributed network has no discrete lines, but it can be described with the help of a continuous time constant spectrum. The physical meaning of this idea is that in a general response any time constant can occur in some amount, some density, so that a density spectrum may suitably represent it.

We define the spectrum function by first introducing a new, logarithmic variable for the time and the time constants:

$$z = \ln t \quad \zeta = \ln \tau \tag{42.42}$$

Let us consider a DRC one-port, the response of which contains numerous exponentials having different time constants and magnitudes. The **time constant density** is defined as

$$R(\zeta) = \lim_{\Delta\zeta \to 0} \frac{\text{sum of magnitudes between } \zeta \text{ and } \zeta + \Delta\zeta}{\Delta\zeta} \tag{42.43}$$

From this definition directly follows the fact that the step-function response can be composed from the time constant density:

$$v(t) = \int_{-\infty}^{\infty} R(\zeta)[1 - \exp(-t/\exp(\zeta))]\, d\zeta \tag{42.44}$$

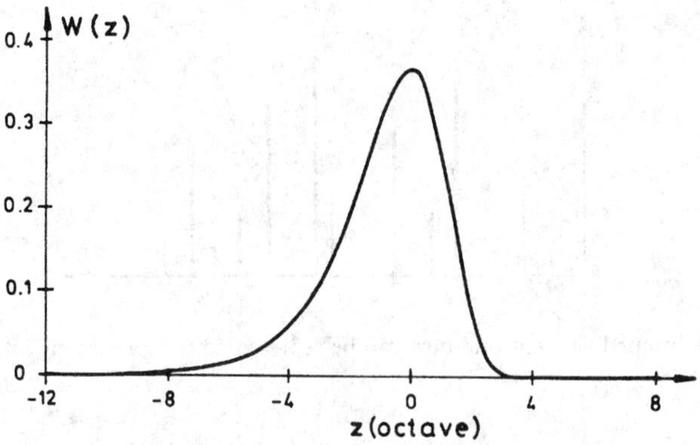

FIGURE 42.13 The shape of the $W(z)$ function.

This integral is obviously the generalization of the summation in (42.41). If the $R(\zeta)$ density function consists of discrete lines (Dirac-δ pulses), (42.41) is given back.

Using the logarithmic time variable in the integral of (42.44)

$$v(z) = \int_{-\infty}^{\infty} R(\zeta)[1 - \exp(-\exp(z - \zeta))]\, d\zeta \tag{42.45}$$

a convolution-type differential equation is obtained. Differentiating both sides with respect to z, we obtain

$$\frac{d}{dz} v(z) = R(z) * W(z) \tag{42.46}$$

where

$$W(z) = \exp(z - \exp(z)) \tag{42.47}$$

is a fixed weighting function whose shape is shown in Fig. 42.13, and $*$ is the symbol of the convolution operation.

It can be proved that the area under the function $W(z)$ is equal to unity

$$\int_{-\infty}^{\infty} W(z)\, dz = 1 \tag{42.48}$$

This means that

$$\int_{-\infty}^{\infty} R(z)\, dz = v(t \to \infty) = R_0 \tag{42.49}$$

where R_0 is the zero-frequency value of the impedance. In other words, the finite step-function response guarantees that the time-constant density has finite integral.

Generalization of Pole–Zero Representation

The task now is to substitute the pole–zero pattern of the lumped network with a continuous (eventually excepting some discrete points) function to describe the general distributed parameter network.

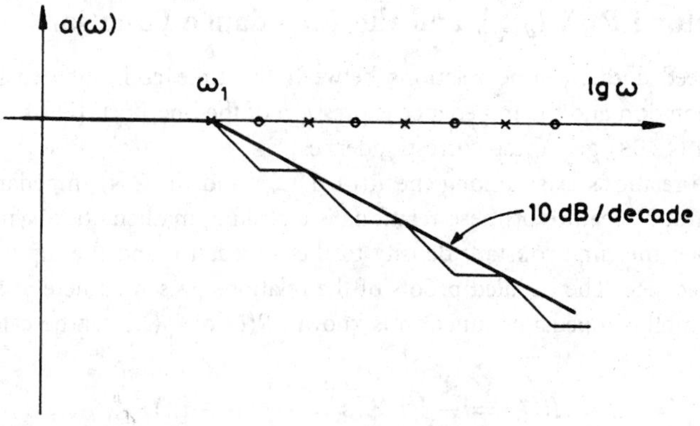

FIGURE 42.14 The 10 dB/decade decay of a DRC line amplitude plot can be approximated with an alternative sequence of poles and zeroes.

As emphasized above the Bode plot of a distributed parameter network frequently shows regions with a 10 dB/decade slope. Fig. 42.14 shows such an amplitude diagram. Using poles and zeroes we can only approximate this behavior. If we place to point ω_1 a pole, the Bode plot turns to the decay of 20 dB/decade, which is too steep. If a zero is placed, the diagram returns to the zero-slope. However, if we *alternate* poles and zeroes in a manner that the *mean value* of the slope should give the prescribed one, then any slope can be approximated. (For the above-mentioned case, if the zeroes are situated exactly midway between the adjacent poles, then the mean slope is 10 dB/decade.) The suitability of the approximation depends on the density of poles and zeroes and can be improved by increasing the density. In this case the network-specific information is not carried by the number of poles and zeroes (their number tends to infinity), but by the *relative* position of the zeroes between the adjacent poles.

An alternative interpretation is also possible. The pair of a neighboring pole and zero constitute a *dipole*. The "intensity" of that dipole depends on the distance between the pole and the zero. If they coincide and cancel each other, then the intensity is equal to zero. If the zero is situated at the maximal distance from the pole (i.e., it is at the next pole), the intensity reaches its maximal value. We choose this to be the unity.

For later convenience we turn to a logarithmic variable on the negative σ-axis:

$$\Sigma = \ln(-\sigma) \tag{42.50}$$

Let us investigate a $\Delta\Sigma$ interval of the logarithmic Σ-axis bounded by two adjacent poles. The distance between the left-hand pole and the inner zero is $\delta\Sigma$. Now, suppose that the density of the poles tends to infinity; i.e., $\Delta\Sigma$ becomes infinitely small. In this case the *dipole intensity function* is

$$I_d(\Sigma) = \lim_{\Delta\Sigma \to 0} \frac{\delta\Sigma}{\Delta\Sigma} \tag{42.51}$$

Considering that the poles and zeroes of an RC port-impedance alternate directly it follows that $0 \leq I_d \leq 1$. For an infinite, distributed RC two-pole the dipole intensity generally has regions in which the I_d value is between 0 and 1. For example, if the Bode plot shows a slope of 10 dB/decade, the value of I_d equals 0.5. This occurs in the case of an infinite, uniform RC line. For discrete circuits the I_d function has only two possible values: 0 or 1.

Relations among $R(\zeta)$, $I_d(\Sigma)$, and the Impedance Function

Obviously we needed one-to-one relations between the time constant density or the dipole intensity representation and the impedance expression of the one-port. [For the lumped element case (42.38) and (42.39) give these correspondences].

Rather simple relations exist among the $R(\zeta)$, $I_d(\Sigma)$, and the $\mathbf{Z}(\mathbf{s})$ impedance function (see below). An interesting feature of these relations is a striking mathematical symmetry: the same expression couples the time constant density to the impedance and the dipole intensity to the logarithmic impedance. The detailed proofs of the relations presented here are given in [3].

If the $\mathbf{Z}(\mathbf{s})$ complex impedance function is known, $R(\zeta)$ or $I_d(\Sigma)$ can be calculated as

$$R(\zeta) = \frac{1}{\pi}\,\mathrm{Im}\,\mathbf{Z}(\mathbf{s} = -\exp(-\zeta)) \tag{42.52}$$

$$I_d(\Sigma) = \frac{1}{\pi}\,\mathrm{Im}(\ln\,\mathbf{Z}(\mathbf{s} = -\exp(\Sigma))) \tag{42.53}$$

If the $R(\zeta)$ or $I_d(\Sigma)$ function is known

$$\mathbf{Z}(\mathbf{S}) = R_0 - \int_{-\infty}^{\infty} R(-x)\,\frac{\exp(\mathbf{S}-x)}{1+\exp(\mathbf{S}-x)}\,dx \tag{42.54}$$

$$\ln\,\mathbf{Z}(\mathbf{S}) = \ln\,R_0 - \int_{-\infty}^{\infty} I_d(x)\,\frac{\exp(\mathbf{S}-x)}{1+\exp(\mathbf{S}-x)}\,dx \tag{42.55}$$

where $\mathbf{S}$ is the complex-valued logarithm of the complex frequency:

$$\mathbf{S} = \ln\,\mathbf{s} \tag{42.56}$$

Using the integral Eq. (42.54) and (42.55), however, we must keep at least one from the two conditions:

1. $\mathbf{s}$ is not located on the negative real axis
2. If $\mathbf{s}$ is located on the negative real axis, then at this point, and in a $\varepsilon \to 0$ neighborhood, $R(\zeta)$ or $I_d(\sigma)$ must be equal to 0

Note that (42.52) to (42.55) are closely related to the Cauchy integral formula of the complex function theory. Substituting (42.52) into (42.54) and exploiting some inherent properties of the RC impedance functions after some mathematics, the Cauchy integral results. The same is true for (42.53) and (42.55).

An important feature of the transformations of (42.52) and (42.54) is that they are linear. This means that the $\mathbf{Z}(\mathbf{s}) \leftrightarrow R(\zeta)$ transformation and the summation are interchangeable.

42.4 Inverse Problem for Distributed RC Circuits

Equation (42.46) offers a direct way to determine the time constant density from the (measured or calculated) response function so that it is a method for the identification of RC one-ports. Using (42.46) for a measured time domain response function the time constant density of the one-port impedance can be determined. By using this method, equivalent circuits can be constructed easily. This possibility is of considerable practical importance [5]; however, only approximate results can be obtained because the calculation leads to the inverse operation of the convolution ("deconvolution"). Unfortunately for the weighting function of (42.47) deconvolution can be done only approximately.

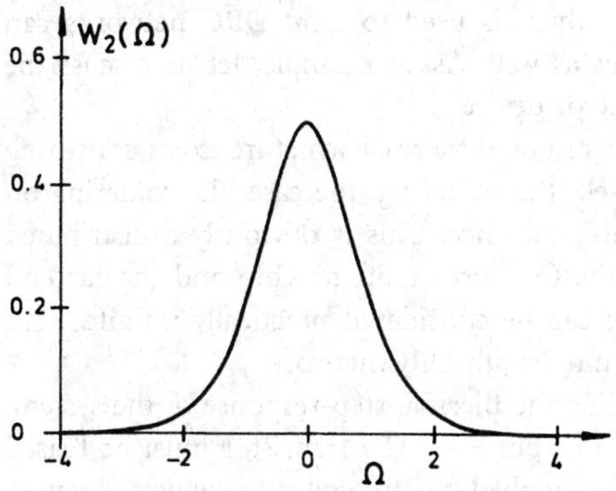

FIGURE 42.15 The shape of the $W_2(\Omega)$ function.

There is a possibility for identification in the frequency domain as well. Ref. [3] shows that a direct convolution relation exists between the Bode diagram of the impedance and the dipole intensity. Introducing the $\Omega = \ln \omega$ notation for the frequency axis, the amplitude Bode diagram can be written in the form of:

$$m(\Omega) = \ln |\mathbf{Z}(\Omega)| \tag{42.57}$$

It can be proved from (42.55) [3] that

$$m(\Omega) = -\int I_d * W_2 + \ln Z(\omega = 0) \tag{42.58}$$

where $\int I_d$ is the integral function of the dipole intensity

$$\int I_d(x) = \int_{-\infty}^{x} I_d(\Sigma)\, d\Sigma \tag{42.59}$$

and W_2 is a new weighting function:

$$W_2(\Omega) = \frac{2\exp(2\Omega)}{(1+\exp(2\Omega))^2}. \tag{42.60}$$

The latter function is plotted in Fig. 42.15.

Note that a similar convolution-type relation exists between $\int I_d(x)$ and the Bode phase plot as well.

Example 42.3 A tight analogy exists between electrical conductance and heat flow. Heat-conducting media, which can be characterized with distributed heat resistance and distributed heat capacitance, behave similarly to the electrical DRC networks. The analogous quantities are as follows:

Voltage → Temperature
Current → Power flow
Resistance → Thermal resistance
Capacitance → Heat capacitance

In the simplest model law 1-V voltage corresponds to 1°C, 1-A current to 1-W power, etc., but different mapping can be applied as well.

The described analogy means that the tool set which is used to treat DRC networks can be applied to the calculation of heat-flow problems as well. As an example, let us discuss the thermal identification of the package of semiconductor devices.

Between an IC chip and its ambient a complex distributed thermal structure exists consisting of many elements. The main parts are the chip itself, the soldering, the case, the soldering on the printed board, the printed board itself, and the ambience. This is obviously a distributed thermal RC network, the input-port of which is the top surface of the chip and the far end is the ambience ("the world"). Thus, the structure can be considered practically infinite. This means, that we have to examine a nonuniform infinite-length DRC network.

Investigations in the time domain require recording the thermal step-response of the system. A probe transistor on the IC chip can be used for this purpose. The transistor must be biased by constant emitter current, while a voltage-step is applied to the collector, which means a step function excitation in the dissipation as well. The V_{BE} voltage of the transistor must be measured because this is proportional to the temperature. The thermal response function can be obtained in this manner. Such a thermal response is shown in Fig. 42.61(a). The time range of the measurement is strikingly wide: 9 decades, from 1 μs to 1000 s. This is indispensable because the thermal time constants of the heat-flow structure vary over a wide range.

According to (42.46), after numerical derivation of the step-response and by a consecutive deconvolution, the time constant density function $R(z)$ can be obtained (see Fig. 42.16(b)). Because of the quantization noise and measuring error the deconvolution operation can be done only approximately with $1 \div 1.5$ octave resolution. A suitable algorithm is discussed in [6]. Figure 42.16(b) shows in the $\tau = 1$ to 1000 ms interval time constants spread over a relative wide range. This refers to the distributed structure of the chip and the package. At $\tau \approx 200$ s a relatively sharp disjunct time constant appears: this can be identified as having originated from the heat capacitance of the whole package and the case-ambience heat resistance.

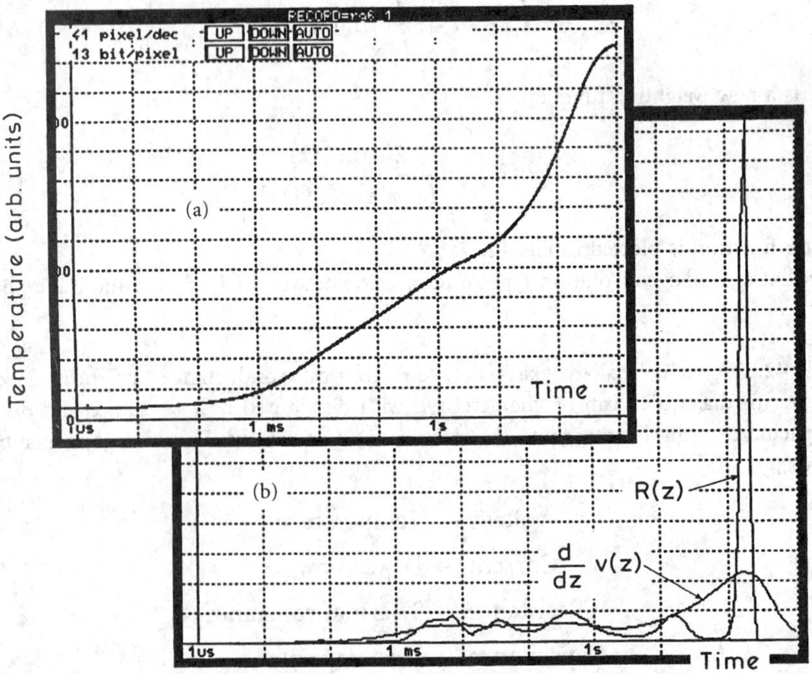

FIGURE 42.16 Thermal identification of an IC package. (a) Thermal response between 1 μs and 1000 s. (b) The $R(z)$ time constant distribution function.

Splitting the resultant time constant spectrum into $\Delta\tau$ time slots each of these slots can be approximated by a Dirac-δ spectrum line proportional in height to the appropriate slot area. These give the data of a lumped element approximation according to (42.39). Now, the equivalent circuit of the heat-flow structure can be generated either in Foster or in Cauer normal form.

Using the Cauer approximation of the DRC line we can calculate the approximate $K(\rho)$ function. From the $K(\rho)$ function the heat-conducting cross-section areas, heat flow path length, etc. can be derived. This means that geometric and physical data of the heat-flow structure can be extracted and checked with the help of an electrical measurement. For more details, see [5].

·ferences

[1] M. S. Ghausi and J. J. Kelly, *Introduction to Distributed Parameter Networks*, New York: Holt, Rinehart and Winston, 1968.

[2] E. N. Protonotarios and O. Wing, "Theory of nonuniform RC lines, part I," *IEEE Trans. Circuit Theory*, vol. 14, pp. 2–12, Mar. 1967.

[3] V. Székely, "On the representation of infinite-length distributed RC one-ports," *IEEE Trans. Circuits Syst.*, vol. 38, pp. 711–719, July 1991.

[4] R. L. Pritchard, *Electrical Characteristics of Transistors*, New York: McGraw-Hill, 1967.

[5] V. Székely and Tran Van Bien, "Fine structure of heat flow path in semiconductor devices: a measurement and identification method," *Solid-State Electr.*, vol. 31, pp. 1363–1368, Sept. 1988.

[6] T. J. Kennett, W. V. Prestwich, and A. Robertson, "Bayesian deconvolution. I. Convergent properties," *Nucl. Instr. Methods*, no. 151, pp. 285–292, 1978.

43

Synthesis of Distributed Circuits

T. K. Ishii
Marquette University, Wisconsin

43.1 Generic Relations

The starting procedure for the synthesis of distributed circuits is the same as for the conventional synthesis of lumped parameter circuits. If a one-port network is to be synthesized, then a desired driving point immittance $H(s)$ must be defined first, where

$$s = \sigma + j\omega \tag{43.1}$$

is the complex frequency, σ is the damping coefficient of the operating signal, and ω is the operating angular frequency. If a two-port network is to be synthesized, then a desired transmittance $T(s)$ must be defined first.

According to conventional principles of network synthesis [1], for the one-port network, $H(s)$ is represented by

$$H(s) = \frac{P(s)}{Q(s)} = \frac{a_n s^n + a_{n-1}s^{n-1} + \cdots + a_1 s + a_0}{b_m s^m + b_{m-1}s^{m-1} + \cdots + b_1 s + b_0} \tag{43.2}$$

where a_n and b_m are constants determined by the network parameters, $Q(s)$ is a driving function, and $P(s)$ is the response function. For a two-port network

$$T(s) = \frac{P(s)}{Q(s)} = \frac{a_n s^n + a_{n-1}s^{n-1} + \cdots + a_1 s + a_0}{b_m s^m + b_{m-1}s^{m-1} + \cdots + b_1 s + b_0} \tag{43.3}$$

Both $H(s)$ and $T(s)$ should be examined for realizability [1] before proceeding.

0-8493-8341-2/95/$0.00+$.50

If the summation of even-order terms of $P(s)$ is $M_1(s)$ and the summation of odd-order terms of $P(s)$ is $N_1(s)$, then

$$P(s) = M_1(s) + N_1(s) \tag{43.4}$$

Similarly,

$$Q(s) = M_2(s) + N_2(s) \tag{43.5}$$

For a one-port network the driving point impedance is synthesized by [1]

$$Z(s) = \frac{N_1(s)}{M_2(s)} \tag{43.6}$$

or

$$Z(s) = \frac{M_1(s)}{N_2(s)} \tag{43.7}$$

For a two-port network [1], if $P(s)$ is even, the transadmittance is

$$y_{21} = \frac{P(s)/N_2(s)}{1 + [M_2(s)/N_2(s)]} \tag{43.8}$$

the open-circuit transfer admittance is

$$y_{21} = \frac{P(s)}{N_2(s)} \tag{43.9}$$

and the open-circuit output admittance is

$$y_{22} = \frac{M_2(s)}{N_2(s)} \tag{43.10}$$

If $P(s)$ is odd,

$$y_{21} = \frac{P(s)/N_2(s)}{1 + [M_2(s)/N_2(s)]} \tag{43.11}$$

$$y_{21} = \frac{P(s)}{M_2(s)} \tag{43.12}$$

and

$$y_{22} = \frac{N_2(s)}{M_2(s)} \tag{43.13}$$

In both cases

$$y_{11} = \frac{y_{21}(s)}{n} \tag{43.14}$$

where n is the current-ratio transfer function from port 1 to port 2. From these y- or z-parameters, the required values for the network components, i.e., L, C, and R, can be determined [1].

In high-frequency circuits the L, C, and R may be synthesized using distributed circuit components. The synthesis of distributed components in microstrip line and circuits is the emphasis of this chapter.

43.2 Synthesis of a Capacitance

If the required capacitive impedance is $-jX_C$ Ω, the capacitance is

$$C = \frac{1}{\omega X_C} \tag{43.15}$$

where

$$X_C > 0 \tag{43.16}$$

and ω is the operating angular frequency. In a distributed circuit the capacitance C is often synthesized using a short section of a short-circuited transmission line of negligibly small transmission line loss. If the characteristic impedance of such a transmission line is Z_0, the operating transmission line wavelength is λ_l, and the length of the transmission line is l in meters, then [2]

$$jX_C = jZ_0 \tan \frac{2\pi l}{\lambda_l} \tag{43.17}$$

where

$$\frac{\lambda_l}{4} < l < \frac{\lambda_l}{2}$$

or, more generally,

$$\frac{n\lambda_l}{2} + \frac{\lambda_l}{4} < l < \frac{n\lambda_l}{2} + \frac{\lambda_l}{2} \tag{43.18}$$

and n is an integer. Detailed information on Z_0 and λ_l is given in Chapter 39. Combining (43.15) and (43.17),

$$C = \frac{1}{\omega Z_0 \tan \dfrac{2\pi l}{\lambda_l}} \tag{43.19}$$

In practical synthesis the transmission line length must be determined. The design equation is

$$l = \frac{\lambda_l}{2\pi} \tan^{-1}\left(\frac{1}{\omega C Z_0}\right) \tag{43.20}$$

Thus, the capacitance C can be synthesized using a section of a short-circuited transmission line.

If an open-circuited transmission line is used instead, then

$$l = \frac{\lambda_l}{2}\left[\frac{1}{\pi} \tan^{-1} \frac{1}{\omega C Z_0} + \frac{1}{2}\right] \tag{43.21}$$

Equation (43.19) is valid provided that (43.20) is used for l or,

$$l = \frac{\lambda_l}{2} \tan^{-1} \omega C Z_0 \tag{43.22}$$

where

$$\frac{n\lambda_l}{2} < l < \frac{\lambda_l}{4} + \frac{n\lambda_l}{2} \tag{43.23}$$

If an open-circuited transmission line section is used instead of the short-circuited transmission line section, just add $\lambda_l/4$ to the line length.

3.3 Synthesis of an Inductance

If the required inductive impedance is $+jX_L\ \Omega$, the inductance is

$$L = \frac{X_L}{\omega} \tag{43.24}$$

where

$$X_L > 0 \tag{43.25}$$

and ω is the operating angular frequency. In a distributed circuit the inductance L is often synthesized using a short section of a short-circuited transmission line of negligibly small transmission line loss. If the characteristic impedance of such a transmission line is Z_0, the operating transmission line wavelength is λ_l, and the length of transmission line is l in meters, then [2]

$$jX_L = jZ_0 \tan \frac{2\pi l}{\lambda_l} \tag{43.26}$$

where

$$0 < l < \frac{\lambda_l}{4} \tag{43.27}$$

or, more generally,

$$\frac{n\lambda_l}{2} < l < \frac{n\lambda_l}{2} + \frac{\lambda_l}{4} \tag{43.28}$$

and n is an integer. Detailed information on Z_0 and λ_l is given in Chapter 39.

Combining (43.24) and (43.26)

$$L = \frac{Z_0}{\omega} \tan 2\pi \frac{l}{\lambda_l} \tag{43.29}$$

In practical synthesis the transmission line length must be designed. The design equation is

$$l = \frac{\lambda_l}{2\pi} \tan^{-1} \frac{\omega L}{Z_0} \tag{43.30}$$

Thus, the inductance L can be synthesized using a section of a short-circuited transmission line.

If an open-circuited transmission line is used instead, then

$$l = \frac{\lambda_l}{2} \left(\frac{1}{\pi} \tan^{-1} \frac{\omega L}{Z_0} + \frac{1}{2} \right) \tag{43.31}$$

Equation (43.29) is valid for the open-circuited transmission line provided that (43.31) is used for l or

$$l = \frac{\lambda_l}{2\pi} \tan^{-1} \frac{Z_0}{\omega L} \tag{43.32}$$

where

$$\frac{n\lambda_l}{2} + \frac{\lambda_l}{4} < l < \frac{\lambda_l}{2} + \frac{n\lambda_l}{2} \tag{43.33}$$

If an open-circuited transmission line is used instead of the short-circuited transmission line section, just add $\lambda_l/4$ to the short-circuited transmission line design.

43.4 Synthesis of a Resistance

A distributed circuit resistance can be synthesized using a lossy transmission line of length l with the line end short or open circuited. When the line end is short circuited, the input impedance of the line of length l and characteristic impedance $\dot{Z}_0$ is [2]

$$\dot{Z}_i = \dot{Z}_0 \tanh \dot{\gamma} l \tag{43.34}$$

where the propagation constant is

$$\dot{\gamma} = \alpha + j\beta \tag{43.35}$$

α is the attenuation constant, and β is the phase constant of the line. Assuming Z_0 is real, the input impedance becomes a pure resistance, i.e., the reactance becomes zero, when

$$\beta l = \frac{2\pi l}{\lambda_l} = \frac{\pi}{2}, \pi, \frac{3}{2}\pi, 2\pi, \ldots \tag{43.36}$$

When

$$\beta l = (2n + 1)\frac{\pi}{2} \tag{43.37}$$

and where

$$n = 0, 1, 2, 3, \ldots \tag{43.38}$$

the short-circuited transmission line is at antiresonance. When

$$\beta l = 2n\,\frac{\pi}{2} \tag{43.39}$$

and where

$$n = 1, 2, 3, \ldots \tag{43.40}$$

the transmission line is at resonance.

When the transmission line is at antiresonance, the input impedance is [2]

$$\dot{Z}_i = \dot{Z}_0 \frac{1 + \varepsilon^{-\alpha(2n+1)\frac{\lambda_l}{2}}}{1 - \varepsilon^{-\alpha(2n+1)\frac{\lambda_l}{2}}} \tag{43.41}$$

If

$$\dot{Z}_0 \approx Z_0 \tag{43.42}$$

the input resistance is

$$R_i \approx Z_0 \frac{1 + \varepsilon^{-\alpha(2n+1)\frac{\lambda_l}{2}}}{1 - \varepsilon^{-\alpha(2n+1)\frac{\lambda_l}{2}}} \tag{43.43}$$

where

$$l = (2n+1)\frac{\lambda_l}{4} \tag{43.44}$$

When the transmission line section is at resonance, the input resistance is

$$R_i \approx Z_0 \frac{1 - \varepsilon^{-\alpha n \lambda_l}}{1 + \varepsilon^{-\alpha n \lambda_l}} \tag{43.45}$$

if (43.41) holds, where

$$l = n\frac{\lambda_l}{2} \tag{43.46}$$

From transmission line theory, α and β can be determined from the transmission line model parameters as [2]

$$\alpha = \frac{\omega LG + BR}{2\beta} \tag{43.47}$$

$$\beta = \left\{ \frac{\omega LB - RG + \sqrt{(RG - \omega LB)^2 + (\omega LG + BR)^2}}{2} \right\}^{\frac{1}{2}} \tag{43.48}$$

where L is the series inductance per meter, G is the shunt conductance per meter, B is the shunt susceptance per meter, and R is the series resistance per meter of the transmission line section.

The characteristic impedance of a lossy line is [2]

$$\dot{Z}_0 = R_0 + jX_0 \tag{43.49}$$

$$R_0 = \frac{\{RG + \omega LB + \sqrt{(RG + \omega LB)^2 + (\omega LG - BR)^2}\}^{\frac{1}{2}}}{\sqrt{2}\sqrt{G^2 + B^2}} \tag{43.50}$$

$$X_0 = \frac{1}{2R_0} \cdot \frac{\omega LG - BR}{G^2 + B^2} \tag{43.51}$$

or simply from (43.34)

$$l = \frac{1}{\dot{\gamma}} \tanh^{-1} \frac{R_i}{\dot{Z}_0} \tag{43.52}$$

to determine l for desired R_i for synthesis.

In practical circuit synthesis convenient, commercially available, surface-mountable chip resistors are often used. Integrated circuit resistors are monolithically developed. Therefore, the technique described here is seldom used.

43.5 Synthesis of Transformers

An impedance $\dot{Z}_1$ can be transformed into another impedance $\dot{Z}_2$ using a lossless or low-loss transmission line of length l and characteristic impedance Z_0 [2]:

$$\dot{Z}_2 = Z_0 \frac{\dot{Z}_1 + jZ_0 \tan \dfrac{2\pi l}{\lambda_l}}{Z_0 + j\dot{Z}_1 \tan \dfrac{2\pi l}{\lambda_l}} \tag{43.53}$$

where λ_l is the wavelength on the transmission line. Solving (43.53) for l,

$$l = \frac{\lambda_l}{2\pi} \tan^{-1} \frac{Z_0(\dot{Z}_2 - \dot{Z}_1)}{j(Z_0^2 - \dot{Z}_1\dot{Z}_2)} \tag{43.54}$$

In (43.53) if

$$l = \frac{\lambda_l}{4} \tag{43.55}$$

then

$$\dot{Z}_2 = Z_0^2/\dot{Z}_1 \tag{43.56}$$

This is a quarter wavelength transmission line transformer. A low-impedance $\dot{Z}_1$ is transformed into a new high-impedance $\dot{Z}_2$, or vice versa. A capacitive $\dot{Z}_1$ is transformed into an inductive $\dot{Z}_2$, or vice versa. However, Z_0 is usually real, which restricts the available transformed impedances.

An admittance $\dot{Y}_1$ can be transformed into another admittance $\dot{Y}_2$ using a lossless or low-loss transmission line of length l and characteristic admittance $\dot{Y}_0$ [2]

$$\dot{Y}_2 = Y_0 \frac{\dot{Y}_1 + jY_0 \tan \dfrac{2\pi l}{\lambda_l}}{Y_0 + j\dot{Y}_1 \tan \dfrac{2\pi l}{\lambda_l}} \tag{43.57}$$

Solving (43.56) for l,

$$l = \frac{\lambda_l}{2\pi} \tan^{-1} \frac{Y_0(\dot{Y}_2 - \dot{Y}_1)}{j(Y_0^2 - \dot{Y}_1\dot{Y}_2)} \tag{43.58}$$

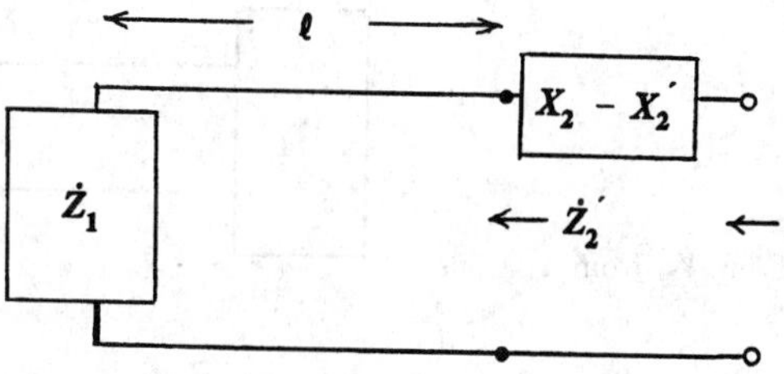

FIGURE 43.1 Synthesis of a $\dot{Z}_1$ to any $\dot{Z}_2$ transformer.

In (43.57) if

$$l = \frac{\lambda_l}{4} \tag{43.59}$$

then

$$\dot{Y}_2 = Y_0^2/\dot{Y}_1 \tag{43.60}$$

This is a one quarter-wavelength transmission line transformer. A low-admittance $\dot{Y}_1$ is transformed into a new high-admittance $\dot{Y}_2$, or vice versa. A capacitive $\dot{Y}_1$ is transformed into an inductive $\dot{Y}_2$, or vice versa. However, Y_0 is usually real, which restricts the available transformed admittances.

The transforming method (43.53) and (43.57) cannot transform $\dot{Z}_1$ to every possible value of $\dot{Z}_2$, especially when the value of Z_0 is given. In practice, this is often the case. If this is the case and $\dot{Z}_1$ is complex, use (43.53) first to transform $\dot{Z}_1$ to $\dot{Z}_2'$, where

$$\dot{Z}_2' = R_2' + jX_2' \tag{43.61}$$

and

$$R_2' = R_2 \tag{43.62}$$

where

$$\dot{Z}_2 = R_2 + jX_2 \tag{43.63}$$

Then synthesize $X_2 - X_2'$ and add it in series at the input of the transmission line, as shown in Fig. 43.1. Then

$$\begin{aligned} \dot{Z}_2 &= \dot{Z}_2' + j(X_2 - X_2') \\ &= R_2' + jX_2' + jX_2 - jX_2' \\ &= R_2' + jX_2 \\ &= R_2 + jX_2 \end{aligned} \tag{43.64}$$

Thus, $\dot{Z}_1$ can be transformed to any $\dot{Z}_2$ desired.

In a similar manner, to transform a shunt admittance $\dot{Y}_1$ to any other shunt admittance $\dot{Y}_2$, first transform $\dot{Y}_1$ to $\dot{Y}_2'$ so that

$$\dot{Y}_2' = G_2' + jB_2' \tag{43.65}$$

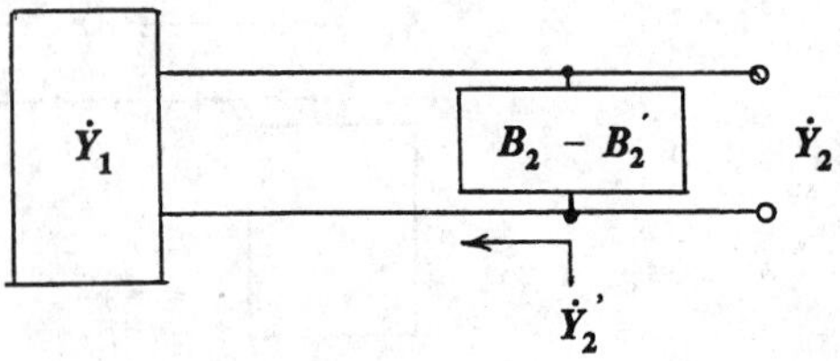

FIGURE 43.2 Synthesis of any $\dot{Y}_2$, from $\dot{Y}_1$ transformation.

and

$$G_2' = G_2 \tag{43.66}$$

where

$$\dot{Y}_2 = G_2 + jB_2 \tag{43.67}$$

Then synthesize $B_2 - B_2'$ and add it in shunt at the input of the transmission line as shown in Fig. 43.2. Then

$$\begin{aligned} \dot{Y}_2 &= \dot{Y}_2' + j(B_2 - B_2') \\ &= G_2' + jB_2' + jB_2 - jB_2' \\ &= G_2' + jB_2 = G_2 + jB_2 \end{aligned} \tag{43.68}$$

Thus, $\dot{Y}_1$ can be transformed to any $\dot{Y}_2$ desired.

43.6 Synthesis Examples

Distributed circuits can be synthesized using a number of transmission line sections. The transmission line sections include waveguides, coaxial lines, two-wire lines, microstrip lines and coplanar waveguides. In this section distributed circuit synthesis examples of distributed circuits using microstrip lines are presented for convenience. Similar techniques can be utilized for other types of transmission lines.

Series *L-C* Circuit

A series *L-C* circuit as shown in Fig. 43.3 is considered to be a phase delay line in distributed line technology. The series impedance of an *L-C* circuit is $j(\omega L - 1/\omega C)$. If the load is a resistance of R_L Ω, then the phase delay in the output voltage is

$$\Delta\theta = \tan^{-1} \frac{\omega L - \dfrac{1}{\omega C}}{R_L} \tag{43.69}$$

The phase "delay" should be interpreted algebraically. Depending on the size of ωL and ωC, $\Delta\theta$ can be either + or −. If it is delay, the sign of $\Delta\theta$ is − and if it is advance, the sign of $\Delta\theta$ must be + in Fig. 43.3.

If the phase constant of the microstrip line is

$$\beta = \frac{2\pi}{\lambda_l} \tag{43.70}$$

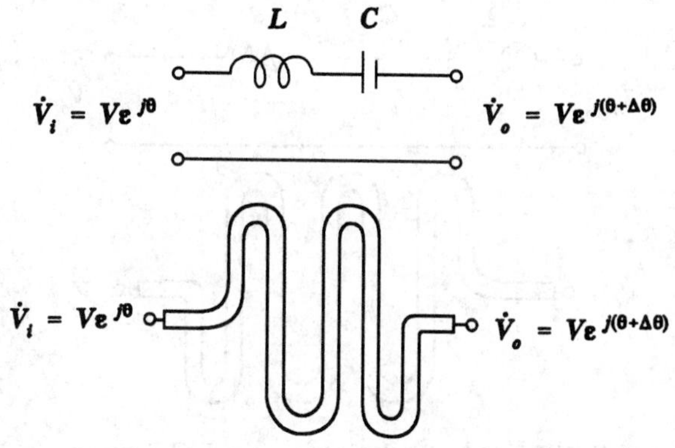

FIGURE 43.3 Synthesis of a series *L*-*C* circuit using a microstrip meander line.

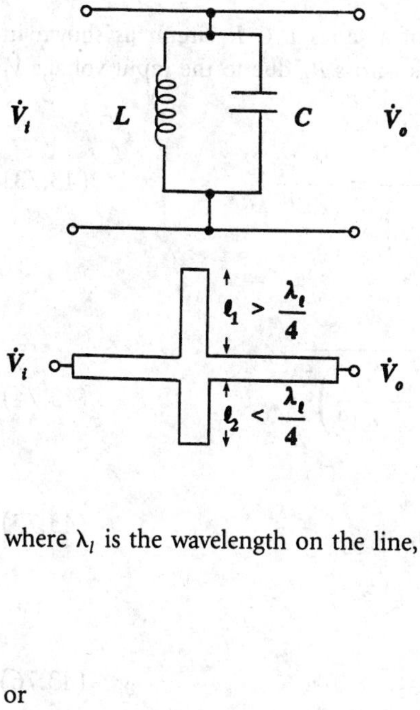

FIGURE 43.4 Synthesis of a parallel *L*-*C* circuit using a microstrip line circuit.

where λ_l is the wavelength on the line, then the synthesizing equation is

$$\Delta\theta = \beta l \tag{43.71}$$

or

$$l = \frac{\Delta\theta}{\beta} = \frac{\lambda_l}{2\pi}\tan^{-1}\frac{\omega L - \dfrac{1}{\omega C}}{R_L} \tag{43.72}$$

Usually the delay line takes the form of a meander line in microstrip line, as shown in Fig. 43.3.

Parallel *L*-*C* Circuit

A parallel *L*-*C* circuit as shown in Fig. 43.4 can be synthesized using distributed circuit components. In this figure the open-circuited microstrip line length l_1 represents inductive shunt admittance and the synthesis equation is given by (43.31). In Fig. 43.4 another open-circuited microstrip line length l_2 represents capacitive shunt admittance. The synthesis equation for this part is given by (43.21).

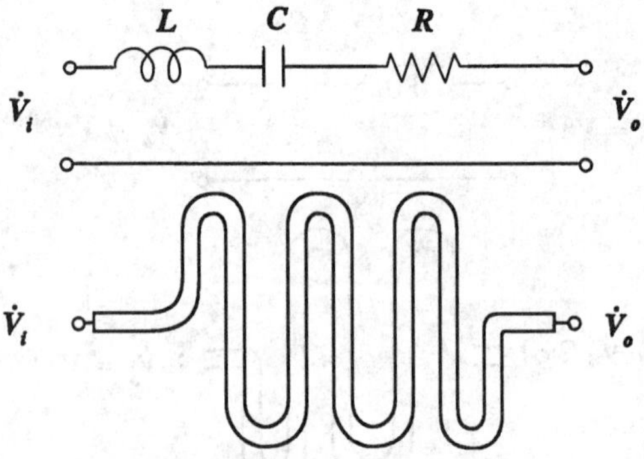

FIGURE 43.5 Synthesis of a series *L-C-R* circuit.

Series *L-C-R* Circuit

If the load resistance connected to the output circuit of a series *L-C-R* circuit as shown in Fig. 43.5 is R_L, then the output voltage $\dot{V}_0$ of this network across R_L due to the input voltage $\dot{V}_i$ is

$$V_0 = \frac{R_L \dot{V}_i}{R_L + R + j\left(\omega L - \dfrac{1}{\omega C}\right)} = \frac{R_L \dot{V}_i}{Z\varepsilon^{j\phi}} \tag{43.73}$$

where

$$Z = \sqrt{(R_L + R)^2 + \left(\omega L - \frac{1}{\omega C}\right)^2} \tag{43.74}$$

$$\phi = \tan^{-1}\frac{\omega L - \dfrac{1}{\omega C}}{R_L + R} \tag{43.75}$$

Thus,

$$\dot{V}_0 = \frac{R_L}{Z}\dot{V}_i \varepsilon^{-j\phi} \tag{43.76}$$

In a distributed circuit, such as the microstrip line shown in Fig. 43.5,

$$V_0 = \dot{V}_i \varepsilon^{-\alpha l} \varepsilon^{-j\beta l} \tag{43.77}$$

where α is the attenuation constant, β is the phase constant, and l is the total length of the microstrip line.

Then, comparing (43.76) with (43.75),

$$\varepsilon^{-\alpha l} = \frac{R_L}{Z} \tag{43.78}$$

or

$$\alpha l = \ln \frac{Z}{R_L} \tag{43.79}$$

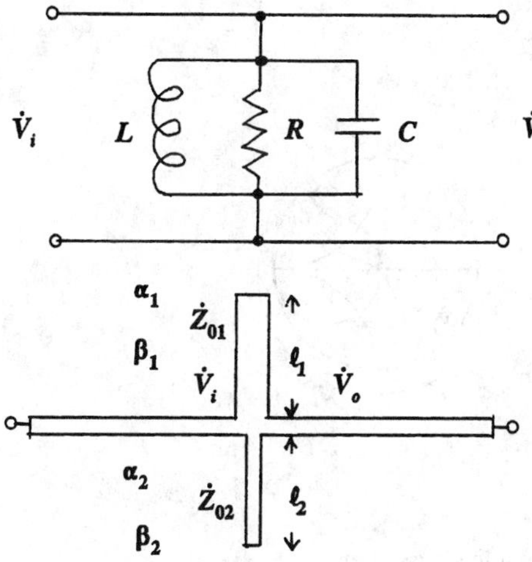

FIGURE 43.6 Synthesis of a parallel *L-C-R* circuit.

and

$$\beta l = \phi \tag{43.80}$$

Parallel *L-C-R* Circuit

A parallel *L-C-R* circuit can be synthesized using the microstrip line, as shown in Fig. 43.6. In this figure the microstrip line of length l_1, characteristic impedance $\dot{Z}_{01}$, attenuation constant α_1, and phase constant β_1 represents a parallel circuit of inductance L and $2R$. Another microstrip line of length l_2, characteristic impedance $\dot{Z}_{02}$, attenuation constant α_2, and phase constant β_2 represents a parallel circuit of capacitance C and $2R$.

The input admittance of the open-circuited lossy microstrip line is

$$\dot{Y}_i = \frac{1}{\dot{Z}_{01}} \tanh(\alpha_1 + j\beta_1)l_1 \tag{43.81}$$

In a parallel L-$2R$ circuit, the parallel admittance is

$$Y_i = \frac{1}{2R} + \frac{1}{j\omega L} \tag{43.82}$$

The synthesis equation for the inductive microstrip line is

$$(\alpha_1 + j\beta_1)l_1 = \tanh^{-1} \dot{Z}_{01}\left(\frac{1}{2R} - j\frac{1}{\omega L}\right) \tag{43.83}$$

Similarly, for the capacitive microstrip line,

$$(\alpha_2 + j\beta_2)l_2 = \tanh^{-1} \dot{Z}_{02}\left(\frac{1}{2R} + j\omega C\right) \tag{43.84}$$

To synthesize this distributed circuit (43.82) and (43.83) must be solved. These are transcendental equations of an excess number of unknowns to be determined by trial and error. Therefore, a digital computer is needed. In many cases lossless lines are used for L and C and a shunt chip resistor is used for R to avoid complications.

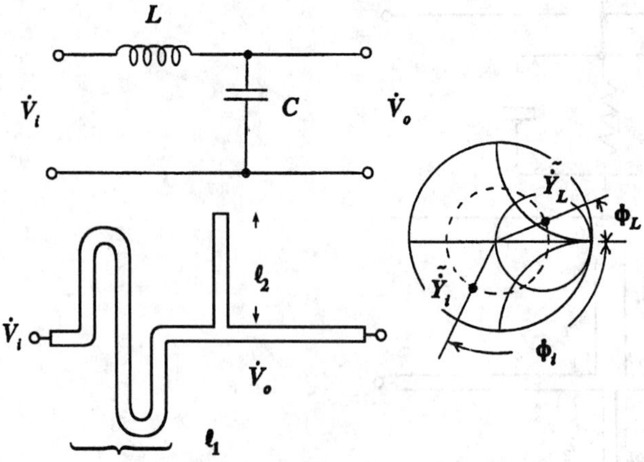

FIGURE 43.7 Synthesis of a low-pass filter.

Low-Pass Filters

If the load resistance of a low-pass filter, as shown in Fig. 43.7 is R_L, then the admittance across R_L is

$$\dot{Y}_L = \frac{1}{R_L} + j\omega C \tag{43.85}$$

If this is synthesized using the microstrip line shown in Fig. 43.7, then the normalized admittance of the microstrip line is

$$\tilde{\dot{Y}}_L \equiv \frac{\dot{Y}_L}{Y_0} = \frac{\dfrac{1}{R_L} + j\omega C}{Y_0} \tag{43.86}$$

where Y_0 is the characteristic admittance of the microstrip line. $\tilde{\dot{Y}}_L$ should be plotted on a Smith chart as shown in Fig. 43.7, and its angle of reflection coefficient ϕ_L must be noted. The input admittance of this circuit is

$$\begin{aligned}\dot{Y}_i &= \frac{1}{j\omega L + \dfrac{R_L \dfrac{1}{j\omega C}}{R_L + \dfrac{1}{j\omega C}}} \\ &= \frac{R_L + j\{\omega C R_C^2(1-\omega^2 LC) - \omega L\}}{R_L^2(1-\omega^2 LC)^2 + (\omega L)^2}\end{aligned} \tag{43.87}$$

This admittance is normalized as

$$\tilde{\dot{Y}}_i = \dot{Y}_i / Y_0 \tag{43.88}$$

and it must be plotted on the Smith chart (Fig. 43.7). The phase angle of the reflection coefficient ϕ_i at this point must be noted. Therefore, the synthesis equation is

$$\beta_1 l_1 = \phi_L + \phi_i \tag{43.89}$$

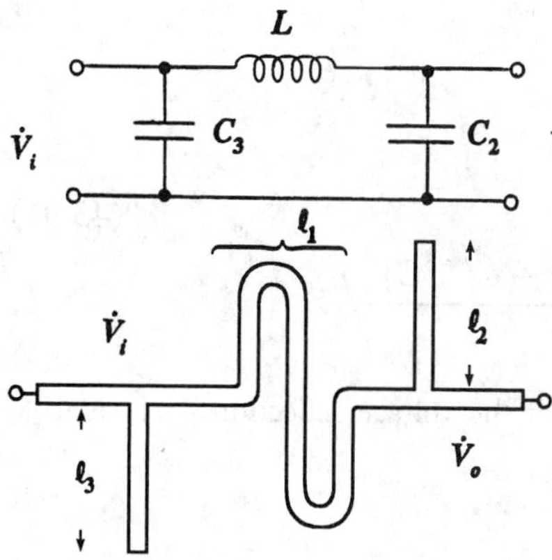

FIGURE 43.8 Synthesis of a π-network low-pass filter.

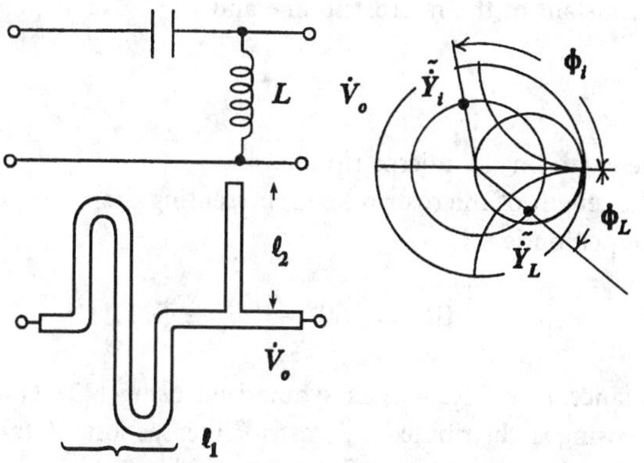

FIGURE 43.9 Synthesis of a high-pass filter.

As seen from Fig. 43.7, the meander line of length l_1, representing the inductance L, and a straight microstrip line of length l_2, representing the capacitance C, are joined to synthesize a microwave low-pass filter. The synthesis of the shunt capacitance C in Fig. 43.7 is accomplished using an open-circuited microstrip line of the phase constant β_2. The length of the microstrip line l_2 is determined from (43.21). If the low-pass filter takes on a π-shape, the microstrip version of synthesis will be as shown in Fig. 43.8. The microstrip line stub l_3 can be synthesized in the same way as the capacitive stub l_2.

High-Pass Filters

If a high-pass filter consists of a capacitance C and an inductance L, as shown in Fig. 43.9, then normalized admittance across the load resistance R_L is

$$\tilde{\dot{Y}}_L = \dot{Y}_L/Y_0 = \left(\frac{1}{R_L} - j\frac{1}{\omega L}\right) \Big/ Y_0 \tag{43.90}$$

where Y_0 is the characteristic admittance of the microstrip line to be used for synthesis. This must be plotted on the Smith chart, as shown in Fig. 43.9, and the angle of reflection coefficient ϕ_L should be noted.

Normalized admittance at the input is

$$\tilde{\dot{Y}}_i \equiv \dot{Y}_i/Y_0 = \frac{1}{\dfrac{1}{j\omega C} + \dfrac{(j\omega L)R_L}{j\omega L + R_L}} \Big/ Y_0$$
$$= \frac{(\omega^2 LC)^2 R_L + j\omega\{R_L^2 C(1-\omega^2 LC) + \omega^2 L^2 C\}}{R_L^2(1-\omega^2 LC)^2 + (\omega L)^2} \Big/ Y_0 \tag{43.91}$$

Plot this on the Smith chart (Fig. 43.9) and the angle of the voltage reflection coefficient ϕ_i should be noted.

The phase delay of the distributed microstrip line is

$$\dot{v}_0 = \dot{v}_i \varepsilon^{-j\beta l_1} \tag{43.92}$$

where β is a phase constant of the microstrip line and

$$\beta = 2\pi/\lambda_l \tag{43.93}$$

where λ_l is the wavelength on the microstrip line.

In (43.91) l_1 is the length of microstrip line representing the C-section of the low-pass filter. Then the synthesis equation is

$$\beta l_1 = 360° - (\phi_i + \phi_L) \tag{43.94}$$

The shunt inductance L in Fig. 43.9 is synthesized using (43.31). Thus, a high-pass filter can be synthesized using a distributed microstrip line section. If the low-pass filter is in a π-network form, then microstrip line configuration will be at the designed frequency, as shown in Fig. 43.10.

Bandpass Filters

A bandpass filter may be a series connection of an inductor L and a capacitor of capacitance C, as shown in Fig. 43.11. This figure is identical to Fig. 43.3. Therefore, the microstrip line synthesis equation is (43.72).

If the bandpass filter is in the shape of a π-network (Fig. 43.12(a)), then the series L_1C_1, represented by the meander microstrip line of length l_5, can be designed using (43.72). Both of

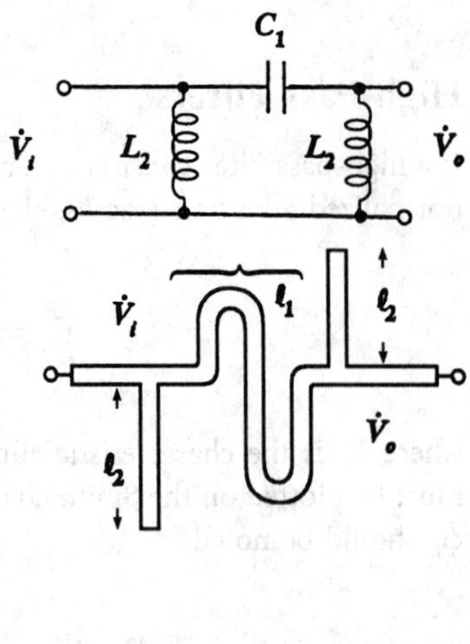

FIGURE 43.10 Synthesis of a π-network high-pass filter.

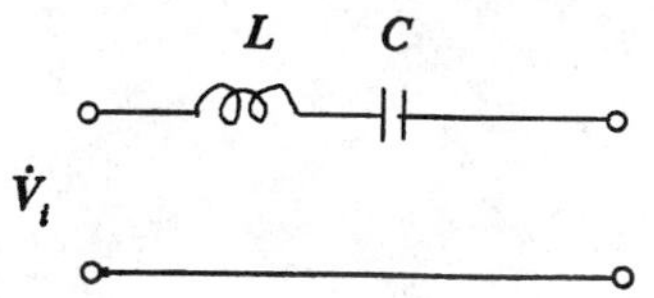

FIGURE 43.11 Configuration of a series L-C bandpass filter.

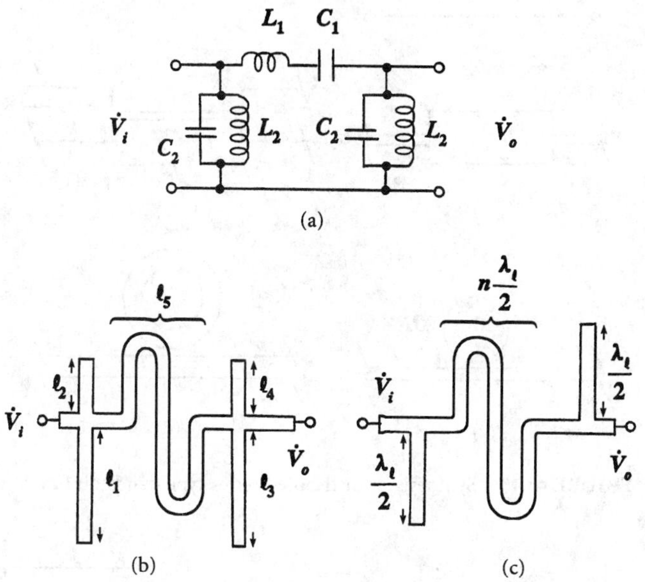

FIGURE 43.12 Synthesis of a distributed microstrip line π-network bandpass filter.

the shunt L_2 and C_2 arms are identical to Fig. 43.4. Therefore, this part is synthesized by utilizing the technique in the section, "Parallel L-C Circuit", which results in the configuration shown in Fig. 43.12(b). An alternative distributed version of Fig. 43.12(a) is shown in Fig. 43.12(c). In this figure the technique in λ_l is the microstrip line wavelength at the center of the desired passband. The frequency bandwidth is determined by the quality factor of the resonating element.

Bandstop Filters

Various types of bandstop filters are available. A schematic diagram of a series type bandstop filter is shown in Fig. 34.13. It is basically a resonant circuit inserted in series with the transmission line. The center frequency of the bandstop is the resonant frequency of the resonant circuit. The loaded quality factor of the resonator determines the bandstop-width.

In the distributed microstrip line the resonator is a section of a half-wavelength microstrip line resonator of either low impedance (Fig. 43.13(a)) or high impedance (Fig. 43.13(b)), or a dielectric resonator placed in proximity to the microstrip line, as shown in (Fig. 43.13(c)) or a ring resonator placed in proximity to the microstrip line (Fig. 43.13(d)). The resonant frequency of the resonators is the center frequency of the bandstop. The bandwidth is determined by the loaded Q and coupling of the resonator.

A bandstop filter can also be synthesized in microstrip line if the filter is a parallel type (Fig. 43.14). Simply attaching a one quarter-wavelength open-circuited stub will create a bandstop filter. The center of the bandstop has a wavelength of λ_l (Fig. 43.14).

A bandstop filter may take a π-network form, as shown in Fig. 43.15, by the combination of the $\lambda_l/2$ resonator and $\lambda_l/4$ open-circuited stubs. The frequency bandwidth of the filter depends on the quality factor of the resonating elements.

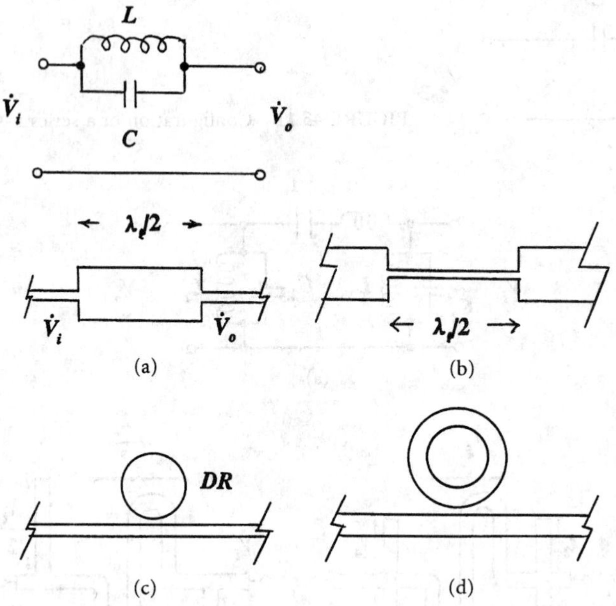

FIGURE 43.13 Synthesis of distributed series type bandstop filters.

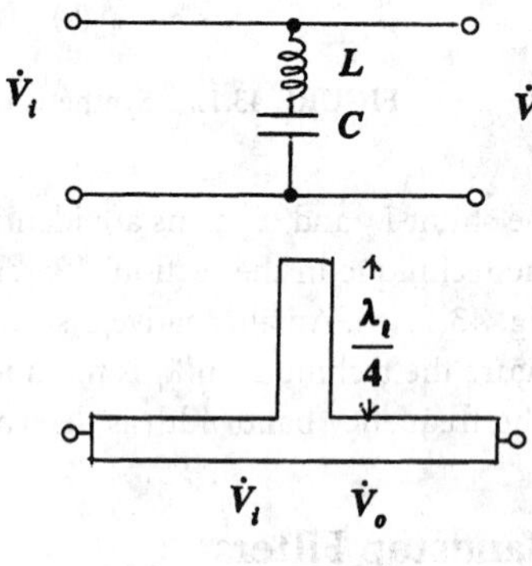

FIGURE 43.14 Synthesis of distributed parallel type bandstop filters.

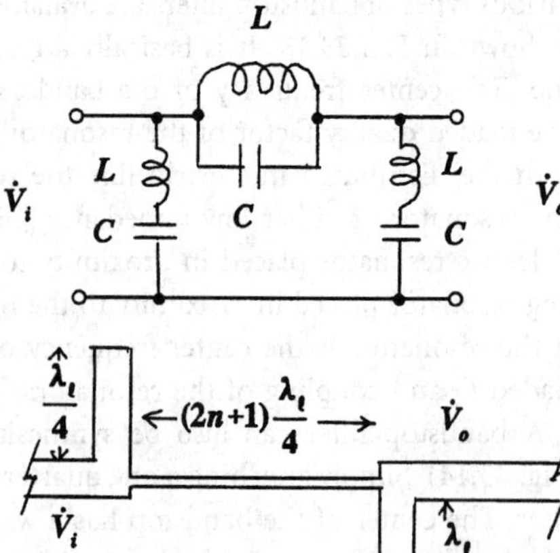

FIGURE 43.15 Synthesis of distributed π-network bandstop filters.

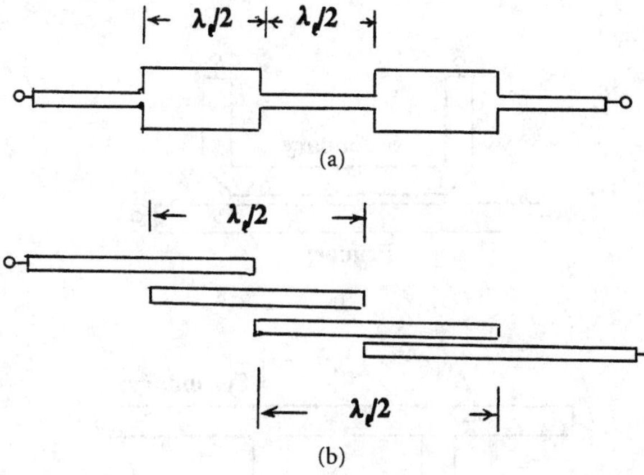

FIGURE 43.16 Microstrip line filters. (a) Bandstop high-low filter. (b) Bandpass coupled resonator filter.

Further Comments on Distributed Circuit Filters

If the unit sections of filters previously described are treated as lumped parameter filters, they can be cascaded into a number of stages to obtain desired filter characteristics, as exemplified by Chebyshev and Butterworth filters. The center frequencies of the stages can be staggered to obtain the desired frequency bandwidth [5].

A radiofrequency choke is a special case of a bandstop filter. Cascading of alternate $\lambda_l/2$ sections of high and low impedance (Fig. 43.16(a)) is common. This is called the high-low filter. A microstrip line bandpass filter can be made using a number of $\lambda_l/2$ microstrip line resonators, as shown in Fig. 43.16(b) [5].

3.7 Synthesis of Couplers

Generic Relations

A coupler is a circuit with coupling between a primary circuit and a secondary circuit. In distributed circuit technology it often takes the form of coupling between two transmission line circuits. The coupling may be done by simply placing the secondary transmission line in proximity to the primary transmission line, as shown in Fig. 43.17(a) and (c), or both the primary and the secondary transmission lines are physically connected, as shown in Fig. 43.17(b). These examples of distributed coupler circuits are shown in microstrip line forms. No lumped parameter equivalent circuit exists for these distributed coupler circuits. Therefore, the previously described synthesis techniques cannot be used.

Proximity Couplers

Coupling of two transmission lines is accomplished simply by existing proximity to each other. If the transmission lines are microstrip lines (Fig. 43.17(a) or Fig. 43.18), the coupler is also called an edge coupler.

The coupling factor (CF) of a coupler is defined as [3]

$$\mathrm{CF} = \frac{P_4}{P_1} \tag{43.95}$$

where P_1 is the microwave power fed to port 1 and P_4 is the microwave power output at port 4. If the power coupling factor per meter of the line edge is $k(x)$ in the coupling region of length

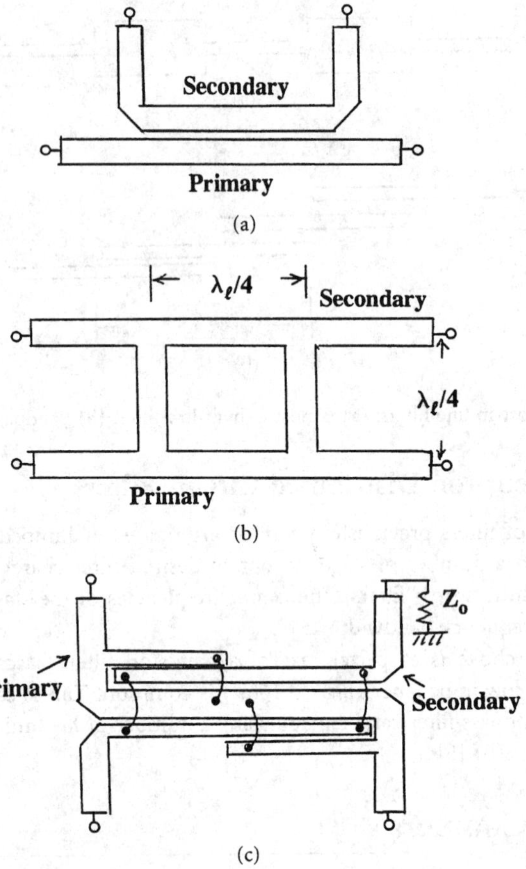

FIGURE 43.17 Distributed circuit couplers. (a) Proximity coupler. (b) Quarter-wavelength coupler. (c) Lange coupler.

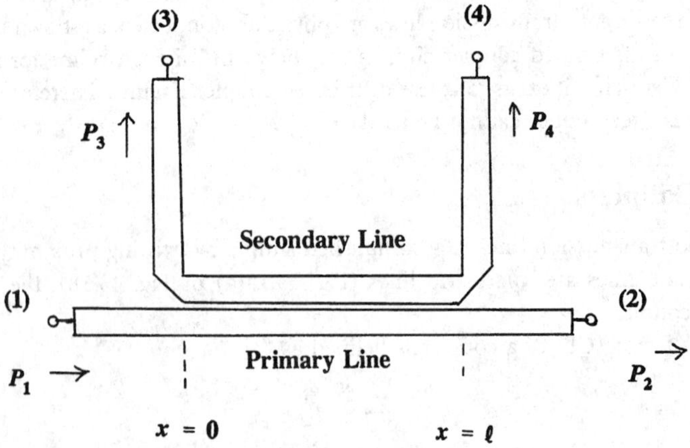

FIGURE 43.18 Microstrip line proximity coupler.

l, then [3]

$$\text{CF} = \left| \int_0^l k(x)\varepsilon^{-j\beta x}\, dx \right| \tag{43.96}$$

This can be expressed in decibels as

$$\text{CF(dB)} = 10 \log \frac{P_4}{P_1} = 10 \log \left| \int_0^l k(x)\varepsilon^{-j\beta x}\, dx \right| \tag{43.97}$$

The directivity of a coupler is defined as [3]

$$\text{dir} = \frac{P_3}{P_4} \tag{43.98}$$

$$\text{dir} = \frac{1}{\text{CF}} \left| \int_0^l k(x)\varepsilon^{-j2\beta x}\, dx \right| \tag{43.99}$$

The decibel expression of directivity, then, is

$$\begin{aligned}\text{dir(dB)} &= 10 \log_{10} \frac{P_3}{P_4} \\ &= 10 \log_{10} \frac{1}{\text{CF}} \left| \int_0^l k(x)\varepsilon^{-j2\beta x}\, dx \right|\end{aligned} \tag{43.100}$$

Generally, $P_3 \neq P_4$; therefore, the coupling has a directional property. This is the reason that this type of coupler is termed the directional coupler.

Insertion loss (IL) of a coupler is defined as

$$\text{IL} = \frac{P_2}{P_1} = \frac{P_1 - (P_3 + P_4)}{P_1} \tag{43.101}$$

Inserting (43.95) and (43.98) into (43.101),

$$\text{IL} = 1 - \text{CF}(1 + \text{dir}) \tag{43.102}$$

or

$$\text{IL(dB)} = 10 \log\{1 - \text{CF}(1 + \text{dir})\} \tag{43.103}$$

A complete analytical synthesis of microstrip line couplers is not available. However, computer software based on semiempirical equations is commercially available [5].

Quarter-Wavelength Couplers

The coupling factor of a quarter-wavelength coupler (Fig. 43.19) is defined as [3]

$$\text{CF} = \frac{P_4}{P_1} \tag{43.104}$$

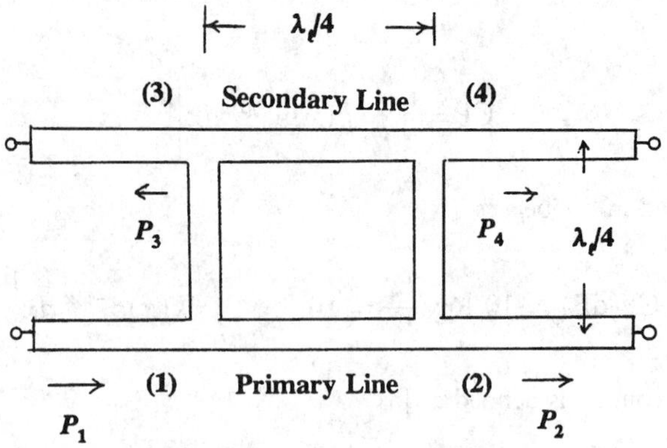

FIGURE 43.19 Quarter-wavelength coupler.

The amount of P_4 or coupling factor can be controlled by the width of microstrip lines connecting the secondary line to the primary line.

$$\mathrm{CF[dB]} = 10 \log_{10} P_4/P_1 \tag{43.105}$$

The directivity is

$$\mathrm{dir} = \frac{P_3}{P_4} \tag{43.106}$$

or

$$\mathrm{dir[dB]} = 10 \log \frac{P_3}{P_4} \tag{43.107}$$

At the precise design operating frequency, P_3 should be zero. The insertion loss for this coupler is

$$\mathrm{IL} = \frac{P_1 - (P_3 + P_4)}{P_1} = 1 - \mathrm{CF}(1 + \mathrm{dir}) \tag{43.108}$$

Lange Couplers

A schematic diagram of a generic Lange coupler [4] is shown in Fig. 43.20. In this figure, the primary transmission line is from port (1) to port (2). The secondary transmission line is from port (3) to port (4). In Lange couplers both the primary and the secondary lines are coupled by edge coupling or proximity coupling. It should be noted that the transmission lines are not physically connected to each other by bonding wires. The bond wires on both ends form connections of the secondary transmission line from port (3) to port (4). The bond wires in the middle of the structure are the electrical potential equalizing bonding in the primary transmission line. Adjusting the values of l_1 and l_2 creates a variety of coupling factors. One of the objectives of Lange couplers is to increase the coupling factor between the primary lines and the secondary line. By comparing the structure to that of a straight edge-coupled coupler, as shown in Fig. 43.18, the Lange coupler shown in Fig. 43.20 has more length of coupling edge. Therefore, a higher coupling factor is obtainable [5].

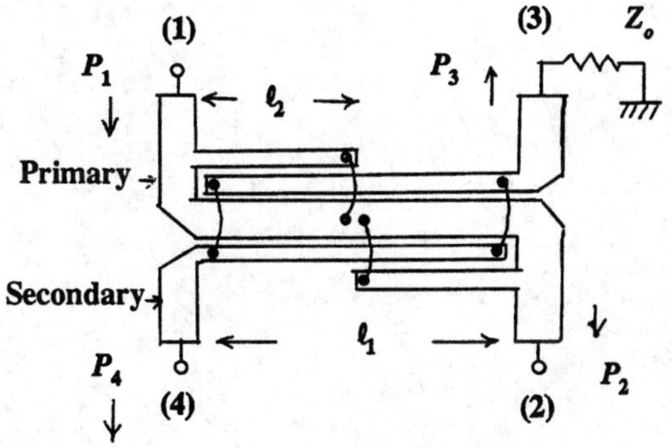

FIGURE 43.20 Lange coupler.

ferences

[1] F. F. Kuo, *Network Analysis and Synthesis*, New York: Wiley, 1962.

[2] T. K. Ishii, *Microwave Engineering*, San Diego: Harcourt Brace Jovanovich, 1989.

[3] F. J. Tischer, *Mikrowellen-Messtechnik*, Berlin: Springer-Verlag, 1958.

[4] J. Lange, "Interdigitated stripline quadrature hybrid," *IEEE Trans. MTT*, vol. 17, no. 11, pp. 1150, 1151, Dec. 1969.

[5] V. F. Fusco, *Microwave Circuit*, Englewood Cliffs, NJ: Prentice-Hall, 1987.

IX

Stability Analysis

Ruey-wen Liu
University of Notre Dame

44

Stability Tests for Polynomials

Peter Bauer
University of Notre Dame

44.1 Introduction

The problem of determining stability for dynamical systems is of key importance in many disciplines of science and engineering. It is therefore not surprising that this subject started to attract attention during the Industrial Revolution in the mid-19th century. Although some of the work performed in this time period was of a purely theoretical nature (e.g., the important contributions by Cauchy and Sturm), many of the research efforts were motivated by problems stemming from practical applications.

One of the first persons to analytically study the problem of stability for a physical system was George Biddell Airy. In 1840 he worked on controlling the speed of an astronomical telescope via a flyball governor. The telescope had to turn in order to compensate for the rotation of the Earth. He treated this problem using differential equations and was probably the first scientist to approach the stability problem of a physical system from a mathematical viewpoint.

In the 1860s James Clark Maxwell was also interested in the stability of motion of dynamic systems. He treated the problem of speed governor-regulated plants by linearizing the corresponding differential equations. Maxwell identified the conditions which had to be imposed on the roots of the arising polynomials in order to guarantee stability. Unfortunately, he was unable to develop a method which allowed the stability of a general order polynomial to be tested. (He did, however, succeed in formulating conditions for polynomials up to the third degree.) At this time, Maxwell urged mathematicians to address this problem and find tests for polynomials of degree four and higher. In the 1870s the Englishman, Edward J. Routh, solved the problem by determining the number of roots which are in the right-half plane by using a coefficient table scheme. He won the 1877 Adams Prize for his paper entitled "Stability of a Given State of

0-8493-8341-2/95/$0.00+$.50

Motion". (It might be interesting to note that Maxwell was a member of the judging committee.) Approximately 20 years later, Hurwitz encountered the same type of problem, being unaware of the solution found earlier by Routh. Hurwitz worked at the Swiss Federal Institute of Technology (ETH) on a problem which arose in the work of the engineer Dr. A. Stodola: The task was to regulate the speed of a high-pressure water turbine. Hurwitz solved the problem by presenting conditions in terms of determinant inequalities. These conditions are identical to those found by Routh, if the polynomial contains no right-half plane zeroes. For a more elaborate treatise of the history of this subject the reader is referred to [4].

Today this criterion is known as the Routh-Hurwitz stability criterion. It is widely used in many disciplines of engineering. A number of modified but related criteria were developed later, some of which are presented in the following subsections.

Fundamental concepts and resulting tests for stability of linear time-invariant systems are formulated. All discussions are limited to polynomials or their corresponding differential (difference) equations. The presentation of the Nyquist criterion is omitted because it is treated separately in Chapter 46.

44.2 Root Clustering and the Principle of Argument

The following notation is used in this chapter:

$\mathcal{R}$:	set of real numbers
$\mathcal{C}$:	set of complex numbers
s:	Laplace transform variable
t:	time
$D^i\{\}$:	differential operator $d^i(\cdot)/dt^i$
$H(s)$:	transfer function (continuous time)
$Y(s)$:	output signal transform (continuous time)
$X(s)$:	input signal transform (continuous time)
$u(t)$:	unit step signal (continuous time)
$h(t)$:	impulse response (continuous time)
$\delta(t)$:	Dirac impulse function
$\Delta \arg\{f(jw)\}\|_a^b$:	phase change of $f(jw)$: $\arg f(jb) - \arg f(ja)$
Re{ }:	real part
Im{ }:	imaginary part
$\text{Re}\{s\} = \sigma$:	real part of s
$\text{Im}\{\} = \omega$:	imaginary part of s
$f^e(s)$:	even part of the polynomial $f(s)$
$f^o(s)$:	odd part of the polynomial $f(s)$
H_m:	$m \times m$ dimensional Hurwitz matrix
Δ_i:	determinants of the principle minors of H_m
z:	z-transform variable
$u(n)$:	unit step (discrete time)
$h(z)$:	symmetric part of the polynomial $f(z)$
$g(z)$:	antisymmetric part of the polynomial $f(z)$

In this section linear time-invariant systems of the form

$$\begin{aligned} a_0 y(t) + a_1 D\{y(t)\} + \cdots + a_m D^m\{y(t)\} \\ = b_0 x(t) + b_1 D\{x(t)\} + \cdots + b_n D^n\{x(t)\}, \quad n \leq m \end{aligned} \tag{44.1}$$

are considered. An equivalent representation is obtained, if the differential equation is transformed to the s-domain via the one-sided Laplace transform. Formally this corresponds to substituting $y(t)$ by $Y(s)$, $x(t)$ by $X(s)$, and D^i by s^i. From (44.1), the following transfer function representation is obtained:

$$H(s) = \frac{Y(s)}{X(s)} = \frac{b_0 + b_1 s + \cdots + b_n s^n}{a_0 + a_1 s + \cdots + a_m s^m}, \qquad n \le m \tag{44.2}$$

In subsequent discussions it is assumed that numerator and denominator in (44.2) are mutually coprime, i.e., they do not contain common factors. If the transfer function contains only real coefficients, the poles will occur in the form of conjugate complex pairs and real poles. After factoring the denominator polynomial into first-order terms, the following partial fraction expansion of the transfer function is obtained:

$$H(s) = A + \sum_{k=1}^{L} \left(\frac{A_{k1}}{s - s_k} + \frac{A_{k2}}{(s - s_k)^2} + \cdots + \frac{A_{km_k}}{(s - s_k)^{m_k}} \right) \tag{44.3}$$

where $A = b_m / a_m$, $\sum_{k=1}^{L} m_k = m$.

The problem of stability can now be approached through the time domain by using the inverse Laplace transform for the terms of the partial fraction expansion

$$\frac{1}{(s - s_k)^n} \leftrightarrow \frac{1}{(n-1)!} t^{n-1} e^{s_k t} u(t), \qquad n \ge 1 \tag{44.4}$$

Because the inverse transform of a constant is the so-called Dirac function, it contributes to the impulse response only at $t = 0$. Therefore, only the terms inside the summation of (44.3) determine stability. The Laplace transform pair (44.4) shows that for exponential stability, i.e., exponential convergence to zero of all partial responses, one requires $\mathrm{Re}\{s_k\} < 0$. If s_k is a real pole, the response affiliated with this pole converges to zero exponentially and aperiodically. If s_k is a complex pole, the response will be a complex oscillation which converges to zero exponentially. Therefore, if all poles satisfy $\mathrm{Re}\{s_k\} < 0$, $k = 1, \ldots, L$, all partial responses will converge to zero, and hence the total impulse response will do so.

If $\mathrm{Re}\{s_k\} = 0$ for at least one k, then from the transform pair (44.4) it is clear that a response portion exists, which will not converge to zero. The special case $\mathrm{Re}(s_k) = 0$ and $n = 1$ produces a bounded nonzero response and is, therefore, not exponentially or asymptotically stable. In the literature this case is sometimes referred to as "marginal stability".

If $\mathrm{Re}\{s_k\} > 0$, the transform pair (44.4) shows that the response portion affiliated with the pole s_k diverges exponentially.

Theorem 44.1: *Assume the numerator and denominator of the rational transfer function*

$$H(s) = \frac{b_0 + b_1 s + \cdots + b_n s^n}{a_0 + a_1 s + \cdots + a_m s^m}, \qquad n \le m$$

are mutually coprime. Then the impulse response of $H(s)$ is asymptotically (and exponentially) stable if and only if (iff) all poles s_k satisfy

$$\mathrm{Re}(s_k) < 0. \tag{44.5}$$

□

The above theorem requires that all roots of the characteristic polynomial are strictly in the open left-half s-plane. The theorem also indicates that asymptotic and exponential stability are equivalent for the class of linear time-invariant systems.

An equivalent time domain statement for stability of the transfer function 44.2 is

$$\int_0^\infty |h(t)|^p \, dt < \infty \quad \text{for any } p \geq 1 \tag{44.6}$$

Theorem 44.1 also shows that the stability boundary for linear time-invariant continuous systems is given by the imaginary axis or equivalently $\text{Re}\{s\} = 0$. The phase changes along this boundary can be used to determine stability without knowledge of the pole locations. The following theorem uses this principle and is the basis for many results on stability of linear systems [9].

Theorem 44.2: (Cremer-Leonhard-Michailov): *The polynomial*

$$f(s) = a_m s^m + \cdots + a_1 s + a_0, \quad a_i \in \Re,\ i = 1, \ldots, m$$

has all its roots in the open left-half s-plane, iff $f(j\omega)$ has a change of argument of $m\pi/2$ for ω changing from 0 to ∞, i.e.,

$$\Delta \arg\{f(j\omega)\}|_0^\infty = \arg\{f(+j\infty)\} - \arg\{f(0)\} = \frac{m\pi}{2}. \tag{44.7}$$

□

This theorem is best illustrated using a graphical argument, as shown in Fig. 44.1. Because all polynomial coefficients are real, the zeroes are either on the real line or they occur in conjugate complex pairs. If all zeroes are in the open left-half plane, then it can be seen from Fig. 44.1 that each real zero contributes a phase change of $\pi/2$ and each conjugate pair contributes a phase change of π, as ω tends from 0 to ∞. Consequently, the total phase change equals $m\pi/2$. In the literature a polynomial which has all roots in the open left-half plane is often referred to as a Hurwitz polynomial.

Similarly, if not all the zeroes are inside the open left-half plane, the phase change resulting from ω changing between 0 and $+\infty$ must be less than $m\pi/2$.

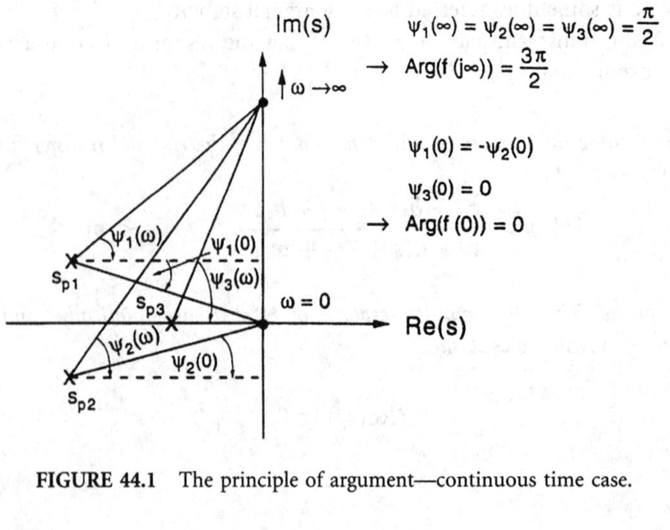

FIGURE 44.1 The principle of argument—continuous time case.

Fact. If $f(s) = a_m s^m + \cdots + a_1 s + a_0$, $a_i \in \Re$, $i = 1, \ldots, m$ has all its roots in the open left-half plane, then the phase function $\arg(f(j\omega))$ is continuous in ω and monotonically increasing [9].

Theorem 44.2 is based on a result commonly known as the principle of argument. In its general version it applies to a large class of functions [10].

Theorem 44.3: *Consider a function $f(s)$, which is holomorphic inside a closed contour C and has no zeroes at any point on the contour. Let $\Delta\{\arg(f(s))\}|_c$ denote the variation of $\arg(f(s))$ for s tracing the contour C once. Then the following condition holds:*

$$\Delta\{\arg(f(s))\}|_c = 2\pi m, \tag{44.8}$$

where m is the number of zeroes of $f(s)$ inside C. □

In section 44.7, this theorem is applied directly to the discrete time case. This theorem can be generalized to other regions of zero clustering in the s-plane. The following result is particularly important if one needs to determine whether a transfer function satisfies a certain margin of stability constraint.

Theorem 44.4: *The polynomial*

$$f(s) = a_m s^m + \cdots + a_1 s + a_0, \quad a_i \in \Re, i = 1, \ldots, m$$

has all its roots in $\mathrm{Re}\{s\} < \sigma$ *iff* $f(\sigma + j\omega)$ *has a change of argument of* $m\pi/2$ *for* ω *changing from 0 to* ∞, *i.e.,*

$$\Delta \arg\{f(\sigma + j\omega)\}|_{\omega=\infty} = \arg\{f(\sigma + j\infty)\} - \arg\{f(\sigma)\} = \frac{m\pi}{2}. \tag{44.9}$$

□

The proof of this theorem can be obtained from a geometric argument similar to that of Theorem 44.2. Further generalizations to sectors in the complex s-plane can be obtained in a similar fashion.

In the complex coefficient case the principle of argument can also be applied to determine whether all roots are in the stable region. The change of argument now must be evaluated for ω ranging from $-\infty$ to $+\infty$, since it is not guaranteed that complex poles occur in conjugate complex pairs.

Theorem 44.5: *The polynomial*

$$f(s) = a_m s^m + \cdots + a_1 s + a_0, \quad a_i \in \mathscr{C}, \ i = 1, \ldots, m$$

has all its roots in $\mathrm{Re}\{s\} < \sigma$, *iff* $f(j\omega)$ *has a change of argument of* $m\pi$ *for* ω *changing from* $-\infty$ *to* $+\infty$, *i.e.,*

$$\Delta \arg\{f(\sigma + j\omega)\}\Big|_{\omega=-\infty}^{\omega=+\infty} = \arg\{f(\sigma + j\infty)\} - \arg\{f(\sigma - j\infty)\} = m\pi \tag{44.10}$$

□

This is the complex coefficient counterpart to Theorem 44.4. The result corresponding to Theorem 44.2 can be obtained by setting $\sigma = 0$ in the theorem above. Similar to the real coefficient case, if $f(s)$ has all its roots in $\mathrm{Re}\{s\} < \sigma$, then $\arg\{f(\sigma + j\omega)\}$ is a continuous and monotonically increasing function of ω, assuming σ remains constant.

44.3 The Interlacing Property

A polynomial

$$f(s) = a_m s^m \cdots + a_1 s + a_0, \quad a_i \in \Re,\ i = 1, \ldots, m$$

can be decomposed into even and odd parts:

$$f^e(s) = a_0 + a_2 s^2 + a_4 s^4 + \cdots \tag{44.11}$$

$$f^o(s) = s(a_1 + a_3 s^2 + a_5 s^4 + \cdots) \tag{44.12}$$

and

$$f(s) = f^e(s) + f^o(s) \tag{44.13}$$

In particular, if $f(s)$ is of even degree $m = 2k$, $k \geq 1$, then one can write for (41.11) and (44.12):

$$f^e(s) = a_0 + a_2 s^2 + \cdots + a_{2k} s^{2k} \tag{44.14}$$

$$f^o(s) = s(a_1 + a_3 s^2 + a_5 s^4 + \cdots + a_{2k-1} s^{2k-2}) \tag{44.15}$$

For the odd degree case, i.e., $m = 2k + 1$, $k \geq 1$, one obtains accordingly

$$f^e(s) = a_0 + a_2 s^2 + a_4 s^4 + \cdots + a_{2k} s^{2k} \tag{44.16}$$

$$f^o(s) = s(a_1 + a_3 s^2 + a_5 s^4 + \cdots + a_{2k+1} s^{2k}) \tag{44.17}$$

Definition (Interlacing Property):

1. The even degree case $m = 2k$: $f(s)$ satisfies the interlacing property iff:
   - i. The coefficients a_{2k} and a_{2k-1} have the same sign.
   - ii. All the zeroes $s_{e,i}$ of $f^e(s)$ and $s_{o,i}$ of $f^o(s)$ are located on the imaginary $j\omega$-axis in the s-plane. The k roots of $f^e(s)$ interlace with the $k-1$ roots of $f^o(s)$ on the positive half of the imaginary axis in the following form:

   $$0 = j\omega_{o,0} < j\omega_{e,1} < j\omega_{o,1} < \cdots < j\omega_{e,k-1} < j\omega_{o,k-1} < j\omega_{e,k} \tag{44.18}$$

   where $s_{e,i} = j\omega_{e,i}$ is the i-th root of $f^e(s)$ and $s_{o,i} = j\omega_{o,i}$ is the i-th root of $f^o(s)$.
2. The odd degree case $m = 2k + 1$: $f(s)$ satisfies the interlacing property iff:
   - i. The coefficients a_{2k} and a_{2k+1} have the same sign.
   - ii. All the zeroes $s_{e,i}$ of $f^e(s)$ and $s_{o,i}$ of $f^o(s)$ are located on the imaginary $j\omega$-axis in the s-plane. The k roots of $f^e(s)$ interlace with the k roots of $f^o(s)$ on the positive half of the imaginary axis in the following form:

   $$0 = j\omega_{o,0} < j\omega_{e,1} < \cdots < j\omega_{e,k-1} < j\omega_{o,k-1} < j\omega_{e,k} < j\omega_{o,k}. \tag{44.19}$$

The interlacing property as defined above requires the zeroes of the even and odd part of the polynomial $f(s)$ to interlace on the positive half of the $j\omega$-axis. Because the coefficients of $f^e(s)$ and $f^o(s)$ are real, all zeroes occur in conjugate complex pairs and therefore also interlace on the negative half of the imaginary axis.

The following theorem, commonly known as the Hermite-Biehler theorem, is a key result in the stability theory of linear continuous time systems.

Theorem 44.6: (Hermite-Biehler): *A real polynomial*

$$f(s) = a_m s^m + \cdots + a_1 s + a_0, \quad a_i \in \Re,\ i = 1, \ldots, m$$

has all its roots s_i in the open left-half plane iff it satisfies the interlacing property. □

A rigorous proof is far beyond the scope of this text. (An especially accessible proof from first principles can be found in [5]). A brief interpretation of the result and its connection to the principle of argument (i.e., Theorem 44.2) is in order at this point:

Condition (i) of the interlacing property is a necessary condition for all zeroes to be in the left-half plane and is discussed in detail in the final section. In fact, it will be shown, that all coefficients must be of the same sign.

Condition (ii) is best explained using Theorem 44.2 and Fig. 44.1. If $f(s)$ is a Hurwitz polynomial of degree m, i.e., it has all zeroes in the open left-half plane, then the total phase change of $f(j\omega)$ is $m\pi/2$ as it tends from 0 to ∞. The trajectory of $f(j\omega)$ starts on the real line for $\omega = 0$. (If all coefficients are positive, the trajectory starts on the positive real line; if all coefficients are negative, it starts on the negative real line.) Because the argument of $f(j\omega)$ increases monotonically with ω, the trajectory of $f(j\omega)$ will circle the origin $m/4$ times by intersecting first with the imaginary axis, then with the real axes, then again with the imaginary axes, etc., until for $\omega \to \infty$ the trajectory will also tend to ∞ parallel to one of the axes. Therefore, the intersection of the trajectory with the real and imaginary axes will alternate or equivalently the zeroes of the imaginary and real part of $f(j\omega)$ will alternate. Because the real part of $f(j\omega)$ is given by $f^e(j\omega)$ and the imaginary parts by $f^o(j\omega)/j$, the zeroes of the real and imaginary parts of $f(j\omega)$ for real ω become the imaginary zeroes of the even and odd parts of $f(s)$, which therefore also interlace. The above argument can be used to show that a Hurwitz polynomial must have the interlacing property.

A similar result holds for polynomials with complex coefficients, i.e.,

$$f(s) = c_0 + c_1 s + \cdots + c_m s^m, \quad c_i \in \mathscr{C},\ i = 1, \ldots, m$$

As in the real coefficient case, one can show that the real and imaginary parts of $f(j\omega)$ satisfy an interlacing property that is similar to the one defined for the real coefficient case. However, real and imaginary parts of $f(j\omega)$ no longer correspond to the even and odd parts of $f(s)$ because the real and imaginary parts of $f(s)$ for $s = j\omega$ are given by:

$$\begin{aligned} f_R(s) &= a_0 + jb_1 s + a_2 s^2 + jb_3 s^3 + \cdots \\ f_I(s) &= jb_0 + a_1 s + jb_2 s^2 + a_3 s^3 + \cdots \end{aligned}$$

where $c_i = a_i + jb_i$; $a_i, b_i \in \Re$, $i = 1, \ldots, m$.

The principle of argument can be applied to regions other than the open left-half plane. Therefore, interlacing properties also can be formulated for more general regions of zero locations. In fact, any region S which has the property that the phase of the polynomial evaluated along the boundary δS of S increases mononically and undergoes a certain net change along this boundary, allows the formulation of an interlacing property. This topic is revisited in a later section on the discrete time case.

44.4 Degree Reduction and the Routh Array

A fundamental result on stability preserving degree reduction is presented for continuous time polynomials. This result allows us to check whether all zeroes of a polynomial are in the open left-half plane. This method uses a consecutive reduction in the degree of the polynomial until it is of degree two, in which case stability is easily determined.

Fact. Let $f(s)$ be a polynomial of degree $m > 0$ and assume that all the coefficients of $f(s)$ are positive:

$$f(s) = a_0 + a_1 s + \cdots + a_m s^m \qquad a_i > 0, \quad i = 0, \ldots, m. \tag{44.20}$$

Then, the polynomial $q(s)$, as defined below, is of degree $m - 1$. For $m = 2k$ (even case):

$$q(s) = f(s) - \frac{a_{2k}}{a_{2k-1}} sf^o(s). \tag{44.21}$$

For $m = 2k + 1$ (odd case)

$$q(s) = f(s) - \frac{a_{2k+1}}{a_{2k}} sf^e(s). \tag{44.22}$$

□

The above fact is easily verified by realizing that in the even case, the highest degree term of $(a_{2k}/a_{2k-1})sf^o(s)$ is equal to $a_{2k}s^{2k}$, and in the odd case, the highest degree term of $(a_{2k+1}/a_{2k})sf^e(s)$ is equal to $a_{2k+1}s^{2k+1}$.

With the above stated fact, the following key result can be stated.

Theorem 44.7: *Assume $f(s)$ has only positive coefficients. Then $f(s)$ has all its roots in the open left-half plane, iff $q(s)$ has all its roots in the open left-half plane.* □

A detailed proof of this theorem can be found in [5]. The proof is based on the interlacing property and makes use of the fact that depending on whether $f(s)$ is of even or odd degree, the resulting polynomial $q(s)$ will have the same odd or even part as $f(s)$.

The last section of the chapter shows that a necessary condition for a polynomial to be Hurwitz is that all its coefficients must have the same sign. Therefore, the formulation of Theorem 44.7 does not restrict the class of polynomials for which the result applies in any way. If the coefficients are all negative, one can simply multiply the whole polynomial by (-1) and the above theorem is directly applicable.

The following test algorithm can now be formulated:

1. Initialize the polynomial $q^i(s)$ for $i = 0$:

$$q^{(0)}(s) := f(s)$$

2. Check all coefficients of $q^i(s)$ for positivity:
   - If all coefficients of $q^i(s)$ are positive and q^i is of degree higher than two, proceed to step 3.
   - If not all coefficients of $q^i(s)$ are positive, terminate the test, as the polynomial is unstable.
   - If all coefficients of $q^i(s)$ are positive and q^i is of degree two, terminate the test, because the polynomial is stable.

3. Generate $q^{i+1}(s)$:

$$q^{(i+1)}(s) := q^{(i)}(s) - \frac{a_{2k}}{a_{2k-1}} sq^{(i)o}(s), \quad m = 2k \text{ (even)}$$

$$q^{(i+1)}(s) := q^{(i)}(s) - \frac{a_{2k+1}}{a_{2k}} sq^{(i)e}(s), \quad m = 2k+1, \text{ (odd)}.$$

4. Set $i = i + 1$. Go to step 2.

The above test terminates under two conditions:

1. If the polynomial contains nonpositive coefficients: In this case the polynomial is unstable because if the original polynomial $f(s)$ has only positive coefficients, the positivity of the coefficients of all $q^i(s)$ is a necessary condition for stability.
2. If the polynomial is of degree two and has positive coefficients only: In the case of a degree two polynomial the positivity condition on the coefficients is actually necessary and sufficient for stability. Therefore, if the degree of $f(s)$ is reduced to two, stability is simply decided by inspecting the sign of the coefficients (see section 44.8).

The above outlined algorithm actually generates the Routh table. This is best seen if the first four rows of the Routh table are constructed and compared to the even and odd parts of $q^{(0)}$ and $q^{(1)}$ in the algorithm. Without loss of generality, assume $f(s)$ is even.

a_{2k}	a_{2k-2}	$\cdots$	a_4	a_2	a_0
a_{2k-1}	a_{2k-3}	$\cdots$	a_3	a_1	
b_{2k-2}	b_{2k-4}	$\cdots$	b_2	b_0	
$\vdots$					

where

$$b_{2k-2} = \frac{a_{2k-1}a_{2k-2} - a_{2k}a_{2k-3}}{a_{2k-1}} = a_{2k-2} - \frac{a_{2k}}{a_{2k-1}} \cdot a_{2k-3}$$

$$b_{2k-4} = \frac{a_{2k-1}a_{2k-4} - a_{2k}a_{2k-5}}{a_{2k-1}} = a_{2k-4} - \frac{a_{2k}}{a_{2k-1}} a_{2k-5}$$

$$\vdots$$

$$b_2 = \frac{a_{2k-1}a_2 - a_{2k}a_1}{a_{2k-1}} = a_2 - \frac{a_{2k}}{a_{2k-1}} a_1$$

$$b_0 = a_0.$$

If we follow the algorithm, the following polynomials are obtained:

$$f(s) = q^{(0)}(s) = a_{2k}s^{2k} + a_{2k-1}s^{2k-1} + \cdots + a_1 s + a_0$$
$$q^{(0)o}(s) = a_{2k-1}s^{2k-1} + \cdots + a_1 s$$
$$q^{(0)e}(s) = a_{2k}s^{2k} + \cdots + a_2 s^2 + a_0.$$

From step 3 we obtain:

$$q^{(1)}(s) = a_{2k-1}s^{2k-1} + \left(a_{2k-2} - \frac{a_{2k}}{a_{2k-1}} s_{2k-3}\right)s^{2k-2} + \cdots + a_1 s + a_0$$

and

$$q^{(1)o}(s) = a_{2k-1}s^{2k-1} + \cdots + a_1 s$$
$$q^{(1)e}(s) = \left(a_{2k-2} - \frac{a_{2k}}{a_{2k-1}} a_{2k-3}\right)s^{2k-2} + \cdots + \left(a_2 - \frac{a_{2k}}{a_{2k-1}} a_1\right)s^2 + a_0.$$

Obviously, the algorithm produces the Routh table.

Adding one new line to the table corresponds to creating $q^{(i+1)}(s)$ from $q^{(i)}(s)$, because $q^{(i+1)}(s)$ and $q^{(i)}(s)$ have either the odd or the even part polynomial in common. Therefore, $q^{(i)}(s)$ and $q^{(i+1)}(s)$ always share a common line in the Routh table. From step 2 in the algorithm it is also apparent that all entries of the Routh table must be positive for the polynomial to be stable.

If one is interested in the number of unstable roots of the polynomial, the Routh table can also provide this information. The number of sign changes in the first column of the Routh array is equal to the number of roots in the right-half plane. Therefore, nonpositive values in the first column indicate instability. It can also happen that the Routh array terminates prematurely by producing a zero in the first column. Several possibilities exist to continue the table in spite of this. Because in this case the polynomial is not Hurwitz stable, this topic will not be discussed further. (A detailed treatment of the singular cases can be found in [11].)

44.5 Hurwitz Criterion

As seen in the previous section, the Routh array can be constructed by using a determinant formulation in terms of the polynomial coefficients. The Routh array indicates stability if and only if all coefficients in the table are positive. This occurs if and only if the first column of the table has only positive entries. Therefore, the positivity condition of the first column can be expressed as a set of determinant inequalities, which are known as the Hurwitz inequalities or the Hurwitz criterion [11]. Although Hurwitz's criterion does not allow us to determine the number of unstable roots (if any), cases exist in which the Hurwitz criterion is easier to apply than the Routh table. Especially for polynomials with uncertain parameters, the Hurwitz criterion is often more convenient.

The Hurwitz criterion is best explained through the Hurwitz matrix. Consider the polynomial

$$f(s) = a_0 + a_1 s + \cdots + a_m s^m, \quad a_i > 0,\ i = 0, \ldots, m. \tag{44.23}$$

The Hurwitz matrix is defined as:

$$H_m = \begin{bmatrix} a_{m-1} & a_{m-3} & a_{m-5} & \cdots \\ a_m & a_{m-2} & a_{m-4} & \cdots \\ 0 & a_{m-1} & a_{m-3} & \cdots \\ 0 & a_m & a_{m-2} & \cdots \\ \vdots & 0 & a_{m-1} & \cdots \\ \vdots & \vdots & a_m & \cdots \\ \vdots & \vdots & 0 & \cdots \\ \vdots & \vdots & 0 & \cdots \\ 0 & 0 & \vdots & \cdots \end{bmatrix} \tag{44.24}$$

H_m is constructed from the coefficients of $f(s)$ by following the above-indicated form and by appending all rows and columns by zeroes to obtain a matrix of dimension $m \times m$. The m

principle minors of H_m are given by:

$$\Delta_1 = |a_{m-1}|$$
$$\Delta_2 = \begin{vmatrix} a_{m-1} & a_{m-3} \\ a_m & a_{m-2} \end{vmatrix}$$
$$\Delta_3 = \begin{vmatrix} a_{m-1} & a_{m-3} & a_{m-5} \\ a_m & a_{m-2} & a_{m-4} \\ 0 & a_{m-1} & a_{m-3} \end{vmatrix} \tag{44.25}$$
$$\vdots$$

Now the Hurwitz criterion is simply given by [11]:

$$\Delta_i > 0, \quad i = 1, \dots, m \tag{44.26}$$

Theorem 44.8: *The polynomial*

$$f(s) = a_0 + a_1 s + \cdots + a_m s^m \quad a_m > 0$$

has all roots in the open left-half plane iff the m Hurwitz determinants Δ_i, $i = 1, \dots, m$ are positive. □

This result is illustrated by the following example:

$$f(s) = s^4 + 3s^3 + 20s^2 + 4s + 1$$
$$H_4 = \begin{bmatrix} 3 & 4 & 0 & 0 \\ 1 & 20 & 1 & 0 \\ 0 & 3 & 4 & 0 \\ 0 & 1 & 20 & 1 \end{bmatrix}$$
$$\Delta_1 = |3| = 3 > 0$$
$$\Delta_2 = \begin{vmatrix} 3 & 4 \\ 1 & 20 \end{vmatrix} = 56 > 0$$
$$\Delta_3 = \begin{vmatrix} 3 & 4 & 0 \\ 1 & 20 & 1 \\ 0 & 3 & 4 \end{vmatrix} > 0$$
$$\Delta_4 > 0$$

Because all principle minors are positive, $f(s)$ is Hurwitz stable.

44.6 Other Related Stability Criteria

Criteria other than those by Hurwitz, Routh, and Michailov-Cremer-Leonhard are briefly discussed. The Nyquist criterion is omitted because it is treated separately in Chapter 46.

The Liénard-Chipart Criterion [11]

Recall that the Hurwitz criterion requires that all determinants of the m principle minors must be positive. The two requirements that all coefficients a_i, $i = 1, \ldots, m$ and all determinants Δ_i, $i = 1, \ldots, m$ must be positive for the roots to be in the open left-half plane, are not independent of each other. In the work of Liénard and Chipart each of the following four conditions were proved to be necessary and sufficient for stability of a polynomial $f(s)$:

$$a_0 > 0, a_2 > 0, a_4 > 0, \ldots \tag{44.27}$$

$$\Delta_1 > 0,\ \Delta_3 > 0,\ \Delta_5 > 0, \ldots \tag{44.28}$$

or

$$a_0 > 0, a_2 > 0, a_4 > 0, \ldots \tag{44.29}$$

$$\Delta_2 > 0, \Delta_4 > 0, \Delta_6 > 0, \ldots \tag{44.30}$$

or

$$a_0 > 0, a_1 > 0, a_3 > 0, a_5 > 0, \ldots \tag{44.31}$$

$$\Delta_1 > 0, \Delta_3 > 0, \Delta_5 > 0, \ldots \tag{44.32}$$

or

$$a_0 > 0, a_1 > 0, a_3 > 0, \ldots \tag{44.33}$$

$$\Delta_2 > 0, \Delta_4 > 0, \Delta_6 > 0, \ldots \tag{44.34}$$

Therefore, the Liénard-Chipart criterion reduces the number of determinants to be checked for positivity by approximately 50%. It simplifies the Hurwitz test significantly.

The Schur-Cohn Criterion for Continuous Time Systems [9]

Theorem 44.7 stated a result on stability-preserving degree reduction for continuous time polynomials. This method led to the formulation of the Routh array. The following result also allows degree reduction and can be viewed as the continuous time counterpart of the Schur-Cohn criterion, which is discussed briefly in Section 44.7.

Theorem 44.9: *Define*

$$p(s) = f(s) - \frac{f(-1)}{f(+1)} f(-s). \tag{44.35}$$

Then $f(s)$ *is Hurwitz stable iff* $p(s)$ *is Hurwitz stable and* $|f(-1)/f(+1)| < 1$. □

Because

$$p(-1) = f(-1) - \frac{f(-1)}{f(1)} f(1) = 0$$

$p(s)$ has a zero at $s = -1$ and its factorization contains the term $(s + 1)$. Therefore, the new polynomial $p(s)/(s + 1)$ is reduced in degree and is also stable, if $f(s)$ is stable. The same procedure can be applied to $p(s)/(s + 1)$.

This scheme can be continued until the resulting polynomial is of degree two. The proof of Theorem 44.9 is based on the interlacing property. If $f(s)$ is assumed to be stable, then it has the interlacing property, as does $p(s)$ because

$$p(s) = f(s) - \frac{f(-1)}{f(1)} f(-s) = f^e(s)\left(1 - \frac{f(-1)}{f(1)}\right) + f^o(s)\left(1 + \frac{f(-1)}{f(1)}\right). \quad (44.36)$$

If $|f(-1)/f(1)| < 1$, the coefficients in the even and odd parts of $f(s)$ and $p(s)$ have the same sign. Furthermore, the zeroes of the even and odd parts of $f(s)$ and $p(s)$ are identical. Hence, if $f(s)$ is Hurwitz stable, then $p(s)$ is Hurwitz stable. (A similar argument can be used to show that if $p(s)$ is Hurwitz stable, then $f(s)$ is also Hurwitz stable.)

The Continued Fraction Expansion Criterion [1]

Consider the polynomial $f(s)$ of degree m:

$$f(s) = f^e(s) + f^o(s)$$

where $f^e(s)$ and $f^o(s)$ are the even and odd parts of $f(s)$, respectively. Then form the transfer function:

$$H(s) = \frac{f^e(s)}{f^o(s)} \quad \text{if } m \text{ is even}$$
$$H(s) = \frac{f^o(s)}{f^e(s)} \quad \text{if } m \text{ is odd.}$$

If one expresses $H(s)$ in the form of a continued fraction expansion, i.e.,

$$H(s) = \beta_1 s + \cfrac{1}{\beta_2 s + \cfrac{1}{\beta_3 s + \cfrac{1}{\cdots \cfrac{1}{\beta_m s}}}} \quad (44.37)$$

the Hurwitz stability of $f(s)$ can be determined.

Theorem 44.10: *The polynomial $f(s)$ has all its roots in the open left-half plane iff all coefficients β_ν, $\nu = 1, \ldots, m$ obtained by the continued fraction expansion in (44.37) are positive.* □

The above theorem applies to polynomials with a positive leading coefficient a_m. It can be shown that the coefficients $\beta_1, \beta_2, \ldots, \beta_m$ are actually the ratios of two consecutive elements in the first column of the Routh array. Therefore, this formulation is equivalent to the Routh-Hurwitz criterion and also can be used to determine the number of zeroes outside the open left-half plane.

The Meerov Criterion of Aperiodic Stability [11]

The following theorem is useful in determining whether the roots of a polynomial are all real and negative. Such a system shows an aperiodic response behavior, i.e., the impulse response only consists of exponentially decaying response portions.

Theorem 44.11. (Meerov Criterion): *Assume the zeroes of $f(s)$ are all simple. Then the zeroes of $f(s)$ are all real and negative iff the following two conditions hold: (a) the coefficients of $f(s)$ are all positive, and (b) the Hurwitz determinants for $\tilde{f}(s) = f(s^2) + sf'(s^2)$ are all positive, where $f'(s^2)$ denotes the derivative of $f(s^2)$ with respect to s.*

44.7 The Discrete Time Case

Conditions for Schur stability, i.e., stability of discrete time polynomials, are summarized here.

Consider the discrete time transfer function

$$H(z) = \frac{b_n z^n + b_{n-1} z^{n-1} + \cdots + b_0}{a_m z^m + a_{m-1} z^{m-1} + \cdots + a_0} \tag{44.38}$$

where z is the transform variable of the z-transform, and a_i, $i = 0, \ldots, m$ and b_j, $j = 0, \ldots, n$ are real coefficients. Also assume that the numerator and the denominator of $H(z)$ are mutually coprime. Then the following theorem holds.

Theorem 44.12: *The impulse response of the transfer function $H(z)$ is asymptotically (and exponentially) stable iff $|z_i| < 1$, $i = 1, \ldots, m$, where the z_i are the poles of the transfer function $H(z)$.* □

Similar to the continuous time case, this result becomes obvious if the transfer function is written in terms of first-order partial fractions. The z-transform correspondence

$$\frac{A}{(z - z_i)^\nu} \leftrightarrow z_i^{n-\nu} u(n - \nu) \frac{(n-1)\cdots(n-\nu+1)}{(\nu-1)!} A \tag{44.39}$$

then shows that all partial responses will tend to zero asymptotically (and exponentially) if and only if all poles z_i satisfy $|z_i| < 1$, $i = 1, \ldots, m$. Consequently, determining the stability of a linear time-invariant discrete time system requires one to check the root locations of the denominator polynomial with respect to the unit disk.

Although the general formulation of the principle of argument (see Theorem 44.3) can be directly applied to the discrete time polynomial, a slightly altered version will be introduced for the case of real coefficient polynomials. In this special situation it is sufficient to check the phase change along the upper or lower half circle because the roots occur in pairs symmetric to the real line.

Theorem 44.13: (Principle of Argument–Discrete Time Case): *The real coefficient polynomial*

$$f(z) = a_m z^m + a_{m-1} z^{m-1} + \cdots + a_0$$

has all its roots in the open unit disk iff $f(e^{j\omega})$ has a change of argument of $m\pi$ for ω changing from 0 to π. □

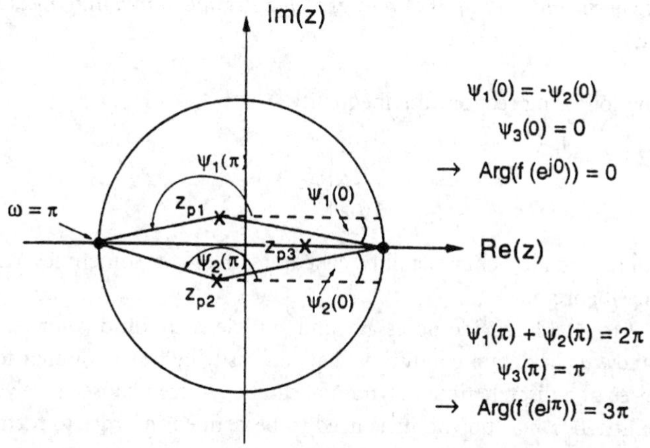

FIGURE 44.2 The principle of argument—discrete time case.

The above theorem is best illustrated by a simple figure. Figure 44.2 shows that if ω varies from 0 to π, each zero inside the unit circle and on the real line contributes a phase change of π, whereas each conjugate complex pair contributes to a phase change of 2π. Hence, the total phase change will be equal to $m\pi$.

If the interlacing property in the discrete time domain were formulated in terms of the real and imaginary part of $f(e^{j\omega})$, a result very similar to the continuous time case could be deduced from Theorem 44.13.

Stability of $f(z)$ would then require that the two functions

$$\text{Re}\{f(e^{j\omega})\} = a_m \cos(m\omega) + \cdots + a_1 \cos \omega + a_0 \tag{44.40}$$

$$\text{Im}\{f(e^{j\omega})\} = a_m \sin(m\omega) + \cdots + a_1 \sin \omega \tag{44.41}$$

have interlacing zeroes on the interval $[0, \pi]$.

The discrete counterpart of the Hermite-Biehler theorem is formulated using the symmetric and antisymmetric parts of $f(z)$ [12].

The symmetric part $h(z)$ of $f(z)$ is defined as

$$h(z) = \frac{1}{2}\left[f(z) + z^m f\left(\frac{1}{z}\right)\right] \tag{44.42}$$

whereas the antisymmetric part $g(z)$ of $f(z)$ is defined as

$$g(z) = \frac{1}{2}\left[f(z) - z^m f\left(\frac{1}{z}\right)\right] \tag{44.43}$$

(Note that the coefficients of $h(z)$ are symmetric with respect to the "center-exponent" $m/2$, whereas the coefficients of $g(z)$ are antisymmetric with respect to $m/2$.)

Any polynomial $f(z)$ can be decomposed in the form

$$f(z) = g(z) + h(z)$$

Theorem 44.14: (Schüssler's Theorem): *The polynomial*

$$f(z) = a_m z^m + \cdots + a_1 z + a_0$$

has all roots in the open unit disk iff $h(z)$ and $g(z)$ have simple alternating roots on the unit circle and $|a_0|/|a_m| < 1$. □

As shown in the following section, the inequality

$$\left|\frac{a_0}{a_m}\right| < 1 \tag{44.44}$$

is a necessary condition for Schur stability. Theorem 44.14 is sometimes referred to as the discrete interlacing theorem.

Theorems 44.12 to 44.14 provide necessary and sufficient conditions for Schur stability of a polynomial and allow us to obtain conditions that can easily be tested. Similar to the continuous time case, the degree of a discrete time polynomial can be reduced consecutively while preserving stability or instability. Two new polynomials need to be defined in order to formulate the degree reduction procedure:

$$q(z) = z^m f\left(\frac{1}{z}\right) = a_0 z^m + a_1 z^{m-1} + \cdots + a_{m-1} z + a_m \tag{44.45}$$

$$r(z) = z^{-1}\left(f(z) - \frac{a_0}{a_m} q(z)\right) \tag{44.46}$$

The polynomial $r(z)$ is always of degree $m - 1$ (or smaller) because

$$\frac{a_0}{a_m} q(z) = \frac{a_0^2}{a_m} z^m + \frac{a_0 a_1}{a_m} z^{m-1} + \cdots + a_0 \tag{44.47}$$

$$f(z) = a_m z^m + a_{m-1} z^{m-1} + \cdots + a_0 \tag{44.48}$$

and hence

$$\begin{aligned} r(z) &= z^{-1}\left(f(z) - \frac{a_0}{a_m} q(z)\right) \\ &= z^{m-1}\left(a_m - \frac{a_0^2}{a_m}\right) + z^{m-2}\left(a_{m-1} - \frac{a_0 a_1}{a_m}\right) + \cdots \\ &\quad \cdots + \left(a_1 - \frac{a_0 a_{m-1}}{a_m}\right). \end{aligned} \tag{44.49}$$

The following key theorem on degree reduction of discrete time polynomials can now be formulated.

Theorem 44.15: (Schur-Cohn-Jury): *Assume*

$$f(z) = a_m z^m + a_{m-1} z^{m-1} + \cdots + a_1 z + a_0$$

satisfies

$$\left|\frac{a_0}{a_m}\right| < 1$$

Then $f(z)$ is Schur stable iff $r(z)$ is Schur stable. □

A simple proof of the above theorem can be found in [5]. A consecutive application of Theorem 41.15 allows us to determine Schur stability of the polynomial $f(z)$. The following stability test algorithm arises:

1. $i = 0$. Set $f^{(i)}(z) := f(z)$.
2. If $|a_0^{(i)}/a_{m-i}^{(i)}| < 1$ and degree $(f^{(i)}(z)) > 1$, go to step 3.
   If $|a_0^{(i)}/a_{m-i}^{(i)}| > 1 \Rightarrow$ instability.
   If $|a_0^{(i)}/a_{m-i}^{(i)}| < 1$ and degree $(f^{(i)}(z)) = 1 \Rightarrow$ stability.
3. Generate

$$f^{(i+1)}(z) := z^{-1}\left(f^{(i)}(z) - \frac{a_0^{(i)}}{a_{m-i}^{(i)}} z^{m-i} f^{(i)}\left(\frac{1}{z}\right)\right)$$

4. $i := i + 1$. Return to step 2.

$a_{m-i}^{(i)}$ is the highest degree coefficient of $f^{(i)}(z)$. The condition $|a_0^{(i)}/a_{m-i}^{(i)}| < 1$ is necessary and sufficient for Schur stability of a first-order polynomial. This inequality is used as a termination condition for stability.

The procedure outlined above generates the Jury-Marden table, which is given below [7].

$a_m^{(0)}$	$a_{m-1}^{(0)}$	$\cdots$	$a_1^{(0)}$	$a_0^{(0)}$
$a_0^{(0)}$	$a_1^{(0)}$	$\cdots$	$a_{m-1}^{(0)}$	$a_m^{(0)}$
$a_{m-1}^{(1)}$	$a_{m-2}^{(1)}$	$\cdots$	$a_0^{(1)}$	
$a_0^{(1)}$	$a_1^{(1)}$	$\cdots$	$a_{m-1}^{(1)}$	
$a_{m-2}^{(2)}$	$a_{m-3}^{(2)}$	$\cdots$		
$a_0^{(2)}$	$a_1^{(2)}$	$\cdots$		
$\vdots$	$\vdots$			
$a_0^{(m)}$				

where

$$a_i^{(j)} = a_{i+1}^{(j-1)} - \frac{a_0^{(j-1)}}{a_{m-j+1}^{(j-1)}} a_{m-i-j}^{(j-1)}. \tag{44.50}$$

The relationship between the above table and the test algorithm is best understood by comparing the coefficients in (44.49) to (44.50). The Jury-Marden table scheme can be remembered more easily in the following determinant representation, which constructs row 3 from rows 1 and 2:

$$a_{m-1}^{(1)} = \frac{1}{a_m^{(0)}} \begin{vmatrix} a_m^{(0)} & a_0^{(0)} \\ a_0^{(0)} & a_m^{(0)} \end{vmatrix}$$

$$a_{m-2}^{(1)} = \frac{1}{a_m^{(0)}} \begin{vmatrix} a_m^{(0)} & a_1^{(0)} \\ a_0^{(0)} & a_{m-1}^{(0)} \end{vmatrix}$$

$$\vdots \qquad \vdots$$

$$a_0^{(1)} = \frac{1}{a_m^{(0)}} \begin{vmatrix} a_m^{(0)} & a_{m-1}^{(0)} \\ a_0^{(0)} & a_1^{(0)} \end{vmatrix}$$

Row 4 is then produced by rewriting row 3 in reverse order. The same determinant scheme is then used to construct row 5 from rows 3 and 4, etc.

A number of other methods are available which allow us to test Schur stability of polynomials. The Schur-Cohn criterion, the Schur-Cohn-Fujiwara criterion, and the Inners formulation are probably the best known. For further reading on these and other Schur stability criteria, the reader is referred to [6, 7].

44.8 Some Simple Stability Checks and Rules

A number of simple sufficient conditions as well as necessary conditions for stability are introduced here. These conditions are often helpful for quick checks by inspection.

The Continuous Time Case

One of the most useful necessary conditions for Hurwitz stability is obtained from the sign of the coefficients.

Theorem 44.16: *If the polynomial*

$$f(s) = a_m s^m + \cdots + a_1 s + a_0 \tag{44.51}$$

is Hurwitz stable, then either

$$a_i > 0, \quad i = 0, \ldots, m \tag{44.52}$$

or

$$a_i < 0, \quad i = 0, \ldots, m. \tag{44.53}$$

In other words, all coefficients must be of the same sign if the polynomial is stable. Hence, missing coefficients are also not permissible. Theorem 44.16 is easily obtained from two facts: (a) $f(s)$ can be factored in first- and second-order factors with real coefficients. (b) Hurwitz stable first- or second-order factors must have coefficients of the same sign. (In fact, the opposite is also true: If the coefficients all have the same sign, then a first or second order factor is Hurwitz stable.)

An important sufficient condition on Hurwitz stability of polynomial $f(s)$ can be found in the work if Lipatov and Sokolov [8]. Because this condition is explicitly stated in Chapter 49 of this handbook, it is not restated here.

Explicit, necessary, and sufficient conditions for Hurwitz stability of low-order polynomials are presented below:

- $m = 1$: $f(s) = a_1 s + a_0$
  Hurwitz stability, iff
  $a_1 > 0,\ a_0 > 0$ or $a_1 < 0,\ a_0 < 0$.
- $m = 2$: $f(s) = a_2 s^2 + a_1 s + a_0$
  Hurwitz stability, iff
  $a_2 > 0,\ a_1 > 0,\ a_0 > 0$ or $a_2 < 0,\ a_1 < 0,\ a_0 < 0$.
- $m = 3$: $f(s) = a_3 s^3 + a_2 s^2 + a_1 s + a_0$
  Hurwitz stability, iff
  $a_0/a_2 < a_1/a_3$ and $a_0 > 0,\ a_1 > 0,\ a_2 > 0$ or
  $a_0/a_2 < a_1/a_3$ and $a_0 < 0,\ a_1 < 0,\ a_2 < 0$.

- $m = 4$: $f(s) = a_4 s^4 + a_3 s^3 + a_2 s^2 + a_1 s + a_0$
  Hurwitz stability, iff
  $a_3 a_2 a_1 - a_0 a_3^2 - a_4 a_1^2 > 0$ and $a_0 > 0$, $a_2 > 0$, $a_3 > 0$ or
  $a_3 a_2 a_1 - a_o a_3^2 - a_4 a_1^2 < 0$ and $a_0 < 0$, $a_2 < 0$, $a_3 < 0$.

The condition for $m = 2$ was used as a termination condition for the stability test algorithm in Section 44.4. Obviously, a partial evaluation of the tests discussed in Section 44.4 to 44.6 can also serve as necessary conditions, e.g., the evaluation of the low-order determinants Δ_1, Δ_2.

The Discrete Time Case

Consider the discrete time polynomial

$$f(z) = a_m z^m + \cdots + a_1 z + a_0 \tag{44.54}$$

Then the necessary condition of Theorem 41.17 holds true.

Theorem 41.17: *If $f(z)$ is Schur stable, then*

$$\frac{|a_0|}{|a_m|} < 1 \tag{44.55}$$

□

This necessary condition follows easily from factoring $f(z)$:

$$f(z) = a_m \prod_{i=1}^{m} (z - z_{0i})$$

where z_{0i} are the polynomial zeroes. Then

$$f(0) = a_0 = a_m \prod_{i=1}^{m} (-z_{0i})$$

and hence

$$\left| \frac{a_0}{a_m} \right| = \prod_{i=1}^{m} |z_{0i}|. \tag{44.56}$$

If the polynomial is stable, then $|z_i| < 1$, $i = 1, \ldots, m$ and the result follows.

For the first-order case, this necessary condition is also sufficient. (This fact was used as a termination condition of the Schur stability test in the previous section.)

If $a_m > 0$ in (44.54), the following two conditions provide a fast initial check:

$$f(1) > 0 \tag{44.57}$$

$$(-1)^m f(-1) > 0 \tag{44.58}$$

(If $a_m < 0$, the above condition can be applied after multiplying $f(z)$ with a negative number.)

Another useful set of necessary conditions is stated below [2].

Theorem 41.18: *If $f(z)$ is Schur stable, then*

$$|a_{m-1}| < \sum_{\substack{i=0 \\ i\neq m-1}}^{m-1} |a_i|$$

$$|a_{m-2}| < \sum_{\substack{i=1 \\ i\neq m-2}}^{m} |a_i| \tag{44.59}$$

$$\vdots$$

$$|a_1| < \sum_{\substack{i=0 \\ i\neq 1}}^{m} |a_i|.$$

□

This theorem essentially states that polynomials which contain a coefficient that is large in magnitude relative to the remaining coefficients tend to be Schur unstable. The following sufficient condition for Schur stability is related to Theorem 44.18.

Theorem 44.19: *The polynomial $f(z)$ is Schur stable if*

$$|a_m| > \sum_{i=0}^{m-1} |a_i|. \tag{44.60}$$

□

The above sufficient condition follows from [3] and is also sufficient for stability of time-variant polynomials which describe discrete time systems.

A number of other sufficient and/or necessary criteria for stability of discrete time systems exist [6, 7] and are too numerous to be listed here. Some of these conditions can be found in Chapter 49.

References

[1] H. Baher, *Synthesis of Electrical Networks*, New York: John Wiley & Sons, 1984.

[2] P. Bauer, "Robust stability of linear systems," Lecture Notes, Department of Electrical Engineering, Notre Dame, IN: University of Notre Dame, Fall 1992.

[3] P. Bauer, M. Mansour, and J. Duran, "Stability of polynomials with time-variant coefficients," *IEEE Trans. Circuits Syst.*, vol. 40, no. 6, pp. 423–426, June 1993.

[4] S. Bennett, *A History of Control Engineering 1800-1930*, London: IEE, 1979.

[5] H. Chapellat, M. Mansour, and S. P. Bhattacharyya, "Elementary proofs of some classical stability criteria," *IEEE Trans. Educ.*, vol. 33, no. 3, pp. 232–239, Aug. 1990.

[6] E. I. Jury, *Inners and Stability of Dynamical Systems*, New York: Wiley-Interscience, 1974.

[7] E. I. Jury, *Theory and Application of the z-Transform Method*, New York: Wiley, 1964.

[8] A. V. Lipatov and N. I. Sokolov, "Some sufficient conditions for stability and instability of continuous linear stationary systems," *Autom. Remote Control*, vol. 39, pp. 1285–1291, 1979.

[9] M. Mansour, "Robust stability in systems described by rational functions," in *Control of Dynamic Systems*, C. T. Leondes, Ed., vol. 51. Orlando, FL: Academic Press, in press.

[10] E. G. Phillips, *Functions of a Complex Variable with Applications*, Edinburgh: Oliver & Boyd, 1963.

[11] B. Porter, *Stability Criteria for Linear Dynamical Systems*, New York: Academic Press, 1968.

[12] H. W. Schüssler, "A stability theorem for discrete systems," *IEEE Trans. Acoust., Signal Speech Proc.*, ASSP-24, pp. 87–89, 1976.

45

Root Locus

Peter Bauer
University of Notre Dame

45.1 Introduction

The previous chapter discussed the problem of stability under the assumption that the characteristic polynomial of the linear time-invariant system is known exactly. In practice, this is seldom the case and one usually must deal with the problem of coefficient uncertainty. In some cases the designer has control over a number of system parameters, and the question of stability needs to be answered for a family of polynomials which contains infinitely many members.

This chapter addresses the problem of stability of the characteristic polynomial in the presence of one uncertain parameter. Basically, two approaches to this problem exist:

1. Root locus: The trajectories of the roots which result from moving one free parameter is constructed.
2. D-partitioning: Sets of polynomials are constructed which have all roots inside a certain specified subregion of the complex plane. (These sets are typically characterized in the parameter space.)

The method of root locus is an attractive tool if a single unknown parameter enters the polynomial in linear form. Root maps can then be constructed as a function of this parameter. In the multiparameter case D-partitioning is the more useful concept because the root loci are no longer 1-D curves in the complex plane and visualization of the root loci is often impossible.

The root locus method is based on the fact that the roots of a polynomial depend continuously on its parameters [3]. For low-order polynomials an exact dependence of the roots on the parameters can be obtained in explicit form. In the case of higher order polynomials one cannot find explicit solutions to the root locus. However, a set of rules can be derived which aid in the construction of root loci. This topic is the main focus of Section 45.2.

A typical application of the root locus method is found in control systems with a variable gain parameter. This is illustrated by the following example.

Consider the feedback system in Fig. 45.1, which contains a variable feedback gain λ. The transfer function of this system is given by

$$H(s) = \frac{Y(s)}{X(s)} = \frac{\dfrac{2}{s(s+1)}}{1 + \lambda \dfrac{2}{s(s+1)}} = \frac{2}{s^2 + s + 2\lambda} \tag{45.1}$$

0-8493-8341-2/95/$0.00 + $.50

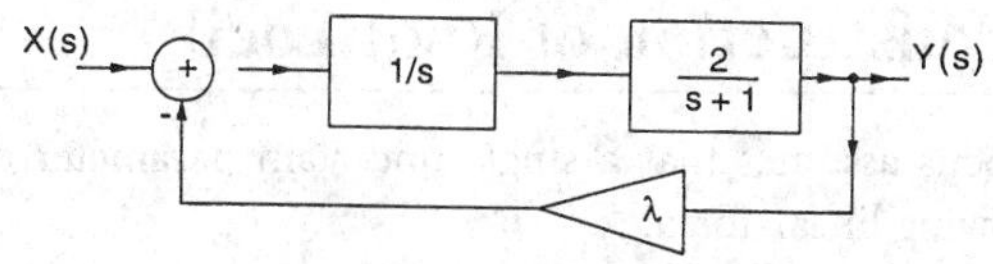

FIGURE 45.1 Feedback system with a variable gain λ.

In this simple case the system poles can be explicitly written as a function of the uncertain parameter λ:

$$s_{p_{1,2}} = -\frac{1}{2} \pm \sqrt{\frac{1}{4} - 2\lambda} \tag{45.2}$$

The resulting root locus is shown in Fig. 45.2.

Figure 45.2 shows that for λ approaching zero from $-\infty$, the root locus approaches $s = -1$ from the left and $s = 0$ from the right along the real axis. The root pair converges to $s = -\frac{1}{2}$ for $\lambda \to \frac{1}{8}$, such that for $\lambda = \frac{1}{8}$ a root of multiplicity two is created at $s = -\frac{1}{2}$. If λ is further increased the root pair "separates" and tends to $-\frac{1}{2} \pm j\infty$ for $\lambda \to +\infty$. The root locus plot also shows that the system remains stable, if $\lambda > 0$. (Of course, this condition also could have been obtained from a result in Chapter 44.8, which states that for a second-order polynomial to be Hurwitz stable it is necessary and sufficient that all coefficients be positive.)

The following notation is used throughout the chapter:

λ:	uncertain parameters $\lambda \in R$
$f(s)$, $q(s)$, $p(s)$:	polynomials with real coefficients
s_{0_i}:	zero of a transfer function
s_{p_i}:	pole of a transfer function
k_0:	degree of the numerator
k_p:	degree of the denominator
$\underline{\lambda}$, $\overline{\lambda}$:	lower/upper boundary of an interval $[\underline{\lambda}, \overline{\lambda}]$
$\underline{f}(s)$, $\overline{f}(s)$:	vertex polynomials of an edge where $\underline{f}(s)$ corresponds to $\underline{\lambda}$ and $\overline{f}(s)$ corresponds to $\overline{\lambda}$
$h(s^2)$:	even part of a continuous time polynomial
$sg(s^2)$:	odd part of a continuous time polynomial
$p^e(s)$:	some even polynomial
$p^o(s)$:	some odd polynomial
$h(z)$:	symmetric part of a discrete time polynomial
$g(z)$:	antisymmetric part of a discrete time polynomial

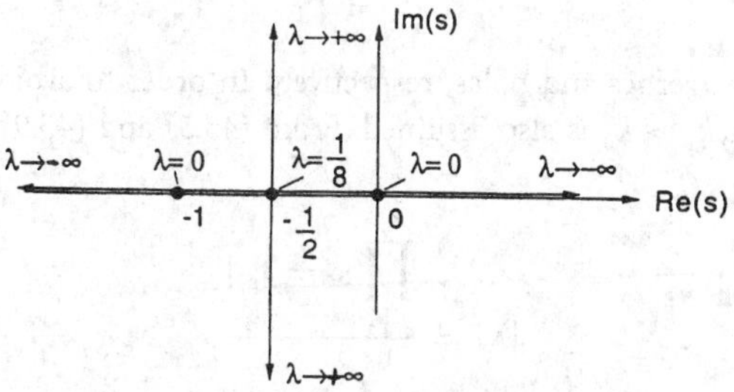

FIGURE 45.2 Root locus of a second-order system.

45.2 Rules for the Construction of Root Loci

The method of root locus assumes that a single uncertain parameter enters the characteristic polynomial in the following linear form:

$$f(s) = p(s) + \lambda q(s) \tag{45.3}$$

Such polynomials typically arise in feedback systems with a variable gain λ. In order to find the roots of (45.3), one must solve the equation

$$1 + \lambda \frac{q(s)}{p(s)} = 0 \tag{45.4}$$

Equation (45.4) yields the following magnitude and phase conditions

$$\left| \lambda \frac{q(s)}{p(s)} \right| = 1 \tag{45.5}$$

and

$$\arg\left(\lambda \frac{q(s)}{p(s)} \right) = (2m+1)\pi. \tag{45.6}$$

For $\lambda > 0$, (45.6) becomes

$$\arg\left(\frac{q(s)}{p(s)} \right) = (2m+1)\pi. \tag{45.7}$$

For $\lambda < 0$, one obtains

$$\arg\left(\frac{q(s)}{p(s)} \right) = 2m\pi. \tag{45.8}$$

If s lies on a root locus, conditions (45.5) to (45.8) hold. Because $q(s)/p(s)$ is a rational transfer function it can be factored in terms of complex first-order terms

$$\frac{q(s)}{p(s)} = \frac{\prod_{i=1}^{k_0} (s - s_{0_i})}{\prod_{i=1}^{k_p} (s - s_{p_i})} \tag{45.9}$$

where s_{0_i} and s_{p_i} are the zeroes and poles, respectively. In order to avoid a degree drop of $f(s)$ for a certain value of $\lambda, k_p > k_0$ is also assumed. From (45.5) and (45.9) it then follows that

$$|\lambda| = \frac{\prod_{i=1}^{k_p} |s - s_{p_i}|}{\prod_{i=1}^{k_0} |s - s_{0_i}|}. \tag{45.10}$$

With (45.9) and (45.7, 45.8) the phase relationships can be written as

$$\sum_{i=1}^{k_0} \arg(s - s_{0_i}) - \sum_{i=1}^{k_p} \arg(s - s_{p_i}) = (2m+1) \quad \text{for } \lambda > 0 \qquad (45.11)$$

$$\sum_{i=1}^{k_0} \arg(s - s_{0_i}) - \sum_{i=1}^{k_p} \arg(s - s_{p_i}) = 2m\pi \quad \text{for } \lambda < 0 \qquad (45.12)$$

The above equations are the key to the derivation of the rules which allow the construction of the root loci. These rules are stated next. Complete proofs can be found in [6], [8], and many others.

1. *The number of root loci:* The number of root loci is equal to the degree of the characteristic polynomial $f(s)$.
2. *Symmetry:* If $f(s)$ contains only real coefficients all root loci are symmetric with respect to the real axis.
3. *Poles of $q(s)/p(s)$:* The poles of $q(s)/p(s)$ are points on the root locus and are found by setting $\lambda = 0$.
4. *Zeroes of $q(s)/p(s)$:* The zeroes of $q(s)/p(s)$ are points on the root locus and are obtained for $\lambda \to \pm\infty$.
5. *Asymptotic behavior of root loci:* For $\lambda \to +\infty$, the root loci asymptotically converge to $k_p - k_0$ rays with angles

$$\frac{(2\mu+1)\pi}{k_p - k_0}, \qquad \mu = 0, 1, \ldots, k_p - k_0 - 1$$

to the real axis. For $\lambda \to -\infty$, the root loci asymptotically converge to $k_p - k_0$ rays with angles

$$\frac{2\mu\pi}{k_p - k_0}, \qquad \mu = 0, 1, \ldots, k_p - k_0 - 1$$

to the real axis.
6. *Intersection of Asymptotes:* The set of asymptotes obtained for $\lambda \to +\infty$ and the set of asymptotes obtained for $\lambda \to -\infty$ intersect on the real line at a point given by

$$\sigma = \left(\sum_{i=1}^{k_p} s_{p_i} - \sum_{i=1}^{k_0} s_{0_i} \right) \Big/ (k_p - k_0)$$

7. *Root loci on the real line:* If $q(s)/p(s)$ has at least one real zero or pole, then a root locus exists which coincides with the real line. In addition, if the sum of the number of poles and the number of zeroes of $q(s)/p(s)$, which are to the right of any particular point of the real axis is odd, then the root locus of this point corresponds to a positive value of λ. If the sum of the number of poles and zeroes of $q(s)/p(s)$, which are to the right of any particular point of the real axis is even, then the root locus of this point corresponds to a negative value of λ.
8. *Branching Points:* Branching points occur at values of s, λ for which $f(s)$ has a multiple root. Furthermore, at a branching point

$$\frac{d\lambda}{ds} = 0$$

holds. The number of branches originating from this point for the same value of λ is equal to the multiplicity of the root.

These eight rules are the most useful construction rules. From (45.10–45.12) a number of additional properties can be derived which are usually very difficult to check. An analysis for complex values of λ can be found in [6]. Rules 1 through 8 are illustrated by using the simple introductory example of Section 45.1:

$$f(s) = s^2 + s + 2\lambda \quad \text{and hence} \quad \frac{q(s)}{p(s)} = \frac{2}{s^2 + s}$$

1. The number of root loci is equal to the degree of $f(s)$, which is two.
2. The two root loci must be symmetric with respect to the real axis.
3. For $\lambda = 0$, the root locus coincides with the roots of $p(s)$ at $s = 0$ and $s = -1$.
4. Since $\frac{q(s)}{p(s)}$ has two zeroes at infinity, the root locus tends to ∞ for $\lambda \to \pm\infty$.
5. $k_p - k_0 = 2$. Therefore, for $\lambda \to +\infty$, the root loci asympotically converge to two rays with angles $\pi/2$ and $3\pi/2$ to the real axis. (In this case, the root locus actually coincides with the rays.) For $\lambda \to -\infty$, the root loci asymptotically converge to two rays with angles 0 and π to the real axis, i.e., the real axis itself (see rule 7). Again, in this case the root locus coincides with the two rays.
6. All the asymptotes intersect at a point on the real line given by

$$\sigma = -\tfrac{1}{2}$$

7. Because $p(s)$ has two roots on the real line at $s = -1$ and $s = 0$, the whole real line must coincide with a root locus. Observe that to the right of the point $s = 0$, the corresponding value of λ is negative because the total number of poles and zeroes of $q(s)/p(s)$ to the right is even and equal to zero. Between -1 and 0, the value of λ is positive because the only pole or zero to the right is $s = 0$. To the left of -1 the value of λ is again negative because we have two poles to the right.
8. A multiple root of $f(s)$ corresponds to a branching point. For $\lambda = \frac{1}{8}$, one obtains

$$f(s) = (s + \tfrac{1}{2})^2$$

and hence $s = -\frac{1}{2}, \lambda = \frac{1}{8}$ correspond to a branching point. With

$$\lambda = -\frac{s^2 + s}{2}$$

one obtains

$$\left.\frac{d\lambda}{ds}\right|_{s=-\frac{1}{2}} = -s - \left.\frac{1}{2}\right|_{s=-\frac{1}{2}} = 0$$

For the same value of λ, two branches leave the point $s = -\frac{1}{2}$. The resulting root locus is shown in Fig. 45.2.

Although in this example only Hurwitz stability was considered, the method of root locus is directly applicable to test stability with respect to any region D.

In many situations explicit knowledge of the root location is not necessary and information on stability is often sufficient. In this case the task often can be formulated as a robust stability problem with one free parameter. If the relevant range of λ is known *a priori* in the form of an interval, stability of an edge polynomial which corresponds to a line in the coefficient space needs to be checked. This problem is investigated in the following section.

45.3 Stability of Polynomials with One Free Parameter

In this section the following problem is considered:
Determine the stability of the following set of polynomials

$$f(s) = p(s) + \lambda q(s), \quad \lambda \in [\underline{\lambda}, \overline{\lambda}] \tag{45.13}$$

The above task occurs not only in determining the stability of feedback systems with variable gain, but it recently became of vital importance for checking stability of uncertain polynomials in general. This is due to the edge theorem [1], which states that all roots of a polytope of polynomials are inside a simply connected region D in the complex plane, if and only if (iff) all the roots of the exposed edges of the polytope are in D (see Theorem 49.2 in Chapter 49).

Equation (45.13) describes an edge polynomial in which the two extreme polynomials $\underline{f}(s) = p(s) + \underline{\lambda} q(s)$ and $\overline{f}(s) = p(s) + \overline{\lambda} q(s)$ correspond to the two end points (vertices) of the edge. Two approaches can be readily applied to this problem and result in necessary and sufficient conditions.

The first method [2] is based on the Hurwitz matrices for $f(s)$. The only requirement of this approach is to test the Hurwitz stability of one extreme polynomial and nonpositive realness of the eigenvalues of a matrix which is constructed from the Hurwitz matrices of $\underline{f}(s)$ and $\overline{f}(s)$. This condition is explicitly stated as Theorem 49.3 in this handbook. It also has a discrete counterpart which can be found in Theorem 49.5.

The second method requires us to find zeroes of a polynomial in ω^2 and is also applicable to the continuous and discrete time cases. First, the continuous time case is addressed [9].

Theorem 45.1: *Assume*

$$\underline{f}(s) = p(s) + \underline{\lambda} q(s) \tag{45.14}$$

and

$$\overline{f}(s) = p(s) + \overline{\lambda} q(s) \tag{45.15}$$

are Hurwitz stable. Then the edge polynomial

$$f(s) = p(s) + \lambda q(s), \quad \lambda \in [\underline{\lambda}, \overline{\lambda}] \tag{45.16}$$

is Hurwitz stable iff

$$\frac{p(j\omega^*) + \overline{\lambda} q(j\omega^*)}{p(j\omega^*) + \underline{\lambda} q(j\omega^*)} > 0 \tag{45.17}$$

for every solution $\omega = \omega^*$ *of*

$$\underline{h}(-\omega^2)\overline{g}(-\omega^2) - \overline{h}(-\omega^2)\underline{g}(-\omega^2) = 0 \tag{45.18}$$

where

$$\underline{f}(s) = \underline{h}(s^2) + s\underline{g}(s^2) \tag{45.19}$$

$$\overline{f}(s) = \overline{h}(s^2) + s\overline{g}(s^2) \tag{45.20}$$

□

This method still requires a computer-aided approach because the real zeroes of a polynomial of arbitrary degree in ω^2 must be determined.

Theorem 45.1 has the following discrete time counterpart [7].

Theorem 45.2: *Assume*

$$\underline{f}(z) = p(z) + \underline{\lambda}q(z) \tag{45.21}$$

and

$$\overline{f}(z) = p(z) + \overline{\lambda}q(z) \tag{45.22}$$

are Schur stable. Then the edge polynomial

$$f(z) = p(z) + \lambda q(z), \quad \lambda \in [\underline{\lambda}, \overline{\lambda}] \tag{45.23}$$

is the Schur stable iff

$$\frac{p(z^*) + \overline{\lambda}q(z^*)}{p(z^*) + \underline{\lambda}q(z^*)} > 0 \tag{45.24}$$

for every $z = z^*$ *solving*

$$\begin{aligned} z^n[(p(z^{-1}) + \underline{\lambda}q(z^{-1}))(p(z) + \overline{\lambda}q(z)) \\ -(p(z) + \underline{\lambda}q(z))(p(z^{-1}) + \overline{\lambda}q(z^{-1}))] = 0 \end{aligned} \tag{45.25}$$

where $|z^*| = 1$. □

Theorems 45.1, 45.2, and 49.3 (Chapter 49) provide necessary and sufficient conditions for the stability of a general polynomial edge. For special cases, it is possible to formulate some simple conditions. These conditions are all based on Kharitonov's results [Kharitonov, 1979] and some discrete counterparts obtained by Hollot and Bartlett [4].

Theorem 45.3: *The edge polynomial*

$$f(s) = p(s) + \lambda q(s), \quad \lambda \in [\underline{\lambda}, \overline{\lambda}] \tag{45.26}$$

where $q(s) = as^i$ *is Hurwitz stable iff the two extreme polynomials*

$$\underline{f}(s) = p(s) + \underline{\lambda}q(s)$$

and

$$\overline{f}(s) = p(s) + \overline{\lambda}q(s)$$

are stable. □

Theorem 45.3 essentially states that edges parallel to the coefficient axis in the parameter space can be checked simply by testing the end points for stability. Using Kharitonov's theorem, it can be shown that the same statement, i.e., Theorem 45.3, is true for polynomials of the form

$$f(s) = p(s) + \lambda q(s), \quad \lambda \in [\underline{\lambda}, \overline{\lambda}]$$

and

$$q(s) = \sum_{\nu=1}^{\nu_{max}} a_{4\nu+\nu_0} s^{4\nu+\nu_0}, \quad \nu_0 + 4\nu_{max} < \text{degree } (p(s)), \quad \nu_0 = 0, 1, 2, 3 \tag{45.27}$$

Furthermore, if $q(s)$ is of the form

$$q(s) = \sum_{\nu=1}^{\nu_{max}} a_{2\nu+\nu_0} s^{2\nu+\nu_0}$$
$$\nu_0 + 2\nu_{max} < \text{degree } (p(s)), \quad \nu_0 = 0, 1. \tag{45.28}$$

only two extreme Kharitonov polynomials arise. Checking these two polynomials is a *sufficient* condition for stability of the edge.

In the most general case, in which $q(s)$ has no special properties, four Kharitonov polynomials can be obtained by treating the uncertain polynomial $f(s)$ as an interval polynomial. (This implies overbounding the uncertainty by a hyperrectangle, which contains the edge.) If the arising four Kharitonov polynomials are stable, the edge is stable.

The following is a result in which Hurwitz stability of one polynomial on the edge implies Hurwitz stability of the entire edge polynomial.

Theorem 45.4: (*a*) *The set of polynomials*

$$p_\lambda(s) = p^e(s) + \lambda p^0(s), \quad \lambda \in (0, \infty) \tag{45.29}$$

is Hurwitz stable iff

$$p_{\lambda_0}(s) = p^e(s) + \lambda_0 p^0(s) \tag{45.30}$$

is Hurwitz stable where λ_0 *is some positive real number.*
(*b*) *The set of polynomials*

$$f_\lambda(s) = \lambda f^e(s) + f^o(s), \quad \lambda \in (0, \infty) \tag{45.31}$$

is Hurwitz stable iff

$$f_{\lambda_0}(s) = \lambda_0 f^e(s) + f^o(s) \tag{45.32}$$

is Hurwitz stable, where λ_0 *is some positive real number.* □

For discrete time polynomials the available results are not as powerful as for the continuous time case. The following theorem is the discrete time counterpart of Theorem 45.3.

Theorem 45.5: *The edge polynomial of degree* k_p

$$f(z) = p(z) + \lambda q(z), \quad \lambda \in [\underline{\lambda}_i, \overline{\lambda}_i] \tag{45.33}$$

where

$$q(z) = az^i, \quad i \leq \begin{cases} \dfrac{k_p}{2} & \text{for } k_p \text{ even} \\ \dfrac{k_p - 1}{2} & \text{for } k_p \text{ odd} \end{cases} \tag{45.34}$$

is Schur stable iff the two extreme polynomials

$$\underline{f}(z) = p(\lambda) + \underline{\lambda} q(z)$$
$$\overline{f}(z) = p(z) + \overline{\lambda} q(z)$$

are Schur stable (k_p is the degree of $p(z)$). □

The above result follows directly from the work of Hollot and Bartlett [4]. Unfortunately, this result is restricted to edges, created by coefficients which belong to the lower powers of z.

A simple, sufficient condition can be obtained from Hollot and Bartlett's result for general polynomials $\tilde{q}(z)$, which satisfy

$$\text{degree } (\tilde{q}(z)) \leq \frac{k_p}{2} \quad \text{for } k_p \text{ even,} \tag{45.35}$$

$$\text{degree } (\tilde{q}(z)) \leq \frac{k_p - 1}{2} \quad \text{for } k_p \text{ odd,} \tag{45.36}$$

where k_p is the degree of $f(z)$. By viewing the edge polynomial $f(z)$ as an interval polynomial in which all coefficients of $\tilde{q}(z)$ are uncertain and independent of each other, a sufficient condition for Schur stability of the edge is given by Schur stability of all $2^{\text{degree}(q(z))+1}$ vertex polynomials. (This again implies overbounding of the uncertainty by a hyperrectangle, which contains the edge.)

For results on Schur, Hurwitz, or general D-stability of polynomials with more than one uncertain parameter, the reader is referred to Chapter 49.

References

[1] A. C. Bartlett, C. V. Hollot, and H. Lin, "Root locations of an entire polytope of polynomials: it suffices to check the edges," *Math. Control, Signals Syst., 1*, pp. 61–71, 1988.

[2] S. Bialas, "A necessary and sufficient condition for the stability of convex combinations of stable polynomials and matrices," *Bull. Pol. Acad. Sci., Tech. Sci.*, 33, pp. 473–480, 1985.

[3] H. Chapellat, M. Mansour, and S. P. Bhattacharyya, "Elementary proofs of some classical stability criteria," *IEEE Trans. Educ.*, vol. 33, no. 3, pp. 232–239, Aug. 1990.

[4] C. V. Hollot and A. C. Bartlett, "Some discrete-time counterparts to Kharitonov's stability criterion for uncertain systems," *IEEE Trans. Autom. Cont.*, vol. 31, pp. 355, 356, 1986.

[5] V. L. Kharitonov, "Asymptotic stability of an equilibrium position of a family of systems of linear differential equations," Differential Equations, vol. 14, pp. 1483–1485, 1979.

[6] A. M. Krall, *Stability Techniques for Continuous Linear Systems*, New York: Gordon & Breach, 1967.

[7] F. J. Kraus, M. Mansour, and E. I. Jury, "Robust Scur-stability of interval polynomials," *Proc. 28th IEEE Conf. Decision Control*, pp. 1908–1910, 1983.

[8] B. Porter, *Stability Criteria for Linear Dynamical Systems*, New York: Academic Press, 1968.

[9] E. Zeheb, "Necessary and sufficient conditions for root clustering of a polytope of polynomials in a simply connected domain," *IEEE Trans. Autom. Contr.*, vol. 34, pp. 986–990, 1989.

46

The Nyquist Criterion*

Charles E. Rohrs
Tellabs Research Center
University of Notre Dame

46.1 Development of the Nyquist Theorem

The Nyquist criterion is a graphical method and deals with the loop gain transfer function, i.e., the open-loop transfer function. The graphical character of the Nyquist criterion is probably one of its most appealing features.

Consider the controller configuration shown in Fig. 46.1. The loop gain transfer function is given simply by $G(s)$. The closed-loop transfer function is given by

$$\frac{Y(s)}{R(s)} = \frac{G(s)}{1+G(s)} = \frac{K_G N_G(s)/D_G(s)}{1+K_G N_G(s)/D_G(s)}$$
$$= \frac{K_G N_G(s)}{D_G(s)+K_G N_G(s)} = \frac{K_G N_G(s)}{D_k(s)}$$

where $N_G(s)$ and $D_G(s)$ are the numerator and denominator of $G(s)$, N, K_G is a constant gain, and $D_k(s)$ is the denominator of the closed-loop transfer function. The closed-loop poles are equal to the zeroes of the function

$$1+G(s) = 1+\frac{K_G N_G(s)}{D_G(s)} = \frac{D_G(s)+K_G N_G(s)}{D_G(s)} \tag{46.1}$$

Of course, the numerator of (46.1) is just the closed-loop denominator polynomial, $D_k(s)$, so that

$$1+G(s) = \frac{D_k(s)}{D_G(s)} \tag{46.2}$$

In other words, we can determine the stability of the closed-loop system by locating the zeroes of $1+G(s)$. This result is of prime importance in the following development.

*Much of the material of this chapter is taken from Rohrs, C. E., J. L. Melsa and D. G. Schultz, *Linear Control Systems*, McGraw-Hill, 1993. It is used with permission.

0-8493-8341-2/95/$0.00+$.50

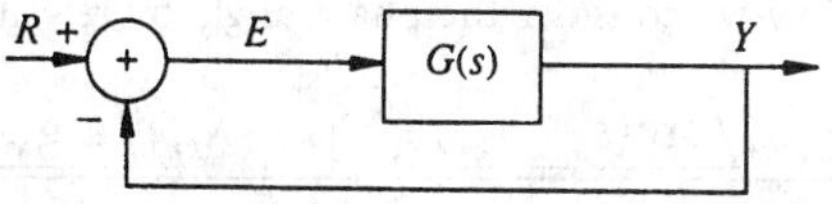

FIGURE 46.1 A control loop showing the loop gain $G(s)$.

For the moment, let us assume that $1 + G(s)$ is known in factored form so that we have

$$1 + G(s) = \frac{(s + \lambda_{k1})(s + \lambda_{k2}) \cdots (s + \lambda_{kn})}{(s + \lambda_1)(s + \lambda_2) \cdots (s + \lambda_n)} \tag{46.3}$$

Obviously, if $1 + G(s)$ were known in factored form, there would be no need for the use of the Nyquist criterion because we could simply observe whether any of the zeroes of $1 + G(s)$, poles of $Y(s)/R(s)$, lie in the right half of the s-plane. In fact, the primary reason for using the Nyquist criterion is to avoid this factoring. Although it is convenient to think of $1 + G(s)$ in factored form at this time, no actual use is made of that form.

Let us suppose that the pole–zero plot of $1 + G(s)$ takes the form shown in Fig. 46.2(a). Consider next an arbitrary closed contour, such as that labeled Γ in Fig. 46.2(a), which encloses one and only one zero of $1 + G(s)$ and none of the poles. Associated with each point on this contour is a value of the complex function $1 + G(s)$. The value of $1 + G(s)$ for any value of s on Γ may be found analytically by substituting the appropriate complex value of s into the function. Alternatively, the value may be found graphically by multiplying the distances of s on Γ to the zeroes and dividing by the distances to the poles.

If the complex value of $1 + G(s)$ associated with every point on the contour Γ is plotted, another closed contour Γ' is created in the complex $(1 + G(s))$-plane, as shown in Fig. 46.2(b). The function $1 + G(s)$ is said to map the contour Γ in the s-plane into the Γ' contour in the $(1 + G(s))$-plane. What we wish to demonstrate is that if a zero is enclosed by the contour Γ, as in Fig. 46.2(a), the contour Γ' encircles the origin of the $(1 + G(s))$-plane in the same sense that the Γ contour encircles the zero in the s-plane. In the s-plane the zero is encircled in the clockwise direction; hence, we must show that the origin of the $(1 + G(s))$-plane is also encircled in the clockwise direction. This result is known as the **principle of the argument.**

The key to the principle of the argument rests in considering the value of the function $1 + G(s)$ at any point s as simply a complex number. This complex number has a magnitude and a phase angle. Because the contour Γ in the s-plane does not pass through a zero, the

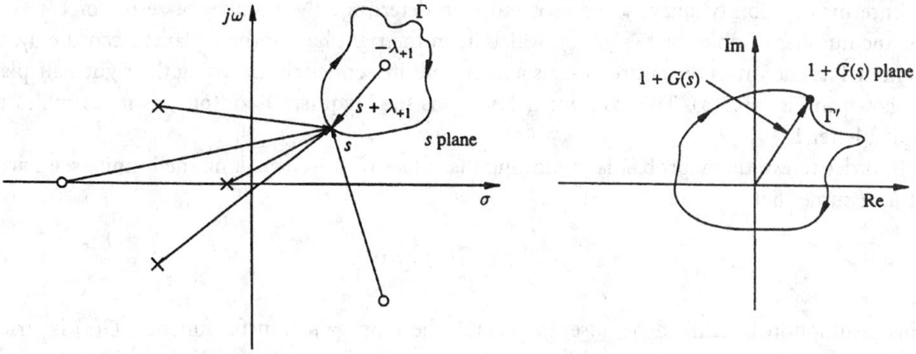

(a) (b)

FIGURE 46.2 (a) Pole–zero plot of $1 + G(s)$ in the s-plane; (b) plot of the Γ' contour in the $(1 + G(s))$-plane.

magnitude is never zero. Now we consider the phase angle by rewriting (46.3) in polar form:

$$1+G(s) = \frac{|s+\lambda_{k1}\underline{/\arg(s+\lambda_{k1})} \cdots |s+\lambda_{kn}|\underline{/\arg(s+\lambda_{kn})}}{|s+\lambda_1|\underline{/\arg(s+\lambda_1)} \cdots |s+\lambda_n|\underline{/\arg(s+\lambda_n)}}$$
$$= \frac{|s+\lambda_{k1}|\cdots|s+\lambda_{kn}|}{|s+\lambda_1|\cdots|s+\lambda_n|}\underline{/\arg(s+\lambda_{k1})+\cdots}$$
$$\underline{/+\arg(s+\lambda_{kn})-\arg(s+\lambda_1)-\cdots-\arg(s+\lambda_n)} \tag{46.4}$$

We assume that the zero encircled by Γ is at $s=-\lambda_{k1}$. Then the phase angle associated with this zero changes by a full $-360°$ as the contour Γ is traversed clockwise in the s-plane. Because the argument or angle of $1+G(s)$ includes the angle of this zero, the argument of $1+G(s)$ also changes by $-360°$. As seen from Fig. 46.2(a), the angles associated with the remaining poles and zeroes make no net change as the contour Γ is traversed. For any fixed value of s, the vector associated with each of these other poles and zeroes has a particular angle associated with it. Once the contour has been traversed back to the starting point, these angles return to their original value; they have not been altered by $\pm 360°$ simply because these poles and zeroes are not enclosed by Γ.

In a similar fashion, we could show that if the Γ contour were to encircle two zeroes of $1+G(s)$ in the clockwise direction on the s-plane, the Γ' contour would encircle the origin of the $(1+G(s))$-plane twice in the clockwise direction. On the other hand, if the Γ contour were to encircle only one pole and no zero of $1+G(s)$ in the *clockwise* direction, then the contour Γ' would encircle the origin of the $(1+G(s))$-plane once in the counterclockwise direction. This change in direction comes about because angles associated with poles are accompanied by negative signs in the evaluation of $1+G(s)$, as indicated by (46.4). In general, the following conclusion can be drawn. **The net number of clockwise encirclements by Θ' of the origin in the $(1+G(s))$-plane is equal to the difference between the number of zeroes n_z and the number of poles n_p of $1+G(s)$ encircled in the clockwise direction by Θ.**

The above result means that the difference between the number of zeroes and the number of poles enclosed by any closed contour Γ may be determined simply by counting the net number of clockwise encirclements of the origin of the $(1+G(s))$-plane by Γ'. For example, if we find that Γ' encircles the origin three times in the clockwise direction and once in the counterclockwise direction, then n_z-n_p must be equal to $3-1=2$. Therefore, in the s-plane, Γ must encircle two zeroes and no poles, three zeroes and one pole, or any other combination such that n_z-n_p is equal to 2.

In terms of stability analysis, the problem is to determine the number of zeroes of $1+G(s)$, i.e., the number of poles of $Y(s)/R(s)$, which lie in the right half of the s-plane. Accordingly, the contour Γ is chosen as the entire $j\omega$-axis and an infinite semicircle enclosing the right-half-plane as shown in Fig. 46.3(a). This contour is known as the **Nyquist D-contour**, as it resembles the capital letter D.

In order to avoid any problems in plotting the values of $1+G(s)$ along the infinite semicircle, let us assume that

$$\lim_{|s|\to\infty} G(s) = 0$$

This assumption is justified because, in general, the loop gain transfer function $G(s)$ is strictly proper. With this assumption, the entire infinite semicircle portion of Γ maps into the single point $s=+1+j0$ on the $(1+G(s))$-plane.

The mapping of Γ therefore involves simply plotting the complex values of $1+G(s)$ for $s=j\omega$ as ω varies from $-\infty$ to $+\infty$. For $\omega \geq 0$, Γ' is nothing more than the polar plot of the frequency response of the function $1 + G(s)$. The values of $1 + G(j\omega)$ for negative values of ω are the mirror image of the values of $1 + G(j\omega)$ for positive values of ω reflected about the real axis. The Γ' contour, therefore, may be found by plotting the frequency response $1 + G(s)$ for positive ω and then reflecting this plot about the real axis to find the plot for negative ω. The Γ' plot is always symmetrical about the real axis of the $(1 + G)$-plane. Care must be taken to establish the direction that the Γ' plot is traced as the D-contour moves up the $j\omega$-axis, around the infinite semicircle, and back up the $j\omega$-axis from $-\infty$ toward 0.

From the Γ' contour in the $(1 + G(s))$-plane, as shown in Fig. 46.3(b), the number of zeroes of $1 + G(s)$ in the right half of the s-plane may be determined by the following procedure. The net number of clockwise encirclements of the origin by Γ' is equal to the number of zeroes minus the number of poles of $1 + G(s)$ in the right half of the s-plane. Note that we must know the number of poles of $1 + G(s)$ in the right-half-plane if we are to ascertain the exact number of zeroes in the right-half-plane, and therefore determine stability. This requirement usually poses no problem because the poles of $1 + G(s)$ correspond to the poles of the loop gain transfer function. In (46.2) the denominator of $1 + G(s)$ is $D_G(s)$, which is usually described in factored form. Hence, the number of zeroes of $1 + G(s)$ or the number of poles of $Y(s)/R(s)$ in the right-half-plane may be found by determining the net number of clockwise encirclements of the origin by Γ' and then adding the number of poles of the loop gain located in the right-half s-plane.

At this point the reader may revolt. Our plan for finding the number of poles of $Y(s)/R(s)$ in the right-half s-plane involves counting encirclements in the $(1 + G(s))$-plane and observing the number of loop gain poles in the right-half s-plane. Yet we were forced to start with the assumption that all the poles and zeroes of $1 + G(s)$ are known, so that the Nyquist contour can be mapped by the function of $1 + G(s)$. Admittedly, we know the poles of this function because they are the poles of the loop gain, but we do not know the zeroes; in fact, we are simply trying to find how many of these zeroes lie in the right-half s-plane.

What we do know are the poles and zeroes of the loop gain transfer function $G(s)$. Of course, this function differs from $1 + G(s)$ only by unity. Any contour that is chosen in the s-plane and mapped through the function $G(s)$ has exactly the same shape as if the contour were mapped through the function $1 + G(s)$, except that it is displaced by one unit. Figure 46.4 is typical of such a situation. In this diagram the -1 point of the $G(s)$-plane is the origin of the $(1 + G(s))$-plane. If we now map the boundary of the right-half s-plane, through the mapping

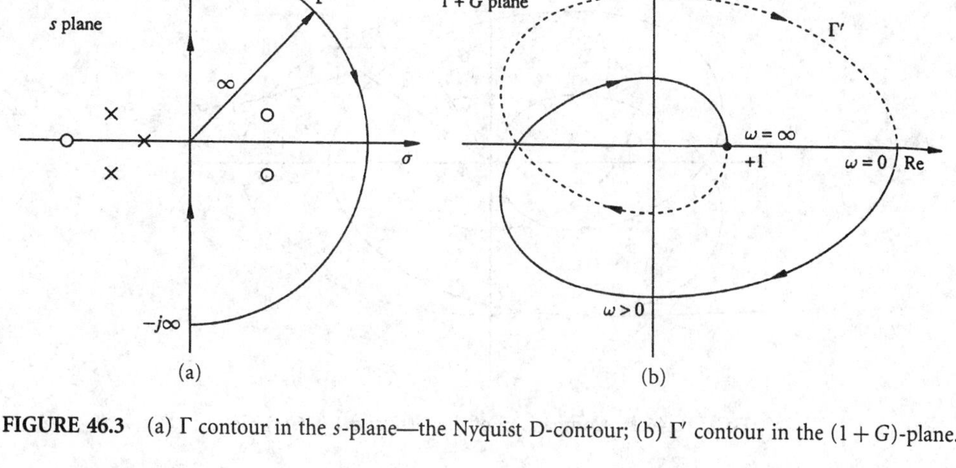

FIGURE 46.3 (a) Γ contour in the s-plane—the Nyquist D-contour; (b) Γ' contour in the $(1 + G)$-plane.

function $G(s)$, which we often know in pole–zero form, information concerning the zeroes of $1 + G(s)$ may be obtained by counting the encirclements of the -1 point. The important point is that by plotting the open-loop frequency-response information, we may reach stability conclusions regarding the closed-loop system.

As mentioned previously, contour Γ of Fig. 43.3a is referred to as the Nyquist D-contour. The map of the Nyquist D-contour through $G(s)$ is called the *Nyquist diagram* of $G(s)$. There are three parts to the Nyquist diagram. The first part is the polar plot of the frequency response of $G(s)$ from $\omega = 0$ to $\omega = \infty$. The second part is the mapping of the infinite semicircle around the right-half-plane. If $G(s)$ is strictly proper, this part maps entirely into the origin of the $G(s)$-plane. The third part is the polar plot of the negative frequencies, $\omega = -\infty$ to $\omega = 0$. The map of these frequencies forms a mirror image in the $G(s)$-plane about the real axis from the first part.

In terms of the Nyquist diagram of $G(s)$, the Nyquist stability criterion may be stated as follows:

The Nyquist Theorem The closed-loop system is stable if and only if (iff) the net number of clockwise encirclements of the points $s = -1 + j0$ by the Nyquist diagram of $G(s)$ plus the number of poles of $G(s)$ in the right-half-plane is zero.

Notice that while the net number of clockwise encirclements is counted in the first part of the Nyquist criterion only the number of right half-plane poles of $G(s)$ is counted in the second part. Right-half-plane zeroes of $G(s)$ are not part of the formula in determining stability using the Nyquist criterion.

Because the Nyquist diagram involves the loop gain transfer function $G(s)$, a good approximation of the magnitude and phase of the frequency response plot can be obtained by using the Bode diagram straight-line approximations for the magnitude and for the phase. The Nyquist plot can then be obtained by transferring the magnitude and phase information to a polar plot. If a more accurate plot is needed, the exact magnitude and phase may be determined for a few values of ω in the range of interest. However, in most cases, the approximate plot is accurate enough for practical problems.

An alternative procedure for obtaining the Nyquist diagram is to plot accurately the poles and zeroes of $G(s)$ and obtain the magnitude and phase by graphical means. In either of these methods, the fact that $G(s)$ is known in factored form is important. Even if $G(s)$ is not known in factored form, the frequency-response plot can still be obtained by simply substituting the values $s = j\omega$ into $G(s)$ or by frequency-response measurements on the actual system.

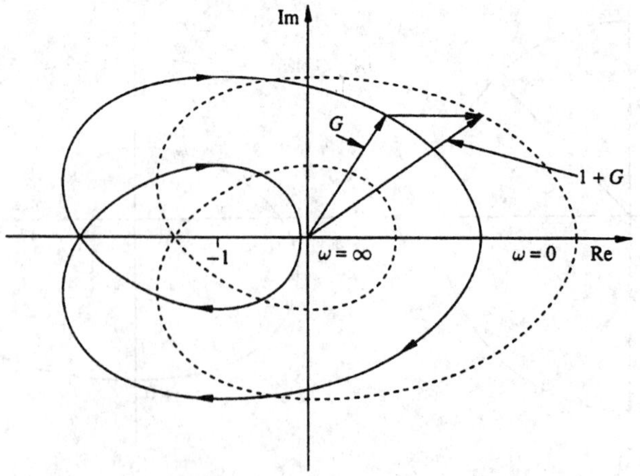

FIGURE 46.4 Comparison of the $G(s)$ and $1 + G(s)$ plots.

Of course, computer programs that produce Nyquist plots are generally available. However, the ability to plot Nyquist plots by hand helps designers know how they can affect such plots by adjusting compensators.

46.2 Examples of the Nyquist Theorem[1]

Example 46.1. To illustrate the use of the Nyquist criterion, let us consider the simple first-order system shown in Fig. 46.5(a). For this system the loop gain transfer function takes the following form:

$$G(s) = KG_P(s) = \frac{K}{s+10} = \frac{50}{s+10}$$

The magnitude and phase plots of the frequency response of $KG_P(s)$ are shown. From these plots the Nyquist diagram of $KG_P(s)$ may be easily plotted, as shown in Fig. 46.6. For example, the point associated with $\omega = 10$ rad/s is found to have a magnitude of $K/(10\sqrt{2})$ and a phase angle of $-45°$. The point at $\omega = -10$ rad/s is the mirror image of the value at $\omega = 10$ rad/s.

From Fig. 46.6 we see that the Nyquist diagram can never encircle the $s = -1 + j0$ point for any positive value of K, and therefore the closed-loop system is stable for all positive values of K. In this simple example it is easy to see that this result is correct because the closed-loop transfer function is given by

$$\frac{Y(s)}{R(s)} = \frac{K}{s+10+K}$$

For all positive values of K, the pole of $Y(s)/R(s)$ is in the left-half-plane.

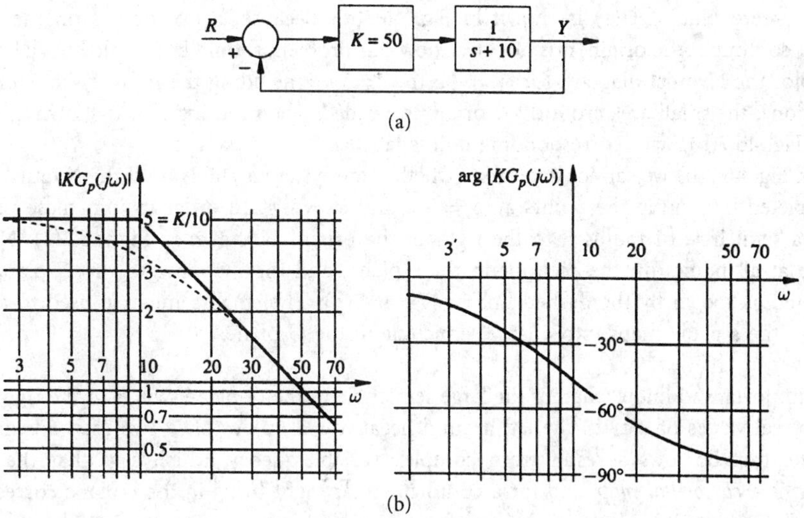

FIGURE 46.5 Simple first-order example. (a) Block diagram; (b) magnitude and phase plots.

[1]Throughout this section we assume that the gain $k > 0$. If $k < 0$, all the theory holds with the critical point shifted to $+1 + j0$.

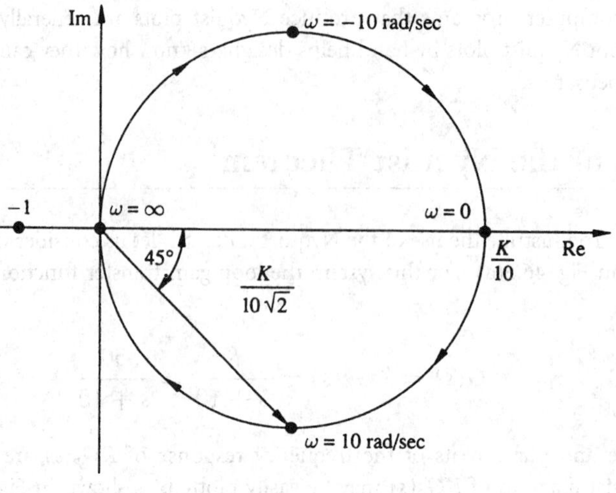

FIGURE 46.6 Nyquist diagram for Example 46.1.

In the above example, $G(s)$ remains finite along the entire Nyquist contour. This is not always the case, even though we have assumed that $G(s)$ approaches zero as $|s|$ approaches infinity. If a pole of $G(s)$ occurs on the $j\omega$-axis, as often happens at the origin because of an integrator in the plant, a slight modification of the Nyquist contour is necessary. The method of handling the modification is illustrated in the following example.

Example 46.2. Consider a system whose loop transfer function is given by

$$G(s) = \frac{(2K/7)[(s+3/2)^2 + (\sqrt{5}/2)^2]}{s(s+2)(s+3)}$$

The pole–zero plot of $G(s)$ is shown in Fig. 46.7(a). Because a pole occurs on the standard Nyquist contour at the origin, it is not clear how this problem should be handled. As a beginning, let us plot the Nyquist diagram for $\omega = +\varepsilon$ to $\omega = -\varepsilon$, including the infinite semicircle; when this is done, the small area around the origin is avoided. The resulting plot is shown as the solid line in Fig. 46.7(b), with corresponding points labeled.

From Fig. 46.7(b) we cannot determine whether the system is stable until the Nyquist diagram is completed by joining the points at $\omega = -\varepsilon$ and $\omega = +\varepsilon$. In order to join these points, let us use a semicircle of radius ε to the right of the origin, as shown in Fig. 46.7(a). Now $G(s)$ is finite at all points on the contour in the s-plane, and the mapping to the G-plane can be completed as shown by the dashed line in Fig. 46.7(b). The small semicircle used to avoid the origin in the s-plane maps into a large semicircle in the G-plane.

It is important to know whether the large semicircle in the G-plane swings to the right around positive real values of s or to the left around negative real values of s. Two methods determine this. The first borrows a result from complex variable theory, which says that the Nyquist diagram is a *conformal map*, and for a conformal map right turns in the s-plane correspond to right turns in the $G(s)$-plane. Likewise, left turns in the s-plane correspond to the left turns in the $G(s)$-plane.

The second method of determining the direction of the large enclosing circle on the $G(s)$-plane comes from a graphical evaluation of $G(s)$ on the circle of radius ε in the s-plane. The magnitude is very large here due to the proximity of the pole. The phase at $s = -\varepsilon$ is slightly larger than $+90°$, as seen from the solid line of the Nyquist plot. The phase contribution from

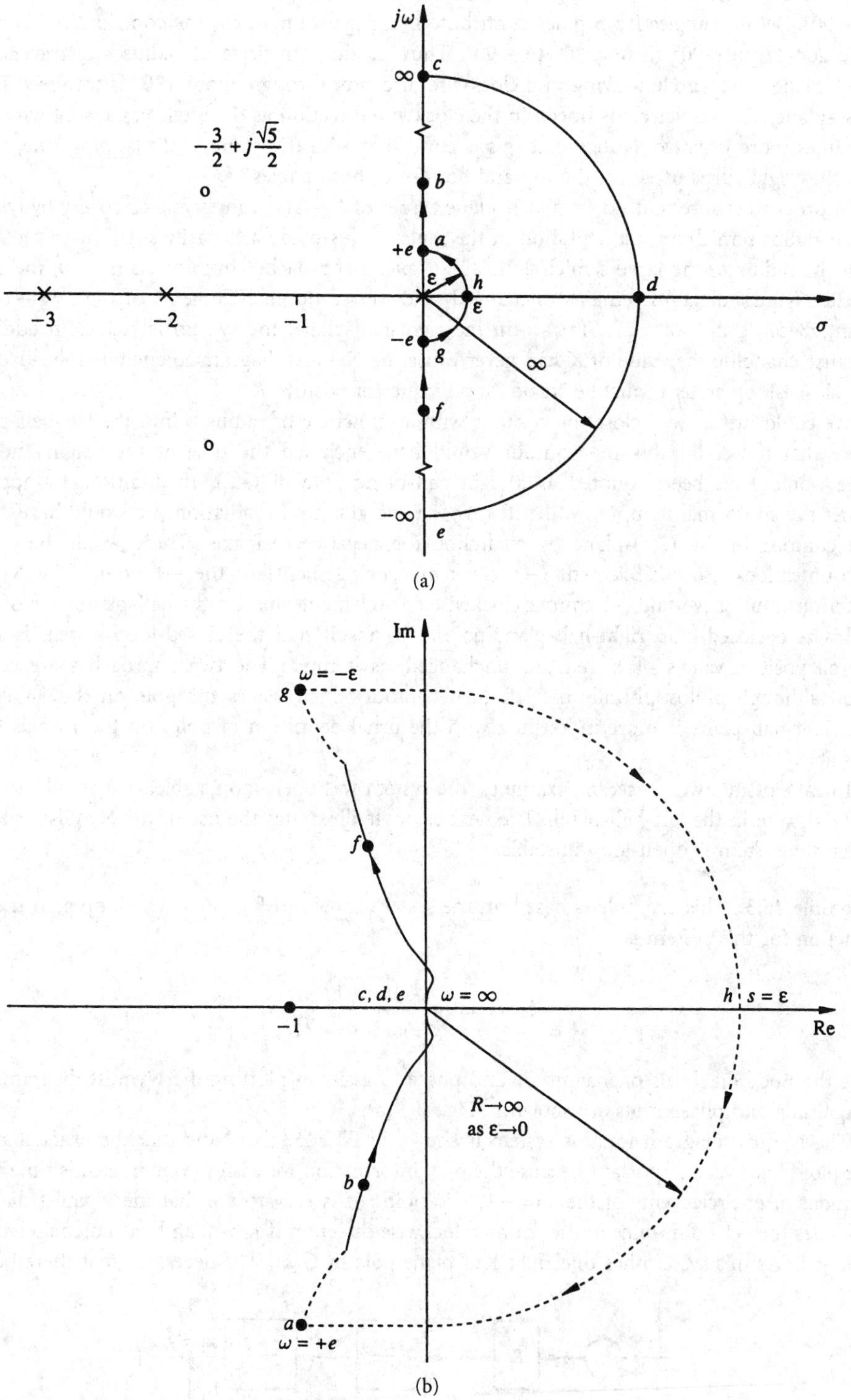

FIGURE 46.7 Example 46.2. (a) Pole–zero plot; (b) Nyquist diagram.

all poles and zeroes, except the pole at the origin, does not change appreciably as the circle of the radius ε is traversed. The angle from the pole at the origin changes from $-90°$ through $0°$ to $+90°$. Because angles from poles contribute in a negative manner, the contribution from the pole goes from $+90°$ through $0°$ to $-90°$. Thus, as the semicircle of radius ε is traversed in the s-plane a semicircle moving in a clockwise direction through about 180° is traversed in the $G(s)$-plane. The semicircle is traced in the clockwise direction as the angle associated with $G(s)$ becomes more negative. Notice that this is consistent with the conformal mapping rule, which matches right turns of 90° at the top and bottom of both circles.

In order to ensure that no right-half-plane zeroes of $1 + G(s)$ can escape discovery by lying in the ε-radius semicircular indentation in the s-plane, ε is made arbitrarily small, with the result that the radius of the large semicircle in the G-plane approaches infinity. As $\varepsilon \to 0$, the shape of the Nyquist diagram remains unchanged, and we see no encirclements of the $s = -1 + j0$ point. Because no poles of $G(s)$ are in the right-half-plane, the system is stable. In addition, because changing the value of K can never cause the Nyquist diagram to encircle the -1 point, the closed-loop system must be stable for all values of positive K.

We could just as well close the contour with a semicircle of radius ε into the left-half-plane. Note that if we did this the contour would have encircled the pole at the origin and this pole would have been counted as a right-half-plane pole of $G(s)$. In addition, by applying either the conformal mapping with left turns or the graphical evaluation, we would have closed the contour in the $G(s)$-plane by encircling the negative real axis. There would have been 1 counterclockwise encirclement (-1 clockwise encirclement) of the -1 point. The Nyquist criterion would have said -1 counterclockwise encirclement plus 1 right-half-plane pole of $G(s)$ yields zero closed-loop right-half-plane poles. The result that the closed-loop system is stable for all positive values of K remains unchanged, as it must. The two approaches are equally good although philosophically the left turn contour which places the pole on the $j\omega$-axis in the right-half-plane is more in keeping with the usual definition of poles on the $j\omega$-axis being unstable.

In each of the two preceding examples, the system was open-loop stable; that is, all the poles of $G(s)$ were in the left-half-plane. The next example illustrates the use of the Nyquist criterion when the system is open-loop unstable.

Example 46.3. This example is based on the system shown in Fig. 46.8. The loop gain transfer function for this system is

$$G(s) = \frac{K(s+1)}{(s-1)(s+2)}$$

Use the Bode diagrams of magnitude and phase to assist in plotting the Nyquist diagram. The magnitude and phase plots are shown in Fig. 46.9(a).

The Nyquist diagram for this system is shown in Fig. 46.9(b). Note that the exact shape of the plot is not very important because the only information we wish to obtain at this time is the number of encirclements of the $s = -1 + j0$ point. It is easy to see that the Nyquist diagram encircles the -1 point once in the counterclockwise direction if $K > 2$ and has no encirclements if $K < 2$. As this system has one right-half-plane pole in $G(s)$, it is necessary that there be one

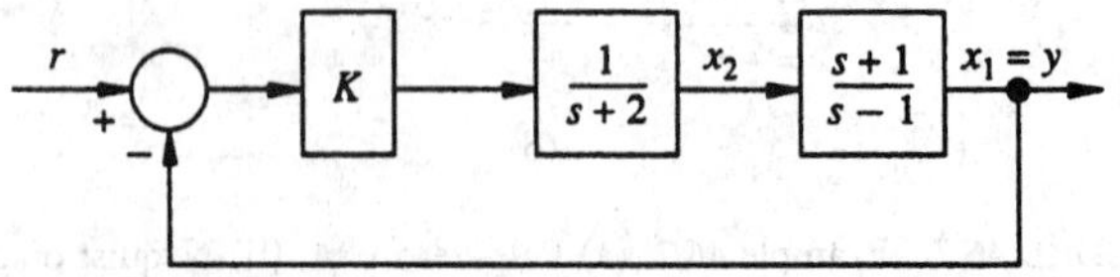

FIGURE 46.8 Example 46.3.

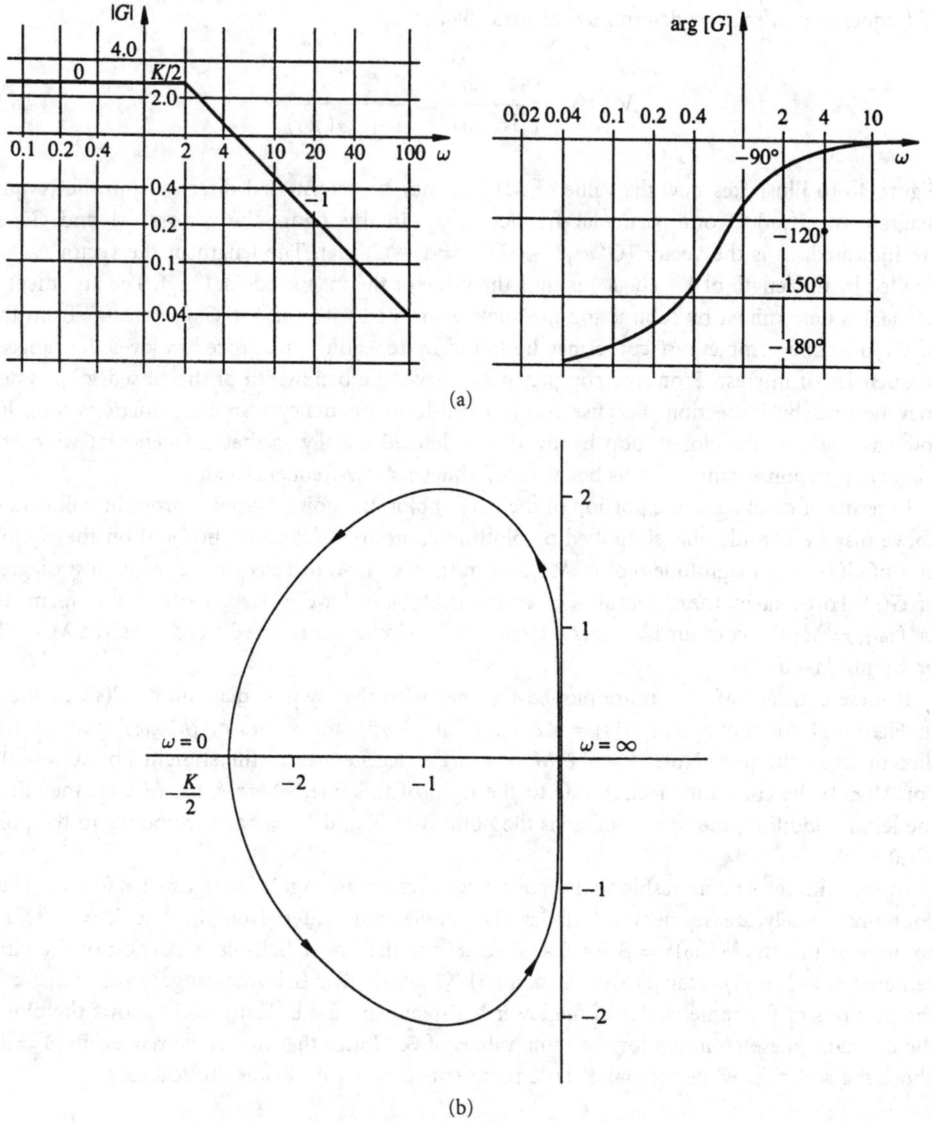

FIGURE 46.9 Example 46.3. (a) Magnitude and phase plots; (b) Nyquist diagram.

counterclockwise encirclement if the system is to be stable. Therefore, this system is stable iff $K > 2$.

46.3 Closed-Loop Response and Nyquist Diagrams

The Nyquist theorem can be used to determine if a closed-loop system is stable. However, many designs are possible which result in closed-loop systems that are stable, but have highly oscillatory and thus unsatisfactory responses to inputs and disturbances. Systems that are oscillatory are often said to be relatively less stable than systems that are more highly damped.

We must start with a system in the G configuration (Fig. 46.1). The key to extracting information about the closed-loop system is to determine the frequency response function of the closed-loop system, often referred to as the M curve. The M curve is, of course, a function of frequency and may be determined analytically as

$$M(j\omega) = \frac{Y(j\omega)}{R(j\omega)} = \frac{G(j\omega)}{1 + G(j\omega)} \tag{46.5}$$

Figure 46.10 illustrates how the value of $M(j\omega_1)$ may be determined directly from the Nyquist diagram of $G(j\omega)$ at one particular frequency, ω_1. In this figure the vectors -1 and $G(j\omega_1)$ are indicated, as is the vector $(G(j\omega_1) - (-1)) = 1 + G(j\omega_1)$. The length of the vector $G(j\omega_1)$ divided by the length of $1 + G(j\omega_1)$ is thus the value of the magnitude $M(j\omega_1)$. The argument of $M(j\omega_1)$ is determined by subtracting the angle associated with the $1 + G(j\omega_1)$ vector from that of $G(j\omega_1)$. The complete M curve may be found by repeating this procedure over the range of frequencies of interest. From the completed M curve, the bandwidth of the closed-loop system may be read by inspection. Because most closed-loop frequency response functions look like low-pass systems the closed-loop bandwidth is defined usually as that frequency at which the frequency response function falls below 0.707 times its low-frequency value.

In terms of the magnitude portion of the $M(j\omega)$ plot, the point-by-point procedure illustrated above may be considerably simplified by plotting contours of constant $|M(j\omega)|$ on the Nyquist plot of $G(s)$. The magnitude plot of $M(j\omega)$ can then be read directly from the Nyquist diagram of $G(s)$. Fortunately, these contours of constant $|M(j\omega)|$ have a particularly simple form. For $|M(j\omega)| = M$, the contour is simply a circle. These circles are referred to as constant M-circles or simply M-circles.

If these constant M-circles are plotted together with the Nyquist diagram of $G(s)$, as shown in Fig. 46.11 for the system $G(s) = 42/s(s + 2)(s + 15)$, the values of $|M(j\omega)|$ may be read directly from the plot. Note that the $M = 1$ circle degenerates to the straight line $X = -0.5$. For $M < 1$, the constant M-circles lie to the right of this line, whereas, for $M > 1$, they lie to the left. In addition, the $M = 0$ circle is the point $0 + j0$, and $M = \infty$ corresponds to the point $-1.0 + j0$.

In an entirely similar fashion, the contours of constant $\arg(M(j\omega))$ can be found. These contours actually are segments of circles. The circles are centered on the line $X = -\frac{1}{2}$. The contour of the $\arg(M(j\omega)) = \beta$ for $0 < \beta < 180°$ is the upper half-plane portion of the circle centered at $-\frac{1}{2} + j1/(2\tan\beta)$ with a radius $|1/(2\sin\beta)|$. For β in the range $-180° < \beta < 0°$, the portions of the same circles in the lower half-plane are used. Figure 46.12 shows the plot of the constant phase contours for the same values of β. Notice that one circle represents $\beta = 45°$ above the real axis, while the same circle represents $\beta = -135°$ below the real axis.

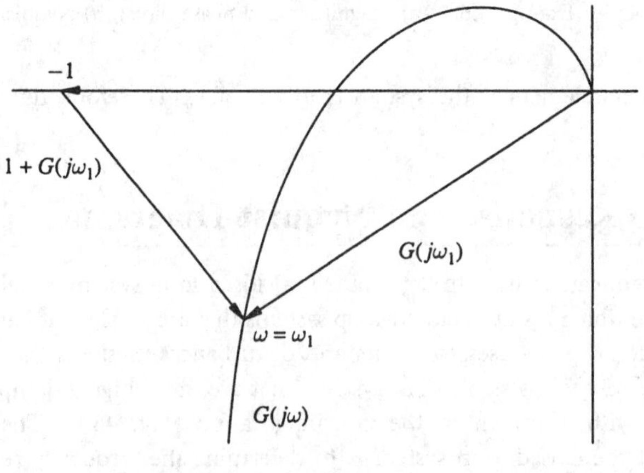

FIGURE 46.10 Graphical determination of $M(j\omega)$.

By using these constant magnitude and constant phase contours, it is possible to read directly the complete closed-loop frequency response from the Nyquist diagram of $G(s)$. In practice it is common to dispense with the constant phase contours because it is the magnitude of the closed-loop frequency response that provides the most information about the closed-loop system's transient response. In fact, it is common to simplify the labor further by considering only one point on the magnitude plot, namely, the point at which M is maximum. This point of peak magnitude is referred to as M_p, and the frequency at which the peak occurs is ω_p. The point M_p may be easily found by considering the contours of increasingly larger values of M until the contour is found that is just tangent to the plot of $G(s)$. The value associated with this contour is then M_p, and the frequency at which the M_p contour and $G(s)$ touch is ω_p. In the plot of $G(s)$ shown in Fig. 46.11, for example, the value of M_p is 1.1 at the frequency $\omega_p \approx 1.1$ rad/s.

One of the primary reasons for determining M_p and ω_p, in addition to the obvious saving of labor as compared to the determination of the complete frequency response, is the close correlation of these quantities with the behavior of the closed-loop system. In particular, for the simple second-order closed-loop system,

$$\frac{Y(s)}{R(s)} = \frac{\omega_n^2}{s^2 + 2s\zeta\omega_n + \omega_n^2} \tag{46.6}$$

the values of M_p and ω_p completely characterize the system. In other words, for this second-order system, M_p and ω_p specify the damping ratio, ζ, and the natural frequency, ω_n, the only parameters of the system. The following equations relate the maximum point of the frequency

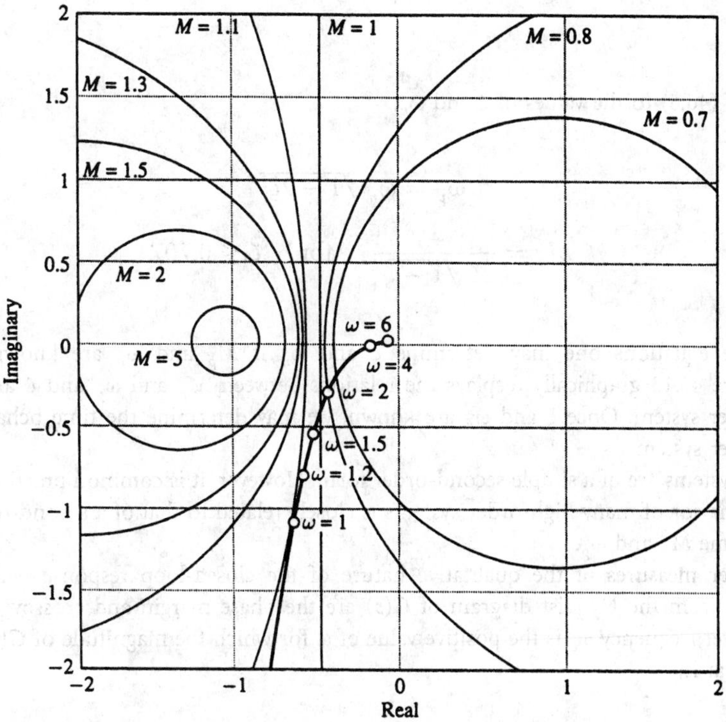

FIGURE 46.11 Constant M-contours.

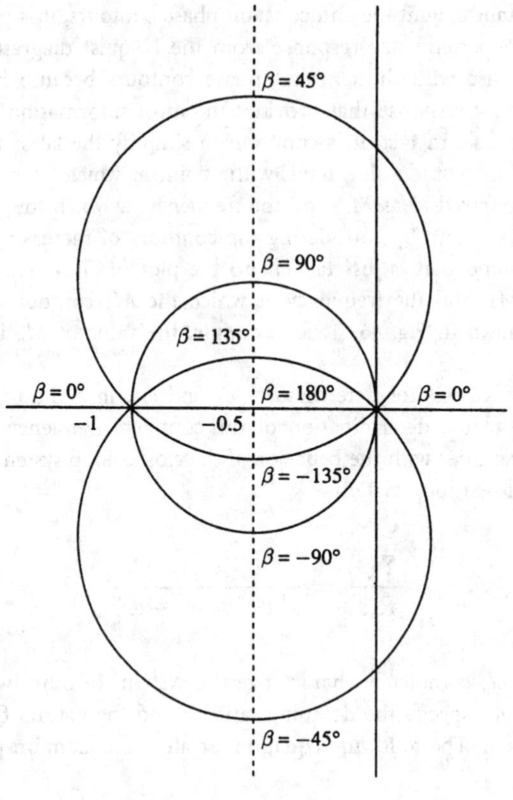

FIGURE 46.12 Constant phase contours.

response of (46.6) to the values of ζ and ω_n.

$$\omega_p = \omega_n\sqrt{1 - 2\zeta^2} \tag{46.7}$$

$$M_p = \frac{1}{2\zeta\sqrt{1 - \zeta^2}} \quad \text{for} \quad \zeta \leq 0.707 \tag{46.8}$$

From these equations one may determine ζ and ω_n if M_p and ω_p are known, and vice versa. Figure 46.13 graphically displays the relations between M_p and ω_p and ζ and ω_n for a second order system. Once ζ and ω_n are known, we may determine the time behavior of this second-order system.

Not all systems are of a simple second-order form. However, it is common practice to assume that the behavior of many high-order systems is closely related to that of a second-order system with the same M_p and ω_p.

Two other measures of the qualitative nature of the closed-loop response which may be determined from the Nyquist diagram of $G(s)$ are the phase margin and crossover frequency. The crossover frequency ω_c is the positive value of ω for which the magnitude of $G(j\omega)$ is equal to unity; that is,

$$|G(j\omega_c)| = 1 \tag{46.9}$$

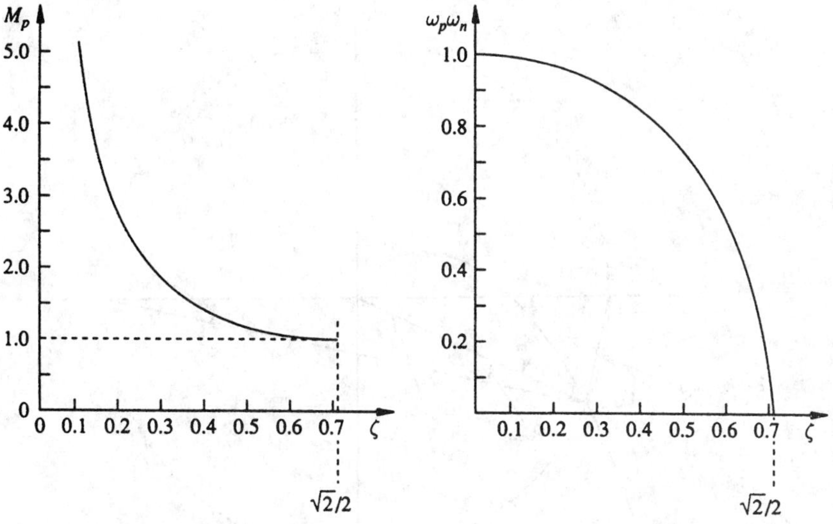

FIGURE 46.13 Plots of M_p and ω_p/ω_n vs. ζ for a simple second-order system.

The phase margin ϕ_m is defined as the difference between the argument of $G(j\omega_c)$ (evaluated at the crossover frequency) and $-180°$. In other words, if we define β_c as

$$\beta_c = \arg(G(j\omega_c)) \tag{46.10}$$

the phase margin is given by

$$\phi_m = \beta_c - (-180°) = 180° + \beta_c \tag{46.11}$$

While it is possible for a complicated system to possess more than one crossover frequency, most systems are designed to possess just one. The phase margin takes on a particularly simple and graphic meaning in the Nyquist diagram of $G(s)$. Consider, for example, the Nyquist diagram shown in Fig. 46.14. We see that the phase margin is simply the angle between the negative real axis and the vector $G(j\omega_c)$. The vector $G(j\omega_c)$ may be found by intersecting the $G(s)$ locus with the unit circle. The frequency associated with the point of intersection is ω_c.

It is possible to determine ϕ_c and ω_c more accurately directly from the Bode plots of the magnitude and phase of $G(s)$. The appropriate value of ω for which the magnitude crosses unity is ω_c. The phase margin is then determined by inspection from the phase plot by noting the difference between the phase shift at ω_c and $-180°$. Consider, for example, the Bode magnitude and phase plots shown in Fig. 46.15 and for the $G(s)$ function of Fig. 46.11. In time constant form this transfer function is

$$G(s) = \frac{1.4}{s(1 + s/2)(1 + s/15)}$$

From Fig. 46.15 we see that $\omega_c = 1.1$ and $\phi_m = 60°$.

The value of the magnitude of the closed-loop frequency response at ω_c can be derived from ϕ_m. We call this value M_c. Often the closest point to the -1 point on a Nyquist plot occurs at a frequency which is close to ω_c. This means that M_c is often a good approximation to M_p. A geometric construction shown in Fig. 46.14 shows that a right triangle exists with a hypotenuse

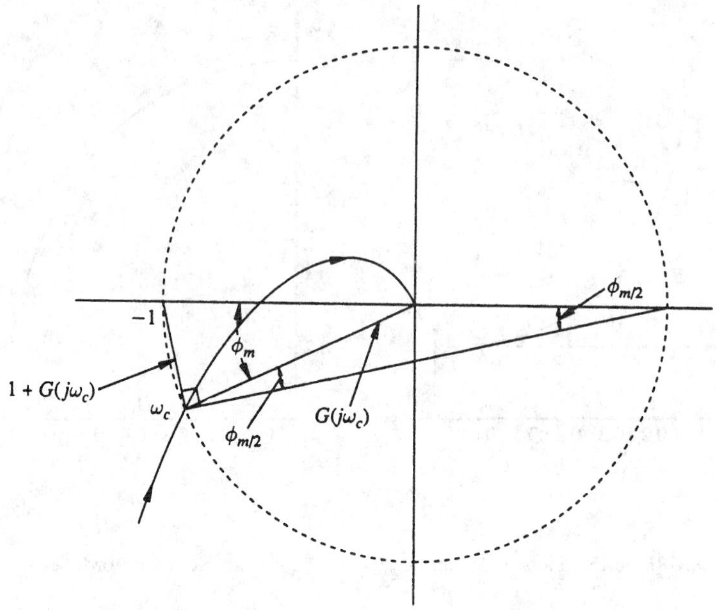

FIGURE 46.14 Definition of phase margin.

of 2, one side of length $|1 + G(j\omega_c)|$, and the opposite angle of $\phi_m/2$, where ϕ_m is the phase margin. From this construction, we see

$$\sin \phi_m/2 = \frac{|1 + G(j\omega_c)|}{2} \tag{46.12}$$

because at

$$\omega = \omega_c \quad |G(j\omega_c)| = 1 \tag{46.13}$$

$$M_c = \frac{|G(j\omega_c)|}{|1 + G(j\omega_c)|} = \frac{1}{2 \sin \phi_m/2} \tag{46.14}$$

Thus, an oscillatory characteristic in the closed-loop time response can be identified by a large peak in the closed-loop frequency response, which, in turn, can be identified by a small phase margin and the corresponding large value of M_c. Unfortunately, the correlation between response and phase margin is somewhat poorer than the correlation of closed-loop time response and the M peak. This lower reliability of the phase margin measure is a direct consequence of the fact that ϕ_m is determined by considering only one point, ω_c on the G plot, whereas M_p is found by examining the entire plot, to find the maximum M. Consider, for example, the two Nyquist diagrams shown in Fig. 46.16. The phase margin for these two diagrams is identical; however, it is obvious that the closed-loop step response resulting from closing the loop gain of Fig. 46.16(b) is far more oscillatory and underdamped than the closed-loop step response resulting from closing the loop gain of Fig. 46.16(a).

In other words, the relative ease of determining ϕ_m as compared to M_p has been obtained only by sacrificing some of the reliability of M_p. Fortunately, systems such as that of Fig. 43.16(b) are fairly rare, and the phase margin provides a simple and effective means of estimating the closed-loop response from the $G(j\omega)$ plot.

A system such as that shown in Fig. 46.16(b) could be identified as a system having a fairly large M_p by checking another parameter, the gain margin. The gain margin is easily determined

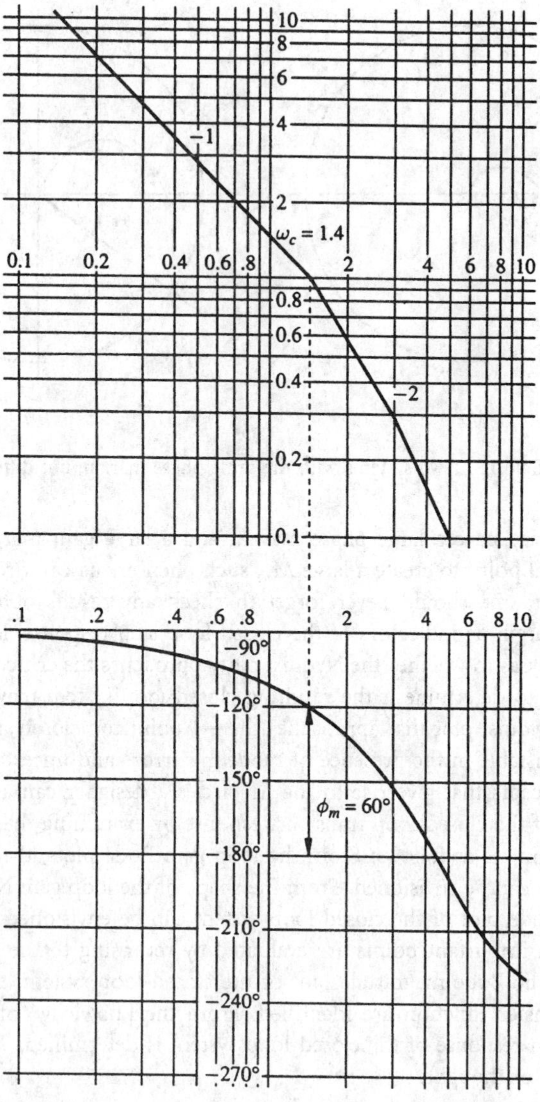

FIGURE 46.15 Magnitude and phase plots of $G(s)$.

from the Nyquist plot of the system. The gain margin is defined as the ratio of the maximum possible gain for stability to the actual system gain. If a plot of $G(s)$ for $s = j\omega$ intercepts the negative real axis at a point $-a$ between the origin and the critical -1 point, then the gain margin is simply

$$\text{Gain margin} = GM = \frac{1}{a}$$

If a gain $\geq 1/a$ were placed in series with $G(s)$, the closed-loop system would be unstable.

While the gain margin does not provide very complete information about the response of the closed-loop system, a small gain margin indicates a Nyquist plot which approaches the -1 point closely at the frequency at which the phase shift is 180°. Such a system will have a large M_p and an oscillatory closed-loop time response independent of the phase margin of the system.

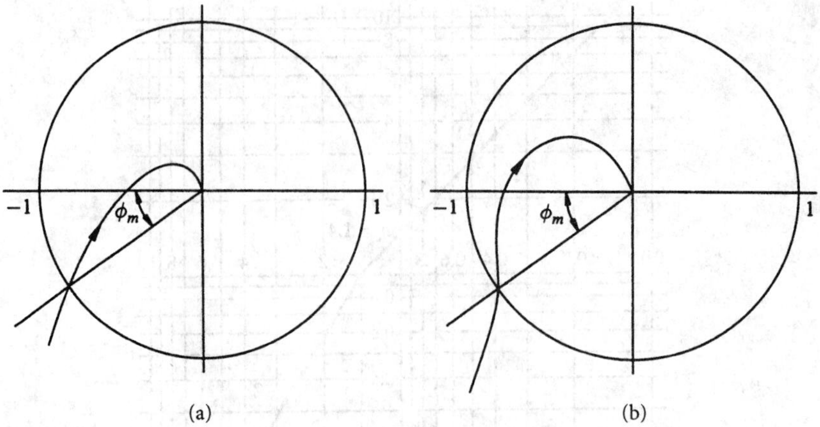

FIGURE 46.16 Two systems with the same phase margin, but different M_p.

While a system may have a large phase margin and a large gain margin and still get close enough to the critical point to create a large M_p, such phenomena can occur only in high-order loop gains. However, one should never forget to check any results obtained by using phase margin and gain margin as indicators of the closed-loop step response, lest an atypical system slip by. A visual check as to whether the Nyquist plot approaches the critical -1 point too closely should be sufficient to determine if the resulting closed-loop system may be too oscillatory. It is also true that a Nyquist plot that approaches the -1 point too closely also indicates a system that may become unstable in the presence of modeling errors and uncertainty.

By using the concepts that give rise to the M-circles a designer can arrive at a pretty good feel for the nature of the closed-loop transient response by examining the loop gain Bode plots. The chain of reasoning is as follows: From the loop gain Bode plots, the shape of the Nyquist plot of the loop gain can be envisioned. From the shape of the loop gain Nyquist plot, the shape of the Bode magnitude plot of the closed-loop system can be envisioned using the concepts of this section. Certain important points are evaluated by returning to the loop gain Bode plots. From the shape of the Bode magnitude plot of the closed-loop system, the dominant poles of the closed-loop transfer function are identified. From the knowledge of the dominant poles, the shape of the step response of the closed-loop system is determined. The following example illustrates this chain of thought.

Example 46.4. Consider the loop gain transfer function

$$G(s) = \frac{80}{s(s+1)(s+10)}$$

The Bode plots for this loop gain are given in Fig. 46.17.

The Nyquist plot can be envisioned from the Bode plots. The Nyquist plot begins far down the negative imaginary axis because the Bode plot has large magnitude and $-90°$ phase at low frequency. It swings to the left as the phase lag increases, and then spirals clockwise toward the origin, cutting the negative real axis and approaching the origin from the direction of the positive imaginary axis, i.e., from the direction associated with $-270°$ phase. From the Bode plot it is determined that the Nyquist plot does not encircle the -1 point because the Bode plot shows that the magnitude crosses unity (0 dB) before the phase crosses $-180°$.

From the Bode plot it can be seen that the Nyquist plot passes very close to the -1 point near the crossover frequency. In this case $\omega_p \approx \omega_c$, and the phase margin is a key parameter to establish how large is M_p, the peak in the closed-loop frequency magnitude plot. The crossover

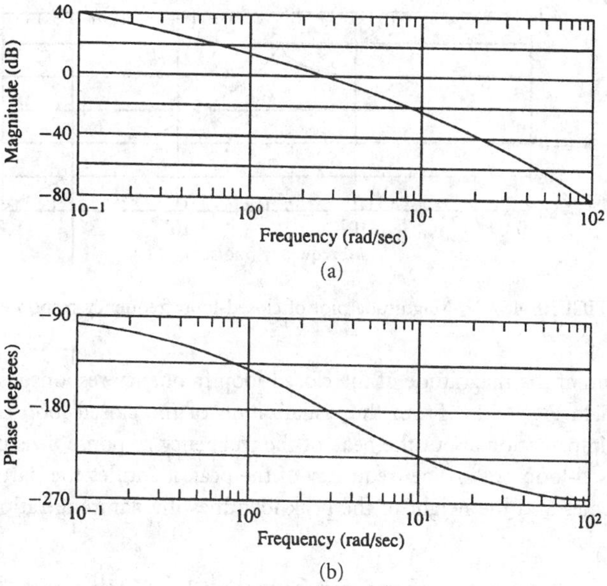

FIGURE 46.17 Bode plots of loop gain (a) magnitude plot; (b) phase plot.

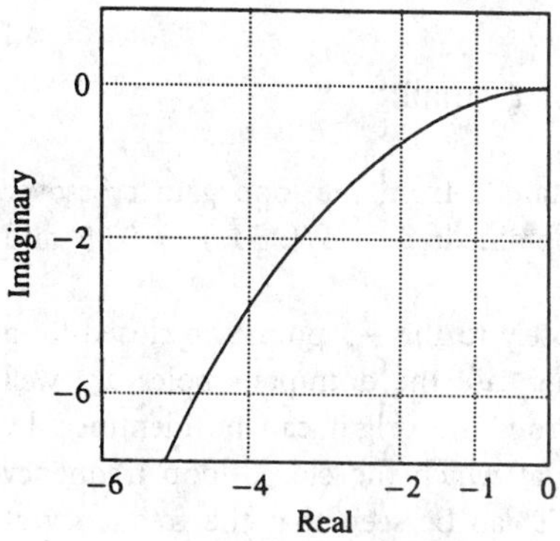

FIGURE 46.18 Nyquist plot of loop gain.

frequency is read from the Bode magnitude plot as $\omega_c = 2.5$ rad/s and the phase margin is read from the Bode phase plot as $\phi_m = 6°$.

Our visualization of the Nyquist plot is confirmed by the diagram of the actual Nyquist plot shown in Fig. 46.18. The magnitude of the closed-loop frequency response for this system can be envisioned using the techniques learned in this section. At low frequencies $G(s)$ is very large, the distance from the origin to the Nyquist plot is very nearly the same as the distance from the -1 point to the Nyquist plot, and the closed-loop frequency response has magnitude near 1. As the Nyquist plot of the loop gain approaches -1, the magnitude of the closed-loop frequency response function increases to a peak. At higher frequencies the loop gain becomes small and the closed-loop frequency response decreases with the loop gain because the distance from the -1 point to the loop gain Nyquist plot approaches unity. Thus, the closed-loop frequency response starts near 0 dB, peaks as the loop gain approaches -1, and then falls off.

The key point occurs when the loop gain approaches the -1 point and the closed-loop frequency response peaks. The closest approach to the -1 point occurs at a frequency very close to the crossover frequency, which has been established as $\omega_c = 2.5$ rad/s. The height of the peak can be established using the phase margin which has been established as $\phi_m = 6°$ and (46.14). The height of the peak should be very close to $(2\sin(\phi_m/2))^{-1} = 9.5 = 19.6$ dB.

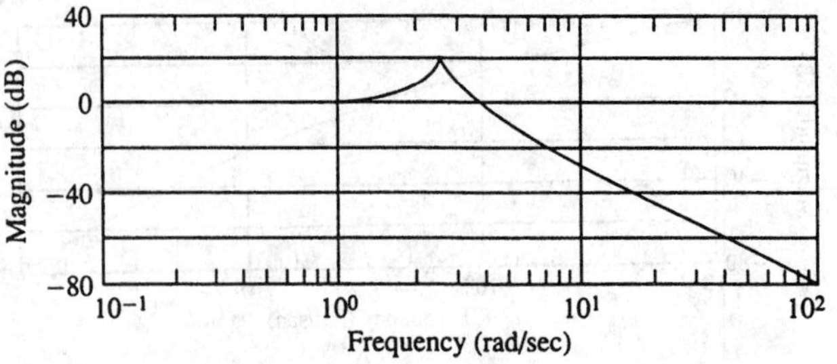

FIGURE 46.19 Magnitude plot of closed-loop frequency response.

Our visualization of the magnitude of the closed-loop frequency response is confirmed by the actual plot shown in Fig. 46.19. From the visualization of the closed-loop frequency response function and the information about the peak of the frequency response, it is possible to identify the dominant closed-loop poles. The frequency of the peak identifies the natural frequency of a pair of complex poles and the height of the peak identifies the damping ratio. More precisely,

$$M_p = \frac{1}{2\zeta\sqrt{1-\zeta^2}} \approx \frac{1}{2\zeta} \quad \text{for } \zeta \text{ small}$$

and

$$\omega_p = \omega_n\sqrt{1-2\zeta^2} \approx \omega_n \quad \text{for } \zeta \text{ small}$$

Using the approximations for ω_p and M_p that are obtained from the loop gain crossover frequency and phase margin, the following values are obtained: $\zeta \approx 1/(2M_p) \approx 0.05$ and $\omega_n \approx \omega_p \approx \omega_c \approx 2.5$ rad/s.

If the Nyquist plot of a loop gain does not pass too closely to the -1 point the closed-loop frequency response does not exhibit a sharp peak. In this case the dominant poles are well damped or real. The distance of these dominant poles from the origin can be identified by the system's bandwidth, which is given by the frequency at which the closed-loop frequency response begins to decrease. From the M-circle concept it can be seen that the frequency at which the closed-loop frequency response starts to decrease is well approximated by the crossover frequency of the loop gain.

Having established the position of the dominant closed-loop poles, it is easy to describe the closed-loop step response. The step response has a percent overshoot given by

$$\text{PO} = 100e^{\left(-\zeta\pi/\sqrt{1-\zeta^2}\right)} \approx 85\%$$

The period of the oscillation is given

$$T_d = \frac{2\pi}{\omega_d} = \frac{2\pi}{\omega_n\sqrt{1-\zeta^2}} \approx 2.5$$

The first peak in the step response occurs at a time equal to half of the period of oscillations, or about 1.25. The envisioned step response is confirmed in the plot of the actual closed-loop step response shown in Fig. 14.20.

The method of the previous example may seem to create a long way to go in order to get an approximation to the closed-loop step response. Indeed, it is much simpler to calculate the

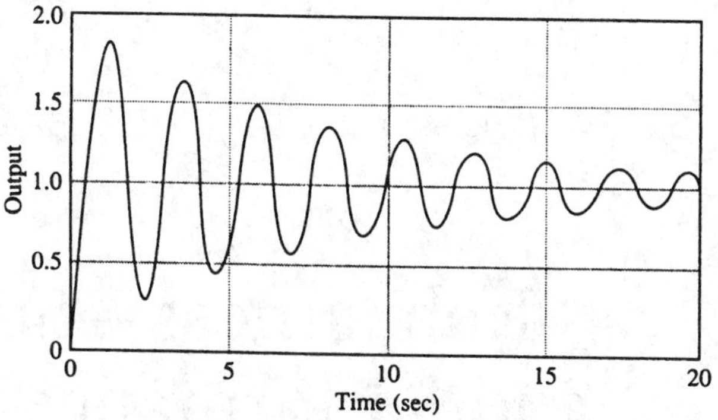

FIGURE 46.20 Closed-loop step response.

closed-loop transfer function directly from the loop gain transfer function. The importance of the logic in the example is not to create a computational method; the importance lies in the insight that is achieved in predicting problems with the closed-loop transient response by examining the Bode plot of the loop gain. The essence of the insight can be summarized in a few sentences: Assume that the Nyquist plot of the loop gain indicates a stable closed-loop system. If the Nyquist plot of the loop gain approaches the -1 point too closely, the closed-loop system's transient response characteristics are oscillatory. The speed of the transient response of the closed-loop system is usually indicated by the loop gain crossover frequency. Detailed information about the loop gain Nyquist plot is available in the loop gain Bode plots. In particular, the crossover frequency and the phase margin can be read from the Bode plots.

Any information that can be wrenched out of the Bode plots of the loop gain is critically important for two reasons. First, the Bode plots are a natural place to judge the properties of the feedback loop. When the magnitude of the loop gain is large, desirable feedback properties such as good disturbance rejection and good sensitivity reduction are obtained. When the magnitude of the loop gain is small these properties are not enhanced. The work of this section completes the missing information about transient response which can be read from the loop gain Bode plots. Second, it is the Bode plot that a designer is able to manipulate directly using series compensation techniques. It is important to be able to establish the qualities of the Bode plots which produce desirable qualities in a control system because only then can the Bode plots be manipulated to attain the desired qualities.

References

[1] H. W. Bode, *Network Analysis and Feedback Amplifier Design*, New York: Van Nostrand, 1945.

[2] R. V. Churchill, J. W. Brown, and R. F. Verhey, *Complex Variables and Applications*, New York: McGraw-Hill, 1976.

[3] Horowitz, I. M., *Synthesis of Feedback Systems*, New York: Academic, 1963.

[4] Nyquist, H., "Regeneration Theory," *Bell System Tech. J.*, vol. 11, pp. 126–146, 1932.

[5] Rohrs, C. E., J. L. Melsa, and D. G. Schultz, *Linear Control Systems*, New York: McGraw-Hill, 1993.

47

Robustness—Stability and Performance Under Modeling Uncertainty*

Charles E. Rohrs
Tellabs Research Center
University of Notre Dame

47.1 Modeling Uncertainty

Imagine that you are a manufacturer of position control systems consisting of a DC motor with a shaft and gears leading to a platform. After manufacturing a number of platform systems, you instruct technicians to measure the frequency response of each system to see how well it matches the model your engineers derived for the system. The following patterns may arise.

Assuming you have a reasonably precise manufacturing operation, the frequency response for the different systems may match the model fairly well at low frequencies. There may be some discrepancies due to slightly different parameters among systems, but the discrepancies should be within an acceptably small level.

As the frequency increases, you may find that the frequency responses start to deviate from the model and even from each other. Phase measurements of different systems at individual frequencies may vary by 180°. At higher frequencies the technician is likely to be unable to get the system to settle down enough to give you an accurate measurement of the frequency response. The technician may only be able to give a range for the magnitude of the response and may not be able to provide reliable phase information at all. The reason for these problems in obtaining high frequency measurements is unmodeled dynamics.

This section discusses a method of extracting the important and obtainable features of modeling uncertainties. The goal is to produce a simple model which is reasonably accurate for all the systems that have been and will be manufactured. Thus, we need to model a class of systems, not just one system. One way to accomplish this is to think about a nominal plant model in conjunction with a perturbation model. The range of frequency responses of a class of plants may be displayed graphically on Bode plots as in Fig. 47.1(a) and (b). Figure 47.1(a) shows the usual characteristics of increasing magnitude uncertainty with increasing frequency and Fig. 47.1(b) shows total phase uncertainty above some frequency.

*Much of the material used in this chapter is taken from Rohrs, C. E., J. L. Melsa, and D. G. Schultz, *Linear Control Systems*, McGraw-Hill, 1993. It is used with permission.

0-8493-8341-2/95/$0.00+$.50

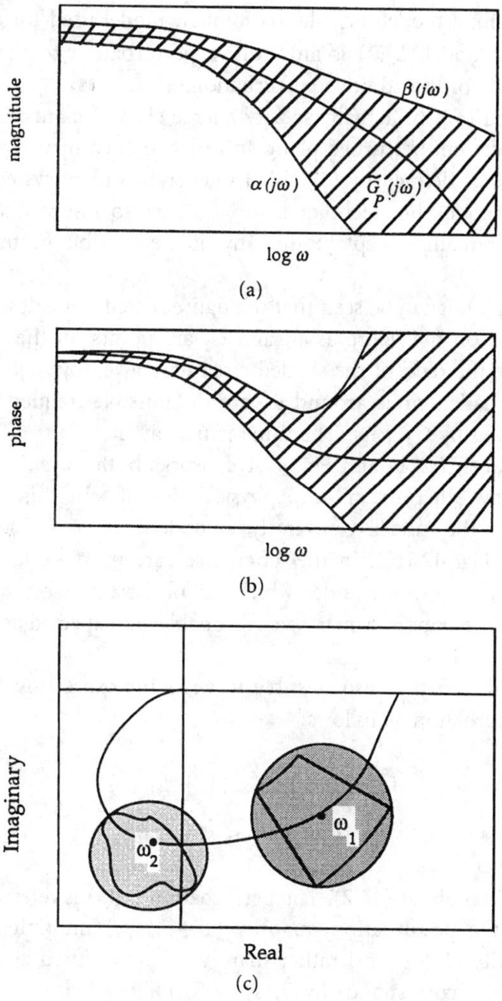

FIGURE 47.1 Areas of the Bode and Nyquist plots within which a perturbed plant model may lie. (a) Bode magnitude areas; (b) Bode phase areas; (c) Polar plot areas.

A second way of viewing the uncertainty of the frequency response at each frequency is by using the polar plot of Fig. 47.1(c). The area of the drawing around the point labeled ω_1 represents the situation in which a range of magnitude responses and a range of phase responses are observed. The resulting uncertainty region is a wedge in the polar plane. The actual frequency response lies somewhere in this wedge. It may be possible by closer observation to find a different irregular shape for the uncertainty region. This is represented by the area in Fig. 47.1(c) surrounding ω_2. To simplify the mathematical model in either case the uncertainty region is enclosed in a circle. The center of the circle is used as the nominal plant model's frequency response and the radius of the circle is maximum magnitude of an additive perturbation. This uncertainty model can be described mathematically as follows:

$$\tilde{G}_p(s) \approx G_p(s) + L_a(s) \tag{47.1a}$$

$$|L_a(j\omega)| < l_a(j\omega) \tag{47.1b}$$

where $G_p(s)$ is the nominal model, i.e., the basic plant model used for the controller designer, $\tilde{G}_p(s)$ is the perturbed model, $L_a(s)$ is an additive perturbation to the model, and $l_a(s)$ is a bound on the magnitude of the additive perturbation.

Equations (47.1a) and (47.1b) actually specify a large class of plants. Each different perturbation, $L_a(s)$, defines a different perturbed plant. In order to encompass a class of plants we allow a class of perturbations as defined by (47.1b). In general, control system design is carried out on the nominal plant model, but techniques are available to assure that a controller designed for the nominal plant remains acceptable for any of the possible perturbed plants defined by (47.1a) and (47.1b).

Referring to Fig. 47.1(c), it can be seen that the figure matches the description if the frequency response points marked on the figure as ω_1 and ω_2 are points on the nominal plant response function, $G_P(j\omega)$, while the radii of the shaded circles are given by $l_a(j\omega_1)$ and $l_a(j\omega_2)$. Then all the points within the shaded circle around ω_1 are the possible frequency responses around ω_1 for the set of plants given by $\tilde{G}_P(j\omega_1)$. The same is true at ω_2.

The connection between (47.1) and Fig. 47.1(c) works both ways. The model of (47.1) may have been developed by other means and physical insights. In this case Fig. 47.1(c) is the proper way to interpret the model. Conversely, a model such as (47.1) may be derived using data represented as in Fig. 47.1(c). In the latter case care must be taken that the number of plants observed and the conditions under which the observations are made are varied enough to encompass all plant responses which can reasonably be expected to be encountered while operating the plant.

With a little manipulation it is also possible to write the expression for the perturbed plant as a multiplicative perturbation as follows:

$$\tilde{G}_p(s) = G_p(s)(1 + L_m(s)) \tag{47.2a}$$

$$|L_m(j\omega)| < l_m(j\omega) \tag{47.2b}$$

We must note a few things about (47.2). The perturbation $L_m(s)$ is referred to as a multiplicative perturbation because the perturbation quantity $(1 + L_m(s))$ multiplies the plant in a series connection. The quantity $(1 + L_m(s))$ rather than just $L_m(s)$ is used as the multiplier so that a small perturbation, $L_m(s)$, corresponds to $\tilde{G}_p(s) \approx G_p(s)$.

Equating $\tilde{G}_p(s)$ in (47.1) and (47.2) shows us that the multiplicative perturbation is a relative perturbation, having the interpretation of a percentage change in the plant model.

$$L_m(s) = \frac{L_a(s)}{G_p(s)} \tag{47.3}$$

The relationship of the phase plot of Fig. 47.1(b) and (47.2) is subtle. This much can be said: When $|l_m(j\omega)| > 1$ the phase of $\tilde{G}_p(j\omega)$ is completely uncertain, i.e., given any arbitrary angle an $L_m(j\omega)$ exists with magnitude slightly larger than 1 such that the $\arg(\tilde{G}_P(j\omega))$ matches the given angle (see Fig. 47.2).

We now look at typical shapes of $l_m(j\omega)$ that may be expected to be found in practice. We discuss two commonly occurring modeling problems and investigate how these problems are captured by (47.1) and (47.2).

Unmodeled Dynamics

Some dynamics are always part of the plant but are excluded from the nominal plant model. These dynamics are excluded for various reasons: the nominal model must remain simple enough to be computationally tractable, the unmodeled dynamics may be so highly sensitive to parameters that their models are inaccurate for most of the systems built, or it is just impossible

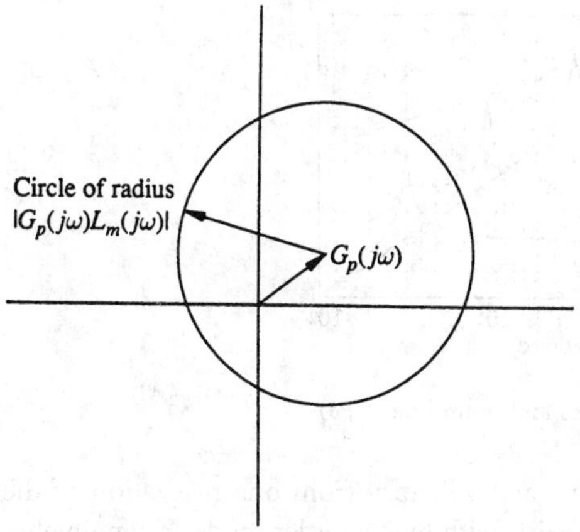

FIGURE 47.2 Possible complex numbers for $\arg(G_p(j\omega) + G_p(j\omega)L_m(j\omega))$.

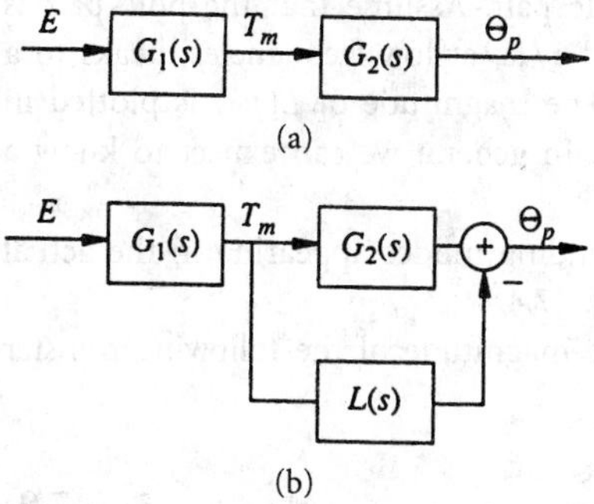

FIGURE 47.3 Block diagrams for a system without and with unmodeled dynamics.

to model all dynamics (Fig. 47.3). In a mechanical system we can expect poorly damped bending modes to be typical of dynamics that are too difficult to model accurately.

Example 47.1. Let the nominal plant be given by

$$G_p(s) = G_1(s)G_2(s) \tag{47.4}$$

The perturbed plant is given by

$$\begin{aligned}\tilde{G}_p(s) &= G_1(s)[G_2(s) - L(s)] \\ &= G_1(s)G_2(s) - G_1(s)L(s) \\ &= G_1(s)G_2(s)\left[1 - \frac{L(s)}{G_2(s)}\right]\end{aligned} \tag{47.5}$$

From (47.5) the additive and multiplicative perturbations can be defined, respectively, as:

$$L_a(s) = -G_1(s)L(s) \tag{47.6}$$

$$L_m(s) = \frac{-L(s)}{G_2(s)} \tag{47.7}$$

Equations (47.6) and (47.7) satisfy (47.3). For this example, $L(s)$ may be given by

$$L(s) = \frac{1/J}{s^2 + (\beta/J)s + K_s/J} \tag{47.8}$$

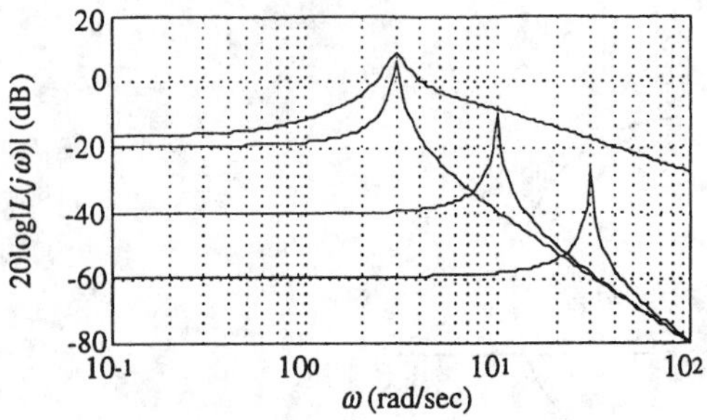

FIGURE 47.4 Bending modes and bounding $|l(j\omega)|$.

The parameters β and K_s are difficult to determine and will vary from one realization of the system to another. Also, a series of bending modes exists with higher-order modes having higher frequencies. The form of (47.8) is that of a complex pole pair. Assume that the pole pair is poorly damped so that $L(j\omega)$ has a small value of around $1/K_s$ at low frequencies, peaks to a value of $(\beta\sqrt{K_s/J})^{-1}$ at $\omega = \sqrt{K_s/J}$, and then falls off. The magnitude of $L(j\omega)$ is plotted in Fig. 47.4 for $J = 1$, $\beta = 0.15$, and $K_s = 9$, 100, and 900. In general we can expect to know a minimum value of K_s but little else about it.

To account for the possibility of any subset of these bending modes appearing in the actual plant, we can bound in magnitude all of the curves in Fig. 47.4.

For this example an appropriate bound is given by the magnitude of the following transfer function:

$$l'(s) = 4\,\frac{(s+1)^2(s^2+3s+9)}{(s+3)^3(s^2+06s+9)} \tag{47.9}$$

The magnitude of $l'(j\omega)$ is plotted along with the possible perturbations in Fig. 47.4.

Now there is a bound on $|L(j\omega)|$

$$|L(j\omega)| < |l'(j\omega)| \tag{47.10}$$

Combining (47.7) and (47.10) produces the desired bound on the possible multiplicative perturbations

$$|L_m(j\omega)| < l_m(j\omega) \tag{47.11}$$

where

$$l_m(j\omega) = \frac{|l'(j\omega)|}{|G_2(j\omega)|} \tag{47.12}$$

The quantity $l_m(j\omega)$ is plotted in Fig. 47.5 for a typical system using the following $G_2(j\omega)$

$$G_2(j\omega) = \frac{1}{s(s+.1)}$$

While the bounds on additive perturbations may level off or decrease at high frequencies, the bounds on multiplication perturbations tend to increase as frequency increases. The reasons for this increase can be seen in (47.3). The plant divides the additive perturbation to form the multiplicative perturbation. Because plant magnitudes roll off (decreases in magnitude with increasing frequency), if the additive perturbation levels off, the multiplication perturbation increases with increasing frequency.

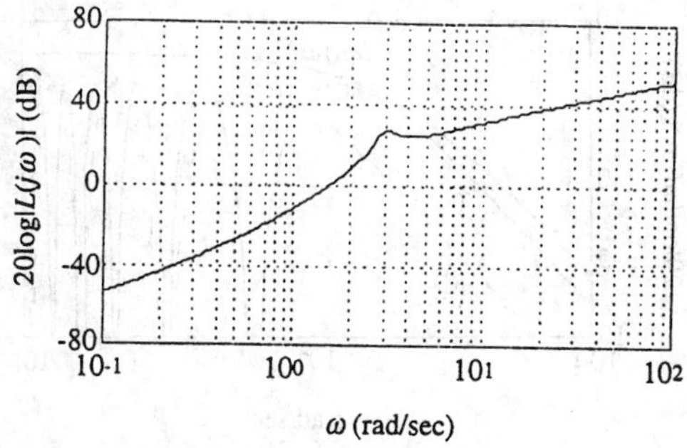

FIGURE 47.5 The bound $l_m(j\omega)$.

Time Delays

Some time delays always occur in the actuation of control commands. In process control systems long time delays may be involved with material flow. In mechanical and aerospace systems nonoscillatory unmodeled dynamics such as those associated with the movement of actuators can be thought of as delays in command actuation. The primary effect of certain nonlinearities can be summarized as a time delay. The model of a time delay is complex in that it does not yield a rational transfer function. Also time delays are inherent limitations to control systems which are not easily overcome. If a time delay is in the system, that system cannot react to a command input in a time frame any faster than the length of the delay. Time delays are often left out of the nominal plant model and thus become part of the possible perturbations to the plant model.

Consider the system of Fig. 47.6, in which an actuation delay is connected with a plant. The impulse response for a delay of T seconds is given as follows

$$h(t) = \delta(t - T) \tag{47.13}$$

The transfer function is given by the transform of the impulse response. We denote the transform of the delay by $L(s)$

$$L(s) = e^{-sT} \tag{47.14}$$

Notice that $L(s)$ is not a rational function, i.e., a ratio of polynomials in s. No finite poles or zeroes exist.

The multiplicative perturbation can be derived from the perturbed plant as follows.

$$\tilde{G}_p(s) = e^{-sT} G_p(s) = (1 + L_m(s))G_p(s) \tag{47.15}$$

$$L_m(s) = e^{-sT} - 1 \tag{47.16}$$

The magnitude of $L_m(s)$ is plotted in Fig. 47.7 for various values of delay T. An upper bound for the magnitude of all possible multiplicative perturbations is achieved by the magnitude of

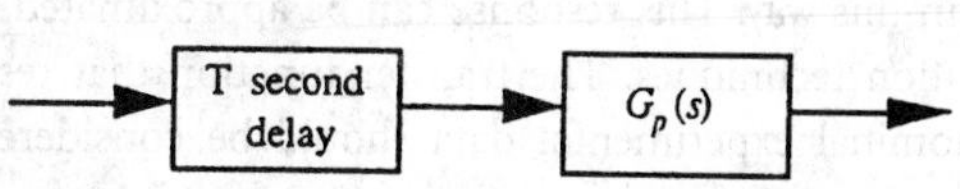

FIGURE 47.6 A system with a time delay.

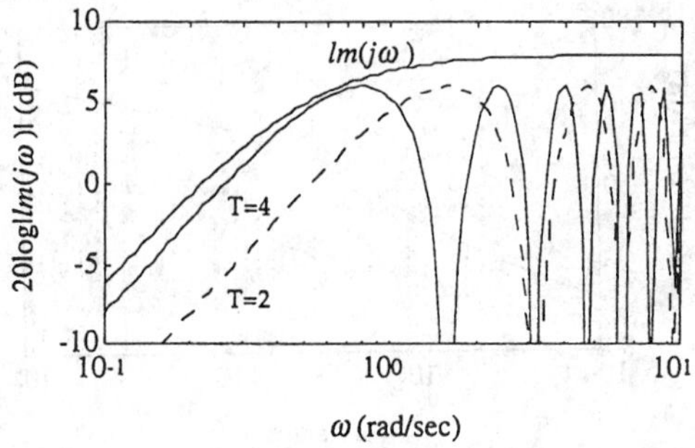

FIGURE 47.7 Magnitude of $l_m(j\omega)$ for a time delay perturbation.

the following transfer function

$$l_m(s) = \left| \frac{2.5s}{s + \dfrac{2}{T_{\max}}} \right| \tag{47.17}$$

where $T_{\max}$ is the maximum expected delay. Of interest is the frequency at which the magnitude of $l_m(j\omega)$ becomes > 1. A direct calculation indicates that $l_m(j\omega)$ becomes > 1 when $\omega \approx 1/T_{\max}$.

In summary, (47.1) and (47.2) provide a mechanism to allow the control designer to work with a simple nominal model, while accounting for the remainder of the dynamics in a less specific, less detailed manner. This structure allows the designer to ascertain whether a controller design is able to function adequately for a whole class of plants. It is, however, important to realize that the ultimate goal of the controller is to work well on the physical system that is to be controlled. This is different from the controller working well on any of the mathematical models within the class defined by (47.1) or (47.2).

Creating a Perturbation Model

The following procedure is given as a sensible method of moving from a set of data which is the result of a number of identification experiments to plant models of the form given in (47.1) and (47.2). First, it should be realized that the data coming from an identification experiment are never as clean as the "experimental" plots that appear in textbooks. Most systems are not purely linear and do not yield a pure sinusoidal output when excited by a sinusoid. Also, identification experiments ought to be made over the entire range of the expected operating environment of the system, including changes of temperature and system wear. Lastly, in some applications a large number of systems may need to be tested in the identification process because each realization of the plant design may vary somewhat from the others, and a controller should work acceptably on each realization.

When a reasonably complete set of identification experiments is performed the response of the plant at each frequency is likely to fill a region on the polar or Nyquist plot, as shown in Fig. 47.1(c). At each frequency, the corresponding region on the plot should be encompassed by a circle of as small a radius as possible. The center of this circle can then be taken as the nominal experimental response at that frequency. The entire nominal experimental frequency response can be plotted in this way. This response can be approximated with a transfer function using standard identification techniques. The transfer function that results from applying that approximation to the nominal experimental data should be considered the nominal transfer function. The frequency response of the nominal transfer function at each frequency should be plotted on the original polar plots of the raw data regions. The nominal frequency response is not likely to match the center of the original circles exactly because of the approximation process used in determining the nominal transfer function. At each frequency, new circles should be drawn, with the response of the nominal transfer function at the center and with a radius large enough to encompass the entire uncertainty region at that frequency. A new Bode magnitude plot should be made, with the magnitude at each frequency being equal to the radius of the circle at that frequency. Identification methods can then be used to find a transfer function, $l_a(s)$, whose magnitude does not simply approximate this magnitude plot, but overbounds it. The transfer function, $l_a(s)$, is the bound on the additive perturbation. One may need to increase the magnitude of this bound further to capture reasonably likely effects that, for some reason, are not present in the data of the identification experiments. The multiplicative perturbation bound can be computed using (47.3).

Equations (47.1) and (47.2) were developed to account for as much of the discrepancy between the physical system and mathematical model as is mathematically tractable. However, the final implementation of a controller requires a leap of faith on the part of the control designer. The control designer must be convinced that the final design will indeed work on the physical system. This confidence comes from experience in using the design tools and in developing an understanding of the real limitations of mathematical models.

We stop momentarily to warn the reader that as this is being written, the subject of how best to account for modeling inaccuracies is the subject of much of the research in control theory. Approaches to this problem that differ from those presented in this section are being studied. In particular, work is being done on the effect of perturbations in the parameters of the plant transfer function. (These kinds of perturbation are often called structured perturbations.) We choose to emphasize perturbations to the frequency response of the plant because we feel that the additive or multiplicative perturbations (often called unstructured perturbations) lead clearly to important design principles.

47.2 Robust Stability

One notion of relative stability is that one system is considered less stable than another if its time response is more oscillatory and less highly damped. The relative stability decreases as the Nyquist plot of the loop gain transfer function approaches the -1 point. Besides an oscillatory time response, a system whose Nyquist plot passes close to the -1 point suffers from another problem. If such a system is even slightly mismodeled, the Nyquist plot can be easily perturbed in such a way that the number of encirclements of the -1 point changes without changing the number of right-half-plane poles in the loop gain transfer function. The Nyquist theorem indicates that the number of right-half-plane poles in the perturbed closed-loop system changes so that if the nominal closed-loop system is stable, the perturbed closed-loop system is unstable.

It is important that a control system which is designed to be stable and perform well when used in conjunction with a nominal plant model still works well when used in conjunction with an actual physical plant. The output of the physical plant can be expected to behave in a manner similar to but not exactly the same as the nominal model. A controller design that works well with a large set of plant models is said to be robust. It is apparent from the preceding discussion that a design with a loop gain Nyquist plot that passes close to the -1 point is not robust. In this section this notion of robustness is formalized using the models of plant uncertainty developed in the previous section. First, conditions assuring that a design remains stable in the face of plant perturbations is developed. Clearly, maintaining stability for expected modeling errors is an absolute requirement. In addition, it is desirable to maintain adequate performance in the face of modeling errors. After the question of stability robustness is resolved in this section the question of robust performance is addressed in the following section.

Earlier we presented techniques for describing a set of possible plant models using a nominal plant model and a perturbation transfer function with a known magnitude bound. Consider first the description using an additive perturbation

$$\tilde{G}_p(s) = G_p(s) + L_a(s) \tag{47.18a}$$

where $L_a(s)$ is itself a stable transfer function containing no right-half-plane poles. A bound, $l_a(j\omega)$, on the magnitude of $L_a(j\omega)$ is known, i.e.,

$$|L_a(j\omega)| < l_a(j\omega) \tag{47.18b}$$

but otherwise $L_a(s)$ is a completely unknown transfer function. We are interested in the stability of the G-configuration system of Fig. 47.8. Let $G(s) = G_C(s)G_P(s)$ be the nominal loop gain, and $\tilde{G}(s) = G_C(s)\tilde{G}_P(s)$ be the perturbed loop gain.

Assume that the nominal design is stable, i.e., that the closed-loop system is stable when $L_a(s) = 0$. The robust stability question is formulated as follows: What conditions must be placed on $G(s)$ so that the configuration of Fig. 47.8 remains stable for all $\tilde{G}_p(s)$ satisfying (47.18)?

The robust stability question can be answered using the same tool that was used to answer the nominal stability and relative stability questions: the Nyquist diagram. Consider Fig. 47.9, which contains a typical Nyquist plot of $G(s)$ and $\tilde{G}(s)$. From the definitions above we can see that

$$\tilde{G}(s) = G_C(s)(G_P(s) + L_a(s)) = G(s) + G_C(s)L_a(s) \tag{47.19}$$

The plot of $\tilde{G}(j\omega)$ can be obtained from the plot of $G(j\omega)$ at each value of ω by adding a vector corresponding to $G_C(j\omega)L_a(j\omega)$ to $G(j\omega)$. Indeed, if each possible $L_a(j\omega)$ satisfying (47.9b) is used in turn, the set of all possible $\tilde{G}(j\omega_0)$ at the frequency ω_0 is given by the interior of a circle centered at $G(j\omega_0)$ with radius $G_C(j\omega_0)l_a(j\omega_0)$. This circle is shown in Fig. 47.9, assuming that the $L_a(j\omega)$ chosen for display has the maximum magnitude. The key observation for the desired result is as follows: Each $L_a(s)$ is assumed to be stable itself, so that the number of

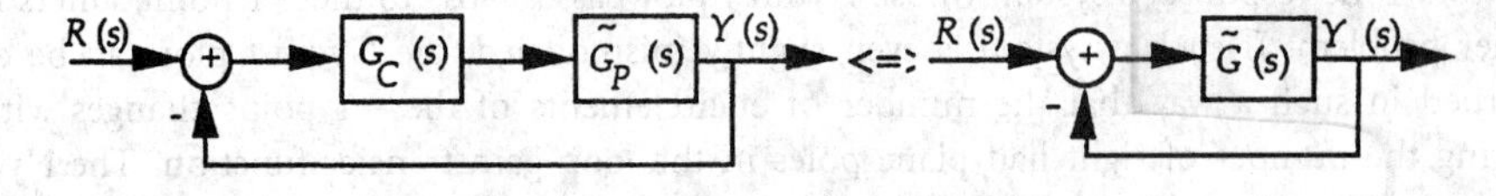

FIGURE 47.8 The perturbed G-configuration.

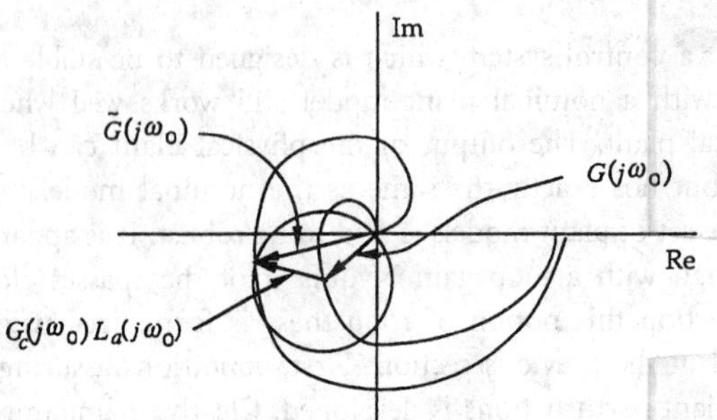

FIGURE 47.9 Nyquist plots of $\tilde{G}(s)$ and $G(s)$.

right-half-plane poles of the loop gain, $\tilde{G}(s)$, is unchanged from the number of right-half-plane poles of the nominal loop gain $G(s)$. Therefore, the closed-loop stability assumed for the configuration in Fig. 47.8 with $L_a(s) = 0$ is maintained for nonzero $L_a(s)$ if and only if the number of encirclements of the -1 point is unchanged in going from $G(j\omega)$ to $\tilde{G}(j\omega)$. One way to assure that the number of encirclements remains unchanged is to ensure that at every single frequency $l_a(j\omega)$ is small enough so that $\tilde{G}(j\omega)$ cannot reach the -1 point.

At each ω, the distance between $G(j\omega)$ and the -1 point is given by

$$|G(j\omega) - (-1)| = |1 + G(j\omega)|$$

Therefore, if at every ω we have the condition

$$|G_C(j\omega)|l_a(j\omega) < |1 + G(j\omega)|$$

or

$$l_a(j\omega) < \frac{|1 + G(j\omega)|}{|G_C(j\omega)|} = |G_C^{-1}(j\omega) + G_P(j\omega)| \tag{47.20}$$

then the perturbed system, $\tilde{G}(j\omega)$, is stable because the perturbation in the Nyquist plot cannot change the number of encirclements.

From (47.20) we see that, as expected, a measure of the robustness of a control system is given by the distance the Nyquist plot of the loop gain transfer function maintains from the -1 point. This distance is expressible as the magnitude of the return difference transfer function, a quantity that should be kept large for such performance requirements as reference input tracking, low parameter sensitivity, and disturbance rejection.

Equation (47.20) provides the condition to guarantee maintenance of stability in the face of stable additive perturbations satisfying (47.18). It is interesting to see what can be said if (47.20) is violated. If (47.20) is violated, the closed-loop system remains stable for many of the possible $L_a(s)$ satisfying (47.18); however, a guarantee no longer exists that stability is maintained for every perturbation satisfying (47.18). Indeed, if $l_a(j\omega)$ is continuous in ω and if, for some ω_0 and $\varepsilon > 0$,

$$l_a(j\omega_0) \geq \frac{|1 + G(j\omega_0)|}{|G_C(j\omega)|} + \varepsilon \tag{47.21}$$

then a perturbation $L_a(j\omega)$ exists satisfying (47.18) which causes instability. The destabilizing perturbation is constructed as follows: Pick $L_a(s)$ stable so that

$$|L_a(j\omega_0)| = \frac{|1 + G(j\omega_0)|}{|G_C(j\omega_0)|} + \varepsilon/2$$

and

$$\arg(L_a(j\omega_0)) = \arg(1 + G(j\omega_0)) - \arg G_C(j\omega_0) + 180^\circ$$

The ensuing Nyquist plot of $\tilde{G}(j\omega)$ has a different number of encirclements of the -1 point than the Nyquist plot of $G(j\omega)$, and the closed-loop stability is lost.

It is often easier to express a class of possible plant models using a multiplicative perturbation rather than an additive perturbation. This is particularly true for stability robustness conditions because the condition of (47.20) has the compensator appearing separately from the loop gain.

The robust stability condition for a multiplicative perturbation is a function only of the loop gain. Assume a set of possible plant models is given as follows:

$$\tilde{G}_P(s) = G_P(s)(1 + L_m(s)) \tag{47.22}$$

where $L_m(s)$ is itself a stable transfer function containing no right-half-plane poles. A bound, $l_m(j\omega)$, on the magnitude of $L_m(j\omega)$ is known, i.e.,

$$|L_m(j\omega)| < l_m(j\omega) \tag{47.23}$$

Then (47.22) and (47.23) are equivalent to (47.18a, b) if the following identification is made:

$$L_a(s) = G_P(s)L_m(s) \tag{47.24}$$

However, the equivalent expression for the loop gain is

$$\tilde{G}(s) = G_C(s)G_P(s)(1 + L_m(s)) = G(s)(1 + L_m(s)) = G(s) + G(s)L_m(s) \tag{47.25}$$

The Nyquist plot is perturbed from $G(s)$ to $\tilde{G}(s)$ by the addition of $G(s)L_m(s)$. The condition for robustness in the multiplicative perturbation setting is obtained by observing again that if at each frequency, the distance that the Nyquist plot of $G(j\omega)$ can be perturbed is less than the distance to the -1 point, a nominally stable closed-loop system is guaranteed to remain stable. Thus, if the closed loop system of Fig. 47.8 is stable for $L_m(s) = 0$, if each $L_m(s)$ is stable, and if for all ω,

$$|l_m(j\omega)G(j\omega)| < |1 + G(j\omega)| \tag{47.26}$$

then the closed-loop system of Fig. 47.8 remains stable. Equation (47.8) can be manipulated into the following equivalent forms.

$$l_m(j\omega) < \frac{|1 + G(j\omega)|}{|G(j\omega)|} \tag{47.27}$$

$$\left|\frac{G(j\omega)}{1 + G(j\omega)}\right| < \frac{1}{l_m(j\omega)} \tag{47.28}$$

or

$$l_m(j\omega)\left|\frac{G(j\omega)}{1 + G(j\omega)}\right| < 1 \tag{47.29}$$

As with the robustness condition for the additive perturbation formulation, something can be said if (47.28) is violated. If (47.28) is violated, then a multiplicative perturbation that satisfies (47.23) causes the perturbed closed-loop system of Fig. 47.8 to be unstable.

The robustness condition of (47.28) usually poses a constraint on the allowable bandwidth of a control system, as can be seen by the following argument. Usually a multiplicative perturbation is small at low frequencies and grows to be larger than unity at higher frequencies. Let ω_1 be the frequency, where $l_m(j\omega_1) = 2$. Assume that for $\omega > \omega_1$, $l_m^{-1}(j\omega) < 1/2$. Now notice that the left-hand side of (47.28) is simply the nominal closed-loop transfer function. It is usually desirable to keep the closed-loop transfer function close to unity for as large a range of frequencies as possible. However, the constraint of (47.28) dictates that the magnitude of the nominal closed-loop transfer function be $< 1/2$ and the nominal loop gain be < 1 for all $\omega > \omega_1$. Thus, the maximum bandwidth of the system is limited to $< \omega_1$ if the controller is to result in a stable closed-loop system for all possible models as given by (47.22) and (47.23).

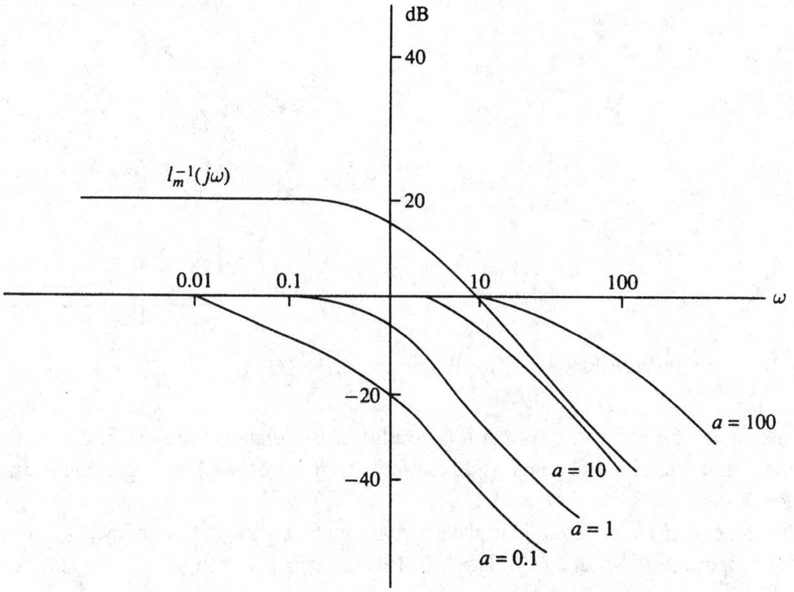

FIGURE 47.10 Bode magnitude plots of (47.31) and (47.34) for various a.

Example 47.2. Let a collection of possible plant models be given by (47.22) and (47.23) with

$$G_p(s) = \frac{1}{s} \tag{47.30}$$

and

$$l_m(s) = \frac{1}{10}\,|(s+1)| \tag{47.31}$$

A series compensator given by

$$G_C(s) = a, \quad \text{a constant} \tag{47.32}$$

makes

$$G(s) = \frac{a}{s} \tag{47.33}$$

By straight calculation, the closed-loop transfer function is

$$\frac{G(s)}{1+G(s)} = \frac{a}{s+a} \tag{47.34}$$

so a single closed-loop pole exists at $s = -a$ and the closed-loop bandwidth covers $\omega = 0$ to $\omega = a$.

Figure 47.10 plots $|l_m^{-1}(j\omega)|$ as derived from (47.31) and the closed-loop frequency response for various values of a. From the figure and a few calculations we can see that the robustness condition of (47.28) is satisfied for all $a < 10$. Thus, the maximum bandwidth is roughly equal to the frequency where $l_m^{-1}(j\omega) = 1$.

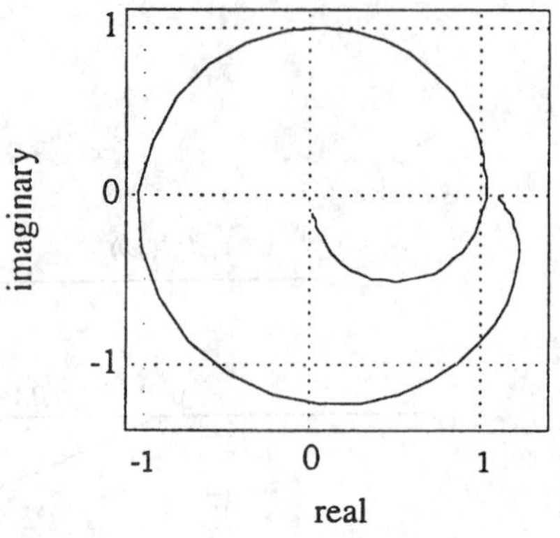

FIGURE 47.11 Perturbed loop gain for Example 43.6.

Now, take $a = 11$ so that the robustness condition is violated. We now find a multiplicative perturbation that satisfies the magnitude bound given by (47.31) and produces an unstable closed-loop system.

In order to construct a destabilizing perturbation, first pick a frequency, ω_0, at which the robustness constraint of (47.28) is violated. In this example, we can pick $\omega_0 = 100$ because

$$\left|\frac{G(j100)}{1+G(j100)}\right| = \left|\frac{11}{j100+11}\right| + \frac{11}{\sqrt{10121}} \geq \frac{1}{l_m(j100)} = \frac{10}{\sqrt{10001}}$$

The concept is to pick $l_m(j\omega_0)$ so that the equivalent additive perturbation given by $l_m(j\omega_0)G(j\omega_0)$ moves the Nyquist plot of $G(j\omega)$ to and past the -1 point at the frequency ω_0. Thus, we pick the magnitude of $L_m(j\omega_0)$ to be slightly larger than $|1 + G(j100)|/|G(j100)|$, while remaining smaller than $l_m(j\omega_0)$, and we pick the phase of the $L_m(j\omega_0)$ to point the vector $L_m(j\omega_0)G_m(j\omega_0)$ in the Nyquist plane from $G(j\omega_0)$ toward -1. This can be achieved as follows:

$$\arg(L_m(j\omega_0)) = \arg(1 + G(j\omega_0)) - 180^\circ - \arg(G(j\omega_0))$$

Performing these calculations the phase of the destabilizing $L_m(j\omega_0)$ for this example is -96° while 9.5 is a destabilizing magnitude. Once we know what $L_m(j\omega_0)$ should equal we only need to fit a transfer function which also satisfies the upper bound of (47.23) around that point. For example, one destabilizing $L_m(s)$ is given by

$$L_m(s) = \frac{0.95(s+1)}{(0.001s+10)} \frac{(-0.0103s+1)^2}{(0.0103s+1)^2}$$

First note that $L_m(s)$ is stable and that

$$|L_m(s)| < \frac{0.95|j\omega + 1|}{10} < l_m(j\omega)$$

Then we can draw the Nyquist plot for the loop gain $G(s)(1 + L_m(s))$ as in Fig. 47.11 and see that the resulting closed-loop system is unstable.

We have seen in this section that the stability of the nominal system can be guaranteed for a system with a stable multiplicative perturbation if (47.29) is satisfied. Such a system is said to possess robust stability with respect to the perturbation in question. Multiplicative perturbations tend to get large at high frequencies. Equation (47.29) requires that the nominal loop gain is made small at high frequencies.

47.3 Performance and Robustness

In the previous section we developed the constraint of (47.28) which characterizes a collection of loop gain transfer functions for which perturbed closed-loop systems remain stable. If such a constraint is satisfied and the actual plant's behavior is adequately described by one of the models in the collection of models given by (47.22) and (47.23), the implementation of the control system produces a stable closed-loop response. However, even though that response is stable it may produce unacceptably bad and oscillatory responses. We need some way to assure that the performance aspects required for the system and met by the nominal design are also met when the controller is implemented on the actual plant. We cannot assure 100% performance on the actual plant as even the collection of plants described by (47.22) and (47.23) cannot perfectly model the actual plant. We can arrive at a constraint that will assure that some performance measure is achieved for any plant that is a member of the collection of plants described by (47.22) and (47.23).

We must first decide on a performance measure that suitably describes such diverse control system performance requirements as possessing desirable transient responses to command inputs, quickly and completely rejecting certain disturbances, and providing a small sensitivity of the response to small parameter changes. One way to abstract these desires into a single performance specification is to realize that all these performance objectives can be met if we can make the return difference transfer function large enough. While the return difference cannot capture every aspect of performance that might be specified, a controller with a large return difference over a broad frequency band can be expected to perform well in most common tests of control system performance. With that in mind we can write down an interesting general performance requirement for a control system.

We can say that a control system in the G-configuration performs adequately if

$$|1+\tilde{G}(j\omega)| > p(j\omega) \tag{47.35}$$

where $p(j\omega)$ is some specified function of frequency. We now show an example which demonstrates how $p(j\omega)$ may be derived from typical performance requirements.

Example 47.3. One performance aspect of interest in a control system is how well disturbances are rejected. A typical specification would be that a control system must reject all constant output disturbances completely as time goes to infinity, and, in addition, it must attenuate the effect of all output disturbances of frequency < 1 rad/s so that the output is disturbed by $< 1\%$ of the magnitude of the disturbance. The effect of an output disturbance on the plant output is

$$Y(s) = \frac{1}{1+\tilde{G}(s)} D(s) \tag{47.36}$$

The disturbance rejection requirement above can be translated to the form of (47.35) by requiring

$$p(j\omega) > 100 \quad \text{for} \quad \omega < 1 \tag{47.37}$$

and

$$\lim_{\omega \to 0} p(j\omega) = \infty \tag{47.38}$$

If (47.35) and (47.37) are satisfied, then

$$|Y| \le \frac{1}{|p(j\omega)|}|D| \le .01|D| \quad \text{for} \quad \omega < 1$$

A second common specification involves the sensitivity of the closed-loop transfer function to changes in the plant transfer function. A specification may read that for all input frequencies < 10 rad/s the closed loop transfer function's magnitude response should not differ from a nominal design by $> 1\%$ of the percentage change in the plant transfer function's response. The usual measure of sensitivity is

$$S_{\tilde{G}_P}^{M} = \frac{\delta M(s)/M(s)}{\delta G(s)/G(s)} = \frac{1}{1+\tilde{G}(s)} \tag{47.39}$$

The sensitivity requirement is met if

$$p(j\omega) > 100 \quad \text{for} \quad \omega < 10 \tag{47.40}$$

Specifications on the transient response of the system translate into specifications on $p(j\omega)$ less directly than do specifications on disturbance rejection and sensitivity reduction. In order to translate typical step response information into a specification on $p(j\omega)$, two intermediate steps are used. First, the transient response specifications are translated into desired dominant closed-loop pole positions. Then, the desired closed-loop pole positions are translated into a desired closed-loop frequency response. Finally, the closed-loop frequency response is translated into a specification on the return difference function or, equivalently, $p(j\omega)$.

Suppose that there is a requirement to produce a closed-loop step response which has $< 20\%$ overshoot and a 5% settling time of < 5.5 s. These specifications are satisfied by a closed-loop transfer function whose dominant behavior is characterized by a single pole pair with damping ratio $\zeta = .47$ and natural frequency $\omega_n = 1.2$. The closed-loop frequency response for such a transfer function has magnitude very close to unity for all frequencies < 1.2 rad/s and falls off for higher frequencies. The peak of the magnitude plot must be < 1.2.

In the G-configuration the magnitude of the closed-loop frequency response, $|\tilde{M}_c(j\omega)|$, is related to the magnitude of the loop gain, $|\tilde{G}(j\omega)|$, and the magnitude of the return difference $|1+\tilde{G}(j\omega)|$, by the expression

$$|\tilde{M}_c(j\omega)| = \frac{|\tilde{G}(j\omega)|}{|1+\tilde{G}(j\omega)|} \tag{47.41}$$

If $|1+\tilde{G}(j\omega)|$ is much greater than 1, then $|\tilde{G}(j\omega)|$ is much greater than 1, and $|\tilde{M}_c(j\omega)|$ is very close to 1. This logic can be quantified using the following inequalities

$$|1+\tilde{G}(j\omega)| - 1 < |\tilde{G}(j\omega)| < |1+\tilde{G}(j\omega)| + 1 \tag{47.42}$$

Dividing inequalities (47.42) by $|1+\tilde{G}(j\omega)|$, and using (47.41) we can arrive at an expression relating the magnitude of the closed-loop frequency response to the magnitude of the return difference

$$1 - \frac{1}{|1+\tilde{G}(j\omega)|} < |\tilde{M}_c(j\omega)| < 1 + \frac{1}{|1+\tilde{G}(j\omega)|} \tag{47.43}$$

If the magnitude of the return difference is > 5 for $\omega < 1.2$, the magnitude of the closed-loop frequency response remains between 0.8 and 1.2 for those frequencies. Inequalities (47.43) provide only a rough guideline to what is needed to achieve a certain closed-loop response. In order to meet the step response specifications, not only must the magnitude of the closed-loop response be kept near 1 for the appropriate frequencies ($\omega < 1.2$ in this case), but also peaks in the magnitude of the closed-loop frequency response at higher frequencies must be avoided. This requires that the return difference be kept from being too small at any frequency.

The guidelines needed to produce an adequate closed-loop frequency response become clearer by looking at Fig. 47.12. This figure shows a polar frequency plot complete with M-circles for a typical loop gain transfer function. From the M-circles, the magnitude of the closed-loop frequency response can be read off at any frequency. The plot of Fig. 47.12 would be an acceptable loop gain for this example because the peak M-circle reached is $M = 1.1$ near $\omega = 1.2$ and the plot moves through M-circles of smaller values for higher frequencies.

In this typical example the guidelines of inequalities (47.43) are conservative in that it is not required to have the return difference be > 5 for all $\omega \leq 1.2$. The guidelines are somewhat lacking as mentioned above in that they do not guarantee that a peak in the closed-loop frequency will not occur at a higher frequency, causing oscillations.

In spite of the shortcomings in precise guarantees, it is still useful to provide a guideline for adequate transient response of the closed-loop system by bounding the return difference from below. For many control systems, the loop gain transfer function decreases monotonically, producing a smooth polar frequency plot similar to Fig. 47.12. For these systems the range of frequencies in which the magnitude of the return difference remains > 1 provides a good estimate of the bandwidth of the controller. The bigger the frequency range over which the magnitude of the return difference stays > 1, the wider the bandwidth of the system and the faster the system can respond to step inputs. In addition, the magnitude of the return difference should be kept as large as possible over all frequencies to guard against high-frequency oscillations.

The example showed that the concept of providing a performance measure for a control system by bounding the magnitude of the return difference as in (47.35) works well for specifications involving disturbance rejection and sensitivity reduction. The bound of (47.35) also provides a guideline for transient response specifications. This lack of preciseness for transient response specification is acceptable because the transient response of a robustly stable control loop which responds almost fast enough usually can be modified to precisely meet specifications by using a prefilter on the command input.

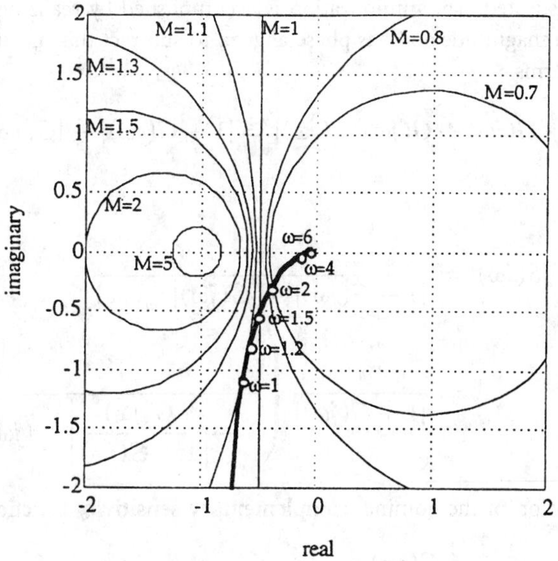

FIGURE 47.12 Nyquist plot with M-circles.

Equation (47.35) can be rewritten as

$$\left|\frac{1}{1+\tilde{G}(j\omega)}\right| < \frac{1}{p(j\omega)} \tag{47.44}$$

or

$$p(j\omega)\left|\frac{1}{1+\tilde{G}(j\omega)}\right| < 1 \tag{47.45}$$

The function $\tilde{S}(j\omega) = (1+\tilde{G}(j\omega))^{-1}$ is called the **sensitivity function** of the perturbed control system because the response of the control system is relatively insensitive to parameter changes and disturbances if this function is small.

Notice that the performance requirement of (47.35) is written as a function of $\tilde{G}(j\omega)$. It is desirable to design a controller based upon the nominal model of the plant, $G(j\omega)$, that meets the performance requirement of (47.35) for any plant $\tilde{G}(j\omega)$ in the collection given by the multiplicative perturbation model of (47.22) and (47.23).

In order to assure that the sensitivity function of the perturbed control system, $\tilde{S}(j\omega)$, is small for any allowable perturbation, $L_m(j\omega)$, it is useful to find how large $\tilde{S}(j\omega)$ can become when facing the most damaging perturbation allowable. The goal is to find the maximum value of $\tilde{S}(j\omega)$ as $L_m(j\omega)$ is allowed to vary over all allowable perturbations:

$$\begin{aligned}\max_{|L_m(j\omega)|<l_m(j\omega)} |\tilde{S}(j\omega)| &= \max_{|L_m(j\omega)|<l_m(j\omega)} \frac{1}{|1+\tilde{G}(j\omega)|} \\ &= \max_{|L_m(j\omega)|<l_m(j\omega)} \frac{1}{|1+G(j\omega)+G(j\omega)L_m(j\omega)|} \\ &= \frac{1}{\min\limits_{|L_m(j\omega)|<l_m(j\omega)} |1+G(j\omega)+G(j\omega)L_m(j\omega)|}\end{aligned} \tag{47.46}$$

Clearly, minimizing the denominator maximizes the expression of (47.46) where various definitions have been substituted. The minimization is accomplished by realizing that the worst case $L_m(j\omega)$ has maximal magnitude and has phase aligned to subtract this maximal magnitude away from the first two terms

$$\min_{|L_m(j\omega)|<l_m(j\omega)} |1+G(j\omega)+G(j\omega)L_m(j\omega)| = |1+G(j\omega)| - |G(j\omega)l_m(j\omega)| \tag{47.47}$$

Thus,

$$\begin{aligned}\max_{|L_m(j\omega)|<l_m(j\omega)} |\tilde{S}(j\omega)| &= \frac{1}{|1+G(j\omega)| - |G(j\omega)|\, l_m(j\omega)} \\ &= \frac{1}{|1+G(j\omega)|}\left(\frac{1}{1-\left|\dfrac{G(j\omega)}{1+G(j\omega)}\right| l_m(j\omega)}\right)\end{aligned} \tag{47.48}$$

The following definition of the nominal **complementary sensitivity function**, $T(j\omega)$, is

$$T(j\omega) = \frac{G(j\omega)}{1+G(j\omega)} = 1 - \frac{1}{1+G(j\omega)} = 1 - S(j\omega) \tag{47.49}$$

The complementary sensitivity function equals one minus the sensitivity function. (The complementary sensitivity function is also equal to the closed-loop response function for the G-configuration.) A final expression for the sensitivity under the worst-case perturbation in terms of the nominal sensitivity and complementary sensitivity functions can be written as follows:

$$\max_{|L_m(j\omega)|<l_m(j\omega)} |\tilde{S}(j\omega)| = |S(j\omega)|\left(\frac{1}{1-|T(j\omega)|\,l_m(j\omega)}\right) \tag{47.50}$$

If the worst-case perturbed sensitivity function remains $< p^{-1}(j\omega)$, then the sensitivity functions for all allowable perturbation are $< p^{-1}(j\omega)$. The robust performance condition of (47.44) is satisfied if

$$|S(j\omega)|\left(\frac{1}{1-|T(j\omega)|\,l_m(j\omega)}\right) < \frac{1}{p(j\omega)} \tag{47.51}$$

Equation (47.51) can be rewritten as

$$|S(j\omega)|p(j\omega) + |T(j\omega)|\,l_m(j\omega) < 1 \tag{47.52}$$

or, more explicitly,

$$\left|\frac{1}{1+G(j\omega)}\right|p(j\omega) + \left|\frac{G(j\omega)}{1+G(j\omega)}\right|l_m(j\omega) < 1 \tag{47.53}$$

Equation (47.52) provides some interesting information. First note that keeping the second term < 1 matches (47.29) and guarantees stability robustness. Keeping the first term < 1 matches (47.44) with $G(j\omega)$, replacing $\tilde{G}(j\omega)$, and this provides for acceptable nominal performance, i.e., acceptable performance would result if the plant actually responded like $G_P(j\omega)$. When mismodeling occurs, as represented by $l_m(j\omega)$, the nominal design must exceed the performance specification by enough of a margin to account for modeling error. Alternatively, the nominal design must not only allow enough stability margin to ensure stability, but must allow a greater margin to maintain performance.

It is perhaps even more interesting to view (47.52) as a weighted tradeoff between two terms. The first term contains the magnitude of the nominal sensitivity function, weighted by the performance requirement, which is large at frequencies at which good performance is required. The second term contains the magnitude of the complementary sensitivity function weighted by the bound on the modeling error, which is large at frequencies at which the plant is not well modeled.

Because by (47.49), the sensitivity function and the complementary sensitivity sum to 1, they cannot both be small at the same frequency. Thus, by using (47.52) it can be seen that good control system performance can only be maintained at frequencies at which the plant is well modeled. The modeling error quantified by $l_m(j\omega)$ is usually large at high frequencies. The complementary sensitivity function is then required to be small at high frequencies. A small complementary sensitivity function means a sensitivity function very near 1, dictating that the achievable performance function, $p(j\omega)$, be somewhat < 1 at high frequencies. The resulting implication that the magnitude of the sensitivity function cannot be kept smaller than 1 for all frequencies means poor performance at some frequencies in the areas of disturbance rejection, sensitivity reduction, and reference input tracking. Luckily, in most situations a large performance bound is required only for low-frequency reference inputs and disturbances. Similar logic then dictates that $l_m(j\omega)$, the modeling error, be small at low frequencies. If the control designer is asked to produce strong performance results at frequencies where the modeling error is large, the reply must be that it cannot be done. Either the performance requirements must be relaxed at those frequencies or a more accurate model must be obtained at those frequencies. Much of control design is involved in judiciously squeezing the tradeoff between the two conflicting requirements of good performance and robustness to modeling errors.

References

[1] Bode, H. W.: *Network Analysis and Feedback Amplifier Design*, New York: Van Nostrand, 1945.

[2] Doyle, J. C., B. A. Francis, and A. R. Tannenbaum, *Feedback Control Theory*, New York: Macmillan, 1992.

[3] Doyle, J. C. and G. Stein: "Multivariable Feedback Design: Concepts for a Classical/Modern Synthesis," *IEEE Trans. Automatic Control*, Vol. AC-26 (1) pp. 4–16, February 1981.

[4] Freudenburg, J. S., and Looze, D. P.: "Right half-plane poles and zeros and design trade-offs in feedback systems," *IEEE Trans. Automatic Control*, Vol. AC-30, pp. 555–556, 1985.

[5] Freudenburg, J. S., and Looze, D. P.: *Frequency Domain Properties of Scalar and Multivariable Feedback Systems*, Berlin: Springer-Verlag, 1988.

[6] Horowitz, I.M., *Synthesis of Feedback Systems*, New York: Academic, 1963.

[7] Nyquist, H.: "Regeneration Theory," *Bell System Tech J*, Vol. 11, pp. 126–147, 1932.

[8] Rohrs, C. E., J. L. Melsa, and D. G. Schultz, *Linear Control Systems*, New York: McGraw-Hill, 1993.

[9] Stein G., *Lecture Notes for 6.232*, MIT, Cambridge, MA 1988.

[10] Stein, G. and N. R. Sandell, Jr., *Classical and Modern Methods for Control System Design*, Cambridge, MA: MIT, 1979.

[11] Zames, G.: "On the Input-Output Stability of Time-Varying Nonlinear Feedback Systems," *IEEE Trans. Automatic Control*, Vol. AC-11, pp. 228–238 and pp. 464–476, 1966.

48

Controller Design Using Nyquist-Bode Techniques*

Charles E. Rohrs
Tellabs Research Center
University of Notre Dame

48.1 General Principles for Designing Series Compensators Using Frequency Response Techniques

We consider series compensators of the form given in Fig. 48.1 where $G_p(s)$ is a model of the plant and $G_C(s)$ is the compensator transfer function which is to be designed. The loop gain is given by

$$G(s) = G_C(s)G_P(s) \tag{48.1}$$

Much information is available concerning performance, stability, and robustness of the closed-loop system from the Bode and Nyquist plots of the loop gain, $G(s)$ (see Chapter 47). The series compensator is particularly convenient because we can start with the Bode plot of the plant model $G_p(s)$ and modify that plot as we add new pole and zero factors to $G(s)$ by including them in the compensator. In this way we can build the Bode diagram of $G(s)$ into a desired shape. This process is sometimes referred to as the loop shaping method of design.

In Chapter 47 we saw that it is possible to define a function $p(j\omega)$ such that many aspects of good performance are satisfied if the following is satisfied.

$$|1 + \tilde{G}(j\omega)| > p(j\omega) \tag{48.2}$$

The set of transfer functions represented by $\tilde{G}(j\omega)$ is the set of loop gains in the configuration of Fig. 48.1 when perturbations are present. Let a set of plant models be characterized by

$$\tilde{G}_P(s) = G_P(s)(1 + L_m(s)) \tag{48.3}$$

*Much of the material of this chapter is taken from Rohrs, C. E., J. L. Melsa, and D. G. Schultz, *Linear Control Systems*, McGraw-Hill, 1993. It is used with permission.

-8493-8341-2/95/$0.00+$.50

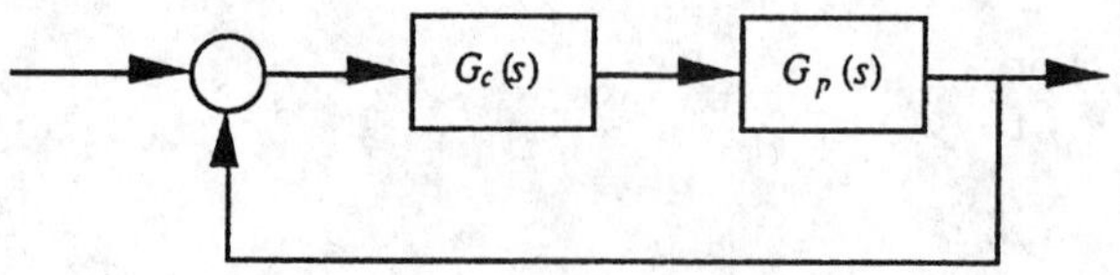

FIGURE 48.1 A series compensator.

$$|L_m(j\omega)| < l_m(j\omega)| \tag{48.4}$$

Using a series compensator the set of loop gains is given by

$$\tilde{G}(s) = G_C(s)\tilde{G}_P(s) \tag{48.5}$$

The nominal loop gain is given by

$$G(s) = G_C(s)G_P(s) \tag{48.6}$$

and using (48.3) to (48.6), we find that

$$\tilde{G}(s) = G(s)(1 + L_m(s)) \tag{48.7}$$

with (48.4) still satisfied. As shown in Chapter 47, the controller is robustly stable and has the performance qualities specified by (48.2) if the following is satisfied for all ω.

$$\left|\frac{G(j\omega)}{1+G(j\omega)}\right| l_m(j\omega) + \left|\frac{1}{1+G(j\omega)}\right| p(j\omega) < 1 \tag{48.8}$$

Satisfying (48.8) requires a tradeoff between the magnitude of the nominal sensitivity function,

$$S(j\omega) = (1 + G(j\omega))^{-1} \tag{48.9}$$

and the magnitude of the nominal complementary sensitivity function.

$$T(j\omega) = G(j\omega)(1 + G(j\omega))^{-1} = 1 - S(j\omega) \tag{48.10}$$

The manner in which the tradeoff is made is determined by the functions $l_m(j\omega)$ and $p(j\omega)$. This tradeoff can be accomplished by manipulating the loop gain transfer function $G(j\omega)$ by adding poles and zeroes to a series compensator. We now develop guidelines as to how the loop gain transfer function should be manipulated so that (48.8) can be satisfied and an acceptable controller can result.

The first thing to notice from (48.8) is that if, for some particular frequency, ω_0, $l_m(j\omega_0) > 1$ and $p(j\omega_0) > 1$, then (48.8) cannot be satisfied. If $l_m(j\omega_0) > 1$ and $p(j\omega_0) > 1$, then

$$\begin{aligned}
&\left|\frac{G(j\omega_0)}{1+G(j\omega_0)}\right| l_m(j\omega_0) + \left|\frac{1}{1+G(j\omega_0)}\right| p(j\omega_0) \\
&\quad > \left|\frac{G(j\omega_0)}{1+G(j\omega_0)}\right| + \left|\frac{1}{1+G(j\omega_0)}\right| \\
&\quad \geq \left|\frac{G(j\omega_0)}{1+G(j\omega_0)} + \frac{1}{1+G(j\omega_o)}\right| = 1
\end{aligned}$$

This fact can be interpreted as showing that aggressive performance objectives cannot be met for frequencies where the system is not well modeled. In general, $l_m(j\omega)$, which represents the modeling error, is small for low frequencies and grows large at high frequencies. To accommodate this, $p(j\omega)$, which represents performance requirements, is made acceptably low at high frequencies where the model is uncertain. Large performance constraint is required and attainable for low-frequency inputs where the model is well known. A typical plot of $l_m(j\omega)$ and $p(j\omega)$ is shown in Fig. 48.2. The frequency axis splits into three regions:

1. Low frequencies where $p(j\omega)$ is large and $l_m(j\omega)$ is small, ($l_m^{-1}(j\omega)$ is large)
2. High frequencies, where $p(j\omega)$ is small and $l_m(j\omega)$ is large, ($l_m^{-1}(j\omega)$ is small)
3. Transition frequencies where $p(j\omega)$ and $l_m(j\omega)$ are both fairly close to 1

The requirements on $G(j\omega)$ so that (48.8) is satisfied can now be developed fairly easily for the high and low frequency sections, while the requirements on $G(j\omega)$ for the transition frequency section are more difficult to determine.

At low frequencies $p(j\omega)$ is much larger than one. In this case (48.8) dictates that the magnitude of the sensitivity function, $S(j\omega)$ must be much less than 1. This situation can be accomplished only if the magnitude of the loop gain, $G(j\omega)$, is made much larger than 1. When $|G(j\omega)|$ is much larger than 1 we can make the following approximations in (48.8). If

$$|G(j\omega)| \gg 1 \tag{48.11a}$$

then

$$\left|\frac{1}{1+G(j\omega)}\right| \approx \frac{1}{|G(j\omega)|} \tag{48.11b}$$

and

$$\left|\frac{G(j\omega)}{1+G(j\omega)}\right| \approx 1 \tag{48.11c}$$

If the approximations are used in (48.8) the following approximate requirement results:

$$\frac{1}{|G(j\omega)|}p(j\omega) < 1 - l_m(j\omega) \tag{48.11d}$$

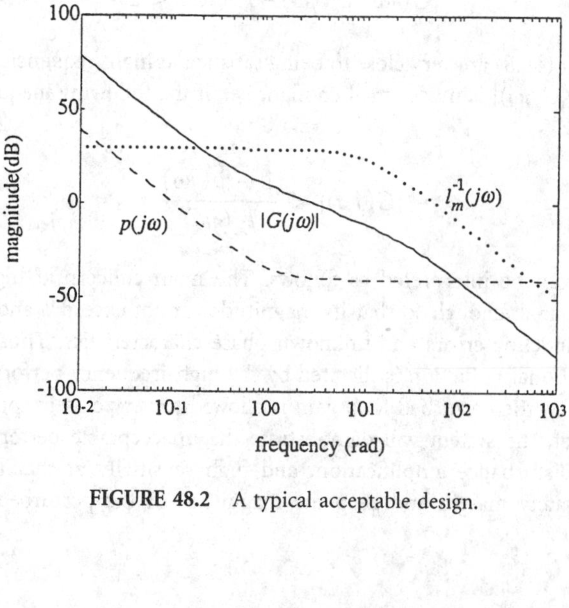

FIGURE 48.2 A typical acceptable design.

Thus, the requirements of (48.8) are very close to being satisfied at low frequencies (where $p(j\omega) \gg 1$ and $l_m(j\omega) \ll 1$) if $|G(j\omega)|$ is made large enough, i.e., if the following inequality, obtained from (48.11d) is satisfied:

$$|G(j\omega)| > \frac{p(j\omega)}{1 - l_m(j\omega)} \tag{48.12}$$

Inequality (48.12) can be interpreted as follows. At low frequencies, large loop gains are required to meet performance requirements. The nominal loop gain must be made slightly larger than the performance requirement dictates so that the performance is achieved in the face of the small amount of modeling error expected at low frequencies.

At high frequencies, $l_m(j\omega)$ is much larger than 1. In this case (48.8) dictates that magnitude of the complementary sensitivity function be much less than 1. However, for the magnitude of the complementary sensitivity function to be much less than 1, the loop gain must be much less than 1:

$$\left|\frac{G(j\omega)}{1 + G(j\omega)}\right| \ll 1 \quad \text{implies} \quad |G(j\omega)| \ll 1 \tag{48.13}$$

When the loop gain is much less than 1, the following approximations can be made. If

$$|G(j\omega)| \ll 1 \tag{48.14a}$$

then

$$\left|\frac{1}{1 + G(j\omega)}\right| \approx 1 \tag{48.14b}$$

and

$$\left|\frac{G(j\omega)}{1 + G(j\omega)}\right| \approx |G(j\omega)| \tag{48.14c}$$

If these approximations are used in (48.8) the following approximation requirement results:

$$|G(j\omega)| l_m(j\omega) + p(j\omega) < 1 \tag{48.14d}$$

The requirements of (48.8) are very close to being satisfied at high frequencies (where $p(j\omega) \ll 1$ and $l_m(j\omega) \gg 1$) if $|G(j\omega)|$ is made small enough, i.e., if the following inequality obtained from (48.14d) is satisfied:

$$|G(j\omega)| < \frac{1 - p(j\omega)}{l_m(j\omega)} \tag{48.15}$$

Inequality (48.15) can be interpreted as follows. The main concern at high frequencies is to keep the loop gain small enough so that its magnitude cannot exceed 1 and cause instability in the face of large modeling errors and unknown phase characteristics. Thus, $|G(j\omega)|$ should be $< l_m^{-1}(j\omega)$. An additional reduction is dictated by the high-frequency performance requirement. If, at some frequency, the perturbed loop gain is allowed to very closely approach the -1 point on the Nyquist plot, the system will demonstrate the unacceptable performance attributes of large oscillations, disturbance amplification, and high sensitivity at that frequency. Thus, we must allow for a safety margin and keep the magnitude of the perturbed loop gain below a threshold which is somewhat < 1. The nominal loop gain should be kept somewhat < $l_m^{-1}(j\omega)$ to provide for not only robust stability but also robust performance at high frequencies.

The following strategy for manipulating the magnitude of the loop gain of a control design is now taking shape. Figure 48.2 shows a plot of $p(j\omega)$, $l_m^{-1}(j\omega)$ and $|G(j\omega)|$ for a possibly acceptable controller design. The loop gain is made larger than $p(j\omega)$ at low frequencies and smaller than $l_m^{-1}(j\omega)$ for high frequencies. There is a safety margin above $p(j\omega)$ at low frequencies to account for possible modeling errors and the fact that approximations were used in attaining (48.12). A safety margin exists below $l_m^{-1}(j\omega)$ at high frequencies to ensure reasonable performance and to account for the approximations used in developing (48.15). Notice that the loop gain need not be above $p(j\omega)$ at high frequencies or below $l_m^{-1}(j\omega)$ at low frequencies.

What qualities should $G(j\omega)$ possess through the transition frequencies? This question is more difficult because the assumptions leading to the approximations (48.11) and (48.14) are not valid. We know that $|G(j\omega)|$ must change from being large at low frequencies to being small at high frequencies. Somewhere in the transition band $|G(j\omega)|$ must pass through the point ω_c where

$$|G(j\omega_c)| = 1 \tag{48.16}$$

This frequency, ω_c, is defined in the control literature as the crossover frequency of the control system. The phase margin, ϕ_m, of the control system is defined as the angle between the polar representation of $G(j\omega_c)$ and the negative real axis

$$\phi_m = 180° + \arg(G(j\omega_c)) \tag{48.17}$$

Much of what we wish to learn about how we should make $G(j\omega)$ behave in the transition frequencies can be seen by studying the control system around ω_c.

It is desirable to create a loop gain with as large a phase margin as possible. A small phase margin means that the Nyquist plot of the loop gain passes very close to the -1 point and indicates two problems. The first problem is that the closed-loop system responds to input and disturbances with poorly damped oscillations at the same frequency as the crossover frequency if the phase margin is small. The second problem is that if ϕ_m is small, a small modeling error in the plant at the crossover frequency may cause the Nyquist plot of the loop again to pass extremely close to the -1 point or actually encircle the -1 point, causing instability. Thus, a nominal performance problem and a robustness problem are associated with a small phase margin.

The combination of a nominal performance and robustness problem can be seen by examining (48.8) at ω_c. Because

$$|G(j\omega_c)| = 1 \tag{48.18a}$$

then

$$\left|\frac{G(j\omega_c)}{1+G(j\omega_c)}\right| = \frac{1}{|1+G(j\omega_c)|} \tag{48.18b}$$

so that (48.8) becomes

$$\left|\frac{1}{1+G(j\omega_c)}\right|(l_m(j\omega_c) + p(j\omega_c)) < 1 \tag{48.18c}$$

At ω_c the magnitude of the sensitivity function which relates to performance and the magnitude of complementary sensitivity function which relates to robustness are both made small by making the magnitude of the return difference large. In fact, (48.18) and (48.8) will both be satisfied at the crossover frequency if the magnitude of the return difference is made larger than the sum of $l_m(j\omega_c)$ and $p(j\omega_c)$:

$$|1 + G(j\omega_c)| > l_m(j\omega_c) + p(j\omega_c) \tag{48.19}$$

However, at the crossover frequency, the magnitude of the return difference can be expressed as a function of the phase margin:

$$|1 + G(j\omega_c)| = 2\left|\sin\left(\frac{\phi_m}{2}\right)\right| \tag{48.20}$$

The largest the return difference can be at the crossover frequency is 2. This maximum occurs when the phase margin is 180°, i.e., when the $\arg(G(j\omega_c)) = 0$. Any phase lag in the loop gain is associated with a decrease in the return difference and an associated drop in the robust performance measure.

Attaining a large phase margin is difficult. In practice, a phase margin of 60° is usually considered to be large. From (48.20) a 60° phase corresponds to a return difference equal to 1 at the crossover frequency. At frequencies higher than the crossover frequency, the return difference is usually slightly smaller than 1. As ω gets large the loop gain approaches zero so that the return difference approaches 1.

In general, $l_m(j\omega)$ grows as ω grows. Equation (48.19) indicates that the crossover frequency should be chosen to be below the frequency where $l_m(j\omega)$ first becomes larger than 1. The crossover frequency is absolutely required to be at a frequency where $l_m(j\omega) < 2$. The crossover frequency, ω_c, marks the boundary between frequencies where the loop gain is large, producing enhanced performance and frequencies where the loop gain is small, producing less desirable performance. Thus, the crossover frequency is often used to describe the bandwidth of a control system. The bandwidth of a control system is limited by the frequencies where an accurate model of the plant can be attained. If perfect modeling were attainable so that $l_m(j\omega)$ were always zero, then any bandwidth could be attained. Such ideal situations, however, exist only in textbooks.

Unmodeled dynamics make $l_m(j\omega)$ large at high frequencies and dictate that the complementary sensitivity function, and thus the loop gain, be rolled off to small values at high frequencies. The desire to enhance performance requires that the return difference, and thus the loop gain, be large at low frequencies. These two desires are fairly easily accomplished. The skill of the control designer comes in making appropriate tradeoffs as the loop gain goes from large values to small values through the transition frequencies.

Two desires conflict in the transition region. We would like the cutoff between low and high frequencies to be as sharp as possible so that the low-frequency region, where performance is enhanced, becomes as large as possible given the high-frequency constraints imposed by modeling uncertainties. The second desire is to keep the amount of phase lag near crossover frequency as small as possible so that the loop gain avoids getting too close to the -1 point or creating extra encirclements.

The fact that these desires conflict can be seen by recalling that the phase plot and the slope of the magnitude plot of a transfer function are related. Assume that the loop gain transfer function of the system is minimum phase. (The situation is worse for nonminimum phase systems.) If a system's magnitude plot has a slope of −40 dB/decade for a decade on either side of the crossover frequency, then the phase shift at the crossover frequency is very close to −180°. Larger negative slopes in the magnitude plot correspond to greater phase lags. In order to achieve a reasonably large phase margin, the magnitude plot must slope gradually through the crossover frequency. If a phase margin of 60° is required, the magnitude plot must have only a -20 dB/decade slope for about 2/3 of a decade on either side of the crossover frequency. Although we would like to transition from high loop gain at low frequencies to low loop gain at high frequencies quickly, the need to maintain a reasonable degree of robust stability and performance throughout the transition region dictates that the transition occur more slowly.

We have now seen the overall desires of how to shape the frequency response of the loop gain transfer function in order to achieve an acceptable level of controller performance and robustness. If a feasible performance specification, $p(j\omega)$, and modeling uncertainty function, $l_m(j\omega)$, are specified, an acceptable controller is one in which the sensitivity function and complementary sensitivity function satisfy (48.8).

48.2 A Design Example Using a Lead-Lag Compensator[1]

The process of designing a control system is demonstrated by the use of a simple, but realistic design problem. The design problem concerns attitude stabilization of satellites. The satellite is geostationary (moves along its orbit at the same rate that the Earth rotates about its axis). Hence, the satellite is stationary with respect to points on Earth. The controller's job is to point the communication antenna at selected receiving areas to within an absolute accuracy of 0.01°. The actual pointing angle is measured with a sensor connected to the communication link itself. Using wave interference techniques, the system can measure the antenna's pointing attitude directly to within approximately 0.01°.

We will consider the control design problem for one axis of motion for this satellite, namely the pitch axis. This axis is normal to the orbit plane and hence, motion about it represents east–west pointing errors of the communication system. Such motions are uncoupled from other attitude motions if all motions are sufficiently small.

Referring to Fig. 48.3(b) we define the variables of interest. The angle θ_p is the pointing angle which is to be controlled. The control is actuated by a reaction wheel, a device that produces torques by changing the speed of a rotating wheel through the use of a DC motor. The control variable V is the voltage driving the motor. The transfer function between the control input, V, and the torque, T, produced by the motor is as follows:

$$\frac{T(s)}{V(s)} = \frac{s}{s+0.01} \tag{48.21}$$

Because essentially no friction exists in space the transfer function between the torque, T, and the angular position, θ_p, is a simple double integrator (with appropriate scaling). The relationship between the control input, V, and the angular position, θ_p is as follows:

$$\frac{\theta_p(s)}{V(s)} = \frac{1}{10^3 s(s+0.01)} \tag{48.22}$$

The angle of the satellite is not sensed instantaneously; dynamics are associated with the sensors. The measured angle, θ_m, is related to the actual angle, θ_p, by the following transfer function:

$$\frac{\theta_m(s)}{\theta_p(s)} = \frac{10}{s+10} \tag{48.23}$$

[1]Much of this material relies on material from Stein and Sandel, *Notes on Control Systems,* MIT, Cambridge, MA, 1978. It is used with permission.

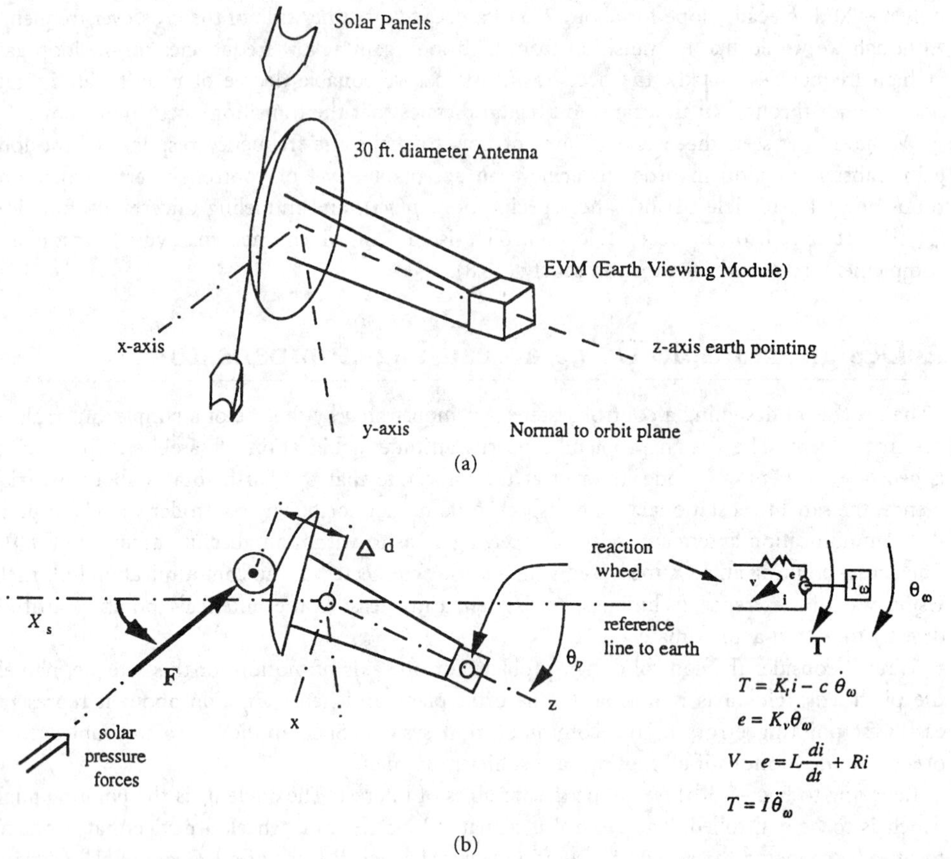

FIGURE 48.3 Diagrams and equations for satellite control design: (a) general diagram (b) variables and equations.

As the sensor follows the pointing angle much faster than the pointing angle is expected to change, we consider the actual and the measured angle to be practically identical and consider the measured angle, $\theta_m(s)$, as our variable both to be fed back and to be controlled. The plant transfer function becomes

$$\frac{\theta_m(s)}{V(s)} = \frac{10}{10^3 s(s + 0.01)(s + 10)} \tag{48.24}$$

The pointing control is to be kept accurate in the face of torque disturbances caused by the pressure of solar radiation on the solar collector panels. The torques are cyclical in nature, with the period equal to the 24 hours it takes the Earth to rotate; the frequency is then equal to 7×10^{-5} rad/s. The maximum torque is 10^{-5} ft. lb. The maximal disturbance is given by

$$d'(t) = 10^{-5} \sin(7 \times 10^{-5} t) \tag{48.25}$$

As the disturbance is itself a torque on the spacecraft it enters the plant at the same place as the control torque, as shown in Fig. 48.4(a). As it is easier to treat disturbances as output disturbances, the disturbance is reflected to the output as in Fig. 48.4(b). The transfer function

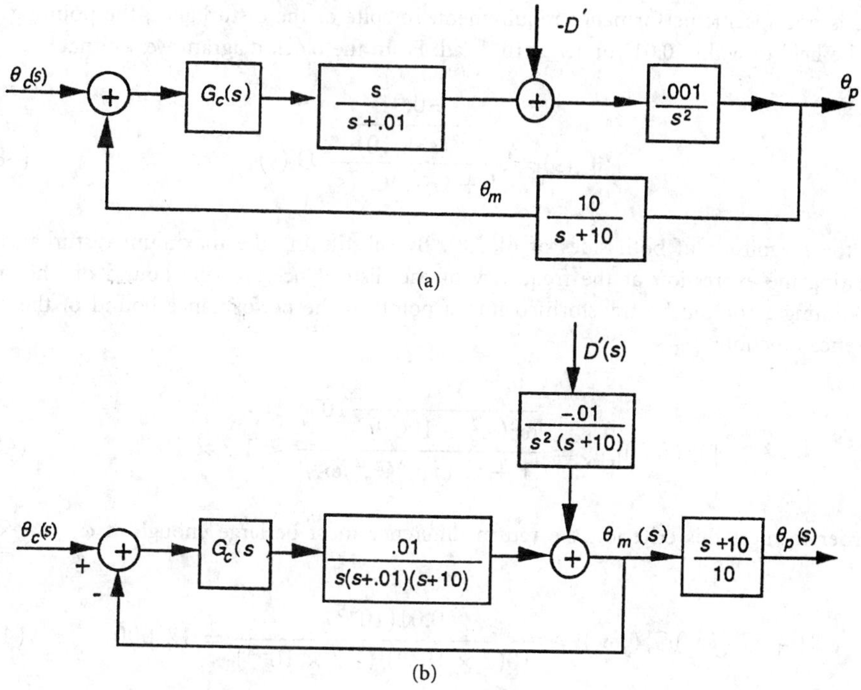

FIGURE 48.4 Block diagrams of the plant: (a) original (b) revised.

from the disturbance $D'(s)$ to $\theta_m(s)$ is given by

$$\frac{\theta_m(s)}{D'(s)} = \frac{-(0.001)(10)}{s^2(s+10)} \tag{48.26}$$

The block diagram manipulations between Fig. 48.4(a) and Fig. 48.4(b) also demonstrate the decision to consider $\theta_m(s)$ as the output to be controlled because it is available for feedback. The resulting loop is in the G-configuration, with the inverse of the sensor dynamics appended in series. Again, this filter can be ignored because it passes with little change signals of frequency < 10 rad/s and the bandwidth of the control loop should be much less than that.

The model is fairly accurate for low frequencies when the satellite is moved slowly enough to avoid exciting any bending modes or flexible resonances. Because the satellite material must be light in order to be launched economically, the satellite's structural members are flexible and oscillate if the satellite is torqued hard or fast enough to create bending. The fundamental bending mode comes from the beams connecting the earth viewing module to the dish antenna. The oscillations are poorly damped due to the lack of atmosphere in space. The resonant frequency occurs at $\omega = 1$ rad/s and produces a resonant peak on the magnitude plot of up to 30 dB along with a sudden phase shift of 180°. Because bending modes at harmonics of 1 rad/s and nonlinearities which become more important at higher frequencies occur, all these model inaccuracies are to be covered by a magnitude bounded multiplicative perturbation

$$\tilde{G}_p(s) = G_p(s)(1 + L_m(s)) \tag{48.27a}$$

$$|L_m(j\omega)| < l_m(j\omega) = \left| \frac{\left(\dfrac{j\omega}{-0.03} + 1 \right)^2}{30} \right| = \frac{1 + \dfrac{\omega^2}{0.0009}}{30} \tag{48.27b}$$

There is one specific performance requirement. In spite of the disturbance, the pointing angle must be held to within 0.01° or 1.7×10^{-4} rad. From the block diagram we see that

$$\theta_m(s) = \frac{\dfrac{-0.01}{s^2(s+10)}}{1 + G_c(s)G_p(s)} D'(s) \tag{48.28}$$

Take the magnitude of both sides of (48.28). By substituting the maximum disturbance and evaluating the expression at the frequency of the disturbance, ω_e, the bound on the size of the pointing error can be transformed into a point on the performance bound of the return difference function:

$$|\theta_m(j\omega_e)|_{\max} = \frac{\dfrac{0.01}{\omega_e^2(\omega_e^2+100)^{1/2}}10^{-5}}{|1 + G_c(j\omega_e)G_p(j\omega_e)|} < 1.7 \times 10^{-4} \tag{48.29}$$

In order to meet this criterion the return difference must be large enough at $\omega_e = 7 \times 10^{-5}$ rad/s:

$$|1 + G_c(j\omega_e)G_p(j\omega_e)| \geq \frac{0.01(10^{-5})}{10(7 \times 10^{-5})^2(1.7 \times 10^{-4})} = 12,000 \tag{48.30}$$

The other performance requirements are general in nature. The return difference should never be much smaller than 1, so that the sensitivity remains low and the responses to disturbances are not too oscillatory or amplifying at any frequency. The bandwidth should be as large as the model accuracy allows, so the response to any disturbance occurs as quickly as possible. The response to a step change in command input, θ_c, should be smooth with little or no overshoot. These requirements can be translated into guidelines for the shape of the loop. Rather than establish a complete performance bounding function before designing the system we try to achieve as good performance as seems sensible and then examine the resulting worst case performance. The one specific requirement we have is from (48.30):

$$p(j\omega_e) = 12,000 \tag{48.31}$$

Otherwise, we expect $p(j\omega)$ to slope down to a value slightly < 1 near crossover and the return to approach 1 asymptotically.

The guiding equation for robust performance is (48.8), repeated here as (48.32):

$$\left|\frac{G(j\omega)}{1 + G(j\omega)}\right| l_m(j\omega) + \left|\frac{1}{1 + G(j\omega)}\right| p(j\omega) < 1 \tag{48.32}$$

In the previous section we made approximations and turned these into conditions on the loop gain for low frequencies and high frequencies, as in (48.12) and (48.13). Equation (48.12), repeated here as (48.33), is accurate at low frequencies when $l_m(j\omega) \ll 1$ and $p(j\omega) \gg 1$:

$$|G(j\omega)| > \frac{p(j\omega)}{1 - l_m(j\omega)} \tag{48.33}$$

The performance requirement given by (48.31) occurs at $\omega_e = 7 \times 10^{-5}$ rad/s. At this frequency, $l_m(j\omega_e) = 1/30$. We extend the requirement of (48.33) for lower frequencies and mark it as an objective for the loop gain on the Bode plot of Fig. 48.5.

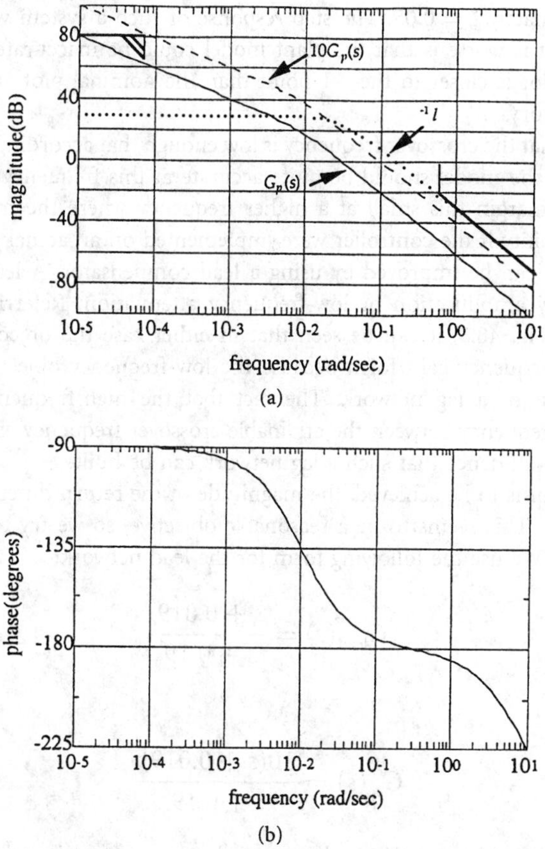

FIGURE 48.5 Bode plots of $G_p(s)$ and $10G_p(s)$: (a) magnitude plots (b) phase plot.

Equation (48.15) repeated here as (48.34), is accurate at high frequencies when $l_m(j\omega) \gg 1$ and $p(j\omega) \ll 1$:

$$|G(j\omega)| < \frac{1 - p(j\omega)}{l_m(j\omega)} \tag{48.34}$$

The approximation is fairly accurate when $l_m(j\omega) > 10$. The curve of $l_m^{-1}(j\omega)$ is sketched as an objective for the loop gain magnitude in Fig. 48.5. The area where $l_m^{-1}(j\omega) < 0.1$ is marked off in a distinct way, as the loop gain must stay below this objective. Although there is no specific performance requirement here, we need to leave a safety margin so that not only stability but also reasonable performance is maintained in the face of perturbations.

Figure 48.5 contains the Bode plots of the plant $G_p(s)$ and $10G_p(s)$. The effect of a simple proportional compensator is seen here. With the additional gain both the low- and high-frequency objectives for the loop gain are met. However, severe problems occur at intermediate frequencies. The crossover frequency is $\omega_c = 0.1$ and the phase margin is only 5°. This is clearly unacceptably small because it implies

$$|1 + G(j\omega)| = 2 \sin \frac{\phi_m}{2} = 0.09 \tag{48.35}$$

If the control design were stopped here, the closed-loop frequency response would contain a peak at ω_c with height $1/0.09 \approx 11$ or 21 dB. This would indicate closed-loop poles with a damping of approximately $\zeta = 0.05$. The step response of such a system would have overshoot and oscillation. What is worse is that the plant model could be inaccurate at $\omega_c = 0.1$ so that the actual Nyquist plot is closer to the -1 point than the nominal plot and the system would be even more oscillatory.

Notice, however, that the crossover frequency is low enough that according to the multiplicative perturbation bound, the model should be quite accurate at this frequency and stability at least is assured. A phase margin this small at a higher frequency where the model is less accurate could result in instability if the controller were implemented on an actual satellite.

The phase margin can be improved by using a lead compensator. A lead network produces either high-frequency amplification or low-frequency attenuation. Referring to the loop gain plots of $10G_p(s)$ in Fig. 48.5, it can be seen that in either case the objectives are missed. We use the unity high-frequency gain form because the low-frequency objective can be recovered by the later addition of a lag network. The fact that the high-frequency constraint allows about 3 decades of frequency between the attainable crossover frequency and the low-frequency constraint provides confidence that such a lag network can be built.

If 60° phase margin is to be achieved, the magnitude of the return difference at the crossover frequency must be 1. This seems to be a reasonable objective, so we try to get 55° of lead out of the lead network. We use the following form for the lead network:

$$G_{\text{lead}}(s) = \frac{s + 0.019}{s + 0.19} \tag{48.36}$$

so that

$$G_{c_1}(s) = \frac{10(s + 0.019)}{s + 0.19} \tag{48.37}$$

The Bode plots of the old loop gain and the new loop gain are given in Fig. 48.6. The new crossover frequency is $\omega_m = 0.06$ with 65° phase margin. The high-frequency gain is unaffected, while the response at low frequency has been attenuated. The next step is to append a lag network to the compensator in order to raise the low-frequency gain while leaving the plots at and above crossover frequency nearly unaffected.

We now design the lag network to meet the low-frequency performance constraint. The key part of the lag network design is to avoid causing problems at the crossover frequency. Because the phase margin of the lead compensated system is 65°, we can afford to place the zero of the lag network only a decade below the crossover frequency, creating 6° phase lag at the crossover frequency. The equation for the lag element is

$$G_{\text{lag}}(s) = \frac{s + 0.005}{s} \tag{48.38}$$

The compensator is now

$$G_{c_2}(s) = \frac{10(s + 0.005)}{s}\,\frac{(s + 0.019)}{(s + 0.19)} \tag{48.39}$$

The Bode plots of the loop gains using compensators $G_{c_1}(s)$ of (48.37) and $G_{c_2}(s)$ of (48.39) are given in Fig. 48.7. It can be seen that the lag compensator raises the low frequency gain while leaving the frequencies higher than the crossover frequency unaffected.

The final loop gain using $G_{c_2}(s)$ as a compensator looks very good. Both high- and low-frequency objectives are met with enough margin to expect adequately robust performance. The phase margin of 60° also indicates adequately robust performance.

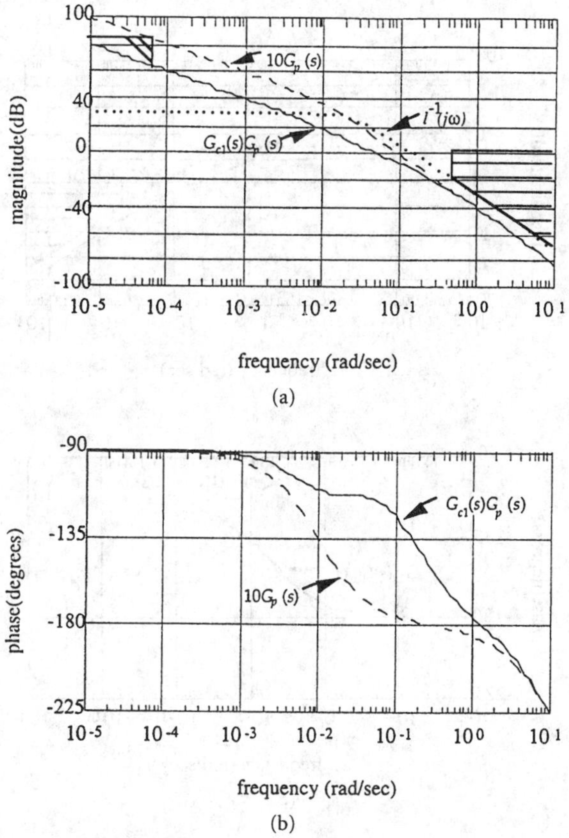

FIGURE 48.6 Bode plots of $10G_p(s)$ an $G_{c_1}(s)G_p(s)$: (a) magnitude plots (b) phase plots.

It is time now to visually check the sensitivity function and the complementary sensitivity function. In addition, the closed-loop step response should be checked.

The plots of Fig. 48.8 show the sensitivity function and the complementary sensitivity function for the loop with the controller $G_{c_2}(s)$. These plots can be used to check the robust performance condition of (48.8). First note that the specific low-frequency disturbance rejection requirement is easily met, as seen by the plot of the sensitivity function. This ascends from very small values at low frequency to a maximum value of about 2 dB at $\omega = 0.1$ and returns toward unity at high frequencies. The plot of the magnitude of the complementary sensitivity function is near unity for low frequencies, has a peak of about 1 dB at $\omega = 0.03$, and falls of rapidly, remaining at least 10 dB below the $l_m^{-1}(j\omega)$ curve. This 10-dB gap indicates that not only does the system remain stable in the face of all allowable perturbations, but reasonable performance is also maintained.

The sensitivity function that results when the system is confronted with the worst-case allowable perturbation is computable using (47.50). In Fig. 48.9 the nominal sensitivity function and the worst-case sensitivity function are plotted. Even when the worst-case perturbation is used the maximal sensitivity is about 4 db = 1.6. This indicates that the system performs robustly for the mismodeling considered.

The last measure of the design to check is the closed-loop response to a step change in command angle θ_c. This step response is plotted in Fig. 48.10. Despite no indications of a complex pole pair from the flat closed-loop frequency response curve (i.e., the complementary sensitivity function curve) of Fig. 48.9, the closed-loop step response displays some overshoot. The overshoot is a common problem with the lead compensator. The overshoot arises from the

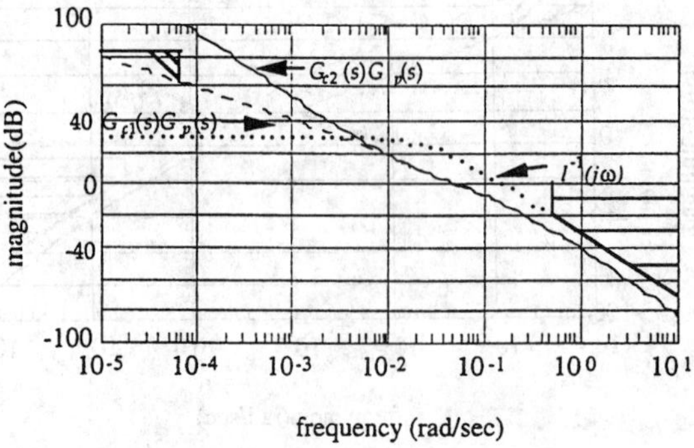

(a)

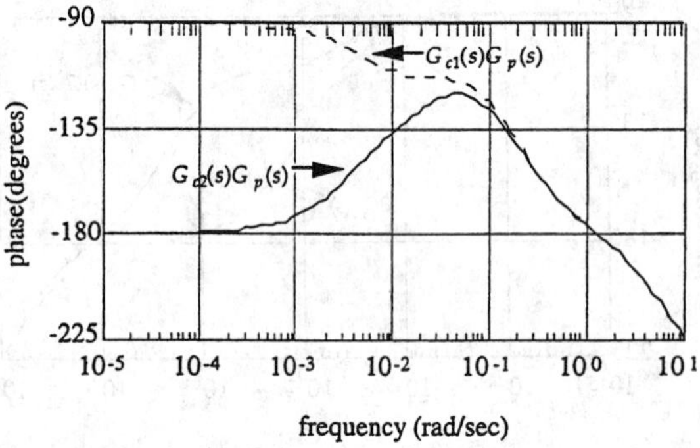

(b)

FIGURE 48.7 Bode plots of $G_{c_2}(s)G_p(s)$: (a) magnitude plots (b) phase plots.

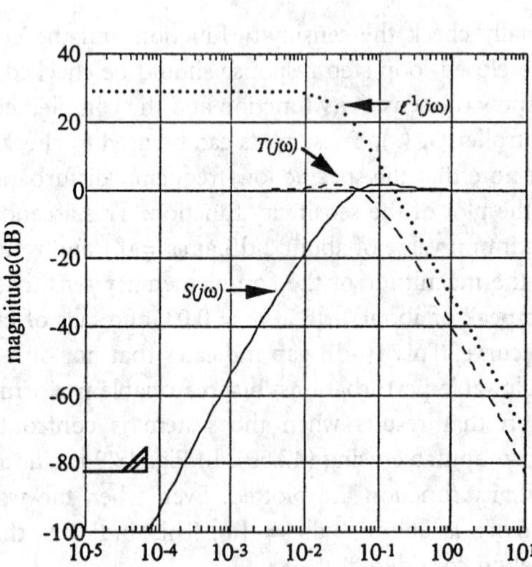

FIGURE 48.8 Sensitivity function, $S(j\omega)$, and complementary sensitivity function, $T(j\omega)$, plots for loop gain $G_{c_2}(s)G_p(s)$.

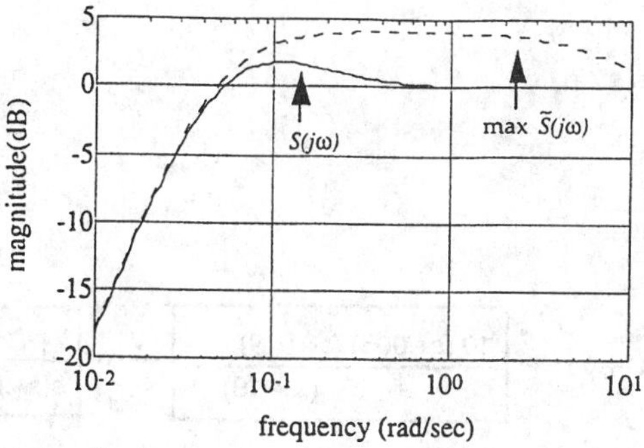

FIGURE 48.9 Nominal, $S(j\omega)$, and worst-case perturbed, $\max \tilde{S}(j\omega)$, sensitivity functions.

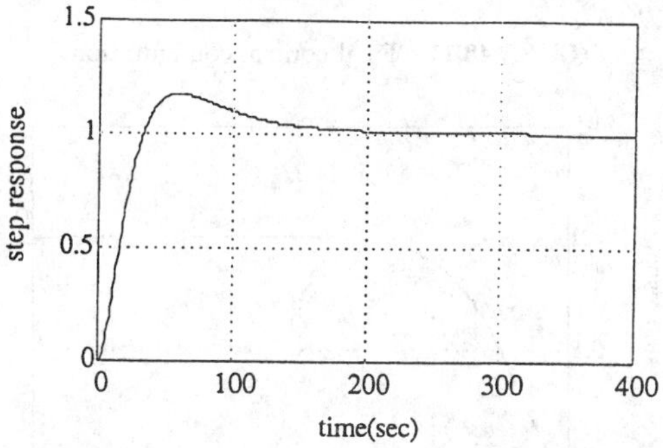

FIGURE 48.10 Step response of closed-loop system.

fact that the zero of the lead network remains closer to the origin than the nearest closed-loop pole. The cause of the overshoot is more clearly shown by examining the closed-loop transfer function:

$$M(s) = 0.1 \frac{(s+0.005)(s+0.019)}{(s+0.00517)(s+0.033)(s+0.05)(s+0.11)(s+10)} \tag{48.40}$$

The system has a pole–zero pair from the lag network which essentially cancels at $s = -0.005$. The next element is a zero closer to the origin than the first dominant pole. This is the cause of the overshoot. Because the zero is part of the compensator, it will not appear in the transfer function from either an input disturbance or an output disturbance. Because the overshoot affects only the closed-loop transient response, it is easily remedied with a prefilter which cancels the offending zero.

$$P(s) = \frac{0.019}{s+0.019}$$

The resulting control configuration is given in Fig. 48.11, where block diagram manipulation has been used to separate the sensor dynamics from the plant. The final step response appears in Fig. 48.12. It is smooth. The time constant of approximately 20 s agrees with the loop gain bandwidth of 0.05 rad/s shown in Fig. 48.7. The response can be made faster if desired by further manipulation of the prefilter.

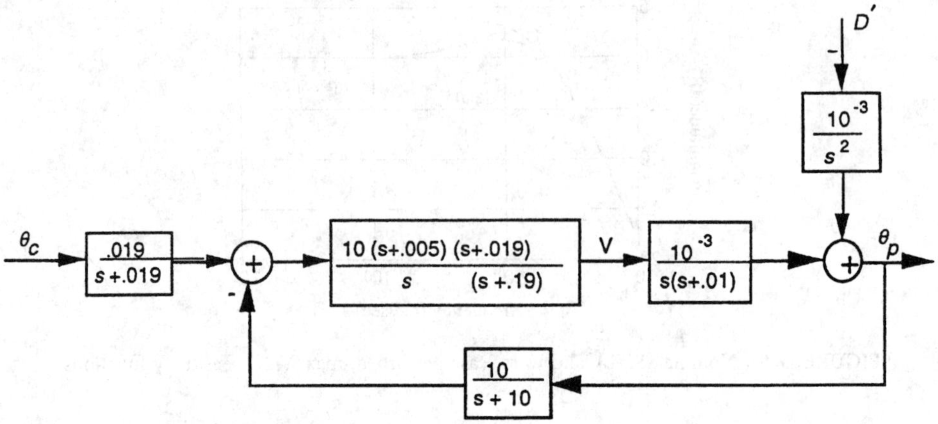

FIGURE 48.11 Final control configuration.

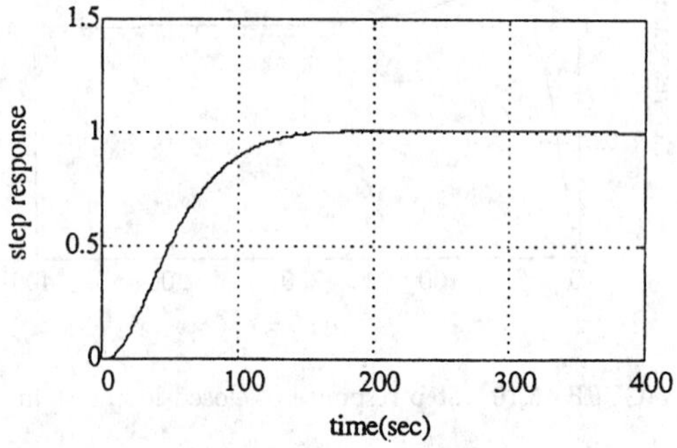

FIGURE 48.12 Step response of closed-loop system with prefilter.

48.3 Controlling Unstable Plants[1]

One of the major advantages of designing closed-loop control systems is that an unstable plant can be stabilized and also meet performance criteria. In this section we consider control systems designs starting with plants that contain poles in the right-half-plane, i.e., unstable plants. It is interesting that the key technique used to stabilize loops with an unstable plant is a lead network, the same lead network that is used to achieve an improved margin of stability in loops with stable plants. Indeed, there is little difference between improving the performance of a stable loop gain and creating acceptable performance with an initially unstable loop gain, except for urgency inherent in stabilizing an otherwise unstable system.

Compare a controller placed in series with two plants, one stable and one unstable. Consider the stable plant

$$G_{p_1}(s) = \frac{100}{(s+0.1)(s+1)(s+10)} \tag{48.41}$$

and the unstable plant

$$G_{p_2}(s) = \frac{100}{(s+0.1)(s-1)(s+10)} \tag{48.42}$$

[1]The author is indebted to Doug Looze for crystalizing the ideas of this section.

If these plants are placed in feedback loops with constant gain controllers the two root-loci of Fig. 48.13 results. The two loci are very similar except for the location of the $j\omega$-axis, which indicates that the loop with the stable plant remains stable for small gains, while the loop with the unstable plant is unstable for all constant gain controllers.

We know that the zero of a lead network pulls the poles of the plant of (48.41) further into the left-half-plane and allows a higher gain while retaining stability. The same effect occurs when a lead network is used in series with the unstable plant of (48.42). Consider the lead compensator

$$G_c(s) = \frac{K(s+1)}{s+10} \tag{48.43}$$

The root loci of the combination of this lead compensator in series with each of the plants is shown in Fig. 48.14. In Fig. 48.14(b), it can be seen that for intermediate values of the gain K, the closed-loop system is stabilized.

To further determine the behavior of the system with the unstable plant, the Bode and Nyquist plots of the system must be examined. In the compensator of (48.43), let $K = 3$. We examine the Bode plots of the plant, $G_{p_2}(s)$, given in (48.42), and the combination of the plant, $G_{p_2}(s)$, with the series compensator, $G_c(s)$, given by (48.43):

$$G_c(s)G_{p_2}(s) = \frac{300(s+1)}{(s+0.1)(s-1)(s+10)^2} \tag{48.44}$$

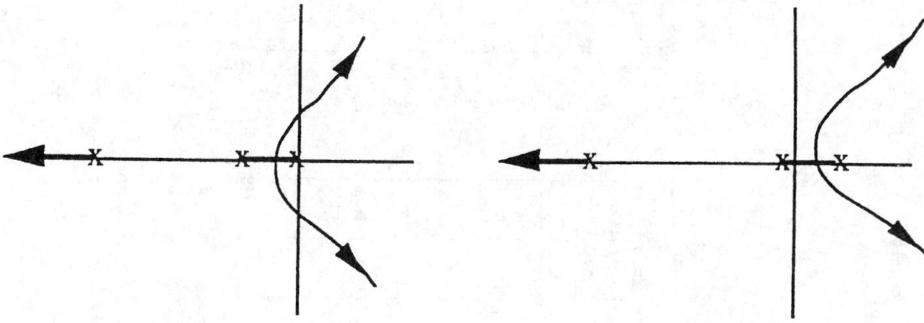

(a) (b)

FIGURE 48.13 Root loci from a stable (a) and an unstable (b) plant.

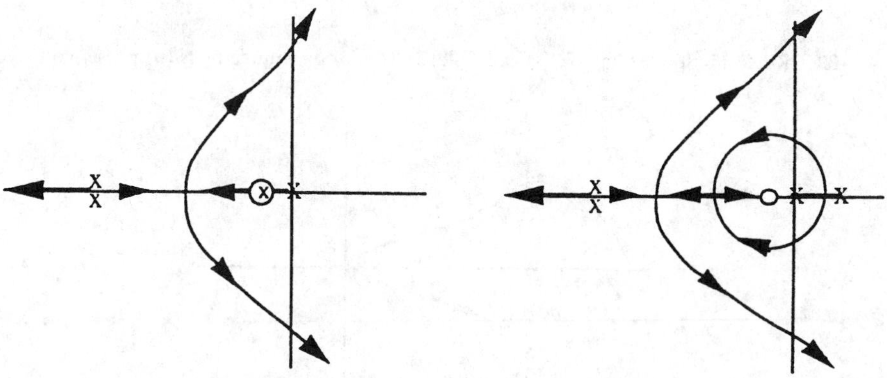

(a) (b)

FIGURE 48.14 (a) The stable plant with lead network. (b) The unstable plant with lead network.

The Bode plots are given in Fig. 48.15. The usual effect of the lead network can be seen in these Bode plots. From the Bode plot, the Nyquist plots can be drawn. The entire Nyquist plot of the unstable plant, $G_{p_2}(s)$ is sketched (not to scale) in Fig. 48.16. The Nyquist plot contains one clockwise encirclement of the -1 point. Because the plant has one right-half-plane pole, the Nyquist analysis indicates that a unity feedback controller in a closed-loop system with this plant produces two unstable closed-loop poles. This observation is in agreement with the root locus of Fig. 48.13(b).

The entire Nyquist plot of $G_c(s)G_{p_2}(s)$ is sketched (not to scale) in Fig. 48.17. The lead network has pulled the portion of the Nyquist plot corresponding to intermediate positive frequency values below the negative real axis on the Nyquist plot. This produces a counterclockwise encirclement of the -1 point and a stable closed-loop system. Thus, we see with frequency domain analysis that an unstable plant can be stabilized with the addition of a lead compensator. We should note that further compensation may be placed in series with $G_c(s)G_{p_2}(s)$ to further improve the performance and robustness of the system. Once the Nyquist plot has the correct number of encirclements, a further lead network can pull the plot further away from the -1 point, giving a better phase margin and less oscillatory response.

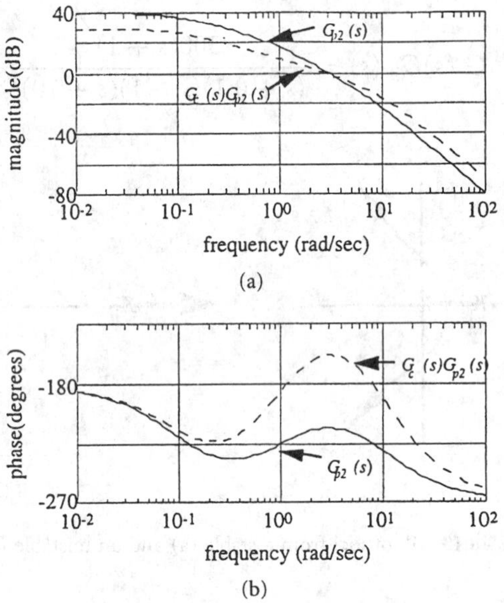

FIGURE 48.15 Bode plots $G_{p_2}(s)$ and $G_c(s)G_{p_2}(s)$: (a) magnitude plots (b) phase plots.

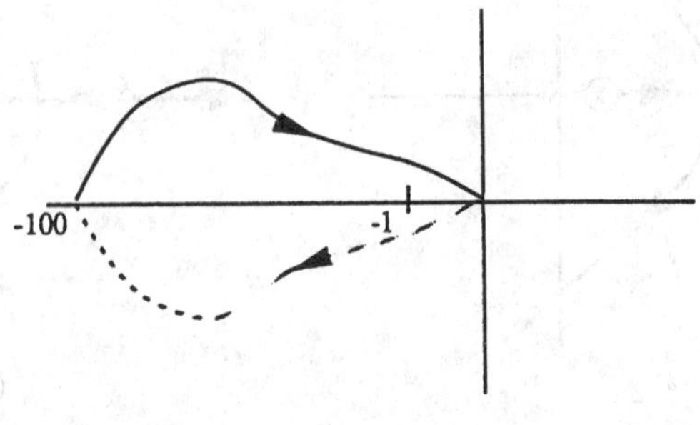

FIGURE 48.16 The Nyquist plot of $G_{p_2}(s)$.

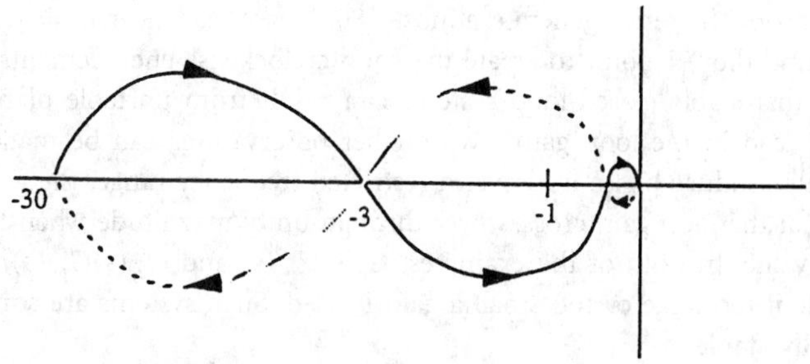

FIGURE 48.17 The Nyquist plot of $G_c(s)G_{p_2}(s)$.

It is interesting to see the Nyquist plots before and after a plant with two unstable poles is stabilized. Let the plant be given by

$$G_{p_3}(s) = \frac{10}{(s+0.1)(s-1)^2} \tag{48.45}$$

A sketch of the entire Nyquist plot (not to scale) of $G_{p_3}(s)$ given by (48.45) is shown in Fig. 48.18. There are no encirclements of the -1 point indicating two unstable closed-loop poles.

It is not clear how one could possibly create two counterclockwise encirclements starting from this plot. However, if enough phase lead can be created, the Nyquist plot of Fig. 48.19 can result. This plot has the required two counterclockwise encirclements to create a stable closed-loop system. The transfer function used to sketch Fig. 48.19 is

$$G_{c_2}(s)G_{p_3}(s) = \frac{6(s+0.1)^2}{(s+10)^2}\,\frac{10}{(s+0.1)(s-1)^2} \tag{48.46}$$

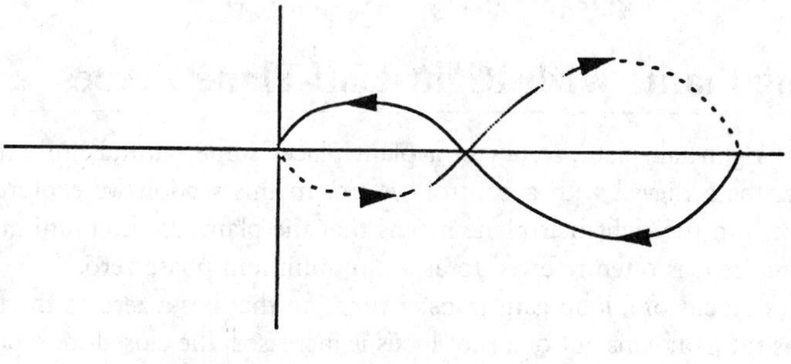

FIGURE 48.18 Nyquist plot of $G_{p_3}(s)$.

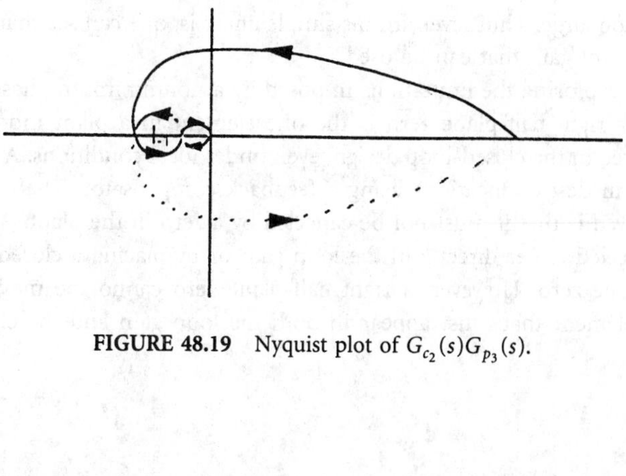

FIGURE 48.19 Nyquist plot of $G_{c_2}(s)G_{p_3}(s)$.

The two lead networks in series generate almost 180° phase lead near $\omega = 1$. This brings the Nyquist plot around the -1 point to create the counterclockwise encirclements.

We have seen that stable closed-loop systems can result from unstable plants by the introduction of phase lead in the loop gain. Two further observations can be made. First, because lead networks only produce phase lead over a restricted frequency range, the gain of the system must be set so that the loop gain crosses over through unity magnitude when the phase lead is near its greatest value. In both of the examples, $G_c(s)G_{p_2}(s)$ and $G_{c_2}(s)G_{p_3}(s)$, the closed-loop system is unstable if too large or too small a gain is used. Such systems are sometimes referred to as **conditionally stable.**

Second, recall that on a Bode phase plot right-half-plane poles produce the same effect as left-half-plane zeroes, i.e., they produce phase lead. In general, so much phase lead is needed to create counterclockwise encirclements on the Nyquist plot that both the lead of a series compensator and the lead produced by unstable poles themselves are needed near the crossover frequency.

There is a subtle indication here. Because the lead contribution from the unstable poles usually must be allowed to develop before the crossover frequency, it is a general rule of thumb that the bandwidth of control systems that stabilize unstable plants cannot be much less than the frequency associated with the highest frequency unstable pole. Unstable plant poles must be actively, not passively controlled. The area of frequency where the return difference is > 1 is the area of active control. The area of active control should include the frequencies around any unstable pole.

The need for the bandwidth to be large enough to actively control the highest frequency unstable poles means that the frequency range where little modeling uncertainty exists must extend beyond the frequency of the highest frequency unstable pole. If such is the case the loop gain can still be rolled off to meet the magnitude constraint given by the unmodeled dynamics at high frequencies after attaining the necessary bandwidth to stabilize the loop.

Now that we have seen how right-half-plane poles affect controller designs, it is natural to investigate how right-half-plane zeroes affect controller designs.

48.4 Controlling Plants with Right-Half-Plane Zeroes

The presence of right-half-plane zeroes in a plant places some natural limitations upon what performance can be achieved with a control system. In this section we explore some of these limitations. A zero in the right-half-plane means that the plant is a nonminimum phase plant. A right-half-plane zero is often referred to as a nonminimum phase zero.

Picture the root locus of a loop gain transfer function that has a zero in the right-half-plane. We know that as the gain constant of a root locus is increased, the closed-loop poles move either toward the loop gain zeroes or toward infinity. If the loop gain has a right-half-plane zero, a closed-loop pole must approach that zero and create an unstable closed-loop pole if the gain constant is made too large. Thus, even in this simple analysis, one can see that a right-half-plane zero limits the control gain that can be used.

Another way of exploring the limitations imposed by a nonminimum phase zero in the plant is to realize that a right-half-plane zero is the one element in a plant model that cannot be effectively eliminated in the closed-loop design, even under ideal conditions. A designer can place closed-loop poles in desired locations using a feedback compensator. The only restrictions on the pole to be moved is that it must not be canceled by a zero in the plant. Also, left-half-plane zeroes can be canceled either directly in the loop gain or by placing a closed-loop pole on top of the left-half-plane zero. However, a right-half-plane zero cannot be moved or canceled. It is the one plant element that must appear in both the loop gain and the closed-loop transfer function.

Because the final loop gain must be zero at the location of a right-half-plane zero, the type of return difference transfer function and sensitivity function that can be achieved are restricted by the fact that each of these functions must equal 1 when evaluated at a right-half-plane plant zero. Arguments from the mathematical theory of complex variables can be used with this fact to show that when the plant contains a right-half-plane zero, the sensitivity function evaluated along the $j\omega$-axis must meet certain conditions.[1] These conditions indicate that if the sensitivity is made < 1 at some frequencies it must become significantly larger than 1 at other frequencies. The closer the right-half-plane zero is to the $j\omega$-axis, the more severe are the tradeoff restrictions. Also, if a right-half-plane zero is close to a right-half-plane pole, the tradeoff restrictions indicate that the sensitivity function must be much larger than 1 for some frequencies. Of course, many of the goals of designing closed-loop control systems can be translated into the objective of achieving a large return difference, and thus a small sensitivity function. Right-half-plane zeroes are natural obstructions to this objective.

Let us examine how a nonminimum phase zero affects attempts to design an appropriate loop gain for a control system. The objectives of the design effort were explained previously. A designer would like a large loop gain at small frequencies with a sharp transition to a small loop gain at high frequencies. The sharpness of the transition from large loop gain to small loop gain is limited by the fact that large negative slopes on the magnitude plot are associated with large amounts of phase lag. Large amounts of phase lag cause stability, performance, and robustness problems near the crossover frequency. The control engineer is constantly trying to negate the effect of phase lag. We have seen the importance of phase lead networks. In the previous section we saw that phase lead is even more important if unstable poles are present in the loop gain.

The effect of a nonminimum phase zero upon the Bode plot of a loop gain is a deadly combination of an increase in magnitude in conjunction with an increase of phase lag. Remember in the transition frequency region, the control engineer is trying to sharply decrease the magnitude while minimizing the phase lag. The nonminimum phase zero works in direct opposition to the objectives of the control designer and it cannot be eliminated by cancellation. No wonder that plants with nonminimum phase zeroes are the most difficult to control!

Notice that the effect of a nonminimum phase zero is minimal if the zero occurs at a frequency that is much higher than the crossover frequency. In the high frequency range the magnitude of the loop gain is small enough to keep the sensitivity function near 1 independent of the phase. Therefore, the extra phase lag from a nonminimum phase zero in this frequency range has little effect.

An example follows which shows the difficulty presented by a nonminimum phase zero.

Example 48.1. Consider a plant described by the following transfer functions

$$G_{p_1}(s) = G_{p_0}(s)G_{AP}(s) \tag{48.47}$$

where

$$G_{p_0}(s) = \frac{0.1}{s\left(\frac{s}{0.01}+1\right)\left(\frac{s}{10}+1\right)} \tag{48.48}$$

[1]See Feudenberg, J. S. and D. P. Looze: "Right half-plane poles and zeros and design tradeoffs in feedback systems," *IEEE Trans. Automatic Control*, Vol. AC-30, pp. 555–565, 1985.

and

$$G_{AP}(s) = \frac{(-s+0.01)}{(s+0.01)} = \frac{\left(\frac{-s}{0.01}+1\right)}{\left(\frac{s}{0.01}+1\right)} \tag{48.49}$$

The transfer function $G_{AP}(s)$ is the same as the plant transfer function of the satellite which was examined in an earlier section. Let us examine $G_{AP}(s)$ and see how it affects the overall transfer function.

The transfer function, $G_{AP}(s)$, contains a right-half-plane zero with a pole located in the left-half-plane at the mirror image of the zero. Notice that the pole and the zero in this arrangement have exactly equal and opposite effects on the Bode magnitude plot. Indeed, the magnitude plot of $G_{AP}(s)$ is equal to unity at all frequencies:

$$|G_{AP}(j\omega)|^2 = \left(\frac{0.01-j\omega}{0.01+j\omega}\right)\left(\frac{0.01+j\omega}{0.01-j\omega}\right) = 1 \tag{48.50}$$

It is said that $G_{AP}(s)$ is an all-pass transfer function because, interpreted as a filter, it passes all frequencies equally. Figure 48.20 shows the Bode plots of $10G_{p_1}(s)$ and $10G_{p_0}(s)$. Unlike the magnitude plot, the phase plot of $G_{p_1}(s)$ is greatly affected by the all-pass section with its nonminimum phase zero. The left-half-plane pole and the right-half-plane zero of (48.49) each add about 45°/decade of phase lag to the phase plot in the area between $\omega = 0.001$ and $\omega = 0.1$. Thus, by the point $\omega = 0.1$ the plant $G_{p_1}(s)$ has almost 180° more phase lag than the plant $G_{p_0}(s)$. This can be seen in Fig. 48.20.

Recall that a major difficulty in the design of the satellite system is to achieve enough phase lead through the crossover frequency range so that a good phase margin and transient response result. With the plant $G_{p_1}(s)$, this problem is exacerbated by the fact that almost 180° of additional phase lead is needed if the crossover frequency is to remain near $\omega = 0.1$.

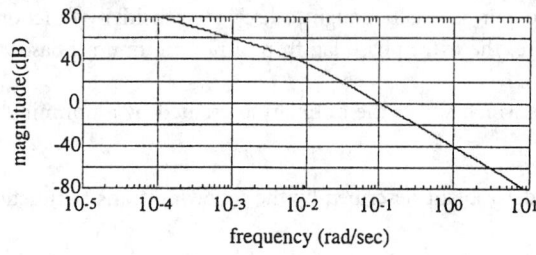

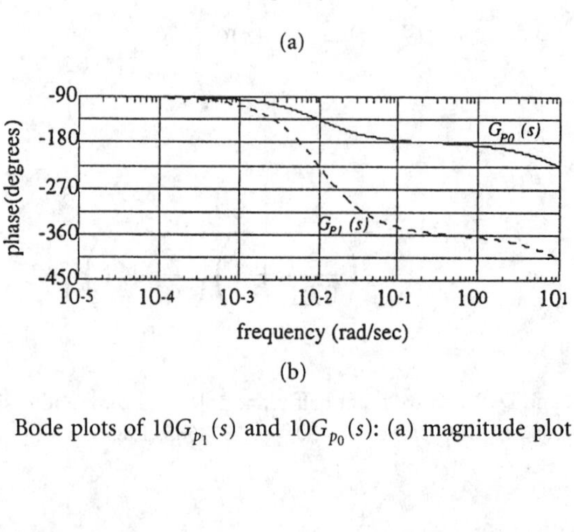

FIGURE 48.20 Bode plots of $10G_{p_1}(s)$ and $10G_{p_0}(s)$: (a) magnitude plot (b) phase plots.

One can try to use a series compensator to negate the effects of the all-pass section of the plant. The pole at $s = -0.01$ can be directly canceled by a zero. Unfortunately, the zero at $s = 0.01$ cannot be canceled. However, its effect on the phase plot can be completely countered by adding another zero at $s = -0.01$. The magnitude plot is greatly affected. For the moment, ignore the fact that a compensator with two zeroes and no poles is unrealizable. We add more poles later. Examine the behavior of

$$G_{L_1}(s) = G_{p_1}(s)G_{C_1}(s) \tag{48.51}$$

with

$$G_{C_1}(s) = 0.01\left(\frac{s}{0.01} + 1\right)^2 \tag{48.52}$$

The Bode plots of $G_{L_1}(s)$ and $10G_{p_0}(s)$ are given in Fig. 48.21. The phase plot of $G_{p_0}(s)$ has been recovered by $G_{L_1}(s)$. However, now the magnitude plot is much flatter. (The difference between $G_{p_0}(s)$ and $G_{L_1}(s)$ is two zeroes: the nonminimum phase zero at $s = +0.01$ and the zero at $s = -0.01$, which restores the phase plot.)

The magnitude of $G_{L_1}(s)$ at $\omega = 0.1$ is 1.005, while the value of the magnitude of $G_{L_1}(s)$ at $\omega = 10$ is 0.7. If this system is to meet the kind of performance constraints that the loop in the earlier section needed to meet, a number of poles need to be added well below $\omega = 0.7$ in order to provide enough roll off in the magnitude plot to meet the high frequency constraint. These poles will cause phase lag in the range $\omega = 0.01$ to $\omega = 0.1$. In addition, a multiple order phase lag is necessary to raise the magnitude over the low-frequency performance constraint. This will further add to the phase lag near crossover. Consider the transfer function

$$G_{L_2}(s) = G_{L_1}(s)G_R(s)G_{\text{lag}}(s) \tag{48.53}$$

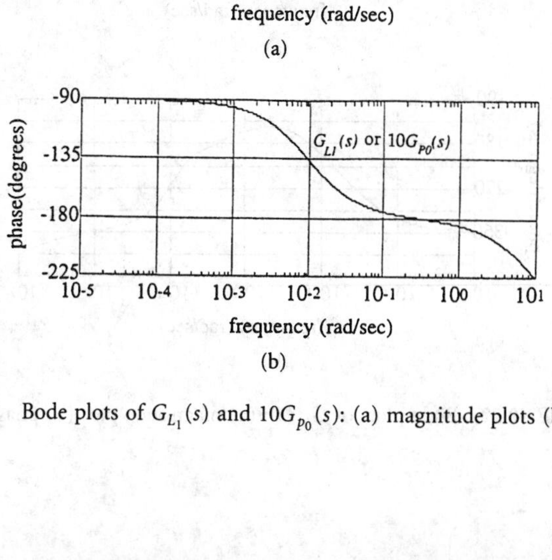

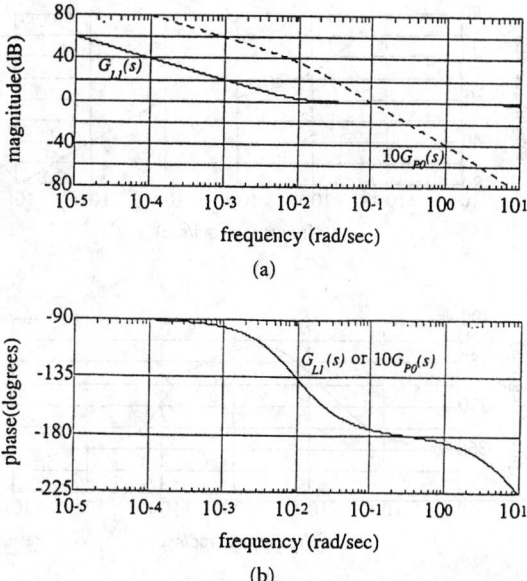

FIGURE 48.21 Bode plots of $G_{L_1}(s)$ and $10G_{p_0}(s)$: (a) magnitude plots (b) phase plots.

where

$$G_R(s) = \frac{1}{(10s+1)^2} \tag{48.54}$$

is used for rolloff and

$$G_{\text{lag}}(s) = \frac{(s+0.001)^2(s+0.0003)}{s^3} \tag{48.55}$$

is a third-order lag network.

The Bode plots of $G_{L_2}(s)$ are given in Fig. 48.22. Notice that the system now meets the magnitude plot objectives with some margin for added robustness. Notice, however, that magnitude plot near crossover is very flat and that the phase lag is $> 180°$ with a large slope at crossover. The phase lag at the crossover frequency cannot be decreased because the increase in the slope of the magnitude plot which accompanies a phase lead network causes the crossover frequency to move out to the area where the phase lag is too large to be overcome. The bandwidth must be decreased. A reasonable phase margin can be achieved by decreasing the crossover frequency without the benefit of a lead network. Simply allowing

$$G_{L_3}(s) = \frac{G_{L_2}(s)}{3} \tag{48.56}$$

lowers the magnitude plot about 10 dB. The final crossover frequency occurs at $\omega = 0.004$, with a phase margin of approximately 30°. A polar plot of $G_{L_3}(s)$ for $\omega \geq 10^{-3}$ is shown in Fig. 48.23. (Use this plot and the transfer function $G_{L_3}(j\omega)$ to draw the entire Nyquist plot of $G_{L_3}(j\omega)$ and convince yourself that the closed-loop design is stable. Be careful how you treat the indentation around the four poles at the origin.) The plot of Fig. 48.23 indicates that the minimum value of the return difference for this design is one half or -6 dB.

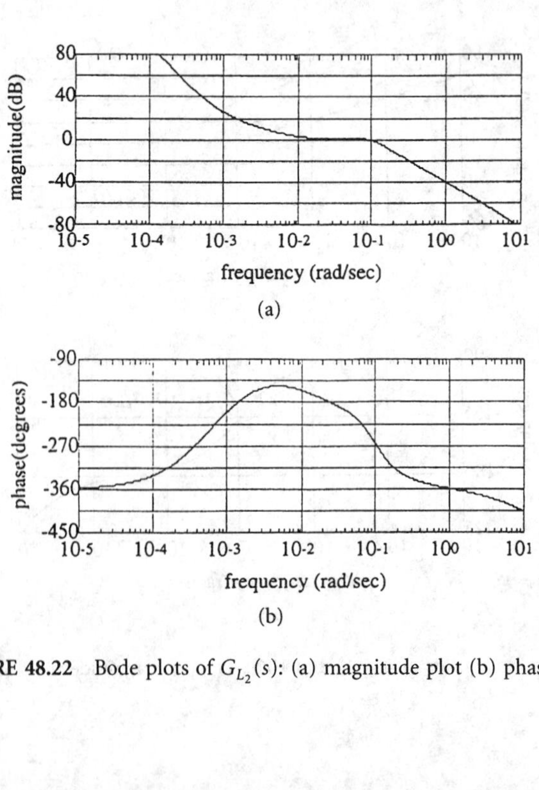

FIGURE 48.22 Bode plots of $G_{L_2}(s)$: (a) magnitude plot (b) phase plot.

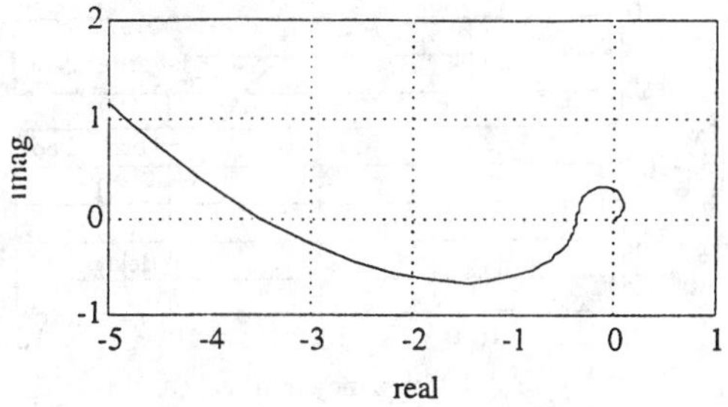

FIGURE 48.23 Polar plot of $G_{L_3}(s)$.

The key difference between the design here starting with a nonminimum phase plant and a design starting with a minimum phase plant is that the nonminimum phase zero in this design forces a reduction of bandwidth. Indeed, the bandwidth is reduced to $\omega_c = 0.004$. This is below the frequency of $\omega = 0.01$ associated with the nonminimum phase zero. It is fortunate that the low-frequency performance specification in this problem is loose enough that it was possible to meet the specification, even with the reduced bandwidth.

It is a general rule that nonminimum phase zeroes restrict the bandwidth of a control system. The bandwidth is usually restricted to be less than the frequency of the lowest frequency nonminimum phase zero so that the phase lag from all the nonminimum phase zeroes is absorbed when the magnitude is low. Couple this rule of thumb with the guideline of the last section, which says that the bandwidth of a system is usually larger than the frequency associated with an unstable pole. It can be seen that a right-half-plane pole at a frequency higher than the right-half-plane zero is a devastating combination. In such a situation something should be done to physically redesign the plant and eliminate either the pole or the zero.

It is important to understand how right-half-plane zeroes physically arise so that they can be eliminated or avoided if plant redesign is possible. Two common physical situations manifest themselves as right-half-plane zeroes in a system model. Right-half-plane zeroes arise due to time delays in a system and they arise when the measurement of important variables cannot or does not occur. We address time delays here.

The impulse response for a system which consists of a time delay of T seconds is

$$h(t) = \delta(t - T) \tag{48.57}$$

The system function is given by

$$H(s) = e^{-sT} \tag{48.58}$$

Taking the magnitude and phase of the frequency response we get

$$\begin{aligned} |H(j\omega)| &= 1 \quad \text{for all} \quad \omega \\ \arg(H(j\omega)) &= -\omega T \end{aligned} \tag{48.59}$$

A delay system is an all-pass system with a phase lag that increases linearly with frequency. A Bode phase plot of the system with a delay of 1 s is shown in Fig. 48.24. Notice that the phase lag which increases linearly with ω becomes an exponential curve on the semilog Bode phase plot. One can see that it is extremely difficult to overcome this phase lag over a substantial frequency range, as lead networks produce a phase increase that is only linear on the Bode plot, and this increase only impacts a small frequency range.

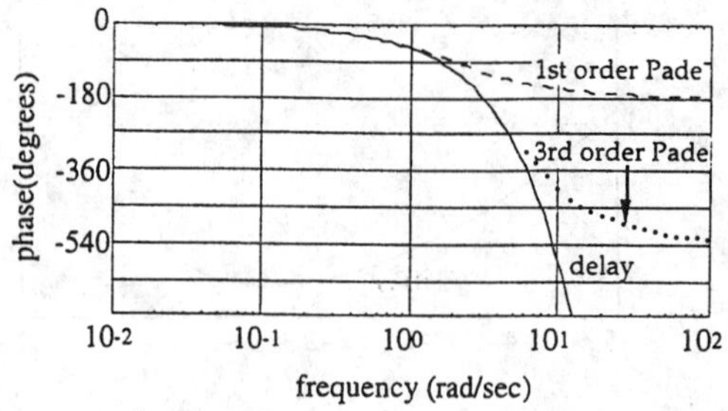

FIGURE 48.24 Bode phase plots of a time delay and approximations.

The frequency response of the delay can be approximated over a finite frequency range by an all-pass network consisting of right-half-plane zeroes and mirror image left-half-plane poles. The all-pass network matches the magnitude plot of delay exactly because both systems are unity magnitude for all frequencies. The phase lag of the all-pass network using poles and zeroes can match the phase lag of the delay over a limited frequency range. This frequency range can be extended by adding more poles and zeroes. The total phase lag from each pole–zero combination is 180°. The phase lag from the delay increases without bound.

A reasonable position of the zeroes and poles can be found by creating a series expansion for the two transfer functions and equating as many terms as possible. Such an approximation is called a Padé approximation. We assume that the number of zeroes equals the number of poles and call the number of poles the order of the approximation. We now demonstrate the technique on a first-order Padé approximation

$$e^{-sT} = 1 - sT + \frac{s^2T^2}{2} - \frac{s^3T^3}{6} + \cdots$$

$$\frac{a-s}{a+s} = 1 - \frac{2}{a}s + \frac{2}{a^2}s^2 + \cdots$$

Equating the first two terms gives $a = 2/T$. Higher order Padé approximations can match more terms and approximate the transfer function of a delay over a wider frequency range. The Bode phase plots of a first-order and a third-order Padé approximation of a 1-s time delay are shown along with the actual Bode phase plot of a 1-s time delay in Fig. 48.24. The first-order approximation is given by

$$G_{D_1}(s) = \frac{-0.5s+1}{0.5s+1} \tag{48.60}$$

The third-order approximation is given by

$$G_{D_3}(s) = \frac{-0.0083s^3 + 0.1s^2 - 0.5s + 1}{0.0083s^3 + 0.1s^2 + 0.5s + 1} = \frac{-1((s-3.67)^2 + (3.5)^2)(s-4.64)}{((s+3.67)^2 + (3.5)^2)(s+4.64)} \tag{48.61}$$

We have seen that time delays cause phase lags which can be modeled with nonminimum phase zeroes. We should note that the Bode plot and Nyquist plot frequency response analysis can be performed directly using the frequency response of the time delay transfer function, e^{-sT}, in series with other elements of the loop gain. The phase lag limits the achievable bandwidth of the system. Time delays are to be avoided in the design of plants that must be controlled. In process control systems this usually means placing valves on inputs very close to the reaction tank and using measurements that do not require much time to produce.

References

[1] Bode, H. W.: *Network Analysis and Feedback Amplifier Design*, New York: Van Nostrand, 1945.

[2] Doyle, J. C., B. A. Francis, and A. R. Tannenbaum, *Feedback Control Theory*, New York: Macmillan, 1992.

[3] Doyle, J. C. and G. Stein: "Multivariable Feedback Design: Concepts for a Classical/Modern Synthesis," *IEEE Trans. Automatic Control*, Vol. AC-26 (1) pp. 4–16, February 1981.

[4] Freudenburg, J. S., and Looze, D. P.: "Right half-plane poles and zeros and design trade-offs in feedback systems," *IEEE Trans. Automatic Control*, Vol. AC-30, pp. 555–556, 1985.

[5] Freudenburg, J. S., and Looze, D. P.: *Frequency Domain Properties of Scalar and Multivariable Feedback Systems*, Berlin: Springer-Verlag, 1988.

[6] Horowitz, I.M., *Synthesis of Feedback Systems*, New York: Academic, 1963.

[7] Nyquist, H.: "Regeneration Theory," *Bell System Tech J*, Vol. 11, pp. 126–147, 1932.

[8] Rohrs, C. E., J. L. Melsa, and D. G. Schultz, *Linear Control Systems*, New York: McGraw-Hill, 1993.

[9] Stein G., *Lecture Notes for 6.232*, Cambridge, MA: MIT, 1988.

[10] Stein, G. and N. R. Sandell, Jr., *Classical and Modern Methods for Control System Design*, Cambridge, MA: MIT, 1979.

49

Stability Robustness

Wu-Sheng Lu
University of Victoria, Canada

49.1 Stability Robustness as Related to Problems in Circuits and Filters

System stability is one of the basic requirements in most design tasks encountered in engineering. In the context of circuits and filters a typical design scenario usually involves selecting design parameters such that the roots of the denominator polynomial of the system's transfer function are inside the left-half-plane, if the system is of continuous time, or inside the unit circle, if the system is of discrete time, so that the system designed will be stable. As highlighted in the preceding chapters, a great deal of research on stability-related issues over the last 60 years has led to powerful stability criteria that can be incorporated into many design methods to ensure, at least in theory, the stability of the circuits or filters designed. On the other hand, however, even if a system is known to stable, the finite precision of hardware or software implementation of the system (e.g., a circuit or a filter) gives rise to a fundamental question: is the *implemented* system also stable? Although the question arises at the implementation stage, it has become increasingly clear that a satisfactory resolution of this issue can be provided at the system design stage by a stability robustness analysis. To further address the issue, it is worthwhile to stress that the imperfection of the system hardware or software implementation can be related to a broad range of parameter uncertainties. These include drifting and fluctuation of the values of the components used in a circuit or filter due to environmental or operating condition changes; external perturbations or disturbances from power supply or other parts of the system; noise

0-8493-8341-2/95/$0.00+$.50

generated by various sensors used in a circuit; and computation errors generated due to finite precision of computing devices such as operational amplifiers in an analog circuit or filter, and fixed-point or floating-point arithmetic units in a microprocessor which implements a digital circuit or filter. Having mentioned these, it is also important to recognize that no matter how advanced the techniques and devices adopted in the implementation are and will be, parameter uncertainty always exists. Thus we see that a fundamental and lasting need exists in engineering practice to study the following problem: Given a stable (nominal) system, which could be one from a preliminary design, and relating the values of the (physical) parameters that are used to describe the system to a point q in the "parameter space", develop computationally feasible techniques for determining the largest region surrounding q such that the system remains stable as long as the point associated with the system is within that region.

Ever since the work of V. L. Kharitonov on robust Hurwitz stability, a considerable body of research has been aimed, in one way or another, at the above-stated robust stability problem. Excellent survey articles [3], [20], and the papers included in [14] have provided a comprehensive exposition of recent developments in this field of research. The rest of the chapter is a concise and organized presentation of several representative results on robust stability. These include the celebrated Kharitonov theorem for the continuous time systems, and its counterpart for the discrete time systems; the edge theorem; simplified criteria for lower-order polynomials; easy-to-apply sufficient conditions; bounds of stability radii for state-space models; and an edge theorem for time delay systems. Whenever possible, the computational aspects of a theoretical result is addressed in the hope that the usefulness of the theorem also could be appreciated in terms of numerical feasibility.

49.2 Stability Robustness of Hurwitz Polynomials

Hurwitz Polynomials

A polynomial

$$p(s) = a_0 + a_1 s + \cdots + a_n s^n \tag{49.1}$$

with real coefficients a_i $(i = 0, 1, \ldots, n)$ is said to be *Hurwitz* or *Hurwitz stable* if all the roots of $p(s)$ have negative real parts.

Kharitonov's Theorem

Suppose the coefficients of $p(s)$ in (49.1) vary over known finite intervals, i.e.,

$$a_i^- \leq a_i \leq a_i^+ \qquad 0 \leq i \leq n \tag{49.2}$$

then (49.1) and (49.2) describe an interval polynomial family, denoted by $\mathcal{P}_n$.

Theorem 49.1 (Kharitonov, 1978): *All polynomials in $\mathcal{P}_n$ are Hurwitz if and only if (iff) the following four polynomials*

$$\begin{aligned}
K_1(s) &= a_0^- + a_1^- s + a_2^+ s^2 + a_3^+ s^3 + a_4^- s^4 + a_5^- s^5 + a_6^+ s^6 + \cdots \\
K_2(s) &= a_0^+ + a_1^+ s + a_2^- s^2 + a_3^- s^3 + a_4^+ s^4 + a_5^+ s^5 + a_6^- s^6 + \cdots \\
K_3(s) &= a_0^+ + a_1^- s + a_2^- s^2 + a_3^+ s^3 + a_4^+ s^4 + a_5^- s^5 + a_6^- s^6 + \cdots \\
K_4(s) &= a_0^- + a_1^+ s + a_2^+ s^2 + a_3^- s^3 + a_4^- s^4 + a_5^+ s^5 + a_6^+ s^6 + \cdots
\end{aligned}$$

are Hurwitz [these four polynomials $K_i(s)$ $(i = 1, 2, 3, 4)$ *are often referred to as Kharitonov polynomials].*

In other words, the stability of the entire polynomial family $\mathcal{P}_n$ can be verified by applying, for example, Routh's stability test four times. In addition to [22], new proofs of this theorem have been provided by a number of researchers. The reader is referred to [33] for an elegant and elementary proof of the Kharitonov theorem.

For low-order polynomials with $n = 2, 3, 4$, and 5, the criterion in Kharitonov's theorem can be simplified [2]. For the sake of simplicity, we consider monic polynomials of order n, i.e., $a_n \equiv 1$. The simplified criterion can be stated as follows:

- For $n = 2$ all polynomials in $\mathcal{P}_2$ are Hurwitz iff

$$a_0^- > 0 \quad \text{and} \quad a_1^- > 0$$

- For $n = 3$, assume $a_i^- > 0$ $(i = 0, 1, 2)$. Then all polynomials in $\mathcal{P}_3$ are Hurwitz iff

$$p_1(s) = s^3 + a_2^- s^2 + a_1^- s + a_0^+$$

  is Hurwitz.
- For $n = 4$, assume $a_i^- > 0$ $(i = 0, 1, 2, 3)$. Then all polynomials in $\mathcal{P}_4$ are Hutwitz iff

$$p_1(s) = s^4 + a_3^- s^3 + a_2^- s^2 + a_1^+ s + a_0^+$$

  and

$$p_2(s) = s^4 + a_3^+ s^3 + a_2^- s^2 + a_1^- s + a_0^+$$

  are Hurwitz.
- For $n = 5$, assume $a_i^- > 0$ $(i = 0, 1, 2, 3, 4)$. Then all polynomials in $\mathcal{P}_5$ are Hurwitz iff

$$p_1(s) = s^5 + a_4^- s^4 + a_3^- s^3 + a_2^+ s^2 + a_1^+ s + a_0^-$$
$$p_2(s) = s^5 + a_4^+ s^4 + a_3^- s^3 + a_2^- s^2 + a_1^+ s + a_0^+$$

  and

$$p_3(s) = s^5 + a_4^+ s^4 + a_3^+ s^3 + a_2^- s^2 + a_1^- s + a_0^+$$

  are Hurwitz.

In other words, for polynomials of order n with $n = 2, 3, 4, 5$, one only needs to check the stability of $(n - 2)$ Kharitonov polynomials in order to verify the Hurwitz property of the entire family $\mathcal{P}_n$.

Because the perturbations of polynomials in the Kharitonov theorem are assumed to be independent of each other, the polynomial family $\mathcal{P}_n$ characterized by (49.1) and (49.2) is associated with a hyperrectangular region in the coefficient space with edges parallel to the coordinate axis. The Hurwitz property of the polynomials at four corners of the hyperrectangle offers the necessary and sufficient condition for the entire polynomial family $\mathcal{P}_n$ being Hurwitz. However, if the coefficient perturbations are not independent, then the Kharitonov theorem could become conservative [4, 6]. This limitation of the Kharitonov theorem is of particular importance from an engineering point of view, as a perturbation parameter often enters into more than one coefficient in many circuits/systems analysis and design problems. In this regard the result of Bartlett et al. [6], known as the edge theorem, describes an important generalization of the Kharitonov's result to deal with the polynomial family where the coefficients depend affine linearly on a set of underlying physical parameters. To describe the edge theorem, the concepts of D-stability, polytope, and its edges need to be introduced.

D-Stability, Polytope, and Its Edges

Polynomial $p(s)$ is said to be D-stable if all the roots of $p(s)$ are in region D. As an example, with D being the open left-half of the complex plane, and D-stable polynomial is a Hurwitz polynomial.

The polytope P, generalized by nth-order, monic, real-coefficient polynomials $p_1(s)$, $p_2(s)$, $\ldots, p_m(s)$, is a family of polynomials defined by

$$P = \left\{ p(s):\ p(s) = \sum_{i=1}^{m} \lambda_i p_i(s) \quad \text{with} \quad \lambda_i \in [0, 1], \sum_{i=1}^{m} \lambda_i = 1 \right\} \tag{49.3}$$

where the polynomials $p_i(s)$, $i = 1, \ldots, m$, are called the vertices of P. To explain how this polytope can be related to a polynomial family encountered in a stability robustness study in certain engineering scenarios, let us consider an nth-order polynomial:

$$p(s) = a_0(r) + a_1(r)s + \cdots + a_{n-1}(r)s^{n-1} + s^n \tag{49.4}$$

with each coefficient $a_i(r)$ depending affine linearly on k physical parameters $r_1, r_2, \ldots, r_k$ which may vary independently over a given, finite uncertainty range $r_i^- \leq r_i \leq r_i^+$, $i = 1, \ldots, k$. The set of polynomials generated by varying the values of $r_1, \ldots, r_k$ over their uncertainty range is the convex hull of the 2^k polynomials $p_1(s), \ldots, p_{2^k}(s)$, which are obtained by setting parameters $r = [r_1 \cdots r_k]$ in (49.4) to all possible extreme points. Evidently, this convex hull is the polytope P defined in (49.3), with $m = 2^k$ (note that m will be $< 2^k$ in case $r_i^- = r_i^+$ for some i's). An edge of P is the convex combination of any two vertices. For instance, the set E_{ij} with $1 \leq i$, $j \leq m$ defined by

$$E_{ij} = \{p(s):\ p(s) = \lambda p_i(s) + (1 - \lambda)p_j(s), \quad \lambda \in [0, 1]\} \tag{49.5}$$

is an edge of P. An exposed edge of P is an edge which can be expressed as the intersection of P and some supporting hyperplane of P.

Theorem 49.2, The Edge Theorem [6]: *With a simply connected domain D, all polynomials in polytope P are D-stable iff polynomials on all exposed edges are D-stable.*

Two important features of the edge theorem are

1. As the exposed edges are of one dimension, the reduction in computation complexity achieved by the theorem is quite significant.
2. By considering different (simply connected) domain D, the edge theorem becomes a powerful tool in the study of a wide range of stability robustness problems, including the root sensitivity problem for discrete time systems.

D-Stability of Polynomials on an Edge

From an implementation point of view, one might want to determine the D-stability of the edges instead of identifying all exposed edges and then determining their D-stability. As is seen from (49.5), an edge is simply a convex combination of two vertices, which can be expressed as $p(s) = \lambda(p_i(s) + Kp_j(s))$, with $K = (\lambda - 1)/\lambda$ varying from 0 to ∞. Consequently, the D-stability of $p(s)$ can readily be verified by checking the root locus of $p(s)$ vs. K to see if the entire locus remains within region D [4].

An easy-to-apply algebraic criterion for the Hurwitz invariance of the polynomials on an edge is given in [8] and [15]. Let D be the open left-half-plane and $H(p)$ be the $n \times n$ *Hurwitz testing matrix* [associated with $p(s)$ in (49.1)] defined by

$$H(p) = \begin{bmatrix} a_{n-1} & a_{n-3} & a_{n-5} & \cdots & \cdots \\ a_n & a_{n-2} & a_{n-4} & \cdots & \cdots \\ 0 & a_{n-1} & a_{n-3} & a_{n-5} & \cdots \\ 0 & a_n & a_{n-2} & a_{n-4} & \cdots \\ & & \cdots & \cdots & \\ & & & & a_0 \end{bmatrix} \tag{49.6}$$

For example, with $n = 5$ and $n = 6$, we have

$$H(p) = \begin{bmatrix} a_4 & a_2 & a_0 & 0 & 0 \\ a_5 & a_3 & a_1 & 0 & 0 \\ 0 & a_4 & a_2 & a_0 & 0 \\ 0 & a_5 & a_3 & a_1 & 0 \\ 0 & 0 & a_4 & a_2 & a_0 \end{bmatrix}$$

and

$$H(p) = \begin{bmatrix} a_5 & a_3 & a_1 & 0 & 0 & 0 \\ a_6 & a_4 & a_2 & a_0 & 0 & 0 \\ 0 & a_5 & a_3 & a_1 & 0 & 0 \\ 0 & a_6 & a_4 & a_2 & a_0 & 0 \\ 0 & 0 & a_5 & a_3 & a_1 & 0 \\ 0 & 0 & a_6 & a_4 & a_2 & a_0 \end{bmatrix}$$

respectively. Theorem 49.3 gives a necessary and sufficient condition for the Hurwitz invariance of polynomials on an edge.

Theorem 49.3 [8] [44]: *The convex combination of two nth-order, real polynomials* $p_0(s)$ *and* $p_1(s)$ *(not necessarily monic) is Hurwitz invariant if* $p_0(s)$ *is Hurwitz and* $H^{-1}(p_0)H(p_1)$ *has no eigenvalues in* $(-\infty, 0]$ *(the nonpositive part of the real axis).*

Further Generalizations

For polynomials with coefficient structures more complicated than affine linearity, e.g., polynomials whose coefficients are *multilinear* functions or *polynomial* functions of independent perturbation parameters, stability on the exposed edges does not in general guarantee the Hurwitz invariance of the whole polynomial family [1]. Useful results are available for robust stability of such polynomial families [13, 39, 40], where a "domain splitting" algorithm and its improved versions are proposed to evaluate the largest stability margin.

49.3 Stability Robustness of Schur Polynomials

Schur Polynomials

A polynomial

$$p(z) = a_0 z^n + a_1 z^{n-1} + \cdots + a_n \tag{49.7}$$

with real coefficients a_i $(i = 0, 1, \ldots, n)$ is said to be *Schur* or *Schur stable* if all the roots of $p(z)$ are within the unit circle.

A Discrete Analog of the Weak Form of the Kharitonov Theorem

The extensions of the Kharitonov theorem and the edge theorem to the discrete time case are not straightforward. Consider, for example, the fourth-order polynomial [10]

$$p(z, a_1) = z^4 + a_1 z^3 + \tfrac{3}{2} z^2 - \tfrac{1}{3}$$

for $a_1 \in [-17/8, 17/8]$. It can be verified that both $p(z, -17/8)$ and $p(z, 17/8)$ are Schur stable, but $p(z, 0)$ is not. It was believed that the difficulty was caused by the incompatibility between the stability region of the polynomial, which is the *open unit disk*, and the region of coefficient perturbations, which is usually rectangular [20]. However, an analog of the weak form of the Kharitonov theorem for discrete time systems has been obtained by Kraus et al. [24]. To describe the result, we first state the weak form of the Kharitonov theorem, where the term *corner polynomial* is used to mean a polynomial $p(s)$ with $a_i \in \{a_i^-, a_i^+\}$, $0 \le i \le n$.

Theorem 49.4 [22]: *All polynomials in $\mathscr{P}_n$, which is defined by (49.1) and (49.2), are Hurwitz iff all 2^{n+1} corner polynomials are Hurwitz.*

Now let us consider a discrete time polynomial family $\mathscr{D}_n$, in which for each $i \neq n/2$, a_i, and a_{n-i} vary inside a region of the form depicted in Fig. 49.1, where A_{ij} for $j = 1, 2, 3, 4$ are called the corner points for coefficients a_i and a_{n-i}. If n is even, $a_{n/2}$ varies in an interval $[a_{n/2}^-, a_{n/2}^+]$. The following theorem is a discrete analog of Theorem 49.4 for $\mathscr{D}_n$.

Theorem 49.5 [24]: *All polynomials in $\mathscr{D}_n$ are Schur stable iff every member of the finite set of $p(z)$, defined by every possible combination of corner points (and interval end points, in case $i = n/2$), is Schur stable.*

In addition to the above result, a discrete analog of the strong Kharitonov theorem, i.e., Theorem 49.1, is given in [24].

Like the continuous time case, simplified criteria for robust Schur stability of low-order discrete time polynomials were developed. The reader is referred to [23] for the details.

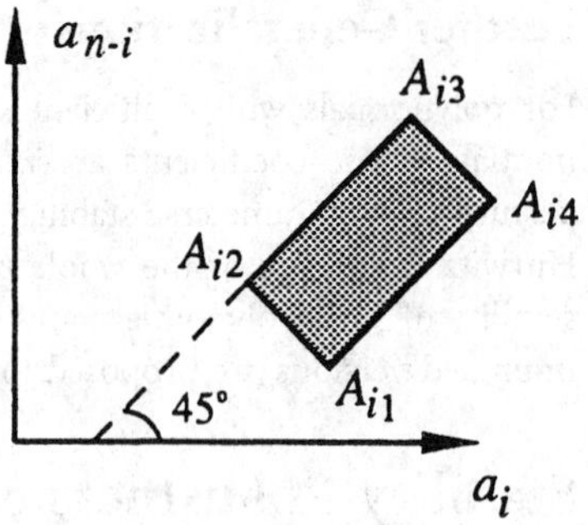

FIGURE 49.1 Region of uncertainty for parameters (a_i, a_{n-i}).

The Edge Theorem for Schur Invariance

As commented in Section 49.2, the edge theorem can be used for studying robust stability of discrete time systems by taking the unit disk as the region D. To implement the theorem, one is concerned with developing feasible criteria for the Schur invariance of the polynomials on a specified edge. Recall that the continuous time counterpart of this problem was elegantly solved by Bialas [8] and Fu and Barmish [22], as stated in Theorem 49.3. The next theorem provides an easy-to-use necessary and sufficient condition for the Schur invariance of the polynomials on an edge, where the Mobiüs transformation

$$z = T(s) \equiv \frac{s+1}{s-1} \tag{49.8}$$

which maps the open left-half-plane onto the open unit disk, is used. To state the theorem, let $p(z)$ be given by (49.7) and define the *transformed polynomial* $p_T(s)$ of $p(z)$ by

$$\begin{aligned} p_T(s) = p[T(s)](s-1)^n &= \sum_{i=0}^{n} a_i(s+1)^{n-i}(s-1)^i \\ &\equiv b_0 s^n + b_1 s^{n-1} + \cdots + b_n \end{aligned} \tag{49.9}$$

Then $n \times n$ Hurwitz testing matrix (49.6), associated with $p_T(s)$ in (49.9), is given by

$$H(p_T) = \begin{bmatrix} b_1 & b_3 & b_5 & \cdots & \cdots \\ b_0 & b_2 & b_4 & \cdots & \cdots \\ 0 & b_1 & b_3 & b_5 & \cdots \\ 0 & b_0 & b_2 & b_4 & \cdots \\ & & \cdots & \cdots & \\ & & & & b_n \end{bmatrix}$$

Theorem 49.6 [7]: *Given two nth-order real polynomials $p_0(z)$ and $p_1(z)$ (not necessarily monic), let $p_{T_0}(s)$ and $p_{T_1}(s)$ be the transformed polynomials of $p_0(z)$ and $p_1(z)$, respectively. Then all polynomials on the edge $E = \{p(z)\colon p(z) = (1-\gamma)p_0(z) + \gamma p_1(z),$ for $0 \le \gamma \le 1\}$ is Schur stable iff* (i) *$p_{T_0}(s)$ and $p_{T_1}(s)$ are nth-order polynomials,* (ii) *the leading coefficients of $p_{T_0}(s)$ and $p_{T_1}(s)$ have the same sign, and* (iii) *$p_{T_0}(s)$ is Hurwitz, and $H^{-1}(p_{T_0})H(p_{T_1})$ has no eigenvalues on $(-\infty, 0]$.*

Sufficient Conditions for Robust Schur Stability

Consider polynomial

$$f(s) = b_0 s^n + b_1 s^{n-1} + \cdots + b_n \tag{49.10}$$

with $b_k > 0$, $k = 0, 1, \ldots, n$. In [27] it was shown that $f(s)$ is Hurwitz stable if the $n - 2$ conditions

$$b_{k-1} b_{k+2} \leq 0.4655\, b_k b_{k+1} \qquad k = 1, 2, \ldots, n-2$$

are satisfied. Assume that each coefficient b_i in (49.10) varies in an interval $b_k^- \leq b_k \leq b_k^+$, with $b_k^- > 0$. It is easy, then, to verify, based on the result of Lipatov and Sokolov, that this family of $f(s)$ is Hurwitz stable if

$$b_{k-1}^+ b_{k+2}^+ \leq 0.4655\, b_k^- b_{k+1}^- \qquad k = 1, 2, \ldots, n-2 \tag{49.11}$$

are satisfied [11]. By using a bilinear transformation, this result leads to a sufficient condition for robust Schur stability, as stated in Theorem 49.7.

Theorem 49.7 [11]: *Consider the polynomial family of*

$$p(z) = a_0 z^n + a_1 z^{n-1} + \cdots + a_n \tag{49.12}$$

where $a_k \in [a_k^-, a_k^+]$, $k = 0, 1, \ldots, n$. *Define*

$$q_{ij} = \sum_{k=0}^{i-1} (-1)^k \binom{j-1}{k} \binom{n+1-j}{i-k-1}$$

for $i = 1, 2, \ldots, n+1$, *and* $j = 1, 2, \ldots, n+1$,

$$b_k^+ = \sum_{j=1}^{n+1} q_{k+1,j} \hat{a}_{j-1} \qquad k = 0, \ldots, n$$

where

$$\hat{a}_{j-1} = \begin{cases} a_{j-1}^- & \text{if } q_{k+1,j} < 0 \\ a_{j-1}^+ & \text{if } q_{k+1,j} > 0 \end{cases}$$

and

$$b_k^- = \sum_{j=1}^{n+1} q_{k+1,j} \tilde{a}_{j-1} \qquad k = 0, \ldots, n$$

where

$$\tilde{a}_{j-1} = \begin{cases} a_{j-1}^+ & \text{if } q_{k+1,j} < 0 \\ a_{j-1}^- & \text{if } q_{k+1,j} > 0 \end{cases}$$

The family of polynomials (49.12) is Schur stable if

$$b_k^+ \geq b_k^- > 0 \qquad k = 0, \ldots, n$$

and

$$b_{i-1}^+ b_{i+2}^+ \leq 0.4655\, b_i^- b_{i+1}^- \qquad i = 1, \ldots, n-2$$

Schur Stability of Interval Matrices: Sufficient Conditions

Let $A^+ = (a_{ij}^+)$ and $A^- = (a_{ij}^-)$ be $n \times n$ real matrices, with $a_{ij}^- \leq a_{ij}^+$ for all i, j. An interval matrix $[A]$, often denoted by $[A] = [A^-, A^+]$, is actually a family of all matrices (a_{ij}) satisfying $a_{ij}^- \leq a_{ij} \leq a_{ij}^+$ for all i, j. $[A]$ is said to be *Schur* or *Schur stable* if the spectrum of $[A]$, denoted by $\sigma[A]$, is contained inside the unit circle. A matrix $A = (a_{ij}) \in [A]$ is called a corner of $[A]$ if $a_{ij} = a_{ij}^+$ or a_{ij}^- for all i, j.

One of the sufficient conditions for the Schur stability of an interval matrix is the following theorem devised by Perron and Frobenius [43].

Theorem 49.8 (Perron and Frobenius): *Let $[A] = [A^-, A^+]$ with $A^- = (a_{ij}^-)$, $A^+ = (a_{ij}^+)$. $[A]$ is Schur stable if matrix $C_{\max}$ is Schur stable where $C_{\max} = (c_{ij})$ is defined by* $c_{ij} = \max(|a_{ij}^-|, |a_{ij}^+|)$.

This theorem is easy to apply, but it sometimes appears to be fairly conservative. For instance, consider the example given in [19]

$$A^- = \begin{bmatrix} 0 & 1 & 0 \\ -0.4 & -0.5 & 1 \\ 0 & -0.5 & -1.35 \end{bmatrix} \qquad A^+ = \begin{bmatrix} 0 & 1 & 0 \\ -0.3 & -0.4 & 1 \\ 0 & -0.4 & -1.1 \end{bmatrix}$$

which gives

$$C_{\max} = \begin{bmatrix} 0 & 1 & 0 \\ 0.4 & 0.5 & 1 \\ 0 & 0.5 & 1.35 \end{bmatrix}$$

It can be shown that the interval matrix $[A^-, A^+]$ is Schur stable [19], but one of the eigenvalues of $C_{\max}$ is 1.8094, hence $C_{\max}$ is not Schur stable.

To describe the results of [19], write an interval matrix as

$$\begin{aligned} [A] &= [A^-, A^+] \\ &= \left\{ A_0 + \sum_{i=1}^{p} \gamma_i A_i \colon \; \gamma_i \in [\gamma_i^-, \gamma_i^+],\ \gamma_i^- < \gamma_i^+ < \infty,\ i = 1, \ldots, p \right\} \end{aligned} \tag{49.13}$$

where $A_0 \in R^{n \times n}$, and each A_i $(i = 1, \ldots, p)$ has one and only one nonzero entry.

Taking the 3×3 interval matrix discussed above as an example, we have

$$A_0 = \begin{bmatrix} 0 & 1 & 0 \\ 0 & 0 & 1 \\ 0 & 0 & 0 \end{bmatrix} \qquad A_1 = \begin{bmatrix} 0 & 0 & 0 \\ 1 & 0 & 0 \\ 0 & 0 & 0 \end{bmatrix} \qquad A_2 = \begin{bmatrix} 0 & 0 & 0 \\ 0 & 1 & 0 \\ 0 & 0 & 0 \end{bmatrix}$$

$$A_3 = \begin{bmatrix} 0 & 0 & 0 \\ 0 & 0 & 0 \\ 0 & 1 & 0 \end{bmatrix} \qquad A_4 = \begin{bmatrix} 0 & 0 & 0 \\ 0 & 0 & 0 \\ 0 & 0 & 1 \end{bmatrix}$$

$$\gamma_1 \in [-0.4, \quad -0.3] \qquad \gamma_2 \in [-0.5, \quad -0.4]$$
$$\gamma_3 \in [-0.5, \quad -0.4] \qquad \gamma_4 \in [-1.35, \quad -1.1]$$

The matrices with each $\gamma_i = \gamma_i^-$ or γ_i^+ are called the exposed vertices of $[A]$ (i.e., a corner of $[A]$), and the set of all exposed vertices is denoted by E°. Obviously, $[A]$ in (49.13) has a total of 2^p vertices. Next, we define the set of the coefficients of the characteristic polynomials of $[A]$ by

$$\phi([A]) = \left\{ (a_1, \ldots, a_n)\colon \det(zI - A) = \det\left(zI - A_0 - \sum_{i=1}^{p} \gamma_i A_i \right) \right.$$
$$\left. = z^n + a_1 z^{n-1} + \cdots + a_n \right\}$$

As A_i has only one nonzero element, set ϕ is separately affine in $\gamma_1, \ldots, \gamma_p$; i.e., ϕ is affine when any $p - 1$ parameters are fixed. Hence, by the mapping theorem [45], the convex hull of $\phi([A])$, denoted by $\text{conv}\{\phi([A])\}$, coincides with $\text{conv}\{\phi(E_i^\circ),\ i = 1, \ldots, 2^p\}$. Further, we define the root space of a given polynomial family Ω by

$$R(\Omega) = \{z\colon\ z^n + a_1 z^{n-1} + \cdots + a_n = 0,\ (a_1, \ldots, a_n) \in \Omega\}$$

In our case, Ω is $\text{conv}\{\phi([A])\}$, which is a polytope as it can be generated by a finite set of elements. It follows from [6] that the boundary of $R(\Omega)$ is contained in the root space of all the exposed edges of Ω. These observations lead immediately to the following theorem.

Theorem 49.9 [19]: *The spectrum of* $[A]$, $\sigma[A]$, *is contained in the root space of conv*$\{\phi[A]\}$. *Furthermore, the boundary of this root space is contained in the root space of the exposed edges of conv*$\{\phi([A])\}$. *Therefore, the root space of these exposed edges form a boundary for* $\sigma[A]$.

Because $\phi([A])$ is in general not convex, what Theorem 49.9 provides is a sufficient condition. It is much less conservative than that of Perron and Frobenius and is computationally feasible, as it reduces to a finite number of single-variable searches. Take the above 3×3 interval matrix again as an example. To apply Theorem 49.9, we compute the root space of all pairwise convex combinations of the 16 vertex polynomials. Thus, for each $i, j \in \{1, 2, \ldots, 16\}$, the root space $R\{(1-\gamma)E_i^\circ + \gamma E_j^\circ$, for $0 \leq \gamma \leq 1\}$ needs to be determined. As shown in [19], these root spaces as the boundary of $\sigma[A]$ are inside the unit circle and, therefore, the interval matrix $[A]$ is Schur stable.

As an additional note to the above theorem, this method can be applied to continuous time systems and claim an interval matrix is Hurwitz stable if the root space of the exposed edges are in the open left-half-plane.

49.4 Lyapunov and Other Non-Kharitonov Approaches: State-Space Models

Stability Radius for Unstructured Perturbations

In addition to the stability problem of interval matrices, which was addressed in the preceding section, the l^2 version of the matrix stability problem is as follows. Let A be a matrix that is Hurwitz (Schur) stable. The problem is to find the largest number $r > 0$ such that $A + \Delta$ is Hurwitz (Schur) stable for *all* Δ satisfying $\bar{\sigma}(\Delta) < r$. Because no structure of Δ is specified, this r is called the stability radius for *unstructured* perturbations.

Stability Radius for Unstructured Complex Perturbations

If Δ is allowed to be complex, then the exact value of r for Hurwitz stability of $A + \Delta$ is given by the following theorem.

Theorem 49.10 [18, 26, 32, 35, 42]: *Let A be Hurwitz stable and r_c be defined by*

$$r_c = \inf\{\bar{\sigma}(\Delta)\colon\ \Delta \in C^{n\times n},\ A + \Delta \text{ unstable}\} \tag{49.14a}$$

Then

$$r_c = \inf_{w\in R} \underline{\sigma}(j\omega I - A) \tag{49.14b}$$

where $j = \sqrt{-1}$, I is the identity matrix of dimension n, R is the set of all real numbers, and $\underline{\sigma}(\cdot)$ denotes the smallest singular value [41] of the matrix involved.

For discrete time systems, we have Theorem 49.11.

Theorem 49.11: *Let A be Schur stable and R_c be defined by*

$$R_c = \inf\{\bar{\sigma}(\Delta)\colon\ \Delta \in C^{n\times n},\ A + \Delta \text{ unstable}\} \tag{49.15a}$$

Then

$$R_c = \inf_{\omega\in[0,2\pi]} \underline{\sigma}(e^{j\omega} I - A) \tag{49.15b}$$

Numerical Evaluation of r_c and R_c

An efficient method for the evaluation of the radii r_c in (49.14b) and R_c in (49.15b) is the bisection method described in [12]. The following two theorems are the keys from which the bisection algorithms are derived.

Theorem 49.12 [12]: *Define*

$$H(\sigma) = \begin{bmatrix} A & -\sigma I \\ \sigma I & -A^H \end{bmatrix} \tag{49.16}$$

where A^H is the conjugate transpose of A. $H(\sigma)$ has an eigenvalue whose real part is zero iff $\sigma \geq r_c$.

Now if α is a lower bound and γ is an upper bound of r_c, then the above theorem suggests that the bounds can be improved by choosing the midpoint of interval $[\alpha, \gamma]$ as σ and checking to see if $H(\sigma)$ has a purely imaginary eigenvalue. The bisection procedure continues until the length of the interval [lower bound, upper bound] is no larger than a given tolerance. Similarly, an algorithm can be developed to evaluate R_c in (49.15b) using the following theorem.

Theorem 49.13 [12]: *For $A \in C^{n\times n}$ a real number V is such that $V \geq R_c$ and for $V \geq \sigma \geq R_c$, the $2n \times 2n$ matrix pencil*

$$F(\sigma) - \lambda G(\sigma) = \begin{bmatrix} -\sigma I & A \\ I & 0 \end{bmatrix} - \lambda \begin{bmatrix} 0 & I \\ A^H & -\sigma I \end{bmatrix} \tag{49.17}$$

has a generalized eigenvalue of magnitude 1. Furthermore, if $\sigma < R_c$, then (49.17) has no generalized eigenvalue of magnitude 1.

Stability Radius for Unstructured Real Perturbations

The robust stability problem for a stable matrix subject to *real* perturbations is obviously more realistic. Technically, however, finding a real stability radius is more difficult than finding a complex stability radius. A solution to this problem is provided by [37]. To describe its main result, let us partition the complex plane C into two disjoint subsets, C_g and C_b, i.e., $C = C_g \cup C_b$, such that C_g is open. A matrix is said to be *stable* if its eigenvalues are in C_g. Note that the above stability concept is consistent with the Hurwitz or Schur stability if C_g is the open left-half-plane or unit disk. Given a real and stable A, its stability radius for unstructured real perturbations is defined as

$$r = \inf\{\bar{\sigma}(\Delta)\colon \Delta \in R^{n\times n},\ A + \Delta \text{ has an eigenvalue on } \partial C_g\}$$

where ∂C_g is the boundary of C_g, and the following theorem reduces the problem of evaluating r to a feasible optimization problem.

Theorem 49.14 [37]:

$$r = \left\{ \sup_{s\in\partial C_g} \mu_r(sI - A)^{-1} \right\}^{-1} \tag{49.18}$$

with

$$\mu_r(M) = \inf_{\gamma\in(0,1]} \sigma_2 \begin{bmatrix} \operatorname{Re} M & -\gamma \operatorname{Im} M \\ \gamma^{-1} \operatorname{Im} M & \operatorname{Re} M \end{bmatrix} \tag{49.19}$$

where the function to be minimized is a unimodal function on (0, 1].

Concerning the computation involved, first note that standard algorithms are available for minimizing a unimodal function on a finite interval [30]. Also note that a purely linear algebraic algorithm is given in [37] for constructing a worst real perturbation Δ such that $A + \Delta$ is unstable with smallest possible $\bar{\sigma}(\Delta)$.

Some special circumstances exist in which the evaluation of r can be simplified considerably. As a matter of fact, it is shown in [36] that if C_g is the left-half-plane (i.e., Hurwitz stability), then

$$r = \begin{cases} \min\{-\mathrm{Re}(\lambda_i(A)),\ i = 1, 2, \ldots, n\} & \text{if } A \text{ is normal} \\ \min\{\underline{\sigma}(A),\ -\frac{1}{2}tr(A)\} & \text{if } A \in R^{2\times 2} \end{cases}$$

and that if C_g is the open unit disk (i.e., Schur stability) and A is normal, then $r = \min\{1 - |\lambda_i(A)|,\ i = 1, 2, \ldots, n\}$.

Stability Bounds Obtained Using Lyapunov Approaches

For continuous time systems, the following matrix equation, known as the Lyapunov equation, has been of use in many stability-related problems

$$A^{\mathrm{T}}P + PA = -Q \tag{49.20}$$

where A is the system matrix of a continuous time system and Q is a given positive definite matrix. It is well known that A is Hurwitz stable iff (49.20) has a unique, positive definite solution P [21]. For a Hurwitz stable A, the positive definite solution of (49.20) can be expressed as

$$P = \int_0^{\infty} e^{A^{\mathrm{T}}t} Q e^{At}\, dt \tag{49.21}$$

or as

$$P = \frac{1}{2\pi} \int_{-\infty}^{\infty} (j\omega I - A)^{-H} Q (j\omega I - A)^{-1}\, d\omega \tag{49.22}$$

Methods for the evaluation of P that are more efficient than using integrals (49.21) or (49.22) exist (e.g., [5, 17]). In the next theorem the largest eigenvalue of P from (49.20) with $Q = 2I$ is related to a lower bound of r_c.

Theorem 49.15 [34]: *Let A be Hurwitz and r_c be defined by (49.14a). Then*

$$r_c \geq \frac{1}{\bar{\lambda}(P)} \tag{49.23}$$

where $\bar{\lambda}(P)$ is the largest eigenvalue of P, satisfying (49.20) with $Q = 2I$.

A Lyapunov equation often used for stability analysis of discrete time systems has the form of

$$A^{\mathrm{T}}PA - P = -Q \tag{49.24}$$

where A is the system matrix of a discrete time system and Q is a given positive-definite matrix. It is well known that A is Schur stable iff (49.24) has a unique, positive definite solution P. Equation (49.24) is linear in P, and for a Schur stable A, its positive definite solution can be expressed as

$$P = \sum_{k=0}^{\infty} (A^{\mathrm{T}})^k Q A^k \tag{49.25}$$

or

$$P = \frac{1}{2\pi}\int_{-\pi}^{\pi} (e^{j\omega}I - A)^{-H}Q(e^{j\omega}I - A)^{-1}\,d\omega \tag{49.26}$$

Efficient methods for solving (49.24) are available [17, 29].

The following theorem is a discrete time counterpart of Theorem 49.15.

Theorem 49.16 [38]: *Let A be Schur stable and R_c be defined by (49.15a). Then*

$$R_c \geq \frac{1}{\bar{\lambda}(P) + \bar{\lambda}^{1/2}(P)\bar{\lambda}^{1/2}(P - I)} \tag{49.27}$$

where P is the positive definite solution of (49.24), with $Q = I$.

Stability Robustness Bounds for Structured Perturbations

When the parameter perturbation Δ exhibits certain structures, improved stability robustness bounds can be obtained. Here, we first consider those $\Delta = (\delta_{ij})$, with each δ_{ij} bounded by constant ε_{ij}, i.e.,

$$|\delta_{ij}| \leq \varepsilon_{ij} \tag{49.28}$$

In [34] it is shown that if $A \in R^{n\times n}$ is Hurwitz, then $A + \Delta$, with Δ satisfying (49.28), remains Hurwitz if

$$\varepsilon \equiv \max_{i,j}\{\varepsilon_{ij}\} < \frac{1}{n\bar{\lambda}(P)} \tag{49.29}$$

where P is the positive definite matrix satisfying (49.20) with $Q = 2I$. In [44], (49.29) is improved as

$$\varepsilon < \frac{1}{\bar{\sigma}[|P|U]_s} \tag{49.30}$$

where $|P|$ is the matrix whose (i, j)th entry is the absolute value of the (i, j)th entry of P, $U = (U_{ij})$, with $U_{ij} = \varepsilon_{ij}/\varepsilon$, and $[\cdot]_s$ denotes the symmetric part of the matrix involved, i.e., $[B] = (B + B^{\mathrm{T}})/2$.

In [46] improved bounds are derived by considering perturbations Δ with the structure

$$\Delta = \sum_{i=1}^{m} k_i E_i \tag{49.31}$$

where E_i are known constant matrices, and k_i are uncertain parameters.

Define

$$P_i = (E_i^{\mathrm{T}}P + PE_i)/2 \qquad i = 1, 2, \ldots, m$$
$$P_e = [P_1\ P_2\ \cdots\ P_m]$$

where P is the solution of (49.20); with $Q = 2I$, we have the following theorem.

Theorem 49.17 [46]: *Let A be Hurwitz and Δ be given by (49.31). Then $A + \Delta$ is Hurwitz if*

$$\sum_{i=1}^{m} k_i^2 < \frac{1}{\bar{\sigma}^2(P_e)} \tag{49.32}$$

or

$$\sum_{i=1}^{m} |k_i| \, \bar{\sigma}(P_i) < 1 \tag{49.33}$$

or

$$|k_j| < \frac{1}{\bar{\sigma}\left(\sum_{i=1}^{m} |P_i|\right)} \qquad \text{for } j = 1, 2, \ldots, m \tag{49.34}$$

49.5 Time Delay and Two-Dimensional Systems

Stability Robustness of Time Delay Systems

The dynamical behavior of circuits that contain time delay components such as delay lines can often be described by delay-differential equations of the form

$$F\dot{x}(t) = \sum_{i=0}^{l} A_i x(t - \tau_i) \tag{49.35}$$

where $x(t) \in R^{n\times 1}$, $A_i \in R^{n\times n}$, $F \in R^{n\times n}$ and nonsingular, and $0 = \tau_0 < \tau_1 < \cdots < \tau_l$. The characteristic equation of (49.35) is

$$p(s) = a_{00}s^n + \sum_{i=1}^{n} \left(\sum_{k=0}^{N} a_{ik} e^{-h_k s} \right) s^{n-i} \tag{49.36}$$

where

$$a_{ik} = \alpha_{ik} + j\beta_{ik} \qquad \alpha_{ik},\ \beta_{ik} \in R$$

are constants, $a_{00} \neq 0$, and $0 = h_0 < h_1 < \cdots < h_N$ are related to $\{\tau_i\}$. System (49.35) is said to be D-stable if the zeroes of $p(s)$ are contained inside region D in the complex plane. In particular, $p(s)$ is said to be *stable* if $p(s)$ is D-stable, with D being the open left-half-plane, in which case system (49.35) is exponentially stable. Note that because of the time delay terms in (49.35), $p(s)$ is no longer a polynomial in s in the normal sense, but a polynomial in s, each of whose coefficients is a linear combination of exponential functions of s. We call $p(s)$ in (49.36) a *quasipolynomial.*

Assume the parameters a'_{ik}s in (49.36) are subject to certain variations, and we are interested in determining the D-stability of (49.35) under such parameter variations. One way to address this problem is to investigate the D-stability of the family of quasipolynomials

$$p = \left\{ p(s) = a_{00}s^n + \sum_{i=1}^{n} \left(\sum_{k=0}^{N} a_{ik} e^{-h_i s} \right) s^{n-i} : h_{00} \neq 0, (a_{00}, a_{10}, \ldots, a_{0N}) \in Q \right\}$$

for some region $Q \subset C^{nN+n+1}$ that characterizes the parameter variations.

A special family of quasipolynomials is the polytope generated by the convex combinations of r nth-order quasipolynomials $p_1(s)$, $p_2(s), \ldots, p_r(s)$. Each has a nonzero leading coefficient a_{00}, i.e.,

$$P = \text{conv}\{p_1(s),\ p_2(s), \ldots, p_r(s)\} \tag{49.37}$$

Similar to the nondelay case, we can define the edges of P and note that an edge of P is a convex combination of a pair of quasipolynomials in P, i.e., $\text{conv}\{p_i(s),\ p_j(s)\}$ for some i, j. The end points of an edge are called vertices, and the vertices of P are called vertices quasipolynomials.

Theorem 49.18, An Edge Theorem for Delay Systems [16]: *Family P in (49.37) is stable iff all the edges of P are stable.*

Denote the edges of P by $E_1, E_2, \ldots, E_t$ and the vertices quasipolynomials by $p_{k_0}(s)$ and $p_{k_1}(s)$ for $1 \le k \le t$. The next theorem may be viewed as a graphical implementation of Theorem 49.18.

Theorem 49.19 [16]: *Family P in (49.37) is stable iff the following two conditions hold for every* E_k, $1 \le k \le t$: *(i) the frequency response plot of* $p_{k_0}(j\omega)/(j\omega + 1)^n$ *does not encircle the origin, and (ii) the frequency response plot of* $p_{k_1}(j\omega)/p_{k_0}(j\omega)$ *does not cross* $(-\infty, 0]$ *(the nonpositive part of the real axis).*

Stability Robustness of Discrete Two-Dimensional Systems

The study of two-dimensional (2-D) systems has been motivated largely by the problems encountered in the analysis and design of 2-D digital filters. As a result, the stability robustness problem is primarily concerned with *discrete* 2-D systems.

Let

$$H(z_1, z_2) = \frac{a(z_1, z_2)}{b(z_1, z_2)}$$

be the transfer function of a 2-D discrete system, where $a(z_1, z_2)$ and $b(z_1, z_2)$ are polynomials in z_1^{-1}, z_2^{-1} with real coefficients, and

$$b(z_1, z_2) = \sum_{i=1}^{n_1} \sum_{j=0}^{n_2} b_{ij} z_1^{-i} z_2^{-j} \qquad b_{00} \neq 0 \tag{49.38}$$

The stability analysis of a discrete 2-D system is far more complex than its 1-D counterpart because of the following reasons. First, the concept of roots (or zeroes) of polynomial $b(z_1, z_2)$ is related to *algebraic curves* rather than isolated and finitely many points, as in the 1-D case, or isolated and infinitely many points, as in the time delay case. As a result the identification of the "roots" of a 2-D polynomial is computationally considerably more expensive than that of a 1-D polynomial. In addition, zero coprime and factor coprime are no longer the same concept, and the zeroes of the numerator polynomial on the bi-unit circle $T^2 = \{(z_1, z_2)\colon |z_1| = 1,\ |z_2| = 1\}$ could play a role when any of these zeroes turn out to belong to the class of nonessential singularities of the second kind [9].

For the sake of simplicity we consider transfer function $H(z_1, z_2)$, which has no nonessential singularities of the second kind on T^2. The stability analysis of $H(z_1, z_2)$ can then be carried out by considering the issue for a simplified transfer function:

$$\hat{H}(z_1, z_2) = \frac{1}{b(z_1, z_2)} \tag{49.39}$$

which always admits a minimal realization in Roesser state space as

$$\begin{bmatrix} x^h(i+1, j) \\ x^v(i, j+1) \end{bmatrix} = \begin{bmatrix} A_1 & A_2 \\ A_2 & A_4 \end{bmatrix} \begin{bmatrix} x^h(i, j) \\ x^v(i, j) \end{bmatrix} + \begin{bmatrix} b_1 \\ b_2 \end{bmatrix} u(i, j) \equiv Ax(i, j) + bu(i, j)$$

$$y(i, j) = [c_1 \quad c_2] \begin{bmatrix} x^h(i, j) \\ x^v(i, j) \end{bmatrix} \equiv cx(i, j)$$

where $A_1 \in R^{n_1 \times n_2}$, $A_2 \in R^{n_2 \times n_2}$ [23]. It is known that (49.39) is stable iff

$$\det \begin{bmatrix} I - z_1^{-1} A_1 & -z_1^{-1} A_2 \\ -z_2^{-1} A_3 & I - z_2^{-1} A_4 \end{bmatrix} \neq 0$$

for $(z_1^{-1}, z_2^{-1}) \in \{(z_1^{-1}, z_2^{-1}): |z_1^{-1}| \leq 1, |z_2^{-1}| \leq 1\}$. As in the 1-D case, the stability radius of a stable 2-D system for *complex* perturbations is defined by

$$R_C^{(2)} = \inf\{\bar{\sigma}(\Delta): \Delta \in C^{(n_1+n_2)\times(n_1+n_2)}, A + \Delta \text{ unstable}\} \tag{49.40}$$

The following theorem gives the exact value of $R_C^{(2)}$.

Theorem 49.20 [28]:

$$R_C^{(2)} = \inf_{(\omega_1, \omega_2)\in\Omega} \underline{\sigma}[I(e^{j\omega_1}, e^{j\omega_2}) - A] \tag{49.41}$$

where $I(e^{j\omega_1}, e^{j\omega_2}) = e^{j\omega_1} I \oplus e^{j\omega_2} I$ *and* $\Omega = \{(\omega_1, \omega_2): -\pi \leq \omega_1, \omega_2 \leq \pi\}$.

A grid-type method was also proposed in [28] for the numerical evaluation of $R_C^{(2)}$ in (49.41).

References

[1] J. Ackermann, H. Z. Hu, and D. Kaesbauer, "Robustness analysis: a case study," *IEEE Trans. Autom. Control,* vol. 35, pp. 352–356, Mar. 1990.

[2] B. D. O. Anderson, E. I. Jury, and M. Mansour, "On robust Hurwitz polynomials," *IEEE Trans. Autom. Control,* vol. 32, pp. 909–912, Oct. 1987.

[3] B. R. Barmish, "New tools for robustness analysis," in *Proc. IEEE Conf. Decision and Control,* pp. 1–6, Austin, TX, Dec. 1988.

[4] B. R. Barmish, "A generalization of Kharitonov's four-polynomial concept for robust stability problems with linearly dependent coefficient perturbations," *IEEE Trans. Autom. Control,* vol. 34, pp. 157–165, Feb. 1989.

[5] R. H. Bartels and G. W. Stewart, "Solution of the equation $AX + XB = C$," *Commun. ACM,* vol. 15, pp. 820–826, 1972.

[6] A. C. Bartlett, C. V. Hollot, and L. Huang, "Root locations of an entire polytope of polynomials: it suffices to check the edges," *Math. Control Signals Syst.,* vol. 1, pp. 61–71, 1988.

[7] A. C. Bartlett and C. V. Hollot, "A necessary and sufficient condition for Schur invariance and generalized stability of polytopes of polynomials," *IEEE Trans. Autom. Control,* vol. 33, pp. 575–578, June 1988.

[8] S. Bialas, "A necessary and sufficient condition for the stability of convex combinations of stable polynomials and matrices," *Bull. Pol. Acad. Sci., Tech. Sci.*, vol. 33, pp. 473–480, 1985.

[9] N. K. Bose, *Applied Multidimensional Systems Theory*, New York: Van Nostrand Reinhold, 1982.

[10] N. K. Bose and E. Zeheb, "Kharitonov's theorem and stability test of multidimensional digital filters," *IEEE Proc.*, vol. 133, part G, pp. 187–190, Aug. 1986.

[11] N. K. Bose, E. I. Jury, and E. Zeheb, "On robust Hurwitz and Schur polynomials," *IEEE Trans. Autom. Control*, vol. 33, pp. 1166–1168, Dec. 1988.

[12] R. Byers, "A bisection method for measuring the distance of a stable matrix to the unstable matrices," *SIAM J. Sci. Stat. Comput.*, vol. 9, pp. 875–881, Sept. 1988.

[13] R. R. E. de Gaston and M. G. Safonov, "Exact calculation of the multiloop stability margin," *IEEE Trans. Autom. Control.*, vol. 33, pp. 156–171, Feb. 1988.

[14] P. Dorato and R. K. Yedavalli (Eds.), *Recent Advances in Robust Control*, New York: IEEE Press, 1990.

[15] M. Fu and B. R. Barmish, "Stability of convex and linear combinations of polynomials and matrices arising in robustness problems," in Proc. Conf. Information Sciences and Systems, Johns Hopkins Univ., Baltimore, MD., 1987.

[16] M. Fu, A. W. Olbrot, and M. P. Polis, "Robust stability for time-delay systems: The edge theorem and graphical tests," IEEE Trans. Autom. Control, vol. 34, pp. 813–820, Aug. 1989.

[17] S. J. Hammarling, "Numerical solution of the stable non-negative definite Lyapunov equation," *IMA J. Numer. Anal.*, vol. 2, pp. 303–323, 1982.

[18] D. Hinrichsen and A. J. Pritchard, "Stability radii of linear systems," *Syst. Control Lett.*, vol. 7, pp. 1–10, Feb. 1986.

[19] C. V. Hollot and A. C. Bartlett, "On the eigenvalues of interval matrices," in *Proc. IEEE Conf. Decision and Control*, pp. 794–799, Los Angeles, Dec. 1987.

[20] E. I. Jury, "Robustness of discrete systems: a review," in *Proc. IFAC 11th Triennial World Cong.*, pp. 197–202, Tallinn, Estonia, USSR, 1990.

[21] T. Kailath, *Linear Systems*, Englewood Cliffs, NJ: Prentice-Hall, 1980.

[22] V. L. Kharitonov, "Asymptotic stability of a family of systems of linear differential equations," *Differential'nye Uravneniya*, vol. 14, pp. 2086–2088, 1978; English translation in *Differential Equations*, vol. 14, pp. 1483–1485, 1979.

[23] F. J. Kraus, B. D. O. Anderson, E. I. Jury, and M. Mansour, "On robustness of low-order Schur polynomials," *IEEE Trans. Circuits Syst.*, vol. 35, pp. 570–577, May 1988.

[24] F. J. Kraus, B. D. O. Anderson, and M. Mansour, "Robust Schur polynomial stability and Kharitonov's theorem," *Int. J. Control*, vol. 47, pp. 1213–1225, 1988.

[25] S. Y. Kung, B. C. Levy, M. Morf, and T. Kailath, "New results in 2-D systems theory. II. 2-D state-space models—realization and notions of controllability, observability, and minimality," *Proc. IEEE*, vol. 65, pp. 945–961, June 1977.

[26] W. H. Lee, "Robustness analysis for state-space models," Report TP-151, Alphatech Inc., 1982.

[27] A. V. Lipatov and N. I. Sokolov, "Some sufficient conditions for stability and instability of continuous linear stationary systems," translated from *Autom. Telemekh.*, no. 9, pp. 30–37, 1978; in *Autom. Remote Control*, vol. 39, pp. 1285–1291, 1979.

[28] W.-S. Lu, "Stability robustness of two-dimensional discrete systems and its computation," *IEEE Trans. Circuits Syst.*, vol. 36, pp. 285–287, Feb. 1989.

[29] W.-S. Lu, H.-P. Wang, and A. Antoniou, "An efficient method for the evaluation of the controllability and observability gramians of 2-D digital filters and systems. II.," *IEEE Trans. Circuits Syst.*, vol. 39, pp. 695–704, Oct. 1992.

[30] D. G. Luenberger, *Linear and Nonlinear Programming*, 2nd ed., Reading, MA: Addison-Wesley, 1984.

[31] M. Mansour, F. J. Kraus, and B. D. O. Anderson, "Strong Kharitonov theorem for discrete systems," in *Proc. IEEE Conf. Decision and Control*, pp. 106–111, Austin, TX, Dec. 1988.

[32] J. F. Martin, "State space measures for stability robustness," *IEEE Trans. Autom. Control*, vol. 32, pp. 509–512, June 1987.

[33] R. J. Minnichelli, J. J. Anagnost, and C. A. Desoer, "An elementary proof of Kharitonov's stability theorem with extensions," *IEEE Trans. Autom. Control*, vol. 34, pp. 995–998, Sept. 1989.

[34] R. V. Patel and M. Toda, "Quantitative measures of robustness of multivariable system," in *Proc. Jt. Am. Control Conf.*, paper TD8-A, San Francisco, 1980.

[35] L. Qiu and E. J. Davison, "New perturbation bounds for the robust stability of linear state space models," in *Proc. IEEE Conf. Decision and Control*, pp. 751–755, Athens, Dec. 1986.

[36] L. Qiu and E. J. Davison, "A new method for the stability robustness determination of state space models with real perturbations," in *Proc. IEEE Conf. Decision and Control*, pp. 538–543, Austin, TX, Dec. 1988.

[37] L. Qiu, B. Bernhardsson, A. Rantzer, E. J. Davison, P. M. Young, and J. C. Doyle, "On the real structured stability radius," in *Proc. 12th IFAC World Cong.*, vol. 8, pp. 71–78, Sydney, July 1993.

[38] M. E. Sezer and D. D. Šiljak, "Robust stability of discrete systems," *Int. J. Control.*, vol. 48, pp. 2055–2063, 1988.

[39] A. Sideris and R. R. E. de Gaston, "Multivariable stability margin calculation with uncertain correlated parameters," in *Proc. IEEE Conf. Decision and Control*, pp. 766–771, Athens, Dec. 1986.

[40] A. Sideris and R. S. Peña, "Fast computation of the multivariable stability margin for real interrelated uncertain parameters," *IEEE Trans. Autom. Control*, vol. 34, pp. 1272–1276, Dec. 1989.

[41] G. W. Stewart, *Introduction to Matrix Computations*, New York: Academic Press, 1973.

[42] C. Van Loan, "How near is a stable matrix to an unstable matrix," *Contemp. Math.*, vol. 47, pp. 465–477, 1985.

[43] R. S. Varga, *Matrix Iterative Analysis*, Englewood Cliffs, NJ: Prentice-Hall, 1962.

[44] R. K. Yedavalli, "Improved measures of stability robustness for linear state space models," *IEEE Trans. Autom. Control*, vol. 30, pp. 577–579, June 1985.

[45] L. A. Zadeh and C. A. Desoer, *Linear System Theory*, New York: McGraw-Hill, 1963.

[46] K. Zhou and P. P. Khargonekar, "Stability robust bounds for linear state-space models with structured uncertainty," *IEEE Trans. Autom. Control*, vol. 32, pp. 621–623, July 1987.

50

Robust Stability of Circuits with Multiple Uncertain Parameters

Li Qiu
Hong Kong University of Science and Technology

50.1 Introduction

Consider the circuit shown in Fig. 50.1(a). Figure 50.1(b) shows its graph and a selected proper tree marked by thick lines. The Kirchhoff's current law equations for the two fundamental cut sets with capacitors are

$$C_1\frac{dv_{C_1}}{dt} = -G_1 v_{C_1} - \alpha v_{C_1} - i_L - G_2(v_{C_1} - v_{C_2})$$
$$C_2\frac{dv_{C_2}}{dt} = \alpha v_{C_1} + i_L + G_2(v_{C_1} - v_{C_2})$$

The Kirchhoff's voltage law equation for the fundamental loop with an inductor is

$$L\frac{di_L}{dt} = v_{C_1} - v_{C_2}$$

A state-space equation of the circuit is given by

$$\begin{bmatrix} \frac{dv_{C_1}}{dt} \\ \frac{dv_{C_2}}{dt} \\ \frac{di_L}{dt} \end{bmatrix} = \begin{bmatrix} -C_1^{-1}(G_1+G_2+\alpha) & C_1^{-1}G_2 & -C_1^{-1} \\ C_2^{-1}(G_2+\alpha) & -C_2^{-1}G_2 & C_2^{-1} \\ L^{-1} & -L^{-1} & 0 \end{bmatrix} \begin{bmatrix} v_{C_1} \\ v_{C_2} \\ i_L \end{bmatrix} \tag{50.1}$$

-8493-8341-2/95/$0.00 + $.50

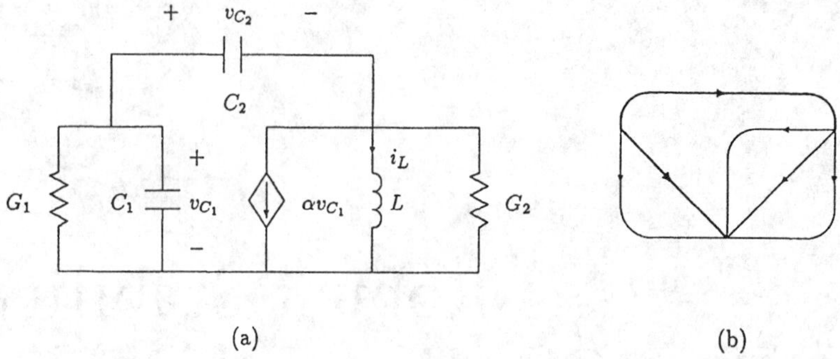

FIGURE 50.1 (a) An active circuit whose stability is of concern. (b) Its graph.

We know that this circuit is stable if all eigenvalues of the 3×3 matrix on the right-hand side of (50.1) have negative real part. Assume that the capacitances of the capacitors and the inductance of the inductor are exactly given as $C_1 = C_2 = 1F$ and $L = 1H$. Also assume that the conductances, G_1 and G_2, of the resistors and the gain α of the voltage-controlled current source are somewhat uncertain. What we know about them is the following:

$$G_1 = G_{10} + \delta G_1, \qquad G_2 = G_{20} + \delta G_2, \qquad \alpha = \alpha_0 + \delta\alpha$$
$$G_{10} = G_{20} = 1S, \qquad \alpha_0 = 2S$$
$$|\delta G_1| \leq 10\% G_{10}, \qquad |\delta G_2| \leq 10\% G_{20}, \qquad |\delta\alpha| \leq 10\% \alpha_0$$

Here, we call G_{10}, G_{20}, and α_0 "nominal values" and call δG_1, δG_2, and $\delta\alpha$ "perturbations". A natural stability analysis problem is to determine if the circuit is stable for all possible values of G_1, G_2, and α.

With the numerical values substituted, the matrix on the right-hand side of (50.1) becomes

$$\begin{bmatrix} -(4 + \delta G_1 + \delta G_2 + \delta\alpha) & 1 + \delta G_2 & -1 \\ 3 + \delta G_2 + \delta\alpha & -(1 + \delta G_2) & 1 \\ 1 & -1 & 0 \end{bmatrix}$$

To facilitate the analysis, we rewrite this matrix in a form so that the perturbation parameters are isolated in a single matrix.

$$\begin{bmatrix} -4 & 1 & -1 \\ 3 & -1 & 1 \\ 1 & -1 & 0 \end{bmatrix} + \begin{bmatrix} -1 & -1 \\ 0 & 1 \\ 0 & 0 \end{bmatrix} \begin{bmatrix} \delta G_1 & 0 \\ \delta\alpha & \delta G_2 \end{bmatrix} \begin{bmatrix} 1 & 0 & 0 \\ 1 & -1 & 0 \end{bmatrix}$$

Our problem then becomes determining whether all matrices in the set

$$\left\{ \begin{bmatrix} -4 & 1 & -1 \\ 3 & -1 & 1 \\ 1 & -1 & 0 \end{bmatrix} + \begin{bmatrix} -1 & -1 \\ 0 & 1 \\ 0 & 0 \end{bmatrix} \begin{bmatrix} \delta G_1 & 0 \\ \delta\alpha & \delta G_2 \end{bmatrix} \begin{bmatrix} 1 & 0 & 0 \\ 1 & -1 & 0 \end{bmatrix} : \right. \qquad (50.2)$$
$$\left. |\delta G_1| \leq 0.1, |\delta G_2| \leq 0.1, |\delta\alpha| \leq 0.2 \right\}$$

are stable.

From this example, we draw two observations:

1. In stability analysis of circuits, the stability of an uncertain matrix, equivalent to a set of matrices mathematically, is often of concern. The set of matrices almost always contains an infinite number of matrices and is often characterized by a large number of uncertain parameters. The problem of determining the stability of a set of matrices is called the *robust stability problem.*
2. An uncertain matrix can be put into the form $A + B\,\Delta C$, where $A \in \mathbb{R}^{n\times n}$ is the nominal matrix, i.e., the value of the matrix when the perturbation is zero; $\Delta \in \mathbb{R}^{m\times p}$ is the perturbation matrix which is only known to belong to a set $\boldsymbol{\Delta} \subset \mathbb{R}^{m\times p}$ containing the origin. $B \in \mathbb{R}^{n\times m}$ and $C \in \mathbb{R}^{p\times n}$ are structural matrices which are used to control the way that the elements of perturbation Δ enter the elements of A. In general, the set Δ may be difficult to characterize because Δ may depend on uncertain parameters in a sophisticated way. In the example given above, $\boldsymbol{\Delta}$ is said to be an interval matrix, i.e., a set of matrices whose elements are independent and take values in intervals. Of course, $\boldsymbol{\Delta}$ in the example is not just an ordinary interval matrix, but an interval matrix centered at the origin.

To obtain a complete solution to the robust stability problem of $A + B\,\Delta C$ when Δ is an arbitrary set containing the origin is an extremely difficult task. By a complete solution, we mean an algorithm which returns an affirmative answer when all members of the set are stable and returns a negative answer when at least one member is not stable. In particular, for the seemingly simple case that $\boldsymbol{\Delta}$ is an interval matrix, the robust stability problem is NP hard, which implies that it is unlikely that a polynomial time algorithm exists to solve this problem. In the following we consider a special case when the set $\boldsymbol{\Delta}$ is given by

$$\{\Delta \in \mathbb{R}^{m\times p}: \|\Delta\| \leq \delta\} \tag{50.3}$$

where $\|\cdot\|$ is a matrix norm which is used to measure the magnitude of a matrix. When the matrix norm is chosen properly, the robust stability problem does have a complete solution. In applications where $\boldsymbol{\Delta}$ is not given by (50.3), the complete solution for (50.3) can be used to obtain a conservative solution in the following way: Choose the smallest $\delta > 0$ so that $\boldsymbol{\Delta} \subset \{\Delta \in \mathbb{R}^{m\times p}: \|\Delta\| \leq \delta\}$; determine if $A + B\,\Delta C$ is stable for all $\Delta \in \{\Delta \in \mathbb{R}^{m\times p}: \|\Delta\| \leq \delta\}$. If the answer is affirmative, then $A + B\,\Delta C$ is stable for all $\Delta \in \boldsymbol{\Delta}$. If the answer is negative, no conclusion can be drawn.

50.2 Matrix Norms and Singular Value Decomposition

The magnitude of a matrix is measured by a matrix norm. Let us focus on matrices in $\mathbb{R}^{m\times n}$. A matrix norm is a function $\|\cdot\|: \mathbb{R}^{m\times n} \to [0, \infty)$ which satisfies the following conditions:

1. $\|A\| = 0$ if and only if $A = 0$ (positivity)
2. $\|\alpha A\| = |\alpha|\,\|A\|$ for all $A \in \mathbb{R}^{m\times n}$ and $\alpha \in \mathbb{R}$ (homogeneity)
3. $\|A + B\| \leq \|A\| + \|B\|$ for all $A, B \in \mathbb{R}^{m\times n}$ (the triangle inequality)

There are many possible matrix norms. One used commonly with many numerical and analytic advantages is given by the largest singular value. Now let us see what the singular values of a matrix are.

A square matrix $U \in \mathbb{R}^{n\times n}$ is said to be unitary if $U'U = I$. A matrix (not necessarily square) is said to be diagonal if all (i, j)th elements for $i \neq j$ are zero. We denote a diagonal matrix by $\mathrm{diag}(d_1, d_2, \ldots, d_{\min\{m,n\}})$ where $d_i, i = 1, 2, \ldots, \min\{m, n\}$, are the diagonal, i.e., the (i, i)th, elements. This notation does not give complete information on the dimension of the diagonal matrix, which is assumed to be given by the context.

Theorem 50.1: (*Singular Value Decomposition*) *For any matrix* $A \in \mathbb{R}^{m\times n}$, *there exist unitary matrices* $U \in \mathbb{R}^{m\times m}$, $V \in \mathbb{R}^{n\times n}$ *and a diagonal matrix* $S = \text{diag}(s_1, s_2, \ldots, s_{\min\{m,n\}}) \in \mathbb{R}^{m\times n}$ *with* $s_1 \geq s_2 \geq \cdots \geq s_{\min\{m,n\}} \geq 0$ *such that* $A = USV'$.

The diagonal elements s_i, $i = 1, 2, \ldots, \min\{m, n\}$, of S are called the *singular values* of A and are denoted by $\sigma_i(A)$.

If we define $\|A\| = \sigma_1(A)$, then we can prove that this $\|\cdot\|$ indeed satisfies the three conditions required for a matrix norm. A reader can try this out as an exercise. Actually, conditions 1 and 2 are easily verified, but condition 3 is a little involved. This particular matrix norm is called the *spectral norm.* Standard references on singular values and matrix norms are [2, 4].

Singular values have applications in many engineering and scientific disciplines. Many mathematical software packages include singular value decomposition as a standard feature. For example, in MATLAB, the following command returns the decomposition:

$$[U, S, V] = svd(A)$$

To obtain the spectral norm of A, one can simply pick up the largest one from the singular values of A, or execute the following command in MATLAB:

$$norm(A)$$

50.3 Unimodal Functions and One-Dimensional Optimization

Consider a continuous function $f\colon [a, b] \to \mathbb{R}$. A point $x^* \in [a, b]$ is called a **local minimum point** of f if there is an $\epsilon > 0$ such that $f(x^*) \leq f(x)$ for all $x \in [a, b]$ with $|x - x^*| < \epsilon$. A point $x^* \in [a, b]$ is called a **global minimum point** of f if $f(x^*) \leq f(x)$ for all $x \in [a, b]$. The value of the function $f(x^*)$ at a local minimum point or a global minimum point x^* is called a **local minimum** or the **global minimum** of f, respectively. The function f is said to be *unimodal* if every local minimum is a global minimum. There are many subtleties in the concept of unimodality. For details consult, e.g [1].

It is well known that a continuous function on a closed interval has a global minimum. One property of a unimodal function is that its global minimum can be found easily. A commonly used method for unimodal function minimization is the golden section search. The idea is to reduce successively the length of the interval containing a global minimum point (often referred to as interval of uncertainty) until the length of the interval becomes so small that it is indistinguishable from a point in a practical sense. Assume we want to minimize a unimodal function $f\colon [a, b] \to \mathbb{R}$. Let us choose α, β such that $a < \alpha < \beta < b$. Then we have the following three cases:

CASE 1: $f(\alpha) > f(\beta)$. In this case any global minimum point must be contained in $[\alpha, b]$.
CASE 2: $f(\alpha) < f(\beta)$. In this case any global minimum point must be contained in $[a, \beta]$.
CASE 3: $f(\alpha) = f(\beta)$. In this case a global minimum point exists in $[\alpha, \beta]$. Although $[\alpha, \beta]$ is shorter than $[a, \beta]$ and $[\alpha, b]$, we usually do not take this advantage. Instead, we only reduce the search interval to either $[a, \beta]$ or $[\alpha, b]$.

Note that we need to evaluate the function twice to carry out the reduction here. However, if we continue to reduce $[a, \beta]$ or $[\alpha, b]$, the point α or β is in (a, β) or $(\alpha, b]$, respectively, and its function value already has been evaluated. Hence, we only need to choose one additional point and evaluate the function at this point. This also applies to any subsequent reduction. If we carry out this reduction iteratively, we will be able to reduce the interval of uncertainty to an arbitrarily small length. The golden section search method provides a good way to choose the points at which the function is evaluated so that the speed of interval reduction is fast, while the number of function evaluation is small.

Initially, the golden section method sets

$$a_1 = a, \qquad b_1 = b, \qquad \alpha_1 = a_1 + 0.312(b_1 - a_1), \qquad \beta_1 = a_1 + 0.618(b_1 - a_1)$$

At iteration k, let the search interval be $[a_k, b_k]$. Then the new search interval $[a_{k+1}, b_{k+1}]$ is determined as follows. If $f(\alpha_k) > f(\beta_k)$, set

$$[a_{k+1}, b_{k+1}] = [\alpha_k, b_k], \qquad \alpha_{k+1} = \beta_k, \qquad \beta_{k+1} = a_{k+1} + 0.618(b_{k+1} - a_{k+1})$$

If $f(\alpha_k) \le f(\beta_k)$, set

$$[a_{k+1}, b_{k+1}] = [a_k, \beta_k], \qquad \alpha_{k+1} = a_{k+1} + 0.312(b_{k+1} - a_{k+1}), \qquad \beta_{k+1} = \alpha_k$$

Continue the iteration until $b_k - a_k$ is less than a predetermined error tolerance. Then any point in $[a_k, b_k]$ can be taken as a global minimum point.

In what follows, we need to deal with a continuous function $f\colon (a, b] \to \mathbb{R}$. Such a function is said to be unimodal if it is unimodal in every closed subinterval in $(a, b]$. In general, such a function may not have a global minimum, but it always has an infimum. In a practical sense, the infimum of a continuous unimodal function $f\colon (a, b] \to \mathbb{R}$ can be considered equal to the global minimum of the same function restricted to $[a + \epsilon, b]$ where ϵ is a positive number smaller than the error tolerance. Note that in carrying out the golden section search, we can treat a function on $(a, b]$ in the same way as a function on $[a, b]$; i.e., we do not need to form $a + \epsilon$ because the function is never evaluated at the end points of the interval during the golden section search process.

The same argument as above applies to functions on $[a, b)$ or (a, b), but we do not need this in the following development.

Golden section search is also a standard feature in many mathematical software packages. In MATLAB the command *fmin* implements golden section search with some enhancement to accelerate the convergence. In order to use it, one needs to create a function file called *fun.m* so that *fun*(x) returns the function value $f(x)$ at any $x \in (a, b)$ and the following command returns the global minimum point:

$$xstar = fmin('fun', a, b)$$

50.4 Real Stability Radius

Let $\mathbb{C}_g$ be an open subset of the complex plane $\mathbb{C}$. Typically, $\mathbb{C}_g = \{s \in \mathbb{C}\colon \mathrm{Re}\, s < 0\}$ or $\mathbb{C}_g = \{s \in \mathbb{C}\colon |s| < 1\}$. A matrix is said to be stable if its eigenvalues are contained in $\mathbb{C}_g$. Given $(A, B, C) \in \mathbb{R}^{n\times n} \times \mathbb{R}^{n\times m} \times \mathbb{R}^{p\times n}$, we define the **real stability radius** of (A, B, C) as

$$r_{\mathbb{R}}(A, B, C) := \inf\{\sigma_1(\Delta)\colon \Delta \in \mathbb{R}^{m\times p} \quad \text{and} \quad A + B\,\Delta C \quad \text{is unstable}\}$$

The motivation for this definition is as follows: Suppose A is a known stable matrix, Δ is an unknown perturbation whose spectral norm can be estimated, and B and C are given matrices determining the structure that Δ enters A. If $\sigma_1(\Delta)$ is less than $r_{\mathbb{R}}(A, B, C)$, then the perturbated matrix $A + B\,\Delta C$ is stable. Furthermore, $r_{\mathbb{R}}(A, B, C)$ is the best possible such bound.

To compute the real stability radius, we need the following preliminary result.

Theorem 50.2: *For each $M \in \mathbb{C}^{p\times m}$,*

$$\sigma_2\left(\begin{bmatrix} \operatorname{Re} M & -\gamma \operatorname{Im} M \\ \gamma^{-1}\operatorname{Im} M & \operatorname{Re} M \end{bmatrix}\right)$$

is an unimodal function of γ on $(0, 1]$.

Let us define a function $\mu_{\mathbb{R}}: \mathbb{C}^{p\times m} \to [0, \infty)$ by

$$\mu_{\mathbb{R}}(M) = \inf_{\gamma\in(0,1]}\sigma_2\left(\begin{bmatrix} \operatorname{Re} M & -\gamma \operatorname{Im} M \\ \gamma^{-1}\operatorname{Im} M & \operatorname{Re} M \end{bmatrix}\right)$$

Then, Theorem 50.2 implies that $\mu_{\mathbb{R}}(M)$ can be computed easily by using, for example, the golden section search.

The following formula for the real stability radius was devised by Qiu et al. [3].

Theorem 50.3: *For $(A, B, C) \in \mathbb{R}^{n\times n} \times \mathbb{R}^{n\times m} \times \mathbb{R}^{p\times n}$ with A stable,*

$$r_{\mathbb{R}}(A, B, C) = \left\{\sup_{s\in\partial\mathbb{C}_g} \mu_{\mathbb{R}}[C(sI - A)^{-1}B]\right\}^{-1}$$

Based on Theorem 50.3, the real stability radius can by computed in two steps:

1. For each s in (a dense grid of) $\partial\mathbb{C}_g$, compute $\mu_{\mathbb{R}}[C(sI - A)^{-1}B]$.
2. Find the supremum of $\mu_{\mathbb{R}}[C(sI - A)^{-1}B]$ over s by a search over all s in (a dense grid of) $\partial\mathbb{C}_g$. This search can be done in a brute force way. Usually, we plot the function $\mu_{\mathbb{R}}[C(sI - A)^{-1}B]$ and find the supremum by inspection.

50.5 Computational Examples

Example 1. Assume $\mathbb{C}_g = \{s \in \mathbb{C}: \operatorname{Re} s < 0\}$. Find $r_{\mathbb{R}}(A, B, C)$ for

$$A = \begin{bmatrix} 79 & 20 & -30 & -20 \\ -41 & -12 & 17 & 13 \\ 167 & 40 & -60 & -38 \\ 33.5 & 9 & -14.5 & -11 \end{bmatrix} \qquad B = \begin{bmatrix} 0.2190 & 0.9347 \\ 0.0470 & 0.3835 \\ 0.6789 & 0.5194 \\ 0.6793 & 0.8310 \end{bmatrix}$$

$$C = \begin{bmatrix} 0.0346 & 0.5297 & 0.0077 & 0.0668 \\ 0.0535 & 0.6711 & 0.3834 & 0.4175 \end{bmatrix}$$

The matrix A is stable because it has eigenvalues $-1 \pm j10$ and $-1 \pm j1$. We plot $\mu_{\mathbb{R}}[C(j\omega I - A)^{-1}B]$ in Fig. 50.2. Its maximal value is 1.9450. The maximum occurs at $\omega = 1.3800$. We obtain $r_{\mathbb{R}}(A, B, C) = 0.5141$.

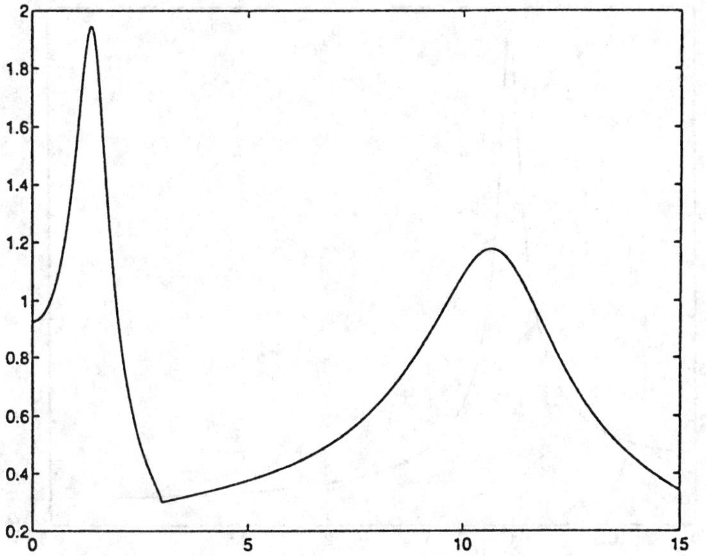

FIGURE 50.2 For Example 1, plot $\mu_{\mathbb{R}}[C(j\omega I - A)^{-1}B]$ vs. ω.

Example 2. Assume $\mathbb{C}_g = \{s \in \mathbb{C}: |s| < 1\}$. Find $r_{\mathbb{R}}(A, B, C)$ for

$$A = \begin{bmatrix} 4.7527 \times 10^{-1} & 7.5787 \times 10^{-1} & 7.9939 \\ -4.1523 \times 10^{-2} & 8.9051 \times 10^{-1} & 7.5787 \times 10^{-1} \\ -7.5787 \times 10^{-2} & -4.1523 \times 10^{-2} & 4.7527 \times 10^{-1} \end{bmatrix}$$

$$B = \begin{bmatrix} 8.0086 \times 10^{-2} & 4.2994 \times 10^{-2} \\ -1.4704 \times 10^{-3} & 9.4791 \times 10^{-2} \\ -4.2994 \times 10^{-3} & -1.4704 \times 10^{-3} \end{bmatrix} \qquad C = \begin{bmatrix} 1 & 0 & 0 \\ 0 & 1 & 0 \end{bmatrix}$$

The eigenvalues of A are $5.1076 \times 10^{-1} \pm j8.0192 \times 10^{-1}$ and 8.1953×10^{-1}, which implies that A is stable. We plot $\mu_{\mathbb{R}}[C(e^{j\theta}I - A)^{-1}B]$ in Fig. 50.3. Its maximal value is 9.6395×10^{-1}. The maximum occurs at $\theta = 1.0053$. We obtain $r_{\mathbb{R}}(A, B, C) = 1.0374$.

Example 3. Here, we consider the circuit example given in the introduction. We have

$$A = \begin{bmatrix} -4 & 1 & -1 \\ 3 & -1 & 1 \\ 1 & -1 & 0 \end{bmatrix} \qquad B = \begin{bmatrix} -1 & -1 \\ 0 & 1 \\ 0 & 0 \end{bmatrix} \qquad C = \begin{bmatrix} 1 & 0 & 0 \\ 1 & -1 & 0 \end{bmatrix}$$

Because the circuit is a continuous time system, it follows that $\mathbb{C}_g = \{s \in \mathbb{C}: \operatorname{Re} s < 0\}$. The eigenvalues of A are -4.3652 and $-0.3174 \pm j0.3583$, so A is stable. We plot $\mu_{\mathbb{R}}[C(j\omega I - A)^{-1}B]$ in Fig. 50.4. Its maximal value is 1.0397. The maximum occurs at $\omega = 0.3980$. We obtain $r_{\mathbb{R}}(A, B, C) = 0.9618$. This means that $A + B\,\Delta C$ is stable for all $\Delta \in \{\Delta \in \mathbb{R}^{2\times 2}: \sigma_1(\Delta) < 0.9618\}$. However, what we are concerned with is if all the matrices in set (50.2) are stable. Note that

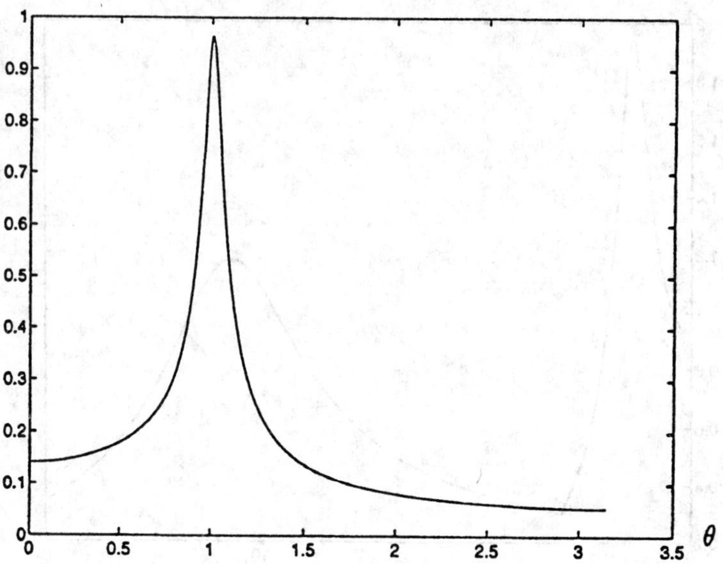

FIGURE 50.3 For Example 2, plot $\mu_{\mathbb{R}}[C(e^{j\theta}I - A)^{-1}B]$ vs. θ.

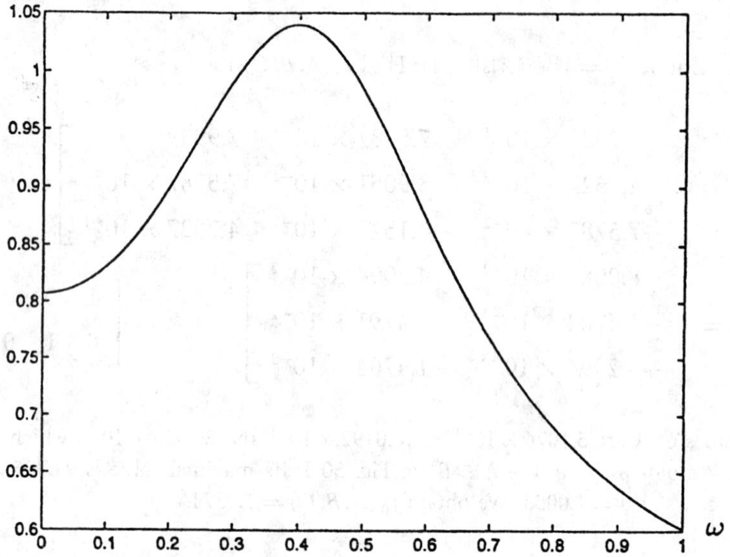

FIGURE 50.4 For Example 3, plot $\mu_{\mathbb{R}}[C(j\omega I - A)^{-1}B]$ vs. ω.

for all matrices

$$\Delta = \begin{bmatrix} \delta G_1 & 0 \\ \delta\alpha & \delta G_2 \end{bmatrix}$$

with $|\delta G_1| \leq 0.1$, $|\delta G_2| \leq 0.1$, and $|\delta\alpha| \leq 0.2$, we have

$$\sigma_1(\Delta) \leq \sigma_1\left(\begin{bmatrix} 0.1 & 0 \\ 0.2 & 0.1 \end{bmatrix}\right) = 0.2414$$

Therefore, all matrices in (50.2) are indeed stable.

References

[1] M. S. Bazaraa, H. D. Sherali, and C. M. Shetty. *Nonlinear Programming—Theory and Algorithms*, 2nd ed. New York: John Wiley & Sons, 1993.

[2] R. A. Horn and C. R. Johnson, *Matrix Analysis*, Cambridge, U.K.: Cambridge University Press, 1985.

[3] L. Qiu, B. Bernhardsson, A. Rantzer, E. J. Davison, P. M. Young, and J. C. Doyle, "A formula for computation of the real stability radius," *Automatica*, vol. 37, no. 7, 1995.

[4] G. W. Stewart and J. G. Sun, *Matrix Perturbation Theory*, San Diego: Academic Press, 1990.

X

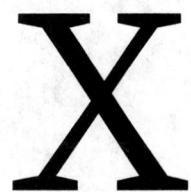

Computer-Aided Design and Optimization

S. M. Kang
University of Illinois at Urbana-Champaign

51

Modeling of Circuit Performances

S. M. Kang
University of Illinois, Urbana-Champaign

A. Dharchoudhury
University of Illinois, Urbana-Champaign

51.1 Introduction

The domain of computer-aided design and optimization for circuits and filters alone is very broad, especially since the design objectives are usually multiple. In general, the circuit design can be classified into

- electrical design
- physical design

Physical design deals with concrete geometrical parameters while electrical design deals with the electrical performances of the physical object. Strictly speaking, both design aspects are inseparable since any physical design is a realization of a particular electrical design. On the other hand, electrical design without proper abstraction of the corresponding physical design can be futile in the sense that its implementation may not meet the design goals. Thus, electrical and physical design should go hand in hand. Most literature deals with both designs using weighted abstractions. For example, in the design of a digital or analog filter, various effects of physical layout are first captured in the parasitic models for interconnects and in the electrical performance models for transistors. Then, electrical design is performed using such models to predict electrical performances. In physical design optimization, geometrical parameters are chosen such that the predicted electrical performances can meet the design objectives. For example, in the timing-driven physical layout of integrated circuits, circuit components will be laid out, placed, and interconnected to meet timing requirements. In order to make

-8493-8341-2/95/$0.00+$.50

the optimization process computationally efficient, performance models are used during the optimization process instead of simulation tools. In this chapter, the focus of design optimization is on the electrical performance. The computer-aided electrical design can be further classified into performance analysis and optimization. Thus, this chapter first discusses the modeling of electrical performances, symbolic and numerical analysis techniques, followed by optimization techniques for circuit timing, and yield enhancement with transistor sizing and statistical design techniques.

51.2 Circuit Performance Measures

The performance of a particular circuit can be measured in many ways according to application and design goals. In essence, each performance measure is an indication of the circuit output which the designer is interested in for a particular operating condition. The circuit performance, in general, is dependent on the values of the designable parameters such as the transistor sizes, over which the designer has some degree of control, and the values of the noise parameters such as the fluctuation in power supply voltage or the threshold voltages of the transistors, which are random in nature. Also, the performance measure used is dependent on the type of the circuit. The performance measures of digital circuits are usually quite different from those of analog circuits or filters. Some of the important performance measures of digital circuits are

- signal propagation delay from an input to an output
- rise and fall times of signals
- clock signal skew in the circuit
- power dissipation
- manufacturing yield
- failure rate (reliability)

Important performance measures for analog circuits are

- small-signal gain and bandwidth in frequency domain
- sensitivities to various nonideal factors (noise)
- slew rate of operational amplifiers
- power dissipation
- manufacturing yield
- failure rate (reliability)

The modeling of digital circuit performances is usually more difficult to derive than the modeling of analog circuit performances, mainly due to the nonlinear nature of digital circuits. Thus, simple transistor models have been used often to obtain simple qualitative models for propagation delays, rise and fall times, and power dissipation. On the other hand, due to the small-signal nature, the device models required by analog designers need to be much more accurate. Another important performance measure is the manufacturing yield which determines the product cost. The yield depends on both the quality of manufacturing process control and the quality of the design. A challenging design problem is how to come up with a design which can be robust to nonideal manufacturing conditions and thereby produce a good manufacturing yield. The model for failure rate is difficult to derive due to several complex mechanisms that determine the lifetime of chips. In such cases, empirical models which are not physical, have been introduced.

In order to develop models for analog circuits, transistors in the circuit are first DC-biased at the proper biasing points. Then, small signal circuit models are developed at those DC operating points. The resulting linear circuits can be analyzed by either using mathematical symbols (symbolic analysis) or by using simulators.

51.3 Input-Output Relationships

The circuit functionality depends on the input-output relationships and thus circuit specifications and test specifications are made on the basis of the input-output relationships. For modeling purposes, the input-output relationships can be classified into static relationships and dynamic relationships.

Memoryless (Static) Relationship

A given input-output pair is said to have a memoryless relationship if it can be described statically without the use of any ordinary or partial differential equation. It is important to note that the choice of input and output variables can determine the nature of the relationship. For instance, the relationship between input charge q and the output voltage v in a 1 F capacitor is static. On the other hand, the relationship between an input current i and the output voltage v, in the same capacitor is nonstatic since the voltage is dependent on the integral of the input current. Ideally, basic circuit elements such as resistor, capacitor, inductor, and source should be characterized by using static relationships, so-called constitutive relationships, with proper choice of input and output variables. However, for circuits and systems design, since their input-output relationships are specified in terms of specific variables of interest, such choice of variables cannot be made arbitrarily. Often, both input and output variables are voltages, especially in digital circuits.

Dynamic Relationships

A given input-output pair is said to have a dynamic relationship if the modeling of the relationship requires a differential equation. In most cases, since the circuits contain parasitic capacitors or inductors or both, the input-output relationship is dynamic. For instance, the input-output relationship of an inverter gate with capacitive load C_{load} can be described as

$$C_{load}\frac{dV_{out}}{dt} = i(V_{out}, V_{in}, V_{DD}) \tag{51.1}$$

where V_{out} and V_{in} denote the output and input voltages, i denotes the current through the pull-up or pull-down transistor, and V_{DD} denotes the power supply voltage. Although the relationship is implicit in this case, (51.1) can be solved either analytically or numerically to find the relationship. Equation (51.1) can be generalized by including noise effects due to variations in the circuit fabrication process.

51.4 Dependency of Circuit Performances on Circuit Parameters

Circuit performances are functions of circuit parameters, which are usually not known explicitly. In order to design and analyze analog or digital circuits, the dependency of the circuit performances on the various circuit parameters needs to be modeled.

The actual value of a circuit parameter in a manufactured circuit is expected to be different from the nominal or target value due to inevitable random variations in the manufacturing processes and in the environmental conditions in which the circuit is operated. For example, the actual channel width W of an MOS transistor can be decomposed into a nominal component W^0 and a statistically varying component ΔW, i.e., $W = W^0 + \Delta W$. The nominal component is under the control of the designer and can be set to a particular value. Such a nominal component is said to be **designable** or **controllable**, e.g., W^0. The statistically varying component, on the other hand, is not under the control of the designer and is random in nature. It is called the **noise** component, and it represents the uncontrollable fluctuation of a circuit parameter about its designable component, e.g., ΔW. For certain circuit parameters, e.g., device model parameters (like the threshold voltages of MOS transistors) and operating conditions (like temperature or power supply voltage), the nominal values are not really under the control of the circuit designer and are set by the processing and operating conditions. For these circuit parameters, the nominal and random components are together called the noise component. In general, therefore, a circuit parameter x_i can be expressed as follows:

$$x_i = d_i + s_i \tag{51.2}$$

where d_i is the designable component and s_i is the noise component. It is common to group all the designable components to form the set of **designable parameters**, denoted by **d**. Similarly, all the noise parameters are grouped together to form the **noise parameters**, denoted by the random vector **s**. Vectorially, (51.2) can be rewritten as

$$\mathbf{x} = \mathbf{d} + \mathbf{s} \tag{51.3}$$

A circuit performance measure is a function of the circuit parameters. For example, the propagation delay of a CMOS inverter circuit is a function of the nominal channel lengths and widths of the NMOS and PMOS transistors. It is also a function of the threshold voltages of the transistors and the power supply voltage. Thus, in general, a circuit performance is a function of the designable as well as the noise parameters:

$$y = y(\mathbf{x}) = y(\mathbf{d}, \mathbf{s}) \tag{51.4}$$

If this function is known explicitly in terms of the designable and noise parameters, then optimization methods can be applied directly to obtain a design which is optimal in terms of the performances or the manufacturing yield. More often than not, however, the circuit performances cannot be expressed explicitly in terms of the circuit parameters. In such cases, the value of a circuit performance for given values of circuit parameters can be obtained by circuit simulation. Circuit simulations are computationally expensive, especially if the circuit is large and transient simulations are required. A compact performance model in terms of the circuit parameters is an efficient alternative to the circuit simulator. Two factors determine the utility of the performance model. First, the model should be computationally efficient to construct and evaluate so that substantial computational savings can be achieved. Second, the model should be accurate.

Figure 51.1 shows the general procedure for constructing the model for a performance measure y in terms of the circuit parameters $\mathbf{x}$. The model-building procedure consists of four steps. In the first step, m *training points* are selected from the $\mathbf{x}$-space. The ith training point is denoted by $\mathbf{x}_i$, $i = 1, 2, \ldots, m$. In the second step, the circuit is simulated at these m training points and the values of the performance measure are obtained from the circuit simulation results as $y(\mathbf{x}_1)$, $y(\mathbf{x}_2), \ldots, y(\mathbf{x}_m)$. In the third step, a preassigned function of y in terms of $\mathbf{x}$ is "fitted" to the data. In the fourth and final step, the model is validated for accuracy. If the model accuracy is deemed inadequate, the modeling procedure is repeated with a larger number of training points or with different models.

The model is called the **response surface model** (RSM) [1] of the performance and is denoted by $\hat{y}(\mathbf{x})$. The computational cost of modeling depends on the number of training points m, and the procedure of fitting the model to the data. The accuracy of the RSM is quantified by

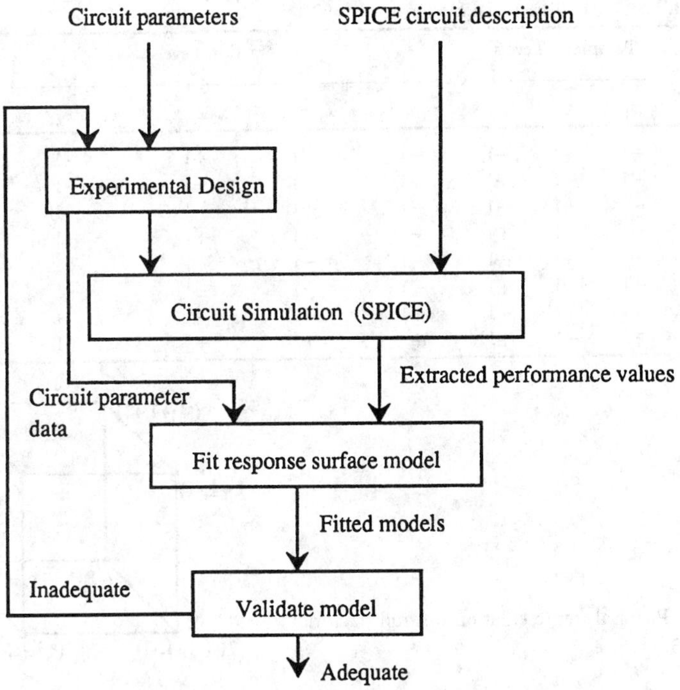

FIGURE 51.1 Performance modeling procedure.

computing error measures which quantify the "goodness of fit". The accuracy of the RSM is greatly influenced by the manner in which the training points are selected from the **x**-space. Design of experiment (DOE) [2] techniques are a systematic means of obtaining the training points such that the maximum amount of information about the model can be obtained from these simulations or experimental runs. In the next section, we review some of the most commonly used experimental design techniques in the context of performance optimization and statistical design. To illustrate certain points in the discussion below, we will assume that the RSM of the performance measure y is the following quadratic polynomial in terms of the circuit parameters:

$$\hat{y}(\mathbf{x}) = \hat{\alpha}_0 + \sum_{i=1}^{n} \hat{\alpha}_i x_i + \sum_{i=1}^{n} \sum_{j=1}^{n} \hat{\alpha}_{ij} x_i x_j \tag{51.5}$$

where the $\hat{\alpha}$'s are the coefficients of the model. Note, however, that the discussion is valid for all RSMs in general.

51.5 Design of Experiments

Factorial Design

In this experimental design, each of the circuit parameters $x_1, x_2, \ldots, x_n$ is quantized into two levels or settings, which are denoted by -1 and $+1$. A full factorial design contains all possible combinations of levels for the n parameters, i.e., it contains 2^n training points or experimental runs. The design matrix for the case of $n = 3$ is shown in Table 51.1 and the design is pictorially described in Fig. 51.2. The value of the circuit performance measure for the kth run is denoted by y_k.

TABLE 51.1 Full Factorial Design for $n = 3$

Run	Parameter Levels			Interaction Levels				y
	x_1	x_2	x_3	$x_1 \times x_2$	$x_1 \times x_3$	$x_2 \times x_3$	$x_1 \times x_2 \times x_3$	
1	−1	−1	−1	+1	+1	+1	−1	y_1
2	−1	−1	+1	+1	−1	−1	+1	y_2
3	−1	+1	−1	−1	+1	−1	+1	y_3
4	−1	+1	+1	−1	−1	+1	−1	y_4
5	+1	−1	−1	−1	−1	+1	+1	y_5
6	+1	−1	+1	−1	+1	−1	−1	y_6
7	+1	+1	−1	+1	−1	−1	−1	y_7
8	+1	+1	+1	+1	+1	+1	+1	y_8

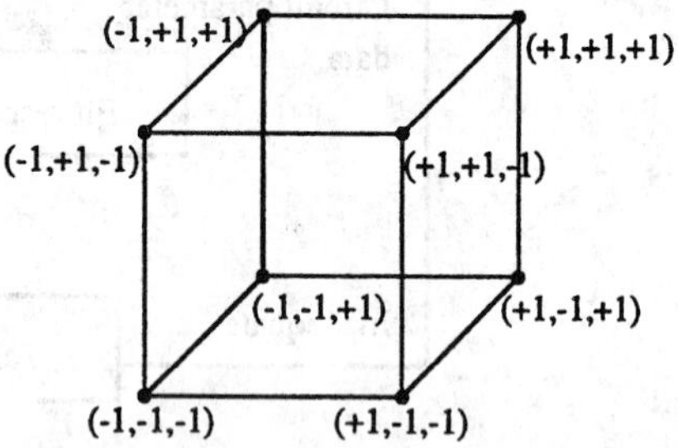

FIGURE 51.2 Pictorial representation of full factorial design for $n = 3$.

Much information about the relationship between the circuit performance y and the circuit parameters x_i, $i = 1, 2, \ldots, n$ can be obtained from a full factorial experiment. The **main** or **individual effect** of a parameter quantifies how much a parameter affects the performance singly. The main effect of x_i, denoted by v_i, is given by

$$v_i = \frac{1}{2^n} \sum_{k=1}^{2^n} x_{ik} \times y_k \tag{51.6}$$

where x_{ik} is the value of x_i in the kth run, and y_k is the corresponding performance value. Thus, the main effect is the difference between the average performance value when the parameter is at the high level (+1) and the average performance value when the parameter is at the low level (−1). The main effect of x_i is the coefficient of the x_i term in the polynomial RSM of (51.5). Similarly, the **interaction effect** of two or more parameters quantifies how those factors jointly affect the performance. The two-factor interaction effect is computed as the difference between the average performance value when both factors are at the same level and the average performance value when they are at different levels. The two-factor interaction effect of parameters x_i and x_j, denoted by $v_{i \times j}$, $i \neq j$, is defined as follows:

$$\begin{aligned} v_{i \times j} &= \frac{1}{2}\{v_{i|j=+1} - v_{i|j=-1}\} \\ &= \frac{1}{2^n} \sum_{k=1}^{2^n} x_{ik} \times x_{jk} \times y_k \end{aligned} \tag{51.7}$$

The two-factor interaction effect of x_i and x_j, $v_{i \times j}$, is the coefficient of the $x_i x_j$ terms in the quadratic RSM of (51.5). Note that $v_{i \times j}$ is the same as $v_{j \times i}$. Moreover, higher order multifactor interaction effects can be computed recursively as before. Note that the interaction columns of Fig. 51.2 are obtained by the aggregated "sign-multiplication" given in (51.7) above.

Thus, the full factorial design allows us to estimate the coefficients of all the first order and cross-factor second order coefficients in the RSM of (51.5). However, it does not allow us to estimate the coefficients of the pure quadratic terms x_i^2. Moreover, the number of experimental runs increases exponentially with the number of circuit parameters; this may become impractical for a large number of circuit parameters. In many cases, high-order multifactor interactions are small and can be ignored. In such cases, it is possible to reduce the size of the design by systematically eliminating some interactions and considering only some of the runs. The accuracy with which the main effects and the low-order interaction effects are estimated does not have to be compromised. Such designs are called **fractional factorial designs**. The most popular class of fractional factorial designs are the 2^{n-p} designs, where a $1/2^p$th fraction of the original 2^n design is used. One of the half fractions of the 2^3 design shown in Table 51.1, i.e., a 2^{3-1} design is shown in Table 51.2.

We observe that some of the columns of Table 51.2 are now identical. It can be seen from (51.6) and (51.7) that we cannot distinguish between effects that correspond to identical columns. Such effects are said to be **confounded** or **aliased** with each other. From Table 51.2, we see that ν_3, the main effect of x_3, and $\nu_{1\times2}$, the interaction effects of x_1 and x_2, are confounded with each other. Moreover the three-factor interaction effect $\nu_{1\times2\times3}$ is confounded with the grand average of the performance (corresponding to a column of 1's). Confounding, however, is not really a problem, since in most applications, high-order interaction effects are negligible. For example, in the quadratic RSM of (51.5), only main effects and two-factor interaction effects are important, and it can be assumed that all higher-order interaction effects are absent.

There is a systematic way of obtaining fractional factorial designs by first isolating a set of **basic** factors (circuit parameters) and then defining the remaining **nonbasic** factors in terms of the interaction of the basic factors. These interactions that define the nonbasic factors in terms of the basic factors are called **generators**. The set of basic factors and the generator functions completely characterize a fractional factorial design. For example, the generator function used in the example of Table 51.2 is

$$x_3 = x_1 \times x_2 \tag{51.8}$$

Note that, once the generator function is known for a fractional factorial design, the confounding pattern is also completely known. Moreover, note that the generator functions are not unique. The generator functions $x_1 = x_2 \times x_3$ and $x_2 = x_1 \times x_3$ also give rise to the design of Table 51.2. This can be understood by realizing that a column corresponding to a parameter sign-multiplied by itself results in a column of all 1's, i.e., the **identity** column, denoted by I. Thus, multiplying both sides of (51.8) by x_2, we get $x_3 \times x_2 = x_1 \times x_2 \times x_2 = x_1 \times I = x_1$, which is also a generator function. Thus, an appropriate set of generator functions must be defined.

Another important concept related to confounding is called the **resolution** of a design, which indicates the clarity with which individual effects and interaction effects can be evaluated in a design. A *resolution III* design is one in which no two main effects are confounded with each other (but they may be confounded with two-factor and/or higher-order interaction effects). An

TABLE 51.2 Half Fraction of a Full Factorial Design for $n = 3$

Run	Parameter Levels			Interaction Levels				y
	x_1	x_2	x_3	$x_1 \times x_2$	$x_1 \times x_3$	$x_2 \times x_3$	$x_1 \times x_2 \times x_3$	
1	−1	−1	+1	+1	−1	−1	+1	y_1
2	−1	+1	−1	−1	+1	−1	+1	y_2
3	+1	−1	−1	−1	−1	+1	+1	y_3
4	+1	+1	+1	+1	+1	+1	+1	y_4

important class of resolution *III* designs is the **saturated** fractional factorial designs. Suppose that the number of factors (parameters), n, can be expressed as $n = 2^q - 1$, for some integer q. Then, a resolution *III* design can be obtained by first selecting q basic factors and then saturating the 2^q design by assigning a nonbasic factor to each of the possible combinations of the q basic factors. There are $p = n - q$ nonbasic factors and generator functions, and these designs are called 2_{III}^{n-p} designs, since $2^q = n + 1$ runs are required for n factors. Note that the fractional factorial design of Table 51.2 is a 2_{III}^{n-p} design, with $n = 3$ and $p = 1$. In the saturated design for $n = 2^q - 1$ factors, the confounding patterns for the basic and nonbasic factors are the same, which implies that any of the q factors may be chosen as the basic factors. However, if $n + 1$ is not an integral power of 2, then q is chosen as $q = \lceil \log_2(n + 1) \rceil$. Next, we define q basic factors and generate a 2^q design. Next, we assign $p = n - q$ nonbasic factors to p generators and form the 2_{III}^{n-p} design. Note that, in this case, only p of the $2^q - q - 1$ available generators are used, and the confounding pattern depends on the choice of the q basic factors and the p generators. The number of runs in the saturated design is considerably less than 2^n.

Another important class of designs commonly used is called *resolution V* designs. In these designs, no main effect or two-factor interaction effect is confounded with another main effect or two-factor interaction effect. Such designs are denoted by 2_V^{n-p} designs, where as before, n is the number of factors and p is the number of generators. These designs can be used to estimate the coefficients of all but the pure quadratic terms in the RSM of (51.5). Since the number of such coefficients (including the constant term) is $C = 1 + n + n(n - 1)/2$, there must be at least C runs. Note that C is still substantially less than 2^n. Another important advantage of the factorial designs is that they are orthogonal, which implies that the model coefficients can be estimated with minimum variances (or errors). An algorithmic method for deriving resolution*V* designs is given in [3].

Central Composite Design

As mentioned above, one of the problems of factorial designs in regard to the RSM of (51.5) is that the coefficients of the pure quadratic terms cannot be estimated. These coefficients can be estimated by using a **central composite design.** A central composite design is a combination of a "cube" and a "star" subdesign. Each factor in a central composite design takes on five values: 0, ± 1 and $\pm\gamma$. Figure 51.3 shows the central composite design for the case of $n = 3$. The cube subdesign, shown in dotted lines in Fig. 51.3, is a fractional factorial design with the factor levels set to ± 1. This design is of resolution *V* so that linear and cross-factor quadratic terms in the RSM of (51.5) can be estimated. The star subdesign, shown in solid lines in Fig. 51.3, is used to

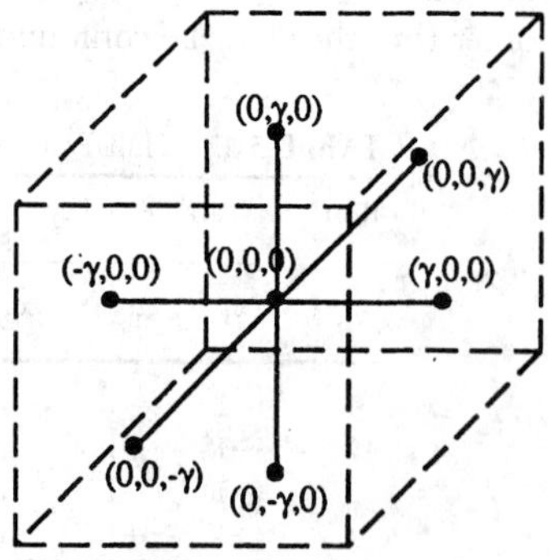

FIGURE 51.3 Central composite design for $n = 3$.

estimate the pure quadratic terms, x_i^2, in the RSM, and consists of

- One *center* point, where all the parameters are set to 0
- $2n$ *Axial* points, one pair for each parameter by setting its value to $-\gamma$ and $+\gamma$ and setting all other parameters to 0

The parameter γ (chosen by the user) is selected so that the composite plan satisfies the **rotatability** property [1]. The main advantage of the central composite design is that all the coefficients of (51.5) can be estimated using a reasonable number of simulations.

Taguchi's Orthogonal Arrays

Taguchi's method using **orthogonal arrays** (OA) [4] is another popular experimental design technique. An orthogonal array is a special kind of fractional factorial design. These orthogonal arrays can be classified into two types. The first category of OAs corresponds to two-level designs, i.e., the parameters are quantized to two levels each, while the second corresponds to three-level designs, i.e., the parameters are quantized to three levels each. These arrays are often available as tables in books which discuss Taguchi techniques [4]. As an example, we show the L18 OA in Table 51.3. The number in the designation of the array refers to the number of experimental runs in the design. The L18 OA belongs to the second category of designs, i.e., each parameter has three levels. In the Taguchi technique, the experimental design matrix for the controllable or designable parameters is called the inner array, while that for the noise parameters is called the outer array. The L18 design of Table 51.3 can be used as an **inner array** as well as an **outer array**.

Latin Hypercube Sampling

The factorial, central composite and Taguchi experimental designs described above set the parameters to certain levels or quantized values within their ranges. Therefore, most of the parameter space remains unsampled. It is therefore desirable to have a more "space-filling" sampling strategy. The most obvious method of obtaining a more complete coverage of the parameter space is to perform **random** or **Monte Carlo** sampling [5]. In random sampling,

TABLE 51.3 Taguchi's L18 Orthogonal Array

Run	Parameter Levels							
1	1	1	1	1	1	1	1	1
2	1	1	2	2	2	2	2	2
3	1	1	3	3	3	3	3	3
4	1	2	1	1	2	2	3	3
5	1	2	2	2	3	3	1	1
6	1	2	3	3	1	1	2	2
7	1	3	1	1	1	3	2	3
8	1	3	2	3	2	1	3	1
9	1	3	3	1	3	2	1	2
10	2	1	1	3	3	2	2	1
11	2	1	2	1	1	3	3	2
12	2	1	3	2	2	1	1	3
13	2	2	1	2	3	1	3	2
14	2	2	2	3	1	2	1	3
15	2	2	3	1	2	3	2	1
16	2	3	1	3	2	3	1	2
17	2	3	2	1	3	1	2	3
18	2	3	3	2	1	2	3	1

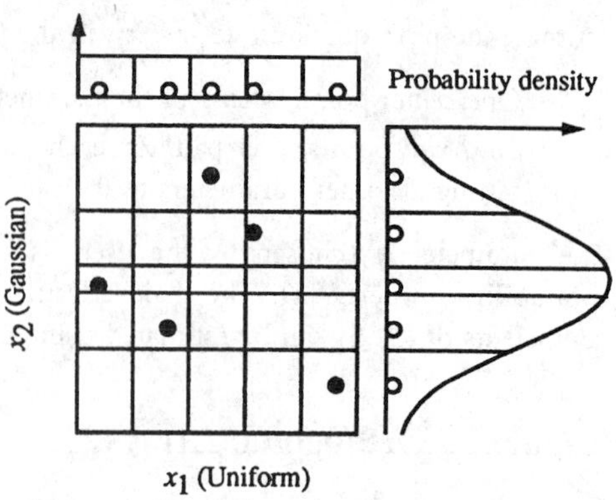

FIGURE 51.4 Latin hypercube sampling for a uniform and a Gaussian random variable.

each parameter has a particular statistical distribution and the parameters are considered to be statistically independent. The distribution of the circuit parameters can often be obtained from test structure measurements followed by parameter extraction. Often, however, based on previous experience or reasonable expectation, the circuit parameters are considered to be normal (or Gaussian). Since the designable parameters are deterministic variables, they are considered to be independent and uniformly distributed for the purposes of random sampling. Random samples are easy to generate and many inferences can be drawn regarding the probability distribution of the performances. The problem with random sampling, however, is that a large sample is usually required to estimate quantities with sufficiently small errors. Moreover, circuit parameters are usually not statistically independent. In particular, the noise parameters are statistically correlated because of the sequential nature of semiconductor wafer processing. Further, a random sample may not be very space filling either. This can be understood by considering the bell-shaped Gaussian density curve. In random sampling, values near the peak of the bell-shaped curve would be more likely to occur in the sample since such values have higher probability of occurrence. In other words, values away from the central region would not be well represented in the sample.

Latin hypercube sampling (LHS) [6, 7] is a sampling strategy that alleviates this problem to a large extent. For each parameter x_i, all portions of its distribution are represented by sample values. If S is the desired size of the sample, then the range of each x_i, $i = 1, 2, \ldots, n$ is divided into S non-overlapping intervals of equal marginal probability $1/S$. Each such interval is sampled once with respect to the probability density in that interval to obtain S values for each parameter. Next, the S values for one parameter are randomly paired with the S values for another parameter, and so on. Figure 51.4 illustrates this process for the case of two circuit parameters x_1 and x_2, where x_1 is a uniform random variable, and x_2 is a Gaussian random variable, and $S = 5$. The marginal probability in an interval is defined as the area under the probability density curve for that interval. Therefore, intervals with equal probability are intervals with equal areas under the probability density curve. For the uniformly distributed x_1, equal probability intervals are also of equal length. For x_2, however, the intervals near the center are smaller (since the density there is higher) than the intervals away from the center (where the density is lower). Figure 51.4 shows that $S = 5$ values are chosen from each region for x_1 and x_2 (shown as open circles). These values are then randomly paired so as to obtain the sample points (shown as filled dots). If the circuit parameters are correlated, then the pairing of the sample values can be controlled so that the sample correlation matrix is close to the user-provided correlation matrix for the circuit parameters. Note that LHS will provide a more uniform coverage of the parameter space than other experimental design techniques. Moreover, a sample of any size can be easily generated and all probability densities are handled.

51.6 Least-Squares Fitting

After the experimental design has been used to select the training points in the parameter space, the circuit is simulated at the training points and the values of the performance measure are extracted from the simulation results. Let S denote the number of experimental runs, and y_k, $k = 1, \ldots, S$ denote the corresponding performance values. The aim in least squares fitting is to determine the coefficients in the RSM so that the fitting error is minimized. Let C denote the number of coefficients in the RSM. For example, the value of C for the RSM of (51.5) is $C = (n+1)(n+2)/2$, where n is the number of circuit parameters. There are **interpolation-based** methods which can be used to determine the coefficients when there are a smaller number of data points than coefficients, i.e., when $S < C$. One such method is called **maximally flat quadratic interpolation** and is discussed later in this chapter. We will restrict our discussion to the case of $S \geq C$. If $S = C$, then there are as many data points as coefficients, and simple simultaneous equation solving will provide the values of the coefficients. In this case, the model interpolates the data points exactly, and there is no fitting error to be minimized. When $S > C$, the coefficients can be determined in such a way that the fitting error is minimized. This method is called **least squares fitting** (in a numerical analysis context) or **linear regression** (in a statistical context). The error measure is called the sum of squared errors and is given by

$$\epsilon = \sum_{i=1}^{S} (y_k - \hat{y}_k)^2 \tag{51.9}$$

where y_k denotes the simulated performance value and $\hat{y}_k$ denotes the model-predicted value at the kth data point. Least squares fitting can then be stated as:

$$\min_{\hat{\alpha}_i} \epsilon = \sum_{i=1}^{S} (y_k - \hat{y}_k)^2 \tag{51.10}$$

where $\hat{\alpha}_i$ represents the (unknown) coefficients of the RSM. The error ϵ is used to determine the adequacy of the model. If this error is too large, the modeling procedure should be repeated with a larger number of training points or a different sampling strategy, or a different model of the performance may have to be hypothesized.

From a statistical point of view (linear regression), the actual performance value is considered to be

$$y_i = \hat{y}_i + e_i, \quad i = 1, 2, \ldots, S \tag{51.11}$$

where the e_i denotes the random error in the model. In least squares fitting, this random error is implicitly assumed to be independent and identically distributed normal random variables with zero mean and constant variance. In particular, we assume that

$$\begin{aligned} E(e_i) &= 0 \\ \mathrm{Var}(e_i) &= \sigma^2 \\ \text{and} \quad \mathrm{Cov}(e_i, e_j) &= 0, \quad i \neq j \end{aligned} \tag{51.12}$$

If the RSM of the performance measure is linear, then it can be written as

$$\hat{y}(\mathbf{x}) = \sum_{i=1}^{C} \alpha_i f_i(\mathbf{x}) \tag{51.13}$$

where $f_i(\mathbf{x})$ are known basis functions of the circuit parameters and C denotes the number of basis functions. Note that the RSM is said to be linear (and least squares fitting can be used) if it is linear in terms of the model coefficients α_i. The basis functions $f_i(\mathbf{x})$ need not be linear themselves. For instance, the basis functions used in the quadratic RSM of (51.5) are 1, x_i and $x_i x_j$.

Let $\mathbf{y}$ and $\mathbf{e}$ be S-vectors: $\mathbf{y} = [y_1, \ldots, y_S]^T$ and $\mathbf{e} = [e_1, \ldots, e_S]^T$, where S is the number of data points. Also, let us define the model coefficients as $\alpha = [\alpha_1, \ldots, \alpha_C]$. Finally, let $\mathbf{X}$ denote the following $S \times C$ matrix

$$X = \begin{bmatrix} f_1(\mathbf{x}_1) & \cdots & f_C(\mathbf{x}_1) \\ \vdots & \vdots & \vdots \\ f_1(\mathbf{x}_S) & \cdots & f_C(\mathbf{x}_S) \end{bmatrix} \tag{51.14}$$

Then, (51.11) can be vectorially written as

$$\mathbf{y} = \mathbf{X}\alpha + \mathbf{e} \tag{51.15}$$

We choose α to minimize the function

$$RSS(\alpha) = (\mathbf{y} - \mathbf{X}\alpha)^T(\mathbf{y} - \mathbf{X}\alpha) \tag{51.16}$$

The least squares estimate $\hat{\alpha}$ of α is given by the following formula:

$$\hat{\alpha} = (\mathbf{X}^T\mathbf{X})^{-1}\mathbf{X}^T\mathbf{y} \tag{51.17}$$

The function (51.16) evaluated at $\hat{\alpha}$ is called the **residual sum of squares**, abbreviated RSS, and is given by

$$RSS = (\mathbf{y} - \mathbf{X}\hat{\alpha})^T(\mathbf{y} - \mathbf{X}\hat{\alpha}) \tag{51.18}$$

It can be shown that an estimate of the variance of the random error is

$$\hat{\sigma}^2 = \frac{RSS}{S - C} \tag{51.19}$$

and that $\hat{\alpha}$ is unbiased, i.e., $E(\hat{\alpha}) = \alpha$, and the variance of the estimate is

$$\mathrm{var}(\hat{\alpha}) = \hat{\sigma}^2(\mathbf{X}^T\mathbf{X})^{-1} \tag{51.20}$$

Another figure of merit of the linear regression is called the **coefficient of determination,** denoted by R^2, and given by

$$R^2 = 1 - \frac{RSS}{\sum_{i=1}^{S}(y_i - \bar{y})^2} \tag{51.21}$$

where $\bar{y}$ is the sample average of the y_i's, $i = 1, \ldots, S$. The coefficient of determination measures the proportion of variability in the performance that is explained by the regression model. Values of R^2 close to 1 are desirable.

51.7 Variable Screening

Variable screening is used in many performance modeling techniques as a preprocessing step to select the significant circuit parameters that are used as independent variables in the performance model. Variable screening can be used to substantially reduce the number of significant circuit parameters that must be considered, and thereby reduce the complexity of the analysis. The significant circuit parameters are included in the performance model, while the others are ignored. Several variable screening approaches have been proposed; in this chapter, we discuss a simple yet effective screening strategy that was proposed in [8].

The variable screening method of [8] is based on the two-level fractional factorial designs discussed earlier. We assume that the main effects [given by (51.6)] of the circuit parameters dominate over the interaction effects and this assumption enables us to use a resolution *III* design. Suppose that there are n original circuit parameters, denoted by $x_1, \ldots, x_n$. Then, we can use the simulation results from the fractional factorial design to compute the following quantities for each of the circuit parameters x_i, $i = 1, \ldots, n$:

1. The main effect v_i computed using (51.6)
2. The *high deviation* d_i^h from the nominal or center point, computed by

$$d_i^h = \frac{2}{n_r} \sum_{k \in K_i^+} y_k - y_c \tag{51.22}$$

   where n_r is the number of simulations, K_i^+ is the set of indices where x_i is $+1$, and y_c is the performance value at the center point, and
3. The *low deviation* d_i^l from the center point computed using a formula similar to (51.22), with K_i^+ replaced by K_i^-.

Note that only two of the above quantities are independent, since the third can be obtained as a linear combination of the other two. A statistical significance test is now used to determine if x_i is significant. To this end, we assume that the quantities v_i, d_i^h, and d_i^l are sampled from a normal distribution with a sample size of n (since there are n circuit parameters). Denoting any one of the above quantities v_i, d_i^h or d_i^l by δ_i, the hypothesis for the significance test is

$$\begin{aligned} &\text{null hypothesis } H_0\text{:} \quad |\delta_i| = 0, \quad \text{i.e., } x_i \text{ is insignificant} \\ &\qquad H_0 \text{ is rejected if } |\delta_i|/\sigma_\delta > t_{\alpha_s/2,n-1} \end{aligned} \tag{51.23}$$

In (51.23) above, a two-sided t-test is used [9] with significance level α_s and σ_δ denotes the estimate of the standard deviation of δ_i. $t_{\alpha_s/2,n-1}$ denotes the value of the t statistic with significance level $\alpha_s/2$ and degrees of freedom $n-1$. We declare a parameter x_i to be insignificant if it is accepted in the tests for at least two of v_i, d_i^h, and d_i^l. This screening procedure is *optimistic* in the sense that an insignificant parameter may be wrongly deemed significant from the above significance tests. The values of the t-variable can be read off from standard t-distribution tables available in many statistics text books.

51.8 Stochastic Performance Models

As stated earlier, the objective of performance modeling is to obtain an inexpensive surrogate to the circuit simulator. This performance model can then be used in a variety of deterministic and statistical optimization tasks in lieu of the expensive circuit simulator. The circuit simulator is a computer model, which is deterministic in the sense that replicate runs of the model with identical inputs will always result in the same values for the circuit performances. This lack of random error makes computer-based experiments fundamentally different from physical experiments. The experimental design and response surface modeling techniques presented above have the tacit assumption that the experiments are physical and the difference between the observed performance and the regression model behaves like white noise. In the absence of random error, the rationale for least squares fitting of a response surface is not clear even though it can be looked upon solely as curve fitting. **Stochastic performance modeling** [10, 11] has been used recently to model the outputs (in our case, the circuit performances) of computer codes or models (in our case, the circuit simulator). The basic idea is to consider the performances as realizations of a stochastic process and this allows one to compute a predictor of the response at untried inputs and allows estimation of the uncertainty of prediction.

The model for the deterministic performance y treats it as a realization of a stochastic process $Y(\mathbf{x})$ which includes a regression model:

$$Y(\mathbf{x}) = \sum_{i=1}^{k} \alpha_i f_i(\mathbf{x}) + Z(\mathbf{x}) \qquad (51.24)$$

The first term in (51.24) is the simple regression model for the performance and the second term represents the systematic departure of the actual performance from the regression model as a realization of a stochastic process. This model is used to choose a design that predicts the response well at untried inputs in the parameter space. Various criteria based on functionals of the mean square error matrix are used to choose the optimal design. Once the design is selected, the circuit simulator is exercised at those design points, and the circuit performance values are extracted. Next, a linear (in the model coefficients $\hat{\alpha}_i s$) predictor is used to predict the values of the performance at untried points. This predicted value can be shown to be the sum of two components: (1) the first component corresponds to the first term in (51.24) and uses the generalized least squares estimate of the coefficients α, and (51.2) the second component is a smoothing term, obtained by interpolating the residuals at the experimental design points. The main drawback of the stochastic performance modeling approach is the excessive computational cost of obtaining the experimental design and of predicting the performance values.

51.9 Other Performance Modeling Approaches

The most popular performance modeling approach combines experimental design and least squares-based response surface modeling techniques described above [8, 12–15]. A commonly used assumption valid for MOS circuits that has been used frequently is based on the existence of **four critical noise parameters** [16–18]. These four critical parameters are the flat-band voltage, the gate-oxide thickness, the channel length reduction, and the channel width reduction.

Several other performance modeling approaches have been proposed, especially in the context of statistical design of integrated circuits. One such approach is based on **self-organizing methods** [19] and is called the **group method of data handling** (GMDH) [20]. An interesting application of GMDH for performance modeling is presented in [21, 22]. Another interesting approach is called **maximally flat quadratic interpolation** (MFQI) [23]. A modeling approach combining the advantages of MFQI and GMDH is presented in [24]. It has been tacitly assumed in the discussion so far that a single response surface model is sufficiently accurate to approximate the circuit performances over the entire circuit parameter space. This may not be true for many circuits. In such cases, a **piecewise** modeling approach may be used advantageously [8]. The circuit parameter space is divided into smaller regions, in each of which a single performance model is adequate. The continuous performance models in the various regions are then *stitched* together to preserve continuity across the regions.

51.10 Example of Performance Modeling

Below, we show an application of performance modeling applied to the clock skew of a clock distribution circuit. Consider the clock driver circuit shown in Fig. 51.5. The top branch has three inverters and the lower branch has two inverters to illustrate the problem of signal skew present in many clock trees. We define the clock skew to be the difference between the times at which *CLK* and its complement cross a threshold of 0.5 V_{DD}. There are two skews, as shown in Fig. 51.6, one corresponding to the rising edge of the *CLK* signal, ΔS_r, and another corresponding to the falling edge of *CLK*, ΔS_f. The performance measure of interest is the signal skew ΔS defined as the larger of the two skews.

The designable circuit parameters of interest in this example are the nominal channel widths of each of the two transistors in the second and third inverters in the top branch and the last inverter in the lower branch. The designable parameters are denoted by W_1, W_2, W_3, W_4, W_5, and W_6 and are marked in Fig. 51.5. The noise parameters of interest are the random variations in the channel widths and lengths of all nMOS and pMOS transistors in the circuit, and the common gate oxide thickness of all the transistors. These are denoted by ΔW_n (channel width variation of nMOS transistors), ΔL_n (channel length variation of nMOS transistors), ΔW_p (channel width variation of pMOS transistors), ΔL_p (channel length variation of pMOS transistors), and t_{ox} (gate oxide thickness of all nMOS and pMOS transistors). We hypothesize a linear response surface model for the clock skew in terms of the designable and noise parameters of the circuit. To this end, we generate an experimental design of size 30 using Latin hypercube sampling. The circuit is simulated at each of these 30 training points, and the value of the clock skew ΔS is extracted from the simulation results. Then, we use regression to fit the following linear

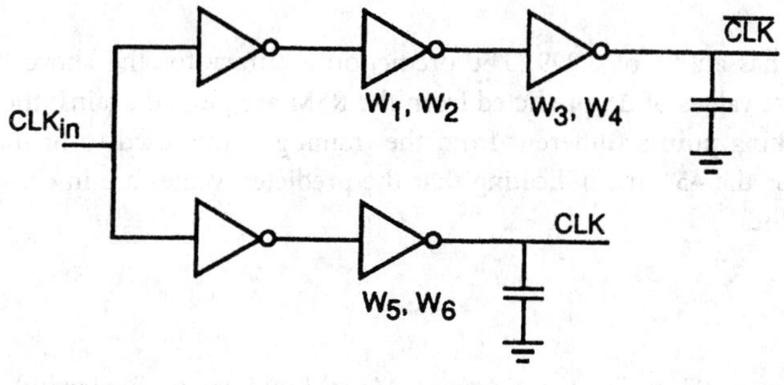

FIGURE 51.5 Clock distribution circuit.

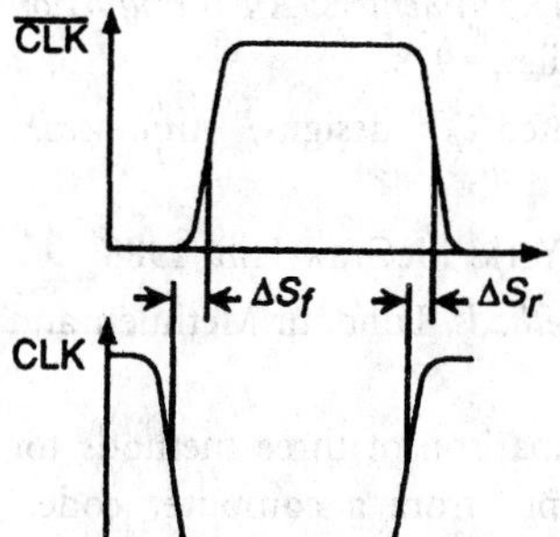

FIGURE 51.6 Definition of rising and falling clock skews.

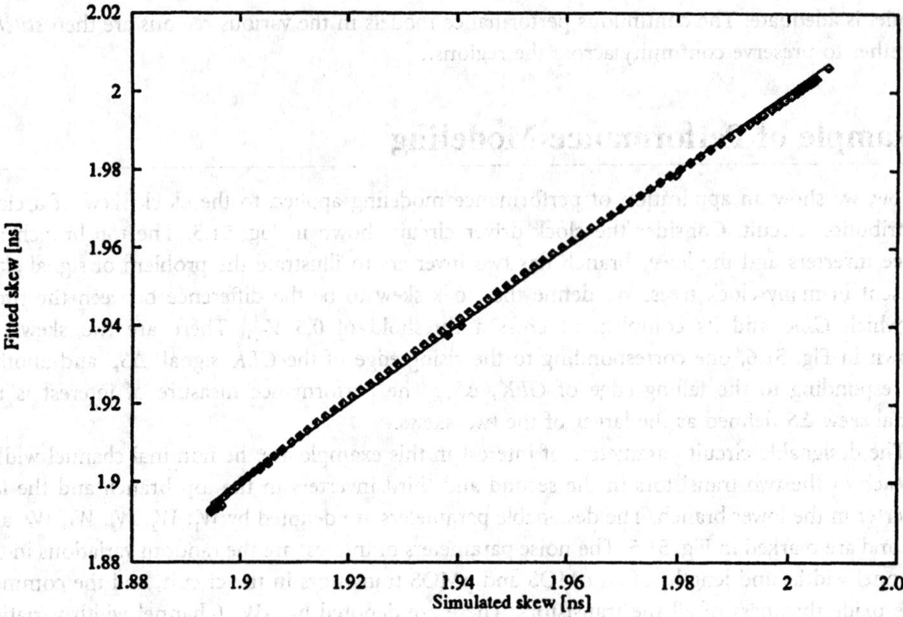

FIGURE 51.7 Plot of predicted versus simulated values for RSM of (51.25).

response surface model:

$$\begin{aligned}\Delta S = 1.93975 - 0.0106W_1 + 0.0121W_2 + 0.0045W_3 - 0.1793W_4 \\ + 0.0400W_5 + 0.0129W_6 - 0.0149\,\Delta W_n - 0.0380\,\Delta W_p \\ - 0.0091\,\Delta L_n + 0.0115\,\Delta L_p + 0.161t_{ox}\end{aligned} \tag{51.25}$$

The above RSM has an R^2 of 0.999. The prediction accuracy for the above RSM is shown in Fig. 51.7 where the values of ΔS predicted from the RSM are plotted against the simulated values for a set of **checking points** (different from the training points used to fit the RSM). Most of the points lie near the 45° line indicating that the predicted values are in close agreement with the simulated values.

References

[1] G. E. P. Box and N. R. Draper, *Empirical Model Building and Response Surfaces*, New York: Wiley, 1987.

[2] G. E. P. Box, W. G. Hunter, and J. S. Hunter, *Statistics for Experimenters: An Introduction to Design, Data Analysis, and Model Building*, New York: Wiley, 1978.

[3] N. R. Draper and T. J. Mitchell, "The construction of saturated 2_R^{k-p} designs," *Ann. Math. Stat.*, vol. 38, pp. 1110–1126, Aug. 1967.

[4] P. J. Ross, *Taguchi Techniques for Quality Engineering*, New York: McGraw-Hill, 1988.

[5] J. M. Hammersley and D. C. Handscomb, *Monte Carlo Methods*, London: Methuen and Co. Ltd., 1964.

[6] M. D. McKay, R. J. Beckman, and W. J. Conover, "A comparison of three methods for selecting values of input variables in the analysis of output from a computer code," *Technometrics*, vol. 21, pp. 239–245, May 1979.

[7] R. L. Iman and W. J. Conover, "A distribution-free approach to inducing rank correlations among input variables," *Comm. Stat.*, vol. B11, no. 3, pp. 311–334, 1982.

[8] K. K. Low and S. W. Director, "An efficient methodology for building macromodels of IC fabrication processes," *IEEE Trans. Computer-Aided Design*, vol. 8, pp. 1299–1313, Dec. 1989.

[9] D. C. Montgomery and E. A. Peck, *Introduction to Linear Regression Analysis*, New York: Wiley, 1982.

[10] J. Sacks, W. J. Welch, T. J. Mitchell, and H. P. Wynn, "Design and analysis of computer experiments," *Stat. Sci.*, vol. 4, no. 4, pp. 409–435. 1989.

[11] M. C. Bernardo, R. Buck, L. Liu, W. A. Nazaret, J. Sacks, and W. J. Welch, "Integrated circuit design optimization using a sequential strategy," *IEEE Trans. Computer-Aided Design*, vol. 11, pp. 361–372, March 1992.

[12] T. K. Yu, S. M. Kang, I. N. Hajj, and T. N. Trick, "Statistical modeling of VLSI circuit performances," Proc. IEEE Int. Conf. Computer-Aided Design, pp. 224–227, Santa Clara, CA, Nov. 1986.

[13] T. K. Yu, S. M. Kang, I. N. Hajj, and T. N. Trick, "Statistical performance modeling and parametric yield estimation of MOS VLSI," *IEEE Trans. Computer-Aided Design*, vol. CAD-6, pp. 1013–1022, Nov. 1987.

[14] T. K. Yu, S. M. Kang, W. Welch, and J. Sacks, "Parametric yield optimization of CMOS analog circuits by quadratic statistical performance models," *Int. J. Circuit Theory Appl.*, vol. 19, pp. 579–592, Nov. 1991.

[15] A. Dharchoudhury and S. M. Kang, "An integrated approach to realistic worst-case design optimization of MOS analog circuits," Proc. 29th ACM/IEEE Design Automation Conference, pp. 704–709, June 1992.

[16] P. Cox, P. Yang, S. S. Mahant-Shetti, and P. Chatterjee, "Statistical modeling for efficient parametric yield estimation," *IEEE J. Solid State Circuits*, vol. SC-20, pp. 391–398, Feb. 1985.

[17] P. Yang and P. Chatterjee, "Statistical modeling of small geometry MOSFETs," IEEE Int. Electron Device Meeting, pp. 286–289, Dec. 1982.

[18] P. Yang, D. E. Hocevar, P. Cox, C. Machala, and P. Chatterjee, "An integrated and efficient approach for MOS VLSI statistical circuit design," *IEEE Trans. Computer-Aided Design*, vol. 5, pp. 5–14, Jan. 1986.

[19] S. J. Farlow, Ed., *Self-Organizing Methods in Modeling: GMDH Type Algorithms*, New York: Marcel Dekker, 1984.

[20] A. G. Ivakhnenko, "The group method of data handling, a rival of the method of stochastic approximation," *Soviet Automatic Control*, vol. 13, no. 3, pp. 43–55, 1968.

[21] A. J. Strojwas and S. W. Director, "An efficient algorithm for parametric fault simulation of monolithic IC's," *IEEE Trans. Computer-Aided Design*, vol. 10, pp. 1049–1058, Aug. 1991.

[22] S. Ikeda, M. Ochiai, and Y. Sawaragi, "Sequential GMDH and its application to river flow prediction," *IEEE Trans. on Systems, Man and Cybernetics*, vol. SMC-6, pp. 473–479, July 1976.

[23] R. M. Biernacki and M. A. Styblinski, "Statistical circuit design with a dynamic constraint approximation scheme," Proc. IEEE ISCAS, pp. 976–979, San Jose, CA, 1986.

[24] M. A. Styblinski and S. A. Aftab, "Efficient circuit performance modeling using a combination of interpolation and self organizing approximation techniques," Proc. IEEE ISCAS, pp. 2268–2271, New Orleans, 1990.

52

Symbolic Analysis Methods

Marwan M. Hassoun
Iowa State University

52.1 Introduction

The late '80s and the early '90s saw a heightened interest in the development of symbolic analysis techniques and their applications to integrated circuits. This resulted in the current generation of symbolic analysis methods which include several software implementations like ISAAC [9], SCAPP [13], ASAP [7], EASY [35], SYNAP [33], SAPEC [24], SCYMBAL [21], GASCAP [19], and SSPICE [43]. This generation is characterized by an emphasis on the production of usable software packages and a large interest in the application of the methods to the area of analog circuit design. This includes extensive efforts to reduce the number of symbolic expressions generated by the analysis through hierarchical methods and approximation techniques [8, 34, 42]. This generation also includes attempts at applying symbolic analysis to analog circuit synthesis [10], parallel processor implementations [16, 30, 42], and time domain analysis [2, 14, 23].

Chapter 21 addressed the details of traditional symbolic analysis techniques as they apply to linear circuit theory. This section addresses symbolic analysis as a computer-aided design tool and its application to integrated circuit design. Modern symbolic analysis is mainly based on a signal flowgraph (SFG) approach or a modified nodal analysis (MNA) approach. The basic SFG methodologies were discussed in Chapter 21 because the methods are part of traditional symbolic analysis techniques. The methodology addressed in this section is based on the concept of modified nodal analysis [17] as it applies to symbolic analysis.

Traditional symbolic circuit analysis is performed in the frequency domain where the results are in terms of the frequency variable s. The main goal of performing symbolic analysis on a circuit in the frequency domain is to obtain a symbolic transfer function of the form

$$H(s, \mathbf{X}) = \frac{N(s, \mathbf{X})}{D(s, \mathbf{X})}, \quad \mathbf{X} = [x_1, x_2, \ldots, x_n], \quad n \le n_{all} \tag{52.1}$$

0-8493-8341-2/95/$0.00 + $.50

The expression is a function of the complex frequency variable s, and the variables x_1 through x_n representing the variable circuit elements, where n is the number of variable circuit elements and n_{all} is the total number of circuit elements. Hierarchical symbolic analysis approaches are based on a decomposed form of Eq. (52.1) [13, 15, 37]. This hierarchical representation is referred to as a **sequence of expressions** representation to distinguish it from the **single expression** representation of Eq. (52.1). The major advantage of having a symbolic expression in single expression form is the insight that can be gained by observing the terms in both the numerator and the denominator. The effects of the different terms can, perhaps, be determined by inspection. This process is very valid for the cases where there are relatively few symbolic terms in the expression.

The heart of any circuit simulator, **symbolic** or **numeric**, is the circuit analysis methodology used to obtain the mathematical equations that characterize the behavior of the system. The circuit must satisfy these equations. These equations are of three kinds, Kirchhoff's Current Law (KCL), Kirchhoff's Voltage Law (KVL), and the Branch Relationship equations (BR) [6]. KCL and KVL are well known and have been discussed earlier in this text. The definition of BR is: for independent branches, **a branch relationship defines the branch current in terms of the branch voltages**. The most common independent branches are resistors, capacitors, and inductors. For the case of dependent branches, the branch current or voltage is defined in terms of other branches' currents or voltages. The four common types of dependent branches are the four types of dependent sources: voltage controlled voltage source (VCVS), voltage controlled current source (VCCS), current controlled voltage source (CCVS), and current controlled current source (CCCS). A linear element would produce a BR equation that is linear, and a nonlinear element would produce a nonlinear BR equation. The scope of this section is linear elements. If a nonlinear element is to be included in the circuit, a collection of linear elements could be used to closely model the behavior of that element around a certain frequency range [11]. BR equations are not used on their own to formulate a system of equations to characterize the circuit. They are, however, used in conjunction with KCL or KVL equations to produce a more complete set of equations to characterize the circuit. Hybrid analysis [6] uses a large set of BR equations in its formulation.

A system of equations to solve for the circuit variables: a collection of node voltages and branch currents, does not need to include all of the above equations. Several analysis methods are based on these equations or a subset thereof. Mathematically speaking, an equal number of circuit unknowns and linearly independent equations is required in order for a solution for the system to exist. The most common analysis methods used in circuit simulators are

1. A modification of nodal analysis as used in the numeric circuit simulators SPICE2 [31] and SLIC [20], and in the formulation of the symbolic equations in [1]. Nodal analysis [6] uses KCL to produce the set of equations that characterizes the circuit. Its advantage is its relatively small number of equations which results in a smaller matrix which is very desirable in computer implementations. Its limitation, however, is that it allows only for node-to-datum voltages as variables. No branch current variables are allowed. Also, the only type of independent power sources it allows are independent current sources. However, voltage sources are handled by using their Norton equivalent. A further limitation to this method is that only voltage-controlled current sources are allowed, none of the other dependent sources are allowed in the analysis. To overcome these limitations, the MNA method was proposed in [17]. It allowed branch current variables in the analysis, which in turn led to the ability to include voltage sources and all four types of controlled sources in the analysis. The MNA method is described in more detail in the next section because it is used in symbolic analysis after further modification to it. These modifications are necessary in order for the analysis to accept ideal op amps,

and to produce a compact circuit characterization matrix. This matrix is referred to as the reduced MNA (RMNA) matrix.

2. Hybrid analysis [6] is used in the numeric simulators ASTAP [41] and ECAP2 [4]. The method requires the selection of a tree and its associated co-tree. It uses cutset analysis (KCL) and loop analysis (KVL) to formulate the set of equations. The equations produced solve for the branch voltages and branch currents. The method is able to accommodate voltage sources and all four types of dependent sources. It utilizes a much larger matrix than the nodal analysis methods; however, it is a very sparse matrix (lots of zero entries). The main drawback is that the behavior of the equations is highly dependent on the tree selected. A bad tree could easily result in an ill-conditioned set of equations. Some methods of tree selection can guarantee a well-conditioned set of hybrid equations [31]. Therefore, special rules must be observed in the tree selection process. That constitutes a large additional overhead on the analysis.
3. Sparse tableau analysis [12] has been used in symbolic analysis to employ the parameter extraction method (Chapter 2). It uses the entire set of circuit variables: node voltages, branch voltages, branch currents for the symbolic formulation in addition to capacitor charges, and inductor fluxes in the case of numeric simulations. This results in a huge set of equations and in turn a very large matrix but a very sparse one. This method uses all the above stated equation formulation methods, KCL, KVL, and the BR. The entire system of equations is solved for all the circuit variables. The disadvantage of the method is its inherent problem of ill-conditioning [31].

It has been found that the sparse tableau method only produces a very minor improvement in the number of mathematical operations over the nodal analysis methods despite its large overhead. This stems from a large matrix and the large and rigid set of variables that have to be solved for. The MNA method gives the choice of branch current variables to be solved for and equations are added accordingly. The tableau method would always have a fixed size matrix and produce unneeded information to the user.

For the purpose of symbolic analysis, as highlighted in Chapter 21, the entire analysis is performed in the frequency domain with the aid of the complex frequency variable s. The Laplace transform representation of the value of any circuit element is therefore used in the formulation. Also, since a modification of the nodal admittance matrix is used, all elements are represented by their admittance values.

52.2 Symbolic Modified Nodal Analysis

The initial step of MNA is to formulate the nodal admittance matrix $\mathbf{Y}$ [6] from the circuit. The circuit variables considered here are all the node-to-datum voltages, referred to simply as node voltages; there are n_v of them. They are included in the variable vector $\mathbf{V}$. So $\mathbf{V}$ has the dimensions of $(n_v \times 1)$. The vector $\mathbf{J}$, also of dimension $(n_v \times 1)$, represents the values of all independent current sources in the circuit. The ith entry of $\mathbf{J}$ represents a current source entering node i. The nodal linear system of equations can be represented in the following matrix form:

$$\mathbf{YV} = \mathbf{J} \tag{52.2}$$

Row i of $\mathbf{Y}$ represents the KCL equation at node i. $\mathbf{Y}$ is constructed by writing KCL equations at each node except for the datum node. The ith equation then would state that the sum of all currents leaving node i is equal to zero. The equations are put into the matrix form of Eq. (52.2). The following example illustrates the process.

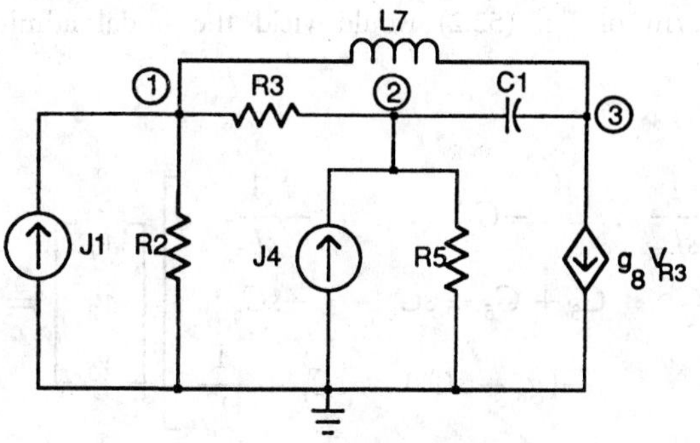

FIGURE 52.1 MNA example circuit.

Example 52.1 Consider the circuit in Fig. 52.1. Collecting the node voltages would produce the following **V**:

$$\mathbf{V} = \begin{bmatrix} v_1 \\ v_2 \\ v_3 \end{bmatrix} \tag{52.3}$$

Considering all the current sources in the circuit, the vector **J** becomes:

$$\mathbf{J} = \begin{bmatrix} J_1 \\ J_4 \\ 0 \end{bmatrix} \tag{52.4}$$

Notice the last entry of **J** is a zero because no independent current sources are connected to it. Now writing KCL equations at the three nondatum nodes produces:

$$G_2 v_1 + G_3(v_1 - v_2) + \frac{1}{sL_7}(v_1 - v_3) = J_1 \tag{52.5}$$

$$G_5 v_2 + G_3(v_2 - v_1) + sC_6(v_2 - v_3) = J_4 \tag{52.6}$$

$$sC_6(v_3 - v_2) + \frac{1}{sL_7}(v_3 - v_1) + g_8 V_{R_3} = 0 \tag{52.7}$$

Substituting $V_{R_3} = v_1 - v_2$ in the above three equations and rearranging their variables results in:

$$\left(G_2 + G_3 + \frac{1}{sL_7}\right) v_1 - G_3 v_2 - \frac{1}{sL_7} v_3 = J_1 \tag{52.8}$$

$$-G_3 v_1 + (G_3 + G_5 + sC_6) v_2 - sC_6 v_3 = J_4 \tag{52.9}$$

$$\left(g_8 + \frac{1}{sL_7}\right) v_1 - (g_8 + sC_6) v_2 + \left(\frac{1}{sL_7} + sC_6\right) v_3 = 0 \tag{52.10}$$

Now realizing the form of Eq. (52.2) would yield the nodal admittance matrix of this circuit **Y**.

$$\begin{bmatrix} G_2 + G_3 + \dfrac{1}{sL_7} & -G_3 & -\dfrac{1}{sL_7} \\ -G_3 & G_3 + G_5 + sC_6 & -sC_6 \\ g_8 - \dfrac{1}{sL_7} & -(g_8 + sC_6) & sC_6 + \dfrac{1}{sL_7} \end{bmatrix} \begin{bmatrix} v_1 \\ v_2 \\ v_3 \end{bmatrix} = \begin{bmatrix} J_1 \\ J_4 \\ 0 \end{bmatrix} \quad (52.11)$$

An automatic technique to construct the nodal admittance matrix is the element stamp method [17]. The method constitutes going through each branch of the circuit and adding its contribution to the nodal admittance matrix in the appropriate positions. It is an easy way to illustrate the impact of each element on the matrix. Resolving Fig. 52.1 using the element stamps would produce exactly the same **Y** matrix, **V** vector, and **J** vector as in Eq. (52.11).

The modified nodal analysis technique introduced in [17] expands on the above nodal analysis in order to readily include independent voltage sources and the three other types of controlled sources: VCVC, CCCS, and CCVS. This is done by introducing some branch currents as variables into the system of equations which in turn allows for the introduction of any branch current as an extra system variable. Each extra current variable introduced would need an extra equation to solve for it. The extra equations are obtained from the BR equations for the branches whose currents are the extra variables. The effect on the nodal admittance matrix is the deletion of any contribution in it due to the branches whose currents have been declared as variables. This matrix is referred to as $\mathbf{Y}_n$. The addition of extra variables and a corresponding number of equations to the system results in the need to append extra rows and extra columns to $\mathbf{Y}_n$. The augmented $\mathbf{Y}_m$ matrix is referred to as the MNA matrix. The new system of equations, in matrix form, is

$$\mathbf{V} = \begin{bmatrix} \mathbf{Y}_n & \mathbf{B} \\ \mathbf{C} & \mathbf{D} \end{bmatrix} \begin{bmatrix} \mathbf{V} \\ \mathbf{I} \end{bmatrix} = \begin{bmatrix} \mathbf{J} \\ \mathbf{E} \end{bmatrix} \quad (52.12)$$

where **I** is a vector of size n_i whose elements are the extra branch current variables introduced, **E** is the independent voltage source values, and **C** and **D** correspond to BR equations for the branches whose currents are in **I**.

According to [17] the MNA current variables should include all branch currents of independent voltage sources, controlled voltage sources, and all controlling currents. Further modifications of this procedure, like reduced MNA formulations (RMNA) [13], compacted MNA formulations (CMNA) [9], and supernode analysis (SNA) [36] relax these constraints and result in fewer MNA equations.

Example 52.2 As an example of an MNA matrix formulation, consider the circuit of Fig. 52.2. The extra current variables are the branch current of the independent voltage source (branch 1) and the CCVS (branch 5). They are referred to as i_1 and i_5 respectively. Using element stamps,

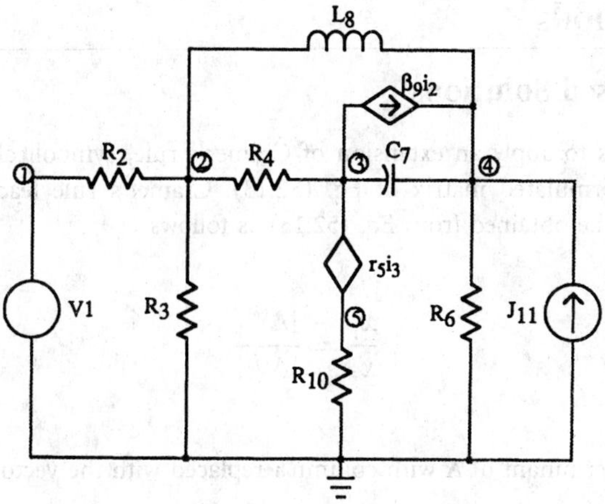

FIGURE 52.2 Circuit of example 52.2.

the MNA system of equations becomes:

$$\begin{array}{c} 1 \\ 2 \\ 3 \\ 4 \\ 5 \\ x_1 \\ x_2 \end{array}\begin{bmatrix} G_2 & -G_2 & 0 & 0 & 0 & 1 & 0 \\ -G_2 & G_2+G_3+G_4+\dfrac{1}{sL_8} & -G_4 & \dfrac{-1}{sL_8} & 0 & 0 & 0 \\ \beta_9 G_2 & -G_4-\beta_9 G_2 & G_4+sC_7 & -sC_7 & 0 & 0 & 1 \\ -\beta_9 G_2 & \dfrac{-1}{sL_8}+\beta_9 G_2 & -sC_7 & G_6+sC_7+\dfrac{1}{sL_8} & 0 & 0 & 0 \\ 0 & 0 & 0 & 0 & G_{10} & 0 & -1 \\ 1 & 0 & 0 & 0 & 0 & 0 & 0 \\ 0 & -r_5 G_3 & 1 & 0 & -1 & 0 & 0 \end{bmatrix}$$

$$\begin{bmatrix} v_1 \\ v_2 \\ v_3 \\ v_4 \\ v_5 \\ i_1 \\ i_5 \end{bmatrix} = \begin{bmatrix} 0 \\ 0 \\ 0 \\ J_{11} \\ 0 \\ V_1 \\ 0 \end{bmatrix}$$

The above equation is in the following linear matrix form

$$\mathbf{Ax} = \mathbf{b} \tag{52.13}$$

where **A** is a symbolic matrix of dimension $n \times n$, **x** is a vector of circuit variables of length n, and **b** is a symbolic vector of constants of length n. Of course, n is the number of circuit variables comprised of currents, voltages, charges or fluxes. The analysis proceeds by solving Eq. (52.13) for **x**.

52.3 Solution Methods

Determinant-Based Solutions

Here the basic idea is to apply an extension of Cramer's rule, symbolically, to find a transfer function from the formulated matrix of Eq. (52.13). Cramer's rule leads to the fact that a transfer function can be obtained from Eq. (52.13) as follows

$$\frac{x_i}{x_j} = \frac{|\mathbf{A}^{(i)}|}{|\mathbf{A}^{(j)}|} \tag{52.14}$$

where $|\mathbf{A}^{(i)}|$ is the determinant of $\mathbf{A}$ with column i replaced with the vector $\mathbf{b}$, and similarly for $|\mathbf{A}^{(j)}|$.

Several symbolic algorithms are based on the concept of using Cramer's rule and calculating the determinants of an MNA matrix. Most notable is the program ISAAC [9] which uses a recursive determinant-expansion algorithm to calculate the determinants of Eq. (52.14). Although there are many other algorithms to calculate the determinant of a matrix, like elimination algorithms and nested minors algorithms [10], recursive determinant-expansion algorithms were found the most suitable for sparse linear matrices [40]. These algorithms are cancellation-free given that all the matrix entries are different which is a very desirable property in symbolic analysis. The determinant of $\mathbf{A}$ is calculated by

$$\begin{aligned} |\mathbf{A}| &= \sum_{j=1}^{n} (-1)^{i+j} a_{ij} |\mathbf{M}_{ij}| \qquad \text{for } (n > 1) \\ &= 1 \qquad \text{for } (n = 1) \end{aligned} \tag{52.15}$$

where row i is an arbitrary row of matrix $\mathbf{A}$, $n \times n$ is the dimension of $\mathbf{A}$ and $|\mathbf{M}_{ij}|$ is a minor of $\mathbf{A}$ of dimension $(n-1) \times (n-1)$ with row i and column j removed. $|\mathbf{M}_{ij}|$ is calculated by using Eq. (52.15) recursively. As an example, the determinant of a 3×3 matrix is calculated by this method as

$$\begin{aligned} \begin{vmatrix} a_{11} & a_{12} & a_{13} \\ a_{21} & a_{22} & a_{23} \\ a_{31} & a_{32} & a_{33} \end{vmatrix} &= a_{11} \begin{vmatrix} a_{22} & a_{23} \\ a_{32} & a_{33} \end{vmatrix} + a_{12} \begin{vmatrix} a_{21} & a_{23} \\ a_{31} & a_{33} \end{vmatrix} + a_{13} \begin{vmatrix} a_{21} & a_{22} \\ a_{31} & a_{32} \end{vmatrix} \\ &= a_{11}a_{22}a_{33} - a_{11}a_{32}a_{23} - a_{12}a_{21}a_{33} + a_{12}a_{31}a_{23} \\ &\quad + a_{31}a_{21}a_{32} - a_{31}a_{31}a_{22} \end{aligned} \tag{52.16}$$

Here, i was chosen to be, arbitrarily, row 1. The expression is cancellation-free here because all the matrix entries are unique. Note that for matrices with duplicated entries, the algorithm is not cancellation-free.

Example 52.3 As an example, the transfer function v_3/v_1 for the circuit in Fig. 52.1 is to be calculated. Equation (52.11) shows its MNA formulation. Using Eq. (52.14) and the recursive application of Eq. (52.15) results in

$$\frac{v_3}{v_1} = \frac{\begin{vmatrix} G_2 + G_3 + \dfrac{1}{sL_7} & -G_3 & J_1 \\ -G_3 & G_3 + G_5 + sC_6 & J_4 \\ g_8 - \dfrac{1}{sL_7} & -g_8 - sC_6 & 0 \end{vmatrix}}{\begin{vmatrix} J_1 & -G_3 & -\dfrac{1}{sL_7} \\ J_4 & G_3 + G_5 + sC_6 & -sC_6 \\ 0 & -g_8 - sC_6 & sC_6 + \dfrac{1}{sL_7} \end{vmatrix}}$$

$$= \frac{\left(g_8 + \dfrac{1}{SL_7}\right)(-J_4G_3 - J_1G_3 - J_1G_5 - J_1sC_6) + (g_8 + sC_6)\left(J_4G_2 + J_4G_3 + \dfrac{J_4}{sL_7} + J_1G_3\right)}{(g_8 + sC_6)\left(-sC_6J_1 + \dfrac{J_4}{sL_7}\right) + \left(sC_6 + \dfrac{1}{sL_7}\right)(J_1G_3 + J_1G_5 + J_1sC_6 + J_4G_3)} \tag{52.17}$$

Close inspection of the above expression will reveal that it is not cancellation-free. This is due to the nonuniqueness of the matrix entries. For instance, in the numerator the terms $g_8G_3J_4$ cancel out and the terms $g_8G_3J_1$ also cancel out. In the denominator the terms $s^2C_6^2J_1$ cancel out.

Parameter Reduction Solutions

These methods use basic linear algebra techniques, applied symbolically, to find the solution to Eq. (52.13). The goal here is to reduce the size of the system of equations by manipulating the entries of matrix **A** down to a 2 × 2 size. The two variables remaining in **x** are the ones for which the transfer function is to be calculated. The manipulation process, which in simple terms is solving for the two desired variables in terms of the others, is done using methods like Gaussian elimination [32]. This methodology is used in the symbolic analysis program SCAPP [13] which uses a successive application of a modified Gaussian elimination process to produce the transfer function of the solution. The process is applied to MNA matrices and the result is a reduced MNA matrix referred to as RMNA ($\mathbf{Y}_R$). The process of reducing a circuit variable, whether a node voltage or branch current, is referred to as variable suppression.

The modification process to the MNA is the suppression of all node voltages and current variables that are not input or output variables. They are referred to as internal variables. The system of equations that describes the circuit in terms of the external variables (circuit inputs and outputs) can be written as

$$\begin{bmatrix} \mathbf{Y}_{Rn} & \mathbf{B}_R \\ \mathbf{C}_R & \mathbf{D}_R \end{bmatrix} \begin{bmatrix} \mathbf{V}_e \\ \mathbf{I}_e \end{bmatrix} = \begin{bmatrix} \mathbf{J}_R \\ \mathbf{E}_R \end{bmatrix} \begin{bmatrix} \mathbf{J}_e \\ 0 \end{bmatrix} \tag{52.18}$$

Where $\mathbf{Y}_{Rn}$ is the reduced node admittance submatrix, $\mathbf{B}_R$ is the reduced contributions of the external currents to KCL equations, $\mathbf{C}_R$ and $\mathbf{D}_R$ represent the reduced BR equations, V_e is the vector of external node variables and $\mathbf{I}_e$ is the vector of external current variables. The ith entry of the $\mathbf{J}_e$ vector represents the currents entering the circuit through the ith node. The $\mathbf{J}_R$ and the $\mathbf{E}_R$ vectors represent the contributions of the independent current and voltage sources, respectively, internal to the circuit.

The process of suppressing an internal node or an internal branch current, in mathematical terms, means solving for the variable in terms of the other system variables. A process to solve for a variable in terms of the other system variables can be done by using a single step of the Gauss elimination method [32]. Each step of the Gauss elimination method consists of reducing a system of n linear equations in n unknowns to a system of $n-1$ linear equations in $n-1$ unknowns by using one of the equations to eliminate one of the unknowns from the remaining $n-1$ equations. The best way to illustrate the method is to consider an example.

Example 52.4 Consider the following system of three linear equations and three unknowns:

$$\begin{aligned} (1)\quad & b_{11}x_1 + b_{12}x_2 + b_{13}x_3 = l_1 \\ (2)\quad & b_{21}x_1 + b_{22}x_2 + b_{23}x_3 = l_2 \\ (3)\quad & b_{31}x_1 + b_{32}x_2 + b_{33}x_3 = l_3 \end{aligned} \tag{52.19}$$

The general matrix form for a linear system of equations can be written as:

$$\mathbf{BX} = \mathbf{L} \tag{52.20}$$

Therefore Eqs. (52.19) can be written as follows:

$$\begin{matrix}1\\2\\3\end{matrix}\begin{bmatrix} b_{11} & b_{12} & b_{13} \\ b_{21} & b_{22} & b_{23} \\ b_{31} & b_{32} & b_{33} \end{bmatrix}\begin{bmatrix} x_1 \\ x_2 \\ x_3 \end{bmatrix} = \begin{bmatrix} l_1 \\ l_2 \\ l_3 \end{bmatrix} \tag{52.21}$$

Now the process of eliminating equation x_2 from the system requires the use of one of the three equations to do so. Without any loss of generality and in order to maintain a certain symmetry in the process, the second equation is chosen. The process is equivalent to eliminating the second row and second column of **B** and the second entry in both **X** and **L**.

First x_2 must be eliminated from the first equation. This is done by multiplying the second equation by b_{12}/b_{22} and subtracting it from the first equation. The next step is to eliminate x_2 from the third equation. This is done by multiplying the second equation by b_{32}/b_{22} and subtracting it from the third equation. The result is

$$\begin{aligned} (1)\quad & \left(b_{11} - \frac{b_{12}}{b_{22}}b_{21}\right)x_1 + \left(b_{13} - \frac{b_{12}}{b_{22}}b_{23}\right)x_3 = l_1 - \frac{b_{12}}{b_{22}}l_2 \\ (2)\quad & \left(b_{31} - \frac{b_{32}}{b_{22}}b_{21}\right)x_1 + \left(b_{33} - \frac{b_{32}}{b_{22}}b_{23}\right)x_3 = l_3 - \frac{b_{32}}{b_{22}}l_2 \end{aligned} \tag{52.22}$$

What has happened here is that the second equation was used to express x_2 in terms of x_1 and x_2. In matrix form the process of suppressing the second row and second column of **B** and the second entry in both **X** and **L** to produce $\mathbf{B}_R$, $\mathbf{X}_R$, and $\mathbf{L}_R$, respectively, can be expressed as:

$$\mathbf{B}_R = \mathbf{B}_{cr} - \frac{1}{b_{22}}\mathbf{C}_2\mathbf{R}_2 \tag{52.23}$$

$$\mathbf{L}_R = \mathbf{L}_{cr} - \frac{1}{b_{22}}\mathbf{C}_2\mathbf{R}_2 \tag{52.24}$$

where $\mathbf{C}_2$ is the second column of $\mathbf{B}$ with b_{22} removed, $\mathbf{R}_2$ is the second row of $\mathbf{B}$ with b_{22} removed, $\mathbf{B}_{cr}$ is $\mathbf{B}$ with the second column and the second row removed, and $\mathbf{L}_{cr}$ is $\mathbf{L}$ with the second entry removed. $\mathbf{X}_R$ is simply $\mathbf{X}$ with x_2 removed since it has been suppressed by Eqs. (52.23) and (52.24).

The new system of linear equations can be written in general terms as :

$$\mathbf{B}_R\mathbf{X}_R = \mathbf{L}_R \tag{52.25}$$

and for this specific example as:

$$\begin{matrix} 1 \\ 3 \end{matrix}\begin{bmatrix} b_{11} - \frac{b_{12}}{b_{22}} b_{21} & b_{31} - \frac{b_{32}}{b_{22}} b_{21} \\ b_{31} - \frac{b_{12}}{b_{22}} b_{23} & b_{33} - \frac{b_{32}}{b_{22}} b_{23} \end{bmatrix} \begin{bmatrix} x_1 \\ x_3 \end{bmatrix} = \begin{bmatrix} l_1 - \frac{b_{12}}{b_{22}} l_2 \\ l_3 - \frac{b_{32}}{b_{22}} l_2 \end{bmatrix} \tag{52.26}$$

The above procedure can be generalized for any system of linear equations in the form of Eq. (52.20) to be reduced to the form of Eq. (52.25) where the variable x_j is to be suppressed. The suppression equations become:

$$\mathbf{B}_R = \mathbf{B}_{cr} - \frac{1}{b_{jj}} \mathbf{C}_j\mathbf{R}_j \tag{52.27}$$

$$\mathbf{L}_R = \mathbf{L}_{cr} - \frac{1}{b_{jj}} \mathbf{C}_j l_j \tag{52.28}$$

where $\mathbf{C}_j$ is the jth column of $\mathbf{B}$ with b_{jj} removed, $\mathbf{R}_j$ is the jth row of $\mathbf{B}$ with b_{jj} removed, $\mathbf{B}_{cr}$ is $\mathbf{B}$ with the jth column and the jth row removed, and the $\mathbf{L}_{cr}$ is $\mathbf{L}$ with the jth entry removed. $\mathbf{X}_R$ is simply $\mathbf{X}$ with x_j removed since it has been suppressed by Eqs. (52.27) and (52.28).

The effect of Eq. (52.27) on each element of $\mathbf{B}$, or in other words the effect of reducing the variable x_j on any element of $\mathbf{B}$, can be expressed as:

$$b_{pq_R} = b_{pq} - \frac{b_{pj}}{b_{jj}} b_{jq} \quad p \neq j \text{ and } q \neq j \tag{52.29}$$

The effect on the members of $\mathbf{L}$ can be expressed as:

$$l_{p_R} = l_p - \frac{b_{pj}}{b_{jj}} l_j \quad p \neq j \tag{52.30}$$

The matrix entry b_{jj} is referred to as the pivot, the entry b_{pj} is referred to as the column pivot, and the entry b_{jq} is referred to as the row pivot. Each column pivot is unique for each row in $\mathbf{B}$ and each row pivot is unique for each column in $\mathbf{B}$.

A simple example will serve to illustrate and easily verify the formulation of the RMNA matrix.

Example 52.5 Consider the circuit in Fig. 52.3. Nodes 1 and 3 are the external nodes. The assumption is that all current variables are to remain internal; in this case i_2. It must be noted that it makes sense that node 3 is the input terminal while node 1 is the output terminal.

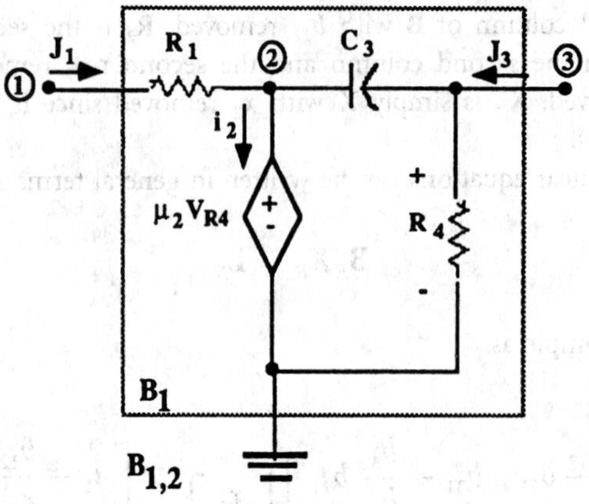

FIGURE 52.3 Circuit of example 52.5.

Therefore, any excitations will come via terminal 3. Using element stamps the MNA matrix for this circuit is expressed as:

$$\begin{array}{c} \\ 1 \\ 2 \\ 3 \\ x_1 \end{array}\begin{array}{c} \begin{array}{cccc} v_1 & v_2 & v_3 & i_2 \end{array} \\ \begin{bmatrix} G_1 & -G_1 & 0 & 0 \\ -G_1 & G_1 + sG_3 & -sC_3 & 1 \\ 0 & -sC_3 & sC_3 + G_4 & 0 \\ 0 & 1 & -\mu_2 & 0 \end{bmatrix} \end{array} \qquad (52.31)$$

To produce the RMNA matrix in terms of v_1 and v_3 only, all the other variables must be suppressed. The anticipated result is a (2 × 2) RMNA matrix. Notice that an attempt to reduce the current variable i_2 first would cause a problem because the pivot is zero. Therefore, it is deferred to the end. A discussion of the reasons can be found in [13].

So the first step is to suppress the internal node 2. The pivot is $G_1 + sC_3$. Performing Eq. (52.27) the MNA matrix of Eq. (52.31) yields:

$$\begin{bmatrix} G_1 & 0 & 0 \\ 0 & sC_3 + G_4 & 0 \\ 0 & -\mu_2 & 0 \end{bmatrix} - \frac{1}{G_1 + sC_3}\begin{bmatrix} -G_1 \\ -sC_3 \\ 1 \end{bmatrix}[-G_1 \quad -sC_3 \quad 1]$$

$$= \begin{array}{c} \\ 1 \\ 3 \\ x_1 \end{array}\begin{bmatrix} v_1 & v_3 & i_2 \\ \dfrac{G_1 sC_3}{G_1 + sC_3} & -\dfrac{G_1 sC_3}{G_1 + sC_3} & \dfrac{G_1}{G_1 + sC_3} \\ -\dfrac{G_1 sC_3}{G_1 + sC_3} & \dfrac{G_1 sC_3}{G_1 + sC_3} + G_4 & \dfrac{sC_3}{G_1 + sC_3} \\ \dfrac{G_1}{G_1 + sC_3} & -\mu_2 + \dfrac{sC_3}{G_1 + sC_3} & -\dfrac{1}{G_1 + sC_3} \end{bmatrix} \qquad (52.32)$$

Notice what happened here: the suppression of node 2, which is one of the nodes of the branch whose current is a variable, produced a fill in the pivot position for row x_1. This now allows for the suppression of the internal current variable i_2. Equation (52.27) is applied again to the matrix of Eq. (52.32) and results in:

$$\begin{bmatrix} \dfrac{G_1 sC_3}{G_1 + sC_3} & -\dfrac{G_1 sC_3}{G_1 + sC_3} \\ -\dfrac{G_1 sC_3}{G_1 + sC_3} & \dfrac{G_1 sC_3}{G_1 + sC_3} + G_4 \end{bmatrix} + (G_1 + sC_3) \begin{bmatrix} \dfrac{G_1}{G_1 + sC_3} \\ \dfrac{sC_3}{G_1 + sC_3} \end{bmatrix} \begin{bmatrix} \dfrac{G_1}{G_1 + sC_3} & -\mu_2 + \dfrac{sC_3}{G_1 + sC_3} \end{bmatrix} \tag{52.33}$$

After canceling some terms and some mathematical manipulation of the matrix of Eq. (52.33), the resulting RMNA system becomes:

$$\begin{matrix} 1 \\ 3 \end{matrix} \begin{bmatrix} G_1 & -G_1 \mu_2 \\ 0 & sC_3 + G_4 - sC_3 \mu_2 \end{bmatrix} \begin{bmatrix} v_1 \\ v_3 \end{bmatrix} = \begin{bmatrix} 0 \\ J_3 \end{bmatrix} \tag{52.34}$$

The transfer function is then readily available from the first equation, which corresponds to KCL at the output node.

52.4 Ideal Operational Amplifiers

Neither the nodal analysis nor MNA approaches account for ideal operational amplifiers (op amps) in their equation formulation. The way op amps are usually handled in those analyses is by modeling each op amp by a voltage-controlled voltage source with a large gain in order to attempt to characterize the infinite gain of the ideal op amp.

The ideal op amp is a 4-terminal device for which one of the terminals is always assumed to be grounded. Op amps are used extensively in the building of a large number of analog circuits, especially analog filters, where symbolic circuit simulators have found extensive application in finding transfer functions for the filters. This section illustrates the expansion of the element set of the MNA approach and the RMNA matrix to include ideal op amps.

The Nullator, the Norator, and the Nullor

Before the characterization of the ideal op amp is attempted, the concepts of the nullator, the norator, and the nullor are explored [5]. They are not real elements. They are tools to introduce some mathematical constraints into a circuit. They are used as an aid to the development of insight into the behavior of ideal devices like the ideal op amp.

The symbol for the nullator is illustrated in Fig. 52.4. A nullator is defined as follows:

Definition A nullator is a two-terminal element defined by the constraints

$$v_1 - v_2 = 0 \tag{52.35}$$

$$i = 0 \tag{52.36}$$

The symbol for the norator is illustrated in Fig. 52.5. A norator is defined as follows.

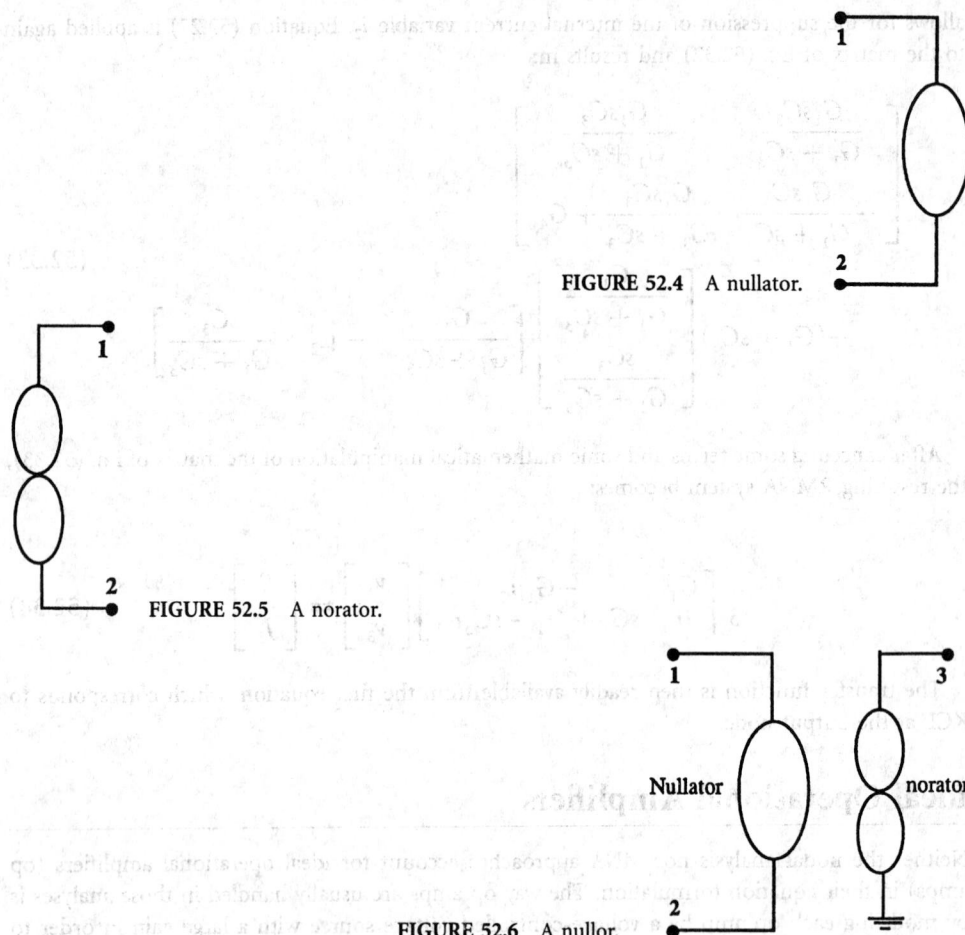

FIGURE 52.4 A nullator.

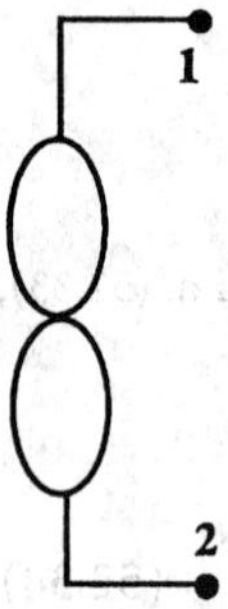

FIGURE 52.5 A norator.

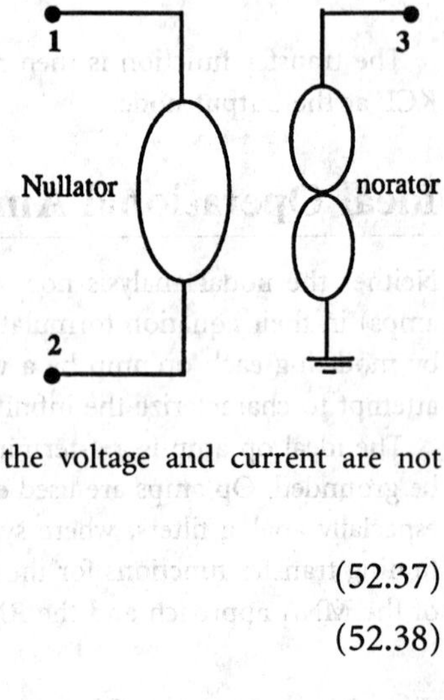

FIGURE 52.6 A nullor.

Definition A norator is a two-terminal element for which the voltage and current are not constrained. That is

$$v_1 - v_2 = arbitrary \tag{52.37}$$

$$i = arbitrary \tag{52.38}$$

A norator in a circuit introduces freedom from some constraints on the nodes it is connected to.

The combination of a nullator and a norator to produce a 4-terminal block as shown in Fig. 52.6 is referred to as a nullor. The equations characterizing this 4-terminal block are represented then by Eqs. (52.35) through (52.38).

The Ideal Op Amp Model

The two characteristics of an ideal op amp are

1. An arbitrary current that the output node supplies
2. The zero voltage differential between its output nodes

This can be modeled using a nullor (Fig. 52.6), with terminal 4 of the nullor grounded.

The rules for writing the KCL equations and finding the MNA matrix $\mathbf{Y_{op}}$ for a circuit with ideal op amps is done by the following procedure:

1. Remove the nullor (the op amp) from the circuit leaving n nodes (plus the reference node).
2. Write the MNA matrix $\mathbf{Y_m}$ for this circuit.
3. For the nullator between nodes p and q, delete column q of $\mathbf{Y_m}$ and add column q to column p. If q is the reference node, simply delete column q and the qth entry in the voltage variable vector $\mathbf{V}$.
4. For the norator between nodes l and k, delete row l of $\mathbf{Y_m}$ and add row l to row k. If k is the reference node, simply delete row l and the lth entry in the right-hand side current vector $\mathbf{J}$.

The result is $\mathbf{Y_{op}}$.

Example 52.6 Consider the 4-terminal block and its nullor model in Fig. 52.7. Following the above procedure, first the $\mathbf{Y_m}$ matrix is built:

$$\begin{matrix} & 1 & 2 & 3 & 4 & 5 \\ 1 \\ 2 \\ 3 \\ 4 \\ 5 \end{matrix}\begin{bmatrix} y_1+y_2 & -y_2 & 0 & -y_1 & 0 \\ -y_2 & y_2+y_3 & 0 & 0 & -y_3 \\ 0 & 0 & 0 & 0 & 0 \\ -y_1 & 0 & 0 & y_1 & 0 \\ 0 & -y_3 & 0 & 0 & y_3 \end{bmatrix} \cdot \begin{bmatrix} v_1 \\ v_2 \\ v_3 \\ v_4 \\ v_5 \end{bmatrix} = \begin{bmatrix} 0 \\ 0 \\ i_3 \\ i_4 \\ i_5 \end{bmatrix} \tag{52.39}$$

By rule 3 above, now column 1 is deleted and added to column 3 as follows:

$$\begin{matrix} & 2 & 3 & 4 & 5 \\ 1 \\ 2 \\ 3 \\ 4 \\ 5 \end{matrix}\begin{bmatrix} -y_2 & y_1+y_2 & -y_1 & 0 \\ y_2+y_3 & -y_2 & 0 & -y_3 \\ 0 & 0 & 0 & 0 \\ 0 & -y_1 & y_1 & 0 \\ -y_3 & 0 & 0 & y_3 \end{bmatrix} \cdot \begin{bmatrix} v_2 \\ v_3 \\ v_4 \\ v_5 \end{bmatrix} = \begin{bmatrix} 0 \\ i_3 \\ i_4 \\ i_5 \end{bmatrix} \tag{52.40}$$

By rule 4 deleting row 2, which is the KCL equation at the output of the op amp, which is a norator terminal, (52.41) is produced:

$$\begin{matrix} & 2 & 3 & 4 & 5 \\ 1 \\ 3 \\ 4 \\ 5 \end{matrix}\begin{bmatrix} -y_2 & y_1+y_2 & -y_1 & 0 \\ 0 & 0 & 0 & 0 \\ 0 & -y_1 & y_1 & 0 \\ -y_3 & 0 & 0 & y_3 \end{bmatrix} \cdot \begin{bmatrix} v_2 \\ v_3 \\ v_4 \\ v_5 \end{bmatrix} = \begin{bmatrix} 0 \\ i_3 \\ i_4 \\ i_5 \end{bmatrix} \tag{52.41}$$

The result here is four equations in four unknowns: the node voltages. To produce the RMNA system of equations that characterize the 4-terminal block, that is, to complete the terminal block analysis, the internal node 2 must be suppressed. Since node 2 is the output terminal of an op amp, where the KCL equation has been deleted because of the arbitrary current entering that terminal, row 1 is used to reduce node 2 voltage variable, which always will be the KCL at the negative terminal of the op amp. The process of modeling the op amp using the nullor produces enough equations to solve for all the system variables. So, the output terminals of

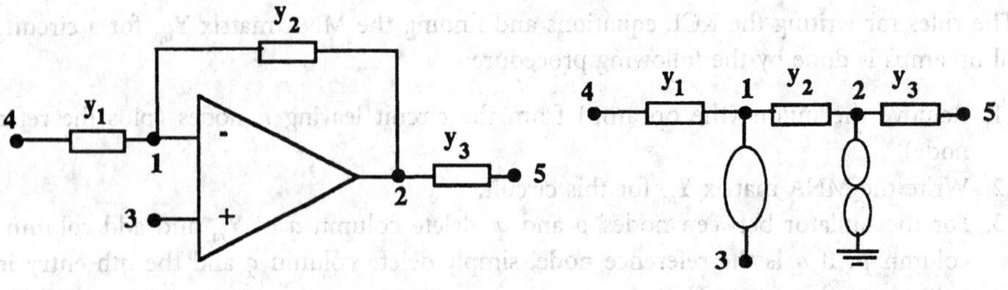

FIGURE 52.7 An op amp circuit.

op amps are always special nodes. They are flagged in the analysis so that when they are to be suppressed, the row corresponding to their negative input terminal is used to suppress them.

If the output terminal of the op amp is to remain a system variable throughout the analysis, then the row corresponding to the negative terminal of the op amp is carried along throughout the entire analysis.

Using Eq. (52.27) to reduce node 2 produces:

$$\begin{array}{c} \\ 3 \\ 4 \\ 5 \end{array}\begin{array}{c} \begin{array}{ccc} 3 & 4 & 5 \end{array} \\ \begin{bmatrix} 0 & 0 & 0 \\ -y_1 & y_1 & 0 \\ 0 & 0 & y_3 \end{bmatrix} \end{array} - \frac{1}{-y_2}\begin{bmatrix} 0 \\ 0 \\ -y_3 \end{bmatrix} \cdot [y_1 + y_2 \quad -y_1 \quad 0]$$

$$= \begin{array}{c} \\ 3 \\ 4 \\ 5 \end{array}\begin{array}{c} \begin{array}{ccc} V_3 & V_4 & V_5 \end{array} \\ \begin{bmatrix} 0 & 0 & 0 \\ -y_1 & y_1 & 0 \\ -\frac{y_3}{y_2}(y_1 + y_2) & \frac{y_3}{y_2} y_1 & y_3 \end{bmatrix} \end{array} \tag{52.42}$$

The above RMNA matrix is the result of the block analysis of the 4-terminal block of the op amp circuit, Fig. 52.7. The hierarchical middle block analysis can then proceed normally if this terminal block is part of a larger circuit.

In the computer implementation in SCAPP [13] of the op amp analysis, KCL equations at nodes corresponding to op amp output terminals are never built. That is, the rows corresponding to that output node are never included in the MNA or RMNA matrix at all. This saves on the overhead of building and then discarding that row. This is done by setting a special flag at every node that is an output of an op amp.

52.5 Applications of Symbolic Analysis

There are many specific applications for which symbolic analysis algorithms have been developed over the years. The following is an attempt to categorize the uses of symbolic analysis methods. For a given application, some overlap of these categories might exist. The goal here is to give a general idea of the applications of symbolic analysis methods. It must be noted that most of these applications cover both s-domain and z-domain analyses.

1. Frequency response evaluation

This is an obvious application of having the symbolic transfer function stated in terms of the circuit variable parameters. The process of finding the frequency response curve over a frequency range for a given circuit involves the repetitive evaluation of Eq. (52.1) with all the parameters numerically specified and sweeping the frequency over the desired range. A numerical simulator would require a simulation run for each frequency point.

2. Circuit response optimization [11, 26]

This process involves the repetitive evaluation of the symbolic function generated by a symbolic simulator. The response of the circuit is repetitively evaluated by substituting different values for the circuit parameters in the equation until a desired numerical response is achieved. The concept, of course, requires a good deal of management in order to reduce the search space and the number of evaluations needed. Such a method for filter design by optimization is discussed in [3, 37] and a method for solving piecewise resistive linear circuits is discussed in [6]. The idea here, is that for a given circuit topology, only one run through the symbolic circuit simulator is necessary.

3. Sensitivity analysis [27]

Sensitivity analysis is the process of finding the effect of the circuit performance due to changes in an element value. The normalized sensitivity of a transfer function H with respect to an element x is given as:

$$S_x^H = \frac{\partial H}{\partial x}\frac{x}{H} \tag{52.43}$$

The above expression can be found symbolically and then evaluated for the different circuit parameter values.

4. Semiconductor device characterization [23]

This process involves the repeated comparison of measured data from fabricated devices with the simulation results using the mathematical models for these devices. The goal of the process is to update the device models to reflect the measured data. The model parameters are incrementally adjusted and the evaluation process is repeated until the difference between the measured and the simulated results is minimized. Such approaches are reported in [22, 23].

5. Statistical Analysis [38]

A widely used statistical analysis method is through Monte Carlo simulations. The circuit behavior has to be evaluated many times in order to evaluate the statistical variation of a circuit output in response to parameter mismatches due to, for instance, integrated circuits process variations.

6. Fault diagnosis of analog circuits [29]

The process reported in [29] takes measurements from the faulty fabricated circuit and compares it to simulation results. The process is continuously repeated with the parameter values in the simulations changed until the faulty element is detected.

7. Insight into circuit operation

The insight that can be provided by obtaining the symbolic transfer function versus its numerical counterpart is very evident. The simple example in Chapter 21 illustrates this powerful application. The effect of the different elements on the behavior of the circuit can be observed by inspecting the symbolic expression. This, of course, is possible if the number of symbolic parameters is small, that is the circuit is small. Insight, however, can also be obtained by observing an approximate symbolic expression which reduces the number of symbolic terms to a manageable figure approximation technique [8, 34, 44].

8. Education [19]

TABLE 52.1 Comparison of Some Symbolic Simulation Programs [11]

	ISSAC	ASAP	SYNAP	SAPEC	SSPICE	SCYMBAL	SCAPP	GASCAP
Analysis domains	s & z	s	dc & s	s	s	z	s	s
Primitive elements	Complete	Complete	Complete	Complete	Complete	—	Complete	Limited
Small-signal linearization	Yes	Yes	Yes	No	Yes	No	Yes	No
Mismatching	Yes	Yes	Yes	No	No	No	No	No
Approximation	Yes	Yes	Yes	No	Yes	No	No	No
Weakly non-linear analysis	Yes	No	No	No	No	No	No	No
Hierarchical analysis	No	No	No	No	Limited	No	Yes	No
Pole/zero extraction	No	Limited	No	No	Limited	No	No	No
Graphical interface	No	Yes	No	Yes	No	No	Yes	Yes
Formulation	CMNA[1]	SFG[2]	MNA[3]	MNA[3]	Y[4]	SFG[2]	RMNA[5] & SFG[2]	—
Language	LSIP	C	C++	LISP	C	FORTRAN	C	C

[1]CMNA: Compact Modified Nodal Analysis.
[2]SFG: Signal Flowgraph.
[3]MNA: Modified Nodal Analysis.
[4]Y: Admittance Matrix.
[5]RMNA: Reduced Modified Nodal Analysis.

Symbolic analysis is most helpful for students as a supplement to linear circuit analysis courses. These courses require the derivation of expressions for circuit impedances, gains, and transfer functions. A symbolic simulator can serve as a check for the correctness of the results in addition to aiding instructors in verifying solutions and making up exercises and examples.

52.6 Symbolic Analysis Software Packages

Several stand-alone symbolic simulators available for public use exist today. A comparison of these software packages was reported in [11]. The software packages that were compared are ISAAC [9], SCAPP [13], ASAP [7], SYNAP [33], SCYMBAL [21], GASCAP [19], SAPEC [28], and SSPICE [43]. The comparison made by [11] was based on the functionality of these simulators and the findings are summarized in Table 52.1.

References

[1] G. E. Alderson and P. M. Lin, "Computer generation of symbolic network functions—a new theory and implementation," *IEEE Trans. on Circuit Theory*, vol. CT-20, pp. 48–56, Jan. 1973.

[2] B. Alspaugh and M. Hassoun, "A mixed symbolic and numeric method for closed-form transient analysis," 1993 IEEE European Conference on Circuit Theory and Design, Davos, Switzerland, pp. 1687–1692, Sept. 1993.

[3] S. Bass, "The application of a fast symbolic analysis routine in a network optimization program," Proceedings of the *Midwest Symposium on Circuit Theory*, 1972.

[4] F. H. Branin, G. R. Hogsett, R. L. Lunde, and L. E. Kugel, "ECAP II—a new electronic circuit analysis program," *IEEE J. Solid-State Circuits*, vol. SC-6, pp. 146–165, August 1971.

[5] L. T. Bruton, *RC-Active Circuits*, Englewood Cliffs, NJ: Prentice-Hall, 1980.

[6] L. O. Chua, and P. M. Lin, *Computer Aided Analysis of Electronic Circuits—Algorithms and Computational Techniques*, Englewood Cliffs, NJ: Prentice-Hall, 1975.

[7] F. V. Fernandez, A. Rodriguez-Vazquez, and J. L. Huertas, "An advanced symbolic analyzer for the automatic generation of analog circuit design equations," Proceedings of the IEEE ISCAS, Singapore, pp. 810–813, June 1991.

[8] F. V. Fernandez, J. Martin, A. Rodriguez-Vazquez, and J. L. Huertas, "On simplification techniques for symbolic analysis of analog integrated circuits," Proceedings of the IEEE ISCAS, San Diego, CA, pp. 1149–1152, May 1992.

[9] G. Gielen, H. Walscharts, and W. Sansen, "ISSAC: a symbolic simulator for analog integrated circuits," *IEEE J. Solid-State Circuits*, vol. SC-24, pp. 1587–1597, Dec. 1989.

[10] G. Gielen and W. Sansen, *Symbolic Analysis for Automated Design of Analog Integrated Circuits*, Kluwer Academic, MA, 1991.

[11] G. Gielen, P. Wambacq, and W. Sansen, "Symbolic analysis methods and applications for analog circuits: a tutorial overview," *Proceedings of the IEEE*, vol. 82, no. 2, pp. 287–301, Feb. 1994.

[12] G. D. Hachtel, et al., "The sparse tableau approach to network and design," *IEEE Trans. on Circuit Theory*, vol. CT-18, pp. 101–113, Jan. 1971.

[13] M. M. Hassoun and P. M. Lin, "A new network approach to symbolic simulation of large-scale networks," *Proc. of the 1989 IEEE International Symp on Ckts and Sys*, pp. 806–809, May 1989.

[14] M. M. Hassoun and J. E. Ackerman, "Symbolic simulation of large-scale circuits in both frequency and time domains," Proc. of the 33rd IEEE Midwest Symposium on Circuits and Systems, Calgary, Alberta, pp. 707–710, Aug. 1990.

[15] M. Hassoun and K. McCarville, "Symbolic analysis of large-scale networks using a hierarchical signal flowgraph approach," *Int. J. Analog Integrated Circuits and Signal Processing*, vol. 3, no. 1, pp. 31–42, Jan. 1993.

[16] M. Hassoun and P. Atawale, "Hierarchical symbolic circuit analysis of large-scale networks on multi-processor systems," Proceedings of the 1993 IEEE ISCAS, Chicago, pp. 1651–1654, May 1993.

[17] C. Ho, A. E. Ruehli, and Brennan, "The modified nodal approach to network analysis," *IEEE Trans. on Circuits and Systems*, vol. CAS-25, pp. 504–509, June 1975.

[18] J. J. Hsu and C. Sechen, "Low-frequency symbolic analysis of large analog integrated circuits," Proceedings of CICC, 1993, pp. 14.7.1–14.7.4.

[19] L. P. Huelsman, "Personal computer symbolic analysis programs for undergraduate engineering courses," IEEE International Symp. on Circuits and Systems, Portland, OR, pp. 798–801, May 1989.

[20] T. E. Idleman, F. S. Jenkins, W. J. McCalla, and D. O. Pederson, "SLIC—a simulator for linear integrated circuits," *IEEE J. of Solid-State Circuits*, vol. SC-6, pp. 188–204, August 1971.

[21] A. Konczykowska and M. Bon, "Automated design software for switched-capacitor ICs with symbolic simulator SCYMBAL," *Proceedings of the Design Automation Conference*, pp. 363–368, 1988.

[22] A. Konczykowska and M. Bon, "Symbolic simulation for efficient repetitive analysis and artificial intelligence techniques in CAD," *IEEE International Symp. on Circuits and Systems*, Portland, OR, pp. 802–805, 1989.

[23] A. Konczykowska, P. Rozes, M. Bon, and W. Zuberek, "Parameter extraction of semiconductor devices electrical models using symbolic approach," *Alta Frequenza Rivista Di Elettronica*, vol. 5, no. 6, pp. 3–5, Nov. 1993.

[24] A. Liberatore and S. Manetti, "SAPEC—A personal computer program for symbolic analysis of electric circuits," IEEE Int. Symp. on Circuits and Systems, Helsinki, pp. 897–900, June 1988.

[25] A. Liberatore, S. Manetti, M. Piccirilli and A. Reatti, "Simulation of switching power converters using symbolic techniques," *Alta Frequenza Rivista Di Elettronica*, vol. 5, No. 6, pp. 3–5, Nov. 1993.

[26] P. M. Lin, *Symbolic Network Analysis*, Elsevier Science, Amsterdam, 1991.

[27] P. M. Lin, "Sensitivity Analysis of Large Linear Networks Using Symbolic Programs," *IEEE Int. Symp. on Circuits and Systems*, San Diego, CA, pp. 1145–1148, May 1992.

[28] S. Manetti, "New approaches to automatic symbolic analysis of electronic circuits," Proceedings of the Institute of Elec. Eng., pp. 22–28, Feb. 1991.

[29] S. Manetti and M. Piccirilli, "Symbolic simulators for the fault diagnosis of nonlinear analog circuits," *Int. J. Analog Integrated Circuits and Signal Processing*, vol. 3, no. 1, pp. 59–72, Jan. 1993.

[30] T. Matsumoto, T. Sakabe, and K. Tsuji, "On parallel symbolic analysis of large networks and systems," IEEE Int. Symp. on Circuits and Systems, Chicago, pp. 1647–1650, May 1993.

[31] L. W. Nagel, "SPICE2: A Computer Program to Simulate Semiconductor Circuits," Memo ERL-M520, Electronics Res. Lab., University of California at Berkeley, May 1975.

[32] B. Noble and J. Daniel, *Applied Linear Algebra*, 2nd ed., Englewood Cliffs, NJ: Prentice-Hall, 1977.

[33] S. Seda, M. Degrauwe, and W. Fichtner, "A symbolic analysis tool for analog circuit design automation," 1988 Int. Conference on Computer-Aided Design, Santa Clara, CA, pp. 488–491, 1988.

[34] S. Seda, M. Degrauwe, and W. Fichtner, "Lazy-expansion symbolic expression approximation in SYNAP," 1992 Int. Conference on Computer-Aided Design, Santa Clara, CA, pp. 310–317, 1992.

[35] R. Sommer, "EASY—an experimental analog design system framework," Int. Workshop on Symbolic Methods and Applications to Circuit Design, Paris, Oct. 1991.

[36] R. Sommer, D. Ammermann, and E. Hennig, "More efficient algorithms for symbolic network analysis: supernodes and reduced loop analysis," *Int. J. Analog Integrated Circuits and Signal Processing*, vol. 3, no. 1, Jan. 1993.

[37] J. A. Starzyk and A. Konczykowska, "Flowgraph analysis of large electronic networks," *IEEE Trans. on Circuits and Systems*, vol. CAS-33, pp. 302–315, March 1986.

[38] M. Styblinski, X. Sun, K. M. Opalska, L. J. Opalski, "Symbolic Time Delay Modeling for Statistical Optimization," International Workshop on Symbolic Methods and Applications to Circuit Design, October 1991.

[39] G. Temes, *Introduction to Circuit Synthesis and Design*, New York: McGraw-Hill, 1977.

[40] P. Wang, "On the expansion of sparse symbolic determinants," Proc. Int. Conference on System Sciences, Honolulu, 1977.

[41] W. T. Weeks, A. J. Jiminez, G. W. Mahoney, D. Mehta, H. Qassemzadeh, and T. R. Scott, "Algorithms for ASTAP–a network analysis program," *IEEE Trans. on Circuit Theory*, vol. CT-20, pp. 628–634, Nov. 1973.

[42] E. Wehrhahn, "Symbolic analysis on parallel computers," European Conference on Circuit Theory and Design, Davos, Switzerland, pp. 1693–1698, 1993.

[43] G. Wie et al., "SSPICE—a symbolic SPICE program for linear active circuits," Proceedings of the Midwest Symposium on Circuits and Systems, 1989.

[44] P. Wombacq, G. Gielen, and W. Sansen, "A cancellation free algorithm for the symbolic simulation of large analog circuits," *IEEE Int. Symp. on Circuits and Systems*, San Diego, CA, pp. 1157–1160, May 1992.

53

Numerical Analysis Methods

Andrew T. Yang
University of Washington

53.1 Equation Formulation

The method by which circuit equations are formulated is essential to a computer-aided circuit analysis program. It affects significantly the set-up time, the programming effort, the storage requirement, and the performance of the program.

A linear time-invariant circuit with n nodes and b branches is completely specified by its network topology and branch constraints. The fundamental equations that describe the equilibrium conditions of a circuit are the Kirchhoff current law equations (KCL), the Kirchhoff voltage law equations (KVL), and the equations which characterize the individual circuit elements. Two methods are popular: the Sparse Tableau approach and the Modified Nodal approach.

Implications of KCL, KVL, and the Element Branch Characteristics

Given an example with $n = 4$ and $b = 6$ (as shown in Fig. 53.1) we can sum the branch currents leaving each node to zero and obtain:

$$\begin{bmatrix} -1 & 1 & 1 & 0 & 0 & 0 \\ 0 & 0 & -1 & 1 & -1 & 0 \\ 0 & 0 & 0 & 0 & 1 & 1 \\ 1 & -1 & 0 & -1 & 0 & -1 \end{bmatrix} \begin{bmatrix} i_{b_1} \\ i_{b_2} \\ i_{b_3} \\ i_{b_4} \\ i_{b_5} \\ i_{b_6} \end{bmatrix} = \begin{bmatrix} 0 \\ 0 \\ 0 \\ 0 \\ 0 \\ 0 \end{bmatrix}$$

or,

$$A_a \times i_b = 0$$

0-8493-8341-2/95/$0.00+$.50

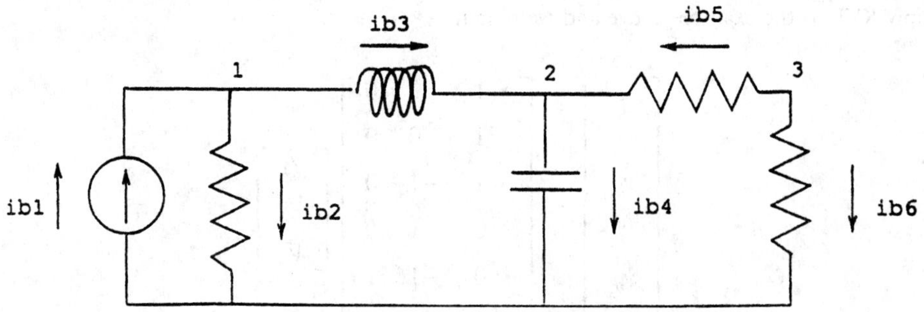

FIGURE 53.1

A_a is an $n \times b$ incidence matrix which contains +1, −1, and 0 entries. i_b is the branch current. Note that A_a is linearly dependent since the entries in each column add up to zero. A unique set of equations can be obtained by defining a datum node and eliminating its corresponding row of A_a. Hence, KCL results in $n - 1$ equations and b unknowns. It implies

$$A \times i_b = 0 \tag{53.1}$$

where A is called the *reduced* incidence matrix.

KVL results in b equations with $b + n - 1$ unknowns. It implies

$$v_b = A^T \times V_n \tag{53.2}$$

where A^T is the transpose of the reduced incidence matrix, v_b is the branch voltage, and V_n is the node-to-datum (nodal) voltage. Define the convention as follows:

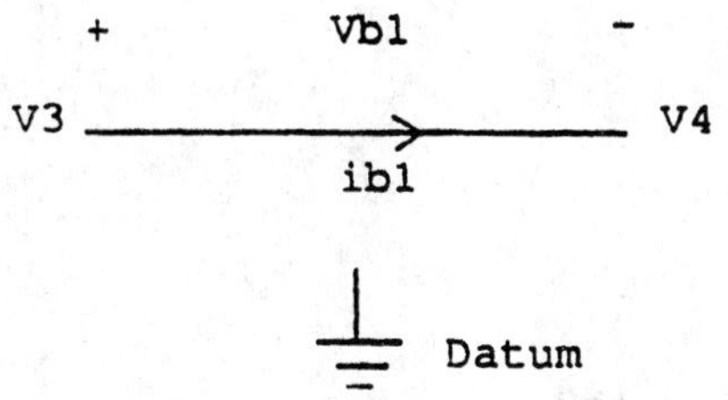

FIGURE 53.2

One can sum the voltages around the loop using KVL

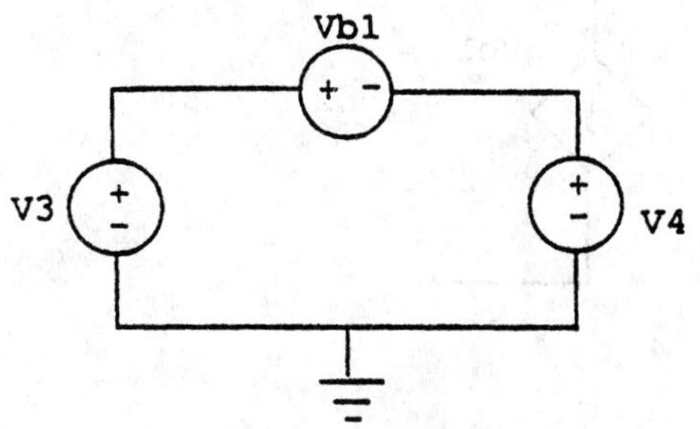

FIGURE 53.3

or,

$$v_{b_1} = V_3 - V_4,$$

Apply KVL to the example above and we obtain

$$\begin{bmatrix} v_{b_1} \\ v_{b_2} \\ v_{b_3} \\ v_{b_4} \\ v_{b_5} \\ v_{b_6} \end{bmatrix} = \begin{bmatrix} -1 & 0 & 0 \\ 1 & 0 & 0 \\ 1 & -1 & 0 \\ 0 & 1 & 0 \\ 0 & -1 & 1 \\ 0 & 0 & 1 \end{bmatrix} \begin{bmatrix} V_1 \\ V_2 \\ V_3 \end{bmatrix}$$

Or,

$$v_b = A^T \times V_n$$

where v_b and V_n are unknowns.

The element characteristic results in generalized branch constraints equations in the form of

$$Y_b \cdot v_b + Z_b \cdot i_b = S_b \tag{53.3}$$

For example
Resistor:

$$\frac{1}{R} \cdot v_b - i_b = 0$$

Voltage source:

$$v_b = E$$

Voltage-Controlled Current Source (VCCS):

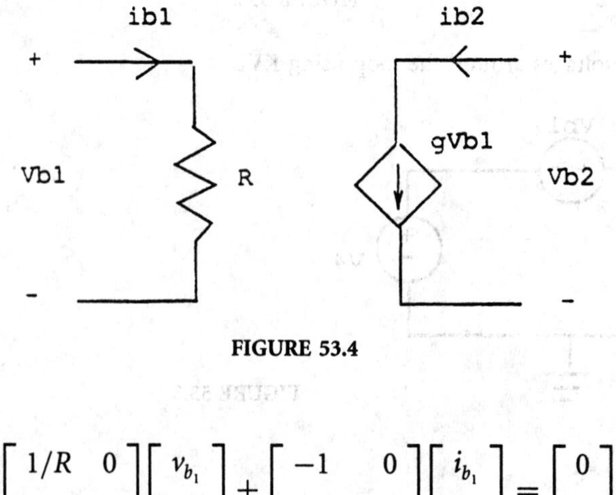

FIGURE 53.4

$$\begin{bmatrix} 1/R & 0 \\ g & 0 \end{bmatrix} \begin{bmatrix} v_{b_1} \\ v_{b_2} \end{bmatrix} + \begin{bmatrix} -1 & 0 \\ 0 & -1 \end{bmatrix} \begin{bmatrix} i_{b_1} \\ i_{b_2} \end{bmatrix} = \begin{bmatrix} 0 \\ 0 \end{bmatrix}$$

Sparse Tableau Formulation

The Sparse Tableau method simply combines Eqs. (53.1), (53.2), and (53.3). The equation and unknown count can be summarized as follows:

	Number of Equations	Number of Unknowns
(1.1)	$n-1$	b
(1.2)	b	$b+n-1$
(1.3)	b	0
Total	$2b+n-1$	$2b+n-1$

The Sparse Tableau in the matrix form is shown below.

$$\begin{bmatrix} 0 & A & 0 \\ A^T & 0 & -I \\ 0 & Z_b & Y_b \end{bmatrix} \begin{bmatrix} V_n \\ i_b \\ v_b \end{bmatrix} = \begin{bmatrix} 0 \\ 0 \\ S_b \end{bmatrix} \tag{53.4}$$

Eq. 53.4 is a system of linear equations.

Note that the matrix formulated by the Sparse Tableau approach is typically sparse with many zero value entries. The main advantage is its generality (i.e., all circuit unknowns including branch current/voltage and node-to-datum voltage can be obtained in one pass). Since the number of equations is usually very large, an efficient sparse linear system solver is essential.

Nodal Analysis

The number of equations in Eq. 53.4 can be reduced significantly by manipulating Eqs. 53.1, 53.2, and 53.3. Motivated by the fact that the number of branches in a circuit is generally larger than the number of nodes, nodal analysis attempts to reduce the number of unknowns to V_n. As we will see, this is achieved by a loss of generality (i.e., not all types of linear elements can be processed).

We can eliminate the branch voltages v_b by substituting Eq. 53.2 into Eq. 53.3. This yields

$$Y_b A^T V_n + Z_b i_b = S_b \tag{53.5}$$

$$A \times i_b = 0 \tag{53.6}$$

Combining Eqs. (5) and (6) to eliminate the branch currents, i_b, we obtain a set of equations with V_n as unknowns.

$$A \cdot Z_b^{-1} \cdot (-Y_b A^T V_n + S_b) = 0 \tag{53.7}$$

Since the Z_b matrix may be singular, not all elements can be processed. For example, a voltage-controlled voltage source, as shown in Fig. 53.5, can be cast in the form of Eq. 53.3.

$$\begin{bmatrix} 0 & 0 \\ u & -1 \end{bmatrix} \begin{bmatrix} v_{b_1} \\ v_{b_2} \end{bmatrix} + \begin{bmatrix} 1 & 0 \\ 0 & 0 \end{bmatrix} \begin{bmatrix} i_{b_1} \\ i_{b_2} \end{bmatrix} = \begin{bmatrix} 0 \\ 0 \end{bmatrix}$$

Note that Z_b is singular and it cannot be inverted. Consider a special case where Eq. 53.3 can be expressed as

$$Y_b v_b - i_b = 0 \tag{53.8}$$

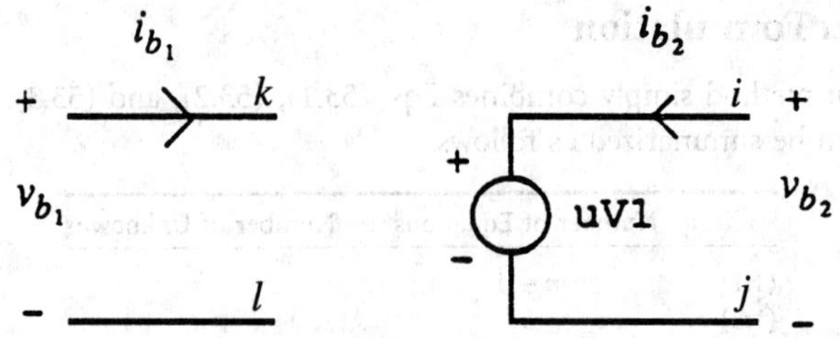

FIGURE 53.5

Or, $-Z_b$ is a unit matrix and $S_b = 0$. This condition is true if the circuit consists of

- Resistor
- Capacitor
- Inductor
- Voltage-controlled current source

For example,

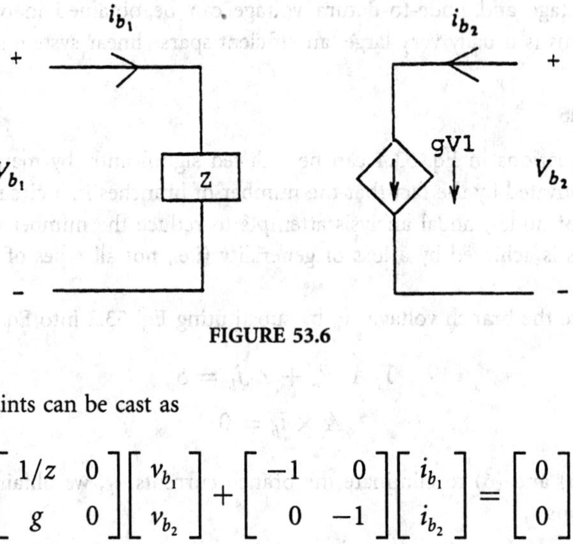

FIGURE 53.6

The branch constraints can be cast as

$$\begin{bmatrix} 1/z & 0 \\ g & 0 \end{bmatrix} \begin{bmatrix} v_{b_1} \\ v_{b_2} \end{bmatrix} + \begin{bmatrix} -1 & 0 \\ 0 & -1 \end{bmatrix} \begin{bmatrix} i_{b_1} \\ i_{b_2} \end{bmatrix} = \begin{bmatrix} 0 \\ 0 \end{bmatrix}$$

and, $z = 1/j\omega C$ for the capacitor, $z = j\omega L$ for an inductor, $z = R$ for a resistor. For this type of circuit, the nodal analysis results in $n - 1$ equations with V_n as unknowns, or,

$$AY_bA^T \times V_n = 0$$

A circuit of this type contains no excitation (voltage/current source). The current sources can be included in the formulation by letting the corresponding edges of the current sources be numbered last. Hence, we can partition the reduced incidence matrix A into

$$A = [A_b \,|A_J] \tag{53.9}$$

where A_J corresponds to the subincidence matrix of the current source branches. Then, Eqs. 53.1, 53.2, 53.8 can be expressed as

$$A_b i_b + A_J J = 0 \tag{53.10}$$

$$v_b = A_b^T V_n \tag{53.11}$$

$$i_b = Y_b v_b \tag{53.12}$$

Rearranging these equations yields

$$A_b Y_b A_b^T \times V_n = -A_J J \tag{53.13}$$

where J is a vector containing the current source values. Voltage sources can also be included in the formulation by simple source transformation which requires moderate preprocessing. The nodal analysis approach, therefore, can be applied to formulate equations for circuits consisting of

- Resistor
- Capacitor
- Inductor
- Voltage source (with preprocessing)
- Current source
- Voltage-controlled current source

Nodal analysis, however, when applied to an inductor, can cause numerical problems when the frequency is low.

$$\left(\frac{1}{j\omega L}\right)(v_{b1}) + (-1)(i_{b1}) = 0$$

Or,

$$\omega \to 0, \quad \frac{1}{j\omega L} \to \infty$$

To sum up, the nodal analysis must be extended to process the following linear elements (without preprocessing).

- Inductor
- Voltage source
- Current-controlled current source
- Current-controlled voltage source
- Voltage-controlled voltage source

Modified Nodal Analysis

A set of selfconsistent modifications to the nodal analysis are proposed and the resultant formulation is called the modified nodal analysis (MNA). The MNA resolves the limitations of the nodal analysis method while preserving its advantages. In this section, we present the basic theory of MNA.

Divide all types of elements into 3 groups.

Group 1: elements that satisfy

$$i_b = Y_b v_b$$

such as resistor, capacitor, and VCCS.

Group 2: elements that satisfy

$$Y_b v_b + Z_b i_b = S_b$$

such as voltage source, inductor, VCVS, CCVS, CCCS

Group 3: current source only.

Apply the partitioning technique to Group 1 and Group 2 elements. We can write

$$[A_1 \, | A_2] \begin{bmatrix} i_1 \\ i_2 \end{bmatrix} = 0 \Rightarrow A_1 i_1 + A_2 i_2 = 0$$

$$\begin{bmatrix} v_1 \\ v_2 \end{bmatrix} = \begin{bmatrix} A_1^T \\ A_2^T \end{bmatrix} V_n \Rightarrow v_1 = A_1^T V_n, \quad v_2 = A_2^T V_n$$

$$i_1 = Y_1 v_1$$

$$Y_2 \cdot v_2 + Z_2 \cdot i_2 = S_2$$

Eliminating v_1, i_1, v_2 from the equations above, we derive a system of linear equations with V_n, i_2 as unknowns.

$$A_1 Y_1 A_1^T V_n + A_2 i_2 = 0 \quad (53.14)$$

$$Y_2 A_2^T V_n + Z_2 i_2 = S_2 \quad (53.15)$$

Casting them in matrix form gives

$$\begin{bmatrix} A_1 Y_1 A_1^T & A_2 \\ Y_2 A_2^T & Z_2 \end{bmatrix} \begin{bmatrix} V_n \\ i_2 \end{bmatrix} = \begin{bmatrix} 0 \\ S_2 \end{bmatrix} \quad (53.16)$$

Finally, we apply the partitioning technique to include the current source (Group 3).

$$\begin{bmatrix} A_1 Y_1 A_1^T & A_2 \\ Y_2 A_2^T & Z_2 \end{bmatrix} \begin{bmatrix} V_n \\ i_2 \end{bmatrix} = \begin{bmatrix} -A_J J \\ S_2 \end{bmatrix} \quad (53.17)$$

or,

$$Y_n \times x = J_n \quad (53.18)$$

where Y_n is the node admittance matrix, J_n is the source vector, and x is the unknown vector.

Implementing Eq. 53.18 by matrix multiplication is difficult. Stamping methods have been developed to stamp in the entries of Y_n, J_n *element by element*. It is a very efficient way to implement Eq. 53.18 into a network analysis program.

Nodal Formulation by Stamps

In this section, we developed the stamping rules for Group 1 and Group 3 elements only. Given a circuit consisting of Group 1 and Group 3 elements only:

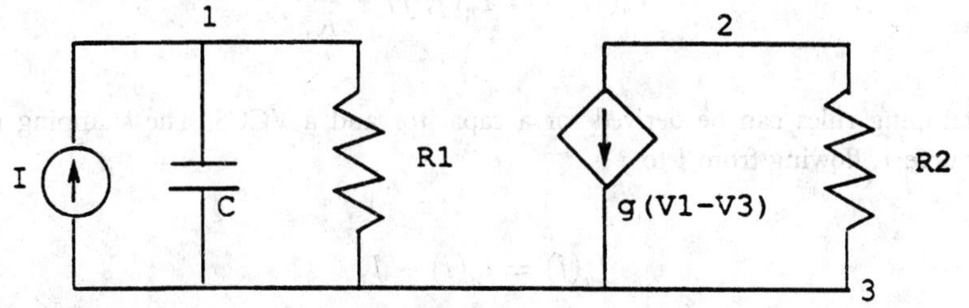

FIGURE 53.7

We write the nodal equations at node 1, 2, and 3:

$$-I + j\omega C(V_1 - V_3) + 1/R_1(V_1 - V_3) = 0 \qquad \text{(node 1)}$$
$$g(V_1 - V_3) + 1/R_2(V_2 - V_3) = 0, \qquad \text{(node 2)}$$
$$I + j\omega C(V_3 - V_1) + 1/R(V_3 - V_1) - g(V_1 - V_3) + 1/R_2(V_3 - V_2) = 0 \qquad \text{(node 3)}$$

Cast them in matrix form:

$$\begin{bmatrix} j\omega C + 1/R_1 & 0 & -j\omega C - 1/R_1 \\ g & 1/R_2 & -1/R_2 - g \\ -j\omega C - 1/R_1 - g & -1/R_2 & j\omega C + 1/R_1 + 1/R_2 + g \end{bmatrix} \begin{bmatrix} V_1 \\ V_2 \\ V_3 \end{bmatrix} = \begin{bmatrix} I \\ 0 \\ -I \end{bmatrix}$$

Note that for the L.H.S., we can write

$$\begin{bmatrix} 1/R_1 & 0 & -1/R_1 \\ 0 & 0 & 0 \\ -1/R_1 & 0 & 1/R_1 \end{bmatrix} + \begin{bmatrix} j\omega C & 0 & -j\omega C \\ 0 & 0 & 0 \\ -j\omega C & 0 & j\omega C \end{bmatrix} + \begin{bmatrix} 0 & 0 & 0 \\ g & 0 & -g \\ -g & 0 & g \end{bmatrix}$$
$$+ \begin{bmatrix} 0 & 0 & 0 \\ 0 & 1/R_2 & -1/R_2 \\ 0 & -1/R_2 & 1/R_2 \end{bmatrix}$$

Therefore, the sampling rule for a resistor is

$$Y_n(i, i) = Y_n(i, i) + \frac{1}{R},$$
$$Y_n(j, j) = Y_n(j, j) + \frac{1}{R},$$
$$Y_n(i, j) = Y_n(i, j) - \frac{1}{R},$$
$$Y_n(j, i) = Y_n(j, i) - \frac{1}{R}$$

If node i is grounded, the corresponding row and column can be eliminated from the node admittance matrix. We then obtained only

$$Y_n(j, j) = Y_n(j, j) + \frac{1}{R}$$

Similar stamping rules can be derived for a capacitor and a VCCS. The stamping rule for a current source J, flowing from i to j is

$$J_n(i) = J_n(i) - J,$$
$$J_n(j) = J_n(j) + J$$

Modified Nodal Formulation by Stamps

Given a circuit including Group 2 elements only:

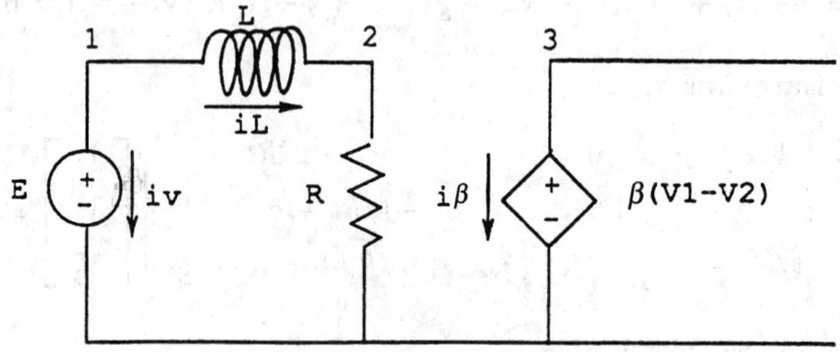

FIGURE 53.8

Let us first define auxiliary branch current unknowns i_v, i_L, and i_β for each type of element in Group 2. From nodal analysis, we obtain the following nodal equations:

$$i_v + i_L = 0, \qquad \text{(node 1)}$$
$$-i_L + 1/R(V_2 - V_4) = 0, \qquad \text{(node 2)}$$
$$i_\beta = 0, \qquad \text{(node 3)}$$
$$-i_v + 1/R(V_4 - V_2) - i_\beta = 0 \qquad \text{(node 4)}$$

For each auxiliary unknown, one auxiliary equation must be provided.

$$V_1 - V_4 = E,$$
$$V_1 - V_2 = j\omega L i_L,$$
$$V_3 - V_4 = \beta(V_1 - V_2)$$

Case these equations in matrix form,

$$\begin{bmatrix} 0 & 0 & 0 & 0 & 1 & 1 & 0 \\ 0 & 1/R & 0 & -1/R & 0 & -1 & 0 \\ 0 & 0 & 0 & 0 & 0 & 0 & 1 \\ 0 & -1/R & 0 & 1/R & -1 & 0 & -1 \\ 1 & 0 & 0 & -1 & 0 & 0 & 0 \\ 1 & -1 & 0 & 0 & 0 & -j\omega L & 0 \\ -\beta & \beta & 1 & -1 & 0 & 0 & 0 \end{bmatrix} \begin{bmatrix} V_1 \\ V_2 \\ V_3 \\ V_4 \\ i_v \\ i_l \\ i_\beta \end{bmatrix} = \begin{bmatrix} 0 \\ 0 \\ 0 \\ 0 \\ E \\ 0 \\ 0 \end{bmatrix}$$

Hence, the following stamping rules are derived.

Voltage source:

L.H.S.:

$$\begin{array}{c|ccc} & i & j & i_v \\ \hline i & 0 & 0 & 1 \\ j & 0 & 0 & -1 \\ i_v & 1 & -1 & 0 \end{array}$$

R.H.S.:

$$\begin{array}{c|c} i & 0 \\ j & 0 \\ i_v & E \end{array}$$

Inductor:

L.H.S.:

$$\begin{array}{c|ccc} & i & j & i_L \\ \hline i & 0 & 0 & 1 \\ j & 0 & 0 & -1 \\ i_L & 1 & -1 & -j\omega L \end{array}$$

Note that the numerical problem associated with an inductor is avoided when $\omega \to 0$.

VCVS (Figure 53.5)

L.H.S.:

$$\begin{array}{c|ccccc} & i & j & k & l & i_v \\ \hline i & 0 & 0 & 0 & 0 & 1 \\ j & 0 & 0 & 0 & 0 & -1 \\ i_v & 1 & -1 & -\mu & \mu & 0 \end{array}$$

The stamping rule for a CCVS and a CCCS can be developed following the same arguments. For DC analysis, set $\omega = 0$. In SPICE, the MNA is employed to formulate network equations. To probe a branch current, a user needs to insert a zero value voltage source between the adjacent nodes. The solution i_v is then the branch current. The implementation of MNA can be summarized as follows.

1. Implement an input parser to read in a circuit description file. A "free" input format does not restrict a user to follow column entry rules.
2. Build an internal node table which has a one-to-one relationship with user-defined nodes. Hence, a user needs not number the network nodes consecutively.
3. Number the auxiliary nodes for group 2 elements last. For each group 2 element, one extra node is needed.
4. Solve the system of linear equations to find output voltages and currents.

53.2 Solution of Linear Algebraic Equations

The methods of solving a set of linear equations are basic to all computer-aided network analysis problems. If the network is linear, the equations are linear. Nonlinear networks lead to a system of nonlinear equations which can be linearized about some operating point. Transient analysis involves solving these linearized equations at many iteration points. Frequency domain analysis (small-signal AC analysis) requires the repeated solution of linear equations at specified frequencies.

The discussion in this section will be an introduction to the direct solution method based on *LU* decomposition, a variant of Gaussian elimination. This method is frequently used because of its efficiency, robustness, and ease of implementation. More advanced topics such as the general sparse matrix techniques are not discussed.

Consider a set of n linear algebraic equations of the form:

$$Ax = b \tag{53.19}$$

where A is an $n \times n$ nonsingular real matrix, x and b are n-vectors. For the system to have a unique solution, A must be nonsingular (i.e., the determinant of A must not be 0). Equation 53.19 can be solved efficiently by first factorizing A into a product of two matrices L and U, which are respectively lower and upper triangular. This so-called *LU* decomposition method for solving a system of linear equations is similar to the Gaussian elimination except b is not required in advance. All operations performed on b using the Gaussian elimination are eliminated to save computation cost. The procedures are expressed as follows:

1. Step 1: Factorization/Decomposition
2. Step 2: Forward Substitution
3. Step 3: Backward Substitution

Factorization

We factor A into a product $L \cdot U$, where L is a lower triangular matrix and U is an upper triangular matrix.

$$A = LU \tag{53.20}$$

Either L or U can have a diagonal of ones. The factors of A, being upper and lower triangular, can be stored in one matrix B, i.e., $B = L + U - I$. In practice, B is stored in place of A to save memory storage. There are two widely used algorithms for factorization: (1) Crout's algorithm (setting the diagonal elements of U to one) and (3) Doolittle's algorithm (setting the diagonal elements of L to one). In the following, we will use a 4×4 matrix to illustrate the direct finding of L and U by the Crout's Algorithm:

$$\begin{bmatrix} l_{11} & 0 & 0 & 0 \\ l_{21} & l_{22} & 0 & 0 \\ l_{31} & l_{32} & l_{33} & 0 \\ l_{41} & l_{42} & l_{43} & l_{44} \end{bmatrix} \begin{bmatrix} 1 & u_{12} & u_{13} & u_{14} \\ 0 & 1 & u_{23} & u_{24} \\ 0 & 0 & 1 & u_{34} \\ 0 & 0 & 0 & 1 \end{bmatrix} = \begin{bmatrix} a_{11} & a_{12} & a_{13} & a_{14} \\ a_{21} & a_{22} & a_{23} & a_{24} \\ a_{31} & a_{32} & a_{33} & a_{34} \\ a_{41} & a_{42} & a_{43} & a_{44} \end{bmatrix}$$

Multiplying the two matrices on the left-hand-side of the equations above gives

$$\begin{bmatrix} l_{11} & l_{11}u_{12} & l_{11}u_{13} & l_{11}u_{14} \\ l_{21} & l_{21}u_{12}+l_{22} & l_{21}u_{13}+l_{22}u_{23} & l_{21}u_{14}+l_{22}u_{24} \\ l_{31} & l_{31}u_{12}+l_{32} & l_{31}u_{13}+l_{32}u_{23}+l_{33} & l_{31}u_{14}+l_{32}u_{24}+l_{33}u_{34} \\ l_{41} & l_{41}u_{12}+l_{42} & l_{41}u_{13}+l_{42}u_{23}+l_{43} & l_{41}u_{14}+l_{42}u_{24}+l_{43}u_{34}+l_{44} \end{bmatrix}$$

$$= \begin{bmatrix} a_{11} & a_{12} & a_{13} & a_{14} \\ a_{21} & a_{22} & a_{23} & a_{24} \\ a_{31} & a_{32} & a_{33} & a_{34} \\ a_{41} & a_{42} & a_{43} & a_{44} \end{bmatrix}$$

The solution sequences, indicated by the superscripts, for solving the 16 unknowns ($n = 4$) are

$$\begin{matrix} l_{11}^{1} & u_{12}^{5} & u_{13}^{6} & u_{14}^{7} \\ l_{21}^{2} & l_{22}^{8} & u_{23}^{11} & u_{24}^{12} \\ l_{31}^{3} & l_{32}^{9} & l_{33}^{13} & u_{34}^{15} \\ l_{41}^{4} & l_{42}^{10} & l_{43}^{14} & l_{44}^{16} \end{matrix}$$

or,

1st column: $l_{11} = a_{11}, \quad l_{21} = a_{21}, \quad l_{31} = a_{31}, \quad l_{41} = a_{41}$

1st row: $u_{12} = \dfrac{a_{12}}{l_{11}}, \quad u_{13} = \dfrac{a_{13}}{l_{11}}, \quad u_{14} = \dfrac{a_{14}}{l_{11}}$

2nd column: $l_{22} = a_{22} - l_{21}u_{12}, \quad l_{32} = a_{32} - l_{31}u_{12}, \quad l_{42} = a_{42} - l_{41}u_{12}$

2nd row: $u_{23} = \dfrac{a_{23} - l_{21}u_{13}}{l_{22}}, \quad u_{24} = \dfrac{a_{24} - l_{21}u_{14}}{l_{22}}$

3rd column: $l_{33} = a_{33} - l_{31}u_{13} - l_{32}u_{23}, \quad l_{43} = a_{43} - l_{41}u_{13} - l_{42}u_{23}$

3rd row: $u_{34} = \dfrac{a_{34} - l_{31}u_{14} - l_{32}u_{24}}{l_{33}}$

4th column: $l_{44} = a_{44} - l_{41}u_{14} - l_{42}u_{24} - l_{43}u_{34}$

Note that l_{11}, l_{22}, and l_{33} are elements by which we divide and they are called *pivots.* Division by a zero pivot is not allowed. We now derive the Crout's algorithm for LU decomposition based on the solution procedures described above.

1. $l_{j1} = a_{j1}$, $j = 1, 2, \ldots, n$ (1st column)
2. $u_{1j} = a_{1j}/l_{11}$, $j = 2, \ldots, n$ (1st row)
3. At the kth step, for column k,
   $l_{jk} = a_{jk} - l_{j1} \cdot u_{1k} - l_{j2} \cdot u_{2k} - \cdots = a_{jk} - \sum_{m=1}^{k-1} l_{jm} \cdot u_{mk} \quad j = k, \ldots, n$
4. At the kth step, for row k,
   $u_{kj} = (1/l_{kk})(a_{kj} - l_{k1}u_{1j} - l_{k2}u_{2j} - \cdots) = (1/l_{kk})\left(a_{kj} - \sum_{m=1}^{k-1} l_{km}u_{mj}\right), \; j = k+1, \ldots, n$

The algorithm can be summarized in a compact form:

1. Set $k = 1$
2. $l_{jk} = a_{jk} - \sum_{m=1}^{k-1} l_{jm} \cdot u_{mk}$, $j = k, \ldots, n$
3. If $k = n$, stop
4. $u_{kj} = (1/l_{kk})\left(a_{kj} - \sum_{m=1}^{k-1} l_{km}u_{mj}\right)$, $j = k+1, \ldots, n$
5. $k = k + 1$, go to step (2)

Forward Substitution

Once A has been factored into L and U, the system of equations is written as follows:

$$LUx = b$$

Define an auxiliary vector y, which can be solved by

$$L \cdot y = b \tag{53.21}$$

Due to the special form of L, the auxiliary vector y can be solved very simply:

$$\begin{bmatrix} l_{11} & 0 & 0 & 0 \\ l_{21} & l_{22} & 0 & 0 \\ l_{31} & l_{32} & l_{33} & 0 \\ l_{41} & l_{42} & l_{43} & l_{44} \end{bmatrix} \begin{bmatrix} y_1 \\ y_2 \\ y_3 \\ y_4 \end{bmatrix} = \begin{bmatrix} b_1 \\ b_2 \\ b_3 \\ b_4 \end{bmatrix}$$

Starting from the first equation we write the solution as follows:

$$y_1 = \frac{b_1}{l_{11}}$$

$$y_2 = \frac{(b_2 - l_{21}y_1)}{l_{22}}$$

$$y_3 = \frac{(b_3 - l_{31}y_1 - l_{32}y_2)}{l_{33}}$$

$$y_4 = \frac{(b_4 - l_{41}y_1 - l_{42}y_2 - l_{43}y_3)}{l_{44}}$$

and, in general

$$y_i = \left(b_i - \sum_{j=1}^{i-1} l_{ij}y_i\right) / l_{ii}, \qquad i = 1, \ldots, n$$

This is called the *forward substitution* process.

Backward Substitution

Once y has been solved, we can proceed to solve for the unknown x by

$$Ux = y \tag{53.22}$$

Again, due to the special form of U, the unknown vector x can be solved very simply:

$$\begin{bmatrix} 1 & u_{12} & u_{13} & u_{14} \\ 0 & 1 & u_{23} & u_{24} \\ 0 & 0 & 1 & u_{34} \\ 0 & 0 & 0 & 1 \end{bmatrix} \begin{bmatrix} x_1 \\ x_2 \\ x_3 \\ x_4 \end{bmatrix} = \begin{bmatrix} y_1 \\ y_2 \\ y_3 \\ y_4 \end{bmatrix}$$

Starting from the first equation we write the solution as follows:

$$x_4 = y_4$$
$$x_3 = y_3 - u_{34}x_4$$
$$x_2 = y_2 - u_{23}x_3 - u_{24}x_4$$
$$x_1 = y_1 - u_{12}x_2 - u_{13}x_3 - u_{14}x_4$$

and, in general,

$$x_i = y_i - \sum_{j=i+1}^{n} u_{ij}x_j, \quad i = n, n-1, \ldots, 1$$

This is called the *backward substitution* process.

Pivoting

If in the process of factorization the pivot (l_{kk}) is zero, it is then necessary to interchange rows and possibly columns to put a nonzero entry in the pivot position so that the factorization can proceed. This is known as *pivoting*.
If $a_{11} = 0$

1. Need to find another row i which has $a_{i1} \neq 0$. This can always be done. Otherwise, all entries of column 1 are zero and hence, $\det |A| = 0$. The solution process should then be aborted.
2. Interchange row i and row 1. Note that this must be done for both A and b.

If $l_{kk} = 0$

1. Find another row r ($r = k+1, \ldots, n$) which has

$$l_{rk} = a_{rk} - \sum_{m=1}^{k-1} l_{rm}u_{mk} \neq 0$$

2. Interchange row r and row k in A and b.

Pivoting is also carried out for numerical stability (i.e., minimize machine round-off error). For example, one would search for the entry with the maximum absolute value of l_{rR} in columns below the diagonal and perform row interchange to put that element on the diagonal. This is called partial pivoting. Complete pivoting involves searching for the element with the maximum absolute value in the unfactorized part of the matrix and moving that particular element to the diagonal position by performing both row and column interchanges. Complete pivoting is more complicated to program than partial pivoting. Partial pivoting is used more often.

Computation Cost of LU Factorization

In this section, we derive the multiplication/division count for the LU decomposition process. As a variation, we derive the computation cost for the Doolittle's algorithm (setting the diagonal elements of L to one).

$$\begin{bmatrix} 1 & 0 & 0 & 0 \\ l_{21} & 1 & 0 & 0 \\ l_{31} & l_{32} & 1 & 0 \\ l_{41} & l_{42} & l_{43} & 1 \end{bmatrix} \begin{bmatrix} u_{11} & u_{12} & u_{13} & u_{14} \\ 0 & u_{22} & u_{23} & u_{24} \\ 0 & 0 & u_{33} & u_{34} \\ 0 & 0 & 0 & u_{44} \end{bmatrix} = \begin{bmatrix} a_{11} & a_{12} & a_{13} & a_{14} \\ a_{21} & a_{22} & a_{23} & a_{24} \\ a_{31} & a_{32} & a_{33} & a_{34} \\ a_{41} & a_{42} & a_{43} & a_{44} \end{bmatrix}$$

L.H.S. after multiplication:

$$\begin{bmatrix} u_{11} & u_{12} & u_{13} & u_{14} \\ l_{21}u_{11} & l_{21}u_{12} + u_{22} & l_{21}u_{13} + u_{23} & l_{21}u_{14} + u_{24} \\ l_{31}u_{11} & l_{31}u_{12} + l_{32}u_{22} & l_{31}u_{13} + l_{32}u_{23} + u_{33} & l_{31}u_{14} + l_{32}u_{24} + u_{34} \\ l_{41}u_{11} & l_{41}u_{12} + l_{42}u_{22} & l_{41}u_{13} + l_{42}u_{23} + l_{43}u_{33} & l_{41}u_{14} + l_{42}u_{24} + l_{43}u_{34} + u_{44} \end{bmatrix}$$

$$\begin{aligned} &\text{Column 1:} \quad (l_{21}, l_{31}, l_{41})u_{11} \\ &\text{Column 2:} \quad (l_{21}, l_{31}, l_{41})u_{12} + (l_{32}, l_{42})u_{22} \\ &\text{Column 3:} \quad (l_{21}, l_{31}, l_{41})u_{13} + (l_{32}, l_{42})u_{23} + l_{43}u_{33} \\ &\text{Column 4:} \quad (l_{21}, l_{31}, l_{41})u_{14} + (l_{32}, l_{42})u_{24} + l_{43}u_{34} \end{aligned} \tag{53.23}$$

Let the symbol $\langle\cdot\rangle$ denote the number of nonzero elements of a matrix or vector. Or,

$\langle U\rangle$: number of nonzeros of U

$\langle L\rangle$: number of nonzeros of L

$\langle A_{i\cdot}\rangle$: number of nonzeros in row i of matrix A

$\langle A_{\cdot j}\rangle$: number of nonzeros in column j of matrix A

From Eq. 53.23, the total number of nonzero multiplications and divisions for LU factorization is given by

$$(\langle L_{\cdot 1}\rangle - 1)(\langle U_{1\cdot}\rangle) + (\langle L_{\cdot 2}\rangle - 1)(\langle U_{2\cdot}\rangle) + (\langle L_{\cdot 3}\rangle - 1)(\langle U_{3\cdot}\rangle) \tag{53.24}$$

Let α be the total number of multiplications and divisions. Express Eq. 53.24 as a summation:

$$\alpha = \sum_{k=1}^{n} (\langle L_{\cdot k}\rangle - 1)(\langle U_{k\cdot}\rangle) \tag{53.25}$$

Or, for $n = 4$,

$$\alpha = 3 \times 4 + 2 \times 3 + 1 \times 2$$

If L and U are full, Eq. 53.25 can be simplified as follows:

$$\alpha = \sum_{k=1}^{n} (n-k)(n-k+1)$$

$$\alpha = \sum_{k=1}^{n} (n^2 + k^2 - 2kn + n - k)$$

$$\alpha = n_3 + \sum_{k=1}^{n} k^2 - 2n \times \frac{n(n+1)}{2} + n^2 - \frac{n(n+1)}{2} \tag{53.26}$$

$$\alpha = \frac{n(n+1)(2n+1)}{6} - \frac{n(n+1)}{2}$$

$$\alpha = \frac{n^3 - n}{3}$$

Total number of mul/div for forward substitution is equal to the total number of nonzeros for L or $\langle L \rangle$. Total number of mul/div for backward substitution is equal to the total number of nonzeros for U or $\langle U \rangle$. Let β be the mul/div count for the forward and backward substitutions,

$$\beta = \langle L \rangle + \langle U \rangle$$

It follows that

$$\beta = n^2 \tag{53.27}$$

if L and U are full. Combining (27) and (26), we obtain the computation cost of solving a system of linear algebraic equations using direct LU decomposition.

$$\text{Total} = \alpha + \beta$$

$$\text{Total} = \alpha + \beta = \frac{n^3}{3} + n^2 - \frac{n}{3} \tag{53.28}$$

References

[1] Hachtel, et al., "The sparse tableau approach to network analysis and design," *IEEE Trans. Circuit Theory*, vol. CT-18, pp. 101–113, Jan. 1971.

[2] L. W. Nagel, "SPICE2: A Computer Program to Simulate Semiconductor Circuits," Memo No. ERL-M520, Electronic Research Laboratory, University of California, Berkeley, May 1975.

[3] C. W. Ho, et al., "The modified nodal approach to network analysis," *IEEE Trans. Circuits and Systems*, vol. CAS-22, June 1975.

[4] J. Vlach and K. Singhal, *Computer Methods for Circuit Analysis and Design*, New York: Van Nostrand Reinhold, 1983.

[5] K. S. Kundert, *Advances in CAD for VLSI: Circuit Analysis, Simulation and Design*, Vol. 3, Part 1, Chap. 6, Amsterdam: North-Holland, 1986.

54

Design by Optimization

Sachin S. Sapatnekar
Iowa State University

54.1 Introduction

In many integrated circuit (IC) design situations, a designer must make complex tradeoffs between conflicting behavioral requirements, dealing with functions that are almost always nonlinear. The number of parameters involved in the design process may be large, necessitating the use of algorithms that provide qualitatively good solutions in a computationally efficient manner.

The theory and utilization of optimization algorithms in computer-aided design (CAD) of integrated circuits are illustrated here. The form of a general nonlinear optimization problem is first presented, and some of the commonly used methods for optimization are overviewed. It is frequently said that setting up an optimization problem is an art, while (arguably) solving it is an exact science. To provide a flavor for both of these aspects, case studies on the following specific design problems are examined:

Transistor Sizing: The delay of a digital circuit can be tuned by adjusting the sizes of transistors within it. By increasing the sizes of a few selected transistors from the minimum size, significant improvements in performance are achievable. However, one must take care to ensure that the area overhead incurred in increasing these sizes is not excessive. The area-delay tradeoff here is the transistor size optimization problem.

Design Centering: The values of design parameters of a circuit are liable to change due to manufacturing variations. This contributes to a deviation in the behavior of the circuit from the norm and may lead to dysfunctional circuits that violate the performance parameters that they were designed for. The problem of design centering attempts to

0-8493-8341-2/95/$0.00 + $.50

ensure that under these variations, the probability of a circuit satisfying its performance specifications is maximized.

54.2 Optimization Algorithms

Nonlinear Optimization Problems

The "standard form " of a **constrained nonlinear optimization problem** is

$$\begin{aligned} &\textit{minimize} \quad f(\mathbf{x}): \mathbf{R}^n \rightarrow \mathbf{R} \\ &\textit{subject to} \quad \mathbf{g}(\mathbf{x}) \leq \mathbf{0} \\ &\mathbf{g}: \mathbf{R}^n \rightarrow \mathbf{R}^m, \mathbf{x} \in \mathbf{R}^n \end{aligned} \tag{54.1}$$

representing the minimization of a function f of n variables under constraints specified by inequalities determined by functions $\mathbf{g} = [g_1 \cdots g_m]^T$. f and g_i are, in general, nonlinear functions, so that the linear programming problem is a special case of the above. The parameters $\mathbf{x}$ may, for example, represent circuit parameters, and $f(\mathbf{x})$ and $g_i(\mathbf{x})$ may correspond to circuit performance functions. Note that "$\geq$" inequalities can be handled under this paradigm by multiplying each side by -1, and equalities by representing them as a pair of equalities. The maximization of an objective function $f(\mathbf{x})$ can be achieved by minimizing $-f(\mathbf{x})$.

The set $\mathcal{F} = \{\mathbf{x} \mid \mathbf{g}(\mathbf{x}) \leq \mathbf{0}\}$ that satisfies the constraints on the nonlinear optimization problem is known as the feasible set, or the feasible region. If $\mathcal{F}$ is empty (nonempty), then the optimization is said to be *unconstrained (constrained).*

Several mathematical programming techniques can be used to solve the optimization problem above; some of these are outlined here. For further details, the reader is referred to a standard text on optimization, such as [1]. Another excellent source for optimization techniques and their applications to IC design is a survey paper by Brayton *et al.* [2].

The above formulation may not directly be applicable to real-life design problems, where, often, multiple conflicting objectives must be optimized. In such a case, one frequently uses techniques that map on the problem to the form in Eq. (54.1) (see Section 54.2).

Basic Definitions

In any discussion on optimization, it is essential to understand the idea of a convex function and a convex set, since these have special properties, and it is desirable to formulate problems as convex programming problems, wherever it is possible to do so without an undue loss in modeling accuracy. (Unfortunately, it is not always possible to do so!)

Definition. A set C in $\mathbf{R}^n$ is said to be a **convex set** if, for every $\mathbf{x}_1, \mathbf{x}_2 \in C$, and every real number $\alpha, 0 \leq \alpha \leq 1$, the point $\alpha\mathbf{x}_1 + (1-\alpha)\mathbf{x}_2 \in C$.

This definition can be interpreted geometrically as stating that a set is convex if, given two points in the set, every point on the line segment joining the two points is also a member of the set. Examples of convex and nonconvex sets are shown in Fig. 54.1.

Two examples of convex sets that will be referred to later are the following geometric bodies:

1. An **ellipsoid** $E(\mathbf{x}, \mathcal{B}, \mathbf{r})$ centered at point $\mathbf{x}$ is given by the equation

$$\{\mathbf{y} \mid (\mathbf{y}-\mathbf{x})^T \mathcal{B} (\mathbf{y}-\mathbf{x}) \leq \mathbf{r}^2\} \tag{54.2}$$

If $\mathcal{B}$ is a scalar multiple of the unit matrix, then the ellipsoid is called a *hypersphere.*

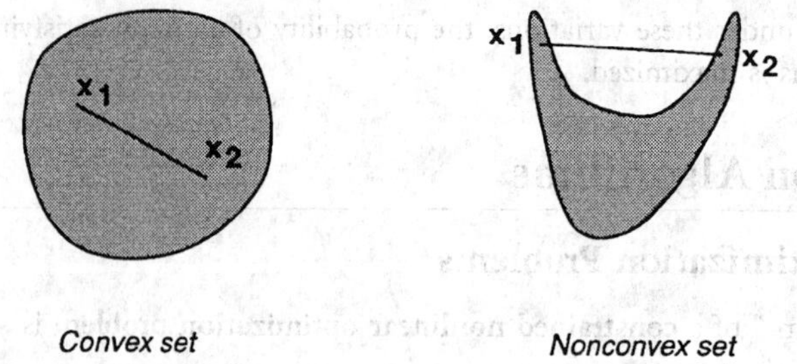

FIGURE 54.1 Convex sets

2. A (convex) **polytope** is defined as an intersection of half spaces, and is given by the equation

$$\mathcal{P} = \{\mathbf{x} \mid A\mathbf{x} \geq \mathbf{b}\}, A \in \mathbf{R}^{m \times n}, \mathbf{b} \in \mathbf{R}^m \tag{54.3}$$

corresponding to a set of m inequalities $\mathbf{a}_i^T\mathbf{x} \geq b_i, \mathbf{a}_i \in \mathbf{R}^n$.

Definition. The **convex hull** of m points, $\mathbf{x}_1, \ldots, \mathbf{x}_m \in \mathbf{R}^n$, denoted $co\{\mathbf{x}_1, \ldots, \mathbf{x}_m\}$, is defined as the set of points $\mathbf{y} \in \mathbf{R}^n$ such that

$$\mathbf{y} = \sum_{i=1}^{m} \alpha_i \mathbf{x}_i; \quad \alpha_i \geq 0 \ \forall\, i, \quad \sum_{i=0}^{m} \alpha_i = 1 \tag{54.4}$$

The convex hull is the smallest convex set that contains the m points. An example of the convex hull of five points in the plane is shown by the shaded region in Fig. 54.2. If the set of points $\mathbf{x}_i$ is of finite cardinality (i.e., m is finite), then the convex hull is a bounded polytope. Hence, a polytope is also often described as the convex hull of its vertices.

Definition. A function f defined on a convex set Ω is said to be a **convex function** if, for every $\mathbf{x}_1, \mathbf{x}_2 \in \Omega$, and every $\alpha, 0 \leq \alpha \leq 1$,

$$f(\alpha\mathbf{x}_1 + (1-\alpha)\mathbf{x}_2) \leq \alpha f(\mathbf{x}_1) + (1-\alpha) f(\mathbf{x}_2) \tag{54.5}$$

f is said to be **strictly convex** is the equality in Eq. (5) is strict for $0 < \alpha < 1$.

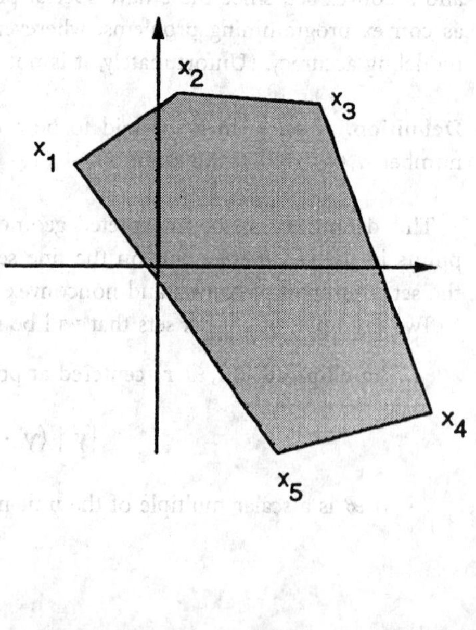

FIGURE 54.2 The convex hull of five points.

Geometrically, a function is convex if the line joining two points on its graph is always above the graph. Examples of convex and nonconvex functions on $\mathbf{R}^n$ are shown in Fig. 54.3.

Definition. A function g defined on a convex set Ω is said to be a **concave function** if the function $f = -g$ is convex. The function g is **strictly concave** if $-g$ is strictly convex.

Definition. The **convex programming problem** is stated as follows:

$$\text{minimize} \quad f(\mathbf{x}) \tag{54.6}$$

$$\text{such that} \quad \mathbf{x} \in S \tag{54.7}$$

where f is a convex function and S is a convex set. This problem has the property that any local minimum of f over S is a global minimum.

Definition. A **posynomial** is a function h of a positive variable $\mathbf{x} \in \mathbf{R}^n$ that has the form

$$h(\mathbf{x}) = \sum_j \gamma_j \prod_{i=1}^{n} x_i^{\alpha_{ij}} \tag{54.8}$$

where the exponents $\alpha_{ij} \in \mathbf{R}$ and the coefficients $\gamma_j > 0$.

For example, the function $3.7x_1^{1.4}x_2^{\sqrt{3}} + 1.8x_1^{-1}x_3^{2.3}$ is a posynomial in the variables x_1, x_2, x_3. Roughly speaking, a posynomial is a function that is similar to a polynomial except that the coefficients γ_j must be positive, and an exponent α_{ij} could be any real number, and not necessarily a positive integer, unlike a polynomial. A posynomial has the useful property that it can be mapped onto a convex function through an elementary variable transformation, $(x_i) = (e^{z_i})$ when $x_i > 0\ \forall\ i$. Such functional forms are useful since in the case of an optimization problem where the objective function and the constraints are posynomial, the problem can easily be mapped onto a convex programming problem.

Constrained Optimization Methods

Most problems in integrated circuit design involve the minimization or maximization of a cost function subject to certain constraints. In this section, a few prominent techniques for constrained optimization are presented. The reader is referred to [1] for details on unconstrained optimization.

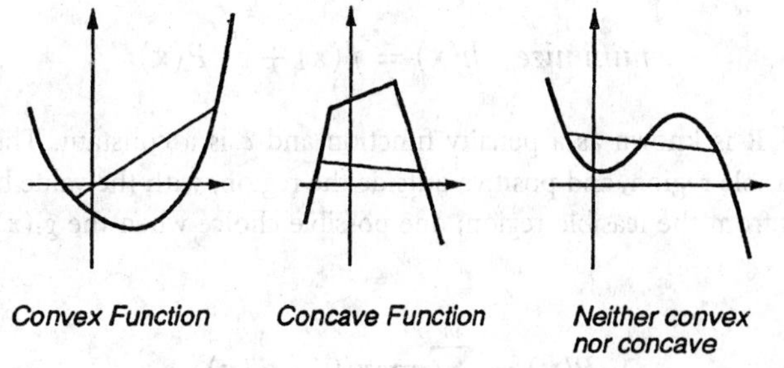

FIGURE 54.3 Convex functions.

Linear Programming

Linear programming is a special case of nonlinear optimization, and is the convex programming problem where the objective and constraints are all linear functions. The problem is stated as

$$\begin{aligned} &\textit{minimize} && \mathbf{c}^T\mathbf{x} \\ &\textit{subject to} && A\mathbf{x} \leq \mathbf{b} \\ &\textit{where} && \mathbf{c}, \mathbf{x} \in \mathbf{R}^n, \quad \mathbf{b} \in \mathbf{R}^m, \quad A \in \mathbf{R}^{m \times n} \end{aligned} \tag{54.9}$$

It can be shown that any solution to a linear program must necessarily occur at a vertex of the constraining polytope. The most commonly used technique for solution of linear programs, the simplex method [1], is based on this principle. The computational complexity of this method can show an exponential behavior for pathological cases, but for most practical problems, it has been observed to grow linearly with the number of variables and sublinearly with the number of constraints. Algorithms with polynomial time worst-case complexity do exist; these include Karmarkar's method and the Shor-Khachiyan ellipsoidal method. The computational complexity of the latter, however, is often seen to be impractical from a practical standpoint.

Some examples of computer-aided design (CAD) problems that are posed as linear programs include placement, gate sizing, clock skew optimization, layout compaction, etc. In several cases, the structure of the problem can be exploited to arrive at graph-based solutions, for example, in layout compaction.

Lagrange Multiplier Methods

These methods are closely related to the first-order **Kuhn-Tucker** necessary conditions on optimality, which state that given an optimization problem of the form in Eq. (54.1), if f and $\mathbf{g}$ are differentiable at $\mathbf{x}^*$, then there is a vector $\boldsymbol{\lambda} \in \mathbf{R}^m, (\boldsymbol{\lambda})_i \geq \mathbf{0}$, such that

$$\nabla f(\mathbf{x}^*) + \boldsymbol{\lambda}^T \nabla \mathbf{g}(\mathbf{x}^*) = 0 \tag{54.10}$$

$$\boldsymbol{\lambda}^T \mathbf{g}(\mathbf{x}^*) = \mathbf{0} \tag{54.11}$$

These correspond to $m+1$ equations in $m+1$ variables; the solution to these provides the solution to the optimization problem. The variables $\boldsymbol{\lambda}$ are known as the Lagrange multipliers.

Note that since $g_1(\mathbf{x}^*) \leq 0$, and because of the nonnegativity constraint on the Lagrange multipliers, $\boldsymbol{\lambda}$, it follows from Eq. (11) that $(\boldsymbol{\lambda})_i = 0$ for inactive constraints (constraints with $g_i(\mathbf{x}) < 0$).

Penalty Function Methods

These methods convert the constrained optimization problem in Eq. (54.1) into an equivalent unconstrained optimization problem, since such problems are easier to solve than constrained problems, as

$$\textit{minimize} \quad h(\mathbf{x}) = f(\mathbf{x}) + c \cdot P(\mathbf{x}) \tag{54.12}$$

where $P(\mathbf{x})\colon \mathbf{R}^n \to \mathbf{R}$ is known as a penalty function and c is a constant. The value of $P(\mathbf{x})$ is zero within the feasible region, and positive outside the region, with the value becoming larger as one moves farther from the feasible region; one possible choice when the $g_i(\mathbf{x})$'s are continuous is given by

$$P(\mathbf{x}) = \sum_{i=1}^{m} \max(0, -g_i(\mathbf{x})) \tag{54.13}$$

For large c, it is clear that the minimum point of Eq. (12) will be in a region where P is small. Thus, as c is increased, it is expected that the corresponding solution point will approach the feasible region and minimize f. As $c \rightarrow \infty$, the solution of the penalty function method converges to a solution of the constrained optimization problem.

In practice, if one were to begin with a high value of c, one may not have very good convergence properties. The value of c is increased in each iteration until c is high and the solution converges.

Method of Feasible Directions

The method of feasible directions is an optimization algorithm that improves the objective function without violating the constraints. Given a point $\mathbf{x}$, a direction $\mathbf{d}$ is feasible if there is a step size $\bar{\alpha} > 0$ such that $\mathbf{x} + \alpha \cdot \mathbf{d} \in \mathcal{F} \ \forall\ 0 \leq \alpha \leq \bar{\alpha}$, where $\mathcal{F}$ is the feasible region. More informally, this means that one can take a step of size up to $\bar{\alpha}$ along the direction $\mathbf{d}$ without leaving the feasible region. The method of feasible direction attempts to choose a value of α in a feasible direction $\mathbf{d}$ such that the objective function f is minimized along the direction, and α is such that $\mathbf{x} + \alpha \cdot \mathbf{d}$ is feasible.

One common technique that uses the method of feasible directions is as follows. A feasible direction at $\mathbf{x}$ is found by solving the following linear program:

$$\begin{aligned} &\textit{minimize} && \epsilon \\ &\textit{subject to} && \langle \nabla f(\mathbf{x}) \cdot \mathbf{d} \rangle \leq \epsilon && (54.14) \\ & && \langle \nabla g_i(\mathbf{x}) \cdot \mathbf{d} \rangle \leq \epsilon && (54.15) \\ &\textit{and} && \textit{normalization requirements on } \mathbf{d} \end{aligned}$$

where the second set of constraints is chosen for all $g_i \geq -b$, where b serves to incorporate the effects of near-active constraints to avoid the phenomenon of jamming (also known as zigzagging) [1]. The value of b is brought closer to 0 as the optimization progresses. One common method that is used as a normalization requirement is to set $\mathbf{d}^T\mathbf{d} = 1$. This constraint is nonlinear and nonconvex, and is not added to the linear program as an additional constraint; rather, it is exercised by normalizing the direction $\mathbf{d}$ after the linear program has been solved. An appropriate step size in this direction is then chosen by solving a one-dimensional optimization problem.

Feasible direction methods are popular in finding engineering solutions because the value of $\mathbf{x}$ at each interation is feasible, and the algorithm can be stopped at any time without waiting for the algorithm to converge, and the best solution found so far can be used.

Vaidya's Convex Programming Algorithm

As mentioned earlier, if the objective function and all constraints in Eq. (54.1) are convex, the problem is a convex programming problem, and any local minimum is a global minimum. The first large-scale practical implementation of this algorithm is described in [3].

Initially, a polytope $\mathcal{P} \in \mathbf{R}^n$ that contains the optimal solution, $\mathbf{x}_{opt}$, is chosen. The objective of the algorithm is to start with a large polytope, and in each iteration, to shrink its volume while keeping the optimal solution, $\mathbf{x}_{opt}$, within the polytope, until the polytope becomes sufficiently small. The polytope $\mathcal{P}$ may, for example, be initially selected to be an n-dimensional box described by the set

$$\{\mathbf{x} \mid x_{min} \leq x_i \leq x_{max}\} \tag{54.16}$$

where x_{min} and x_{max} are the minimum and maximum values of each variable, respectively. The algorithm proceeds iteratively as follows.

STEP 1. A **center** $\mathbf{x}_c$ deep in the interior of the polytope $\mathscr{P}$ is found.

STEP 2. An **oracle** is invoked to determine whether the center $\mathbf{x}_c$ lies within the feasible region $\mathscr{F}$. This may be done by verifying that all of the constraints of the optimization problem are met at $\mathbf{x}_c$.

If the point $\mathbf{x}_c$ lies outside $\mathscr{F}$, it is possible to find a *separating hyperplane* passing through $\mathbf{x}_c$ that divides $\mathscr{P}$ into two parts, such that $\mathscr{F}$ lies entirely in the part satisfying the constraint

$$\mathbf{c}^T\mathbf{x} \geq \beta \tag{54.17}$$

where $\mathbf{c} = -[\nabla g_p(\mathbf{x})]^T$ is the negative of the gradient of a violated constraint, g_p, and $\beta = \mathbf{c}^T\mathbf{x}_c$. The separating hyperplane above corresponds to the tangent plane to the violated constraint.

If the point $\mathbf{x}_c$ lies within the feasible region $\mathscr{F}$, then there exists a hyperplane (17) that divides the polytope into two parts such that $\mathbf{x}_{opt}$ is contained in one of them, with

$$\mathbf{c} = -[\nabla f(\mathbf{x})]^T \tag{54.18}$$

being the negative of the gradient of the objective function, and β being defined as $\beta = \mathbf{c}^T\mathbf{x}_c$ once again.

STEP 3. In either case, the constraint (17) is added to the current polytope to give a new polytope that has roughly half the original volume.

STEP 4. Go to Step 1; the process is repeated until the polytope is sufficiently small.

Example. The algorithm is illustrated by using it to solve the following problem

$$\textit{minimize} \quad f(x_1, x_2) \quad \textit{such that} \quad (x_1, x_2) \in S \tag{54.19}$$

where S is a convex set and f is a convex function. The shaded region in Fig. 54.4(a) is the set S, and the dotted lines show the level curves of f. The point $\mathbf{x}_{opt}$ is the solution to this problem.

1. The expected solution region is bounded by a rectangle, as shown in Fig. 54.4(a).
2. The center, $\mathbf{x}_c$, of this rectangle is found.
3. The oracle is invoked to determine whether $\mathbf{x}_c$ lies within the feasible region. In this case, it can be seen that $\mathbf{x}_c$ lies outside the feasible region. Hence, the gradient of the constraint function is used to construct a hyperplane through $\mathbf{x}_c$, such that the polytope is divided into two parts of roughly equal volume, one of which contains the solution $\mathbf{x}_{opt}$. This is illustrated in Fig. 54.4(b), where the shaded region corresponds to the updated polytope.
4. The process is repeated on the new smaller polytope. Its center lies inside the feasible region; hence, the gradient of the objective function is used to generate a hyperplane that further shrinks the size of the polytope, as shown in Fig. 54.4(c).
5. The result of another iteration is illustrated in Fig. 54.4(d). The process continues until the polytope has been shrunk sufficiently.

Other Methods

The compact nature of this handbook makes it unfeasible to enumerate or describe all of the methods that are used for nonlinear optimization. Several other methods, such as Newton's and modified Newton/quasi-Newton methods, conjugate gradient methods etc., are often useful in engineering optimization. For these and more, the reader is referred to a standard text on optimization, such as [1].

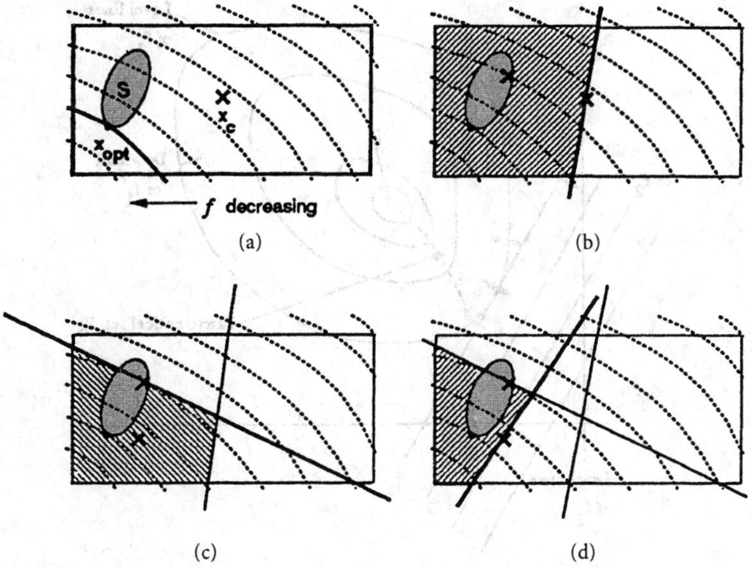

FIGURE 54.4 Example illustrating the convex programming algorithm.

Multicriterion Optimization and Pareto Criticality

Most integrated circuit design problems involve tradeoffs between multiple objectives. In cases where one objective can be singled out as the most important one, and a reasonable constraint set can be defined in terms of the other objectives, the optimization problem can be stated using the formulation in Eq. (54.1). This is convenient since techniques for the solution of a problem in this form have been extensively studied, and a wide variety of optimization algorithms is available.

Let $\mathbf{f}$ be a vector of design objectives that is a function of the design variables $\mathbf{x}$, where

$$\mathbf{f}(\mathbf{x}): \mathbf{R}^n \rightarrow \mathbf{R}^m = (f_1(\mathbf{x}), f_2(\mathbf{x}), \ldots, f_m(\mathbf{x})) \tag{54.20}$$

It is extremely unlikely in a real application that all of the f_i's will be optimal at the same point, and hence one must trade off the values of the f_i's in search for the best design point.

In this context, we note that at a point $\mathbf{x}$, we are interested in taking a step δ in a direction $\mathbf{d}$, $\|\mathbf{d}\| = 1$, so that

$$f_i(\mathbf{x} + \delta \cdot \mathbf{d}) \leq f_i(\mathbf{x}) \ \forall\ 1 \leq i \leq m \tag{54.21}$$

A **Pareto critical point** is defined as a point $\mathbf{x}$ where no such small step of size less than δ exists in any direction. If a point is Pareto critical for *any* step size from the point $\mathbf{x}$, then, $\mathbf{x}$ is a **Pareto point**. The notion of a Pareto critical point is, therefore, similar to that of a local minimum, and that of a Pareto point to a global minimum. In computational optimization, one is concerned with the problem of finding a local minimum since, except in special cases, it is the best that one can be guaranteed of finding without an exhaustive search. If the set of all Pareto critical points is P_c, and the set of Pareto points is P, then clearly $P \subset P_c$. In general, there could be an infinite number of Pareto points, but the best circuit design must necessarily occur at a Pareto point $\mathbf{x} \in P$.

In Fig. 54.5, the level curves of two objective functions are plotted in $\mathbf{R}^2$. f_1 is nonlinear and has a minimum at P^*. f_2 is linear and decreases as both x_1 and x_2 decrease. The Pareto critical

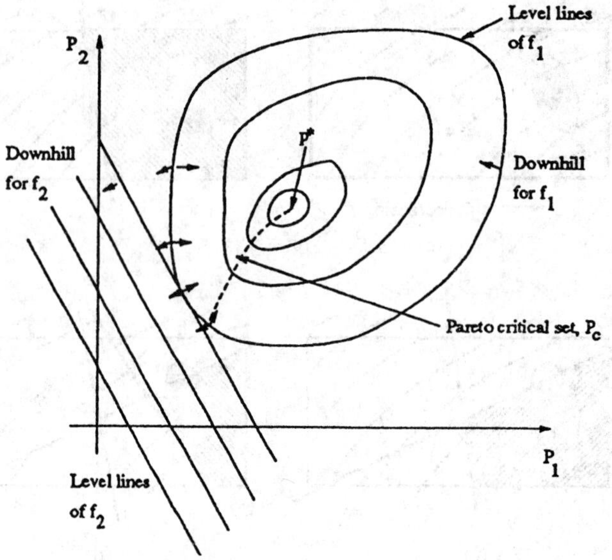

FIGURE 54.5 Exact conflict at a Pareto critical point.

set, P_c, is given by the dashed curve. At a few of the points, the unit normal to the level lines of f_1 and f_2, i.e., the negative gradients of f_1 and f_2 is shown. From the figure, it can be seen that if the unit normals at the point $\mathbf{x}$ are not equal and opposite, then the unit normals will have a common downhill direction allowing a simultaneous decrease in f_1 and f_2, and hence, $\mathbf{x}$ would not be a Pareto critical point. Therefore, a Pareto critical point is one where the gradients of f_1 and f_2 are opposite in direction, i.e., $\lambda \nabla f_1 = -\nabla f_2$, where λ is some scale factor.

In higher dimensions, a Pareto critical point is characterized by the existence of a set of weights, $w_i > 0 \ \forall \ 1 \le i \le m$, such that

$$\sum_{i=1}^{m} w_i \nabla f_i = \mathbf{0} \tag{54.22}$$

Some of the common methods that are used for multicriterion optimization are

Weighted sums optimization: The multiple objectives, $f_1(\mathbf{x}), \ldots, f_m(\mathbf{x})$ are combined as:

$$F(\mathbf{x}) = \sum_{i=1}^{m} w_i f_i(\mathbf{x}) \tag{54.23}$$

where $w_i > 0 \ \forall \ i = 1 \cdots m$, and the function $F(\mathbf{x})$ is minimized.

At any local minimum point of $F(\mathbf{x})$, the relation in Eq. (22) is seen to be valid, and hence, $\mathbf{x} \in P_c$. In general, $P \neq P_c$, but it can be shown that when each f_i is a convex function, then $P = P_c$; if so, it can also be shown that all Pareto points can be obtained by optimizing the function F in Eq. (23). However, for nonconvex functions, there are points $\mathbf{x} \in P$ that cannot be obtained by the weighted sum optimization since Eq. (22) is only a necessary condition for the minimum of F. A characterization of the Pareto points that can be obtained by this technique is provided in [2].

In practice, the w_i's must be chosen to reflect the magnitudes of the f_i's. For example, if one of the objectives is a voltage quantity whose typical value is a few volts, and another is a capacitor value that is typically a few picofarads, the weight corresponding to the capacitor value would be roughly 10^{12} times that for the voltage, in order to ensure that each objective has a reasonable contribution to the objective function value. The designer may further weigh the relative importance of each objective in choosing the w_i's. This objective may be combined with additional constraints to give a formulation of the type in Eq. (54.1).

Minmax optimization: The following objective function is used for Eq. (54.1)

$$\textit{minimize} \quad F(\mathbf{x}) = \max_{1 \le i \le m} w_i f_i(\mathbf{x}) \tag{54.24}$$

where the weights $w_i > 0$ are chosen as in the case of weighted sums optimization.

The above can equivalently be written as the following constrained optimization problem:

$$\begin{aligned} &\textit{minimize} \quad r \\ &\textit{subject to} \quad w_i f_i(\mathbf{x}) \le r \end{aligned} \tag{54.25}$$

Minimizing the objective function described by Eq. (24) with different sets of w_i values can be used to obtain all Pareto points [2].

Since this method can, unlike the weighted-sums optimization method, be used to find *all* Pareto critical points, it would seem to be a more natural setting for obtaining Pareto points than the weighted sum minimization. However, when the f_i's are convex, the weighted sums approach is preferred since it is an unconstrained minimization, and is computationally easier than a constrained optimization. It must be noted that when the f_i's are not all convex, the minmax objective function is nonconvex, and finding all local minima is a nontrivial process for any method.

54.3 The Transistor Sizing Problem for CMOS Digital Circuits

Problem Description

Circuit delays in integrated circuits often have to be reduced to obtain faster response times. A typical CMOS digital integrated circuit consists of multiple stages of combinational logic blocks that lie between latches that are clocked by system clock signals. For such a circuit, delay reduction must ensure that valid signals are produced at each output latch of a combinational block, before any transition in the signal clocking the latch. In other words, the worst-case input-output delay of each combinational stage must be restricted to be below a certain specification.

Given the circuit topology, the delay of a combinational circuit can be controlled by varying the sizes of transistors in the circuit. Here, the size of a transistor is measured in terms of its channel width, since the channel lengths of MOS transistors in a digital circuit are generally uniform. In coarse terms, the circuit delay can usually be reduced by increasing the sizes of certain transistors in the circuit. Hence, making the circuit faster usually entails the penalty of increased circuit area. The area-delay trade-off involved here is, in essence, the problem of transistor size optimization.

Three formal statements of the problem are stated below:

1. Minimize *Area* subject to *Delay* $\le T_{spec}$ (54.26)
2. Minimize *Delay* subject to *Area* $\le A_{spec}$ (54.27)
3. Minimize *Area* $\cdot$ $[\textit{Delay}]^c$ (54.28)

where c is a constant. Of all of these, the first form is perhaps the most useful practical form, since a designer's objective is typically to meet a timing constraint dictated by a system clock.

Delay Modeling

We examine the delay modeling procedure used in TILOS at the transistor, gate, and circuit levels. Most existing transistor-sizing algorithms use minor variations on this theme.

Transistor Level Model

An MOS transistor is modeled as a voltage-controlled switch with an on-resistance, R_{on}, between drain and source, and three grounded capacitances, C_d, C_s, and C_g, at the drain, source, and gate terminals, respectively, as shown in Fig. 54.6. The behaviors of the resistance and capacitances associated with a MOS transistor of channel width x are modeled as

$$R_{on} \propto 1/x \tag{54.29}$$

$$C_d, C_s, C_g \propto x \tag{54.30}$$

Gate Level Model

At the gate level, delays are calculated in the following manner. For each transistor in a pull-up or pull-down network of a complex CMOS gate, the largest resistive path from the transistor to the gate output is computed, as well an the largest resistive path from the transistor to a supply rail. Thus, for each transistor, the network is transformed into an RC line, and its Elmore time constant [4] is computed and is taken to be the gate delay.

While finding the Elmore delay, the capacitances that lie between the switching transistor and the supply rail are assumed to be at the voltage level of the supply rail at the time of the switching transition and do not contribute to the Elmore delay. For the example in Fig. 54.7, the capacitance at node n_1 is ignored while computing the Elmore delay, the expression for which is

$$(R_1 + R_2)C_2 + (R_1 + R_2 + R_3)C_3 \tag{54.31}$$

Each R_i is inversely proportional to the corresponding transistor size, x_i, and each C_i is some constant (for wire capacitance) plus a term proportional to the width of each transistor whose gate, drain, or source is connected to node i. Thus, Eq. (31) can be rewritten as

$$(A/x_1 + A/x_2)(Bx_2 + Cx_3 + D) + (A/x_1 + A/x_2 + A/x_3)(Bx_3 + E)$$

which is a posynomial function (see Section 54.2) of x_1, x_2, and x_3.

Circuit Level Model

At the circuit level, the PERT technique, which will be described shortly, is used to find the circuit delay. The procedure is best illustrated by means of a simple example. Consider the circuit in Fig. 54.8, where each box represents a gate and the number within the box represents its delay. We assume that the worst-case arrival time for a transition at any primary input, i.e., at the inputs to boxes A, B, C, and D, is 0.

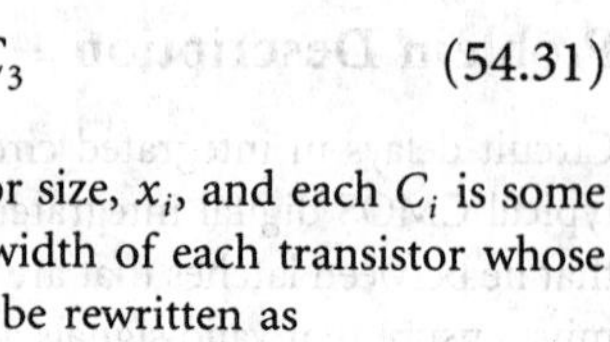

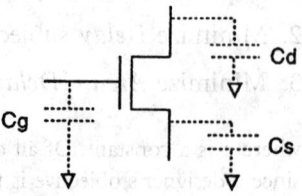

FIGURE 54.6 An RC transistor model.

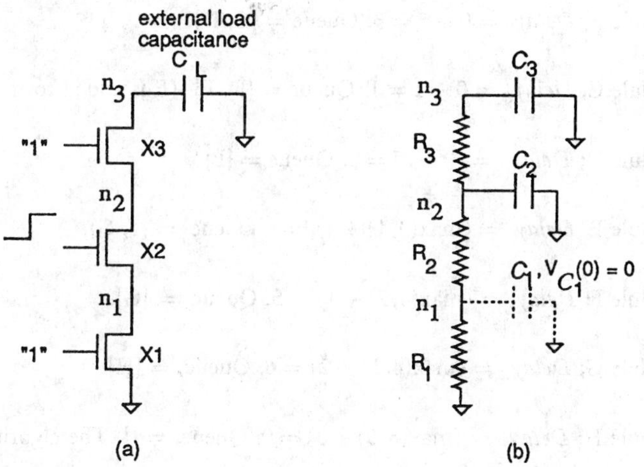

FIGURE 54.7 (a) A sample pull-down network. (b) Its RC representation.

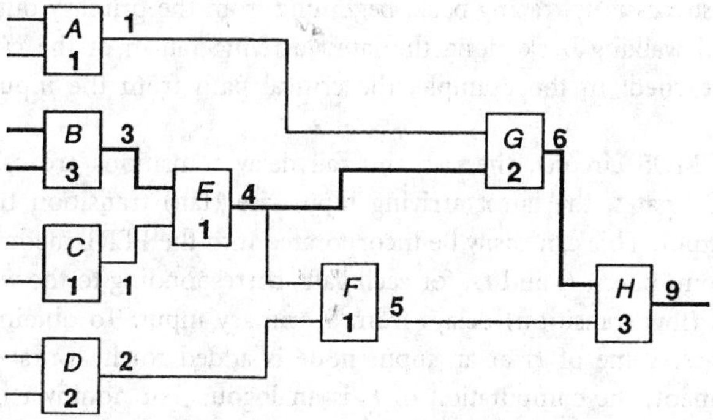

FIGURE 54.8 The PERT technique.

A component is said to be **ready for processing** when the signal arrival time information is available for all of its inputs. Initially, since signal arrival times are known only at the primary inputs, only those components that are fed soley by primary inputs are ready for processing. These are placed in a queue and are scheduled for processing. In the iterative process, the component at the head of the queue is scheduled for processing. Each processing step consists of

- Finding the latest arriving input to the component, which triggers the output transition. This involves finding the maximum of all worst-case arrival times of inputs to the component.
- Adding the delay of the component to the latest arriving input time to obtain the worst-case transition time at the output.
- Checking all of the components that the current component fans out to, to find out whether they are ready for processing. If so, the component is added to the tail of the queue.

The iterations end when the queue is empty. In the example, the algorithm is executed as follows:

STEP 1. Initially, Queue = {A, B, C, D}.

STEP 2. Schedule A; $Delay_A = 0 + 1 = 1$. Queue = {B, C, D}.

STEP 3. Schedule B; $Delay_B = 0 + 3 = 3$. Queue = {C, D}.

STEP 4. Schedule C; $Delay_C = 0 + 1 = 1$. Queue = {D, E}. (E is added to the queue).

STEP 5. Schedule D; $Delay_D = (0 + 2) = 2$. Queue = {E}.

STEP 6. Schedule E; $Delay_E = (\max(3, 1) + 1) = 4$. Queue = {F, G}.

STEP 7. Schedule F; $Delay_F = (\max(4, 2) + 1) = 5$. Queue = {G}.

STEP 8. Schedule G; $Delay_G = (\max(4, 1) + 2) = 6$. Queue = {H}.

STEP 9. Schedule H; $Delay_H = (\max(6, 5) + 3) = 9$. Queue = {}. The algorithm terminates.

The worst-case delays at the output of each component are shown in Fig. 54.8. The **critical path**, defined as the path between an input and an output with the maximum delay, can now easily be found by successively tracing back, beginning from the primary output with the latest transition time, and walking back along the latest arriving fan-in of the current gate, until a primary input is reached. In the example, the critical path from the input to the output is B-E-G-H.

In the case of CMOS circuits, the rise and fall delay transitions are calculated separately. For inverting CMOS gates, the latest arriving input rise (fall) transition triggers a fall (rise) transition at the output. This can easily be incorporated into the PERT method described above, by maintaining two numbers, t_r and t_f, for each gate, corresponding to the worst-case rise (high transition) and fall (low transition) delays from a primary input. To obtain the value of t_f at an output, the largest value of t_r at an input node is added to the worst-case fall transition time of the component; the computation of t_r is analogous. For noninverting gates, t_r and t_f are obtained by adding the rise (fall) transition time to the worst-case input rise (fall) transition time.

Since each gate delay is a posynomial, the circuit delay found by the PERT technique is a sum of gate delays, the circuit delay is also a posynomial function of the transistor sizes. In general, the path delay can be written as

$$\sum_{i,j=1}^{n} a_{ij}\frac{x_i}{x_j} + \sum_{i=1}^{n}\frac{b_i}{x_i} + K \qquad (54.32)$$

The Area Model

The exact area of a circuit cannot easily be represented as a function of transistor sizes. This is unfortunate, since a closed functional form facilitates the application of optimization techniques. As an approximation, the following formula is used by many transistor-sizing algorithms, to estimate the active circuit area.

$$Area = \sum_{i=1}^{n} x_i \qquad (54.33)$$

where x_i is the size of the ith transistor and n is the number of transistors in the circuit. In other words, the area is approximated as the sum of the sizes of transistors in the circuit which, from the definition Eq. (54.8), is clearly a posynomial function of the x_i's.

The Sensitivity-Based TILOS Algorithm

Steps in the Algorithm

The algorithm that was implemented in TILOS (TImed LOgic Synthesizer) was the first to recognize the fact that the area and delay can be represented as posynomial functions of the transistor sizes. The algorithm proceeds as follows:

An initial solution is assumed where all transistors are at the minimum allowable size. In each iteration, a static timing analysis is performed on the circuit, as explained earlier, to determine the critical path for the circuit.

Let N be the primary output node on the critical path. The algorithm then walks backward along the critical path, starting from N. Whenever an output node of a gate, $Gate_i$, is visited, TILOS examines the largest resistive path between V_{DD} (ground) and the output node if $Gate_i$'s t_r (t_f) causes the timing failure at N. This includes

- The **critical transistor**, i.e., the transistor whose gate terminal is on the critical path. In Fig. 54.7, X2 is the critical transistor.
- The **supporting transistors**, i.e., transistors along the largest resistive path from the critical transistor to the power supply (V_{DD} or ground). In Fig. 54.7, X1 is a supporting transistor.
- The **blocking transistors**, i.e., transistors along the highest resistance path from the critical transistor to the logic gate output. In Fig. 54.7, X3 is a blocking transistor.

TILOS finds the sensitivity, which is the reduction in circuit delay per increment of transistor size, for each critical, blocking, and supporting transistor. (The procedure of sensitivity computation is treated in greater detail shortly.) The size of the transistor with the greatest sensitivity is increased by multiplying it by a constant, BUMPSIZE, a user-settable parameter that defaults to 1.5. The above process is repeated until all constraints are met, implying that a solution is found, or the minimum delay state has been passed, and any increase in transistor sizes would make it slower instead of faster, in which case TILOS cannot find a solution.

Note that since in each iteration, exactly one transistor size is changed, the timing analysis method can employ incremental simulation techniques to update delay information from the previous iteration. This substantially reduces the amount of time spent in critical path detection.

Sensitivity Computation

Figure 54.9 illustrates a configuration in which the critical path extends back along the gate of the upper transistor, which is the critical transistor. The sensitivity for this transistor is calculated as follows: set all transistor sizes, except x, to the size of the critical transistor. R is the total resistance of an RC chain driving the gate and C is the total capacitance of an RC chain being driven by the configuration. The total delay, $D(x)$, of the critical path is

$$D(x) = K + RC_u x + \frac{R_u C}{x} \tag{54.34}$$

where R_u and C_u are the resistance and capacitance of a unit-sized transistor, K is a constant that depends on the resistance of the bottom transistor, the capacitance in the driving RC chain and the resistance in the driven RC chain. The sensitivity, $D'(x)$, is then

$$D'(x) = RC_u - \frac{R_u C}{x^2} \tag{54.35}$$

The sensitivities of supporting transistors and blocking transistors can similarly be calculated.

Note that increasing the size of a transistor with negative sensitivity only means that the delay along the current critical path can be reduced by changing the size of this transistor, and does

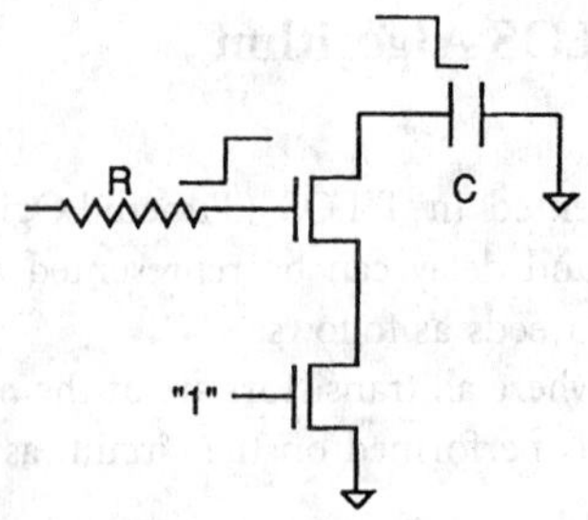

FIGURE 54.9 Sensitivity calculations in TILOS.

not necessarily mean that the circuit delay can be reduced; the circuit delay is the maximum of all path delays in the circuit, and a change in the size of this transistor could increase the delay along some other path, making a new path critical. This is the rationale behind increasing the size of the most sensitive transistor by only a small factor.

From an optimization viewpoint, the procedure of bumping up the size of the most sensitive transistor could be looked upon in the following way. Let the ith transistor (out of n transistors) be the one with the maximum sensitivity. Define $\mathbf{e}_i \in \mathbf{R}^n$ as $(\mathbf{e}_i)_j = 0$ is $i \neq j$ and $(\mathbf{e}_i)_j = 1$ if $i = j$. In each iteration, the TILOS optimization procedure works in the n-dimensional space of the transistor sizes, chooses $\mathbf{e}_i$ as the search direction, and attempts to find the solution to the problem by taking a small step along that direction.

Transistor Sizing Using the Method of Feasible Directions

Shyu *et al.* [12] proposed a two-stage optimization approach to solve the transistor sizing problem. The delay estimation algorithm is identical to that used in TILOS.

The algorithm can be summarized in the following pseudo-code:

```
Use TILOS to size the entire circuit;
While (TRUE) {
   Select G_1,...,G_k, the k most critical paths,
   and X = {x_i}, the set of design parameters
   Solve the optimization problem
```

$$\begin{aligned} &\textit{minimize} && \textstyle\sum_{x_i \in X} x_i \\ &\textit{such that} && G_i(X) \le T \;\forall\, i = 1, \ldots, k \\ &\textit{and} && x_i \ge \textit{minsize} \;\forall\, x_i \in X \end{aligned}$$

```
If all constraints are satisfied, exit    }
```

In the first stage, the TILOS heuristic is used to generate an initial solution. The heuristic finds a solution that satisfies the constraints, and only the sized-up transistors are used as design parameters. Although TILOS is not guaranteed to find an optimal solution, it can serve as an initial guess solution for an iterative technique.

In the second stage of the optimization process, the problem is converted into a mathematical optimization problem, and is solved by a Method of Feasible Directions (MFD) algorithm described in Section 54.2 using the feasible solution generated in the first stage as an initial guess. To reduce the computation, a sequence of problems with a smaller number of design parameters is solved.

At first, the transistors on the worst-delay paths (usually more than one) are selected as design parameters. If, with the selected transistors, the optimizer fails to meet the delay constraints, and some new paths become the worst-delay paths, the algorithm augments the design parameters with the transistors on those paths, and restarts the process. However, while this procedure reduces the run time of the algorithm, one faces the risk of finding a suboptimal solution since only a *subset* is the design parameters is used in each step.

The MFD optimization method proceeds by finding a search direction **d**, a vector in the n-dimensional space of the design parameters, based on the gradients of the cost function and some of the constraint functions. Once the search direction has been computed, a step along this direction is taken, so that the decrease in the cost and constraint functions is large enough. The computation stops when the length of this step is sufficiently small.

This algorithm has the feature that once the feasible region (the set of transistor sizes where all delay constraints are satisfied) is entered, all subsequent improvements will remain feasible, and the algorithm can be terminated at any time with a feasible solution.

Practical Implementation Aspects

The Generalized Gradient For convergence, the MFD requires that the objective and constraint functions be continuously differentiable. However, since the circuit delay is defined as the maximum of all path delays, the delay constraint functions are usually not differentiable. To illustrate that the maximum of two continuously differentiable functions, $g_1(x)$ and $g_2(x)$, need not be differentiable, consider the example in Fig. 54.10. The maximum function, shown by the bold lines, is nondifferentiable at x_0.

To cope with the nondifferentiability of the constraint functions, a modification of the MFD is used, which employs the concept of the *generalized gradient*. The idea is to use a convex combination of the gradients of the active, or nearly active constraints near a discontinuity.

Scaling It is important to scale the gradients of the cost and path delay functions since the use of their unscaled values may produce poor descent search directions due to the large difference in magnitude of the gradients of the constraint and objective functions. When a gradient has a magnitude that is much smaller than other gradients, it dominates the search direction. In such a case, the descent direction is unjustly biased away from the other constraints and/or the cost function.

Lagrangian Multiplier Approaches

As can be seen from the approaches studied so far, the problem of transistor sizing can be formulated as a constrained nonlinear programming problem. Hence, the method of Lagrangian multipliers, described in Section 54.2 is applicable. Early approaches that used Lagrangian multipliers rely on the user to provide critical path information, which may be impractical since

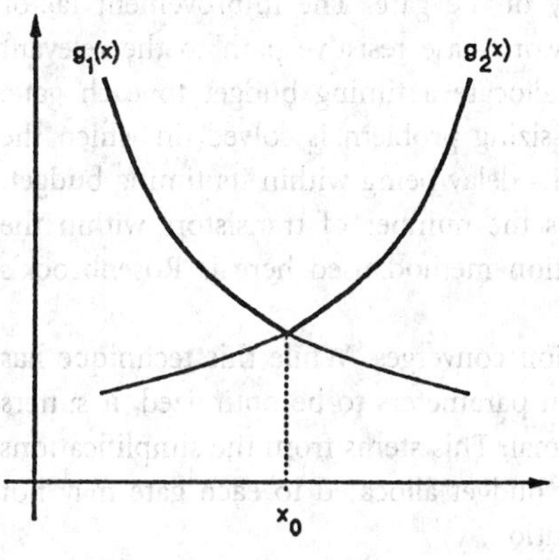

FIGURE 54.10 Nondifferentiability of the *max* function.

critical paths are liable to change as sizing progresses. An alternative solution to transistor size optimization using Lagrangian multipliers was presented by Marple [13]. This technique uses a different area model and employs the idea of introducing intermediate variables to reduce the number of delay constraints from an exponential number to a number that is linear in the circuit size.

This technique begins with a prespecified layout, and performs the optimization using an area model for that layout. While such an approach has the disadvantage that it may not result in the minimal area over *all* layouts, it still maintains the feature that the area and delay constraints are posynomials. Apart from the delay constraints, there also exist some area constraints, modeled by constraint graphs that are commonly used in layout compaction. These constraints maintain the minimum spacing between objects in the final layout, as specified by design rules.

The delay of the circuit is modeled by a delay graph, $D(V, E)$ where V is the set of nodes (gates) in D, and E is the set of arcs (connections among gates) in D. This is the same graph on which the PERT analysis is to be carried out. Let m_i represent the worst-case delay at the output of gate i, from the primary inputs. Then for each gate, the delay constraint is expressed as

$$m_i + d_j \leq m_j \tag{54.36}$$

where gate $i \in$ fan-in(gate j), and d_j is the delay of gate j. Thus, the number of delay constraints is reduced from a number that could, in the worst case, be exponential in $|V|$, to one that is linear in $|E|$, using $|V|$ additional variables.

Two-Step Optimization

Since the number of variables in the transistor-sizing problem, which equals the number of transistors in a combinational segment, is typically too large for most optimization algorithms to handle efficiently, many algorithms choose a simpler route by performing the optimization in two steps. Examples of algorithms that use this idea to solve the transistor-sizing problem are iCOACH, MOSIZ, and CATS.

In the first step in MOSIZ, each gate is mapped onto an equivalent macromodeling primitive, such as an inverter. The transistor-sizing problem on this simplified circuit is then solved. Note that the number of variables is substantially reduced when each gate is replaced by a simple primitive, with fewer transistors. The delay of each equivalent inverter, with the transistor sizes obtained above, is taken as the **timing budget** for the gate represented by that inverter, and the gate is optimized under the timing budget.

iCOACH uses macromodels for timing analysis of the circuit, and has the capability of handling dynamic circuits. The optimizer employs a heuristic to estimate an **improvement factor** for each gate, which is related to the sensitivity of the gate. The improvement factor depends on the fan-in count, fan-out count, and the worst-case resistive path to the relevant supply rail. The improvement factor is then used to allocate a timing budget to each gate. In the second step, for each gate, a smaller transistor-sizing problem is solved, in which the area-delay product of the gate is minimized, subject to its delay being within its timing budget. The number of variables for each such problem equals the number of transistors within the gate, which is typically a small number. The optimization method used here is Rosenbrock's rotating coordinate scheme [5].

The two steps are repeated iteratively until the solution converges. While this technique has the obvious advantage of reducing the number of design parameters to be optimized, it suffers from the disadvantage that the solution may be nonoptimal. This stems from the simplifications introduced by the timing budget allocation; the timing budget allocated to each gate may not be the same as the delay of the gate for the optimal solution.

The Convex Programming-Based Approach

The chief shortcoming of most of the approaches above is that the simplifying assumptions that are made to make the optimization problem more tractable may lead to a suboptimal solution. The algorithm in iCONTRAST solves the underlying optimization problem exactly.

The objective of the algorithm is to solve the transistor-sizing problem in Eq. (54.28), where both the area and the delay are posynomial functions of the vector **x** of transistor sizes. The procedure described below may easily be extended to solve the formulations in Eqs. (54.29) and (54.30) as well; however, these formulations are not as useful to the designer. The variable transformation, $(x_i) = (e^{z_i})$ maps the problem in (54.28) to

$$\begin{aligned} &\textit{minimize} \quad Area(\mathbf{z}) = \sum_{i=1}^{n} e^{z_i} \\ &\textit{subject to} \quad D(\mathbf{z}) \le T_{spec} \end{aligned} \tag{54.37}$$

The delay of a circuit is defined to be the maximum of the delays of all paths in the circuit. Hence, it can be formulated as the maximum of posynomial functions of **x**. This is mapped by the above transformation onto a function $D(\mathbf{z})$ that is a maximum of convex functions; a maximum of convex functions is also a convex function. The area function is also a posynomial in **x**, and is transformed into a convex function by the same mapping. Therefore, the optimization problem defined in (54.28) is mapped to a *convex programming* problem, i.e., a problem of minimizing a convex function over a convex constraint set. Due to the unimodel property of convex functions over convex sets, any local minimum of (54.28) is also a global minimum.

Vaidya's convex programming method described in Section 54.2 is then used to find the unique global minimum of the optimization problem.

Gradient Calculations

In an iteration of Vaidya's convex programming algorithm, when the center $\mathbf{z}_c$ of a polytope lies within the feasible region S, the gradient of the area function is required to generate the new hyperplane passing through the center. The gradient of the area function (Eq. (54.37)) is given by

$$\nabla Area(\mathbf{z}) = [e^{z_1}, e^{z_2}, \ldots, e^{z_n}] \tag{54.38}$$

In the case when the center $\mathbf{z}_c$ lies outside the feasible region S, the gradient of the critical path delay function $D_{critpath}(\mathbf{z}_c)$ is required to generate the new hyperplane that is to be added. Note that transistors in the circuit can contribute to the kth component of the gradient of the delay function in either of two ways:

1. If the kth transistor is a critical, supporting or blocking transistor (as defined in Section 54.3)
2. If the kth transistor is a capacitive load for some critical transistor.

Transistors that satisfy neither of these requirements make no contribution to the gradient of the delay function.

Concluding Remarks

The list of algorithms presented above is among the prominent ones used for transistor sizing. For more information, the reader is referred to [6] and sources such as the International

Conference on Computer-Aided Design, the Design Automation Conference, and the *IEEE Transactions on Computer-Aided Design.*

The TILOS algorithm is found to be fast and gives near-optimal solutions under loose timing constraints. As the timing constraints become tight, the heuristic becomes less optimal, and more rigorous methods must be used. The MFD algorithm, which was compared directly with TILOS, was found to give small (up to 5%) improvements over TILOS. The convex programming-based algorithm, iCONTRAST, solves the underlying convex programming problem exactly, and finds the optimal solution. However, the run times of iCONTRAST are not as small as those of TILOS. No comparative data for other algorithms described herein have been presented in the literature.

The reader must be cautioned here that the actual optimization problem in transistor sizing is not exactly a posynomial programming problem. The use of Elmore delay models (which are accurate within about 20%) to approximate the circuit delay, and the use of approximate area models allows the problem to be formulated as a convex program, and hence although one may solve this optimization problem exactly, one still has to live with the inaccuracies of the modeling functions. In practice, in most cases, this is not a serious problem.

54.4 The Design Centering Problem

Problem Description

While manufacturing a circuit, it is inevitable that process variations will cause design parameters, such as component values, to waver from their nominal values. As a result, the manufactured circuit may no longer meet some behavioral specifications, such as requirements on the delay, gain and bandwidth, that is has been designed to satisfy. The procedure of design centering attempts to select the nominal values of design parameters so as to ensure that the behavior of the circuit remains within specifications, with the greatest probability. In other words, the aim of design centering is to ensure that the manufacturing yield is maximized.

The values of n design parameters may be ordered as an n-tuple that represents a point in $\mathbf{R}^n$. A point is **feasible** if the corresponding values for the design parameters satisfy the behavioral specifications on the circuit. The feasible region (or the region of acceptability), $R_f \subset \mathbf{R}^n$, is defined as the set of all design points for which the circuit satisfies all behavioral specifications.

The random variations in the values of the design parameters are modeled by a probability density function, $\Phi(\mathbf{z}): \mathbf{R}^n \rightarrow [0, 1]$, with a mean corresponding to the nominal value of the design parameters. The **yield** of the circuit, $\mathbf{Y}$, as a function of the mean, $\mathbf{x}$, is given by

$$\mathbf{Y}(\mathbf{x}) = \int_{R_f} \Phi_{\mathbf{x}}(\mathbf{z})\, d\mathbf{z} \tag{54.39}$$

The **design center** is the point $\mathbf{x}$ at which the yield, $\mathbf{Y}(\mathbf{x})$, is maximized. There have traditionally been two approaches to solving this problem: one based on geometrical methods, and another based on statistical sampling. In addition, several methods that hybridize these approaches also exist. We will now provide an exposition of geometrical approaches to the optimization problem of design centering, and leave the discussion of other techniques to Chapter 55.

A common assumption made by geometrical design centering algorithms is that R_f is a convex bounded body. Geometrical algorithms recognize that the evaluation of the integral (54.39) is computationally difficult, and generally proceed as follows: the feasible region in the space of design parameters, i.e., the region where the behavioral specifications are satisfied, is approximated by a known geometrical body, such is a polytope or an ellipsoid. The center of this body is then approximated, and is taken to be the design center.

The Simplicial Approximation Method

Outline of the Method

The simplicial approximation method is a method for approximating a feasible region by a polytope and finding its center. This method proceed in the following steps:

1. Determine a set of $m \geq n + 1$ points on the boundary of R_f.
2. Find the convex hull (see Section 54.2) of these points and use this polyhedron as the initial approximation to R_f. In the two-dimensional example in Fig. 54.11(a), the points 1, 2, and 3 are chosen in Step (1), and their convex hull is the triangle with the vertices 1, 2, and 3. Set $k = 0$.
3. Inscribe the largest n-dimensional hypersphere in this approximating polyhedron and take its center as the first estimate of the design center. This process involves the solution of a linear program. In Fig. 54.11(a), this is the hypersphere C^0.
4. Find the midpoint of the largest face of the polyhedron, i.e., the face in which the largest $(n - 1)$-dimensional hypersphere can be inscribed. In Fig. 54.11(a), the largest face is 2-3, the face in which the largest $(n - 1)$-dimensional hypersphere can be inscribed.
5. Find a new boundary point on R_f by searching along the outward normal of the largest face found in Step (4) extending from the midpoint of this face. This is carried out by performing a line search. In Fig. 54.11(a), point 4 is thus identified.
6. Inflate the polyhedron by forming the convex hull of all previous points, plus the new point generated in Step (5). This corresponds to the quadrilateral 1, 2, 3, 4 in Fig. 54.11(a).
7. Find the center of the largest hypersphere inscribed in the new polyhedron found in Step (6). This involves the solution of a linear program. Set $k = k + 1$, and go to Step (4). In Fig. 54.11(a), this is the circle C^1.

Further iterations are shown in Fig. 54.11(b). The process is terminated when the sequence of radii of the inscribed hypersphere converges.

Inscribing the Largest Hypersphere in a Polytope

Given a polytope specified by Eq. (54.3), if the $\mathbf{a}_i$'s are chosen to be unit vectors, then the distance of a point $\mathbf{x}$ from each hyperplane of the polytope is given by $r = \mathbf{a}_i^T\mathbf{x} - b_i$.

The center $\mathbf{x}$ and radius r of the largest hypersphere that can be inscribed within the polytope $\mathcal{P}$ are then given by the solution of the following linear program:

$$\begin{aligned} &\textit{minimize} \quad r \\ &\textit{subject to} \quad a_i^T\mathbf{x} - r \geq b_i \end{aligned} \tag{54.40}$$

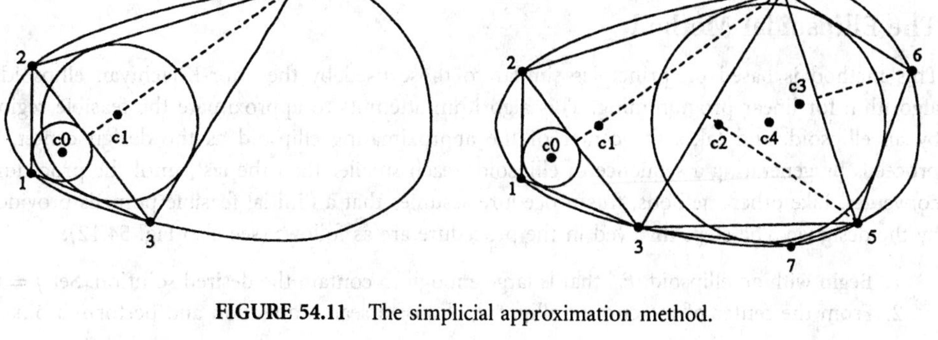

FIGURE 54.11 The simplicial approximation method.

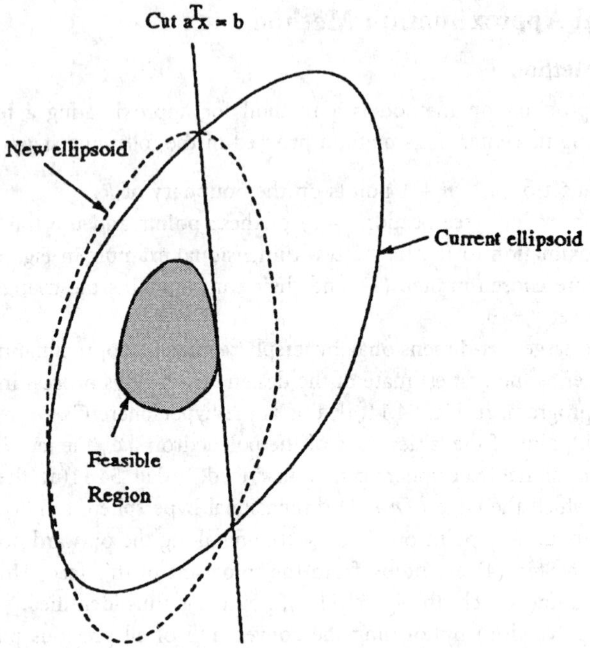

FIGURE 54.12 The ellipsoidal method.

Since the number of unknowns of this linear program is typically less than the number of constraints, it is more desirable to solve its dual [1]. A similar technique can be used to inscribe the largest hypersphere in a face of the polytope.

This method has also been generalized for the inscription of maximal norm bodies, to handle joint probability density functions with (nearly-)convex level contours.

Elongated Feasible Regions

If the design centering procedure outlined earlier is applied to a rectangular feasible region, the best possible results may not be obtained by inscribing a hypersphere. For elongated feasible regions, it is more appropriate to determine the design center by inscribing an ellipsoid rather than a hypersphere. Simplicial approximation handles this problem by scaling the axes so that the lower and upper bounds for each parameter differ by the same magnitude and one may inscribe the largest ellipsoid by inscribing the largest hypersphere in a transformed polytope. This procedure succeeds in factoring in reasonably the fact that feasible regions may be elongated; however, it considers only a limited set of ellipsoids, which have their axes aligned with the coordinate axis, as candidates for inscription within the polytope.

The Ellipsoidal Method

This method is based on principles similar to those used by the Shor-Khachiyan ellipsoidal algorithm for linear programming. This algorithm attempts to approximate the feasible region by an ellipsoid, and takes the center of the approximating ellipsoid as the design center. It proceeds by generating a sequence of ellipsoids, each smaller than the last, until the procedure converges. Like other methods, this procedure assumes that an initial feasible point is provided by the designer. The steps involved in the procedure are as follows (see also Fig. 54.12):

1. Begin with an ellipsoid, E_0, that is large enough to contain the desired solution. Set $j = 0$.
2. From the center of the current ellipsoid, choose a search direction and perform a binary

search to identify a boundary point along that direction. One convenient set of search directions is the parameter directions, searching along the ith, $i = 1, 2, \cdots, n$ in a cycle, and repeating the cycle, provided the current ellipsoid center is feasible. If not, a linear search is conducted along a line from the current center to the given feasible point.

3. A supporting hyperplane [1] at the boundary point can be used to generate a smaller ellipsoid, E_{j+1}, that is guaranteed to contain the feasible region R_f, if R_f is convex. The equation of E_{j+1} is provided by an update procedure.
4. Increment j, and go to Step (1) unless the convergence criterion is met. The convergence criterion is triggered when the volume is reduced by less a given factor, ϵ. Upon convergence, the center of the ellipsoid is taken to be the design center.

Convexity-Based Approaches

Introduction

Here, the feasible region is first approximated by a polytope in the first phase. Next, two geometrical approaches to find the design center are proposed. In the first, the properties of polytopes are utilized to inscribe the largest ellipsoid within the approximating polytope. The second method proceeds by formulating the design centering problem as a convex programming problem, assuming that the variations in the design parameters are modeled by Gaussian probability distributions and use Vaidya's convex programming algorithm described in Section 54.2 to find the solution.

Feasible Region Approximation

The feasible region, $R_f \subset \mathbf{R}^n$, is approximated by a polytope given by Eq. (54.3) in this step. The algorithm begins with an initial feasible point, $\mathbf{z}_0 \in R_f$. An n-dimensional box, namely, $\{\mathbf{z} \in \mathbf{R}^n \mid z_{min} \leq z_i \leq z_{max}\}$, containing R_f is chosen as the initial polytope $\mathcal{P}_0$. In each iteration, n orthogonal search directions, $\mathbf{d}_1, \mathbf{d}_2 \cdots \mathbf{d}_n$ are chosen (possible search directions include the n coordinate directions). A binary search is conducted from $\mathbf{z}_0$ to identify a boundary point $\mathbf{z}_{\mathrm{b}i}$ of R_f, for each direction $\mathbf{d}_i$. If $\mathbf{z}_{\mathrm{b}i}$ is relatively deep in the interior of $\mathcal{P}$, then the tangent plane to R_f at $\mathbf{z}_{\mathrm{b}i}$ is added to the set of constraining hyperplanes in Eq. (54.3). A similar procedure is carried out along the direction $-\mathbf{d}_i$. Once all of the hyperplanes have been generated, the approximate center of the new polytope is calculated. Then $\mathbf{z}_0$ is reset to be this center, and the above process is repeated.

Therefore, unlike the simplicial approximation which tries to expand the polytope outward, this method starts with a large polytope and attempts to add constraints to shrink it inward. The result of polytope approximation on an ellipsoidal feasible region is illustrated in Fig. 54.13.

Algorithm I: Inscribing the Largest Hessian Ellipsoid

For a polytope given by Eq. (54.3), the log-barrier function is defined as

$$F(\mathbf{z}) = -\sum_{i=1}^{m} \log_e(\mathbf{a}_i^T \mathbf{z} - b_i) \tag{54.41}$$

The Hessian ellipsoid centered at a point $\mathbf{x}$ in the polytope $\mathcal{P}$, is defined as the ellipsoid $E(\mathbf{x}, \mathcal{H}(\mathbf{x}), 1)$ (see Eq. (54.2)), where $\mathcal{H}(\mathbf{x}_c)$ is the Hessian [1] of the log-barrier function above, and is given by

$$\mathcal{H}(\mathbf{z}) = \nabla^2 F(\mathbf{z}) = \sum_{i=1}^{m} \frac{\mathbf{a}_i \mathbf{a}_i^T}{(\mathbf{a}_i^T \mathbf{z} - b_i)^2} \tag{54.42}$$

This is known to be a good approximation to the polytope locally around $\mathbf{x}$.

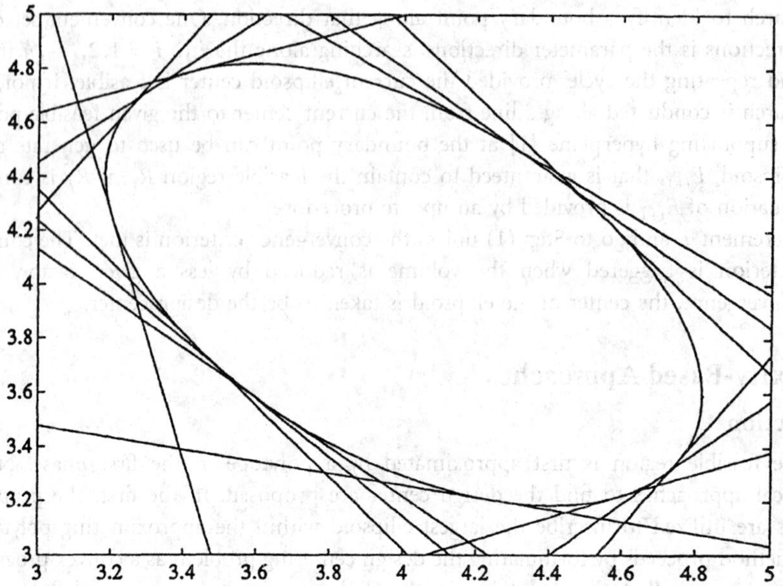

FIGURE 54.13 Polytope approximation for the convexity-based methods.

Hence, the goal is to find the largest ellipsoid in the class $\mathbf{E}(\mathbf{x}, \mathcal{H}(\mathbf{x}), \mathbf{r})$ that can be inscribed in the polytope, and its center $\mathbf{x}_c$. The point $\mathbf{x}_c$ will be taken to be the computed design center. An iterative process is used to find this ellipsoid. In each iteration, the Hessian at the current point $\mathbf{x}_k$ is calculated, and the largest ellipsoid $\mathbf{E}(\mathbf{x}, \mathcal{H}(\mathbf{x}_k), \mathbf{r})$ is inscribed in the polytope. The inscription of this ellipsoid is equivalent to inscribing a hypersphere in a transformed polytope. The process of inscribing a hypersphere in a polytope is explained in this section.

Algorithm II: The Convex Programming Approach

When the probability density functions that represent variations in the design parameters are Gaussian in nature, the design centering problem can be posed as a convex programming problem.

The joint Gaussian probability density function of n independent random variables $\mathbf{z} = (z_1, \cdots, z_n)$, with mean $\mathbf{x} = (x_1, \cdots, x_n)$ and variance $\sigma = (\sigma_1, \cdots, \sigma_n)$ is given by

$$\Phi_{\mathbf{x}}(\mathbf{z}) = \frac{1}{(2\pi)^{n/2}\sigma_1\sigma_2\cdots\sigma_n} \exp\left[\sum_{i=0}^{i=n} -\frac{(z_i - x_i)^2}{2\sigma_i^2}\right] \tag{54.43}$$

This is known to be a log-concave function of $\mathbf{x}$ and $\mathbf{z}$. Also, note that the arbitrary covariance matrices can be handled, since a symmetric matrix may be converted into a diagonal form by a simple linear (orthogonal) transformation. The design centering problem is now formulated as

$$\begin{aligned} &\textit{maximize} \quad \mathbf{Y}(\mathbf{x}) = \int_{\mathcal{P}} \Phi_{\mathbf{x}}(\mathbf{z}) d\mathbf{z} \\ &\textit{such that} \quad \mathbf{x} \in \mathcal{P} \end{aligned} \tag{54.44}$$

where $\mathcal{P}$ is the polytope approximation to the feasible region R_f. It is a known fact that the integral of a log-concave function over a convex region is also a log-concave function. Thus, the yield function $\mathbf{Y}(\mathbf{x})$ is log-concave, and the above problem reduces to a problem of maximizing a log-concave function over a convex set. Hence, this can be transformed into a convex programming problem. The convex programming algorithm in Section 54.2 is then applied to solve the optimization problem.

Concluding Remarks

The above list of algorithms is by no means exhaustive, but provides a general flavor for how optimization methods are used in geometrical design centering. The reader is referred to [7–10] and to Chapter 55 for further information about statistical design. In conclusion, it is appropriate to list a few drawbacks associated with geometrical methods:

Limitations of the Approximating Bodies. In the case of ellipsoidal approximation, certain convex bodies cannot be approximated accurately, because an ellipsoid is symmetrical about any hyperplane passing through its center, and is inherently incapable of producing a good approximation to a body that has a less symmetric structure. A polytope can provide a better approximation to a convex body than an ellipsoid since any convex body can be thought of as a polytope with an infinite number of faces. However, unlike the ellipsoidal case, calculating the exact center of a polytope is computationally difficult and one must resort to approximations.

Statistical Effects are Ignored. Methods such as simplicial approximation, ellipsoidal approximation and Algorithm I of the convexity-based methods essentially approximate the feasible region by means of an ellipsoid, and take the center of that ellipsoid to be the design center, regardless of the probability distributions that define variations in the design parameters. However, the design center could be highly dependent on the exact probability distributions of the variables, and would change according to these distributions.

Nonconvexities. Real feasible regions are seldom convex. While in many cases, they are "nearly convex," there are documented cases where the feasible region is not very well-behaved. In a large number of cases of good designs, since the joint probability density function of the statistical variables decays quite rapidly from the design center, a convex approximation does not adversely affect the result. However, if the nominal design has a very poor yield, a convex approximation will prove to be inadequate.

The Curse of Dimensionality. Geometrical methods suffer from the so-called "curse of dimensionality," whereby the computational complexity of the algorithm increases greatly with the number of variables. However, as noted in [11], for local circuit books within a die, performance variations in digital MOS circuits depend only on four independent statistical variables. Moreover, for this class of circuits, the circuit performance can be modeled with reasonable accuracy by linear functions. For such circuits, the deterministic (i.e., nonstatistical) algorithm in [11] is of manageable complexity.

eferences

[1] D. G. Luenberger, *Linear and Nonlinear Programming,* 2nd ed., Reading, MA: Addison-Wesley, 1984.

[2] R. K. Brayton, G. D. Hachtel, and A. L. Sangiovanni-Vincentelli, "A survey of optimization techniques for integrated-circuit design," *Proceedings of the IEEE,* vol. 69, pp. 1334–1362, Oct. 1981.

[3] S. S. Sapatnekar, V. B. Rao, P. M. Vaidya, and S. M. Kang, "An exact solution to the transistor sizing problem for CMOS circuits using convex optimization," *IEEE Trans. on Computer-Aided Design,* vol. 12, pp. 1621–1634, Nov. 1993.

[4] J. Rubenstein, P. Penfield, and M. A. Horowitz, "Signal delay in RC tree networks," *IEEE Trans. on Computer-Aided Design*, vol. CAD-2, pp. 202–211, July 1983.

[5] H. H. Rosenbrock, "An automatic method for finding the greatest or least value of a function," *Computer Journal*, vol. 3, pp. 175–184, Oct. 1960.

[6] S. S. Sapatnekar and S. M. Kang, *Design Automation for Timing-Driven Layout Synthesis.* Boston, MA: Kluwer Academic Publishers, 1993.

[7] S. W. Director, P. Feldmann, and K. Krishna, "Statistical integrated circuit design," *IEEE Journal of Solid-State Circuits*, vol. 28, pp. 193–202, March 1993.

[8] M. D. Meehan and J. Purviance, *Yield and Reliability in Microwave Circuit and System Design*, Boston, MA: Artech House, 1993.

[9] S. W. Director, W. Maly, and A. J. Strojwas, *VLSI Design for Manufacturing: Yield Enhancement*, Boston, MA: Kluwer Academic Publishers, 1990.

[10] R. Spence and R. S. Soin, *Tolerance Design of Integrated Circuits*, Reading MA: Addison-Wesley, 1988.

[11] D. E. Hocevar, P. F. Cox, and P. Yang, "Parametric yield optimization for MOS circuit blocks," *IEEE Trans. on Computer-Aided Design*, vol. 7, pp. 645–658, June 1988.

[12] J. M. Shyu, A. Sangiovanni-Vincentelli, J. P. Fishburn and A. E. Dunlop, "Optimization-Based Transistor Sizing," *IEEE J. Solid-State Circuits*, vol. 23, pp. 400–409, April 1988.

[13] D. Marple and A. El Gamal, "Optimal Selection of Transistor Sizes in Digital VLSI Circuits," Stanford Conference on VLSI, pp. 151–172, 1987.

55

Statistical Design Optimization

ʟaciej A. Styblinski
xas A&M University

55.1 Introduction

Manufacturing process variations and environmental effects (such as temperature) result in the variations of the values of circuit elements and parameters. Statistical methods of circuit design optimization take those variations into account and apply statistical (or statistical/deterministic) optimization techniques to obtain an "optimal" design. Statistical design optimization belongs to a general area of statistical circuit design.

55.2 Problems and Methodologies of Statistical Circuit Design

A broad class of problems exists in this area: statistical **analysis** involves studying the effects of element variations on circuit performance. It applies statistical techniques, such as Monte Carlo simulation [34] and the **variance propagation method** [39], to estimate variability of performances. **Design centering** attempts to find a center of the acceptability region [12] such that manufacturing yield is maximized. **Direct methods** of yield optimization use yield as the objective function and utilize various statistical (or mixed statistical/deterministic) algorithms to find the yield maximum in the space of designable circuit/process parameters. **Design centering**

ı493-8341-2/95/$0.00 + $.50

and **tolerance assignment** (used mostly for discrete circuits) attempt to find the design center, with simultaneous optimal assignment of circuit element tolerances, minimizing some suitable cost function and providing 100% yield (worst-case design) [4, 6]. To solve this problem, mostly deterministic algorithms of nonlinear programming are used. Worst-case design is often too pessimistic and too conservative, leading to substantial overdesign. This fact motivates the use of **statistical techniques**, which provide a much more realistic estimation of the actual performance variations and lead to superior designs. Stringent requirements of the contemporary Very Large Scale Integration (VLSI) design prompted a renewed interest in the practical application of these techniques. The most significant philosophy introduced recently in this area is statistical Design for Quality (DFQ). It was stimulated by the practical appeal of the DFQ methodologies introduced by Taguchi [30], oriented toward "on-target" design with performance variability minimization. In what follows, mostly the techniques of manufacturing yield optimization and their design for quality generalization are discussed.

55.3 Underlying Concepts and Techniques

Circuit Variables, Parameters, and Performances

Designable Parameters

Designable parameters, represented by the n-dimensional vector[1] $x = (x_1, \ldots, x_n)$, are used by circuit designers as "decision" variables during circuit design and optimization. Typical examples are nominal values of passive elements, nominal MOS transistor mask dimensions, process control parameters, etc.

Random Variables

The t-dimensional vector of random variables (or "noise" parameters in Taguchi's terminology [30]) is denoted as $\theta = (\theta_1, \ldots, \theta_t)$. It represents statistical R, L, C element variations, disturbances or variations of manufacturing process parameters, variations of device model parameters such as t_{ox} (oxide thickness), V_{TH} (threshold voltage), and environmental effects such as temperature, supply voltages, etc. Usually, θ represents **principal random variables**, selected to be **statistically independent** and such that all other random parameters can be related to them through some **statistical models**. Probability density function (p.d.f.) of θ parameters will be denoted as $f_\theta(\theta)$.

Circuit (Simulator) Variables

These variables represent parameters and variables used in circuit, process, or system simulators such as SPICE. They are represented as the c-dimensional vector $e = (e_1, \ldots, e_c)$. Specific examples of e variables are: R, L, C elements, gate widths W_j and lengths L_j of MOS transistors, device model parameters, or, if a process simulator is used, process-related control, and physical and random parameters available to the user. The e vector contains only those variables that are **directly related** to the x and θ vectors.[2] This relationship is, in general, expressed as

$$e = e(x, \theta) \tag{55.1}$$

The p.d.f. of θ is transformed into $f_e(e)$, the p.d.f. of e. This p.d.f. can be **singular**, i.e., defined in a certain **subspace** of the e-space (see examples below). Moreover, it can be very complicated, with highly nonlinear statistical dependencies between different parameters, so it is very difficult to represent it directly as a p.d.f. of e. In the majority of cases, the analytic form of $f_e(e)$ is not known. For that reason techniques of statistical modeling are used (see the next section).

[1] Vectors are denoted by lower case letters without subscripts or superscripts.
[2] This means that, e.g., some SPICE parameters will always be *fixed*.

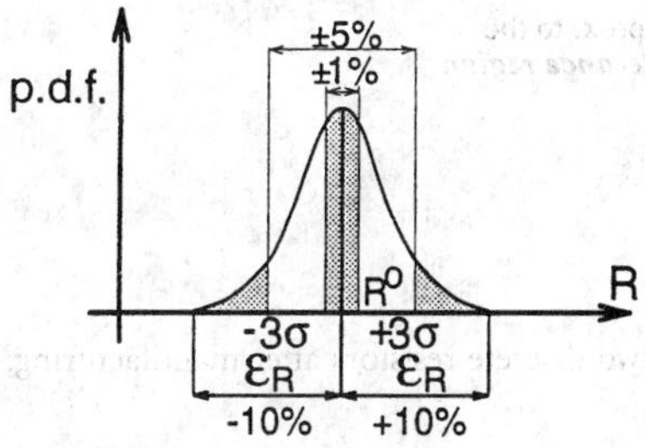

FIGURE 55.1 A typical probability density function (p.d.f.) of a discrete resistor before and after "binning."

Circuit Performances

The vector of circuit performance function (or simply performances) is defined as the m-dimensional vector $y = (y_1, \ldots, y_m)$. Its elements can be gain, bandwidth, slew rate, signal delay, and circuit response for a single frequency or time, etc. Each of the performances y_j is a function of the vector of circuit elements e: $y_j = y_j(e) = y_j(e(x, \theta))$. These transformations are most often not directly known in analytical form and a circuit simulator (such as SPICE) must be used to find the values of y_j's corresponding to the given values of x and θ. The overall simulator time required for the determination of all the performances y can be substantial for large circuits. This is the major limiting factor for the practical application of statistical circuit design techniques. To circumvent this problem, new **statistical macromodeling** techniques are being introduced [31] (see example in last subsection of 55.4).

Statistical Modeling of Circuit (Simulator) Variables

Statistical modeling is the process of finding a suitable transformation $e = e(x, \theta)$, such that given the distribution of θ, the distribution of e can be generated. The transformation $e = e(x, \theta)$ can be described by closed form analytical formulas or by a computer algorithm.

For **discrete active RLC circuits** (e.g., such as a common emitter amplifier), vector e is composed of two parts: the first part contains the **actual** values of statistically perturbed RLC elements: i.e., $e_i = x_i + \theta_i$ for those elements. In this formula, θ represents **absolute** element spreads and its expected (average) value, $E\{\theta\} = 0$ and x_i is the **nominal** value of e_i, often selected as the expected value of e_i. This implies that the variance of e_i, $\text{var}\{e_i\} = \text{var}\{\theta_i\}$; $E\{e_i\} = x_i$ and the distribution of e_i is the same as that of θ_i, with the expected value shifted by x_i. Alternatively, if θ_i represents **relative** element spreads, $e_i = x_i(1 + \theta_i)$, where $E\{\theta_i\} = 0$. Therefore, $E\{e_i\} = x_i$; $\text{var}\{e_i\} = x_i^2 \,\text{var}\{\theta_i\}$, i.e., the standard deviations σ_{e_i} and σ_{θ_i} are related: $\sigma_{e_i} = x_i \sigma_{\theta_i}$, or $\sigma_{e_i}/E\{e_i\} = \sigma_{e_i}/x_i = \sigma_{\theta_i}$. This means that with fixed σ_{θ_i}, the **relative** standard deviation of e_i is constant, as it is often the case in practice, where standard deviations of RLC elements are described in percents of the element nominal values. Both forms of e_i indicate that each e_i is directly associated with its corresponding θ_i and x_i, and that there is one-to-one mapping between e_i and θ_i. These dependencies are important, since many of the yield optimization algorithms were developed assuming that $e_i = x_i + \theta_i$. A typical p.d.f. for discrete elements is shown in Fig. 55.1, before "binning" into different categories (the whole curve) and after binning into ±1%, ±5%, and ±10% resistors (the shaded and white areas: e.g., the ±10% resistors will have the distribution characterized by the external shaded areas, with a ±5% "hole" in the middle).

Usually, passive discrete elements are statistically independent[3] as shown in Fig. 55.2. The cross-section shown is often called a **level set**, a norm body [12], or a **tolerance body**. It is defined as a set of element values for which the element p.d.f. is larger than a prescribed value.

[3]But, for instance, if R_L (R_C) is a loss resistance of an inductor L (capacitor C) then L and R_L (C and R_C) are correlated.

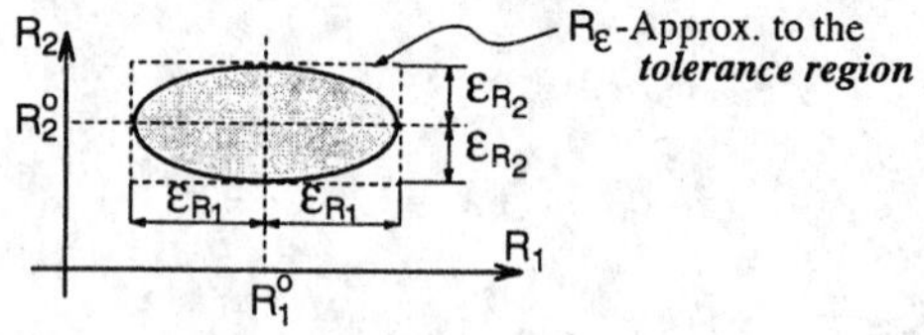

FIGURE 55.2 A level set (cross section) of a p.d.f. function for two discrete resistors after manufacturing.

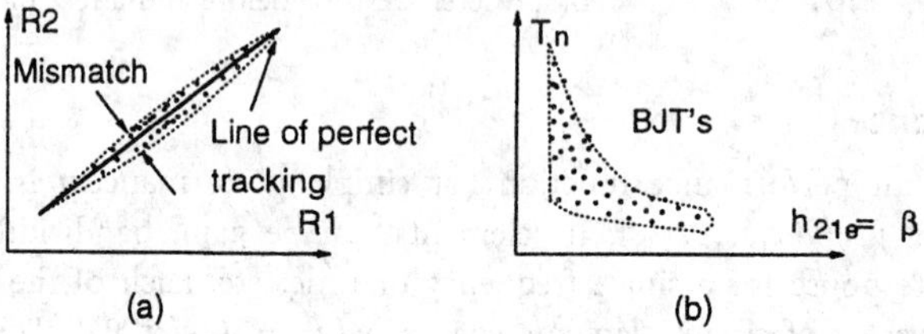

FIGURE 55.3 Dependencies between passive IC elements and between device model parameters: (a) Linear, between two resistors R1 and R2; (b) nonlinear, between the base transit time T_n and the current gain $h_{21e} = \beta$ for bipolar junction transistors.

This value is selected such that the probability of the element values falling into the tolerance body is equal to e.g., 95%, (i.e., the tolerance body represents 95% of the entire element population). Alternatively, the tolerance body can be represented as a (hyper-)box shown in the figure, with the sides equal to $2\varepsilon_i$ ($2\varepsilon_{R_i}$ in the figure), called the **tolerance region** and denoted by R_ε. Figure 55.2 also shows that the dependence $e_i = x_i + \theta_i$ is equivalent in this case to $R_i = R_{NOM,i} + \Delta R_i$, where ($x_i = R_{NOM,i}$; $\theta = \Delta R_i$).

The second part of e is composed of the parameters representing active device (e.g., BJT, MOS) model parameters. They are usually strongly correlated within each device model (same applies to IC's), but typically, there are no correlations between device model parameters of **different** devices. Each of the device model parameters e_d is related through a specific model $e_d = e_d(x, \theta)$ to the vector of θ parameters,[4] representing **principal random variables**, which are themselves often some device model parameters (such as oxide thickness t_{ox} of MOS transistors), and/or are some **dummy** random variables. For example, in the BJT empirical statistical model introduced in [3], the base transient time T_n is modeled as follows:

$$e_d = e_d(x_d, \theta_1, \theta_2) \equiv T_n(\beta, X_{r5}) = \left(a + \frac{b}{\sqrt{\beta}}\right)(1 + cX_{r5}) \tag{55.2}$$

i.e., it is the function of the current gain $\theta_1 = \beta$ (the principal random variable, affecting the majority of the BJT model parameters) and $\theta_2 = X_{r5}$ (a dummy random variable, uniformly distributed in the interval $[-1, 1]$ independent of β and having no physical meaning); a, b, c are empirically selected constants [see Fig. 55.3(b)].

In the discrete circuit case, each designable parameter x_i has a random variable θ_i added to it (i.e., x and θ are in the **same space** in which the p.d.f. $f_e(e) = f_\theta(x + \theta)$ is defined.

For **integrated circuits**, the passive elements (R's, C's), active device dimensions (such as W's and L's for MOS transistors), and other transistor model parameters are strongly correlated [see Fig. 55.3(a)]. Because of this, the distribution of θ parameters is limited to a certain **subspace** of the entire e-space, i.e., $f_e(e(x, \theta))$ is singular, and the formula $e_i = x_i + \theta_i$ does not hold. As an example, consider a subset of all e parameters representing the gate lengths $x_1 = L_1$,

[4]Observe that these models are parametrized by x: e.g., the MOS transistor model parameters e_d will also depend on the device length L and width W.

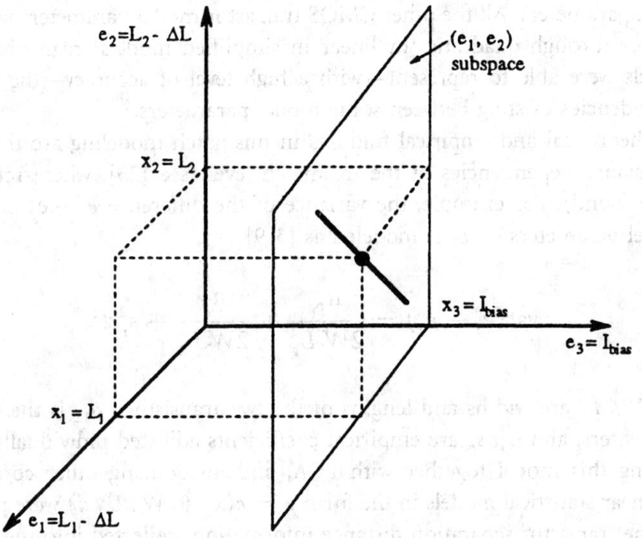

FIGURE 55.4 Singular distribution $f_e(e_1, e_2, e_3)$ in 3-dimensional e-space, represented by the thick line in the (e_1, e_1)-subspace (plane). This is due to the perfect matching of $\Delta L \equiv \Delta L_1 \equiv \Delta L_2$ values.

$x_2 = L_2$ (designable parameters) of two MOS transistors T_1, T_2 and an independent designable parameter $x_3 \equiv I_{\text{bias}}$, representing the transistor bias current. Assume also (as is frequently the case) that $\theta \equiv \Delta L$, the technological gate length reduction (a **common** random parameter) changes the same way for T_1 and T_2, i.e., there is ideal matching of $\Delta L = \Delta L_1 = \Delta L_2$,[5] i.e., $e_1 \equiv x_1 + \theta \equiv L_1 - \Delta L$, $e_2 \equiv x_2 + \theta \equiv L_2 - \Delta L$, $e_3 \equiv x_3 \equiv I_{\text{bias}}$. The only random variable in this model is ΔL, as shown in Fig. 55.4. The p.d.f. $f_e(e)$ is in this case defined only in the **one-dimensional** subspace of the e-space (i.e., all the realization of the vector e are located on the thick line in Fig. 55.4). The major consequence of this difference is that, in general, many yield optimization algorithms developed for discrete circuits (i.e., using the explicit assumption that $e_i = e_i + \theta_i$), cannot be used for IC design.

Statistical modeling for IC's is concerned with representing **global** variations and parameter **mismatches**[6] occurring between different devices on the same chip. In the last example, mismatch can be modeled by two additional independent **local mismatch variables** ε_{L_1}, ε_{L_2}, representing small random local deviations of ΔL_1 and ΔL_2 from the ideal value ΔL, i.e., $e_1 \equiv x_1 + \theta + \varepsilon_i \equiv L_1 - \Delta L - \varepsilon_{L_1}$, $e_2 \equiv x_2 + \theta + \varepsilon_i \equiv L_2 - \Delta L - \varepsilon_{L_2}$. With this model, ΔL represents a global or **common factor**, affecting both channel lengths, and ε_{L_1}, ε_{L_2} represent *specific* (local) **variations.** The model just presented is a simple (simple, one-factor) case of the Factor Analysis (FA) [20] model of correlations between e_1 and e_2.

In [8], FA analysis, together with Principal Component Analysis (PCA)[7] and nonlinear regression were used for the determination of linear and nonlinear statistical models for CMOS transistor parameters. The following common factors $F_1, \ldots, F_8$ were identified: t_{ox} (oxide thickness common to n- and p-type transistors) $N_{SUB,n}$, $N_{SUB,p}$ (n- and p-type substrate doping), ΔL_n, ΔL_p (length reduction), ΔW_n, ΔW_p (width reduction—for narrow transistors only), and XJ_p (p-type junction depth). These variables were causing about 96% of the total

[5]Such a model is commonly used for digital IC's [9], since it is of sufficient accuracy for digital applications.

[6]Element mismatches are very important for analog IC's.

[7]Principal component analysis involving coordinate rotation and leads to *uncorrelated principal components.*

variability of all parameters. All the other CMOS transistor model parameters were related to the $F_1, \ldots, F_8$ factors through quadratic [or linear in simplified models] regression formulas. The resulting models were able to represent—with a high level of accuracy—the strong nonlinear statistical dependencies existing between some model parameters.

The major theoretical and empirical findings in mismatch modeling are the device area (or length) and distance dependencies of the mismatch level (see [23], where references to other authors can be found). For example, the variance of the difference $e_1 - e_2$ between two MOS transistor model parameters e_1, e_2 is modeled as [3, 9]

$$\mathrm{var}(e_1 - e_2) = \frac{a_p}{2W_1L_1} + \frac{a_p}{2W_2L_2} + s_p^2 d_{12}^2 \tag{55.3}$$

where W_1, L_1, W_2, L_2 are widths and lengths of the two transistors, d_{12} is the distance between the transistor centers, and a_p, s_p are empirical coefficients adjusted individually for each model parameter. Using this model together with PCA, and introducing other concepts, two quite sophisticated linear statistical models in the form $e_i = e(x_i, \theta, W_i, L_i, d)$ were proposed in [23]. They include the transistor separation distance information, collected into the vector d, in two different forms. The models, constructed from on-chip measured data, were used for practical yield optimization. θ parameters were divided into two groups: a group of **correlated** random variables responsible for the **common** part of each parameter variance and **correlations** between model parameters of each *individual* transistor, and the second group of **local** (mismatched related) random variables, responsible for mismatches between **different** transistors. Additional dependencies, related to transistor spacing and device area related coefficients, maintain proper mismatch dependencies.

Acceptability Regions

The **acceptability region** A_y is defined as a region in the space of performance parameters y (y-space), for which all inequality and equality constraints imposed on y are fulfilled. In the majority of cases, A_y is a hyperbox, i.e., all the constraints are of the form: $S_j^L \le y_j \le S_j^L$; $j = 1, \ldots, m$, where S_j^L, S_j^U are the (designer defined) lower and upper bounds imposed on y_j, called also **designer's specifications**. More complicated specifications, involving some relations between y_j parameters, or S_j^L, S_j^U bounds can also be defined.

For the simplest case of the y-space box-constraints, the **acceptability region** A in the e-space is defined as such a set of e vectors in the c-dimensional space, for which all inequalities $S_j^L \le y_j(e) \le S_j^U$, $j = 1, \ldots, m$ are fulfilled. Illustration of this definition is shown in Fig. 55.5. It can be interpreted as the mapping of the acceptability region A_y from the y-space into the e-space. Acceptability regions A can be very complicated: they can be non-convex and can contain internal infeasible regions (or "holes"), as shown in Fig. 55.6 for a simple active RC filter [25].

For discrete circuits, A is normally represented in the e-space, due to the simple relationship $e_i = x_i + \theta_i$ (or $e_i = x_i(1 + \theta_i)$), between e_i, x_i, and θ_i. For integrated circuits e is related to x, and θ through the statistical model, x and θ are in different spaces (or subspaces), the dimension of θ is lower than the dimension of e, and the p.d.f. $f_e(e)$ is singular and usually unknown in analytic form. For these reasons, it is more convenient to represent A in the joint (x, θ)-space, as shown in Fig. 55.7. For a fixed x, A can be defined in the θ-space and labeled as $A_\theta(x)$, since it is parametrized by the actual values of x, as shown in Fig. 55.7. The shape and location of $A_\theta(x)$ change with x, as shown. For a fixed x, $A_\theta(x)$ is defined as such a region in the t-dimensional θ space, for which all the inequalities $S_j^L \le y_j(e(x, \theta)) \le S_j^U$; $j = 1, \ldots, m$ are fulfilled.

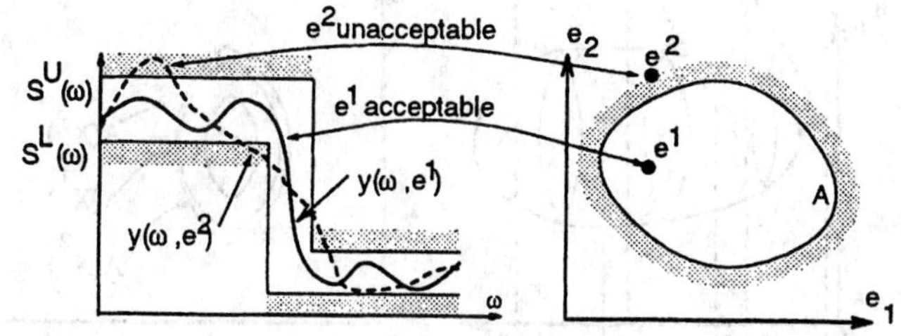

FIGURE 55.5 Illustration of the mapping of the $S^L(\omega)$, $S^U(\omega)$ constraints imposed on $y(\omega, e)$ into the e-space of circuit parameters.

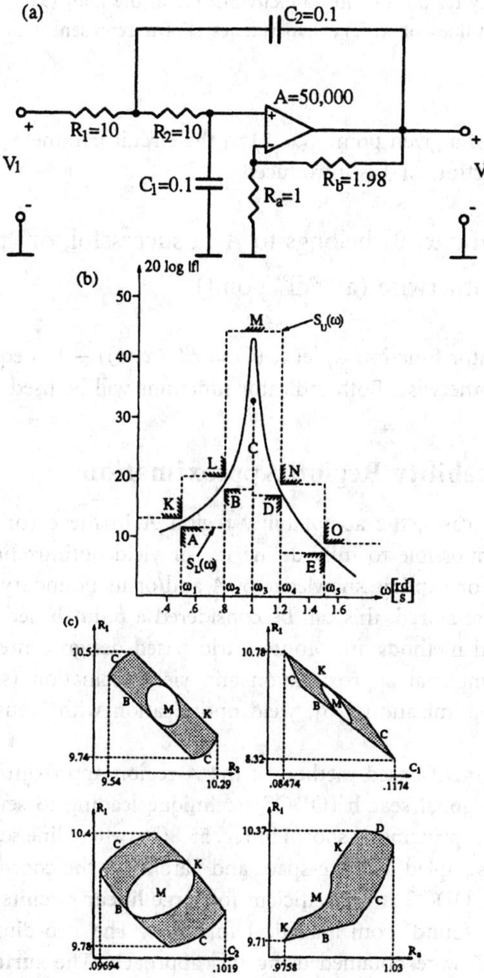

FIGURE 55.6 (a) A Sallen-Key active filter; (b) the lower and upper bounds imposed in the filter frequency response; (c) two-dimensional cross-sections of the acceptability region A. Capital letters in (b) and (c) indicate the correspondence of constraints to the boundaries of A.

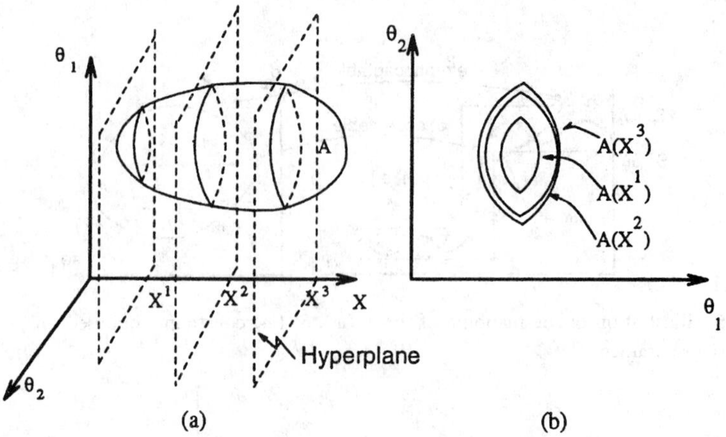

FIGURE 55.7 Acceptability region for integrate circuits: (a) in the joint (x, θ)-space, (b) in the θ-space, parametrized by different values of x. The hyperplanes shown represent t-dimensional subspaces of θ parameters.

In order to recognize if a given point $e(x, \theta)$ in the circuit parameter space belongs to A [or $A_\theta(x)$] an **indicator function** $\phi(\cdot)$ is introduced

$$\phi(e(x, \theta)) = \begin{cases} 1 & \text{if } e(x, \theta) \text{ belongs to } A \text{ (a successful, or "pass" point)} \\ 0 & \text{otherwise (a "fail" point)} \end{cases} \tag{55.4}$$

A complementary indicator function $\phi_F(e(x, \theta)) = \phi(e(x, \theta)) - 1$ is equal to 1 if $e(x, \theta)$ does not belong to A and 0 otherwise. Both indicator functions will be used in what follows.

Methods of Acceptability Region Approximation

Except for some simple cases, the acceptability region A in the e (or the joint (x, θ))-space is unknown and it is impossible to **fully** define it. For yield optimization and other statistical design tasks an **implicit** or **explicit** knowledge of A and/or its boundary is required. If only the points belonging to A are stored, this can be considered a **point-based** "approximation" to A. Some of the point-based methods are Monte Carlo based design centering, centers of gravity method, point-based simplicial approximation and yield evaluation (see below), "parametric sampling-based" yield optimization [36], yield optimization with "reusable" points [39], and others.

The **acceptability segment**-based method of the A-region approximation was called in [27] a one-dimensional orthogonal search (ODOS) technique leading to several yield optimization methods [25, 46]. Its basic principle is shown in Fig. 55.8(b), where line segments passing through the points e^i randomly sampled in the e-space and parallel to the coordinate axes, are used for the approximation of A. ODOS is very efficient for **large linear circuits**, since the intersections with A can be **directly** found from analytical formulas. The two-dimensional cross-sections of A, shown in Fig. 55.6 were obtained using this approach. The **surface** integral-based yield and yield gradient estimation and optimization method proposed in [13] also use the segment approximation to the **boundary** of A. A variant of this method is segment approximation in one direction, as shown in Fig. 55.8(c). This method was then extended to plane and hyperplane approximation to A in [42] [Fig. 55.8(d)]. In another approach, called "radial exploration of space" in [54], the segments approximating A are in **radial directions**, as shown in Fig. 55.8(e).

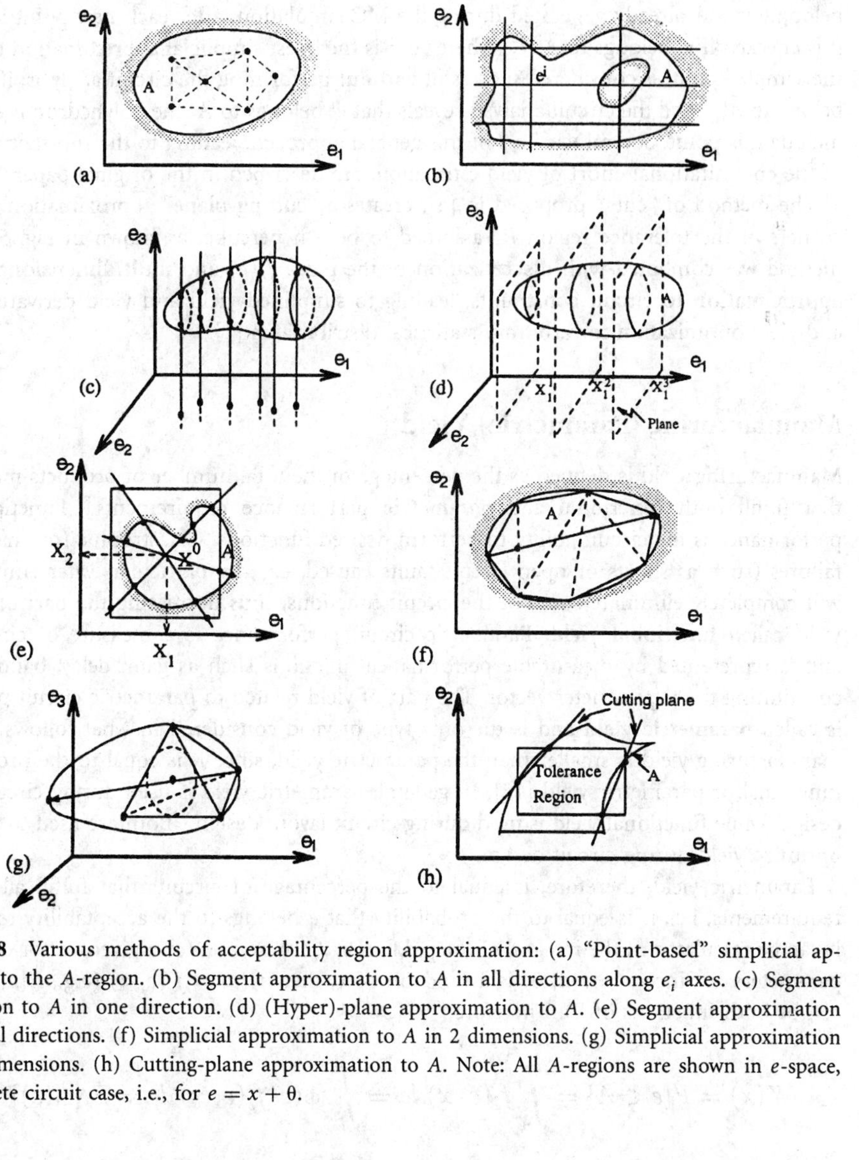

FIGURE 55.8 Various methods of acceptability region approximation: (a) "Point-based" simplicial approximation to the A-region. (b) Segment approximation to A in all directions along e_i axes. (c) Segment approximation to A in one direction. (d) (Hyper)-plane approximation to A. (e) Segment approximation to A in radial directions. (f) Simplicial approximation to A in 2 dimensions. (g) Simplicial approximation to A in 3 dimensions. (h) Cutting-plane approximation to A. Note: All A-regions are shown in e-space, for the discrete circuit case, i.e., for $e = x + \theta$.

The techniques just described rely on the fact that "segment" yields (calculated in the e subspace) can be, for some special cases [27,54], calculated more efficiently than using a standard Monte Carlo method. This leads to higher efficiency and accuracy of yield estimation, in comparison to the point-based yield estimation.

The **simplicial approximation** proposed in [10] (described in more detail in Section 54.4), is based on approximating the boundary of A in the e-space, by a polyhedron, i.e., by the union of those partitions of a set of c-dimensional hyperplanes which lie inside of the boundary of A or on it. The boundary of A-region is assumed to be **convex** [see Fig. 55.8(f)(g)]. The approximating polyhedron is a convex hull of points. Simplicial approximation is obtained by locating points on the boundary of A, by a systematic expansion of the polyhedron. The search for next vertex is always performed in the direction passing through the center of the largest face of the polyhedron already existing. In the MC oriented-"point-based" version of the method [11] [see Fig. 55.8(a)] subsequent simplicial approximations $\tilde{A}_i$ to A are not based on the points located on the boundary of A (which is computationally expensive) but on the points e^i

belonging to A already generated during the MC simulation; after each new point is generated, it is checked if e^i belongs to $\tilde{A}^{i-1}$, (where $\tilde{A}^{i-1}$ is the latest simplicial approximation to A); if yes, the sampled point is considered successful **without** performing the circuit analysis; if e^i does not belong to $\tilde{A}^{i-1}$ and the circuit analysis reveals that e^i belongs to A, the polyhedron is expanded to include this point. Several versions of this general approach, leading to the substantial reduction of the computational effort of yield estimation, are described in the original paper.

The method of "cuts" proposed in [5], creates a "cutting-plane" approximation to A in the corners of the tolerance region R_ε assumed to be a hypercube, as shown in Fig. 55.8(h). The method was combined with **discretization** of the p.d.f. $f_e(e)$ and **multidimensional quadratic approximation** to circuit constraints, leading to simplified yield and yield derivative formulas and yield optimization for arbitrary statistical distributions [1].

Manufacturing (Parametric) Yield

Manufacturing yield is defined as the percentage of the total number of products manufactured that fulfill both **functional** and **parametric** performance requirements.[8] Functional circuit performance is the circuit ability to perform desired functions. Catastrophic (or "hard") circuit failures (such as shorts or open circuit faults caused, e.g., by particular wafer contamination) will completely eliminate some of the circuit functions, thus decreasing the part of the overall yield called **functional yield.** Parametric circuit performance is a measure of circuit quality and is represented by measurable performance functions such as gain, delay, bandwidth, etc., constituting the y-parameter vector. The part of yield related to parametric circuit performance is called **parametric yield** and is the only type of yield considered in what follows. The actual manufacturing yield is smaller than the parametric yield, since it is equal to the product of the functional or parametric yield [12]. In general, parametric yield is used during circuit electrical design, while functional yield is used during circuit layout design.[9] Both are used to **predict** and **optimize** yield during circuit design.

Parametric yield, therefore, is equal to the percentage of circuits that fulfill all parametric requirements, i.e., it is equal to the probability that e belongs to the acceptability region A. So, it can be calculated as the integral of the p.d.f. of e, $f_e(e)$ over A, for a given vector of designable parameters x. Since $e = e(x, \theta)$ is a function of x, then $f_e(e) = f_e(e, x)$ (e.g., $E\{e\}$ and $\text{var}\{e\}$ can be both functions of x). Therefore,[10]

$$Y(x) = P\{e \in A\} = \int_A f_e(e, x)de = \int_{R^c} \phi(e) f_e(e, x)\,de = E_e\{\phi(e)\} \quad (55.5)$$

where $P\{\cdot\}$ denotes probability, $\phi(e)$ is the indicator function (4), and $E_e\{\cdot\}$ is expectation with respect to the random variable e. The above formula is useful if $f_e(e, x)$ is a nonsingular p.d.f., which is usually the case for discrete circuits, for which $e_i = x_i + \theta_i$, (or $e_i = x_i(1 + \theta_i)$). In a general case, however, (e.g., for integrated circuits) the p.d.f. $f_e(e)$ is not known, since it has to be obtained from a complicated transformation $e = e(x, \theta)$, given the p.d.f. $f_\theta(\theta)$ of θ. Therefore, it is more convenient to integrate directly in the θ-space. Since parametric yield is also the probability that θ belongs to $A_\theta(x)$ (the acceptability region in the θ-space for any fixed

[8] For more detailed yield definitions, involving different types of yield, e.g., design yield, wafer yield, probe yield, processing yield, etc. see [12].

[9] Layout design (i.e., transistor spacing, location, size) has also influence on parameter variations and mismatch, as discussed in Section 55.3.

[10] Multiple integration performed below is over the acceptability region A, or over the entire c-dimensional space R^c of real numbers. $\in$ means "belongs to."

x), yield becomes

$$Y(x) = P\{\theta \in A_\theta(x)\} = \int_{A_\theta(x)} f_\theta(\theta)\, d\theta$$
$$= \int_{R^t} \phi(e(x, \theta)) f_\theta(\theta)\, d\theta = E_\theta\{\phi(e(x, \theta))\} \tag{55.6}$$

Formula (55.6) is general, and is valid for both discrete and integrated circuits. An unbiased estimator of $E_\theta\{\phi(e(x, \theta))\} \equiv E_\theta\{\phi(\theta)\}$ (for fixed x), is the arithmetic mean, based on N points θ^i, sampled in θ-space with the p.d.f. $f_\theta(\theta)$, for which the function $\phi(\theta^i)$ is calculated (this involves circuit analyses). Thus, the yield estimator $\hat{Y}$ is expressed as

$$\hat{Y} = \frac{1}{N} \sum_{i=1}^{N} \phi(\theta^i) = \frac{N_S}{N} \tag{55.7}$$

where N_S is the number of successful trials, i.e., the number of circuits for which $\theta \in A_\theta(x)$ (all circuit constraints are fulfilled). Integral (55.6) is normally calculated using Monte Carlo (MC) simulations [34] and (7). The MC method is also used to determine statistical parameters of the p.d.f. $f_y(y)$ of $y = y(x, \theta)$. In order to sample the θ parameters with p.d.f. $f_\theta(\theta)$, special numerical procedures, called **random number generators**, are used. The basic MC algorithm is as follows

1. Set $i = 0$, $N_S = 0$ (i is the current index of a sampled point and N_S is the total number of successful trials).
2. Substitute $i = i + 1$, generate the i-th realization of θ: $\theta^i = (\theta_1^i, \ldots, \theta_t^i)$, with the p.d.f. $f_\theta(\theta)$.
3. Calculate the i-th realization of $y^i = (y_1^i, \ldots, y_m^i) = y(x, \theta^i)$, with the aid of an appropriate circuit analysis program, and store the results.
4. Check if all circuit constraints are fulfilled, i.e., if $S^L \le y^i \le S^U$; if yes, set $N_S = N_S + 1$.
5. If $i \ne N$, go to (2); otherwise:
   Find the yield estimator $\hat{Y} = N_S/N$. If needed, find also some statistical characteristics of y-parameters (e.g., create histograms of y, find statistical moments of y, etc.).

To generate θ^i's with the p.d.f. $f_\theta(\theta)$, the **uniformly** distributed random numbers are generated first and then transformed to $f_\theta(\theta)$. The most typical random number generator (r.n.g.), generating a sequence of pseudo-random, uniformly distributed integers θ_k in the interval $[0, M)$ (M is an integer), is a multiplicative r.n.g., using the formula [34]: $\theta_{k+1} = c\theta_k (\mathrm{mod}\, M)$ where c is an integer constant and $\theta_k (\mathrm{mod}\, M)$ denotes a remainder from dividing θ_k by M. The initial value θ_0 of θ_k is called the "seed" of the r.n.g., and, together with c, should usually be chosen very carefully, to provide good quality of the random sequence generated. Several other r.n.g.'s are used in practice [34]. The r_k numbers in the $[0, 1)$ interval are obtained from $r_k = \theta_k/M$. Distributions other than uniform are obtained through different transformations of the uniformly distributed random numbers, such as the inverse of the cumulative distribution function [34]. To generate **correlated normal** variables $\theta = (\theta_1, \ldots, \theta_n)$, with a given covariance matrix K^θ, the transformation $\theta = CZ$ is used, where $Z = (z_1, z_2, \ldots, z_n)$ is the vector of independent normal variables with $E\{z_i\} = 0$, $\mathrm{var}\{z_i\} = 1$, $i = 1, \ldots, n$, and C is a matrix obtained from the so-called Cholesky decomposition of the covariance matrix K^θ, such that $K^\theta = CC^t$, where t denotes transposition. C is usually lower or upper triangular and can be easily constructed from a given matrix K^θ [34].

The yield estimator $\hat{Y}$ is a random variable, since performing different, independent MC simulations, one can expect different values of $\hat{Y} = N_S/N$. As a measure of $\hat{Y}$ variations, variance or standard deviation of $\hat{Y}$ can be used. It can be shown [7] that the standard deviation of $\hat{Y}$, is equal to $\sigma_{\hat{Y}} = \sqrt{Y(1-Y)/N}$, i.e., it is proportional to $1/\sqrt{N}$. Hence, to decrease the error of $\hat{Y}$ 10 times, the number of samples has to be increased 100 times. This is a major drawback of the MC method. However, the accuracy of the MC method (measured by $\sigma_{\hat{Y}}$) is **independent** of the dimensionality of the θ-space, which is usually a drawback of other methods of yield estimation. One of the methods of variance reduction of the $\hat{Y}$ estimator is **importance sampling** [7, 34, 39]. Assume that instead of sampling θ with the p.d.f. $f_\theta(\theta)$, some other p.d.f. $g_\theta(\theta)$ is used. Then

$$Y = \int_{R^t} \phi(e(\theta)) \frac{f_\theta(\theta)}{g_\theta(\theta)} g_\theta(\theta)\, d\theta \equiv E\left\{\phi(e(\theta))\frac{f_\theta(\theta)}{g_\theta(\theta)}\right\} \tag{55.8}$$

where $g_\theta(\theta) \neq 0$ if $\phi(\theta) = 1$. Yield Y can now be estimated as

$$\hat{Y} = \frac{1}{N}\sum_{i=1}^{N} \phi(e(\theta^i))\frac{f_\theta(\theta^i)}{g_\theta(\theta^i)} \tag{55.9}$$

sampling N points θ^i with the p.d.f. $g_\theta(\theta)$. The variance of this estimator is $\text{var}\{\hat{Y}\} = E\{[\phi(\theta)f_\theta(\theta)/g_\theta(\theta) - Y]^2\}/N$. If it is possible to choose $g_\theta(\theta)$ such that it mimics (or is similar to) $\phi(\theta)f_\theta(\theta)/Y$, the variability of $[\phi(\theta)f_\theta(\theta)/g_\theta(\theta) - Y]$ is reduced, and thus the variance of $\hat{Y}$. This can be accomplished if some **approximation** to $\phi(\theta)$, i.e., to the acceptability region A is known. Some possibilities of using importance sampling techniques were studied, e.g., in [16]. One of such methods, called parametric sampling was used in [36], and other variants of important sampling were used in [2, 40] for yield optimization. There are several other methods of variance reduction, such as the method of control variates, correlated sampling, stratified sampling, antithetic variates, and others [7, 34, 39]. Some of them have been used for statistical circuit design [7, 39].

55.4 Statistical Methods of Yield Optimization

The objective of the yield optimization is to find a vector of designable parameters $x = x_{opt}$, such that $Y(x_{opt})$ is maximized. This is illustrated in Fig. 55.9 for the case of discrete circuits where $e_1 = x_1 + \theta_1$, $e_2 = x_2 + \theta_2$.[11] Case (a) corresponds to low initial yield proportional to the weighted (by the p.d.f. $f_e(e)$) area (hypervolume, in general), represented by the dark shaded part of the tolerance body shown. Case (b) corresponds to optimized yield, obtained by shifting the nominal point (x_1, x_2) to the vicinity of the **geometric** center of the acceptability region A. Because of this geometric property, approximate yield maximization can often be accomplished using methods called **deterministic** or **geometrical design centering.** They solve the yield maximization problem **indirectly** (since yield is not the objective function optimized), using a geometrical concept of **maximizing the distance** from x_{opt} to the boundary of A. The best known method of this class [10, 12] inscribes the largest hypersphere (or other norm-body) into the **simplicial approximation** to the boundary of A, shown in Fig. 55.8(g). This approach is described in more detail in Chapter 54.4, together with other geometrical design centering methods.

There is also a class of methods which can be called "**performance-space oriented** design centering methods." They attempt to maximize the scaled distances of the performances y_j from the lower S_j^U and/or upper S_j^U specifications, leading to approximate design centering. Some of

[11]Alternatively, the model $e_i = x_i(1 + \theta_i)$ can be used, in which case the size of the tolerance body (see Section 55.3) will increase proportionally to x_i.

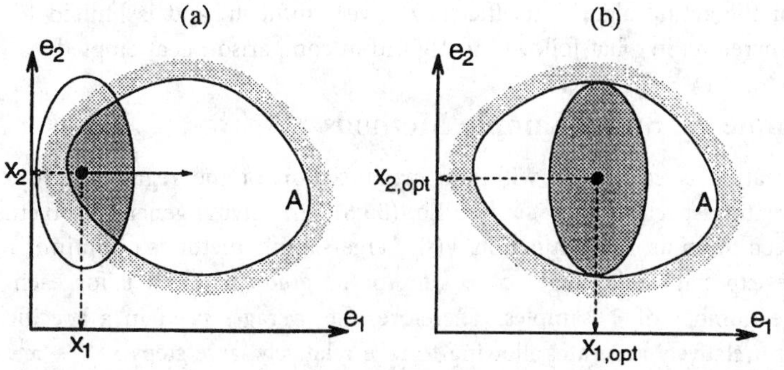

FIGURE 55.9 Interpretation of yield maximization for discrete circuits: (a) initial (low) yield (b) optimized yield.

these methods are also used for variability minimization and performance tuning, which are the most important tenets of design for quality [30], to be briefly discussed later in this section.

The major feature of **statistical** yield optimization methods, referred to as **statistical design centering,** is statistical sampling in either θ-space only or in both θ and x-spaces. Sampling can be also combined with some geometrical approximation to the A-region, such as the segment, or radial-segment approximation.

In dealing with various statistical methods of yield optimization, the type of the transformation $e = e(x, \theta)$ from the θ-space to the circuit-parameter space e, has to be considered. The early, mostly heuristic yield optimization methods were based on the simple additive model $e_i = x_i + \theta_i$, valid for discrete circuits. Because of that, the majority of these methods cannot be used for IC circuit or manufacturing process optimization, where θ and x are in different spaces, or subspaces, and the distribution of θ is defined over some subspace of the e-space, i.e., it is **singular.**

The type of the statistical yield optimization algorithm to be used in practice strongly depends on the type of the circuit to be considered and the information available, namely: whether the transformation $e = e(x, \theta)$ is a simple one: $e_i = x_i + \theta_i$ (or $e_i = x_i(1 + \theta_i)$) (for discrete circuits) or general: $e = e(x, \theta)$ (for IC's); whether the values of $y_j(x, \theta)$ for given x and θ only or also the **derivatives** of y_j with respect to x_k and/or θ_s are available from the circuit simulator; whether some approximation $\tilde{y}(x, \theta)$ to $y(x, \theta)$ is available—either with respect to θ only (for a fixed x), or with respect to both x and θ; whether the analytical form of $f_\theta(\theta)$ is known and $f_\theta(\theta)$ is differentiable w.r.t. θ, or only samples θ^i of θ are given (obtained from a numerical algorithm or from measurements), so, analytical forms of $f_\theta(\theta)$ and its derivatives w.r.t. θ are not known.

Different combinations of the cases listed above require different optimization algorithms. The more general a given algorithm is, the larger number of cases it is able to cover, but simultaneously it can be less efficient than specialized algorithms covering only selected cases. An ideal algorithm would be the one that is **least restrictive** and could use the **minimum necessary** information, i.e., it could handle the most difficult case characterized by: the general transformation $e = e(x, \theta)$, with the values of $y = y(x, \theta)$ only available (generated from a circuit simulator) but without derivatives, no approximation to $y(x, \theta)$ available, and unknown analytic form of $f_\theta(\theta)$ (only the samples of θ given). Moreover, under these circumstances such an algorithm should be reasonably efficient even for large problems, and with the presence of some additional information, should become more efficient. The selected yield optimization algorithms discussed below fulfill the criteria of the algorithm "optimality" to a quite different level of satisfaction. It has to be stressed that due to different assumptions made during the development of different algorithms and the statistical nature of the results, an entirely fair evaluation of the **actual** algorithm efficiency is very difficult, and is limited to some specific cases only. Therefore, in what follows, no algorithm comparison is attempted.

Large-Sample vs. Small-Sample Methods

Yield optimization is concerned with the maximization of the **regression function** $Y(x) = E_\theta\{\phi(x, \theta)\}$ with respect to (w.r.t.) x [see Eq. (55.6)]. In solving general problems of this type, $\phi(\cdot)$ is replaced by an arbitrary function $w(\cdot)$. **Large-sample** methods of optimizing $E_\theta\{w(x, \theta)\}$ calculate the expectation (average) of w (and/or its gradient) w.r.t. θ for each $x^0, x^1, x^2, \ldots$ from a **large** number of θ^i samples. Therefore, the averages used in a specific optimization procedure are relatively accurate, allowing to take relatively large steps $x^{k+1} - x^k$. On the other hand, small-sample methods use just a few (very often just one) samples of $w(x, \theta^i)$ for any given point x, and make relatively **small** steps in the x-space, but they utilize also a special **averaging procedure**, which calculates the average of w or its gradient over a certain number of steps. So, in this case, the averaging in θ-space and progression in the x-space are **combined**, while in the large-sample methods they are **separated.** Both techniques have proven convergence under certain (different) conditions. The majority of yield optimization methods belong to the large-sample category (but some can be modified to use a small number of samples per iteration). A class of small-sample yield optimization methods was proposed in [50] and is based on the well-known techniques of **stochastic approximation** [34], to be discussed later in this section.

Methods Using Standard Deterministic Optimization Algorithms

The most natural method of yield optimization would be to estimate the yields $Y(x^0), Y(x^1), \ldots$ from a large number of samples (as described in the previous section) for each $x^0, x^1, \ldots$ of the sequence $\{x^k\}$ generated by a standard, nonderivative deterministic search algorithm, such as the simplex method of Nelder and Mead, Powell, or other algorithms discussed in [14]. This is very appealing, since most of the conditions for algorithm's "optimality" are fulfilled: it would work for any $e = e(x, \theta)$ and only the **values** of $y = y(x, \theta)$ and the **samples** of θ would be required. However, if no approximation to $y = y(x, \theta)$ was available, the method would require tens of thousands of circuit analyses, which would be prohibitively expensive. Moreover, if the number of samples per iteration was reduced to increase efficiency, the optimizer would be receiving a highly noise-corrupted information leading to poor algorithm convergence or divergence, since standard optimization algorithms work poorly with noisy data (special algorithms, able to work under **uncertainty**—such as stochastic approximation algorithms—have to be used).

If some approximating functions $\hat{y} = \hat{y}(x^k, \theta)$ are available **separately** for each x^k, a large number of MC analyses can be cheaply performed, reducing the statistical error. In practice, such an approach is most often too expensive, due to the high cost of obtaining the approximating formulas, if the number of important θ parameters is large. The approximating functions $\hat{y}_j(x, \theta)$ for each y_j can be also created in the *joint* (x, θ) space [56,61]. In [45], an efficient new approximating methodology was created, highly accurate for a relatively large range of the x_i values. However, also in this case the dimension of the joint space (x, θ) cannot be too large, since the cost of **obtaining** the approximating functions $\hat{y}_j(x, \theta)$ becomes itself prohibitively high. Because of these difficulties, several **dedicated** yield optimization methods have been developed, for which the use of function approximation is not required. Some of these methods are described in what follows.

Large-Sample Heuristic Methods for Discrete Circuits

These methods have been developed mostly for discrete circuits, for which $e_i = x_i + \theta_i$. Only $y^i = y(x + \theta^i)$ function values are required, for the samples θ^i obtained in an arbitrary way.

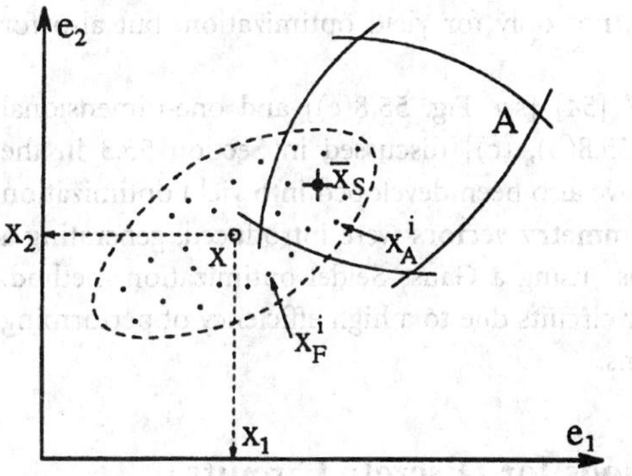

FIGURE 55.10 Interpretation of the original centers of gravity method.

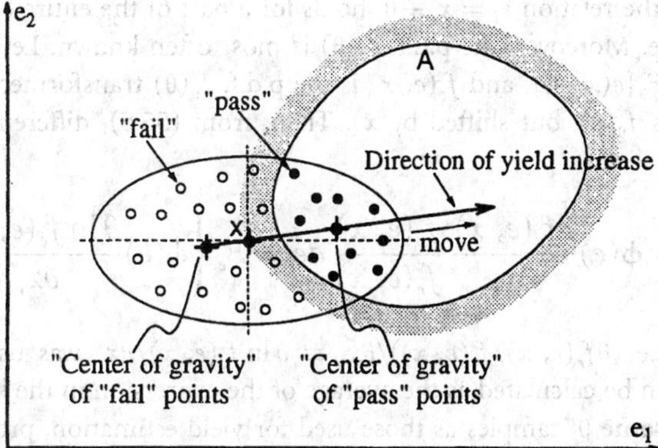

FIGURE 55.11 Interpretation of the modified centers of gravity method.

Approximation in the e-space can be constructed to increase efficiency, but for discrete circuits the number of θ parameters can be large (proportional to the number of active devices, since no correlations between different devices exist), so the use of approximation is most often not practical. The most typical representative of this class is the centers of gravity method [39]. The method is based on a simple observation that if $\overline{x}_S$ is the center of gravity of "pass" points (as shown in Fig. 55.10), defined as $\overline{x}_S = (x_A^1 + x_A^2 + \cdots + x_A^{N_A})/N_A$, where N_A is the number of points x_A^i falling into the A-region, then a step from x to $\overline{x}_S$ will improve yield. In [37], also the center of gravity of the "fail" points $\overline{x}_F = (x_F^1 + x_F^2 + \cdots + x_F^{N_F})/N_F$ was defined, and the direction of yield increase taken as going from $\overline{x}_F$ through $\overline{x}_A$, as shown in Fig. 55.11. Moving in this direction with the step-size equal to $\mu(\overline{x}_S - \overline{x}_F)$, where $\mu \approx 0.2 - 2$ (often taken as $\mu = 1$) leads to a sequence of optimization steps, which is stopped if $\|\overline{x}_S - \overline{x}_F\|$ is less than a predefined small constant. This is based on the property (proved for a class of p.d.f.'s in [43]) that, under some conditions, at the yield maximum $\|\overline{x}_S - \overline{x}_F\| = 0$ (and $\overline{x}_S = \overline{x}_F = \hat{x}$, where $\hat{x}$ is the point of the yield maximum). It was also shown in [43] that for the normal p.d.f. $f_\theta(\theta)$ with zero correlations and all standard deviations $\sigma_{\theta_i} = \sigma_\theta$, $i = 1, \ldots, t$, equal, the "centers-of-gravity" direction coincides with the yield gradient direction. However, with correlations and σ_{θ_i}'s not equal, the two directions can be quite different. Various schemes, aimed at the reduction of the total required number of analyses were developed based on the concepts of "re-usable" points [39].

In [18, 19] the original centers of gravity method was significantly improved, introducing a concept of "Gaussian adaptation" of the covariance matrix of the sampled points of θ^i, such that they (temporarily) adopt to the shape of the acceptability region, leading to higher (optimal) efficiency of the algorithm. The method was successfully used on large industrial design examples, involving as many as 130 designable parameters, not only for yield optimization, but also for standard function minimization.

The methods of "radial exploration of space" [54] [see Fig. 55.8(e)] and one-dimensional orthogonal searches (ODOS) [26, 27] [see Fig. 55.8(b), (c)] discussed in Section 55.3 in the context of Acceptability Region approximation, have also been developed into yield optimization methods: in the "radial exploration" case the **asymmetry vectors** were introduced, generating a direction of yield increase, and in the ODOS case using a Gauss-Seidel optimization method. Both techniques were especially efficient for linear circuits due to a high efficiency of performing circuit analyses in radial and orthogonal directions.

Large-Sample, Derivative-Based Methods for Discrete Circuits

For discrete circuits, the relation $e_i = x_i + \theta_i$ holds for a part of the entire e vector, so, x_i and θ_i are in the same space. Moreover, the p.d.f. $f_\theta(\theta)$ is most often known. Let x denote the vector of expectations $x \equiv E_\theta\{e(x, \theta)\}$, and $f_e(e, x)$ is the p.d.f. $f_\theta(\theta)$ transformed to the e-space (i.e., of the same shape as $f_\theta(\theta)$, but shifted by x). Then, from (55.5), differentiating w.r.t. x_i, one obtains

$$\frac{\partial Y(x)}{\partial x_i} = \int_{R^n} \phi(e) \frac{\partial f_e(e, x)}{\partial x_i} \frac{f_e(e, x)}{f_e(e, x)} de = E_e \left\{ \phi(e) \frac{\partial \ln f_e(e, x)}{\partial x_i} \right\} \tag{55.10}$$

where the equivalence $(\partial f_e(e, x)/f_e(e, x))/\partial x_i \equiv \partial \ln f_e(e, x)/\partial x_i$ was used. Therefore, yield derivatives w.r.t. x_i can be calculated as the **average** of the expression in the braces of Eq. (55.10), calculated from **the same** θ^i samples as those used for yield estimation, provided that the p.d.f. $f_e(e, x)$ is **differentiable** w.r.t. x (e.g., the normal or log-normal p.d.f.'s are differentiable, but the uniform p.d.f. is not). Notice that instead of sampling with the p.d.f. $f_e(e, x)$, some other (better) p.d.f. $g_e(e, x)$ can be used as in the **importance sampling** yield estimation [see Eq. (55.8)]. Then

$$\frac{\partial Y(x)}{\partial x_i} = E_e \left\{ \phi(e) \frac{\partial f_e(e, x)}{\partial x_i} \left(\frac{f_e(e, x)}{g_e(e, x)} \right) \right\} \tag{55.11}$$

where sampling is performed with the p.d.f. $g_e(e, x) \neq 0$. This technique was used in [2, 36, 51] (to be discussed below). Consider the multivariate normal p.d.f., with the positive definite covariance matrix K

$$f_e(e) = \frac{1}{(2\pi)^{t/2}\sqrt{\det K}} \exp\left[-\frac{1}{2}(e - x)^t K^{-1} (e - x) \right] \tag{55.12}$$

where $e - x \equiv \theta$ (discrete circuits), and $\det K$ is the determinant of K. Then, it can be shown that the yield gradient $\nabla_x Y(x)$ is expressed by

$$\nabla_x Y(x) = E\{\phi(e) K^{-1}(e - x)\} = Y(x) K^{-1}(\overline{x}_S - x) \tag{55.13}$$

where $\overline{x}_S$ is the center of gravity of "pass" points. If yield $Y(x)$ is a continuously differentiable function of x, then the necessary condition for the yield maximum is $\nabla_x Y(\hat{x}) = 0$, which combined with Eq. (55.13) means that the stationary point $\hat{x}$ for the yield function (the yield maximum if $Y(x)$ is also concave) is $\hat{x} = \overline{x}_S$, the center of gravity of the pass points. This result justifies (under the assumptions stated above) the centers of gravity method of yield optimization (since its objective is to make $\hat{x} = \overline{x}_S \equiv \overline{x}_F$).

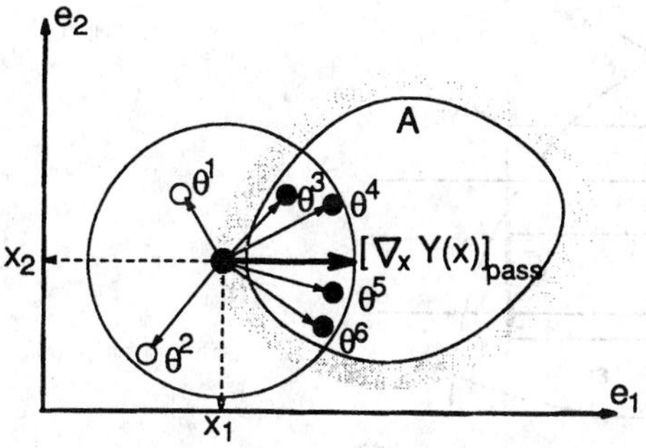

FIGURE 55.12 Interpretation of the yield gradient formula for normal p.d.f., with no correlations and $\sigma_{\theta_1} = \sigma_{\theta_2} = 1$. For the "pass" points (black dots) $[\nabla_x Y(x)]_{pass} \approx (\theta^3 + \theta^4 + \theta^5 + \theta^6)/4$; for the "fail" points (white dots) $[\nabla_x Y(x)]_{fail} = (-\theta^1 - \theta^2)/2$. The two estimators can be combined together (equal weighting assumed): $\nabla_x Y(x) = (-\theta^1 - \theta^2 + \theta^3 + \theta^4 + \theta^5 + \theta^6)/6$. It is clearly seen that the two yield gradient estimators coincide with the center-of-gravity of the "pass" and "fail" points, respectively.

For $K = \text{diag}\{\sigma_{e_1}^2, \ldots, \sigma_{e_t}^2\}$, i.e., with zero correlations, the yield gradient w.r.t x is expressed as

$$\nabla_x Y(x) = E_e\left\{\phi(e)\left[\frac{\theta_1}{\sigma_{\theta_1}^2}, \ldots, \frac{\theta_t}{\sigma_{\theta_t}^2}\right]^t\right\} \tag{55.14}$$

where $\sigma_{\theta_i} \equiv \sigma_{e_i}$ was used instead of σ_{e_i}. It can be readily shown that for all $\sigma_{\theta_1} = \cdots = \sigma_{\theta_t} = \sigma_\theta$ equal, the yield gradient direction coincides with the center-of-gravity direction [43] (Fig. 55.12).

Since all higher order derivatives of the normal p.d.f. exist, all higher order yield derivatives can also be estimated from the same sampled points θ^i, as those used for yield estimation. The yield gradient can also be calculated from the "fail" points simply using the $\phi_F(\cdot) = \phi(\cdot) - 1$ indicator function in all the expressions above. Then, the two resulting estimators can be combined as one joint average, as it was done in the $\overline{x}_S$, $\overline{x}_F$-based centers of gravity method. Actually, there exists an optimal **weighted** combination of the two estimators for any given problem, resulting in the minimum variability of the gradient, but it is difficult to precisely determine in practice. A general rule is that at the beginning of optimization, when x is far away from $\hat{x}$ (the optimal point), the yield gradient estimator based on the "pass" points should be more heavily weighted; the opposite is true at the end of optimization, when the "fail" points carry more precise gradient information. An interpretation of the yield gradient formula (55.14) is shown in Fig. 55.12, for the case where $\sigma_{\theta_1} = \sigma_{\theta_2} = 1$.

In the majority of practical applications of large-sample derivative methods, it was assumed that $f_e(e)$ was normal. A typical iteration step is made in the gradient direction

$$x^{k+1} = x^k + \alpha_k \nabla_x Y(x^k) \tag{55.15}$$

where α_k is most often selected empirically, since yield maximization along the gradient direction is too expensive, unless some approximating functions $\hat{y} = \hat{y}(x + \theta)$ are used (normally, this is not the case for the class of methods discussed). Since the number of points θ^i sampled for each x^k, is large, the main difference between various published algorithms is how to most efficiently use the information already available. The three methods to be discussed (introduced almost at the same time [2, 36, 51]), utilize for that purpose some form of **importance sampling**, discussed in Section 55.3.

In [36], a "parametric sampling" technique was proposed, in which the θ^i points were sampled with the p.d.f. $g_e(e, x)$ in a broader range than for the original p.d.f. (i.e., all the σ_{θ_i}'s were artificially increased). All points sampled were stored in a database, and the gradient-direction steps made according to (55.15). The importance sampling-based gradient formula (55.11) was used in subsequent iterations within the currently available database. Then, a new set of points was generated and the whole process repeated.

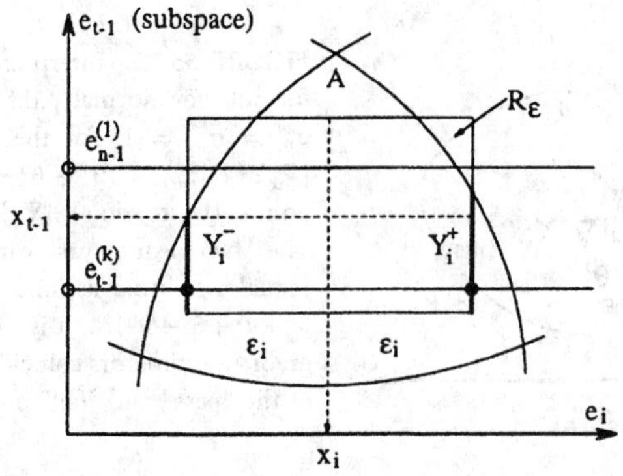

FIGURE 55.13 Yield and yield derivative estimation for uniform distribution. Y^+ and Y^- denote symbolically the "yields" calculated on the $t-1$ dimensional faces of the tolerance hypercube R_ε. The $e_{t-1}^{(k)}$ points are sampled in the $t-1$ dimensional subspaces of e_{t-1} parameters, with the uniform p.d.f. $f_{e_{t-1}}(e_{t-1})$.

The methods developed in [51] and [2] were also based on the importance sampling concept, but instead of using the gradient steps as in (55.15), the yield gradient and **Hessian**[12] matrices were calculated and updated within a given database. Then, a more efficient Newton's direction was taken in [51] or a specially derived and efficient "yield prediction formula" used in [2]. In order to deal with the singularity or nonpositive definiteness of the Hessian matrix (which is quite possible due to the randomness of data and the behavior of $Y(x)$ itself), suitable Hessian **corrections** were implemented using different kinds of the Hessian matrix decomposition (Cholesky-type in [51] and eigenvalue decomposition in [2]).

As it was the case for the heuristic methods, the methods just discussed are relatively insensitive to the dimensionality of x and θ spaces.

For the **uniform** p.d.f., centered at $e = x$ (see Fig. 55.13) and defined within a hyperbox $x_i - \varepsilon_i \le e_i \le x_i + \varepsilon, i = 1, \ldots, t$, where ε_i are element tolerances (see Fig. 55.2), the yield gradient formula (55.10) cannot be used, since the uniform p.d.f. is nondifferentiable w.r.t. x_i. It can be shown [26, 42, 43] that yield can be calculated by sampling in the $t-1$ dimensional subspace of the e-space, represented by $e_{t-1} \equiv (e_1, \ldots, e_{i-1}, e_{i+1}, \ldots, e_t)$ and analytical integration in the one-dimensional subspace e_i, as shown in Fig. 55.13. Using this approach, it can be further proved [43] that the yield derivatives w.r.t. x_i are expressed by

$$\frac{\partial Y(x)}{\partial x_i} = \frac{1}{2\varepsilon_i}(Y_i^+ - Y_i^-) \tag{55.16}$$

where Y_i^+, Y_i^- are "yields" calculated on the faces of the $t-1$ tolerance hyperbox R_ε, corresponding to $x_i + \varepsilon_i$ and $x_i - \varepsilon_i$, respectively. Calculation of these "yields" is very expensive, so in [43][13] different algorithms improving efficiency were proposed. In [53], an approximate method using efficient 3-level orthogonal array (OA) sampling on the faces of R_ε was proposed, in which (due to specific properties of OA's) **the same** sampled points were utilized on different faces (actually one third of **all** sampled points were available for a single face). This has lead to substantial computational savings and faster convergence.

[12]Matrix of second derivatives.

[13]In [43] general formulas for yield gradient calculation for *truncated* p.d.f.'s were derived.

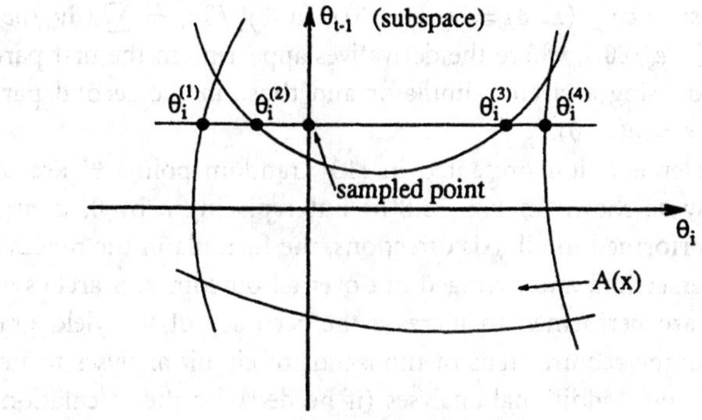

FIGURE 55.14 Interpretation of "local yield" calculation along a parallel.

A Large-Sample, Derivative-Based Method for Integrated Circuits

In this case, which is typical of IC yield optimization, yield gradient calculations cannot be performed in the e-space, as was the case for discrete circuits. In general, yield gradient could be calculated by differentiating the $\phi(e(x,\theta)) \equiv \phi(x,\theta)$ term in the θ-space-based yield formula (55.6), derived for the general case, where the general transformation $e = e(x,\theta)$ is used. Differentiation of $\phi(x,\theta)$ is, however, not possible in the traditional sense, since $\phi(x,\theta)$ is a nondifferentiable unit step function determined over the acceptability region A. One possible solution was proposed in [15]. In what follows, a related, but more general method proposed in [13] is discussed.

It was first shown in [13] that yield can be evaluated as a **surface-integral** rather than the **volume-integral** (as it is normally done using the MC method). To understand this, observe that yield can be evaluated by sampling in the $t-1$ subspace of the t-dimensional θ-subspace, as shown in Fig. 55.14, evaluating "local yields" along the lines parallel to the θ_i axis, and averaging all the local yields.[14] Each "local yield" can be evaluated using the values of the cumulative (conditional) distribution function[15] along each parallel, at the points of its intersection with the boundary of the acceptability region A. This process is equivalent to calculating the surface integral over the values of the cumulative density function calculated (with appropriate signs) on the boundary of the acceptability region.

The next step in [13] was to differentiate the "surface-integral" based yield formula w.r.t. x_i, leading to the following yield gradient formula

$$\nabla Y(x) = E_{\theta_{t-1}}\left\{\sum_k f_{\theta_i}(\theta_i^{(k)}) \cdot \left.\frac{\nabla_x y_a(x,\theta)}{|\partial y_a(x,\theta)/\partial\theta_i|}\right|_{x,\theta_{t-1},\theta_i^{(k)}}\right\} \tag{55.17}$$

where summation is over all intersection points $\theta_i^{(k)}$ of the parallel shown in Fig. 55.14 with the boundary of $A_\theta(x)$; y_a is that **specific** performance function $y_j(x,\theta)$ which, out of all the other performances, actually determines the boundary of the acceptability region at the $\theta_i^{(k)}$ intersection point.[16] The gradient $\nabla_x y_a(x,\theta)$, and the derivative $\partial y_a(x,\theta)/\partial\theta_i$ have to be calculated for every fixed sampled point (x, θ_{t-1}), at each intersecting point $\theta_i^{(k)}$ shown in Fig. 55.14. Observe that the derivative calculations in (55.17) can (and often will have to) be

[14]Identical technique was used in the previous section [42, 43] to derive the yield derivative formula (16) (see Fig. 55.13).

[15]If θ_i's are independent, it is the *marginal* p.d.f. of θ_i, as it was assumed in [13].

[16]At this point, a specific equation $y_a(x, \theta_{t-1}, \theta_i^{(k)}) = S_a$ (where $S_a \equiv S_a^L$ or $S_a \equiv S_a^U$ are lower and upper bounds on y_a, respectively) must be *solved* to find $\theta_i^{(k)}$.

performed in two steps: since $y_a(x, \theta) = y_a(e(x, \theta))$, so, $\partial y_a/\partial x_p = \sum_s(\partial y_a/\partial e_s)(\partial e_s/\partial x_p)$, and $\partial y_a/\partial\theta_i = \sum_s(\partial y_a/\partial e_s)(\partial e_s/\partial\theta_i)$, where the derivatives appearing in the first parentheses of both formulas are calculated using a circuit simulator and those in the second parentheses from a given statistical model $e = e(x, \theta)$.

In the practical implementation presented in [13], random points θ^r are sampled in the θ space, in the same way as shown in Fig. 55.8(b) but replacing e by θ, then searches for the intersections $\theta_i^{(k)}$ are performed in **all** axis directions, the formula in the braces of Eq. (55.17) is calculated for each intersection, and averaged out over all outcomes. Searches along the parallel lines in **all** directions are performed to increase the accuracy of the yield gradient estimator. Observe that this technique requires tens of thousands of circuit analyses to iteratively find the intersection points $\theta_i^{(k)}$, plus additional analyses (if needed) for the calculation of the gradient and the derivatives in (55.17). This problem has been circumvented in [13] by constructing approximating functions $\hat{y} = \hat{y}(x, \theta)$ w.r.t. θ for each x, together with approximating functions for all the derivatives.[17] Due to a high level of **statistical** accuracy obtained in evaluating both yield and its gradients, an efficient, gradient-based deterministic optimization algorithm, based on sequential quadratic programming was used, requiring a small number of iterations (from 5 to 11 for the examples discussed in [13]). The gradient $\nabla_x y_a(x, \theta)$ was either directly obtained from a circuit simulator, or (if not available) using (improved) finite difference estimators. The method showed to be quite efficient for a moderate size of the θ-space (10–12 parameters).

The resulting yield optimization method is independent of the form of $e = e(x, \theta)$, the derivatives of y_j w.r.t. to both x_k and θ_s are required and the **analytical** form of $f_\theta(\theta)$ and its cumulative function distribution must be known. The method cannot practically work without constructing the approximating functions $\tilde{y} = y(x, \theta)$ (approximating functions in the joint space (x, θ) could also be used, if available).

Small-Sample Stochastic Approximation-Based Methods

Standard methods of nonlinear programming perform poorly in solving problems with statistical errors in calculating the objective functions (i.e., yield) and its derivatives.[18] One of the methods **dedicated** to the solution of such problems is the stochastic approximation (SA) approach [33] developed for solving the regression equations, and then adopted to the unconstrained and constrained optimization by several authors. These methods are aimed at finding a minimum (maximum) of a function corrupted by noise (a regression function). The SA methods were first applied to yield optimization and statistical design centering in [49, 50]. The theory of SA methods is well established, so its application to yield optimization offers the theoretical background missing e.g., in the heuristic methods of Section 55.3. As compared to the large-sample methods, the SA algorithms to be discussed use a **few** (or **just one**) randomly sampled points per iteration, which is compensated for by a large number of iterations exhibiting a trend toward the solution. The method tends to bring large initial improvements with a small number of circuit analyses, efficiently utilizing the high content of the deterministic information present at the beginning of optimization.

In 1951, in their pioneering work, Robins and Monro [33] proposed a scheme for finding a root of a regression function, which they named the **stochastic approximation** procedure. The problem was to find a zero of a function, whose "noise corrupted" values could be observed only, namely $G(x) = g(x) + \theta(x)$ where $g(x)$ is a given (unknown) function of x, θ is a random variable, such that $E\{\theta\} = 0$, and $\text{var}\{\theta\} \leq L < \infty$. Therefore, a zero of the regression function

[17]Low degree polynomials were used with very few terms, generated from a *stepwise regression* algorithm.

[18]This was the reason a high level of accuracy was required while calculating both yield and its derivatives using the method described in the previous section (and *consistency* between the yield and gradient estimators), otherwise, the deterministic optimization algorithm would often diverge, as observed in [13].

$g(x) \equiv E\{G(x)\} = 0$ was to be found. The SA algorithm proposed in [33] works as follows: given a point $x^{(k)}$, set the next point as $x^{(k+1)} = x^{(k)} - a_k G(x^{(k)})$. The sequence of $\{x^{(k)}\}$ points converges to the solution $\hat{x}$ under some conditions (one of them is, for instance, that a_k can change according to the harmonic sequence $\left\{1, \frac{1}{2}, \frac{1}{3}, \frac{1}{4}, \ldots\right\}$). Assuming that $G(x) \equiv \xi^k$ is a "noise corrupted" observation of the **gradient** of a regression function $f(x) = E_\theta\{w(x, \theta)\}$, the algorithm can be used to find a stationary point (e.g. a maximum) of $f(x)$, since this is equivalent to finding $\hat{x}$ such that $E\{G(\hat{x})\} = E\{\xi(\hat{x})\} = \nabla_x Y(\hat{x}) = 0$. For yield optimization $f(x) \equiv Y(x) = E\{\phi(x, \theta)\}$. The simplest scheme that can be used, if the yield gradient estimator is available is

$$x^{k+1} = x^k + \tau_k \xi^k \tag{55.18}$$

where $\xi^k \equiv \widehat{\nabla_x Y}(x^k)$ is an estimate of the yield gradient, based on **one** (or more) points θ^i sampled with the p.d.f. $f_\theta(\theta)$ and $\tau_k > 0$ is the step length coefficient selected such that the sequence: $\{\tau_k\} \to 0$, and $\sum_{k=0}^{\infty} \tau_k = \infty$, $\sum_{k=0}^{\infty} \tau_k^2 < \infty$ (e.g., the harmonic series $\{1, \frac{1}{2}, \frac{1}{3}, \ldots\}$ fulfills these conditions). For the convergence with probability one, it is also required that the conditional expectation $E\{\xi^k \mid x^1, x^2, \ldots, x^k\} = \nabla Y(x^k)$. The algorithm (55.18) is similar to the steepest ascent algorithms of nonlinear programming, so, it will be slowly convergent for ill-conditioned problems. A faster algorithm was introduced in [35] and used for yield optimization in [49, 50]. It is based on the following iterations

$$x^{k+1} = x^k + \tau_k d^k \tag{55.19}$$

$$d^k = (1 - \rho_k) d^{k-1} + \rho_k \xi^k, \quad 0 \le \rho_k \le 1 \tag{55.20}$$

where ξ^k is a (one- or more-point) estimator of $\nabla_x Y(x^k)$ and $\{\tau_k\} \to 0$, $\{\rho_k\} \to 0$ are nonnegative coefficients. $d^{(k)}$ is a convex combination of the previous (old) direction d^{k-1} and the new gradient estimate ξ^k, so the algorithm is an analog of a more efficient, conjugate gradient method. Formula (2) provides **gradient averaging.** The ρ_k coefficient controls the "memory" or "inertia" of the search direction d^k, as shown in Fig. 55.15. If ρ_k is small, the "inertia" of the algorithm is large, i.e. the algorithm tends to follow the previous gradient directions. For convergence with probability one, the same conditions must hold as those for (55.18).

The coefficients τ_k and ρ_k automatically determined based on some heuristic statistical algorithms proposed in [35]. Several other enhancements were used to speed up the convergence of the algorithm, especially in its much refined version proposed in [35]. For solving the yield

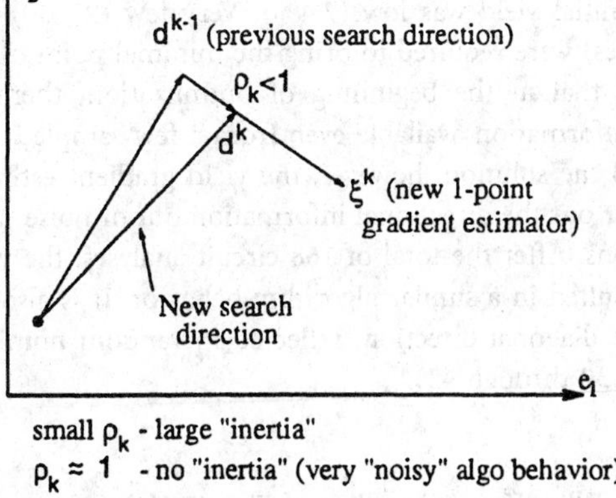

FIGURE 55.15 An illustration of the gradient averaging equation.

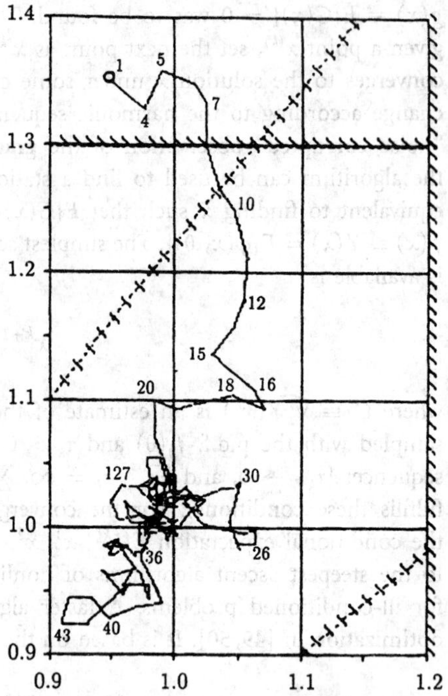

FIGURE 55.16 A typical trajectory of the stochastic approximation algorithm (Example 1).

optimization problems, the one- or two-point yield gradient estimator is found from (55.10) in general, and from (55.13) or (55.14) for the normal p.d.f.

Example 1

A simple two-dimensional case will be considered to illustrate the algorithm properties. The acceptability region (for a voltage divider) [49] is defined in the two dimensional space (e_1, e_2) by the inequalities:

$$0.45 \leq \frac{e_2}{e_1 + e_2} \leq 0.55; \quad 0 \leq e_1 \leq 1.2; \quad 0 \leq e_2 \leq 1.3 \qquad (55.21)$$

where $e_i = x_i + \theta_i$ and the p.d.f. of θ_i is normal with $E\{\theta_1\} = E\{\theta_2\} = 0$, and $\text{cov}\{\theta_1, \theta_2\} = 0$ (no correlations).

A typical algorithm trajectory is shown in Fig. 55.16. The one-sample yield gradient formula (55.14) was used. The initial yield was low (7.4%). Very few (25–30) iterations (equal to the number of circuit analyses) were required to bring the minimal point close to the final solution. This is due to the fact that at the beginning of optimization, there is a high contents of **deterministic** gradient information available even from a few sampled points, so the algorithm progress is fast. Close to the solution, however, the yield gradient estimator is very noisy, and the algorithm has to filter out the directional information out of noise, which takes the majority of the remaining iterations. After the total of 168 circuit analyses, the yield increases to 85.6%. Other starting points resulted in a similar algorithm behavior. It is also observed that the yield optimum is "flat" in the diagonal direction, reflected in random nominal point movement in this direction (iterations 30 through 43).

Example 2

Parametric yield for the Sallen-Key active filter of Fig. 55.6(a) (often used in practice as a test circuit) with the specifications on its frequency response shown in Fig. 55.6(b) was optimized (recall, that the shape of the acceptability region for this case was very complicated, containing internal non-feasible regions ("holes")). All R,C variables were assumed designable. The relative standard deviations were assumed equal to 1% for each element, the p.d.f. was normal with 0.7 correlation coefficient between the like elements (R's *or* C's) and zero correlations between different elements. The initial yield was 6.61%. Using the SA algorithm as in the previous example, 50 iterations (equal to the number of circuit analyses) brought yield to 46.2%; the next 50 iterations to 60%, while the remaining 132 iterations increased yield to only 60.4% (again the initial convergence was very fast). The algorithm compared favorably with the Hessian matrix-based, large-sample method discussed previously [51], which required about 400 circuit analyses to obtain the same yield level, and whose initial convergence was also much slower. It has to be stressed, however, that the results obtained are **statistical** in nature and it is difficult to draw strong general conclusions. One observation (confirmed also by other authors) is that the SA-type algorithms provide, in general, fast initial convergence into the **neighborhood** of the optimal solution, as it was shown in the examples just investigated.

Small-Sample Stochastic Approximation Methods for Integrated Circuits

The methods of yield gradient estimation for discrete circuits cannot be used for IC's because of the form of the $e = e(x, \theta)$ transformation, as discussed at the beginning of Section 55.4. Previously in this section, a complicated algorithm was described for gradient estimation and yield optimization in such situations. In this section, a simple method based on **random perturbations** in x-space, proposed in [47, 48], is described. It is useful in its own merit, but especially in those cases, where the conditions for the application of the yield gradient formula (55.17) are hard to meet, namely: the cost of constructing the approximating functions $\hat{y} = \hat{y}(x, \theta)$ for fixed x is high and calculating the gradient of y w.r.t. x is also expensive (which is the case, e.g., if the number of important x and θ parameters is large), and the analytical form of $f_\theta(\theta)$ is not available (as it is required in (55.17)).

The method applications go beyond yield optimization: a general problem is to find a minimum of a general regression function $f(x) = E_\theta\{w(x, \theta)\}$, using the SA method, in the case where the gradient estimator of $f(x)$ **is not directly available**. Several methods have been proposed to estimate the gradient **indirectly**, all based on adding some **extra perturbations** to x parameters. Depending on their nature, size, and the way the perturbations are **changed** during the optimization process, different interpretations result, and different problems can be solved. In the simplest case, some extra deterministic perturbations (usually double-sided) are added **individually** to each x_k (one-at-a-time), while random sampling is performed in the θ-space, and the estimator of the derivative of $f(x)$ w.r.t. x_k is estimated from the difference formula $\hat{\xi}_k = \widehat{\partial f(x)/\partial x_k} = (1/N)\sum_{i=1}^{N}\{[w(x + a_k e_k, \theta_1^i) - w(x - a_k e_k, \theta_2^i)]/(2a_k)\}$, where $a_k > 0$ is the size of the perturbation step and e_k is the unit vector along the x_k coordinates axis. Usually $\theta_1^i \equiv \theta_2^i$ to reduce the variance of the estimator. Normally, $N = 1$. Other approaches use **random direction** derivative estimation by sampling points randomly on: a unit sphere of radius a [34], randomly at the vertices of a hypercube in the x-space [38], or at the points generated by orthogonal arrays [53], commonly used in the design of experiments. Yet another approach, dealing with **nondifferentiable** p.d.f.'s was proposed in [55]. In what follows, the random perturbation approach resulting in **convolution function smoothing** is described for a more general case, where a *global* rather than a local minimum of $f(x)$ is to be found.

The multi-extremal regression function $f(x)$ defined above, can be considered a superposition of an uni-extremal function (i.e., having just one minimum) and other multi-extremal functions that add some deterministic "noise" to the uni-extremal function (which itself has also some "statistical noise"—due to θ—superimposed on it). The objective of convolution smoothing can be visualized as "filtering out" both types of noise and performing minimization on the "smoothed" uni-extremal function (or on a family of these functions), in order to reach

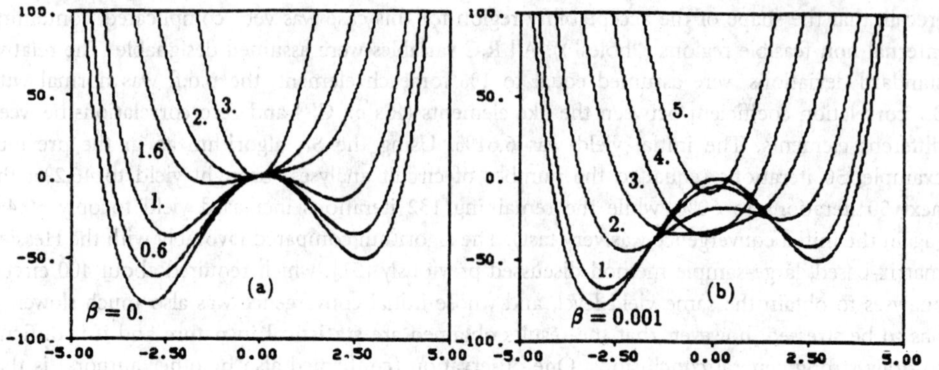

FIGURE 55.17 Smoothed functional $f(x, \beta)$ for different β's using: (a) Gaussian kernel (b) uniform kernel.

the global minimum. Since the minimum of the smoothed uni-extremal function does not, in general, coincide with the global function minimum, a sequence of minimization runs is required with the amount of smoothing eventually reduced to zero in the neighborhood of the global minimum. The smoothing process is performed by averaging $f(x)$ over some region of the n dimensional parameter space x using a proper weighting (or smoothing) function $\hat{h}(x)$, defined below. Let the n-dimensional vector η denote a vector of random perturbations; it is added to x to create the convolution function [34]

$$\tilde{f}(x,\beta) = \int_{R^n} \hat{h}(\eta,\beta) f(x-\eta)\, d\eta$$
$$= \int_{R^n} \hat{h}(x-\eta,\beta) f(\eta)\, d\eta = E_\eta\{f(x-\eta)\} \tag{55.22}$$

where $\tilde{f}(x, \beta)$ is the **smoothed approximation** to the original multi-extremal function $f(x)$, and the kernel function $\hat{h}(\eta, \beta)$ is the p.d.f. used to sample η. Note that $\tilde{f}(x, \beta)$ can be interpreted as an **averaged** version of $f(x)$ weighted by $\hat{h}(\eta, \beta)$. Parameter β controls the dispersion of $\hat{h}$, i.e., the degree of $f(x)$ **smoothing** (e.g., β can control the standard deviations of $\eta_1 \ldots \eta_n$). $E_\eta\{f(x-\eta)\}$ is the expectation with respect to the random variable η. Therefore, an unbiased estimator $\hat{f}(x, \beta)$ of $\tilde{f}(x, \beta)$ is the average: $\hat{f}(x, \beta) = (1/N)\sum_{i-1}^{N} f(x - \eta^i)$, where η is sampled with the p.d.f. $\hat{h}(\eta, \beta)$. The kernel function $\hat{h}(\eta, \beta)$ should have certain properties discussed in [34], fulfilled by several p.d.f.'s, e.g., the Gaussian and uniform. For the function $f(x) = x^4 - 16x^2 + 5x$, which has two distinct minima, the smoothed functionals, obtained using (55.22), are plotted in Fig. 55.17 for different values of $\beta \to 0$, for: (a) Gaussian and (b) uniform kernels. As seen, smoothing is able to eliminate the local minima of $\tilde{f}(x, \beta)$ if β is sufficiently large. If $\beta \to 0$ then $\tilde{f}(x, \beta) \to f(x)$.

The objective now is to solve the following optimization problem: minimize the smoothed functional $\tilde{f}(x, \beta)$ with $\beta \to 0$ as $x \to \hat{x}$, where $\hat{x}$ is the global minimum of the original function $f(x)$. The modified optimization problem can be written as: minimize $\tilde{f}(x, \beta)$ w.r.t. x, with $\beta \to 0$ as $x \to \hat{x}$. Differentiating (55.22) and using variable substitution, the gradient formula is obtained

$$\nabla_x \tilde{f}(x,\beta) = \int_{R^n} \nabla_\eta \hat{h}(\eta,\beta) f(x-\eta)\, d\eta = \frac{1}{\beta}\int_{R^n} \nabla_\eta h(\eta) f(x-\beta\eta)\, d\eta, \tag{55.23}$$

where $\hat{h}(\cdot)$ is as defined above, and $h(y)$ is a normalized version of $\hat{h}(y)$ (obtained if $\beta = 1$). For normalized multinormal p.d.f. with zero correlations, the gradient of $\tilde{f}(x, \beta)$ is

$$\begin{aligned}\nabla_x \tilde{f}(x,\beta) &= \frac{-1}{\beta}\int_{R^n} \eta f(x-\beta\eta)h(\eta)\,d\eta = \frac{-1}{\beta}E_\eta\{\eta f(x-\beta\eta)\} \\ &= \frac{-1}{\beta}E_{\eta,\theta}\{\eta w(x-\beta\eta,\theta)\}\end{aligned} \tag{55.24}$$

where sampling is performed in x-space with the p.d.f. $h(\eta)$, and in θ-space with the p.d.f. $f_\theta(\theta)$; $E_{\eta,\theta}$ denotes expectation w.r.t. *both* η and θ, and it was taken into account for $f(x)$ is a noise corrupted version of $w(x, \theta)$, i.e., $f(x) = E_\theta\{w(x, \theta)\}$. The unbiased **single-sided** gradient estimator is therefore

$$\hat{\nabla}_x \tilde{f}(x,\beta) = \frac{-1}{\beta}\frac{1}{N}\sum_{i=1}^{N} \eta^i w(x-\beta\eta^i,\theta^i) \tag{55.25}$$

In practice, a **double-sided** estimator [34] of smaller variance is used

$$\hat{\nabla}_x \tilde{f}(x,\beta) = \frac{1}{2\beta}\frac{1}{N}\sum_{i=1}^{N} \eta^i[w(x+\beta\eta^i,\theta_1^i) - w(x-\beta\eta^i,\theta_2^i)]. \tag{55.26}$$

Normally, $N = 1$ for best overall efficiency. Statistical properties of these two estimators (such as their variability) were studied in [52]. To reduce variability, *the same* $\theta_1^i = \theta_2^i$ are usually used in Eq. (55.26) for positive and negative $\beta\eta^i$ perturbations. For yield optimization, $w(\cdot)$ is simply replaced by the indicator function $\phi(\cdot)$. For multi-extremal problems, β values should be originally relatively large and then systematically reduced to some small number rather than to zero. For single-extremal problems (this might be the case for yield optimization) it is often sufficient to perform just a *single* optimization with a relatively small value of β, as it was done in the examples to follow.

Case Study: Process Optimization for Manufacturing Yield Enhancement

The object of the work presented in [48] was to investigate how to modify the MOS control process parameters together with a **simultaneous** adjustment of transistor widths and lengths to maximize parametric yield.[19] To make it possible, both process/device simulator(s) (such as FABRICS, SUPREM, PISCES, etc.) and a circuit simulator must be used. In what follows, FABRICS [12, 22, 24] is used as a process/device simulator. IC technological process parameters are statistical in nature, but variations of some of the parameters (e.g., times of operations, implant doses, etc.) might have small relative variations, and some parameters are common to several transistors on the chip. Because of that, the transformation $e = e(x, \theta)$ (where θ are now process related random variables), is such that standard methods of yield optimization developed for discrete circuits cannot be used. Let $Z_1 = z_1 + \xi_1$ be the process control parameters (doses, times, temperatures), where ξ_1 are random parameter variations, and z_1 are deterministic **designable** parameters. Let $Z_2 = z_2 + \xi_2$ be the designable layout dimensions, where z_2 are designable and ξ_2 are random variations (common to several transistors on a chip). $P = p + \psi$ are process physical parameters (random, nondesignable) such as diffusivities, impurity concentrations, etc. (as above, p is the nominal value and ψ is random). All random

[19]This might be important, e.g., for the process refinement and IC cell *redesign*.

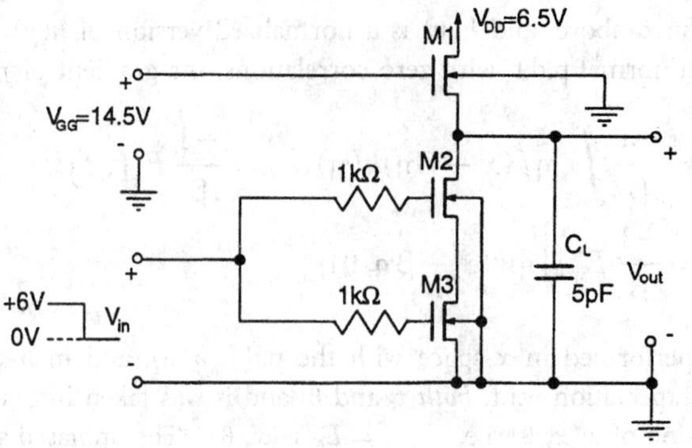

FIGURE 55.18 NMOS NAND gate of Example 1.

perturbations are collected into the vector of random parameters θ, called also the vector of **process disturbances:** $\theta = (\xi_1, \xi_2, \psi)$. The vector of designable parameters $x = (z_1, z_2)$ is composed of the process z_1 and layout z_2 designable parameters. Therefore, x and θ are in **different** spaces (sub-spaces). There are also other difficulties: the analytical form of the p.d.f. of θ is most often not known, since θ parameters are hierarchically generated from a numerical procedure [12], the derivatives of the performances y w.r.t. x and θ are not known from FABRICS and can only be estimated by finite differences, and the θ and x spaces are very large (see below). So, creating approximation and/or finding derivatives using finite differences is expensive. Because of these difficulties, the smoothed-functional approach discussed in this section will be used, as shown in the following example.

Example 1

The objective is to maximize parametric yield for the NMOS NAND gate shown in Fig. 55.18 [48], by automated adjustment of process and layout parameters. Specifications are: $V_0 = V_{out}(t = 0) \leq 0.7$ V, $V_{out}(t_1 = 50 \text{ ns}) > 6.14$ V, circuit area $\leq 2500\ \mu\text{m}^2$. There are 45 designable parameters: all 39 technological process parameters, 6 transistor dimensions, and about 40 noise parameters, so it is a large problem suitable for the use of the SA-based random perturbation method described above.

The initial yield was $Y = 20\%$. After the first optimization using the method of random perturbations with 2% relative perturbations of each of the designable parameters involving the total of 110 FABRICS/SPICE analyses, yield increased to 100% and the nominal area decreased to 2138 μm^2. Then, specs were tightened to $V_{out}(t_2 = 28 \text{ ns}) > 6.14$ V with the same constraints on V_0 and area, causing yield to drop to 10.1%. After 60 FABRICS/SPICE analyses, using the perturbation method, yield increased to: $Y = 92\%$, and area= 2188 μm^2. These much improved results produced the nominal circuit responses shown in Fig. 55.19. Several technological process parameters were changed during optimization in the range between 0.1 to 17%: times of oxidation, annealing, drive-in, partial pressure of oxygen, and others, while the transistor dimensions changed in the range between 0.8% to 6.3%. The cost of obtaining these results was quite reasonable: the total of 170 FABRICS/SPICE analyses, in spite of the large number of optimized and noise parameters. Other examples are discussed in [48].

Generalized Formulation of Yield, Variability, and Taguchi Circuit Optimization Problems

Parametric yield is not the only criterion that should be considered during statistical circuit design. Equally, and often more important is minimization of performance variability caused by various manufacturing and environmental disturbances. Variability minimization has been

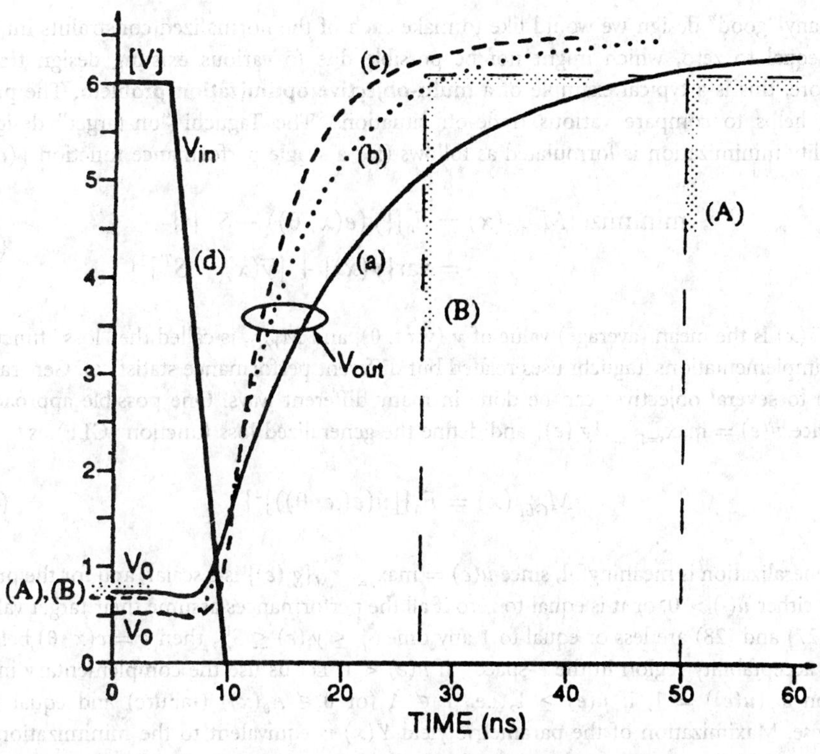

FIGURE 55.19 Nominal transient responses for Example 1: (a) Initial. (b) After first optimization. (c) After tightening the specs and second optimization.

an important issue in circuit design for many years [17] (see [17] for earlier references). It leads (indirectly) to parametric yield improvement. Circuits characterized by low performance variability are regarded as high quality products. Most recently, variability minimization has been re-introduced into practical industrial design due to the work of Taguchi [30]. He has successfully popularized the notion of "Off-line Quality Control" through an intensive practical implementation of his strategy of designing products with low performance variability, tuned to the "target" values S_j^T of the performance functions y_j.

In what follows, a generalized approach proposed in [44] is discussed, in which a broad range of various problems, including yield optimization and Taguchi's variability minimization as special cases can be solved, using the SA approach and the general gradient formulas developed previously in this section.

It is convenient to introduce the M dimensional vector $g(e)$ of **scaled** constraints, composed of the vectors $g^L(e)$, and $g^U(e)$, defined as

$$g_k^L(e) = \frac{S_k^T - y_k(e)}{S_k^T - S_k^L}, \quad k = 1, \ldots, M_L \tag{55.27}$$

$$g_t^U(e) = \frac{y_t(e) - S_t^T}{S_t^U - S_t^T}, \quad t = 1, \ldots, M_U \tag{55.28}$$

where $e = e(x, \theta)$, $M = M_L + M_U \leq 2$ m (note that in general $M \leq 2$ m, since some of the lower or upper specs might not be defined). These constraints are linear functions of $y_i(\cdot)$; they have important properties: $g^L = g^U = 0$ if $y = S^T$, $g^L = 1$ if $y = S^L$, and $g^U = 1$ if $y = S^U$. For $S^L < y < S^U$ and $y \neq S^T$, either g^L or g^U is greater than zero, but never both.

For any "good" design we would like to make **each** of the normalized constraints introduced above equal to zero, which might not be possible due to various existing design trade-offs. Therefore, this is a typical example of a **multi-objective optimization problem.** The proposed scaling helps to compare various trade-off situations. The Taguchi "on-target" design with variability minimization is formulated as follows (for a **single** performance function $y(e)$)

$$\begin{aligned}\text{minimize}\{M_{TAG}(x) &= E_\theta\{[y(e(x,\theta)) - S^T]^2\}\\ &= \text{var}\{y(x)\} + [\overline{y}(x) - S^T]^2\}\end{aligned} \tag{55.29}$$

where $\overline{y}(x)$ is the mean (average) value of y (w.r.t. θ), and M_{TAG}, is called the "loss" function (in actual implementations Taguchi uses related but different performance statistics). Generalization of (29) to **several** objectives can be done in many different ways. One possible approach is to introduce $u(e) = \max_{s=1,\ldots,M}\{g_s(e)\}$ and define the generalized loss function (GLF) as

$$M_{GLF}(x) = E_\theta\{[u(e(x,\theta))]^2\} \tag{55.30}$$

This generalization is meaningful, since $u(e) = \max_{s=1,\ldots,M}\{g_s(e)\}$ is a scalar, and for the proposed scaling either $u(\cdot) > 0$, or it is equal to zero if all the performances assume their target values S_i^T. Since (27) and (28) are less or equal to 1 any time $S_i^L \le y_i(e) \le S_i^U$, then $e = e(x,\theta)$ belongs to A (the acceptability region in the e-space), if $u(e) < 1$. Let us use the complementary indicator function $\phi_F(u(e)) = 1$, if $u(e) \ge 1$, i.e., $e \notin A$ (or $\theta \notin A_\theta(x)$) (failure) and equal to zero otherwise. Maximization of the parametric yield $Y(x)$ is equivalent to the minimization of the probability of failures $F(x) = 1 - Y(x)$, which can be formulated as the following minimization process, w.r.t. x

$$\text{minimize}\left\{\begin{aligned}F(x) &= P\{\theta \notin A_\theta(x)\}\\ &= \int_{R^t}\phi_F(u(x,\theta))f_\theta(\theta)\,d\theta = E_\theta[\phi_F(u(x,\theta))]\end{aligned}\right\} \tag{55.31}$$

The scalar "step" function $\phi_F(\cdot)$ (whose argument is also a scalar function $u(e) = \max_{1,\ldots,M}\{g_s(e)\}$), can now be generalized into a scalar weight function $w(\cdot)$, in the same spirit as in the Zadeh fuzzy set theory [62, 63]. For further generalization, the original p.d.f. $f_\theta(\theta)$ used in (31) is parametrized, multiplying θ by the **smoothing parameter** β to control the dispersion of e, which leads to the following optimization problem, utilizing a **generalized measure** $M_w(x,\beta)$

$$\text{minimize}\{M_w(x,\beta) = E_\theta\{w(u(e(x,\beta\theta)))\}\} \tag{55.32}$$

where $u(e(x,\beta\theta)) = \max_{s=1,\ldots,M}\{g_s(e(x,\beta\theta))\}$. If $0 \le w(\cdot) \le 1$, $M_w(\cdot)$ corresponds to the **probability measure of a fuzzy event** introduced by Zadeh [63]. The choice of $w(\cdot)$ and β leads to different optimization problems and different algorithms for yield/variability optimization. The standard yield optimization problem results if $w(\alpha) = \phi_F(\alpha)$ and $\beta = 1$ [Eq. (55.31)] [Fig. 55.20(a)]. Yield can be also approximately optimized **indirectly**, using a "smoothed" (e.g., sigmoidal) membership function $w(\alpha)$ [Fig. 55.20(e)]. For variability minimization, we have a whole family of possible approaches: (1) The generalized Taguchi approach with $w(\alpha) = \alpha^2$ [see (30)] and $\beta = 1$ [see Fig. 55.20(c)]. (2) If $w(\alpha) = \alpha$ and $\beta = 1$ is kept **constant**, we obtain a **statistical** mini-max problem, since the expected value of the max function is used; this formulation will also lead to performance variability reduction. (3) If $w(\alpha)$ is piecewise

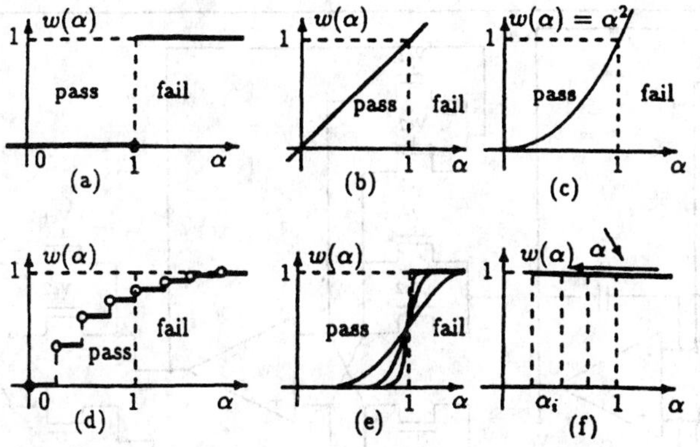

FIGURE 55.20 Various cases of the weight function $w(\alpha)$.

constant [Fig. 55.20(d)], the approach is equivalent to the **income index maximization** with separate **quality classes**, introduced in [28], and successfully used there for the increase of the percentage of circuits belonging to the best classes {i.e., those characterized by small values of $u(e)$); this approach also reduces variability of circuit performance functions. For all the cases, optimization is performed using the proposed SA approach. The smoothed functional gradient formulas (55.25) or (55.26) have to be used, in general, since the "discrete-circuit" type gradient formulas are valid only for a limited number of situations.

Example 1 Performance Variability Minimization for a MOSFET-C Filter

The objective of this example [31] was to reduce performance variability and to tune to target values the performances of the MOSFET-C filter shown in Fig. 55.21. The MOSFETS are used in groups composed of four transistors implementing equivalent resistors with improved linearity. Design specifications are f_0, the center frequency, with the S^L, S^T, S^U values specified as {45, 50, 55} kHz; H_0, the voltage gain at f_0, with the specs {15, 20, 25} dB; and Q, the pole quality factor, with the specs {8, 10, 12}. The Taguchi-like "on-target" design and tuning was performed by minimizing the generalized measure (55.32) with $w(x) = x^2$, using the SA approach with convolution smoothing, since the designable and random parameters are in different (sub)spaces. The circuit has the total number of 90 transistors so its **direct** optimization using transistor-level simulation would be too costly. Therefore, in [31] the operational amplifiers (op-amps) shown in Fig. 55.21 were modeled by a **statistical macromodel**, representing the most relevant op-amp characteristics: DC gain A_0, output resistance R_0, and the -20 dB/dec frequency roll-off. Transistor model parameters were characterized by a statistical model developed in [8], based on 6 **common factors** (see Section 55.3): t_{ox} (oxide thickness), ΔL_n, ΔL_p (length reduction), $N_{SUB,n}$, $N_{SUB,p}$ (substrate doping) and x_{jp} (junction depth). All other transistor model parameters were calculated from the 6 common factors using second-order regression formulas developed in [8]. The major difficulty was to create **statistical** models of the op-amp macromodel parameters A_0, R_0, and $f_{3\,\text{dB}}$ as the functions of the 6 common factors listed above. A special extraction procedure (similar in principle to the model of factor analysis [20]) was developed and the relevant models created. Perfect matching between the transistor model parameter of the 22 transistors of each of the op-amps was assumed for simplicity[20] (correlational dependencies **between** individual parameters for **individual** transistors were maintained using the statistical mode of [8], described above). Moreover, perfect matching between the three macromodels (for the 3 op-amps) was also assumed. Mismatches between the threshold voltages and K_p (gain) coefficients of all the transistors in Fig. 55.21 were taken into account, introducing 48

[20]A more sophisticated model, taking the mismatches into account, was later proposed in [32].

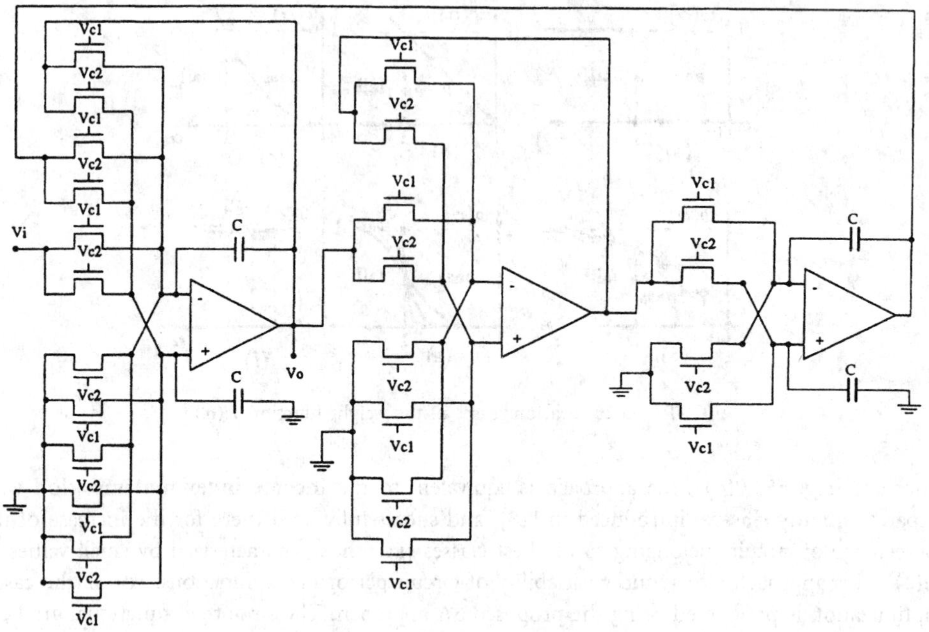

FIGURE 55.21 The MOSFET-C bandpass filter optimized in Example 1.

additional noise parameters. So, including the 6 global noise parameters discussed above, the overall number of θ parameters was 54. Moreover, it was found that even if a large number of noise parameters had small individual effect, their **total** contributions was significant, so, no meaningful reduction of the dimension of the θ-space was justified. Because of that, it was difficult to use approximating models in the θ-space, and even more difficult in the joint (x, θ) space of $7 + 54 = 61$ parameters. Therefore, no approximation was used.

The Monte Carlo studies showed that the proposed statistical macromodel provided quite reasonably statistical accuracy for f_0 (less than 1% errors for both the mean and the standard deviation), and less than 9.5% errors for both, H_0 and Q. The macromodel-based analysis was about 30 times faster than using the full device-level SPICE analysis. Twenty-five designable parameters were selected, including transistor channel length of the "resistor-simulating transistors," 4 capacitors, and one of the gate voltages. Due to the designable parameter tracking for transistor quadruples, the actual vector of optimized parameters was reduced to 7: $x = (V_{G_2}, C_1, C_2, L_2, L_3, L_4, L_5)$, where V_{G_2} denotes the gate voltage, and C_i, L_j are capacitor and channel length values, respectively. The original yield was 43.5% and relative standard deviations for H_0, f_0, and Q, were to 3.98%, 7.24%, and 22.06% respectively, and all nominal values were close to their target values. After SA optimization with convolution smoothing involving 5% perturbations for each of the x parameters, and the total of about 240 circuit analyses, yield increased to 85.0%, and H_0, f_0, and Q relative standard deviations were reduced to 3.87%, 6.35%, and 13.91%, respectively. Therefore, the largest variability reduction (for Q) was about 37%, with simultaneous significant yield increase.

The above example demonstrates a typical approach that has to be taken in the case of large analog circuits: the circuit has to be (hierarchically) macromodeled first[21] and suitable *statistical* macromodels have to be created, including mismatch modeling.

[21] Behavioral models can also be used.

55.5 Conclusion

Several different techniques of **statistical** yield optimization were presented. Other approaches to yield estimation and optimization can be found, e.g., in [21] (process optimization), and in [15, 57, 59, 60, 61]. It was also shown that yield, variability minimization, Taguchi design, and other approaches can be generalized into one methodology called design for quality[22] (DFQ). DFQ is a quickly growing area [30], where the major objective is not to maximize the parametric yield as the only design criterion (yield is often only vaguely defined), but to minimize the performance variability around the designer specified **target performance values**. This has been a subject of research for many years using a sensitivity-based approach, and most recently, using the Taguchi methodology, based on some of the techniques of Design of Experiments rather than on the use of sensitivities. To increase the efficiency of the (mostly manual) Taguchi techniques [30], some automated methods started to appear, such as the generalized methodology described previously, or the automated approach based on capability indices C_p/C_{pk} (used extensively in process quality control) proposed in [41].

Statistical design optimization is still an active research area, but several mature techniques have been already developed and practically applied to sophisticated industrial IC design. This is of great importance to the overall manufacturing cost reduction, circuit quality improvement, and shortening of the overall IC design cycle.

References

[1] H. L. Abdel-Malek and J. W. Bandler, Yield optimization for arbitrary statistical distributions: part I—theory, *IEEE Trans. Circuit Syst.*, CAS-27(4), pp. 245–253, April 1980.

[2] K. J. Antreich and R. K Koblitz, Design centering by yield prediction, *IEEE Trans. Circuits Systems*, CAS-29, pp. 88–95, Feb. 1982.

[3] P. Balaban and J. J. Golembeski, Statistical analysis for practical circuit design, *IEEE Trans. on Circuits and Systems*, CAS-22(2), pp. 100–108, Feb. 1975.

[4] J. W. Bandler, Optimization of design tolerance using nonlinear programming, *J. Optimization Theory and Applications*, vol. 14, p. 99, 1974; also in Proc. Princeton Conf. Information Sciences and Systems, p. 655, Princeton, NJ, Feb. 1972.

[5] J. W. Bandler and H. L. Abdel-Malek, Optimal centering, tolerancing and yield determination via updated approximations and cuts, *IEEE Trans. Circuits and Systems*, CAS-25, pp. 853–871, 1978.

[6] J. W. Bandler, P. C. Liu, and H. Tromp, A nonlinear programming approach to optimal design centering, tolerancing and tuning, *IEEE Trans. CAS*, CAS-23, p. 155, March 1976.

[7] P. W. Becker and F. Jensen, *Design of Systems and Circuits for Maximum Reliability or Maximum Production Yield*, New York: McGraw-Hill, 1977.

[8] J. Chen and M. A. Styblinski, A systematic approach of statistical modeling and its application to CMOS circuits, in Proc. IEEE Int'l. Symposium on Circuits and Systems '93, pp. 1805–1808, Chicago, May 1993.

[9] P. Cox, P. Yang, S. S. Mahant-Shetti, and P. Chatterjee, Statistical modeling for efficient parametric yield estimation of MOS VLSI circuits, *IEEE Trans. on Electron Devices*, ED-32, pp. 471–478, Feb. 1985.

[10] S. W. Director and G. D. Hatchel, The simplicial approximation to design centering, *IEEE Trans. Circuits Syst.*, CAS-24(7), pp. 363–372, July 1977.

[22]Other approaches to yield generalization/Design for Quality were presented in [13, 28, 58].

[11] S. W. Director and G. D. Hatchel, A point basis for statistical design, in *IEEE Proc.*, New York, 1978. ISCAS-78.

[12] S. W. Director, W. Maly, and A. J. Strojwas, *VLSI Design for Manufacturing: Yield Enhancement*, Boston: Kluwer Academic Publishers, 1990.

[13] P. Feldman and S. W. Director, Integrated circuit quality optimization using surface integrals, *IEEE Trans. on CAD*, 12(12), pp. 1868–1879, Dec. 1993.

[14] P. H. Gill, W. Murray, and M. H. Wright, *Practical Optimization*, San Diego: Academic Press, 1981.

[15] D. E. Hocevar, P. F. Cox, and P. Yang, Parametric yield optimization for MOS circuit blocks, *IEEE Trans. on Computer-Aided Design*, 7(6), pp. 645–658, June 1988.

[16] D. E. Hocevar, M. R. Lightner, and T. N. Trick, A study of variance reduction techniques for estimating circuit yields, *IEEE Trans. on Computer-Aided Design*, CAD-2(3), pp. 180–192, July 1983.

[17] A. Ilumoka, N. Maratos, and R. Spence, Variability reduction: statistically based algorithms for reduction of performance variability of electrical circuits, IEEE Proc., vol. 129, pt. G(4), pp. 169–180, Aug. 1982.

[18] G. Kjellstrom and L. Taxen, Stochastic optimization in system design, *IEEE Trans. Circuits Systems*, CAS-28, pp. 702–715, July 1981.

[19] G. Kjellstrom, L. Taxen, and P. O. Lindberg, Discrete optimization of digital filters using Gaussian adaptation and quadratic function minimization, *IEEE Trans. on Circuits and Systems*, CAS-34(10), pp. 1238–1242, Oct. 1987.

[20] D. N. Lawley and A. E. Maxwell, *Factor Analysis as a Statistical Method*, New York: Elsevier, 1971.

[21] K. K. Low and S. W. Director, An efficient methodology for building macromodels of IC fabrication processes, *IEEE Trans. on Computer-Aided Design*, 8(12), pp. 1299–1313, Dec. 1989.

[22] W. Maly and A. J. Strojwas, Statistical simulation of IC manufacturing process, *IEEE Trans. on CAD*, CAD-1, July 1982.

[23] C. Michael and M. Ismail, *Statistical Modeling for Computer-Aided Design of MOS VLSI Circuits*, Boston: Kluwer Academic Publishers, 1993.

[24] S. R. Nassif, A. J. Strojwas, and S. W. Director, FABRICS II. *IEEE Trans. on CAD*, CAD-3, pp. 40–46, Jan. 1984.

[25] J. Ogrodzki, L. Opalski, and M. A. Styblinski, Acceptability regions for a class of linear networks, in Proc. IEEE Int. Symp. on Circuits and Systems, Houston, TX, May 1980.

[26] J. Ogrodzki and M. A. Styblinski, Optimal tolerancing, centering, and yield optimization by One-Dimensional Orthogonal Search (ODOS) technique, in Proc. Eur. Conf. Circ. Theory and Design (ECCTD), vol. 2, pp. 480–485, Warsaw, Poland, Sept. 1980.

[27] L. Opalski, M. A. Styblinski, and J. Ogrodzki, An orthogonal search approximation to acceptability regions and its application to tolerance problems, in Proc. Conf. SPACECAD, Bologna, Italy, Sept. 1979.

[28] L. J. Opalski and M. A. Styblinski, Generalization of yield optimization problem: maximum income approach, *IEEE Trans. on Comp. Aided Design of ICAS*, CAD-5(2), pp. 346–360, April 1986.

[29] M. J. M. Pelgrom, A. C. J. Duinmaijer, and A. P. G. Welbers, Matching properties of MOS transistors, *IEEE J. of Solid State Circuits*, 24, pp. 1334–1362, Oct. 1989.

[30] M. S. Phadke, *Quality Engineering Using Robust Design*, Englewood Cliff, NJ: Prentice-Hall, 1989.

[31] M. Qu and M. A. Styblinski, Hierarchical approach to statistical performance improvement of CMOS analog circuits, in SRC TECHCON '93, Atlanta, GA, Sept. 1993.

[32] M. Qu and M. A. Styblinski, Statistical characterization and modeling of analog functional blocks, in Proc. IEEE Int. Symp. on Circuits and Systems, London, May–June 1994.

[33] H. Robins and S. Monro, A stochastic approximation method, *Annal. Math. Stat.*, 22, pp. 400–407, 1951.

[34] R. Y. Rubinstein, *Simulation and the Monte Carlo Method*, New York: John Wiley & Sons, 1981.

[35] A. Ruszczynski and W. Syski, Stochastic approximation algorithm with gradient averaging for constrained problems, *IEEE Trans. Automatic Control*, AC-28, pp. 1097–1105, Dec. 1983.

[36] K. Singhal and J. F. Pinel, Statistical design centering and tolerancing using parametric sampling, *IEEE Trans. Circuits Syst.*, CAS-28, pp. 692–702, July 1981.

[37] R. S. Soin and R. Spence, Statistical exploration approach to design centering, in *Proc. Inst. Elect. Eng.*, vol. 127, part G, pp. 260–262, 1980.

[38] J. C. Spall, Multivariate stochastic approximation using a simultaneous perturbation gradient approximation, *IEEE Trans. on Automatic Control*, 37(3), 1992.

[39] R. Spence and R. S. Soin, *Tolerance Design of Electronic Circuits*, Electronic Systems Engineering Series, Reading, MA, Addison-Wesley, 1988.

[40] W. Strasz and M. A. Styblinski, A second derivative Monte Carlo optimization of the production yield, in Proc. European Conf. Circuit Theory and Design, vol. 2, pp. 121–131, Warsaw, Poland, Sept. 1980.

[41] M. A. Styblinski and S. A. Aftab, IC variability minimization using a new C_p and C_{pk} based variability/performance measure, in Proc. IEEE Int. Symp. on Circuits and Systems, London, May–June 1994.

[42] M. A. Styblinski, Estimation of yield and its derivatives by Monte Carlo sampling and numerical integration in orthogonal subspaces, in Proc. Eur. Conf. Circ. Theory and Design (ECCTD), vol. 2, pp. 474–479, Warsaw, Poland, Sept. 1980.

[43] M. A. Styblinski, Problems of yield gradient estimation for truncated probability density functions, *IEEE Trans. on Comp. Aided Design of ICAS*, CAD-5(1), pp. 30–38, Jan. 1986 (special issue on statistical design of VLSI circuits).

[44] M. A. Styblinski, Generalized formulation of yield, variability, minimax and Taguchi circuit optimization problems, *Microelectron Reliab.*, 34(1), 31–37, 1994.

[45] M. A. Styblinski and S. A. Aftab, Combination of interpolation and self organizing approximation techniques—a new approach to circuit performance modeling, *IEEE Trans. on Computer-Aided Design*, 12(11), pp. 1775–1785, Nov. 1993.

[46] M. A. Styblinski, J. Ogrodzki, L. Opalski, and W. Strasz, New methods of yield estimation and optimization and their application to practical problems (invited paper), in Proc. Int. Symp. on Circuits and Systems, Chicago, IL, 1981.

[47] M. A. Styblinski and L. J. Opalski, A random perturbation method for IC yield optimization with deterministic process parameters, in Proc. IEEE Int. Symp. on Circuits and Systems, pp. 977–980, Montreal, May 7–10, 1984.

[48] M. A. Styblinski and L. J. Opalski, Algorithms and software tools for IC yield optimization based on fundamental fabrication parameters, *IEEE Trans. on Comp. Aided Design of ICAS*, CAD-5(1), pp. 79–89, January 1986 (special issue on statistical design of VLSI circuits).

[49] M. A. Styblinski and A. Ruszczynski, Stochastic approximation approach to production yield optimization, in Proc. 25th Midwest Symp. on Circuits and Systems, Houghton, MI, Aug. 30–31, 1982.

[50] M. A. Styblinski and A. Ruszczynski, Stochastic approximation approach to statistical circuit design, *Electron. Lett.*, 19(8), 300–302, April 14, 1983.

[51] M. A. Styblinski and W. Strasz, A second derivative Monte Carlo optimization of the production yield, in *Proc. ECCTD'80*, vol. 2, pp. 121–131, Warsaw, Sept. 1980.

[52] M. A. Styblinski and T.-S. Tang, Experiments in nonconvex optimization: stochastic approximation with function smoothing and simulated annealing, *Neural Networks J.*, 3(4), 1990.

[53] M. A. Styblinski and J. C. Zhang, Orthogonal array approach to gradient based yield optimization, in Proc. Int. Symp. on Circuits and Systems, pages 424–427, New Orleans, LA, May 1990.

[54] K. S. Tahim and R. Spence, A radial exploration algorithm for the statistical analysis of linear circuits, *IEEE Trans. on CAS*, CAS-27(5), pp. 421–425, May 1980.

[55] T.-S. Tang and M. A. Styblinski, Yield optimization for non-differentiable density functions using convolution techniques, *IEEE Trans. on CAD of IC and Systems*, 7(10), pp. 1053–1067, 1988.

[56] W. J. Welch, T.-K. Yu, S. M. Kang, and J. Sacks, Computer experiments for quality control by parameter design, *J. of Quality Technol.*, 22(1), pp. 15–22, Jan. 1990.

[57] P. Yang, D. E. Hocevar, P. F. Cox, C. Machala, and P. K. Chatterjee, An integrated and efficient approach for MOS VLSI statistical circuit design, *IEEE Trans. CAD of VLSI Circ. Syst.*, CAD-5, pp. 5–14, Jan, 1986.

[58] D. L. Young, J. Teplik, H. D. Weed, N. T. Tracht, and A. R. Alvarez, Application of statistical design and response surface methods to computer-aided VLSI device design II: desirability functions and Taguchi methods, *IEEE Trans. on Computer-Aided Design*, 10(1), pp. 103–115, Jan. 1991.

[59] T.-K. Yu, S. M. Kang, I. N. Hajj, and T. N. Trick, Statistical performance modeling and parametric yield estimation of MOS VLSI, *IEEE Trans. CAD of VLSI Circ. Syst.*, CAD-6(6), pp. 1013–1022, Nov. 1987.

[60] T. K. Yu, S. M. Kang, I. N. Hajj, and T. N. Trick, iEDISON: an interactive statistical design tool for MOS VLSI circuits, in IEEE Int. Conf. on Computer-Aided Design, ICCAD-88, pp. 20–23, Santa Clara, CA, Nov. 7–10, 1988.

[61] T. K. Yu, S. M. Kang, J. Sacks, and W. J. Welch, Parametric yield optimization of MOS integrated circuits by statistical modeling of circuit performances, Technical Rep. 27, Dept. of Statistics, University of Illinois, Champaign, IL, July 1989.

[62] L. A. Zadeh, Fuzzy sets, *Inform. Control*, 8, pp. 338–353, 1965.

[63] L. A. Zadeh, Probability measures of fuzzy events, *J. Math. Anal. Appl.*, 23, pp. 421–427, 1968.

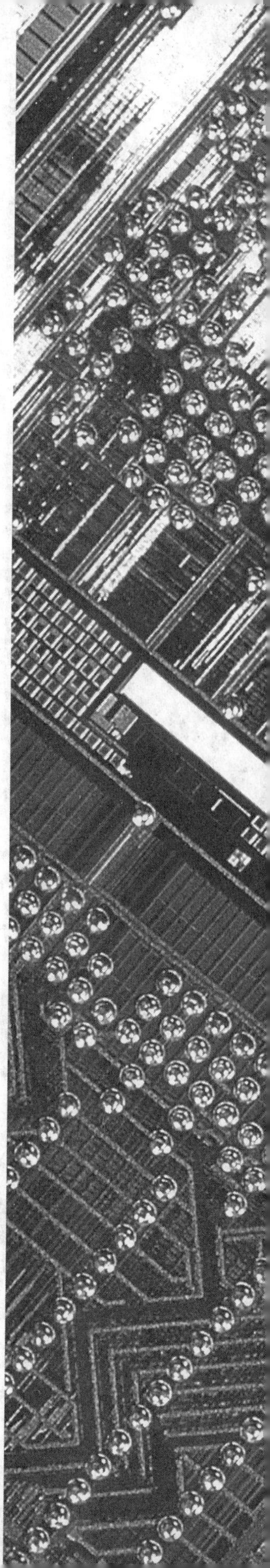

XI

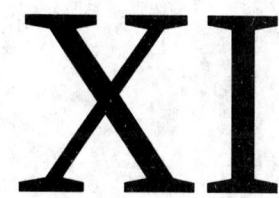

Analog Integrated Circuits

John Choma, Jr.
University of Southern California

56

Monolithic Device Models

B. M. Wilamowski
University of Wyoming

John Choma, Jr.
University of Southern California

Stephen I. Long
University of California Santa Barbara, CA

Nhat M. Nguyen
Hewlett Packard Company

Martin A. Brooke
Georgia Institute of Technology

56.1 Bipolar Junction Transistor

B. M. Wilamowski

The bipolar transistor is the most commonly used discrete active device. It has the highest transconductance among all transistors. Therefore, it can drive a reactive load with very high frequencies. The MOS transistor has a high switching speed only in integrated circuits, where parasitic capacitances are very small. The bipolar transistor also has some disadvantages. It requires a large area on the surface of integrated circuits. It has a relatively small input impedance and the switching speed is very much limited by the storage time of injected minority carriers.

In recent years the complexity of electronic circuits has increased significantly. Computer tools aid the design process. In a case of integrated circuits, computer-aided design is a necessity. Obviously the accuracy of design depends on the modeling accuracy. The knowledge of transistor models, and the understanding of model parameters is therefore essential.

Ebers–Moll Model

The bipolar transistor has three terminals: emitter, base, and collector. It consists of two junctions: the forward-biased emitter–base junction and the reverse-biased collector–base junction. Minority carriers injected from emitter to base are then extracted from the base by the reverse-biased

0-8493-8341-2/95/$0.00 + $.50

collector junction. Therefore, the collector current is proportional to the emitter–base current. In the forward-active mode, the current–voltage characteristic of the emitter junction is described by the well-known diode equation

$$I_E = I_{EF} = I_{E0}\left[\exp\left(\frac{V_{BE}}{V_T}\right) - 1\right] \tag{56.1}$$

where I_{E0} is the emitter saturation current and $V_T = kT/q$ is the thermal potential (about 25 mV at room temperature). The collector current is always smaller than the emitter current $I_{CF} = \alpha_F I_{EF}$, where α_F is the forward current gain, which is smaller than unity.

The positions of the emitter and collector can be switched. The base–collector junction can become the forward-biased junction and the base–emitter junction can become the reverse-biased junction. In this reverse-active mode the collector current can be expressed by

$$I_C = I_{ER} = I_{C0}\left[\exp\left(\frac{V_{BC}}{V_T}\right) - 1\right] \tag{56.2}$$

where I_{C0} is the collector saturation current. In a similar way, $I_{CR} = \alpha_R I_{ER}$, where α_R is the reverse current gain. When both junctions, the base–emitter and the base–collector, are forward biased, transistor operates in saturation mode and an equivalent transistor model is composed of two transistors operating in forward- and reverse-active modes, as Fig. 56.1 illustrates. The forward transistor operation is described by (56.1), and the reverse transistor operation is described by (56.2). From Kirchoff's current law one can write $I_C = I_{CF} - I_{ER}$, $I_E = I_{EF} - I_{ER}$, and $I_B = I_{BF} + I_{CR}$. Using (56.1) and (56.2) the emitter and collector currents can be described as

$$\begin{aligned} I_E &= a_{11}\left(\exp\frac{V_{BE}}{V_T} - 1\right) - a_{12}\left(\exp\frac{V_{BC}}{V_T} - 1\right) \\ I_C &= a_{21}\left(\exp\frac{V_{BE}}{V_T} - 1\right) - a_{22}\left(\exp\frac{V_{BC}}{V_T} - 1\right) \end{aligned} \tag{56.3}$$

which are known as the Ebers–Moll equations [1]. The Ebers–Moll coefficients a_{ij} are given as

$$a_{11} = I_{E0} \qquad a_{12} = \alpha_R I_{C0} \qquad a_{21} = \alpha_F I_{E0} \qquad a_{22} = I_{C0} \tag{56.4}$$

The Ebers–Moll coefficients are a very strong function of the temperature

$$a_{ij} = K_x T^m \exp\frac{V_{go}}{V_T} \tag{56.5}$$

where K_x is proportional to the junction area and independent of the temperature, $V_{go} = 1.21$ V is the potential gap in silicon (referenced to 0 K), and m is a material constant with a value between 2.5 and 4. When the transistor saturates, the current injection through the collector junction may activate parasitic transistors, where base acts as emitter, collector as base, and substrate as collector. In typical integrated circuits, bipolar transistors must not operate in saturation. Therefore, for the integrated bipolar transistor the Ebers–Moll equations can be simplified to the form

$$\begin{aligned} I_E &= a_{11}\left(\exp\frac{V_{BE}}{V_T} - 1\right) \\ I_C &= a_{21}\left(\exp\frac{V_{BE}}{V_T} - 1\right) \end{aligned} \tag{56.6}$$

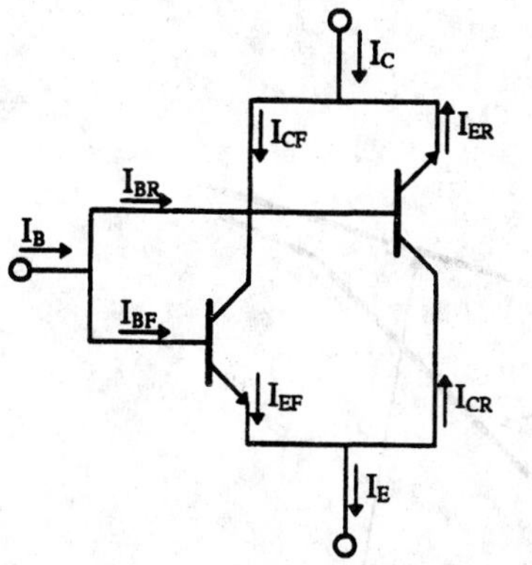

FIGURE 56.1 The equivalent circuit of the bipolar transistor for forward, reverse, and saturated modes.

where $a_{21}/a_{11} = \alpha_F$. This equation corresponds to the circuit diagram as shown in Fig. 56.1, but without the reverse transistor.

Circuit-Level Gummel–Poon Model

In real bipolar transistors the current–voltage characteristics are more complex than those described by the Ebers–Moll equations. Typical current–voltage characteristics of the bipolar transistor, plotted in semilogarithmic scale, are shown in Fig. 56.2. At small base–emitter voltages, due to the generation–recombination phenomena, the base current is proportional to

$$I_{BL} \propto \exp \frac{V_{BE}}{2V_T} \tag{56.7}$$

Also, due to the base conductivity modulation at high-level injections, the collector current for larger voltages can be expressed by the similar relation

$$I_{CH} \propto \exp \frac{V_{BE}}{2V_T} \tag{56.8}$$

Note, that the collector current for wide range is given by

$$I_C = I_s \exp \frac{V_{BE}}{V_T} \tag{56.9}$$

The saturation current is a function of device structure parameters

$$I_s = \frac{qAn_i^2 V_T \mu_B}{\int_0^{w_B} N_B(x)\,dx} \tag{56.10}$$

where $q = 1.6 \cdot 10^{-19}$ C is the electron charge, A is the emitter–base junction area, n_i is the intrinsic concentration ($n_i = 1.5 \cdot 10^{10}$ at 300 K), μ_B is the mobility of the majority carriers in the transistor base, w_B is the effective base thickness, and $N_B(x)$ is the distribution of impurities in the base. Note that the saturation current is inversely proportional to the total impurity dose in the base. In the transistor with the uniform base, the saturation current is given by

$$I_s = \frac{qAn_i^2 V_T \mu_B}{w_B N_B} \tag{56.11}$$

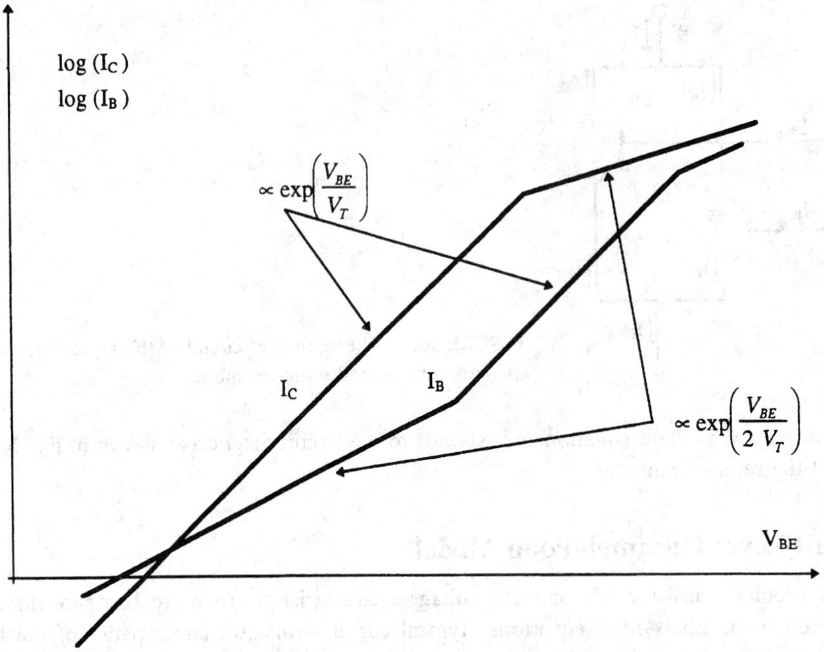

FIGURE 56.2 Collector and base currents as a function of base–emitter voltage.

When a transistor operates in the reverse-active mode (emitter and collector are switched) then the current of such biased transistor is given by

$$I_E = I_s \exp \frac{V_{BC}}{V_T} \tag{56.12}$$

Note that the I_s parameter is the same for forward and reverse mode of operation. The Gummel–Poon transistor model [2] was derived from the Ebers–Moll model using the assumption that $a_{12} = a_{21} = I_s$. For the Gummel–Poon model (56.3) is simplified to the form

$$\begin{aligned} I_E &= I_s\left(\frac{1}{\alpha_F}\exp\frac{V_{BE}}{V_T} - \exp\frac{V_{BC}}{V_T}\right) \\ I_C &= I_s\left(\exp\frac{V_{BE}}{V_T} - \frac{1}{\alpha_R}\exp\frac{V_{BC}}{V_T}\right) \end{aligned} \tag{56.13}$$

These equations require only three coefficients, while the Ebers–Moll requires four. The saturation current I_s is constant for a wide range of currents. The current gain coefficients α_F and α_R have values smaller, but close to unity. Often, instead of using the current gain as $\alpha = I_C/I_E$, the current gain β as a ratio of the collector current to use the base current $\beta = I_C/I_B$ is used. The mutual relationships between α and β coefficients are given by

$$\alpha_F = \frac{\beta_F}{\beta_F + 1} \quad \beta_F = \frac{\alpha_F}{1 - \alpha_F} \quad \alpha_R = \frac{\beta_R}{\beta_R + 1} \quad \beta_R = \frac{\alpha_R}{1 - \alpha_R} \tag{56.14}$$

The Gummel–Poon model was implemented in SPICE [3] and other computer programs for circuit analysis. To make the equations more general, the material parameters η_F and η_R were introduced

$$I_C = I_s\left[\exp\frac{V_{BE}}{\eta_F V_T} - \left(1+\frac{1}{\beta_R}\right)\exp\frac{V_{BC}}{\eta_R V_T}\right] \tag{56.15}$$

The values of η_F and η_R vary from one to two.

Current Gains of Bipolar Transistors

The transistor current gain β is limited by two phenomena: base transport efficiency and emitter injection efficiency. The effective current gain β can be expressed as

$$\frac{1}{\beta} = \frac{1}{\beta_I} + \frac{1}{\beta_T} + \frac{1}{\beta_R} \tag{56.16}$$

where β_I is the transistor current gain caused by emitter injection efficiency, β_T is the transistor current gain caused by base transport efficiency, and β_R is the recombination component of the current gain. As one can see from (56.16), smaller values of β_I, β_T, and β_R dominate. The base transport efficiency can be defined as a ratio of injected carriers into the base, to the carriers that recombine within the base. This ratio is also equal to the ratio of the minority carrier lifetime, to the transit time of carriers through the base. The carrier transit time can be approximated by an empirical relationship

$$\tau_{\text{transit}} = \frac{w_B^2}{V_T \mu_B (2+0.9\eta)} \qquad \eta = \ln\left(\frac{N_{BE}}{N_{BC}}\right) \tag{56.17}$$

where μ_B is the mobility of the minority carriers in base, w_B is the base thickness, N_{BE} is the impurity doping level at the emitter side of the base, and N_{BC} is the impurity doping level at the collector side of the base. Therefore, the current gain due to the transport efficiency is

$$\beta_T = \frac{\tau_{\text{life}}}{\tau_{\text{transit}}} = (2+0.9\eta)\left(\frac{L_B}{w_B}\right)^2 \tag{56.18}$$

where $L_B = \sqrt{V_T \mu_B \tau_{\text{life}}}$ is the diffusion length of minority carriers in the base.

The current gain β_I due to the emitter injection efficiency is given

$$\beta_I = \frac{\mu_B \int_0^{w_E} N_{E\text{eff}}(x)\,dx}{\mu_E \int_0^{w_B} N_B(x)\,dx} \tag{56.19}$$

where μ_B and μ_E are minority carrier mobilities in the base and in the emitter, respectively, $N_B(x)$ is impurity distribution in the base, and $N_{E\text{eff}}$ is the effective impurity distribution in the emitter.

The recombination component of current gain β_R is caused by the different current–voltage relationship of base and collector currents, as can be seen in Fig. 56.2. The slower base current increase is due to the recombination phenomenon within the depletion layer of the base–emitter junction. Since the current gain is the ratio of the collector current to the base current, the relation for β_R can be found as

$$\beta_R = K_{R0} I_C^{1-(1/\eta_R)} \tag{56.20}$$

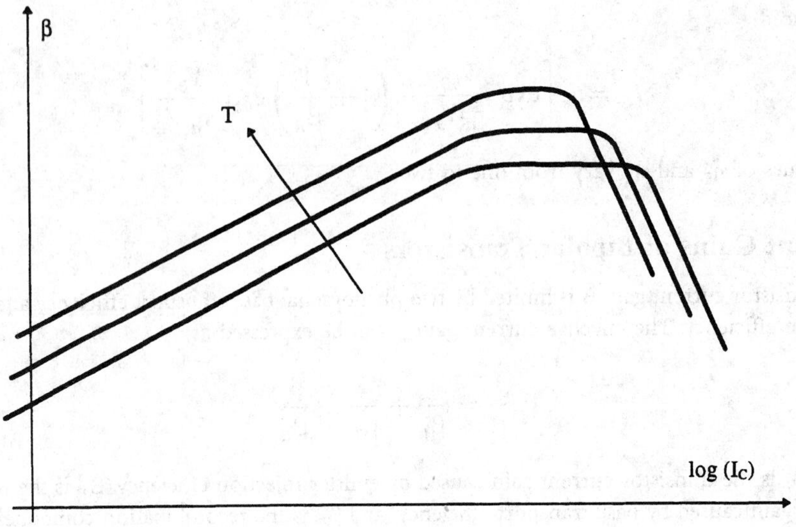

FIGURE 56.3 The current gain β as the function of the collector current.

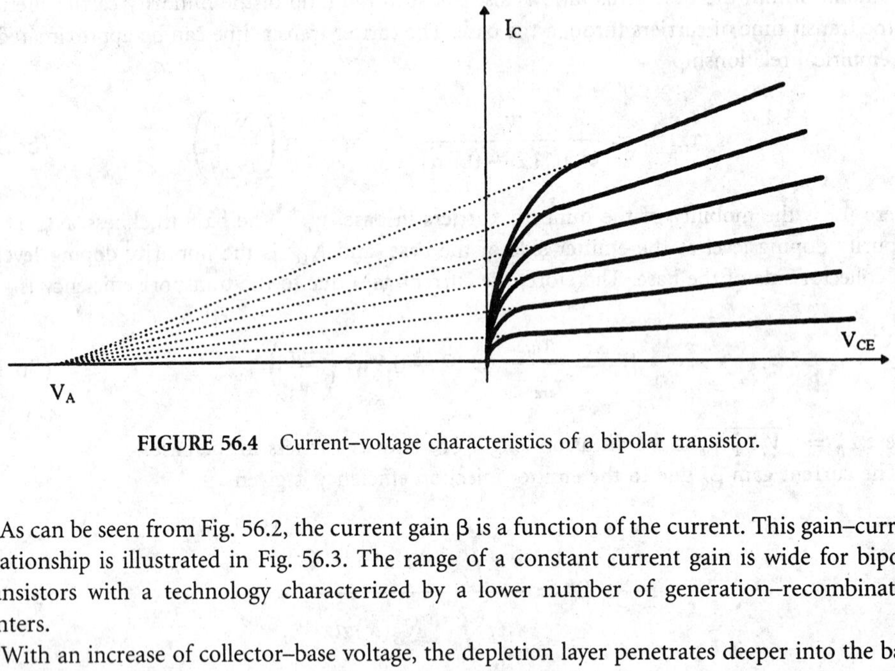

FIGURE 56.4 Current–voltage characteristics of a bipolar transistor.

As can be seen from Fig. 56.2, the current gain β is a function of the current. This gain–current relationship is illustrated in Fig. 56.3. The range of a constant current gain is wide for bipolar transistors with a technology characterized by a lower number of generation–recombination centers.

With an increase of collector–base voltage, the depletion layer penetrates deeper into the base. Therefore, the effective thickness of the base decreases. This leads to an increase of transistor current gain with applied collector voltages. Figure 56.4 illustrates this phenomenon, which is known as Early's effect. The extensions of transistor characteristics (dotted lines in Fig. 56.4) are crossing the voltage axis at the point $-V_A$, where V_A is known as the Early voltage. The current gain β, as a function of collector voltage, is usually expressed using the relation

$$\beta = \beta_o\left(1 + \frac{V_{CE}}{V_A}\right) \tag{56.21}$$

A similar equation can be defined for the reverse mode of operation.

High-Current Phenomena

The concentration of minority carriers increases with the rise of transistor currents. When the concentration of moving carriers exceeds a certain limit, the transistor property degenerates. Two phenomena are responsible for this limitation. The first is related to the high concentration of moving carriers (electrons in the npn transistor) in the base–collector depletion region. This is known as the Kirk effect. The second phenomenon is caused by a high level of carriers injected into the base. When the concentration of injected minority carriers in the base exceeds the impurity concentration there, then the base conductivity modulation limits the transistor performance.

To understand the Kirk effect, consider the npn transistor in forward-active mode with the base–collector junction reverse biased. The depletion layer consists of the negative lattice charge of the base region and the positive lattice charge of the collector region. Boundaries of the depletion layer are such that the total positive and negative charges are equal. When a collector current carrying negatively charged electrons flows through the junction, the effective negative charge on the base side of junction increases. Also, the positive lattice charge of the collector side of the junction is compensated by the negative charge of moving electrons. This way, the collector–base space–charge region moves toward the collector, resulting in a thicker effective base. With a large current level, the thickness of the base may be doubled or tripled. This phenomenon, known as the Kirk effect, becomes very significant when the charge of moving electrons exceeds the charge of the lightly doped collector N_C. The threshold current for the Kirk effect is given by

$$I_{\max} = qAv_{\text{sat}}N_C \tag{56.22}$$

where v_{sat} is saturation velocity for electrons ($v_{\text{sat}} = 10^7$ cm/s for silicon).

The conductivity modulation in the base, or high-level injection, starts when the concentration of injected electrons into the base exceeds the lowest impurity concentration in the base $N_{B\min}$. This occurs for the collector current $I_{\max}$ given by

$$I_{\max} < qAN_{B\max}v = \frac{qAV_T\mu_B N_{B\max}(2 + 0.9\eta)}{w_B} \tag{56.23}$$

The above equation is derived using (56.17), for estimation of the base transient time.

The high-current phenomena are significantly enlarged by the current crowding effect. The typical cross section of a bipolar transistor is shown in Fig. 56.5. The horizontal flow of the base current results in the voltage drop across the base region under the emitter. This small voltage difference on the base–emitter junction causes a significant difference in the current densities at the junction. This is due to the very nonlinear junction current–voltage characteristics. As a result, the base–emitter junction has very nonuniform current distribution across the junction. Most of the current flows through the part of the junction closest to base contact. For transistors with larger emitter areas, the current crowding effect is more significant. This nonuniform transistor current distribution makes the high-current phenomena, such as the base conductivity modulation and the Kirk effect, start for smaller currents than given by (56.22) and (56.23). The current crowding effect is also responsible for the change of the effective base resistance with a current. As a base current increases, the larger part of emitter current flows closer to the base contact, and the effective base resistance decreases.

Small-Signal Model

Small-signal transistor models are essential for ac circuit design. The small-signal equivalent circuit of the bipolar transistor is shown in Fig. 56.6(a). The lumped circuit shown in Fig. 56.6(a)

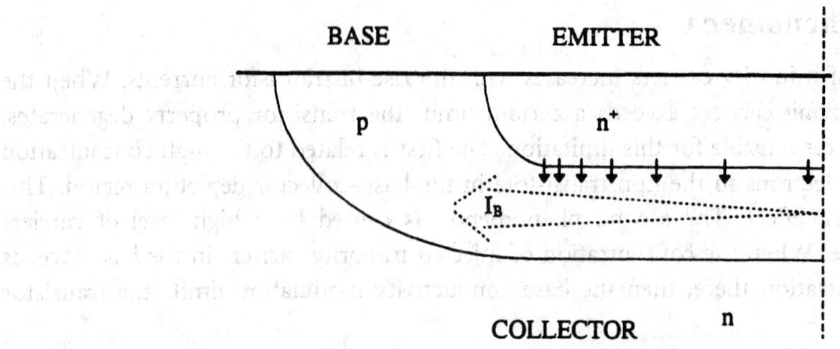

FIGURE 56.5 Current crowding effect.

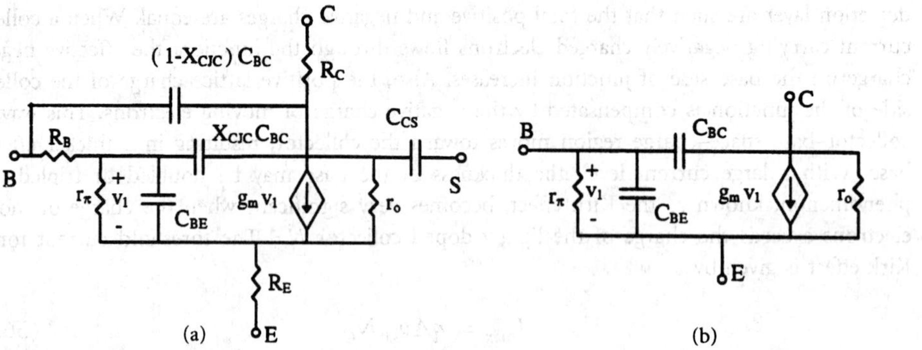

FIGURE 56.6 Bipolar transistor equivalent diagrams: (a) SPICE model, (b) simplified model.

is only an approximation. In real transistors, resistances and capacitances have a distributed character. For most design tasks, this lumped model is adequate, or even the simple equivalent transistor model shown in Fig. 56.6(b) can be considered. The small-signal resistances r_π and r_o are inversely proportional to the transistor currents, and the transconductance g_m is directly proportional to the transistor currents

$$r_\pi = \frac{\eta_F V_T}{I_B} = \frac{\eta_F V_T \beta_F}{I_C} \qquad r_o = \frac{V_A}{I_C} \qquad g_m = \frac{I_C}{\eta_F V_T} \tag{56.24}$$

where η_F is the forward emission coefficient, ranging from 1.0 to 2.0, and V_T is the thermal potential ($V_T = 25$ mV at room temperature). Equations similar to (56.24) can be written for the reverse transistor operation as well.

The series base, emitter, and collector resistances R_B, R_E, and R_C, respectively, are usually neglected for simple analysis [Fig. 56.6(b)]. However, for high-frequency analysis it is essential to use at least the base series resistance R_B. The series emitter resistance R_E usually has a constant bias-independent value. The collector resistance R_C may significantly vary with the biasing current. The value of the series collector resistance may be lowered by one or two orders of magnitude if the collector junction becomes forward biased. A large series collector resistance may force the transistor into the saturation mode. Usually however, when the collector–emitter voltage is large enough, the effect of collector resistance is not significant. The SPICE model assumes a constant value for the collector resistance R_C.

The series base resistance R_B may significantly limit the transistor performance at high frequencies. Due to the current crowding effect and the base conductivity modulation, the series base resistance is a function of the collector current I_C [4].

$$R_B = R_{B\min} + \frac{R_{B0} - R_{B\min}}{0.5 + \sqrt{0.25 + \dfrac{i_C}{I_{KF}}}} \tag{56.25}$$

where I_{KF} is β_F high-current roll-off current, R_{B0} is the base resistance at very small currents, and $R_{B\min}$ is the minimum base resistance at high currents. Another possible approximation of the base series resistance R_B as a function of the base current I_B is [4]

$$R_B = 3(R_{B0} - R_{B\min})\frac{\tan z - z}{z \tan^2 z} + R_{B\min} \qquad z = \frac{\sqrt{1 + \dfrac{1.44 I_B}{\pi^2 I_{RB}}} - 1}{\dfrac{24}{\pi^2}\sqrt{\dfrac{I_B}{I_{RB}}}} \tag{56.26}$$

where I_{RB} is the base current for which the base resistance falls halfway to its minimum value.

The base–emitter capacitance C_{BE} is composed of two terms: the diffusion capacitance, which is proportional to the collector current, and the depletion capacitance, which is a function of the base–emitter voltage V_{BE}. The C_{BE} capacitance is given by

$$C_{BE} = \tau_F \frac{I_C}{\eta_F V_T} + C_{JE0}\left(1 - \frac{V_{BE}}{V_{JE0}}\right)^{-m_{JE}} \tag{56.27}$$

where V_{JE0} is the base–emitter junction potential, τ_F is the base transit time for forward direction, C_{JE0} is base–emitter zero-bias junction capacitance, and m_{JC} is the base–emitter grading coefficient.

The base–collector capacitance C_{BC} is given by a similar expression as (56.27). In the case when the transistor operates in forward-active mode, it can be simplified to

$$C_{BE} = C_{JC0}\left(1 - \frac{V_{BC}}{V_{JC0}}\right)^{-m_{JC}} \tag{56.28}$$

where V_{JC0} is the base–collector junction potential, C_{JC0} is the base–collector zero-bias junction capacitance, and m_{jc} is the base–collector grading coefficient.

In the case when the bipolar transistor is in the integrated form, the collector–substrate capacitance C_{CS} has to be considered

$$C_{CS} = C_{JS0}\left(1 - \frac{V_{CS}}{V_{JS0}}\right)^{-m_{JS}} \tag{56.29}$$

where V_{JS0} is the collector–substrate junction potential, C_{JS0} is the collector–substrate zero-bias junction capacitance, and m_{JS} is the collector–substrate grading coefficient.

When the transistor enters saturation, or it operates in reverse-active mode, (56.27) and (56.28) should be modified to

$$C_{BE} = \tau_F \frac{I_S \exp\left(\dfrac{V_{BE}}{\eta_F V_T}\right)}{\eta_F V_T} + C_{JE0}\left(1 - \frac{V_{BE}}{V_{JE0}}\right)^{-m_{JE}} \tag{56.30}$$

$$C_{BC} = \tau_R \frac{I_s \exp\left(\dfrac{V_{BC}}{\eta_R V_T}\right)}{\eta_R V_T} + C_{JC0}\left(1 - \frac{V_{BC}}{V_{JC0}}\right)^{-m_{JC}} \tag{56.31}$$

Technologies

Bipolar technology was used to fabricate the first integrated circuits more than 30 years ago. A similar standard bipolar process is still used. In recent years, for high-performance circuits and for BiCMOS technology, the standard bipolar process was modified by using thick selective silicon oxidation instead of p-type isolation diffusion. Also, the diffusion process was substituted by the ion implantation process.

Standard Bipolar Process

The structure of the typical integrated bipolar transistor is shown in Fig. 56.7. The process starts with n^+ buried layer diffusion into the p-type substrate. For this process, arsenic (As) or antimony (Sb) impurities are used. These impurities diffuse very slowly and this process usually requires many hours, at temperatures around 1250°C. Slowly diffusing impurities are used so that once the buried layer is formed, it is not affected by subsequent high temperature processes. On top of the substrate, with the diffused buried layer, the single crystal epitaxial n-type layer is grown. Typical concentration of the impurities in the epitaxial layer is about 10^{16} cm^{-3}, and its thickness is 5–10 μm. A thicker epitaxial layer is required for higher voltage integrated circuits. To form the collector islands, deep p-type isolation diffusion is carried on. The p-type substrate and the p-type isolation layer are connected to the most negative potential in the circuit. This way, the collector–substrate (isolation) junction is always reverse biased, and transistor collectors are electrically isolated. Next, the p-type base and n^+-type emitter layers are diffused into collector islands. The ohmic contact to the collector is also formed out of the n^+ emitter layer. Contact formation and metal connections conclude the process. The typical impurity profile of a bipolar transistor is shown in Fig. 56.8. The emitter doping level is much higher than the base doping, so large current gains are possible [see (56.19)]. The base is narrow and it has an impurity gradient, so the carrier transit time through the base is short [see (56.17)]. Collector concentration near the base–collector junction is low; therefore, the transistor has a large breakdown voltage, large Early voltage V_{AF}, and collector–base depletion capacitance is low. High impurity concentration in the buried layer leads to small collector series resistance. The emitter strips have to be as narrow as technology allows, to reduce the base series resistance and the current crowding effect. If large emitter area is required, many narrow emitter strips interlaced with base contacts have to be used in a single transistor. Special attention has to be taken during the circuit design, so the base–collector junction is not forward biased. If the base–collector junction is forward biased, then the parasitic pnp transistors activate. This leads to undesired circuit operation. Thus, the integrated bipolar transistors must not operate in reverse or in saturation modes.

Lateral and Vertical pnp Transistors

Standard bipolar technology is oriented for fabrication of npn transistors with the structure shown in Fig. 56.7. Using the same process, other circuit elements can be fabricated as well. For example, using the base layer, integrated resistors can be obtained, as shown in Fig. 56.7. Typical sheet resistance of the base layer is about 150–200 Ω/square. Resistors with required resistances are composed of the proper number of squares. For small resistors, an emitter layer with a sheet resistance of about 5 Ω/square can be used. Large resistors can be made of the base layer under the emitter, with a sheet resistance of 2–5 kΩ/square.

The standard bipolar process allows also for fabrication of pnp transistors, as shown in Fig. 56.9. The lateral transistor shown in Fig. 56.9(a) uses the base p-type layer for both emitter

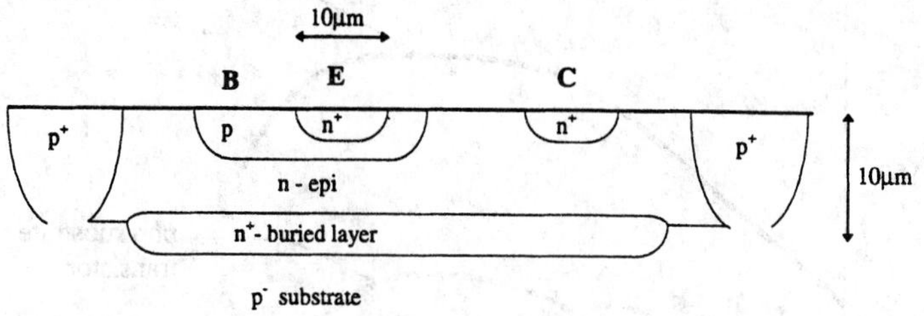

FIGURE 56.7 Standard bipolar structure.

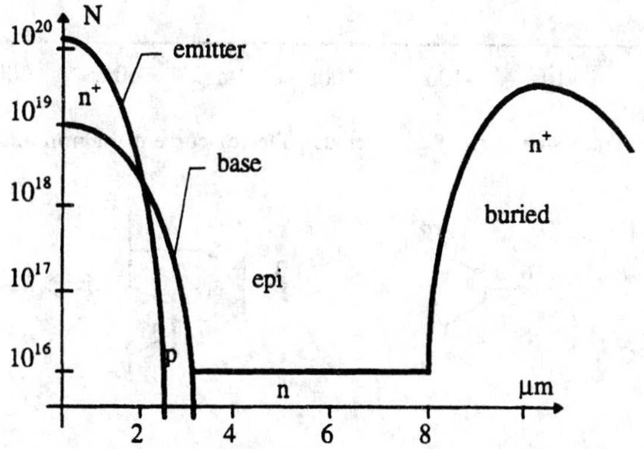

FIGURE 56.8 Cross section of the standard bipolar transistor.

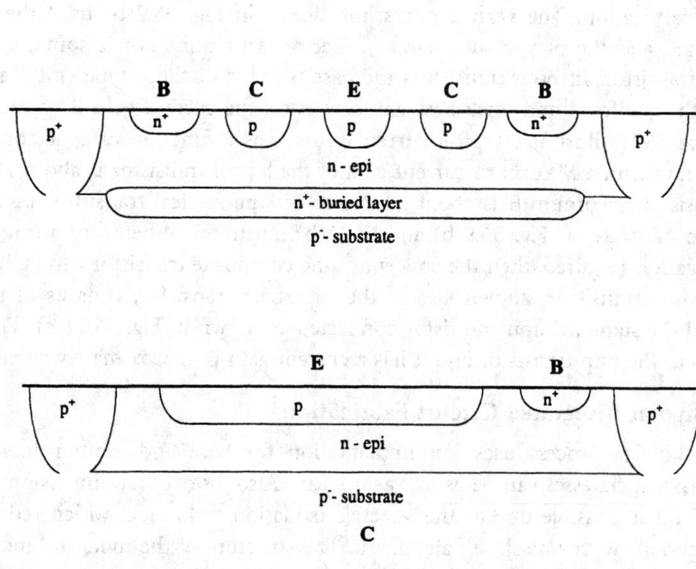

FIGURE 56.9 Integrated pnp transistors: (a) lateral pnp transistor, (b) substrate pnp transistor.

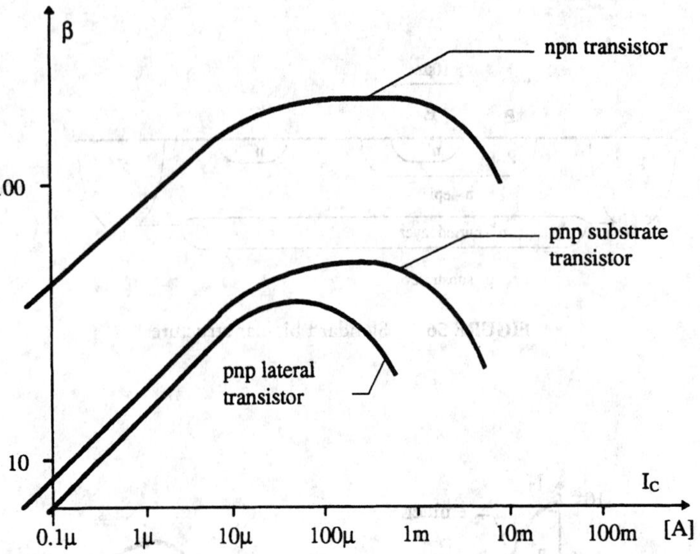

FIGURE 56.10 Transistor current gain versus collector current for npn and pnp transistors.

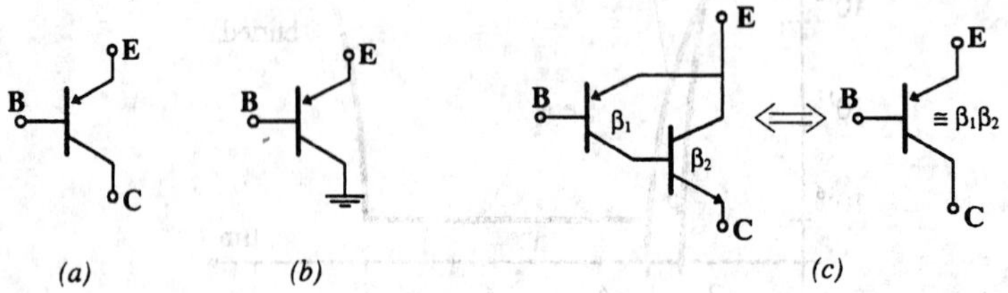

FIGURE 56.11 Integrated pnp transistors: (a) lateral transistor, (b) substrate transistor, (c) composed transistor.

and collector fabrication. The vertical transistor shown in Fig. 56.9(b) used the p-type base layer for emitter, and the p-type substrate as collector. This transistor is sometimes known as the substrate transistor. In both transistors the base is made of the n-type epitaxial layer. Such transistors with a uniform and thick base are slow. Also, the current gain β of such transistors is small. Figure 56.10 illustrates typical current gains β as a function of collector currents for npn and pnp transistors. Maximum current gain of the lateral transistor is about 30, and in the substrate transistor maximum β is about 50. Note that the vertical transistor has the collector shorted to the substrate, as Fig. 56.9(b) and 56.11(b) illustrates. When a pnp transistor with a large current gain is required, then the concept of the composite transistor can be implemented. Such a composite transistor, known also as the superbeta transistor, consists of a pnp lateral transistor, and the standard npn transistor connected as shown in Fig. 56.11(c). The composed transistor acts as the pnp transistor and it has a current gain β approximately equal to $\beta_{pnp}\beta_{npn}$.

Advanced Bipolar Integrated Circuit Fabrication

The modern bipolar process uses ion implantation for base and emitter fabrication. The emitter and base thicknesses can be as low as 0.1 μm. Also, horizontal dimensions are reduced significantly. This is possible due to the selective oxidation technique, which reduces parasitic capacitances and allows for mask self alignment. The structure of the modern bipolar transistor is shown in Fig. 56.12, and its impurity profile in Fig. 56.13.

The popular MOS integrated circuits are very fast only if the parasitic capacitances are very small. Indeed, inside integrated circuits these capacitances are on the order of fF(10^{-15} F) and

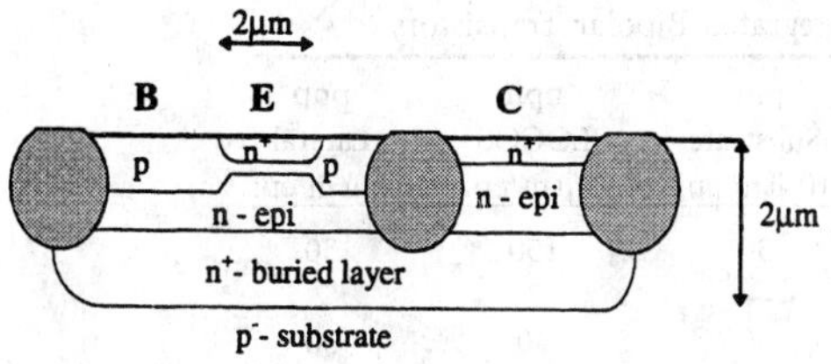

FIGURE 56.12 Structure of modern npn transistor fabricated using ion implantation and selective oxidation process.

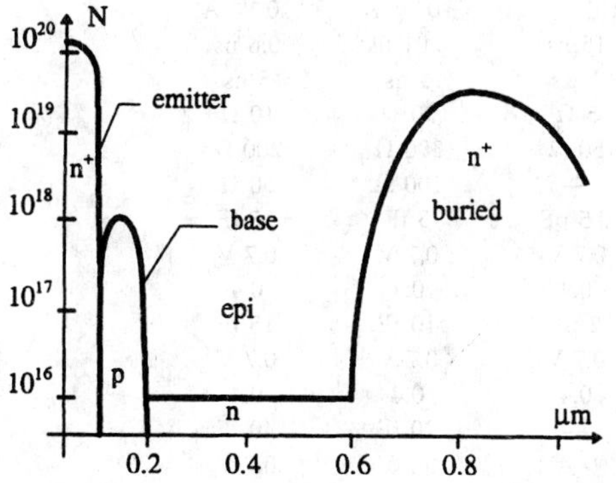

FIGURE 56.13 Impurity profile in modern bipolar transistor.

the switching times of transistors operating with a single μA range are in single ns range. The problem arises when signals have to be transmitted outside the integrated circuits, where loading capacitances are in the pF range (1000 times larger). With the same device construction the switching time will be reduced by a factor of 1000. To overcome this problem a special chain of power MOS transistors is used in output buffers. This is a costly solution. Occasionally those output buffers consume more space of integrated circuits than the functional circuitry. Another approach is to use BiCMOS technology, where bipolar transistors are used as output buffers. In this case, the high current driving capabilities of the bipolar transistors are utilized. The bipolar transistor in BiCMOS technology has a similar structure, as shown in Fig. 56.12. In BiCMOS technology, both bipolar and MOS transistors are fabricated on the same wafer. Once the bipolar transistors are introduced into the CMOS process, they are then used for another circuitry other than buffers.

Typical Parameters of Integrated Bipolar Transistors

Various types of transistors can be fabricated using integrated circuit technologies. Typical parameters of some bipolar integrated transistors are shown in Table 56.1.

Model Parameters

It is essential to use proper transistor models in computer-aided design tools. The accuracy of simulation results depends on the model accuracy, and on the values of the model parameters used. In this section the thermal sensitivity and second-order effects in the transistor model are discussed. The SPICE bipolar transistor model parameters are also discussed.

Thermal Sensitivity

All parameters of the transistor model are temperature dependent. Some parameters are very strong functions of temperature. To simplify the model description, the temperature dependence of some parameters are often neglected. In this section, the temperature dependence of the transistor model is described based on the model of the SPICE program [3]–[5]. Deviations from the actual temperature dependence will also be discussed. The temperature dependence of

TABLE 56.1 Typical Parameters of Integrated Bipolar Transistors

	npn Small 10 μm epi	pnp Lateral 10 μm epi	pnp Substrate 10 μm epi	npn LOCOS 1 μm epi	pnp Lateral 1 μm epi
β_F	200	20	30	150	50
β_R	2	0.3	—	2	2
V_{AF}	100	30	30	30	30
V_{AR}	50	30	—	30	30
I_s	10^{-15} A	10^{-15} A	10^{-14} A	10^{-17} A	10^{-17} A
τ_F	0.25 ns	30 ns	15 ns	0.01 ns	0.6 ns
τ_R	200 ns	3 μs	1 μs	5 ns	5 ns
r_E	2 Ω	5 Ω	5 Ω	30 Ω	10 Ω
r_B	200 Ω	100 Ω	50 Ω	300 Ω	200 Ω
r_C	100 Ω	5 Ω	—	100 Ω	50 Ω
C_{JE}	1.5 pF	0.5 pF	0.5 pF	5 fF	15 fF
V_{JE}	0.8 V	0.7 V	0.7 V	0.7 V	0.7 V
m_{JE}	0.33	0.4	0.4	0.4	0.4
C_{JC}	0.7 pF	2 pF	2 pF	10 fF	15 fF
V_{JC}	0.7 V	0.7 V	0.7 V	0.7 V	0.7 V
m_{JC}	0.5	0.4	0.4	0.4	0.4
C_{JS}	3 pF	4 pF	—	20 fF	40 fF
V_{JS}	0.6	0.7 V	—	0.7 V	0.7 V
m_{JS}	0.4	0.4	—	0.4	0.4
BV_{CE}	30 V	30 V	30 V	8 V	15 V
BV_{BE}	7 V	50 V	50 V	20 V	20 V
BV_{BC}	50 V	50 V	50 V	6 V	20 V

junction capacitance is given by

$$C_J(T) = C_J \left\{1 + m_J \left[4.010^{-4}(T - T_{\text{NOM}}) + \left(1 - \frac{V_J(T)}{V_J}\right)\right]\right\} \tag{56.32}$$

where T_{NOM} is the nominal temperature, which is specified in the SPICE program in the .OPTIONS statement. The junction potential $V_J(T)$ is a function of temperature

$$V_J(T) = V_J \frac{T}{T_{\text{NOM}}} - 3V_T \ln\left(\frac{T}{T_{\text{NOM}}}\right) - E_G(T) + E_G \frac{T}{T_{\text{NOM}}} \tag{56.33}$$

The value of 3 in the multiplication coefficient of the above equation is from the temperature dependence of the effective state densities in the valence and conduction bands. The temperature dependence of the energy gap is computed in the SPICE program from

$$E_G(T) = E_G - \frac{7.0210^{-4}T^2}{T + 1108} \tag{56.34}$$

The transistor saturation current as a function of temperature is calculated as

$$I_s(T) = I_s \left(\frac{T}{T_{\text{NOM}}}\right)^{X_{TI}} \exp\left[\frac{E_G(T - T_{\text{NOM}})}{V_T T_{\text{NOM}}}\right] \tag{56.35}$$

where E_G is the energy gap at the nominal temperature. The junction leakage currents I_{SE} and I_{SC} are calculated using

$$I_{SE}(T) = I_{SE} \left(\frac{T}{T_{\text{NOM}}}\right)^{X_{TI}-X_{TB}} \exp\left[\frac{E_G(T - T_{\text{NOM}})}{\eta_E V_T T_{\text{NOM}}}\right] \tag{56.36}$$

and

$$I_{SC}(T) = I_{SC}\left(\frac{T}{T_{\text{NOM}}}\right)^{X_{TI}-X_{TB}} \exp\left[\frac{E_G(T - T_{\text{NOM}})}{\eta_C V_T T_{\text{NOM}}}\right] \tag{56.37}$$

The temperature dependence of the transistor current gains β_F and β_R are modeled in the SPICE program as

$$\beta_F(T) = \beta_F\left(\frac{T}{T_{\text{NOM}}}\right)^{XTB} \qquad \beta_R(T) = \beta_R\left(\frac{T}{T_{\text{NOM}}}\right)^{XTB} \tag{56.38}$$

The SPICE model does not give accurate results for the temperature relationship of the current gain β at high currents. For high current levels the current gain decreases sharply with the temperature, as can be seen from Fig. 56.3. Also, the knee current parameters IKF, IKR, and IKB are temperature dependent, and this is not implemented in the SPICE program.

Second-Order Effects

The current fain β is sometimes modeled indirectly by using different equations for the collector and base currents [4], [5]

$$\begin{aligned} I_C = {} & \frac{I_S(T)}{Q_b}\left(\exp\frac{V_{BE}}{\eta_F V_T} - \exp\frac{V_{BC}}{\eta_R V_T}\right) \\ & - \frac{I_S(T)}{\beta_R(T)}\left(\exp\frac{V_{BC}}{\eta_R V_T} - 1\right) - I_{SC}(T)\left(\exp\frac{V_{BC}}{\eta_C V_T} - 1\right) \end{aligned} \tag{56.39}$$

where

$$Q_b = \frac{1 + \sqrt{1 + 4Q_X}}{2\left(1 - \frac{V_{BC}}{V_{AF}} - \frac{V_{BE}}{V_{AR}}\right)} \tag{56.40}$$

$$Q_X = \frac{I_s(T)}{I_{KF}}\left(\exp\frac{V_{BE}}{\eta_F V_T} - 1\right) + \frac{I_s(T)}{I_{KR}}\left(\exp\frac{V_{BC}}{\eta_R V_T} - 1\right) \tag{56.41}$$

and

$$\begin{aligned} I_B = {} & \frac{I_S}{\beta_F}\left(\exp\frac{V_{BE}}{\eta_F V_T} - 1\right) + I_{SE}\left(\exp\frac{V_{BE}}{\eta_E V_T} - 1\right) \\ & + \frac{I_S}{\beta_R}\left(\exp\frac{V_{BC}}{\eta_R V_T} - 1\right) + I_{SC}\left(\exp\frac{V_{BC}}{\eta_C V_T} - 1\right) \end{aligned} \tag{56.42}$$

where I_{se} is the base–emitter junction leakage current, I_{sc} is the base–collector junction leakage current, η_E is the base–emitter junction leakage emission coefficient, and η_C is the base–collector junction leakage emission coefficient.

The forward transit time τ_F is a function of biasing conditions. In the SPICE program the τ_F parameter is computed using

$$\tau_F = \tau_{F0}\left[1 + X_{TF}\left(\frac{I_{CC}}{I_{CC} + I_{TF}}\right)^2 \exp\frac{V_{BC}}{1.44V_{TF}}\right] \quad I_{CC} = I_s\left(\exp\frac{V_{BE}}{\eta_F V_T} - 1\right) \tag{56.43}$$

At high frequencies the phase of the collector current shifts. This phase shift is computed in the SPICE program in the following way:

$$I_C(\omega) = I_C \exp(j\omega P_{TF}\tau_F) \tag{56.44}$$

where P_{TF} is a coefficient for excess phase calculation.

Noise is usually modeled as the thermal noise for parasitic series resistances, and as shot and flicker noise for collector and base currents

$$\overline{i_R^2} = \frac{4kT\,\Delta f}{R} \tag{56.45}$$

$$\overline{i_B^2} = \left(2qI_B + \frac{K_F I_B^{A_F}}{F}\right)\Delta f \tag{56.46}$$

$$\overline{i_C^2} = 2qI_C\,\Delta f \tag{56.47}$$

where K_F and A_F are the flicker noise coefficients. More detailed information about noise modeling is given in the bipolar noise section of this text (Chapter 58.2).

SPICE Model of the Bipolar Transistor

The SPICE model of bipolar transistor uses similar or identical equations as described in this chapter [3]–[5]. Table 56.2 below shows the parameters of the bipolar transistor model and its relation to the parameters used in this chapter.

The SPICE (Simulation Program with Integrated Circuit Emphasis [3]) was developed primarily for the analysis of integrated circuits. During the analysis it is assumed that the temperatures of all circuit elements are the same. This is not true for power integrated circuits, where the junction temperatures may differ by 30 K or more. This is obviously not true for circuits composed of discrete elements, where the junction temperatures may differ by 100 K or more. These temperature effects, which can significantly affect the analysis results, are not implemented in the SPICE program.

Although the SPICE bipolar transistor model used more than 40 parameters, many features of the bipolar transistor are not included in the model. For example, the reverse junction characteristics are described by (56.36) and (56.37). This model does not give accurate results. In the real silicon junction the leakage current is proportional to the thickness of the depletion layer, which is proportional to $V^{1/m}$. Also, the SPICE model of the bipolar transistor assumes that there are no junction breakdown voltages. A more accurate model of the reverse junction characteristics is described in the diode section of this text. The reverse transit time τ_R is very important in order to model the switching property of the lumped bipolar transistor, and it is a strong function of the biasing condition and temperature. Neither phenomena are implemented in the SPICE model.

Relative Machability

The device parameters of transistors fabricated on the same chip are almost the same. This feature is used for unique integrated circuit design. For various technological processes, the distribution of device parameters are different. For example, the layer's resistances in the typical diffusion process may vary by ±20 percent. When the ion implantation process is used, those variations are below ±5 percent, or occasionally, even below ±2 percent. From the circuit designer's point of view, the relative machability is still more important. The circuit parameters usually depend on the resistor to resistor ratio or the ratio of the device areas. Special care should be taken during the device layout design. For example, the integrated resistor shown

TABLE 56.2 Parameters of SPICE Bipolar Transistor Model

Name Used	Equations	SPICE Name	Parameter Description	Unit	Typical Value	SPICE Default
I_s	(56.10), (56.11)	IS	Saturation current	A	10^{-15}	10^{-16}
I_{SE}	(56.36), (56.39)	ISE	B–E leakage saturation current	A	10^{-12}	0
I_{SC}	(56.37), (56.39)	ICS	B–C leakage saturation current	A	10^{-12}	0
β_F	(56.14), (56.16), (56.21)	BF	Forward Current gain	—	100	100
β_R	(56.14), (56.16), (56.21)	BF	Reverse current gain	—	0.1	1
η_F	(56.15), (56.24), (56.30), (56.31), (56.39)–(56.41)	NF	Forward current emission coefficient	—	1.2	1.0
η_R	(56.15), (56.24), (56.30), (56.31), (56.39)–(56.42)	NR	Reverse current emission coefficient	—	1.3	1.0
η_E	(56.42)	NE	B–E leakage emission coefficient	—	1.4	1.5
η_C	(56.39), (56.42)	NC	B–C leakage emission coefficient	—	1.4	1.5
V_{AF}	(56.21), (56.40)	VAF	Forward Early voltage	V	200	∞
V_{AR}	(56.21), (56.40)	VAR	Reverse Early voltage	V	50	∞
I_{KF}	(56.22), (56.23), (56.41)	IKF	β_F high-current roll-off corner	A	0.05	∞
I_{KR}	(56.22), (56.23), (56.41)	IKR	β_R high-current roll-off corner	A	0.01	∞
I_{RB}	(56.26)	IRB	Current where base resistance falls by half	A	0.1	∞
R_B	(56.25), (56.26)	RB	Zero base resistance	Ω	100	0
$R_{B\,min}$	(56.25), (56.26)	RBM	Minimum base resistance	Ω	10	RB
R_E	Fig. 56.6	RE	Emitter series resistance	Ω	1	0
R_C	Fig 56.6	RC	Collector series resistance	Ω	50	0
C_{JE0}	(56.27)	CJE	B–E zero-bias depletion capacitance	F	10^{-12}	0
C_{JC0}	(56.28)	CJC	B–C zero-bias depletion capacitance	F	10^{-12}	0
C_{JS0}	(56.29)	CJS	Zero-bias collector-substrate capacitance	F	10^{-12}	0
V_{JE0}	(56.27)	VJE	B–E built-in potential	V	0.8	0.75
V_{JC0}	(56.28)	VJC	B–C built-in potential	V	0.7	0.75
V_{JS0}	(56.29)	VJS	Substrate junction built-in potential	V	0.7	0.75
m_{JE}	(56.27)	MJE	B–E junction exponential factor	—	0.33	0.33
m_{JC}	(56.28)	MJC	B–C junction exponential factor	—	0.5	0.33
m_{JS}	(56.29)	MJS	Substrate junction exponential factor	—	0.5	0
X_{CJC}	Fig. 56.6	XCJC	Fraction of B–C capacitance connected to internal base node (see Fig. 56.6)	—	0.5	0
τ_F	(56.17), (56.28), (56.30), (56.42)	TF	Ideal forward transit time	s	10^{-10}	0
τ_R	(56.31)	TR	Reverse transit time	s	10^{-8}	0
X_{TF}	(56.43)	XTF	Coefficient for bias dependence of τ_F		—	0
V_{TF}	(56.43)	VTF	Voltage for τ_F dependence on V_{BC}	V		∞
I_{TF}	(56.43)	ITF	Current where $\tau_F = f(I_C, V_{BC})$ starts	A		0
P_{TF}	(56.44)	PTF	Excess phase at freq = $1/(2\pi\tau_F)$ Hz	deg		0
X_{TB}	(56.38)	XTB	Forward and reverse beta temperature exponent			0
E_G	(56.34)	EG	Energy gap	eV	1.1	1.11
X_{TI}	(56.35)–(56.37)	XTI	Temperature exponent for effect on I_s	—	3.5	3
K_F	(56.46)	KF	Flicker-noise coefficient	—		0
A_F	(56.46)	AF	Flicker-noise exponent	—		1
F_C	—	FC	Coefficient for the forward-biased depletion capacitance formula	—	0.5	0.5
T_{NOM}	(56.32)–(56.38)	TNOM	Nominal temperature specified in .OPTION statement	K	300	

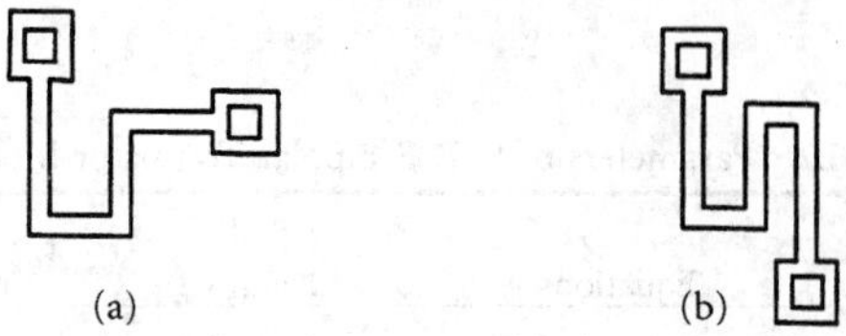

FIGURE 56.14 Two resistor topologies: (a) sensitive to contacts misalignment; (b) nonsensitive to contacts misalignment.

in Fig. 56.14(a) is very sensitive to the misalignment of the contact mask, while the resistor shown in Fig. 56.14(b) is not. With small mask misalignment, the increase of resistance near one contact is compensated by the decrease of resistance near another contact. This concept can be extended for the transistor layout design. If the ratio of emitter areas of two transistors should be, for example, three, then it is recommended to fabricate three identical transistors and connect them in parallel. Any attempt to fabricate a transistor with an area three times larger than another will lead to meaningful device mismatching. This is due not only to edge effects, but also due to the current crowding effect and other phenomena.

The SPICE program has the ability to model both absolute and relative parameter distributions. It can be done by using the *LOT* and *DEV* parameters in the .MODEL statement. For example if $LOT = 10$ percent and $DEV = 1$ percent, then parameters for all devices may vary by ±10 percent, but relative variation from device to device will be ±1 percent. This feature makes the SPICE program very suitable for integrated circuit simulation. The Monte Carlo analysis implemented in the PSPICE program makes the integrated circuit design even easier.

References

[1] J. J. Ebers and J. M. Moll, "Large signal behavior of bipolar transistors," *Proc. IRE*, vol. 42, pp. 1761–1772, Dec. 1954.

[2] H. K. Gummel and H. C. Poon, "An integral charge-control model of bipolar transistors," *Bell Syst. Tech. J.*, vol. 49, pp. 827–852, May 1970.

[3] L. W. Nagel and D. O. Pederson, "SPICE (Simulation program and integrated circuit emphasis)," Univ. California, Berkeley, ERL Memo ERL M382, Apr. 1973.

[4] P. Antognetti and G. Massobrio, *Semiconductor Device Modeling with SPICE*, New York: McGraw-Hill, 1988.

[5] A. Vladimiresku, *The SPICE Book*, New York: Wiley, 1994.

[6] A. S. Grove, *Physics and Technology of Semiconductor Devices*, New York: Wiley, 1967.

[7] S. M. Sze, *Physics of Semiconductor Devices*. 2nd ed., New York: Wiley, 1981.

[8] G. W. Neudeck, *The PN Junction Diode*, vol. II, Modular Series on Solid-State Devices, Reading, MA: Addison-Wesley, 1983.

[9] R. S. Muller and T. I. Kamins, *Device Electronics for Integrated Circuits*, 2nd ed., New York: Wiley, 1986.

[10] E. S. Yang, *Microelectronic Devices*, New York: McGraw-Hill, 1988.

[11] B. G. Streetman, *Solid State Electronic Devices*, 3rd ed. Englewood Cliffs, NJ: Prentice Hall, 1990.

[12] D. A. Neamen, *Semiconductor Physics and Devices*, Irwin 1992.

56.2 MOSFET Technology Devices

John Choma, Jr.

Introduction

The most widely used monolithic semiconductor, particularly in digital and in hybrid analog–digital circuit applications, is the silicon **metal–oxide–semiconductor field-effect transistor**, or MOSFET. The MOSFET differs from its competitive active element, the **bipolar junction transistor** (BJT), in three fundamental respects. First, BJT's are current-controlled devices in that their output (collector) current is controlled by their base current. Under normal operation, this base current is manifested by hole electron recombination between the background majority charge resident in the base region and the minority charge injected across a forward-biased emitter–base junction. On the other hand, MOSFETs are voltage-controlled elements. Their output (drain) current is controlled by an input voltage that modulates the conductivity of a narrow channel of mobile majority charges whose directed transport comprises the output current. As a result of this different current conduction mechanism, the MOSFET features an input impedance that is significantly larger than that of its bipolar counterpart. Moreover, the MOSFET boasts a linearity between its controlling input and controlled output variables that is better than the I/O linearity of a BJT. The latter feature is exploited in state-of-the-art stereo applications for which the active element of choice in the power output stages is the MOS field-effect transistor.

A second difference between the MOSFET and the BJT is that the MOSFET is a majority carrier device; that is, the mobile charges comprising the output current are majority charges in the channel region in which they are transported. In contrast, bipolar currents materialize principally from minority charge transported across a narrow base region. The immediate consequence of this difference is that the output current in a MOSFET displays a negative temperature coefficient, whereas the output current of a BJT increases with increasing temperature.

Finally, and primarily because of the material structure that achieves voltage-controlled current outputs, the cross-section geometry of a MOSFET is simpler than that of a BJT, and its surface area is smaller. These fabrication characteristics, its ability to function at low power levels, and its inherently negative temperature coefficient make the MOSFET ideally suited in digital very large scale integrated (VLSI) electronic applications for which given signal-processing specifications may require tens or even hundreds of thousands of transistors.

Although the MOS transistor is synonymous with digital signal processing, there is an escalating interest in applying MOS technologies to service a variety of analog operating requirements. Aside from low design, development, and fabrication costs, and the general inability of MOSFETs to compare favorably to bipolar transistors with respect to forward gain and bandwidth, there is at least one other engineering issue that motivates this interest. This issue derives from the burgeoning use of microprocessors and other types of computer-based controllers that are used to optimize the analog responses of automotive, consumer, and industrial electronics. The logic variables produced by these control units require interfacing to their analog inputs and their resultantly processed analog outputs. Until recently, these interface circuits were realized as bipolar cells implemented peripherally to the digital control circuits. But manufacturing costs abate and system reliability improves when peripheral analog interface circuitry is supplanted by monolithic circuits incorporated on the digital circuit chip in the same VLSI device technology that accomplishes requisite digital signal processing.

There are four types of MOSFETs: the *N-channel* and *P-channel enhancement-mode* MOSFETs and the *N-channel* and *P-channel depletion-mode* MOSFETs. A fifth type of MOSFET is *complementary MOS*, or *CMOS*. Although commonly referred to as a CMOS "device," CMOS is actually a composite connection of N-channel and P-channel enhancement-mode MOSFETs.

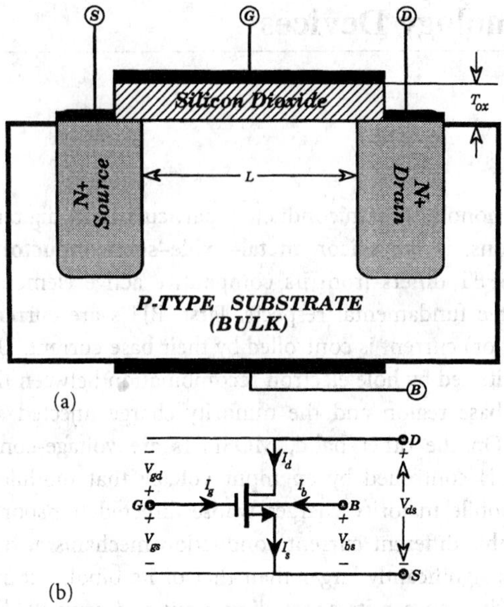

FIGURE 56.15 (a) Simplified cross section of an N-channel enhancement-mode MOSFET. The diagram is not drawn to scale. (b) Electrical schematic diagram of an N-channel enhancement-mode MOSFET.

This section addresses the circuit level models that emulate the static and dynamic volt–ampere characteristics of the foregoing five types of MOS devices. Additionally, illustrations of the applications of these models to simple circuit applications are provided.

Enhancement-Mode MOSFET

There are two types of enhancement-mode MOSFETs, which are also known as **insulated gate field-effect transistors** or *IGFETs*: the *N-channel type* and the *P-channel type.* As shown in the simplified cross section of Fig. 56.15(a) the N-channel enhancement mode MOSFET consists of two heavily doped N-type semiconductor regions that are diffused or implanted into a relatively lightly doped P-type *substrate* or *bulk.* The heavily doped regions, which are indicated in gray shading as the N+ *source* and N+ *drain,* have average concentrations of the order of 10^{19} atoms/cm^3, maximum lateral widths of a few microns, and depths in the range of a micron. In contrast, the P-type substrate is doped to an average concentration of the order of 10^{14} atoms/cm^3, has a vertical thickness that can be hundreds of microns, and a net width that also approaches hundreds of microns. Metallic or polysilicon electrical contacts, shown as black areas in the figure, are made to the source and drain regions on the surface of the device, as well as to the substrate bulk. These contacts are externally accessible on the *source* (S), *drain* (D), and *substrate,* or *bulk,* (B) terminals.

The region between the inner sidewalls of the source and drain diffusions in the immediate neighborhood of the device surface is referred to as the *channel* of the MOSFET. This channel, whose length is indicated as L in Fig. 56.15(a), is typically 2 to 20 microns long. In very-high-speed integrated circuit (VHSIC) technology, L can be significantly smaller than 2 μm. *Submicron MOSFET technology* refers to MOSFETs having $L < 1$ μm. A layer of *silicon dioxide* (SiO_2), depicted as the crosshatched area in Fig. 56.15(a) is thermally grown atop this surface channel, and a metallic or *polysilicon* electrical contact is made to the top of the oxide layer. This contact is the external *gate* terminal (G), to which voltage is applied to control the current flowing in the MOSFET drain lead. The thickness T_{ox}, of the oxide layer is in the range of a few hundred to a few thousand angstroms (an angstrom is 10^{-8} cm, or 10^{-4} μm). If W represents the depth of the gate contact (in a direction perpendicular to the face of the page on which the MOSFET cross sectional diagram appears), LW represents the *gate area*, or *active channel area*, of the device. The performance of many MOSFET circuits is strongly dependent on this area, as well as on the *gate aspect ratio* W/L, which is a designable parameter whose value can be as small as one to as large as a few hundred.

In Fig. 56.15(a), a small part of the gate oxide, and hence gate metal, interfacial area overlaps both the source and drain regions. These *overlap areas* give rise to *overlap capacitances* between both the gate and source and the gate and drain. Since overlap capacitances degrade the switching speed and frequency response of a MOSFET, a monolithic process known as **self-aligning gate technology** [1] has been developed to minimize these overlapping areas.

The electrical schematic symbol of the N-channel enhancement mode MOSFET is given in Fig. 56.15(b). The four terminals (D), (S), (G), and (B), in the schematic diagram correspond to the drain, source, gate, and substrate bulk contacts delineated in the companion diagram of Fig. 56.15(a). The schematic symbol also defines the positive reference directions of four MOSFET currents. These are the *drain current* I_d, the *source current* I_s, the *gate current* I_g, and the *bulk current* I_b. The drain, gate, and bulk currents are positive when they flow into the N-channel MOSFET, while the source current is positive when it flows out of the transistor. The four indicated MOSFET currents are not independent electrical variables, for by KCL,

$$I_s = I_d + I_g + I_b \tag{56.48}$$

The schematic also defines the positive reference polarities of four MOSFET terminal voltages. In particular, V_{ds} is the *drain to source voltage*, V_{bs} is the *bulk to source voltage*, V_{gs} is the *gate to source voltage*, and finally, V_{gd} is the *gate to drain voltage*. By KVL,

$$V_{ds} = V_{gs} - V_{gd} \tag{56.49}$$

The cross-section diagram and corresponding electrical schematic symbol for a P-channel enhancement mode MOSFET are given in Fig. 56.16. The N+ diffusions or implants for the source and drain in the N-channel transistor become P+ regions in the P-channel device. In contrast to the N-channel transistor, the P-channel device uses an N-type substrate. Otherwise, the cross sections for these two transistor types are identical. The only change in the electrical schematic is that the arrow in the source lead of the P-channel transistor is pointed in a direction that is opposite to that shown in the electrical schematic of the N-channel MOSFET.

The positive reference directions for all P-channel device currents and terminal voltages are reversed with respect to their individual orientations in the N-channel circuit diagram. Specifically, the positive direction of the drain, gate, and substrate bulk currents is out of the P-channel transistor, and positive source current flows in to the device. Likewise, the positive reference polarities of all terminal voltages are reversed; that is, the *source to drain voltage* V_{sd}, the *source to bulk voltage* V_{sb}, the *source to gate voltage* V_{sg}, and the *drain to gate voltage* V_{dg}, are positive electrical variables when their measured polarities mirror those indicated in Fig. 56.16(b). Neither (56.48) nor (56.49) is affected by these polarity reversals.

Static Characteristic Equations

The Ebers–Moll and Gummel–Poon models for a bipolar junction transistor provide a single set of algebraic equations that governs the volt–ampere relationships for each of three BJT currents over all possible regimes of transistor operation. Unfortunately, no single equation set is available to describe the static volt–ampere relationships of a MOSFET. Instead, separate equations are written for the three MOSFET operating regimes: namely, the *cutoff*, the *ohmic* (also known as *triode*), and the *saturation* (also known as *linear*) regimes.

Before discussing these static volt–ampere equations, it should be noted that for low signal frequencies, the current I_g flowing in the gate lead of either an N-channel or a P-channel

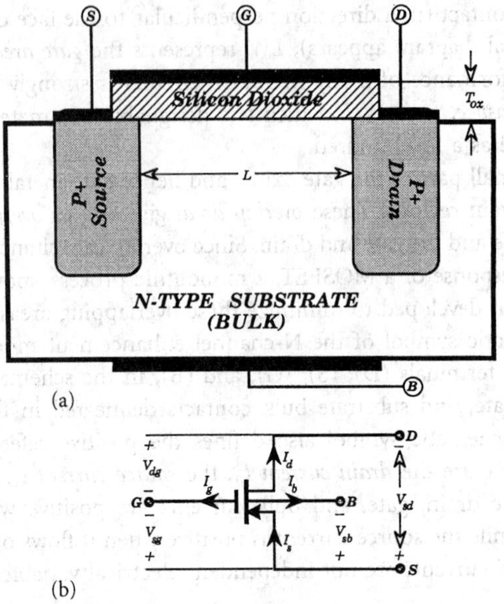

FIGURE 56.16 (a) Simplified cross section of a P-channel enhancement-mode MOSFET. The diagram is not drawn to scale. (b) Electrical schematic diagram of a P-channel enhancement-mode MOSFET.

MOSFET is zero because the silicon dioxide layer lying beneath the gate metal is an almost perfect insulator. Moreover, the substrate current, I_b, is zero because in the majority of MOSFET applications, the PN junctions formed between the substrate bulk and both the drain and source regions are reverse biased. In N-channel devices, this requirement is satisfied by connecting the P-type substrate to the most negative potential afforded by the circuit in which the MOSFET is embedded. For P-channel units, the N-type substrate is connected to the most positive of available circuit node potentials. In either case, since the substrate current, I_b, is the superposition of the bulk–drain junction leakage current, the bulk–source leakage current, and the current flowing between substrate and gate, $I_b \approx 0$. Since I_g and I_b are zero in each of three MOSFET operating regimes, (56.48) shows that the drain current I_d equates to the source current I_s at least at the low signal frequencies for which the capacitive charging currents associated with PN junction depletion layers and oxide capacitances can be ignored.

Cutoff Regime. A MOSFET is said to be cutoff when its drain current I_d is zero. In an N-channel enhancement mode MOSFET, cutoff is realized if the gate-source voltage V_{gs} is at most a positive voltage, say V_{hn}, which is termed the gate to source threshold voltage or simply, the threshold voltage of the MOSFET. The threshold voltage is effectively a turn on voltage for the MOSFET, in the sense that V_{gs} must exceed V_{hn} to cause measurable current conduction through the device. As such, its engineering significance is similar to that of V_γ, the emitter–base junction turn-on voltage for a BJT. But unlike the 700 or so mV that typifies V_γ, V_{hn} is in the range of slightly less than a volt to a few volts. Also unlike V_γ, which is independent of BJT junction voltages and currents, V_{hn} is a function of the bulk–source voltage V_{bs}. However, in the first-order static model presented here, V_{hn} is taken to be a constant, independent of MOSFET terminal voltages and lead currents.

Cutoff in a P-channel transistor is defined analogously. Specifically, cutoff, in the sense of $I_d = 0$ in the schematic representation of Fig. 56.16(b), requires that $V_{sg} \leq V_{hp}$, where V_{hp} is the positive *source to gate threshold voltage* of the P-channel device.

Ohmic (Triode) Regime. An N-channel MOSFET operates in its ohmic, or triode, regime if its gate–source voltage satisfies the constraint $V_{gs} \geq V_{hn}$ and simultaneously, its drain–source voltage is such that $V_{ds} < (V_{gs} - V_{hn})$. In this regime, the drain current I_d is given by [2]

$$I_d = K_n\left(\frac{W}{L}\right)V_{ds}\left(V_{gs} - V_{hn} - \frac{V_{ds}}{2}\right) \tag{56.50}$$

Note that for given drain–source and gate–source voltages, the drain current is proportional to the gate aspect ratio, W/L. In an actual design setting, a desired drain current is achieved by altering this ratio through an appropriate choice of the gate depth W.

The constant K_n in (56.50) is termed a **transconductance coefficient**. This parameter is in the range of tens to hundreds of micromhos per volt and is given by

$$K_n = \mu_n C_{ox} \tag{56.51}$$

where μ_n is the *surface mobility of electrons* in the channel. Because carrier mobility diminishes with increasing operating temperatures, the transconductance coefficient, and thus the drain current, decreases with increasing temperature. Finally, the parameter C_{ox} is the *density of the gate oxide capacitance.* It has dimensions of farads per square meter and is related to the oxide thickness T_{ox} by the expression

$$C_{ox} = \frac{\epsilon_{ox}}{T_{ox}} \tag{56.52}$$

where ϵ_{ox} is the *dielectric constant* of silicon dioxide (345.2 fF/cm).

For P-channel transistors, equations of the form of (56.50) and (56.51) prevail, but μ_n in (56.51) is replaced by μ_p, the *surface mobility of holes* in the channel. In particular,

$$I_d = K_p\left(\frac{W}{L}\right)V_{sd}\left(V_{sg} - V_{hp} - \frac{V_{sd}}{2}\right) \tag{56.53}$$

with the understanding that

$$K_p = \mu_p C_{ox} = \mu_p\left(\frac{\epsilon_{ox}}{T_{ox}}\right) \tag{56.54}$$

The triode region of the static MOSFET characteristic curve is more commonly referenced as the ohmic regime because for small drain–source voltages, the MOSFET behaves electrically as a linear, voltage-controlled, two terminal resistance. This contention is confirmed by (56.50), which produces a differential drain–source conductance $(1/R_{ds})$, of

$$\frac{\partial I_d}{\partial V_{ds}} \triangleq \frac{1}{R_{ds}} = K_n\left(\frac{W}{L}\right)(V_{gs} - V_{hn} - V_{ds}) = \frac{I_d}{V_{ds}}\left(\frac{V_{gs} - V_{hn} - V_{ds}}{V_{gs} - V_{hm} - \frac{V_{ds}}{2}}\right) \tag{56.55}$$

For $V_{ds} \ll (V_{gs} - V_{hn})$, the drain–source voltage, V_{ds}, and the drain current, I_d, conform to the simple Ohm's law relationship

$$V_{ds} \approx R_{ds} I_d \tag{56.56}$$

where R_{ds} is logically viewed as the effective resistance presented between the drain and source terminals of a conducting MOSFET operated at small drain–source voltages. It is interesting to note that the value of this drain–source resistance is controllable by the gate–source voltage, V_{gs}. Specifically, R_{ds} varies inversely with V_{gs}. Accordingly, the possibility of using a MOSFET as a voltage-variable resistance in the design of an electronically controlled attenuator is suggested. It is also interesting to observe that R_{ds} varies directly with the length L of the channel whose resistance is represented by R_{ds} and inversely with W, which is the cross-sectional width of the subject channel. These geometric dependencies mirror the geometric resistance properties of conventional passive resistors.

Saturated (Linear) Regime. An N-channel MOSFET enters the saturation, or linear, region of its static characteristic curves if $V_{gs} > V_{hn}$ and $V_{ds} \geq (V_{gs} - V_{hn})$. The corresponding drain current expression derives from (56.50) by equating V_{ds} to $(V_{gs} - V_{hn})$, with the result that

$$I_d = \frac{K_n}{2}\left(\frac{W}{L}\right)(V_{gs} - V_{hn})^2 \tag{56.57}$$

To first order, the saturated value of drain current is independent of the drain–source voltage V_{ds} and is a function of only the gate–source voltage V_{gs}. In short, the saturated MOSFET functions as a voltage-controlled current source, with V_{gs} serving as the controlling electrical variable for the controlled variable, I_d. This voltage-controlled current is the fundamental mechanism allowing for voltage gain from the input (gate–source) port to the output (drain–source) port. Accordingly, MOSFETs earmarked for high gain, nominally linear, signal-processing applications are biased in their saturated regimes. Note that MOSFET saturation has a distinctly different connotation that does saturation in a BJT. In the latter device, the ability to provide forward gain in the saturation regime is sharply degraded by charge injection and charge storage phenomena.

The difference voltage $(V_{gs} - V_{hn})$, is defined as the **drain saturation voltage**, V_{dss}; that is,

$$V_{dss} \triangleq V_{gs} - V_{hn} \tag{56.58}$$

As is highlighted by (56.55), the drain saturation voltage is significant in that for $V_{ds} = V_{dss}$, the static volt–ampere characteristic in the ohmic domain has zero slope and therefore, the ohmic value of drain current is maximized. By (56.50), note that this maximum drain current, say I_{dss}, which is termed the **drain saturation current**, is identical to the saturated drain current given by (56.57). Thus, in the saturated regime of operation, the drain current is alternatively expressible as

$$I_d = \frac{K_n}{2}\left(\frac{W}{L}\right)(V_{gs} - V_{hn})^2 \equiv \frac{K_n}{2}\left(\frac{W}{L}\right)V_{dss}^2 \triangleq I_{dss} \tag{56.59}$$

Equations (56.57)–(56.59) remain applicable for P-channel transistors, provided that V_{gs} is replaced by the source–gate voltage, V_{sg} and V_{hn} is supplanted by V_{hp}, the source-gate threshold voltage. In (56.58), the drain–source saturation voltage becomes the source–drain saturation voltage V_{sds}.

In summary, the simplified static volt–ampere equations for an N-channel enhancement mode MOSFET are

$$I_d = \begin{cases} 0, & V_{gs} \leq V_{hn} \\ K_n\left(\frac{W}{L}\right)V_{ds}\left(V_{gs} - V_{hn} - \frac{V_{ds}}{2}\right), & V_{gs} > V_{hn} \text{ and } V_{ds} < (V_{gs} - V_{hn}) \\ \frac{K_n}{2}\left(\frac{W}{L}\right)(V_{gs} - V_{hn})^2, & V_{gs} > V_{hn} \text{ and } V_{ds} \geq (V_{gs} - V_{hn}) \end{cases} \tag{56.60}$$

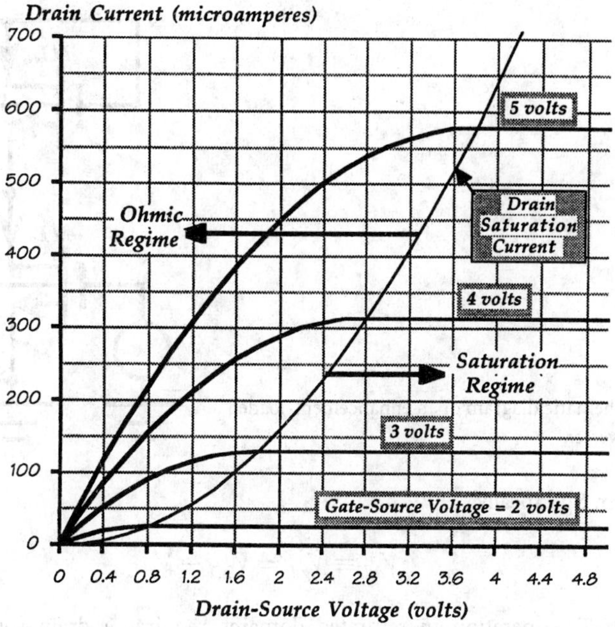

FIGURE 56.17 Static common source volt–ampere characteristic curve for an N-channel enhancement-mode MOSFET having $(K_n W/L) = 80$ μmho/V and a threshold voltage V_{hn} of 1.2 V.

These equations, which collectively comprise the common drain static characteristic curves, are plotted in Fig. 56.17 for the case of an N-channel MOSFET having $V_{hn} = 1.2$ V and $K_n(W/L) = 80$ μmho/V. The superimposed plot of (56.59), which is shown dashed in Fig. (56.17), is a locus of volt–ampere points that separate the ohmic regime on its left from the saturated regime on its right. Observe that for very small drain–source voltages, the ohmic regime curves are nominally linear, as predicted earlier. On the other hand, constant drain currents, whose individual values are determined by gate–source voltages in accordance with (56.60), are evidenced in the saturation regime.

Example 1. The common source inverter is a canonic cell for both analog and digital MOSFET applications. In the enhancement-loaded common source inverter of Fig. 56.18, the N-channel driver transistor *MD* has a gate aspect ratio of (W_d/L) and a threshold voltage of V_{hnd}. On the other hand, the N-channel load transistor *ML* has a gate aspect ratio of (W_l/L) and a threshold voltage of V_{hnl}. Both transistors have identical channel properties and accordingly, both devices share the same transconductance coefficient K_n. The input voltage V_i is presumed nonnegative, thereby ensuring that the grounded substrates of both transistors conduct negligible current. Since the gate current flowing into a MOSFET is zero at low signal frequencies, no internal source resistance is included in the input source voltage branch.

Use the simple static model of an enhancement-mode MOSFET to determine the relationship of the output voltage V_o to the applied input drive V_i. Investigate this transfer relationship from the viewpoint of setting criteria that allow using the circuit as either a linear amplifier or as a logical inverter.

Solution.

1) The load transistor, *ML*, has a drain–source voltage, say V_{dsl}, that is identical to its gate–source voltage, V_{gsl}, since the drain and gate terminals are electrically incident with

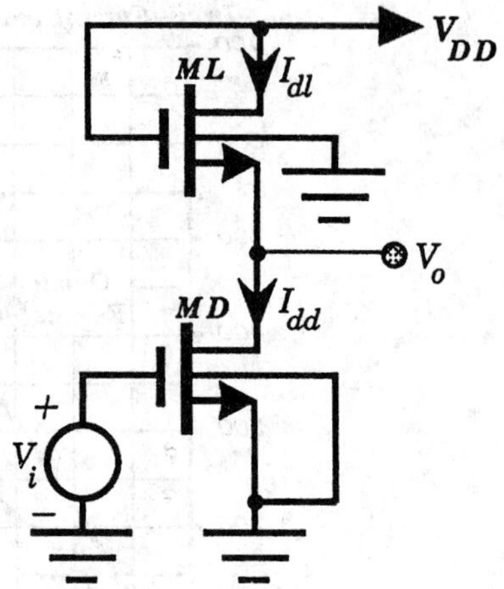

FIGURE 56.18 Schematic diagram of an enhancement-loaded common source inverter.

one another. Specifically,

$$V_{dsl} \equiv V_{gsl} = V_{DD} - V_o$$

Since MOSFET operation in saturated domains requires a drain–source voltage that exceeds the corresponding gate–source voltage, less a threshold voltage drop, identical drain–source and gate–source voltages ensure volt–ampere operation in the saturation region. Thus, transistor *ML* is always saturated and for $V_{gsl} > V_{hnl}$, the current I_{dl} flowing through the load transistor in Fig. 56.18 is

$$I_{dl} = \frac{K_n}{2}\left(\frac{W_l}{L}\right)(V_{DD} - V_o - V_{hnl})^2$$

Observe from the first equation above that since V_{gsl} must exceed V_{hnl} in order for *ML* to conduct, V_o must correspondingly be smaller than $(V_{DD} - V_{hnl})$. This statement is corroborated by the second of the foregoing two equations, which confirms that $I_{dl} = 0$ when $V_o = (V_{DD} - V_{hnl})$. It follows that the maximum possible output voltage, say V_{max}, is

$$V_{max} = V_{DD} - V_{hnl}$$

whence

$$I_{dl} = \frac{K_n}{2}\left(\frac{W_l}{L}\right)(V_{max} - V_o)^2$$

2) The gate–source voltage V_{gsd} of the driver transistor, *MD*, is the input voltage V_i, while the drain–source voltage, V_{dsd}, of this device is the output voltage, V_o, of the circuit. For $V_i > V_{hnd}$ and $V_o \geq (V_i - V_{hnd}) > 0$, *MD* is saturated, and its drain current, I_{dd}, is

$$I_{dd} = \frac{K_n}{2}\left(\frac{W_d}{L}\right)(V_i - V_{hnd})^2$$

But since the substrates and gates of either MOSFET conduct zero static current, $I_{dd} \equiv I_{dl}$, which implies that for $V_0 \geq (V_i - V_{hnd}) > 0$,

$$V_o = V_{max} - \sqrt{\lambda_{dl}}(V_i - V_{hnd})$$

where, subject to the assumptions of identical transconductance coefficients and channel lengths,

$$\lambda_{dl} = \frac{W_d/L}{W_l/L} = \frac{W_d}{W_l}$$

is the ratio of the gate aspect ratio of the driver to the gate aspect ratio of the load transistor.

3) Several noteworthy points underscore the foregoing analysis.

   a) Despite the fact that the currents of both saturated transistors are quadratic functions of their respective gate–source voltages, the expression derived for the output voltage, V_o, is a linear function of the circuit input voltage, V_i. This is to say that as long as $V_o \geq (V_i - V_{hnd}) > 0$, the enhancement-loaded common source inverter functions as a linear signal processor. This observation confirms an original contention that I/O linearity in MOSFET technology circuits requires that all pertinent MOSFETs be biased in the saturation regime of their static volt–ampere characteristic curves.

   b) Since the circuit at hand operates linearly for $V_o \geq (V_i - V_{hnd}) > 0$, a small-signal voltage gain, say A_v, can be meaningfully defined. Returning to the output voltage expression, this gain is seen to be

$$A_v = \frac{dV_0}{dV_i} = -\sqrt{\lambda_{dl}}$$

   where the negative algebraic sign suggests phase inversion between the small-signal input voltage and its corresponding small-signal output response. A laudable attribute of the common source inverter is that its voltage gain depends on only the ratio of gate aspect geometries, which is accurately controlled during processing.

   c) The voltage range, $V_o \geq (V_i - V_{hnd}) > 0$, required for I/O linearity translates to an input voltage requirement of

$$V_{hnd} < V_i \leq V_{hnd} + \frac{V_{\max}}{1 + \sqrt{\lambda_{dl}}}$$

   and a corresponding output voltage range of

$$V_{\max} > V_o \geq \frac{V_{\max}}{1 + \sqrt{\lambda_{dl}}}$$

   The last result suggests that the optimum output bias voltage, say V_{oQ}, in the sense of maximum output signal swing commensurate with linear circuit operation is

$$V_{oQ} = \frac{1}{2}\left(\frac{\sqrt{\lambda_{dl}}}{1 + \sqrt{\lambda_{dl}}}\right)V_{\max}$$

   Using the previously derived output voltage expression, this disclosure corresponds to an input bias, V_{iQ} given by

$$V_{iQ} = V_{hnd} + \frac{1}{2\sqrt{\lambda_{dl}}}\left(\frac{2 + \sqrt{\lambda_{dl}}}{1 + \sqrt{\lambda_{dl}}}\right)V_{\max}$$

4) While the load transistor is always saturated, the driver *MD* slips into its ohmic regime for $V_o < (V_i - V_{hnd})$ where

$$I_{dd} = K_n\left(\frac{W_d}{L}\right)V_o\left(V_i - V_{hnd} - \frac{V_o}{2}\right)$$

Since $I_{dd} \equiv I_{dl}$, where I_{dl} is given in step 1 above, the output voltage is now prescribed by

$$K_n\left(\frac{W_d}{L}\right)V_o\left(V_i - V_{hnd} - \frac{V_o}{2}\right) = K_n\left(\frac{W_l}{L}\right)(V_{max} - V_o)^2$$

which produces the quadratic relationship

$$(1 + \lambda_{dl})V_o^2 - 2[V_{max} + \lambda_{dl}(V_i - V_{hnd})]V_o + V_{max}^2 = 0$$

An approximate solution for the resultantly small output voltage V_o is

$$V_o \approx \frac{V_{max}}{2\left[1 + \lambda_{dl}\left(\frac{V_i - V_{hnd}}{V_{max}}\right)\right]}$$

Unlike the linear output response corresponding to saturation of *MD*, the present result shows that the output voltage decreases nonlinearly with increasing input voltage V_i.

5) When used as a logical inverter, the logic "1" or "high" level for the circuit in Fig. 56.18 is V_{max}, as defined in step 1. It follows that when the input voltage is V_{max}, the logic "0" or "low" voltage V_{min} is established at the output port. From the output result derived in the preceding step,

$$V_{min} \approx \frac{V_{max}}{2\left[1 + \lambda_{dl}\left(\frac{V_{max} - V_{hnd}}{V_{max}}\right)\right]}$$

The proper operation of a digital inverter requires that if a logic high input, $V_i = V_{max}$, produces a logic low output, $V_o = V_{min}$, the identical logic low input, $V_i = V_{min}$, must result in an output response whose value mirrors the defined logic high; that is $V_o = V_{max}$. Since an output voltage that is numerically equal to V_{max} is established only if zero current flows through the load transistor, and thus through the driver transistor as well, design care must be exercised to ensure that the logic "0" voltage V_{min} is smaller than the threshold voltage V_{hnd} of the driver. Using the preceding relationship, $V_{min} < V_{hnd}$ implies the requirement

$$\lambda_{dl} > \frac{\dfrac{V_{max}}{2V_{hnd}} - 1}{1 - \dfrac{V_{hnd}}{V_{max}}}$$

that is, the subject inverter must be capable of sufficient gain. Figure 56.19 summarizes the foregoing disclosures by sketching V_o as a function of V_i.

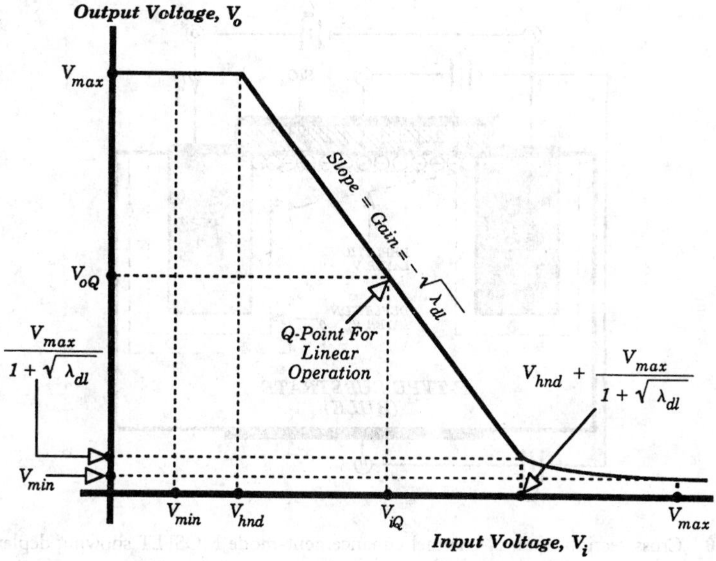

FIGURE 56.19 Static transfer characteristic of an N-channel enhancement-loaded common source inverter.

Engineering Description of Device Operation

A circuit analysis based on the model defined by (56.60) is useful only to the extent that a first-order estimate of MOSFET circuit performance is required. A more accurate estimate of circuit responses demands the use of more sophisticated MOSFET models. Before discussing these refinements, it is instructive to articulate several engineering and physical concepts that underlie the simplified device representation.

To the foregoing end, consider an N-channel enhancement-mode MOSFET operated under the usual case of nonpositive bulk–source biasing V_{bs}, a small drain–source voltage V_{ds}, and a gate–source voltage V_{gs}, that is allowed to vary from zero to a voltage that is larger than the gate–source threshold potential V_{hn}. Figure 56.20 shows the simplified cross section of the device, together with the depletion layers corresponding to the indicated biasing.

For the case of $V_{ds} = V_{bs} = 0$, depletion layers are formed around all bulk–source and bulk–drain junctions. These depletion layers are the zero bias transition regions that establish device equilibrium. For equally doped source and drain diffusions, the depletion layers are uniformly thick, and because the bulk is much more lightly doped than is either the drain or the source diffusion, most of the depletion layer thickness extends into the bulk. If V_{bs} is made negative, the reverse biasing across the bulk–source and bulk–drain junctions increases, and the depletion layer thickness increases uniformly. If V_{ds} is also increased, the depletion region about the drain becomes thicker than the depletion layer about the source, since the net reverse bias at the bulk–drain junction perimeter is $V_{ds} - V_{bs}$, while at the bulk–source junction, the reverse bias is only $-V_{bs}$.

The Depletion Condition. If the gate–source voltage V_{gs}, is allowed to increase to a positive value that is smaller than the gate–source threshold potential, and if the bulk–source voltage V_{bs}, is at most 0 V, no drain current, I_d, can flow even if the drain–source voltage, V_{ds}, is large. In order to understand this fact, it is fruitful to paint a qualitative electrostatic picture corresponding to the biasing, $V_{bs} < 0$, $V_{ds} > 0$, and $0 < V_{gs} < V_{hn}$. This biasing circumstance defines the so-called *depletion condition* in an N-channel enhancement mode MOSFET.

The immediate effect of $V_{gs} > 0$ is to establish a vertical electric field from the gate metal (or polysilicon), through the oxide layer, and toward the silicon, as shown by the directed lines in

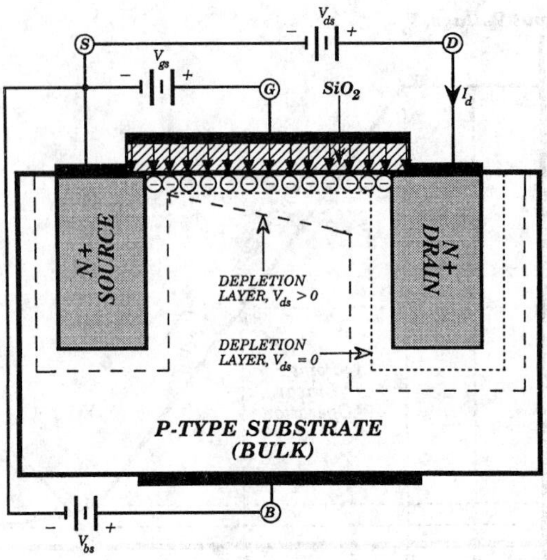

FIGURE 56.20 Cross section of an N-channel enhancement-mode MOSFET showing depletion layers for zero and positive drain-source voltages V_{ds}. The applied bulk source voltage V_{bs} is presumed to be negative or zero, and the voltage V_{gs} applied from gate to source is positive. The encircled negative signs symbolize immobile ionized acceptor atoms. The vertical arrows in the oxide layer represent the electric-field induced by applied gate-source potential. The diagram is not drawn to scale.

the diagram of Fig. 56.20. Electric field lines must terminate on negative charges. Since the only substantive source of such charges in an N-channel enhancement-mode MOSFET is ionized acceptor impurities in the bulk, the electric field corresponding to positive V_{gs} induces a depletion region beneath the gate in the MOSFET channel between source and drain. Equivalently, the electrostatic force associated with positive V_{gs} drives the free holes originally present in the interfacial channel region deeper into the bulk, thereby leaving only immobile ionized impurity atoms (depicted as encircled negative signs in the figure) in the channel beneath the gate. The resultant *channel depletion layer,* or *surface depletion layer,* complements the depletion layers already established around the perimeters of the source and drain diffusions. The bulk side boundaries of these transition layers are illustrated in Fig. 56.20 for the cases of $V_{ds} = 0$ and $V_{ds} > 0$.

The reason underlying zero drain current response to even relatively large drain–source voltages is now apparent. In particular, appreciable drain current is realized only if a correspondingly large electron concentration is transported from the source to the drain. Only two paths are conceivable for this charge transport. One is through the channel depletion region, and the other is through the substrate bulk lying beneath the surface depletion layer. But no free electrons are present in the surface depletion region. Furthermore, no free electrons can be injected from the source to the bulk, thence into the drain, because the junction perimeters of both the source and drain regions are reverse biased in the depletion mode of operation. Hence, I_d is essentially zero in the depletion mode of operation.

The Inversion Condition. If the thin interfacial channel region described in the preceding subsection were to contain a large concentration of free electrons, in addition to immobile acceptor ions, a drain current would flow in response to the application of positive drain–source voltage. Positive V_{ds} serves to attract channel electrons in the direction of source to drain, thereby effecting a drain current path from drain, through the channel, to the source. The transport of electrons through the channel in response to a lateral electric field set up in the direction of drain to source by the application of positive V_{ds} is, in fact, the principle mechanism for current flow in an N-channel MOSFET. The problem now is to understand how an appreciable electron concentration is induced in the interfacial channel that is heavily populated by ionized bulk impurities.

Boltzmann statistics and the mass action law provide the answer to the foregoing query. The electric field at the surface of the MOSFET produces a potential, say V_y, at the interface of oxide and silicon. If an allowance is made for a potential drop, say V_{ox}, from the interface of gate metal (or polysilicon) and gate oxide to the interface of gate oxide and silicon surface, this induced interfacial potential subscribes to the KVL relationship

$$V_{gs} = V_{ox} + V_y \tag{56.61}$$

For analytical expedience, the electrical effects of work function differences [3] between gate lead and gate metal (or polysilicon) and between gate metal (or polysilicon) and gate silicon dioxide are presumed to have been incorporated into the voltage V_{ox}. The aforementioned interfacial potential V_y, defines the free-electron concentration in the interfacial channel region in accordance with

$$n = n_o e^{V_y/V_T} = \frac{N_{iB}^2}{p_o} e^{V_y/V_T} \tag{56.62}$$

where V_T is the Boltzmann voltage, N_{iB} is the intrinsic carrier concentration in the substrate bulk p_o, the equilibrium concentration of holes in the bulk, is very nearly equal to the average bulk impurity concentration N_A, and

$$n_o = \frac{N_{iB}^2}{p_o} \approx \frac{N_{iB}^2}{N_A} \tag{56.63}$$

represents the equilibrium concentration of free electrons in the bulk.

Equation (56.62) establishes the mechanism for achieving large free-electron concentrations in the interfacial channel region. It shows that n rises dramatically with the potential V_y. Suppose that V_y is allowed to increase to a voltage that forces n to equal the intrinsic carrier concentration, N_{iB}. This particular value of V_y is symbolized as V_F and is termed the **Fermi potential** of the substrate bulk. From (56.62) and (56.63)

$$V_F = V_T \ln\left(\frac{N_A}{N_{iB}}\right) \tag{56.64}$$

which is typically of the order of 300 mV. It is crucial to understand that at an interfacial potential of V_F, the channel immediately beneath the oxide–silicon interface is intrinsic; that is, it is neither P-type, as it was prior to the application of the Fermi potential at the interface, nor is it N-type, which is mandated for significant drain currents. Further increases in V_y, which are realized by increasing the gate–source voltage V_{gs}, result in corresponding increases in electron concentration. When V_y is slightly greater than V_F, the interfacial channel is said to experience *weak inversion* in the sense that there are now more electrons than there are holes in the interfacial channel region.

There is a practical limit to the magnitude of V_y. This limit is the interfacial potential corresponding to $n = N_A$, since only one hole can be relinquished by, and hence only one electron can be attracted to, each bulk impurity atom. From (56.62) and (56.63), $n = N_A$ implies $V_y = 2V_F$. This value of interfacial potential is essentially constant, and when it is achieved, the N-channel enhancement mode MOSFET is said to be in *strong inversion*.

Although drain currents can be observed for interfacial potentials in the weak inversion range, $V_F < V_y < 2V_F$, the condition for substantial current flow, in the sense of that predicted by the simplified static model of (56.60), is generally taken to be strong inversion; that is $V_y = 2V_F$. The gate–source voltage V_{gs} corresponding to the strong inversion requirement is the threshold voltage V_{hn}. By (56.61)

$$V_{hn} = V_{ox} + 2V_F \tag{56.65}$$

However, $V_{gs} > V_{hn}$ ensures strong inversion, and hence a substantive supply of free-channel electrons, at only the source end of the channel. In order for strong inversion to prevail throughout the entire source to drain channel, it is also necessary that the gate–drain voltage V_{gd} exceed the threshold potential V_{hn}. Since

$$V_{gd} = V_{gs} - V_{ds} \tag{56.66}$$

$V_{gd} > V_{hn}$ implies

$$V_{ds} < V_{gs} - V_{hn} \equiv V_{dss} \tag{56.67}$$

Thus, the biasing constraints $V_{gs} > V_{hn}$ and $V_{ds} < V_{dss}$, which ensure MOSFET operation in the ohmic regime of its static characteristic curves, are equivalent to establishing the criteria for strong inversion of the entire interfacial channel region.

Figure 56.21 depicts the channel inversion situation for $V_{ds} = 0$ and for $0 < V_{ds} < V_{dss}$, with the bulk–source voltage V_{bs} taken as zero in both cases. For $V_{ds} = 0$, the diagram in Fig. 56.21(a) shows that the depletion layers around the perimeters of the source, drain, and inverted channel are uniformly thick, as is the channel inversion layer itself. This uniformity reflects identical reverse biases imposed across both the bulk–source and bulk–drain junctions as a result of zero drain–source voltage. The uniformly thick inverted channel volume can be perceived as a resistive semiconductor bar, for which the resistance is R_{ds}, as defined by (56.55). From a physical perspective, it also derives from

$$R_{ds} = \frac{\rho_c L}{A_{\text{eff}}} \tag{56.68}$$

where ρ_c is the average resistivity of the inverted channel, and A_{eff} is the effective cross-sectional area pierced by electrons in transit from the N+ source region to the N+ drain region. This area is the thickness of the inverted channel multiplied by W, the depth of the channel, measured in a direction that is perpendicular to the face of the page in Fig. 56.21(a). Since the free-channel charge concentration, n, is predominantly electrons, the resistivity in (56.68) is

$$\rho_c \approx \frac{1}{q\mu_n n} \tag{56.69}$$

with μ_n representing the mobility of channel electrons. The Ohm's law relationship $I_d = V_{ds}/R_{ds}$ therefore suggests that the ohmic regime drain current, I_d, is proportional to the electron mobility, μ_n, as well as to the geometrical factor, (W/L). The latter contention reinforces the previously discussed direct dependence of drain current on the gate aspect ratio.

The situation corresponding to $V_{ds} > 0$ is depicted in Fig. 56.21(b). The thickness of the depletion layer about the perimeter of the bulk–drain junction is now larger than the depletion layer thickness about the bulk–source junction. This enhanced layer thickness stems from a reverse bias about the bulk–drain junction that is larger than that across the bulk–source junction by an amount V_{ds}. Additionally, the inversion layer nearer the drain end of the channel is thinner than it is near the source end because the gate–drain voltage supporting inversion at

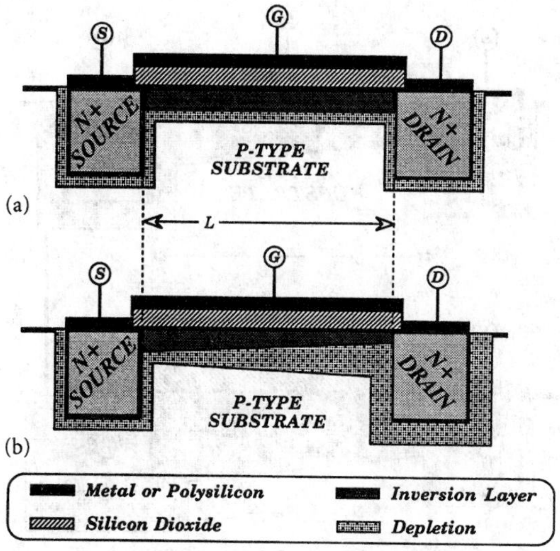

FIGURE 56.21 Simplified cross sections of an N-channel enhancement-mode MOSFET. The substrate contact is now shown, and it is assumed that the bulk-source voltage V_{bs} is zero. (a) The case of $V_{gs} > V_{hn}$ and $V_{ds} = 0$. (b) The case of $V_{gs} > V_{hn}$ and $0 < V_{ds} < V_{dss}$.

the drain site is smaller by the same amount, V_{ds}, than is the gate-source voltage that induces an inversion layer near the source. This nonuniform inversion layer suggests that the resistance R_{ds} associated with the resistive inversion region is no longer independent of the drain–source voltage. It follows that the resultant drain current $I_d = V_{ds}/R_{ds}$, is dependent on V_{ds} as is borne out by the second of the static characteristic equations in (56.60).

There is yet another important aspect of positive V_{ds}. In general, the current flowing through a semiconductor volume is the superposition of drift and diffusion components. When the free-carrier concentration that supports current flow has no charge gradient in the plane of charge transport, the current is exclusively a drift component proportional to the electric field established in the charge transport plane. Such a situation prevails for small V_{ds}, which is to say the drain current arising from the application of small drain–source voltages is approximately proportional to the lateral field established in the direction of drain to source by positive V_{ds}. Further increases in the drain–source voltage incur a more pronounced gradient to the inversion layer charge. Accordingly, the diffusion component of net drain current progressively increases with increasing V_{ds}. Despite increases in this diffusion current component, the simplified static volt–ampere relationships of (56.60) account for only drift current effects in the interfacial channel of a MOSFET.

Pinchoff and Saturation. As discussed in the preceding section, the thickness of the inversion layer narrows nearer the drain site for increasing values of the drain–source voltage V_{ds}. If V_{ds} is increased to a value that equals the drain saturation voltage V_{dss}, the inversion layer thickness vanishes at the drain end of the channel since the interfacial potential at that end becomes zero. Figure 56.22(a) depicts this so-called channel *pinchoff condition.* Since the inversion layer thickness is zero at the drain end of the channel when $V_{ds} = V_{dss}$, the differential resistance there is infinitely larger, which corroborates the channel resistance conclusions drawn earlier in conjunction with (56.55).

If V_{ds} is increased beyond V_{dss}, the second equation in (56.60) is impertinent to the volt–ampere characteristics of an N-channel enhancement mode MOSFET because the channel inversion region no longer embraces the entire metallurgical channel length L, as depicted in Fig. 56.22(b). Stated in other terms, the gradient of the channel charge profile increases for

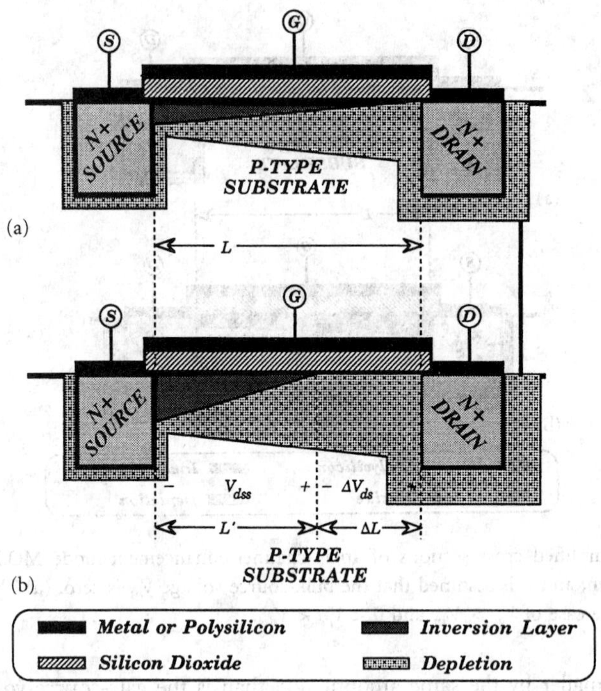

FIGURE 56.22 Simplified cross sections of an N-channel enhancement-mode MOSFET. The substrate contact is not shown, and it is assumed that the bulk–source voltage V_{bs} is zero. (a) The case of $V_{gs} > V_{hn}$ and $V_{ds} \equiv V_{dss}$. (b) The case of $V_{gs} > V_{hn}$ and $V_{ds} = V_{dss} + \Delta V_{ds}$.

$V_{ds} > V_{dss}$ to such an extent that drift is no longer the predominant charge transport mechanism supporting the flow of drain current. The diminished channel length contention can be argued as follows. The application of a drain–source voltage V_{ds} establishes a lateral electric field directed through the channel region from the drain to the source. This electric field induces a channel voltage, say V_x, with respect to the source at any position x within the length of the channel. If V_{dss} is the value of drain–source voltage that precisely establishes the inversion charge profile depicted in Fig. 56.22(a), $V_{ds} > V_{dss}$ suggests that the total electron charge embodied by this profile must be contained in a cross-sectional channel area whose length, say L', is smaller than L. In particular, if V_{ds} is written as

$$V_{ds} = V_{dss} + \Delta V_{ds} \tag{56.70}$$

V_{dss} is the potential dropped over the length, L', of the channel, while ΔV_{ds} is the voltage dropped over the channel segment ΔL, as indicated in Fig. 56.22(b).

Note that the channel region of length ΔL comprises a depletion layer whose intrinsic electric field is a function of the difference voltage, ΔV_{ds}. Intuitively, a drain current dependence on ΔL, and hence on ΔV_{ds}, can be suspected since the large lateral electric field associated with large ΔL (and thus, large V_{ds}) encourages the transport of inversion layer electrons into the drain region. This intuitively deduced dependence of I_d on ΔL is ignored in the third equation of (56.60) for the static model of an N-channel enhancement-mode MOSFET.

Refinements to the Simple MOSFET Model

Several refinements to the simple static model of a MOSFET are typically invoked in either manual or computer-aided analyses of MOS technology circuits. Among these are channel length modulation corrections to the drain saturation current, incorporation of threshold voltage dependence on bulk–source bias, incorporation of series lead resistances, and inclusion of charge storage phenomena in the carrier transport mechanism that underlies MOSFET operation. These refinements are discussed below with respect to only N-channel devices. With appropriate symbolic changes, they apply equally well to P-channel MOSFETs.

Channel Length Modulation. As discussed earlier, the effective channel length of a MOSFET decreases by an amount ΔL when the transistor operates in the saturated regime of its static characteristic curves. Recall from (56.59) that the drain current flowing in an N-channel MOSFET is I_{dss} when the drain–source voltage V_{ds} is V_{dss}. Since I_{dss} is inversely proportional to the channel length L, the drain current I_d for $V_{ds} > V_{dss}$ must be larger than I_{dss} by a factor of $[L/(L - \Delta L)]$; that is,

$$I_d\Big|_{V_{ds} \geq V_{dss}} = I_{dss}\left(\frac{L}{L - \Delta L}\right) = \frac{\frac{K_n}{2}\left(\frac{W}{L}\right)(V_{gs} - V_{hn})^2}{1 - \frac{\Delta L}{L}} \tag{56.71}$$

In this semiquantitative relationship, the factor $\Delta L/L$ relates empirically to the drain–source voltage as

$$\frac{\Delta L}{L} = \frac{1}{1 + \frac{V_\lambda}{V_{ds} - V_{dss}}} \tag{56.72}$$

so that in the saturated domain,

$$I_d\Big|_{V_{ds} \geq V_{dss}} = \frac{K_n}{2}\left(\frac{W}{L}\right)(V_{gs} - V_{hn})^2\left(1 - \frac{V_{ds} - V_{dss}}{V_\lambda}\right) \tag{56.73}$$

In this expression, V_λ, termed the **channel length modulation voltage**, is of the order of tens of volts. Instead of the zero slope in the saturated common drain static volt–ampere characteristics that is indicated in Fig. 56.17, the effect of a finite channel length modulation voltage is to incur a nonzero slope I_{dss}/V_λ in each of these curves.

Example 2. In the circuit of Fig. 56.23, all transistors have a transconductance coefficient of 200 μmho/V. Transistor $M2$ has a threshold voltage of 2.3 V, a channel length modulation voltage of 25 V, and a gate aspect ratio that is five times larger than that of transistor $M1$. In turn, $M1$ is characterized by a threshold voltage of 2.0 V, and a channel length modulation voltage of 20 V. Transistor $M3$ has a gate aspect ratio of 10, a gate–source threshold voltage of 1.8 V, and a channel length modulation voltage of 45 V. If the power line voltage, V_{DD}, is 5.2 V, calculate the current, I_{d3}, flowing in the drain of transistor $M3$ for $V_{CC} = 5.2$ V and $V_{CC} = 10.4$ V.

Solution.

1) Both transistors $M1$ and $M2$ operate in their saturation regions since the gate and drain terminals of each of these devices are electrically tied together. Moreover, since the substrates of all transistors are grounded and no negative voltages are applied to the

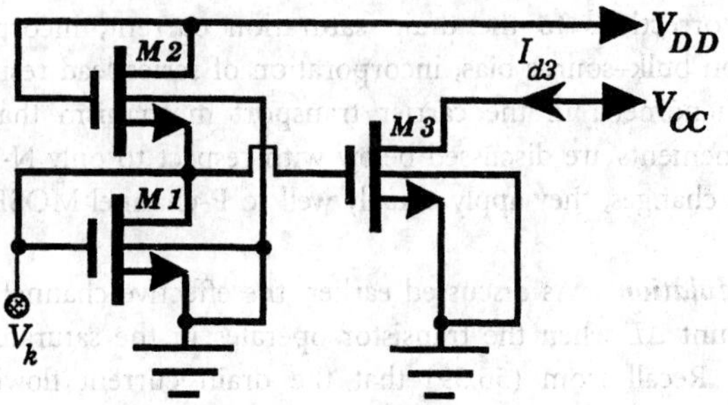

FIGURE 56.23 Schematic diagram of an enhancement-mode N-channel MOSFET constant current sink.

circuit, the currents conducted by transistors $M1$ and $M2$ are identical. Accordingly, from (56.74),

$$\frac{K_n}{2}\left(\frac{W_2}{L}\right)(V_{gs2} - V_{hn2})^2\left(1 + \frac{V_{ds2} - V_{dss2}}{V_{\lambda 2}}\right) = \frac{K_n}{2}\left(\frac{W_1}{L}\right)(V_{gs1} - V_{hn1})^2\left(1 + \frac{V_{sd1} - V_{dss1}}{V_{\lambda 1}}\right)$$

In this expression, the gate–source voltage of $M2$, V_{gs2}, is $(V_{DD} - V_k) \equiv V_{ds2}$, the drain–source voltage of transistor $M2$, and V_{gs1}, the gate–source voltage of transistor $M1$, which is also the drain–source voltage of $M1$, is the voltage V_k, indicated in the schematic diagram. Moreover, from (56.67), the drain–source saturation voltage, V_{dss2}, of $M2$ is $(V_{gs2} - V_{hn2}) \equiv (V_{DD} - V_k - V_{hn2})$, while the drain–source saturation voltage V_{dss1} of transistor $M1$ is $(V_{gs1} - V_{hn1}) \equiv (V_k - V_{hn1})$. Using these observations and introducing a parameter h such that

$$h = \sqrt{\frac{1 + \dfrac{V_{hn1}}{V_{\lambda 1}}}{\left(\dfrac{W_2/L}{W_1/L}\right)\left(1 + \dfrac{V_{hn2}}{V_{\lambda 2}}\right)}}$$

the foregoing current relationship yields

$$V_k = \frac{V_{DD}}{h+1} - \frac{V_{hn2} - hV_{hn1}}{h+1}$$

For the stipulated device parameters and circuit variables, V_k is 2.62 V. Note that the $M1$–$M2$ subcircuit functions as a voltage divider that serves to transform the power line voltage V_{DD} into a voltage V_k that is suitable for biasing the current sinking transistor $M3$.

2) For transistor $M3$ in Fig. 56.23, the gate–source voltage is V_k, the drain–source voltage is V_{CC}, and the drain–source saturation voltage is $V_k - V_{hn3}$. Since $V_{CC} > (V_k - V_{hn3})$, at even the smallest value of V_{CC} of interest, $M3$ is always saturated. Thus,

$$I_{d3} = \frac{K_n}{2}\left(\frac{W_3}{L}\right)(V_k - V_{hn3})^2\left(1 + \frac{V_{CC} - V_k + V_{hn3}}{V_{\lambda 3}}\right)$$

For $V_{CC} = 5.2$ V, I_{d3} is 740.0 μA, while $V_{CC} = 10.4$ V delivers $I_{d3} = 817.9$ μA. It follows that the effect of a finite channel length modulation voltage in transistor $M3$ is to increase the drain current flowing in this device by 10.5% when V_{CC} doubles its original 5.2 V value.

Substrate Bulk Effect on Threshold Voltage. In an N-channel enhancement mode MOSFET, the gate-source threshold voltage, V_{hn}, is not a constant, as inferred by (56.65). Instead, V_{hn} is modulated by the applied bulk–source voltage, V_{bs}. This *threshold modulation*, which is commonly called **bulk** or **body effect**, can be understood qualitatively by studying the effect of a strong reverse bias applied from the substrate to the source. In particular, a large negative value of V_{bs} serves to deplete the bottom part of the substrate in the immediate neighborhood of the bulk contact. As a result, free electrons in this neighborhood are displaced toward the oxide–semiconductor interface. Since interfacial depletion precedes the channel inversion caused by a gate–source bias that is at least as large as the threshold potential, these displaced free electrons must be pushed back down into the body of the substrate before channel inversion can occur. The threshold voltage expression given by (56.65) accounts for only equilibrium (zero net reverse bias across the boundary separating the inversion layer from the substrate) concentrations of free substrate electrons. Thus, the actual threshold voltage must be that given by (56.65) plus a correction to allow for the additional energy required to displace substrate electrons that are no longer in equilibrium because of nonzero V_{bs}. This correction is minimal for lightly doped substrates. Substrate bias-induced threshold voltage modulation is also minimized by thin oxides. Thin oxide layers promote strong vertical field intensities for even small gate–source biases, thereby expediting interfacial channel depletion.

A suitable expression for the corrected threshold voltage, say V_{hnc}, is

$$V_{hnc} = V_{hn} + \sqrt{2V_\theta(2V_F - 2V_T - V_{bs})} \tag{56.74}$$

where V_F is defined by (56.64). With

$$L_a = \sqrt{\frac{\epsilon_s V_T}{qN_A}} \tag{56.75}$$

representing the *Debye length* of electrons in a substrate having an average impurity concentration of N_A, the parameter V_θ, which is known as the **body effect voltage**, is

$$V_\theta = \frac{(qN_A L_a/C_{ox})^2}{V_T} = \frac{qN_A \epsilon_s}{C_{ox}^2} \tag{56.76}$$

In (56.75) and (56.76), ϵ_s is the dielectric constant of silicon [1.053 pF/cm], while C_{ox} is the density of gate oxide capacitance given by (56.52).

In (56.74), the effect of a bulk–source reverse bias $V_{bs} < 0$ is to increase the effective threshold voltage V_{hnc}. The amount of such increase depends on the value of the body effect voltage V_θ. Confirming earlier contentions, the amount of requisite threshold correction is small if V_θ is small, which requires small substrate concentration and/or large C_{ox} (corresponding to small oxide thickness T_{ox}).

Equation (56.74) is easily modified to embrace P-channel transistors by replacing V_{hnc} by V_{hpc}, the corrected source to gate threshold voltage of a P-channel enhancement mode MOSFET. Additional changes entail replacing V_{hn} by V_{hp}, substituting V_{sb}, the source to substrate bias, for V_{bs}, and finally, replacing N_A in (56.75) and (56.76) by N_D, the average donor concentration in the N-type substrate of a P-channel device.

It is often convenient to express the effective threshold voltage of an enhancement mode MOSFET in terms of the zero bias threshold voltage; that is, the threshold voltage prevailing with zero bias applied between bulk and source. From (56.74), the zero bias threshold voltage, say V_{hno}, is

$$V_{hno} = V_{hn} + 2\sqrt{V_\theta(V_F - V_T)} \tag{56.77}$$

It follow that

$$V_{hnc} = V_{hno} + 2\sqrt{V_\theta(V_F - V_T)}\left[\sqrt{1 - \frac{V_{bs}}{2(V_F - V_T)}} - 1\right] \tag{56.78}$$

Ohmic Resistances. Although ohmic resistances generally exert negligible effects on the volt–ampere characteristics of MOSFETs, they nonetheless materialize in the bulk, drain, and source regions. The topological significance of these parasitic resistances is abstracted by the symbolic schematic diagram of the N-channel enhancement-mode MOSFET shown in Fig. 56.24.

The resistance of the heavily doped drain region is r_{dd}, and its source region counterpart is r_{ss}. Both of these resistances are of the order of a few ohms. Unfortunately, they are difficult to calculate satisfactorily because their respective net values incorporate the contact resistances formed at the surface of the drain and source. Moreover, they are spreading resistances, since the currents in the drain and source do not flow in one direction. The resistance, r_{bb}, is a spreading resistance associated with the bulk substrate. Because of the relatively light impurity concentration in the bulk, r_{bb} is considerably larger than either r_{dd} or r_{ss}. Under static conditions, however, the voltage dropped across r_{bb} is about the same as, or even less than, the voltages dropped across the internal drain and source resistances, since the substrate–source and substrate–drain junctions are reverse biased. These junctions are represented respectively by the diodes, *DBS* and *DBD*, which have their own internal ohmic resistances, r_{bs} and r_{bd}, respectively.

The immediate effect of incorporating regional ohmic resistances into the MOSFET model is to separate internal drain, source, and bulk nodes from the externally accessible drain, source,

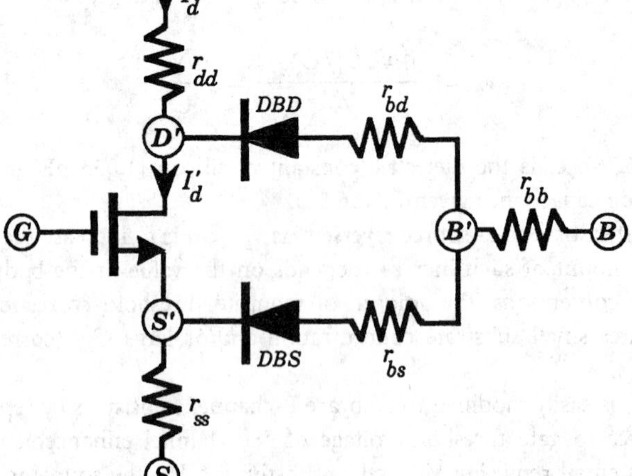

FIGURE 56.24 Schematic diagram of an N-channel enhancement-mode MOSFET with internal ohmic resistances and bulk-drain and bulk-source diodes explicitly delineated. For P-channel devices, the diode connections are reversed, as are the indicated reference directions of the currents, I_d and I'_d.

and bulk terminals. Thus, the drain–source (V_{ds}), gate–source (V_{gs}), bulk–source (V_{bs}), bulk–drain (V_{bd}), and gate–drain (V_{gd}) voltages exploited previously in all modeling relationships and discussions must now be interpreted as the voltages appearing from internal drain (D′) to internal source (S′), gate to internal source, internal bulk (B′) to internal source, internal bulk to internal drain, and gate to internal drain terminals, respectively. Likewise, the drain current, I_d, in all preceding modeling expressions must be viewed as the channel current I'_d in Fig. 56.24. Note that the channel current differs from the measurable drain current by an amount that equals the small reverse bias current that flows through the drain–bulk PN junction diode.

MOSFET Capacitances. The frequency response of MOS analog circuits and the switching speed of MOS digital circuits are limited by the time required to charge and discharge capacitances that appear between internal device nodes, from interconnecting metallization to the circuit ground plane, and between interconnecting electrical lines. Figure 56.25 illustrates the topological interconnection of intrinsic device capacitances.

The capacitors, C_{bd} and C_{bs}, represent *transition capacitances* associated with the PN bulk–source and bulk–drain junctions. Both of these elements have the voltage function form,

$$C = \frac{C_0}{\left[1 - \dfrac{V}{V_j - 2V_T}\right]^m} \tag{56.79}$$

where m is the *grading coefficient*, C_0 is the *zero-bias value of the transition capacitance*, and V_j is *built-in potential* of the pertinent junction. When computing C_0, care must be taken to include

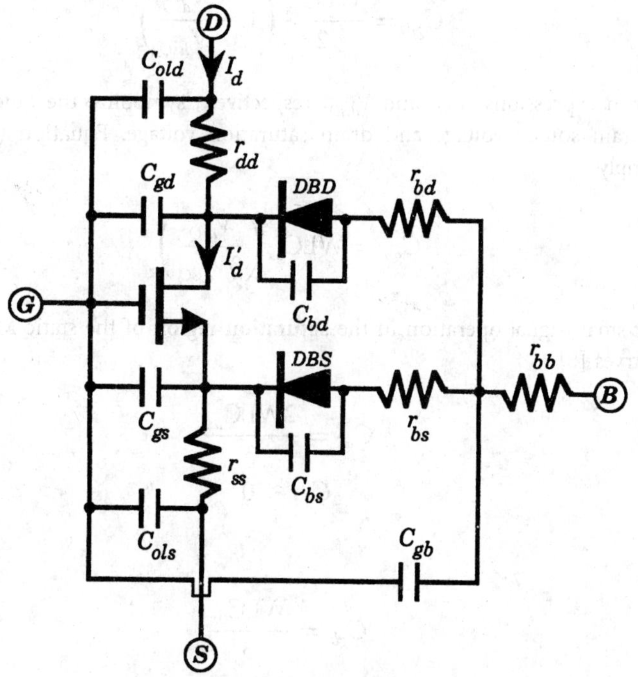

FIGURE 56.25 Schematic diagram of an N-channel enhancement mode MOSFET with internal ohmic resistances, bulk-drain and bulk-source diodes, and internal device capacitances explicitly delineated. The diodes *DBD* and *DBS* appear as ideal elements in the senses of zero internal capacitances and resistances. For P-channel devices, the diode connections are reversed, as are the indicated reference directions of the currents I_d and I'_d.

the areas of the drain and source diffusion sidewalls, as well as the areas of the bottoms of these diffused or implanted regions.

The capacitors, C_{old} and C_{ols}, are *overlap capacitances* between the gate and drain and the gate and source, respectively. These capacitances are zero when the subject MOSFET is fabricated in self-aligning gate technology.

To first order, the sum of gate–bulk (C_{gb}), gate–drain (C_{gd}), and gate-source (C_{gs}) capacitances is the net effective capacitance of the gate oxide layer; that is,

$$\mathrm{WLC}_{ox} = C_{gb} + C_{gd} + C_{gs} \tag{56.80}$$

The partitioning of WLC_{0x} into three component capacitances entails a complex calculation. Such partitioning must ensure consistency between capacitive charging currents and the charge storage and transport implications of the static volt–ampere relationships of the MOSFET mode [4]–[6]. For manual analysis purposes, the following approximate relationships have proven to be effective.

1. For linear small-signal operation in the **ohmic region** of the static MOSFET characteristic curves [7],

$$C_{gs} = \frac{\mathrm{WLC}_{ox}}{2}\left(1 + \frac{V_{dsQ}}{3V_{dssQ}}\right) \tag{56.81}$$

and

$$C_{gd} = \frac{\mathrm{WLC}_{ox}}{2}\left(1 - \frac{V_{dsQ}}{V_{dssQ}}\right) \tag{56.82}$$

In these two expressions, V_{dsQ} and V_{dssQ}, respectively, symbolize the quiescent values of internal drain–source voltage and drain saturation voltage. Equation (56.80) through (56.82) imply

$$C_{gb} = \mathrm{WLC}_{ox}\left(\frac{V_{dsQ}}{3V_{dssQ}}\right) \tag{56.83}$$

2. For linear small-signal operation in the **saturation region** of the static MOSFET characteristic curves [8]

$$C_{gs} = \frac{2\mathrm{WLC}_{ox}}{3} \tag{56.84}$$

$$C_{gd} \approx 0 \tag{56.85}$$

and

$$C_{gb} = \frac{\mathrm{WLC}_{0x}}{3} \tag{56.86}$$

Note that despite $C_{gd} \approx 0$, the net effective gate–drain capacitance is not zero; rather, it is C_{old}, the gate–drain overlap capacitance.

Other High-Order Effects. At risk of oversimplification, most of the deviations between the observed electrical behavior of MOSFETs and the response predicted by either the simplified model or any of its embellished versions result from processing and manufacturing attempts aimed toward improving the electrical performance of these devices. For example, a continuing design issue is increased switching speed. In MOSFETs, the most commonly invoked mechanism for improved response speed is reduction in channel length. But there are several serious drawbacks to a progressively decreased channel length, most of which stem from the commensurate increase in the strength of the lateral electric field induced by drain–source bias.

One drawback of large channel fields is the possible saturation of the nominal velocity at which majority carriers (electrons in N-channel devices) are transported from the source to the drain [9]. In particular, at low field strengths, the carrier velocity increases almost linearly with the strength of the electric field in which the carrier is immersed. The relevant proportionality constant between carrier velocity and small electric field strengths is, in fact, the carrier mobility. But at high fields, this carrier velocity approaches a constant (known as the **saturated limited velocity**), thereby actually limiting the ability of a MOSFET to respond quickly to applied excitation. Moreover, *carrier velocity saturation* decreases the transconductance coefficient and increases the thermal sensitivity of static drain current [10].

A second ramification of the high electric fields associated with short channel lengths is the production of *hot carriers* [11]. Hot carriers differ from the routine majority carriers that populate the channel in that they absorb an amount of energy from the high lateral channel field that is large enough for them to overcome the potential barrier at the silicon dioxide–silicon interface and thus be injected into the gate oxide layer. This injection alters the voltage drop across the gate oxide and hence, the threshold voltage. The resultant threshold shift causes device turn on voltage perturbations that are especially troublesome in logic cells that are operated in noisy environments or in logic cells for which the logic swing (different between the logical "1" and logical "0" voltages) is small. The trapping of injected charge in the oxide layer also degrades the operational reliability of a MOSFET through a phenomenon known as *oxide wearout* [12], and it increases the vulnerability of the device to electromagnetic and nuclear radiation [13].

Yet another effect of high channel fields is *impact ionization*, or *avalanche multiplication*, in the high-field drain end of the channel [14]. The *secondary carrier emission* associated with this phenomenon can, like hot carrier production, lead to oxide charging. It can also induce substrate currents, thereby significantly modifying the static and dynamic behavior predicted by the simplified model and its first-order refinements [15].

A more insidious problem materializes when impact ionization occurs in the bulk. Secondary carriers emitted in the bulk can diffuse large distances owing to the relatively light doping concentration of the substrate. These diffusing carriers can perturb the electrical operation of proximate devices on the same chip, thereby limiting the packing densities of junction isolated circuits [16]. Yet another substrate problem is *punchthrough*, wherein the depletion layers about the source and drain regions coalesce in the channel [17]. Punchthrough is an extreme condition that results from careless application of excessive drain–source voltage.

Finally, thin gate oxides are desirable from the viewpoint that they diminish the sensitivity of threshold voltage to bulk–source voltage. But the strong vertical electric fields encouraged by thin oxides also decrease the lateral, source to drain, mobility of free channel charges [18], since such fields tend to repel carriers toward the substrate bulk. As a result, the effective mobility of channel charges, which is known as *surface mobility*, can be markedly diminished from the concentration-dependent mobility experienced by current carriers under equilibrium conditions. This decrease in carrier mobility degrades the speed of MOSFET response, as well as reducing the static drain current predicted by the simplified model.

Small-Signal Equivalent Circuit

The development of the small-signal MOSFET model parallels the small-signal modeling of any nonlinear element. In particular, the nonlinear volt–ampere branch relationships of a MOSFET are expanded into a Taylor series about the quiescent operating point, and only the linear terms of these expansions are retained. For the model depicted in Fig. 56.25, the only nonlinearity, assuming that the bulk–drain and bulk–source diodes are reverse biased, is the internal drain current I'_d, which is a function of the internal gate–source, internal drain–source, and internal bulk–source voltages V_{gs}, V_{ds}, and V_{bs}, respectively. Thus,

$$I'_d \approx I'_{dQ} + \left.\frac{\partial I'_d}{\partial V_{gs}}\right|_Q (V_{gs} - V_{gsQ}) + \left.\frac{\partial I'_d}{\partial V_{ds}}\right|_Q (V_{ds} - V_{dsQ}) + \left.\frac{\partial I'_d}{\partial V_{bs}}\right|_Q (V_{bs} - V_{bsQ}) \tag{56.87}$$

If the signal components of drain current, gate–source voltage, drain–source voltage, and bulk–source voltage are denoted by

$$\begin{aligned} i'_{da} &= I'_d - I'_{dQ} \\ v_{ga} &= V_{gs} - V_{gsQ} \\ v_{da} &= V_{ds} - V_{dsQ} \\ v_{ba} &= V_{bs} - V_{bsQ} \end{aligned} \tag{56.88}$$

$$i'_{da} = g_{mf} v_{ga} + g_{mb} v_{ba} + \frac{v_{da}}{r_o} \tag{56.89}$$

Assuming, as is usually the case for linear signal processing applications, that the MOSFET is biased in its saturated domain, (56.73), (56.78), (56.87), and (56.89) yield a *forward transconductance* g_{mf} of

$$g_{mf} \stackrel{\Delta}{=} \left.\frac{\partial I'_d}{\partial V_{gs}}\right|_Q = \sqrt{2K_n\left(\frac{W}{L}\right)I_{dQ}} \times \left[\frac{V_\lambda + V_{dsQ} - \left(\frac{3}{2}\right)V_{dssQ}}{\sqrt{V_\lambda(V_\lambda + V_{dsQ} - V_{dssQ})}}\right] \tag{56.90}$$

where I_{dQ}, V_{dsQ}, and V_{dssQ} are the quiescent values of the drain current, drain–source voltage, and drain saturation voltage, respectively. Observe that while the small-signal forward transconductance of a bipolar junction transistor increases linearly with the quiescent collector current, the small-signal forward transconductance of a MOSFET increases with only the square root of the quiescent drain current. Observe further that g_{mf} rises with the square root of gate aspect geometry and that for a large channel length modulation voltage V_λ,

$$g_{mf}\Big|_{V_\lambda \to \infty} \approx \sqrt{2K_n\left(\frac{W}{L}\right)I_{dQ}} \tag{56.91}$$

The equations used to arrive at (56.90) produce a *bulk transconductance* g_{mb} of

$$g_{mb} \stackrel{\Delta}{=} \left.\frac{\partial I'_d}{\partial V_{bs}}\right|_Q = \lambda_b g_{mf} \tag{56.92}$$

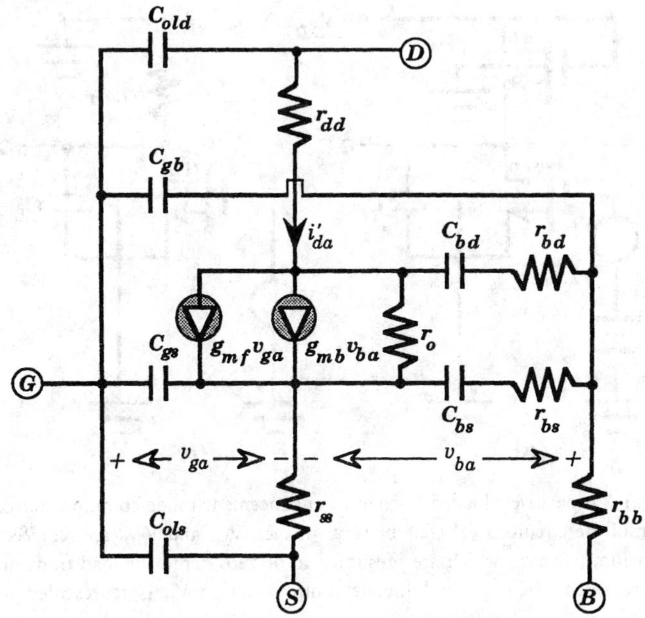

FIGURE 56.26 Small-signal high-frequency equivalent circuit for N-channel and P-channel enhancement-mode MOSFETs. All indicated capacitances are evaluated at the quiescent operating point of the subject MOSFET.

where

$$\lambda_b = \sqrt{\frac{V_\theta/2}{2(V_F - V_T) - V_{bsQ}}} \tag{56.93}$$

Note that the *threshold modulation factor* λ_b and hence the small-signal bulk transconductance, is small if V_θ, which is a measure of the influence exerted on the threshold voltage by bulk–source biasing, is small. Finally, the small-signal *channel resistance* r_o derives from

$$\frac{1}{r_o} \triangleq \left.\frac{\partial I'_d}{\partial V_{ds}}\right|_Q = \frac{I_{dQ}}{V_\lambda + V_{dsQ} - V_{dssQ}} \tag{56.94}$$

Equation (56.94) shows that the drain–source channel resistance is infinitely large for an infinitely large channel length modulation voltage V_λ.

Equation (56.89) and Fig. 56.25 give rise to the small-signal equivalent circuit of Fig. 56.26. In this model the intrinsic drain–bulk and source–bulk diodes are replaced by open circuits on the assumption that the bulk–drain and bulk–source junctions are back biased. The resultant equivalent circuit is applicable to N-channel, as well as to P-channel enhancement-mode MOSFETs.

Example 3. The MOSFET inverter of Fig. 56.27(a) is biased at its input port by a voltage V_{GG} and design care is exercised to ensure that the driver transistor *MD* operates in its saturated regime. Specifically, if V_{oQ} symbolizes the quiescent component of the net output voltage V_o, $V_{oQ} > (V_{GG} - V_{hnd})$, where V_{hnd} is the threshold voltage of the driver. The driver transistor has a gate aspect ratio of (W_D/L), while the gate aspect ratio of the load device *ML* is W_L/L; otherwise, the two transistors in the circuit are identical. Derive expressions for the small-signal

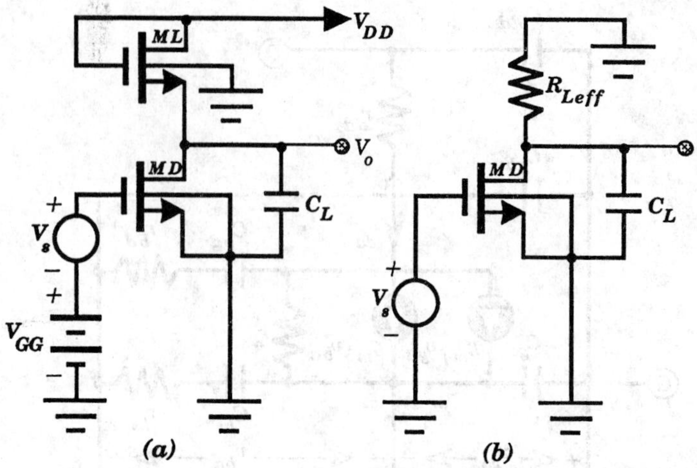

FIGURE 56.27 (a) A capacitively loaded N-channel enhancement-mode common source inverter. (b) AC schematic diagram of the circuit in (a). The battery voltages, V_{GG} and V_{DD}, are replaced by short circuits (or by their internal impedances, which are presumed to be zero here), the load transistor *ML* is replaced by an effective load resistance R_{Leff}, and the net output voltage V_o, is represented by its small-signal component V_{os}.

low-frequency voltage gain $A_v = V_{os}/V_s$ of the circuit, and the small-signal driving point output resistance R_{out}, faced by the load capacitance C_L. Additionally, give an expression for the 3-dB bandwidth, $B_{3\,dB}$, of the circuit, assuming that capacitance C_L is the dominant energy storage element in the neighborhood of the circuit 3-dB bandwidth.

Solution.

1) The first step toward solving this problem is to recognize that at low signal frequencies, the load transistor *ML* functions as a two terminal linear resistance, say R_{Leff}, as is suggested by the AC schematic diagram in Fig. 56.27(b). This resistance is the ratio V_x/I_x in the load transistor AC schematic diagram drawn in Fig. 56.28(a), for which the low-frequency small-signal equivalent circuit is the structure depicted in Fig. 56.28(b). An inspection of the latter diagram reveals

$$v_{ga} = v_{ba} = -V_x + r_{ssl}I_x$$

and

$$V_x = (r_{ssl} + r_{ddl})I_x + r_{ol}[I_x + g_{mfl}v_{ga} + g_{mbl}v_{ba}]$$

Combining these two relationships and recalling (56.92), the effective resistance R_{Leff} presented to the drain of the driver transistor *MD* by the load transistor *ML* is

$$R_{L\,\text{eff}} \triangleq \frac{V_x}{I_x} = r_{ssl} + \frac{r_{ddl} + r_{ol}}{1 + (1 + \lambda_{bl})g_{mfl}r_{ol}}$$

where λ_{bl} is the threshold modulation factor of the load transistor. Since the source and drain ohmic resistances, r_{ssl} and r_{ddl}, respectively, are very small and since the drain–source

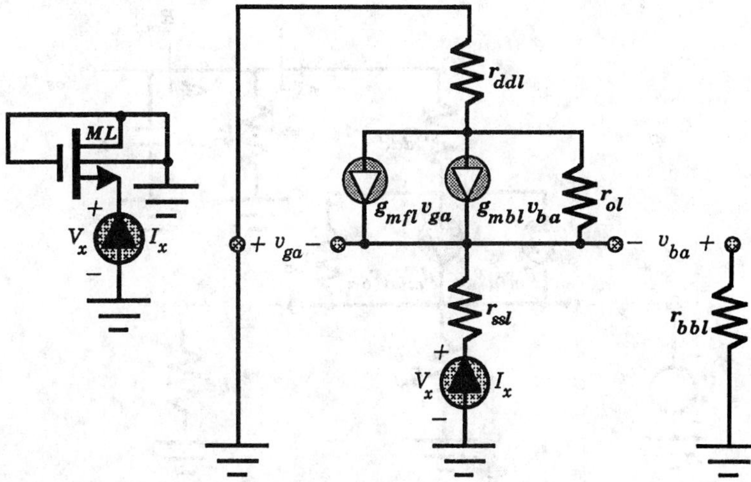

FIGURE 56.28 (a) AC schematic diagram of the load transistor in the amplifier of Fig. 56.27(a). (b) Small-signal low-frequency equivalent model of the circuit in (a). The subscript "*l*," appended to the symbology for the conventional small-signal parameters of a MOSFET signifies that the subject parameter corresponds to the load transistor.

channel resistance, r_{ol}, is typically at least in the mid-tens of thousands of ohms, R_{Leff} can be approximated as

$$R_{Leff} \approx \frac{1}{(1+\lambda_{bl})g_{mfl}}$$

2) With the load transistor modeled as a two terminal load resistance, the AC schematic diagram of the inverting amplifier in Fig. 56.27(a) becomes the structure in Fig. 56.27(b), for which the small-signal model in the diagram is drawn in Fig. 56.29(a). Noting in the latter diagram that

$$v_{ga} = V_s + \left(\frac{r_{ssl}}{R_{Leff}}\right)V_{os}$$

and

$$v_{ba} = \left(\frac{r_{ssl}}{R_{Leff}}\right)V_{os}$$

a straightforward low-frequency circuit analysis produces a small-signal voltage gain A_v of

$$A_v = \frac{V_{os}}{V_s} = -\frac{g_{mfd}R_{Leff}}{1+(1+\lambda_{bd})g_{mfd}r_{ssd} + \dfrac{R_{Leff}+r_{ddd}+r_{ssd}}{r_{od}}}$$

Observe a 180° phase inversion, as expected, between the source excitation and resultant output voltage response. Note further that for a large drain–source channel resistance, r_{od}, and small ohmic resistances, r_{ddd} and r_{ssd}, in the drain and source leads, respectively, the voltage gain relationship collapses to the simple expression

$$A_v \approx -g_{mfd}R_{Leff}$$

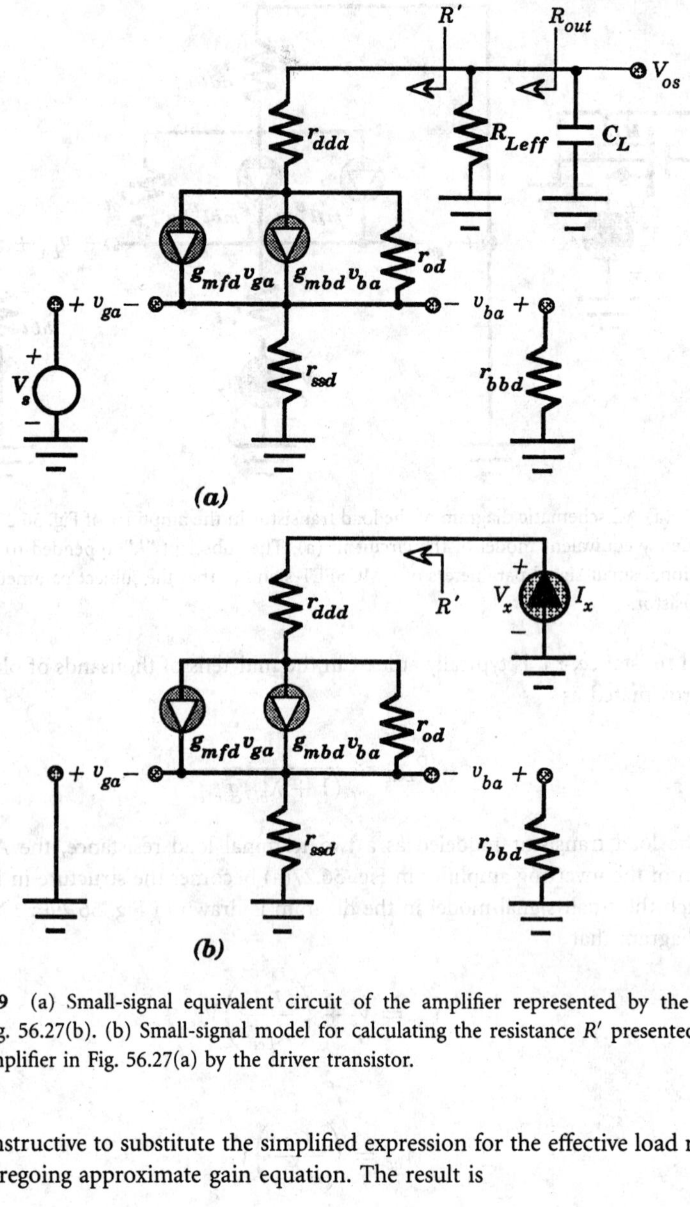

FIGURE 56.29 (a) Small-signal equivalent circuit of the amplifier represented by the AC schematic diagram in Fig. 56.27(b). (b) Small-signal model for calculating the resistance R' presented to the output port of the amplifier in Fig. 56.27(a) by the driver transistor.

3) It is instructive to substitute the simplified expression for the effective load resistance into the foregoing approximate gain equation. The result is

$$A_v \approx -\frac{g_{mfd}}{(1+\lambda_{bl})g_{mfl}}$$

The approximate gain relationship exploits the assumption of infinitely large drain–source channel resistance, and hence infinitely large channel length modulation voltage. Thus, (56.91) can be used for the transconductances that appear in the preceding gain equation. But since the load and driver transistors conduct the same quiescent drain current and since the two transistors are identical except for differing gate aspect ratios,

$$A_v \approx -\left(\frac{1}{1+\lambda_{bl}}\right)\sqrt{\frac{W_d/L}{W_l/L}}$$

The last result agrees well with the results of the voltage gain analysis performed in conjunction with Example 1. In that example, an analogous inverter was addressed analytically by applying a simple MOSFET model that inherently ignores both channel length and threshold modulation phenomena. In fact, the present result agrees exactly with the relevant results of Example 1 if in the present case, threshold modulation is ignored to constrain λ_{bl} to zero.

4) Since the load capacitance, C_L, in the inverting amplifier of Fig. 56.27(a) is assumed to be the dominant energy storage element at signal frequencies lying in the immediate neighborhood of the circuit 3-dB bandwidth, C_L is the only capacitive element that appears in the corresponding small-signal model of Fig. 56.29(a). With only one capacitance embedded in a low-pass network, the 3-dB bandwidth (in units of radians per second) is simply the inverse of the time constant established by the capacitance. Thus,

$$B_{3\,\text{dB}} = \frac{1}{R_{\text{out}}C_L} = \frac{1}{(R' \| R_{L\text{eff}})C_L}$$

where R_{out} is the driving point resistance facing C_L, and R' in Fig. 56.29(a) is the resistance seen to the left of the effective load resistance $R_{L\text{eff}}$. The resistance R' is the V_x/I_x ratio in the model of Fig. 56.29(b), which can be shown to be

$$R' = r_{ddd} + r_{ssd} + [1 + (1 + \lambda_{bl})g_{mfd}r_{ssd}]r_{od}$$

Clearly, the resistance R' is larger than the relatively large drain–source channel resistance, r_{od}. It follows that R' is significantly larger than $R_{L\text{eff}}$, whence the approximate 3-dB bandwidth is

$$B_{3\,\text{dB}} \approx \frac{1}{R_{L\text{eff}}C_L} \approx \frac{(1 + \lambda_{bl})g_{mfl}}{C_L}$$

Complementary MOS (CMOS)

Complementary MOS, or *CMOS*, is an actively loaded enhancement-driven MOSFET network. Its topology, which is depicted in Fig. 56.30, consists of an N-channel enhancement mode MOSFET (MN) whose drain terminal is connected to the drain terminal of a P-channel enhancement mode MOSFET (MP). Because the gate terminals of both the N-channel and the P-channel devices are indicated as electrically tied together, the response of a CMOS cell coalesces the response afforded by an N-channel amplifier driving a P-channel enhancement load with the response delivered by a P-channel amplifier driving an N-channel enhancement load. Many of the advantages of CMOS technology in digital logic applications are made transparent by an analysis of the static transfer characteristic of the CMOS cell.

Static Transfer Characteristic

The P-channel transistor MP in Fig. 56.30 is saturated when its source–drain voltage $V_{sdp} = (V_{DD} - V_o)$ is at least as large as its source–gate voltage $V_{sgp} = (V_{DD} + V_i)$, less one source–gate threshold voltage, say V_{hp}. Thus, MP is saturated for output voltages that satisfy

$$V_o \leq V_i = V_{hp} \tag{56.95}$$

Analogous reasoning leads to

$$V_o \geq V_i - V_{hn} \tag{56.96}$$

as the saturation requirement for the N-channel device, MN.

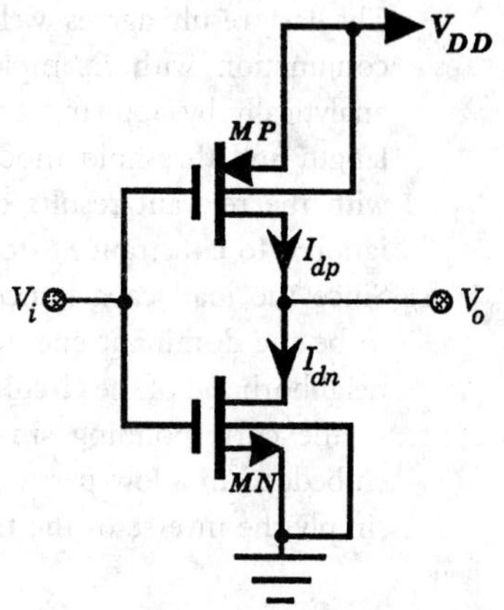

FIGURE 56.30 The circuit schematic diagram of a simple CMOS cell.

Consider first the case of $V_i < V_{hn}$, which corresponds to cutoff of transistor *MN*, or $I_{dn} = 0$ in Fig. 56.30. Since V_i is small, (56.95) is not satisfied, and transistor *MP* operates in its ohmic regime. Using the simplified model for the static volt–ampere characteristics of an enhancement mode P-channel MOSFET, it follows that

$$I_{dp} = K_p\left(\frac{W_p}{L_p}\right)(V_{DD} - V_o)\left[(V_{DD} - V_i) - V_{hp} - \frac{(V_{DD} - V_o)}{2}\right] \tag{56.97a}$$

where K_p is the transconductance coefficient of *MP* and (W_p/L_p) is the gate aspect ratio of *MP*. Since $I_{dp} = I_{dn} = 0$, $V_o = V_{DD}$ for all $V_i < V_{hn}$. As diagrammed in the static transfer characteristic of Fig. 56.31, the logic "1" level is the full power supply voltage V_{DD}. This logic result contrasts with that of the enhancement-loaded enhancement-driven inverter, for which the logic "1" level is one load threshold voltage below the power line voltage.

For output voltages in the neighborhood of V_{DD}, (56.95) is violated, while (56.96) is satisfied; that is, *MP* remains in its ohmic regime, but *MN* is saturated. Accordingly, with K_n and (W_n/L_n) denoting the transconductance coefficient and gate aspect ratio, respectively, of the N-channel MOSFET, respectively, $I_{dp} = I_{dn}$ implies

$$\begin{aligned} I_{dp} &= K_p\left(\frac{W_p}{L_p}\right)(V_{DD} - V_o)\left[(V_{DD} - V_i) - V_{hp} - \frac{(V_{DD} - V_o)}{2}\right] \\ &= \left(\frac{K_n}{2}\right)\left(\frac{W_n}{L_n}\right)(V_i - V_{hn})^2 = I_{dn} \end{aligned} \tag{56.97b}$$

If V_{ic} is the input voltage at which V_o falls from V_{DD} to the level $V_i + V_{hp} = V_{ic} + V_{hp}$, the last result delivers

$$V_{ic} = \frac{V_{DD} - V_{hp} + \sqrt{\eta_{np}}\,V_{hn}}{1 + \sqrt{\eta_{np}}} \tag{56.98}$$

where

$$\eta_{np} = \left(\frac{K_n}{K_p}\right)\left(\frac{W_n/L_n}{W_p/L_p}\right) \tag{56.99}$$

Thus for $V_i < V_{ic}$, *MP* operates in its ohmic regime, *MN* is in saturation, and the static transfer characteristic is derived implicitly from (56.97b).

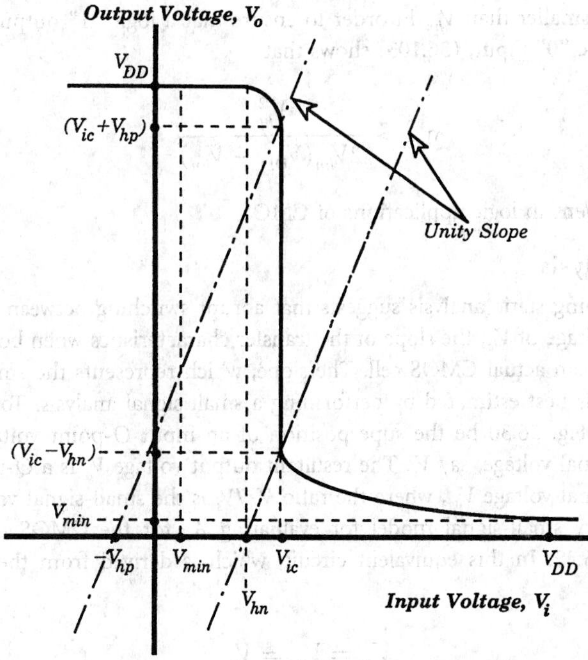

FIGURE 56.31 Approximate static forward transfer characteristic of the CMOS cell in Fig. 56.30.

When the output voltage V_o simultaneously satisfies (56.95) and (56.96) both *MP* and *MN* are saturated. It follows that since $I_{dp} = I_{dn}$,

$$\left(\frac{K_p}{2}\right)\left(\frac{W_p}{L_p}\right)(V_{DD} - V_i - V_{hp})^2 = \left(\frac{K_n}{2}\right)\left(\frac{W_n}{L_n}\right)(V_i - V_{hn})^2 \tag{56.100}$$

which is independent of V_o. Moreover, this relationship is satisfied for $V_i = V_{ic}$. Thus, for $(V_i - V_{hn}) < V_o < (V_i + V_{hp})$, $V_i \equiv V_{ic}$, implying, as shown in Fig. 56.31, an infinitely large incremental slope in the static transfer characteristic at $V_i = V_{ic}$. In practice, the magnitude of this slope is large, but not infinitely large, owing to channel length modulation and body effects.

For $V_i > V_{ic}$, which corresponds to $V_o < (V_i - V_{hn})$, transistor *MN* operates in its ohmic regime, while transistor *MP* is saturated. The resultant static characteristic derives from

$$\left(\frac{K_p}{2}\right)\left(\frac{W_p}{L_p}\right)(V_{DD} - V_i - V_{hp})^2 = \left(\frac{K_n}{2}\right)\left(\frac{W_n}{L_n}\right)V_o\left(V_i - V_{hn} - \frac{V_o}{2}\right) \tag{56.101}$$

If $V_{\min}$ denotes the logic "0" output voltage corresponding to a logic "1" input voltage of $V_i = V_{DD}$, (56.101) produces

$$V_{\min} = (V_{DD} - V_{hn})\sqrt{1 - \frac{1}{\eta_{np}}\left(\frac{V_{hp}}{V_{DD} - V_{hn}}\right)^2} \tag{56.102}$$

which can be approximated as

$$V_{\min} \approx \frac{1}{2\eta_{np}}\left[\frac{V_{hp}^2}{V_{DD} - V_{hn}}\right] \tag{56.103}$$

Since V_{min} must be smaller than V_{hn} in order to ensure that a logic "1" output results from an application of a logic "0" input, (56.103) shows that

$$\eta_{np} > \frac{V_{hp}^2}{2V_{hn}(V_{DD} - V_{hn})} \tag{56.104}$$

is a design requirement in logic applications of CMOS.

Small-Signal Analysis

Although the foregoing static analysis suggests that abrupt switching between logic states takes place at an input voltage of V_{ic}, the slope of the transfer characteristics when both transistors are saturated is finite in an actual CMOS cell. The slope, which represents the small-signal voltage gain of the circuit, is best estimated by performing a small signal analysis. To this end, let the input voltage V_i in Fig. 56.30 be the superposition of an input Q-point voltage, say V_{iQ}, and an applied small-signal voltage, say V_s. The resultant output voltage V_o is a Q-point component V_{oQ}, and a small-signal voltage V_{os}, where the ratio V_{os}/V_s is the small-signal voltage gain A_v.

The low-frequency small-signal model for evaluating A_v for the CMOS cell in Fig. 56.30 is offered in Fig. 56.32. In this equivalent circuit, which is derived from the model given in Fig. 56.26

$$V_{sn} = V_{sp} = V_s \tag{56.105}$$

The partitioning of the input signal voltage into two identical components V_{sn} and V_{sp} reflects the simultaneous application of the input signal V_s to the gates of both the N-channel and the P-channel transistors. Since the small-signal equivalent circuit is linear, this voltage decomposition allows superposition theory to be applied in the evaluation of the voltage gain. In particular,

$$V_{os} = A_{vn}V_{sn} + A_{vp}V_{sp} = (A_{vn} + A_{vp})V_s \tag{56.106}$$

where A_{vn} is the circuit voltage gain due solely to the application of V_{sn} to the N-channel transistor. Similarly, A_{vp} is the voltage gain attributed exclusively to the application of V_{sp} to the gate of the P-channel device.

The N-channel transistor gain, A_{vn}, can be computed by setting the signal source, V_{sp}, to zero. With $V_{sp} = 0$, the P-channel transistor acts as an effective load on the drain terminal of the N-channel transistor. The effective resistance of this P-channel load, say R_{Lp}, derives from an analysis of the circuit given in Fig. 56.33. In this model

$$V_a = V_b = -I_x r_{ssp} \tag{56.107}$$

whence

$$R_{Lp} = \frac{V_x}{I_x} = r_{ddp} + r_{ssp} + [1 + (1 + \lambda_{bp})g_{mfp}r_{ssp}]r_{op} \tag{56.108}$$

Given R_{Lp} as the effective resistance terminating the drain of the N-channel transistor, the voltage gain, A_{vn}, follows from a consideration of the equivalent circuit provided in Fig. 56.33(b). It can be shown that

$$A_{vn} = -\frac{g_{mfn}R_{Lp}}{1 + (1 + \lambda_{bn})g_{mfn}r_{ssn} + \dfrac{R_{Lp} + r_{ddn} + r_{ssn}}{r_{on}}} \tag{56.109}$$

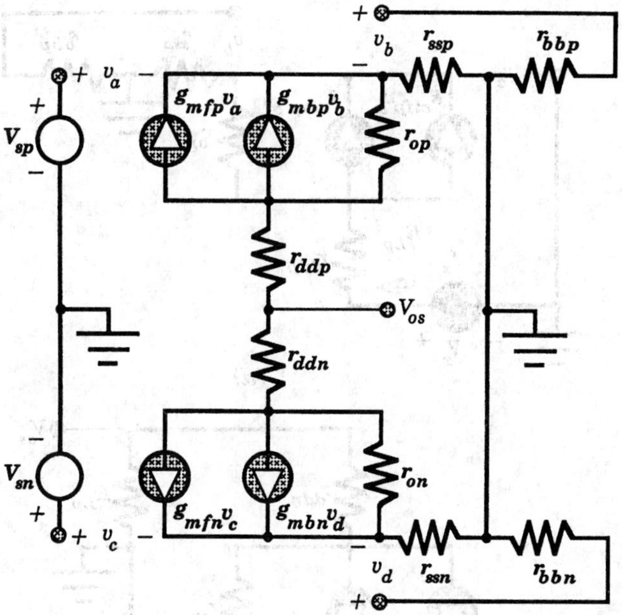

FIGURE 56.32 Small-signal low-frequency equivalent circuit of the CMOS cell in Fig. 56.30.

The gain, A_{vp}, attributed to the P-channel transistor can be found by repeating the foregoing calculations for the case of $V_{sn} = 0$. In this case, the N-channel transistor loads the drain of the P-channel device by a resistance, say R_{Ln}. By symmetry, R_{Ln} is given by (56.108) with the subscript "p," replaced by the subscript, "n," to connote N-channel device parameters. Similarly, A_{vp} is derived from (56.109), with the subscript "n" supplanted by subscript "p." The result is

$$A_{vp} = -\frac{g_{mfp}R_{Ln}}{1 + (1 + \lambda_{bp})g_{mfp}r_{ssp} + \dfrac{R_{Ln} + r_{ddp} + r_{ssp}}{r_{op}}} \tag{56.110}$$

where

$$R_{Ln} = r_{ddn} + r_{ssn} + [1 + (1 + \lambda_{bn})g_{mfn}r_{ssn}]r_{on} \tag{56.111}$$

The substitution of (56.109) and (56.110) into (56.106) leads directly to an expression for the desired small-signal voltage gain. This gain is the slope of the static transfer characteristics curve at a quiescent operating point that constrains the operation of both transistors in the CMOS cell to their saturated regimes.

The CMOS cell voltage gain expression can be simplified by assuming that the internal series drain and source resistances are negligible and the channel length modulation voltages are large in both the P-channel and the N-channel transistors. Accordingly,

$$R_{Ln} \approx r_{on} \approx \frac{V_{\lambda n}}{I_{dQ}}$$

and (56.112)

$$R_{Lp} \approx r_{op} \approx \frac{V_{\lambda p}}{I_{dQ}}$$

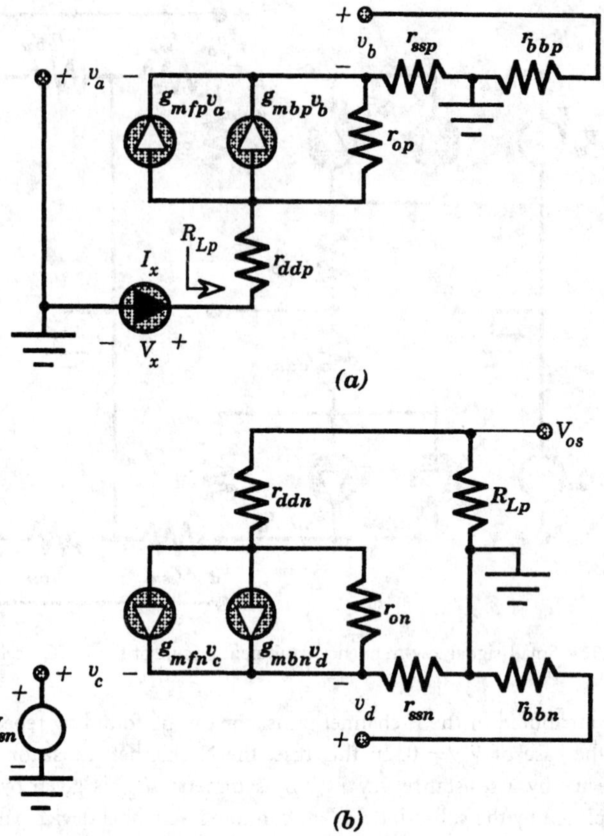

FIGURE 56.33 (a) Small-signal low-frequency equivalent circuit used to evaluate the effective resistance R_{Lp} that loads the drain of the N-channel transistor MN in the CMOS amplifier of Fig. 56.30. (b) Equivalent circuit for evaluating the voltage gain due solely to the N-channel transistor in the CMOS cell.

where I_{dQ} is the quiescent drain current flowing through both transistors and $V_{\lambda n}$ and $V_{\lambda p}$ represent the channel length modulation voltages of the N-channel and P-channel transistors, respectively. Thus

$$A_v = \frac{V_{os}}{V_s} \approx -(g_{mfn} + g_{mfp})(r_{on} \| r_{op}) \tag{56.113}$$

and using (56.112),

$$A_v = -\sqrt{\frac{2K_n(W_n/L_n)}{I_{dQ}}} \left[\frac{1 + \sqrt{\dfrac{K_p(W_p/L_p)}{K_n(W_n/L_n)}}}{\dfrac{1}{V_{\lambda n}} + \dfrac{1}{V_{\lambda p}}} \right] \tag{56.114}$$

Interestingly, (56.114) shows that the voltage gain magnitude is inversely proportional to the square root of the quiescent drain current flowing on both devices of the CMOS cell.

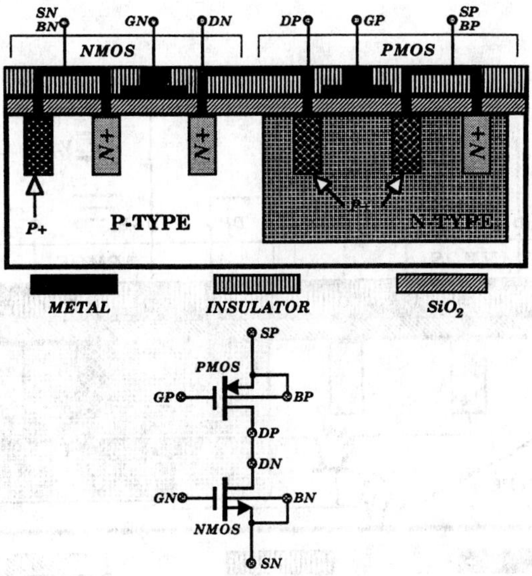

FIGURE 56.34 Simplified cross section of a monolithic CMOS transistor.

Cross Section of Monolithic CMOS

In principle, the CMOS cell shown in Fig. 56.30 can be realized by interconnecting an individual N-channel (NMOS) enhancement-mode MOSFET and an individual P-channel (PMOS) enhancement-mode transistor. But both MOSFETs can be fabricated monolithically on the same chip, thereby explaining why a CMOS cell is often referred to as CMOS "device."

The simplified cross section of a monolithic CMOS transistor is shown in Fig. 56.34. The left-hand side of this structure comprises the N-channel component of the CMOS cell. This half structure consists of gate oxide, surface metallization for the source, drain, gate, and bulk electrical contacts for the NMOS transistor, and a P-type substrate into which a P^+ and two N^+ regions are diffused or implanted. The two N^+ diffusions form the source and drain regions of the NMOS transistor, while the P^+ diffusion establishes a low resistance contact to the substrate bulk. The right-hand side of the structure is the PMOS transistor. It has a lightly doped N-type well that is diffused into the P-type bulk. Into this N-type region, which forms the substrate of the PMOS transistor, two P^+ regions are diffused for the PMOS source and drain. An N^+ diffusion or implant into the lightly doped N-type region ensures a low resistance electrical contact to the PMOS bulk.

The ability to synthesize CMOS as a monolithic interconnection of its NMOS and PMOS components is advantageous from both fabrication and circuit design perspectives. However, potentially serious problems plague the resultant structure [19]. The source of most of these problems is the unavoidable appearance of bipolar transistors and semiconductor resistors beneath the surface of the CMOS cell. To crude first order, these bulk parasitic elements are formed topologically as shown in Fig. 56.35. The emitter of the PNP bipolar junction transistor in this diagram is the P^+ drain/source diffusion of the P-channel MOSFET, its base is the N-type PMOS well, and its collector is the P-type tub that serves as the substrate for the NMOS device. The NPN transistor has an emitter which is the N^+ drain/source diffusion of the NMOS transistor, its base is the P-type tub, and its collector is formed of the N-type well that is used as the PMOS substrate.

The electrical equivalent of the circuit topology abstracted in Fig. 56.35 is the schematic offered in Fig. 56.36. Under most operational circumstances, the surface current, I_d, flowing through the CMOS device is significantly larger than the bulk current, I_{bulk}, flowing through the

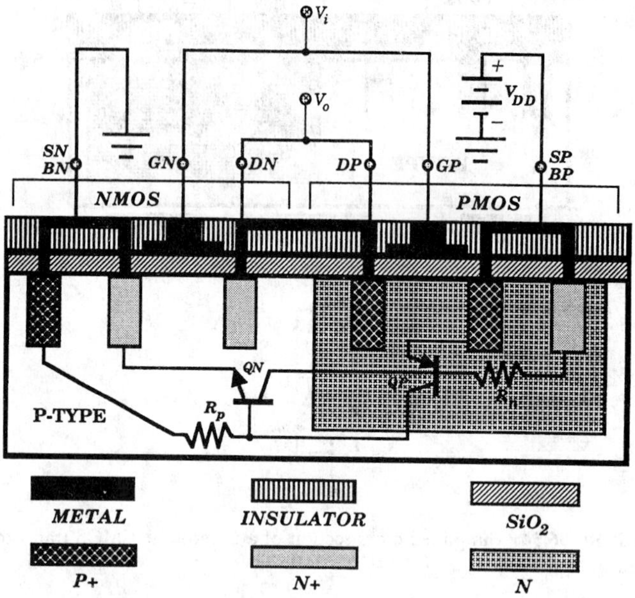

FIGURE 56.35 Simplified topological representation of parasitic bulk transistors and resistors in the cross section of monolithic CMOS. Also shown is the biasing and the I/O terminals for normal inverter operation of the CMOS cell.

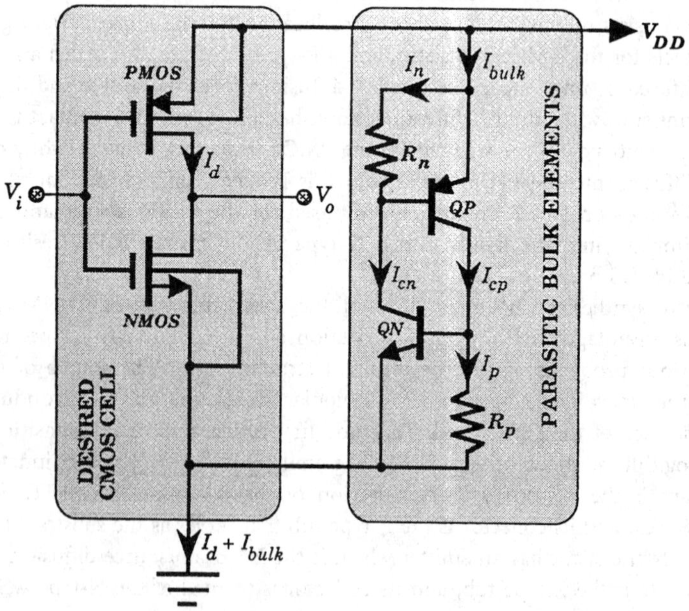

FIGURE 56.36 Equivalent circuit of electrical cross section depicted in Fig. 56.35.

bulk parasitic elements. One way of ensuring $I_d \gg I_{\text{bulk}}$ is to preclude the bipolar devices from turning on, which requires that resistors R_n and R_p be so small as to disallow the development of an appreciable forward bias across the emitter–base junctions of either BJT. Small R_n and R_p require large doping concentrations in the P-type substrate and in the N-type well, respectively. Unfortunately, these large substrate concentrations have the undesirable effect of increasing the threshold voltages of, and exacerbating the body effect in, both MOS transistors.

Short of achieving an ideal goal of $R_n = R_p = 0$, there is no practical way of precluding turn on of both the parasitic bipolar devices, because the bipolar subcircuit on the right-hand side of Fig. 56.36 is a positive feedback network. The instability mechanism can be illustrated qualitatively as follows. Assume that both of the parasitic bipolar units conduct slightly. Let the collector current of transistor *QN* be I_{cn} and that of *QP* be I_{cp}. The collector current, I_{cn}, produces a base current, I_{cn}/β_n, in the base of *QN*, where β_n is the static forward current transfer ratio, or static beta of *QN*. Similarly, the base current flowing in transistor *QP* is I_{cp}/β_p, with β_p symbolizing the static beta of *QP*. The resultant current, I_n, flowing in the resistance, R_n, is

$$I_n = I_{cn} - \frac{I_{cp}}{\beta_p} \tag{56.115}$$

Likewise, R_p conducts a current I_p of

$$I_p = I_{cp} - \frac{I_{cn}}{\beta_n} \tag{56.116}$$

Let the original collector current I_{cn} increase be a small amount, say ΔI_{cn}. This perturbation is arguably the result of a change in circuit operating temperature, impacting electromagnetic radiation, or a spurious signal that superimposes with the power line voltage V_{DD}. Then I_n can be expected to change by an amount ΔI_n such that

$$\Delta I_n = \Delta I_{cn} - \frac{\Delta I_{cp}}{\beta_p} > 0 \tag{56.117}$$

The current change ΔI_{cp} in the *QP* collector current is positive, since positive ΔI_n increases the forward bias across the emitter–base junction of transistor *QP*. But an increase, ΔI_{cp}, in the *QP* collector current, I_{cp}, results in a change, say ΔI_p, in the current flowing through the resistance R_p. If, in addition to satisfying (56.117),

$$\Delta I_p = \Delta I_{cp} - \frac{\Delta I_{cn}}{\beta_n} > 0 \tag{56.118}$$

a further increase in the collector current of *QN* materializes, owing to the fact that an enhanced current through R_p increases the forward bias across the emitter–base junction of the NPN transistor, *QN*. This additional increase, which superimposes with the originally incurred increase in the collector current of *QN*, propagates through the bulk bipolar subcircuit in a fashion that is similar to the foregoing scenario. It therefore causes further increases in the *QN* (and in the *QP*) collector current.

The process described above continues until both bipolar transistors saturate, which is the so-called **latchup** condition. Since saturated bipolar transistors develop only small voltages across their collector–emitter terminals, the latchup condition in monolithic CMOS is tantamount to a very low resistance path between the power supply line and circuit ground. The resultantly large power supply current, most of which is the saturation current supplied to the bulk parasitic elements, causes significant self heating of the bulk. Aside from the fact that this self heating supports still further increases in bulk bipolar collector currents, it can damage the semiconductor bulk, thereby leading to catastrophic failure of the entire CMOS structure.

It is important to note that two conditions must be satisfied if latchup is to occur. These two conditions, which are defined by (56.117) and (56.118), can be combined into the single latchup condition,

$$\beta_n \beta_p > 1 \tag{56.119}$$

Recalling that the beta of bipolar devices is inversely proportional to base width, the likelihood of latchup can be minimized by designing the CMOS cell so that its N-type substrate well is sufficiently wide and/or its P-type substrate is also wide. Unfortunately, these width guidelines do not complement ongoing demands to downsize integrated devices. This observation explains why latchup is an especially perplexing problem in many state of the art VLSI architectures that feature minimal geometry devices. Another way to reduce bipolar beta is to increase base region doping. This increased doping decreases the resistances R_n and R_p, but only at the expense of increasing NMOS and PMOS threshold voltages and increasing the MOSFET threshold voltage dependence on bulk–source biasing.

The foregoing difficulties with respect to minimizing the risk of CMOS latchup compel creative engineering solutions to the latchup problem. One of these solutions involves selective light doping of the P^+ and/or N^+ drain/source regions with gold or other inert impurities. Such doping has the effect of reducing the minority carrier diffusion lengths to which bipolar transistor betas are dependent. Another solution places a lightly doped trench of semiconductor material around the PMOS transistor. This *guard ring* inserts a large resistance in series with the collectors of both transistors *QN* and *QP* in the bulk parasitic subcircuit. The resultantly high net collector resistances inhibit collector current increases, thereby reducing the chance of latchup. In some applications, the guard ring is an oxide trench that serves to open circuit the collectors of *QN* and *QP*.

Depletion-Mode MOSFET

As is the case with enhancement-mode transistors, there are two types of *depletion mode* MOSFETs: the **N-channel device** and the **P-channel transistor.** Figure 56.37(a) displays the cross section of the N-channel depletion mode MOSFET, and Fig. 56.37(b) is the corresponding circuit schematic symbol. The P-channel counterparts to the N-channel depletion-mode transistor are offered in Fig. 56.38. A comparison of these two figures to the device cross sections offered in Figs. 56.15(a) and 56.16(a) reveals that the cross sections of depletion-mode MOSFETs differ from those of enhancement-mode transistors only in that depletion-mode transistors have a semiconductor impurity layer diffused or implanted in the source to drain spacing immediately beneath the gate oxide. This thin channel region has an impurity concentration that is greater than the substrate dopant concentration and is typically of the order of 10^{16} atoms/cm^3.

Static Volt–Ampere Curves

The cross section of the N-channel depletion mode MOSFET differs from that of the N-channel enhancement mode MOSFET by a thin N-type layer, which is diffused or implanted in the source–drain channel of the depletion mode device. An understanding of the electrical effect of this channel is facilitated by recalling that in N-channel enhancement mode transistors, a concentration of free electrons must be induced in the source–drain channel in order for a drain current to flow. The vehicle for inducing such a channel of free electrons is the gate–source voltage, V_{gs}. In particularly, V_{gs} must exceed the gate–source threshold voltage, V_{hnc}, to allow an enhancement MOSFET to operate in either the ohmic or saturated regimes of its

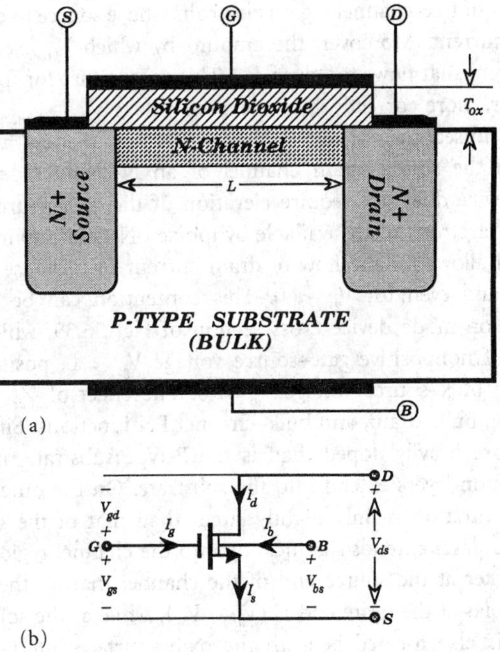

FIGURE 56.37 (a) Simplified cross section of an N-channel depletion-mode MOSFET. The diagram is not drawn to scale. (b) Electrical schematic symbol of an N-channel depletion-mode MOSFET.

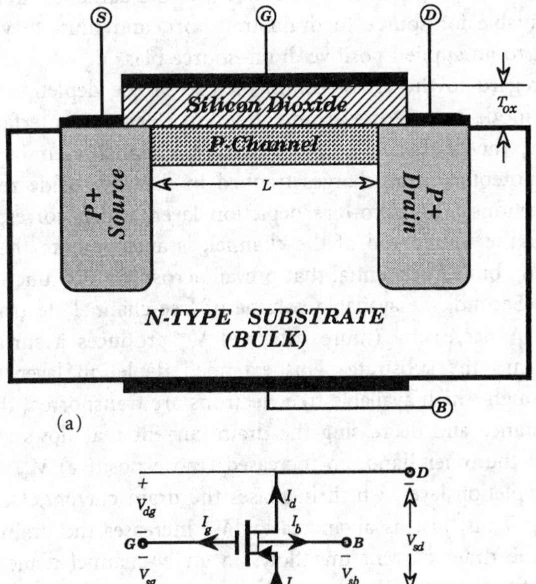

FIGURE 56.38 (a) Simplified cross section of a P-channel depletion-mode MOSFET. The diagram is not drawn to scale. (b) Electrical schematic symbol of a P-channel depletion-mode MOSFET.

static characteristic curves. Thus, V_{hnc} is the minimum gate–source voltage commensurate with establishing the supply of free-channel region electrons whose source to drain transport comprises the transistor drain current. Moreover, the amount by which V_{gs} exceeds V_{hnc} determines the amount of drain current that flows for given V_{ds}. This means that for fixed V_{ds}, the drain current is controlled by V_{gs} or, more correctly, by the excess voltage $V_{gs} - V_{hnc}$.

In contrast, the diffused or implanted N-type channel renders a copious supply of free electrons available in the source–drain channel of an N-channel depletion mode MOSFET. Consequently, this device does not require elevation of the gate–source voltage to a threshold level. In fact, the free electrons made available by ionized N-type impurity atoms in the diffused or implanted channel allows for the flow of drain current in response to positive drain-source voltage for $V_{gs} = 0$ and even for $V_{gs} < 0$. This contention can be understood by studying the N-channel depletion-mode device cross section of Fig. 56.39, which is drawn to illustrate the electrical effects of nonpositive gate–source voltage $V_{gs} < 0$, positive drain–source voltage $V_{ds} > 0$, and negative bulk–source voltage $V_{bs} < 0$. The effect of $V_{bs} < 0$ is a depletion layer about the bulk–source, bulk–drain, and bulk–channel PN junctions. Since the source and drain regions are much more heavily doped than is the P-type substrate, most of the bulk–source and bulk–drain depletion layers extend into the substrate. On the other hand, the fact that the channel dopant concentration is only slightly larger than that of the substrate means that the bulk-channel depletion layer extends significantly into the channel region. The net bulk-channel depletion layer is thicker at the source end of the channel than at the drain end, because the reverse bulk-channel bias at the drain end is $(V_{ds} - V_{bs})$, while at the source end, it is only $-V_{bs}$.

A depletion layer is also formed beneath the oxide–surface interface, since negative gate–source voltage serves to repel free-channel electrons away from the interface and deeper into the channel. This interfacial depletion region is thicker at the drain site, where the net interface-channel reverse bias is the gate–drain voltage, $V_{gd} = (V_{gs} - V_{ds})$, than at the source end, where the interface-channel reverse bias is only the gate–source voltage, V_{gs}. As a result, free-channel electrons are confined to the trapezoidal region shown as the darkened area in Fig. 56.39. These free electrons are available for source to drain transport, and thus, they give rise to a drain current, I_d, in respect to an applied positive drain–source bias, V_{ds}.

Several points in regard to the cross-section picture of the depletion layers formed in an N-channel depletion-mode transistor warrant attention. First, an interface-channel depletion layer is established at the source end of the N-type channel even if $V_{gs} \equiv 0$ because of metal–oxide contact potential, stray charges trapped in the gate oxide region, and interfacial oxide-surface imperfections. This zero-bias depletion layer, which corresponds to a zero-bias interfacial potential at the source end of the channel, is analogous to the zero-bias transition layer and corresponding built-in potential that prevail across the PN junction of a conventional semiconductor diode. Second, the available volume of free-channel electrons is controllable by the gate–source bias. A decreasing (more negative) V_{gs} produces a surface depletion region that extends deeper into the substrate. This extended depletion layer reduces the channel cross-section area through which available free electrons are transported, thereby increasing the effective channel resistance and decreasing the drain current that flows as a result of applied drain–source bias. On the other hand, an increased (more positive) V_{gs} narrows the depth of the surface-channel depletion layer, which increases the drain current, I_d, for a given value of drain–source bias, V_{ds}. Thus, just as an increasing V_{gs} increases the drain current flowing for fixed V_{ds} in NMOS, the drain current that flows in an N-channel depletion mode MOSFET (DNMOS) biased at a fixed positive drain–source voltage is a monotonically increasing function of gate–source bias.

A third important point is that the bulk–source bias, V_{bs}, like the gate–source bias, V_{gs}, modulates the drain current. An increasingly negative V_{bs} makes the bulk-channel depletion layer protrude further into the channel, thereby diminishing the cross-section channel area pierced by transported free electrons and thus, the drain current that flows for a given V_{ds}. Accordingly,

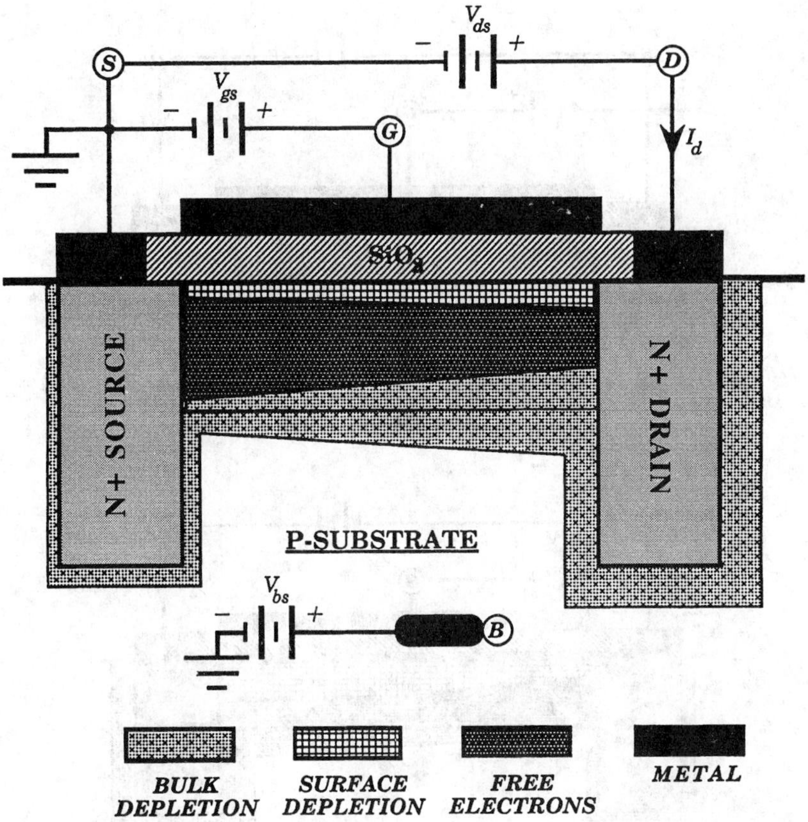

FIGURE 56.39 Cross section of an N-channel depletion mode transistor. The indicated depletion layers correspond to $V_{ds} > 0$, $V_{gs} < 0$, and $V_{bs} < 0$. The diagram is not drawn to scale.

a body effect induced by substrate reverse bias, not unlike the body effect in enhancement-mode transistors, materializes in depletion-mode devices.

A fourth point is that for the depletion condition illustrated in Fig. 56.39, the free electrons available for transport in the source–drain channel are isolated from the oxide–semiconductor interface by the surface depletion region induced by $V_{gs} < 0$. These electrons are not as affected by the vertical electric fields arising from gate–source bias as are those free electrons that prevail in enhancement mode devices. Thus, even if the doping concentration in the N-type channel of a DNMOS device is identical to the P-type substrate doping concentration in an NMOS transistor, the average mobility of free electrons in a depletion-mode MOSFET is larger than that of free electrons in an enhancement-mode unit. Since the transit time, which is the average time required for the transport of electrons from the source to the drain, is inversely related to electron mobility, the electrical responses of DNMOS circuits are potentially faster than those of comparable NMOS circuits.

For given V_{ds} and V_{bs}, it is clear that there exists a negative value of gate–drain voltage, say $V_{gd} = (-V_{pnc})$, such that the surface depletion layer coalesces with the channel side of the bulk-channel depletion layer at the drain end of the N-type channel. This condition, which is diagrammed in Fig. 56.40(a), is referred to as **channel pinchoff** and corresponds to a gate–source voltage of

$$V_{gs} = V_{gs} + V_{ds} = -V_{pnc} + V_{dss} \qquad (56.120)$$

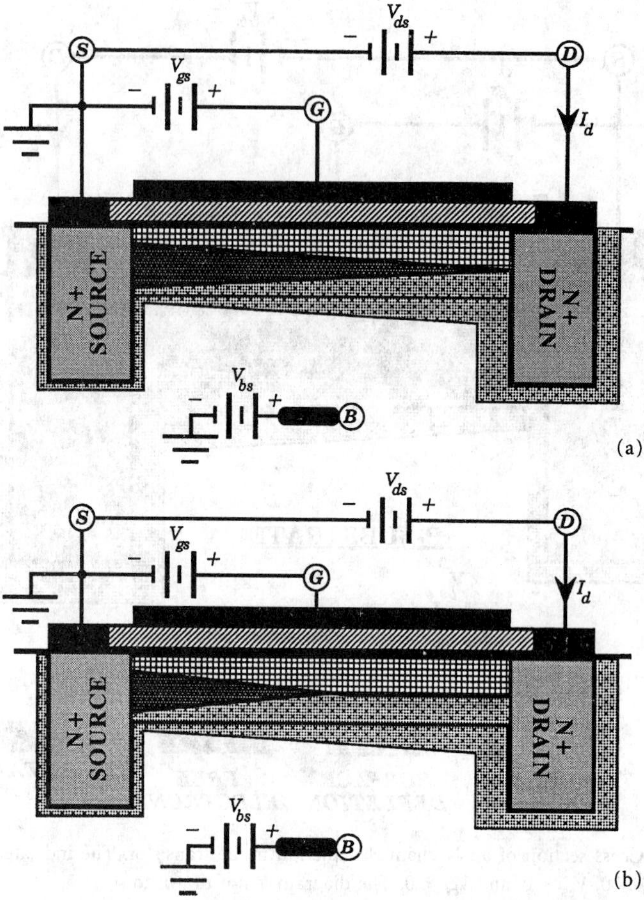

FIGURE 56.40 Cross section of N-channel depletion-mode MOSFET. The diagram is not drawn to scale. (a) Pinchoff at drain end of channel, $V_{gd} = -V_{pnc}$. (b) Operation beyond pinchoff, $V_{gd} \leq -V_{pnc}$.

where V_{pnc}, a positive number, is termed the **channel pinchoff voltage.** The parameter V_{dss} is the drain saturation voltage; it is the particular drain–source voltage commensurate with channel pinchoff at the drain site. For $V_{gd} < (-V_{pnc})$, which corresponds to $V_{ds} > (V_{gs} + V_{pnc}) \equiv V_{dss}$, the point of channel pinchoff moves closer to the source end of the channel, as indicated in Fig. 56.40(b). Channel length modulation is therefore observed in an N-channel depletion mode transistor, just as it is in enhancement-mode technology. A first-order static model of the drain current increase arising from channel length modulation is identical to the model invoked for channel length modulation in NMOS. This is to say that the effect of channel length modulation is given by (56.73), subject to replacing V_{hn} by $(-V_{pmc})$ and interpreting V_{dss} as $(V_{gs} + V_{pnc})$.

Progressive increases in V_{ds} above V_{dss} move the channel pinchoff point closer to the source end of the channel. When the gate–source bias V_{gs} decreases to $(-V_{pnc})$, the pinchoff point is at the source, and the entire channel is said to be pinched off. Under this operating condition, no drain current can flow, since the entire source to drain channel is depleted of free electrons. It follows that a necessary condition for nonzero drain current, I_d, in DNMOS technology is $V_{gs} > (-V_{pnc})$, which suggests viewing $(-V_{pnc})$ as an effective threshold voltage for DNMOS. To first order, the dependence of pinchoff voltage V_{pnc} on the bulk–source bias is the threshold voltage function given by (56.78), provided V_{hno} in (56.78) is replaced by V_{pno}, the pinchoff voltage for zero bulk–source bias. Moreover, the substrate Debye length implicit to the variable V_θ is taken to be the average Debye length in the N-type channel of DNMOS.

The static volt–ampere characteristic of an N-channel depletion-mode MOSFET derives approximately from the static equations for enhancement-mode transistors [20]. In particular,

$$I_d = \begin{cases} 0, & V_{gs} \le -V_{pnc} \\ K_n\left(\frac{W}{L}\right)V_{ds}\left(V_{gs} + V_{pnc} - \frac{V_{ds}}{2}\right), & V_{gs} > -V_{pnc} \text{ and } \\ & V_{ds} < (V_{gs} + V_{pnc}) \\ \frac{K_n}{2}\left(\frac{W}{L}\right)(V_{gs} + V_{pnc})^2\left(1 + \frac{V_{ds} - V_{dss}}{V_\lambda}\right), & V_{gs} > -V_{pnc} \text{ and } \\ & V_{ds} \ge (V_{gs} + V_{pnc}) \end{cases} \quad (56.121)$$

where I_d is taken as a current flowing into the drain terminal of an N-channel transistor. Because of internal drain, source, and substrate resistances, V_{gs}, V_{ds}, and V_{bs} (on which V_{pnc} is dependent), respectively, are internal gate–source, internal drain–source, and internal bulk–source voltages, respectively. For P-channel depletion-mode MOSFETs, (56.121) remains applicable provided I_d is interpreted as a current flowing out of the drain terminal, V_{gs} is replaced by the source–gate voltage V_{sg}, V_{ds} is supplanted by the source–drain voltage V_{sd}, and V_{bs} is replaced by the bulk–source voltage V_{sb}. The commentary offered earlier in regard to second-order effects in enhancement-mode technology apply to depletion-mode technology as well.

Depletion-Loaded Inverter

Depletion-mode transistors, like enhancement-mode MOSFETs, can be used as load devices in inverters designed for either digital or analog signal-processing applications. The simplified schematic diagram of a typical enhancement-driven depletion-loaded inverter is given in Fig. 56.41. As the following analyses confirm, the inverter offers a logic swing capability that extends to the power supply voltage, V_{DD}. In linear signal processors, the inverter in Fig. 56.41 also delivers a small-signal voltage gain that is larger than the gain afforded by an enhancement-driven enhancement-loaded inverter.

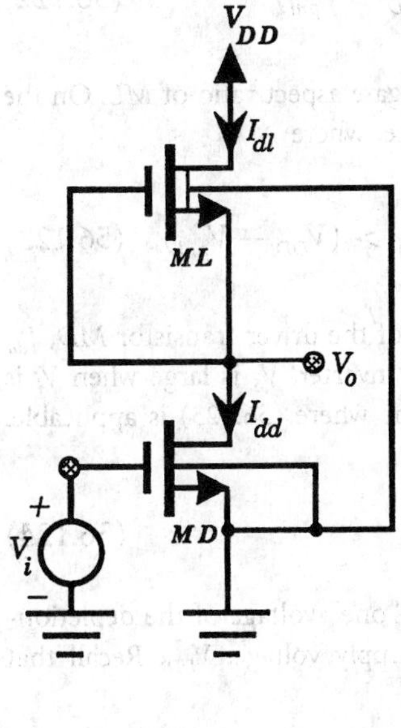

FIGURE 56.41 Schematic diagram of an enhancement-driven depletion-loaded N-channel MOSFET inverter.

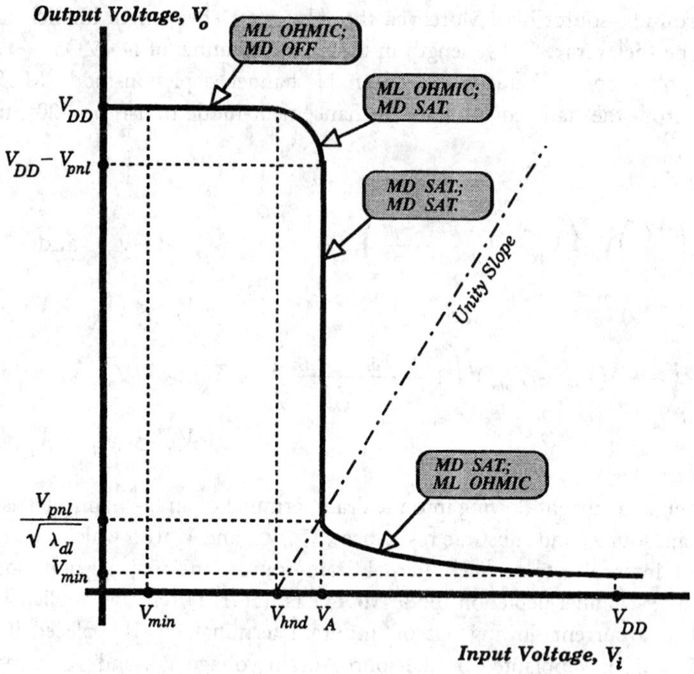

FIGURE 56.42 Static transfer characteristic of an enhancement-driven depletion-loaded common-source MOSFET inverter. The characteristic exploits the simplified MOSFET model equations.

Static Transfer Analysis. The static transfer characteristic of the depletion loaded, enhancement driver inverter is depicted in Fig. 56.42. Its derivation commences with the observation that the gate–source voltage of *ML* is zero. Accordingly, *ML* in the circuit of Fig. 56.41 is saturated if $(V_{DD} - V_o) > V_{pnl}$, where V_o is the indicated output voltage and V_{pnl} is the pinchoff voltage of *ML*. (For simplicity, body effect, as well as channel length modulation, series internal device resistances, and other second-order effects are ignored.) With *ML* saturated, the simplified static model of a MOSFET gives for its drain current I_{dl}

$$I_{dl} = \frac{K_{nl}}{2}\left(\frac{W_l}{L_l}\right)V_{pnl}^2, \quad \text{for} \quad V_o \le (V_{DD} - V_{pnl}) \tag{56.122}$$

where K_{nl} is the transconductance coefficient and (W_l/L_l) is the gate aspect ratio of *ML*. On the other hand, $V_o > (V_{DD} - V_{pnl})$ constrains *ML* to its ohmic regime, where

$$I_{dl} = \frac{K_{nl}}{2}\left(\frac{W_l}{L_l}\right)(V_{DD} - V_o)\left[V_{pnl} - \frac{V_{DD} - V_o}{2}\right] \quad \text{for } V_o > (V_{DD} - V_{pnl}) \tag{56.123}$$

If the input voltage V_i is smaller than the threshold voltage V_{hnd} of the driver transistor *MD*, I_{dd} in Fig. 56.41 is zero. Moreover, since the circuit operates as an inverter, V_o is large when V_i is small. Thus, $V_i < V_{hnd}$ forces *ML* to conduct in its ohmic regime where (56.123) is applicable. Since I_{dl} is identical to I_{dd},

$$V_o = V_{DD}, \quad \text{for } V_i < V_{hnd} \tag{56.124}$$

Within the context of simplified MOSFET modeling, the logical "one" voltage of the depletion-loaded enhancement-driven inverter is seen to be the power supply voltage, V_{DD}. Recall that in the enhancement-loaded enhancement-driven inverter considered earlier, the logical "one" voltage is one load threshold voltage below the supply voltage.

As V_i rises above the threshold voltage, V_{hnd}, the driver begins to conduct a drain current, I_{dd}. If V_i is such that V_o, which is the drain–source voltage developed across *MD*, is larger than $(V_i - V_{hn})$, *MD* is saturated, whence

$$I_{dd} = \frac{K_{nd}}{2}\left(\frac{W_d}{L_d}\right)(V_i - V_{hnd})^2 \quad \text{for } V_o \geq V_i - V_{hnd} \tag{56.125}$$

with K_{nd} denoting the transconductance coefficient of the driver, and W_d/L_d its gate aspect ratio. As long as V_o, in addition to exceeding $V_i - V_{hnd}$, remains above $V_{DD} - V_{pnl}$, *ML* continues operating in its ohmic domain. Since $I_{dd} = I_{dl}$, (56.123) and (56.125) provide

$$(V_{DD} - V_o)\left[V_{pnl} - \frac{V_{DD} - V_o}{2}\right] = \frac{\lambda_{dl}}{2}(V_i - V_{hnd})^2$$
$$\text{for } V_o \geq (V_i - V_{hnd}) \text{ and } V_o > (V_{DD} - V_{pnl}) \tag{56.126}$$

where

$$\lambda_{dl} \triangleq \frac{K_{nd}}{K_{nl}}\left(\frac{W_d/L_d}{W_l/L_l}\right) \tag{56.127}$$

Equation (56.126) implicitly defines the static forward transfer characteristic for the indicated range of output voltages. The subject relationship shows that V_o reduces to $V_{DD} - V_{pnl}$ when V_i rises to a value, say V_A, given by

$$V_A = V_{hnd} + \frac{V_{pnl}}{\sqrt{\lambda_{dl}}} \tag{56.128}$$

When V_o falls below $V_{DD} - V_{pnl}$, *ML* becomes saturated. But as long as V_o remains above $V_i - V_{hnd}$, *MD* remains saturated. The pertinent transfer characteristic derives from equating (56.122) and (56.125), so that

$$V_{pnl}^2 = \lambda_{dl}(V_i - V_{hnd})^2$$
$$\text{for } V_o > (V_i - V_{hnd}) \text{ and } V_o \leq (V_{DD} - V_{pnl}) \tag{56.129}$$

This relationship is independent of the output voltage V_o and additionally, it is satisfied for only one value of the input voltage; namely, $V_i = V_A$, as defined by (56.128). It follows that the circuit at hand, biased at $V_i = V_A$, delivers a theoretically infinitely large magnitude of small-signal gain, subject to the assumption that channel length and threshold voltage modulation phenomena can be ignored.

When V_o falls below $V_i - V_{hnd}$, the load transistor remains saturated, while the driver enters its saturation regime. Under this condition, the pertinent device volt–ampere equations deliver a static transfer characteristic that derives from the solution of the quadratic expression,

$$\lambda_{dl}V_o^2 - 2\lambda_{dl}(V_i - V_{hnd})V_o + V_{pnl}^2 = 0 \tag{56.130}$$

The logical "0" output, say V_{min}, corresponding to a logical "1" input of $V_i = V_{DD}$ can be determined by applying the quadratic formula to this relationship. To first order,

$$V_{min} = \frac{V_{pnl}^2}{2\lambda_{dl}(V_{DD} - V_{hnd})} \tag{56.131}$$

Because a logical "0" input must produce a logical "1" output, and recalling that the logical "1" output of V_{DD} is produced whenever $V_i < V_{hnd}$, (56.131) implies the design requirement,

$$V_{DD} > V_{hnd} + \frac{V_{pnl}^2}{2\lambda_{dl} V_{hnd}} \tag{56.132}$$

Small-Signal Analysis. Recall that the static volt–ampere equations of a depletion-mode MOSFET differ from those of its enhancement-mode counterpart only in that the threshold voltage V_{hnc} of the enhancement-mode transistor is replaced by $(-V_{pnc})$, where V_{pnc} symbolizes the pinchoff voltage, corrected for body effect, of the depletion-mode device. It follows that the topology of the low-frequency model for a depletion-mode MOSFET is identical to that of the enhancement-mode transistor. Indeed, the high-frequency model depicted in Fig. 56.26 remains applicable to depletion-mode units, provided V_{hnc} is replaced by $(-V_{pnc})$ in (56.90)–(56.94), which analytically define the small-signal model parameters g_{mf}, g_{mb}, and r_o. The discussion in regard to capacitances and various high-order effects submitted in the section on refinements to the simple MOSFET model remain generally applicable as well.

Example 4. The input voltage, V_i, in the enhancement-driven depletion-loaded MOSFET inverter of Fig. 56.41 is of the form $V_i = (V_{iQ} + V_s)$, where V_{iQ} is a constant voltage that biases both transistors in their saturated domains. The voltage V_s is a small amplitude, low-frequency sinusoidal signal. Derive a general expression for the low-frequency small-signal voltage gain $A_V = V_{os}/V_s$, and compare the results to those obtained earlier in conjunction with the enhancement-driven enhancement-loaded MOSFET inverter. Do not ignore the electrical effects of internal series resistances, bulk-induced threshold and pinchoff voltage modulation, and channel length modulation.

Solution.

1) The schematic diagram in Fig. (56.41) indicates that at low frequencies, transistor *ML* imposes a resistive load on the drain of the driver transistor, *MD*. This contention follows from the observations that V_{DD} is grounded for small-signal considerations and that the only signal exciting transistor *ML* is applied simultaneously to its gate and source terminals by the drain of *MD*. Thus, for small input signals, *ML* behaves as a two terminal linear resistor of resistance, say R_{Leff}. Accordingly, the pertinent small-signal equivalent circuit is identical to the equivalent circuit shown in Fig. 56.29, save for the fact that no load capacitor is currently considered. From the results of Example 3, the voltage gain is of the form

$$A_v = \frac{V_{os}}{V_s} = -\frac{g_{mfd} R_{Leff}}{1 + (1 + \lambda_{bd}) g_{mfd} r_{ssd} + \dfrac{R_{Leff} + r_{ddd} + r_{ssd}}{r_{od}}}$$

The only remaining task is the determination of an expression for the resistance R_{Leff}.

2) The equivalent circuit pertinent to finding R_{Leff} is offered in Fig. 56.43, in which the drain, to which V_{DD} is applied in the original circuit schematic diagram, is grounded, and the gate is connected directly to the source. Let a signal current, I_x, be applied as indicated from ground to the source–gate terminal. The desired resistance, which is the ratio of the indicated voltage, V_x, to the applied current, I_x, is

$$R_{Leff} = \frac{(r_{ssl} + r_{ddl}) + [1 + (1 + \lambda_{bl}) g_{mfl} r_{ssl}] r_{ol}}{1 + \lambda_{bl} g_{mfl} r_{ol}}$$

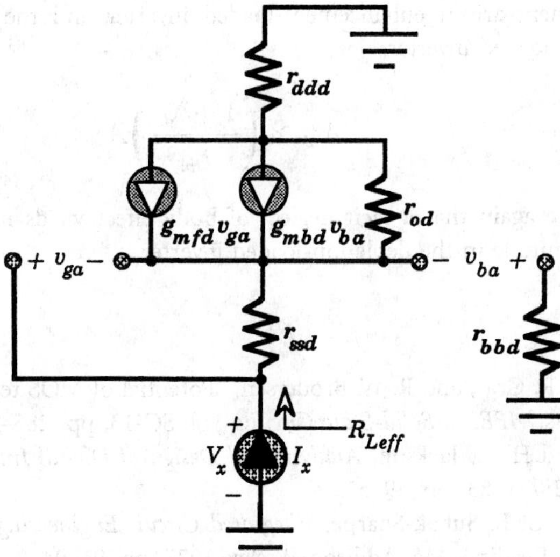

FIGURE 56.43 Equivalent circuit used to determine the effective load resistance R_{Leff} presented to the enhancement-mode driver by the depletion-mode load in the inverter of Fig. 56.41.

3) Several important aspects of the foregoing analysis warrant attention.

   a. If all model parameters in both the gain and the effective resistance expressions are evaluated for MOSFET operation in the saturated regime, the voltage gain obtained is the slope at $V_i = V_A$ of the static transfer characteristic curve plotted in Fig. 56.42. Although the voltage gain is not infinitely large, as is implied by the aforementioned transfer characteristic, its magnitude approaches infinity if $r_{ol} \rightarrow \infty$ and if $\lambda_{bl} \rightarrow 0$. But $r_{ol} \rightarrow \infty$ implies negligible channel length modulation, while $\lambda_{bl} \rightarrow 0$ corresponds to the neglect of threshold body phenomena. These simplifying approximations are implicit to the plot of Fig. 56.42.
   b. If the internal series resistances are very small, and if the channel length modulation resistance is very large in both the load and driver transistors,

$$R_{\text{Leff}} \approx \frac{1}{\lambda_{bl} g_{mfl}}$$

and

$$A_v \approx -g_{mfd} R_{\text{Leff}}$$

The resultant voltage gain collapses to

$$A_v \approx -\frac{g_{mfd}}{\lambda_{bl} g_{mfl}}$$

   c. A comparison of the approximate gain expression found above to the corresponding gain approximation for the enhancement-loaded inverter considered in Example 3 indicates that the gain afforded by the depletion-loaded inverter is slightly larger. In particular, if A_{ve} and A_{vd}, respectively, denote the approximate voltage gains of the enhancement-driven enhancement-loaded inverter and the enhancement-driven depletion-loaded inverter,

$$A_{vd} \approx \left(\frac{1+\lambda_{bl}}{\lambda_{bl}}\right) A_{ve}$$

Note once again that a tacit neglect of body effect yields infinitely large voltage gain magnitude in the depletion-loaded inverter.

References

[1] D. A. Hodges, P. R. Gray, and R. W. Brodersen, "Potential of MOS technologies for analog integrated circuits," *IEEE J. Solid-State Circuits*, vol. SC-13, pp. 285–294, June 1978.

[2] D. A. Hodges and H. G. Jackson, *Analysis and Design of Digital Integrated Circuits*, New York: McGraw-Hill, 1983, pp. 49–52.

[3] A. B. Glaser and G. E. Subak-Sharpe, *Integrated Circuit Engineering: Design, Fabrication, and Applications*, Reading, Ma: Addison-Wesley, 1977, pp. 91–94.

[4] D. E. Ward and R. W. Dutton, "A charge-oriented model for MOS transistor capacitances," *IEEE J. Solid-State Circuits*, vol. SC-13, pp. 703–708, Oct. 1978.

[5] S.-Y. Oh, D. E. Ward, and R. W. Dutton, "Transient analysis of MOS transistors," *IEEE J. Solid-State Circuits*, vol. SC-15, pp. 636–643, Aug. 1980.

[6] B. J. Sheu, D. L. Scharfetter, C. Hu, and D. O. Pederson, "A compact IGFET charge model," *IEEE Trans. Circuits Syst.*, vol. CAS-31, pp. 745–748, Aug. 1984.

[7] J. E. Meyer, "MOS models and circuit simulation," *RCA Review*, vol. 32, pp. 42–63, Mar. 1971.

[8] S. Liu and L. W. Nagel, "Small-signal MOSFET models for analog circuit design," *IEEE J. Solid-State Circuits*, vol. SC-17, pp. 983–998, Aug. 1982.

[9] F. N. Trofimenkoff, "Field-dependent mobility analysis of the field-effect transistor," *Proc. IEEE*, vol. 53, pp. 1765–1766, Nov. 1965.

[10] M. H. White, F. Van de Wiele, and J. P. Lambot, "High-accuracy MOS models for computer-aided design," *IEEE Trans. Electron Devices*, vol. ED-27, pp. 899–906, May 1980.

[11] P. E. Cottrell, R. R. Troutman, and T. H. Ning, "Hot electron emission in N-channel IGFET's," *IEEE Trans. Electron Devices*, vol. ED-26, pp. 520–523, Apr. 1979.

[12] K.-L. Chen *et. al.*, "Reliability effects on MOS transistors due to hot carrier injection," *IEEE Trans. Electron Devices*, vol. ED-32, pp. 386–393, Feb. 1985.

[13] T. H. Ning, C. M. Osborn, and H. N. Yu, "Effect of electron trapping on IGFET characteristics," *J. Electron. Mater.*, vol. 6, p. 65, 1977.

[14] P. K. Chatterjee, "VLSI dynamic NMOS design constraints due to drain induced primary and secondary impact ionization," *IEDM Tech. Digest*, p. 14, 1979.

[15] T. Tsuchiya and S. Nakajima, "Emission mechanism and bias-dependent emission efficiency of photons induced by drain avalanche in Si MOSFETs," *IEEE Trans. Electron Devices*, vol. ED-32, pp. 405–412, Feb. 1985.

[16] J. R. Brews, *et. al.*, "Generalized Guide for MOSFET Miniaturization," *IEEE Electron Dev. Lett.*, vol. EDL-1, p. 2, 1980.

[17] C.-Y. Wu, W.-Z. Hsiao, and H.-H. Chen, "A simple punchthrough voltage model for short-channel MOSFETs with single channel implantation in VLSI," *IEEE Trans. Electron Devices*, vol. ED-32, pp. 1704–1707, Sept. 1985.

[18] C. G. Sodini, P. K. Ko, and J. L. Moll, "The effect of high fields on MOS devices and circuit performance, *IEEE Trans. Electron Devices*, vol. ED-31, pp. 1386–1393, Oct. 1984.

[19] R. R. Troutman, *Latchup In CMOS Technology: The Problem and Its Cure.* Boston: Kluwer Academic, 1986.

[20] M.-W. Chiang, J. Choma, Jr., and C. Kao, "A simulation method to completely model the various transistor *I–V* operational modes of long channel depletion MOSFETs," *IEEE Trans. Computer-Aided Design*, vol. CAD-4, pp. 322–328, July 1985.

56.3 JFET Technology Transistors

Stephen I. Long

Both the MOSFET and the junction FET or JFET are used in analog IC design, usually in conjunction with bipolar transistors. In this section, the mechanisms responsible for the operation of the JFET will be described, models presented, and fabrication technologies discussed. Performance of integrated and discrete silicon JFET's will be compared to each other and to compound semiconductor FET devices.

Introduction

The JFET consists of a conductive channel with source and drain contacts whose conductance is controlled by a gate electrode. The channel can be fabricated in either conductivity type, n or p, and both normally on (depletion-mode-type devices) and normally off (enhancement-mode-type) devices are possible. The circuit symbols typically used for JFET's are shown in Fig. 56.44 along with the bias polarities of active region operation for these four device possibilities. For analog circuit applications, the depletion mode is almost exclusively utilized because it provides a larger range of input voltage and therefore greater dynamic range. In silicon, both p- and n-channel JFET's are used, but when compound semiconductor materials such as GaAs or InGaAs are used to build the FET, n-channel devices are used almost exclusively.

When fabricated with silicon, the JFET is used in analog IC processes for its high input impedance, limited by the depletion capacitance and leakage current of a reverse-based pn junction. When the JFET's are used at the input stage, an op-amp with low input bias current, at least at room temperature, can be built. Fortunately, a p-channel JFET can be fabricated with a standard bipolar process with few additional process steps. This enables inexpensive BiFET

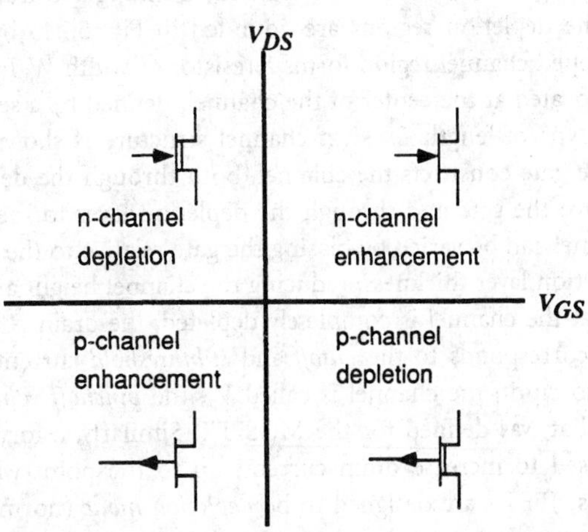

FIGURE 56.44 The circuit symbols typically used for JFET's are shown with the bias polarities for active region operation.

processes to be employed for such applications. Unfortunately, the simple process modifications required for integrating JFET's and BJT's are not consistent with the requirements for high-performance devices. Short gate lengths and high channel doping levels are generally not possible. So the transconductance per channel width and the gain–bandwidth product of JFET's integrated with a traditional analog BJT process are not very good. The short-circuit current gain–bandwidth product (f_T) is about 50 MHz for an integrated p-channel JFET. The MOSFETs in a BiCMOS process are much better devices, however a BiCMOS process does not often include both NPN and PNP BJT's needed for high-performance analog circuits.

Discrete silicon JFET's are available with much better performance because they can be fabricated with a process optimized for the JFET. Typical applications are for low-noise amplifiers up to the VHF/UHF range. Noise figures less than 0.1 dB can be obtained at low frequencies with high source impedances and 2 dB at high frequencies at the noise matching input condition with high-performance discrete silicon JFET's. The low input gate current I_G, which can be in the picoamp range, causes the shot noise (proportional to $\sqrt{I_G}$) component to be very low. The input equivalent noise current of the JFET is mainly due to input referred channel (Johnson) noise. This property gives very low noise performance when presented with a high source impedance. In this case, the JFET is often superior to a BJT for noise. For low source impedances, the BJT is generally better.

Compound semiconductor materials such as GaAs and InGaAs are used to fabricate JFET-like devices called MESFET's (metal-semiconductor FET) and HEMT's (high electron mobility transistor). The reason for using these materials is superior performance at high frequencies. These devices are unequaled for gain–bandwidth, ultralow noise, and power amplification at frequencies above 10 GHz and up to 150 GHz. Integrated analog microwave circuits are fabricated with these devices and are commercially available for use in low noise receiver and power amplifier applications. Some representative results will be summarized in the section on GaAs MESFET and HEMT technologies.

JFET Static I–V Characteristics

The JFET differs in structure and in the details of its operation from the MOSFET discussed in the previous section. Figure 56.45 shows an idealized cross section of a JFET. The channel consists of a doped region, which can be either p on n type, with source and drain contacts at each end. The channel is generally isolated from its surrounding substrate material by a reverse biased pn junction. The depletion regions are bounded in Fig. 56.45 by dashed lines and are unshaded. The thin, doped channel region forms a resistor of width W into the page and height d. A gate electrode is located at the center of the channel, defined by a semiconductor region of opposite conductivity type of length L. An n-channel structure is shown here for purposes of illustration. The p-type gate constricts the channel, both through the depth of the diffusion or implant used to produce the gate and through the depletion layer formed at the p–n junction. The height of the channel can be varied by biasing the gate relative to the source (V_{GS}). A reverse bias increases the depletion layer thickness, reducing the channel height and the drain current. If V_{GS} is large enough that the channel is completely depleted, the drain current will become very small. This condition corresponds to the *cutoff* and *subthreshold* current regions of operation, and the V_{GS} required to cutoff the channel is called V_P, the *pinchoff voltage.* V_P corresponds to the threshold voltage that was defined for the MOSFET. Similarly, a forward bias between gate and channel can be used to increase drain current, up to the point where the gate junction begins to conduct. Most JFET's are designed to be *depletion mode* (normally on); drain current can flow when $V_{GS} = 0$ *and* they are normally operated with a reverse-biased gate junction. It is also possible, however, to fabricate *enhancement-mode* JFET's by use of a thinner or more lightly doped channel.

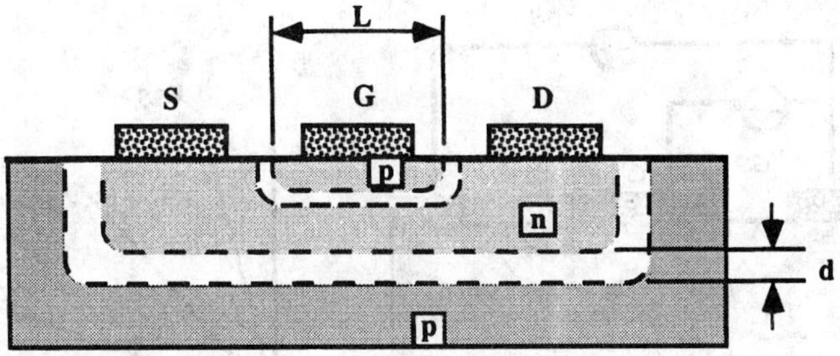

FIGURE 56.45 Idealized cross section of a JFET. The depletion regions are bounded with dashed lines and are unshaded.

The pinchoff voltage is a sensitive function of the doping and thickness of the channel region. It can be found if the channel doping profile $N(x)$ is known through Poisson's equation. For a nonuniform profile,

$$V_P = V_{BI} - \frac{q}{\varepsilon}\int_0^d xN(x)\,dx \tag{56.133}$$

For uniform doping, $N(x) = N_D$ and the familiar result in (56.134) shows that the pinchoff voltage depends on the square of the thickness. This result shows that very precise control of profile depth is needed if good matching and reproductibility of pinchoff voltage is to be obtained [11].

$$V_P = V_{BI} - \frac{qN_D d^2}{2\varepsilon} \tag{56.134}$$

JFET Operating Regions

The static current–voltage characteristics of the JFET can be categorized by the five regions of operation shown in Fig. 56.46 for an n-channel device. The mechanisms that produce these

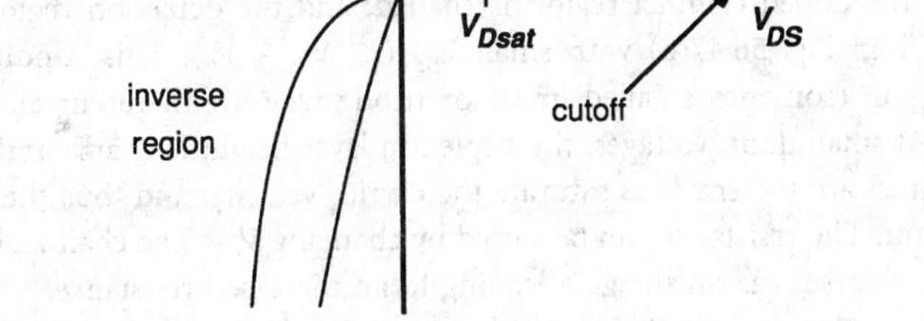

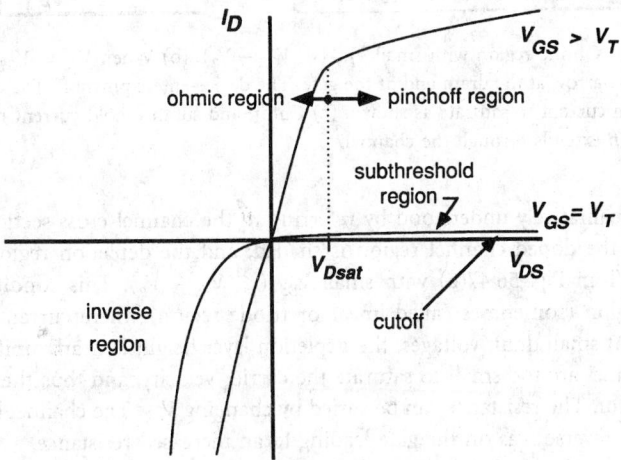

FIGURE 56.46 The static current–voltage characteristics of the JFET can be categorized by five regions of operation. An n-channel device is shown in this illustration with grounded source.

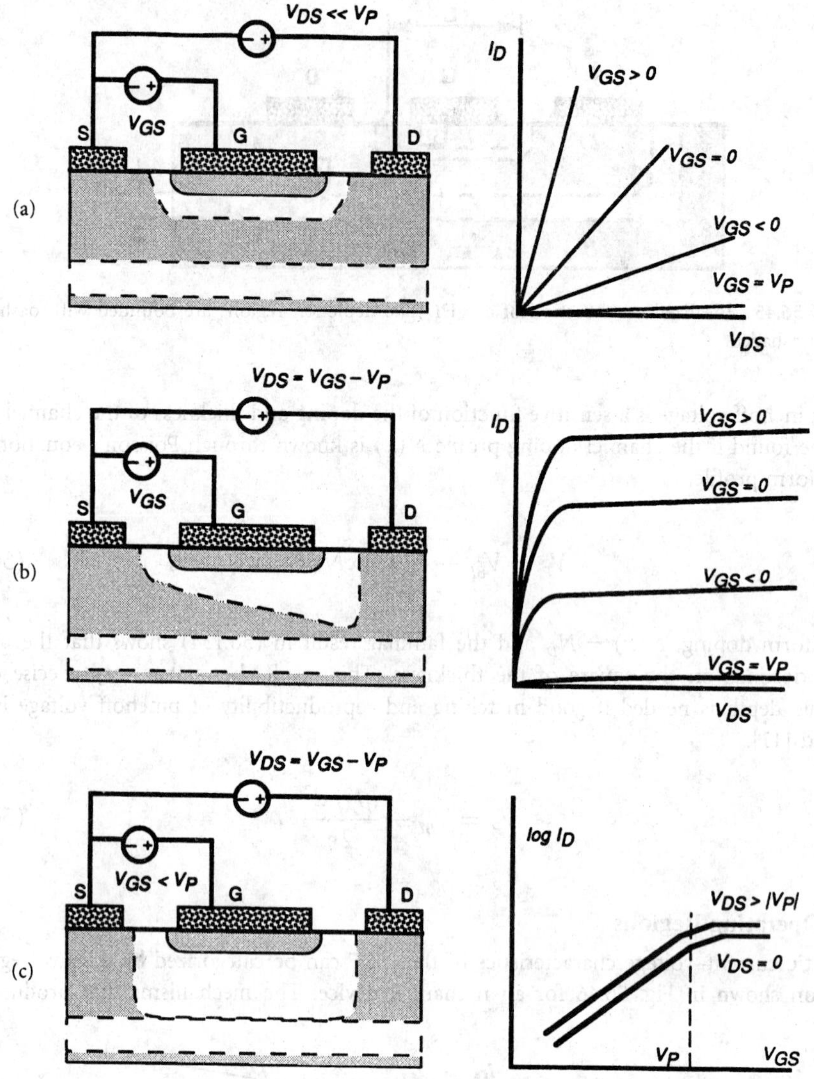

FIGURE 56.47 (a) Ohmic region with small V_{DS} ($\ll V_{GS} - V_P$). (b) When $V_{DS} = V_{GS} - V_P$, the channel height will become narrow at the drain end of the gate. The device enters pinchoff. The constricted channel will cause the drain current to saturate as shown. (c) Cutoff and subthreshold current regions occur when the depletion region extends through the channel.

regions can be qualitatively understood by referring to the channel cross sections in Fig. 56.47. In these figures, the doped channel region is shaded, and the depletion region is white. First, consider the JFET in Fig. 56.47(a) with small V_{DS} ($\ll V_{GS} - V_P$). This condition corresponds to the ohmic region (sometimes called linear or triode region) where current and voltage are linearly related. At small drain voltages, the depletion layer height is nearly uniform, the electric fields in the channel are too small to saturate the carrier velocity, and thus the channel behaves like a linear resistor. The resistance can be varied by changing V_{GS}. The channel height is reduced by increasing the reverse bias on the gate leading to an increased resistance.

As V_{DS} increases, the depletion layer thickness grows down the length of the channel, as shown in Fig. 56.47(b). This occurs because the drain current causes a voltage increase along the channel as it flows through the channel resistance. Since the depletion layer thickness is governed by the gate-to-channel voltage (V_{GC}), there is an increasing reverse bias that leads to constriction of the channel at the drain end of the gate. Ideally, when $V_{DS} = V_{GS} - V_P$, then $V_{GC} = V_P$, and the channel height will approach zero (pinchoff). The constricted channel will cause the drain current to saturate as shown. Further increases in V_{DS} do not cause the drain current to increase since the channel has already constricted to a minimum height and the additional potential is accommodated by lateral extension of the depletion region at the drain end of the gate. This region of operation is generally described as the **pinchoff region** (rather than the saturation region in order to avoid confusion with BJT saturation). The height of the channel is not actually zero but is limited by the mobile channel charge, which travels at saturated drift velocity in this high field region.

If $V_{GS} < 0$, then the initial channel height at the source is reduced, I_D is less, and the pinchoff region occurs at a smaller drain voltage $V_{DS} = V_{GS} - V_P$. The saturation of drain current can also occur at smaller V_{DS} if the gate length is very small. In this case, the electric field in the channel is large, and the carrier velocity will saturate before the channel can reach pinchoff. Velocity saturation will also limit drain current.

The *subthreshold* region of operation, shown in Fig. 56.47(c) is defined when small drain currents continue to flow even though $V_{GS} \leq V_P$. While technically this gate bias should produce cutoff, some small fraction of the electrons from the source region will have sufficient energy to overcome the potential barrier caused by the gate depletion region and will drift into the drain region and produce a current. Since the energy distribution is exponential with potential, the current flow in this region varies exponentially with V_{GS}.

The *inverse* region occurs when the polarity of the drain bias is reversed. This region is of little interest for the JFET since gate-to-drain conduction of the gate diode limits the operation to the linear region only.

Channel Length Modulation Effect

A close look at the I–V characteristic in the pinchoff region shows that the incremental conductivity or slope of this region is not equal to zero. There is some finite slope that is not expected from the simple velocity saturation or pinchoff models. Channel length modulation is one explanation for this increase; the position under the gate where pinchoff or velocity-saturation first occurs moves toward the source as V_{DS} increases. This is due to the expansion of the drain side depletion region at large V_{DS}. Figure 56.48 illustrates this point. Here, a channel cross section is shown for $V_{DS} = V_{GS} - V_P$ in Fig. 56.48(a) and for $V_{DS} \gg V_{GS} - V_P$ in Fig. 56.48(b). While pinchoff always occurs when the gate-to-channel voltage is V_P, the higher drain voltage causes the location of this point (L) to move closer to the source end of the channel (L'). Since the electric field in this region E is roughly proportional to $(V_{GS} - V_P)/L$ where L is now a function of V_{DS} and V_{GS}, then the current must increase as the channel length decreases due to increasing carrier velocity ($v = \mu E$). If the channel length is short, velocity saturation may cause the drain current to saturate. In this case, the velocity saturation point moves closer to the source as drain voltage is increased. Since the length has decreased, less gate-to-channel voltage is needed to produce the critical field for velocity saturation. Less voltage implies a wider channel opening, hence more current.

Temperature Effects

There are two mechanisms that influence the drain current of the JFET when temperature is changed [4], [12]. First, the pinchoff voltage becomes more negative (for n-channel) with increase in temperature, therefore requiring lower V_{GS} to cut off the channel or to enter the pinchoff region. Therefore, when the device is operating in the pinchoff region, and $V_{GS} - V_P$ is small, the drain current will increase with temperature. This effect is caused by the decrease in the built-in voltage of the gate-to-channel junction with increasing temperature. Second, the carrier mobility and saturated drift velocity decreases with temperature. This causes a reduction

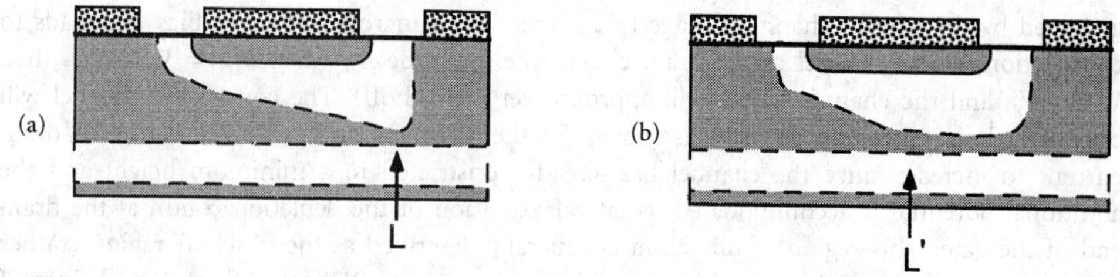

FIGURE 56.48 A channel cross section is shown for $V_{DS} = V_{GS} - V_P$ in (a) and for $V_{DS} \gg V_{GS} - V_P$ in (b). While pinchoff always occurs when the gate-to-channel voltage is V_P, the higher drain voltage causes the location of this point (L) to move closer to the source end of the channel (L').

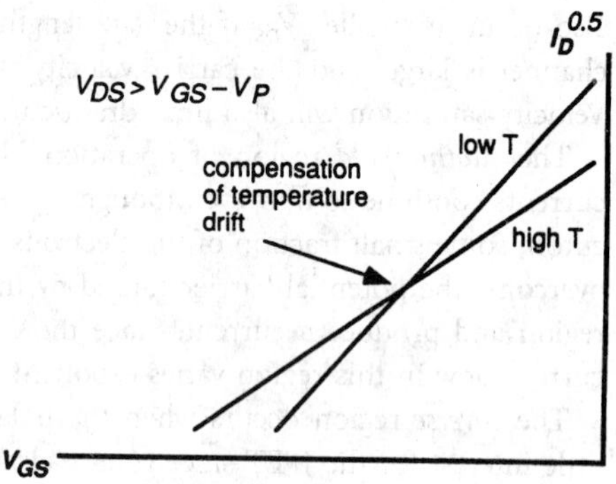

FIGURE 56.49 Effect of temperature on the drain current in the pinchoff region.

in drain current that is in opposition to the first effect. This effect dominates for large $V_{GS} - V_P$. Therefore, there is a V_{GS} value for which the drain current is exactly compensated by the two effects. This is illustrated qualitatively in Fig. 56.49.

The gate current is also affected by temperature, as it is the reverse current of a pn junction. The current increases roughly by a factor of 2 for each 10°C increase in temperature. At high temperatures, the input current of a JFET input stage may become comparable to that of a well-designed BJT input stage of an op-amp, thus losing some of the benefit of the mixed BJT–JFET circuit design.

JFET Models

Most applications of the JFET in Analog IC's employ the pinchoff region of operation. It is this region that provides power gain and buffer (source follower) capability for the device, so the models for the JFET presented below will concentrate on this region. It will also be assumed that the gate–source junction will not be biased into forward conduction. Although forward conduction is simple to model using the ideal diode equation within the FET equivalent circuit models, this bias condition is not useful for the principal analog circuit applications of the JFET and will also be avoided in the discussion that follows.

Large-Signal Model: Drain Current Equations

Equations modeling the large signal JFET $I_D - V_{GS}$ characteristic can be derived for the two extreme cases of FET operation in the pinchoff region. A gradually decreasing channel height and mobility limited drift velocity in the channel are appropriate assumptions for very long gate length FET's. A fixed channel height at pinchoff with velocity saturation limited drift velocity are more suitable for short gate lengths.

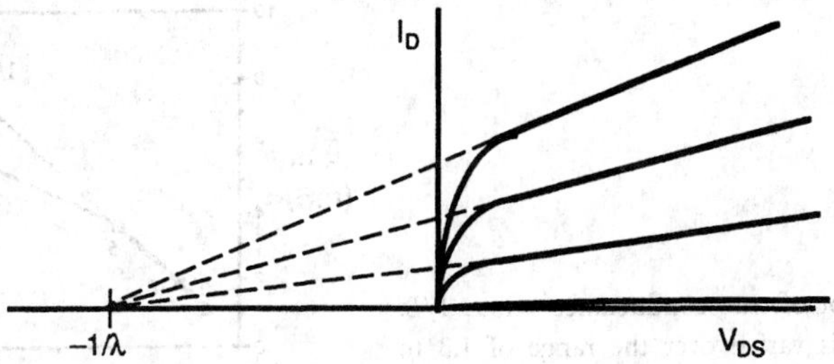

FIGURE 56.50 The channel length modulation parameter λ is defined by the extrapolation of the drain current in saturation to $I_D = 0$.

The square-law transfer characteristic [2] given by (56.135) provides a good

$$I_D = I_{DSS}\left(1 - \frac{V_{GS}}{V_P}\right)^2 (1 + \lambda V_{DS}) \tag{56.135}$$

approximation to measured device characteristics in the case of long gate length (> 5 μm) or very low electrical fields in the channel $(V_{GS} - V_P)/L < E_{sat}$. In both cases, the channel height varies slowly and the velocity remains proportional to mobility. E_{sat} is the critical field for saturation of drift velocity, about 3.5 kV/cm for GaAs and 20 kV/cm for Si. I_{DSS} is defined as the drain current in the pinchoff region when $V_{GS} = 0$. The first two terms of the equation are useful for approximate calculation of dc biasing. The third term models the finite drain conductance caused by the channel length modulation effect. The parameter λ in this term is derived from the intercept of the drain current when extrapolated back to zero as shown in Fig. 56.50.

Equation (56.135) is also used to represent the pinchoff region in the SPICE JFET model. It is parameterized in a slightly different form as shown below in (56.136).

$$I_D = \beta(V_{GS,i} - V_{TO})^2(1 + \lambda V_{DS}) \tag{56.136}$$

These equations are the same if $V_{TO} = V_P$, and

$$\begin{aligned} \beta &= \frac{I_{DSS}}{V_P^2} \\ V_{GS,i} &= V_{GS} - I_D R_S \\ V_{DS,i} &= V_{DS} - I_D R_S - I_D R_D \end{aligned} \tag{56.137}$$

The pinchoff region is defined for $V_{DS,i} \geq V_{GS,i} - V_{TO}$ as is usual for the gradual channel approximation. R_S and R_D are the parasitic source and drain resistances associated with the contacts and the part of the channel that is outside of the gate junction. These resistances will reduce the internal device voltages below the applied terminal voltages as shown in (56.137).

For shorter gate length devices, improved models have been proposed and implemented in SPICE3 and some of the many commercial SPICE products, often in the MESFET model. The Raytheon or Statz model [10] is frequently used for this purpose. This model modifies the drain current dependence on V_{GS} by adding a velocity saturation model parameter b in the denominator as shown in (56.138).

$$I_D = \left[\frac{\beta(V_{GS,i} - V_{TO})^2}{1 + b(V_{GS,i} - V_{TO})}\right](1 + \lambda V_{DS,i}) \tag{56.138}$$

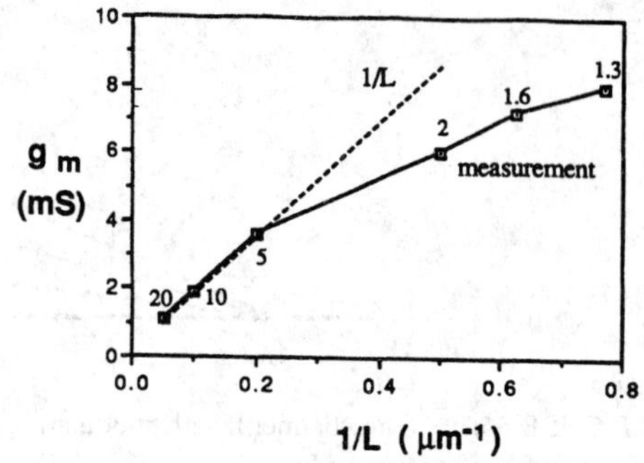

FIGURE 56.51 Plot of transconductance versus $1/L$. The gate length L is varied over the range of 1.3 to 20 μm (lengths shown next to data points). $1/L$ scaling is observed for gate lengths greater than 5 μm.

This added term allows the drain current to be nearly square law in V_{GS} for small $V_{GS} - V_{TO}$, but it becomes almost linear when V_{GS} is large, effectively emulating the rapid rise in transconductance followed by saturation that is typical in short channel devices. Although the specific behavior of the drain current is sensitive to the vertical doping profile in the channel [15], (56.138) is flexible enough to accommodate most short channel device characteristics with uniform or nonuniform channel doping. Another feature of short gate length FET's that this model predicts adequately is a saturation of I_D at $V_{DS,i} < V_{GS,i} - V_{TO}$. This early transition into the pinchoff region is also a consequence of velocity saturation and is widely observed.

Width and Length Scaling

Classical FET theory suggests that the drain current should scale linearly with channel width W and inversely with gate length L. In practice, this applies adequately for JFET's with long gate lengths ($L \geq 5$ μm) and wide channels ($W \geq 5$ μm) but breaks down below these approximate dimensions. In the case of the width, the drain current of a narrow channel JFET will be larger that expected by a linear scaling due to the lateral diffusion that occurs during implantation or diffusion. The channel may be wider than the mask dimension by possibly 0.5 μm for this reason. If the process design rules allow use of very narrow channels, experimental data should be used to determine the effective channel width.

Scaling of drain current with gate length is less straightforward. For very short gate lengths, at the limit of velocity saturation, no increase in drain current is predicted by (56.139) unless the channel depth d is also scaled. An increase in channel doping is required to reduce d. Since channel doping in a given process is constant, the drain current will scale less sensitively than $1/L$ if gate lengths below 5 μm are used. This effect is illustrated in Fig. 56.51 for a GaAs MESFET, a type of JFET to be described in the section on JFET technologies, where the transconductance is plotted against $1/L$. The gate length L is varied over the range of 1.3–20 μm (lengths shown next to data points). $1/L$ scaling is observed for gate lengths greater than 5 μm.

$$\beta = \frac{\varepsilon v_{\text{sat}} W}{d} \tag{56.139}$$

In practice, data measured on devices covering a range of L is needed in order to predict length scaling accurately. Note that the SPICE2 or SPICE3 JFET and MESFET models do not attempt to scale device parameters for gate length, so the length scaling must be accomplished manually through use of parameter sets optimized for each length.

Device Capacitances

There will be capacitances associated with the pn junctions comprising the JFET. Figure 56.45 shows that there are junctions between the gate and channel and between the channel and substrate. In addition, for some silicon JFET devices as discussed in the section on JFET technologies, there will also be a junction between the gate and substrate. Each of these capacitances are voltage dependent and should be considered in a device model if it is to be accurate at high frequencies.

The gate-to-channel capacitance is split into two components, a gate-to-source capacitance C_{gs} and a gate-to-drain capacitance C_{gd}. The V_{GS} and V_{GD} dependence of these capacitances is often approximated by the depletion capacitance of a pn junction; eqs. (56.140) and (56.141) are typically used to predict variation of capacitance in the indicated bias range. For $V_{GS} < V_P$ the capacitance rapidly drops to a small value limited by parasitic capacitances and is insensitive to either V_{GS} or V_{DS}. V_{BI} is the apparent built-in voltage of the junction. In some models, the parameter m is made adjustable to accommodate nonuniform doping profile effects. Typically, m =0.5 is used to represent a uniformly doped channel.

$$C_{gs} = \frac{C_{gso}}{(1 - V_{GS}/V_{BI})^m} \tag{56.140}$$

$$C_{gd} = \frac{C_{gdo}}{(1 - V_{GD}/V_{BI})^m} \tag{56.141}$$

While this model is severely in error in the linear region [6], if operation of the device is limited to the pinchoff region it can provide an adequate prediction. For the GaAs MESFET, C_{gd} will be of the order of 15 percent of the total gate capacitance; C_{gs} about 85 percent. For the Si JFET, C_{gd} is about 25 percent; C_{gs} is about 75 percent. This formulation is found in both the SPICE JFET and MESFET models; however, the MESFET model implementation [10] includes the pinchoff effects and is considerably more accurate in the linear region.

The parasitic gate-to-substrate capacitance C_{gss} found in certain Si JFET technologies can also be modeled by functions like (56.140) or (56.141). However, it is not included explicitly as part of the SPICE JFET model. If it is necessary, the diode model can be used to simulate C_{gss} by connecting an external diode from gate-to-substrate.

Small-Signal Model

The small-signal model for the JFET in the pinchoff region is shown in Fig. 56.52. The voltage-dependent current source models the transconductance g_m as a constant that can be derived from the drain current equations above from

$$g_m = \frac{\partial I_D}{\partial V_{GS}} \tag{56.142}$$

The square-law current model (56.135) predicts a linearly increasing g_m with V_{GS}

$$g_m = -\frac{2I_{DSS}}{V_P}\left(1 - \frac{V_{GS}}{V_P}\right) \tag{56.143}$$

whereas a model that includes some velocity saturation effects such as (56.138) would predict a saturation in g_m.

The small-signal output resistance r_o models the channel length modulation effect. This is also derived from the drain current equations through

$$r_o^{-1} = \frac{\partial I_D}{\partial V_{DS}} \tag{56.144}$$

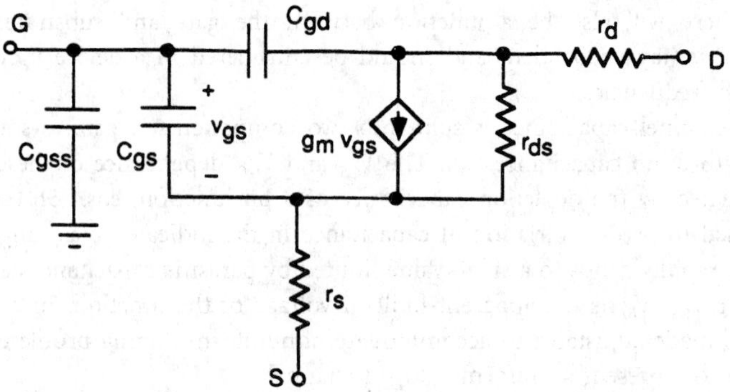

FIGURE 56.52 The small-signal model for the JFET in the pinchoff region.

For both models, r_o is determined by

$$r_o = \frac{1}{I_D \lambda} \tag{56.145}$$

The small-signal values of the nonlinear, voltage dependent C_{gs}, C_{gd}, and C_{gss} are also shown in Fig. 56.52. Parasitic source and drain resistances R_S and R_D can also be included, as shown. If they are not included in the small-signal model, the effect of these parasitics can sometimes be produced in the intrinsic FET model by reducing the intrinsic g_m of the device.

The short-circuit current gain–bandwidth product f_T defined in (56.146) is a high-frequency figure of merit for transistors. It is inversely proportional to the transit time τ of the channel charge, and it is increased by reducing the gate length. Reduced L also reduces the gate capacitance and increases transconductance. The FET channel material also affects f_T as higher drift velocity leads to higher g_m.

$$f_T = \frac{g_m}{2\pi(C_{gss} + C_{gs} + C_{gd})} = \frac{1}{\tau} \tag{56.146}$$

SPICE Model Parameter Summary

In Tables 56.3 and 56.4, a brief summary of the basic set of parameters used in all SPICE JFET and MESFET models is provided. These parameters were defined in the section on the large-signal model. Note that some SPICE versions will include other parameters designed to improve the accuracy or versatility of the model. These parameters scale directly or inversely in channel width W for most model implementations. If the gate length is changed, generally a new set of parameters extracted for the new gate length will be required. The parameters β, C_{gso}, C_{gdo}, and I_S scale directly with W and R_S, R_D inversely. Other parameters are not affected by width to the first order.

JFET Technologies

The IC fabrication technology used to make JFET's depends primarily on the material. Discrete Si JFET's are available that provide f_T above 500 MHz and very low input rms noise currents through optimizing the channel design and minimizing parasitic capacitances, resistances, and gate diode leakage currents. However, a silicon IC process is rarely designed to optimize the performance of the JFET; rather, the JFET is made to accommodate an existing bipolar process with as few modifications as possible [2]. Then, the extra circuit design flexibility and performance benefits of a relatively inexpensive mixed FET/BJT process (often called BiFET) can be obtained

TABLE 56.3 Summary of SPICE JFET Model Parameters and Typical Values for a p-Channel Ion-Implanted Si JFET with $W/L = 25$ [2]

Parameter	Units	Typical Value
Pinchoff voltage (V_P)	Volts	1.0
Transconductance coefficient (β)	Amps/Volt^2	3.0e-4
Channel length modulation (λ)	Volt^{-1}	−0.01
Zero-bias *GS* junction capacitance (C_{gso})	Farad	2e-12
Zero-bias *GD* junction capacitance (C_{gdo})	Farad	2e-12
Zero-bias *GB* junction capacitance (C_{gss})	Farad	4e-12
Junction grading coefficient (GS and GD)		0.33
Junction grading coefficient (GB)		0.5
Gate junction built-in voltage (V_{BI})	Volt	0.5
Gate junction saturation current (I_S)	Amp	1e-10

TABLE 56.4 Summary of SPICE MESFET Model Parameters and Typical Values for a 1 μm Gate Length GaAs MESFET Normalized to a 1 μm Channel Width [6]

Parameter	Units	Typical Value
Pinchoff voltage (V_P)	Volts	−1.0
Transconductance coefficient (β)	Amps/Volt^2	1.0e-4
Doping tail extending parameter (b)	Volt^{-1}	0.3
Channel length modulation (λ)	Volt^{-1}	0.05
Saturation voltage parameter (α)	Volt^{-1}	2.5
Source resistance (R_S)	Ohm	900
Drain resistance (R_D)	Ohm	900
Zero-bias *GS* junction capacitance (C_{gso})	Farad	1.2e-15
Zero-bias *GD* junction capacitance (C_{gdo})	Farad	2.8e-15
Gate junction built-in voltage (V_{BI})	Volt	0.8
Gate junction saturation current (I_S)	Amp	5e-16

with small incremental cost. On the other hand, gallium arsenide IC processes exist only for the improved performance that is possible with these devices. Thus, the process technology is carefully designed to optimize device performance for bandwidth, noise, or power and to minimize the parasitic capacitances and inductances associated with passive components [9].

Silicon JFET Technologies

In principle, it would be possible to build p-channel Si JFET's in a standard analog BJT process without additional mask steps if the base diffusion has suitable doping and thickness to give a useful pinchoff voltage when overlaid with the emitter diffusion. Unfortunately, this is usually not the case, since the emitter diffusion is too shallow, and the pinchoff voltage resulting from this approach would be too high (positive in the case of the p-channel device). Therefore, the channel of the JFET must be made thinner either through the use of an additional diffusion or by providing the channel and gate with ion implantations.

The double diffused JFET process was the earliest approach to be used, since the channel depth could be reduced through the use of only one additional diffusion step. A cross-sectional view is shown in Fig. 56.53. Here, the depth of the gate diffusion is extended by an extra predeposition and drive-in diffusion step carried out prior to the emitter diffusion steps. Therefore, a narrower base region is obtained allowing the pinchoff voltage of the JFET to be independent of the requirements of the BJT. While this simple process modification is inexpensive to implement, the absolute control and matching of pinchoff voltage is not accurate enough for low input offset voltage in JFET differential amplifier pairs. In addition, the forced compatibility with the BJT process requires use of the collector layer under the channel. This forms a lower gate electrode that is by necessity less heavily doped than the channel. Therefore, the depletion region at this

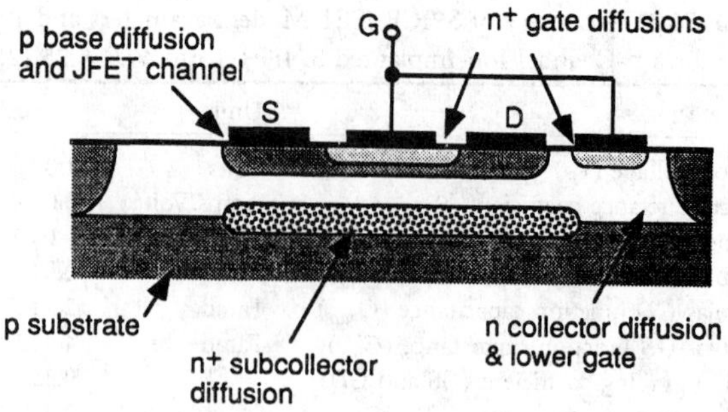

FIGURE 56.53 Cross section of a double diffused silicon JFET (not to scale).

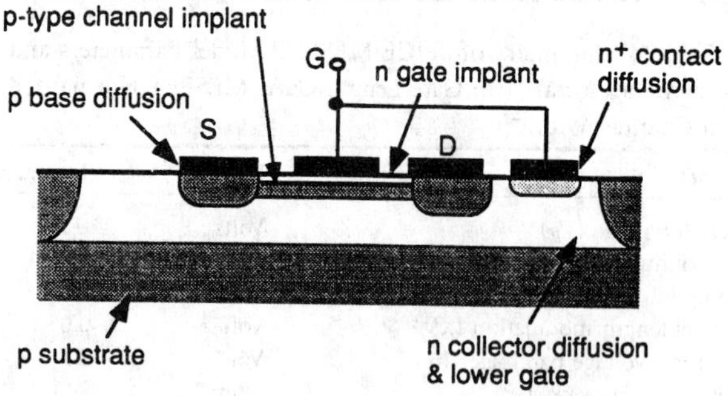

FIGURE 56.54 Cross section of an ion implanted silicon JFET (not to scale).

interface extends primarily into the gate, and the lower gate is rather ineffective in contributing to the total transconductance of the JFET. It does add the parasitic capacitance C_{gss} to the device at the collector to substrate junction, limiting frequency response. Finally, due to the relatively high base doping level, the drain breakdown voltage is low.

Because of the performance compromises inherent in this approach, it has been largely superceded by the ion implantation method. Figure 56.54 illustrates the cross section of an ion-implanted JFET. In order to gain better control of the pinchoff voltage and transconductance, both the channel and the gate are formed by ion implantation. This allows use of a channel that is more lightly doped than the base diffusion enabling the collector "gate" layer to more effectively contribute to the transconductance of the device. In addition, the predeposition of channel and gate charge is much more repeatable with ion implantation, so device matching and reproducibility of pinchoff voltage is greatly improved. The f_T will be improved by the larger g_m per unit width and the slightly reduced gate capacitances, and the drain breakdown voltage will be increased as is often needed for an analog IC process. However, low channel doping is not a good recipe for a high-frequency transistor with short gate length, so the f_T of these devices is still only 50 MHz or so.

GaAs MESFET and HEMT Technologies

High-performance GaAs MESFET devices are constructed with a metal–semiconductor junction gate instead of a diffused or implanted pn junction gate. The metal gate forms a Schottky barrier diode directly on an n-type channel as shown in Fig. 56.55 and allows the channel height to be varied in the same manner as the JFET. No gate dielectric or diffusion is necessary. The channel is formed either by ion implantation of an n-type dopant into a high resistivity (semiinsulating) GaAs substrate or a lightly p-type substrate or by growth of the channel region with molecular beam epitaxy. These MESFET's are used as the primary active device in analog microwave and mm-wave monolithic integrated circuits (sometimes

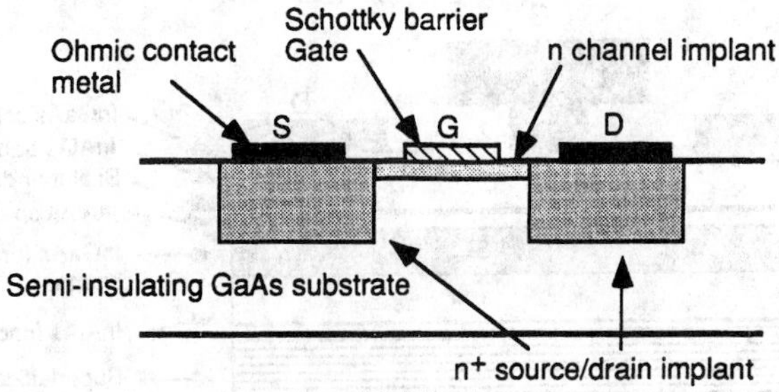

FIGURE 56.55 Cross section of a GaAs MESFET (not to scale).

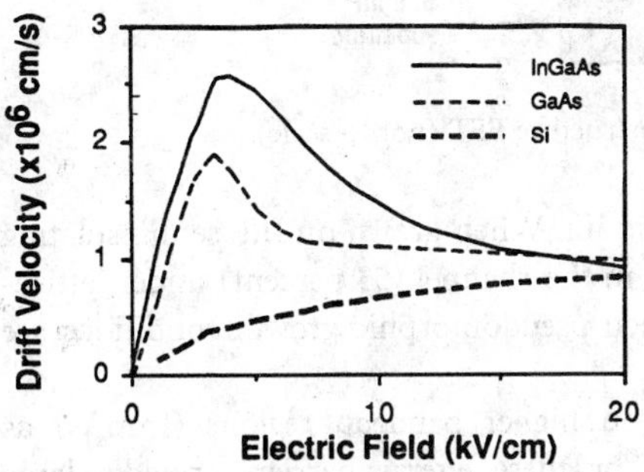

FIGURE 56.56 Electron velocity versus electric field for InGaAs (53 percent In) [5], GaAs [8], and Si [3].

called MMIC's). Since analog MESFET circuits are used for low noise, wide bandwidth, or high power amplification up to 50 GHz, gate lengths are often quite small; 0.25 μm is not unusual. The high performance of these devices comes through a combination of factors. First, the mobility and saturated drift velocity of n-GaAs are five times and two times larger than n-Si, respectively. In addition, the peak drift velocity is reached at much lower electric field for GaAs [8] as illustrated in Fig. 56.56. The parasitic substrate capacitance is quite small due to the high resistivity provided by semiinsulating GaAs. These factors lead to a much higher f_T (20–50 GHz) and lower parasitic resistances (f_{max}) at lower operating voltages for the GaAs MESFET than the Si JFET. These are desirable qualities for high performance microwave circuits, and the high electron velocity at low bias voltage can lead to reduced power as well. Also, patterning submicrometer gate lengths with metal or silicide gate materials is relatively straightforward, and gate resistance can be reduced through plating thick gate stripes with mushroom cross sections. Passive interconnections can have low losses through the use of plated air bridge interconnections suspended over the surface of the MMIC, and plated-through substrate vias can be fabricated when needed to reduce parasitic inductances.

MESFET MMICC's are available commercially from several sources. Custom MMIC design and fabrication can also be supported by several GaAs IC foundries.

Further reduction in noise figure and increase in bandwidth has been demonstrated by the use of heterojunction MESFET's, sometimes called HEMT's (high electron mobility transistors) or P-HEMT's (pseudomorphic HEMT's). This device achieves its improved performance mainly through the high mobility, undoped InGaAs channel material. The electron velocity versus electric field of $In_{0.53}Ga_{0.47}As$ is compared with GaAs and Si in Fig. 56.56, where it can be seen that higher drift velocity is obtained in $In_{0.53}Ga_{0.47}As$ [5] than either GaAs [8] or Si [3]. The higher the In concentration in the InGaAs, the higher the mobility and velocity and the lower the noise. InGaAs cannot be lattice matched to GaAs substrates, but about 20 percent In can be tolerated if the InGaAs channel is very thin. In this case the mismatch is accommodated by elastic strain rather than by the formation of lattice defects. This strained condition is called

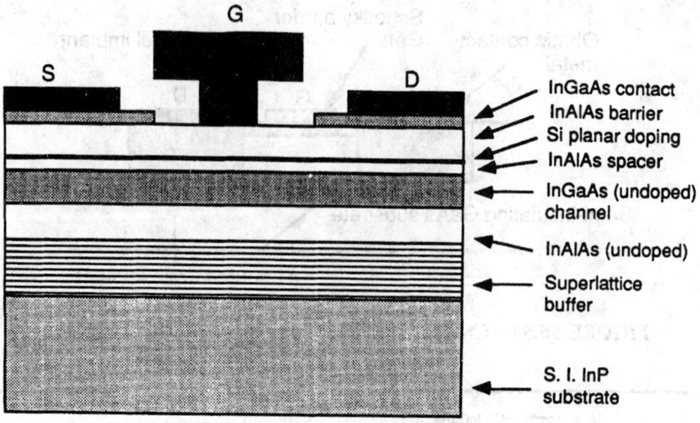

FIGURE 56.57 Cross section of a heterostructure FET (not to scale).

pseudomorphic: a forced lattice match due to elastic strain. When grown on InP semiinsulating substrates, much higher In concentration can be used in the channel (53 percent) under lattice-matched conditions, and even more In is possible when pseudomorphic growth conditions are used.

In addition, the channel layer is confined between higher bandgap regions (InAlAs) as illustrated in Fig. 56.57. The confinement provided by these energy barriers provides large channel electron sheet concentrations, also improving g_m.

Finally, the gate barrier heterojunction also enables good Schottky gate characteristics to be obtained even though the channel material itself has a low bandgap and would otherwise provide a poor barrier height if the metal were in direct contact.

Excellent performance of HEMT or P-HEMT devices and MMIC's at microwave and millimeter wave frequencies has been reported. Table 56.5 summarizes some representative results.

TABLE 56.5 Summary of Performance of HEMT and P-HEMT Devices and MMIC's. *G* refers to the Power Gain Associated with the Frequency, Noise, or Wide Band Condition cited.

Device Technology	Frequency (GHz)	NF (dB)	*G* (dB)	Reference
InP P-HEMT	95	1.3	8.2	[1]
(Discrete MIC)	141		7.3	
(three-stage MMIC)	94	3.5	21	[13]
InP HEMT	5–100		5.5	[7]
(Distributed amplifier)				
GaAs P-HEMT (two-stage MMIC)	112–115	5.5	10	[14]

References

[1] P. D. Chow, K. Tan *et al.*, "*W*-band and *D*-band low noise amplifiers using 0.1 micron pseudomorphic InAlAs/InGaAs/InP HEMT's," IEEE MTT Symp., Albuquerque, NM, 1992.

[2] P. R. Gray and R. Meyer, *Analysis and Design of Analog Integrated Circuits*, New York: Wiley, 1993.

[3] C. Jacobini, C. Canali *et al.*, "A review of some charge transport properties of silicon," *Solid-State Electron.* vol. 20, pp. 77, 1977.

[4] S. J. Lee and C. P. Lee, "Temperature effect on low threshold voltage ion-implanted GaAs MESFET's," *Electron. Lett.* vol. 17, no. 20, pp. 760–761, 1981.

[5] M. A. Littlejohn, K. W. Kim *et al.*, "High-field transport in InGaAs and related heterostructures," in *Properties of Lattice-Matched and Strained Indium Gallium Arsenide*. London: Inspec, IEE, 1993, pp. 107–116.
[6] S. I. Long and S. E. Butner, *Gallium Arsenide Digital Integrated Circuit Design*. New York: McGraw-Hill, 1990.
[7] R. Majidi-Ahy, C. Nishimoto *et al.*, "5-100 Ghz InP coplanar waveguide MMIC distributed amplifier," *IEEE Trans. Microwave Theory Tech.*, vol. 38, pp. 1986–1991, Dec. 1990.
[8] F. Pozela and A. Reklaitis, "Electron transport properties in GaAs at high electric fields," *Solid-State Electron.* vol. 23, pp. 927–933, 1980.
[9] R. A. Pucel, "Design considerations for monolithic microwave circuits," *IEEE Trans. Microwave Theory Tech.*, vol. 29, pp. 513–534, June 1981.
[10] H. Statz, P. Newman *et al.*, "GaAs FET device and circuit simulation in SPICE," *IEEE Trans. Electron*, vol. 34, pp. 160–169, Feb. 1987.
[11] S. M. Sze, *Physics of Semiconductor Devices*, New York: Wiley-Interscience, 1981.
[12] G. W. Taylor, H. M. Darley *et al.*, "A Device Model for an Ion-Implanted MESFET." *IEEE Trans. Electron*, vol. ED-26, pp. 172–179, 1979.
[13] H. Wang, T. Ton *et al.*, "An ultralow noise *W*-band monolithic three-stage amplifier using 0.1 μm pseudomorphic InGaAs/GaAs HEMT technology," IEEE MTT Symposium, Albuquerque, NM, 1992.
[14] H. Wang, T. Ton *et al.*, "A *D*-band monolithic low noise amplifier," IEEE GaAs IC Sym. Miami Beach, FL, 1992.
[15] R. E. Williams and D. W. Shaw, "Graded channel FET's: Improved linearity and noise figure," *IEEE Trans. Electron.* vol. 25, pp. 600–605, June 1978.

56.4 Passive Components

Nhat M. Nguyen

Resistors

Resistors available in monolithic form are classified in general as semiconductor resistors and thin-film resistors. Semiconductor structures include diffused, pinched, epitaxial, and ion-implanted resistors. Commonly used thin-film resistors include tantalum, nickel–chromium (Ni–Cr), cermet (Cr–SiO), and tin oxide (SnO_2). Diffused, pinched, and epitaxial resistors can be fabricated along with other circuit elements without any additional processing steps. Ion-implanted and thin-film resistors require additional processing steps for monolithic integration but offer lower temperature coefficient, smaller absolute value variation, and superior high-frequency performance.

Resistor Calculation. The simplified structure of a uniformly doped resistor of length L, width W, and thickness T is shown in Fig. 56.58. The resistance is

$$R = \frac{1}{\sigma}\frac{L}{WT} = \left(\frac{\rho}{T}\right)\frac{L}{W} = R_{\square}\frac{L}{W} \tag{56.147}$$

where σ and ρ are conductivity and resistivity of the sample, respectively, and $R_{\square}$ is referred to as the *sheet resistance.* From the theory of semiconductor physics, the conductivity of a semiconductor sample is

$$\sigma = q(\mu_n n + \mu_p p) \tag{56.148}$$

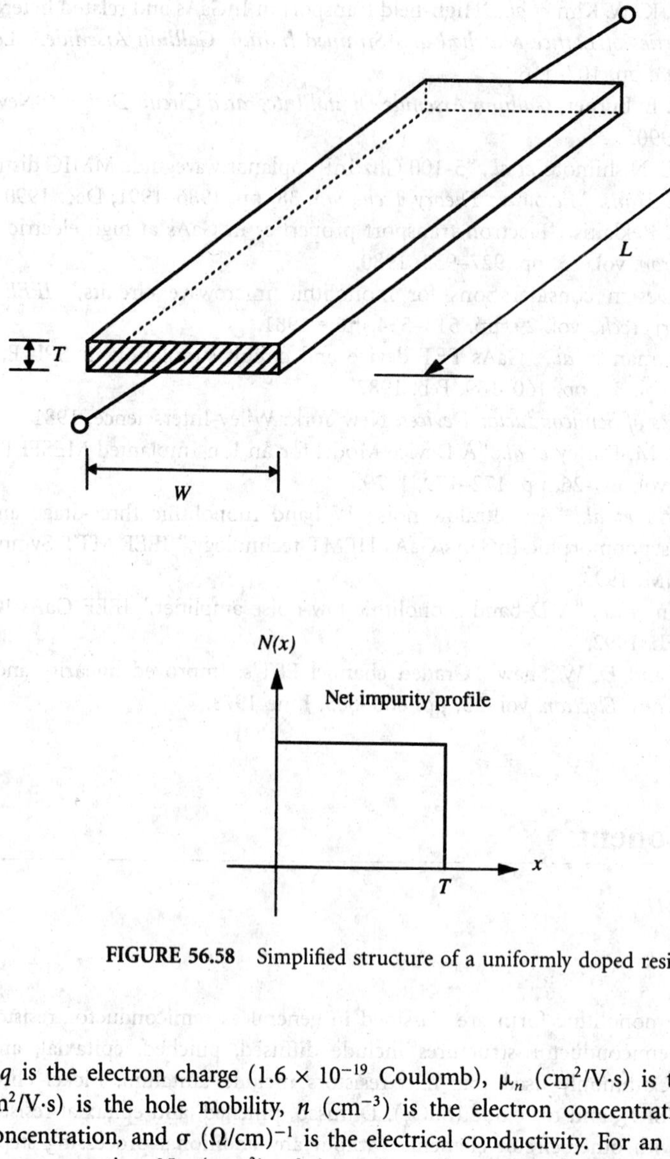

FIGURE 56.58 Simplified structure of a uniformly doped resistor.

where q is the electron charge (1.6×10^{-19} Coulomb), μ_n (cm^2/V·s) is the electron mobility, μ_p (cm^2/V·s) is the hole mobility, n (cm^{-3}) is the electron concentration, p (cm^{-3}) is the hole concentration, and σ $(\Omega/\text{cm})^{-1}$ is the electrical conductivity. For an n-type doped sample with a concentration N_D (cm^{-3}) of donor impurity atoms, the electron concentration n is approximately equal to N_D. Given the *mass-action law* $np = n_i^2$, the conductivity of an n-type doped sample is approximated by

$$\sigma = q\left(\mu_n N_D + \mu_p \frac{n_i^2}{N_D}\right) \approx q\mu_n N_D \qquad (56.149)$$

where n_i (cm^{-3}) is the *intrinsic* concentration. For a p-type doped sample, the conductivity is

$$\sigma = q\left(\mu_n \frac{n_i^2}{N_A} + \mu_p N_A\right) \approx q\mu_p N_A \qquad (56.150)$$

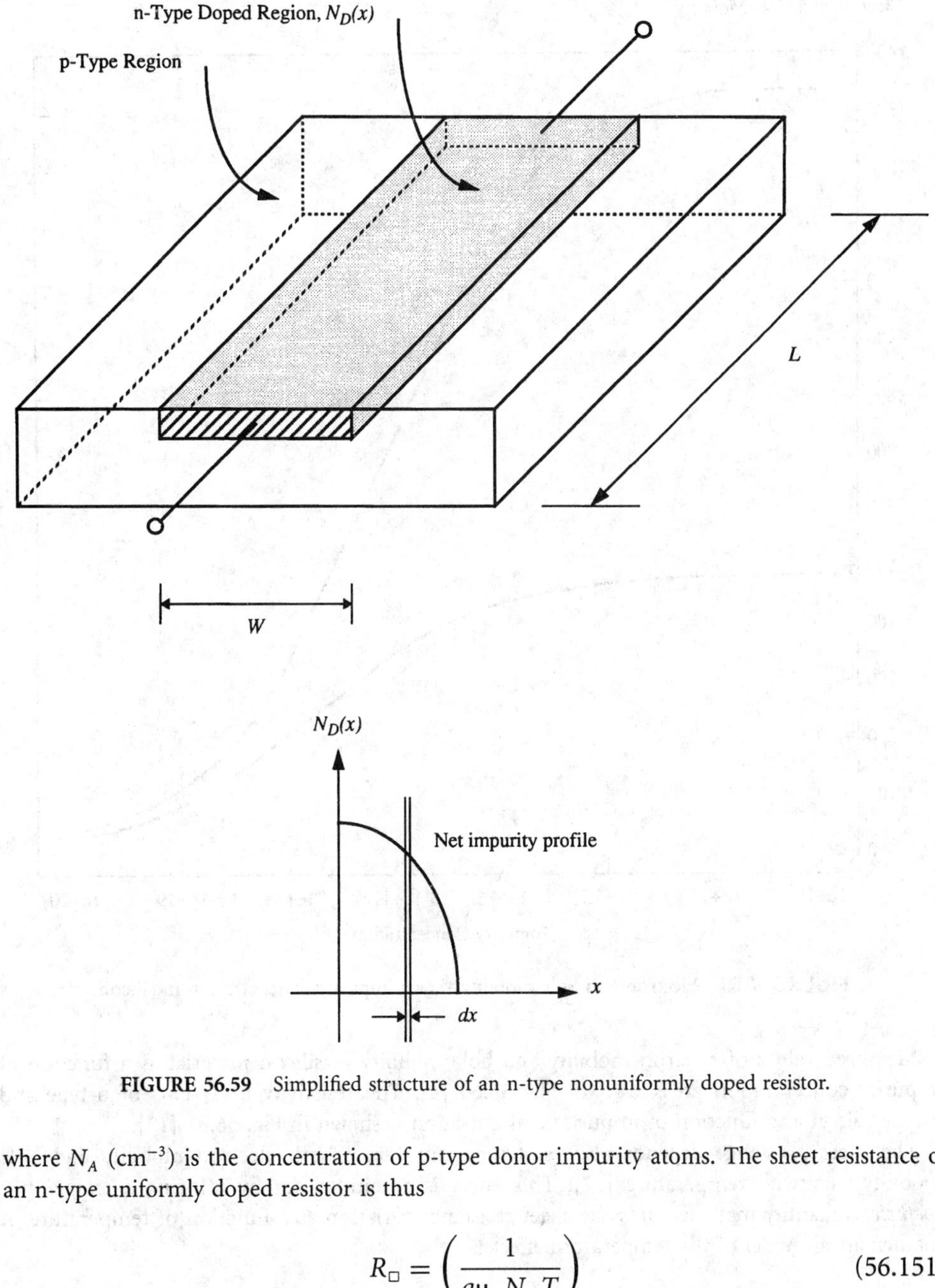

FIGURE 56.59 Simplified structure of an n-type nonuniformly doped resistor.

where N_A (cm^{-3}) is the concentration of p-type donor impurity atoms. The sheet resistance of an n-type uniformly doped resistor is thus

$$R_{\square} = \left(\frac{1}{q\mu_n N_D T} \right) \tag{56.151}$$

For an n-type nonuniformly doped resistor as shown in Fig. 56.59, where n-type impurity atoms are introduced into the p-type region by means of a high-temperature diffusion process, the sheet resistance [7] is

$$R_{\square} = \left[\int_0^{x_j} q\mu_n N_D(x)\, dx \right]^{-1} \tag{56.152}$$

where x_j is the distance from the surface to the edge of the junction depletion layer.

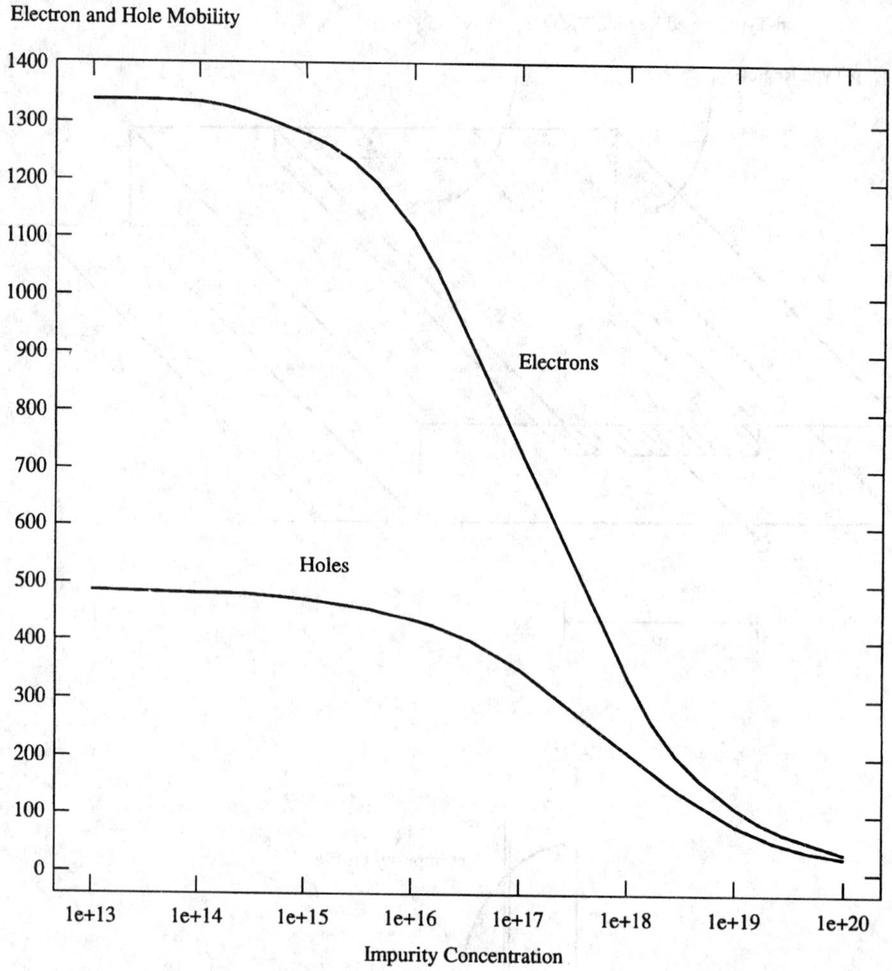

FIGURE 56.60 Electron and hole mobility versus impurity concentration in silicon.

Measured values of electron mobility and hole mobility in silicon material as a function of impurity concentration are shown in Fig. 56.60 [4]. The resistivity ρ (Ω-cm) of n-type and p-type silicon as a function of impurity concentration is shown in Fig. 56.61 [12].

The sheet resistance depends also on temperature since both electron mobility and hole mobility vary with temperature [17]. This effect is accounted for by utilizing a *temperature coefficient* quantity that measures the sheet resistance variation as a function of temperature. A mathematical model of the temperature effect is

$$R_{\square}(T) = R_{\square}(T_o)[(T - T_o)TC] \tag{56.153}$$

where T_o is the room temperature and TC is the temperature coefficient.

Diffused Resistors

In MOS (metal–oxide–semiconductor) technology, the diffused layer forming the source and drain of the MOS transistors can be used to form a diffused resistor. In silicon bipolar technology, the available diffused layers are base diffusion, emitter diffusion, active base region, and epitaxial layer.

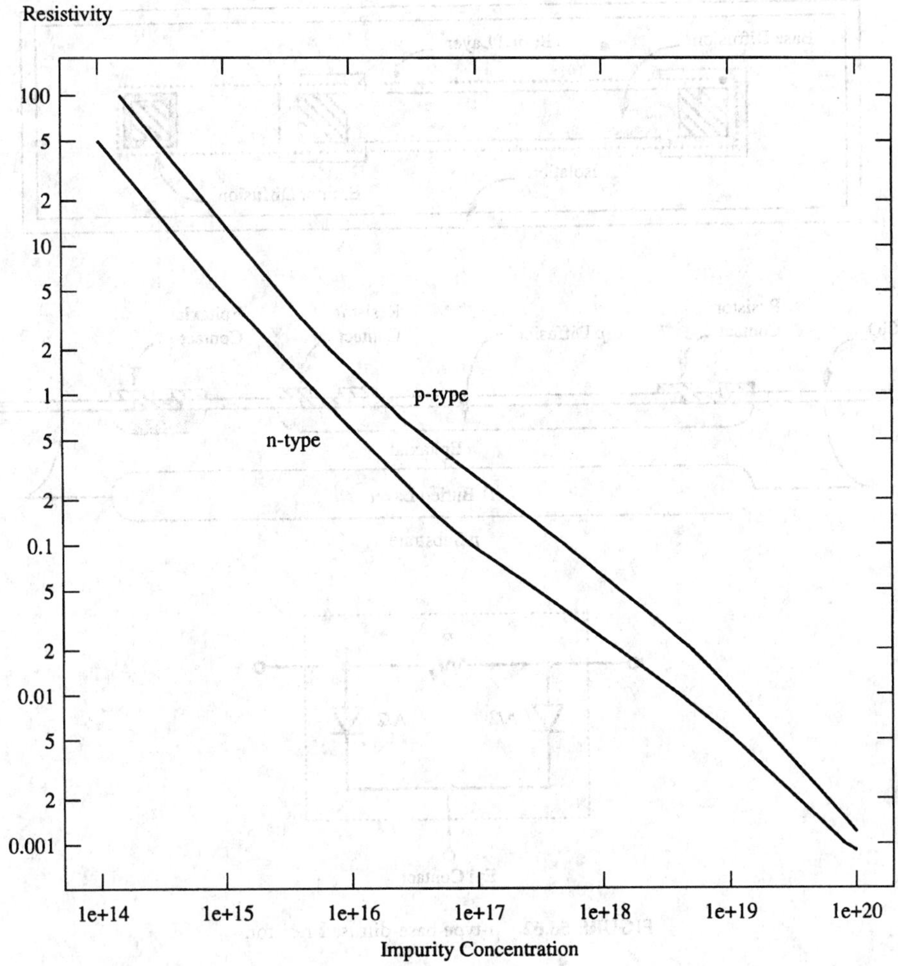

FIGURE 56.61 Resistivity of p-type and n-type silicon versus impurity concentration.

Base Diffused Resistors. The structure of a typical base diffused resistor is shown in Fig. 56.62, where the substrate material is assumed of p-type silicon material. The diffused resistor is formed by using the p-type base diffusion of the npn transistors. The resistor contacts are formed by etching selected windows of the SiO_2 passivation layer and depositing thin films of conductive metallic material. The isolation region can be formed with either a p-type doped junction or a trench filled with SiO_2 dielectric material. The pn junction formed by the p-type resistor and the n-type epitaxial (epi) layer must be reverse biased in order to eliminate the undesired dc current path through the pn junction. The impedance associated with a forward-biased pn junction is low and thus would also cause significant ac signal loss. To ensure this reverse bias constraint the epi region must be connected to a potential that is more positive than either end of the resistor contacts. Connecting the epi region to a relatively higher potential also eliminates the conductive action due to the parasitic pnp transistor formed by the p-type resistor, the n-type epi region, and the p-type substrate. When the base diffused resistor is fabricated along with other circuit elements to form an integrated circuit, the epitaxial contact is normally connected to the most positive supply of the circuit.

The resistance of a diffused resistor is given by (56.147), where the diffused sheet resistance is between 100 and 200 $\Omega/\square$. Due to the lateral diffusion of impurity atoms, the effective

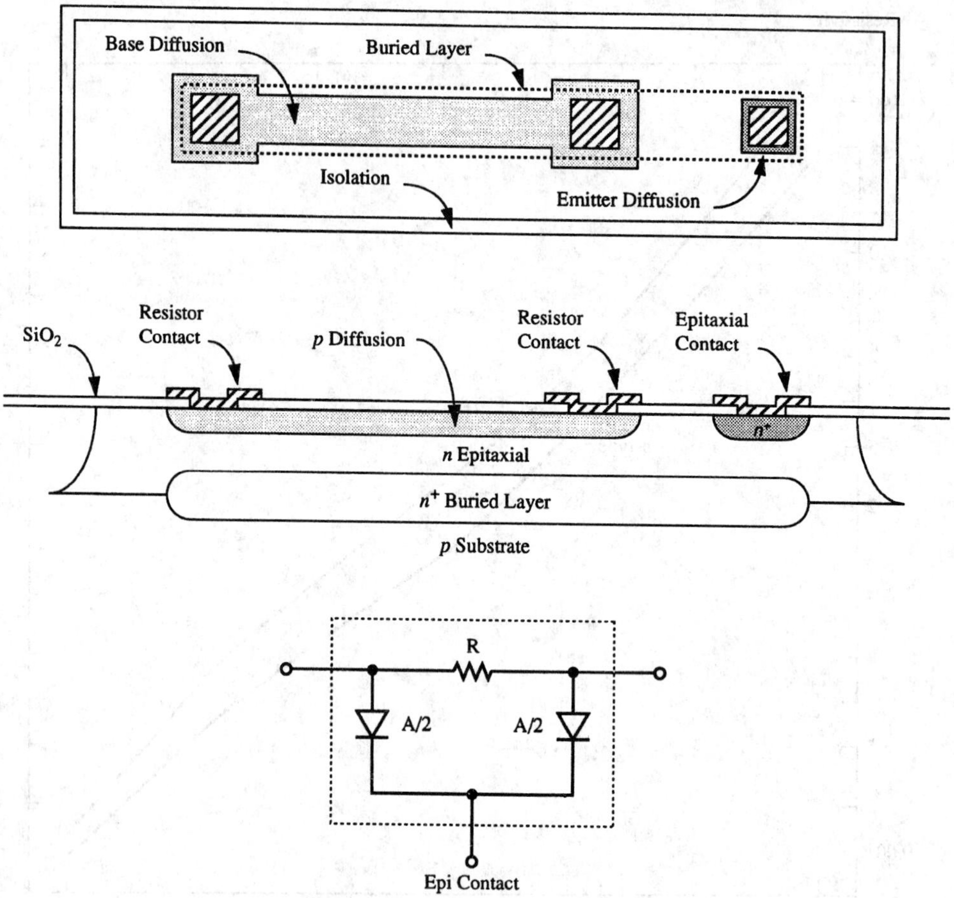

FIGURE 56.62 p-type base-diffused resistor.

cross-sectional area of the resistor is larger than the width determined by photomasking. This lateral or side diffusion effect can be accounted for by replacing the resistor width W by an effective width W_{eff}, where $W_{\text{eff}} \geq W$. The resistance from the two resistor contacts must also be accounted for, especially for small values of L/W [3]. Base-diffused resistors have a typical temperature coefficient between +1500 and +2000 ppm/°C.

The maximum allowable voltage for the base-diffused resistor of Fig. 56.62 is limited by the breakdown voltage between the p-type base diffusion and the n-type epi. This voltage equals the breakdown voltage BV_{CBO} of the collector-base junction of the npn transistor and typically causes an *avalanche breakdown* mechanism across the base–epi junction. As the applied voltage approaches the breakdown voltage, a large leakage current flows from the epi region to the base region and can cause excessive heat dissipation.

For analog integrated circuit applications where good matching tolerance between adjacent resistors is required, the resistor width should be made as large as possible. Base-diffused resistors with 50-μm resistor widths can achieve a matching tolerance of ±0.2 percent. The minimum resistor width is limited by photolithographic consideration with typical values between 3 and 5 μm. Also, in order to avoid the self-heating problem of the resistor it is important to ensure a minimum resistor width for a given dc current level, with a typical value of about 3 μm for every 1 mA of current.

With respect to high-frequency performance, the reverse-biased pn junction between the p-type base diffusion and the n-type epi contributes a distributed depletion capacitance which in turn causes an impedance roll-off at 20 dB/decade. This capacitance depends on the voltage applied across the junction and the junction impurity-atom dopings. For most applications the electrical lumped model as shown in Fig. 56.62 is adequate for characterizing this capacitive effect where the effective pn junction area is divided equally between the two diodes. Figure 56.63 shows a normalized impedance response as a function of the *RC* distributed stage. The frequency at which impedance value is reduced by 3 dB is given by

$$f_{-3\,\mathrm{dB}} = \begin{cases} \left(\dfrac{1}{2\pi}\right)\dfrac{2.0}{RC} & N = 1 \quad \text{(Circuit model of Fig. 56.62)} \\ \left(\dfrac{1}{2\pi}\right)\dfrac{2.32}{RC} & N = 2 \\ \left(\dfrac{1}{2\pi}\right)\dfrac{2.42}{RC} & N = 3 \\ \left(\dfrac{1}{2\pi}\right)\dfrac{2.48}{RC} & N = 4 \end{cases} \tag{56.154}$$

Emitter Diffused Resistors. Emitter-diffused resistors are formed by using the heavily doped n^+ emitter diffusion layer of the npn transistors. Due to the high doping concentration, the sheet resistance can be as low as 2 to 10 $\Omega/\square$ with a typical absolute value tolerance of ± 20 percent.

Figure 56.64 shows an emitter-diffused resistor structure where an n^+ diffusion layer is formed directly on top of the n-type epitaxial region and the ohmic contacts are composed of conductive metal thin films. Since the resistor body and the epi layer are both n-type doped, they are electrically connected in parallel but the epi layer is of much higher resistivity due to its lower concentration doping, and thus the effective sheet resistance of the resistor structure is determined solely by the n^+ diffusion layer. The pn junction formed between the p-type substrate and the n-type epi region must always be reverse biased, which is accomplished by connecting the substrate to a most negative potential. Because of the common n-type epi layer, each resistor structure of Fig. 56.64 requires a separate isolation region.

Figure 56.65 shows another emitter diffused resistor structure where the n^+ diffusion layer is situated within a p-type diffused well. Several such resistors can be fabricated in the same p-type well or in the same isolation region because the resistors are all electrically isolated. The p-type well and the n^+ diffusion region form a pn junction that must always be reverse biased for electrical isolation. In order to eliminate the conductive action due to the parasitic npn transistor formed by the n-type resistor body, the p-type well, and the n-type epi, the junction potential across the well contact and the epi contact must be either short circuited or reverse biased. The maximum voltage that can be applied across the emitter-diffused resistor of Fig. 56.65 is limited by the breakdown voltage between the n^+ diffusion and the p-type well. This voltage equals the breakdown voltage BV_{EBO} of the emitter–base junction of the npn transistor, with typical values between 6 and 8 V.

Pinched Resistors

The active base region for the npn transistor can be used to construct pinched resistors with typical sheet resistance range from 2 to 10 $K\Omega/\square$. These high values can be achieved due to a thin cross-sectional area through which the resistor current traverses. The structure of a p-type base-pinched resistor is shown in Fig. 56.66, where the p-type resistor body is "pinched"

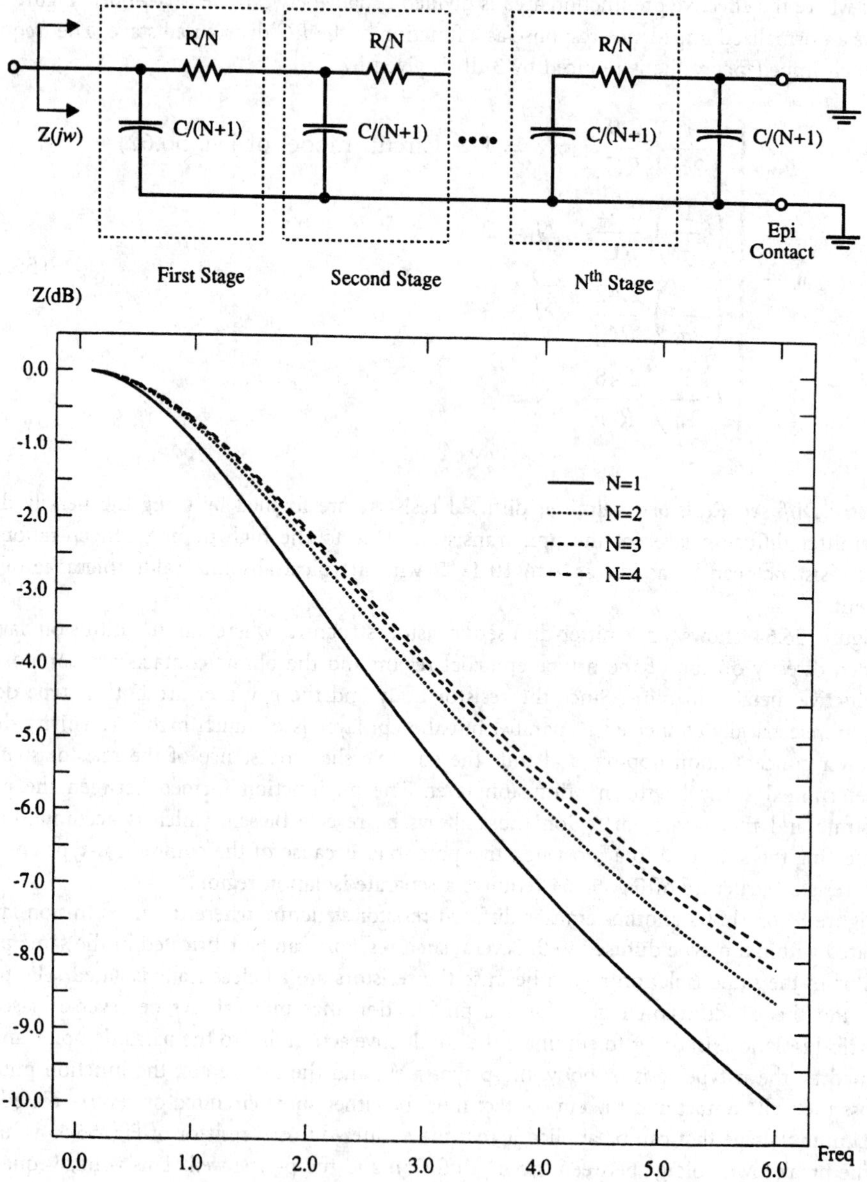

FIGURE 56.63 Normalized frequency response of a diffused resistor for $N = 1, 2, 3, 4$. The epi contact and one end of the resistor are grounded.

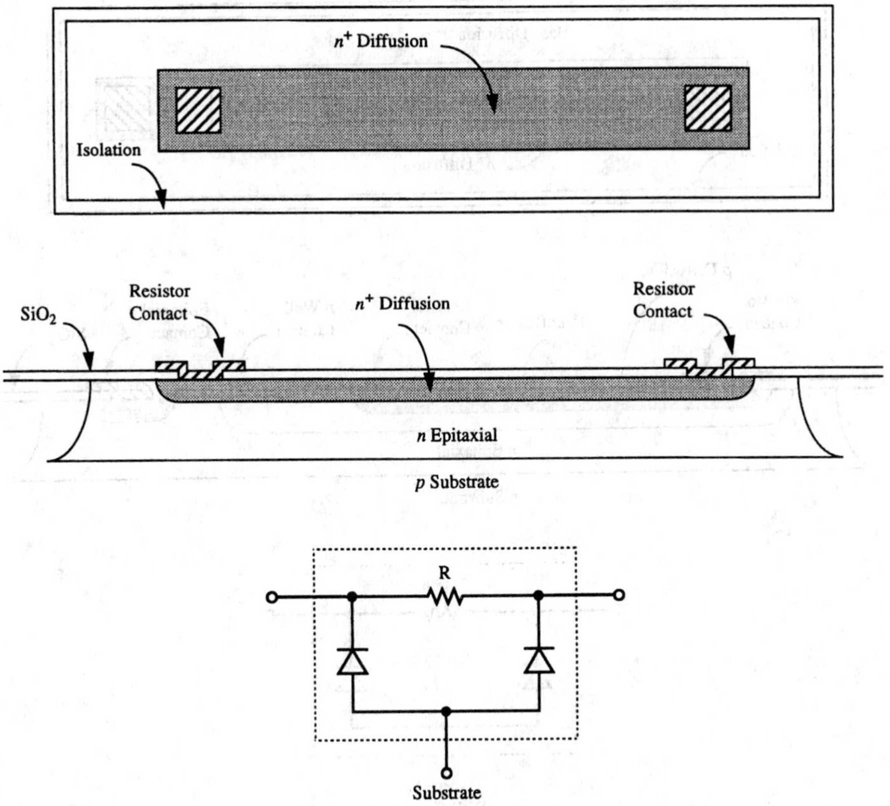

FIGURE 56.64 n-type emitter-diffused resistor I.

between the n^+ diffusion layer and the n-type epitaxial layer. The n^+ diffusion layer overlaps the p-type diffusion layer and is therefore electrically connected to the n-type epi. In many aspects the base-pinched resistor behaves like a p-channel JFET, in which the active base region functions as the p channel, the two resistor contacts assume the drain and source, and the n^+ diffusion and the epi constitute the n-type gate. When the pn junction formed between the active base and the surrounding n^+ diffusion and n epi is subject to a reverse bias potential, the carrier-free depletion region increases and extends into the active base region, effectively reducing the resistor cross section and consequently increasing the sheet resistance. Since the carrier-free depletion region varies with reverse bias potential, the pinched resistance is voltage controlled and is nonlinear.

Absolute values for the base-pinched resistors can vary as much as ±50 percent due to large process variation in the fabrication of the active base region. The maximum voltage that can be applied across the base-pinched resistor of Fig. 56.66 is restricted by the breakdown voltage between the n^+ diffusion layer and the p-type base diffusion. The breakdown voltage has a typical value around 6 V.

Epitaxial Resistors

Large values of sheet resistance can be obtained either by reducing the effective cross-sectional area of the resistor structure or by using a low doping concentration that forms the resistor body. The first technique is used to realize the pinched resistor while the second is used to realize the epitaxial resistor. Figure 56.67 shows an epitaxial resistor structure where the resistor is formed with a lightly doped epitaxial layer. For an epi thickness of 10 μm and a doping concentration

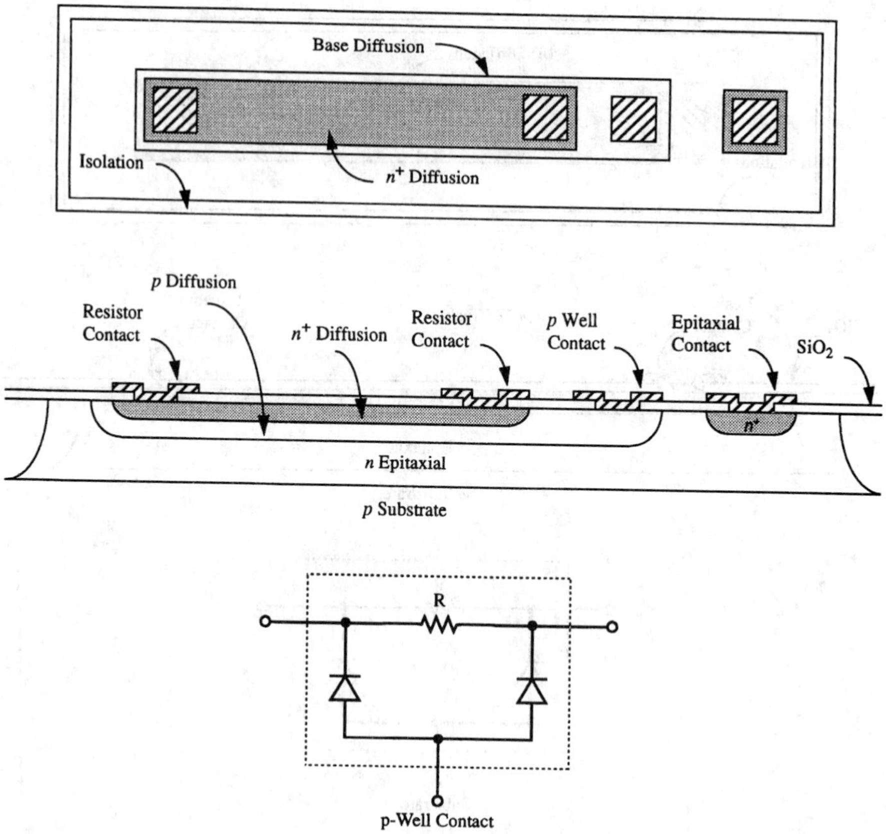

FIGURE 56.65 n-type emitter-diffused resistor II.

of 10^{15} donor atoms/cm^3, this structure achieves a resistivity of 5 Ω-cm and an effective sheet resistance of 5 KΩ/□. The temperature coefficient of the epitaxial resistor is relatively high with typical values around +3000 ppm/°C. This large temperature variation is a direct consequence of the hole and electron mobilities undergoing more drastic variations against temperature at particularly low doping concentrations [13]. The maximum voltage that can be applied across the epitaxial resistor is significantly higher than that for the pinched resistor. This voltage is set by the breakdown voltage between the n-type epi and the p-type substrate which varies inversely with the doping concentration of this pn junction.

Epitaxial Pinched Resistors. By putting a p-type diffusion plate on top of the epitaxial resistor of Fig. 56.67, even larger sheet resistance value can be obtained. The p-type diffusion plate overlaps the epi region and is electrically connected to the substrate through the p-type isolation. The epi layer is thus pinched between the p-type diffusion plate and the p-type substrate. When the n-type epi and the surrounding p-type regions is subject to a reverse bias potential, the junction depletion width extends into the epi region and effectively reduces the cross-sectional area. Typical sheet resistance values are between 4 and 5 KΩ/□. The epitaxial-pinched resistor behaves like an n-channel JFET, in which the effective channel width is controlled by the substrate voltage.

Ion-Implanted Resistors

Ion implantation is an alternative technique beside diffusion for inserting impurity atoms into a silicon wafer [17]. Commonly used impurities for implantation are the p-type boron

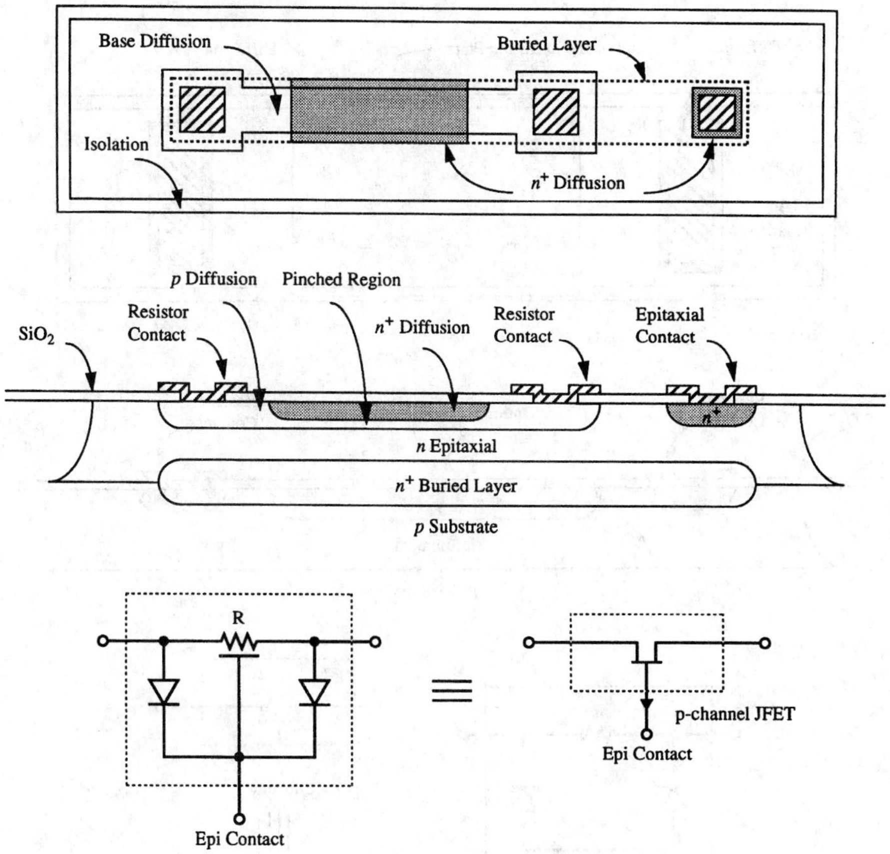

FIGURE 56.66 p-type base-pinched resistor.

atoms. The desired impurity atoms are first ionized and then accelerated to a high energy by an electric field. When a beam of these high-energy ions is directed at the wafer, the ions penetrate into exposed regions of the wafer surface. The penetration depth depends on the velocity at contact and is typically between 0.1 and 0.8 μm. The exposed regions of the wafer surface are defined by selectively etching a thick thermally grown SiO_2 layer that covers the wafer and functions as a barrier against the implanted ions. Unique characteristics of the ion-implantation technique include a precise control of the impurity concentration, uniformly implanted layers of impurity atoms, and no lateral diffusion. The structure of a p-type ion-implanted resistor is shown in Fig. 56.68, where the p-type diffused regions at the contacts are used to achieve good ohmic contacts to the implanted resistor. The pn junction formed between the p-type implanted region and the n-type epitaxial layer must be reverse biased for electrical isolation. By connecting the epi region to a potential relatively more positive than the substrate potential, the conductive action due to the parasitic pnp transistor formed by the p-type implanted, the n-type epi, and the p-type substrate is also eliminated. Ion-implanted resistors exhibit relatively tight absolute value tolerance and excellent matching. Absolute value tolerance down to ±3 percent and matching tolerance of ±2 percent are typical performance.

Table 56.6 provides a summary of the typical characteristics for the diffused, pinched, epitaxial, and ion-implanted resistors.

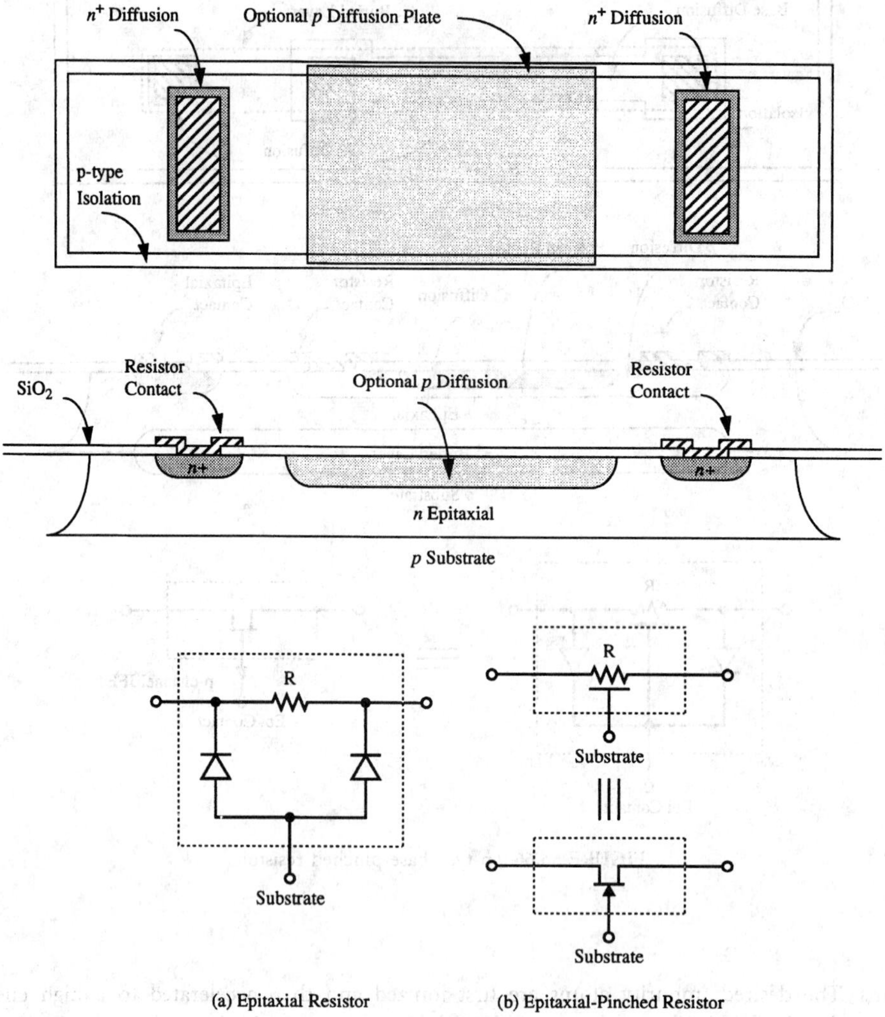

FIGURE 56.67 n-type epitaxial and epitaxial-pinched resistors.

TABLE 56.6 Typical Properties of Semiconductor Resistors

Resistor Type	Sheet ρ (Ω/□)	Absolute Tolerance (%)	Matching Tolerance (%)	Temperature Coefficient (ppm/°C)
Base diffused	100–200	±20	±2 (5 μm wide) ±0.2 (50 μm wide)	+1500 to +2000
Emitter diffused	2–10	±20	±2	+600
Base pinched	2–10 K	±50	±10	+2500
Epitaxial	2–5 K	±30	±5	+3000
Epitaxial pinched	4–10 K	±50	±7	+3000
Ion implanted	100–1000	±3	±2 (5 μm wide) ±0.15 (50 μm wide)	Controllable to ±100

Source: P. R. Gray and R. G. Meyer, *Analysis and Design of Analog Integrated Circuits.* New York: Wiley, 1984, p. 119.

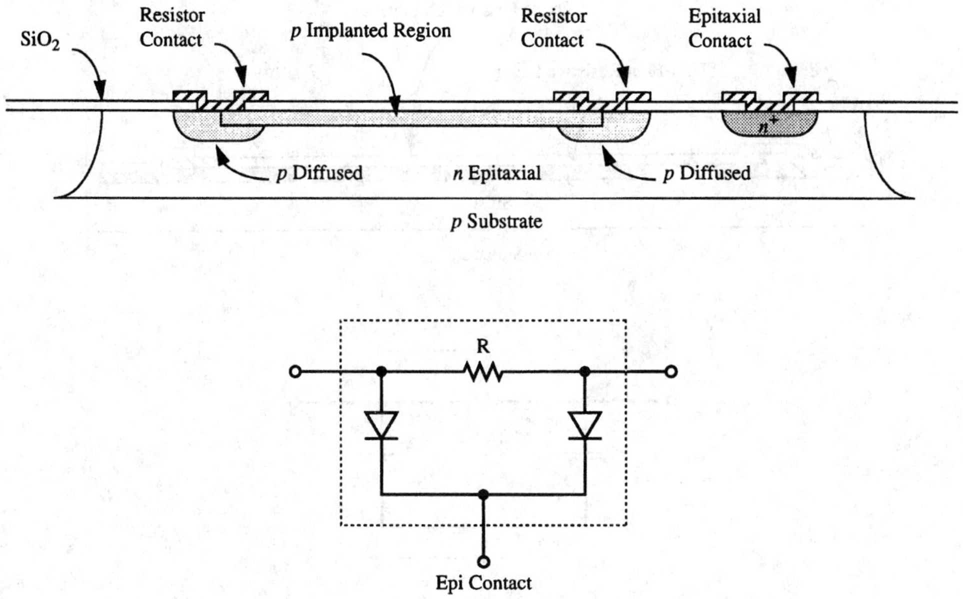

FIGURE 56.68 p-type ion-implanted resistor.

Thin-Film Resistors

Compared with diffused resistors, thin-film resistors offer advantages of a lower temperature coefficient, a smaller absolute value variation, and an excellent high-frequency characteristic. Commonly used resistive thin films are tantalum, nickel–chromium (Ni–Cr), cermet (Cr–SiO), and tin oxide (SnO_2). A typical thin-film resistor structure is shown in Fig. 56.69, where a thin-film resistive layer is deposited on top of a thermally grown SiO_2 layer and a thin-film conductive metal layer is used to form the resistor contacts. The oxide layer functions as an insulating layer for the resistor. Various CVD (chemical vapor deposition) techniques can be used to form the thin films [8]. The oxide passivation layer deposited on top of the resistive film and the conductive film protects the device surface from contamination. The electrical lumped model as shown in Fig. 56.69 is adequate to characterize the high-frequency performance of the resistor. The parallel-plate capacitance formed between the thin-film resistive and the substrate is divided equally between the two capacitors. Table 56.7 provides a summary of the characteristics for some commonly used thin-film resistors.

Capacitors

Monolithic capacitors are widely used in analog and digital integrated circuits for functions such as circuit stability, bandwidth enhancement, ac signal coupling, impedance matching, and charge

TABLE 56.7 Typical Characteristics of Thin-Film Resistors

Resistor Type	Sheet ρ (Ω/□)	Absolute Tolerance (%)	Matching Tolerance (%)	Temperature Coefficient (ppm/°C)
Ni–Cr	40–400	±5	±1	±100
Ta	10–1000	±5	±1	±100
SnO_2	80–4000	±8	±2	0–1500
Cr–SiO	30–2500	±10	±2	±50–±150

Source: A. B. Grebene, *Bipolar and MOS Analog Integrated Circuit Design.* New York: Wiley, 1984, p. 155.

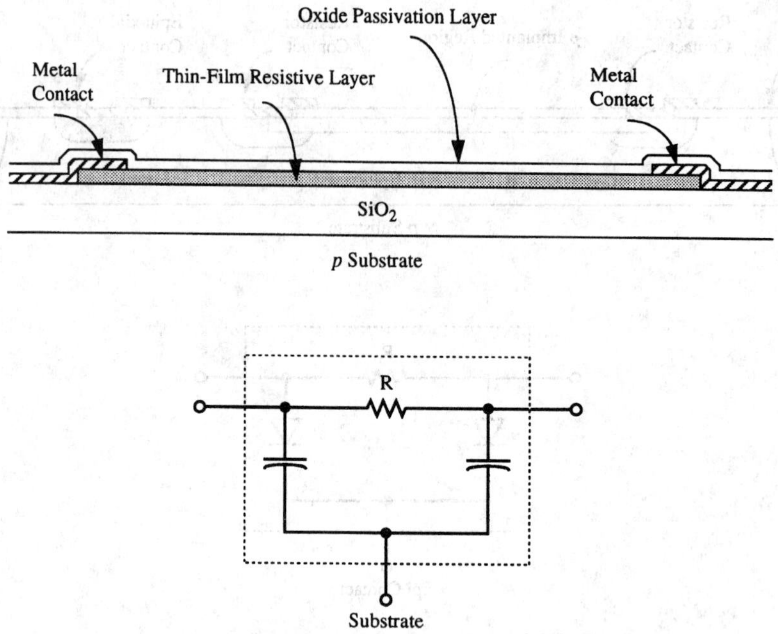

FIGURE 56.69 Thin-film resistor.

storage cells. Capacitor structures available in monolithic form include pn junction, MOS, and polysilicon capacitors. pn junctions under reverse-biased conditions exhibit a nonlinear voltage-dependent capacitance. MOS and polysilicon capacitors, on the other hand, closely resemble the linear parallel-plate capacitor structure as shown in Fig. 56.70. If the insulator thickness T of the parallel-plate structure is small compared with the plate width W and length L, the electric field between the plates is uniform (fringing field neglected). Under this condition the capacitance can be calculated by

$$C = \frac{\kappa \varepsilon_0}{T} WL \tag{56.155}$$

where κ is the relative dielectric constant of the insulating material and ε_0 is the permittivity constant in vacuum (8.854×10^{-14} F/cm).

Junction Capacitors

The structure of an *abrupt* pn junction is shown in Fig. 56.71, where the doping is assumed uniform throughout the region on both sides. The acceptor impurity concentration of the p region is N_A atoms/cm^3 and the donor impurity concentration of the n region is N_D atoms/cm^3. When the two regions are brought in contact, mobile holes from the p region diffuse across the junction to the n region and mobile electrons diffuse from the n to the p region. This diffusion process creates a *depletion* region that is essentially free of mobile carriers (*depletion approximation*) and contains only fixed acceptor and donor ions. Ionized acceptor atoms are negatively charged and ionized donor atoms are positively charged. In equilibrium the diffusion process is balanced out by a drift process that arises from a *built-in* voltage ψ_o across the junction. This voltage is positive from the n region relative to the p region and is given by [17]

$$\psi_o = \frac{kT}{q} \ln \frac{N_A N_D}{n_i^2} \tag{56.156}$$

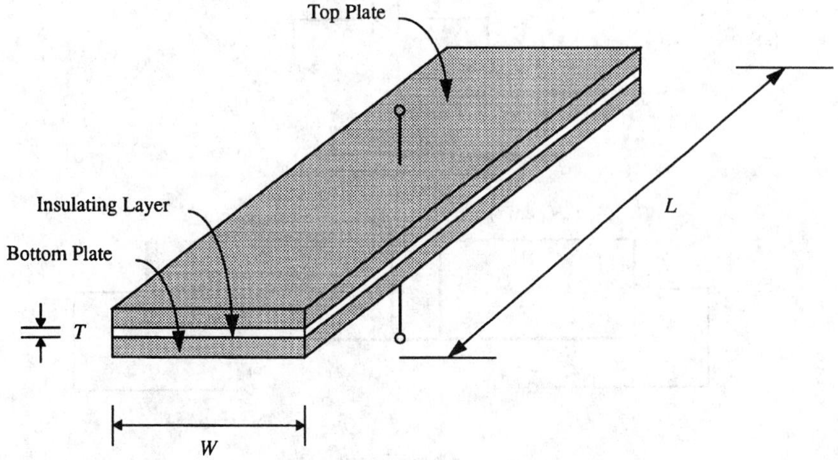

FIGURE 56.70 Structure of a parallel-plate capacitor.

where k is the Boltzmann constant (1.38×10^{-23} V·Coulomb/K), T is the temperature in Kelvin (K), q is the electron charge (1.60×10^{-19} C), and n_i (cm^{-3}) is the *intrinsic* carrier concentration in a pure semiconductor sample. For silicon at 300 K, $n_i \approx 1.5 \times 10^{10}\ \text{cm}^{-3}$.

When the pn junction is subject to an applied reverse bias voltage V_R, the drift process is augmented by the external electric field and more mobile electrons and holes are pulled away from the junction. Because of this effect, the depletion width W_d and consequently the charge Q on each side of the junction vary with the applied voltage. A junction capacitor can thus be defined to correlate this charge–voltage relationship. The Poisson's equation relating the junction voltage $\phi(x)$ to the electric field $\xi(x)$ and the total charge Q is

$$\frac{d^2\phi(x)}{dx^2} = -\frac{d\xi(x)}{dx} = -\frac{q}{\varepsilon_S}(p - n + N_D - N_A) \approx \begin{cases} \dfrac{qN_A}{\varepsilon_S} & -x_p < x < 0 \\ -\dfrac{qN_D}{\varepsilon_S} & 0 < x < x_n \end{cases} \tag{56.157}$$

where ε_S ($11.8\varepsilon_0 = 1.04 \times 10^{-12}$ F/cm) is the permittivity of the silicon material. The first integral of (56.157) yields the electric field as

$$\xi(x) = \begin{cases} -\dfrac{qN_A}{\varepsilon_S}(x + x_p) & -x_p < x < 0 \\ -\dfrac{qN_D}{\varepsilon_S}(x_n + x) & 0 < x < x_n \end{cases} \tag{56.158}$$

The electric field is shown in Fig. 56.71, where the maximum field strength occurs at the junction edge. This value is given by

$$|\xi_{\max}| = \frac{qN_A}{\varepsilon_S}x_p = \frac{qN_D}{\varepsilon_S}x_n$$

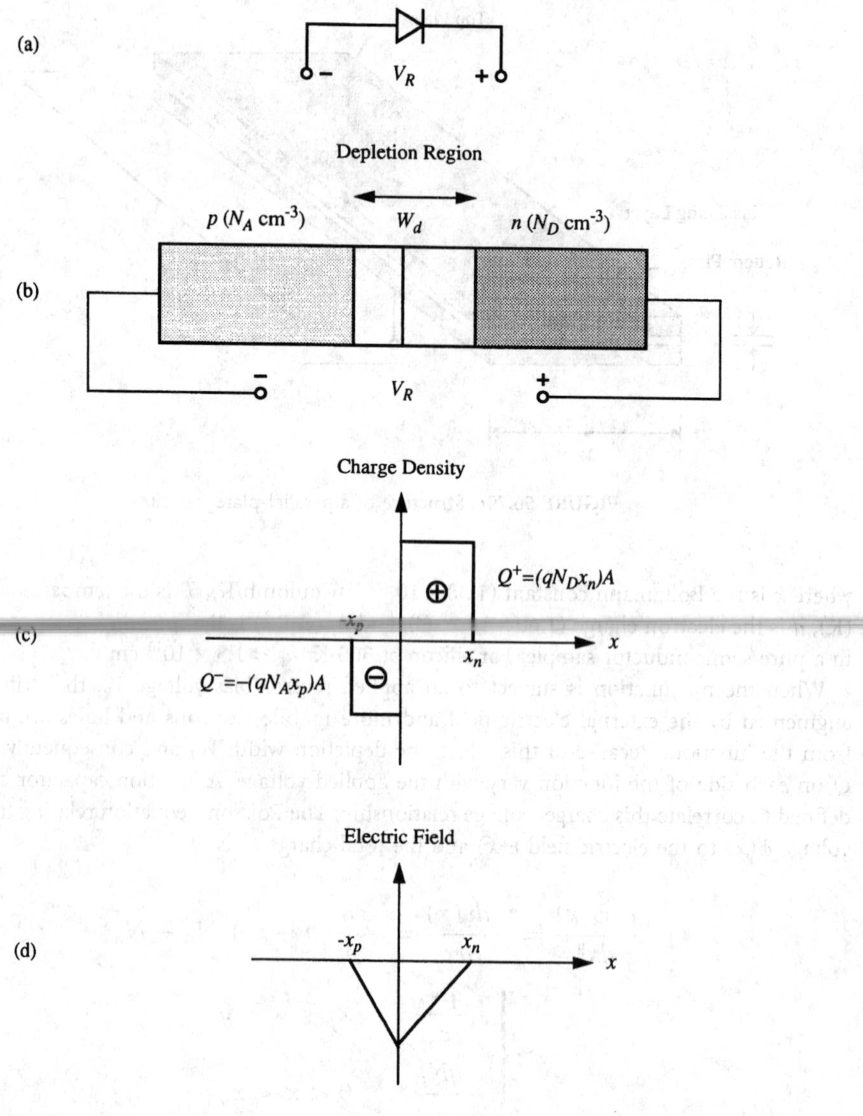

FIGURE 56.71 Abrupt p–n junction: (a) p–n junction symbol; (b) depletion region; (c) charge density within the depletion region; (d) electric field.

The partial depletion width x_p on the p region and the partial depletion width x_n on the n region can then be related to the depletion width W_d as

$$x_p + x_n = W_d$$

$$x_p = \frac{N_D}{N_A + N_D} W_d$$

$$x_n = \frac{N_A}{N_A + N_D} W_d$$

Taking the second integral of (56.157) yields the junction voltage

$$\phi(x) = \begin{cases} \dfrac{qN_A}{\varepsilon_S}\left(\dfrac{x_p^2}{2} + x_p x + \dfrac{x^2}{2}\right) & -x_p < x < 0 \\ \dfrac{qN_D}{\varepsilon_S}\left(\dfrac{x_n x_p}{2} + x_n x - \dfrac{x^2}{2}\right) & 0 < x < x_n \end{cases} \tag{56.159}$$

where the voltage at x_p is arbitrarily assigned to be zero. The total voltage $\psi_o + V_R$ can be expressed as

$$\psi_o + V_R = \phi(x_n) = \frac{qN_D}{2\varepsilon_S}\left(1 + \frac{N_D}{N_A}\right)x_n^2$$

Finally, the depletion width W_d and the total charge Q in terms of the total voltage across the junction can be derived to be

$$W_d = \left[\frac{2\varepsilon_S}{q}(\psi_o + V_R)\left(\frac{1}{N_A} + \frac{1}{N_D}\right)\right]^{1/2}$$
$$|Q| = A(qN_A x_p) = A(qN_D x_n) = A\left[2q\varepsilon_S(\psi_o + V_R)\left(\frac{1}{N_A} + \frac{1}{N_D}\right)^{-1}\right]^{1/2} \tag{56.160}$$

The junction capacitance is thus

$$C_j = \left|\frac{dQ}{dV_R}\right| = A\left[\frac{q\varepsilon_S}{2}\left(\frac{1}{\psi_o + V_R}\right)\left(\frac{1}{N_A} + \frac{1}{N_D}\right)^{-1}\right]^{1/2}$$
$$= \frac{C_{jo}}{\left(1 + \dfrac{V_R}{\psi_o}\right)^{1/2}} \tag{56.161}$$

where A is the effective cross-sectional junction area and C_{jo} is the value of C_j for $V_R = 0$. If the doping concentration in one side of the pn junction is much higher than that in the other, the depletion width and the junction capacitance can be simplified to

$$W_d = \left[\frac{2\varepsilon_S}{qN_L}(\psi_o + V_R)\right]^{1/2} \tag{56.162}$$

$$C_j = A\left[\frac{\varepsilon_S q N_L}{2}\left(\frac{1}{\psi_o + V_R}\right)\right]^{1/2} \tag{56.163}$$

where N_L is the concentration of the lightly doped side. Figure 56.72 displays the junction capacitance per unit area as a function of the total voltage $\psi_o + V_R$ and the concentration on the lightly doped side of the junction [3].

In silicon bipolar technology the base–emitter, the base–collector, and the collector–substrate junctions under reverse bias are often utilized for realizing a junction capacitance. The collector–substrate junction has only a limited use since it can only function as a shunt capacitor due to the substrate being connected to an ac ground.

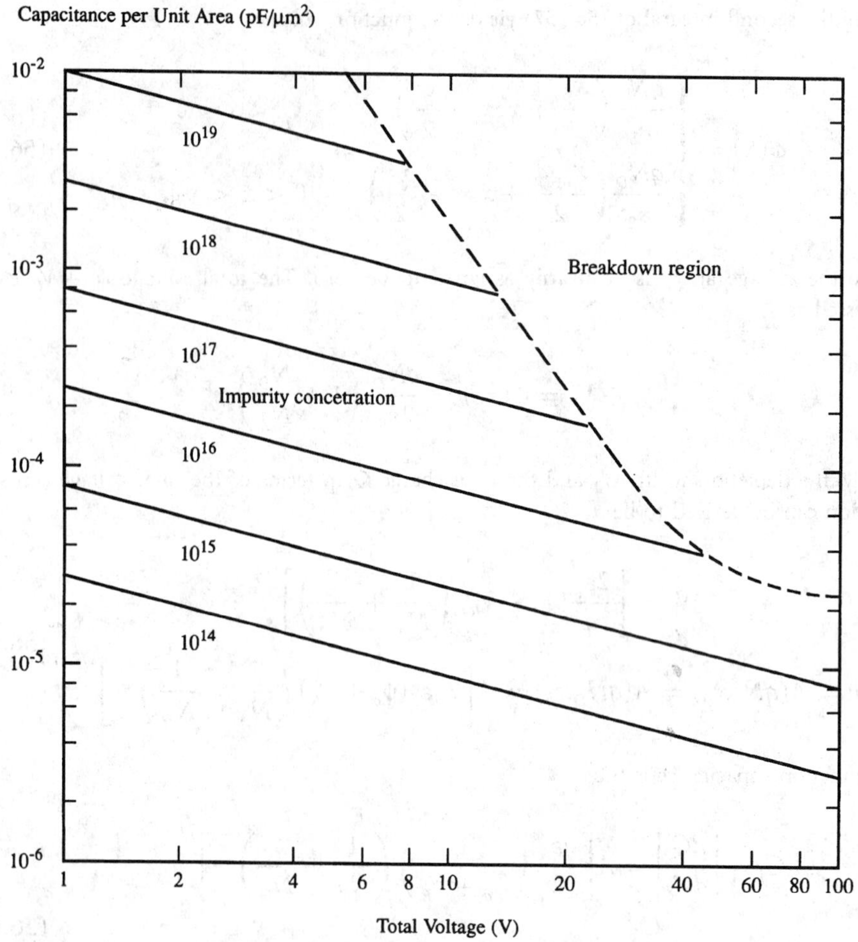

FIGURE 56.72 Junction capacitance as a function of the total voltage and the concentration on the lightly doped side.

Base–Collector Junction Capacitor. A typical base–collector capacitor structure is shown in Fig. 56.73 together with an equivalent lumped circuit model. A heavily doped n^+ buried layer is used to minimize the series resistance R_C. For the base–collector junction to operate in reverse bias, the n-type collector must be connected to a voltage relatively higher than the voltage at the p-type base. The junction breakdown voltage is determined by BV_{CBO} of the npn transistor, which has a typical value between 25 and 50 V.

Base–Emitter Junction Capacitor. Figure 56.74 shows a typical base–emitter capacitor structure where the parasitic junctions D_{BC} and D_{SC} must always be in reverse bias. The base–emitter junction achieves the highest capacitance per unit area among the base–collector, base–emitter, and collector–substrate junctions due to the relatively higher doping concentrations in the base and emitter regions. For the base–emitter junction to operate in reverse bias, the n-type emitter must be connected to a voltage relatively higher than the voltage at the p-type base. The breakdown voltage of the base–emitter junction is relatively low, determined by the BV_{EBO} of the npn transistor, which has a typical value of about 6 V.

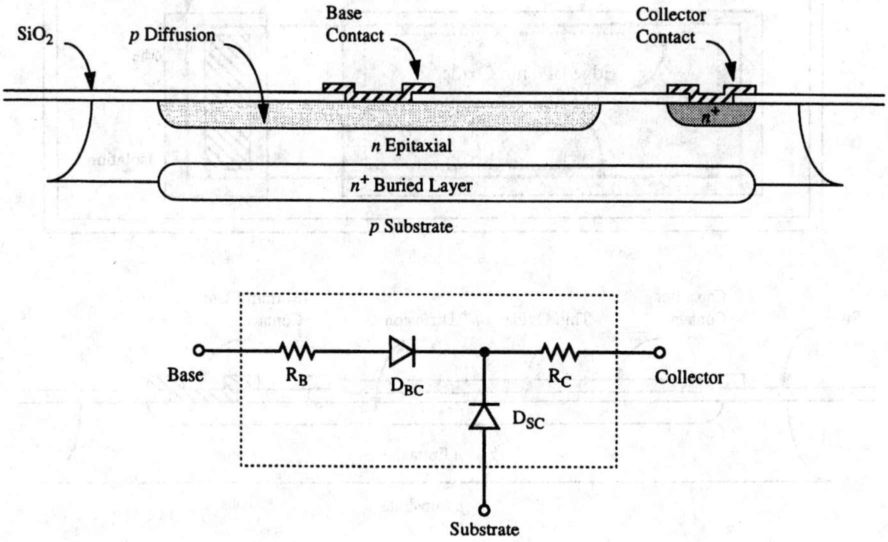

FIGURE 56.73 Base–collector junction capacitor.

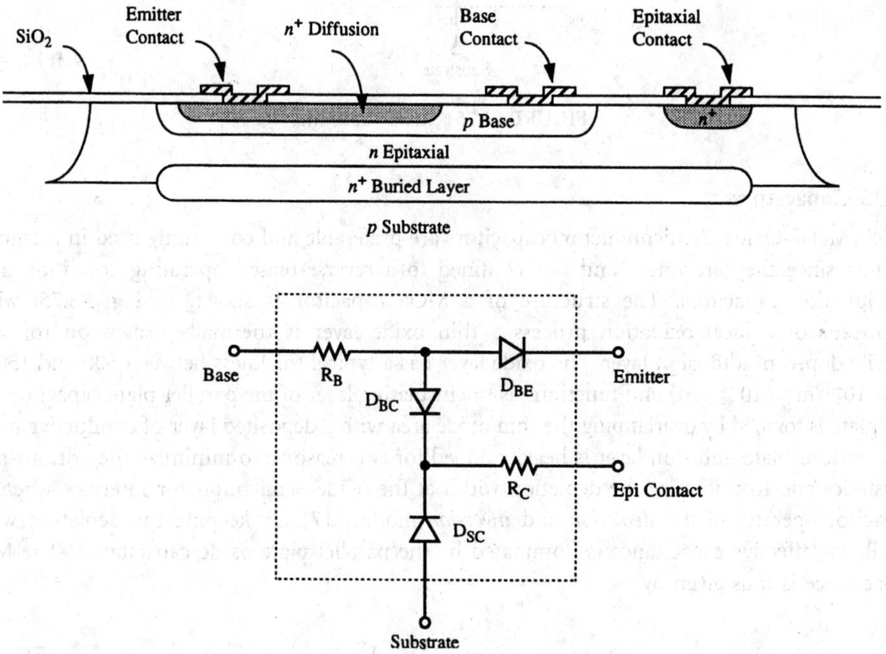

FIGURE 56.74 Base–emitter junction capacitor.

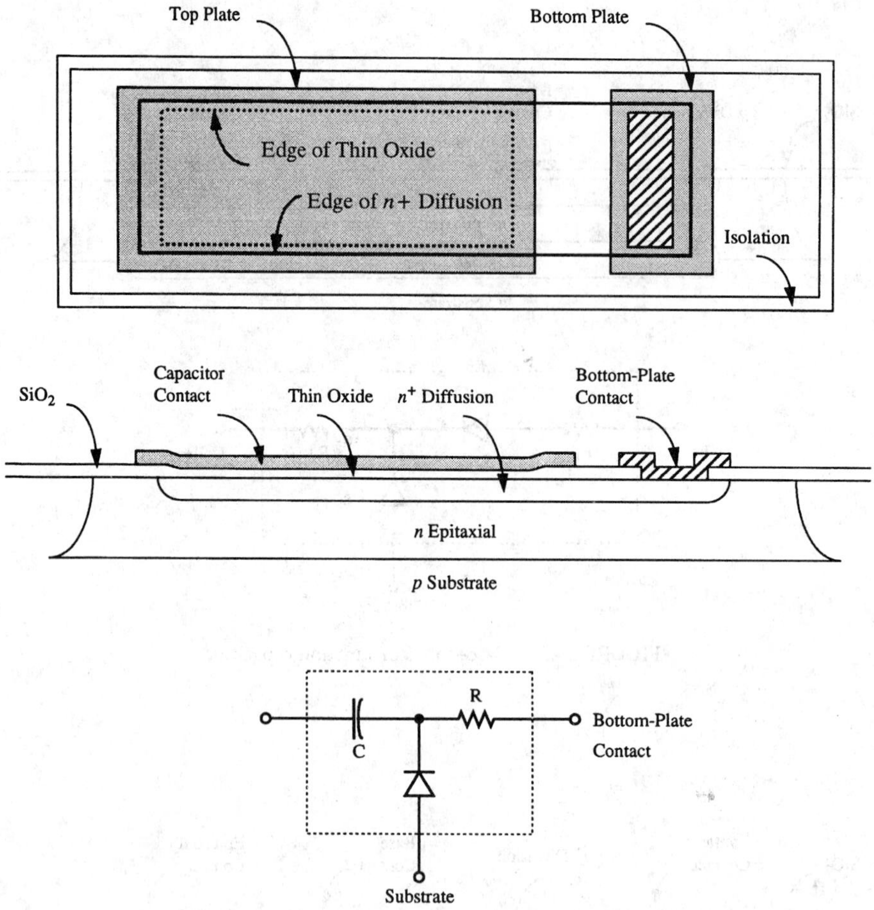

FIGURE 56.75 MOS capacitor.

MOS Capacitors

MOS (Metal–Oxide–Semiconductor) capacitors are preferable and commonly used in integrated circuits since they are linear and not confined to a reverse-biased operating condition as in the junction capacitors. The structure of a MOS capacitor is shown in Fig. 56.75, where by means of a local oxidation process a thin oxide layer is thermally grown on top of a heavily doped n^+ diffusion layer. The oxide layer has a typical thickness between 500 and 1500 Å (Å$= 10^{-10}$m $= 10^{-4}$ μm) and functions as the insulating layer of the parallel-plate capacitor. The top plate is formed by overlapping the thin oxide area with a deposited layer of conductive metal. The bottom-plate diffusion layer is heavily doped for two reasons: to minimize the bottom-plate resistance and to minimize the depletion width at the oxide–semiconductor interface when the capacitor operates in the *depletion* and *inversion* modes [17]. By keeping the depletion width small, the effective capacitance is dominated by the parallel-plate oxide capacitance. The MOS capacitance is thus given by

$$C = \frac{\kappa_{ox}\varepsilon_0}{T} A \qquad (56.164)$$

where κ_{ox} is the relative dielectric constant of SiO_2 (2.7 to 4.2), ε_0 is the permittivity constant, T is the oxide thickness, and A is the area defined by the thin oxide layer.

In practice a thin layer of silicon nitride (Si_3N_4) is often deposited on the thin oxide layer and is used to minimize the charges inadvertently introduced in the oxide layer during oxidation and subsequent processing steps. These *oxide charges* are trapped within the oxide and can cause detrimental effect to the capacitor characteristic [17]. The silicon nitride assimilates an additional insulating layer and effectively creates an additional capacitor in series with the oxide capacitor. The capacitance for such a structure can be determined by an application of *Gauss's law*. It is given by

$$C = \frac{\varepsilon_0}{\left(\frac{T_{ni}}{\kappa_{ni}}\right) + \left(\frac{T_{ox}}{\kappa_{ox}}\right)} A \tag{56.165}$$

where T_{ni} and T_{ox} are the thickness of the silicon nitride and oxide layers, respectively, and κ_{ni} (2.7 to 4.2) and κ_{ni} (3.5 to 9) are the relative dielectric constant of oxide and silicon nitride, respectively. In the equivalent circuit model of Fig. 56.75, the parasitic junction between the p-type substrate and the n-type bottom plate must always be reverse biased. The bottom-plate contact must be connected to a voltage relatively higher than the substrate voltage.

Polysilicon Capacitors

Polysilicon capacitors are conveniently available in MOSFET technology, where the gate of the MOSFET transistor is made of polysilicon material. Polysilicon capacitors also assimilate the parallel-plate capacitor. Figure 56.76 shows a typical structure of a polysilicon capacitor, where a thin oxide is deposited on top of a polysilicon layer and serves as an insulating layer between the top-plate metal layer and the bottom-plate polysilicon layer. The polysilicon region is isolated from the substrate by a thick oxide layer that forms a parasitic parallel-plate capacitance between the polysilicon layer and the substrate. This parasitic capacitance must be accounted for in the equivalent circuit model. The capacitance of the polysilicon capacitor is determined by either (56.164) or (56.165) depending on whether a thin silicon nitride is used in conjunction with the thin oxide.

Inductors

Planar inductors have been implemented using a variety of substrates such as standard PC boards, ceramic and sapphire hybrids, monolithic GaAs [24], and more recently monolithic silicon [18]. In the early development of silicon technology, planar inductors were investigated [26], but the prevailing lithographic limitations and relatively large inductance requirements (for low-frequency applications) resulted in excessive silicon area and poor performance. Reflected losses from the conductive silicon substrate were a major contribution to low inductor Q. Recent advances in silicon IC processing technology have achieved fabrication of metal width and metal spacing in the low micrometer range and thus allow many more inductor turns per unit area. Also, modern oxide-isolated processes with multilayer metal options allow thick oxides to help isolate the inductor from the silicon substrate. Practical applications of monolithic inductors in low-noise amplifiers, impedance matching amplifiers, filters and microwave oscillators in silicon technologies have been successfully demonstrated [19], [20].

Monolithic inductors are especially useful in high-frequency applications where inductors of a few nano-Henrys of inductance are sufficient. Inductor structures in monolithic form include strip, loop, and spiral inductors. Rectangular and circular spiral inductors are by far the most commonly used structures.

Rectangular Spiral Inductors

The structure of a rectangular spiral inductor is shown in Fig. 56.77, where the spiral loops are formed with the top metal layer M_2 and the connector bridge is formed with the bottom metal

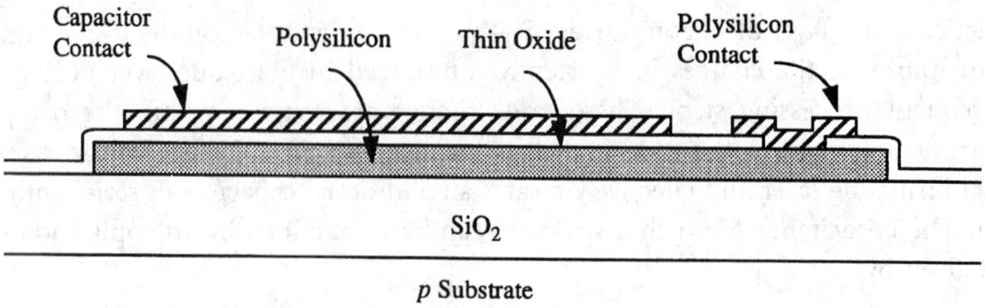

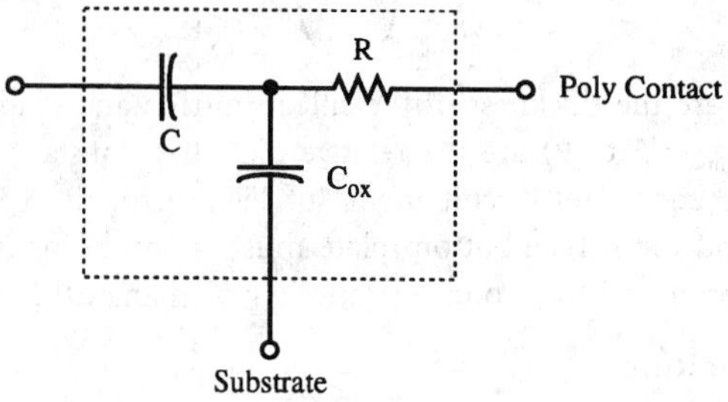

FIGURE 56.76 Polysilicon capacitor.

layer M_1. Using the top metal layer to form the spiral loops has the advantage of minimizing the parasitic metal-to-substrate capacitance. The metal width is denoted by W and the metal spacing is denoted by S. The total inductance is given by

$$L_T = \sum_{i=1}^{4N} L_S(i) + 2 \cdot \sum_{i=1}^{4N-1} \sum_{j=i+1}^{4N} L_M(ij) \tag{56.166}$$

where N is the number of turns, $L_S(i)$ is the *self inductance* of the rectangular metal segment i and $L_M(ij)$ is the *mutual inductance* between metal segments i and j. The self inductance is due to the magnetic flux surrounding each metal segment. The mutual inductance is due to the magnetic flux coupling around every two parallel metal segments and has a positive value if the currents applied to the metal conductors flow in the same direction and a negative value otherwise. Perpendicular metal segments have negligible mutual inductance.

The self inductance and mutual inductance for straight rectangular conductors can be determined by the *geometric mean distance* method [10], in which the conductors are replaced by equivalent straight filaments whose basic inductive characteristics are well known.

Self Inductance. The self inductance for the rectangular conductor of Fig. 56.78 depends on the conductor length L, the conductor width W, and the conductor thickness T. The static self inductance is given by [9], [10].

$$L_S = 2L\left[\ln\left(\frac{2L}{GMD}\right) - 1.25 \pm \left(\frac{AMD}{L}\right) + \left(\frac{\mu_r}{4}\right)\zeta\right] (\text{nH}) \tag{56.167}$$

where μ_r is the relative permeability constant of the conductor, **GMD** is the geometric mean distance, **AMD** is the arithmetic mean distance, and ζ is a frequency-dependent parameter that equals 1 for direct and low-frequency alternating currents and approaches 0 for very

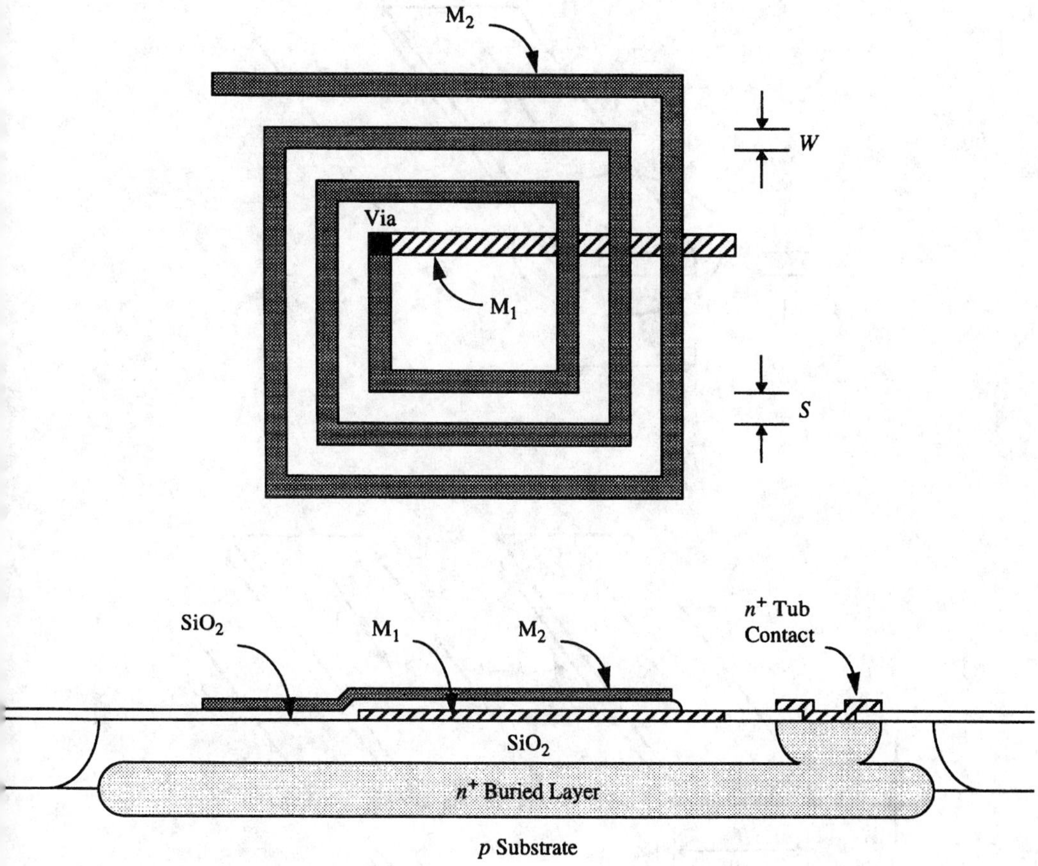

FIGURE 56.77 Rectangular spiral inductor.

high-frequency alternating currents. The *AMD* and *GMD* for the rectangular conductor of Fig. 56.78 are

$$AMD = \left(\frac{W+T}{3}\right)$$

$$GMD = \begin{cases} 0.22313 \cdot (W+T) & T \to 0 \\ 0.22360 \cdot (W+T) & T = W/2 \\ 0.223525 \cdot (W+T) & T \to W \end{cases} \tag{56.168}$$

The rectangular dimensions L, W, and T are normalized to the centimeter in the above expressions.

Mutual Inductance. The mutual inductance for the two parallel rectangular conductors of Fig. 56.78 depends on the conductor length L, the conductor width W, and the conductor thickness T, and the distance D separating the conductor centers. The static mutual inductance is [10]

$$L_M = 2L\alpha \text{ (nH)} \tag{56.169}$$

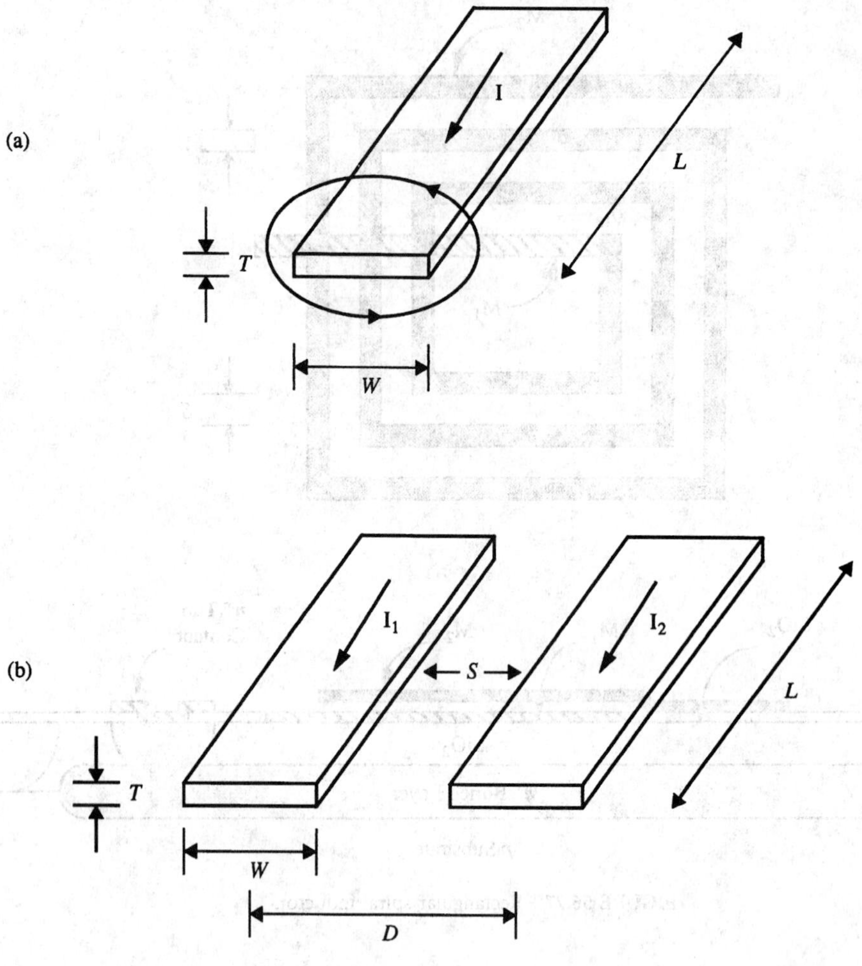

FIGURE 56.78 Calculation of (a) self inductance and (b) mutual inductance for parallel rectangular conductors.

where

$$\alpha = \ln\left[\left(\frac{L}{GMD}\right) + \left[1 + \left(\frac{L}{GMD}\right)^2\right]^{1/2}\right] - \left[1 + \left(\frac{GMD}{L}\right)^2\right]^{1/2} + \left(\frac{GMD}{L}\right)$$

and

$$GMD = \exp(\ln D - \beta) \tag{56.170}$$

$$\beta = \begin{cases} \frac{1}{12}\left(\frac{D}{W}\right)^{-2} + \frac{1}{60}\left(\frac{D}{W}\right)^{-4} + \frac{1}{168}\left(\frac{D}{W}\right)^{-6} + \frac{1}{360}\left(\frac{D}{W}\right)^{-8} \\ +\frac{1}{660}\left(\frac{D}{W}\right)^{-10} + \cdots \\ 0.1137 \quad \text{for} \quad D = W \end{cases}$$

The GMD closed-form expression (56.170) is valid for rectangular conductors with small thickness-to-width ratios T/W. As the thickness T approaches the width W, the *GMD* approaches the distance D and the *GMD* is no longer represented by the above closed-form expression. Figure 56.79 shows plots of the self inductance and the mutual inductance as expressed in (56.167) and (56.169), respectively. The conductor dimensions are given in μm (μm $= 10^{-4}$ cm).

For the inductor structure of Fig. 56.77 it is important to emphasize that since the spiral loops are of nonmagnetic metal material, the total inductance depends only on the geometry of the conductors and not on the current strength. At high-frequencies, especially those above the *self-resonant* frequency of the inductor, the *skin effect* due to current crowding toward the surface and the propagation delay as the current traverses the spiral must be fully accounted for [16], [22]. The ground-plane effect due to the inductor image must also be considered regardless of the operation frequency.

An equivalent lumped model for the rectangular spiral inductor of Fig. 56.77 is shown in Fig. 56.80. This model consists of the total inductance L_T, the accumulated metal resistance R_S, the coupling capacitance C_{CP} between metal segments due to the electric fields in both the oxide region and the air region, the parasitic capacitances C_{IN} and C_{OUT} from the metal layers to the buried layer [2], [11], [15], and the buried-layer resistance R_P. Since the spiral structure of Fig. 56.77 is not symmetrical, the parasitic capacitors C_{IN} and C_{OUT} are not the same, though the difference is relatively small. The self-resonant frequency can be approximated using the circuit model of Fig. 56.80 with one side of the inductor being grounded. For simplicity, let $C_{\text{IN}} = C_{\text{OUT}} \equiv C_P$ and neglect the relatively small coupling capacitor C_{CP}, the self-resonant frequency is given by

$$f_R = \frac{1}{2\pi}\frac{1}{\sqrt{L_T C_P}}\left[\frac{1 - R_S^2\left(\dfrac{C_P}{L_T}\right)}{1 - R_P^2\left(\dfrac{C_P}{L_T}\right)}\right]^{1/2} \tag{56.171}$$

Transformer Structures. Transformers are often used in high-performance analog integrated circuits that require conversions between single-ended signals and differential signals. In monolithic technology, transformers can be fabricated using the basic structure of the rectangular spiral inductor. Figure 56.81 shows a planar interdigitated spiral transformer that requires only two metal layers M_1 and M_2. The structure of Fig. 56.82, on the other hand, requires three layers of metal for which the top metal layer M_3 is used for the upper spiral, the middle metal layer M_2 is used for the lower spiral, and the bottom metal layer M_1 is used for the two connector bridges. This structure can achieve a higher inductance per unit than that of Fig. 56.81 due to a stronger magnetic coupling between the upper spiral and the lower spiral through a relatively thin oxide layer separating metal layers M_2 and M_3. An equivalent lumped model is shown in Fig. 56.83. In addition to all the circuit elements of the two individual spiral inductors, there are also a magnetic coupling factor k and a coupling capacitance C_C between the primary and secondary coils.

Circular Spiral Inductors

The structure of a concentric circular spiral inductor is shown in Fig. 56.84 where the circular loops share the same center point. The top metal layer M_2 is used for the circular conductors and the bottom metal layer M_1 is used for the connector bridge. The metal width is denoted by

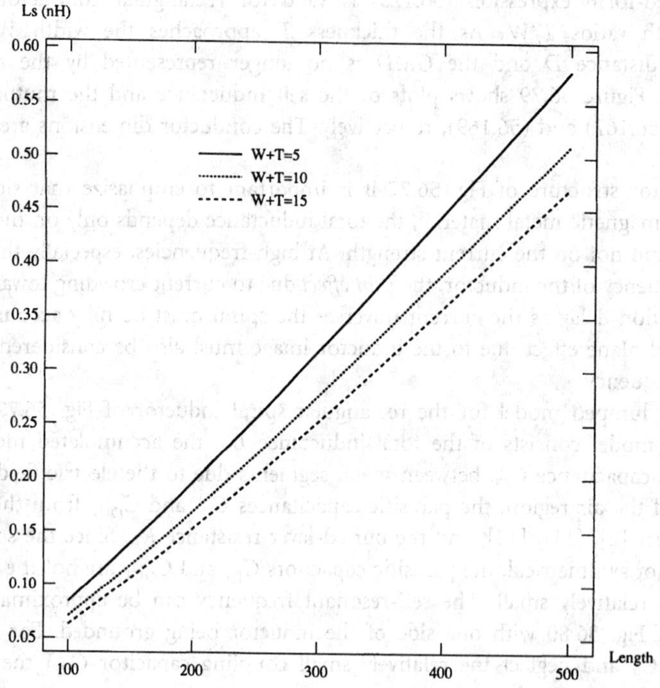

(a)

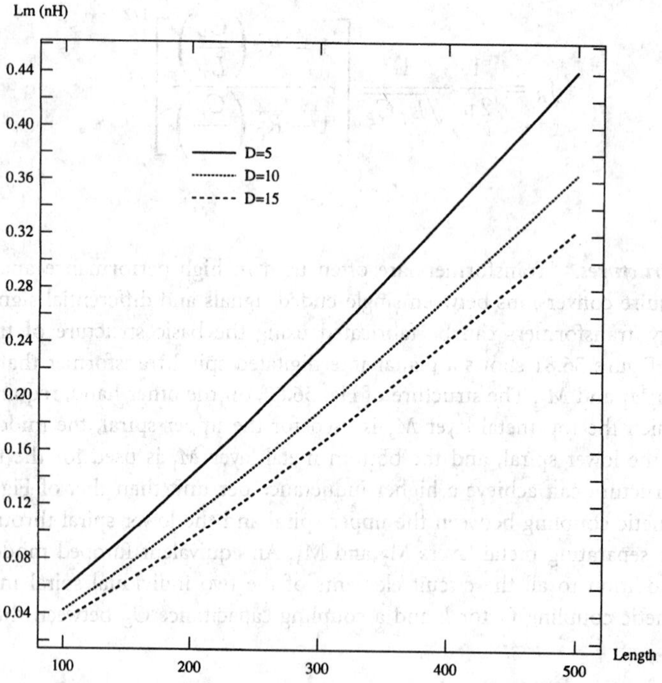

(b)

FIGURE 56.79 (a) Self inductance as a function of width, thickness and length for rectangular conductors. (b) Mutual inductance as a function of distance and length for rectangular conductors ($W = 5$, $T = 0$).

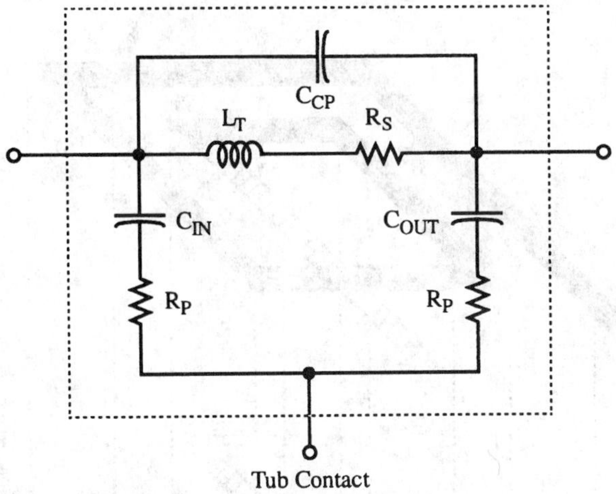

FIGURE 56.80 Electrical model for the spiral inductor.

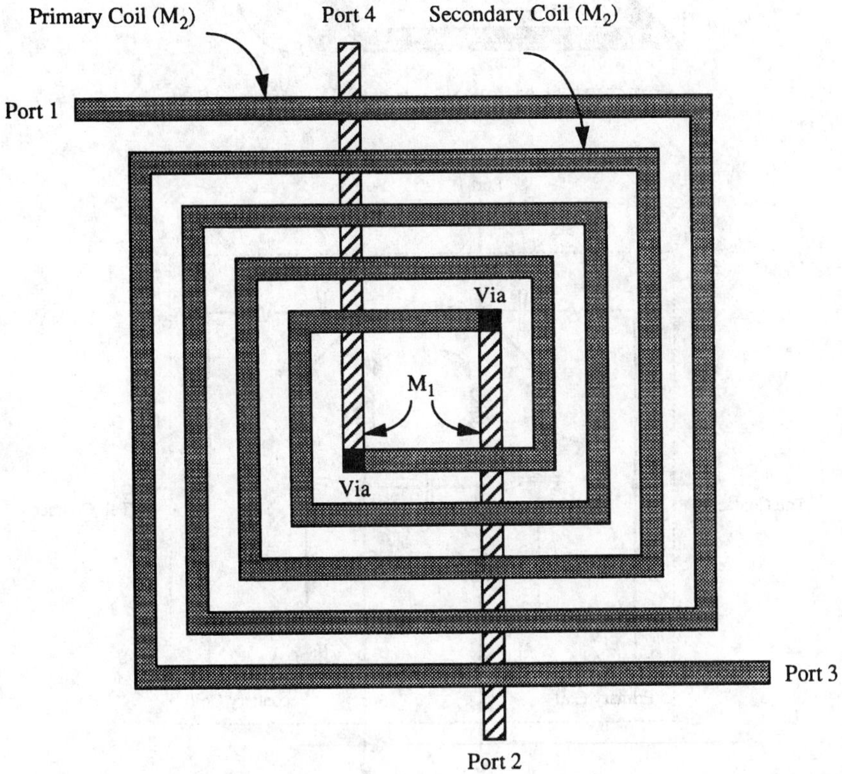

FIGURE 56.81 Rectangular spiral transformer I.

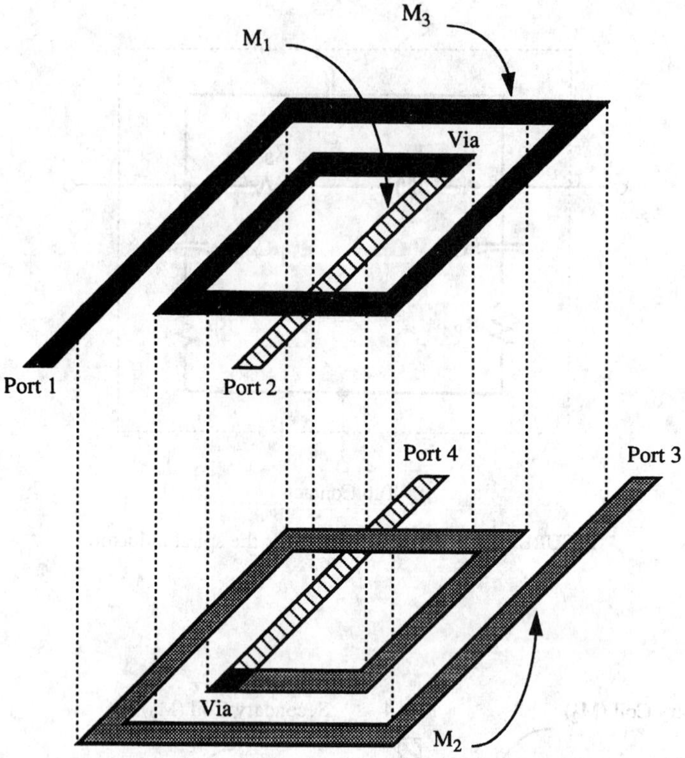

FIGURE 56.82 Rectangular spiral transformer II.

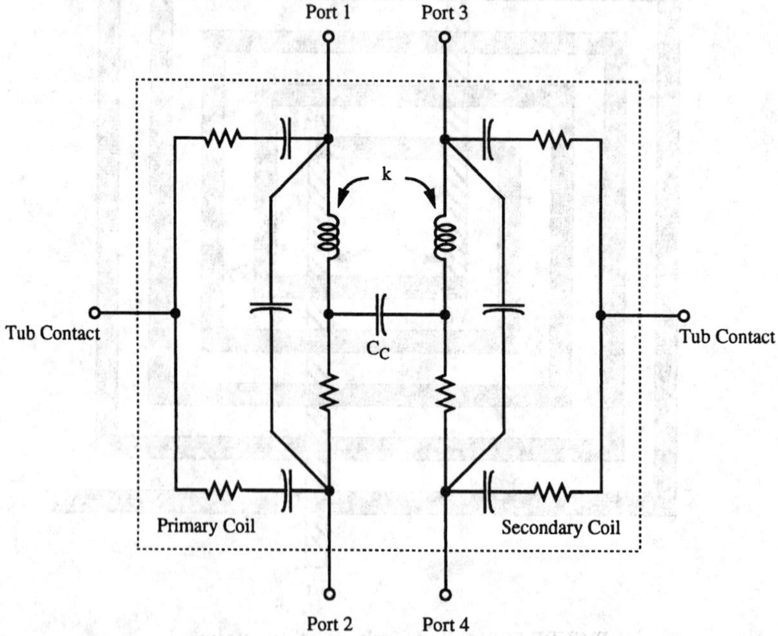

FIGURE 56.83 Electrical model for the spiral transformer.

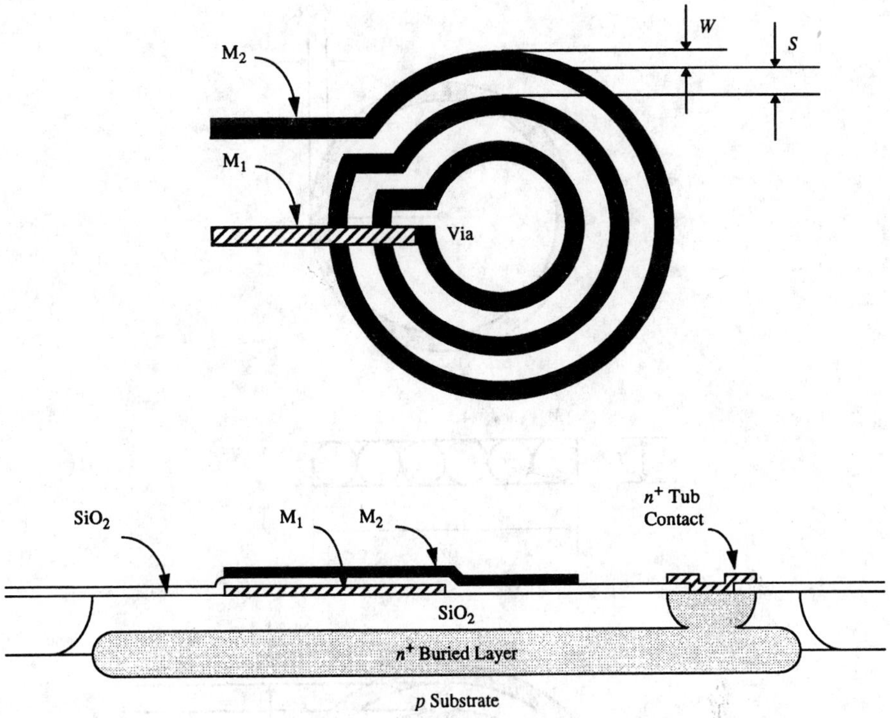

FIGURE 56.84 Concentric circular spiral inductor.

W and the spacing between two adjacent loops is denoted by S. The total inductance is given by

$$L_T = \sum_{i=1}^{N} L_S(i) + 2 \cdot \sum_{i=1}^{N-1} \sum_{j=i+1}^{N} L_M(ij) \tag{56.172}$$

where N is the number of circular turns, $L_S(i)$ is the *self inductance* of the circular conductor i and $L_M(ij)$ is the *mutual inductance* between conductors i and j.

Self Inductance. Consider the single circular conductor of Fig. 56.85(a) that has a radius R and a width W. A current I applied to this conductor produces a magnetic flux encircled by the loop and another magnetic flux inside the conductor itself. The inductance associated with the former and the latter magnetic flux component is referred to as the *external* self inductance and the *internal* self inductance, respectively. The external self inductance characterizing the change in the encircled magnetic flux to the change in current is [25].

$$L_S = \mu(2R - \delta)\left[\left(1 - \frac{k^2}{2}\right)K(k) - E(k)\right] \text{ (nH)} \tag{56.173}$$

where

$$k^2 = \frac{4R(R-\delta)}{(2R-\delta)^2} \tag{56.174}$$

and μ is the permeability of the conductor (equals 4π nH/cm for nonmagnetic conductors), and δ is one-half the conductor width W. $K(k)$ and $E(k)$ are the *complete elliptic integrals* of

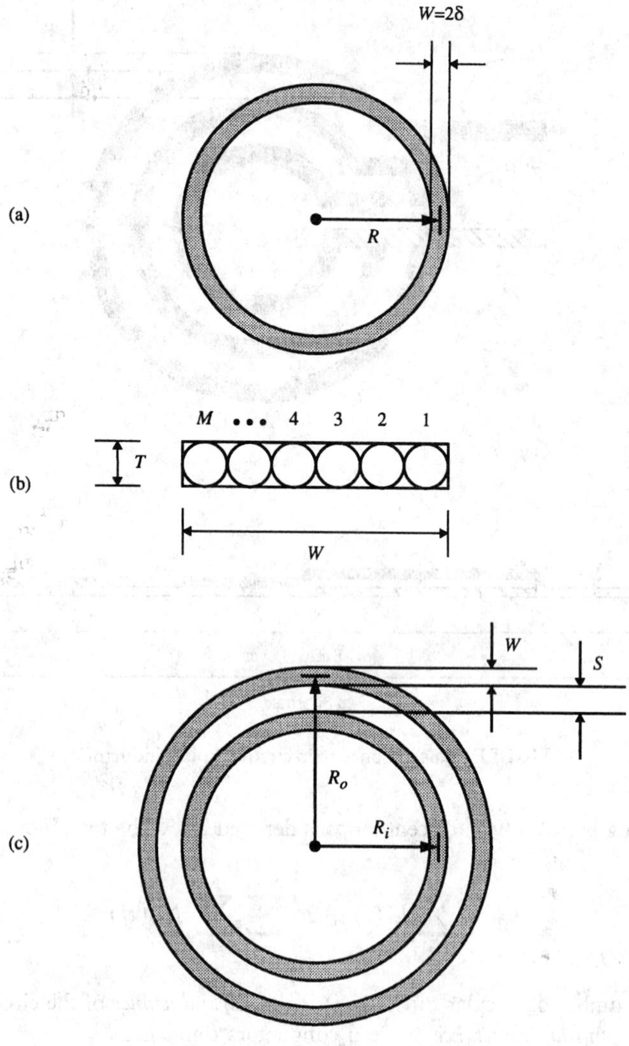

FIGURE 56.85 Calculation of self inductance and mutual inductance for circular conductors. (a) External self inductance; (b) internal self inductance; (c) mutual inductance.

the first and second kind, respectively, and are given by

$$K(k) = \int_0^{\pi/2} \frac{d\phi}{\sqrt{1 - k^2 \sin^2 \phi}}$$

$$E(k) = \int_0^{\pi/2} \sqrt{1 - k^2 \sin^2 \phi}\, d\phi$$

The internal self inductance is determined based on the concept of magnetic field energy. As shown in Fig. 56.85(b), the flat circular conductor is first approximated by an M number of round circular conductors [14] that are electrically in parallel and each conductor has a diameter equal to the thickness T of the flat conductor. The internal self inductance of each

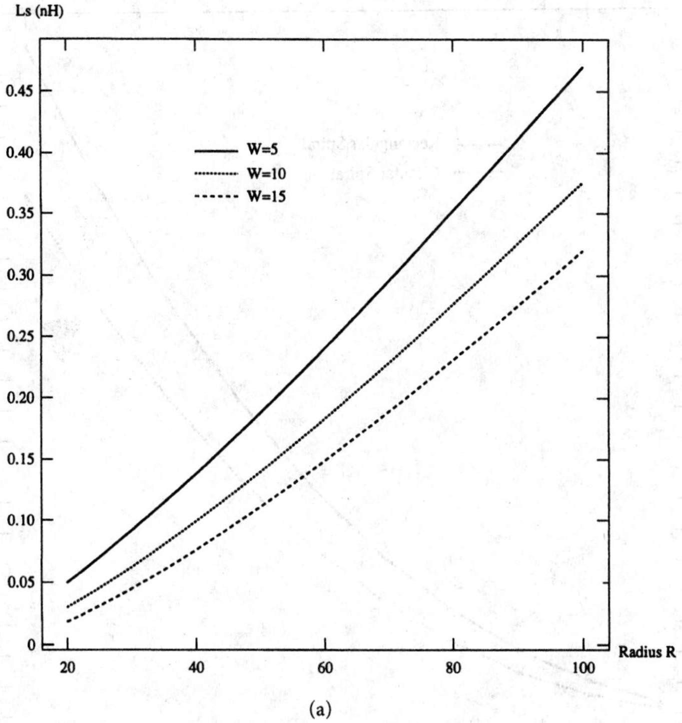

(a)

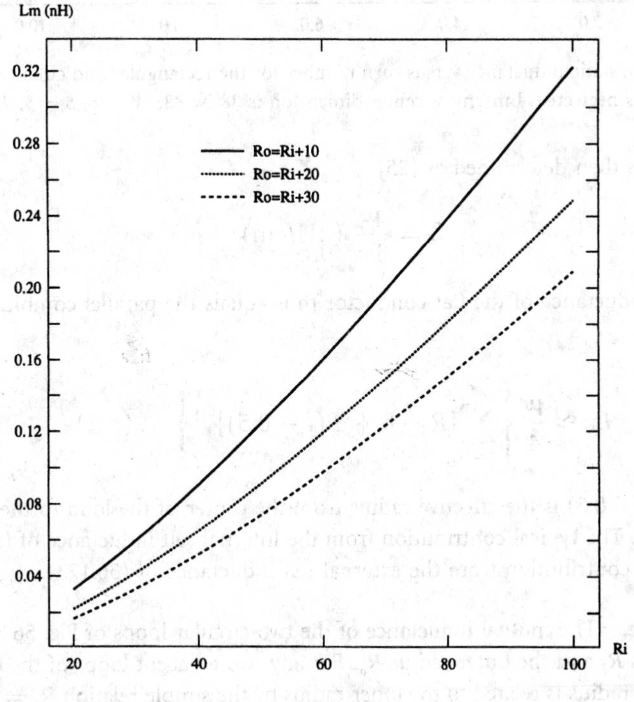

(b)

FIGURE 56.86 (a) External self inductance as a function of radius and width for circular conductors. (b) Mutual inductance as a function of radii R_i and R_o for circular conductors.

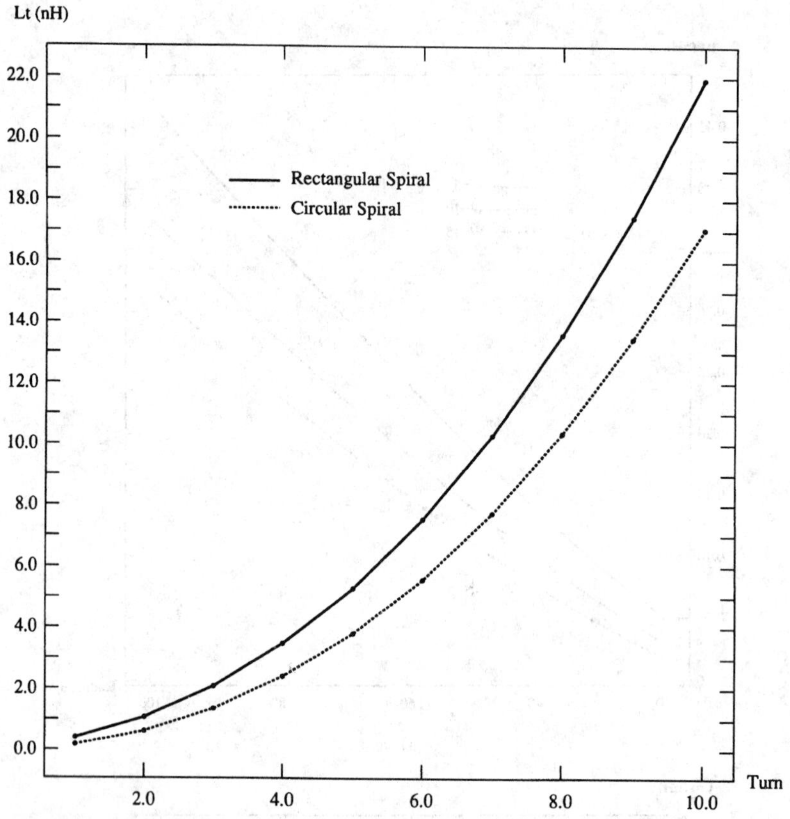

FIGURE 56.87 Total static inductance versus turn number for the rectangular and circular inductors. The ground-plane effect is neglected. Innermost center dimension is 88 by 88, $W = 6$, $S = 3$, $T = 1.2$.

round conductor is then determined as [25]

$$L = \frac{\mu}{8\pi} \text{ (nH/cm)}$$

The internal self inductance of the flat conductor thus equals the parallel combination of these M components

$$L_S \approx \frac{\mu}{4} \left\{ \sum_{i=1}^{M} [R - \delta + T(i - 0.5)]^{-1} \right\}^{-1} \text{ (nH)} \tag{56.175}$$

where $R - \delta + T(i - 0.5)$ is the effective radius from the center of the loop to the center of the round conductor i. The typical contribution from the internal self inductance of (56.175) is less than 5 percent the contribution from the external self inductance of (56.173).

Mutual Inductance. The mutual inductance of the two circular loops of Fig. 56.85(c) depends on the inner radius R_i and the outer radius R_o. For any two adjacent loops of the circular spiral inductor, the outer radius is related to the inner radius by the simple relation $R_o = R_i + (W + S)$. The mutual inductance is determined based on the *Neumann's* line integral given as follows:

$$L_M = \frac{\mu}{4\pi} \int_C \int_C \frac{d\mathbf{l}_1 \cdot d\mathbf{l}_2}{D}$$

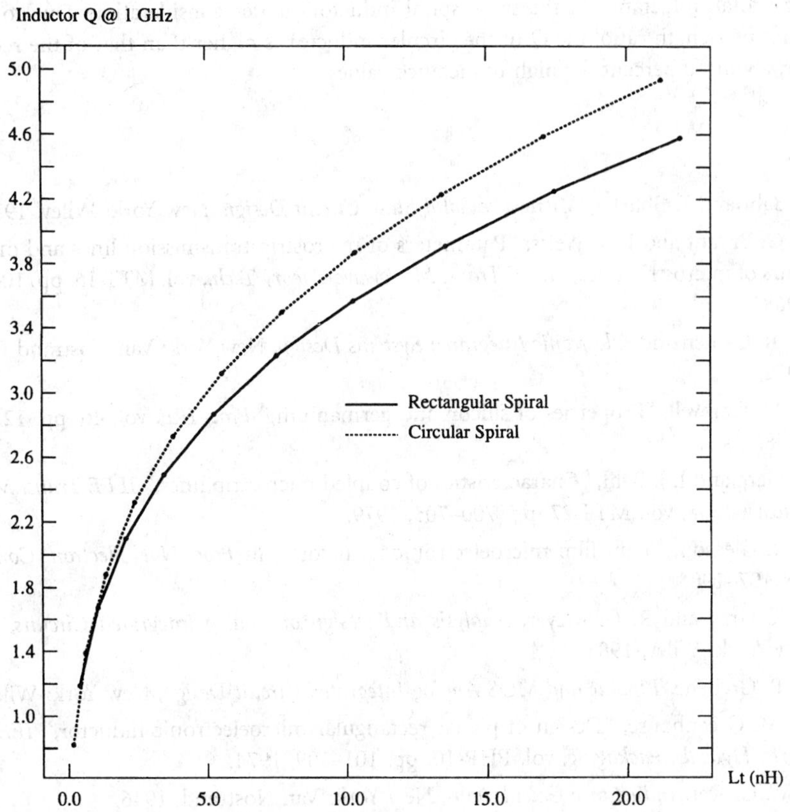

FIGURE 56.88 Inductor Q versus total inductance for the rectangular and circular inductors. Metal sheet resistance is 25 m$\Omega/\square$. Innermost dimension is 88 by 88, $W = 6$, $S = 3$, $T = 1.2$.

where $d\mathbf{l}_1 \cdot d\mathbf{l}_2$ represents the dot product of the differential lengths and D is the distance separating the differential l_1 vector and l_2 vector. The static mutual inductance [25] is

$$L_M = \mu\sqrt{R_i R_o}\left[\left(\frac{2}{k} - k\right)K(k) - \frac{2}{k}E(k)\right] \text{ (nH)} \tag{56.176}$$

where

$$k^2 = \frac{4R_i R_o}{(R_i + R_o)^2} \tag{56.177}$$

Fig. 56.86 shows plots of the external self inductance and the mutual inductance as expressed in (56.173) and (56.176), respectively. The conductor dimensions are given in μm.

As in the rectangular spiral inductor, the ground-plane effect and the retardation effect of the circular spiral inductor must be fully accounted for. The circuit model of Fig. 56.80 can be used to characterize the electrical behavior of the circular inductor.

A comparison between the rectangular spiral of Fig. 56.77 and the circular spiral of Fig. 56.84 is shown in Fig. 56.87, where the total inductance L_T is plotted against the turn number N. Both inductors have the same innermost dimension, the same conductor width, space, and thickness. The dimensions are given in μm, and the ground-plane effect and the retardation effect are not considered. For a given turn number, the rectangular spiral yields a higher inductance per semiconductor area than the circular spiral. Figure 56.88 shows a plot of the inductor Q

versus the total inductance of the same spiral inductors under consideration. Due to a higher inductance per length ratio, the Q of the circular inductor is higher than that of the rectangular inductor, about 10 percent for high inductance values.

References

[1] I. Bahl and P. Bhartia, *Microwave Solid State Circuit Design*, New York: Wiley, 1988.

[2] T. G. Bryant and J. A. Weiss, "Parameters of microstrip transmission lines and of coupled pairs of microstrip lines," *IEEE Trans. Microwave Theory Tech.*, vol. MTT-16, pp. 1021–1027, 1968.

[3] H. R. Camenzind, *Electronic Integrated Systems Design*, New York: Van Nostrand Reinhold, 1972.

[4] E. M. Conwell, "Properties of silicon and germanium," *Proc. IRE*, vol. 46, pp. 1281–1300, 1958.

[5] R. Garg and I. J. Bahl, "Characteristics of coupled microstrip lines," *IEEE Trans. Microwave Theory Tech.*, vol. MTT-27, pp. 700–705, 1979.

[6] F. R. Gleason, "Thin-film microelectronic inductors," in *Proc. Nat. Electron. Conf.*, 1964, pp. 197–198.

[7] P. R. Gray and R. G. Meyer, *Analysis and Design of Analog Integrated Circuits*, 2nd ed., New York: Wiley, 1984.

[8] A. B. Grebene, *Bipolar and MOS Analog Integrated Circuit Design*, New York: Wiley, 1984.

[9] H. M. Greenhouse, "Design of planar rectangular microelectronic inductor," *IEEE Trans. Parts, Hybrids, Packaging*, vol. PHP-10, pp. 101–109, 1974.

[10] F. W. Grover, *Inductance Calculations*, New York: Van Nostrand, 1946.

[11] E. Hammerstad and O. Jensen, "Accurate models for microstrip computer-aided design," *IEEE MTT-S Dig.*, pp. 407–409, 1980.

[12] J. C. Irwin, "Resistivity of bulk silicon and of diffused layers in silicon," *Bell Syst. Tech. J.*, vol. 41, pp. 387–410, 1962.

[13] C. Jacoboni, C. Canali, G. Ottaviani, and A. A. Quaranta, "A review of some charge transport properties of silicon," *Solid State Electron.*, vol. 20, 1977.

[14] R. L. Kemke and G. A. Burdick, "Spiral inductors for hybrid and microwave applications," in *Proc. Electron. Components Conf.*, 1974, pp. 152–161.

[15] M. Kirschning and R. H. Jansen, "Accurate wide-range design equations for the frequency-dependent characteristics of parallel-coupling microstrip lines," *IEEE Trans. Microwave Theory Tech.*, vol. MTT-32, pp. 83–90, 1984.

[16] D. Krafcsik and D. Dawson, "A close-form expression for representing the distributed nature of the spiral inductor," *IEEE MTT-S Dig.*, pp. 87–92, 1986.

[17] R. S. Muller and T. I. Kamins, *Device Electronics for Integrated Circuits*, 2nd ed., New York: Wiley, 1986.

[18] N. M. Nguyen and R. G. Meyer, "Si IC-compatible inductors and LC passive filters," *IEEE J. Solid-State Circuits*, vol. 25, pp. 1028–1031, 1990.

[19] ——, "A Si bipolar monolithic RF bandpass amplifier," *IEEE J. Solid-State Circuits*, vol. 27, pp. 123–127, 1992.

[20] ——, "A 1.8-GHz monolithic LC voltage-controlled oscillator," *IEEE J. Solid-State Circuits*, vol. 27, pp. 444–450, 1992.

[21] ——, "Start-up and frequency stability in high-frequency oscillators," *IEEE J. Solid-State Circuits*, vol. 27. pp. 810–820, 1992.

[22] M. Parisot, Y. Archambault, D. Pavlidis, and J. Magarshack, "Highly accurate design of spiral inductors for MMIC's with small size and high cut-off frequency characteristics," *IEEE MTT-S Digest*, pp. 106–110, 1984.

[23] E. Pettenpaul, H. Kapusta, A. Weisgerber, H. Mampe, J. Luginsland, and I. Wolff, "CAD models of lumped elements on GaAs up to 18 GHz," *IEEE Trans. Microwave Theory Tech.*, vol. 36, pp. 294–304, 1988.

[24] R. A. Pucel, "Design considerations for monolithic microwave circuits," *IEEE Trans. Microwave Theory Tech.*, vol. MTT-29. 513–534, 1981.

[25] S. Ramon, J. R. Whinnery, and T. V. Duzer, *Fields and Waves in Communication Electronics*, 2nd ed., New York: Wiley, 1984.

[26] R. M. Warner, Jr., and J. N. Fordemwalt, *Integrated Circuits*, New York: McGraw-Hill, 1965.

56.5 Chip Parasitics in Analog Integrated Circuits

Martin A. Brooke

The parasitic elements in electronic devices and interconnect limit the performance of all integrated circuits. No amount of improvement in device performance or circuit design can completely eliminate these effects. Thus, as circuit speeds increase, unaccounted for interconnect parasitics become a more and more common cause of analog integrated circuit design failure. Hence the causes, characterization, and modeling of significant interconnect parasitics are essential knowledge for good analog integrated circuit design [1]–[4].

Interconnect Parasitics

The parasitics due to the wiring used to connect devices together on chip produce a host of problems. Unanticipated feedback through parasitic capacitances can cause unwanted oscillation. Mismatch due to differences in interconnect resistance contribute to unwanted offset voltages. For very-high-speed integrated circuits, the inductance of interconnects is both a useful tool and a potential cause of yield problems.

Even the interactions between interconnect lines are both important and very difficult to model. So too are the distributed interactions of resistance, capacitance, and (in high-speed circuits) inductance that produce transmission line effects.

Parasitic Capacitance

Distributed Capacitance of integrated circuit lines is perhaps the most important of all integrated circuit parasitics. It can lower the bandwidth of amplifiers, alter the frequency response of filters, and cause oscillations.

Physics. Every piece of integrated circuit interconnect has capacitance to the substrate. In the case of silicon circuitry, the substrate is conductive and connected to an ac ground, thus there is a capacitance to ground from every circuit node due to the interconnect. Figure 56.89 illustrates this substrate capacitance interconnect parasitic. The capacitance value will depend on the total area of the interconnect, and on the length of edge associated with the interconnect. This edge effect is due to the nonuniformity of the electric field at the interconnect edges. The nonuniformity of the electric field at edges is such that the capacitance value is larger for a given area of interconnect near the edge than elsewhere.

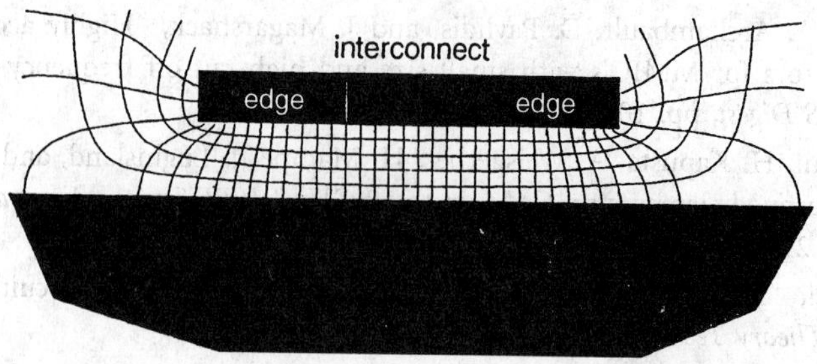

FIGURE 56.89 Substrate capacitance. The electric field distorts at the edges, making the capacitance larger there than elsewhere.

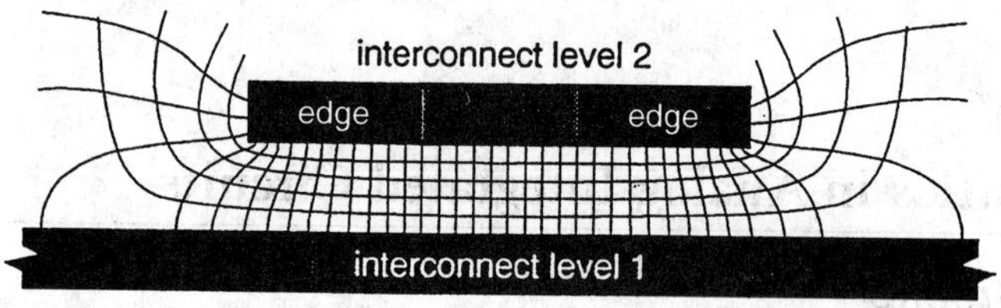

FIGURE 56.90 Overlap capacitance. The bottom interconnect level will have edges into and out of the page with distorted electric field similar to that shown for the top level of interconnect.

In addition to the substrate capacitance, all adjacent pieces of an interconnect will have capacitance between them. This capacitance is classified into two forms, overlap capacitance, and parallel line capacitance (also known as proximity capacitance). Overlap capacitance occurs when two pieces of interconnect cross each other, while parallel line capacitance occurs when two interconnect traces run close to each other for some distance.

When two lines cross each other, the properties of the overlapping region will determine that size of the overlap capacitance. The electric field through a cross section of two overlapping lines is illustrated in Fig. 56.90. The electric field becomes nonuniform near the edges of the overlapping region, producing an edge-dependent capacitance term. The capacitance per unit area at the edge is always greater than elsewhere and, if the overlapping regions are small, the edge capacitance effect can be significant.

The size of parallel line capacitance depends on the distance for which the two lines run side by side and on the separation of the lines. Since parallel line capacitance occurs only at the edges of an interconnect, the electric field that produces it is very nonuniform. This particular nonuniformity, as illustrated in Fig. 56.91, makes the capacitance much smaller for a given area of interconnect than either overlap or substrate capacitance. Thus two lines must run parallel for some distance for this capacitance to be important. The nonuniformity of the electric field makes the dependence of the capacitance on line separation highly nonlinear, as a result the capacitance value decreases much more rapidly with separation than it would if it depended linearly on the line separation.

Modeling. In the absence of significant interconnect resistance effects, all of the parasitic capacitances can be modeled with enough accuracy for most analog circuit design applications by dissecting the interconnect into pieces with similar capacitance characteristics and adding up the capacitance of each piece to obtain a single capacitance term. For example, the dissected view

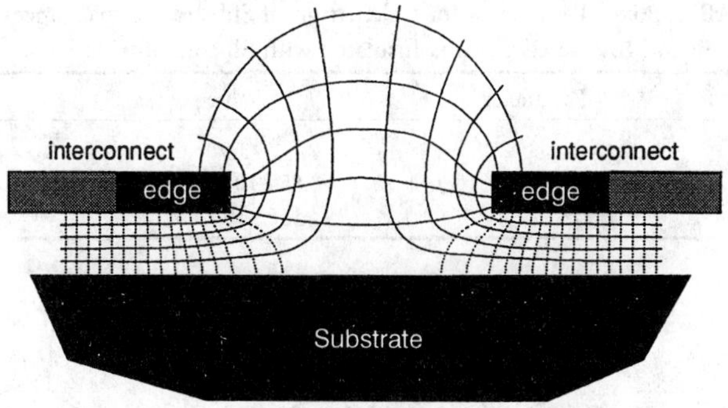

FIGURE 56.91 Parallel line capacitance. Only the solid field lines actually produce line-to-line capacitance, the dashed lines form substrate capacitance.

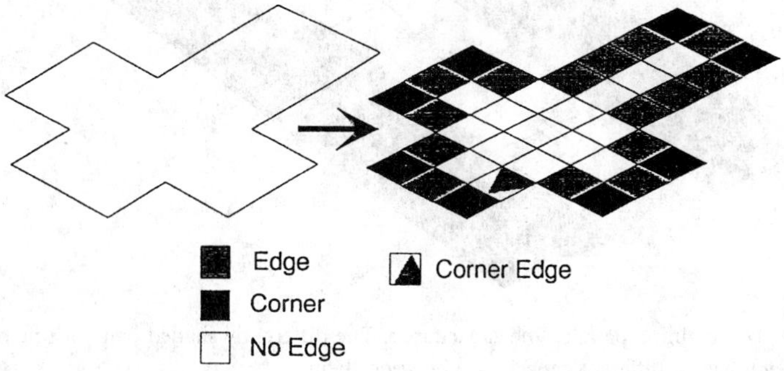

FIGURE 56.92 Determining substrate capacitance. The capacitance of each square in the dissected interconnect segment is summed.

of a piece of interconnect with substrate capacitance is shown in Fig. 56.92. The interconnect has been dissected into squares that fall into three classes: two types of edges, and one center type. The capacitance to the substrate for each of these squares in parallel and thus the total capacitance of the interconnect segment is simply the sum of the capacitance of each square. If the substrate capacitance contribution of each square has been previously measured or calculated, the calculation of the total interconnect segment substrate capacitance involves summing each type of squares capacitance multiplied by the number of squares of that type in the segment.

The accuracy of this modeling technique depends solely on the accuracy of the models used for each type of square. For example, in Fig. 56.92, the accuracy could be improved by adding one more type of edge square to those that are modeled. One of these squares has been shaded differently in the figure and is called the corner edge square.

For the nonedge pieces of the interconnect the capacitance is approximately a parallel-plate capacitance and can be computed from (56.178).

$$C = \frac{A \cdot \varepsilon_r \cdot \varepsilon_0}{t} \tag{56.178}$$

In (56.178), A is the area of the square or piece of interconnect, t is the thickness of the insulation layer beneath the interconnect, ε_r is the relative dielectric constant of the insulation material, and ε_0 is the dielectric constant of free space. For silicon integrated circuits insulated with silicon dioxide the parameters are given in Table 56.8.

The capacitance of edge interconnect pieces will always be larger than nonedge pieces. The amount by which the edge capacitance increases will depend on the ratio of the size of the

TABLE 56.8 Parameters for Calculation of Substrate Capacitance in Silicon Integrated Circuits Insulated with Silicon Dioxide

Parameter	Value
ε_r	3.9
ε_0	$8.854 \cdot 10^{-12}$ F/m
t	$1\text{–}5 \cdot 10^{-6}$ m

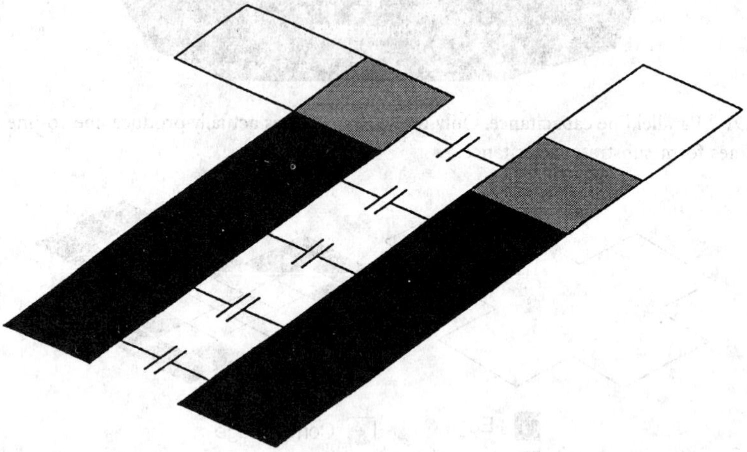

FIGURE 56.93 Determining parallel line capacitance. The differently shaded pairs of squares are different types and will each have a different capacitance between them.

piece of interconnect and the thickness of the insulation layer beneath the interconnect. If the interconnect width is significantly larger than the thickness of the insulation then edge effects are probably small and can be ignored. However, when thin lines are used in integrated circuits the edge effects are usually significant. The factor by which the edge capacitance can increase over the parallel-plate approximation can easily be as high as 1.5 for thin lines.

The modeling of overlap capacitance is handled in the same fashion as substrate capacitance. The region where interconnect lines overlap is dissected into edges and nonedges and the value of capacitance for each type of square summed up to give a total capacitance between the two circuit nodes associated with each piece of interconnect that overlaps. The area of overlap between the two layers of interconnect can be used as A in (56.178), while that separation between the layers can be used as t. The strong distortion of the electric fields will increase the actual value above this idealized computed value by a factor that depends on the thickness of the lines. This factor can be as high as 2 for thin lines.

Parallel line capacitance can also be handled in a manner similar to that used for substrate and overlap capacitance. However, we must now locate *pairs* of edge squares, one from each of the adjacent interconnect lines. In Fig. 56.93 one possible pairing of the squares from adjacent pieces of interconnect is shown. The capacitance for each type of pair of squares is added together, weighted by the number of pairs of each type to get a single capacitance that connects the circuit nodes associated with each interconnect line.

The effect on the capacitance of the spacing between pairs must be either measured or computed for each possible spacing, and type of pair of squares. One approach to this is to use a table of measured or computed capacitances and separation distances. The measured parallel line capacitance between silicon integrated circuit lines for a variety of separations is presented in Fig. 56.94. From the figure we see that the capacitance value decreases exponentially with line

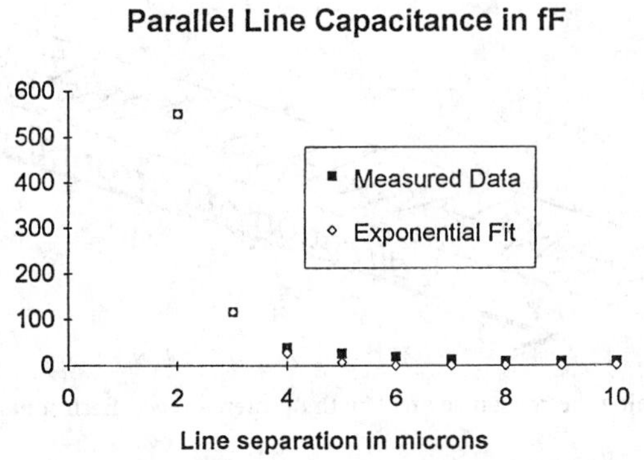

FIGURE 56.94 Parallel line capacitance measured from a silicon integrated circuit. The diamonds are an exponental fit to the data [using (56.179)]. The fit is excellent at short separations when the capacitance is largest.

separation. Thus an exponential fit to measured or simulated data is good choice for computing the capacitance [7], [8].

Equation (56.179) can be used to predict the parallel line capacitance C for each type of pair of edge squares. L is the length of the edge of the squares, and the parameters C_c and S_d are computed or fit to measured coupling capacitance data like that in Fig. 56.94.

$$C = C_c \cdot L \cdot e^{-(s/S_d)} \tag{56.179}$$

Effects on Circuits. The effects that parasitic capacitances are likely to produce in circuits range from parametric variations, such as reduced bandwidth, to catastrophic failures, such as amplifier oscillation. Each type of parasitic capacitance produces a characteristic set of problems. Being aware of these typical problems will ease diagnosis of actual, or potential, parasitic capacitance problems.

Substrate capacitance usually causes lower than expected bandwidth in amplifiers and lowering of the poles in filters. The capacitance is always to ac ground and thus increases device and circuit capacitances to ground. Thus, circuit nodes that have a dominant effect on amplifier bandwidth, or filter poles, should be designed to have as little substrate capacitance as possible. Another, more subtle, parametric variation that can be caused by substrate capacitance is frequency-dependent mismatch. For example, if the parasitic capacitance to ground is different between the two inputs of a differential amplifier, then, for fast transient signals, the amplifier will appear unbalanced. This could limit the accuracy high-speed comparators, and is sometimes difficult to diagnose since the error only occurs at high speeds.

Overlap and parallel line capacitance can cause unwanted ac connections to be added to a circuit. These connections will produce crosstalk effects and can result in unstable amplifiers. The output interconnect and input interconnect of high-gain or high-frequency amplifiers must thus be kept far apart at all times. Care must be taken to watch for series capacitances of this type. For example, if the output and input interconnect of an amplifier both cross the power supply interconnect, unwanted feedback can result if the power supply line is not well ac grounded. This is a very common cause of integrated circuit amplifier oscillation. Because of the potential for crosstalk between parallel or crossing lines, great care should also be taken to keep weak (high-impedance) signal lines away from strong (low-impedance) signal lines.

Parasitic Resistance

For analog integrated circuit designers, the second most important interconnect parasitic is resistance. This unexpected resistance can cause both parametric problems, such as increased

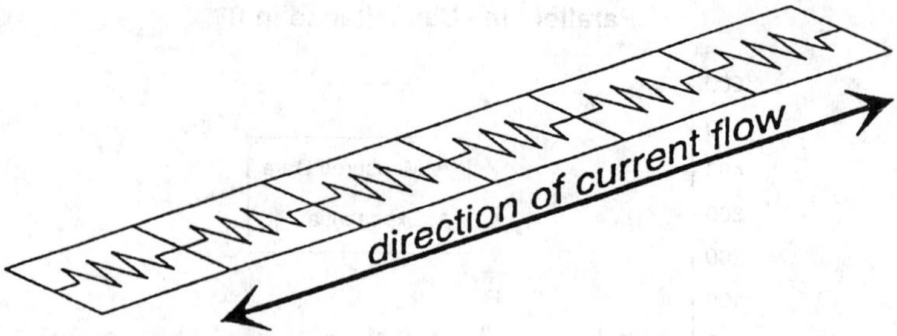

FIGURE 56.95 Determining the resistance of a length of interconnect. Each square has the same resistance regardless of size.

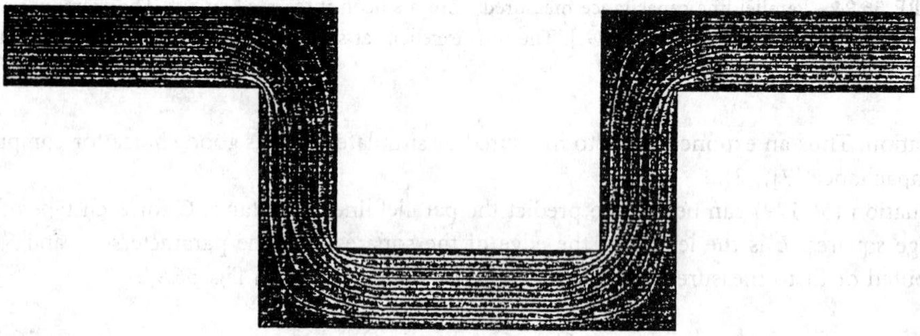

FIGURE 56.96 Current flow in a complex interconnect geometry.

offset voltages, and catastrophic problems such as amplifier oscillation (for example, poorly sized power supply lines can cause resistive positive feedback paths in high gain amplifiers called "ground loops"). To make matters worse, the resistivity of integrated circuit interconnect has been steadily increasing as the line widths of circuits have decreased.

Physics. Except for superconductors, all conductors have resistance. A length of interconnect used in an integrated circuit is no exception. The resistance of a straight section of interconnect is easily found by obtaining the resistance per square for the particular interconnect layer concerned, and then adding up the resistance of each of the series of squares that makes up the section. This procedure is illustrated in Fig. 56.95.

For more complicated interconnect shapes the problem of determining the resistance between two points in the interconnect is also more complex. The simplest approach is to cut the interconnect up into rectangles and assume each rectangle has a resistance equal to the resistance per square of the interconnect material times the number of full and partial squares that will fit along the direction of current flow in the rectangle [5]. This scheme works whenever the direction of current flow is clear; however, for corners and intersections of interconnect the current flow is in fact quite complex. Figure 56.96 shows the kind of current flow that can occur in an interconnect section with complex geometry.

Modeling. To account for the effects of complex current flows the resistance of complex interconnect geometries must be determined by measurement or simulation. One simple empirical approach is to cut out sections of resistive material in the same shape as the interconnect shape to be modeled, and then measure the resistance. The resistance for other materials can be found by multiplying by the ratio of the respective resistances per square of the two materials.

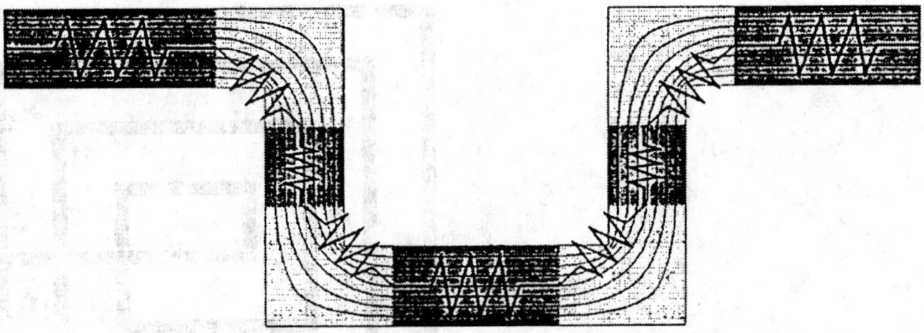

FIGURE 56.97 The process of breaking a complex geometry into subgeometries, constructing the equivalent connected resistance, and forming a single resistance for an interconnect section. In this example only two subgeometries are used: a corner subgeometry and a basic rectangular subgeometry.

Once the resistance has been found for a particular geometry it can be used for any linear scaling of that geometry. For most types of integrated circuit interconnect all complex geometries can be broken up into relatively few important subgeometries. If tables of the resistance of these subgeometries for various dimensions and connection patterns are obtained, the resistance of quite complex shapes can be accurately calculated by connecting the resistance of each subgeometry together and calculating the resistance of the connected resistances. This calculation can usually be performed quickly by replacing series and parallel connected resistor pairs with their equivalents. The process of breaking a complex geometry into subgeometries, constructing the equivalent connected resistance, and forming a single resistance for an interconnect section is illustrated in Fig. 56.97.

Effects on Circuits. The resistance of interconnect can have both parametric and catastrophic effects on circuit performance. Even small differences in the resistance on either input side of a differential amplifier can lead to increased offset voltage. Thus, when designing differential circuits care must be taken to make the interconnect identical on both input sides, as this ensures that the same resistance is present in both circuits.

The resistance of power supply interconnect can lead to both parametric errors in the voltages supplied and catastrophic failure due to oscillation. If power supply voltages are assumed to be identical in two parts of a circuit and, due to interconnect resistance, there is a voltage drop from one point to the next, designs that rely on the voltages being the same may fail. In high-gain and feedback circuits the resistance of the ground and power supply lines may become an unintentional positive feedback resistance which could lead to oscillation. Thus output and input stages for high-gain amplifiers will usually require separate ground and power supply interconnects. This ensures that no parasitic resistance is in a feedback path.

When using resistors provided in an integrated circuit process, the extra resistance provided by the interconnect may cause inaccuracies in resistor values. This would be most critical for small resistance values. The only solution in this case is to accurately compute the interconnect resistance. Since most resistance layers provided in analog integrated circuit processes are just a form of high resistivity interconnect, the methods described here for accurately computing the resistance of interconnect are also useful for predicting the resistance of resistors to be fabricated.

Parasitic Inductance

In high-speed integrated circuits the inductance of long lines of interconnect becomes significant. In integrated circuit technologies that have an insulating substrate, such as gallium arsenide (GaAs) and silicon on insulator (SOI), reasonably high-performance inductive devices can be made from interconnect. In technologies with conductive substrates, resistive losses in the

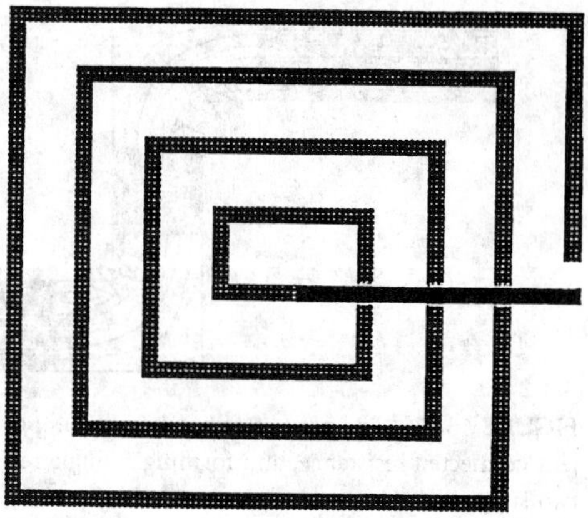

FIGURE 56.98 Spiral inductance used in insulated substrate integrated circuits for gigahertz frequency operation.

substrate restrict the application of interconnect inductance. High-frequency circuits are often tuned using interconnect inductance and capacitance (*LC*) to form a narrow bandpass filter or tank circuit, and *LC* transmission lines, or stubs, made from interconnect are useful for impedance matching. There is much similarity between this use of parasitic inductance and the design of microstripline-printed circuit boards. The major difference being that inductance does not become significant in integrated circuit interconnect until frequencies in the gigahertz are reached.

In order to make a good interconnect inductance, there are two requirements. First, there must not be any resistive material within range of the magnetic field of the inductance. If this occurs then induced currents flowing in the resistive material will make the inductor have high series resistance (low Q factor). This would make narrow bandwidth bandpass filters difficult to make using the inductance, and make transmission lines made from the interconnect lossy. The solution is to have an insulating substrate, or to remove the substrate from beneath the inductor.

The second requirement for large inductance is to form a coil or other device to concentrate the magnetic field lines. Within the confines of current integrated circuit manufacturing, spiral inductors, like that illustrated in Fig. 56.98 are the most common method used to obtain useful inductances.

Transmission Line Behavior

There are two types of transmission line behavior that are important in integrated circuits, *RC* transmission lines and *LC*/*RLC* transmission lines. For gigahertz operation inductive transmission lines are important. These can be lossy *RLC* transmission lines if a conductive substrate such as silicon is used, or nearly lossless *LC* transmission lines if an insulating substrate such as gallium arsenide is used. The design of inductive transmission lines is very similar to designing microstripline-printed circuit boards. At lower frequencies of 10 to 1000 MHz resistive capacitive (*RC*) transmission lines are important for long low resistivity interconnect lines or short high resistivity lines.

RC transmission lines are of concern to analog circuit designers working in silicon integrated circuits. When used correctly, an interconnect can behave as though it were purely capacitive in nature. However, when a higher resistivity interconnect layer, such as polysilicon or diffusion is used, the distributed resistance and capacitance can start to produce transmission line effects at relatively short distances. Similarly, for very long signal distribution lines or power supply lines, if they are not correctly sized, transmission line behavior ensues.

Physics. One method for modeling distributed transmission line interconnect effects is lumped equivalent modeling [6]. This method is useful for obtaining approximate models of complex

TABLE 56.9 The Maximum Length of Minimum Width Polysilicon Line that can be Modeled with a Single Lumped *RC T* or Π Network and Remain 10 percent Accurate

Frequency (MHz)	Length (micrometers)
10	1262
100	399
1000	126

geometries quickly, and is the basis of accurate numerical finite element simulation techniques. For analog circuit designers the conversion of interconnect layout sections into lumped equivalent models also provides an intuitive tool to understanding distributed transmission line interconnect behavior.

To be able to model a length of interconnect as a lumped *RC* equivalent, the error between the impedance of the interconnect when correctly treated as a transmission line, and when replaced with the lumped equivalent, must be kept low. If this error is e, then it can be shown that the maximum length of interconnect that can be modeled as a simple *RC T* or Π network is given in (56.180). In the equation R is the resistance per square of the particular type of interconnect used, C is the capacitance per unit area, and ω is the frequency of operation in radians per second.

$$D \leq \sqrt{\frac{3 \cdot e}{\omega \cdot R \cdot C}} \tag{56.180}$$

This length can be quite short. Consider the case of a polysilicon interconnect line in a 1.2 μm CMOS process that has a resistance per square of 40 Ω, a capacitance per unit are of 0.1 fF/μm^2. For an error e of 10 percent the maximum line length of minimum width line that can be treated as a lumped T or Π network for various frequencies is given in Table 56.9. Longer interconnect lines than this must be cut up into lengths less than or equal to the length given by (56.180).

Modeling. The accurate modeling of distributed transmission line effects in integrated circuits is best performed with lumped equivalent circuits. These circuits can be accurately extracted by dissecting the interconnect geometry into lengths that are, at most, as long as the length given by (56.180). These lengths are then modeled by either a T or Π lumped equivalent *RC* network. The extraction of the resistance and capacitance for these short interconnect sections can now follow the same procedures as were described in the previous sections on parasitic interconnect resistance and capacitance. The resulting *RC* network is then an accurate transmission line model of the interconnect. Fig. 56.99 shows an example of this process.

Effects on Circuits. Several parametric and catastrophic problems can arise due to unmodeled transmission line behavior. Signal propagation delays in transmission lines are longer than predicted by a single lumped capacitance and resistance model of interconnect. Thus ignoring the effects of transmission lines can result in slower circuits than expected. If the design of resistors for feedback networks results in long lengths of the resistive interconnect used to make the resistors, these resistors may in fact be *RC* transmission lines. The extra delay produced by the transmission line may well cause oscillation of the feedback loops using these resistors. The need for decoupling capacitors in digital and analog circuit power supplies is due to the *RC* transmission line behavior of the power supply interconnect. Correct modeling of the *RC*

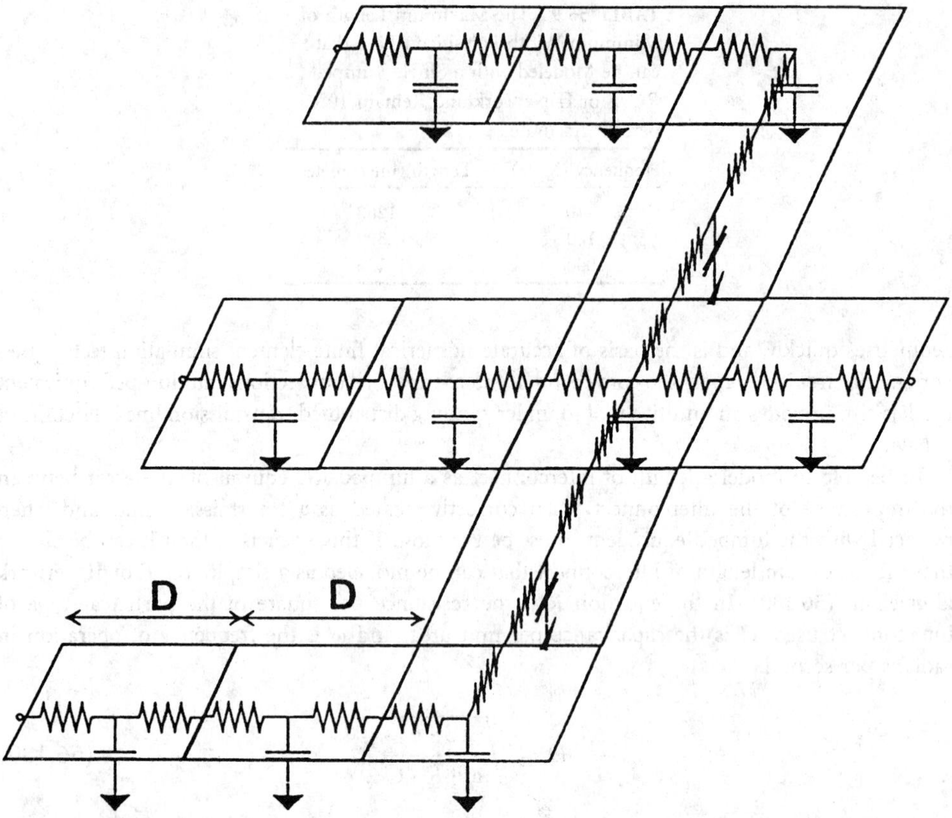

FIGURE 56.99 The extraction of an accurate *RC* transmission line model for resistive interconnect. The maximum allowable length *D* is computed from (56.180).

properties of the power distribution interconnect is needed to see whether fast power supply current surges will cause serious changes in the supply voltage or not.

Nonlinear Interconnect Parasitics

There are a number of types of interconnect that can have nonlinear parasitics. These nonlinear effects are a challenge to model accurately because the effect can change with the operating conditions of the circuit. A conservative approach is to model the effects as constant at the worst likely value they can attain. This is adequate for predicting parameters, like circuit bandwidth, that need only exceed a specification value. If the specifications call for accurate prediction of parasitics then large nonlinear parasitics are generally undesirable and should be avoided.

Most nonlinear interconnect parasitics are associated with depletion or inversion of the semiconductor substrate. A diffusion interconnect is insulated from conducting substrates such as silicon by a reversed biased diode. This diode's depletion region width varies with the interconnect voltage and results in a voltage-dependent capacitance to the substrate. For example, the diffusion interconnect in Fig. 56.100 has voltage dependent capacitance to the substrate due to a depletion region. The capacitance value depends on the depletion region thickness, which depends on the voltage difference between the interconnect and the substrate.

$$C = C_0 \cdot \left(1 - \left(\frac{V_S}{\phi_B}\right)\right)^M \qquad (56.181)$$

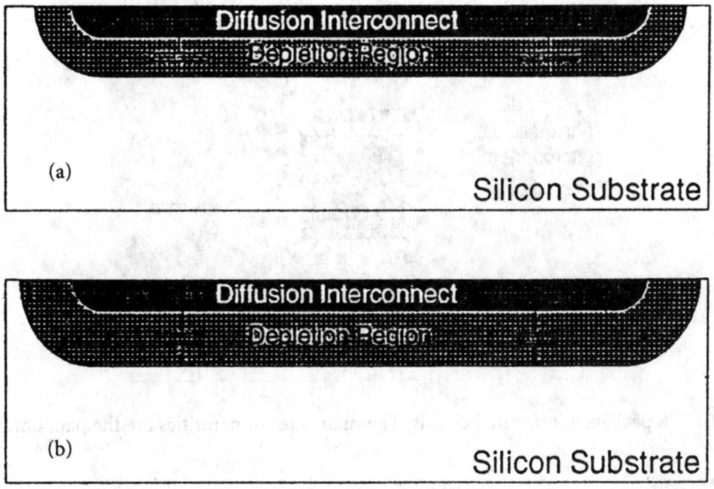

FIGURE 56.100 Diffusion interconnect has a voltage-dependent capacitance produced by the depletion region between the interconnect and the substrate. At low voltage difference between the interconnect and substrate (a) the capacitance is large. However, the capacitance decreases for larger voltage differences (b).

The typical equation for depletion capacitance is given in (56.181). In this equation V_S is the voltage from the interconnect to the substrate, ϕ_B is the built-in potential of the semiconductor junction, M is the grading coefficient of the junction, while C_0 is the zero-bias capacitance of the junction. Since the capacitance is less than C_0 for reverse bias and the junction would not insulate for forward bias, we can assume that the capacitance is always less than C_0 and use C_0 as a conservative estimate of C. Because of the uncertainty in the exact structure of most semiconductor junctions ϕ_B and M are usually fit to measured capacitance versus voltage (CV) data.

Another common nonlinear parasitic occurs when metal interconnect placed over a conducting semiconductor substrate creates inversions at the semiconductor surface. This inversion layer increases the substrate capacitance of the interconnect and is voltage dependent. To prevent this most silicon-integrated circuit manufacturers place an inversion-preventing implant on the surface of the substrate. The depletion between the substrate and n-type or p-type wells diffused into the substrate also creates a voltage-dependent capacitance. Thus use of the well as a high resistivity interconnect for making high value resistors will require consideration of a nonlinear capacitance to the substrate.

Pad and Packaging Parasitics

All signals and supply voltages that exit an integrated circuit must travel across the packaging interconnections. Just like the on-chip interconnect, the packaging interconnect has parasitic resistance, capacitance, and inductance. However, some of the packaging materials are significantly different in properties and dimension to those used in the integrated circuit, thus there are major differences in the importance of the various types of parasitics. Fig. 56.101 shows a typical packaged integrated circuit. The chief components of the packaging are the pads on the chip, the wire or bump bond used to connect the pad to the package, and then the package interconnect.

The pads used to attach wire bonds or bump bonds to integrated circuits are often the largest features on an integrated circuit. The typical pad is 100 μm on a side and has a capacitance of 100 fF. In addition, there are often protection diodes used on pads that will add a small nonlinear component to the pad capacitance.

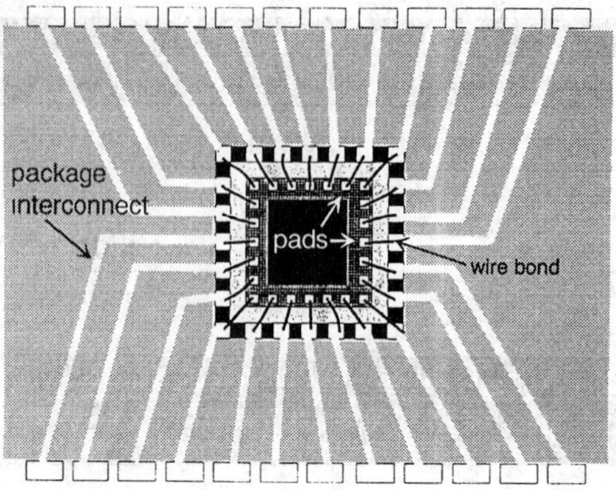

FIGURE 56.101 A packaged integrated circuit. The main sites of parasitics are the pad, bond, and package interconnect.

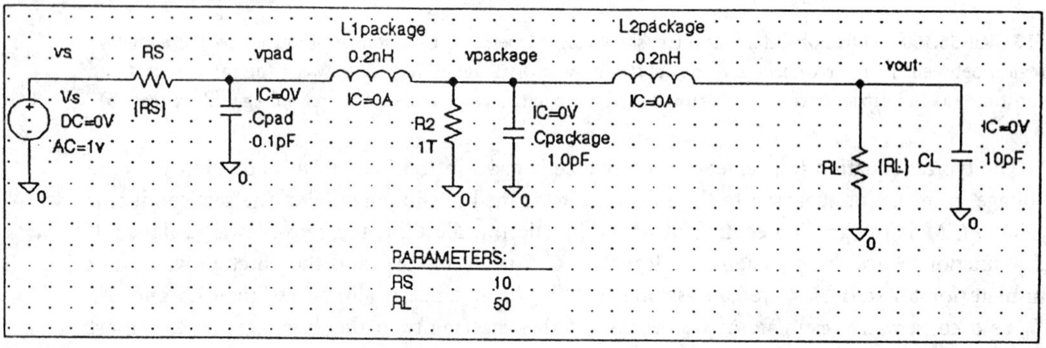

FIGURE 56.102 The circuit model of a high-frequency package output and associated parasitics. Cpad is the pad capacitance. L1package, Cpackage, R2, and L2package model the package interconnect. *RS* is the source resistance of the circuit. *RL* and *CL* are the external load.

The wire bonds that attach the pads to the package are typically very low resistivity and have negligible capacitance. Their major contribution to package parasitics is inductance. Typically the package interconnect inductance is greater than the wire bond inductance; however, when wire bonds are used to connect two integrated circuits directly together, then the wire bond inductance is significant.

Often the dominant component of package parasitics comes from the packaging interconnect itself. Depending on the package, there is inductance, capacitance to ground, and parallel line capacitance produced by this interconnect. Carefully made high-frequency packages do not exhibit much parallel line capacitance (at the expense of much capacitance to ground due to shielding), but in low-frequency packages with many connections this can become a problem.

Typical inductance and capacitance values for a high-speed package capable of output bandwidths around 5 GHz are incorporated into a circuit model for the package parasitics in Fig. 56.102. When simulated with a variety of circuit source resistances (*RS*) this circuit reaches maximum bandwidth without peaking when the output resistance is 4 Ω. At lower output resistance, Fig. 56.103 shows that considerable peaking in the output frequency response occurs.

Parasitic Measurement

The major concern when measuring parasitics is to extract the individual parasitic values independently from measured data. This is normally achieved by exaggerating the effect that

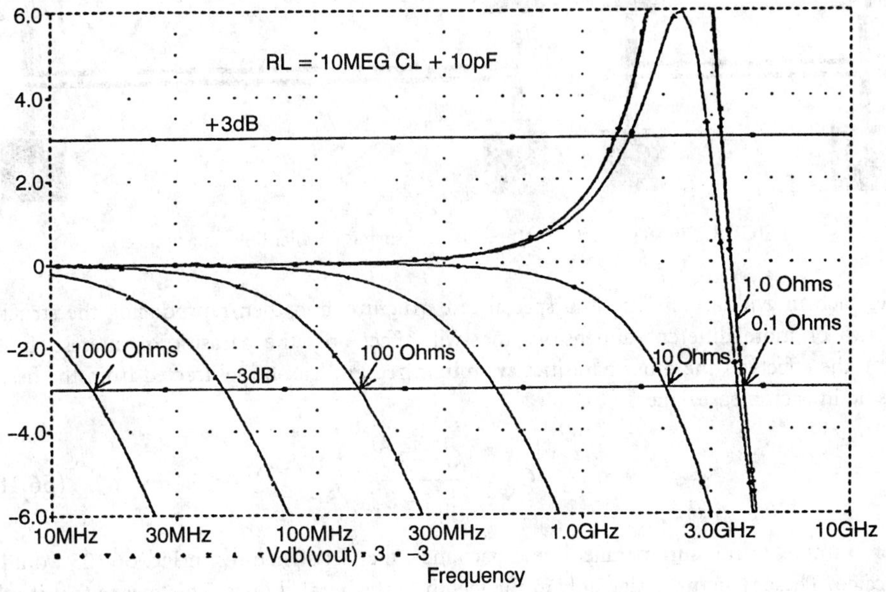

FIGURE 56.103 The PSPICE™ ac simulation of circuit in Fig. 56.102 when the load resistance *RL* is 10 MΩ. This shows how the package inductance causes peaking for sufficiently low output resistance. In this case peaking occurs for *RS* below 4 Ω and at about 2 GHz.

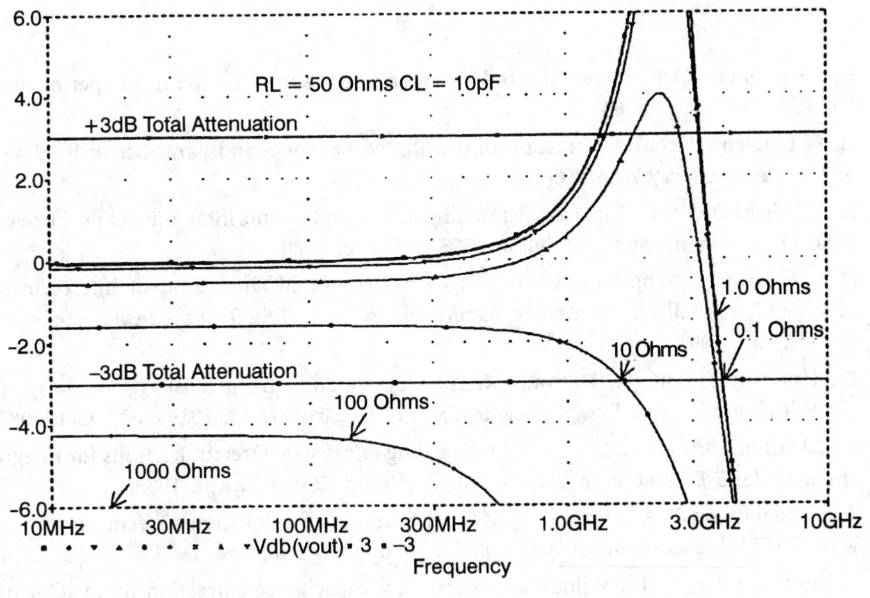

FIGURE 56.104 The PSPICE™ ac simulation of circuit in Fig. 56.102 when the load is 50 Ω. The package inductance still causes peaking for *RS* below 4 Ω.

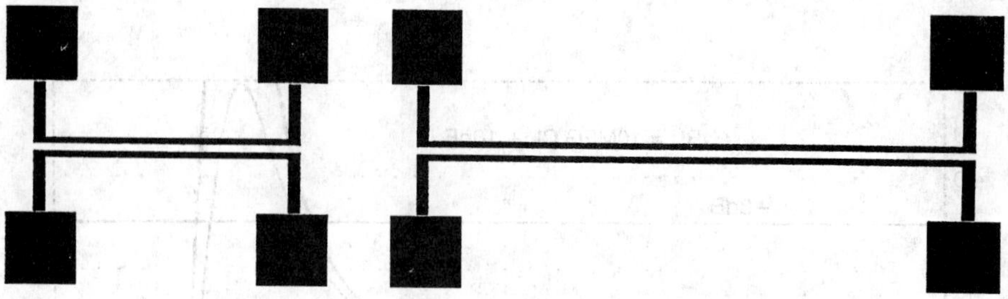

FIGURE 56.105 Test structures for measuring parallel line capacitance.

causes each individual parasitic in a special test structure, and then reproducing the structure with two or more different dimensions that will affect only the parasitic of interest. In this fashion the effects of the other parasitics are minimized and can be subtracted from the desired parasitic in each measurement.

$$C_P = \frac{C_1 - C_2}{L_1 - L_2} \tag{56.182}$$

For example, to measure parallel line capacitance, the test structures in Fig. 56.105 would be fabricated. These structures vary only in the length of the parallel lines. This means that if other parasitic capacitance ends up between the two signal lines used to measure the parasitic, then it will be a constant capacitance that can be subtracted from both measurements. The parallel line capacitance will vary in proportion to the variation of length between the two test structures. Thus the parallel line capacitance per unit length can be found from (56.182). In this equation C_P is the parallel line capacitance per unit length, C_1 and C_2 are the capacitances measured from each test structure, and L_1 and L_2 are the length of the two parallel interconnect segments.

References

[1] D. L. Carter and D. F. Guise, "Effects of interconnections on submicron chip performance," *VLSI DESIGN*, pp. 63–68, 1984.

[2] H. B. Lunden, "Detailed extraction of distributed networks and parasitics in IC designs," in *Proc. Euro. Conf. Circuit Theory, Design*, 1989, 84–88.

[3] R. A. Sainati and T. J. Moravec, "Estimating high speed circuit interconnect performance," *IEEE Trans. Circuits Syst.*, vol. 36, pp. 533–541, Apr. 1989.

[4] D. S. Gao, A. T. Yang, and S. M. Kang, "Modeling and simulation of interconnection delays and crosstalks in high-speed integrated circuits," *IEEE Trans. Circuits Syst.*, vol. 37, pp. 1–8, Jan. 1990.

[5] M. Horowitz and R. W. Dutton, "Resistance extraction from mask layout data," *IEEE Trans. Computer-Aided Design Integrat. Circuits Syst.*. vol. 7, pp. 1029–1037, Oct. 1988.

[6] R. J. Antinone and G. W. Brown, "The modeling of resistive interconnections for integrated circuits," *IEEE J. Solid-State Circuits*, vol. SC-18, pp. 200–203, Apr. 1983.

[7] A. E. Ruehli and P. A. Brennan, "Capacitance models for integrated circuit, metalization wires," *IEEE J. Solid-State Circuits*, vol, SC-10, pp. 530–536, Dec. 1975.

[8] S. Mori, I. Suwa, and J. Wilmore, "Hierarchical capacitance extraction in an IC artwork verification system," in *Proc. IEEE Int. Conf. Computer-Aided Design*, 1984, pp. 266–268.

57

Analog Circuit Cells

. V. Noren
niversity of Idaho
hn Choma, Jr.
Trujillo
niversity of Southern
lifornia
. G. Haigh
niversity College of London,
nited Kingdom
. Redman-White
niversity of Southampton
nited Kingdom
. Akbari-Dilmaghani
niversity College of London,
nited Kingdom
I. Ismail
hio State University
.-C. Huang
ıtung Inst. of Technology
ıipei, Taiwan
:.-C. Hung
hio State University
. Saether
he Norwegian Institute of
echnology, Norway

57.1 Bipolar Biasing Circuits

K. V. Noren

Fundamental in the design of bipolar integrated circuits is the establishment of the bias currents and voltages at which the subcircuits of the design are to operate. This of course determines the small-signal parameters of transistors meant to be operated as small-signal linear devices and the dc voltage levels used in direct-coupled circuits, such as op amps. The subcircuits that perform biasing include current sources, voltage references, and level shifters. Typically, it is desired that the BJT design be robust, and independent of a variety of external conditions that can affect circuit performance. These factors include variations in supply voltages, in temperature, and process parameters. Desires to improve on these issues have led to many refinements and developments based upon simple ideas and basic building blocks. We present in this section some of the fundamental biasing building blocks that are used in bipolar integrated circuit technology and the refinements of them that have evolved over time.

Common BJT Biasing Circuits

Current Mirrors and Sources

The current mirror is a circuit that reproduces a reference current at one or more locations in the circuit. A simple current mirror is depicted in Fig. 57.1. Since $V_{BE1} = V_{BE2}$, $I_{OUT} \approx I_{REF}$, and the reference current is effectively mirrored to another part of the circuit. In order to begin to classify and characterize current mirrors, we must first define an ideal current mirror. An ideal current mirror can be characterized as follows.

3493-8341-2/95/$0.00 + $.50

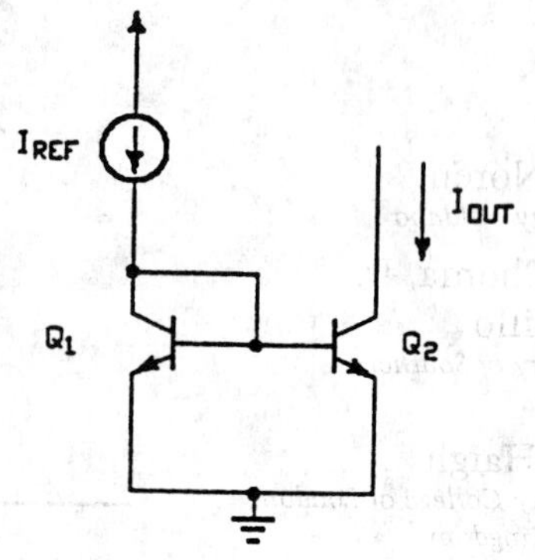

FIGURE 57.1 Simple current mirror.

1. It reproduces the reference current, exactly.
2. It does not vary with loading (infinite output resistance).
3. It is insensitive to power supply variations.
4. It is insensitive to temperature.

Simple Current Mirror. For the simple current mirror, including the effects of finite β (base–current drain) for matched transistors, the transfer function is

$$\frac{I_{OUT}}{I_{REF}} \approx \frac{1}{1 + 2/\beta} \tag{57.1}$$

The error due to β results from the fact that the I_{REF} must supply current to the bases of both Q_1 and Q_2. From the small-signal equivalent circuit of Fig. 57.1, it is seen that the output resistance of the simple current source is simply r_o of the transistor Q_2.

If the mirror in Fig. 57.1 is configured with N multiple outputs, the transfer function becomes

$$\frac{I_{OUT}}{I_{REF}} \approx \frac{1}{1 + (N+1)/\beta} \tag{57.2}$$

Wilson Current Mirror. A current source that improves the output resistance and β sensitivity to the transistors is the Wilson current mirror shown in Fig. 57.2 [1].

An analysis of the circuit shows the need to still supply base current to Q_1 and Q_2. This is supplied by the emitter current of Q_3, which draws the current from I_{REF}, but reduced by a

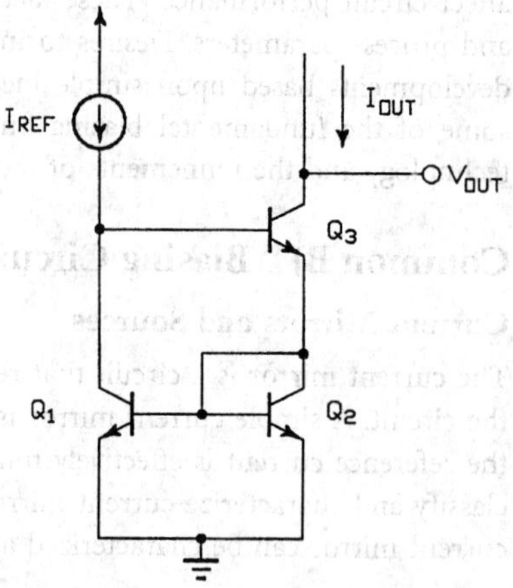

FIGURE 57.2 Wilson current mirror.

factor of β over the drain from I_{REF} from the original current source. A rigorous analysis will show, neglecting the effects of r_o,

$$\frac{I_{OUT}}{I_{REF}} = \frac{1}{1 + 2/(\beta^2 + 2\beta)} \approx \frac{1}{1 + 2/\beta^2} \tag{57.3}$$

supporting the intuitive analysis. It should be noted that the equation also depends on the betas of the transistors being matched as well. An improvement in output resistance can be seen by noting the circuit has negative feedback. For example, considering matched transistors and the effects of output voltage for all of the transistors, an increase in V_{OUT} gives rise to an increase in output current. This, in turn, causes an increase in I_{C2}. Since Q_1 and Q_2 themselves form a simple current mirror, I_{C1} also increases, which forces a decrease in I_{B3}, since I_{REF} is taken to be constant, thus forcing a reduction in I_{OUT}. Changes in I_{OUT} due to changes in V_{OUT} have been effectively reduced due to the feedback. Inserting the small-signal model for the transistors and solving for R_o gives

$$R_o \approx \beta r_o/2 \tag{57.4}$$

The Wilson current mirror can also be extended to N multiple outputs by placing extra transistors across Q_3. It can be shown that for N devices

$$\frac{I_{OUT}}{I_{REF}} \approx \frac{1}{N} \frac{1}{1 + 2/\beta^2} \tag{57.5}$$

dividing the current between the devices with about the same error as the $N = 1$ case.

Widlar Current Mirror. In bipolar integrated circuit design, it is sometimes desirable to create low currents levels, on the order of μA, for example [2]. With the simple current mirror or the Wilson current mirror, this must be done by creating very low reference currents and may require large values of resistances. This is undesirable. The current mirror depicted in Fig. 57.3 is capable of producing a small output current, from a nominal valued reference current.

To analyze the circuit, it is assumed that I_{REF} is responsible for determining the voltage of V_{BE1}. In the simple current mirror, all of V_{BE1} appears across V_{BE2}. In the Widlar current mirror, only a fraction of V_{BE1} can be placed across V_{BE2}, since a portion must be placed across R_1 as well.

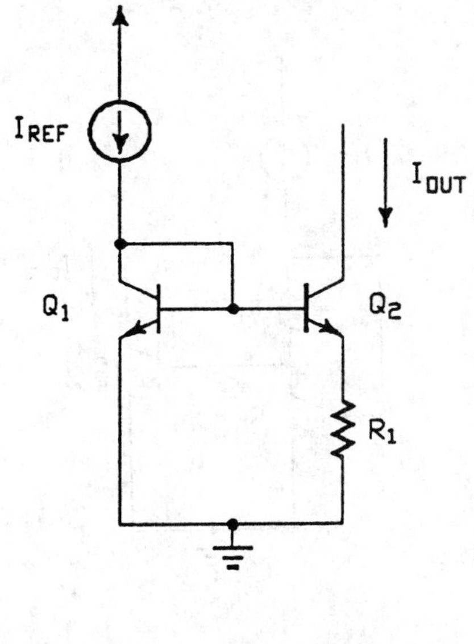

FIGURE 57.3 Widlar current mirror.

Thus, with proper selection of R_1, it is expected that we are capable of producing lower current levels due to the lower V_{BE2}, achieved by voltage dividing V_{BE1}. It can be shown, neglecting the effects of β, that

$$I_{\text{OUT}}R_1 = V_T \ln\left(\frac{I_{\text{REF}}}{I_{\text{OUT}}}\right) \tag{57.6}$$

For example, to create an I_{OUT} of 5 μA, from a reference of 1 mA, with $V_T = 26$ mV, we have $R_1 = 27$ kΩ. If the same 5 μA were desired from the simple current mirror, and we assume that the emitter areas of Q_1 and Q_2 are equal, we would need to generate a reference current of 5 μA. To do this, I_{REF} is replaced by a resistor R_{REF} tied to the positive supply voltage V_{CC}, for example. Suppose for this example that V_{CC} is 5 V. Then, the voltage drop across R_{REF} is $V_{CC} - V_{BE}$. Taking $V_{BE} = 0.7$ V, gives a value of $R_{\text{REF}} = (5 - 0.7)/5$ μm $= 860$ kΩ. This, of course, takes up much more chip real estate than for $R_1 = 27$ kΩ and is undesirable.

If the effects of β are included, as in the simple current mirror, it is necessary to draw the equivalent of 2 base currents, here I_{B2} although being less than I_{B1}. It can be expected that (57.2) gives an upper bound for beta errors for the Widlar current mirror (the upper bound being that of $I_{\text{OUT}} = I_{\text{REF}}$) and errors for the lower bound being $1/(1 + 1/\beta)$, supplying base current to only 1 transistor. Thus, the errors due to base current drain are on the same order as that of the simple current mirror.

For the output resistance, R_1 provides negative feedback to help increase the output resistance. This is identical to emitter degeneration in a common-emitter amplifier, and in fact, when using the small-signal equivalent circuit to derive the output resistance, the topology is identical to this amplifier, neglecting a collector resistance. Thus, it is easily verified that

$$R_o \approx r_o(1 + g_m R_1 \| r_\pi) \tag{57.7}$$

Low-Bias Current Mirror. An alternative to the Widlar current source that also provides a low output current is the current source shown in Fig. 57.4 [3], [4].

Again, V_{BE1} is generated and forced by I_{REF}. Using KVL,

$$V_{BE2} = V_{BE1} - I_{\text{REF}}R_1 \tag{57.8}$$

Thus, again, the voltage generated at V_{BE2} must be smaller than V_{BE1} and effectively only a fraction of I_{REF} is mirrored. A more exact relationship for the output current and reference

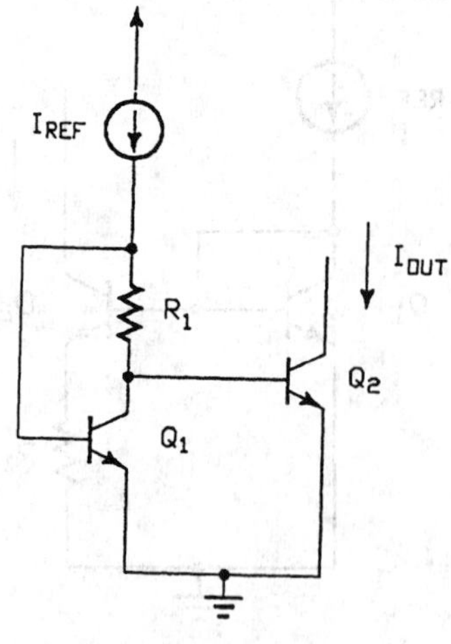

FIGURE 57.4 Low-bias current source.

current can be derived from (57.8), and this is

$$I_{OUT} = I_{REF} \exp\left(\frac{-I_{REF}R_1}{V_T}\right) \tag{57.9}$$

and

$$R_1 = \frac{-V_T \ln\left(\frac{I_{OUT}}{I_{REF}}\right)}{I_{REF}} \tag{57.10}$$

In fact, this current source is capable of supplying even lower currents than the Widlar current source for a given I_{REF} and an upper bound for R_1. Consider the same example as was given for the Widlar current source. With $I_{REF} = 1$ mA, $I_{OUT} = 5$ μA, and $V_T = 26$ mV, we find that $R_1 = 137.75\ \Omega$. This is a substantial decrease in resistance from the Widlar current mirror example.

This current mirror is also called a g_m-compensated mirror, and derives this name from its performance when R_1 is selected to be $1/g_{m_1}$ [5]. To understand the operation, consider the effect of a small change in I_{REF}. Neglecting base current, the change in voltage for V_{BE1} is $\Delta I_{REF} r_{e1}$. The change in voltage across R_1 is $\Delta I_{REF} R_1 = \Delta I_{REF}/g_{m_1} \approx \Delta I_{REF} r_{e1}$. This of course means, as seen by (57.8), the net change in voltage across the base–emitter of Q_2 is about 0 V.

The basic current mirrors can be extended to complementary bipolar technology (CBT) as well. Figure 57.5 gives an example of a current mirror that can be found in CBT. A problem that arises in the current mirrors used in bipolar technology results in the fact that pnp and npn transistors have a different gummel number [5]. This results in different base–emitter voltages (magnitudes) for the same collector currents. The effects of this can be deduced from Fig. 57.5, where now ΔV_{BE} errors must be considered. Thus, care must be taken when biasing complementary bipolar designs. The issue of balancing and matching is a fundamental problem in this technology [5].

Supply Independent Biasing. For many current mirrors, I_{REF} is determined by a resistor tied to the positive power supply. In the Widlar current mirror example, $I_{REF} = (V_{CC} - V_{BE})/R_{REF}$. The reference current is directly proportional to the supply voltage. In many situations, this is undesirable. One solution is to replace V_{CC} by one of the many available circuits that provide a voltage reference that is independent of supply voltage. Another alternative is shown in Fig. 57.6.

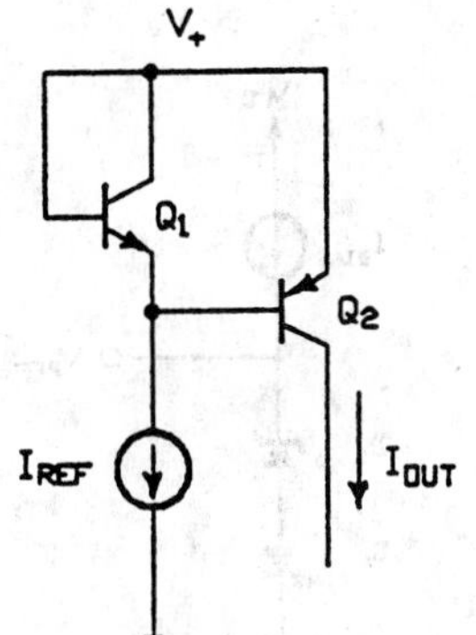

FIGURE 57.5 Current mirror for complementary bipolar design.

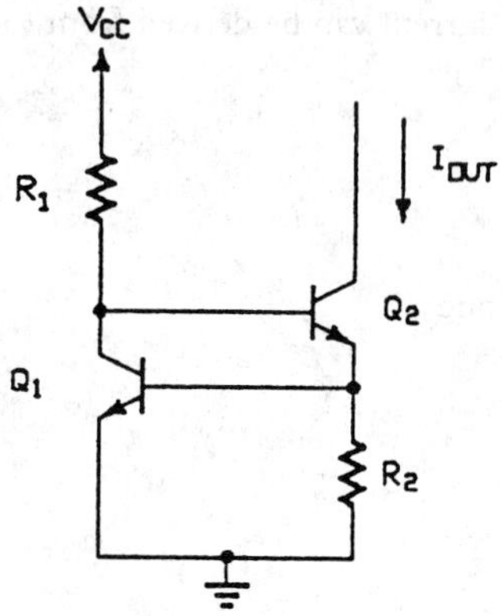

FIGURE 57.6 Current source that has reduced supply voltage dependency.

A current is generated through R_2 via the V_{BE1} that is forced across it. Neglecting finite betas for the transistors, this current is approximately I_{OUT}. A complete derivation yields

$$I_{OUT} = \frac{V_T}{R_2} \ln\left(\frac{V_{CC}}{R_1 I_S}\right) \tag{57.11}$$

This shows the output current to have logarithmic variation with V_{CC}, an improvement over the linear relationship found in other current mirrors.

Voltage references

Also fundamental to biasing in BJT circuits is the voltage reference. As with the current sources, we may define an ideal voltage source in order to have an adequate way of evaluating voltage references. An ideal voltage reference can be characterized as follows.

1. It does not vary with loading (zero output resistance).
2. It is insensitive to power supply variations.
3. It is insensitive to temperature.

The simplest voltage reference is the zener diode reference, but is it well known that a zener diode exhibits temperature dependence and has a fairly high equivalent output resistance. Some of the deviations include placing a diode series with a zener, strings of diodes, and the common-collector stage. Figure 57.7 gives an example. Here it is assumed, but not always true, that V_{DZ} has a positive coefficient and V_{D1} has a negative temperature coefficient and provides some cancellation of the coefficients. There are literally hundreds of deviations based on this fundamental principle.

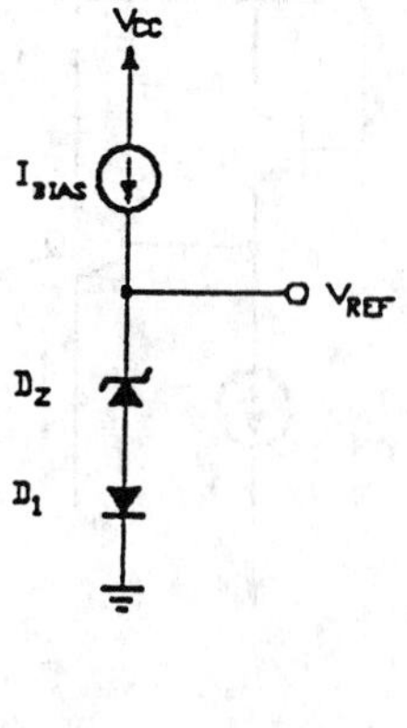

FIGURE 57.7 Zener-based voltage references.

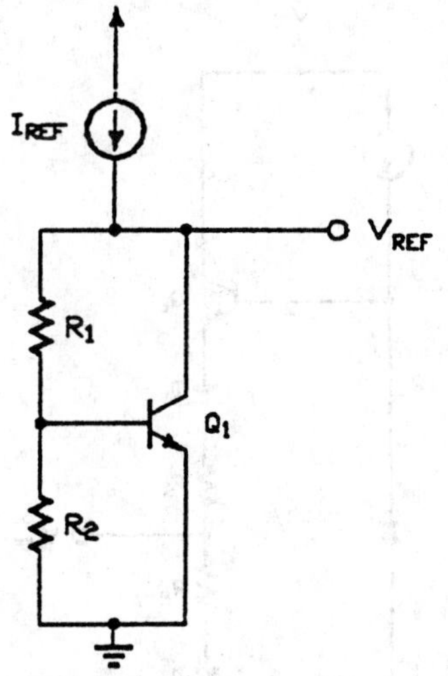

FIGURE 57.8 V_{BE} multiplier circuit.

A circuit that produces a voltage that is an arbitrary multiple of V_{BE} is the V_{BE} multiplier circuit shown in Fig. 57.8.

The circuit works on the principle that a current equal to V_{BE}/R_2 is generated and, neglecting base current, flows through R_1. Thus, the voltage across R_1 is $(V_{BE}/R_2)R_1$ and the total voltage can be written as

$$V_{\text{REF}} = V_{BE}(1 + R_2/R_1) \tag{57.12}$$

Thermal Desensitization

Many applications for biasing circuits demand that their performance remain constant through a wide range of temperatures. Thus, many circuits for temperature insensitive biasing have emerged. Ideally, a temperature insensitive output, voltage or current, would depend on a temperature insensitive element. Since all semiconductor components exhibit variation with temperature, most of the schemes for temperature independence involve some form of cancellation technique or compensation [6]. Instead of eliminating all sensitivity, the design techniques strive to minimize the errors. For example, a common solution is to place devices with positive temperature coefficients in series with devices with negative temperature coefficients, scaling some of these coefficients if necessary, to provide nearly zero sensitivity to temperature for an output is taken across the devices. Figs. 57.7 and 57.9 are examples.

Consider the circuit in Fig. 57.9. The reference voltage can be written as

$$V_{\text{REF}} = \frac{R_2 V_{DZ} + V_{BE}(R_1 - 2R_2)}{R_1 + R_2} \tag{57.13}$$

where $V_{BE1} = V_{D1} = V_{D2} = V_{BE}$. By taking the partial derivative of V_{REF} with respect to temperature and setting this equal to zero, a relationship by which a zero temperature coefficient for the reference voltage can be attained and this is

$$\frac{R_1 - 2R_2}{R_2} = -\frac{TC_{VDZ}}{TC_{VBE}} \tag{57.14}$$

where TC_{VZ} and TC_{VBE} are the temperature coefficients of the zener diode and base-emitter voltage.

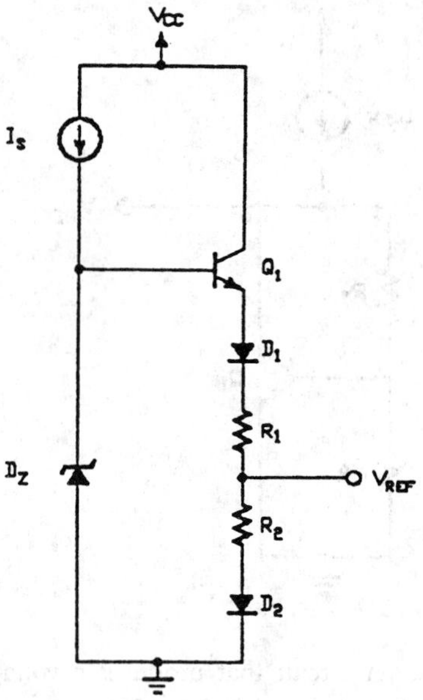

FIGURE 57.9 Simple zener-biased circuit with temperature coefficient cancellation scheme.

An alternative to this is one of the many variations of bandgap references available. A simple bandgap reference is depicted in Fig. 57.10.

Bandgap circuits also operate on a principle of cancellation of temperature coefficients. Generally, a voltage is developed that is a scaled value of V_T. This scaled value has a well-defined temperature coefficient that is the scaling constant times the temperature coefficient of V_T, which is positive. This voltage is added to a base–emitter voltage that has a negative temperature coefficient. The scaling factor is chosen so that the sum of the temperature coefficients is zero. The output is then taken across the two voltages to produce a voltage with a temperature coefficient of approximately zero. Generally, the output voltage has the form

$$V_{REF} = V_{BE} + KV_T \tag{57.15}$$

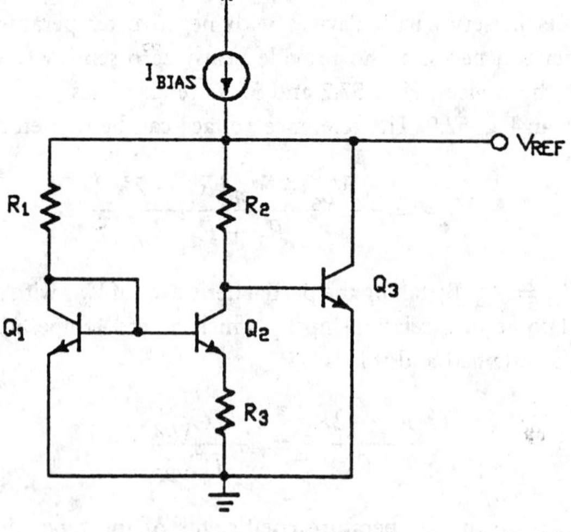

FIGURE 57.10 Simple bandgap reference.

If the temperature coefficient of V_{BE} is taken to be -2 mV/°C and the temperature coefficient of V_T is taken to be $k/q \approx +0.085$ mV/°C, this results in a value for K of about 23.52. This gives a value for V_{OUT} of about 0.7 V + 23.52(25.9 mV) = 1.3 V, close to the bandgap voltage of silicon, and gives rise to the nomenclature of bandgap references.

For the circuit in Fig. 57.10, we assume that Q_1 and Q_2 operate at different current densities. This is done either by operating Q_1 and Q_2 at different collector current levels, with matched emitter areas, or by operating them at the same collector currents with emitter areas being mismatched. A voltage ΔV_{BE} is developed across R_3 and then current through R_3 and R_2, neglecting the effects of β, is Δ_{BE}/R_3. This gives $V_{REF} = \Delta V_{BE}R_2/R_3 + V_{BE3}$. Since $\Delta V_{BE} = V_T \ln(J_1/J_2)$, this gives the constant K as being $R_2/R_3 \ln(J_1/J_2)$.

An improved bandgap reference is depicted in Fig. 57.11. This reference forms a basic building block for several commercial voltage references.

In this circuit, I_{C1} and I_{C2} are forced to be equal by the high-gain amplifier operating with negative feedback. The current densities of Q_1 and Q_2 are made unequal by sizing the emitter areas of Q_1 and Q_2 differently. In this case, $A_1 = nA_2$. This means $\Delta V_{BE} = V_T \ln(J_1/J_2) = V_T \ln(n)$. The voltage drop across R_2 is $\Delta V_{BE}R_2/R_1$. Finally, $V_{REF} = V_{BE} + V_T \ln(n)R_2/R_1$. Thus, in this case, $K = R_2/R_1 \ln(n)$.

Process and Modeling Parameter Desensitization

It is given that fabricated devices exhibit deviation from some "nominal" value assumed in the fabrication process. In designing process-insensitive and model-parameter-insensitive devices, it is sought to minimize the effects of variation-process- and model-parameters-desensitization techniques. A good example of this is the simple current mirror with emitter degeneration shown in Fig. 57.12.

If I_{REF} and R_1 are such that the voltage drop across V_{BE1} is small in comparison, then the dominant voltage at V_B is $I_{REF}R_1$. Likewise, if R_2 and I_{OUT} are such that the voltage drop across V_{BE2} can be neglected, then we have $I_{REF}R_1 \approx I_{OUT}R_2$ and

$$I_{OUT} \approx \frac{R_2}{R_1} I_{REF} \tag{57.16}$$

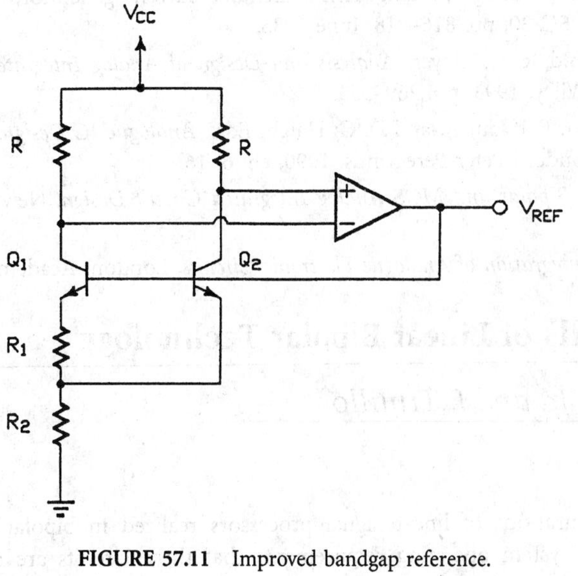

FIGURE 57.11 Improved bandgap reference.

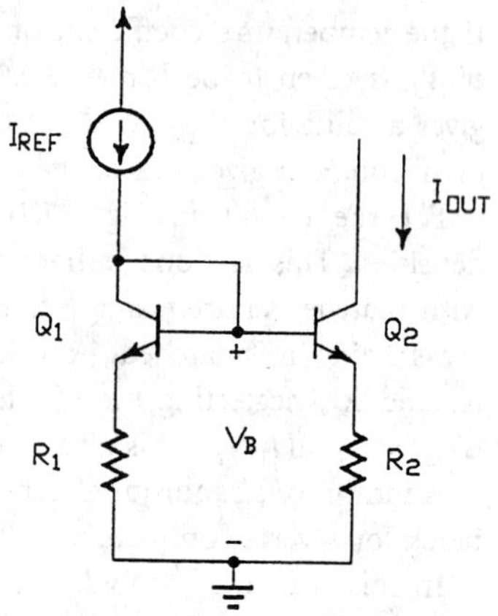

FIGURE 57.12 Simple current mirror with emitter degeneration.

For the simple current mirror, neglecting β and considering the possibility of mismatched emitter areas, we have

$$I_{\text{OUT}} \approx \frac{A_2}{A_1} I_{\text{REF}} \tag{57.17}$$

Since resistors can be matched to ±.1 percent and NPN matching for transistors can be as poor as ±1 percent, the current mirror with emitter degeneration is less susceptible to processing errors.

There is a wealth of information available in the literature. Further depth into bias circuit behavior is provided in the accompanying references for this section.

References

[1] G. R. Wilson, "A monolithic junction FET-NPN operational amplifier," *IEEE J. Solid-State Circuits*, vol. SC-3, pp. 341–343, Dec. 1968.

[2] R. J. Widlar, "Some circuit design techniques for linear integrated circuits," *IEEE Trans. Circuit Theory*, vol. CT-12, pp. 586–590, Dec. 1965.

[3] C. Kwok, "Low-voltage peaking complementary current generator," *IEEE J. Solid-State Circuits*, vol. SC-20, pp. 816–818, June 1985.

[4] P. R. Gray and R. G. Meyer, *Analysis and Design of Analog Integrated Circuits*, 3rd ed., New York: Wiley, 1993, pp. 269–353.

[5] C. Toumazou, F. J. Lidgey, and D. G. Haigh, Eds., *Analogue IC Design: The Current-Mode Approach*, London: Peter Peregrinus, 1990, ch. 6, 16.

[6] A. Grebene, *Bipolar and MOS Analog Integrated Circuit Design*, New York: Wiley, 1984, ch. 4.

[7] J. Davidse, *Integration of Analogue Electronic Circuits*, London: Academic, 1979, ch. 4.

57.2 Canonic Cells of Linear Bipolar Technology

John Choma, Jr. and J. Trujillo

Introduction

The circuit configurations of linear signal processors realized in bipolar technology are as diverse as are the system operating requirements that these circuits are designed to satisfy.

Despite topological diversity, most practical open-loop linear bipolar circuits are derived from interconnections of surprisingly few basic subcircuits. These subcircuits include the diode-connected bipolar junction transistor (BJT), the common-emitter amplifier, the common-base amplifier, the common-emitter–common base cascode, the emitter follower, the Darlington connection, and the balanced differential pair. Because these open-loop subcircuits underpin linear bipolar circuit technology, they are rightfully termed the *canonic cells* of linear bipolar circuit design.

By examining the low-frequency performance characteristics of the canonic cells of linear bipolar technology, this section achieves two objectives. First, the forward gain, the driving point input resistance, and the driving point output resistance are catalogued for each canonic circuit. This information produces Thévenin and Norton I/O port equivalent circuits that expedite the analysis and design of multistage electronics. Second, the forthcoming work establishes a basis for prudent circuit design in that all analytical results are studied by highlighting the attributes and uncovering the limitations of each cell. The understanding that resultantly accrues paves the way toward systematic design procedures that yield optimal circuit architectures capable of circumventing observed subcircuit shortcomings.

Small-Signal Model

The fundamental tool exploited in the analyses that follow is the low-frequency small-signal equivalent circuit of a bipolar junction transistor shown in Fig. 57.13(a). This equivalent circuit, which applies to NPN and PNP discrete component and monolithic transistors, is derived from the low-frequency large-signal NPN BJT model offered in Fig. 57.13(b) [1], [2]. As is depicted in Fig. 57.13(c), the PNP large-signal transistor model is topologically identical to its NPN counterpart. The only difference between the two models is a reversal in the direction of all controlled current sources and branch currents and a reversal in polarity of all assigned branch and port voltages.

The large-signal models in Fig. 57.13(b) and (c) are simplified to reflect transistor biasing that assures nominally linear device operation for all values of applied input signal voltages. A necessary condition for linear operation is that the internal emitter–base junction voltage v_e be at least as large as the threshold voltage, say v_γ, of the junction for all time; that is

$$v_e(t) \geq v_\gamma \quad \text{for all time } t \tag{57.18}$$

For silicon transistors, v_γ is typically in the neighborhood of 700 to 750 mV. A second condition underlying transistor operation in its linear regime is that the internal base–collector junction voltage v_c is never positive, that is

$$v_c(t) \leq 0 \quad \text{for all time } t \tag{57.19}$$

In the models of Fig. 57.13(b) and (c), r_b represents the **effective base resistance** of a BJT, r_c is its net *internal* **collector resistance**, and r_e is the net *internal* **emitter resistance.** All three resistances, and particularly r_b, decrease monotonically with increasing quiescent base and collector currents, I_{BQ} and I_{CQ}, respectively [3], [4]. The collector resistance also decreases with increases in the intrinsic collector–emitter voltage v_x. Large base, collector, and emitter resistances conduce reduced circuit gain, diminished gain–bandwidth product, and increased electrical noise. In view of these observations and in the interest of formulating a mathematically tractable analysis that produces conservative estimates of bipolar circuit performance, these resistances are usually interpreted as constants equal to their respective low current, low voltage values.

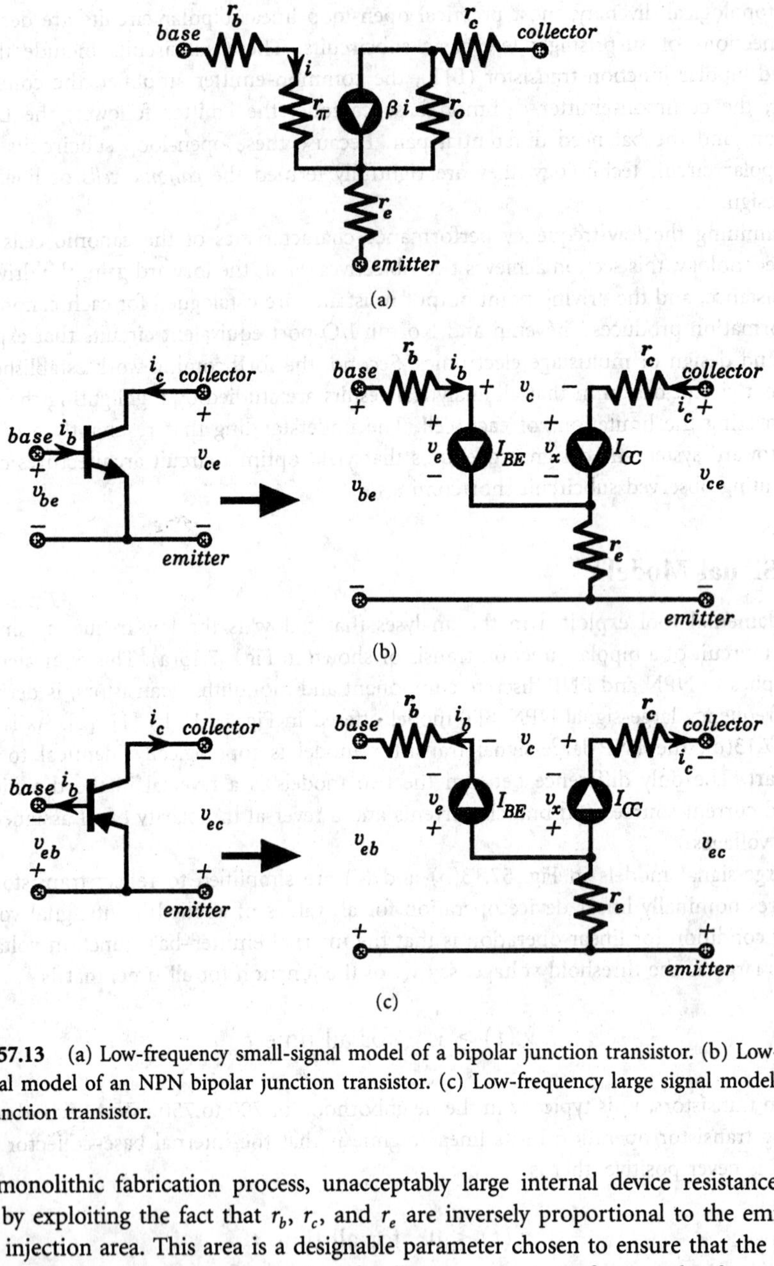

FIGURE 57.13 (a) Low-frequency small-signal model of a bipolar junction transistor. (b) Low-frequency large-signal model of an NPN bipolar junction transistor. (c) Low-frequency large signal model of a PNP bipolar junction transistor.

In a monolithic fabrication process, unacceptably large internal device resistances can be reduced by exploiting the fact that r_b, r_c, and r_e are inversely proportional to the emitter–base junction injection area. This area is a designable parameter chosen to ensure that the transistor in question conducts the proper density of collector current. Unfortunately, the engineering price potentially paid for a reduction of device resistances through increases in junction area is circuit response speed, since the capacitances associated with transistor junctions are directly proportional to device injection area.

The current I_{BE} in Fig. 57.13(b) and (c) is given approximately by

$$I_{BE} = \frac{A_E J_S}{\beta_F} e^{v_e/n_f V_T} \tag{57.20}$$

where A_E is the aforementioned *emitter–base junction injection area*, J_S, is the *density of transistor saturation current*, and β_F is the *forward short-circuit current transfer ratio*. Moreover, n_f is the *injection coefficient of the emitter–base junction,* v_e is the *internal junction voltage* serving to forward bias the emitter–base junction, and

$$V_T = \frac{kT_j}{q} \tag{57.21}$$

is the **Boltzmann voltage.** In the last expression, k is *Boltzmann's constant* [1.38 (10^{-23}) J/K], T_j is the *absolute temperature of the emitter–base junction,* and q is the *magnitude of electron charge* [1.6 (10^{-19}) C].

The current I_{CC} is derived from [5]

$$I_{CC} = A_E J_S e^{v_e/n_f V_T}\left(1 - \frac{I_{CC}}{I_{KF}}\right)\left(1 + \frac{v_x}{V_{AF}}\right) \tag{57.22}$$

where I_{KF}, which is proportional to A_E, is the **forward knee current** of the transistor [6], and V_{AF}, which is independent of A_E, is the *forward Early voltage* [7]. Note that the base current i_b is the current I_{BE}, while the collector current i_c is I_{CC}. Thus, the *static* **common-emitter current gain** (often referred to as the **DC beta**), h_{FE}, of a bipolar junction transistor is

$$h_{FE} = \frac{i_c}{i_b} = \frac{I_{CC}}{I_{BE}} = \beta_F\left(1 - \frac{i_c}{I_{KF}}\right)\left(1 + \frac{v_x}{V_{AF}}\right) \tag{57.23}$$

which is functionally dependent on both the collector current and the intrinsic collector–emitter voltage.

Unlike the base, collector, and emitter resistances, the resistance r_π in the small-signal model of Fig. 57.13(a) is not an ohmic branch element. It is a mathematical resistance that arises from the Taylor series expansion of the current I_{BE} about the **quiescent operating point,** or *Q-point,* of the transistor. In particular, r_π, which is known as the **emitter–base junction diffusion resistance,** derives from

$$\frac{1}{r_\pi} = \left.\frac{\partial I_{BE}}{\partial v_e}\right|_Q \tag{57.24}$$

where it is understood that the indicated derivative is evaluated at the Q-point of the device. This Q-point is unambiguously defined by the *zero signal,* or *static,* values of the base current I_{BQ}, the collector current I_{CQ}, and the internal collector–emitter voltage V_{XQ}. Using (57.20), (57.23), and the fact that $i_b \equiv I_{BE}$,

$$r_\pi = \frac{h_{FE} n_f V_T}{I_{CQ}} \tag{57.25}$$

The inverse dependence of r_π on quiescent collector current renders r_π large at low collector current biases.

Similarly, r_o, the *forward Early resistance,* derives from

$$\frac{1}{r_o} = \left.\frac{\partial I_{CC}}{\partial v_x}\right|_Q \tag{57.26}$$

It can be shown that

$$r_o = \frac{V_{XQ} + V_{AF}}{I_{CQ}\left(1 - \dfrac{I_{CQ}}{I_{KF}}\right)} \tag{57.27}$$

Like r_π, r_o is also large for low-level biasing.

Finally, the parameter β, which is the *low-frequency small-signal common-emitter short current gain* (often more simply referred to as the **AC beta**) of the transistor, is

$$\beta = g_m r_\pi \tag{57.28}$$

where g_m, the *forward transconductance* of a BJT is

$$g_m = \left.\frac{\partial I_{CC}}{\partial v_e}\right|_Q \tag{57.29}$$

From (57.22), (57.23), (57.25), and (57.28),

$$\beta = h_{FE}\left(1 - \frac{I_{CQ}}{I_{KF}}\right) \tag{57.30}$$

To the extent that $I_{CQ} \ll I_{KF}$, β is nominally independent of both Q-point collector current and emitter–base junction injection area.

Single-Input–Single-Output Canonic Cells

Diode-Connected Transistor

The simplest of the single-input–single-output, or *single-ended* canonic cells for linear bipolar circuits is the *diode-connected transistor* offered in Fig. 57.14(a). This transistor connection emulates the volt–ampere characteristics of a conventional PN junction diode. It can therefore be used in rectifier, voltage regulator, dc level shifting, and other applications that exploit conventional diodes. But unlike a conventional PN junction diode, the diode-connected transistor proves especially useful in current mirror biasing schemes. These and other similar circuits require that the base–emitter terminal voltage, v, of the diode track the base–emitter terminal voltage of a second, presumably identical transistor, over wide variations in junction operating temperatures.

If the voltages dropped across the internal base, collector, and emitter resistances are small, the intrinsic emitter–base junction voltage, v_e, is approximately the indicated terminal voltage, v. Moreover, the intrinsic base–collector junction voltage, v_c, is essentially zero. It follows that for $v > v_\gamma$, the transistor in the diode connection operates in its linear regime.

In the subject diagram, the terminal voltage v is depicted as a superposition of a static voltage, V_Q, and a signal component, v_s. The resultant diode current, i, is a superposition of a quiescent current, I_Q, and a signal current, i_s. The quiescent components of diode voltage and current arise from static power supplied to the diode circuit to ensure that the diode-connected device operates in its linear regime. On the other hand, the signal components are established by a time-varying signal applied to the input port of the circuit in which the diode-connected transistor is embedded. In order to achieve reasonably linear processing of the applied input signal, the value of V_Q must be such as to ensure that $v = V_Q + v_s > v_\gamma$ for all values of the time-varying signal voltage v_s. Since v_s can be positive or negative at any instant of time, the requirement $V_Q + v_s > v_\gamma$ mandates that the amplitude of v_s be sufficiently small.

The immediate impact of the small-signal condition corresponding to the linearity requirement $V_Q + v_s > v_\gamma$ is that the small-signal volt–ampere characteristics of the diode are linear. And since the diode is a two-terminal element, these characteristics can be modeled at low-signal frequencies by a simple resistance, say R_d, as suggested by the single-element macromodel offered in Fig. 57.14(b). The resistance in the latter figure can be determined by using the small-signal transistor model of Fig. 57.13(a) to construct the small-signal equivalent circuit of the diode-connected transistor shown in Fig. 57.14(c). In this figure, the ratio of the test

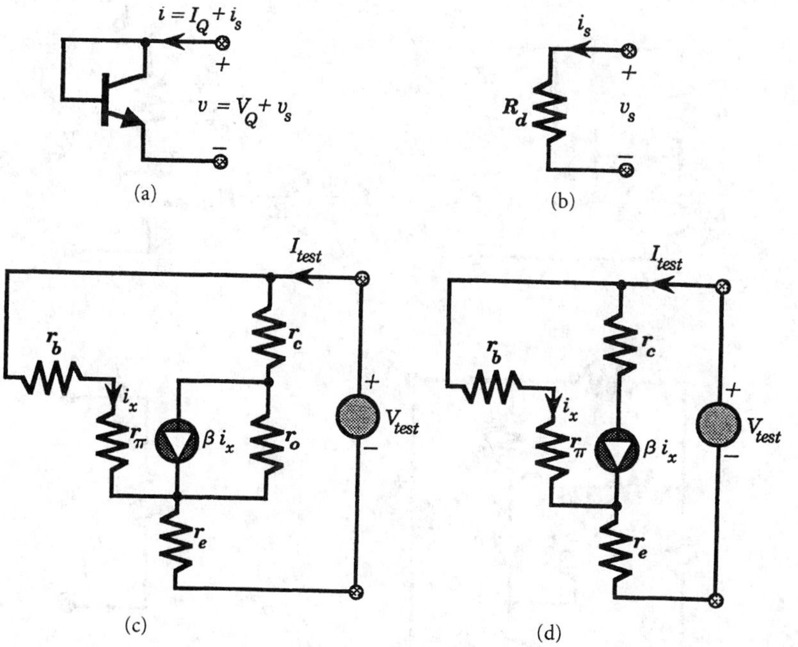

FIGURE 57.14 (a) Diode-connected bipolar junction transistor. (b) The small-signal equivalent circuit of the diode in (a). (c) Low-frequency small-signal model of the diode-connected transistor in (a). The ratio $V_{\text{test}}/I_{\text{test}}$ is the small-signal resistance R_d presented at the terminals of the diode-connected transistor. (d) The model in (c) approximated for the case of very large Early resistance.

voltage, V_{test}, to the test current, I_{test}, is the desired resistance, R_d. A straightforward KVL analysis confirms that

$$R_d = \frac{V_{\text{test}}}{I_{\text{test}}} = r_e + \frac{(r_o + r_c)\|(r_b + r_\pi)}{1 + \dfrac{\beta r_o}{r_o + r_c + r_b + r_\pi}} \tag{57.31}$$

Typically, r_o is 25 kΩ or larger, r_c is smaller than 75 Ω, r_b is of the order of 100 Ω, and r_π is in the range of 1 kΩ for a minimal geometry device. It follows that $r_o \gg (r_c + r_b + r_\pi)$, and R_d can be approximated as

$$R_d \approx r_e + \frac{r_b + r_\pi}{\beta + 1} \tag{57.32}$$

Note that this terminal resistance is of the order of the low tens of ohms. For example, if $r_b = 100\ \Omega$, $r_\pi = 1.2$ kΩ, $r_e = 1\ \Omega$, and $\beta = 100$, $R_d = 13.9\ \Omega$. It is instructive to note that the approximation, $r_o \gg (r_c + r_b + r_\pi)$, collapses the model in Fig. 57.14(c) to the structure in Fig. 57.14(d), from which (57.32) follows immediately.

A variation of the diode scheme is the so-called V_{BE} *multiplier* depicted in Fig. 57.15(a). This circuit finds extensive use in regulator and level shifting applications that require either a series interconnection of more than one diode or a circuit branch voltage drop whose requisite value is a nonintegral multiple of the base–emitter terminal voltage of a single diode.

The circuit under consideration establishes a static terminal voltage, V_Q, whose value is a designable multiple of the static base–emitter terminal voltage, V_{BEQ}. To confirm this contention,

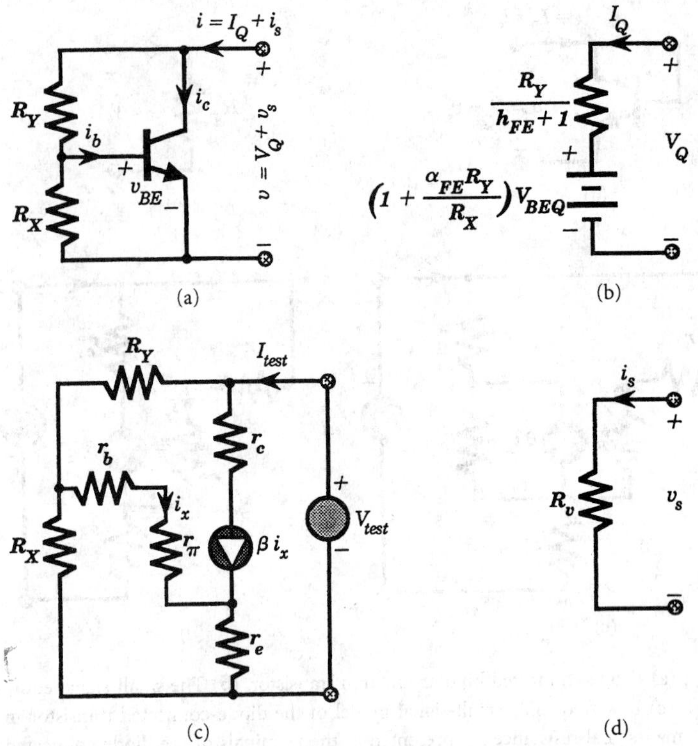

FIGURE 57.15 (a) Schematic diagram of V_{BE} multiplier. (b) DC macromodel of the multiplier in (a). (c) Low-frequency small-signal equivalent circuit of the Y_{BE} multiplier. (d) The small-signal equivalent resistance at the terminals of the V_{BE} multiplier.

observe that for static operating conditions, V_Q is

$$V_Q = R_Y(I_Q - I_{CQ}) + V_{BEQ} \tag{57.33}$$

where the voltage component V_{BEQ} of the net base–emitter terminal voltage v_{be} is

$$V_{BEQ} = R_X\left[I_Q - \left(\frac{h_{FE}+1}{h_{FE}}\right)I_{CQ}\right] \tag{57.34}$$

The current, I_{CQ}, is the static component of the net collector current, i_c, and h_{FE} is the collector current to base current transfer ratio defined by (57.23). An elimination of I_{CQ} from the foregoing two expressions leads to

$$V_Q = \left(1 + \frac{\alpha_{FE}R_Y}{R_X}\right)V_{BEQ} + \left(\frac{R_Y}{h_{FE}+1}\right)I_Q \tag{57.35}$$

where

$$\alpha_{FE} = \frac{h_{FE}}{h_{FE}+1} \tag{57.36}$$

is known as the **static common-base current gain** (often referred to as the **DC alpha**) of a BJT. Equation (57.35) suggests that the static electrical behavior of the V_{BE} multiplier approximates a battery, whose voltage is controllable by the resistive ratio R_Y/R_X.

The internal resistance of this effective battery is inversely dependent on $(h_{FE} + 1)$, and is therefore small. The macromodel in Fig. 57.15(b) reflects the foregoing electrical interpretation. Note that for $R_Y = 0$ and R_X infinitely large, the circuit in Fig. 57.15(a) collapses to the diode-connected transistor of Fig. 57.14(a), and V_Q understandably reduces to V_{BEQ}, the quiescent base–emitter terminal voltage of a diode-connected transistor.

For $V_Q + v_s > v_\gamma$, the transistor in the V_{BE} multiplier operates linearly. Accordingly, the pertinent small-signal terminal characteristics emulate a resistance, say R_v, which can be determined by applying the model of Fig. 57.13(a) to the circuit in Fig. 57.15(a). The resultant equivalent circuit, simplified to reflect the realistic assumption of large r_o, is shown in Fig. 57.15(c) while Fig. 57.15(d) postulates the small-signal macromodel. An analysis of the circuit in Fig. 57.15(c) reveals that

$$R_v \approx R_X \parallel R_d + R_Y\left[1 - \frac{\alpha R_X}{R_X + R_d}\right] \tag{57.37}$$

where

$$\alpha = \frac{\beta}{\beta + 1} \tag{57.38}$$

is the *low-frequency small-signal common-base short-circuit current gain* (more simply referred to as the **ac alpha**) of the transistor, and R_d is the resistance given by (57.31). For $R_Y = 0$, $R_X = \infty$ reduces R_v to the expected result, $R_v \approx R_d$. Note further that for $\beta \gg 1$ (which makes $\alpha \approx 1$) and $R_d \ll R_X$, R_v is essentially the small-signal resistance presented at the terminals of a diode-connected transistor.

Common Emitter Amplifier

The most commonly used single-ended canonic gain cell is the *common-emitter amplifier*, whose NPN and PNP AC schematic diagrams are shown in Figs. 57.16(a) and (b), respectively. The AC schematic diagram delineates only the signal paths of a circuit. Thus, the biasing subcircuits required for linear operation of the transistors are not shown, thereby affording topological and analytical simplification. This simplification is accomplished without loss of engineering generality, for the results produced by an analysis of the AC schematic diagram reveal all salient performance traits of the common-emitter configuration.

The common-emitter amplifier is distinguished by the facts that signal is applied to the base of the transistor, and the resultant response is extracted as either the voltage, V_{OS}, or the current, I_{OS}, at the collector port. The effective load resistance terminating the collector to ground is indicated as R_{LT}, while the signal source is represented as a traditional Thévenin equivalent circuit. Alternatively, a Norton representation of the input source can be used, with the understanding that the Norton equivalent signal current, say I_{ST}, is simply the ratio of the Thévenin signal voltage, V_{ST}, to the Thévenin source resistance, R_{ST}.

The common-emitter amplifier is capable of large magnitudes of voltage and current gains, moderately large input resistance, and very large driving point output resistance. An analytical confirmation of these contentions begins by drawing the small-signal equivalent circuit of the amplifier. This structure is given in Fig. 57.16(c) and is valid for either the NPN or the PNP versions of the amplifier. An analysis of the small-signal model yields a voltage gain,

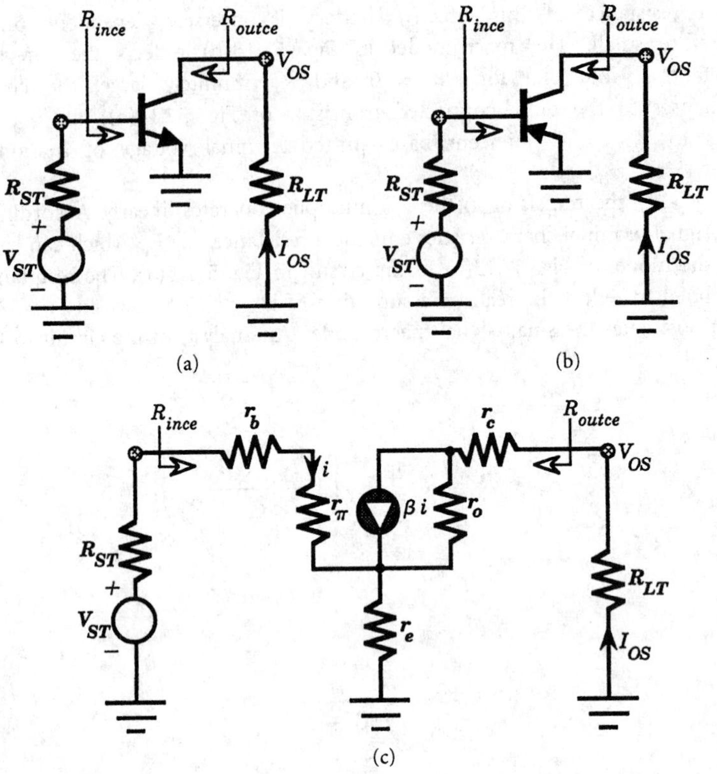

FIGURE 57.16 (a) AC schematic diagram of an NPN common-emitter amplifier. (b) AC schematic diagram of a PNP common-emitter amplifier. (c) Small-signal low-frequency equivalent circuit of the common-emitter amplifier.

$A_{vce} = V_{OS}/V_{ST}$, of

$$A_{vce} = -\left\{ \frac{\left(\beta - \dfrac{r_e}{r_o}\right)\left(\dfrac{r_o}{r_o + r_c + r_e + R_{LT}}\right) R_{LT}}{R_{ST} + r_b + r_\pi + \left(\dfrac{\beta r_o}{r_o + r_c + R_{LT}} + 1\right)[r_e \| (r_o + r_c + R_{LT})]} \right\} \quad (57.39)$$

This relationship can be simplified by exploiting the fact that the internal emitter resistance r_e of a transistor is small. Thus, $\beta \gg r_e/r_o$ and $r_e \ll (r_o + r_c + R_{LT})$, thereby implying

$$A_{vce} \approx -\frac{\beta_{\text{eff}} R_{LT}}{R_{ST} + r_b + r_\pi + (\beta_{\text{eff}} + 1) r_e} \quad (57.40)$$

where

$$\beta_{\text{eff}} \triangleq \beta \left[\frac{r_o}{r_o + r_c + R_{LT}} \right] \quad (57.41)$$

is an attenuated version of the AC beta for the utilized transistor. This effective beta approximates β itself, since $r_o \gg (r_c + R_{LT})$ is typical.

In concert with earlier arguments, (57.40) confirms a diminished magnitude of gain for large internal device resistances. Note also that phase inversion, as inferred by the negative sign in either (57.39) or (57.40) prevails between the Thévenin source voltage, V_{ST}, and the voltage signal response, V_{OS}. Finally, observe that large magnitudes of voltage gain are possible in the common-emitter orientation when β_{eff} is sufficiently large.

The driving point input resistance, R_{ince}, of the common-emitter amplifier can be determined as the ratio V_x/I_x for the test structure depicted in Fig. 57.17(a). It is easily shown that

$$R_{ince} = r_b + r_\pi + (\beta_{eff} + 1)[r_e \| (r_o + r_c + R_{LT})] \tag{57.42}$$

Since $r_e \ll (r_o + r_c + R_{LT})$, (57.42) collapses to

$$R_{ince} \approx r_b + r_\pi + (\beta_{eff} + 1)r_e \tag{57.43}$$

Similarly, the driving point output resistance, R_{outce}, is derived as the V_x/I_x ratio of the equivalent circuit offered in Fig. 57.17(b). In particular,

$$R_{outce} = r_c + r_e \| (r_\pi + r_b + R_{ST}) + \left(\frac{\beta r_e}{r_e + r_\pi + r_b + R_{ST}} + 1 \right) r_o \tag{57.44}$$

Since the model resistance r_π varies inversely with collector bias current, R_{ince} is moderately large when the common-emitter transistor is biased at low currents. On the other hand, R_{outce} is very large since (57.44) confirms $R_{outce} > r_o$.

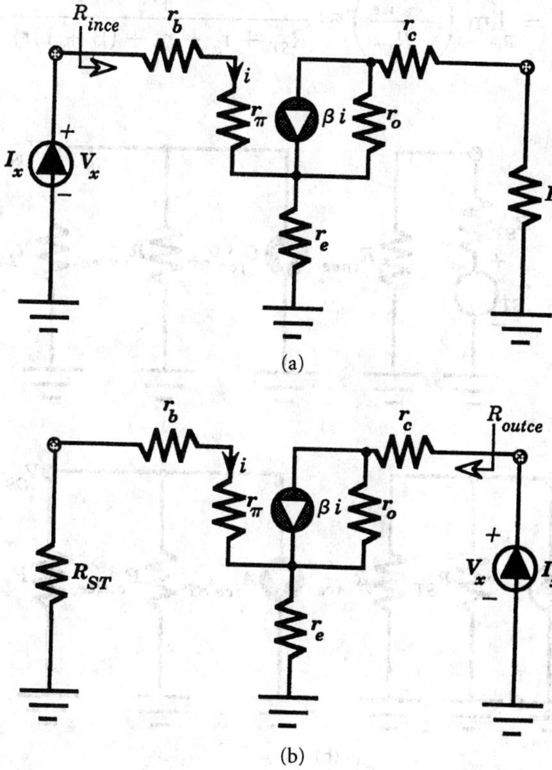

FIGURE 57.17 (a) Small-signal test structure used to determine the driving point input resistance of the common-emitter amplifier. (b) Small-signal test structure used to determine the driving point output resistance of the common-emitter amplifier.

When the foregoing results are simplified to reflect the practical special case of a very large forward Early resistance, r_o, the cumbersome small-signal equivalent circuit of Fig. 57.16(c) reduces to a *small-signal macromodel* useful for design-oriented circuit analysis of multistage amplifiers. To this end, note that a large r_o produces a driving point common-emitter input resistance that is independent of the terminating load resistance. Such independence implies no internal feedback from the output to input ports. It follows that the small-signal volt–ampere characteristics at the input port of a common-emitter amplifier can be modeled approximately by a simple resistance of value, R_{ince}, as defined by (57.43). On the other hand, the large driving point output resistance R_{outce} suggests that a prudent output port model of a common emitter stage is a Norton equivalent circuit. The Norton, or short-circuit, output current is proportional to the applied input signal voltage, V_{ST}, as depicted in Fig. 57.18(a). Alternatively, it can be expressed as a proportionality of the Norton input signal current, I_{ST}, as suggested in Fig. 57.18(b). In the former figure, the Norton current is

$$G_{fce}V_{ST} = \lim_{R_{LT}\to 0} I_{OS} = \lim_{R_{LT}\to 0}\left(-\frac{V_{OS}}{R_{LT}}\right) = \lim_{R_{LT}\to 0}\left(-\frac{A_{vce}V_{ST}}{R_{LT}}\right) \tag{57.45}$$

Subject to the assumption of large r_o,

$$G_{fce} = \lim_{R_{LT}\to 0}\left(-\frac{A_{vce}}{R_{LT}}\right) \approx \frac{\beta}{R_{ST} + r_b + r_\pi + (\beta + 1)r_e} \tag{57.46}$$

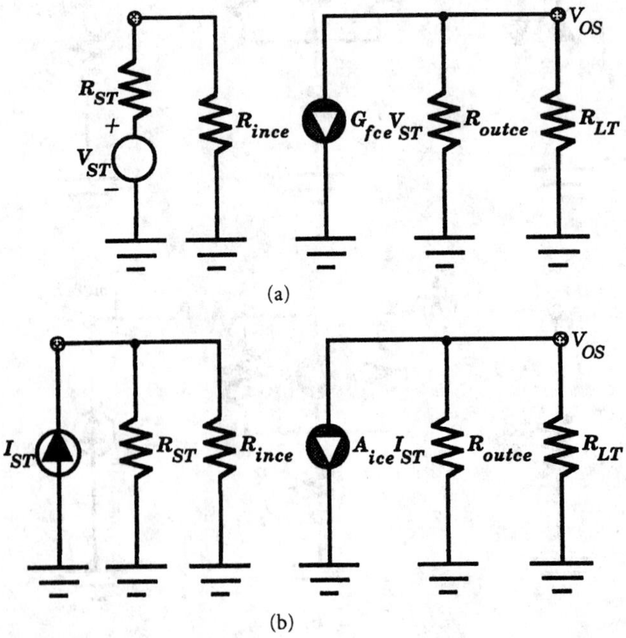

FIGURE 57.18 (a) Small-signal macromodel of a common-emitter amplifier in which the Norton output port circuit uses a voltage-controlled current source. (b) Small-signal macromodel of a common-emitter amplifier in which the Norton output port circuit uses a current-controlled current source.

Recalling (57.28), this effective forward transconductance of the amplifier can be expressed in terms of the transconductance, g_m of the transistor utilized in the amplifier. Specifically,

$$G_{fce} \approx \frac{g_m}{1 = g_m r_e + \dfrac{r_e + r_b + R_{ST}}{r_\pi}} \tag{57.47}$$

In the macromodel of Fig. 57.18(a), R_{outce} is very large by virtue of large r_o. Accordingly, the parallel combination of R_{outce} and R_{LT} is essentially R_{LT}, thereby implying an approximate common-emitter voltage gain of

$$A_{vce} \approx -G_{fce} R_{LT} \tag{57.48}$$

For the alternative macromodel in Fig. 58.18(b), the Norton current is

$$A_{ice} I_{ST} = A_{ice}\left(\frac{V_{ST}}{R_{ST}}\right) = \lim_{R_{LT}\to 0} I_{OS} = G_{fce} V_{ST} \tag{57.49}$$

Since $V_{ST} = R_{ST} I_{ST}$, it follows that A_{ice} is, for large r_o,

$$A_{ice} = G_{fce} R_{ST} \approx \beta\left[\frac{R_{ST}}{R_{ST} + r_b + r_\pi + (\beta + 1) r_e}\right] \tag{57.50}$$

Note that the Norton current proportionality A_{ice}, which is, in fact, the approximate ratio of the indicated output current, I_{OS}, to the Norton source current, I_{ST}, in the common-emitter configuration, is always smaller than β.

Example 1. Transistor Q1 in the amplifier depicted in Fig. 57.19(a) is fundamentally a common-emitter configuration since input signal is applied to its base terminal and the output voltage signal response is extracted at its collector. The amplifier uses coupling capacitors C_i and C_o at its input and output ports. The input coupling capacitor, C_i, blocks the flow of static current in the source signal branch consisting of the series interconnection of the voltage, V_S, and the Thévenin source resistance, R_S. Accordingly, C_i precludes R_S from affecting the biasing of both transistors used in the amplifier. Similarly, the output coupling capacitor, C_o, blocks the flow of static current in the external load resistance, R_L. Thus, both C_i and C_o can be viewed as open circuits for dc considerations. But simultaneously, these capacitors can be rendered transparent for AC situations by choosing them sufficiently large so that they emulate short circuits at the lowest frequency, say f_l, of signal-processing interest. In this problem, it is tacitly assumed that C_i and C_o, which are perfect DC open circuits, behave as good approximations of AC short circuits.

The subject amplifier utilizes a diode-connected transistor (Q2) for temperature compensation of the static collector current conducted by transistor Q1. For simplicity, assume that these two transistors have identical small-signal parameters of $r_b = 90\ \Omega$, $r_c = 55\ \Omega$, $r_e = 1.5\ \Omega$, $r_\pi = 970\ \Omega$, $r_o = 42\ \text{k}\Omega$, and $\beta = 115$. Let the indicated circuit parameters be $R_1 = 2.2\ \text{k}\Omega$, $R_2 = 1.3\ \text{k}\Omega$, $R_{EE} = 75\ \Omega$, $R_{CC} = 3.9\ \text{k}\Omega$, $R_L = 1.0\ \text{k}\Omega$, and $R_S = 300\ \Omega$. Assuming that these circuit variables ensure linear operation of both devices, determine the small-signal voltage gain, $A_v = V_{OS}/V_S$, the driving point input resistance, R_{in}, and the driving point output resistance, R_{out}, of the amplifier. Finally, calculate the requisite minimum values of the input and output coupling capacitors, C_i and C_o, such that the lowest frequency f_l of interest is 500 Hz.

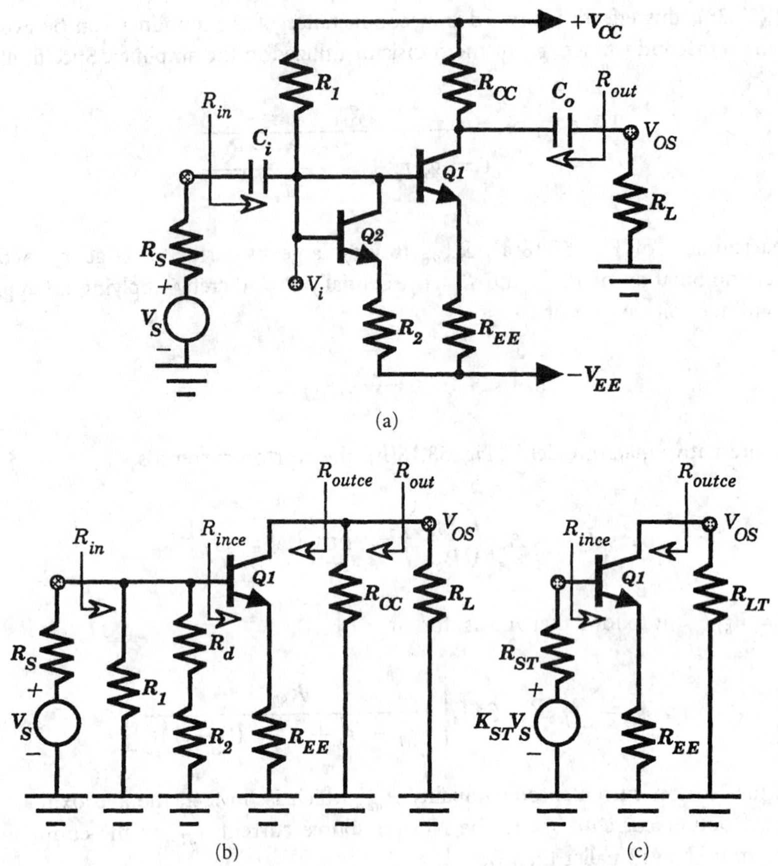

FIGURE 57.19 (a) Common-emitter amplifier with capacitively coupled input and output signal ports. (b) AC schematic diagram of the amplifier in (a). (c) Simplified AC schematic diagram of the amplifier in (a).

Solution.

1) The first step of the solution process entails drawing the AC schematic diagram of the subject amplifier. By casting this diagram in the form of the canonic cell shown in Fig. 57.16(a), the gain and resistance expressions provided above can be exploited directly to assess the small-signal performance of the circuit at hand. Such a solution tack maximizes design-oriented understanding by avoiding the algebraic tedium implicit to an analysis of the entire small-signal equivalent circuit of the amplifier.

To the foregoing end, observe that transistor $Q2$ operates in its linear regime as a diode. It can therefore be viewed as the two terminal resistance R_d, given by (57.31) or (57.32). Since the Early resistance is large, the latter expression can be used to arrive at $R_d = 10.64\ \Omega$.

Since the power supply voltage, V_{EE}, is presumed ideal in the sense that it contains no signal component, the resultant series combination of R_d and R_2 returns the base of transistor $Q1$ to ground, as depicted in Fig. 57.19(b). Similarly, R_1 appears in shunt with the series interconnection of R_d and R_2, since V_{CC}, like V_{EE}, is also presumed to be an ideal constant (zero signal component) source of voltage. The AC schematic diagram of the input port is completed by noting that the AC short-circuit nature of the coupling capacitance C_i effectively connects the Thévenin representation of the signal source directly between the base of $Q1$ and ground.

At the output port of the amplifier, R_{CC} connects between the collector and ground since, as already exploited, V_{CC} is an AC short circuit. Moreover, the external load resistance, R_L, shunts R_{CC}, as shown in Fig. 57.19(b), because C_o behaves as an AC short circuit. The AC schematic diagram is completed by inserting the emitter degeneration resistance R_{EE} as a series element between ground and the emitter of transistor $Q1$.

2) The diagram in Fig. 57.19(b) can be straightforwardly collapsed to the simplified topology of Fig. 57.19(c). In the latter circuit, the effective load resistance, R_{LT} is

$$R_{LT} = R_{CC} \| R_L = 795.9\ \Omega$$

At the input port, the Thévenin resistance seen by the base of $Q1$ is

$$R_{ST} = R_1 \| (R_d + R_2) \| R_S = 219.7\ \Omega$$

while the corresponding Thévenin signal voltage can be expressed as $K_{ST} V_S$, where the voltage divider K_{ST} is

$$K_{ST} = \frac{R_1 \| (R_d + R_2)}{R_1 \| (R_d + R_2) + R_S} = 0.733$$

The implication of this calculation is that insofar as the active transistor $Q1$ is concerned, the biasing resistances R_1 and R_2 cause a loss of more than 25 percent of the applied input signal.

3) The resultant AC schematic diagram in Fig. 57.19(c) is virtually identical to the canonic topology in Fig. 57.16(a). Indeed, if the circuit resistance R_{EE} is absorbed into $Q1$, where it appears in series with the internal emitter resistance r_e, the diagram is identical to the AC schematic diagram of the canonic common-emitter cell. Since $(r_e + R_{EE}) = 76.5\ \Omega$ is better than 560 times smaller than the resistance sum, $(r_o + r_c + R_{LT}) = 42.85\ \text{k}\Omega$, and $\beta = 115$ is more than 63,000 times larger than the resistance ratio $(r_e + R_{EE})/r_o = 0.00182$, the simplified expression in (57.40) can be used to evaluate the voltage gain $V_{OS}/K_{ST} V_S$ in Fig. 57.19(c). From (57.41), the effective small-signal beta is $\beta_{\text{eff}} = 112.7$. Then, with r_e replaced by $(r_e + R_{EE}) = 76.5\ \Omega$, (57.40) gives

$$A_{vce} = \frac{V_{OS}}{K_{ST} V_S} = -8.99\ \text{V/V}$$

It follows that the actual voltage gain of the amplifier in Fig. 57.19(a) is

$$A_v = \frac{V_{OS}}{V_S} = K_{ST} A_{vce} = -6.59\ \text{V/V}$$

A better design, in the sense of achieving an adequate desensitization of circuit transfer characteristics with respect to parametric uncertainties, entails the use of a slightly larger emitter degeneration resistance R_{EE} selected to ensure that $(\beta_{\text{eff}} + 1)(r_e + R_{EE}) \approx \beta_{\text{eff}} R_{EE} \gg (R_{ST} + r_b + r_\pi)$. For such a design, (57.40) produces

$$A_v \approx -\frac{K_{ST} R_{LT}}{R_{EE}}$$

which is nominally independent of transistor parameters.

4) With r_e replaced by $(r_e + R_{EE}) = 76.5\ \Omega$, (57.43) gives for the input resistance seen looking into the base of transistor Q1 in Fig. 57.19(c), $R_{ince} = 9.76\ \text{k}\Omega$. It follows that the driving point input resistance seen by the source circuit in Fig. 57.19(b) is

$$R_{\text{in}} = R_1 \| (R_d + R_2) \| R_{ince} = 757.6\ \Omega$$

5) The output resistance, R_{outce}, seen looking into the collector of transistor Q1 in the diagram of Fig. 57.19(c) is derived from (57.44). With r_e replaced by $(r_e + R_{EE}) = 76.5\ \Omega$, $R_{outce} = 2.49$ MΩ. The resultant driving point output resistance seen by the load circuit in Fig. 57.19(b) is

$$R_{\text{out}} = R_{CC} \| R_{outce} = 3{,}894\ \Omega$$

The circuit output resistance is only a scant 6 Ω smaller than the collector biasing resistance R_{CC}, owning to the large value of R_{outce}. In turn, the value of the latter resistance is dominated by the last term on the right-hand side of (57.44), which is proportional to the large forward Early resistance, r_o.

6) The input coupling capacitance, C_i, can be calculated with the help of the input port macromodel of Fig. 57.20(a). In this model, the subcircuit to the right of C_i in Fig. 57.19(a) is replaced by its Thévenin equivalent circuit, which consists of the driving point input resistance, R_{in}, calculated above. The voltage transfer function of this input port is

$$\frac{V_i(j\omega)}{V_S(j\omega)} = \left(\frac{R_{\text{in}}}{R_{\text{in}} + R_S}\right)\left[\frac{j\omega(R_{\text{in}} + R_S)C_i}{1 + j\omega(R_{\text{in}} + R_S)C_i}\right]$$

An inspection of the foregoing relationship confirms that the dynamical effect of C_i on the voltage transfer function response is minimized if $\omega\,(R_{\text{in}} + R_S)C_i \gg 1$. At $\omega = 2\pi f_l$, the value of C_i that makes the left-hand side of this inequality equal to one is $C_i = 0.30\ \mu\text{F}$. Observe that at $\omega_l(R_{\text{in}} + R_S)C_i = 1$,

$$\left|\frac{V_i(j\omega_l)}{V_S(j\omega_l)}\right| = \left(\frac{R_{\text{in}}}{R_{\text{in}} + R_S}\right)\left|\frac{j}{1+j}\right| = \frac{1}{\sqrt{2}}\left(\frac{R_{\text{in}}}{R_{\text{in}} + R_S}\right)$$

that is, the magnitude of the input port voltage transfer function is a factor of the square root of two, or 3 dB, below the transfer function value realized at signal frequencies that are significantly higher than f_l. If this 3-dB attenuation is acceptable, $C_i = 0.30\ \mu\text{F}$ is appropriate to the design requirement.

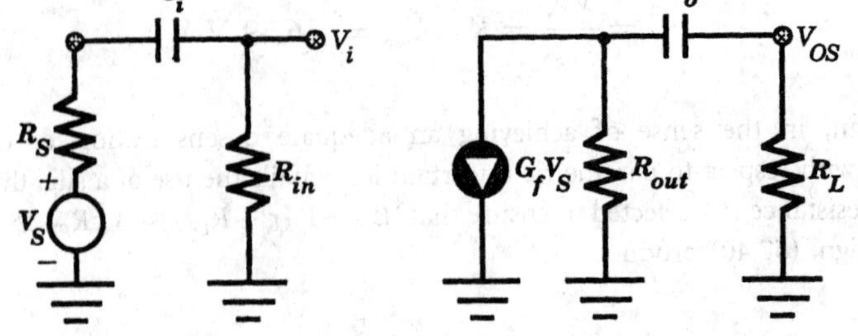

FIGURE 57.20 (a) Input port macromodel used in the calculation of the input coupling capacitor C_i. (b) Output port macromodel used to calculate the output coupling capacitor C_o.

7) The output coupling capacitance, C_o, is calculated analogously by exploiting the macromodel concepts overviewed in Fig. 57.18(a). To this end, the output port macromodel is offered in Fig. 57.20(b), where the effective forward transconductance, G_f, is such that $-G_f(R_{out}\|R_L) = A_v$, as calculated in step 3. The voltage gain is seen to be

$$A_v(j\omega) = \frac{V_{OS}(j\omega)}{V_S(j\omega)} = -[G_f(R_{out}\|R_L)]\left[\frac{j\omega(R_{out}+R_L)C_o}{1+j\omega(R_{out}+R_L)C_o}\right]$$

which has an algebraic form that is similar to the foregoing transfer relationship for the amplifier input port. Thus, C_o is

$$C_o \geq \frac{1}{2\pi f_l(R_{out}+R_L)} = 0.065\ \mu\text{F}$$

Since the 0.30 μF input capacitor and the 0.065 μF output capacitor establish identical input and output port left-half-plane poles at the same frequency (f_l), the resultant attenuation at f_l is actually larger than 3 dB. If this enhanced attenuation is unacceptable, the smaller of the two coupling capacitances can be made larger by a factor of three or so, thereby translating the associated pole frequency downward by a factor of three. In the present case, a plausible value of C_o is $C_o \geq (3)(0.065\ \mu\text{F}) = 0.2\ \mu\text{F}$.

It should be noted that the requisite two coupling capacitances are orders of magnitude too large for monolithic realization. Accordingly, if the subject amplifier is an integrated circuit, C_i and C_o are necessarily off-chip elements.

Example 2. When very large magnitudes of voltage gains are required, the output port of a common-emitter configuration can be terminated in an active load, as opposed to the passive resistive load encountered in the preceding example. Consider, for example, the complementary, NPN–PNP transistor amplifier whose schematic diagram appears in Fig. 57.21(a). The subcircuit containing transistors Q1 and Q2 is identical to that of the amplifier in Fig. 57.19(a). Indeed, for the purpose of this example, let the resistances, R_1, R_2, R_{EE}, and R_S, as well as the small-signal parameters of transistors Q1 and Q2, remain at the values respectively stipulated for them in the preceding example. In the present diagram, the PNP transistor Q3, along with its peripheral biasing resistances R_3, R_4, and R_5, supplants the resistance R_{CC} in the previously addressed common-emitter unit. Since no signal is applied to the base of Q3, the subcircuit consisting of Q3, R_3, R_4, and R_5 serves only to supply the appropriate biasing current to the collector of transistor Q1. To the extent that this static current is invariant with temperature and the voltage signal response, V_{OS}, established at the collector of Q1, the Q1 load circuit functions as a nominally constant current source. As a result, the effective load resistance, indicated as R_L in the subject figure, seen by the collector of Q1 is very large. In view of the absence of an external load appended to the output port, the resultant voltage gain of the amplifier is commensurately large.

Let $R_3 = 1.8$ kΩ, $R_4 = 3.3$ kΩ, and $R_5 = 100$ Ω. Moreover, let the small-signal parameters of the PNP transistor be $r_{bp} = 40\ \Omega$, $r_{cp} = 70\ \Omega$, $r_{ep} = 9\ \Omega$, $r_{\pi p} = 1100\ \Omega$, $r_{op} = 30$ kΩ, and $\beta_p = 60$. Assuming linear operation of all devices, determine the small-signal voltage gain $A_v = V_{OS}/V_S$, the driving point input resistance R_{in}, and the driving point output resistance R_{out} of the amplifier. As in the preceding example, the input coupling capacitance C_i can be presumed to act as an ac short circuit for the signal frequencies of interest.

Solution.

1) The ac schematic diagram of the amplifier in Fig. 57.21(a) is given in Fig. 57.21(b), where the PNP transistor load subcircuit is represented as an effective two terminal load resistance,

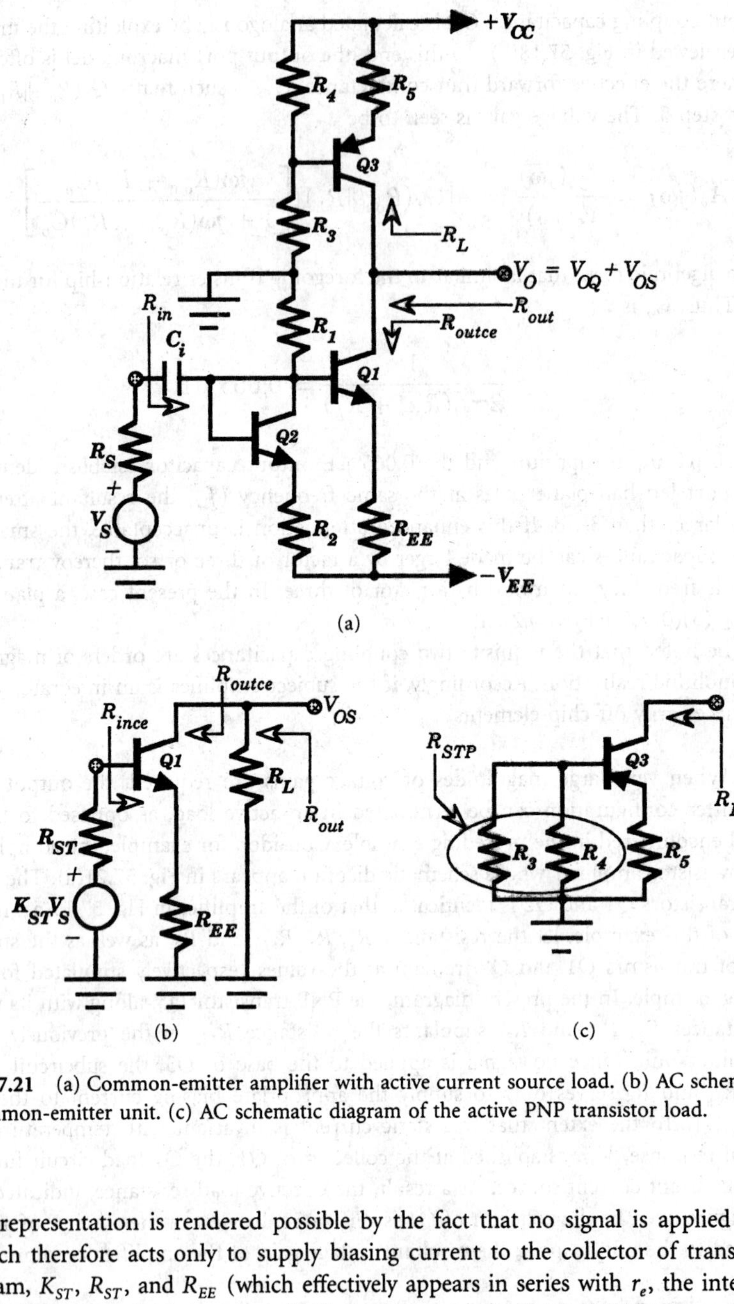

FIGURE 57.21 (a) Common-emitter amplifier with active current source load. (b) AC schematic diagram of the common-emitter unit. (c) AC schematic diagram of the active PNP transistor load.

R_L. This representation is rendered possible by the fact that no signal is applied to the PNP load, which therefore acts only to supply biasing current to the collector of transistor $Q1$. In the diagram, K_{ST}, R_{ST}, and R_{EE} (which effectively appears in series with r_e, the internal emitter resistance of $Q1$) remain the same as in Example 1, namely, $K_{ST} = 0.733$, $R_{ST} = 219.7\ \Omega$, and $R_{EE} = 75\ \Omega$.

2) The AC schematic diagram of the $Q3$ subcircuit alone appears in Fig. 57.21(c). A comparison of this figure with that shown in Fig. 57.21(b) suggests that the subject diagram represents a PNP common-emitter amplifier under zero signal conditions. In particular, the Thévenin source resistance seen by the base of the PNP unit is $R_{STP} = R_3 \parallel R_4 = 1165\ \Omega$, while the emitter degeneration resistance of this subcircuit is $R_5 = 100\ \Omega$. It follows that the effective AC load resistance R_L terminating the collector port of $Q1$ is the driving point output resistance of a common-emitter stage. With $r_c \triangleq r_{cp} = 70\ \Omega$, $r_b \triangleq r_{bp} = 40\ \Omega$, $r_e \triangleq (r_{ep} + R_5) = 109\ \Omega$,

$r_\pi \triangleq r_{\pi p} = 1.1\ \text{k}\Omega$, $R_{ST} \triangleq R_{STP} = 1165\ \Omega$, $r_o \triangleq r_{op} = 30\ \text{k}\Omega$, and $\beta \triangleq \beta_p = 60$, (57.44) yields $R_{outce} \triangleq R_L = 111.5\ \text{k}\Omega$.

3) The voltage gain, input resistance, and output resistance of the actively loaded common-emitter amplifier can now be computed. For $r_c \triangleq r_{cn} = 55\ \Omega$, $r_b \triangleq r_{bn} = 90\ \Omega$, $r_e \triangleq (r_{en} + R_{EE}) = 76.5\ \Omega$, $r_\pi \triangleq r_{\pi n} = 970\ \Omega$, $R_{ST} \triangleq R_{STN} = 219.7\ \Omega$, $r_o \triangleq r_{on} = 42\ \text{k}\Omega$, $R_{LT} \triangleq R_L = 111.5\ \text{k}\Omega$, and $\beta \triangleq \beta_n = 115$, (57.41) gives an effective NPN transistor beta of $\beta_{\text{eff}} = 31.46$, and (57.40) yields a voltage gain of $A_{vce} = V_{OS}/K_{ST}V_{ST} = -931.9$. It follows that the small-signal voltage gain of the stage at hand is $A_v = V_{OS}/V_S = K_{ST}A_{vce} = -682.6$ V/V. It should be noted that this voltage gain is the ratio of only the signal component, V_{OS}, of the net output voltage, V_O (which contains a quiescent component of V_{OQ}) to the source signal voltage, V_S.

4) From (57.42), the driving point input resistance seen looking into the base of transistor $Q1$ in Fig. 57.19(b) is $R_{ince} = 3.54\ \text{k}\Omega$. Then,

$$R_{\text{in}} = R_1 \| (R_d + R_2) \| R_{ince} = 666.7\ \Omega$$

This input resistance differs slightly from the corresponding calculation in the preceding example owing to the reduction in the effective forward AC beta caused by the large active load resistance.

5) The resistance R_{outce} seen looking into the collector of transistor $Q1$ remains the same as calculated in Example 1, namely $R_{outce} = 2.49\ \text{M}\Omega$. It follows that the driving point output resistance of the amplifier under investigation is

$$R_{\text{out}} = R_L \| R_{outce} = 106.7\ \text{k}\Omega$$

This large output resistance means that the actively loaded common-emitter configuration is a relatively poor voltage amplifier. In particular, an output buffer is mandated to couple virtually any practical external load resistance to the amplifier output port. In addition to reducing the output resistance, such a properly designed and implemented output buffer can reliably establish and stabilize the quiescent output voltage, V_{OQ}.

Common Base Amplifier

The second of the canonic linear bipolar gain cells is the *common-base amplifier*, whose NPN and PNP AC schematic diagrams and corresponding small-signal equivalent circuit appear in Fig. 57.22(a) and (b), respectively. As is confirmed below, the input resistance, R_{incb}, of this stage is very small and the output resistance, R_{outcb}, is very large. Accordingly, the common-base unit comprises a relatively poor voltage amplifier in the sense that its voltage gain, though potentially large, is a sensitive function of both the Thévenin source resistance R_{ST} and the Thévenin load resistance R_{LT}.

Although the common-base amplifier is not well suited for general voltage gain applications, it is an excellent *current buffer*, which is ideally characterized by zero input resistance, infinitely large output resistance, and unity current gain. When used for current buffering purposes, the common-base amplifier rarely appears as a stand alone single-stage amplifier, since signal excitations, particularly at the input and the output ports of an electronic system are invariably formatted as voltages. Instead, it is invariably used in conjunction with an input voltage to current converter and/or an output current to voltage converter to achieve desired system performance characteristics.

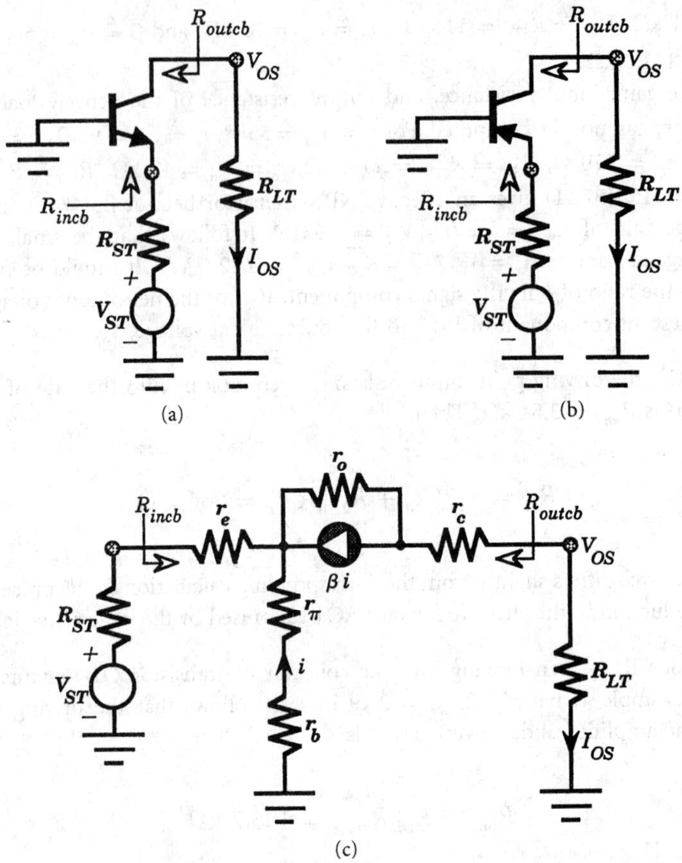

FIGURE 57.22 (a) AC schematic diagram of an NPN common-base amplifier. (b) AC schematic diagram of a PNP common-base amplifier. (c) Small-signal low-frequency equivalent circuit of the common-base amplifier.

The small-signal analysis of the common-base stage is considerably simplified if the assumption of large r_o is exploited at the outset. To this end, the equivalent circuit shown in Fig. 57.22(b) reduces to the structure of Fig. 57.23(a). In the latter equivalent circuit, observe a signal emitter current i_{es} that relates to the indicated signal base current i in accordance with the KCL constraint

$$i_{es} = -(\beta + 1)i \tag{57.51}$$

The signal component of the output current I_{OS} is therefore expressible as

$$I_{OS} = -\beta_i = \left(\frac{\beta}{\beta + 1}\right) i_{es} = \alpha i_{es} \tag{57.52}$$

where (57.51) and (57.52) are used. The last result suggests the alternative model in Fig. 57.23(b), which is a slightly more convenient version of the model in Fig. 57.23(a) in that the current-controlled current source, αi_{es}, is dependent on the signal input port current i_{es}, as opposed to the signal current, i, that flows in the grounded base lead.

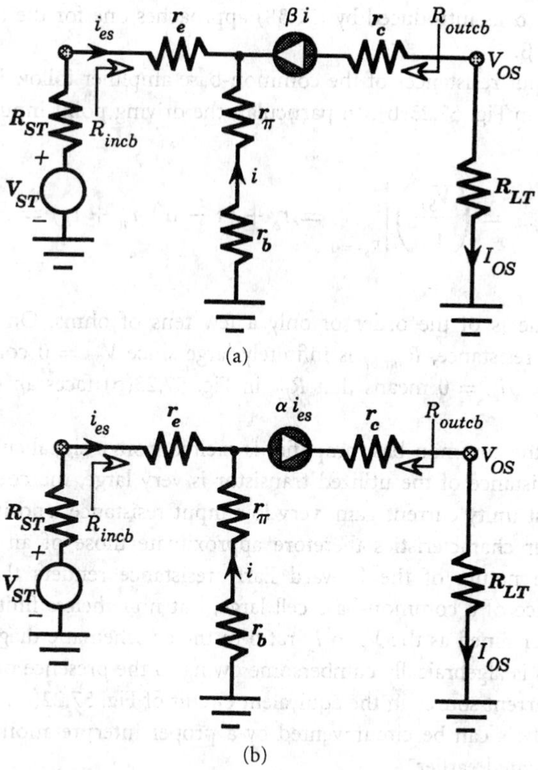

FIGURE 57.23 (a) The equivalent circuit of Fig. 57.22(c), simplified for the case of very large Early resistance r_o. (b) Modification of the circuit in (a) in which the current-controlled current source is rendered dependent on the input signal current i_{es}.

By inspection of the equivalent circuit in Fig. 57.23(b), the small-signal voltage gain A_{vcb} of the common-base cell is

$$A_{vcb} = \frac{\alpha R_{LT}}{R_{ST} + r_e + (1-\alpha)(r_\pi + r_b)} = \frac{\alpha R_{LT}}{R_{ST} + R_d} \tag{57.53}$$

where (57.38) is used once again and R_d is the diode resistance defined by (57.32). In contrast to the common-emitter cell, the common-base stage has no voltage gain phase inversion. But like the common-emitter configuration, the common-base voltage gain is directly proportional to the effective load resistance. It is also almost inversely proportional to the effective source resistance, given that the diode resistance R_d is small.

Although the voltage gain is vulnerable to uncertainties in the terminating load and source resistances, the common-base current gain, A_{icb}, is virtually independent of R_{LT} and R_{ST}. This contention follows from the fact that A_{icb}, which is the ratio of I_{OS} to the Norton equivalent source current, V_{ST}/R_{ST}, is

$$A_{icb} = \left(\frac{R_{ST}}{R_{LT}}\right) A_{vcb} = \frac{\alpha R_{ST}}{R_{ST} + R_d} \tag{57.54}$$

which is independent of R_{LT} (to the extent that the Early resistance r_o can indeed be ignored). Since the signal in a current-drive amplifier is likely to have a large source resistance, $R_{ST} \gg R_d$, which implies $A_{icb} \approx \alpha$, independent of R_{LT} and R_{ST}. Note that this approximate current gain is essentially unity, since α as introduced by (57.38) approaches one for the typically encountered circumstance of large β.

The input and output resistances of the common-base amplifier follow immediately from an analysis of the model in Fig. 57.23(b). In particular, the driving point input resistance, R_{incb}, is

$$R_{incb} = \left(\frac{V_{ST}}{i_{es}}\right)\bigg|_{R_{ST}=0} = r_e + (1-\alpha)(r_\pi + r_b) = R_d \tag{57.55}$$

whose numerical value is of the order of only a few tens of ohms. On the other hand, the driving point output resistance, R_{outcb}, is infinitely large since $V_{ST} = 0$ constrains i_{es}, and thus αi_{es}, to zero. In turn, $\alpha i_{es} = 0$ means that R_{LT} in Fig. 57.23(b) faces an open circuit, whence $R_{outcb} = \infty$.

To the extent that the common-base amplifier is excited from a signal current source and that the forward Early resistance of the utilized transistor is very large, the common-base amplifier is seen to have almost unity current gain, very low input resistance, and infinitely large output resistance. Its transfer characteristics therefore approximate those of an ideal current buffer. Of course, the finite nature of the forward Early resistance renders the observable driving point output resistance of a common-base cell large, but nonetheless finite. The actual output resistance can be determined as the V_x to I_x ratio in the ac schematic diagram of Fig. 57.24(a). The requisite analysis is algebraically cumbersome owing to the presence of r_o in shunt with the current-controlled current source in the equivalent circuit of Fig. 57.22(c). Fortunately, however, an actual circuit analysis can be circumvented by a proper interpretation of cognate common emitter results formulated earlier.

In order to demonstrate the foregoing contention, consider Fig. 57.24(b), which depicts the AC schematic diagram for determining the driving point output resistance R_{outce} of a common-emitter amplifier. The only difference between the two AC schematic diagrams in Fig. 57.24 is the topological placement of the effective source resistance, R_{ST}. In the common-base stage, this source resistance is in series with the emitter of a transistor whose base is grounded. On the other hand, R_{ST} appears in the common emitter configuration as an element in series with the base of a transistor whose emitter is grounded. It follows that (57.44) can be used to deduce an expression for R_{outcb}, provided that in (57.44) R_{ST} is set to zero and r_e is replaced by $(r_e + R_{ST})$.

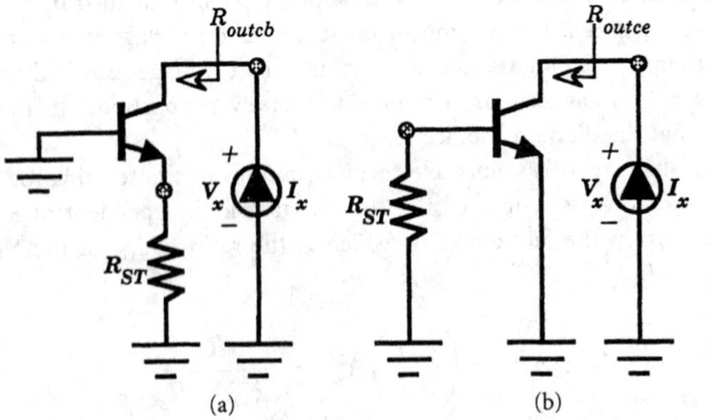

FIGURE 57.24 (a) AC schematic diagram appropriate to the computation of the driving point output resistance of a common-base amplifier. (b) AC schematic diagram pertinent to computing the driving point output resistance of a common-emitter amplifier.

The result is

$$R_{outcb} = r_c + (r_e + R_{ST}) \| (r_\pi + r_b) + \left[\frac{\beta(r_e + R_{ST})}{r_e + R_{ST} + r_\pi + r_b} + 1 \right] r_0 \qquad (57.56)$$

For large R_{ST} and large r_o, (57.46) reduces to

$$R_{outcb} \approx (\beta + 1) r_o \qquad (57.57)$$

which is an extremely large output resistance.

The common-base stage is generally used in conjunction with a common-emitter amplifier to form the *common-emitter–common base cascode,* whose schematic diagram is shown in Fig. 57.25. In this application, the common-emitter stage formed by transistor $Q1$, the emitter degeneration resistance, R_{EE}, and the biasing elements, R_1 and R_2, serves as a transconductor that converts the input signal voltage, V_S, to a collector current whose signal component is i_{1s}. Note that such conversion is encouraged by the fact that the effective load resistance, R_{Leff}, terminating the collector of Q1, is the presumably low input resistance of the common-base stage formed by transistor Q2 and the biasing resistances, R_3 and R_4. Since the current gain of a common-base stage is essentially unity, Q2 translates the signal current in its emitter to an almost identical signal current flowing through the collector load resistance, R_L. The latter element acts as a current to voltage converter to establish the signal component, V_{OS}, of the net output voltage V_O.

The analysis of the common-emitter–common-base cascode begins by representing the collector port of the common-emitter configuration by its Norton equivalent circuit. Assuming that

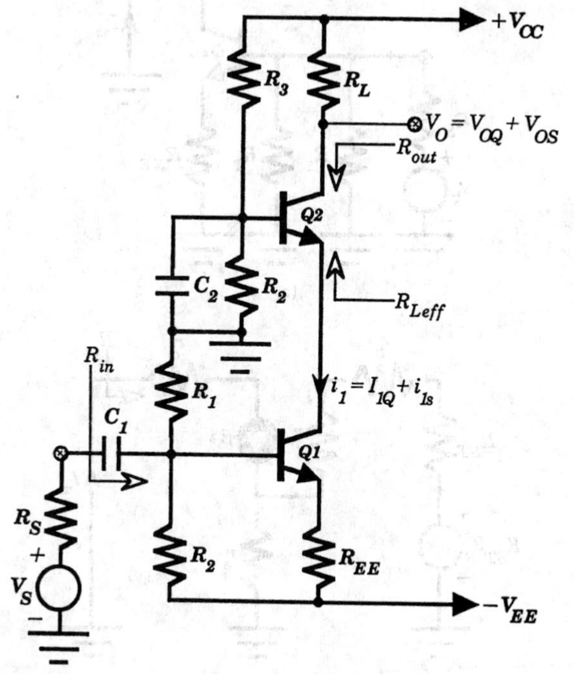

FIGURE 57.25 Schematic diagram of a common-emitter–common-base cascode. The common-emitter stage formed by transistor Q1 and its peripheral elements acts as a voltage to current converter. Transistor Q2 and its associated biasing elements functions as a current amplifier, while the load resistance, R_L, acts as a current to voltage converter.

the input coupling capacitor, C_1, is sufficiently large to enable its replacement by a short circuit over the signal frequency range of interest, the pertinent AC schematic diagram is the circuit in Fig. 57.26(a), where I_{ns} symbolizes the Norton, or short-circuit signal current conducted by the collector of transistor $Q1$. The corresponding small-signal equivalent circuit appears in Fig. 57.26(b), where the Early resistance is tacitly ignored, the effective source resistance, R_{ST}, seen by the base of $Q1$ is

$$R_{ST} = R_S \| R_1 \| R_2 \tag{57.58}$$

and the voltage divider K_{ST} is

$$K_{ST} = \frac{R_1 \| R_2}{R_1 \| R_2 + R_S} \tag{57.59}$$

Using the model in Fig. 57.26(b) it is a simple matter to show that

$$I_{ns} = \beta i = G_{ns} V_S \tag{57.60}$$

where G_{ns}, which can be termed the **Norton transconductance** of the common-emitter stage, is

$$G_{ns} = \frac{\beta K_{ST}}{R_{ST} + r_b + r_\pi + (\beta + 1) r_e} \tag{57.61}$$

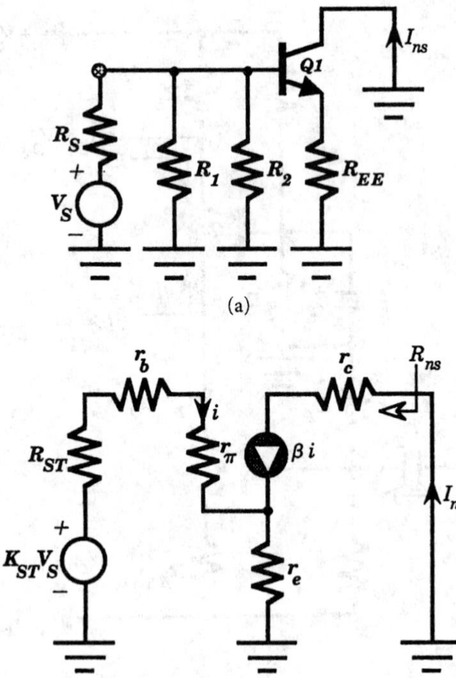

FIGURE 57.26 (a) AC schematic diagram used to calculate the Norton equivalent output circuit of the common-emitter subcircuit in the cascode configuration of Fig. 57.25. (b) Small-signal model of the AC circuit in (a).

The Norton output resistance R_{ns} is infinitely large by virtue of the assumption of infinitely large Early resistance.

The foregoing results permit drawing the AC schematic diagram of the common-base component of the common-emitter–common-base cascode in the topological form depicted in Fig. 57.27(a). From the corresponding small-signal equivalent circuit in Fig. 57.27(b), which assumes that the capacitor, C_2, behaves as an AC short circuit and which once again ignores transistor Early resistance, the voltage gain, say A_v, follows immediately as

$$A_v = \frac{V_{OS}}{V_S} = -\alpha G_{ns} R_L = -\frac{\alpha\beta K_{ST} R_L}{R_{ST} + r_b + r_\pi + (\beta+1) r_e} \tag{57.62}$$

The driving point output resistance R_{out}, like the Norton output resistance of the common-emitter stage, is infinitely large. In fact, R_{out} is a good approximation of infinity. Its numerical value approaches $(\beta + 1)r_o$, since the terminating resistance, R_{ns}, seen in the emitter circuit of transistor Q2 is of the order of r_o.

Equation (57.62) is similar in form to (57.40) which defines the voltage gain of a simple common-emitter amplifier. A careful comparison of the two subject relationships suggests that the voltage gain of the common-emitter–common-base cascode of Fig. 57.25 is equivalent to the voltage gain achieved by a simple common-emitter stage, whose output port is loaded in an effective load resistance of αR_L. Although α is close to unity, but nonetheless always less than one, an effective load of αR_L implies that the voltage gain of the cascode is slightly less

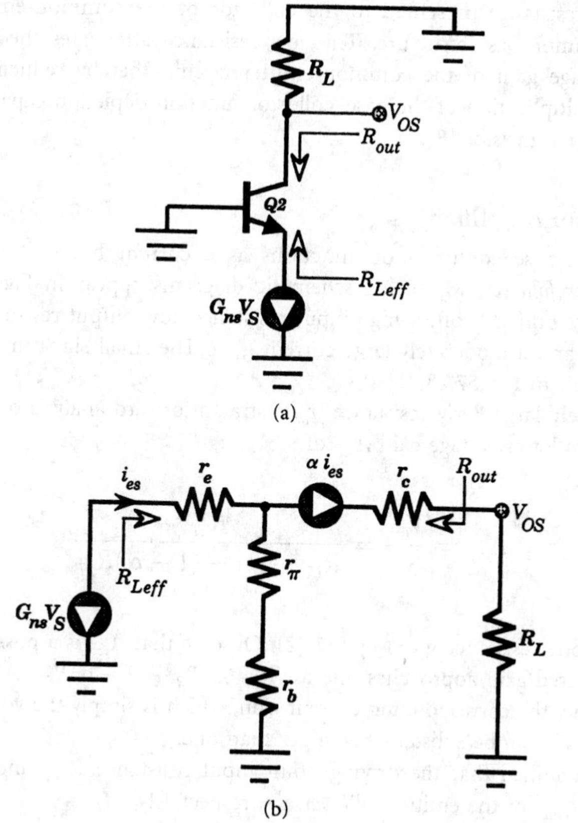

FIGURE 57.27 (a) The effective AC schematic diagram of the common-base component of the common-emitter–common-base cascode of Fig. 57.25. (b) Small-signal model of the AC circuit in (a).

than that achieved by the common-emitter stage alone, provided, of course, that the transistors utilized in both configurations have identical small-signal parameters. A question therefore arises as to the prudence of incorporating common-base signal processing in conjunction with a common-emitter unit stage.

In fact, no practical purpose is served by a common-emitter–common-base cascode if the load resistance R_L driven by the amplifier is very small. But if the load resistance imposed on the output port of a simple common-emitter amplifier is large, as it is when the load itself is realized actively, as per Example 2, the effective transistor beta defined by (57.41) is appreciably smaller than the actual small-signal beta. The result is a degraded common-emitter amplifier voltage gain. In this situation, the insertion of a common-base stage between the output port of the common-emitter amplifier and R_L, as diagrammed in Fig. 57.25, increases the degraded gain of the common-emitter amplifier alone by restoring the effective beta of the common emitter transistor to a value that approximates the actual small-signal beta of the transistor. This observation follows from the fact that the effective load resistance, $R_{L\,\text{eff}}$, seen by the collector of the common emitter transistor in the cascode topology is of the order of only a small diode resistance. It follows from (57.41) that β_{eff} for $R_{LT} = R_{L\,\text{eff}}$ is likely to be significantly larger than the value of β_{eff} that derives from the load condition, $R_{LT} = R_L$.

The reason for using common-base circuit technology in conjunction with a common-emitter amplifier is circuit broadbanding. In particular, a carefully designed common-emitter–common-base cascode configuration displays a 3-dB bandwidth and a gain–bandwidth product that are significantly larger than the bandwidth and gain–bandwidth product afforded by a common-emitter stage of comparable gain. The primary reason underlying this laudable attribute is the low effective load resistance presented to the collector of the common-emitter stage by the emitter of the common-base structure. This low resistance attenuates the magnitude of the phase-inverted voltage gain of the common-emitter circuit, thereby reducing the deleterious effects of Miller multiplication of the base–collector junction depletion capacitance implicit to the common-emitter transistor [8].

Common Collector Amplifier

While the common-base configuration functions as a current buffer, the *common-collector amplifier*, or *emitter follower*, whose AC schematic diagrams appear in Fig. 57.28(a) and (b) operates as a voltage buffer. It offers high input resistance, low output resistance, a voltage gain approaching unity, and a moderately large current gain. The small-signal model of the emitter follower is the circuit in Fig. 57.28(c).

Assuming infinitely large Early resistance, r_o, a straightforward analysis of the subject model reveals an emitter-follower voltage gain A_{vcc} of

$$A_{vcc} = \frac{V_{OS}}{V_S} = \frac{R_{LT}}{R_{LT} + R_d + (1 - \alpha)R_{ST}} \tag{57.63}$$

where R_d is the diode resistance given by (57.32). Observe that A_{vcc} is a positive less than unity number. The indicated gain approaches one for $R_{LT} \gg R_d + (1 - \alpha)R_{ST}$. Although the voltage gain is less than one, the corresponding current gain, which is simply the voltage gain scaled by a factor of (R_{ST}/R_{LT}), can be substantially larger than one.

It is simple to confirm that the driving point input resistance R_{incc} and the driving point output resistance R_{outcc} of the emitter follower are, respectively,

$$R_{incc} = r_b + r_\pi + (\beta + 1)(r_e + R_{LT}) = (\beta + 1)(R_d + R_{LT}) \tag{57.64}$$

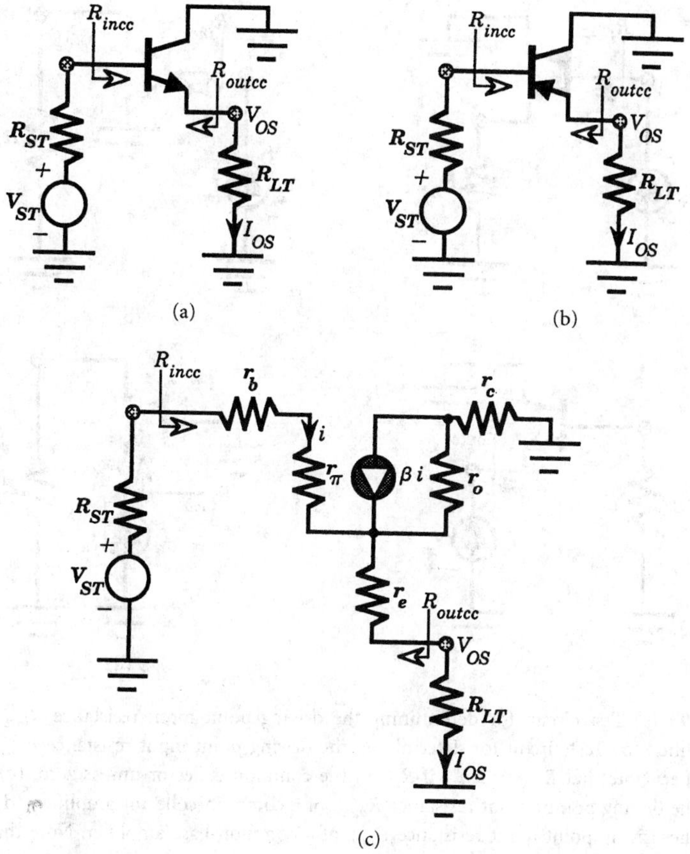

FIGURE 57.28 (a) AC schematic diagram of an NPN common-collector amplifier. (b) AC schematic diagram of a PNP common-collector amplifier. (c) Small-signal low-frequency equivalent circuit of the common-collector amplifier, assuming that the Early resistance is sufficiently large to ignore.

and

$$R_{outcc} = r_e + \frac{r_\pi + r_b + R_{ST}}{(\beta + 1)} = R_d + (1 - \alpha)R_{ST} \tag{57.65}$$

It is interesting and instructive to note that the driving point input resistance R_{incc}, of an emitter follower is of the same form as the driving point input resistance R_{ince}, of a common-emitter stage. Indeed, if R_{LT} in the emitter follower is zero, $R_{incc} \equiv R_{ince}$. This result is reasonable in view of the fact that for both the emitter-follower and common-emitter configurations, the input resistance is, as suggested by the test circuits in Figs. 57.29(a) and (b), the Thévenin resistance presented to the source circuit by the base of the subject transistor. Moreover, the common-collector output resistance R_{outcc} mirrors the driving point input resistance R_{incb} for the common-base amplifier. In fact, $R_{incb} \equiv R_{outcc}$ if the base of a common-base amplifier is terminated to ground in a resistance of value R_{ST}. Once again, the latter observation is intuitively correct, for, as depicted in Figs. 57.29(c) and (d), the emitter of the transistor comprises the input terminal for the common-base amplifier and the output terminal of an emitter follower.

An inspection of (57.63) confirms a common-collector voltage gain that tends toward unity for progressively larger Thévenin load resistances, R_{LT}. Correspondingly, the driving point input resistance of an emitter follower increases dramatically with increasing R_{LT}. These observations

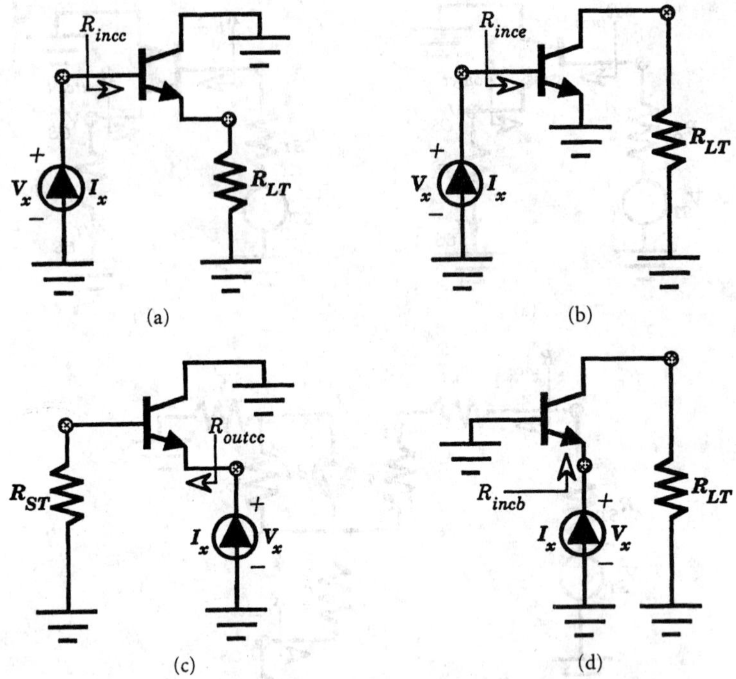

FIGURE 57.29 (a) Test circuit for determining the driving point input resistance R_{incc} of a common-collector amplifier. (b) Test circuit for determining the driving point input resistance R_{ince} of a common-emitter amplifier. Note that $R_{incc} = R_{ince}$ if R_{LT} in the common collector unit is zero. (c) Test circuit for determining the driving point output resistance R_{outcc} of a common-collector amplifier. (d) Test circuit for determining the driving point input resistance R_{incb} of a common-base amplifier. Note that $R_{outcc} = R_{incb}$ if R_{ST} in the common-collector unit is zero.

are often pragmatically exploited by supplanting the passive load in the schematic diagram of Fig. 57.28(a) with an active load, as suggested in Fig. 57.30(a). Since the effective AC load resistance must be large to achieve a near unity voltage gain, this active load must function as a sink of nominally constant current. To this end, the active load in question is shown conducting a net current that consists of a static current component I_{CS} and a signal current component I_{OS}, where the constant current sink nature of the load implies $I_{CS} \gg I_{OS}$.

The subject active load can be represented by its Norton equivalent circuit, as diagrammed in Fig. 57.30(b), where I_{CS} is depicted as a constant current source and I_{OS} is made to flow through a resistance, R_{cs}. The latter branch element represents the dynamic resistance presented to the emitter-follower output port by the two terminal active termination. Note that $R_{cs} = \infty$ yields $I_{OS} = 0$, which implies an active load that behaves as an ideal constant current sink. The corresponding AC schematic diagram, which is offered in Fig. 57.30(c), is derived from Fig. 57.30(b) by setting I_{CS} to zero, since I_{CS} itself is a constant current that is devoid of any signal component. Additionally, the biasing sources, V_{CC}, V_{EE}, and E_{SS} are presumed ideal and are therefore set to zero as well.

The AC schematic diagram in Fig. 57.30(c) is identical to that in Fig. 57.28(a), subject to the proviso that $R_{LT} = R_{cs}$. But since R_{cs} is presumably a large resistance, substituting $R_{LT} = R_{cs}$ into (57.63) to evaluate the voltage gain of the actively loaded emitter follower is at least theoretically inappropriate because the subject gain equation is premised on the assumption of a large Early resistance, r_o; specifically (57.63) reflects the assumption $R_{LT} \ll r_o$. A voltage gain expression, more accurate than (57.63), derives from an analysis of the model in Fig. 57.28(c). If r_o is

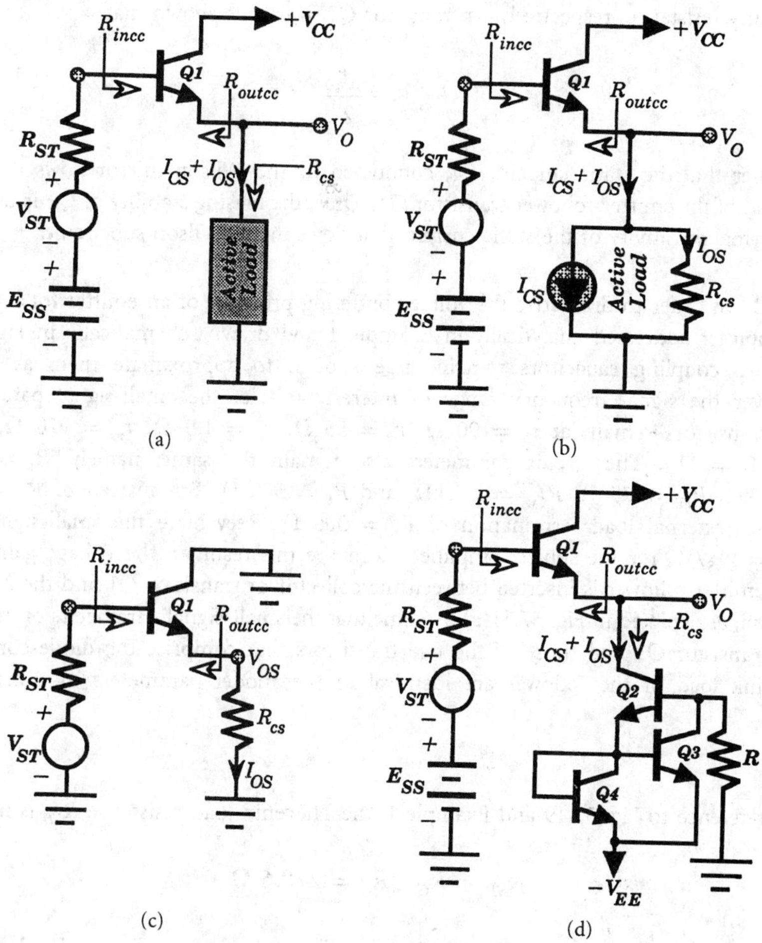

FIGURE 57.30 (a) Emitter follower with active load that conducts a static current I_{CS} and a signal component I_{OS}. (b) The Norton equivalent circuit of the active load. (c) AC schematic diagram of the actively loaded emitter follower. (d) Wilson current mirror realization of the active load.

included in this analysis, but if r_e and r_c are sufficiently small to justify their neglect, the result for $R_{LT} = R_{cs}$ is

$$A_{vcc} = \frac{V_{OS}}{V_S} \approx \frac{r_o \| R_{cs}}{(r_o \| R_{cs}) + R_d + (1-\alpha)R_{ST}} \tag{57.66}$$

whose algebraic form collapses to (57.63) if $R_{cs} \ll r_o$. Similarly, the revised expression for the driving point input resistance is

$$R_{incc} \approx r_b + r_\pi + (\beta + 1)(r_o \| R_{cs}) \tag{57.67}$$

The output resistance R_{outcc} remains as stipulated by (57.65).

The active load appearing in Fig. 57.30(a) can be realized as any one of a variety of NPN current sources [9]. Figure 57.30(d) offers an example of such a realization in the form of the Wilson current mirror formed of transistors Q2, Q3, and Q4, and the current setting resistor, R [10]. This subcircuit establishes an extremely high dynamic resistance between the collector of transistor Q2 and the signal ground. In particular, if β_2 and r_{o2} symbolize the AC beta and forward Early resistance, respectively, of transistor Q2, it can be shown that

$$R_{cs} \approx \frac{\beta_2 r_{o2}}{2} \qquad (57.68)$$

Note further that the static current, I_{CS}, conducted by the Wilson mirror flows through the emitter lead of the emitter-follower transistor Q1. Thus, the biasing stability of Q1 is determined by the thermal sensitivity of the static current that flows in the Wilson subcircuit.

Example 3. In order to dramatize the voltage buffering property of an emitter follower, return to the amplifier addressed analytically in Example 1 and drawn schematically in Fig.57.19(a). Let the two coupling capacitors remain large enough to approximate them as AC short circuits over the signal frequency range of interest, and let the small-signal parameters of the two transistors remain at $r_b = 90\ \Omega$, $r_c = 55\ \Omega$, $r_e = 1.5\ \Omega$, $r_\pi = 970\ \Omega$, $r_o = 42$ kΩ, and $\beta = 115$. The circuit parameters also remain the same; namely, $R_1 = 2.2$ kΩ, $R_2 = 1.3$ kΩ, $R_{EE} = 75\ \Omega$, $R_{CC} = 3.9$ kΩ, and $R_S = 300\ \Omega$. But instead of $R_L = 1.0$ kΩ, consider an external load termination of $R_L = 300\ \Omega$. Reevaluate the small-signal voltage gain, $A_v = V_{OS}/V_S$, for the subject amplifier. Compare this result to the voltage gain achieved when an emitter follower is inserted between the collector of transistor Q1 and the 300 Ω load termination, as depicted in Fig. 57.31(a). Assume that the small-signal parameters of the emitter-follower transistor Q3 and those of the two transistors that comprise the diode-compensated current sink load of the follower are identical to the model parameters of transistors Q1 and Q2.

Solution.

1) With reference to Fig. 57.19 and Example 1, the Thévenin load resistance R_{LT} is now

$$R_{LT} = R_{CC} \| R_L = 278.6\ \Omega$$

The parameters R_{ST} and K_{ST} in Fig. 57.19(c) remain unchanged at the previously computed values of $R_{ST} = 219.7\ \Omega$ and $K_{ST} = 0.733$. Then, ignoring the effects of the finite, but large, Early resistance r_o, the voltage gain is

$$A_v = \frac{V_{OS}}{V_S} \approx -\frac{\beta K_{ST} R_{LT}}{R_{ST} + r_b + r_\pi + (\beta + 1)(r_e + R_{EE})}$$

whence $A_v = -2.31$ V/V.

2) Consider now the amplifier modification shown in Fig. 57.31(a). Transistor Q3 functions as an emitter follower to buffer the terminating load resistance R_L effectively seen by the gain stage formed of transistor Q1 and its peripheral elements. Transistors Q3 and Q4 form a diode-compensated current sink that comprises the active load presented to the emitter-follower output port under static operational conditions. To the extent that r_o can be tacitly ignored, this current sink comprises an infinitely large dynamic resistance. Accordingly, the AC schematic diagram seen to the right of the collector of transistor Q1 is the structure identified in Fig. 57.31(b).

3) The source circuit that drives the base of transistor Q3 in Fig. 57.31(b) is the Thévenin equivalent circuit established at the collector of transistor Q1 in Fig. 57.31(a). The signal voltage associated with this source circuit is the open circuit voltage developed at the Q1 collector; that is, it is the voltage at the Q1 collector with the load formed of transistor Q3 and its peripheral

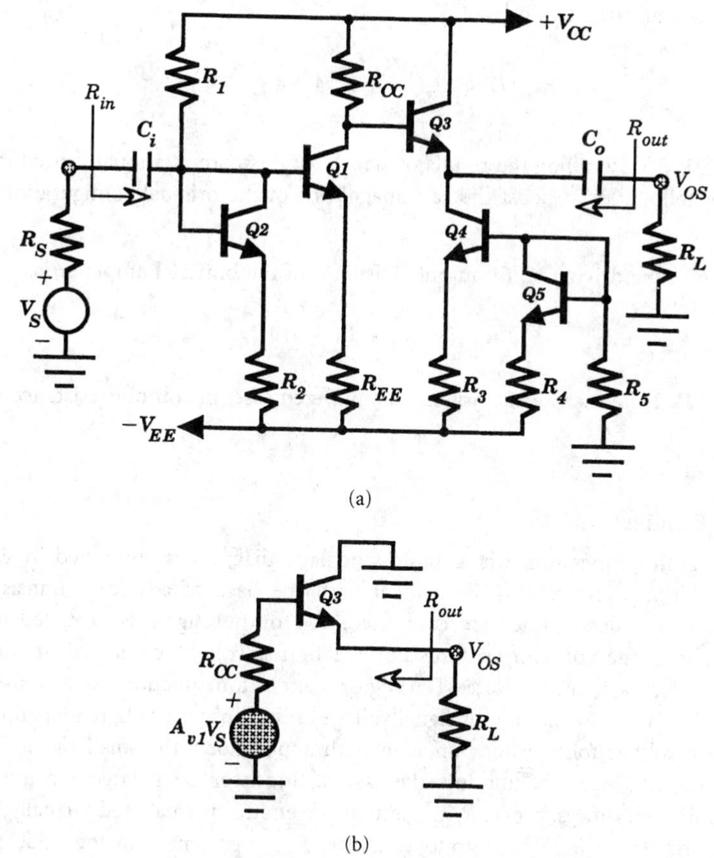

FIGURE 57.31 (a) The amplifier of Fig. 57.19, but with an emitter-follower buffer inserted between the gain stage and the terminating load resistance. (b) AC schematic diagram of the amplifier in (a).

elements removed. Since the circuit to the left of the base of transistor $Q3$ is a linear network, this Thévenin voltage is necessarily proportional to the input signal V_S. The indicated constant of proportionality in Fig. 57.31(b), A_{v1}, can rightfully be termed the open-circuit voltage gain of the first stage of the subject amplifier. This is to say that A_{v1} is A_v, as determined in step 1) above, but with R_L removed and therefore, R_{LT} set equal to R_{CC}. It follows that

$$A_{v1} = -\frac{\beta K_{ST} R_{CC}}{R_{ST} + r_b + r_\pi + (\beta + 1)(r_e + R_{EE})}$$

or $A_{v1} = -32.38$ V/V. Since r_o is taken to be infinitely large, the resistance seen looking into the collector of transistor $Q1$ is likewise infinitely large. As a result, the Thévenin resistance associated with the source circuit in the AC diagram of Fig. 57.31(b) is $R_{CC} = 3.9\ \text{k}\Omega$.

4) Recalling (57.63), the voltage gain of the circuit in Fig. 57.19(b) is

$$A_{vcc} = \frac{V_{OS}}{A_{v1} V_S} = \frac{R_L}{R_L + R_d + (1 - \alpha) R_{CC}} = 0.871\ \text{V/V}$$

The resultant overall circuit gain is

$$A_v = \frac{V_{OS}}{V_S} = A_{v1}A_{vcc}$$

or $A_v = -28.21$ V/V. Recalling the results of step 1) of this computational procedure, the effect of the emitter follower is to boost the gain magnitude of the original configuration by a factor of about 12.2.

5) From (57.65), the driving point output resistance of the buffered amplifier is

$$R_{\text{out}} = R_d + (1 - \alpha)R_{CC}$$

or $R_{\text{out}} = 44.3\ \Omega$. Note that for the original nonbuffered case, the output resistance is $R_{CC} = 3.9$ kΩ.

Darlington Connection

In the **Darlington connection**, whose basic schematic diagram is abstracted in Fig. 57.32(a), the emitter of one transistor $Q1$ is incident with the base of a second transistor $Q2$ and the two transistor collector leads are connected. The output signal is extracted as either the current flowing in the collector of transistor $Q2$ or the voltage developed at the emitter of Q2. In the former case, the indicated Darlington connection functions as a transconductance amplifier. In the latter case, an output signal voltage at the emitter of Q2 renders the connection functional as a voltage follower, or buffer. In both applications, the small-signal driving point input resistance, R_{ind}, seen looking into the base of transistor Q1 is large. On the other hand, the driving point output resistance, R_{outed}, seen at the emitter is small and virtually independent of the source resistance R_S. The output resistance, R_{outcd}, presented to the node at which the two transistor collectors are incident is large. At the expense of forward transconductance, R_{outcd} can be enhanced by returning the collector of Q1 to the $+V_{CC}$ bus, instead of to the collector of transistor Q2. For a nonzero collector load resistance, R_{LC}, this alternate connection, which is diagrammed in Fig. 57.32(b) eliminates Miller multiplication of the base–collector junction capacitance of transistor Q1, thereby resulting in an improved transconductance frequency response.

A fundamental problem that plagues both of the foregoing Darlington connections is the fact that the static emitter current conducted by Q1 is identical to the static base current drawn by Q2. Accordingly, the emitter current of Q1 is likely to be much smaller than the biasing current commensurate with optimal gain–bandwidth product in this device. Moreover, this emitter current cannot be predicted accurately since it is inversely proportional to the Q2 static beta, whose numerical value is an unavoidable uncertainty. This poor biasing translates into an unreliable delineation of the static and small signal parameters for Q1. In turn, potentially significant uncertainties shroud the forward transfer and driving point resistance characteristics of the Darlington configuration.

To remedy the situation at hand, an additional current path, usually directed to signal ground, is provided at the junction of the Q1 emitter and the Q2 base, as suggested in Fig. 57.32(c) and (d). The appended current path can be a simple two terminal resistance, although care must be exercised to ensure that this resistance is sufficiently large to avoid seriously compromising the large driving point input resistance afforded by the basic Darlington connection in either of the preceding diagrams. Since large resistance and realistic biasing currents may prove to be conflicting design requirements, the appended current path is often an active current sink, such as the Wilson mirror load explored earlier in conjunc-

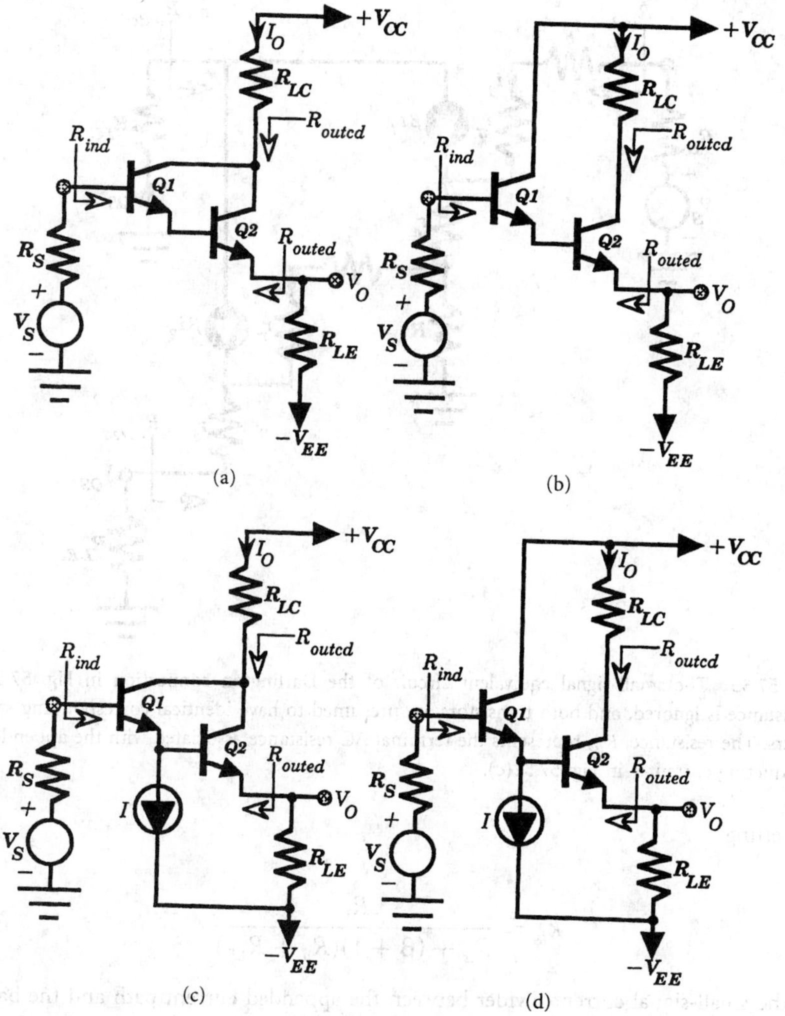

FIGURE 57.32 (a) The basic Darlington connection. (b) Alternative Darlington connection for wideband transconductance response. (c) Darlington connection with input transistor current compensation. (d) Alternative Darlington connection with input transistor current compensation.

tion with the common-collector amplifier. Note in the latter two diagrams that the current, indicated as I, conducted by the appended passive or active current path is essentially the emitter current of transistor Q1, provided that I is much larger than the base currents of Q2.

The small-signal bipolar junction transistor equivalent circuit of Fig. 57.13(a) can be used to deduce the transfer and driving point resistance characteristics of any of the Darlington connections depicted in Fig. 57.32. An analysis is provided herewith for only the configuration in Fig. 57.32(c), since this topology is the most commonly encountered Darlington circuit and the others are amenable to very straightforward analyses. To this end, the model for the subject structure is offered in Fig. 57.33, where it is assumed that both transistors are biased so that their corresponding small-signal parameters are nominally identical. Moreover, the Early resistance of each transistor is presumed to be sufficiently large to warrant its neglect, and a resistance, R_{is} is included to account for the terminal resistance of the appended current path discussed

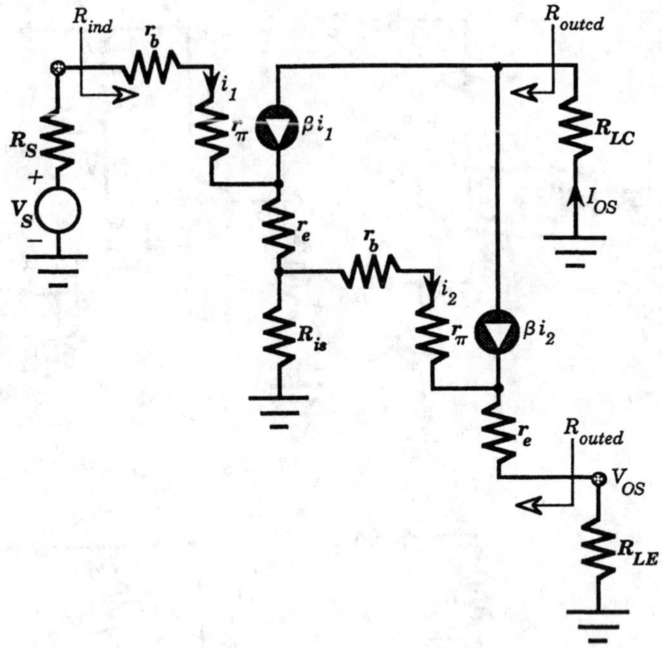

FIGURE 57.33 The small-signal equivalent circuit of the Darlington connection in Fig. 57.32(c). The Early resistance is ignored, and both transistors are presumed to have identical corresponding small-signal parameters. The resistance, R_{is}, represents the terminal AC resistance associated with the appended current path conducting current I in Fig. 57.32(c).

above. Letting

$$k_{is} \triangleq \frac{R_{is}}{R_{is} + (\beta + 1)(R_d + R_{LE})} \tag{57.69}$$

denote the small-signal current divider between the appended current path and the base circuit of transistor Q2, it can be shown that the driving point input resistance R_{ind} is

$$R_{ind} = (\beta + 1)[R_d + (\beta + 1)k_{is}(R_d + R_{LE})] \tag{57.70}$$

For large ac beta,

$$R_{ind} \approx (\beta + 1)^2 k_{is}(R_d + R_{LE}) \tag{57.71}$$

which is maximal for $k_{is} \approx 1$. From (57.69), the latter constraint mandates that the appended current path be designed so that its small-signal terminal resistance satisfies the inequality $R_{is} \gg (\beta + 1)(R_d + R_{LE})$.

The voltage gain, A_{vd}, from the signal source to the emitter port is

$$A_{vd} = \frac{V_{OS}}{V_S} = \frac{(\beta + 1)^2 k_{is} R_{LE}}{R_S + (\beta + 1)R_d + (\beta + 1)^2 k_{is}(R_d + R_{LE})} \tag{57.72}$$

where V_{OS} is the signal component of the net output voltage V_O. Equation (57.52) reduces to

$$A_{vd} \approx \frac{R_{LE}}{R_d + R_{LE}} \tag{57.73}$$

for large ac beta. The corresponding driving point output resistance, R_{outed}, is

$$R_{outed} = R_d + \frac{R_d}{(\beta + 1)k_{is}} + \frac{R_S}{(\beta + 1)^2 k_{is}} \approx R_d \tag{57.74}$$

At the collector port, the driving point output resistance R_{outcd} is infinitely large to the extent that the Early resistance r_o of both transistors can be ignored. For finite r_o this resistance is of the order of, and slightly larger than, $(r_o/2)$. Finally, the model in Fig. 57.33 yields a forward transconductance, G_{fd}, from the signal source to the collector port of

$$G_{fd} = \frac{I_{OS}}{V_S} = \frac{\beta[1 + (\beta + 1)k_{is}]}{R_S + (\beta + 1)R_d + (\beta + 1)^2 k_{is}(R_d + R_{LE})} \tag{57.75}$$

where I_{OS} is the signal component of the net output current I_O. Equation (57.75) collapses to

$$G_{fd} \approx \frac{\alpha}{R_d + R_{LE}} \tag{57.76}$$

for large AC beta.

Differential Amplifier

The **differential amplifier** is a four-port network, as suggested in Fig. 57.34(a). Source signals represented by the voltages, V_{S1} and V_{S2}, which have Thévenin resistances of R_{S1} and R_{S2}, respectively, are applied to the two amplifier input ports. The two output ports are terminated in three load resistances. Two of these loads, R_{L1} and R_{L2}, are **single-ended terminations** in that they provide a signal path to ground from each of the two output terminals. A third load resistance, R_{LL}, is differentially connected between the two output terminals. In response to the two applied source signals, two *single-ended output* voltages, V_{O1} and V_{O2}, are generated across R_{L1} and R_{L2}, and a *differential output voltage*, V_{DO}, is established across R_{LL}. This third output response is the difference between V_{O1} and V_{O2}; that is,

$$V_{DO} = V_{O1} - V_{O2} \tag{57.77}$$

The salient features of a differential amplifier are unmasked by the concepts of *differential- and common-mode* excitation and response. To this end, let the *differential input source voltage*, V_{DI}, be defined as

$$V_{DI} \triangleq V_{S1} - V_{S2} \tag{57.78}$$

and let the *common-mode input voltage*, V_{CI}, be

$$V_{CI} \triangleq \frac{1}{2}(V_{S1} + V_{S2}) \tag{57.79}$$

The differential input voltage is seen as the difference between the two applied source excitations. On the other hand, the common-mode input voltage is the arithmetic average of the two source voltages.

When solved for V_{S1} and V_{S2}, (57.78) and (57.79) give

$$V_{S1} = V_{CI} + \frac{1}{2}V_{DI} \tag{57.80a}$$

$$V_{S2} = V_{CI} - \frac{1}{2}V_{DI} \tag{57.80b}$$

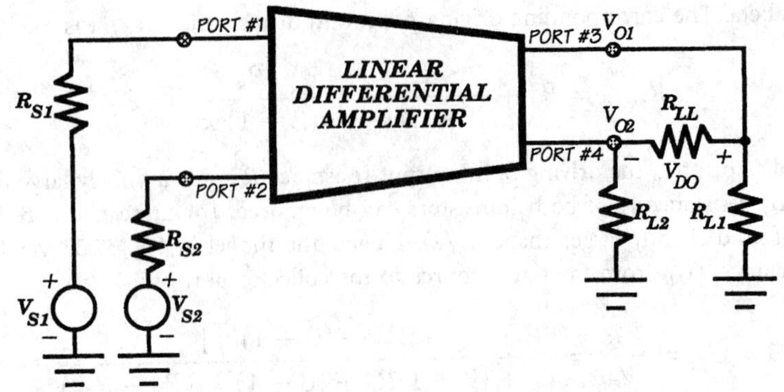

(a)

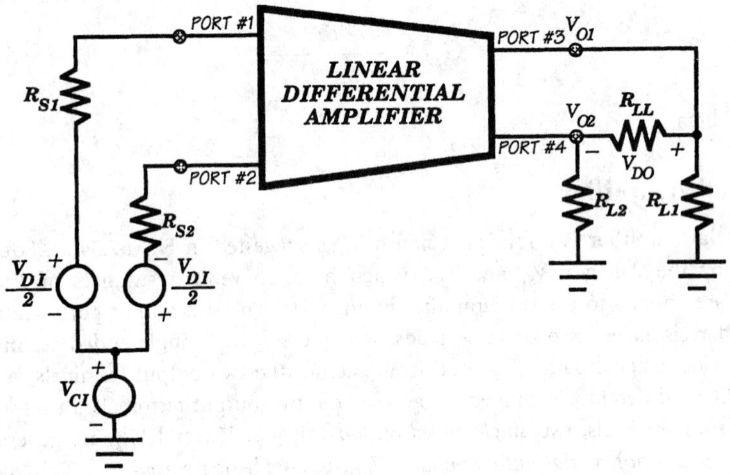

(b)

FIGURE 57.34 (a) System level diagram of a differential amplifier. (b) System level diagram that depicts the electrical implications of the common-mode and the differential-mode input source voltages.

The preceding two expressions allow the diagram of Fig. 57.34(a) to be drawn in the form shown in Fig. 57.34(b). This alternative representation underscores the fact that the Thévenin voltage applied to either input port is the superposition of a common-mode source voltage and a component proportional to the differential-mode source voltage. The common-mode component raises both of the open-circuit input terminals to a voltage that lies above ground by an amount, V_{CI}. Superimposed with V_{CI} at the open circuit terminals of port 1 is a differential-mode voltage, $V_{DI}/2$. Simultaneously, a voltage of $-V_{DI}/2$ superimposes with V_{CI} at the open-circuit terminals of port 2.

The fact that two general source excitations applied to a four-port system can be separated into a voltage component that appears only differentially across the two system input ports and a single-ended common-mode voltage component that is simultaneously incident with both of the system input ports makes it possible to achieve signal discrimination in a differential circuit. In particular, a differential amplifier can be designed so that it amplifies the differential component of two source signals while rejecting (in the sense of amplifying with near zero gain) their common-mode component. Signal discrimination is useful whenever an electronic system must process low-level electrical signals that are contaminated by spurious inputs, such as the voltage ramifications of electromagnetic interference or the biasing perturbations induced by temperature. If the two input ports of a differential amplifier are geometrically proximate and have matched driving point input impedances, these spurious excitations impact the two input ports identically. The undesired inputs are therefore common-mode excitations that can be rejected by a differential amplifier that is well designed in the sense of producing output port responses that are sensitive to only differential inputs.

If the differential amplifier in Fig. 57.34 is linear, superposition theory gives

$$V_{O1} = A_{11}V_{S1} + A_{12}V_{S2} \tag{57.81a}$$

$$V_{O2} = A_{21}V_{S1} + A_{22}V_{S2} \tag{57.81b}$$

where the A_{ij} are constants, independent of V_{S1} and V_{S2}. When (57.80a) and (57.80b) are inserted into the last two relationships, the single-ended output voltages are expressible as

$$V_{O1} = (A_{11} + A_{12})V_{CI} + (A_{11} - A_{12})\frac{V_{DI}}{2} \tag{57.82a}$$

$$V_{O2} = (A_{22} + A_{21})V_{CI} - (A_{22} - A_{21})\frac{V_{DI}}{2} \tag{57.82b}$$

It follows that the differential output voltage is

$$V_{DO} = (A_{11} - A_{22} + A_{12} - A_{21})V_{CI} + (A_{11} + A_{22} - A_{12} - A_{21})\frac{V_{DI}}{2} \tag{57.83}$$

Since the common-mode output voltage is

$$V_{CO} \triangleq \frac{1}{2}(V_{O1} + V_{O2}) \tag{57.84}$$

(57.82a) and (57.82b) yield

$$V_{CO} = \left(\frac{A_{11} + A_{22} + A_{12} + A_{21}}{2}\right)V_{CI} + \left(\frac{A_{11} - A_{22} - A_{12} + A_{21}}{2}\right)\frac{V_{DI}}{2} \tag{57.85}$$

The ability of a differential amplifier to process differential excitations is measured by the **differential-mode voltage gain**, A_D. This performance index is defined as the ratio of the differential output voltage to the differential input voltage, under the condition of zero common-mode input voltage. From (57.83),

$$A_D \triangleq \left[\frac{V_{DO}}{V_{DI}}\right]\Bigg|_{V_{CI}=0} = \frac{A_{11} + A_{22} - A_{12} - A_{21}}{2} \tag{57.86}$$

On the other hand, the *common-mode voltage gain*, A_C, is a measure of the common-mode signal rejection characteristics of a differential amplifier. It is the ratio of the common-mode output voltage to the common-mode input voltage, under the condition of zero differential input voltage. Using (57.85),

$$A_C \triangleq \left[\frac{V_{CO}}{V_{CI}}\right]\Bigg|_{V_{DI}=0} = \frac{A_{11} + A_{22} + A_{12} + A_{21}}{2} \tag{57.87}$$

A measure of the degree to which a differential amplifier rejects common-mode excitation is the **common-mode rejection ratio** ρ which is the ratio of the differential-mode voltage gain to the common-mode voltage gain. From (57.86) and (57.87)

$$\rho \stackrel{\Delta}{=} \frac{A_D}{A_C} = \frac{A_{11} + A_{22} - A_{12} - A_{21}}{A_{11} + A_{22} + A_{12} + A_{21}} \tag{57.88}$$

A common-mode gain of zero indicates that no common-mode output results from the application of common-mode input signals. Therefore, a practical design goal is the realization of a differential amplifier that has the largest possible magnitude of common-mode rejection ratio.

Balanced Differential Amplifier

Most differential amplifiers are *balanced.* Two operating requirements are satisfied by balanced differential systems. First, with zero common-mode input voltage, the two single-ended output voltages are mutually phase inverted, but otherwise identical. By (57.82a) and (57.82b), the balance requirement implies the parametric constraint,

$$A_{11} - A_{12} = A_{22} - A_{21} \tag{57.89}$$

Second, equal single-ended output voltages result when the differential-mode input voltage is zero. Using (57.82a) and (57.82b) once again, this stipulation requires

$$A_{11} + A_{12} = A_{22} + A_{21} \tag{57.90}$$

Equations (57.89) and (57.90) combine to deliver the balanced operating requirement

$$\left.\begin{aligned} A_{11} &= A_{22} \\ A_{12} &= A_{21} \end{aligned}\right\} \tag{57.91}$$

From (57.86) through (57.88), the differential-mode voltage gain, the common-mode voltage gain, and the common-mode rejection ratios of a balanced differential amplifier are

$$A_D = A_{11} - A_{12} \tag{57.92}$$

$$A_C = A_{11} + A_{12} \tag{57.93}$$

$$\rho = \frac{A_{11} - A_{12}}{A_{11} + A_{12}} \tag{57.94}$$

Moreover, the single-ended output voltages in (57.82a) and (57.82b) become

$$V_{O1} = A_C V_{CI} + A_D \frac{V_{DI}}{2} = \frac{A_D}{2}\left(1 + \frac{2V_{CI}}{\rho V_{DI}}\right) V_{DI} \tag{57.95a}$$

$$V_{O2} = A_C V_{CI} - A_D \frac{V_{DI}}{2} = -\frac{A_D}{2}\left(1 - \frac{2V_{CI}}{\rho V_{DI}}\right) V_{DI} \tag{57.95b}$$

which give rise to a differential response of

$$V_{DO} = V_{O1} - V_{O2} = A_D V_{DI} \tag{57.96}$$

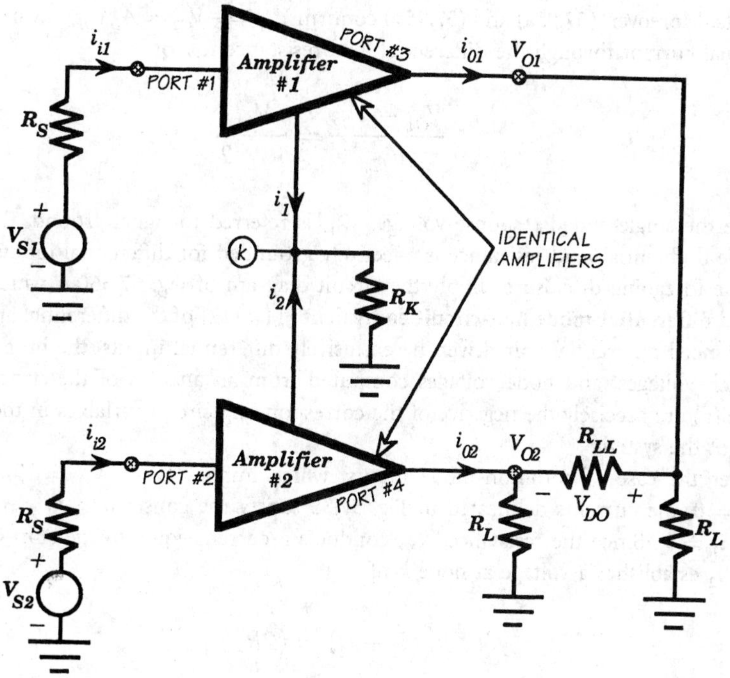

FIGURE 57.35 Generalized system diagram of a balanced differential system. The topology is an AC schematic diagram in that requisite biasing subcircuits of either amplifier are not shown.

Equations (57.95a) and (57.95b) show that a balanced differential amplifier having a very large common-mode rejection ratio produces single-ended outputs that are nominally phase-inverted versions of one another and approximately independent of the common-mode input voltage. On the other hand, the differential output voltage of a balanced system is independent of the common-mode input signal, regardless of the value of the common-mode rejection ratio.

Figure 57.35 depicts the most straightforward way to implement balance in a differential configuration. In this abstraction, two identical single-ended amplifiers, such as those discussed in earlier subsections, are interconnected to establish signal flow paths between single-ended input and single-ended output ports. This topology boasts integrated circuit practicality, since it exploits the inherent ability of a mature monolithic fabrication process to produce well-matched equivalent components. The two single-ended amplifiers in the subject figure are topologically identical, and they incorporate matched active devices that are biased at the same quiescent operating points. Thus, amplifiers 1 and 2 have small-signal two-port equivalent circuits that are reflective of one another. In order to ensure balanced operation, the single-ended output ports of each amplifier are terminated to ground in equal load resistances R_L. Similarly, the Thévenin source resistances are equivalent. Observe that balance implies that the upper half of the circuit in Fig. 57.35 is a mirror image of the lower half of the system schematic diagram. This interpretation begets the common reference to a balanced differential amplifier as a **differential pair.**

The balance condition entails the following engineering constraints.

1. Under the case of differential-mode excitation, which implies $V_{S1} = -V_{S2} = V_{DI}/2$ and hence, $V_{CI} = 0$, the currents indicated in Fig. 57.35 are such that $i_{i1} = -i_{i2}$, $i_{o1} = -i_{o2}$, and $i_1 = -i_2$. Since the resistance, R_K, conducts a current equal to the sum of i_1 and i_2, $i_1 = -i_2$ clamps node k to signal ground potential for exclusively differential-mode

inputs. Moreover, (57.95a) and (57.95b) confirm $V_{O1} = -V_{O2} = A_D V_{DI}/2$, which produces a signal current through the differential load resistance R_{LL} of

$$\frac{V_{O1} - V_{O2}}{R_{LL}} = \frac{V_{O1}}{R_{LL}/2}$$

Since the single-ended response voltage, V_{O1}, is referred to signal ground, the midpoint of the differential load resistance is effectively grounded for differential inputs.

The foregoing disclosures imply the circuit diagram of Fig. 57.36(a), which is the so-called **differential-mode half-circuit equivalent** [11], [12] of the differential amplifier. For a balanced differential pair driven by exclusively differential inputs, the branch currents, branch voltages, and node voltages computed from an analysis of the structure in Fig. 57.36(a) are precisely the negative of the corresponding circuit variables in the remaining half of the system.

2. Under the case of common-mode inputs, which implies $V_{S1} = V_{S2} = V_{CI}$ and hence, $V_{DI} = 0$, the currents delineated in Fig. 57.35 satisfy the constraints, $i_{i1} = i_{i2}$, $i_{o1} = i_{o2}$, and $i_1 = i_2$. Since the resistance, R_K, conducts a current equal to the sum of i_1 and i_2, $i_1 = i_2$ establishes a voltage at node k of

$$R_K(i_1 + i_2) = (2R_K)i_1$$

that is, the signal voltage developed at node k corresponds to an amplifier 1 current of

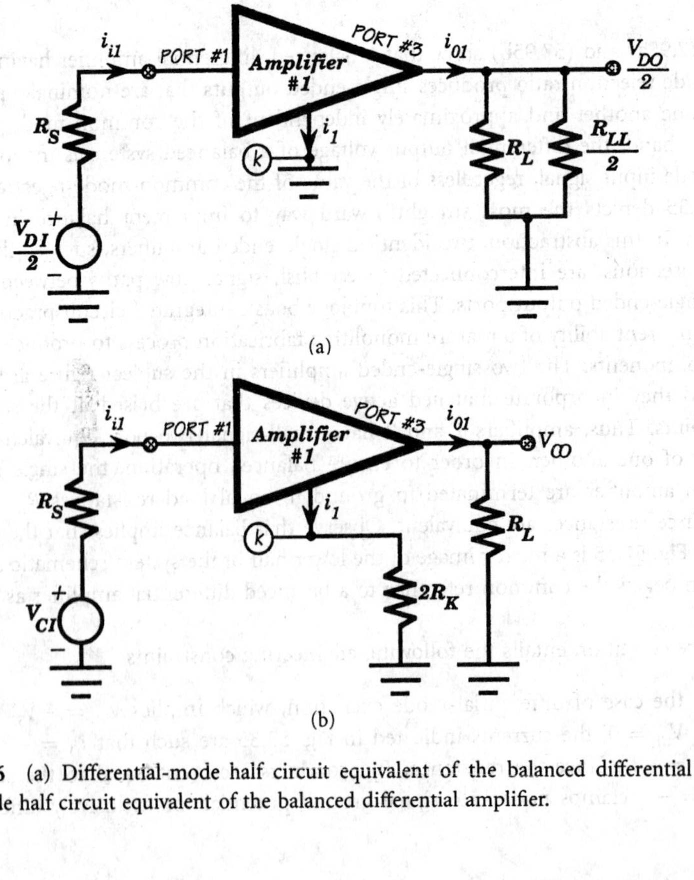

FIGURE 57.36 (a) Differential-mode half circuit equivalent of the balanced differential amplifier. (b) Common-mode half circuit equivalent of the balanced differential amplifier.

i_1 flowing through a resistance whose value is twice R_K. Additionally, $V_{O1} = V_{O2} = V_{CO}$, which means that no signal current flows through R_{LL} under exclusively common-mode excitation.

It follows that the *common-mode half-circuit equivalent* of the differential amplifier is as drawn in Fig. 57.36(b). For a balanced differential pair driven by exclusively common-mode input signals, the branch currents, branch voltages, and nodal voltages computed for the structure in Fig. 57.36(b) are identical to the corresponding circuit variables in the other half of the circuit.

Thévenin Equivalent I/O Circuits

Because the two input ports of a differential amplifier electrically interact with one another, the Thévenin equivalent circuit seen by the two signal sources in Fig. 57.34(a) is itself a two-port network. The branch elements of the Thévenin model are defined in terms of a *differential-mode input resistance*, R_{DI}, and a *common-mode input resistance*, R_{CI}.

Consider the test circuit of Fig. 57.37(a), which is configured to formulate the Thévenin equivalent input circuit. This structure is analogous to that of Fig. 57.34(a), except that the linear differential unit is presumed balanced. Furthermore, the original source circuits are supplanted by the test voltages, V_{t1} and V_{t2}, which establish the input port currents I_{t1} and I_{t2}. Because no signals other than V_{t1} and V_{t2} are applied to the differential pair, the Thévenin equivalent circuit seen between ports 1 and 2 is a resistance, say R_{XI}. Similarly, a second resistance, R_{XX}, is introduced to terminate port 1 to ground. System balance implies that a resistance of the same value terminates the second input port. The hypothesized Thévenin equivalent input circuit is given in Fig. 57.37(b).

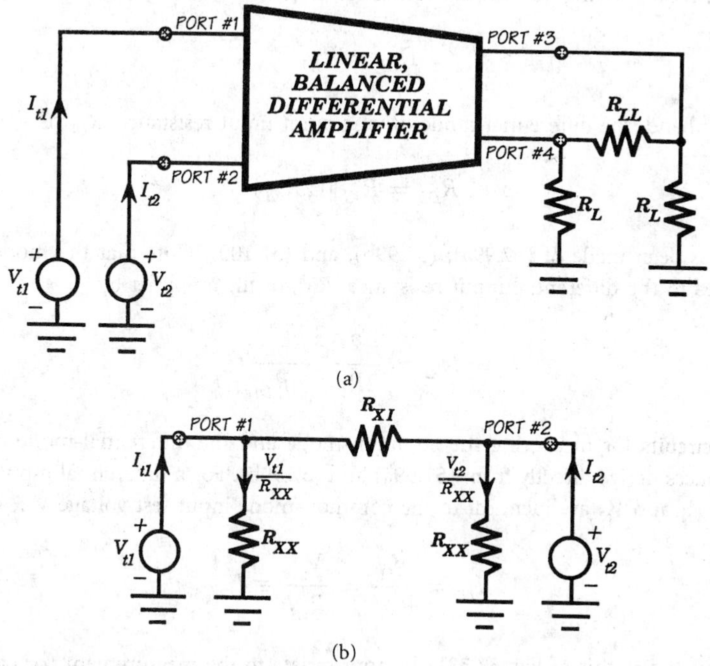

FIGURE 57.37 (a) Test circuit used to evaluate the Thévenin input equivalent circuit of a balanced differential amplifier. Note the connection of balanced loads at ports 3 and 4. (b) Hypothesized Thévenin equivalent input circuit.

The application of Kirchhoff's current and voltage laws to the model in Fig. 57.37(b) produces

$$I_{t1} - \frac{V_{t1}}{R_{XX}} = \frac{V_{t2}}{R_{XX}} - I_{t2} \tag{57.97a}$$

$$V_{t1} - V_{t2} = R_{XI}\left(I_{t1} - \frac{V_{t1}}{R_{XX}}\right) \tag{57.97b}$$

If the test voltages V_{t1} and V_{t2}, are decomposed into their differential (V_{Dt}) and common mode (V_{Ct}) components in accordance with

$$V_{t1} = V_{Ct} + \frac{V_{Dt}}{2} \tag{57.98a}$$

$$V_{t2} = V_{Ct} - \frac{V_{Dt}}{2} \tag{57.98b}$$

Equations (57.97a) and (57.97b) lead to

$$I_{t1} = \frac{V_{Ct}}{R_{XX}} + \frac{V_{Dt}}{R_{XI}\|(2R_{XX})} \tag{57.99a}$$

$$I_{t2} = \frac{V_{Ct}}{R_{XX}} - \frac{V_{Dt}}{R_{XI}\|(2R_{XX})} \tag{57.99b}$$

Equations (57.99a) and (57.99b) implicitly define the common-mode and differential-mode components of the currents I_{t1} and I_{t2} resulting from the test voltages V_{t1} and V_{t2}. Accordingly, the common-mode driving point input resistance, R_{CI}, is

$$R_{CI} = R_{XX} \tag{57.100}$$

On the other hand, the differential-mode driving point input resistance, R_{DI}, is

$$R_{DI} = R_{XI}\|(2R_{CI}) \tag{57.101}$$

where use has been made of (57.99a), (57.99b), and (57.100). Note that the model resistance, R_{XI}, is related to the differential input resistance, R_{DI}, of the amplifier by

$$R_{XI} = \frac{2R_{CI}R_{DI}}{2R_{CI} - R_{DI}} \tag{57.102}$$

The test circuits for measuring the common-mode and the differential-mode driving point input resistances derive directly from (57.99a) and (57.99b). For a differential input test voltage, V_{Dt} of zero, V_{t1} and V_{t2} are identical to the common-mode input test voltage V_{Ct}, whence

$$I_{t1} = \frac{V_{t1}}{R_{XX}} = \frac{V_{t1}}{R_{CI}} \equiv I_{t2} \tag{57.103}$$

It follows that the circuit of Fig. 57.38(a) is appropriate to the measurement (or calculation) of R_{CI} as the Ohm's law ratio of the common-mode test voltage to the resultant common-mode test current. Observe that half circuit analysis measures apply, wherein R_{CI} is the resistance seen looking into port 1, with port 3 terminated to ground in the resistance R_L. For differential

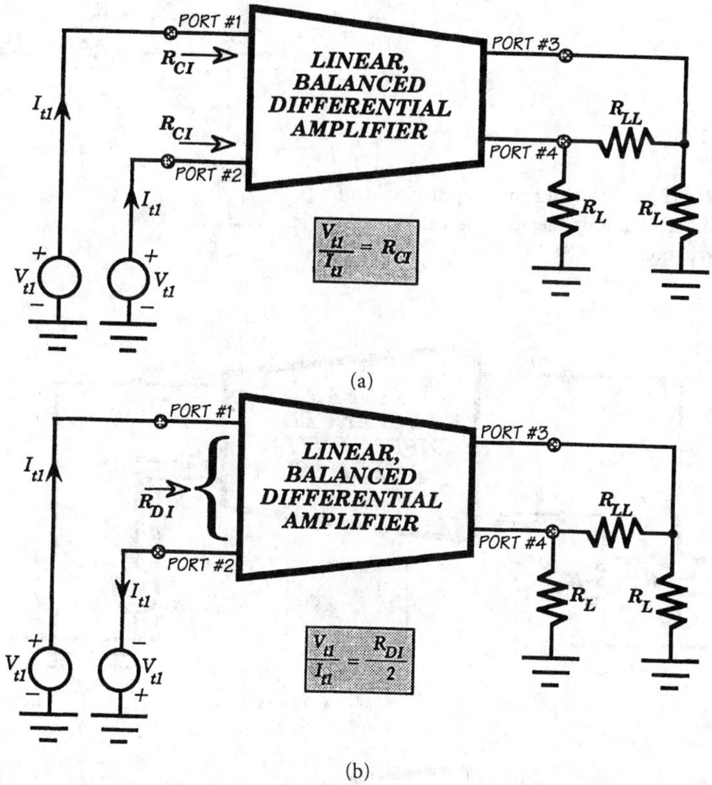

FIGURE 57.38 (a) Test circuit used to evaluate the driving point common-mode input resistance of a linear, balanced differential pair. (b) Test circuit used to evaluate the driving point differential-mode input resistance of a linear, balanced differential pair.

testing, $V_{t1} = -V_{t2} = V_{Dt}/2$, whence

$$I_{t1} = \frac{2V_{t1}}{R_{XI}\|(2R_{CI})} = \frac{2V_{t1}}{R_{DI}} = -I_{t2} \tag{57.104}$$

The pertinent test cell is shown in Fig. 57.38(b). For half-circuit analysis, care should be exercised to recognize that the ratio of the test voltage, V_{t1}, to the corresponding test current, I_{t1}, is one-half of the driving point differential input resistance, R_{DI}. In addition, the subcircuit connecting port 2 to port 4 must be removed, and port 3 must be terminated to ground by the shunt interconnection of the resistances, R_L and $R_{LL}/2$.

Just as a Thévenin model can be constructed for the input ports of a balanced differential pair, a Thévenin equivalent circuit can be developed for the output ports. Under zero input conditions, this output model, which is presented in Fig. 57.39, is topologically identical to the equivalent circuit in Fig. 57.37(b). In a fashion that reflects the computation of the resistance parameters for the input equivalent circuit, Fig. 57.40(a) is the test circuit for evaluating the driving point common-mode output resistance, R_{CO}. Figure 57.40(b) is the test structure for calculating the driving point differential-mode output resistance, R_{DO}. Following (57.102), R_{XO} in Fig. 57.39 is given by

$$R_{XO} = \frac{2R_{CO}R_{DO}}{2R_{CO} - R_{DO}} \tag{57.105}$$

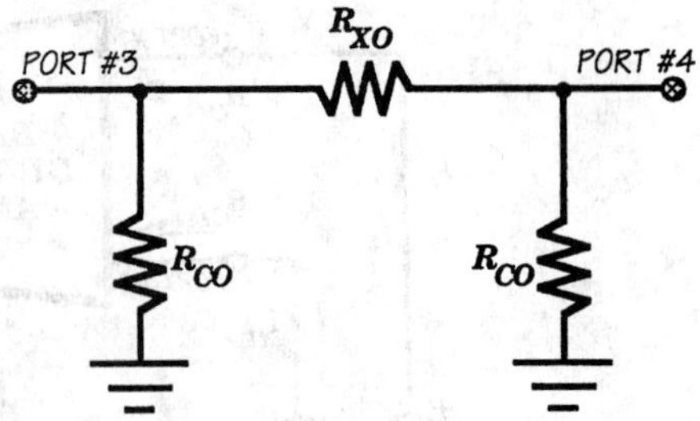

FIGURE 57.39 Thévenin equivalent output circuit of the balanced differential amplifier for the case of zero input signal excitation.

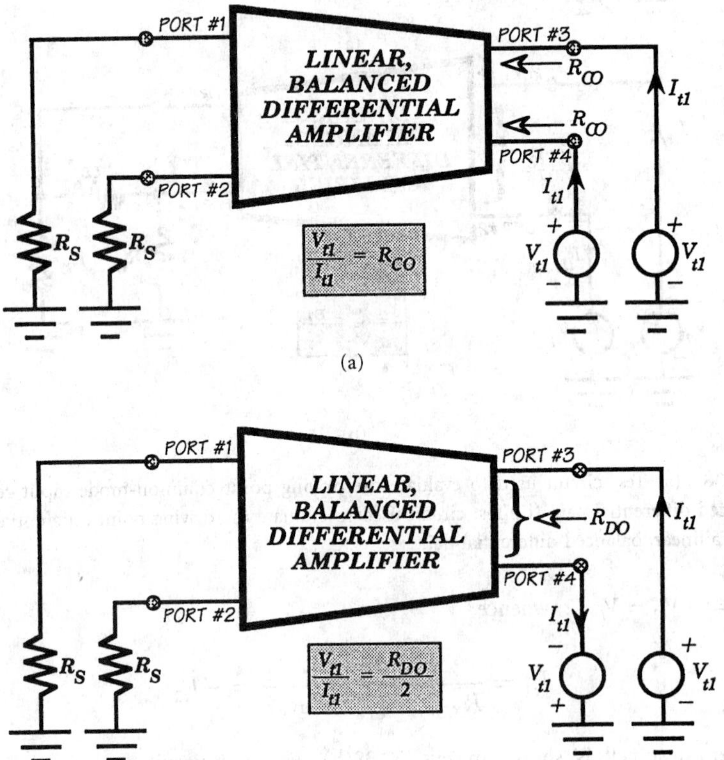

(a)

(b)

FIGURE 57.40 (a) Test circuit used to evaluate the driving point common-mode output resistance of a linear, balanced differential pair. (b) Test circuit used to evaluate the driving point differential-mode output resistance of a linear, balanced differential pair.

There are two electrically interactive output ports in the differential system of Fig. 57.34(a). Thus, two Thévenin voltage V_{th1} and V_{th2}, each of which is linearly dependent on the differential-mode and the common-mode components of the applied source signals must be evaluated. These Thévenin output responses derive from open circuited load conditions, as indicated in Fig. 57.41(a). With $R_{S1} = R_{S2} \triangleq R_S$, balance prevails, and V_{th1} and V_{th2} are characterized by differential and common-mode components, analogous to the characterization of the terminated outputs, V_{O1} and V_{O2}.

The Thévenin voltages in question derive from an analysis of the equivalent circuit in Fig. 57.41(b), which represents the model of Fig. 57.39 modified to account for nonzero source excitation. The proportionality constants, k_c and k_d, are related to the previously determined

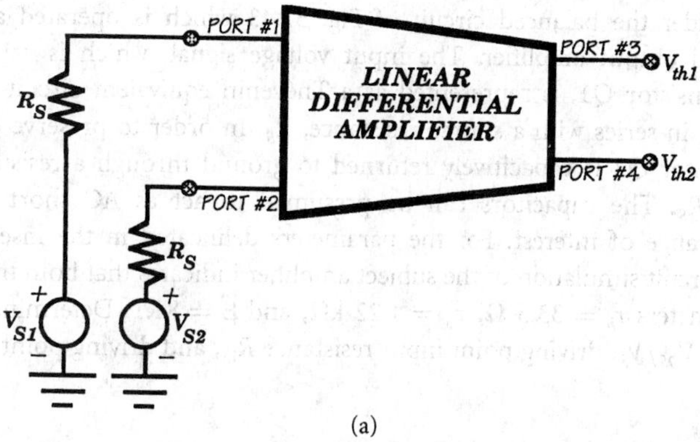

(a)

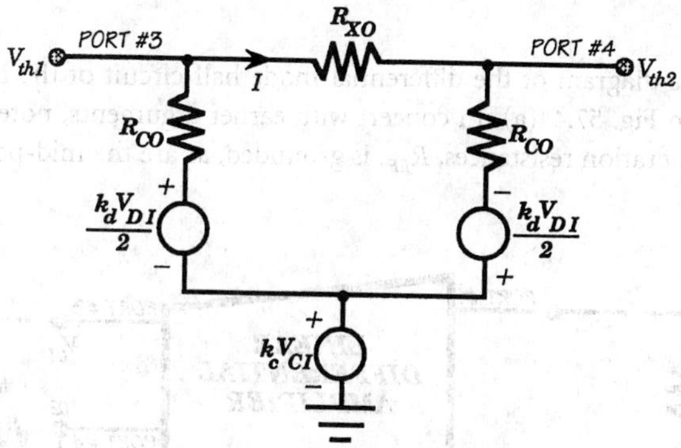

(b)

FIGURE 57.41 (a) System schematic diagram used to define the Thévenin voltages at the output ports of a balanced differential amplifier. (b) Thévenin equivalent circuit for the output ports of a balanced differential pair.

common-mode and differential-mode voltage gains. It is a simple matter to confirm that

$$V_{\text{th1}}, V_{\text{th2}} = k_c V_{CI} \pm \left(\frac{k_d R_{XO}}{2R_{CO} + R_{XO}}\right)\frac{V_{DI}}{2} \tag{57.106}$$

The first term on the right-hand side of this relationship is the open-circuit common-mode output voltage, while the second term is the open-circuit differential-mode output voltage. It follows that k_c represents the open-circuit common-mode voltage gain, A_{CO}; that is,

$$k_c = \lim_{\substack{R_L \to \infty \\ R_{LL} \to \infty}} (A_C) \triangleq A_{CO} \tag{57.107}$$

On the other hand, the open-circuit differential-mode gain, A_{DO}, is

$$\frac{k_d R_{XO}}{2R_{CO} + R_{XO}} = \lim_{\substack{R_L \to \infty \\ R_{LL} \to \infty}} (A_D) \triangleq A_{DO} \tag{57.108}$$

Using (57.52), the Thévenin model parameter, k_d, in the last expression can be cast as

$$k_d = \left(\frac{2R_{CO}}{R_{DO}}\right) A_{DO} \tag{57.109}$$

Fig. 57.42 summarizes the foregoing modeling results [13].

Example 4. Consider the balanced circuit of Fig. 57.43 which is operated as a single-ended input–single-ended output amplifier. The input voltage signal, which is capacitively coupled to the base of transistor Q1, is represented as a Thévenin equivalent circuit consisting of the source voltage, V_S, in series with a source resistance, R_S. In order to preserve electrical balance, the base of transistor Q2 is capacitively returned to ground through a resistance whose value is also equal to R_S. The capacitors can be presumed to act as AC short circuits over the signal frequency range of interest. For the parameters delineated in the inset to Fig. 57.43, a computer-aided circuit simulation of the subject amplifier indicates that both transistors have the small-signal parameters $r_b = 33.5\ \Omega$, $r_\pi = 1.22\ \text{k}\Omega$, and $\beta = 81.1$. Determine the small-signal voltage gain $A_v = V_{OS}/V_S$, driving point input resistance R_{in}, and driving point output resistance R_{out}.

Solution.

1) The AC schematic diagram of the differential-mode half circuit of the balanced amplifier in Fig. 57.43 is shown in Fig. 57.44(a). In concert with earlier arguments, note that the junction of the two emitter degeneration resistances, R_{EE}, is grounded, as are the mid-point of the resistance,

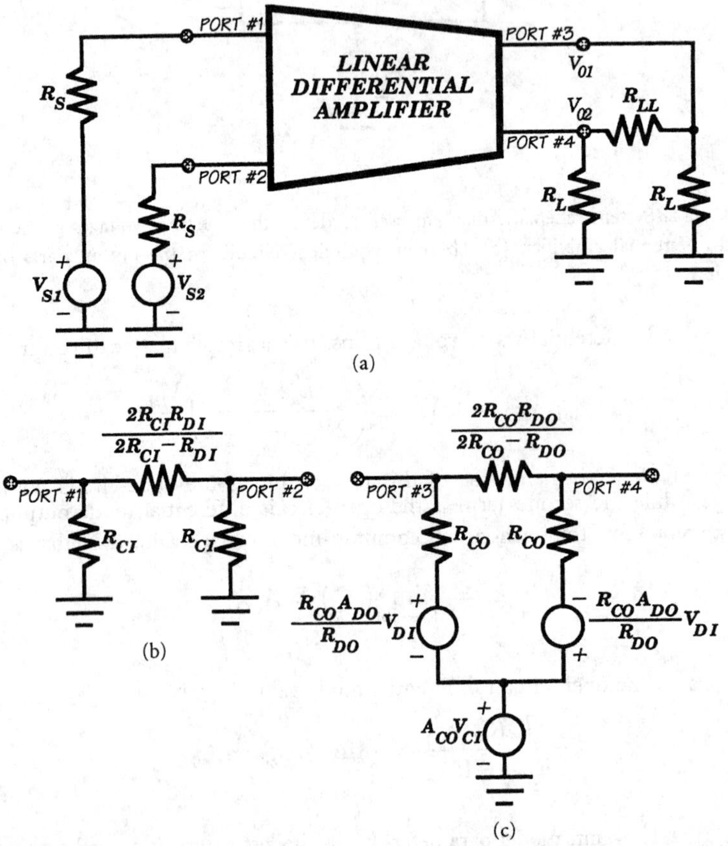

FIGURE 57.42 (a) System schematic diagram of a linear, balanced differential amplifier. (b) Thévenin equivalent input circuit. (c) Thévenin equivalent output circuit. The parameters, A_{DO} and A_{CO}, represent the open-circuit values of the differential- and common-mode voltage gains, respectively, of the balanced pair in (a).

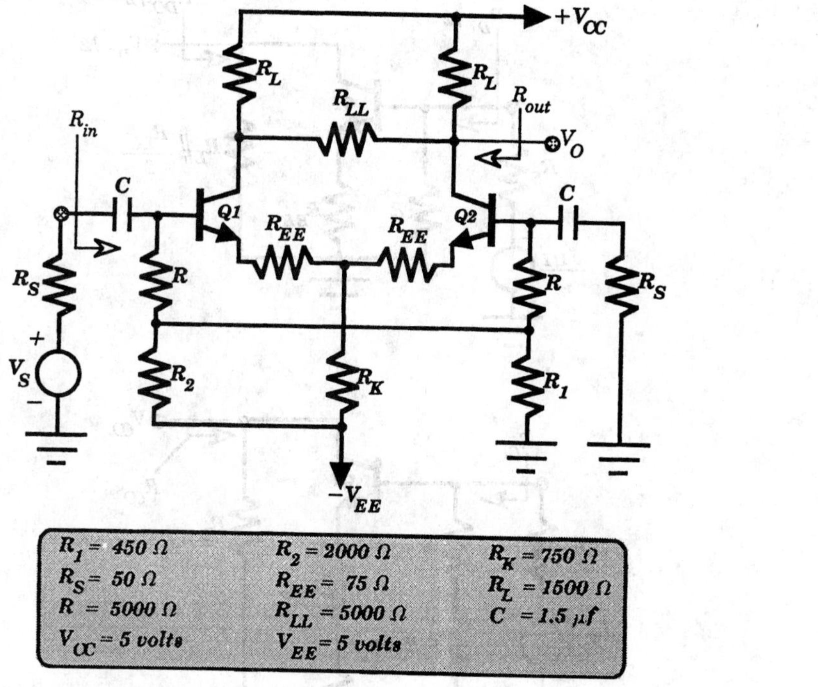

FIGURE 57.43 A balanced bipolar differential amplifier used in a single-ended input–single-ended output mode.

R_{LL}, and the node at which R_1, R_2, and the two resistances labeled R are incident. Using the bipolar model of Fig. 57.13(a), with r_o and r_e ignored, the voltage gain of this structure is the differential-mode voltage gain of the differential pair. Moreover, the driving point input resistance of the circuit at hand is one-half of the differential input resistance of the original pair, while its driving point output resistance is one-half of the differential-mode output resistance. Analysis confirms

$$A_D = \frac{V_{DO}/2}{V_{DI}/2} = -\frac{\beta\left(\dfrac{R}{R+R_S}\right)\left(R_L \| \dfrac{R_{LL}}{2}\right)}{(R_S \| R) + r_b + r_\pi + (\beta+1)R_{EE}}$$

$$\frac{R_{DI}}{2} = R \,\|\, [r_b + r_\pi + (\beta+1)R_{EE}]$$

$$\frac{R_{DO}}{2} = R_L \| \frac{R_{LL}}{2}$$

Numerically, $A_D = -10.09$, $R_{DI} = 5971\ \Omega$, and $R_{DO} = 1875\ \Omega$.

2) The AC schematic diagram of the pertinent common-mode half circuit is given in Fig. 57.44(b). The input signal voltage is now the common-mode input voltage, V_{CI}, which produces the common-mode output response, V_{CO}. Using the bipolar model of Fig. 57.13(a), with r_o and

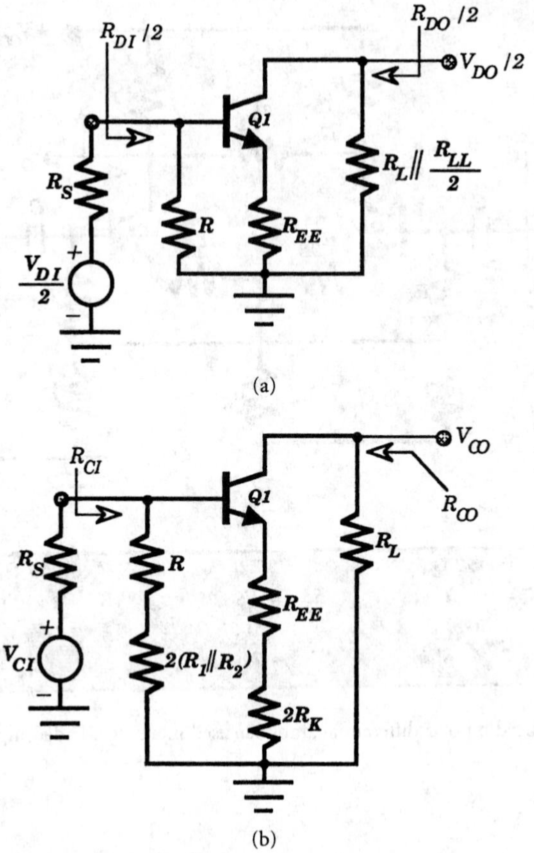

FIGURE 57.44 (a) Differential-mode half-circuit ac equivalent schematic of the differential amplifier shown in Fig. 57.43. (b) Common-mode half-circuit ac equivalent schematic of the differential amplifier shown in Fig. 57.43.

r_e ignored, it is easily shown that

$$A_C = \frac{V_{CO}}{V_{CI}} = -\frac{\beta\left(\dfrac{R + 2(R_1 \| R_2)}{R + 2(R_1 \| R_2) + R_S}\right) R_L}{\{R_S \| [R + 2(R_1 \| R_2)]\} + r_b + r_\pi + (\beta + 1)(R_{EE} + 2R_K)}$$

$$R_{CI} = [R + 2(R_1 \| R_2)] \| [r_b + r_\pi + (\beta + 1)(R_{EE} + 2R_K)]$$

$$R_{CO} = R_L$$

Numerically, $A_C = -923.3\ (10^{-3})$, $R_{CI} = 5493\ \Omega$, and $R_{CO} = 1500\ \Omega$. The common-mode rejection ratio $\rho = A_D/A_C = 10.93$, is small owing to the relatively small value of the resistance R_K.

3) Since the output voltage is extracted at the collector of transistor $Q2$, (57.95b) is the applicable equation for determining the output signal voltage, V_{OS}. With only a single source voltage, V_S, applied, the differential input voltage is V_S, and the common-mode input voltage is $V_S/2$. It follows that

$$V_{OS} = A_C V_{CI} - \left(\frac{A_D}{2}\right) V_{DI} = \left(\frac{A_C - A_D}{2}\right) V_S$$

whence a voltage gain of

$$A_v = \frac{V_{OS}}{V_S} = \frac{A_C - A_D}{2}$$

These analyses give $A_v = 4.583$.

4) In order to evaluate the driving point input and output resistances, the parameters, R_{XI} and R_{XO} must be calculated. From (57.102) and (57.105), $R_{XI} = 13.08$ kΩ, and $R_{XO} = 5.0$ Ω.

The two-port model for calculating the driving point input resistance R_{in} is given in Fig. 57.42(b). Recall that a circuit resistance, whose value is numerically equal to the internal signal source resistance, R_S, is connected between ground and the node to which the base of transistor Q2 is incident. By inspection,

$$R_{in} = R_{CI} \| [R_{XI} + (R_{CI} \| R_S)]$$

or $R_{in} = 3.87$ kΩ.

The output port model that emulates the driving point output resistance is given in Fig. 57.42(c). This model is analogous to that of Fig. 57.42(a), except that no external loads are connected between signal ground and the node to which the collector of transistor $Q1$ is incident. Clearly,

$$R_{out} = R_{CO} \| (R_{XO} + R_{CO})$$

which produces $R_{out} = 1219$ Ω.

References

[1] J. J. Ebers and J. L. Moll, "Large-signal behavior of junction transistors," *Proc. IRE*, vol. 42, pp. 1761–1772, Dec. 1954.

[2] H. K. Gummel and H. C. Poon, "An integral charge-control model of bipolar transistors," *Bell System Tech. J.*, vol. 49, pp. 115–120, May–June 1970.

[3] H. N. Ghosh, "A distributed model of the junction transistor and its application in the prediction of the emitter–base diode characteristic, base impedance, and pulse response of the device." *IEEE Trans. Electron Devices*, vol. ED-12, pp. 513–531, Oct. 1965.

[4] J. R. Hauser, "The effects of distributed base potential on emitter current injection density and effective base resistance for stripe transistor geometries," *IEEE Trans. Electron Devices*, vol. ED-11, pp. 238–242, May 1965.

[5] P. R. Gray and R. G. Meyer, *Analysis and Design of Analog Integrated Circuits*, New York: Wiley, 1977, pp. 16–19.

[6] C. T. Kirk, "A theory of transistor cut-off frequency (f_T) at high current densities," *IEEE Trans. Electron Devices*, vol. ED-9, pp. 164–174, Mar. 1962.

[7] J. M. Early, "Effects of space–charge layer widening in junction transistors," *Proc. IRE*, vol. 46, pp. 1141–1152, Nov. 1952.

[8] A. S. Sedra and K. C. Smith, *Microelectronic Circuits*, New York: Holt, Rinehart and Winston, 1987, pp. 52–57 and 639–642.

[9] A. B. Grebene, *Bipolar and MOS Analog Integrated Circuit Design*, New York: Wiley-Interscience, 1984, pp. 170–182.

[10] G. R. Wilson, "A monolithic junction FET-NPN operational amplifier," *IEEE J. Solid-State Circuits*, vol. SC-3, pp. 341–348, Dec. 1968.

[11] E. J. Angelo, *Electronic Circuits*, New York: McGraw-Hill, 1970, ch. 4.

[12] A. B. Grebene, *Bipolar and MOS Analog Integrated Circuit Design*, New York: Wiley-Interscience, 1984, pp. 217–224.
[13] S. A. Witherspoon and J. Choma, Jr., "The analysis of balanced linear differential circuits," *IEEE Trans. Education*, vol. 38, pp. 40–50, Feb. 1995.

57.3 MOSFET Biasing Circuits

D. G. Haigh, B. Redman-White, and R. Akbari-Dilmaghani

Introduction

CMOS technology is finding a very wide range of applications in analog and analog–digital mixed-mode circuit implementations in addition to its traditional role in digital circuits. In mixed-mode circuits compatibility of analog circuits with digital VLSI is important, particularly in cost-sensitive areas and situations where low power consumption is required. Such CMOS circuits require a range of biasing circuits and it is this topic that is the main subject of this section, although the subject is mentioned elsewhere in this text, where particular designs are covered.

The requirements for biasing circuits in CMOS circuit design can be divided into the requirements for voltage and for current sources. These sources can be further subdivided into two additional categories, high precision and noncritical. High-precision voltage or current sources are essential components in data converters, both analog-to-digital and digital-to-analog and the precision required depends on the overall target precision of the data converter. For a large number of bits, the precision required could be very great indeed and would need to be maintained over a specified temperature and supply voltage range and in the presence of on-chip, chip-to-chip and wafer-to-wafer component parameter variations. Precision sources are also required in other applications such as dc pedestals for video signals in video systems.

Noncritical voltage sources are generally required for setting up an internal analog ground or for biasing the gates of FETs in common-gate configuration, as in a cascode FET. In these cases, the sensitivity of overall circuit performance parameters to the bias voltage would generally not be high and moderate precision circuit techniques would be acceptable. The main considerations would be to maintain correct operation, especially in terms of signal headroom, over all variations in process, power supply, and temperature conditions. The requirement for current sources for biasing CMOS analog and mixed-mode circuits is very considerable. The reason for this is that most circuits, such as an operational amplifier, consist of several stages, each of which requires biasing. In discrete circuits the tendency is to use resistors for biasing, for reasons of cost and the poor sample-to-sample tolerances on discrete active device parameters. In integrated circuit implementation, on the other hand, relatively well-matched devices on the same chip are available and the use of current source biasing minimizes gain loss due to loading effects. Furthermore, using a multioutput current mirror to supply different parts of the circuit allows stabilization of the current against temperature and power supply voltage variations to be performed at one location only (on or off the chip) and the stabilized current can be distributed throughout the chip or subcircuit using current mirror circuits. This also produces significant immunity to localized power supply fluctuation and noise. Full or partial stabilization of bias currents with operating and environmental changes is desirable in order to minimize the range of operating current for which a design must be specified, allowing higher design performance targets to be achieved.

In many cases, CMOS circuits have their power supplies derived from an off-chip bipolar regulator (with its own internal bandgap). In these situations, a reasonable voltage reference can sometimes be obtained from the power supply voltage via a potential divider. A voltage obtained in this way can be applied to an external low tolerance resistor to obtain a current reference of moderate precision. The value of realizing a reference on chip is that the cost of the external reference can be avoided. For example, in battery supplied equipment, a large degree of supply voltage immunity is required in the presence of a widely varying battery voltage. It is possible to use an on-chip voltage regulator or an on-chip switched-mode power supply. In CMOS technology, high-precision voltage references are difficult to design, although a reference with good power supply rejection is possible. BiCMOS technology overcomes many of the problems experienced with CMOS technology since well-controlled bipolar devices for very high-precision references are available on-chip.

Many references to biasing appear in this text under the heading of the circuit concerned and only some general guidelines and principles, together with some example circuits will be given here. We begin this section on CMOS biasing circuits by considering the devices available for biasing in a CMOS integrated circuit including parasitically realized bipolar junction devices. Some useful simplified models of these devices and a brief examination of the variability of the relevant model parameters will be presented. We then consider different types of references and biasing circuits. Since voltage and current references are closely interrelated, they are dealt with concurrently. The material is presented according to a gradually increasing level of sophistication and achievable precision, starting with simple circuits with only minor supply voltage, temperature, and process independence and leading to fully curvature-compensated bandgap references. This is followed by a consideration of references based on less usual devices that may not be available or usable in every process but that offer potentially attractive solutions. We then illustrate the application of some biasing techniques in the context of simple operational amplifier circuits. Finally, we consider the biasing of amplifiers for very low supply voltages, where rail-to-rail optimized performance is required, and dynamic biasing techniques. The topic of CMOS biasing circuits is sufficiently large to warrant an entire book and we include a list of references that will help to provide the reader with more detailed information.

Device Types and Models for Biasing

Devices

The principal devices of CMOS technology are the enhancement-mode N-channel and P-channel MOSFET, which are shown schematically in Fig. 57.45(a) and (b) for N-well and P-well technologies, respectively. Currently, N-well technology is more widely available than P-well. Depletion-mode devices are not routinely provided in CMOS (in contrast with NMOS), but they are available in some processes [23] and they can be used to realize reference circuits [4], [5], [23]. They are produced by an additional implantation under the gate region (N-type for NMOS and P-type for PMOS). The NMOS depletion-mode symbol is shown schematically in Fig. 57.46.

In most processes, the MOSFET gate material is highly N-doped polysilicon. In some processes [8], P-doped gates are available, and these can be used to realize reference circuits. The symbol of an N-channel enhancement MOSFET with P-type doped base is shown in Fig. 57.47.

In order to realize temperature and process desensitized biasing of CMOS circuits, the special properties of bipolar junction transistors (BJTs) are advantageous. BJT devices can be realized in CMOS technology as parasitic devices with certain restrictions. Two classes of such device are available, namely vertical and lateral. For the vertical device [13], the restrictions are that an N-well process can realize PNP devices with the collectors connected to the substrate (most negative supply rail) and that a P-well process can realize NPN devices with the collectors connected to the substrate (most positive supply rail). The realization of these vertical devices in the case of N-well and P-well processes is shown schematically in Fig. 57.48(a) and (b), respectively. The devices can have typical current gains of around 100 and may have high leakage, and certain precautions have to be taken in the layout. One problem is that the control

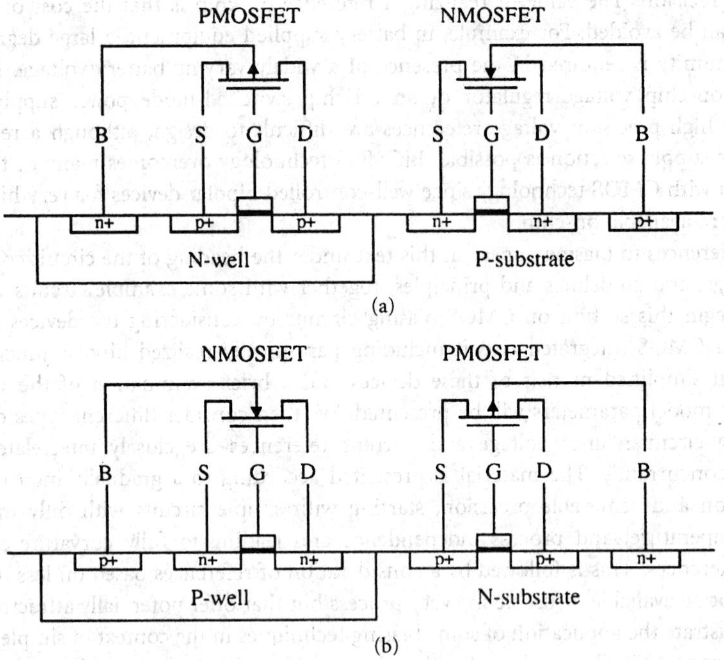

FIGURE 57.45 Realization of NMOS and PMOS FETs in CMOS technology. (a) N-well process. (b) P-well process.

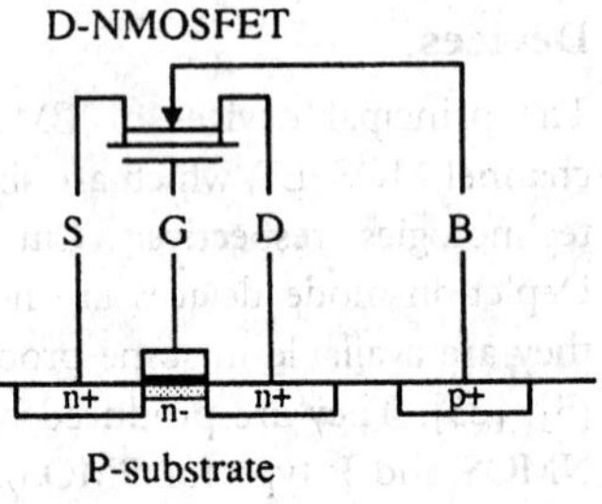

FIGURE 57.46 Realization of NMOS depletion-mode FET in N-well CMOS technology.

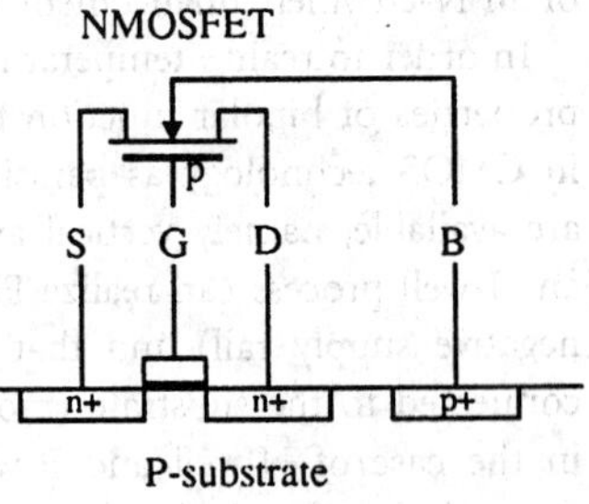

FIGURE 57.47 NMOS enhancement-mode FET with P-doped polysilicon gate.

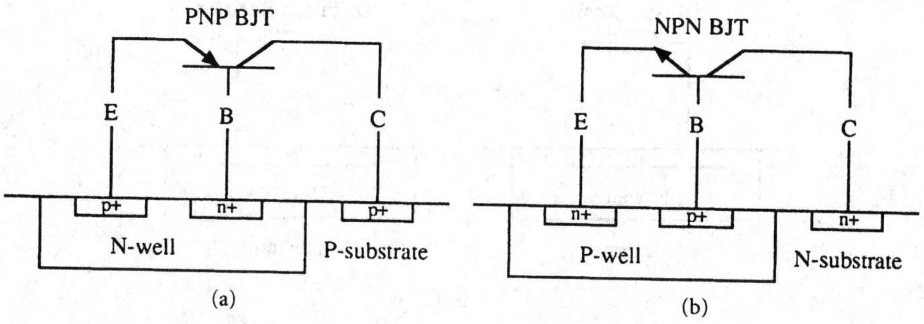

FIGURE 57.48 Realization of vertical BJT devices in CMOS technology. (a) N-well process. (b) P-well process.

on the parameters of these devices in production is minimal. Nevertheless, such devices are adequate to realize moderate to high precision bias and reference circuits.

The restriction on the collector connections of the vertical BJT devices in CMOS technology is effectively removed in lateral BJT devices [16]. The realization of these devices in the case of N-well and P-well processes is illustrated in Fig. 57.49(a) and (b), respectively. It should be noted that β_f for lateral bipolar transistors is not related to α_f in the usual way due to substrate currents. For both vertical and lateral parasitic bipolar devices, it is often the case that the foundry will not provide detailed characterization and models, and also that the parameters of the devices will not be well controlled.

Apart from MOSFETs and BJTs, the remaining component needed to realize bias circuits are resistors and in some cases capacitors. In the case of precision bias circuits, the realized variable (voltage or current) would be dependent on resistor ratios rather than on absolute values. On-chip resistors may be realized using polysilicon, or as N-well or P-well diffusion, as illustrated in Fig. 57.50. Diffused resistors are sensitive to substrate potential and thus they are not suitable for precision potential dividers. In some cases, voltages or currents on a chip that are required to have high stability are referred to an off-chip highly stable and accurate discrete resistor. In some advanced processes, film resistors (usually nichrome) are available and they have excellent temperature stability, a wide range of values and can sometimes be laser-trimmed.

Capacitors, both on-chip and external, are used for decoupling bias and reference voltages and are especially valuable where low noise is critically important. On-chip capacitors also provide the basic components used for dynamic biasing.

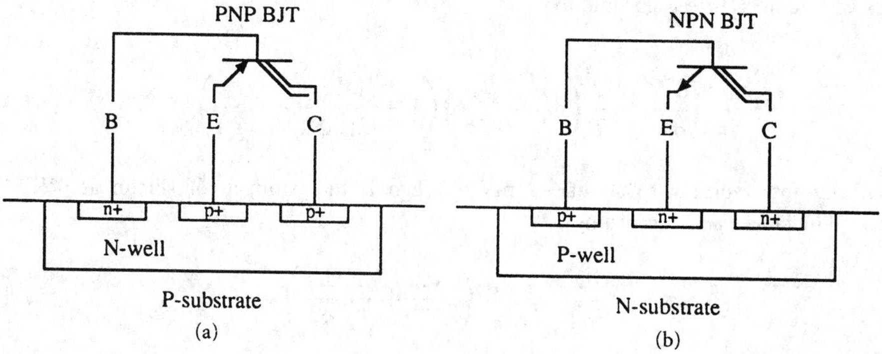

FIGURE 57.49 Realization of lateral BJT devices in CMOS technology. (a) N-well process. (b) P-well process.

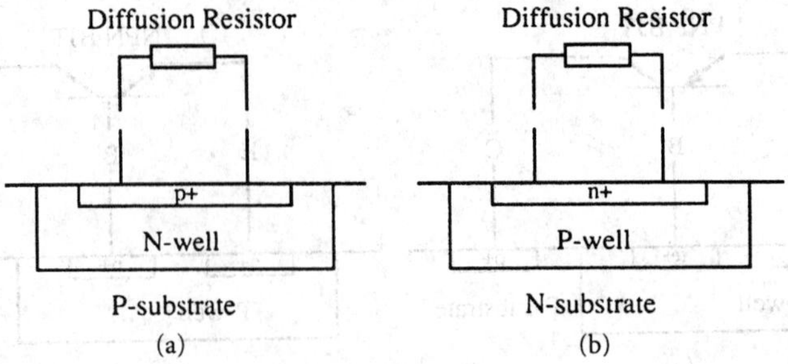

FIGURE 57.50 Realization of diffusion resistors in CMOS technology. (a) N-well process. (b) P-well process.

Device Models and Parameter Variability

The MOSFETs in Fig. 57.45 may be very approximately described by [13], [15]

$$I_d = \beta(V_{gs} - V_t)^2(1 + \lambda V_{ds}) \tag{57.110}$$

where

$$\beta = \mu C_{ox} W/L \tag{57.111}$$

and

$$V_t = V_{to} + \gamma[(2\phi_{fb} - V_{bs})^{0.5} - 2\phi_{fb}^{0.5}] \tag{57.112}$$

with

$$\phi_{fb} = \left(\frac{kT}{q}\right)\ln\left(\frac{N_{\text{sub}}}{n_i}\right) \tag{57.113}$$

and

$$\mu \propto T^{-\eta} \tag{57.114}$$

The parameters in (57.110)–(57.114) are defined in Table 57.1. As a result of (57.110)–(57.114), MOSFET parameters show considerable temperature dependence. In particular, threshold voltage varies with temperature according to

$$\frac{\partial V_t}{\partial T} = \frac{2}{T}\left(\phi_F - \frac{E_{go}}{2q}\right)\left(1 + \frac{\gamma}{2(2\phi_{fb} - V_{bs})}\right) \tag{57.115}$$

which amounts typically to about -2 mV/°C (Ego is the bandgap of silicon at 0°K) [15]. Transconductance varies according to

$$\frac{\partial g_m}{\partial T} = \frac{g_m}{2}\left(-\frac{\eta}{T} + \frac{1}{I_d}\frac{\partial I_d}{\partial T}\right) \tag{57.116}$$

The temperature and process dependencies of MOS devices have led to the exploitation of some properties of combinations of less usual MOS devices. It has been observed that the

TABLE 57.1 Device Parameters

Symbol	Parameter
I_d	Drain current
β	Transconductance parameter
V_{gs}	Gate–source voltage
V_t	Threshold voltage
λ	Output conductance parameter
V_{ds}	Drain–source voltage
μ	Mobility
C_{ox}	Oxide capacitance per unit area
W	Gate width
L	Gate length
V_{to}	Zero substrate bias threshold voltage
γ	Body factor
ϕ_{fb}	Fermi potential
V_{bs}	Substrate–source voltage
k	Boltzmann's constant
T	Absolute temperature in degrees K
q	Electronic charge
N_{sub}	Substrate doping density
n_i	Intrinsic carrier density
η	Mobility temperature coefficient
V_{FB}	Flat-band voltage
Q_{SS}	Surface charge per unit area
ϕ_P	Bulk potential
Q_d	Charge per unit area in inversion layer
ϕ_{bi}	Channel to substrate built-in potential
Q_i	Implanted charge per unit area
C_{impl}	Capacitance defined by implanted channel depth
ϕ_G	Bandgap voltage for silicon
ϕ_{GO}	Bandgap voltage for silicon at 0 K

difference between the threshold voltages of an enhancement- and depletion-mode FET pair is relatively temperature independent. The threshold voltages for enhancement- and depletion-mode MOSFETs may be written

$$V_{t(\text{enh})} = V_{FB} - \frac{Q_{SS}}{C_{ox}} + 2|\phi_P| + \frac{|Qd|}{C_{ox}} \tag{57.117}$$

$$V_{t(\text{depl})} = V_{FB} - \frac{Q_{SS}}{C_{ox}} + \phi_{bi} + (|Q_d| - |Q_i|)\left(\frac{1}{C_{ox}} + \frac{1}{C_{\text{impl}}}\right) \tag{57.118}$$

where the meaning of the parameters is given in Table 57.1 [4]. Many of the parameters in (57.117) and (57.118) show considerable temperature dependence. The difference between the threshold voltages is given by

$$\begin{aligned} V_{t(\text{diff})} &= V_{t(\text{enh})} - V_{t(\text{depl})} \\ &= 2|\phi_P| - \phi_{bi} - |Q_i|\left(\frac{1}{C_{ox}} + \frac{1}{C_{\text{impl}}}\right) \end{aligned} \tag{57.119}$$

assuming that $1/C_{ox} \gg 1/C_{\text{impl}}$, which is true in practice. In practice, it is also the case that $2|\phi_P| \approx \phi_{bi}$. Since the implanted charge Q_i is controllable and independent of temperature to first order [4], the threshold voltage difference exhibits temperature independence to first order.

The difference between the threshold voltages of two N-channel FETs with polysilicon gates of opposite doping polarity (P and N) also shows relative insensitivity to temperature variations [8]. To a first approximation, the threshold voltage difference is given by

$$\Delta V_G = \phi_G = 1.12\ \text{V} \quad \text{(room temperature)} \tag{57.120}$$

which is the bandgap voltage for silicon [8]. A more detailed analysis [1] gives

$$\Delta V_G(T) = \phi_{GO} - \frac{\alpha T^2}{T + \beta} \tag{57.121}$$

where $\alpha = 7.02 \cdot 10^{-4}$ V/K, $\beta = 1109$ K and the meaning of the remaining parameters is given in Table 57.1. In practice, the degree of temperature independence obtained provides useful reference circuits [8]. The exploitation of both the enhancement–depletion FET threshold difference and the N-P doped polysilicon gate threshold voltage difference for the design of references will be described later.

The strong temperature dependence of conventional MOS device parameters means that for stable biasing circuits, MOS devices are mainly useful where the critical variable depends on a ratio of parameters of similar devices. Even in this case, the matching is not as good as for bipolar devices. The matching of the gate–source voltages of two similar devices with nominally identical drain currents is inversely proportional to the square root of the gate area and is typically of the order of 10 mV, which limits the minimum offset voltage of a CMOS op amp. Since op amps form key components in many voltage and current reference circuits, this is a serious limitation.

In contrast to the rather complex dependence of MOS device parameters with temperature, the situation in the case of BJT devices is relatively straightforward [15], [17], [26]. The BJT may be described by

$$I_c = I_s e^{V_{be} q/kT} \tag{57.122}$$

where the additional parameters are defined in Table 57.1. Equation (57.122) may alternatively be written

$$V_{be} = \frac{kT}{q} \ln \frac{I_c}{I_s} \tag{57.123}$$

For two devices with an emitter area ratio of A

$$A = \frac{I_{s1}}{I_{s2}} \tag{57.124}$$

We have

$$\begin{aligned} \Delta V_{be} &= V_{be1} - V_{be2} \\ &= \frac{kT}{q} \ln \frac{I_{c1}}{I_{c2}} \frac{1}{A} \\ &= V_T \ln \frac{1}{A} \quad (\text{for } I_{c1} = I_{c2}) \end{aligned} \tag{57.125}$$

Thus the difference between the V_{be}'s of two BJTs with different current densities is proportional to the thermal voltage V_T, which is proportional to absolute temperature (PTAT). The positive temperature coefficient of V_T can be effectively used to cancel the negative temperature coefficient of V_{be} [13]. This is referred to as the bandgap principle.

Resistors are also key elements in MOS biasing circuits and they may be realized using diffusion, polysilicon and, in some advanced processes, using film techniques. Polysilicon and diffused resistors suffer from a high temperature coefficient that is positive for diffusion. The resistivity of gate polysilicon is typically rather low at about 20 Ω/square and its initial value tolerance is quite high. Film resistors have a very low temperature coefficient.

Voltage and Current Reference and Bias Circuits

Supply-Voltage-Referenced Voltage and Current References

When the supply voltages to a chip are well-regulated off-chip, then a voltage reference, acceptable in some cases, can be realized by a simple potential divider from the power supply voltage, as shown in Fig. 57.51. An external decoupling capacitor may be used if needed and the voltage dividing elements may be resistors [Fig. 57.51(a)] or MOSFETs [Fig. 57.51(b)]. If the power supply voltages are well-controlled off-chip, using an external regulator circuit, then a simple current reference can be realized by the arrangement in Fig. 57.52, where the reference current is defined by applying a well-defined fraction of the controlled supply voltage to a well-controlled external resistor.

A very simple biasing circuit providing multiple current sources and sinks, as required in CMOS analog signal-processing circuits is shown in Fig. 57.53(a) [13]. Assuming large FET W/L ratios, the voltage across the resistor is approximately $V_{dd} - V_{ss} - 2V_t$. The current in R is

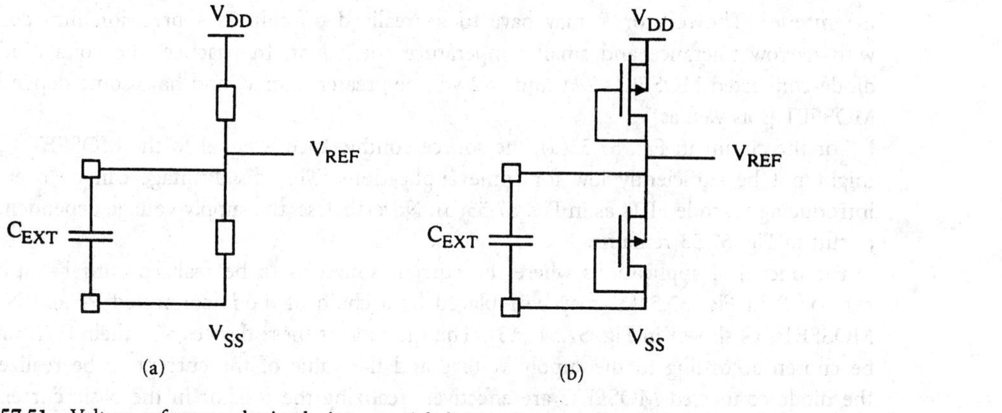

FIGURE 57.51 Voltage reference obtained via potential divider from regulated power supply: (a) using resistors; (b) using MOSFETs.

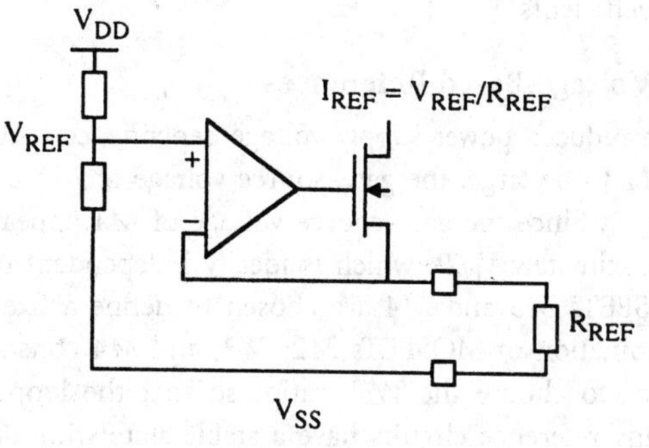

FIGURE 57.52 Current source realized using potential divider.

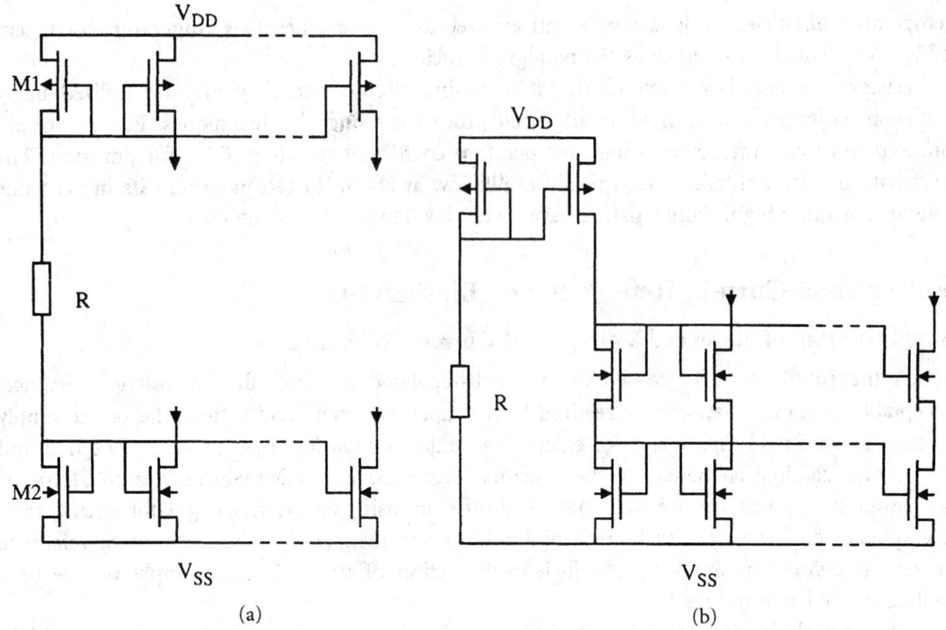

FIGURE 57.53 Resistor/current mirror bias circuits: (a) simple; (b) cascode.

mirrored in the output MOSFETs whose W/L ratios may be chosen to provide required current magnitudes. The resistor R may have to be realized off-chip as a precision film component with narrow tolerance and small temperature coefficient. In practice, the voltage across the diode-connected MOSFETs $M1$ and $M2$ will be greater than V_t and have some dependence on MOSFET β as well as V_t.

For the circuit in Fig. 57.53(a), the source conductance is equal to the MOSFET g_{ds}, which might not be sufficiently low for some applications. This disadvantage can be overcome by introducing cascode FETs as in Fig. 57.53(b). Nevertheless, the supply voltage dependence of the circuit in Fig. 57.53 remains.

For uncritical applications where the current source is to be realized entirely on-chip, the resistor R in Fig. 57.53(a) may be replaced by a chain of diode-connected P- and N-channel MOSFETs, as shown in Fig. 57.54 [13]. The number of these devices and their W/L ratios may be chosen according to the supply voltage and the value of the current to be realized. Since the diode-connected MOSFETs are effectively realizing the resistor in the basic current source of Fig. 57.53(a), the power supply voltage dependence of the current remains. In addition, the effective resistance realized depends on MOSFET β and V_t, both of which have large tolerances and high temperature coefficients.

MOSFET Threshold Voltage-Based References

A current reference with reduced power supply voltage dependence is shown in Fig. 57.55 [13]. By choosing W/L for $M1$ to be large, the gate–source voltage of $M1$ can be made close to the device threshold voltage V_t. Since the gate–source voltage of $M1$ appears across the resistor R, the current in R is approximately V_t/R, which is ideally independent of power supply voltage. The W/L ratios of MOSFETs $M3$ and $M4$ are chosen to define a fixed ratio for the currents in $M1$ and R. The combination of MOSFETs $M2$, $M3$, and $M4$ constitute a positive feedback loop and it is important to choose the W/L ratios so that the loop gain is less than unity to avoid oscillation. Many reference circuits have a stable state with all currents zero. In such cases, it is necessary to provide a start-up circuit [13] to prevent the reference circuit locking

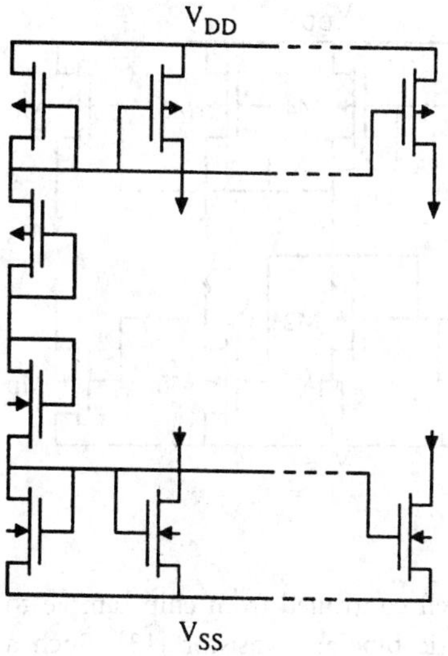

FIGURE 57.54 FET/current mirror bias circuit.

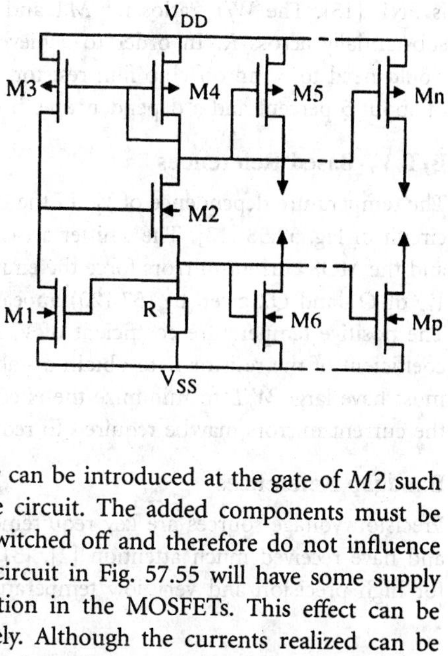

FIGURE 57.55 V_t-referenced current bias circuit.

into an undesired operating point. A source follower can be introduced at the gate of $M2$ such that in this condition a current is injected into the circuit. The added components must be such that in the normal operating point, they are switched off and therefore do not influence operation. In practice, the currents realized by the circuit in Fig. 57.55 will have some supply voltage dependence due to channel length modulation in the MOSFETs. This effect can be reduced by introducing cascode devices appropriately. Although the currents realized can be made substantially independent of supply voltage, the dependence on resistance R remains. For high precision and temperature independence, R may need to be realized as an off-chip film resistor. It must be borne in mind that the device threshold voltage on which the current depends is rather variable (typically 0.5–0.8 V) and also rather temperature dependent. Solution of this problem requires the introduction of alternative techniques based on BJT or unconventional CMOS devices, which will be discussed.

An alternative circuit to that in Fig. 57.55 is shown in Fig. 57.56 [25]. This circuit regulates the MOSFET drain currents with the result that MOSFET transconductance is proportional to $1/R$. This circuit also relies on positive feedback and care must be taken in the design to avoid instability. It has been shown in [21] that a practical stable design results from the choice $(W/L)_4 = (W/L)_3$ and $(W/L)_2 = 4(W/L)_1$, giving $g_m = 1/R$. In processes where R can be realized as a film resistor on-chip, this circuit can stabilize transconductances to within 3 percent over a 100°C temperature range [25].

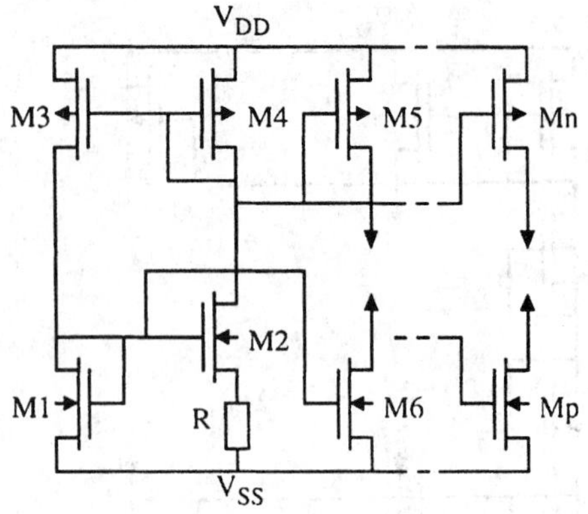

FIGURE 57.56 g_m-R tracking current reference circuit.

BJT V_{be}-Based References

The problem that MOSFET threshold voltage is not very well controlled from chip sample to chip sample leads to the idea of using the V_{be} of a parasitic bipolar transistor [13]. Such a V_{be}-referenced circuit is shown in Fig. 57.57 for the case of an N-well process, where the BJT is PNP [13]. The W/L ratios for $M1$ and $M2$ are made large so that the V_{be} of $T1$ appears substantially across R. In order to achieve high precision and temperature independence, R would need to be an off-chip film resistor. However, the V_{be} has a process-dependent tolerance of about 5 percent and a dependence with temperature of about −2 mV/°C.

BJT V_T-Based References

The temperature dependence of V_{be} in the circuit of Fig. 57.57 can be overcome in the V_T-based circuit of Fig. 57.58 [13]. The emitter areas of the BJTs $Q1$ and $Q2$ are scaled in the ratio $1 : n$ and the MOS current mirrors force the emitter currents to be equal. The difference between the V_{be} of $Q1$ and $Q2$ given by (57.120) appears across the resistor R, hence defining the current. The positive temperature coefficient of V_T can be used to counteract the positive temperature coefficient of the resistor R to obtain a stable current. In the circuit of Fig. 57.58, $M1$ and $M2$ must have large W/L to minimize the effect of MOSFET process variability. Also, cascoding of the current mirrors may be required to reduce the effect of channel length modulation.

Bandgap References

Precision voltage sources are key requirements for the realization of precision data converters and have received much attention [2], [3], [7], [10], [12]–[14], [20], [22]. The requirements for high precision and very low temperature and supply voltage dependence have led to the

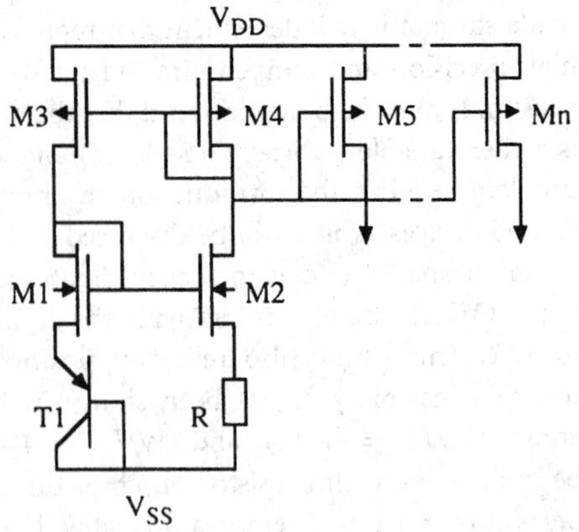

FIGURE 57.57 V_{be}-referenced current bias circuit.

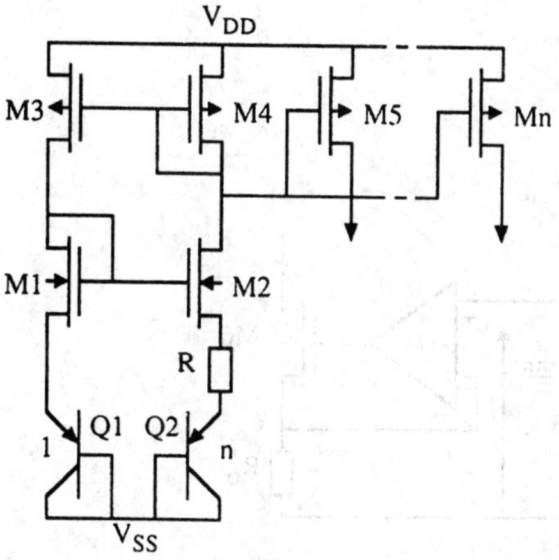

FIGURE 57.58 V_T-referenced current bias circuit.

development of the bandgap principle [2], [3], [7], [10], [13]. The bandgap principle was originally developed for bipolar technology and a typical architecture is shown in Fig. 57.59(a). As described above, the difference between the V_{be}'s of two BJTs with different current densities is proportional to the thermal voltage V_T and is PTAT. In Fig. 57.59(a), the difference between the V_{be}'s appears across $R3$ and in scaled form across $R2$. Thus the output voltage is equal to the V_{be} of $Q1$ plus the scaled version of ΔV_{be}. Thus $R2/R3$ may be chosen so that the opposite temperature coefficients of V_{be} and ΔV_{be} cancel. The ratio $R1/R2$ determines the ratio of the currents in $Q1$ and $Q2$. The circuit in Fig. 57.59(a) is incompatible with implementation using CMOS technology with vertical parasitic bipolar devices because the collectors are not grounded. This can be overcome using the architecture in Fig. 57.59(b). However, there is the further problem that the offset voltage of the operational amplifier is multiplied by the internal gain of the feedback loop and added to the output. Offset is worse for CMOS than for bipolar operational amplifiers. This problem has been overcome in various ways. In [12] and [22], use is made of a discrete time offset compensated differential amplifier, which can have very low offset. Another approach is to make use of lateral bipolar devices, which do not suffer from the topological restrictions of their vertical counterparts [16]. Thus the architecture of Fig. 57.59(a) or an equivalent topology may be implemented.

A typical example of a current reference based on a bandgap voltage reference is shown in Fig. 57.60 [13]. The current in the resistor xR is V_T-referenced as in Fig. 57.58 and therefore has a negative temperature coefficient (the BJTs are vertical parasitic devices). This current is

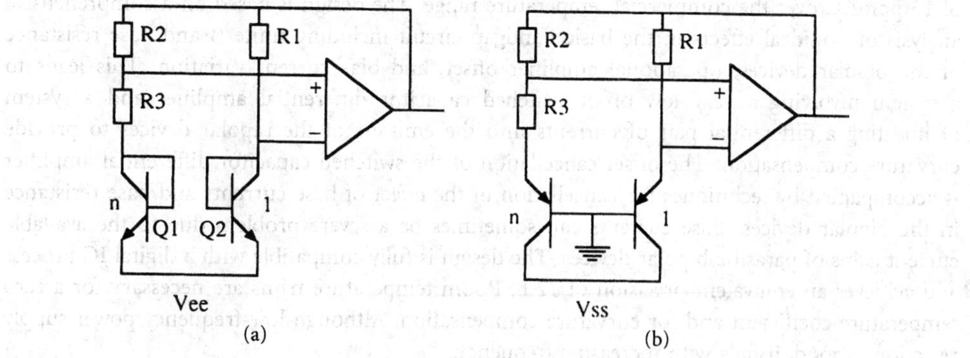

FIGURE 57.59 Basic bandgap circuits. (a) Classical bandgap circuit. (b) Modified form with grounded-collector PNP transistors, assuming an N-well process.

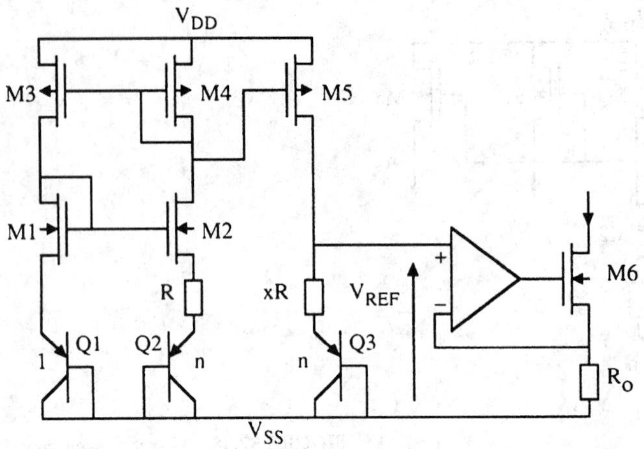

FIGURE 57.60 Bandgap current bias circuit.

converted to a voltage and weighted by the resistor xR before being added to the V_{be} of $Q3$, which has a negative temperature coefficient. The parameter x is chosen to obtain an overall zero temperature coefficient for the output current, which is given by V_{ref}/R_o. Clearly, R_o needs to be a high-precision resistor and could be external to the chip. The current mirrors need to be very well matched as any offset is amplified. The operational amplifier needs to have a low offset voltage since this is added to the reference voltage. Further current mirroring may be used to change the sign of the current or to increase the permissible range of the output voltage, referred to as compliance.

Curvature-Compensated Bandgap References

The bandgap reference principle can provide a zero temperature coefficient at a single temperature, leaving a temperature dependence that is dominated by a second-order temperature dependence. Very sophisticated techniques have been developed [12], [24] to eliminate this second-order dependence to leave a typically much smaller third-order dependence. This technique is referred to as curvature compensation. An example of a curvature-compensated current reference [24] is shown in Fig. 57.61. This circuit can achieve precisions of the order of 5 ppm/°C for supply voltages over 5 to 15 V.

Discrete Time Bandgap References

The voltage reference in [12] provides curvature compensation and achieves a drift of the order of 13 ppm/°C over the commercial temperature range. The design is based on a comprehensive analysis of nonideal effects in the basic bandgap circuit including finite β and base resistance of the bipolar devices, operational amplifier offset, and bias current variation. This leads to a system involving a very low offset switched capacitor differential amplifier and a system of injecting a differential pair of currents into the emitters of the bipolar devices to provide curvature compensation. The offset cancellation of the switched capacitor differential amplifier is accompanied by techniques for cancellation of the effect of base currents and base resistance in the bipolar devices. Base currents can sometimes be a severe problem due to the available current gains of parasitic bipolar devices. The design is fully compatible with a digital IC process and achieves an equivalent precision of 12 b. Room temperature trims are necessary for a zero temperature coefficient and for curvature compensation. Although low-frequency power supply rejection is good, it falls with increasing frequency.

In [22], a floating voltage reference for signal-processing applications with a good power supply rejection ratio of at least 85 dB maintained up to 500 MHz is realized. Over a temperature

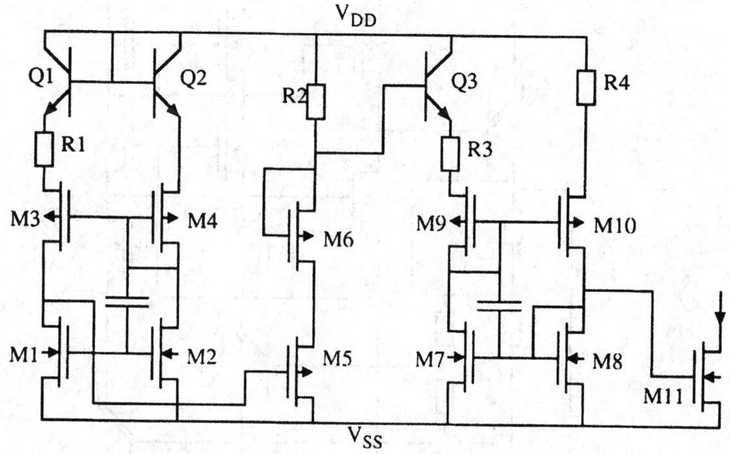

FIGURE 57.61 Curvature-compensated current bias circuit.

range of −40 to +85°C, voltage dependence is 40 ppm/°C and supply voltage dependence ±5 percent. The circuit has the important advantage that trimming is not required.

Voltage and Current References Based on Less Usual Devices

Use of Device in Subthreshold Region

An alternative approach to current reference making use of MOSFETs in the subthreshold region is reported in [19]. The principle of the approach is illustrated in Fig. 57.62(a), where for thermal stabilization the voltage source is required to be PTAT. The PTAT voltage source is realized as a cascade of 5 of the PTAT voltage sources shown in Fig. 57.62(b), which rely on the subthreshold mode operation of the devices. In practice, cascoding of the current mirrors and current sources is required and a start-up circuit is needed. A current accuracy of 3 percent with temperature stability of 3 percent over 0–80°C can be achieved with this approach [19].

Voltage Reference Circuits Using Lateral Bipolar Devices

The circuit diagram of a bandgap voltage reference making use of lateral bipolar devices is shown in Fig. 57.63 [16]. The circuit is designed to be insensitive to low β and α of the bipolar

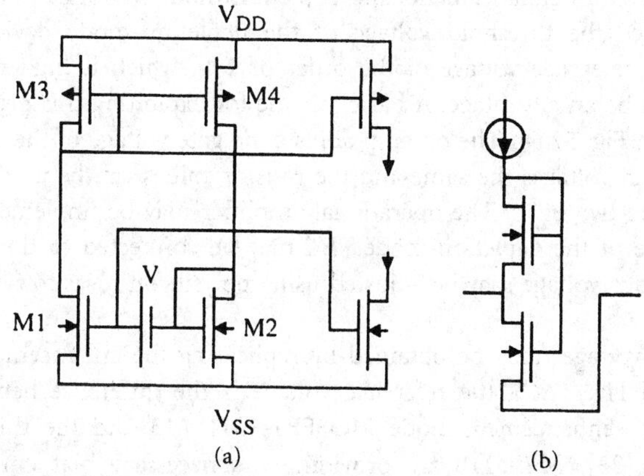

FIGURE 57.62 MOS current bias circuit based on weak inversion operation: (a) basic circuit; (b) voltage source cell.

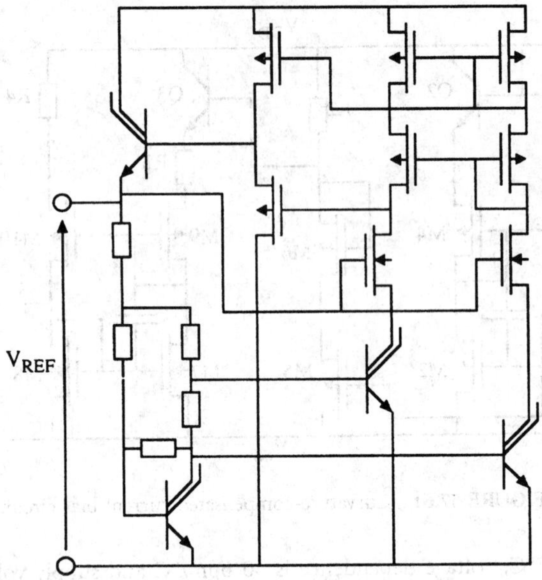

FIGURE 57.63 Voltage reference using lateral bipolar devices.

devices. It is also insensitive to offsets and mismatch. A single trim at room temperature is required and a high power supply rejection ratio, at least at low frequencies, is obtained. The output voltage is stable to within 2 mV over a wide temperature range.

Voltage References Based on Enhancement-Depletion Mode Threshold Voltage Difference

The topology restrictions and imperfections of the BJT devices available in CMOS technology have led to the development of alternative techniques for designing references without needing bipolar devices. In one technique, the fact that the difference between the threshold voltage of depletion-mode and enhancement-mode devices is relatively temperature independent has been exploited [4], [5], [23].

It has been shown in the section on device models and parameter variability that the difference in threshold voltages of an enhancement- and depletion-mode MOSFET is relatively insensitive to temperature. Since the threshold voltage of the depletion-mode device is negative, this approach leads to a reference voltage of the order of 2 V, which is higher than the bandgap voltage, and this can be an advantage. A basic scheme for exploiting this principle for a voltage reference is shown in Fig. 57.64. The op amp adjusts the gate voltage of the enhancement-mode FET to keep the drain voltages the same and the resistor values can be used to adjust the ratio of the currents in the two FETs. The operational amplifier may be implemented at device level [4]. The gate voltage of the depletion-mode FET may be connected to the output of a buffer amplifier whose output voltage may be adjusted using polysilicon fuses to typically 3.15 V±0.02 V [5].

Higher reference voltages may be obtained by replicating the enhancement- and depletion-mode MOSFETs. In Fig. 57.65, the reference voltage is the difference between the threshold voltages of the three enhancement-mode MOSFETs *M*1–*M*3 and the three depletion-mode MOSFETs *M*4–*M*6 [23]. *M*7–*M*10 are providing the necessary bias currents. In [23], the variation of V_{REF} with temperature is 1.5 mV/°C, which is useful for many biasing situations and the reference voltage is of the order of 3 V in spite of low threshold voltage devices.

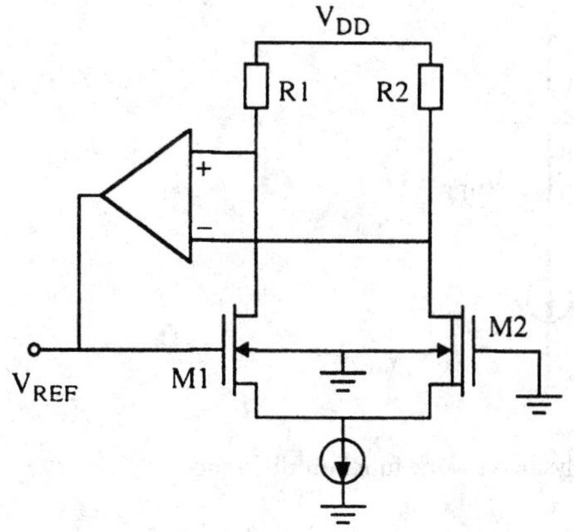

FIGURE 57.64 Basic reference based on enhancement–depletion threshold difference.

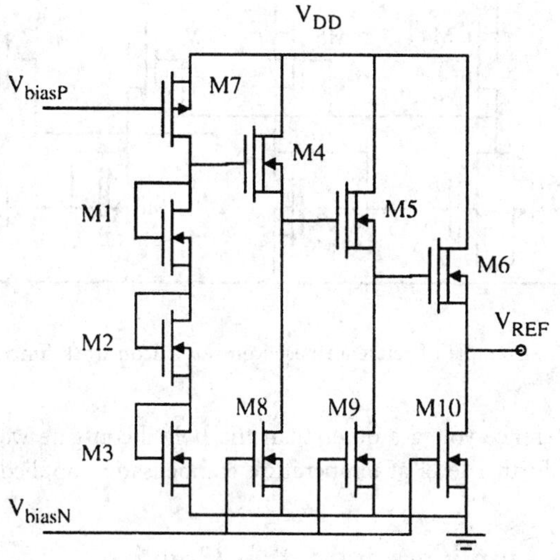

FIGURE 57.65 High-output enhancement–depletion threshold difference reference.

Voltage References Based on N- and P-Doped Polysilicon Gate Threshold Voltage Difference

In CMOS technology, the gate material is usually polysilicon with N-type doping. In some processes, selective doping to provide P-type doping of the polysilicon gate is also available and the presence of both types of doping has been exploited for reference circuit design [8].

The basic principle of a voltage reference based on the difference between the threshold voltages of N- and P-doped polysilicon gate MOSFETs is illustrated in Fig. 57.66. $M1$ has a P-doped gate and a higher threshold voltage than $M2$. $M1$ and $M2$ are in different P-wells but have the same effective dimensions and bias currents.

A full transistor-level implementation of the basic circuit in Fig. 57.66 is shown in Fig. 57.67 [8]. $M1$ and $M2$ are the reference MOSFETs. $M3$ has a very long channel and its current is the same as that in $M1$ by virtue of the current mirror $M4 : M5$. Thus the current in $M1$ adjusts itself to the crosspoint of the characteristics of $M1$ and $M3$. $M7$, $M8$, and $M9$ ensure that the currents in $M1$ and $M2$ are identical. $M6$ is a start-up device. When the power supply is switched on, $M6$ comes on but within 1 ms is switched off by the reverse leakage resistance of the polysilicon diode D. In [8], $M1$ and $M2$ can have a W/L of 100 μm/20 μm; supply voltage sensitivity of $< 10^{-3}$ is obtained for V_{DD} between 2 and 9 V [8]. Digital tuning using

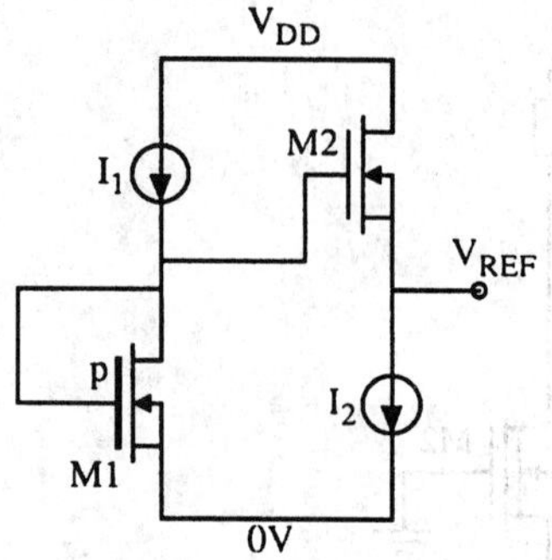

FIGURE 57.66 Basic reference based on polysilicon work function difference.

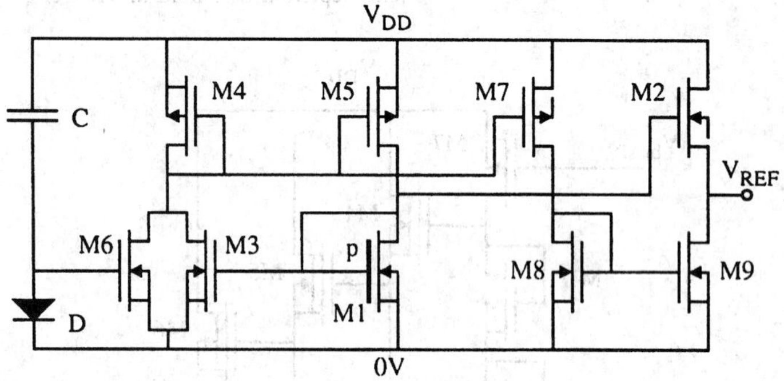

FIGURE 57.67 Example of reference based on polysilicon work function difference.

polysilicon fuses to reference voltages other than the polysilicon gate work function difference can be obtained and a further level of temperature compensation applied [8].

Biasing of Simple Amplifiers and Other Circuits

Simple Amplifiers

In traditional two-stage amplifier design, the bias for the whole circuit is easily set up from one reference current and no critical voltage differences have to be set up. It is only necessary to ensure that the operating currents, and hence the transconductance, of each device have a required value. A typical example of the biasing of a two-stage amplifier is shown in Fig. 57.68 [13]. The ratio of currents between the first- and second-stage controls the separation of the poles and also the systematic offset. The internal biasing circuit consists simply of a set of current mirrors.

Cascode Amplifiers

In cascode amplifiers [13], the idea is to raise the amplifier output impedance in order to increase the gain. An important requirement, especially in a low supply voltage environment, is to obtain maximum output voltage swing, or compliance, in the cascode amplifier. This requirement makes the biasing of the cascode devices critical.

A simple amplifier designed for capacitive loads is shown in Fig. 57.69. Only a single current mirror from the main current reference is needed to control all the bias currents. For reasonable low-frequency gains of say > 60 dB, the output impedance must be made high while keeping component parameters practical. Use of very long channel output FETs is undesirable because of poor bandwidth and the chip area requirement. Therefore, the cascode technique, as shown in Fig. 57.70, is the ideal solution. This raises output impedance by approximately $g_{m(M11)}r_{ds(M11)}$ and $g_{m(M12)}r_{ds(M12)}$ on each side. However, the maximum output voltage swing, or compliance, of the circuit has now been reduced by at least the saturation voltages of $M11$ and $M12$. The cascode devices must be biased so that the voltage across $M8$ and $M9$ are just above V_{dsat}.

The usual way of achieving this is to arrange that a current is passed through an FET with a scaled width so that one obtains a voltage $V_{TN} + V_{Dsat(M6)} + V_{Dsat(M12)}$ for the N-channel side and $V_{TP} + V_{Dsat(M8)} + V_{Dsat(M11)}$ for the P-channel side. If the saturation voltage of the driver FET ($M6$) and the cascode FET ($M12$) are the same, then this requires a bias FET with a current density four times higher. In reality the body effect in $M12$ and tolerance considerations mean that this factor will have to be somewhat higher than 4 [13].

Hence, there is a requirement for more replicas of the incoming reference current. The whole scheme develops to the configuration shown in Fig. 59.71, where the circuit is represented in its simplest ideal form. Note that all FET scaling is applied to device widths; the lengths are the same throughout. In practice, this circuit would have a large offset due to the unequal drain voltages in the various current mirrors, and balancing dummy FETs would be needed.

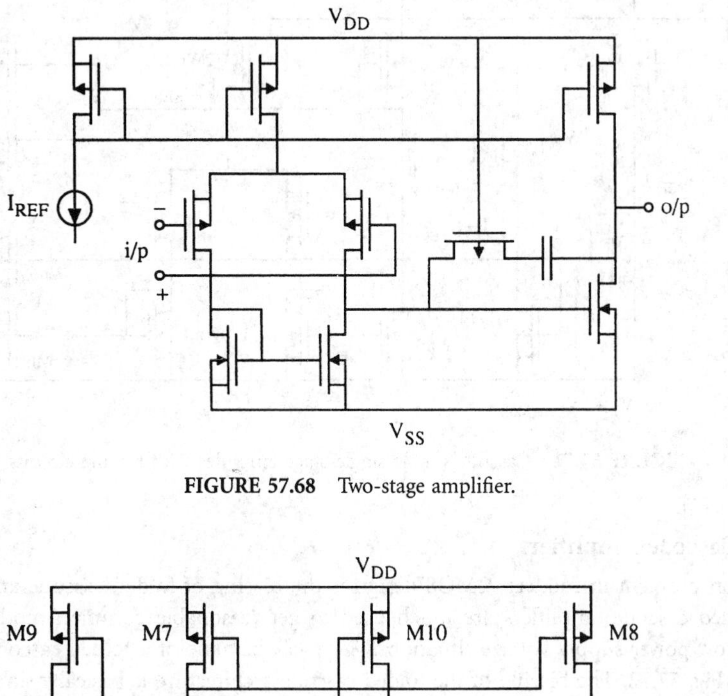

FIGURE 57.68 Two-stage amplifier.

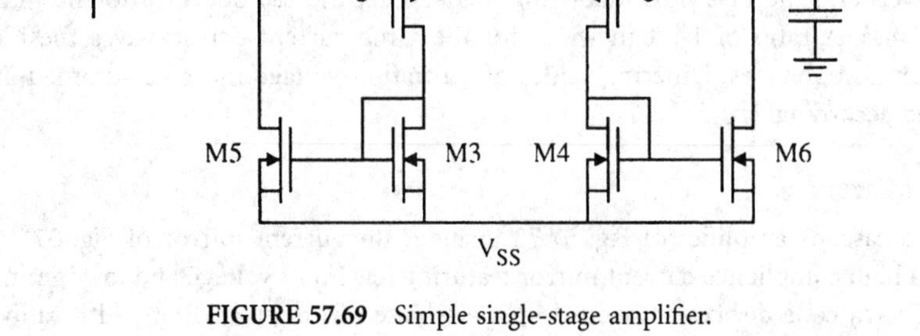

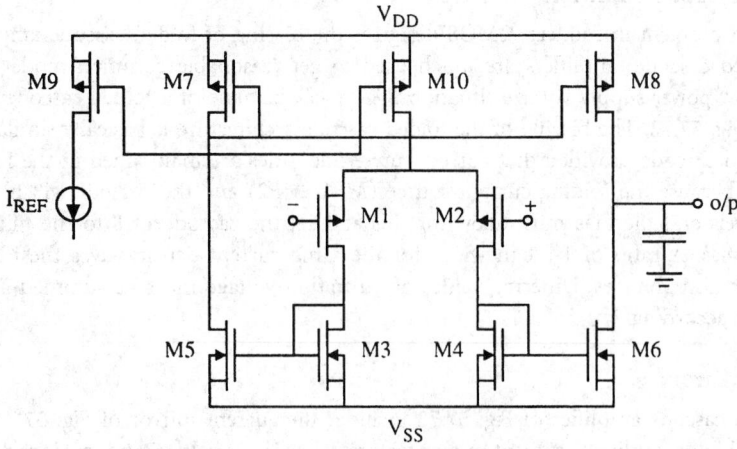

FIGURE 57.69 Simple single-stage amplifier.

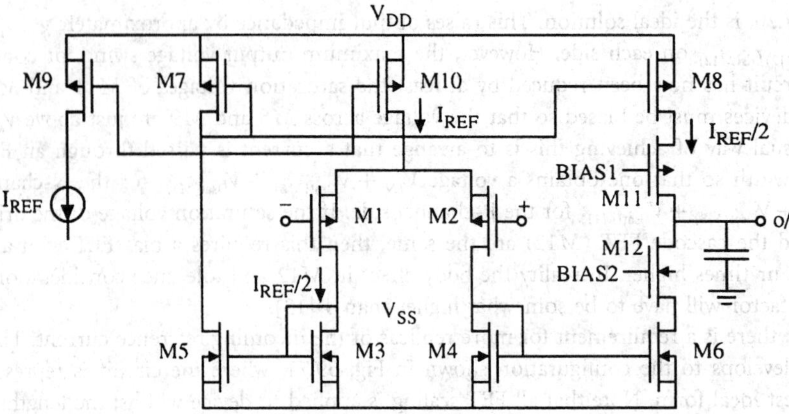

FIGURE 57.70 Simple single-stage amplifier with cascoded output FETs.

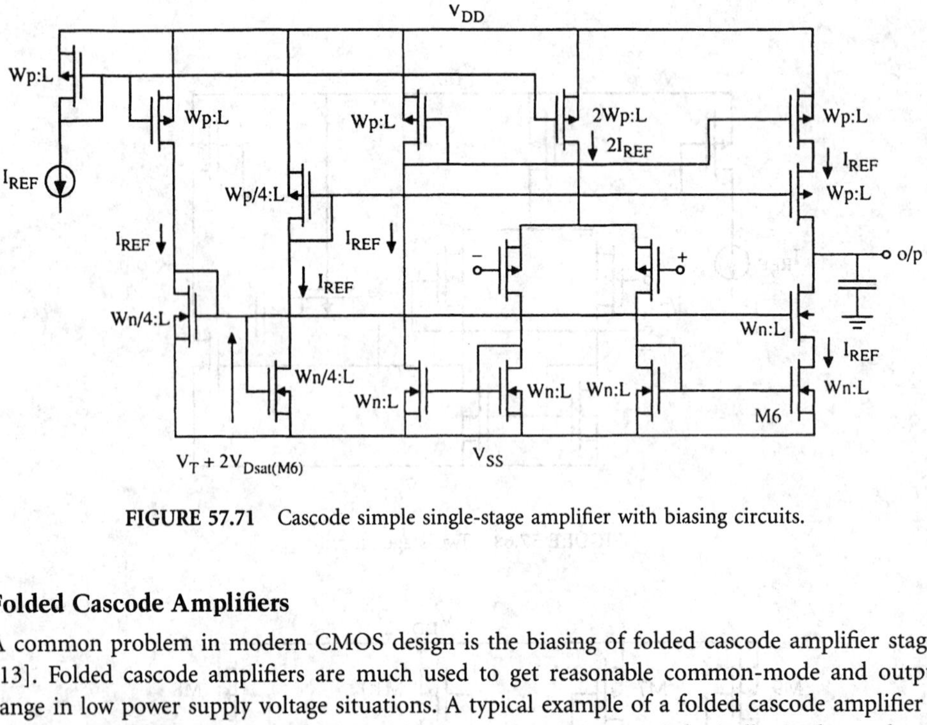

FIGURE 57.71 Cascode simple single-stage amplifier with biasing circuits.

Folded Cascode Amplifiers

A common problem in modern CMOS design is the biasing of folded cascode amplifier stages [13]. Folded cascode amplifiers are much used to get reasonable common-mode and output range in low power supply voltage situations. A typical example of a folded cascode amplifier is shown in Fig. 57.72. The biasing of the folded cascode architecture is basically similar to that of a nonfolded cascode provided that correct current densities are maintained in the FETs. This is important because the folding current source (*MF*1, *MF*2) and the cascode will have different current levels and the bias must allow for this and set the cascode FET to the minimum safe operating bias. A ratio of 1 : 4 in width for the same current density gives the ideal bias for equal saturation voltages. Differing values of saturation voltage must be summed and the bias FET scaled accordingly.

Current Mirrors

The folded cascode amplifier of Fig. 57.72 includes the current mirror of Fig. 57.73 [18]. This circuit is a high compliance current mirror featuring low input voltage and low minimum output voltage. The cascode devices embedded in it are biased from an FET of width ratio running at

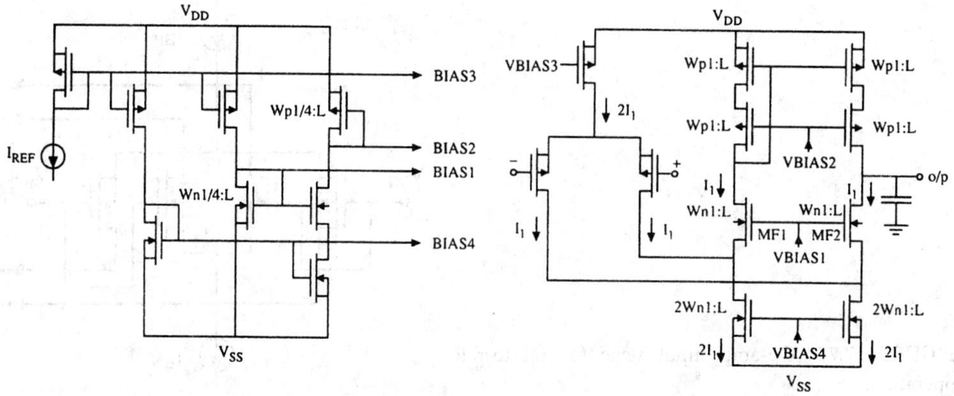

FIGURE 57.72 Folded cascode amplifier with biasing circuits.

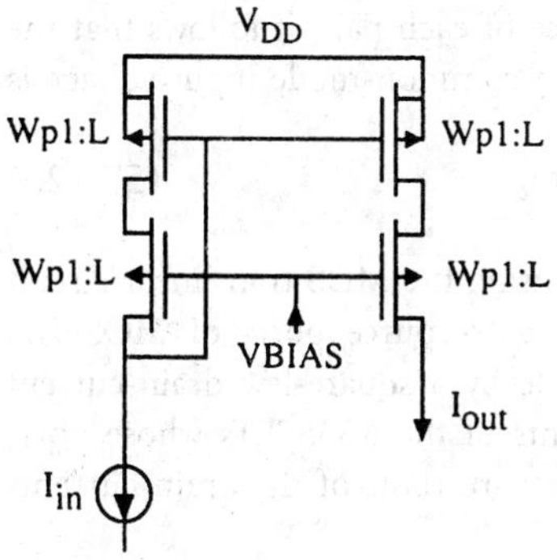

FIGURE 57.73 High-compliance current mirror.

the same quiescent current as the mirror. A width ratio of 1 : 4 is predicted by simple theory for equal saturation voltages in the mirror.

Biasing of Circuits with Low Power Supply Voltage

The topic of low voltage analog MOS circuits is an important one because of the requirement for battery operation and also for compatibility with advanced digital IC processes with low power supply voltages. In order to maintain a reasonably high dynamic range in low supply voltage analog circuits, it is essential that the circuits operate with signal swings that are a very large fraction of the total supply voltage. Since operational amplifiers are key components in analog circuits, this has led to the design of operational amplifiers with "rail-to-rail" input common-mode voltage and output voltage capability. The design of such an input stage requires new approaches to biasing and we shall give brief details of an example of one such circuit [27].

A conventional differential pair of N- or P-channel MOSFETs would not provide sufficient input voltage common-mode range. This is because the input FET pair and the FET realizing the tail current source would tend to come out of saturation at one extreme of input common-mode voltage (negative for NMOS, positive for PMOS). This is overcome by combining an NMOS and PMOS differential pair as shown in Fig. 57.74. However, this circuit has the disadvantage that the effective transconductance varies widely with common-mode input voltage since in the middle range, both differential pairs are conducting but at the extremes, only one differential pair is conducting. This produces a common-mode voltage dependence of amplifier dynamic performance and makes it difficult to optimize the dynamic performance for all input conditions.

An elegant solution to this problem is reported in [27]. Assuming that the drain current of a MOSFET can be described by a square-law relationship, then transconductance is proportional

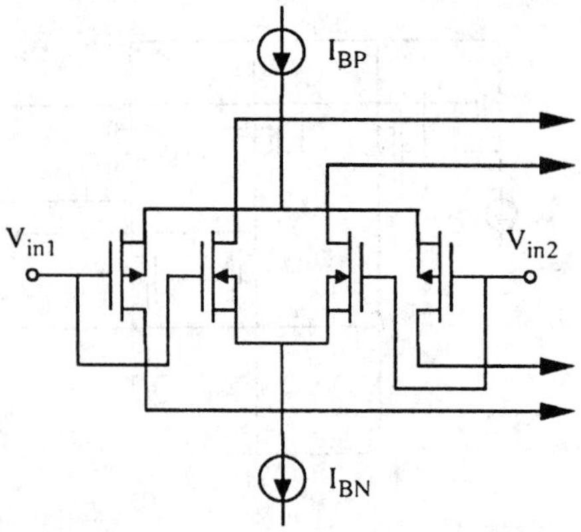

FIGURE 57.74 Op-amp input stage for rail-to-rail operation.

to the square root of bias current. Since the overall effective transconductance of the input stage in Fig. 57.74 is the sum of the effective transconductance of each pair, it follows that the condition when the overall transconductance is independent of common-mode input voltage is

$$\sqrt{I_{BN}} + \sqrt{I_{BP}} = \text{constant} \tag{57.126}$$

Bias currents satisfying this relationship can be implemented using the MOS translinear circuit principle [27]. This principle applies to circuits where the gate–source ports of MOSFETs form a closed loop. Assuming that the devices are describable by a square-law drain-current relationship, the sum of the square roots of the drain currents of the MOSFETs whose ports are connected in a clockwise fashion equals the sum of the square roots of the drain currents of the counterclockwise-connected MOSFETs.

The application of the basic idea to implement bias currents according to (57.126) is shown in Fig. 57.75. Since the clockwise-connected MOSFETs $M1$ and $M2$ have a constant drain current I_o, the translinear principle implies that the drain current in $M3$ and $M4$ satisfy (57.126). The development of the schematic bias circuit in Fig. 57.75, the input stage in Fig. 57.74 and a class AB output stage into a fully operational amplifier is described in [27]. The circuit operates with a minimum power supply of 2.5 V.

Dynamic Biasing

Dynamic biasing is a technique that is applicable to amplifiers in sampled data systems, such as switched capacitor filters and data converters [6], [9]. Such amplifiers are required to meet two key requirements. These are fast settling time in order to allow high switching rates and high

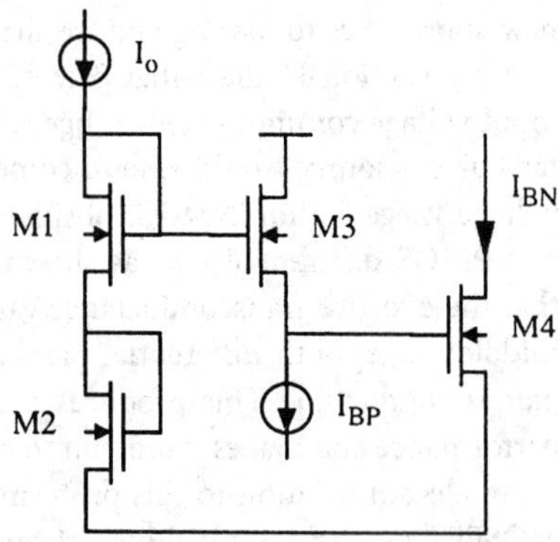

FIGURE 57.75 Bias circuit based on MOS translinear principle.

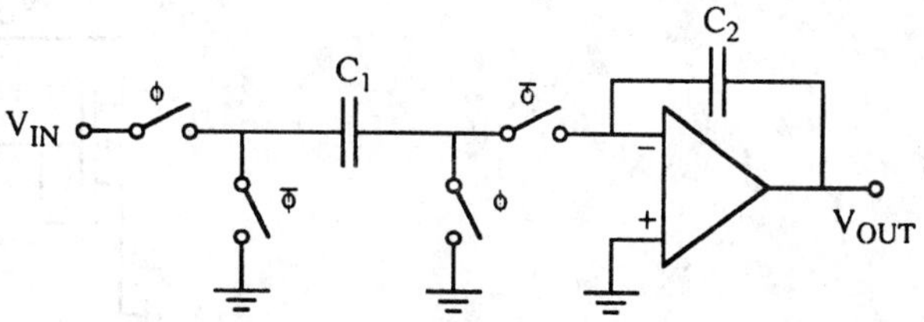

FIGURE 57.76 Typical switched capacitor integrator.

gain in order to obtain precision performance [13]. Fast settling time is obtained for maximum effective device transconductance and device transconductance g_m is approximately proportional to $\sqrt{I_B}$, where I_B is bias current. Gain is given by g_m/g_o, where g_o is output conductance. Since g_o is proportional to bias current, gain is *inversely* proportional to $\sqrt{I_B}$. Thus maximum settling time requires a high bias current and maximum gain requires a low bias current. The dynamic bias technique reconciles these two requirements.

Figure 57.76 shows a typical switched capacitor integrator, which is a basic building block for implementing high-order switched capacitor systems. The switches are controlled by two-phase nonoverlapping clock signals ϕ and $\bar{\phi}$. The operational amplifier would generally have a first stage comprising a differential MOSFET pair with constant current source bias. The equivalent of this with dynamic biasing is shown in Fig. 57.77, where the constant current source has been replaced by the combination of capacitor C and switches $S1$ and $S2$. We refer to the integrator of Fig. 57.76. During phase ϕ, the capacitor C_1 is being charged up to the input voltage and the amplifier is inactive. Meanwhile, in the dynamically biased amplifier of Fig. 57.77, the capacitor C is being discharged. In phase $\bar{\phi}$, capacitor C_1 in Fig. 57.76 is connected to the input of the amplifier, whose output voltage is required to change to absorb the incoming charge. At the same time, capacitor C in Fig. 57.77 is connected to the differential pair and immediately starts to conduct a high current. The high current through the amplifier MOSFETs provides a high effective amplifier slew rate and fast initial settling time, although the gain of the amplifier during this initial part of the clock phase is low. As time progresses, capacitor C becomes charged and the current in the amplifier MOSFETs reduces. This increases the gain of the amplifier leading to a high precision of the amplifier output voltage. Eventually, the amplifier current falls to zero with the output voltage at this required level. If, as would usually be the case, it is required to sample the amplifier output voltage in both phases, then the dynamic current source comprising capacitor C and the two switches $S1$ and $S2$ in Fig. 57.77 would need to be duplicated with opposite switch phasing. This technique considerably increases the gain available from an amplifier since the effective gain depends on the low bias current condition and is very high. Dynamic biasing may, however, be easily applied to both stages of a two-stage amplifier if required [9]. Also, efficient schemes are available for the dynamic biasing of several amplifiers in a circuit.

Dynamic biasing is well worth considering in sampled date applications, such as switched capacitor filters and data converters. It can maximally exploit a given low power consumption to obtain good dynamic circuit performance. A variant of this approach [11] is adaptive biasing in which the input differential signal is sensed and the bias current is increased for large differential input signals to speed up the slewing response.

Conclusions

The task of designing voltage and current references and bias circuits is an important one. The requirements are very diverse, ranging from high precision, as required in data converters, to moderate, as required in general biasing situations. In this section, there has been space

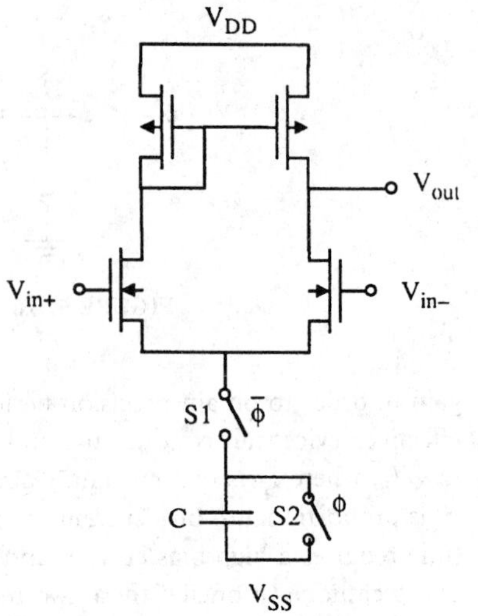

FIGURE 57.77 Simple op amp with dynamic biasing.

sufficient only to refer to some outstanding work in the area and briefly discuss some of the main principles. It is hoped that the reader will consult the references for more detailed information.

References

[1] S. M. Sze, *Physics of Semiconductor Devices*, New York: Wiley-Interscience, 1969.

[2] R. J. Widlar, "New developments in IC voltage regulators," *IEEE J. Solid-State Circuits*, vol. SC-6, pp. 2–7, Feb. 1971.

[3] Y. P. Tsividis, "A CMOS voltage reference," *IEEE J. Solid-State Circuits*, vol. SC-13, pp. 774–778, Dec. 1978.

[4] R. A. Blauschild, P. A. Tucci, R. S. Muller, and R. G. Meyer, "A new NMOS temperature-stable voltage reference," *IEEE J. Solid-State Circuits*, vol. SC-13, pp. 767–773, Dec. 1978.

[5] M. E. Hoff, J. Huggins, and B. M. Warren, "An NMOS telephone Codec for transmission and switching applications," *IEEE J. Solid-State Circuits*, vol. SC-14, pp. 47–50, Feb. 1979.

[6] M. A. Copelend and J. M. Rabaey, "Dynamic amplifier for MOS technology," *Electron. Lett.*, vol. 15, pp. 301, 302, May 1979.

[7] E. A. Vittoz and O. Neyroud, "A low voltage CMOS bandgap reference," *IEEE J. Solid-State Circuits*, vol. SC-14, pp. 573–577, June 1979.

[8] H. I. Oguey and B. Gerber, "MOS voltage reference based on polysilicon gate work function difference," *IEEE J. Solid-State Circuits*, vol. SC-15, pp. 264–269, June 1980.

[9] B. J. Hosticka, "Dynamic CMOS amplifiers," *IEEE J. Solid-State Circuits*, vol. SC-15, pp. 887–894, Oct. 1980.

[10] R. Ye and Y. Tsividis, "Bandgap voltage reference sources in CMOS technology," *Electron. Lett.*, vol. 18, no. 1, pp. 24, 25, Jan. 7, 1982.

[11] M. C. Degruwe, J. Rijmenants, E. A. Vittoz, and H. J. de Man, "Adaptive biasing CMOS amplifiers," *IEEE J. Solid-State Circuits*, vol. SC-17, pp. 522–528, Oct. 1980.

[12] B.-S. Song and P. R. Gray, "A precision curvature-compensated CMOS bandgap reference," *IEEE J. Solid-State Circuits*, vol. SC-18, pp. 634–643, Dec. 1983.

[13] P. R. Gray and R. G. Meyer, *Analysis and Design of Analog Integrated Circuits*, New York: Wiley, 1984, pp. 730–737.

[14] J. Michejda and S. K. Kim, "A precision CMOS bandgap reference," *IEEE J. Solid-State Circuits*, vol. SC-19, pp. 1014–1021, Dec. 1984.

[15] B. J. Hosticka, K.-G. Dalsab, D. Krey, and G. Zimmer, "Behavior of analog MOS integrated circuits at high temperatures," *IEEE J. Solid-State Circuits*, vol. SC-20, pp. 871–874, Aug. 1985.

[16] M. G. K. R. Degrauwe, O. N. Leuthold, E. A. Vittoz, H. J. Oguey, and A. Descombes, "CMOS voltage references using lateral bipolar transistors," *IEEE J. Solid-State Circuits*, vol. SC-20, pp. 1151–1157, Dec. 1985.

[17] S. L. Lin and C. A. T. Salama, "A $V_{be}(T)$ model with application to bandgap reference design," *IEEE J. Solid-State Circuits*, vol. SC-20, pp. 1283–1285, Dec. 1985.

[18] A. J. J. Boudewijns, "Amplifier arrangement," U.S. Patent 4893090, granted Jan. 8, 1990, submitted Sept. 1988.

[19] W. M. Sansen, F. O. Eynde, and M. Steyaert, "A CMOS temperature-compensated current reference," *IEEE J. Solid-State Circuits*, vol. SC-23, pp. 821–824, June 1988.

[20] M. Ferro, F. Salerno, and R. Castello, "A floating CMOS bandgap voltage reference for differential applications," *IEEE J. Solid-State Circuits*, vol. SC-24, pp. 690–697, June 1989.

[21] J. M. Steininger, "Understanding wideband MOS transistors," *IEEE Circuits Devices Mag.*, vol. 6, pp. 26–31, May 1990.

[22] G. Nicollini and D. Senderowicz, "A CMOS bandgap reference for differential signal processing," *IEEE J. Solid-State Circuits*, vol. SC-21, pp. 41–50, Jan. 1991.

[23] K. Ishibashi and K. Sasaki, "A voltage down converter with submicroampere standby current for low power static RAMs," *IEEE J. Solid-State Circuits*, vol. SC-27, pp. 920–925, June 1992.

[24] C.-Y. Wu and S.-Y. Chin, "High precision curvature-compensated CMOS band-gap voltage and current references," *J. Analog Integrat. Circuits Signal Process.*, (Kluwer), vol. 2, no. 3, pp. 207–215, Sept. 1992.

[25] S. D. Willingham and K. W. Martin, "A BiCMOS low distortion 8 MHz low pass filter," *IEEE J. Solid-State Circuits*, vol. SC-28, pp. 1234–1245, Dec. 1993.

[26] J. Choma, Jr., "Temperature stable voltage controlled current source," *IEEE Trans. Circuits Syst. I*, vol. 41, pp. 405–411, May 1994.

[27] J. H. Botma, R. Jiegerink, S. L. J. Gierkink, and R. F. Wassenaar, "Rail-to-rail constant Gm input stage and class AB output stage for low-voltage CMOS op amps," *Analog Integrat. Circuits Signal Process.*, (Kluwer), vol. 6, no. 2, pp. 121–133, Sept. 1994.

57.4 Canonical Cells of MOSFET Technology

M. Ismail, S.-C. Huang, C.-C. Hung, and T. Saether

Analog integrated circuits have long been designed in technologies other than CMOS. But modern analog and mixed-signal VLSI applications in areas such as telecommunications, smart sensors, battery-operated consumer electronics, and artificial neural computation require CMOS analog design solutions. In recent years, analog CMOS circuit design has shown signs of dramatic change. Field programmable analog arrays and modular analog VLSI circuits [1] are representatives of emerging analog design philosophies leading to a whole new generation of analog circuit and layout design methodologies.

This section discusses basic cells used in contemporary CMOS analog integrated circuits. The performance of a CMOS (bipolar) circuit can often be improved further by incorporating a limited number of bipolar (CMOS) transistors on the same substrate. The resulting circuits are called BiCMOS circuits. BiCMOS circuits that are predominantly CMOS will also be discussed. First, we discuss primitive analog cells. These cells may or may not require device matching for proper operation. Second, we introduce modern and simple circuit techniques to mitigate nonideal effects and significantly improve circuit performance, and finally, we discuss basic voltage amplifier circuits. The presented cells will help in the systematic design of analog integrated circuits and could constitute an efficient analog VLSI cell library. Throughout this section, MOS transistors are assumed to be biased in strong inversion.

Matched Device Pairs

Figure 57.78 shows basic MOS transistor pairs [2] operating in the saturation region, where only NMOS transistor pairs are shown. Figure 57.78(a) shows a differential pair with no direct connection between the two transistors. The resultant differential pair is characterized by the difference in the drain currents (using the simple square-law equation); that is

$$\begin{aligned} I_{a1} - I_{a2} &= \frac{K}{2}(V_{G1} - V_{S1} - V_T)^2 - \frac{K}{2}(V_{G2} - V_{S2} - V_T)^2 \\ &= \frac{K}{2}[(V_{G1} - V_{G2}) - (V_{S1} - V_{S2})] \\ &\quad \times [(V_{G1} + V_{G2}) - (V_{S1} + V_{S2}) - 2V_T] \end{aligned} \tag{57.127}$$

where $K(= \mu C_{ox} W/L)$ and V_T are the transconductor parameter and the threshold voltage of the transistors respectively. Figure 57.78(b) is a common-source or source-coupled differential pair, a special case of circuit (a) with $V_{S1} = V_{S2} = V_S$, and the differential current is

$$I_{b1} - I_{b2} = \frac{K}{2}(V_{G1} - V_{G2})[(V_{G1} + V_{G2}) - 2V_S - 2V_T] \tag{57.128}$$

Figure 57.78(c) is a common-gate differential pair with $V_{G1} = V_{G2} = V_G$ in circuit (a), and the differential current is obtained as

$$I_{c1} - I_{c2} = -\frac{K}{2}(V_{S1} - V_{S2})[2V_G - (V_{S1} + V_{S2}) - 2V_T] \tag{57.129}$$

Differential pairs are essential building blocks of circuits such as op amps, differential difference amplifiers (DDAs) and operational transconductance amplifiers (OTAs). Several linear V–I converters built by these cells have been developed.

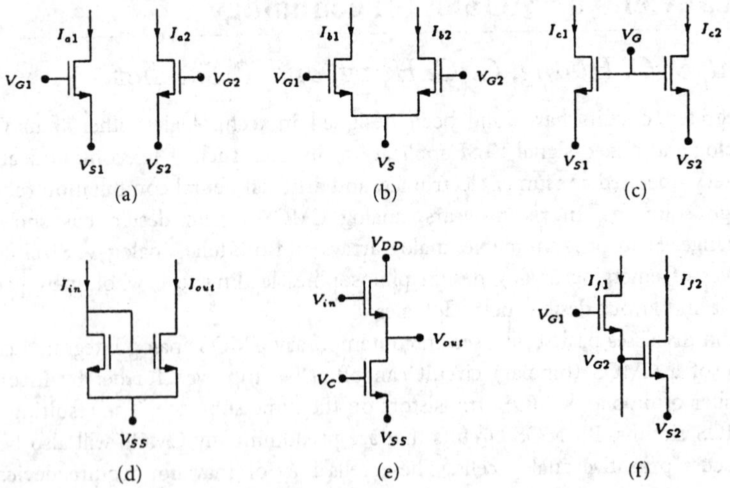

FIGURE 57.78 Matched primitive cells operating in the saturation region.

Current mirrors are usually used as loads for amplifier stages. Moreover, current mirrors are essential building blocks in modern current-mode analog integrated circuits. Figure 57.78(d) shows a well-known simple current mirror. Ideally, the input current I_{in} is equal to the output current I_{out} for matched transistors. In practice, a nonunity I_{out} to I_{in} ratio occurs due to finite output resistance resulting from channel length modulation effects. The output resistance can be increased by Wilson or cascode current mirrors at the expense of a limited output swing, which is not desired in low-voltage applications. A regulated current mirror can improve both the output resistance and swing but increase circuit complexity. Detail analysis and comparison are discussed in [3].

A voltage follower is shown in Fig. 57.78(e). Since the same current flows in both transistors, their gate–source voltages are the same. That is

$$V_{\text{in}} - V_{\text{out}} = V_C - V_{SS} \tag{57.130}$$

and therefore

$$V_{\text{out}} = V_{\text{in}} - V_C + V_{SS} \tag{57.131}$$

Alternatively, a transistor pair can be arranged as the circuit shown in Fig. 57.78(f), which is used as a basic cell for composite MOSFET (COMFET) circuits [4]. The differential current is given by

$$\begin{aligned} I_{f1} - I_{f2} &= \frac{K}{2}(V_{G1} - V_{G2} - V_T)^2 - \frac{K}{2}(V_{G2} - V_{S2} - V_T)^2 \\ &= \frac{K}{2}(V_{G1} - V_{S2} - 2V_T)(V_{G1} - 2V_{G2} + V_{S2}) \end{aligned} \tag{57.132}$$

With proper biasing, linear V–I conversion can be achieved by this transistor cell.

Transistor pairs operating in the triode region are found mostly in simulating linear transconductors and resistors, e.g., those in MOSFET-C filters. Figure 57.79 shows three popular examples, where the nonlinear terms in the drain current equations are canceled. A simple drain current equation in the triode region is

$$I_D = K\left[(V_G - V_T)(V_D - V_S) - \frac{1}{2}(V_D^2 - V_S^2)\right] \tag{57.133}$$

$$= \frac{K}{2}(V_G - V_S - V_T)^2 - \frac{K}{2}(V_G - V_D - V_T)^2 \tag{57.134}$$

Equation (57.134) gives another form of the triode current equation. In some cases, circuit analysis can be performed more easily with this form than using (57.133). The resulting current equations of the equivalent "MOS resistors" are now obtained.

For the circuit shown in Fig. 57.79(a), a two-transistor transconductor, the current difference is given by [5]

$$\begin{aligned} I_a - I_a^{'} &= K\left[(V_C - V_T)(V_X - V_Y) - \frac{1}{2}(V_X^2 - V_Y^2)\right] \\ &\quad - K\left[(V_C - V_T)(-V_X - V_Y) - \frac{1}{2}(V_X^2 - V_Y^2)\right] \\ &= 2K(V_C - V_T)V_X \end{aligned} \tag{57.135}$$

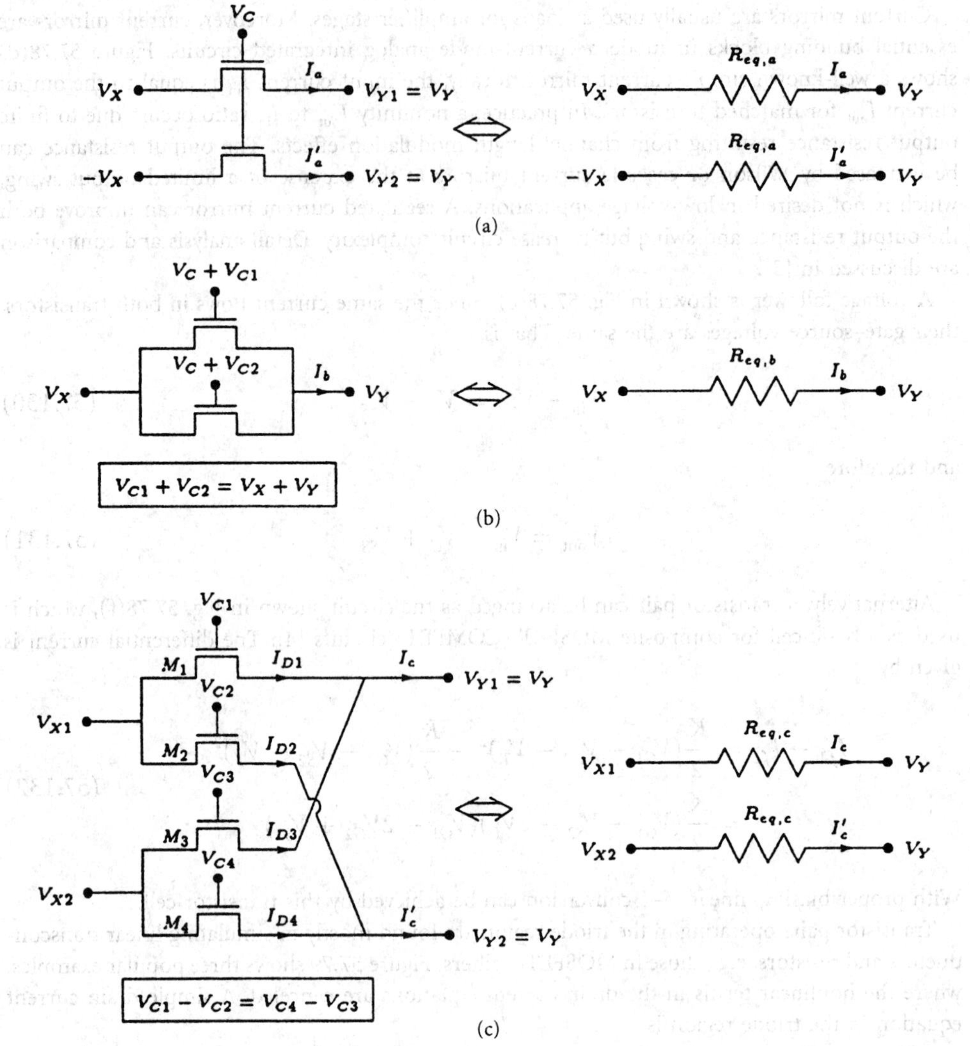

FIGURE 57.79 Matched primitive cells operating in the triode region.

and the equivalent resistance is obtained as

$$R_{\text{eq},a} = \frac{2V_X}{I_a - I_a'} = \frac{1}{K(V_C - V_T)} \tag{57.136}$$

The circuit shown in Fig. 57.79(b) realizes a floating resistor [5], where

$$\begin{aligned} I_b &= K[(V_C + V_{C1} - V_T)(V_X - V_Y) - \frac{1}{2}(V_X^2 - V_Y^2)] \\ &\quad + K[(V_C + V_{C2} - V_T)(V_X - V_Y) - \frac{1}{2}(V_X^2 - V_Y^2)] \\ &= K[(2V_C - 2V_T + V_X + V_Y)(V_X - V_Y) - (V_X^2 - V_Y^2)] \\ &= 2K(V_C - V_T)(V_X - V_Y) \end{aligned} \tag{57.137}$$

and the equivalent resistance is

$$R_{eq,b} = \frac{V_X - V_Y}{I_b} = \frac{1}{2K(V_C - V_T)} \tag{57.138}$$

An implementation with $V_{C1} = V_{C2} = (V_X + V_Y)/2$ has been described in [6].

The circuit shown in Fig. 57.79(c), a four-transistor transconductor [7], gives the following current equation:

$$\begin{aligned} I_c - I_c' &= K[(V_{C1} - V_T)(V_{X1} - V_Y) - \frac{1}{2}(V_{X1}^2 - V_Y^2)] \\ &\quad - K[(V_{C2} - V_T)(V_{X1} - V_Y) - \frac{1}{2}(V_{X1}^2 - V_Y^2)] \\ &\quad + K[(V_{C3} - V_T)(V_{X2} - V_Y) - \frac{1}{2}(V_{X2}^2 - V_Y^2)] \\ &\quad - K[(V_{C4} - V_T)(V_{X2} - V_T) - \frac{1}{2}(V_{X2}^2 - V_Y^2)] \end{aligned} \tag{57.139}$$

$$\begin{aligned} &= K(V_{C1} - V_{C2})(V_{X1} - V_{X2}) \\ &= K(V_{C4} - V_{C3})(V_{X1} - V_{X2}) \end{aligned} \tag{57.140}$$

and

$$R_{eq,c} = \frac{V_{X1} - V_{X2}}{I_c - I_c'} = \frac{1}{K(V_{C1} - V_{C2})} = \frac{1}{K(V_{C4} - V_{C3})} \tag{57.141}$$

Note that I_c and I_c' are taken at the V_Y nodes. It is very interesting to know that nonlinearity cancellation is also achieved with the four transistors operating in the saturation region [2].

Figure 57.79(a) and (c) are usually used together with op amps to simulate resistors, where the virtual short property of op amps makes $V_{Y1} = V_{Y2}$.

Unmatched Device Pairs

Figure 57.80 shows primitive cells that do not require matching, unless specified. Figure 57.80(a) and (b) are parallel and series composite NMOS transistors, respectively, which are very useful in laying out very wide or long transistors, respectively. The equivalent device transconductance parameter K_{eq} is calculated as follows:

For the parallel composite transistor, the drain current is written as

$$\begin{aligned} I_{Dp} &= \frac{K_{eq,p}}{2}(V_G - V_S - V_T)^2 \\ &= \frac{K_1}{2}(V_G - V_S - V_T)^2 + \frac{K_2}{2}(V_G - V_S - V_T)^2 \end{aligned} \tag{57.142}$$

That is,

$$K_{eq,p} = K_1 + K_2 \tag{57.143}$$

or alternatively

$$\left(\frac{W}{L}\right)_{eq,p} = \left(\frac{W}{L}\right)_1 + \left(\frac{W}{L}\right)_2 \tag{57.144}$$

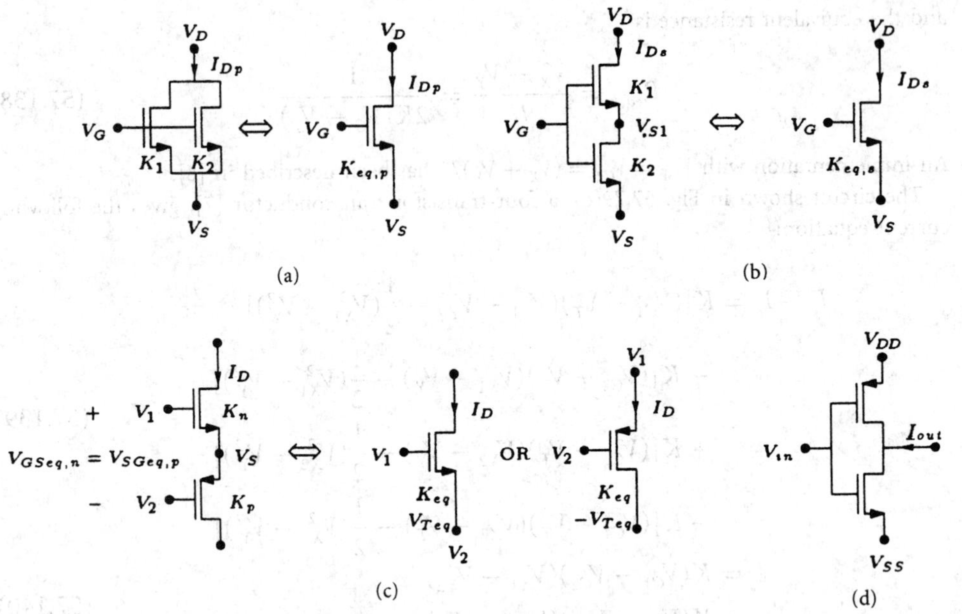

FIGURE 57.80 Unmatched primitive cells.

Using the same channel length L, it can be simplified as

$$W_{eq,p} = W_1 + W_2 \tag{57.145}$$

As a result, a wider transistor can be realized by parallel connection of two or more narrower transistors.

For the series composite transistor, note that the lower transistor is always operating in the triode region due to the requirement $V_G - V_{S1} > V_T$, to turn on the upper transistor. The resultant drain current is given by

$$I_{Ds} = \frac{K_{eq,s}}{2}(V_G - V_S - V_T)^2 \tag{57.146}$$

$$= \frac{K_1}{2}(V_G - V_{S1} - V_T)^2 \tag{57.147}$$

$$= \frac{K_2}{2}[(V_G - V_S - V_T)^2 - (V_G - V_{S1} - V_T)^2] \tag{57.148}$$

From the above equations, we have

$$(V_G - V_S - V_T)^2 = \frac{2I_{Ds}}{K_{eq,s}} \tag{57.149}$$

$$(V_G - V_{S1} - V_T)^2 = \frac{2I_{Ds}}{K_1} \tag{57.150}$$

Substituting the above equations into (57.148), we obtain

$$I_{Ds} = \frac{K_2}{2}\left(\frac{2I_{Ds}}{K_{eq,s}} - \frac{2I_{Ds}}{K_1}\right) \tag{57.151}$$

That is,

$$\frac{1}{K_{\text{eq},s}} = \frac{1}{K_1} + \frac{1}{K_2} \tag{57.152}$$

or

$$\left(\frac{L}{W}\right)_{\text{eq},s} = \left(\frac{L}{W}\right)_1 + \left(\frac{L}{W}\right)_2 \tag{57.153}$$

Similarly, with fixed channel width W, the above equation is simply obtained as

$$L_{\text{eq},s} = L_1 + L_2 \tag{57.154}$$

which indicates that the equivalent transistor can be used to realize a long-channel device with shorter channel ones.

Figure 57.80(c) is a CMOS composite transistor, which can be seen as equivalent to either an NMOS or a PMOS transistor operating in the saturation region. In contrast, the composite transistors shown in Fig. 57.80(a) and (b) can operate in both saturation and triode regions. The main advantage of the equivalent composite transistor shown in Fig. 57.80(c) is that both their equivalent gate and source nodes have high input impedances, which is desired in some circuits. The equivalent K and V_T are obtained by the equations of the gate–source voltages.

$$\begin{aligned} V_1 - V_2 &= V_{GSn} + V_{SGp} \\ &= \sqrt{\frac{2I_D}{K_n}} + \sqrt{\frac{2I_D}{K_p}} + V_{Tn} - V_{Tp} \\ &= \sqrt{\frac{2I_D}{K_{\text{eq}}}} + V_{T\text{eq}} \end{aligned} \tag{57.155}$$

which give

$$\frac{1}{\sqrt{K_{\text{eq}}}} = \frac{1}{\sqrt{K_n}} + \frac{1}{\sqrt{K_p}} \tag{57.156}$$

and

$$V_{T\text{eq}} = V_{Tn} - V_{Tp} \tag{57.157}$$

Finally, Fig. 57.80(d) shows a CMOS inverter, which could be used as a transconductor [8]. Its output current is

$$\begin{aligned} I_{\text{out}} &= \frac{K_n}{2}(V_{\text{in}} - V_{SS} - V_{Tn})^2 - \frac{K_p}{2}(V_{DD} - V_{\text{in}} + V_{Tp})^2 \\ &= a(V_{\text{in}} - V_{Tn})^2 + bV_{\text{in}} + c \end{aligned} \tag{57.158}$$

where

$$a = \frac{1}{2}(K_n - K_p)$$
$$b = -K_n V_{SS} + K_p(V_{DD} - V_{Tn} + V_{Tp})$$
$$c = \frac{K_n}{2}(2V_{SS}V_{Tn} + V_{SS}^2) + \frac{K_p}{2}[V_{Tn}^2 - (V_{DD} + V_{Tp})^2]$$

Composite Transistors

The body effect of a transistor is due to nonzero source to bulk voltage (V_{SB}), which widens the depletion region between the source and bulk and therefore increases the absolute value of its threshold voltage. The threshold voltage (referred to the source) is dependent on V_{SB} and is given by

$$V_{Tn} = V_{Tno} + \gamma(\sqrt{2|\phi_F| + V_{SB}} - \sqrt{2|\phi_F|}) \quad \text{for NMOS}$$
$$V_{Tp} = V_{Tpo} - \gamma(\sqrt{2|\phi_F| + V_{BS}} - \sqrt{2|\phi_F|}) \quad \text{for PMOS}$$

where $2|\phi_F|$ is the potential required for strong inversion and γ is the body effect parameter.

Usually, bulk regions of an NMOS transistor and a PMOS transistor are tied to the most negative voltage (V_{SS}) and the most positive voltage (V_{DD}) respectively to turn off the parasitic diodes associated with source-bulk and drain-bulk. In some cases, bulk regions are directly connected to transistor sources ($V_{SB} = 0$) to eliminate the body effect; e.g., in the follower of Fig. 57.78(e), the bulk must be connected to the source in each transistor to ensure equal threshold voltages. This is achieved by putting each device in a separated well, which must be the p-well for NMOS devices. However, separate wells require large layout areas. Besides, unless twin-tub processes are used, only one type of transistor (either NMOS or PMOS depending on the process) can be connected this way.

Due to the body effect, the equivalent threshold voltage of a CMOS composite transistor would be large (two threshold voltages plus extra voltage resulting from the body effect), which would render it unsuitable for low-voltage applications. The equivalent threshold voltage could be reduced by replacing one of the MOS transistors with a BJT, as shown in Fig. 57.81 [9]. For the stacked composite BiCMOS transistors, the equivalent threshold voltage is given by $V_{T\text{eq}} \approx |V_T| + 0.7$ V, where V_T is the threshold voltage of the NMOS or PMOS transistor, and 0.7 V is the BJT turn-on voltage V_{BE}, which is not subject to body effects. It can be further reduced by the folded arrangement as shown in Fig. 57.81(b), where $V_{T\text{eq}} \approx |V_T| - 0.7$ V. An all-MOS folded composite transistor can be implemented in a similar manner as shown in Fig. 57.82 [10], where $K_2 \gg K_1$. As a result,

$$K_{\text{eq}} \approx K_1 \tag{57.159}$$

$$V_{T\text{eq}} = V_T - V_{GS2} \approx -\sqrt{\frac{2I}{K_2}} \tag{57.160}$$

where $V_{GS2} \approx \sqrt{2I/K_2} + V_T$. SPICE simulation results for the N-type folded transistors ($W_1 = L_1 = L_2 = 3$ μm) operating in the saturation region ($V_{DG} = 0$) with various W_2 are compared with a single NMOS depletion transistor ($W = L = 3$ μm) with various V_T as shown in Fig. 57.83. It can be seen that a smaller K_2 results in a smaller $V_{T\text{eq}}$ (more negative), but with a

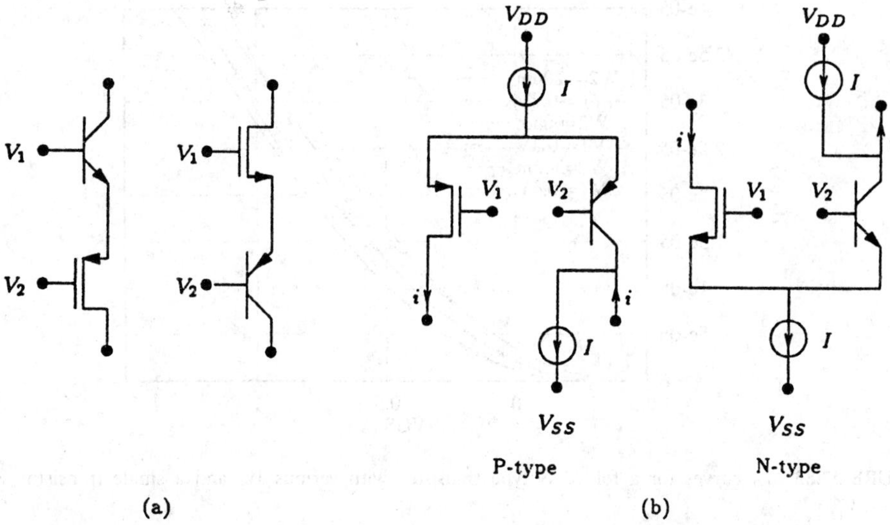

FIGURE 57.81 BiCMOS composite transistors: (a) stacked version and (b) folded version.

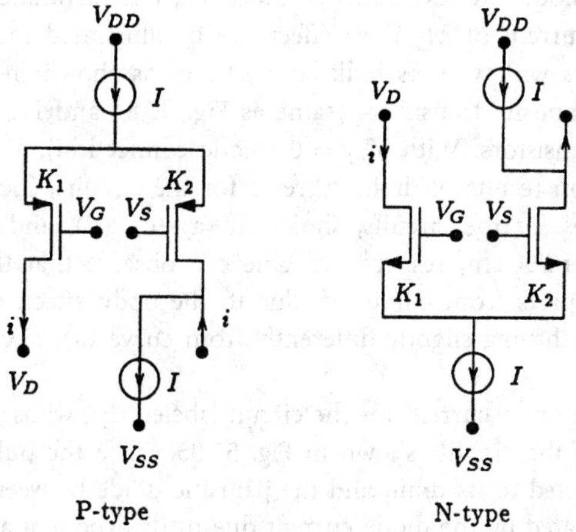

FIGURE 57.82 MOS folded composite transistors.

larger K_2 the composite transistor behaves more like a single transistor having a smaller $|V_{Teq}|$, which could be useful in low-voltage applications.

Forcing $V_{SB} = 0$ to eliminate the body effect is usable only for MOS circuits conducting currents in a single direction. For the circuits shown in Fig. 57.79, since resistors operate bidirectionally, the bulk regions of each transistor must be connected to the rail to assure that parasitic diodes are turned off when currents flow in either direction. In fact, the transistors are operating in the triode region symmetrically between drain and source and biased at $V_{DS} \approx 0$. It would, however, result in nonzero V_{SB} and increase the threshold voltage, which increases the equivalent resistance, in Fig. 57.79(a) and (b), but introduces nonlinearities. To overcome this problem, one may configure two transistors into one composite transistor as shown in Fig. 57.84(a), where the bulks of the transistors are interconnected to node V_{S1}. Due to symmetry, this composite transistor can be operated in either direction. Its physical cross section is shown in Fig. 57.84(b), where the diodes represent the p–n junctions composed by bulk and source/drain nodes. This configuration is equivalent to Fig. 57.84(c), and one can find that the parasitic diode connected between V_{S1} and V_S would turn on when $V_{S1} - V_S$ is larger than the turn-on

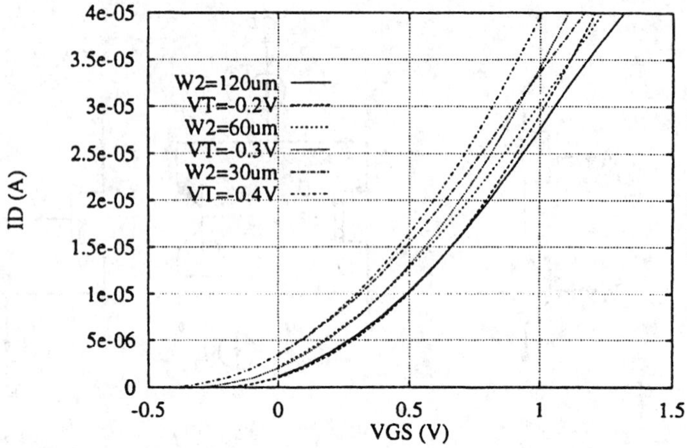

FIGURE 57.83 I_D curves for a folded N-type transistor with various W_2 and a single transistor with various V_T.

voltage of the parasitic diode. However, this is undesired, but fortunately the diode current is restricted by the drain current of M_2. This effect can be illustrated more clearly through the comparison of transistors with various bulk connections, as shown in Fig. 57.85, where Fig. 57.85(d) and (e) are composite transistors [same as Fig. 57.84 and Fig 57.80(b), respectively], which simulate single transistors. With $V_{DG} = 0$ (diode connection), $V_S = 0$ and $V_{SS} = -5$ V, Fig. 57.86 gives simulation results of drain currents for the circuits, shown in (a) and (c)–(e), where the transistor sizes for the circuits, shown in (a)–(c), (d), and (e) are 20 μm/3 μm, 30 μm/3 μm, and 36 μm/3 μm, respectively. One can observe that the curve (c) shown in Fig. 57.86 completely departs from curve (a), due to the body effect. Curve (e) fits perfectly to curve (a). Although behaving slightly differently from curve (a), circuit (d) approximates a single transistor as well.

Figure 57.87 shows the drain current for the circuit labeled (b), whose current is much larger than those of the rest of the circuits shown in Fig. 57.85. Since the bulk of the circuit shown in Fig. 57.85(b) is connected to its drain and the parasitic diode between the drain and bulk is on, the current is dominated by the diode current due to its exponential nature. By using it as in Fig. 57.85(d), the diode current is limited to the current level of a MOS transistor. This can be seen from Fig. 57.88, showing V_{S1} of the circuit (d) saturated at a voltage ~ 0.7 V, which is close to the turn-on voltage of a p–n junction diode. The transistor sizes of the circuits labeled (d) and (e) are adjusted to achieve the same K value of the single transistor (a). According to (57.152), $1/K_{eq,s} = 1/K_1 + 1/K_2$. That is, to achieve $K_{eq,s} = K$, $K_e (= K_1 = K_2)$ is given by

$$K_e = 2K \tag{57.161}$$

For circuit (d), K_d is obtained by rewriting (57.148), which follows that

$$\begin{aligned} I_{Ds} &= \frac{K}{2}(V_G - V_S - V_T)^2 \\ &= \frac{K_d}{2}(V_G - V_{S1} - V_T)^2 \\ &= \frac{K_d}{2}[(V_G - V_S - V_T)^2 - (V_G - V_{S1} - V_T)^2] + I_s e^{(V_{S1}-V_S)/U_T} \end{aligned}$$

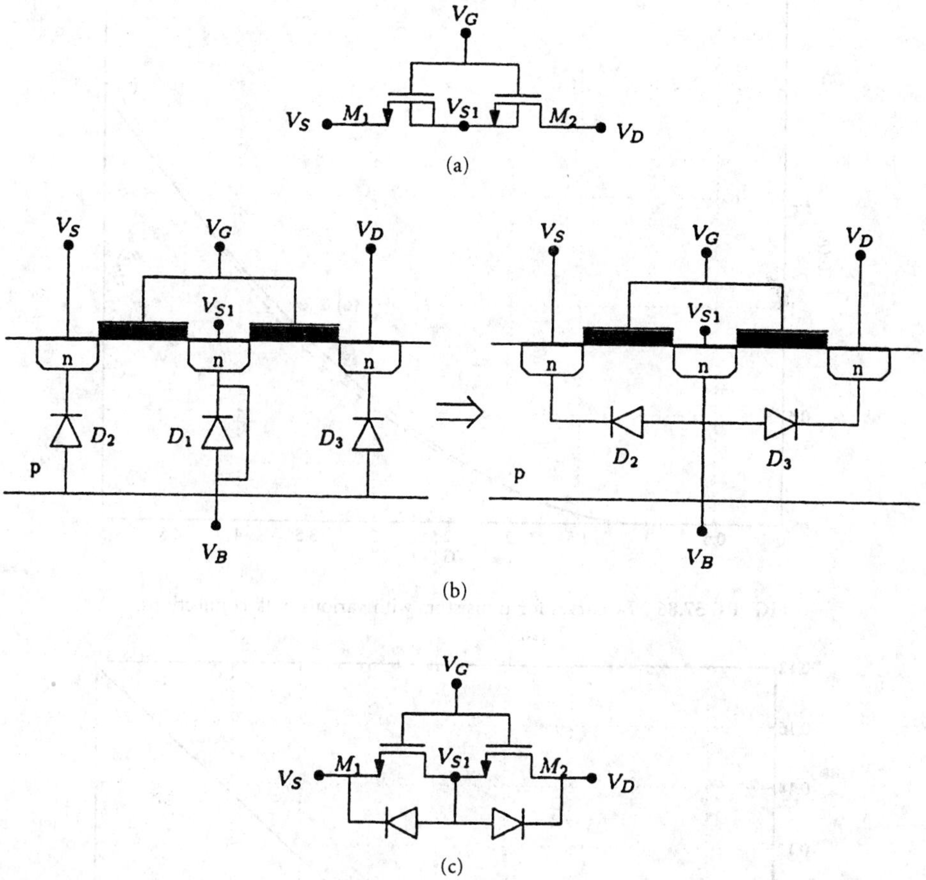

FIGURE 57.84 [10] (a) Composite bidirectional transistor with reduced body effect. (b) Cross-sectional view of the physical device, where the short connection across D_1 is, in effect, placing D_2 and D_3 back-to-back between V_D and V_S and forming a parasitic symmetrical bipolar device. (c) Equivalent circuit of (a).

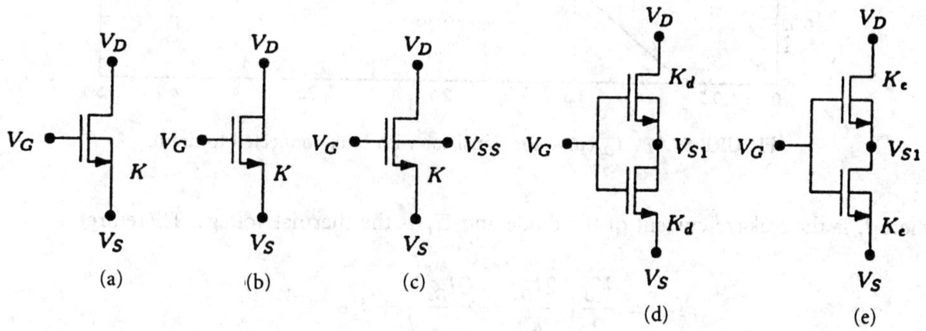

FIGURE 57.85 Transistors with various bulk connections.

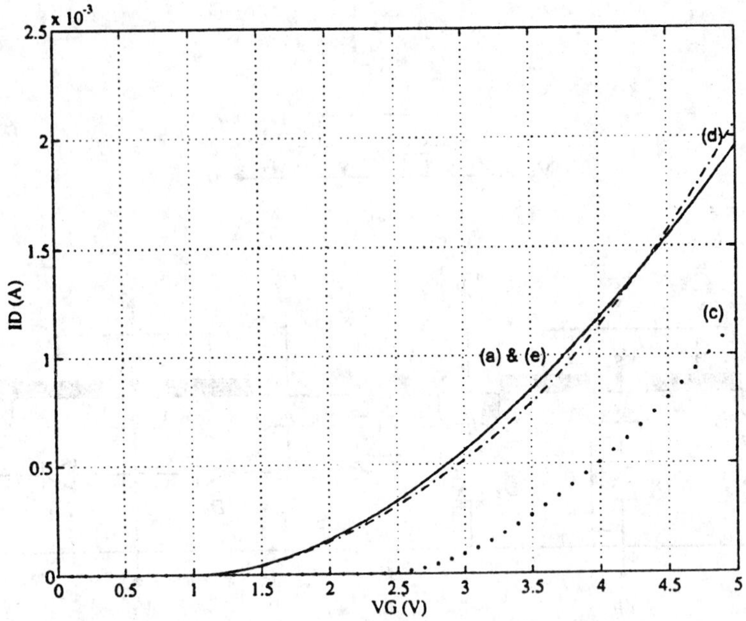

FIGURE 57.86 I_D curves for transistors with various bulk connections.

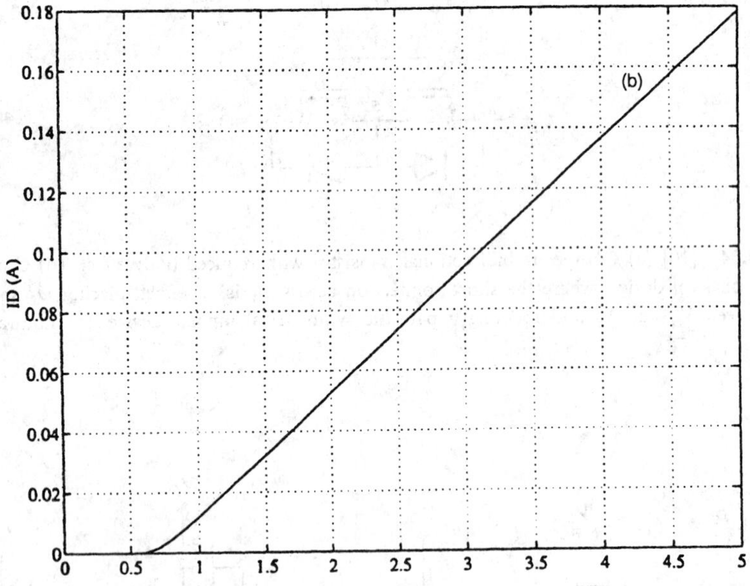

FIGURE 57.87 I_D curve for transistor with bulk connected to drain.

where I_s is the leakage current of the diode and U_T is the thermal voltage. Therefore,

$$\begin{aligned} I_{Ds} &= \frac{K_d}{2}\left(\frac{2I_{Ds}}{K} - \frac{2I_{Ds}}{K_d}\right) + I_s e^{(V_{S1}-V_S)/U_T} \\ &= I_{Ds}\left(\frac{K_d}{K} - 1\right) + I_s e^{(V_{S1}-V_S)/U_T} \end{aligned} \tag{57.162}$$

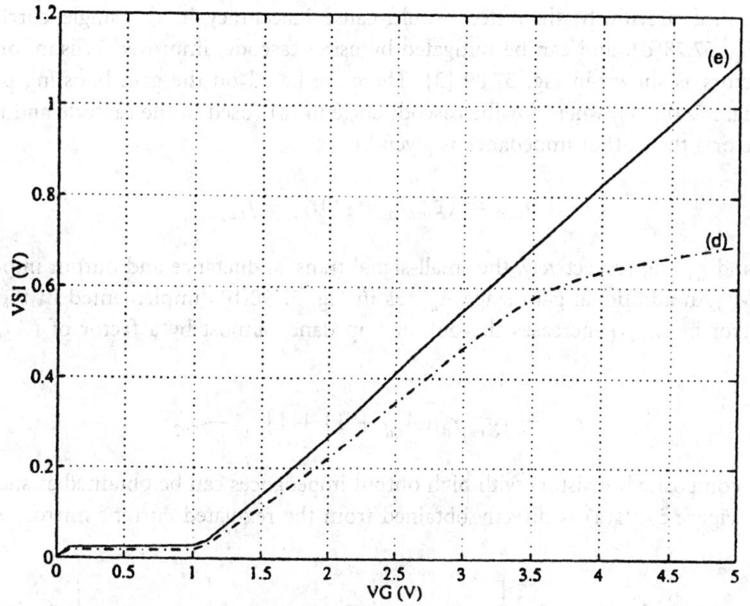

FIGURE 57.88 V_{S1} curves for composite transistors, labeled (d) and (e).

Divided by I_{Ds}, the above equation becomes

$$\begin{aligned} 1 &= \frac{K_d}{K} - 1 + \frac{I_s}{I_{Ds}} e^{(V_{S1}-V_S)/U_T} \\ 2K &= K_d + K\frac{I_s}{I_{Ds}} e^{(V_{S1}-V_S)/U_T} \end{aligned} \tag{57.163}$$

and hence

$$K_d < 2K \tag{57.164}$$

due to the parasitic diode. A SPICE level 2 model is used in the simulation and its higher order effects result in using a device size of 36 μm/3 μm($K_e \neq 2K$), instead of 40 μm/3 μm.

Super MOS Transistors

The channel length modulation effect models the channel shortening effect in the saturation region due to the increase in the depletion width near the drain when increasing V_{DS}. It is modeled as

$$I_D = \frac{K}{2}(V_{GS} - V_T)^2(1 + \lambda V_{DS}) \tag{57.165}$$

where λ is the channel length modulation parameter. The effect results in a finite output impedance of a transistor, since the output impedance is given by

$$r_o = \left(\frac{\partial I_D}{\partial V_{DS}}\right)^{-1} = \frac{1}{\lambda \frac{K}{2}(V_{GS} - V_T)^2} \simeq \frac{1}{\lambda I_D} \tag{57.166}$$

As mentioned previously, this effect would cause inaccuracy in the single current mirror shown in Fig. 57.78(d), and can be mitigated by using cascode, improved Wilson, or regulated current mirrors as shown in Fig. 57.89 [3]. These are based on the gain-boosting principle as shown in Fig. 57.90 [11], where for the cascode stage in (a) (used in the cascode and the Wilson current mirrors) the output impedance is given by

$$r_{o,a} = (g_{m2}r_{o2} + 1)r_{o1} + r_{o2} \tag{57.167}$$

where g_{mi} and r_{oi} are, respectively, the small-signal transconductance and output impedance for transistor M_i. An additional gain stage A_{dd}, as in Fig. 57.90(b) (implemented in the regulated current mirror by M_{add}) increases the output impedance almost by a factor of $(A_{dd} + 1)$ and gives

$$r_{o,b} = [g_{m2}r_{o2}(A_{dd} + 1) + 1]r_{o1} + r_{o2} \tag{57.168}$$

As a result, composite transistors with high output impedances can be obtained as shown in Fig. 57.91 [10]. Figure 57.91(a) is directly obtained from the regulated current mirror, where M_{N1}

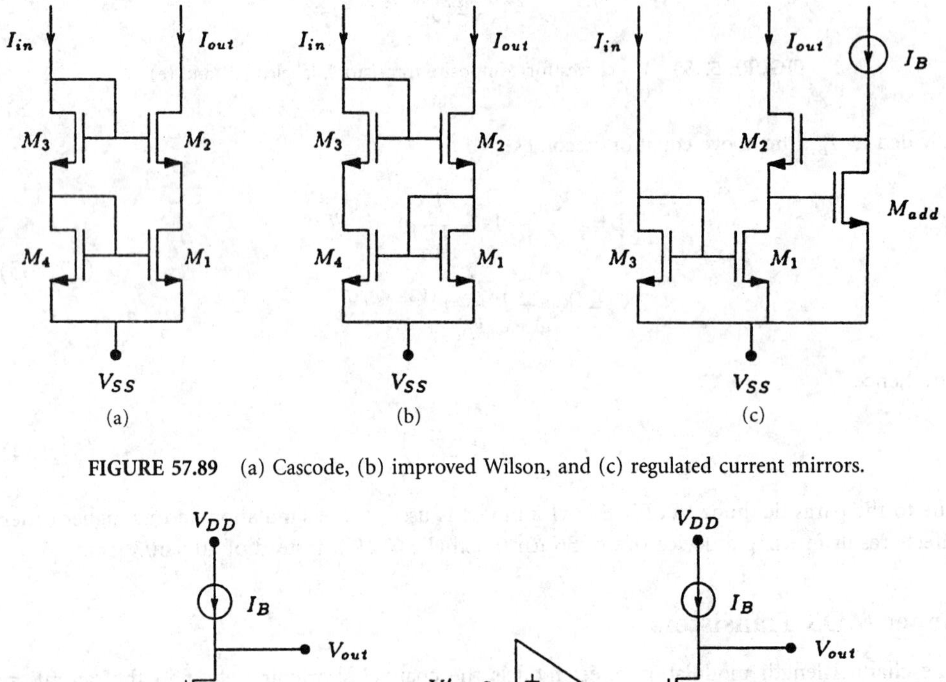

FIGURE 57.89 (a) Cascode, (b) improved Wilson, and (c) regulated current mirrors.

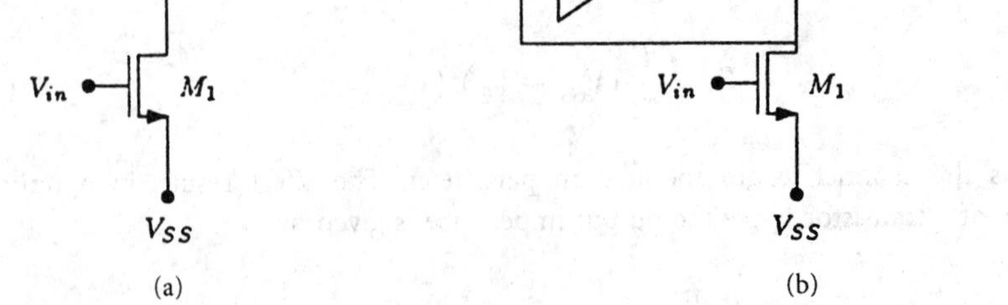

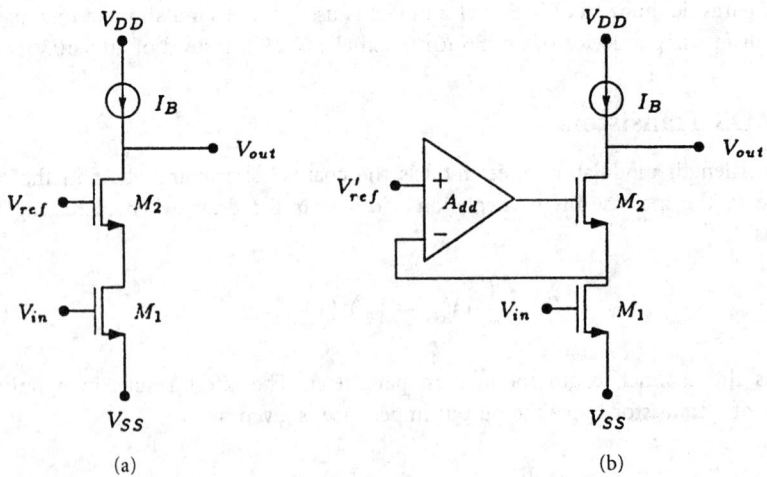

FIGURE 57.90 (a) Cascode stage and (b) cascode stage with an additional gain stage.

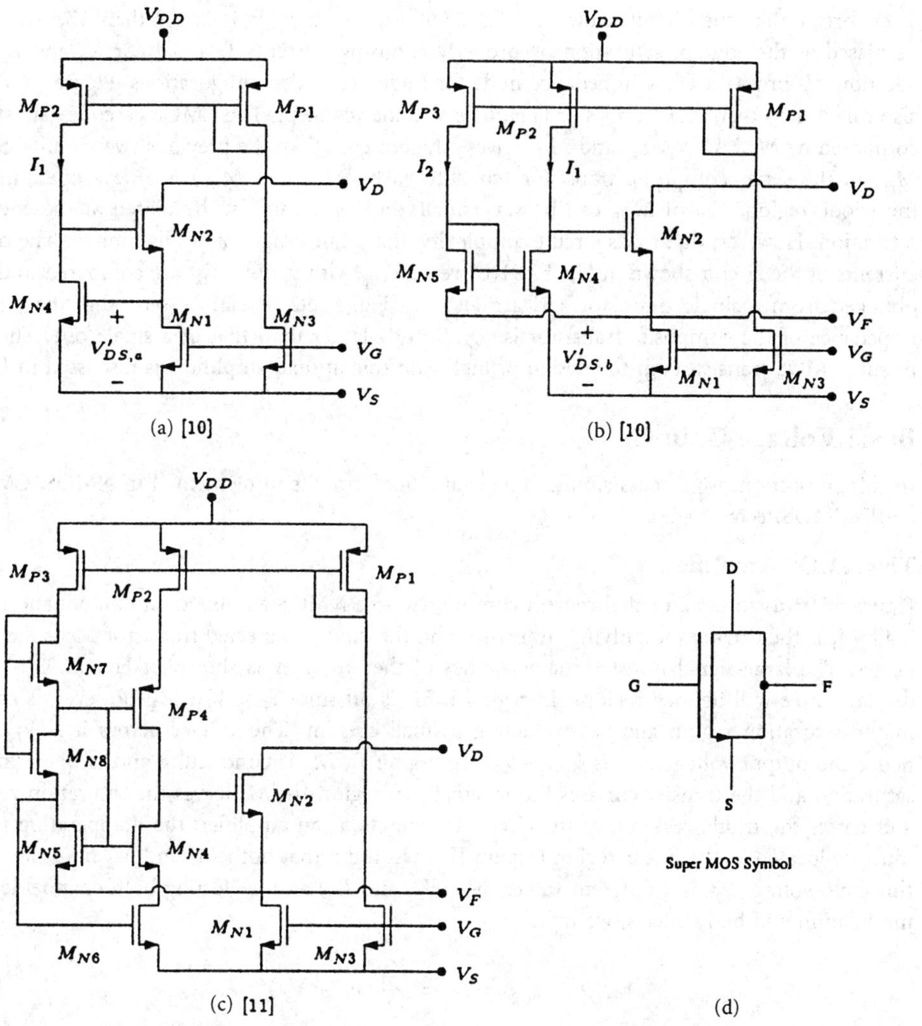

FIGURE 57.91 Composite super NMOS transistors. *Source:* (a), (b) S.-C. Huang, "Systematic design solutions for analog VLSI circuits," Ph.D dissertation, Dep. Electr. Eng., Ohio State Univ., 1994. (c) K. Bult and G. J. Geelen, "The CMOS gain-boosting technique," *J. Analog Integrat. Circuits Signal Process.*, vol. 1, 1991.

and M_{N2} are cascoded and M_{P2} and M_{N4} compose an additional gain stage. The drain–source voltage of M_{N1} biased by I_1 is given by

$$V_{DS,a}^{'} = \sqrt{\frac{2I_1}{K_4}} + V_T \tag{57.169}$$

Figure 57.91(b), a modified version of (a), employs the biasing technique in [12], biasing V_{DS} for a triode-mode V–I converter. The resultant $V_{DS,b}^{'}$ is given by

$$V_{DS,b}^{'} = \sqrt{\frac{2I_1}{K_4}} - \sqrt{\frac{2I_2}{K_5}} \tag{57.170}$$

Therefore, unlike the circuit shown in Fig. 57.90(a), where $V'_{DS,a}$ is larger than V_T, M_{N1} can be biased at the edge of saturation by properly choosing currents I_1 and I_2 or K_4 and K_5. In addition, V_F provides a low impedance node for folded cascode configurations. Figure 57.91(c), also called the super-MOS transistor [11], uses a similar concept. The CMOS cascode gain stage, composed by M_{P2}, M_{P4}, M_{N4}, and M_{N6}, gives a higher gain than the previous two circuits. Since M_{N8} in the series composite transistor (constituted by M_{N7} and M_{N8}) is always operating in the triode region, V_{DS} of M_{N8} can be very small, and M_{N1} can also be biased at the edge of saturation. However, due to its circuit complexity, the input range for V_{GS} is limited. The drain currents of the circuit shown in Fig. 57.91(b) versus V_{DS} with various V_{GS} are compared to those obtained from a single transistor and are given in Fig. 57.92. It can be seen that the output impedance of the composite transistor is significantly larger than that of a single one. The use of super MOS transistors in the design of high-gain operational amplifiers is discussed in [11].

Basic Voltage Gain Cells

In this subsection, we discuss simple voltage amplifier circuits implemented in NMOS, CMOS, and BiCMOS technologies.

The NMOS Amplifier

Figure 57.93(a) shows an enhancement common-source NMOS amplifier with an enhancement load. M_1 is the driving (amplifying) transistor and the diode-connected transistor M_2 is the load device. The large-signal transfer characteristics of the amplifier is shown in Fig. 57.93(b) and displays three well-defined regions. In region I, M_1 is off since $v_I < V_{T1}$. M_2, however, is always in the saturation region and is conducting a small current. The voltage across it is V_{T2} and hence the output voltage, v_O, is $V_{DD} - V_{T2}$. In region II, M_1 is conducting and is operating in saturation and the transfer curve is linear. Finally in region III, M_1 leaves the saturation region and enters the triode region. For the circuit to operate as an amplifier, the dc operating point must be located in the linear region (region II). Assuming that both M_1 and M_2 have the same threshold voltage V_T, but different values of K (K_1 and K_2) and neglecting both channel length modulation and body effects, we write

$$i_{D1} = i_{D2} = i_D = \frac{K_1}{2}(v_I - V_T)^2 \tag{57.171}$$

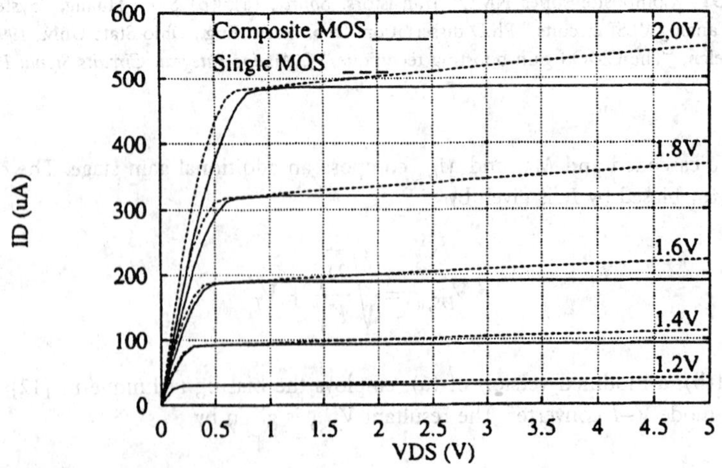

FIGURE 57.92 Simulated I_D curves for the high output impedance composite transistor shown in Fig. 57.91, (b) and for a single transistor.

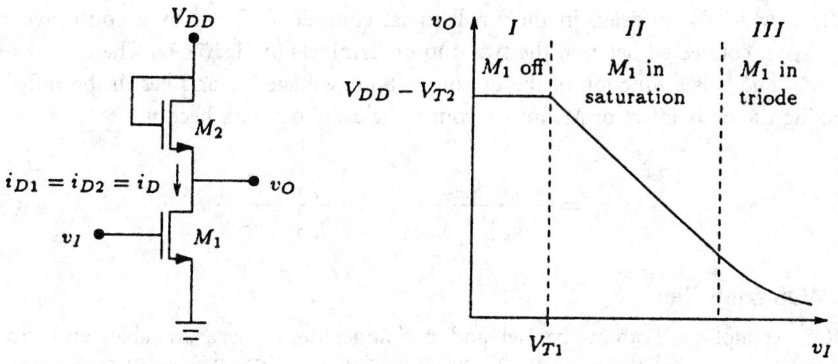

(a) (b)

FIGURE 57.93 (a) The NMOS amplifier. (b) Transfer characteristics.

and

$$\begin{aligned} i_D &= \frac{K_2}{2}(v_{GS2} - V_T)^2 \\ &= \frac{K_2}{2}(V_{DD} - v_O - V_T)^2 \end{aligned} \tag{57.172}$$

Combining (57.171) and (57.172) and with some manipulations, we obtain

$$v_O = \left(V_{DD} - V_T + \sqrt{\frac{K_1}{K_2}}V_T\right) - \sqrt{\frac{K_1}{K_2}}v_I \tag{57.173}$$

which is a linear equation between v_O and v_I. This is obviously the equation of the straight-line portion (region II) of the transfer curve.

The first term in (57.173) represents the dc component of the output voltage V_O. The second term represents the small-signal component and thus the ac small-signal gain of the amplifier A_v is

$$A_v = \frac{v_o}{v_i} = -\sqrt{\frac{K_1}{K_2}} = -\sqrt{\frac{(W/L)_1}{(W/L)_2}} \tag{57.174}$$

The small-signal equivalent circuit of the amplifier in Fig. 57.93(a) is shown in Fig. 57.94. Since D_2 and G_2 are connected in M_2, the voltage across the controlled current-source $g_{m2}v_{gs2}$ is v_{gs2}. Therefore, the controlled current-source can be represented by a resistance $1/g_{m2}$. Since $v_{gs1} = v_i$, we obtain the voltage gain as follows:

$$A_v = \frac{v_o}{v_i} = -\frac{g_{m1}}{g_{m2} + 1/r_{o1} + 1/r_{o2}} \tag{57.175}$$

Now, if r_{o1} and r_{o2} are much larger than $(1/g_{m2})$, the gain reduces to $A_v \simeq -g_{m1}/g_{m2}$, which can easily be shown to lead to the expression in (57.174). Note that the gain can also be determined by inspection from the circuit in Fig. 57.93(a) as $-g_{m1}$ multiplied by the equivalent small-signal resistance seen at the drain of M_1, which is $(1/g_{m2} \| r_{o1} \| r_{o2})$.

Practically, M_1 and M_2 share the same substrate, which is normally connected to the most negative supply voltage in the circuit (ground in this case). It follows that M_2 suffers from body effect, which is modeled in the small-signal equivalent circuit by a controlled current-source $g_{mb2}v_{bs2}$ connected between the two output terminals in Fig. 57.94, where $v_{bs2} = v_{gs2}$ and $g_{mb2} = \chi g_{m2}$ and χ is a function of the dc source-body voltage V_{SB} and lies in the range 0.1–0.3 [13]. Taking the body effect of M_2 into account, the amplifier gain becomes

$$A_v = -\frac{g_{m1}}{g_{m2} + g_{mb2}} = -\frac{g_{m1}}{g_{m2}}\frac{1}{1+\chi} \tag{57.176}$$

The CMOS Amplifier

In CMOS technology, both n-channel and p-channel devices are available, and are usually fabricated in a way that eliminates the body effect. The basic CMOS amplifier is shown in Fig. 57.95. Here M_2 and M_3 in Fig. 57.95(c) are a pair of PMOS devices operating as a current source active load and implement the current source I_{B1} in Fig. 57.95(a). M_2 is biased in the saturation region and when M_1 is operating in the saturation region, the small-signal voltage gain will be equal to $-g_{m1}$, multiplied by the total resistance seen between the output and ground which is $(r_{o1} \| r_{o2})$.

Cascode versions of the amplifier as shown previously in Fig. 57.90 can be used to boost the gain significantly. For instance, the cascode amplifier in Fig. 57.90(a) has a gain equal to the effective transconductance $-g_{meff}$, multiplied by $r_{o,a}$ given by (57.167), where g_{meff} is given by [14]

$$g_{meff} = g_{m1}\frac{g_{m2}r_{o1} + \dfrac{r_{o1}}{r_{o2}}}{g_{m2}r_{o1} + \dfrac{r_{o1}}{r_{o2}} + 1} \tag{57.177}$$

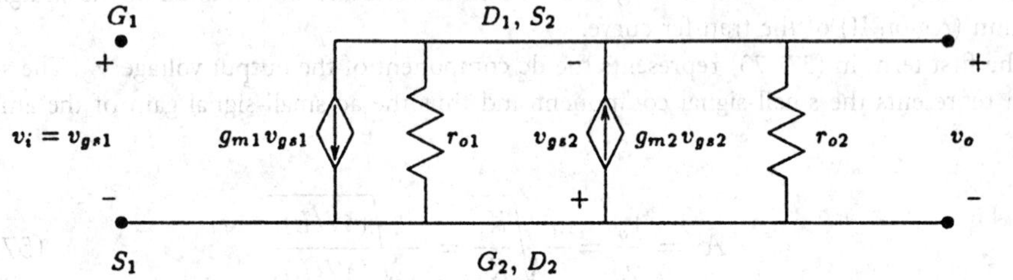

FIGURE 57.94 Small-signal equivalent circuit of the NMOS amplifier.

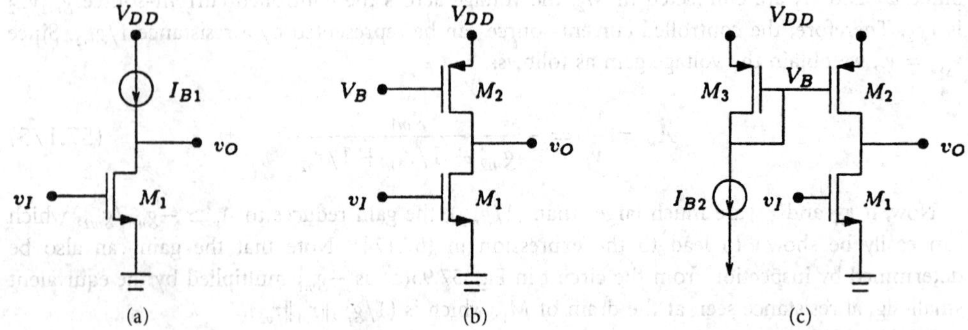

FIGURE 57.95 The CMOS amplifier: (a) basic circuit, (b) and (c) CMOS implementations.

BiCMOS Amplifiers

A BiCMOS technology combines bipolar and CMOS transistors on a single substrate. A bipolar transistor has the advantage over a MOS of a much higher transconductance (g_m) for the same dc bias current. Also, bipolar transistors have better high-frequency performance than their MOS counterparts. On the other hand, the practically infinite input resistance at the gate of a MOSFET makes it possible to design amplifiers with extremely high input resistance and an almost zero input bias current. For these reasons, there has been an increasing interest in BiCMOS technologies for implementing high-performance integrated circuits. While most BiCMOS processes offer high-quality NMOS, PMOS, and NPN transistors, advanced BiCMOS processes offer PNP transistors as well.

Figure 57.96 shows three basic folded-cascode single-ended high-performance BiCMOS amplifiers [15]. The main features of these amplifiers are a high gain–bandwidth product and extremely high dc input impedance. The high gain is achieved by cascoding transistors in the signal path [M_1, Q_2 in Fig. 57.96(a), M_1, M_2 in Fig. 57.96(b), and M_1, Q_2, Q_3 in Fig. 57.96(c)]. The high bandwidth is achieved by exploiting the exponential nature of the current–voltage characteristics of bipolar transistors. For instance, let us consider the amplifier circuit in Fig. 57.96(a). The internal node in the signal path at the emitter of Q_2 has an extremely low impedance. This is due to the fact that while the emitter current can change significantly with the input signal, the emitter voltage remains almost constant. The same argument can be made about the internal nodes in the signal paths of the other two amplifiers [the base of Q_3 in Fig. 57.96(b) and the emitters of Q_2 and Q_3 in Fig. 57.96(c)]. The low impedance of internal nodes in the signal path places nondominant poles at very high frequencies.

The Differential Amplifier

The amplifier circuits discussed previously are of the single-ended type. A single-ended amplifier has both input and output voltage signals referred to ground. In most IC applications, a differential amplifier is utilized. In this case the amplifier has a differential input and may also have a differential output, in which case it is called a fully differential amplifier. It is usually easy and straightforward to convert single-ended amplifiers to differential architectures.

The most widely used differential amplifier is based on the common-source or common-gate differential pairs shown respectively in Fig. 57.78(b) and (c). The common-source pair is shown here again in Fig. 57.97. The only difference here is that the circuit is biased by a constant current source I that is usually implemented using a current-mirror circuit [see Fig. 57.78(d)].

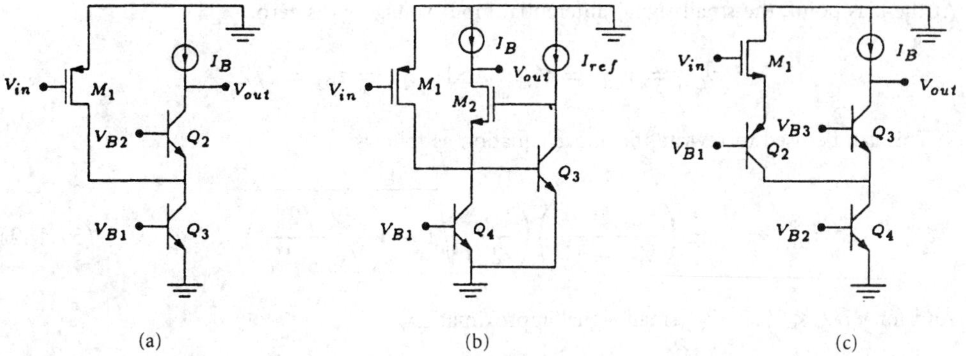

FIGURE 57.96 BiCMOS basic amplifier circuits: (a) common-source, common-base, (b) common-source, common-gate with active-feedback, and (c) common-drain, common-base, common-base.

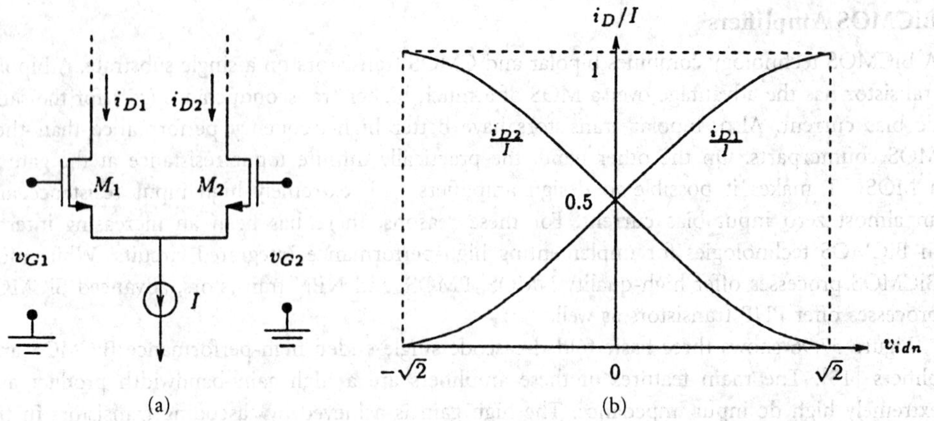

FIGURE 57.97 The MOS differential pair: (a) the circuit and (b) the transfer characteristic.

Assuming that M_1 and M_2 are identical and neglecting both channel length modulation and body effects, we write

$$i_{D1} = \frac{K}{2}(v_{GS1} - V_T)^2 \tag{57.178}$$

$$i_{D2} = \frac{K}{2}(v_{GS2} - V_T)^2 \tag{57.179}$$

Taking the square root of both sides in each of the two equations above and defining the differential input as $v_{id} = v_{GS1} - v_{GS2}$, we get

$$\sqrt{i_{D1}} - \sqrt{i_{D2}} = \sqrt{\frac{K}{2}}v_{id} \tag{57.180}$$

But since the current-source bias imposes the constraint that $i_{D1} + i_{D2} = I$, one can easily show that

$$i_{D1,2} = \frac{I}{2} \pm \sqrt{KI}\left(\frac{v_{id}}{2}\right)\sqrt{1 - \frac{(v_{id}/2)^2}{(I/K)}} \tag{57.181}$$

At the bias point, the small-signal differential input voltage v_{id} is zero,

$$v_{GS1} = v_{GS2} = V_{GS} \quad \text{and} \quad i_{D1} = i_{D2} = I/2$$

This can be used to rewrite the above equation as follows:

$$i_{D1,2} = \frac{I}{2} \pm \left(\frac{I}{V_{GS} - V_T}\right)\left(\frac{v_{id}}{2}\right)\sqrt{1 - \left(\frac{v_{id}/2}{V_{GS} - V_T}\right)^2} \tag{57.182}$$

And for $v_{id}/2 \ll V_{GS} - V_T$ (small-signal approximation),

$$i_{D1,2} \simeq \frac{I}{2} \pm \left(\frac{I}{V_{GS} - V_T}\right)\left(\frac{v_{id}}{2}\right) \tag{57.183}$$

The differential pair tranconductance g_m, defined as $g_m = (i_{D1} - i_{D2})/v_{id}$ is then given by $I/(V_{GS} - V_T)$. We recall that a single MOS transistor biased at a drain current I_D has a transconductance $2I_D/(V_{GS} - V_T)$. Thus, we see that each transistor in the pair has a transconductance $2(I/2)/(V_{GS} - V_T)$, which is equal to the differential pair transconductance, g_m. Equation (57.182) and (57.183) indicate that for small-signal inputs, the current in M_1 increases by i_d and that in M_2 decreases by i_d. From (57.182) we can find v_{id} at which current steering between M_1 and M_2 occurs. That is $i_{D1} = I$ and $i_{D2} = 0$ or vice versa for negative v_{id}. Equating the second term in (57.182) to $I/2$, we get

$$|v_{id}|_{\max} = \sqrt{2}(V_{GS} - V_T) \tag{57.184}$$

Figure 57.97(b) shows plots of the normalized currents i_{D1}/I and i_{D2}/I versus the normalized differential input voltage $v_{idn} = v_{id}/(V_{GS} - V_T)$.

A simple CMOS differential amplifier is shown in Fig. 57.98, where the PMOS pair is used as an active load. The small-signal current i is given by $g_m(v_{id}/2)$, where $g_m = I/(V_{GS} - V_T)$. The small-signal output voltage is given by $v_o = 2i(r_{o2} \| r_{o4})$ and the voltage gain is $A_v = v_o/v_{id} = g_m(r_{o2} \| r_{o4})$.

When $v_{id} = 0$, the bias current I does not actually split equally between M_1 and M_2. This is due to mismatches in K, ΔK, and V_T, ΔV_T, which contribute to a dc offset voltage that is usually larger than that in differential amplifiers implemented with bipolar transistors. For instance, modern silicon-gate MOS technologies have ΔV_T as high as 2 mV [13]. Note that ΔV_T has no counterpart in bipolar junction transistors.

The Folded-Cascode Operational Amplifier

The folded-cascode operational amplifier (op amp) is a basic building block in modern analog integrated circuits. Figure 57.99 shows two folded-cascode op amp circuits in CMOS and BiCMOS technologies [15]. Actually, several combinations of bipolar and CMOS devices could be used in the design of this amplifier. Here, we assume that PNP bipolar transistors are not available, which implies a BiCMOS process having less complexity. The input common-source MOS pair is of the PMOS type. The cascode common-gate or common-base pair [M_5 and M_6 in Fig. 57.99(a) and Q_5 and Q_6 in Fig. 57.99(b)] is "folded" and, therefore, is implemented with devices of the opposite type to that used in the input pair. This is unlike the basic "unfolded" cascode amplifier in Fig. 57.90, where both input and cascode devices are of the same type.

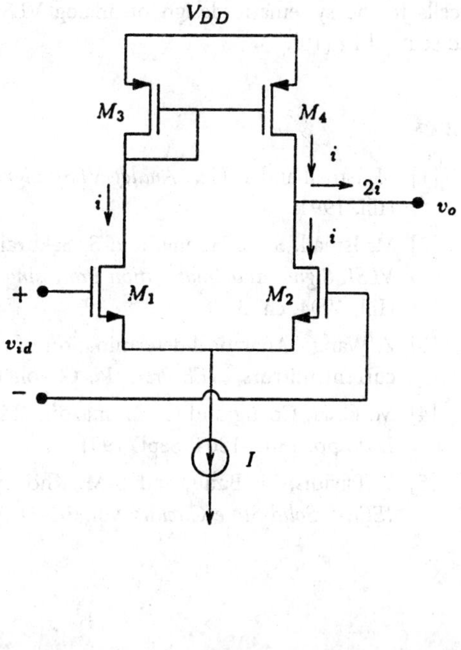

FIGURE 57.98 The simple differential-input, single-ended output CMOS amplifier.

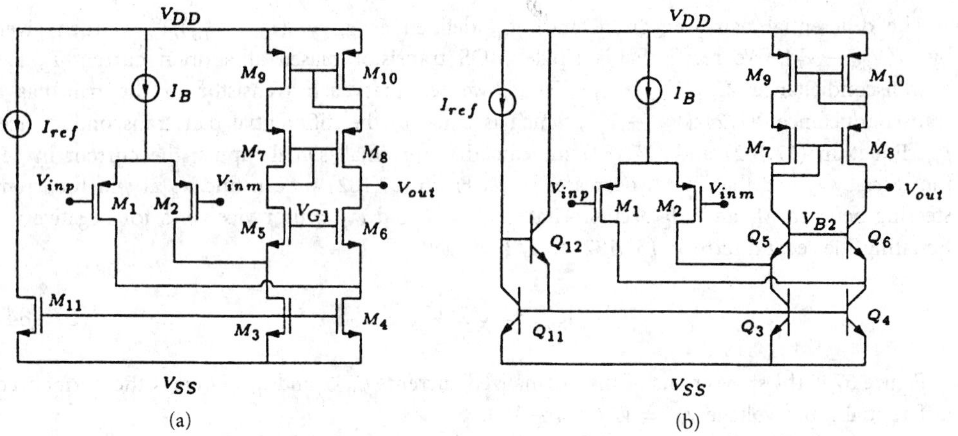

FIGURE 57.99 Folded-cascode amplifiers: (a) common-source, common-gate and (b) common-source, common-base.

The greater values of transconductance associated with the common-base bipolar devices in the BiCMOS op amp place the nondominant parasitic poles at much higher frequencies. Note that the BiCMOS op amp circuit is based on the single-ended amplifier shown in Fig. 57.96(a). The BiCMOS op amp combines the increased bandwidth with the advantages of an MOS input stage; namely a nearly infinite input impedance, a zero input bias current, and a higher slew rate [13].

Conclusion

Analog design is more complicated and less systematic than digital design and involves many trade-offs to meet certain design specifications. It strongly relies on human heuristics. The transfer of these human experiences into a computer-aided design environment is essential to the success of analog design in the context of very large scale integration (VLSI) of both analog and mixed analog/digital integrated circuits. This transfer, however, requires the development of systematic approaches to the analysis and design of analog integrated circuits. To this end, understanding the basic operations of the analog cells discussed here is critical. The use of these cells in the systematic design of analog VLSI systems, such as filters and data converters, is discussed in [10].

References

[1] M. Ismail and T. Fiez, *Analog VLSI: Signal and Information Processing*, New York: McGraw-Hill, 1994.

[2] M. Ismail, S.-C. Huang, and S. Sakurai, "Continuous-time signal processing," in *Analog VLSI: Signal and Information Processing*, M. Ismail and T. Fiez, Eds., New York: McGraw-Hill, 1994, ch. 3.

[3] Z. Wang, "Analytical determination of output resistance and DC matching errors in MOS current mirrors," *IEE Proc.: Pt. G*, vol. 137, pp. 397–404, Oct. 1990.

[4] M. C. H. Cheng and C. Toumazou, "Linear composite MOSFETs (COMFETs)," *Electron. Lett.*, pp. 1802–1804, Sept. 1991.

[5] Y. Tsividis, M. Banu, and J. M. Khoury, "Continuous-time MOSFET-C filters in VLSI," *IEEE J. Solid-State Circuits*, vol. SC-21, pp. 15–30, Feb. 1986.

[6] M. Banu and Y. Tsividis, "Floating voltage-controlled resistors in CMOS technology," *Electron. Lett.*, vol. 18, pp. 678, 679, July 1982.

[7] M. Ismail, "Four-transistor continuous-time MOS transconductor," *Electron. Lett.*, vol. 23, pp. 1099, 1100, Sept. 1987.

[8] B. Nauta, "A CMOS transconductance-C filter technique for very high frequencies," *IEEE J. Solid-State Circuits*, vol. 27, pp. 142–153, Feb. 1992.

[9] J. Ramirez-Angulo, "Applications of composite BiCMOS transistors," *Electron. Lett.*, vol. 27, pp. 2236–2238, Nov. 1991.

[10] S.-C. Huang, "Systematic design solutions for analog VLSI circuits," Ph.D. dissertation, Dep. Electr. Eng., Ohio State Univ., Columbus, OH, 1994.

[11] K. Bult and G. J. Geelen, "The CMOS gain-boosting technique," *J. Analog Integrat. Circuits Signal Process.*, vol. 1, pp. 119–135, 1991.

[12] U. Gatti, F. Maloberti, and G. Torelli, "A novel CMOS linear transconductance cell for continuous-time filters," in *Proc. IEEE Int. Symp. Circuits Syst.* (ISCAS), 1990, pp. 1173–1176.

[13] A. S. Sedra and K. C. Smith, *Microelectronic Circuits*, 3rd ed., Philadelphia: Holt, Rinehart and Winston, Series in Electr. Eng., 1991, chs. 5, 6, and 10.

[14] K. Bult, "Basic CMOS circuit techniques," in *Analog VLSI: Signal and Information Processing*, M. Ismail and T. Fiez, Eds., New York: McGraw-Hill, 1994, ch. 2.

[15] S. R. Zarabadi, M. Ismail, and F. Larsen, "Basic BiCMOS circuit techniques," in *Analog VLSI: Signal and Information Processing*, M. Ismail and T. Fiez, Eds., New York: McGraw-Hill, 1994, ch. 5.

58

High-Performance Analog Circuits

Chris Toumazou
Imperial College of Science, Technology, and Medicine

Alison Payne
Imperial College of Science, Technology, and Medicine

John Lidgey
Oxford Brookes University

B. M. Wilamowski
University of Wyoming

58.1 Broadband Bipolar Networks

Chris Toumazou, Alison Payne, and John Lidgey

Introduction

Numerous textbooks have presented excellent treatments of the design and analysis of broadband bipolar amplifiers. This chapter is concerned with techniques for integrated circuit amplifiers, and is written mainly as a tutorial aimed at the practicing engineer.

For broadband bipolar design it is first important to identify the key difference between lumped and distributed design techniques. Basically when the signal wavelengths are close to the dimensions of the integrated circuit, then characteristic impedances become significant, lines become lossy, and we essentially need to consider the circuit in terms of transmission lines. At lower frequencies where the signal wavelength is much larger than the dimensions of the circuit, the design can be considered in terms of lumped components, allowing some of the more classical low-frequency analog circuit techniques to be applied. At intermediate frequencies we enter the realms of hybrid lumped/distributed design. Many RF designs fall into this category, although every day we see new technologies and circuit techniques developed that increase the frequency range for which lumped approaches are possible. In broadband applications, IC's are generally designed without the use of special microwave components, so broadband techniques are very similar to those employed at lower frequencies. However, several factors still have to be considered in RF design: all circuit parasitics must be identified and included to ensure accurate simulation; feedback can generally only be applied locally as phase shifts per stage are significant; the cascading of several local feedback stages is difficult since ac coupling is often impractical; the NPN bipolar transistor is the main device used in silicon, since it has potentially a higher f_t

0-8493-8341-2/95/$0.00 + $.50

than PNP bipolar or MOSFET devices; active PNP loads are generally avoided due to their poor frequency and noise performance and so resistive loads are used instead.

The frequency performance of an RF or broadband circuit will depend on the frequency capability of the devices used, and no amount of good design can compensate for transistors with an inadequate range. As a rule designs are kept as simple as possible, since at high frequencies all components have associated parasitics.

Miller's Theorem

It is important to describe at the outset a very useful approximation that will assist in simplifying the high-frequency analysis of some of the amplifiers to be described. The technique is known as Miller's theorem and will be briefly discussed here. A capacitor linking input to output in an inverting amplifier results in an input referred shunt capacitance that is multiplied by the voltage gain of the stage, as shown in Fig. 58.1. This increased input capacitance is known as the Miller capacitance.

It is straightforward to show that the input admittance looking into the inverting input of the amplifier is approximately $Y_{in} = j\omega C_f(1 + A)$. The derivation assumes the inherent poles within the amplifier are at a sufficiently high frequency so that the frequency response of the circuit is dominated by the input of the amplifier. If this is not the case then Miller's approximation should be used with caution as will be discussed later. From the above model it is apparent that the Thévenin input signal source sees an enlarged capacitance to ground. Miller's approximation is often a useful way of simplifying circuit analysis by assuming that the input dominant frequency is given by the simple lowpass *RC* filter in Fig. 58.1. However, the effect is probably one of the most detrimental in broadband amplifier design, affecting both frequency performance and/or stability.

Bipolar Transistor Modeling at High Frequencies

In this section we consider the high-frequency small-signal performance of the bipolar transistor. The section assumes the reader has some knowledge of typical device parameters, and has some

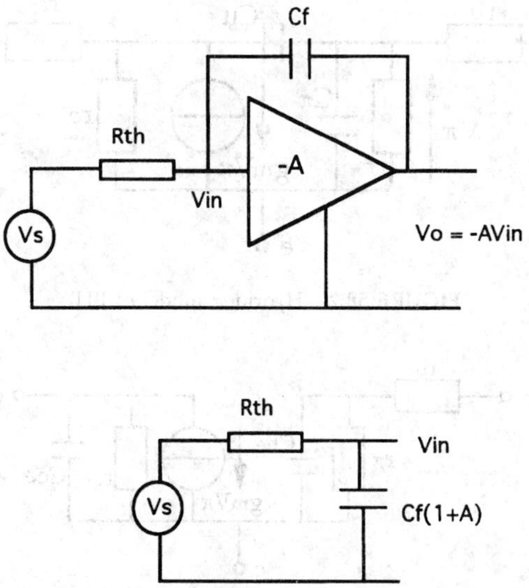

FIGURE 58.1 Example of the Miller effect.

familiarity with the technology. For small-signal analysis, the simplified hybrid-π model shown in Fig. 58.2 is used, where

r_b = base series resistance
r_c = collector series resistance
r_π = dynamic base–emitter resistance
r_o = dynamic collector–emitter resistance
C_π = base–emitter junction capacitance
C_μ = collector–base junction capacitance
C_{cs} = collector–substrate capacitance
g_m = small-signal transconductance.

At low frequencies the Miller approximation allows the hybrid-π model to be simplified to the circuit shown in Fig. 58.3, where the net input capacitance now becomes $C_{be} = C_\pi + C_\mu(1 - A_v)$, the net output capacitance becomes $C_{ce} = C_\mu(1 - 1/A_v)$, where A_v is the voltage gain given by $A_v = (V_{ce}/V_{be}) \approx -g_m R_1$ where R_1 is the collector load resistance. r_C and C_{cs} have been neglected. Thus $C_{be} \approx C_\pi + g_m R_1 C_\mu$ and $C_{ce} \approx C_\mu$. The output capacitance C_{ce} is often neglected from the small-signal model. The approximation $A_v = -g_m R_1$ assumes that $r_\pi \gg r_b$, and that the load is purely resistive. At high frequencies, however, we cannot neglect the gain roll-off due to C_π and C_μ, and even at frequencies as low as 5 percent of f_t the Miller approximation can introduce significant errors.

A simplified hybrid-π model that takes the high-frequency gain roll-off into account is shown in Fig. 58.4. C_μ is now replaced by an equivalent current source $sC_\mu(V_\pi - V_{ce})$.

A further modification is to split the current source between the input and output circuits as shown in Fig. 58.5.

Finally, the input and output component terms can be rearranged leading to the modified equivalent circuit shown in Fig. 58.6, which is now suitable for broadband design. From Fig. 58.6 the transconductance $(g_m - sC_\mu)$ shows the direct transmission of the input signal through C_μ. The input circuit current source $(sC_\mu V_{ce})$ shows the feedback from the output to the input via C_μ. Depending on the phase shift between V_{ce} and V_{be}, this feedback can cause high-frequency oscillation. At lower frequencies $sC_\mu \ll g_m$ and $V_{ce}/V_\pi \approx -g_m R_1$, which is identical to the Miller

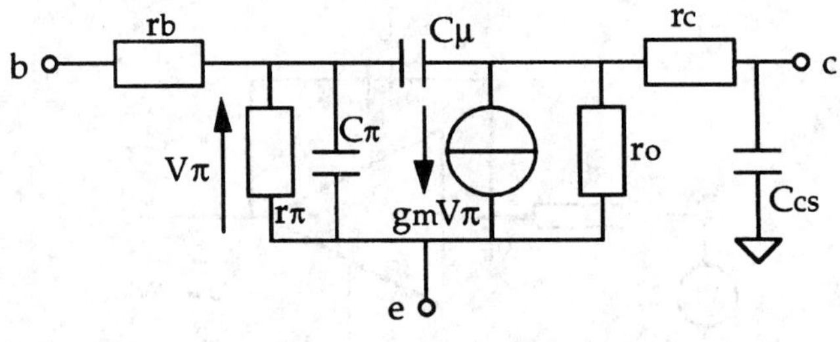

FIGURE 58.2 Hybrid π model of BJT.

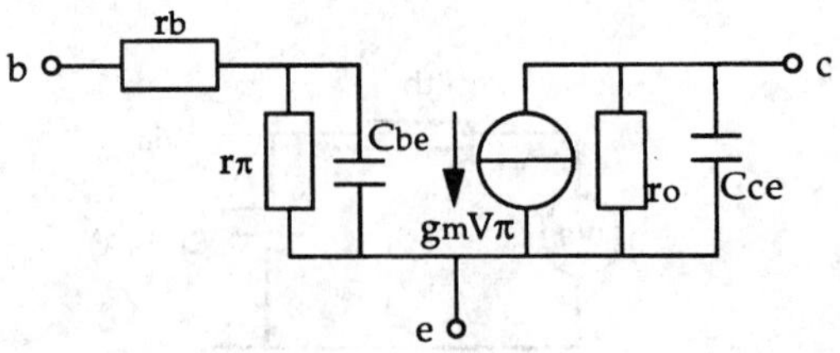

FIGURE 58.3 Simplified Miller-approximated hybrid π model of BJT.

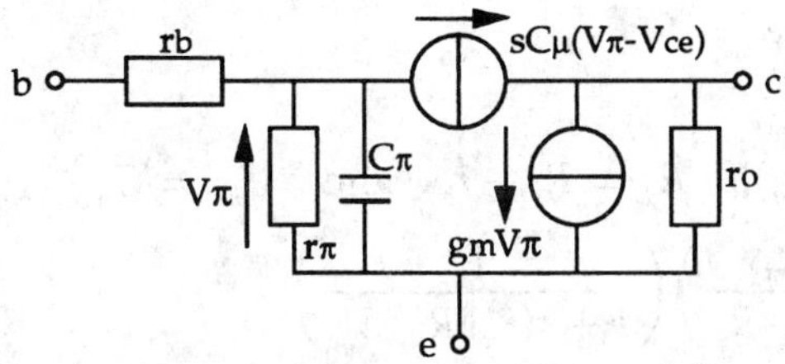

FIGURE 58.4 Simplified high-frequency model.

approximation. The model of Fig. 58.6 is the most accurate for broadband amplifier design, particularly at high frequencies.

Single-Gain Stages

Consider now the high-frequency analysis of single-gain stages.

Common-Emitter (CE) Stage

Figure 58.7 shows a common-emitter amplifier with load R_l and source R_s. External biasing components are excluded from the circuit.

First analysis using the Miller approximation yields the small-signal high-frequency model

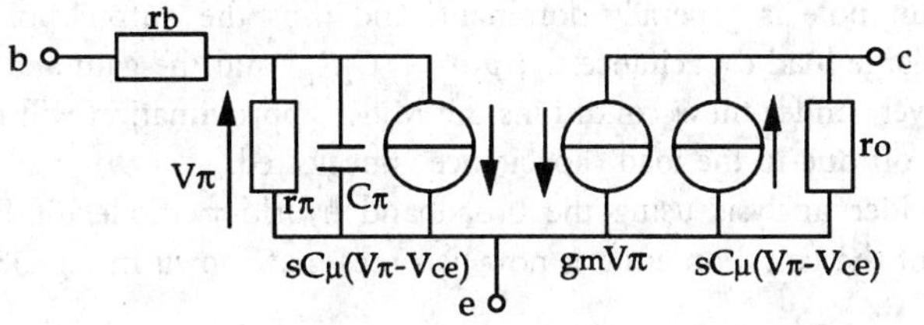

FIGURE 58.5 Split current sources.

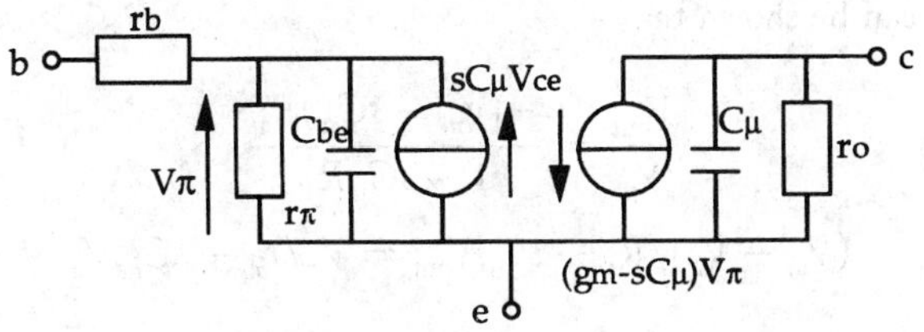

FIGURE 58.6 Modified equivalent circuit.

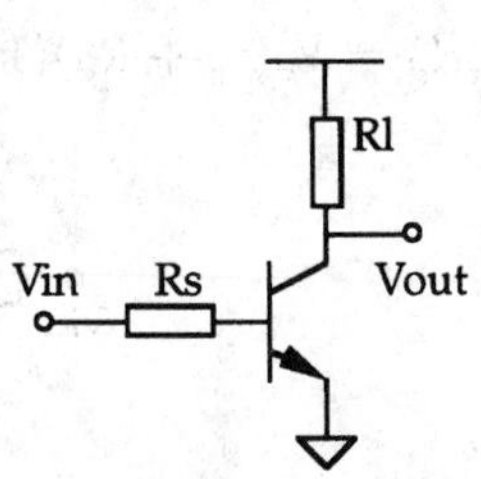

FIGURE 58.7 Common-emitter amplifier.

shown in Fig. 58.8, where

$$R_{1'} = (R_1 \| r_o), \quad R_{s'} = R_s + r_b \quad \text{and} \quad C_{be} = C_\pi + g_m R_{1'} C_\mu$$

$$\frac{V_\pi}{V_{\text{in}}} = \left(\frac{r_\pi}{r_\pi + R_{s'}}\right)\left(\frac{1}{1 + s(r_\pi \| R_{s'})C_{be}}\right) \tag{58.1}$$

$$\frac{V_{\text{out}}}{V_\pi} = \frac{-g_m R_{1'}}{1 + sC_\mu R_{1'}},$$

and thus

$$\frac{V_{\text{out}}}{V_{\text{in}}} = -\left(\frac{g_m R_{1'} r_\pi}{r_\pi + R_{s'}}\right)\left(\frac{1}{(1 + sC_\mu R_{1'})(1 + s(r_\pi \parallel R_{s'})C_{be})}\right) \tag{58.2}$$

This approximate analysis shows:

- "Ideal" voltage gain $= -g_m R_1$.
- Input attenuation caused by $R_{s'}$ in series with r_π.
- Input circuit pole p_1 at $s = 1/C_{be}(r_\pi \parallel R_{s'}) \approx 1/C_{be}R_{s'}$.
- Output attenuation caused by r_o in parallel with R_1.
- Output circuit pole p_2 at $s = 1/C_\mu R_{1'}$.

The input circuit pole is generally dominant, and thus the output pole p_2 can often be neglected. With a large load capacitance C_1, $p_2 \approx 1/C_1 R_{1'}$, and the gain and phase margin will be reduced. However, under these conditions the Miller approximation will no longer be valid, since the gain roll-off due to the load capacitance is neglected.

If we now consider analysis using the broadband hybrid-π model of Fig. 58.6, then the equivalent model of the common emitter now becomes that shown in Fig. 58.9, where

$$C_{be} = C_\pi + C_\mu, \quad R_{s'} = R_s + r_b \quad \text{and} \quad R_{1'} = R_1 \parallel r_o$$

From the model it can be shown that

$$\frac{V_{\text{out}}}{V_\pi} = \frac{-(g_m - sC_\mu)R_{1'}}{1 + sC_\mu R_{1'}} \tag{58.3}$$

$$(V_{\text{in}} - V_\pi)/R_{s'} + sC_\mu V_{\text{out}} = V_\pi/r_\pi + sC_{be}V_\pi \tag{58.4}$$

and

$$V_{\text{in}} r_\pi + V_{\text{out}} sC_\mu r_\pi R_{s'} = V_\pi (r_\pi + R_{s'})(1 + sC_{be}(r_\pi \parallel R_{s'})) \tag{58.5}$$

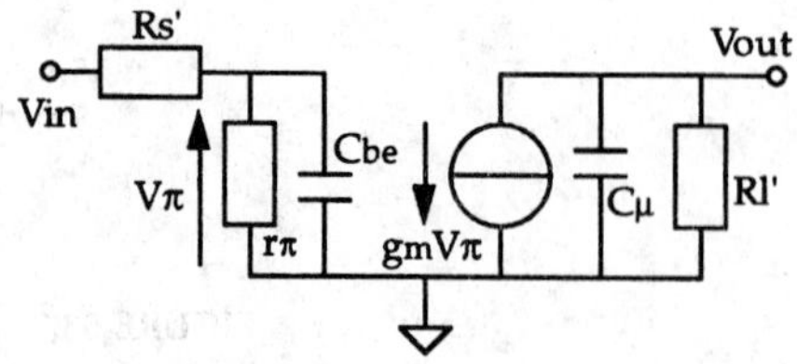

FIGURE 58.8 High-frequency model of the common emitter.

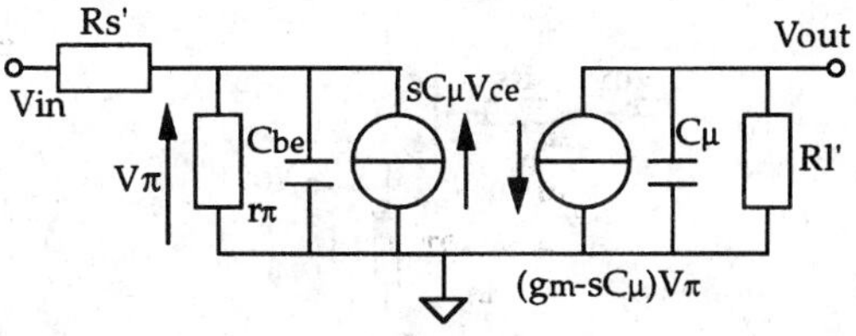

FIGURE 58.9 Equivalent circuit model of the common emitter.

Rearranging these equations yields

$$\frac{V_{\text{out}}}{V_{\text{in}}} = -\left(\frac{g_m R_{1'} r_\pi}{r_\pi + R_{s'}}\right) \times \left(\frac{1 - sC_\mu/g_m}{(1 + sC_{be}R_{s'})(1 + sC_\mu R_{1'}) + sC_\mu g_m R_{1'} R_{s'} - s^2 C_\mu^2 R_{s'} R_{1'}}\right) \tag{58.6}$$

This analysis shows that there is a right-hand-plane (RHP) zero at $s = 1/(C_\mu r_e)$, which is not predicted by the Miller approximation. Assuming $r_\pi \gg R_{s'}$ and $C_\pi \gg C_\mu$, the denominator can be written as

$$1 + s(R_{s'}(C_\pi + C_\mu g_m R_{1'}) + C_\mu R_{1'}) + s^2 C_\mu C_\pi R_{1'} R_{s'} \tag{58.7}$$

which can be described by the second-order characteristic equation

$$1 + s(1/p_1 + 1/p_2) + s^2/p_1 p_2 \tag{58.8}$$

By comparing coefficients in (58.7) and (58.8) the sum of the poles is the same as that obtained in (58.2) using the Miller approximation, but the pole product $p_1 p_2$ is greater. This means that the poles are farther apart than predicted by the Miller approximation. In general, the Miller approximation should be reserved for analysis at frequencies of operation well below f_t, and for situations where the capacitive loading is not significant. The equivalent circuit of Fig. 58.9 therefore gives a more accurate result for high-frequency analysis. For a full understanding of RF behavior, computer simulation of the circuit including all parasitics is essential.

Since the CE stage provides high current and voltage gain, oscillation may well occur. Therefore, care must be taken during layout to minimize parasitic coupling between the input and output. The emitter should be at ground potential for ac signals, and any lead inductance from the emitter to ground will generate phase-shifted negative feedback to the base, which can result in instability.

Common-Collector (CC) Stage

The common collector or emitter follower shown in Fig. 58.10 is a useful circuit configuration since it generally serves to isolate a high gain stage from a load. The high-frequency performance of this stage must be good enough so as not to degrade the frequency performance or stability of the complete amplifier. An equivalent high-frequency small-signal model of the common collector is shown in Fig. 58.11.

The following set of equations can be derived from Fig. 58.11:

$$(V_{\text{in}} - V_b)/R_{s'} = V_\pi/r_\pi + sC_{be}V_\pi + sC_\mu V_{out}, \quad V_b = V_{out} + V_\pi \tag{58.9}$$

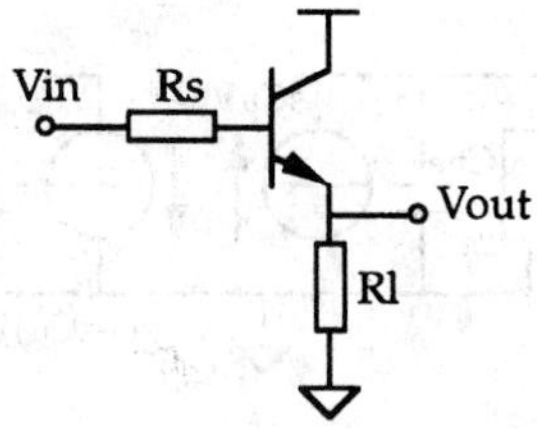

FIGURE 58.10 Common-collector amplifier.

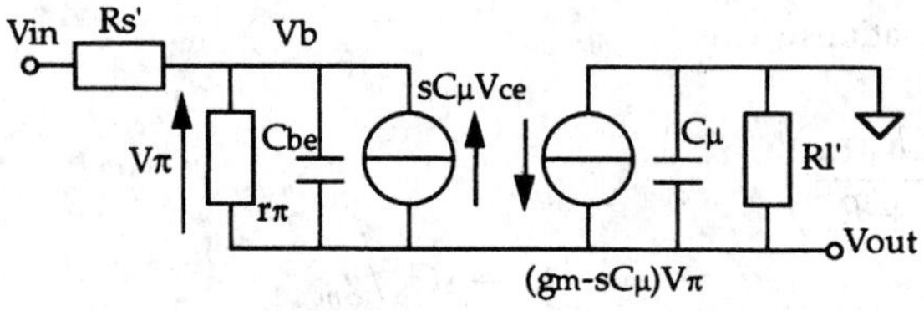

FIGURE 58.11 Equivalent circuit of the common collector.

and

$$V_\pi/r_\pi + sC_{be}V_\pi + sC_\mu V_{out} + (g_m - sC_\mu)V_\pi - sC_\mu V_{out} - V_{out}/R_{l'} = 0 \tag{58.10}$$

Rearranging these equations yields

$$\frac{V_{out}}{V_{in}} = \frac{R_{l'}(1+g_m r_\pi + sC_\pi r_\pi)}{(R_{s'}+r_\pi)(1+s(R_{s'}\|r_\pi)C_{be})+R_{l'}(1+sC_\mu R_{s'})(1+g_m r_\pi + sC_\pi r_\pi)} \tag{58.11}$$

The above expression can be simplified by assuming $r_\pi \gg R_{s'}$, $g_m r_\pi \gg 1$, $C_\pi \gg C_\mu$ to,

$$\frac{V_{out}}{V_{in}} = \left(\frac{r_\pi}{r_\pi + R_{s'}}\right)\left(\frac{1 + sC_\pi/g_m}{(1 + sC_\mu R_{s'})(1 + sC_\pi/g_m) + (1 + sC_\pi R_{s'})/g_m R_{l'}}\right) \tag{58.12}$$

This final transfer function indicates the presence of a left-half-plane (LHP) zero at $s = (g_m/C_\pi) = \omega_t$. The denominator can be rewritten as approximately

$$(1 + 1/g_m R_{l'}) + s(C_\mu R_{s'} + C_\pi/g_m + C_{be}R_{s'}/g_m R_{l'}) + s^2 C_\mu C_\pi R_{s'}/g_m \tag{58.13}$$

which simplifies to

$$1 + s(C_\pi r_e + C_\mu R_{s'} + (C_\mu + C_\pi)r_e R_{s'}/R_{l'}) + s^2 C_\mu C_\pi R_{s'} R_{l'} \tag{58.14}$$

Assuming a second-order characteristic form of $1 + s(1/p_1 + 1/p_2) + s^2/p_1 p_2$, if $p_1 \ll p_2$, the above reduces to $1 + s/p_1 + s^2/p_1 p_2$. If $(R_{s'}/R_{l'}) \ll 1$, then $p_1 \approx 1/(C_\pi r_e)$, and this dominant pole will be approximately canceled by the zero. The frequency response will then be limited by the nondominant pole $p_2 \approx 1/C_\mu R_{s'}$.

The frequency response of a circuit containing several stages is thus rarely limited by the CC stage, due to this dominant pole–zero cancellation. For this analysis to be valid, $R_{s'} \ll R_{l'}$. As $R_{s'}$ increases the poles will move closer together, and the pole–zero cancellation will degrade. In practice, the CC stage is often used as a buffer, and is thus driven from a high source resistance into a low value load resistance.

A very important parameter of the common-collector stage is output impedance. It is generally assumed that the output impedance of a common collector is low, also that there is good isolation between a load and the amplifying stage, and that any amount of current can be supplied to the load. Furthermore, it is assumed that capacitive loads will not degrade the frequency performance since the load will be driven by an almost short circuit. While this may be the case at low frequencies it is a different story at high frequencies. Consider the following high-frequency analysis. We first assume that the small-signal model shown in Fig. 58.12 is valid.

From the Fig. 58.12, the output impedance can be approximated as

$$\frac{V_{\text{out}}}{I_{\text{out}}} = \frac{Z_\pi + R_{s'}}{1 + g_m Z_\pi} \tag{58.15}$$

where $Z_\pi = (r_\pi \parallel C_{be})$ and $R_{s'} = R_s + r_b$. At very low frequencies ($\omega \to 0$):

$$R_{\text{out}} = \frac{r_\pi + R_{s'}}{1 + g_m r_\pi} \approx 1/g_m + R_{s'}/g_m r_\pi \approx r_e + R_{s'}/\beta \tag{58.16}$$

At very high frequencies ($\omega \to \infty$):

$$R_{\text{out}} = \frac{1/sC_{be} + R_{s'}}{1 + g_m/sC_{be}} \approx R_{s'} \tag{58.17}$$

If $r_e > R_{s'}$, then the output impedance decreases with frequency, that is, Z_{out} is capacitive. If $R_{s'} > r_e$, then Z_{out} increases with frequency and so Z_{out} appears inductive. It is usual for an emitter follower to be driven from a high source resistance, thus the output impedance appears to be inductive and can be modeled as shown in Fig. 58.13, where

$$R_1 = r_e + R_{s'}/\beta, \quad R_2 = R_{s'}, \quad L = R_{s'}/\omega_t$$

The inductive behavior of the CC stage output impedance must be considered in broadband design since any capacitive loading on this stage could result in peaking or instability. The transform from base resistance to emitter inductance arises because of the 90° phase shift between base and emitter currents at high frequencies, due principally to C_π. This transform property can be used to advantage to simulate an on-chip inductor by driving a CC stage from a high source resistance. Similarly, by loading the emitter with an inductor we can increase the effective base series resistance $R_{s'}$ without degrading the noise performance of the circuit. A capacitive load will also be transformed by 90° between the base and emitter;

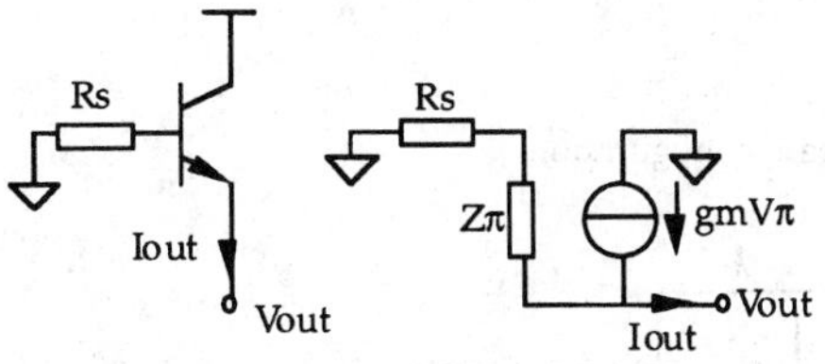

FIGURE 58.12 Equivalent circuit of the CC output stage.

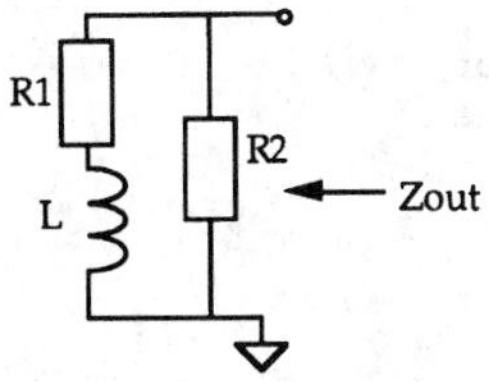

FIGURE 58.13 Equivalent high-frequency model of CC output stage.

for example, a capacitive loading on the base can look like a negative resistance at the emitter.

Common-Base (CB) Stage

The common-base amplifier shown in Fig. 58.14 offers the highest frequency performance of all the single-stage amplifiers. When connected as a unity gain current buffer, the CB stage operates up to the f_t of the transistor.

Using the simplified hybrid π model of Fig. 58.3, it follows that

$$\frac{I_{\text{out}}}{I_{\text{in}}} \approx \frac{\beta}{\beta + 1} \quad \text{where} \quad \beta = \frac{\beta_o}{1 + s/\omega_o} \tag{58.18}$$

$$\frac{I_{\text{out}}}{I_{\text{in}}} \approx \frac{a_o}{1 + s/\omega_t} \quad \text{where} \quad a_o = \beta_o/(\beta_0 + 1) \quad \text{and} \quad \omega_t = \beta_o\omega_o \tag{58.19}$$

The CB stage thus provides wideband unity current gain. Note that the input impedance of the CB stage is the same as the output impedance of the CC stage, and thus can appear inductive if the base series resistance is large.

In many situations the CB stage is connected as a voltage amplifier, an example of this being the current-feedback amplifier, which will be discussed in a later section. Consider the following high-frequency analysis of the CB stage being employed as a voltage gain amplifier. Figure 58.15 shows the circuit together with a simplified small-signal model. From the equivalent model the gain of the circuit can be approximated as

$$\frac{V_{\text{out}}}{V_{\text{in}}} = \frac{kR_1}{R_s}\left(\frac{1 - sC_\mu/g_m}{1 + s(C_\pi/g_m)(kR_{s'}/R_s)}\right) \tag{58.20}$$

where

$$R_{s'} = R_s + r_b, \quad \text{and} \quad k \approx \frac{R_s}{R_s + 1/g_m}$$

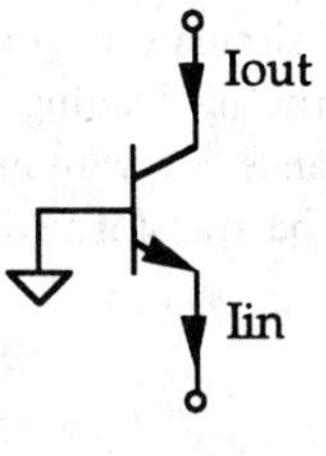

FIGURE 58.14 Common-base configuration.

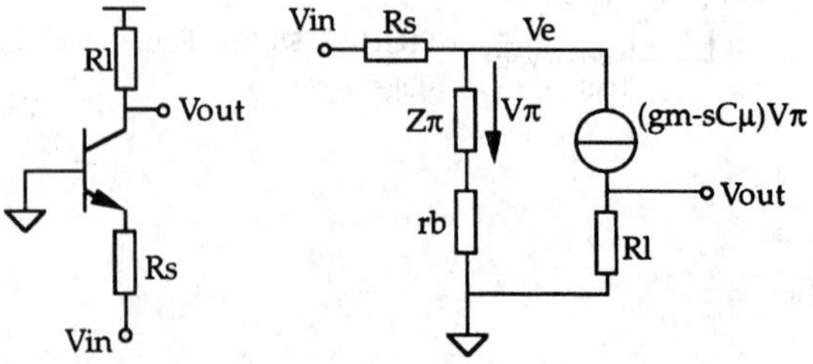

FIGURE 58.15 CB stage as a voltage amplifier.

If $R_s \gg 1/g_m$, then $k \approx 1$ and so

$$\frac{V_{\text{out}}}{V_{\text{in}}} = \frac{R_1}{R_s}\left(\frac{1 - sC_\mu/g_m}{1 + s(C_\pi/g_m)(1 + r_b/R_s)}\right) \tag{58.21}$$

Thus it can thus be seen that the circuit has a right-hand-plane zero at $s = 1/(r_e C_\mu)$, since $r_e = 1/\text{gm}$ and a pole at $1/C_\pi r_e(1 + r_b/R_s) = \omega_t/(1 + r_b/R_s)$. Note that in the case of a current source drive ($R_s \gg r_b$), the pole is at the ω_t of the transistor. However, this does assume that the output is driven into a short circuit. Note also the excellent isolation between the input and output circuits, since there is no direct path through C_μ and so no Miller effect.

Neutralization of C_μ

Many circuit techniques have been developed to compensate for the Miller effect in amplifiers and hence extend the frequency range of operation. The common-emitter stage provides the highest potential power gain, but the bandwidth of this configuration is limited since the amplified output voltage effectively appears across the collector–base junction capacitance resulting in the Miller capacitance multiplication effect. This bandwidth limiting due to C_μ can be overcome by using a two-transistor amplifying stage such as the CE–CB cascode stage or the CC–CE cascade. Consider now a brief qualitative description of each in turn. The circuit diagram of the CE–CB cascode is shown in Fig. 58.16.

The CE transistor Q_1 provides high current gain of approximately β and a voltage gain of $A_{v1} \approx -g_{m1}R_1 = -g_{m1}r_{e2}$, which in magnitude will be close to unity. Therefore, the Miller multiplication of C_μ is minimized, and the bandwidth of Q_1 is maximized. The CB transistor Q_2 provides a voltage gain $A_{v2} \approx R_1/r_{e2}$. The total voltage gain of the circuit can be approximated as $A_v \approx g_{m1}R_1$, which is equal to that of a single CE stage. The total frequency response is given by the cascaded response of both stages. Since both transistors exhibit wideband operation, then the dominant poles of each stage may be close in frequency. As a result, the total phase shift through the cascode configuration is likely to be greater than that obtained with a single device, and care should be taken when applying negative feedback around the pair.

Consider now the CC–CE stage of Fig. 58.17. In this case voltage gain is provided by the CE stage transistor Q_2 and is $A_{v2} \approx -g_{m2}R_1$. This transistor is being driven from the low output impedance of Q_1 and so the input pole frequency of this device ($\approx 1/C_{be2}R_{s2}$) is maximized. The CC stage transistor Q_1 is effectively a buffer that isolates C_μ of Q_2 from the source resistance R_s. The low-frequency voltage gain of this circuit is reduced when compared with a single-stage configuration because the input signal effectively appears across two base–emitter junctions.

The two-transistor configurations help to maintain a wideband frequency response by isolating the input and output circuits. In integrated circuit design, another method of neutralizing the effect of C_μ is possible when differential gain stages are used.

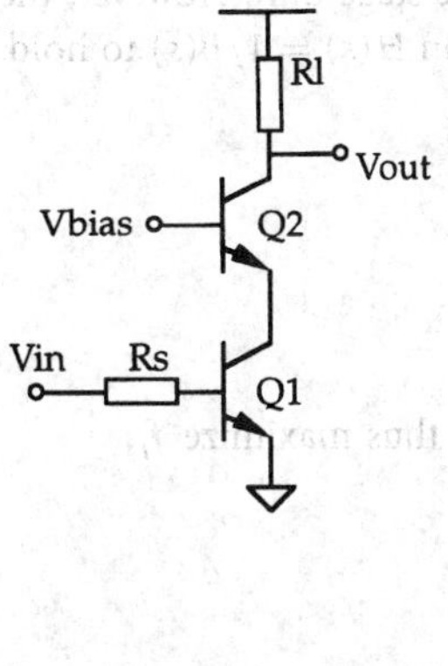

FIGURE 58.16 CE–CB cascode.

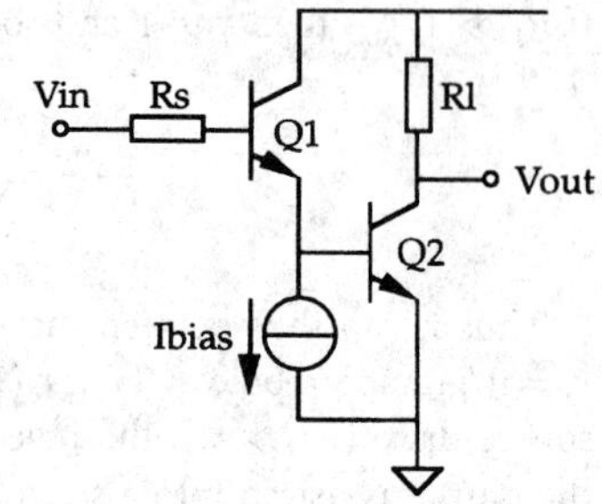

FIGURE 58.17 CC–CE stage.

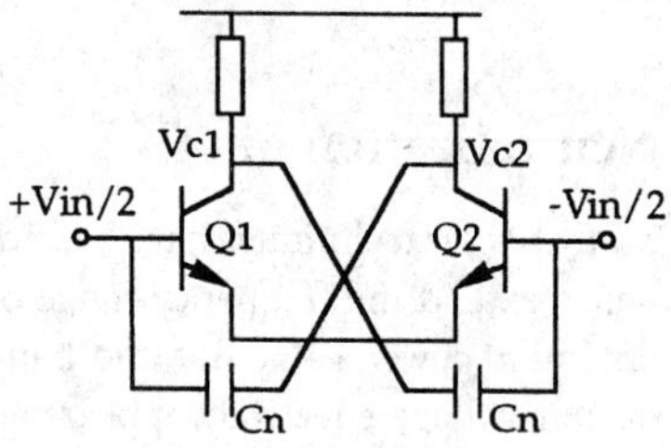

FIGURE 58.18 Differential gain stage.

For example, Fig. 58.18 shows a section of a differential input amplifier. If the inputs are driven differentially, then the collector voltages V_{c1} and V_{c2} will be 180° out of phase. The neutralization capacitors C_n thus inject a current into the base of each transistor that is equal and opposite to that caused by the intrinsic capacitance C_μ. Consequently, the neutralization capacitors should be equal to C_μ in order to provide good signal cancellation, and so they may be implemented from the junction capacitance of two dummy transistors with identical geometries to Q_1 and Q_2 as shown in Fig. 58.19.

Negative Feedback

Negative feedback is often employed around high gain stages to improve the frequency response. In effect, the gain is reduced in exchange for a wider, flatter bandwidth. The transfer function of a closed-loop system can be written

$$H(s) = \frac{A(s)}{1 + A(s)B(s)} \tag{58.22}$$

where $A(s)$ is the open-loop gain and $B(s)$ is the feedback fraction. If the open-loop gain $A(s)$ is large, then $H(s) \approx 1/B(s)$. In RF design, compound or cascaded stages can produce excessive phase shifts that result in instability when negative feedback is applied. To overcome this problem it is generally accepted to apply local negative feedback around a single stage only. However, the open-loop gain of a single stage is usually too low for the approximation $H(s) = 1/B(s)$ to hold.

RF Bipolar Transistor Layout

When laying out RF transistors, the aim is to

- Minimize C_μ and C_π.
- Minimize base width to reduce the forward transit time t_f and thus maximize f_t.
- Minimize series resistance r_b and r_c.

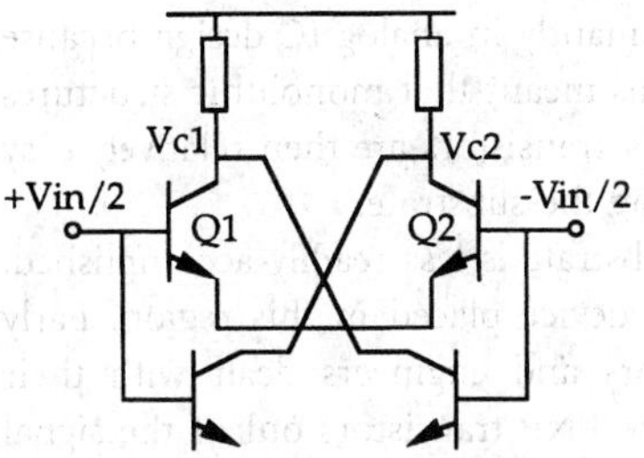

FIGURE 58.19 Implementation of neutralization capacitors.

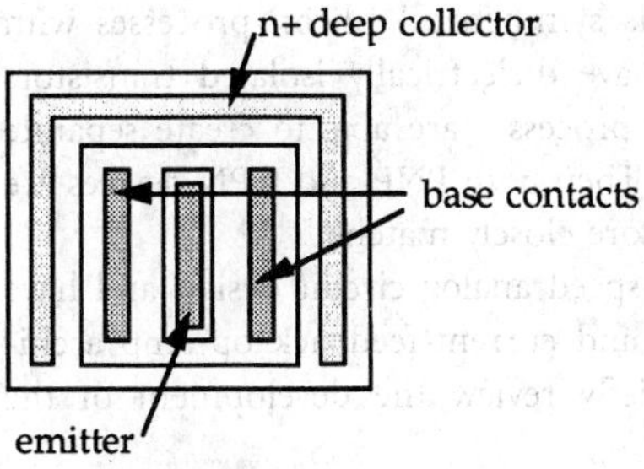

FIGURE 58.20 Stripe geometry.

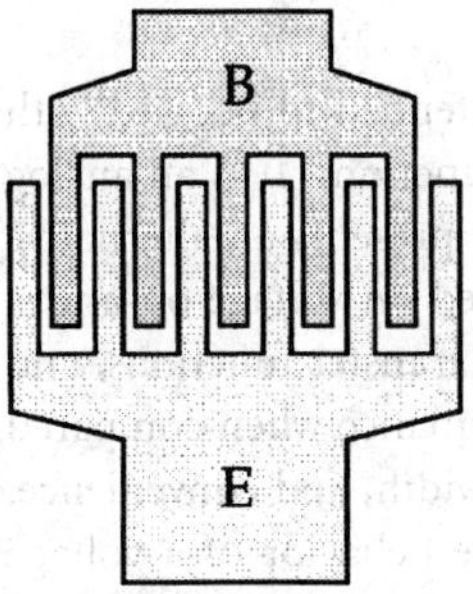

FIGURE 58.21 Transistor layout with interleaving fingers.

To minimize junction capacitance, the junction area must be reduced; however, this will tend to increase the series resistance. Transistors are generally operated at fairly high currents to maximize f_t. However, if the emitter gets too crowded, then the effective value of β will be reduced. The requirements given above are generally best met by using a stripe geometry of the type shown in Fig. 58.20.

The stripe geometry maximizes the emitter area-to-periphery ratio, which reduces emitter crowding while minimizing the junction capacitance. The length of the emitter is determined by current-handling requirements. The base series resistance is reduced by having two base contacts and junction depths are minimized to reduce capacitance. The buried layer, or deep collector, reduces the collector series resistance. High-power transistors are produced by paralleling a number of transistors with interleaving "fingers," as shown in Fig. 58.21. This preserves the frequency response of the stripe geometry while increasing the total current-handling capability.

Bipolar Current-Mode Broadband Circuits

Recently there has been strong interest in applying so-called current-mode techniques to electronic circuit design. Considering the signal operating parameter as a current and driving into low impedance nodes has allowed the development of a wealth of circuits with broadband properties. Many of the following circuit and system concepts date back several years; it is progress in integrated circuit technology that has given a renewed impetus to "practical" current-mode techniques.

The NPN bipolar transistor, for example, is used predominantly in analog IC design because electron mobility is greater than hole mobility in silicon. This means that monolithic structures are typically built on P-type substrates, because vertical NPN transistors are then relatively easy to construct and to isolate from each other by reverse biasing the substrate.

Fabricating a complementary PNP device on a P-type substrate is less readily accomplished. An N-type substrate must be created locally and the PNP device placed in this region. Early bipolar processes created PNP devices as lateral transistors and engineers dealt with their inherently poor, low-frequency characteristics by keeping the PNP transistors out of the signal path whenever possible.

However, high-speed analog signal processing demands symmetrical silicon processes with fully complementary BJTs. Newer, advanced processes have dielectrically isolated transistors rather than reversed-biased PN junction isolation. These processes are able to create separate transistors, each situated in a local semiconductor region. Then both PNP and NPN devices are vertical and their performance characteristics are much more closely matched.

Dielectric isolation processes have revolutionized high-speed analog circuit design and have been key in making high-performance current-conveyor and current-feedback op-amp architectures practical. In the following sections we will briefly review the development of the current-conveyor and current-feedback op-amp.

The Current Conveyor

The current conveyor is a versatile broadband analog amplifier that is intended to be used with other circuit components to implement many analog signal-processing functions. It is an analog circuit building block in much the same way as a voltage op-amp, but it presents an alternative method of implementing analog systems that traditionally have been based on voltage op-amps. This alternative approach leads to new methods of implementing analog transfer functions, and in many cases the conveyor-based implementation offers improved performance when compared to the voltage op-amp-based implementation in terms of accuracy, bandwidth, and convenience. Circuits based on voltage op-amps are generally easy to design since the behavior of a voltage op-amp can be approximated by a few simple design rules. This is also true for current conveyors, and once the appropriate design rules are understood, the application engineer is able to design conveyor-based circuits just as easily.

The first generation current conveyor or CCI was proposed by Smith and Sedra in 1968 [1] and the more versatile second-generation current conveyor, or CCII, was introduced by the same two authors in 1970 [2], as an extension of the CCI. The CCII is without doubt the more valuable and adaptable building block of the two, and we will concentrate mostly on this device. Figure 58.22(a) shows the voltage–current describing matrix for the CCII, while Fig. 58.22(b) shows the schematic normally used for the CCII with the power supply connections omitted.

The voltage at the low-impedance input node X follows that at the high-impedance input node Y, while the input current at node X is mirrored or "conveyed" to the high-impedance output node Z. The $\pm$ sign indicates the polarity of the output current with respect to the input current; by convention, a positive sign indicates that both the input and output currents simultaneously flow into or out of the device, thus Fig. 58.22(b) illustrates a CCII+. For the first-generation conveyor, or CCI, the input current at node X was reflected to input Y, that is, the two inputs had equal currents. In the case of the second-generation conveyor input, Y draws

$$\begin{bmatrix} I_Y \\ V_X \\ I_Z \end{bmatrix} = \begin{bmatrix} 0 & 0 & 0 \\ 1 & 0 & 0 \\ 0 & \pm 1 & 0 \end{bmatrix} \begin{bmatrix} V_Y \\ I_X \\ V_Z \end{bmatrix}$$

(a)

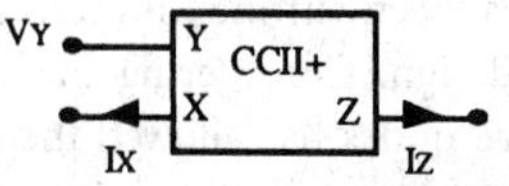

(b)

FIGURE 58.22 The CCII current conveyor. (a) I–V describing matrix. (b) Schematic.

no current, and this second generation, or CCII formulation, has proved to be much more adaptable and versatile than its first-generation predecessor. Because of the combined voltage and current following properties, CCII's may be used to synthesize a number of analog circuit functions that are not so easily or accurately realizable using voltage op-amps.

Some of these application areas are shown in Fig. 58.23. As current-conveyors become more readily available and circuit designers become more familiar with the versatility of this device it is certain that further ingenious uses will be devised.

The Ideal Transistor and the Current-Conveyor. So far a transistor-level realization of the CCII has not been discussed. The current–voltage transfer relationship for the CCII+ is given by

$$V_X = V_Y, \quad I_Y = 0 \quad \text{and} \quad I_Z = I_X \tag{58.23}$$

These equations show that there is a simple voltage-following action between input node Y and output node X, and that there is a simple current-following action between input node X and output node Z. Also, these characteristic equations tell us that the impedance relationship for the ideal current conveyor are

$$Z_{INY} = \infty, \quad Z_X = 0 \quad \text{and} \quad Z_{OUTZ} = \infty. \tag{58.24}$$

Figure 58.24 shows a schematic representation of a CCII− built with a single BJT and on reflection it is clear that the current conveyor is effectively an ideal transistor, with infinite β and infinite g_m.

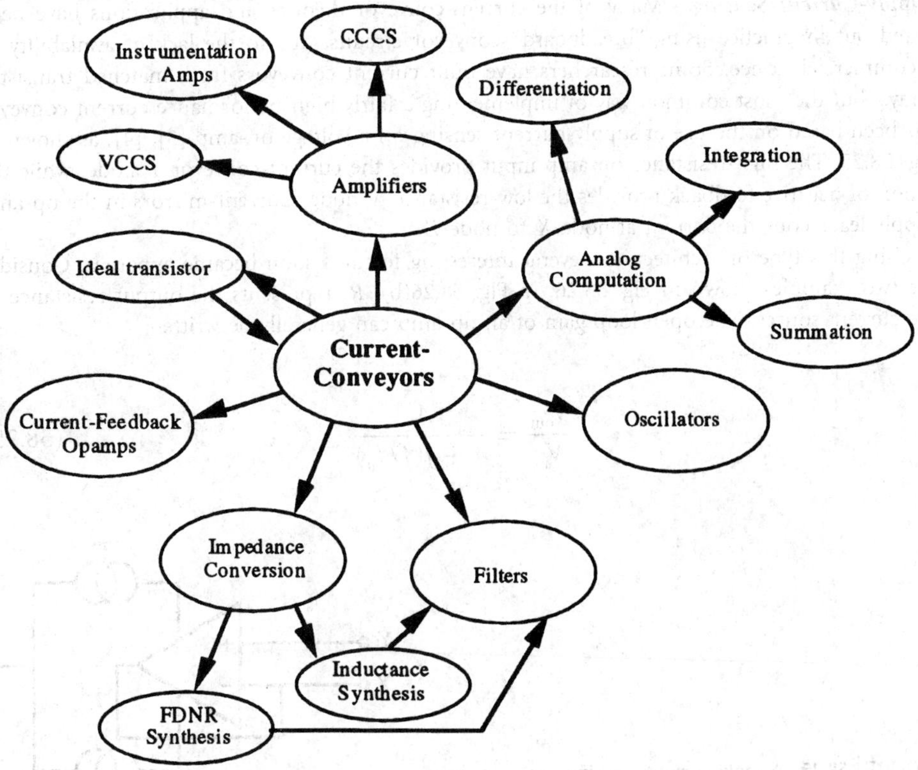

FIGURE 58.23 Current-conveyor applications.

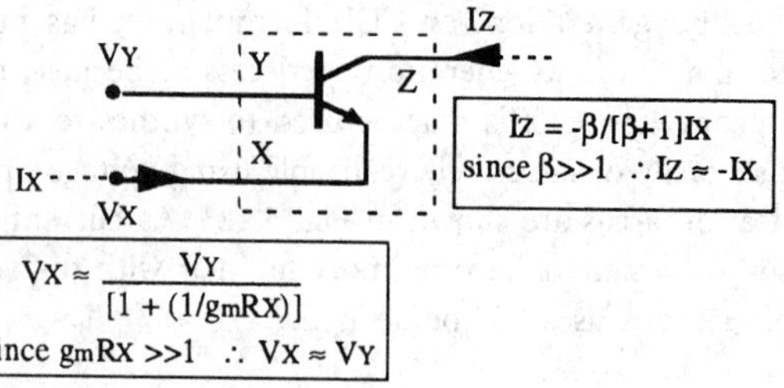

FIGURE 58.24 Single BJT CCII−.

Driving into the base of a BJT gives almost unity voltage gain from input base to output emitter, with high input impedance and low output impedance, and driving into the emitter of a BJT gives almost unity current gain from emitter input to collector output, with low input impedance and high output impedance. Drawing the comparison further, the high-input-impedance Y node corresponds to the base (or gate) of a transistor, the low-input-impedance X node corresponds to the emitter (or source) of a transistor, and the high-output-impedance Z node corresponds to the collector (or drain) of a transistor. Clearly, one transistor cannot function alone as a complete current conveyor since an unbiased single transistor at best can only handle unipolar signals and the high-accuracy unity voltage and unity current gain required for a high-performance current conveyor cannot be obtained. However, the generic relationship between the current conveyor and an ideal transistor is valid, and it provides valuable insight into the development and operation of monolithic current conveyors described in the next section.

Supply-Current Sensing. Many of the current-conveyor theories and applications have been tested out in practice using "breadboard" conveyor circuits, due to the lack of availability of a commercial device. Some researchers have built current conveyors from matched transistor arrays, but the most common way of implementing a fairly high-performance current conveyor has been based on the use of supply-current sensing on a voltage op-amp [3], [4], as shown in Fig. 58.25. The high-resistance op-amp input provides the current-conveyor Y node, while the action of negative feedback provides the low-resistance X node. Current-mirrors in the op-amp supply leads copy the current at node X to node Z.

Using this type of architecture, several interesting features soon became apparent. Consider the two examples shown in Fig. 58.26. In Fig. 58.26(b), R_s represents the output resistance of the current source. The open-loop gain of an op-amp can generally be written

$$\frac{V_{\text{out}}}{V_{\text{in}}} = \frac{A_o}{1 + j(f/f_o)} \tag{58.25}$$

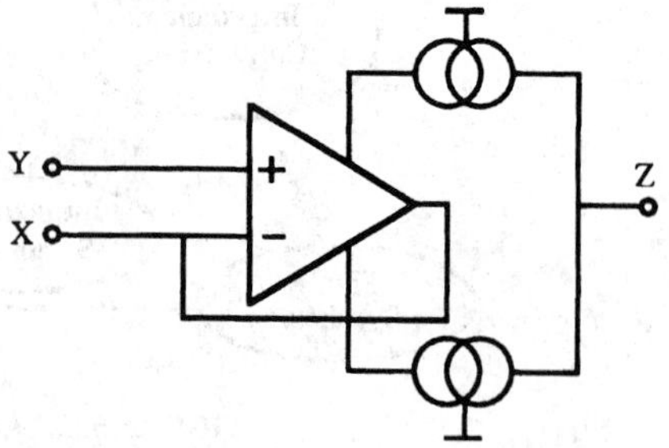

FIGURE 58.25 Supply-current sensing on a voltage op-amp.

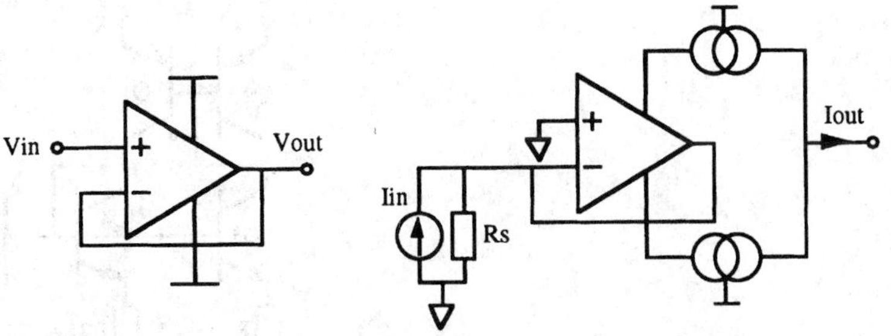

FIGURE 58.26 (a) Voltage follower. (b) Current follower.

where A_o is the open-loop dc gain magnitude and f_o is the open-loop −3 dB bandwidth. Since $A_o \gg 1$, the transfer function of the voltage follower of Fig. 58.26(a) can be written as

$$\frac{V_{\text{out}}}{V_{\text{in}}} \approx \frac{1}{1 + j(f/GB)} \tag{58.26}$$

where $GB = A_o f_o$. From (58.26), the −3 dB bandwidth of the closed-loop voltage follower is equal to the open-loop gain-bandwidth product or GB of the op-amp. If the op-amp is configured instead to give a closed-loop voltage gain K, it is well known that the closed-loop bandwidth correspondingly reduces by the factor K.

The transfer function for the current-follower circuit of Fig. 58.26(b), as shown in [4], is given by

$$\frac{I_{\text{out}}}{I_{\text{in}}} \approx \lambda \frac{1 + j(f/GB)}{1 + j(f/kGB)} \tag{58.27}$$

where λ is the current transfer ratio of the current mirrors and $k = (R_s + r_o/A_o)/(R_s + r_o)$, and r_o represents the output resistance of the op-amp. Since $A_o \gg R_s \gg r_o$, then $k \approx 1$, and the pole and zero in (58.27) almost cancel. The current-follower circuit thus operates well above the gain–bandwidth product GB of the op-amp, and the −3-dB frequency of this circuit will be determined by higher frequency parasitic poles within the current mirrors.

This "extra" bandwidth is achieved because the op-amp is being used with input and output nodes held at virtual ground. The above example is generic in the development of many of the circuits that follow. It demonstrates that reconfiguring a circuit topology to operate with current signals can often result in a superior frequency performance.

First-Generation Current Conveyors. Smith and Sedra's original paper presenting the first-generation CCI current conveyor showed a transistor-level implementation based on discrete devices, shown in Fig. 58.27. Assuming that transistors Q_3–Q_5 and resistors R_1–R_3 are matched, then to first order the currents through these matched components will be equal. Transistors Q_1 and Q_2 are thus forced to have equal currents, and equal V_{be}s. Input nodes X and Y therefore track each other in both voltage and current. In practice there will be slight differences in the collector currents in the different transistors, due to the finite β of the devices. These differences can be reduced, for example, by using more elaborate current mirrors. The polarity of the output current at node Z can be inverted easily by using an additional mirror stage, and the entire circuit can also be inverted by replacing NPN transistors with PNPs, and vice versa. Connecting two complementary current conveyors, as shown in Fig. 58.28 results in a class AB circuit capable of bipolar operation. Note that in practice this circuit may require additional components to guarantee start-up.

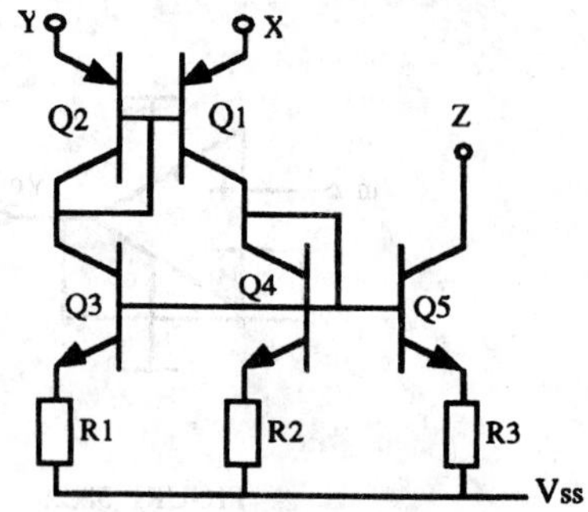

FIGURE 58.27 First-generation current conveyor (CCI).

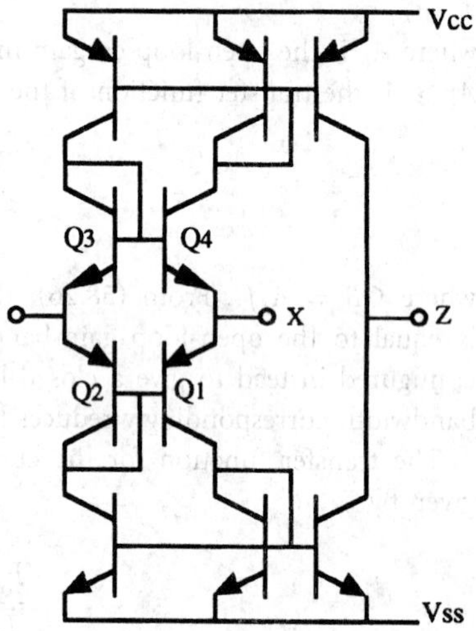

FIGURE 58.28 Class *AB* current conveyor.

An integrated current conveyor based on the architecture shown in Fig. 58.27 is commercially available as the PA630 [5], and the basic topology of this device is shown in Fig. 58.29. An NPN Wilson mirror (Q_1–Q_3) and a PNP Wilson mirror (Q_4–Q_6) are used to provide the current and voltage following properties between inputs X and Y, similar to the circuit of Fig. 58.27. Taking a second output from the PNP current mirror to provide the Z output would destroy the base-current compensation scheme of the Wilson mirror. Therefore, a second NPN Wilson mirror (Q_7–Q_9) is used to perform a current-splitting action and so the combined emitter current of Q_7 and Q_8 is divided in two, with one half being shunted via Q_9 to the supply rail, and the other half driving an output PNP Wilson mirror (Q_{10}–Q_{12}). This results in an output current at node Z that to first order is virtually equal to that at the X and Y inputs. Q_{13} is included to ensure that the device always starts up when turned on. The complete architecture of the PA630 CCI also includes frequency compensation to ensure stability, and modified output current mirrors that use the "wasted" collector current of Q_9 to effectively double the output resistance at node Z. A full description of the architecture and operation of this device can be found in [6].

The current-conveyor architecture shown in Fig. 58.29 includes both NPN and PNP transistors in the signal path, and thus the bandwidth and current-handling capability of this device will be poor if only lateral PNPs are available. The development of complementary bipolar processes, with vertical PNP as well as NPN transistors, has made possible the implementation of high-performance integrated circuit current conveyors.

Second-Generation Current Conveyors. A second generation current conveyor (CCII) can also be simply implemented on a complementary bipolar process, by replacing the diode at the CCI

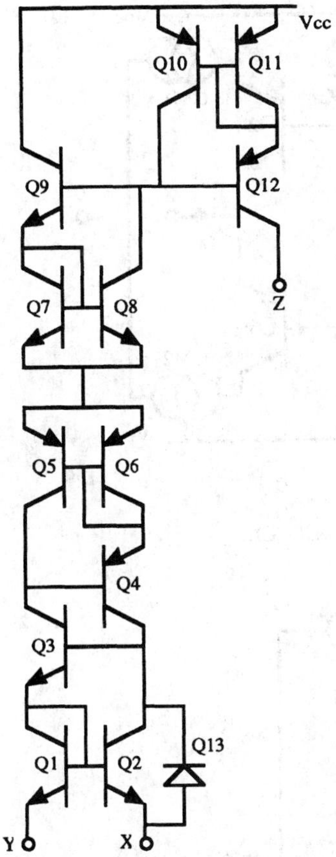

FIGURE 58.29 Simplified PA630 current conveyor (CCI).

Y input with a transistor, and taking the input from the high resistance base terminal, as shown in Fig. 58.30(a). This can be extended to a class AB version, as shown in Fig. 58.30(b). Referring to Fig. 58.30(b), transistors Q_1–Q_4 act as a voltage buffer that transfers the voltage at node Y to node X. The current source and sink ($I_{B1} = I_{B2} = I_B$) provide the quiescent bias current for these input transistors. Any input current (I_x) at node X is split between Q_2 and Q_3, and is copied by current mirrors CM_1 and CM_2 to the output node Z. This CCII architecture forms the basis of the commercially available CCII01 current conveyor [7]. As we shall see later, it is also used as the basic input stage of the current-feedback op-amp, which has emerged as a high-speed alternative to the more conventional voltage op-amp [8].

The simple CCII architecture of Fig. 58.30(b) will clearly exhibit a quiescent voltage offset between nodes X and Y due to the mismatch between the V_{be}s of the NPN and PNP transistors Q_1/Q_2 and Q_3/Q_4, since

$$\begin{aligned} V_Y - V_X &= V_{BE}(p) - V_{BE}(n) \\ &= V_T \ln(I_{sp}/I_{sn}) \end{aligned} \tag{58.28}$$

where I_{sp} and I_{sn} are the reverse saturation currents of the PNP and NPN transistors, respectively, and V_T is the thermal voltage. This process-dependent voltage offset can be reduced by including additional matching diodes in the input stage, as shown in Fig. 58.31. Referring to this diagram,

$$\begin{aligned} V_Y - V_X &= V_{BE}(Q_1) + V_{D_2} - V_{BE}(Q_2) - V_{D_1} \\ V_Y - V_X &= [V_{BE}(Q_1) - V_{D_1}] - [V_{BE}(Q_2) - V_{D_2}] \end{aligned} \tag{58.29}$$

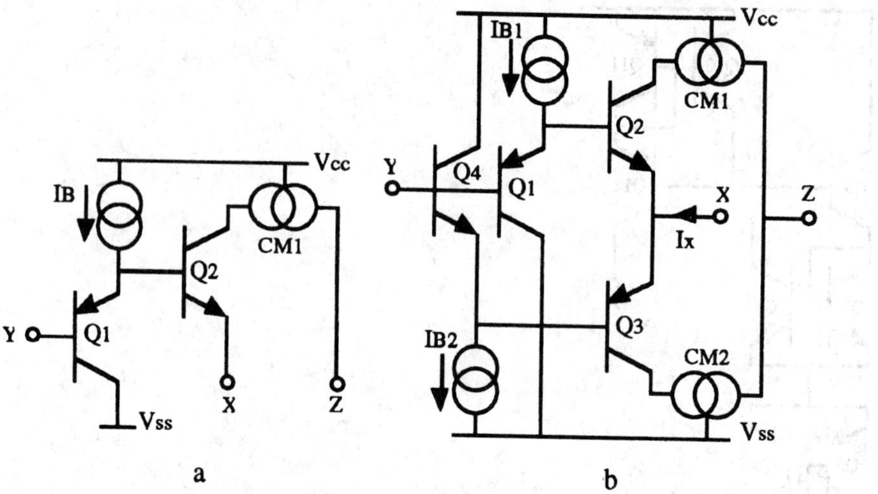

FIGURE 58.30 (a) Class *A* CCII. (b) Class *AB* CCII.

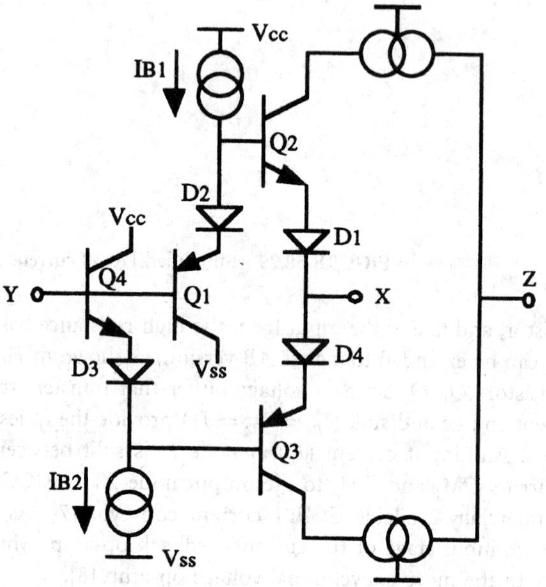

FIGURE 58.31 CCII with input matching diodes.

Inclusion of these diodes clearly reduces the quiescent input voltage offset, provided that D_1 is matched to Q_1, D_2 is matched to Q_2, etc. However, the addition of diodes D_1 and D_2 has several disadvantages. First, the input voltage dynamic range of the circuit will be reduced by the forward voltage across the additional diode. Second, the small-signal input resistance seen looking into node X will be double that for the basic architecture given in Fig. 58.30(b). This nonzero input resistance at node X (R_x) will compromise the performance of the current conveyor, especially in applications where a nonzero input voltage is applied at node Y. The effect of the small-signal input resistance R_x is to produce a signal-dependent voltage offset V_d between nodes X and Y, where

$$V_d = R_x I_x \tag{58.30}$$

Since the value of R_x is determined by the small-signal resistance $(r_{e2} + r_{d2})$ in parallel with $(r_{e3} + r_{d3})$, its value could be reduced by increasing the value of the quiescent bias current I_B. However, an increase in bias current will lead to an increase in the total power consumption, as well as a possible increase in offsets, and so is certainly not an ideal solution. Further techniques for CCII implementation are discussed in [14].

The above conveyor is typical of commercial conveyor architectures [7], which are generally built on a high-speed dielectric isolation (fully complementary) bipolar process. Such devices feature an equivalent slew rate of some 2000V/μs and a bandwidth of around 100 MHz.

Until high-performance current conveyors are widely available, these devices will continue to be used in research laboratories rather than in the applications arena. Process technologies and design techniques have now advanced to the stage where the implementation of an integrated current conveyor is both desirable and viable, and a whole host of applications are waiting for its arrival.

Current-Feedback Operational Amplifier

In this section the design and development of a high-gain wide-bandwidth transimpedance or current-feedback operational amplifier is considered. The design of conventional operational amplifiers has remained relatively unchanged since the introduction of the commercial operational amplifier in 1965. Recently, a new amplifier architecture, called a current-feedback operational amplifier, has been introduced. This amplifier architecture is basically a transimpedance amplifier, or a current-controlled voltage source, while the classical voltage-feedback operational amplifier is a voltage-controlled voltage source.

There are two major advantages of the current-feedback operational amplifier compared to its voltage-feedback counterpart. First, the closed-loop bandwidth of the current-feedback amplifier is larger than that of classical voltage-feedback design for comparable open-loop voltage gain. Second, the current-feedback operational amplifier is able to provide a constant closed-loop bandwidth for closed-loop voltage gains up to about 10. A further advantage of the current-feedback architecture is an almost unlimited slew rate due to the class-*AB* input drive, which does not limit the amount of current available to charge up the compensation capacitor as is the case in the conventional voltage-feedback op-amp. This high-speed performance of the current-feedback operational amplifier is extremely useful for analog signal processing applications within video and telecommunication systems.

The generic relationship between the CCII+ and the current-feedback op-amp is extremely close and several of the features offered by the CCII are also present in the current-feedback op-amp. The basic structure of the current-feedback op-amp is essentially that of a CCII+ with the Z node connected directly to an output voltage follower, as shown in Fig. 58.32. Any current flowing into the low impedance inverting input is conveyed to the gain node (Z_T), and the resulting voltage is buffered to the output. Z_T is thus the open-loop transimpedance gain of the current-feedback op-amp, which in practice is equal to the parallel combination of the CCII+ output impedance, the voltage buffer input impedance and any additional compensation capacitance at the gain node. Generally, in current-feedback op-amps, the gain node is not connected to an external pin, and so the Z node of the CCII+ cannot be accessed.

Current-Feedback Op-Amp Architecture. In the following sections we review the basic theory and design of the current-feedback op-amp and will identify the important features and mechanisms that result in broadband performance. We will begin by reviewing the voltage-feedback op-amp and comparing it with the current-feedback op-amp in order to see the differences clearly.

A schematic of the classical voltage-feedback op-amp comprising a long-tail pair input stage is shown in Fig. 58.33(a), which contrasts a typical current-feedback architecture, which is shown

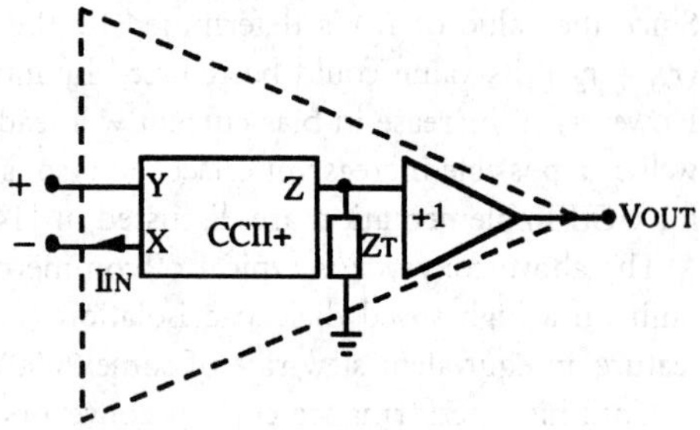

FIGURE 58.32 Current-feedback op-amp structure.

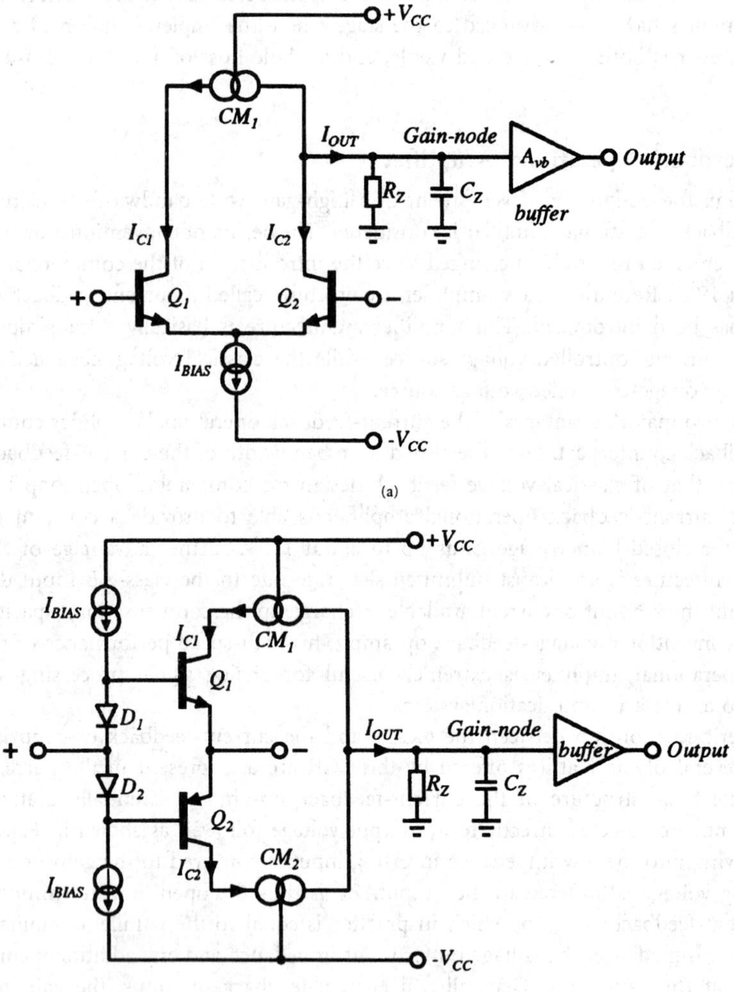

FIGURE 58.33 (a) Simplified classic voltage-feedback op-amp architecture. (b) Typical current-feedback op-amp architecture.

in Fig. 58.33(b). In both circuits, current mirrors are represented by two interlocking circles with an arrow denoting the input side of the mirror.

The current-feedback op-amp of Fig. 58.33(b) shows that the noninverting input is a high-impedance input that is buffered to a low-impedance inverting terminal via a class *AB* complementary common-collector stage (Q_1, Q_2, D_1, D_2). Note that this classical input buffer architecture is used here for simplicity. In practice a higher performance topology such as that described in Fig. 58.31 would more likely be employed. The noninverting input is a voltage input; this voltage is then buffered to the inverting low-impedance current input to which feedback is applied. In contrast, both the noninverting and inverting input of the voltage-feedback op-amp are high-impedance voltage inputs at the bases of transistor Q_1 and Q_2.

In both architectures the collector currents of Q_1 and Q_2 are transferred by the current mirror to a high-impedance node represented by resistance R_Z and capacitance C_Z. This voltage is then transferred to the output by voltage buffers that have a voltage gain A_{vb}, providing the necessary low output impedance for current driving. In the case of the current-feedback op-amp, the output buffer is usually the same topology as the input buffer stage shown in the Fig. 58.33(b), but with slightly higher output current bias levels and larger output devices to provide an adequate output drive capability. Ideally, the bias current I_{CQ1} and I_{CQ2} will be canceled at the gain node giving zero offset current.

Differential-Mode Operation of the Current-Feedback Op-Amp. A schematic diagram of the current-feedback op-amp with a differential input voltage applied at the noninverting and inverting input is shown in Fig. 58.34.

The positive input voltage is applied to the base of transistor Q_1 (NPN) via D_1, and the negative input voltage is applied to the emitter of Q_1, causing the V_{BE} of Q_1 to increase and the V_{BE} of Q_2 to reduce. I_{C1} will therefore increase by an amount ΔI and so I_{C2} will decrease by the same amount $-\Delta I$. A net current of $2\Delta I$ is therefore sourced out of the high-impedance node (Z) giving rise to a positive voltage ($2\Delta IZ$). This voltage is then buffered to the output.

With negative feedback applied around the current-feedback op-amp, the low-impedance inverting input will sense the current "feedback" from the output via the feedback network. This feedback current flowing into the inverting input is given by

$$i_{\text{in}-} = I_{C2} - I_{C1} \tag{58.31}$$

The difference between the collector current I_{C1} and I_{C2}, $i_{\text{in}-}$, will thus be driven into gain node Z, giving rise to the output voltage

$$V_{\text{out}} = Zi_{\text{in}-} \tag{58.32}$$

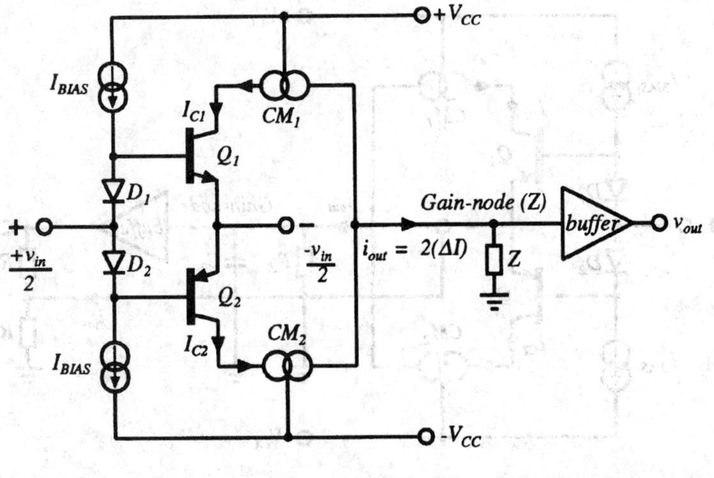

FIGURE 58.34 Current-feedback op-amp with differential input voltage applied.

It is clear that the output voltage is dependent on the current that flows into the inverting input, hence the amplifier has a high open-loop transimpedance gain Z.

Closed-Loop Noninverting Operation of the Current-Feedback Op-Amp. A schematic diagram of the current-feedback op-amp connected with negative feedback as a noninverting amplifier is shown in Fig. 58.35. For a positive input voltage v_{in}, the output voltage v_{out} will swing in the positive direction and the inverting input current i_{in-} will flow out:

$$i_{in-} = \frac{v_{in-}}{R_1} - \frac{(v_{out} - v_{in-})}{R_2} \tag{58.33}$$

The input stage is simply a voltage follower and so ideally, $v_{in+} = v_{in-} = v_{in}$. Because $v_{out} = Zi_{in-}$, then substituting for v_{in-} and i_{in-} in (58.33) yields

$$\frac{v_{out}}{Z} = \frac{v_{in}}{R_1} - \frac{(v_{out} - v_{in})}{R_2} \tag{58.34}$$

rearranging for v_{out}/v_{in}

$$v_{out}\left(\frac{1}{R_2} + \frac{1}{Z}\right) = v_{in}\left(\frac{1}{R_1} + \frac{1}{R_2}\right) \tag{58.35}$$

$$\frac{v_{out}}{v_{in}} = \left(1 + \frac{R_2}{R_1}\right)\left(\frac{1}{1 + \dfrac{R_2}{Z}}\right) \tag{58.36}$$

This result shows that the closed-loop noninverting gain of the current-feedback op-amp is similar to that of a classical voltage-feedback op-amp. From (58.36), the open-loop transimpedance gain Z must be as large as possible to give good closed-loop gain accuracy. Since v_{out}/Z represents the error current i_{in-} then maximizing the Z term will minimize the inverting error current. Note that at this stage it is only the R_2 term in the denominator of the second term in (58.36) that sets the bandwidth of the amplifier; the gain-setting resistor R_1 has no effect on the closed-loop bandwidth.

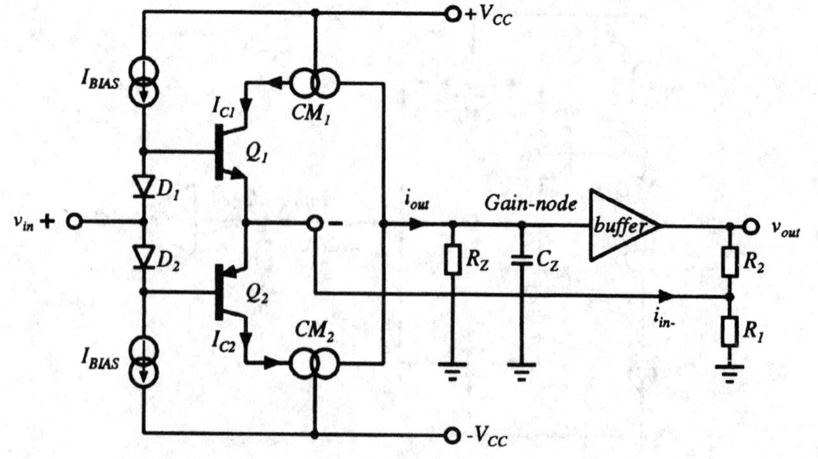

FIGURE 58.35 Noninverting current-feedback op-amp.

Closed-Loop Inverting Operation of Current-Feedback Op-Amp. A current-feedback op-amp connected as an inverting amplifier is shown in Fig. 58.36. The low-impedance inverting input samples the input current and drives the output until the voltage at its terminal is at a virtual ground because of negative feedback. Ideally the closed-loop gain is given by

$$A_{CL} = -\frac{R_2}{R_1} \tag{58.37}$$

From Fig. 58.36, application of Kirchhoff current laws to the current i_1, $i_{\text{in}-}$, and i_2 gives

$$i_{\text{in}-} + i_2 = i_1$$

$$i_{\text{in}-} - \frac{v_{\text{out}}}{R_2} = \frac{v_{\text{in}}}{R_1}$$

because $v_{\text{out}}/Z = -i_{\text{in}-}$, then

$$-\frac{v_{\text{out}}}{Z} - \frac{v_{\text{out}}}{R_2} = \frac{v_{\text{in}}}{R_1}$$

which can be rearranged as

$$\frac{v_{\text{out}}}{v_{\text{in}}} = -\frac{R_2}{R_1}\left(\frac{1}{1+\frac{R_2}{Z}}\right) \tag{58.38}$$

Again, the high-Z term is required to provide good closed-loop gain accuracy.

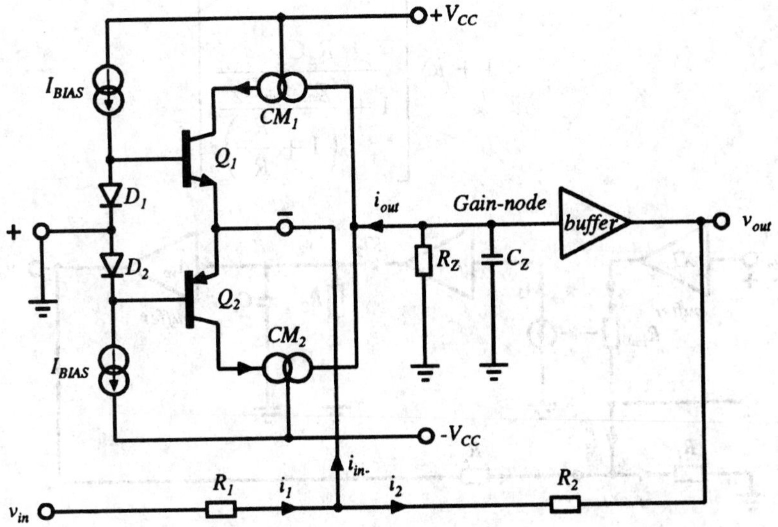

FIGURE 58.36 Inverting current-feedback op-amp amplifier.

More Detailed Analysis of the Current-Feedback Op-Amp. A simplified macromodel of the current-feedback architecture configured as a noninverting amplifier is shown in Fig. 58.37. The input stage is represented by a semi-ideal voltage buffer to the inverting input. The output resistance of the input stage buffer R_{inv} is included since it has a significant effect on the bandwidth of the amplifier, as will be shown later. The current that flows out from the inverting terminal i_3 is transferred to the gain node, which is represented by R_Z and C_Z, via a current mirror that has a current gain K. The voltage at the gain node is transferred to the output in the usual way by a voltage buffer, with voltage gain A_{vb}. The net transfer function is given by

$$\frac{v_{\text{out}}}{v_{\text{in}}} = \frac{1+\dfrac{R_2}{R_1}}{1+j\omega C_Z\left[\dfrac{R_{\text{inv}}\left(1+\dfrac{R_2}{R_1}\right)+R_2}{A_{vb}K}\right]} \tag{58.39}$$

Hence the pole frequency is also given by

$$f_{-3\,\text{dB}} = \frac{A_{vb}K}{2\pi C_Z\left[R_{\text{inv}}\left(1+\dfrac{R_2}{R_1}\right)+R_2\right]} \tag{58.40}$$

(A full derivation of this transfer function is given in Appendix A.)

To compare this result to the classical voltage-mode op-amp architecture, a simplified schematic diagram of the voltage-feedback op-amp configured as a noninverting amplifier is shown in Fig. 58.38.

Again from a full analysis, given in Appendix B, the transfer function obtained is

$$\frac{v_{\text{out}}}{v_{\text{in}}} = \frac{1+\dfrac{R_2}{R_1}}{\left[1+j\omega\dfrac{\dfrac{R_z C_Z}{g_m A_{vb} R_Z}}{1+\dfrac{1}{\left(1+\dfrac{R_2}{R_1}\right)}}\right]} \tag{58.41}$$

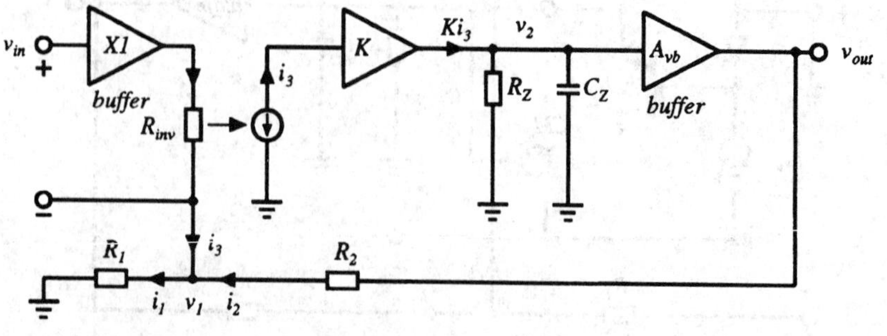

FIGURE 58.37 Inverting amplifier with current-feedback op-amp macromodel.

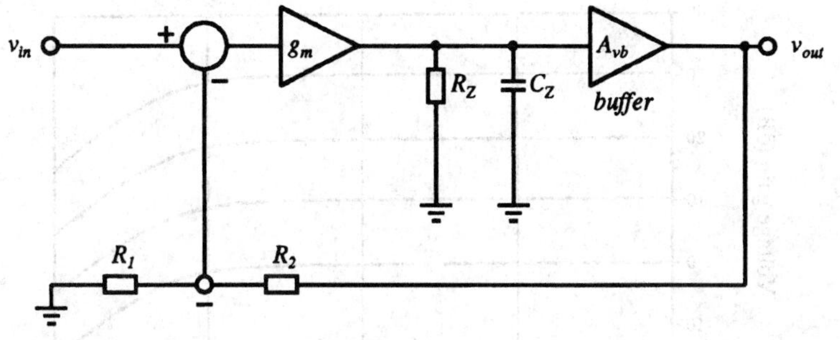

FIGURE 58.38 Noninverting amplifier with voltage-feedback op-amp macromodel.

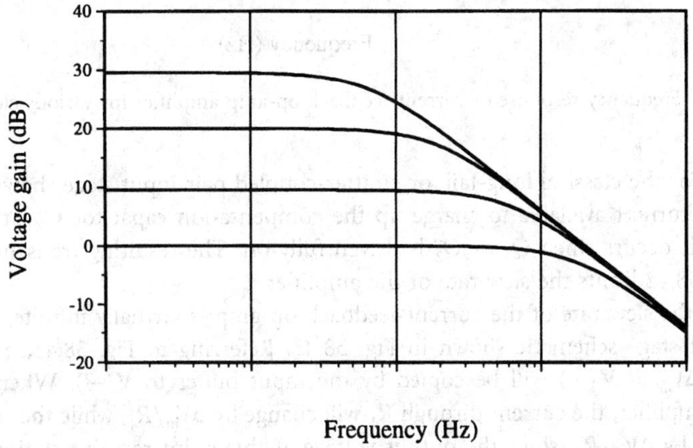

FIGURE 58.39 Frequency response of voltage-feedback op-amp amplifier for various closed-loop gains.

The pole frequency is given by

$$f_{-3\,\mathrm{dB}} = \frac{1 + \dfrac{g_m A_{vb} R_Z}{\left(1 + \dfrac{R_2}{R_1}\right)}}{2\pi R_Z C_Z} \tag{58.42}$$

Pole Frequency Comparison. If one compares the closed-loop pole frequency Eqs. (58.40) and (58.42) for the current-feedback and voltage-feedback op-amp, respectively, it is clear that the bandwidth of the voltage-feedback op-amp is dependent on the closed-loop gain $(1 + R_2/R_1)$ resulting in the well-known constant gain–bandwidth product $f_{\mathrm{MAX}} = (A_v)_{CL} f_T$. This means that an increase in the closed-loop gain results in a decrease in the bandwidth by the same factor as illustrated in Fig. 58.39. In contrast, the pole frequency of the current-feedback op-amp is directly dependent on R_2 and can be set almost independently of the closed-loop gain. Thus the closed-loop bandwidth is almost independent of closed-loop gain as shown in Fig. 58.40, assuming that R_{inv} is close to zero. Intuitively this is the case since the feedback error current that is set by the feedback resistor R_2 is the current available to charge up the compensation capacitor. However, if one considers (58.40) in some detail it can be seen that for high closed-loop gains and a nonzero R_{inv}, then the R_{inv} term starts to dictate and so the bandwidth will become more dependent on the closed-loop gain.

Slew Rate of the Current-Feedback Op-Amp. As mentioned earlier, one other advantage of the current-feedback op-amp over the classical voltage-feedback op-amp is the high slew rate

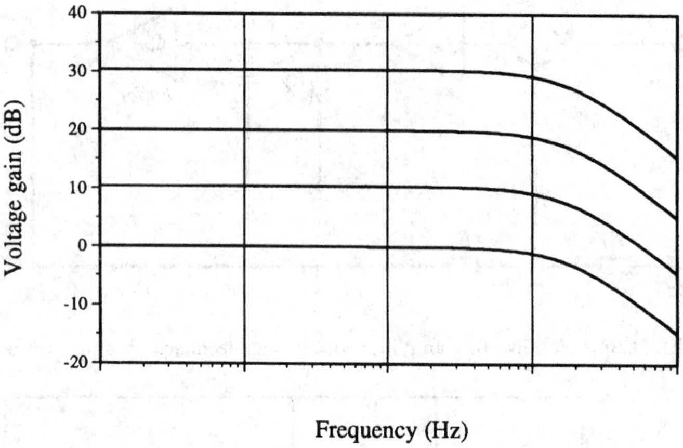

FIGURE 58.40 Frequency response of current-feedback op-amp amplifier for various closed-loop gains.

performance. For the classical long-tail, or emitter-coupled pair input stage shown in Fig. 58.41, the maximum current available to charge up the compensation capacitor C_Z at the gain node is I_{BIAS}, and this occurs when Q_1 or Q_2 is driven fully on. The resulting transconductance plot shown in Fig. 58.42 limits the slew rate of the amplifier.

In contrast, the slew rate of the current-feedback op-amp is virtually infinite, as can be seen from the input stage schematic shown in Fig. 58.43. Referring to Fig. 58.43, a change in the input voltage ΔV_{in} at $V(+)$ will be copied by the input buffer to $V(-)$. When connected as noninverting amplifier, the current through R_1 will change by $\Delta V_{in}/R_1$, while the current through R_2 will change by $\Delta V_{in}/R_2$, since the output voltage at this point remains stationary. The total change in current through R_1 and R_2 must be supplied by the internal input buffer, and will be $\Delta I(-) = \Delta V_{in}((R_2 + R_1)/(R_2 \cdot R_1))$. This large input error current causes a rapid change in the output voltage, until V_{out} is again at the value required to balance the circuit once more, and reduce $I(-)$ to zero. The larger the input voltage slew rate, the larger the change in input error current, and thus the faster the output voltage slew rate. Current-feedback op-amps theoretically have no slew-rate limit. A typical current-feedback op-amp will exhibit a slew rate of between 500 and 2000 V/μS.

An analysis of this input stage (see Appendix C) shows that the transconductance follows a sinh(x) type function, as shown in Fig. 58.44. In theory, this characteristic provides nearly

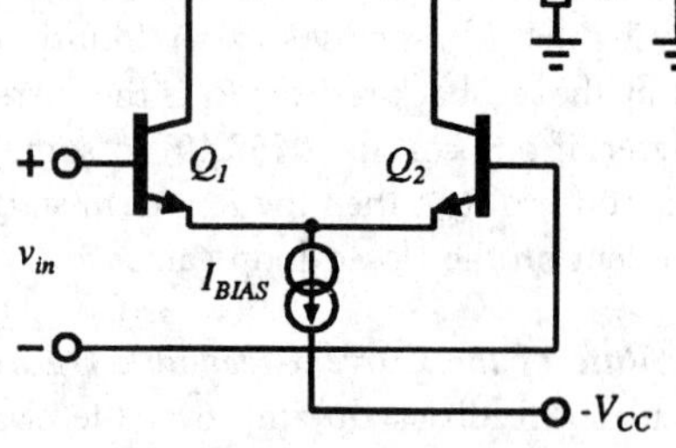

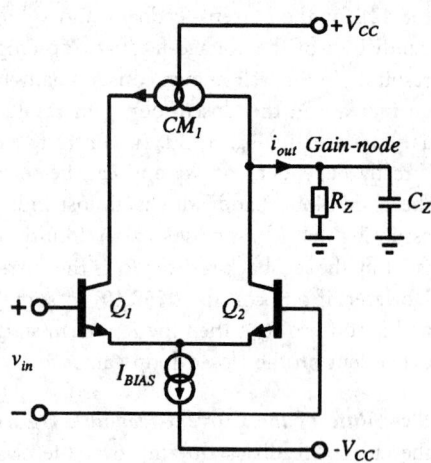

FIGURE 58.41 Long-tail pair input stage.

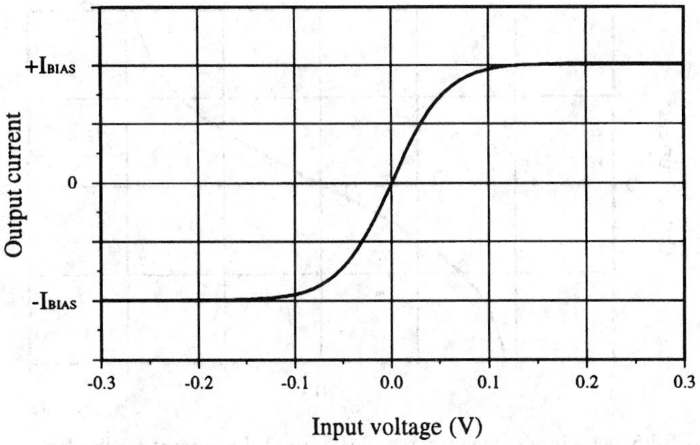

FIGURE 58.42 Long-tail pair input transconductance.

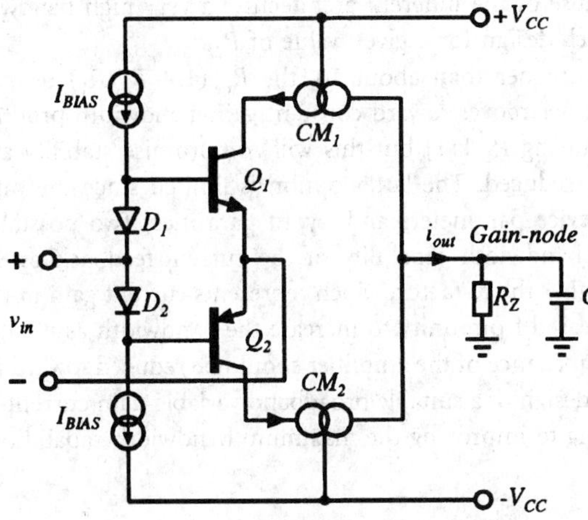

FIGURE 58.43 Current-feedback op-amp input stage.

unlimited slew-rate capability [9]. However, in practice a maximum slew rate will be limited by the maximum current drive into the gain node, which depends on the power dissipation of the circuit, the ability of power supply to deliver sufficient current, and the current-handling capability of the current mirrors.

Wideband and High-Gain Current-Feedback Op-Amp. Previously, we have shown that the bandwidth of the current-feedback op-amp is almost independent of the closed-loop gain setting. Therefore, the closed-loop gain–bandwidth product *GB* increases linearly with the closed-loop gain. However, the bandwidth of the practical current-feedback op-amp starts decreasing with high gain as a result of the finite inverting-input impedance [10], as shown by (58.40). This is because for high gain, $R_{inv}(1 + R_2/R_1) > R_2$, and so the $R_{inv}(1 + R_2/R_1)$ term dominates the expression for closed-loop bandwidth, resulting in a direct conflict between gain and bandwidth.

At low gains when $R_2 > R_{inv}(1 + R_2/R_1)$, the closed-loop pole frequency is determined only by the compensation capacitor and the feedback resistor R_2. Thus the absolute value of the feedback resistor R_2 is important, unlike the case of the voltage-feedback op-amp. Usually, the manufacturer specifies a minimum value of R_2 that will maximize bandwidth but still ensure

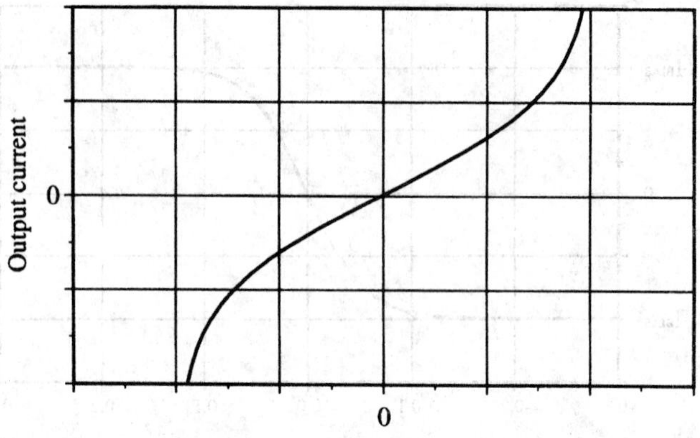

FIGURE 58.44 Input-stage transconductance of the current-feedback op-amp.

stability. Note that because of the inherent architecture a very high bandwidth can be achieved with the current-feedback design for a given value of R_2.

In practice, for gains higher than about 10, the $R_{inv}(1 + R_2/R_1)$ term in (58.40) becomes dominant and the amplifier moves toward constant gain–bandwidth product behavior. The *GB* can be increased by reducing R_2 [11] but this will compromise stability and/or bandwidth, or alternatively, C_Z can be reduced. The latter option is limited since the minimum value of C_Z is determined by the device parameters and layout parasitics. Two possible ways of improving the high-gain constant bandwidth capability of the current-feedback op-amp can be seen by inspection of (58.40). Either the K factor, which represents current gain in the current mirrors at the Z-node can be increased from unity to increase the bandwidth as it rolls off with high gain, or the inverting input impedance of the amplifier should be reduced toward zero. In the following section we consider the design of a suitable broadband variable-gain current-mirror circuit with a possible application being to improving the maximum bandwidth capability of current-feedback op-amps.

Basic Current Mirror. A typical current-feedback op-amp circuit is shown in Fig. 58.45. It includes a complementary common-collector input stage (Q_1–Q_4) and a similar output buffer (Q_5–Q_8), with linking cascode current mirrors setting the Z-node impedance (Q_{12}–Q_{14}, Q_9–Q_{11}). The cascoded mirror provides unity current gain. Any attempt to increase the current gain via emitter degeneration usually results in much poorer current-mirror bandwidth. Consider now the development of a suitable broadband, variable gain current mirror.

A schematic diagram of a simple Widlar current mirror and its small-signal equivalent circuit are shown in Figs. 58.46 and 58.47, respectively. For simplicity, we will assume that the impedance of the diode-connected transistor Q_1 is resistive and equal to R_D. The dc transfer function of the mirror is derived in Appendix D and is given by

$$\frac{I_{out}}{I_{in}} = \frac{\beta}{\beta + 2} \tag{58.43}$$

and the −3-dB bandwidth is given by

$$f_{-3\,dB} = \frac{1}{2\pi C_\pi \left\{ \dfrac{r_{\pi 2}(r_{bb2} + R_D)}{r_\pi + r_{bb2} + R_D} \right\}} \tag{58.44}$$

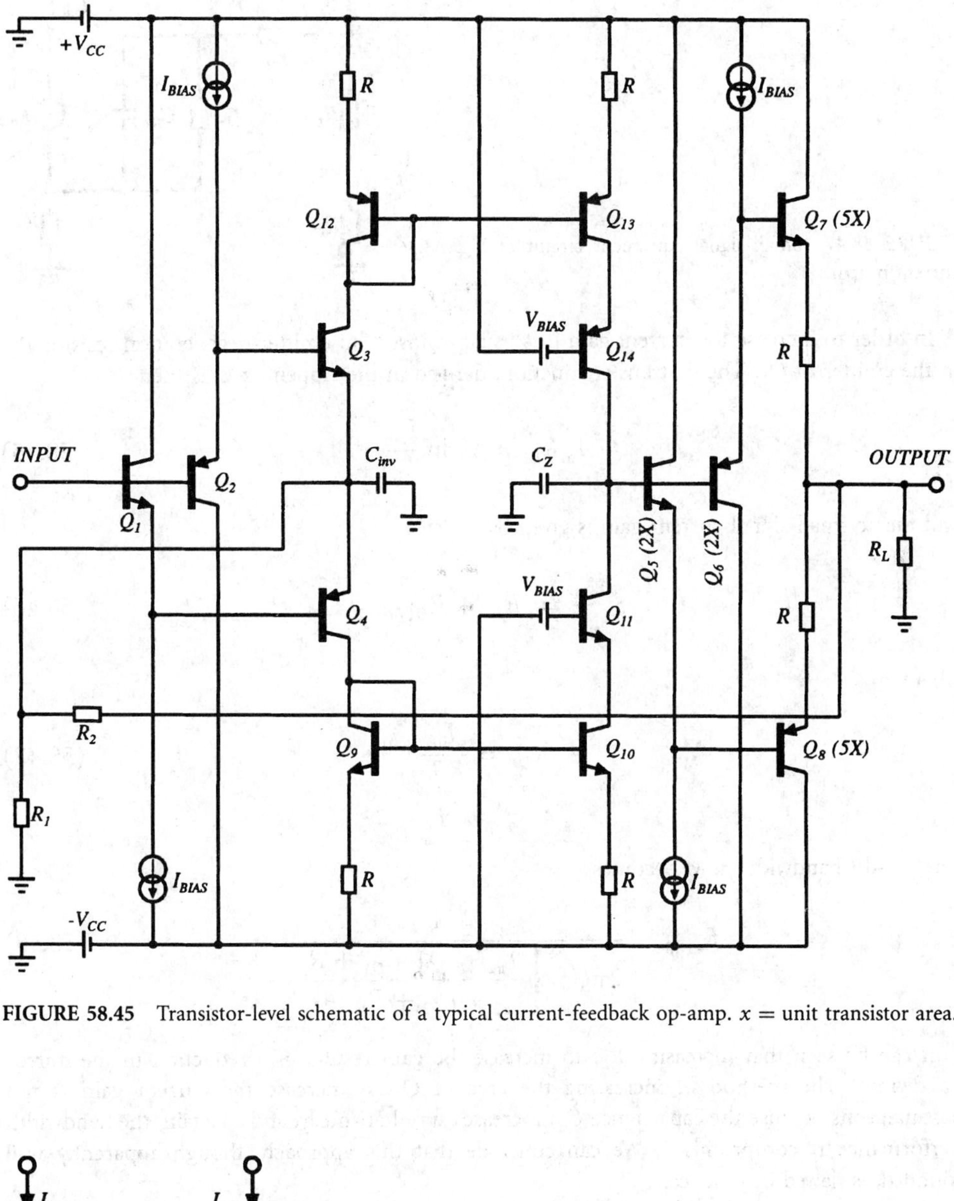

FIGURE 58.45 Transistor-level schematic of a typical current-feedback op-amp. x = unit transistor area.

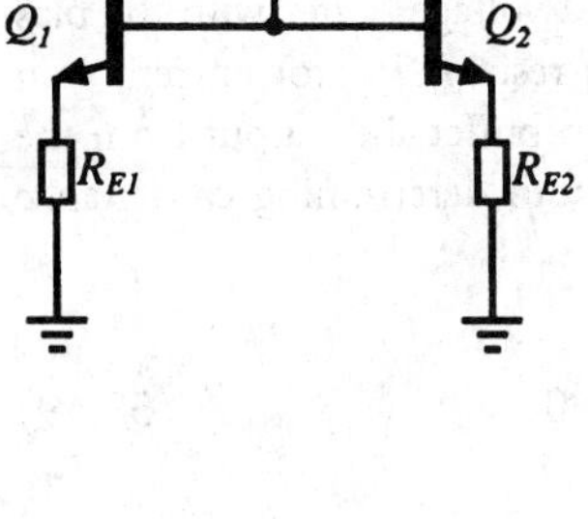

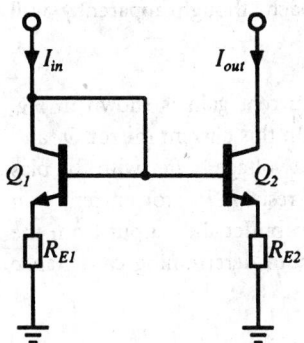

FIGURE 58.46 Simple Widlar current mirror with emitter degeneration.

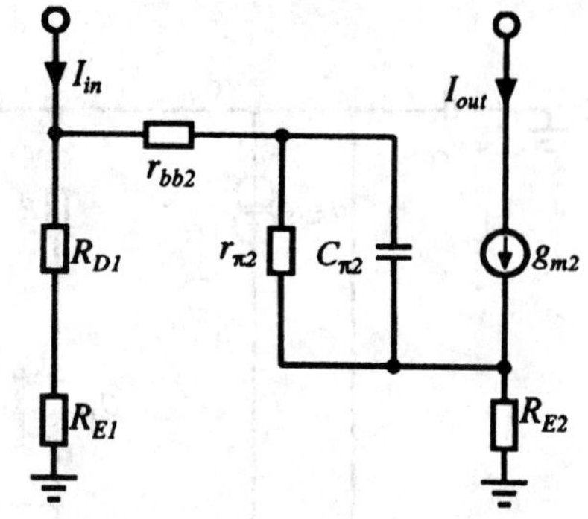

FIGURE 58.47 Small-signal equivalent circuit of Fig. 58.46 current mirror.

In order to increase the current gain it is usual to insert an emitter-degeneration resistor R_{E1} in the emitter of Q_1. The dc transfer function, derived in the Appendix E, is then

$$I_{\text{in}} R_{E1} = V_T \ln \frac{I_{\text{out}}}{I_{\text{in}}} \tag{58.45}$$

and the ac small-signal current gain is given by

$$\frac{i_{\text{out}}}{i_{\text{in}}} = (R_{E1} + R_{D1}) g_{m2} \tag{58.46}$$

where

$$R_{D1} = \frac{I_{\text{in}}}{\dfrac{KT}{q}} \tag{58.47}$$

The −3-dB bandwidth now becomes

$$f_{-3\,\text{dB}} = \frac{1}{2\pi C_{\pi 2} \left\{ \dfrac{r_{\pi 2}(r_{bb2} + R_{D1} + R_{E1})}{r_{\pi 2} + r_{bb2} + R_{D1} + R_{E1}} \right\}} \tag{58.48}$$

It can be seen that increasing R_{E1} to increase the gain results in a reduction in the mirror bandwidth. The method of increasing the area of Q_2 to increase the current gain is not advantageous because the capacitance $C_{\pi 2}$ increases simultaneously, and so again, the bandwidth performance is compromised. We can conclude that this approach, though apparently well founded, is flawed in practice.

Improved Broadband Current Mirror. A current mirror with current gain is shown in Fig. 58.48 and the small-signal equivalent circuit is shown in Fig. 58.49. In this current mirror Q_1 and Q_2 are connected as diodes in series with R_{E1}. Q_3 is connected as a voltage buffer with the bias current source I_{EQ3}. Q_4 is the output transistor with degeneration resistor R_{E4} for current gain setting. The basic idea is to introduce the common collector Q_3 to buffer the output from the input and hence isolate gain setting resistor R_{E4} from the bandwidth determining capacitance of the input. The dc transfer function is given by

$$I_{\text{in}} R_{E1} - I_{\text{out}} R_{E4} + V_T \ln \frac{I_{\text{in}}^2}{I_{CQ3} I_{\text{out}}} = 0 \tag{58.49}$$

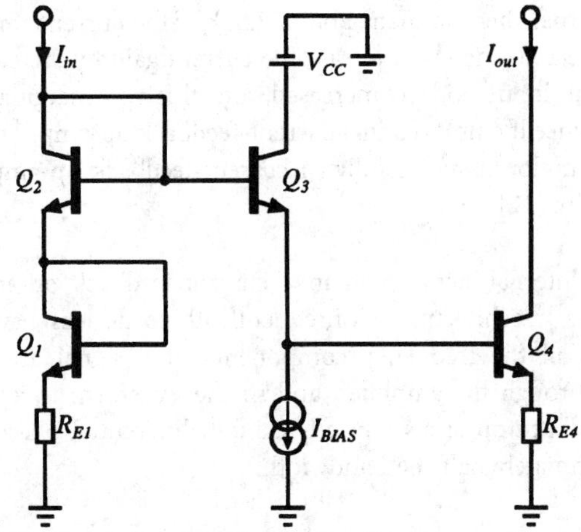

FIGURE 58.48 Improved current mirror with current gain.

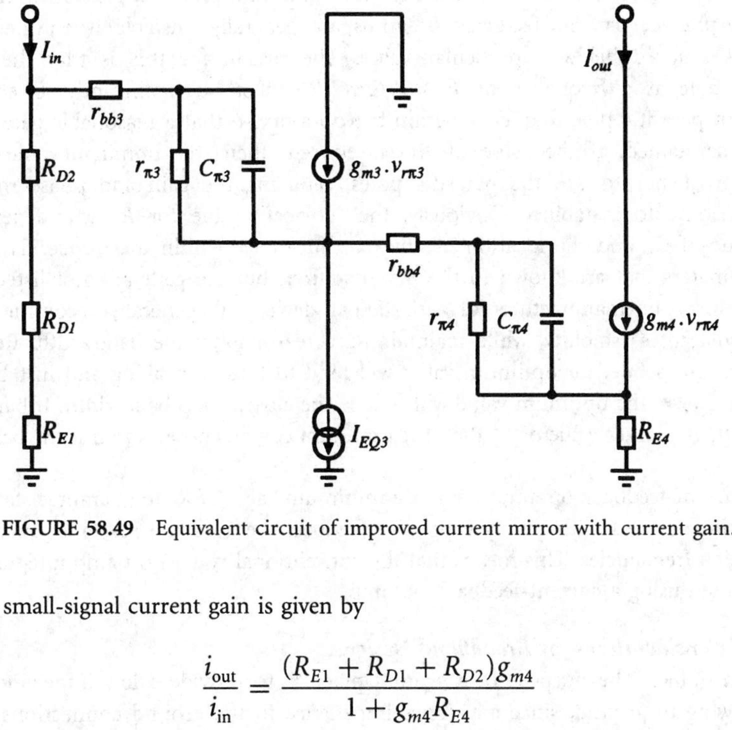

FIGURE 58.49 Equivalent circuit of improved current mirror with current gain.

and the ac small-signal current gain is given by

$$\frac{i_{\text{out}}}{i_{\text{in}}} = \frac{(R_{E1} + R_{D1} + R_{D2})g_{m4}}{1 + g_{m4}R_{E4}} \tag{58.50}$$

and the −3-dB bandwidth now becomes

$$f_{-3\,\text{dB}} = \frac{1}{2\pi C_{\pi4}\left(\dfrac{r_{\pi4}R_x}{r_{\pi4} + R_x}\right)} \tag{58.51}$$

where

$$R_x = r_{bb4} + \frac{r_{\pi3} + r_{bb3} + R_{D1} + R_{D2} + R_{E1}}{\beta_3} \tag{58.52}$$

It can be seen clearly that the dominant pole (58.51) of the current mirror with current gain is now only slightly decreased when we increase the current gain by increasing R_{E1}. However the nondominant pole at the input node is increased, and this will marginally effect the resultant overall stability performance if employed in a current-feedback op-amp. This current mirror with current gain has been employed successfully in current-feedback op-amp design for increased gain--bandwidth capability [12].

Phase Linearity. The internal signal path in a current-feedback op-amp is very linear due largely to the symmetrical architecture. Consequently these devices have a very linear phase response. Furthermore, all the frequency components of a signal are delayed by the same amount when passing through the amplifier, and so the waveform is reproduced accurately at the output. Current-feedback op-amps typically exhibit differential phase error of around $\pm 1°$ at frequencies of approximately half the bandwidth.

Choosing the Value of R_2. From (58.40), we can see that for a fixed value of C_z, a smaller feedback resistor R_2 will give a higher closed-loop bandwidth. It might be expected that the maximum bandwidth would be obtained with the minimum feedback resistance; that is, with $R_2 = 0$. In practice, current-feedback op-amps are generally unstable when their feedback resistance is reduced below a particular value. The reason for this is that the dominant closed-loop pole at a frequency of $f \approx 1/2\pi C_z R_2$ must be significantly lower than any nondominant parasitic pole frequency within the op-amp, so that a reasonable gain and phase margin is maintained. If the value of R_2 is reduced, then this dominant pole will move upward in frequency toward the parasitic poles, reducing the gain and phase margin, and eventually leading to instability. Obviously, the "correct" value for R_2 will depend on the internal value of C_z and the location of any parasitic poles within the device. These are the sort of parameters that are known to the manufacturer, but are generally not listed in a data sheet. Therefore, the manufacturer of a particular device will generally recommend a value of R_2 that guarantees stability, while maintaining a reasonably wide bandwidth. Reducing R_2 below this recommended or optimum value will tend to lead to peaking and instability, while increasing R_2 above the optimum value will reduce the closed-loop bandwidth. If band limiting is required, then a larger value of R_2 than the optimum can be chosen to limit the bandwidth as required.

Since a current-feedback op-amp requires a minimum value of R_2 to guarantee stability, these devices cannot be used with purely capacitive feedback because the reactance of a capacitor reduces at high frequencies. This means that the conventional voltage op-amp integrator cannot be implemented using a current-feedback op-amp.

Practical Considerations for Broadband Designs.

1. *Ground Planes.* The purpose of a ground plane is to provide a low impedance path for currents flowing to ground, since any series impedance in the ground connections will mean that not all ground nodes are at the same potential. In addition, the inductance of a printed circuit track is approximately inversely proportional to the track width, and so the use of thin tracks can result in inductive ground loops, leading to ringing or even oscillations. The use of an unbroken ground plane on one side of the circuit board can minimize the likelihood of inductive loops within the circuit. However, any particularly sensitive ground-connected nodes in the circuit should be grounded as physically close together as is possible.

2. *Bypass Capacitors.* Power supply lines often have significant parasitic inductance and resistance. Large transient load currents can therefore result in voltage spikes on the power supply lines, which can couple onto the signal path within the device. Bypass capacitors are therefore used to lower the impedance of the power supply lines at the point of load, and thus short out the effect of the supply line parasitics. The type of bypass capacitor

to use is determined by the application and frequency range of interest. High-speed op-amps work best when their power supply pins are decoupled with radio-frequency-quality capacitors.

Manufacturers often recommend using a composite large–small parallel bypass capacitor with something like a 4.7-uF tantalum capacitor on all supply pins, with a parallel 100-nF ceramic to ensure good capacitive integrity at higher frequencies, where the tantalum becomes inductive. However, a note of caution here: this large–small double capacitor technique relies on the large capacitor having sufficiently high ESR so that at resonance the two capacitors do not create a high-*Q* parallel filter. In surface-mount designs a single bypass capacitor may well be better than two due to the inherent high-*Q* of surface-mount capacitors.

All bypass capacitor connections should be minimized, since track lengths will simply add more series inductance and resistance to the bypass path. The capacitor should be positioned right next to the power supply pin, with the other lead connected directly to the ground plane.

3. *Sensitive Nodes.* Certain nodes within a high-frequency circuit are often sensitive to parasitic components. A current-feedback op-amp, for example, is particularly sensitive to parasitic capacitance at the inverting input, since any capacitance at this point combines with the effective resistance at that node to form a second nondominant pole in the feedback loop. The net result of this additional pole is a reduced phase margin, leading to peaking and even instability. Clearly, great care must be taken during layout to reduce track lengths, etc., at this node. In addition, the stray capacitance to ground at $V(-)$ can be reduced by putting a void area in the ground plane at this point. If the op-amp is used as an inverting amplifier, then the potential of the inverting input is held at virtual ground, and any parasitic capacitance will have less effect. Consequently, the current-feedback op-amp is more stable when used in the inverting rather than the noninverting configuration.

4. *Unwanted Oscillations.* Following the guidelines given above should ensure that your circuit is well behaved. If oscillations still occur, a likely source is unintentional positive feedback due to poor layout. Output signal paths and other tracks should be kept well away from the amplifier inputs to minimize signal coupling back into the amplifier. Input track lengths should also be kept as short as possible for this same reason.

Broadband Amplifier Stability

Operational amplifiers are generally designed with additional on-chip frequency compensation capacitance in place. This is done to present the applications engineer with an op-amp that is simple to use in negative feedback, with minimal chance of unstable operation. In theory all will be well, but there are three main reasons that op-amps become unstable in the real world of analog electronic circuit design. This section outlines the three main causes for unstable operation of broadband amplifiers and shows practical ways of avoiding these pitfalls.

Op-Amp Internal Compensation Strategy

Before dealing with specific stability problems in broadband amplifiers and how to solve them, we will look briefly at the internal frequency compensation strategy used in op-amp design. Generally, op-amps can be classified into two groups, those with two high-voltage gain stages and those with only one stage. The two-stage design provides higher open-loop gain but relatively low bandwidth, while the higher speed single-stage amplifier provides lower open-loop gain but much higher usable bandwidth. Insight into the internal op-amp architecture and the type of compensation used will give the designer valuable information on how to tame the unstable op-amp.

Review of the Classical Feedback System

Analyzing the classical feedback system in Fig. 58.50 gives the well-known expression for the closed-loop gain, A_c:

$$A_c = A/[1 + B \cdot A] \tag{58.53}$$

where A is the open-loop gain of the amplifier and B the feedback fraction. $T = B \cdot A$ is referred to as the loop-gain, and the behavior of T over frequency is a key parameter in feedback system design. Clearly, if $T \gg 1$ or $A \gg A_c$, then the closed-loop gain is virtually independent of the open-loop gain A, thus

$$A_c \approx B^{-1} \tag{58.54}$$

This is the most important and desirable feature of negative feedback systems. However, the system will not necessarily be stable since, at higher frequencies, phase lag in the open-loop gain A may cause the feedback to become positive.

Stability Criteria

Though negative feedback is desirable, it results in potential instability when the feedback becomes positive. The loop-gain T is the best parameter to test whether an amplifier is potentially unstable. The phase margin Φ_M is a common feature of merit used to indicate how far the amplifier is from becoming an oscillator:

$$\Phi_M = 180^\circ + \Phi(|BA| = 1) \tag{58.55}$$

When $\Phi_M = 0^\circ$, the phase of the loop gain, $T = B \cdot A$ is exactly -180° for $|B \cdot A| = 1$. The closed-loop gain A_c will become infinite and we have got an oscillator! Clearly, what is required is that $\Phi_M > 0$ and generally the target is to make $\Phi_M \geq 45^\circ$ for reasonably stable performance. However, excessive Φ_M is undesirable if settling time is an important parameter in a particular application.

An op-amp is a general purpose part and so the IC designer strives to produce a maximally versatile amplifier by ensuring that even with 100 percent feedback, the amplifier circuit will not become unstable. This is done by maintaining a $\Phi_M > 0$ for 100 percent feedback, that is, when $B = 1$. If the feedback network B is taken to be purely resistive, then any additional phase lag in the loop gain must come from the open-loop amplifier A. Tailoring the phase response of A so that the phase lag is less than 180° up to the point at which $|A| < 1$ or 0 dB, ensures that the amplifier is "unconditionally stable"; that is, with any amount of resistive feedback, stable operation is "guaranteed."

Most open-loop op-amps, whether single-stage or two-stage, will exhibit a two-pole response. The separation of these two poles whether at low frequency or high frequency will have a major effect on the stability of the system and it is the op-amp designer's objective to locate these

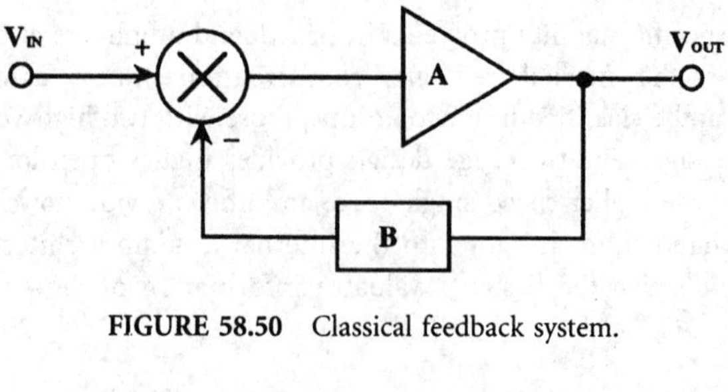

FIGURE 58.50 Classical feedback system.

open-loop poles to best advantage to achieve maximum bandwidth, consistent with versatile and stable performance.

Two-Stage Op-Amp Architecture

A schematic of the standard two-stage op-amp topology is shown in Fig. 58.51. The input differential pair T_1/T_2 provides high gain, as does the second gain stage of T_3/T_4 Darlington pair common emitter. A high-voltage gain is achieved with this structure, so the output stage is usually a unity voltage gain common-collector output buffer to provide a useful load current drive capability.

The amplifier structure in Fig. 58.51 has two internal high-impedance nodes, node X and node Y. These high impedance nodes are responsible for introducing two dominant poles into the frequency response and their relative location is critical in determining the stability of the amplifier. Each pole contributes a low-pass filter function to the open-loop gain expression of the form

$$[1 + jf/f_P]^{-1} \tag{58.56}$$

Each pole introduces 45° of phase lag at the pole frequency f_P and an additional 45° at $f \approx 10 \cdot f_P$. With a two-pole amplifier the open-loop gain A is given by

$$A = A_0/[1 + jf/f_{P1}][1 + jf/f_{P2}] \tag{58.57}$$

where A_0 is the dc open-loop gain and f_{P1} and f_{P2} are the two-pole frequencies. A typical plot of A versus f is shown in Fig. 58.52(a). At low frequencies, where $f \ll f_{P1}$ the gain is flat, and at f_{P1} the gain begins to fall at a rate increasing to -20 dB/decade. The roll-off steepens again at f_{P2} to a final gradient of -40 dB/decade.

It is generally the case that $f_{P1} \ll f_{P2}$ as shown in Fig. 58.52(a). Turning our attention to the phase plot in Fig. 58.52(a), at $f = f_{P1}$ the output lags the input by 45°, and as the frequency rises toward f_{P2} the phase lag increases through 135° at f_{P2} to 180° at $f \approx 10 \cdot f_{P2}$. To ensure unconditionally stable performance, the second pole must be sufficiently far from the first so that the phase margin is large enough.

Figure 58.53 shows curves of the dc value of open-loop gain A_0 versus the ratio N of the pole frequencies ($N = f_{P2}/f_{P1}$) for different values of phase margin. For a given value of $A_0 = 1000$ or +60 dB, the ratio of the pole frequencies must be $N \approx 700$ to obtain a phase margin of 45°.

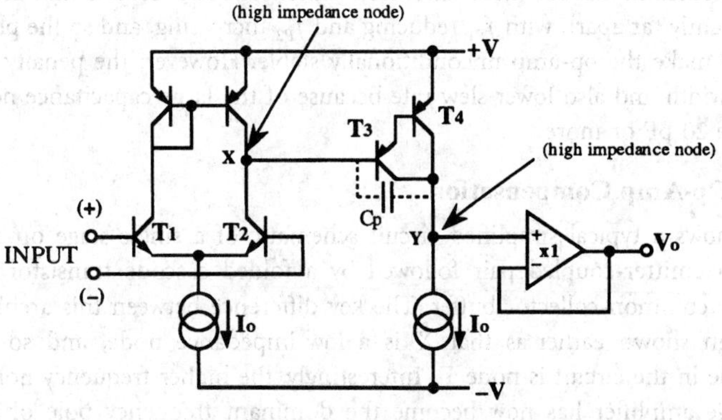

FIGURE 58.51 Architecture of the standard two-stage op-amp.

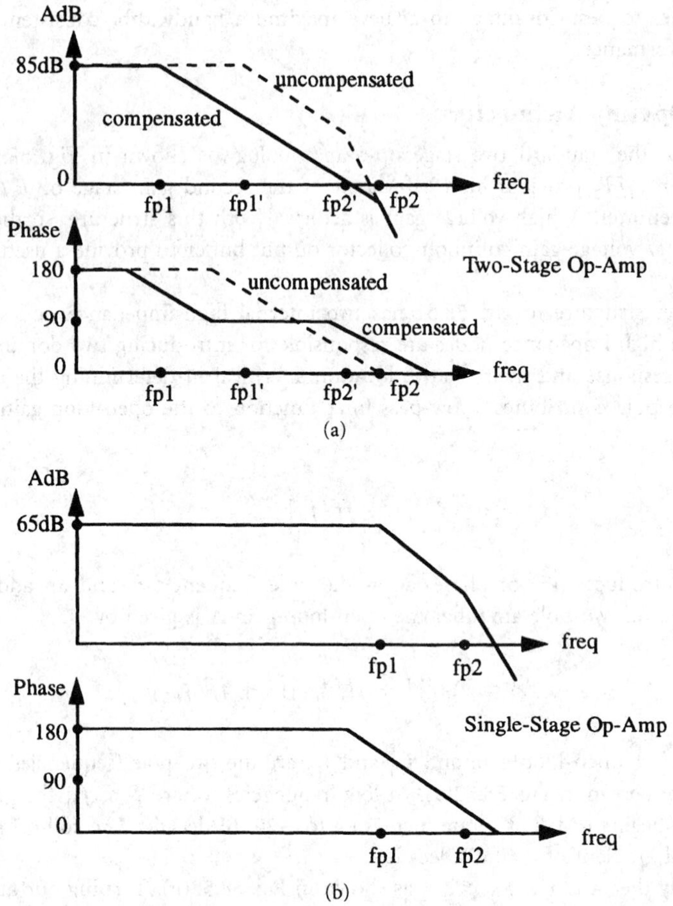

FIGURE 58.52 Pole frequency and phase response for (a) two-stage op-amp and (b) single-stage op-amp.

Miller Compensation and Pole Separation

Without any added compensation capacitance the two open-loop poles of the op-amp are invariably too close to make the amplifier unconditionally stable. The most common compensation method is to add a capacitor between the base and collector of the Darlington pair, shown as C_P in Fig. 58.51. This is known as Miller compensation because this strategy makes use of the Miller capacitance multiplication effect discussed earlier. The net result is that the two poles now become significantly far apart, with f_{P1} reducing and f_{P2} increasing, and so the phase margin can be increased to make the op-amp unconditionally stable. However, the penalty of this method is poorer bandwidth and also lower slew rate because of the large capacitance needed, which in practice may be 20 pF or more.

Single-Stage Op-Amp Compensation

Figure 58.54 shows a typical simplified circuit schematic of a single-stage op-amp. The input is a differential emitter-coupled pair followed by a folded cascode transistor and an output complementary common-collector buffer. The key difference between this architecture and the two-stage design shown earlier is that X is a low-impedance node, and so the only high-impedance node in the circuit is node Y. Interestingly, the higher frequency nondominant pole of the two-stage amplifier has now become the dominant frequency pole of the single-stage design, as indicated by the second set of curves in Fig. 58.52(b), which leads to several advantages.

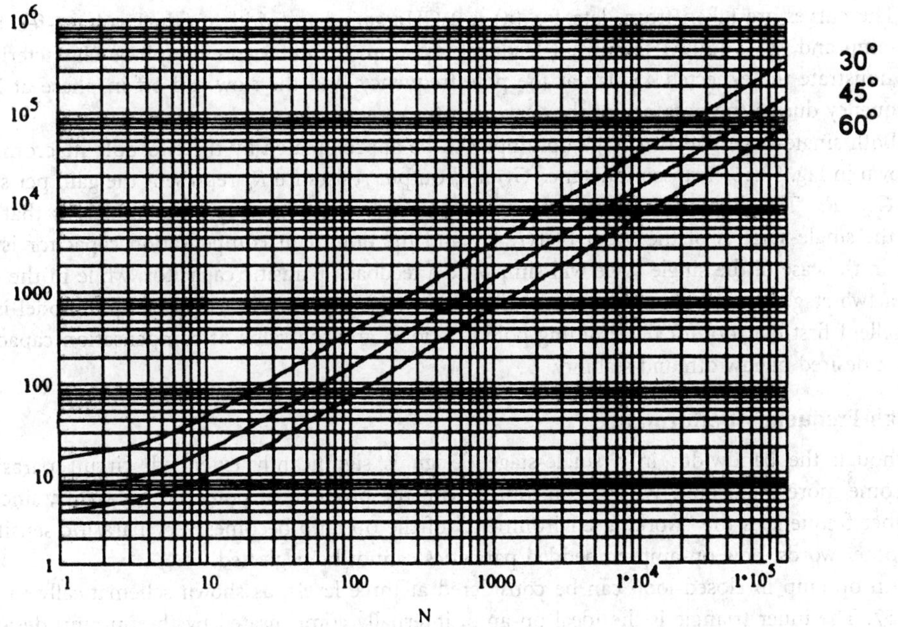

FIGURE 58.53 Low-frequency gain A_0 versus N $(= f_{P2}/f_{P1})$ for a two-pole amplifier.

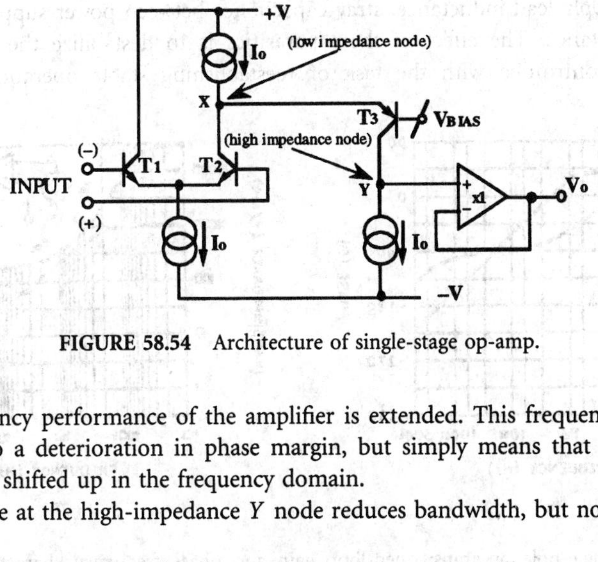

FIGURE 58.54 Architecture of single-stage op-amp.

1. The frequency performance of the amplifier is extended. This frequency extension does not lead to a deterioration in phase margin, but simply means that the phase margin problem is shifted up in the frequency domain.
2. Capacitance at the high-impedance Y node reduces bandwidth, but now improves phase margin.
3. A single value of a few pF's of grounded capacitor at Y will now act as a satisfactory compensation capacitor, unlike the large Miller capacitor required in the two-stage design.
4. The slewing capability of this single-stage structure is very good as a result of the much smaller compensation capacitor.
5. Clearly, it is much more straightforward to develop a stable amplifier for high-frequency applications if it has essentially only one voltage gain stage and so high-frequency op-amp designers generally opt for a single gain stage architecture.

Grounded Capacitor Compensation

Typical A_{OL} versus f responses of two single-stage op-amps are shown in Fig. 58.55, indicating one high-frequency pole and its proximity to the nondominant pole.

The curves are taken from data for (a) a 2-GHz gain–bandwidth product voltage-feedback op-amp and (b) a 150-MHz current-feedback op-amp. In both cases the phase characteristics demonstrate the expected 45° lag at the pole frequency, and the slow roll-off in phase at high frequency due to the presence of the very-high-frequency poles.

Both single-stage and two-stage op-amps can be approximated by the two-pole macromodel shown in Fig. 58.56. Transconductance G_M and output resistance R_0 represent the gain per stage of $G_M \cdot R_0$. The difference between the two-stage and single-stage op-amp models is that R_{01} of the single-stage is of the order of $[G_M]^{-1}$ and the dominant compensation capacitor is C_2. C_P in the case of the single stage will simply be a feedback parasitic capacitor, while in the case of a two-stage it will be the dominant Miller compensating capacitor. This simple model is an excellent first-cut tool for determining pole locations, and the value of compensation capacitor for a desired bandwidth and stability.

High-Frequency Performance

Although the bandwidth in a single-stage design is significantly extended, circuit parasitics become more important. We are confronted with the problem of potential instability, since at higher frequencies the "working environment" of the op-amp becomes very parasitic sensitive; in other words, now op-amp-embedded parasitics cannot be neglected.

An op-amp in closed-loop can be considered at three levels, as shown schematically in Fig. 58.57. The inner triangle is the ideal op-amp, internally compensated by the op-amp designer for stable operation using the circuit techniques outlined earlier. High-frequency amplifiers are sensitive to parasitics of the surrounding circuit. The key parasitics within the outer triangle include power supply lead inductance, stray capacitance between power supply pins, and input to ground capacitance. The effect of these parasitics is to destabilize the amplifier, and so the designer is confronted with the task of reestablishing stable operation. The approach

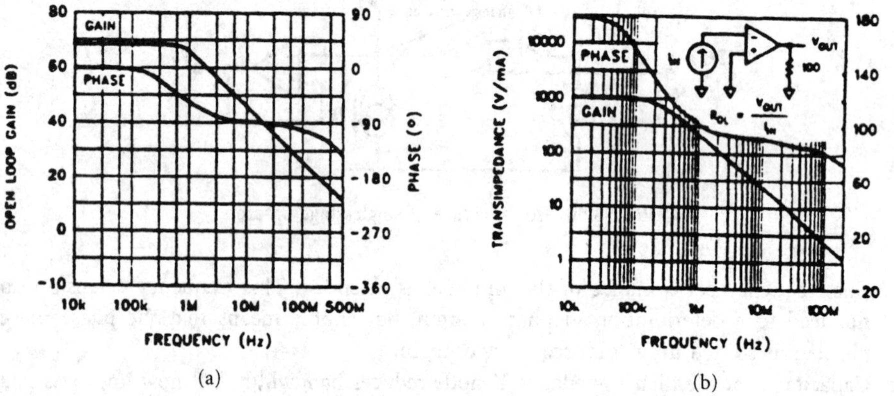

FIGURE 58.55 Single-pole op-amps; open-loop gain and phase frequency characteristics. (a) Voltage feedback. (b) Current feedback.

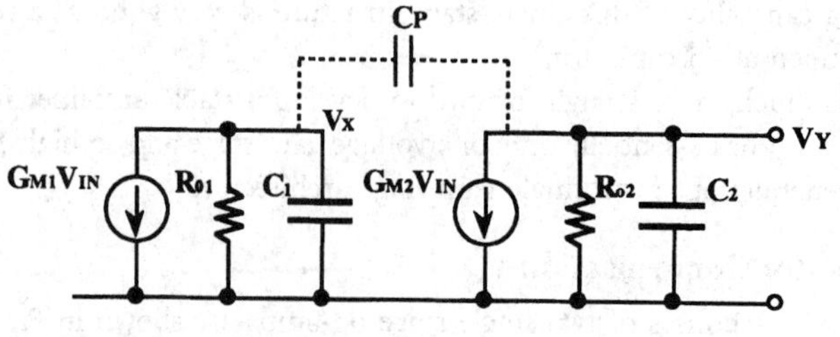

FIGURE 58.56 Partial equivalent circuit of two-pole op-amp.

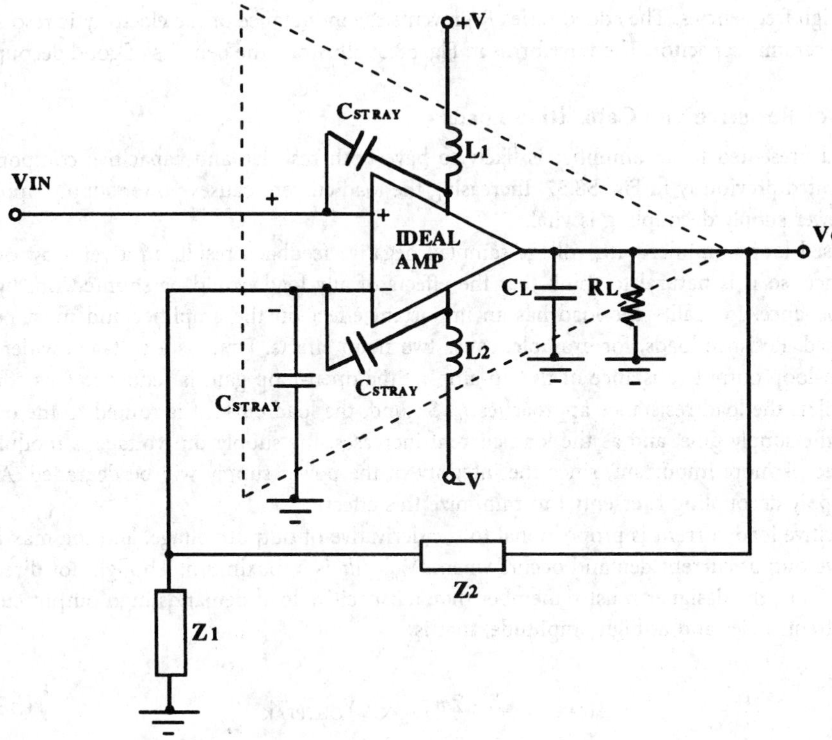

FIGURE 58.57 Real feedback amplifier.

needed to achieve this parallels the work of the op-amp designer. The parasitics almost always introduce additional extrinsic nondominant poles, which need to be compensated. The task of compensation cannot be attempted without considering the outer or third level, which includes the closed-loop gain defining components together with the load impedance. Again, stray reactance associated with these components will modify the loop gain, and so to guarantee stable operation of the closed-loop amplifier it is necessary to compensate the complete circuit.

Power Supply Impedance

In this section we will consider the ways in which the impedance of the power supply can affect the frequency response of the amplifier. First, some important rules:

1. There is no such thing as an ideal zero-impedance power supply.
2. Real power supplies have series *R–L* impedance and at high frequencies the inductance matters most.
3. Power supply inductance causes "bounce" on the power supply voltage, generating unwanted feedback via parasitic capacitive links to the inputs. Power supply "bounce" increases with increasing load current.
4. Supply decoupling capacitors act as "short term local batteries" to maintain power supply integrity, and it is important that they are placed as close as possible to the power supply pins of the op-amp.

Large electrolytic capacitors are fine at low frequencies but are inductive at high frequencies. Figure 58.58 shows commonly used decoupling circuitry. Small-sized tantalum electrolytics are preferred, while a parallel ceramic capacitor with low series inductance takes over the decoupling role at high frequencies. The added series R prevents the inductance of the electrolytic resonating with the ceramic capacitor. The waveforms in Fig. 58.59 illustrate the benefits of good decoupling.

Effects of Resistive and Capacitive Loads

The load presented to an amplifier is likely to have both resistive and capacitive components, as illustrated previously in Fig. 58.57. Increasing the load current causes power supply ripple, so good power supply decoupling is vital.

A closed-loop amplifier with voltage-sampled negative feedback results in a very-low output impedance, so it is natural to think that the effects of any load would be shunted out by this low impedance. In reality the load has an important effect on the amplifier and must not be overlooked. Resistive loads, for example, cause two main effects. First, as a voltage divider with the open-loop output resistance of the op-amp r_0, the open-loop gain is reduced. This effect is small unless the load resistance approaches r_0. Second, the load current is routed to the output pin via the supply pins, and as the load current increases, the supply pin voltage is modulated. This effect is more important, since the integrity of the power supply will be degraded. Again, good supply decoupling is essential to minimize this effect.

Capacitive load current is proportional to the derivative of output voltage, and the maximum capacitive output current demand occurs when dV_{OUT}/dt is a maximum. Though not directly a stability issue, the designer must remember than a capacitive load demands high output current at high frequencies and at high amplitude, that is

$$I_{MAX} = C_L \cdot 2\pi f_{MAX} \cdot V_{OUTPEAK} \tag{58.58}$$

Figure 58.60 illustrates the effect of load capacitance on the loop gain.

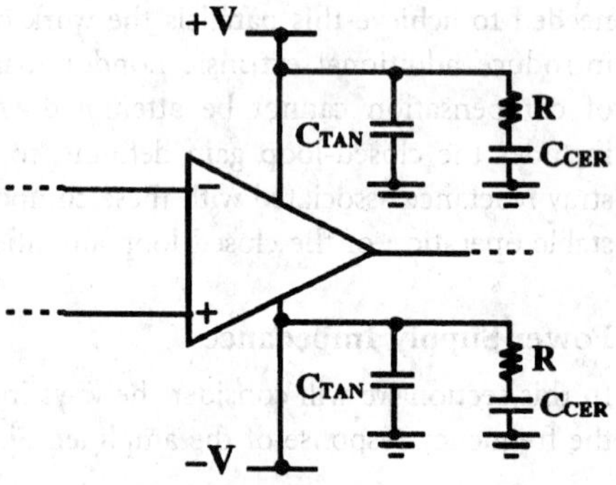

FIGURE 58.58 Supply decoupling circuitry. *Note:* C_{CER} = ceramic capacitor and C_{TAN} = tantalum.

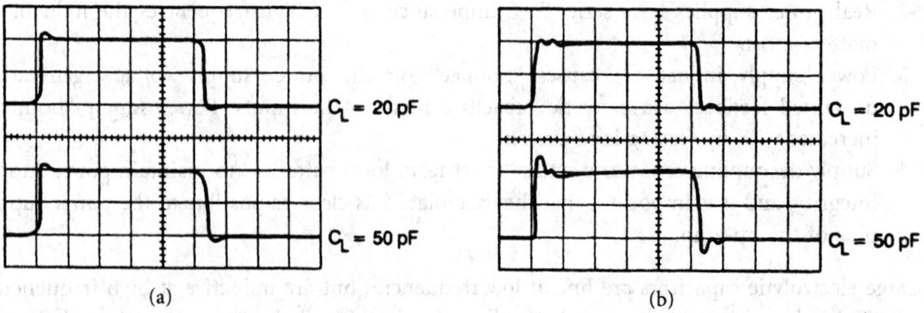

FIGURE 58.59 High-speed voltage buffer: (a) with and (b) without supply decoupling.

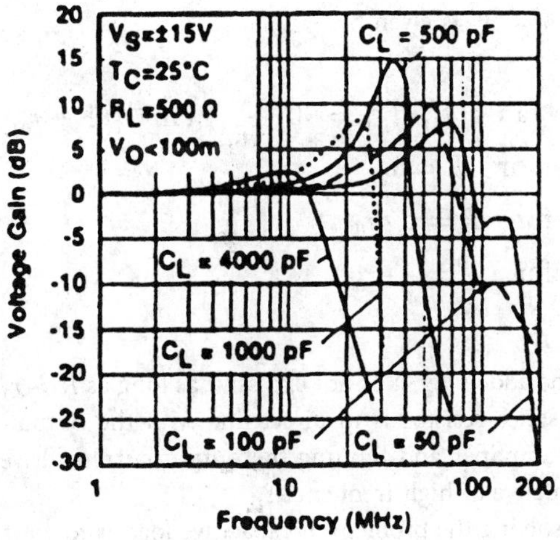

FIGURE 58.60 Load capacitance causes gain peaking.

C_L together with the equivalent output resistance of the op-amp, adds an additional pole into the loop gain of the form

$$V_F/V_{\mathrm{OUT}} = B = 1/[1 + jf/f_L] \quad \text{where} \quad f_L = 1/2\pi r_0 \cdot C_L \tag{58.59}$$

The load resistance has a minor influence on the loop gain compared to the effects of load capacitance by slightly reducing the value of dc open-loop gain by factor K, where $K = R_L/[r_0 + R_L]$, as described above. Since the effective output resistance reduces to $r_{0'} = r_0//R_L$, then f_L changes to $f_{L'} = 1/2\pi r_{0'} C_L$.

Neutralizing the Phase Lag

To compensate for high-frequency phase lag, the simplest technique is to add a series resistance R between the output of the op-amp and the load connection point, as shown in Fig. 58.61.

The series resistor adds a zero into the V_F/V_{OUT} equation, which changes to

$$V_F/V_{\mathrm{OUT}} = K \cdot [1 + jf/f_Z]/[1 + jf/f_P] \tag{58.60}$$

where $K = [R + R_L]/[r_0 + R + R_L]$, $f_P = 1/[2\pi(r_0 + R)//R_L \cdot C_L]$ and $f_Z = 1/[2\pi R_L//R \cdot C_L] = f_P \cdot [1 + r_0/R]$, so clearly, $f_P < f_Z$.

The phase lag introduced by the pole is compensated by the phase lead of the zero at higher frequencies. The maximum phase lag is limited if the zero is close to the pole, almost eliminating the effects of the load capacitor. Maximum phase lag in V_F/V_{OUT} occurs at $f = f_M$, where f_M is given by

$$f_M = [f_P \cdot f_Z]^{1/2} = f_P \cdot (1 + r_0/R)^{1/2} \tag{58.61}$$

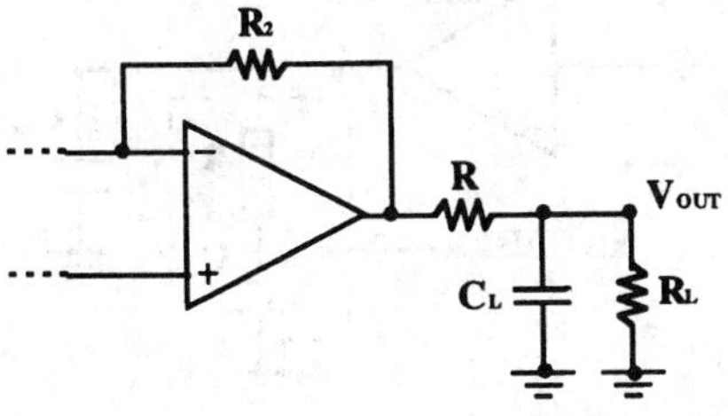

FIGURE 58.61 Load capacitance neutralization.

and at f_M the phase lag $\Phi = \Phi'$ is given by

$$\begin{aligned}
\Phi' &= 90^\circ - 2 \cdot \tan^{-1}[f_M/f_P] = 90^\circ - 2 \cdot \tan^{-1}[(1 + r_0/R)^{1/2}] \\
\Phi' &\approx -19.5^\circ \quad \text{for} \quad R = r_0 \\
\Phi' &\approx -8.2^\circ \quad \text{for} \quad R = 2 \cdot r_0 \\
\Phi' &\approx -6.4^\circ \quad \text{for} \quad R = 3 \cdot r_0
\end{aligned} \tag{58.62}$$

These values show that the added lag Φ' is not excessive as long as $R > r_0$. The disadvantage with this method is that the series resistor is in direct line with the output current, increasing the output resistance of the amplifier and limiting the output current drive capability. The output impedance also goes inductive at high frequencies.

An alternative way of solving the problem of capacitive load is to view the closed-loop output resistance of the op-amp as being inductive, since the closed-loop output impedance of the op-amp is essentially the open-loop output resistance divided by the loop gain. As the loop gain falls with frequency, the output impedance rises, and thus appears inductive. Adding a load capacitor generates a resonant circuit. The solution is to "spoil" the Q of the resonator, therefore minimizing the added phase lag of C_L.

Adding a so-called series R–C "snubber," as in Fig. 58.62, effects a cure. The resistor R is ac coupled by the capacitor at high frequencies and spoils the Q. Effectively, C_L resonates with the inductive output impedance, and at this frequency leaves the R–C snubber as a "new" load. The equivalent circuit is therefore close to the previous compensation method shown in Fig. 58.61, but with the added advantage that now the load current is not carried by the series resistance. To select the snubber component values, make $R = 1/2\pi f_0 C$, where f_0 is the resonant frequency, which can simply be determined experimentally from the amplifier without the snubber in place. The value of the series capacitance is a compromise: too big and it will increase the effective load capacitance. Choosing $C = C_L$ works reasonably well in practice.

Inverting Input Capacitance to Ground

With most broadband bipolar op-amps, parasitic capacitance to ground adds an additional pole (and hence phase lag) into the feedback path, which threatens stability. Stray capacitance C_1 at the inverting input pin (shown previously in Fig. 58.57) modifies B and adds phase lag in the loop-gain T, compromising stability.

Solving for B with C_1 taken into account will clarify the problem. It is simple to show that

$$B = V_F/V_{\text{OUT}} = Z_1/[Z_1 + Z_2] \tag{58.63}$$

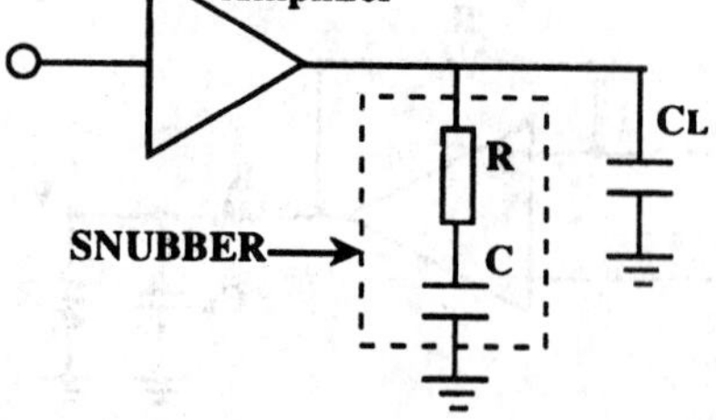

FIGURE 58.62 Snubber cures capacitive load peaking.

where $Z_1 = R_1/[1 + j\omega R_1 C_1]$ and $Z_2 = R_2$. Substituting we get

$$B = K/[1 + jf/f_C] \tag{58.64}$$

where $K = R_1[R_1 + R_2]$ and $f_C = 1/[2\pi C_1 R_1/R_2]$.

The additional pole at $f = f_C$ will now give the circuit a very undesirable three-pole loop gain, which could cause significant gain peaking, as shown in Fig. 58.63. f_C could be made high by choosing relatively low values of $R_1//R_2$ but the additional pole can be eliminated by adding a feedback capacitor C_2 across resistor R_2 to give pole–zero cancellation.

$$Z_1 = R_1/[1 + j\omega R_1 C_1] \quad \text{and} \quad Z_2 = R_2/[1 + j\omega R_2 C_2] \tag{58.65}$$

If $R_1 C_1 = R_2 C_2$, then $B = Z_1/[Z_1 + Z_2] = R_1/[R_1 + R_2]$, making B frequency independent. The design equation for C_2 is then

$$C_2 = C_1 \cdot R_1/R_2 \tag{58.66}$$

If the open-loop phase margin Φ_M needs to be increased for the desired value of closed-loop gain, and the inverting capacitance C_1 has its inevitable high-frequency influence, then the optimum solution for C_2 would be to locate the zero on the second pole of the loop-gain response following the procedure given above.

Conclusions

This chapter has hopefully served to illustrate some of the modern techniques the practicing engineer will encounter when designing broadband bipolar amplifiers. The section has focused mainly upon key generic building blocks and methodologies for broadband design. Many circuits and design techniques have not been covered, but analysis techniques described should serve as a foundation for the analysis of other broadband designs. Furthermore, comprehensive analytical treatment of many alternative broadband bipolar circuits can be found in the texts [6] and [13]–[15].

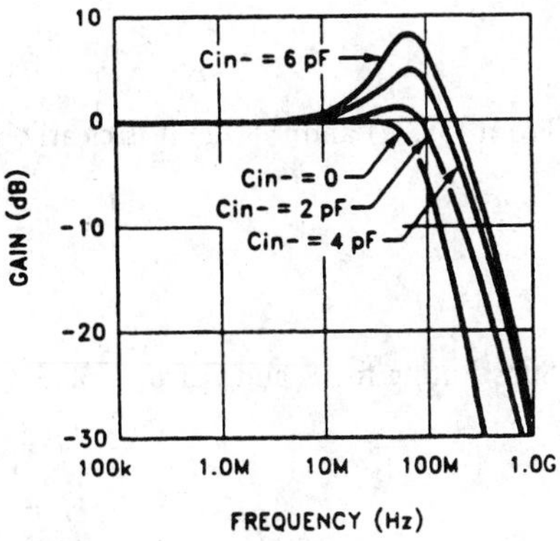

FIGURE 58.63 Stray input capacitance causes gain peaking.

Appendices

A: Transfer Function and Bandwidth Characteristic of Current-Feedback Operational Amplifier

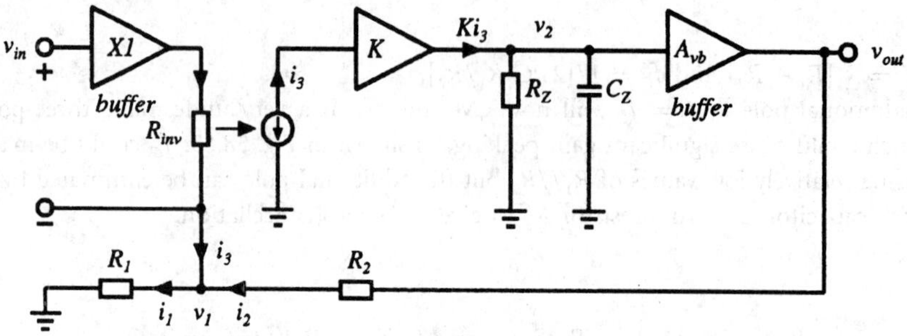

$$-i_1 + i_2 + i_3 = 0 \tag{58.67}$$

$$i_1 = \frac{v_1}{R_1} \tag{58.68}$$

$$i_2 = \frac{v_{out} - v_1}{R_2} \tag{58.69}$$

$$i_3 = \frac{v_{in} - v_1}{R_{inv}} \tag{58.70}$$

$$v_2 = \frac{Ki_3R_Z}{1 + j\omega R_Z C_Z} \tag{58.71}$$

$$v_{out} = A_{vb} v_2 \tag{58.72}$$

Substituting (58.68)–(58.70) into (58.67) yields

$$-\frac{v_1}{R_1} + \frac{v_{out} - v_1}{R_2} + \frac{v_{in} - v_1}{R_{inv}} = 0$$

Rearranging for v_1 gives

$$v_1 = \frac{\dfrac{v_{in}R_2}{R_{inv}} + v_{out}}{1 + \dfrac{R_2}{R_1} + \dfrac{R_2}{R_{inv}}}$$

From (58.71) and (58.72) it is clearly seen that

$$v_{out} = \frac{A_{vb}Ki_3R_Z}{1 + j\omega R_Z C_Z} \tag{58.73}$$

Substituting for i_1 and i_2 from (58.68) and (58.69) into (58.67) gives

$$i_3 = v_1\left(\frac{1}{R_1} + \frac{1}{R_2}\right) - \frac{v_{out}}{R_2}$$

Substitute for v_1:

$$i_3 = \left[\frac{\dfrac{v_{in}R_2}{R_{inv}} + v_{out}}{1 + \dfrac{R_2}{R_1} + \dfrac{R_2}{R_{inv}}}\right]\left(\frac{1}{R_1} + \frac{1}{R_2}\right) - \frac{v_{out}}{R_2}$$

Substitute for i_3 from (58.73):

$$\frac{v_{out}(1 + j\omega R_Z C_Z)}{A_{vb}KR_Z} = \left\{\left[\frac{\dfrac{v_{in}R_2}{R_{inv}} + v_{out}}{1 + \dfrac{R_2}{R_1} + \dfrac{R_2}{R_{inv}}}\right]\left(\frac{1}{R_1} + \frac{1}{R_2}\right) - \frac{v_{out}}{R_2}\right\}$$

rearranging

$$v_{out}\left[\frac{(1 + j\omega R_Z C_Z)}{A_{vb}KR_Z} - \frac{\left(\dfrac{1}{R_1} + \dfrac{1}{R_2}\right)}{\left(1 + \dfrac{R_2}{R_1} + \dfrac{R_2}{R_{inv}}\right)} + \frac{1}{R_2}\right] = \frac{\dfrac{v_{in}R_2}{R_{inv}}\left(\dfrac{1}{R_1} + \dfrac{1}{R_2}\right)}{\left(1 + \dfrac{R_2}{R_1} + \dfrac{R_2}{R_{inv}}\right)}$$

$$\frac{v_{out}}{v_{in}} = \frac{1 + \dfrac{R_2}{R_1}}{\dfrac{R_{inv}\left(1+\dfrac{R_2}{R_1}+\dfrac{R_2}{R_{inv}}\right)(1+j\omega R_Z C_Z)}{A_{vb}KR_Z} - R_{inv}\left(\dfrac{1}{R_1}+\dfrac{1}{R_2}\right) + \dfrac{R_{inv}\left(1+\dfrac{R_2}{R_1}+\dfrac{R_2}{R_{inv}}\right)}{R_2}}$$

$$\frac{v_{out}}{v_{in}} = \frac{1 + \dfrac{R_2}{R_1}}{\dfrac{R_{inv}\left(1 + \dfrac{R_2}{R_1}\right) + R_2}{A_{vb}KR_Z} + \dfrac{\left(R_{inv}\left(1 + \dfrac{R_2}{R_1}\right) + R_2\right)j\omega R_Z C_Z}{A_{vb}KR_Z} + 1}$$

Factorize the denominator

$$\frac{v_{\text{out}}}{v_{\text{in}}}=\frac{1+\dfrac{R_2}{R_1}}{\left[1+\dfrac{R_{\text{inv}}\left(1+\dfrac{R_2}{R_1}\right)+R_2}{A_{vb}KR_Z}\right]\left[1+\dfrac{\dfrac{\left(R_{\text{inv}}\left(1+\dfrac{R_2}{R_1}\right)+R_2\right)j\omega R_Z C_Z}{A_{vb}KR_Z}}{1+\dfrac{R_{\text{inv}}\left(1+\dfrac{R_2}{R_1}\right)+R_2}{A_{vb}KR_Z}}\right]}$$

$$\frac{v_{\text{out}}}{v_{\text{in}}}=\frac{1+\dfrac{R_2}{R_1}}{\left[1+\dfrac{R_{\text{inv}}\left(1+\dfrac{R_2}{R_1}\right)+R_2}{A_{vb}KR_Z}\right]\left[1+j\omega C_Z\left\{\dfrac{R_{\text{inv}}\left(1+\dfrac{R_2}{R_1}\right)+R_2}{A_{vb}K+\dfrac{R_{\text{inv}}\left(1+\dfrac{R_2}{R_1}\right)+R_2}{R_Z}}\right\}\right]}$$

If we assume that R_Z is very large, then

$$\frac{R_{\text{inv}}\left(1+\dfrac{R_2}{R_1}\right)+R_2}{R_Z}\approx 0$$

and the transfer function becomes

$$\frac{v_{\text{out}}}{v_{\text{in}}}=\frac{1+\dfrac{R_2}{R_1}}{1+j\omega C_Z\left[\dfrac{R_{\text{inv}}\left(1+\dfrac{R_2}{R_1}\right)+R_2}{A_{vb}K}\right]}$$

The pole frequency is given by

$$f_{-3\,\text{dB}}=\frac{A_{vb}K}{2\pi C_Z\left[R_{\text{inv}}\left(1+\dfrac{R_2}{R_1}\right)+R_2\right]}$$

The gain–bandwidth product is given by

$$GBW=\frac{A_{vb}K\left[1+\dfrac{R_2}{R_1}\right]}{2\pi C_Z\left[R_{\text{inv}}\left(1+\dfrac{R_2}{R_1}\right)+R_2\right]}$$

B: Transfer Function and Bandwidth Characteristic of Voltage-Feedback Operational Amplifier

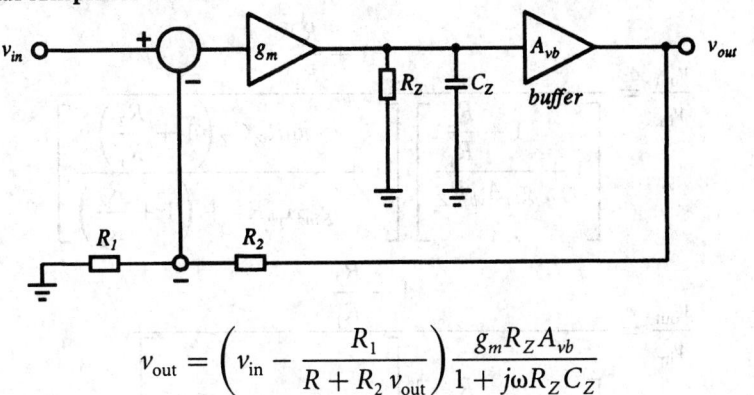

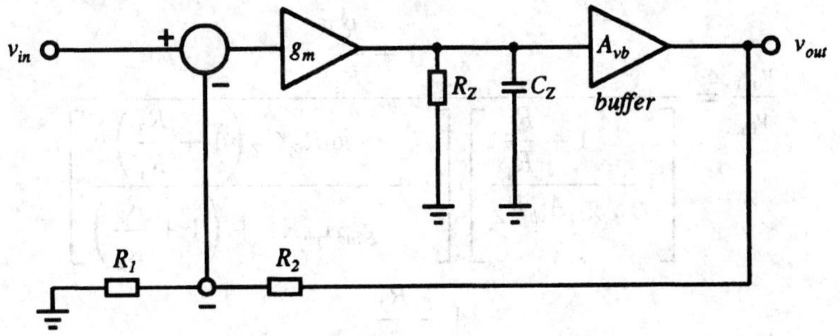

$$v_{out} = \left(v_{in} - \frac{R_1}{R + R_2\, v_{out}}\right)\frac{g_m R_Z A_{vb}}{1 + j\omega R_Z C_Z} \tag{58.74}$$

$$v_{out}\left[1 + \frac{R_1 g_m A_{vb} R_Z}{(R_1+R_2)(1 + j\omega R_Z C_Z)}\right] = v_{in}\frac{g_m A_{vb} R_Z}{1 + j\omega R_Z C_Z}$$

$$\frac{v_{out}}{v_{in}} = \frac{\dfrac{g_m A_{vb} R_Z}{1 + j\omega R_Z C_Z}}{1 + \dfrac{R_1 g_m A_{vb} R_Z}{(R_1+R_2)(1 + j\omega R_Z C_Z)}}$$

Multiply the numerator and denominator by $(1 + j\omega R_Z C_Z)/g_m A_{vb} R_Z$

$$\frac{v_{out}}{v_{in}} = \frac{1}{\dfrac{1 + j\omega R_Z C_Z}{g_m A_{vb} R_Z} + \dfrac{R_1}{(R_1+R_2)}}$$

$$\frac{v_{out}}{v_{in}} = \frac{1 + \dfrac{R_2}{R_1}}{\left[\dfrac{1 + j\omega R_Z C_Z}{g_m A_{vb} R_Z}\right]\left[1 + \dfrac{R_2}{R_1}\right] + 1}$$

$$\frac{v_{out}}{v_{in}} = \frac{1 + \dfrac{R_2}{R_1}}{\dfrac{1 + \dfrac{R_2}{R_1}}{g_m A_{vb} R_Z} + \dfrac{j\omega R_Z C_Z\left(1 + \dfrac{R_2}{R_1}\right)}{g_m A_{vb} R_Z} + 1}$$

get $1 + [1 + (R_2/R_1)/g_m A_{vb} R_Z]$ out of the denominator

$$\frac{v_{out}}{v_{in}} = \frac{1 + \dfrac{R_2}{R_1}}{\left[1 + \dfrac{1 + \dfrac{R_2}{R_1}}{g_m A_{vb} R_Z}\right]\left[1 + \dfrac{\dfrac{j\omega R_Z C_Z\left(1 + \dfrac{R_2}{R_1}\right)}{g_m A_{vb} R_Z}}{1 + \dfrac{1 + \dfrac{R_2}{R_1}}{g_m A_{vb} R_Z}}\right]}$$

multiply the denominator bracket by $g_m A_{vb} R_Z/[1 + (R_2/R_1)]$

$$\frac{v_{\text{out}}}{v_{\text{in}}} = \frac{1 + \dfrac{R_2}{R_1}}{\left[1 + \dfrac{1 + \dfrac{R_2}{R_1}}{g_m A_{vb} R_Z}\right]\left[1 + \dfrac{j\omega R_Z C_Z\left(1 + \dfrac{R_2}{R_1}\right)}{g_m A_{vb} R_Z + \left(1 + \dfrac{R_2}{R_1}\right)}\right]}$$

$$\frac{v_{\text{out}}}{v_{\text{in}}} = \frac{1 + \dfrac{R_2}{R_1}}{\left[1 + \dfrac{1 + \dfrac{R_2}{R_1}}{g_m A_{vb} R_Z}\right]\left[1 + \dfrac{j\omega R_Z C_Z}{1 + \dfrac{g_m A_{vb} R_Z}{\left(1 + \dfrac{R_2}{R_1}\right)}}\right]}$$

assuming that $g_m A_{vb} R_Z$ is much larger than $1 + R_2/R_1$, then

$$\frac{v_{\text{out}}}{v_{\text{in}}} = \frac{1 + \dfrac{R_2}{R_1}}{1 + j\omega\left[\dfrac{R_Z C_Z}{1 + \dfrac{g_m A_{vb} R_Z}{\left(1 + \dfrac{R_2}{R_1}\right)}}\right]}$$

The pole frequency is given by

$$f_{-3\,\text{dB}} = \frac{1 + \dfrac{g_m A_{vb} R_Z}{\left(1 + \dfrac{R_2}{R_1}\right)}}{2\pi R_Z C_Z}$$

The gain–bandwidth product is given by

$$GBW = \frac{\left(1 + \dfrac{R_2}{R_1}\right)\left[1 + \dfrac{g_m A_{vb} R_Z}{\left(1 + \dfrac{R_2}{R_1}\right)}\right]}{2\pi R_Z C_Z}$$

C: Transconductance of the Current-Feedback Op-Amp Input Stage

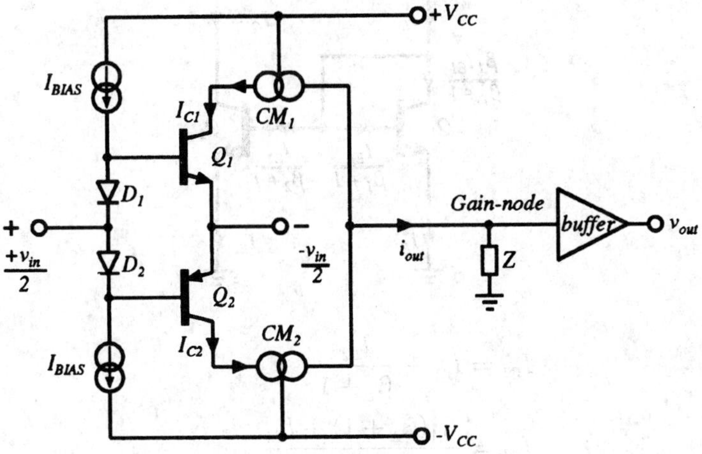

$$v_{\text{in}} = v_1 - v_2$$

$$i_{\text{out}} = I_{C1} - I_{C2}$$

$$I_{C1} = I_{S1} e^{\frac{V_{BE1}}{V_T}}$$

$$I_{C2} = I_{S2} e^{\frac{V_{BE2}}{V_T}}$$

$$V_{BE1} = V_{DQ1} + v_{\text{in}}$$

$$V_{BE2} = v_{\text{in}} - V_{DQ2}$$

$$I_{C1} = I_{S1} e^{\left(\frac{V_{DQ1}}{V_T} + \frac{v_{\text{in}}}{V_T}\right)}$$

$$I_{C2} = I_{S2} e^{\left(\frac{V_{DQ1}}{V_T} - \frac{v_{\text{in}}}{V_T}\right)}$$

$$I_{C1} = I_{CQ1} e^{\frac{v_{\text{in}}}{V_T}}$$

$$I_{C2} = I_{CQ2} e^{-\frac{v_{\text{in}}}{V_T}}$$

Assuming matched transistors then, $I_{CQ1} = I_{CQ2} = I_{CQ}$

$$i_{\text{out}} = I_{C1} - I_{C2} = I_{CQ}\left[e^{+\left(\frac{v_{\text{in}}}{V_T}\right)} - e^{-\left(\frac{v_{\text{in}}}{V_T}\right)}\right]$$

$$\frac{i_{\text{out}}}{I_{CQ}} = y = [e^x - e^{-x}] = 2\sinh(x)$$

where $x = +v_{\text{in}}/V_T$.

D: Transfer Function of Widlar Current Mirror

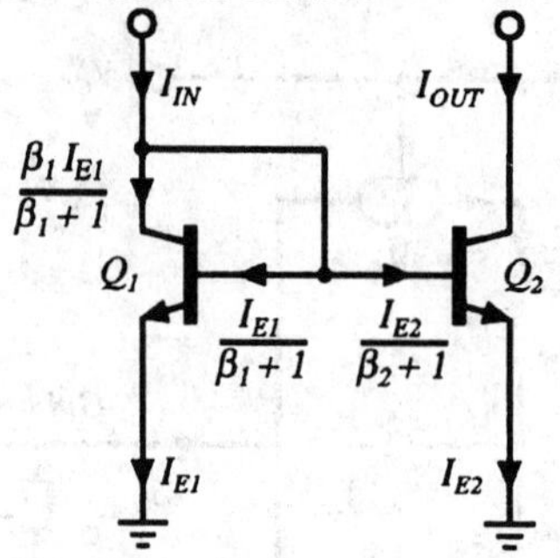

$$I_{\text{IN}} = I_{E1} + \frac{I_{E2}}{\beta_2 + 1}$$

$$I_{\text{IN}} = \frac{I_{E1}(\beta_2 + 1) + I_{E2}}{\beta_2 + 1}$$

$$I_{\text{OUT}} = \beta_2 I_{B2}$$

$$I_{\text{OUT}} = \frac{\beta_2 I_{E2}}{\beta_2 + 1}$$

$$\frac{I_{\text{OUT}}}{I_{\text{IN}}} = \frac{(\beta_2 I_{E2})(\beta_2 + 1)}{(\beta_2 + 1)[I_{E1}(\beta_2 + 1) + I_{E2}]}$$

$$\frac{I_{\text{OUT}}}{I_{\text{IN}}} = \frac{\beta_2 I_{E2}}{I_{E1}(\beta_2 + 1) + I_{E2}}$$

$$\frac{I_{\text{OUT}}}{I_{\text{IN}}} = \frac{1}{\dfrac{I_{E1}(\beta_2 + 1)}{I_{E2}\beta_2} + \dfrac{1}{\beta_2}}$$

For

$$\frac{I_{E1}}{I_{E2}} = \frac{I_{S1}\left(\dfrac{\beta_1 + 1}{\beta_1}\right)e^{\frac{V_{BE1}}{V_T}}}{I_{S2}\left(\dfrac{\beta_2 + 1}{\beta_2}\right)e^{\frac{V_{BE2}}{V_T}}}$$

Then, as $V_{BE1} = V_{BE2}$,

$$\frac{I_{E1}}{I_{E2}} = \frac{I_{S1}\left(\dfrac{\beta_1 + 1}{\beta_1}\right)}{I_{S2}\left(\dfrac{\beta_2 + 1}{\beta_2}\right)}$$

$$\frac{I_{\text{OUT}}}{I_{\text{IN}}} = \frac{1}{\dfrac{I_{S1}(\beta_1 + 1)}{I_{S2}\beta_1} + \dfrac{1}{\beta_2}}$$

Assume $\beta_1 = \beta_2 = \beta$, $I_{S1} = I_{S2}$. Then

$$\frac{I_{\text{OUT}}}{I_{\text{IN}}} = \frac{\beta}{\beta + 2}$$

E: Transfer Function of Widlar Current Mirror with Emitter Degeneration Resistors

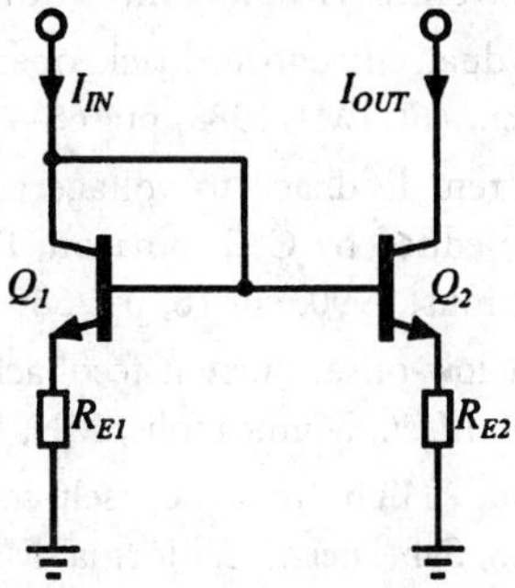

Assuming that $\beta \gg 1$, then

$$V_{BE1} + I_{\text{IN}}R_1 = V_{BE2} + I_{\text{out}}R_2$$

$$I_{\text{out}} = \frac{I_{\text{IN}}R_1}{R_2} + \frac{(V_{BE1} - V_{BE2})}{R_2}$$

$$\frac{I_{\text{OUT}}}{I_{\text{IN}}} = \frac{R_1}{R_2} + \frac{(V_{BE1} - V_{BE2})}{I_{\text{IN}}R_2}$$

$$\frac{I_{\text{OUT}}}{I_{\text{IN}}} = \frac{R_1}{R_2} + \frac{V_T \ln\left(\frac{I_{\text{IN}}}{I_{S1}} \frac{I_{S2}}{I_{\text{OUT}}}\right)}{I_{\text{IN}}R_2}$$

$$\frac{I_{\text{OUT}}}{I_{\text{IN}}} = \frac{R_1}{R_2} + \frac{V_T\left(\ln \frac{I_{\text{IN}}}{I_{\text{OUT}}} + \frac{\Delta V_{BE}}{V_T}\right)}{I_{\text{IN}}R_2}$$

$$\frac{I_{\text{OUT}}}{I_{\text{IN}}} = \frac{R_1}{R_2} + \frac{V_T\left(\ln \frac{I_{\text{IN}}}{I_{\text{OUT}}}\right)}{I_{\text{IN}}R_2} + \frac{\Delta V_{BE}}{I_{\text{IN}}R_2}$$

Assuming that the term $V_T[\ln(I_{\text{IN}}/I_{\text{OUT}})]/I_{\text{IN}}R_2$ is small compared to the other terms, then

$$\frac{I_{\text{OUT}}}{I_{\text{IN}}} = \frac{R_1}{R_2} + \frac{\Delta V_{BE}}{I_{\text{IN}}R_2}$$

eferences

[1] K. C. Smith and A. S. Sedra, "The current conveyor—A new circuit building block," *Proc. IEEE*, vol. 56, pp. 1368–1369, Aug. 1968.

[2] A. Sedra and K. C. Smith, "A Second Generation Current-Conveyor and its Applications," *IEEE Trans. Circuit Theory*, vol. CT-17, pp. 132–134, 1970.

[3] B. Wilson, "High performance current conveyor implementation," *Electron. Lett.*, vol. 20, no. 24, pp. 990–991, 1984.

[4] C. Toumazou, F. J. Lidgey, and C. Makris, "Extending voltage-mode op-amps to current-mode performance," *Proc. IEE: Pt. G,* vol. 137, no. 2, pp. 116–130, 1990.

[5] PA630 Data Sheet, Photronics Co., Ottawa, P.Q., Canada.

[6] C. Toumazou, F. J. Lidgey, and D. Haigh, Eds., *Analogue IC Design—The Current-Mode Approach,* Exeter, England: Peter Peregrinus, 1990.

[7] CCII01 Data Sheet, LTP Electronics, Headington, Oxford, England.

[8] D. F. Bowers, "A precision dual current-feedback operational amplifier," in *Proc. IEEE Bipolar Circuits Technol. Meet., (BCTM),* 1988, pp. 68–70.

[9] D. F. Bowers, "Applying current feedback to voltage amplifier," in *Analogue IC Design: The Current-Mode Approach,* edited by C. Toumazou, F. J. Lidgey and D. G. Haigh, Eds. Exeter, England: Peter Peregrinus, 1990, ch. 16, pp. 569–595.

[10] I. A. Koullias, "A wideband low-offset current-feedback op amp design," in *Proc. IEEE 1989 Bipolar Circuits Technol. Meet.,* Minneapolis, MN, Sept. 18–19, 1989, pp. 120–123.

[11] A. Payne and C. Toumazou, "High frequency self-compensation of current feedback devices," in *Proc. IEEE ISCAS,* San Diego, California, May 10–13, 1992, pp. 1376–1379.

[12] T. Vanisri and C. Toumazou, "Wideband and high gain current-feedback op-amp," *Electron. Lett.,* vol. 28, no. 18, pp. 1705–1707, Aug. 27, 1992.

[13] A. Grebene, *Bipolar and MOS Analog Integrated Circuit Design,* New York: Wiley, 1984.

[14] C. Toumazou, Ed., *Circuits and Systems Tutorials,* New York: IEEE ISCAS, 1994.

[15] *High Performance Analog Integrated Circuits.* Élantec Data Book, 1994.

58.2 Bipolar Noise

B. M. Wilamowski

Bipolar transistors and other electronic devices generate internal electrical noise. This limits the device operation at a small-signal range. There are a few different sources of noise, such as thermal noise, shot noise, flicker noise or $1/f$ noise, burst noise or "popcorn noise," and avalanche noise [1], [6].

Thermal Noise

Thermal noise is created by random motion of charge carriers due to the thermal excitation [1]. This noise is sometimes known as the Johnson noise. The thermal motion of carriers creates a fluctuating voltage on the terminals of each resistive element. The average value of this voltage is zero, but the power on its terminals is not zero. The internal noise voltage source or current source is described by Nyquist equation

$$\overline{v_n^2} = 4kTR\,\Delta f \qquad \overline{i_n^2} = \frac{4kT\,\Delta f}{R} \tag{58.75}$$

where k is the Boltzmann constant, T is absolute temperature, and $4kT$ is equal to $1.66 \times 10^{-20}\,\text{V}\cdot\text{C}$ at room temperature. The thermal noise is proportional to the frequency bandwidth Δf. It can be represented by the voltage source in series with resistor R, or by the current source in parallel to the resistor R. The maximum noise power can be delivered to the load when $R_L = R$. In this case maximum noise power in the load is $kT\,\Delta f$. The noise power density $dP_n/df = kT$, and it is independent of frequency. Thus, the thermal noise is the white noise. The rms noise voltage and the rms noise current are proportional to the square root of the frequency bandwidth Δf. The thermal noise is associated with every physical resistor in the circuit. In a bipolar transistor, the thermal noise is generated mainly by series base, emitter, and collector resistances.

Shot Noise

Shot noise is associated with the carrier injection through the pn junction. In each forward biased junction, there is a potential barrier that can be overcome by the carriers with higher thermal energy. This is a random process and the noise current is given by

$$\overline{i_n^2} = 2qI\,\Delta f \tag{58.76}$$

where q is the electron charge and I is the forward junction current. The shot noise depends on the thermal energy of carriers near the potential barrier. Similar to the thermal noise, the shot noise has constant and frequency-independent noise power density. It is also the white type of noise. Shot noise is usually considered as a current source connected in parallel to the small-signal junction resistance.

Flicker—1/*f* Noise

Flicker noise in bipolar transistors is associated mainly with generation-recombination centers [2]–[4]. Free carriers are randomly trapped and released by these centers. This is a relatively slow process and it cannot be seen at high frequencies. The flicker noise is always associated with a current and is approximated by

$$\overline{i_n^2} = K_F I^{A_F} \frac{\Delta f}{f} \tag{58.77}$$

where K_F is the flicker-noise coefficient and A_F is the flicker-noise exponent. Both K_F and A_F are device dependent. With modern technology the number of trapping centers can be significantly lowered, thus the effect of flicker noise is meaningfully reduced. The $1/f$ nature of the flicker noise is such that sometimes this noise component is considered to be responsible for the long term device parameter fluctuation.

Other Types of Noise

The burst or "popcorn" noise is another type of noise at low frequencies [3], [4]. This noise is not fully understood, but it seems that it is related to the heavy-metal ion contamination. The burst noise looks, on an oscilloscope, like a square wave with the constant magnitude, but with random pulse widths. It has significant effect at low frequencies. In audio amplifiers the burst noise sounds as random shoots, which are similar to the sound associated with making popcorn. Obviously, bipolar transistors with large burst noise must not be used in audio amplifiers and in other analog circuitry. The burst noise is often approximated by

$$\overline{i_n^2} = K_B \frac{i_D^{A_B}}{1 + \left(\frac{f}{f_B}\right)^2} \Delta f \tag{58.78}$$

where K_B, A_B, and f_B are experimentally chosen parameters, which usually vary from one device to another. Furthermore, a few different sources of the burst noise can exist in a single transistor. In such a case, each noise source should be modeled by a separate Eq. (58.78) with different parameters (usually different corner frequency f_B).

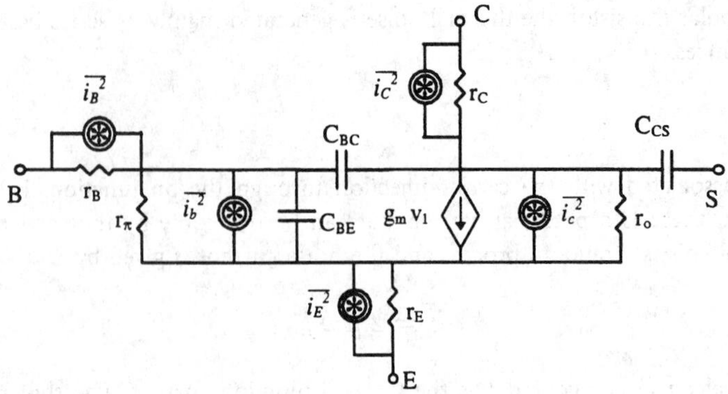

FIGURE 58.64 Equivalent diagram of the bipolar transistor, which includes noise sources.

The avalanche noise is another noise component, which can be found in bipolar transistors. For large reverse voltages on the collector junction, the collector current can be multiplied by the avalanche phenomenon. Carriers in the collector–base junctions gain energies in high electrical field, then losing this energy during collision with the crystal lattice. If the energy gained between collisions is large enough, then during collision another pair of carriers (electron and hole) can be generated. This way the collector current can be multiplied. This is a random process and obviously the noise source is associated with the avalanche carrier generation. The magnitude of the avalanche noise is usually much larger than any other noise component. Fortunately, the avalanche noise exists only in the pn junction biased with a voltage close to the breakdown voltage. The avalanche phenomenon is often used to build the noise sources [5].

Noise Characterization

Many different methods are used in literature for noise characterization. Sometimes the noise is characterized by an equivalent noise resistance, sometimes by an equivalent noise temperature, sometimes by an equivalent rms noise voltage or current or sometimes by a noise figure.

Equivalent Noise Voltage and Current

The equivalent noise voltage or current is the most commonly used method for modeling the noise in semiconductor devices. The equivalent diagram of the bipolar transistor, including various noise components, is shown in Fig. 58.64. The noise components are given by

$$\overline{i_B^2} = \frac{4kT\,\Delta f}{r_B} \qquad \overline{i_E^2} = \frac{4kT\,\Delta f}{r_E} \qquad \overline{i_C^2} = \frac{4kT\,\Delta f}{r_C}$$

$$\overline{i_c^2} = 2qI_C\,\Delta f \tag{58.79}$$

$$\overline{i_b^2} = 2qI_B\,\Delta f + K_F\frac{I_B^{A_F}}{f}\Delta f + K_B\frac{I_B^{A_B}}{1+\left(\dfrac{f}{f_B}\right)^2}\Delta f$$

Thermal noise is associated with physical resistors only, such as base, emitter, and collector series resistances. The small-signal equivalent resistances, such as r_π and r_o, do not exhibit thermal noise. The shot noise is associated with both collector and base currents. It was found experimentally that the flicker and the burst noise are associated with the base current. The typical noise characteristic of a bipolar transistor is shown in Fig. 58.65. The corner frequency of the flicker noise can vary from 100 Hz to 1 MHz.

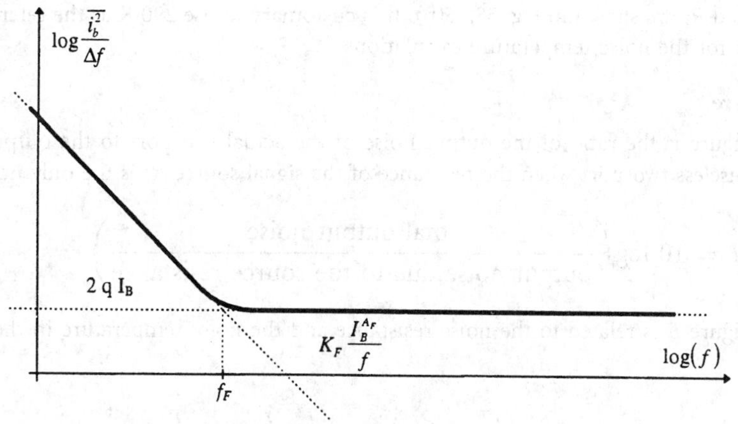

FIGURE 58.65 Bipolar Noise as a function of frequency.

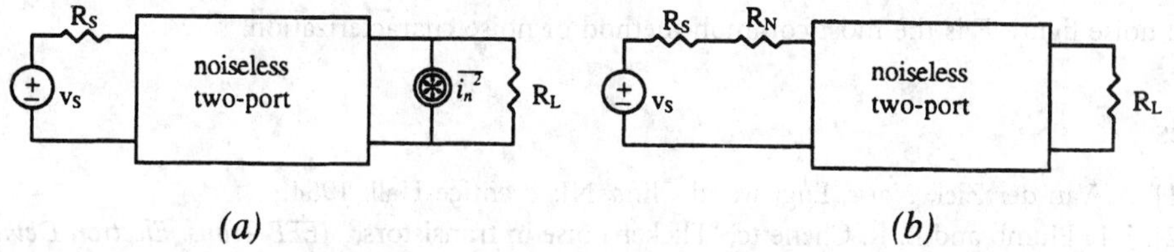

FIGURE 58.66 Noise characterization for two-ports: (a) using the noise source at the output, (b) using noise resistance R_N at the input.

Equivalent Noise Resistance and Noise Temperature

The noise property of a two-port element can be described by a noise current source connected in parallel to the output terminals, as Fig. 58.66(a) shows. Knowing that noise current can be expressed as the shot noise of the dc device current [(58.76)], the two-port noise can be expressed by means of an equivalent dc noise current

$$I_{eq} = \frac{\overline{i_n^2}}{2q\,\Delta f} \tag{58.80}$$

Another way to model the two-port noise in the two-port is to use the thermal noise at the input. This can be done using an additional "noisy" resistor connected to the input, as Fig. 58.66(b) shows:

$$R_n = \frac{\overline{v_{n1}^2}}{4kT\,\Delta f} = \frac{\overline{v_{n2}^2}}{A_v^2 4kT\,\Delta f} \tag{58.81}$$

where A_v is the voltage gain of the two-port, and v_{n1}^2 and v_{n2}^2 are equivalent noise voltage sources at the input and the output, respectively. The equivalent noise resistance is not a very convenient way to represent the noise property of the two-port. This additional resistance R_n must not be on the circuit diagram for small-signal analysis. To overcome this difficulty the concept of the equivalent noise temperature was introduced. This is a temperature increment of the source resistance required to obtain the same noise magnitude at the output if this source resistance is the only noise source. The noise temperature can be calculated from the simple formula

$$T_n = \frac{R_n}{R_s} 290 \text{ K} \tag{58.82}$$

Where R_n and R_s are shown in Fig. 58.66(b). It is customary to use 290 K as the reference room temperature for the noise temperature calculations.

Noise Figure

The noise figure is the ratio of the output noise of the actual two-port to the output noise of the ideal noiseless two-port when the resistance of the signal source R_s is the only noise source.

$$F = 10\log\left(\frac{\text{total output noise}}{\text{output noise due to the source resistance}}\right) \tag{58.83}$$

The noise figure F is related to the noise resistance and the noise temperature in the following way:

$$F = 10\log\left(1 + \frac{R_n}{R_s}\right) = 10\log\left(1 + \frac{T_n}{290\text{ K}}\right) \tag{58.84}$$

The noise figure F is the most common method of noise characterization.

References

[1] A. Van der Ziel, *Noise*, Englewood Cliffs, NJ: Prentice-Hall, 1954.

[2] J. L. Plumb and E. R. Chenette, "Flicker noise in transistors," *IEEE Trans. Electron Devices*, vol. ED-10, pp. 304–308, Sept. 1963.

[3] R. C. Jaeger and A. J. Broderson, "Low-frequency noise sources in bipolar junction transistors," *IEEE Trans. Electron Devices*, vol. ED-17, pp. 128–134, Feb. 1970.

[4] R. G. Meyer, L. Nagel, and S. K. Lui, "Computer simulation of $1/f$ noise performance of electronic circuits," *IEEE J. Solid-State Circuits*, vol. SC-8, pp. 237–240, June 1973.

[5] R. H. Haitz, "Controlled noise generation with avalanche diodes," *IEEE Trans. Electron Devices*, vol. ED-12, pp. 198–207, Apr. 1965.

[6] P. R. Gray and R. G. Mayer, *Analysis and Design of Analog Integrated Circuits*, 3rd ed., New York: Wiley, 1993.

XII

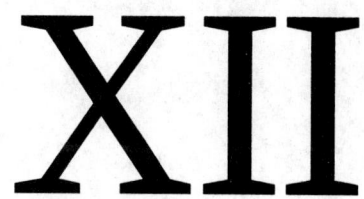

Digital and Analog VLSI

John Choma, Jr.
University of Southern California

59

Physical Design Automation[1]

Naveed Sherwani
Western Michigan University

59.1 Introduction

In the last three decades integrated circuit (IC) fabrication technology has evolved from integration of a few transistors in **small scale integration** (SSI) to integration of several million transistors in **very large scale integration** (VLSI). This phenomenal progress has been made possible by automating the process of design and fabrication of VLSI chips.

Integrated circuits consist of a number of electronic components, built by layering several different materials in a well-defined fashion on a silicon base called a **wafer**. The designer of an IC transforms a circuit description into a geometric description, which is known as a **layout**. A layout consists of a set of planar geometric shapes in several layers. The layout is then checked to ensure that it meets all the design requirements. The result is a set of design files in a particular unambiguous representation, known as an **intermediate form**, that describes the layout. The

[1]The material presented in this chapter was adapted from the author's book *Algorithms for VLSI Physical Design Automation*, Kluwer Academic Publishers, 1993, with editorial changes for clarity and continuity. Copyright permission has been obtained from Kluwer Academic Publishers.

0-8493-8341-2/95/$0.00+$.50

design files are then converted into pattern generator files, which are used to produce patterns called **masks** by an optical pattern generator. During fabrication, these masks are used to pattern a silicon wafer using a sequence of photolithographic steps. The component formation requires very exacting details about geometric patterns and separation between them. The process of converting the specifications of an electrical circuit into a layout is called the **physical design**. It is an extremely tedious and error-prone process because of the tight tolerance requirements and the minuteness of the individual components. Currently, the smallest geometric feature of a component can be as small as 0.35 μm (1 $\mu m = 1.0 \times 10^{-6}$ m). However, it is expected that the feature size can be reduced to 0.1 μm within 5 years. This small feature size allows fabrication of as many as 4.5 million transistors on a 25 $\times$ 25 mm maximum size chip. Due to the large number of components and the exacting details required by the fabrication process, the physical design is not practical without the help of computers. As a result, almost all phases of physical design extensively use computer-aided design (CAD) tools and many phases have already been partially or fully automated. This automation of the physical design process has increased the level of integration, reduced the turnaround time, and enhanced chip performance.

Very large scale integration physical design automation is essentially the study of algorithms related to the physical design process. The objective is to study optimal arrangements of devices on a plane (or in a three-dimensional space) and efficient interconnection schemes between these devices to obtain the desired functionality. Because space on a wafer is very expensive real estate, algorithms must use the space very efficiently to lower costs and improve yield. In addition, the arrangement of devices plays a key role in determining the performance of a chip. Algorithms for physical design must also ensure that all the rules required by the fabrication are observed and that the layout is within the tolerance limits of the fabrication process. Finally, algorithms must be efficient and should be able to handle very large designs. Efficient algorithms not only lead to fast turnaround time, but also permit designers to iteratively improve the layouts.

With the reduction in the smallest feature size and increase in the clock frequency, the effect of electrical parameters on physical design will play a more dominant role in the design and development of new algorithms.

In this section we present an overview of the fundamental concepts of VLSI physical design automation. Different design styles are discussed in Section 59.4. Section 59.2 discusses the design cycle of a VLSI circuit. In Section 59.3 different steps of the physical design cycle are discussed. The rest of the sections discuss each step of the physical design cycle.

59.2 Very Large Scale Integration Design Cycle

Starting with a formal specification, the VLSI design cycle follows a series of steps and eventually produces a packaged chip. A flow chart representing a typical design cycle is shown in Fig. 59.1.

1. **System Specification:** The specifications of the system to be designed are exactly specified here. This necessitates creating a high-level representation of the system. The factors to be considered in this process include performance, functionality, and the physical dimensions. The choice of fabrication technology and design techniques are also considered. The end results are specifications for the size, speed, power, and functionality of the VLSI system to be designed.
2. **Functional Design:** In this step, behavioral aspects of the system are considered. The outcome is usually a timing diagram or other relationships between subunits. This information is used to improve the overall design process and to reduce the complexity of the subsequent phases.

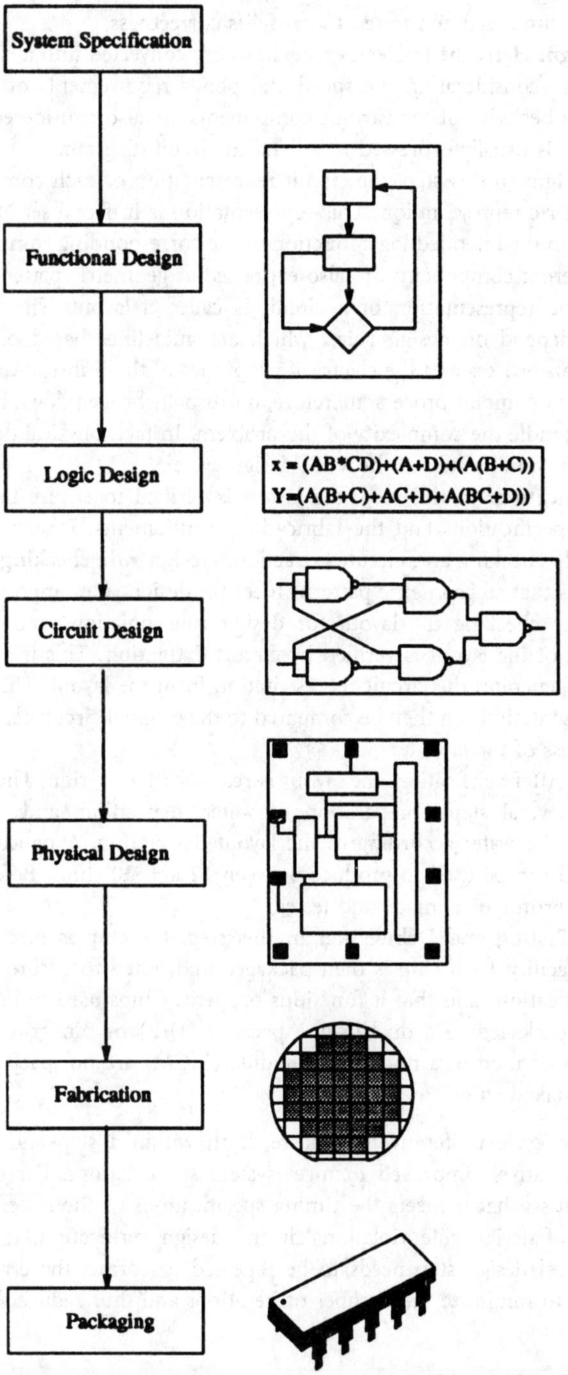

FIGURE 59.1 Design process steps.

3. **Logic Design:** In this step, the functional design is converted into a logical design, typically represented by boolean expressions. These expressions are minimized to achieve the smallest logic design which conforms to the functional design. This logic design of the system is simulated and tested to verify its correctness.
4. **Circuit Design:** Here, the boolean expressions are converted into a circuit representation by taking into consideration the speed and power requirements of the original design. The electrical behavior of the various components are also considered in this phase. The circuit design is usually expressed in a detailed circuit diagram.
5. **Physical Design:** In this step, the circuit representation of each component is converted into a geometric representation. This representation is in fact a set of geometric patterns which perform the intended logic function of the corresponding component. Connections between different components are also expressed as geometric patterns. As stated earlier, this geometric representation of a circuit is called a layout. The exact details of the layout also depend on design rules, which are guidelines based on the limitations of the fabrication process and the electrical properties of the frabrication materials. Physical design is a very complex process; therefore, it is usually broken down into various substeps in order to handle the complexity of the problem. In fact, physical design is arguably the most time-consuming step in the VLSI design cycle.
6. **Design Verification:** In this step, the layout is verified to ensure that the layout meets the system specifications and the fabrication requirements. Design verification consists of design rule checking and circuit extraction. **Design rule checking** (DRC) is a process which verifies that all geometric patterns meet the design rules imposed by the fabrication process. After checking the layout for design rule violations and removing them the functionality of the layout is verified by **circuit extraction**. This is a reverse engineering process and generates the circuit representation from the layout. This reverse engineered circuit representation can then be compared to the original circuit representation to verify the correctness of the layout.
7. **Fabrication:** After verification, the layout is ready for fabrication. The fabrication process consists of several steps: preparation of wafer, deposition, and diffusion of various materials on the wafer according to the layout description. A typical wafer is 10 cm in diameter and can be used to produce between 12 and 30 chips. Before the chip is mass produced, a prototype is made and tested.
8. **Packaging, Testing and Debugging:** In this step, the chip is bricated and diced in a fabrication facility. Each chip is then packaged and tested to ensure that it meets all the design specifications and that it functions properly. Chips used in printed circuit boards (PCBs) are packaged in a dual-in-line package (DIP) or pin grid array (PGA). Chips which are to be used in a multi-chip module (MCM) are not packaged because MCMs use bare or naked chips.

The VLSI design cycle is iterative in nature, both within a step and between steps. The representation is iteratively improved to meet system specifications. For example, a layout is iteratively improved so that it meets the timing specifications of the system. Another example may be detection of design rule violations during design verification. If such violations are detected, the physical design step needs to be repeated to correct the error. The objective of VLSI CAD tools is to minimize the number of iterations and thus reduce the time-to-market.

59.3 Physical Design Cycle

Physical design cycle converts a circuit diagram into a layout. This is accomplished in several steps such as partitioning, floorplanning, placement, routing, and compaction, as shown in Fig. 59.2.

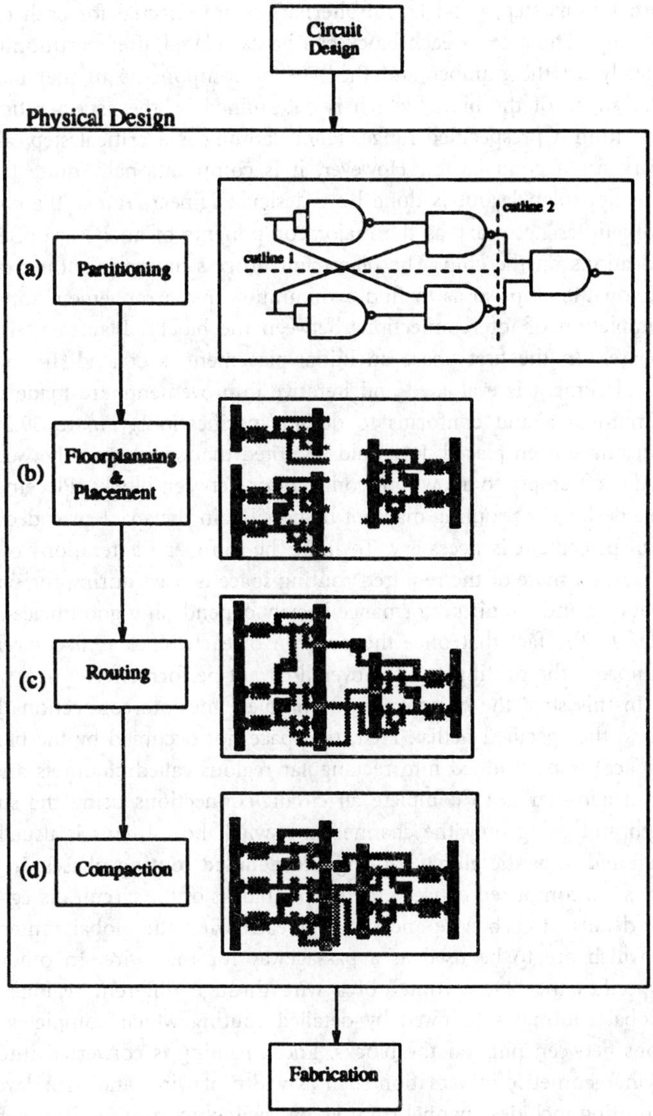

FIGURE 59.2 Physical design cycle.

1. **Partitioning:** The complex task of chip layout is divided into several smaller tasks. A chip may contain several million transistors. Layout of the entire circuit cannot be handled due to the limitation of memory space as well as computation power available. Therefore, it is normally partitioned by grouping the components into blocks (subcircuits/modules). The actual partitioning process considers many factors such as size of the blocks, number of blocks, and number of interconnections between the blocks. The output of partitioning is a set of blocks along with the interconnections required between blocks. The set of interconnections required is referred to as a **netlist**. Figure 59.2(a) shows that the input circuit has been partitioned into three blocks. In large circuits the partitioning process is hierarchical and at the topmost level a chip may have between 5 and 25 blocks. Each module is then partitioned recursively into smaller blocks.

2. **Placement:** In this step, good layout alternatives are selected for each block, as well as the entire chip. The area of each block can be calculated after partitioning and is based approximately on the number and the type of components in that block. The actual rectangular shape of the block, which is determined by the aspect ratio may, however, be varied within a prespecified range. Floorplanning is a critical step, as it sets up the groundwork for a good layout. However, it is computationally quite hard. Very often the task of floorplan layout is done by a design engineer, rather than by a CAD tool. This is sometimes necessary as the major components of an IC are often intended for specific locations on the chip. The placement process determines the exact positions of the blocks on the chip, so as to find a minimum area arrangement for the blocks that allows completion of interconnections between the blocks. Placement is typically done in two phases. In the first phase an initial placement is created. In the second phase the initial placement is evaluated and iterative improvements are made until the layout has minimum area and conforms to design specifications. Figure 59.2(b) shows that three blocks have been placed. It should be noted that some space between the blocks is intentionally left empty to allow interconnections between blocks. Placement may lead to unroutable design, i.e., routing may not be possible in the space provided. Thus, another iteration of placement is necessary. To limit the number of iterations of the placement algorithm, an estimate of the required routing space is used during the placement phase. A good routing and circuit performance heavily depend on a good placement algorithm. This is due to the fact that once the position of each block is fixed, very little can be done to improve the routing and the overall circuit performance.
3. **Routing:** In this step, the objective is to complete the interconnections between blocks according to the specified netlist. First, the space not occupied by the blocks (called the **routing space**) is partitioned into rectangular regions called **channels** and **switchboxes.** The goal of a router is to complete all circuit connections using the shortest possible wire length and using only the channel and switch boxes. This is usually done in two phases, referred to as the **global routing** and **detailed routing** phases. In global routing, connections are completed between the proper blocks of the circuit disregarding the exact geometric details of each wire and pin. For each wire, the global router finds a list of channels which are to be used as a passageway for that wire. In other words, global routing specifies the "loose route" of a wire through different regions in the routing space. Global routing is followed by detailed routing which completes point-to-point connections between pins on the blocks. Loose routing is converted into exact routing by specifying geometric information such as width of wires and their layer assignments. Detailed routing includes channel routing and switchbox routing. Routing is a very well-studied problem and several hundred articles have been published about all its aspects. Because almost all problems in routing are computationally hard, the researchers have focused on heuristic algorithms. As a result, experimental evaluation has become an integral part of all algorithms and several benchmarks have been standardized. Due to the nature of the routing algorithms, complete routing of all the connections cannot be guaranteed in many cases. As a result, a technique called **rip-up and reroute** is used, which basically removes troublesome connections and reroutes them in a different order. The routing phase of Fig. 59.2(c) shows that all the interconnections between three blocks have been routed.
4. **Compaction:** In this step, the layout is compressed in all directions such that the total area is reduced. By making the chip smaller, wire lengths are reduced, which in turn reduces the signal delay between components of the circuit. At the same time, a smaller area may imply more chips can be produced on a wafer, which in turn reduces the cost of manufacturing. However, the expense of computing time mandates that extensive compaction is used only when large quantities of ICs are produced. Compaction must

ensure that no rules regarding the design and fabrication process are violated during the process. The final diagram in Fig. 59.2(d) shows the compacted layout.

Physical design, like VLSI design, is iterative in nature, and many steps such as global routing and channel routing are repeated several times to obtain a better layout. In addition, the quality of results obtained in a step depends on the quality of solutions obtained in earlier steps. For example, a poor quality placement cannot be "cured" by high quality routing. As a result, earlier steps have more influence on the overall quality of the solution. In this sense partitioning, floorplanning, and placement problems play a more important role in determining the area and chip performance, as compared to routing and compaction. Because placement may produce an "unroutable" layout, the chip might need to be replaced or repartitioned before another routing is attempted. In general, the whole design cycle may be repeated several times to accomplish the design objectives. The complexity of each step varies depending on the design constraints as well as the design style used.

59.4 Design Styles

Even after decomposing physical design into several conceptually easier steps, it has been shown that each step is computationally very hard. Market requirements demand a quick time-to-market and high yield. As a result, restricted models and design styles are used in order to reduce the complexity of physical design. This practice began in the late 1960s and led to the development of several restricted design styles [8].

The most general form of layout is called the **full-custom** design style, in which the circuit is partitioned into a collection of subcircuits according to some criteria such as functionality of each subcircuit. In this design style each subcircuit is called a functional block or simply a block. The full-custom design style allows functional blocks to be of any size. Figure 59.3 shows an example of a very simple circuit with few blocks. Internal routing in each block is not shown for the sake of clarity. Blocks can be placed at any location on the chip surface without restrictions. In other words, this style is characterized by the absence of any constraints on the physical design process. This design style allows for very compact designs. However, the process of automating a full-custom design style has a much higher complexity than other restricted models. For this reason it is used only when final design must have a minimum area and designing time is less of a factor. The automation process for a full-custom layout is still a topic of intensive research. Some phases of physical design of a full-custom chip may be done manually to optimize the layout. Layout compaction is a very important aspect in full-custom. The rectangular solid boxes around the boundary of the circuit are called **I-O pads**. Pads are used to complete interconnections between chips or interconnections between chip and the board. The space not occupied by blocks is used for routing of interconnecting wires. Initially all the blocks are placed within the chip area, with the objective of minimizing the total area. However, enough space must be left between the blocks to complete the routing. Usually several metal layers are used for routing interconnections. Currently, two metal layers are common for routing and the three-metal layer process is gaining acceptance, as the fabrication costs become more feasible. The routing area needed between the blocks becomes increasingly smaller as more routing layers are used. This is because some routing is done on top of the transistors in the additional metal layers. If all the routing can be done on top of the transistors, the total chip area is determined by the area of the transistors.

In a hierarchical design of circuit each block in full-custom design may be very complex and may consist of several subblocks, which in turn may be designed using the full-custom design style or other design styles. It is easy to see that because any block is allowed to be placed anywhere on the chip, the problem of optimizing area and interconnection of wires becomes difficult. Full-custom design is very time consuming; thus, the method is inappropriate for very

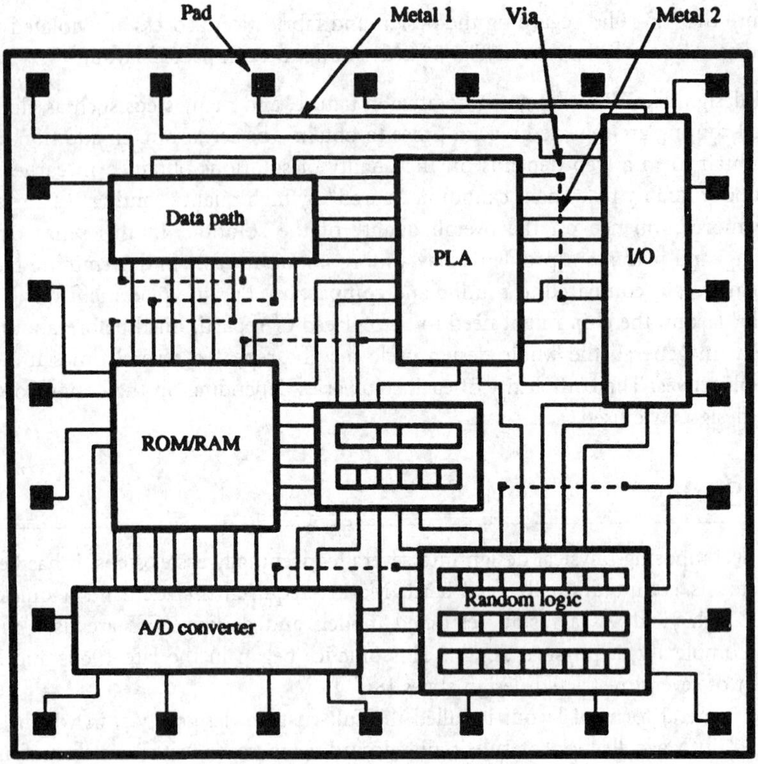

FIGURE 59.3 Full-custom structure.

large circuits, unless performance is of utmost importance. Full-custom is usually used for the layout of microprocessors.

A more restricted design style is called the **standard cell design** style. The design process in standard cell design style is simpler than a full-custom design style. Standard cell methodology considers the layout to consist of rectangular cells of the same height. Initially, a circuit is partitioned into several smaller blocks, each of which is equivalent to some predefined subcircuit (cell). The functionality and the electrical characteristics of each predefined cell are tested, analyzed, and specified. A collection of these cells is called a **cell library**, usually consisting of 200 to 400 cells. Terminals on cells may be located either on the boundary or in the center of the cells. Cells are placed in rows and the space between two rows is called a **channel**. These channels are used to perform interconnections between cells. If two cells to be interconnected lie in the same row or in adjacent rows, then the channel between the rows is used for interconnection. However, if two cells to be connected lie in two nonadjacent rows, then their interconnection wire passes through the empty space between any two cells, or **feedthrough.**

Standard cell design is well suited for moderate-size circuits and medium production volumes. Physical design using standard cells is simpler as compared to full-custom and efficient using modern design tools. The standard cell design style is also widely used to implement the "random logic" of the full-custom design, as shown in Fig. 59.3. While standard cell designs are developed more quickly, a substantial initial investment is needed in the development of the cell library, which may consist of several hundred cells. Each cell in the cell library is "handcrafted" and requires a highly skilled design engineer. Each type of cell must be created with several transistor sizes. Each cell must then be tested by simulation and its performance must be characterized. A standard cell design usually takes more area than a full-custom or a hand-crafted design.

However, as more metal layers become available for routing, the difference in area between the two design styles will gradually be reduced.

The **gate array** design style is a simplified version of the standard cell design style. Unlike the cells in standard cell designs, all the cells in gate array are identical. The entire wafer is prefabricated with an array of identical gates or cells. These cells are separated by both vertical and horizontal spaces called vertical and horizontal channels. The circuit design is modified such that it can be partitioned into a number of identical blocks. Each block must be logically equivalent to a cell on the gate array. The name "gate array" signifies the fact that each cell may simply be a gate, such as a three-input NAND gate. Each block in the design is mapped or placed onto a prefabricated cell on the wafer during the partitioning/placement phase, which is reduced to a block-to-cell assignment problem. The number of partitioned blocks must be less than or equal to that of the total number of cells on the wafer. Once the circuit is partitioned into identical blocks, the task is to make the interconnections between the prefabricated cells on the wafer using horizontal and vertical channels to form the actual circuit. The uncommitted gate array is taken into the fabrication facility and routing layers are fabricated on top of the wafer. The completed wafer is also called a **customized wafer**.

This simplicity of gate array design is gained at the cost of rigidity imposed upon the circuit both by the technology and the prefabricated wafers. The advantage of gate arrays is that the steps involved for creating any prefabricated wafer are the same, and only the last few steps in the fabrication process actually depend on the application for which the design will be used. Hence, gate arrays are cheaper and easier to produce than full-custom or standard cell. Similar to standard cell design, gate array is also a nonhierarchical structure. The gate array architecture is the most restricted form of layout. This also means it is the simplest for algorithms to work with. For example, the task of routing in gate array is to determine if a given placement is routable. The routability problem is conceptually simpler as compared to the routing problem in standard cell and full-custom design styles.

The choice of design style for a particular circuit depends on many factors such as functionality of the chip, time-to-market, and the volume of chips to be manufactured. Full-custom is typically reserved for high-performance, high-volume chips, while standard cells are used for moderate performance, where the cost of full-custom cannot be justified. Gate arrays are typically used for low-performance, low-cost applications. A design style may be applicable to the entire chip or a block of the chip.

Irrespective of the choice of the design style, all steps of the physical design cycle need to be carried out. However, the complexity, the effectiveness, and the algorithms used differ considerably depending on the design style. The following sections discuss algorithms for different phases of the physical design cycle.

59.5 Partitioning

As stated earlier, the basic purpose of partitioning is to simplify the overall design process. The circuit is decomposed into several subcircuits to make the design process manageable. This section considers the partitioning phase of the physical design cycle, study constraints, and objective functions for this problem and presents efficient algorithms.

Given a circuit, the partitioning problem is to decompose it into several subcircuits subject to constraints while optimizing a given objective function. The constraints for the partitioning problem include area constraints and terminal constraints. Area constraints are user specified for design optimization and terminal count depends on the area and aspect ratio of the block. In particular, the terminal count for a partition is given by the ratio of the perimeter of the partition to the terminal pitch. The minimum spacing between two adjacent terminals is called **terminal pitch** and is determined by the design rules. The objective functions for a partitioning problem include the minimization of the number of nets that cross the partition boundaries, and the minimization of the maximum number of times a path crosses the partition boundaries. The constraints and the objective functions used in the partitioning problem vary depending upon the partitioning level and the design style used. The actual objective function and constraints chosen for the partitioning problem may also depend on the specific problem. The number of nets which connect a partition to other partitions cannot be greater than the terminal count of the partition. In addition, the number of nets cut by partitioning should be minimized to simplify the routing task. The minimization of the number of nets cut by partitioning is one of the most important objectives in partitioning.

A disadvantage of the partitioning process is that it may degrade the performance of the final design. During partitioning, critical components should be assigned to the same partition. If such an assignment is not possible, then appropriate timing constraints must be generated to keep the two critical components close together. Usually several components, forming a critical path, determine the chip performance. If each component is assigned to a different partition, the critical path may be too long. Minimizing the length of critical paths improves system performance.

After a chip has been partitioned, each of the subcircuits must be placed on a fixed plane and the nets between all the partitions must be interconnected. The placement of the subcircuits is done by the placement algorithms and the nets are routed by using routing algorithms.

Classification of Partitioning Algorithms

The partitioning algorithms also may be classified based on the nature of the algorithms, of which two types exist: deterministic algorithms and probabilistic algorithms. **Deterministic algorithms** produce repeatable or deterministic solutions. For example, an algorithm which makes use of deterministic functions will always generate the same solution for a given problem. On the other hand, the **probabilistic algorithms** are capable of producing a different solution for the same problem each time they are used, as they make use of some random functions.

The partitioning algorithms also may be classified on the basis of the process used for partitioning. Thus, we have group migration algorithms, simulated annealing and evolution-based algorithms, and other partitioning algorithms.

The **group migration algorithms** [9, 17] start with some partitions, usually generated randomly, and then move components between partitions to improve the partitioning. The group migration algorithms are quite efficient. However, the number of partitions must be specified, which is usually not known when the partitioning process starts. In addition, the partitioning of an entire system is a multilevel operation, and the evaluation of the partitions obtained by the partitioning depends on the final integration of partitions at all levels, from the basic subcircuits to the whole system. An algorithm used to find a minimum cut at one level may sacrifice the quality of cuts for the following levels. The group migration method is a deterministic method which is often trapped at a local optimum and cannot proceed further.

The **simulated annealing/evolution algorithms** [3, 10, 18, 26] carry out the partitioning process by using a cost function, which classifies any feasible solution, and a set of moves, which allows movement from solution to solution. Unlike deterministic algorithms, these algorithms accept moves which may adversely affect the solution. The algorithm starts with a random solution and as it progresses, the proportion of adverse moves decreases. These degenerate moves act as a safeguard against entrapment in local minima. These algorithms are computationally intensive as compared to group migration and other methods.

Among all the partitioning algorithms, the group migration and simulated annealing/evolution has been the most successful heuristics for partitioning problems. The group migration and simulated annealing/evolution methods are most widely used, and extensive research has been carried out in these two types of algorithms.

The Kernighan-Lin Partitioning Algorithm

The Kernighan-Lin (K-L) algorithm is a bisectioning algorithm. It starts by initially partitioning the graph $G = (V, E)$ into two subsets of equal size. Vertex pairs are exchanged across the bisection, if the exchange improves the cutsize. The above procedure is carried out iteratively until no further improvement can be achieved.

The basic idea of the K-L algorithm is illustrated with the help of an example before presenting the algorithm formally. Consider the example given in Fig. 59.4(a). The initial partitions are

$$A = \{1, 2, 3, 4\}$$

$$B = \{5, 6, 7, 8\}$$

Notice that the initial cutsize is 9. The next step of the K-L algorithm is to choose a pair of vertices whose exchange results in the largest decrease of the cutsize or results in the smallest increase, if no decrease is possible. The decrease of the cutsize is computed using gain values $D(i)$ of vertices v_i. The gain of a vertex v_i is defined as

$$D(i) = inedge(i) - outedge(i)$$

where *inedge(i)* is the number of edges of vertex i that do not cross the bisection boundary and *outedge(i)* is the number of edges that cross the boundary. The amount by which the cutsize decreases, if vertex v_i changes over to the other partition, is represented by $D(i)$. If v_i and v_j are exchanged, the decrease of the cutsize is $D(i) + D(j)$. In the example given in Fig. 59.4, a suitable vertex pair is (3, 5), which decreases the cutsize by 3. A tentative exchange of this pair is made. These two vertices are then locked. This lock on the vertices prohibits them from taking part in any further tentative exchanges. The above procedure is applied to the new partitions, which gives a second vertex pair of (4, 6). This procedure is continued until all the vertices are locked. During this process, a log of all tentative exchanges and the resulting cutsizes is stored. Table 59.1 shows the log of vertex exchanges for the given example. Note that the partial sum of cutsize decrease $g(i)$ over the exchanges of first i vertex pairs is given in the table; e.g., $g(1) = 3$ and $g(2) = 8$. The value of k for which $g(k)$ gives the maximum value of all $g(i)$ is determined from the table. In this example $k = 2$ and $g(2) = 8$ is the maximum partial sum. The first k pairs of vertices are actually exchanged. In the example the first two vertex pairs (3, 5) and (4, 6) are actually exchanged, resulting in the bisection shown in Fig. 59.4(b). This completes an iteration and a new iteration starts. However, if no decrease of cutsize is possible during an iteration, the algorithm stops. Figure 59.5 presents the formal description of the K-L algorithm.

The procedure INITIALIZE finds initial bisections, and initializes the parameters in the algorithm. The procedure IMPROVE tests if any improvement has been made during the last

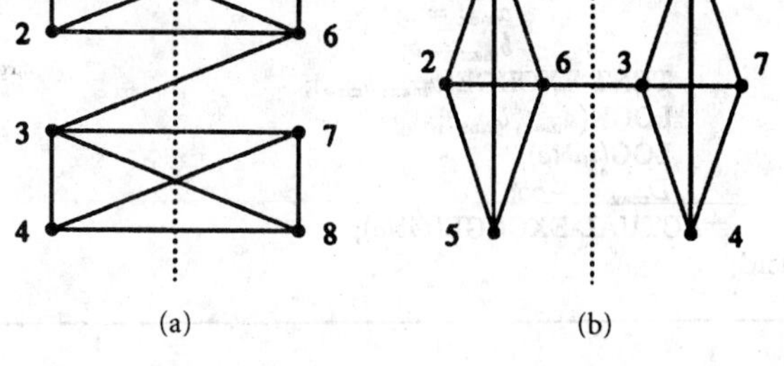

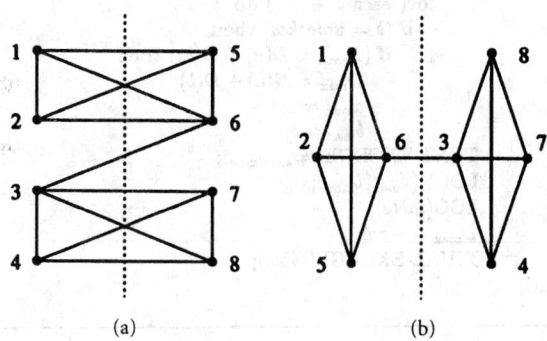

FIGURE 59.4 A graph bisected by K-L algorithm. (a) Initial bisections. (b) Final bisections.

TABLE 59.1 Log of the Vertex Exchanges

i	Vertex Pair	$g(i)$	$\sum_{j=1}^{i} g(i)$	Cutsize
0	—	—	—	9
1	(3, 5)	3	3	6
2	(4, 6)	5	8	1
3	(1, 7)	−6	2	7
4	(2, 8)	−2	0	9

iteration, while the procedure UNLOCK checks if any vertex is unlocked. Each vertex has a state of either *locked* or *unlocked*. Only those vertices whose status is *unlocked* are candidates for the next tentative exchanges. The procedure TENT-EXCHGE tentatively exchanges a pair of vertices. The procedure LOCK locks the vertex pair, while the procedure LOG stores the log table. The procedure ACTUAL-EXCHGE determines the maximum partial sum of $g(i)$, selects the vertex pairs to be exchanged, and fulfills the actual exchange of these vertex pairs.

The time complexity of the K-L algorithm is $O(n^3)$. The K-L algorithm is, however, quite robust. It can accommodate additional constraints, such as a group of vertices requiring to be in a specified partition. This feature is very important in layout because some blocks of the circuit are to be kept together due to the functionality. For example, it is important to keep all components of an adder together. However, the K-L algorithm has several disadvantages. For example, the algorithm is not applicable for hypergraphs, it cannot handle arbitrarily weighted graphs and the partition sizes must be specified before partitioning. Finally, the complexity of the algorithm is considered too high even for moderate size problems.

```
Algorithm KL

Begin
    INITIALIZE();
    while( IMPROVE(table) = TRUE ) do
    (* if an improvement has been made during last iteration,
    the process is carried out again. *)
      while ( UNLOCK(A) = TRUE ) do
      (* if there exists any unlocked vertex in A,
      more tentative exchanges are carried out. *)
        for ( each a ∈ A ) do
          if (a = unlocked) then
            for( each b ∈ B ) do
              if (b = unlocked) then
                if (D_max < D(a) + D(b)) then
                  D_max = D(a) + D(b);
                  a_max = a;
                  b_max = b;
        TENT-EXCHGE(a_max, b_max);
        LOCK(a_max, b_max);
        LOG(table);
        D_max = -∞;
      ACTUAL-EXCHGE(table);
End;
```

FIGURE 59.5 Algorithm K-L.

59.6 Other Partitioning Algorithms

In order to overcome the disadvantages of the K-L algorithm, several extensions of the K-L algorithm such as the Fiduccia-Mattheyses algorithm, the Goldberg and Burstein algorithm, the component replication algorithm, and the ratio cut algorithm were developed. In the class of probabilistic and iterative algorithms simulated annealing and evolution-based algorithms have been developed for partitioning. For details on these partitioning algorithms refer to Chapter 4 of [29].

59.7 Placement

The placement phase follows the partitioning phase of the physical design cycle. After the circuit has been partitioned, the area occupied by each block (subcircuit) can be calculated and the number of terminals (pins) required by each block is known. Partitioning also generates the netlist which specifies the connections between the blocks. The layout is completed by arranging the blocks on the layout surface and interconnecting their pins according to the netlist. The arrangement of blocks is done in the placement phase, while interconnection is completed in the routing phase. In the placement phase blocks are assigned a specific shape and are positioned on a layout surface in such a fashion that no two blocks overlap and enough space is left on the layout surface to complete the interconnections between the blocks. The blocks are positioned so as to minimize the total area of the layout. In addition, the locations of pins on each block are also determined.

The input to the placement phase is a set of blocks, the number of terminals for each block, and the netlist. If the layout of the circuit within a block has been completed, then the dimensions of the block are also known. The blocks for which the dimensions are known are called **fixed blocks** and the blocks for which dimensions are yet to be determined are called **flexible blocks**. Thus, during the placement phase, we need to determine an appropriate shape for each block (if shape is not known), the location of each block on the layout surface, and determine the locations of pins on the boundary of the blocks. The problem of assigning locations to the fixed blocks on a layout surface is called the **placement problem**. If some or all of the blocks are flexible, the problem is called the **floorplanning problem**. Hence, the placement problem is a restricted version of the floorplanning problem. The terminology is slightly confusing as floorplanning problems are placement problems as well, but these terminologies have been widely used and accepted. It is desirable that the pin locations are identified at the same time the block locations are fixed. However, due to the complexity of the placement problem, the problem of identifying the pin locations for the blocks is solved after the locations of all the blocks are known. This process of identifying pin locations is called **pin assignment**.

The placement phase is crucial in the overall physical design cycle because an ill-placed layout cannot be improved by high-quality routing. In other words, the overall quality of the layout in terms of area and performance is mainly determined in the placement phase.

Classification of Placement Algorithms

The placement algorithms can be classified on the basis of the input to the algorithms, the nature of output generated by the algorithms, and the process used by the algorithms.

Depending on the input, the placement algorithms can be classified into two major groups: **constructive placement** and **iterative improvement** methods. The input to the constructive placement algorithms consists a set of blocks along with the netlist. The algorithm finds the locations of blocks. On the other hand, iterative improvement algorithms start with an initial placement. These algorithms modify the initial placement in search of a better placement. These algorithms are typically used in an iterative manner until no improvement is possible.

The nature of output produced by an algorithm is another way of classifying the placement algorithms. Some algorithms generate the same solution when presented with the same problem; i.e., the solution produced is repeatable. These algorithms are called **deterministic placement algorithms.** Algorithms that function on the basis of fixed connectivity rules (or formulas) or determine the placement by solving simultaneous equations are deterministic and always produce the same result for a particular placement problem. Some algorithms, on the other hand, work by randomly examining configurations and may produce a different result each time they are presented with the same problem. Such algorithms are called **probabilistic placement algorithms.**

The classification based on the process used by the placement algorithms is perhaps the best way of classifying these algorithms. Two important classes of algorithms come under this classification: simulation-based algorithms and partitioning-based algorithms. **Simulation-based algorithms** simulate some natural phenomenon while **partitioning-based algorithms** use partitioning for generating the placement. The algorithms which use clustering and other approaches are classified under "other" placement algorithms.

Simulated Annealing Placement Algorithm

Simulated annealing is one of the most well-developed methods available [2, 10–12, 16, 19, 23, 25–28]. The simulated annealing technique has been successfully used in many phases of VLSI physical design, e.g., circuit partitioning. A detailed description of the application of this method to partitioning may be found in Chapter 4 of [29]. Simulated annealing is used in placement as an iterative improvement algorithm. Given a placement configuration, a change to that configuration is made by moving a component or interchanging locations of two components. In the case of the simple pairwise interchange algorithm it is possible that an achieved configuration has a cost higher than that of the optimum, but no interchange can cause a further cost reduction. In such a situation the algorithm is trapped at a local optimum, and cannot proceed further. Actually, this happens quite often when this algorithm is used in real-life examples. Simulated annealing avoids getting stuck at a local optimum by occasionally accepting moves that result in a cost increase.

In simulated annealing all moves that result in a decrease in cost are accepted. Moves that result in an increase in cost are accepted with a probability that decreases over the iterations. The analogy to the actual annealing process is heightened with the use of a parameter called **temperature** T. This parameter controls the probability of accepting moves which result in an increased cost. Additional moves are accepted at higher values of temperature than at lower values. The acceptance probability can be given by $e^{-\Delta C/T}$, where ΔC is the increase in cost. The algorithm starts with a very high value of temperature, which gradually decreases so that moves that increase cost have a lower probability of being accepted. Finally, the temperature reduces to a very low value which causes only moves that reduce cost to be accepted. In this way the algorithm converges to an optimal or near-optimal configuration.

In each stage the configuration is shuffled randomly to obtain a new configuration. This random shuffling could be achieved by displacing a block to a random location, an interchange of two blocks, or any other move which can change the wire length. After the shuffle, the change in cost is evaluated. If a decrease in cost occurs, the configuration is accepted; otherwise, the new configuration is accepted with a probability that depends on the temperature. The temperature is then lowered using some function which, for example, could be exponential in nature. The process is stopped when the temperature has dropped to a certain level. The outline of the simulated annealing algorithm is shown in Fig. 59.6.

The parameters and functions used in a simulated annealing algorithm determine the quality of the placement produced. These parameters and functions include the cooling schedule consisting of initial temperature *(init_temp)*, final temperature *(final_temp)*, and the function

```
Algorithm Simulated-Annealing

Begin
    temp = INIT-TEMP;
    place = INIT-PLACEMENT;
    while (temp > FINAL-TEMP) do
       while (inner_loop_criterion = FALSE) do
          new_place = PERTURB(place);
          ΔC = COST(new_place) - COST(place);
          if (ΔC < 0) then
             place = new_place;
          else if (RANDOM(0, 1) > e^(ΔC/T)) then
             place = new_place;
       temp = SCHEDULE(temp);
End;
```

FIGURE 59.6 The simulated annealing algorithm.

used for changing the temperature (SCHEDULE), *inner_loop_criterion*, which is the number of trials at each temperature, the process used for shuffling a configuration (PERTURB), acceptance probability (F), and the cost function (COST). A good choice of these parameters and functions can result in a good placement in a relatively short time.

Other Placement Algorithms

Several other algorithms which simulate naturally occurring processes have been developed for routing. The simulated evolution algorithm is analogous to the natural process of mutation of species as they evolve to better adapt to their environment. The force directed placement explores the similarity between the placement problem and the classical mechanics problem of a system of bodies attached to springs. The partitioning-based placement techniques include Breuer's algorithm and the terminal propagation algorithm. Several other algorithms such as the cluster growth algorithm, the quadratic assignment algorithm, the resistive network optimization algorithm, and the Branch and Bound algorithm also exist. For more details on these algorithms, refer to Chapter 5 of [29].

59.8 Routing

The exact locations of circuit blocks and pins are determined in the placement phase. A netlist is also generated which specifies the required interconnections. Space not occupied by the blocks can be viewed as a collection of regions. These regions are used for routing and are called **routing regions**. The process of finding the geometric layouts of all the nets is called **routing**. Each routing region has a capacity, which is the maximum number of nets that can pass through that region. The capacity of a region is a function of the design rules and dimensions of the routing regions and wires. Nets must be routed within the routing regions and must not violate the capacity of any routing region. In addition, nets must not short circuit; that is, nets must not intersect each other. The objective of routing is dependent on the nature of the chip. For general purpose chips, it is sufficient to minimize the total wire length. For high-performance chips, total wire length may not be a major concern. Instead, we may want to minimize the longest wire to minimize the delay in the wire and therefore maximize its performance. Usually routing involves special treatment of such nets as clock nets, power, and ground nets. In fact, these nets are routed separately by special routers.

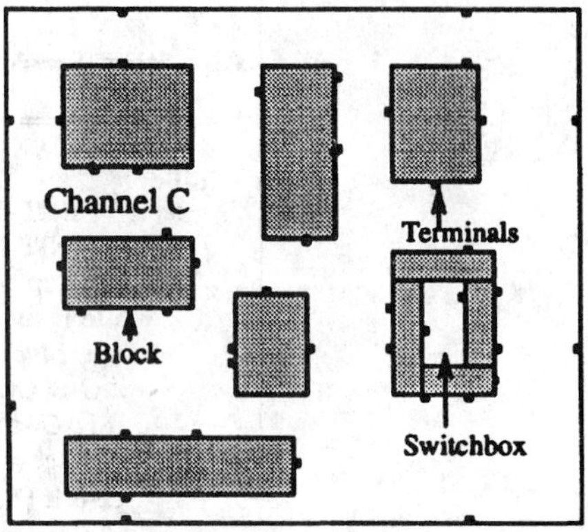

FIGURE 59.7 Layout of circuit blocks and pins after placement.

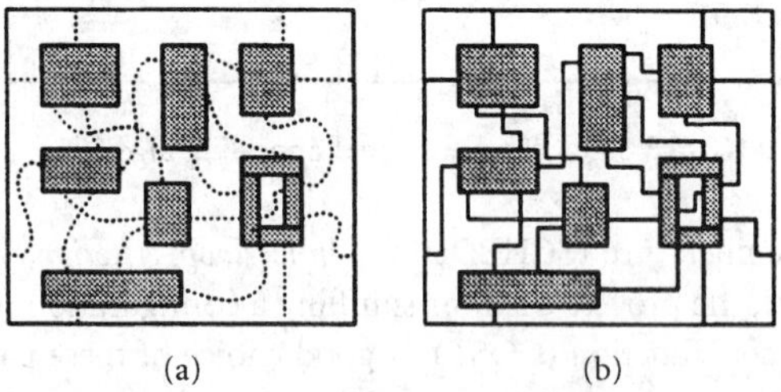

FIGURE 59.8 (a) Global routing. (b) Detailed routing.

Channels and switchboxes are the two types of routing regions. A **switchbox** is a rectangular area bounded on all sides. A **channel** is a rectangular area bounded on two opposite sides by the blocks. The capacity of a channel is a function of the number of layers (l), height (h) of the channel, wire width (w), and wire separation (s); i.e., $Capacity = (l \times h)/(w + s)$. For example, if for channel C (Fig. 59.7), $l = 2$, $h = 18\lambda$, $w = 3\lambda$, $s = 3\lambda$, then the capacity is $(2 \times 18)/(3 + 3) = 6$.

A VLSI chip may contain several million transistors. As a result, tens of thousands of nets must be routed to complete the layout. In addition, several hundred routes are possible for each net. This makes the routing problem computationally hard. In fact, even when the routing problem is restricted to channels, it cannot be solved in polynomial time; i.e., the channel routing problem is NP-complete [30]. Therefore, routing traditionally is divided into two phases. The first phase is called **global routing** and generates a "loose" route for each net. In fact, it assigns a list of routing regions to each net without specifying the actual geometric layout of wires. The second phase, which is called **detailed routing**, finds the actual geometric layout of each net within the assigned routing regions (see Fig. 59.8(b)). Unlike global routing, which considers the entire layout, a detailed router considers just one region at a time. The exact layout is produced for each wire segment assigned to a region, and vias are inserted to complete the layout. Detailed routing includes channel routing and switchbox routing. Another approach to routing is called **area routing**, which is a single-phase routing technique. However, this technique is computationally infeasible for general VLSI circuits and is typically used for specialized problems.

59.9 Classification of Global Routing Algorithms

Two approaches solve the global routing problem: the sequential and the concurrent.

1. **Sequential Approach:** As the name suggests, nets are routed one by one. However, once a net has been routed it may block other nets which are yet to be routed. As a

result, this approach is very sensitive to the order in which the nets are considered for routing. Usually, the nets are sequenced according to their criticality, perimeter of the bounding rectangle, and number of terminals. The criticality of a net is determined by the importance of the net.

2. **Concurrent Approach:** This approach avoids the ordering problem by considering routing of all the nets simultaneously. The concurrent approach is computationally hard and no efficient polynomial algorithms are known, even for two-terminal nets. As a result, integer programming methods have been suggested. The corresponding integer program is usually too large to be employed efficiently. Hence, hierarchical methods that work from the top down are employed to partition the problem into smaller subproblems, which can be solved by integer programming.

59.10 Classification of Detailed Routing Algorithms

Many ways are possible for classifying the detailed routing algorithms. The algorithms could be classified on the basis of the routing models used. Some routing algorithms use grid-based models, while some other algorithms use the gridless model. The gridless model is more practical as all the wires in a design do not have the same width. Another possible classification scheme could be to classify the algorithms based on the strategy they use. Thus, we could have greedy routers or hierarchical routers to name two. We classify the algorithms based on the number of layers used for routing. Single-layer routing problems frequently appear as subproblems in other routing problems which deal with more than one layer. Two-layer routing problems have been thoroughly investigated because until recently, due to limitations of the fabrication process, only two metal layers were allowed for routing. A third metal layer is now allowed, thanks to improvements in the fabrication process, but it is expensive compared to the two-layer metal process. Several multilayer routing algorithms also were developed recently, which can be used for routing MCMs, which have up to 32 layers.

Lee's Algorithm for Global Routing

This algorithm, which was developed by Lee in 1961 [20], is the most widely used algorithm for finding a path between any two vertices on a planar rectangular grid. The key to the popularity of Lee's maze router is its simplicity and its guarantee of finding an optimal solution, if one exists.

The exploration phase of Lee's algorithm is an improved version of the breadth-first search. The search can be visualized as a wave propagating from the source. The source is labeled '0' and the wavefront propagates to all the unblocked vertices adjacent to the source. Every unblocked vertex adjacent to the source is marked with a label '1'. Then, every unblocked vertex adjacent to vertices with a label '1' is marked with a label '2', and so on. This process continues until the target vertex is reached or no further expansion of the wave can be carried out. An example of the algorithm is shown in Fig. 59.9. Due to the breadth-first nature of the search, Lee's maze router is guaranteed to find a path between the source and the target, if one exists. In addition, it is guaranteed to be the shortest path between the vertices.

The input to Lee's algorithm is an array B, the source, s, and target, t, vertex. $B[v]$ denotes if a vertex v is blocked or unblocked. The algorithm uses an array, L, where $L[v]$ denotes the distance from the source to the vertex v. This array will be used in the procedure RETRACE that retraces the vertices to form a path, P, which is the output of Lee's algorithm. Two linked lists, *plist* (propagation list) and *nlist* (neighbor list), are used to keep track of the vertices on the wavefront and their neighbor vertices, respectively. These two lists are always retrieved from tail to head. We also assume that the neighbors of a vertex are visited in counterclockwise order, that is, top, left, bottom, and then right.

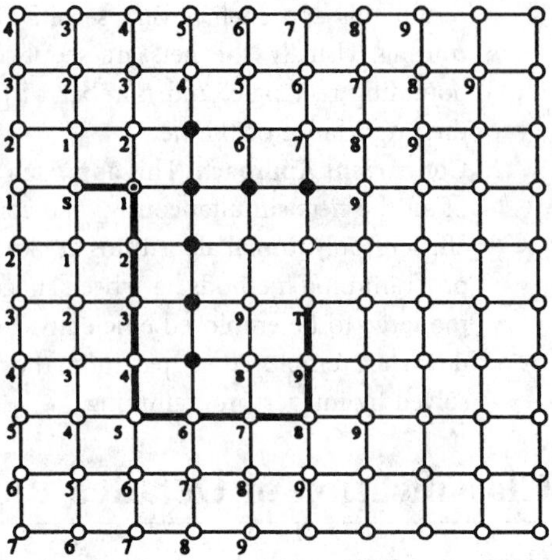

FIGURE 59.9 A net routed by Lee's algorithm.

The formal description of Lee's algorithm appears in Fig. 59.10. The time and space complexity of Lee's algorithm is $O(h \times w)$ for a grid of dimension $h \times w$.

Lee's routing algorithm requires a large amount of storage space and its performance degrades rapidly when the size of the grid increases. Numerous attempts have been made to modify the algorithm to improve its performance and reduce its memory requirements.

Lee's algorithm requires up to $k + 1$ bits per vertex, where k bits are used to label the vertex during the exploration phase and an additional bit is needed to indicate whether the vertex is blocked. For an $h \times w$ grid, $k = \log_2(h \times w)$. Aker [1] noticed that in the retrace phase of Lee's

```
Algorithm Lee-Router (B, s, t, P)

Input: B, s, t
Output: P

Begin
    plist = s;
    nlist = φ;
    temp = 1;
    path_exists = FALSE;
    while plist ≠ φ do
      for each vertex v_i in plist do
        for each vertex v_j neighboring v_i do
        if B[v_j] = UNBLOCKED then
          L[v_j] = temp;
          INSERT(v_j,nlist);
          if v_j = t then
            path_exists = TRUE;
            exit while;
      temp=temp + 1;
      plist = nlist;
      nlist = φ;
    if path_exists = TRUE then RETRACE (L, P);
    else path does not exist;
End;
```

FIGURE 59.10 Algorithm LEE-ROUTER.

algorithm, only two types of neighbors of a vertex need to be distinguished; vertices toward the target and vertices toward the source. This information can be coded in a single bit for each vertex. The vertices in wavefront, L, are always adjacent to the vertices in wavefront $L - 1$ and $L + 1$. Thus, during wave propagation, instead of using a sequence $1, 2, 3, \ldots$, the wavefronts are labeled by a sequence such as $0, 0, 1, 1, 0, 0, \ldots$. The predecessor of any wavefront is labeled differently from its successor. Thus, each scanned vertex is either labeled '0' or '1'. Besides these two states, additional states ('block' and 'unblocked') are needed for each vertex. These four states of each vertex can be represented by using exactly two bits, regardless of the problem size. Compared to Acker's scheme, Lee's algorithm requires at least 12 bits per vertex for a grid size of 2000×2000.

It is important to note that Acker's coding scheme only reduces the memory requirement per vertex. It inherits the search space of Lee's original routing algorithm, which is $O(h \times w)$ in the worst case.

Greedy Channel Router for Detailed Routing

Assigning the complete trunk of a net or a two-terminal net segment of a multiterminal net severely restricts LEA and dogleg routers. Optimal channel routing can be obtained if for each column it can be guaranteed that only one horizontal track per net exists. Based on this observation, one approach to reduce channel height could be to route nets column by column trying to join split horizontal tracks (if any) that belong to the same net as much as possible.

Based on the above observation and approach, Rivest and Fiduccia [24] developed the **greedy channel router**. This makes fewer assumptions than LEA and dogleg routers. The algorithm starts from the leftmost column and places all the net segments of a column before proceeding to the next right column. In each column the router assigns net segments to tracks in a greedy manner. However, unlike the dogleg router, the greedy router allows the doglegs in any column of the channel, not necessarily where the terminal of the doglegged net occurs.

Given a channel routing problem with m columns, the algorithm uses several steps while routing a column. In the fist step the algorithm connects any terminal to the trunk segment of the corresponding net. This connection is completed by using the first empty track, or the track that already contains the net. In other words, minimum vertical segment is used to connect a trunk to a terminal. For example, net 1 in Fig. 59.11(a) in column 3 is connected to the same net. The second step attempts to collapse any **split nets** (horizontal segments of the same net present on two different tracks) using a vertical segment as shown in Fig. 59.11(b). A split net will occur when two a terminals of the same net are located on different sides of the channel and cannot be connected immediately because of existing vertical constraints. This step also brings a terminal connection to the correct track if it has stopped on an earlier track. If two sets of split nets overlap, the second step will only be able to collapse one of them.

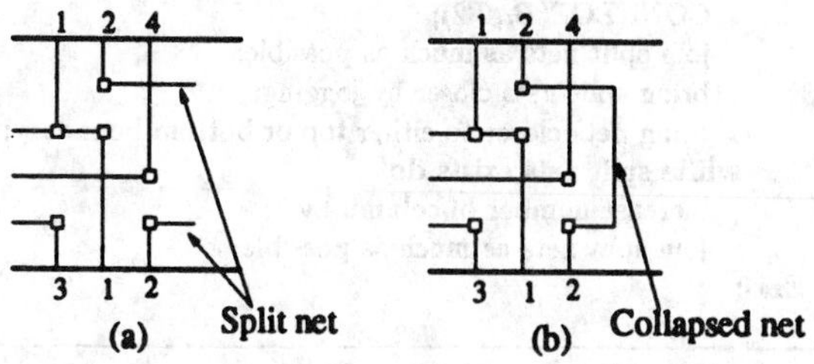

FIGURE 59.11 (a) A split net. (b) The collapsed split net.

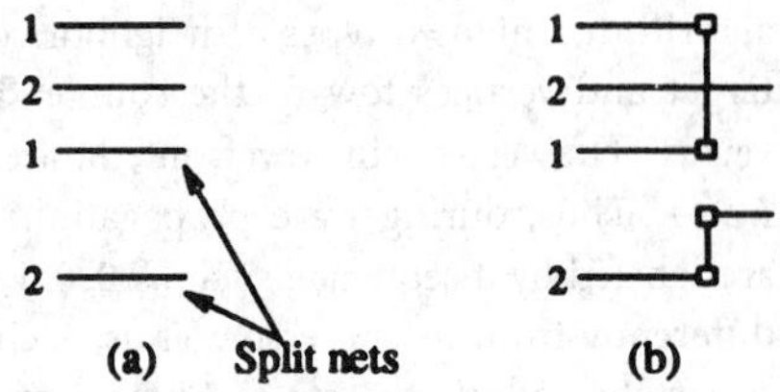

FIGURE 59.12 (a) Reducing the distance between split nets. (b) The result.

In the third step the algorithm tries to reduce the range or the distance between two tracks of the same net. This reduction is accomplished by using a dogleg, as shown in Fig. 59.12(a) and (b). The fourth step attempts to move the nets closer to the boundary which contains the next terminal of that net. If the next terminal of a net being considered is on the top (bottom) boundary of the channel, then the algorithm tries to move the net to the upper (lower) track. In case no track is available, the algorithm adds extra tracks and the terminal is connected to this new track. After all five steps have been completed, the trunks of each net are extended to the next column and the steps are repeated. A detailed description of the greedy channel routing algorithm is in Fig. 59.13.

The greedy router sometimes gives solutions which contain an excessive number of vias and doglegs. It does, however, have the capability of providing a solution even in the presence of cyclic vertical constraints. The greedy router is more flexible in the placement of doglegs due to fewer assumptions about the topology of the connections. An example routed by the greedy channel router is shown in Fig. 59.14.

```
Algorithm Greedy-Channel-Router (N)

Input: N
Begin
   d = DENSITY(N);
  (* calculate the lower bound of channel density *)
    insert d tracks to channel;
    for i = 1 to m do
       T1 = GET-EMPTY-TRACK;
       if T1 = 0 then
          ADD-TRACK(T1);
          ADD-TRACK(T2);
       else
          T2 = GET-EMPTY-TRACK;
          if T2 = 0 then
             ADD-TRACK(T2);
       CONNECT(T_i, T1);
       CONNECT(B_i, T2);
       join split nets as much as possible;
       bring split nets closer by jogging;
       bring nets closer to either top or bottom boundary;
    while split nets exists do
       increase number of column by 1;
       join split nets as much as possible;
End;
```

FIGURE 59.13 Algorithm GREEDY-CHANNEL-ROUTER.

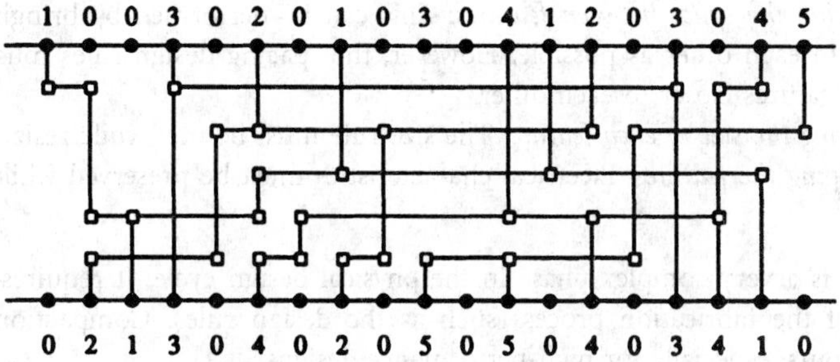

FIGURE 59.14 Channel routed using a greedy router.

Other Routing Algorithms

Soukup proposed an algorithm which basically cuts down the search time of Lee's algorithm by exploring in the direction toward the target, without changing the direction until it reaches the target or an obstacle. An alternative approach to improve upon the speed was suggested by Hadlock. One class of algorithms is called the line-probe algorithms. The basic idea of the line-probe algorithms is to reduce the memory requirement by using line segments instead of grid nodes in the search. Several algorithms based on Steiner trees have been developed. The main advantage of these algorithms is that they can be used for routing multiterminal nets. For further details on these global routing algorithms, refer to Chapter 6 of [29].

Extensive research has been carried out in the area of detailed routing and several algorithms exist for channel and switchbox routing. There are LEA-based algorithms which use a reserved layer model and do not allow any vertical constraints or doglegs. The YACR2 algorithm can handle vertical constraint violations. The net merge channel router works for two-layer channel routing problems and it exploits the graph theoretic structure of channel routing problems. Glitter is an algorithm for gridless channel routing and can handle nets of varying widths. The hierarchical channel router divides the channel routing problem into smaller problems, each of which is solved in order to generate the complete routing for the nets in the channel. Several algorithms such as the extended net merge channel routing algorithm, HVH routing from the HV solution, and the hybrid HVH-VHV channel routing algorithm exist for the three-layer channel routing problem. For further details on these detailed routing algorithms, refer to Chapter 7 of [29].

9.11 Compaction

After completion of detailed routing, the layout is functionally complete. At this stage, the layout is ready to be used to fabricate a chip. However, due to nonoptimality of placement and routing algorithms, some vacant space is present in the layout. In order to minimize the cost, improve performance, and yield, layouts are reduced in size by removing the vacant space without altering the functionality of the layout. This operation of layout area minimization is called layout compaction.

The compaction problem is simplified by using symbols to represent primitive circuit features, such as transistors and wires. The representation of layout using symbols is called a **symbolic layout**. Special languages [4, 21, 22] and special graphic editors [13, 14] are available to describe symbolic layouts. To produce the actual masks, the symbolic layouts are translated into actual geometric features. Although a feature can have any geometric shape, in practice only rectangular shapes are considered.

The goal of compaction is to minimize the total layout area without violating any design rules. The area can be minimized in three ways:

1. *By reducing the space between features:* This can be performed by bringing the features as close to each other as possible. However, the spacing design rules must be met while moving features closer to each other.
2. *By reducing the size of each feature:* The size rule must be met while resizing the features.
3. *By reshaping the features:* Electrical characteristics must be preserved while reshaping the feature.

Compaction is a very complex phase in the physical design cycle. It requires understanding many details of the fabrication process such as the design rules. Compaction is critical for full-custom layouts, especially for high-performance designs.

Classification of Compaction Algorithms

Compaction algorithms can be classified in two ways. The first classification scheme is based on the direction of movements of the components (features): *one-dimensional* (1-D) and *two-dimensional* (2-D). In **1-D compaction** components are moved only in the x- or the y-direction. As a result, either the x- or the y-coordinate of the components is changed due to the compaction. If the compaction is done along the x-direction then it is called *x-compaction.* Similarly, if the compaction is done along the y-direction, then it is called *y-compaction.* In **2-D compaction** the components can be moved in both the x- and y-directions simultaneously. As a result, in 2-D compaction both x- and y-coordinates of the components are changed at the same time in order to minimize the layout area.

The second approach to classify the compaction algorithms is based on the technique for computing the minimum distance between features. In this approach we have two methods, **constraint graph-based compaction** and **virtual grid-based compaction.** In the constraint graph method the connections and separations rules are described using linear inequalities which can be modeled using a weighted directed graph (constraint graph). This constraint graph is used to compute the new positions for the components. On the other hand, the virtual grid method assumes the layout is to be drawn on a grid. Each component is considered attached to a grid line. The compaction operation compresses the grid along with all components placed on it, keeping the grid lines straight along the way. The minimum distance between two adjacent grid lines depends on the components on these grid lines. The advantage of the virtual grid method is that the algorithms are simple and can be easily implemented. However, the virtual grid method does not produce compact layouts as does the constraint graph method.

In addition, compaction algorithms can be classified on the basis of the hierarchy of the circuit. If compaction is applied to different levels of the layout, it is called **hierarchical compaction**. Any of the above-mentioned methods can be extended to hierarchical compaction. A variety of hierarchical compaction algorithms have been proposed for both constraint graph and virtual grid methods. Some compaction algorithms actually "flatten the layout" by removing all hierarchy and then perform compaction. In this case it may not be possible to reconstruct the hierarchy, which may be undesirable.

Shadow-Propagation Algorithm for Compaction

A widely used and one of the best known techniques for generating a constraint graph is the **shadow-propagation** used in the CABBAGE system [15]. The "shadow" of a feature is propagated along the direction of compaction. The shadow is caused by shining an imaginary light from behind the feature under consideration (see Fig. 59.15). Usually the shadow of the feature is extended on both sides of the features in order to account for diagonal constraints. This leads to greater than minimal Euclidean spacings because an enlarged rectangle is used to account for corner interactions. (See shadow of feature in Fig. 59.15.)

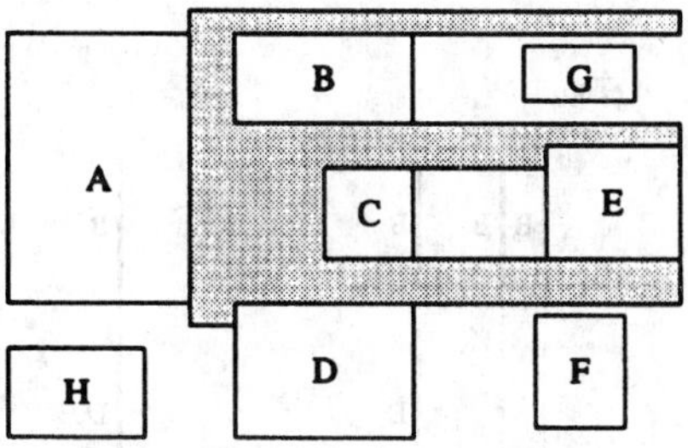

FIGURE 59.15 Example of shadow propagation.

When the shadow is obstructed by another feature, an edge is added to the graph between the vertices corresponding to the propagating feature and the obstructing feature. The obstructed part of the shadow is then removed from the front and no longer propagated. The process is continued until all of the shadow has been obstructed. This process is repeated for each feature in the layout. The algorithm SHADOW-PROPAGATION, given in Fig. 59.16, presents an overview of the algorithm for x-compaction of a single feature from left to right.

The SHADOW-PROPAGATION routine accepts the list of components *(Comp_list)*, which is sorted on the x-coordinates of the left corner of the components and the component *(component)* for which the constraints are to be generated. The procedure, INITIALIZE-SCANLINE, computes the total length of the interval in which the shadow is to be generated. This length includes the design rule separation distance. The y-coordinate of the top and the bottom of this interval are stored in the global variables, *top* and *bottom*, respectively. The procedure, GET-NXT-COMP, returns the next component *(curr_comp)* from *Comp_list*. This component is then removed from the list. Procedure LEFT-EDGE returns the vertical interval of component, *curr_comp*. If this interval is within the *top* and *bottom* then *curr_comp* may have a constraint with *component*. This check is performed by the procedure IN-RANGE. If the interval for *curr_component* lies within *top* and *bottom* and if this interval is not already contained within one of the intervals in the interval set, $\mathcal{I}$, then the component lies in the shadow of *component*, and hence a constraint must be generated. Each interval represents the edge at which the shadow is blocked by a component. The constraint is added to the constraint graph by the procedure ADD-CONSTRAINT. The procedure UNION inserts the interval corresponding to *curr_comp* in the interval set at the

```
Algorithm Shadow-Propagation (Comp_list, component)

Input: Comp_list, component
Begin
    INITIALIZE-SCANLINE( component );
    I = φ;
    while( (LENGTH-SCANLINE( I ) < (top − bottom))
        and ( Comp_list ≠ φ) )
        curr_comp = GET-NXT-COMP( Comp_list );
        Ii = LEFT-EDGE( curr_comp );
        if( IN-RANGE( Ii, top, bottom ) )
            I' = UNION( Ii, I );
            if( I' ≠ I )
                ADD-CONSTRAINT( component, curr_comp );
                I = UNION( Ii, I );
End;
```

FIGURE 59.16 Shadow-propagation algorithm.

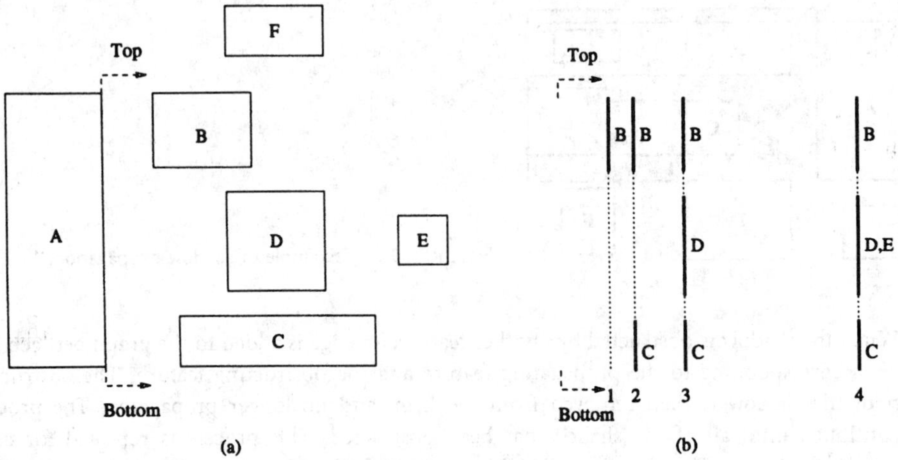

FIGURE 59.17 Interval generation for shadow propagation.

appropriate position. This process is carried out until the interval set completely covers the interval from *top* to *bottom* or no more components are in *Comp_list*. Figure 59.17(a) shows the layout of components. The constraint for component *A* with other components is being generated. Figure 59.17(b) shows the intervals in the interval set as the shadow is propagated. From Fig. 59.17(b) it is clear that the constraints will be generated between components *A* and components *B*, *C*, and *D* in that order. As component *F* lies outside the interval defined by *top* and *bottom* it is not considered for constraint generation. The interval generated by component *E* lies within one of the intervals in the interval set. Hence, no constraint is generated between components *A* and *E*.

Other Compaction Algorithms

Several algorithms such as constraint graph-based compaction algorithms, scanline algorithm, and grid-based compaction algorithms exist for the 1-D compaction problem. An algorithm based on a simulation of the zone-refining process also was developed. This compactor is considered a $1\frac{1}{2}$-D compactor, as the key idea is to provide enough lateral movements to blocks during compaction to resolve interferences. For further details on these algorithms, refer to Chapter 10 of [29].

59.12 Summary

The sheer size of the VLSI circuit, the complexity of the overall design process, the desired performance of the circuit, and the cost of designing a chip dictate that the whole design process must be automated. Also, the design process must be divided into different stages because of the complexity of the entire process. Physical design is one of the steps in the VLSI design cycle. In this step each component of a circuit is converted into a set of geometric patterns which achieves the functionality of the component. The physical design step can be divided further into several substeps. All the substeps of the physical design step are interrelated. Efficient and effective algorithms are required to solve different problems in each of the substeps. Despite significant research efforts in this field, the CAD tools still lag behind the technological advances in fabrication. This calls for the development of efficient algorithms for physical design automation.

eferences

[1] S. B. Aker, "A Modification of Lee's Path Connection Algorithm," *IEEE Trans. Comput.*, pp. 97, 98, Feb. 1967.

[2] P. Bannerjee and M. Jones, "A Parallel Simulated Annealing Algorithm for Standard Cell Placement on A Hypercube Computer," *Proc. IEEE Int. Conf. Comput. Design*, p. 34, 1986.

[3] A. Chatterjee and R. Hartley, "A New Simultaneous Circuit Partitioning and Chip Placement Approach Based on Simulated Annealing," *Proc. Design Autom. Conf.*, pp. 36–39, 1990.

[4] P. A. Eichenberger, Fast Symbolic Layout Translation for Custom VLSI Integrated Circuits, Ph.D. thesis, Stanford University, Stanford, CA, 1986.

[5] A. El Gamal et al., "An Architecture for Electrically Configurable Gate Arrays," *IEEE JSSC*, 24(2), pp. 394–398, Apr. 1989.

[6] H. Hseih, et al., "A 9000-Gate User-Programmable Gate Array," *Proc. 1988 CICC*, pp. 15.3.1–15.3.7, May 1988.

[7] S. C. Wong et al. "A 5000-Gate CMOS EPLD with Multiple Logic and Interconnect Arrays," *Proc. 1989 CICC*, pp. 5.8.1–5.8.4, May 1989.

[8] M. Feuer, "VLSI Design Automation: An Introduction," *Proc. IEEE*, 71(1), pp. 1–9, Jan. 1983.

[9] C. M. Fiduccia and R. M. Mattheyses, "A Linear-Time Heuristics for Improving Network Partitions," *Proc. 19th Design Autom. Conf.*, pp. 175–181, 1982.

[10] J. Greene and K. Supowit, "Simulated Annealing Without Rejected Moves," *Proc. Int. Conf. Comput. Design*, pp. 658–663, Oct. 1984.

[11] L. K. Grover, "Standard Cell Placement Using Simulated Sintering," *Proc. 24th Design Autom. Conf.*, pp. 56–59, 1987.

[12] B. Hajek, "Cooling Schedules for Optimal Annealing," *Oper. Res.*, pp. 311–329, May 1988.

[13] D. D. Hill, "ICON: A Toll for Design at Schematic, Virtual-Grid and Layout Levels," *IEEE Design Test*, 1(4), pp. 53–61, 1984.

[14] M. Y. Hsueh, "Symbolic Layout and Compaction of Integrated Circuits," Technical Report UCB/ERL M79/80, Electronics Research Laboratory, University of California, Berkeley, 1979.

[15] M. Y. Hsueh and D. O. Pederson, Computer-Aided Layout of LSI Circuit Building-Blocks, Ph.D. thesis, University of California, Berkeley, Dec. 1979.

[16] M. D. Huang, F. Romeo, and A. Sangiovanni-Vincentelli, "An Efficient General Cooling Schedule for Simulated Annealing," *Proc. IEEE Int. Conf. CAD*, pp. 381–384, 1986.

[17] W. Kernighan and S. Lin, "An Efficient Heuristic Procedure for Partitioning Graphs," *Bell Syst. Tech. J.*, 49, pp. 291–307, 1970.

[18] S. Kirkpatrick, G. D. Gellat, and M. P. Vecchi, "Optimization by Simulated Annealing," *Science*, 220, pp. 671–680, May 1983.

[19] J. Lam and J. Delosme, "Performance of a New Annealing Schedule," *Proc. 25th Design Autom. Conf.*, pp. 306–311, 1988.

[20] C. Y. Lee, "An Algorithm for Path Connections and Its Applications," *IRE Trans. Electr. Comput.*, 1961.

[21] T. M. Lin and C. A Mead, "Signal Delay in General RC Networks," *IEEE Trans. CAD*, CAD-3, No. 4, pp. 331–349, Oct. 1984.

[22] J. M. Da Mata, "Allenda: A Procedural Language for the Hierarchical Specification of VLSI Layout," *Proc. 22nd Design Autom. Conf.*, pp. 183–189, 1985.

[23] T. Ohtsuki, *Partitioning, Assignment and Placement*, Amsterdam: North-Holland, 1986.

[24] R. Rivest and C. Fiduccia, "A Greedy Channel Router," *Proc. 19th ACM/IEEE Design Autom. Conf.*, pp. 418–424, 1982.

[25] F. Romeo and A. Sangiovanni-Vincentelli, "Convergence and Finite Time Behavior of Simulated Annealing," *Proc. 24th Conf. Decision Control*, pp. 761–767, 1985.

[26] F. Romeo, A. S. Vincentelli, and C. Sechen, "Research on Simulated Annealing at Berkeley," *Proc. IEEE Int. Conf. Comput. Design*, pp. 652–657, 1984.

[27] C. Sechen and K. W. Lee, "An Improved Simulated Annealing Algorithm for Row-Based Placement," *Proc. IEEE Int. Conf. CAD*, pp. 478–481, 1987.

[28] C. Sechen and A. Sangiovanni-Vincentelli, "The Timber Wolf Placement and Routing Package," *IEEE J. Solid-State Circ.*, SC-20, pp. 510–522, 1985.

[29] N. Sherwani, *Algorithms for VLSI Physical Design Automation*, Boston: Kluwer Academic Publishers, 1992.

[30] T. G. Szymanski, "Dogleg Channel Routing is NP-Complete," *IEEE Trans. CAD*, CAD-4, pp. 31–41, Jan. 1985.

60

Design Automation Technology

llen M. Dewey
M Microelectronics
ughkeepsie, NY

0.1 Introduction

The field of *design automation* technology, also commonly called **computer-aided design** (CAD) or **computer-aided engineering** (CAE), involves developing computer programs to conduct portions of product design and manufacturing on behalf of the designer. Competitive pressures to produce in less time and use fewer resources, new generations of products having improved function and performance are motivating the growing importance of design automation. The increasing complexities of microelectronic technology, shown in Fig. 60.1, illustrate the importance of relegating portions of product development to computer automation [1, 3]. Advances in microelectronic technology enable over 1 million devices to be manufactured on an integrated circuit substrate smaller in size than a postage stamp, yet the ability to exploit this capability remains a challenge. Manual design techniques are unable to keep pace with product design cycle demands and are being replaced by automated design techniques [2, 4].

Figure 60.2 summarizes the historical development of design automation technology and computer programs. Design automation computer programs are also called **applications** or **tools.** Design automation efforts started in the early 1960s as academic research projects and captive industrial programs, focusing on individual tools for physical and logical design. Later developments extended logic simulation to more detailed *circuit* and *device* simulation and more abstract *functional* simulation. Starting in the mid to late 1970s, new areas of test and synthesis emerged and commercial design automation products appeared. Today, the electronic design automation industry is an international business with a well-established and expanding technical base [5]. The electronic design automation technology base will be examined by presenting an

93-8341-2/95/$0.00 + $.50

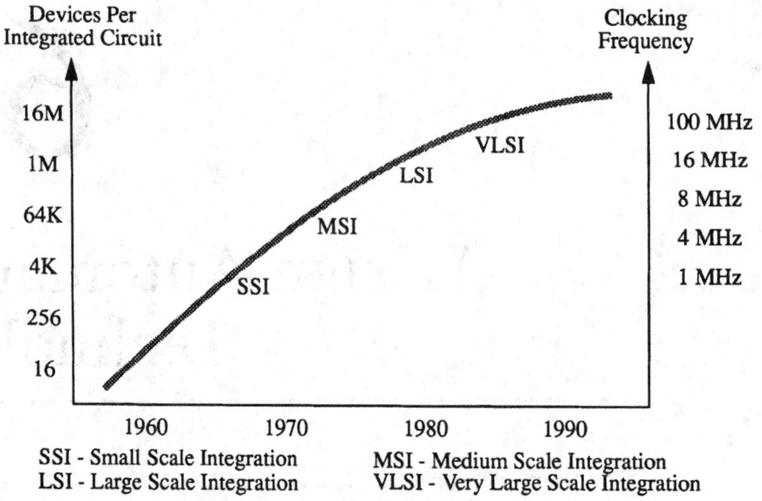

FIGURE 60.1 Complexity of microelectronic technology.

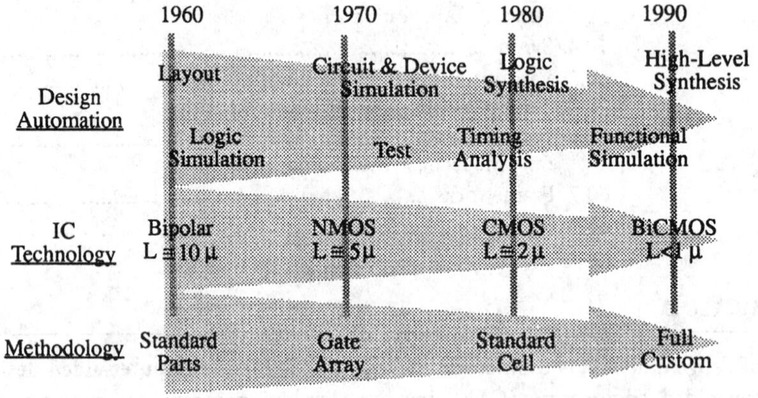

FIGURE 60.2 The development of design automation technology.

overview of the following topical areas:

- Design entry
- Conceptual design
- Synthesis
- Verification
- Testing
- Frameworks

60.2 Design Entry

Design entry, also called **design capture**, is the process of communicating with a design automation system. Design entry involves describing a system design to a design automation system, invoking applications to analyze and/or modify the system design, and querying the results. In short, design entry is how an engineer "talks" to a design automation application and/or system.

Any sort of communication is composed of two elements: language and mechanism. Language provides common semantics; mechanism provides a means by which to convey the common semantics. For example, two people communicate via a language such as English, French, or German, and a mechanism, such as a telephone, electronic mail, or facsimile transmission. For design, a digital system can be described in many ways, involving different perspectives or *abstractions.* An **abstraction** is a model for defining the behavior or semantics of a digital system, i.e., how the outputs respond to the inputs. Figure 60.3 illustrates several popular levels of abstractions and the trends toward higher levels of design entry abstraction to address greater levels of complexity [10, 12].

The physical level of abstraction involves geometric information that defines electrical devices and their interconnection. Geometric information includes the shaped objects and where objects are placed relative to each other. For example, Fig. 60.4 shows the geometric shapes defining a simple complementary metal oxide-semiconductor (CMOS) inverter. The shapes denote different materials, such as aluminum and polysilicon, and connections, called *contacts* or *vias.*

The design entry mechanism for physical information involves textual and graphic techniques. With textual techniques, geometric shape and placement are described via an artwork description language, such as Caltech Intermediate Form or Electronic Design Intermediate Form. With

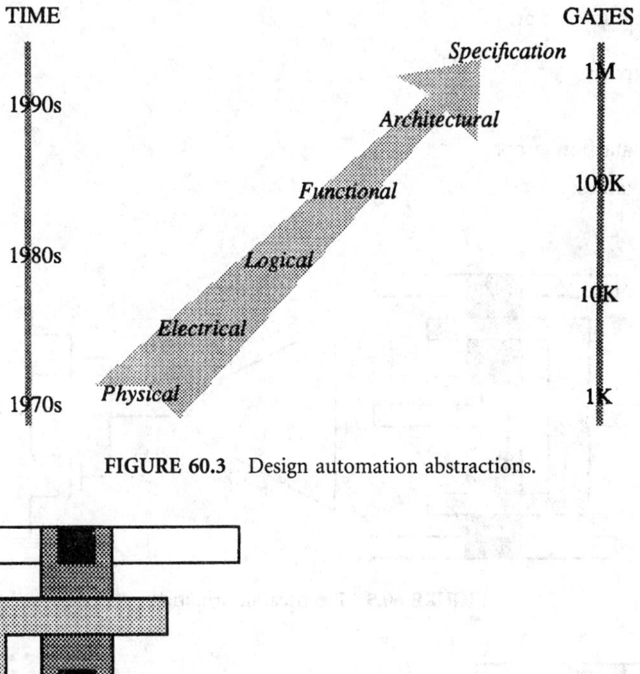

FIGURE 60.3 Design automation abstractions.

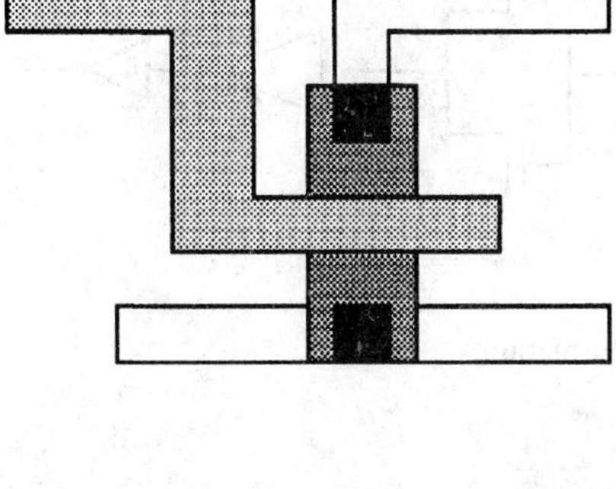

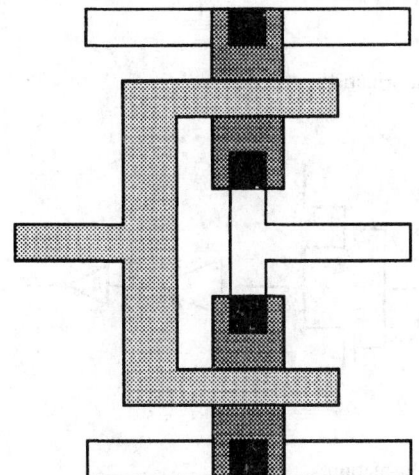

FIGURE 60.4 Physical information.

graphical techniques, geometric shape and placement are described by drawing the objects on a display terminal [13].

The electrical level abstracts geometric information into corresponding electrical devices, such as capacitors, transistors, and resistors. For example, Fig. 60.5 shows the electrical symbols denoting a CMOS inverter. Electrical information includes the device behavior in terms of terminal or pin current and voltage relationships. Device behavior also may be defined in terms of manufacturing parameters, such as resistances or chemical compositions.

The logical level abstracts electrical information into corresponding logical elements, such as and gates, or gates, and inverters.Logical information includes truth table and/or characteristic switching algebra equations and active level designations. For example, Fig. 60.6 shows the logical symbol for a CMOS inverter. Notice how the amount of information decreases as the level of abstraction increases.

Design entry mechanisms for electrical and logical abstractions are commonly collectively called **schematic capture** techniques. Schematic capture defines hierarchical structures, commonly called **netlists,** of components. A designer creates instances of components supplied from a library of predefined components and connects component pins or ports via wires [9, 11].

The functional level abstracts logical elements into corresponding computational units, such as registers, multiplexers, and arithmetic logic units (ALUs). The architectural level abstracts functional information into computational algorithms or paradigms. Some examples of common computational paradigms are

- State diagrams
- Petri nets
- Control/data flow graphs
- Function tables

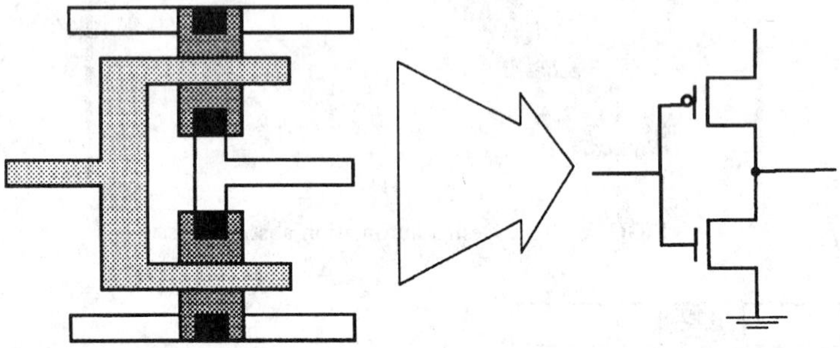

FIGURE 60.5 Electrical information.

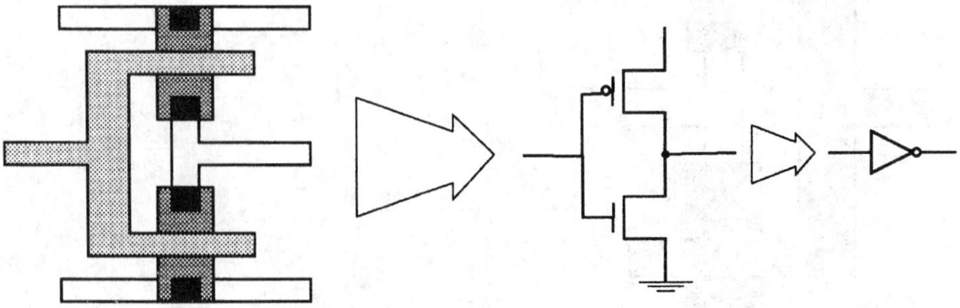

FIGURE 60.6 Logical information.

- Spreadsheets
- Binary decision diagrams

These higher levels of abstraction support a more expressive "higher bandwidth" communication interface between an engineer and design automation programs. An engineer can focus his or her creative, cognitive skills on concept and behavior, rather than the complexities of detailed implementation. Associated design entry mechanisms typically use hardware description languages with a combination of textual and graphic techniques [6].

Figure 60.7 shows a simple state diagram. The state diagram defines three states, denoted by circles. State-to-state transitions are denoted by labeled arcs; state transitions depend on the present state and the input X. The output, Z, per state is given within each state. Because the output is dependent on only the present state, the digital system is classified as a **Moore finite state machine.** If the output is dependent on the present state and input, then the digital system is classified as a **Mealy finite state machine.** The state table corresponding to the state diagram in Fig. 60.7 is given in Table 60.1. A hardware description language model written in VHDL of the Moore finite state machine is given in Fig. 60.8. The VHDL model, called a **design entity**, uses a "date flow" description style to describe the state machine [7, 8]. The entity statement defines the interface, i.e., the ports. The ports include two input signals, X and CLK, and an output signal, Z. The ports are of type BIT, which specifies that the signals may only carry the values '0' or '1'. The architecture statement defines the I-O transform via two concurrent signal assignment statements. The internal signal STATE holds the finite state information and is driven by a guarded, conditional concurrent signal assignment statement that executes when the associated block expression (CLK = '1' **and not** CLK'STABLE) is true, which is only on the rising edge of the signal CLK. STABLE is a predefined attribute of the signal CLK; CLK'STABLE is true if CLK has not changed value. Thus, if **not** CLK'STABLE is true, meaning that CLK has just changed value, and CLK = '1', then a rising transition has occurred on CLK. The

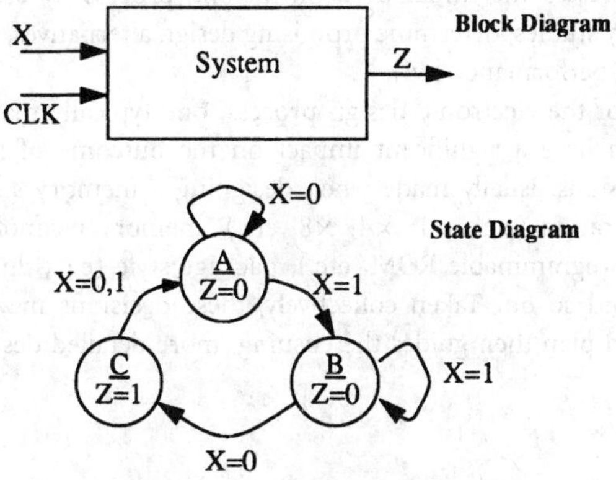

FIGURE 60.7 State diagram.

TABLE 60.1 State Table

	Next State		
Present State	$X = 0$	$X = 1$	Output
A	A	B	0
B	C	B	0
C	A	A	1

```
entity MOORE_MACHINE is
  port (X, CLK : in BIT;  Z : out BIT);
end MOORE_MACINE;
architecture FSM of MOORE_MACHINE is
  type STATE_TYPE is (A, B, C);
  signal STATE : STATE_TYPE := A;
begin
  NEXT_STATE:
  block (CLK='1' and not CLK'STABLE)
  begin
    STATE <= guarded B when (STATE=A and X='1') else
                     C when (STATE=B and X='0') else
                     A when (STATE=C) else
                     STATE;
  end block NEXT_STATE;

  with STATE select
    Z <= '0' when A,
         '0' when B,
         '1' when C;
end FSM;
```

FIGURE 60.8 VHDL model.

output signal Z is driven by a nonguarded, selected concurrent signal assignment statement that executes anytime STATE changes value.

60.3 Conceptual Design

Figure 60.9 shows that the conceptual design task generally follows the design entry task. Conceptual design addresses the initial decision-making process by assisting the designer to conduct initial feasibility studies, determine promising design alternatives, and obtain preliminary indication of estimated performance [16].

During early stages of the electronic design process, one typically makes a number of high-level decisions that can have a significant impact on the outcome of a design. For example, consider the early decisions usually made when designing a memory subsystem. The designer must choose a configuration (e.g., ×1, ×4, ×8, etc.), memory technology (e.g., static RAM, dynamic RAM, mask programmable ROM, etc.), package style (e.g., dual-in-line, chip carrier, pin grid array, etc.), and so on. Taken collectively, these decisions may be referred to as the **design plan**. The design plan then guides the ensuing, more detailed design process, acting as a high-level road map.

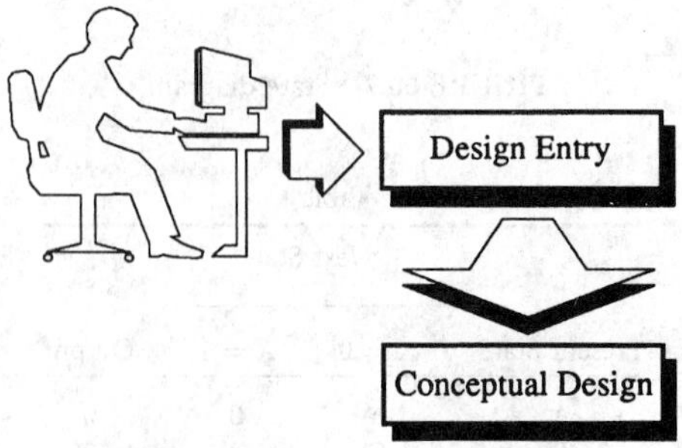

FIGURE 60.9 Design process.

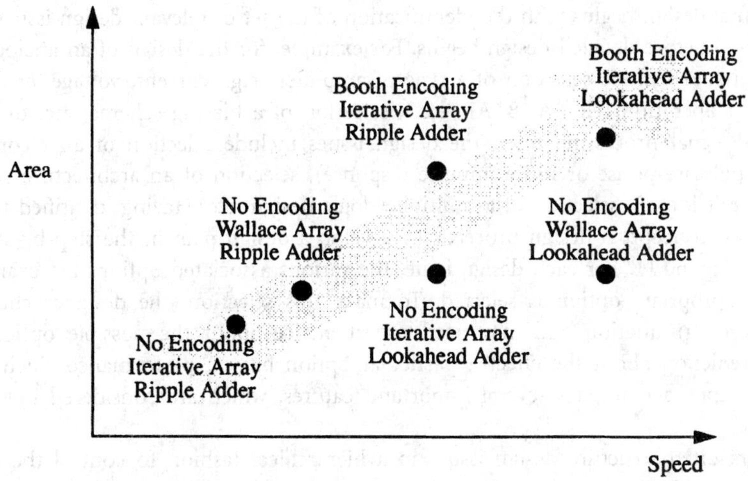

FIGURE 60.10 Performance tradeoffs.

In general, many different plans may result in designs that satisfy the same set of functional specifications, but have varying tradeoffs in terms of performances, such as area or power dissipation. For example, Fig. 60.10 shows relative area and speed performance tradeoffs for six combinational multiplier designs. Figure 60.10 shows the typical trend that increasing speed implies increasing area. More importantly, the chart indicates how various multiplier designs relate to each other and which designs present the best area/speed tradeoffs for a particular application.

Given this large set of possible design plans to consider, it is easy to imagine getting hopelessly lost in analyzing the various design plans and keeping track of the tradeoffs and decision interdependencies. Choosing the most appropriate design plan is important because it is usually not economically feasible to choose an alternative plan and repeat the detailed design process if the final design does not meet the desired specifications. To aid the designer during the early stages of the design process in maximizing his or her chances that certain high-level decisions will result in a design that meets the desired specifications, conceptual design tools assist the designer in analyzing possible ramifications of high-level design decisions before actually undertaking the detailed design and fabrication process.

During conceptual design, the engineer has a global view of the design process in considering the key decisions associated with each lower level of abstraction. Thus, the engineer can undertake a breadth-first design methodology. Costly errors can be avoided by investing time at the outset in thoroughly thinking through a particular plan, the result being a more complete and credible specification that better guides the ensuing actions. In addition, the unnecessary use of detailed design CAD tools is minimized because conceptual design acts as a "coarse filter" eliminating infeasible design plans.

Design Planning

To discuss conceptual design, it is convenient to introduce the following terms:

- *Design Issue*—Pertinent aspect of a design for which a decision must be made.
- *Design Option*—Possible choice that can be made for a design issue.
- *Design Decision*—Particular option chosen for a design issue.
- *Design Plan*—Set of design decisions which identifies a high-level approach for implementing the design.

Conceptual design begins with the identification of the set of relevant design issues that must be considered before detailed design begins. For example, for the design of an analog amplifier, the design issues include selection of a type of amplifier (e.g., current, voltage, etc.), selection of a class of operation (e.g., A, B, AB, etc.), selection of a biasing scheme, etc.For the design of a digital signal processing filter, the design issues include selection of an algorithm (e.g., infinite impulse response or finite impulse response), selection of an architecture (e.g., direct, cascade, parallel, etc.), selection of a multiplier logic design, etc. Having identified the relevant design issues, conceptual design proceeds to develop a design plan in the step-by-step fashion outlined in Fig. 60.11. For each design issue, the various associated options are examined, and the most appropriate option is selected. To make this selection, the designer chooses some method of discrimination and may need expert advice about the possible options or may desire a prediction about the effect a particular option has on performance. Such a conceptual design approach implies several important features, which are considered in more detail below.

Designers often structure design issues in a hierarchical fashion to control the complexity of the conceptual design process. Complex design issues are decomposed into sets of design subissues or *subplans* [22]. For example, an expanded, hierarchical listing of candidate design issues for digital signal processing filters is given below [14, 20]:

1. Selection of a filter algorithm
2. Selection of a polynomial approximation technique
3. Selection of a filter architecture
   - Selection of a filter structure
   - Selection of a pipelining strategy
4. Selection of a logic design
   - Selection of a multiplier logic design
     - Selection of an encoding technique
     - Selection of an array design
     - Selection of a final adder design
   - Selection of an adder logic design
5. Selection of a fabrication technology
6. Selection of a layout design style

The selection of a multiplier logic design can be decomposed into three design subissues: selection of an encoding technique, selection of a combinational array design, and selection of a final adder design. Notice that many details about digital filter design are not listed because conceptual design is typically not concerned with every design issue required to eventually fully design and fabricate the initial specification. Hierarchical design issues provide for a

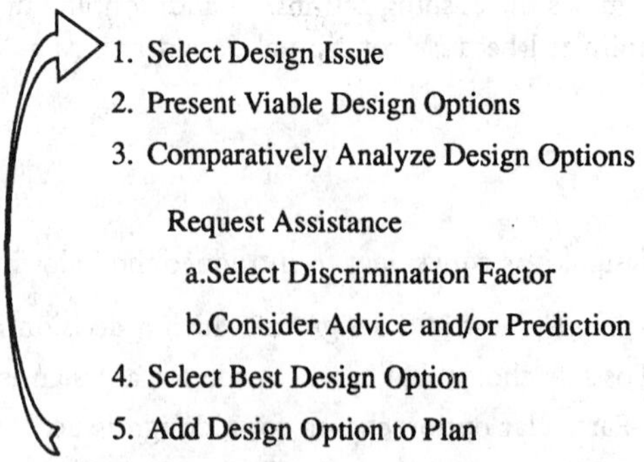

FIGURE 60.11 Constructing a design plan.

more intuitive representation of the conceptual design process, enabling the designer to better comprehend the overall magnitude and composition of the design domain.

Each design issue has associated commonly used options or selections. For example, the possible digital filter polynomial approximation techniques are

1. Equiripple
2. Frequency sampling
3. Windowing
4. Beta
5. Elliptic
6. Butterworth
7. Chebyshev

During conceptual design, the order in which the designer examines the various issues may not be completely arbitrary because the options available for a design issue may depend on which options were chosen for other design issues. Hence, a partial ordering may be imposed on addressing design issues. As an example, equiripple, frequency sampling, windowing, and beta digital filter approximations apply only to the finite impulse response algorithm, while elliptic, Butterworth, and Chebyshev digital filter approximations apply only to the infinite impulse response algorithm. Hence, the polynomial approximation technique depends on the choice of the filter algorithm and thus, the choice of a filter algorithm must precede the choice of a polynomial approximation technique. Similarly, if the designer chooses a bipolar logic family, then the Manchester carry adder logic design is not a viable option because this option requires pass transistor technology that exists only in metal oxide semiconductor-related logic families. Hence, the adder's logic design depends on the choice of the logic family and thus, the choice of a logic family must precede the choice of a logic design style.

Conceptual design supports such ordering of design issues via *ordering* constraints, enforcing that the more pervasive design issues be addressed first [23]. In specifying the interdependencies between design issues, as many or as few ordering constraints may be imposed as is warranted by the particular design domain. At one extreme, a strict sequential ordering of design issues may be required. At the other extreme, no ordering may occur, signifying that the design issues have no interdependencies and that each issue may be examined independently of the others. Partial orderings may also occur in which some design issues typically arise that may be examined and some design issues that may not be examined, owing to interdependencies. Hence, ordering constraints provide a general mechanism within the conceptual design methodology for encoding various design strategies, such as top-down, bottom-up, or meet-in-the-middle.

In addition to interdependencies between design issues (ordering constraints), there may be interdependencies between options. For the digital adder design example, in which the designer investigates the logic design style issue, only the options that are consistent with the selected logic family option are viable. Interdependencies between design options are supported via *consistency* constraints. Using consistency constraints, a system planner helps to determine, for a particular design issue, which options are consistent with earlier decisions. The designer is not presented with an option (step 2 in Fig. 60.11) that is inconsistent with his or her earlier decisions. As such, the designer is prevented from developing an incorrect or inconsistent plan.

Decision Assistance

It may be the situation for complex conceptual designs that the designer consults an expert on a particular issue. The expert offers assistance in understanding the relative advantages and disadvantages of the available options in order to select the most appropriate option(s). Two general forms of assistance, *advice* and *prediction*, are available.

Advice describes the qualitative advantages and/or disadvantages that should be considered before accepting or rejecting a design option. On the other hand, prediction describes the quantitative advantages and/or disadvantages. Prediction is a valuable tool for reaching down through several levels of the design hierarchy to obtain quantitative insights into the implications of high-level design decisions. Such advice is given in lieu of the more expensive route of actually executing the detailed design process in order to determine the effects of a decision on design performance. While the predicted performance is only an approximation, it is often sufficient to allow the designer to choose a particular option, or at least to rule out an option.

Different types of prediction techniques are summarized in the taxonomy presented in Fig. 60.12. Heuristic models use general guidelines or empirical studies to model the design process [15]. Analytical models employ a more rigorous, mathematical approach to model the design process. Some such models may be based on probability theory, e.g., wirability theory, and capture the particulars of a design activity in a set of random variables [17, 18], while other analytical models may be based on graph and information theory, e.g., very large scale integration (VLSI) complexity theory [19, 24, 25]. Finally, simulation models include the judicious partial execution of a design activity on critical portions of the design. The use of previous designs serves to capture the experience or "corporate-history" factor used by an expert designer in attempting to predict the performance of a new function based on previous similar designs.

As an example, Fig. 60.13 shows how performance estimates for digital filters can be obtained in a hierarchical, incremental fashion, employing a variety of prediction models that cooperate to yield the desired results. The prediction task is performed by incrementally translating the initial specifications into a series of complexity measures. Each succeeding complexity measure is a refinement of the performance estimate, starting with a performance estimate in terms of the filter order and ending with the performance estimate in terms of the physical units of area, speed, and power dissipation. Each prediction model draws its required input information from higher-level prediction models, the initial functional specifications, and the applicable decisions from the conceptual plan. Incrementally generating a prediction allows the flexibility to use different types of prediction techniques as the situation warrants to model the implications of different parts of the design plan.

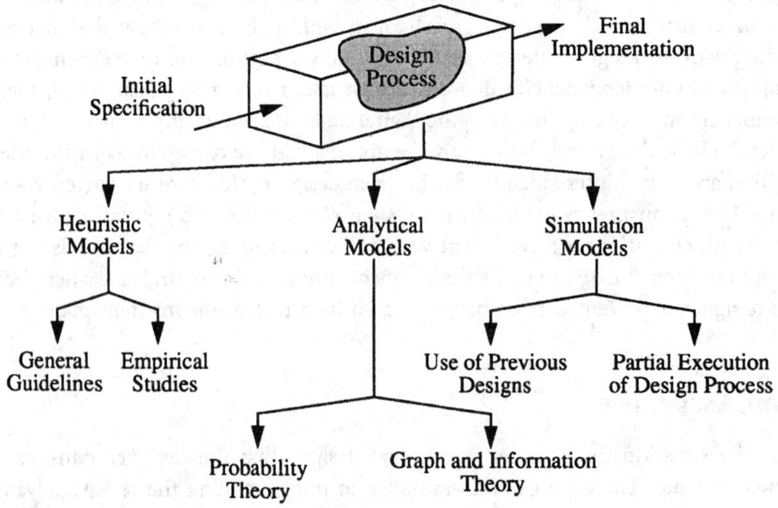

FIGURE 60.12 Types of prediction techniques.

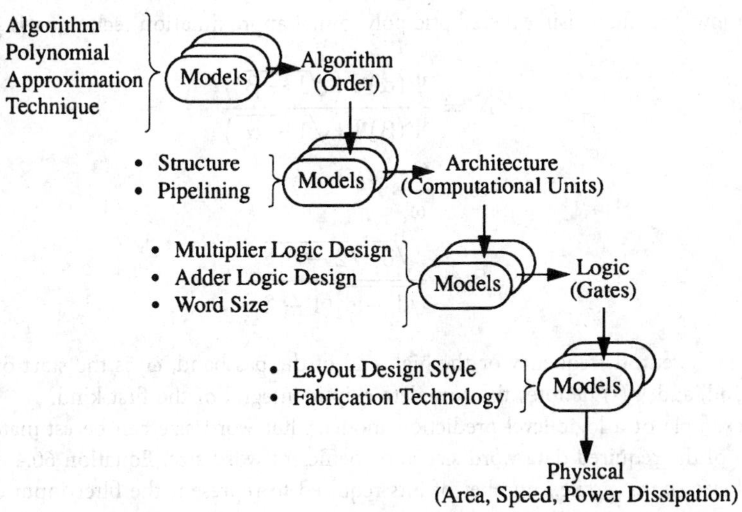

FIGURE 60.13 Hierarchical structure of filter performance prediction.

The starting point in the prediction methodology is the initial functional specifications. For digital filters, functional specifications are usually given in terms of the desired frequency response, involving the following:

- The frequency bands of importance
- The widths of the transition bands between the desired frequency bands.
- The levels of amplification or attenuation per frequency band
- The allowable tolerance or deviation from the specified levels of amplification and/or attenuation

An extensive set of prediction models has been developed for predicting the performance of linear, shift-invariant digital filters having the general I-O behavior given by the difference equation

$$y(n) = \sum_{k=1}^{N} a_k(n-k) + \sum_{k=0}^{M} b_k x(n-k) \tag{60.1}$$

The variables $x(n)$ and $y(n)$ denote the values of signals $x(t)$ and $y(t)$ at time t_n. As an example of an algorithm-level prediction model, the order of a low-pass filter using the equiripple polynomial approximation technique is given by the empirical model shown in Equation 60.2 [21].

$$N \approx \frac{\left\{ \begin{array}{l} [5.309 \times 10^{-3}(\log_{10}\delta_1)^2 + 0.07114\log_{10}\delta_1 - 0.4761]\log_{10}\delta_2 \\ \quad +[-2.66 \times 10^{-3}(\log_{10}\delta_1^2) - 0.5941\log_{10}\delta_1 - 0.4278] \end{array} \right\}}{\Delta F} - [0.51244(\log_{10}\delta_1 - \log_{10}\delta_2) + 11.01217]\Delta F + 1 \tag{60.2}$$

where N is the filter order, δ_1 is the passband tolerance, δ_2 is the stopband tolerance, and ΔF is the width of the transition band. As another example of an algorithm prediction model, the order of a low-pass filter using the elliptic polynomial approximation technique is given by

$$N = \frac{\Psi(\alpha)\Psi\left(\sqrt{1-\beta^2}\right)}{\Psi(\beta)\Psi\left(\sqrt{1-\alpha^2}\right)}$$

$$\alpha = \frac{\omega_p}{\omega_s} \tag{60.3}$$

$$\beta = \frac{\sqrt{\delta_1}\sqrt{2-\delta_1\delta_2}}{\sqrt{1-\delta_2^2(1-\delta_1)}}$$

where ω_p is the cutoff frequency or the high end of the passband, ω_s is the start or low end of the stopband, and $\Psi(\cdot)$ denotes the complete elliptic integral of the first kind.

As an example of a logic level prediction model, filter word size can be estimated to be the maximum of the required data word size and coefficient word size. Equation 60.4 estimates the data word size ($data_{WS}$), i.e., number of bits required to represent the filter input data samples to ensure an adequate signal:noise ratio:

$$data_{WS} = \left\lceil \frac{SNR_q - 4}{6} \right\rceil \tag{60.4}$$

where $\lceil \cdot \rceil$ denotes the "smallest integer not less than" and SNR_q denotes the root mean square signal:quantization noise ratio in decibels. Equation 60.5 estimates the coefficient word size ($coeff_{WS}$), i.e., number of bits required to represent the filter coefficients to ensure that finite precision coefficient quantization and arithmetic round-off errors do not appreciably deteriorate filter frequency response:

$$coeff_{WS} = \left\lceil \log_2\left(\frac{\omega_s + \omega_p}{\omega_s - \omega_p}\right) + \log_2\left(\frac{1}{\min(\delta_1, \delta_2)}\right) - \frac{1}{2}\log_2\left(\frac{\omega_S}{\omega_s - \omega_p}\right) + 3 \right\rceil \tag{60.5}$$

where ω_S denotes the sampling frequency.

Exploring Alternative Designs

After analyzing the viable options, the final steps in Fig. 60.11 involve making the design decision. Sometimes designers want to delay making a decision during a conceptual design session until other aspects of the design can be explored. If more than one option for a particular design issue looks equally viable, then the designer may elect to consider all of the viable options. For example, returning to digital filters, the designer may wish to investigate the implications of using both the lookahead carry and block lookahead carry logic design styles for the adder. Later in the conceptual design session, when the competing adder design plans have been more fully developed and their relative merits more clearly understood, the designer may have the information necessary to select one logic design style over the other. Such an approach is often referred to as **nonlinear planning** or adopting a **least-commitment strategy**.

Multiple, alternative plans support delaying design decisions. New plans are created each time more than one option is chosen for a design issue. For example, if the designer wishes to investigate both adder logic designs, then the planner records the decision by generating two plans. One design plan uses the lookahead carry adder logic design, and the other plan uses the block lookahead carry logic design. Having generated alternative plans, the designer can then further develop each plan independently by requesting that the planner switch the focus of the conceptual design session between the competing plans. It should be noted that a consequence of creating alternative plans is that the conceptual design process may not always yield one plan. Rather, the conceptual design process may yield multiple plans that appear roughly equally promising. The conceptual design methodology is not intended to always possess the ability to definitively differentiate between alternative plans due to the general nature of the plans or the approximate nature of the predictions. In cases in which the conceptual design process yields more than one candidate plan, the designer could invoke more detailed synthesis and analysis design automation tools to further investigate and resolve the plans.

In a lengthy conceptual design session a designer may want to reconsider and possibly change an earlier decision; this situation is supported via backtracking. When the designer requests a reconsideration of an earlier decision for a particular design issue, the planner must back up the state of the conceptual design session and reactivate the associated viable options. The designer may then reexamine the options and possibly select a different option. The backtracking process also involves checking the rest of the plan to see if any of the designer's other, earlier decisions might be affected by the new decision. If any other design decisions depend on the decision that the designer wants to change, then these decisions must by similarly undone. In this manner only the appropriate part of the conceptual design session is backed up; the decisions that are not dependent on the desired change are left unaffected.

Applications

Conceptual design involves well-defined, algorithmic-based knowledge and ill-defined, heuristic-based knowledge. For instance, maintaining alternative plans involves primarily algorithmic-based knowledge. For each plan, the available design issues are determined in accordance with the ordering constraints. For each available design issue, the viable options are determined in accordance with the decisions made thus far and the consistency constraints. As the conceptual design session progresses, new design decisions are recorded, or possibly, previous design decisions are changed.

In contrast, providing advice involves primarily heuristic-based knowledge. Based on the state of the plan, the appropriate piece of advice or prediction model is identified, guidance is formulated, and the results are supplied to the designer. Advice and predictions serve as knowledge sources working to solve the conceptual design "problem". The knowledge sources opportunistically invoke themselves to address an applicable portion of the advice or prediction request, taking their input from the system plan(s), and posting their results back to the system plan(s) in an attempt to refine and develop a conceptual design.

Domain-dependent information concerning design issues, options, tradeoffs, and interdependencies is typically specified as part of the knowledge acquisition process involved in initially constructing a system planner, and is separate from domain-independent information to facilitate creating planners for different design tasks and moving existing planners to new design tasks. If alternative plans are developed during conceptual design, multiple versions of the domain-dependent information are correspondingly created. Each version of the domain-dependent information, also called **contexts, worlds**, or **belief spaces**, contains a plan's state information, such as which design issues have been addressed, which options have been chosen, and which advice has been given. When a designer changes the focus of the conceptual design session between alternative plans, a context switch is performed between the associated versions of the domain-dependent information. An analogous situation is an operating system that allows multiple jobs. Each time the computer switches to a new job, the state of the current job must be saved, and the state of the new job must be restored.

Clearly, for complex design domains, the amount of domain-dependent information can be sizable. To limit the search time required to find applicable options or appropriate advice, the domain-dependent information can be segmented or partitioned by design issue. Such an organization allows for dynamically limiting the portion of information searched. For instance, in searching for appropriate advice, only the advice that is associated with the current design issue is made accessible. Thus, any search is confined to the advice associated with the current design issue; a search is never executed over the entire domain-dependent information.

Conceptual design is an emerging area of design automation that demonstrates the potential for significantly improving our design capability. Conceptual design involves studying a design process, modeling the process as a hierarchy of design issues, delineating the options, and identifying the associated interdependencies and discriminating characteristics. A conceptual design plan realizes the initial specifications by delineating the important design decisions. Such a methodology emphasizes spending more time at the initial stage of the design process in obtaining a more credible system specification. In addition to beginning an expensive design process with a careful definition of what is required of the design, conceptual design emphasizes the need for a general plan for how to realize the design.

Associated applications play a supporting role of managing and presenting the large and complex amount of information typically involved in conceptual design to assist the designer in making the decisions which constitute the conceptual design plan. Plans are developed incrementally and interactively. In a breadth-first fashion conceptual design guides the designer through a myriad of alternative plans, allowing the designer to explore differences and seek optimal solutions. The advantage of conceptual design is that expensive design resources are not wasted in trying plans that eventually prove unsatisfactory. Applications also provide performance estimation, which can be useful in coordinating between groups involved with a design. For example, an early estimate of chip area can be used to check with the manufacturing group that the design will be within acceptable fabrication yield margins or will adhere to any packaging or printed circuit board size restrictions. An early estimate of chip performance also can be used to coordinate with marketing and management in projecting potential applications, probable costs, and future business.

60.4 Synthesis

Figure 60.14 shows that the synthesis task generally follows the conceptual design task. The designer describes the desired system via design entry, generates a conceptual plan of major

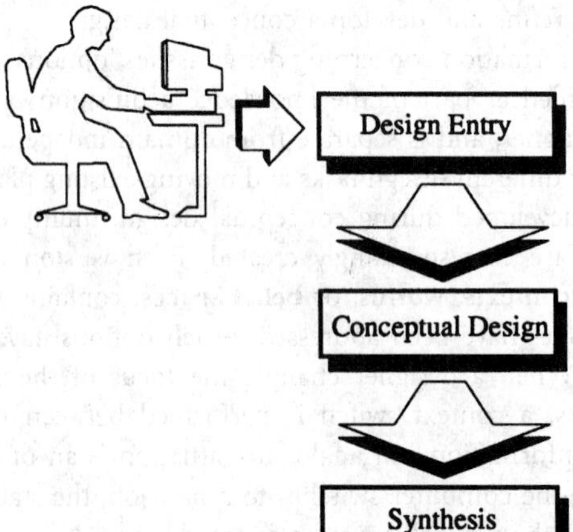

FIGURE 60.14 The design process.

design decisions, and then invokes synthesis design automation programs to assist in generating the required implementation [32].

Synthesis translates or transforms a design at one level of abstraction to another, more detailed level of abstraction. The more detailed level of abstraction may be only an intermediate step in the entire design process or it may be the final implementation. Synthesis programs that yield a final implementation are sometimes called **silicon compilers** because the programs generate sufficient detail to proceed directly to silicon fabrication [26, 29].

Like design abstractions, synthesis techniques can be hierarchically categorized, as shown in Fig. 60.15. The higher levels of synthesis offer the advantage of less complexity, but also the disadvantage of less control over the final implementation.

Algorithmic synthesis also called **behavioral synthesis**, addresses "multicycle" behavior, which means behavior that spans more than one *control step.* A control step equates to a clock cycle of a synchronous, sequential digital system, i.e., a state in a finite state machine controller or a microprogram step in a microprogrammed controller. Algorithmic synthesis typically accepts sequential design descriptions using a procedural or imperative modeling style. Such descriptions define the I-O transform, but provide little information about the parallelism of the implementation [28, 30].

Partitioning decomposes the design description into smaller behaviors. Partitioning is an example of a *high-level transformation* that modifies the initial sequential design description to optimize a hardware implementation.High-level transformations include several common software programming compiler optimizations, such as loop unrolling, subprogram in-line expansion, constant propagation, and common subexpression elimination.

Resource allocation associates behaviors with hardware computational units and scheduling determines the order in which behaviors execute. Behaviors that are mutually exclusive can potentially share computational resources.Allocation is performed using a variety of graph clique covering or node coloring algorithms. Allocation and scheduling are interdependent and different synthesis strategies perform allocation and scheduling in different ways. Sometimes scheduling is performed first, followed by allocation; sometimes allocation is performed first, followed by scheduling; and sometimes allocation and scheduling are interleaved.

Scheduling assigns computational units to control steps, thereby determining which behaviors execute in which clock cycles. At one extreme, all computational units can be assigned to a single control step, exploiting maximum concurrency. At the other extreme, computational units can be assigned to individual control steps, exploiting maximum sequentiality. Figure 60.16

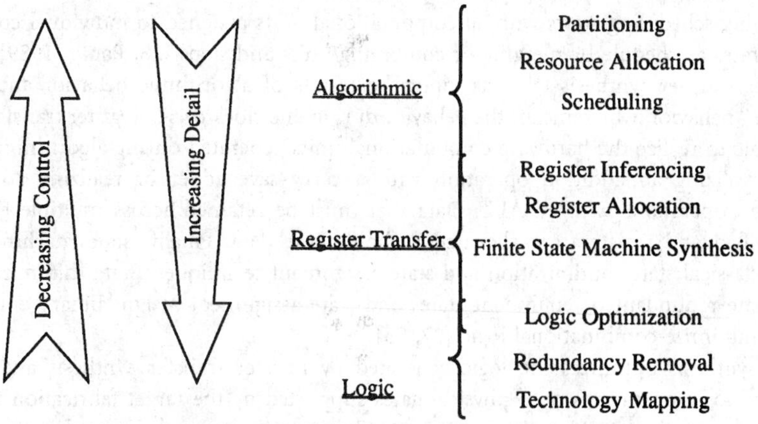

FIGURE 60.15 Taxonomy of synthesis techniques.

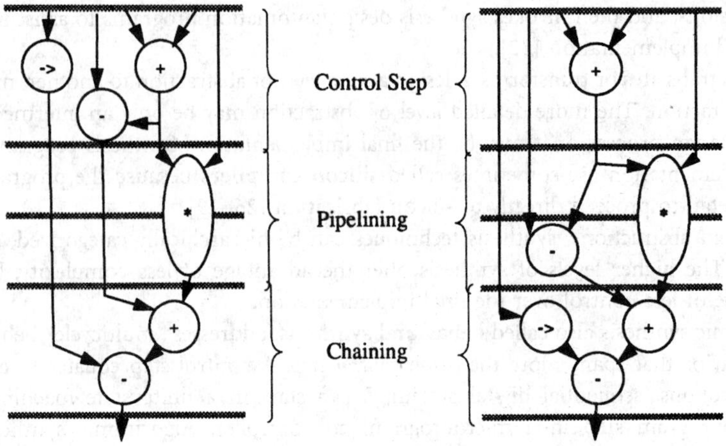

FIGURE 60.16 Scheduling.

illustrates two possible schedules of the same behavior. *Pipelining* schedules an operation across multiple control steps, whereas *chaining* combines multiple operations within a single control step.

Several popular scheduling algorithms are

- As soon as possible (ASAP)
- As late as possible (ALAP)
- List scheduling
- Force directed scheduling
- Control step splitting/merging

As soon as possible and as late as possible scheduling algorithms order the computational units based on data dependencies. List scheduling is based on ASAP and ALAP scheduling, but can consider additional global constraints, such as a maximum number of control steps. Force-directed scheduling computes the probabilities of computational units being assigned to control steps and attempts to evenly distribute computation activity among all control steps. Control step splitting scheduling starts with all computational units assigned to one control step and generates a schedule by splitting the computational units into multiple control steps. Control step merging scheduling starts with all computational units assigned to individual control steps and generates a schedule by merging or combining units and steps [28, Paulin 1989]

Register transfer synthesis takes as input the results of algorithmic behavior and addresses "per-cycle" behavior, which means the behavior during one clock cycle. Register transfer synthesis selects logic to realize the hardware computational units generated during algorithmic synthesis, such as realizing an addition operation with a carry–save adder or realizing addition and subtraction operations with an ALU. Data that must be retained across multiple clock cycles are identified and registers are allocated to hold these data. Finally, state machine synthesis involves classical state minimization and state assignment techniques. State minimization seeks to eliminate redundant or equivalent states and state assignment assigns binary encodings for states to minimize combinational logic [27, 33].

Logic synthesis optimizes the logic generated by register transfer synthesis and maps the minimized logic operations onto physical gates supported by the target fabrication technology. Technology mapping considers the foundry cell library and associated electrical restrictions, such as fanin/fanout limitations.

60.5 Verification

Figure 60.17 shows that the verification task generally follows the synthesis task. The verification task checks the correctness of the function and performance of a design to ensure that an intermediate or final implementation faithfully realizes the initial, desired specification.

Several common types of verification are

- Timing analysis
- Simulation
- Emulation
- Formal verification

These types of verification are examined in more detail in the following sections [42].

Timing Analysis

As the name implies, timing analysis checks that the overall design satisfies operating speed requirements and individual signals within a design satisfy transition requirements. Common **signal transition requirements**, also called **timing hazards**, include *rise* and *fall* times, *propagation delays, clock times, race conditions, glitch detection,* and *setup* and *hold* times. Set and hold timing checks are illustrated in Fig. 60.18. For synchronous sequential digital systems, memory devices (level-sensitive latches or edge-sensitive flip-flops) require that the data and control signals obey setup and hold timings to ensure correct operation, i.e., that the memory device correctly and reliably stores the desired data. The control signal is typically a clock signal and Fig. 60.18 assumes that a control signal transition, rising or falling, triggers activation of the memory device. The data signal carrying the information to be stored in the memory device must be stable for a period equal to the setup time prior to the control signal transition to ensure the correct value is sensed by the memory device. Also, the data signal must be stable for a period

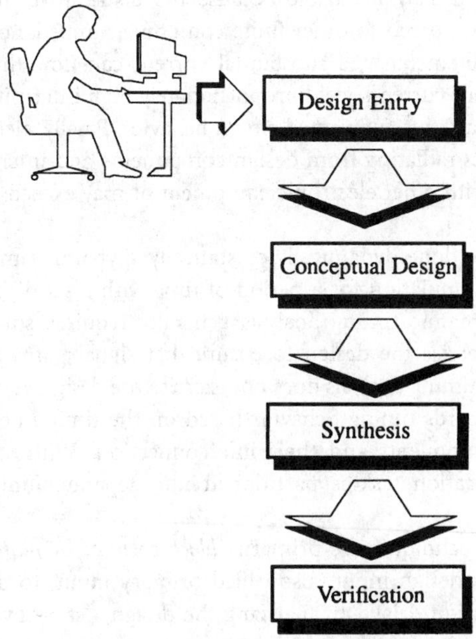

FIGURE 60.17 Design process.

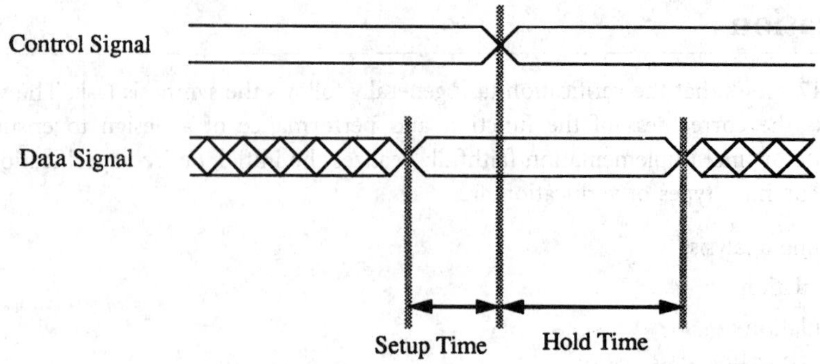

FIGURE 60.18 Signal transition requirements.

equal to the hold time after the control signal transition to ensure the memory device has enough time to store the sensed value.

Another class of timing transition requirements, commonly called **signal integrity checks,** include *reflections, crosstalk, ground bounce,* and *electromagnetic interference.* Signal integrity checks are typically required for high-speed designs operating at clock frequencies above 75 MHz. At such high frequencies, the transmission line behavior of wires must be analyzed. A wire must be properly terminated, i.e., connected, to a port having an impedance matching the wire characteristic impedance to prevent signal reflections. Signal reflections are portions of an emanating signal that "bounce back" from the destination to the source. Signal reflections reduce the power of the emanating signal and can damage the source. Crosstalk refers to unwanted reactive coupling between physically adjacent signals, providing a connection between signals that are supposed to be electrically isolated. Crosstalk causes information carried on a signal to interfere or corrupt information carried on a neighboring signal. Ground bounce is another signal integrity problem. Because all conductive material has a finite impedance, a ground signal network does not in practice offer the same electrical potential throughout an entire design. These potential differences are usually negligible because the distributive impedance of the ground signal network is small compared to other finite component impedances. However, when many signals switch value simultaneously, a substantial current can flow through the ground signal network. High intermittent currents yield proportionately high intermittent potential drops, i.e., ground bounces, which can cause unwanted circuit behavior. Finally, electromagnetic interference refers to signal harmonics radiating from design components and interconnects. This harmonic radiation may interfere with other electronic equipment or may exceed applicable environmental safety regulatory limits [38].

Timing analysis can be done dynamically or statically. Dynamic timing analysis exercises the design via simulation or emulation for a period of time with a set of input stimuli and records the timing behavior. Dynamic timing analysis generally requires simulating many input test vectors to extensively exercise the design to ensure that signal paths are sufficiently identified and characterized. Static timing analysis does not exercise the design via simulation or emulation. Rather, static analysis records timing behavior based on the timing behavior (e.g., propagation delay) of the design components and their interconnection. With static timing analysis, the complexity of the verification task is partitioned into separate functional and performance checks.

Static timing analysis techniques are primarily *block oriented* or *path oriented.* Block-oriented timing analysis generates design input, also called primary input, to design output, also called primary output, propagation delays by analyzing the design, "stage-by-stage" and summing up the individual stage delays. All devices driven by primary inputs constitute stage 1; all devices driven by the outputs of stage 1 constitute stage 2, and so on. Starting with the first stage,

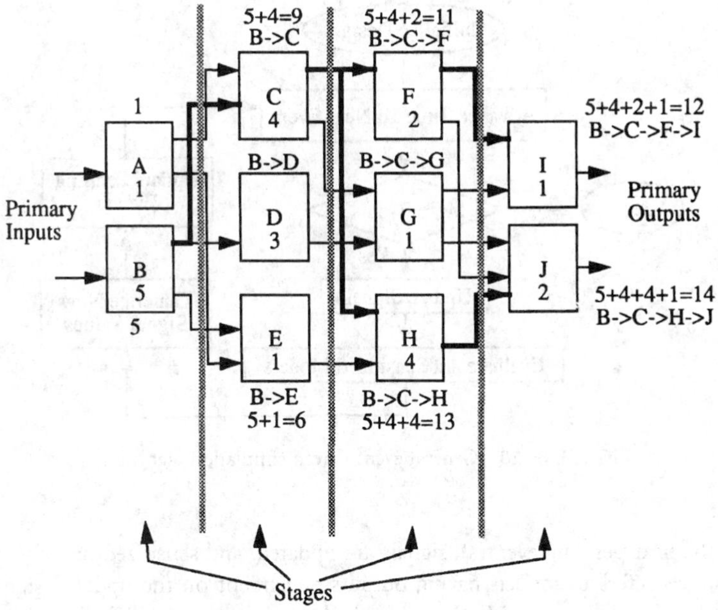

FIGURE 60.19 Block oriented static timing analysis.

all devices associated with a stage are annotated with worst-case delays. A worst-case delay is the propagation delay of the device plus the delay of the last input to arrive at the device, i.e., the signal path with the longest delay leading up to the device inputs. For example, the device labeled "H" in stage 3 in Fig. 60.19 is annotated with the worst-case delay of 13, representing the device propagation delay of 4 and the delay of the last input to arrive through devices "B" and "C" of 9 [39]. When the devices associated with the last stage, i.e., the devices driving the primary outputs, are processed, the accumulated worst-case delays record the longest delay from primary inputs to primary outputs, also called the critical paths. The critical path for each primary output is highlighted in Fig. 60.19.

Path oriented timing analysis generates primary input to primary output propagation delays by traversing all possible signal paths one at a time. Thus, finding the critical path via path oriented timing analysis is equivalent to finding the longest path through a directed acyclic graph, where devices are graph vertices and interconnections are graph edges [41].

A limitation of static timing analysis concerns detecting *false violations* or *false paths.* False violations are signal timing conditions that may occur due to the static structure of the design, but do not occur due to the dynamic nature of the design's response to actual input stimuli.

To account for realistic variances in component timing due to manufacturing tolerances, aging, or environmental effects, timing analysis often provides stochastic or statistical checking capabilities. Statistical timing analysis uses random number generators based on empirically observed probabilistic distributions to determine component timing behavior. Thus, statistical timing analysis describes design performance and the likelihood of the design performance.

Simulation

Simulation exercises a design over a period of time by applying a series of input stimuli and generating the associated output responses. The general event-driven, or schedule-driven, simulation algorithm is diagrammed in Fig. 60.20. An event is a change in signal value. Simulation starts by initializing the design; initial values are assigned to all signals. Initial values include starting values and pending values which constitute future events. Simulation time is

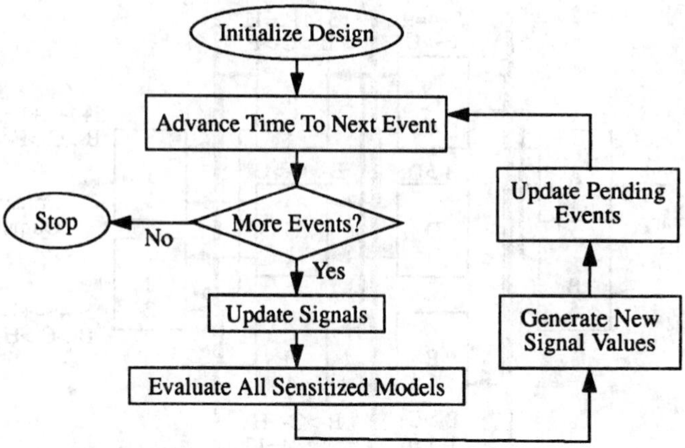

FIGURE 60.20 General event-driven simulation algorithm.

advanced to the next pending event(s), signals are updated, and sensitized models are evaluated. Sensitized models refers to models having outputs dependent on the updated signals [35, 40]. The process of evaluating the sensitized models yields new, potentially different values for signals, i.e., a new set of pending events. These new events are added to the list of pending events, time is advanced to the next pending event(s), and the simulation algorithm repeats. Each pass through the loop of evaluating sensitized models at a particular time step is called a simulation cycle (see Fig. 60.20). Simulation ends when the design yields no further activity, i.e., no more pending events exist to process.

Logic simulation is a computational intensive task for large, complex designs. Prior to committing to manufacturing, processor designers often simulate bringing up or "booting" the operating system. Such simulation tasks require sizable simulation computational resources. As an example, consider simulating 1 s of a 200-K gate, 20-MHz processor design. Assuming that on average only 10% of the total 200 K gates are active or sensitized on each processor clock cycle, Equation 60.6 shows that simulating 1 s of actual processor time equates to 400 billion events.

$$\begin{aligned} 400\text{ B events} &= (20\text{ M clock cycles})(200\text{ K gates})(10\%\text{ activity}) \\ 140\text{ h} &= (400\text{ B events})\left(\frac{50\text{ instructions}}{\text{event}}\right)\left(\frac{50\text{ M instructions}}{\text{s}}\right) \end{aligned} \tag{60.6}$$

Assuming that on average a simulation program executes 50 computer instructions per event on a computer capable of processing 50 million instructions per second, Equation 60.6 also shows that processing 400 billion events requires 140 h or just short of 6 days. This simple example demonstrates the sizable computational properties of simulation. Figure 60.21 shows how simulation computation scales with design complexity.

To address the growing computational demands of simulation, several simulation acceleration techniques have been introduced. Schedule- or event-driven simulation (explained above) can be accelerated by removing layers of interpretation and running simulation as a native executable image; such an approach is called **complied, scheduled driven simulation**. Schedule-driven simulation can be accelerated also by using more efficient event management schemes. In a conventional, central event management scheme all events are logged into a time-ordered list. As simulation time advances, pending events become actual events and the corresponding sensitized devices are executed to compute the response events. On the other hand, in a dataflow event

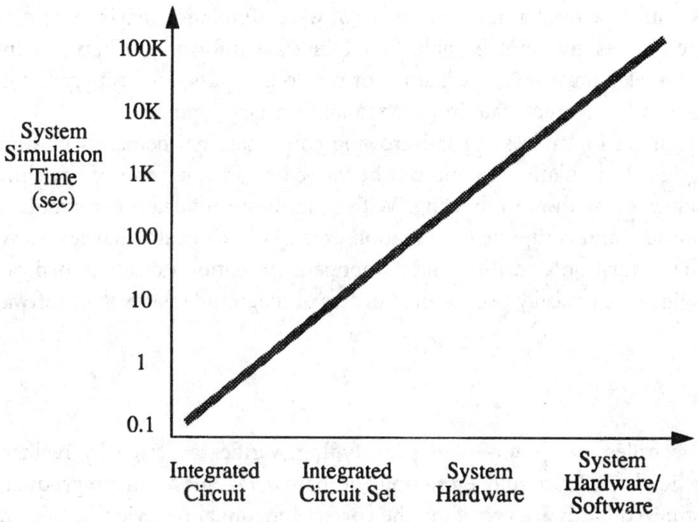

FIGURE 60.21 Simulation requirements.

management scheme events are "self-empowered", active agents that flow through networks and trigger device evaluations without registering with a central time-order list and dispatcher [37].

Instead of evaluating a device in a stimulus-response manner, *cycle-driven* simulation avoids the overhead of event queue processing by evaluating all devices at regular intervals of time. Cycle-driven simulation is efficient when a design exhibits a high degree of concurrency, i.e., a large percentage of the devices are active per simulation cycle. Based on the staging of devices, the devices are *ranked-ordered* to determine the order in which they are evaluated at each time step to ensure the correct causal behavior yielding the proper ordering of events. For functional verification, logic devices are often assigned zero-delay and memory devices are assigned unit-delay. Thus, any number of stages of logic devices may execute between system clock periods.

Another simulation acceleration technique is *message-driven* simulation, also called **parallel** or **distributed** simulation. Device execution is divided among several processors, and the device simulations communicate event activity via messages. Messages are communicated using a conservative or an optimistic strategy. Optimistic message passing strategies, such as *time warp* and *lazy cancellation*, make assumptions about future event activity to advance local device simulation. If the assumptions are correct, the processors operate more independently and better exploit parallel computation.However, if the assumptions are incorrect, then local device simulations may be forced to "roll back" to synchronize local device simulations [34, 36].

Schedule-driven, cycle-driven, and message-driven simulation are software-based simulation acceleration techniques. Simulation also can be accelerated by relegating certain simulation activities to dedicated hardware. For example, *hardware modelers* can be attached to simulators to accelerate the activity of device evaluation. As the name implies, hardware modeling uses actual hardware devices instead of software models to obtain stimulus-response information. In a typical scenario the hardware modeler receives input stimuli from the software simulator. The hardware modeler then exercises the device and sends the output response back to the software simulator. Using actual hardware devices reduces the expense of generating and maintaining software models and provides an environment to support application software development. However, the hardware device must exist, which means hardware modeling has limited use in the initial stages of design in which hardware implementations are not available. Also, it is sometimes difficult for a slave hardware modeler to preserve accurate real time device operating response characteristics within a master nonreal time software simulation environment. For example, some hardware devices may not be able to retain state information between invocations, so the hardware modeler must save the history of previous inputs and reapply them to bring the hardware device to the correct state in order to apply a new input.

Another technique for addressing the growing computational demands of simulation is via *simulation engines.* A simulation engine can be viewed as an extension of the simulation acceleration techniques of hardware modeling. With a hardware modeler, the simulation algorithm executes in software, and component evaluation executes in dedicated hardware. With a simulation engine, the simulation algorithm and component evaluation execute in dedicated hardware. Simulation engines are typically two to three orders of magnitude faster than software simulation [43].

Emulation

Emulation, also called **computer-aided prototyping**, verifies a design by realizing the design in "preproduction" hardware and exercising the hardware. The term preproduction hardware means nonoptimized hardware providing the correct functional behavior, but not necessarily the correct performance. That is, emulation hardware may be slower, require more area, or dissipate more power than production hardware. Presently, preproduction hardware commonly involves some form of *programmable logic devices*, typically *field-programmable gate arrays.* Programmable logic devices provide generic combinational and sequential digital system logic that can be programmed to realize a wide variety of designs [44].

Emulation begins by partitioning the design; each design segment is realized by a programmable logic device. The design segments are then interconnected to realize a preproduction implementation and exercised with a series of input test vectors.

Emulation offers the advantage of providing prototype hardware early in the design cycle to check for errors or inconsistencies in initial functional specifications. Problems can be isolated and design modifications can be accommodated easily by reprogramming the logic devices. Emulation can support functional verification at a computational rate much greater than conventional simulation. However, emulation does not generally support performance verification because, as explained above, prototype hardware typically does not operate at production clock rates.

60.6 Test

Figure 60.22 shows that the test task generally follows the verification task. Although the verification and test tasks both seek to check for correct function and performance, verification focuses on a model of the design before manufacturing, whereas test focuses on the actual hardware after manufacturing. Thus, the primary objective of test is to detect a faulty device by applying input test stimuli and observing expected results [47, 53].

The test task is difficult because designs are growing in complexity; more components provide more opportunity for manufacturing defects. Test is also challenged by new microelectronic fabrication processes having new failure modes which again provides more opportunity for manufacturing defects. New microelectronic fabrication processes also offer higher levels of integration with fewer access points to probe internal electrical nodes. To illustrate the growing demands of test, Table 60.2 shows the general proportional relationship between manufacturing defects and required fault coverage, i.e., quality of testing. Figure 60.23 shows the escalating costs of testing equipment.

Testing involves three general testing techniques or strategies: *functional, parametric,* and *fault.* Functional testing checks that the hardware device realizes the correct I-O digital system behavior. Parametric testing checks that the hardware device realizes the correct performance

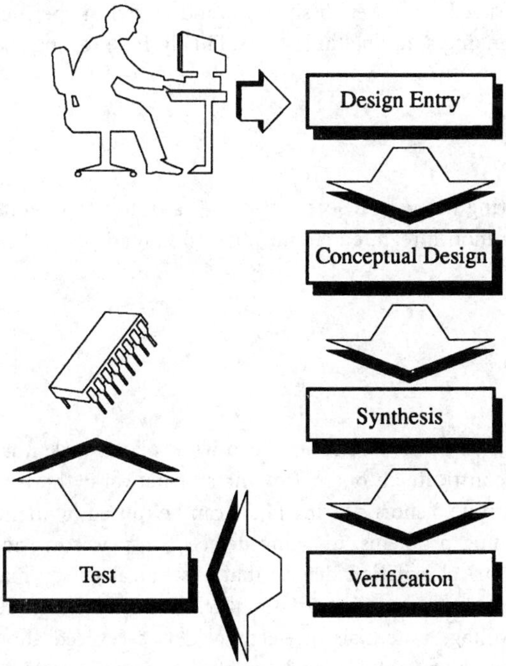

FIGURE 60.22 Design process.

TABLE 60.2 Fault Coverage and Defect Rate

Microelectronic Fabrication Process	Defect Rate 70% Fault Coverage	Defect Rate 90% Fault Coverage	Defect Rate 99% Fault Coverage
2 μm 90% Yield	3%	1%	0.1%
1.5 μm 50% Yield	19%	7%	0.7%
1 μm 10% Yield	50%	21%	2%

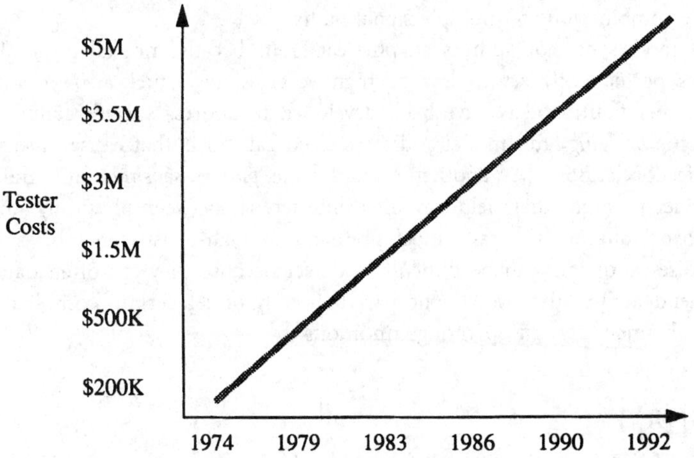

FIGURE 60.23 Integrated circuit tester equipment cost.

specifications, such as speed or power dissipation, and electrical specifications, such as voltage polarities and current sinking/sourcing limitations. Finally, fault testing checks for manufacturing defects or "faults".

Fault Modeling

A fault is a manufacturing or aging defect that causes a device to operate incorrectly or to fail. A sample listing of common integrated circuit physical faults is given below:

- Wiring faults
- Dielectric faults
- Threshold faults
- Soft faults

Wiring faults are unwanted opens and shorts. Two wires or networks that should be electrically connected but are not constitute an open. Two wires or networks that should not be electrically connected but are constitute a short. Wiring faults can be caused by manufacturing defects such as metallization or etching problems, or aging defects, such as corrosion or electromigration. Dielectric faults are electrical isolation defects that can be caused by masking defects, material impurities or imperfections, and electrostatic discharge. Threshold faults occur when the turn-on and turn-off voltage potentials of electrical devices exceed allowed ranges. The faulty devices cannot be properly operated, which results in component failure. Soft faults occur when radiation exposure temporarily changes electrical charge distributions. Such changes can alter circuit voltage potentials, which can, in turn, change logical values, also called "dropping bits". Radiation effects are called "soft" faults because the hardware is not permanently damaged [54].

To simplify the task of fault testing, the physical faults described above are translated into logical faults. Typically, a single logical fault covers several physical faults. A popular logical fault model is the *single stuck line* fault model. The single stuck line fault model considers faults where any single signal line or wire is permanently set to a logic 0, "stuck-at-0", or a logic 1, "stuck-at-1". These signal or interconnect faults are assumed to be time invariant.

Building on the single stuck line fault model, the *multiple stuck line* fault model considers multiple signal wire stuck-at-0/stuck-at-1 faults. The multiple stuck line fault model is more expressive and can cover more physical faults than the single stuck line model. However, fault testing for the multiple stuck line fault model is more difficult because of the exponential growth in the possible combinations of multiple signal faults.

Stuck fault models do not address all physical faults because not all physical faults result in signal lines permanently set to low or high voltages, i.e., stuck-at-0 or stuck-at-1 logic faults. Thus, other fault models have been developed to address specific failure mechanisms. For example, the *bridging* fault model addresses electrical shorts that cause unwanted coupling or spurious feedback loops. As another example, the *pattern-sensitive* fault model addresses wiring and dielectric faults that yield unwanted interference between physically adjacent signals. Pattern-sensitive faults are generally most prevalent in high-density memories incorporating low signal : noise ratios and can be difficult to detect because they are often data-pattern and data-rate dependent. In other words, the part fails only under certain combinations of input stimuli and only under certain operating conditions.

Fault Testing

Having identified and categorized the physical faults that may cause device malfunction or failure and determined how the physical faults relate to logical faults, the next task is to develop tests to detect these faults. When the tests are generated by a computer program, this activity is called **automatic test program generation**. Examples of fault testing techniques are listed below:

- Stuck-at techniques
- Scan techniques
- Signature techniques
- Ad hoc techniques
- Coding techniques
- Electrical monitoring techniques

The following paragraphs review these testing strategies.

Basic stuck-at techniques generate input stimuli for fault testing combinational digital systems. Three of the most popular stuck-at fault testing techniques are the D algorithm, the path oriented decision making (Podem) algorithm, and the fan algorithm. These algorithms first identify a circuit fault (e.g., stuck-at-0 or stuck-at-1) and then try to generate an input stimulus that detects the fault and makes the fault visible as an output. Detecting a fault is often called **fault sensitization** and making a fault visible is often called **fault propagation**. To illustrate this process, consider the simple combinational design in Fig. 60.24 [46, 49, 50]. The design is defective because a manufacturing defect has caused the output of the `and` gate to be permanently tied to ground, i.e., stuck-at-0, using a positive logic convention. To sensitize the fault, the inputs A and B should both be set to 1, which should force the `and` gate output to a 1 for a good circuit. To propagate the fault, the inputs C and D should both be set to 0, which will force the `xor` gate output to 1, again for a good circuit. Thus if A = 1, B = 1, C = 0, and D = 0 in Fig. 60.24, then a good circuit would yield a 1, but the defective circuit yields a 0 which detects the stuck-at-0 fault at the `and` gate output.

Sequential automatic test program generation is a more difficult task than combinational automatic test program generation because exercising or sensitizing a particular circuit path to detect the presence of a possible manufacturing fault may require a sequence of input test vectors. One technique for testing sequential digital systems is called scan fault testing. Scan fault testing is called a **design-for-testability** technique because it modifies or constrains the design in a manner that facilitates fault testing. Scan techniques impose a logic design discipline that all state registers be connected in one or more chains to form "scan rings", as shown in Fig. 60.25 [48]. During normal device operation, the scan rings are disabled and the registers serve as conventional memory (state) storage elements. During test operation, the scan rings are enabled and stimulus test vectors are shifted into the memory elements to set the state of the digital system. The digital system is exercised for one clock cycle and then the results are shifted out of the scan ring to record the response.

The principal advantage of scan design-for-testability is that the scan ring decouples stages of combinational logic, thereby transforming a sequential design into effectively a combinational

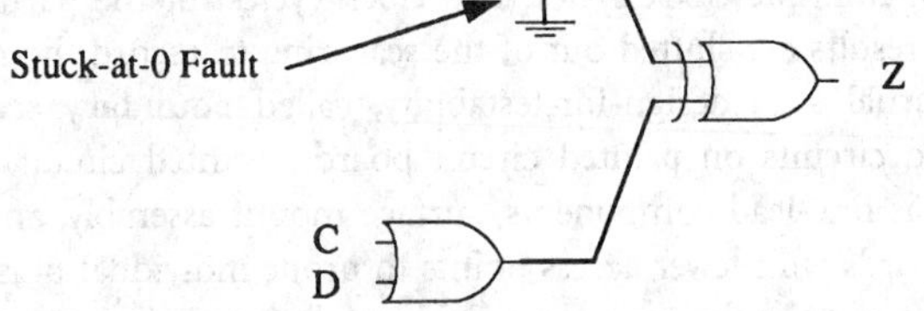

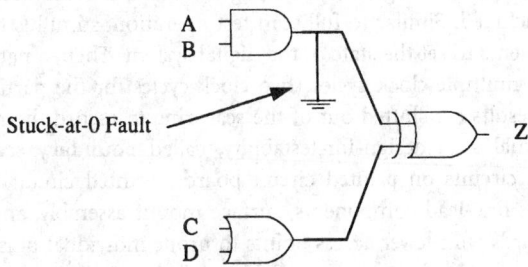

FIGURE 60.24 Combinational logic stuck-at fault testing.

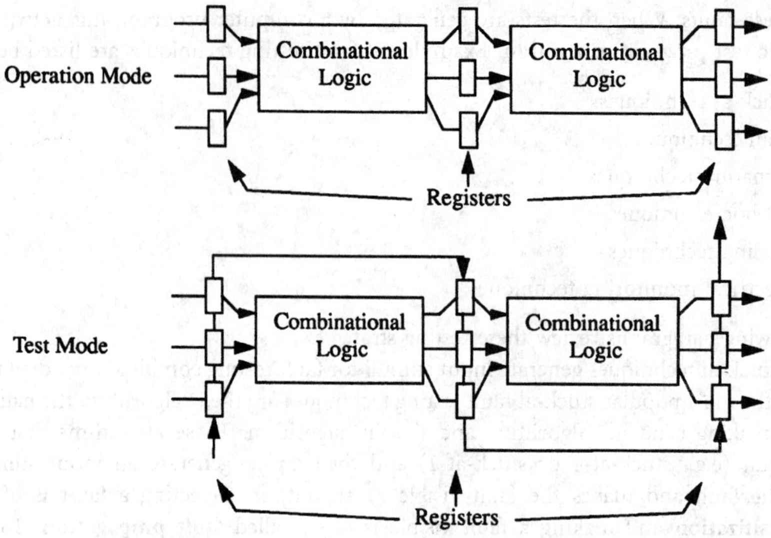

FIGURE 60.25 Scan-based design-for-testability.

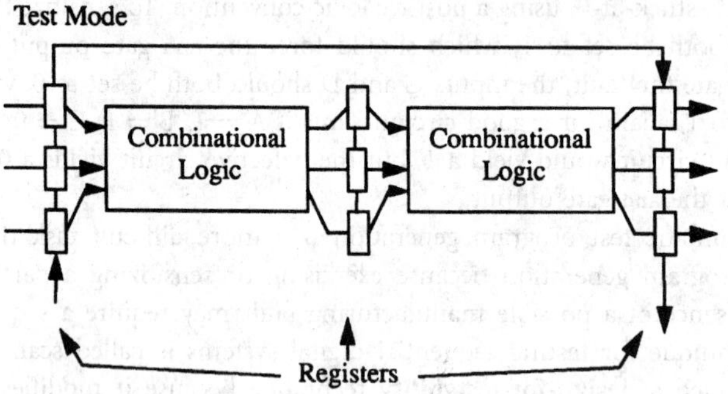

FIGURE 60.26 Partial scan-based design-for-testability.

design and correspondingly a difficult sequential test task into a simpler combinational test task. However, fault tests must still be generated for the combinational logic, and shifting in stimuli and shifting out response requires time. Also, scan paths require additional hardware resources that can impose a performance (speed) penalty. To address these limitations, "partial" scan techniques have been developed that offer a compromise between testability and performance. Instead of connecting every register into scan chains, Fig. 60.26 shows that a partial scan selectively connects a subset of registers into scan chains. Registers in critical performance circuit paths are typically excluded and registers providing control and observability to portions of the design are typically included. Similar to full scan test operation, stimulus test vectors are shifted into the memory elements to set the state of the digital system. Then, a partial scan test exercises the digital system for multiple clock cycles (two clock cycles for the partial scan shown in Fig. 60.26) and then the results are shifted out of the scan ring to record the response.

A variation of partial scan design-for-testability, called **boundary scan**, has been defined for testing integrated circuits on printed circuit boards. Printed circuit board manufacturing developments, such as fine-lead components, surface mount assembly, and multichip modules, yield high-density boards with fewer access points to probe individual pins. Such printed circuit boards are difficult to test. As the name implies, boundary scan imposes a design discipline on printed circuit board components, typically integrated circuits, such that the I-O pins on the

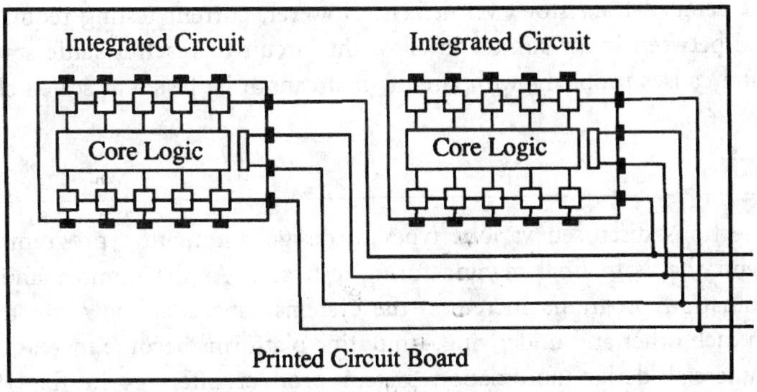

FIGURE 60.27 Boundary scan.

integrated circuits can be connected into scan chains. Figure 60.27 shows that each integrated circuit configured for boundary scan contains scan registers between the I-O pins and the core logic to enable the printed circuit board test bus to control and observe the behavior of individual integrated circuits [51].

Another design-for-testability technique is signature analysis, also called **built-in-self-test.** Signature testing techniques use additional logic, typically linear feedback shift registers, to automatically generate pseudorandom test vectors. The output responses are compressed into a single vector and compared to a known good vector. If the output response vector does not exactly match the known good vector, then the design is considered faulty. Matching the output response vector and a known good vector does not guarantee correct hardware; however, if enough pseudorandom test vectors are exercised, then the chances are acceptably small of obtaining a false positive result. Signature analysis is often used to test memories [45].

Ad hoc testing techniques selectively insert test hardware and access points into a design to improve observability and controllability. Typical candidates for additional access points include storage elements (set and reset controls), major system communication buses, and feedback loops. Ad hoc testing techniques, also called behavioral testing techniques, can avoid the performance penalty of more structured, logical fault testing techniques, such as scan testing, and more closely mimics the actions of an expert test engineer. However, the number of access points is often restricted by integrated circuit or printed circuit board I-O pin limitations.

Coding test techniques encode signal information so that errors can be detected and possibly corrected. Although often implemented in software, coding techniques can also be implemented in hardware. For example, a simple coding technique called **parity checking** is often implemented in hardware. Parity checking adds an extra bit to multibit data. The parity bit is set such that the total number of logic 1's in the multibit data and parity bit is either an even number (even parity) or an odd number (odd parity). An error has occurred if an even parity encoded signal contains an odd number of logic 1's or an odd parity encoded signal contains an even number of logic 1's. Coding techniques are used extensively to detect and correct transmission errors on system buses and networks, storage errors in system memory, and computational errors in processors [52].

Finally, the electrical monitoring testing technique, also called current/voltage testing, relies on the simple observation that an out-of-range current or voltage often indicates a defective or bad part. Possibly a short or open is present, causing a particular I-O signal to have the wrong voltage or current. Current testing (I_{ddq} *testing*) is particularly useful for digital systems using CMOS integrated circuit technology. Normally, CMOS circuits yield very low static or quiescent currents. However, physical faults, such as gate oxide defects, can increase static current by several orders of magnitude. Such a substantial change in static current is straightforward to detect. The principle advantages of current testing are that the tests are simple and the fault models address detailed transistor-level defects. However, current testing requires that enough time be allotted between input stimuli to allow the circuit to reach a static state, which slows down testing and causes problems with circuits that cannot be tested at scaled clock rates.

60.7 Frameworks

The previous sections discussed various types of design automation programs ranging from initial design entry tasks to final manufacturing test tasks. As the number and sophistication of design automation programs increases, the systems aspects of how the tools should be integrated with each other and underlying computing platforms become increasingly important. The first commercial design automation system product offerings in the 1970s consisted primarily of design automation programs and an associated computing platform "bundled" together in turnkey systems. These initial product offerings offered little flexibility to "mix-and-match" design automation programs and computing products from different sources to take advantage of state-of-the-art capabilities and construct a design automation system tailored to particular end-user requirements. In response to these limitations, vendor offerings in the 1980s started to address *open systems.* Vendors "unbundled" software and hardware and introduced interoperability mechanisms to enable a design automation program to execute on different vendor systems. The interoperability mechanisms are collectively called **frameworks** [56].

Frameworks support design automation systems potentially involving many users, application programs, and host computing environments or platforms. Frameworks manage the complexities of a complete design methodology by coordinating and conducting the logistics of design automation programs and design data. Supporting the entire design cycle, also called **concurrent engineering**, improves productivity by globally optimizing the utilization of resources and the minimization of design errors and associated costs.

A general definition of a framework is given below:

> A CAD framework is a software infrastructure that provides a common operating environment for CAD tools. A framework should enable users to launch and manage tools; create, organize, and manage data; graphically view the entire design process; and perform design management tasks such as configuration and version management. Among the key elements of a CAD framework are platform-independent graphics and user interfaces, inter-tool communications, and design data and process management services. (CAD Framework Initiative, CFI)[1]

Figure 60.28 illustrates that a framework is essentially a domain-specific (i.e., electronic systems) layer of operating system software that facilitates "plug-compatible" design automation programs. Framework services are specific to electronic systems design and are mapped into the more generic services provided by general purpose computing platforms [55].

User interface services provide a common and consistent "look-and-feel" to application programs. User interface services include support for consistent information display styles using menus and/or windows. These services also include support for consistent command styles using programming function keys and/or mouse buttons.Application program services provide support for program interactive/batch invocation and normal/abnormal termination. Application services also provide support for program-to-program communication and process management. Process management implements a design methodology that defines application-to-application and application-to-data dependencies. Process management describes a general sequencing of design automation programs and provisions for iterations. In other words, process management ensures a designer is working with the right tool at the right time with the right data.

[1]CAD Framework Initiative is a consortium of companies established to define and promote industry standards for design automation system interoperability.

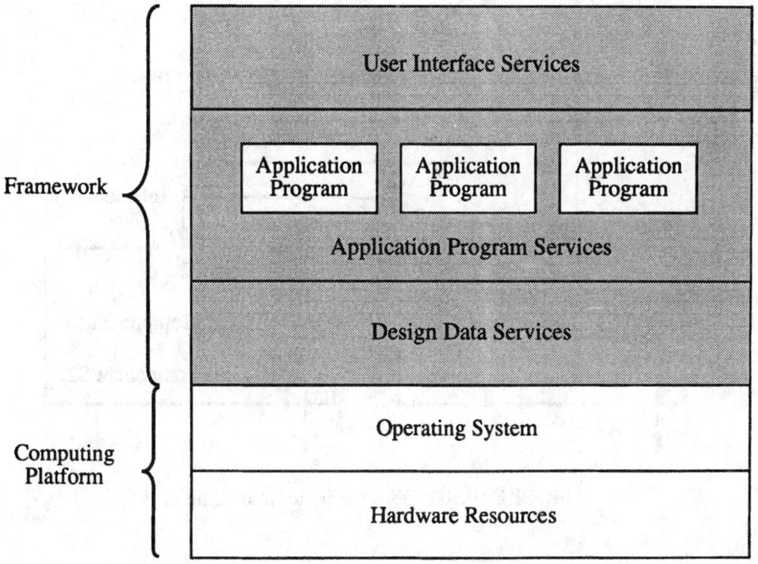

FIGURE 60.28 Framework architecture.

Finally, a framework provides design data services to support access, storage, configuration, and integrity operations. Figure 60.29 shows relationships among *relational, network*, and *object-oriented* data management schemes. Due to the comparatively large size of design data and the length of design operations, framework data services are evolving toward object-oriented paradigms [57–59].

Object-oriented paradigms match data structures and operations with design automation objects and tasks, respectively. Figure 60.30 shows that hardware devices, components, and products become natural choices for software "objects".

The hierarchical relationships between hardware units become natural choices for software "inheritance". Finally, design automation tasks and subtasks that operate on hardware units, such as display or simulate, become natural choices for software "methods".

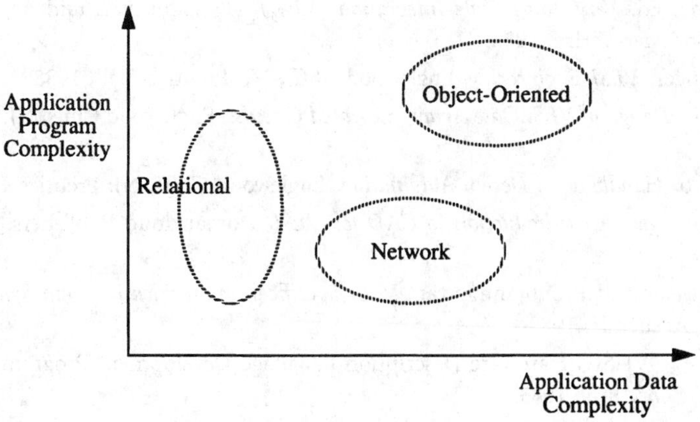

FIGURE 60.29 Data management technology.

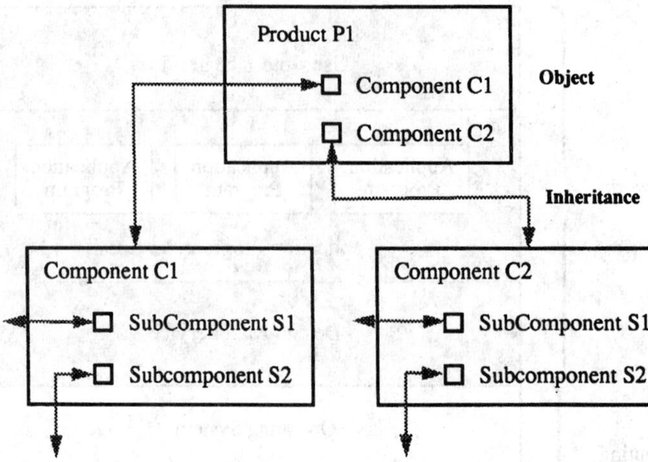

FIGURE 60.30 Object-oriented modeling.

60.8 Summary

Design automation technology offers the potential of serving as a powerful fulcrum in leveraging the skills of a designer against the growing demands of electronic system design and manufacturing. Design automation programs help to relieve the designer of the burden of tedious, repetitive tasks that can be labor intensive and error prone.

Design automation technology can be broken down into several topical areas, such as design entry, conceptual design, synthesis, verification, testing, and frameworks. Each topical area has developed an extensive body of knowledge and experience.

Design entry defines a desire specification. Conceptual design refines the specification into a design plan. Synthesis refines the design plan into an implementation. Verification checks that the implementation faithfully realizes the desired specification. Testing checks that the manufactured part performs functionally and parametrically correctly. Finally, frameworks enable individual design automation programs to operate collectively and cohesively within a larger computing system environment.

References

[1] D. Barbe, Ed. *Very Large Scale Integration (VLSI)—Fundamentals and Applications*, New York: Springer-Verlag, 1980.

[2] T. Dillinger, *VLSI Engineering*, Englewood Cliffs, NJ: Prentice-Hall, 1988.

[3] E. Hollis, *Design of VLSI Gate Array Integrated Circuits*, Englewood Cliffs, NJ: Prentice-Hall, 1987.

[4] S. Sapiro, *Handbook of Design Automation*, Englewood Cliffs, NJ: Prentice-Hall, 1986.

[5] S. Trimberger, *An Introduction to CAD for VLSI*, Domancloud Publishers, San Jose, CA, 1990.

[6] G. Birtwistle and P. Subrahmanyan, *VLSI Specification, Verification, and Synthesis*, Boston: Kluwer Academic, 1988.

[7] A. Dewey, "VHSIC Hardware Description Language Development Program," *Proc. Design Autom. Conf.*, June 1983.

[8] A. Dewey, "VHDL: Towards a Unified View of Design," *IEEE Design Test Comput.*, June 1992.

[9] J. Douglas-Young, *Complete Guide to Reading Schematic Diagrams*, Englewood Cliffs, NJ: Prentice-Hall, 1988.

[10] C. Liu, *Elements of Discrete Mathematics*, New York: McGraw-Hill, 1985.

[11] M. Pechet, Ed., *Handbook of Electrical Package Design*, New York: Marcel Dekker, 1991.

[12] J. Peterson, *Petri Net Theory and Modeling of Systems*, Englewood Cliffs, NJ: Prentice-Hall, 1981.

[13] B. Spinks, *Introduction to Integrated Circuit Layout*, Englewood Cliffs, NJ: Prentice-Hall, 1985.

[14] N. Bose, *Digital Filters: Theory and Applications*, New York: North-Holland, 1985.

[15] X. Chen and M. Bushnell, "A Module Area Estimator For VLSI Layout," *Proc. Design Autom. Conf.*, June 1988.

[16] A. Dewey, *Principles of VLSI Systems Planning: A Framework for Conceptual Design*, Boston: Kluwer Academic, 1990.

[17] W. Donath, "Wiring Space Estimation for Rectangular Gate Arrays," *Proc. Int. Conf. VLSI*, 1981.

[18] W. Heller, "Wirability—Designing Wiring Space for Chips and Chip Packages," *IEEE Design Test Comput.*, Aug. 1984.

[19] C. Leiserson, Area-Efficient VLSI Computation, Ph.D. dissertation, Pittsburgh: Carnegie Mellon University, 1981.

[20] A. Oppenheim and R. Schafer, *Digital Signal Processing*, Englewood Cliffs, NJ: Prentice-Hall, 1975.

[21] L. Rabiner, O. Hermann, and D. Chan, "Practical Design Rules for Optimum Finite Impulse Response Low-Pass Digital Filters," *Bell Syst. Tech.J.*, July–Aug. 1973.

[22] E. Sacerdoti, "Planning in a Hierarchy of Abstraction Spaces," *Artif. Intell.*, Sept. 1974.

[23] M. Stefik, "Planning with Constraints," *Artif. Intell.*, Feb. 1980.

[24] C. Thompson, "Area-Time Complexity for VLSI," *Proc. Caltech Conf.VLSI*, Jan. 1979.

[25] L. Valiant, "Universality Considerations in VLSI Circuits," *IEEE Trans. Comput.*, Feb. 1981.

[26] R. Ayres, *VLSI: Silicon Compilation and the Art of Automatic Microchip Design*, Englewood Cliffs, NJ: Prentice-Hall, 1983.

[27] R. Brayton et al., *Logic Minimization Algorithms for VLSI Synthesis*, Boston: Kluwer Academic, 1992.

[28] R. Camposano and W. Wolfe, *High-Level VLSI Synthesis*, Boston: Kluwer Academic, 1991.

[29] D. Gajski, Ed., *Silicon Compilation*, Reading, MA: Addison-Wesley, 1988.

[30] D. Gajski et al., *High-Level Synthesis—Introduction to Chip andSystem Design*, Boston: Kluwer Academic, 1992.

[31] P. Paulin and J. Knight, "Force-Directed Scheduling for the Behavioral Synthesis of ASIC's," *IEEE Design Test Comput*,, Oct. 1989.

[32] B. Preas, M. Lorenzetti, and B. Ackland, Eds., *Physical Design Automation of VLSI Systems*, New York: Benjamin Cummings, 1988.

[33] T. Sasao, Ed., *Logic Synthesis and Optimization*, Boston: Kluwer Academic, 1993.

[34] R. Bryant, "Simulation on Distributed Systems," *Proc. Int. Conf.Distributed Systems*, 1979.

[35] J. Butler, Ed., *Multiple-Valued Logic in VLSI Design*, New York: IEEE Computer Society Press, 1991.

[36] K. Chandy and J. Misra, "Asynchronous Distributed Simulation Via a Sequence of Parallel Computations," *Commun. ACM*, Apr. 1981.

[37] W. Hahn and K. Fischer, "High Performance Computing For Digital Design Simulation," *VLSI85*, Amsterdam: Elsevier, 1985.

[38] R. McHaney, *Computer Simulation: A Practical Perspective*, New York: Academic Press, 1991.

[39] T. McWilliams and L. Widdoes, "SCALD—Structured Computer Aided Logic Design," *Proc. Design Autom. Conf.*, June 1978.

[40] U. Pooch, *Discrete Event Simulation: A Practical Approach*, Boca Raton, FL: CRC Press, 1993.

[41] T. Sasiki et al., "Hierarchical Design and Verification for Large Digital Systems," *Proc. Design Autom. Conf.*, June 1978.

[42] J. Sifakis, Ed., *Automatic Verification Methods for Finite State Systems*, New York: Springer-Verlag, 1990.

[43] S. Takasaki, F. Hirose, and A. Yamada, "Logic Simulation Engines in Japan," *IEEE Design Test Comput.*, Oct. 1989.

[44] S. Walters, "Computer-Aided Prototyping For ASIC-Based Synthesis," *IEEE Design Test Comput.*, June 1991.

[45] V. Agrawal, C. Kime and K. Saluja, "A Tutorial on Built-In Self Test," *IEEE Design Test Comput.*, June 1993.

[46] M. Breuer and A. Friedman, *Diagnosis and Reliable Design of Digital Systems*, New York: Computer Science Press, 1976.

[47] A. Buckroyd, *Computer Integrated Testing*, New York: John Wiley & Sons, 1989.

[48] E. Eichelberger and T. Williams, "A Logic Design Structure for LSI Testability," *Proc. Design Autom. Conf.*, June 1977.

[49] H. Fujiwara and T. Shimono, "On the Acceleration of Test Generation Algorithms," *IEEE Trans. Comput.*, Dec. 1983.

[50] P. Goel, "An Implicit Enumeration Algorithm to Generalize Tests for Combinational Logic Circuits," *IEEE Trans. Comput.*, Mar. 1981.

[51] K. Parker, "The Impact of Boundary Scan on Board Test," *IEEE Design Test Comput.*, Aug. 1989.

[52] W. Peterson and E. Weldon, *Error-Correcting Codes*, Cambridge, MA: MIT Press, 1972.

[53] M. Weyerer and G. Goldemund, *Testability of Electronic Circuits*, Englewood Cliffs, NJ: Prentice-Hall, 1992.

[54] G. Zobrist, *VLSI Fault Modeling and Testing Technologies*, New York: Ablex Publishing, 1993.

[55] T. Barnes, *Electronic CAD Frameworks*, Boston: Kluwer Academic, 1992.

[56] D. Bedworth, M. Henderson and P. Wolfe, *Computer-Integrated Design and Manufacturing*, New York: McGraw-Hill, 1991.

[57] R. Gupta and E. Horowitz, Eds., *Object-Oriented Databases with Applications to CASE Networks, and VLSI CAD*, Englewood Cliffs, NJ: Prentice-Hall, 1990.

[58] W. Kim, *Object-Oriented Concepts, Databases, and Applications*, Reading, MA: Addison-Wesley, 1989.

[59] E. Nahouranii and F. Petry, Eds., *Object-Oriented Databases*, IEEE Computer Society Press, Los Alamitos, CA, 1991.

61

Computer-Aided Analysis

Gregory Rollins
echnology Modeling
ssociates, Inc.

eter Bendix
echnology Modeling
ssociates, Inc.

51.1 Analog Circuit Simulation

J. Gregory Rollins

Introduction

Computer-aided simulation is a powerful aid during the design or analysis of electronic circuits and semiconductor devices. The first part of this chapter focuses on analog circuit simulation. The second part covers simulations of semiconductor processing and devices. While the main emphasis is on analog circuits, the same simulation techniques may, of course, be applied to digital circuits (which are, after all, composed of analog circuits). The main limitation will be the size of these circuits because the techniques presented here provide a very detailed analysis of the circuit in question and, therefore, would be too costly in terms of computer resources to analyze a large digital system.

The most widely known and used circuit simulation program is SPICE (simulation program with integrated circuit emphasis). This program was first written at the University of California at Berkeley by Laurence Nagel in 1975. Research in the area of circuit simulation is ongoing at many universities and industrial sites. Commercial versions of SPICE or related programs are available on a wide variety of computing platforms, from small personal computers to large mainframes. A list of some commercial simulator vendors can be found in the Appendix.

It is possible to simulate virtually any type of circuit using a program like SPICE. The programs have built-in elements for resistors, capacitors, inductors, dependent and independent voltage and current sources, diodes, MOSFETs, JFETs, BJTs, transmission lines, transformers, and even transformers with saturating cores in some versions. Found in commercial versions are libraries of standard components which have all necessary parameters prefitted to typical

3493-8341-2/95/$0.00+$.50

specifications. These libraries include items such as discrete transistors, op amps, phase-locked loops, voltage regulators, logic integrated circuits (ICs) and saturating transformer cores.

Computer-aided circuit simulation is now considered an essential step in the design of integrated circuits, because without simulation the number of "trial runs" necessary to produce a working IC would greatly increase the cost of the IC. Simulation provides other advantages, however:

- The ability to measure "inaccessible" voltages and currents. Because a mathematical model is used all voltages and currents are available. No loading problems are associated with placing a voltmeter or oscilloscope in the middle of the circuit, with measuring difficult one-shot wave forms, or probing a microscopic die.
- Mathematically ideal elements are available. Creating an ideal voltage or current source is trivial with a simulator, but impossible in the laboratory. In addition, all component values are exact and no parasitic elements exist.
- It is easy to change the values of components or the configuration of the circuit. Unsoldering leads or redesigning IC masks are unnecessary.

Unfortunately, computer-aided simulation has it own problems:

- Real circuits are distributed systems, not the "lumped element models" which are assumed by simulators. Real circuits, therefore, have resistive, capacitive, and inductive parasitic elements present besides the intended components. In high-speed circuits these parasitic elements are often the dominant performance-limiting elements in the circuit, and must be painstakingly modeled.
- Suitable predefined numerical models have not yet been developed for certain types of devices or electrical phenomena. The software user may be required, therefore, to create his or her own models out of other models which are available in the simulator. (An example is the solid-state thyristor which may be created from a NPN and PNP bipolar transistor).
- The numerical methods used may place constraints on the form of the model equations used.

The following sections consider the three primary simulation modes: DC, AC, and transient analysis. In each section an overview is given of the numerical techniques used. Some examples are then given, followed by a brief discussion of common pitfalls.

DC (Steady-State) Analysis

DC analysis calculates the state of a circuit with fixed (non-time varying) inputs after an infinite period of time. DC analysis is useful to determine the operating point (Q-point) of a circuit, power consumption, regulation and output voltage of power supplies, transfer functions, noise margin and fanout in logic gates, and many other types of analysis. In addition DC analysis is used to find the starting point for AC and transient analysis. To perform the analysis the simulator performs the following steps:

1. All capacitors are removed from the circuit (replaced with opens).
2. All inductors are replaced with shorts.
3. Modified nodal analysis is used to construct the nonlinear circuit equations. This results in one equation for each circuit node plus one equation for each voltage source. Modified nodal analysis is used rather than standard nodal analysis because an ideal voltage source or inductance cannot be represented using normal nodal analysis. To represent the voltage sources, loop equations (one for each voltage source or inductor), are included as well as the standard node equations. The node voltages and voltage source currents, then,

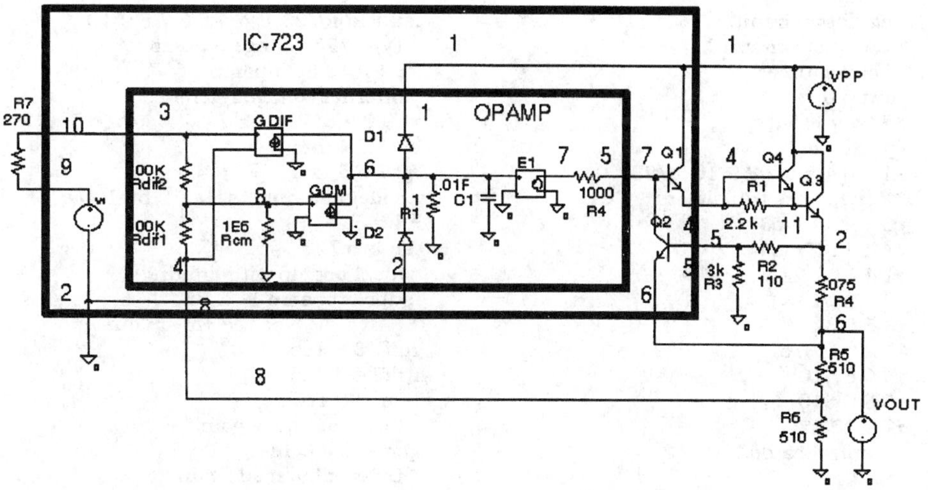

FIGURE 61.1 Regulator circuit to be used for DC analysis, created using PSPICE.

represent the quantities which are solved for. These form a vector **x**. The circuit equations can also be represented as a vector $\mathbf{F}(\mathbf{x}) = \mathbf{0}$.

4. Because the equations are nonlinear, Newton's method (or a variant thereof) is used to solve the equations. Newton's method is given by the following equations:

$$J = \left.\frac{\partial \mathbf{F}}{\partial \mathbf{x}}\right|_{\mathbf{x}^i}$$

$$\mathbf{x}^{i+1} = \mathbf{x}^i - J^{-1} \cdot \mathbf{F}(\mathbf{x}^i)$$

Here, if $\mathbf{x}^i$ is an estimate of the solution, $\mathbf{x}^{i+1}$ is a better estimate. The equations are used iteratively, and hopefully the vector **x** converges to the correct solution. The square matrix J of partial derivatives is called the Jacobian of the system. Most of the work in calculating the solution is involved in calculating J and its inverse J^{-1}. It may take as many as 100 iterations for the process to converge to a solution. Parameters control this process in most simulation programs (see the .OPTIONS statement in SPICE). For example, the maximum number of iterations allowed and error limits which must be satisfied before the process is considered to be converged, but normally the default limits are appropriate.

Example 61.1. Simulation Voltage Regulator: We shall now consider simulation of the type 723 voltage regulator IC, shown in Fig. 61.1. We wish to simulate the IC and calculate the sensitivity of the output IV characteristic and verify that the output current follows a "fold-back" type characteristic under overload conditions.

The IC itself contains a voltage reference source and operational amplifier. Simple models for these elements are used here rather than representing them in their full form, using transistors to illustrate model development. The use of simplified models can also greatly reduce the simulation effort. (For example, the simple op amp used here requires only eight nodes and ten components, yet realizes many advanced features.)

Note in Fig. 61.1 that the numbers next to the wires represent the circuit nodes. These numbers are used to describe the circuit to the simulator. In most SPICE-type simulators the nodes are represented by numbers, with the ground node being node zero. Referring to Fig. 61.2, the 723 regulator and its internal op amp are represented by subcircuits. Each subcircuit

```
Regulator circuit.
* Complete circuit *
* Load source*
vout 6 0
* Power input *
vpp 1 0 11
x1 1 0 4 5 6 7 8 9 10 ic723
* Series Pass transistors *
q3 1 4 11 mq3
q4 1 11 2 mq4
r1 4 11 2.2k
r2 5 2 110
r3 5 0 3k
r4 2 6 0.075
r5 6 8 510
r6 8 0 510
r7 9 10 270
* Control cards *
.op
.model mq3 npn(is=1e-9 bf=30
+ br=5 ikf=50m)
.model mq4 npn(is=1e-6 bf=30
+ br=5 ikf=10)
.dc vout 1 5.5 .01
.plot dc i(vout)
.probe
.subckt ic723 1 2  4 5 6 7 8 9 10
* Type 723 voltage regulator *
x1 1 2 10 8 7 opamp
* Internal voltage reference *
vr 9 2 2.5
q1 3 7 4 mm
q2 7 5 6 mm
.model mm npn (is=1e-12 bf=100
+ br=5)
.ends ic723
* Ideal opamp with limiting
.subckt opamp 1   2   3   4   5
*                vcc vee +in -in out
rdif1 3 8 1e5
rdif2 4 8 1e5
rcm  8 0 1e6
* Common mode gain *
gcm 6 0 8 0 1e-1
* Differential mode gain *
gdif 6 0 4 3 100
r1 6 0 1
* Single pole response *
c1 6 0 .01
d1 6 1 ideal
d2 2 6 ideal
e1 7 0 6 0 1
rout 5 7 1e3
.model ideal d (is=1e-6 n=.01)
.ends opamp
```

FIGURE 61.2 SPICE input listing of regulator circuit shown in Fig. 61.1.

has its own set of nodes and components. Subcircuits are useful for encapsulating sections of a circuit or when a certain section needs to be used repeatedly (see next section).

The following properties are modeled in the op amp:

1. Common mode gain
2. Differential mode gain
3. Input impedance
4. Output impedance
5. Dominant pole
6. Output voltage clipping

The input terminals of the op amp connect to a "T" resistance network, which sets the common and differential mode input resistance. Therefore, the common mode resistance is RCM + RDIF = 1.1E6 and the differential mode resistance is RDIF1 + RDIF2 = 2.0E5.

Dependent current sources are used to create the main gain elements. Because these sources force current into a 1-Ω resistor, the voltage gain is Gm*R at low frequency. In the differential mode this gives (GDIF*R1 = 100). In the common mode this gives (GCM*R1*(RCM/(RDIF1 + RCM = 0.0909). The two diodes D1 and D2 implement clipping by preventing the voltage at node 6 from exceeding VCC or going below VEE. The diodes are made "ideal" by reducing the ideality factor n. Note that the diode current is $I_d = I_s[\exp(V_d/(nV_t)) - 1]$, where V_t is the thermal voltage (0.026 V). Thus, reducing n makes the diode turn on at a lower voltage.

A single pole is created by placing a capacitor (C1) in parallel with resistor R1. The pole frequency is therefore given by 1.0/(2*π*R1*C1). Finally, the output is driven by the voltage-controlled voltage source E1 (which has a voltage gain of unity), through the output resistor R4. The output resistance of the op amp is therefore equal to R4.

To observe the output voltage as a function of resistance, the regulator is loaded with a voltage source (VOUT) and the voltage source is swept from 0.05 to 6.0 V. A plot of output voltage vs. resistance can then be obtained by plotting VOUT vs. VOUT/I(VOUT) (using PROBE in this case; see Fig. 61.3). Note that for this circuit, even though a current source would seem a more natural choice, a voltage source must be used as a load rather than a current source because the output characteristic curve is multivalued in current. If a current source were used it would not be possible to easily simulate the entire curve. Of source, many other interesting quantities can be plotted; for example, the power dissipated in the pass transistor can be approximated by plotting IC(Q3)*VC(Q3).

Several restrictions exist as to what constitutes a valid circuit, and in most cases the simulators will complain if the restrictions are violated:

1. All nodes must have a DC path to ground.
2. Voltage sources must not be connected in a loop.
3. Current sources may not be connected in series.
4. Each node must be connected to at least two elements.

For these simulations PSPICE was used running on an IBM PC. The simulation took < 1 min of CPU time.

Pitfalls. Many SPICE users forget that the first line in the input file is used as the title. Therefore, if the first line of the file is a circuit component, the component name will be used as the title and the component will not be included in the circuit. Convergence problems are sometimes experienced if "difficult" bias conditions are created. An example of such a condition is if a diode is placed in the circuit backwards, resulting in a large forward bias voltage, SPICE will have trouble resolving the current. Another difficult case is if a current source were used instead of a voltage to bias the output in the previous example. If the user then tried to increase the output current above 10 A, SPICE would not be able to converge because the regulator will not allow such a large current.

AC Analysis

AC analysis uses phasor analysis to calculate the frequency response of a circuit. The analysis is useful for calculating the gain, 3 dB frequency input and output impedance, and noise of a circuit as a function of frequency, bias conditions, temperature, etc.

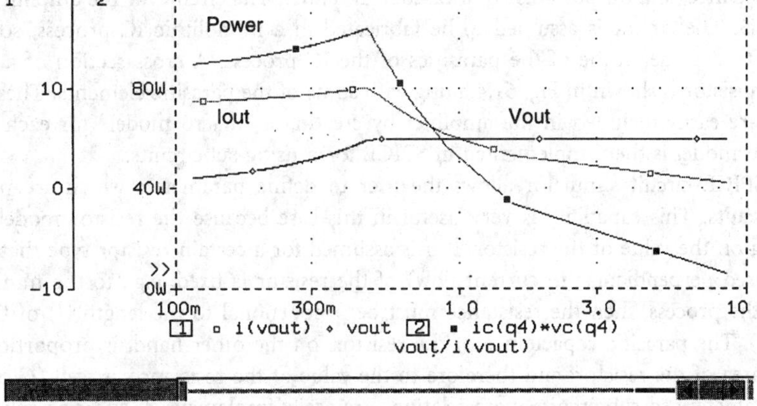

FIGURE 61.3 Output characteristics of regulator circuit using PSPICE.

Numerical Method

1. A DC solution is performed to calculate the Q-point for the circuit.
2. A linearized circuit is constructed at the Q point. To do this, all nonlinear elements are replaced by their linearized equivalents. For example, a nonlinear current source $I = aV_1^2 + bV_2^3$ would be replaced by a linear voltage controlled current source $I = V_1(2aV_{1q}) + V_2(3bV_{2q}^2)$.
3. All inductors and capacitors are replaced by complex impedances, and conductances evaluated at the frequency of interest.
4. Nodal analysis is now used to reduce the circuit to a linear algebraic complex matrix. The AC node voltages may now be found by applying an excitation vector (which represents the independent voltage and current sources) and using Gaussian elimination (with complex arithmetic) to calculate the node voltages.

AC analysis does have limitations and the following types of nonlinear or large signal problems cannot be modeled:

1. Distortion due to nonlinearities such as clipping, etc.
2. Slew rate-limiting effects
3. Analog mixers
4. Oscillators

Noise analysis is performed by including noise sources in the models. Typical noise sources include thermal noise in resistors $I_n^2 = 4kT\,\Delta f/R$, and shot $I_n^2 = 2qI_d\,\Delta f$, and flicker noise in semiconductor devices. Here, T is temperature in Kelvins, k is Boltzmann's constant, and Δf is the bandwidth of the circuit. These noise sources are inserted as independent current sources, $\text{In}_j(f)$ into the AC model. The resulting current due to the noise source is then calculated at a user-specified summation node(s) by multiplying by the gain function between the noise source and the summation node $A_{js}(f)$. This procedure is repeated for each noise source and then the contributions at the reference node are root mean squared (RMS) summed to give the total noise at the reference node. The equivalent input noise is then easily calculated from the transfer function between the circuit input and the reference node $A_{is}(f)$. The equation describing the input noise is therefore:

$$I_i = \frac{1}{A_{is}(f)}\sqrt{\sum_j [A_{js}(f)\text{In}_j(f)]^2}$$

Example 61.2. Cascode Amplifier with Macro Models: Here, we find the gain, bandwidth, input impedance, and output noise of a cascode amplifier. The circuit for the amplifier is shown in Fig. 61.5. The circuit is assumed to be fabricated in a monolithic IC process, so it will be necessary to consider some of the parasitics of the IC process. A cross-section of a typical IC bipolar transistor is shown in Fig. 61.4 along with some of the parasitic elements. These parasitic elements are easily included in the amplifier by creating a "macro model" for each transistor. The macro model is then implemented in SPICE form using subcircuits.

The PSPICE circuit simulator allows the user to define parameters which are passed into the subcircuits. This capability is very useful in this case because the resistor model will vary, depending on the value of the resistor. It if is assumed for a certain resistor type that the width w (measured perpendicular to current flow) of the resistor is fixed, e.g., to the minimum line width of the process, then the resistance must be proportional to the length (l) of the resistor ($R \propto l/w$). The parasitic capacitance of the resistor, on the other hand, is proportional to the junction area of the resistor, and therefore to the value of the resistance as well ($C \propto lw \propto R$). Using parameterized subcircuits these relations are easily implemented, and one subcircuit can be used to represent many different resistors (see Fig. 61.6). Here we represent the capacitance

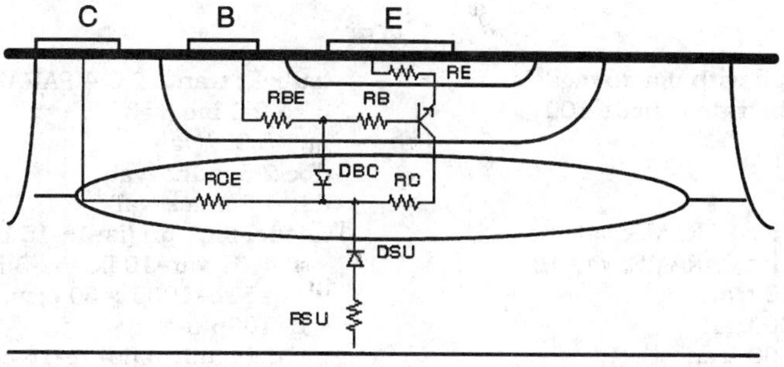

FIGURE 61.4 BJT cross-section with macro model elements.

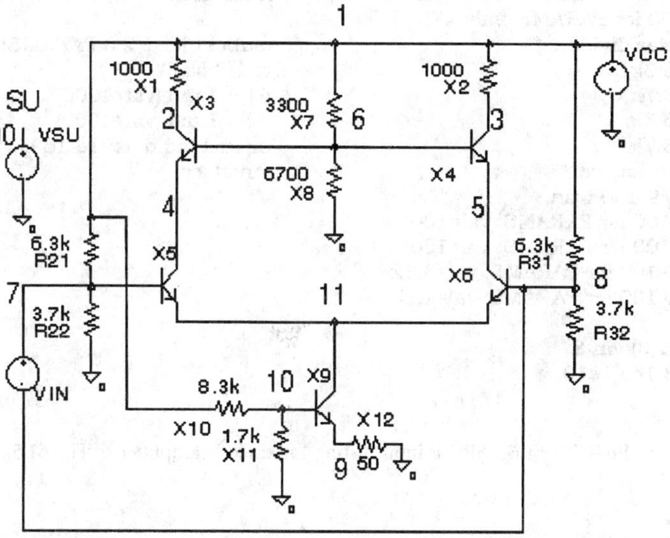

FIGURE 61.5 Cascode amplifier for AC analysis, created using PSPICE.

using 2 diodes one at each end of the resistor. This was done because the resistor junction capacitance is voltage dependent.

The input to the circuit is a voltage source (VIN), applied differentially to the amplifier. The output will be taken differentially across the collectors of the two upper transistors at nodes 2 and 3. The input impedance of the amplifier can be calculated as VIN/I(VIN) or because VIN = 1.0 just as 1/I(VIN). These quantities are shown plotted using PROBE in Fig. 61.7. It can be seen that the gain of the amplifier falls off at high frequency as expected. The input impedance also drops because parasitic capacitances shunt the input.

It is also requested in Fig. 61.6 that noise analysis be performed at every 20th frequency point. A portion of the noise printout is shown in Fig. 61.8. It can be seen that the simulator calculates the noise contributions of each component in the circuit at the specified frequencies and displays them. The total noise at the specified summing node (differentially across nodes 2 and 3) in this case is also calculated as well as the equivalent noise referenced back to the input. This example took < 1 min on an IBM PC.

Pitfalls. Many novice users will forget that AC analysis is a linear analysis. They will, for example, apply a 1-V signal to an amplifier with 5-V power supplies and a gain of 1000 and be surprised when SPICE tells them that the output voltage is 1000 V. Of course, the voltage generated in a simple amplifier must be less than the power supply voltage, but to examine

```
Cascode amp with macro models.
* P type substrate is node 100
vcc 1 0 10
vsu 100 0 0
vin 7 8 ac 1
x1 1 2 100 icr PARAMS: val=1k
x2 1 3 100 icr PARAMS: val=1k
x3 2 6 4 100 tran
x4 3 6 5 100 tran
x5 4 7 11 100 tran
x6 5 8 11 100 tran
* cascode base bias divider *
x7 1 6 100 icr PARAMS: val=3.3k
x8 6 0 100 icr PARAMS: val=6.7k
* input bias dividers *
r21 1 7 6.3k
r22 7 0 3.7k
r31 1 8 6.3k
r32 8 0 3.7k
* Current Source *
x9 11 10 9 100 tran
x10 9 0 100 icr PARAMS: val=100
x11 9 0 100 icr PARAMS: val=100
x12 1 10 100 icr PARAMS: val=8.3k
x13 10 0 100 icr PARAMS: val=1.7k
.op
.noise v(2,3) vin 8
.ac dec 8 1e6 1e10
.probe

.subckt tran 1 2 3 4 PARAMS: val=1
q1 5 2 3 mq {val}
rc 1 5 20/{val}
dbc 2 5 mdbc {val}
dsc 4 5 mdcs {val}
.model mq npn (is=1e-15 bf=100 br=5
+  vaf=30 var=10 ikf=5e-3 ikr=1e-4
+  re=5 rb=100 rc=50 cje=.1p cjc=.05p
+ tf=100p tr=1n)
.model mdbc d (is=1e-16 cjo=.05p)
.model mdcs d (is=1e-16 cjo=.01p
+ rs=1000)
.ends tran

.subckt icr 1 2 3 PARAMS: val=1000
rr 1 2 {val}
d1 3 1 md {val/1000}
d2 3 1 md {val/1000}
.model md d (is=1e-16 cjo=10f rs=1k)
.ends icr
```

FIGURE 61.6 SPICE input listing for cascode amplifier of Fig. 61.5.

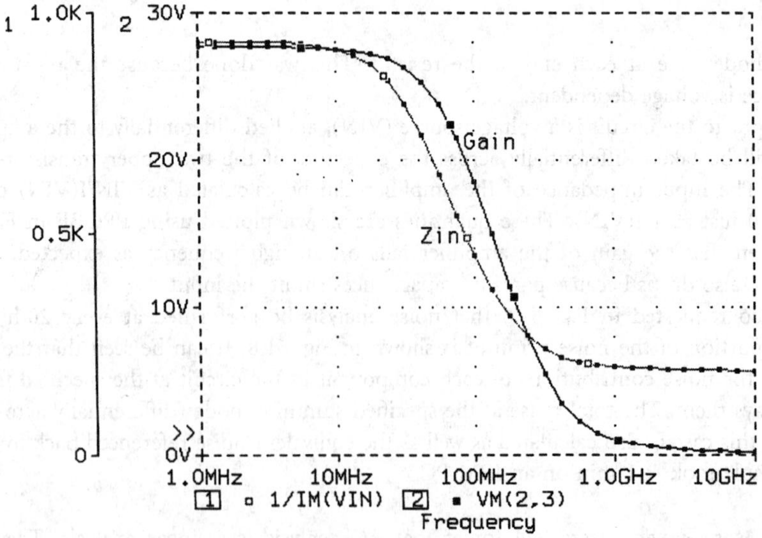

FIGURE 61.7 Gain and input impedance of cascode amplifier.

```
****    NOISE ANALYSIS                 TEMPERATURE =   27.000 DEG C
   FREQUENCY =   1.000E+09 HZ

**** DIODE SQUARED NOISE VOLTAGES (SQ V/HZ)

            x1.d1       x1.d2       x2.d1       x2.d2       x3.dsc      x4.dsc
RS       0.000E+00   0.000E+00   0.000E+00   0.000E+00   6.098E-21   6.098E-21
ID       0.000E+00   0.000E+00   0.000E+00   0.000E+00   2.746E-24   2.746E-24
FN       0.000E+00   0.000E+00   0.000E+00   0.000E+00   0.000E+00   0.000E+00
TOTAL    0.000E+00   0.000E+00   0.000E+00   0.000E+00   6.100E-21   6.100E-21

**** TRANSISTOR SQUARED NOISE VOLTAGES (SQ V/HZ)

            x3.q1       x4.q1       x5.q1       x6.q1       x7.q1
RB       1.519E-19   1.519E-19   1.522E-16   1.522E-16   0.000E+00
RC       3.185E-20   3.185E-20   7.277E-20   7.277E-20   0.000E+00
RE       8.056E-21   8.056E-21   7.418E-18   7.418E-18   0.000E+00
IB       2.190E-18   2.190E-18   3.390E-18   3.390E-18   0.000E+00
IC       8.543E-17   8.543E-17   1.611E-16   1.611E-16   0.000E+00
FN       0.000E+00   0.000E+00   0.000E+00   0.000E+00   0.000E+00
TOTAL    8.782E-17   8.782E-17   3.242E-16   3.242E-16   0.000E+00


**** TOTAL OUTPUT NOISE VOLTAGE            = 8.554E-16 SQ V/HZ
                                           = 2.925E-08 V/RT HZ

    TRANSFER FUNCTION VALUE:
      V(2,3)/vin                           = 9.582E+00
    EQUIVALENT INPUT NOISE AT vin          = 3.052E-09 V/RT HZ
```

FIGURE 61.8 Noise analysis results for cascode amplifier.

such clipping effects, transient analysis must be used. Likewise, selection of a proper Q point is important. If the amplifier is biased in a saturated portion of its response and AC analysis is performed, the gain reported will be much smaller than the actual large signal gain.

Transient Analysis

Transient analysis is the most powerful analysis capability of a simulator because the transient response is so hard to calculate analytically. Transient analysis can be used for many types of analysis, such as switching speed, distortion, basic operation of certain circuits like switching power supplies. Transient analysis is also the most CPU intensive and can require 100 or 1000 times the CPU time as a DC or AC analysis.

Numerical Method

In a transient analysis time is discretized into intervals called time steps. Typically the time steps are of unequal length, with the smallest steps being taken during portions of the analysis when the circuit voltages and currents are changing most rapidly. The capacitors and inductors in the circuit are then replaced by voltage and current sources based on the following procedure.

The current in a capacitor is given by $I_c = CdV_c/dt$. The time derivative can be approximated by a difference equation:

$$I_c^k + I_c^{k-1} = 2C\,\frac{V_c^k - V_c^{k-1}}{t^k - t^{k-1}}$$

In this equation the superscript k represents the number of the time step. Here, k is the time step we are presently solving for and $(k - 1)$ is the previous time step. This equation can be solved to give the capacitor current at the present time step.

$$I_c^k = V_c^k(2C/\Delta t) - V_c^{k-1}(2C/\Delta t) - I_c^{k-1}.$$

Here, $\Delta t = t^k - t^{k-1}$, or the length of the time step. As time steps are advanced, $V_c^{k-1} \rightarrow V_c^k$; $I_c^{k-1} \rightarrow I_c^k$. Note that the second two terms on the right hand side of the above equation are dependent only on the capacitor voltage and current from the previous time step, and are therefore fixed constants as far as the present step is concerned. The first term is effectively a conductance ($g = 2C/\Delta t$) multiplied by the capacitor voltage, and the second two terms could be represented by an independent current source. The entire transient model for the capacitor therefore consists of a conductance in parallel with two current sources (the numerical values of these are, of course, different at each time step). Once the capacitors and inductors have been replaced as indicated, the normal method of DC analysis is used. One complete DC analysis must be performed for each time point. This is the reason that transient analysis is so CPU intensive. The method outlined here is the trapezoidal time integration method and is used as the default in SPICE.

Example 61.3. Phase-Locked Loop Circuit: Figure 61.9 shows the phase-locked loop circuit. The first analysis considers only the analog multiplier portion (also called a phase detector). The second analysis demonstrates the operation of the entire PLL. For the first analyses we wish to show that the analog multiplier does indeed multiply its input voltages. To do this, sinusoidal signals of 0.5 V amplitude are applied to the inputs, one at 2 MHz, the second at 3 MHz. The analysis is performed in the time domain and a Fourier analysis is used to analyze the output. Because the circuit functions as a multiplier, the output should be

$$V_{\text{out}} = AV_a \sin(2e6\pi t) V_b \sin(3e6\pi t) = AV_a V_b[\cos(1e6\pi t) - \cos(5e6\pi t)]$$

The 1 and 5 MHz components are of primary interest. Feedthrough of the original signal will also occur, which will give 2 and 3 MHz components in the output. Other forms of nonlinear distortion will give components at higher frequencies. The SPICE input deck is shown in Fig. 61.10. The output of the Fourier analysis is shown in Fig. 61.11. It can be seen that

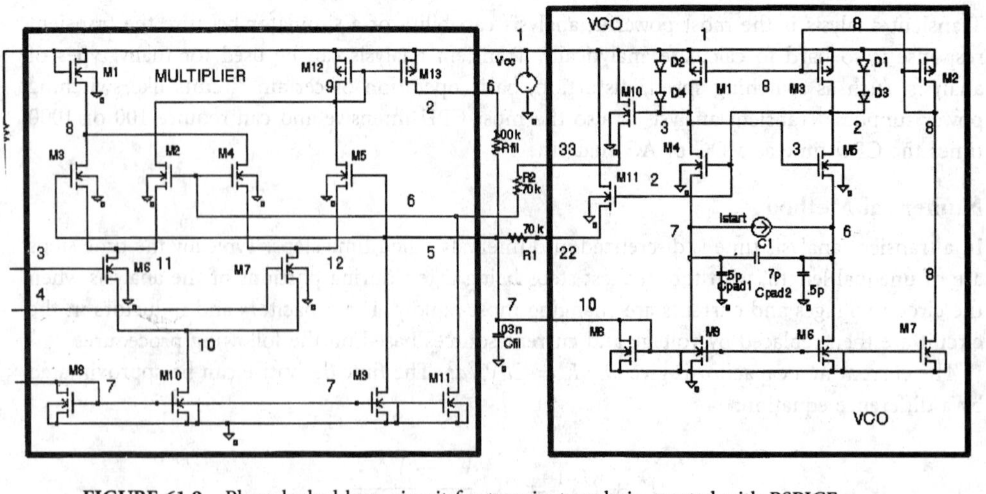

FIGURE 61.9 Phase-locked loop circuit for transient analysis, created with PSPICE.

```
Analog Multiplier Circuit.

Vcc 1 0 10.0
Rbias 1 2 800k
x1 1 9 3 4 5 6 2 10 mult
rdum 9 0 1meg

rin1 4 7 10k
vbias 7 0 2
vin 3 4 sin(0 .5 2meg 0 0)

rin2 5 8 10k
vbias2 8 0 5.0
vin2 5 6 sin(0 .5 3meg 0 0)

.tran 5n 3u 0 5n
.four 1meg 20 v(10)
.op
.probe
.options defl=4u defas=200p
defad=200p

.subckt mult 1  2   3  4  5  6  7   8
*  Pwr Iout In1 In2 Pin1 Pin2 Ibias
Vout
* load resistor *
m1 1 1 8 0 nmos w=30u l=4u
* Upper diff pairs *
m2 9 5 11 0 nmos w=60u l=4u
m3 8 6 11 0 nmos w=60u l=4u
m4 8 5 12 0 nmos w=60u l=4u
m5 9 6 12 0 nmos w=60u l=4u
* Lower diff pairs *
m6 11 3 10 0 nmos w=60u l=4u
m7 12 4 10 0 nmos w=60u l=4u
* Drive Current Mirror*
m8 7 7 0 0 nmos w=60u l=4u
m9 5 7 0 0 nmos w=60u l=4u
m10 10 7 0 0 nmos w=60u l=4u
m11 6 7 0 0 nmos w=60u l=4u
* Output curent mirror *
m12 9 9 1 1 pmos w=60u l=4u
m13 2 9 1 1 pmos w=60u l=4u

.model nmos nmos (level=2 tox=5e-8
+ nsub=2e15 tpg=1 vto=.9 uo=450
+ ucrit=8e4 uexp=.15 cgso=5.2e-10
+ cgdo=5.2e-10)
.model pmos pmos (level=2 tox=5e-8
+nsub=2e15 tpg=-1 vto=-.9 uo=200
+ ucrit=8e4 uexp=.15 cgso=5.2e-10
+ cgdo=5.2e-10)
.ends mult
```

FIGURE 61.10 SPICE input file for analog multiplier portion of PLL.

FOURIER COMPONENTS OF TRANSIENT RESPONSE V(10)

DC COMPONENT = 8.125452E+00

NO	FREQ HZ	FOURIER COMP.	NORMALIZED COMP.	PHASE (DEG)	NORMALIZED PHASE (DEG)
1	1.000E+06	9.230E-02	1.000E+00	8.973E+01	0.000E+00
2	2.000E+06	6.462E-03	7.002E-02	-1.339E+01	-1.031E+02
3	3.000E+06	3.207E-02	3.474E-01	6.515E+01	-2.458E+01
4	4.000E+06	1.097E-02	1.189E-01	-1.604E+02	-2.501E+02
5	5.000E+06	8.229E-02	8.916E-01	-1.074E+02	-1.971E+02
6	6.000E+06	1.550E-02	1.680E-01	-1.683E+02	-2.580E+02
7	7.000E+06	3.695E-02	4.004E-01	7.984E+01	-9.886E+00
8	8.000E+06	7.943E-03	8.606E-02	1.302E+02	4.044E+00
9	9.000E+06	1.962E-02	2.126E-01	-1.149E+02	-2.046E+02

FIGURE 61.11 Results of transient and Fourier analyses of analog multiplier.

the components at 1 and 5 MHz dominate and are a factor of 3 larger than the next largest component. The phase of the 1 and 5 MHz component is also offset by approximately 90° from the input (at 0°), as expected.

Simulation of the entire PLL will now be performed. The SPICE input deck is shown in Fig. 61.12. The phase detector and voltage-controlled oscillator are modeled in separate subcircuits. The phase detector, or multiplier, subcircuit is the same as that of Fig. 61.10 and is omitted from Fig. 61.12 for brevity. Examine the VCO subcircuit and note the PULSE-type current source ISTART connected across the capacitor. The source gives a current pulse 0.3E-6 s wide at the start of the simulation to start the VCO running. To start a transient simulation SPICE first computes a DC operating point (to find the initial voltages V_c^{k-1} on the capacitors). As this DC point is a valid, although not necessarily stable, solution, an oscillator will remain at this point indefinitely unless some perturbation is applied to start the oscillations. Remember, this is an ideal mathematical model and no noise sources or asymmetries exist that would start a real oscillator—it must be done manually. The capacitor C1 would have to be placed off-chip, and bond pad capacitance (CPAD1 and CPAD2) have been included at the capacitor nodes. Including the pad capacitances is very important if a small capacitor C1 is used for high-frequency operation.

In this example, the PLL is to be used as a FM detector circuit and the FM signal is applied to the input using a single frequency FM voltage source. The carrier frequency is 600 kHz and the modulation frequency is 60 kHz. Figure 61.13 shows the input voltage and the output voltage of the PLL at the VCO output and at the phase detector output. It can be seen that after a brief starting transient, the PLL locks onto the input signal and that the phase detector output has a strong 60-kHz component. This example took 251 s on a Sun SPARC-2 workstation (3046 time steps, with an average of 5 Newton iterations per time step).

Pitfalls. Occasionally SPICE will fail and give the message "Timestep too small in transient analysis", which means that the process of Newton iterations at certain time steps could not be

```
Phase locked loop circuit.
Vcc 1 0 10.0
Rbias 1 2 800k
rin1 4 5 100k
vbias 5 0 2
Rfil 6 7 100k
Cfil 6 0 .03n
x1 1 6 3 4 18 19 2 10 mult
r1 8 18 70k
r2 9 19 70k
vsens 7 17 0
x2 1 8 9 17 vco
vin 3 4 sffm(0 1 600k 2 60k)
.tran .05u 60u
.probe
.options acct defl=4u defas=200p
+ defad=200p

.subckt vco 1  22   33  10
*         Pwr out1 out2 Ictl

* P current mirror
m1 2 8 1 1 pmos w=10u l=4u
m2 8 8 1 1 pmos w=10u l=4u
m3 3 8 1 1 pmos w=10u l=4u

* Oscillator
m4 3 2 7 0 nmos w=10u l=4u
m5 2 3 6 0 nmos w=10u l=4u

* N current mirror
m6 6 10 0 0 nmos w=20u l=4u
m7 8 10 0 0 nmos w=4u l=4u
m8 10 10 0 0 nmos w=20u l=4u
m9 7 10 0 0 nmos w=20u l=4u

* source follower buffers
m10 1 3 33 0 nmos w=80u l=4u
m11 1 2 22 0 nmos w=80u l=4u

* Frequency setting capacitor *
c1   6 7 7pf
cpad6 6 0 .5pf
cpad7 7 0 .5pf

* Diode swing limiters *
d1 1 5 md
d2 1 4 md
d3 5 2 md
d4 4 3 md
* Pulse to start VCO *
istart 6 0 pulse (0 400u .1us .01us
+.01us  .3us 100)
```

FIGURE 61.12 SPICE input listing for phase-locked loop circuit.

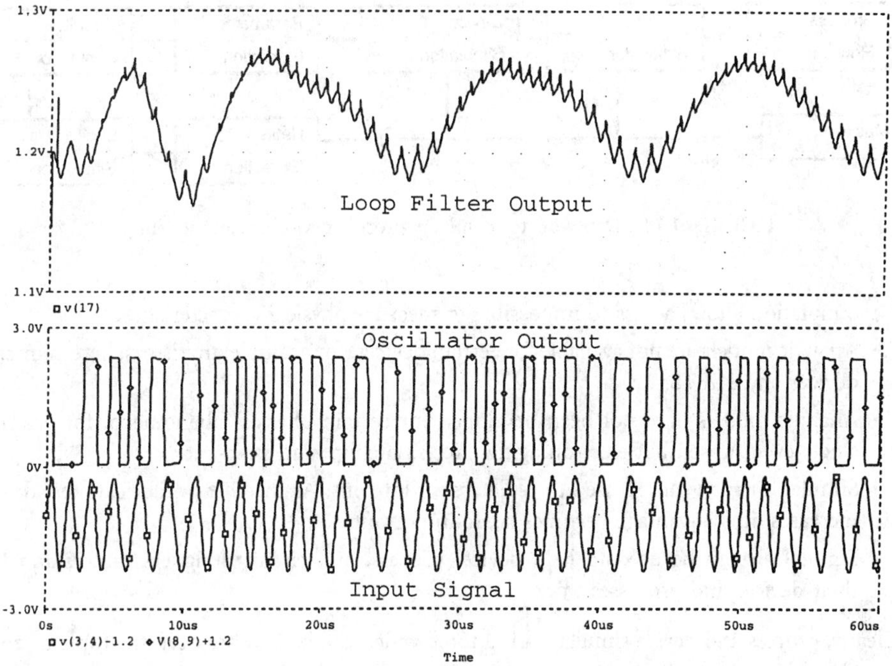

FIGURE 61.13 Transient analysis results of PLL circuit, created using PSPICE.

made to converge. One of the most common causes of this is the specification of a capacitor with a value that is much too large, for example, specifying a 1-F capacitor instead of a 1 pF capacitor (an easy mistake to make by not adding the "p" in the value specification). Unfortunately, we usually have no way to tell which capacitor is at fault from the type of failure generated other than to manually search the input deck.

Other transient failures are caused by MOSFET models. Some models contain discontinuous capacitances (with respect to voltage) and others do not conserve charge. These models can vary from version to version so it is best to check the user's guide.

Process and Device Simulation

Process and devices simulation are the steps that precede analog circuit simulation in the overall simulation flow (see Fig. 61.14). The simulators are also different in that they are not measurement driven as are analog circuit simulators. The input to a process simulator is the sequence of process steps performed (times, temperatures, gas concentrations) as well as the mask dimensions. The output from the process simulator is a detailed description of the solid-state device (doping profiles, oxide thickness, junction depths, etc.). The input to the device simulator is the detailed description generated by the process simulator (or via measurement). The output of the device simulator is the electrical characteristics of the device (IV curves, capacitances, switching transient curves).

Process and device simulation are becoming increasingly important and widely used during the integrated circuit design process. A number of reasons exist for this:

- As device dimensions shrink, second-order effects can become dominant. Modeling of these effects is difficult using analytical models.
- Computers have greatly improved, allowing time-consuming calculations to be performed in a reasonable amount of time.

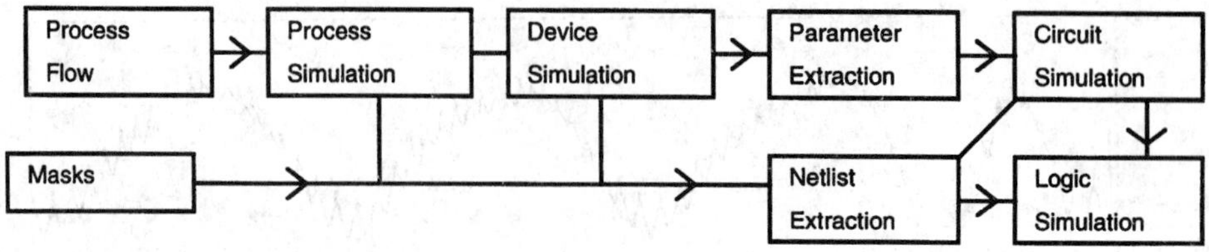

FIGURE 61.14 Data flow for complete process-device-circuit modeling.

- Simulation allows access to impossible to measure physical characteristics.
- Analytic models are not available for certain devices, for example, thyristors, heterojunction devices and IGBTS.
- Analytic models have not been developed for certain physical phenomena, for example, single event upset, hot electron aging effects, latchup, and snap-back.
- Simulation runs can be used to replace split lot runs. As the cost to fabricate test devices increases, this advantage becomes more important.
- Simulation can be used to help device, process, and circuit designers understand how their devices and processes work.

Clearly, process and device simulation is a topic which can be and has been the topic of entire texts. The following sections attempt to provide an introduction to this type of simulation, give several examples showing what the simulations can accomplish, and provide references to additional sources of information.

Process Simulation

Integrated circuit processing involves a number of steps which are designed to deposit (deposition, ion implantation), remove (etching), redistribute (diffusion), or transform (oxidation) the material of which the IC is made. Most process simulation work has been in the areas of diffusion, oxidation, and ion implantation; however, programs are available that can simulate the exposure and development of photo-resist, the associated optical systems, as well as gas and liquid phase deposition and etch.

A number of programs are available (either from universities or commercial vendors) which can model silicon processing. The best known program is SUPREM-IV, which was developed at Stanford University (Stanford, CA). SUPREM-IV is capable of simulating oxidation, diffusion, and ion implantation, and has simple models for deposition and etch. In the following section a very brief discussion of the governing equations used in SUPREM will be given along with the results of an example simulation showing the power of the simulator.

Diffusion

The main equation governing the movement of electrically charged impurities (acceptors in this case) in the crystal is the diffusion equation:

$$\frac{\partial C}{\partial t} = \nabla \cdot \left(D\nabla C - \frac{DqC_a}{kT}\mathbf{E} \right)$$

Here, C is the concentration (#/cm^3) of impurities, C_a is the number of electrically active impurities (#/cm^3), q is the electron charge, k is Boltzmann's constant, T is temperature in degrees Kelvin, D is the diffusion constant, and E is the built-in electric field. The built-in electric field E in (V/cm) can be found from:

$$\mathbf{E} = -\frac{kT}{q}\frac{1}{n}\boldsymbol{\nabla} n$$

In this equation n is the electron concentration (#/cm^3), which in turn can be calculated from the number of electrically active impurities (C_a). The diffusion constant (D) is dependent on many factors. In silicon the following expression is commonly used:

$$D = F_{IV}\left(D_x + D_+\frac{n_i}{n} + D_-\frac{n}{n_i} + D_=\left(\frac{n}{n_i}\right)^2\right)$$

The four D components represent the different possible charge states for the impurity: (x) neutral, (+) positive, (−) negative, (=) doubly negatively charged. n_i is the intrinsic carrier concentration, which depends only on temperature. Each D component is in turn given by an expression of the type

$$D = A\ \exp\left(-\frac{B}{kT}\right)$$

Here, A and B are experimentally determined constants, different for each type of impurity ($x, +, -, =$). B is the activation energy for the process. This expression derives from the Maxwellian distribution of particle energies and will be seen many times in process simulation. It is easily seen that the diffusion process is strongly influenced by temperature. The term F_{IV} is an enhancement factor which is dependent on the concentration of interstitials and vacancies within the crystal lattice (an interstitial is an extra silicon atom which is not located on a regular lattice site; a vacancy is a missing silicon atom which results in an empty lattice site) $F_{IV} \propto C_I + C_V$. The concentration of vacancies, C_V, and interstitials, C_I, are in turn determined by their own diffusion equation:

$$\frac{\partial C_v}{\partial t} = +\nabla \cdot D_V \cdot \nabla C_V - R + G$$

In this equation D_V is another diffusion constant of the form $A\exp(-B/kT)$. R and G represent the recombination and generation of vacancies and interstitials. Note that an interstitial and a vacancy may recombine and in the process destroy each other, or an interstitial and a vacancy pair may be simultaneously generated by knocking a silicon atom off its lattice site. Recombination can occur anywhere in the device via a bulk recombination process $R = A(C_V C_I)\exp(-B/kT)$. Generation occurs where there is damage to the crystal structure, in particular at interfaces where oxide is being grown or in regions where ion implantation has occurred, as the high-energy ions can knock silicon atoms off their lattice sites.

Oxidation

Oxidation is a process whereby silicon reacts with oxygen (or with water) to form new silicon dioxide. Conservation of the oxidant requires the following equation:

$$\frac{dy}{dt} = \frac{\mathbf{F}}{N}$$

Here, F is the flux of oxidant (#/cm^2/s), N is the number of oxidant atoms required to make up a cubic centimeter of oxide, and dy/dt is the velocity with which the Si–SiO_2 interface moves into the silicon. In general the greater the concentration of oxidant (C_0), the faster the growth of the oxide and the greater the flux of oxidant needed at the Si–SiO_2 interface. Thus, $F = k_s C_0$.

The flux of oxidant into the oxide from the gaseous environment is given by:

$$F = h(HP_{ox} - C_0)$$

Here H is a constant, P is the partial pressure of oxygen in the gas, and C_0 is the concentration of oxidant in the oxide at the surface and h is of the form $A\exp(-B/kT)$. Finally, the moment of the oxidant within the already existing oxide is governed by diffusion: $\mathbf{F} = D_0 \nabla C$. When all these equations are combined, it is found that (in the one-dimensional case) oxides grow linearly $dy/dt \propto t$ when the oxide is thin and the oxidant can move easily through the existing oxide. As the oxide grows thicker $dy/dt \propto \sqrt{t}$ because the movement of the oxidant through the existing oxide becomes the rate-limiting step.

Modeling two-dimensional oxidation is a challenging task. The newly created oxide must "flow" away from the interface where it is being generated. This flow of oxide is similar to the flow of a very thick or viscous liquid and can be modeled by a creeping flow equation:

$$\nabla^2 V \propto \nabla P$$

$$\nabla \cdot V = 0$$

V is the velocity at which the oxide is moving and P is the hydrostatic pressure. The second equation results from the incompressibility of the oxide. The varying pressure P within the oxide leads to mechanical stress, and the oxidant diffusion constant D_0 and the oxide growth rate constant k_s are both dependent on this stress. The oxidant flow and the oxide flow are therefore coupled because the oxide flow depends on the rate at which oxide is generated at the interface and the rate at which the new oxide is generated depends on the availability of oxidant, which is controlled by the mechanical stress.

Ion Implantation

Ion implantation is normally modeled in one of two ways. The first involves tables of moments of the final distribution of the ions which are typically generated by experiment. These tables are dependent on the energy and the type of ion being implanted. The second method involves Monte-Carlo simulation of the implantation process. In Monte-Carlo simulation the trajectories of individual ions are followed as they interact with (bounce off) the silicon atoms in the lattice. The trajectories of the ions, and the recoiling Si atoms (which can strike more Si atoms) are followed until all come to rest within the lattice. Typically several thousand trajectories are simulated (each will be different due to the random probabilities used in the Monte-Carlo method) to build up the final distribution of implanted ions. Monte-Carlo has advantages in that the damage to the lattice can be calculated during the implant process. This damage creates interstitials and vacancies which affect impurity diffusion, as was seen earlier. Monte-Carlo can also model channeling, which is a process whereby the trajectories of the ions align with the crystal planes, resulting in greater penetration of the ion than in a simple amorphous target. Monte-Carlo has the disadvantage of being CPU intensive.

Process simulation is always done in the transient mode using time steps as was done with transient circuit simulation. Because partial differential equations are involved, rather than ordinary differential equations, spatial discretization is needed as well. To numerically solve the problem, the differential equations are discretized on a grid. Either rectangular or triangular grids in one, two, or three dimensions are commonly used. This discretization process results in the conversion of the partial differential equations into a set of nonlinear algebraic equations. The nonlinear equations are then solved using a Newton method in a way very similar to the method used for the circuit equations in SPICE.

Example 61.4. NMOS Transistor: In this example the process steps used to fabricate a typical NMOS transistor will be simulated using SUPREM-4. These steps are

1. Grow initial oxide (30 min at 1000 K)
2. Deposit nitride layer (a nitride layer will prevent oxidation of the underlying silicon)
3. Etch holes in nitride layer
4. Implant P+ channel stop (boron dose = 5e12, energy = 50 keV)
5. Grow the field oxide (180 min at 1000 K wet O_2)
6. Remove all nitride
7. Perform P channel implant (boron dose = 1e11, energy = 40 keV)
8. Deposit and etch polysilicon for gate
9. Oxidize the polysilicon (30 min at 1000 K, dry O_2)
10. Implant the light doped drain (arsenic dose = 5e13 energy = 50 keV)
11. Deposit sidewall space oxide
12. Implant source and drain (arsenic, dose = 1e15, energy = 200 keV)
13. Deposit oxide layer and etch contact holes
14. Deposit and etch metal

The top 4 μm of the completed structure, as generated by SUPREM-4, is shown in Fig. 61.15. The actual simulation structure used is 200 μm deep to allow correct modeling of the diffusion of the vacancies and interstitials. The gate is at the center of the device. Notice how the edges of the gate have lifted up due to the diffusion of oxidant under the edges of the polysilicon (the polysilicon, as deposited in step 8, is flat). The dashed contours show the concentration of dopants in both the oxide and silicon layers. The short dashes indicate N-type material, while the longer dashes indicate P-type material. This entire simulation requires about 30 min on a Sun SPARC-2 workstation.

Device Simulation

Device simulation uses a different approach from that of conventional lumped circuit models to determine the electrical device characteristics. Whereas with analytic or empirical models all characteristics are determined by fitting a set of adjustable parameters to measured data, device simulators determine the electrical behavior by numerically solving the underlying set of differential equations. The first of these equations is the Poisson equation, which describes the

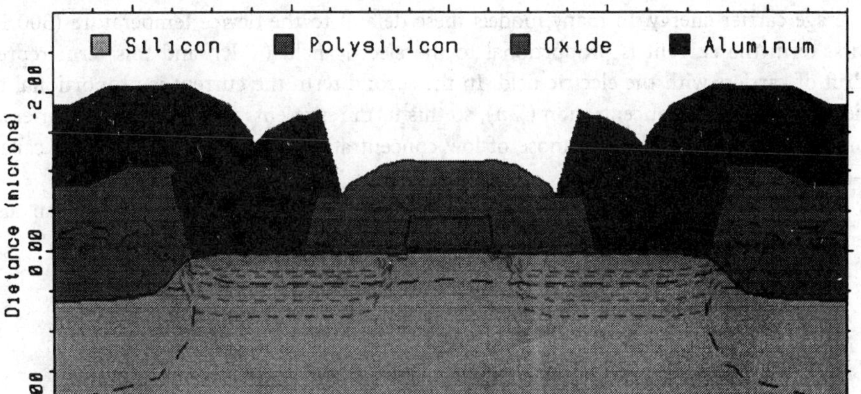

FIGURE 61.15 Complete NMOS transistor cross section generated by process simulation, created with TMA SUPREM-4.

electrostatic potential within the device

$$\nabla \cdot \varepsilon \cdot \nabla \Psi = q(N_a^- - N_d^+ - p + n - Q_f)$$

N_d and N_a are the concentration of donors and acceptors, i.e., the N- and P-type dopants. Q_f is the concentration of fixed charge due, for example, to traps or interface charge. The electron and hole concentrations are given by n and p, respectively, and Ψ is the electrostatic potential.

A set of continuity equations describes the conservation of electrons and holes:

$$\frac{\partial n}{\partial t} = \left(\frac{1}{q} \nabla \cdot \mathbf{J}_n - R + G\right)$$

$$\frac{\partial p}{\partial t} = \left(-\frac{1}{q} \nabla \cdot \mathbf{J}_p - R + G\right)$$

In these equations R and G describe the recombination and generation rates for the electrons and holes. The recombination process is influenced by factors such as the number of electrons and holes present as well as the doping and temperature. The generation rate is also dependent upon the carrier concentrations, but is most strongly influenced by the electric field, with increasing electric fields giving larger generation rates. Because this generation process is included, device simulators are capable of modeling the breakdown of devices at high voltage. $\mathbf{J}_n$ and $\mathbf{J}_p$ are the electron and hole current densities (in amperes per square centimeter). These current densities are given by another set of equations

$$\mathbf{J}_n = q\mu\left(-n \nabla \Psi + \frac{kT_n}{q} \nabla n\right)$$

$$\mathbf{J}_p = q\mu\left(-p \nabla \Psi - \frac{kT_p}{q} \nabla p\right)$$

In this equation k is Boltzmann's constant, μ is the carrier mobility, which is actually a complex function of the doping, n, p, electric field, temperature, and other factors. In silicon the electron mobility will range between 50 and 1000 and the hole mobility will normally be a factor of 2 smaller. In other semiconductors such as gallium arsenide the electron mobility can be as high as 5000. T_n and T_p are the electron and hole mean temperatures, which describe the average carrier energy. In many models these default to the device temperature (300 K). In the first term the current is proportional to the electric field ($\nabla \Psi$), and this term represents the drift of carriers with the electric field. In the second term the current is proportional to the gradient of the carrier concentration (∇n), so this term represents the diffusion of carriers from regions of high concentration to those of low concentration. The model is therefore called the drift-diffusion model.

In devices in which self-heating effects are important, a lattice heat equation can also be solved to give the internal device temperature:

$$\sigma(T)\frac{\partial T}{\partial t} = H + \nabla \cdot \lambda(T) \cdot \nabla T$$

$$H = -(\mathbf{J}_n + \mathbf{J}_p) \cdot \nabla \Psi + H_R$$

where H is the heat generation term, which includes resistive (Joule) heating as well as recombination heating, H_u. The terms $\sigma(T)$, $\lambda(T)$ represent the specific heat and the thermal conductivity of the material (both temperature dependent). Inclusion of the heat equation is essential in many power device problems.

As with process simulation partial differential equations are involved, therefore, a spatial discretization is required. As with circuit simulation problems, various types of analysis are available:

- Steady state (DC), used to calculate characteristic curves of MOSFETs, BJTs diodes, etc.
- AC analysis, used to calculate capacitances, Y-parameters, small signal gains, and S-parameters.
- Transient analysis used for calculation of switching and large signal behavior, and special types of analysis such as radiation effects.

Example 61.5. NMOS IV Curves: The structure generated in the previous SUPREM-IV simulation is now passed into the device simulator and bias voltages are applied to the gate and drain. Models were included which account for Auger and Shockley Reed Hall recombination, doping and electric field-dependent mobility, and impact ionization. The set of drain characteristics obtained is shown in Fig. 61.16. Observe how the curves bend upward at high V_{ds} as the device breaks down. The $V_{gs} = 1$ curve has a negative slope at $I_d = 1.5\text{e-}4\text{A}$ as the device enters snap-back. It is possible to model this type of behavior because impact ionization is included in the model.

Figure 61.17 shows the internal behavior of the device with $V_{gs} = 3$ V and $I_d = 3\text{e-}4\text{A}$. The filled contours indicate impact ionization, with the highest rate being near the edge of the drain right beneath the gate. This is to be expected because this is the region in which the electric field is largest due to the drain depletion region. The dark lines indicate current flow from the source to the drain. Some current also flows from the drain to the substrate. This substrate current consists of holes generated by the impact ionization. The triangular grid used in the simulation can be seen in the source, drain, and gate electrodes. A similar grid was used in the oxide and silicon regions.

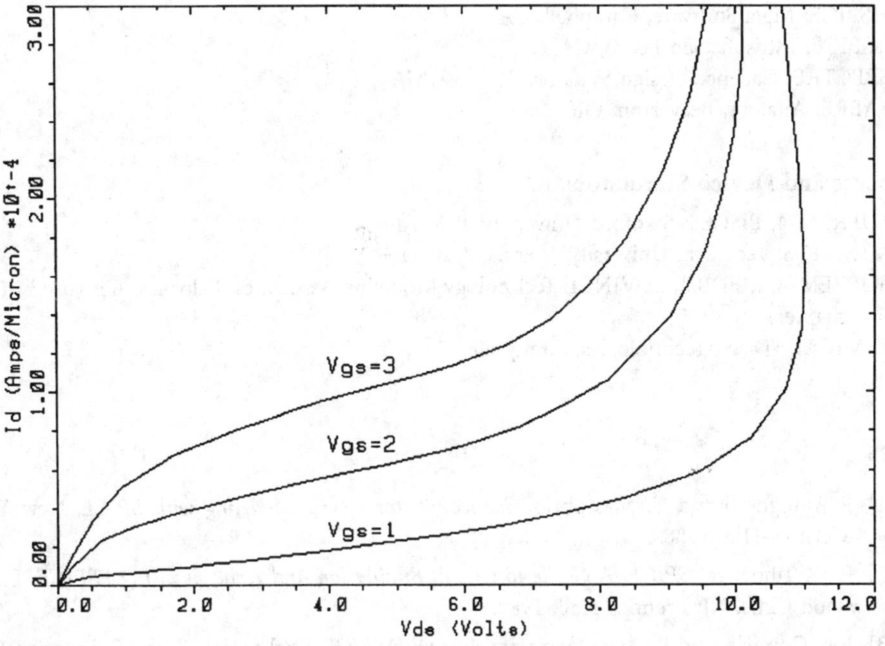

FIGURE 61.16 I_d vs. V_{ds} curves generated by device simulation, created with TMA MEDICI.

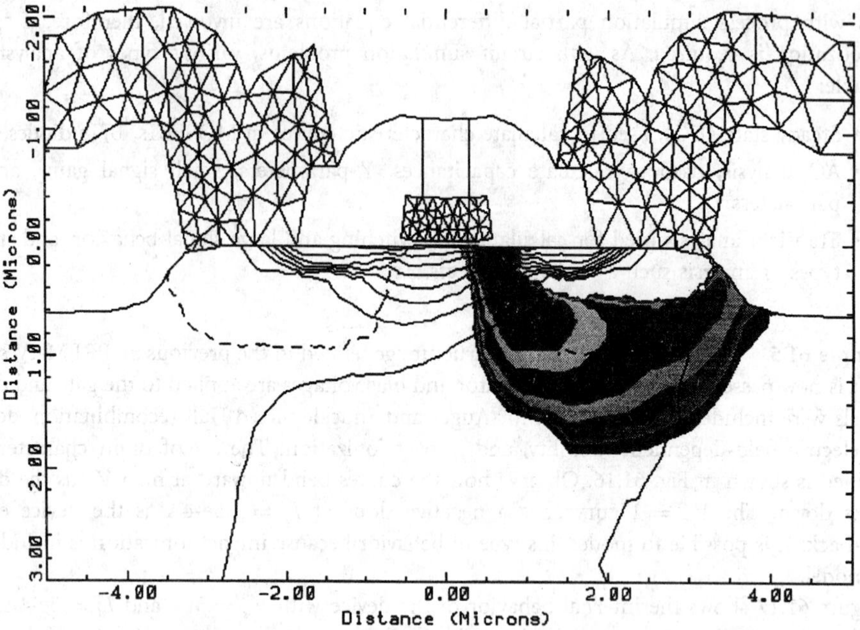

FIGURE 61.17 Internal behavior of MOSFET under bias, created with TMA MEDICI.

Appendix

Circuit Analysis Software

SPICE2, SPICE3: University of California, Berkeley
PSPICE: MicroSim Corporation, Irvine, CA (used in this chapter)
HSPICE: Meta Software, Campbell, CA
IsSPICE: Intusoft, San Pedro, CA
SPECTRE: Cadence Design Systems, San Jose, CA
SABRE: Analogy, Beaverton, OR

Process and Device Simulators

SUPREM-4, PISCES: Stanford University, Palo Alto, CA
MINIMOS: Technical University, Vienna, Austria
SUPREM-4, MEDICI, DAVINCI: Technology Modeling Associates, Palo Alto CA (used in this chapter)
SEMICAD: Dawn Technologies, Sunnyvale, CA

References

[1] P. Antognetti and G. Massobrio, *Semiconductor Device Modeling with SPICE*, New York: McGraw-Hill, 1988.

[2] P. W. Tuinenga, *SPICE, A Guide to Circuit Simulation and Analysis Using PSPICE*, Englewood Cliffs, NJ: Prentice-Hall, 1988.

[3] J. A. Connelly and P. Choi, *Macromodeling with SPICE*, Englewood Cliffs, NJ: Prentice-Hall, 1992.

[4] S. Selberherr, *Analysis and Simulation of Semiconductor Devices*, Berlin: Springer-Verlag, 1984.

[5] R. Dutton and Z. Yu, *Technology CAD, Computer Simulation of IC Process and Devices*, Boston: Kluwer Academic, 1993.

61.2 Parameter Extraction for Analog Circuit Simulation

Peter Bendix

Introduction

Definition of Device Modeling

We use various terms such as device characterization, parameter extraction, optimization, and model fitting to address an important engineering task. In all of these, we start with a mathematical model that describes the transistor behavior. The model has a number of parameters which are varied or adjusted to match the IV (current-voltage) characteristics of a particular transistor or set of transistors. The act of determining the appropriate set of model parameters is what we call device modeling. We then use the model with the particular set of parameters that represent our transistors in a circuit simulator such as SPICE[1] to simulate how circuits with our kinds of transistors will behave. Usually the models are supplied by the circuit simulator we chose. Occasionally we may want to modify these models or construct our own models. In this case we need access to the circuit simulator model subroutines as well as the program that performs the device characterization.

Most people believe the above description covers device modeling. However, we feel this is a very narrow definition. One can obtain much more information by characterizing a semiconductor process than just providing models for circuit simulation. If the characterization is done correctly, it can provide detailed information about the fabrication process. This type of information is useful for process architects as an aid in developing future processes. In addition, the wealth of information from the parameters obtained can be used by process and product engineers to monitor wafers routinely. This provides much more information than what is usually obtained by doing a few crude electrical tests for process monitoring. Finally, circuit designers can use device characterization as a means of optimizing a circuit for maximum performance. Therefore, instead of just being a tool used by specialized device physicists, device characterization is a tool for process architects, process and product engineers, and circuit designers. It is a tool that can and should be used by most people concerned with semiconductor manufacture and design.

Steps Involved in Device Characterization

Device characterization begins with a test chip. Without the proper test chip structures, proper device modeling cannot be done from measured data. A good test chip for MOS technology would include transistors of varying geometries, gate oxide capacitance structures, junction diode capacitance structures, and overlap capacitance structures. This would be a minimal test chip. Additional structures might include ring oscillators and other circuits for checking the AC performance of the models obtained. It is very important that the transistors be well designed and their geometries be chosen appropriate for the technology as well as the desired device model. For bipolar technology modeling, the layout of structures used to get S-parameter

[1]SPICE is a circuit simulation program from the Department of Electrical Engineering and Computer Science at the University of California at Berkeley.

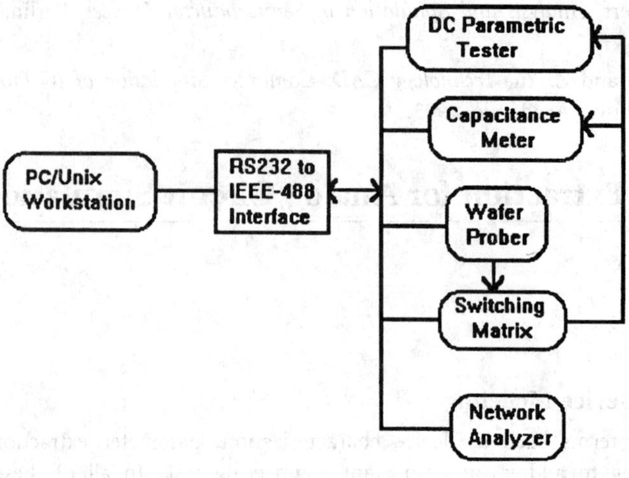

FIGURE 61.18 Typical hardware measuring equipment for device characterization.

measurements is very critical. Although a complete test chip description is beyond the scope of this book, be aware that even perfect device models cannot correct for a poor test chip.

Next we need data that represent the behavior of a transistor or set of transistors of different sizes. These data can come from direct measurement or they can be produced by a device simulator such as PISCES.[2] Typically, one does voltage sweeps on various nodes of the transistor and measures the output current at some or all of these nodes. For example, in an MOS transistor, one might sweep the gate voltage and measure the resulting drain current, holding the drain and bulk voltages fixed. The equipment used to do DC device characterization measurements is usually a DC parametric tester. This equipment usually has a set of SMUs (source/measure units) that can force voltage and measure current, or force current and measure voltage. The measuring equipment can be manually run or controlled by a computer. In either case the measured data must be put onto a hard disk.

For oxide capacitors, junction diodes, and overlap capacitors, a capacitance or inductance, capacitance, resistance (LCR) meter is used to do the measurements. For junction capacitance models, a junction diode (source to bulk, for example) has a reverse bias sweep done and the capacitance is measured at each bias point. For oxide and overlap capacitances, only a single capacitance measurement at one bias voltage is required.

In a more automated environment, switching matrices, semiautomated probers, and temperature-controlled chucks are added. The same equipment is used for bipolar devices. In addition, network analyzers are often used for *S*-parameter characterization to obtain the related AC transit time parameters of bipolar transistors. A typical collection of measuring equipment might look like Fig. 61.18.

It is also possible to use a device simulator like PISCES, or a combination of a process simulator like SUPREM-IV[3] coupled to a device simulator, to provide the simulated results. The process recipe steps are fed into SUPREM-IV, which produces a device structure and associated profiles of all dopants; this information is then fed into the device simulator along with a description of voltage sweeps to be applied to various device electrodes. The output of the device simulator is a set of IV characteristics which closely represent what would have been

[2]PISCES is a process simulation program from the Department of Electrical Engineering at Stanford University, Stanford, CA.

[3]SUPREM-IV is a process simulation program from the Department of Electrical Engineering at Stanford University, Stanford, CA.

measured on a real device. The benefits of using simulation over measurement are that no expensive measurement equipment or fabricated wafers are necessary. This can be very helpful when trying to predict the device characteristics of a new fabrication process before any wafers have been produced.

Once the measured (or simulated) data are available, parameter extraction software is used to find the best set of model parameter values to fit the data. This can be done by extraction, optimization, or a combination of the two. Extraction is simpler and quicker, but not as accurate as optimization. In extraction simplifying assumptions are made about the model equations and parameter values are extracted using slope and intercept techniques, for example. In optimization general least-squares curve fitting routines are employed to vary a set or subset of parameters to fit the measured data. Extraction and optimization are covered in more detail in the following sections.

Least-Squares Curve Fitting (Analytical)

We begin this section by showing how to do least-squares curve fitting by analytical solutions, using a simple example to illustrate the method. We then discuss least-squares curve fitting using numerical solutions in the next section. We can only find analytical solutions to simple problems. The more complex ones must rely on numerical techniques.

Assume a collection of measured data, $m_1, \ldots, m_n$. For simplicity, let these measured data values be functions of a single variable, v, which was varied from v_1 through v_n, measuring each m_i data point at each variable value v_i, i running from 1 to n. For example, the m_i data points might be drain current of an MOS transistor, and the v_i might be the corresponding values of gate voltage. Assume that we have a model for calculating simulated values of the measured data points, and let these simulated values be denoted by $s_1, \ldots, s_n$. We define the least-squares, root mean square (RMS) error as

$$\text{Error}_{\text{rms}} = \left[\frac{\sum_{i=1}^{n} \{\text{weight}_i (s_i - m_i)\}^2}{\sum_{i=1}^{n} \{\text{weight}_i m_i\}^2} \right]^{1/2} \tag{61.1}$$

where a weighting term is included for each data point. The goal is to have the simulated data match the measured data as closely as possible, which means we want to minimize the RMS error. Actually, what we have called the RMS error is really the relative RMS error, but the two terms are used synonymously. There is another way of expressing the error, called the absolute RMS error, defined as follows:

$$\text{Error}_{\text{rms}} = \left[\frac{\sum_{i=1}^{n} \{\text{weight}_i (s_i - m_i)\}^2}{\sum_{i=1}^{n} \{\text{weight}_i m_{\text{min}}\}^2} \right]^{1/2} \tag{61.2}$$

where we have used the term m_{min} in the denominator to represent some minimum value of the measured data. The absolute RMS error is usually used when the measured values approach zero to avoid problems with small or zero denominators in (61.1). For everything that follows, we consider only the relative RMS error. The best result is obtained by combining the relative RMS formula with the absolute RMS formula by taking the maximum of the denominator from (61.1) or (61.2).

We have a simple expression for calculating the simulated data points, s_i, in terms of the input variable, v, and a number of model parameters, $p_1, \ldots, p_m$. That is,

$$s_i = f(v_i, p_1, \ldots, p_m) \tag{61.3}$$

where f is some function. Minimizing the RMS error function is equivalent to minimizing its square. Also, we can ignore the term in the denominator of (61.1) as concerns minimizing, because it is a normalization term. In this spirit, we can define a new error term,

$$\text{Error} = (\text{Error}_{\text{rms}})^2 \left[\sum_{i=1}^{n} \{\text{weight}_i m_i\}^2 \right] \tag{61.4}$$

and claim that minimizing Error is equivalent to minimizing Error_{rms}. To minimize Error, we set all partial derivatives of it with respect to each model parameter equal to zero; that is, write

$$\frac{\partial(\text{Error})}{\partial p_j} = 0, \quad \text{for } j = 1, \ldots, m \tag{61.5}$$

Then solve the above equations for the value of p_j.

Least-Squares Curve Fitting. Analytic Example: Write a simple expression for the drain current of an MOS transistor in the linear region in terms of a single variable, the gate voltage, V_{gs}, and also in terms of two model parameters, β and V_{th};

$$I_{ds} = \beta(V_{gs} - V_{th})V_{ds} \tag{61.6}$$

We denote the measured drain current by I_{meas}. In terms of our previous notation we have the following substitutions:

$s \rightarrow I_{ds}$
$v \rightarrow V_{gs}$
$p_1 \rightarrow \beta$
$p_2 \rightarrow V_{th}$

We have two conditions to satisfy for minimizing the error:

$$\frac{\partial(\text{Error})}{\partial \beta} = 0 \tag{61.7}$$

$$\frac{\partial(\text{Error})}{\partial V_{th}} = 0 \tag{61.8}$$

Equations (61.7) and (61.8) imply, respectively,

$$\sum_{i=1}^{n} [\beta(V_{gs_i} - V_{th})V_{ds} - I_{\text{meas}_i}](V_{gs_i} - V_{th}) = 0 \tag{61.9}$$

$$\sum_{i=1}^{n} [\beta(V_{gs_i} - V_{th})V_{ds} - I_{\text{meas}_i}] = 0 \tag{61.10}$$

Solving (61.9) and (61.10) is straightforward but tedious. The result is

$$V_{th} = \frac{[\text{Term}_1\text{Term}_2 - \text{Term}_3\text{Term}_4]}{[n\text{Term}_1 - \text{Term}_2\text{Term}_4]} \tag{61.11}$$

$$\beta = \left(\frac{1}{V_{ds}}\right)\frac{[n\text{Term}_1 - \text{Term}_2\text{Term}_4]}{[n\text{Term}_3 - \text{Term}_2^2]}, \tag{61.12}$$

where n is the number of data points that i is summed over, and

$$\text{Term}_1 = \sum_{i=1}^{n} V_{gs_i} I_{\text{meas}_i} \tag{61.13}$$

$$\text{Term}_2 = \sum_{i=1}^{n} V_{gs_i} \tag{61.14}$$

$$\text{Term}_3 = \sum_{i=1}^{n} V_{gs_i}^2 \tag{61.15}$$

and

$$\text{Term}_4 = \sum_{i=1}^{n} I_{\text{meas}_i} \tag{61.16}$$

Thus, analytical solutions can become quite messy, even for the simplest model expressions. In fact, it is usually not possible to solve the system of equations analytically for more complicated models. Then we must rely on numerical solutions.

One further word of caution is required. One may try comparing analytical solutions to those obtained by numerical techniques. They almost always will not match. The reason is that the weighting functions used to calculate Error_{rms} are different for the two cases. In order to compare the two techniques, one must know the exact algorithm that the numerical least-squares curve fitting routine is using and be able to match the weighting functions. Often weighting is implicit in the numerical least-squares curve fitting algorithm that is not explicitly obvious.

Least-Squares Curve Fitting (Numerical)

For almost all practical applications we are forced to do least-squares curve fitting numerically, because the analytic solutions as previously discussed are not obtainable in closed form. What we are calling least-squares curve fitting is more generally known as nonlinear optimization. Many fine references on this topic are available. We highlight only the main features here, and paraphrase parts of [2] in what follows.

Basically, we want to solve the optimization problem whose goal is to minimize the function $F(\mathbf{x})$, where F is some arbitrary function of the vector $\mathbf{x}$. If extra conditions exist on the components of $\mathbf{x}$, we are then solving a constrained nonlinear optimization problem. When using least-squares fitting for model parameter optimization, we usually constrain the range of the fitting parameters. This type of constraint is a very simple one and is a special case of what is called nonlinear constrained optimization. This or the unconstrained approach are the types normally used for model parameter optimization.

For simplicity, let F be a function of a single variable, x, instead of the vector, $\mathbf{x}$. We state without proof that a necessary condition for a minimum of $F(x)$ to exist is that

$$F' = \frac{\partial F}{\partial x} = 0 \tag{61.17}$$

This means we solve the minimization problem if we can solve for the zeroes of F'. This is exactly what we did when we went through the least-squares analytical method.

One numerical approach to solving for the zeroes of a function, f, is Newton's method. This is an iterative method in which we write

$$x_{k+1} = x_k - \frac{f(x_k)}{f'(x_k)} \tag{61.18}$$

Steepest Descents Method. Next consider the more general case in which F is now a function of many variables, $\mathbf{x}$; i.e., F is multivariate. We want to consider numerical techniques for finding the minimum of F by iteration. We also would like to take iteration steps that decrease the value of F at each step. This imposes what is called a descent condition:

$$F(\mathbf{x}_{k+1}) < F(\mathbf{x}_k) \tag{61.19}$$

where k is the iteration number. Solving the minimization problem amounts to choosing a search direction for the vector $\mathbf{x}$ such that at each iteration F decreases until some convergence criterion is satisfied with $\mathbf{x}_k$ as the solution. Let us write F as a Taylor series expanded around the point $\mathbf{x}$, with $\mathbf{p}$ a vector of unit length and h a scalar

$$F(\mathbf{x} + h\mathbf{p}) = F(\mathbf{x}) + hg^{\mathrm{T}}(\mathbf{x})\mathbf{p} + \tfrac{1}{2}h^2\mathbf{p}^{\mathrm{T}}G(\mathbf{x})\mathbf{p} + \cdots \tag{61.20}$$

where in (61.20), $g(\mathbf{x})$ is the gradient (first derivative in the $\mathbf{p}$ direction) of $F(\mathbf{x})$ and $G(\mathbf{x})$ is the curvature (second derivative in the $\mathbf{p}$ direction) of $F(\mathbf{x})$. The upper case T means matrix transpose.

We want to choose a direction and step length for the next iteration in which

$$\mathbf{x}_{k+1} = \mathbf{x}_k + h_k\mathbf{p}_k \tag{61.21}$$

where in (61.21), $\mathbf{p}$ is a unit vector in the search direction and h is the step size. In the method of steepest descents we set

$$\mathbf{p}_k = -g(\mathbf{x}_k) \tag{61.22}$$

That is, the method of steepest descents chooses the search direction to be opposite the first derivative in the direction $\mathbf{x}_k$.

Newton's Method. The method of steepest descents is a first derivative method. Newton's method is a second derivative method (this is not to be confused with the simple Newton's method discussed previously). Referring to (61.20), the Newton direction of search is defined to be the vector $\mathbf{p}$, which satisfies

$$G(\mathbf{x}_k)\mathbf{p}_k = -g(\mathbf{x}_k) \tag{61.23}$$

Finding $\mathbf{p}_k$ from the above equation involves calculating second derivatives to compute G.

Gauss-Newton Method. We want to modify Newton's method so that second derivative calculations are avoided. To do this, consider the so-called least-squares problem in which F can be written as

$$F(\mathbf{x}) = \sum [f_i(\mathbf{x})]^2 \tag{61.24}$$

In this special case we can write

$$g(\mathbf{x}) = J^{\mathrm{T}}(\mathbf{x}) f(\mathbf{x}) \tag{61.25}$$

$$G(\mathbf{x}) = J^{\mathrm{T}}(\mathbf{x}) J(\mathbf{x}) + Q(\mathbf{x}) \tag{61.26}$$

where J is the Jacobian of f and

$$Q(\mathbf{x}) = \sum f_i(\mathbf{x}) G_i(\mathbf{x}) \tag{61.27}$$

where G_i is the so-called Hessian of f. Substituting (61.25) and (61.26) into (61.23) gives

$$[J^{\mathrm{T}}(\mathbf{x}_k) J(\mathbf{x}_k) + Q_k]\mathbf{p}_k = -J^{\mathrm{T}}(\mathbf{x}_k) f(\mathbf{x}_k) \tag{61.28}$$

If we throw away the Q term, (61.28) becomes

$$[J^{\mathrm{T}}(\mathbf{x}_k) J(\mathbf{x}_k)]\mathbf{p}_k = -J^{\mathrm{T}}(\mathbf{x}_k) f(\mathbf{x}_k) \tag{61.29}$$

a condition for finding the search direction, $\mathbf{p}_k$, which does not involve second derivatives. Solving (61.29) is tantamount to solving the linear least-squares problem of minimizing

$$F(\mathbf{x}) = \sum [j_i(\mathbf{x}_k)\mathbf{p}_k + f_i(\mathbf{x}_k)]^2 \tag{61.30}$$

When the search direction, $\mathbf{p}_k$, is found from (61.29), we call this the Gauss-Newton method.

Levenberg-Marquardt Algorithm. Both the method of steepest descents and the Gauss-Newton method have advantages and disadvantages. Briefly stated, the method of steepest descents will move toward the correct solution very quickly, but due to the large steps it takes, it can jump past the solution and not converge. The Gauss-Newton method, however, moves more slowly toward the correct solution, but because it tends to take smaller steps, it will converge better as its approaches the solution. The Levenberg-Marquardt algorithm was designed to combine the benefits of both the steepest descents and Gauss-Newton methods.

The Levenberg-Marquardt search direction is found by solving

$$[J^{\mathrm{T}}(\mathbf{x}_k) J(\mathbf{x}_k) + \lambda_k I]\mathbf{p}_k = -J^{\mathrm{T}}(\mathbf{x}_k) f(\mathbf{x}_k) \tag{61.31}$$

where λ_k is a non-negative scalar and I is the identity matrix. Note that (61.31) becomes (61.29) when λ_k is zero, so that the Levenberg-Marquardt search direction becomes the Gauss-Newton in this case; as $\lambda_k \to \infty$, $\mathbf{p}_k$ becomes parallel to $-J^{\mathrm{T}}(\mathbf{x}_k) f(\mathbf{x}_k)$, which is just $-g(\mathbf{x}_k)$, the direction of steepest descents. Generally speaking, in the Levenberg-Marquardt algorithm the value of λ_k is varied continuously, starting with a value near the steepest descents direction (large λ_k) and moving toward smaller values of λ_k as the solution is approached.

Extraction (as Opposed to Optimization)

The terms "extraction" and "optimization" are, unfortunately, used interchangeably in the semiconductor industry; however, strictly speaking, they are not the same. By optimization, we mean using generalized least-squares curve fitting methods such as the Levenberg-Marquardt algorithm to find a set of model parameters. By extraction, we mean any technique that does not use general least-squares fitting methods. This is a somewhat loose interpretation of the term extraction, but perhaps the following discussion will justify it.

Suppose we have measured the drain current of an MOS transistor in the linear region at zero back-bias (the bulk node bias) and we want to find the threshold voltage by extraction (as opposed to optimization). We could plot the drain current vs. gate voltage, draw a tangent line at the point where the slope of I_{ds} is a maximum, and find the V_{gs} axis intercept of this tangent line. This would give use a crude value of the threshold voltage, V_{th} (neglecting terms such as $V_{ds}/2$). This would be a graphic extraction technique; it would not involve using least-squares optimization.

We could also apply the above graphic technique in an equivalent algorithmic way, without drawing any graphs. Let us write a linear region equation for the drain current

$$I_{ds} = \beta(V_{gs} - V_{th})V_{ds}[1 + Ke^{-\alpha(V_{gs}-V_{th})}] \tag{61.32}$$

In (61.32), the exponential term is the subthreshold contribution to the drain current; it becomes negligible for $V_{gs} \gg V_{th}$. To find the value of V_{gs} where the maximum slope occurs, we set the derivative of I_{ds} with respect to V_{gs} equal to zero and solve for V_{gs}. The solution is

$$V_{gs}(\text{maximum slope}) = V_{th} + \left[\frac{1+K}{2\alpha K}\right] \tag{61.33}$$

where in obtaining (61.33) we used the approximation

$$e^{-\alpha(V_{gs}-V_{th})} \cong 1 - \alpha(V_{gs} - V_{th}) \tag{61.34}$$

If we rewrite (61.33) for V_{th} in terms of V_{gs}(maximum slope) we have

$$V_{th} = V_{gs}(\text{maximum slope}) - \left[\frac{1+K}{2\alpha K}\right] \tag{61.35}$$

Therefore, the approach is to use a measurement search routine that seeks the maximum slope, $\Delta I_{ds}/\Delta V_{gs}$, where the Δ is calculated as numerical derivatives, and having found V_{gs}(maximum slope) through measurement, use (61.35) to calculate V_{th} from V_{gs}(maximum slope).

We could continue by including the effects of mobility degradation, modifying (61.32) to

$$I_{ds} = \left[\frac{\beta}{1+\theta(V_{gs} - V_{th})}\right](V_{gs} - V_{th})V_{ds}[1 + Ke^{-\alpha(V_{gs}-V_{th})}] \tag{61.36}$$

where we have now included the effects of mobility degradation with the θ term. The extraction analysis can be done again by including this term, with a somewhat different solution obtained using suitable approximations.

The main point is that we write the equations we want and then solve them by whatever approximations we choose, as long as these approximations allow us to get the extracted results in closed form. This is parameter extraction.

Extraction vs. Optimization

Extraction has the advantage of being much faster than optimization, but it is not always as accurate. It is also much harder to supply extraction routines for models that are being developed. Each time you make a change in the model, you must make suitable changes in the corresponding extraction routine. For optimization, however, no changes are necessary other than the change in the model itself, because least-squares curve fitting routines are completely general. Also, if anything goes wrong in the extraction algorithm (and no access to the source code is available), almost nothing can be done to correct the problem. With optimization, one can always change the range of data, weighting, upper and lower bounds, etc. A least-squares curve fitting program can be steered toward a correct solution.

Novices at device characterization find least-squares curve fitting somewhat frustrating because a certain amount of user intervention and intuition is necessary to obtain the correct results. These beginners prefer extraction methods because they do not have to do anything. However, after being burned by extraction routines that do not work, a more experienced user will usually prefer the flexibility, control, and accuracy that optimization provides.

Commercial software is available that provides both extraction and optimization together. The idea here is to first use extraction techniques to make reasonable initial guesses and then use these results as a starting point for optimization, because optimization can give very poor results if poor initial guesses for the parameters are used. Nothing is wrong with using extraction techniques to provide initial guesses for optimization, but for an experienced user this is rarely necessary, assuming that the least-squares curve fitting routine is robust (converges well) and the experienced user has some knowledge of the process under characterization. Software that relies heavily on extraction may do so because of the nonrobustness of its optimizer.

These comments apply when an experienced user is doing optimization locally, not globally. For global optimization (a technique we do not recommend), the above comparisons between extraction and optimization are not valid. The following section contains more detail about local vs. global optimization.

Strategies: General Discussion

The most naive way of using an optimization program would be to take all the measured data for all devices, put them into one big file, and fit to all these data with all model parameters simultaneously. Even for a very high quality, robust optimization program the chances of this method converging are slight. Even if the program does converge, it is almost certain that the values of the parameters will be very unphysical. This kind of approach is an extreme case of global optimization. We call any optimization technique that tries to fit with parameters to data outside their region of applicability a global approach. That is, if we try to fit to saturation region data with linear region parameters such as threshold voltage, mobility, etc., we are using a global approach. In general, we advise avoiding global approaches, although in the strategies described later, sometimes the rules are bent a little.

Our recommended approach is to fit subsets of relevant parameters to corresponding subsets of relevant data in a way that makes physical sense. For example, in the MOS level 3 model, VT0 is defined as the threshold voltage of a long, wide transistor at zero back-bias. It does not make sense to use this parameter to fit to a short channel transistor, or to fit at nonzero back-bias values, or to fit to anywhere outside the linear region. In addition, subsets of parameters should be obtained in the proper order so that those obtained at a later step do not affect those obtained at earlier steps. That is, we would not obtain saturation region parameters before we have obtained linear region parameters because the values of the linear region parameters would influence the saturation region fits; we would have to go back and reoptimize on the saturation region parameters after obtaining the linear region parameters. Finally, never use optimization to obtain a parameter value when the parameter can be measured directly. For example, the

MOS oxide thickness, TOX, is a model parameter, but we would never use optimization to find it. Always measure its value directly on a large oxide capacitor provided on the test chip. The recommended procedure for proper device characterization follows:

1. Have all the appropriate structures necessary on your test chip. Without this, the job cannot be performed properly.
2. Always measure whatever parameters are directly measurable. Never use optimization for these.
3. Fit the subset of parameters to corresponding subsets of data, and do so in physically meaningful ways.
4. Fit parameters in the proper order so that those obtained later do not affect those obtained previously. If this is not possible, iteration may be necessary.

Naturally, a good strategy cannot be mounted if one is not intimately familiar with the model used. There is no substitute for learning as much about the model as possible. Without this knowledge, one must rely on strategies provided by software vendors, and these vary widely in quality.

Finally, no one can provide a completely general strategy applicable to all models and all process technologies. At some point the strategy must be tailored to suit the available technology and circuit performance requirements. This not only requires familiarity with the available device models, but also information from the circuit designers and process architects.

MOS DC Models

Available MOS Models

A number of MOS models have been provided over time with the original circuit simulation program, SPICE. In addition, some commercially available circuit simulation programs have introduced their own proprietary models, most notably HSPICE.[4] This section is concentrated on the standard MOS models provided by UC Berkeley's SPICE, not only because they have become the standard models used by all circuit simulation programs, but also because the proprietary models provided by commercial vendors are not well documented and no source code is available for these models to investigate them thoroughly.

MOS Levels 1, 2, and 3. Originally, SPICE came with three MOS models known as level 1, level 2, and level 3. The level 1 MOS model is a very crude first-order model that is rarely used. The level 2 and level 3 MOS models are extensions of the level 1 model and have been used extensively in the past and present [11]. These two models contain about 15 DC parameters each and are usually considered useful for digital circuit simulation down to 1 μm channel length technologies. They can fit the drain current for wide transistors of varying length with reasonable accuracy (about 5% RMS error), but have very little advanced fitting capability for analog application. They have only one parameter for fitting the subthreshold region, and no parameters for fitting the derivative of drain current with respect to drain voltage, G_{ds} (usually considered critical for analog applications). They also have no ability to vary the mobility degradation with back-bias, so the fits to I_{ds} in the saturation region at high back-bias are not very good. Finally, these models do not interpolate well over device geometry; e.g., if a fit is made to a wide-long device and a wide-short device, and then one observes how the models track for lengths between these two extremes, they usually do not perform well. For narrow devices they can be quite poor as well. Level 3 has very little practical advantage over level 2, although the level 2 model is proclaimed to be more physically based, whereas the level 3 model

[4]HSPICE is a commercially available, SPICE-like circuit simulation program from Meta Software, Campbell, CA.

is called semiempirical. If only one can be used, perhaps level 3 is slightly better because it runs somewhat faster and does not have quite such an annoying kink in the transition region from linear to saturation as does level 2.

Berkeley Short-Channel Igfet Model (BSIM). To overcome the many short-comings of level 2 and level 3, the BSIM and BSIM2 models were introduced. The most fundamental difference between these and the level 2 and 3 models is that BSIM and BSIM2 use a different approach to incorporating the geometry dependence [3, 6]. In level 2 and 3 the geometry dependence is built directly into the model equations. In BSIM and BSIM2 each parameter (except for a very few) is written as a sum of three terms

$$\text{Parameter} = \text{Par}_0 + \frac{\text{Par}_\text{L}}{\text{L}_\text{eff}} + \frac{\text{Par}_\text{W}}{\text{W}_\text{eff}}, \tag{61.37}$$

where Par_0 is the zero-order term, Par_L accounts for the length dependence of the parameter, Par_W accounts for the width dependence, and L_eff and W_eff are the effective channel width and length, respectively. This approach has a large influence on the device characterization strategy, as discussed later. Because of this tripling of the number of parameters and for other reasons as well, the BSIM model has about 54 DC parameters and the BSIM2 model has over 100.

The original goal of the BSIM model was to fit better than the level 2 and 3 models for submicron channel lengths, over a wider range of geometries, in the subthreshold region, and for nonzero back-bias. Without question, BSIM can fit individual devices better than level 2 and level 3. It also fits the subthreshold region better and it fits better for nonzero back-biases. However, its greatest shortcoming is its inability to fit over a large geometry variation. This occurs because (31.37) is a truncated Taylor series in $1/\text{L}_\text{eff}$ and $1/\text{W}_\text{eff}$ terms, and in order to fit better over varying geometries, higher power terms in $1/\text{L}_\text{eff}$ and $1/\text{W}_\text{eff}$ are needed. In addition, no provision was put into the BSIM model for fitting G_{ds}, so its usefulness for analog applications is questionable. Many of the BSIM model parameters are unphysical, so it is very hard to understand the significance of these model parameters. This has profound implications for generating skew models (fast and slow models to represent the process corners) and for incorporating temperature dependence. Another flaw of the BSIM model is its wild behavior for certain values of the model parameters. If model parameters are not specified for level 2 or 3, they will default to values that will at least force the model to behave well. For BSIM, not specifying certain model parameters, setting them to zero, or various combinations of values can cause the model to become very ill-behaved.

BSIM2. The BSIM2 model was developed to address the shortcomings of the BSIM model. This was basically an extension of the BSIM model, removing certain parameters that had very little effect, fixing fundamental problems such as currents varying the wrong way as a function of certain parameters, adding more unphysical fitting parameters, and adding parameters to allow fitting G_{ds}. BSIM2 does fit better than BSIM, but with more than twice as many parameters as BSIM, it should. However, it does not address the crucial problem of fitting large geometry variations. Its major strengths over BSIM are fitting the subthreshold region better, and fitting G_{ds} better. Most of the other shortcomings of BSIM are also present in BSIM2, and the large number of parameters in BSIM2 makes it a real chore to use in device characterization.

BSIM3. Realizing the shortcomings of BSIM2, UC Berkeley recently introduced the BSIM3 model. This is an unfortunate choice of name because it implies BSIM3 is related to BSIM and BSIM2. In reality, BSIM3 is an entirely new model that in some sense is related more to level 2 and 3 than BSIM or BSIM2. The BSIM3 model abandons the length and width dependence approach of BSIM and BSIM2, preferring to go back to incorporating the geometry dependence directly into the model equations, as do level 2 and 3. In addition, BSIM3 is a more physically based model, with about 30 fitting parameters (the model has many more parameters, but the majority of these can be left untouched for fitting), making it more manageable, and it has abundant parameters for fitting G_{ds}, making it a strong candidate for analog applications.

It is an evolving model, so perhaps it is unfair to criticize it at this early stage. Its greatest shortcoming is, again, the inability to fit well over a wide range of geometries. It is hoped that future modifications will address this problem. In all fairness, however, it is a large order to ask a model to be physically based, have not too many parameters, be well behaved for all default values of the parameters, fit well over temperature, fit G_{ds}, fit over a wide range of geometries, and still fit individual geometries as well as a model with over 100 parameters, such as BSIM2. Some of these features were compromised in developing BSIM3.

Proprietary Models. A number of other models are available from commercial circuit simulator vendors, the literature, etc. Some circuit simulators also offer the ability to add a researcher's own models. In general, we caution against using proprietary models, especially those which are supplied without source code and complete documentation. Without an intimate knowledge of the model equations, it is very difficult to develop a good device characterization strategy. Also, incorporating such models into device characterization software is almost impossible. To circumvent this problem, many characterization programs have the ability to call the entire circuit simulator as a subroutine in order to exercise the proprietary model subroutines. This can slow program execution by a factor of 20 or more, seriously impacting the time required to characterize a technology. Also, if proprietary models are used without source code, the circuit simulator results can never be checked against another circuit simulator. Therefore, we want to stress the importance of using standard models. If these do not meet the individual requirements, the next best approach is to incorporate a proprietary model whose source code one has access to. This requires being able to add the individual model not only to circuit simulators, but also to device characterization programs; it can become a very large task.

MOS Level 3 Extraction Strategy in Detail

The strategy discussed here is one that we consider to be a good one, in the spirit of our earlier comments. Note, however, that this is not the only possible strategy for the level 3 model. The idea here is to illustrate basic concepts so that this strategy can be refined to meet particular individual requirements.

In order to do a DC characterization, the minimum requirement is one each of the wide-long, wide-short, and narrow-long devices. We list the steps of the procedure and then discuss them in more detail.

STEP 1. Fit the wide-long device in the linear region at zero back-bias, at V_{gs} values above the subthreshold region, with parameters VT0 (threshold voltage), U0 (mobility), and THETA (mobility degradation with V_{gs}).

STEP 2. Fit the wide-short device in the linear region at zero back-bias, at V_{gs} values above the subthreshold region, with parameters VT0, LD (length encroachment), and THETA. When finished with this step, replace VT0 and THETA with the values from step 1, but keep the value of LD.

STEP 3. Fit the narrow-long device in the linear region at zero back-bias, at V_{gs} values above the subthreshold region, with parameters VT0, DW (width encroachment), and THETA. When finished with this step, replace VT0 and THETA with the values from step 1, but keep the value of DW.

STEP 4. Fit the wide-short device in the linear region at zero back-bias, at V_{gs} values above the subthreshold region, with parameters RS and RD (source and drain series resistance).

STEP 5. Fit the wide-long device in the linear region at all back-biases, at V_{gs} values above the subthreshold region, with parameter NSUB (channel doping affects long channel variation of threshold voltage with back-bias).

STEP 6. Fit the wide-short device in the linear region at all back-biases, at V_{gs} values above the subthreshold region, with parameter XJ (erroneously called the junction depth; affects short-channel variation of threshold voltage with back-bias).

STEP 7. Fit the narrow-long device in the linear region at zero back-bias, at V_{gs} values above the subthreshold region, with parameter DELTA (narrow channel correction to threshold voltage).

STEP 8. Fit the wide-short device in the saturation region at zero back-bias (or all back-biases) with parameters VMAX (velocity saturation), KAPPA (saturation region slope fitting parameter), and ETA (V_{ds} dependence of threshold voltage).

STEP 9. Fit the wide-short device in the subthreshold region at whatever back-bias and drain voltage is appropriate (usually zero back-bias and low V_{ds}) with parameter NFS (subthreshold slope fitting parameter). One may need to fit with VT0 also and then VT0 is replaced after this step with the value of VT0 obtained from step 1.

This completes the DC characterization steps for the MOS level 3 model. One would then go on to do the junction and overlap capacitance terms (discussed later). Note that this model has no parameters for fitting over temperature, although temperature dependence is built into the model that the user cannot control.

In Step 1 VT0, U0, and THETA are defined in the model for a wide-long device at zero back-bias. They are zero-order fundamental parameters without any short or narrow channel corrections. We therefore fit them to a wide-long device. It is absolutely necessary that such a device be on the test chip. Without it, one cannot obtain these parameters properly. The subthreshold region must be avoided also because these parameters do not control the model behavior in subthreshold.

In Step 2 we use LD to fit the slope of the linear region curve, holding U0 fixed from step 1. We also fit with VT0 and THETA because without them the fitting will not work. However, we want only the value of LD that fits the slope, so we throw away VT0 and THETA, replacing them with the values from step 1.

Step 3 is the same as step 2, except that we are getting the width encroachment instead of the length.

In Step 1 the value of THETA that fits the high V_{gs} portion of the wide-long device linear region curve was found. Because the channel length of a long transistor is very large, the source and drain series resistances have almost no effect here, but for a short-channel device, the series resistance will also affect the high V_{gs} portion of the linear region curve. Therefore, in step 4 we fix THETA from step 1 and use RS and RD to fit the wide-short device in the linear region, high V_{gs} portion of the curve.

In Step 5 we fit with NSUB to get the variation of threshold voltage with back-bias. We will get better results if we restrict ourselves to lower values of V_{gs} (but still above subthreshold) because no mobility degradation adjustment exists with back-bias, and therefore the fit may not be very good at higher V_{gs} values for the nonzero back-bias curves.

Step 6 is just like step 5, except we are fitting the short-channel device. Some people think that the value of XJ should be the true junction depth. This is not true. The parameter XJ is loosely related to the junction depth, but XJ is really the short-channel correction to NSUB. Do not be surprised if XJ is not equal to the true junction depth.

Step 7 uses DELTA to make the narrow channel correction to the threshold voltage. This step is quite straightforward.

Step 8 is the only step that fits in the saturation region. The use of parameters VMAX and KAPPA is obvious, but one may question using ETA to fit in the saturation region. The parameter ETA adjusts the threshold voltage with respect to V_{ds}, and as such one could argue that ETA should be used to fit measurements of I_{ds} sweeping V_{gs} and stepping V_{ds} to high values. In doing so, one will corrupt the fit in the saturation region, and usually we want to fit the saturation region better at the expense of the linear region.

Step 9 uses NFS to fit the slope of the log(I_{ds}) vs. V_{gs} curve. Often, the value of VT0 obtained from step 1 will prevent one from obtaining a good fit in the subthreshold region. If this happens, try fitting with VT0 and NFS, but replacing the final value of VT0 with that from step 1 at the end, keeping only NFS from this final step.

The above steps illustrate the concepts of fitting relevant subsets of parameters to relevant subsets of data to obtain physical values of the parameters, as well as fitting parameters in the proper order so that those obtained in the later steps will affect those obtained in earlier steps minimally. Please refer to Figs. 61.19 and 61.20 for how the resulting fits typically appear (all graphs showing model fits are provided by the device modeling software package Aurora, from Technology Modeling Associates, Inc., Palo Alto, CA).

An experienced person may notice that we have neglected some parameters. For example, we did not use parameters KP and GAMMA. This means KP will be calculated from U0, and GAMMA will be calculated from NSUB. In a sense U0 and NSUB are more fundamental parameters than KP and GAMMA. For example, KP depends on U0 and TOX; GAMMA

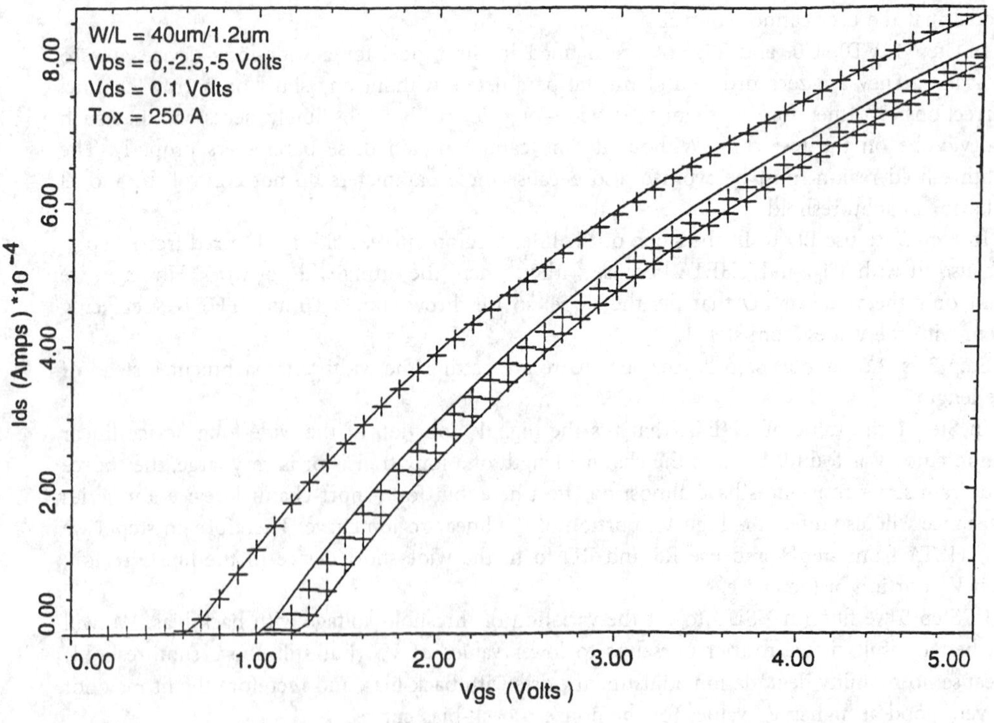

FIGURE 61.19 Typical MOS level 3 linear region measured and simulated plots at various V_{bs} values for a wide-short device.

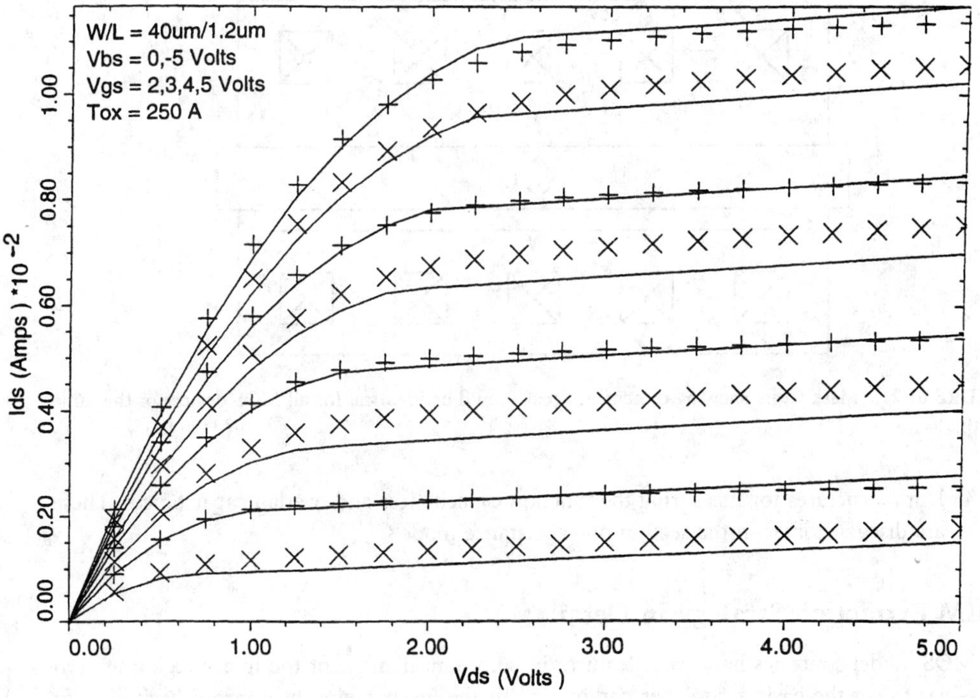

FIGURE 61.20 Typical MOS level 3 saturation region measured and simulated plots at various V_{gs} and V_{bs} values for a wide-short device.

depends on NSUB and TOX. If one is trying to obtain skew models, it is much more advantageous to analyze statistical distributions of parameters that depend on a single effect than those that depend on multiple effects. KP will depend on mobility and oxide thickness; U0 is therefore a more fundamental parameter. We also did not obtain parameter PHI, so it will be calculated from NSUB. The level 3 model is very insensitive to PHI, so using it for curve fitting is pointless. This illustrates the importance of being very familiar with the model equations. The kind of judgments described here cannot be made without such knowledge.

Test Chip Warnings. The following hints will greatly assist in properly performing device characterization.

1. Include a wide-long device; without this, the results will not be physically correct.
2. All MOS transistors with the same width should be drawn with their sources and drains identical. No difference should be seen in the number of source/drain contacts, contact spacing, source/drain contact overlap, poly gate to contact spacing, etc. In Fig. 61.21, c is the contact size, cs is the contact space, cov is the contact overlap of diffusion, and c2p is the contact to poly spacing. All these dimensions should be identical for devices of different L but identical W. If not, extracting the series resistance will become more difficult.
3. Draw devices in pairs. That is, if the wide-long device is W/L = 20/20, make the wide-short device the same width as the wide-long device; e.g., make the short device 20/1, not 19/1. If the narrow-long device is 2/20, make the narrow-short device of the same width; i.e., make it 2/1, not 3/1, and similarly for the lengths. (Make the wide-short and the narrow-short devices have the same length.)

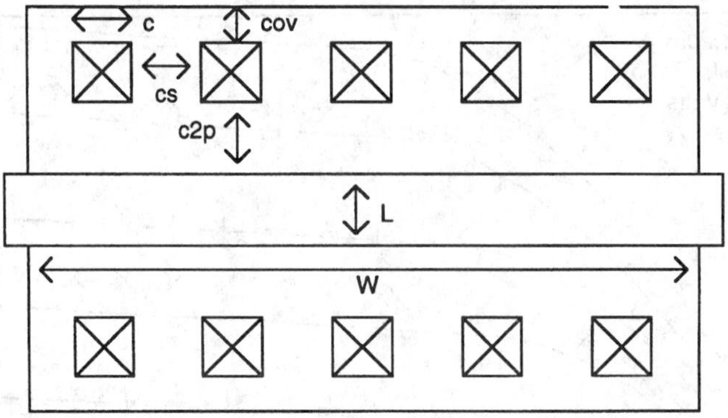

FIGURE 61.21 Mask layers showing dimensions that should be identical for all MOS devices of the same width.

4. Draw structures for measuring the junction capacitances and overlap capacitances. These are discussed later in the section on capacitance models.

BSIM Extraction Strategy in Detail

All MOS model strategies have basic features in common; namely, fit the linear region at zero back-bias to get the basic zero-order parameters, fit the linear region at nonzero back-bias, fit the saturation region at zero back-bias, fit the saturation region at nonzero back-bias, and then fit the subthreshold region. It is possible to extend the type of strategy we covered for level 3 to the BSIM model, but that is not the way BSIM was intended to be used.

The triplet sets of parameters for incorporating geometry dependence into the BSIM model, (61.37), allow an alternate strategy. We obtain sets of parameters without geometry dependence by fitting to individual devices without using the Par_L and Par_W terms. We do this for each device size individually. This produces sets of parameters relevant to each individual device. So, for device number 1 of width W(1) and length L(1) we would have a value for the parameter VFB which we will call VFB(1); for device number n of width W(n) and length L(n) we will have VFB(n). To get the Par_0, Par_L, and Par_W terms, we fit to the parameters themselves, rather than the measured data. That is, to get VFB_0, VFB_L, and VFB_W we fit to the "data points" VFB(1),...,VFB(n) with parameters VFB_0, VFB_L and VFB_W using (61.37), where L_{eff} and W_{eff} are different for each index, 1 through n.

Note that as L and W become very large, the parameters must approach Par_0. This suggests that we use the parameter values for the wide-long device as the Par_0 terms and only fit the other geometry sizes to get the Par_L and Par_W terms. For example, if we have obtained VFB(1) for our first device which is our wide-long device, we would set $\text{VFB}_0 = \text{VFB}(1)$, and then fit to VFB(2),...,VFB(n) with parameters VFB_L and VFB_W, and similarly for all the other triplets of parameters. In order to use a general least-squares optimization program in this way the software must be capable of specifying parameters as targets, as well as measured data points.

We now list a basic strategy for the BSIM model:

STEP 1. Fit the wide-long device in the linear region at zero back-bias, at V_{gs} values above the subthreshold region, with parameters VFB (flatband voltage), MUZ (mobility), and U0 (mobility degradation), with DL (length encroachment) and DW (width encroachment) set to zero.

STEP 2. Fit the wide-short device in the linear region at zero back-bias, at V_{gs} values above the subthreshold region, with parameters VFB, U0, and DL.

STEP 3. Fit the narrow-long device in the linear region at zero back-bias, at V_{gs} values above the subthreshold region, with parameters VFB, U0, and DW.

STEP 4. Refit the wide-long device in the linear region at zero back-bias, at V_{gs} values above the subthreshold region, with parameters VFB, MUZ, and U0, now that DL and DW are known.

STEP 5. Fit the wide-short device in the linear region at zero back-bias, at V_{gs} values above the subthreshold region, with parameters VFB, RS, and RD. When finished, replace the value of VFB with the value found in step 4.

STEP 6. Fit the wide-long device in the linear region at all back-biases, at V_{gs} values above the subthreshold region, with parameters K1 (first-order body effect), K2 (second-order body effect), U0, and X2U0 (V_{bs} dependence of U0).

STEP 7. Fit the wide-long device in the saturation region at zero back-bias with parameters U0, ETA (V_{ds} dependence of threshold voltage), MUS (mobility in saturation), U1 (V_{ds} dependence of mobility), and X3MS (V_{ds} dependence of MUS).

STEP 8. Fit the wide-long device in the saturation region at all back-biases with parameter X2MS (V_{bs} dependence of MUS).

STEP 9. Fit the wide-long device in the subthreshold region at zero back-bias and low V_{ds} value with parameter N0; then fit the subthreshold region nonzero back-bias low V_{ds} data with parameter NB; and finally fit the subthreshold region data at higher V_{ds} values with parameter ND. Or, fit all the subthreshold data simultaneously with parameters N0, NB, and ND.

Repeat steps 6 through 10 for all the other geometries, with the result of sets of geometry-independent parameters for each different size device. Then follow the procedure described previously for obtaining the geometry-dependent terms Par_0, Par_L, and Par_W.

In the above strategy we have omitted various parameters either because they have minimal effect or because they have the wrong effect and were modified in the BSIM2 model. Because of the higher complexity of the BSIM model over the level 3 model, many more strategies are possible than the one just listed. One may be able to find variations of the above strategy that suit the individual technology better. Whatever modifications are made, the general spirit of the above strategy probably will remain.

Some prefer to use a more global approach with BSIM, fitting to measured data with Par_L and Par_W terms directly. Although this is certainly possible, it is definitely not a recommended approach. It represents the worst form of blind curve fitting, with no regard for physical correctness or understanding. The BSIM model was originally developed with the idea of obtaining the model parameters via extraction as opposed to optimization. In fact, UC Berkeley provides software for obtaining BSIM parameters using extraction algorithms, with no optimization at all. As stated previously, this has the advantage of being relatively fast and easy. Unfortunately, it does not always work. One of the major drawbacks of the BSIM model is that certain values of the parameters can cause the model to produce negative values of G_{ds} in saturation. This is highly undesirable, not only from a modeling standpoint, but also because of the convergence problems it can cause in circuit simulators. If an extraction strategy is used that does not guarantee non-negative G_{ds}, very little can be done to fix the problem when G_{ds} becomes negative. Of course, the extraction algorithms can be modified, but this is difficult and time consuming. With optimization strategies, one can weight the fitting for G_{ds} more heavily and thus force the model to produce non-negative G_{ds}. We, therefore, do not favor extraction strategies for BSIM, or anything else. As with most things in life, minimal effort provides minimal rewards.

BSIM2 Extraction Strategy

We do not cover the BSIM2 strategy in complete detail because it is very similar to the BSIM strategy, except more parameters are involved. The major difference in the two models is the inclusion of extra terms in BSIM2 for fitting G_{ds} (refer to Fig. 61.22, which shows how badly BSIM typically fits $1/G_{ds}$ vs. V_{ds}). Basically, the BSIM2 strategy follows the BSIM strategy for the extraction of parameters not related to G_{ds}. Once these have been obtained, the last part of the strategy includes steps for fitting to G_{ds} with parameters that account for channel length modulation and hot electron effects. The way this proceeds in BSIM2 is to fit I_{ds} first, and then parameters MU2, MU3, and MU4 are used to fit to $1/G_{ds}$ vs. V_{ds} curves for families of V_{gs} and V_{bs}. This can be a very time consuming and frustrating experience, because fitting to $1/G_{ds}$ is quite difficult. Also, the equations describing how G_{ds} is modeled with MU2, MU3, and MU4 are very unphysical and the interplay between the parameters makes fitting awkward. The reader is referred to Fig. 61.23, which shows how BSIM2 typically fits $1/G_{ds}$ vs. V_{ds}. BSIM2 is certainly better than BSIM, but it has its own problems fitting $1/G_{ds}$.

BSIM3 Comments

The BSIM3 model is very new and will undoubtedly change in the future [5]. We will not list a BSIM3 strategy here, but focus instead on the features of the model that make it appealing for analog modeling.

BSIM3 has terms for fitting G_{ds} that relate to channel length modulation, drain-induced barrier lowering, and hot electron effects. They are incorporated completely differently from the G_{ds} fitting parameters of BSIM2. In BSIM3 these parameters enter through a generalized Early

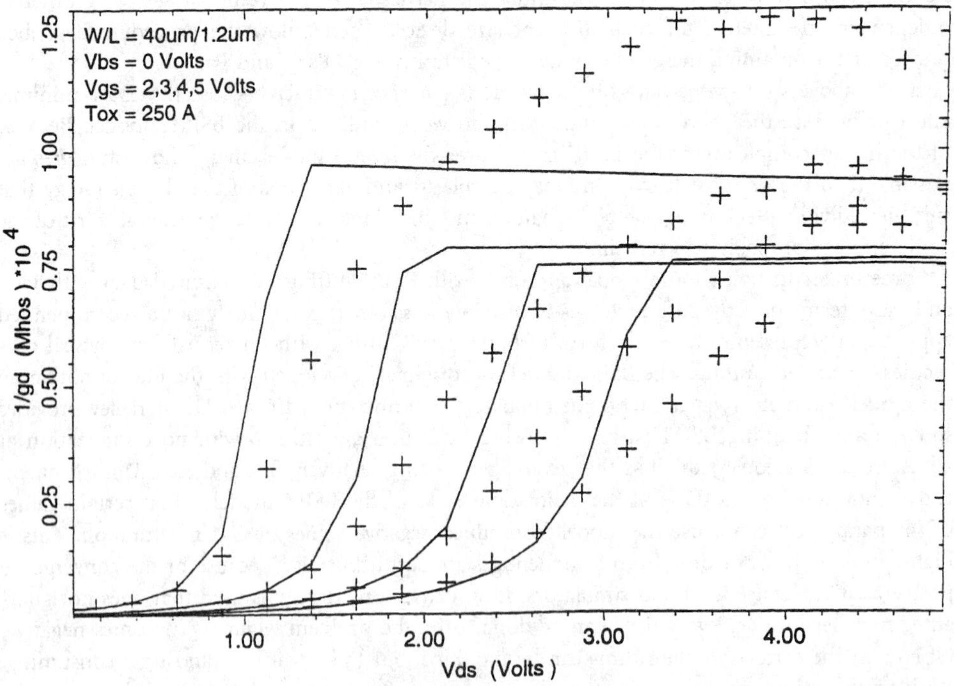

FIGURE 61.22 Typical BSIM $1/G_d$ vs. V_{ds} measured and simulated plots at various V_{gs} values for a wide-short device.

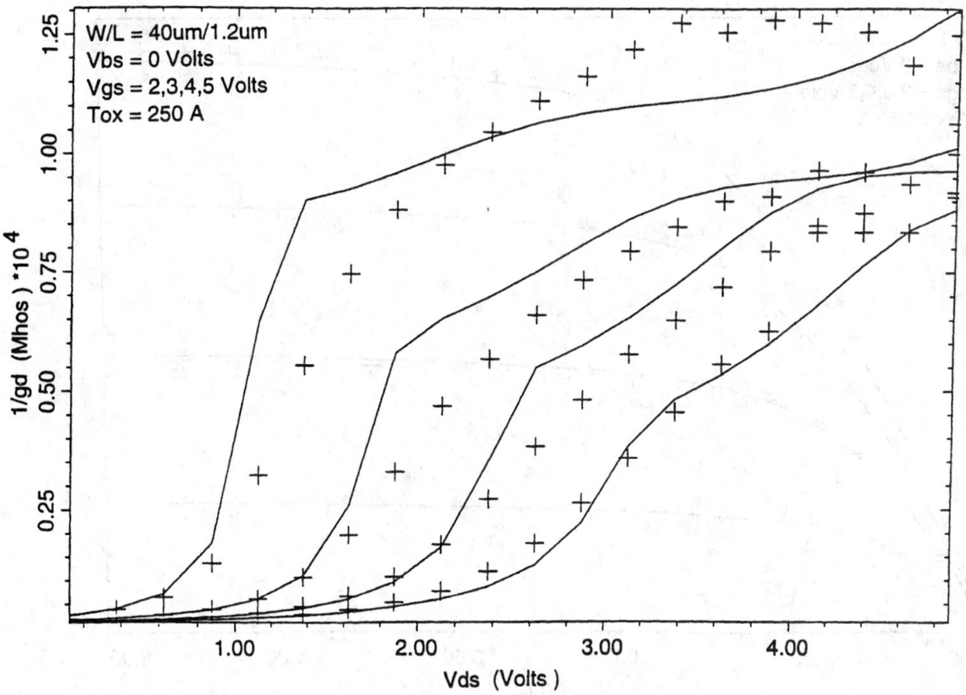

FIGURE 61.23 Typical BSIM2 $1/G_d$ vs. V_{ds} measured and simulated plots at various V_{gs} values for a wide-short device.

voltage relation, with the drain current in saturation written as

$$I_{ds} = I_{d\,\mathrm{sat}}\left[1 + \frac{(V_{ds} - V_{d\,\mathrm{sat}})}{V_A}\right] \tag{61.38}$$

where V_A is a generalized Early voltage made up of three terms as

$$\frac{1}{V_A} = \frac{1}{V_{\mathrm{ACLM}}} + \frac{1}{V_{\mathrm{ADIBL}}} + \frac{1}{V_{\mathrm{AHCE}}} \tag{61.39}$$

with the terms in (61.39) representing generalized Early voltages for channel length modulation (CLM), drain-induced barrier lowering (DIBL), and hot carrier effects (HCE). This formulation is more physically appealing than the one used in BSIM2, making it easier to fit $1/G_{ds}$ vs. V_{ds} curves with BSIM2. Figures 61.24 and 61.25 show how BSIM3 typically fits I_{ds} vs. V_{ds} and $1/G_{ds}$ vs. V_{ds}.

Most of the model parameters for BSIM3 have physical significance so they are obtained in the spirit of the parameters for the level 2 and 3 models. The incorporation of temperature dependence is also easier in BSIM3 because the parameters are more physical. All this, coupled with the fact that about 30 parameters exist for BSIM3 as compared to over 100 for BSIM2, makes BSIM3 a logical choice for analog design. However, BSIM3 is evolving, and shortcomings to the model may still exist that may be corrected in later revisions.

Which MOS Model To Use?

Many MOS models are available in circuit simulators, and the novice is bewildered as to which model is appropriate. No single answer exists, but some questions must be asked before

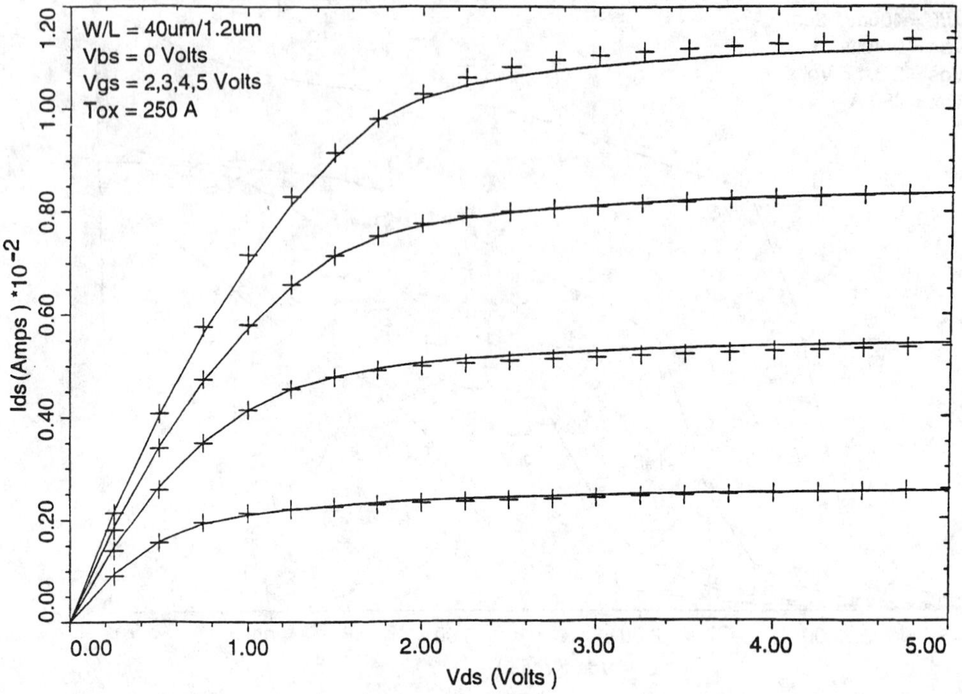

FIGURE 61.24 Typical BSIM3 saturation region measured and simulated plots at various V_{gs} values for a wide-short device.

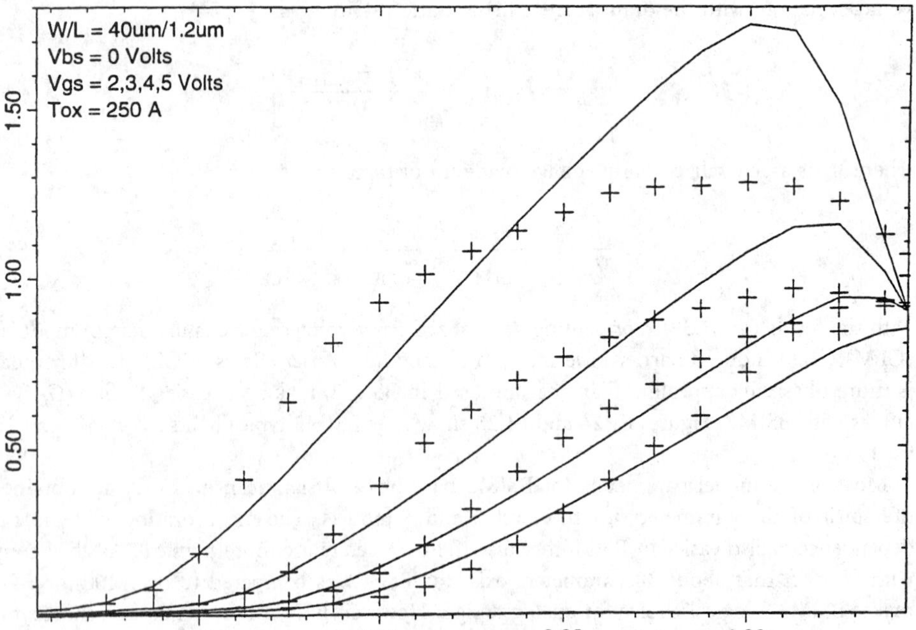

FIGURE 61.25 Typical BSIM3 $1/G_d$ vs. V_{ds} measured and simulated plots at various V_{gs} values for a wide-short device.

making a choice:

1. What kind of technology am I characterizing?
2. How accurate a model do I need?
3. Do I want to understand the technology?
4. How important are the skew model files (fast and slow parameter files)?
5. How experienced am I? Do I have the expertise to handle a more complicated model?
6. How much time can I spend doing device characterization?
7. Do I need to use this model in more than one circuit simulator?
8. Is the subthreshold region important?
9. Is fitting G_{ds} important?

Let us approach each question with regard to the models available. If the technology is not submicron, perhaps a simpler model such as level 3 is capable of doing everything needed. If the technology is deep submicron, then use a more complicated model such as BSIM, BSIM2, or BSIM3. If high accuracy is required, then the best choice is BSIM3, mainly because it is more physical than all the other models and is capable of fitting better.

For a good physical understanding of the process being characterized, BSIM and BSIM2 are not good choices. These are the least physically based of all the models. The level 2 and 3 models have good physical interpretation for most of the parameters, although they are relatively simple models. BSIM3 is also more physically based, with many more parameters than level 2 or 3, so it is probably the best choice.

If meaningful skew models need to be generated, then BSIM and BSIM2 are very difficult to use, again, because of their unphysical parameter sets. Usually, the simplest physically based model is the best for skew model generation. A more complicated physically based model such as BSIM3 may also be difficult to use for skew model generation.

If the user is inexperienced, none of the BSIM models should be used until the user's expertise improves. Our advice is to practice using simpler models before tackling the harder ones.

If time is critical, the simpler models will definitely be much faster for use in characterization. The more complicated models require more measurements over wider ranges of voltages as well as wider ranges of geometries. This, coupled with the larger number of parameters, means they will take some time with which to work. The BSIM2 model will take longer than all the rest, especially if the G_{ds} fitting parameters are to be used.

The characterization results may need to be used in more than one circuit simulator. For example, if a foundry must supply models to various customers, they may be using different circuit simulators. In this case proprietary models applicable to a single circuit simulator should not be used. Also, circuit designers may want to check the circuit simulation results on more than one circuit simulator. It is better to use standard Berkeley models (level 2, level 3, BSIM, BSIM2, and BSIM3) in such cases.

If the subthreshold region is important, then level 2 or level 3 cannot be used, and probably not even BSIM. BSIM2 or BSIM3 must be used instead. These two models have enough parameters for fitting the subthreshold region.

If fitting G_{ds} is important, BSIM2 and BSIM3 are, again, the only choices. None of the other models have enough parameters for fitting G_{ds}.

Finally, if a very unusual technology is to be characterized, none of the standard models may be appropriate. In this case commercially available specialized models or the user's own models must be used. This will be a large task, so the goals must justify the effort.

Bipolar DC Model

The standard bipolar model used by all circuit simulators is the Gummel-Poon model, often called the BJT (bipolar junction transistor) model [4]. Most circuit simulator vendors have added various extra parameters to this model, which we will not discuss, but they all have the basic parameters introduced by the original UC Berkeley SPICE model.

We now list a basic strategy for the BJT model:

STEP 1. Fit a Gummel plot (log(IC) and log(IB) vs. VBE) for the IC curve in the straight line portion of the curve (low to middle VBE values) with parameters IS and NF. Note that it is best if the Gummel plots are done with VCB = 0.

STEP 2. Fit plots of IC vs. VCE, stepping IB, in the high VCE region of the curves with parameters VAF, BF, and IKF.

STEP 3. Fit plots of IC vs. VCE, stepping IB, in the low VCE region of the curves with parameter RC.

STEP 4. Fit a Gummel plot for the IB curve for VBE values in the mid- to high range with parameter BF.

STEP 5. Fit a Gummel plot for the IB curve for low VBE values with parameters ISE and NE.

STEP 6. Try to obtain parameter RE by direct measurement or by simulation using a device simulator such as PISCES. It is also best if RB can be obtained as a function of current by measurement on special structures, or by simulation. One may also obtain RB by *S*-parameter measurements (discussed later). Failing this, RB may be obtained by fitting to the Gummel plot for the IB curve for VBE values in the mid- to high range with parameters RB, RBM, and IRB. This is a difficult step to perform with an optimization program.

STEP 7. Fit a Gummel plot for the IC curve for high VBE values with parameter IKF.

STEP 8. At this point all the DC parameters have been obtained for the device in the forward direction. Next, obtain the reverse direction data by interchanging the emitter and collector to get the reverse Gummel and IC vs. VCE data.

Repeat Step 2 on the reverse data to get VAR, BR, IKR.
Repeat Step 4 on the reverse data to get BR.
Repeat Step 5 on the reverse data to get ISC, NC.
Repeat Step 7 on the reverse data to get IKR.

Fitting the reverse direction parameters is very frustrating because the BJT model is very poor in the reverse direction. After seeing how poor the fits can be, one may decide to obtain only a few of the reverse parameters.

MOS and Bipolar Capacitance Models

MOS Junction Capacitance Model

The MOS junction capacitance model accounts for bottom wall (area component) and side wall (periphery component) effects. The total junction capacitance is written as a sum of the two terms

$$\mathrm{CJ(total)} = \frac{A(\mathrm{CJ})}{1 + [(VR/\mathrm{PB})]^{\mathrm{MJ}}} + \frac{P(\mathrm{CJSW})}{1 + [(VR/\mathrm{PB})]^{\mathrm{MJSW}}} \qquad (61.40)$$

where A is the area and P is the periphery of the junction diode capacitor, and VR is the reverse bias (a positive number) across the device. The parameters of this model are CJ, CJSW, PB, MJ, and MJSW.

It is very easy to find parameter values for this model. Typically, one designs special large junction diodes on the test chip, large enough so that the capacitances being measured are many tens of picofarads. This means the diodes have typical dimensions of hundreds of microns. Two junction diodes are needed—one with a large area:periphery ratio, and another with a small area:periphery ratio. This is usually done by drawing a rectangular device for the large area:periphery ratio, and a multifingered device for the small area:periphery ratio (see Fig. 61.26).

The strategy for this model consists of a single step, fitting to all the data simultaneously with all the parameters. Sometimes it is helpful to fix PB to some nominal value such as 0.7 or 0.8, rather than optimize on PB, because the model is not very sensitive to PB.

BJT Junction Capacitance Model

The BJT model is identical to the MOS, except it uses only the bottom wall (area term), not the side wall. Three sets of BJT junction diodes need to be characterized: emitter-base, collector-base, and collector-substrate. Each of these has three model parameters to fit. For example, for the emitter-base junction diode, the parameters are CJE, VJE, and MJE (CJE corresponds to CJ, VJE to PB, and MJE to MJ of the MOS model). Similarly, for the collector-base junction diode, we have CJC, VJC, and MJC. Finally, for the collector-substrate junction diode, we have CJS, VJS, and MJS.

These parameters are all fit in a single strategy step on the relevant data. Again, it is best if large test structures can be drawn rather than the values measured directly on a transistor, because the transistor junction diodes have such small areas that the measurements are prone to accuracy problems.

MOS Overlap Capacitance Model

In addition to the junction capacitance model, the MOS models include terms for the overlap or Miller capacitance. This represents the capacitance associated with the poly gate overlap over the source and drain regions. As for the junction diodes, special test chip structures are necessary for obtaining the overlap capacitance terms CGS0, CGD0, and CGB0. These structures usually consist of large, multifingered devices that look just like the junction capacitance periphery diodes, except that where the junction diodes are diffusion over well structures, the overlap capacitors are poly over diffusion.

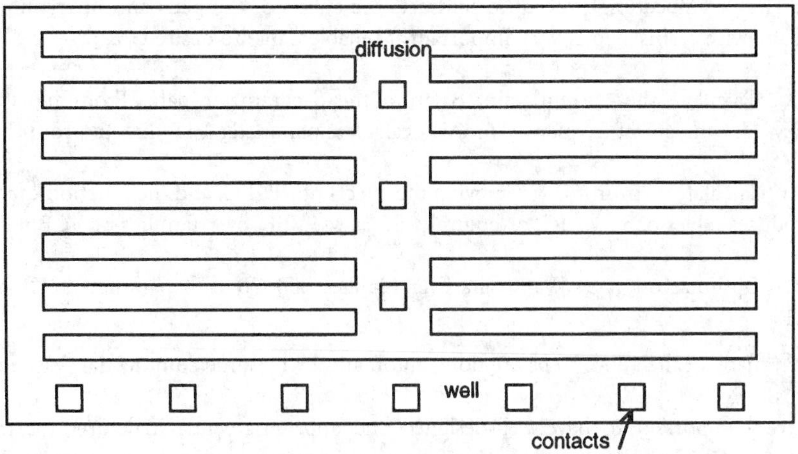

FIGURE 61.26 Mask layers showing the periphery multifingered junction diode structure.

Bipolar High Frequency Model

The SPICE BJT model has five parameters for describing the high-frequency behavior of bipolar devices. These are parameters TF, XTF, ITF, VTF, and PTF, all of which are associated with the base transit time, or equivalently, the cutoff frequency. Parameter TF is the forward transit time, XTF and ITF modify TF as a function of VBE (or IC), and VTF modifies TF as a function of VCB. Parameter PTF represents the excess phase at the cutoff frequency. The expression used by SPICE for the transit time is

$$\tau_f = \mathrm{TF}\left[1 + \mathrm{XTF}\left\{\frac{\mathrm{IF}}{\mathrm{IF}+\mathrm{ITF}}\right\}^2 e^{[\mathrm{VBC}/\{(1.44)(\mathrm{VTF})\}]}\right] \qquad (61.41)$$

where

$$\mathrm{IF} = \mathrm{IS}[e^{\{(q)(V_{be})\}/\{(\mathrm{NF})(k)(\mathrm{T})\}} - 1]. \qquad (61.42)$$

SPICE uses the transit time to calculate the diffusion capacitance of the emitter-base junction in forward bias.

Obtaining the bipolar high frequency parameters is very complicated, time-consuming, and difficult. It also requires very expensive measuring equipment, including a network analyzer for measuring S-parameters, and a quality high-frequency wafer prober with high-frequency probes. Having measured the S-parameters, after suitable data manipulation and optimization it is possible to find the high-frequency SPICE parameters.

It is beyond the scope of this book to cover high-frequency bipolar theory completely. We will, however, list the basic steps involved in obtaining the high frequency bipolar transit time parameters:

STEP 1. Measure the S-parameters over an appropriate frequency range to go beyond the cutoff frequency, f_T. Do these measurements over a large range of IC values and stepping over families of VCB values as well. Be sure to include VCB = 0 in these data. Be sure that the network analyzer being used is well calibrated. These measurements can be done in either common emitter or common collector mode. Each has advantages and disadvantages.

STEP 2. De-embed the S-parameter measurements by measuring dummy pad structures that duplicate the layout of the bipolar transistors, including pads and interconnect, but with no devices. The de-embedding procedure subtracts the effects of the pads and interconnects from the actual devices. This step is very important if reliable data are desired [10].

STEP 3. Calculate the current gain, β, from the S-parameter data. From plots of β vs. frequency, find the cutoff frequency, f_T. Also calculate parameter PTF directly from these data.

STEP 4. Calculate τ_f from f_T by removing the effects of RE, RC, and the junction capacitances. This produces tables of τ_f vs. IC, or equivalently, τ_f vs. VBE, over families of VCB.

STEP 5. Optimize on τ_f vs. VBE data over families of VCB with parameters TF, XTF, ITF, and VTF.

Note that many of the above steps are done automatically by some commercial device modeling programs.

It is very important to have well-designed and appropriate test structures for measuring S-parameters. The pad layout and spacing is very critical because of the special size and shape of the high-frequency probes (see Fig. 61.27). The test chip structures must be layed out separately

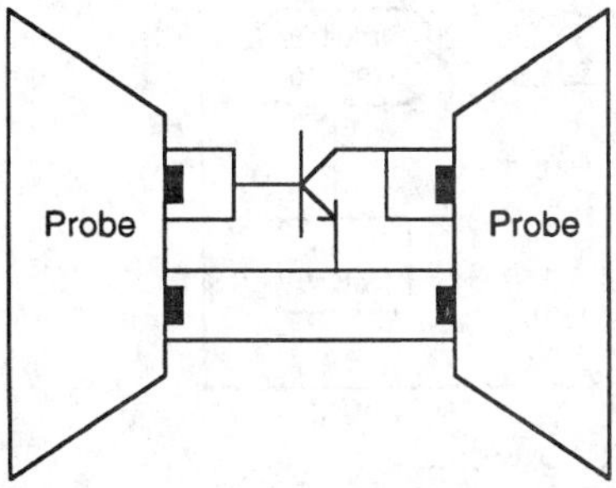

FIGURE 61.27 Typical bipolar transistor test chip layout for measuring *S*-parameters with high-frequency probes. This shows the common emitter configuration.

for common emitter and common collector modes if measurements are necessary for both these modes.

It is also possible to use the *S*-parameter measurements to obtain the base resistance as a function of collector current. From this information one can optimize to obtain parameters RB, RBM, and IRB. However, the base resistance obtained from *S*-parameters is not a true DC base resistance, and the model is expecting DC values of base resistance.

Miscellaneous Topics

Skew Parameter Files

This chapter discussed obtaining model parameters for a single wafer, usually one that has been chosen to represent a typical wafer for the technology being characterized. The parameter values obtained from this wafer correspond to a typical case. Circuit designers also want to simulate circuits with parameter values representing the extremes of process variation, the so-called fast and slow corners, or skew parameter files. These represent the best and worst case of the process variation over time.

Skew parameter values are obtained usually by tracking a few key parameters, measuring many wafers over a long period of time. The standard deviation of these key parameters is found and added to or subtracted from the typical parameter values to obtain the skew models. This method is extremely crude and will not normally produce a realistic skew model. It will almost always overestimate the process spread, because the various model parameters are not independent—they are correlated.

Obtaining realistic skew parameter values, taking into account all the subtle correlations between parameters, is more difficult. In fact, skew model generation is often more an art than a science. Many attempts have been made to utilize techniques from a branch of statistics called multivariate analysis [1]. In this approach principal component or factor analysis is used to find parameters that are linear combinations of the original parameters. Only the first few of these new parameters will be kept; the others will be discarded because they have less significance. This new set will have fewer parameters than the original set and therefore will be more manageable in terms of finding their skews. The user sometimes must make many choices in the way the common factors are utilized, resulting in different users obtaining different results.

Unfortunately, a great deal of physical intuition is often required to use this approach effectively. To date, we have only seen it applied to the simpler MOS models such as level 3. It is not known if this is a viable approach for a much more complicated model such as BSIM2 [7].

Macro Modeling

Sometimes a device is designed and manufactured that cannot be modeled by a single transistor model. In such cases one may try to simulate the behavior of a single device by using many basic device elements together, representing the final device. For example, a single real bipolar

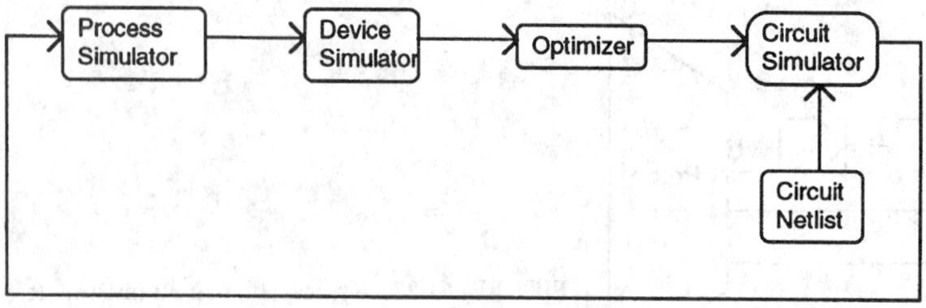

FIGURE 61.28 Block diagram of a TCAD system.

device might be modeled using a combination of several ideal bipolar devices, diodes, resistors, and capacitors. This would be a macro model representing the real bipolar device.

Macro modeling usually uses least-squares optimization techniques. One chooses model parameter values to represent the ideal devices in the technology, and then optimizes on geometry size, capacitance size, resistance size, etc., to obtain the final macro model. This is similar to the optimization techniques used to find model parameters for a standard model, but in this case we are calling on an entire circuit simulator rather than a model subroutine to provide the output characteristics of our macro model.

Obviously, macro modeling can be very computation intensive. A complex macro model can take a very long time to optimize. Also, it may be impractical to use a macro model for every real device in the circuit. For example, if the circuit in question has 100,000 real transistors and each of these is being represented by a macro model with 10 components, the number of circuit nodes introduced by the macro model ($> 10 \times 10,000$) might be prohibitive in terms of simulation time. Nevertheless, for critical components, macro modeling can provide tremendous insight.

Modeling in a TCAD Environment

Programs that do optimization are part of a larger set of software that is sometimes called TCAD (technology computer-aided design). Typically, TCAD encompasses software for process simulation (doping distributions), device simulation (electrical characteristics), device modeling, and circuit simulation. Other elements of TCAD can include lithography simulation, interconnect modeling (capacitance, resistance, and inductance), etc.

In the past each of the software components of TCAD were used as stand-alone programs, with very little communication between the components. The trend now is to incorporate all these pieces into a complete environment that allows them to communicate with each other seamlessly, efficiently, and graphically, so that the user is not burdened with keeping track of file names, data, etc. In such a system one can easily set up split lots and simulate from process steps up through circuit simulation. One will also be able to close the loop, feeding circuit simulation results back to the process steps and run optimization on an entire process to obtain optimal circuit performance characteristics (see Fig. 61.28).

The future of device modeling will surely be pushed in this direction. Device modeling tools will be used not only separately, but within a total TCAD environment, intimately connected with process, device, interconnect, and circuit simulation tools.

References

[1] W. R. Dillon and M. Goldstein, *Multivariate Analysis Methods and Applications*, New York: John Wiley & Sons, 1984.

[2] P. E. Gill, W. Murray, and M. Wright, *Practical Optimization*, Orlando, FL: Academic Press, 1981.

[3] J. S. Duster, J.-C. Jeng, P. K. Ko, and C. Hu, "User's Guide for BSIM2 Parameter Extraction Program and The SPICE3 with BSIM Implementation," Electronic Research Laboratory, Berkeley: University of California, 1988.

[4] I. Getreu, *Modeling the Bipolar Transistor*, Beaverton, OR: Tektronix, 1976.

[5] J.-H. Huang, Z. H. Liu, M.-C. Jeng, P. K. Ko, and C. Hu, "BSIM3 Manual," Berkeley: University of California, 1993.

[6] M.-C. Jeng, P. M. Lee, M. M. Kuo, P. K. Ko, and C. Hu, "Theory, Algorithms, and User's Guide for BSIM and SCALP," Version 2.0, Electronic Research Laboratory, Berkeley: University of California, 1987.

[7] J. A. Power, A. Mathewson, and W. A. Lane, "An Approach for Relating Model Parameter Variabilities to Process Fluctuations," *Proc. IEEE Int. Conf. Microelectronic Test Struct.*, vol. 6, Mar. 1993.

[8] W. H. Press, B. P. Flannery, S. A. Teukolsky, and W. T. Vetterling, *Numerical Recipes in C*, Cambridge, U.K.: Cambridge University Press, 1988.

[9] B. J. Sheu, D. L. Scharfetter, P. K. Ko, and M.-C. Jeng, "BSIM: Berkeley Short-Channel IGFET Model for MOS Transistors," *IEEE J. Solid-State Circuits*, vol. SC-22, no. 4, Aug. 1987.

[10] P. J. van Wijnen, *On the Characterization and Optimization of High-Speed Silicon Bipolar Transistors*, Beaverton, OR: Cascade Microtech, Inc., 1991.

[11] A. Vladimirescu and S. Liu, "The Simulation of MOS Integrated Circuits Using SPICE2," memorandum no. UCB/ERL M80/7, Berkeley: University of California, 1980.

Other recommended publications which are useful in device characterization are

L. W. Nagel, "SPICE2: A Computer Program to Simulate Semiconductor Circuits," memorandum no. ERL-M520, Berkeley: University of California, 1975.

G. Massobrio and P. Antognetti, *Semiconductor Device Modeling with SPICE*, New York: McGraw-Hill, 1993.

62

Digital Circuits

John P. Uyemura
Georgia Institute of Technology

Bing J. Sheu
University of Southern California

Robert C. Chang
University of Southern California

62.1 MOS Logic Circuits

John P. Uyemura

Introduction

MOS-based technology has become the default standard for high-density logic designs. There are several reasons for this, the most obvious being that MOSFETs can be made with side dimensions of < 1 μm (10^{-6} m), allowing for complex logic functions to be constructed in small areas. The section is an investigation of the basics of designing and characterizing logic gates in an MOS technology.

MOSFET Models for Digital Circuits

The properties of digital logic gates are derived from a large-signal analysis of the circuits. Because transistor characteristics are intrinsically nonlinear, accurate analytic modeling becomes quite complicated, and closed-form solutions can be difficult to come by. To overcome this problem, simplified MOSFET models are used to estimate the circuit operation in first-cut designs. Once the basic operation is established, computer simulations are used to obtain more accurate information.

Square-law models are useful for understanding the operation of MOS logic circuits. Consider first an *n*-channel, enhancement-mode MOSFET that has a threshold voltage $V_{Tn} > 0$. As shown in Fig. 62.1, the primary device voltages are V_{DS}, V_{GS}, and V_{SB}. The value gate-source voltage V_{GS} relative to the threshold voltage V_{Tn} determines if drain current I_D flows. If $V_{GS}, < V_{Tn}$ then $I_D \approx 0$, establishing the condition of **cutoff**. Elevating the gate-source voltage to a value $V_{GS}, > V_{Tn}$ places the MOSFET into the **active region** where I_D will be nonzero if a drain-source voltage V_{DS} is applied; the value of I_D depends on the values of the device voltages.

To describe active operation, we introduce the drain-source saturation voltage $V_{DS,\text{sat}}$ defined by

$$V_{DS,\text{sat}} = V_{GS} - V_{Tn} \tag{62.1}$$

0-8493-8341-2/95/$0.00+$.50

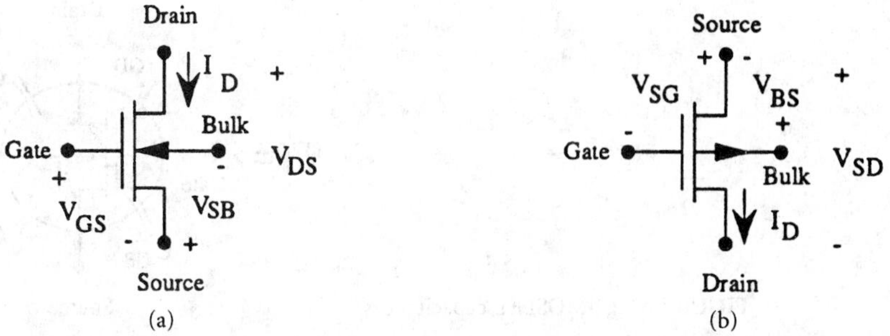

FIGURE 62.1 MOSFET symbols. (a) n-channel MOSFET. (b) p-channel MOSFET.

The threshold voltage is affected by the source-bulk (body) voltage V_{SB} by

$$V_{Tn} = V_{T0n} + \gamma(\sqrt{2|\phi_F| + V_{SB}} - \sqrt{2|\phi_F|}) \tag{62.2}$$

where V_{T0n} is the nFET zero-body-bias threshold voltage, γ is the body bias (or, body effect) coefficient, and ϕ_F is the bulk Fermi potential. When $V_{DS} \le V_{DS,\text{sat}}$, the MOSFET is nonsaturated with

$$I_D \approx \left(\frac{\beta_n}{2}\right)[2(V_{GS} - V_{Tn})V_{DS} - V_{DS}^2] \tag{62.3}$$

In this equation β_n is the device transconductance given by $\beta_n = k'_n(W/L)$, with k'_n the process transconductance in units of $[A/V^2]$, W the channel width, and L the channel length; the width-to-length (W/L) is called the **aspect ratio** of the transistor. The process transconductance is given by $k'_n = \mu_n C_{\text{ox}}$, where μ_n is the electron surface mobility and C_{ox} the oxide capacitance per unit area. For an oxide layer with thickness t_{ox}, the MOS capacitance per unit area is calculated from

$$C_{\text{ox}} = \frac{\varepsilon_{\text{ox}}}{t_{\text{ox}}} \tag{62.4}$$

where ε_{ox} is the oxide permittivity. In the current technologies t_{ox} is smaller than about 300 Å. If $V_{DS} \ge V_{DS,\text{sat}}$, the MOSFET is saturated with

$$I_D \approx \left(\frac{\beta_n}{2}\right)(V_{GS} - V_{Tn})^2 \tag{62.5}$$

This ignores several effects, most notably that of **channel-length modulation**, but is still a reasonable approximation for estimating basic performance parameters.

The structure of the MOSFET gives rise to several parasitic capacitances that tend to dominate the circuit performance. Two types of capacitors are contained in the basic model shown in Fig. 62.2. The contributions C_{GS} and C_{GD} are due to the MOS layering of the gate-oxide-semiconductor, which is the origin of the field effect. The total gate capacitance C_G is calculated from

$$C_G = C_{\text{ox}}WL \tag{62.6}$$

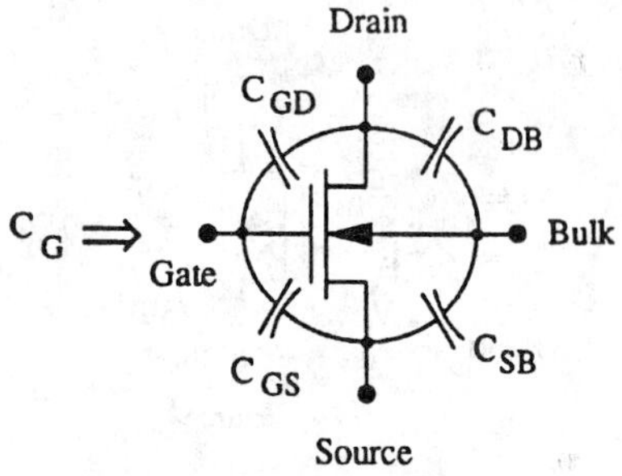

FIGURE 62.2 MOSFET capacitances.

and the gate-source and gate-drain contributions can be approximated to first order by

$$C_{GD} \approx \frac{C_G}{2} \approx C_{GS} \tag{62.7}$$

Capacitors C_{DB} and C_{SB} are depletion contributions from the reverse-biased *pn* junction at the drain and source. These are nonlinear, voltage-dependent elements that decrease with increasing reverse voltage.

A *p*-channel MOSFET (pMOS or pFET) is the electrical complement of an *n*-channel device. An enhancement-mode pFET is defined to have a negative threshold voltage, i.e., $V_{Tp} < 0$. It is common to use device voltage of V_{SG}, V_{SD}, and V_{BS}, as shown in Fig. 62.1 to describe the operation. Cutoff occurs if $V_{SG} < |V_{Tp}|$, while the device is active if $V_{SG} \geq |V_{Tp}|$. The saturation voltage of the pFET is defined by

$$V_{SD,\,\text{sat}} = V_{SG} - |V_{Tp}| \tag{62.8}$$

With $V_{SG} \geq |V_{Tp}|$ and $V_{SG} < V_{SD,\,\text{sat}}$, the transistor is nonsaturated with

$$I_D \approx \left(\frac{\beta_p}{2}\right)[2(V_{SG} - |V_{Tp}|)V_{SD} - V_{SD}^2] \tag{62.9}$$

For the pFET, β_p is the device transconductance $\beta_p = k'_p(W/L)$, where $k'_p = \mu_p C_{\text{ox}}$ is the process transconductance, and (W/L) is the aspect ratio of the device. In complementary metal oxide-semiconductor inverters (CMOS) nFETs and pFETs are used in the same circuit, and it is important to note that $k'_n > k'_p$ due to the fact that the electron mobility is larger than the hole mobility, typically by a factor of 2 to 2.5.

It is often convenient to use the simplified MOSFET symbols shown in Fig. 62.3. The polarity of the transistor (nMOS or pMOS) is made explicit by the absence or presence of the gate inversion "bubble", as shown. These symbols do not show the bulk electrode explicitly, but it is important to remember that all nFETs have their bulks connected to the lowest voltage in the circuit (usually ground), while all pFET bulks are connected to the highest voltage (usually the power supply V_{DD}).

In digital circuit design it is useful to model MOSFETs as voltage-controlled switches, as shown in Fig. 62.4. The MOSFET switches are controlled by a gate input signal G, which is taken to be a Boolean variable. Employing a positive logic convention, $G = 0$ corresponds to a low voltage (below V_{Tn}), while $G = 1$ is a high voltage. The operation of the FET switches is straightforward. An input of $G = 0$ places the nFET into cutoff, corresponding to an OPEN switch; $G = 1$ implies active operation, and the switch is CLOSED. The pFET has a complementary behavior, with $G = 0$ giving a CLOSED switch, and $G = 1$ giving an OPEN switch.

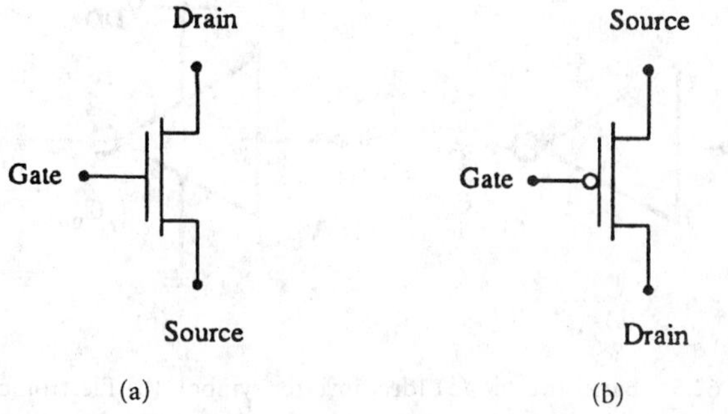

FIGURE 62.3 Simplified MOSFET symbols. (a) nMOSFET; (b) pMOSFET.

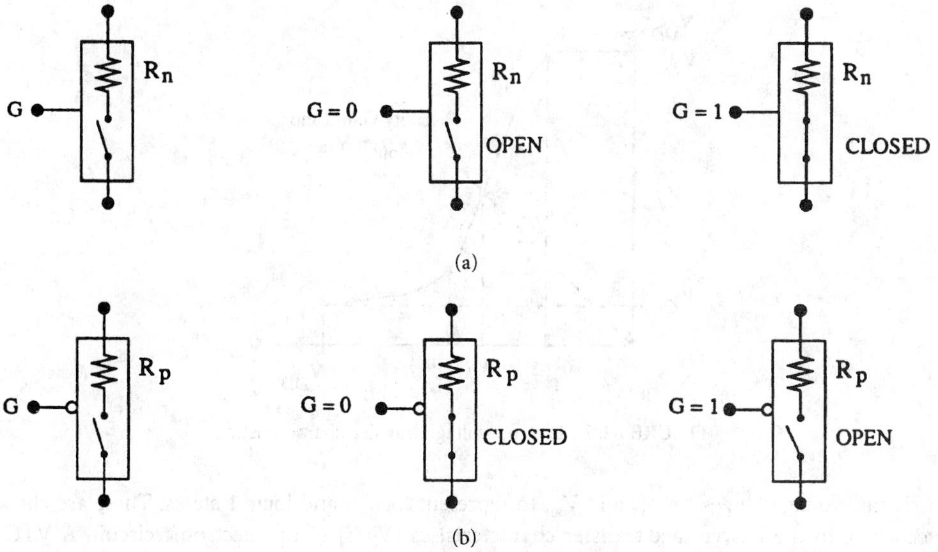

FIGURE 62.4 MOSFET switching models. (a) nFET switch model; (b) pFET switch model.

The switch models include parasitic drain-to-source resistances R_n and R_p, which are usually estimated using

$$
\begin{aligned}
R_n &= \frac{1}{k'_n(\mathrm{W/L})_n(V_{DD} - V_{Tn})} \\
R_p &= \frac{1}{k'_p(\mathrm{W/L})_p(V_{DD} - |V_{Tp}|)}
\end{aligned}
\tag{62.10}
$$

These equations illustrate the general dependence that the drain-source resistance R is inversely proportional to the aspect ratio (W/L). However, the MOSFET is at best a nonlinear resistor, so that these are only rough estimates. It is important to note the MOSFET parasitic capacitances C_{GS}, C_{GD}, C_{SB}, and C_{DB} must be included in the switching models when performing transient analysis.

The Digital Inverter

An ideal digital inverter is shown in Fig. 62.5. In terms of the Boolean variable A the inverter accepts A and produces the complement $\overline{A}$. Electronic implementation of the inverter requires

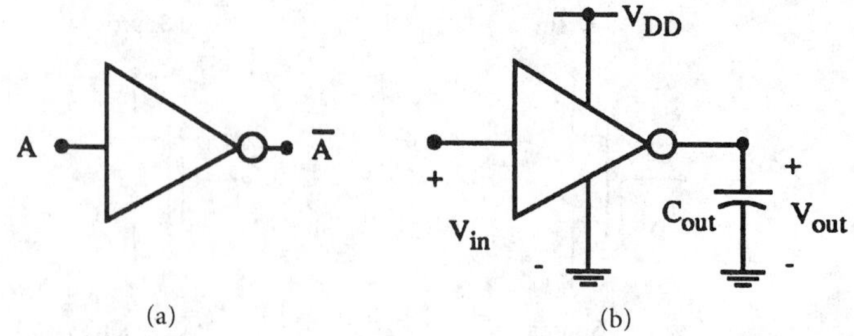

FIGURE 62.5 Basic inverter. (a) Ideal inverter symbol. (b) Electronic parameters.

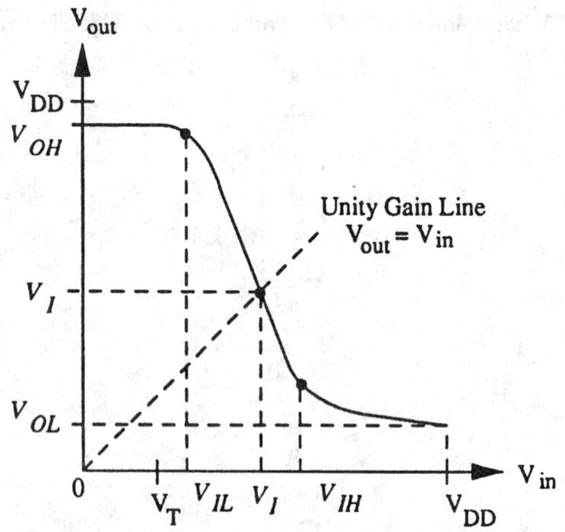

FIGURE 62.6 Inverter voltage transfer characteristics.

assigning voltage ranges for V_{in} and V_{out} to represent logic 0 and logic 1 states. These are chosen according to the DC **voltage transfer characteristics** (VTC) of the electronic circuit. A VTC is simply a plot of the output voltage V_{out} as a function of V_{in}; a general VTC is shown in Fig. 62.6.

Consider first the output voltage V_{out}. The maximum value of V_{out} is denoted by V_{OH}, and is called the output high voltage. This is used to represent an ideal logic 1 output voltage. Conversely, the smallest value of V_{out} is denoted as V_{OL}, and is called the output low voltage. V_{OL} is the output logic 0 voltage. The logic swing of the inverter is then defined by $(V_{OH} - V_{OL})$. The range of input voltage V_{in} used to represent logic 0 and logic 1 input states is usually determined by points on the VTC at which the slope has a value of $(dV_{out}/dV_{in}) = -1$. Logic 0 voltages are those with values between 0 V and V_{IL}, the input low voltage. Similarly, voltages in the range from the input voltage V_{IH} to V_{OH} represent logic 1 input levels. The intersection of the VTC with the unity gain line defined by $V_{out} = V_{in}$ gives the inverter threshold voltage V_I; this represents the switching point of the circuit. The numerical values of V_{OH}, V_{OL} V_{IL}, V_{IH}, and V_I are determined by the circuit and topology and the characteristics of the devices used in the circuit.

The transient characteristics of the gate are defined by two basic transition times. Figure 62.7 shows V_{in} and V_{out}, and the most important output switching intervals. The input voltage has been taken as an ideal step-like pulse. In a more realistic situation the input voltage is better approximated by a ramp or an exponential. The idealized pulse is used here because it allows a comparison among various circuits.

The most important switching properties of the inverter are the low-to-high time t_{LH}, and the high-to-low time t_{HL} as shown in Fig. 62.7. These represent the minimum response times of the inverter. Note that these time intervals are usually defined between the 10 and 90% voltages

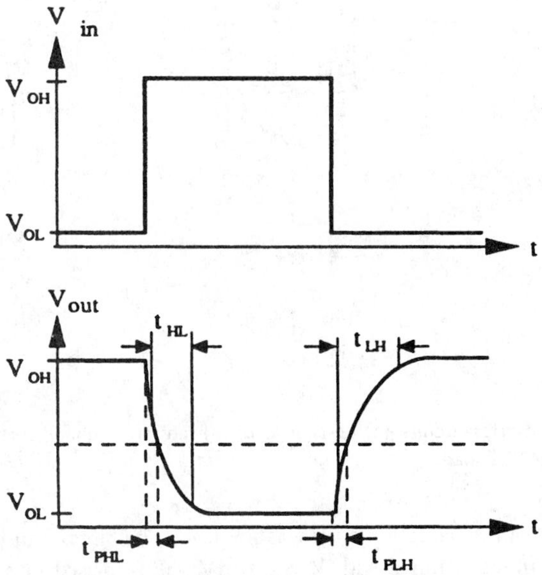

FIGURE 62.7 Inverter switching times.

instead of the full logic swing. The maximum switching frequency is computed from

$$f_{\max} = \frac{1}{t_{LH} + t_{HL}} \tag{62.11}$$

The propagation delay t_P for the gate is the average time required for a change in the input to be seen at the output. It is computed using the time intervals t_{PHL} and t_{PLH} shown in the diagram from

$$t_P = \left(\frac{1}{2}\right)(t_{PHL} + t_{PLH}) \tag{62.12}$$

Note that the transition times for this parameter are measured to the 50% voltage.

nMOS Logic Gates

Early generations of MOS logic circuits were based on a single type of MOSFET. The Intel 4004, for example, used only pMOS transistors, while subsequent microprocessor chips such as the Intel 8088, the Zilog Z80, and the Motorola 6800 used only *n*-channel MOSFETs. Although all current MOS-based designs are implemented in CMOS, which employs both nMOS and pMOS devices, it is worthwhile to examine nMOS-only logic circuits. This provides an introduction to the basic characteristics of MOS logic circuits, many of which are used in even the most advanced CMOS techniques.

Several types of inverter circuits can be constructed using *n*-channel MOSFETs. Three configurations are shown in Fig. 62.8. Each circuit uses a switching transistor MD, known as the **driver**, which is controlled by the input voltage V_{in}. The VTC is determined by the **load** device that connects the drain of MD to the power supply V_{DD}. The MD can be viewed as the switched "pull-down" device, while the load serves as the "pull-up" device. In Fig. 62.8(a) a

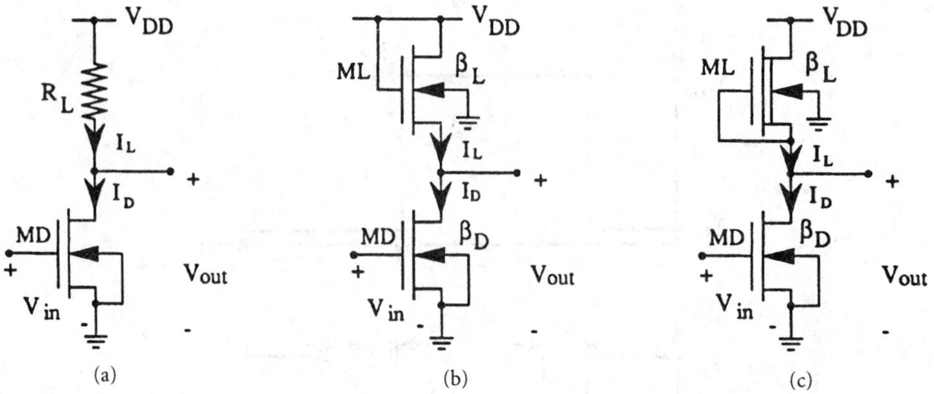

FIGURE 62.8 nMOS inverter circuits. (a) Resistor load; (b) saturated enhancement mode MOSFET load; (c) depletion mode MOSFET load.

simple linear resistor with a value R_L is used as the load. The circuits in part (b) of the figure use an enhancement-mode (defined with $V_{Tn} > 0$) nMOSFET biased into saturation, while (c) has a depletion-mode nMOSFET (where $V_{Tn} < 0$) as an active load. Active loads provide better switching characteristics due to the nonlinearity of the device. In addition, MOSFETs are much smaller than resistors and process variations are not as critical because the circuit characteristics depend on the ratio of driver-to-load dimensions.

Although the three nMOS inverter circuits are similar in structure, they have distinct switching properties. Consider the output voltage swing. Circuits (a) and (c) of the figure both have $V_{OH} \approx V_{DD}$, but the active load in (b) gives

$$V_{OH} = V_{DD} - V_{TL} \tag{62.13}$$

with the theshold voltage computed from

$$V_{TL} = V_{T0n} + \gamma(\sqrt{2|\phi_F| + V_{OH}} - \sqrt{2|\phi_F|}) \tag{62.14}$$

This is referred to as a **threshold voltage loss**, and is due to the fact that the load must have a minimum gate-source voltage of $V_{GSL} = V_{TL}$ to be biased into the active mode. Obviously, $V_{OH} < V_{DD}$ for this circuit.

The value of $V_{OL} > 0$ is determined by a ratio of driver parameters to load parameters. In circuit (a), this ratio is given by $R_L\beta_D$, which is inversely proportional to V_{OL}. This means a small V_{OL} requires that both the load resistance and the driver dimensions are large. In circuits (b) and (c), V_{OL} is set by the driver-to-load ratio $\beta_R = (\beta_D/\beta_L) = (W/L)_D/(W/L)_L$; increasing β_R decreases V_{OL}. For the depletion MOSFET load circuit in (c), the design equation is given by

$$\beta_R = \frac{|V_{TL}|^2}{2(V_{DD} - V_{TD})V_{OL} - V_{OL}^2} \tag{62.15}$$

A condition of $\beta_R > 1$ is generally required to achieve a functional inverter, implying that the driver MOSFET is always larger than the load device. Also, note that it is not possible to achieve $V_{OL} = 0$ because this requires an infinite driver-to-load ratio.

The transient switching characteristics are obtained by including the output capacitance C_{out} as shown in Fig. 62.5 at the output. C_{out} consists of the input gate capacitance seen looking

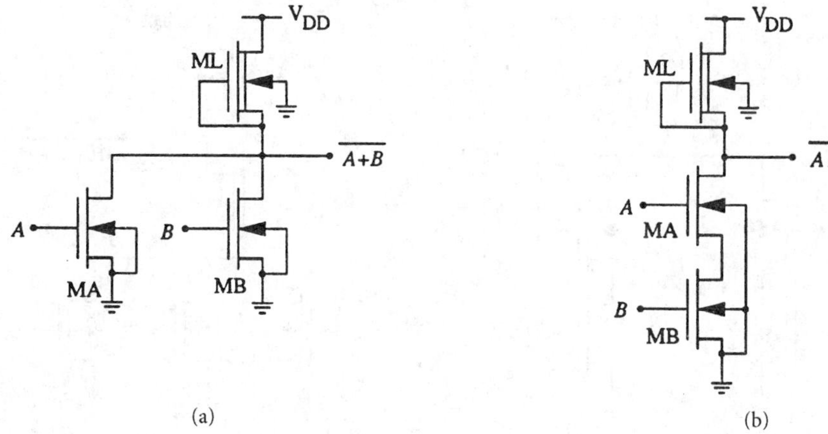

FIGURE 62.9 nMOS NOR and NAND gates. (a) Two-input NOR gate; (b) two-input NAND gate.

in the MOSFET of the next stage, and also has parasitic contributions from the MOSFETs and interconnects. By using a switch model for the driver, it is seen that the transient characteristics are determined by the time required to charge and discharge C_{out}. The high-to-low time t_{HL} represents the time it takes to discharge the capacitor through the driver MOSFET with a device transconductance value of β_D. A rough estimate is obtained using the RC time constant such that $t_{HL} \approx 2R_D C_{\text{out}}$, with R_D the equivalent resistance. Similarly, the low-to-high time t_{LH} is the time interval needed to charge C_{out} through the load device. With respect to Fig. 62.8 circuit (c) has the best transient response such that $t_{LH} \approx 2R_L C_{\text{out}}$, where R_L represents the equivalent resistance of the load MOSFET. nMOS circuits in the mid-1980s had inverter transition times on the order of a few nanoseconds. Because the DC design requires that $\beta_R = (\beta_D/\beta_L) > 1$, and the drain-source resistance of a MOSFET is inversely proportional to β, these circuits exhibit nonsymmetrical switching times with $t_{LH} > t_{HL}$. The propagation delay times can be estimated using $t_{PHL} \approx R_D C_{\text{out}}$ and $t_{PLH} \approx R_L C_{\text{out}}$ because these are measured relative to the 50% voltage levels.

MOS-based logic design allows one to easily construct other logic functions using the inverter circuit as a guide. For example, adding another driver MOSFET in parallel gives the NOR operation, while adding a series connected driver yields the NAND operation; these are shown in Fig. 62.9.

Complex logic gates for AOI (AND-OR-INVERT) and OAI (OR-AND-INVERT) canonical logic functions can be constructed using the simple rules

- nMOSFETs (or groups) in parallel provide the NOR operation
- nMOSFETs (or groups) in series provide the NAND operation

Examples are provided in Fig. 62.10. It should be noted that this type of circuit structuring is possible because the drain and source are interchangeable. The main problem that arises in design complex nMOS logic gates is that the circuit requires large driver-to-load ratios to achieve small V_{OL} values. The switching FET arrays collectively act like a driver network that must be designed to have a large overall effective β-value. Although parallel-connected MOSFETs are not a problem, the pull-down resistance of series-connected MOSFETs can be large unless the individual aspect ratios are increased. Satisfying this condition requires additional chip area, decreasing the logic density.

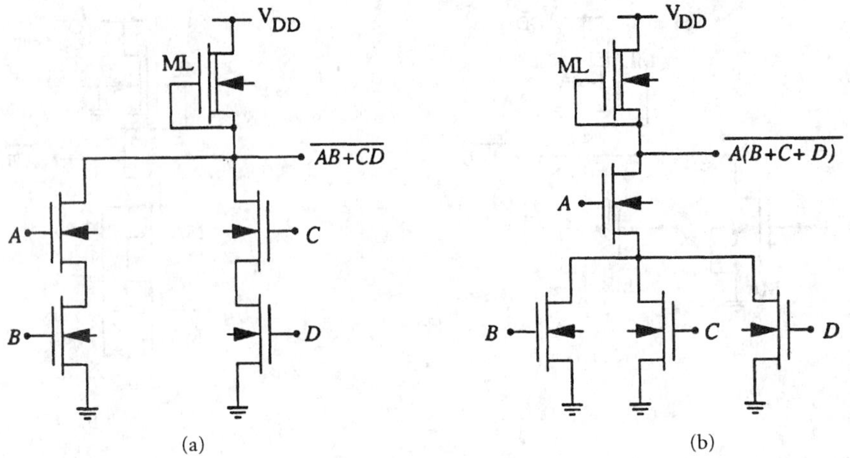

FIGURE 62.10 nMOS AOI logic gates.

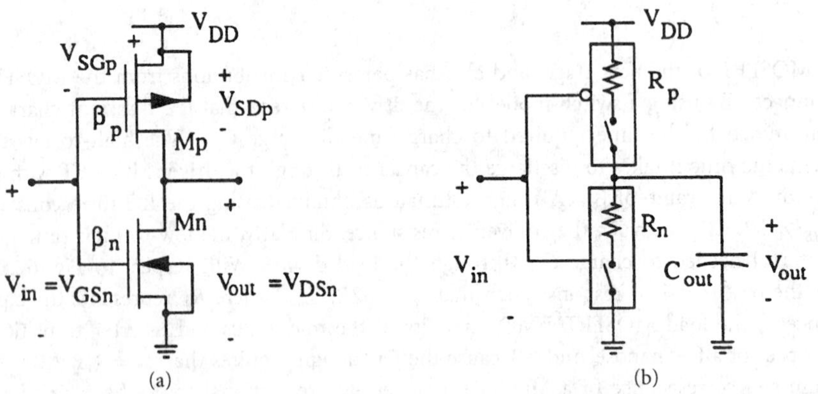

FIGURE 62.11 CMOS inverter. (a) Circuit; (b) switch model.

CMOS Inverter

A CMOS inverter is shown in Fig. 62.11. This circuit uses a pair of transistors, one nMOS and one pMOS, connected with their gates together. When $V_{in} < V_{Tn}$, the pFET is active and the nFET is in cutoff. Conversely, when $V_{in} > (V_{DD} - V_{Tn})$, the nFET is active while the pFET is in cutoff. The two MOSFETs are said to form a complementary pair.

The complementary arrangement of the MOSFETs gives the circuit a full rail-to-rail output range, i.e., $V_{OL} = 0$ V and $V_{OH} = V_{DD}$. The devices are connected in such a way that terminal voltages satisfy

$$\begin{aligned} V_{GSn} + V_{SGp} &= V_{DD} \\ V_{DSn} + V_{SDp} &= V_{DD} \end{aligned} \tag{62.16}$$

Note in particular the relationship between the gate-source voltages. Increasing the current in one transistor automatically decreases the current through the other. This provides the VTC with a very sharp transition, as shown in Fig. 62.12. Moreover, the shape of the VTC is almost insensitive to the power supply value V_{DD}, which allows CMOS circuits based on this construction to be used with a range of values. The minimum value of V_{DD} is set by the device threshold voltages, and is usually estimated as being about 3 V_T. This is based on the input switching

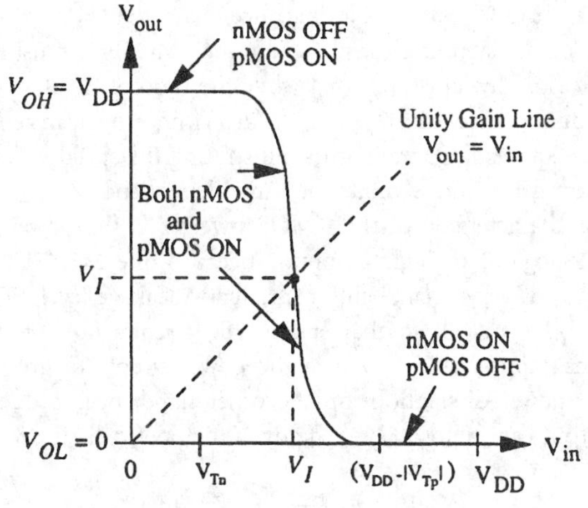

FIGURE 62.12 CMOS VTC.

voltage V_{in}, and allows one V_T to switch the nFET, one V_T to switch the pFET, and one V_T for separation. Currently, V_T values equal ~ 0.75 V, so that the minimum V_{DD} is about 2.3 V. Because V_T is set in the fabrication, the minimum power supply used in low-voltage designs depends upon the process specifications. The maximum value of the power supply voltage is limited by the reverse breakdown voltages of the drain-bulk junctions. This is typically around 14 to 17 V.

Because the structure of the CMOS circuit automatically gives a full-rail output logic swing, the DC design of the gate centers around setting the inverter threshold voltage V_I. At this point, both FETs are saturated, and equating currents gives the expression

$$V_I = \frac{\sqrt{\beta_n/\beta_p}\, V_{Tn} + (V_{DD} - |V_{Tp}|)}{1 + \sqrt{\beta_n/\beta_p}} \tag{62.17}$$

This equation shows that V_I can be set by adjusting the ratio β_n/β_p. If $\beta_n = \beta_p$, and $V_{Tn} \approx |V_{Tp}|$, then $V_I \approx (V_{DD}/2)$. Increasing this ratio decreases the inverter switching voltage. If the nFET and pFET are of equal size, then $\beta_n > \beta_p$ (as $k'_n > k'_p$), and $V_I < (V_{DD}/2)$.

The transient characteristics are obtained by analyzing the charge and discharge current flow paths through the transistors. By using the switch model in Fig. 62.11(b), the primary time constants are

$$\tau_n = R_n C_{out} = \frac{C_{out}}{\beta_n(V_{DD} - V_{Tn})}$$
$$\tau_p = R_p C_{out} = \frac{C_{out}}{\beta_p(V_{DD} - |V_{Tp}|)} \tag{62.18}$$

Analyzing the transitions with a step input voltage yields

$$t_{HL} = \tau_n \left[\frac{2(V_{Tn} - V_0)}{(V_{DD} - V_{Tn})} + \ln\left(\frac{2(V_{DD} - V_{Tn})}{V_0} - 1 \right) \right]$$
$$t_{LH} = \tau_p \left[\frac{2(|V_{Tp}| - V_0)}{(V_{DD} - |V_{Tp}|)} + \ln\left(\frac{2(V_{DD} - |V_{Tp}|)}{V_0} - 1 \right) \right] \tag{62.19}$$

where $V_0 = 0.1\ V_{DD}$ is the 10% voltage. Noting once again that $k'_n > k'_p$, equal size transistors will give $t_{LH} > t_{HL}$. To obtain symmetrical switching, the pMOSFET must have an aspect ratio of $(W/L)_p = (k'_n/k'_p)(W/L)_n$. This illustrates that while the ratio of β-values sets the DC switching voltage V_I, the individual choices for β_n and β_p determine the transient switching times. In general, fast switching requires large transistors, illustrating the speed vs. area tradeoff in CMOS design. The propagation delay time exhibits the same dependence.

Another interesting characteristic of the CMOS inverter is the power dissipation. Consider an inverter with stable logic 0 or logic 1 inputs. Because one MOSFET is in cutoff, the DC power supply current I_{DD} is very small, being restricted to leakage levels. The standby DC power dissipation is $P_{DC} = I_{DD}V_{DD} \approx 0$, so that static logic circuits do not dissipate much power under static conditions. Appreciable I_{DD} from the power supply to ground flows only during a transition. Dynamic power dissipation, on the other hand, occurs due to the charging and discharging of the output capacitance C_{out}. The dynamic power dissipation can be estimated by

$$P_{Dynamic} = C_{out}V_{DD}^2 f \tag{62.20}$$

where f is the switching frequency of the signal. Qualitatively, this is understood by noting that this is just twice the average stored energy multiplied by the frequency. This illustrates the important result that the power dissipation of a CMOS circuit increases with the switching frequency.

Static CMOS Logic Gates

Static logic gates are based on the inverter. The term "static" means that the output voltages (logic levels) are well defined as long as the inputs are stable. The nFET rules discussed for nMOS logic gates still apply to CMOS. However, static logic gates provide an nFET and a pFET for every input. Proper operation requires that rules be developed for the pMOSFET array as follows:

- pMOSFETs (or groups) in parallel provide the NAND operation
- pMOSFETs (or groups) in series provide the NOR operation

When these rules are compared to the nMOS rules, it is seen that the nFET and pFET arrays are logical duals of one another (i.e., OR goes to AND, and vice versa).

An N-input static CMOS logic gate requires 2N transistors. NAND and NOR gates are shown in Fig. 62.13 using the rules; this type of logic is termed series-parallel, for obvious reasons. Examples of complex logic gates are shown in Fig. 62.14. Note in particular the circuit in Fig. 62.14(b). This implements the XOR function by means of

$$A \oplus B = \overline{AB + \overline{A}\,\overline{B}} = A\overline{B} + \overline{A}B \tag{62.21}$$

Reductions of this type are often performed to work the AOI or OAI equation into a more familiar form.

As seen from these examples, the logic function is determined by the placement of the nFETs and pFETs in their respective arrays. Electrically, the design problem centers around choosing the aspect ratios to achieve acceptable switching times. Because a MOSFET has a parasitic drain-source resistance that varies as (1/β), series-connected transistor chains exhibit larger time constants than parallel-connected arrangements. Recalling that $R_n < R_p$ shows that for equal size devices, series chains of nFETs are preferable to the same number of series-connected pFETs. Consequently, NAND gates are used more frequently than NOR gates, and AOI logic functions with a small number of OR operations are better. It is also possible to expand to transistor

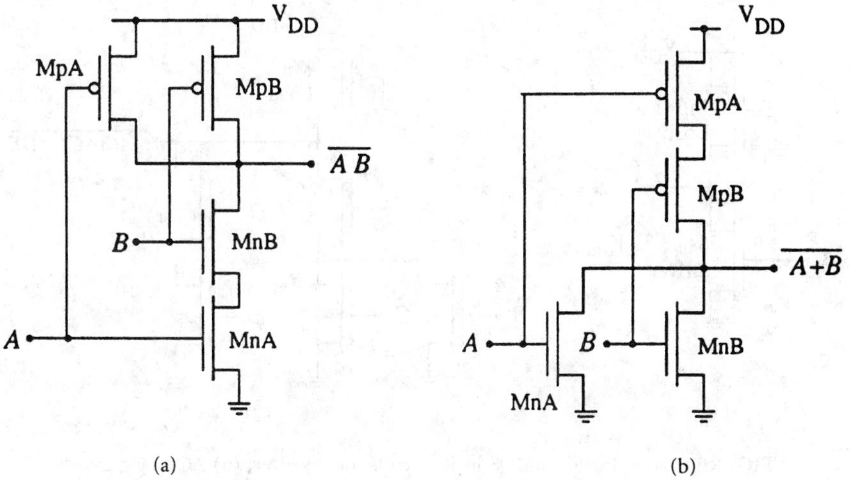

FIGURE 62.13 CMOS (a) NAND and (b) NOR gates.

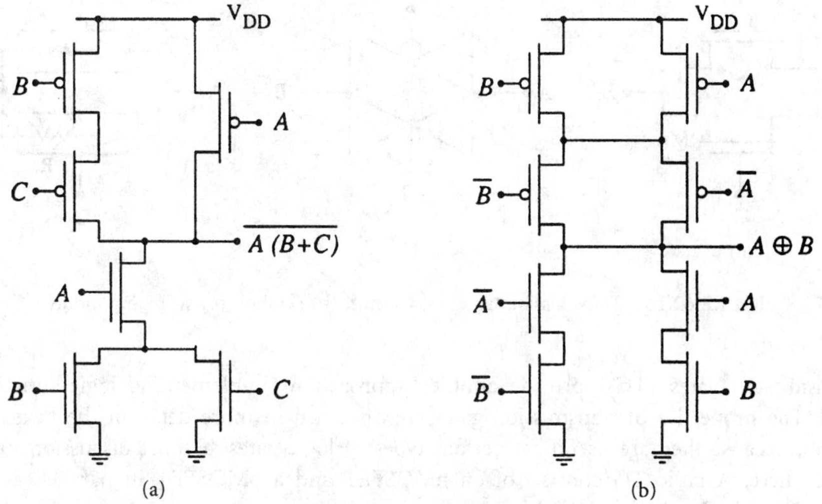

FIGURE 62.14 CMOS AOI logic examples. (a) AOI gate; (b) XOR circuit.

arrays that are not of the series-parallel type, such as a delta configuration, but it is difficult to devise general design guidelines for these circuits.

Canonical CMOS static logic design is based on using pairs of nMOS and pMOS transistors. In modern very large scale integration (VLSI) design the design complexity is limited by the interconnect (as opposed to the number of transistors), so that the need to connect every input to two transistors may result in problems in the chip layout. Pseudo-nMOS circuits provide an alternative to standard CMOS circuits. These logic gates implement logic using nFET arrays; however, the pMOS array is replaced by a single *p*-channel MOSFET that acts as a load device. Figure 62.15 shows an inverter and an AOI circuit implement based on pseudo-nMOS structuring. In both circuits the load pMOSFET is biased active with $V_{SGp} = V_{DD}$ by grounding the gate. Although the circuits are simplified, two main problems arise with this type of circuit. First, the output low voltage V_{OL} is determined by the driver-to-load ratio $(\beta_n/\beta_p) > 1$, so that large driver nFETs are required. Second, if the input voltage is high, then the circuit dissipates DC power. Despite these drawbacks, pseudo-nMOS circuits may be useful in certain situations.

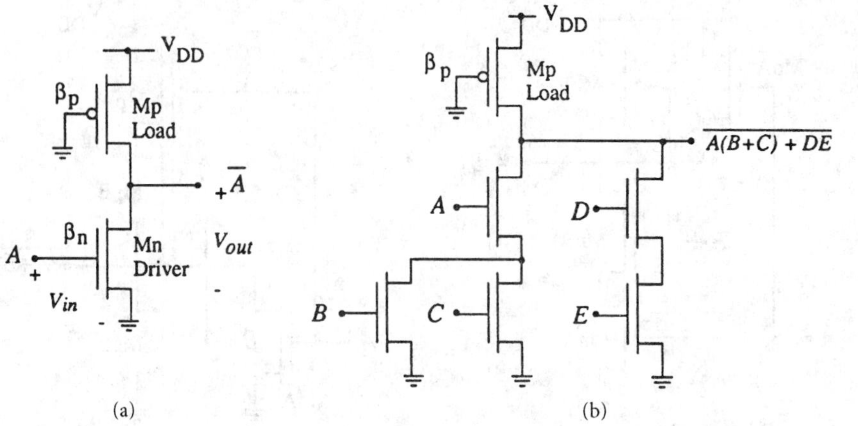

FIGURE 62.15 Pseudo-nMOS logic circuits. (a) Inverter; (b) AOI logic gate.

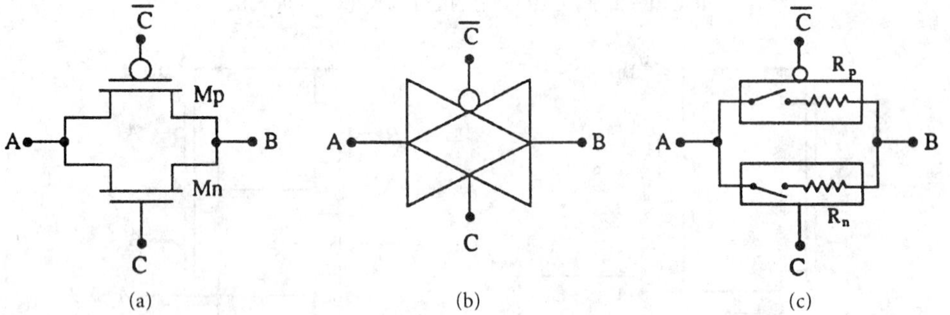

FIGURE 62.16 Transmission Gate. (a) Circuit; (b) symbol; (c) switching model.

Transmission gates (TGs) provide another approach to implementing logic functions in CMOS. The properties of transmission gates are discussed in more detail in the next section. However, because they are useful for certain types of logic gates, a short discussion has been included here. A basic TG consists of an nMOSFET and a pMOSFET in parallel, as shown in Fig. 62.16(a); the symbol in (b) represents the composite structure. Transmission gates act like voltage-controlled switches: logically, a condition of $C = 0$ gives an open switch, while $C = 1$ gives a closed switch. Transmission gates can pass the full range of voltages (from 0 V to V_{DD}) in either direction; this is not possible with a single device, due to the threshold voltage characteristic discussed earlier in this section.

Figure 62.17 illustrates a simple 2 : 1 multiplexer (MUX) with two input lines, D_0 and D_1, and a control bit S. When $S = 0$, the upper TG is closed, and the output is $F = D_0$. Conversely, $S = 1$ closes the bottom TG, so $F = D_1$. The operation of this circuit is expressed by

$$F = \overline{S}D_0 + SD_1 \tag{62.22}$$

The circuit can be expanded easily to create larger multiplexers. For example, an 8 : 1 requires three select bits, and each of the eight lines will be a switching network using three TGs. Several other TG-based logic functions are popular in CMOS design. Figure 62.18 shows the exclusive-OR (XOR) and exclusive-NOR (XNOR) circuits. The primary drawbacks of TG-based logic circuits are that (1) the TG does not have a connection to the power supply, and acts as a parasitic RC element to the stage that drives it, and (2) the chip layout may become large,

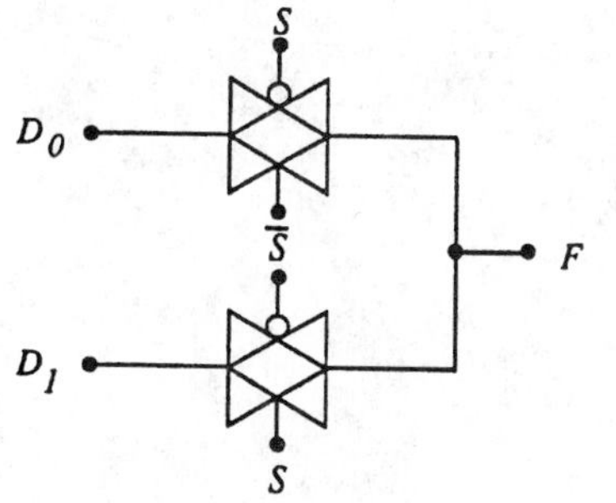

FIGURE 62.17 TG-based 2 : 1 multiplexer.

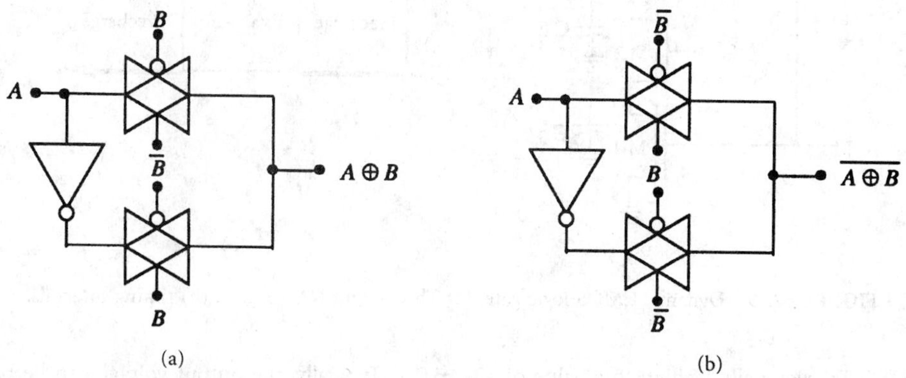

FIGURE 62.18 TG-based (a) XOR and (b) XNOR logic gates.

complicated, or both. In particular, (1) implies that TG circuits may be slower than equivalent functions designed using basic static CMOS techniques.

Dynamic CMOS Logic Gates

Dynamic CMOS logic gates are characterized as having outputs that are valid only for a limited time interval. Although this property inherently makes the circuit design more challenging, dynamic logic circuits can potentially achieve fast switching speeds. In general, dynamic circuits use parasitic capacitors in the MOS circuit to store charge Q. Because $Q = CV$, the presence or absence of charge corresponds to a logic 1 or logic 0 level, respectively. MOSFETs are used as voltage-controlled switches to "steer" the charge on and off the logic nodes. Several dynamic logic families have appeared in the literature, each having distinct characteristics. We now merely touch on some characteristics of a basic circuit which illustrates the important points.

Consider the dynamic logic circuit shown in Fig. 62.19 for a three-input NAND gate. The transistors labeled MP and MN are controlled by the clock $\phi(t)$, and provide synchronization of the data flow. Note the presence of capacitors C_{out}, C_1, C_2, and C_3. These represent parasitic capacitances due to the transistors and interconnect, and are crucial to the operation.

The circuit is controlled by the timing provided by $\phi(t)$. When $\phi = 0$, MP is ON and MN is OFF. During this time, C_{out} is charged to a voltage $V_{out} = V_{DD}$, which is called a **precharge event.** When ϕ changes to a level $\phi = 1$, MP is driven into cutoff, but MN is biased ON; the operation of the circuit during this time is termed a **conditional discharge event.** If the inputs are set to $(A, B, C) = (1, 1, 1)$, then all three logic transistors, MA, MB, and MC, are ON, and C_{out} can discharge through these transistors and MN to a final voltage of $V_{out} = 0$ V. If at least one input is a logic 0, then C_{out} does not have a direct discharge path to ground. Ideally, V_{out} would stay at V_{DD}. However, charge leakage occurs across the reverse-biased drain-bulk *pn* junctions in the

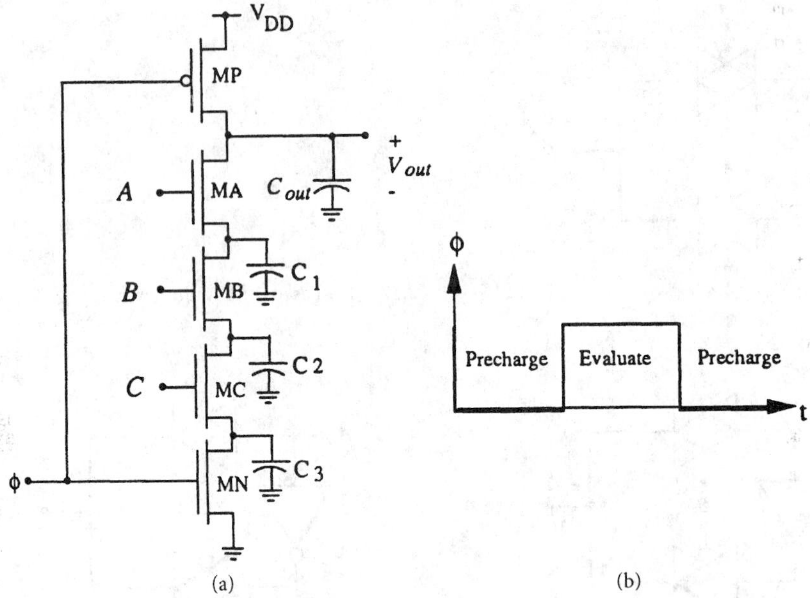

FIGURE 62.19 Dynamic CMOS logic gate. (a) Three-input NAND gate; (b) timing intervals.

MOSFETs, eventually leading to a value of $V_{out} = 0$ V. Typically, the output voltage can be held only for a few milliseconds, thus leading to the name "dynamic circuit".

Another problem that arises in dynamic logic circuits is that of **charge sharing.** Consider the three-input NAND gate with inputs of $(A, B, C) = (0, X, X)$ during the precharge, where X is a do not care condition. The total charge transferred to the circuit from the power supply is

$$Q_T = C_{out} V_{DD} \tag{62.23}$$

Now suppose that the inputs are switched to $(A, B, C) = (1, 1, 0)$ during the evaluation phase. MOSFETs MA and MB are ON, but MC is OFF, blocking the discharge path. Charge sharing occurs because the charge originally stored on C_{out} is now shared with C_1 and C_2. After the transients have decayed, the three capacitors are in parallel. Ignoring any threshold drop, they will share the same final voltage V_f such that

$$Q_T = (C_{out} + C_1 + C_2)V_f \tag{62.24}$$

Equating the two expressions for charge gives

$$V_f = \frac{C_{out}}{C_{out} + C_1 + C_2} V_{DD} \leq V_{DD} \tag{62.25}$$

To ensure that the output voltage remains at a logic 1 high voltage, the capacitors must satisfy the relation

$$C_{out} \gg C_1 + C_2 \tag{62.26}$$

The capacitance values are proportional to the sizes of the contributing regions, so that the performance is closely tied to the layout of the chip.

eferences

The material in this section is quite general. The references listed below are books in the field of digital MOS integrated circuits that provide further reading on the topics discussed here.

[1] L. A. Glasser and D. W. Dobberpuhl, *The Design and Analysis of VLSI Circuits*, Reading, MA: Addison-Wesley, 1985.

[2] H. Haznedar, *Digital Microelectronics*, Reading, MA: Addison-Wesley, 1991.

[3] J. P. Uyemura, *Circuit Design for CMOS VLSI*, Norwell, MA: Kluwer Academic, 1992.

[4] J. P. Uyemura, *Fundamentals of MOS Digital Integrated Circuits*, Reading, MA: Addison-Wesley, 1988.

62.2 Transmission Gates

Robert C. Chang and Bing J. Sheu

A signal propagates through a transmission gate (TG) in a unique manner. In conventional logic gates the input signal is applied to the gate terminal of an MOS transistor and the output signal is produced at the drain or the source terminal. In a TG the input signal propagates between the source and the drain terminals through the transistor channel, while the gate voltage is held at a constant value. The TG is turned off if the voltage applied to the gate terminal is below the threshold voltage. The TG approach can be used in digital data processing to implement special switching functions with high performance as well as a small transistor count [1]. It also can be used in analog signal processing to act as a compact voltage-controlled resistor.

Digital Processing

Single Transistor Version

A TG can be constructed by a single nMOS or pMOS transistor, as shown in Fig. 62.20. For an nMOS TG to pass a signal V_{in} to the output terminal, the selection signal S is set to the logic 1 value, i.e., the gate voltage V_G is set to a high voltage value V_{DD}. If the input signal is also the V_{DD} value, the output voltage V_{out} is determined by [2],

$$V_{\text{out}}(t) = (V_{DD} - V_{thn})\left[\frac{t/\tau_{nc}}{1 + (t/\tau_{nc})}\right] \tag{62.27}$$

where V_{thn} is the threshold voltage of the nMOS transistor with the body effect and τ_{nc} is the charging time constant which can be expressed as

$$\tau_{nc} = \frac{2C_{\text{out}}}{\mu_n C_{\text{OX}}(\text{W/L})(V_{DD} - V_{thn})} \tag{62.28}$$

Here, μ_n is the carrier mobility, C_{OX} is the per-unit-area capacitance value, and W/L is the transistor aspect ratio. If time t goes to ∞, then V_{out} will approach $V_{DD} - V_{thn}$, which indicates that a threshold voltage loss occurs in the signal from the input node to the output node. This is due to the fact that V_{GS} must be greater than the threshold voltage to turn on the nMOS transistor. Owing to this voltage reduction, an nMOS TG can only transmit a "weak" logic 1 value. However, a logic 0 can be transmitted by an nMOS TG without penalty. In order

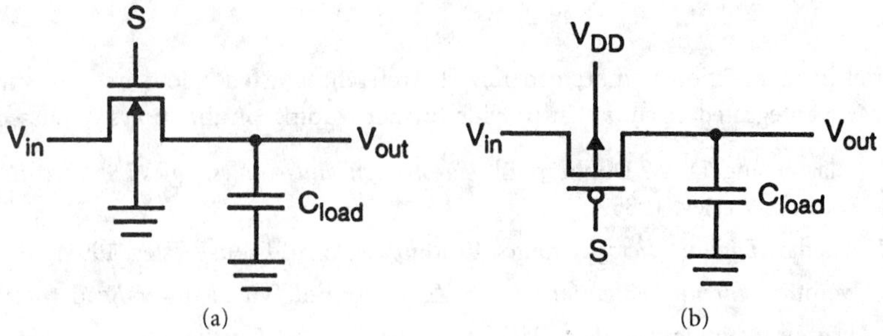

FIGURE 62.20 (a) nMOS TG. (b) pMOS TG.

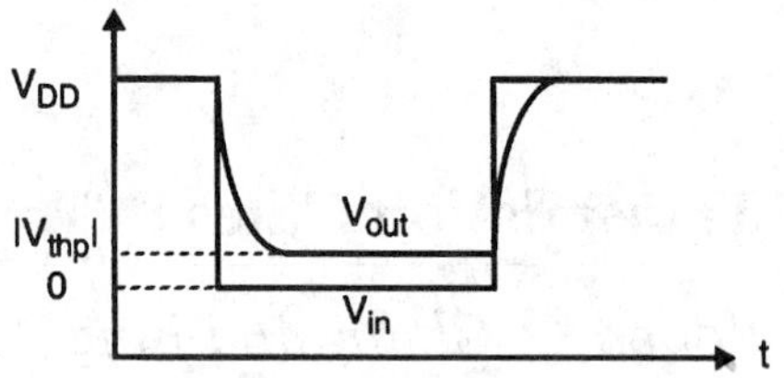

FIGURE 62.21 Characteristics of nMOS TG.

to analyze this case, we set $V_{in} = 0$ and $V_{out}(t = 0) = V_{DD} - V_{thn}$. The output voltage V_{out} is determined by

$$V_{out}(t) = (V_{DD} - V_{thn})\left[\frac{2e^{-(t/\tau_{nd})}}{1 + e^{-(t/\tau_{nd})}}\right] \tag{62.29}$$

where the discharge time constant can be expressed as

$$\tau_{nd} = \frac{C_{out}}{\mu_n C_{OX}(W/L)(V_{DD} - V_{thn})} \tag{62.30}$$

Notice that V_{out} will approach zero as time goes to infinity. Input-output (I-O) characteristics of an nMOS transmission gate are shown in Fig. 62.21.

The schematic diagram of a pMOS TG is shown in Fig. 62.20(b). For a pMOS transmission gate to pass a signal V_{in} to the output terminal, the selection signal S is set to the logic 0 value. To transmit a logic 0 value with the initial V_{out} value being V_{DD}, the expression for V_{out} is given as

$$V_{out}(t) = |V_{thp}| + \frac{V_{DD} - |V_{thp}|}{1 + (V_{DD} - |V_{thp}|)(t/2\tau_{pd})} \tag{62.31}$$

where τ_{pd} is the discharging time constant for the pMOS TG. As time goes to infinity, V_{out} will approach $|V_{thp}|$, so that the pMOS TG can only transmit a "weak" logic 0 value. On the other hand, the pMOS transmission gate can perfectly transmit a logic 1 value. To analyze this case, we set $V_{in} = V_{DD}$ and assume the initial V_{out} value as $|V_{thp}|$. The expression for V_{out} is given as

$$V_{out}(t) = V_{DD} - (V_{DD} - |V_{thp}|)\left[\frac{2e^{-(t/\tau_{pc})}}{1 + e^{-(t/\tau_{pc})}}\right] \tag{62.32}$$

where τ_{pc} is the charging time constant for the pMOS TG. The output voltage will approach V_{DD} as time goes to ∞. The transfer characteristics of the pMOS transmission gate is shown in Fig. 62.22.

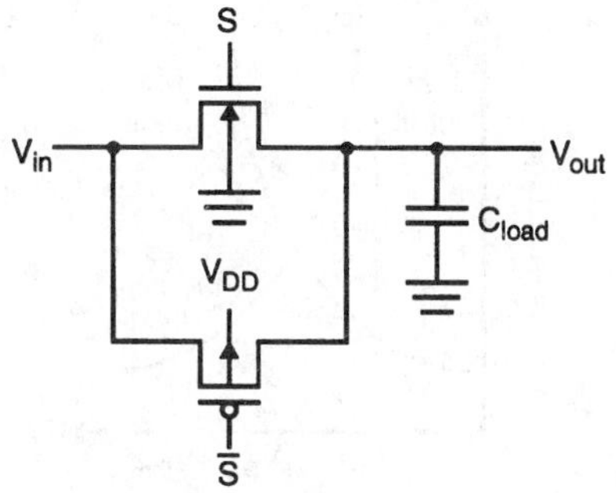

FIGURE 62.22 Characteristics of pMOS TG.

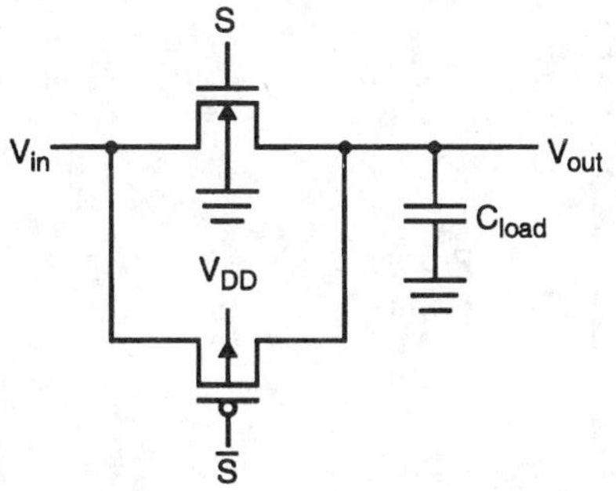

FIGURE 62.23 CMOS TG.

Complementary Transistor Version

Figure 62.23 shows the schematic diagram of a complementary transistor version of the TG which can be constructed by combining the characteristics of nMOS and pMOS TGs. The CMOS TG can transmit both the logic 0 and logic 1 values without any degradation. The voltage transmission properties of the single transistor and CMOS TGs are summarized in Table 62.1. The overall behavior of the CMOS TG can be described as follows. When the selection signal S is low, both the nMOS and pMOS transistors are cut off. The output voltage V_{out} will remain at a high impedance state. When the selection signal S is high, both the nMOS and pMOS transistors are turned on and the output voltage will be equal to the input voltage.

Three regions of operation exist for a CMOS TG. In region 1 $V_{in} < |V_{thp}|$, then nMOS transistor is in the triode region and the pMOS transistor is in the cutoff region. Because the pMOS transistor is turned off, the total current, I_{tot}, is supplied by the nMOS transistor and I_{tot} decreases as V_{in} increases. In region 2 $|V_{thp}| < V_{in} < V_{DD} - V_{thn}$; both the nMOS and pMOS transistors are in the triode region. In this region the nMOS transistor current decreases and the pMOS transistor current increases as V_{in} increases. Thus, I_{tot} is approximately a constant value. In region 3 $V_{in} > V_{DD} - V_{thn}$, the nMOS transistor is turned off and the pMOS transistor is in the triode region. The plot of the TG on-resistance is shown in Fig. 62.24.

TABLE 62.1 Transmission gate characteristics

V_{out} V_{in} Type	$V_{in} = 0$ (logic 0)	$V_{in} = V_{DD}$ (logic 1)
nMOS	0	$V_{DD} - V_{thn}$
pMOS	$\|V_{thp}\|$	V_{DD}
CMOS	0	V_{DD}

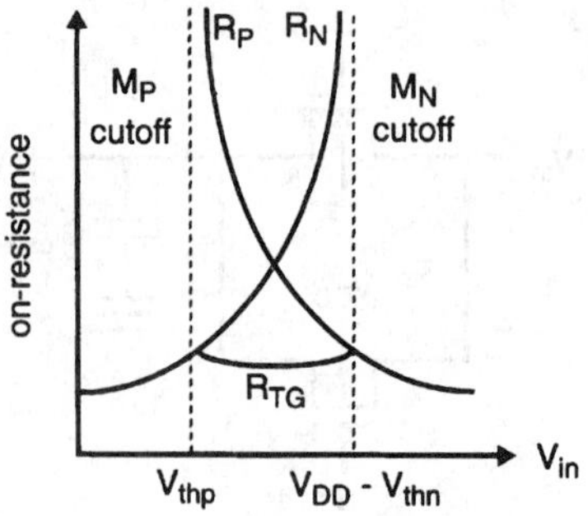

FIGURE 62.24 TG resistances.

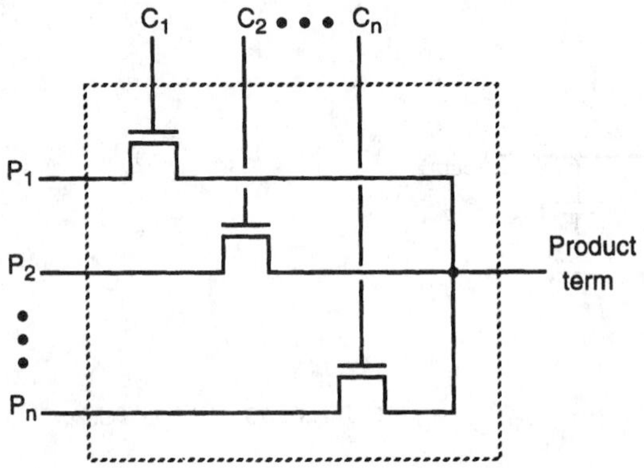

FIGURE 62.25 Model for pass transistor logic.

Pass-Transistor Logic

Pass-transistor logic is a family of logic which is composed of TG. Methods for deriving pass-transistor logic using nMOS TGs have been reported [3]. Figure 62.25 shows the schematic diagram of the pass-transistor logic in which a set of pass signals, $P_i's$, are applied to the sources of the nMOS transistors and another set of control signals, $C_i's$, are applied to the gates of the nMOS transistors.

The desired logic function F can be expressed as $F = C_1 \cdot P_1 + C_2 \cdot P_2 + \cdots + C_n \cdot P_n$. When $C_i's$ are high, $P_i's$ are transmitted to the output node. $P_i's$ can be logic 0, logic 1, true, or complement of the ith input variable X_i, or the high-impedance state Z. Constructing a Karnaugh map can help one to design the pass-transistor circuit. The pass function rather than the desired output values is put to the corresponding locations in the Karnaugh map. Then any variables that may act as a control variable or a pass variable are grouped.

For example, consider the design of a two-input XOR function. The truth table and the modified Karnaugh map of the XOR function are given in Tables 62.2 and 62.3, respectively. By grouping the A column when B is 0, and the $\overline{A}$ column when B is 1, the function can be expressed as

$$F = \overline{B} \cdot A + B \cdot \overline{A} \tag{62.33}$$

where the B is a control variable and A is a pass variable.

Figure 62.26(a) and (b) show the schematic diagrams of nMOS and CMOS implementations of the XOR function. When the control variable B is with a logic 0 value, the pass variable A is transmitted to the output. When the control variable B is with a logic 1 value, the pass variable $\overline{A}$ is transmitted to the output. Another implementation of the XOR function is shown in Fig. 62.26(c).

TABLE 62.2 Truth table of XOR function

A	B	A ⊕ B	Pass function
0	0	0	$A + B$
0	1	1	$\overline{A} + B$
1	0	1	$A + \overline{B}$
1	1	0	$\overline{A} + \overline{B}$

TABLE 62.3 Modified Karnaugh map for XOR function

		B			
		0		1	
A	0	A		$\overline{A}$	
			B		B
	1	A		$\overline{A}$	
			$\overline{B}$		$\overline{B}$

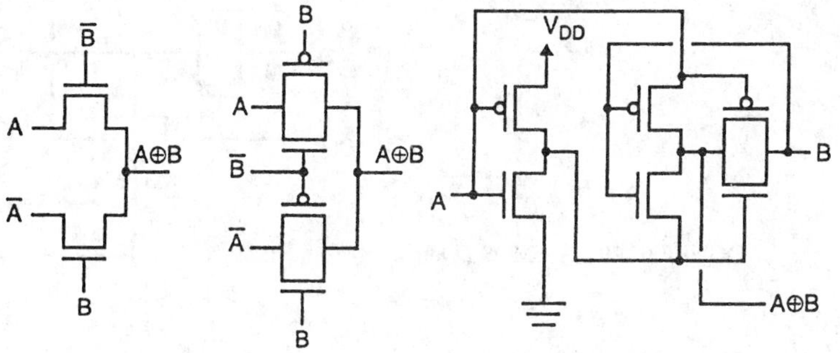

(a) (b) (c)

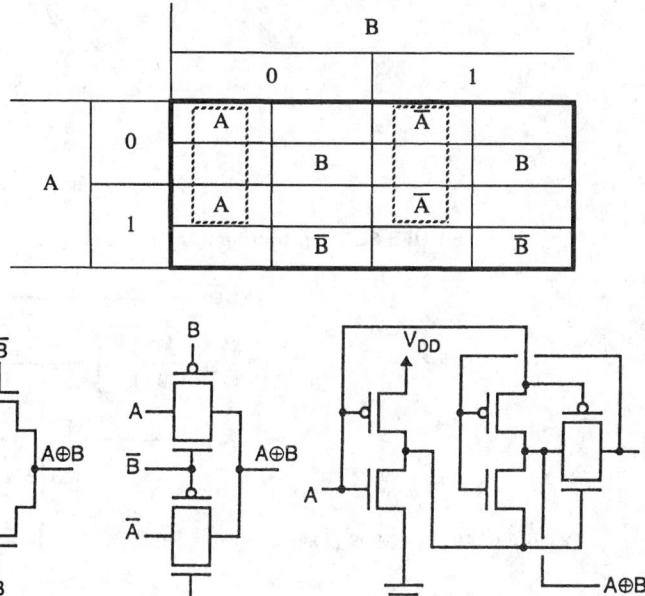

FIGURE 62.26 XOR gates. (a) nMOS version. (b) Complementary version I. (c) Complementary version II.

It is not permitted to have groupings that transmit both true and false values of the input variable to the output simultaneously. The final expression must contain all the cells in the Karnaugh map. Note that the *p*-transistor circuit is the dual of the *n*-transistor circuit. Thus, the *p*-pass function must be constructed when a complementary version is required. In addition, the pass variable with logic 0 value is transmitted by the nMOS network in a complementary implementation while the pass variable with logic 1 value is transmitted by the pMOS network.

The OR function can be constructed by one pMOS transistor and one CMOS TG, as shown in Fig. 62.27. When the input signal *A* is with the logic 0 value, the CMOS TG is turned on and the input signal *B* is passed to the output node. On the other hand, if the input signal *A* is with the logic 1 value, the pMOS TG is turned on and the logic 1 value of input signal *A* is transmitted to the output node. Because the pMOS TG can propagate a "strong" logic 1 value, it is not necessary to use another CMOS TG.

Transmission gates can be used to construct a multiplexer which selects and transmits one of the inputs to the output. Figure 62.28 shows the circuit schematic diagram of a two-input multiplexer which is composed of CMOS TGs. The output function of the two-input

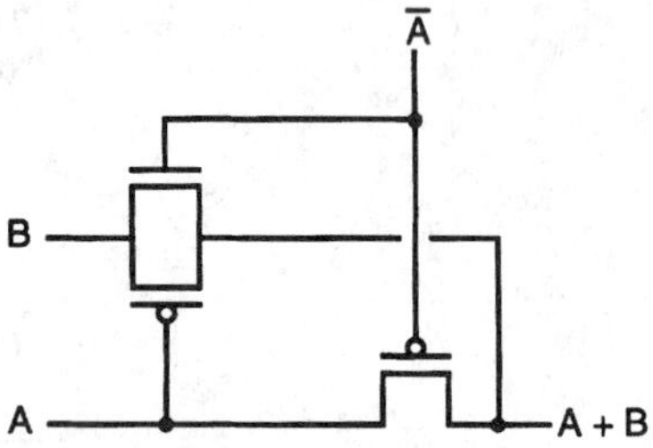

FIGURE 62.27 OR gates.

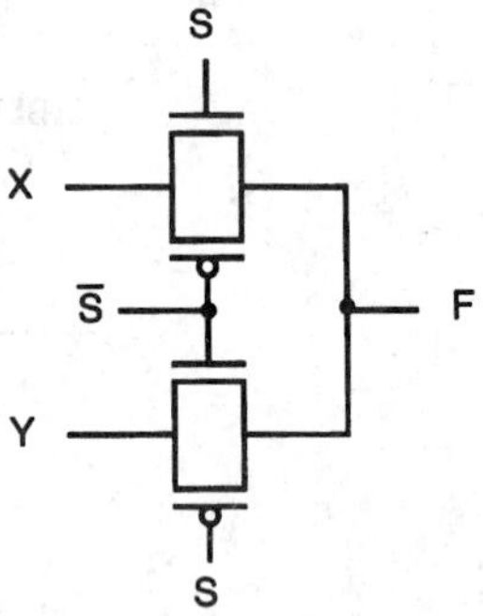

FIGURE 62.28 A two-input multiplexer.

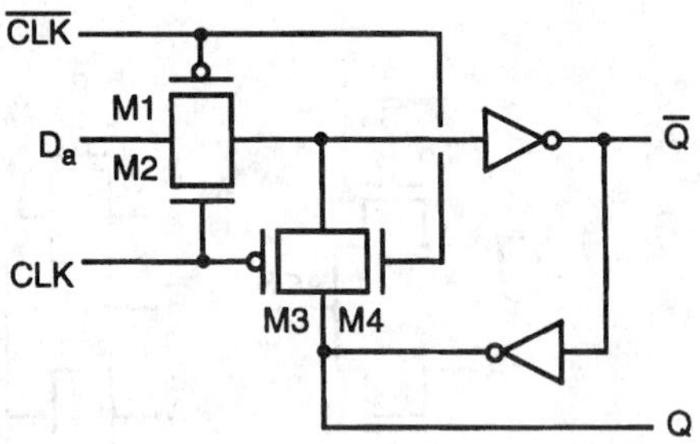

FIGURE 62.29 A CMOS *D* latch.

multiplexer is

$$F = X \cdot S + Y \cdot \overline{S} \tag{62.34}$$

If the selection signal *S* is at a logic 1 value, the input signal *X* is transmitted to the output. On the other hand, if the selection signal *S* is at a logic 0 value, the input signal *Y* is transmitted to the output. Multiplexers are important components in CMOS data manipulation structures and memory elements.

A basic *D* latch can be constructed by two TGs and two inverters, as shown in Fig. 62.29. When the *CLK* signal is at a logic 0 value, pass transistors M1 and M2 are turned off so that the input signal D_a cannot be transmitted to the outputs *Q* and $\overline{Q}$. In addition, pass transistors M3 and M4 are turned on so that a feedback path around the inverter pair is established and the current state of *Q* is stored. When the *CLK* signal is at a logic 1 value, M1 and M2 are turned on and M3 and M4 are turned off. Thus, the output signal *Q* is set to the input signal D_a and $\overline{Q}$ is set to $\overline{D}_a$. Because the output signal *Q* will follow the change of input signal D_a when the *CLK* signal is high, this circuit is a positive level-sensitive *D* latch. A positive edge-trigger register or so-called *D* flip-flop can be designed by combining one positive level-sensitive *D* latch and one negative level-sensitive *D* latch. By cascading *D* flip-flops, a shift register can be constructed.

Transmission gates can be used in the design of memory circuits. A typical random access memory (RAM) architecture consists of one row/word decoder, one column/bit decoder, and

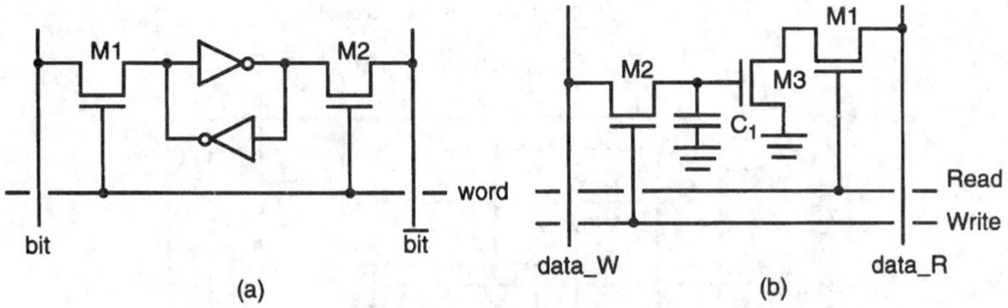

FIGURE 62.30 (a) SRAM cell. (b) DRAM cell.

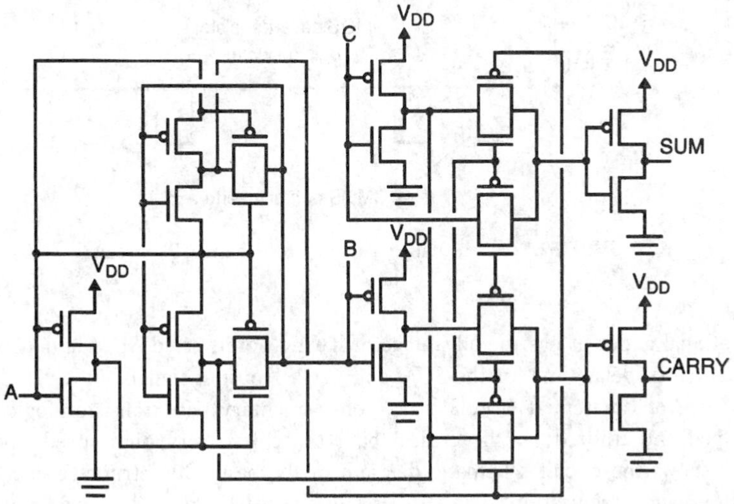

FIGURE 62.31 TG adder.

memory cells. The memory cells used in RAMs can be categorized into static cells and dynamic cells. Memory data/charges are stored on the latches in static cells, while on the capacitors in dynamic cells. The static random access memories (SRAMs) are not forced to include the refresh circuitry and are faster than the dynamic random access memories (DRAMs). However, the size of SRAM cells is much larger than that of DRAM cells. The most commonly used circuit in the design of SRAM cells is the six-transistor circuit shown in Fig. 62.30. Four transistors are used to form two cross-coupled inverters. The other two transistors, M1 and M2, are TGs to control the read/write operation of the memory cell. If the word line is not selected, the data stored on the latch will not change as long as the leakage current is small. If the word line is selected, the transistors M1 and M2 are turned on. Through the bit and $\overline{bit}$ lines, data can be written into the latch or the stored data can be read out by the sense amplifier. Transmission gates can also be found in the four-transistor DRAM cell circuit, as shown in Fig. 62.30(b). When the *Read* line is selected, pass transistor M1 is turned on and the data stored on the capacitor C_1 are read out. When the *Write* line is selected, pass transistor M2 is turned on and the data from *data_W* line are written into the cell.

Figure 62.31 shows the circuit schematic diagram of a TG adder which consists of four transmission gates, four inverters, and two XOR gates [4]. The *SUM* output, which represents $A \oplus B \oplus C$, is constructed by a multiplexer controlled by $A \oplus B$ and its complement. Notice that when $A \oplus B$ is false, the *CARRY* output equals A or B. Otherwise, *CARRY* output takes the value of input signal C. Although the TG adder has the same number of transistors as

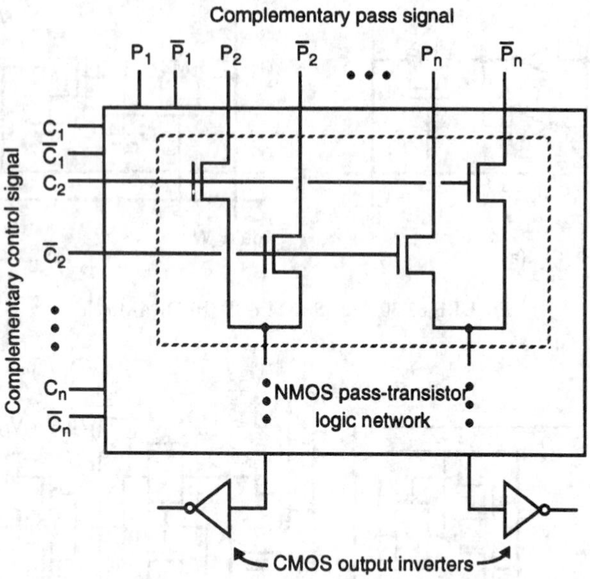

FIGURE 62.32 Schematic structure of the basic CPL circuit.

the combinational adder, it has the advantage of having noninverted *SUM* and *CARRY* output signals and an equal delay time for the *SUM* and *CARRY* output signals.

Another form of differential CMOS logic, complementary pass-transistor logic (CPL), has been developed and utilized on the critical path to achieve very high speed operation [5]. Figure 62.32 shows the circuit schematic diagram of the basic CPL structure using an nMOS pass-transistor logic organization. The CPL is constructed by an nMOS pass-transistor logic network, complementary inputs and outputs, and CMOS output inverters. As the nMOS pass transistor will transmit a logic 1 signal with one threshold voltage reduction, the output signals must be amplified by the CMOS inverters which can shift the logic threshold voltage and drive a large capacitive load. One attractive feature of the CPL design is that complementary outputs are generated by the simple four-transistor circuits. Because inverters are not required in CPL circuits, the number of critical-path gate stages can be reduced.

Figure 62.33 shows the schematic diagrams of four basic CPL circuit modules: an AND/NAND module, an OR/NOR module, an XOR/XNOR module, and a wired-AND/NAND module [5, 6]. By combining these four circuit modules, arbitrary Boolean functions can be constructed. These modules have an identical circuit schematic and are distinguished by different arrangements of input signals. This property of CPL is quite suitable for master-slice design.

The schematic diagram of a CPL full adder is shown in Fig. 62.34. Both the circuitry to produce the *SUM* output signal and the circuitry to produce the *CARRY* output signal are constructed from basic CPL modules. The *SUM* circuitry consists of two XOR/XNOR modules, while the *CARRY* circuitry consists of three wired-AND/NAND modules. The CMOS output inverters are fixed "overhead" because they are required whether the circuit has one, two, or many inputs. Thus, designing with a complex Boolean function in a CPL gate is preferred to minimize the delay time and overall device count.

Figure 62.35 shows the block diagram of a 16×16 bit multiplier, which is constructed by using a parallel multiplication architecture. A carry-look-ahead (CLA) adder and a Wallace-tree adder array are used to minimize the critical-path gate stages. The number of transistors in the CPL multiplier is less than that in a full CMOS counterpart [7].

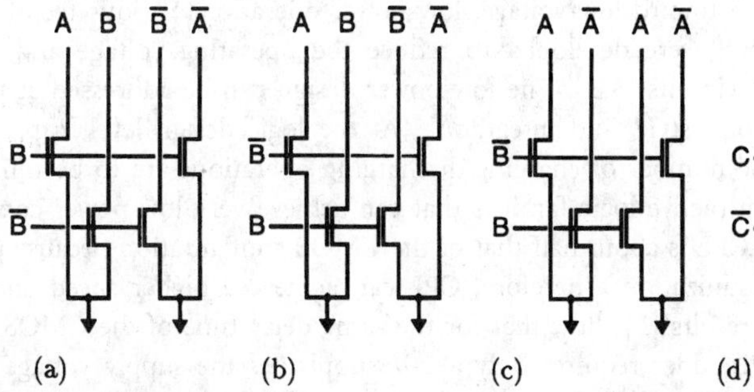

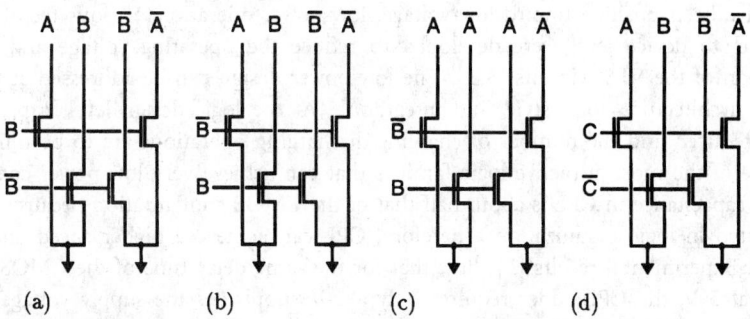

FIGURE 62.33 CPL circuit modules. (a) AND/NAND. (b) OR/NOR. (c) XOR/XNOR. (d) wire-AND/NAND.

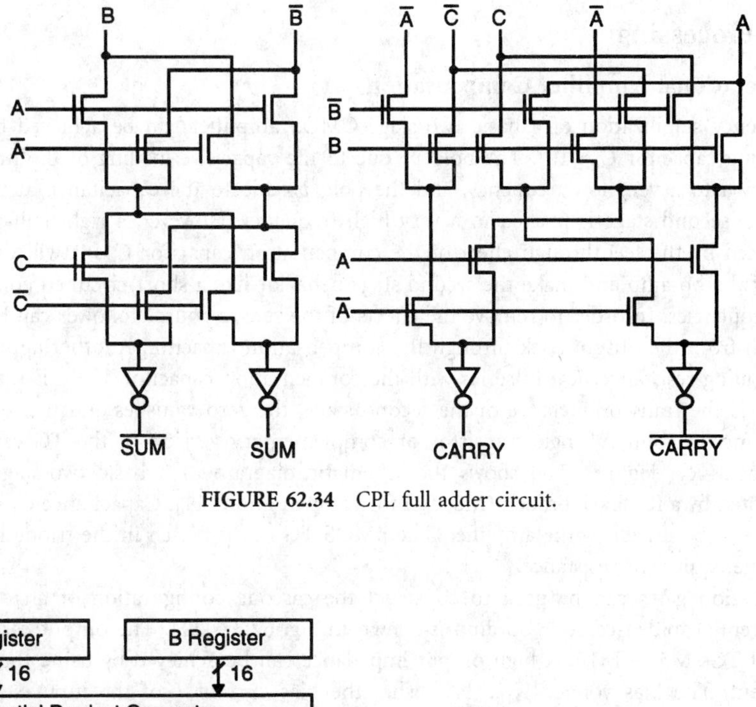

FIGURE 62.34 CPL full adder circuit.

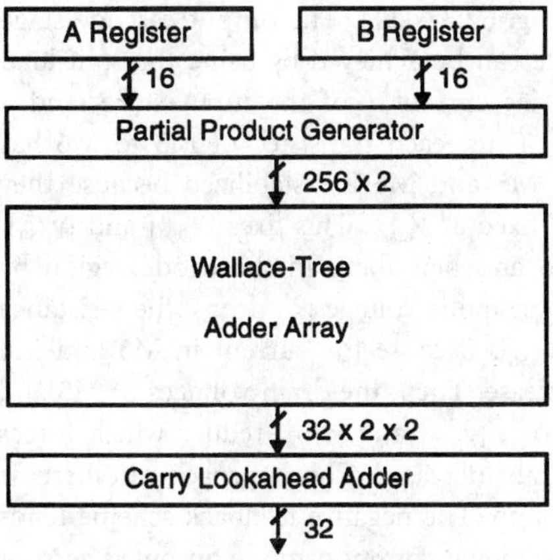

FIGURE 62.35 Block diagram of the 16×16 bit multiplier.

Due to the continued device miniaturization and the recent drive of portable systems, VLSI systems have been pushing toward low-voltage, low-power operation. Various techniques from system level to device level were developed to reduce the operating voltage and the power consumption of the VLSI circuits [8, 9]. The low-power design can be addressed at four levels: algorithm, architecture, logic style, and integration. At the logic design level, capacitive loads are to be reduced and the number of charging/discharging operations are to be minimized. A CPL is one of the most attractive logic families that can achieve very low power consumption. The input capacitance in CPL is about half that of the CMOS configuration because pMOS can be eliminated in logic organization. Therefore, CPL can achieve a higher speed and dissipate less power. Experimental results [5] show that for the same delay time of the CMOS full adder operating at 5 V, the CPL adder requires only a 2-V supply. As the supply voltage decreases, the delay time will increase, but the power-delay product will decrease. Hence, it is desirable to operate at the slowest allowable speed to reduce power dissipation. Experimental results indicate that performance of CPL logic style is better than the conventional CMOS logic style from the viewpoint of power consumption.

Analog Processing

MOS Operational Amplifier Compensation

The frequency stabilization of a basic two-stage CMOS amplifier can be achieved by using a pole-splitting capacitor C_C [10]. The pole p_1 due to the capacitive loading of the first stage is pushed down to a very low frequency, and the pole p_2 due to the capacitance at the output node of the second stage is pushed to a very high frequency. However, a right-half-plane zero is introduced by the feedthrough effect of the compensation capacitor C_C. It will degrade the stability of the op amp and make the second stage behavior like a short-circuited configuration at high frequencies. In order to remove the effects of the zero, a source follower can be inserted in the path from the output back through the compensation capacitor. Another approach is to insert a nulling resistance, R_Z, in series with the compensation capacitor. If R_Z is set to $1/g_{m2}$, where g_{m2} is the transconductance of the second stage, the zero vanishes and the feedthrough effect is cancelled out. A single transistor or complementary version of the TG can be used to implement R_Z. Figure 62.36 shows the schematic diagram of a basic two-stage op amp supplemented by a feedback branch (M8, C_C) for compensation [11]. Capacitance C_L is the load capacitance to be driven by the amplifier. The pMOS TG M8 is biased in the triode region and provides the equivalent resistance.

Transmission gates can be used to construct the cascode configuration of an op amp. A fully differential folded-cascode op amp is shown in Fig. 62.37 [11]. The output cascode stage consists of TGs M5 to M10. A high output impedance can be achieved by using the split-load arrangement. The bias voltage V_{bias1} establishes the bias current I of the input stage, and a bias current I_0 in the output transistors M7 to M12. Thus, each transistor of M3 to M6 has a stabilized current $I_0 + I/2$. The source voltages of M5 and M6 are stabilized because they conduct stabilized currents and their gate voltages are fixed at V_{bias2}. This fixes $|v_{DS3}|$ and $|v_{DS4}|$. Let transistors M3 and M4 have the same W/L ratio and bias them in the triode region by choosing a suitable value for V_{bias2}. If the output common-mode voltage $v_{0,c}$ drops, the resistance of M3 and M4 reduces, which increases $|v_{GS5}|$ and $|v_{GS6}|$. Because the current in M5 and M6 remains unchanged, $|v_{DS5}|$ and $|v_{DS6}|$ are forced to decrease. Then, the drain voltages of M5 and M6 increase, which increases $|v_{GS7}|$ and $|v_{GS8}|$. Therefore, $|v_{DS7}|$ and $|v_{DS8}|$ reduce which forces v_0^+ and v_0^- to rise. The common-mode voltage $v_{0,c}$ is thus increased. This approach can increase the common-mode rejection ratio (CMRR) of the op amp. The negative feedback scheme tends to keep $v_{0,c}$ at a constant value. It means that the small-signal common-mode output is zero or a very small value. Thus, a high CMRR is achieved.

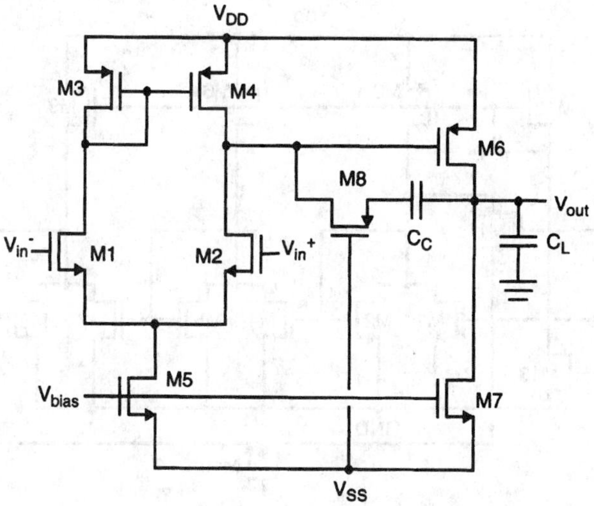

FIGURE 62.36 CMOS op amp.

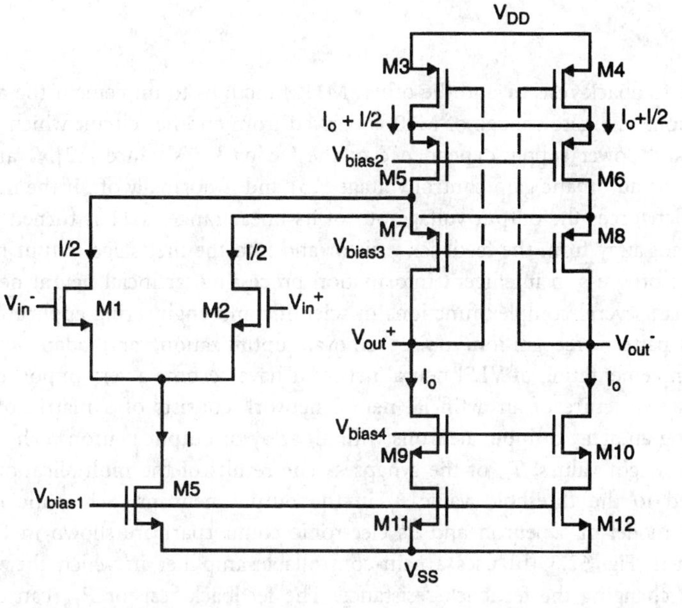

FIGURE 62.37 A fully differential CMOS op amp with stabilized DC output level.

Transimpedance Compensation

The optical receiver is an important component of the optical fiber communication system. One of the basic modules in a high performance optical receiver is the low-noise preamplifier. Several approaches are available to design the preamplifier. One approach is to use the transimpedance design which can avoid the equalization and limited dynamic range problems by using negative feedback. Transmission gates can be used to provide the feedback resistance for a transimpedance amplifier.

A complete preamplifier circuit schematic is given in Fig. 62.38 [12]. This circuit consists of three gain stages and two TGs. Each gain stage is composed of a pMOS current source achieving a common-source amplification with a folded nMOS load. One TG transistor, M10,

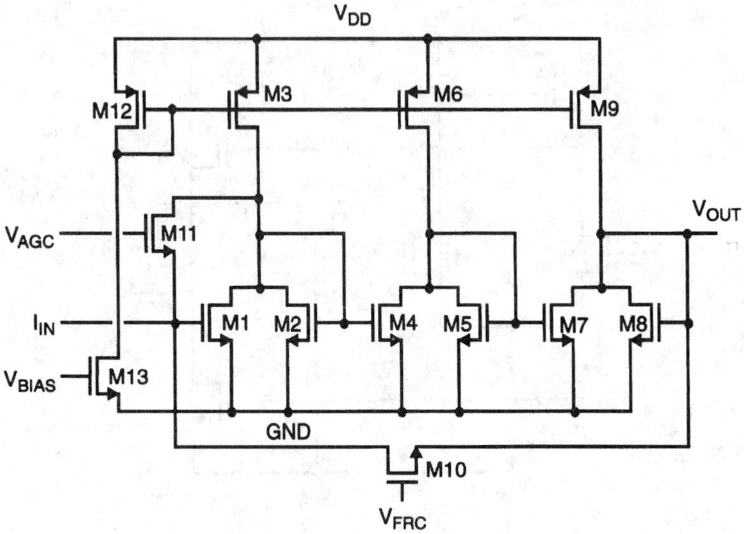

FIGURE 62.38 Circuit schematic of a preamplifier.

functions as a feedback resistor and the other, M11, functions to implement the automatic gain control function. The gate voltage of M10 is derived from another circuit which minimizes the temperature and power supply dependence of the feedback resistance [12]. Transistor M11 is controlled by the automatic gain control voltage [13] and is normally off. If the input current to the preamplifier forces the output voltage out of its linear range, M11 is turned on and begins to shunt current away from the feedback resistor and into the first stage output node.

With recent progress in intelligent information processing, artificial neural networks can be used to perform several complex functions in scientific and engineering applications, including classification, pattern recognition, noise removal, optimization, and adaptive control [14]. Design and implementation of VLSI neural networks have become a very important engineering task. The basic structure of an artificial neural network consists of a matrix of synapse cells interconnecting an array of input neurons with an array of output neurons. The inputs, V_i, are multiplied by weight values, T_i, of the synapses. The results of the multiplication are summed and compared to the threshold value θ_i in the output neurons. Schematic diagrams of a mathematical model of a neuron and its electronic counterpart are shown in Fig. 62.39. The circuit shown in Fig. 62.39(b) uses a gain-controllable amplifier in which the voltage gain is controlled by changing the feedback resistance. The feedback resistor R_{FB} can be constructed by the TG structure so that feedback resistance can be adjusted by the gain-control voltage V_{GC} [15].

Continuous-Time Filters

Resistors are important components in the construction of continuous time filters [16]. However, the implementation of resistors by integrator circuit (IC) fabrication technologies was found to be lacking in several areas of performance. The TG can be used to realize active resistance. For example, a double-MOS differential configuration, shown in Fig. 62.40, is used to implement a differential AC resistor [17].

This circuit consists of four nMOS transmission gates. Not only can it linearize the AC resistor, but it can also eliminate the effects of the bulk-source voltage [18]. To determine the AC resistance, assume that all the transistors are matched and are biased in the triode region.

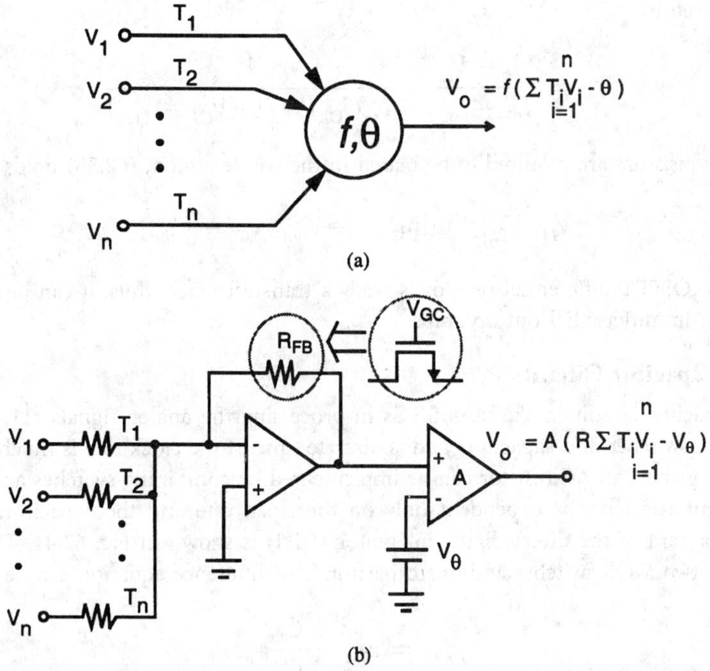

FIGURE 62.39 Neuron and synapse operation. (a) Mathematical model. (b) Analog circuit model with adjustable gain.

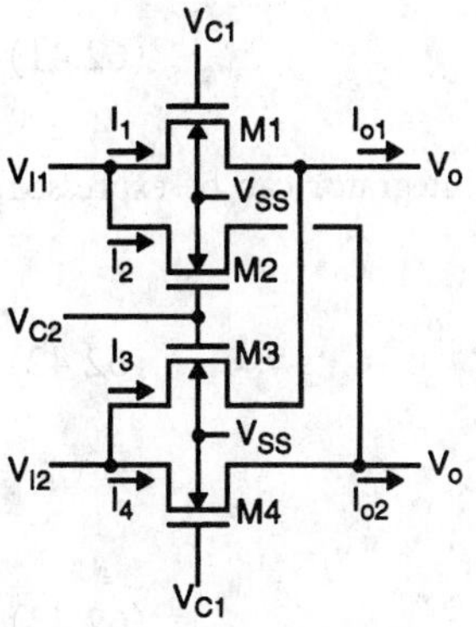

FIGURE 62.40 Double-MOS implementation of a differential AC resistor.

The currents I_{01} and I_{02} can be expressed as

$$\begin{aligned} I_{01} &= I_1 + I_3 \\ &= \mu_n C_{OX}(W/L)[(V_{C1} - V_0 - V_{thn})(V_{I1} - V_0) - (1/2)(V_{I1} - V_0)^2] \\ &\quad + \mu_n C_{OX}(W/L)[(V_{C2} - V_0 - V_{thn})(V_{I2} - V_0) - (1/2)(V_{I2} - V_0)^2] \end{aligned} \tag{62.35}$$

$$\begin{aligned} I_{02} &= I_2 + I_4 \\ &= \mu_n C_{OX}(W/L)[(V_{C2} - V_0 - V_{thn})(V_{I1} - V_0) - (1/2)(V_{I1} - V_0)^2] \\ &\quad + \mu_n C_{OX}(W/L)[(V_{C1} - V_0 - V_{thn})(V_{I2} - V_0) - (1/2)(V_{I2} - V_0)^2] \end{aligned} \tag{62.36}$$

Equations (62.35) and (62.36) can be combined to determine the differential current

$$I_{01} - I_{02} = \mu_n C_{OX}(W/L)[(V_{C1} - V_{C2})(V_{I1} - V_{I2})] \tag{62.37}$$

Thus r_{ac} is given by

$$r_{ac} = \frac{V_{I1} - V_{I2}}{I_{01} - I_{02}} = \frac{1}{\mu_n C_{OX}(W/L)(V_{C1} - V_{C2})} \tag{62.38}$$

Because all transistors are required to be biased in the triode region, (62.38) holds when

$$V_{I1}, V_{I2} \leq \min[V_{C1} - V_{thn}, V_{C2} - V_{thn}] \tag{62.39}$$

The double-MOSFET differential resistor is really a transresistance, thus, it can be applied only to differential-in, differential-out op amps.

Switched-Capacitor Circuits

Switched-capacitor circuits make use of TGs in processing the analog signals [11, 17, 19]. This approach uses switches and capacitors and is discrete time. If the clock rate is much higher than the signal frequency, an AC resistor can be implemented by combining switches and capacitors. The equivalent resistance is dependent only on the clock rate and the capacitor. The circuit schematic diagram of the direct digital integrator (DDI) is shown in Fig. 62.41. The resistance is realized by two MOS switches and one capacitor. The difference equation can be expressed as

$$v_{0,n+1} = v_{0,n} - \frac{C_S}{C_I} v_{in}. \tag{62.40}$$

After taking the z-transform, the new expression becomes

$$z \cdot V_0(z) = V_0(z) - \frac{C_S}{C_I} V_{in}(z) \tag{62.41}$$

By rearranging the various terms, the transfer function of the DDI integrator can be expressed as

$$\frac{V_0(z)}{V_{in}(z)} = -\frac{C_S}{C_I} \cdot \frac{z^{-1}}{1 - z^{-1}} \tag{62.42}$$

By setting $z = e^{j\omega T}$, the frequency response can be determined:

$$\frac{V_0}{V_{in}}(j\omega) = -\frac{C_S}{C_I}\frac{1}{j\omega T} \cdot \frac{\omega T/2}{\sin(\omega T/2)} \cdot e^{-j\omega T/2}. \tag{62.43}$$

where T is the period of the clock. In (62.43) the first term corresponds to an ideal integrator, the second term contributes to the magnitude error, and the third term is the phase error. Because the frequency response of the ideal integrator is $-[j\omega R_{eq} C_I]^{-1}$, the equivalent resistance value is determined by

$$R_{eq} = \frac{T}{C_S} \tag{62.44}$$

A ladder network can be constructed by cascading the DDI integrators. In the ladder network all cascaded stages sample the input signal at clock Φ_1 and transform the signal at clock Φ_2, where Φ_1 and Φ_2 are nonoverlapping clock signals. This clocking scheme induces the extra half-cycle phase delay. This phase error can cause extra peaking in frequency response and generate cyclic response. In order to remove the excess phase, other integrators, such as lossless

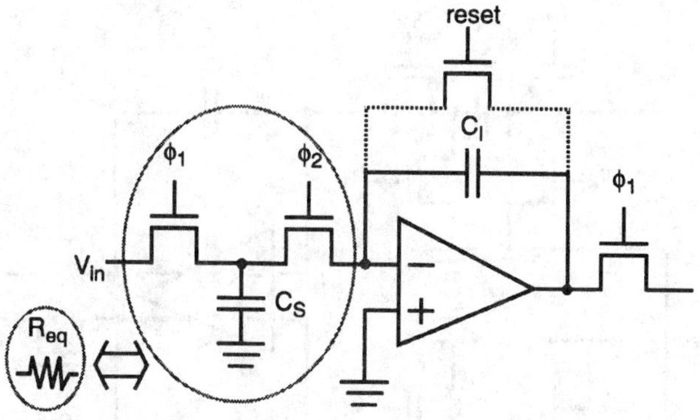

FIGURE 62.41 Direct digital integrator.

digital integrators (LDI) or bilinear integrators, can be used. In the LDI ladder network the odd-number stages sample the input signal at clock Φ_1 and transform the signal at clock Φ_2, while the even-number stages sample the input signal at clock Φ_2 and transform the signal at clock Φ_1. Thus, the frequency response of an LDI integrator can be expressed by

$$\frac{V_0}{V_{\text{in}}}(j\omega) = -\frac{C_S}{C_I}\frac{1}{j\omega T} \cdot \frac{\omega T/2}{\sin(\omega T/2)} \tag{62.45}$$

Figure 62.42 shows the circuit schematic diagram of the bottom-plate differential-input LDI. Output of an LDI integrator is more insensitive to parasitic components.

Figure 62.43 shows the circuit schematic diagram of a differential bilinear integrator. The transfer function of the bilinear integrator is

$$\frac{V_0^+ - V_0^-}{V_{\text{in}}^+ - V_{\text{in}}^-} = \frac{C_S}{C_I} \cdot \frac{1 + z^{-1}}{1 - z^{-1}} \tag{62.46}$$

As the output of the bilinear integrator does not change during clock Φ_1, it can be used to feed another identical integrator.

Transmission gates can be used to initialize the switched-capacitor circuits. For example, capacitor C_I in Fig. 62.41 is to perform the integration function and to be reset or discharged before operation. An nMOS TG can be put in parallel with the capacitor C_I. Before normal operation, the TG is turned on and the capacitor C_I is discharged so that the initial capacitor voltage value is reset to zero.

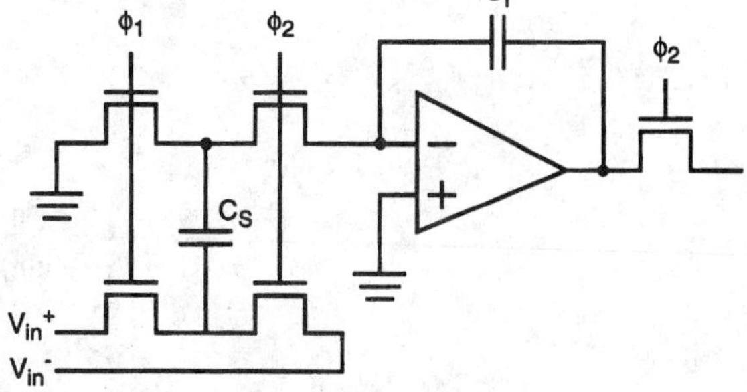

FIGURE 62.42 Bottom-plate differential-input lossless digital integrator.

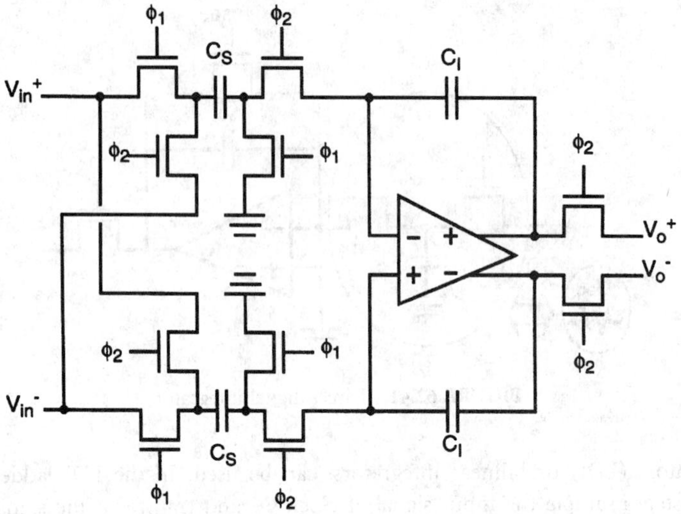

FIGURE 62.43 Differential bilinear integrator.

The accuracy of switched-capacitor circuits is disturbed by charge injection when the controlling switch turns off [20]. The turn-off of an MOS switch consists of two phases. The gate voltage is higher than the transistor threshold voltage V_{th} during the first phase. A conduction channel extends from the source to the drain of the transistor. As the gate voltage decreases, mobile carriers exit through both the drain and the source terminals and the channel conduction decreases. During the second phase, the gate voltage is smaller than V_{th} and the conduction channel no longer exists. The coupling between the gate and the data-holding node is only through the gate-to-diffusion overlap capacitance. The following analysis is focused on the switch charge injection due to the first phase of the switch turn-off.

Figure 62.44 shows the circuit schematic corresponding to the general case of switch charge injection. Capacitance C_L is the lumped capacitance at the data-holding node. Capacitance C_S could be the lumped capacitance associated with the amplifier output node, while resistance R_S could be the output resistance of an op amp. Let C_G represent the total gate capacitance of the switch, including both the channel capacitance and gate-to-drain/gate-to-source overlap capacitances. Kirchhoff's current law at node A and node B requires

$$C_L \frac{dv_L}{dt} = -i_d + \frac{C_G}{2} \frac{d(V_G - v_L)}{dt} \tag{62.47}$$

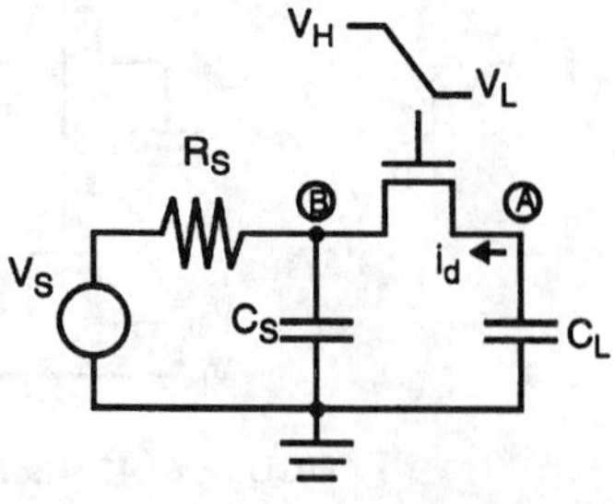

FIGURE 62.44 Circuit for analysis of switch charge injection.

and

$$\frac{v_S}{R_S} + C_S \frac{dv_S}{dt} = i_d + \frac{C_G}{2} \frac{d(V_G - v_S)}{dt} \tag{62.48}$$

where v_L and v_S are the error voltages at the data-holding node and the signal-source node, respectively. Gate voltage is assumed to decrease linearly with time from the turn-on value V_H:

$$V_G = V_H - \alpha t \tag{62.49}$$

where α is the falling rate. When the transistor is biased in the strong inversion region,

$$i_d = \beta(V_{HT} - \alpha \cdot t)(v_L - v_S) \tag{62.50}$$

where

$$\beta = \mu C_{OX} \frac{W}{L} \tag{62.51}$$

and

$$V_{HT} = V_H - V_S - V_{thn} \tag{62.52}$$

Here, V_{thn} is the transistor effective threshold voltage, including the body effect. For small-geometry transistors, narrow- and short-channel effects should be considered in determining the V_{thn} value. Under the condition $|dV_G/dt| \gg |dv_L/dt|$ and $|dv_S/dt|$, (62.47) and (62.48) can be simplified to

$$C_L \frac{dv_L}{dt} = -\beta(V_{HT} - \alpha t)(v_L - v_S) - \frac{C_G}{2}\alpha \tag{62.53}$$

and

$$\frac{v_S}{R_S} + C_S \frac{dv_S}{dt} = \beta(V_{HT} - \alpha t)(v_L - v_S) + \frac{C_G}{2}\alpha \tag{62.54}$$

No closed-form solution to this set of equations can be found. Numerical integration can be employed to find final results. Analytical solutions to special cases are given below.

Figure 62.45(a) shows the circuit schematic diagram of the case, with only a voltage source at the signal-source node. Because $C_S \gg C_L$, v_S can be approximated as zero and the governing equation reduces to

$$C_L \frac{dv_L}{dt} = -\beta(V_{HT} - \alpha t)v_L - \frac{C_G}{2}\alpha \tag{62.55}$$

When the gate voltage reaches the threshold condition, the error voltage at the data-holding node is

$$v_L = -\sqrt{\frac{\pi \alpha C_L}{2\beta}} \left(\frac{C_G}{2C_L}\right) erf\left(\sqrt{\frac{\beta}{2\alpha C_L}} V_{HT}\right) \tag{62.56}$$

Notice that the value of the error function $erf(\cdot)$ can be found from mathematical tables.

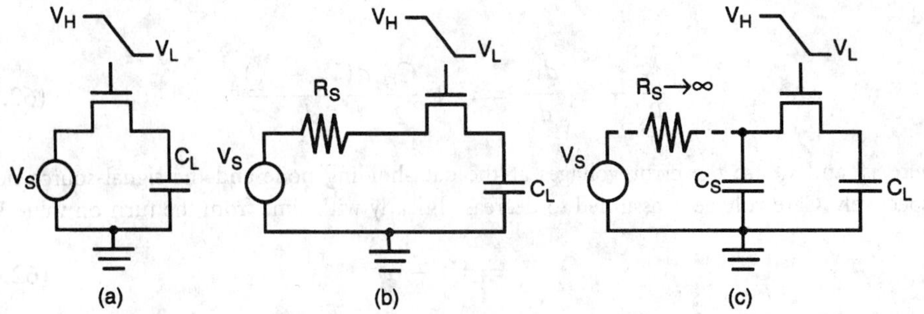

FIGURE 62.45 Special cases of switch charge injection. (a) No source resistance and capacitance. (b) No source capacitance. (c) Infinitely large source resistance.

Another special case is when the source capacitance is negligibly small, as is shown in Fig. 62.45(b). The governing equations reduce to

$$C_L \frac{dv_L}{dt} = -\beta(V_{HT} - \alpha t)(v_L - v_S) - \frac{C_G}{2}\alpha \tag{62.57}$$

and

$$\frac{v_S}{R_S} = \beta(V_{HT} - \alpha t)(v_L - v_S) + \frac{C_G}{2}\alpha \tag{62.58}$$

When the gate voltage reaches the threshold condition, the error voltage at the data-holding node is

$$\begin{aligned} v_L = -\frac{\alpha C_G}{2C_L} \exp\left(-\frac{V_{HT}}{\alpha C_L R_S}\right) \cdot \int_0^{V_{HT}/\alpha} & [\beta R_S(V_{HT} - \alpha\theta) + 1]^{1/C_L \beta R_S^2 \alpha} \\ \cdot \exp\left(\frac{\theta}{C_L R_S}\right)\left(2 - \frac{1}{1 + \beta R_S(V_{HT} - \alpha\theta)}\right) d\theta & \end{aligned} \tag{62.59}$$

If a time constant $R_S C_S$ is much larger than the switch turn-off time, then the channel charge will be shared between C_S and C_L, as shown in Fig. 62.45(c). For the case of a symmetrical transistor and $C_S = C_L$, half of the channel charge will be deposited to each capacitor. Otherwise the following equations can be used to find the results:

$$C_L \frac{dv_L}{dt} = -\beta(V_{HT} - \alpha t)(v_L - v_S) - \frac{C_G}{2}\alpha \tag{62.60}$$

and

$$C_S \frac{dv_S}{dt} = \beta(V_{HT} - \alpha t)(v_L - v_S) + \frac{C_G}{2}\alpha \tag{62.61}$$

We can multiply (62.61) by the ratio C_L/C_S and then subtract the result from (62.60) to obtain

$$C_L \frac{d(v_L - v_S)}{dt} = -\beta(V_{HT} - \alpha t)\left(1 + \frac{C_L}{C_S}\right)(v_L - v_S) - \frac{\alpha C_G}{2}\left(1 - \frac{C_L}{C_S}\right) \tag{62.62}$$

When the gate voltage reaches the threshold condition, the amount of voltage difference between the data-holding node and the signal-source node becomes

$$v_L - v_S = -\sqrt{\frac{\pi\alpha C_L}{2\beta(1 + C_L/C_S)}}\left(\frac{C_G(1 - C_L/C_S)}{2C_L}\right) \cdot erf\left(\sqrt{\frac{\beta(1 + C_L/C_S)}{2\alpha C_L}}V_{HT}\right) \tag{62.63}$$

References

[1] N. Weste and K. Eshraghian, *Principles of CMOS VLSI Design*, 2nd ed. Reading, MA: Addison-Wesley, 1993.

[2] J. P. Uyemura, *Fundamentals of MOS Digital Integrated Circuits*, Reading, MA: Addison-Wesley, 1988.

[3] D. Radhakrishnan, S. R. Whitaker, and G. K. Maki, "Formal design procedures for pass transistor switching circuits," *IEEE J. Solid State Circuits*, vol. 20, no. 2, pp. 531–536, Apr. 1985.

[4] Y. Suzuki, K. Odagawa, and T. Abe, "Clocked CMOS calculator circuitry," *IEEE J. Solid State Circuits*, vol. 8, no. 6, pp. 734–739, Dec. 1973.

[5] K. Yano, T. Yamanaka, T. Nishida, M. Saito, K. Shimohigashi, and A. Shimizu, "A 3.8-ns CMOS 16×16-b multiplier using complementary pass-transistor logic," *IEEE J. Solid State Circuits*, vol. 25, no. 2, pp. 388–395, Apr. 1990.

[6] T. Kengaku, Y. Shimazu, T. Tokuda, and O. Tomisawa, *IECE Jpn.*, 2–83, 1987.

[7] Y. Oowaki et al., "A 7.4ns CMOS 16×16 multiplier," in *ISSCC Dig. Tech. Papers*, pp. 52, 53, 1987.

[8] A. P. Chandrakasan, S. Sheng, and R. W. Brodersen, "Low-power CMOS digital design," *IEEE J. Solid-State Circuits*, vol. 27, no. 4, pp. 473–484, Apr. 1992.

[9] K. Shimohigashi and K. Seki, "Low-voltage ULSI design," *IEEE J. Solid-State Circuits*, vol. 28, no. 4, pp. 408–413, Apr. 1993.

[10] P. G. Gray and R. G. Meyer, "MOS operational amplifier design—a tutorial overview," *IEEE J. Solid-State Circuits*, vol. 17, no. 6, pp. 969–982, Dec. 1982.

[11] R. Gregorian and G. C. Temes, *Analog MOS Integrated Circuits for Signal Processing*, New York: John Wiley & Sons, 1986.

[12] D. M. Pietruszynski, J. M. Steininger, and E. J. Swanson, "A 50-Mbit/s CMOS monolithic optical receiver," *IEEE J. Solid-State Circuits*, vol. 23, no. 6, pp. 1426–1433, Dec. 1988.

[13] G. Williams, U.S. Patent 4,574,249, Mar. 4, 1986.

[14] P. K. Simpson, "Foundations of neural networks," in *Artificial Neural Networks: Paradigms, Applications, and Hardware Implementations*, E. Sánchez-Sinencia and C. Lau, Eds. New York: IEEE Press, 1992, pp. 3–24.

[15] S. M. Gowda, B. J. Sheu, J. Choi, C.-G. Hwang, and J. S. Cable, "Design and characterization of analog VLSI neural network modules," *IEEE J. Solid-State Circuits*, vol. 28, no. 3, pp. 301–313, Mar. 1993.

[16] M. Ismail, S. V. Smith, and R. G. Beale, "A new MOSFET-C universal filter structure for VLSI," *IEEE J. Solid-State Circuits*, vol. 23, no. 2, pp. 183–194, Feb. 1988.

[17] R. E. Geiger, P. E. Allen, and N. R. Strader, *VLSI Design Techniques for Analog and Digital Circuits*, New York: McGraw-Hill, 1990.

[18] M. Banu and Y. Tsividis, "Fully integrated active RC filters in MOS technology," *IEEE J. Solid-State Circuits*, vol. 18, no. 6, pp. 644–651, Dec. 1983.

63

Digital Systems

F. Gail Gray
Virginia Polytechnic Institute and State University

Wayne D. Grover
TR Labs, University of Alberta

Josephine C. Chang
University of Southern California

Bing J. Sheu
University of Southern California

Roland Priemer
University of Illinois at Chicago

Kung Yao
University of California

Flavio Lorenzelli
University of California

63.1 Programmable Logic Devices

F. Gail Gray

Programmable logic devices (PLDs) allow circuit designers to implement logic circuits with fewer chips relative to standard gate-level designs based on primitive gates and flip flops. The term "programmable logic device" is used to refer to the class of moderately complex single chip devices, in which the function of the device can be programmed by the user. Included are such devices as the programmable logic array (PLA), programmable array logic (PAL), and programmable read-only memories (PROMs). Not discussed is the more complex field programmable gate array (FPGA). The use of PLDs usually lowers layout and unit production costs. Because most commercial vendors provide software design aids for mapping designs to their specific chips, initial design costs are low. Perhaps the greatest advantage of PLD designs is *flexibility*. Design changes do not require physical changes to the printed circuit board as long

0-8493-8341-2/95/$0.00+$.50

TABLE 63.1 Complexity Ladder of Logic Devices

Device	Complexity of a Single Chip	Restrictions on Realizable Functions	Initial Design and Cost of Design Changes	Unit Production Cost
SSI discrete gate chip	Lowest	None	High	High
MSI chip		Very highly	High	High
PLD[a]		Highly	Low	Moderate
Gate array[a]		Moderately	Moderate	Low
LSI (PROM)[a]		None	Very low	High
Custom chip	Highest	None	Very high	Very low

[a]Field programmable.

as the revised functions still fit onto the same PLD. The low cost of design revisions make PLDs very attractive for prototype design and low volume production. Designers often move up the design ladder once proven designs move into high-volume production.

The discussion begins by comparing PLD devices to other logic devices and by providing a brief overview of device technologies used in commercial PLDs. By comparing the internal structures of PLAs, PALs, and PROMs, the capabilities and limitations of each type of PLD are described. Because the PAL is currently the most popular PLD device, design methodology for both combinational and sequential PAL devices is described. By emphasizing the differences between designing with PALs and designing with standard logic gates, practical insights are provided about PLD design. This section concludes with a discussion of features of PLD development systems that will help the reader select a system suitable for a specific design environment.

Table 63.1 shows the position of PLDs on the complexity ladder of device types. Programmable logic devices fall between MSI chips and gate arrays in device complexity. In the "Restrictions.." column, the range of realizations for various device types are compared. Discrete gates can implement any function if enough gates are available. MSI chips implement very specialized functions such as shift registers, multiplexers, decoders, etc. The table compares PLDs, gate arrays, and PROMs relative to the range of functions that can be implemented on a single chip. A PROM chip with n address inputs can implement any combinational function of n variables. A PLD chip with n inputs can implement only a subset of the combinational functions of n variables. Fortunately, practical functions tend to be in the set of possible functions, but not always. Gate arrays can implement a larger subset of functions, but certainly not all. Of course, we can implement any function with a custom chip. Custom chips are preferred for large-volume production because of the very low unit production costs. However, initial design costs and the cost of design changes are very high for custom chip design. Also, the design of custom chips require highly trained personnel and a large investment in equipment. The low design cost and low cost of design changes make PLDs a good choice for lower volume production and for prototype development.

Device Technologies

Companies produce PLD devices in different technologies to meet varying design and market demands. Technologies are organized into two categories. **Process technology** refers to the underlying semiconductor structure which affects device speed, power consumption, device density, and cost. **Programming technology** refers to the physics of chip programming and affects ease of programming and the ability to reprogram chips with errors.

Process Technologies

The dominant technologies in PLD devices are bipolar and complementary metal oxide semiconductor inverter (CMOS). Bipolar devices are faster (as low as 3 ns), less expensive to manufacture, but consume more power than CMOS devices. The higher power requirements of bipolar devices limit the gate density. Typical CMOS devices, therefore, achieve much higher gate densities than bipolar devices. The power consumption of CMOS devices depends on the application because a CMOS device only consumes power when it is switching state. The amount of power consumed increases with the speed of switching. Therefore, the total amount of power consumed depends on the frequency and speed of state changes in the device. Some devices have programmable power standby activation that puts the device in a lower power consumption configuration if no input signal changes for a predefined amount of time. The device then responds to the next input change much slower than normal but switches back to the faster speed configuration and maintains the faster speed as long as input changes continue to occur frequently. When programmed in the *standby power mode*, power consumption is reduced on the average at the expense of response time of the device. When programmed to operate in the *turbo mode*, the device stays in the faster configuration at all times. The result is higher power consumption, but faster response time. The mode of operation is selectable by the user to match the requirements of an application.

To take advantage of the higher densities of CMOS devices and still be compatible with bipolar devices many CMOS PLDs have special driver circuits at the input and output pins to allow pin compatibility with popular bipolar devices such as the commonly used TTL (Transistor Transistor Logic) devices.

ECL (Emitter Coupled Logic) is a very high speed technology used in some PLDs. Although ECL has the highest speed of the popular technologies, the power consumption is very high, which severely limits the gate density.

Security is another issue that is related to process technology. Many PLDs have a programmable option that prevents reading the program. Because the software provided by most manufacturers allows the user to read the program in the chip in order to verify correct programming, it is extremely easy to copy designs. To prevent illegal copying of patented designs, one simply blows the *security fuse*, which permanently prevents anyone from reading the program by normal means. However, the program in most bipolar circuits can easily be read by removing the case and examining the programmed fuses under a microscope. The CMOS PLDs are much more secure because it is virtually impossible to determine the program by examining the circuit.

Programming Technologies

The programming technologies used in PLDs are virtually the same as the programming technologies available for read-only memory (ROM). Programming technologies are divided into two broad categories: mask programmable devices and field programmable devices.

In *mask programmable* technologies identical base chips are produced en masse. The final metallization step is simply omitted. A mask programmable PLD chip is programmed by performing a final metal deposition step that selects the programming options. Clearly, this step must be performed at the manufacturers' plant. The user makes entries on an order form that specify how the chip is to be programmed and sends it to the manufacturer. The manufacturer must then prepare one or more production masks prior to making the chip. Mask programmable devices incur a high setup cost to make the first device, but unit costs are typically less than half of that for field programmable technologies. The usual practice is to use field programmable devices for prototype work and implement only proven designs in mask programmable technologies when a large production volume is required. Many PLDs are available in both mask programmable and field programmable versions, which make the conversion easy and reliable.

Field programmable technologies can be programmed by the user directly. Specialized equipment is needed. Modern programming devices can actually program both ROM and PLD devices. The programmer is typically controlled by a small computer (PC) and uses files prepared in standard format (JEDEC) by software provided by the manufacturer or written by software vendors. Such software can include elegant features such as a programming language (ABEL), truth table input, equation input, or state machine input. Selection of a chip vendor should include careful evaluation of the support software for programming the chip.

Field programmable PLD technologies can be classified into three broad categories: fusible link PLDs, ultraviolet (UV) erasable PLDs (EPLDs), and electrically erasable PLDs (EEPLDs). Field programmable ROMs come in analogous forms: fusible link ROMs (PROMs), ultraviolet erasable ROMs (EPROMs), and electrically erasable ROMs (EEPROMs).

Fusible link PLDs typically utilize bipolar process technology. The programmer blows selected fuses in the device. Because higher than normal voltages and currents are required to blow the fuses, programming fusible link PLDs can be quite stressful for the device. Overheating is a common problem. However, this technology is quite well developed and the design of programming devices is sufficiently mature that reliable results can be expected as long as directions are carefully followed. Fusible link technologies provide the convenience of in-site programming, which reduces the time required to develop designs and the time required to make design changes. The tradeoff involves at least a twofold increase in per unit cost and a significant reduction in device density relative to mask programmable devices because the fuses take up considerable chip space. A fusible link PLD can be programmed only once because the blown fuses cannot be restored.

Ultraviolet erasable PLDs have a window on the top of the chip. Programming the chip involves storing charges at internal points in the circuit that control switch settings. The charges can be dissipated by shining UV light through the window on the chip. Therefore, EPLDs provide the convenience of reprogramming as a design evolves. On the downside, EPLDs cost at least three times as much per chip as mask programmable PLDs and operate at much slower speeds. Because EPLDs typically utilize CMOS technology, they are slower than fusible link PLDs but require less power. Therefore, EPLDs are often used in development work, with the final design being implemented in either fusible link technology (for faster speed) or mask programmable technology (for faster speed and lower density). In spite of the fact that EPLDs cost more than fusible link PLDs, the reprogramming feature eventually results in a lower cost for development than using fusible link PLDs. This technology requires an additional piece of hardware to erase the chips.

Electrically erasable PLDs provide the convenience of reprogramming without the need to erase the previous program because the chip is programmed by setting the states of flip flops inside the device. It is therefore not necessary to purchase an erasing device. The reprogramming also requires less time to accomplish. Of course, EEPLD chips cost more and have a lower gate density than EPLD chips.

Notation

Programmable logic devices typically have many logic gates with a large number of inputs. Also, PLDs often have many gates that have the same set of inputs. For example, the Intel 5C031 has 74 AND gates, each with the same 36 gate inputs. Obviously, such a complex circuit using standard AND gate symbols would be extremely complex and difficult to read. Figure 63.1 shows the conventional diagram for an eight-input AND gate. Clearly, a similar diagram for a 36-input AND gate would be very cumbersome. Figure 63.2 shows the same eight-input AND gate in PLD notation. The eight parallel wires that actually occur as inputs to the AND gate

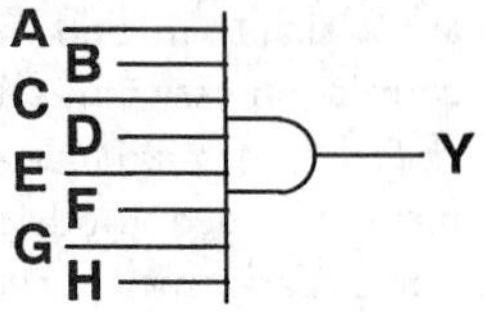

FIGURE 63.1 Conventional diagram for an eight-input AND gate.

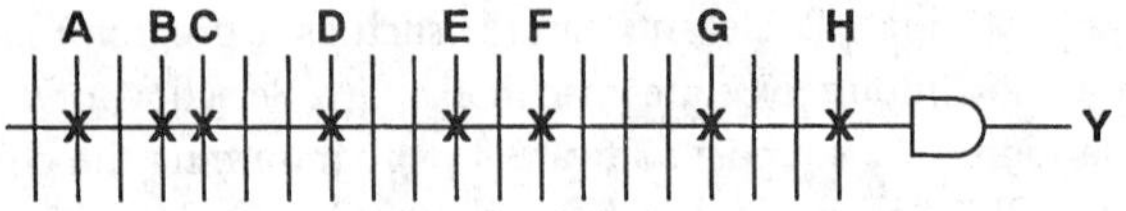

FIGURE 63.2 PLD notation for an eight-input AND gate.

are represented by a single horizontal line in PLD notation. The actual inputs to the AND gate are drawn perpendicular to the single line. There may be more signal lines than just the eight needed for this gate. An X is placed at the intersection of the single line with each of the perpendicular lines that provide actual inputs to the AND gate. Keep in mind that the single horizontal line actually represents eight parallel wires that are not physically connected to each other.

Programmable Logic Array

Figure 63.3 shows that the basic PLA consists of a programmable AND array followed by a programmable OR array. Vertical lines in the AND array represent the input variables (A, B, C, D). Because each input drives many AND gates, an internal buffer provides high current signals in both true and complemented format to each AND gate. Initially, there is a connection from each input variable and its complement to each AND gate. In this example circuit each AND gate initially has eight inputs (A, $\overline{A}$, B, $\overline{B}$, C, $\overline{C}$, D, $\overline{D}$). Each AND gate input line contains a fuse or electronic switch. We program the chip by blowing the fuses in lines that are not needed, or by programming the electronic switches. After programming, the X's are removed from the lines that are disconnected. For example, in the programmed chip of Fig. 63.4, the upper AND gate implements product term ($\overline{A} \cdot \overline{C} \cdot \overline{D}$).

In the OR array of Fig. 63.3, there is an initial connection from each AND gate output to an input on each OR gate. Again, the single vertical line connected on the input side of each OR gate represents all six wires. Each of the input lines to the OR gates also contains a fuse or programmable switch. Fig. 63.4 shows that after programming, output X connects to product terms $\overline{A} \cdot \overline{C} \cdot \overline{D}$, $B \cdot D$, and $C \cdot \overline{D}$.

The number of product lines on a chip limits the range of functions that fit onto the chip. The PLA chip in Fig. 63.3 can implement any three functions of the same four variables as long as the *total number of required product terms* is less than six. However, there are 80 different product terms involving four variables.

In order to fit functions onto the chip, designers must be able to simplify multiple output functions using gate sharing whenever possible. Finding a minimal gate implementation of multiple output functions with gate sharing is a very complex task. The goal is to minimize the total number of gates used. The size of gates does not matter. For example, whether an AND gate has four inputs or two inputs is not important. All that changes is the number of fuses that are blown. This differs dramatically from the minimization goals when discrete gates are used. For discrete gate minimization, a four-input gate costs more than a two-input gate. Therefore, the classical minimization programs need to be modified to reflect the different goals for PLA development. Three parameters determine the capacity of a PLA chip. Let n be the number of inputs, p be the number of product terms, and m be the number of outputs. Then the PLA

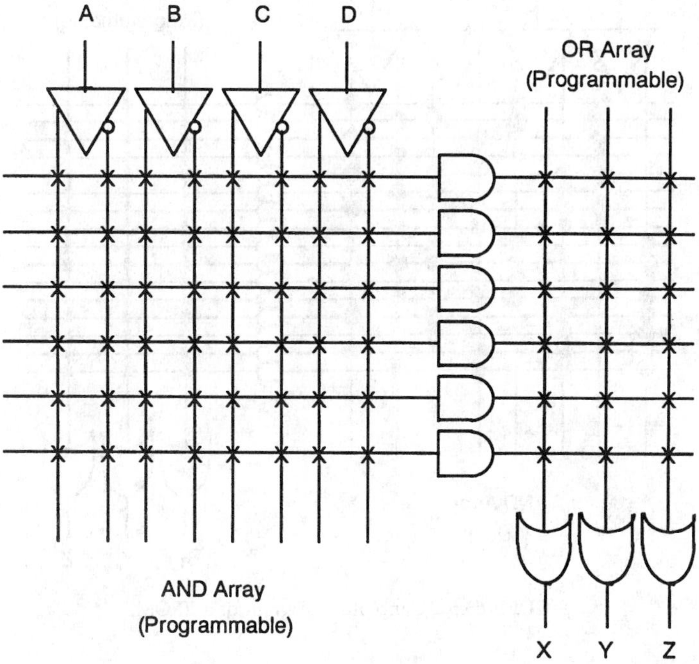

FIGURE 63.3 Basic architecture for a PLA.

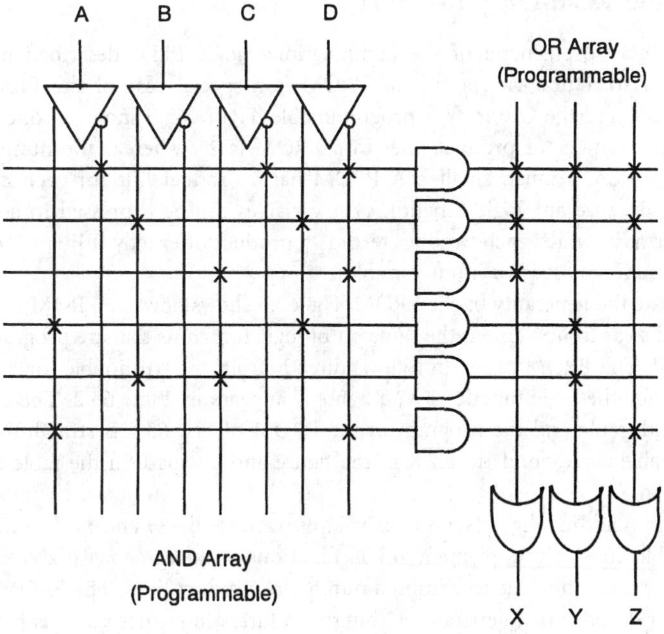

FIGURE 63.4 An example of a programmed PLA.

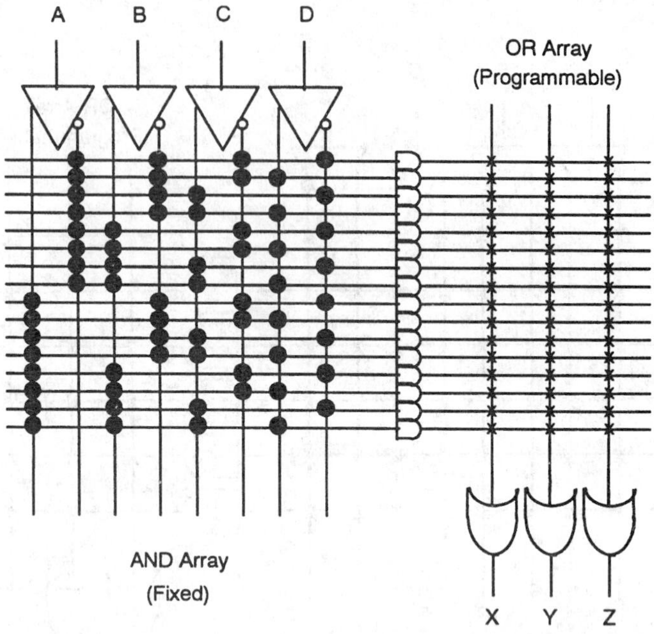

FIGURE 63.5 Conceptual diagram of a PROM.

chip can implement any m functions of the same n variables that require a total of p or fewer product terms. The device complexity is proportional to $(m + n)p$.

Programmable Read-Only Memory

The PROM is the most general of the combinational logic PLDs described in this chapter. However, from a structural viewpoint, the PROM is a special case of the PLA in which the AND array is fixed and the OR array is programmable. Figure 63.5 shows a conceptual diagram of a PROM. The number of product lines in a PROM is 2^n, whereas the number of product lines in a typical PLA is much smaller. A PROM has a product line for each combination of input variables. Because any logic function of n variables can be expressed in a canonical sum of minterms form, in which each product term is a product of exactly n literals, the PROM can implement any function of its n input variables.

To demonstrate the generality of the PROM, Fig. 63.6 shows how the PROM of Fig. 63.5 must be programmed so as to implement the same set of logic functions that are programmed into the PLA of Fig. 63.4. The PROM program follows directly from the truth table for a logic function. The truth table for the logic functions X, Y, and Z appears in Table 63.2. The correspondence between the truth table and the program in the PROM of Fig. 63.6 is straightforward. A logic 1 in the truth table corresponds to an X in the figure and a logic 0 in the table corresponds to the absence of an X.

A PROM with n address lines (serving as n input variable lines) and m data lines (serving as m output variable lines) can implement any m functions of the same n variables. Unlike a PLA, a PROM has no restrictions due to a limited number of product lines. The PROM contains an n input, 2^n output decoder that generates 2^n internal address lines that serve as product lines. As the decoder grows exponentially in size with n, the cost of a PROM also increases rapidly with n. The justification for a PLA is to reduce the cost of the PROM decoder by providing fewer product terms as many practical functions require significantly fewer than 2^n product terms. As

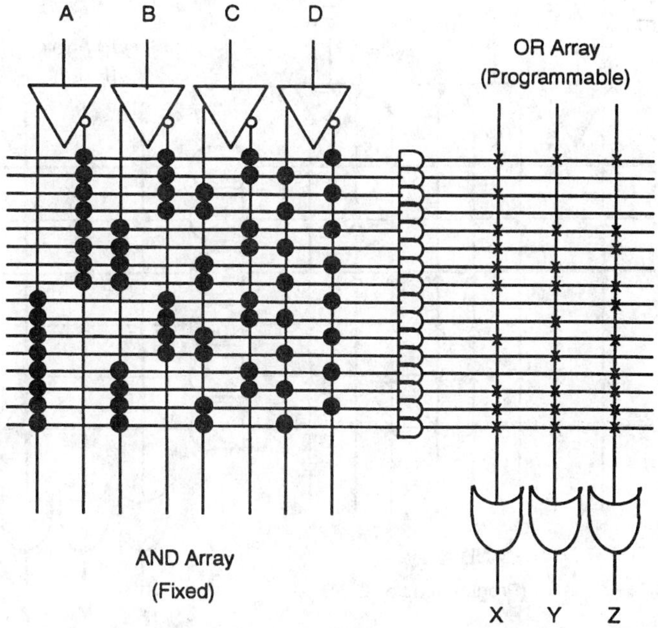

FIGURE 63.6 An example of a programmed PROM.

TABLE 63.2 Truth Table for the Logic Functions Implemented in the PROM

ABCD	*XYZ*	*ABCD*	*XYZ*
0000	111	1000	001
0001	000	1001	010
0010	100	1010	101
0011	000	1011	010
0100	111	1100	001
0101	101	1101	111
0110	110	1110	111
0111	111	1111	111

a result, some n variable functions will not fit onto a PLA chip with n input variables whereas all n variable functions will fit onto a PROM with n address lines.

Programmable Array Logic

Programmable array logic is the most popular form of PLD today. Lower price, higher gate densities, and ease of programming all tend to make the PAL more popular than the PLA. On the negative side, the range of functions that can fit onto a chip with the same number of inputs, outputs, and product lines is less for a PAL than for a PLA. The reason will be immediately obvious when we study the architecture of the PAL.

Figure 63.7 shows the basic architecture of a PAL. The PAL architecture is a special case of the PLA architecture in which the OR array is fixed. The filled circles in the OR array indicate permanent connections. Only the AND array is programmable. Compare this PAL architecture to the PLA architecture in Fig. 63.3. Because the OR array is not programmable, it is immediately evident that the range of functions that fit onto the PAL is less than the range of functions that fit onto the PLA. In the PLA the product terms can be divided among the three outputs in any way desired, and product terms that are used in more than one output can share the same

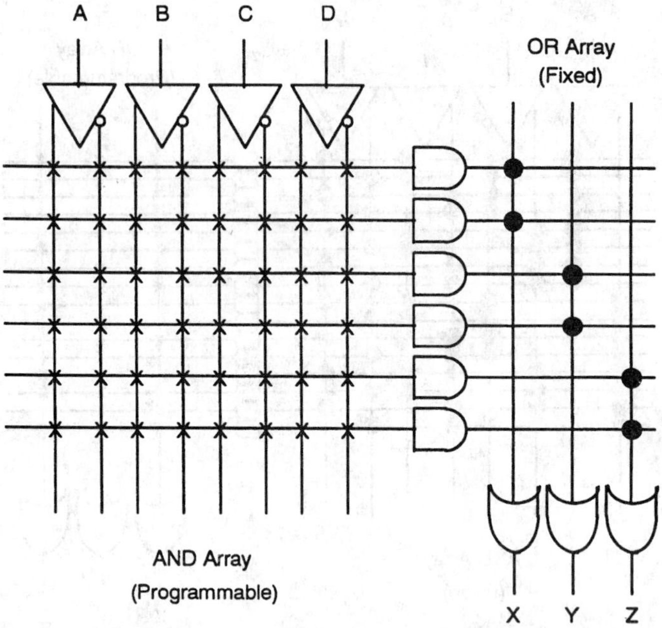

FIGURE 63.7 Basic architecture of a PAL.

product line. In the PAL each output is limited to a fixed number of product terms. In this case all outputs are limited to two product terms. Additionally, if two output functions both require the same product term in a PAL, two different product lines must be used.

Consider the three functions implemented on the PLA in Fig. 63.4. Because the three functions require a total of nine product terms, they will not fit onto the PAL of Fig. 63.7. However, any function that would fit onto this PAL would obviously fit onto the PLA because the OR array in the PLA can be programmed to be identical to the OR array of the PAL. Figure 63.8 shows an example of three functions that fit onto this PAL. Note that two different product lines must be used to provide the same product term ($\overline{A} \cdot \overline{C} \cdot \overline{D}$) to outputs X and Y.

In order to describe the range of applications for a PAL, we must know the number of inputs, n, the number of outputs, m, and the number of product lines that are permanently connected to each output OR gate. The PAL described here has four inputs, $n = 4$, three outputs, $m = 3$, and has two product lines connected to each OR gate. This PAL is described as a 2-2-2 PAL with four input variables. Many PALs have the same number of product terms permanently connected to each output. In this case the three parameters, n, m, and p, completely describe the size of the PAL. For PALs, the parameter p usually represents the number of product terms per output instead of the total number of product terms, as was the case for PLAs.

The minimization algorithm for multiple output PALs is significantly less complex than the minimization algorithm for a PLA because gate sharing is eliminated as a possibility by the fact that the OR array is not programmable. This means that each output function can be minimized independently. Minimizing a single output function is much less complex than minimizing a set of output functions where gate sharing must be considered.

Combinational Logic Programmable Logic Devices

The programmability of the AND and OR array provide a convenient means to classify combinational logic PLD types. The classification of Table 63.3 illustrates comparative features of combinational logic PLD devices. Even though PLAs have the most general structure (i.e.,

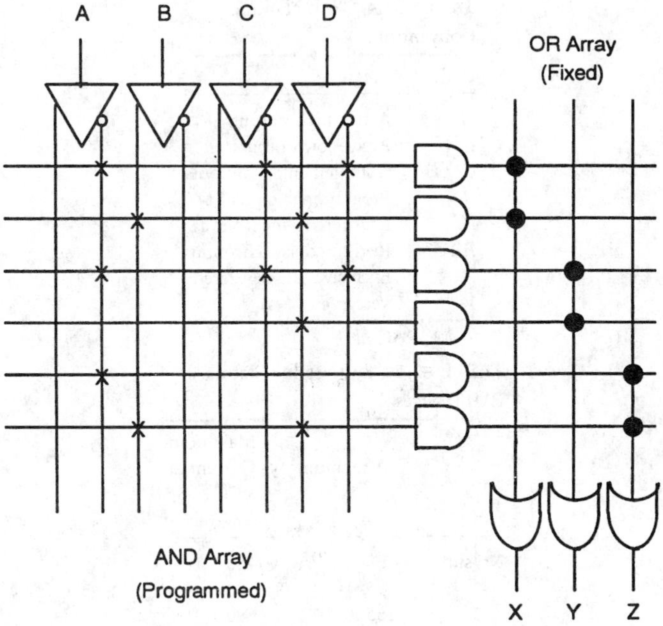

FIGURE 63.8 An example of a programmed PAL.

TABLE 63.3 Classification of Combinational Logic PLDs

AND Array	OR Array	Device	Typical No. of Product Terms per Output Gate
Fixed	Mask programmable	ROM	2^n
Fixed	Field programmable	PROM, EPROM, EEPROM	2^n
Field programmable	Fixed	PAL	2–10
Field programmable	Field programmable	PLA	50–150

both the AND and OR arrays are programmable), the number of functions that fit onto the chips are limited by the number of product terms per output. Both ROMs and PROMs have fixed AND arrays, but all possible product terms are provided. Therefore, PROMs and ROMs are the most general device from a functional viewpoint. Applications are only limited by the size and cost of available devices. Although PALs are the most restrictive devices from both the structural and functional viewpoints, the lower cost relative to PROMs and PLAs, higher gate densities relative to PLAs, and wider variety of available chip types have contributed to a rapid rise in the popularity of PAL devices. For this reason, the remainder of this discussion is concentrated on PAL devices.

Naming Convention for Programmable Array Logic

Fortunately, the PAL vendors have developed a naming convention for PAL devices that describes important parameters associated with the device. The general form for a name is *PALqXpS*. Parameter q represents the number of pairs of vertical lines in the AND array. It is the number of input variables to the AND array but not necessarily the number of input pins on the chip. Parameter X represents a type of output. Parameter p is the number of outputs of type X. Parameter S provides information about the process technology used to construct the chip.

Table 63.4 shows some of the letter codes used in the X field along with their corresponding meanings. For example, the PAL 16L8 chip has 16 pairs of inputs into the AND array and has

TABLE 63.4 PAL Naming Conventions

Code	Meaning
H	Active high outputs
L	Active low outputs
C	Complementary outputs
P	Programmable polarity
R	Registered synchronous
RA	Registered asynchronous
X	Exclusive OR registered
V	Versatile

TABLE 63.5 Suffixes for Standard PAL Devices

Suffix	Maximum Delay (ns)	Maximum Current at VCC = 5 V (mA)
No suffix	35	180
A	25	180
A-2	35	90
A-4	55	50
B	15	180
B-2	25	90
B-4	35	55
D	10	180
E	7.5	180

eight active low combinational outputs. The PAL16R4 has 16 pairs of inputs into the AND array and four registered outputs. Not indicated in the part number is the fact that the PAL16R4 also has four active low combinational outputs. The *X* field can contain multiple codes. For example, the PAL16RP8 has eight registered outputs with programmable polarity. Code *V* is used when the individual cells have many programmable features. In addition to output polarity, many devices allow the user to select whether the output is to be combinational or registered. The Intel chip described later in this chapter is an example of a *V*-type chip.

The suffix field, *S*, provides information about the speed and power consumption of the device. Table 63.5 shows a list of typical suffixes used in "standard" lines of PAL devices that use TTL low power Schottky technology. The letter represents the delay of the basic technology. For example, suffix *A* represents a device with a maximum delay of 25 ns and a maximum current of 180 mA at 5 V. The number following the dash represents a reduced power version of the device. For example, $A - 2$ means the basic technology is type *A*, but the power is reduced to one half that of a normal type *A* device. $A - 4$ means that the power of a type *A* device has been reduced by a factor of four. Of course, if power is reduced, the delay must increase.

Many manufacturers add suffix letters for a variety of purposes. For example, Texas Instruments adds an additional suffix to indicate the temperature rating for the device. For example, the Texas Instruments PAL16L8AC is a 25-ns device with maximum load current of 180 mA that operates over a temperature range from 0° to 70° C. The PAL168A-2M is a half-power version that is rated for a maximum delay of 35 ns, a maximum current of 95 mA, and a temperature range from −55° to +125° C.

The letter C after the PAL designation indicates a CMOS device with very low current requirements. These chips have very low operating current requirements. A *Z* suffix, for "zero power", indicates the presence of a standby mode of operation in which the chip enters a special standby mode when no input has changed for a time. When an input changes, the chip must change back to normal mode before responding to the input change. This feature saves power over time, but reduces the response time when in standby mode. For example, the PALC20R8Z is a CMOS version of the standard PAL20R8 that has the special low power standby mode of operation.

Designing with Combinational Logic Programmable Array Logic Devices

Determining whether a function will fit onto a PAL device is a complex procedure. To demonstrate the difficulty, the types of functions that can fit onto a PAL16L8 chip were examined. Figure 63.9 shows that the PAL16L8 chip has 8 output pins and 16 pairs of vertical input lines to the AND array. The L indicates that the eight outputs are all active low combinational logic outputs. The most important information not provided by the device name is the number of product terms per output, which is seven for this device. An additional product term provides an output enable for each output pin, therefore, eight product lines are available per output pin (a total of 64 product lines). The number of primary input pins is ten (pin numbers 1 to 9, and 11). In terms of our definitions, it would seem that $n = 10$, $m = 8$, and $p = 7$. However, this simplistic analysis significantly understates the capacity of this chip (see below).

A simplistic analysis would say that the PAL16L8 chip can implement any eight functions of the ten input variables as long as each function requires no more than seven product terms. As far as it goes, this statement is correct. However, it significantly understates the capacity of the chip because it does not take into account the fact that six of the output pins are internally connected as inputs to the AND array (pins 13 to 18). This is the source of the additional six inputs to the AND array. These internal feedback connections significantly expand the capacity of the chip.

Consider the following logic function

$$X = A\overline{B}C+B\overline{C}D+\overline{A}E+D\overline{E}F+\overline{A}C+\overline{D}+\overline{F}GH+F\overline{G}I+BE\overline{H}+\overline{C}H+\overline{I}J+\overline{B}E\overline{J}+\overline{D}H$$

It appears that this logic function will not fit onto the PAL16L8 chip because it requires 13 product terms and each output only has seven product lines. However, if not all chip outputs are needed for the application, we can use one of the feedback lines to fit this function onto the chip.

First, the function X is partitioned:

$$X = A\overline{B}C + B\overline{C}D + \overline{A}E + D\overline{E}F + \overline{A}C + \overline{D} + Y$$

$$Y = \overline{F}GH + F\overline{G}I + BE\overline{H} + \overline{C}H + \overline{I}J + \overline{B}E\overline{J} + \overline{D}H$$

As Y has only 7 product terms, function Y is mapped to the macro-cell connected to pin 18 and the feedback line from pin 18 is used to connect Y to a pair of vertical lines in the AND array. Function Y is now available to all other cells as an input variable. Function X also has seven product terms; thus, X is mapped to the macro-cell connected to pin 17. One of the product terms in X is the single variable Y. Figure 63.10 shows the final implementation. To obtain the needed product terms for output X, two macro-cells were used in the array. As a result, pin 18 is no longer available as an output unless function Y can be used outside the chip.

Two practical matters need to be considered. First, some signals must now pass through the AND array twice as they proceed from an input to an output due to the feedback path. Therefore, the delay of the chip is now twice what it was when the feedback path was not utilized. Second, the outputs are inverted by the buffer; therefore, $\overline{X}$ is actually obtained on pin

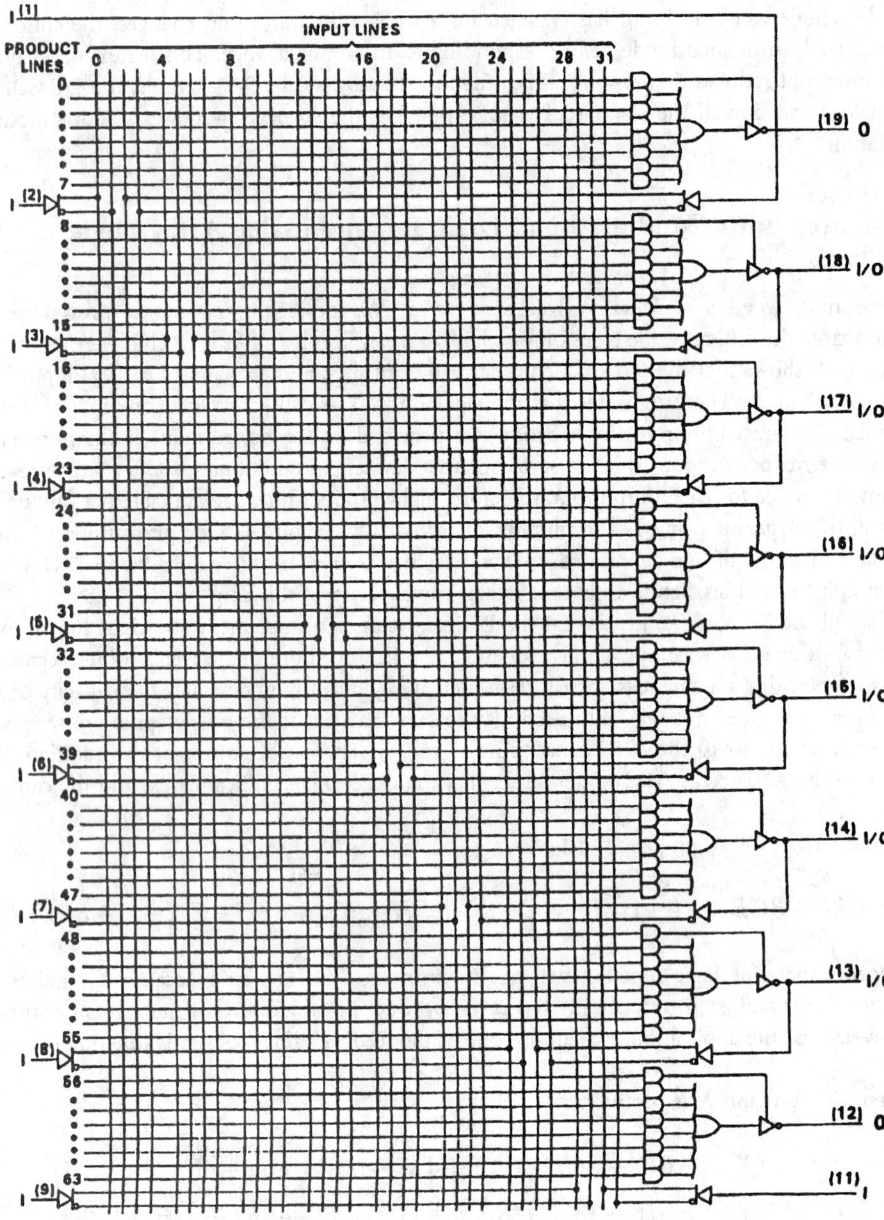

FIGURE 63.9 Logic diagram of the PAL16L8 chip. (Courtesy of Texas Instruments.)

17. If X is specified to be *active low*, then the output on pin 17 is exactly what is needed. If the output X is specified to be active high, than an inverter is required. Alternatively, $\overline{X}$ could be implemented. Because PALs are available with both active low and active high outputs, a designer should select an appropriate PAL to eliminate the need for inverters.

Another feature that adds to the flexibility of the chip is that pins 13 to 18 can be used either as inputs or as outputs. For example, the enable for the output buffer at pin 15 can be permanently disabled. Because pin 15 is connected directly into the AND array, it is no different from any other input, e.g., pin 2. Of course, by permanently disabling the buffer at pin 15, the OR array connected to the buffer is also disconnected. In order to use pin 15 as an input,

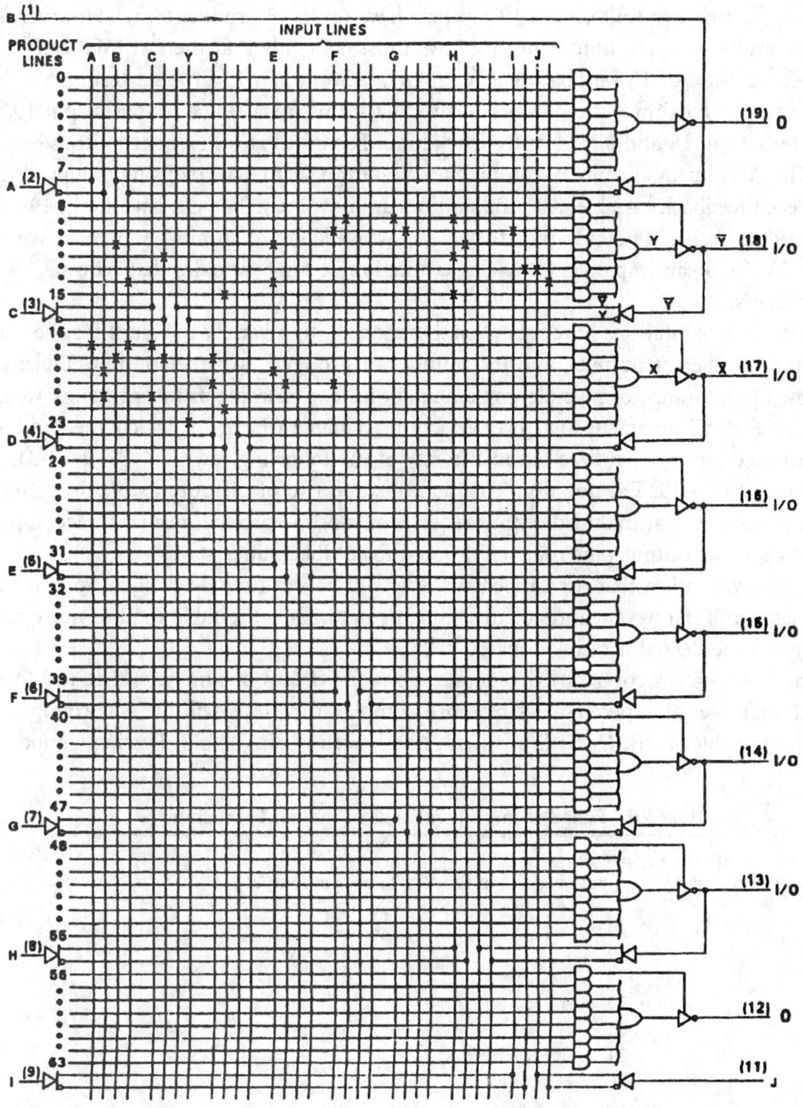

FIGURE 63.10 Implementation of function with 13 product terms on PAL16L8 chip.

the macro-cell connected to pin 15 cannot be used. However, dual-use pins and feedback lines extend the range of applications for the chip dramatically. For example, suppose we need only one output and select pin 19 for that use. Then pins 13 to 18 are available either as inputs or as feedback lines. We can, therefore, fit a wide range of functions onto the chip with varying numbers of inputs and product terms.

Assuming that we need only the one output on pin 19, then we could have up to 16 inputs (pins 1 to 9, 11, 13 to 18). We could, of course, use only the seven product lines in the macro-cell connected to pin 19 to implement our function. Thus, we conclude that the PAL16L8 chip can implement any single function of 16 variables that requires no more than seven product terms.

If more product terms but not as many input variables are needed, then we can connect pin 11 to pin 12. This connects the output on pin 12 back into the AND array in the same way as any other input variable. The output in pin 19 can pick up the seven product terms on pin 12 as an input. This takes up one of the product lines for pin 19, but six product lines remain.

Therefore, the single output on pin 19 can now have up to 13 product terms. However, pin 11 is no longer available as an input. Therefore, we must conclude that the PAL16L8 can implement any single function of 15 variables that requires no more than 13 product terms.

If we want to maximize the number of product terms for a single output at pin 19, then we can use pins 13 to 18 and pin 11 as feedback lines. Each feedback line contributes seven product terms. The AND array for pin 19 can pick up all 49 product terms by using one product line to pick up each feedback variable. All product lines are now busy. The OR gate at pin 19 then sums all 49 product terms. The only pins that are now available for inputs are 1 to 9. We conclude that the PAL16L8 can implement any single function of nine variables that requires 49 or fewer product terms.

Clearly, implementations have many combinations with a variety of values for the number of inputs, the number of outputs, and the number of product terms per output. Tables 63.6 and 63.7 show the full range of possible implementations. For example, from the tables, we note that the PAL16L8 can implement any three functions of ten variables in which the product terms are distributed among the three outputs in any of the following ways: 7-7-37, 7-13-31, 7-19-25, 13-13-25, or 13-19-19. The notation 7-7-37 means that two of the outputs require at most seven product terms and that the third output requires at most 37 product terms. To accomplish these results, five of the output pins must be devoted to feedback lines.

Any implementation that uses a feedback line will have a time delay equal to twice that of a single macro-cell. However, none of the implementations in the table require more than twice the normal time delay for a single macro-cell.

Although the tables cover broad generalizations for classes of functions that will fit onto the PAL16L8 chip, certain special types of more complex functions will fit. For example, suppose that input variables *A*, *B*, *C*, *D*, *E*, *F*, *G*, *H*, and *I* occupy pins 1 to 9. Further, suppose that we

TABLE 63.6 Range of Implementations of the PAL16L8 Chip

No. of Outputs	No. of Inputs	No. of Product Terms per Output
1	16	7
1	15	13
1	14	19
1	13	25
1	12	31
1	11	37
1	10	43
1	9	49
2	16	7-7
2	15	7-13
2	14	7-19, 13-13
2	13	7-25, 13-19
2	12	7-31, 13-25, 19-25
2	11	7-37, 13-31, 19-25
2	10	7-43, 13-37, 19-31, 25-25
3	15	7-7-7
3	14	7-7-13
3	13	7-7-19, 7-13-13
3	12	7-7-25, 7-13-19, 13-13-13
3	11	7-7-31, 7-13-25, 7-19-19, 13-13-19
3	10	7-7-37, 7-13-31, 7-19-25, 13-13-25, 13-19-19
4	14	7-7-7-7
4	13	7-7-7-13
4	12	7-7-7-19, 7-7-13-13
4	11	7-7-7-25, 7-7-13-19, 7-13-13-13
4	10	7-7-7-31, 7-7-13-25, 7-7-19-19, 7-13-13-19, 13-13-13-13

TABLE 63.7 Range of Implementations of the PAL16L8 Chip

No. of Outputs	No. of Inputs	No. of Product Terms per Output
5	13	7-7-7-7-7
5	12	7-7-7-7-13
5	11	7-7-7-7-19, 7-7-7-13-13
5	10	7-7-7-7-25, 7-7-7-13-19, 7-7-13-13-13
6	12	7-7-7-7-7-7
6	11	7-7-7-7-7-13
6	10	7-7-7-7-7-19, 7-7-7-7-13-13
7	11	7-7-7-7-7-7-7
7	10	7-7-7-7-7-7-13
8	10	7-7-7-7-7-7-7-7

implement functions S, T, V, W, X, Y, and Z using the macro-cells connected to pins 12, 13, 14, 15, 16, 17, and 18, respectively. Further suppose that we connect pin 12 to pin 11 so that all of these functions are connected to a pair of vertical input lines in the AND array. Thus, all of these functions are now available to the single output P at pin 19. This approach allows many very complex logic functions to fit onto the chip. For example, let each of the functions S, T, V, W, X, Y, and Z be a sum of product expression involving the nine input variables with at most seven product terms. S might be

$$\begin{aligned} S = {} & A\overline{B}CD\overline{E}F\overline{G}HI + \overline{A}B\overline{C}DE\overline{F}GH\overline{I} + \overline{B}CD\overline{E}F\overline{G}HI + \overline{F}G\overline{H}I \\ & + \overline{A}H\overline{I} + \overline{D}EF\overline{G}HI + ABCDEFGHI \end{aligned}$$

Variables T, V, W, X, Y, and Z could be of similar complexity. Then output P might be

$$P = A\overline{B}CD\overline{E}FG\overline{H}IS\overline{T}VW\overline{X}Y\overline{Z} + B\overline{C}DEF\overline{G}\,\overline{S}T\overline{V}WXY\overline{Z} + \cdots$$

where P has at most seven such product terms. The delay of this implementation is still twice the delay of one basic macro-cell.

Let us consider embedded factors. Each equation has at most seven product terms involving the listed variables:

$$\begin{aligned}
S &= f(A-I) = \overline{A}BC\overline{D}E\overline{F}GH\overline{I} + A\overline{B}\,\overline{C}DEF\overline{G}HI + \cdots \\
T &= f(S, A-I) = AB\overline{C}D\overline{E}FG\overline{H}IS + \overline{A}BCD\overline{E}\,\overline{F}GHI\overline{S} + \cdots \\
V &= f(S, T, A-I) = \overline{C}DE\overline{F}GH\overline{I}\,\overline{S}T + BC\overline{D}HIS\overline{T} + \cdots \\
W &= f(S, T, V, A-I) = \overline{A}BCD\overline{E}F\overline{G}HI\overline{S}T\overline{V} + \overline{D}E\overline{F}HIS\overline{T}V + \cdots \\
X &= f(S, T, V, W, A-I) = \overline{A}\,\overline{B}CDE\overline{F}G\overline{H}I\overline{S}TV\overline{W} + \overline{E}T\overline{V}W + \cdots \\
Y &= f(S, T, V, W, X, A-I) = ABC\overline{D}EFG\overline{H}IS\overline{T}VW\overline{X} + \overline{D}H\overline{I}S\overline{T}V\overline{W}X + \cdots \\
Z &= f(S, T, V, W, X, Y, A-I) = A\overline{B}CD\overline{E}FGH\overline{I}\,\overline{S}T\overline{V}WX\overline{Y} + \overline{F}H\overline{I}WXY + \cdots \\
P &= f(S, T, V, W, X, Y, Z, A-I) = A\overline{B}CD\overline{E}FG\overline{H}IS\overline{T}VW\overline{X}Y\overline{Z} \\
&\quad + B\overline{C}DEF\overline{G}\,\overline{S}T\overline{V}WXY\overline{Z} + \cdots
\end{aligned}$$

The delay of this implementation is eight times the delay of a single macro-cell because an input signal change might have to propagate serially through all of the macro-cells on its way to the output at pin 19.

These examples demonstrate that very complex functions can fit onto the chip. Determining the optimal way to factor the equations is a very complex issue. Chip manufacturers and third-party vendors provide software packages that aid in the fitting process.

Designing with Sequential Programmable Array Logic Devices

The concept of registered outputs extends the range of PAL devices to include sequential circuits. Figure 63.11 shows the logic diagram for the PAL16R4 chip. Again, this chip has 16 pairs of inputs to the AND array. The R4 part of the designation means that the chip has four outputs connected directly to D-type flip flops, i.e., the outputs are *registered*. Let us add another parameter, k, to designate the number of flip flops on the chip. An examination of Fig. 63.11 indicates that the PAL16R4 also has four combinational outputs with feedback connections to the AND array. These pins are I/O pins because they can be used as inputs if the OR output to the pins are permanently disabled. All outputs are active low. The chip has eight input pins.

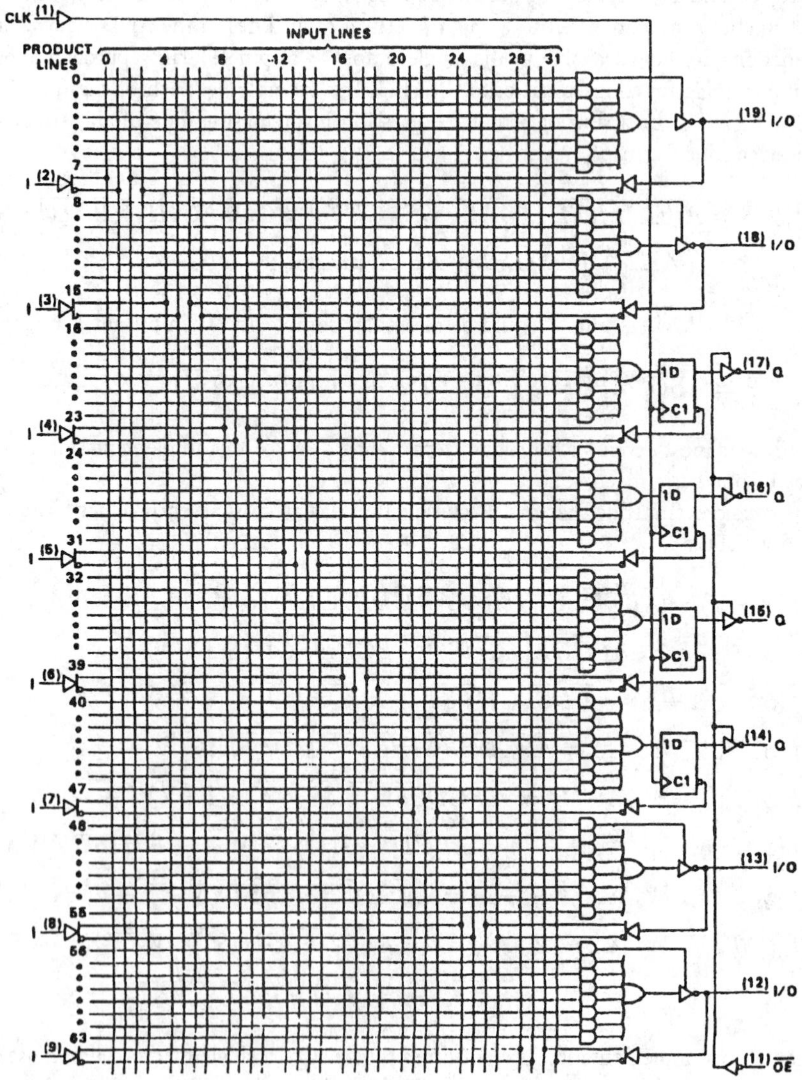

FIGURE 63.11 Logic diagram of the PAL16R4 chip.

Using our parameter system, the PAL16R4 has $n = 8$, $m = 4$, $k = 4$, $p = 7$ for combinational outputs and $p = 8$ for registered outputs.

Because the four registered outputs are also connected back into the AND array, this chip can implement a sequential circuit, with the registered outputs serving as state variables. Therefore, this chip can implement any eight-input, four-output, sequential circuit that needs no more than four-state variables (16 states) and no more than seven product terms for each output or eight product terms for each state variable. Separate pins provide an enable for the state variables (pin 11) and a clock for the flip flops (pin 1). Thus, the state variables are also available at output pins.

By an analysis similar to that used in the previous section, we can determine the different ways in which the product terms can be distributed among the combinational output and state variable equations. Table 63.8 shows the range of basic possibilities. For example, the table indicates that the PAL16R4 chip can implement any single output, eight-input, sequential circuit that requires no more than four state variables (16 states) and in which the available product terms may be divided among the outputs and state variables in seven different distributions. The notation (7)-(8-8-8-26) means that the single output can have up to seven product terms, that one state variable can have up to 26 product terms, and that the other three state variables can have up to eight product terms each.

Designing with Programmable Array Logic Devices with Programmable Macro-Cell Outputs

The PAL16R4 chip has limited application potential because the outputs from pins 14 to 17 *must* be registered. Most newer chips allow a user to decide whether to have registered or combinational outputs at each pin and also allow the user to select either active high or active low outputs.

The Intel 5C031 chip demonstrates this additional flexibility. Figure 63.12 shows the architecture of the 5C031 chip. Unfortunately, Intel does not use the conventional naming conventions. If it did, the name of this chip would be PAL18V8 because 18 pairs of inputs go to the AND array and the eight outputs are under the control of the architecture control block, i.e., the outputs are Versatile. Each of eight macro-cells contains a normal PAL AND array (labeled PLA array in Fig. 63.12) and an I/O architecture control block. Each PAL AND array provides eight product terms permanently connected as inputs to an OR gate and an additional product term

TABLE 63.8 Range of Basic Implementations of the PAL16R4 Chip

No. of Outputs	No. of Inputs	No. of State Variables	No. of Product Terms per (Combinational Output)-(State Variable)
1	11	4	(7)-(8-8-8-8)
1	10	4	(7)-(8-8-8-14), (13)-(8-8-8-8)
1	9	4	(7)-(8-8-8-20), (7)-(8-8-14-14), (13)-(8-8-8-14), (19)-(8-8-8-8)
1	8	4	(7)-(8-8-8-26), (7)-(8-8-14-20), (7)-(8-14-14-14), (13)-(8-8-8-20), (13)-(8-8-14-14), (19)-(8-8-8-14), (25)-(8-8-8-8)
2	10	4	(7-7)-(8-8-8-8)
2	9	4	(7-7)-(8-8-8-14), (7-13)-(8-8-8-8)
2	8	4	(7-7)-(8-8-8-20), (7-7)-(8-8-14-14), (7-13)-(8-8-8-14), (7-19)-(8-8-8-8), (13-13)-(8-8-8-8)
3	9	4	(7-7-7)-(8-8-8-8)
3	8	4	(7-7-7)-(8-8-8-14), (7-7-13)-(8-8-8-8)
4	8	4	(7-7-7-7)-(8-8-8-8)

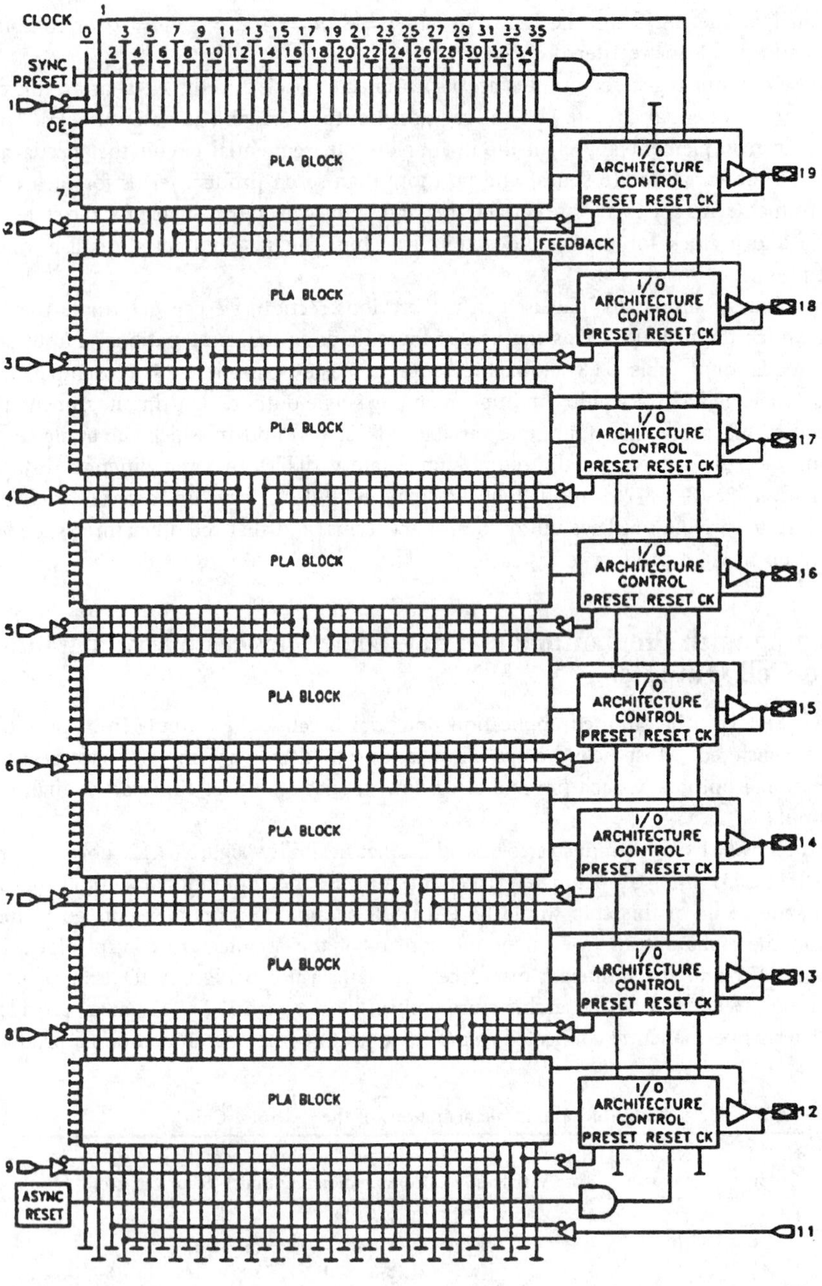

FIGURE 63.12 Architecture of the Intel 5C031. (Courtesy of Intel Corporation.)

that enables the output buffer. Figure 63.13 shows that this array is similar in form to those for PAL chips described earlier in this chapter. Figure 63.12 shows that 18 vertical pairs of input lines go to the AND array. Ten of these pairs are connected directly to input pins 1 to 9 and pin 11. The other eight pairs are feedback lines from the architecture control blocks of the eight macro-cells. Each macro-cell is associated with a bidirectional pin (pins 12 to 19) that can be used as an input pin, as an output pin, or as a bidirectional bus pin. If used as a bidirectional bus pin, the designer must control the output enable using a product term from the AND array.

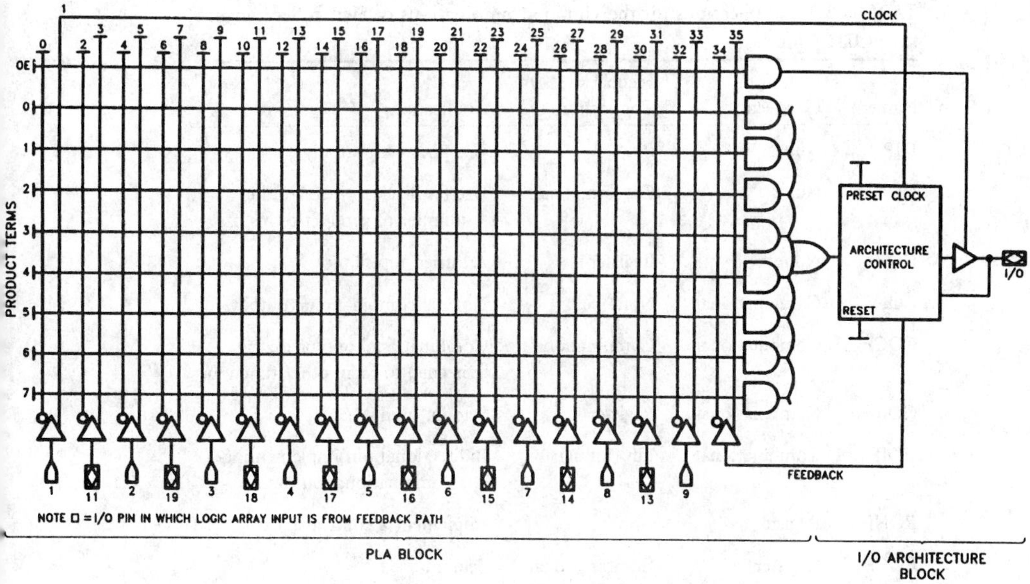

FIGURE 63.13 Typical AND array for Intel 5C031 chip. (Courtesy of Intel Corporation.)

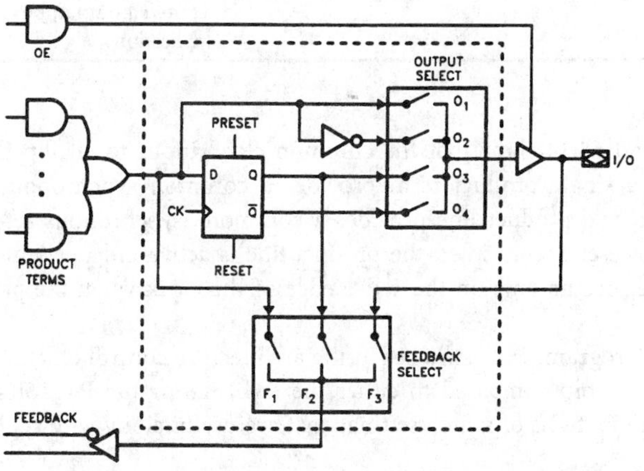

FIGURE 63.14 Macro-cell architecture of the 5C031 chip. (Courtesy of Intel Corporation.)

The architecture control block in each macro-cell provides designer control over the signal that is connected to the bidirectional pin and feedback line associated with that macro-cell. Figure 63.14 shows that the architecture control block contains a *D* flip-flop, an inverter, two sets of programmable switches, and an output buffer with an enable. The signal controlled by the buffer enable (OE) can be either the direct output of the combinational AND array (either active high or active low) or the data value stored in the *D* flip-flop (either active high or active low). If the enable is active, the buffer steers the selected signal to the pin. An inactive enable causes the buffer to enter the high impedance state, which effectively disconnects the buffer from the pin. The pin can, therefore, be used as a bus or as an input.

The feedback signal can be the combinational output of the AND array, the data value stored in the flip flop, or the pin signal. Therefore, the feedback line can be used to expand the number of product terms, to produce a state variable for a sequential circuit, to provide an additional input for the chip, or to implement a bidirectional pin.

TABLE 63.9 Applications for the Unique Combinations of Switch Settings in the 5C031 Chip

Name	Output Connection	Feedback Connection	Application
INP	None	Pin	Use pin as input only
NOCF	None	Combinational	Use feedback to increase the number of product terms
NORF	None	Register	A buried flip flop
CONF	Combinational	None	Normal combinational output
COCF	Combinational	Combinational	Combinational output that is needed to form other functions
CORF	Combinational	Register	Seldom used
COIF	Combinational	Pin (input)	Bidirectional pin implementing a combinational output
RONF	Register	None	Implement a control signal
ROCF	Register	Combinational	Seldom used
RORF	Register	Register	Typical state machine
ROIF	Register	Pin (input)	Bidirectional pin with registered output; bus applications

Figures 63.12 and 63.14 show that the common clock input to all flip flops comes directly from pin 1, that a single product term provides a common synchronous preset for all flip flops, and that a single product line provides a common asynchronous reset for all flip flops. The asynchronous reset occurs when the product line is active independent of clock state. The synchronous preset occurs only on the active edge of the clock when the preset product line is active.

The two sets of programmable switches in the architecture control block significantly increase the flexibility of the chip compared to either the PAL16L8 or the PAL16R4. Table 63.9 shows the 11 unique combinations of switch settings that are useful along with typical applications for each setting.

If each macro-cell is programmed to be COIF, then the 5C031 is pin compatible with the PAL16L8 and has combinational logic functional capabilities similar to those of Table 63.6 and 63.7, except that the 5C031 has eight product terms per macro-cell instead of seven. As a result, all of the numbers in the product terms section of the tables will be increased slightly. The number of inputs and outputs are the same in all cases except for the single output case in which the 5C031 has one more input than the PAL16L8. The extra input arises because pin 12 has no feedback connection in the PAL16L8 and it does in the 5C031. For example, the 5C031 can implement any single output combinational logic function with at most 13 inputs (instead of 12) that requires no more than 36 product terms (instead of 31) and can implement any three-output combinational logic function with at most ten inputs in which the number of product terms per output is no more than 15, 22, and 22, respectively, instead of 13-19-19.

Similarly, by programming four of the 5C031 marco-cells as RORF and four as COIF, the 5C031 will emulate the PAL16R4 chip. Therefore, Table 63.8 describes possible 5C031 implementations of sequential circuits with exactly four state variables if the number of product terms for combinational outputs is increased to account for the fact that the 5C031 has eight

TABLE 63.10 Sequential Circuits that Fit onto the 5C031 Chip

$k+m$	n	No. of Product Terms per Output or State Variable
2	15	8-8
2	14	8-15
2	13	8-22, 15-15
2	12	8-29, 15-22
2	11	8-36, 15-29, 22-22
2	10	8-43, 15-36, 22-29
2	9	8-50, 15-43, 22-36, 29-29
3	14	8-8-8
3	13	8-8-15
3	12	8-8-22, 8-15-15
3	11	8-8-29, 8-15-22, 15-15-15
3	10	8-8-36, 8-15-29, 8-22-22, 15-15-22
3	9	8-8-43, 8-15-36, 8-22-29, 15-15-29, 15-22-22
4	13	8-8-8-8
4	12	8-8-8-15
4	11	8-8-8-22, 8-8-15-15
4	10	8-8-8-29, 8-8-15-22, 8-15-15-15
4	9	8-8-8-36, 8-8-15-29, 8-8-22-22, 8-15-15-22, 15-15-15-15
5	12	8-8-8-8-8
5	11	8-8-8-8-15
5	10	8-8-8-8-22, 8-8-8-15-15
5	9	8-8-8-8-29, 8-8-8-15-22, 8-8-15-15-15
6	11	8-8-8-8-8-8
6	10	8-8-8-8-8-15
6	9	8-8-8-8-8-22, 8-8-8-8-15-15
7	10	8-8-8-8-8-8-8
7	9	8-8-8-8-8-8-15
8	9	8-8-8-8-8-8-8-8

product terms per cell instead of seven. For example, the 5C031 can implement any sequential circuit with 2 outputs, up to 8 inputs, up to 15 product terms per output (instead of 13) and up to 8 product terms per state variable.

The 5C031 is much more versatile than either the PAL16L8 or the PAL16R4. Because pins 12 to 19 can be used as inputs, combinational outputs, or registered outputs, the following inequalities describe the possibilities.

$$k, m \geq 0$$
$$1 \leq k + m \leq 8$$
$$n \leq 17 - (k + m)$$

Any pin that is not used as an input, output, or state variable may be used to increase the number of product terms available for the outputs or state variables. Table 63.10 shows how these limits affect implementations of sequential circuits. For example, the 5C031 chip can implement any 9-input ($n = 9$) sequential circuit with 1 output ($m = 1$), 3 state variables ($k = 3$) (up to 8 states), that requires no more than 15 product terms for any output or state variable.

Programmable Logic Device Development Systems

From the previous discussion, it is clear that fitting designs to PLD chips is a complex process. Manufacturers of PLDs and third-party vendors market software packages that help engineers map designs onto chips. Selecting a package appropriate for a particular design environment is a critical decision that will significantly affect the productivity of the design group.

1. Design entry
2. Behavioral simulation
3. PLD synthesis
4. Device simulation
5. Compilation
6. Programming the PLD chip
7. Testing the programmed chip

FIGURE 63.15 Design steps supported by PLD development systems.

Three types of PLD development systems are available: user-designed packages, vendor-designed packages, and universal packages. Because these programs are very complex and require many years of effort to develop, most design groups lack the time and resources to develop their own. Many vendors provide design aids that are specific to a particular product line. Great danger is inherent in becoming dependent upon one vendor's products because new products in this field appear frequently. Clearly, a universal design package that supports a wide variety of product lines is most desirable. A variety of PLD development systems with different features, capabilities, and prices are available.

Figure 63.15 shows the basic design steps supported by PLD development systems. **Design entry** refers to the use of an editor to create a source file that specifies the *functional behavior* of the device. Some development systems support *behavioral simulation* at this level. The step **PLD synthesis** refers to the process of mapping the design to specific PLD chips after verification of correct functional behavior. Most PLD development systems support *device simulation* at this level to verify that the PLD design functions correctly. The user then *compiles* the design into a standard format language to provide information to the programming device. Next, the PLD programmer reads the compiled design and programs the PLD chip. After programming, the chip needs to be tested to verify correct programming. Each of these steps are now discussed in more detail.

Design Entry

It is essential for a PLD development system to have a variety of design entry modes. Many vendors market the more complex design entry modes as optional features. This section describes some of the more common design entry modes and their value to PLD designers.

Equations are the most frequently used method of design entry. A PLD design system must support Boolean equation entry because the AND-OR arrays on PLD chips directly implement Boolean equations. Almost all PLD designers will use Boolean equation entry extensively. Some PLD tools support other operators, such as arithmetic or relational operators. If these are provided as part of a behavioral description language in combination with automated synthesis tools, they can be very useful. In isolation their value for PLD design is minimal because most PLD chips cannot directly implement these higher-level operations.

Truth table entry allows specification of a combinational logic function by defining the output for each of the 2^n input combinations. This form is particularly valuable if *don't care* entries exist. Truth table entry is most commonly used for functions with a small number of input variables that are not easily described by Boolean equations. Code converters, decoders, and lookup tables are examples. A good PLD design tool will support truth table entry.

Symbolic state machine entry is crucial for PLD designers because PLD chips with registered outputs are often used to implement state machine controllers. Current tools have features described as state machine entry that vary dramatically in form and usefulness. Before selecting a tool, the specifics of the state machine entry format should be carefully investigated. The most useful formats allow symbolic representation of states and specification of state transitions using some form of conditional statement such as *if_then_else*, or *case* statements. Relational operators are also useful in this context. The tool should perform automated state assignment and should fit the state variable equations to the target PLD chip.

State diagrams using graphics are useful, but not essential for PLD design. This feature is mainly a convenience, provided that symbolic state machine entry is available.

Schematic entry is a widely accepted way to describe logic systems. It us useful for PLD development only if it is combined with a powerful partitioning and mapping tool that can fit the circuit onto PLD chips. The reason is that most design paradigms that produce schematic diagrams use Boolean equations to specify the schematic. The Boolean equations could be input directly to the PLD development system, omitting the derivation of a schematic from the equations. Schematic entry is useful to convert existing gate-level designs into PLD implementations.

Hardware description language (HDL) entry is potentially the most useful of all methods. This entry method is not currently available in most PLD packages. However, HDL development systems often have PLD options. These options are fairly new and should be carefully examined for maturity before investing in such a system because the cost of an HDL development system is much more than the cost of a simple PLD development system. An HDL development system that adequately supports PLD design provides great flexibility for design entry. The better systems can automatically map a high-level behavioral language representation, such as VHDL or Verilog, directly onto a variety of PLDs.

Synthesis

Programmable logic device synthesis is the process of transforming a given description of a device into an equivalent description that is suitable for a PLD implementation. For example, the process of transforming a symbolic state machine description into a set of Boolean equations that fit onto a specific PLD chip is an example of PLD synthesis. The power of the synthesis algorithms in a PLD development system is perhaps the most important feature of the system. The following processes are essential in a PLD development system.

Logic minimization is obviously an essential process in a PLD development system because the number of product terms per output gate on PAL chips is severely limited. The goal of logic minimization for PLD designs is to reduce the number of product terms, not the size of the product terms. Classical logic minimization algorithms use cost functions that reward reduction of the number of gate inputs. This is important for TTL gate implementations, for example, because an eight-input gate costs about four times as much as a four-input gate. In PLD designs the number of gate inputs does not matter. Each product term consumes one product line in the chip. A one-literal product term, such as X, costs exactly the same as a ten-literal product term, such as $A\overline{B}CD\overline{E}F\overline{G}HIJ$. Therefore, traditional logic minimization programs, such as Espresso, need to be modified for PLD development. If product terms can be shared among different outputs, then multiple output minimization is necessary. However, for most PAL devices, the product terms cannot be shared, therefore, single-output minimization algorithms are sufficient. Single-output minimization algorithms are much less complex and take much less time to execute than multiple output minimization algorithms. Therefore, systems that do single-output minimization result in higher productivity. Therefore, one must be careful of systems that advertise well-known traditional logic minimization algorithms to market their products, especially if multiple-output minimization is stressed.

Equation factoring, which is sometimes called *multiple-level minimization,* is essential in order to fit large functions onto PLD chips using multiple cells combined with feedback lines inside the chips. This feature is missing from most vendor PLD development systems. However, in order to provide effective automated PLD synthesis, this operation is absolutely necessary. In most current PLD development systems the designer must interact with the synthesis program to implement multiple-level minimization. Such interaction requires extensive skill from the user of the software package.

Device fitting algorithms are needed to map designs to a single PLD chip. Such algorithms often try both product of sums and sum of product forms, or both true and complemented signal forms, for example. These algorithms also attempt to make use of internal feedback paths to increase the number of product terms available in an output function.

If a design requires more than one PLD chip, then a *partitioning algorithm* must be able to decompose the design into blocks that can map to individual PLD chips. Unfortunately, efficient partitioning algorithms are in the realm of current research. Look for them to appear soon.

Hardware description language synthesis is another goal of current research. A tool that can convert a high-level behavioral description, in VHDL, for example, into a PLD implementation would be very valuable. Such tools are currently under development. Initial products are just starting to appear. Look for these to improve dramatically in the near future.

Simulation of Programmable Logic Device Designs

All good PLD development systems include some form of simulation capability. The simulators vary widely in scope, user interface, and general usefulness.

Behavioral simulation allows high-level design descriptions to be simulated independent of implementation. Behavioral simulators verify the I/O behavior of the device. Correct behavioral simulation verifies the correctness of the algorithms prior to mapping the design to specific hardware components. Only top of the line development systems currently support simulation at this level.

Device simulators verify the function of the design after mapping the design to a specific PLD chip, but before actually programming the chip. This is the most common type of simulator in current PLD development systems. A device simulator will construct a software model of the target PLD architecture, map the design to that architecture, then simulate the behavior of the specific PLD. The better simulators will provide timing information as well as functional information.

Generating Test Vectors for Programmable Logic Device Designs

It is very difficult to decide what test vectors to apply to a device in order to verify that it is functioning correctly. Many development tools provide software to help generate test vectors for the device. Test vectors can be produced at any level in the design process. It is an open problem as to how to use test vectors derived at one level of representation at a lower level.

Programming Programmable Logic Device Devices

Most PLD development systems interface to a programming device via a standard format file, such as JEDEC or EDIF. The information in these files also can be used to interface to other specialized tools, such as simulators or testers.

Verification of the Programmed Device

After programming, the device needs to be tested for correct performance. This step usually requires a specialized piece of testing equipment that can input test vectors at full operating speed and capture the output sequence for verification. Such equipment is often expensive and specific to a particular line of products.

An Example Design to Illustrate a Programmable Logic Device Development System

To illustrate the use of automated software, Fig. 63.16 shows how a design is represented in the Intel PLDshell/PLDasm development system. The first few lines consists of a header that

```
Logic Optimizing Compiler Utilization Report            pingpong.rpt

; HEADING INFORMATION
;
TITLE         PROJECT 5, FALL 1993
PATTERN       PDS
REVISION      1
AUTHOR        F. GAIL GRAY
COMPANY       VIRGINIA TECH

OPTIONS
  TURBO = OFF
  SECURITY = OFF

CHIP  PONG 5C031

PIN   1   CLK
PIN   2   INIT
PIN   3   DIR
PIN   [4:9]   X[0:5]
PIN   19   TEMP   CMBFBK
PIN   18   Q0
PIN   [17:12]   Z[0:5]

 EQUATIONS

TEMP = /Q0 * /INIT * Z4 * X3
     + /Q0 * /INIT * Z5 * X4
     + /Q0 * /INIT * Z0 * X5
     + /Q0 * /INIT * Z1 * X0
     + /Q0 * /INIT * Z2 * X1
     + /Q0 * /INIT * Z3 * X2
TEMP.TRST = VCC

/Q0.D := /INIT * Z5 * X0 * /TEMP
       + /INIT * Z0 * X1 * /TEMP
       + /INIT * Z1 * X2 * /TEMP
       + /INIT * Z2 * X3 * /TEMP
       + /INIT * Z3 * X4 * /TEMP
       + /INIT * /TEMP * Z4 * X5
       + /Q0 * /INIT * /TEMP
       + INIT * /TEMP * /DIR
 Q0.CLKF = CLK
 Q0.RSTF = GND
 Q0.SETF = GND
 Q0.TRST = VCC

Z0.D := INIT * X0
      + Q0 * Z5 * /INIT
      + /Q0 * Z1 * /INIT
Z0.CLKF = CLK
Z0.RSTF = GND
Z0.SETF = GND
Z0.TRST = VCC

Z1.D := INIT * X1
      + Q0 * Z0 * /INIT
      + /Q0 * Z2 * /INIT
Z1.CLKF = CLK
Z1.RSTF = GND
Z1.SETF = GND
Z1.TRST = VCC

Z2.D := INIT * X2
      + Q0 * Z1 * /INIT
      + /Q0 * Z3 * /INIT
Z2.CLKF = CLK
Z2.RSTF = GND
Z2.SETF = GND
Z2.TRST = VCC

Z3.D := INIT * X3
      + Q0 * Z2 * /INIT
      + /Q0 * Z4 * /INIT
Z3.CLKF = CLK
Z3.RSTF = GND
Z3.SETF = GND
Z3.TRST = VCC
```

FIGURE 63.16 Example of design entry for a PLD development system.

```
Z4.D := INIT * X4
    + Q0 * Z3 * /INIT
    + /Q0 * Z5 * /INIT
Z4.CLKF = CLK
Z4.RSTF = GND
Z4.SETF = GND
Z4.TRST = VCC

Z5.D := INIT * X5
    + Q0 * Z4 * /INIT
    + /Q0 * Z0 * /INIT
Z5.CLKF = CLK
Z5.RSTF = GND
Z5.SETF = GND
Z5.TRST = VCC

SIMULATION

  VECTOR X := [ X5 X4 X3 X2 X1 X0 ]
  VECTOR Z := [ Z5 Z4 Z3 Z2 Z1 Z0 ]
  TRACE_ON CLK INIT DIR X Z Q0
  SETF /CLK INIT /DIR /X5 X4 /X3 /X2 /X1 /X0
  CLOCKF CLK
  SETF /INIT /X5 /X4 X3 /X2 X1 /X0
  FOR J := 1 TO 15 DO
    BEGIN
    CLOCKF CLK
    END
  SETF /CLK INIT DIR /X5 /X4 /X3 /X2 X1 /X0
  CLOCKF CLK
  SETF /INIT /X5 /X4 X3 /X2 /X1 /X0
  CLOCKF CLK
  FOR J := 1 TO 15 DO
    BEGIN
    CLOCKF CLK
    END
  SETF /CLK INIT DIR /X5 /X4 /X3 X2 /X1 /X0
  CLOCKF CLK
  SETF /INIT /X5 X4 /X3 /X2 X1 /X0
  CLOCKF CLK
  FOR J := 1 TO 15 DO
    BEGIN
    CLOCKF CLK
    END
  TRACE_OFF
```

FIGURE 63.16 (continued)

describes general information about the design such as its title, author, and company. In the OPTIONS section both TURBO and SECURITY are turned off. If the TURBO option is on, the device is optimized for maximum speed. Turning TURBO off optimizes the design for minimum power consumption, accompanied by a reduction in speed. Turning SECURITY on prevents the contents of the chip from being read. This feature allows companies to protect their designs from being read easily. The CHIP section contains the target chip specification (5C031) and the definition of PIN signals. The logic of the design is entered in equation format in the EQUATION section.

Note that in order to make the design fit onto an Intel 5C031, the logic equations are factored into a multiple-level format. The subfunction TEMP is factored out of the original sum of products form. This subfunction, along with its complement, is then made available to the flip flop logic equation for Q0.D. The software package is intelligent enough to know that the complement form of Q0 is less complex than the true form. Once the subfunction TEMP is specified, the software package is able to map the function onto the 5C031 chip. However, it cannot find the necessary subfunction TEMP using only the original sum of products form for Q0. In this case, which is typical, the designer must interact with the software to obtain the optimum design.

Finally, the SIMULATION section contains commands to the built-in simulator that will simulate the function of the PLD chip prior to programming the chip. Fig. 63.17 shows the

```
Logic Optimizing Compiler Utilization Report          pingpong.rpt

***** Design implemented successfully

        5C031
        -----
   CLK -| 1   20|- Vcc
   INIT -| 2   19|- TEMP
   DIR -| 3   18|- Q0
    X0 -| 4   17|- Z0
    X1 -| 5   16|- Z1
    X2 -| 6   15|- Z2
    X3 -| 7   14|- Z3
    X4 -| 8   13|- Z4
    X5 -| 9   12|- Z5
   GND -|10   11|- Gnd
        -----

CMOS Device: ground unused inputs and I/Os   Gnd = unused input or I/O pin.
RESERVED = Leave pins unconnected on board.  N.C. = unconnected pins

                                             pingpong.rpt
**OUTPUTS**

  Name Pin Resource MCell PTerms |  Reset  | Preset

  TEMP 19   COCF     1     6/8   |    -    |   -
    Q0 18   RORF     2     8/8   |    -    |   -
    Z0 17   RORF     3     3/8   |    -    |   -
    Z1 16   RORF     4     3/8   |    -    |   -
    Z2 15   RORF     5     3/8   |    -    |   -
    Z3 14   RORF     6     3/8   |    -    |   -
    Z4 13   RORF     7     3/8   |    -    |   -
    Z5 12   RORF     8     3/8   |    -    |   -

                                             pingpong.rpt
**INPUTS**

  Name Pin Resource MCell PTerms |  Reset  | Preset

  INIT  2   INP       -      -   |    -    |   -
   DIR  3   INP       -      -   |    -    |   -
    X0  4   INP       -      -   |    -    |   -
    X1  5   INP       -      -   |    -    |   -
    X2  6   INP       -      -   |    -    |   -
    X3  7   INP       -      -   |    -    |   -
    X4  8   INP       -      -   |    -    |   -
    X5  9   INP       -      -   |    -    |   -
   CLK  1   INP       -      -   |    -    |   -

**UNUSED RESOURCES**

   Name Pin Resource MCell PTerms

    -   11   INPUT    -     -

**PART UTILIZATION**

8/8 MacroCells (100%), 50% of used Pterms Filled
9/10 Input Pins (90%)
   PTerms Used 50%

**RESOURCE MNEMONICS**
INP  = Pin Input to Logic Array
COCF = Comb.      pin Output, Comb.     Feedback
RORF = D-Register pin Output, Register  Feedback

Macrocell Interconnection Cross Reference              pingpong.rpt
```

FIGURE 63.17 Report file generated by software package.

```
FEEDBACKS:            M M M M M M M M
                      0 0 0 0 0 0 0 0
                      1 2 3 4 5 6 7 8
TEMP . COCF @M1 -> .  *  .  .  .  .  .  . @19
Q0 ....... RORF @M2 -> *  *  *  *  *  *  *  * @18
Z0 ....... RORF @M3 -> *  *  .  *  .  .  .  * @17
Z1 ....... RORF @M4 -> *  *  *  .  *  .  .  . @16
Z2 ....... RORF @M5 -> *  *  .  *  .  *  .  . @15
Z3 ....... RORF @M6 -> *  *  .  .  *  .  *  . @14
Z4 ....... RORF @M7 -> *  *  .  .  .  *  .  * @13
Z5 ....... RORF @M8 -> *  *  *  .  .  .  *  . @12


INPUTS:

CLK ..... INP  @1  ->  .  *  *  *  *  *  *  *
INIT ..... INP @2  ->  *  *  *  *  *  *  *  *
DIR ...... INP @3  ->  .  *  .  .  .  .  .  .
X0 ....... INP @4  ->  *  *  *  .  .  .  .  .
X1 ....... INP @5  ->  *  *  .  *  .  .  .  .
X2 ....... INP @6  ->  *  *  .  .  *  .  .  .
X3 ....... INP @7  ->  *  *  .  .  .  *  .  .
X4 ....... INP @8  ->  *  *  .  .  .  .  *  .
X5 ....... INP @9  ->  *  *  .  .  .  .  .  *
                       T Q Z Z Z Z Z Z
                       E 0 0 1 2 3 4 5
                       M
                       P

. = not connected      x = no connection possible
* = signal feeds cell  ? = error, unable to fit
```

FIGURE 63.17 (continued)

report file generated by the software package. The PIN assignments are displayed, followed by a list of the configuration mode for each macro-cell. Utilization information indicates what resources are used and what resources are still free. The number of unused product terms in each macro-cell is also provided. Finally, the interconnection cross-reference section lists the dependencies of each signal in the array. This information is useful in debugging the circuit and in predicting the effects of input and variable changes on outputs and state variables.

References

[1] *Programmable Logic Data Book*, Dallas, TX: Texas Instruments.

[2] *Programmable Logic*, Mt. Prospect, IL: Intel Corporation.

[3] C. Alford, *Programmable Logic Designer's Guide*, Indianapolis, IN: Howard W. Sams & Company, 1989.

[4] H. Katz, *Contemporary Logic Design*, Redwood City, CA: Benjamin/Cummings, 1984.

[5] L. Pappas, *Digital Design*, St. Paul, MN: West Publishing, 1994.

[6] D. Pellerin and M. Holley, *Practical Design Using Programmable Logic*, Englewood Cliffs, NJ: Prentice-Hall, 1991.

[7] J. F. Wakerly, *Digital Design, Principles & Practices*, 2nd ed. Englewood Cliffs, NJ: Prentice-Hall, 1994.

[8] *PLDshell/PLDasm User's Guide, V2.1*, Santa Clara, CA: Intel Corporation, 1992.

53.2 Clocking Schemes

Wayne D. Grover

Introduction

Advances in VLSI processing technology, particularly CMOS, have resulted in submicron processes with roughly fourfold increases in circuit densities and two- to threefold increases in speed over the preceding 1.2 μm processes. Consequently, CMOS applications at speeds of 200 MHz are now feasible. New design challenges must be mastered to realize systems at these speeds. In particular, clocking-related issues of skew, delay, power dissipation, and switching noise are potentially design-limiting factors. For example, at 200 MHz, 1 ns of skew represents 20% of the clock period. Yet to get the most advantage from available process speed we would like to keep skew under 10% of the clock period, thus setting a target of 500 ps of skew for dies that may be as large as 2 cm on a side.

In large synchronous designs the clock net is also typically the largest contributor to on-chip power dissipation and electrical noise generation, particularly "ground bounce", which reduces noise margin. Ground bounce is a rise in ground potential due to surges of current returning through a nonzero (typically inductive) ground path impedance. A typical CMOS flip flop consumes 12 μW/MHz. At 200 MHz a system with 2 K flip flops will consume 4.8 W for clocking alone. At the board and shelf level, clock distribution networks also can be a source of electromagnetic emissions, and may require considerable delay tuning for optimization of the clock distribution network.

In the past multiphase clocking schemes and dynamic logic structures helped minimize transistor count, but this is now less important than achieving low skew, enhancing routability, controlling clock-related switching noise, and providing effective CAD tools for clock net synthesis and documentation. For these reasons, a shift has occurred toward single-phase clocking and fully static logic in all but the largest custom designs today. In addition, phase-feedback control schemes using phase-locked loops (PLLs) are becoming common, as are algorithmic clock-tree synthesis methods.

This chapter focuses on the issues and alternatives for on-chip and multichip clocking, with the primary emphasis on CMOS technology. We first review the fundamental nature and sources of skew and the requirements for the clocking of storage elements. We then outline and compare a number of "open-loop" clock distribution approaches, such as the single-buffer, clock trunk, clock ring, H-tree, and balanced clock tree approaches. Phase-locked loop synchronization methods and PLL-based clock generation are then outlined. In closing, we look at future technologies and developments for high-speed clocking. The concepts and methods of this chapter apply to many circuit technologies on- and off-chip. However, we emphasize CMOS because CMOS processes (including bi-CMOS) presently represent the vast majority of digital VLSI designs and are expected to do so into the next century.

Asynchronous, self-timed, and wavefront array systems are outside the scope of this chapter. These approaches aim to minimize the need for low-skew synchronous clocking. However, truly asynchronous modules tend to require a large overhead in logic for interaction with each other, so that speed, size, and power often suffer relative to synchronous design. Nonetheless, self-timing can be an effective approach for random-access memory and read-only memory (RAM and ROM) cells, to which considerable optimization effort can be invested for reuse in many designs. Self-timed methods should be considered the alternative to fully synchronous design in large, highly modularized systems, particularly where well-defined autonomous modules have relatively infrequent interactions. The main issues in self-timed systems are the possibly high delay required to avoid metastability problems between self-timed modules, and the circuit costs of the synchronization protocol for intermodule communication.

Clocking Principles

Most of us accept the clocked nature of digital systems without question, but what, fundamentally, is the reason for clocking? Any digital system can be viewed either as a pipeline or as a finite state machine (FSM) architecture, as outlined in Fig. 63.18. In the pipelined architecture clocked sections are cascaded, each section comprising an asynchronous combinational logic block followed by a latch or storage element that samples and holds the logic state at the clock instant. In the FSM the only difference is that the next state input and the system outputs are determined by the asynchronous logic block, and the sampled next state value(s) are fed back into the combinational logic. The FSM can therefore be conceptually unfolded and also represented in a pipeline fashion. The fundamental reason for clocking digital systems is seen in this pipelined abstraction of a digital system: it is to bring together, and retain coordination amongst, asynchronously evolved intermediate results. With physical delays that are temperature, process, and input dependent in combinational logic, we need to create *agreed-upon time instants* at which all analog voltages in a system are valid when interpreted as boolean logic states. Clocking deals with delay uncertainty in logic circuit paths by holding up the fast signals and waiting for slower signals so that both are valid before they are again combined or interact with each other. Without this coordination, purely asynchronous logic would develop severe propagation path differences, and be slow in repetitive operations. Ultimately, a valid state would evolve, but all inputs would have to be stable for the entire time required for this evolution. On the other hand, when the overall combinational logic function is appropriately partitioned between clocked storage latches, system speed can approach the limit given by the delay of a single gate because each logic sub-block is reused in each clock period.

From this we obtain several insights: (1) only the storage elements of a digital system need become loads on the clock net (assuming static logic gates); (2) the system cannot be clocked faster than the rate set by the slowest combinational signal path delay between clocked storage elements; (3) any uncertainty in clock timing (skew) is indistinguishable from uncertainty in the settling time of the intervening combinational logic, (4) for a logic family to work, its storage elements must: (i) at no time be transparent (i.e., simultaneously connect input to output), (ii) have a setup time less than $(T - t_{\text{clk}-Q})$ where T is the clock period, and $t_{\text{clk}-Q}$ is the clock-to-Q output delay of the same type of flop, (iii) have a hold time less than their clock-to-output delay. The last points may be better appreciated by considering that the combinational logic block may be null, i.e., a zero-delay wire, such as in a shift-register.

An implication of [4(i)] is that two-phase nonoverlapping clocks, or an equivalent sequencing process, is fundamental for the storage elements of a digital system. This may sound unusual to readers who have already designed entire systems with SSI and MSI parts, or in gate-array design systems, without having seen anything but single-phase edge-triggered flip flops, latches, counters, etc. However, at least two clock phases (or clock-enabling phases) are internally required in any clocked storage device. An analogy is of a ship descending in elevation through a lock [27]. During the first clock phase sluice gates "charge" the lock up to the incoming water level and open the input gate to bring a ship in. Throughout this phase it is essential that the output gate is closed, or water will race destructively right though the lock. Only when the ship

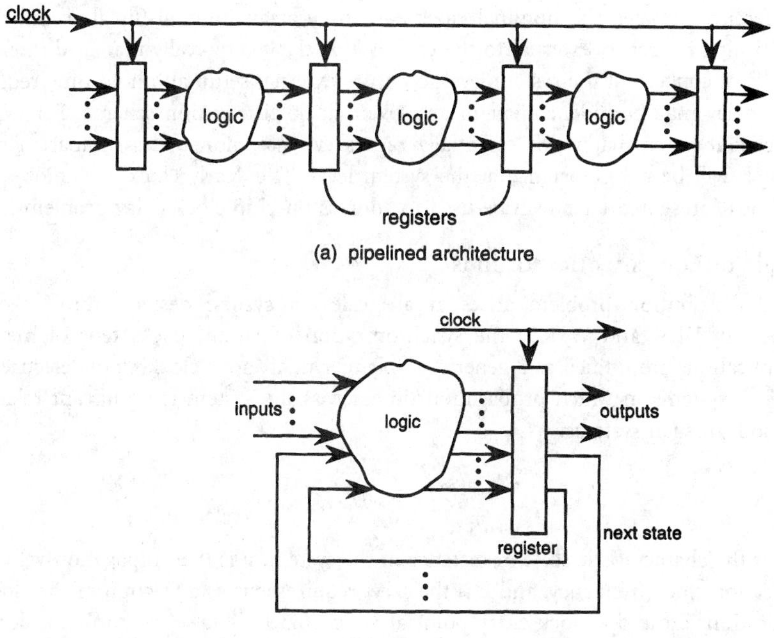

FIGURE 63.18 Architecture of digital systems.

is entirely in the lock and the input gate is closed (isolating the input) can the output sluice gates be opened to equalize the water level to that on the outgoing side, allowing the ship to leave (producing a new output). Similarly, in a flip flop or latch the currently stored value, which appears at the output, must be isolated from the input while the input evolves to its next value.

Skew and Delay

Clock *skew* is defined most generally as the difference in time between the actual and the desired instant of the active clock edge at a given clocked storage element. In the majority of designs in which the desired instant of clocking is the same at all storage elements, skew is the maximum difference in clock waveform timing at different latches. Clock skew is of concern because it ultimately leads to the violation of setup or hold times within latches, or to clock race problems in multiphase clocking. Furthermore, from a design viewpoint, uncertainty in clock timing must be treated as equivalent to actual clock skew. Skew or timing uncertainty are therefore equivalent to an increase in critical path logic delay. In either case the clock period must be extended to ensure valid logic levels and proper setup/hold requirements relative to the clock time.

To illustrate the equivalence of skew (either actual or design timing uncertainty) to a loss of system speed, consider a 200-MHz process used in a design which has 25% skew, (i.e., actual clock edge dispersion, or, equivalently, uncertainty in clock timing is 1.25 ns). If a competitor uses the same process at 200 MHz and achieves 5% skew (0.25 ns), then the latter design has 20% more of each 5 ns clock cycle for settling combinational logic paths. Alternatively, for the same logic functions, the low-skew system could be clocked at 250 MHz with the same timing margins as the high-skew system. Skew, therefore, represents a loss of performance relative to basic process capabilities developed at great expense. However, skew reduction costs relatively little, and is in the logic designer's control, not the process developer's, and yet it is directly equivalent to a basic enhancement in process speed.

Skew is usually of primary concern on-chip, or within any module that is designed on the presumption of a uniform clock phase throughout the module. Clock *delay*, on the other hand, is the difference between the nominal clock edge time at an internal flip flop and the system clock, or timing reference, external to the chip. While skew is typically managed internal to the die, delay is of concern at the system level to ensure external setup and hold time requirements. Skew and delay may be independent in any given clock distribution scheme. For example, an on-chip clocking tree that yields essentially zero skew may, nonetheless, impart a high clock delay, which will be of importance at the system level. The "early clock" technique and some PLL methods (presented later) can be used to address on-chip clock delay problems.

Isochronic or Equipotential Regions

The clock distribution problem arises at all scales of system design, from on-chip clock distribution in VLSI and WSI to the synchronization of circuit packs tens of meters apart. These applications are unified as a generic problem: synchronous clocking of "electrically large" systems, i.e., systems in which propagation time across the system is significant relative to the clock period. In such systems:

$$D/v > k/f_{\mathrm{app}} \tag{63.1}$$

where D is the characteristic scale or distance of the system, v is the propagation velocity, f_{app} is the application clock frequency, and k is the skew requirement as a fraction of the clock period. For all locations around a clock entry point at which (63.1) is false, we can consider events to be essentially simultaneous (or, equivalently, the region is a single electrical node), and the clock can be distributed within such regions without delay equalization. The region over which clock can be distributed without any significant skew is also known as an equipotential [27] region, or an isochronic region [1]. In this chapter we are concerned only with cases in which (63.1) is true, but it is implicit that the clocked end nodes may be either clocked loads directly or a buffer that feeds a local isochronic region.

The diameter (and shape) of an isochronic region on-chip depends on the wire type employed for interconnection. To control skew on chip, we need to consider delay differences due both to wire lengths and to the lumped capacitive effects of the driven loads. Where the RC time constant of the wiring interconnect τ_w is much less than the RC combination of the driving source resistance and the lumped capacitance of N clocked loads on the net ($\tau_{\mathrm{net}} = R_s C_{\mathrm{gate}} N = N\tau_g$), we can consider all points on a net to be isochronic, meaning that the net acts like one electrical node characterized by the total lumped capacitance of gates on the net. Wires on a chip are often modeled as distributed R_0C_0 sections, where R_0 and C_0 are the resistance and capacitance per unit length, respectively (see Fig. 63.19). In such cases the propagation delay for a wire of length l follows the diffusion equation [31]

$$\tau_w = R_0 C_0 l^2/2 \tag{63.2}$$

Therefore, if we consider a net of length l, with N standard loads, we can consider the net to be isochronic if $\tau_w \ll N\tau_g$. From (63.2) this implies:

$$l \ll \sqrt{\frac{2NR_s C_{\mathrm{gate}}}{R_0 C_0}} \tag{63.3}$$

This relationship provides a guideline for the maximum length over which wire delays may be neglected relative to gate-charging delays. Based on typical values for a 1-μm process, $\tau_g < 500$ ps, isochronic regions for lightly loaded lines ($N = 1$) are up to 10,000 λ for lines in third layer metal, 5000 and 8000 λ for first and second layer metal, respectively, and 200 λ for polysilicon wires, where λ is the minimum feature size of the process [31]. This illustrates

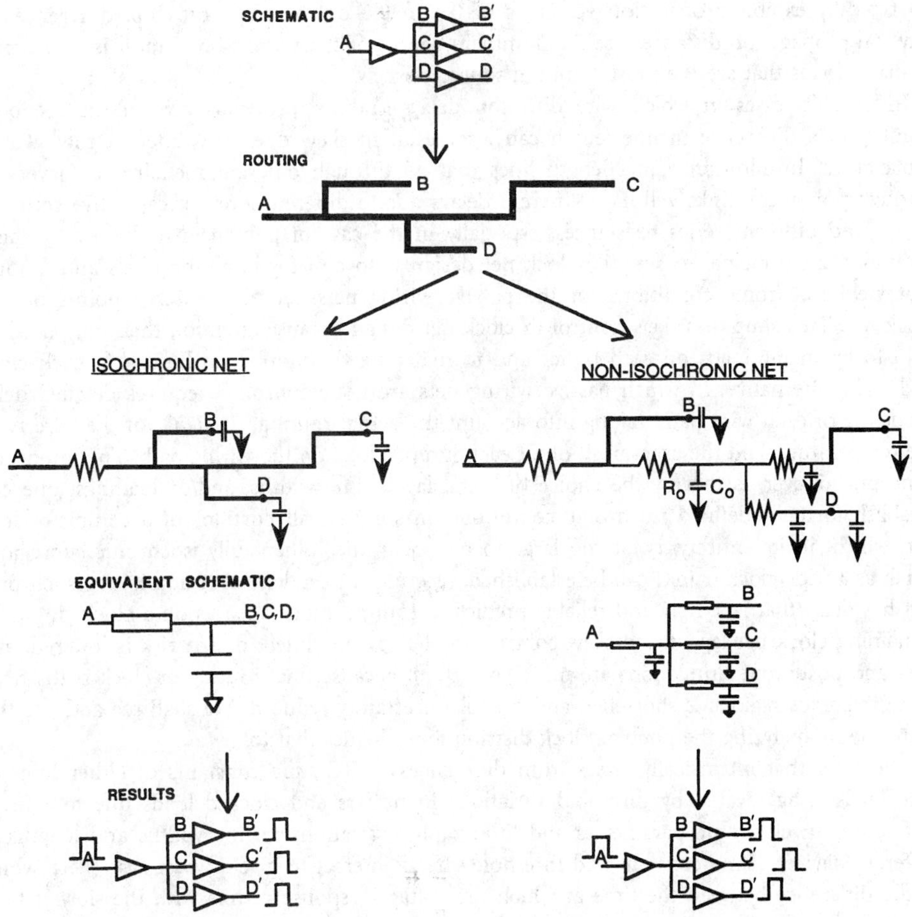

FIGURE 63.19 Isochronic and nonisochronic passive wiring nets.

the importance of distributing clock within metal layers to the greatest extent possible. Even a few short polysilicon links may introduce sufficient series resistance to drastically reduce the isochronic region for the given clock line. This also illustrates that if clock is distributed in metal layers, and is always buffered before exceeding the isochronic distance, it will be primarily differences in lumped capacitive loads and not wire lengths that determine clock signal delays, and hence relative clock skews.

Nature of Skew On-Chip

The concept of isochronic regions helps us understand the nature of clock skews in VLSI and helps explain why on-chip skews may be greater than those between off-chip points that are physically many more times distant. A key realization is that signals do not propagate at the "speed of light". If they did, absolute delays across even the largest chips (2 cm edges) would be subnanosecond and the isochronic diameter would easily encompass an entire die at clock speeds up to 200 MHz. Rather, on-chip propagation delay depends much more on the time needed for output drivers to charge the total lumped capacitance associated with all the gate inputs of the driven net. In other words, fanout and driver current abilities have more to do with delay than path lengths. This is especially true when clock distribution is exclusively via metal layers, as is the norm in a modern design. On the other hand, off-chip, we route signals via impedance-controlled coaxial cables or microstrip lines, or via optical fiber, and these media do typically exhibit propagation velocities of 0.6 to 0.8 c. Therefore, off-chip, differences in physical propagation distances are the dominant source of skew, while on-chip, it is imbalances in driver loads that are the most common source of skew.

In on-chip cases in which wire diffusion delays and lumped capacitive effects are both significant, a difference in line length can also result in skew due to a different total wiring capacitance. In addition, equal length lines that go through different metallization layers or through polysilicon links will have different delays due to different levels of capacitive coupling to V_{SS} and different series resistances, especially in the case of polysilicon links. Accordingly, an important principal to simplify clock net design is to aim for buffering levels and fanouts that yield isochronic conditions for the passive wiring nets between buffered points on the clock net. This simplifies skew control in clock net design because attention then only need be paid to balancing loads on each buffer and to matching the number of buffers in each clock path. The alternative, in which passive wiring nets are not isochronic, requires detailed delay modeling of each wire path, taking into account the actual routing, the R_0C_0 of the wire type, the temperature, and the exact position of each lumped load on the wiring path. The important concept, however, is that by the choice of metal layers, line widths, and/or loadings, one can establish formally defined isochronic conditions on some or all portions of a complete clock net, which, in its entirety, is far too large to be isochronic. When fully isochronic subregions (such as a wide clock trunk) can be established, or even when a defined region is not isochronic but has delay that is simply and reliably predicted from position (such as on a clock ring), the remaining clock net layout and skew control problem is simplified, design risk is lowered, and pre- and postlayout simulations are more consistent, because final routing of clock paths from these reference regions is shortened and overall uncertainty reduced. We shall see and use this principle in analyzing the popular clock distribution schemes that follow.

The skew that intrinsically arises from differences in RC time constants of either lines or gate loads is aggravated by threshold variations in buffers and clocked loads due to minute differences in electronic parameters and lithographic variations in line widths and lengths at different devices. Time-constant and threshold effects interact to give a worst-case skew, which is the difference between the time at which the voltage response of line with the slowest time constant, τ_{max}, crosses the threshold of the logic element with the *highest* threshold, $V_{T\,max}$, until switching of the device with the lowest threshold driven by the line with fastest RC time constant. Taking the difference of the earliest and latest switching times we have [32]:

$$\delta = \tau_{min} \ln\left(1 - \frac{V_{T\,min}}{V_{DD}}\right) - \tau_{max} \ln\left(1 - \frac{V_{T\,max}}{V_{DD}}\right) \tag{63.4}$$

Equation (63.4) implies that a clock system design in which buffered electrical segments of the clock net have 10% variation in τ about τ_{nom}, and 10% variation of V_T about $V_{DD}/2$, will have an estimated skew of at least 17% of τ_{nom}.

Single-Phase Clocking

Clocks ultimately always drive a storage register or latch of some type. The form of clock signal(s) required in a system therefore depends on the type of latch or flip flop element used and on properties of the combinational logic circuits used. True single-phase clocking is the most complex clocking principle with which to design systems, and has traditionally not been used, although recent work has assessed some truly single-phase logic families [2]. The reason for caution with single-phase clocking is that invalid states may be passed to the output in two ways, as shown in Fig. 63.20: (a) if the combinational logic (CL) delay is *less than* T_h (i.e., too fast) or (b) the CL delay is *greater than* $T_C - t_{charge}$ (i.e, too slow). In other words, a two-sided (min and max) constraint on logic path delay exists for single-phase clocking [27]. This means that although attractive to minimize total interconnect, buffer counts, and interphase skew is

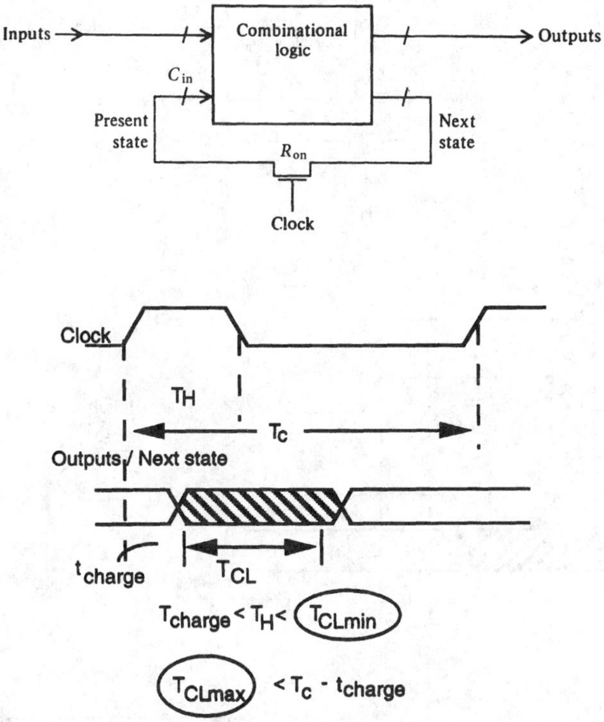

FIGURE 63.20 In single-phase clocking the combinational logic path must be neither too slow nor too fast.

avoided, truly single-phase clocking involves a greater design risk and timing analysis complexity. Because of this, the most common overall clocking scheme is single-phase clock *distribution* with local *generation* of a two-phase clock.

Two-Phase Clocking

With two nonoverlapping clock phases, we can eliminate one of the risks of single-phase clocking, that of a logic path being *too fast.* On the first phase the input is made transparent and allowed to affect the CL, charging the inputs through $R_{on}C_{in}$ in time t_{charge}. During this time the CL outputs are isolated from the input latch. On the second phase the new CL output values are stored by the second phase latch while the input latch is opaque, isolating inputs from the new values until the next phase one clock time. A nonoverlapping period between phases ensures that at no time does direct transparency occur from input to output. With two-phase nonoverlapping clocks, as shown in Fig. 63.21(a), we need to ensure only that the maximum delay in the CL is less than $T_C - t_{charge} - T_3 - t_{preset}$. It is essential that the nonoverlapping interval, T_2, be greater than zero, but T_3 can be arbitrarily short. When present, however, T_3 acts as an extra timing margin against skew.

It is obviously desirable to make T_2 as small as possible, but we can do so only when distributing two-phase clock directly if the interphase skew is less than T_2. In the worst case the interphase skew may be twice the skew of each of the two clock phase nets individually. Skew, therefore, necessitates at least a 1 : 1 derating in speed for two-phase clocking in addition to the basic loss of clock cycle time for logic settling, to ensure correct operation of storage devices. If skews in the two clock phase nets are uncorrelated, however, the extra penalty could be as high as 2 : 1. Every nanosecond of skew in the clock net for each phase then not only reduces the basic critical path logic timing margin by 1 ns, but also adds 2 ns to the T_2 requirement.

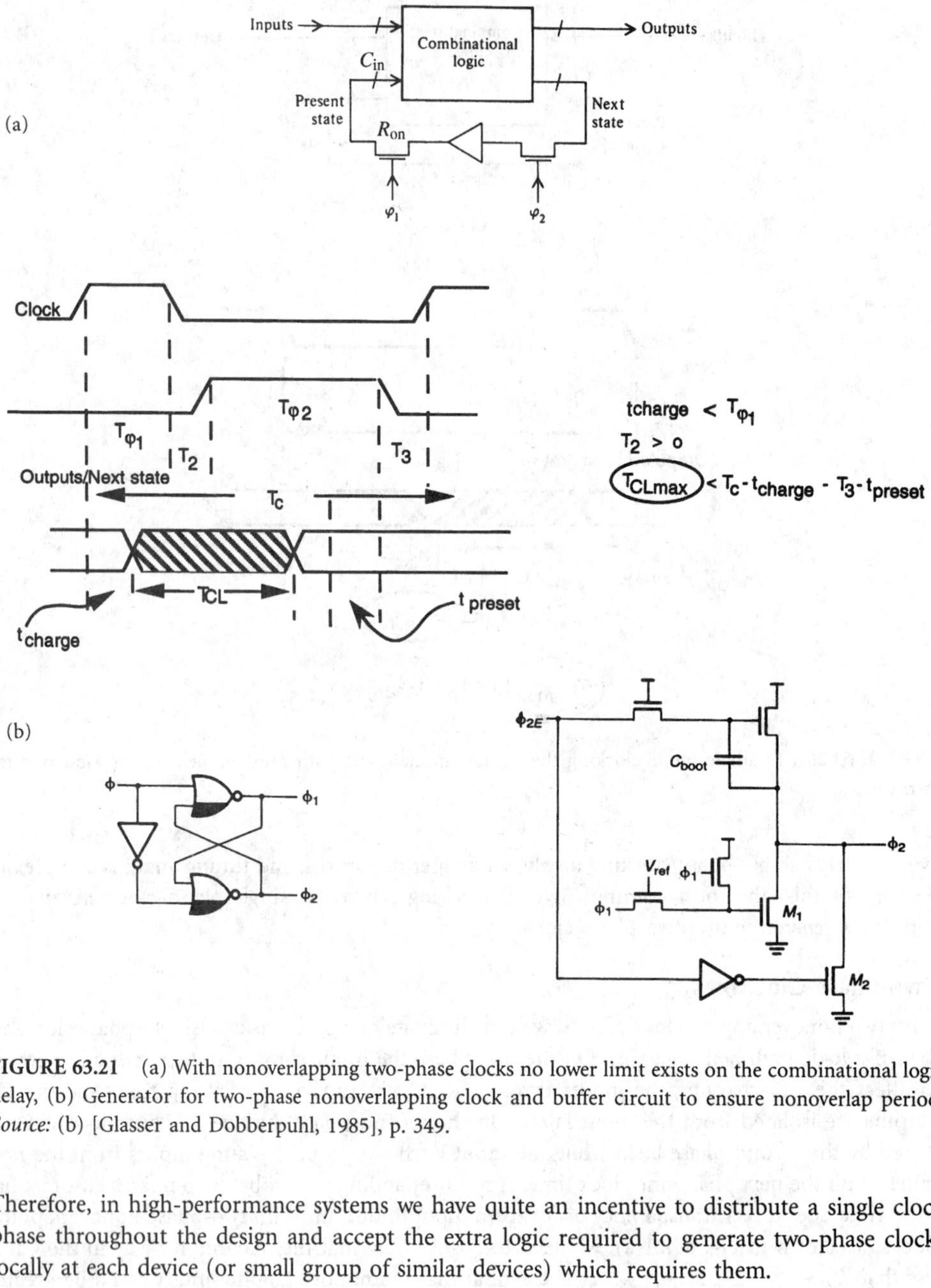

FIGURE 63.21 (a) With nonoverlapping two-phase clocks no lower limit exists on the combinational logic delay, (b) Generator for two-phase nonoverlapping clock and buffer circuit to ensure nonoverlap period. *Source:* (b) [Glasser and Dobberpuhl, 1985], p. 349.

Therefore, in high-performance systems we have quite an incentive to distribute a single clock phase throughout the design and accept the extra logic required to generate two-phase clocks locally at each device (or small group of similar devices) which requires them.

Two-Phase Clock Generator Circuit

The canonical form of circuit to generate the local two-phase nonoverlapping clocks from a single phase clock is shown in Fig. 63.21(b). The feedback of φ2 into NOR 1 ensures that φ2 must be low before φ1 can go high after the single-phase φ_{in} input has gone low, and vice versa. A special clock buffer circuit is shown in Fig. 63.21(b), which helps ensure that a period of nonoverlap exists in the presence of the threshold variations in the driven loads [12]. It does this by using transistor M1 to clamp the φ2 output low until far into the fall of φ1. φ2 is held low until φ1 has fallen below ($V_{ref} - V_{thresh}$) to finally cut off M1. V_{ref} might be set at 2 V in a 5-V process, thereby ensuring that φ1 is well below the logic threshold of all clocked loads

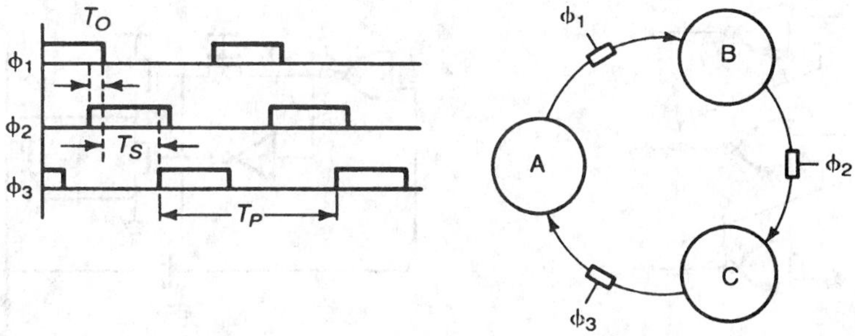

FIGURE 63.22 Principle of multiphase clocking. Outputs are isolated from inputs by other stages of nonactive logic even though any two active clock waveforms may overlap. *Source:* [Glasser and Dobberpuhl], 1985, p. 352.

before φ2 begins to rise, while at the same time minimizing but guaranteeing the existence of a nonoverlap period, which is lost processing time.

Multiple-Phase Overlapping Clocks

Generating and/or distributing nonoverlapping clocks with a minimal T_2 can be quite difficult in large systems. An alternative is to define three or more primary functional logic steps and use a similar number of overlapping clock phases. In this case the multiple stages of clocking removes the need for the guaranteed nonoverlap period in two-phase clocking. Let us consider three-phase overlapping clocking. The principles generalize to any higher number of phases.

In three-phase clocking the middle phase can be though of as providing the nonoverlap time which ensures time isolation between input and output activation for the module enabled on each clock phase. In fact, each phase plays a similar isolating role with respect to operations performed on its adjacent phases. Figure 63.22 illustrates the concept. The number of phases and the role for each phase typically reflects some natural cycle or step sequence of the basic system being designed; for example: bus input, add to accumulator, bus output.

In WSI systems, in which uncertainty in clock delays, circuit speeds, and interconnect impedances may be high [10], overlapping clock logic can give high tolerance to clock skew, and is compatible with self-timing in selected subcircuits. Three-phase overlapping clocking has a distinct advantage: no hazard exists unless all three clock phases overlap in time. In the presence of severe clock skew this can be a major advantage. Although called overlapping clocks, the circuits still function if successive phases do not actually overlap, although speed is sacrificed if overlap is lost.

Overlapping Clock Phase Generator

Figure 63.23 illustrates a circuit for generating three-phase overlapping clocks. Phase overlap is ensured because it is the onset of each phase that kills it predecessor. A Johnston counter comprised of three static *D*–flip flops generates the phase-enabling signals which sequence the actual generator stage, which is comprised of the three cross-coupled NOR gates. A deliberately limited positive-going drive ability of the enable input ensures that the Johnston counter exercises underlying rate and sequence control while the output waveforms are determined by the interactions between the actual clock phase signals themselves. While the enable inputs to each NOR are logically sufficient to drive the output high when the other input is low, they are arranged not to be able to drive the NOR output low on their own when the enable signal returns high. The output of phase i therefore stays high after its own enable signal has disappeared (gone high) until the phase $i + 1$ output is also high in response to the low-going phase $i + 1$ enable. Figure 63.23 shows this logic and a NORing clock buffer circuit in which the phase $i + 1$ signal is necessary to assist in returning the phase i output to zero.

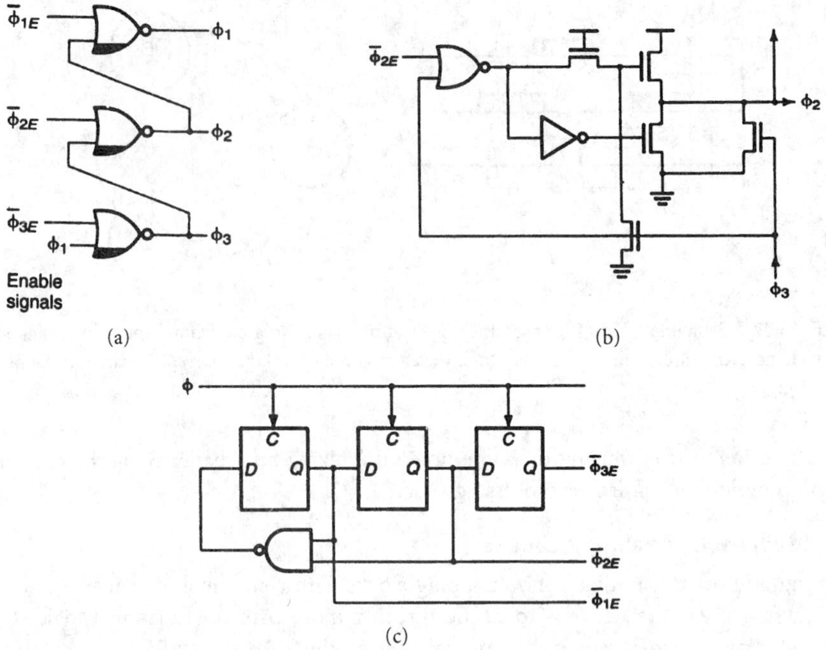

FIGURE 63.23 Three-phase clock generator logic and buffer design to ensure overlap. *Source:* [Glasser and Dobberpuhl], 1985, (a) p. 348, (b), (c) p. 354.

Clocking Latches

A latch is a storage element which is *level* sensitive to the clock waveform. The latch output conforms to the input that is present while the clock waveform is at its active level for that latch, and then continues to hold that value when the clock level falls below the active level. Latches have setup and hold time requirements analogous to those in the flip flops that follow. Circuit designs for high and low active latches are given in [31]. By combining latches of opposite active polarity, with logic between the stages, there can be two logic operations per clock period. In this type of operation, however, skew adds to the needed clock period as usual, but in addition any imbalance in the clock *duty cycle* requires a further margin because the minimum duty cycle half-width must remain greater than the worst-case logic delay. The clock edges also must be kept sharp enough that transparency never occurs between two successive latches working on opposite clock phases simultaneously, or that some minimum logic delay always exists between latches that exceeds the possible overlap time. The recent DEC ALPHA mircoprocessor is a two-phase latch machine in which both phases drive latches that are active in the respective phases permitting logic evaluation twice per cycle. This is one case in which to control the very high transistor count, two-phase clock is distributed globally rather than generated at each module. The entire clock net on each phase is driven from a single large buffer placed at the center of the die where a PLL is also fabricated to advance the on-chip clocking phase relative to external bus timing, thereby compensating for buffer delay.

Two principles for maintaining 50% clock duty cycle in a latch machine are (1) whenever generating or phase-locking to a system clock, do so at twice the rate needed, then divide by two. This results in a 50/50 clock waveform, regardless of the original clock source waveform. (2) When repeatedly buffering clock in a chain, or when distributing clock through a hierarchy of clock buffers, use inverting clock buffers at each stage. Inverting the clock at every buffering stage inherently counteracts the effects of different rise and fall times in buffers. Otherwise, these can accumulate to extend or shorten the ON period of the clock waveform. For example, if a noninverting buffer has greater fall time than rise time, a clock path transiting several of these buffers will develop a broadened ON period. This effect is self-compensating in a chain of inverting buffers.

Clocking Flip Flops

Flip flops are more complex storage circuits than latches, but have no dependency on clock duty cycle because they are sensitive only during an active edge of the clock waveform. A rising edge *D*–flip flop (for instance) updates its output to match its *D* input on the rising edge of the clock waveform. The *Q* output retains the updated value thereafter, regardless of further changes in the input or clock waveform (with the exception of another rising clock transition). A latch is a more fundamental circuit element than the *D*–flip flop in that edge-triggered behavior is attained only by implementing two latches and generating a pair of two-phase nonoverlapping clock pulses internally, in response to the active clock transition at the edge-triggered input.

For instance, Fig. 63.24 shows a typical *D*–flip flop in which inverters *I*1 and *I*2 generate the internal two-phase clock signals for level-sensitive latches *L*1 and *L*2. In specialized applications it may be advantageous to design a custom module within which multiphase clocks are distributed directly, without incurring greatly increased skew problems. For example, an error-correcting codec ASIC prototype for 45 Mb/s digital communications includes a 2613 stage tapped delay line comprised of seven-gate single-phase *D*–flip flop modules. The use of single-phase clock flip flops in this context is relatively expensive, but appropriate for fast validation of the system design. For cost- and power-reduced production in volume, a more detailed latch-based design using directly distributed two-phase nonoverlapping clocks may be worthwhile. In general, while it is most common to conduct system design based on the single-phase clocking model, two-phase or multiphase clocking may be advantageous within specialized substructures.

Although edge triggered, a minimum clock pulse width of about 1 to 1.5 ns is still typically required to deliver enough switching energy on the clock line. For correct operation (specifically, to avoid uncertain outputs due to metastability) of an edge-triggered flip flop, data must be stable at the input(s) for a minimum setup time before the clock edge, and the data must remain stable at the input for the hold time, after the clock edge. The time until the *D*–flip flop output is valid after the clock edge occurs is the clock-to-*Q* delay. For hazard-free transfer of data from one stage to another with *D*–flip flops, without assuming a minimum logic delay constraint between stages, the clock-to-*Q* delay must exceed the hold time. Typical values for a range of *D*–flip flop types in a 1.5-μm CMOS process are $t_{\text{setup}} = 0.8$ to 1.5 ns, $t_{\text{hold}} = 0.2$ to 0.4 ns, and $t_{\text{clk}-Q} = 1.3$ to 3.5 ns for *Q* output fanouts of 1 to 16, respectively. With the extra input logic delays in a JK flip flop, many JK flip flop cell implementations exhibit $t_{\text{hold}} = 0.0$ ns. By comparing magnitudes of typical setup and hold time requirements, it is apparent that skew is more likely to cause a setup time violation on critical delay logic paths than it is to result in a hold time violation.

Role of Clocks in Dynamic Logic

Clock signals are also used to implement a variety of logic gate functions in a dynamic circuit style, i.e., based on short-term charge storage, not static logic. This typically involves precharging on one phase and logic evaluation steered by the inputs on the second phase. The "Domino" logic approach combines a dynamic NMOS gate with a static CMOS buffer [8]. In "NORA" (no-race) logic dynamic logic blocks are combined with clocked CMOS latch stages. A variety of other dynamic logic circuits, using up to four clock signals to structure the precharge and to evaluate timing, are covered by [12] and [31]. In all of these gate level circuit implementations the clocking-related issues are ultimately manifestations of the basic principles already seen for two-phase clocking; i.e., of never simultaneously enabling a direct path from input (or

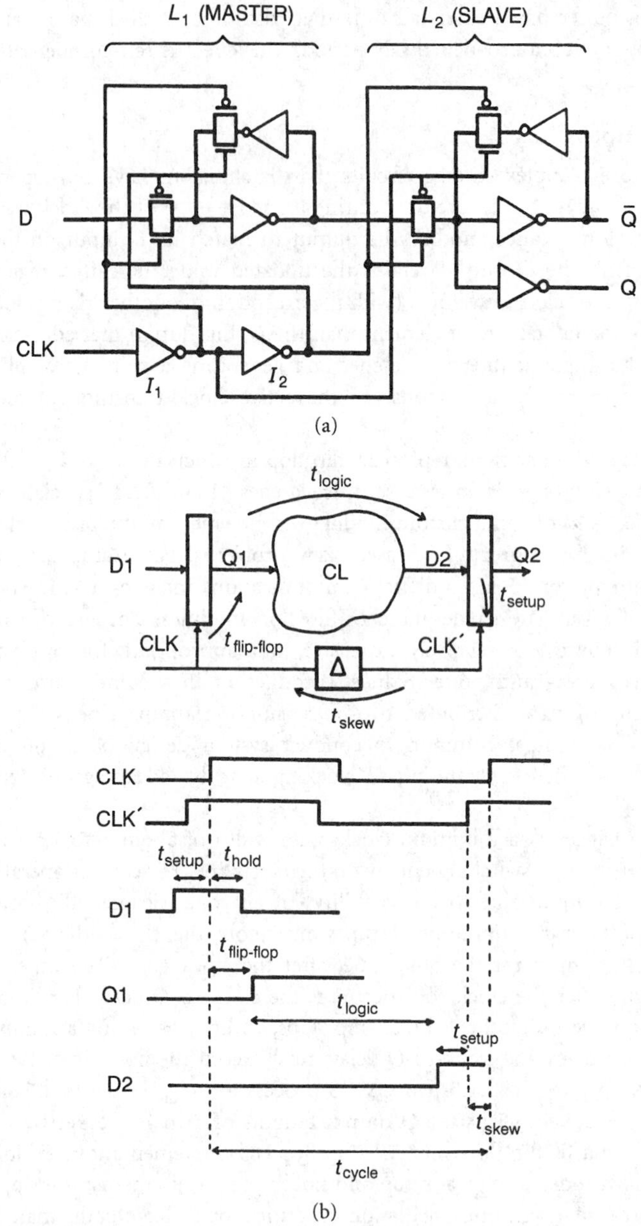

FIGURE 63.24 (a) Two-phase latch structure of a typical CMOS positive edge-triggered *D* flip flop. (b) Setup, hold, and delay times for a *D*-flip flop showing how skew is equivalent to a shorter clock period and threatens setup time margin. *Source:* (b) [Bakoglu 1990], p. 345.

precharge source) to output. These logic styles were developed to reduce transistor counts. However, modern designers will most often be faced with a greater challenge in managing system-level problems of skew in a single clock phase distributed to static registers than the challenge of reducing transistor count. For standard cell and gate array designs, fully static logic and self-contained single-phase static registers are usually the only options provided. Static logic with single-phase edge-triggered *D*–flip flops are recommended for most designs today.

Synchronizers and Metastability

Many systems need to sample external inputs which may be timed independently of the synchronous system clock, such as switch-based control inputs, keyboard states, or external process states in a real-time controller. The external state needs to be synchronized with the system time base for processing. Metastability leading to synchronizer failure is a fundamental possibility that can never be entirely eliminated.

Figure 63.25 shows a basic synchronizer circuit. Our concern is that it is possible for the synchronizer output to take an arbitrarily long time to settle to one or the other valid logic states if the input signal voltage is sampled in the intermediate voltage range, i.e., $V_{iL} < v_{\text{in}}(t) < V_{iH}$. In this range it is possible to find the input voltage at a value that leaves the cross-coupled latches internal to a flip flop in an intermediate state, with insufficient positive feedback to snap the output to either high or low valid states. System noise or quantum fluctuation will ultimately perturb such a precarious balance point and the output runs to one direction or the other, but it can take an arbitrarily long time for the synchronizer to reach a valid output state. As shown in Fig. 63.25(b), some flip flop outputs may also tend to rise at least halfway toward the positive output level before deciding the input was really a zero. This glitch may trigger edge-sensitive circuits following the synchronizer.

Fortunately, the probability of an indeterminate latch output falls exponentially with the time T after sampling the possibility indeterminate input:

$$P(t > T) = f_{\text{clk}} f_{\text{in}}\, \Delta e^{-T/\tau_{sw}} \tag{63.5}$$

where f_{clk} is the synchronous sampling frequency, f_{in} is the frequency of external transitions to be synchronized, Δ is the time taken for the input voltage in transition to cross from V_{iL} to V_{iH} (or vice versa), and τ_{sw} is the time constant characterizing the bandwidth of the latch device.

Having recognized the strict possibility of a metastable logic state resulting from synchronizer input, the designer can address the issue in several practical ways:

1. Use a high gain fast comparator to minimize Δ by minimizing the voltage range V_{iL} to V_{iH}.
2. Ensure or specify fast transitions in external sensors or other devices to be sampled, if design control extends to them.
3. If there is no real-time penalty from an additional clock period of input response delay, the synchronizing latch should be followed by one or two more identical synchronizer latch stages, thereby increasing T in (63.5) to reduce the chance of a metastable state being presented to internal circuitry to an acceptably low probability.

The effect of input metastability on the system also should be analyzed for its impact. If it is extremely crucial to avoid a metastability hazard, then phase locking the external system to the system clock may be considered, or if the system and external timebases are free running but well characterized (for example, in terms of a static frequency offset or known phase modulation), then the anticipated times of synchronization hazard may be calculated and avoided or otherwise resolved.

As a practical matter, the way in which design software handles metastability should be considered. Potentially metastable conditions should be flagged as a warning to the user, but not necessarily treated as a violation prohibited by the design environment. Some applications, particularly in VLSI for telecommunications, need design support for plesiochronous (near-synchronous), phase-modulated, or jittered signal environments. This implies test vector support to represent clocks interacting through logic at slightly different long-term or instantaneous free-running frequencies with design and simulation rules that permit the metastable conditions inherent as such clocks walk relative to one another. Circuit simulations must be allowed to

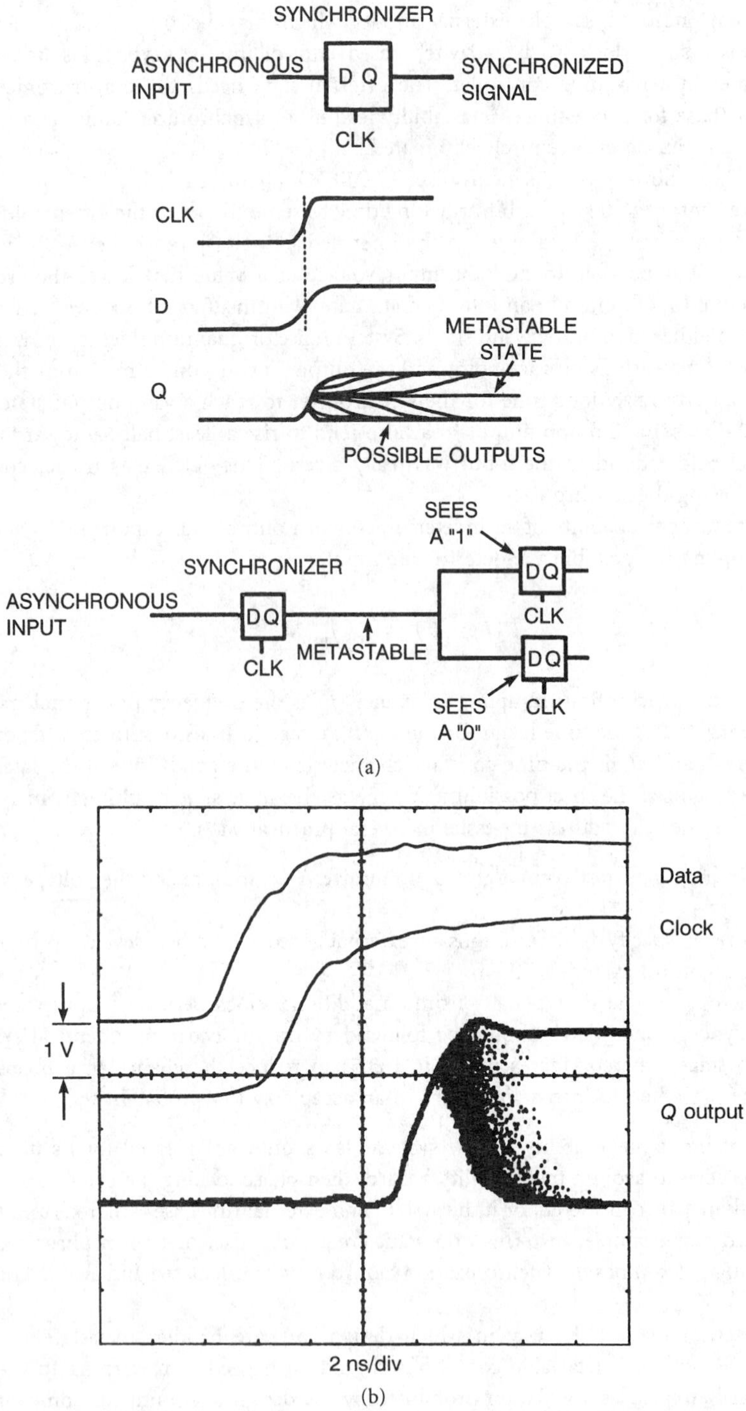

FIGURE 63.25 (a) Metastability in a synchronizer circuit. *Source:* [Bakoglu 1990], p. 357. (b) Experimental illustration of metastability in 74F/74 (TTL) *D*-flip flop showing output rise before indeterminate final response (10-s point accumulation). *Source:* [Johnson and Graham, 1993]. p. 130.

continue with a random value resulting from the simulated "synchronizer failure" to be useful in such applications.

Controlled Introduction of Skew

Skew is not necessarily all bad. In fact, from the viewpoint of the system power supply, and power- and ground-related noise current surges, it is undesirable to have all logic transitions occurring exactly simultaneously. In a CMOS IC with 20 K register stages at 0.1 pF load each and a 1-ns clock rise time, 10 A of peak current can be drawn by the clock net. This can present a serious $L\, dI/dt$ problem through power and ground pins and can even lead to electromigration problems for the metallic clock lines. Chip clocking strategies should take this into account early in the design by seeking ways to deliberately stagger or slightly disperse the timing of some modules with respect to others. Also, the system clock waveform may not necessarily need the fastest possible rise time. Consistent with avoiding slow-clock problems, and controlling threshold-related skew in buffers, the clock edge should not be made faster than this as an end in itself. Excessively fast clock edges only aggravate power and ground noise problems as well as ringing and potentially causing electromagnetic radiation problems in the chip-to-chip interconnect. These principles motivate the widely used 10 K ECL logic family, which is based on the much-faster 100 K series with explicit measures to *slow down* the rise and fall times of the basic 100 K logic gates.

When considering random skew, it may or may not be beneficial to pursue skew reduction below a certain level in the design. In the case of a microprocessor or a design in which the fastest possible IC speed is always useful skew reduction does mean a performance improvement. In some other applications, however, the clock speed is set by the application. For instance, a VLSI circuit for a telecommunications multiplexer may be required to operate at a standard line rate of 45 MHz. In this case there may be no premium for a design that can perform the same functions at a higher speed. A working design with skew of 5 to 7 ns (out of a 22-ns clock period) may then be *more* desirable than a functionally equivalent design with 0.5 ns skew because dI/dT effects are eased by distributing the total switching current over time in the former. This principle may be important in practise as automated clock synthesis tools become more widely used and effective at achieving low skew, possibly creating unnecessary system level noise and EMI emission problems. Future clock-related CAD tools should possibly aim to disperse clock timing at various loads while satisfying a target worst-case skew, rather than absolutely minimizing skew.

Strictly speaking, skew can also be beneficial when allowed to build up in a controlled way in certain regular logic structures. For instance, by propagating clock in the opposite direction to data in a shift register, one enhances the effective setup time of the data transfer from register to register. In general, however, it is not feasible or advisable to try to design every clock path with a desired (non-simultaneous) clocking time at each register, taking into account the individual logic paths of signals leading each clocked latch input. Especially when designing larger systems mediated by CAD tools for placement, routing, and delay estimation, the most practical and low-risk approach is to consider any deviations from a common nominal clock time as undesired skew. Indeed, for any one latch, timing margin may be enhanced by the actual skew that arises, but with thousands of logic paths, it is impossible to analyze the relative data and clock timing for each latch. Only one instance in which the skew works against the assumed timing margin is enough to fail a design. Therefore, the "customized skew" approach is recommended only for small and very high speed specialized circuit design.

Clock Signal Manipulation

As a matter of design discipline, some commercial ASIC and cell-based layout systems may prohibit a designer from directly gating or manipulating a clock signal. Any needed clock qualification is done through defined enable or reset inputs on register structures. As in software development, in which structured design disciplines have been developed, gating the clock may riskier than its apparent efficiency warrants. In addition, clock gating within random logic designs can interfere with test pattern generation. The risk also exists of creating clock glitches or even logical lockups when clocked logic functions decode conditions that gap its own clock. On the other hand, in high performance and in large system-level designs, clock gating for power down and *clock tuning* may be unavoidable.

Wagner [30] discusses clock pulse-width manipulation, defining four canonical subcircuits that can be used to "chop", "shrink", or "stretch" the clock waveform for either delay tuning or duty cycle maintenance. The effect of these circuits on the positive pulse portion of a clock waveform is shown in Fig. 63.26, where AND gates have delay d_a, OR gates have delay d_0, inverters have delay d_i, and the delay elements have delay D. Aside from a single gate delay, the chopper and stretchers leave the rising edge unaltered and tune the trailing edge. These can be used to maintain a balanced clock duty cycle or to tune the nonoverlap period in two-phase clocking. The shrinker delays the rising edge of the clock as might be helpful to specifically delay clocking a latch or a flip flop that is known to follow a particularly long logic path delay. This is not generally a preferred design approach, especially when manufacturing repeatability and temperature dependence of delay elements are considered.

By *clock gating* we mean selectively removing or masking active phases or edges from the clock signal at one or more latches. One valid reason to gate the clock in CMOS is to reduce power consumption. Many circuit designs possess large modules or subsystems which it makes sense to stop cycling in certain application states. Gating the clock off is therefore the simplest form of power management, because CMOS has negligible power dissipation in a static state. However, even for this simple use of clock gating, the main issue is avoiding glitches when gating the clock.

Before gating any clock, the designer should see if gating can be avoided with an alternate design style. For example, if it is desired to hold the data on one register for a number of cycles while other registers on the same clock proceed, a preferred approach is to use a 2:1 mux at the register input. Rather than gate the clock, the mux is steered to select the register's own output for those cycles in which gating would have occurred. Ultimately, if it is appropriate to gate out one or more clock pulses, a recommended way of doing so in rising edge active logic is to OR out the undesired clock edges, decoding the clock gapping conditions on the same clock polarity as the one being qualified (see Fig. 63.27). A natural tendency seems to be to AND out the gapped clock edge and/or to decode the gapping condition on $\overline{\text{CLK}}$, but these approaches are more apt to generate clock line glitch than the OR-based approach. In the AND approach the gating line returns high at the same time as the falling edge after the last-gapped active edge. In the case of minimum delay through the gapping logic the risk is that both AND inputs are momentarily above threshold.

Minimizing Delay Relative to an External Clock

In a large system skew can build up between clock and data at the system level, even if the clock is skew-free everywhere within the ICs because data paths through ICs can build up delay relative to the system clock. For instance, if an ECL or TTL system clock line is distributed to a large CMOS IC, then the system clock must be level shifted and a large clock buffer may be required in each IC to drive its internal clock net. The delay through the on-chip clock interface and buffer can mean that even if the chip timing is internally skew-free, the on-chip clock is significantly delayed relative to the external system timing. Short of using the phase-lock methods described later, a simple technique to minimize this form of system-level skew is either to retime chip outputs with a separate copy of the system clock that has not gone through the internal clock buffer, or, if electrically compatible, to use the external system clock to directly retime the output signals from each IC (Fig. 63.28). This is called the *early clock* concept. Note

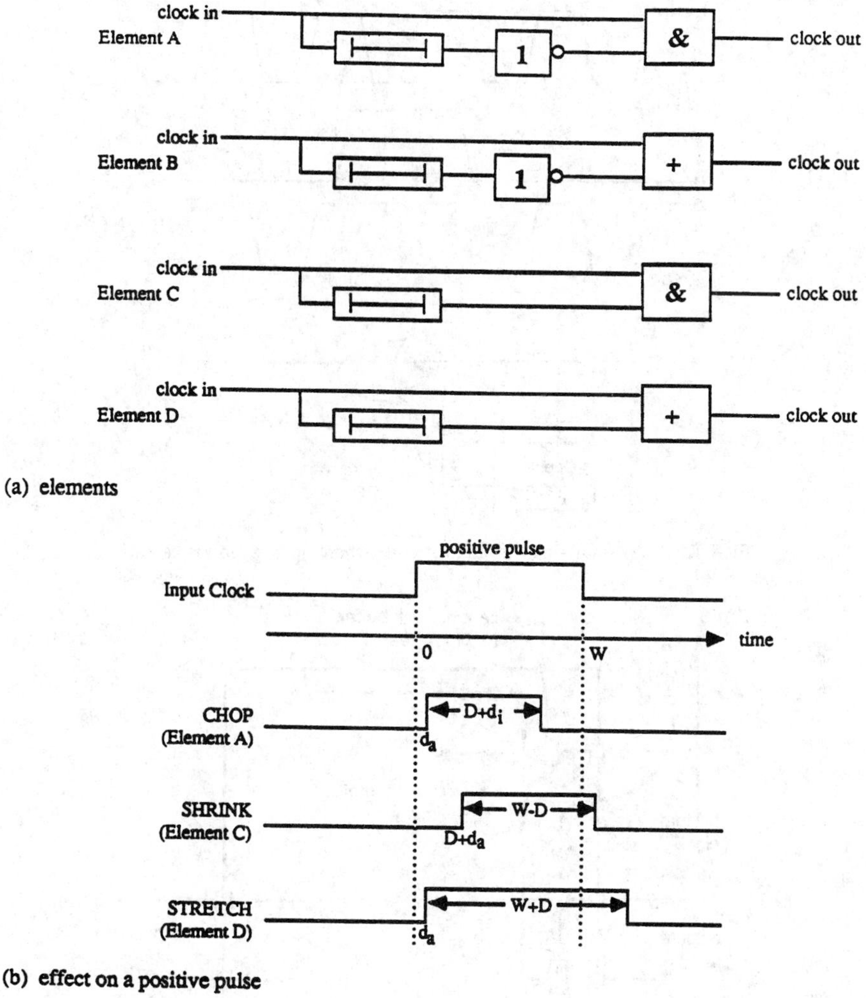

FIGURE 63.26 Standard circuits for chopping, stretching, and shrinking a clock waveform to adjust duty cycle or timing margins. (*Source:* Adapted from [30], p. 15.)

that this assumes an adequate timing margin exists in the final stage of internal logic to permit the relatively early sampling of logic states.

Clock Distribution Schemes

Single-Driver Configurations

Often a single on-chip clock buffer is the simplest and best approach to clock distribution. A single adequately sized clock buffer is typically located at the perimeter to drive the entire clock net of the chip, as shown in Fig. 63.29. This approach can also perform well in large systems if the clock is distributed in a low R_0C_0 routing layer, such as third-layer metal. The main advantage regarding skew is that no matter how the single-clock driver delays or otherwise responds to the external clock input, the output waveform is electrically common to all loads. No intervening buffers and no separate passive net segments develop skew. Moreover, if the global clock net comprises an electrically isochronic region ($NC_{gate} \gg l^2RC_{wire}$) in which clock loads are reasonably uniformly distributed, the clock net voltage rises virtually simultaneously

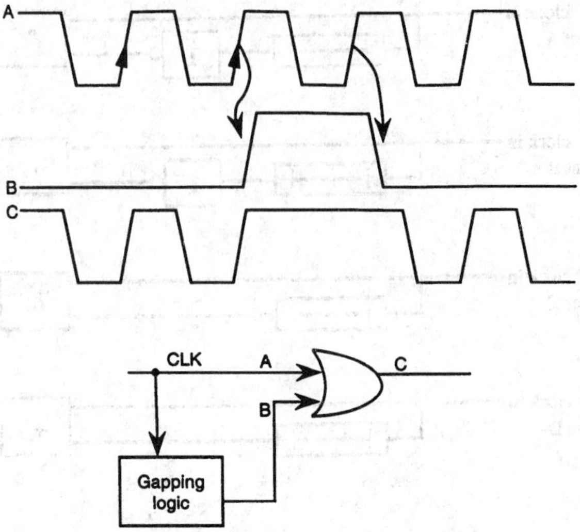

FIGURE 63.27 OR-ing out a clock edge when clock gating is essential.

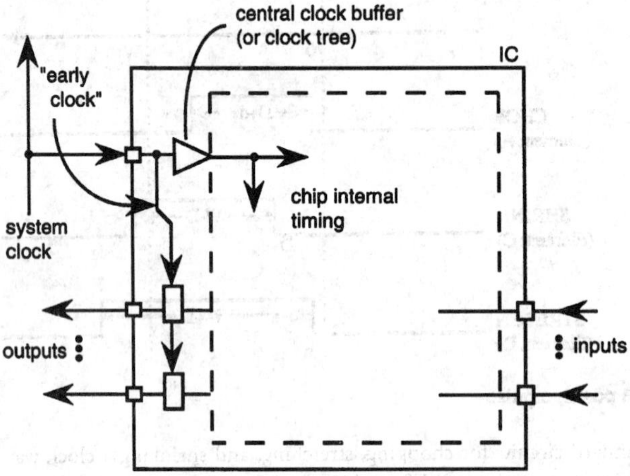

FIGURE 63.28 The "early clock" technique for reducing chip delay relative to external system timing.

at all points on the charging clock net, resulting in extremely low skew. Often this leads to lower skew than in a buffered clock fanout tree. There is also only one global clock wiring net, simplifying documentation.

On the other hand, skew can be larger and more routing dependent with a single buffer (than with some following schemes) when clock routing lengths vary significantly and wiring capacitance and resistance are significant. In such cases an isochronic net is not an accurate model, and performance depends on the way clocked loads are distributed on arms branching from the medial clock node. It is possible for neighboring flip flops to be connected to the central driver via quite different path lengths, making prelayout simulation relatively uncertain and requiring considerable postlayout tuning. Another caution is that even if no actual skew is present because a single waveform is common to all loads, the rise time of the clock waveform may show considerable loading effects, so that threshold-dependent skew arises in the clocked loads. Finally, the potential for conducted switching noise problems is high with the single-buffer configuration because the entire clock net capacitance is charged in typically under 1 ns. Power

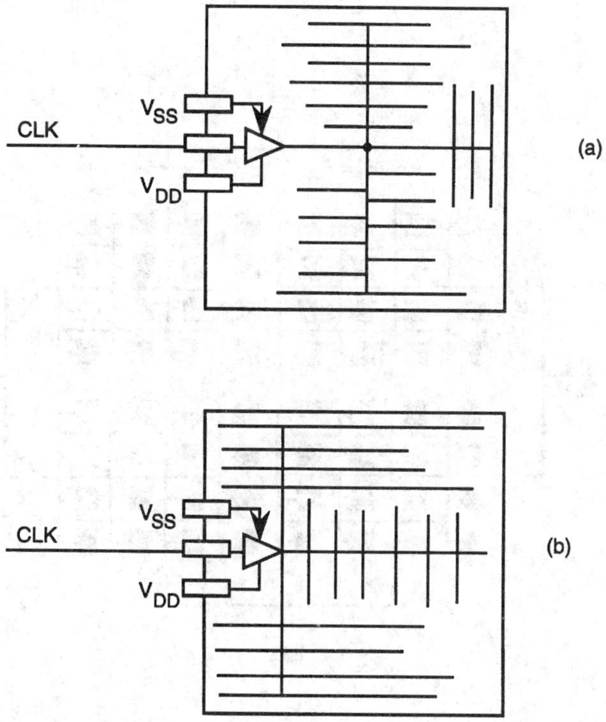

FIGURE 63.29 Single clock buffer placed in the I–O perimeter with dedicated power, ground pins (a) branching from a medial point on die (current density on line to medial point may be high), (b) branching immediately (skew may be high).

supply decoupling and ground bound problems and even current density (electromigration limits) considerations may need to be given special attention if using this configuration. It is usually recommended that the single clock buffer be physically adjacent to the clock input pin, and the clock pin should be flanked by dedicated power and ground pins that feed only the clock driver. This principle, which applies in general to clock input buffers in all clocking schemes, keeps the clock switching noise out of the core power and ground bus lines. Also, the delay through a large clock buffer may be considerable, so a lightly loaded "early clock" can be picked off from the buffer input and used to retime chip output registers, or a PLL may advance the phase to the internal clock buffer so as to align internal and external timing regardless of the buffer delay.

An interesting central clock buffer design, which also has attributes of the clock trunk scheme which follows, is reported in [9]. Here, the area for central clock driver fabrication is a strip across the center along one axis of the die. External clock is fed from one pin on the same axis as the buffer, and internal clock lines radiate systematically away from the central linear distributed buffer. Data flow is also highly structured and arranged to progress away from the central driver strip, further minimizing clock-to-data skew.

Four-Quadrant Clocking Approaches

In the quadrant-oriented approach we may use up to four clock pads and four smaller clock drivers placed at the centers of the die edge, preferably also with dedicated power and ground from flanking pins. There may be one, two, or four external clock pins. Figure 63.30(a) shows a quadrant scheme tested in [24]. In Fig. 63.30(b) a single-pin quadrant-oriented scheme in 1-μm two-layer metal CMOS achieved 0.6 ns skew amongst 400 registers. A four-pin quadrant

(a)

(b)

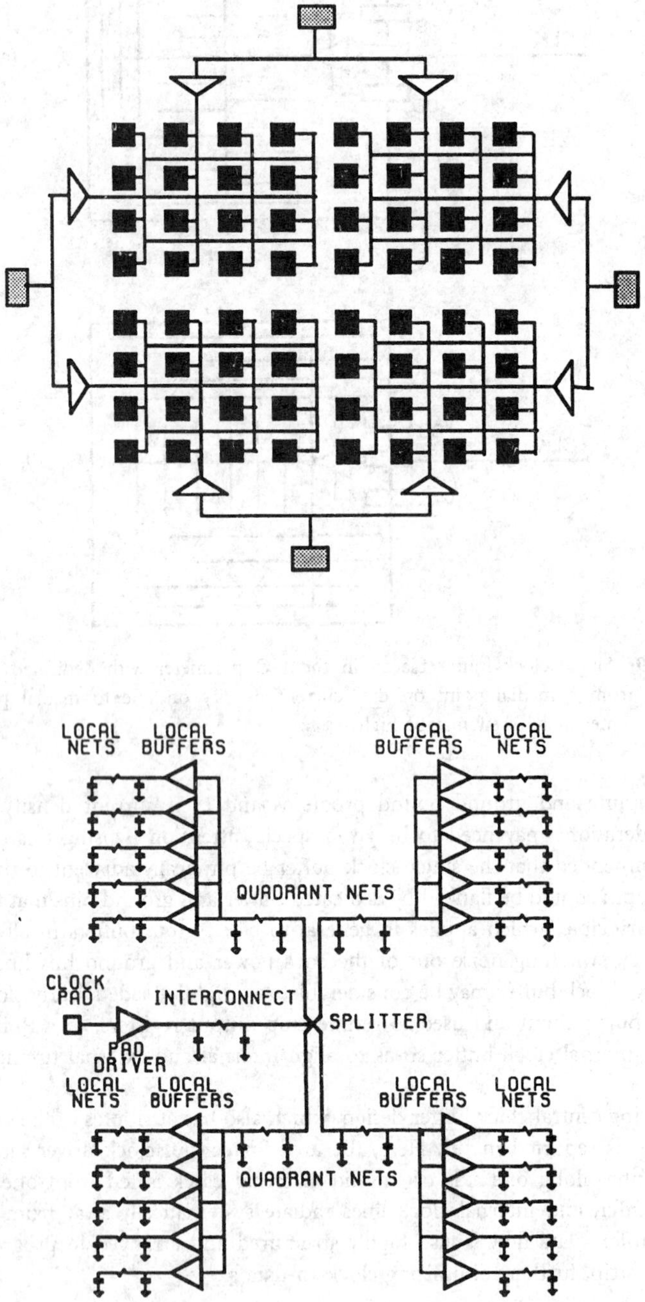

FIGURE 63.30 Quadrant-oriented clock distribution schemes: (a) 4 pins, 2 parallel buffers per quadrant. *Source:* [Nigam and Keezer, 1993], (b) Single-pin quadrant scheme with 2 buffer levels. *Source:* [Boon et al., 1989].

approach was successfully used in [29] to develop a 90-MHz CMOS CPU. If more than one pin is used for clocking, pin count goes up but absolute delay through the clock buffers can be reduced. The maximum wiring RC delay on the internal clock net and the peak current and $L\, dI/dt$ effects through any one pin and bonding wire inductance may all be reduced in this case. Total loads on each of the four internal clock nets should be well balanced and/or the drivers sized for the actual number of loads in each quadrant. In many cases reduction of each clocked area to one fourth of the die area can result in isochronic regions for which no further design attention other than total load balancing is required for clock routing within each region. The quadrant approach can also reduce the clock routing problem when the clock shares only two metallization layers available for all signal routing. Two other considerations apply to the quadrant schemes: (1) external tracking to multiple clock pins should be laid out with delay balancing in mind; (2) skew should be particularly considered on data paths which must cross from quadrant to quadrant, bridging timing across clock subnetworks.

Symmetric and Generalized Clock Buffer Trees

A symmetric or regular clock buffer tree [Fig. 63.31(a)] has equal fanouts from buffers at the same level, equivalent (or isochronic) passive interconnect paths at each stage, identical buffer types at each level, and equal groups of loads at each leaf of the tree. This ideal can be approximated if loads are regularly distributed and layout and routing can be controlled to ensure equal interconnect delays from each buffer to its subtending buffers. The remaining skew will primarily be due to threshold variation in buffers and terminal loads.

A more general view of the clock buffer tree that arises with irregular layouts and routing is that the buffer tree has unequal interconnect delays and differing numbers of loads at each buffer. Figure 63.31(b) illustrates the electrical model of such a clock tree. (R and C values are all different.) The basic approach to skew control in such a generalized buffer tree is to size each buffer specifically for the individual loads it drives and the delay of the interconnect path to its input from the preceding level buffer. In practice this means that generalized clock buffer trees may be the most handcrafted (and/or custom-tuned) of all designs, especially in a large system-level clock tree that extends down from a master clock source through multiple shelves, backplanes, connectors, circuit packs, and individual ICs. The system-level hierarchical clock tree design for the VAX 8800 is a good example described by Samaras [25]. Here, a two-phase clock was distributed to 20 large circuit packs over a 21-in. backplane with a global skew of 7.5 ns, for operation at about 50 MHz (37% skew).

Some basic methods and principles are, however, identifiable for designing generalized buffer trees so that it is not all just careful tuning and handcrafting:

1. Inverting buffers at each level will preserve clock duty cycle better than a tree of noninverting buffers.
2. Total *delay* (root to leaves) of the clock net is theoretically minimized when driving primarily capacitive loads, by a fanout ratio of $e = 2.718\ldots$ at each level of the tree [23]. In practice this implies a fanout ratio of about $n = 3$, with appropriately sized buffers, if delay is to be minimized. However, $n = 3$ can lead to relatively deep trees of many buffers in which skew can build-up from threshold and time-constant variations as in (63.4).
3. If identical buffers are used within each level, then the design aim is to make sure that the load of further buffers and/or interconnect RC load is delay-equivalent for each buffer. Dummy loads may be required to balance out portions of the clock tree whose total fanout is not needed. At the end-nodes of the tree equal numbers of standard loads should be grouped together on each leaf node, within a locally isochronic region.
4. If the tree is deep, the skew amongst members of one local clock group at the bottom of the tree may be considerably smaller than the skew between groups of clocked loads

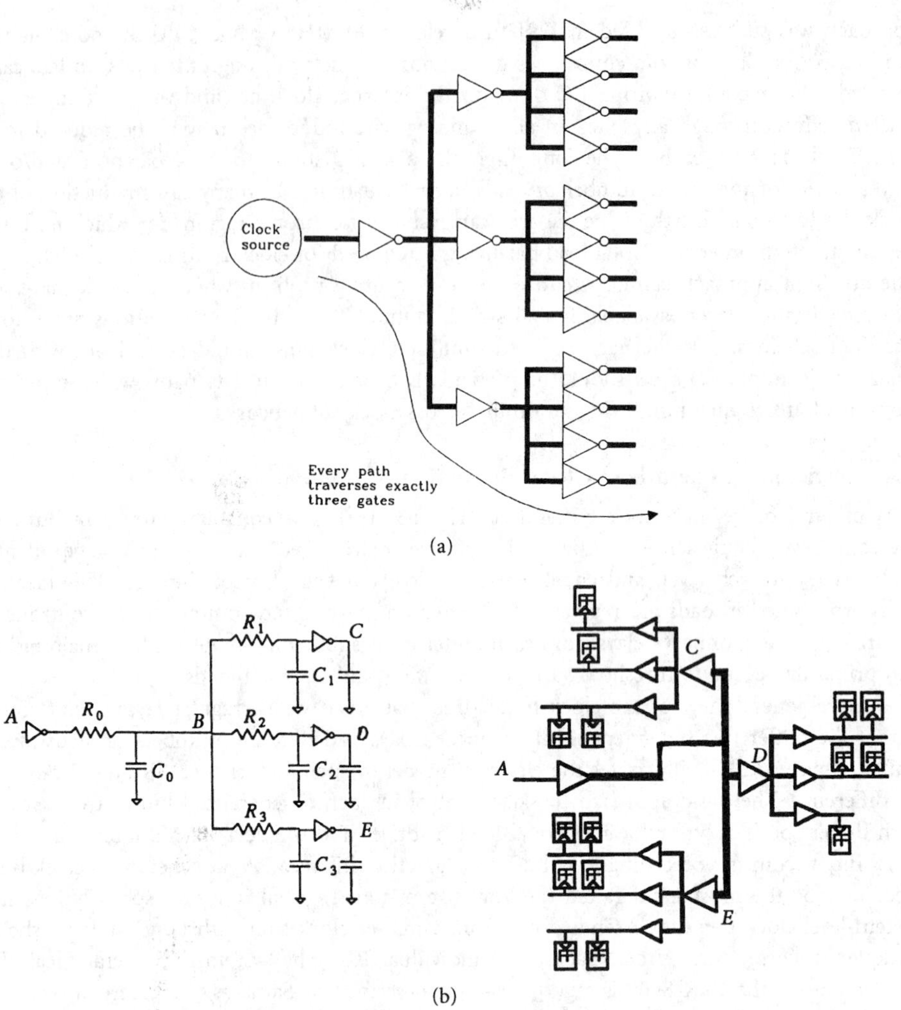

FIGURE 63.31 (a) Idealized symmetric buffer clock tree. *Source:* [Johnson and Graham, 1993], p. 348, (b) Generalized clock buffer tree where interconnects and loads are not identical.

at different end-nodes of the tree. If the inter- and intragroup skews are separately characterized, however, logic path delays can be designed to take advantage of the low intragroup skew and to allow greater timing margin on intergroup paths [25].

5. The choice of clock buffer type for use in any clock tree should take into account the effects of power supply sensitivity. For example, the *bootstrapped clock buffer* of Fig. 63.32(a) can provide a very sharp rise time, although with relatively high delay through the buffer. Sharp rise times minimize skew due to switching threshold variations in the following buffers or clocked loads. The output switching time of the bootstrapped clock buffer is, however, relatively sensitive to supply voltage. On the other hand, the *phase correcting buffer* of Fig. 63.32(b) is very tolerant to supply variations, but is not as fast in its output rise time. This leads to a mixed buffer strategy in which the bootstrapped buffer is used in the relatively small population of buffers in the first few stages of a clock tree. Here, special attention can be paid to ensuring well-equalized power voltages. The phase-correcting buffer is more appropriate in later stages, nearer the more numerous individual loads, amongst which on-chip or wafer supply levels may exhibit more IR voltage drop variation.

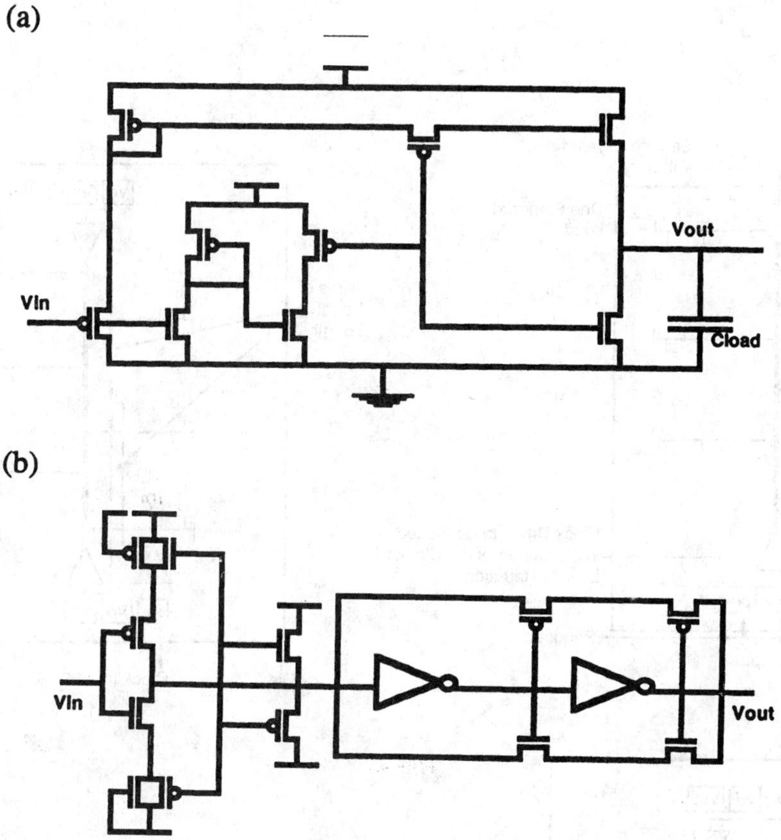

FIGURE 63.32 (a) Bootstrapped clock buffer for use in first levels of a clock tree; (b) phase-correcting clock buffer for use deeper in clock tree. *Source:* [Fried 1986].

Algorithms for the generalized clock tree design problem are also emerging. The algorithm in [28] can generate a buffer design based on custom transistor sizing to drive each heterogenous load, all with the same nominal delay. A still more general optimization approach is under development by Cirit [5]. The sizes of all buffers in the tree are jointly optimized with respect to unequal interconnect RC totals and unequal loads that each drives, as in Fig. 63.31(b). Thus, the minimum *total* tree delay is found for which all paths from root to leaf of the tree also have *equal* delay. The procedure is intended to be iterated along with placement and routing alternatives until a set of feasible buffer sizes and acceptable total delay is found.

Clock Trunk Schemes

The clock trunk concept is gaining popularity and is now supported within several CAD systems for CMOS processes with two or more metallization layers. Three variants of clock trunk structures are shown in Fig. 63.33. An input clock signal is buffered (its input pad is at the center of one side of the chip edge) and is routed either to the midpoint, or one or both ends of the internal clock "trunk". The trunk itself is a metal line specially widened for low resistance, thereby making delay and the $R_0[C_{load} + C_0]$ rise time particularly small on the trunk portion of the overall clock net. As long as the lumped capacitive loads (C_{load}) dominate the trunk's C_0, the time constant of the trunk drops as it is widened. C_0 can be kept particularly low by forming the trunk in third-layer metal. The idea is to size and drive the clock trunk so that an isochronic region is created down the middle of the die, reducing all remaining clock path

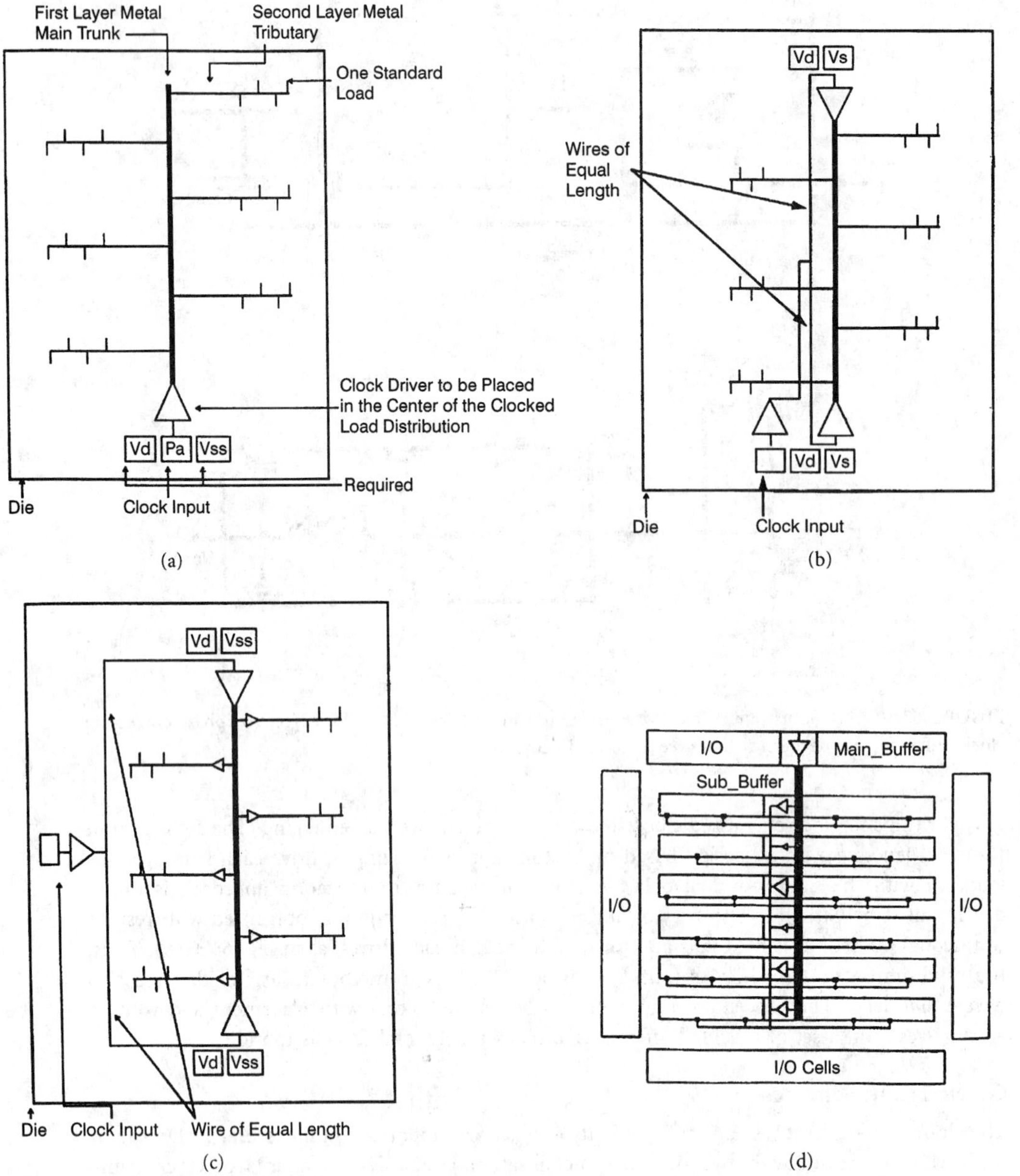

FIGURE 63.33 The clock trunk concept: (a) single-ended unbuffered clock trunk, (b) double-ended unbuffered, (c) buffered clock trunk, (d) clock trunk with shorted branch buffer outputs. *Sources:* (a)–(c): [LSI Logic 1992]. (d): [Saigo et al., 1990].

distances to not more than half the die diameter and setting up a situation in which the final distribution of clock from the isochronic trunk to the loads is in line with one routing axis. This means that branch routing from the trunk to loads can be exclusively contained within one metal layer and travel in a shortest direct line from trunk to load. This is highly desirable for processes in which all horizontal and vertical routing are dedicated to one metal layer or the other. Overall layout and routing is simplified and clock paths are predictable and uniform in routing and wiring type. If, however, the total fanout is very small, the wide metal trunk may add more to trunk capacitance than it decreases R_0 in the $R_0[C_{\text{load}} + C_0]$ product for the trunk. This is undesirable because for an isochronic trunk, we want C on the trunk to be dominated by its loads, not by the distributed capacitance of the trunk. In practice therefore, the trunk scheme is typically recommended for fanouts of over 50. Below this, a single buffer scheme is recommended.

By designing with the following principles for single-ended, double-ended, and buffered clock trunks, clock nets of up to 2000 fanouts can achieve < 1.5 ns skew in 1.0- or 0.7-μm CMOS gate array technology [21]. When the clock fanout is between 50 and 500 unit loads, a single-ended trunk scheme, as shown in Fig. 63.33(a) provides a good tradeoff among skew, area, and delay. A single clock driver input buffer is used to drive the trunk line, which is typically realized by six first-layer metal lines in parallel with metal filled in. Clock trunk sizing for a given fanout must set a minimum width to take current density limits into account, given that all of the clock net current flows through the trunk if the tributaries are not buffered. Tributaries of nominally constant fanout branch out in second-layer metal. To the extent possible, macro-cells and hard-coded mega-cells should be laid out with the clock trunk in mind, ideally permitting the clock trunk to be located in the middle of the logic to which it fans out.

The tributary branches may or may not be buffered with their own clock drivers, depending on the total number of loads to be driven. Buffering primarily has the effect of reducing overall delay, rise time, and total current density in the trunk, by allowing a smaller trunk driver. From a purely skew-oriented view, however, it is better not to have the secondary drivers as long as the trunk is isochronic and the size and delay of the main driver is acceptable. When using local buffering, it is important that the branch loads on all tributaries be balanced, more so than when using an unbuffered trunk. Local buffering is, therefore, primarily a way of distributing the total buffering load so that no one buffer needs to be extremely large.

For layout software simplicity, the main trunk may be constrained to use vertical (or horizontal) routing channels only. The main trunk (or each of possibly several main trunks) should be placed as close as possible to the centroid of area of all the loads which it drives. Layout or floorplanning software for commercial ASIC design can typically assist the designer in clock trunk placement by visualizing the spatial distribution of clock loads.

When a design has between 500 and 2000 clock fanouts, a double-ended clock trunk, as in Fig. 63.33(b) or (c) is recommended. The double-ended clock drivers are internal buffers which use the I/O slots of two pins that will not be used externally thereafter. Both single-ended and double-ended clock trunk schemes use only one external clock input pin, with adjacent pins providing an AC ground and switching noise isolation by powering the drivers with dedicated V_{ss} and V_{dd} pins for the clock buffers. In general, clock input pins should always be surrounded by nondriven pins to minimize the possibility of crosstalk coupling into the clock waveform. Clock pins also should be chosen so that minimal internal routing is required between the predriver associated with the input pad and the clock trunk drivers. As a clock trunk design is laid out, the spike current draw from V_{ss} due to simultaneous switching of large fanouts on the clock net should be assessed and considered in determining how many V_{ss} pins are needed in a particular design.

In the double-ended trunk scheme some care must be taken to ensure that an equal-length path can be routed from the midpoint of the trunk, where the clock input branches to the trunk drivers at both ends of the clock trunk, and that a direct routing from the side of the die to

TABLE 63.11 Fanouts of Clock Trunk Schemes in 1-μm CMOS Gate Arrays (Maximum Number of Clock Loads Driven with < 1.5 ns Skew)

Clock Frequency (MHz)	Single-ended Trunk	Double-ended Trunk	Double-ended Trunk with Local Buffering
50	500	2000	>2000
60	450	1500	>1500
70	400	1200	>1200

Source: Clock Scheme for One Micron Technologies, Rev. 1.1, LSI Logic Application Note, LSI Logic Corp., Aug. 1992.

the branch point at the center of the die is feasible. Particularly when preconfigured macro-cell functions have been placed, the metallization layer needed to bring the clock predriver into the middle of the die may be blocked. This leads to the recommendation that the clock input pin be placed on the side of the die that is in line with the clock trunk [see Fig. 63.33(c)]. The line to the branch point and the two branch lines can then use the same metallization layer as the clock trunk and can be automatically provided for as part of the routing channel width reserved by the clock trunk layout software.

The two double-ended arrangements will have a basic skew given by the (nontrunk) R_0C_0 delay across one half of the chip's dimension (typically under 300 ps), plus skew due to any imbalance in buffer loads and thresholds. An advantage of the buffered clock trunk is that the capacitive load of the clock tree is distributed somewhat in both time and position across multiple buffer stages, reducing the current spikes occurring during a clock edge and their impact on ground bounce and injected power supply noise. On the other hand, a total of three or four buffer stages associated with this structure (for low skew in large applications) may cause high delay between the clock edge used to latch incoming and outgoing data on chip and the external system clock. Clock net fanouts achievable using the clock trunk scheme in commercial gate arrays are summarized in Table 63.11.

In even larger dies with fanouts of over 3K flip flops, multiple symmetrically driven double-ended clock trunks can be established to control the maximum distance of any point from a clock trunk. For example, with two trunks placed one fourth of the die width in from the sides, no load is over one fourth the die diameter from a trunk, and the loading of branch buffers is half that required with one trunk. Branch lines from different trunks should not be connected together where they meet in the middle of the die. These points are far enough away in terms of delay from their common driving points that joining them could cause power-wasting buffer output fights. In a further variant on the buffered clock trunk, buffers have their outputs ganged (i.e., shorted) by an additional vertical metal line parallel to the trunk, close to the buffer outputs. The effect of shorting the branch buffers is to equalize the propagation delay through the trunk and distribute the capacitance per buffer more uniformly. This has been found to reduce skew considerably if the branch buffers were not equally loaded and also reduced skew (although less so) in the balanced buffer case [26].

Clock Ring Configuration

The clock ring approach shown in Fig. 63.34 combines aspects of the clock trunk, quadrant, and the single large buffer approaches to achieve a combination of moderately low skew, and moderately low delay without the possibly high routing-dependent skew of the pure single-buffer scheme. The ring approach also simplifies overall clock and signal routing conflicts in a two-layer metal process. The external clock is buffered at entry with a moderate- to large-scale buffer which drives a clock ring that follows the die perimeter. The ring is not a widened trunk because typically less than 50 other buffers are driven off the ring, not the entire clock net. Therefore,

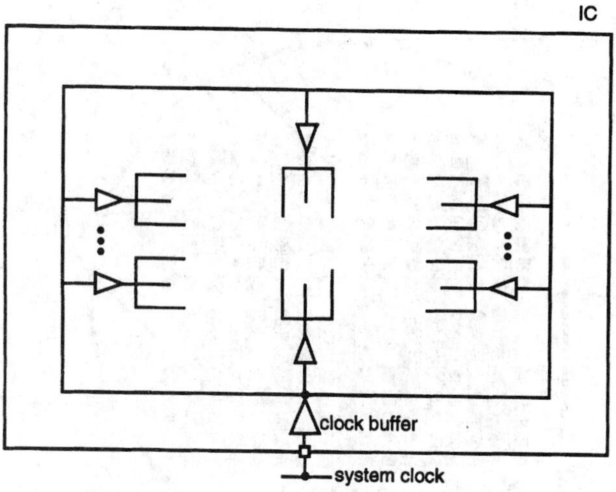

FIGURE 63.34 The clock ring concept.

with relatively low C_{load} on the ring, it is not widened. The extra capacitance of a widened ring would be relatively high (on a square die, the total ring length will be four times the length of a corresponding central trunk) and only increase skew and delay. The ring drives secondary buffers sized to drive balanced groups of flip flops in the core of the chip. The aim of the ring on a large die is not to create a wholly isochronic perimeter, although this could be approached by driving the ring at multiple symmetric locations. Rather, the ring establishes a relatively low-skew reference perimeter from which any interior clock load can be reached either by a purely vertical path or purely horizontal path (i.e., in a direct line of a single metal layer) no longer than one half the die size. The worst case routing distance from the ring is the same as in the clock trunk scheme, but two thirds of all locations are within half that distance in the ring, whereas only one half of all uniformly distributed loads are within half of the maximum distance of a central trunk. In practice with good load balance on the secondary drivers, clock skew of 0.8 to 1.0 ns has been obtained in a designs of up to 30K gates. If the secondary drivers are well balanced, skew in this architecture will depend primarily on the R_0C_0 delay from the ring driving point to its far side, around the periphery, typically 0.8 ns for a die of 350 to 400 mil. Relatively low chip delay is obtained by using the clock signal on the ring as an "early clock" with which to time I/O latches. The ring is electrically closed as this helps distribute the subbuffer capacitance and equalize delays, especially if driven at two opposing points.

H-Trees

The H-Tree is an area-efficient regular structure most suited to clock distribution in systems in which the synchronized modules are identical in size and placed in a regular array. Figure 63.35 illustrates a 256-module H-tree tested on a 4-in. wafer by Keezer and Jain [17]. The scheme balances the R_0C_0 delay through the clock network by geometric symmetry so that the delay is nominally constant from the root to any leaf node. Loads are clocked only at leaf nodes of the tree. The minimum feature size of the process can be assumed to set the line width of the H-tree at its leaf nodes and each preceding level has progressively wider lines to maintain constant current density and to minimize impedance mismatch effects (at wafer scale and above) when no branching buffers exist. The H-tree is driven by a buffer at its root and may or may not have additional buffers at branching points. ASIC manufacturers have been able to achieve skew below 500 ps at fanouts of > 5000 with experimental H-tree layouts in third-layer metal [22].

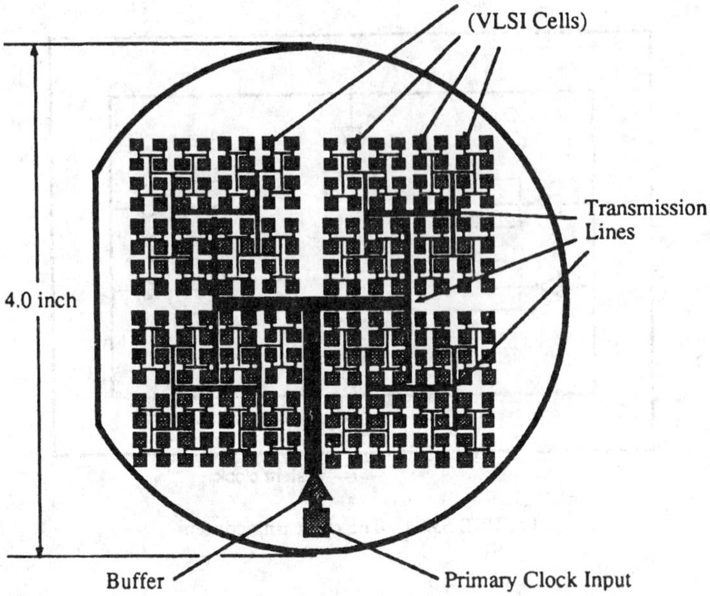

FIGURE 63.35 The H-tree concept illustrated in the form of a 256-cell passive H-tree for wafer-scale integration. *Source:* [Keezer and Jain, 1992].

The H-tree approach is most practical only if an entire layer in a multilayer PCB or a third or fourth metallization layer in CMOS can be dedicated for H-tree clock distribution. Otherwise, the H-tree may encounter (or cause) a large number of routing blockages or require poly links which will disrupt the H-tree performance. In addition, many VLSI designs include memory cells or other hardcoded cells that are incompatible with the ideal symmetry of the H-tree. However, if a suitable layer is available for H-tree layout, it may be applied to random-logic designs by considering each leaf of the node as a clock supply point for all clocked loads within an isochronic region around each leaf node. The whole die is then tiled with overlapping isochronic regions, each fed from out of the plane by a leaf of the overlying H-tree. Each leaf of the H-tree might also use a variably sized buffer to drive the particular number of clocked loads of the random logic layout that fall in its particular leaf node zone.

Kung and Gal-Ezer [18] have given an expression for the time constant of an H-tree, which characterizes the total delay from root node to leaf:

$$\tau_H = 1.43N^3\left(3 - \frac{2}{N}\right)R_0C_0 \tag{63.6}$$

where an $N \times N$ array of leaf nodes is driven. Absolute delay rises as N^3 for large N. This in itself does not limit the clocking speed because, at least theoretically, more than one clock pulse could be propagating toward the leaf nodes within the H-tree. As a practical matter, however, the delay of the H-tree expressed in (63.6) is essentially a rise time effect on the clock waveform. A slow rising edge out of the H-tree can lead to significant skew due to threshold variations in the leaf node buffers. These considerations apply to the on-chip context in which the H-tree clock network is dominated by RC diffusion delay effects. Equation (63.6) also describes an unbuffered H-tree. By placing buffers at selected levels of the overall tree, total propagation delay through the tree will increase, but the bandwidth of the tree may be preserved by effectively reducing N to the portions of the overall tree between buffers, in (63.6). In contrast to the rapid bandwidth fall-off on-chip, at the multi-chip module level of system integration an H-tree may be designed from an impedance-controlled transmission line standpoint to obtain very high clock bandwidth. In experiments of this type Bakoglu [3] has achieved 30 ps of skew at 2 GHz with a 16 leaf H-tree covering a 15 × 15 cm wafer area.

In an H-tree the total clock path length doubles each time one moves up two levels from the leaf nodes toward the root. Based on this, Kugelmass and Steiglitz [19] have shown that given σ_b and σ_w as the standard deviation of buffer delay (if present) and wire delay, respectively, the total delay of an H-tree considering buffers and wires has variance:

$$\sigma^2 = \sigma_b^2 \log_2(N) = \sigma_w^2 2(\sqrt{N} - 1) \tag{63.7}$$

and that the average case skew between any two leaf nodes is bounded by:

$$E[\text{skew}] = \sigma_w 4\sqrt{(\sqrt{(N-1)}\ \ln(N))} \tag{63.8}$$

where N is large and wire length effects dominate. Average case (or expected) skew is the maximum difference between clock times, averaged over many design trials, not the average clock time difference in any one design. Kugelmass and Steiglitz [19] also gives results for the probability that a sample value of the skew exceeds the mean skew by a given factor in either an H-tree or a binary tree, based on assumptions that all wire length and buffer delay variables are independent and identically distributed:

$$P(\text{skew} > E[\text{skew}] + a) \leq \left[1 + \left(\frac{a}{E[\text{skew}]}\right)^2 48\,(\ln N)^2/\pi^2\right]^{-1} \tag{63.9}$$

where a is the amount of time by which the mean skew is exceeded. These expressions may be used to estimate skew-limited production yield at a given target clock speed.

Delay, Skew, and Rise Time Comparison

The five clock distribution schemes described so far were studied in a unified, experimental way by Nigam and Keezer [24] using HSPICE simulations. They compared each scheme on a 5-in. wafer holding an 8 × 8 grid of modules to be clocked. Each module presented at a total load of 2 pF and the interconnect R and C values were taken for a typical 2-μm double-metal CMOS process. Clock distribution lines were 10 μm wide for the buffer tree and the H-tree, except for its trunk, which was 40 μm. The clock trunk schemes used a 20-μm trunk width. All interconnect was modeled as distributed RC, with transmission line delay effects included. The results are tabulated in Table 63.12 and give an excellent overview of the relative characteristics of each method. The H-tree has essentially no skew, but has the highest delay and slowest clock edge, which can translate into skew due to threshold variations in the loads. The clock trunk has good skew and moderate delay. The best overall performance is achieved by the four-quadrant scheme, essentially by virtue of reducing the clocking area to one fourth of the overall size of the other clock networks.

Balanced Binary Trees

A balanced binary tree (BBT) is an unbuffered clock distribution tree in which each branch node has exactly two subtending nodes, and the delay from the root to all leaf nodes is made constant by placing branch points at the "balance point" of the two subtending trees from any node. Balanced binary trees are not simply clock buffer trees with fanouts of two. The significance of the BBT is that constant delay is achieved through multiple levels, without any buffers, and the BBT can be constructed by a fairly simple algorithm. Passive BBTs also may be used in practice to implement delay-equalized fanout networks between the active buffer levels of a larger buffered clock tree. The BBT concept should not be confused with the buffered clock

TABLE 63.12 Comparative Performance of Clock Distribution Networks (8 × 8 Array of Loads Clocked at 31 MHz and Constant Total Power, (650 mW)

Scheme	Delay (ns)	Skew (ns)	Rise/Fall Time (ns)
3-level symmetric buffer tree	7	3	12.5
Single buffer H-tree	15	~0.0	38
Clock trunk with branch buffers	13	4	14.2
Clock trunk with ganged branch buffers	14.2	2	16
4-pin-quadrant scheme, 2 buffers per quadrant	4.3	1.3	9

Source: N. Nigam and D. C. Keezer, "A Comparative Study of Clock Distribution Approaches for WSI," *Proc. IEEE 1993 Int. Conf. WSI*, pp. 243–251.

tree concept in general, however. The key is that the generalized buffer clock tree does not have path delay equivalence if its buffers are removed, whereas the BBT has this property.

The basic ideas and methods of generalized balanced tree synthesis are explained in [6]. The clock tree that results has two branches at every node. Clocked loads appear only at the leaves of the tree. The line lengths at each level can be different than those at other levels of the tree, and the two line segments that are children of any node also can be of unequal lengths. The key, however, is that at each branch the total distance from the branch point to any subtending leaf via one outgoing path is equal to that in the other outgoing direction.

Figure 63.36 illustrates the basic procedure for BBT synthesis. The process works from the bottom up, by considering all leaf node positions, i.e., clock entry points for modules or isochronic local regions. Leaf nodes are subjected to a generalized matching or pairing process in which each node is paired with the other node closest to it in the Manhattan street length sense within the available routing channels. A first-level branch point is then defined at the midpoint on each line joining the paired leaf nodes. A similar pairing of the first-level branch points then defines a new set of line segments, each of which has two leaf nodes symmetrically balanced at its ends. Second-level branch points are then defined at the RC balance point on each line joining first-level branch points, and so on. Each iteration consists of finding the minimum path length matching of pairs of branch points from the previous iteration and defining the next set of branch points at the time-constant balance points on the line between just-matched lower-level branch points. Assuming leaf nodes present equal loads, the first-level branch points are the midpoints of the pair-matching line segments. After that, the balance point for level i is the point on the line segment joining matched level $(i - 1)$ branch points at which the total RC wiring time constants (simply total length if all wires and layers are identical) to the leaf nodes in the right and left directions are equal. This is repeated until only two branch points remain to be matched. A line is routed between them and driven by the clock signal at the final balance point, which defines the root of the BBT.

Clock trees developed in this way are in one sense highly structured and symmetric in that the total delay from root to any leaf is nominally constant, like the H-tree. Unlike the H-tree, however, the actual layout is irregular, allowing the BBT to accommodate the actual placement of cells and modules and to cope with the limited and irregular routing channels available in designs that do not use a completely regular layout.

Skew in BBTs has been considered theoretically and experimentally. Kugelmass and Steiglitz [19] showed that in a BBT with independent variations in the delay at each stage of the tree, with σ_0^2 variance, the expected skew is fairly tightly bounded by:

$$E[\text{skew}] \leq \frac{4\sigma_0}{\sqrt{2 \ln 2}} \ln N \qquad (63.10)$$

where N is the number of leaf nodes of the tree.

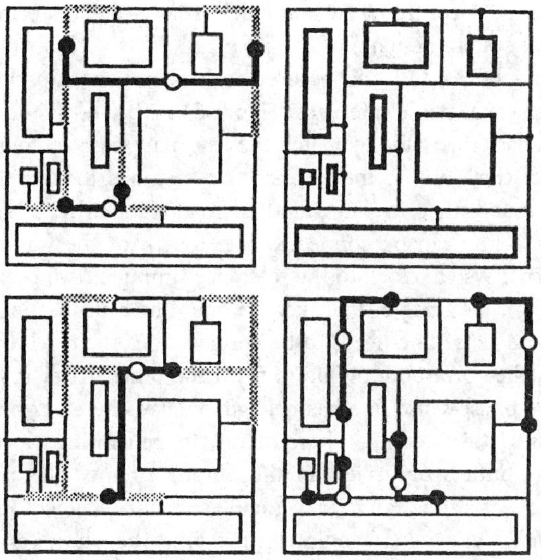

FIGURE 63.36 BBT synthesis in an 8-terminal net. Solid dots are roots of subtrees in the previous level; hollow dots are roots of new subtrees computed at the current level. *Source:* [Cong et al., 1991].

Using the above expressions we can compare the H-tree to a BBT. The comparison shows that when the regular structure of an H-tree is feasible, it is of relative merit for large fanouts because the expected skew grows more slowly ($O(N^{1/4}(\ln N)^{1/2})$) than the BBT tree in which expected skew grows as $O(\ln N)$. For comparison, assuming 10,000 leaf nodes and the same σ_w per unit wiring length, the expected skew of the H-tree is about one half that expected of the BBT. This outcome is primarily because the BBT must be deeper (have more levels) than the H-tree for the same fanout.

Experimentally, Cong et al. [6] produced a large sample of design trials of 16- and 32-node BBT clock trees, synthesized on a 1000 × 1000 grid. It was shown that the BBT resulted in less than 2% of the skew from a corresponding minimum spanning tree (MST) for clock distribution to the same loads, even though the BBT had 24 to 77% more total routing length than the MST. The MST benchmark characterizes the skew that would typically result if the clock was routed as an ordinary signal net, with no special concern about skew.

Balanced clock tree synthesis is now supported by at least one gate array provider [22] in combination with a three-level clock buffer tree hierarchy. Skew of < 500 ps is achieved in 0.5 μm designs of up to 13,440 clocked loads. By using appropriately sized buffers and wire width at each level of the balanced tree, clock rise time is typically 0.8 to 0.9 ns at the terminal nodes. The clock tree compiler is invoked after floorplanning. The compiler takes into account the resistance and capacitance of different wire types, the length and width of wires, and the input capacitance of clock pins and buffers. Up to three active buffering levels can be used, with fanouts of up to 64, 14, and 15, respectively, from buffers at each level. The fanout subnet driven by each buffer is laid out as a (passive) BBT, so that the leaves of one balanced tree are the buffers that act as the roots of further passive binary balanced subtrees driven by the next buffering level. At the lowest level, local buffers each drive up to 15 loads via a final four-stage passive BBT.

Clocking Schemes Involving Phase-Locked Loops

A PLL is a negative-feedback control system in which the phase (and, implicitly, frequency) of a voltage-controlled oscillator (VCO) or phase shifter is brought into alignment, or to a predefined static phase offset, with respect to the phase of a periodic reference signal. To date, the application of PLLs is to control skew and clock delay problems primarily at the multichip and interboard levels of system design.

Figure 63.37(a) shows how a PLL can be used to lock the on-chip clock phase at a selected point on-chip, to an external phase reference. Figure 63.37(b) shows the mid-trunk phase on a single-ended clock trunk being made to match the external phase reference. Here, the feedback line from the middle of the trunk to the PLL input is assumed to have negligible delay in itself because it is a metal line with only one standard load. Similarly, in the double-ended clock trunk scheme, the sense line can be connected one fourth of the way along the clock trunk. This will lock the internal system clock to the reference timing at two points on the clock trunk, as shown in Fig. 63.37(c), reducing overall skew to one fourth of that in the single-ended clock trunk scheme. The phase-sense line needs to be connected to the clock trunk at only one point because, by symmetry, the corresponding point from the other driver is similarly phase locked.

In general, when the phase-sense line has negligible delay, the clock phase at the sense point is driven into lock with the reference phase. Thus, in generating clock signals we can null out the delays of large buffers or drivers in the output circuits as well as their process and temperature-dependent variations, and, in general, coordinate clock and data phases at the inputs to another chip at any remote point by bringing the phase-sense line back from the actual point where the phase-controlled relationship is desired. In this way even the delay of an off-chip driver can be cancelled out by including it within the PLL feedback loop. If delay in the phase-sensing feedback path is not negligible, then its effect is to advance the phase at the desired control point. The feedback delay can be compensated by a matching delay in the forward path from the VCO to the phase-sensing point, or at the PLL input. An inverter in the feedback signal path is also a convenient way to cause a 180° phase shift between reference and VCO without requiring any loop delay.

With the addition of a frequency divider (divide by N) in the feedback path, as in Fig. 63.37(d), the VCO operates at N times the frequency of the reference clock input. For $N = 2^n$ frequency multiplication, the feedback divider can be a simple ripple counter of n toggle flip flop stages. For other multipliers a synchronous counter is usually used. The delay-matching element at the PLL phase detector input compensates for delay from the feedback path divider. On-chip frequency multiplication can ease a number of system-level design problems. The overall system clock rate need not be equal to that of the fastest chip in the system. Transmission line effects across the relatively long distances of PCBs or backplanes can be reduced by operating at a lower clock frequency outside of the system ICs. The lower frequency of system reference distribution may also reduce power, and usually assists in meeting radiated emission specifications for electronic equipment. Because a PLL regenerates the clock in each IC, a considerable amount of clock edge slew rate control can be used on the external system clock, further easing EMI and power supply switching noise problems. The difficulty of retaining clock waveform integrity getting on- and off-chip at high frequencies through inductive packaging and bonding leads is also eased for the same reason.

PLLs for CMOS

A block diagram of a PLL is shown in Fig. 63.38(a). The VCO exhibits a positive monotonic frequency of oscillation in response to a control voltage, characterized by the slope of its frequency vs. voltage curve. The loop filter, $H(f)$, is of a general low-pass characteristic, often of an all-poles design to avoid any jitter peaking (or AC gain) in the closed loop transfer function of the PLL. The loop filter must provide a DC coupled path between the phase detector and VCO. The phase (and/or frequency) detector compares the VCO output phase to the input reference phase and generates an output signal that is either of a DC nature or has a DC component that is proportional to the phase difference between the reference and feedback signal. The phase detector is characterized by the rate of change in the DC component of its output vs. phase input difference in volts/radian.

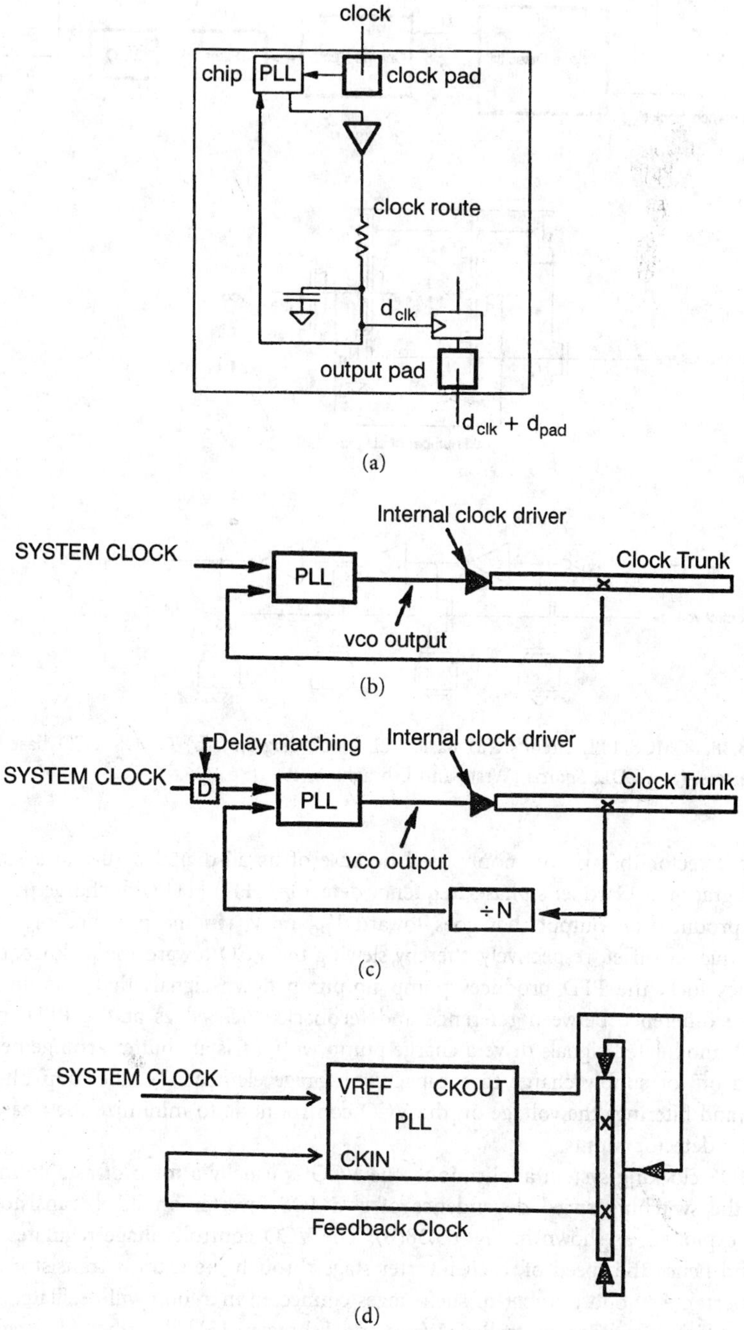

FIGURE 63.37 PLLs for skew and delay control: (a) cancelling internal buffer and clock net delay, (b) halving the skew in a single-ended clock trunk, (c) reducing double-ended clock trunk skew to one fourth of the single-ended trunk, (d) on-chip frequency multiplication. *Sources* (a): [Weste and Eshraghian, 1993], p. 335. (b)–(d): LSI Logic Corp., *Phase-Locked Loop Application Note*, Nov. 1991.

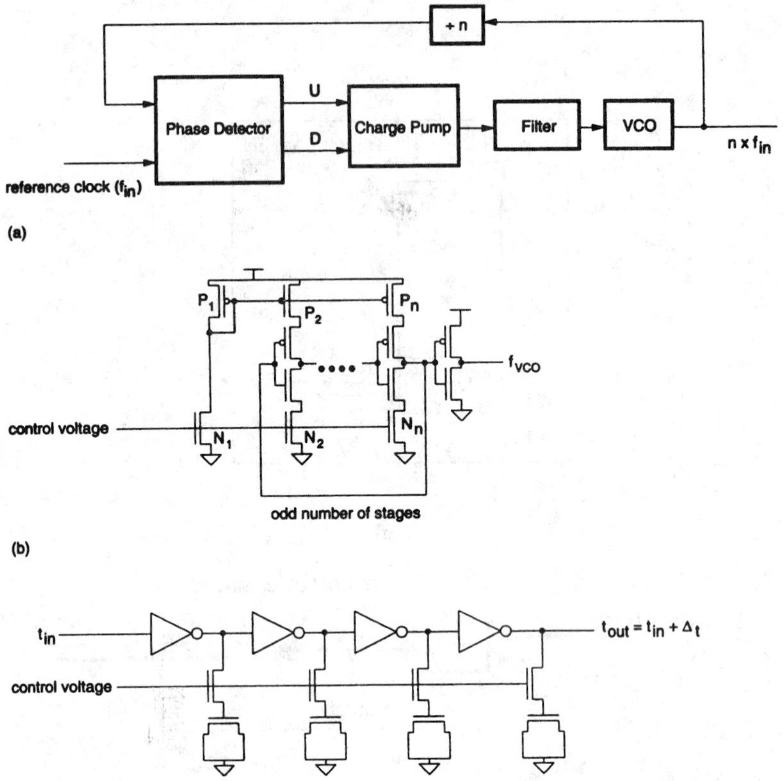

FIGURE 63.38 CMOS PLL circuits: (a) basic PLL block diagram, (b) CMOS VCO based on current starved inverters, (c) VCDL. *Source:* [Weste and Eshraghian, 1993], p. 336.

A phase detector that is commonly used because of its all-digital nature and suitability for CMOS integration is Gardner's phase-frequency detector (PFD) [11] with charge pump outputs. The PFD produces an output that goes toward V_{DD} or V_{SS} in the presence of a negative or positive frequency offset, respectively, thereby slewing the VCO toward the lock frequency. Once in frequency lock, the PFD produces pump up/pump down signals that vary in proportion to the time difference between reference and feedback clock edges at the PFD input. These pulse-width modulated signals drive a charge pump with a tristate buffer arrangement to either hold, bleed off, or supply charge to a capacitive storage element (i.e, the loop filter), thereby adjusting (and filtering) the voltage on the VCO control node to minimize the phase difference at the phase detector inputs.

For CMOS clocking system applications, the VCO is usually a form of astable multivibrator in which the switching speed dependence of a CMOS inverter on its n-transistor pulldown current is exploited, as shown in Fig. 63.38(b). The VCO control voltage regulates the current flow in, and hence the speed of, each inverter stage through the extra n-transistor stage added to each inverter. Any odd number of such stages connected in a loop will oscillate, but now the relaxation period is voltage controlled. Weste and Eshragian [31] describe a 13-stage "current-starved inverter" VCO based on this approach. A related VCO design is based on varying the load capacitance seen by each inverter (in a chain of inverters) by applying the VCO control voltage to an n-MOSFET in series with the gate of another transistor configured as a capacitive load [16]. An on-chip RC loop filter can also be constructed from a CMOS transmission gate biased as a resistor and MOS gates used as capacitors (source and drain both connected to V_{SS}) [31].

If frequency multiplication is desired, a PLL with true VCO is required. Otherwise, many PLL applications can use a voltage-controlled delay line (VCDL) in conjunction with a "raw-clock" input signal, as in Fig. 63.38(c). All the phase-lock feedback principles are the same, except we phase-shift the raw-clock input as required, rather than controlling the VCO oscillation phase. This eliminates the risk of the PLL ever failing to lock in and may be simpler to fabricate. On the other hand, system design must take into account a more limited range of phase-shifting ability (a full half-cycle of phase control range may require a lot of delay stages) and to make sure initial delays are nominally centered within the positive-only delay control range of the VCDL. Another more subtle point is that while a VCO introduces a perfect integrator ($1/s$ term) in the PLL closed loop response, a VCDL does not. A VCDL, therefore, should not be simply substituted for a VCO without revisiting the closed loop response characteristics for noise bandwidth and possible jitter peaking.

Special power, grounding, and testing considerations apply when a PLL is used. A PLL is basically a linear circuit, so noise is especially important. Particularly when a frequency multiplying PLL is used, the VCO power supply should be well decoupled from system noise, and the input phase reference should be highly stable, as the PLL output clock will have N times the reference's phase noise. Noise voltages coupled into the analog PFD output and LPF signal path are similarly converted into phase noise that is N times worse than in a $\times 1$ PLL. Leadless on-chip decoupling capacitors are recommended as are dedicated power and ground pins for the PLL. The R and C components for the loop filter are often off-chip. In this case it is important that they are connected (depending on the $H(s)$ configuration) to the same analog ground reference as the VCO and PFD. The VCO output, or a divided down version of it, should be brought to an external pin for lock-in validation and as an aid in possible global system clock tuning. For testability, several other separate pins are typically required for independent access to input and output of the PFD, LPF, and VCO each. An on-chip PLL can require up to six or more pinouts.

Anceau's PLL Scheme

Anceau [1] developed a PLL-based approach for large systems in which modules are well-defined, relatively independent, and could be entirely self-timed if not for the need to avoid metastability in communication with other modules. Anceau recognized two natural system scales which are isochronic below different maximum frequencies. One is a global region encompassing the entire system, with a clock period determined by propagation distance delays, or, on-chip, by RC diffusion delays. The isochronic rate for this scale defines a slower clock rate for a system-wide communication bus. The second type of clocking region is smaller local regions which can run at full speed and are characterized by critical logic path delays and lumped capacitive loads within modules, not distance-dependent delays. Each smaller region will be free to operate in an almost self-timed mode.

The clocking style within each module (e.g., logic type and number of clock phases) can be as appropriate for the individual modules. Skew at the highest clock speeds in the system need be considered only within each module, except that timing must be controlled when reading the common data bus to avoid metastability. This is done by reference to the active edge of the slower-rate communications clock (comm_clk), formed by dividing down the master module clock frequency. The rising edge of comm_clk strobes the enabled driver data onto the bus. All other nondriving modules in the comm_clk cycle read the bus on an internal clock edge that is kept away in time from this transition in comm_clk, for metastability avoidance. Figure 63.39 illustrates the overall scheme. A PLL phase locks the module clock at a predefined angle relative to the comm_clk, thus keeping the raw module clock away from the transition times in the lower rate communication clock. The read timing is then safe because it is always preceded by the comm_clk transition on which new data were strobed to the bus. A monostable triggered by the comm_clk edge can be used inside each module as a delay generator to prohibit any bus read

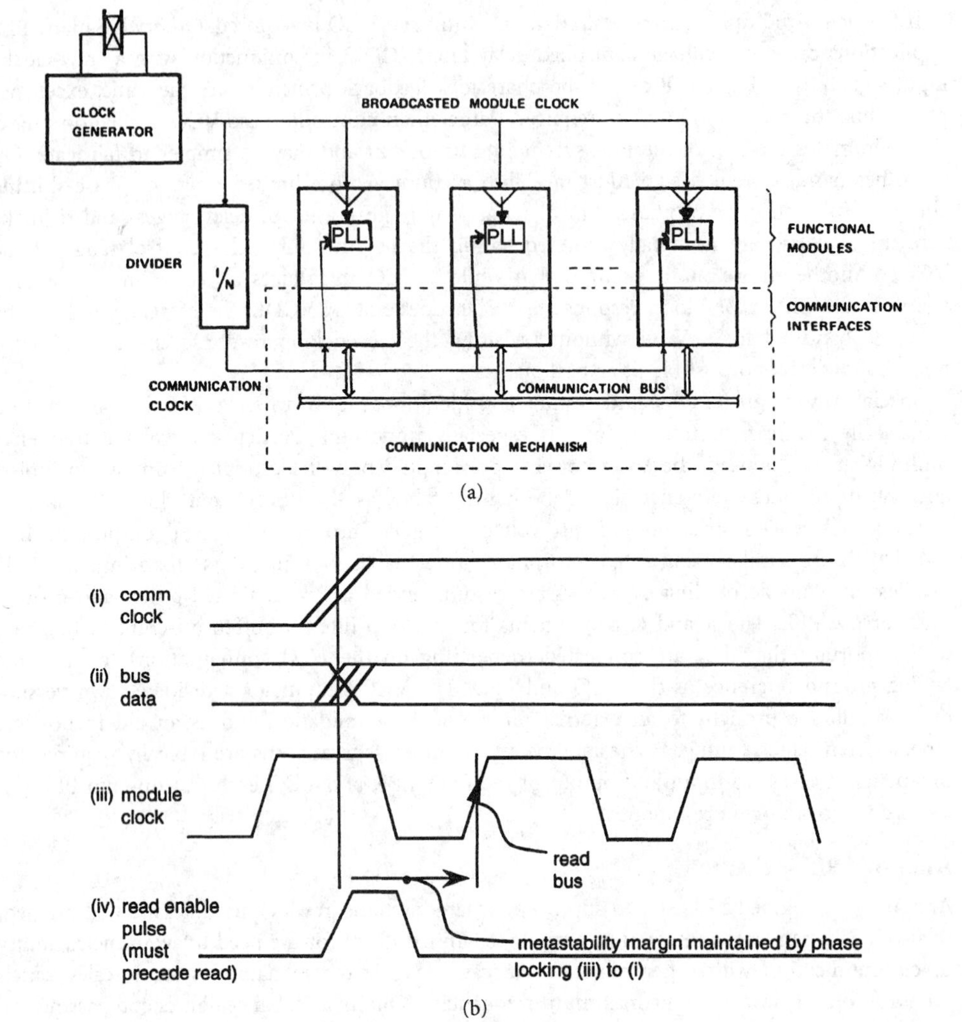

FIGURE 63.39 Anceau's scheme for metastability avoidance: (a) system architecture, (b) interface timing. (*Source:* Adapted from [1].)

in the metastable region. This way, as long as modules write to the bus only on the comm_clk edge, other modules that read the bus will never do so at a moment when the bus data are still in transition.

Grover's Interval-Halving PLL Scheme

A new PLL-based approach to clock distribution in large systems was recently developed for full-speed synchronization of all clocks in a large system [13, 14]. In this scheme any number of nonisochronic points arbitrarily located on a single- or double-conductor reference line independently derive clock that is in absolute phase-lock to a common system-wide reference time. The central principle is that the time between appearances of an isolated pulse traveling down and back on a reference line is the same regardless of the point of observation, as shown in Fig. 63.40(a). This figure plots the trajectory in space-time of an isolated pulse that travels from a site at one end of a line and is returned at the end of the line to its origin (where it is electrically terminated). Figure 63.40(a) is drawn for the most general case of physically separate go and return conductors, looped at the right-hand end ($x = D$), but the space-time

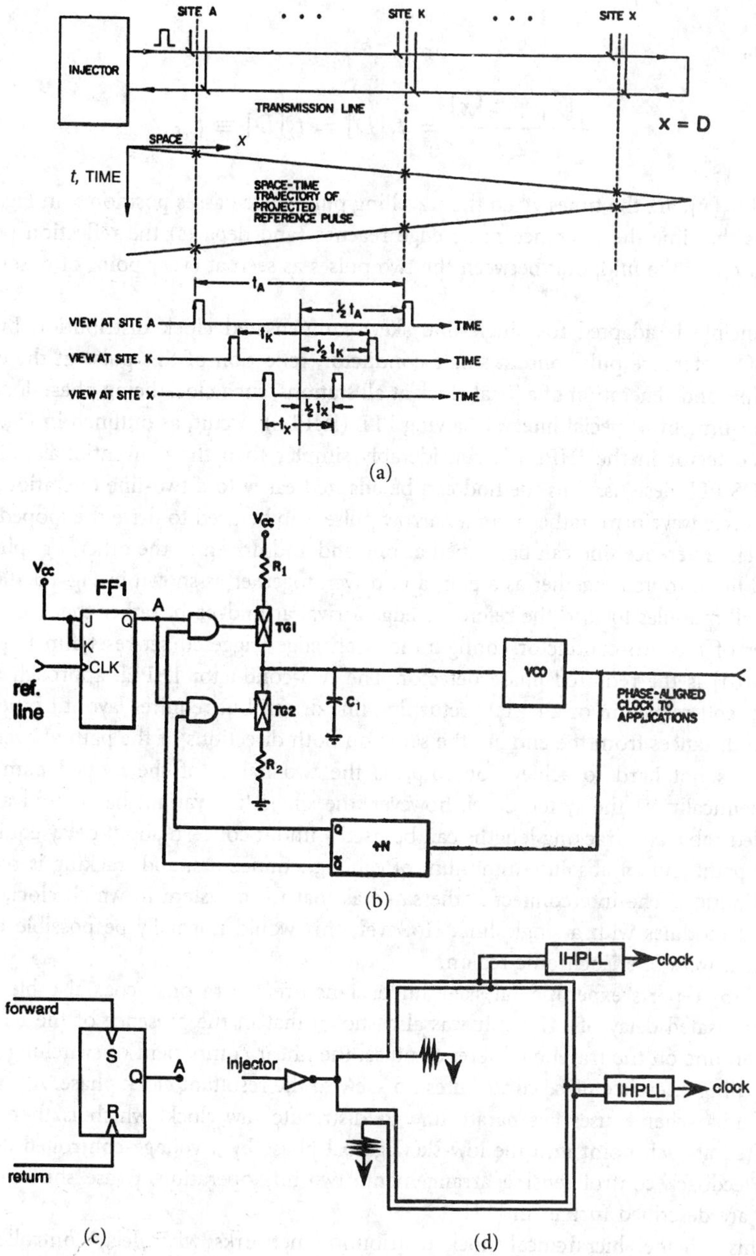

FIGURE 63.40 Grover's interval-halving clock distribution scheme: (a) interval-halving reference-time principle, (b) IHPLL circuit for single-line skew-compensated clock distribution, (c) phase detector for 2-line operation, (d) driving two lines rather than one looped line. *Source* (a) (b): [Grover, 1994].

trajectory of the isolated pulse is identical if the line is a single conductor open circuited at the end and driven by an impedance-matched source. Equivalently, a tristate buffer can terminate and regenerate the returning pulse at the end of the line for on-chip use. In either case it is evident that the instant in time that is halfway between the outgoing and returning pulse edges is the same for all points of observation on the line, regardless of the propagation velocity of the line, i.e.,

$$\frac{t_1[x] + t_2[x]}{2} = t_1[D] = t_2[D] \equiv t_{\text{ref}} \qquad (63.11)$$

where $t_1[x]$, $t_2[x]$, are the times when the travelling pulse edge passes position x in Fig. 63.40(a), and $t[D]$ is the time the reference pulse edge reaches (and departs) the reflection point. This time, called t_{ref}, is the midpoint between the two pulses as seen at every point of observation on the line.

This principle is adapted for single-line skew-compensated clock distribution by periodic injection of a reference pulse onto a single conductor, reflection of this pulse at the end of the reference line, and generation of a local clock at all stations, such clock being phase-locked to the interval mid-time by a special interval-halving PLL (IHPLL) circuit, as outlined in Fig. 63.40(b). The phase detector in the IHPLL is considerably simpler than the conventional PFD used in many CMOS PLL designs. This method can be adapted easily to a two-line operation, in which a full duty cycle waveform, rather than a narrow pulse, can be used to drive the looped reference line path. The reference line can be looped at one end and driven at the other, or split into two terminated lines, routed together as a pair, and driven together as shown in Fig. 63.40(c). In the latter case all modules lock to the reference edge arrival at mid-path, rather than at its end.

In either of the two-conductor configurations an edge-triggered set-reset flip flop function [Fig. 63.40(d)] is the required phase detector. The two-conductor IHPLL approach avoids the need for an end-reflection or a tristate returning line driver, but requires layout of the reference line so that distances from the end are the same on both directions of the path at every tapping point. This is not hard to achieve on-chip, as the two halves of the looped path could be laid out identically. At the system level, however, the single-line variant has the advantage that uncontrolled cable and tracking lengths can be used without concern about delay equivalence in the return path, and an absolute minimum of cabling, connectors, and tracking is required for clock distribution. The interconnect is the same as that for a system in which clock is directly wired to all modules with a single line. However, this would normally be possible only if the whole system were one isochronic region.

Grover [13] reports experimental skew under 1 ns over 30 m on a coaxial cable which has an uncompensated delay of 147 ns. It was also shown that in the presence of the effects of the transmission line on the traveling reference pulse, the linear component of switching time error on the traveling reference pulse contributes no skew to the resultant clock phase. A phase-shifter variant of this scheme uses a separate line to distribute raw-clock, which is then adaptively phase shifted at each point into the low-skew global phase by a voltage-controlled delay under the same feedback control sensing arrangement. Two-line operation, phase shifter, and other variations are described further in [14].

With this scheme, hierarchical clock distribution networks with delay-controlled cabling, delay-tuning, and numerous temperature- and load-dependent intermediate buffers may be replaced by one conductor with arbitrary routing. Both EMI and conducted noise are reduced by buffer driver elimination and because of the reduced average power of the reference pulse compared to the full clock signal. It may also be possible to add new clock-deriving taps in service, offering a growth path that is not limited by a predesigned clock–tree fanout limit. In many applications hybrids of reduced-depth clock trees, fanning out from skew-compensated roots on a single-line clock system of this type, may give the best combination of techniques.

Anceau's and Grover's schemes are similar in that a reference line is distributed to all modules and a PLL generates a local clock at each module. However, in the Anceau scheme modules do not run phase synchronously. Actual skew between modules remains arbitrarily high at the module clock frequency because the comm_clk line is set slow enough to be isochronic over the whole system. Actual delay in comm_clk, which is significant at the higher module rate, is not compensated at modules. Each module derives only enough information to coordinate its bus accesses with other modules, at the slower comm_clk frequency. In Grover's scheme, however, truly synchronous full-speed global clocking of all modules is achieved by returning the signal on the reference line (by reflection or looping) and exploiting the interval-halving time-reference principle and IH-PLL to cancel global skew. Gate-to-gate interaction on any clock cycle is feasible between modules in this case, as opposed to interaction only through the metastable-avoidance bus interface protocol.

Clock Tuning in Large Systems

In large systems clock tuning at the chip, circuit pack, shelf, and rack levels of physical equipment may be required for the highest performance. Circuits to permit clock tuning can be a tapped delay-line circuit with a programmable selector, or (on a circuit card) a printed-in set of loops to be shunted out as needed by a suitcase jack, or a voltage-controlled varactor clock delay buffer. All of these are described in [15]. In general, to aid in the tuning process, one pin on each IC should be devoted to give external observability of the worst-case (if known) clock phase from inside the IC. This way the tuning process can compensate for the delays through I/O pads and clock buffers in large ICs.

Tuning begins by designing cable lengths, tracking, and connectors so that clock paths have nominally equivalent delays. The active tuning process then measures and adjusts relative delay starting from the master clock source to predefined levels of tuning points (TPs) electrically farther from the master clock source, denoted by TP_0. The delay measurement and adjustment repeats through lower level tuning points until the clock in every IC is tuned. In going from the first to successive tuning points, it is preferable to refer delay measurements directly back to TP_0 each time. This may, however, be physically unmanageable, in which case delay tuning to level TP_n can be relative to TP_{n-1}, although overall error relative to TP_0 will be higher in the relative tuning scheme. For systems that must grow in service, operational (i.e., in-service) signals should be the basis of the delay measurement, not requiring off-line signals or patterns.

One convenient way of indirectly measuring delay between points that are not easily accessed simultaneously for oscilloscope measurements is to make an oscillator out of the signal path to be measured. If the number of inversions in the path between TPs is odd, then a multiplexer can be switched to loop the tuning point signal at TP_n back to the TP_{n-1} driving point. A frequency counter can then measure the oscillation frequency, providing data to support automatic clock delay adjustments at the subordinate tuning point.

One mainframe computer used an automatic tuning scheme in which a clock phase-shifter chip produced multiple, slightly time-shifted copies of the clock on each system PCB. Individually selected delayed clock instances were then supplied to each IC on the board through a programmable crosspoint matrix IC. Each clock-receiving IC also provided several internal clock observation outputs to support delay measurements down to the gate level. After automatically measuring the delays of the observable internal clocks in up to 30 ICs per board, the on-board clock selection matrix was programmed, giving each IC its best clock phase for overall system timing margins and minimal skew [30].

Future Directions

Current-Steered Logic

One way of reducing power supply noise injection from clocking is with current-steered logic. Experiments on differential current mode flip flops in CMOS predict very short setup times (300 to 500 ps). Such devices would be very quiet electrically and much less susceptible to varying load effects on delay than on conventional CMOS. On the other hand, such devices might be about twice as large as conventional CMOS flip flops and require more power. Using the bi-CMOS ability to integrate ECL type structures with CMOS may, however, be part of the solution for clocking very high speed medium-scale integrated devices.

Reduced Voltage Swing

Another potential method for reducing clocked load power consumption is to reduce voltage swings. Some experiments indicate a significant reduction in clock-related power and ground noise, but skew and delay objectives are more difficult to meet as the devices slow down in response to lower switching voltage swings.

Mixed Technology

Here, clock speed increases are envisaged by using current mode logic circuits selectively to implement critical timing paths in otherwise all-CMOS systems. Net power reductions may also be obtained with ECL-based high-speed serial-multiplexed interfaces to replace wide buses which have many parallel CMOS drivers.

$\overline{Q}$ Elimination

Most logic families provide flip flops with both Q and $\overline{Q}$ outputs as standard cells. An approach that could potentially halve the clock-related switching current and power is to eliminate the $\overline{Q}$ output buffer and develop corresponding logic synthesis tools to utilize inverted inputs and other logic means to assemble logic functions without the $\overline{Q}$ outputs from flip flops. In one experiment of which the author is aware $\overline{Q}$ buffers were removed, halving overall flip flop power consumption, at the cost of only a 5% degradation in clock-to-Q delay.

Dedicated Layer for Clock Distribution

A number of workers are advocating or are already using dedicated third-layer metal for clock distribution. This affects process cost, but the advantages can be significant in high-performance applications. Third-layer metal is lower than other layers in resistance and capacitance. By moving the clock net, which is the largest single net in many designs, out of the other layers, routability of all other signals is improved and floorplanning simplified. Moreover, the clock tree can avoid uncertain delays due to unpredictable routing or due to polysilicon links in series when routing in fewer shared signal layers. In addition, noise due to clocking can be more easily isolated in the third-layer metal approach.

Optoelectronic Clock Distribution

Optical clock distribution takes advantage of the three-dimensional nature of imaging optics to remove all but the last-stage buffering levels of the clock distribution tree from the plane of the circuit, thereby eliminating multiple stages of buffering and metallization for clock routing. Figure 63.41 is an overview of the basic idea proposed by Clymer and Goodman [7]. The optical clock signal is generated off-chip and drives a laser diode at the top of the figure. The optical beam is expanded onto a transmission hologram which focuses the light intensity onto predefined locations where optical detectors are fabricated into the wafer or die. The optical signal is detected, amplified, and used to drive a local clock generator–buffer which supplies a local isochronous region. The optical path length differences are not equalized in this scheme as the optical path velocity is so high as to make the all optical path delays negligible as compared to the diffusion and lumped capacitive delays that determine the clock rate of the electronic system. When sources, detector, and packaging for this type of approach are developed, the potential exists for very low skew–high speed clock distribution, with greater on-chip densities by eliminating most clock routing. One of the main challenges is in attaining uniform response times from the optical detector-amplifier combination (which tend to be sensitive to feature size variations) and the development of sources in the optical wavelength range for photodetectors that can be fabricated within the conventional CMOS circuit environment.

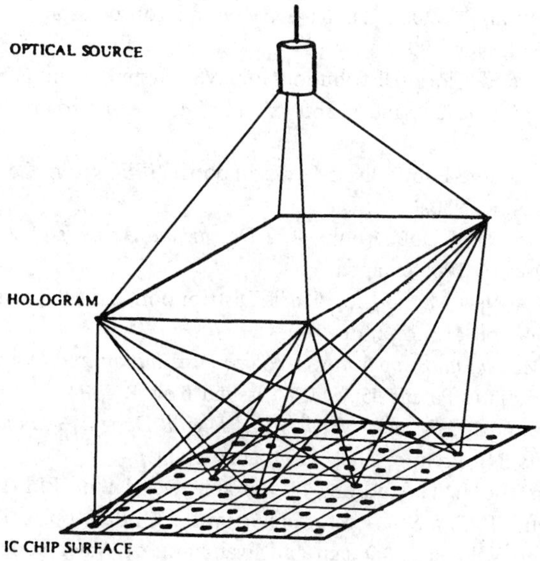

FIGURE 63.41 Clymer and Goodman's concept for wafer-scale optical clock distribution. *Source:* [Clymer and Goodman, 1986].

Reconfigurable Clock Nets

In WSI systems, where a single short in a very large clock net may disable an entire wafer-level system, Fried [10] advocates methods of restructuring a clock net to enhance yield, primarily through the addition of a controllable tristate output stage to clock buffers within the clock distribution network. This way failed portions of the clock net can be isolated, or, with redundant interconnect and buffers, they may clocked by an alternate path. In particular, tristate buffers may be programmed on or off to select clock for each module from redundant connections to the central clock net, or to simply isolate failed clock net subregions from the drivers of unfailed portions.

References

[1] F. Anceau, "A Synchronous Approach for Clocking VLSI Systems," *IEEE J. Solid-State Circuits*, vol. SC-17, no. 1, pp. 51–56, Feb. 1982.

[2] M. Afghahi and C. Svensson, "A Unified Single-Phase Clocking Scheme for VLSI Systems," *IEEE J. Solid-State Circuits*, vol. 25, no. 1, pp. 225–233, Feb. 1990.

[3] H. B. Bakoglu, *Circuits, Interconnections and Packaging for VLSI*. Reading, MA: Addison-Wesley, 1990, chapter 8.

[4] S. Boon, S. Butler, R. Byrne, B. Setering, M. Casalanda, and A. Scherf, "High Performance Clock Distribution for CMOS ASICS," *Proc. IEEE 1989 Custom Integrated Circuits Conf.*, pp. 15.4.1–15.4.5, 1989.

[5] M. A. Cirit, "Clock Skew Elimination in CMOS VLSI," *Proc. IEEE Int. Symp. Circuits Systems*, 1990, pp. 861–864.

[6] J. Cong, A. Kahng, and G. Robins, "On Clock Routing for General Cell Layouts," *Proc. 4th IEEE Int. ASIC Conf.*, pp. 14.5.1–14.5.4, Sept. 1991.

[7] B. D. Clymer and J. W. Goodman, "Optical Clock Distribution to Silicon Chips," *Opt. Engin.*, vol. 25, no. 10, pp. 1103–1108, Oct. 1986.

[8] R. H. Cramback, C. M. Lee, and H. S. Law, "High-Speed Compact Circuits with CMOS," *IEEE J. Solid-State Circuits*, vol, SC-17, pp. 614–619, June 1982.

[9] D. Dobberpuhl et al., "A 200 MHz 64-b CMOS microprocessor," *IEEE JSSC*, vol. 27, no. 11, pp. 1555–1567, Nov. 1992.

[10] J. Fried, "Power and Clock Distribution for WSI Systems," in *Proc. IFIP Workshop on Wafer Scale Integration*, G. Saucier and J. Trilhe, Eds. Amsterdam: North-Holland, 1986, pp. 127–141.

[11] F. M. Gardner, "Charge-Pump Phase-Locked Loops," *IEEE Trans. Commun.*, vol. COM-28, pp. 1849–1858, Nov. 1980.

[12] L. A. Glasser and D. W. Dobberpuhl, *The Design and Analysis of VLSI Circuits.* Reading, MA: Addison-Wesley, 1985, chapter 6.

[13] W. D. Grover, "A New Method for Clock Distribution," *IEEE Trans. Circuits & Systems, Part I*, Feb. 1994, vol. 41, no. 2, pp. 149–160.

[14] W. D. Grover, *Method and Apparatus for Clock Distribution and Distributed Clock Synchronization*, United States Patent #5,361,277, Issued Nov. 1, 1994.

[15] M. W. Johnson and M. Graham, *High-Speed Digital Design: A Handbook of Black Magic.* Englewood Cliffs, NJ: Prentice-Hall, 1993, chapter 11.

[16] M. G. Johnson and E. L. Hudson, "A Variable Delay Line PLL for CPU-Coprocessor Synchronization," *IEEE J. Solid-State Circuits*, vol. 23, no. 5, pp. 1218–1223, Oct. 1988.

[17] D. C. Keezer and V. K. Jain, "Design and Evaluation of Wafer Scale Clock Distribution," *IEEE Int. Conf. WSI*, 1992, pp. 168–175.

[18] S.-Y. Kung and R. J. Gal-Ezer, "Synchronous vs Asynchronous Computation in Very Large Scale Integrated (VLSI) Array Processors," *Proc. SPIE*, vol. 341, pp. 53–64, 1982.

[19] S. D. Kugelmass and K. Steiglitz, "An Upper Bound on Expected Clock Skew in Synchronous Systems," *IEEE Trans. Comput.*, vol. 39, no. 12, pp. 1475–1477, Dec. 1990.

[20] LSI Logic Corp., *Phase-Locked Loop Application Note*, LSI Logic Application Note, Nov. 1991.

[21] LSI Logic Corp., *Clock Scheme for One Micron Technologies*, Rev. 1.1, LSI Logic Application Note, Aug. 1992.

[22] LSI Logic Corp., *Clock Distribution Schemes for 300K Technologies*, Rel. 2.0, LSI Logic Application Note, May 1993.

[23] A. M. Moshen and C. A. Mead, "Delay-Time Optimization for Driving and Sensing of Signals on High-Capacitance Paths of VLSI Systems," *IEEE Trans. Electron Devices*, ED-26, pp. 540–548, 1979.

[24] N. Nigam and D. C. Keezer, "A Comparative Study of Clock Distribution Approaches for WSI," *Proc. IEEE 1993 Int. Conf. WSI*, pp. 243–251, 1993.

[25] W. A. Samaras, "The CPU Clock System in the VAX 8800 Family," *Digital Tech. J.*, no. 4, pp. 34–40, Feb. 1987.

[26] T. Saigo, S. Watanabe, Y. Ichikawa, S. Takayama, T. Umetsu, K. Mima, T. Yamamoto, J. Santos, and J. Buurma, "Clock Skew Reduction Approach for Standard Cell," *Proc. IEEE 1990 Custom Integrated Circuits Conf.*, pp. 16.4.1–16.4.4, 1990.

[27] C. L. Seitz, "System Timing," in *Introduction to VLSI Systems*, C. Mead and L. Conway, Eds. Reading, MA: Addison-Wesley, 1980, chapter 7.

[28] J. Shyu, A. Sangiovanni-Vincentelli, J. Fishburn, and A. Dunlop, "Optimization-Based Transistor Sizing," *IEEE J. Solid-State Circuits*, vol. 23, no. 2, pp. 400–409, Apr. 1988.

[29] D. Tanksalvala et al., "A 90 MHz RISC CPU Designed for Sustained Performance," *IEEE Solid-State Circuits Conf.*, pp. 52–53, Feb. 1990.

[30] K. D. Wagner, *A Survey of Clock Distribution Techniques in High Speed Computer Systems*, Report CRC 86-20. Stanford, CA: Stanford University Center for Reliable Computing, Dec. 1986.

[31] N. Weste and K. Eshraghian, *Principles of CMOS VLSI Design: A Systems Perspective*, 2nd ed. Reading, MA: Addison-Wesley, 1993, pp. 317–335 (clocking strategies), pp. 334–336 (PLL methods), pp. 685–689.

[32] D. F. Wann and M. A. Franklin, "Asynchronous and Clocked Control Structures for VLSI Based Interconnection Networks," *IEEE Trans. Comput.*, vol. C-32, No. 3, pp. 284–293, Mar. 1983.

63.3 MOS Storage Circuits

Josephine C. Chang and Bing J. Sheu

In a large digital system, a sequence of operations must be performed for a particular function. The results of each operation depends on the results of previous operations. Therefore, the outputs of a logic circuit block typically depends not only on present input signals, but also on the history of the inputs. A combinational logic circuit becomes more useful if it is combined with memory elements. To construct a sequential system, the most common and straightforward way is to employ a central clock to synchronize the sequence of operations.

Instead of using memory elements in a sequential system, we can use dynamic logic circuits to store temporary data. With the building blocks of inverters and transmission gates, the MOS transistors can be used as dynamic storage components to store data temporarily on the device capacitances. Dynamic storage is widely used in MOS technologies because of the simplicity of the required circuitry. Because a memory element such as a static circuit latch occupies a large area and consumes power, elimination of latches has a positive effect on circuit density and power consumption. However, the disadvantages of dynamic logic gates include high transient power disturbances and less noise margins in some applications [1].

Dynamic logic circuits design is based on the synchronized movement of charge through the MOS circuit. A typical capacitance value associated with a logic gate is on the order of a few femtofarads, which means the amount of charge $Q = CV$ dynamically stored on the capacitance is on the order of femtocoulombs. Therefore, perturbations from ideal behavior can become critical to the operation of a circuit.

Dynamic Charge Storage

The MOS technologies have two attractive features that lead to an efficient way to store data momentarily. These two features are the extremely high input impedance of MOS transistor and the ability of a MOS transistor to function as a nearly ideal electrical switch. In order to store the charge on a capacitive node, the node must be isolated from both the power supply and ground. Various types of storage nodes can be realized in CMOS technologies. For example, charge can be stored at a node between sources (or drains) of two MOS transistors such as nMOS–nMOS, pMOS–nMOS, and nMOS–pMOS [2]; or the source (or the drain) terminal of one MOS transistor connected to the gate terminal of a second MOS transistor. Because the stored charge will leak away over time, this circuit is termed *dynamic storage circuit.*

Figure 63.42 shows the schematic diagrams of three combinations of source-drain connection. The distinction among the three connection types comes from the difference in voltage transmission levels for nMOS and pMOS gates.

Dynamic charge storage requires clocking the data at a sufficiently high rate so that the charge on the various nodes does not lead away significantly. Typically, this requires a minimum refresh rate of 500 Hz to 1 kHZ, corresponding to a charge storage time of about 1 to 2 ms.

nMOS–nMOS

An nMOS transistor is perfect for transmitting logic 0 signals, but imperfect for transmitting logic 1 signals due to the threshold voltage loss through the transistor. The voltage level of V_x

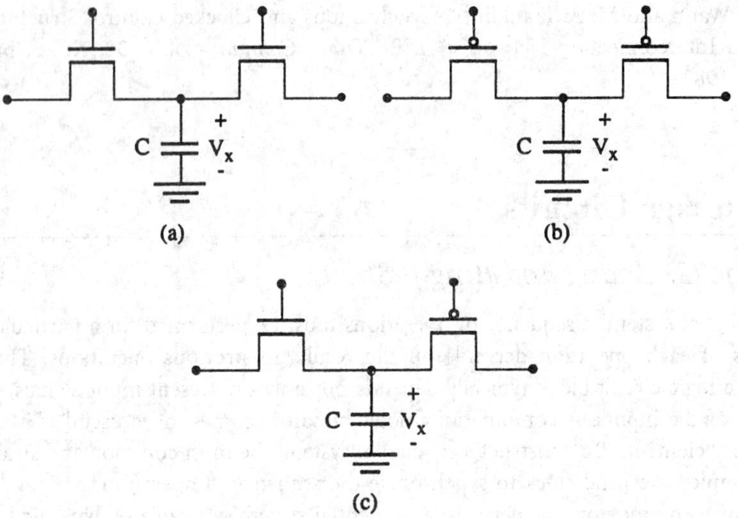

FIGURE 63.42 MOSFET source-drain connection storage nodes. (a) nMOS–nMOS; (b) pMOS–pMOS; (c) nMOS–pMOS.

which can be stored on the capacitor C is therefore limited by

$$0 \leq V_x \leq (V_{DD} - V_{th,n}) = V_{\max} \tag{63.12}$$

where

$$V_{th,n} = V_{th0,n} + \gamma_n(\sqrt{2\phi_{Fn} + V_{\max}} - \sqrt{2\phi_{Fn}}) \tag{63.13}$$

Charge storage on an nMOS–nMOS node is affected by the leakage paths through the p-type bulk to the ground. This affects the long-term storage of a logic 1 value.

pMOS–pMOS

A pMOS–pMOS node is the complement storage component of an nMOS–nMOS node. The voltage level of V_x is limited by

$$V_{\min} = |V_{th,p}| \leq V_x \leq V_{DD} \tag{63.14}$$

where

$$V_{th,p} = V_{th0,p} + |\gamma_p|\left(\sqrt{|2\phi_{Fp}| + (V_{DD} - V_{\min})} - \sqrt{2|\phi_{Fp}|}\right) \tag{63.15}$$

Because both p-channel MOS transistors have n-type bulks which are connected to V_{DD}, this type of storage node receives leakage current from the power supply. The logic 1 values can be held indefinitely, but the logic 0 values can only exist for a limited period of time.

nMOS–pMOS

A complementary nMOS–pMOS storage node can benefit from both nMOS and pMOS in transmitting logic 0 and logic 1, respectively. The voltage level which is stored on the capacitance is in the range of

$$0 \leq V_x \leq V_{DD} \tag{63.16}$$

if the maximum input value of V_{DD} is transmitted through the pMOS transistor and the minimum input value of 0 V is transmitted through the nMOS transistor. On the other hand, if the case is reversed, the maximum input value is entered through the nMOS transistor and the minimum input value is entered through the pMOS transistor, then the voltage range is reduced to

$$|V_{th,p}| \leq V_x \leq V_{DD} - V_{th,n} \tag{63.17}$$

This type of operation should be avoided because it greatly reduces the noise margins. In a standard nMOS–pMOS storage node both leakage paths to the power supply and ground exist. The ability to retain logic 0 and logic 1 values depends on which leakage path dominates.

Source–Gate Connection

This type of storage node is the connecting point between the source terminal of a pass transistor and a gate terminal of another MOS transistor [3]. Electrical charge can be temporarily stored on or removed from the gate terminal of the second transistor. When the gate terminal of the pass transistor is at a logic low value, the pass transistor is turned off, and the charge on the gate terminal is isolated. This charge determines the stored logic value. If the stored charge is perfectly isolated, the logic value would be stored indefinitely. In a practical situation the isolation is less than perfect, primarily because of leakage through the reverse-biased diode operation between the source diffusion region of the pass transistor and the substrate. In addition, leakage also can occur through the pass transistor. With the continuous advances in VLSI technologies, subthreshold leakage through the channel of the pass transistor becomes more important due to scale-down in device sizes. Leakage currents alter the node voltage which may lead to a logic error.

Two major problems arise in maintaining the integrity of a stored logic state. First is the parasitic conduction paths in the transistors that lead to charge leakage. Leakage currents alter the node voltage, which may cause a logic error. The second problem is charge sharing, which occurs when two isolated storage nodes become connected by a switching event and must equalize their voltages by redistributing charge. Charge sharing may result in a logic error, or may block logic propagation entirely.

Charge Sharing

Beside charge leakage, a problem called charge sharing may also damage the integrity of a stored logic state. Charge sharing occurs when a dynamic charge-storage node is used to drive another isolated node in a switching network [4]. Typically, when two capacitors with different voltages are connected by a pass transistor, as shown on Fig. 63.43, charge sharing may occur. When the pass transistor is turned on, the voltages on the capacitors equilibrate to some intermediate value. In Fig. 63.43 capacitors C_1 and C_2 are in parallel when the transmission gate is conducting. This forces the voltages across C_1 and C_2 to be equal. If the two capacitors are charged to different initial voltages, charge sharing will occur when the transmission gate turns on. Let the initial voltage charge on C_1 be V_1 and Q_1, and the initial voltage and charge on C_2 be V_2 and Q_2. The initial charge balance equation is

$$Q_1 + Q_2 = C_1V_1 + C_2V_2 \tag{63.18}$$

After the transmission gate turns on, the final charges on C_1 and C_2 become Q_1' and Q_2', respectively, and both capacitors are charged to the same value V'. The final charge balance equation is

$$Q_1' + Q_2' = (C_1 + C_2)V' \tag{63.19}$$

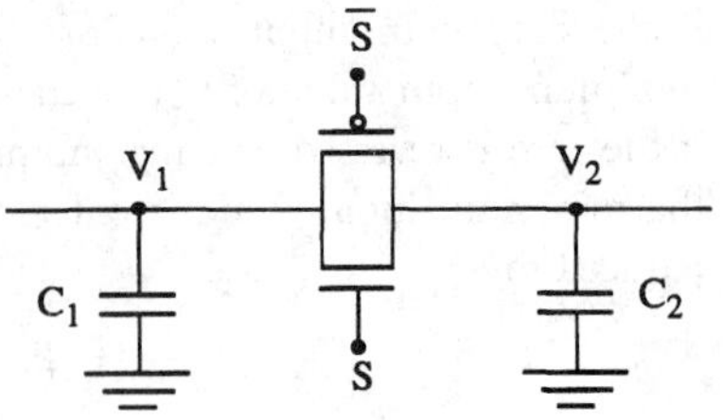

FIGURE 63.43 A charge-sharing-prone structure.

By applying the charge-conservation principle, we can obtain

$$V' = \frac{C_1V_1 + C_2V_2}{C_1 + C_2} \tag{63.20}$$

and

$$Q_1' = \frac{C_1}{C_1 + C_2}(C_1V_1 + C_2V_2) \tag{63.21}$$

A precharged circuit might work incorrectly due to charge-sharing errors, which could occur inside the pulldown network or at the output circuit. To control a precharged circuit, a gated clock can be present only at the input of the bottom transistor, while all other inputs to the gates of transistors in series in the pulldown chain must have a stable signal over the same clock phase to prevent charge-sharing problems. A *sneak path* is created when two pass transistors in series are both turned on at the same time and one is connected to V_{DD} while the other is connected to the ground. Charge can leak through this sneak path.

Shift Register

A frequent use of dynamic storage circuits is the shift register. Shift registers are most often used to provide temporary storage of digital signals. The shift register storage can be used as a simple way to delay the arrival of a signal for a specific number of clock cycles. Shift register storage is also frequently used as the temporary memory for a sequential logic circuit. In general, shift registers provide dense, limited access memory for many applications in digital integrated circuits.

Simple Shift Register

Figure 63.44 shows the schematic diagram of a multistage MOS shift register, with each stage composed of a pass transistor and an inverter [5]. The nonoverlapping clock waveforms Φ_1 and Φ_2 are used. Assume that a logic signal is placed at the input of the first shift register stage while the Φ_1 clock is low and the transmission gate of the first stage is turned off. Next, when the Φ_1 clock goes high, if the signal at the input to the first stage is held constant, it will be propagated to the input of the inverter in the first stage. After a short delay, the output of the first inverter will provide the inverted logic signal to the input of the second shift register stage. At this time, the Φ_2 clock is low and the transmission gate in the second stage will not pass this signal. When the clock values change so that Φ_2 becomes high, the transmission gate of the second stage will propagate the output signal of the first stage to the second inverter, and then the output of the second stage is produced. The signal will be stopped by the transmission gate of the third stage because Φ_1 is low while Φ_2 is high. This sequence continues through the shift register chain as the clock signals alternate, causing the input signal to propagate through the shift register stages. The data are stored on the capacitances associated with the gate terminals of the inverter. The transmission gate acts as the switch that lets charge flow into and out of the capacitors when

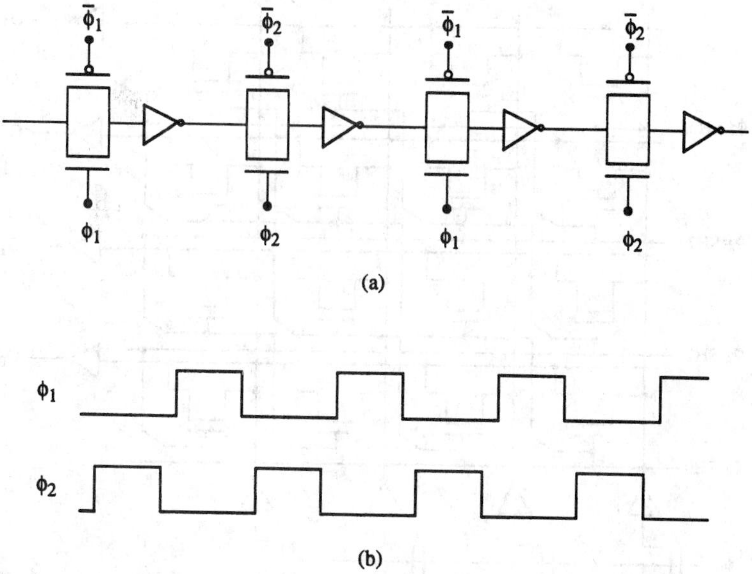

FIGURE 63.44 (a) A four-stage MOS shift register. (b) Nonoverlapping waveforms of Φ_1 and Φ_2.

they are turned on. The charge is trapped at the capacitor when the transmission gate is turned off.

Each time the Φ_1 clock changes to a high value, the shift register input signal will propagate to the gate of the first inverter and the output signal of the first stage will be produced. A sequence of alternating Φ_1 and Φ_2 clock signals will cause an input signal to propagate through the whole structure at the rate of two stages of the shift register for each complete cycle of the clock signals. After N clock cycles, a logic input value will have shifted through $2N$ stages of the shift register chain. When a two-phase clock is used to control a shift register, it is important that the two clock phases do not overlap. If both phases of the clock were high simultaneously, a data value could propagate through multiple stages during the clock overlap time. This would cause erroneous operation of the shift register.

Parallel Shift Register

Several copies of the multistage shift registers can be combined in parallel with the same clock lines to form a parallel shift register to transmit a group of signals in lock-step fashion. Such a parallel shift of 8, 16, or 32 data bits is often used in microprocessor circuits. The basic structure of this set of shift registers demonstrates two principles which are important for the efficient geometrical layout of digital circuits. The data for the shift register flow from left to right while the control signals (Φ_1 and Φ_2 clocks) flow from top to bottom. Such an orthogonal structure of data paths and control signals within a circuit module is widely used to provide a regular organization of logic circuits within a VLSI chip. Physical layout of the shift register stages can be mirrored with respect to the ground and V_{DD} lines. This mirroring technique allows shared power and ground connections and reduces required circuit layout area. It is important to minimize the size of the basic shift register stage because this stage is repeated many times in a large shift register.

Clocked Barrel Shifter

A *barrel shifter* is a wraparound shifter that forms a very useful switch array [6]. The basic layout is shown in Fig. 63.45. The inputs are labeled I_i; the shift controls $\Phi_2 \cdot SHi$, and the outputs O_i. The input lines run horizontally while the output lines run vertically. The operation of the first

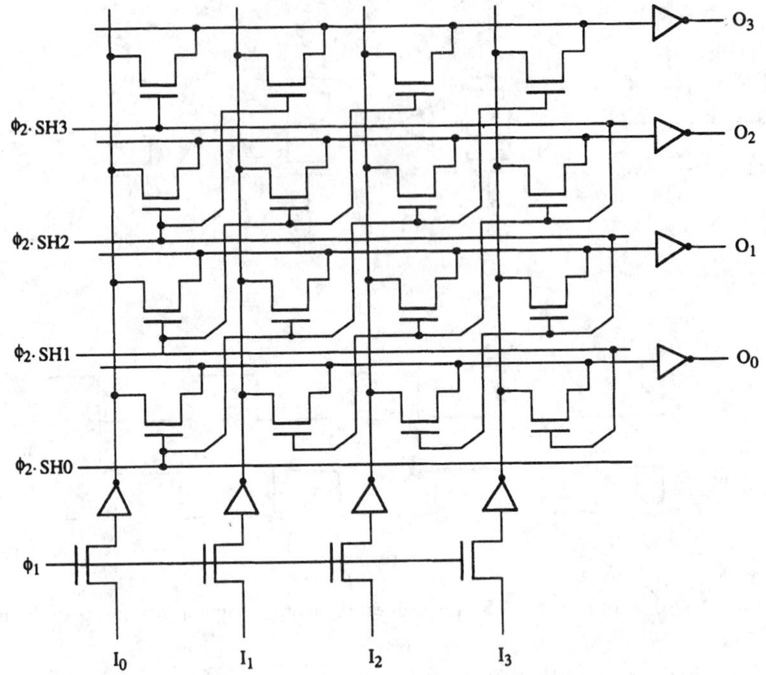

FIGURE 63.45 A four-bit clocked barrel shifter.

shift register stage is the same as explained earlier. In the second stage the four output signals from the four inverters in the first stage can be shifted without changing the order, or each signal can move up one, two, or three locations.

Dynamic CMOS Logic

The dynamic CMOS logic design consists of dynamic circuits based on precharging the output node to a particular level when the clock is at the logic 0 level. During the precharge phase, the inputs to the circuits change. When the clock is at the logic 1 value, the output of the logic gate may be pulled to a complementary value, depending on the input conditions.

The choice of using static or dynamic logic is dependent on many criteria. When low-power performance is desired, it appears that dynamic logic has some inherent advantages in a number of areas including reduced switching activity due to hazards, elimination of short-circuit dissipation, and reduced parasitic node capacitances. Static logic circuits have advantages on charge sharing and precharge operation.

Static circuits design can exhibit spurious transitions due to races. These spurious transitions dissipate extra power over that required to perform the computation. The number of these extra transitions is a function of input patterns, internal state assignment in the logic design, delay skew, and logic depth. Although it is possible with careful logic design to eliminate these transitions, dynamic logic intrinsically does not have this problem because any node can undergo at most one power-consuming transition per clock cycle [7].

Short-circuit currents are found in static CMOS circuits. However, by sizing transistors for equal rise and fall times, the short-circuit component of the total power dissipated can be kept to < 20% of the dynamic switching component. Dynamic logic does not exhibit this problem, except for those cases in which static pullup devices are used to control charge sharing or when clock skew is significant. Dynamic logic typically used fewer transistors to implement a given

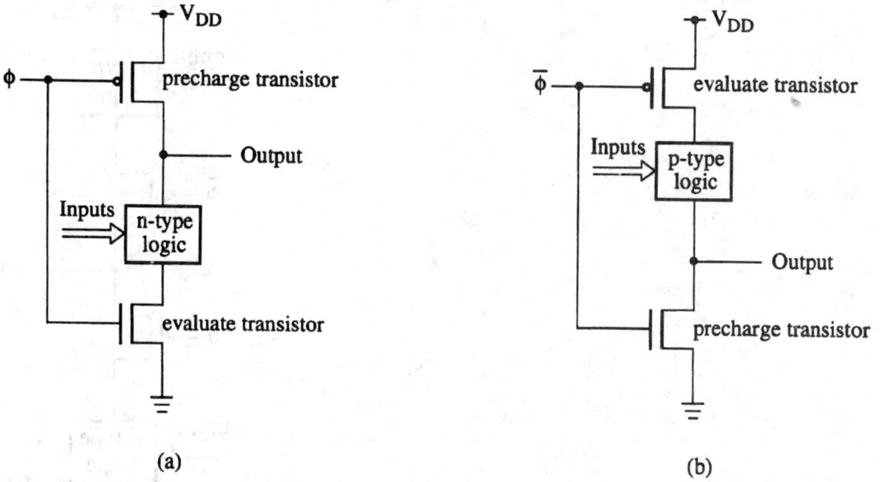

FIGURE 63.46 A precharge-evaluate logic gate.

logic function, which directly reduces the amount of capacitance being switched and thus has a direct impact in the power-delay product.

Precharge-Evaluate Logic

The schematic diagram of a basic precharge-evaluate logic is shown in Fig. 63.46. It consists of an nMOS logic structure whose output node is precharged to V_{DD} by a pMOS precharge transistor; and conditionally discharged by the n-transistor network connected to the ground. Alternatively, and n-transistor precharge to the ground and a pMOS logic structure to conditionally discharge to V_{DD} may be used. A single-phase clock Φ is used for high-speed operation. For the former case, the precharge phase occurs when the clock Φ is low. The path to the ground is activated via the n-transistor network when the clock Φ is high. The input capacitance of this logic gate is the same as a pseudo-nMOS gate which has a single p-transistor, with the gate connected to the ground, as a load device. The pullup time is better than a pseudo-nMOS gate by virtue of the active switch but the pulldown time is increased due to the ground switch.

Clocked CMOS Logic (C^2MOS)

This clocked CMOS logic gates were originally used to build low-power dissipation logic gates. The reasons for the reduced dynamic power dissipation stem mainly from metal–gate CMOS layout considerations. The main use of such logic structures is to form clocked structures that incorporate latches or interface with other dynamic forms of logic structure. The gates have the same input capacitance as regular complementary gates, but larger rise and fall times due to the serially connected clocking transistors.

The schematic diagram of a clocked CMOS logic gate is shown in Fig. 63.47. In this circuit the clocked transistors are placed in series with the transistors in the p- and n-type logic blocks. The primary use of C^2MOS is in dynamic shift registers. In a C^2MOS dynamic shift register the p-type logic block is a p-transistor network and the n-type logic is an n-transistor network. All transistors can normally be chosen as minimum-size devices because each stage is only required to drive the capacitance of an identical shift register stage.

Although the C^2MOS circuit requires the same number of transistors, external connections, and clock phases as the standard CMOS dynamic shift register, the layout is greatly simplified because the source/drain regions of the two p-channel transistors can be merged, and the corresponding regions of the two n-channel transistors can be merged. This feature helps to reduce circuit capacitance, number of contacts, and layout area.

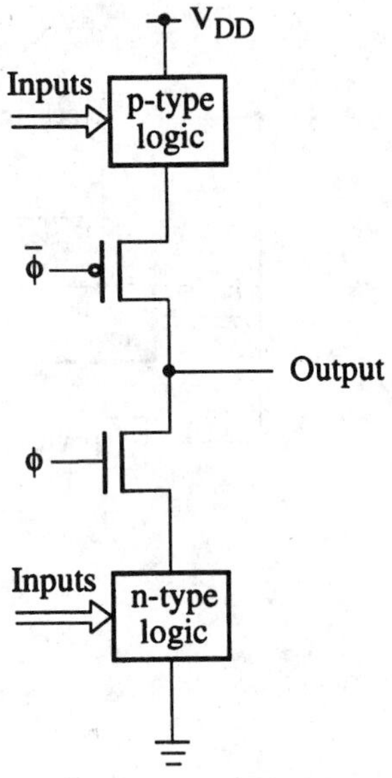

FIGURE 63.47 A clocked CMOS logic gate.

Operation of the C^2MOS circuit is quite simple. The gates of the pMOS pullup transistor and the nMOS pulldown transistor of the inverter are both connected to the input signals. For a valid logic input, one of these transistors is turned off while the other is turned on. Clocked transistors placed in series with the pullup and pulldown transistors serve to connect these transistors to the output node when the clock Φ is high. If the input signals match the n-type logic portion, the output storage node will be discharged. Otherwise, the output storage node will be charged. When the clock Φ is low, the output node will remain in its present state. In contrast to other clocked logic circuits, the output of C^2MOS is available during the entire clock cycle, although it is actively driven only when the clock is high. A C^2MOS circuit is more susceptible to interference from the load circuit attached to the stage because the load capacitance is the storage node for the dynamic charge.

Domino CMOS Logic

The domino logic gate design can provide glitch-free cascades of nMOS logic structures. It is a modification to the clocked CMOS logic gate to allow a single clock to precharge and evaluate a cascaded set of dynamic logic blocks. A domino logic gate consists of two elements: a precharge-evaluate logic stage followed by a static inverter buffer at the output, as shown in Fig. 63.48. The logic gate can be built in two forms: mostly n-transistors and mostly p-transistors. During the precharge phase when clock Φ is low, the output node of the dynamic gate is precharged high, and inverted by the static buffer to provide a logic 0 output for the domino CMOS gate. As subsequent logic stages are driven by this buffer, transistors in subsequent logic blocks will be turned off during the precharge phase. When the gate is evaluated, the node voltage of the logic stage is conditionally pulled down according to the input signal values. If the logic condition of the gate is satisfied, the node voltage is pulled down. It is inverted by the static buffer to provide a logic 1 output. Each gate in sequence can make at most one transition ($1 \rightarrow 0$). Hence, the buffer can only make a transition from ($0 \rightarrow 1$). In a cascaded set of logic blocks each state evaluates and causes the next stage to evaluate in the same manner as a stack of dominos fall. Any number of logic stages may be cascaded, provided that the sequence can evaluate within the given clock phase. A single clock can be used to precharge and evaluate all logic gates within a block.

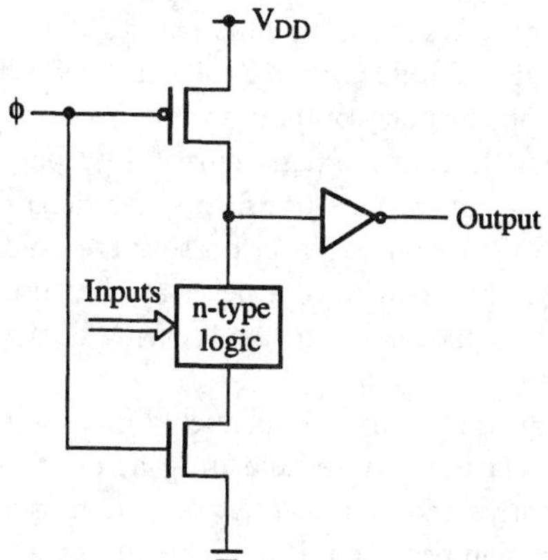

FIGURE 63.48 A domino CMOS logic gate.

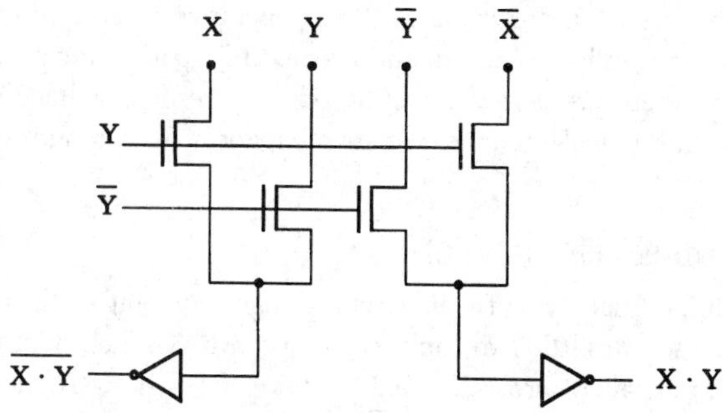

FIGURE 63.49 A CPL AND/NAND cell.

The structure has some limitations. First, only noninverting structures can be constructed. Second, each logic gate must be buffered. Finally, in common with a clocked-CMOS gate, charge redistribution can be a problem. The effects of these problems can be minimized. For example, in complex logic circuits, such as arithmetic logic units, the necessary XOR gates may be implemented conventionally as complementary gates and driven by the last domino circuit. The buffer is often needed to drive large capacitive load and dies not contribute to any extra cost.

Static storage of charge can be realized by a domino logic gate by including a weak *p*-transistor. A weak *p*-transistor is one that has low gain which is realized with a small W/L ratio. It should have a small gain in order not to fight with the pulldown transistors, yet to balance the effects of leakage. This will allow low frequency or static operation when the clock is held high. In this case the pullup speed could be an order of magnitude slower than the pulldown speed. Notice that the precharge transistor may be eliminated if the time between evaluation phases is long enough to allow the weak pullup to charge the output node. A domino logic gate has advantages over a simple precharge-evaluate logic structure. For example, the static buffer provides output-driving capability to either V_{DD} or the ground. In the precharge-evaluate logic gate the output can be favorably driven only to the ground in response to logical conditions, not to V_{DD}. When the logic condition of the precharge-evaluate gate is not satisfied, dynamic charge storage at the output must maintain the logic 1 value. The dynamic logic portion of a domino CMOS gate always has a fanout of 1, thereby simplifying device sizing within the gate structure.

Complementary Pass-Transistor Logic (CPL)

The CPL gate is constructed by using an nMOS pass-transistor network for logic function and eliminating the pMOS latch [8]. It consists of complementary inputs and outputs, nMOS pass transistor logic network, and CMOS output inverters. Figure 63.49 shows the schematic diagram of a CPL AND/NAND cell. The pass-transistors function as pulldown and pullup devices. Thus, the pMOS latch can be eliminated, allowing the advantage of differential circuits to be fully utilized. One attractive feature of the CPL gate is that complementary outputs are produced by the simple four-transistor circuits. Because the logic 1 value level of the pass-transistor outputs is lower than the supply voltage V_{DD} by the threshold voltage of the pass-transistors, the signals must be amplified by the output inverters. In addition, the CMOS output inverters shift the logic threshold voltage and drive the capacitive load. The logic threshold shift is necessary because that of the output inverter is lower than half of the supply voltage, due to the lowering of the logic 1 value.

The CPL gate is attractive because fewer transistors are required to implement important functions. However, a CPL gate has two basic problems. First, the threshold drop across the single-channel pass-transistor results in reduced current drive and hence slower operation at a reduced supply voltage. This is important for low-power design because it is desirable to operate at the lowest possible voltage levels.

Second, because the logic 1 input value at the regenerative inverters is not V_{DD}, the pMOS device in the inverter is not fully turned off, and hence direct-path static power dissipation could be significant. To solve these problems, reduction of the threshold voltage has proven effective, although if taken too far it will incur a cost in dissipation due to subthreshold leakage and reduced noise margins.

Cascade Voltage Switch Logic (CVSL)

The CVSL gate is a differential style of logic circuit design requiring both true and complement signals to be routed to the gates [9]. Two complementary nMOS switch structures are constructed and then connected to a pair of cross-coupled *p* pullup transistors as shown in Fig. 63.50(a). When the inputs switch, node voltages Q and $\overline{Q}$ are either pulled up or down. Positive feedback applied to the *p* pullup transistors causes the gate to switch. The logic trees may be further minimized from the fully differential form using logic minimization algorithms. This version is slower than a conventional complementary gate employing a *p*-tree and an *n*-tree because during the switching action, the *p* pullup transistors must compete with the *n* pulldown tree. The schematic diagram of a dynamic charge-storage version of the CVSL logic gate design is shown in Fig. 63.50(b). It consists of two domino logic gates with complementary input logic trees. The advantage of CVSL gate over a domino logic gate is the capability to generate a complete logic function rather than just the noninverting logic function. However, extra silicon area is needed.

NORA CMOS Dynamic Logic

NORA logic is capable of handling signal race problems in transmission gates [10]. It is based on dynamic CMOS logic, but uses latches instead of transmission gates to control signal flow. In a NORA logic dynamic nMOS and pMOS logic circuits are cascaded into a C^2MOS latch. Figure 63.51 shows the schematic diagrams of both Φ stage and $\overline{\Phi}$ stage. Static inverters are provided at the outputs of dynamic circuits to realize logic inversion. This allows direct implementation of arbitrary functions without modification. In the Φ stage the logic circuit used $\Phi = 0$ for precharge and $\Phi = 1$ for evaluation. The latch accepts data when $\Phi = 1$ and holds the data when $\Phi = 0$. No new data can be accepted during the hold time. The operation in the $\overline{\Phi}$ stage is similar when reversing clock signals are used.

By alternating Φ and $\overline{\Phi}$ clock stages makes NORA chains well suited for pipelined logic. The schematic diagram of a generic structure of a NORA chain is shown in Fig. 63.52 [2]. Logic flows through the chain at a rate set by the clock. The problem of logic races by using transmission gates as latches between logic circuits has been eliminated because of the dynamic C^2MOS latch circuit.

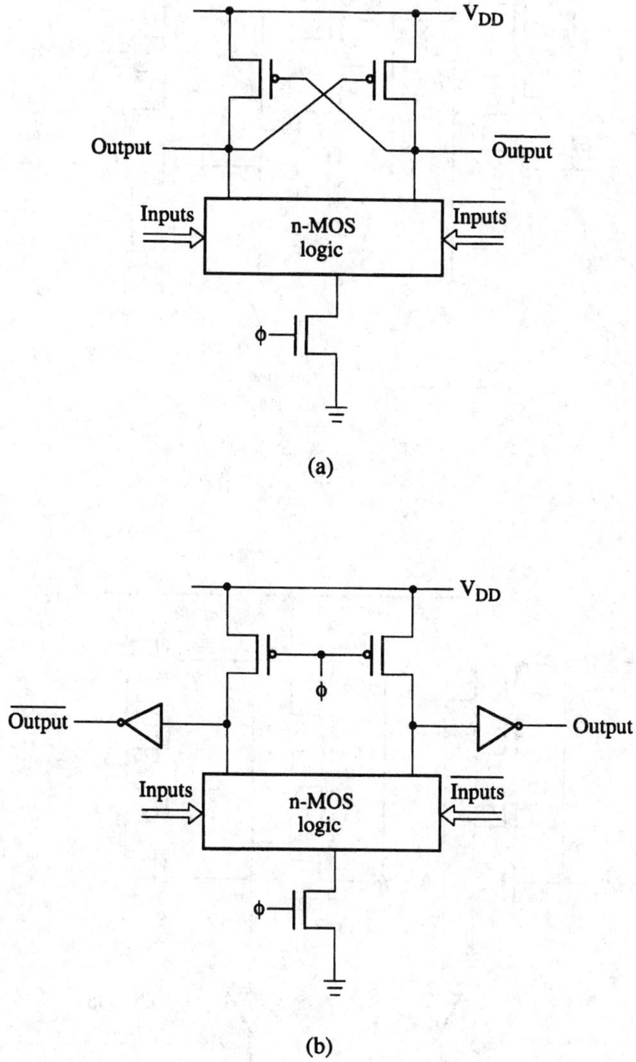

FIGURE 63.50 A CVSL gate. (a) Static version. (b) Dynamic version.

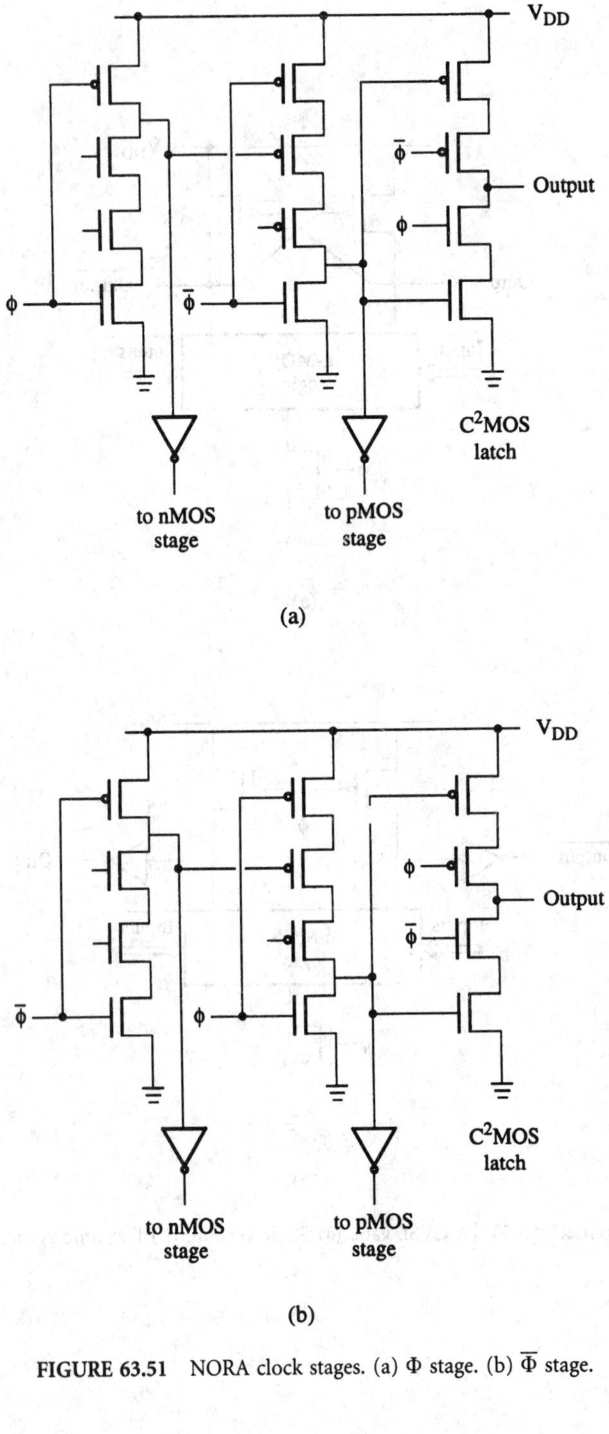

FIGURE 63.51 NORA clock stages. (a) Φ stage. (b) $\overline{\Phi}$ stage.

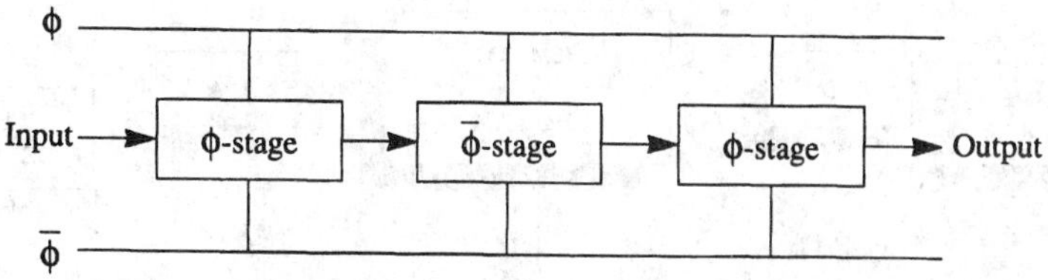

FIGURE 63.52 A NORA chain.

References

[1] N. Wang, *Digital MOS Integrated Circuits*, Englewood Cliffs, NJ: Prentice-Hall, 1989.
[2] J. P. Uyemura, *Circuit Design for CMOS VLSI*, Boston: Kluwer Academic, 1992.
[3] R. L. Geiger, P. E. Allen, and N. R. Strader, *VLSI Design Techniques for Analog and Digital Circuits*, New York: McGraw-Hill, 1990.
[4] A. Mukherjee, *Introduction to nMOS and CMOS VLSI Systems Design*, Englewood Cliffs, NJ: Prentice-Hall, 1986.
[5] L. A. Glasser and D. W. Dobberpuhl, *The Design and Analysis of VLSI Circuits*, Reading, MA: Addison-Wesley, 1985.
[6] E. D. Fabricius, *Introduction to VLSI Design*, New York: McGraw-Hill, 1990.
[7] A. Chandrakasan, S. Sheng, and R. W. Brodersen, "Low-Power CMOS Digital Design," vol. 27, no. 4, pp. 473–484, Apr. 1992.
[8] K. Yano et al., "A 3.8-ns CMOS 16×16-b Multiplier Using Complementary Pass-Transistor Logic," *IEEE J. Solid-State Circuits*, vol. 25, no. 2, pp. 388–395, Apr. 1990.
[9] N. Weste and K. Eshraghian, *Principles of CMOS VLSI Design*, Reading, MA: Addison-Wesley, 1993.
[10] N. F. Goncalves and H. J. De Man, "NORA: A Racefree Dynamic CMOS Technique for Pipelined Logic Structures," *IEEE J. Solid-State Circuits*, vol. 18, no. 3, pp. 261–266, June 1983.

63.4 Microprocessor-Based Design

Roland Priemer

Introduction

During the past two decades microprocessors (μPs) shave become components that are routinely and widely used in machines and systems that engineers design. This is due to their flexibility and ability to perform tasks at a low cost. Because they are programmable, μPs (microprocessors) are used to achieve operations of devices and systems with complexity that we have come to take for granted. The competition amount the numerous manufacturers of microprocessors has brought about a great variety of these machines to increase their suitability in ever-widening fields, more products, and new markets. Moreover, turnaround times have become short enough and costs have come down enough so that system and application engineers have the option of specifying the design of a customized microprocessor to meet their application's specific performance requirements.

This section is intended to introduce the reader to design with a microprocessor. It is assumed that the reader is acquainted with the material that is generally covered in an introductory course on digital systems. The goal is to help the designer who is not experienced with the design

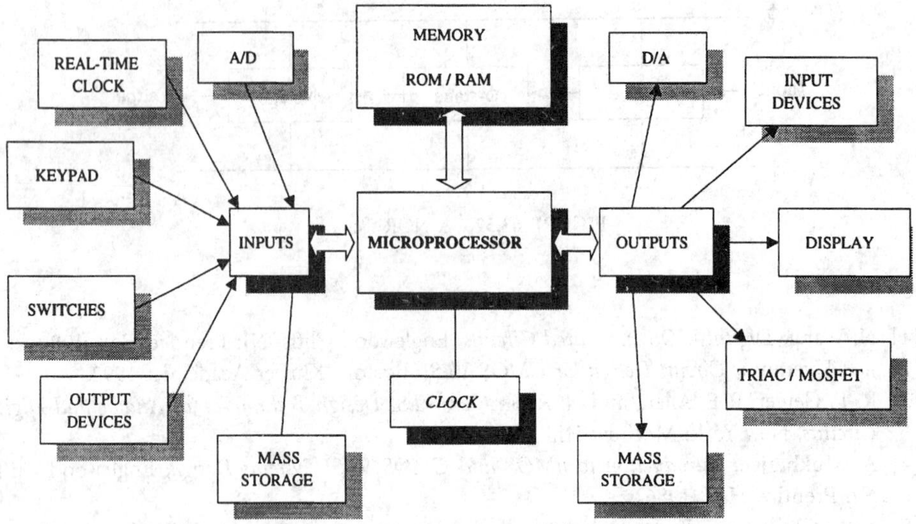

FIGURE 63.53 Conceptual diagram of a microprocessor-based system.

of microprocessor-based systems to come to a basic understanding of what is involved. Two representative and significantly different microprocessors will be used to do this. These are the Zilog Z80, a general-purpose microprocessor, and the Motorola M68HC11, also called a microcontroller. These and similar microprocessors have been used in many diverse kinds of dedicated applications.

The design and development of the hardware and software are two broad aspects to utilizing a microprocessor. Here, the emphasis is on digital hardware design. To understand the role that a microprocessor plays in the design process we need to understand the operation of several major building blocks: memory, architecture of a microprocessor, the system bus and timing of bus signals, and the supportive devices used to interface these building blocks together. At the level of design to be considered here, hardware and software are and should remain inseparable. From both points of view we gain a better understanding of the other. However, due to space limitations, no significant presentation of programming in assembly language is included in the sequel. Instead software is only utilized to see how it influences hardware requirements and how it makes hardware work.

Features of a Microprocessor-Based System

Conceptually, Fig. 63.53 depicts a microprocessor-based system. The inputs to the microprocessor are all binary (can only assume one of two values) or digital signals. However, the origins and meanings of these signals can be very diverse. Inputs can be due to manually activated switch closures from buttons or keypads or they can be due to switches that are embedded within sensors that detect, for example, presence or absence of an object, pressure that is over or under a certain level, temperature that is above or below a particular temperature, presence or absence of an ultrasound or light beam, voltage that is above or below a desired value, etc. An input can come from a real-time clock. Inputs can also be information about the status or readiness of devices that receive outputs.

A single digital signal is called a bit, and it can assume only one of two logic values, i.e., logic 1 or logic 0, which in the circuits that we will use correspond to 5 or 0 volts, respectively. Typically, these circuits are designed to accept as logic 1 any voltage in the range 3.5 to 5, and accept as logic 0 any voltage in the range 0 to 1.5. Voltages in the range 1.5 to 3.5 will produce

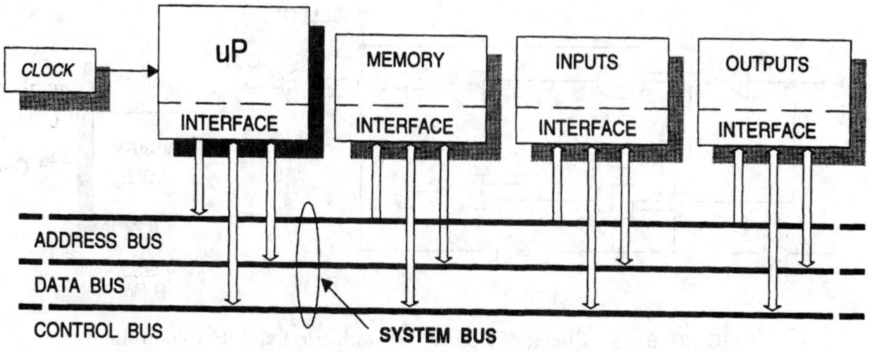

FIGURE 63.54 System block diagram.

indeterminate results. Unless explicitly noted, all signals are voltages as they vary with respect to a common reference.

Taken together, n bits that are labeled, for example, with $d_{n-1}\, d_{n-2} \cdots d_1\, d_0$ can have any one of 2^n different binary combinations or binary assignments. A group of $n = 8$ bits is called a byte, which can assume any one of 256 different binary assignments. Another commonly used quantity is $2^{10} = 1024$, which is denoted by $1K$. Thus, for example, $16K = 2^{14} = 16{,}384$.

Inputs can be analog in origin, and the microprocessor receives the result of sampling and analog to digital (A/D) conversion. Usually, analog signals are electrical signals generated by transducers that sense physical quantities of interest, such as speed of a motor, weight of an object, sound of someone speaking, stress of material under test, temperature in a building, electrocardiogram of a patient, acceleration of some structure, wind direction, oil pressure in an engine; the possibilities are endless. Thus, a microprocessor (the program it is executing) can receive information about time and the state of the world in which it is intended to operate.

The memory module of Fig. 63.53 serves several purposes. First, it contains the binary instruction codes of the program that is executed by the microprocessor. The program instruction codes are another kind of input to the microprocessor. Memory also may be used for temporary storage and later retrieval of data produced by the program as it executes the algorithms that process the inputs.

Figure 63.54 shows a more functional structure than Fig. 63.53 of a microprocessor-based system. An important feature of this structure is the use of a set of common communications paths, called the system bus, between all major modules. To identify any particular device, the microprocessor places a binary number, called an address, on the m bit address bus. Actually, the electronic circuitry that comprises the microprocessor sets the voltages on the m address bus conductors to some combination of 0 and 5 volts. 2^m different addresses can be placed on the address bus. Every other module receives the address, and only the particular device that recognizes its own address, which is assigned and incorporated by the system designer, is supposed to respond. The data bus simultaneously transfers n bits of data from the microprocessor to any device or from any device to the microprocessor. The control bus informs all modules about the kind of activity that is presently occurring or is about to take place on the system bus.

Usually, the different modules in the system have different operating characteristics. It is the purpose of the interfaces to make compatible the modalities of these different modules. It is often interface design that is the focus of an overall hardware design effort. We will work with microprocessors that produce an $m = 16$-bits wide address, which permits $2^{16} = 65{,}536 = 64K$ different addresses, and that use an $n = 8$-bits wide data bus. Also, depending on the kind of control signals provided by a microprocessor and the kind of control signals that are preferred or required on the control bus, a microprocessor may also require an interface circuit to the system bus.

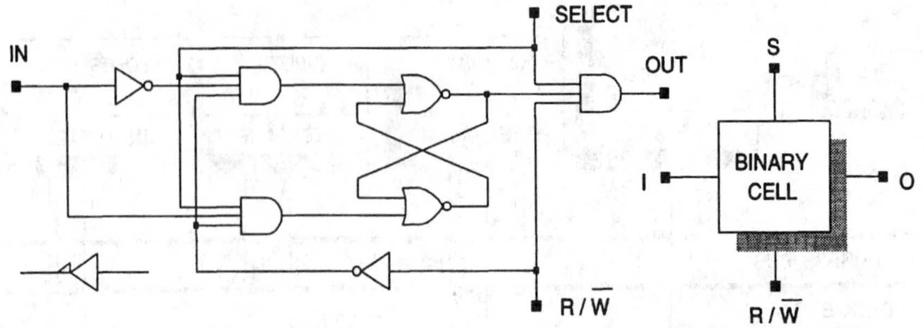

FIGURE 63.55 Circuit of a binary cell with select and R/$\overline{W}$ control.

The different standard system bus definitions are numerous. For example, the public availability of the definition of the IBM PC bus has permitted the connection of innumerable different accessories, produced by many different manufacturers, to this bus. Such accessories were designed to be compatible with the definition of this bus standard. Such standardization contributed to the wide success of the PC. However, in a dedicated application, adherence to a bus standard may not be necessary, particularly when no need exists to interconnect many major modules. Here, we use the particular control signals of the microprocessor to form the system control bus as needed.

Memory

Almost all of the internal and external activity of a microprocessor depends on the ability to store, transfer, and retrieve data. A designer must understand how these operations take place. By way of presenting some memory devices, some terminology and supportive hardware are also introduced.

A circuit for a memory cell is shown in Fig. 63.55. It is also called a binary cell (BC). Here, the activity of the flip flop is determined by the logic signals at the points labeled S, for select, and R/$\overline{W}$, for read or write. The small circles attached to the OR gates and the buffers are called bubbles, and they mean logical complement. Thus, the buffer followed by a bubble is a NOT gate, and the flip flop is made with two NOR gates. Another commonly used notation for logical complement is a small right triangle, as shown in the figure.

Note that we are using positive logic so that logic 1 means that a signal is high (5 volts) and logic 0 means that a signal is low (0 volts). When an operation or activity is enabled by taking a control signal high (low), then we say the control is active high (low). More briefly and without regard to the required level, an operation is enabled by asserting (or activating) its control signal(s).

The read and select control inputs of the binary cell are active high, while the write control input is active low, which is indicated by the bar over the W. To store a bit in the binary cell, the sequence of operations is (1) select the cell, (2) apply an input, and (3) momentarily take the R/$\overline{W}$ control signal low. Thus, the write operation is complete.

A set of binary cells that can be read or written in parallel is called a register, and the set of bits that a register can hold is called a word. A 4 words × 3 bits per word memory module is shown in Fig. 63.56. The $m = 2$ bits input $a_1 a_0$ is decoded with the 2 × 4 decoder, which has its own active low enable control. When enabled, the decoder output selects, according to its inputs, just one register (row of binary cells) for a read or write operation. The binary assignment to $a_1 a_0$ is called the address of the word to be referenced. For a read operation, the OR gates receive the bits of the selected word, while all other OR gate inputs are logic 0. For a write operation all binary cells in a column receive the same input bit, while only the binary cell in the selected row accepts it. With additional columns we can increase the register length

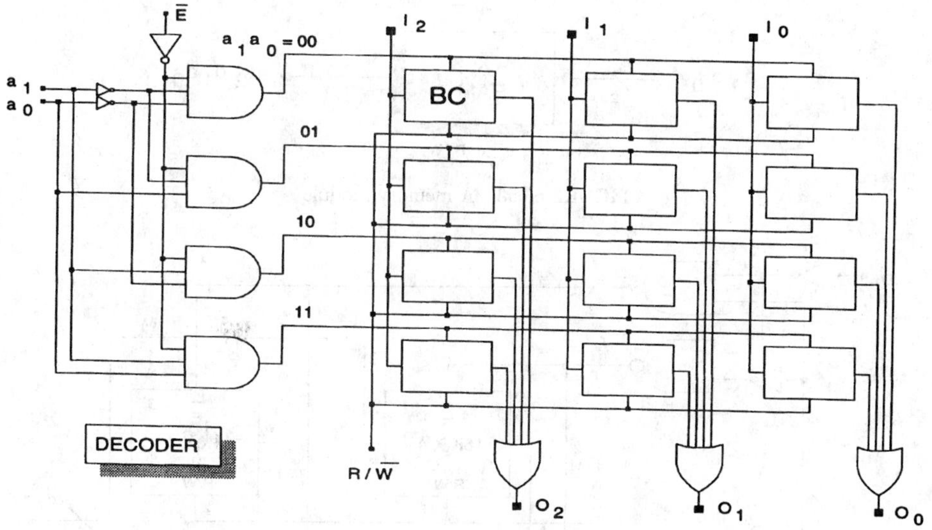

FIGURE 63.56 A multiword read/write memory circuit.

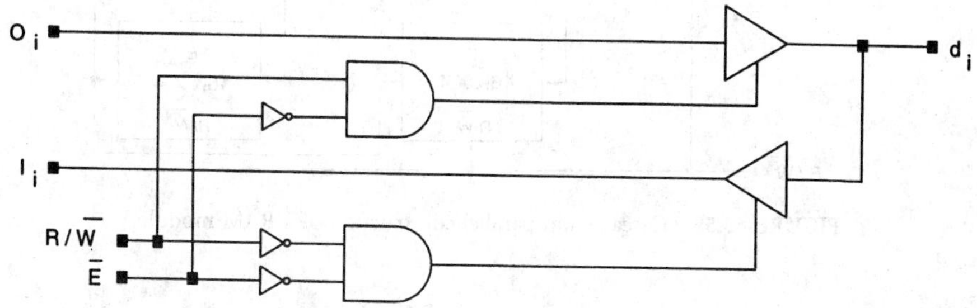

FIGURE 63.57 Unidirectional to bidirectional bus conversion.

to n bits per word, and with an $m \times 2^m$ decoder we can have a memory size of 2^m registers, while only requiring m address bits to specify any particular memory register that is the object of a read or write operation (memory reference).

Figure 63.56 implies that the system data bus must consist of an n bit unidirectional input bus to transfer data from memory to the microprocessor and an n bit unidirectional output bus to transfer data from the microprocessor to memory. Practically, to reduce the number of communication paths, an n bit bidirectional system data bus is preferred. This can be accommodated by connecting to each memory input/output (I/O) bit pair, say the ith pair, an arrangement of tristate buffers as shown in Fig. 63.57, where d_i can be connected to the ith bit of the system data bus. When both buffers are in tristate (high impedance state), which occurs when $\overline{E}$ is high, then d_i is independent of O_i and I_i. Thus the memory module can be electrically disconnected from the system bidirectional data bus. When $\overline{E}$ is asserted, then $R/\overline{W}$ controls the direction of data transfer. Figure 63.58 shows a more concise notation for the entire memory module. Here, the two address lines are grouped into one bus, and the three bidirectional data lines are grouped into another bus.

Because no particular address sequence is required for reading or writing data to this memory, it is called random access memory (RAM). Stored data remains intact as long as power is supplied to the circuit. Therefore, this memory is called static RAM, contrary to another memory type, called dynamic RAM, in which the logic content of a binary cell is not based on the state of a bistable circuit, but is based on the presence or absence of charge on a cell capacitor. Because

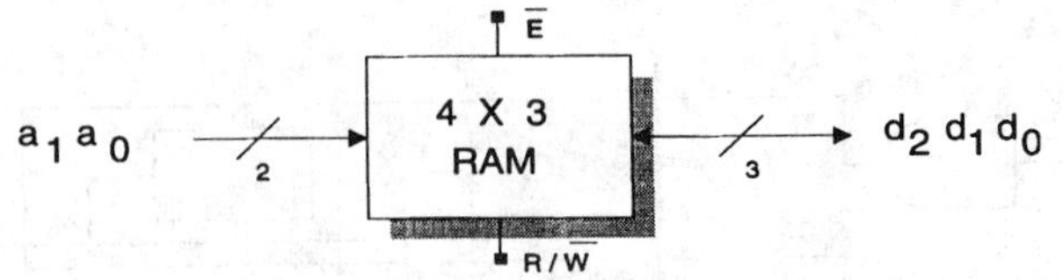

FIGURE 63.58 A memory module.

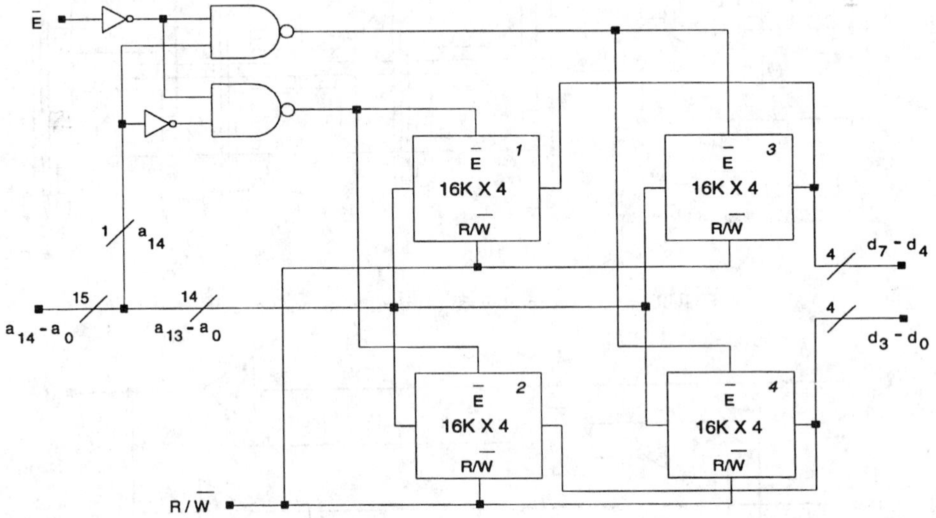

FIGURE 63.59 Cascade and parallel construction of a RAM module.

charge on capacitors dissipates, dynamic RAM must be periodically refreshed, which requires additional control circuitry. However, dynamic RAM can be fabricated to achieve much higher bit densities than of static RAM, resulting in the substantially lower cost of dynamic RAM.

Random access memories in various sizes and package types are available. Commonly available static RAMs range in size from 256×1 to $128K \times 8$ (a 128 K byte RAM), and the function and number of the control signals also varies a little. Commonly available dynamic RAMs range in sized from $16K \times 1$ to $16M \times 1$ (a 16 mega bit RAM). Important parameters for each RAM type are the read and write cycle times, which must be less than the width of read and write time windows allowed by the devices that will reference this memory.

Larger RAM modules are constructed by the cascade and parallel connection of smaller RAMs. This is illustrated in Fig. 63.59, which shows a $32K \times 8$ RAM made from $16K \times 4$ RAMs. If $\overline{E}$ is asserted, which takes the entire module out of tristate, then address bit a_{14} enables either RAMs 1 and 2 or RAMs 3 and 4 to access the data bus. The remaining address bits select a particular 4-bit register within each $16K \times 4$ RAM. Here, RAMs 1 and 3 hold the upper half of each data byte, while RAMs 2 and 4 hold the lower half of each data byte.

Whenever power is removed from the circuits of static or dynamic RAM, the data content is lost, and such memory devices are said to be volatile. To provide software for execution immediately after power is applied to a microprocessor, ROM (read-only memory) is used. Figure 63.60 shows an 8 words $\times 2$ bits per word ROM that has been, for example, programmed with the data given in the table. By opening (indicated with an $\times$) the connections called links, the ROM is programmed. These links remain open through power-down and power-up cycles, and thus, a ROM is said to be nonvolatile. The tristate buffers are controlled by the output enable (OE) signal so that the ROM output can be electrically disconnected from a system data

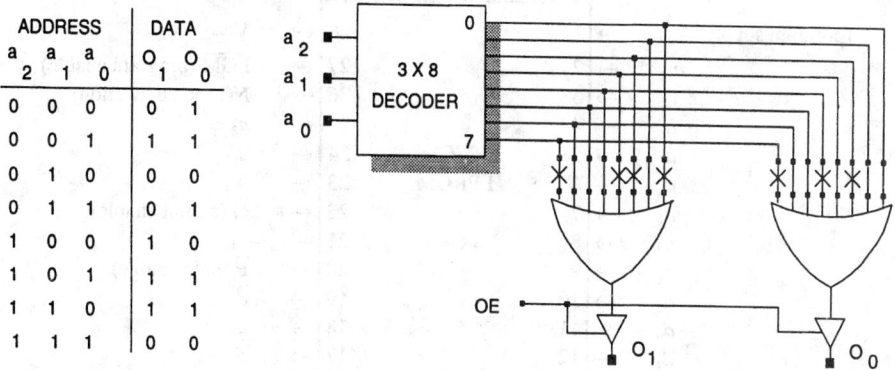

ADDRESS a_2	a_1	a_0	DATA O_1	O_0
0	0	0	0	1
0	0	1	1	1
0	1	0	0	0
0	1	1	0	1
1	0	0	1	0
1	0	1	1	1
1	1	0	1	1
1	1	1	0	0

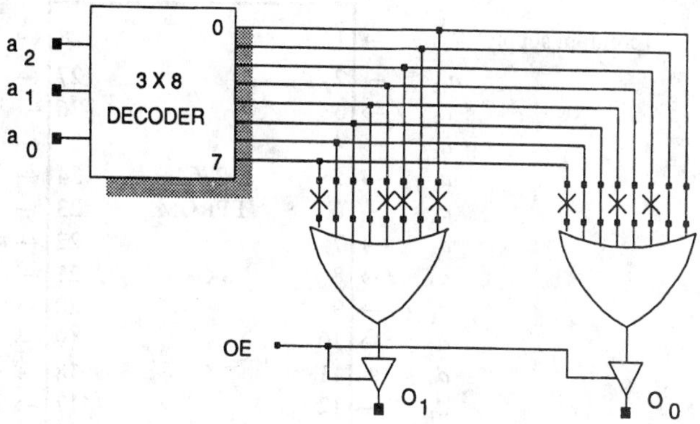

FIGURE 63.60 An 8 words × 2 bits/word programmable ROM.

bus, which may then be used for transfer of data among other devices. An important parameter of a ROM is its read cycle time.

By providing a manufacturer a table of contents, a ROM can be programmed at the time of fabrication. Blank (unprogrammed) ROMs are available that can be one-time programmed (OTP) in the field. These are called PROMs for programmable read-only memory. There are also PROMs that can be erased by exposing the integrated circuit (IC) to a certain level of ultraviolet light for several minutes. Erasing is made possible by using an IC package with a clear window just above the IC. These are called EPROMs for erasable programmable read only memory. After an EPROM has been erased, it can be programmed again. Another EPROM type is the EEPROM, which can be electrically erased by essentially overwriting previously stored data. However, the write cycle time is significantly longer than the write cycle time of conventional RAM. Commonly available PROMs and EPROMs range in size from $2K \times 8$ to $1024K \times 8$ (a 1 mega byte ROM). Commonly available EEPROMS are not as large as other ROMs, and they cost more than other ROM types. Like RAM modules, larger ROM modules can be made by combining smaller ROMs in a manner illustrated in Fig. 63.59.

Figure 63.61 shows an $8K \times 8$ EPROM and the package pin assignment. To place a byte into a particular register, the register address and intended content must be supplied first at the address and data pins, respectively. Then in addition to V_{CC}, the V_{PP} supply must be provided, the chip enable is asserted, and the $\overline{\text{PGM}}$ input is activated for a particular time duration. An instrument called a PROM programmer is used to do this. Some PROM programmers work by connection to a computer such as an IBM PC, and, in conjunction with software running on the PC, it can program a great variety of PROMs, which are selected from a menu of those PROMs supported by the software, with code in a file that it receives from the PC.

A programmed ROM can serve several purposes. Most computers have a ROM that contains program code that is always executed at power-up. Typically, this software is a utility (called a boot) that transfers software from a mass storage device to RAM. Then, after this transfer is complete, or perhaps after some additional ROM resident initializing code has executed, the software in RAM starts to execute. In a dedicated application such as the microprocessor-based system that controls a car engine, it is more likely that ROM will contain all of the application software for use whenever the system is powered up.

ROM is used for other purposes. For example, by connecting three variables to the address inputs of the ROM given in FIgure 63.60, this ROM can, through programming, be used to realize any two Boolean functions of three variables. ROMs are used to hold tables of data. For example, multiplication can be performed through table look-up. If the 4-bit binary codes of two digits that we want to multiply are used to form an 8-bit address of, e.g., a 256×8 ROM, then the upper and lower halves of the retrieved data could each be the 4-bit binary codes of

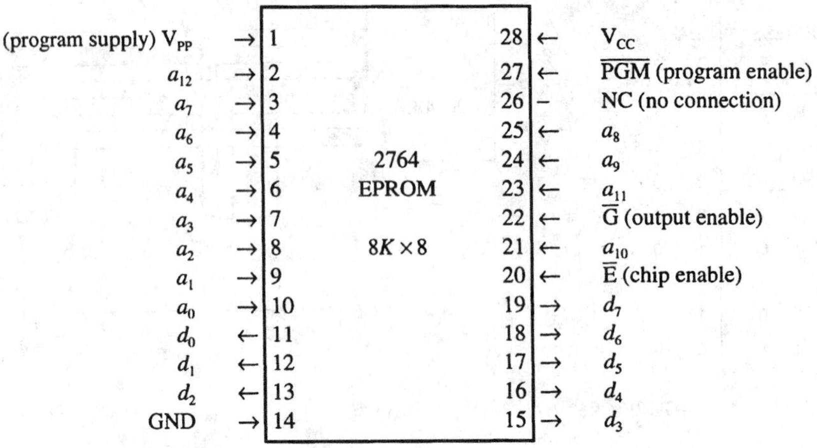

FIGURE 63.61 Pin assignment of a typical (the 2764) EPROM.

the product digits. The idea is that the address is formed with the input(s) of an operation (or argument of a function), and the precomputed and stored data are the output of the operation (or value of the function). Often EEPROMs are used to hold tables of data that are generated during program execution, which must be available after a power-down and power-up cycle. A battery-backed RAM also can be used for such purposes.

Another way to use a ROM is for signal generation. By using a counter to supply an address sequence to a ROM, each bit of the data as it is clocked out of the ROM could be used as a control signal of some process. The frequency of the clock that drives the counter determines real time, and the variation between 0 and 1 of any particular data bit determines the resulting control signal shape. On the other hand, if the data word sequence coming from a ROM is submitted to a D/A converter, then a ROM can be used to generate an arbitrarily shaped analog signal that is repeated each time the counter repeats its count sequence.

The kind of memory that is used, its size, and the addresses to which any particular memory will be responsive must meet the requirements of an application. This information is often given in the form of a system memory map. The size of this map is the range of addresses that a microprocessor can specify. With $m = 16$, an address in binary is denoted by $a_{15}a_{14}\cdots a_1a_0$. The most significant address bit a_{15} splits the entire $64K$ word memory space into two $32K$ word blocks and $0xxx\,xxxx\,xxxx\,xxxx$, where x means don't care, is any address in the lower $32K$ word block, while $1xxx\,xxxx\,xxxx\,xxxx$ is any address in the upper $32K$ word block. Similarly, address bits $a_{15}a_{14}$ together split the entire memory space into four $16K$ word blocks. If $a_0 = 0$, then the address is an even number, while if $a_0 = 1$, then it is an odd number.

Sometimes it is more convenient to use hexadecimal notation. The symbol $ will be placed in front of all numbers written with hexadecimal notation. Thus, $*XXXX* is any 16-bit number in hexadecimal notation, while $*XX* is any 8-bit number. As the most significant hexadecimal address digit changes by $1, the address changes by $4K$.

Suppose a design must include $8K$ bytes of EPROM starting at address $0000, $1K$ bytes of EEPROM starting at address $4000, and $32K$ bytes of RAM starting at address $8000. The ROM, EEPROM, and RAM must be responsive to addresses in the ranges $0000 to $1FFF, $4000 to $43FF, and $8000 to $FFFF, respectively. Because the three most significant address bits split the memory space into eight $8K$ word blocks, the ROM should be enabled when these address bits are $a_{15}a_{14}a_{13} = 000$, and then the remaining address bits $a_{12}\cdots a_0$, covering an $8K$ word space, are the address input to the ROM. Similarly, the EEPROM should be enabled when $a_{15}\cdots a_{10} = 010000$, and the remaining address bits $a_9\cdots a_0$, covering a $1K$ word space, are

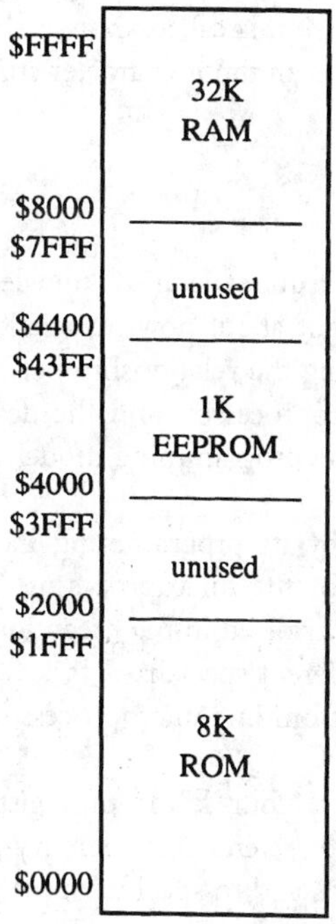

⋮	⋮	⋮	⋮
107BH			
107AH	$01		
1079H	$D3	OUT (01H),	A
1078H	$80	ADD A,	B
1077H	$C0		
1076H	$00		
1075H	$3A	LD A,	(C000H)
1074H			
⋮	⋮	⋮	⋮
$0007			
$0006	00110000		
$0005	00111110	LD A,	30H
$0004	11110000		
$0003	00000000		
$0002	00110001	LD SP,	F000H
$0001	01010110		
← $0000	11101101	IM 1	

FIGURE 63.62 Memory map, sample ROM content, and assembly language source.

the address input to the EEPROM. The RAM should be enabled when $a_{15} = 1$. Thus, the size and position of a memory device determine how to place it in the memory space. The other important aspect of interfacing memory to the system bus and eventually to a microprocessor is that a memory module must be responsive to control signals issued by the microprocessor to present and accept data in certain particular time windows. This will be considered further when we look at the timing of microprocessor control signals.

A memory map for this design is shown in Fig. 63.62. Here, we see the size, kind, and position in the memory space of the microprocessor of actual memory devices. When possible, it is also useful to describe the location and purpose of particular program code modules.

Assuming that the EPROM contains program code, some example instructions are placed at the beginning of this memory. These are binary numbers that are machine instruction codes for the Z80 microprocessor. The Z80 microprocessor is a type of processor that after being reset, fetches instruction codes starting at address $0000. The first instruction, $ED $56, is the Z80 machine code that selects its method 1 for responding to interrupts. Another commonly used notation for hexadecimal is to attach the suffix H. The next instruction, 31H, is the machine code that loads the stack pointer register with an address given by the next two bytes, $F000. These instructions could be the part of a program that initializes the processor. To understand how an instruction is processed, it is useful to study the architecture of a microprocessor.

For programming convenience, programs are usually written using mnemonics of instruction codes that are indicative of instruction activity. The set of mnemonics and associated notational convention for all of the instruction codes that a particular microprocessor can execute form the assembly language of the microprocessor. Different microprocessors with different instruction sets have different assembly languages. Furthermore, different manufacturers adopt different mnemonics for their microprocessor instruction sets, causing another variation among assembly languages. However, in principle, many common attributes remain from one assembly language to another. A program written in assembly language must be converted into machine code for storage and eventual execution. A program that performs this conversion is called an assembler. From the viewpoint of the assembler, an assembly language program is an input character string, and the machine code output is another character string.

Microprocessor Architecture

Employing a microprocessor in a dedicated application does not require detailed knowledge about its internal behavior. However, it is useful to have some insight about how instructions, i.e., their codes, are processed (executed) by hardware. Understanding the relationship between software and hardware can affect the selection of a particular microprocessor, and the design process. Moreover, the hardware designer should also understand the programming model of a microprocessor.

From a programmer's viewpoint, a microprocessor is defined by its programming model and its instruction set, which together comprise the architecture of the microprocessor. The programming model consists of the set of internal registers that are involved in the execution of operations as specified by the instruction set. These registers do different specialized tasks. The purpose, capability, and number of these registers can vary greatly from one microprocessor to another.

Basically, within the programming model, microprocessors have four kinds of registers. Address registers are used to form and hold addresses to be used for referencing memory and other devices to obtain program instruction codes and their operands and to specify source and destination memory locations for data read and write operations. Data registers can be the source of data for an operation or the destination for the result of an operation. Operational registers have associated hardware to perform, for example, logical and arithmetic operations. Finally, status/control registers configure the operation of the microprocessor and support different kinds of conditional instructions.

An operational register possessed by most microprocessors is the accumulator. It is commonly denoted by register A. Figure 63.63 illustrates how an A register functions within a microprocessor. Associated with register A are the arithmetic logic unit (ALU) and the condition code register (CCR) or status register (SR). Inputs to the ALU can come from register A, register B, and the CCR, and its activity is determined by the function select word $f_{k-1} \cdots f_1 f_0$. Register A receives its input from the ALU, and register B receives its input from the n bit internal data bus. The activity of this circuit is determined by the control signals that are applied at all of the points labeled with triangles. For example, by asserting the E (enable) input of register A, the n bit word coming from the ALU is latched (loaded) into register A. The content of register A can be placed on the internal data bus by asserting the control signal of the n tristate buffers connected to its output. The content of the internal n bit data bus can be latched into register B by asserting its E control input.

Figure 63.64 illustrates how the ALU performs its task at the gate (bit) level. Notice how the function select lines determine, as they do for the multiplexers of all the other bits, which

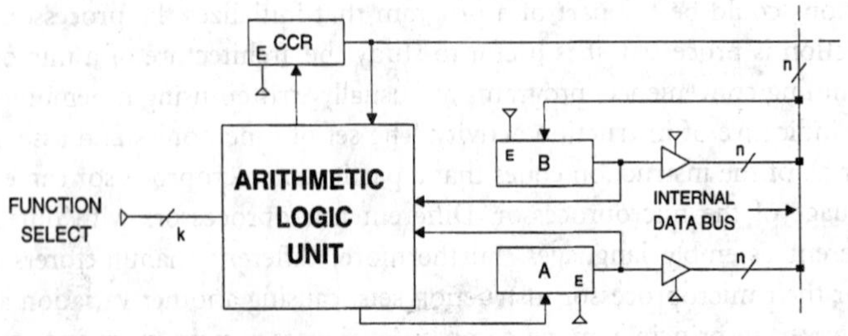

FIGURE 63.63 Register to register transfer activity of an accumulator.

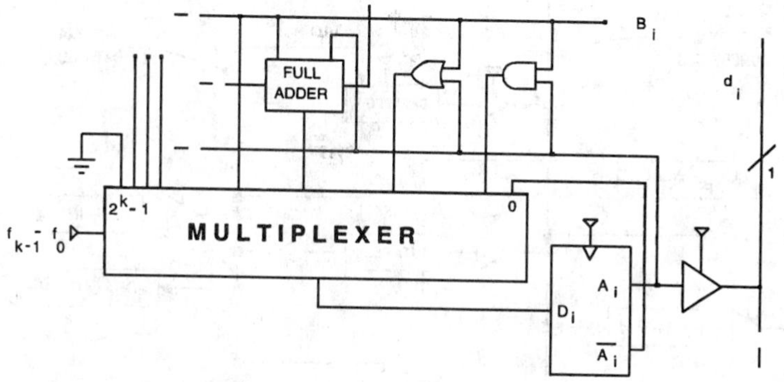

FIGURE 63.64 Bit level activity of an ALU.

multiplexer input is latched into the ith flip flop of the accumulator when its E control input is asserted. For example, if $f_{k-1} \cdots f_0 = 0 \cdots 0$, then the accumulator will be complemented, as if the microprocessor has just read in the machine code for the complement accumulator instruction, which has the Z80 assembler mnemonic CPL. Thus, the machine codes of instructions such as ADD A, B; RLA; INC A; LD A, 80H; etc., eventually determine the binary assignments of these function select lines to accomplish the instruction tasks. As many as 2^k different operations can exist involving the accumulator that can be achieved by this ALU structure.

All of the control bits required in Figures 63.63 and 63.64 come from memory called control ROM or microstore. Figure 63.65 illustrates the data paths of a microprogrammable microprocessor. Each word, consisting of perhaps 16 to 128 bits, in control ROM is called a microinstruction. The busing of microinstructions throughout the microprocessor is not shown in Fig. 63.65. Instead, these interconnections are indicated by labeling with the small triangles.

Figure 63.65 also illustrates a level of design called the register transfer level. Still another level of greater detail is required before a level of IC design detail is reached that shows the interconnection of transistors, resistors, diodes, and conductors. This is the logic gate level. However, at that level, the detail of design would probably be too much and detract from an understanding of microprocessor operation. Here, the intent is to exemplify how hardware processes an instruction code.

Each microinstruction is partitioned into a set of fields, and one of these fields holds the accumulator function select line assignment. Other fields hold data that signify (1) which registers are the source of data for the internal and external buses, (2) which registers are the destination of data on the internal and external buses, (3) register activity control such as clear, increment, decrement, load, and tristate, (4) a j bit address of the next microinstruction, and (5) the external control word that informs external devices of the present microprocessor activity. Each microinstruction coming out of control ROM can cause many activities to take place at the same time within the microprocessor. The other registers in the diagram of Fig. 63.65 perform the following tasks:

MAR—Memory address register. It holds the address that the microprocessor can place on the external address bus, and it is loaded from the internal address bus by asserting its enable input.

PC—Program counter. This register holds the address of the next instruction or instruction operand. It can be cleared, loaded, and incremented by asserting appropriate enable inputs. To fetch program code (either an instruction code or instruction operand(s)), its content is transferred to the MAR. Usually after each time its content has been used to fetch a byte of program code, its content is incremented.

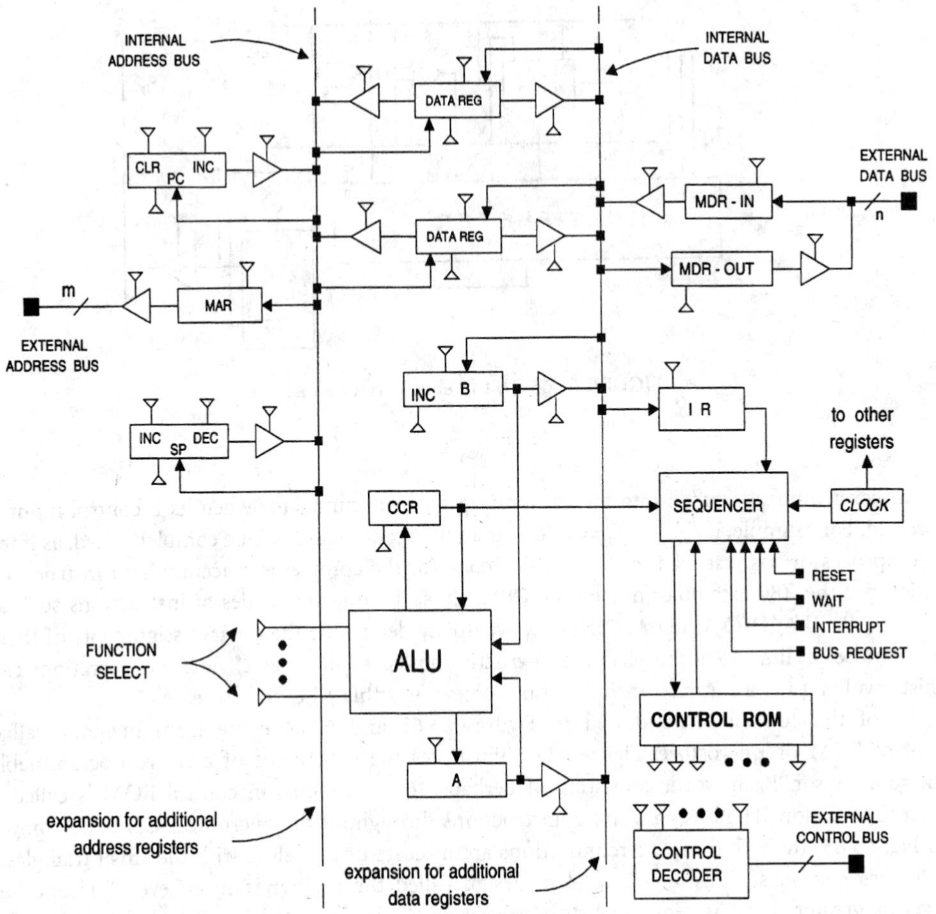

FIGURE 63.65 Data paths of a microprogrammable microprocessor.

SP—Stack pointer. It holds an address that points to RAM that can be used for temporary storage. It can be loaded, incremented, and decremented by asserting appropriate enable inputs. Certain instructions can cause its content to be transferred to the MAR for a memory read or write operation. The RAM that is referenced with the SP is called the stack. Typically, the SP is implicitly decremented before (after) writing to the stack, and it is implicitly incremented after (before) reading from the stack. Thereby, the stack is a last in and first out memory area that is used to support, among other things, subroutine calls.

MDR-IN—Memory data register in. It receives its input from the external data bus, and it can drive the internal data bus.

MDR-OUT—Memory data register out. It receives its input from the internal data bus, and it drives the external data bus.

IR—Instruction register. When the microprocessor is executing an instruction fetch from external memory, this register receives the content of the MDR-IN register, which is then assumed to be an instruction code.

CRR—Condition code register. Each bit in this status register is indicative of the result of some previous microprocessor operation. The meaning of the flags (bits) in the CCR vary from one microprocessor to another. Typically, this register contains: (1) a carry flag that is set or reset depending on whether or not the previous add or subtract instruction produced a carry or required a borrow out of the most significant bit of the arithmetic operation, (2) a zero flag that is set if the previous instruction produced a zero result and reset if the instruction result is not zero, (3) an interrupt enable flag that can be set or reset by an instruction to allow for software control over whether or not the microprocessor can respond to maskable interrupts, and (4) other flags. Usually, a microprocessor's instruction set includes program flow control instructions that are conditioned on these status flags such that if the flag is set (reset), then the instruction is executed, and if the flag is reset (set), then the instruction is not executed (skipped).

Sequencer. The sequencer consists mainly of a ROM and some sequential logic. It uses the instruction code and status register flags to form an address to its own ROM from which it obtains a *j* bit microinstruction address that is applied to the control ROM. Furthermore, in response to asynchronous external control signals, such as reset, interrupt, bus request, wait, and others, it generates addresses of microinstructions that cause activity appropriate for these inputs.

Control decoder. Depending on the input coming from control ROM, the decoder generates external control signals that are intended to be used to inform external devices about the present activities of the microprocessor and to synchronize activities on the external address and data buses.

Data register. This register is used for the temporary storage of data. Usually, there are several registers like it. Some can also drive the lower byte of the internal address bus, while others can also drive the upper byte of the internal address bus. Still others can receive data from the upper or lower byte of the internal address bus. Some may have low level arithmetic/logic capabilities.

The execution of each microprocessor instruction involves the execution of a set of microinstructions, called a microprogram. Thus, a microprocessor's instruction set is determined by the microprograms stored in control ROM. After system reset or the completion of each instruction, unless an external asynchronous input is active, the microprogram that performs an instruction fetch is executed. It uses the content of the program counter (PC) to point to the instruction, and it increments the program counter so that the program counter points to an instruction operand or the next instruction. Then, depending on the instruction code and status register content presented to the sequencer, a particular microprogram is executed to complete execution of an instruction. If an instruction code requires that operands be fetched, then the program counter is further incremented so that after an instruction has executed, the program counter is pointing to the next instruction. Microprograms for program flow control instructions such as JUMP or BRANCH elsewhere cause the program counter to be loaded with the address operand of the instruction. Moreover, microprograms for instructions such as CALL or BRANCH subroutine first cause the stack pointer (SP) to be used for storing in the stack the content of the program counter before loading the program counter with the address operand. Then, the microprogram for an instruction such as RETURN, which is used to terminate a subroutine, causes the stack pointer to be used for retrieving from the stack the address of the instruction following the CALL subroutine instruction, which is then loaded into the program counter.

This architecture can be expanded to include additional address registers and other special-purpose address registers such as index registers, another program counter and stack pointer, additional accumulators and data registers, and so on. The widths of the address and data bus can be increased, contingent upon fabrication issues. Moreover, algorithmic instruction types can be supported because executing the desired activity of an instruction is a matter of writing a microprogram to accomplish all of the required register to register transfer activities.

Design Using a General-Purpose Microprocessor

The Z80 is a general-purpose microprocessor, and it is available in a 40 pin DIP (dual in-line package). It is an 8-bit machine, i.e., its data bus is 8 bits wide. From a software development

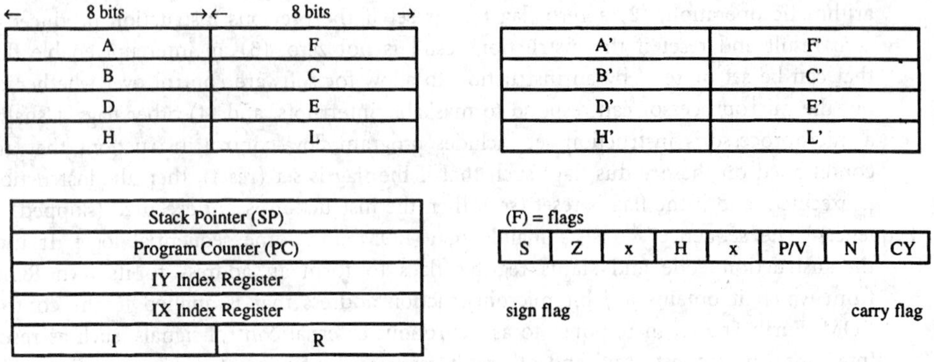

FIGURE 63.66 Z80 programming model.

viewpoint, this machine has numerous features, as an examination of its extensive instruction set and programming model, which is given in Fig. 63.66, will show. It has an accumulator (A), status or flag (F) register, six data registers (B, C, D, E, H, and L) that can be paired for use as address registers (BC, DE, and HL), a stack pointer (SP), a program counter (PC), two index registers (IX and IY), a register (I) that is used by its indirect vectored interrupt processing method, and a 7-bit counter register (R) that can be utilized for dynamic RAM refresh. Furthermore, it can quickly change programming context by swapping with the alternate (primed) register set. Among its over 600 instructions are data transfer, arithmetic, logical and rotate, branch (or jump), stacking, I/O, program control, exchange, block transfer, search, and bit manipulation instructions. Due to space limitations, the information given here is necessarily limited. Complete information can be found in textbooks or the data book from the manufacturer.

The pin assignment of the Z80 is shown in Fig. 63.67. Power is supplied at pins 11 and 29, and a conventional crystal controlled oscillator is used to provide the system clock ϕ at pin 6. If the load of each system address bus bit is only one or two TTL loads, then the outputs $a_{15}a_{14}\cdots a_1a_0$ can drive the system address bus. However, to accommodate a greater load, two unidirectional tristate octal buffers, as shown in Fig. 63.67, can be used. These buffers also have hysteresis, which helps to produce sharper bus signals. Similarly, it is likely that the Z80 data signals $d_7d_6\cdots d_1d_0$ must be buffered to accommodate system data bus loading. A bidirectional buffer, as shown in Fig. 63.67, can be used for this. Z80 control signals must be used to control buffer direction. Also, define the system data bus by terminating it with pull-up resistors. The Z80 control signals perform the following tasks:

$\overline{\text{RESET}}$—This active low input resets the interrupt enable flag, clears registers I and R, causes all control signals to become inactive, sets the interrupt processing method to method 0, and clears the program counter. When this input is released, the Z80 starts to fetch the first instruction code (op-code) from the memory register with address held by the program counter.

$\overline{\text{MREQ}}$—This output becomes active whenever the Z80 is performing an operation, such as op-code fetch or instruction operand fetch, that references an external device with a 16-bit address. It indicates that the address bus holds a valid address. The devices that respond are said to be positioned in the memory space.

$\overline{\text{IORQ}}$—This output becomes active whenever the Z80 is executing either an IN, for input, or an OUT, for output, instruction, both of which reference external devices with an 8-bit address that is placed on $a_7\cdots a_0$. It also indicates that the address bus holds a valid 8-bit address. The devices that respond are said to be positioned in the I/O space, which is separate from the memory space.

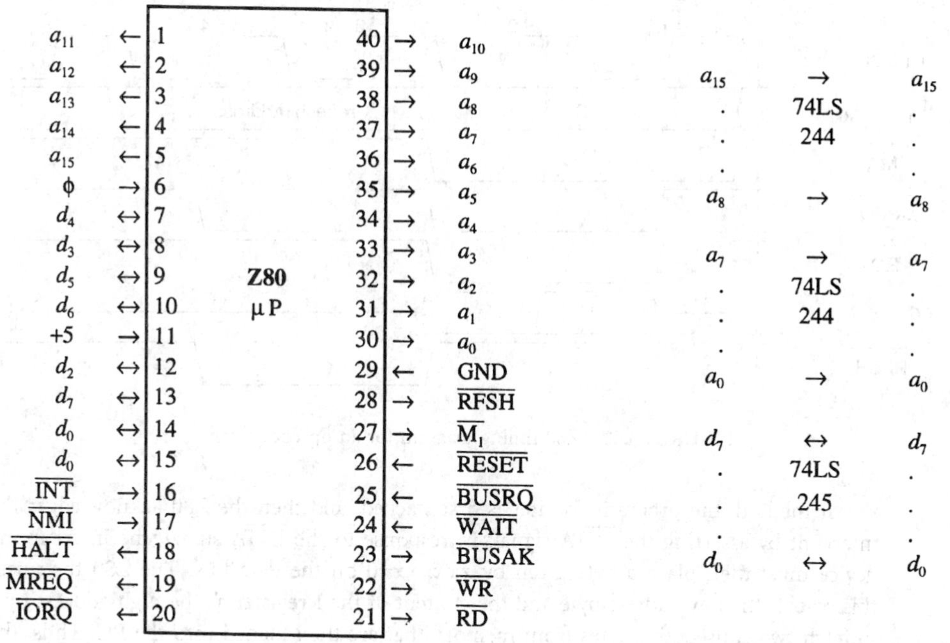

FIGURE 63.67 Pin assignment of the Z80 microprocessor.

$\overline{RD}$—This indicates that an external device must place valid data on the data bus, which the Z80 will soon accept.

$\overline{WR}$—This indicates that the Z80 is driving the data bus with valid data.

$\overline{M_1}$—This output is active while the Z80 is fetching an op-code. The only other time it becomes active is in response to a maskable interrupt.

$\overline{NMI}$—Whenever this nonmaskable interrupt input goes through an active low edge, an internal flip flop is set, and other interrupts are disabled. The interrupt is said to be latched. At the completion of every instruction, this flip flop is checked, and if it is set, then the Z80 first stacks the program counter and then loads the program counter with \$0066, where the first op-code of an interrupt service routine should be located. This way of loading the program counter in response to an interrupt is called a direct interrupt, and it is intended for an unconditional and fast response to, for example, battery low detected, temperature too high detected, or some other urgent situation, because it requires no further action from the interrupting device.

$\overline{INT}$—At the completion of every instruction the Z80 checks this maskable interrupt input. If it is active and interrupts have been enabled with the EI instruction, then interrupt processing commences according to the interrupt method stipulated by the IM i, $i = 0$, 1, or 2, instruction. If $i = 1$, for direct method, the program counter is first stacked, and then it is loaded with 0038H, the address used to obtain the first op-code of an interrupt service routine. If $i = 0$, for vectored method, the Z80 acknowledges the interrupt by asserting the $\overline{IORQ}$ line while the $\overline{M_1}$ signal is active. This unique activity is then interpreted externally to be an interrupt acknowledge signal, denoted by $\overline{INTA}$. In response to the $\overline{INTA}$ signal, the interrupting device then has the opportunity to place the op-code for the one byte Z80 instruction RST, N on the data bus, where N= 0, 1,· · ·, 7. Then, the Z80 stacks the program counter and loads it with an address given by 8N. Thus, depending on N, a 3-bit code embedded within the RST instruction, the first op-code of the interrupt service routine can be located in one of eight locations. If $i = 2$, for indirect

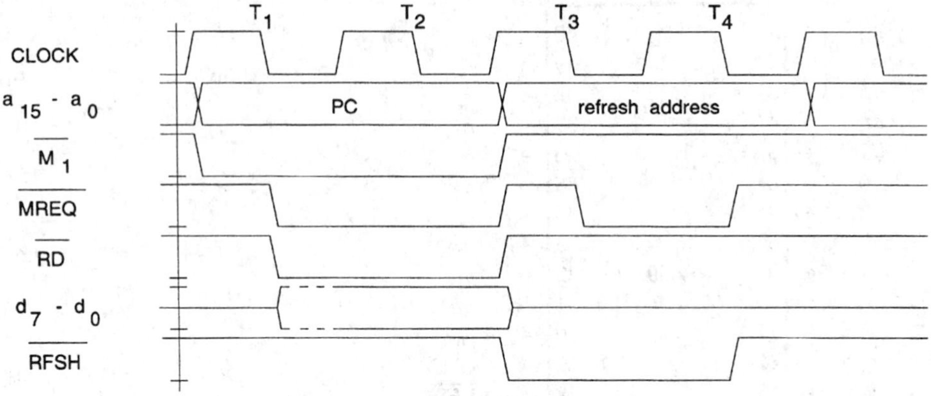

FIGURE 63.68 Z80 timing diagram for an op-code fetch.

vector method, the program counter is first stacked, and then the Z80 acknowledges the interrupt by asserting the $\overline{\mathrm{INTA}}$ signal. In response to the $\overline{\mathrm{INTA}}$ signal, the interrupting device must then place a byte given by *xxxxxxx*0 on the data bus. The Z80 then uses this byte as the low address byte and the content of the I register as the high address byte to fetch two consecutive bytes from memory that are then loaded into the PC. Thus, the I register points to a 256-byte block of memory, where one of 128 interrupt vectors can be selected by the byte provided by the interrupting device.

$\overline{\mathrm{HALT}}$—This output becomes active when the Z80 has stopped fetching additional instructions due to having executed a HALT instruction. The microprocessor can only continue instruction execution upon activation of an interrupt input.

$\overline{\mathrm{WAIT}}$—This input, when active, causes the Z80 to hold constant its address and control signal outputs until this input is released (no longer active). When referenced, a slow memory or I/O device can cause the Z80 to wait by asserting the $\overline{\mathrm{WAIT}}$ input until sufficient time has elapsed to allow the device to respond to the reference.

$\overline{\mathrm{BUSRQ}}$ and $\overline{\mathrm{BUSAK}}$—The first signal is an input that indicates to the Z80 that an external device wants to take control of the system bus. The Z80 completes execution of the present machine cycle, takes its address, data, and tristate control signals into tristate, and acknowledges the request with an active $\overline{\mathrm{BUSAK}}$ signal.

$\overline{\mathrm{RFSH}}$—When active, this indicates the $a_6 \cdots a_0$ holds the content of counter register R, which along with an active $\overline{\mathrm{MREQ}}$, can be used to refresh dynamic memory.

Most of these signals are representative of the kinds of control signals that microprocessors have. Their usefulness becomes more apparent as we look at the timing of microprocessor activity and the impact this has on the design of hardware so that the microprocessor can accomplish read and write data transfers. Figure 63.68 shows a Z80 timing diagram for an op-code fetch, and Fig. 63.69 shows timing diagrams for the other memory and I/O references. For the purpose of studying the timing of events we do not need to know actual binary assignments on the address and data bus lines. Therefore, the convention in these diagrams is intended to indicate when these signals either become relevant for the operation at hand or switch to the present binary assignment, whatever it may be, from some previous binary assignment.

As we look at the timing diagrams, it will also be useful to see how these signals are used. Figure 63.70 shows a schematic of a microcomputer designed according to the memory map of Fig. 63.62. Whenever possible, labeling is used to indicate connections.

Depending on the instruction, the Z80 uses from 4 to 23 clock cycles to execute an instruction. A group of clock cycles within the execution time of an instruction that accomplishes a major activity is called a machine cycle. Each clock cycle within a machine cycle is labeled with

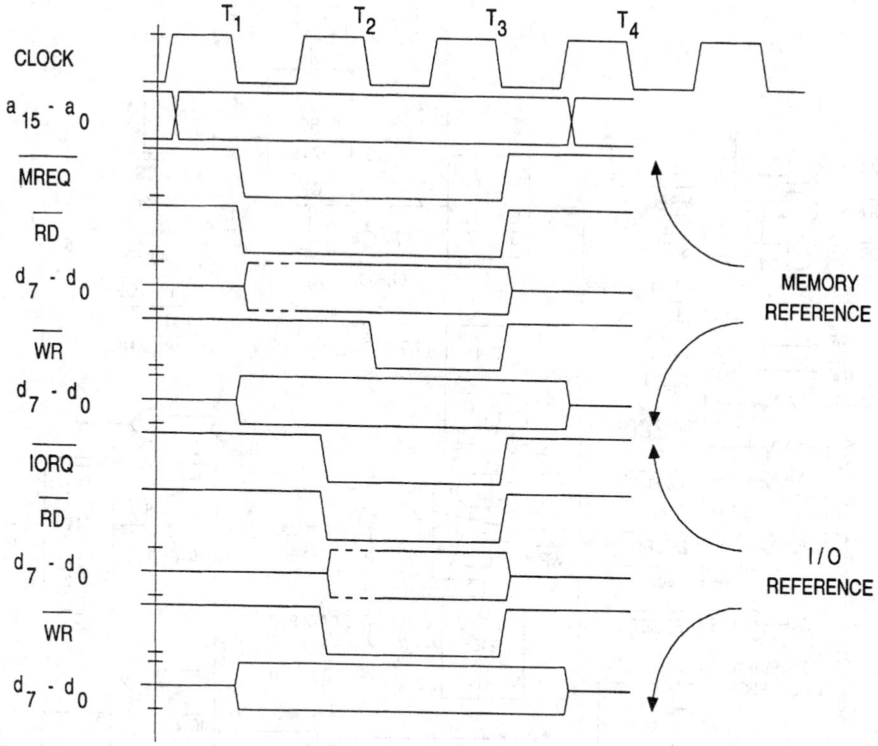

FIGURE 63.69 Z80 timing diagrams.

T_i, $i = 1, 2, \cdots$, and all the activities occur at either leading or trailing clock edges. To better understand such groupings, suppose that after executing many instructions that were retrieved from ROM, the Z80 is about to execute the next few instructions located in memory shown in Fig. 63.62.

The program counter is set to PC=1075H, pointing to the op-code at memory location 1075H. This is the op-code for load accumulator with data using the two bytes that follow the op-code to form an address to point to the data, i.e., LD A, (C000H). This is a three byte instruction, and, according to the Z80 manual, it requires 13 clock cycles, which are grouped into four machine cycles, to execute.

Referring to Fig. 63.68, at the leading edge of the first clock cycle of the first machine cycle of this instruction the Z80 asserts the $\overline{M_1}$ signal, and it drives the external address bus with the address 1075H. At the trailing edge of clock cycle T_1 it asserts the $\overline{MREQ}$ signal and the $\overline{RD}$ signal. At the time that $\overline{RD}$ becomes active the Z80s data bus input register is enabled to follow the content of the external data bus. At the leading edge of clock cycle T_3 the Z80s data bus input register is disabled from following the external data bus, and the content of this register is accepted as an op-code.

During the time interval from the trailing edge of T_1 until just before the leading edge of T_3, which is slightly less than 1.5 clock cycles, external hardware has the opportunity to place the addressed data, which will be processed as an instruction op-code, on the data bus. A designer must ensure that external hardware is fast enough to do this in a timely manner. If it is necessary to use an external device with an access time that is greater than this allotted time, then a counter, called a wait state generator, can be used. Once the slow external device detects (by address and control signal decode) that it is supposed to put valid data on the data bus, then it enables the wait state generator, which should be driven by the system clock, to activate

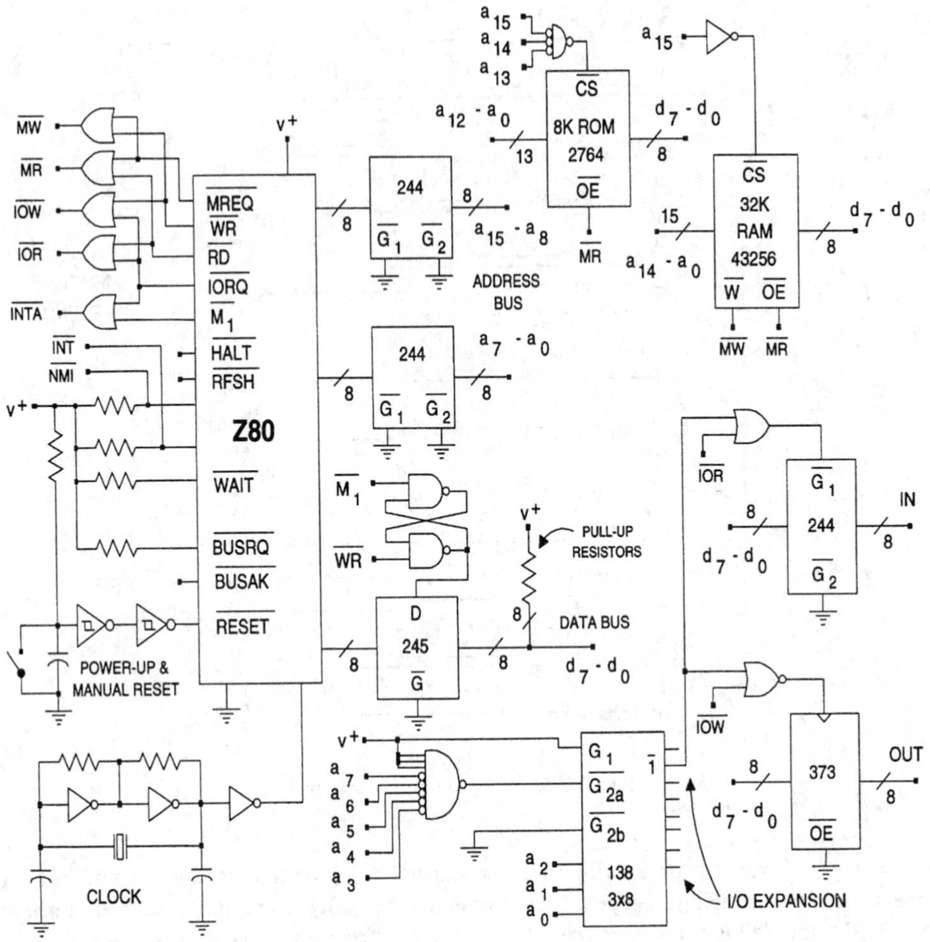

FIGURE 63.70 Z80-based microcomputer.

the $\overline{\text{WAIT}}$ input of the microprocessor for a number of clock cycles that will give the device an opportunity to place data on the data bus. Each clock cycle is called a wait state (labeled with T_w), and the number of wait states needed will depend on the access time of the device being referenced. If several external devices need to generate different numbers of wait states, then the outputs of the wait state generators of all such devices can be OR'd to present just one wait input to the microprocessor. Therefore, the machine can run as fast as each different external device will allow.

During the next two clock cycles the received op-code is interpreted so that the Z80 becomes set to perform two more memory references, called machine cycles M_2 and M_3, to fetch each byte of the address operand. Machine cycle M_1 consists of four clock cycles, and while the Z80 is processing a retrieved op-code during clock cycles T_3 and T_4, the address bus is available for dynamic RAM refresh. During machine cycle M_4 the address C000H, which was obtained during M_2 and M_3, is placed on the external address bus, and execution of this instruction is completed by transferring the content of memory location C000H to the accumulator. Machine cycles M_2 and M_3 each require three clock cycles, as does machine cycle M_4. If whenever this RAM module is referenced, it issues wait states, then the actual number of clock cycles required to execute this instruction will be more than 13 clock cycles.

Notice that the op-code determines the kind and number of additional machine cycles that are necessary to complete execution of an instruction. It also determines how much the program counter must be incremented. Thus, whenever execution of an instruction has been completed, the program counter is pointing to an op-code of the next instruction. Furthermore, virtually all activity involves some kind of synchronized register-to-register transfer.

The op-code at location 1078H is the code for the one byte instruction, ADD A, B. Because this addition requires no additional memory reference, it requires four clock cycles (one machine cycle) to execute. The next instruction, i.e., OUT (01H), A, is a two byte instruction, and it requires three machine cycles to execute. Because this is an OUT instruction, the Z80, during machine cycle M_3, does the following: (1) transfers the address 01H, which it obtained during M_2, to the lower byte of the memory address register to drive the lower byte of the system address bus; (2) transfers the content of the accumulator to the data bus out register at the trailing edge of T_1 to drive the external data bus, which, as shown in Fig. 63.60, incurs a setup delay; (3) activates the $\overline{\text{IORQ}}$ control signal at the leading edge of T_2, which incurs a setup delay; and (4) activates the $\overline{\text{WR}}$ control signal, which also incurs a setup delay, until the trailing edge of T_3. Thus, the I/O device that is supposed to receive the content of the accumulator has slightly less then 1.5 clock cycles during which it must capture the content of the system data bus. By decoding the address, which for an I/O device is called an I/O port address, and the control signals, an I/O interface can enable an external register to start to follow the data bus shortly after $\overline{\text{IORQ}}$ and $\overline{\text{WR}}$ have become active. Then, by the time that the $\overline{\text{WR}}$ signal becomes inactive, the external register should hold the content of the data bus.

Interfacing

The circuitry, or more broadly, the method that makes compatible the operations of devices so that these devices can exchange signals (data or codes), is loosely called an interface. The devices on opposite sides of an interface can be different in many ways. Interfacing two devices can encompass a variety of requirements, such as, (1) impedance matching, (2) voltage or current level conversion, (3) translation of control signal meanings, (4) protocol conversion, (5) signal timing alignment, (6) electrical isolation, (7) exchange of status information, (8) data format conversion, (9) and resolution of other incompatibility issues.

Figure 63.70 gives some examples of the goals of interface design. The unidirectional and bidirectional buffers make compatible the drive (current source) capability of the microprocessor and the drive requirements of the address and data bus. The particular control signals of the Z80 are interfaced to the system control bus with combinational logic to provide drive as well as more explicit control signals. The interfaces between the memory devices and the system bus utilize the address bus to position memory in the desired locations in the memory space. We must be certain that only one device can be a source to the data bus at a time. The interfaces also utilize control bus signals to produce the particular kinds of control signals required by the memory devices and to activate memory at times compatible with microprocessor timing.

Figure 63.70 also shows an input and output port. The interface for this parallel I/O port also utilizes the address bus, and, instead of using control bus signals that go active due to memory reference instructions, it uses control bus signals that go active due to the IN and OUT instructions of the Z80. For output, it decodes the control bus and the lower half of the address bus to enable (clock) the octal latch to capture the content of the data bus at the right time and in response to the intended (its own) address \$01. Thus, by executing an output instruction, like the one located at address \$1079 in Fig. 63.62, the octal latch receives the content of the accumulator. For input, the interface decodes the address and control bus to enable (take out of tristate) the octal buffer, which permits it to drive the data bus with its input at the right time and in response to the intended address. Thus, by executing an input instruction such as IN

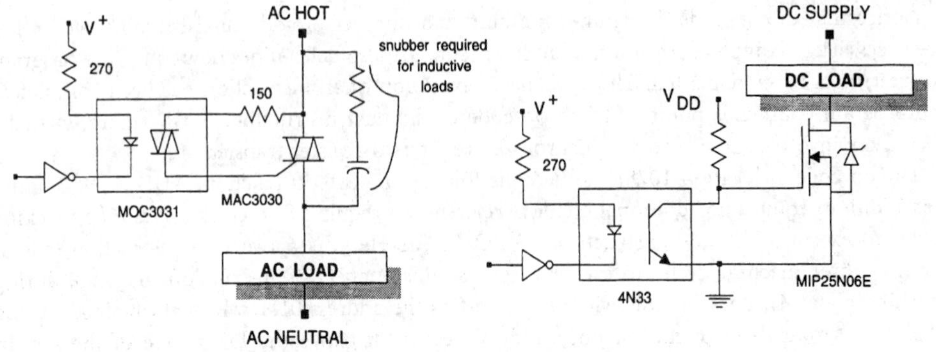

FIGURE 63.71 Circuits for opto-isolated power control.

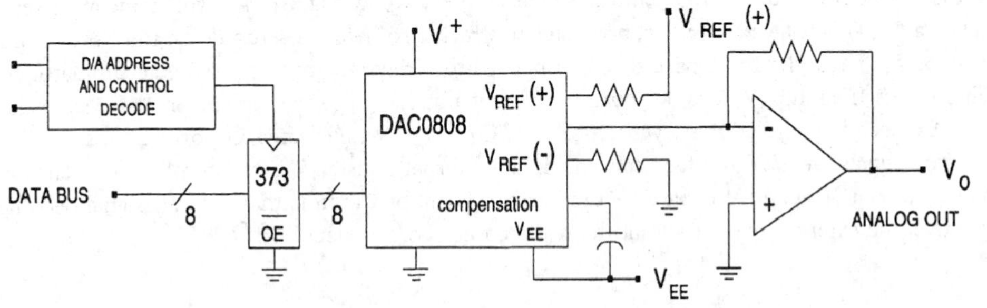

FIGURE 63.72 Output to a D/A converter.

A, (01H), the accumulator will receive the byte at the buffer input. Depending on the address decode logic, the I/O port can be positioned anywhere in the I/O space of the microprocessor.

By using the $\overline{\text{MREQ}}$ control signal, instead of the $\overline{\text{IORQ}}$ control signal, and the entire address bus, I/O ports can also be positioned in the memory space of the microprocessor. The variety of conventional memory move instructions can then be used for I/O. Some microprocessors, such as the M68HC11, do not have I/O space separate from memory space and instructions for I/O in addition to memory move instructions. Instead, no distinction is made between memory and I/O references. This eliminates the need for explicit control signals that interface hardware requires to distinguish memory and I/O references.

One or more parallel I/O ports can serve many different applications. With circuits such as those given in Fig. 63.71, a bit of an output port can turn on and off a light or sound indicator, or DC or AC power supply to a load. With a set of bits, software can produce codes that control a device such as a printer. Because the octal latch in Fig. 63.70 has a tristate output, output enable could be controlled by address and control signal decode from another microprocessor to place the octal latch output on another data bus. Thus, data transfer between two machines is accomplished. Also, software can produce binary time functions to control some process. An analog signal can be generated by outputting to a D/A converter, as shown in Fig. 63.72.

A program can receive information through a parallel input port. This may be the position of a set of switches, as shown on Fig. 63.73, or the output of some kind of an encoder, such as a keypad encoder, rotating shaft position encoder, A/D converter, as shown in Fig. 63.74, or even data from another microprocessor.

For data transfer from one system to another, the sender needs to know if the intended recipient is ready to receive data, and, if data has been sent, then was it received? To facilitate asynchronous transfer of data, consider the output interface shown in Fig. 63.75. This interface is the output port given in Fig. 63.70 and additional circuitry to support ready-to-receive

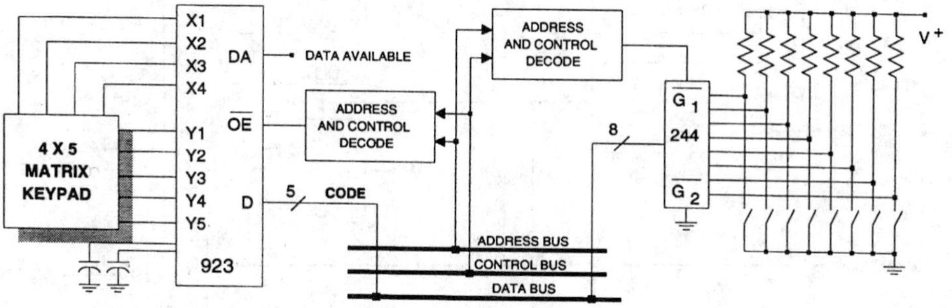

FIGURE 63.73 Circuits to input switch closures.

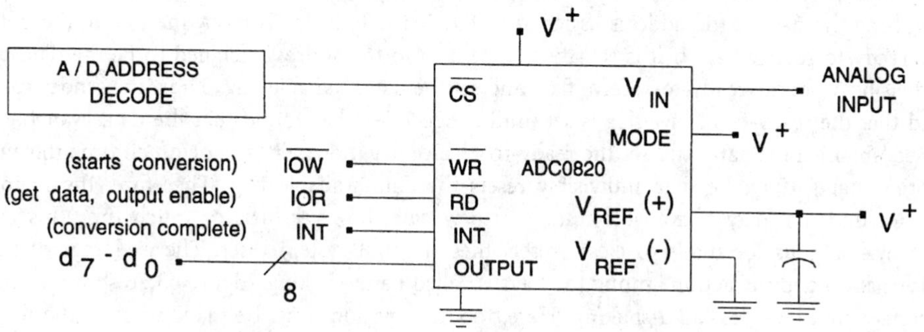

FIGURE 63.74 Write controlled A/D that includes sample and hold.

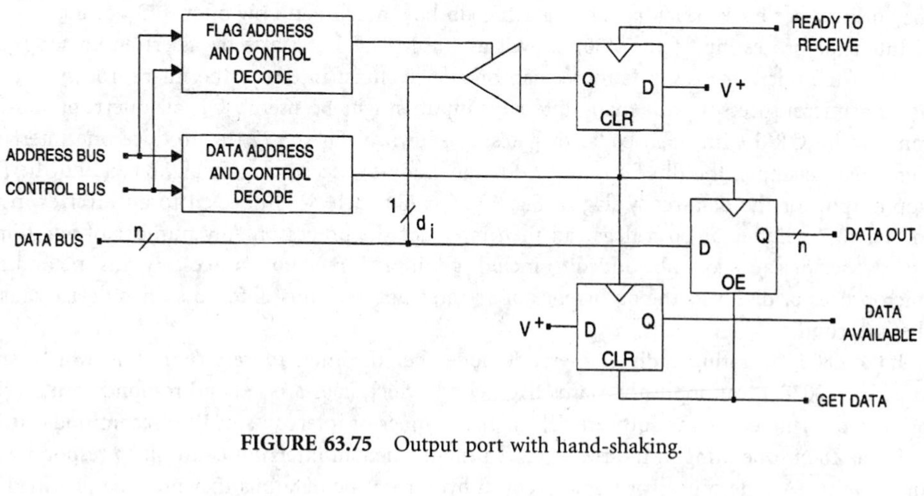

FIGURE 63.75 Output port with hand-shaking.

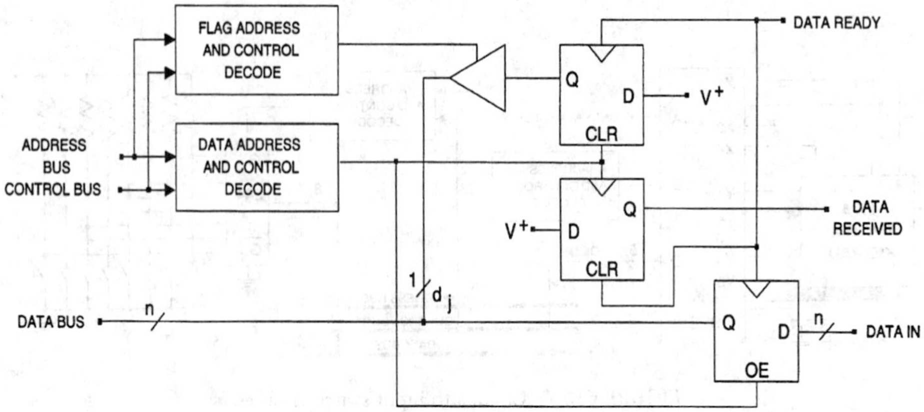

FIGURE 63.76 Input port with hand-shaking.

and data-available flags. The sender can poll (obtain and check) the ready-to-receive flag by inputting the data at the address assigned to flags. If data bit d_i is reset, then continue polling the ready-to-receive flag. If it is set, then write data to the address assigned to be the data port, indivisibly reset the ready-to-receive flag, and set the data-available flag. Thus, the sender cannot find that the ready-to-receive flag is set until the recipient has found that the data-available flag is set, obtained the data, and set the ready-to-receive flag, however little or much time this takes. Notice that getting the data indivisibly resets the data-available flag. Therefore, the recipient cannot find that data is available until it is new data, however little or much time this takes. A provision must be made to clear (reset) these flags at system reset. The exchange of status information and the action among the flags is called hand-shaking. Figure 63.76 shows an input interface that includes hand-shaking. Here, too, a provision must be made to clear the flags at system reset.

To accomplish data transfer, the I/O methods of Figs. 63.75 and 63.76 require that a program must continually poll the flags concerned with I/O. This works as long as the microprocessor is not required to do anything else. If the microprocessor must be used to do other tasks, then I/O can be serviced by either periodically polling I/O status flags or by using I/O status flags to interrupt the microprocessor while it is executing some program. Periodic polling of status flags may or may not be satisfactory. This depends on how much and how often I/O occurs.

Interrupt processing of the Z80, as well as the M68HC11, starts by asserting an interrupt input of the microprocessor. If more than one device must interrupt the microprocessor, then using a microprocessor with many interrupt inputs might be useful. Or, all interrupt sources can be wire OR'd with open collector gates, as shown in Fig. 63.77, to produce one interrupt signal. For example, the flip flop output for the ready-to-receive flag in Fig. 63.75 and the flip flop output for the data-ready flag in Fig. 63.76 could each be connected to an inverter input in Fig. 63.77. Either or both flags can then interrupt the processor. Any number of additional open collector gates could be added to include additional interrupt sources. By this method the microprocessor discovers the occurrence of an interrupt, but must find out which device caused the interrupt.

If the Z80 is operating in direct interrupt mode, then the interrupt service routine, which starts at address 0038H, can input the status flags, check which flags are set, and respond with service priority determined by the software. Thus, any number of interrupts can be accommodated.

If the Z80 is operating in indirect vectored mode, then an interrupt causes it to respond with an active $\overline{\text{INTA}}$. The processor then accepts a byte from the data bus that must be provided by the interrupting device, and this byte, along with the content of the I register, is used as an

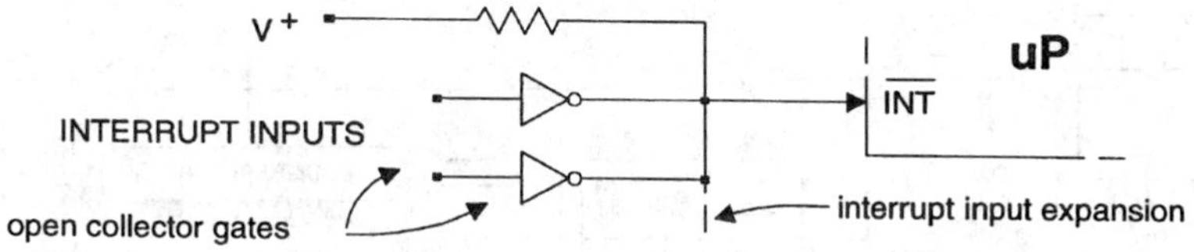

FIGURE 63.77 A wire OR'd circuit.

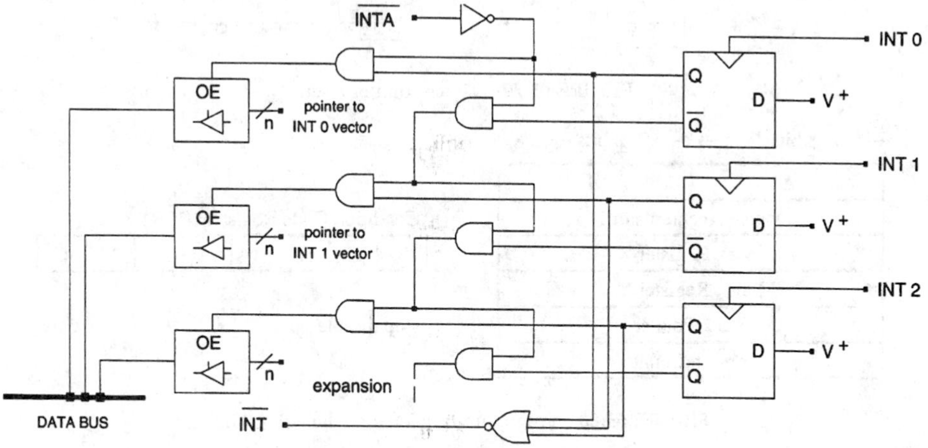

FIGURE 63.78 Priority with a daisy chain.

address to get the service routine address that is then loaded into the program counter. Thus, the I register points to an interrupt vector table, and each interrupting device can point to the address of its own service routine.

If more than one interrupt source exists, then because only one device is allowed to drive the data bus, a priority must be established, in which the interrupting device with highest priority is serviced first. A daisy chain can be used so that hardware determines priority. This is shown in Fig. 63.78. The device connected to the top of the chain has the highest priority. The $\overline{\text{INTA}}$ signal propagates down the chain until is is blocked by a set status flag. This condition is then used to place a response to $\overline{\text{INTA}}$ on the data bus.

In applications such as control it is often necessary that a program keep track of elapsed time. This can be accomplished with a circuit like to the one shown in Fig. 63.79. The 16-bit counter is clocked by the system clock or a clock derived from the system clock. The state of the counter is compared to the output of two registers, and a match, which clears the counter, is then used to interrupt the microprocessor. Thus, through I/O ports, software can set the time interval between interrupts, which can be disabled (a write to the LSB) or enabled (a write to the MSB), and thereafter the microprocessor can be periodically interrupted to perform tasks in real time.

Solutions to interfacing problems are not unique. Variables such as the number of I/O devices, I/O device characteristics, amount of I/O, required response times, hand-shaking, software and hardware tradeoffs, cost of parts and eventual manufacture, etc., all influence interface design. Also, usually numerous options are available to solve a particular problem. All of this requires an understanding of system (microprocessor) bus activity to properly time the occurrence of data transfer events. Then, there are issues of response to asynchronous events and status information about data transfer, polling and interrupt processing, and hardware and software tradeoffs.

Solutions to common I/O problems such as parallel I/O, serial I/O, timer functions, etc., are often available within single IC packages. Typically, manufacturers of microprocessors provide a

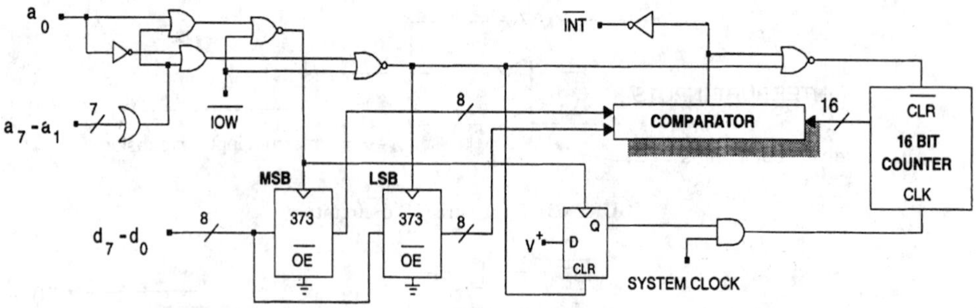

FIGURE 63.79 Timer-counter circuit.

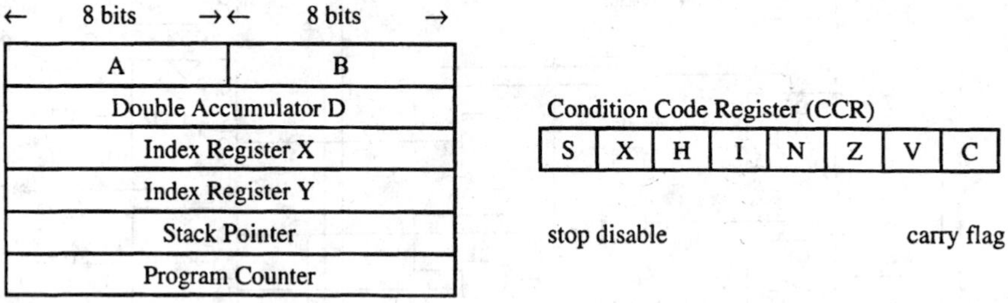

FIGURE 63.80 Programming model of the M68HC11.

family of compatible ICs for each microprocessor. This can simplify the design task and reduce package count. Also, to reduce package count, consideration should be given to combining gate-level hardware into a single package of a programmable logic device (PLD).

Design Using a Microcontroller

The M68HC11 is a microcomputer within a single IC package. It is available in a 48-pin DIP and a 52-pin LCC (leaded chip carrier). In addition to the microprocessor, it can be configured for a variety of resources within the same package. These additional internal resources can be ROM (PROM or EPROM), RAM, parallel I/O ports, serial I/O ports, EEPROM, timer, and an A/D converter. Through a reserved bank of 64 status/control and I/O registers, software can select modes of operation of this additional hardware from a multitude of options. With these additional resources, especially the timer and A/D converter, the device is also called a microcontroller.

The programming model of the M68HC11 is shown in Fig. 63.80. Registers A and B are both accumulators, and they can be used together as accumulator D for 16-bit add, subtract, multiply, and divide instructions. The machine language (op-codes), instruction set, and mnemonics are not the same as for the Z80, and thus the M68HC11 has its own assembly language.

The microprocessor also has different modes of operation. One of four modes is determined by the inputs at two pins labeled MODA and MODB. These modes are MODB = 0, MODA = 0, special bootstrap, (01) special test, (10) normal single chip, and (11) normal expanded. We now consider the single chip and expanded modes of operation.

The M68HC11A8 is a version that has eight analog input channels, and in the normal expanded mode of operation its pin functions are shown in Fig. 63.81, which gives a diagram of a conventional microcomputer. Notice the selection of the normal expanded mode. For convenient reference, 38 of the 52 pins are grouped and labeled as follows: port A = pins 27 to 34, port B = pins 35 to 42, port C = pins 9 to 16, port D = pins 20 to 25, and port E = pins 43 to 50. The remaining pins are used for power supply (V_{SS} = pin 1 and V_{DD} = pin 26), the

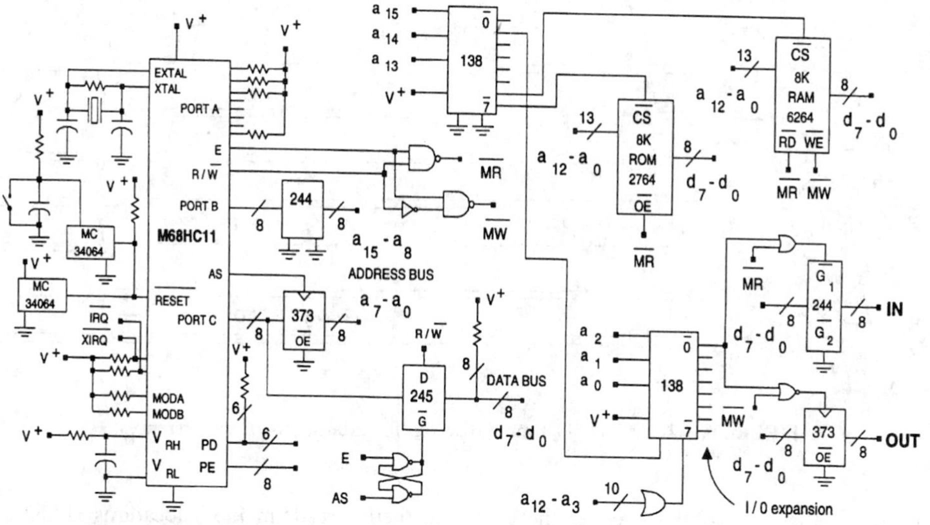

FIGURE 63.81 M68HC11A8-based microcomputer.

crystal for the internal clock circuit (pins 7 and 8), $\overline{\text{RESET}}$ (pin 17), nonmaskable interrupt ($\overline{\text{XIRQ}}$ = pin 18), maskable interrupt ($\overline{\text{IRQ}}$ = pin 19), control bus (AS = pin 4, E clock = pin 5, and R/$\overline{\text{W}}$ = pin 6), and analog input range (V_{RL} = pin 51 to V_{RH} = pin 52).

Port B provides the upper address byte. Port C is the data bus, and, to economize on the package pin count, the lower address byte is time multiplexed with the data bus. The control signal AS, for address strobe, signifies when the data bus contains the lower address byte. The 74HC373 captures the lower address byte when AS is active. Thus, the external system address bus consists of port B and the output of the octal latch. The remainder of the control bus consists of the R/$\overline{\text{W}}$ signal and the E clock signal, which has a frequency equal to one fourth of the system clock frequency.

Because all I/O is memory mapped, the timing of system bus activity is especially straightforward. Figure 63.82 shows M68HC11 timing for all external memory references. After the lower address byte has been captured, the E clock goes high, the data bus is available for data transfer, and the R/$\overline{\text{W}}$ signal determines the direction. The E clock and the R/$\overline{\text{W}}$ signal together control read and write operations. During a read operation, the M68HC11 begins to follow the data bus at the leading edge of the E clock, and it accepts the content of the data bus at the trailing edge of the E clock. Thus, external hardware has slightly less than 2 clock cycles to provide valid data. During a write operation, the microprocessor begins to drive the data bus with valid data at the leading edge of the E clock. Thus, external hardware has slightly less than 2 clock cycles to follow the data bus, and at the trailing edge of the E clock it must accept the content of the data bus. All external memory references require one E clock cycle.

At power-up, an active low reset is accomplished with the MC34064 device in Fig. 63.81. The output of this device switches low whenever its input supply is below a particular limit, and the output goes into tristate whenever its input exceeds a particular limit. Upon reset, the microprocessor initializes internal registers to default conditions, and then it reads memory locations $FFFE and $FFFF to obtain the reset vector that it loads into the program counter. Thus, a designer can locate code that is to be executed at reset almost anywhere in the memory space.

Interrupt processing occurs in a similar way. Just prior to fetching each op-code the microprocessor checks the $\overline{\text{IRQ}}$ input. If it is active, and if the I bit (interrupt mask) in the CCR is reset, then the microprocessor stacks nine bytes from registers in its programming model,

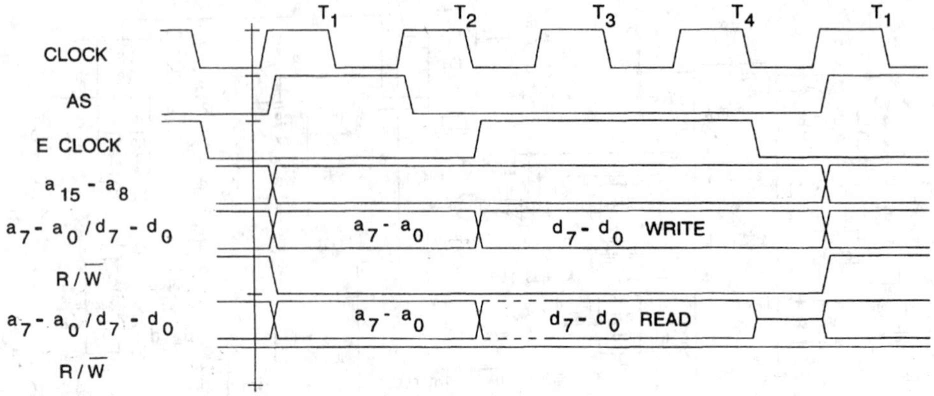

FIGURE 63.82 M68HC11 timing diagram for external memory references.

disables further interrupts by setting the I bit, and then it reads memory locations $FFF2 and $FFF3 to obtain a vector that is loaded into the program counter. The vector is the address of (points to) the first instruction of the $\overline{\text{IRQ}}$ interrupt service routine, which means that a designer can locate this service routine almost anywhere in the memory space.

The response to many interrupting sources can be handled in the same way as is illustrated by Figs. 63.75 to 63.77, where the flags of interrupting devices are wire OR'd into one interrupt input to the microprocessor. Then, upon an interrupt, the service routine must first find out which device caused the interrupt. The control bus does not provide any special means to accomplish this. Moreover, as the service routine responds to each interrupt, a provision must be made to clear each associated interrupting flag.

With the same vectored method, the M68HC11 is designed to accommodate numerous other external and internal sources of interrupts such as internal timer overflow, illegal op-code, software interrupt, nonmaskable interrupt, etc. Each interrupt is associated with two particular memory locations within the vector table, where its vector must be placed. This provides flexibility, and it does not require additional hardware for a reaction by an interrupting device to further control signals.

With the resources and method of interrupt processing of this microprocessor, the memory map of an M68HC11-based system is somewhat fixed. Figure 63.83 shows the memory map of the system given in Fig. 63.81.

Through access to the system bus, additional memory and I/O ports can be added to this system. However, some restrictions exist. Internal address and control signal decoding positions the internal EEPROM block to start at memory location $B600. Because of its long write cycle time, we cannot store data into this memory with simply a regular memory write instruction. Nonvolatile memory must be located at the top end of the memory space to hold the vector table. The positions of the internal RAM and the control register block are programmable. The default locations are given in Fig. 63.83. They can be positioned to start at any 4*K* boundary according to the data stored in the INIT register, which has address $103D after reset.

Many pins of the M68HC11 serve more than one purpose. For example, the mode select inputs are read during reset to determine the mode of operation. Thereafter, MODA is an active low output that becomes active during the first E cycle of each instruction. This is useful for debugging to know when the data bus must hold an op-code. The MODB pin can be used to provide standby power to maintain the content of the internal RAM when V_{DD} is not present.

Port A can serve as three input pins (PA0 to PA2), four output pins (PA3 to PA6), and one pin (PA7) that can be configured for I/O, depending on the data direction control bit labeled DDRA7 in the PACTL register at memory location $1026. From a programmer's viewpoint, port

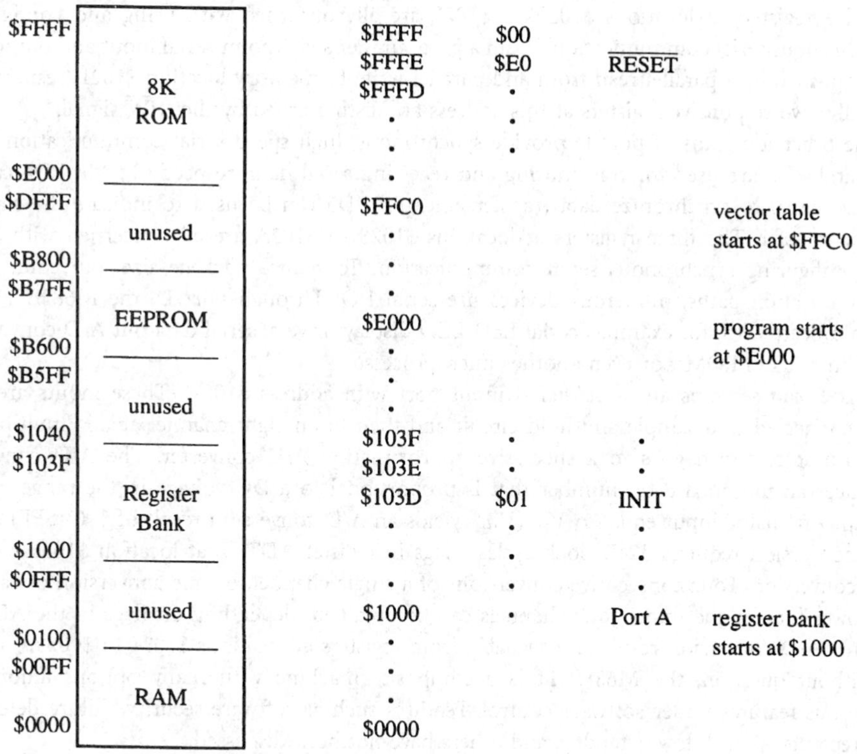

FIGURE 63.83 Memory map and sample ROM content.

A is a memory register with address \$1000, and the register that controls its I/O functionality is another memory register.

The M68HC11 has a free-running 16-bit counter that is clocked by the output of a programmable prescaler, which is clocked by the E clock. The two bytes of the counter can be read through buffers at locations \$100E (high byte) and \$100F (low byte). All M68HC11 timer functions are based on this counter. The 29 registers at locations \$100B to \$1027 are all concerned with using and configuring various kinds of timer functions. Thus, software can initiate a periodic real-time interrupt (RTI), which has an associated RTI vector located at \$FFF0 and \$FFF1.

Another purpose of port A is to provide access to some of these timer functions. For example, pins PA0 to PA2 can be used to measure the edge-to-edge time durations of incoming pulses. The M68HC11 contains circuitry for edge detection, and for each input the edge polarity that is to be detected is programmable. Also, for each input a unique interrupt vector location is associated with edge detection. Thus, to measure the period of a pulse train input at PA0, for example, the counter state must be captured upon an edge detection. For PA0, counter capture is controlled by the IC3F flag in register TFLG1 at location \$1023. Registers at locations \$101A and \$101B receive the counter state. For PA0, edge detection interrupt is enabled with the IC3I flag in register TMSK1 at location \$1022. The vector for this interrupt is located at \$FFEA and \$FFEB. If the I bit in the CCR is reset, then an edge at PA0 will receive interrupt service, which should read the captured counter. Comparing counter states from successive edges of the same polarity can then be used to find the period.

Port D can serve as a general purpose 6-bit I/O port with address \$1008. The direction of each bit is programmable through the DDRD register at location \$1009. When serial communication is enabled, this port provides asynchronous serial input ($R \times D$) at PD0 and serial output ($T \times D$) at PD1. The baud rate is controlled by the contents of the BAUD register at location \$102B.

The five registers at locations $102B to $102F are all concerned with using and configuring asynchronous serial communication. From a programmer's viewpoint serial input and output are accomplished by a parallel read from and parallel write to memory location $102F, respectively. Actually, two responsive registers at this address are distinguished by the $R/\overline{W}$ signal.

The other four pins of port D provide synchronous, high-speed serial communication. Pins PD3 and PD2 are used for transmitting and receiving serial data, respectively. Pin PD4 carries a clock signal to synchronize data transfer, and pin PD5 can be used to indicate the start of a data transfer. The three registers at locations $1028 to $102A are all concerned with using and configuring synchronous serial communication. To reduce package size, pin count, and communication paths, numerous devices use serial I/O. Through port D, the M68HC11 can communicate with, for example, serial LED/LCD display drivers, serial data out A/D converters, serial in/out EEPROMs, or even another microprocessor.

Port E can serve as an 8-bit digital input port with address $100A. These inputs are also each connected to a sample and hold circuit and then to an eight-channel analog multiplexer, the output of which goes to a successive approximation A/D converter. The A/D converter produces an unsigned 8-bit number that is proportional to a DC voltage in the range of V_{RL} to V_{RH}. An analog input equal to V_{RL} (V_{RH}) yields an A/D conversion result of $00 ($FF). Each A/D conversion requires 32 E clock cycles. Flags in register ADTCL at location $1030 control A/D conversion. Four consecutive conversions of a single channel, or one conversion of each of the lower four or the upper four channels can be obtained, depending on flags in the ADCTL register. A/D conversion results are available from registers at locations $1031 to $1034.

Without question, the M68HC11 is a composite machine with many options among its numerous features under software control. Features such as software security, failure detection and recovery, power-down/standby, and others have not been discussed.

In the single chip mode shown in Fig. 63.84 ports B and C and the control signals serve other purposes. Here, port B is an 8-bit output port with address $1004. Pin 6, which is the $R/\overline{W}$ signal in the expanded mode, can now be configured as an output strobe (STRB) that produces a pulse whenever a write to $1004 (port B) occurs. Port C is a general purpose I/O port with address $1003, and its data direction is controlled by the DDRC register at location $1007. The eight pins of port C also go to a register with address $1005. Pin 4, which is the AS signal in the expanded mode, is now an input strobe (STRA). An edge, the polarity of which is programmable, at this input will cause the data at the port C pins to be latched into the register PORTCL with address $1005. Full hand-shaking is implemented, because when data is latched into the PORTCL register, the flag STAF in status register PIOC at location $1002 becomes set, and after both the flag and PORTCL have been read, the flag becomes reset.

In single chip mode one can have as many as 27(12) output bits and 11(26) input bits, depending on data direction control, or fewer I/O bits if some of these pins are used for A/D, serial communication, or timer functions. Moreover, software must reside in the internal $8K$ byte ROM of the M68HC11A8 or the internal $12K$ byte ROM of the M68HC11E9. Like external ROM, this ROM is positioned at the top end of the memory space. It must receive its content at the time of manufacture. With some restrictions, it can be enabled or disabled by the ROMON flag of the CONFIG register at location $103F. This is a special one byte EEPROM so that it will retain its content through power-down and power-up cycles. The M68HC711E9 is an EPROM version of the M68HC11E9. The EPROM can be programmed/erased in the field for development in the single chip mode. There is also an OTP version of the E9. Several other members of the M68HC11 family of microcontrollers have varying amounts of hardware resources.

Design Guidelines

Looking back at Fig. 63.53, we see that its simplicity is deceptive. Nonetheless, in view of Fig. 63.84 or even Fig. 63.70, Fig. 63.53 is not so unrealistic. To use a microprocessor requires

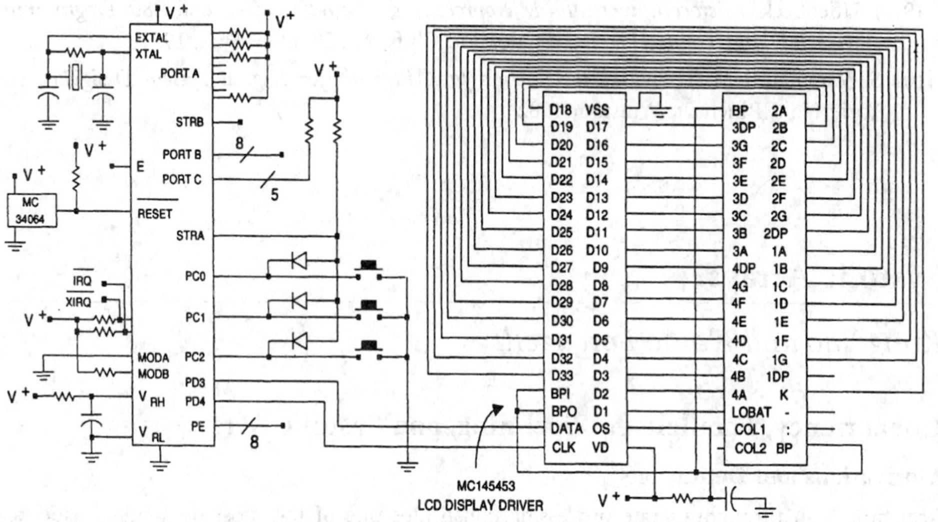

FIGURE 63.84 M68HC11 microcontroller used for a 3-button and 4-digit LCD display device.

an awareness of what may seem like an untold amount of information (facts). This should not and cannot be avoided. Time spent in the beginning learning the details will likely save time and expense that may be spent later making revisions. Here we have only raised a few issues to explore the possibilities.

As we look at different processors we find that while the programming model, instruction set, and hardware resources change, hardware design is concerned mostly with interfacing to achieve electrical, timing, and functional compatibility. Some important principles carry over from one microprocessor to another. It is a component that is fundamentally intended to provide tradeoffs between hardware and software. This is a matter of degree, and will vary from one application to another. Much care must be taken to suitably allocate the hardware-software tradeoffs. Software and hardware development tools such as software simulators, hardware in circuit emulators, logic analyzers, and others are essential. Software will likely carry the greater burden to achieve a product that functions according to a desired theory of operation, and it will also incur the greater development cost.

References

[1] D. M. Auslander and P. Sagues, *Microprocessors For Measurement and Control*, New York: Osborne/McGraw-Hill, 1981.

[2] J. S. Byrd and R. O. Pettus, *Microcomputer Systems: Architecture and Programming*, Englewood Cliffs, NJ: Prentice-Hall, 1993.

[3] M. Cavenor and J. Arnold, *Microcomputer Interfacing: An Experimental Approach Using the Z80*, Englewood Cliffs, NJ: Prentice-Hall, 1989.

[4] F. F. Driscoll, R. F. Coughlin, and R. S. Villanucci, *Data Aquisition and Process Control with the M68HC11 Microcontroller*, Merrill, 1994.

[5] J. D. Greenfield, *The 68HC11 Microcontroller*, Philadelphia: W. B. Saunders, 1992.

[6] H. Haznedar, *Digital Microelectronics*, Menlo Park, CA: Benjamin/Cummings, 1991.

[7] G. H. Miller, *Microcomputer Engineering*, Englewood Cliffs, NJ: Prentice-Hall, 1993.

[8] P. Spasov, *Microcontroller Technology*, Englewood Cliffs, NJ: Prentice-Hall, 1993.

[9] J. Uffenbeck, *Microcomputers and Microprocessors, The 8080, 8085, and Z80, Programming, Interfacing, and Troubleshooting*, Englewood Cliffs, NJ: Prentice-Hall, 1991.

[10] S. Waser and M. J. Flynn, *Introduction to Arithmetic for Digital System Designers*, New York: Holt, Rinehart, Winston, 1982.

63.5 Systolic Arrays

Kung Yao and Flavio Lorenzelli

Concurrency, Parallelism, Pipelining, and Systolic Array

Motivations and Definitions

Real-time high throughput rate processing constitutes one of the most demanding aspects of modern digital signal processing. In order to achieve the desired throughput rate, various forms of concurrent operations are needed. **Concurrency** denotes the ability of a processing system to perform more than one operation at a given time. Concurrency can be achieved through either parallelism or pipelining, or both. **Parallelism** addresses concurrency by replicating some desired processing functions many times. High throughput rate is achieved by having simultaneous operations performed by these functions on different parts of the program. On the other hand, **pipelining** tackles concurrency by breaking some demanding part of the task into many smaller simpler pieces, with many corresponding processing elements (PE), so that processing can be performed in a pipeline manner. This digital pipe is arranged so that it is capable of processing the instructions and data independent of the number of PEs in the pipe. Then, high throughput rate can be achieved by having fast PEs in the pipe. As we shall see, a systolic array can exploit both the parallelism and pipelining capability of some algorithms.

The term **systolic array** was coined by Kung and Leiserson [2] to denote one simple class of concurrent processors, in which processed data move in a regular and periodic manner similar to that of the systolic pumping action of the blood by the heart. The earlier definition of a systolic array by Kung [3] requires: (1) Only a small class of PEs is in the array, with each element in a class performing identical operation; (2) all operations are performed in a synchronous manner independent of the processed data—the only control data broadcast to the PEs is the synchronous clock signal; and (3) the PEs have only nearest-neighbor communications. These regular structure and local communication properties of a systolic array are consistent with efficient modern VLSI designs. Later, various extensions of these assumptions were made: (1) Some of the PEs can perform a limited number of different functions, depending on the presence of some control data; (2) wavefront array allows PEs to start/end/control their own processing tasks, depending on the data; (3) PEs can have communications to few nearby neighbors; wraparound communications among PEs located at the edge of the array are allowed.

Systolic arrays can be designed as linear arrays or two-dimensional rectangular or triangular arrays. In Fig. 63.85(a) consider a uniprocessor system requiring μ time unit to complete a basic operation. If some task requires N such repeated identical operations, the effective throughput rate of this system is given by $r_a = 1/N\mu$. In Fig. 63.85(b) consider a linear array consisting of a single pipe with N such PEs. Then the rate of this linear systolic array is given by $r_b = 1/\mu$. This demonstrates the pipelining aspects of the array. In Fig. 63.85(c) consider a rectangular array consisting of M pipes, with each pipe having N PEs. The rate of this rectangular systolic array is given by $r_c = M/\mu$. This demonstrates both the pipelining and parallelism of the array. While the three models in Fig. 63.85 are overly simple, nevertheless they demonstrate the fact

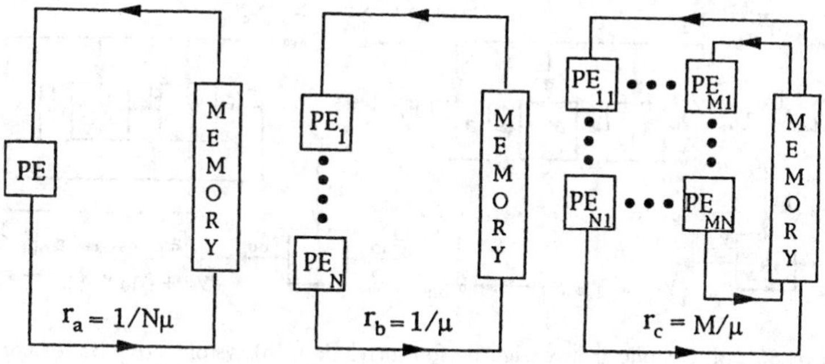

FIGURE 63.85 (a) A uniprocessor system; (b) a linear systolic array; (c) a rectangular systolic array.

that if a given task can be designed for systolic processing, different systolic arrays can yield significantly higher throughput rates as compared to a uniprocessor of a given capability. This is the most basic aspect of systolic processing in which a higher hardware complexity is traded for a higher throughput rate.

Systolic Arrays for Correlation

Consider the linear correlation of a data sequence $\{x_1, x_2, \ldots, x_M\}$ with a weight sequence $\{a_1, a_2, \ldots, a_N\}$ to yield an output sequence $\{y_1, y_2, \ldots, y_{M-N+1}\}$, given by $y_i = a_1x_i + a_2x_{i+1} + \cdots + a_Nx_{i+N-1} = \sum_{j=1}^{N} a_jx_{i+j-1}$, $i = 1, 2, \ldots, M - N + 1$. For the case of $N = 3$ and $M > N$, we have

$$
\begin{aligned}
y_1 &= a_1x_1 + a_2x_2 + a_3x_3 \\
y_2 &= \qquad\qquad a_1x_2 + a_2x_3 + a_3x_4 \\
y_3 &= \qquad\qquad\qquad\qquad a_1x_3 + a_2x_4 + a_3x_5 \\
&\vdots
\end{aligned}
$$

Here, we show two of many possible systolic arrays that can implement the above correlation operations. Design B1 in Fig. 63.86(a) uses three identical PEs to perform the accumulation (multiply and add) operation. Here, the weights a_i are preloaded to the cells and stay throughout the computation. Partial results y_i move systolically from cell to cell. Starting at the third iteration, $y_1, y_2, \ldots$, are outputted from the rightmost cell at the rate of one output per iteration. For each iteration, a x_i is broadcast to all the cells, and a y_i, initialized to zero, enters the leftmost cell. The broadcasted data x_i is marked with an arrow ↓ in Table 63.13. Indeed, by comparison we see y_1, y_2, and y_3, outputted at iteration $T = 3$, 4, and 5, agree with those given from the correlation equations.

In design B2, shown in Fig. 63.86(b), each input x_i is again broadcasted to each cell, each y_i stays at each cell to accumulate terms, while the weights a_i circulate around the cells in the array. A tag bit is associated with a_1 to reset the contents of the accumulator, while a tag bit is associated with a_3 to output the contents of the accumulator after the first two iterations. Data movements in design B2 are shown in Table 63.14. Note that resets occur at cell 1 at iteration 1, cell 2 at iteration 2, cell 3 at iteration 3, cell 1 at iteration 4, etc. Similarly, output y_1 occurs from cell 1 at iteration 3, y_2 from cell 2 at iteration 4, y_3 from cell 3 at iteration 5, etc.

Systolic Array Design Techniques

Systolic array designs, as shown above for the correlation case, can be obtained by ad hoc approaches. More formal procedures for the systematic design of systolic arrays have been

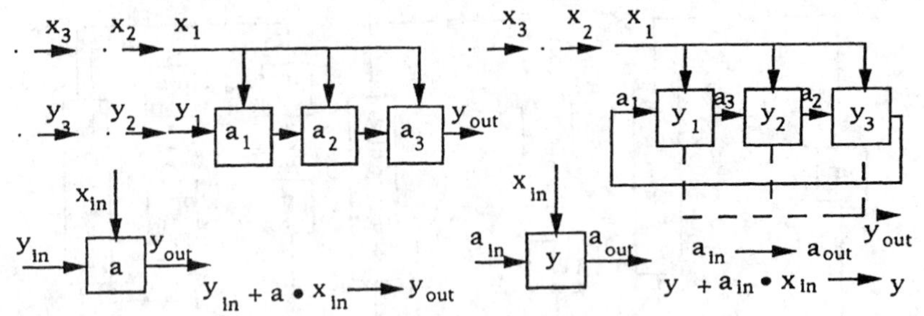

FIGURE 63.86 (a) Systolic array design B1 for correlation; (b) systolic array B2 for correlation.

TABLE 63.13 Data Movement in Design B1

Iteration	Cell 1	Cell 2	Cell 3
$T = 1$	$\overset{\downarrow}{x_1}a_1$	$\overset{\downarrow}{x_1}a_2$	$\overset{\downarrow}{x_1}a_3$
$T = 2$	$\overset{\downarrow}{x_2}a_1$	$a_1a_1 + \overset{\downarrow}{x_2}a_2$	$x_1a_2 + \overset{\downarrow}{x_2}a_3$
$T = 3$	$\overset{\downarrow}{x_3}a_1$	$x_2a_1 + \overset{\downarrow}{x_3}a_2$	$x_1a_1 + x_2a_2 + \overset{\downarrow}{x_3}a_3 \rightarrow y_1$
$T = 4$	$\overset{\downarrow}{x_4}a_1$	$x_3a_1 + \overset{\downarrow}{x_4}a_2$	$x_2a_1 + x_3a_2 + \overset{\downarrow}{x_4}a_3 \rightarrow y_2$
$T = 5$	$\overset{\downarrow}{x_5}a_1$	$x_4a_1 + \overset{\downarrow}{x_5}a_2$	$x_3a_1 + x_4a_2 + \overset{\downarrow}{x_5}a_3 \rightarrow y_3$

TABLE 63.14 Data movement in Design B2

Iteration	Cell 1	Cell 2	Cell 3
$T = 1$	$0 + a_1\overset{\downarrow}{x_1}$	$a_3\overset{\downarrow}{x_1}$	$a_2\overset{\downarrow}{x_1}$
$T = 2$	$a_1x_1 + a_2\overset{\downarrow}{x_2}$	$0 + a_1\overset{\downarrow}{x_2}$	$a_2x_1 + a_3\overset{\downarrow}{x_2}$
$T = 3$	$a_1x_1 + a_2x_2 + a_3\overset{\downarrow}{x_3} \rightarrow y_1$	$a_1x_2 + a_2\overset{\downarrow}{x_3}$	$0 + a_1\overset{\downarrow}{x_3}$
$T = 4$	$0 + a_1\overset{\downarrow}{x_4}$	$a_1x_2 + a_2x_3 + a_3\overset{\downarrow}{x_4} \rightarrow y_2$	$a_1x_3 + a_2\overset{\downarrow}{x_4}$
$T = 5$	$a_1x_4 + a_2\overset{\downarrow}{x_5}$	$0 + a_1\overset{\downarrow}{x_5}$	$a_1x_3 + a_2x_4 + a_3\overset{\downarrow}{x_5} \rightarrow y_3$

proposed by Moldovan [7], Quinton [9], Kung [5], Rao [10], Darte and Delosme [1] and others. All these more formal procedures are collectively referred to as dependence graph mapping techniques for systolic array design.

In this approach an algorithm must be formulated in the **single assignment algorithm** form. Each variable has a unique value during the evaluation of the algorithm. Those variables with multiple values can be converted to single values by vectorizing the variables through the introduction of new indices. As an example, consider the matrix multiplication of $C = AB$, where $A = [a_{ik}]$ is $N_1 \times N_3$, $B = [b_{kj}]$ is $N_3 \times N_2$, and $C = [c_{ij}]$ is $N_1 \times N_2$. A conventional formulation of this algorithm contains the expression, $c_{ij} = c_{ij} + a_{ik}b_{kj}$, for $i = 1$ to N_1, $j = 1$ to N_2, and $k = 1$ to N_3. We note, c_{ij} has multiple values for $k = 1, \ldots, N_3$. We can modify it to have single values by replacing it by the variable c_{ijk}. The above equation for c_{ij} then becomes $c_{ijk} = c_{ij(k-1)} + a_{ik}b_{kj}$, $c_{ij0} = 0$, c_{ijN_3}, $i = 1, \ldots, N_1$, $j = 1, \ldots, N_2$, $k = 1, \ldots, N_3$.

All algorithm variables are assumed to be indexed variables with V variable names, denoted by the generic names of V_m, $1 \leq m \leq V$. In the above matrix–matrix multiplication problem $V = 3$, and we can take $X_1 = c$, $X_2 = a$, and $X_3 = b$. For each variable name, the domain of the index vectors is a subset in a S-dimensional space. This subset is called the algorithm's **index space** and S is its dimension. For most iterative signal processing problems, time is usually one of the index space coordinates. For the above matrix–matrix multiplication problem, we need to propagate a_{ik}

across the j variables as well as b_{kj} over the i variables in order to perform the basic multiplication operation. These a_{ik} and b_{kj} are propagating variables because they involve no computations, but need to be made available at various stages of the computation. In the matrix–matrix problem clearly $S = 3$ and the index space is $\mathbf{S_0} = \{(i, j, k)\colon 1 \le i \le N_1,\ 1 \le j \le N_2,\ 1 \le k \le N_3\}$. Furthermore, the initializations of the new variables are given by $a(i, 0, k) = a_{ik}$, $b(0, j, k) = b_{jk}$, $c(i, j, 0) = 0$, $c(i, j, N_3) = c_{ij}$, and the algorithm is finally given by $a(i, j, k) = a(i, j-1, k)$, $b(i, j, k) = a(i-1, j, k)$, $c(i, j, k) = c(i, j, k-1) + a(i, j, j)b(i, j, k)$, for $(i, j, k) \in \mathbf{S_0}$.

In general, a point (or node) in the index space is called an **index point**. Thus, $X_m(I)$ is the variable X_m defined at the index point I. A dependence graph mapping is a representation of a single assignment algorithm, where the dependencies among the variables are represented by directed arcs among the nodes. A basic property of the class of algorithms of interest is that of **shift-invariance**.

An algorithm is shift-invariant if the dependence graph is regular. That is, if $X(I)$ depends on $Y(J)$, then $X(I+K)$ depends on $Y(J+K)$ for all I, J, and K in the index space. Three well-known shift-invariance algorithms include:

1. Uniform recurrence equations (URE): $X_1(I) = F_1(X_1(I - D_1), \ldots, X_V(I - D_V))$, $X_i(I) = X_i(I - D_i)$, $2 \le i \le V$. Computation occurs only in $F_1(\cdot)$ and propagations in all the other variables. Clearly, the final form of the above matrix–matrix multiplication algorithm is a URE algorithm with $V = 3$, $F_1(\cdot) = c(i, j, k)$, with $X_1(\cdot) = c(\cdot)$, $X_2(\cdot) = a(\cdot)$, $X_3(\cdot) = b(\cdot)$, $I = (i, j, k)$, $D_1 = [0, 0, 1]^{\mathrm{T}}$, $D_2 = [0, 1, 0]^T$, and $D_3 = [1, 0, 0]^T$.
2. Generalized uniform recurrence equations (GURE): $X_m(I) = F_m(X_{m_1}(I - D_{m_1}), \ldots, X_{m_{k(m)}}(I - D_{m_{k(m)}}))$, $1 \le m \le V$, where $m_1, \ldots, m_{k(m)}$ belongs to $\{1, \ldots, m\}$. In GURE we can have computations in all V functions of $F_m(\cdot)$. The number of independent variables, $m_{k(m)}$, depends on each m. The shift index dependence, $I - D_{m_i}$, is fixed for each X_{m_i}.
3. Regular iterative algorithm (RIA): $X_m(I) = F_m(X_{m_1}(I - D_{m_1, m}), \ldots, X_{m_{k(m)}}(I - D_{m_{k(m)}, m}))$, $1 \le m \le V$. Here, the shift index dependency, $I - D_{m_i, m}$, is not fixed but is a function of m_i and m.

Each processor of the systolic array is assumed to have all the necessary computational modules to compute $F_m(\cdot)$. For URE, we need only one such module, but for GURE and RIA, we need V modules. The time required for the computation of $F_m(\cdot)$ is denoted by τ_m, and the minimum time between such computations is denoted by h_m. In most cases, we can set τ_m and h_m to unity. The design of a processor array to perform the algorithm requires spatial and temporal assignments. Each $X_m(I)$ must be assigned to a processor at each integral time slot. The processor **allocation function**, $A(I)$, assigns all variables with the index I to the processors in the array. The **scheduling function**, $S_m(I)$, assigns the start of the computation for the variable $X_m(I)$. The simplest form of scheduling and processor allocations are based on the projection of the high multidimensional dependence graph onto the lower dimensional processor array. Variables represented by nodes in the dependence graph are mapped to processors which perform the computations. The directed arcs of the dependence graphs are transformed to physical communication links in the processor array.

The essence of the allocation function $A(\cdot)$ is thus to return for every index value $I \in \mathbf{S_0}$ a vector which indicates the processor in charge of the computation represented by a point in a lower dimensional space. Analogously, the scheduling function $S_m(\cdot)$ provides the relative start of the execution for the computation indexed by I. These two functions cannot be chosen independently because two computations assigned to the same processor cannot be scheduled for the same time (**compatibility constraint**). Additional details on this constraint are given later. While in principle $A(\cdot)$ and $S_m(\cdot)$ can be any function, we shall consider only **affine functions**, in the sense that $A(I) = A^{\mathrm{T}}I$, $S_m(I) = \lambda^{\mathrm{T}}i + \gamma_m$, where A is a suitable matrix, λ a vector, and γ_m an integral constant.

The dependence graph of an algorithm can be interpreted as a *lattice* embedded in a multidimensional integral space (i.e., a proper bounded subset of $\mathbf{Z}^S$, where $\mathbf{Z}$ is the set of relative integers), enclosed in a convex polyhedron. We assume the lattice to be "dense" in the sense that all the integral points in it correspond to actual computations. The whole procedure of mapping an algorithm onto a systolic-type processor consists of two conceptually different but interdependent operations of using a space transformation and a time transformation. The former actually *projects* the dependence graph onto a lower dimensional structure which then can be mapped one-to-one onto the physical array, while the latter gives the start of the execution of each computation.

For simplicity, consider the projection of the S-dimensional space onto an $(S-1)$-dimensional processor space. The more general problem of projecting the dependence graph onto an $(S-p)$-dimensional space ($p \geq 1$) can be expressed using a similar but more involved notation and is omitted here. Instead of considering allocation functions, we refer to the *projection vector* u, which is orthogonal to the processor space onto which we project. Assume that we have chosen both the projection and the scheduling vectors (u and λ, respectively). For normalization purposes, they are chosen to be coprime vectors, such that the greatest common divisor of their components is 1, and their first nonzero element is positive.

Two sets of constraints must be satisfied by u and λ. Assume nodes I and J are located along a direction parallel to the projection vector u such that $J = I + \alpha u$, $\alpha \in \mathbf{Z}$. Then the computations associated with the two nodes will be projected onto the same processor. Consequently, compatibility constraint requires that they be performed at different times. Analytically, this is equivalent to $|\lambda^T u| \geq \max_{m=1,\dots,V} h_m$, which for $h_m = 1$ simplifies to $|\lambda^T u| > 0$. Thus, for this case λ and u cannot be perpendicular. Furthermore, the quantity $c \triangleq |\lambda^T u|$ represents the number of time slots between successive calculations scheduled on the same processor. $1/c$ is sometimes called the **efficiency** of the processors because the larger the c, the more time the processors can idle. One common approach is to select the projection vector and the scheduling vector so as to achieve the highest efficiency, with c being as close to 1 as possible.

Consider the case in which the variable $X_m(I)$ depends on $X_n(I - D_{nm})$. The **precedence constraint** implies the calculation of $X_n(I - D_{nm})$ must be scheduled so as to be completed before the start of the calculation of $X_m(I)$. Analytically, the precedence constraint is equivalent to $\lambda^T D_{nm} + \gamma_m - \gamma_n \geq \tau_n = 1$, for all $1 \leq m \leq v$ and for all dependences D_{nm}. If the γ constants are chosen to be all equal, the precedence constraint becomes $\lambda^T D_{nm} \geq 1\ \forall m = 1, \dots, V$.

Assume the precedence and compatability constraints are satisfied and λ and u are coprime vectors. Then it is possible to extend both vectors to two unimodular matrices. A matrix with integral entries is called **unimodular** when its determinant is equal to ± 1. This implies that they admit integral inverses. The unimodular extension of coprime vectors is not unique. We will choose U and Λ to be the unimodular extended matrices that have u and λ, respectively, as their first columns. It is possible to show that the columns of any S-dimensional unimodular matrix can constitute a basis for the space $\mathbf{Z}^S$. Moreover, if we denote $\sigma_1, \dots, \sigma_S$ to be the columns of $\Sigma = U^{-T}$, then we have $\sigma_1^T u = 1$ and $\sigma_i^T u = 0$ for all $i = 2, \dots, S$. Therefore, $\{\sigma_2, \dots, \sigma_S\}$ will be a basis of the processor space of the resulting logic array. Similarly, the first column of T (the inverse of Λ^T), t_1, represents the direction in which time increases by one step; i.e., it is the vector defining the hyperplane of the points computed at the same time. The other columns of T (denoted by $t_2, \dots, t_S$) are a basis of such a hyperplane.

If we denote by $\Sigma_+ = [\sigma_2, \dots, \sigma_S]$ the matrix basis of the processor space, the allocation function and the scheduling function have the form $A(I) = \Sigma_+^T I$, $S_m(I) = \lambda^T I + \gamma_m$, $m = 1, \dots, V$. With these elements, we can have the complete description of the final array. The processors are labeled by $A(I) = \Sigma_+^T I$ as I ranges over the index space. The dependences D_{nm}

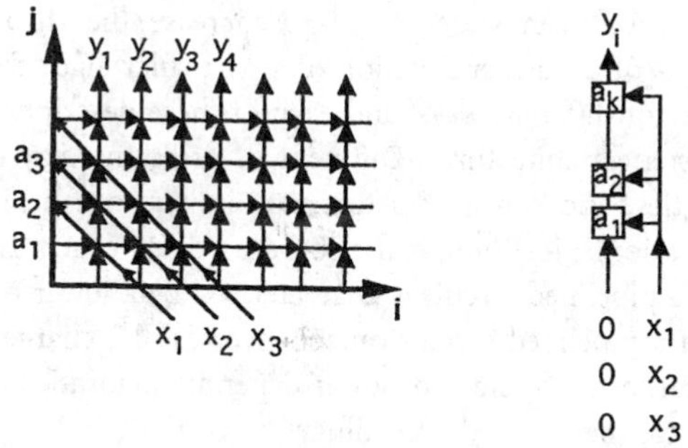

(a) (b)

FIGURE 63.87 (a) Two-dimensional dependence graph; (b) one-dimensional dependence graph.

are mapped onto communication links $\Sigma_+^T D_{nm}$ and the delay registers on such links must be in number equal to $\lambda^T D_{nm} + \gamma_m - \gamma_n - \tau_n = \lambda^T D_{nm} - 1$.

Reconsider the systolic correlation problem using the weights $\{a_1, \ldots, a_k\}$ and the data $\{x_1, \ldots, x_n\}$, as discussed earlier. Recall the correlation is given by $y_i = a_1 x_i + a_2 x_{i+1} + \cdots + a_k x_{i+k-1}$, $1 \le i \le n+1-k$. A recurrence equation formulation of this equation is given by $y(i, j) = y(i, j-1) + w(i, j)x(i, j)$, $y(i, 0) = 0$, $y_i = y(i, k)$; $w(i, j) = w(i-1, j)$, $w(0, j) = a_j$; and $x(i, j) = x(i+1, j-1)$, $x(i, 0) = x_{i-1}$, all with $1 \le i \le n+1-k$, $1 \le j \le k$. A dependence graphical representation of these equations is shown in Fig. 63.87(a).

A URE reformulation of the recurrence equations yields $X_1(I) = y(i, j) = F_1(X_1(I - D_1), X_2(U - D_2), X_3(I - D_3))$, $X_2(I) = w(i, j) = X_2(I - D_2)$, $X_3(I) = x(i, j) = X_3(I - D_3)$, with the index point $I = [i, j]^T$ and displacement vectors $D_1 = [0, 1]^T$, $D_2 = [1, 0]^T$, and $D_3 = [-1, 1]^T$. In particular, consider the URE representation of the B1 design based on the choice of $u = [1, 0]^T$, $\Sigma_+ = [0, 1]^T$, $\lambda = [1, 1]^T$, $\gamma_m = 0$, $m = 1, 2$, and 3, and $\tau_n = h_n = 1$, $n = 1$, 2, and 3. Then the two-dimensional graph of Fig. 63.87(a) is projected onto the one-dimensional graph of Fig. 63.87(b). Specifically, for any index $I = [i, j]^T$, the processor allocation function yields $A(I) = \Sigma_+^T I = j$, $i \le j \le k$, which is a valid projection from two-dimensions to one. On the other hand, the index point for each input data x_l (with variable name of X_3) is given by $I = [i, l - i + 1]^T$, $l = 1, \ldots, n$. Then the scheduling function $S_3(I)$ is given by $S_3(I) = \lambda^T I = [1, 1][i, l - i + 1]^T = i + l - i + 1 = l + 1$, $l = 1, \ldots, n$. This indicates each x_l for the above given I must be available at all the processors at time $l + 1$. Thus, there is no propagation and this means all the x_l must be broadcasted to all the processors. However, the simplistic definition of a systolic design does not allow broadcasting. Indeed, the precedence constraint is **not** satisfied with D_3. That is, $\lambda^T D_3 = [1, 1][-1, 1]^T = 0 \not\geq \tau_n = 1$. Of course, other choices of λ and u generate other forms of systolic array architecture for correlation. For more complicated signal processing tasks such as QR decomposition, recursive least-squares estimation, singular value decomposition, Kalman filtering, etc., the design of efficient systolic array architectures are generally difficult. The dependence graph mapping technique provides a systematic approach to such designs by providing the proper selections of these λ and u vectors.

Digital Filters

The application of digital filtering has spread tremendously in recent years to numerous fields, such as signal processing, digital communications, image processing, and radar processing. It is well known that the sampling rate, which is closely related to the system clock, must be higher than the Nyquist frequency of the signals of interest. It follows that in order to perform real-time filtering operations when high frequency signals are involved, high-speed computing hardware is necessary.

Pipelining techniques have been widely used to increase the throughput of synchronous hardware implementations of a transfer function of a particular algorithm. Most algorithms can be described in a number of different ways, and each of these descriptions can be mapped onto a set of different concurrent architectures. Different descriptions may lead to realizations with entirely different properties, and can have a dramatic impact on the ultimate performance of the hardware implementation. Pipelining can also be used for other than throughput increase. For a fixed sample rate, a pipelined circuit is characterized by a lower power consumption. This is due to the fact that in a pipelined system capacitances can be charged and discharged with a lower power supply. Because the dissipated power depends quadratically on the voltage supply, the power consumption can be reduced accordingly.

An increase in the speed of the algorithm also can be achieved by using *parallelism*. By replicating a portion of the hardware architecture, similar or identical operations can be performed by two or more concurrent circuits, and an intelligent use of this hardware redundancy can result in a net throughput increase, at the expense of area. Note that VLSI technologies favor the design in which individual sections of the layout are replicated numerous times. A regular and modular design can be achieved at relatively low costs. For a fixed sample rate, parallelism can be exploited for a low power design due to the reduced speed requirements on each separate portion of the circuit.

Much work has been done in the field of systolic synthesis of finite and infinite impulse response (FIR/IIR) filters, as can be seen from the literature references. In the following we consider possible strategies which can be used to increase the throughput of the concurrent architectures of the FIR and IIR filters.

Finite Impulse Response Filters

Finite impulse response filters have been largely employed because of certain desirable properties. In particular, they are always stable, and causal FIR filters can possess linear phase. A large number of algorithms have been devised for the efficient implementation of FIR filters, which minimize the number of multipliers, the round-off noise, or the coefficient sensitivity. The generic expression that relates the output $y(n)$ at time n to the inputs $x(n-i)$ at times $(n-i)$, $i = 0, 1, \ldots, q$ is given by

$$y(n) = \sum_{i=0}^{q} a_i x(n-i)$$

where the $\{a_i\}_{i=0}^{q}$ are the FIR filter coefficients. Here, we consider only issues of pipelining and parallelism.

The pipeline rate, or throughput, of implemented nonrecursive algorithms such as FIR filters can be increased without changing the overall transfer function of the algorithm by means of a relatively simple modification of the internal structure of the algorithm. In particular, one set of latches and storage buffers can be inserted across any feed-forward cutset of the data flow graph. Figure 63.88 illustrates the increase of throughput achieved by pipelining in a second-order three-tap FIR filter. The sample rate of the circuit of Fig. 63.88(a) is limited by the throughput of one multiplication *and* two additions. After placing the latches at the locations shown in Fig. 63.88(b), the throughput can be increased to the rate of one multiplication *or* two additions. Pipelining can be used to increase the sample rate in all the cases in which no feedback loops are present. The drawbacks of pipelining are an increased latency and a larger number of latches and buffers.

Parallelism can be used to increase the speed of an FIR filter. Consider Fig. 63.89, in which the three-tap FIR filter of Fig. 63.88 was duplicated. Because at each time instant two input samples are processed and two samples are output, the effective throughput rate is exactly doubled.

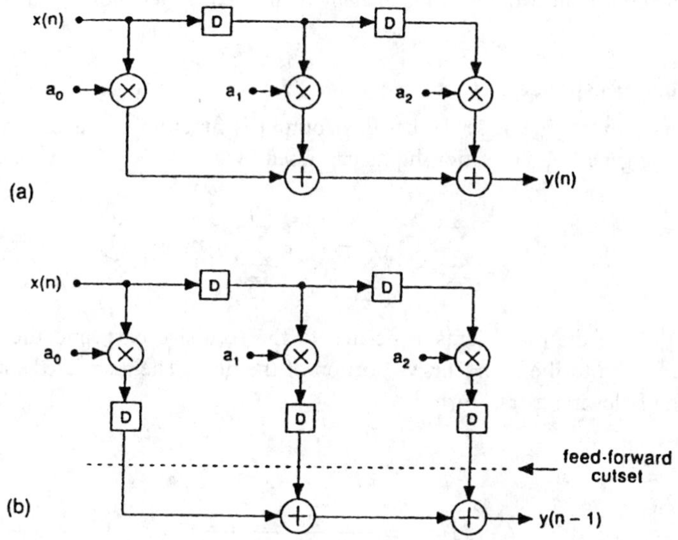

FIGURE 63.88 A three-tap FIR filter. (a) With no pipelining, the throughput is limited by the rate of one multiplication and two additions. (b) With pipelined circuit, the throughput is increased to the rate of one multiplication or two additions.

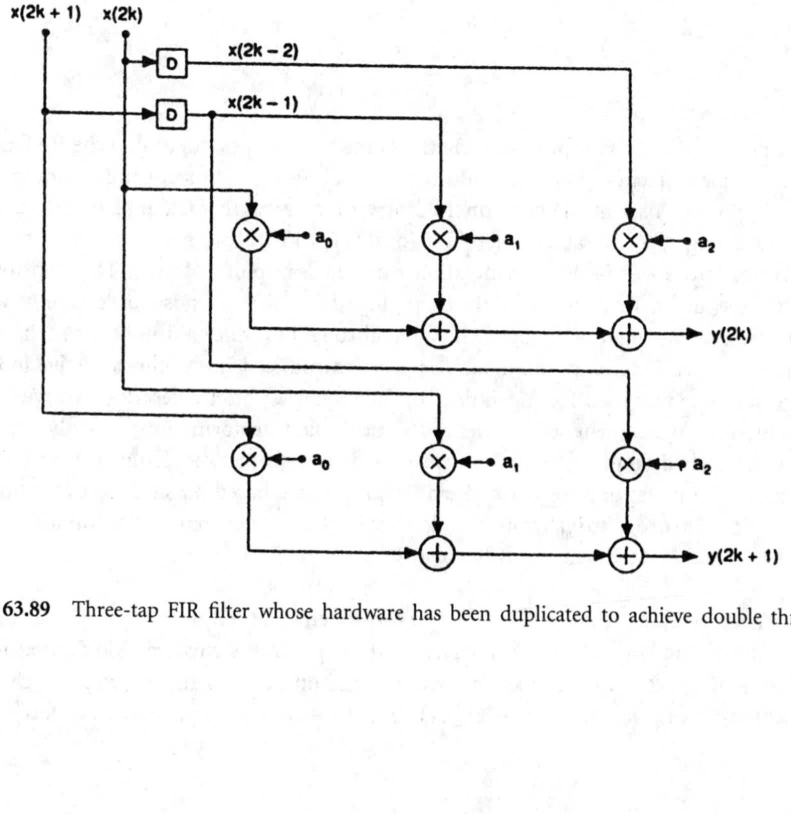

FIGURE 63.89 Three-tap FIR filter whose hardware has been duplicated to achieve double throughput rate.

As can be seen from Fig. 63.89, parallelism leads to speed increase at a considerable hardware cost. For many practical implementations, parallelism and pipelining can be used concomitantly, when either method alone would be insufficient or limited by technology such as I–O, clock rate, etc.

Infinite Impulse Response Filters

These are recursive filters in the sense that their output is function of current inputs as well as past outputs. The generic I–O relationship is expressed by

$$y(n) = \sum_{j=1}^{p} a_j y(n-j) + \sum_{i=0}^{q} b_i x(n-i) \tag{63.22}$$

where the $\{a_j\}_{j=1}^{p}$ are the coefficients associated to the recursive part, and the $\{b_i\}_{i=0}^{q}$ are the coefficients associated to the nonrecursive portion of the filter. The associated transfer function is written as the following z-transform

$$H(z) = \frac{\sum_{i=0}^{q} b_i z^{-i}}{1 - \sum_{j=1}^{p} a_j z^{z-j}}$$

For stability reasons, it is required that all the poles (i.e., the zeroes of the denominator of $H(z)$) be inside the unit circle in the z-plane.

Consider a circuit in which L loops are present, each with latency τ_k, $k = 1, \ldots, L$. The number of latches present in each loop is equal to ν_k, $k = 1, \ldots, L$. Then the throughput period cannot be shorter than

$$T_{\max} \equiv \max_{k=1,\ldots,L} \left\lceil \frac{\tau_k}{\nu_k} \right\rceil$$

The pipeline can be increased by increasing the number of latches internal to the feedback loops. The computational latency associated with the internal feedback prevents one from introducing pipeline simply by inserting latches on feedforward cutsets. In fact, inserting latches in the loop would change the overall transfer function. This difficulty can be overcome by recasting the algorithm into an equivalent formulation from an I–O point of view. The transformations applied to the algorithm, prior to the mapping, have the purpose of creating additional concurrency, thereby increasing the achievable throughput rate. Without ever changing the algorithm's transfer function, additional delays are introduced inside the recursive loop. These delays are subsequently used for pipelining. In the sequel we briefly describe two types of look-ahead techniques, which generate the desired algorithmic transformations, namely the clustered and the scattered look-ahead techniques proposed by Loomis and Sinha [8] and Parhi and Messerschmitt [19], respectively. Look-ahead techniques are based on successive iterations of the basic recursion, in order to generate the desired level of concurrency. The implementation is then based on the iterated version of the algorithm.

Clustered Look-Ahead: In a pth order recursive system, the output at time n is a function of the past output samples $y(n-1)$, $y(n-2), \ldots, y(n-p)$. In the clustered look-ahead technique the recursion is iterated m times so that the current output is a function of the cluster of p consecutive samples $y(n-m)$, $y(n-m-1), \ldots, y(n-m-p)$. The original order-$p$ recursive filter is emulated by a $(p+m)$th filter, where m canceling poles and zeroes have been added. In this way the m delays generated inside the feedback loop can be used to pipeline by m stages.

By iterating 63.22 m times, we can derive the following I–O relationship

$$y(n) = \sum_{j=1}^{p-1}\left[\sum_{k=j+1}^{p} a_k r_{j+m-k}\right] y(n-j-m) + \sum_{j=0}^{m-1}\sum_{k=0}^{q} b_k x(n-k-j)$$

where the coefficients $\{r_i\}$ can be precomputed off-line, and are such that $r_i = \sum_{k=1}^{p} a_k r_{i-k}$, $i > 0$, $r_0 = 1$, and $r_i = 0$, $i = -(p-1), \ldots, -1$. This implementation requires $(p+m)$ multiplications for the nonrecursive part, and p for the recursive part, for a total of $(2p+m)$, which grows linearly with m. The transfer function is equal to

$$H(z) = \frac{\sum_{j=0}^{m-1}\sum_{k=0}^{q} b_k z^{-k-j}}{1 - \sum_{j=1}^{p-1}\left[\sum_{k=j+1}^{p} a_k r_{j+m-k}\right] z^{-j-m}}$$

The clustered look-ahead technique does not guarantee that the resulting filter is stable because it may introduce poles outside the unit circle.

Consider the following simple example with a stable transfer function

$$H(z) = \frac{1}{1 - 1.3z^{-1} + 0.35z^{-2}}$$

with poles at $z = 0.7$ and $z = 0.5$. The two-stage equivalent filter can be obtained by introducing the canceling pole–zero pair at $z = -1.3$, as follows:

$$H(z) = \frac{1 + 1.3z^{-1}}{(1 - 1.3z^{-1} + 0.35z^{-2})(1 + 1.3z^{-1})} = \frac{1 + 0.9z^{-1}}{1 - 1.34z^{-2} + 0.455z^{-3}}$$

Because a pole is found at $z = -1.3$, this transfer function is clearly unstable.

Scattered Look-Ahead: In the scattered look-ahead technique the current output sample, $y(n)$, is expressed in terms of the (scattered) p past outputs $y(n-m)$, $y(n-2m), \ldots, y(n-mp)$. The original order-p filter is now emulated by an order-mp filter. For each pole of the original filter, $(m-1)$ canceling pole–zero pairs are introduced at the same distance from the origin as the original pole. Thus, stability is always assured. The price we must pay is higher complexity, on the order of mp. To best describe the technique, it is convenient to write the transfer function $H(z)$ as a ratio of polynomials, i.e., $H(z) = N(z)/D(z)$. The transformation can be written as follows:

$$H(z) = \frac{N(z)}{D(z)} = \frac{N(z)\prod_{k=1}^{m-1} D(ze^{j(2\pi k/m)})}{\prod_{k=0}^{m-1} D(ze^{j(2\pi k/m)})}$$

Note that the transformed denominator is now a function of z^{-m}.

Consider the example of the previous section. For the scattered look-ahead technique it is necessary to introduce pole–zero pairs at $z = 0.7e^{\pm j(2\pi/3)}$ and $z = 0.5e^{\pm j(2\pi/3)}$. The transformed denominator equals $1 - 0.125z^{-3}$.

The complexity of the nonrecursive part of the transformed filter is $(pm + 1)$ multiplications, while the recursive part requires p multiplications, for a total of $(pm + p + 1)$ pipelined multiplications. Although the complexity is still linear in m, it is much higher than in the clustered look-ahead technique for a large value of p.

Parhi and Messerschmitt [19] presented a technique to reduce the complexity of the nonrecursive portion down to $O(p \log_2 m)$, applicable when m is a power of 2. This technique can be described as follows. Assume that the original recursive portion of the given IIR filter is given by

$$H(z) = \frac{1}{1 - \sum_{j=1}^{p} a_j^{(1)} z^{-j}}$$

An equivalent two-stage implementation of the same filter can be obtained by multiplying numerator and denominator by the polynomial $(1 - \sum_{j=1}^{p} (-1)^j a_j^{(1)} z^{-j})$, which is given by

$$H(z) = \frac{1 - \sum_{j=1}^{p} (-1)^j a_j^{(1)} z^{-j}}{1 - \sum_{j=1}^{p} a_j^{(2)} z^{-j}}$$

where the set of coefficients $\{a_j^{(2)}\}_{j=1}^{p}$ is obtained from the original set $\{a_j^{(1)}\}_{j=1}^{p}$ by algebraic manipulation. By repeating this process $\log_2 m$ times one can obtain an m-stage pipelined implementation, equivalent to the original filter. In this way the hardware complexity only grows logarithmically with the number of pipelining stages.

Bidirectional Systolic Arrays for Infinite Impulse Response Filtering: Lei and Yao [16] showed that many IIR filter structures can be considered as special cases of a general class of systolizable filters, as shown in Fig. 63.90. These filters can be pipelined by rescaling the time so that $z' = z^{1/2}$, and by applying a cutset transformation. This time rescaling causes the hardware factorization to reduce to merely 50%, which is quite inefficient. Lei and Yao [17] later proposed two techniques to improve the efficiency of these bidirectional IIR filters.

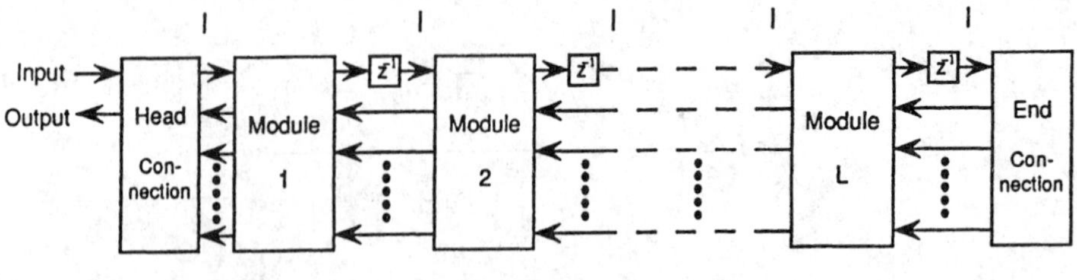

FIGURE 63.90 A general structure of bidirectional IIR filters.

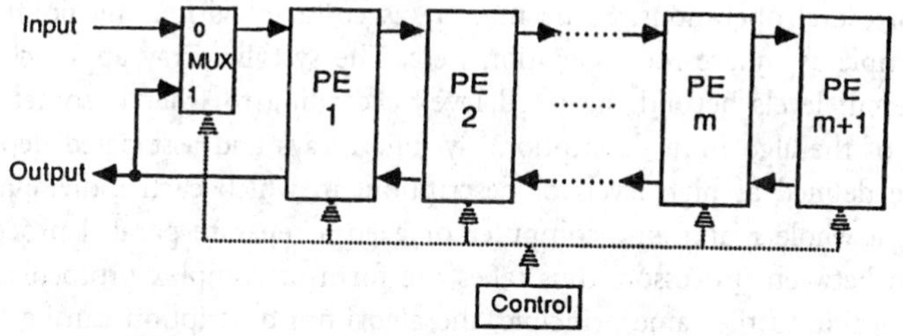

FIGURE 63.91 The overlapped subfilter scheme for IIR filtering.

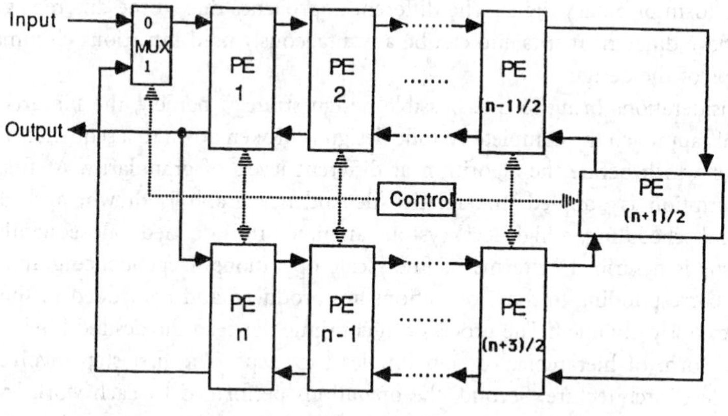

FIGURE 63.92 The systolic ring scheme for IIR filtering.

In the first method ("overlapped subfilter scheme") one makes use of the possibility to factor the numerator and the denominator of the given transfer function. For instance, if

$$H(z) = \frac{N(z)}{D(z)} = \underbrace{\frac{N_a(z)}{D_a(z)}}_{H_a(z)} \cdot \underbrace{\frac{N_b(z)}{D_b(z)}}_{H_b(z)}$$

where $a + b = p$, $a - b = 0$, 1, or 2, and p is the number of modules of the original transfer function. Then the two subfilters, $H_a(z)$ and $H_b(z)$, can be realized on the same systolic array of $a + 1$ modules, as in Fig. 63.91. A multiplexer at the input of the array chooses the incoming data at even time instants, and the data from the output of the first module at odd time instants. The modules alternately perform operations associated to $H_a(z)$ and $H_b(z)$ in such a way as to interleave the operations and have an overall 100% efficiency.

In the second technique ("systolic ring scheme") the number of modules is about half of the order of the original transfer function. The modules of the new structure are arranged as a systolic ring, as in Fig. 63.92. For example, a five-module ring can be used to implement a ten-module IIR filter: module i performs the operations associated to modules i and $(5 + i)$ of the original array, for $i = 1, \ldots, 5$. Note that in the original structure every other module is idle. The resulting ring is therefore 100% efficient.

Systolic Word and Bit-Level Designs

Previous discussions on systolic array designs have taken place at the word level. This is to say that the smallest data or control item exchanged between pairs of processors is constituted by a *word* representable by B bits. Each processor in a word-level system has the capability of performing word-level operations. Some may be as complex as floating point multiplications or others as simple as square root operations, etc. The systolic array approach can be applied at various different levels beyond the word level, according to what is sometimes referred to as **granularity** of the algorithm description. Systolic arrays and associated dependence graphs can, in fact, be defined at high levels of description, in which each individual processor can in principle be a whole mainframe computer or even a separate parallel processor array. The communication between processors thus takes the form of complex protocols and entire data sequences. According to the same principle, the algorithm description can be done also at the lowest level of operation, namely, at the bit level, at which each processor is a simple latched logic gate, capable of performing a logic binary operation. The exchanged data and control also take the form of binary digits. The different approaches due to the different granularity of description have different merits and can be advantageously used in various circumstances or at different steps of the design.

These considerations bring to one possible design strategy, namely, the **hierarchical systolic design**. In this approach the complete systolic design is broken down to a sequence of hierarchical steps, each of which define the algorithm at different levels of granularity. At first, the higher level of description is adopted, the relative dependence graph is drawn, and, after suitable projection and scheduling, a high-level systolic architecture is defined. Subsequently, each high-level processor is described in terms of finer scale operations. Dependence graph and systolic architecture corresponding to these operations are produced and embedded in the higher level structure previously obtained. The process can continue down to the desired level of granularity. The simplest form of hierarchical design implies two steps. The first step involves the design of the word-level architecture. Second, the operations performed by each work-level processor are described at bit level. The corresponding bit-level arrays are then nested into the word-level array, after ensuring that data flows at both levels are fully compatible.

The hierarchical approach has the merits of reducing the complexity of each step of the design. The dependence graphs involved usually have reduced dimensionality (thus, are more manageable), and the procedure is essentially recursive. The drawback of a hierarchical design is that it implicitly introduces somewhat arbitrary boundaries between operations, thereby reducing the set of resulting architectures. An approach which leaves all options open is to consider the algorithm at bit level from the outset. This approach has led to new insights and novel architectures. The price to pay is that the designer must deal with dependence graphs of higher dimensionality. As an example, the dependence graph of the inner product between two N-vectors, $c = \sum_{i=0}^{N-1} a_i b_i$ is two dimensional. If the same inner product is written at bit level, i.e., $c_k = \sum_{i=0}^{N-1} \sum_{j=0}^{B-1} a_{i,j} b_{i,k-j}$+carries, $k = 0, \ldots, B-1$, then it produces a three-dimensional dependence graph.

Examples of the two design procedures applied to the convolution problem are considered below. First, consider the factors that can make bit-level design advantageous:

- **Regularity.** Most bit-level arrays are highly regular. Only relatively simple cells need be designed and tested. The communication pattern is simple and regular. Neighbor-to-neighbor connections allow high packing density and low transmission delays.
- **High pipeline rate.** Because the individual cells have reduced computation time (on the order of the propagation delay through a few gates), the overall throughput can be made very high.
- **Inexpensive fault tolerance.** The use of by-pass circuitry can be made without wasting too much of the silicon area.

It must be borne in mind that bit level arrays realistically cannot be operated in wavefront array mode because the interprocessor hand-shaking protocols would be too expensive as compared to the data exchange. A good clock signal distribution is therefore needed to synchronize the

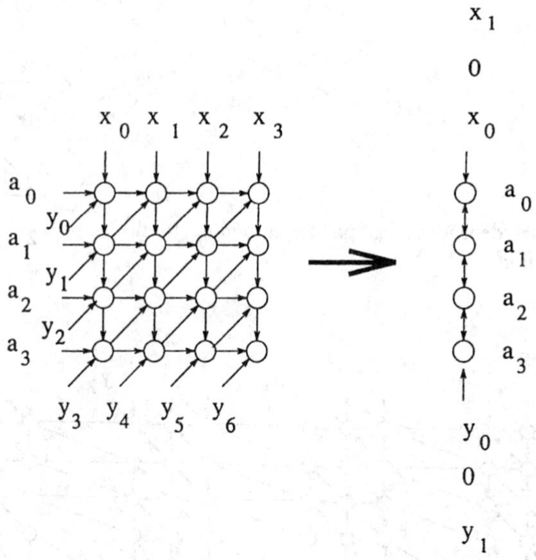

FIGURE 63.93 Word-level dependence graph of an N point convolution operation with one possible systolic realization.

array operations. In systolic arrays, unlike synchronous architectures of a different sort, only the incremental clock skew must be minimized by suitably designing the clock signal distribution lines. This problem may become particularly delicate in bit-level arrays, where the number of processors involved is very high.

Bit-Level Design of a Serial Convolver

Bit-level systolic design was first proposed by McCanny and McWhirter [29]. Subsequently, they and others have applied this technique to various algorithms. As a simple example, consider the bit-level design of a serial convolver. The word-level output of an N point convolver can be written as $y_k = \sum_{i=0}^{N-1} a_i x_{k-i}$, $k = 0, 1, \ldots$, where $\{a_i\}_{i=0}^{N-1}$ is a given set of coefficients and x_i, $i = 0, 1, \ldots$, is a sequence of input data. Coefficients and data values are assumed to be B-bit words. The word-level dependence graph is shown in Fig. 63.93, together with one possible systolic realization. In this case the coefficients are permanently stored in each individual cell. Input and output values are propagated in opposite directions. In each processing element the corresponding coefficient is multiplied by the incoming data value. This product is added to the partial output value and the accumulated result is propagated forward. Each cell performs the simple multiply and add operation expressed by $y_{k,i+1} \leftarrow y_{k,i} + a_i x_{k-i}$, $y_k = y_{k,N}$.

According to the hierarchical approach, one must now proceed to determine the dependence graph corresponding to the bit-level description of the multiply-and-add operation. The complete dependence graph can be subsequently obtained by embedding the finer scale graph into the higher level graph. If both a_i and x_i are B-bit binary numbers, then the jth bit of $y_{k,i}$ can be computed according to $y_{k,i,j} = y_{k,i,j} + s_{i,k,j}$, $s_{i,k,j} \equiv \sum_{l=0}^{B-1} a_{i,l} x_{k-i,j-l} +$ carries, where $a_{i,l}$ and $x_{i,l}$, $l = 0, \ldots, B-1$, represent the lth bit of a_i and x_i. The dependence graph corresponding to this operation is given in Fig. 63.94, where subscripts only indicate the bit position, and $B = 3$. Note that this graph is quite similar to the graph corresponding to a convolver, apart from the carry bits, which are taken care of by the insertion of an additional row of cells.

The combined dependence graph, obtained from the word dependence graph of Fig. 63.93, in which each cell is replaced by the bit-level dependence graph of Fig. 63.94, is given in Fig. 63.95.

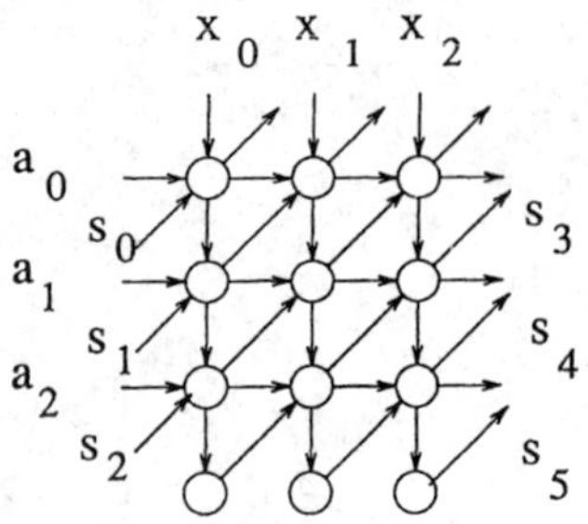

FIGURE 63.94 Bit-level dependence graph corresponding to the multiply-and-add operation.

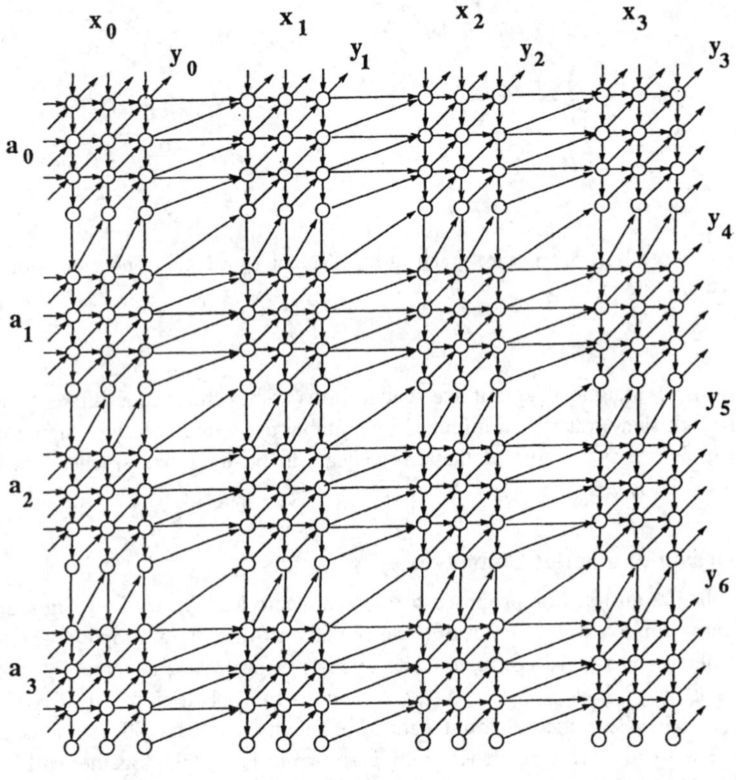

FIGURE 63.95 Bit-level dependence graph for convolution obtained by embedding the bit-level graph into the word-level graph.

The data flows are fully compatible at both word and bit levels. At this point, a full two-dimensional bit-level systolic array can be obtained from the final dependence graph by simply replacing each node with latched full adder cells. Different linear systolic implementations can be obtained by projecting the combined dependence graph along various directions. One possibility is again to keep the coefficients resident in individual cells, and have input data bits and accumulated results propagate in opposite directions. The schematic representation of the systolic array with these features is drawn in Fig. 63.95. Judgment about the merits of different projections involves desired data movement, I–O considerations, throughput rate, latency time, efficiency factor (ratio of idle time to busy time per cell), etc.

As discussed above, the convolution operation can be described at bit level from the very beginning. In this case the expression for the jth bit of the kth output can be expressed

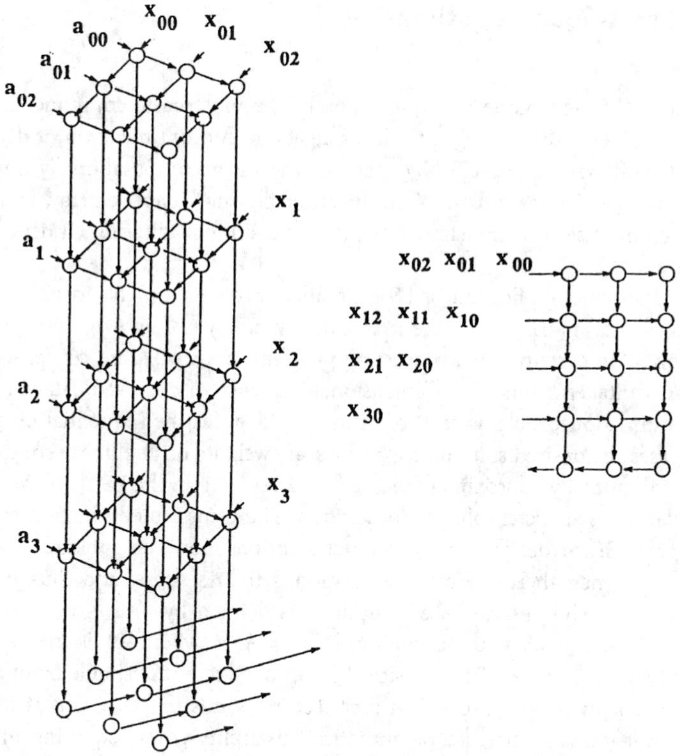

FIGURE 63.96 General 3-dimensional bit-level dependence graph for convolution.

as follows:

$$y_{k,j} = \sum_{i=0}^{N-1} \sum_{l=0}^{B-1} a_{i,l} x_{k-i,j-l} + \text{carries} \tag{63.23}$$

By using this expression as a starting point, one is capable of generating a number of feasible systolic realizations potentially much larger than what is attainable from the two-step hierarchical approach. The reason for this can be simply understood by noting that in this formulation no arbitrary precedence relationship is imposed between the two summations on i and l, whereas earlier we required that the summation on l would always *precede* the summation on i. The result is a fairly complicated three-dimensional dependence graph of size $N \times B \times$ number of inputs, as shown in Fig. 63.96. Observe that the bottom level of the dependence graph corresponds to the summation over l in (62.23). In the same figure a schematic two-dimensional bit-level systolic realization of the algorithm is given, in which the coefficient bits are held in place. Projections along different directions have different characteristics and may be considered preferable in different situations. The choice ultimately must be made according to given design constraints or to efficiency requirements.

The concept of bit-level design, as considered here, can be applied to a large variety of algorithms. Indeed, it has generated a number of architectures, including FIR/IIR filters, arrays for inner product computation, median filtering, image processing, eigenvalue problems, Viterbi decoding, etc.

Recursive Least-Squares Estimation

Least-Squares Estimation

The least-squares (LS) technique constitutes one of the most basic components of all modern signal processing algorithms dealing with linear algebraic and optimization of deterministic and random signals and systems. Specifically, some of the most computationally intensive parts of modern spectral analysis, beam formation, direction finding, adaptive array, image restoration, robotics, data compression, parameter estimation, and Kalman filtering all depend crucially on LS processing.

Regardless of specific application, a LS estimation problem can be formulated as $Ax \approx y$, where the $m \times n$ data matrix A and the $m \times 1$ data vector y are known, and we seek the $n \times 1$ desired solution x. In certain signal processing problems rows of A are composed of sequential blocks of lengths n taken from a one-dimensional sequence of observed data. In other n-sensor multichannel estimation problems each column of A denotes the sequential outputs of a given sensor. In all cases the desired solution x provides the weights on the linear combinations of the columns of A to optimally approximate the observed vector y in the LS sense. When $m = n$ and A is nonsingular, then an exact solution for x exists. The Gaussian elimination method provides an efficient approach for determining this exact solution. However, for most signal processing problems, such as when there are more observations than sensors, and thus $m > n$, then no exact solution exists. The optimum LS solution $\hat{x}$ is defined by $\|A\hat{x} - y\| = \min_x \|Ax - y\|$. The classical approach in LS solution is given by $\hat{x} = A^{+}y$, where A^{+} is the pseudo-inverse of A defined by $A^{+} = (A^{\mathrm{T}}A)^{-1}A^{\mathrm{T}}$. The classical LS approach is not desirable from the complexity, finite precision sensitivity, and processing architecture points of view. This is due to the need for a matrix inversion, the increase of numerical instability from "squaring of the condition number" in performing the $A^{\mathrm{T}}A$ operation, and the block nature of the operation in preventing a systolic update processing and architecture for real-time applications.

The QR decomposition (QRD) approach provides a numerically stable technique for LS solution that avoids the objections associated with the classical approach. Consider a real-valued $m \times n$ matrix A with $m \geq n$ and all the columns are linearly independent (i.e., rank $A = n$). Then, from the QRD, we can find a $m \times m$ orthogonal matrix Q such that $QA = \overline{R}$. The $m \times n$ matrix $\overline{R} = [R^{\mathrm{T}}, 0^{\mathrm{T}}]^{\mathrm{T}}$ is such that R is an $n \times n$ upper triangular matrix (with nonzero diagonal elements) and 0 is an all-zero $(m - n) \times n$ matrix. This upper triangularity of R is used crucially in the following LS solution problem.

Because the l_2 norm of any vector is invariant with respect to an orthogonal transformation, an application of the QRD to the LS problem yields $\|Ax - y\|^2 = \|Q(Ax - y)\|^2 = \|\overline{R}x - f\|^2$, where f is a $m \times 1$ matrix given by $f = Qy = [u^{\mathrm{T}}, v^{\mathrm{T}}]^{\mathrm{T}}$. Denote $e = Ax - y$ as the *residual* of the LS problem. Then, the above LS problem is equivalent to $\|e\|^2 = \|Ax - y\|^2 = \|[Rx, 0x]^{\mathrm{T}} - [u^{\mathrm{T}}, v^{\mathrm{T}}]^{\mathrm{T}}\|^2 = \|Rx - u\|^2 + \|v\|^2$. Because R is a nonsingular upper triangular square matrix, the back substitution procedure of the Gaussian elimination method can be used to solve for the exact solution $\hat{x}$ of $R\hat{x} = u$. Finally, the LS problem reduces to $\min_x \|Ax - y\|^2 = \|A\hat{x} - y\|^2 = \|\overline{R}\hat{x} - f\|^2 = \|R\hat{x} - u\|^2 + \|v\|^2 = \|v\|^2$. For the LS problem, any QRD technique such as the Gram-Schmidt method, the modified-Gram-Schmidt (MGS) method, the Givens transformation, and the Householder transformation is equally valid for finding the matrix R and the vector v. For a systolic implementation, the Givens transformation yields the simplest architecture, but the MGS and Householder transformation techniques are also possible with slight advantages under certain finite precision conditions.

Recursive Least-Squares Estimation

The complexity involved in the computation of the optimum residual $\hat{e}$ and the optimum LS solution vector $\hat{x}$ can become arbitrarily large as the number of samples in the column vectors of A and y increases. In practice we must limit m to some finite number greater than the number of columns n. Two general approaches in addressing this problem are available. In the **sliding window** approach we periodically incorporate the latest observed set of data (i.e., *updating*) and possibly remove an older set of data (i.e., *downdating*). In the **forgetting factor** approach a fixed scaling constant with a magnitude between 0 and 1 is multiplied again the R matrix and thus exponentially forget older data. In either approach, we find the optimum LS solution weight vector $\hat{x}$ in a recursive least-squares manner. As the statistics of the signal change over each window, these $\hat{x}$ vectors change *adaptively* with time. This observation motivates the development of a recursive least-squares solution implemented via the QRD approach. For simplicity, we consider only the updating aspects of the sliding window recursive least-squares problem.

Let m denote the present time of the sliding window of size m. Consider the $m \times n$ matrix $A(m)$, the $m \times 1$ column vector $y(m)$, the $n \times 1$ solution weight column vector $x(m)$, and the $m \times 1$ residual column vector $e(m)$ expressed in terms of their values at time $m-1$ as $A(m) = [\alpha(1), \ldots, \alpha(m)]^{\mathrm{T}} = [A(m-1)^{\mathrm{T}}, \alpha(m)]^{\mathrm{T}}$, $y(m) = [y_1, \ldots, y_m]^{\mathrm{T}} = [y(m-1)^{\mathrm{T}}, y_m^{\mathrm{T}}]^{\mathrm{T}}$, $x(m) = [x_1(m), \ldots, x_n(m)]^{\mathrm{T}}$, and $e(m) = A(m)x(m) - y(m) = [e_1(m), \ldots, e_n(m)]^{\mathrm{T}}$. By applying the orthogonal matrix $Q(m) = [Q_1(m)^{\mathrm{T}}, Q_2(m)^{\mathrm{T}}]^{\mathrm{T}}$ of the QRD of the $m \times n$ matrix $A(m)$, we obtain $Q(m)A(m) = [R(m)^{\mathrm{T}}, 0^{\mathrm{T}}]^{\mathrm{T}} = R_0(m)$ and $Q(m)y(m) = [Q_1(m)^{\mathrm{T}}, Q_2(m)^{\mathrm{T}}]^{\mathrm{T}} y(m) = [u(m)^{\mathrm{T}}, v(m)^{\mathrm{T}}]^{\mathrm{T}}$. The square of the l_2 norm of the residual e is then given by $\epsilon(m) = \|e(m)\|^2 = \|A(m)x(m) - y(m)\|^2 = \|Q(m)(A(m)x(m) - y(m))\|^2 = \|R(m)x(m) - u(m)\|^2 + \|v(m)\|^2$. The residual is minimized by using the back substitution method to find the optimum LS solution $\hat{x}(m)$ satisfying $R(m)\hat{x}(m) = u(m) = [u_1(m), \ldots, u_n(m)]^{\mathrm{T}}$. It is clear that the optimum residual $\hat{e}(m)$ is available after the optimum LS solution $\hat{x}(m)$ is available as seen from $\hat{e}(m) = A(m)\hat{x}(m) - y(m)$. It is interesting to note that it is not necessary to first obtain $\hat{x}(m)$ explicitly and then solve for $\hat{e}(m)$ as shown above. It is possible to use a property of the orthogonal matrix $Q(m)$ in the QRD of A and the vector $y(m)$, to obtain $\hat{e}(m)$ explicitly. Specifically, note $\hat{e}(m) = A(m)\hat{x}(m) - y(m) = Q_1(m)^{\mathrm{T}} R(m)\hat{x}(m) - y(m) = [Q_1(m)^{\mathrm{T}} Q_1(m) - I_m]y(m) = -Q_2(m)^{\mathrm{T}} Q_2(m)y(m) = -Q_2(m)^{\mathrm{T}} v(m)$ This property is used explicitly in the following systolic solution of the last component of the optimum residual.

Recursive QR Decomposition

Consider the recursive solution of the QRD. First, assume the decomposition at step $m-1$ has been completed as given by $Q(m-1)A(m-1) = [R(m-1)^{\mathrm{T}}, 0^{\mathrm{T}}]^{\mathrm{T}}$ by using a $(m-1) \times (m-1)$ orthogonal matrix. Next, define a new $m \times m$ orthogonal transformation $T(m) = [Q(m-1), 0; 0, 1]$. By applying $T(m)$ on the new $m \times n$ data $A(m)$, which consists of the previously available $A(m-1)$ and the newly available row vector $\alpha(m)^{\mathrm{T}}$, we have

$$T(m)A(m) = \begin{bmatrix} Q(m-1) & 0 \\ 0 & 1 \end{bmatrix} \begin{bmatrix} A(m-1) \\ \alpha(m)^{\mathrm{T}} \end{bmatrix} = \begin{bmatrix} Q(m-1)A(m-1) \\ \alpha(m)^{\mathrm{T}} \end{bmatrix}$$

$$= \begin{bmatrix} R(m-1) \\ 0 \\ \alpha(m)^{\mathrm{T}} \end{bmatrix} = R_1(m)$$

While $R(m-1)$ is a $n \times n$ upper triangular matrix, $R_1(m)$ does not have the same form as the desired $R_0(m) = [R(m)^{\mathrm{T}}, 0^{\mathrm{T}}]^{\mathrm{T}}$ where $R(m)$ is upper triangular.

Givens Orthogonal Transformation

Next, we want to transform $R_1(m)$ to the correct $R_0(m)$ form by an orthogonal transformation $G(m)$. While any orthogonal transformation is possible, we will use the Givens transformation approach due to its simplistic systolic array implementation. Specifically, denote $G(m) =$

$G_n(m)G_{n-1}(m)\ldots G_1(m)$, where $G(m)$ as well as each $G_i(m)$, $i = 1,\ldots,n$, are all $m \times m$ orthogonal matrices. Define

$$G_i(m) = \begin{array}{c} \\ 1 \\ \\ i \\ \\ m \end{array} \begin{array}{c} \begin{array}{ccccc} 1 & & i & & m \end{array} \\ \begin{bmatrix} 1 & & 0 & & 0 \\ & 1 & & & \\ 0 & & c_i(m) & & s_i(m) \\ & & & 1 & \\ 0 & & -s_i(m) & & c_i(m) \end{bmatrix} \end{array}, \quad i = 1,\ldots,n$$

as a $m \times m$ identity matrix, except that the (i, i) and (m, m) elements are specified as $c_i(m) = \cos\theta_i(m)$, where $\theta_i(m)$ represents the rotation angle at the ith iteration, the (i, m) element as $s_i(m) = \sin\theta_i(m)$, and the (m, i) element as $-s_i(m)$. By cascading all the $G_i(m)$, $G(m)$ can be reexpressed as

$$G(m) = \begin{bmatrix} k(m) & 0 & d(m) \\ 0 & I_{m-n-1} & 0 \\ h^{\mathrm{T}}(m) & 0 & \gamma(m) \end{bmatrix}$$

where $k(m)$ is $n \times n$, $d(m)$ and $h(m)$ are $n \times 1$, and $\gamma(m)$ is 1×1. In general $k(m)$, $d(m)$, and $h(m)$ are quite involved functions of $c_i(m)$ and $s_i(m)$, but $\gamma(m)$ is given simply as $\gamma(m) = \prod_{i=1}^{n} c_i(m)$ and will be used in the evaluation of the optimum residual.

Use $G(m)$ to obtain $G(m)T(m)A(m) = G(m)R_1(m)$. In order to show the desired property of the n orthogonal transformation operations of $G(m)$, first consider

$$G_1(m)R_1(m) = \begin{bmatrix} c_1(m) & & & & s_1(m) \\ & 1 & & & \\ & & 1 & & \\ & & & 1 & \\ -s_1(m) & & & & c_1(m) \end{bmatrix} \begin{bmatrix} x & x & \cdots & x \\ 0 & x & \cdots & x \\ & & \cdots & x \\ 0 & 0 & \cdots & x \\ x & x & \cdots & x \end{bmatrix}$$

$$= \begin{bmatrix} x & x & \cdots & x \\ 0 & x & \cdots & x \\ & & \cdots & x \\ 0 & 0 & \cdots & x \\ 0 & x & \cdots & x \end{bmatrix}$$

In the above expression an x denotes some nonzero valued element. The purpose of $G_1(m)$ operating on $R_1(m)$ is to obtain a zero at the $(m, 1)$ position without changing the $(m-2) \times n$ submatrix from the second to the $(m-1)$st rows of the r.h.s. of the expression. In general, at the ith iteration, we have

$$G_i(m)\begin{bmatrix} x & x & \cdot & \cdot & \cdot & x \\ & x & \cdot & \cdot & \cdot & x \\ & & & & & x \\ & & \vdots & & & \\ 0 & 0 & \cdot & \cdot & \cdot & 0 \\ 0 & 0 & 0 & x & x & x \end{bmatrix} = \begin{bmatrix} x & x & \cdot & \cdot & \cdot & x \\ & x & \cdot & \cdot & \cdot & x \\ & & & & & x \\ & & & \vdots & & \\ 0 & 0 & \cdot & \cdot & \cdot & 0 \\ 0 & 0 & 0 & 0 & x & x \end{bmatrix}$$
$$\qquad\qquad i-1 \qquad\qquad\qquad\qquad\qquad i$$

The above zeroing operation can be explained by noting that the Givens matrix $G_i(m)$ operates as a $(m-2)\times(m-2)$ identity matrix on all the rows on the right of it except the ith and the mth rows. The crucial operations at the ith iteration on these two rows can be represented as

$$\begin{bmatrix} c & s \\ -s & c \end{bmatrix}\begin{bmatrix} 0 & \cdots & 0 & r_i & r_{i+1} & \cdots & r_n \\ 0 & \cdots & 0 & a_i & a_{i+1} & \cdots & a_n \end{bmatrix}$$
$$i$$
$$= \begin{bmatrix} 0 & \cdots & 0 & r_i^{\mathrm{T}} & r_{i+1}^{\mathrm{T}} & \cdots & r_n^{\mathrm{T}} \\ 0 & \cdots & 0 & 0 & a_{i+1}^{\mathrm{T}} & \cdots & a_n^{\mathrm{T}} \end{bmatrix}$$
$$i$$

For simplicity of notation, we suppress the dependencies of i and m on c and s. Specifically, we want to force $a_i^{\mathrm{T}} = 0$ as given by $0 = a_i^{\mathrm{T}} = -sr_i + ca_i$. In conjunction with $c^2 + s^2 = 1$, this requires $c^2 = r_i^2/(a_i^2 + r_i^2)$ and $s^2 = a_i^2/(a_i^2 + r_i^2)$. Then $r_i^{\mathrm{T}} = cr_i + sa_i = \sqrt{(a_i^2 + r_i^2)}$, $c = r_i/r_i^{\mathrm{T}}$, and $s = a_i/r_i^{\mathrm{T}}$. This shows from the individual results of $G_1(m), G_2(m), \ldots, G_n(m)$, the overall results yield $Q(m)A(m) = G(m)\bar{R}(m) = [R(m)^{\mathrm{T}}, 0^{\mathrm{T}}]^{\mathrm{T}} = R_0(m)$, with $Q(m) = G(m)T(m)$.

Recursive Optimal Residual and LS Solutions

Consider the recursive solution of the last component of the optimum residual $\hat{e}(m) = [\hat{e}_1(m), \ldots, \hat{e}_m(m)]^{\mathrm{T}} = -Q_2(m)^{\mathrm{T}}v(m) = -Q_2(m)^{\mathrm{T}}[v_1(m), \ldots, v_m(m)^{\mathrm{T}}]$. Because $Q_2(m) = [Q_2(m-1), 0; h(m)^{\mathrm{T}}Q_1(m-1), \gamma(m)]$, then $\hat{e}(m) = [\hat{e}_1(m), \ldots, \hat{e}_m(m)]^{\mathrm{T}} = -[Q_2^{\mathrm{T}}(m-1), Q_1^{\mathrm{T}}(m-1)h(m); 0, \gamma(m)][v_1(m), \ldots, v_m(m)]$. Thus, the last component of the optimum residual is given by $\hat{e}_m(m) = -\gamma(m)v_m(m) = -\prod_{i=1}^{n} c_i(m)v_m(m)$, which depends on all the products of the cosine parameters $c_i(m)$ in the Givens QR transformation, and $v_m(m)$ is just the last component of $v(m)$, which is the result of $Q(m)$ operating on $y(m)$.

As considered earlier, the LS solution $\hat{x}$ satisfies the triangular system of equations. After the QR operation on the extended matrix $[A(m), y(m)]$, all the r_{ij}, $j \geq i = 1, \ldots, n$ and u_i, $i = 1, \ldots, n$ are available. Thus, $\{\hat{x}_1, \ldots, \hat{x}_n\}$ can be obtained by using the back substitution method of $\hat{x}_i = (u_i - \sum_{j=i+1}^{n} r_{ij}\hat{x}_j/r_{ij})$, $i = n, n-1, \ldots, 1$. Specifically, if $n = 1$, then $\hat{x}_1 = u_1/r_{11}$. If $n = 2$, then $\hat{x}_2 = u_2/r_{22}$ and $\hat{x}_1 = u_1 - r_{12}\hat{x}_2/r_{11} = u_1/r_{11} - u_2r_{12}/r_{11}r_{22}$. If $n = 3$, then $\hat{x}_3 = u_3/r_{33}$, $\hat{x}_2 = u_2 - r_{23}\hat{x}_3/r_{22} = u_2/r_{22} - r_{23}u_3/r_{22}r_{23}$, and $\hat{x}_1 = u_1 - r_{12}\hat{x}_2 - r_{13}\hat{x}_3/r_{11} = u_1/r_{11} - r_{12}u_2/r_{11}r_{22} + u_3[-r_{13}/r_{11}r_{33} + r_{12}r_{23}/r_{11}r_{22}r_{33}]$.

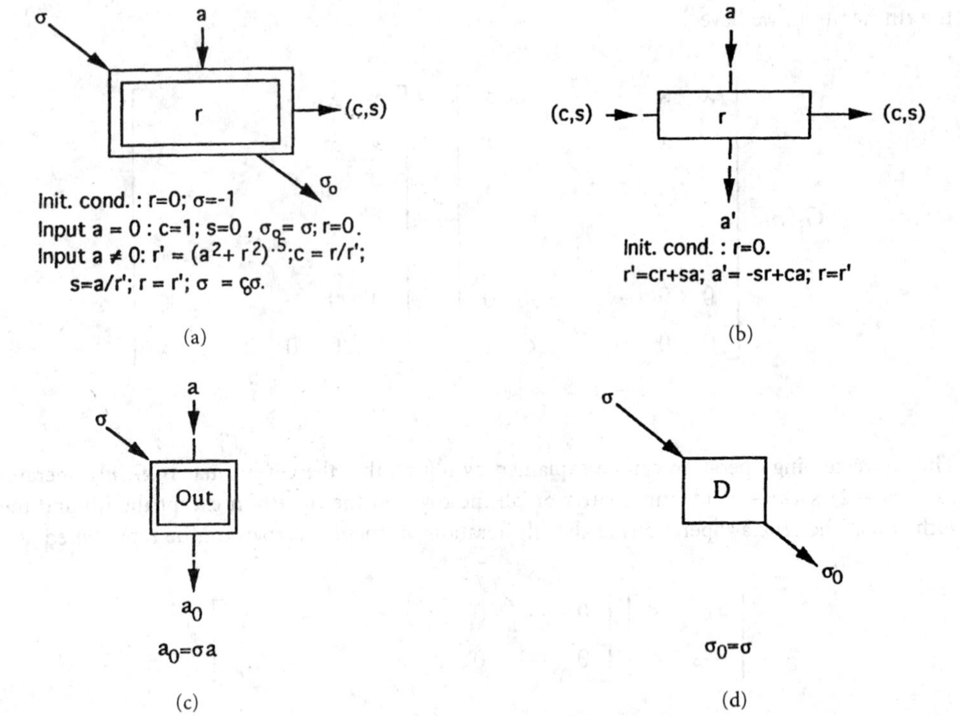

FIGURE 63.97 (a) Boundary cell; (b) internal cell; (c) output cell; (d) delay cell.

Systolic Array Implementation for QR Decomposition and LS Solution

The recursive QRD considered above can be implemented on a two-dimensional triangular systolic array based on the usage of four kinds of processing cells. Fig. 63.97(a) shows the boundary cell for the generation of the sine and cosine parameters, s and c, needed in the Givens rotations. Figure 63.97(b) shows the internal cell for the proper updating of the QRD transformations. Figure 63.97(c) shows the single output cell needed in the generation of the last component of the optimal residual $\hat{e}_m(m)$ as well as the optimal LS solution $\hat{x}(m)$. Figure 63.97(d) shows the delay cell which performs a unit time delay for proper time skewing in the systolic processing of the data.

Figure 63.98 shows a triangular systolic array capable of performing the recursive QRD for the optimal recursive residual estimation and the recursive least-squares solution by utilizing the basic processing cells in Fig. 63.97. In particular, the associated LS problem uses an augmented matrix $[A, y]$ consisting of the $m \times n$ observed data matrix A and the $m \times 1$ observed vector y. The number of processing cells in the triangular array consists of n boundary cells, $n(n+1)/2$ internal cells, one output cell, and n delay cells.

The input to the array in Fig. 63.98 uses the augmented matrix

$$[A, y] = \begin{bmatrix} a_{11} & a_{12} & \cdots & a_{1n} & y_1 \\ a_{21} & a_{22} & \cdots & a_{2n} & y_2 \\ & & \vdots & & \\ a_{m1} & a_{m2} & \cdots & a_{mn} & y_m \end{bmatrix}$$

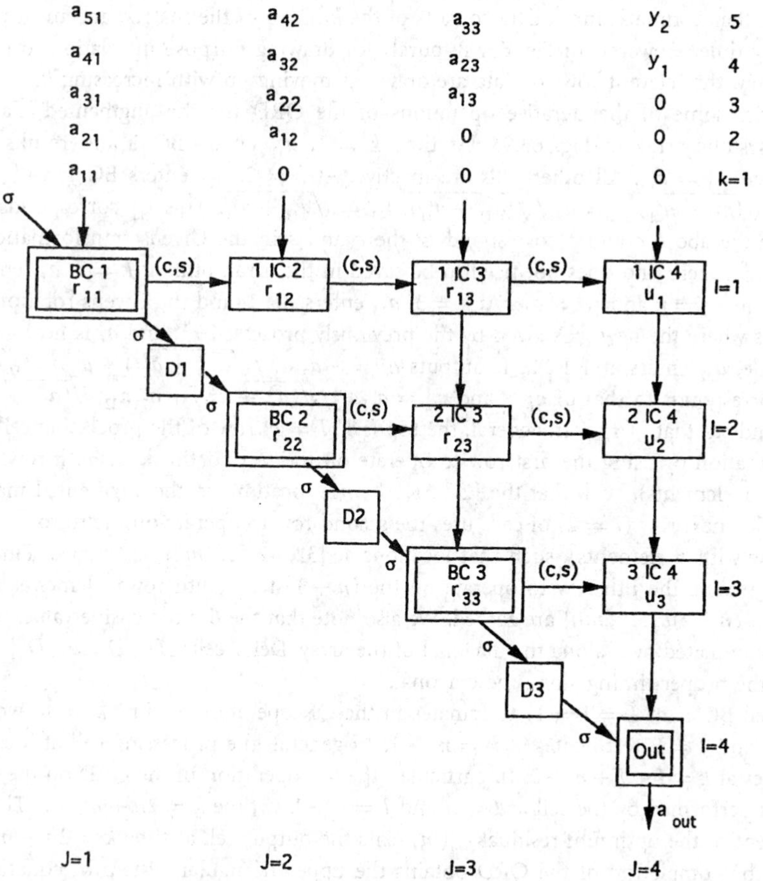

FIGURE 63.98 Triangular systolic array implementation of a $n = 3$ QRD-recursive least-squares solver.

skewed in a manner such that each successive column from left to right is delayed by a unit time as given by

$$
\left[\begin{array}{cccccc}
a_{11} & 0 & \cdots & & 0 & 0 \\
a_{21} & a_{12} & 0 & \cdots & 0 & 0 \\
a_{31} & a_{22} & a_{13} & \cdots & 0 & 0 \\
& & \vdots & & & \\
a_{n1} & a_{(n-1)2} & & & a_{1n} & 0 \\
a_{(n+1)1} & a_{n2} & & & a_{2n} & y_1 \\
& & \vdots & & & \\
a_{m1} & a_{(m-1)2} & & & a_{(m-n+1)n} & y_{m-n} \\
& & \vdots & & & \\
0 & & & & a_{mn} & y_{m-1} \\
0 & & & & 0 & y_m
\end{array}\right]
\begin{array}{c}
k = 1 \\ 2 \\ 3 \\ \vdots \\ n \\ n+1 \\ \vdots \\ m \\ \vdots \\ m+n-1 \\ m+n
\end{array}
$$

We see that at time k, input data consists of the kth row of the matrix, and moves down with increasing time. However, in Fig. 63.98, purely for drawing purpose in relation to the position of the array, the relevant rows of data are drawn as moving up with increasing k.

Consider some of the iterative operations of the QRD for the augmented matrix $[A, y]$ for the systolic array in Fig. 63.98. At time $k = 1$, a_{11} enters BC 1 and results in $c = 0$, $s = 1$, and $r_{11} = a_{11}$. All other cells are inactive. At $k = 2$, a_{21} enters BC 1, with the results $c = a_{11}/\sqrt{(a_{11}^2 + a_{21}^2)}$, $s = a_{21}/\sqrt{(a_{11}^2 + a_{21}^2)}$, $r_{11} = \sqrt{a_{11}^2 + a_{21}^2}$. This r_{11} corresponds to that of r_i^{T}, while the above c and s correspond to the c and s in the Givens transformation. Indeed, the new a_i^{T} is zero and does not need to be saved in the array. Still, at $k = 2$, a_{12} enters 1 IC 2 and outputs $a^{\mathrm{T}} = 0$ and $r_{12} = a_{12}$. At $k = 3$, a_{13} enters BC 1, and the Givens rotation operation continues where the new r_i is given by the previously processed r_i^{T}, and a_i is now given by a_{13}. Meanwhile, a_{22} enters at 1 IC 2. It outputs $a^{\mathrm{T}} = -a_{21}a_{12}/\sqrt{(a_{11}^2 + a_{21}^2)} + a_{22}a_{21}/\sqrt{(a_{11}^2 + a_{21}^2)}$, which corresponds to that of a_{i+1}^{T}, and $r_{12} = a_{11}a_{12}/\sqrt{(a_{11}^2 + a_{21}^2)} + a_{21}a_{22}/\sqrt{(a_{11}^2 + a_{21}^2)}$, which corresponds to that of r_{i+1}^{T}. In general, the top (i.e., $I = 1$) row of the processing cells performs Givens rotation by using the first row to operate on the second, third, ..., mth rows (each row with $n + 1$ elements), such that the $\{21, 31, \ldots, m1\}$ locations in the augmented matrix are all zeroed. The next row ($I = 2$) of cells uses the second row to operate on the third, ..., mth rows (each row with n elements), such that locations at $\{32, 42, \ldots, m2\}$ are zeroed. Finally, at row $I = n$, by using the nth row to operate on the $(n + 1)$st, ..., mth rows, elements at locations $\{(n+1)n, (n+2)n, \ldots, mn\}$ are zeroed. We also note that the desired cosine values in $\gamma(m)$ are being accumulated by c along the diagonal of the array. Delay cells $\{D_1, D_2, \ldots, D_n\}$ are used to provide the proper timing along the diagonal.

The cell BC 1 (at $I = J = 1$) terminates in the QR operation at time $k = m$, while the cell at $I = 1$ and $J = 2$ terminates at $k = m + 1$. In general, the processing cell at location (I, J) terminates at $k = I + J + m - 2$. In particular, the last operation in the QRD on the augmented matrix is performed by the cell at $I = n$ and $J = n + 1$ at time $k = 2n + m - 1$. Then, the last component of the optimum residual $e_m(m)$ exits the output cell at time $k = 2n + m$.

After the completion of the QRD obtains the upper triangular system of equation, we can "freeze" the r_{IJ} values in the array to solve for the optimum LS solution $\hat{x}$ by the back substitution method. Specifically, we can append $[I_n, 0]$, where I_n is a $n \times n$ identity matrix and 0 is an $n \times 1$ vector of all zeroes, to the bottom of the augmented matrix $[A, y]$. Of course, this matrix is skewed as before when used as input to the array. In particular, immediately after the completion of the QR operation at BC 1, we can input the unit value at time $k = m + 1$. This is stage 1 of the back substitution method. Due to skewing, a unit value appears at the $I = 1$ and $J = 2$ cell at stage 3. Finally, at stage $(2n - 1)$, which is time $k = m + 2n - 1$, the last unit value appears at the $I = n$ and $J = 1$ cell. For our example of $n = 3$, this happens at stage 5. The desired LS solution $\hat{x}_1$ appears at stage $(2n + 1)$ (i.e., stage 7 for $n = 3$), which is time $k = 2n + m + 1$, while the last solution $\hat{x}_n$ appears at stage $3n$ (i.e., stage 9 for $n = 3$), which is time $k = 3n + m$. The values of $\{\hat{x}_1, \hat{x}_2, \hat{x}_3\}$ at the output of the systolic array are identical to those given by the back substitution method solution of the LS problem.

Kalman Filtering

Introduction

Kalman filtering (KF) was developed in the late 1950s as a natural extension of the classical Wiener filtering. It has profound influence in the theoretical and practical aspects of estimation and filtering. It is used almost universally for tracking and guidance of aircraft, satellites, GPS, and missiles as well as many system estimation and identification problems. KF is not one unique method, but is a generic name for a class of state estimators based on noisy measurements. KF can be implemented as a specific algorithm on a general-purpose mainframe/mini/microcomputer operating in a batch mode, or it can be implemented on a dedicated system using either DSP, ASIC, or custom VLSI processors in a real-time operating mode.

Classically, an analog or a digital filter is often viewed in the frequency domain having some low-pass, bandpass, high-pass, etc. properties. A KF is different from the classical filter in that it may have multiple inputs and multiple outputs with possibly nonstationary and time-varying characteristics performing optimum states estimation based on the unbiased minimum variance estimation criterion.

In the following discussions we first introduce the basic concepts of KF, followed by various algorithmic variations of KF. Each version has different algorithmic and hardware complexity and implementational implications. Because there are myriad of KF variations, we then consider two simple systolic versions of KF.

Basic Kalman Filtering

The KF model consists of a discrete-time linear dynamical system equation and a measurement equation. A linear discrete-time dynamical system with $n \times 1$ state vector $x(k+1)$, at time $k+1$, is given by $x(k+1) = A(k)x(k) + B(k)u(k) + w(k)$, where $x(k)$ is the $n \times 1$ state vector at time k, $A(k)$ is a $n \times n$ system coefficient matrix, $B(k)$ is an $n \times p$ control matrix, $u(k)$ is a $p \times 1$ deterministic vector, which for some problems may be zero for all k, and $w(k)$ is an $n \times 1$ zero-mean system noise vector with a covariance matrix $W(k)$. The input to the KF is the $m \times 1$ measurement (also called observation) vector $y(k)$, modeled by $y(k) = C(k)x(k) + v(k)$, where $C(k)$ is a $m \times n$ measurement coefficient matrix, and $v(k)$ is a $m \times 1$ zero-mean measurement noise vector with a $m \times m$ positive-definite covariance matrix $V(k)$. The requirement of the positive-definite condition on $V(k)$ is to guarantee the Cholesky (square root) factorization of $V(k)$ for certain KF algorithms. In general, we will have $m \leq n$ (i.e., the measurement vector dimension is less than or equal to that of the state vector dimension). It is also assumed that $w(k)$ is uncorrelated to $v(k)$. That is, $\mathrm{E}\{w(i)v(j)^{\mathrm{T}}\} = 0$. We also assume each noise sequence is white in the sense $\mathrm{E}\{w(i)w(j)^{\mathrm{T}}\} = \mathrm{E}\{v(i)v(j)^{\mathrm{T}}\} = 0$, for all $i \neq j$.

The KF provides a recursive linear estimation of $x(k)$ under the minimum variance criterion based on the observation of the measurement $y(k)$. Let $\hat{x}(k)$ denote the optimum filter state estimate of $x(k)$ given measurements up to and including $y(k)$, while $x_+(k)$ denotes the optimum predicted state estimate of $x(k)$ given measurements up to and including $y(k-1)$. Then the $n \times n$ **optimum estimation error covariance matrix** is given by $P(k) = \mathrm{E}\{(x(k) - \hat{x}(k))(x(k) - \hat{x}(k))^{\mathrm{T}}\}$, while the **minimum estimation error variance** is given by $J(k) = \mathrm{Trace}\{P(k)\} = \mathrm{E}\{(x(k) - \hat{x}(k))^{\mathrm{T}}(x(k) - \hat{x}(k))\}$. The $n \times n$ **optimum prediction error covariance matrix** is given by $P_+(k) = \mathrm{E}\{(x(k) - x_+(k))(x(k) - x_+(k))^{\mathrm{T}}\}$.

The original KF recursively updates the optimum error covariance and the optimum state estimate vector by using two sets of update equations. Thus, it is often called the **covariance KF**. The **time update equations** for $k = 1, 2, \ldots,$ are given by $x_+(k) = A(k-1)\hat{x}(k-1) + B(k-1)u(k-1)$ and $P_+(k) = A(k-1)P(k-1)A^{\mathrm{T}}(k-1) + W(k-1)$. The $n \times n$ **Kalman gain matrix** $K(k)$ is given by $K(k) = P_+(k)C^{\mathrm{T}}(k)[C(k)P_+(k)C^{\mathrm{T}}(k) + V(k)]^{-1}$. The **measurement update equations** are given by $\hat{x}(k) = x_+(k) + K(k)(y(k) - C(k)x_+(k))$ and $P(k) = P_+(k) - K(k)C(k)P_+(k)$. The first equation shows the update relationship of $\hat{x}(k)$ to the predicted state estimate $x_+(k)$, for $x(k)$ based on $\{\ldots, y(k-2), y(k-1)\}$, when the latest observed value $y(k)$ is available. The second equation shows the update relationship between $P(k)$ and $P_+(k)$. Both equations depend on the $K(k)$, which depends on the measurement coefficient matrix $C(k)$ and the statistical property of the measurement noise, covariance matrix $V(k)$. Furthermore, $K(k)$ involves a $m \times m$ matrix inversion.

Other Forms of Kalman Filtering

The basic KF algorithm considered above is called the covariance form of KF because the algorithm propagates the prediction and estimation error covariance matrices $P_+(k)$ and $P(k)$.

Many versions of the KF are possible, characterized partially by the nature of the propagation of these matrices. Ideally, under infinite precision computations, no difference in results is observed among different versions of the KF. However, the computational complexity and the systolic implementation of different versions of the KF are certainly different. Under finite precision computations, especially for small numbers of bits under fixed point arithmetics, the differences among different versions can be significant. In the following discussions we may omit the deterministic control vector $u(k)$ because it is usually not needed in many problems. In the following chol$(\cdot)$, qr$(\cdot)$, and triu$(\cdot)$ stand for Cholesky factor, QR decomposition, and triangular factor, respectively.

1. **Information Filter.** The inverse of the estimation error covariance matrix $P(k)$ is called the information matrix and is denoted by $PI(k)$. A KF can be obtained by propagating the information matrix and other relevant terms. Specifically, the information filter algorithm is given by time updates for $k = 1, 2, \ldots,$ of $L(k) = A^{-T}(k-1)PI(k-1)A^{-1}(k-1) \times [W^{-1}(k-1) + A^{-T}(k-1)PI(k-1)A^{-1}(k-1)]^{-1}$, $d_+(k) = [I - L(k)]A^{-T}(k-1)d(k-1)$, $PI_+(k) = [I - L(k)]A^{-T}(k-1)PI(k-1)A^{-1}(k-1)$. The measurement updates are given by $d(k) = d_+(k) + C^T(k)V^{-1}(k)y(k)PI(k) = PI_+(k) + C^T(k)V^{-1}(k)C(k)$.
2. **Square-Root Covariance Filter (SRCF).** In this form of the KF, we propagate the square root of $P(k)$. In this manner we need to use a lower dynamic range in the computations and obtain a more stable solution under finite precision computations. We assume all three relevant covariance matrices are positive-definite and have the factorized form of $P(k) = S^T(k)S(k)$, $W(k) = S_W^T(k)S_W(k)$, $V(k) = S^T(k)S_V(k)$. In particular, $S(k) = \text{chol}(P(k))$, $S_W(k) = \text{chol}(W(k))$, $S_V(k) = \text{chol}(V(k))$, are the upper triangular Cholesky factorizations of $P(k)$, $W(k)$, and $V(k)$, respectively. The time updates for $k = 1, 2, \ldots,$ are given by $x_+(k) = A(k-1)\hat{x}(k-1)$, $U(k) = \text{triu}(\text{qr}([S(k-1)A^T(k-1); S_W(k-1)]))$, $P_{+S}(k) = U(k)(1\colon n; 1\colon n)$. The measurement updates are given by $P_+(k) = P_{+S}^T(k)P_{+S}(k)$, /, $K(k) = P_+(k)C^+(k)[C(k)P_+(k)C^+(k) + V(k)]^{-1}$, $\hat{x}(k) = x_+(k) + K(k)(y(k) - C(k)x_+(k))$, $Z(k) = \text{triu}(\text{qr}([S_V(k), 0_{mn}; P_{+S}(k)C^+(k), P_{+S}(k)]))$ and $S(k) = Z(k)(m+1\colon m+n, m+1\colon m+n)$.
3. **Square-Root Information Filter (SRIF).** In the SRIF form of the KF we propagate the square root of the information matrix. Just as in the SRCF approach, as compared to the conventional covariance form of the KF, the SRIF approach, as compared to the SRIF approach, needs to use a lower dynamic range in the computations and obtain a more stable solution under finite precision computations. First, we denote $SI(k) = (\text{chol}(P(k)))^{-1}$, $SI_W(k) = (\text{chol}(W(k)))^{-1}$, and $SI_V(k) = (\text{chol}(V(k)))^{-1}$. The time updates for $k = 1, 2, \ldots,$ are given by $U(k) = \text{triu}(\text{qr}([SI_W(k-1), 0_{n\times n}, 0_{n\times 1}; SI(k-1)A^{-1}(k-1), SI(k-1)A^{-1}(k-1), b(k-1)]))$, $P_{+S}(k) = U(k)(n+1\colon 2n, n+1\colon 2n)$ and $b_+(k) = U(k)(n+1\colon 2n, 2n+1)$. The measurement updates are given by $Z(k) = \text{triu}(\text{qr}([P_{+S}(k), b_+(k); SI_V(k)C(k), SI_v(k)y(k)]))SI(k) = Z(k)(1\colon n, 1\colon n)$, and $b(k) = Z(k)(1\colon n, n+1)$. At any iteration, $\hat{x}(k)$ and $P(k)$ are related to $b(k)$ and $SI(k)$ by $\hat{x}(k) = SI(k)b(k)$ and $P(k) = (SI^T(k)SI(k))^{-1}$.

Systolic Matrix Implementation of the KF Predictor

The covariance KF for the optimum state estimate $\hat{x}(k)$ includes the KF predictor $x_+(k)$. In particular, if we are only interested in $x_+(k)$, a relatively simple algorithm for $k = 1, 2, \ldots,$ is given by $K(k) = P_+(k)C^T(k)[C(k)P_+(k)C^T(k) + V(k)]^{-1}$, $x_+(k+1) = A(k)x_+(k) + A(k)K(k)[y(k) - C(k)x_+(k)]$ and $P_+(k+1) = A(k)P_+(k)A^T(k) - A(k)K(k)C(k)P_+(k)A^T(k) + W(k)$. To start this KF prediction algorithm, we use $\hat{x}(0)$ and $P(0)$ to obtain $x_+(1) = A(0)\hat{x}(0)$ and $P_+(1) = A(0)P(0)A^T(0) + W(0)$. The above operations involve matrix inverse; matrix-matrix and matrix-vector multiplications; and matrix and vector additions. Fortunately, the matrix inversion of $\alpha = C(k)P_+(k)C^T(k) + V(k)$ can be approximated by the iteration of

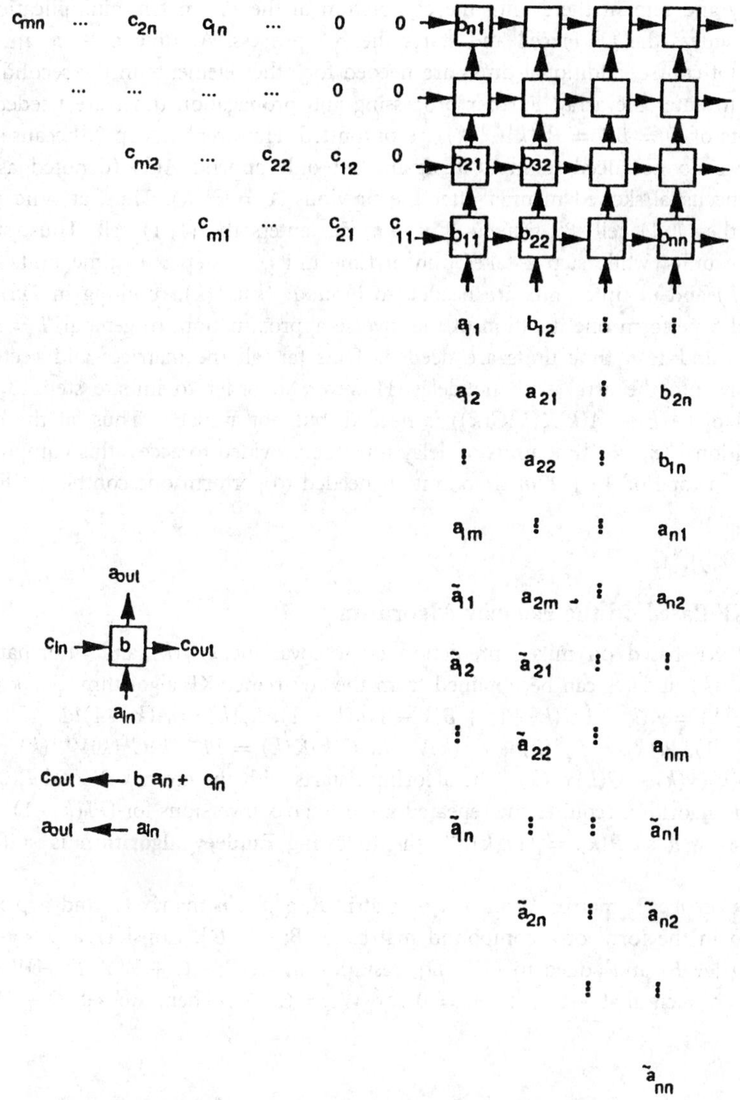

FIGURE 63.99 Systolic matrix multiplication and addition of $B \times A + C$.

$\beta(i+1) = \beta(i)[2I - \alpha\beta(i)]$, $i = 1, \ldots, I$. Here, $\beta(i)$ is the ith iteration estimate of the inverse of the matrix α. While the above equation is not valid for arbitrary α and $\beta(i)$, for KF applications, we can use $I = 4$ because a good initial estimate $\beta(1)$ of the desired inverse is available from the previous step in the KF. Clearly, with the use of the above equation for the matrix inversion, all the operations needed in the KF predictor can be implemented on an orthogonal array using systolic matrix operations of the form $D = B \times A + C$, as shown in Fig. 63.99.

The recursive algorithm of the KF predictor is decomposed as a sequence of matrix multiplications, as shown in Table 63.15. In step 1 the $n \times n$ matrix $P_+(k)$ and the $m \times n$ matrix $C^T(k)$ are denoted as B and A, respectively. The rows of B (starting from the $n, n-1, \ldots, 1$ row) are skewed and inputted to the $n \times n$ array starting at time 1. By time n (as shown in Fig. 63.99, all the elements of the first column of B (i.e., $b_{n1}, \ldots, b_{11}$) are in the first column of the array. At time $n+1, \ldots, 2n-1$, elements of the second to nth columns of B are inputted to the array and remain there until the completion of the BA matrix multiplication. At time $n+1$, a_{11} enters the (1, 1) cell and starts the BA process. At time $n+m$, a_{1m} enters the (1, 1) cell. Of course, additional times are needed for other elements in the second to the nth rows of A to enter the array. Further processing and propagation times are needed before all the elements of $D = BA = P_+(k)C^T(k)$ are outputted. However, in step 2, because B remains as $P_+(k)$, we do not need to input it again, but only append $A(k)$ (denoted as $\tilde{A}$ in Fig. 63.99) in the usual skewed manner after the previous $A = C^T(k)$. Thus, at time $n+m+1$, $\tilde{a}_{11}$ enters the (1, 1) cell. By time $n+m+n$, $\tilde{a}_{1n}$ enters the (1, 1) cell. Thus, step 1 takes $n+m$ time units, while step 2 takes only n time units. In step 3 m time units are needed to load $C(k)$ and m time units are needed to input $P_+(k)C^T(k)$, resulting in $2m$ time units. Steps 4 and 5 perform one iteration of the inverse approximation. In general, $I = 4$ iterations is adequate, and $16m$ time units are needed. Thus far, all the matrices and vectors are fed continuously into the array with no delay. However, in order to initiate step 13, the $(n, 1)$ component of $(A(k) - A(k)K(k)C(k))$ is needed, but not available. Thus, at the end of step 11, an additional $(n-3)$ time units of delay must be provided to access this component. From Table 63.15, a total of $9n + 22m$ time units is needed to perform one complete KF prediction iteration.

Systolic KF Based on the Faddeev Algorithm

A form of KF based on mixed prediction error covariance $P_+(k)$ and information matrix $PI(k) = P^{-1}(k)$ updates can be obtained from the covariance KF algorithm. For $k = 1, 2, \ldots,$ we have $x_+(k) = A(k-1)\hat{x}(k-1) + B(k-1)u(k-1)$, $P_+(k) = A(k-1)PI^{-1}(k-1)A^T(k-1) + W(k-1)$, $PI(k) = P_+^{-1}(k) + C^T(k)V^{-1}(k)C(k)K(k) = PI^{-1}(k)C^T(k)V^{-1}(k)$ and $\hat{x}(k) = x_+(k) + K(k)(y(k) - C(k)x_+(k))$ The algorithm starts with the given $\hat{x}(0)$ and $P(0)$, as usual. Because this algorithm requires the repeated use of matrix inversions for $(PI(k-1))$, $(P_+(k))^{-1}$, $(V(k))^{-1}$, as well as $P(k) = (PI(k))^{-1}$, the following **Faddeev algorithm** is suited for this approach.

Consider an $n \times n$ matrix A, an $n \times m$ matrix B, a $p \times n$ matrix C, and a $p \times m$ matrix D arranged in the form of a compound matrix $[A\ B; -C\ D]$. Consider a $p \times n$ matrix W multiplying $[A\ B]$ and added to $[-C\ D]$, resulting in $[A\ B; -C + WA\ D + WB]$. Assume W is chosen such that $-C + WA = 0$, or $W = CA^{-1}$. Then, we set $D + WB = D + CA^{-1}B$.

TABLE 63.15 Systolic Matrix Operations of a KF Predictor

Step	B	A	C	D	Time
1	$P_+(k)$	$C^T(k)$	0	$P_+(k)C^T(k)$	$n+m$
2	$P_+(k)$	$A^T(k)$	0	$P_+(k)A^T(k)$	n
3	$C(k)$	$P_+(k)C^T(k)$	$V(k)$	$C(k)P_+(k)C^T(k) + V(k) = \alpha$	$2m$
4	α	$-\beta(i)$	$2I$	$2I - \alpha\beta(i)$	$2Im$
5	$\beta(i)$	$2I - \alpha\beta(i)$	0	$\beta(i+1)$	$2Im$
6	$P_+(k)C^T(k)$	β	0	$K(k)$	$n+m$
7	$A(k)$	$K(k)$	0	$A(k)K(k)$	$n+m$
8	$A(k)$	$x_+(k)$	0	$A(k)x_+(k)$	1
9	$-C(k)$	$x_+(k)$	$y(k)$	$y(k) - C(k)x_+(k)$	$m+1$
10	$A(k)K(k)$	$-C(k)$	$A(k)$	$A(k) - A(k)K(k)C(k)$	$2n$
11	$A(k)K(k)$	$y(k) - C(k)x_+(k)$	$A(k)x_+(k)$	$x_+(k+1)$	1
12					$n-3$
13	$A(k) - A(k)K(k)C(k)$	$P_+(k)A^T(k)$	$W(k)$	$P_+(k)$	$2n$

In particular, by picking $\{A, B, C, D\}$ appropriately, the basic matrix operations needed above can be obtained using the Faddeev algorithm. Some examples are given by

$$\begin{matrix} A & I \\ -I & 0 \end{matrix} \Rightarrow D + W B = A^{-1}$$

$$\begin{matrix} I & B \\ -C & 0 \end{matrix} \Rightarrow D + W B = C B$$

$$\begin{matrix} I & B \\ -C & D \end{matrix} \Rightarrow D + W B = D + C B$$

$$\begin{matrix} A & B \\ -I & 0 \end{matrix} \Rightarrow D + W B = A^{-1} B$$

A modified form of the above Faddeev algorithm first triangularizes A with an orthogonal transformation Q, which is more desirable from the finite precision point of view. Then, the nullification of the lower left portion can be performed easily using the Gaussian elimination procedure. Specifically, applying a QRD, $Q[A \;\; B] = [R \;\; QB]$. Then, applying the appropriate W yields

$$\begin{bmatrix} R & QB \\ -C + WQA & D + WQB \end{bmatrix} = \begin{bmatrix} R & QB \\ 0 & D + CA^{-1}B \end{bmatrix} \qquad (63.24)$$

The above mixed prediction error covariance and information matrix KF algorithm can be reformulated as a sequence of Faddeev algorithm operations, as given in Table 63.16. The times needed to perform steps 2, 3, 4, and 6 are clearly just the sum of the lengths of the two matrices in the corresponding steps. Step 1 requires only n time units to input the second row of matrices because $\hat{x}(k-1)$ is already located in the array from the previous iteration (step 8 output) and one time unit to output $x_+(k)$. Due to the form of $[-I \;\; 0]$ in step 4, $C(k)$ of step 5 can be inputted before the completion of $P_+^{-1}(k)$ in step 4. Thus, only n time units are needed in step 5. Similarly, $x_+(k)$ of step 7 can be inputted in step 6. Thus, we need only $m+1$ time units to input $[C(k) \;\; y(k)]$ and complete its operations. In step 8 only $m+1$ time units are needed as in step 1. Thus, a total of $9n + 3m + 3$ time units are needed for the Faddeev algorithm approach to the KF.

Other Forms of Systolic KF and Conclusions

While the operations of a KF can be expressed in many ways, only some of these algorithms are suited for systolic array implementations. For a KF problem with a state vector of dimension n and a measurement vector of dimension m, we have shown the systolic matrix-matrix multiplication implementation of the predictor form of the KF needs $9n + 22m$ time steps for each iteration. A form of KF based on mixed update of prediction error covariance and information matrices is developed based on the Faddeev algorithm using matrix-matrix systolic array implementation. It has a total of $9n + 3m + 3$ time steps per iteration. A modified form of the SRIF algorithm can be implemented as a systolic array consisting of an upper rectangular array of $n(n+1)/2$ internal cells, and a lower n-dimensional triangular array of n boundary cells and $(n-1)^2/2$ internal cells, plus a row of n internal cells, and $(n-1)$ delay cells. It has a total of n boundary cells, $((n-1)^2 + 2n^2 + 2n)/2$ internal cells, and $(n-1)$ delay cells. Its throughput rate is $3n$ time steps per iteration. A modified form of the SRCF algorithm utilizing the Faddeev algorithm results in a modified SRCF form of a KF consisting of a trapezoidal

TABLE 63.16 Faddeev Algorithm Solution to KF

Step	Compound Matrix	$D + WB$	Time
1	$\begin{matrix} I & \hat{x}(k-1) \\ -A(k-1) & B(k-1)u(k-1) \end{matrix}$	$x_+(k)$	$n+1$
2	$\begin{matrix} P^{-1}(k-1) & A^{T}(k-1) \\ -A(k-1) & W(k-1) \end{matrix}$	$P_+(k)$	$2n$
3	$\begin{matrix} V(k-1) & I \\ -C^{T}(k) & 0 \end{matrix}$	$C^{T}(k)V^{-1}(k-1)$	$m+n$
4	$\begin{matrix} P_+(k) & I \\ -I & 0 \end{matrix}$	$P_+^{-1}(k)$	$2n$
5	$\begin{matrix} I & C(k) \\ -C^{T}(k)V^{-1}(k) & P_+^{-1}(k) \end{matrix}$	$P^{-1}(k)$	n
6	$\begin{matrix} P^{-1}(k) & C^{T}(k+1)V^{-1}(k) \\ -I & 0 \end{matrix}$	$K(k)$	$2n$
7	$\begin{matrix} I & x_+(k) \\ C(k) & y(k) \end{matrix}$	$y(k) - C(k)x_+(k)$	$m+1$
8	$\begin{matrix} I & y(k) - C(k)x_+(k) \\ -K(k) & x_+(k) \end{matrix}$	$\hat{x}(k)$	$m+1$

section, a linear section, and a triangular section systolic array. The total of these three sections needs $(n+m)$ boundary cells, n linear cells, and $((m-1)^2 + 2nm + (n-1)^2)/2$ internal cells. Its throughput rate is $3n + m + 1$ time steps per iteration. The operations of both of these systolic KF are quite involved and detailed discussions are omitted here. In practice, in order to compare different systolic KFs, one needs to concern oneself not only with the hardware complexity and the throughput rate, but other factors involving the number of bits needed under finite precision computations, data movement in the array, and I–O requirements are also important.

Eigenvalue and Singular Value Decompositions

Results from linear algebra and matrix analysis have led to many powerful techniques for the solution of a wide range of practical engineering and signal processing problems. Although known for many years, these mathematical tools have been considered too computationally demanding to be of any practical use, especially when the speed of calculation is an issue. Due to the lack of computational power, engineers had to content themselves with suboptimal methodologies of simpler implementation. Only recently, due to the advent of parallel/systolic computing algorithms, architectures, and technologies, have engineers employed these more sophisticated mathematical techniques. Among these techniques are the so-called eigenvalue decomposition (EVD) and the singular value decomposition (SVD). As an application of these methods, we consider the important problem of spatial filtering.

Motivation–Spatial Filtering Problem

Consider a linear array consisting of L sensors uniformly spaced with an adjacent distance d. A number M, $M < L$, of narrowband signals of center frequency f_0, impinging on the array. These signals arrive from M different spatial direction angles $\theta_1, \ldots, \theta_M$, relative to some reference direction. Each sensor is provided with a variable weight. The weighted sensor outputs are then collected and summed. The goal is to compute the set of weights to enhance the estimation of the desired signals arriving from directions $\theta_1, \ldots, \theta_M$.

In one class of beamformation problems one sensor (sometimes referred to as main sensor) receives the desired signal perturbed by interference and noise. The remaining $L-1$ sensors (auxiliary sensors) are mounted and aimed in such a way as to collect only the (uncorrelated) interference and noise components. In this scenario the main sensor gain is to be kept at a fixed value, while the auxiliary weights are adjusted in such a way as to cancel out as much perturbation as possible. Obviously, the only difference of this latter case is that one of the weights (the one corresponding to the main sensor) is kept at a constant value of unity.

Let the output of the ith sensor, $i = 1, \ldots, L$, at discrete time $n = 0, 1, \ldots,$ be given by

$$\overline{x}_i(n) = \Re\{[x_i(n) + v_i(n)]e^{j(2\pi f_0 n)}\}, \qquad x_i(n) = a_i(n)\sum_{k=1}^{M} S_k(n)e^{j2\pi(i-1)d \sin\theta_k/\lambda}$$

where $a_i(n)$ is the antenna gain at time n, S_k is the complex amplitude of the kth signal, inclusive of the initial phase, and λ is the signal wavelength. The vectors $x_i(n)$ and $v_i(n)$ are analytic signal representations. The noise $v_i(n)$ is assumed to be uncorrelated white and Gaussian, of power σ_N^2. In order to avoid the ill effects of spatial aliasing, let us also assume that $d \leq \lambda/2$. The outputs of the sensor array for times $n = 0, 1, \ldots, N$, can be collected in matrix form as follows,

$$\underbrace{X}_{N\times L} = \underbrace{S}_{N\times M}\underbrace{A}_{M\times L} + \underbrace{V}_{N\times L}$$

The matrix A is referred to as the **steering matrix.** In the case in which $a_i(n) = 1$ for all i, the matrix A is Vandermonde and full rank, and its kth row can be expressed as $A_k = (1, e^{j2\pi d \sin\theta_k/\lambda}, \ldots, e^{j2\pi(L-1)d \sin\theta_k/\lambda})$.

The data correlation matrix, $R_X = \mathrm{E}\{X^{\mathrm{H}}X\}$, where $\mathrm{E}\{\cdot\}$ is the ensemble average operator, is equal to $R_X = A^{\mathrm{H}}R_S A + \sigma_N^2 I$, $R_S = \mathrm{E}\{S^{\mathrm{H}}S\}$. We note:

1. The matrix R_S has rank M by definition as does the matrix $A^{\mathrm{H}}R_S A$
2. The rows of A are in the range space of R_X
3. The value σ_N^2 is an eigenvalue of R_X with multiplicity $L-M$, given that $\det(R_X - \sigma_N^2 I) = 0$ and the rank of $A^{\mathrm{H}}R_S A$ is M

The eigenvalue decomposition of R_X can therefore be written as $R_X = V_S\Lambda_S V_S^{\mathrm{H}} + \sigma_N^2 V_N V_N^{\mathrm{H}}$, where V_S is $L \times M$, V_N is $L \times (L-M)$, $V_S^{\mathrm{H}}V_S = I$, $V_N^{\mathrm{H}}V_N = I$, and $V_S^{\mathrm{H}}V_N = 0$. Moreover, we have that $AV_N = 0$. Let $A(\theta)$ be a generic steering vector, defined as $A(\theta) \equiv (1, e^{j2\pi d \sin\theta/\lambda}, \ldots, e^{j2\pi(L-1)d \sin\theta/\lambda})$. Then the function $\mathcal{S}(\theta) = 1/|A(\theta)V_N|^2$ has M poles at the angles $\theta = \theta_k$, $k = 1, \ldots, M$. Alternatively, any linear combination, w, of the columns of V_N is such that $\mathrm{E}\{\|Xw\|_2\} = \min_z \mathrm{E}\{\|Xz\|_2\} = \sigma_N$. In other words, the signals impinging from angular directions $\theta_1, \ldots, \theta_M$ are totally canceled out in the system output.

The desired weighting vector for our spatial filtering problem can consequently be expressed as $w = V_N p$, $p = [p_1, \ldots, p_{L-M}]^{\mathrm{T}}$, for any nonzero vector p. From the above discussion, we see that the solution to the original spatial filtering problem can be obtained from the eigenvalue decomposition of the correlation matrix $R_X = \mathrm{E}\{X^{\mathrm{H}}X\}$. In practice the sample correlation matrix $\hat{R}_X$ is used instead, where the ensemble average is replaced by a suitable temporal average.

The computation of the covariance matrix implies the computation of the matrix product $X^{\mathrm{H}}X$. Some small elements in X are then squared and the magnitude of the resulting element can become comparable or smaller than the machine precision. Rounding errors can often impair and severely degrade the computed solution. In these cases it is better to calculate the desired quantities (correlation eigenvalues and eigenvectors) directly from the data matrix X using the SVD technique as considered below.

Eigenvalue Decomposition of a Symmetric Matrix

Consider an $L \times L$ real symmetric matrix $A = A^{\mathrm{T}}$. In the spatial filtering example above, $A = R_X$. Let

$$G(i, j, \theta) = \begin{array}{c} \\ i \\ \\ j \\ \\ \end{array} \begin{pmatrix} I_{i-1} & & & & \\ & c & & s & \\ & & I_{j-i-1} & & \\ & -s & & c & \\ & & & & I_{L-j} \end{pmatrix}$$

(column labels i and j above the columns containing c, s)

be an **orthogonal Givens rotation matrix**, where $c = \cos\theta$ and $s = \sin\theta$. Pre- or postmultiplication of A by G leaves A unchanged, except for rows (columns) i and j, which are replaced by a linear combination of old rows (columns) i and j. A **Jacobi rotation** is obtained by simultaneous pre- and postmultiplication of a matrix by a Givens rotation matrix, as given by $G(i, j, \theta)^{\mathrm{T}} A G(i, j, \theta)$, where θ is usually chosen in order to zero out the (i, j) and (j, i) entries of A.

The matrix A can be driven toward diagonal form by iteratively applying Jacobi rotations, as given by $A_0 \leftarrow A$, $A_{k+1} \leftarrow G_k^{\mathrm{T}} A G_k$, where G_k is a Givens rotation matrix. A **sweep** is obtained by applying $L(L-1)/2$ Jacobi rotations, each nullifying a different pair of off-diagonal elements, according to a prespecified order. Given the matrix $A_k = (a_{pq}^{(k)})$ at the kth iteration, and a pair of indices (i, j), the value of $\tan\theta$ can be obtained from the following equations

$$u = \frac{a_{jj}^{(k)} - a_{ii}^{(k)}}{2a_{ij}^{(k)}} \qquad \tan\theta = \frac{\mathrm{sign}(u)}{|u| + \sqrt{1+u^2}} \tag{63.25}$$

It is possible to demonstrate that each Jacobi rotation reduces the matrix off-norm. The matrix A_k indeed tends to diagonal form and for all practical purposes it reaches it after $\mathcal{O}(\log L)$ sweeps.

The matrix V of eigenvectors is obtained by applying the same rotations to a matrix initialized to the identity, as follows: $V_0 \leftarrow I$, $V_{k+1} \leftarrow V_k G_k$. A two-dimensional systolic array implementation of the previous algorithm is shown in Fig. 63.100, for the case $L = 8$. At the beginning of iteration k, processor P_{ij} contains elements

$$\begin{pmatrix} a_{2i-1,2j-1}^{(k)} & a_{2i-1,2j}^{(k)} \\ a_{2i,2j-1}^{(k)} & a_{2i,2j}^{(k)} \end{pmatrix}, \quad i, j = 1, 2, \ldots, L/2$$

The diagonal processors compute the rotation parameters and apply the rotation to the four entries they store. Subsequently they propagate the rotation parameters horizontally and vertically to their neighbors, which, upon receiving them, apply the corresponding rotation to their stored entries. After the rotation is applied, each processor swaps its entries with its four neighbors along the diagonal connections. The correct data movement at the edges of the array is also shown in Fig. 63.100. A correct scheduling of operations requires that each processor be idle for two out of three time steps, which translates into an *efficiency* of 33%. Each sweep takes $3(L-1)$ time steps and the number of sweeps can be chosen on the order of $\log L$.

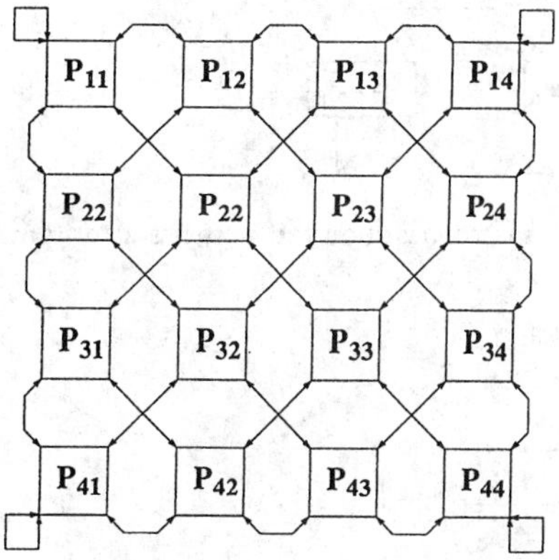

FIGURE 63.100 Systolic array for an EVD of a symmetric matrix based on Jacobi rotations.

Singular Value Decomposition of a Rectangular Matrix via the Hestenes Algorithm

Consider an $N \times L$ real matrix A, $N \geq L$. Its SVD can be written as follows

$$\underbrace{A}_{N\times L} = \underbrace{U}_{N\times L} \underbrace{\Sigma}_{L\times L} \underbrace{V^{\mathrm{T}}}_{L\times L}$$

where U and V have orthonormal columns. The matrix $\Sigma = \mathrm{diag}(\sigma_1, \ldots, \sigma_L)$ is the diagonal matrix of singular values, where $\sigma_1 \geq \sigma_2 \geq \cdots \geq \sigma_L \geq 0$. Consider the following recursion $A_0 \leftarrow A$, $A_{k+1} \leftarrow A_k G_k$, where the Givens rotations are chosen not to zero out entries of A_k, but to orthogonalize pairs of its columns. A sweep is now defined as a sequence of Givens rotations which orthogonalize all $\binom{L}{2}$ pairs of columns exactly once. Observe the similarity with the algorithm described above for the calculation of eigenvalues and eigenvectors. If $G(i, j, \theta)$ is the Givens rotation which orthogonalizes columns i and j of A, then $G(i, j, \theta)^{\mathrm{T}} M G(i, j, \theta)$ is the Jacobi rotation that zeroes out the entries (i, j) and (j, i) of $M = A^{\mathrm{T}}A$. A sweep (as defined here) of rotations applied to the rectangular matrix A corresponds exactly to a sweep (as defined earlier) of rotations applied to the symmetric matrix M.

At any time step, the original matrix A can be expressed as follows:

$$A = A_k V_k^{\mathrm{T}} \qquad V_k = \prod_{i=1}^{k} G_k$$

where V_k has orthonormal columns for any k (by definition of Givens rotations). After a number of sweeps (on the order of $\log L$) the matrix A_k approaches a matrix, W, of orthogonal columns, $A_k \rightarrow W$, $V_k \rightarrow V$. If σ_i is the norm of the ith column of W, $i = 1, \ldots, L$, then we have $W = U\,\mathrm{diag}(\sigma_1, \ldots, \sigma_L)$, $A = U\Sigma V^{\mathrm{T}}$.

This SVD approach based on the Hestenes algorithm can be realized on a Brent-Luk [58] linear systolic array, as shown in Fig. 63.101, for the case $L = 8$. Each processor stores a pair of columns; in particular, the procedure starts by storing columns $2k - 1$ and $2k$ in processor P_k. Each processor computes the rotation parameters which orthogonalize the pair of columns.

Let x and z be the two stored columns. Let ξ and ζ be their norms, and η be their inner product. Then the value of $\tan\theta$ from

$$u = \frac{\zeta - \xi}{2\eta} \qquad \tan\theta = \frac{\mathrm{sign}(u)}{|u| + \sqrt{1 + u^2}}$$

After applying the computed rotation, each processor swaps its pair of columns with the two neighboring processors along the connections shown in Fig. 63.101. The column indices stored

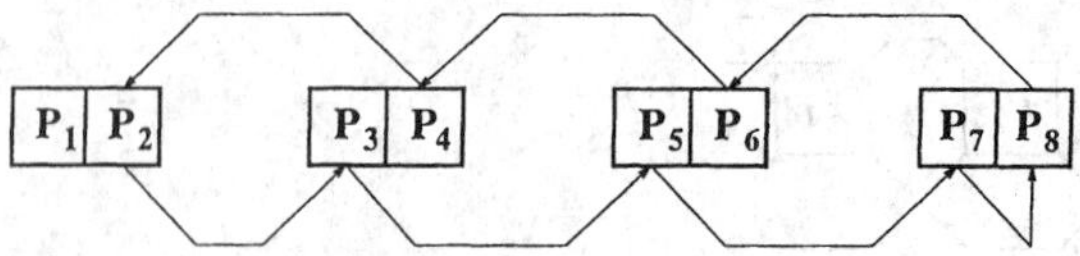

FIGURE 63.101 Linear systolic array for SVD of a rectangular matrix based on the Hestenes algorithm.

TABLE 63.17 Movement of Matrix Columns During One Sweep ($L = 8$)

P_1	P_2	P_3	P_4
(1, 2)	(3, 4)	(5, 6)	(7, 8)
(1, 4)	(2, 6)	(3, 8)	(5, 7)
(1, 6)	(4, 8)	(2, 7)	(3, 5)
(1, 8)	(6, 7)	(4, 5)	(2, 3)
(1, 7)	(8, 5)	(6, 3)	(4, 2)
(1, 5)	(7, 3)	(8, 2)	(6, 4)
(1, 3)	(5, 2)	(7, 4)	(8, 6)
(1, 2)	(3, 4)	(5, 6)	(7, 8)

in each processor at the different steps of a single sweep are given in Table 63.17. Note that all the $\binom{L}{2}$ pairs of indices are generated by using the Brent-Luk scheme. The stopping criteria can be set in advance. A possible criterion is by inspecting the magnitude of the rotating angles. When they are all in absolute value below a given threshold, then the algorithm can stop. More commonly, a predetermined number of sweeps is chosen ahead of time. Observation shows that a number of sweeps on the order of $\log L$ is sufficient for convergence.

Singular Value Decomposition of a Rectangular Nonsymmetric Matrix via the Jacobi Algorithm

The SVD algorithm described in the previous section has the drawback of a somewhat complicated updating procedure. In many signal processing applications continuous updating of the matrix decomposition as new samples are appended to the data matrix is required. Such problems occur in spectral analysis, direction-of-arrival estimation, beam forming, etc. An efficient updating procedure for the SVD of rectangular matrices of growing row size is given by the algorithm described in this section, based on the succession of two basic operations: a QR updating step, followed by a rediagonalization operation. This algorithm is otherwise known as a version of the Kogbetliantz algorithm for triangular matrices.

Given the $m \times L$ data matrix at time m, $A_m = [a_1, \ldots, a_m]^T$, where a_i, $i = 1, \ldots, m$ are the rows of A_m, one defines the exponentially weighted matrix $B_m(\beta)A_m$, where $B_m(\beta)$ is the diagonal forgetting matrix $B_m(\beta) \equiv \text{diag}(\beta^{m-1}, \beta^{m-2}, \ldots, \beta, 1)$, and $0 < \beta \leq 1$ is the forgetting factor. The updating problem is to determine the SVD of the updated weighted matrix $B_{m+1}(\beta)A_{m+1}$, given the SVD at time m,

$$B_m(\beta)A_m = U_m \Sigma_m V_m$$

Often only the singular values and right singular vectors are of interest. This is fortunate, because the left singular matrix grows in size as time increases, while the sizes of Σ_m and V_m remain unchanged.

The algorithm can be summarized as follows. Given the matrices V_m and Σ_m and the new data sample x_{m+1},

$$\Sigma'_m \leftarrow \begin{pmatrix} \beta\Sigma_m \\ x_{m+1}V_m \end{pmatrix} \qquad V'_m \leftarrow V_m$$

the QR updating step

$$\begin{pmatrix} \Sigma'_m \\ 0 \end{pmatrix} \leftarrow Q_{m+1}\Sigma'_m$$

the rediagonalization using permuted Jacobi rotations

for $k = 1, \ldots, l,$
  for $i = 1, \ldots, n-1,\ j = i+1,$
    $\Sigma'_m \leftarrow \Pi_{ij} G(i, j, \theta)\Sigma'_m G(i, j, \phi)^{\mathrm{T}}\Pi_{ij},$
    $V'_m \leftarrow V'_m G(i, j, \phi)^{\mathrm{T}}\Pi_{ij}.$
  end
end
$\Sigma_{m+1} \leftarrow \Sigma'_m,$
$V_{m+1} \leftarrow V'_m.$

In the above algorithm the parameter l determines both the number of computations between subsequent updates and the estimation accuracy at the end of each update step. When l is chosen equal to the problem order L, then one complete sweep is performed. In practice, the l parameter can be chosen as high as $\sim 10L$ (usually for block computations or slow updates) and as small as 1 (for very high updating rates, with understandable degradation of estimation accuracy). The matrices Σ_m and Σ'_m are upper triangular at all times. This is ensured by the application of the permuted left and right Givens rotations in the rediagonalization step. After the application of any Jacobi rotation, the rotated rows and columns are subsequently permuted. This expedient not only preserves the upper triangularity of the Σ-matrices, but also makes it possible for the rotations to be generated on the diagonal and propagated along physically adjacent pairs of rows and columns. All these features make this algorithm a very attractive candidate for a systolic implementation.

A schematic diagram of the systolic array proposed by Moonen et al. [63] is shown in Fig. 63.102, where the triangular array stores the matrices Σ_m and Σ'_m, for all m, and the square array stores the V-matrix. The incoming data samples, x_{m+1}, are input into the V-array, where the vector-by-matrix multiplication $x_{m+1}V_m$ is performed. The output is subsequently fed into the triangular array. As it propagates through the array, the QR updating step is carried on: left rotations are generated in the diagonal elements of the array and propagated through the corresponding rows of the Σ-matrix. One does not need to wait for the completion of the QR step to start performing the Jacobi rotations associated with the diagonalization step. It is known that the two operations (QR update and diagonalization) can be interleaved without compromising the final result. The parameters relative to the left rotations are, as before, propagated along the rows of the triangular matrix, while the right rotation parameters move along the columns of Σ_m, and are passed on to the V-array. Due to the continual modification of the matrix V_m caused by these right rotations, and because of the use of finite precision arithmetic, the computed right singular matrix may deviate from orthogonality. It is also known that in a realistic environment the norm of $V_m V_m^{\mathrm{T}} - I$ grows linearly with m. Reorthogonalization procedures must therefore be included in the overall scheme. A complicated reorthogonalization procedure based on left rotations, which interleaves with the other operations, was described in

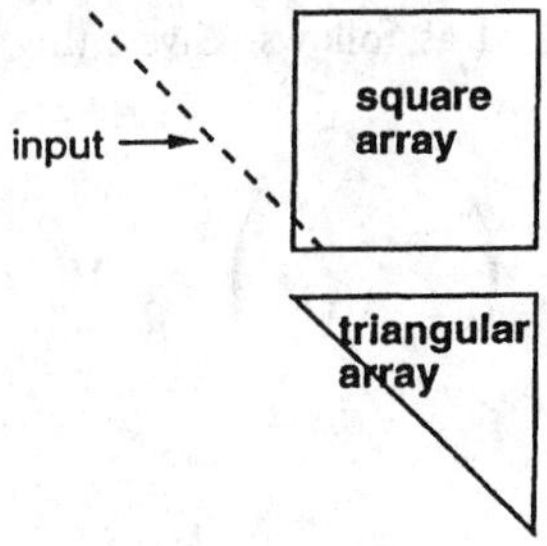

FIGURE 63.102 Two-dimensional systolic array for updating the Jacobi SVD algorithm for a nonsymmetric matrix.

Moonen et al. [63]. An improved reorthogonalization algorithm was proposed by Vanpoucke and Moonen [66], where the matrix V_m is stored in parametrized form, thereby guaranteeing orthogonality at all times. The resulting triangular array and its modes of operation were also described.

References

[1] A. Darte and J. M. Delosme, "Partitioning for Array Processors," Tech. Rep. LIP-IMAG 90-23, Laboratoire de l'Informatique du Parallélisme, Ecole Supérieure de Lyon, Oct. 1990.

[2] H. T. Kung and C. E. Leiserson, "Systolic Arrays (for VLSI)," in *Introduction to VLSI Systems,* C. A. Meads and L. A. Conway, Eds. Reading, MA: Addison-Wesley, 1980, chap. 8.3.

[3] H. T. Kung, "Why systolic architecture?" *Computer*, vol. 15, pp. 37–46, 1982.

[4] S. Y. Kung, K. S. Arun, R. J. Gal-Elzer, and D. V. B. Rao, "Wavefront Array Processor: Language, Architecture, and Applications," *IEEE Trans. Comput.*, vol. 31, pp. 1054–1066, 1982.

[5] S. Y. Kung, *VLSI Array Processing,* Englewood Cliffs, NJ: Prentice-Hall, 1988.

[6] F. Lorenzelli, "Systolic Mapping with Partitioning and Computationally Intensive Algorithms for Signal Processing," Ph.D. thesis, University of California, Los Angeles, 1993.

[7] D. I. Moldovan, "On the Analysis and Synthesis of VLSI Algorithms," *IEEE Trans. on Comput.*, vol. 31, pp. 1121–1126, 1982.

[8] M. Newman, *Integral Matrices,* New York: Academic Press, 1972.

[9] P. Quinton, "The Systematic Design of Systolic Arrays," IRISA Tech. Report 193, Apr. 1983.

[10] S. Rao, "Regular Iterative Algorithms and Their Implementations on Processor Arrays," Ph.D. thesis, Stanford, CA: Stanford University, 1985.

[11] E. Angelidis and J. E. Diamessis, "A Novel Method for Designing FIR Filters with Nonuniform Frequency Samples," *IEEE Trans. Signal Process.*, vol. 42, pp. 259–267, 1994.

[12] S. Chanekar, S. Tantaratana, and L. E. Franks, "Multiplier-Free IIR Filter Realization Using Periodically Time-Variable State-Space Structures, I and II," *IEEE Trans. Signal Process.*, vol. 42, pp. 1008–1027, 1994.

[13] L. A. Ferrari and P. V. Sankar, "Minimum Complexity FIR Filters and Sparse Systolic Arrays," *IEEE Trans. Comput.*, vol. 37, pp. 760–764, 1988.

[14] A. Jayadeva, "A New Systolic Design for Digital IIR Filters," *IEEE Trans. Circuits Syst.*, vol. 37, pp. 653–654, 1990.

[15] S. C. Knowles, J. G. McWhirter, R. F. Woods, and J. V. McCanny, "Bit-Level Systolic Architectures for High Performance IIR Filtering," *J. VLSI and Signal Process.*, vol. 1, pp. 9–24, 1989.

[16] S. M. Lei and K. Yao, "A Class of Systolizable IIR Digital Filters and Its Design for Proper Scaling and Minimum Output Roundoff Noise," *IEEE Trans. Circuits Syst.*, vol. 37, pp. 1217–1230, 1990.

[17] S. M. Lei and K. Yao, "Efficient Systolic Array Implementation of IIR Digital Filtering," *IEEE Trans. Circuits Syst.*, vol. 39, pp. 581–584, 1992.

[18] H. H. Loomis, Jr. and B. Sinha, "High-Speed Recursive Digital Filters," *Circuits, Syst. Signal Process.*, vol. 3, pp. 267–294, 1984.

[19] K. K. Parhi and D. G. Messerschmitt, "Pipeline Interleaving and Parallelism in Recursive Digital Filters. I. Pipelining Using Scattered Look-Ahead and Decomposition," *IEEE Trans. Acoust. Speech Signal Process,*, vol. 37, pp. 1099–1117, 1989.

[20] N. R. Shanbhag and K. K. Parhi, *Pipelined Adaptive Digital Filters*, Boston: Kluwer Academic, 1994.

[21] R. F. Woods and J. V. McCanny, "Design of High Performance IIR Digital Filter Chip," *IEEE Proc. E, Comput. Digital Tech.*, vol. 139, pp. 195–202, 1992.

[22] C. W. Wu and J.-C. Wang, "Testable Design of Bit-Level Systolic Block FIR Filters," *Proc. IEEE Int. Symp. Circuits Syst.*, pp. 1129–1132, 1992.

[23] R. Wyrzykowski and S. Ovramenko, "Flexible Systolic Architecture for VLSI FIR Filters," *IEEE Proc. E, Comput. Digital Tech.*, vol. 139, pp. 170–172, 1992.

[24] L. W. Chang and J. H. Lin, "A Bit-Level Systolic Array for Median Filter," *IEEE Trans. Signal Process.*, vol. 40, pp. 2079–2083, 1992.

[25] J. M. Delosme, "Bit-Level Systolic Array for Real Symmetric and Hermitian Eigenvalue Problems," *J. VLSI Signal Process.*, vol. 4, pp. 69–88, 1992.

[26] R. A. Evans, J. V. McCanny, J. G. McWhirter, A. Wood, and K. W. Wood, "A CMOS Implementation of a Systolic Multibit Convolver Chip," *Proc. VLSI*, pp. 227–235, 1983.

[27] G. Fettweis and H. Meyr, "High-Rate Viterbi Processor: A Systolic Array Solution," *IEEE J. Sel. Topics Commun.*, vol. 8, pp. 1520–1534, 1990.

[28] S. C. Knowles, J. G. McWhirter, R. F. Woods, and J. V. McCanny, "Bit-Level Systolic Architectures for High Performance IIR Filtering," *J. VLSI Signal Process.*, vol. 1, pp. 9–24, 1989.

[29] J. V. McCanny and J. G. McWhirter, "On the Implementation of Signal Processing Function Using One Bit Systolic Arrays," *Elect. Lett.*, vol. 18, pp. 241–243, 1982.

[30] J. V. McCanny, J. G. McWhirter, and S. Y. Kung, "The Use of Data Dependence Graphs in the Design of Bit-Level Systolic Arrays," *IEEE Trans. Acoust., Speech Signal Process.*, vol. 38, pp. 787–793, 1990.

[31] J. V. McCanny, R. F. Woods, and M. Yan, "Systolic Arrays for High Performance Digital Signal Processing," in *Digital Signal Processing: Principles, Devices, and Applications*, N. B. Jones and J. D. M. Watson, Eds. New York: Peter Peregrinus, 1990, pp. 276–302.

[32] C. L. Wang, "An Efficient and Flexible Bit-Level Systolic Array for Inner Product Computation," *J. Chin. Inst. Eng.*, vol. 41, pp. 567–576, 1991.

[33] C. W. Wu, "Bit-Level Pipelined 2-D Digital Filters Image Processing," *IEEE Trans. Circuits Syst. Video Technol.*, vol. 1, pp. 22–34, 1991.

[34] M. G. Bellanger and P. A. Regalia, "The FLS-QR Algorithm for Adaptive Filtering: The Case of Multi-Channel Signals," *Signal Process.*, vol. 22, pp. 115–126, 1991.

[35] J. M. Cioffi, "The Fast Adaptive ROTOR's RLS Algorithm," *IEEE Trans. Acoust. Speech Signal Process.*, vol. 38, pp. 631–653, 1990.

[36] W. M. Gentleman and H. T. Kung, "Matrix Triangularization by Systolic Arrays," *Proc. SPIE, Real-Time Signal Process.*, vol. 298, pp. 298–303, 1981.

[37] S. Haykin, *Adaptive Filter Theory*, 2nd ed. Englewood Cliffs, NJ: Prentice-Hall, 1991.

[38] F. Ling and J. G. Proakis, "A Recursive Modified Gram-Schmidt Algorithm with Applications to Least-Squares Estimation and Adaptive Filtering," *IEEE Trans. Acoust. Speech Signal Process*, vol. 34, pp. 829–836, 1986.

[39] K. J. R. Liu, S. F. Hsieh, K. Yao, and C. T. Chiu, "Dynamic Range, Stability, and Fault-Tolerant Capability of Finite-Precision RLS Systolic Array Based on Givens Rotations," *IEEE Trans. Circuits Systems*, June, pp. 625–636, 1991.

[40] K. J. R. Liu, S. F. Hsieh, and K. Yao, "Systolic Block Householder Transformation for RLS Algorithm with Two-Level Pipelined Implementation," *IEEE Trans. Signal Process.*, vol. 40, pp. 946–958, 1992.

[41] J. G. McWhirter, "Recursive Least-Squares Minimization Using a Systolic Array," *Proc. SPIE, Real-Time Signal Process. VI*, vol. 431, pp. 105–112, 1983.

[42] J. G. McWhirter, "Algorithmic Engineering in Adaptive Signal Processing," *IEEE Proc. F*, vol. 139, pp. 226–232, 1992.

[43] B. Yang and J. F. Böhme, "Rotation-Based RLS Algorithms: Unified Derivations, Numerical Properties, and Parallel Implementations," *IEEE Trans. Acoust. Speech Signal Process.*, vol. 40, pp. 1151–1167, 1992.

[44] M. J. Chen and K. Yao, "On Realization of Least-Squares Estimation and Kalman Filtering by Systolic Arrays," in *Systolic Arrays*, W. Moore, A. McCabe, and R. Urquhart, Eds. Bristol, U.K.: Adam Hilger, 1986, pp. 161–170.

[45] F. Gaston, G. Irwin, and J. McWhirter, "Systolic Square Root Covariance Kalman Filtering," *J. VLSI Signal Process.*, pp. 37–49, 1990.

[46] J. H. Graham and T. F. Kadela, "Parallel Algorithm Architectures for Optimal State Estimation," *IEEE Trans. Comput.*, vol. 34, pp. 1061–1068, 1985.

[47] R. E. Kalman, "A New Approach to Linear Filtering and Prediction Problems," *J. Basic Eng.*, vol. 82, pp. 35–45, 1960.

[48] P. G. Kaminski, "Discrete Square Root Filtering: A Survey of Current Techniques," *IEEE Trans. Autom. Control*, vol. 16, pp. 727–735, 1971.

[49] S. Y. Kung and J. N. Hwang, "Systolic Array Design for Kalman Filtering," *IEEE Trans. Signal Process.*, vol. 39, pp. 171–182, 1991.

[50] R. A. Lincoln and K. Yao, "Efficient Systolic Kalman Filtering Design by Dependence Graph Mapping," in *VLSI Signal Processing III*, R. W. Brodersen and H. S. Moscovitz, Eds. New York, IEEE Press, 1988, pp. 396–407.

[51] J. G. Nash and S. Hansen, "Modified Faddeev Algorithm for Matrix Manipulation," *Proc. SPIE*, vol. 495, pp. 39–46, 1984.

[52] C. C. Paige and M. A. Saunders, "Least Squares Estimation of Discrete Linear Dynamic Systems Using Orthogonal Transformation," *SIAM J. Numer. Anal.*, vol. 14, pp. 180–193, 1977.

[53] G. M. Papadourakis and F. J. Taylor, "Implementation of Kalman Filters Using Systolic Arrays," *Proc. Int. Conf. Acoust., Speech and Signal Process.*, pp. 783–786, 1987.

[54] P. Rao and M. A. Bayoumi, "An Algorithm Specific VLSI Parallel Architecture for Kalman Filter," in *VLSI Signal Processing IV*, H. S. Moscovitz, K. Yao, and R. Jain, Eds. New York: IEEE Press, 1991, pp. 264–273.

[55] T. Y. Sung and Y. H. Hu, "Parallel Implementation of the Kalman Filter," *IEEE Trans. Aero. Electr. Syst.*, vol. 23, pp. 215–224, 1987.

[56] H. Yeh, "Systolic Implementation of Kalman Filters," *IEEE Trans. Acoust., Speech, and Sig. Process.*, pp. 1514–1517, 1988.

[57] E. Biglieri and K. Yao, "Some Properties of Singular Value Decomposition and Their Application to Signal Processing," *Signal Process.*, vol. 18, pp. 277–289, 1989.

[58] R. Brent and F. T. Luk, "The Solution of Singular-Value and Symmetric Eigenvalue Problems on Multiprocessor Arrays," *SIAM J. Sci. Stat. Comput.*, vol. 6, pp. 69–84, 1985.

[59] G. H. Golub and C. F. Van Loan, *Matrix Computations*, 2nd ed., Baltimore: Johns Hopkins University Press, 1989.

[60] S. Haykin, *Adaptive Filter Theory*, 2nd ed. Englewood Cliffs, NJ: Prentice-Hall, 1991.

[61] M. R. Hestenes, "Inversion of Matrices by Biorthogonalization and Related Results," *J. Soc. Ind. Appl. Math.*, vol. 6, pp. 51–90, 1958.

[62] F. T. Luk, "A Triangular Processor Array for Computing Singular Values," *Linear Algebra Appl.*, vol. 77, pp. 259–273, 1986.

[63] M. Moonen, P. Van Dooren, and J. Vandewalle, "A Systolic Array for SVD Updating," *SIAM J. Matrix Anal. Appl.*, vol. 14, pp. 353–371, 1993.

[64] R. O. Schmidt, "A Signal Subspace Approach to Multiple Emitter Location and Spectral Estimation," Ph.D thesis, Stanford, CA: Stanford University, 1981.

[65] G. W. Stewart, "A Jacobi-Like Algorithm for Computing the Schur Decomposition of a Nonhermitian Matrix," *SIAM J. Sci. Stat. Comput.*, vol. 6, pp. 853–864, 1985.

[66] F. Vanpoucke and M. Moonen, "Numerically Stable Jacobi Array for Parallel Subspace Tracking," *Proc. SPIE*, vol. 2296, 1994.

[67] F. D. Van Veen and K. M. Buckley, "Beamforming: A Versatile Approach to Spatial Filtering," *IEEE ASSP Mag.*, vol. 5, pp. 4–24, 1988.

64

Data Converters

Bang-Sup Song
University of Illinois, Urbana-Champaign

Ramesh Harjani
University of Minnesota

64.1 Digital-to-Analog Converters

Bang-Sup Song

Introduction

Digital-to-analog converters (DACs), referred to as decoders in communications terms, are devices by which digital processors communicate with the analog world. Although DACs are used as key elements in analog-to-digital converters (ADCs), they find numerous applications as stand-alone devices from CRT display systems and voice/music synthesizers to automatic test systems, waveform generators, digitally controlled attenuators, process control actuators, and digital transmitters in modern digital communications systems.

The basic function of the DAC is the conversion of input digital numbers into analog waveforms. An N-bit DAC provides a discrete analog output level, either voltage or current, for every level of 2^N digital words, $\{D_i;\ i = 0, 1, 2, \ldots, 2^N - 1\}$, that is applied to the input. Therefore, an ideal voltage DAC generates 2^N discrete analog output voltages for digital inputs varying from 000...00 to 111...11 as illustrated in Fig. 64.1 for the four-bit example. The output has a one-to-one correspondence with the input

$$V_{\text{out}}(D_i) = V_{\text{ref}}\left(\frac{b_N}{2} + \frac{b_{N-1}}{2^2} + \cdots + \frac{b_2}{2^{N-1}} + \frac{b_1}{2^N}\right) \tag{64.1}$$

where V_{ref} is a reference voltage setting the output range of the DAC and $b_N b_{N-1}, \ldots, b_1$ is the binary representation of the input digital word D_i. In the unipolar case, as shown, the reference point is 0 when the digital input D_0 is 000...00, but in bipolar or differential DACs, the reference point is the midpoint of the full scale when the digital input is 100...00, and the range is defined from $-V_{\text{ref}}/2$ to $V_{\text{ref}}/2$. Although purely current-output DACs are possible, voltage-output DACs are common in most applications.

0-8493-8341-2/95/$0.00+$.50

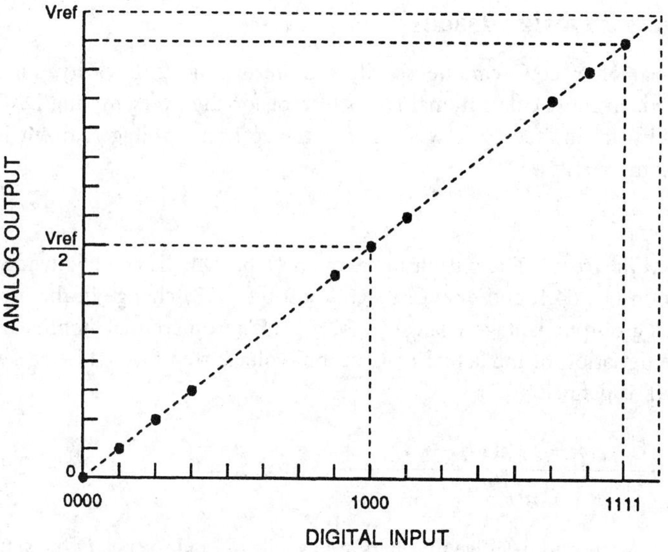

FIGURE 64.1 Transfer characteristics of a unipolar DAC.

Signal-to-Noise Ratio and Dynamic Range

The resolution is a term used to describe a minimum voltage or current that a DAC can resolve. The fundamental limit of a DAC is the quantization noise due to the finite resolution of the DAC. If the input digital word is N bits long, the minimum step that a DAC can resolve is $V_{\mathrm{ref}}/2^N$. If output voltages are reproduced with this minimum step of uncertainty, an ideal DAC should have a maximum signal-to-noise ratio (SNR) of

$$\mathrm{SNR} = \frac{3}{2}2^{2N} \approx 6N + 1.8\ (\mathrm{dB}) \tag{64.2}$$

where SNR is defined as the power ratio of the maximum signal to the inband uncorrelated noise. For example, an ideal 16-bit DAC has an SNR of about 97.8 dB. The spectrum of the quantization noise is evenly distributed up to the Nyquist bandwidth (half the sampling frequency). Therefore, this inband quantization noise decreases by 6 dB when the oversampling ratio is doubled. This implies that when oversampled, the SNR within the signal band can be made lower than the quantization noise limited by (64.2).

The resolution of a DAC is usually characterized in terms of SNR, but the SNR accounts only for the uncorrelated noise. The real noise performance is better represented by TSNR, which is the ratio of the signal power to the total inband noise, including harmonic distortion. Also, a slightly different term is often used in place of the SNR. The useful signal range or dynamic range is defined as the power ratio of the maximum signal to the minimum signal. The minimum signal is defined as the smallest input, for which the TSNR is 0 dB, while the maximum signal is the full-scale input. Therefore, the SNR of nonideal DACs can be lower than the ideal dynamic range because the noise floor can be higher with a large signal present. In practice, DACs are limited not only by the quantization noise, but also by nonideal factors such as noises from circuit components, power supply coupling, noisy substrate, timing jitter, insufficient settling, and nonlinearity, etc.

Basic Converter Design Issues

The performance of a DAC can be specified in terms of its linearity, monotonicity, and conversion speed. In most conventional DACs, except for the oversampling DACs, the linearity and monotonicity are limited by how accurately the reference voltage/current is divided using passive/active components.

Linearity

Differential Nonlinearity. The output range of an N-bit DAC is equally divided into 2^N small units, as shown in Fig. 64.1, and one least significant bit (LSB) change in the input digital word makes the analog output voltage change by $V_{ref}/2^N$. The differential nonlinearity (DNL) is a measure of the deviation of the actual DAC output voltage step from this ideal voltage step for 1 LSB. The DNL is defined as

$$\text{DNL} = \frac{V_{out}(D_{i+1}) - V_{out}(D_i) - V_{ref}/2^N}{V_{ref}/2^N}, \quad \text{for } i = 0, 1, \ldots, 2^N - 2 \text{ (LSB)} \qquad (64.3)$$

and the largest positive and negative numbers are usually quoted to specify the static performance of a DAC.

Integral Nonlinearity. The overall linearity of a DAC can be specified in terms of the integral nonlinearity (INL), which is a measure of deviation of the actual DAC output voltage from the ideal straight line drawn between two endpoints, 0 and V_{ref}. Because the ideal output is $i \times V_{ref}/2^N$ for any digital input D_i, the INL is defined as

$$\text{INL} = \frac{V_{out}(D_i) - i \times V_{ref}/2^N}{V_{ref}/2^N}, \quad \text{for } i = 0, 1, \ldots, 2^N - 1 \text{ (LSB)} \qquad (64.4)$$

and the largest positive and negative numbers are usually quoted to specify the static performance of a DAC.

However, several definitions of INL may result depending on how two endpoints are defined. In some DAC architectures the two endpoints are not exactly 0 and V_{ref}. The nonideal reference point causes an offset error, while the nonideal full-scale range gives rise to a gain error. In most DAC applications these offset and gain errors resulting from the nonideal endpoints do not matter, and the integral linearity can be better defined in a relative measure using a straight line linearity concept rather than the end point linearity in the absolute measure. The straight line can be defined as two endpoints of the actual DAC output voltages or as a theoretical straight line adjusted to best fit the actual DAC output characteristics. The former definition is sometimes called endpoint linearity, while the latter is called best straight line linearity.

Monotonicity

The DAC output should increase over its full range as the digital input word to the DAC increases. That is, the negative DNL should be < -1 LSB for a DAC to be monotonic. Monotonicity is critical in most applications, in particular, in digital control applications. The source of nonmonotonicity is an inaccuracy in binary weighting of a DAC. For example, the most significant bit (MSB) has a weight of one half of the full range. If the MSB weight is smaller than the ideal value, the analog output change can be smaller than the ideal step $V_{ref}/2^N$ when the input digital word changes from 0111...11 to 1000...00 at the midpoint of the DAC range. If this decrease in the output is $>$ 1 LSB, the DAC becomes nonmonotonic. The similar nonmonotonicity can take place when switching the second or lower MSB bits in binary-weighted multi-bit DACs.

Monotonicity is inherently guaranteed if an N-bit DAC is made of 2^N elements for thermometer decoding. However, it is impractical to implement high-resolution DACs using 2^N elements because the number of elements grows exponentially as N increases. Therefore, to guarantee monotonicity in practical applications, DACs have been implemented using either a segmented DAC or an integrator-type DAC. Oversampling interpolative DACs also achieve monotonicity using a pulse-density modulated bitstream converted into analog voltages by a lossy integrator or by a low-pass filter.

Segmented Digital-to-Analog Converters. Applying a two-step conversion concept, a DAC can be made in two levels using coarse and fine DACs. The fine DAC divides one coarse MSB segment into fine LSBs. If one fixed MSB segment is subdivided to generate LSBs, matching among MSB segments creates a nonmonotonicity problem. However, if the next MSB segment is subdivided instead of the fixed segment, the segmented DAC can maintain monotonicity regardless of the MSB matching. This is called **next-segment approach**. Unless the next segment approach is used to make a segmented DAC with a total $M+N$ bits, the MSB DAC should have a resolution of $M+N$ bits for monotonicity, while the LSB DAC requires an N-bit resolution. Using the next-segment approach, an MSB DAC made of 2^M identical elements guarantees monotonicity, although INL is still limited by the MSB matching.

To implement a segmented DAC using two resistor-string DACs, voltage buffers are needed to drive the LSB DAC without loading the MSB DAC. Although the resistor-string MSB DAC is monotonic, overall monotonicity is not guaranteed due to the offsets of the voltage buffers. The use of a capacitor-array LSB DAC eliminates a need for voltage buffers. The most widely used segmented DAC is a current-ratioed DAC, whose MSB DAC is made of identical elements for the next-segment approach, but the LSB DAC is a current divider. A binary-weighted current divider can be used as an LSB DAC, as shown in Fig. 64.2. For monotonicity, the MSB M-bits are selected by a thermometer code, but one of the MSB current sources corresponding to the next segment of the thermometer code is divided by a current divider for fine LSBs.

Integrator-Type Digital-to-Analog Converters. As mentioned, monotonicity is guaranteed only in a thermometer-coded DAC. The thermometer coding of a DAC output can be implemented either by repeating identical DAC elements many times or by repeatedly using the same element. The former requires more hardware, but the latter more time. In the continuous time integrator-type DAC the integrator output is a linear ramp and the time to stop integrating can be controlled by the digital codes. Therefore, monotonicity can be maintained. Similarly, the

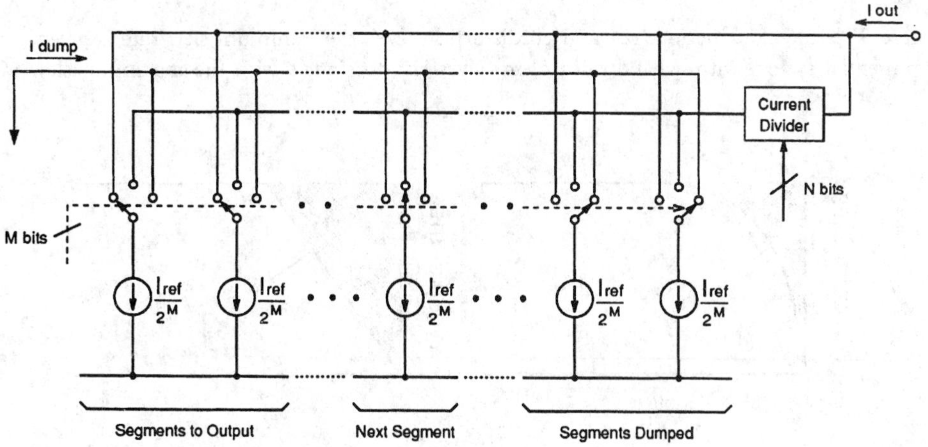

FIGURE 64.2 Segmented DAC for monotonicity.

discrete time integrator can integrate a constant amount of charge repeatedly and the number of integrations can be controlled by the digital codes. The integration approach can give high accuracy, but its disadvantage is its slow speed limiting its applications.

Although it is different in concept, oversampling interpolative DACs modulate the digital code into a bitstream, and its pulse density represents the DAC output. Due to the incremental nature of the pulse density modulation, oversampling DACs are monotonic. A DAC for the pulse-density modulated bitstream is a lossy integrator. The integrator integrates a constant charge if the pulse is high, while it subtracts the same charge if the pulse is low. In principle, it is equivalent to the discrete time integrator DAC, but the output is represented by the average charge on the integrator.

Conversion Speed

The output of a DAC is a sampled-and-held step waveform held constant during a word clock period. Any deviation from the ideal step waveform causes an error in the DAC output. High-speed DACs usually have a current output, but even current-output DACs are either terminated with a 50 to 75 Ω low-impedance load or buffered by a wideband transresistance amplifier. Therefore, the speed of a DAC is limited either by the RC time constant of the output node or by the bandwidth of the output buffer amplifier. Figure 64.3 illustrates two step responses of a DAC when it settles with a time constant of τ and when it slews with a slew rate of S, respectively. The transient errors given by the shaded areas of Fig. 64.3 are h/τ and $h^2/2S$, respectively. This implies that a single time-constant settling of the former case only generates a linear error in the output, which does not affect the DAC linearity, but the slew-limited settling of the buffer generates a nonlinear error. Even in the single-time constant case (the former), the code-dependent time constant in settling can introduce a nonlinearity error because the settling error is a function of the time constant τ. This is true for a resistor-string DAC, which exhibits a code-dependent settling time because the output resistance of the DAC depends on the digital input.

The slew rate limit of the buffer is a significant source of nonlinearity since the error is proportional to the square of the signal, as shown in Fig. 64.3(b). The height and the width of the error term change with the input. The worst-case harmonic distortion (HD) when generating a sinusoidal signal with a magnitude V_0 with a limited slew rate of S is [1]

$$\mathrm{HD}_k = 8\,\frac{\sin^2 \dfrac{\omega T_c}{2}}{\pi k(k^2-4)} \times \frac{V_0}{ST_c}, \quad k = 1, 3, 5, 7, \ldots \tag{64.5}$$

where T_c is the clock period. For a given distortion level, the minimum slew rate is given. Any exponential system with a bandwidth of ω_0 gives rise to signals with the maximum slew rate

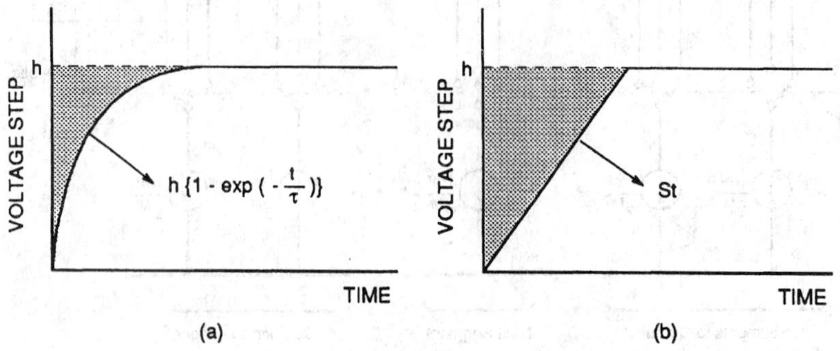

FIGURE 64.3 Errors in step response: (a) settling, and (b) slewing.

of $2\omega_0 V_0$. Therefore, by making $2\omega_0 V_0 > S_{min}$, no slew rate is limited and the DAC system will exhibit no distortion.

Converter Architectures

Many circuit techniques are used to implement DACs, but a few popular techniques used widely today are of the parallel type, in which all bits change simultaneously upon applying an input code word. Serial DACs, on the other hand, produce an analog output only after receiving all digital input data in a sequential form. When DACs are used as stand-alone devices, their output transient behavior limited by glitch, slew rate, word clock jitter, settling, etc. are of paramount importance, but are used as subblocks of ADCs, digital-to-analog converters need only to settle within a given time interval. In stand-alone DAC applications the digital input word made of N bits should be synchronously applied to the DAC with a precise timing accuracy. Thus, the input data latches are used to hold the digital input during the conversion. The output analog sample-and-hold, usually called deglitcher, is often used for the better transient performance of a DAC. The three most popular architectures in integrated circuits are DACs using a resistor string, ratioed current sources, and a capacitor array. The current-ratioed DAC finds the greatest application as a stand-alone DAC, while the resistor-string and capacitor-array DACs are used mainly as ADC subblocks.

Resistor-String Digital-to-Analog Converters

The simplest voltage divider is a resistor string. Reference levels can be generated by connecting 2^N identical resistors in series between V_{ref} and 0. Switches to connect the divided reference voltages to the output can be either a 1-out-of-2^N decoder or a binary tree decoder as shown in Fig. 64.4 for the 3-bit example. Because it requires a good switch, the stand-alone resistor-string DAC is easier to implement using CMOS. However, the lack of switches does not limit the application of the resistor string as a voltage reference divider subblock for ADCs in other process technologies.

Resistor strings are used widely as reference dividers, an integral part of the flash ADC. All resistor-string DACs are inherently monotonic and exhibit good differential linearity. However,

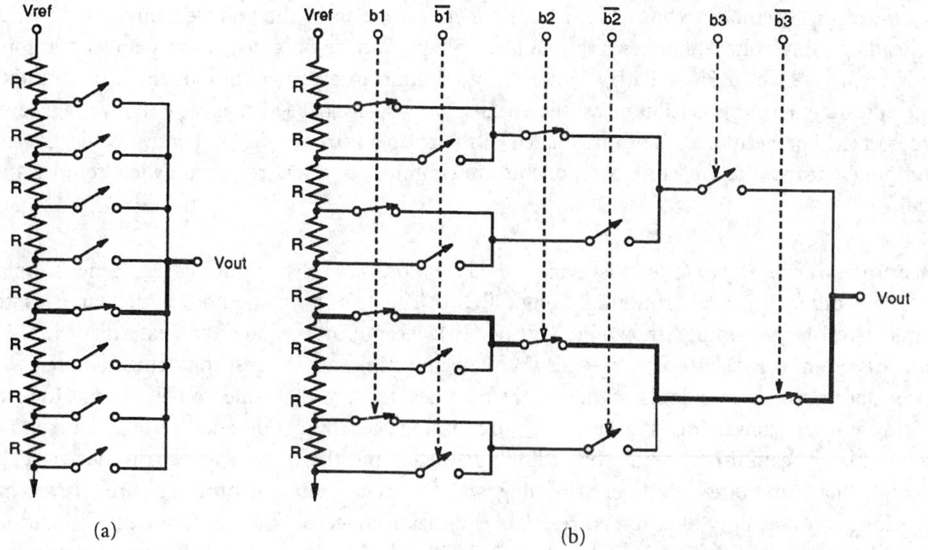

FIGURE 64.4 Resistor-string DAC: (a) with 1-out-of-2^N decoder, and (b) with a binary tree decoder.

TABLE 64.1 Resistors in IC Processes

Resistor Type	Sheet R (Ω/sq.)	Tolerance (%)	10–20 μm Matching (%)	T.C. (ppm/°C)
Diffusion	100–200	±20	±0.2–0.5	1500
Ion implantation	500–1k	±5	±0.1	200
Thin film	1k	±5	±0.1	10–100
Undoped poly	100–500	±20	±0.2	1500

they suffer from a poor integral linearity and also have the drawback that the output resistance depends on the digital input code. This causes a code-dependent settling time when charging the capacitive load of the output bus. The code-dependent settling time has no effect on the reference divider performance as an ADC subblock, but the performance is severely degraded as a stand-alone DAC. This nonuniform settling time problem can be alleviated by adding low-resistance parallel resistors and by compensating the MOS switch overdrive voltages.

In bipolar technology the most common resistors are thin-film resistors made of tantalum, Ni-Cr, or Cr-SiO, which exhibit very low voltage and temperature coefficients. In CMOS either diffusion or undoped poly resistors are common. Four of the most frequently used resistors are listed in Table 64.1. Conventional trimming or adjustment techniques are impractical to be applied to all 2^N resistor elements. The following four methods are often used to improve the integral linearity of resistor-string DACs.

Layout Techniques. The use of large geometry devices and/or careful layout is effective in improving the matching marginally. Large geometry devices reduce the random edge effect, and the layout using a common centroid or geometric averaging can reduce the process gradient effect. However, typical matching of resistors in integrated circuits is still limited to an 8- to 10-bit level due to the mobility and resistor thickness variations. Differential resistor DACs with large feature sizes are reported to exhibit a higher matching accuracy of an 11- to 12-bit level.

Off-Chip Adjustment. It is possible to set tap points of a resistor-string to specified voltages by connecting external voltage sources to them, as shown in Fig. 64.5(a) for the 3-bit example. Simply put, the more taps adjusted, the better the integral linearity obtained. An additional benefit of this method is the reduced RC time constant due to the voltage sources at the taps. Instead of using voltage sources, the required voltages can be obtained using parallel trimming resistors, as shown in Fig. 64.5(b). However, in addition to external components for trimming, fine adjustments and precision measurement instruments are needed to ensure that voltage levels are correct. Furthermore, due to mismatch in the temperature coefficients between the external components and the on-chip components, retrimming is often required when temperature changes.

Postprocess Trimming. The most widely used methods are laser trimming [2], Zener zapping [3], and other electrical trimming using PROM. The trimming method is the same as the parallel resistor trimming shown in Fig. 64.5(b) except for the fact that external trimming resistors are now integrated on the chip. While being trimmed, the resistor string is biased with a constant current. Individual segments are trimmed to have the same voltage drop. However, during normal conversion, the current source is replaced by a reference voltage source. The focused laser beam for trimming has a finite diameter, and the resistor to be trimmed occupies a large chip area. Both the laser trimming and the Zener zapping processes are irreversible. The long-term stability of trimmed resistors is a major concern, although the electrical and the PROM trimming (if PROM is replaced by EPROM) can be repeated. All trimming methods in this category are time consuming and require precision instruments.

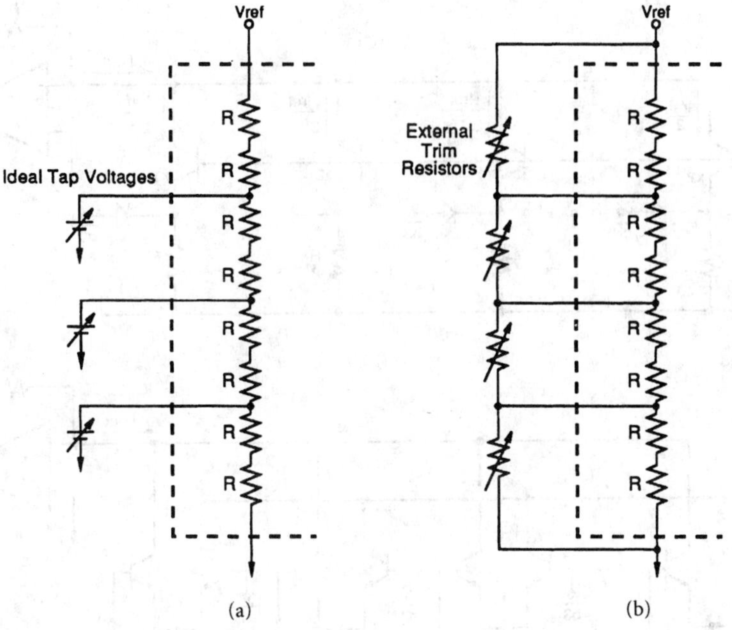

FIGURE 64.5 INL improvements: (a) by external voltage sources, and (b) by parallel resistors.

On-Chip Buffers. The voltage at intermediate taps of the resistor string can be controlled by another resistor string through on-chip unity-gain buffers. This is actually an improved version of the off-chip method. The controlling resistors can be either laser trimmed or electronically controlled by switches. Laser-trimmed controlling resistors have the same problems mentioned earlier. Trimming network can be implemented to electronically control resistor values. In either case buffers with a high open-loop gain, a low output resistance, a large current driving capability, and a wide bandwidth for accurate and fast settling are required.

Current-Ratioed Digital-to-Analog Converters

The most popular stand-alone DACs in use today are current-ratioed DACs, of which the two types are a weighted-current DAC and an R-$2R$ DAC.

Binary-Weighted Current Digital-to-Analog Converters. The weighted current DACs shown in Fig. 64.6 are made of an array of switched binary-weighted current sources and the current summing network. In bipolar technology binary weighting is achieved by ratioed transistors and emitter resistors with binary-related values of R, R/2, R/4, and so on, while in MOS technology, only ratioed transistors are used. One example is a video random access memory DAC in CMOS, which is made of simple PMOS differential pairs with binary-weighted tail currents. Digital-to-analog converters relying on active device matching can achieve an 8-bit level performance with a 0.2 to 0.5% matching accuracy using a 10 to 20 μm device feature size while degeneration with thin-film resistors gives a 10-bit level performance. The current sources are switched on or off by means of switching diodes or emitter-coupled differential pairs (source-coupled pairs in CMOS), as shown in Fig. 64.6. The output current summing is done by a wideband transresistance amplifier, but in high-speed DACs, the output current is used directly to drive a resistor load for a maximum speed but with a limited output swing. The weighted current design has the advantage of simplicity and high speed, but it is difficult to implement a high-resolution DAC because a wide range of emitter resistors and transistor sizes is used and very large resistors cause problems with both temperature stability and speed.

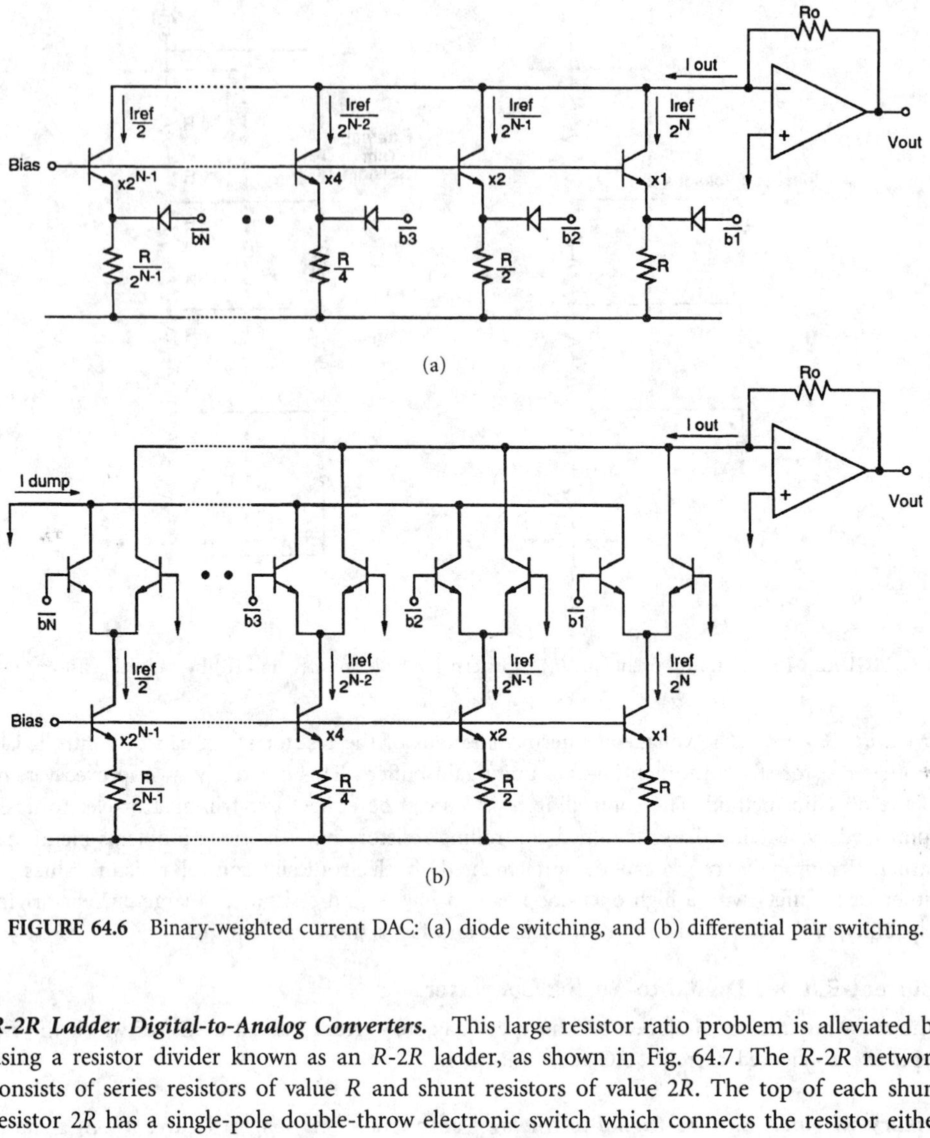

FIGURE 64.6 Binary-weighted current DAC: (a) diode switching, and (b) differential pair switching.

***R*-2*R* Ladder Digital-to-Analog Converters.** This large resistor ratio problem is alleviated by using a resistor divider known as an *R*-2*R* ladder, as shown in Fig. 64.7. The *R*-2*R* network consists of series resistors of value *R* and shunt resistors of value 2*R*. The top of each shunt resistor 2*R* has a single-pole double-throw electronic switch which connects the resistor either to ground or to the output current summing node. The operation of the *R*-2*R* ladder network is based on the binary division of current as it flows down the ladder. At any junction of series resistor *R*, the resistance looking to the right side is 2*R*. Therefore, the input resistance at any junction is *R*, and the current splits into two equal parts at the junction because it sees equal resistances in either direction. The result is binary-weighted currents flowing into each shunt resistor in the ladder. The digitally controlled switches direct the currents to either ground or to the summing node. The advantage of the *R*-2*R* ladder method is that only two values of resistors are used, greatly simplifying the task of matching or trimming and temperature tracking. In addition, for high-speed applications relatively low resistor values can be used. Excellent results can be obtained using laser-trimmed thin-film resistor networks. Because the output of the *R*-2*R* DAC is the product of the reference voltage and the digital input word, the *R*-2*R* ladder DAC is often called a multiplying DAC (MDAC).

Both the weighted-current DAC and the *R*-2*R* DAC can be used as a current divider to make a sub-DAC. To make a segmented DAC for monotonicity based on the next-segment approach, as discussed earlier, the MSB should be made of thermometer-coded equal currents. Once the

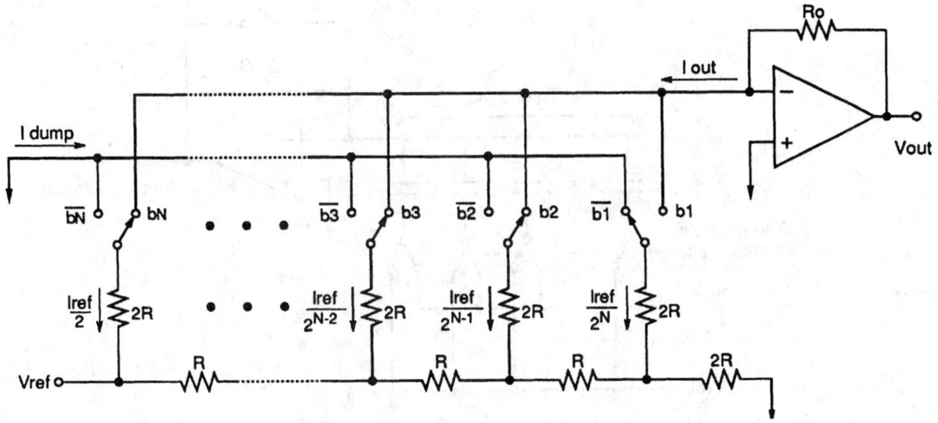

FIGURE 64.7 *R*-2*R* ladder DAC.

MSB is selected, the next segment should be divided further into LSBs as shown in Fig. 64.2. Integral nonlinearity can be improved by dynamically matching or by self-calibrating the MSB current sources as discussed later.

Capacitor-Array Digital-to-Analog Converter. Capacitors made of double-poly or poly-diffusion in MOS technology are considered one of the most accurate passive components comparable to thin-film resistors in bipolar process, both in the matching accuracy and voltage and temperature coefficients [4]. The only disadvantage in the capacitor-array DAC implementation is the use of a dynamic charge redistribution principle. A switched-capacitor counterpart of the resistor-string DAC is a parallel capacitor array of 2^N unit capacitors (C) with a common top plate. The capacitor-array DAC is not appropriate for stand-alone applications without a feedback amplifier virtually grounding the top plate and an output sample-and-hold or deglitcher. The operation of the capacitor-array DAC shown in Fig. 64.8 is based on the thermometer-coded DAC principle, and has a distinct advantage of monotonicity if the system is implemented properly. However, due to the complexity of handling the thermometer-coded capacitor array, a binary-weighted capacitor array is often used, as shown in Fig. 64.9, by grouping unit capacitors in binary ratio values. A common centroid layout of the capacitor array is known to give a 10-bit level matching for this application when the unit capacitor size is over 12 μm × 12 μm. The matching accuracy of the capacitor in MOS technology depends on the geometry sizes of the capacitor width and length and the dielectric thickness.

As a stand-alone DAC, the top plate of the DAC is precharged either to the offset of the feedback amplifier or to the ground. One smallest capacitor is not necessary for this application. However, as a subblock of an ADC, the total capacitor should be $2^N C$, as drawn in Fig. 64.9, and the top plate of the array is usually connected to the input nodes of comparators or high-gain operational amplifiers, depending on the ADC architectures. As a result, the top plate has a parasitic capacitance, but its effect on the DAC performance is negligible. The capacitor-array DAC requires two-phase nonoverlapping clocks for proper operation. Initially, all capacitors should be charged to ground. After initialization, depending on the digital input, the bottom plates are connected either to V_{ref} or to ground. Consider the case in which the top plate is floating without the feedback amplifier. If the charge at the top plate finishes its redistribution, the top plate voltage neglecting the top plate parasitic effect becomes

$$V_0 = \sum_{i=1}^{N} \frac{b_k}{2^i} V_{ref} \tag{64.6}$$

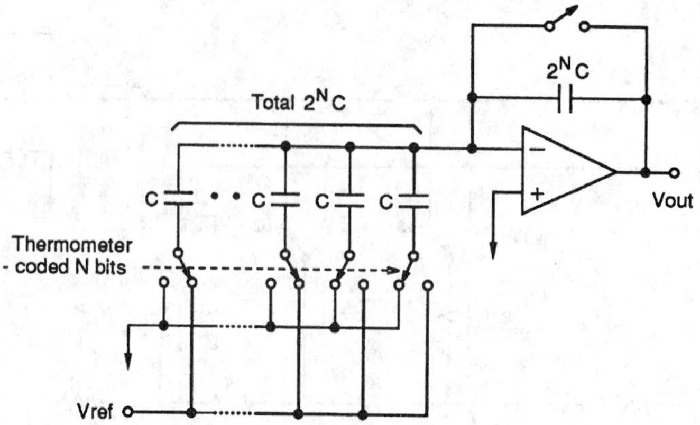

FIGURE 64.8 Thermometer-coded capacitor-array DAC.

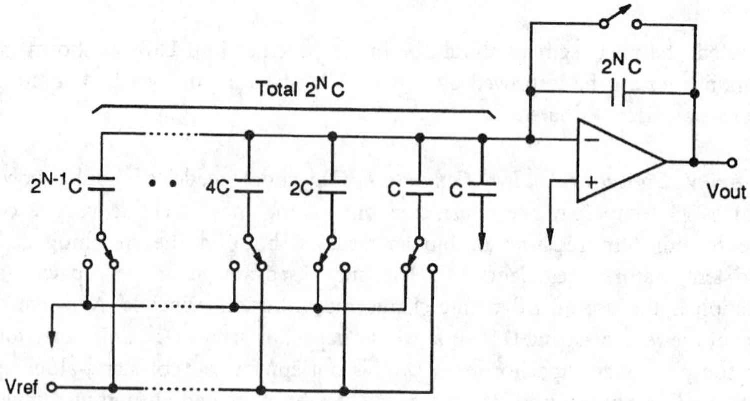

FIGURE 64.9 Binary-weighted capacitor-array DAC.

where $b_N b_{N-1}, \ldots, b_2 b_1$ is the input binary word. For example, switching the MSB capacitor bottom to V_{ref} changes the output voltage by

$$\frac{2^{N-1}C}{\sum_{i=1}^{N} C_i} V_{ref} = \frac{V_{ref}}{2} \tag{64.7}$$

where the capacitor C_i for the ith bit is nominally scaled to $2^{i-1}C$. Therefore, the nonlinearity at the midpoint of the full range is limited by the ratio mismatch of the half sum of the capacitor array to the total sum of the array. Similarly, the nonlinearity at one fourth of the range is limited by the ratio of one fourth of the capacitor array to the total array, and so on.

One important application of the capacitor-array DAC is as a reference DAC for ADCs. As in the case of the *R*-2*R* MDAC, the capacitor-array DAC can be used as an MDAC to amplify residue voltages for multistep ADCs. As shown in Fig. 64.9, if the input is sampled on the bottom plates of capacitors instead of the ground, the voltage amplified is the amplified input voltage minus the DAC output. By varying the feedback capacitor size, the MDAC can be used as an interstage residue amplifier in multistep pipelined ADCs. For example, if the feedback capacitor is C and the digital input is the coarse *N*-bit decision of the sampled analog voltage, the amplifier output is a residue voltage amplified by 2^N for the subsequent LSB conversion.

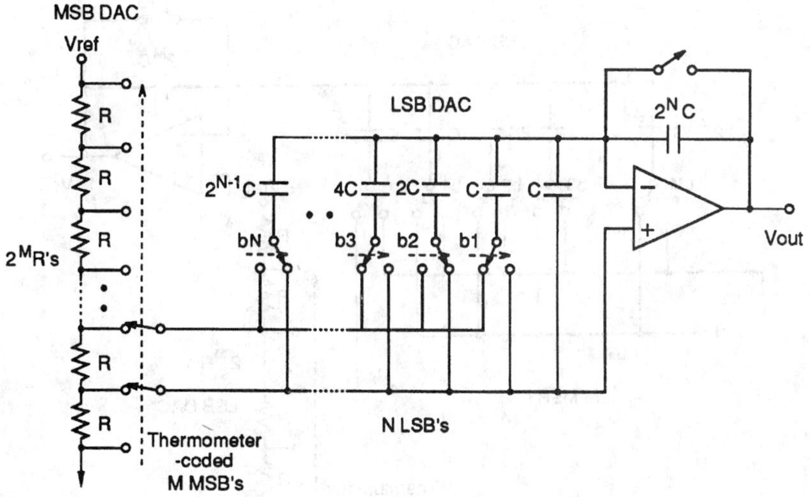

FIGURE 64.10 $R + C$ combination DAC.

$R + C$ or $C + R$ Combination Digital-to-Analog Converters

Both resistor-string and capacitor-array DACs need 2^N unit elements for N-bits, and the number grows exponentially. Splitting arrays into two, one for MSBs and the other for the LSBs, requires a buffer amplifier to interface between two arrays. Although a floating capacitor connects two capacitors arrays, the parasitic capacitance of the floating node is not well controlled. A more logical combination for high-resolution DAC is between resistor and capacitor DACs. This combination does not require any coupling capacitors or interface buffer amplifiers.

In the $R + C$ combination the MSB is set by the resistor string, and next segment of the resistor-string DAC supplies the reference voltage of the LSB capacitor-array DAC, as shown in Fig. 64.10. When the top plate is initialized, all capacitor bottom plates are connected to the lower voltage of the next segment of the resistor-string DAC. During the next clock phase, the bottom plates of capacitors are selectively connected to the higher voltage of the segment if the digital bit is ONE, but stays switched to the lower voltage if ZERO. This segmented DAC approach gives an inherent monotonicity as far as the LSB DAC is monotonic within its resolution. Although INL is poor, the fully differential implementation of this architecture benefits from the lack of the even-order nonlinearity, thereby achieving improved INL. On the other hand, in the $C + R$ combination shown in Fig. 64.11, the operation of the capacitor DAC is the same. The MSB side reference voltage is fixed, but the reference voltage of the smallest capacitor is supplied by the LSB resistor-string DAC. This approach exhibits nonmonotonicity due to the capacitor DAC matching. Both combination DACs are seldom used as stand-alone DACs due to their limited speed, but are used frequently as subblocks of high-resolution ADCs.

Techniques for High-Resolution Digital-to-Analog Converters

Most DACs are made of passive or active components such as resistors, capacitors, or current sources, and their linearity relies on the matching accuracy of those components. Among frequently used DAC components, diffused resistors and transistors are in general known to exhibit an 8-bit level matching while thin-film resistors and capacitors are matched to a 10-bit level. Trimming or electronic calibration is needed in order to obtain a higher linearity than what is achievable with bare component matching. The traditional solutions to this have been the wafer-level trimming methods such as laser trimming and Zener zapping. Although many other promising trimming or matching techniques such as polysilicon fuse trimming, electrical

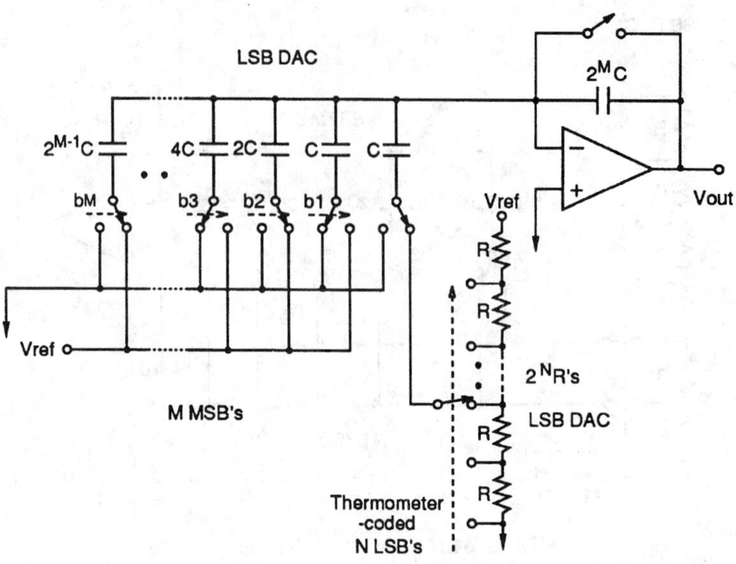

FIGURE 64.11 $C + R$ combination DAC.

trimming techniques using PROM, and large device matching by geometrical averaging have been proposed, conventional factory-set trimming or matching techniques give no flexibility of retrimming. How successfully these techniques can be applied to large-volume production of high-resolution DACs and how the factory-trimmed components will perform over the long term are still in question.

The future trend is toward more sophisticated and intelligent electronic solutions that overcome and complement some of the limitations of conventional trimming techniques. The methods recently developed are dynamic circuit techniques [5] for component matching, switched-capacitor integration [6], electronic calibration [7] of a DAC nonlinearity, and oversampling interpolation techniques [8] which trade speed with resolution. In particular, the oversampling interpolative DACs are used widely in stand-alone applications such as digital audio playback systems or digital communications due to their inherent monotonicity.

Dynamic Matching Techniques

In general, a dynamic element matching to improve the accuracy of the binary ratio is a time-averaging process. For simplicity, consider a simple voltage or current divide-by-two element, as shown in Fig. 64.12. Due to mismatches in the ratio of resistors, transistors, and capacitors, the divided voltage or current is not exactly $V_{ref}/2$ or $I_{ref}/2$, but their sum is V_{ref} or I_{ref}. The dynamic matching concept is to multiplex these two outputs with complementary errors of Δ and $-\Delta$ so that the errors Δ and $-\Delta$ can be averaged out over time while the average value of $V_{ref}/2$ or $I_{ref}/2$ remains. It is in effect equivalent to the suppressed carrier balanced modulation of the error component Δ. The high-frequency energy can be filtered out using a post low-pass filter. This technique relies on the accurate timing of the duty cycle. Any duty cycle error or timing jitter results in inaccurate matching. The residual matching inaccuracy becomes a second-order error proportional to the product of the original mismatch and the timing error.

The application of dynamic element matching to the binary-weighted current DAC is a straightforward switching of two complementary currents. Its application to the binary voltage divider using two identical resistors or capacitors requires exchanging of resistors or capacitors. This can be achieved easily by reversing the polarity of the reference voltage for the divide-by-two case. However, in the general case of N element matching the current division is inherently

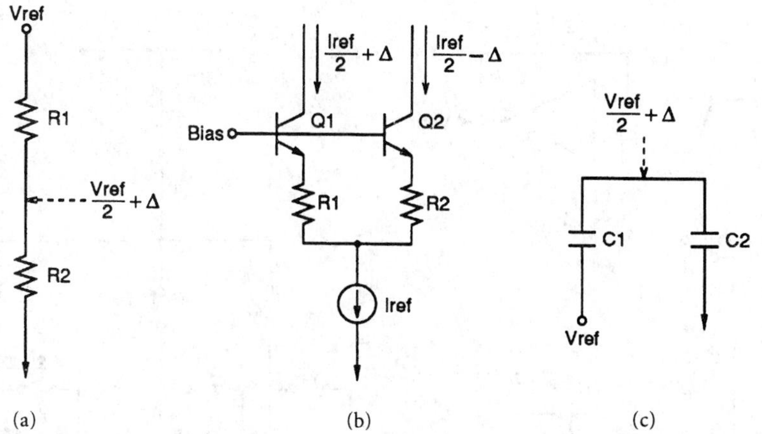

FIGURE 64.12 Divide-by-two elements: (a) resistor, (b) current, and (c) capacitor.

simpler in implementation than the voltage division. In general, to match the N independent elements, a switching network with N inputs and N outputs is required. The function of the switching network is to connect any input out of N inputs to one output with an average duty cycle of $1/N$. The simplest one is a barrel shifter rotating the I-O connections in a predetermined manner [5]. This barrel shifter generates a low-frequency modulated error when N becomes larger because the same pattern repeats every N clocks. A more sophisticated randomizer with the same average duty cycle can distribute the mismatch error over the wider frequency range. The latter technique finds applications as a multi-bit DAC in the multibit noise-shaping sigma-delta data converter, whose linearity relies on the multi-bit DAC.

Voltage or Current Sampling. The voltage or current sampling concept is an electronic alternative to direct mechanical trimming. To sample voltage or current using a voltage or current sampler is equivalent to trimming individual voltage or current sources. The voltage sampler is usually called a sample-and-hold (S/H) circuit, while the current sampler is called a current copier. The voltage is usually sampled on the input capacitor of a buffer amplifier and the current is usually sampled on the input capacitor of a transconductance amplifier such as an MOS transistor gate. Therefore, both voltage and current sampling techniques are ultimately limited by their sampling accuracy.

The idea behind the voltage or current sampling DAC is to use one voltage or current element repeatedly. One example of the voltage sampling DAC is a discrete-time integrator-type DAC with many S/H amplifiers for sampling output voltages. The integrator integrates a constant charge repeatedly, and its output is sampled on a new S/H amplifier every time the integrator finishes an integration as shown in Fig. 64.13(a). This is equivalent to generating equally spaced reference voltages by stacking identical unit voltages [6]. The fundamental problem associated with this sampling voltage DAC approach is the accumulation of the sampling error and noise in generating larger voltages. Similarly, the current sampling DAC can sample a constant current on current sources made of MOS transistors, as shown in Fig. 64.13(b) [7]. Because one reference current is copied on other identical current samplers, the matching accuracy can be maintained as far as the sampling errors are kept constant. It is not practical to make a high-resolution DAC using voltage or current sampling alone. Therefore, this approach is limited to generating MSB DACs for the segmented DAC or for the subranging ADCs.

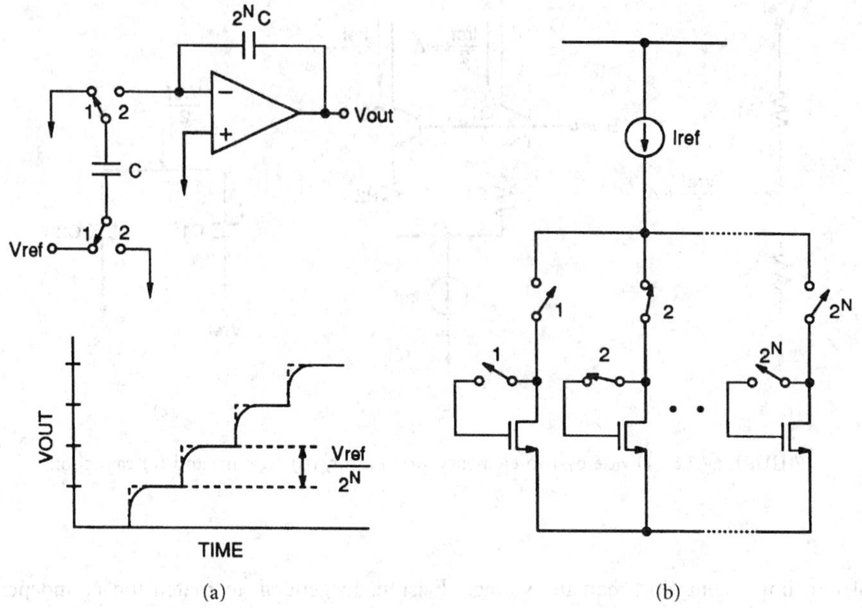

FIGURE 64.13 Voltage and current sampling concepts: (a) integrator, and (b) current copier.

Electronic Calibration Techniques

Electronic calibration is a general term to describe various circuit techniques, which usually predistort the DAC transfer characteristic so that the DAC linearity can be improved. One such technique is a straightforward code mapping, and the other is a self-calibration. The code-mapping calibration is a very limited technique only for the factory because it requires a precision measurement setup and a large digital memory. The self-calibration is to incorporate all the calibration mechanisms and hardware on the DAC as a built-in function so that users can recalibrate whenever calibrations are needed. The self-calibration is based on an assumption that the segmented DAC linearity is limited by the MSB DAC so that only errors of MSBs may be measured, stored in memory, and recalled during normal operation. There are two ways of measuring the MSB errors. In one method individual-bit nonlinearities, usually appearing as component mismatch errors, are measured digitally [9], and a total error, which is called a code error, is computed from individual-bit errors depending on the output code during normal conversion. On the other hand, the other method measures and stores digital code errors directly and eliminates the digital code-error computation during normal operation [10]. The former requires less digital memory during normal conversion while the latter requires fewer digital computations.

Direct Code Mapping. The simplified code mapping of a DAC can be done with a calibration DAC, digital memory, and a precision instrument to measure the DAC output, as shown in Fig. 64.14. The idea is to measure the DAC error using a calibration DAC so that the DAC output corrected by the calibration DAC can produce an ideal DAC output. The input code of the calibration DAC is stored as a code error in digital memory addressed by the DAC input code. This code error is recalled to predistort the DAC output during normal operation. This technique needs a 2^N memory with a word length corresponding to the number of bits of the calibration DAC. This method can correct any kinds of DAC nonlinearities as long as the calibration DAC has an output range wide enough to cover the whole range of nonlinearity. However, the same method can be implemented without the use of a calibration DAC if the

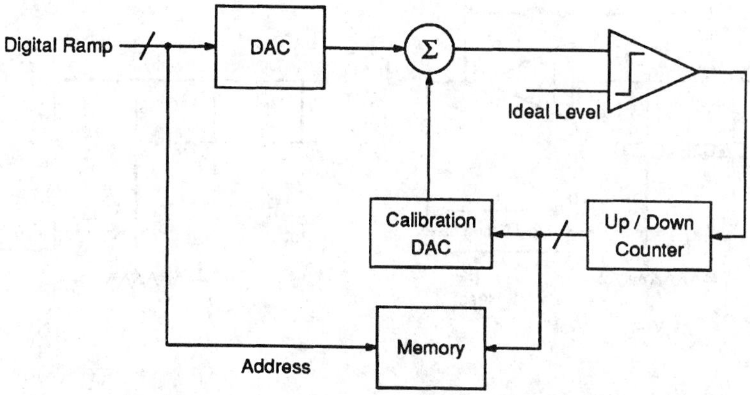

FIGURE 64.14 Code mapping with a calibration DAC.

main DAC is monotonic with extra bits of resolution. In this case the calibration is a simple code mapping, selecting correct input digital codes for correct DAC output voltages among redundant input digital codes.

Self-Calibration for Individual Capacitor Errors. The idea of measuring the individual bit errors using a calibration DAC is to quantize the difference Δ in the divide-by-two elements shown in Fig. 64.12 because the ideal divide ratio is 1 : 2. For example, the MSB should be half of the whole range of a DAC, the second MSB is half of the MSB, and so on. Unless buffer amplifiers are used, the ideal calibration DACs for *R* and *C* DACs are *C* and *R* DACs, respectively. The ratio error measurement cycle of two-bit *C* DAC is illustrated in Fig. 64.15. Errors can be quantized using a successive approximation method, but the up/down converter is shown here for simplicity. Initially, the top plate is charged to the comparator offset and the bottom plates of C_1 and C_2 sample 0 and V_{ref}. At the same time, the bottom plate of C_C samples $V_{\text{ref}}/2$ and the up/down counter is reset to make $V_{\text{cal}} = 0$. In the next clock period the charge is redistributed by swapping 0 and V_{ref} on the bottom plates of C_1 and C_2. Then the top plate residual error V_x will be from the charge conservation

$$V_x = \frac{C_1}{C_1 + C_2 + C_C} V_{\text{ref}} - \frac{C_2}{C_1 + C_2 + C_C} V_{\text{ref}} = \frac{\Delta C}{2C + C_C} V_{\text{ref}} \tag{64.8}$$

if $V_{\text{cal}} = 0$, $C_1 = C + \Delta C/2$, and $C_2 = C - \Delta C/2$. This top plate residual voltage can be nulled out by changing the calibration DAC voltage V_{cal}. The measured calibration voltage is approximately

$$V_{\text{cal}} = -\frac{\Delta C}{C_C} V_{\text{ref}} \tag{64.9}$$

As the actual error V_x is half of the measured value when the V_{ref} is applied to C_1, the actual calibration DAC voltage to be subtracted during normal operation becomes $V_{\text{cal}}/2$. Similarly, the multi-bit calibration can start from the MSB measurement and move down to the LSB side [9].

The extension of this calibration technique to current DACs is straightforward. For example, two identical unipolar currents, I_1 and I_2, can be compared using a voltage comparator and a calibration DAC, as shown in Fig. 64.16. After I_1 is switched in, the calibration DAC finds an equilibrium as a null. Then the difference can be measured by interchanging I_1 and I_2 and finding a new equilibrium. Therefore, the current difference error is obtained as

$$I_{\text{cal}} = I_2 - I_1 \tag{64.10}$$

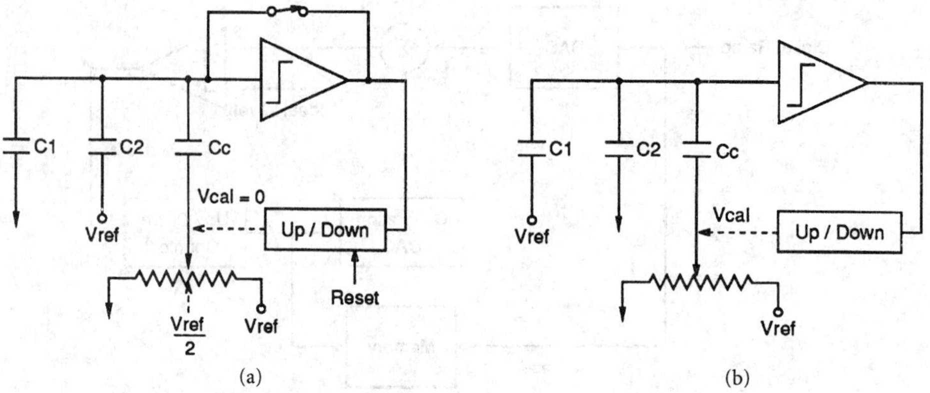

FIGURE 64.15 Capacitor ratio error measurement cycles: (a) initialization, and (b) error quantization.

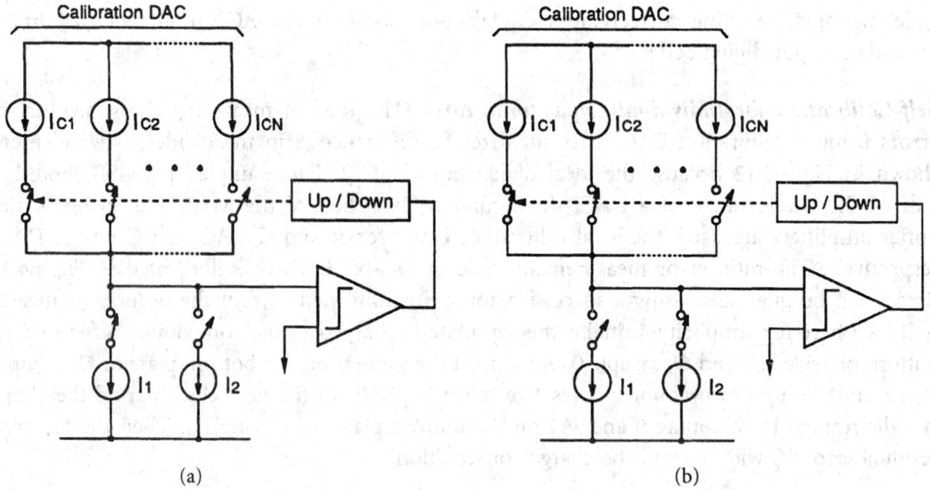

FIGURE 64.16 Current difference measurement cycles: (a) initialization, and (b) error quantization.

During normal operation, half of this value should be added to the DAC output using the same calibration DAC every time the current I_1 is switched to the DAC output. Similarly, the same amount is subtracted if I_2 is switched to the output.

Code-Error Calibration. The code-error calibration is based on the simple fact that the thermometer-coded MSBs of a DAC are made of segments of equal magnitude [10]. Any nonuniform segment will contribute to the overall nonlinearity of a DAC. The segment error between two adjacent input codes is measured by comparing the segment with the ideal segment. Starting from the reference point, 0 or $V_{ref}/2$, the same procedure is repeated until all the segment errors are measured. Therefore, the current code error, Error(j), is obtained by adding the current segment error to the accumulated sum of all the previous segment errors:

$$\text{Error}\,(j) = \sum_{k=1}^{j} \text{Seg}(k) \tag{64.11}$$

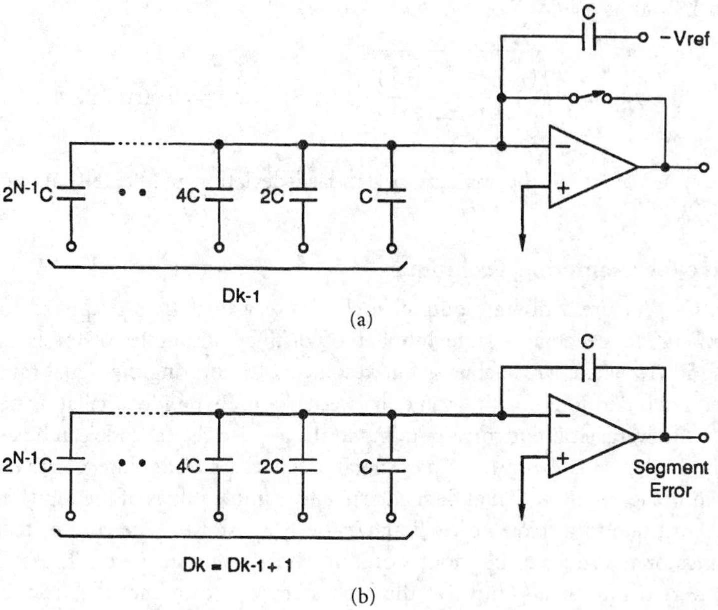

FIGURE 64.17 Code-error measurement cycle: (a) initialization, and (b) error quantization.

where Seg(k) is the kth segment error from the reference point. These measured code errors are stored in memory addressed by digital codes so that they can be subtracted from uncalibrated raw digital outputs during normal conversion.

The segment error measurement of a current-ratioed DAC with thermometer-coded MSBs is similar to the current difference measurement shown in Fig. 64.16. The only difference in measurement is the use of the reference segment current in place of one of the two currents to be compared. That is, each MSB current source is compared to the reference segment. For the capacitive DAC, the kth segment error can be measured in two cycles. After the output of the DAC is initialized to have a negative ideal segment voltage with the input digital code corresponding to $k - 1$, the input code is increased by 1 as shown in Fig. 64.17. Applying digital codes to the capacitor-array DAC means connecting the bottom plates to either V_{ref} or ground depending on the corresponding digital bits. Then the kth segment error is generated at the output and can be measured digitally using subsequent ADC stages or using a calibration DAC as shown in Fig. 64.15.

Digital Truncation Errors. All calibration methods need extra bits of resolution in the error measurements because digital truncation errors are accumulated during code-error computations. For example, if the truncation errors are random, the additions of n digital numbers will increase the standard deviation of the added number by $n^{1/2}$. This accumulated truncation error affects both the DNL and INL of the converter self-calibrated using measured errors of individual bits. On the other hand, if calibrated using measured segment errors, the DNL of the converter is always guaranteed to be within $\pm\frac{1}{2}$ LSB of a target resolution because all segment errors are measured with one extra bit of resolution, but the INL will still be affected by the digital truncation because code errors are obtained by accumulating segment errors. The effect of the digital truncation errors due to n repeated digital additions on the INL can be modeled using uncorrelated and independent random variables, and the standard deviation of INL is calculated in LSB units as

$$\sigma_{\text{INL}} = \sqrt{\frac{(n-i)(i-1)}{12(n-1)}} \text{ (LSB)} \quad \text{for } i = 1, 2, \ldots, n \tag{64.12}$$

For example, when $n = 16$, the maximum standard deviation of the INL at the midpoint is about 0.56 LSB.

Interpolative Oversampling Techniques

Ordinary DACs generate a discrete output level for every digital word applied to their input, and it is difficult to generate a large number of distinct output levels for long words. The oversampling interpolative DAC achieves fine resolution by covering the signal range with a few widely spaced levels and interpolating values between them. By rapidly oscillating between coarse output levels, the average output corresponding to the applied digital code can be generated with reduced noise in the signal band [8]. The general architecture of the interpolative oversampling DAC is shown in Fig. 64.18. A digital filter interpolates sample values of the input signal in order to raise the word rate to a frequency well above the Nyquist rate. The core of the technique is a digital truncator to truncate the input words to shorter output words. These shorter words are then converted into analog form at the high sample rate so that the truncation noise in the signal band may be satisfactorily low. The sampling rate upconversion for this is usually done in stages using two upsampling digital filters. The first filter, usually a two to four times oversampling FIR, is to shape the signal band for sampling rate upconversion and to equalize the passband droop resulting from the second SINC filter for higher-rate oversampling.

A noise-shaping delta-sigma modulator can be built in digital form to make a digital truncator as shown in Fig. 64.19. Using a linearized model, the z-domain transfer function of the modulator is

$$Y(z) = \frac{\alpha H(z)}{1 + \alpha H(z)} X(z) + \frac{1}{1 + \alpha H(z)} Q(z) \tag{64.13}$$

where $Q(z)$ is the quantization noise and α is the quantizer gain. The loop filter $H(z)$ can be chosen so that the quantization noise may be high-pass filtered while the input signal is low-pass filtered. For the first-order modulator, the loop filter is just an integrator with a transfer function of

$$H(z) = \frac{z^{-1}}{1 - z^{-1}} \tag{64.14}$$

while for the second-order modulator, the transfer function is

$$H(z) = \frac{z^{-1}(2 - z^{-1})}{(1 - z^{-1})^2} \tag{64.15}$$

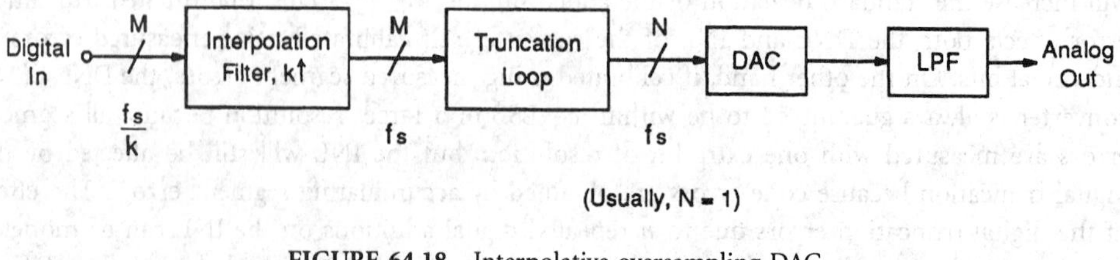

FIGURE 64.18 Interpolative oversampling DAC.

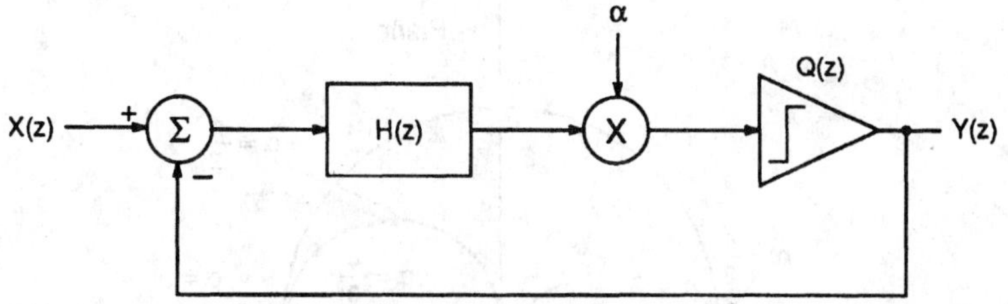

FIGURE 64.19 Delta-sigma modulation as a digital truncator.

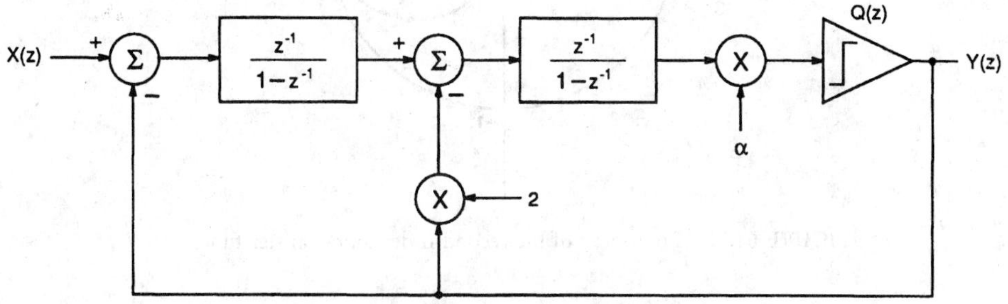

FIGURE 64.20 Second-order one-bit modulator.

However, the standard second-order modulator is implemented, as shown in Fig. 64.20, using a double integration loop. In general, first-order designs tend to produce correlated idling patterns. Second-order designs are vastly superior to first-order designs both in terms of the required oversampling ratio to achieve a particular signal-to-noise ratio as well as in the improved randomness of the idling patterns. However, even the second-order loop is not entirely free of correlated fixed patterns in the presence of small DC inputs. The second-order loop needs dithering to reduce fixed pattern noises, but loops of a higher order than third do not exhibit fixed pattern noises.

Stability. The quantizer gain α plays an important role in keeping the modulator stable. Considering α, the transfer function of the second-order loop shown in Fig. 64.20 becomes

$$Y(z) = \frac{\alpha z^{-2} X(z) + (1 - z^{-1})^2 Q(z)}{1 - 2(1-\alpha)z^{-1} + (1-\alpha)z^{-2}} \tag{64.16}$$

The root locus of the transfer function in the z-domain is shown in Fig. 64.21. As shown, the second-order loop becomes unstable for $\alpha > 4/3$ because one pole moves out of the unit circle. This in turn implies that the signal at the input of the quantizer becomes too large. Most delta-sigma modulators become unstable if the signal to the quantizer exceeds a certain limit. Higher-order modulators tend to be overloaded easily at higher quantizer gain than first- or second-order modulators. Therefore, for the stability reason, the integrator outputs of the loop filter are clamped so that the signal at the input of the quantizer is limited for linear operation. Digital truncators of a higher order than second are feasible in digital circuits because signal levels can be easily detected and controlled. The straightforward third order or higher loop using multiple loops is unstable, but higher order modulators can be built using either the cascaded MASH [11] or the single-bit higher order [12] architecture.

Dynamic Range. In general, for the Nth order loop, the noise falls by $6N + 3$ dB for every doubling of the sampling rate, providing $N + 0.5$ extra bits of resolution. Because the advantage

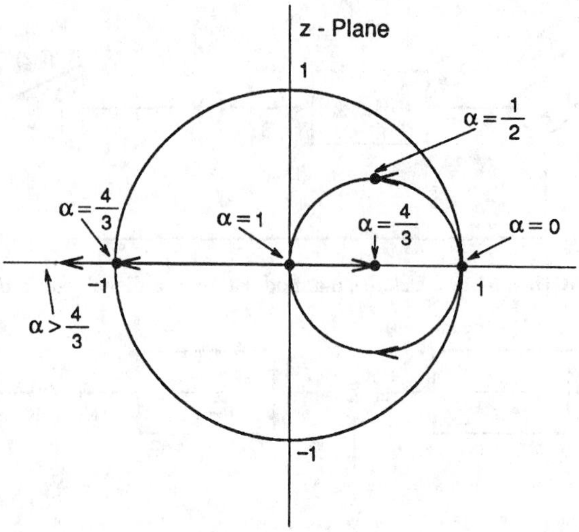

FIGURE 64.21 Root locus of the second-order loop transfer function.

of oversampling begins to appear when the oversampling ratio is > 2, a practically achievable dynamic range by oversampling is approximately

$$DR > (6N + 3)(\log_2 M - 1)\ \text{dB} \tag{64.17}$$

where M is the oversampling ratio. For example, a second-order loop with 256 times oversampling can give a dynamic range of > 105 dB, but the same dynamic range can be obtained using a third-order loop with only 64 times oversampling. The dynamic range is not a limit in the digital modulator. In practice, the dynamic range is limited in the rear-end analog DAC and postfilter.

One-Bit or Multi-Bit DAC. The rear end of the interpolative oversampling DAC is an analog DAC. Because the processing in the interpolation filter and truncator is digital instead of analog, achieving precision is easy. Therefore, the oversampling DAC owes its performance to the rear-end analog part because the conversion of the truncated digital words into analog form takes place in the rear-end DAC. The one-bit quantizer can be easily overloaded and needs clamping to be stable, while multi-bit quantizers are more stable due to their small quantization errors. However, the multi-bit system is limited by the accuracy of the multi-bit DAC. Although the analog techniques such as dynamic matching or self-calibration can improve the performance of the multi-bit DAC, the one-bit DAC is simpler to implement and its performance is not limited by component matching. It is true that a continuous time filter can convert the one-bit digital bitsteam into an analog waveform, but it is difficult to construct an ideal undistorted digital waveform without clock jitter. However, if the bitstream is converted into a charge packet, a high linearity is guaranteed due to the uniformity of the charge packets.

A typical differential one-bit switched-capacitor DAC with one-pole roll-off can be built as shown in Fig. 64.22 using two-phase nonoverlapping clocks 1 and 2. There are many advantages in a fully differential implementation. The dynamic range increases by 3 dB because the signal is doubled (6 dB) but the noise gains by 3 dB. It also rejects most noisy coupling through power supplies or through the substrate as a common-mode signal. Furthermore, the linearity is improved because the even-order nonlinearity components of the capacitors and the op amp are canceled. In the implementation of Fig. 64.22, a resistor as a loss element can be replaced by a capacitor switched in and out at f_c as illustrated. The bandwidths of these filters in both

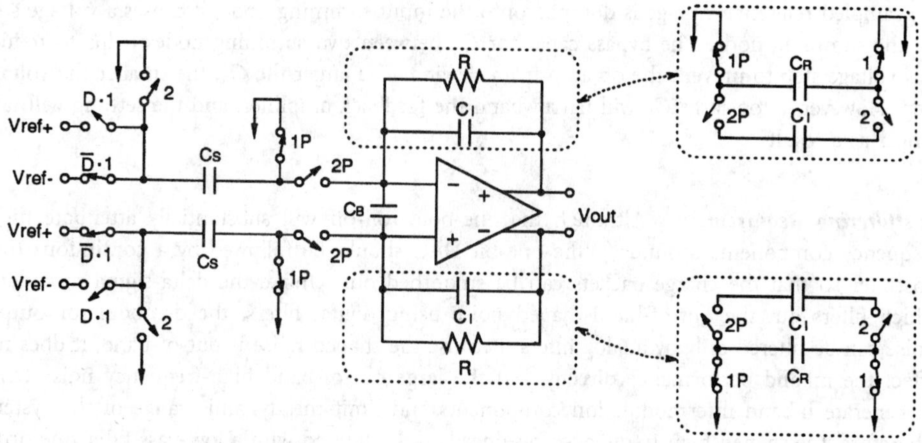

FIGURE 64.22 Switched-capacitor one-bit DAC/filter.

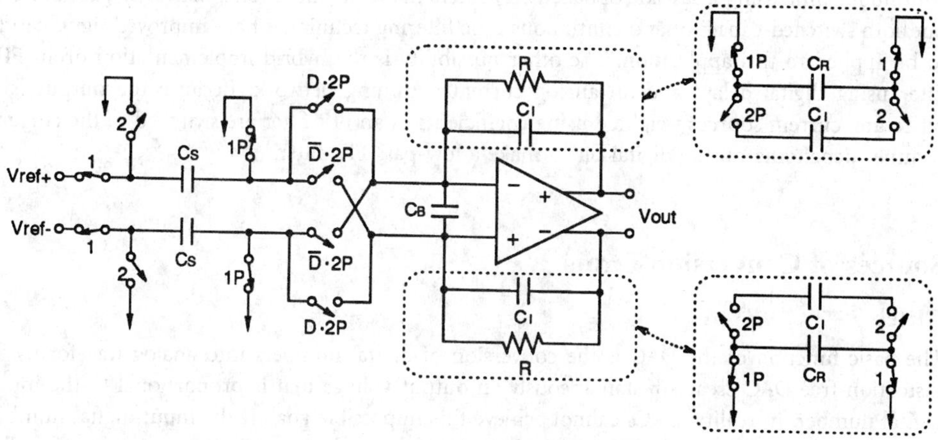

FIGURE 64.23 Alternative one-bit DAC sampling constant V_{ref}.

cases are set by $1/RC_I$ and $f_c C_R/C_I$, respectively. Also, the filter DC gains are defined by $Rf_c C_S$ and C_S/C_I, respectively. The digital bitstream is converted into a charge packet by sampling the reference voltage on the bottom plates of the sampling capacitors (C_S). If the digital bit is ZERO, $-V_{ref}$ is sampled during the clock phase 1 and the charge on C_S is dumped on the lossy integrator during 2. On the other hand, if the digital bit is ONE, V_{ref} is sampled instead. To reduce the input-dependent switch-feedthrough component, the switches connected to the top plates should be turned off slightly earlier than the bottom plate switches using $1p$ and $2p$. Alternatively, a slightly different one-bit DAC is possible by sampling a constant reference voltage by inverting the polarity of the integration depending on the digital bit as shown in Fig. 64.23.

The op amp for this application should have a high DC gain and a fast slew rate. The op amp DC gain requirement is a little alleviated considering the linear open-loop transfer characteristic of most op amps within a limited swing range. As discussed earlier, the slew-limited settling generates an error proportional to the square of the magnitude. Therefore, a nonslewing op amp such as a class AB input op amp performs better for this application. The op amp starts to slew when a larger voltage than its linear input range is applied. When the charge packet of the sampled reference voltage is dumped onto the input summing node, it causes a voltage step on the summing node. The bypass capacitor C_B between two summing nodes helps to reduce this voltage step to prevent the op amp from slewing. The larger the C_B, the smaller the voltage step. However, a too-large C_B will narrowband the feedback amplifier, and the settling will take longer as a result.

Postfiltering Requirement. Although the one-pole roll-off will substantially attenuate high-frequency components around f_c, the one-bit DAC should be followed by a continuous time postfilter so that the charge packets can be smoothed out. Unlike the delta-sigma modulator which filters out the out-of-band-shaped noise using digital filters, the demodulator output noise can be filtered only by analog filters. Because the shaped noise is out-of-band, it does not affect the inband performance directly, but the large out-of-band high-frequency noise tends to generate inband intermodulation components and limit the dynamic range of the system. Therefore, the shaped high-frequency noise needs to be filtered with a low-pass filter one order higher than the order of the modulator. It is challenging to meet this postfiltering requirement with analog filtering techniques. Analog filters for this application are often implemented in continuous time using a cascade of Sallen-Key filters made of emitter follower unity-gain buffers, but both switched-capacitor and continuous time filtering techniques have improved significantly to be applied to this application. The other possibility is the hybrid implementation of an FIR filter using digital delays and an analog current summing network. Because the output is a bitstream, current sources weighted using coefficients of an FIR filter are switched to the current summer depending on the digital bit to make a low-pass FIR filter.

Sources of Conversion Errors

Glitch

The basic function of the DAC is the conversion of digital numbers into analog waveforms. A distortion-free DAC creates instantaneously an output voltage that is proportional to the input digital number. In reality, DACs cannot achieve this impossible goal. If the input digital number changes from one value to a different one, the DAC output voltage always reaches a new value sometime later. For DACs, the shape of the transient response is a function governed in large part by two mechanisms, glitch and slew rate limit. The ideal transient response of a DAC to a step is a single-time constant exponential function, which only generates an error growing linearly with the input signal, as explained in Fig. 64.3. Any other transient responses give rise to errors that have no bearing on the input signal. The glitch impulse is described in terms of a picovolts times seconds or equivalent unit.

Glitches are caused by small time differences between some current sources turning off and others turning on. Take, for example, the major code transition at half scale from 011...11 to 100...00. Here, the MSB current source turns on while all other current sources turn off. The small difference in switching times results in a narrow half-scale glitch, as shown in Fig. 64.24. Such a glitch, for example, can produce distorted characters in CRT display applications. To alleviate both glitch and slew-rate problems related to transients, a DAC is followed by a S/H amplifier, usually called a deglitcher. The deglitcher stays in the hold mode while the DAC changes its output value. After the switching transients have settled, the deglitcher is changed to the sampling mode. By making the hold time suitably long, the output of the deglitcher can be made independent of the DAC transient response. Thus, the distortion during transients can be circumvented by using a fast S/H amplifier. However, the slew rate of the deglitcher is on the same order as that of the DAC, and the transient distortion will still be present, now as an artifact of the deglitcher.

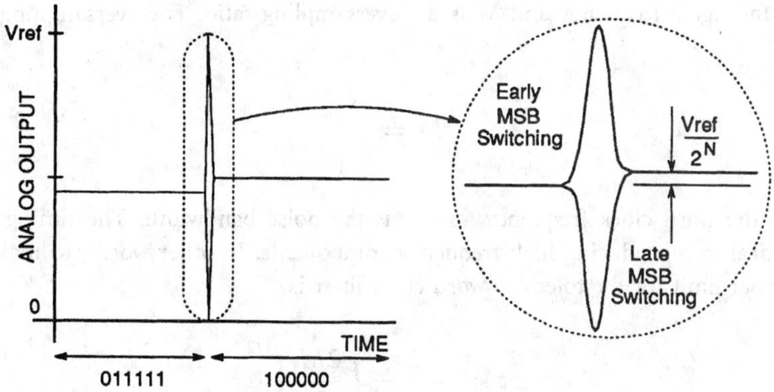

FIGURE 64.24 Glitch impulse at a major carry.

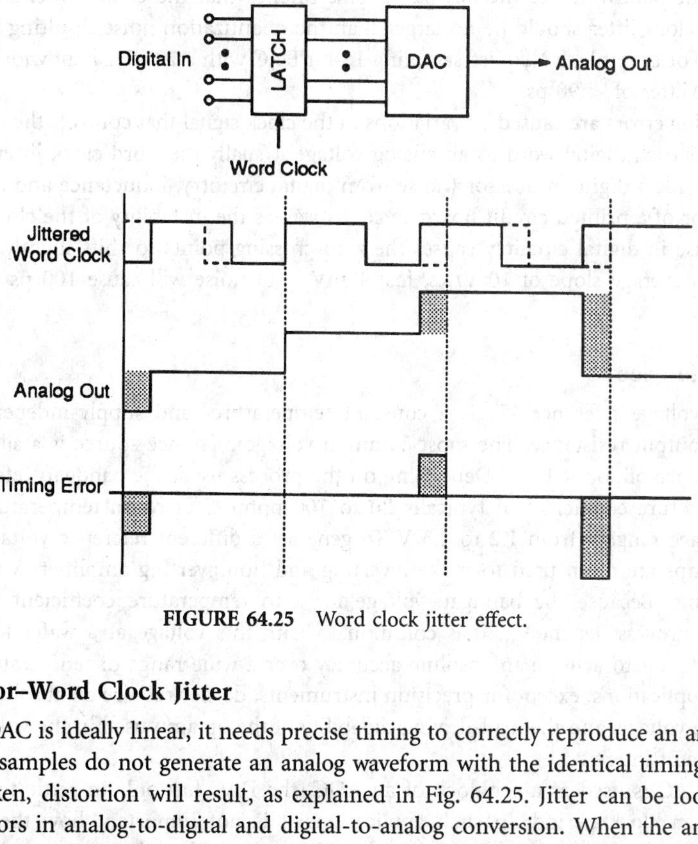

FIGURE 64.25 Word clock jitter effect.

Timing Error–Word Clock Jitter

Although a DAC is ideally linear, it needs precise timing to correctly reproduce an analog output signal. If the samples do not generate an analog waveform with the identical timing with which they were taken, distortion will result, as explained in Fig. 64.25. Jitter can be loosely defined as timing errors in analog-to-digital and digital-to-analog conversion. When the analog voltage is reconstructed using a DAC with timing variations in the word clock, the sample amplitudes, the ONEs and ZEROs are correct, but they come out at the wrong time. Because the right amplitude at the wrong time is the wrong amplitude, a timing jitter in the word clock produces an amplitude variation in the DAC output, causing the waveform to change shape. This in turn introduces either spurious components related to the jitter frequency or raises the noise floor of a DAC, unless the jitter is periodic. If the jitter has a Gaussian distribution with a root mean square jitter of Δt, the worst-case SNR resulting from this random word clock jitter is

$$\mathrm{SNR} = -20 \times \log \frac{2\pi f \Delta t}{M^{1/2}} \tag{64.18}$$

where f is the signal frequency and M is the oversampling ratio. The oversampling ratio M is defined as

$$M = \frac{f_c}{2f_n} \tag{64.19}$$

where f_c is the word clock frequency and f_n is the noise bandwidth. The timing jitter error is more critical in reproducing high-frequency components. In other words, to make an N-bit DAC, an upper limit for the tolerable word clock jitter is

$$\Delta t < \frac{1}{2\pi B 2^N}\left(\frac{2M}{3}\right)^{1/2} \tag{64.20}$$

where B is the bandwidth of the baseband. This implies that the error power induced in the baseband by clock jitter should be no larger than the quantization noise resulting from an ideal N-bit DAC. For example, a Nyquist-sampling 16-bit DAC with a 22-kHz bandwidth should have a word clock jitter of < 90 ps.

These timing errors are caused by variations in the clock signal that controls the time when the DAC converts each digital word to an analog voltage. Usually the word clock jitter results from everything inside a digital processor (noise from digital circuitry, inductance and capacitance of a clock bus or of a printed circuit board trace) as well as the instability of the clock source. For example, noise in digital circuitry causes the zero-crossing points to shift slightly. If the digital signal has an average slope of 10 V/μs, just 1 mV_{rms} of noise will cause 100 ps of root mean square jitter.

Voltage Reference

Ideally, the voltage reference V_{ref} is a constant temperature- and supply-independent voltage with a zero output resistance. The most common voltage reference source is a silicon bandgap voltage reference of about 1.2 V. Depending on the process used, the bandgap reference voltage has a temperature coefficient of typically 20 to 100 ppm/°C at room temperature when it is set to a voltage ranging from 1.2 to 1.3 V. To generate a different reference voltage other than 1.2 V, op amps are often used to make inverting and noninverting amplifiers with trimmable feedback gains. Because the bandgap voltage for zero temperature coefficient is not clearly defined and process dependent, it is common to trim this voltage at a wafer level, and it is extremely difficult to achieve an absolute accuracy over a wide range of temperature. However, most DAC applications, except for precision instruments, do not require an absolute accuracy of the reference voltage, and the load-driving capability of the op amp used in the voltage reference is of paramount interest.

When a DAC is used as a subblock of an ADC, the DAC has only to settle to a final value within a given clock period. In such applications it is not important how the DAC settles. However, the DAC as a stand-alone device should have a single-pole response without a glitch and a slew limit. Except for the current-ratioed DAC, the voltage reference is periodically loaded and disturbed by a switched-in load at the clock rate, and the reference output should be restored immediately so that the DAC can settle fast. When disturbed, the voltage reference needs to settle with a time-constant shorter than that of the DAC with an unlimited slew rate. Figure 64.26 illustrates this situation when the voltage reference periodically refreshes the loading capacitor. If the op amp unity-gain bandwidth is $1/\tau_0$ rad/s, the op amp restores the output with a time constant $(1 + R_2/R_1)\tau_0$ of the feedback amplifier. The large capacitor C_D helps to prevent the op amp from slewing because the voltage dip of $V_{ref}C_L/(C_D + C_L)$ decreases as C_D increases.

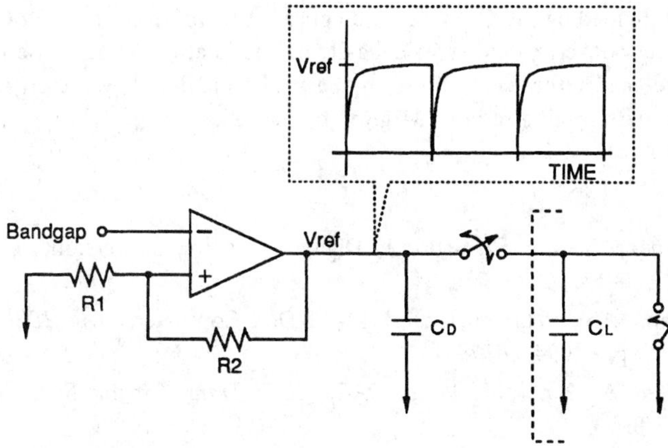

FIGURE 64.26 Voltage reference periodically charging a load capacitor.

Noise

Noise is a fundamental limit in high-resolution DACs. The resolution of a DAC is limited by a quantization noise given by (64.2), unless the circuit contributes higher noise. There are many noise sources in real DACs. The DAC output is corrupted by noises directly coupled from bouncing power supplies, bias lines, and ground, as well as the noise sources such as thermal noise, flicker noise, and shot noise, etc. To reduce the former noise sources, it is necessary to carefully separate analog and digital supplies and to eliminate sets of ground loops using a star-ground configuration. The use of a fully-differential architecture helps to reduce such coupling noises. However, the latter set of noise sources is predictable and can be reduced by a low-noise design.

The dominant noise source of a resistor-string DAC shown in Fig. 64.4 is a white thermal noise of resistors. If the output resistance of the DAC is R_{out}, the root mean square output noise of the DAC is $(4kTR_{out})^{1/2}$ V/Hz$^{1/2}$, where k is the Boltzman constant and T is the absolute temperature. The thermal noise is about 4 nV/Hz$^{1/2}$ if R_{out} is 1 kΩ. The output resistance of this DAC depends on the digital input code. The worst-case R_{out} is one quarter of the total resistance of the string when the output is connected to the center tap. For lower noise, the total resistance should be minimized at the cost of high power consumption. The noise of the voltage reference appears at the output after being divided by the resistor string. The voltage noise of the output buffer, if used, is directly added to this noise.

On the other hand, the noise source of the current-ratioed DAC of Figs. 64.6 and 64.7 is the current noise contributed either by the shot noise of the current source or by the parallel resistor connected to the output node. If the total output current is I_{out} and the total output shunt resistance is R_p, the root mean square output shot noise is $2\ qI_{out}$ A/Hz$^{1/2}$ and the root mean square current noise contributed by the shunt resistor is $4\ kT/R_p$ A/Hz$^{1/2}$, where q is 1.6×10^{-19} C. The shot noise of a 50-μA current and the current noise of a 1-kΩ shunt resistor give the same root mean square noise of 4 pA/Hz$^{1/2}$. The shunt feedback resistance of the transresistance amplifier also contributes to the total noise. If the current DAC is terminated by a low-impedance source of 50 to 75 Ω, the current noise is usually dominated by this termination resistor.

The main source of the capacitor-array DAC shown in Figs. 64.8 and 64.9 is a well-known kT/C noise. The sampled root mean square noise voltage of the capacitor of C is $(kT/C)^{1/2}$ V without regard to the switch-on resistance. The 1-pF sampling capacitor gives a 64-μV sampled noise. This noise is evenly distributed over the Nyquist band (half the sampling frequency), and puts a lower limit on the smallest signal the DAC can handle. Furthermore, the reference and ground noises are divided by the C divider and appear at the output of the DAC. When applied to the oversampling converters either in delta-sigma modulators or in a one-bit DAC, shown in Figs. 64.22 and 64.23, the kT/C noise in the signal band is lower than the Nyquist-rate converters by the oversampling ratio of M given by (64.19).

References

[1] D. M. Freeman, "Slewing Distortion in Digital-to-Analog Conversion," *J. Audio Eng. Soc.*, vol. 25, pp. 178–183, 1977.

[2] R. B. Craven, "An Integrated Circuit 12-Bit D/A Converter," *Dig. IEEE Int. Solid-State Circuits Conf.*, pp. 40, 41, 1975.

[3] D. T. Comer, "A Monolithic 12-Bit DAC," *IEEE Trans. Circuits Syst.*, vol. CAS-25, pp. 504–509, 1978.

[4] J. M. McCreary and P. R. Gray, "All-MOS Charge Redistribution Analog-to-Digital Conversion Techniques. I," *IEEE J. Solid-State Circuits*, vol. SC-10, pp. 371-379, 1975.

[5] R. J. Van de Plassche, "Dynamic Element Matching for High Accuracy Monolithic D/A Converters," *IEEE J. Solid-State Circuits*, vol. SC-11, pp. 795–800, 1976.

[6] M. J. M. Pelgrom and M. Roorda, "An Algorithmic 15-Bit CMOS Digital-to-Analog Converter," *IEEE J. Solid-State Circuits*, vol. SC-23, pp. 1402–1405, 1988.

[7] D. W. J. Groeneveld, H. J. Schouwenaars, H. A. H. Termeer, and C. A. A. Bastiaansen, "A Self-Calibration Technique for Monolithic High-Resolution D/A Converters," *IEEE J. Solid-State Circuits*, vol. SC-24, pp. 1517–1522, 1989.

[8] J. C. Candy and A. N. Huynh, "Double Interpolation for Digital-to-Analog Conversion," *IEEE Trans. Commun.*, vol. 34, pp. 77–81, 1986.

[9] H. S. Lee, D. A. Hodges, and P. R. Gray, "A Self-Calibrating 15 Bit CMOS A/D Converter," *IEEE J. Solid-State Circuits*, vol. SC-19, pp. 813–819, 1984.

[10] S. H. Lee and B. S. Song, "Digital-Domain Calibration of Multistep Analog-to-Digital Converters," *IEEE J. Solid-State Circuits*, vol. SC-27, pp. 1679–1688, 1992.

[11] T. Hayashi, Y. Inabe, K. Uchimura, and T. Kimura, "A Multistage Delta-Sigma Modulator without Double Integration Loop," *Dig. IEEE Int. Solid-State Circuits Conf.*, pp. 182, 183, 1986.

[12] W. L. Lee and C. G. Sodini, "A Topology for Higher-Order Interpolative Coder," *Proc. Int. Symp. Circuits Syst.*, pp. 459–462, 1987.

64.2 Analog-to-Digital Converters

Ramesh Harjani

Introduction

With the increased complexity possible in modern-day integrated circuits analog-to-digital (ADC) and digital-to-analog (DAC) converters have become ubiquitous components of mixed-signal integrated circuits. Analog-to-digital converters transform an analog signal, V_A, into an N-bit digital representation, V_d. Such a converter is said to have a resolution of N bits. The digital signal V_d is an approximation of the original analog signal, V_A, and the maximum error during this conversion process for an N-bit converter is equal to $1/2^N$ of the full-scale value. This error is called the quantization error.

A number of topologies exist for ADC. They can be classified as Nyquist rate converters or oversampled converters. Nyquist rate converters sample the input at the minimum sampling rate, i.e., two times the maximum signal frequency. As the name implies, oversampled converters take more samples than is mandated by the Nyquist criterion to generate extremely high resolution. Nyquist rate converters are usually classified as (1) high-speed, (2) medium-speed, and (3) high-resolution converters. However, the different architectures are better categorized by evaluating the number of clock cycles they use to perform the analog-to-digital conversion. For example, for N-bits of resolution a high-speed converter performs the conversion in one or two clock cycles, a medium-speed converter performs the conversion in N clock cycles, and a high-resolution converter performs the conversion in 2^N clock cycles. Thus, we classified them here as 1-clock converters, N-clock converters and 2^N-clock converters.

Before describing the details of the various ADC architectures we discuss some of the performance characteristics of ADCs in general and describe some test techniques that are used to measure these characteristics.

Analog-to-Digital Converter Test Techniques

Analog-to-digital converters are primarily tested using parametric techniques. That is, a key set of parameters which characterize an ADC are verified. An ADC is characterized by its static and dynamic performance. In static performance we are primarily concerned about the linearity of the I-O transfer characteristics, and in dynamic performance we are concerned about the operation of the converter at full operating speed.

Figure 64.27 shows the transfer characteristics for an ideal 3-bit converter. The dashed line shows the transfer characteristics of an infinite precision converter and the bold line shows the transfer characteristic of a 3-bit version. We note that the least significant bit (LSB) value is equal to $1/2^N$ of the full-scale value. Figures 64.28 to 64.31 show examples of the static performance characteristics of an ADC. A gain error is said to be present when the maximum digital value does not correspond to the full-scale analog value. An offset error corresponds to a horizontal shift in the transfer characteristics. The integral nonlinearity error specifies the maximum deviation of the transfer characteristics from the ideal code center. Differential nonlinearity specifies the deviation of each stepsize from $1/2^N$ of the full-scale value.

Dynamic ADC performance characteristics include signal-to-noise ratio (SNR), effective bits, aperture errors, and input signal bandwidth. During full-speed operation some additional errors become evident because of the finite settling time and bandwidth limitations of the circuits within the ADC. The SNR and effective bits are dynamic ways of measuring the minimum resolution and errors in the transfer characteristics. Different terms are used to specify similar performance parameters largely because different measurement techniques are used to generate them. The input signal bandwidth specifies the maximum frequency of the input signal.

Because of the large number of performance specifications for ADCs quite a few techniques are used to test them as well. These testing techniques also reflect the different types of

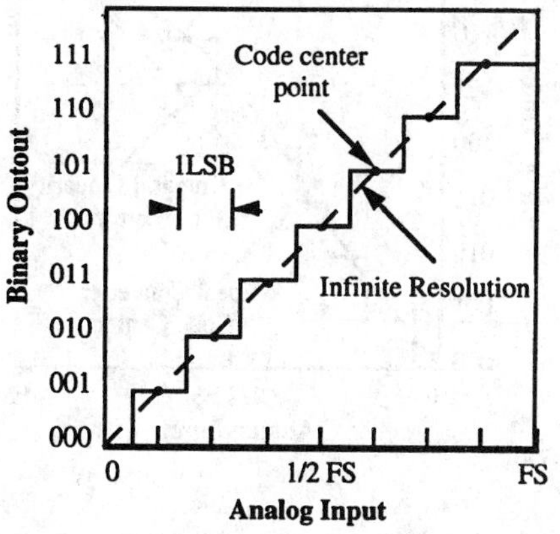

FIGURE 64.27 Transfer characteristics for an ideal 3-bit ADC.

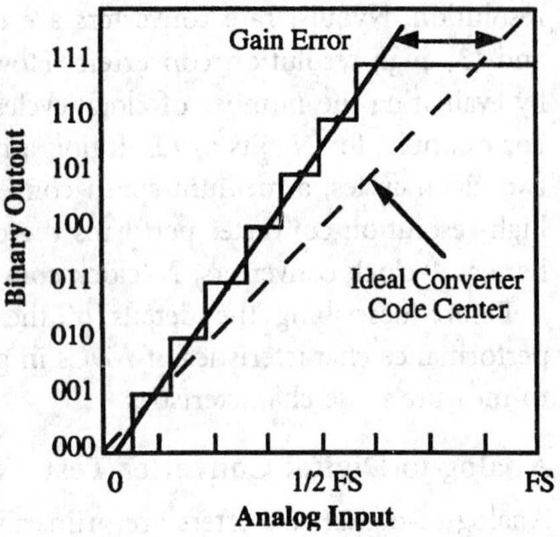

FIGURE 64.28 Gain error in an ADC.

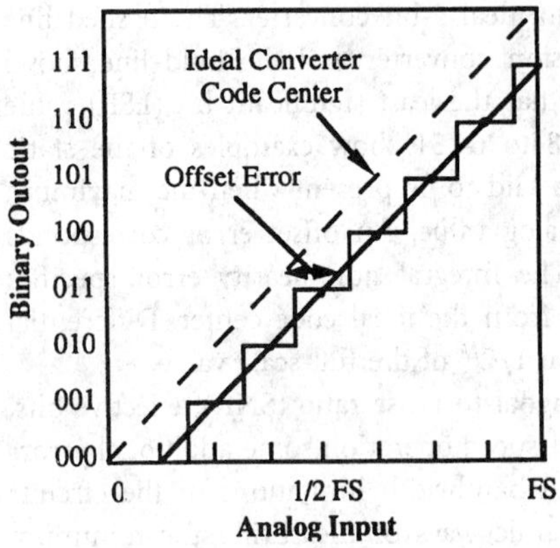

FIGURE 64.29 Offset error in an ADC.

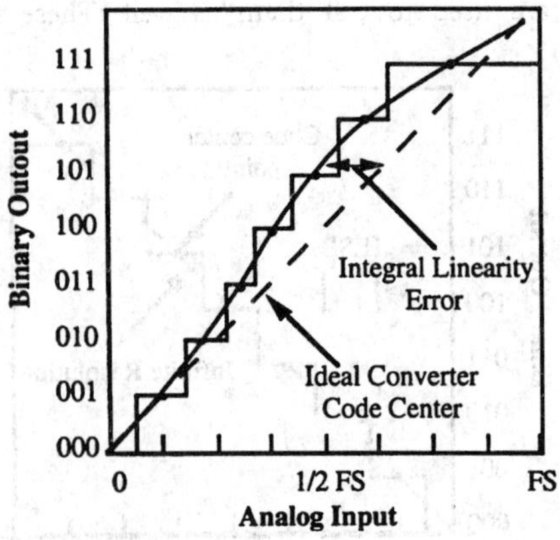

FIGURE 64.30 Integral nonlinearity error.

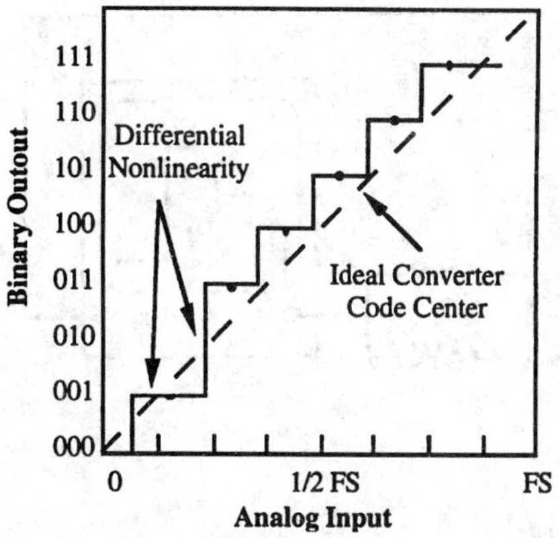

FIGURE 64.31 Differential nonlinearity error.

performance characteristics, i.e., static and dynamic. The more traditional methods of testing ADC are primarily concerned with checking the static characteristics of these converters. Examples of such test techniques include analog difference signal methods, crossplot methods, and servo loop code transition measurement methods [3]. Dynamic techniques to test linearity are based on using a well-known and near-full-scale input signal and evaluating the output code probability over a large input sample size. This technique is called the **code density test** or the **histogram test**. The output code probability for a linear ramp input is uniform for all codes. However, it is difficult to generate extremely accurate ramp signals at high speed, therefore, sine-wave signals are used. Unfortunately, with a sine wave the output code probability is no longer uniform but is instead "cuspshaped". An estimate of the differential nonlinearity can be generated by evaluating the difference between the expected code probability and measured values. The code density test requires a large number of data samples in the range of several hundreds of thousands. Additionally, a small number of large magnitude errors are easily masked in this technique. More recently, frequency domain-based techniques have been used to measure the harmonic content of the converted signal to provide an estimate of the SNR and effective number of bits. In this technique a discrete time Fourier transform (DTFT) is performed on the output data sequence and is used to measure the data converter performance characteristics. Although the number of data samples required are fewer than the code density approach, this approach still requires a few thousand data samples. Additionally, proper windowing functions and synchronized sampling may be required to reduce spectral leakage.

Data converters are relatively time consuming to test. However, rather than explain all the techniques that can be used to test data converters, we concentrate on a single method, the crossplot method, to provide the reader with some insight into the complexity of testing ADCs. For additional details about the other methods, readers are referred to [3]. A block diagram for the crossplot method is shown in Fig. 64.32. In the crossplot technique the output of the lowest two or three bits of an ADC are fed to the Y input of an oscilloscope and a separate triangular *dither* signal is fed to the X input of the oscilloscope. The input to the ADC is generated using a discrete summing amplifier that adds the output of a DAC, which is itself swept through its input space at a much slower rate, and the dither signal, mentioned earlier. This technique generates a staircase waveform on the oscilloscope. Therefore, the linearity of the last two or three bits around a fixed bias voltage, which is generated by the DAC, can be seen easily on an oscilloscope. The primary advantage of this technique is that it uses fairly inexpensive equipment. However, it is extremely time consuming to evaluate all 2^N combinations and the technique provides no information about the integral linearity, offset, and gain errors. In general, the techniques that are used to test ADCs are fairly complex and either

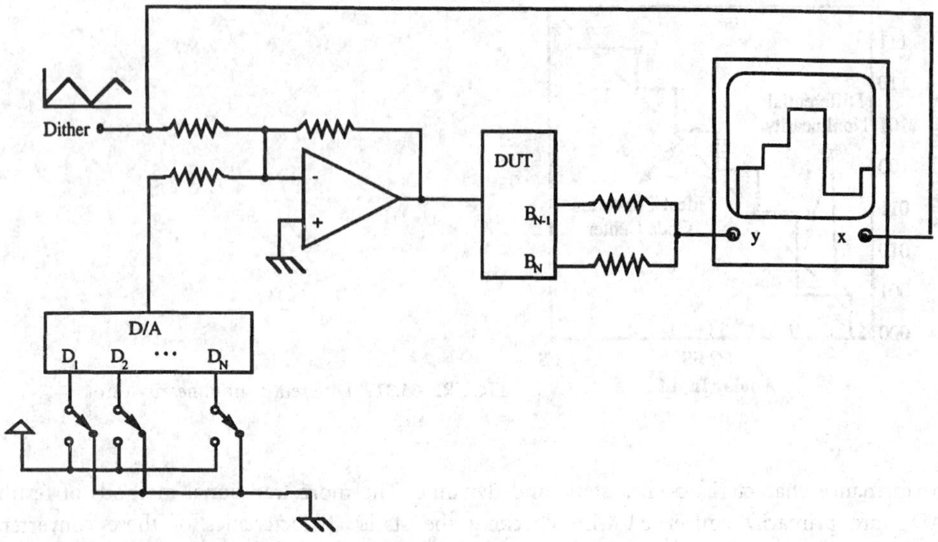

FIGURE 64.32 Crossplot technique used to test ADC static characteristics. (*Source:* M. J. Demler, *High-Speed Analog-to-Digital Conversion.* New York: Acadmic Press, 1991, p. 162.)

require a large sample size, i.e., high cost, or use extremely complex equipment to perform the tests.

Having described the basic characteristics of ADC we now describe some of the different ADC architectures in detail. We first consider Nyquist rate converter topologies and then describe oversampled converters.

Nyquist Rate Converters

As mentioned earlier, Nyquist rate converters are classified as 1-clock converters, N-clock converters, and 2^N-clock converters. Examples of 1-clock converters include the flash architecture, the pipelined architecture [11, 12], and the voltage folding architecture. Examples of N-clock converters include the successive-approximation architecture [9] and the algorithmic architecture [6]. Examples of 2^N-clock converters include the single-slope and dual-slope architectures [4].

1-Clock Converters

Flash Converters. Flash or parallel converters are the highest rate converters and are sometimes also called **video rate converters** because they operate at rates necessary for video signals. Parallel converters are O(1)-clock converters in that they require one or two clock cycles to perform an N-bit conversion. The basic principle of operation is fairly simple and is easily explained with the help of Fig. 64.33. This figure shows a block diagram for a 3-bit flash converter. The input is compared to $(2^3 - 1)$ comparators. The comparison voltage for the ith comparators is set to be equal to $(iV_{FS})/(2^3)$, where V_{FS} is the full-scale voltage. The resulting outputs are then converted from thermometer code to binary code. In general, the time required to perform the analog-to-digital conversion is equal to the comparator resolution time plus the time taken to perform the digital code conversion, i.e., one clock cycle. The reference voltages for the comparators are usually generated using an equally spaced resistor string that is offset by $R/2$, as shown in Fig. 64.34. The primary advantage of providing the $R/2$ offset is that it allows the code center to pass through the origin. The resolution of integrated MOS flash converters is limited to approximately 8 bits. This is primarily due to the matching constraints placed on the resistors and also because of the exponential increase in the number of components ($\approx 2^N$)

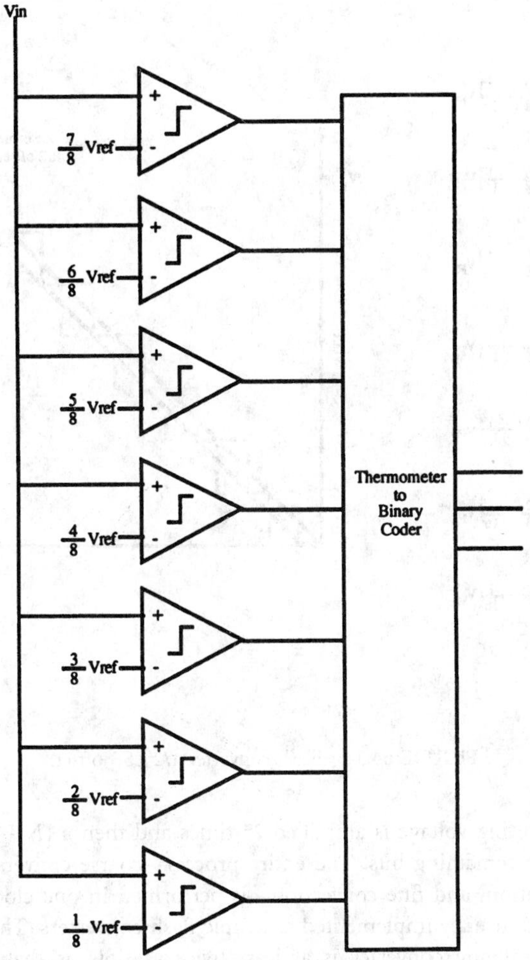

FIGURE 64.33 3-bit flash converter block diagram.

required for higher resolution. The speed for MOS flash converters is usually limited to about 50 MHz. However, if higher speed operation is required, then the process of time interleaving may be employed. Here, M (N-bit) ADCs are operated in parallel, but each is delayed somewhat, as shown in Fig. 64.35 [1]. By starting the conversion operation for A/D_2 before the operation of A/D_1 is complete allows the complete interleaved converter to operate at M times the speed of each individual converter. Although the technique allows for higher operating speed, the number of parallel paths has increased. Therefore, due to size limitations the resolution is usually lower than for simple flash converters.

Subranging Converters. The exponential area penalty ($\propto 2^N$) for flash converters can be mitigated by using subranging techniques without incurring severe speed penalties. Here, instead of deciding all N-bits at one time, only a subset of these is decided during the first clock phase, and the rest are decided in the following clock phases. If two clock phases are used to decide all N-bits, this is called a two-step subranging converter topology.

A block diagram for a two-step subranging converter is shown in Fig. 64.36. In this converter topology the α most significant bits (MSBs) are decided by the coarse ADC, the results of which are then passed on to an α-bit DAC. The output voltage of the DAC is subtracted from the

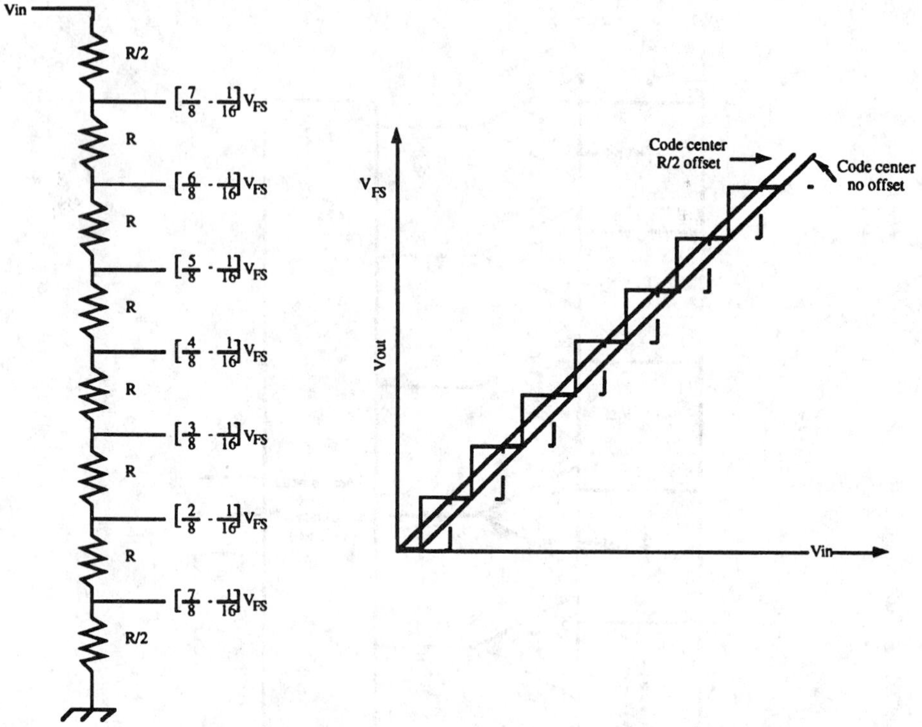

FIGURE 64.34 Flash converter 1/2 LSB offset.

input signal. The resulting voltage is amplified 2^{α} times and then a $(N - \alpha)$-bit fine converter is used to resolve the remaining bits. The entire process—coarse conversion, digital-to-analog subtraction, amplification, and fine conversion—is performed in one clock period. The coarse and fine converters are usually implemented as simple flash converters. Therefore, the operating speed of a pure subranging converter is at least twice as slow as that of a flash converter. However, the total number of comparators (and associated circuits) is reduced from 2^N to $2^{\alpha} + 2^{N-\alpha}$. The savings in area can be substantial. Usually the value of α is selected to be roughly equal to $N/2$. For example, for $N = 10$ and selecting $\alpha = 5$, the number of comparators reduce from $2^{10} = 1024$ to $2^6 = 64$, which is a savings of 93.75%. This subranging process can be extended further to a larger number of sequential stages; however, the total propagation delay usually limits the operation speed fairly significantly.

Pipelined Converters. Instead of operating both the coarse and fine converter in the same clock period a sample-and-hold (S/H) could be added to the circuit in Fig. 64.36, such that when the fine converter is resolving the lower $(N - \alpha)$ bits for the first input sample the coarse converter can begin resolving the upper α bits for the next input sample. This kind of converter can be made to operate at speeds comparable to flash converters and is called a pipelined converter. A block diagram for a two-stage pipelined converter is shown in Fig. 64.37. Because the α MSBs are decided during the previous clock period, the results from the fine and coarse converters need to be synchronized by delaying the output of the coarse converter by a single clock period. As with the subranging converter the number of comparators have decreased from 2^N to $2^{N-\alpha} + 2^{\alpha}$.

Unlike the subranging topology, the pipeline methodology can easily be extended to N sequential stages, each resolving only 1 bit, because of the sample-and-held circuits between each of the stages. The block diagram for such an N-step converter is shown in Fig. 64.38.

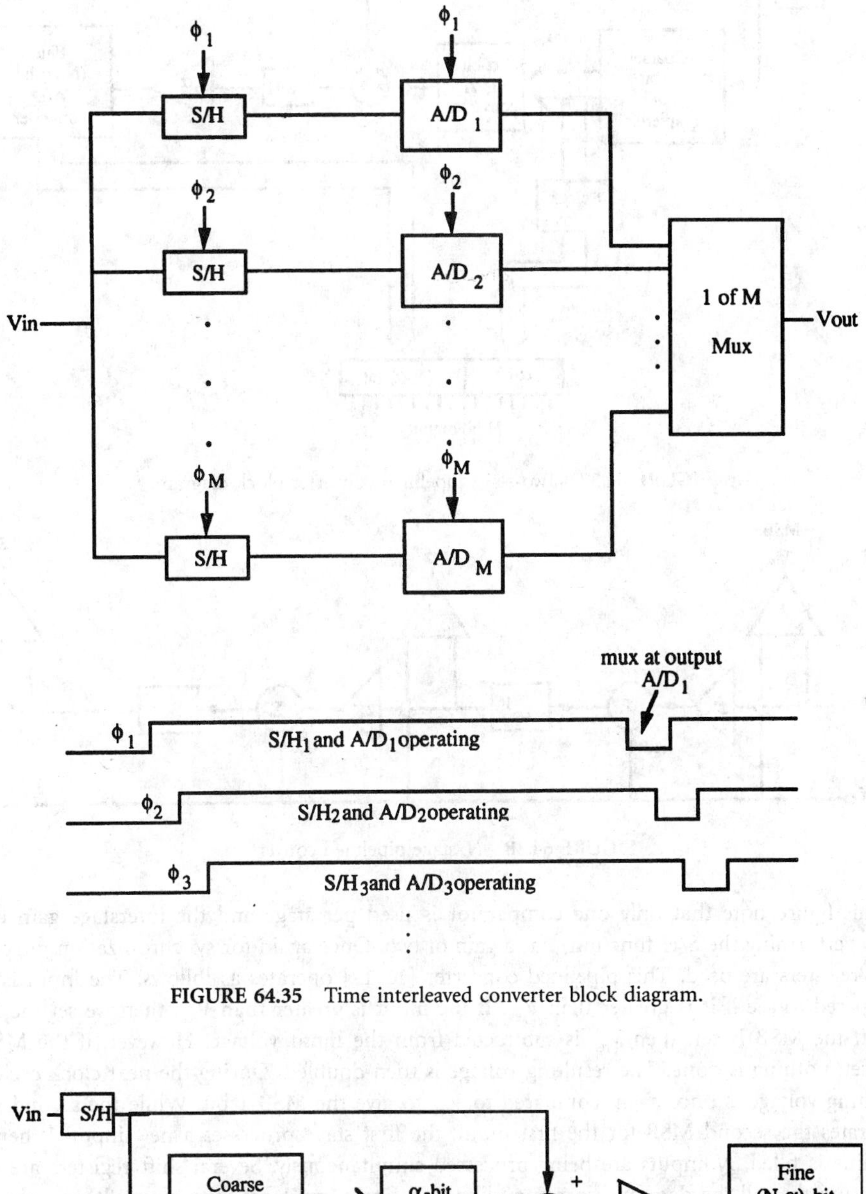

FIGURE 64.35 Time interleaved converter block diagram.

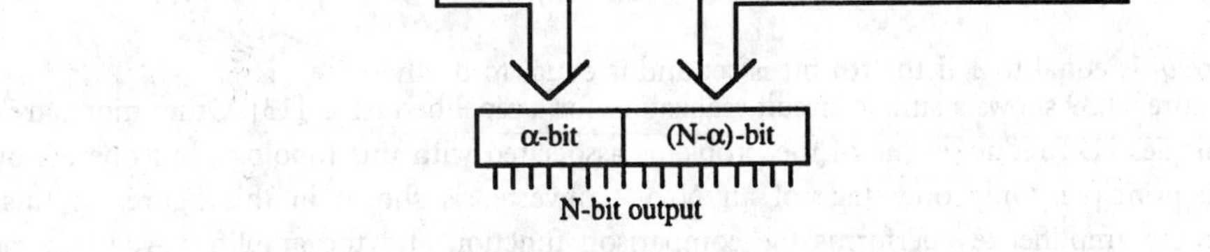

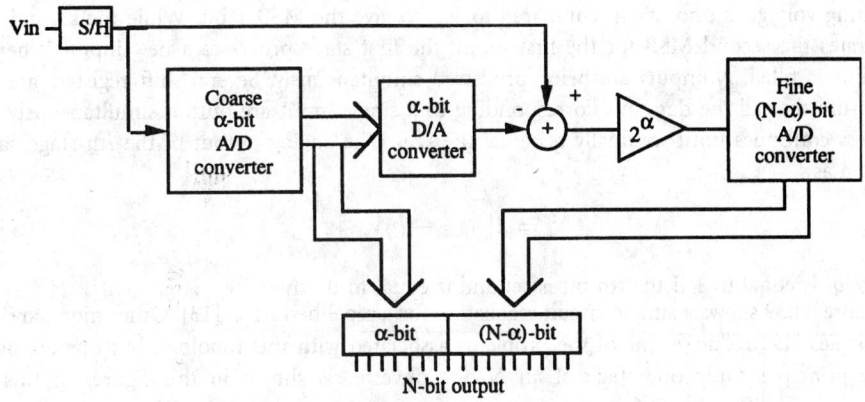

FIGURE 64.36 Subranging converter block diagram.

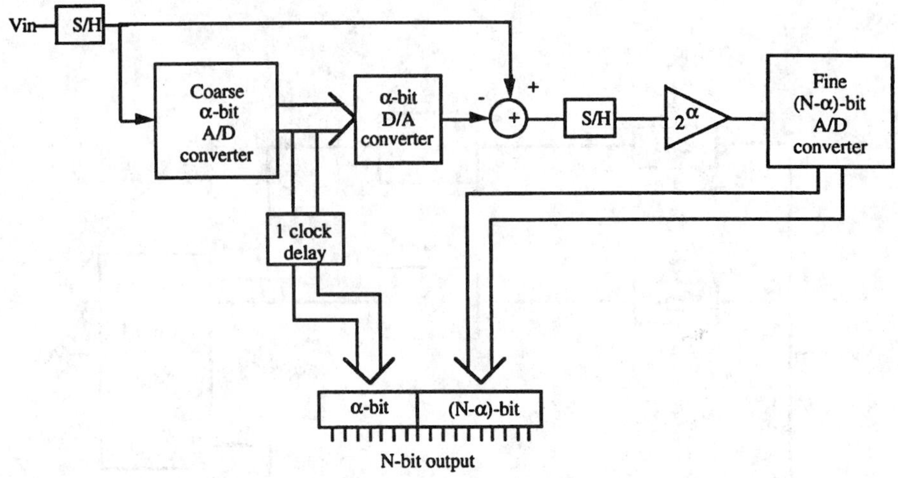

FIGURE 64.37 Two-stage pipelined converter block diagram.

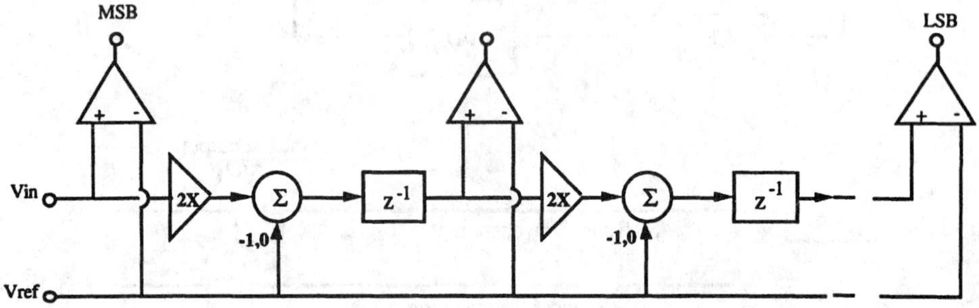

FIGURE 64.38 N-stage pipelined converter.

In this figure note that only one comparator is used per stage and the interstage gain block (also performing the S/H function) has a gain of two. Once again for synchronization purposes, shift registers are used. This pipelined converter [11, 12] operates as follows. The input is first compared to see if it is greater than V_{ref}. If the input is greater than V_{ref}, then we set the MSB bit. If the MSB is set, then V_{ref} is subtracted from the input voltage. However, if the MSB is not set, nothing is done. The resulting voltage is then doubled. During the next clock cycle the resulting voltage is once again compared to V_{ref} to give the MSB-1 bit. While the second stage generates the second MSB for the first input, the first stage processes a new input. When the pipeline is filled, N inputs are being processed simultaneously. Several shift registers are used to ensure that all the data bits corresponding to a single input are output simultaneously. This process continues until we finally generate the LSB. The analog output of the ith stage can be written as

$$V_i = 2[V_{i-1} - b_i V_{ref}]z^{-1} \tag{64.21}$$

where b_i is equal to 1 if the ith bit is set and is equal to 0 otherwise.

Figure 64.39 shows a simple circuit realization for a pipelined ADC [13]. Other more advanced topologies [12] reduce some of the problems associated with this topology, but operate on the same principle. Only one stage of an N-bit converter is shown in this figure. In this first stage the amplifier A_1 performs the comparison function and the amplifier A_2 performs the $2\times$ multiplication and the conditional subtraction function. Both amplifiers, along with the associated capacitors and switches, also implement a S/H function. Let us first concentrate on

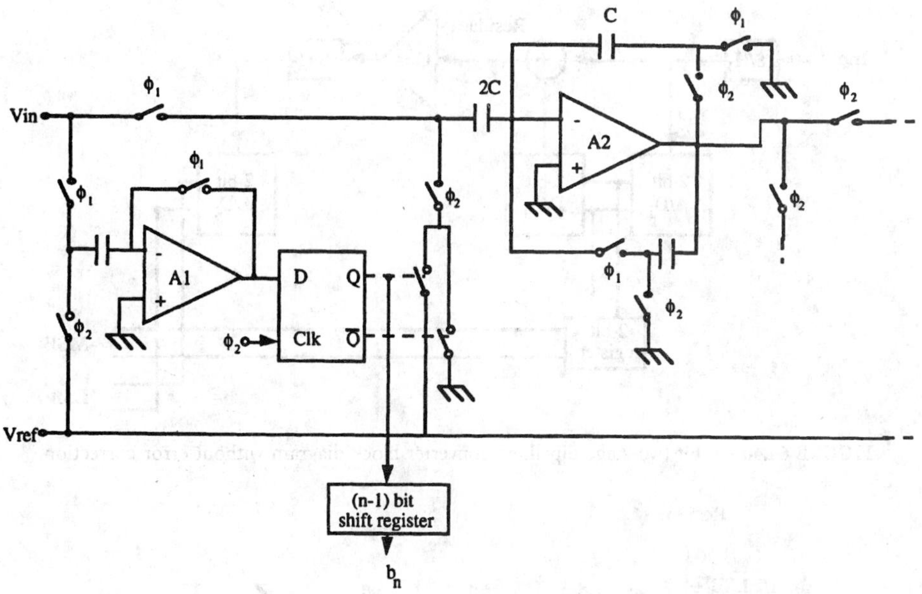

FIGURE 64.39 Circuit realization of an N-stage pipelined converter.

the comparison function. For a sufficiently large gain, during ϕ_1 the capacitor C_h is charged to $V_a - V_{\text{off}}$, where V_{off} is equal to the amplifier offset voltage. During ϕ_2 the reference voltage, V_{ref}, is connected to the positively charged terminal of capacitor C_h and the feedback around the amplifier is removed. During this period the amplifier A_1 acts as a comparator and compares the voltage $[V_{\text{ref}} - (V_a - V_{\text{off}})]$ against V_{off}; i.e., the effect of the amplifier offset voltage is completely canceled. The above discussion is valid only for sufficiently large gain. However, if the gain of amplifier A_1 is not sufficiently large then the charge transfer is not complete and an error is introduced in the comparison; i.e., the effective comparison is not with V_{ref} but with $V_{\text{ref}} + \Delta V$. The error in the comparison is dependent on the input signal voltage. Note that the first stage samples the input at ϕ_1 while the next stage samples the input at ϕ_2. Likewise, the third stage samples the input at ϕ_1 and the fourth stage samples the input at ϕ_2, and so forth.

Digital-Error Correction in Multistep Converters. Pipelined and subranging converters that have an interstage gain > 1 can utilize digital-error correction [14] to improve linearity. We shall illustrate the principle of digital-error correction with the help of a 4-bit two-stage pipelined converter. Each of the stages resolves two bits. Digital-error correction can be used to correct for linearity errors in all except the last stage. Additionally, it is unable to correct for digital-to-analog linearity and op amp settling time errors. Therefore, for our two-stage example we will only be able to correct for errors in the first stage and we shall assume an ideal DAC.

Figure 64.40 shows the block diagram for the 4-bit two-stage pipelined converter without digital-error correction. This circuit is a 4-bit version of Fig. 64.37. The input signal is sample-and-held by S/H_1. The course MSB bits for the overall converter are generated by the first-stage subconverter (A/D_1). The analog value corresponding to these bits is then generated by the first-stage digital-to-analog subconverter (D/A_1). The difference between the input signal and the digital-to-analog output is called the residue. This residue is amplified by the interstage gain stage $(G = 4)$ and passed on to the second-stage analog-to-digital subconverter. The second-stage analog-to-digital subconverter (A/D_1) then generates the lower two bits. Because the second stage is working on the signal after one clock delay, an intermediate delay stage is added to synchronize the outputs of the two stages.

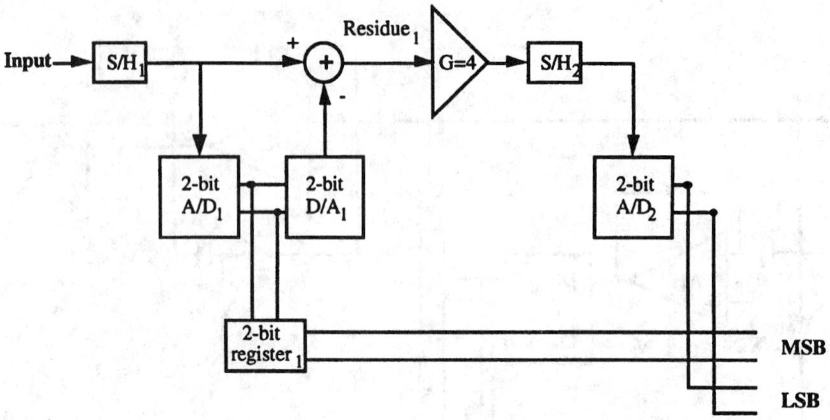

FIGURE 64.40 4-bit two-stage pipelined converter block diagram without error correction.

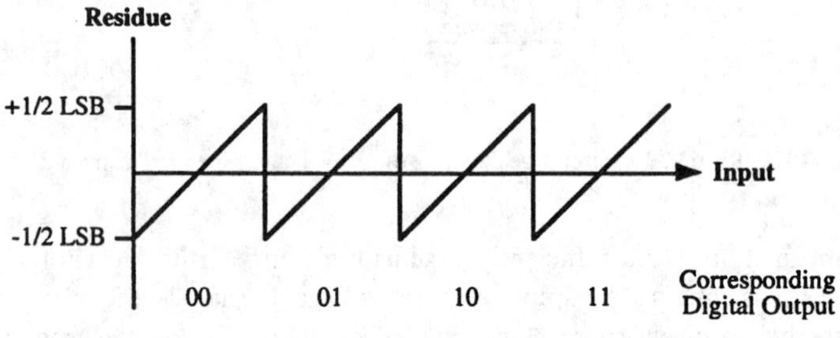

FIGURE 64.41 Ideal subconverter residue.

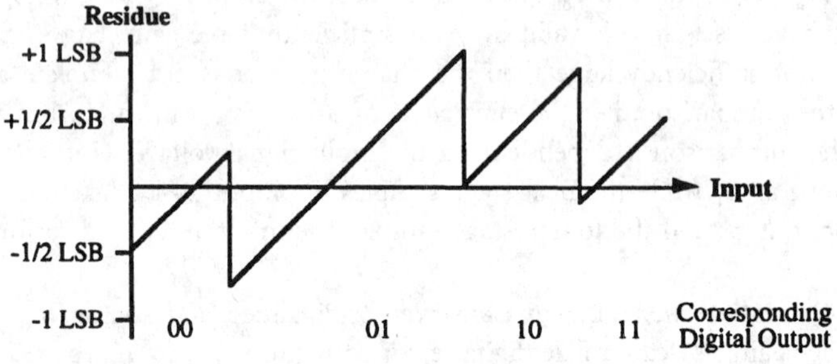

FIGURE 64.42 Nonideal subconverter residue.

The residue for an ideal converter varies from $-\frac{1}{2}$ LSB to $+\frac{1}{2}$ LSB of the first-stage subconverter resolution, as shown in Fig. 64.41. In the case of nonlinearity in the first subconverter the residue will have excursions above and below the $\pm\frac{1}{2}$ LSB value, as shown in Fig. 64.42. For an ideal digital-to-analog conversion, the residue corresponding to each digital code is still accurate and no data have been lost as yet. In the traditional pipelined converter shown in Fig. 64.40 any residue value from the first converter that is greater than $\pm\frac{1}{2}$ LSB of the first stage saturates the second stage and produces errors.

If, however we change the overall converter topology such that the resolution of the second subconverter is increased by 1 bit, i.e., we double the number of levels, and reduce the interstage gain by half, then we can detect when the residue exceeds the $\pm\frac{1}{2}$ LSB levels and correct for its effect digitally. Figure 64.43 shows a block diagram for the pipelined converter in Fig. 64.40 with digital-error correction. Whenever the residue from the first stage exceeds $+\frac{1}{2}$ LSB it implies

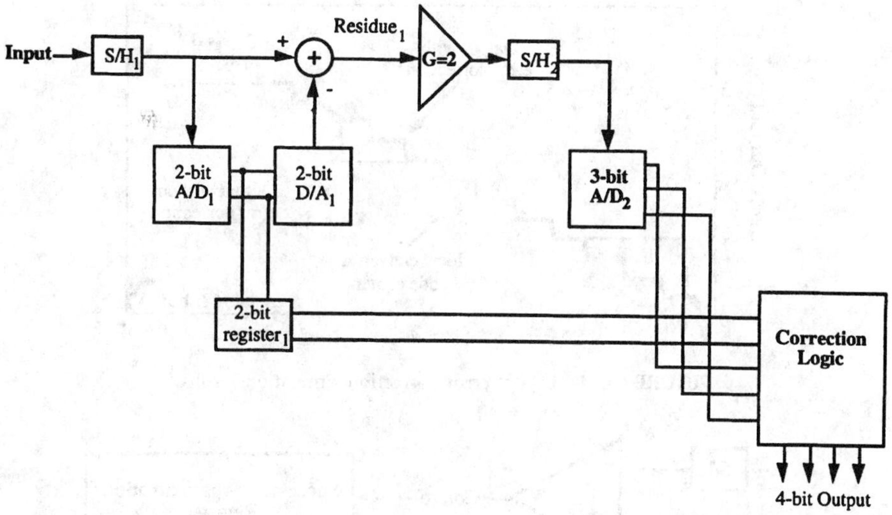

FIGURE 64.43 4-bit two-stage pipelined converter block diagram with error correction.

that the digital output of the first stage subconverter is too small. Likewise, whenever the residue is less than $-\frac{1}{2}$ LSB it implies that the digital output of the first stage is too large. By adding a $\frac{1}{2}$ LSB offset at the input of the first analog-to-digital subconverter and at the output of the first DAC the input to the second subconverter for an ideal first subconverter is restricted between $\frac{1}{4}$ full-scale and $\frac{3}{4}$ full-scale. Any excursion outside this region implies an error in the first analog-to-digital subconverter. The approximate value for this error is measured by the second-stage subconverter and is then subtracted digitally from the final value. Nonlinearities from the second stage are not corrected via this scheme; however, as the interstage gain is > 1 the effect of the nonlinearities in the second stage will have a much lower effect than those resulting from the first stage. Nonlinearities in the DAC can be reduced substantially by utilizing reference feedforward compensation [15]. Here, the reference for the second stage changes dynamically and is obtained by amplifying the first-stage digital-to-analog subconverter segment voltage that corresponds to the most current digital output code of the first-stage analog-to-digital subconverter. Figure 64.44 shows simulation results for the 4-bit two-stage pipelined converter with and without digital error correction. For purposes of clarity the second-stage subconverter is made ideal. In a real converter some nonlinearity would still exist, but would be limited to that introduced by the last stage. Traditionally, even though the resolution of the first subconverter is only 2 bits, it needs to be linear to the overall converter resolution. Digital-error correction can reduce the linearity requirements such that it is commensurate with its resolution.

N-Clock Converters

Both the successive approximation and algorithmic analog-to-digital topologies require N clock cycles to perform an N-bit conversion. They both perform 1 bit of conversion per clock cycle. The successive approximation converter is a subclass of the subranging converter, in which during each clock cycle only 1 bit of resolution is generated. The algorithmic converter is a variation of the pipelined converter, in which the pipeline is folded back into a loop. Both topologies essentially perform a binary search to generate the digital value. However, in the case of the successive approximation converter the binary search is performed on the reference voltage, while in the case of the algorithmic converter the search is performed on the input signal.

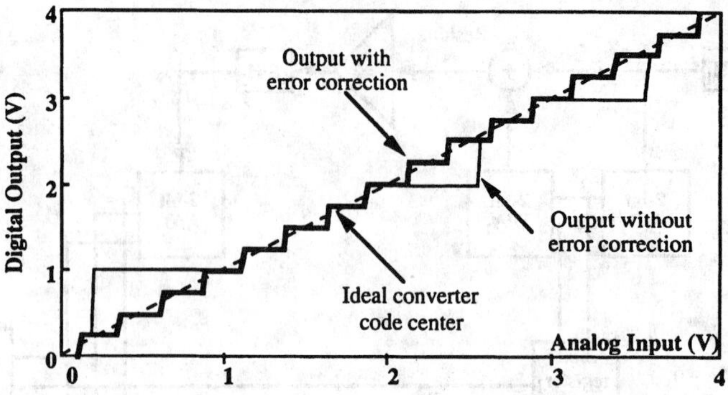

FIGURE 64.44 Digital-error correction simulation results.

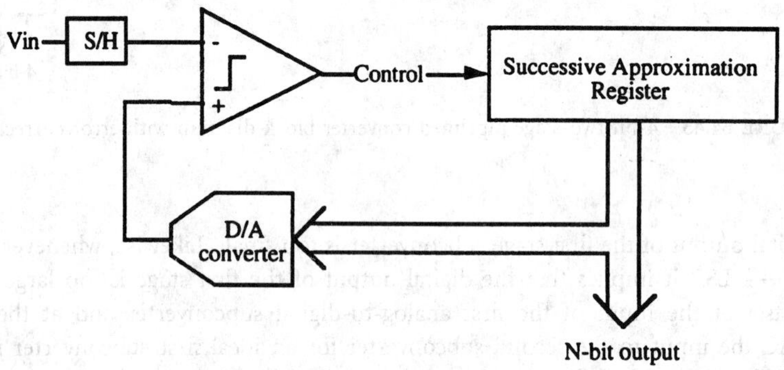

FIGURE 64.45 Successive approximation converter block diagram.

Successive Approximation Converters. A block diagram for the successive approximation converter is shown in Fig. 64.45. Because the conversion requires N clock cycles a S/H version of the input signal is provided to the negative input of the comparator. The comparator controls the digital logic circuit that performs the binary search. This logic circuit is called the successive approximation register (SAR). The output of the SAR is used to drive the DAC that is connected to the positive input of the comparator.

During the first clock period the input is compared to the MSB, i.e., the MSB is temporarily raised high. If the output of the comparator remains high, then the input lies somewhere between 0 and $V_{ref}/2$, and the MSB is reset to 0. However, if the comparator output is low, then the input signal is somewhere between $V_{ref}/2$ and V_{ref}, and the MSB is set high. During the next clock period the MSB-1 bit is evaluated in the same manner. This procedure is repeated such that at the end of N clock periods all N bits have been resolved. Figure 64.46 shows the binary search procedure for a 4-bit converter and shows the comparator output sequence that corresponds to an input equal to 72% of V_{ref}.

The successive approximation converter is one of the most popular topologies in both MOS and bipolar technologies. In MOS technologies the charge-redistribution implementation [9] of the successive approximation methodology is the most commonly used. The circuit diagram of a 4-bit charge redistribution converter is shown in Fig. 64.47. In this circuit the binary weighted capacitors $\{C, C/2, \ldots, C/8\}$ and the switches $\{S_1, S_2, \ldots, S_5\}$ form the 4-bit scaling DAC. For each conversion the circuit operates as a sequence of three phases. During the first phase (sample) switch S_0 is closed and all the other switches $\{S_1, S_2, \ldots, S_6\}$ are connected such that the input voltage V_{in} is sampled onto all the capacitors. During the next phase (hold) S_0 is open and the bottom plates of all the capacitors are connected to ground; i.e., switches

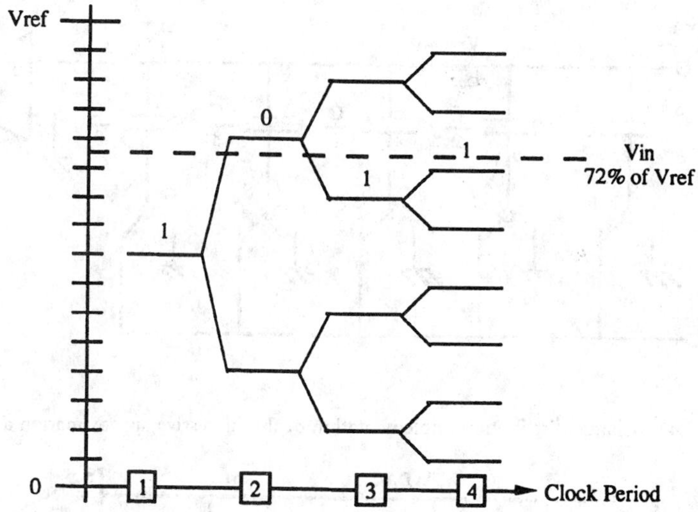

FIGURE 64.46 Binary search process for successive approximation.

$\{S_1, S_2, \ldots, S_5\}$ are switched to ground. The voltage, V_x, at the top plate of the capacitors at this time is equal to $-V_{in}$ and the total charge in all the capacitors is equal to $-2CV_{in}$. The final phase (redistribution) begins by testing the input voltage against the MSB. This is accomplished by keeping the switches $\{S_2, S_3, \ldots, S_5\}$ connected to ground and switching S_1 and S_6 such that the bottom plate of the largest capacitor is connected to V_{ref}. The voltage at the top plate of the capacitor is equal to

$$V_x = \frac{V_{ref}}{2} - V_{in} \tag{64.22}$$

If $V_x > 0$ then the comparator output goes high, signifying that $V_{in} < (V_{ref}/2)$ and switch S_1 is switched back to ground. If the comparator output is low, then $V_{in} > (V_{ref}/2)$ and the switch S_1 is left connected to V_{ref} and the MSB is set high. In a similar fashion the next bit, MSB-1, is evaluated. This procedure is continued until all N bits have resolved. After the conversion process the voltage at the top plate is such that

$$V_x = -V_{in} + \left\{ b_3 \frac{V_{ref}}{2^1} + b_2 \frac{V_{ref}}{2^2} + b_1 \frac{V_{ref}}{2^3} + b_0 \frac{V_{ref}}{2^0} \right\} \tag{64.23a}$$

$$-1\text{LSB} < V_x < 0 \tag{64.23b}$$

where b_i is $\{0, 1\}$ depending upon if bit$_i$ was set to 0 or 1.

One of the advantages of the charge-redistribution topology is that the parasitic capacitance from the switches has little effect on the accuracy. Additionally, the clock feedthrough from switch S_0 only causes an offset and the clock feedthrough from switches $\{S_1, S_2, \ldots, S_5\}$ is input signal independent because they are always connected to either ground or V_{ref}. However, any mismatch in the binary ratios of the capacitors in the array causes nonlinearity, which limits the accuracy to 10 or 12 bits.

Self-Calibration Successive Approximation Converters. Fortunately, self-calibrating [7] techniques have been introduced that correct for errors in the binary ratios of the capacitors. Figure 64.48 shows the block diagram for a successive approximation-based self-calibrating ADC. The circuit consists of an N-bit binary weighted capacitor array main DAC, an M-bit

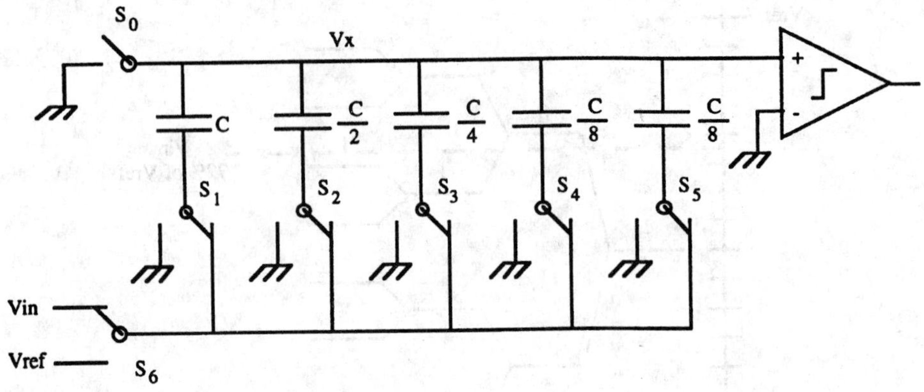

FIGURE 64.47 Charge-distribution implementation of the successive approximation architecture.

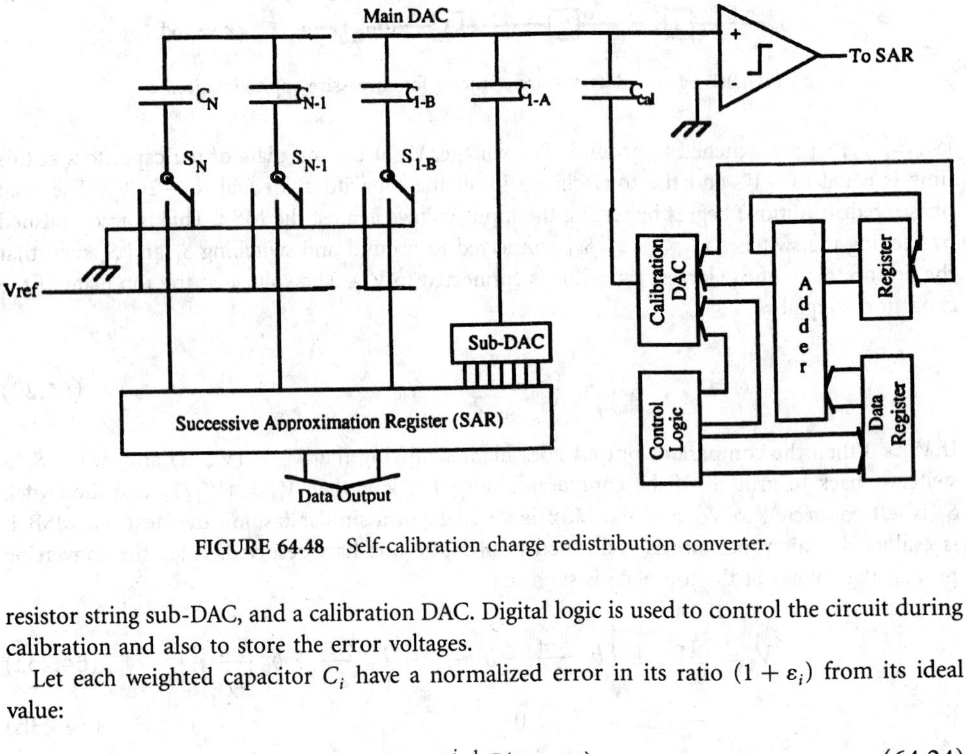

FIGURE 64.48 Self-calibration charge redistribution converter.

resistor string sub-DAC, and a calibration DAC. Digital logic is used to control the circuit during calibration and also to store the error voltages.

Let each weighted capacitor C_i have a normalized error in its ratio $(1 + \varepsilon_i)$ from its ideal value:

$$C_i = 2^{i-1}C(1 + \varepsilon_i) \tag{64.24}$$

Each capacitor contributes an error voltage at the top plate which is equal to

$$V_{e_i} = \frac{V_{\text{ref}}}{2^N} 2^{i-1}\varepsilon_i \quad i = 1B, 2, \ldots, N \tag{64.25}$$

Therefore, the total linearity error is equal to

$$V_{\text{error}} = \sum_{i=1B}^{N} V_{\varepsilon_i} b_i \tag{64.26}$$

where b_i is the logic value of the ith bit.

The calibration cycle begins by measuring the error contribution from the largest capacitor and progressing to the smallest. The error from the MSB capacitor is evaluated by closing S_0 and setting switches $\{S_1, S_2, \ldots, S_5\}$ such that all the capacitors except C_{MSB} are charged to V_{ref}. Next, the switch S_0 is opened and switches $\{S_1, S_2, \ldots, S_5\}$ are switched to connect the bottom plates to ground. Under ideal conditions, i.e., $C_{\text{MSB}} = 2^{N-1}C$, the voltage at the top plate is equal to zero. It should be noted that the total capacitance is equal to $2C$. However, because $C_{\text{MSB}} = 2^{N-1}C(1 + \varepsilon_{\text{MSB}})$, the top plate voltage $V_x = (V_{\text{ref}}/2)\varepsilon_{\text{MSB}}$, such that $V_{x_{\text{MSB}}} = 2V_{\varepsilon_{\text{MSB}}}$. Therefore, the error voltage at the top plate is a direct measure of the corresponding error in the capacitor ratio. A successive approximation search using the sub-DAC is used to measure these voltages. The relationship between the measured residual voltage and the error voltage is equal to

$$V_{\varepsilon_i} = \frac{1}{2}\left\{V_{x_i} - \sum_{j=i+1}^{N} V_{\varepsilon_j}\right\} \tag{64.27}$$

which corresponds to the equivalent error terms on the digital side. These digital correction terms are stored and subsequently added or subtracted during the normal operation cycle. Self-calibration improves the resolution of successive approximation converters to approximately 15 or 16 bits.

Algorithmic Converters. As stated earlier, the algorithmic ADC is formed by modifying a pipelined converter. Here, the pipeline has been closed to form a loop. All N-bits are evaluated by a single stage, therefore implying that a N-bit conversion requires N clock cycles. A block diagram for the algorithmic converter is shown in Fig. 64.49 [6] and consists of a S/H, a $2\times$ amplifier, a comparator, and a reference subtraction circuit. The circuit operates as follows. The input is first sampled and held by setting S_1 to V_{in}. This signal is then multiplied by 2 (by the $2\times$ amplifier). The result of this multiplication, V_0, is compared to V_{ref}. If $V_{0_N} > V_{\text{ref}}$ then the most significant bit, b_N, is set to 1 or it is set to 0. If b_N is equal to 1, then S_2 is connected to V_{ref} such that V_{b_N} is equal to

$$V_{b_N} = 2V_{0_N} - b_N V_{\text{ref}} \quad b_N = \{0, 1\} \tag{64.28}$$

This voltage is then sample-and-held and used to evaluate the MSB-1 bit. This procedure continues until all N-bits are resolved. The general expression for V_0 is equal to

$$V_{0_i} = [2V_{0_{i-1}} - b_i V_{\text{ref}}]z^{-1} \tag{64.29}$$

where b_i is the comparator output for the ith evaluation and z^{-1} implies a delay of one clock period.

A circuit implementation for this ADC topology is shown in Fig. 64.50 [10]. This circuit uses three amplifiers, five ratio-matched capacitors (C_1 to C_5), an arbitrary valued capacitor, C_6, and a comparator. Two amplifiers and the capacitors (C_1 to C_5) form the recirculating register and the gain of two amplifiers. The amplifier A_3 and capacitor C_6 form an offset compensated comparator. The switches controlled by V_3, V_4, and V_5 load the input or selectively subtract the reference voltage. The conversion is started by setting V_1, V_2, and V_3 high. This forces V_x and V_Y to 0 and loads V_{in} into C_1. Then, V_1 is set low and V_5 is set high. Therefore, the charge $V_{\text{in}} * C_1$ is transferred from C_1 to C_2. C_1 is made to be equal to C_2, therefore, $V_x = V_{\text{in}}$ (C_3 is also charged to V_x). Because V_1 has been set low the comparator output goes high if $V_{\text{in}} > 0$, or else it remains low. This determines the MSB. The MSB-1 is determined by setting V_2 low and setting V_1 high. This forces the charge from C_3 to transfer to C_4 ($V_4 = V_x$; C_5 is also charged

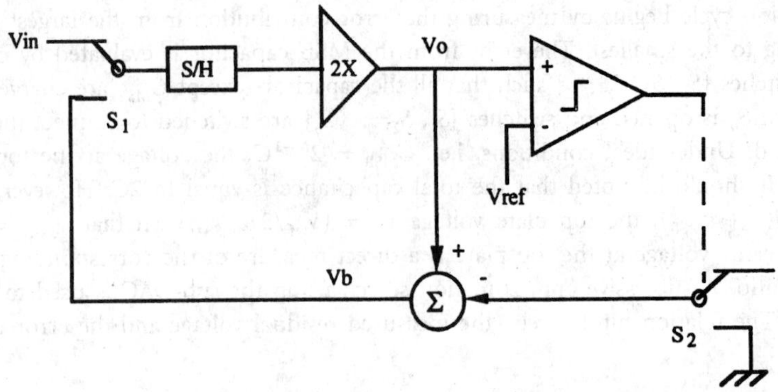

FIGURE 64.49 Algorithmic ADC block diagram.

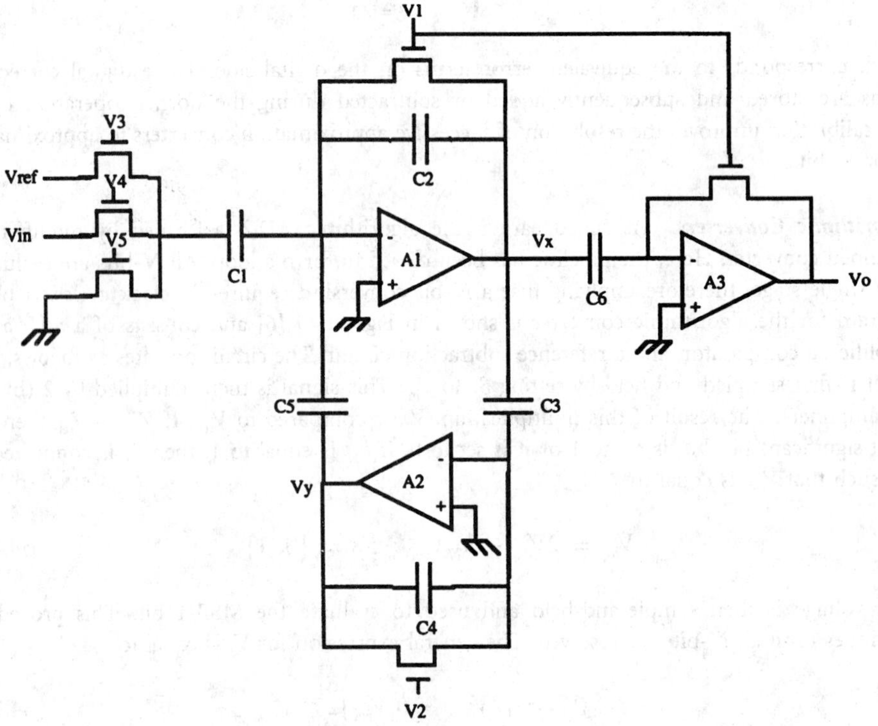

FIGURE 64.50 Example circuit implementation of the algorithmic converter.

to V_4). During the same period C_1 is connected to ground if MSB = 1, or it is connected to V_{ref}. Next, V_2 is set low and V_1 is set high, while C_1 is switched from ground to V_{ref} for V_{ref} to ground. This transfers a charge equivalent to $C_1 \pm V_{ref}$ from C_1 to C_2 and transfers the charge in C_5, $C_5 * V_y$, to C_2. The capacitor C_5 is made to be twice as large as C_2, therefore, the voltage at V_x is equal to $2 * V_{in} \pm V_{ref}$. This process is repeated and the comparator determines bit MSB-1. This circuit has been shown to provide up to 10 bits of resolution at a maximum conversion rate of 200 kHz.

The maximum resolution of the algorithmic converter is limited by the ratio matching of the capacitors, clock feedthrough, capacitor voltage coefficient, parasitic capacitance, and offset voltages. The previous topology solves the problem of parasitic capacitances and amplifier offset voltage, however, its maximum resolution is limited by the ratio matching of the capacitors

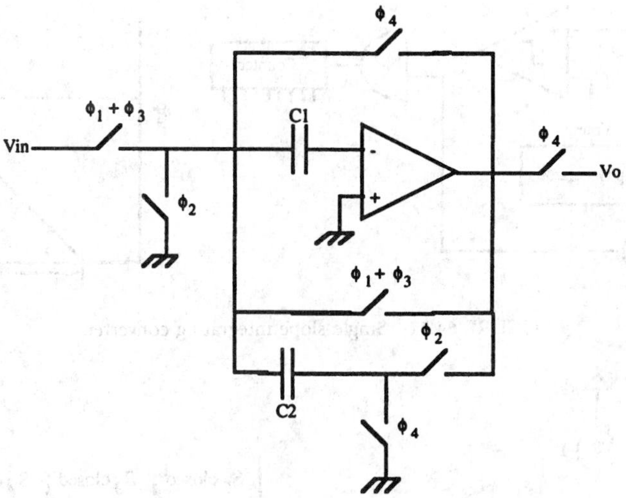

FIGURE 64.51 Ratio-independent multiply-by-two circuit.

that are used to realize the gain of two amplifiers. This problem is partially resolved by using a ratio-independent multiply-by-two algorithm [6] to increase the maximum resolution to the 12-bit level. The ratio-independent multiply-by-two algorithm is easily explained by the circuit shown in Fig. 64.51. During ϕ_1 capacitor C_1 is charged to V_{in}. This charge is then transferred onto C_2 during ϕ_2. The charge on C_2 is equal to $V_{in} * C_1$. During ϕ_3 C_2 is disconnected from the feedback path and V_{in} is once again sampled onto C_1. During ϕ_4 the charge in C_2 is added to C_1. The total charge in C_1 is now equal to $C_1 V_{in} + C_1 V_{in} = 2C_1 V_{in}$ and is completely independent of the value of C_2. Therefore, the voltage at the output at ϕ_4 is equal to $2V_{in}$. The only constraint is that the input voltage be held steady, i.e., S/H during ϕ_1 and ϕ_3.

2^N-Clock Converters

The basic principle of the integrating converter can be explained with the help of Fig. 64.52. A comparator compares the input signal with the output of a ramp voltage generator. The ramp voltage generator is zeroed after each measurement. The output of this comparator is used to gate the clock to an interval counter. The counter output corresponding to the ramp time T_{in} provides an accurate measure of the input voltage. The input voltage V_{in} is equal to $T_{in} * U$, where U is the ramp rate. Because the absolute values of components are not well controlled and also because of the large offset voltages associated with MOS amplifiers and comparators, a calibration or a reference cycle is usually added to calculate the ramp rate and the offset voltage. A simple circuit for a single-slope integrating converter that includes the calibration cycle is shown in Fig. 64.53 [4].

The ramp voltage is generated using a constant current source to charge a capacitor. The ramp voltage, V_{ramp}, is equal to $\int_0^t (I/c)\,dt$, which is equal to $(I\,\Delta t)/c$ for a constant current I. The ramp voltage is compared against the analog ground voltages, V_{in} and V_{ref}, respectively. The addition of the third calibration cycle eliminates any offset errors. The final resolution is dependent only on the linearity of the ramp generator, i.e., the linearity of the current source. In the single-slope approach just described the calibration is done in digital. However, the complete calibration can be performed in analog as well, as in the dual-slope approach. Further improvements include a charge balancing technique [4] that uses an oscillating integration process to keep the voltage across the capacitor closer to zero, thereby reducing the linearity constraints on the ramp generator. The primary advantage of the integrating converter is the small number of precision analog components that are required to generate extremely high

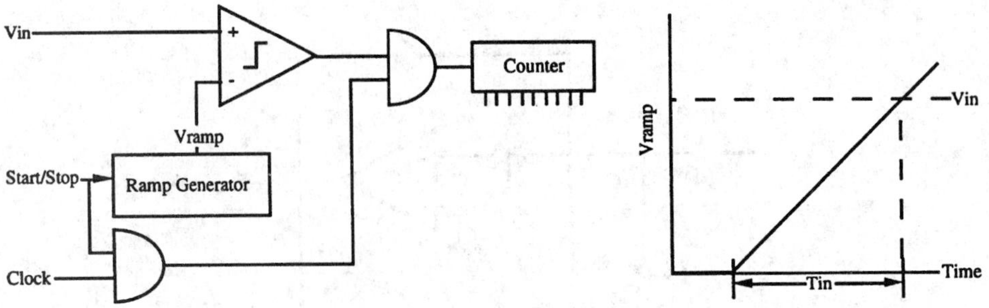

FIGURE 64.52 Single-slope integrating converter.

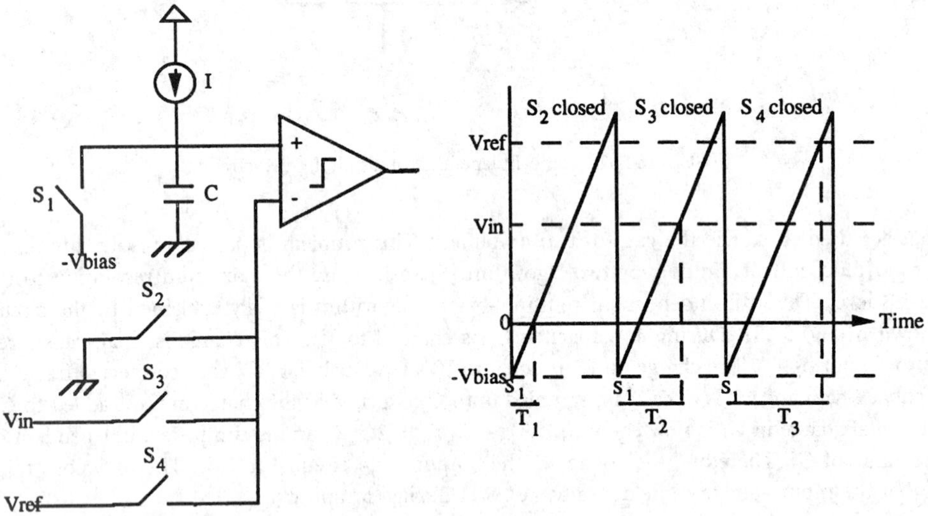

FIGURE 64.53 Single-slope integrating converter with calibration cycle.

resolution. However, the primary disadvantage is the conversion time required. It takes 2^N clock cycles to generate a N-bit conversion.

Oversampled Converters

Oversampling converters have the advantage over Nyquist rate converters in that they do no require very tight tolerances from the analog components and also because they simplify the design of the anti-alias filter. Examples of oversampling converters include the noise-shaping architecture and the interpolative architecture. Our discussion centers around noise-shaping converters.

If the analog input signal V_{in} has a frequency spectrum from 0 to f_0 then $2f_0$ is defined as the Nyquist rate. Oversampling converters sample the input at a rate larger than the Nyquist frequency. If f_s is the sampling rate, then $(f_s)/(2f_0) = \text{OSR}$ is called the oversampling ratio. Oversampling converters use "signal averaging" along with a low-resolution converter to provide extremely high resolution. This technique can best be understood by considering the following example shown in Fig. 64.54.

Let the input be exactly in the middle of V_n and V_{n+1} and let it be sampled a number of times. If, in addition to the input signal, we add some random noise, then for a large number of samples the output would fall on V_n 50% of the time and on V_{n+1} the other 50% of the time. If the signal was a littler closer to V_{n+1}, then the percentage of times the output falls on V_{n+1}

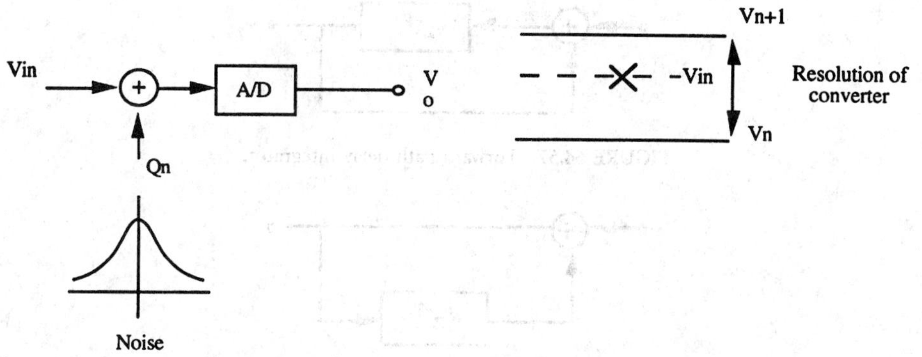

FIGURE 64.54 Higher resolution provided by oversampling.

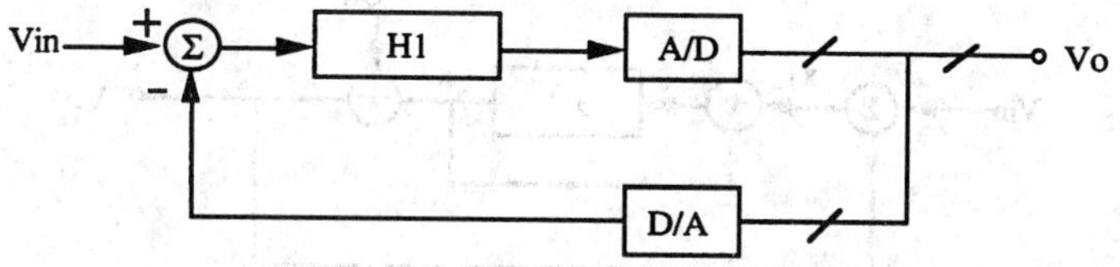

FIGURE 64.55 Noise-shaping oversampling converters.

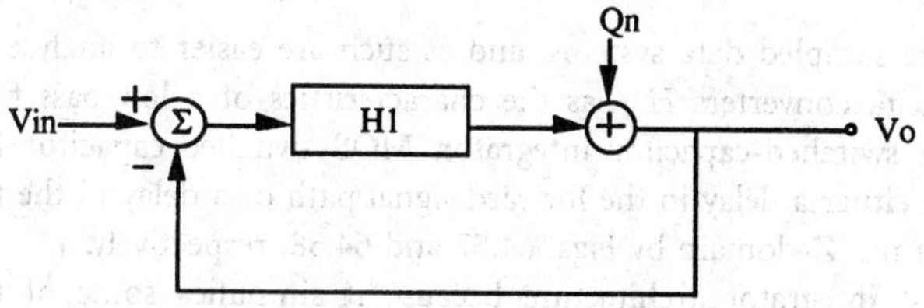

FIGURE 64.56 Linear system model of noise-shaping converter.

would increase. Using this averaging technique we can get a better estimate of the input signal. However, in simple oversampling the resolution only increases by $\sqrt{n}$, where n is the number of samples of V_{in} that are averaged. Therefore, to increase the resolution of the converter by one additional bit we are required to increase the number of samples by 4×.

Noise-shaping converters use feedback to generate the necessary noise and additionally perform frequency shaping of the noise spectrum to reduce the amount of oversampling necessary. This can be illustrated with the help of Fig. 64.55. The output from H_1 is quantized by an N-bit ADC. This digital value is then converted to an analog value by the N-bit DAC. This value is subtracted from the input and the result is sent to H_1. Here, we assume an N-bit converter for simplicity, however, for the special case in which $N = 1$ the noise-shaping converter is called a sigma-delta converter. The quantization process approximates an analog value by a finite-resolution digital value. This step introduces a quantization error, Q_n. Further, if we assume that the quantization error is not correlated to the input, the system can now be modeled as a linear system, as shown in Fig. 64.56. Here we note that the error introduced by the analog-to-digital process is modeled by Q_n. The output voltage for this system can now be written as

$$V_0 = \frac{Q_n}{[1 + H_1]} + \frac{V_{\text{in}} H_1}{[1 + H_1]} \tag{64.30}$$

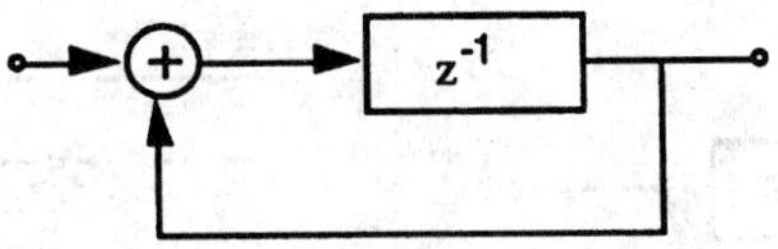

FIGURE 64.57 Forward path delay integrator.

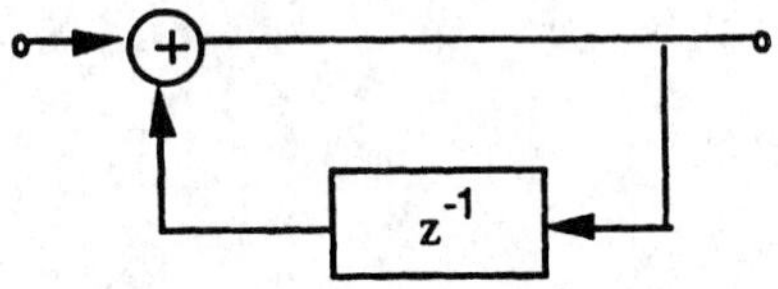

FIGURE 64.58 Feedback path delay integrator.

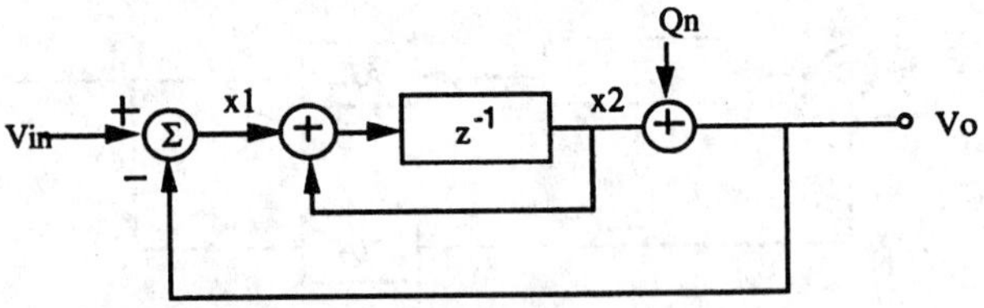

FIGURE 64.59 First-order noise-shaping converter.

Data converters are sampled data systems, and as such are easier to analyse in the Z-domain. For most sigma-delta converters H_1 has the characteristics of a low-pass filter and is usually implemented as a switched-capacitor integrator. MOS switched-capacitor integrators can be implemented with either a delay in the forward signal path or a delay in the feedback path, and can be modeled in the Z-domain by Figs. 64.57 and 64.58, respectively.

We use the first integrator architecture because it simplifies some of the algebra. For a first-order sigma-delta converter H_1 is realized as a simple switched-capacitor integrator, i.e., $H_1 = (z^{-1})/(1 - z^{-1})$. Therefore, Fig. 64.56 can now be drawn as Fig. 64.59. Replacing H_1 by $(z^{-1})/(1 - z^{-1})$ in (64.30) we can write the transfer function for the first-order sigma-delta converter as

$$V_0 = V_{\text{in}} z^{-1} + Q_n(1 - z^{-1}) \tag{64.31}$$

As can be seen from (64.31) the output is a delayed version of the input plus the quantization noise multiplied by the factor $(1 - z^{-1})$. This function has a high-pass characteristic, as shown in Fig. 64.60. We note here that the quantization noise is substantially reduced at lower frequencies and increases slightly at higher frequencies. In this figure f_0 is the input signal bandwidth and $f_s/2 = \pi$ corresponds to the Nyquist rate of the oversampling converter. For simplicity the quantization noise is usually assumed to be white[1] with a spectral density equal to $e_{\text{rms}}\sqrt{2/f_s}$. Therefore, the magnitude of the output noise spectrum can be written as

$$N(f) = e_{\text{rms}}\sqrt{\frac{2}{f_s}}|1 - z^{-1}| = 2e_{\text{rms}}\sqrt{\frac{2}{f_s}}\sin\left(\frac{\pi f}{f_s}\right) \tag{64.32}$$

[1]Quantization noise is clearly not uncorrelated or white for the first-order sigma-delta modulator, but becomes increasingly so for the higher-order systems.

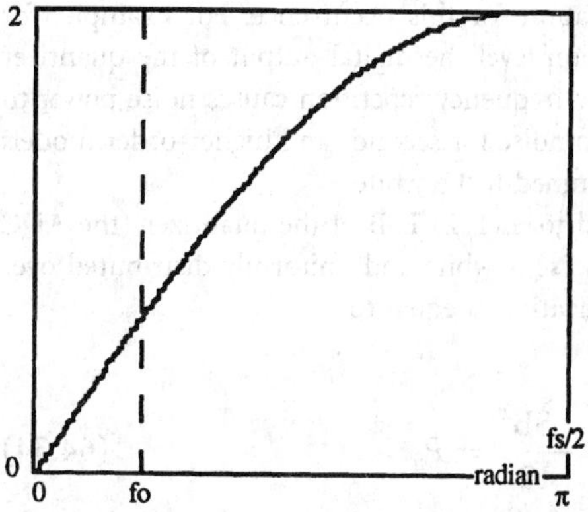

FIGURE 64.60 Magnitude response of the function $(1 - z^{-1})$.

Further, if $f_0 \ll f_s$ we can approximate the root mean square noise in the signal band, $(0 < f < f_0)$, by

$$N_{f_0} \approx e_{\text{rms}} \frac{\pi}{3} \left(\frac{2f_0}{f_s} \right)^{3/2}. \tag{64.33}$$

As the oversampling ratio increases the quantization noise in the signal band decreases; i.e., for a doubling of the oversampling ratio the quantization noise drops by $20 \log(2)^{3/2} \approx 9$ dB. Therefore, for each doubling of the oversampling ratio we effectively increase the resolution of the oversampling converter by an additional 1.5 bits.

The previous analysis was based on the assumption that the quantization noise was not correlated to the input and uniformly distributed across the Nyquist band. We now reexamine these assumptions. The assumption that the quantization noise is not correlated with the input only holds for extremely busy input signals. This is particularly not true for the first-order modulator assumed in the analysis above, such that for extremely low frequency or DC inputs the first-order modulator generates pattern noise (also called tones), as shown in Fig. 64.61. The peaks of the pattern noise occur at input voltages that are integer divisors of the quantization

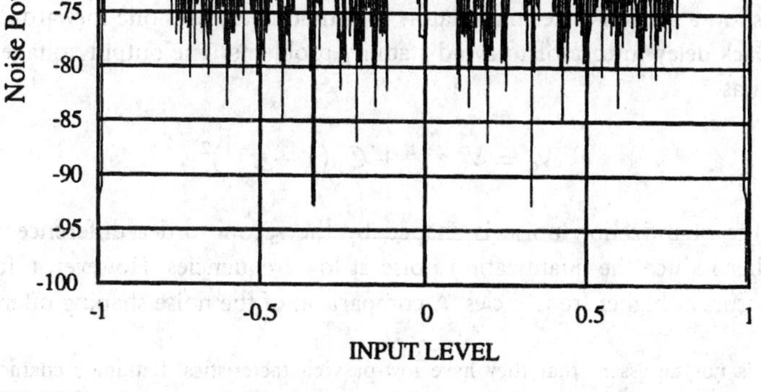

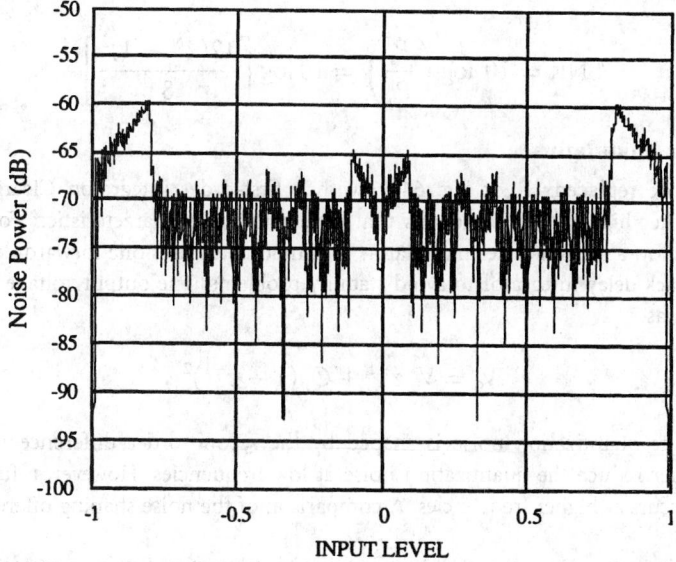

FIGURE 64.61 Pattern noise for a first-order sigma-delta modulator for DC inputs.

step. It is possible to provide a conceptual explanation for this occurrence. For example, for an input that is an integer divisor of the quantization level the digital output of the quantizer repeats itself at an extremely low frequency. This low-frequency repetition causes noise power to be introduced into the signal band. The quantization noise for second- and higher-order models is significantly more uncorrelated and is usually assumed to be white.

The quantization error has a value that is limited to $\pm(1/2)$ LSB of the quantizer (the ADC in Fig. 64.45). If we assume that the quantization noise is white and uniformly distributed over the quantization level, then the average noise quantization is equal to

$$\int_{-\frac{1}{2}\mathrm{LSB}}^{+\frac{1}{2}\mathrm{LSB}} x^2\,dx = \frac{\mathrm{LSB}^2}{12} = P_n \tag{64.34}$$

Because the quantization noise is sampled at the clock frequency f_s, the entire noise power is aliased back into the overall converter Nyquist band $[0-(f_s/2)]$. Therefore, the spectral density of the quantization noise is equal to

$$P_n = \frac{\mathrm{LSB}^2}{12} = \int_0^{f_s/2} n_e(f)^2\,\delta f = n_e(f)^2\frac{f_s}{2} \tag{64.35a}$$

$$n_e(f) = P_n\sqrt{\frac{2}{f_s}} \tag{64.35b}$$

The SNR for an ADC is defined as $10\log(P_s/P_n)$, where P_s is the signal power. The signal power is highly waveform dependent. For example, the P_s for a full-scale sine wave input, $(A/2)\sin(\omega T)$, which is applied to an N-bit quantizer, can be written in terms of the quantization level as

$$\frac{A^2}{8} = \frac{[(2^N-1)\mathrm{LSB}]^2}{8} \tag{64.36}$$

Therefore,

$$\mathrm{SNR} = 10\log\left(\frac{P_s}{P_n}\right) = 10\log\left[\frac{12(2^N-1)^2}{8}\right] \tag{64.37}$$

Higher-Order Modulators

In Fig. 64.59 we replaced H_1 of Fig. 64.56 with a first-order integrator. Clearly, H_1 can be replaced by other higher-order functions that have a low-pass characteristic.[2] For example, in Fig. 64.62 we show a second-order modulator. This modulator uses one forward delay integrator and one feedback delay integrator to avoid stability problems. The output voltage for this figure can be written as

$$V_0 = V_{\mathrm{in}}z^{-1} + Q_n(1-z^{-1})^2 \tag{64.38}$$

Note that the quantization noise is shaped by the second-order difference equation. This serves to further reduce the quantization noise at low frequencies. However, a further increase in the noise occurs at higher frequencies. A comparison of the noise shaping offered by the first-

[2]Actually, it is not necessary that they have low-pass characteristics. Bandpass characteristics may be preferred if the input signal is to be bandlimited.

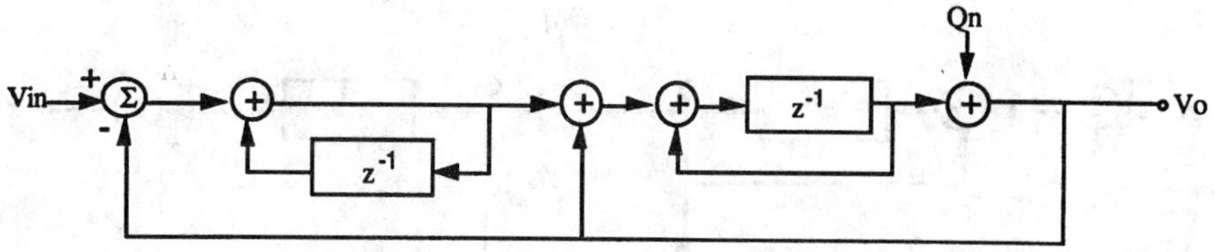

FIGURE 64.62 Second-order modulator block diagram.

and second-order modulators is shown in Fig. 64.63. Once again, assuming that $f_o \ll f_s$ we can write an expression for the root mean square noise in the signal band for the second-order modulator as

$$N_{f_0} \approx e_{\text{rms}} \frac{\pi^2}{\sqrt{5}} \left(\frac{2f_0}{f_s}\right)^{5/2} \tag{64.39}$$

The noise power in the signal bandwidth falls by 15 dB for every doubling of the oversampling ratio. One of the added advantages of the second-order modulator over the first-order modulator is that quantization noise has been shown to be less correlated to the input, therefore, less pattern noise.

From our analysis so far it would seem that increasing the order of the filter would reduce the necessary oversampling ratio for a given resolution. This is true, however, the simple Candy-style modulator (shown in Figs. 64.59 and 64.62) with orders > 2 results in stability problems. This is because for higher-order modulators the later integrator stages are easily overloaded and saturated. This in turn increases the noise in the signal band. However, higher-order modulators can be realized by using a cascade of lower order modulators in the MASH architecture [8]. In the cascaded MASH technique both the digital output and the output of the integrator of each lower-order modulator is passed on to the next module. A second-order MASH architecture using two cascaded first-order sections is shown in Fig. 64.64. It can be shown that the output is equal to

$$Y = z^{-2}X - Q_{n_2}(1 - z^{-1})^2 \tag{64.40}$$

Once again, we note that the quantization noise is multiplied by the second-order difference equation. The sign in front of the noise term is not important. However, for complete cancellation of the quantization noise from the first integrator the gain of the first loop needs to be identical to the gain of the second loop. Therefore, the amplifier gain and capacitor matching become

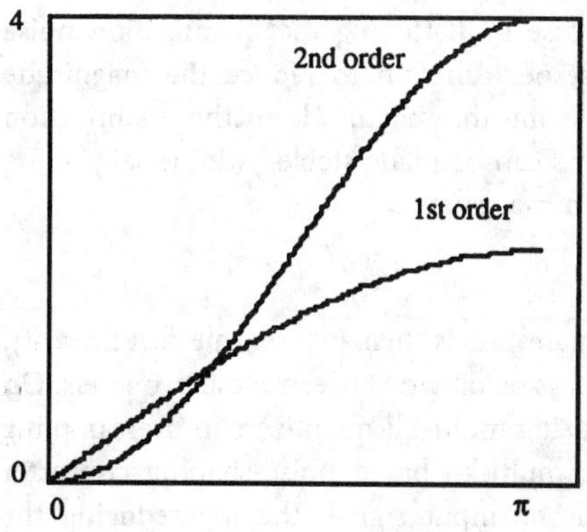

FIGURE 64.63 Noise shaping due to the second-order modulator shown in Fig. 64.62.

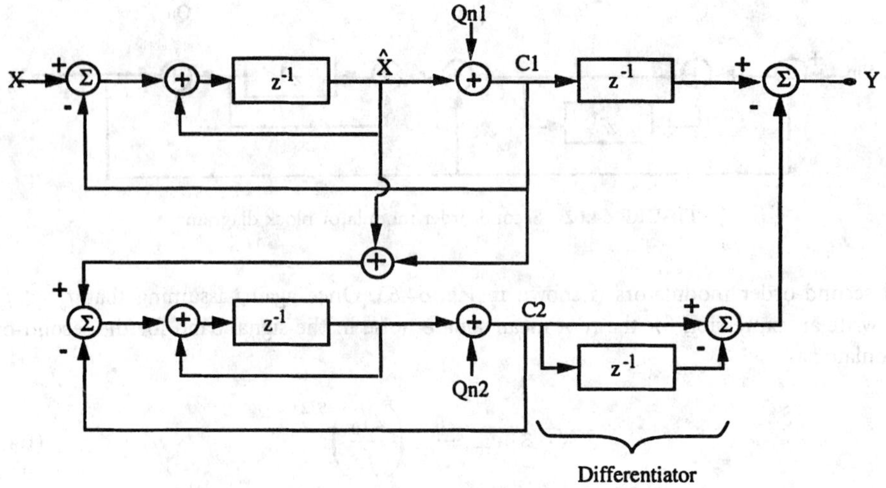

FIGURE 64.64 MASH architecture for a second-order modulator.

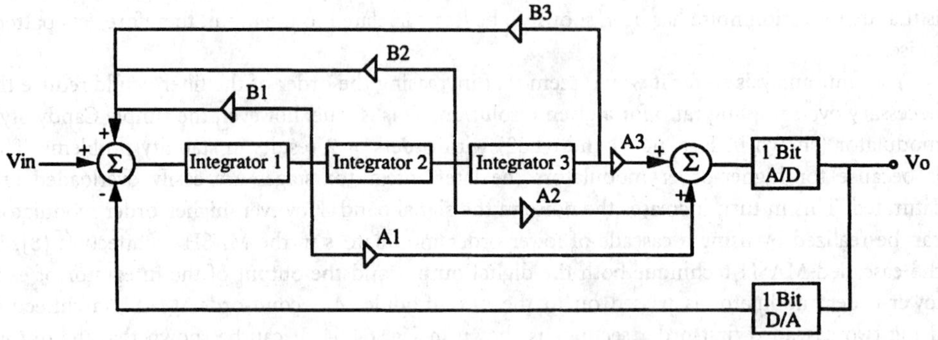

FIGURE 64.65 Finite pole–zero loop filter higher-order modulator.

extremely important. It has been shown that a 1% matching and an op amp gain of 80 dB are sufficient for 16 bits of accuracy [8].

An alternate methodology to stabilize higher-order oversampled coders is the use of finite poles and zeroes for the loop filter [16], H_1 in Fig. 64.55. Up until now all the loop filters have been integrators with poles at DC and zeroes at extremely high frequencies. The loop filter can be realized using additional feedback and feedforward paths as shown in Fig. 64.65. A third-order modulator is shown in this figure. Having finite poles and zeroes serves two purposes: (1) the nonzero poles function to reduce the in-band noise by flattening the quantization noise transfer function at low frequencies, (2) the finite zeroes function to reduce the magnitude of the quantization noise at high frequencies. By reducing the magnitude of the quantization noise at high frequencies, even higher-order modulators can be made stable. Additionally, these modulators have been shown to be devoid of pattern noise artifacts.

Multi-Bit Quantizers

The primary reason for using single-bit or two-level quantizers is their inherent perfect linearity. Because only two levels can exist, a straight line can always be drawn between these two levels. On the other hand a number of advantages are found in using multi-bit quantizers in oversampling converters. The quantization noise generated in the multi-bit-based noise-shaping converter is significantly more "white" and uncorrelated with the input signal, thereby reducing the

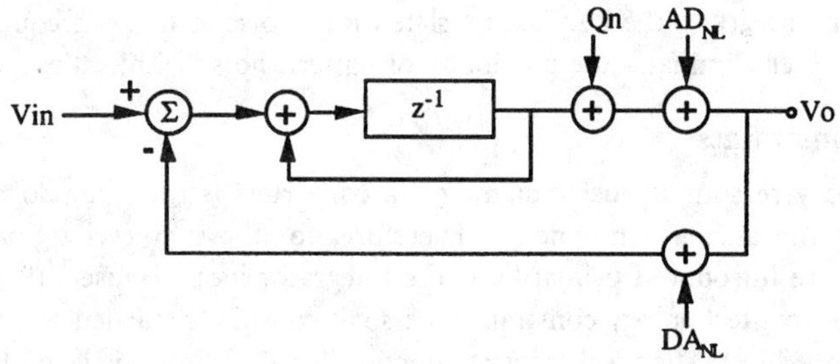

FIGURE 64.66 Model for nonlinearity associated with multi-bit quantizers.

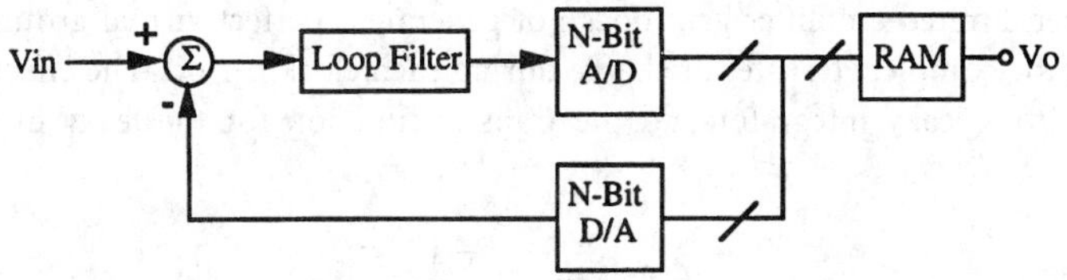

FIGURE 64.67 Digital error correction for multi-bit quantizers.

probability of pattern noise. Additionally, the quantization noise power goes down exponentially as the number of bits in the quantizer increases. However, the primary problem associated with multilevel quantizers is the nonlinearity errors present with the DAC in the modulator loop. This problem can be illustrated with the help of Fig. 64.66.

In Fig. 64.66 the error resulting from the nonlinearity in the multi-bit ADC is included as AD_{NL} and the error resulting from the nonlinearity in the multi-bit DAC is included as DA_{NL}. The output voltage is given by (64.41). Here, note that the analog-to-digital nonlinearity is suppressed by the loop filter, while the digital-to-analog nonlinearity is only subjected to a unit delay. Therefore, any digital-to-analog nonlinearity directly appears in the output. A number of methods have been applied to reduce the effects of nonlinearity associated with multi-bit quantizers. The two most promising methods that have emerged are digital error correction [18] and dynamic element matching [17, 19].

$$V_0 = V_{\text{in}}z^{-1} - DA_{NL}z^{-1} + Q_n(1 - z^{-1}) + AD_{NL}(1 - z^{-1}) \qquad (64.41)$$

A block diagram for digital error correction for multi-bit quantizer-based noise-shaping converters is shown in Fig. 64.67. The random access memory (RAM) and the multi-bit DAC have the same input signal. Because of the high gain in the loop at low frequencies the output of the DAC is almost identical to the input voltage, V_{in}. Now if the digital RAM is programmed to generate the exact digital equivalent of the digital-to-analog output for any digital input, then the RAM output and the digital-to-analog output will be identical to each other. Because the output of the DAC is almost identical to the input voltage, the output voltage will also be the exact digital equivalent of the analog input. The RAM can be programmed by reconfiguring the modulator stages and feeding the system with a multi-bit digital ramp [18].

In the dynamic element matching approach the various analog elements that are used to generate the different analog voltage levels are dynamically swapped around. The various elements can be swapped randomly [17] or in a periodic fashion [19]. The use of random permutations translates the nonlinearity of the DAC into random noise that is distributed throughout the oversampling converter Nyquist range. This method virtually eliminates errors due to nonlinearity, but unfortunately it also increases the noise level in the signal band. In a variation of this basic technique the various analog elements are swapped in a periodic fashion such that the nonlinearity in the DAC is translated into noise at higher frequencies. Individual level averaging further eliminates the possibility of pattern noise within the signal band [19].

Technology Constraints

One of the primary reasons for using sigma-delta converters is that they do not require good matching among the analog components. Therefore, for the two-level sigma-delta converter the nonidealities are introduced primarily by the integrator loop [Hauser, 1991]. To aid in the analysis of the various technology constraints we shall consider a particular implementation of an integrator (Fig. 64.68). The ideal transfer function for this circuit is given by $-1/(1-z^{-1})$. To realize this ideal transfer function the circuit relies on the virtual ground generated at the negative input of the integrator to accomplish complete charge transfer during each clock period. However, limited amplifier gain does not generate a perfect virtual ground, thereby not accomplishing the complete transfer of charge during each clock period. The effect of the limited gain is similar to a leaky integrator and the transfer function for the leaky integrator can be written as

$$H(z) = \frac{-1}{1-\alpha z^{-1}} \tag{64.42a}$$

where

$$\alpha = \frac{1}{1-\dfrac{1}{A}\left(1+\dfrac{C_1}{C_2}\right)} \approx \frac{1}{1-\dfrac{2}{A}} \tag{64.42b}$$

The net effect of finite gain is to increase the modulation noise in the signal band as illustrated by Fig. 64.69 for the first-order modulator. In this figure the X-axis is plotted from 0 to 1 rather than the complete Nyquist band to emphasize the signal band. The noise transfer function has been plotted for a number of amplifier gains. When compared to Fig. 64.60, note the increase in the noise level in the signal band. The effect of finite gain is felt throughout the input signal magnitude range as shown in Fig. 64.70. The graph for Fig. 64.69 was generated using the linearized model for the modulator presented in (64.21) and the graph in Fig. 64.70 was generated using the difference equation method. The difference equation method does not make any assumptions about linearity nor does it assume that the input is uncorrelated with the quantization noise, however, it requires considerably more simulation time. Because of oversampling the bandwidth requirements for the op amps in the integrators are usually

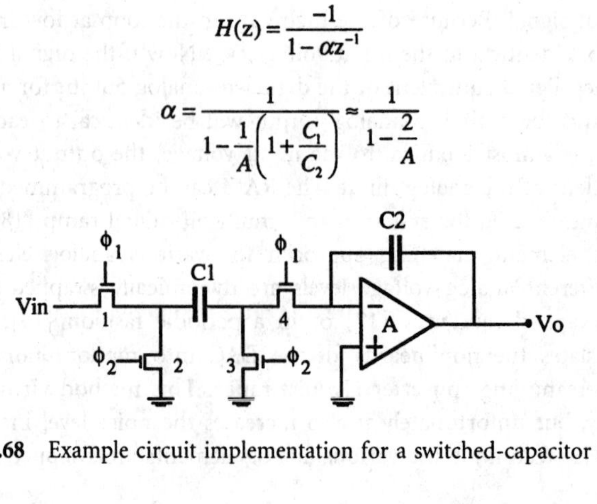

FIGURE 64.68 Example circuit implementation for a switched-capacitor integrator.

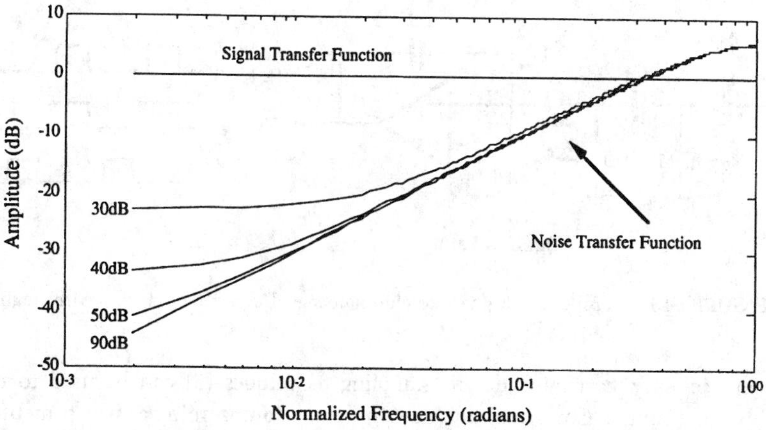

FIGURE 64.69 Effect of finite amplifier gain on noise transfer function using a linear model.

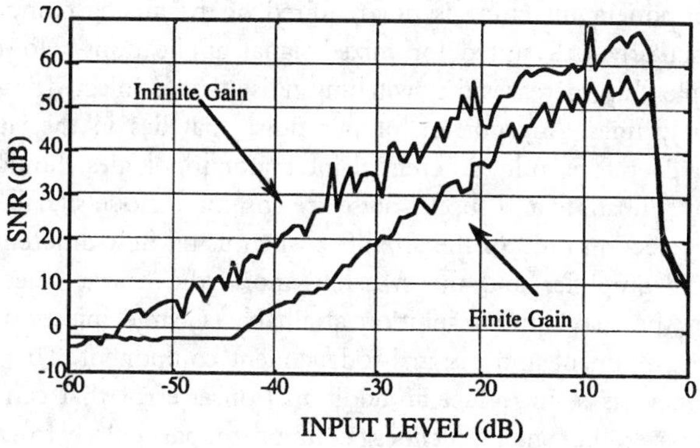

FIGURE 64.70 Effect of finite amplifier gain using difference equation simulations.

large. Unfortunately, it is extremely difficult to realize extremely high gain and extremely high bandwidth amplifiers in MOS. One solution that attempts to mitigate the finite gain effect is to estimate the amount of incomplete charge transfer and compensate for it [5].

Circuit noise provides additional limitations to the maximum resolution realizable by an oversampled converter. The primary noise sources are the thermal noise generated by the switches in the integrator, amplifier noise, charge injection, and clock feedthrough from the switches. Because of sampling the thermal noise associated with the finite on resistance of the switches is aliased back into the Nyquist band of the oversampling converter. The total noise aliased into the baseband for large bandwidth amplifiers is equal to kT/C for each switch pair, where k is the Boltzmann constant, T is the temperature in degrees Kelvin, and C is the value of the sampling capacitor in the integrator. For the parasitic insensitive integrator shown in Fig. 64.68, the total noise from this source is equal to $2kT/C$. This noise is evenly spread across the Nyquist band, but only the fraction $2f_0/f_s$ of this noise appears in the signal band. The rest is filtered out by the digital LPF. Using this constraint, for a full-scale sine wave input the minimum sampling capacitance is given by

$$C_{\min} = 16 \cdot kT \cdot \text{SNR}_{\text{desired}} \tag{64.43}$$

The inband portion of the amplifier noise is also added to the output signal. In general, only the noise of the first amplifier is important for higher-order converters. For MOS amplifiers the flicker noise component is significantly more important as it tends to dominate in the signal

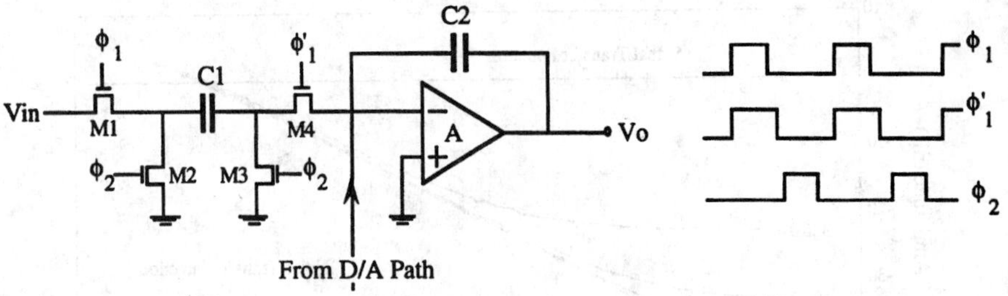

FIGURE 64.71 Proper clock phase to eliminate signal-dependent clock feedthrough.

band. When necessary, correlated double sampling techniques [5] can be used to reduce the effect of this noise source. Correlated double sampling, or autozeroing as it is sometimes called, has the added benefit that it eliminates any amplifier offset voltages. It is usually important to remove this offset voltage only for data acquisition applications.

Because tight component matching is not required of the analog components, sigma-delta converters are particularly well suited for mixed-signal applications. However, having digital circuits on the sample chip increases the switching noise that is injected into the substrate and into the power supply lines. Any portion of this noise that lies in the signal band is added to the input signal. Therefore, fully differential integrator topologies should be used for high-resolution converters. Substrate and supply noise are common-mode signals and are reduced by the common-mode rejection ratio of the amplifier when using fully differential circuits.

In addition to the amplifier and the switching noise the charge injection from switches also sets a limit on the maximum resolution attainable. Charge injection from switches has a signal-dependent component and a signal-independent component. The effect of the signal-independent component is to introduce an additional offset error that can easily be calibrated out, if necessary. However, the signal-dependent component, particularly from the input sampling transistor (transistor M_1 in Fig. 64.71), cannot be distinguished from input signal. This signal-dependent component is highly nonlinear and can be reduced substantially by using proper clock phasing. Signal-dependent charge injection from transistors M_1 and M_2 in Fig. 64.71 can be canceled to first order by delaying the turn off of ϕ_1' slightly [6].

A number of topologies for the digital low-pass filters have been tried. However, it has been shown that simple finite impulse response (sinc) filters are probably the optimal choice. It has been shown that the number of stages of sinc filtering necessary is equal to the modulator order plus 1 [2]. Noise-shaping converters have the ability to provide extremely high resolution. However, care must be used when using simple linear assumptions. Clearly, for the first-order modulator the white noise assumption breaks down. Additionally, it has been shown that the simple linear model overestimates the realizable SNR. For example, the linearized model overestimates the attainable SNR by as much as 14 dB for the second-order modulator [Hauser, 1991].

Acknowledgment

The author wishes to acknowledge the help of his students in completing this manuscript, in particular, that of Feng Wang.

References

[1] W. C. Black, "High Speed CMOS A/D Conversion Techniques," Ph.D. thesis, Berkeley: University of California, 1980.

[2] J. C. Candy and G. C. Temes, *Oversampling Methods for A/D and D/A Conversion.* New York: IEEE Press, 1992.

[3] M. J. Demler, *High-Speed Analog-to-Digital Conversion*, New York: Academic Press, 1991.

[4] P. R. Gray and D. A. Hodges, "All-MOS Analog-Digital Conversion Techniques," *IEEE Trans. Circuits Syst.*, vol. CAS-25(7), pp. 482–489, 1978.

[5] P. J. Hurst and R. A. Levinson, "Delta-Sigma A/Ds with Reduced Sensitivity to Op Amp Noise and Gain," *IEEE Int. Symp. Circuits Syst.*, pp. 254–257, 1989.

[6] P. W. Li, M. J. Chin, P. R. Gray, and R. Castello, "A Ratio-Independent Algorithmic Analog-to-Digital Conversion Technique," *IEEE J. Solid-State Circuits*, vol. SC-19(6), pp. 828–836, 1984.

[7] H. S. Lee, D. A. Hodges, and P. R. Gray, "A Self-Calibrating 15 Bit CMOS A/D Converter," *IEEE J. Solid-State Circuits*, vol. SC-19(6), pp. 813–819, 1983.

[8] Y. M. Matsuya, K. Uchimura, and A. Iwata, "A 16-Bit Oversampling A-to-D Conversion Technology using Triple Integration Noise Shaping," *IEEE J. Solid-State Circuits*, vol. SC-22, pp. 921–929, 1987.

[9] J. L. McCreary and P. R. Gray, "All-MOS Charge Redistribution Analog-to-Digital Conversion Techniques. I," *IEEE J. Solid-State Circuits*, vol. SC-10, pp. 371–379, 1975.

[10] R. H. McCharles, V. A. Saletore, W. C. Black, and D. A. Hodges, "An Algorithmic Analog-to-Digital Converter," *IEEE Int. Solid-State Circuits Conf.*, 1977.

[11] S. Masuda, Y. Kitamura, S. Ohya, and M. Kikuchi, "A CMOS Pipelined Algorithmic A/D Converter," *IEEE Custom Integrated Circuits Conf.*, pp. 559–562, 1984.

[12] G. C. Temes, F. J. Wang, and K. Watanabe, "Novel Pipeline Data Converters," *IEEE Int. Symp. Circuits Systems*, pp. 1943–1946, 1988.

[13] R. Unbehauen and A. Cichocki, *MOS Switched-Capacitor and Continuous-Time Integrated Circuits and Systems.* New York: Springer-Verlag, 1989.

[14] S. H. Lewis, Video-Rate Analog-to-Digital Conversion Using Pipelined Architectures. Ph.D. thesis, Berkeley: University of California, 1987.

[15] S. Sutarja and P. R. Gray, "A Pipelined 13-Bit, 250-ks/s, 5-V Analog-to-Digital Converter," *IEEE J. Solid-State Circuits*, vol. SC-23, pp. 1316–1323, 1988.

[16] K. Chao, S. Nadeem, W. Lee, and C. Sodini, "A Higher Order Topology for Interpolative Modulators for Oversampling A/D Converters," *IEEE Trans. Circuits Syst.*, pp. 309–318, Mar. 1990.

[17] R. Carley, "A Noise-Shaping Coder Topology for 15+ Bit Converters," *IEEE J. Solid-State Circuits*, pp. 267–273, Apr. 1989.

[18] R. Walden, et al., "Architectures for Higher-Order Multibit Sigma-Delta Modulators," *IEEE Int. Symp. Circuits Syst.*, pp. 895–898, 1990.

[19] B. Leung and S. Sutarja, "Multibit Sigma-Delta A/D Converter Incorporating a Novel Class of Dynamic Element Matching Techniques," *IEEE Trans. Circuits Syst.*, vol. CAS-39(1), pp. 35–51, 1992.

urther Information

Max Hauser provides an extremely good overview of oversampling converters in "Principles of Oversampling A/D Conversion," *J. Audio Eng. Soc.*, vol. 39(1/2), pp. 3–26, 1991.

Sources for further reading about Nyquist rate converters include D. J. Dooley, *Data Conversion Integrated Circuits.* New York: IEEE Press, 1980, and R. J. van der Plassche, *Integrated Analog-to-Digital and Digital-to-Analog Converters.* Boston: Kluwer Academic, 1994.

IEEE Journal of Solid-State Circuits, particularly the December issues, and the *IEEE Transactions on Circuits and Systems* are good sources for more recent research on data converters.

XIII

ilter Characteristics

ıwrence P. Huelsman
ıiversity of Arizona

65

General Characteristics of Filters

ndreas Antoniou
iversity of Victoria,
nada

5.1 Introduction

An electrical filter is a system that can be used to modify, reshape, or manipulate the frequency spectrum of an electrical signal according to some prescribed requirements. For example, a filter may be used to amplify or attenuate a range of frequency components, reject or isolate one specific frequency component, and so on. The applications of electrical filters are numerous, for example,

- to eliminate signal contamination such as noise in communication systems
- to separate relevant from irrelevant frequency components
- to detect signals in radios and TV's
- to demodulate signals
- to bandlimit signals before sampling
- to convert sampled signals into continuous-time signals
- to improve the quality of audio equipment, e.g., loudspeakers
- in time-division to frequency-division muliplex systems

3-8341-2/95/$0.00+$.50

- in speech synthesis
- in the equalization of transmission lines and cables
- in the design of artificial cochleas.

Typically, an electrical filter receives an input signal or *excitation* and produces an output signal or *response*. The frequency spectrum of the output signal is related to that of the input by some rule of correspondence. Depending on the type of input, output, and internal operating signals, three general types of filters can be identified, namely, continuous-time, sampled-data, and discrete-time filters.

A **continuous-time** signal is one that is defined at each and every instant of time. It can be represented by a function $x(t)$ whose domain is a range of numbers (t_1, t_2), where $-\infty \leq t_1$ and $t_2 \leq \infty$. A **sampled-data** or **impulse-modulated** signal is one that is defined in terms of an infinite summation of continuous-time impulses (see [1, ch. 6]). It can be represented by a function

$$\hat{x}(t) = \sum_{n=-\infty}^{\infty} x(nT)\delta(t - nT)$$

where $\delta(t)$ is the impulse function. The value of the signal at any instant in the range $nT < t < (n + 1)T$ is zero. The frequency spectrum of a continuous-time or sampled-data signal is given by the Fourier transform.[1]

A **discrete-time** signal is one that is defined at discrete instants of time. It can be represented by a function $x(nT)$, where T is a constant and n is an integer in the range (n_1, n_2) such that $-\infty \leq n_1$ and $n_2 \leq \infty$. The value of the signal at any instant in the range $nT < t < (n + 1)T$ can be zero, constant, or undefined depending on the application. The frequency spectrum in this case is obtained by evaluating the z transform on the unit circle $|z| = 1$ of the z plane.

Depending on the format of the input, output, and internal operating signals, filters can be classified either as **analog** or **digital** filters. In analog filters the operating signals are varying voltages and currents, whereas in digital filters they are encoded in some binary format. Continuous-time and sampled-data filters are always analog filters. However, discrete-time filters can be analog or digital.

Analog filters can be classified on the basis of their constituent components as

- passive *RLC* filters
- crystal filters
- mechanical filters
- microwave filters
- active *RC* filters
- switched-capacitor filters.

Passive RLC filters comprise resistors, inductors, and capacitors. **Crystal** filters are made of piezoelectric resonators that can be modeled by resonant circuits. **Mechanical** filters are made of mechanical resonators. **Microwave** filters consist of microwave resonators and cavities that can be represented by resonant circuits. **Active** *RC* filters comprise resistors, capacitors, and amplifiers; in these filters, the performance of resonant circuits is simulated through the use of feedback or by supplying energy to a passive circuit. **Switched-capacitor** filters comprise resistors, capacitors, amplifiers, and switches. These are discrete-time filters that operate like active filters but through the use of switches the capacitance values can be kept very small. As a result, switched-capacitor filters are amenable to VLSI implementation.

[1]See Chapter 4 by W. K. Jenkins.

This section provides an introduction to the characteristics of analog filters. Their basic characterization in terms of a differential equation is reviewed in Section 65.2 and by applying the Laplace transform, an algebraic equation is deduced that leads to the s-domain representation of a filter. The representation of analog filters in terms of the transfer function is then developed. Using the transfer function, one can obtain the time-domain response of a filter to an arbitrary excitation, as shown in Section 65.3. Some important time-domain responses, i.e., the impulse and step responses, are examined. Certain filter parameters related to the step response, namely, the overshoot, delay time, and rise time, are then considered. The response of a filter to a sinusoidal excitation is examined in Section 65.4 and is then used to deduce the basic frequency-domain representations of a filter, namely, its frequency response and loss characteristic. Some idealized filter characteristics are then identified and the differences between idealized and practical filters are delineated in Section 65.5. Practical filters tend to introduce signal degradation through amplitude and/or delay distortion. The causes of these types of distortion are examined in Section 65.6. In Section 65.7, certain special classes of filters, e.g., minimum-phase and allpass filters, are identified and their applications mentioned. This chapter concludes with a review of the design process and the tasks that need to be undertaken to translate a set of filter specifications into a working prototype.

65.2 Characterization

A linear causal analog filter with input $x(t)$ and output $y(t)$ can be characterized by a differential equation of the form

$$b_n \frac{d^n y(t)}{dt^n} + b_{n-1} \frac{d^{n-1} y(t)}{dt^{n-1}} + \cdots + b_0 y(t) = a_n \frac{d^n x(t)}{dt^n} + a_{n-1} \frac{d^{n-1} x(t)}{dt^{n-1}} + \cdots + a_0 x(t)$$

The coefficients $a_0, a_1, \cdots, a_n$ and $b_0, b_1, \cdots, b_n$ are functions of the element values and are real if the parameters of the filter (e.g., resistances, inductances, etc.) are real. If they are independent of time, the filter is time invariant. The input $x(t)$ and output $y(t)$ can be either voltages or currents. The order of the differential equation is said to be the *order* of the filter.

An analog filter must of necessity incorporate reactive elements that can store energy. Consequently, the filter can produce an output even in the absence of an input. The output on such an occasion is caused by the initial conditions of the filter, namely,

$$\left.\frac{d^{n-1} y(t)}{dt^{n-1}}\right|_{t=0}, \quad \left.\frac{d^{n-2} y(t)}{dt^{n-2}}\right|_{t=0}, \quad \cdots, \; y(0)$$

The response in such a case is said to be the *zero-input response.* The response obtained if the initial conditions are zero is sometimes called the *zero-state response.*

The Laplace Transform

The most important mathematical tool in the analysis and design of analog filters is the Laplace transform. It owes its widespread application to the fact that it transforms differential into algebraic equations that are a lot easier to manipulate. The Laplace transform of $x(t)$ is defined as[2]

$$X(s) = \int_{-\infty}^{\infty} x(t) e^{-st}\, dt$$

[2]See Chapter 3 by J. R. Deller, Jr. for a detailed exposition of the Laplace transform.

where s is a complex variable of the form $s = \sigma + j\omega$. Signal $x(t)$ can be recovered from $X(s)$ by applying the inverse Laplace transform, which is given by

$$x(t) = \frac{1}{2\pi j}\int_{C-j\infty}^{C+j\infty} X(s)e^{st}\, ds$$

where C is a positive constant. A short-hand notation of the Laplace transform and its inverse are

$$X(s) = \mathscr{L}x(t) \quad \text{and} \quad x(t) = \mathscr{L}^{-1}X(s)$$

Alternatively,

$$X(s) \leftrightarrow x(t)$$

A common practice in the choice of symbols for the Laplace transform and its inverse is to use upper case for the s domain and lower case for the time domain.

On applying the Laplace transform to the nth derivative of some function of time $y(t)$, we find that

$$\mathscr{L}\left[\frac{d^n y(t)}{dt^n}\right] = s^n Y(s) - s^{n-1}y(0) - s^{n-2}\left.\frac{dy(t)}{dt}\right|_{t=0} - \cdots - \left.\frac{d^{n-1}y(t)}{dt^{n-1}}\right|_{t=0}$$

Now on applying the Laplace transform to an nth-order differential equation with constant coefficients, we obtain

$$(b_n s^n + b_{n-1}s^{n-1} + \cdots + b_0)Y(s) + \Psi_y(s) = (a_n s^n + a_{n-1}s^{n-1} + \cdots + a_0)X(s) + \Psi_x(s)$$

where $X(s)$ and $Y(s)$ are the Laplace transforms of the input and output, respectively, and $\Psi_x(s)$ and $\Psi_y(s)$ are functions that combine all the initial-condition terms that depend on $x(t)$ and $y(t)$, respectively.

The Transfer Function

An important s-domain characterization of an analog filter is its *transfer function*, as for any other linear system. This is defined as the ratio of the Laplace transform of the response to the Laplace transform of the excitation.

An arbitrary linear, time-invariant, continuous-time filter, which may or may not be causal, can be represented by the convolution integral

$$y(t) = \int_{-\infty}^{\infty} h(t-\tau)x(\tau)\, d\tau = \int_{-\infty}^{\infty} h(\tau)x(t-\tau)\, d\tau$$

where $h(t)$ is the impulse response of the filter. The Laplace transform yields

$$\begin{aligned} Y(s) &= \int_{-\infty}^{\infty}\left[\int_{-\infty}^{\infty} h(t-\tau)x(\tau)\, d\tau\right] e^{-st}\, dt \\ &= \int_{-\infty}^{\infty}\int_{-\infty}^{\infty} h(t-\tau)e^{-st}x(\tau)\, d\tau\, dt \\ &= \int_{-\infty}^{\infty}\int_{-\infty}^{\infty} h(t-\tau)e^{-st}\cdot e^{s\tau}\cdot e^{-s\tau}x(\tau)\, d\tau\, dt \end{aligned}$$

Changing the order of integration, we obtain

$$Y(s) = \int_{-\infty}^{\infty} \int_{-\infty}^{\infty} h(t-\tau)e^{-s(t-\tau)} \cdot x(\tau)e^{-s\tau}\, dt\, d\tau$$

$$= \int_{-\infty}^{\infty} \int_{-\infty}^{\infty} h(t-\tau)e^{-s(t-\tau)}\, dt \cdot x(\tau)e^{-s\tau}\, d\tau$$

Now if we let $t = t' + \tau$, then $dt/dt' = 1$ and $t - \tau = t'$; hence

$$Y(s) = \int_{-\infty}^{\infty} \int_{-\infty}^{\infty} h(t')e^{-st'}\, dt' \cdot x(\tau)e^{-s\tau}\, d\tau$$

$$= \int_{-\infty}^{\infty} h(t')e^{-st'}\, dt' \cdot \int_{-\infty}^{\infty} x(\tau)e^{-s\tau}\, d\tau$$

$$= H(s)X(s)$$

Therefore, the transfer function is given by

$$H(s) = \frac{Y(s)}{X(s)} = \mathscr{L}h(t) \tag{65.1}$$

In effect, the transfer function is equal to the Laplace transform of the impulse response.

Some authors define the transfer function as the Laplace transform of the impulse response. Then through the use of the convolution integral, they show that the transfer function is equal to the ratio of the Laplace transform of the response to the Laplace transform of the excitation. The two definitions are, of course, equivalent.

Typically, in analog filters the input and output are voltages, e.g., $x(t) \equiv v_i(t)$ and $y(t) \equiv v_o(t)$. In such a case the transfer function is given by

$$\frac{V_o(s)}{V_i(s)} = H_V(s)$$

or simply by

$$\frac{V_o}{V_i} = H_V(s)$$

However, on occasion the input and output are currents, in which case

$$\frac{I_o(s)}{I_i(s)} \equiv \frac{I_o}{I_i} = H_I(s)$$

The transfer function can be obtained through network analysis using one of several classical methods,[3] e.g., by using

- Kirchhoff's voltage and current laws
- matrix methods
- flow graphs

[3]See Chapters 17 to 23.

- Mason's gain formula
- state–space methods.

A transfer function is said to be *realizable* if it characterizes a stable and causal network. Such a transfer function must satisfy the following constraints:

1. It must be a rational function of s with real coefficients.
2. Its poles must lie in the left-half s plane.
3. The degree of the numerator polynomial must be equal to or less than that of the denominator polynomial.

A transfer function may represent a network comprising elements with real parameters only if its coefficients are real. The poles must be in the left-half s plane to ensure that the network is stable and the numerator degree must not exceed the denominator degree to assure the existence of a causal network.

65.3 Time-Domain Response

From (65.1)

$$Y(s) = H(s)X(s)$$

Therefore, the time-domain response of a filter to some arbitrary excitation can be deduced by obtaining the inverse Laplace transform of $Y(s)$, i.e.,

$$y(t) = \mathscr{L}^{-1}\{H(s)X(s)\}$$

General Inversion Formula

If

1. the singularities of $Y(s)$ in the finite plane are poles,[4] and
2. $Y(s) \to 0$ uniformly with respect to the angle of s as $|s| \to \infty$ with $\sigma \leq C$, where C is a positive constant, then [3]

$$y(t) = \begin{cases} 0 & \text{for } t < 0 \\ \dfrac{1}{2\pi j}\displaystyle\int_{C-j\infty}^{C+j\infty} Y(s)e^{st}\,ds = \dfrac{1}{2\pi j}\displaystyle\int_{\Gamma} Y(s)e^{st}\,ds & \text{for } t \geq 0 \end{cases} \tag{65.2}$$

where Γ is a contour in the counterclockwise sense make up of the part of the circle $s = Re^{j\theta}$ to the left of line $s = C$ and the segment of the line $s = C$ that overlaps the circle, as depicted in Fig. 65.1; C and R are sufficiently large to ensure that Γ encloses all the finite poles of $Y(s)$.

From the residue theorem [2] and (65.2), we have

$$y(t) = \begin{cases} 0 & \text{for } t < 0 \\ \dfrac{1}{2\pi j}\displaystyle\int_{\Gamma} Y(s)e^{st}\,ds = \sum_{i=1}^{K} \operatorname*{res}_{s=p_i} Y_0(s) & \text{for } t \geq 0 \end{cases}$$

[4]Such a function is said to be meromorphic [2], [3].

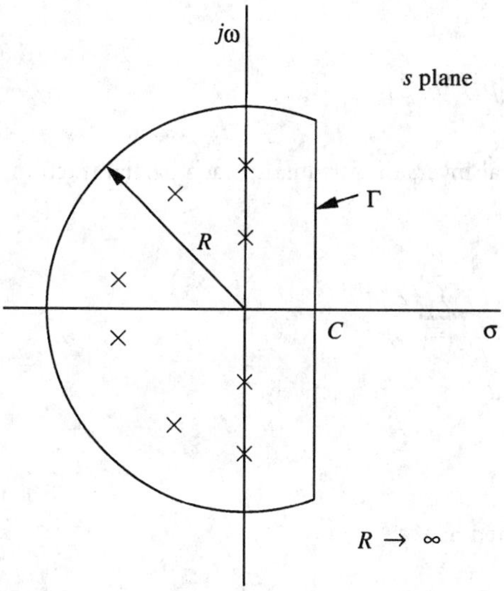

FIGURE 65.1 Contour Γ for the evaluation of the inverse Laplace transform.

where $Y_0(s) = Y(s)e^{st}$ and K is the number of poles in $Y(s)$. If $Y_0(s)$ has a pole p_i of order m_i, the residue can be obtained by using the general formula [2]

$$\operatorname*{res}_{z=p_i} Y_0(s) = \frac{1}{(m_i - 1)!} \lim_{s \to p_i} \frac{d^{m_i-1}}{ds^{m_i-1}} [(s - p_i)^{m_i} Y_0(s)]$$

Note that complex poles yield complex residues. Hence, like the poles of $Y_0(s)$, its residues occur in complex-conjugate pairs. For this reason, $y(t)$ is found to be a real function of t, as can be easily verified.

Condition 1 above may not be satisfied sometimes, for example, if

$$\lim_{s \to \infty} Y(s) = A_0$$

where A_0 is a constant. In such a case, we can express $Y(s)$ as

$$Y(s) = A_0 + Y'(s)$$

where $Y'(s)$ satisfies conditions 1 and 2. Thus

$$y(t) = A_0 \delta(t) + \mathscr{L}^{-1} Y'(s)$$

The inverse Laplace transform of $Y'(s)$ can now be obtained by using the inversion formula.

Inverse by Using Partial Fractions

The simplest way to obtain the time-domain response of a filter is to express $H(s)X(s)$ as a partial-fraction expansion and then invert the resulting fractions individually. If $Y(s)$ has simple poles, we can write

$$Y(s) = A_0 + \sum_{i=1}^{K} \frac{A_i}{s - p_i}$$

where A_0 is a constant and

$$A_i = \lim_{s \to p_i} [(s - p_i) Y(s)]$$

is the residue of pole $s = p_i$. On applying the general inversion formula to each partial fraction, we obtain

$$y(t) = A_0 \delta(t) + u(t) \sum_{i=1}^{K} A_i e^{p_i t}$$

where $\delta(t)$ and $u(t)$ are the impulse function and unit step, respectively.

Impulse and Step Responses

The response of a filter to an impulse $\delta(t)$ designated as

$$y(t) = \mathscr{R}\delta(t) \equiv h(t)$$

where $\mathscr{R}$ is an operator, is of considerable importance. Its absolute integrability guarantees the stability of the filter[5] and its Laplace transform, namely, $H(s)$, is the transfer function as has been shown in the section on the transfer function.

For an Nth-order, causal, linear, and time-invariant filter

$$H(s) = \frac{a_0 + a_1 s + a_2 s^2 + \cdots + a_M s^M}{b_0 + b_1 s + b_2 s^2 + \cdots + b_N s^N}$$

where $M \leq N$.

The step (or unit-step) response is the output of a filter to the signal

$$u(t) = \begin{cases} 1 & \text{for } t \geq 0 \\ 0 & \text{for } t < 0 \end{cases}$$

The Laplace transform of $u(t)$ is $1/s$. Hence the step response of an arbitrary filter is obtained as

$$y(t) = \mathscr{R}u(t) \equiv y_u(t) = \mathscr{L}^{-1}\left[\frac{H(s)}{s}\right]$$

Overshoot, Delay Time, and Rise Time

There are three time-domain parameters of a filter that are usually associated with the step response [4], namely, the overshoot, delay time and rise time. The *overshoot* γ is the difference between the peak value and the asymptotic value of the step response in percent as $t \to \infty$. The *delay time* τ_d is the time required for the step response to reach 50 percent of the asymptotic value. The *rise time* τ_r is the time required for the step response to increase from 10 to 90 percent of the asymptotic value. These three parameters are illustrated in Fig. 65.2, where $K = a_0/b_0$ is a scaling constant that normalizes the asymptotic value of the step response as $t \to \infty$ to unity.

[5] See Chapter 15, contribution of J. Vlach.

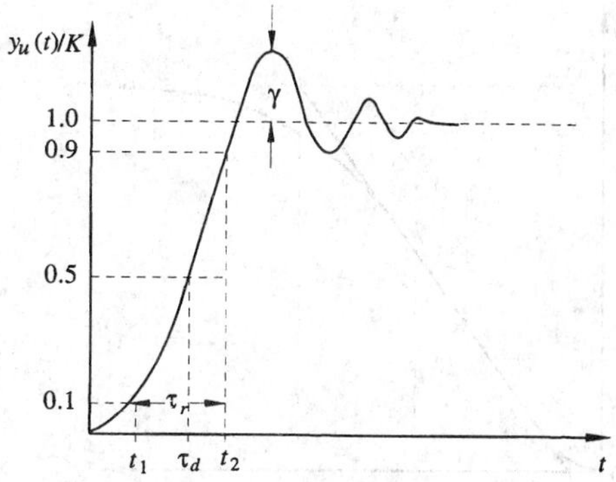

FIGURE 65.2 Overshoot, delay time, and rise time.

The delay and rise times defined in terms of the step response entail quite a bit of computation. Alternative definitions of these parameters that are easier to use have been proposed by Elmore [4]. These are based on the impulse response and give accurate results if the overshoot is small. The delay time is defined as

$$\tau_D = \int_0^\infty th(t)\,dt$$

and the rise time assumes the form

$$\tau_R = \left[2\pi \int_0^\infty (t-\tau_D)^2 h(t)\,dt\right]^{1/2} = \sqrt{2\pi}\left[\int_0^\infty t^2 h(t)\,dt - \tau_D^2\right]^{1/2}$$

The physical interpretation of these parameters is illustrated in Fig. 65.3(a) and (b). If the overshoot is small, say less than 1 percent, then

$$\tau_D \approx \tau_d \quad \text{and} \quad \tau_R \approx \tau_r$$

The simplification brought about by Elmore's definitions can be easily demonstrated. Consider a filter whose step response approaches unity as $t \to \infty$. Such a filter has a transfer function of the form

$$H(s) = \frac{1 + a_1 s + a_2 s^2 + \cdots + a_M s^M}{1 + b_1 s + b_2 s^2 + \cdots + b_N s^N} \tag{65.3}$$

that is, $a_0 = b_0 = 1$. From the definition of the Laplace transform

$$\begin{aligned} H(s) &= \int_0^\infty h(t)e^{-st}\,dt \\ &= \int_0^\infty h(t)\left(1 - st + \frac{s^2t^2}{2!} - \cdots\right)dt \\ &= \int_0^\infty h(t)\,dt - s\int_0^\infty th(t)\,dt + \frac{s^2}{2!}\int_0^\infty t^2h(t)\,dt - \cdots \\ &= \int_0^\infty h(t)\,dt - s\tau_D + \frac{s^2}{2!}\left(\frac{\tau_R^2}{2\pi} + \tau_D^2\right) - \cdots \end{aligned} \tag{65.4}$$

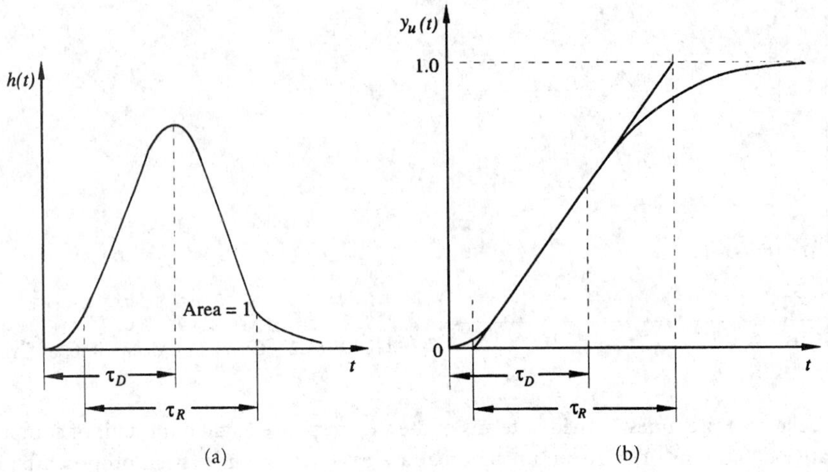

FIGURE 65.3 Physical interpretation of Elmore's definitions of delay and rise times: (a) impulse response $h(t)$, (b) unit-step response $y_u(t)$.

Alternatively, from (65.3), direct division gives

$$H(s) = 1 - (b_1 - a_1)s + (b_1^2 - a_1 b_1 + a_2 - b_2)s^2 + \cdots \tag{65.5}$$

Now by comparing (65.4) and (65.5), we deduce

$$\int_0^\infty h(t)\,dt = 1, \; \tau_D = b_1 - a_1$$

and

$$\tau_R = \{2\pi[b_1^2 - a_1^2 + 2(a_2 - b_2)]\}^{1/2}$$

The above definitions are based on the assumption that the unit-step response approaches unity as $t \to \infty$. If this is not the case, i.e., coefficients a_0 and b_0 are not equal to unity, then we can write

$$H(s) = KH'(s)$$

where $K = a_0/b_0$ and

$$H'(s) = \frac{1 + a'_1 s + a'_2 s^2 + \cdots + a'_M s^M}{1 + b'_1 s + b'_2 s^2 + \cdots + b'_N s^N}$$

Using the coefficients of $H'(s)$ in the formulas for τ_D and τ_R yields approximate values for the delay time and rise time, since these parameters are independent of the absolute value of the step response.

65.4 Frequency-Domain Analysis

The frequency response of an analog filter is deduced by finding its steady-state sinusoidal response, as we shall now demonstrate.

Sinusoidal Response

Consider an Nth-order analog filter characterized by a transfer function $H(s)$. The sinusoidal response of such a filter is

$$y(t) = \mathscr{L}^{-1}[H(s)X(s)]$$

where

$$X(s) = \mathscr{L}[u(t)\sin\omega t] = \frac{\omega}{(s+j\omega)(s-j\omega)} \tag{65.6}$$

The product $H(s)X(s)$ satisfies conditions 1 and 2 imposed on the general inversion formula of (65.2). Hence for $t \geq 0$, we have

$$y(t) = \frac{1}{2\pi j}\int_{\Gamma} Y(s)e^{st}\,ds = \sum \text{res}[H(s)X(s)e^{st}] \tag{65.7}$$

where Γ is a contour enclosing the poles of $H(s)$ and $X(s)$ as in Fig. 65.1.

From (65.6) and (65.7)

$$y(t) = \sum_{i=1}^{N} X(p_i)e^{p_i t} \operatorname*{res}_{s=p_i} H(s) + \frac{1}{2j}[H(j\omega)e^{j\omega t} - H(-j\omega)e^{-j\omega t}] \tag{65.8}$$

If the filter is assumed to be stable, then the poles are in the left-half s plane, i.e., $p_i = \sigma_i + j\omega_i$ with $\sigma_i < 0$.[6] As a consequence

$$\lim_{t\to\infty} e^{p_i t} = \lim_{t\to\infty}(e^{\sigma_i t} \cdot e^{j\omega_i t}) = 0$$

and since the residues of $H(s)$ are finite, the steady-state sinusoidal response is obtained from (65.8) as

$$\tilde{y}(t) = \lim_{t\to\infty} y(t) = \frac{1}{2j}[H(j\omega)e^{j\omega t} - H(-j\omega)e^{-j\omega t}] \tag{65.9}$$

Now from the definition of the Laplace transform

$$H(s) = \int_{-\infty}^{\infty} h(t)e^{-st}\,dt$$

and hence

$$H(-j\omega) = \int_{-\infty}^{\infty} h(t)e^{j\omega t}\,dt = \left[\int_{-\infty}^{\infty} h(t)e^{-j\omega t}\,dt\right]^* = H^*(j\omega) \tag{65.10}$$

[6] See Chapter 15 and 44.

If we write

$$H(j\omega) = M(\omega)e^{j\theta(\omega)} \tag{65.11}$$

where

$$M(\omega) = |H(j\omega)| \quad \text{and} \quad \theta(\omega) = \arg H(j\omega) \tag{65.12}$$

the steady-state sinusoidal response of the filter is obtained from (65.9)–(65.12) as

$$\begin{aligned}\tilde{y}(t) &= \frac{1}{2j}[M(\omega)e^{j\theta(\omega)}e^{j\omega t} - M(\omega)e^{-j\theta(\omega)}e^{-j\omega t}] \\ &= M(\omega)\frac{1}{2j}[e^{j[\omega t+\theta(\omega)]} - e^{-j[\omega t+\theta(\omega)]}] \\ &= M(\omega)\sin[\omega t + \theta(\omega)]\end{aligned}$$

The preceding analysis has shown that the steady-state response of an analog filter to a sinusoid of unit amplitude is a sinusoid of amplitude $M(\omega)$, shifted by an angle $\theta(\omega)$. In effect, for a given frequency ω, the filter introduces a *gain* $M(\omega)$ and a *phase shift* $\theta(\omega)$.

As functions of frequency, $M(\omega)$ and $\theta(\omega)$ are known as the **amplitude** (or **magnitude**) **response** and **phase response** of the filter, respectively. The transfer function evaluated on the imaginary axis, namely, $H(j\omega)$ is the **frequency response** and, as was shown, its magnitude and angle are the amplitude response and phase response, respectively.

Two other quantities of a filter, which are of significant interest, are its *phase* and *group delays*. These are defined as

$$\tau_p(\omega) = -\frac{\theta(\omega)}{\omega} \quad \text{and} \quad \tau_g(\omega) = -\frac{d\theta(\omega)}{d\omega}$$

respectively. For filters, the group delay is the more important of the two. As a function of frequency, $\tau_g(\omega)$ is usually referred to as the delay characteristic.

Graphical Construction

Consider a filter characterized by a transfer function of the form

$$H(s) = H_0\frac{N(s)}{D(s)} = H_0\frac{\prod_{i=1}^{M}(s - z_i)}{\prod_{i=1}^{N}(s - p_i)^{m_i}} \tag{65.13}$$

where H_0 is a constant. The frequency response of the filter is obtained as

$$H(j\omega) = M(\omega)e^{j\theta(\omega)} = \frac{H_0\prod_{i=1}^{M}(j\omega - z_i)}{\prod_{i=1}^{N}(j\omega - p_i)^{m_i}}$$

By letting

$$j\omega - z_i = M_{z_i} e^{j\psi_{z_i}} \tag{65.14}$$

$$j\omega - p_i = M_{p_i} e^{j\psi_{p_i}} \tag{65.15}$$

we obtain

$$M(\omega) = \frac{|H_0| \prod_{i=1}^{M} M_{z_i}}{\prod_{i=1}^{N} M_{p_i}^{m_i}} \tag{65.16}$$

and

$$\theta(\omega) = \arg H_0 + \sum_{i=1}^{M} \psi_{z_i} - \sum_{i=1}^{N} m_i \psi_{p_i} \tag{65.17}$$

where $\arg H_o = \pi$ if H_0 is negative.

The gain and phase shift $M(\omega)$ and $\theta(\omega)$ for some frequency $\omega = \omega_i$ can be determined graphically by using the following procedure:

1. Mark the zeros and poles of the filter in the s plane.
2. Draw the phasor $s = j\omega_i$, where ω_i is the frequency of interest.
3. Draw a phasor of the type in (65.14) for each simple zero of $H(s)$.
4. Draw m_i phasors of the type in (65.15) for each pole of order m_i.
5. Measure the magnitudes and angles of the phasors in steps 3 and 4 and use them in (65.16) and (65.17) to calculate the gain $M(\omega_i)$ and phase shift $\theta(\omega_i)$, respectively.

The amplitude and phase responses of a filter can be determined by repeating the above procedure for frequencies $\omega = \omega_1, \omega_2, \cdots$, in the range 0 to ∞. The procedure is illustrated in Fig. 65.4.

It should be mentioned that the modern approach for the analysis of filters is through the use of the many circuit analysis programs such as SPICE.[7] Nevertheless, the above graphical method is of interest and merits consideration for two reasons. First, it illustrates some of the fundamental properties of filters. Second, it provides a certain degree of intuition about the expected amplitude or phase response of a filter. For example, if a filter has a pole close to the $j\omega$ axis, then as ω approaches the neighborhood of the pole, the magnitude of the phasor from the pole to the $j\omega$ axis decreases rapidly to a very small value and then increases as ω increases above this value. As a result, the amplitude response will exhibit a large peak in the frequency range close to the pole. On the other hand, a zero close to or on the $j\omega$ axis will lead to a notch in the amplitude response when ω is in the neighborhood of the zero.

There are other situations of interest, for example, if the poles of a filter are located in a band of the s plane below the horizontal line $s = \omega_c$ and its zeros are located above this line, then the filter will pass low-frequency and attenuate high-frequency components since $M_{zi} < M_{pi}$ if $\omega > \omega_c$ for all i. Such a filter is said to be a *low-pass* filter. If the zeros are located below the line $s = \omega_c$ and the poles above it, then the filter will pass high-frequency and attenuate low-frequency components, i.e., the filter will be a *high-pass* one.

[7]See Chapter 61 contribution of J. G. Rollins.

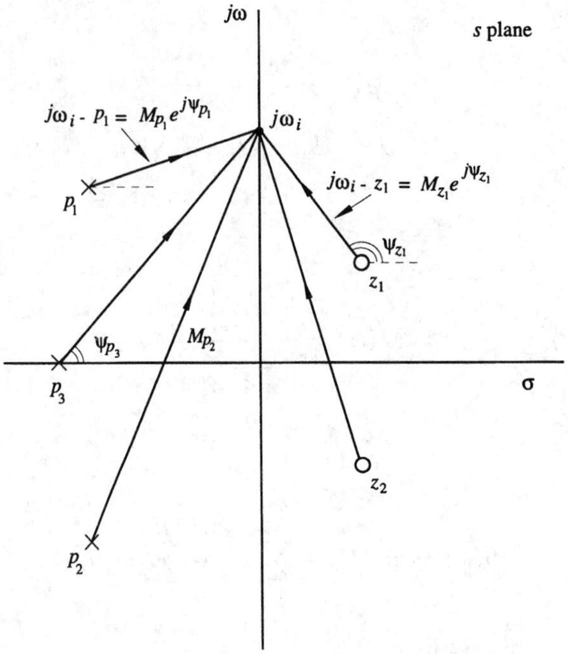

FIGURE 65.4 Graphical method for the evaluation of the frequency response.

Loss Function

Quite often, it is desirable to represent a filter in terms of its loss function. Consider a filter represented by the voltage transfer function

$$\frac{V_o(s)}{V_i(s)} = H(s) = \frac{N(s)}{D(s)}$$

where $V_i(s)$ and $V_o(s)$ are the Laplace transforms of the input and output voltages, respectively, and $N(s)$ and $D(s)$ are polynomials in s. The loss (or attenuation) of the filter in decibels is defined as

$$A(\omega) = 20 \log \left| \frac{V_i(j\omega)}{V_o(j\omega)} \right| = 20 \log \frac{1}{|H(j\omega)|} = 10 \log L(\omega^2) \tag{65.18}$$

where

$$L(\omega^2) = \frac{1}{H(j\omega)H(-j\omega)}$$

$A(\omega)$ as a function of ω is the *loss characteristic.*

With $\omega = s/j$ in (65.18), the function

$$L(-s^2) = \frac{D(s)D(-s)}{N(s)N(-s)}$$

can be formed. This is called the *loss function* of the filter and, as is evident, its zeros are the poles of $H(s)$ and their negatives, whereas its poles are the zeros of $H(s)$ and their negatives.

65.5 Ideal and Practical Filters

An ideal low-pass filter is one that will pass only low-frequency components. Its loss characteristic is given by

$$A(\omega) = \begin{cases} 0 & \text{for } 0 \le \omega < \omega_c \\ \infty & \text{for } \omega_c < \omega < \infty \end{cases}$$

The frequency ranges 0 to ω_c and ω_c to ∞ are the *passband* and *stopband*, respectively. The boundary between the passband and stopband, namely, ω_c, is the *cutoff frequency.* An ideal high-pass filter will pass all components with frequencies above the cutoff frequency and reject all components with frequencies below the cutoff frequency, i.e.,

$$A(\omega) = \begin{cases} \infty & \text{for } 0 \le \omega < \omega_c \\ 0 & \text{for } \omega_c < \omega < \infty \end{cases}$$

Idealized loss characteristics can similarly be identified for bandpass and bandstop filters as

$$A(\omega) = \begin{cases} \infty & \text{for } 0 \le \omega < \omega_{c1} \\ 0 & \text{for } \omega_{c1} < \omega < \omega_{c2} \\ \infty & \text{for } \omega_{c2} \le \omega < \infty \end{cases}$$

and

$$A(\omega) = \begin{cases} 0 & \text{for } 0 \le \omega < \omega_{c1} \\ \infty & \text{for } \omega_{c1} < \omega < \omega_{c2} \\ 0 & \text{for } \omega_{c2} \le \omega < \infty \end{cases}$$

respectively.

Practical filters differ from ideal ones in that the passband loss is not zero, the stopband loss is not infinite, and the transition between passband and stopband is gradual. Practical loss characteristics for low-pass, high-pass, bandpass, and bandstop filters assume the forms

$$A_{\text{LP}}(\omega) \begin{cases} \le A_p & \text{for } 0 \le \omega \le \omega_p \\ \ge A_a & \text{for } \omega_a \le \omega \le \infty \end{cases}$$

$$A_{\text{HP}}(\omega) \begin{cases} \ge A_a & \text{for } 0 \le \omega \le \omega_a \\ \le A_p & \text{for } \omega_p \le \omega \le \infty \end{cases}$$

$$A_{\text{BP}}(\omega) \begin{cases} \ge A_a & \text{for } 0 \le \omega \le \omega_{a1} \\ \le A_p & \text{for } \omega_{p1} \le \omega \le \omega_{p2} \\ \ge A_a & \text{for } \omega_{a2} \le \omega \le \infty \end{cases}$$

and

$$A_{BS}(\omega)\begin{cases}\leq A_p & \text{for} \quad 0 \leq \omega \leq \omega_{p1}\\ \geq A_a & \text{for} \quad \omega_{a1} \leq \omega \leq \omega_{a2}\\ \leq A_p & \text{for} \quad \omega_{p2} \leq \omega \leq \infty\end{cases}$$

respectively, where ω_p, ω_{p1}, and ω_{p2} are passband edges, ω_a, ω_{a1}, and ω_{a2} are stopband edges, A_p is the maximum passband loss, and A_a is the minimum stopband loss. In practice, A_p is determined from the allowable amplitude distortion (see Section 65.6) and A_a is dictated by the allowable adjacent channel interference and the desirable signal-to-noise ratio.

It should be mentioned that in practical filters the cutoff frequency ω_c is not a very precise term. It is often used to identify some hypothetical boundary between passband and stopband such as the 3-dB frequency in Butterworth filters, the passband edge in Chebyshev filters, the stopband edge in inverse-Chebyshev filters, or the geometric mean of the passband and stopband edges in elliptic filters.

If a filter is required to have a piecewise constant loss characteristic (or amplitude response) and the shape of the phase response is not critical, the filter can be fully specified by its band edges, the minimum passband and maximum stopband losses A_p and A_a, respectively.

65.6 Amplitude and Delay Distortion

In practice, a filter can distort the information content of the signal. Consider a filter characterized by a transfer function $H(s)$ and assume that its input and output signal are $v_i(t)$ and $v_o(t)$. The frequency response of the filter is given by

$$H(j\omega) = M(\omega)e^{j\theta(\omega)}$$

where $M(\omega)$ and $\theta(\omega)$ are the amplitude and phase responses, respectively.

The frequency spectrum of $v_i(t)$ is its Fourier transform, namely, $V_i(j\omega)$. Assume that the information content of $v_i(t)$ is concentrated in frequency band B given by

$$B = \{\omega\colon \omega_L \leq \omega \leq \omega_H\}$$

and that its frequency spectrum is zero elsewhere.

Let us assume that the amplitude response is constant with respect to band B, i.e.,

$$M(\omega) = G_0 \quad \text{for} \quad \omega \in B \tag{65.19}$$

and that the phase response is linear, i.e.,

$$\theta(\omega) = -\tau_g\omega + \theta_o \quad \text{for} \quad \omega \in B \tag{65.20}$$

where τ_g is a constant. This implies that the group delay is constant with respect to band B, i.e.,

$$\tau(\omega) = -\frac{d\theta(\omega)}{d\omega} = \tau_g \quad \text{for} \quad \omega \in B$$

The frequency spectrum of the output signal $v_o(t)$ can be obtained from (65.19) and (65.20) as

$$\begin{aligned} V_o(j\omega) &= H(j\omega)V_i(j\omega) = M(\omega)e^{j\theta(\omega)}V_i(j\omega) \\ &= [G_0 e^{-j\omega\tau_g + j\theta_0}]V_i(j\omega) = G_0 e^{j\theta_0}[e^{-j\omega\tau_g}V_i(j\omega)] \end{aligned}$$

and from the time-shifting theorem of the Fourier transform

$$v_o(t) = G_0 e^{j\theta_0} v_i(t - \tau_g)$$

We conclude that is the amplitude response of the filter is flat and its phase response is a linear function of ω (i.e., the delay characteristic is flat) in band B, then the output signal is a delayed replica of the input signal except that a gain G_0 and a constant phase shift θ_0 are introduced.

If the amplitude response of the filter is not flat in band B, then *amplitude distortion* will be introduced since different frequency components of the signal will be amplified by different amounts.

If the delay characteristic is not flat in band B, then *delay* (or *phase*) *distortion* will be introduced since different frequency components will be delayed by different amounts.

Amplitude distortion can be quite objectionable in practice and, consequently, in each frequency band that carries information, the amplitude response is required to be constant to within a prescribed tolerance. The amount of amplitude distortion allowed determines the maximum passband loss A_p.

If the ultimate receiver of the signal is the human ear, e.g., when a speech or music signal is to be processed, delay distortion is quite tolerable. However, in other applications it can be as objectionable as amplitude distortion and the delay characteristic is required to be fairly flat. Applications of this type include data transmission, where the signal is to be interpreted by digital hardware, and image processing, where the signal is used to reconstruct an image that is to be interpreted by the human eye. The allowable delay distortion dictates the degree of flatness in the delay characteristic.

65.7 Minimum-Phase, Nonminimum-Phase, and Allpass Filters

Filters satisfying prescribed loss specifications for applications where delay distortion is unimportant can be readily designed with transfer functions whose zeros are on the $j\omega$ axis or in the left-half s plane. Such transfer functions are said to be **minimum-phase** since the phase response at a given frequency ω is increased if any one of the zeros is moved into the right-half s plane, as will now be demonstrated.

Minimum-Phase Filters

Consider a filter whose zeros z_i for $i = 1, 2, \cdots, M$ are replaced by their mirror images and let the new zeros be located at $z = \bar{z}_i$, where

$$\text{Re}\ \bar{z}_i = -\text{Re}\ z_i \quad \text{and} \quad \text{Im}\ \bar{z}_i = \text{Im}\ z_i$$

as depicted in Fig. 65.5. From the geometry of the new zero-pole plot, the magnitude and angle of each phasor $j\omega - \bar{z}_i$ are given by

$$M_{\bar{z}_i} = M_{z_i} \quad \text{and} \quad \psi_{\bar{z}_i} = \pi - \psi_{z_i}$$

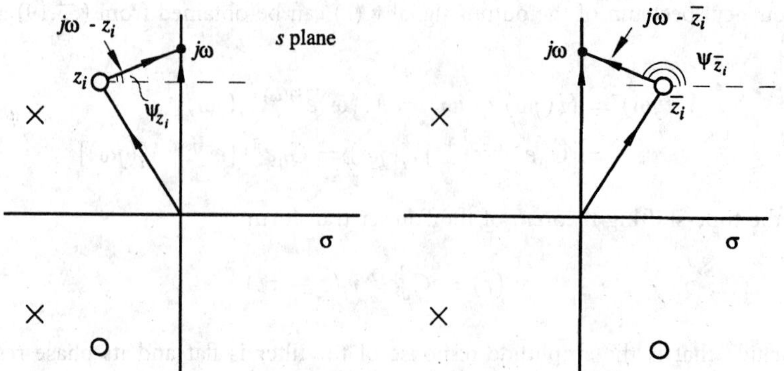

FIGURE 65.5 Zero-pole plots of minimum-phase and corresponding nonminimum-phase filter.

respectively. The amplitude response of the modified filter is obtained from (65.16) as

$$\overline{M}(\omega) = \frac{|H_0| \prod_{i=1}^{M} M_{\bar{z}_i}}{\prod_{i=1}^{N} M_{p_i}^{m_i}} = \frac{|H_0| \prod_{i=1}^{M} M_{z_i}}{\prod_{i=1}^{N} M_{p_i}^{m_i}} = M(\omega)$$

Therefore, replacing the zeros of the transfer function by their mirror images leaves the amplitude response unchanged.

The phase response of the original filter is given by (65.17) as

$$\theta(\omega) = \arg H_0 + \sum_{i=1}^{M} \psi_{z_i} - \sum_{i=1}^{N} m_i \psi_{p_i} \tag{65.21}$$

and since $\psi_{\bar{z}_i} = \pi - \psi_{z_i}$, the phase response of the modifier filter is given by

$$\begin{aligned} \bar{\theta}(\omega) &= \arg\, H_0 + \sum_{i=1}^{M} \psi_{\bar{z}_i} - \sum_{i=1}^{N} m_i \psi_{p_i} \\ &= \arg\, H_0 + \sum_{i=1}^{M} (\pi - \psi_{z_i}) - \sum_{i=1}^{N} m_i \psi_{p_i} \end{aligned} \tag{65.22}$$

that is, the phase response of the modified filter is different from that of the original filter. Furthermore, from (65.21) and (65.22)

$$\bar{\theta}(\omega) - \theta(\omega) = \sum_{i=1}^{M} (\pi - 2\psi_{z_i})$$

and since $-\pi/2 \leq \psi_{z_i} \leq \pi/2$, we have

$$\bar{\theta}(\omega) - \theta(\omega) \geq 0$$

or

$$\bar{\theta}(\omega) \geq \theta(\omega)$$

As a consequence, the phase response of the modified filter is equal to or greater than that of the original filter for all ω.

A frequently encountered requirement in the design of filters is that the delay characteristic be flat to within a certain tolerance within the passband(s) in order to achieve tolerable delay distortion, as was demonstrated in Section 65.6. In these and other filters in which the specifications include constraints on the phase response or delay characteristic, a nonminimum-phase transfer function is almost always required.

Allpass Filters

An **allpass filter** is one that has a constant amplitude response. Consider a transfer function of the type given by (65.13). From (65.10), $H(-j\omega)$ is the complex conjugate of $H(j\omega)$ and hence a constant amplitude response can be achieved if

$$M^2(\omega) = H(s)H(-s)\big|_{s=j\omega} = H_0^2 \frac{N(s)}{D(s)} \times \frac{N(-s)}{D(-s)}\bigg|_{s=j\omega} = H_0^2$$

Hence an allpass filter can be obtained if

$$N(-s) = D(s)$$

that is, the zeros of such a filter must be the mirror images of the poles and vice versa. A typical zero-pole plot for an allpass filter is illustrated in Fig. 65.6. A second-order allpass transfer function is given by

$$H_{AP}(s) = \frac{s^2 - bs + c}{s^2 + bs + c}$$

where $b > 0$ for stability. As above, we can write

$$M^2(\omega) = H_{AP}(s)H_{AP}(-s)\big|_{s=j\omega} = \frac{s^2 - bs + c}{s^2 + bs + c} \times \frac{s^2 + bs + c}{s^2 - bs + c}\bigg|_{s=j\omega} = 1$$

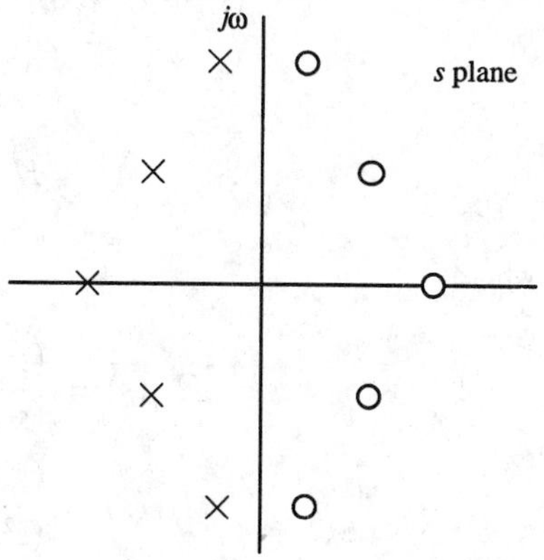

FIGURE 65.6 Typical zero-pole plot of an allpass filter.

Allpass filters can be used to modify the phase responses of filters without changing their amplitude responses. Hence they are used along with minimum-phase filters to obtain nonminimum-phase filters that satisfy amplitude and phase response specifications simultaneously.

Decomposition of Nonminimum-Phase Transfer Functions

Some methods for the design of filters satisfying amplitude and phase response specifications, usually methods based on optimization, yield a nonminimum-phase transfer function. Such a transfer function can be easily decomposed into a product of a minimum-phase and an allpass transfer function, i.e.,

$$H_{\mathrm{N}}(s) = H_{\mathrm{M}}(s)H_{\mathrm{AP}}(s)$$

Consequently, a nonminimum-phase filter can be implemented as a cascade arrangement of a minimum-phase and an allpass filter.

The above decomposition can be obtained by using the following procedure:

1. For each zero in the right-half s plane, augment the transfer function by a zero and a pole at the mirror image position of the zero.
2. Assign the left-half s-plane zeros and the original poles to the minimum-phase transfer function $H_{\mathrm{M}}(s)$.
3. Assign the right-half s-plane zeros and the left-hand s-plane poles generated in step 1 to the allpass transfer function $H_{\mathrm{AP}}(s)$.

This procedure is illustrated in Fig. 65.7. For example, if

$$H_{\mathrm{N}}(s) = \frac{(s^2+4s+5)(s^2-3s+7)(s-5)}{(s^2+2s+6)(s^2+4s+9)(s+2)}$$

then we can write

$$H_{\mathrm{N}}(s) = \frac{(s^2+4s+5)(s^2-3s+7)(s-5)}{(s^2+2s+6)(s^2+4s+9)(s+2)} \times \frac{(s^2+3s+7)(s+5)}{(s^2+3s+7)(s+5)}$$

Hence

$$H_{\mathrm{N}}(s) = \frac{(s^2+4s+5)(s^2+3s+7)(s+5)}{(s^2+2s+6)(s^2+4s+9)(s+2)} \times \frac{(s^2-3s+7)(s-5)}{(s^2+3s+7)(s+5)}$$

or

$$H_{\mathrm{N}}(s) = H_{\mathrm{M}}(s)H_{\mathrm{AP}}(s)$$

where

$$H_{\mathrm{M}}(s) = \frac{(s^2+4s+5)(s^2+3s+7)(s+5)}{(s^2+2s+6)(s^2+4s+9)(s+2)}$$

$$H_{\mathrm{AP}}(s) = \frac{(s^2-3s+7)(s-5)}{(s^2+3s+7)(s+5)}$$

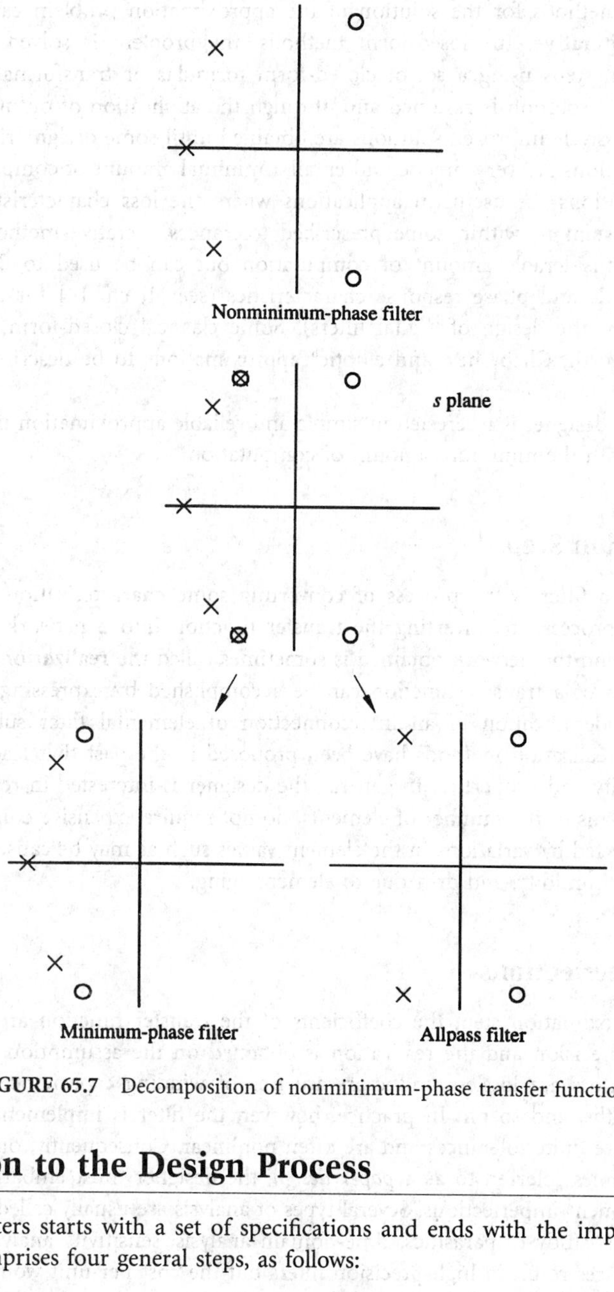

FIGURE 65.7 Decomposition of nonminimum-phase transfer function.

65.8 Introduction to the Design Process

The design of filters starts with a set of specifications and ends with the implementation of a prototype. It comprises four general steps, as follows:

1. approximation
2. realization
3. study of imperfections
4. implementation.

The Approximation Step

The **approximation step** is the process of generating a transfer function that satisfies the desired specifications, which may concern the amplitude, phase, and possibly the time-domain response of the filter.

The available methods for the solution of the approximation problem can be classified as **closed-form** or **iterative.** In closed-form methods, the problem is solved through a small number of design steps using a set of closed-form formulas or transformations. In iterative methods, an initial solution is assumed and, through the application of optimization methods, a series of progressively improved solutions are obtained until some design criterion is satisfied. Closed-form solutions are very precise and entail a minimal amount of computation. However, the available solutions are useful in applications where the loss characteristic is required to be piecewise constant to within some prescribed tolerances. Iterative methods, on the other hand, entail a considerable amount of computation but can be used to design filters with arbitrary amplitude and phase response characteristics (see [1, ch. 14] for the application of these methods for the design of digital filters). Some classical closed-form solutions are the so-called Butterworth, Chebyshev, and elliptic[8] approximations to be described in Chapter 66 by A. M. Davis.

In general, the designer is interested in simple and reliable approximation methods that yield precise designs with the minimum amount of computation.

The Realization Step

The *synthesis* of a filter is the process of converting some characterization of the filter into a network. The process of converting the transfer function into a network is said to be the **realization step** and the network obtained is sometimes called the **realization.**

The realization of a transfer function can be accomplished by expressing it in some form that allows the identification of an interconnection of elemental filter subnetworks and/or elements. Many realization methods have been proposed in the past that lead to structures of varying complexity and properties. In general, the designer is interested in realizations that are economical in terms of the number of elements, do not require expensive components, and are not seriously affected by variations in the element values such as may be caused by variations in temperature and humidity, and drift due to element aging.

Study of Imperfections

During the approximation step, the coefficients of the transfer function are determined to a high degree of precision and the realization is obtained on the assumption that elements are ideal, i.e., capacitors are lossless, inductors are free of winding capacitances, amplifiers have infinite bandwidths, and so on. In practice, however, the filter is implemented with nonideal elements that have finite tolerances and are often nonlinear. Consequently, once a realization is obtained, sometimes referred to as a *paper design*, the designer must embark on the study of the effects of element imperfections. Several types of analysis are usually called for ranging from tolerance analysis, study of parasitics, time-domain analysis, sensitivity analysis, noise analysis, etc. Tight tolerances result in high-precision filters but the cost per unit would be high. Hence the designer is obliged to determine the highest tolerance that can be tolerated without violating the specifications of the filter throughout its working life. Sensitivity analysis is a related study that will ascertain the degree of dependence of a filter parameter, e.g., the dependence of the amplitude response on a specific element. If the loss characteristic of a filter is not very sensitive to a certain capacitance, then the designer would be able to use a less precise and cheaper capacitor, which would, of course, decrease the cost of the unit.

[8]To be precise, the elliptic approximation is not a closed-form method, since the transfer function coefficients are given in terms of certain infinite series. However, these series converge very rapidly and can be treated as closed-form formulas for most practical purposes (see [1, ch. 5]).

Implementation

Once the filter is thoroughly analyzed and found to meet the desired specifications under ideal conditions, a prototype is constructed and tested. Decisions to be made involve the type of components and packaging, and the methods to be used for the manufacture, testing, and tuning of the filter. Problems may often surface at the implementation stage that may call for one or more modifications in the paper design. Then the realization and possibly the approximation may have to be redone.

65.9 Introduction to Realization

Realization tends to depend heavily on the type of filter required. The realization of passive *RLC* filters differs quite significantly from that of active filters which, in turn, is entirely different from the realization of microwave filters.

Passive Filters

Passive *RLC* filters have been the mainstay of communications since the 1920's and, furthermore, they continue to be of considerable importance today for frequencies in the range 100–500 kHz.

The realization of passive *RLC* filters has received considerable attention through the years and it is, as a consequence, highly developed and sophisticated. It can be accomplished by using available filter-design packages such as FILSYN [5] and FILTOR [6]. In addition, several filter-design handbooks and published design tables are available [7]–[10].

The realization of passive *RLC* filters starts with a resistively terminated *LC* two-port network such as that in Fig. 65.8. Then through one of several approaches, the transfer function is used to generate expressions for the z or y parameters of the *LC* two-port. The realization of the *LC* two-port is achieved by realizing the z or y parameters. The realization of passive filters is considered in Section XIV by W. K. Chen.

Active Filters

Since the reactance of an inductor is ωL, increased inductance values are required to achieve reasonable reactance values at low frequencies. For example, an inductance of 1 mH which will present a reactance of 6.28 kΩ at 1 MHz will present only 0.628 Ω at 100 Hz. Thus, as the frequency range of interest is reduced, the inductance values must be increased if a specified impedance level is to be maintained. This can be done by increasing the number of turns on the inductor coil and to some extent by using ferromagnetic cores of high permeability. Increasing the number of turns increases the resistance, the size, and the cost of the inductor. The resistance is increased because the length of the wire is increased [$R = (\rho \times length)/Area$] and hence the Q factor is reduced. The cost goes up because the cost of materials as well as the cost of labor go up, since an inductor must be individually wound. For these reasons, inductors are generally incompatible with miniaturization or microcircuit implementation.

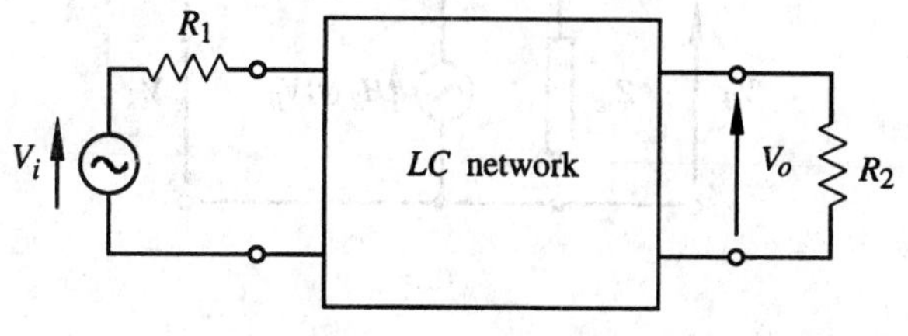

FIGURE 65.8 Passive *RLC* filter.

The above physical problem has led to the invention and development of a class of inductorless filters known collectively as **active filters**. Sensitivity considerations, which will be examined in Chapter 68 by I. Filanovsky, have led to two basic approaches to the design of active filters. In one approach, the active filter is obtained by simulating the inductances in a passive *RLC* filter or by realizing a signal flow graph of the passive *RLC* filter. In another approach, the active filter is obtained by cascading a number of low-order filter sections of some type, as depicted in Fig. 65.9(a), where Z_{o0} is the output impedance of the signal source.

Each filter section is made up of an interconnection of resistors, capacitors, and active elements, and by Thévenin's theorem, it can be represented by its input impedance, open-circuit voltage transfer function, and output impedance as shown in Fig. 65.9(b). The voltage transfer function of the configuration is given by

$$H(s) = \frac{V_o}{V_i}$$

and since the input voltage of section k is equal to the output voltage of section $k-1$, i.e., $V_{ik} = V_{o(k-1)}$ for $k = 2, 3, \cdots, K$, and $V_o = V_{oK}$, we can write

$$H(s) = \frac{V_o}{V_i} = \frac{V_{i1}}{V_i} \times \frac{V_{o1}}{V_{i1}} \times \frac{V_{o2}}{V_{i2}} \times \cdots \times \frac{V_{oK}}{V_{iK}} \tag{65.23}$$

where

$$\frac{V_{i1}}{V_i} = \frac{Z_{i1}}{Z_{o0} + Z_{i1}} \tag{65.24}$$

and

$$\frac{V_{ok}}{V_{ik}} = \frac{Z_{i(k+1)}}{Z_{ok} + Z_{i(k+1)}} H_k(s) \tag{65.25}$$

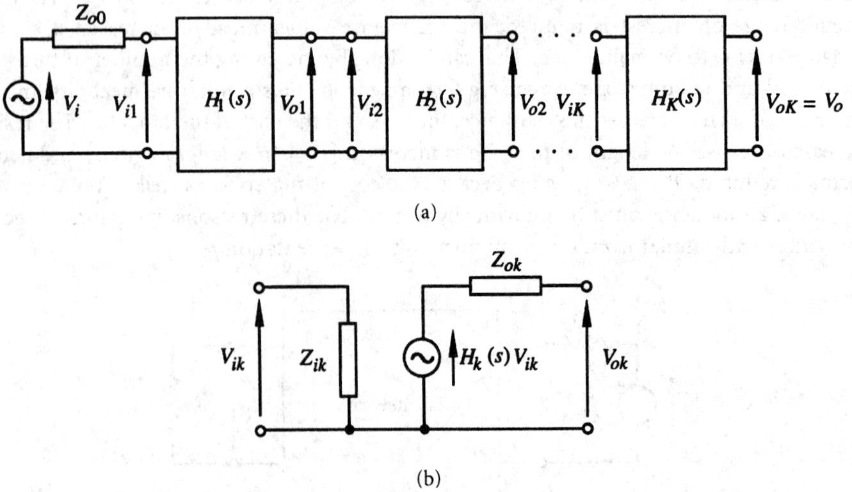

FIGURE 65.9 (a) Cascade realization, (b) Thévenin equivalent circuit of filter section.

is the transfer function of the kth section. From (65.23)–(65.25), we obtain

$$H(s) = \frac{V_o}{V_i} = \frac{Z_{i1}}{Z_{o0} + Z_{i1}} \prod_{k=1}^{K} \frac{Z_{i(k+1)}}{Z_{ok} + Z_{i(k+1)}} H_k(s)$$

Now if

$$|Z_{ik}| \gg |Z_{o(k-1)}|$$

for $k = 1, 2, \cdots, K$, then the loading effect produced by section $k+1$ on section k can be neglected and hence

$$H(s) = \frac{V_o}{V_i} = \prod_{k=1}^{K} H_k(s)$$

Evidently, a highly desirable property in active filter sections is that the magnitude of the input impedance be large and/or that of the output impedance be small since in such a case the transfer function of the cascade structure is equal to the product of the transfer functions of the individual sections.

An arbitrary Nth-order transfer function obtained by using the Butterworth, Bessel, Chebyshev, inverse-Chebyshev, or elliptic approximation can be expressed as

$$H(s) = H_0(s) \prod_{k=1}^{K} \frac{a_{2k}s^2 + a_{1k}s + a_{0k}}{s^2 + b_{1k}s + b_{0k}}$$

where

$$H_0(s) = \begin{cases} \dfrac{a_{10}s + a_{00}}{b_{10}s + b_{00}} & \text{for odd } N \\ 1 & \text{for even } N \end{cases}$$

The first-order transfer function $H_0(s)$ for the case of an odd-order filter can be readily realized using the RC network of Fig. 65.10.

Biquads

From the above analysis, we note that all we need to be able to realize an arbitrary transfer function is a circuit that realizes the biquadratic transfer function

$$H_{BQ}(s) = \frac{a_2s^2 + a_1s + a_0}{s^2 + b_1s + b_0} = \frac{a_2(s+z_1)(s+z_2)}{(s+p_1)(s+p_2)} \tag{65.26}$$

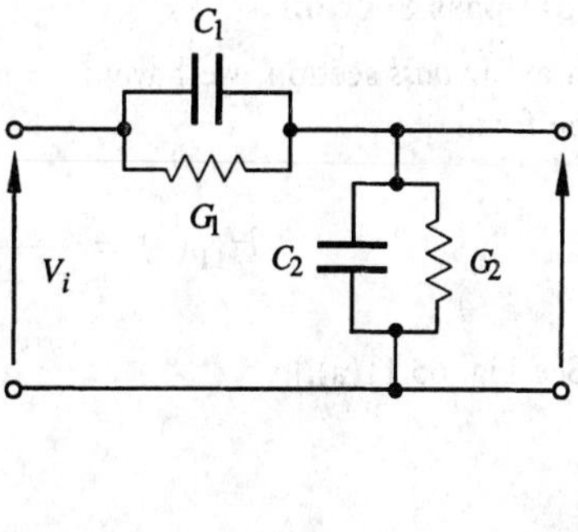

FIGURE 65.10 First-order RC network.

where zeros and poles occur in complex conjugate pairs, i.e., $z_2 = z_1^*$ and $p_2 = p_1^*$. Such a circuit is commonly referred to as a *biquad*.

After some manipulation, the transfer function in (65.26) can be expressed as

$$H_{\text{BQ}}(s) = K \frac{s^2 + (2\,\text{Re}\, z_1)s + (\text{Re}\, z_1)^2 + (\text{Im}\, z_1)^2}{s^2 + (2\,\text{Re}\, p_1)s + (\text{Re}\, p_1)^2 + (\text{Im}\, p_1)^2}$$
$$= K \frac{s^2 + (\omega_z/Q_z)s + \omega_z^2}{s^2 + (\omega_p/Q_p)s + \omega_p^2}$$

where $K = a_2$, ω_z, and ω_p are the zero and pole frequencies, and Q_z and Q_p are the zero and pole *quality factors* (or Q factors for short), respectively. The formulas for the various parameters are as follows:

$$\omega_z = \sqrt{(\text{Re}\, z_1)^2 + (\text{Im}\, z_1)^2}$$
$$\omega_p = \sqrt{(\text{Re}\, p_1)^2 + (\text{Im}\, p_1)^2}$$
$$Q_z = \frac{\omega_z}{2\ \text{Re}\ z_1}$$
$$Q_p = \frac{\omega_p}{2\ \text{Re}\ p_1}$$

The zero and pole frequencies are approximately equal to the frequencies of minimum gain and maximum gain, respectively. The zero and pole Q factors have to do with the selectivity of the filter. A high zero Q factor results in a deep notch in the amplitude response, whereas a high pole Q factor results in a very peaky amplitude response.

The dc gain and the gain as $\omega \to \infty$ in dB are given by

$$M_0 = 20 \log |H_{\text{BQ}}(0)| = 20 \log\left(K \frac{\omega_z^2}{\omega_p^2}\right)$$

and

$$M_\infty = 20 \log |H_{\text{BQ}}(j\infty)| = 20 \log K$$

respectively.

Types of Basic Filter Sections

Depending on the values of the transfer function coefficients, five basic types of filter sections can be identified, namely, low-pass, high-pass, bandpass, notch (sometimes referred to as bandreject), and allpass. These sections can serve as building blocks for the design of filters that can satisfy arbitrary specifications. They are actually sufficient for the design of all the standard types of filters, namely, Butterworth, Chebyshev, inverse-Chebyshev, and elliptic filters.

Low-pass Section

In a *low-pass* section, we have $a_2 = a_1 = 0$ and $a_0 = K\omega_p^2$. Hence the transfer function assumes the form

$$H_{\text{LP}}(s) = \frac{a_0}{s^2 + b_1 s + b_0} = \frac{K\omega_p^2}{s^2 + (\omega_p/Q_p)s + \omega_p^2}$$

[See Fig. 65.11(a).]

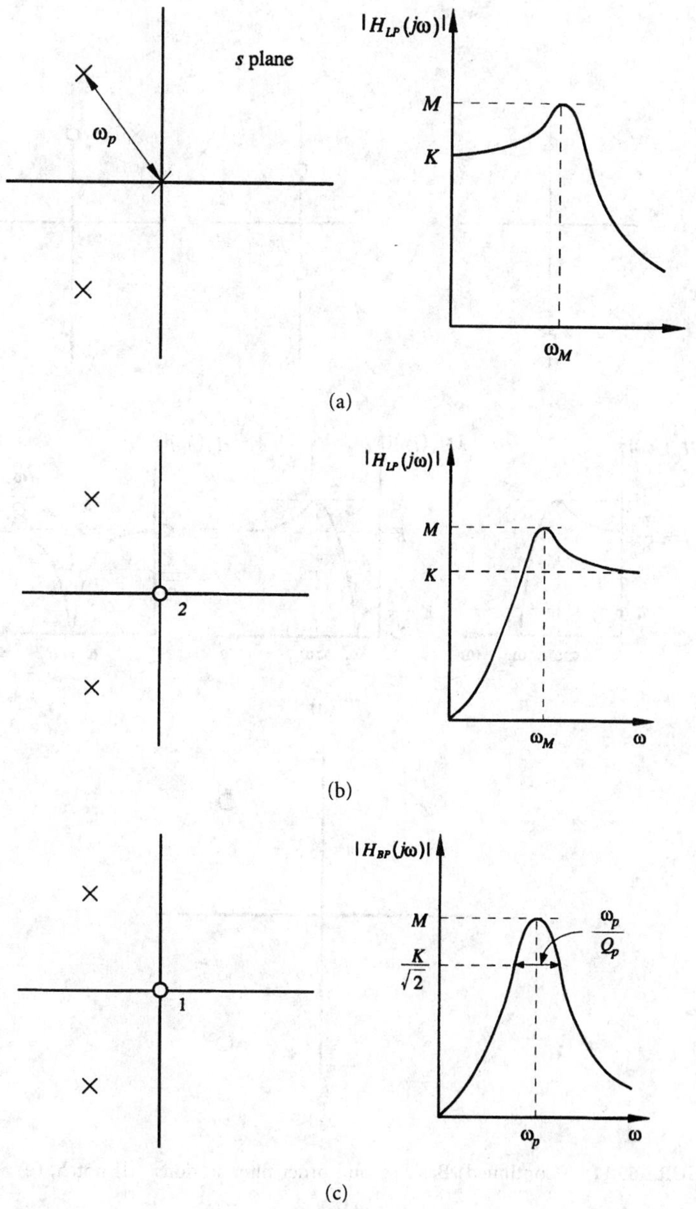

FIGURE 65.11 Basic second-order filter sections: (a) low-pass, (b) high-pass, (c) bandpass.

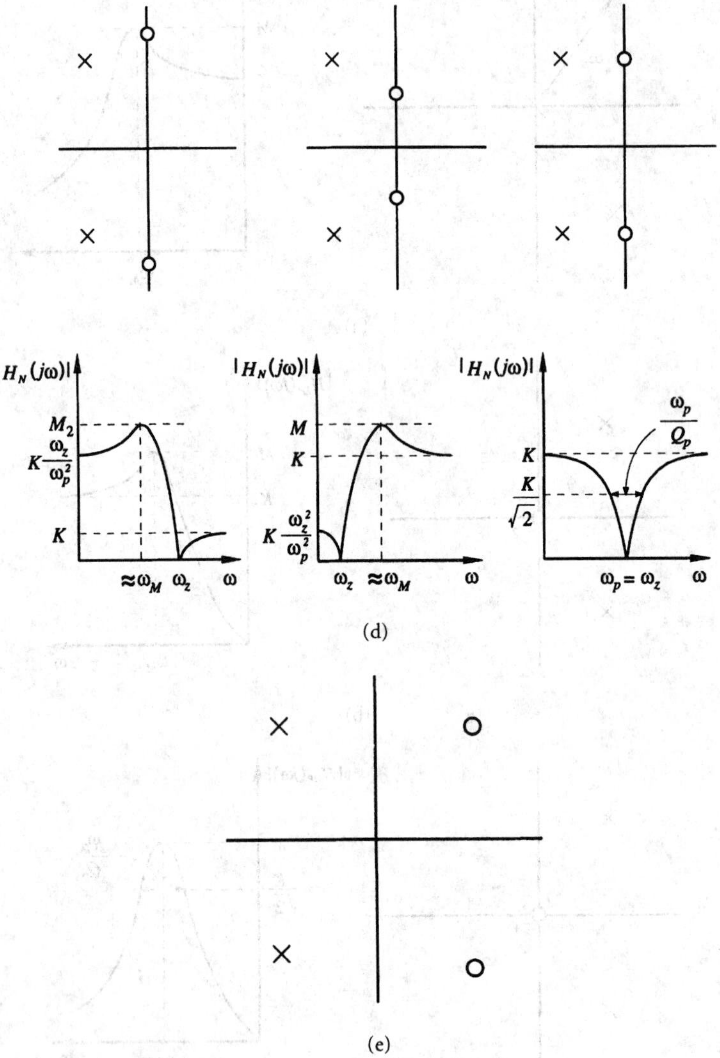

FIGURE 65.11 (Continued) Basic second-order filter sections: (d) notch, (e) allpass.

High-Pass Section

In a *high-pass* section, we have $a_2 = K$ and $a_1 = a_0 = 0$. Hence the transfer function assumes the form

$$H_{HP}(s) = \frac{a_2 s^2}{s^2 + b_1 s + b_0} = \frac{Ks^2}{s^2 + (\omega_p/Q_p)s + \omega_p^2}$$

[See Fig. 65.11(b).]

Bandpass Section

In a *bandpass* section, we have $a_1 = K\omega_p/Q_p$ and $a_2 = a_0 = 0$. Hence the transfer function assumes the form

$$H_{BP}(s) = \frac{a_1 s}{s^2 + b_1 s + b_0} = \frac{K(\omega_p/Q_p)s}{s^2 + (\omega_p/Q_p)s + \omega_p^2}$$

[See Fig. 65.11(c).]

Notch Section

In a *notch* section, we have $a_2 = K$, $a_1 = 0$, and $a_0 = K\omega_z^2$. Hence the transfer function assumes the form

$$H_N(s) = \frac{a_2 s^2 + a_0}{s^2 + b_1 s + b_0} = \frac{K(s^2 + \omega_z^2)}{s^2 + (\omega_p/Q_p)s + \omega_p^2}$$

[See Fig. 65.11(d).]

Allpass Section

In an *allpass* section, we have $a_2 = K$, $a_1 = -K\omega_p/Q_p$, and $a_0 = K\omega_p^2$. Hence the transfer function assumes the form

$$H_{AP}(s) = \frac{a_2 s^2 + a_1 s + a_0}{s^2 + b_1 s + b_0} = \frac{K[s^2 - (\omega_p/Q_p)s + \omega_p^2]}{s^2 + (\omega_p/Q_p)s + \omega_p^2}$$

[See Fig. 65.11(e).]

The design of active and switched-capacitor filters is treated in some detail in Section XV.

References

[1] A. Antoniou, *Digital Filters: Analysis, Design, and Applications*, 2nd ed. New York: McGraw-Hill, 1993.

[2] E. Kreyszig, *Advanced Engineering Mathematics*, 3rd ed. New York: Wiley, 1972.

[3] R. J. Schwarz and B. Friedland, *Linear Systems*, New York: McGraw-Hill, 1965.

[4] R. Schaumann, M. S. Ghausi, and K. R. Laker, *Design of Analog Filters*, Englewood Cliffs, NJ: Prentice Hall, 1990.

[5] G. Szentirmai, "FILSYN—A general purpose filter synthesis program," *Proc. IEEE*, vol. 65, pp. 1443–1458, Oct. 1977.

[6] A. S. Sedra and P. O. Brackett, *Filter Theory and Design: Active and Passive*, Portland, OR: Matrix, 1978.

[7] J. K. Skwirzynski, *Design Theory and Data for Electrical Filters*, London: Van Nostrand, 1965.

[8] R. Saal, *Handbook of Filter Design*, Backnang: AEG Telefunken, 1979.

[9] A. I. Zverev, *Handbook of Filter Synthesis*, New York: Wiley, 1967.

[10] E. Chirlian, *LC Filters: Design, Testing, and Manufacturing*, New York: Wiley, 1983.

66

Approximation

Artice M. Davis
San Jose State University

66.1 Introduction

The approximation problem for filters is illustrated in Fig. 66.1. A filter is often desired to produce a given slope of gain over one or more frequency intervals, to remain constant over other intervals, and to completely reject signals having frequencies contained in still other intervals. Thus, in the example shown in the figure, the desired gain is zero for very low and very high frequencies. The center line, shown dashed, is the nominal behavior and the shaded band shows the permissible variation in the gain characteristic. Realizable circuits must always generate smooth curves and so cannot exactly meet the piecewise linear specification represented by the center line. Thus, the realizable behavior is shown by the smooth, dark curve that lies entirely within the shaded tolerance band.

What type of frequency response function can be postulated that will meet the required specifications and—at the same time—be realizable: constructible with a specified catalog of elements? The answer depends upon the types of elements allowed. For instance, if one allows pure delays with a common delay time, summers, and scalar multipliers, a trigonometric polynomial will work; this, however, will cause the gain function to be repeated in a periodic manner. If this is permissible, one can then realize the filter in the form of an FIR digital filter or as a commensurate transmission line filter—and, in fact, it can be realized in such a fashion that the resulting phase behavior is precisely linear. If one fits the required behavior with a rational trigonometric function, a function that is the ratio of two trigonometric polynomials, an economy of hardware will result. The phase, however, will unfortunately no longer be linear. These issues are discussed at greater length in [1].

Another option would be to select an ordinary polynomial in ω as the approximating function. Polynomials, however, behave badly at infinity. They approach infinity as $\omega \rightarrow \pm\infty$, a highly undesirable situation. For this reason, one must discard polynomials. A rational function of ω, however, will work nicely for the ratio of two polynomials will approach zero as $\omega \rightarrow \pm\infty$ if the degree of the numerator polynomial is selected to be of lower degree than that of the denominator. Furthermore, by the Weierstrass theorem, such a function can approximate any continuous function arbitrarily closely over any closed interval of finite length [2]. Thus, one sees that the rational functions in ω offer a suitable approximation for analog filter design and—in fact—do not have the repetitive nature of the trigonometric rational functions.

0-8493-8341-2/95/$0.00 + $.50

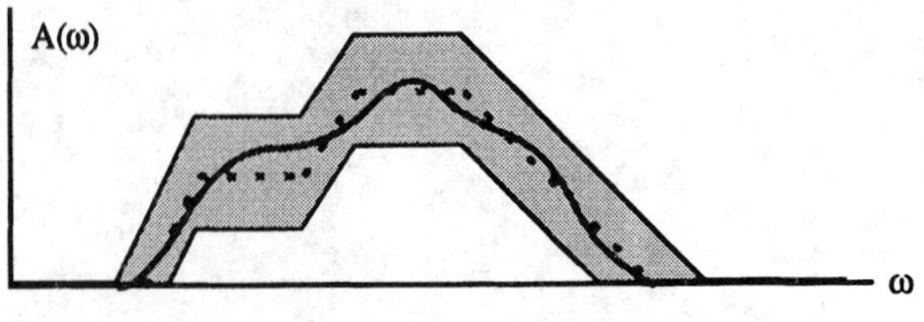

FIGURE 66.1 The general approximation problem.

Suppose, therefore, that the gain function is of the form

$$A(\omega) = \frac{N(\omega)}{D(\omega)} = \frac{a_0 + a_1\omega + a_2\omega^2 + \cdots + a_r\omega^r}{b_0 + b_1\omega + b_2\omega^2 + \cdots + b_q\omega^q} \tag{66.1}$$

where $r \leq q$ for reasons mentioned above. Assuming that the filter to be realized is constrained to be constructable with real[1] elements, one must require that $A(-\omega) = A(\omega)$—that is, that the gain be an event function of frequency. But then, as it is straightforward to show, one must require that all the odd coefficients of both numerator and denominator be zero. This means that the gain is a function of ω^2:

$$A(\omega) = \frac{N(\omega^2)}{D(\omega^2)} = \frac{a_0 + a_1\omega^2 + \cdots + a_m\omega^{2m}}{b_0 + b_1\omega^2 + \cdots + b_n\omega^{2n}} = A(\omega^2) \tag{66.2}$$

The expression has been reindexed and the constants redefined in an obvious manner. The net result is that one must approximate the desired characteristic by the ratio of two polynomials in ω^2; the objective is to determine the numerator and denominator coefficients to meet the stated specifications. Once this is accomplished one must compute the filter transfer function $G(s)$ in order to synthesize the filter [4], [5]. Assuming that $G(s)$ is real (has real coefficients), then its complex conjugate satisfies $G^*(s) = G(s^*)$, from which it follows that $G(s)$ is related to $A(\omega^2)$ by the relationship

$$[G(s)G(-s)]_{s=j\omega} = G(j\omega)G^*(j\omega) = |G(j\omega)|^2 = A^2(\omega^2) \tag{66.3}$$

In fact, it is more straightforward to simply cast the original approximation problem in terms of $A^2(\omega^2)$, rather than in terms of $A(\omega)$. In this case, (66.2) becomes

$$A^2(\omega^2) = \frac{N(\omega^2)}{D(\omega^2)} = \frac{a_0 + a_1\omega^2 + \cdots + a_m\omega^{2m}}{b_0 + b_1\omega^2 + \cdots + b_n\omega^{2n}} \tag{66.4}$$

Thus, one can assume that the approximation process produces $A^2(\omega^2)$ as the ratio of two real polynomials in ω^2. Since (66.3) requires the substitutions $s \rightarrow j\omega$, one also has $s^2 \rightarrow -\omega^2$, and conversely. Thus, (66.3) becomes

$$G(s)G(-s) = A^2(-s^2) \tag{66.5}$$

Though this has been shown to hold only on the imaginary axis, it continues to hold for other complex values of s as well by analytic continuation.[2]

[1]Complex filters are quite possible to construct, as recent work [3] shows.

[2]A function analytic in a region is completely determined by its values along any line segment in that region—in this case, by its values along the $j\omega$ axis.

The problem now is to compute $G(s)$ from (66.5), a process known as the **factorization problem**. The solution is not unique; in fact, the phase is arbitrary—subject only to certain realizability conditions. To see this, just let $G(j\omega) = A(\omega)e^{j\phi(\omega)}$, where $\phi(\omega)$ is an arbitrary phase function. Then, (66.3) implies that

$$G(j\omega)G^*(j\omega) = A(\omega)e^{j\phi(\omega)} \cdot A(\omega)e^{-j\phi(\omega)} = A^2(\omega) \tag{66.6}$$

If the resulting structure is to have the property of minimum phase [6], the phase function is determined completely by the gain function. If not, one can simply perform the factorization and accept whatever phase function results from the particular process chosen. As has been pointed out earlier in this chapter, it is often desirable that the phase be a linear function of frequency. In this case, one must follow the filter designed by the above process with a phase equalization filter, one that has constant gain and a phase characteristic that, when summed with that of the first filter, produces linear phase. As it happens, the human ear is insensitive to phase nonlinearity, so the phase is not of much importance for filters designed to operate in the audio range. For those intended for video applications, however, it is vitally important. Nonlinear phase produces, for instance, the phenomenon of multiple edges in a reproduced picture.

If completely arbitrary gain characteristics are desired, computer optimization is necessary [6]. Indeed, if phase is of great significance, computer algorithms are available for the simultaneous approximation of both gain and phase. These are complex and unwieldy to use, however, so for more modest applications the above approach relying upon gain approximation only suffices. In fact, the approach arose historically in the telephone industry in its earlier days in which voice transmission was the only concern, data and video transmission being unforeseen at the time. Furthermore, the frequency division multiplexing of voice signals was the primary concern; hence, a number of standard desired shapes of frequency response were generated: low-pass, high-pass, bandpass, and band-reject (or notch). Typical but stylized specification curves are shown in Fig. 66.2. This figure serves to define the following parameters: the minimum passband gain A_p, the maximum stopband gain A_s, the passband cutoff frequency ω_p, the stopband cutoff frequency ω_s (the last two parameters are for low-pass and high-pass filters only), the center frequency ω_o, upper passband and stopband cutoff frequencies ω_{pu} and ω_{su} and lower passband and stopband cutoff frequencies ω_{pl} and ω_{sl} (the last four parameters are for the bandpass and band-reject filters only). As shown in the figure, the maximum passband gain is usually taken to be unity. In the realization process, the transfer function scale factor is often allowed to be a free parameter that is resolved in the design procedure. The resulting "flat gain" (frequency independent) difference from unity is usually considered to be of no consequence—as long as it is not too small, thus creating signal-to-noise ratio problems. There is a fifth standard type that we have not shown: the allpass filter. It has a constant gain for all frequencies, but with a phase characteristic that can be tailored to fit a standard specification in order to compensate for phase distortion. Frequency ranges where the gain is relatively large are called **passbands** and those where the gain is relatively small, **stopbands**. Those in between—where the gain is increasing or decreasing—are termed **transition bands**.

In order to simplify the design even more, one bases the design of all other types of filter in terms of only one: the low pass. In this case, one says that the low-pass filter is a **prototype**. The specifications of the desired filter type are transformed to those of an equivalent low-pass prototype, and the transfer function for this filter is determined to meet the transformed specifications. Letting the low-pass frequency be symbolized by Ω and the original by ω, one sets

$$\Omega = f(\omega) \tag{66.7}$$

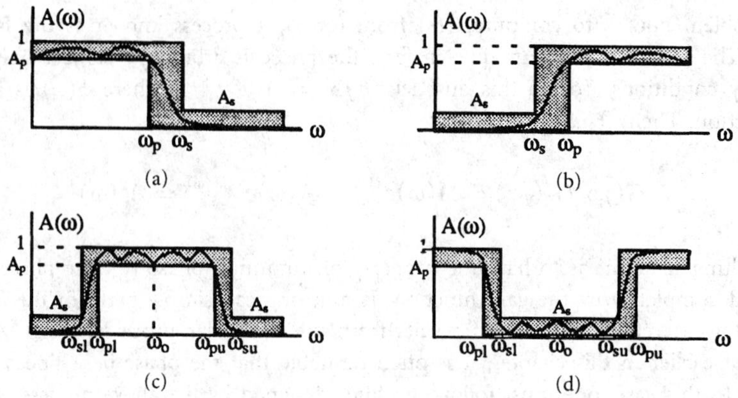

FIGURE 66.2 Catalog of basic filter types: (a) lowpass, (b) highpass, (c) bandpass, (d) bandreject.

The approximation problem is then solved in terms of Ω. Letting

$$p = j\Omega \tag{66.8}$$

one then has $G(p)$, the desired transfer function.

Two approaches are now possible. One is to apply the inverse transformation, letting

$$s = j\omega = jf^{-1}(\Omega) = jf^{-1}(p/j) \tag{66.9}$$

thus obtaining the desired transfer function

$$G(s) = G(j\omega) = G[jf^{-1}(\Omega)] = G[jf^{-1}(p/j)] \tag{66.10}$$

The other consists of designing the circuit to realize the low-pass prototype filter, then transform *each of the elements* from functions of p to functions of s by means of the complex frequency transformation

$$s = j\omega = jf^{-1}(\Omega) = jf^{-1}(p/j) \tag{66.11}$$

As it happens, the transformation $p = f(s)$ has a special form. It can be shown that if f is real for real values of s, thereby having real parameters, and maps the imaginary axis of the p plane into the imaginary axis of the s plane then it must be an odd function; furthermore, if it is to map *positive real rational functions* into those of like kind (necessary for realizability with R, L, and C elements, as well as possibly ideal transformers[3]), it must be a *reactance function.* (See [11] for details.) Since it is always desirable from an economic standpoint to design filters of minimum order, it is desirable that the transformation be of the smallest possible degree. As a result, the following transformations are used:

$$(\text{lpp} \leftrightarrow \text{lp}) \qquad p = ks \tag{66.12}$$

$$(\text{lpp} \leftrightarrow \text{hp}) \qquad p = \frac{k}{s} \tag{66.13}$$

$$(\text{lpp} \leftrightarrow \text{bp}) \qquad p = \frac{s^2 + \omega_o^2}{Bs} \tag{66.14}$$

[3]For active realizations—those containing dependent sources—this condition is not necessary.

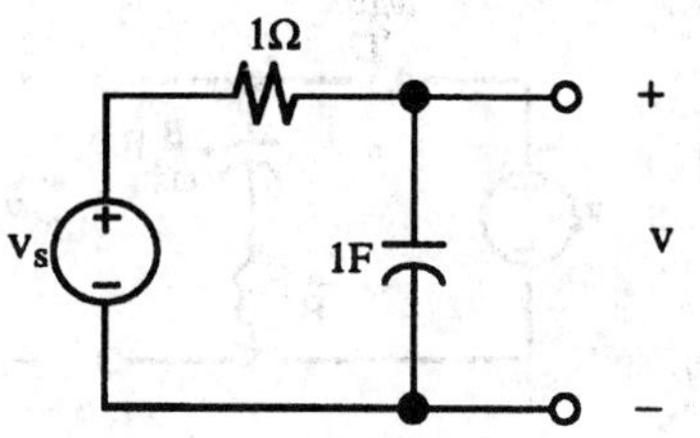

FIGURE 66.3 Low-pass prototype.

$$(\text{lpp} \leftrightarrow \text{br}) \qquad p = \frac{Bs}{s^2 + \omega_o^2} \tag{66.15}$$

where the parameters k, B, and ω_o are real constants to be determined by the particular set of specifications. We have used the standard abbreviations of lpp for *low-pass prototype,* lp for *low pass,* hp for *high pass,* bp for *bandpass,* and br for *bandreject.* Often the letter f is added; for example, one might use the acronym brf for *bandreject filter.* The reason for including the transformation in (66.12) is to allow standardization of the lpp. For instance one can transform from an lpp with, say, a passband cutoff frequency of 1 rad/s to a low-pass filter with a passband cutoff of perhaps 1 kHz.

As a simple example, suppose a bandreject filter were being designed and that the result of the approximation process were

$$H(p) = \frac{1}{p+1} \tag{66.16}$$

Then the br transfer function would be

$$H(s) = [H(p)]_{p=[Bs/(s^2+\omega_o^2)]} = \frac{1}{\dfrac{Bs}{s^2+\omega_o^2} + 1} = \frac{s^2 + \omega_o^2}{s^2 + Bs + \omega_o^2} \tag{66.17}$$

(The parameters ω_o and B would be determined by the bandreject specifications.) As one can readily see, a first-order lpp is transformed into a second-order brf. In general, for bandpass and bandreject design, the object transfer function is of twice the order of the lpp. Since the example is so simple, it can readily be seen that the circuit in Fig. 66.3 realizes the lpp voltage gain function in (66.16). If one applies the transformation in (66.15) the 1-Ω resistor maps into a 1-Ω resistor, but the 1 F capacitor maps into a combination of elements having the admittance

$$Y(p) = p = \frac{1}{\dfrac{s}{B} + \dfrac{\omega_o^2}{Bs}} \tag{66.18}$$

But this is simply the series connection of a capacitor of value B/ω_o^2 farads and an inductor of value $1/B$ henrys. The resulting bandreject filter is shown in Fig. 66.4.

The only remaining "loose end" is the determination of the constant(s) in the appropriate transformation equation selected appropriately from (66.12)–(66.15). This will be done here for the lpp $\leftrightarrow$ bp transformation, (66.14). It is typical, and the reader should have no difficulty working out the other cases. Substituting $p = j\Omega$ and $s = j\omega$ in (66.14), one gets

$$j\Omega = \frac{-\omega^2 + \omega_o^2}{jB\omega} \tag{66.19}$$

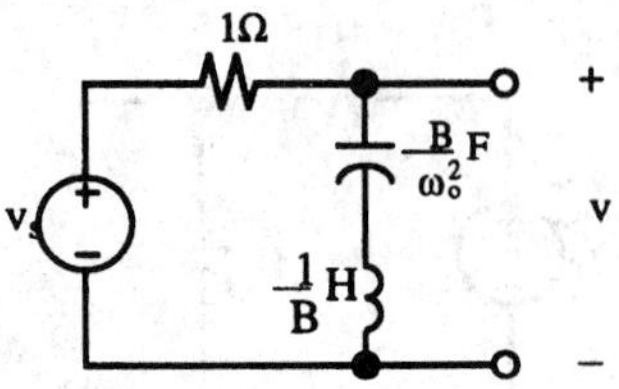

FIGURE 66.4 Resulting bandreject filter.

or

$$\Omega = \frac{\omega^2 - \omega_o^2}{B\omega} = \frac{\omega}{B} - \frac{\omega_o^2}{B\omega} \tag{66.20}$$

This clearly shows that $\omega = \pm\omega_o$ maps into $\Omega = 0$ and $\omega = \pm\infty$ into $\Omega = \pm\infty$. However, as $\omega \to 0+$, $\Omega \to -\infty$ and as $\omega \to 0-$, $\Omega \to +\infty$. Of perhaps more interest is the inverse transformation. Solving (66.20) for ω in terms of Ω, one finds that[4]

$$\omega = \frac{B\Omega}{2} + \sqrt{\left[\frac{B\Omega}{2}\right] + \omega_o^2} \tag{66.21}$$

Now consider pairs of values of Ω, of which one is the negative of the other. Letting ω_+ be the image of Ω with $\Omega > 0$ and ω_- be the image of $-\Omega$, one has

$$\omega_+ = \frac{B\Omega}{2} + \sqrt{\left[\frac{B\Omega}{2}\right] + \omega_o^2} \tag{66.22}$$

and

$$\omega_- = \frac{-B\Omega}{2} + \sqrt{\left[\frac{B\Omega}{2}\right] + \omega_o^2} \tag{66.23}$$

Subtracting, one obtains

$$\omega_+ - \omega_- = B\Omega \tag{66.24}$$

$$\omega_+\omega_- = \omega_o^2 \tag{66.25}$$

Thus, the geometric mean of ω_+ and ω_- is the parameter ω_o; furthermore, the lpp frequencies $\Omega = 1$ rad/s map into points whose difference is the parameter B. Recalling that $A(\Omega)$ has to be an even function of Ω, one sees that the gain magnitudes at these two points must be identical. If the lpp is designed so that $\Omega = 1$ rad/s is the "bandwidth" (single-sided), then the object bpf will have a (two-sided) bandwidth of B rad/s.

An example should clarify things. Figure 66.5 shows a set of bandpass filter gain specifications. Some slight generality has been allowed over those shown in Fig. 66.2 by allowing the maximum stopband gains to be different in the two stopbands.

The graph is semilog: the vertical axis is linear with a dB scale and the horizontal axis is a log scale (base 10). The −0.1-dB minimum passband gain, by the way, is called the *passband ripple* because actual realized response is permitted to "ripple" back and forth between 0 and −0.1 dB. Notice that the frequency has been specified in terms of kHz—often a practical unit. Equation

[4]The negative sign on the radical gives $\omega < 0$ and the above treatment only considers positive ω.

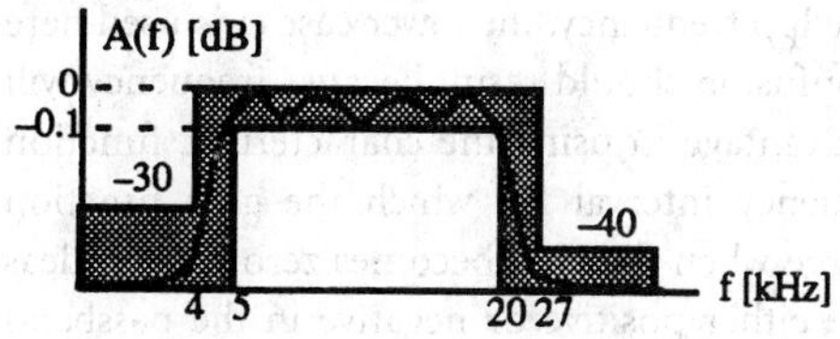

FIGURE 66.5 Bandpass filter specifications.

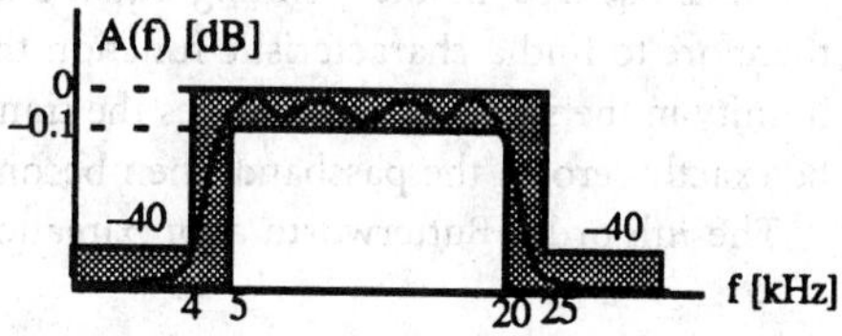

FIGURE 66.6 Bandpass filter specifications.

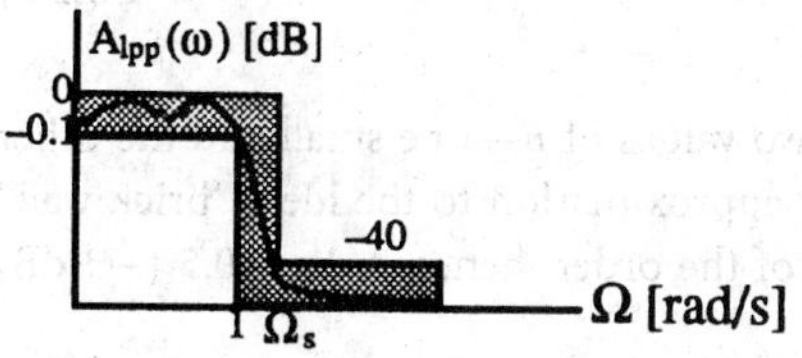

FIGURE 66.7 Bandpass filter specifications.

(66.20), however, can be normalized to any unit without affecting Ω. Thus, by normalizing ω to $2\pi \times 10^3$, one can substitute f in kHz for ω; the parameters ω_o (replaced by f_o symbolically) and B will then be in kHz also.

Now notice that $5 \times 20 = 100 \neq 4 \times 27 = 108$, so the specifications are not geometrically symmetric relative to any frequency. Somewhat arbitrarily choosing $f_o = \sqrt{5 \times 20} = 10$ kHz, one can force the specifications to have geometric symmetry by the following device: simply reduce the upper stopband cutoff frequency from 27 to 25 kHz. Then force the two stopband attenuations to be identical by decreasing the −30-dB lower stopband figure to −40 dB. This results in the modified specifications shown in Fig. 66.6. If one chooses to map the upper and lower passband cutoff frequencies to $\Omega = 1$ rad/s (a quite typical choice, as many filter design catalogs are tabulated under this assumption), one then has

$$B = 20 - 5 = 15 \text{ kHz} \tag{66.26}$$

This fixes the parameters in the transformation and the lpp stopband frequency can be determined from (66.20):

$$\Omega_s = \frac{25^2 - 10^2}{15 \times 25} = \frac{25}{15} - \frac{100}{375} = 1.498 \text{ rad/s} \tag{66.27}$$

The lpp specifications then assume the form shown in Fig. 66.7. Once the lpp approximation problem is solved, one can then transform either the derived transfer function or the synthesized circuit back up to bandpass form since the parameters of the transformation are known.

66.2 The Butterworth LPP Approximation

For performing lpp approximations, it is more convenient to work with the *characteristic function* $k(\omega)$ than with the gain function. It is defined by the equation

$$A^2(\omega) = \frac{1}{1 + K^2(\omega)} \tag{66.28}$$

Although Ω was used in the last subsection to denote lpp frequency, the lower case ω is used here and throughout the remainder of the section. No confusion should result because frequency will henceforth always mean lpp frequency. The main advantage in using the characteristic function is simply that it approximates zero over any frequency interval for which the gain function approximates unity. Further, it becomes infinitely large when the gain becomes zero. These ideas are illustrated in Fig. 66.8. Notice that $K(\omega)$ can be either positive or negative in the passband for it is squared in the defining equation. The basic problem in lpp filter approximation is therefore to find a characteristic function that approximates zero in the passband, approximates infinity in the stopband, and makes the transition from one to the other rapidly. Ideally, it would be exactly zero in the passband, then become abruptly infinity for frequencies in the stopband.

The nth order Butterworth approximation is defined by

$$K(\omega) = \omega^n \tag{66.29}$$

This characteristic function is sketched in Fig. 66.9 for two values of n—one small and the other large. As is easily seen, the larger order provides a better approximation to the ideal "brick wall" lpp response. Notice, however, that $K(1) = 1$ regardless of the order; hence $A(1) = 0.5$ (−3 dB) regardless of the order.

It is conventional to define the *loss function* $H(s)$ to be the reciprocal of the gain function:

$$H(s) = \frac{1}{G(s)} \tag{66.30}$$

Letting $s = j\omega$ and applying (66.28) results in

$$|H(j\omega)|^2 - K^2(\omega) = 1 \tag{66.31}$$

which is one form of *Feldtkeller's equation*, a fundamental equation in the study of filters. The loss approximates unity wherever the characteristic function approximates zero and infinity when the latter approximates infinity.

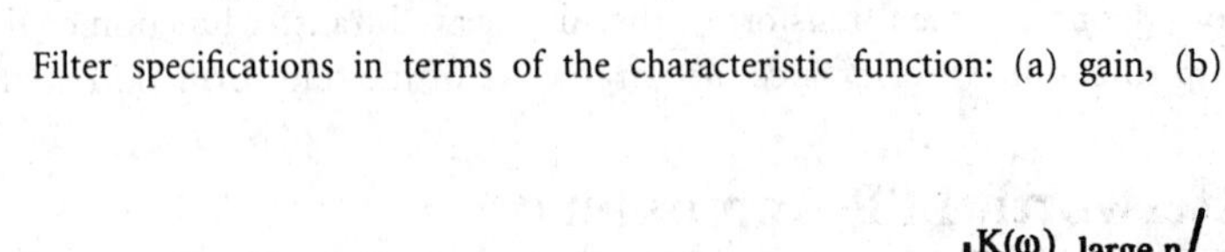

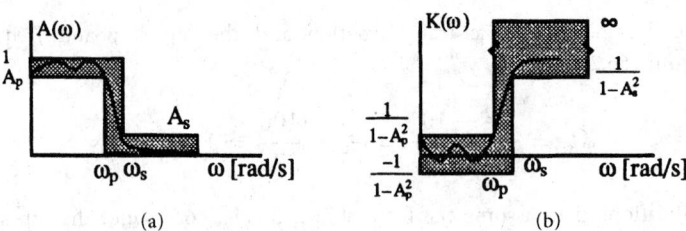

FIGURE 66.8 Filter specifications in terms of the characteristic function: (a) gain, (b) characteristic function.

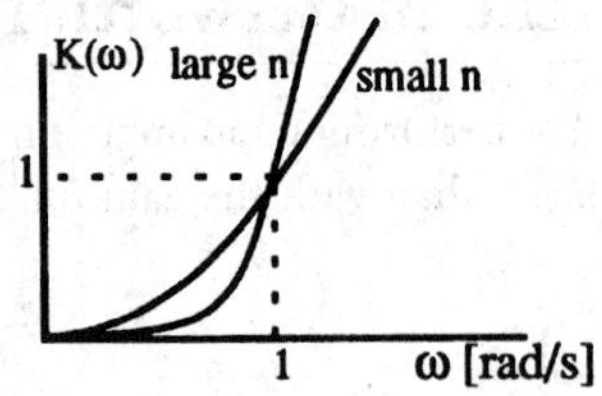

FIGURE 66.9 The Butterworth characteristic function.

The loss function can be used to illustrate a striking property of the Butterworth approximation. Taking the kth derivative of (66.31) one has

$$\frac{d^k|H(j\omega)|^2}{d\omega^k} = \frac{d^k K^2(\omega)}{d\omega^k} = \frac{d^k \omega^{2n}}{d\omega^k} = \frac{(2n)!}{k!}\omega^{2n-k} \tag{66.32}$$

This has the value zero at $\omega = 0$ for $k \leq 2n - 1$. It is the unique polynomial having this property among the set of all monic[5] polynomials of order $2n$ or lower having the value zero at the origin. But this means that the square of the Butterworth characteristic function $K(\omega)$ is the flattest of all such polynomials at $\omega = 0$. Since adding one to $K^2(\omega)$ produces the loss function $|H(j\omega)|^2$, the same is true of it relative to the set of all loss functions having the value unity at the origin. For this reason, the Butterworth approximation is often called the *maximally flat magnitude* (or MFM) approximation.

The passband ripple parameter A_p is always $1/\sqrt{2}$ for a Butterworth lpp; note that if a different ripple parameter is desired, one must treat the corresponding filter as a general low-pass filter, then apply (65.12) in the introduction to Chapter 65. The value of the parameter in that frequency transformation is determined by the requirement that the frequency at which the desired filter assumes the value $1/\sqrt{2}$ map into a lpp passband frequency of 1 rad/s. The required order is determined from the equation

$$\frac{1}{\sqrt{1+\omega_s^{2n}}} \leq A_s \tag{66.33}$$

or, rearranged,

$$n \geq \frac{\log\left[\dfrac{1}{A_s^2} - 1\right]}{2\log(\omega_s)} \tag{66.34}$$

The value of n is, of course, chosen to be the smallest integer greater than the expression on the right-hand side of (66.34).

Since there is only one parameter in the Butterworth approximation, (66.34) completely determines $A(\omega)$. That is, the MFM is a one-parameter approximation. The only remaining item of concern is the determination of $G(s)$, for synthesis of the actual filter requires knowledge of the transfer function. As was pointed out in the last subsection, this is the factorization problem. In general, the solution is given by (66.5), repeated here for convenience as (66.35):

$$G(s)G(-s) = A^2(-s^2) \tag{66.35}$$

In the present case, we have

$$G(s)G(-s) = \frac{1}{1+(-s^2)^n} = \frac{1}{1+(-1)^n s^{2n}} \tag{66.36}$$

How does one find $G(s)$ from this equation? The solution merely lies in applying the restriction that the resulting filter is to be stable. This means that the poles of $G(s)G(-s)$ that lie in the right-half plane must be discarded and the remaining ones assigned to $G(s)$. In this connection, observe that any poles on the imaginary axis must be of even multiplicity since $G(s)G(-s)$ is an even function (or, equivalently, since $A^2(\omega^2) = |G(j\omega)|^2$ is nonnegative). Furthermore,

[5]A monic polynomial is one whose leading coefficient (highest power of ω) is unity.

any even-order poles of $G(s)G(-s)$ on the imaginary axis could only result from one or more poles of $G(s)$, itself, at the same location. But such a $G(s)$ represents a filter that is undesirable because, at best, it is only marginally stable. As will be shown, this situation does not occur for Butterworth filters.

The problem now is merely to find all the poles of $G(s)G(-s)$, then to sort them. These poles are located at the zeros of the denominator in (66.36). Thus, one must solve

$$1 + (-s^2)^n = 0 \tag{66.37}$$

or, equivalently,

$$s^{2n} = (-1)^{n-1} \tag{66.38}$$

Representing s in polar coordinates by

$$s = \rho e^{j\phi} \tag{66.39}$$

one can write (66.38) in the form

$$\rho^{2n} e^{j2n\phi} = e^{j(n-1)\pi} \tag{66.40}$$

This has the solution

$$\rho = 1 \tag{66.41}$$

and

$$\phi = \frac{\pi}{2} + (2k - 1)\frac{\pi}{2n} \tag{66.42}$$

where k is any integer. Of course, only those values of ϕ between 0 and 2π radians are to be considered unique.

As an example, suppose that $n = 2$; that is, one is interested in determining the transfer function of a second-order Butterworth filter. Then the unique values of ϕ determined from (66.42) are $\pi/4$, $3\pi/4$, $5\pi/4$, and $7\pi/4$. All other values are simply these four with integer multiples of 2π added. The last two represent poles in the right-half plane, so are simply discarded. The other two correspond to poles at $s = -0.707 \pm j0.707$. Letting $D(s)$ be the numerator polynomial of $G(s)$, then, one has

$$D(s) = s^2 + \sqrt{2}s + 1 \tag{66.43}$$

The poles of $G(s)G(-s)$ are sketched in Fig. 66.10. Notice that if one assigns the left-half plane poles to $G(s)$, then those in the right-half plane will be those of $G(-s)$.

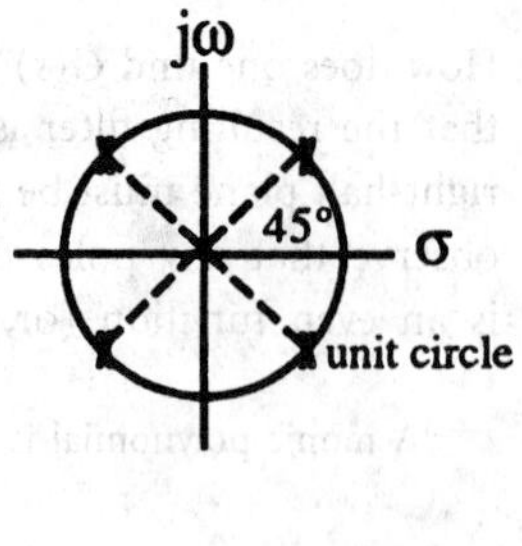

FIGURE 66.10 Poles of the second-order Butterworth transfer function.

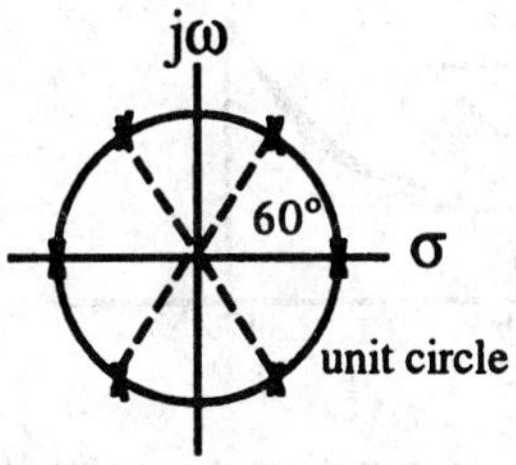

FIGURE 66.11 Poles of the third-order Butterworth transfer function.

Perhaps another relatively simple example is in order. To this end, consider the third-order Butterworth transfer function: $n = 3$. In this case, the polar angles of the poles of $G(s)$ are at $\pi/3$, $2\pi/3$, π, $4\pi/3$, $5\pi/3$, and 2π. This pole pattern is shown in Fig. 66.11. The denominator polynomial corresponds to those in the left-half plane. It is

$$D(s) = (s + 1)(s^2 + s + 1) = s^3 + 2s^2 + 2s + 1 \tag{66.44}$$

The pattern for the general nth order case is similar—all the poles of $G(s)G(-s)$ lie on the unit circle and are equally spaced at intervals of π/n radians, but are offset by $\pi/2$ radians relative to the positive real axis. A little thought will convince one that this means that no poles ever fall on the imaginary axis.

The factorization procedure determines the denominator polynomial in $G(s)$. But what about the numerator? Since the characteristic function is a polynomial, it is clear that $G(j\omega)$—and hence $G(s)$ itself—will have a constant numerator. For this reason, the Butterworth approximation is referred to as an *all-pole* filter. Sometimes it is also called a *polynomial* filter, referring to the fact that the characteristic function is a polynomial. As was mentioned earlier, the constant is usually allowed to float freely in the synthesis process and is determined only at the conclusion of the design process. However, more can be said. Writing

$$G(s) = \frac{a}{D(s)} \tag{66.45}$$

one can apply (66.36) to show that $a^2 = 1$, provided that $|G(0)|$ is to be one, as is the case for the normalized lpp. This implies that

$$a = \pm 1 \tag{66.46}$$

If a passive unbalanced (grounded) filter is desired the positive sign must be chosen. Otherwise, one can opt for either.

66.3 The Chebyshev LPP Approximation

The main advantage of the Butterworth approximation is that it is simple. It does not, however, draw upon the maximum approximating power of polynomials. In fact, a classical problem in mathematics is to approximate a given continuous function on a closed bounded interval with a polynomial of a specified maximum degree. One can choose to define the error of approximation in many ways, but the so-called *minimax* criterion seems to be the most suitable for filter design. It is the minimum value, computed over all polynomials of a specified maximum degree, of the maximum difference between the polynomial values and those of the specified function. This is illustrated in Fig. 66.12. The minimax error in this case occurs at ω_x. It is the largest value of the magnitude of the difference between the function values $f(\omega)$ and those of a given candidate polynomial $p_n(\omega)$. The polynomial of best fit is the one for which this value is the smallest.

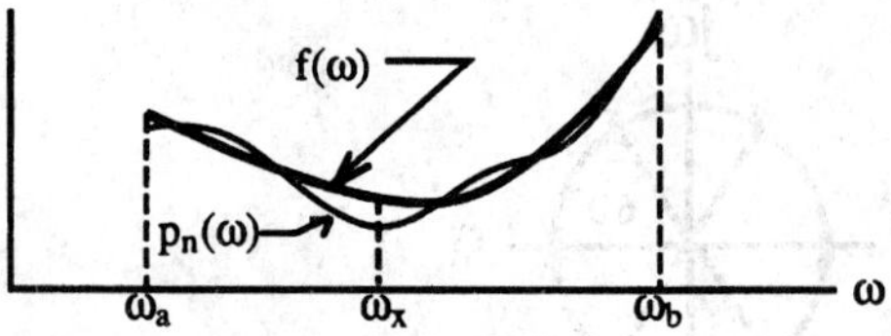

FIGURE 66.12 The minimax error criterion.

The basic lpp approximation problem is to pick the characteristic function to be that polynomial of a specified maximum degree no more than, say, n, which gives the smallest maximum error of approximation *to the constant value 0* over the interval $0 \leq \omega \leq 1$ (arbitrarily assuming that the passband cutoff is to be 1 rad/s). In this special case, the solution is known in closed form: it is the Chebyshev[6] polynomial of degree n. Then $K(\omega)$ is the polynomial $\varepsilon T_n(\omega)$, where

$$T_n(\omega) = \cos[n \cos^{-1}(\omega)] \tag{66.47}$$

and ε is the minimax error (a constant).

It is perhaps not clear that $T_n(\omega)$ is actually a polynomial; however, upon computing the first few by applying simple trigonometric identities one has the results shown in Table 66.1. In fact, again by calling upon simple trigonometric identities, one can derive the general recurrence relation

$$T_n(\omega) = 2\omega T_{n-1}(\omega) - T_{n-2}(\omega); \quad n \geq 2 \tag{66.48}$$

The Chebyshev polynomials have an enormous number of interesting properties and to explore them all would require a complete monograph. Among those of the most interest for filtering applications, however, are these. First, from the recursion relationship (66.48) one can see that $T_n(\omega)$ is indeed a polynomial of order n; furthermore, its leading coefficient is 2^{n-1}. If n is even, $T_n(\omega)$ is an even polynomial in ω and if n is odd, $T_n(\omega)$ is an odd polynomial. The basic definition in (66.47) clearly shows that the extreme values of $T_n(\omega)$ over the interval $0 \leq \omega \leq 1$ are ± 1. Some insight into the behavior of the Chebyshev polynomials can be obtained by making the transformation $\phi = \cos^{-1}(\omega)$. Then $T_n(\phi) = \cos(n\phi)$, a trigonometric function that is quite well known. The behavior of T_{15}, for example, is shown in Fig. 66.13. The basic ideas is this: the Chebyshev polynomial of nth order is merely a cosine of "frequency" $n/4$, which "starts" at $\omega = 1$ and "runs backward" to $\omega = 0$. Thus, it is always 1 at $\omega = 1$ and—as ω goes from 1 rad/s to 0 rad/s backward, it goes through n quarter-periods (or $n/4$ full periods). Thus, at $\omega = 0$ the value of this polynomial will be either 0 or ± 1, depending upon the specific value of n. If n is even, an integral number of half-periods will have been described and the resulting value will be ± 1; if n is odd, an integral number of half-periods plus a quarter-period will have been described and the value will be zero.

TABLE 66.1

n	$T_n(\omega)$
0	1
1	ω
2	$2\omega^2 - 1$
3	$4\omega^3 - 3\omega$
4	$8\omega^4 - 8\omega^2 + 1$

[6] If the name of the Russian mathematician is transliterated from the French, in which the first non-Russian translations were given, it is spelled Tchebychev.

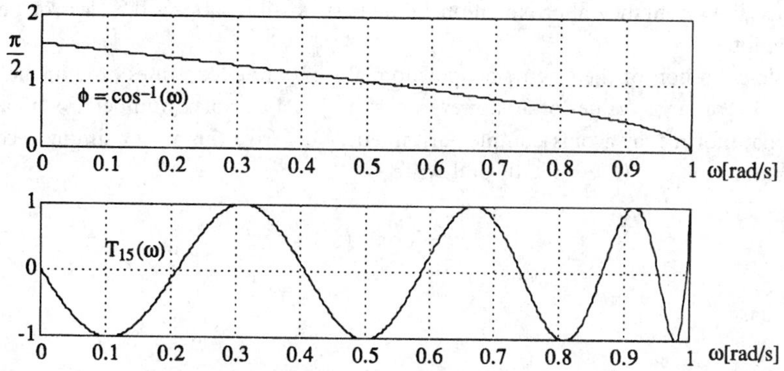

FIGURE 66.13 Behavior of the Chebyshev polynomials.

Based on the foregoing theory, one sees that the best approximation to the ideal lpp characteristic over the passband is, for a given passband tolerance ε, given by

$$A^2(\omega) = \frac{1}{1 + \varepsilon^2 T_n^2(\omega)} \tag{66.49}$$

It is, of course, known as the Chebyshev approximation and the resulting filter as the Chebyshev lpp of order n. The gain magnitude $A(\omega)$ is plotted for $n = 5$ and $\varepsilon = 0.1$ in Fig. 66.14. The passband behavior looks like ripples in a container of water, and since the crests are equally spaced above and below the average value, it is called **equiripple** behavior. In the passband, the maximum value is 1 and the minimum value is

$$A_{\text{min}} = \frac{1}{\sqrt{1 + \varepsilon^2}} \tag{66.50}$$

The passband ripple is usually specified as the peak to peak variation in dB. Since the maximum value is one, that is 0 dB, this quantity is related to the *ripple parameter* ε through the equation

$$\text{passband ripple in dB} = 20 \log \sqrt{1 + \varepsilon^2}. \tag{66.51}$$

The Chebyshev approximation is the best possible among the class of all-pole filters—*over the passband.* But what about its stopband behavior? As was pointed out above, it is desirable that—in addition to approximating zero in the passband—the characteristic function should go to infinity as rapidly as possible in the stopband. Now it is a happy coincidence that the Chebyshev polynomial goes to infinity for $\omega > 1$ faster than any other polynomial of the same

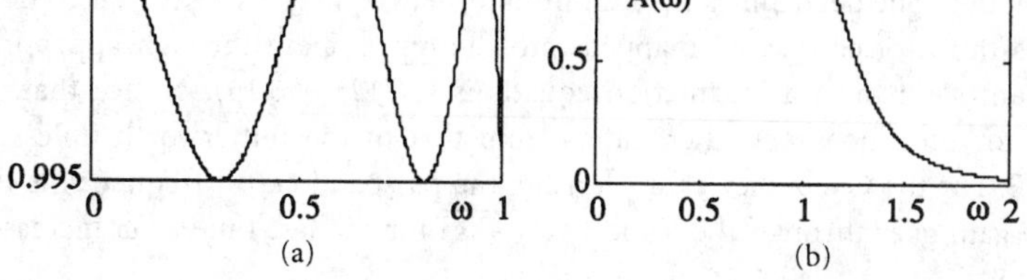

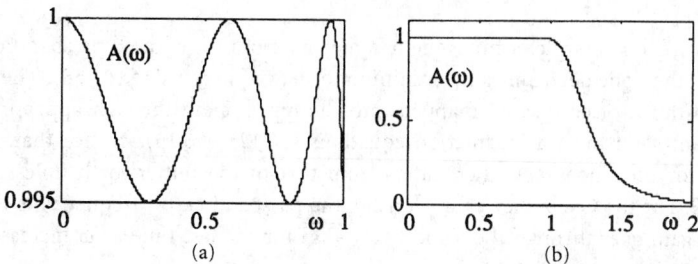

FIGURE 66.14 Frequency response of a fifth-order Chebyshev lpp: (a) passband, (b) overall.

order. Thus, the Chebyshev approximation is the best possible among the class of polynomial, or all-pole, filters.

The basic definition of the Chebyshev polynomial works fine for values of ω in the passband, where $\omega \leq 1$. For larger values of ω, however, $\cos^{-1}(\omega)$ is a complex number. Fortunately, there is an alternate form that avoids complex arithmetic. To derive this form, simply recognize the complex nature of $\cos^{-1}(\omega)$ explicitly and write

$$x = \cos^{-1}(\omega) \tag{66.52}$$

One then has[7]

$$\omega = \cos(x) = \cos[j(jx)] = \cosh(jx) \tag{66.53}$$

so

$$jx = \cosh^{-1}(\omega) \tag{66.54}$$

Thus, one can also write

$$T_n(\omega) = \cos[jn(jx)] = \cosh[n(jx)] = \cosh[n \cosh^{-1}(\omega)] \tag{66.55}$$

This result is used to compute the required filter order. Assuming as usual that A_s is the maximum allowed stopband gain, one uses the square root of (66.49) to get

$$A(\omega_s) = \frac{1}{\sqrt{1 + \varepsilon^2 T_n^2(\omega_s)}} \leq A_s \tag{66.56}$$

Solving, one has

$$T_n(\omega_s) \geq \frac{1}{\varepsilon}\sqrt{\frac{1}{A_s^2} - 1} \tag{66.57}$$

Since $\omega_s \geq 1$, it is most convenient to use the hyperbolic form in (66.55):

$$n \geq \frac{\cosh^{-1}\left[\dfrac{1}{\varepsilon}\sqrt{\dfrac{1}{A_s^2} - 1}\right]}{\cosh^{-1}(\omega_s)} \tag{66.58}$$

To summarize, one first determines the parameter ε from the allowed passband ripple, usually using (66.51); then one determines the minimum order required using (66.58). The original filter specifications must, of course, be mapped into the lpp domain through appropriate choice(s) of the constant(s) in the transformation equations (66.12)–(66.15). Notice that the definition of passband for the Chebyshev filter differs from that of the Butterworth unless the passband ripple is 3 dB. For the Chebyshev characteristic, the passband cutoff frequency is that frequency at which the gain goes through the value $1/\sqrt{1+\varepsilon^2}$ for the last time as ω increases.

[7] Since $\cos(x) = \cosh(jx)$.

The only remaining item is the determination of the transfer function $G(s)$ by factorization. Again, this requires the computation of the poles of $G(s)G(-s)$. Using (66.49) one has

$$G(s)G(-s) = [A^2(\omega)]_{s=j\omega} = \frac{1}{1 + \varepsilon^2 [T_n^2(\omega)]_{s=j\omega}} \tag{66.59}$$

Thus, the poles are at those values of s for which

$$[T_n^2(\omega)]_{s=j\omega} = \frac{-1}{\varepsilon^2} \tag{66.60}$$

or[8]

$$\cos\left[n \cos^{-1}\left(\frac{s}{j}\right)\right] = \pm j \frac{1}{\varepsilon} \tag{66.61}$$

Letting

$$\cos^{-1}(s/j) = a + jb, \text{ there results} \tag{66.62}$$

$$\cos(na)\cosh(nb) - j\sin(na)\sinh(nb) = \pm j\frac{1}{\varepsilon} \tag{66.63}$$

Equating real and imaginary parts, one has

$$\cos(na)\cos(nb) = 0 \tag{66.64a}$$

and

$$\sin(na)\sinh(nb) = \pm\frac{1}{\varepsilon} \tag{66.64b}$$

Since $\cosh(nb) > 0$ for any b, one must have

$$\cos(na) = 0 \tag{66.65}$$

which can hold only if

$$a = (2k+1)\frac{\pi}{2n}; \ k \text{ any integer} \tag{66.66}$$

But, in this case, $\sin[(2k+1)\pi/2] = \pm 1$, so application of (66.64b) gives

$$\sinh(nb) = \pm\frac{1}{\varepsilon} \tag{66.67}$$

One can now solve for b:

$$b = \pm\frac{1}{n}\sinh^{-1}\left[\frac{1}{\varepsilon}\right] \tag{66.68}$$

[8]Since s is complex anyway, nothing is to be gained from using the hyperbolic form.

Equations (66.66) and (66.68) together determine a and b, hence $\cos^{-1}(s/j)$. Taking the cosine of both sides of (66.62) and using (66.65) and (66.68) gives

$$s = \pm\sin\left[(2k+1)\frac{\pi}{2n}\right]\sinh\left(\frac{1}{n}\sinh^{-1}\left[\frac{1}{\varepsilon}\right]\right) + j\cos\left[(2k+1)\frac{\pi}{2n}\right]\cosh\left(\frac{1}{n}\sinh^{-1}\left[\frac{1}{\varepsilon}\right]\right) \tag{66.69}$$

Letting $s = \sigma + j\omega$ as usual, one can rearrange (66.69) into the form

$$\left[\frac{\sigma}{\sinh\left[\frac{1}{n}\sinh^{-1}\left[\frac{1}{\varepsilon}\right]\right]}\right]^2 + \left[\frac{\omega}{\cosh\left[\frac{1}{n}\sinh^{-1}\left[\frac{1}{\varepsilon}\right]\right]}\right]^2 = 1 \tag{66.70}$$

which is the equation of an ellipse in the s-plane with real axis intercepts of

$$\sigma_o = \pm\sinh\left[\frac{1}{n}\sinh^{-1}\left[\frac{1}{\varepsilon}\right]\right] \tag{66.71}$$

and imaginary axis intercepts of

$$\omega_o = \pm\cosh\left[\frac{1}{n}\sinh^{-1}\left[\frac{1}{\varepsilon}\right]\right] \tag{66.72}$$

This is shown in Fig. 66.15.

As an example, suppose that a Chebyshev lpp is to be designed that has a passband ripple of 0.1 dB, a maximum stopband gain of -20 dB, and $\omega_s = 2$ rad/s. Then one can use (66.51) to find[9] the ripple parameter ε:

$$\varepsilon = \sqrt{10^{0.1/10} - 1} = 0.1526 \tag{66.73}$$

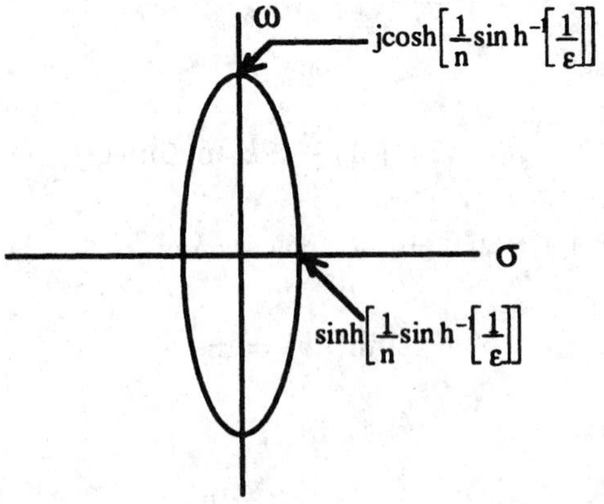

FIGURE 66.15 Pole locations for $G(s)G(-s)$ for the Chebyshev filter.

[9]For practical designs, a great deal of precision is required for higher order filters.

TABLE 66.2

k	Real Part	Imaginary Part
0	0.2642	1.1226
1	0.6378	0.4650
2	0.6378	−0.4650
3	0.2642	−1.1266
4	−0.2642	−1.1266
5	−0.6378	−0.4650
6	−0.6378	0.4650
7	−0.2642	1.1226

Equation (66.58) gives the minimum order required:

$$n \geq \frac{\cosh^{-1}\left[\frac{1}{0.1526}\sqrt{\frac{1}{(0.1)^2}-1}\right]}{\cosh^{-1}(2)} = \frac{\cosh^{-1}(65.20)}{\cosh^{-1}(2)} = 3.70 \tag{66.74}$$

In doing this computation by hand, one often uses the identity

$$\cosh^{-1}(x) = \ln[x + \sqrt{x^2 - 1}] \tag{66.75}$$

which can be closely approximated by

$$\cosh^{-1}(x) = \ln(2x) \tag{66.76}$$

if $x \gg 1$. In the present case, a fourth-order filter is required. The poles are shown in Table 66.2 and graphed in Fig. 66.16.

By selecting the left-half plane poles and forming the corresponding factors, then multiplying them, one finds the denominator polynomial of $G(s)$ to be

$$D(s) = s^4 + 1.8040s^3 + 2.2670s^2 + 2.0257s + 0.8286 \tag{66.77}$$

As for the Butterworth example of the last subsection, one finds the scale factor:

$$k = 0.8286 \tag{66.78}$$

Therefore, the complete transfer function is

$$G(s) = \frac{k}{D(s)} = \frac{0.8286}{s^4 + 1.8040s^3 + 2.2670s^2 + 2.0257s + 0.8286} \tag{66.79}$$

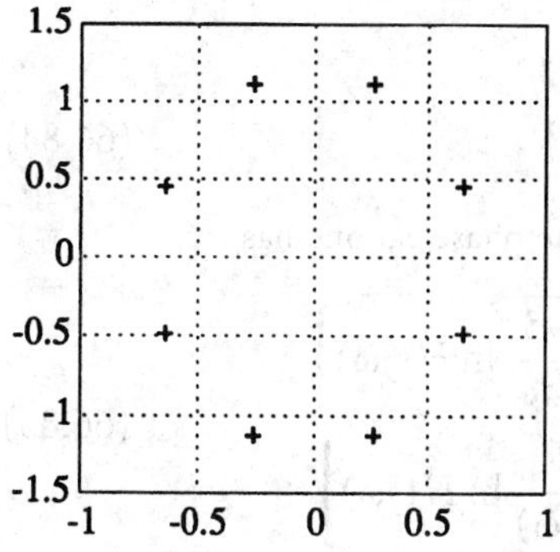

FIGURE 66.16 Pole locations for example filter.

Of course, if completely automated algorithms are not being used to design such a filter as the one in our example, numerical assistance is required. The computation in the preceding example was performed in MATLAB—a convenient package for many computational problems in filter design.

66.4 The Bessel–Thompson LPP Approximation

Thus far, two all-pole approximations have been presented. As was pointed out in the preceding subsection, the Chebyshev is better than the Butterworth—in fact, it is the best all-pole approximation available. So why bother with the Butterworth at all, other than as an item of historical interest? The answer lies in the phase. The Chebyshev filter has a phase characteristic that departs farther from linearity than that of the Butterworth; put differently, its group delay deviates from a constant by a greater amount. Thus, the Butterworth approximation is still a viable approximation in applications where phase linearity is of some importance.

The question naturally arises as to whether there is an lpp approximation that has better phase characteristics than that of the Butterworth. The answer is yes, and that is the topic of this section, which will follow [7]—perhaps the simplest development of all.

Recall that the gain function $G(s)$ is the inversion of the loss function $H(s)$:

$$G(s) = \frac{1}{H(s)} \tag{66.80}$$

Also, recall that it was the loss function $|H(j\omega)| = 1/A(\omega)$ (or, rather, its square $|H(j\omega)|^2$) that was required to have the maximally flat magnitude (MFM) property: at $\omega = 0$,

$$\left[\frac{d^k|H(j\omega)|^2}{d\omega^k}\right]_{\omega=0} = \left[\frac{d^k\omega^{2n}}{d\omega^k}\right]_{\omega=0} = \left[\frac{(2n)!}{k!}\omega^{2n-k}\right]_{\omega=0} = 0 \tag{66.81}$$

for $k = 0, 1, \cdots, 2n-1$. The question to be asked and answered in this section is whether there exists a similar approximation for the group delay $\tau_g(\omega)$: a maximally flat delay (MFD) approximation.

To answer this question, the phase will be written in terms of the loss function $H(j\omega)$. Since the latter quantity can be written in polar form as

$$H(j\omega) = \frac{1}{G(j\omega)} = \frac{1}{A(\omega)e^{j\phi(\omega)}} = \frac{1}{A(\omega)}e^{-j\phi(\omega)} \tag{66.82}$$

the (complex) logarithm is

$$\ln H(j\omega) = -\ln A(\omega) - j\phi(\omega) \tag{66.83}$$

Thus,

$$\phi(\omega) = -\operatorname{Im}\{\ln H(j\omega)\} \tag{66.84}$$

The group delay is merely the negative of the derivative of the phase, so one has

$$\begin{aligned}\tau_g(\omega) &= -\frac{d\phi}{d\omega} = \frac{d}{d\omega}\operatorname{Im}\{\ln H(j\omega)\} = \operatorname{Im}\left\{\frac{d}{d\omega}\ln H(j\omega)\right\} \\ &= \operatorname{Im}\left\{j\frac{d}{d(j\omega)}\ln H(j\omega)\right\} = \operatorname{Re}\left\{\frac{d}{d(j\omega)}\ln H(j\omega)\right\}\end{aligned} \tag{66.85}$$

Recalling that the even part of a complex function $F(s)$ is given by

$$Ev\{F(s)\} = \frac{F(s) + F(-s)}{2} \tag{66.86}$$

one can use the symmetry property for real $F(s)$ (in the present context, $F(s)$ is assumed to be a rational function, so a real $F(s)$ is one with real coefficients) to show that

$$Ev\{F(s)\}_{s=j\omega} = \frac{F(j\omega) + F(-j\omega)}{2} = \frac{F(j\omega) + F^*(j\omega)}{2} = \text{Re}\{F(j\omega)\} \tag{66.87}$$

In this manner, one can analytically extend the group delay function $\tau_g(j\omega)$ so that it becomes a function of s:

$$\tau_g(s) = Ev\left\{\frac{d}{ds}\ln H(s)\right\} = Ev\left[\frac{H'(s)}{H(s)}\right] = \frac{1}{2}\left[\frac{H'(s)}{H(s)} + \frac{H'(-s)}{H(-s)}\right] \tag{66.88}$$

In the present case, it will be assumed that $H(s)$ is a polynomial. In this manner, an all-pole (or polynomial) filter will result. Thus, one can write

$$H(s) = \sum_{k=0}^{n} a_k^n s^k \tag{66.89}$$

The superscript on the (assumed real) coefficients matches the upper limit on the sum and is the assumed filter order. The group delay function is, thus, a real rational function:

$$\tau_g(s) = \frac{N(s)}{D(s)} \tag{66.90}$$

where $N(s)$, the numerator polynomial, and $D(s)$, the denominator polynomial, have real coefficients and are of degrees no greater than $2n - 1$ and $2n$, respectively, by inspection of (66.88). But, again according to (66.88), $\tau_g(s)$ is an even function. Thus,

$$\tau_g(-s) = \frac{N(-s)}{D(-s)} = \tau_g(s) = \frac{N(s)}{D(s)} \tag{66.91}$$

The last equation, however, implies that

$$\frac{N(-s)}{N(s)} = \frac{D(-s)}{D(s)} = 1 \tag{66.92}$$

The last equality is arrived at by the following reasoning. $N(s)$ and $D(s)$ are assumed to have no common factors—any such have already been canceled in the formation of $\tau_g(s)$. Thus, the two functions of s in (66.92) are independent; since they must be equal for all s, they must therefore equal a constant. But this constant is unity, as is easily shown by allowing s to become infinite and noting that the degrees and leading coefficients of $N(-s)$ and $N(s)$ are the same.

The implication of the preceding development is simply that $N(s)$ and $D(s)$ consist of only even powers of s. Looking at $N(s)$, for example, and letting it be written

$$N(s) = \sum_{k=0}^{2n-1} \rho_k s^k \tag{66.93}$$

one has

$$N(-s) = \sum_{k=0}^{2n-1} \rho_k(-s)^k = N(s) = \sum_{k=0}^{m} \rho_k(s^k) \tag{66.94}$$

This, however, implies that the ρ_k are zero for odd k. Hence $N(s)$ consists of only even powers of s. The same is true of $D(s)$ and therefore of $\tau_g(s)$. Clearly, therefore, $\tau_g(\omega)$ will consist of only even powers of ω—that is, it will be a function of ω^2. Now there is to be an MFD approximation one must have by analogy with the MFM approximation in the section on the Butterworth LPP approximation,

$$\frac{d^k \tau_g(\omega)}{d(\omega^2)^k} = 0 \tag{66.95}$$

for $k = 1, 2, \cdots, n-1$. The constraint is, of course, that τ_g must come from a polynomial loss function $H(s)$ whose zeros [poles of $G(s)$] all lie in the left-half plane.

It is convenient to normalize time so that the group delay T, say, at $\omega = 0$ is 1 second; this is equivalent to scaling the frequency variable ω to be ωT. Here, it will be assumed that this has been performed already. There is a slight difference between the MFD and MFM approximations; the latter approximates zero at $\omega = 0$, while the former approximates $T = 1s$ at $\omega = 0$. The two become the same, however, if one considers the function

$$\tau_g(s) - 1 = \frac{P(s)}{2H(s)H(-s)} \tag{66.96}$$

where $P(s)$ has a maximum order of $2n$. The form on the right-hand side of this expression can be readily verified by consideration of (66.88), the basic result for the following derivation. Furthermore, writing $P(s)$ in the form

$$P(s) = \sum_{k=0}^{n} P_k(s^2)^k \tag{66.97}$$

it is readily seen that

$$p_o = p_1 = \cdots = p_{n-1} = 0 \tag{66.98}$$

The lowest order coefficient is clearly zero because $\tau_g(0) = 1$; furthermore, all odd coefficients are zero since τ_g is even. Finally, all other coefficients in (66.98) have to be zero if one imposes the MFD condition in (66.95).[10]

At this stage, one can write the group delay function in the form

$$\tau_g(s) = \frac{N(s)}{D(s)} = 1 + \frac{p_n s^{2n}}{2H(s)H(-s)} = \frac{2H(s)H(-s) + p_n s^{2n}}{2H(s)H(-s)} \tag{66.99}$$

It was pointed out immediately after (66.90), however, that the degree of $N(s)$ is at most $2n-1$. Hence the coefficient of s^{2n} in the numerator of (66.99) must vanish, and one therefore also has

$$2(-1)^n a_n + p_n = 0 \tag{66.100}$$

[10]The derivatives are easy to compute, but this is omitted here for reasons of space.

or, equivalently,

$$p_n = 2(-1)^{n+1} a_n \tag{66.101}$$

Finally, this allows one to write (66.99) in the form

$$\tau_g(s) = \frac{H(s)H(-s) + (-1)^{n+1} a_n s^{2n}}{H(s)H(-s)} \tag{66.102}$$

If one now equates (66.102) and (66.88) and simplifies, there results

$$H'(s)H(-s) + H'(-s)H(s) - 2H(s)H(-s) = 2(-1)^{n+1} a_n s^{2n} \tag{66.103}$$

Multiplying both sides by s^{-2n}, taking the derivative (noting that $(d/ds)H(-s) = -H'(-s)$, one obtains (after a bit of algebra)

$$Ev\{[sH''(s) - 2(s+n)H'(s) + 2nH(s)]H(-s)\} = 0 \tag{66.104}$$

Now if s_o is a zero of this even function, then $-s_o$ will also be a zero. Further, since all of the coefficients in $H(s)$ are real, s_o^* and $-s_o^*$ must also be zeros as well; that is, the zeros must occur in a *quadrantally symmetric* manner. Each zero must belong either to $H(-s)$, or to the factor it multiplies, or to both. The degree of the entire expression in (66.104) is $2n$ and $H(-s)$ has n zeros. Thus, the expression in square brackets must have n zeros. Now here is the crucial step in the logic: if the filter being designed is to be stable, then all n zeros of $H(-s)$ must be in the (open) right-half plane. This implies that the factor in square brackets must have n zeros in the (open) left-half plane. Since that expression has degree n, these zeros can be found from the equation

$$sH''(s) - 2(s+n)H'(s) + 2nH(s) = 0 \tag{66.105}$$

This differential equation can be transformed into that of Bessel; here, however, the solution will be derived directly by recursion. Using (66.89) for $H(s)$, computing its derivatives, reindexing, and using (66.105), one obtains, for $0 \le k \le n-1$,

$$(k+1)ka_{k+1}^n - 2n(k+1)a_{k+1}^n - 2ka_k^n + 2na_k^n = 0 \tag{66.106}$$

This produces the recursion formula

$$a_{k+1}^n = \frac{2(n-k)}{(2n-k)(k+1)} a_k^n; \quad 0 \le k \le n-1 \tag{66.107}$$

or, normalizing the one free constant so that $a_n^n = 1$ and reindexing, gives

$$a_{k+1}^n = \frac{2(n-k)}{(2n-k)(k+1)}; \quad 0 \le k \le n-1 \tag{66.108}$$

The resulting polynomials for $H(s)$ are closely allied with the Bessel polynomials. The first several are given in Table 66.3 and the corresponding gain and group delay characteristics are plotted (using MATLAB) in Fig. 66.17. Notice that the higher the order, the more accurately the group delay approximates a constant and the better the gain approximates the ideal lpp; the latter behavior, however, is fairly poor. A view of the group delay behavior in the passband is shown for the third-order filter in Fig. 66.18.

TABLE 66.3

n	$H(s)$
1	$s+1$
2	s^2+3s+3
3	$s^3+6s^2+15s+15$
4	$s^4+10s^3+45s^2+105s+105$

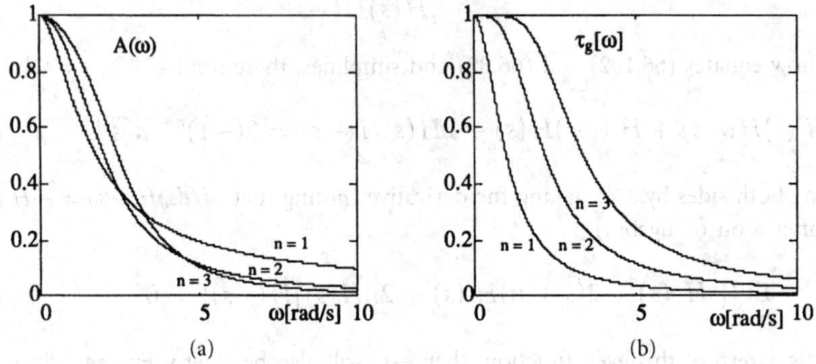

FIGURE 66.17 Gain and group delay characteristics for the Bessel–Thompson filters of orders one through three: (a) gain, (b) group delay.

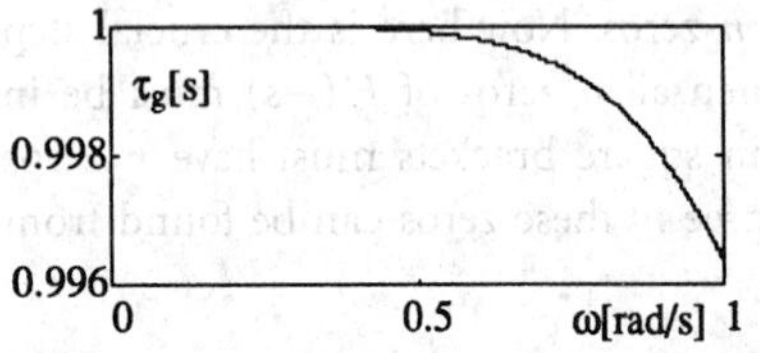

FIGURE 66.18 Group delay of third-order Bessel filter.

66.5 The Elliptic Approximation

The Butterworth approximation provides a very fine approximation to the ideal "brick wall" lpp response at $\omega = 0$, but is poor for other frequencies; the Chebyshev, on the other hand, spreads out the error of approximation throughout the passband and thereby achieves a much better amplitude approximation. This is one concrete application of a general result in approximation theory known as the Weierstrass theorem [2], which asserts the possibility of uniformly approximating a continuous function on a compact set by a polynomial or by a rational function.

A compact set is the generalization of a closed and bounded interval—the setting for the Chebyshev approximation sketched in Fig. 66.19. The compact set is the closed interval $[-1, 1]$,[11] the continuous function to be approximated is the constant 0, and the approximating function is a polynomial of degree n (in the figure n is three). A careful inspection of this figure reveals that the maximum error of approximation occurs exactly four times on a set of four discrete points within the interval and the *sign* of this maximum error alternates from one such point to its adjacent neighbor. This is no coincidence; in fact, the polynomial of best approximation of degree n is characterized by this alternation or *equiripple* property: the error function achieves its maximum value exactly $n+1$ times on the interval of approximation and the signs at these points alternate. If one finds a polynomial by any means that has this property it is the unique polynomial of best approximation.

[11]Here, we are explicitly observing that the characteristic function to be approximated is even in ω.

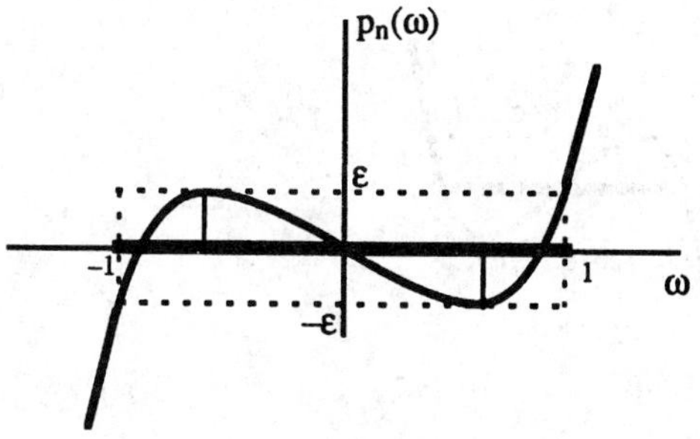

FIGURE 66.19 Chebyshev approximation.

In the present case, one simply notes that the cosine function has the required equiripple behavior. That is, as ϕ varies between $-\pi/2$ and $\pi/2$, $\cos(n\phi)$ varies between its maximum and minimum values of ± 1 a total of $n+1$ times. But $\cos(n\phi)$ is not a polynomial, and the problem (for reasons stated in the introduction) is to determine a *polynomial* having this property. At this point, one observes that if one makes the transformation

$$\phi = \cos^{-1}(\omega) \tag{66.109}$$

then, as ω varies from -1 to $+1$, ϕ varies from $-\pi/2$ to $+\pi/2$. This transformation is, fortunately, one-to-one over this range; furthermore, even more fortunately, the overall function

$$T_n(\omega) = \cos[n \cos^{-1}(\omega)] \tag{66.110}$$

is a polynomial as desired. Of course, the maximum error is ± 1, an impractically large value. This is easily rectified by requiring that the approximating polynomial be

$$p_n(\omega) = \varepsilon T_n(\omega) = \varepsilon \cos[n \cos^{-1}(\omega)] \tag{66.111}$$

In terms of gain and characteristic functions, therefore, one has

$$G(\omega) = \frac{1}{1+K^2(\omega)} = \frac{1}{1+\varepsilon^2 T_n^2(\omega)} \tag{66.112}$$

That is, $p_n(\omega) = \varepsilon T_n(\omega)$ is the best approximation to the characteristic function $K(\omega)$.

Since the gain function to be approximated is even in ω, $K(\omega)$—and, therefore, $p_n(\omega) = \varepsilon T_n(\omega)$—must be either even or odd. As one can recall, the Chebyshev polynomials have this property. For this reason, it is only necessary to discuss the situation for $\omega \geq 0$ for the extension to negative ω is then obvious.

The foregoing sets the stage for the more general problem, which will now be addressed. As noted above, the Weierstrass theorem holds for more general compact sets. Thus, as a preliminary exercise, the approximation problem sketched in Fig. 66.20 will be discussed. That figure shows two closed intervals as the compact set on which the approximation is to occur. The function to be approximated is assumed to be the constant a on the first interval and the constant b on the second.[12] In order to cast this into the filtering context, it will be assumed that the constant a is small and the constant b large. The sharpness of the transition characteristic can then be set to any desired degree by allowing ω_p to approach unity from

[12]One should note that this function is continuous on the compact set composed of the two closed intervals.

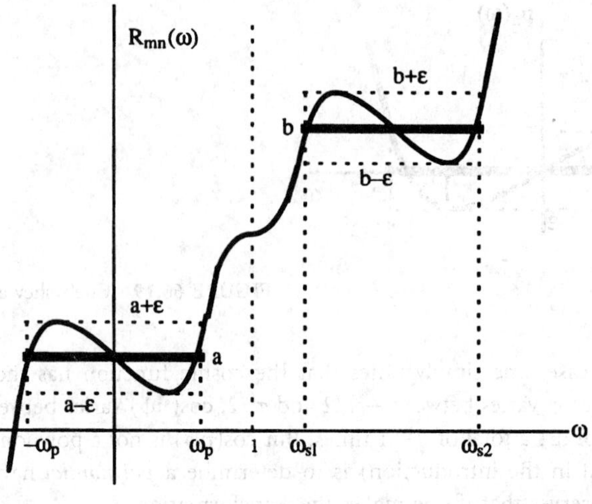

FIGURE 66.20 Chebyshev rational approximation.

below and ω_{s1} to approach unity from above. Notice that the frequency scaling will be different here than for the all-pole case in which the right-hand interval endpoint was allowed to be unity.

In addition to more general compact sets, the Weierstrass theorem allows the approximating function to be a general real rational function, which has been denoted by $R_{mn}(\omega)$ in the figure. This notation means that the numerator and denominator polynomials have maximum degree m and n respectively. Now suppose that one specifies that $b = 1/a$ and suppose $R_{mn}(\omega)$ is a reciprocal function, that is, one having the property that

$$R_{mn}\left(\frac{1}{\omega}\right) = \frac{1}{R_{mn}(\omega)} \tag{66.113}$$

Then if one determines its coefficients such that the equiripple property holds on the interval $-\omega_p \leq \omega \leq \omega_p$, it will also hold on the interval $\omega_{s1} \leq \omega \leq \omega_{s2}$. Of course, this interval is constrained such that $\omega_{s1} = 1/\omega_p$. The other interval endpoint will then extend to $+\infty$ (and, by symmetry, there will also be an interval on the negative frequency axis with one endpoint at $-1/\omega_p$ and the other at $-\infty$). This is, of course, not a compact set—but the approximation theorem continues to hold anyway because it is equivalent to approximating $a = 1/b$ on the low-pass interval $[-\omega_p, \omega_p]$. Thus, one can let $a = 0$ and $b = \infty$ and simultaneously approximate the ideal characteristic function

$$K(\omega) = \begin{cases} 0; & |\omega| \leq \omega_p \\ \infty; & |\omega| \geq \dfrac{1}{\omega_p} \end{cases} \tag{66.114}$$

with the real and even or odd (for realizability) rational function $R_{mn}(\omega)$ having the aforementioned reciprocal property. The Weierstrass theorem equiripple property for such rational functions demands that the total number of error extrema[13] on the compact set be $m + n + 2$. (This assumes the degree of both polynomials to be relative to the variable ω^2.)

[13]There can be degeneracy in the general approximation problem, but the constraints of the problem being discussed here preclude this from occurring.

Based on the preceding discussion, one has the following form for $R_{mn}(\omega)$:

$$R_{2n,2n}(\omega) \equiv R_{2n}(\omega) = \frac{(\omega_1^2 - \omega^2)(\omega_3^2 - \omega^2)\cdots(\omega_{2n-1}^2 - \omega^2)}{(1 - \omega_1^2\omega^2)(1 - \omega_3^2\omega^2)\cdots(1 - \omega_{2n-1}^2\omega^2)} \tag{66.115}$$

$$R_{2n+1,2n}(\omega) \equiv R_{2n+1}(\omega) = \frac{\omega(\omega_2^2 - \omega^2)(\omega_4^2 - \omega^2)\cdots(\omega_{2n}^2 - \omega^2)}{(1 - \omega_2^2\omega^2)(1 - \omega_4^2\omega^2)\cdots(1 - \omega_{2n}^2\omega^2)} \tag{66.116}$$

The first is clearly an even rational function and the latter odd. The problem now is to find the location of the pole and zero factors such that equiripple behavior is achieved in the passband. The even case is illustrated in Fig. 66.21 for $n = 2$. Notice that the upper limit of the passband frequency interval has been taken for convenience to be equal to $\sqrt{k}$; hence, the lower limit of the stopband frequency interval is $1/\sqrt{k}$. Thus,

$$k = \frac{\omega_p}{\omega_s} \tag{66.117}$$

is a measure of the sharpness of the transition band rolloff and is always less than unity. The closer it is to one, the sharper the rolloff. It is an arbitrarily specified parameter in the design. Notice that equiripple behavior in the passband implies equiripple behavior in the stopband—in the latter, the approximation is to the constant whose value is infinity and the minimum value (which corresponds to a maximum value of gain) is $1/\varepsilon$, while the maximum deviation from zero in the passband is ε. The zeroes are all in the passband and are mirrored in poles, or infinite values, in the stopband. The notation ω_0 and ω_2 for the passband frequencies of maximum error has been introduced. In general, their indices will be even for even approximations and odd for odd approximations.

A sketch for the odd case would differ only in that the plot would go through the origin, that is the origin would be a zero rather than a point of maximum error. Notice that the resulting filter will have finite gain unlike the Butterworth or Chebyshev, which continue to rolloff toward zero as ω becomes infinitely large. As noted in the figure the approximating functions are called the *Chebyshev rational functions.*

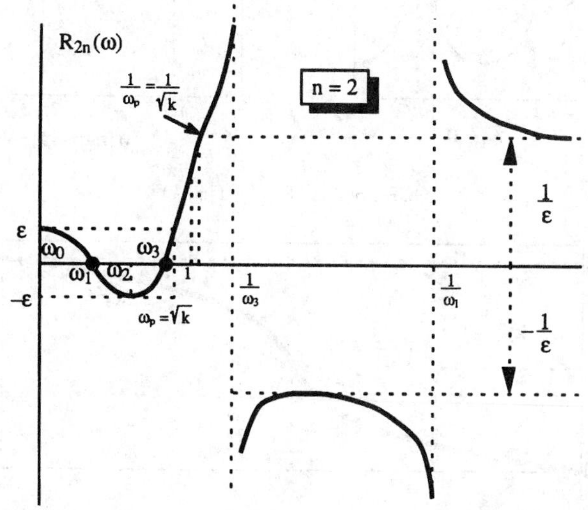

FIGURE 66.21 Typical plot of an even-order Chebyshev rational function.

The procedure [8] now is quite analogous to the Chebyshev polynomial approximation, though rather more complicated. One looks for a continuous, periodic waveform that possesses the desired equiripple behavior. Since the Jacobian elliptic functions are generalizations of the more ordinary sinusoids, it is only natural that they be investigated with an eye toward solving the problem under attack. With that in mind, some of the more salient properties will now be reviewed.

Consider the function

$$I(\phi, k) = \frac{1}{\sqrt{1 - k^2 \sin^2(\phi)}} \tag{66.118}$$

which is plotted for two values of k in Fig. 66.22. If $k = \pm 1$, the peak value is infinite; in this case, $I(\phi, k) = \sec(\phi)$. For smaller values of k the peak value depends upon k with smaller values of k resulting in lower peak values. The peaks occur at odd multiples of $\pi/2$. Note that $I(\phi, k)$ has the constant value one for $k = 0$. Figure 66.23 shows the running integral of $I(\phi, k)$ for $k^2 = 0.99$ and for $k^2 = 0$. For $k^2 = 0$ the curve is a straight line (shown dashed); for other values of k^2, it deviates from a straight line by an amount that depends upon the size of k^2. Observe that the running integral has been plotted over one full period of $I(k, \phi)$—that is, from 0 to π. The running integral is given in analytical form by

$$u(\phi, k) = \int_0^{\phi} I(\alpha, k)\, d\alpha = \int_0^{\phi} \frac{1}{\sqrt{1 - k^2 \sin^2(\alpha)}}\, d\alpha \tag{66.119}$$

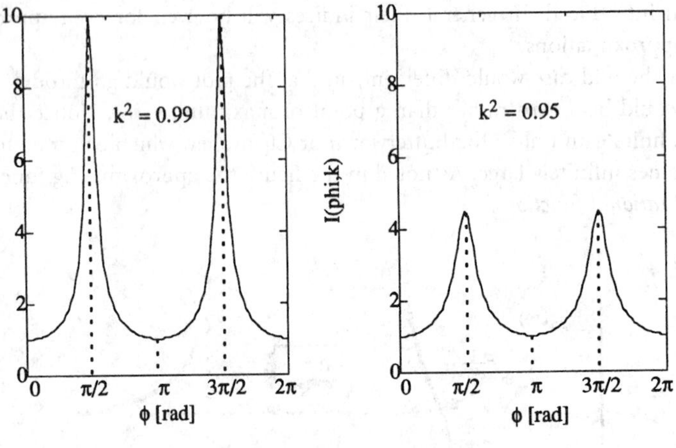

FIGURE 66.22 Plot of $I(\phi, k)$.

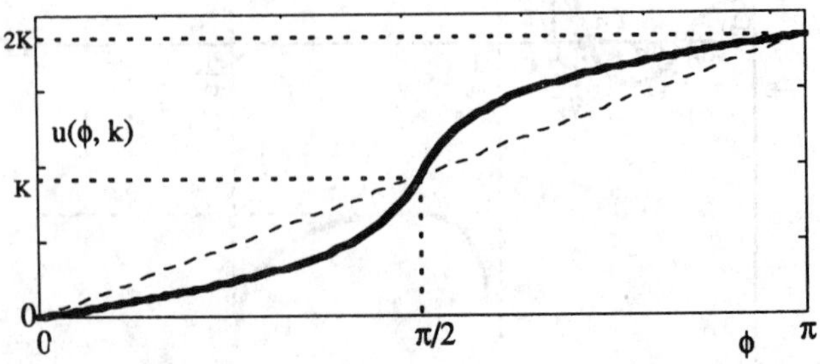

FIGURE 66.23 Running integral of I normalized to K.

The quantity K, shown by the lowest dashed horizontal line in Fig. 66.23, is the integral of $I(\phi, k)$ from 0 to $\pi/2$; that is, it is the area beneath the $I(k, \phi)$ curve in Fig. 66.22 from 0 to $\pi/2$. Thus, it is the area beneath the curve to the left of the first vertical dashed line in that figure. It is given by

$$u\left(\frac{\pi}{2}, k\right) = \int_0^{\pi/2} I(\alpha, k)\, d\alpha = \int_0^{\pi/2} \frac{1}{\sqrt{1 - k^2 \sin^2(\alpha)}}\, d\alpha = K \tag{66.120}$$

and is referred to as the *complete elliptic integral of the first kind.*

The sine generalization sought can now be defined. Since the running integral $u(\phi, k)$ is monotonic, it can be inverted and thereby solved for ϕ in terms of u; for each value of u there corresponds a unique value of ϕ. The **elliptic sine function** is defined to be the ordinary sine function of ϕ:

$$sn(u, k) = \sin(\phi, k) \tag{66.121}$$

Now, inspection of Fig. 66.23 shows that, as ϕ progresses from 0 to 2π, $u(\phi, k)$ increases from 0 to $4K$; since angles are unique only to within multiples of 2π, therefore, it is clear that $sn(u, k)$ is periodic with period $4K$. Hence, K is a quarter-period of the elliptic sine function.

The integral in (66.119) cannot be evaluated in closed form; thus, there is not a simple, compact expression for the required inverse. Therefore, the elliptic sine function can only be tabulated in numerical form or computed using numerical techniques. It is shown for $k^2 = 0.99$ in Fig. 66.24 and compared with a conventional sine function having the same period ($4K$).

Recall, now, the objective of exploring the preceding generalization of the sinusoid: one is looking for a transformation that will convert the characteristic functions given by (66.115) and (66.116) into equivalent waveforms having the equiripple property. As one might suspect (since so much time has been spent on developing it), the elliptic sine function is precisely the transformation desired. The crucial aspect of showing this is the application of a fundamental property of $sn(u, k)$, known as an **addition formula:**

$$sn(u + a, k)sn(u - a, k) = \frac{sn^2(u, k) - sn^2(a, k)}{1 - k^2 sn^2(u, k)sn^2(a, k)} \tag{66.122}$$

The right-hand side of this identity has the same form as one factor in (66.115) and (66.116), that is of one zero factor coupled with its corresponding pole factor. This suggests the transformation

$$\omega = \sqrt{k}\, sn(u, k) \tag{66.123}$$

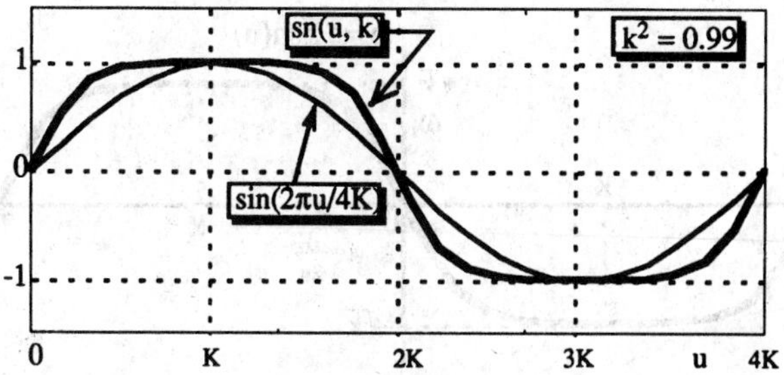

FIGURE 66.24 Plots of the elliptic and ordinary sine functions.

The passband zeros then are given by

$$\omega_i = \sqrt{k}\, sn(u_i, k) \tag{66.124}$$

and the factors mentioned above map into

$$\frac{\omega_i^2 - \omega^2}{1 - \omega_i^2\omega^2} = \frac{k[sn^2(u_i, k) - sn^2(u, k)]}{1 - k^2 sn^2(u_i, k) sn^2(u, k)} = ksn(u + u_i)sn(u - u_i) \tag{66.125}$$

For specificity, the even-order case will be discussed henceforth. The odd-order case is the same if minor notational modifications are made. Thus, one sees that

$$R_{2n}(\omega) = \prod_{i=1,3,\cdots,2n-1} \frac{\omega_i^2 - \omega^2}{1 - \omega_i^2\omega^2} = \prod_{i=1,3,\cdots,2n-1} ksn(u + u_i)sn(u - u_i) \tag{66.126}$$

The u_i are to be chosen. Before doing this, it helps to simplify the preceding expression by defining $u_{-i} = -u_i$ and reindexing. Calling the resulting function $G(u)$, one has

$$G(u) = \prod_{i=-1,-3,\cdots,-(2n-1)}^{i=1,3,\cdots,2n-1} ksn(u + u_i) \tag{66.127}$$

Refer now to Fig. 66.24. Each of the *sn* functions is periodic with period $4K$ and is completely defined by its values over one quarter-period $[0, K)$. Suppose that one defines

$$u_i = i\frac{K}{2n} \tag{66.128}$$

Figure 66.25 shows the resulting transformation corresponding to (66.123) and (66.124). As u progresses from $-K$ to $+K$, ω increases from $-\sqrt{k}$ to $+\sqrt{k}$ as desired. Furthermore, because of the symmetry of $sn(u)$, the set $\{u_i\}$ forms an additive group—adding $K/2n$ to the index of any u_i results in another u_i in the set. This means that $G(u)$ in (66.127) is periodic with period $K/2n$. Thus, as ω increases from $-\sqrt{k}$ to $+\sqrt{k}$, $R_{2n}(\omega)$ achieves $2n+2$ extrema, that is, positive and negative peak values. But this is sufficient for $R_{2n}(\omega)$ to be the Chebyshev rational function of best approximation.

The symmetry of the zero distribution around the peak value of $sn(u)$, that is, around $u = K$, reveals that the peak values of $R_{2n}(\omega)$ occur between the zeros; that is, recalling that in Fig. 66.21 the peak values have been defined by the symbols ω_i for even i, one has

$$\omega_i = \sqrt{k}\, sn\left(\frac{iK}{2n}\right);\ i = 0, 1, 2, \cdots, 2n \tag{66.129}$$

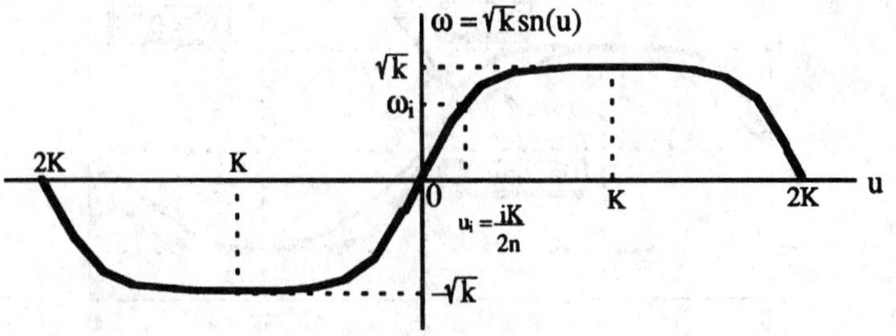

FIGURE 66.25 The elliptic sine transformation.

where the odd indices correspond to the zeros and the even ones to the peak values. Note that $i = 0$ corresponds to $\omega = 0$ and $i = 2n$ to $\omega = \sqrt{k}$. This permits one to compute the minimax error:

$$\varepsilon = |R_{2n}(0)| = (\omega_1\omega_3 \cdots \omega_{2n-1})^2 \tag{66.130}$$

As pointed out above, the analysis for odd-order filters proceeds quite analogously. The only difference lies in the computation of ε. In this case, $R_{2n+1}(0) = 0$; thus, $\omega = 0$ is a zero—not a point of maximum deviation. One can, however, note that the quantity $\varepsilon^2 - R^2_{2n+1}(\omega)$ has double zeros at the frequencies of maximum deviation (except at $\omega = \omega_{2n+1} = \sqrt{k}$) and the same denominator as $R^2_{2n+1}(\omega)$. Hence,

$$\varepsilon^2 - R^2_{2n+1}(\omega) = \frac{(\omega_1^2 - \omega^2)^2(\omega_3^2 - \omega^2)^2 \cdots (\omega_{2n+1}^2 - \omega^2)}{(1 - \omega_2^2\omega^2)^2(1 - \omega_4^2\omega^2)^2 \cdots (1 - \omega_{2n}^2\omega^2)^2} \tag{66.131}$$

Notice here that $\omega_{2n+1} = \sqrt{k}$. Since $R_{2n+1}(0) = 0$, one can evaluate this expression at $\omega = 0$ to get

$$\varepsilon = (\omega_1\omega_3 \cdots \omega_{2n-1})^2\sqrt{k} \tag{66.132}$$

Note that one uses the zero frequencies in the even-order case and the frequencies of maximum deviation in the odd-order case.

As an example, suppose that the object is to design an elliptic filter of order $n = 2$. Further, suppose that the passband ripple cutoff frequency is to be 0.95. Then one has

$$\omega_p = \sqrt{k} = 0.95 \tag{66.133}$$

The quarter period of the elliptic sine function is

$$K = 2.9083 \tag{66.134}$$

Evaluating the zeros and points of maximum deviation of the Chebyshev rational function numerically using (66.129), one obtains the values shown in Table 66.4. Thus, the required elliptic rational function is

$$R_4(\omega) = \frac{(0.3508 - \omega^2)(0.8746 - \omega^2)}{(1 - 0.3508\omega^2)(1 - 0.8746\omega^2)} \tag{66.135}$$

Finally, the maximum error is given by (66.130):

$$\varepsilon = (\omega_1\omega_2)^2 = (0.5923 \times 0.9352)^2 = 0.3078 \tag{66.136}$$

TABLE 66.4

i	ω_i
0	0.0000
1	0.5923
2	0.8588
3	0.9352
4	0.9500

Figure 66.26 shows the resulting gain plot. Observe the transmission zero at $\omega = 1/0.9352 = 1.069$, corresponding to the pole of the Chebyshev rational function located at the inverse of the largest passband zero. Also, as anticipated, the minimum gain in the passband is

$$A_p = \frac{1}{\sqrt{1+\varepsilon^2}} = \frac{1}{\sqrt{1+(0.3078)^2}} = 0.9558 \tag{66.137}$$

and the maximum stopband gain is

$$A_s = \frac{1}{\sqrt{1+\frac{1}{\varepsilon^2}}} = \frac{1}{\sqrt{1+\frac{1}{(0.3078)^2}}} = 0.2942 \tag{66.138}$$

Perhaps a summary of the various filter types is in order at this point. The elliptic filter has more flexibility than the Butterworth or the Chebyshev because one can adjust its transition band rolloff independently of the passband ripple. However, as the sharpness of this rolloff increases, as a more detailed analysis shows, the stopband gain increases and the passband ripple increases. Thus, if these parameters are to be independently specified—as is often the desired approach—one must allow the order of the filter to float freely. In this case, the design becomes a bit more involved. The reader is referred to [9], which presents a simple curve for use in determining the required order. It proceeds from the result that the filter order is the integer greater than or equal to the following ratio:

$$n \geq \frac{f(M)}{f(\Omega)} \tag{66.139}$$

where

$$f(x) = \frac{K(\sqrt{1-x^{-2}})}{K(1/x)} \tag{66.140}$$

Lin [9] presents a curve of this function $f(x)$. K is the complete elliptic integral given in (66.120). The parameters M and Ω are defined by

$$M = \sqrt{\frac{1-10^{-0.1A_s}}{1-10^{-0.1A_p}}} \tag{66.141}$$

and

$$\Omega = 1/k \tag{66.142}$$

Of course, there is still the problem of factorization. That is, now that the appropriate Chebyshev rational function is known, one must find the corresponding $G(s)$ transfer function

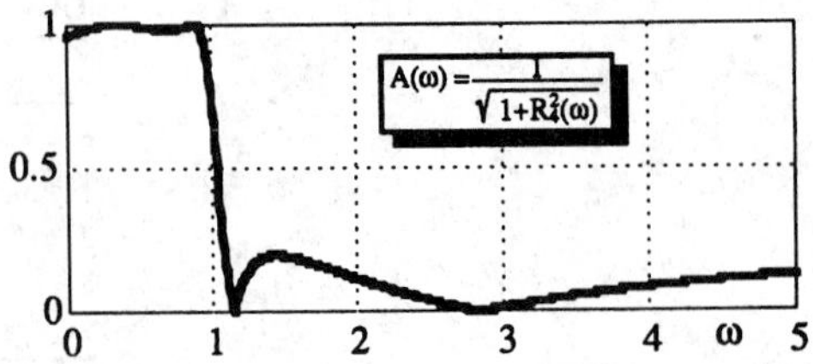

FIGURE 66.26 Gain plot for the example elliptic filter.

TABLE 66.5

Filter Type	Gain Rolloff	Phase Linear
Bessel	Worst	Best
Butterworth	Poor	Better
Chebyshev	Better	Poor
Elliptic	Best	Worst

of the filter. The overall design process is explained in some detail in [10], which develops a numerically efficient algorithm for directly computing the parameters of the transfer function.

The Butterworth and Chebyshev filters are of the all-pole variety, and this means that the synthesis of such filters is simpler than is the realization of elliptic filters, which requires the realization of transmission zeros on the finite $j\omega$ axis.

Finally, a word about phase (or group delay). Table 66.5 shows a rank ordering of the filters in terms of both gain and phase performance. As one can see, the phase behavior is inverse to the gain performance. Thus, the elliptic filter offers the very best standard approximation to the ideal lpp "brickwall" gain behavior, but its group delay deviates considerably from a constant. On the other end of the spectrum, one notes that the Bessel–Thompson filters offers excellent phase performance, but a quite high order is required to achieve a reasonable gain characteristic.

If both excellent phase and gain performance are absolutely necessary, two approaches are possible. One either uses computer optimization techniques to simultaneously approximate gain and phase, or one uses one of the filters described in this section followed by an *allpass* filter, one having unity gain over the frequency range of interest, but whose phase can be designed to have the inverse characteristic to the filter providing the desired gain. This process is known as *phase compensation.*

References

[1] A. M. Davis, "The Approximation Theoretic Foundation of Analog and Digital Filters," IEEE International Symposium on Circuits and Systems, San José, California, May, 1986.

[2] T. J. Rivlin, *An Introduction to the Approximation of Functions*, Dover, 1981.

[3] G. R. Lang and P. O. Brackett, "Complex Analog Filters," in *Proc. Euro. Conf. Circuit Theory, Design*, The Hague, The Netherlands, Aug. 1981, pp. 412–415.

[4] W. K. Chen, *Passive and Active Filters, Theory and Implementations*, New York: Wiley, 1986.

[5] L. P. Huelsman, *Theory and Design of Active RC Circuits*, New York: McGraw-Hill, 1968.

[6] G. Szentirmai, *Computer-Aided Filter Design*, New York: IEEE, 1973.

[7] G. C. Temes and J. W. LaPatra, *Circuit Synthesis and Design*, New York: McGraw-Hill, 1977.

[8] E. A. Guillemin, *Synthesis of Passive Networks*, New York: Wiley, 1957.

[9] P. M. Lin, "Single Curve for Determining the Order of an Elliptic Filter," *IEEE Trans. Circuits Syst.*, vol. 37, no. 9, pp. 1181–1183, Sept. 1990.

[10] A. Antoniou, *Digital Filters: Analysis and Design*, New York: McGraw-Hill, 1979.

[11] A. M. Davis, "Realizability-preserving transformations for digital and analog filters," *J. Franklin Institute*, vol. 311, no. 2, pp. 111–121, Feb. 1981.

67

Frequency Transformations

Jaime Ramirez-Angulo
New Mexico State University

67.1 Low-Pass Prototype

As discussed in Chapter 66, conventiional approximation techniques (Butterworth, Chebyshev, Elliptic, Bessel, etc.) lead to a normalized transfer function denoted *low-pass prototype* (LPP). The LPP is characterized by a passband frequency $\Omega_p = 1.0$ rad/s, a maximum passband ripple A_p (or $A_{\max}$), a minimum stopband attenuation A_s (or $A_{\min}$), and a stopband frequency Ω_s. A_p and A_s are usually specified in dB. Tolerance bounds (also called box constraints) for the magnitude response of an LPP are illustrated in Fig. 67.1(a). The ratio Ω_s/Ω_p is called the **selectivity factor** and it has a value Ω_s for an LPP filter. The passband and stopband edge frequencies are defined as the maximum frequency with the maximum passband attenuation A_p and the minimum frequency with the minimum stopband attenuation A_s, respectively. The passband ripple and the minimum passband attenuation are expressed by

$$A_p = 20\log\left|\frac{K}{H(\omega_p)}\right| \qquad A_s = 20\log\left|\frac{K}{H(\omega_s)}\right| \tag{67.1}$$

where K is the maximum value of the magnitude response in the passband (usually unity). Figure 67.1(b) shows the magnitude response of a Chebyshev LPP transfer function with specifications $A_p = 2$ dB, $A_s = 45$ dB, and $\Omega_s = 1.6$.

Transformation of Transfer Function. Low-pass, high-pass, bandpass and band-reject transfer functions (denoted in what follows LP, HP, BP, and BR, respectively) can be derived from an LPP transfer function through a transformation of the complex frequency variable. For convenience, the transfer function of the LPP is expressed in terms of the complex frequency variable s, where $s = u + j\Omega$ while the transfer functions obtained through the frequency transformation (low-pass, high-pass, bandpass, or band-reject) are expressed in terms of the transformed complex frequency variable $p = \sigma + j\omega$.

0-8493-8341-2/95/$0.00+$.50

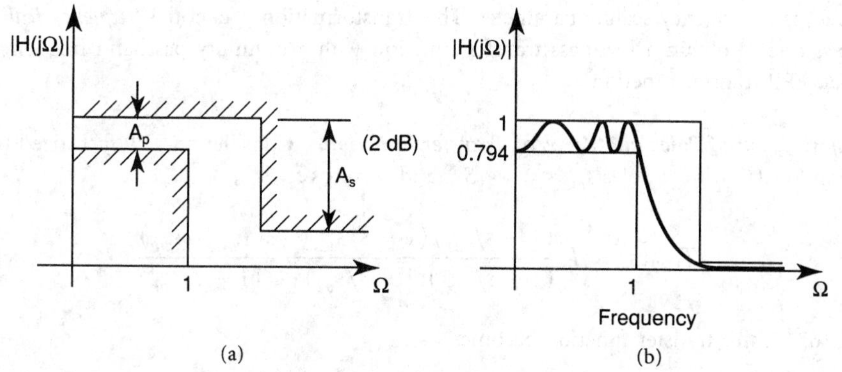

FIGURE 67.1 (a) Tolerance bounds for magnitude response of low-pass prototype. (b) Chebyshev LPP response.

The approximation of a LP, HP, BP, or BR transfer function with passband ripple (s) A_p and stopband attenuation (s) A_s involves three steps:

1. Determination of the stopband edge frequency or selectivity factor Ω_s of an LPP, which can be transformed into the desired LP, HP, BP, or BR filter.
2. Approximation of an LPP transfer function $T_{\text{LPP}}(s)$ with selectivity factor Ω_s and with same passband ripple and stopband attenuation A_p and A_s as the desired LP, HP, BP, or BR filter respectively.
3. Transformation of the LPP transfer function $T_{\text{LPP}}(s)$ into the desired transfer function (LP, HP, BP, or BR) $T(p)$ through a frequency transformation of the form

$$s = f(p) \tag{67.2}$$

Transformation of a network with LPP magnitude response into a low-pass, high-pass, bandpass, or band-rejection network can be done directly on the elements of the network. This procedure is denoted **network transformation.** It is very convenient in practice because element values for double terminated lossless ladder networks with LPP specifications have been extensively tabulated for some common values of A_p, A_s, and Ω_s. Also, a host of personal computer programs have become available in recent years that allow one to determine the component values of LPP ladder networks for arbitrary values A_p, A_s, and Ω_s. In what follows we will study the frequency transformation $s = f(p)$ for each specific type of filter response (LP, HP, BP, and BR). We will show how to calculate the selectivity factor of the equivalent LPP based on box constraint specifications for each type of filter. We will then show how mapping of the imaginary frequency axis from s to p leads to LP, HP, BP, or BR magnitude responses. We will analyze how poles and zeros are mapped from the s-plane to the p-plane for each transformation and finally we will show the element transformations required to directly transform LPP networks into any of the filter types addressed above.

67.2 Frequency and Impedance Scaling

Frequency Scaling

The simplest frequency transformation is a scaling operation expressed by

$$s = \frac{p}{\omega_o} \tag{67.3}$$

where ω_o is a frequency scaling parameter. This transformation is denoted *frequency scaling* and it allows one to obtain a low-pass transfer function with a nonunity passband frequency edge from an LPP transfer function.

Transformation of Poles and Zeros of Transfer Function. Consider an LPP factorized transfer function $T_{\text{LPP}}(s)$ with n poles $s_{p1}, s_{p2}, \cdots, S_{pn}$ and m zeros $s_{z1}, s_{z2}, \cdots, s_{zm}$

$$T_{\text{LPP}}(s) = K\frac{(1 - s/s_{z1})(1 - s/s_{z2})\cdots(1 - s/s_{zm})}{(1 - s/s_{sp1})(1 - s/s_{p2})\cdots(1 - s/s_{pn})} \tag{67.4}$$

Using (67.3), this transfer function becomes

$$T_{\text{LP}}(p) = K\frac{(1 - p/p_{z1})(1 - p/p_{z2})\cdots(1 - p/p_{zm})}{(1 - p/p_{p1})(1 - p/p_{p2})\cdots(1 - p/p_{pn})} \tag{67.5}$$

where poles and zeros (s_{zi} and s_{pj}) of the LPP transfer function $P_{\text{LPP}}(s)$ become simply poles and zeros in $T_{\text{LP}}(p)$, which are related to those of $T_{\text{LPP}}(s)$ by the scaling factor ω_o:

$$p_{pi} = \omega_o s_{pi} \qquad p_{zj} = \omega_o s_{zj} \tag{67.6}$$

To determine the magnitude (or frequency) response of the LP filter we evaluate the magnitude of the transfer function on the imaginary axis (for $s = j\Omega$). The magnitude response of the transformed transfer function $|T_{\text{LP}}(j\omega)|$ preserves a low-pass characteristic as illustrated in Fig. 67.2. The frequency range from 0 to ∞ in the Ω axis is mapped to the range 0 to ∞ in the ω axis. A frequency and its mirror image in the negative axis $\pm\Omega$ is mapped to frequencies $\omega = \pm\omega_o\Omega$ with the same magnitude response: $|T_{\text{LPP}}(j\Omega)| = |T_{\text{LP}}(j\omega_o\Omega)|$. The passband and stopband edge frequencies $\Omega_p = 1$ rad/s and Ω_s of the LPP are mapped into passband and stopband edge frequencies $\omega_p = \omega_o$ and $\omega_s = \omega_o\Omega_s$, respectively. From this, it can be seen that for given low-pass filter specifications ω_s, ω_p the equivalent LPP is determined based on the relation $\Omega_s = \omega_s/\omega_p$, while the frequency scaling parameter ω_o corresponds to the passband edge frequency of the desired LP.

LP Network Transformation. Capacitors and inductors are the only elements that are frequency dependent and that can be affected by a change of frequency variable. Capacitors and inductors in an LPP network have impedances $z_c = 1/sc_n$ and $z_l = sl_n$, respectively. Using (67.3) these become $Z_c(p) = 1/pC$ and $Z_L(p) = pL$, where $C = c_n/\omega_o$ and $L = l_n/\omega_o$. The LPP to LP frequency transformation is performed directly on the network by simply dividing the values of all capacitors and inductors by the frequency scaling factor ω_o. This is illustrated in Fig. 67.3(a). The transformation expressed by (67.3) can be applied to any type of filter and it has the effect of scaling the frequency axis without changing the shape of its magnitude response.

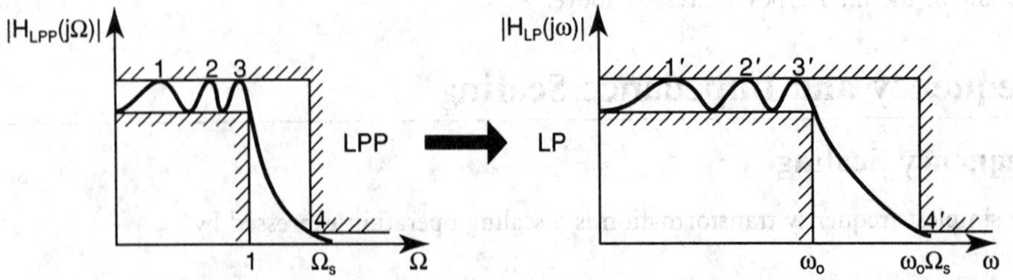

FIGURE 67.2 Derivation of low-pass response from a low-pass prototype by frequency scaling.

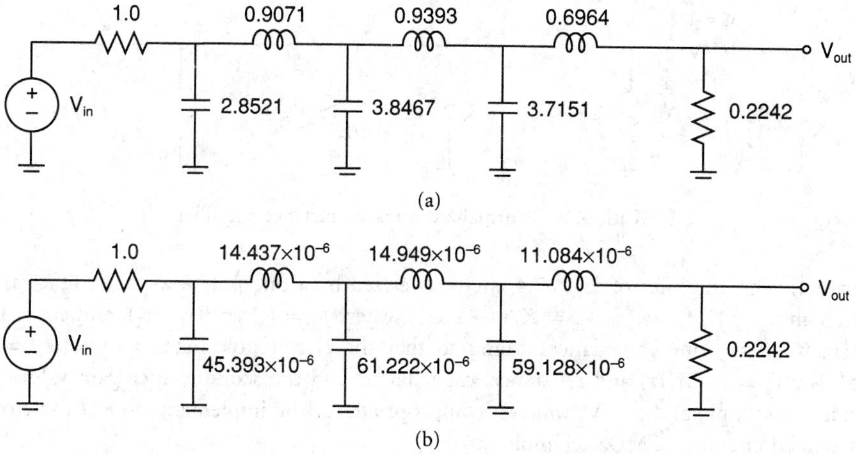

FIGURE 67.3 (a) Low-pass prototype ladder network. (b) LP network with passband frequency $f_p = 10$ kHZ derived from (a).

This is illustrated in Fig. 67.3, where the elements of an LPP with $A_p = 2$ dB, $A_s = 45$ dB, and selectivity $\Omega_s = 1.6$ rad/s are scaled [Fig. 67.3(b)] to transform the network into an LP network with passband and stopband edge frequencies $\omega_p = 2\pi$ 10 krad/s and $\omega_s = 2\pi$ 16 krad/s (or $f_p = 10$ kHz and $f_s = 16$ kHz), respectively.

Impedance Scaling

Dimensionless transfer functions defined by ratios of voltages (V_{out}/V_{in}) or currents (I_{out}/I_{in}) remain unchanged if all impedances of a network are scaled by a common scaling factor "a". On the other hand, transfer functions of the transresistance type (V_{out}/I_{in}) or of the transconductance type (I_{out}/V_{in}) are simply modified by the impedance scaling factor a and $1/a$, respectively. If we denote a normalized impedance by z_n, then the impedance scaling operation leads to an impedance $Z = az_n$. When applied to resistors (r_n), capacitors (c_n), inductors (l_n), transconductance gain coefficients (g_n), and transresistance gain coefficients (r_n) result in the following relations for the elements (R, C, L, g, and r) of the impedance scaled network.

$$\begin{aligned} R &= ar_n \\ L &= al_n \\ C &= \frac{1}{a}c_n \\ g &= \frac{1}{a}g_n \\ r &= ar_n \end{aligned} \tag{67.7}$$

Dimensionless voltage-gain and current-gain coefficients are not affected by impedance scaling. Technologies for fabrication of microelectronic circuits (CMOS, bipolar, BiCMOS monolithic integrated circuits, thin-film and thick-film hybrid circuits) only allow elements values and time constants (or pole and zero frequencies) within certain practical ranges. Frequency and impedance scaling are very useful to scale normalized responses and network elements resulting from standard approximation procedures to values within the range achievable by the implementation technology. This is illustrated in the following example.

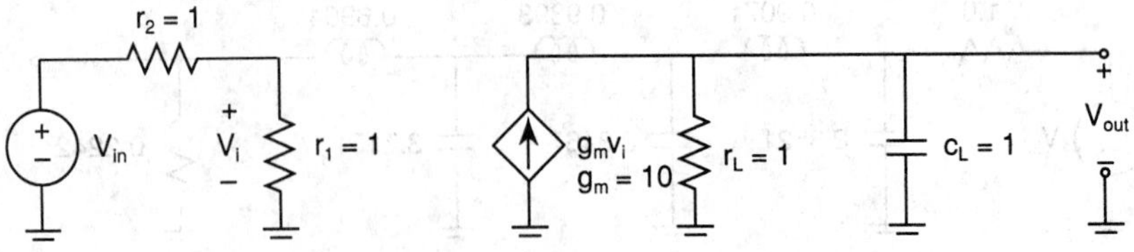

FIGURE 67.4 Normalized transconductance amplifier.

Example 1. The amplifier of Fig. 67.4 is characterized by a one-pole low-pass voltage transfer function given by $H(s) = V_{\text{out}}/V_{\text{in}} = K/(1+s/\omega_p)$, where $K = g_m r_L r_1/(r_1+r_2)$, and $\omega_p = 1/r_L c_L$. Perform frequency and impedance scaling so that the circuit pole takes a value $\omega_p = 2\pi 10$ Mrad/s (or $f_p = 10$ MHz) and resistance, capacitance, and transconductance gain values are in the range of kΩ, pF, and μA/V, which are appropriate for the implementation of the circuit as an integrated circuit in CMOS technology.

Solution. The required location of the pole and range of values for the circuit elements can be achieved using frequency and impedance scaling factors $\omega_o = 2\pi \cdot 10^7$ and $a = 10^4$, respectively. These result in $R_1 = ar_1 = 10$ kΩ, $R_2 = ar_2 = 10$ kΩ, $g = g_m/a = 1000$ μA/V, $R_L = ar_L = 10$ kΩ, and $C_L = c_L/a\omega_o = 1.59$ pF.

67.3 Low-Pass to High-Pass Transformation

The LPP to high-pass transformation is defined by

$$s = \frac{\omega_o^2}{p} \tag{67.8}$$

Using this substitution in the LPP transfer function (67.4) it becomes

$$T_{\text{HP}}(p) = K\frac{p^{n-m}(p-p_{z1})(p-p_{z2})\cdots(p-p_{zm})}{(p-p_{p1})(p-p_{p2})\cdots(p-p_{pn})} \tag{67.9}$$

where the poles and zeros of (67.8) are given by

$$\begin{aligned} p_{zi} &= \frac{\omega_o^2}{s_{zi}} \quad \text{for } i \in \{1,2,\ldots,m\} \\ p_{pj} &= \frac{\omega_o^2}{s_{pj}} \quad \text{for } j \in \{1,2,\ldots,n\} \end{aligned} \tag{67.10}$$

It can be seen that zeros and poles of $T_{\text{HP}}(p)$ are reciprocal to those of $T_{\text{LPP}}(s)$ and scaled by the factor ω_o^2. $T_{\text{HP}}(p)$ has $n-m$ zeros at $s=0$, which can be considered to originate from $n-m$ zeros at ∞ in $T_{\text{LPP}}(s)$.

Let us consider now the transformation of the imaginary axis in s to the imaginary axis in p. For $s = j\Omega$, p takes the form $p = j\omega$ where

$$\omega = -\omega_o^2/\Omega \tag{67.11}$$

From (67.11) it can be seen that positive frequencies in the LPP transform to reciprocal and scaled frequencies of the HP filter. Specifically, the frequency range from 0 to ∞ in Ω maps to the frequency range $-\infty$ to 0 in ω, while the range $-\infty$ to 0 in Ω maps to 0 to ∞ in ω. The passband edge frequency ($\Omega_p = \pm 1$) and the stopband edge frequency $\pm\Omega_s$ of the LPP is mapped to $\omega_p = \mp\omega_o^2$ and $\omega_s = \mp\omega_o^2/\Omega_s$ in the high-pass response. This is illustrated in Fig. 67.5.

The procedure to obtain the specifications of the equivalent LPP given specifications ω_p and ω_s for a HP circuit can be outlined as follows:

1. Calculate the selectivity factor of the LPP according to $\Omega_s = \omega_p/\omega_s$.
2. Approximate an LPP transfer function $T_{\text{LPP}}(s)$ with the selectivity Ω_s and the passband ripple and stopband attenuation of the desire high-pass response.
3. Perform an LPP to HP transformation either by direct substitution $p = \omega_o^2/s$ in $T_{\text{LPP}}(s)$ or by transforming poles and zeros of $T_{\text{LPP}}(s)$ using (67.10).

Network Transformation. Consider a capacitor c_n and an inductor l_n in an LPP network. They have impedances $z_c(s) = 1/sc_n$, $z_l(s) = sl_n$, respectively. Using (67.8) these become impedances $Z_L(p) = pL$ and $Z_c = 1/pC$ in the high-pass network, where $L = 1/\omega_o^2 c_n$ and $C = 1/\omega_o^2 l_n$.

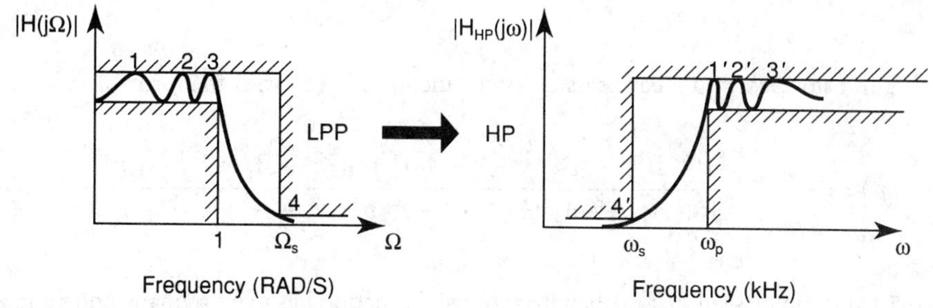

FIGURE 67.5 Transformation of a low-pass into a high-pass response.

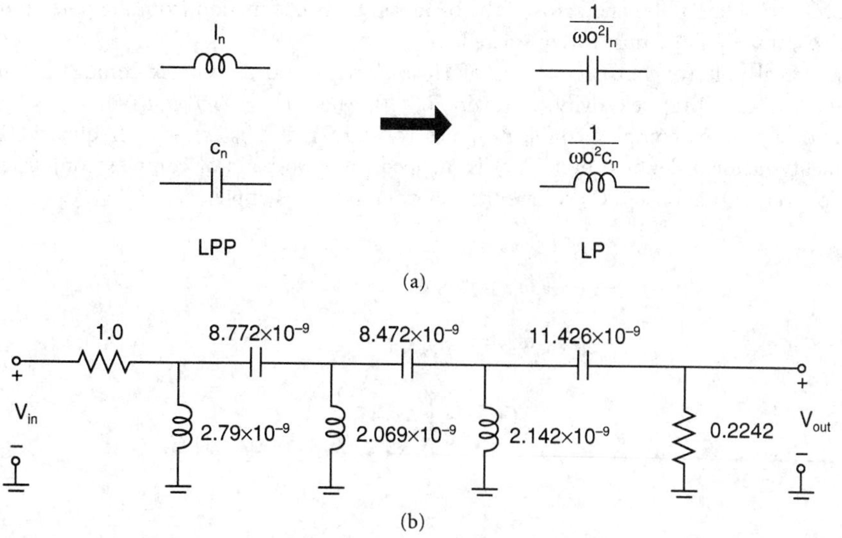

FIGURE 67.6 (a) LPP to high-pass network element transformations. (b) High-pass network derived from LPP of Fig. 67.3(a).

It can be seen that an LPP to HP transformation can be done directly on an LPP network by replacing capacitors by inductors and inductors by capacitors. For illustration, Fig. 67.6 shows a high-pass network with a passband edge frequency $\omega_p = 2\pi\ 20$ Mrad/s or ($fp = 20$ MHz) derived from the LPP network shown in Fig. 67.6(a).

67.4 Low-Pass to Bandpass Transformation

If the LPP transfer function is expressed now as a rational function

$$H_{\text{LPP}}(s) = K\frac{b_o + b_1 s + b_2 s^2 + \cdots + b_m s_m}{a_o + a_1 s + a_2 s^2 + \cdots + a_n s^n} \tag{67.12}$$

then through the substitution

$$s = \frac{1}{\text{BW}}\frac{p^2 + \omega_o^2}{p} \tag{67.13}$$

$H_{\text{LPP}}(s)$ is transformed into a bandpass transfer function $H_{\text{BP}}(p)$ with the form

$$H_{\text{BP}}(p) = K' p^{n-m}\frac{1 + B_1 p + B_2 p^2 + \cdots + B_2 p^{2m-2} + B_1 p^{2m-1} + p^{2m}}{1 + A_1 p + A_2 p^2 + \cdots + A_2 p^{2m-2} + A_1 p^{2m-1} + p^{2m}} \tag{67.14}$$

From (67.14) it can be seen that the bandpass transfer function has twice as many poles and zeros as the LPP transfer function. In addition it has $n - m$ zeros at the origin. The coefficients of the numerator and denominator polynomials are symmetric and are a function of the coefficients of $H_{\text{LPP}}(s)$.

In order to obtain poles and zeros of the bandpass transfer function from the poles and zeros of $T_{\text{LPP}}(p)$, three points must by considered.

First, a real pole (or zero) $s_p = -u_p$ of $H_{\text{LPP}}(s)$ maps into a complex conjugate pair with frequency ω_o and Q (or selectivity) factor in $H_{\text{BP}}(p)$, where $Dq = \omega_o/(U_p BW)$.[1]

Second, a pair of complex conjugate poles (or zeros) of $H_{\text{LPP}}(s)$ with frequency Ω_o and pole-quality factor q denoted by (Ω_o, q) is mapped into two pairs of complex conjugate poles (or zeros) (ω_{o1}, Q) and (ω_{o2}, Q), where the following relations apply:

$$\begin{aligned}\omega_{o1} &= \omega_o M \\ \omega_{o2} &= \frac{\omega_o}{M} \\ Q &= \frac{a}{c}\left(M + \frac{1}{M}\right)\end{aligned} \tag{67.15a}$$

[1]A complex conjugate pole pair can be expressed as $s_p, s_p^* = u_p \pm j\omega_p = \omega_c e^{\pm j\theta} = (\omega_c, Q)$, where the pole quality factor Q is given $Q = \frac{1}{2}\cos\theta$ and $\omega_c = (G_p^2 + \omega_p^2)^{\frac{1}{2}}, \theta = tg^{-1}\frac{\omega_p}{G_p}$.

and the definitions

$$a = \frac{\omega_o}{\text{BW}}$$

$$b = \frac{\Omega_o}{2a}$$

$$c = \frac{\Omega_o}{q}$$

$$M = \sqrt{b^2 + \sqrt{(1+b^2)^2 - \frac{c}{(2a)^2}} + \sqrt{2}b\sqrt{1 + b^2 - \frac{c}{2\Omega_o^2} + \sqrt{(1+b^2)^2 - \frac{c}{(2a)^2}}}} \tag{67.15b}$$

apply.

Narrow-Band Approximation. If the condition $\text{BW}/\omega_o \ll 1$ is satisfied, then following simple transformations known as the narrow band approximation[2] can be used to map directly poles (or zeros) from the s-plane to the p-plane

$$p_p \approx \frac{\text{BW}}{2} s_p + j\omega_o, \qquad p_z \approx \frac{\text{BW}}{2} s_z + j\omega_o \tag{67.16}$$

These approximations are valid only if the transformed poles and zeros are in the vicinity of $j\omega_o$, that is, if $|s_p - \omega_o|/\omega_o << 1$.

Third, in order to obtain poles and zeros of the bandpass transfer function, mapping of complex zeros on the imaginary Ω axis ($s_z, s_z^* = \pm j\Omega_z$) takes place using the same mapping relations discussed next.

Mapping of Imaginary Frequency Axis. Consider a frequency $s = j\Omega$ and its mirror image $s = -j\Omega$ in the LPP. Using (67.13) these two frequencies are mapped into four frequencies: $\pm\omega_1$ and $\pm\omega_2$, where ω_1 and ω_2 are given by

$$\omega_2 = \Omega\frac{\text{BW}}{2} + \sqrt{\omega_o^2 + \left(\Omega\frac{\text{BW}}{2}\right)^2}$$
$$\omega_1 = -\Omega\frac{\text{BW}}{2} + \sqrt{\omega_o^2 + \left(\Omega\frac{\text{BW}}{2}\right)^2} \tag{67.17}$$

from (67.17), the following relations can be derived:

$$\omega_2 - \omega_1 = \text{BW}\,\Omega$$
$$\omega_1\omega_2 = \omega_o^2 \tag{67.18}$$

It can be seen that with the LPP to bandpass transformation *frequencies are mapped into bandwidths.* A frequency Ω and its mirror image $-\Omega$ are mapped into two pairs of frequency points that have ω_o as center of geometry. The interval from 0 to ∞ in the positive Ω axis maps into two intervals in the ω axis: the first from ω_o to $+\infty$ on the positive ω axis *and* the second from $-\omega_o$ to 0 in the negative ω axis. The interval $-\infty$ to 0 on the negative Ω axis maps into

[2] L. P. Huelsman, "An algorithm for the low-pass to bandpass transformation," *IEEE Trans. Education,* vol. E-11, p. 72, Mar. 1968.

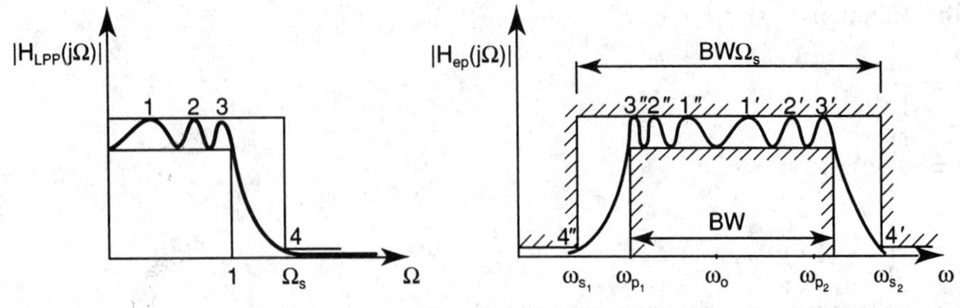

FIGURE 67.7 Low-pass to bandpass transformation.

two intervals: from $-\infty$ to $-\omega_o$ in the negative ω axis and from 0 to ω_o in the positive ω axis. The LPP passband and stopband edge frequencies $\Omega_p = \pm 1$ and $+\Omega_s$ are mapped into passband edge frequencies ω_{p1}, ω_{p2} and into stopband edge frequencies ω_{s1}, ω_{s2} that satisfy

$$\begin{aligned} \omega_{p2} - \omega_{p1} &= BW \\ \omega_{s2} - \omega_{s1} &= BW\Omega_s \end{aligned} \tag{67.19}$$

and

$$\omega_{p1}\omega_{p2} = \omega_{s1}\omega_{s2} = \omega_o^2 \tag{67.20}$$

Figure 67.7 shows mapping of frequency points 1, 2, 3, and 4 in the Ω axis to points 1′, 2′, 3′ and 4′ and 1″, 2″, 3″, 4″ in the ω axis of the bandpass response.

If the bandpass filter specifications do not satisfy (67.20) (which is usually the case) then either one of the stopband frequencies or one of the passband frequencies has to be redefined so that they become symmetric w.r.t. ω_o and an equivalent LPP filter can be specified. For given passband and stopband specifications ω_{p1}, ω_{p2}, ω_{s1}, ω_{s2}, and A_p, A_s, the procedure to determine Ω_s for an equivalent LPP is as follows:

1. Calculate first the parameter ω_o in terms of the passband frequencies according to $\omega_o = \sqrt{\omega_{p1}\omega_{p2}}$
2. Make the stopband frequencies geometrically symmetric with respect to ω_o determined in step 1 by redefining either ω_{s1} or ω_{s2} so that one of these frequencies becomes more constrained.[3] If $\omega_{s1} < \omega_o^2/\omega_{s2}$, then assign ω_{s1} the new value: $\omega_{s1} = \omega_o^2/\omega_{s2}$. Otherwise assign ω_{s2} the new value $\omega_{s2} = \omega_o^2/\omega_{s1}$.
3. Calculate a selectivity factor of the LPP based on the redefined stopband frequency according to $\Omega_s = (\omega_{s2} - \omega_{s1})/(\omega_{p2} - \omega_{p1})$. This expression follows directly from (67.19) and (67.20).
4. Calculate now the parameter ω_o in terms of the stopband frequencies, according to $\omega_o = \sqrt{\omega_{s1}\omega_{s2}}$.
5. Symmetrize the passband frequencies w.r.t. the value of ω_o determined in step 4 by constraining either ω_{p1} or ω_{p2}. If $\omega_{p1} > \omega_o^2/\omega_{p2}$, then assign ω_{p1} the new value: $\omega_{p1} = \omega_o^2/\omega_{p2}$. Otherwise assign ω_{p2} the new value $\omega_{p2} = \omega_o^2/\omega_{p1}$.
6. Calculate the selectivity factor based on the new set of passband frequencies using the same expression as in step 3.

[3]The term "constraint specifications" is used here in the sense of redefining either a stopband frequency or a passband frequency so that one of the transition bands becomes narrower, which corresponds to tighter design specifications.

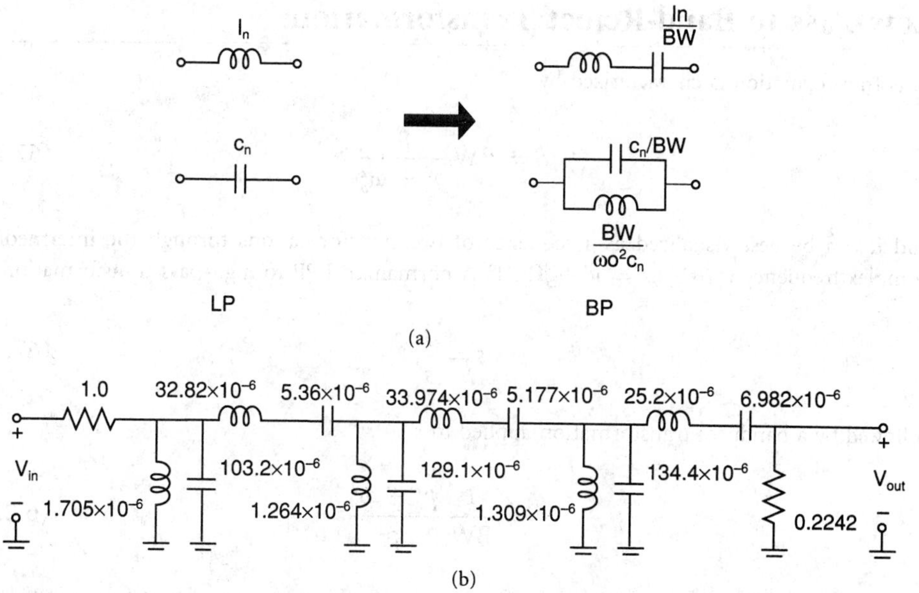

FIGURE 67.8 (a) LPP to bandpass network element transformations. (b) Bandpass network derived from LPP network of Fig. 67.3(a).

7. Select from step 3 or step 6 the maximum selectivity factor Ω_s and determine the transformation parameters ω_o and BW from the values calculated in steps 1 to 3 or in 4 to 6, whichever sequence leads to the maximum Ω_s, which leads to the lowest order n for $T_{LPP}(s)$ and with this to the least expensive filter implementation.

Example 2. Consider the following nonsymmetric specifications for a bandpass filter: $\omega_{s1} = 2\pi$ 9, $\omega_{s2} = 2\pi$ 17, $\omega_{p1} = 2\pi$ 10 and $\omega_{p2} = 2\pi$ 14.4 (all frequencies specified in krad/s). Application of the above procedure leads to a new value for the upper stopband frequency $\omega_{s2} = 16$ and from this to the following parameters: BW $= 2\pi$ (14.4–10) $= 2\pi$ 4.4, $\omega_o^2 = 2\pi$ 9 2π 16 $= 2\pi$ 10 2π 14.4 $= (2\pi)^2$ 144, and $\Omega_s = 16–9/(14.4–10) = 1.59$.

Bandpass Network Transformation. Consider now the transformation of capacitors and inductors in an LPP filter. An inductor in the LPP network has an impedance $z_l(s) = sl_n$. This becomes an impedance $z_s(p) = pL_s + 1/pC_s$, where $L_s = l_n/\mathrm{BW}$ and $C_s = \mathrm{BW}/l_n\omega_o^2$. Now consider a capacitor c_n in the LPP with admittance $Y_c(s) = sc_n$. Using the transformation (67.12), this becomes an admittance $Y_p(p) = pC_p + 1/pL_p$, where $C_p = c_n/\mathrm{BW}$ and $L_p = \mathrm{BW}/c_n\omega_o^2$. This indicates that to transform an LPP network into a bandpass network, inductors in the LPP network are replaced by the series connection of an inductor with value L_s and a capacitor with value C_s and capacitors in the LPP are replaced by the parallel combination of a capacitor C_p and inductor L_p. This is illustrated in Fig. 67.8. As with other transformations, resistors remain unchanged since they are not frequency dependent.

Example 3. Consider the LPP network shown in Fig. 67.3(a) with specifications $Ap = 2$ dB, $As = 45$ dB, and $\Omega_s = 1.6$. Derive a bandpass network using the parameters calculated in Example 2: $\omega_o^2 = (2\pi)^2 144$, BW $= 2\pi$ 4.4 krad/s.

Solution. Straightforward application of the relations shown in Fig. 67.8(a) leads to the network of Fig. 67.8(b).

67.5 Low-Pass to Band-Reject Transformation

This transformation is characterized by

$$S = \mathrm{BW}\frac{p}{p^2 + \omega_o^2} \tag{67.21}$$

and it can be best visualized as a sequence of two transformations through the intermediate complex frequency variable $s' = u' + j\Omega'$. 1) A normalized LPP to high-pass transformation

$$s = \frac{1}{s'} \tag{67.22}$$

followed by a bandpass transformation applied to s'

$$s' = \frac{1}{\mathrm{BW}}\frac{p^2 + \omega_o^2}{p} \tag{67.23}$$

Mapping of the imaginary ω axis to the Ω axis through this sequence of transformations leads to a band-rejection response in ω; frequency points 1, 2, 3, 4 have been singled out to better illustrate the transformation. Figure 67.9(a) shows the magnitude response of the LPP, Fig. 67.9(b) shows the high-pass response obtained through (67.22). This intermediate response is a normalized high-pass response with a passband extending from $\Omega_p' = 1$ to ∞ and a stopband extending from 0 to $\Omega_s' = 1/\Omega_s$. Figure 67.9(c) shows the magnitude response obtained by applying (67.23) to the variable s'. The frequency range from $\Omega' = 0$ to ∞ in Fig. 67.9(b) is mapped into the range from ω_o to ∞ and from $-\omega_o$ to 0. The frequency range from $-\infty$ to 0 in Ω' is mapped into the ranges from 0 to ω_o and from $-\infty$ to $-\omega_o$ in ω as indicated in Fig. 67.9(c). It can be seen that the bandpass transformation applied to a normalized high-pass response creates two passbands with ripple A_p from $\omega = 0$ to ω_{p_1} and from ω_{p2} to ∞ and a stopband with attenuation A_s from ω_{s1} to ω_{s2}. The following conditions are satisfied:

$$\begin{aligned}
\omega_{s1}\omega_{s2} &= \omega_{p1}\omega_{p2} = \omega_o^2 \\
\omega_{s2} - \omega_{s1} &= \mathrm{BW}\,\Omega_s' = \mathrm{BW}/\Omega_s \\
\omega_{p2} - \omega_{p1} &= \mathrm{BW}
\end{aligned} \tag{67.24}$$

From (67.24) if ω_{s1}, ω_{s2}, ω_{p1}, ω_{p2} as well as A_p and A_s are specified for a band-rejection filter, then the selectivity factor of the equivalent LPP is calculated according to $\Omega_s = (\omega_{p2} - \omega_{p1})/(\omega_{s2} - \omega_{s1})$. Similar to the case of bandpass filters, the band-rejection specifications must be made symmetric with respect to ω_o so that an equivalent LPP can be specified. This is done following a similar procedure as for the bandpass transformation by constraining either one of the passband frequencies or one of the stopband frequencies so that the response becomes geometrically symmetric with respect to ω_o. The option leading to the largest selectivity factor (that corresponds in general to the least expensive network implementation) is then selected. In this case constraining design specifications refers to either increase ω_{p1} or ω_{s2} or to decrease ω_{p2} or ω_{s1}.

Example 4. Make following band-rejection filter specifications symmetric so that an equivalent LPP with the lowest Ω_s can be found: $\omega_{s1} = 2\pi\ 10$, $\omega_{s2} = 2\pi\ 14.4$, $\omega_{p1} = 2\pi\ 9$, $\omega_{p2} = 2\pi\ 17$.

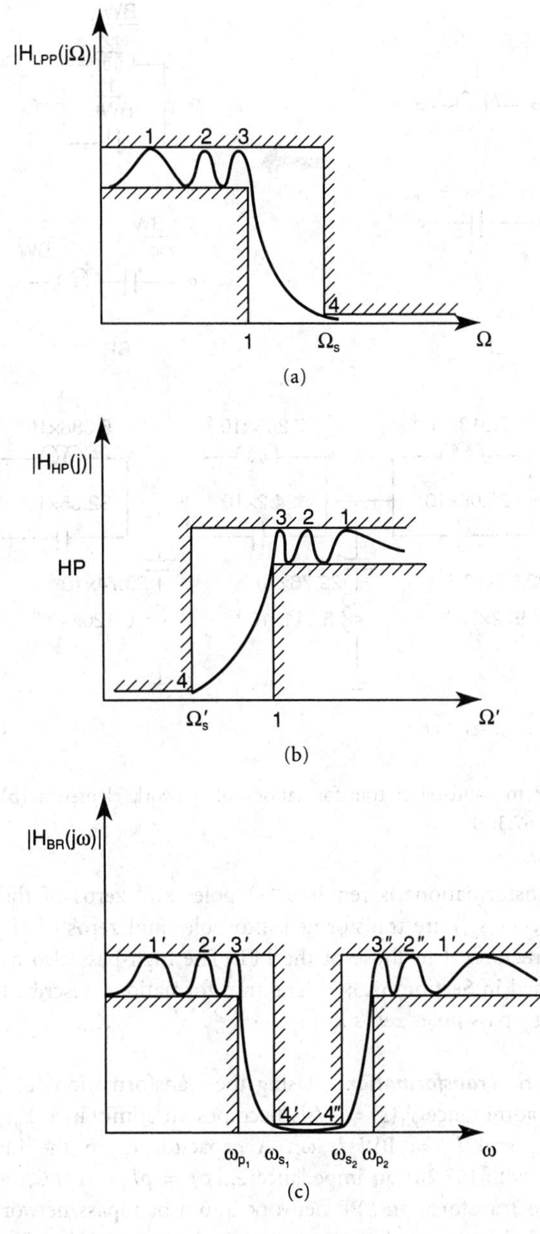

FIGURE 67.9 Low-pass to band-reject transformation: (a) Low-pass response, (b) normalized high-pass response, (c) band-reject response derived from (a) and (b).

Solution. Defining ω_o in terms of the passband frequencies the upper stopband frequency acquires the new value $\omega_{s2} = 2\pi\ 15.3$ and the selectivity factor $\Omega_s = 1.509$ is obtained. If ω_o is defined in terms of the stopband frequencies the upper passband frequency is assigned the new value $\omega_{p2} = 2\pi\ 16$ and the selectivity factor $\Omega_s = 1.59$ is obtained. Therefore, the second option with $\Omega_s = 1.59$ corresponds to the largest selectivity factor and the following transformation parameters result: $\text{BW} = \omega_{p2} - \omega_{p1} = 2\pi(16–9) = 2\pi\ 7$, $\omega_o^2 = \omega_{p1}\ \omega_{p1} = (2\pi)^2\ 144$ is made.

Transformation of Poles and Zeros of the LPP Transfer Function. To determine the poles and zeros of the band-rejection transfer function $H_{\text{BR}}(p)$ starting from those of $H_{\text{LPP}}(s)$, again

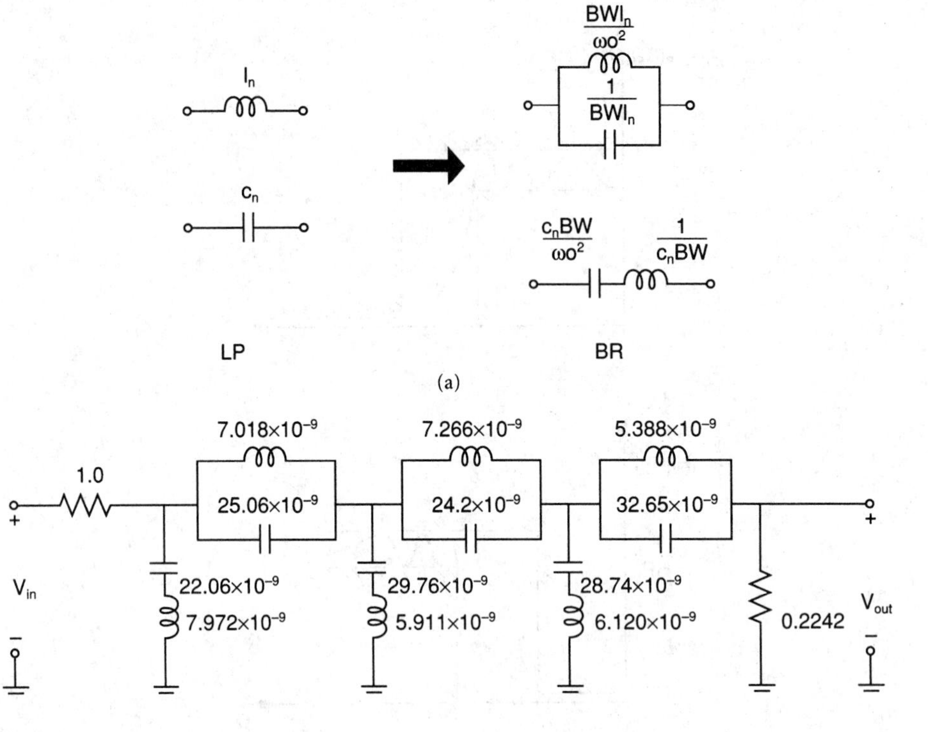

FIGURE 67.10 (a) LPP to band-reject transformations of network elements. (b) Band-reject network derived from LPP of Fig. 67.3(a).

a sequence of two transformations is required: 1) poles and zeros of the LPP (denoted s_{p1}, $s_{p2}, \cdots, s_{pn}$ and $s_{z1}, s_{z2}, \cdots, s_{zn}$) are transformed into poles and zeros of $H_{HP}(s')s'_{p1}, s'_{p2}, \cdots, s'_{z1}$, $s'_{z2}, \cdots, s'_{zm}$ which are reciprocal to those of the LPP. The high-pass also acquires $n - m$ zeros at the origin as explained in Section 67.3; 2) The transformations described in Section 67.4 are then applied to the high-pass poles zeros $s'_{p1}, s'_{p2}, \cdots, s'_{z1}, s'_{z2}, \cdots, s'_{zm}$.

Band-Rejection Network Transformation. Using the transformation (67.21), an inductor in the LPP network with admittance $y_l(s) = 1/sl_n$ becomes an admittance $Y_p(p) = pC_p + 1/pL_p$, where $C_p = 1/\text{BW}\ l_n$ and $L_p = \text{BW}\ l_n/\omega_o^2$. A capacitor c_n in the LPP with impedance $z_c(s) = 1/sc_n$ becomes with (67.20) an impedance $Z_s(p) = pL_s + 1/pC_s$, where $L_s = 1/c_n\text{BW}$ and $C_s = c_n\text{BW}/\omega_o^2$. To transform an LPP network into a bandpass network, capacitors in the LPP network are replaced by a series connection of an inductor with value L_s and a capacitor with value C_s, while inductors in the LPP are replaced by the parallel combination of a capacitor C_p and inductor L_p. This is illustrated in Fig. 67.10(a).

Example 5. Consider the LPP network shown in Fig. 67.3(a). It corresponds to the specifications: $A_p = 2$ dB, $A_s = 45$ dB and $\Omega_s = 1.6$. Transform it into a bandpass network using the following parameters from Example 4: $\omega_o^2 = (2\pi)^2\ 144$, $\text{BW} = 2\pi\ 7$ Mrad/s (units for ω_0 are μrad/s).

Solution. The circuit of Fig. 67.10(b) is obtained applying the transformations indicated in Fig. 67.10(a).

68

Sensitivity and Selectivity

gor M. Filanovsky
Jniversity of Alberta,
'anada

68.1 Introduction

Using sensitivity one can evaluate the change in a filter performance characteristic (bandpass, Q-factor) or in a filter function (input impedance, transfer function) resulting from a change in the nominal value of one or more of the filter components. Hence, sensitivities and based on them, sensitivity measures, can be used to compare different realizations of electric filters that meet the same specifications. Sensitivities can also be used to estimate the spread of the performance characteristic caused by the spread of the element values. In the design of filters one is interested both in choosing realizations that have low sensitivities and in minimizing the sensitivities. This allows use of components with wider tolerances for a given variation or a given spread of the filter characteristic or function.

68.2 Definitions of Sensitivity

Let y be the filter performance characteristic and x be the value of the parameter of a filter element that is causing the characteristic change. The **relative sensitivity** is defined as follows:

$$\mathbf{S}_x^y(y,x) = \frac{\partial y}{\partial x}\frac{x}{y} = \frac{\partial y/y}{\partial x/x} = \frac{\partial(\ln y)}{\partial(\ln x)} \tag{68.1}$$

493-8341-2/95/$0.00 + $.50

It is usually used to establish the approximate relationship between the relative changes $\delta y = \Delta y/y$ and $\delta x = \Delta x/x$. Here, Δy and Δx are absolute changes. The interpretation of relative changes δy and δx depends on the problem at hand. If these relative changes are small, one writes that[1]

$$\delta y \approx \mathbf{S}_x^y(y, x)\, \delta x \tag{68.2}$$

The argument in the parentheses of (68.2), when it does not involve any ambiguity, will usually be omitted, i.e., we will write simply $\mathbf{S}_x^y$. Some simple properties of the sensitivity determined by (68.1) can be established by differentiation only. They are summarized in Table 68.1 [1].

One can also define two *semirelative sensitivities*

$$\mathbf{S}_x(y, x) = x\frac{\partial y}{\partial x} = \frac{\partial y}{\partial x/x} = \frac{\partial y}{\partial(\ln x)} \tag{68.3}$$

which is here frequently denoted by $\mathbf{S}_x(y)$ and

$$\mathbf{S}^y(y, x) = \frac{1}{y}\frac{\partial y}{\partial x} = \frac{\partial y/y}{\partial x} = \frac{\partial(\ln y)}{\partial x} \tag{68.4}$$

which is also denoted by $\mathbf{S}^y(x)$. Both these sensitivities can be used in a way similar to (68.2) to establish the approximate relationships between one relative and one absolute change. Finally, the *absolute sensitivity* $\mathbf{S}(y, x)$ is simply the partial derivative of y with respect to x, i.e., $\mathbf{S}(y, x) = \partial y/\partial x$ and can be used to establish the relationship between absolute changes. The variable x represents the value for any component of the filter. The set of values for all the components will be denoted as $\mathbf{x} = \{x_i\}$, where $i = 1, 2, \cdots, n$.

68.3 Function Sensitivity to One Variable

Let the chosen quantity y be the filter function $F(s, x)$. When it does not involve any ambiguity this function will be denoted as $F(s)$. The element x is chosen as some passive or active element in the circuit realization of the function. The *function sensitivity* is defined as

$$\mathbf{S}_x^{F(s,x)} = \frac{\partial F(s, x)}{\partial x}\frac{x}{F(s, x)} \tag{68.5}$$

Under conditions of sinusoidal steady state, when $s = j\omega$, the function $F(j\omega, x)$ can be represented as

$$F(j\omega, x) = \left|F(j\omega, x)\right| e^{j\arg F(j\omega,x)} = e^{-\alpha(\omega,x)+j\beta(\omega,x)} \tag{68.6}$$

and using the left-hand part of (68.6), one finds that

$$\mathbf{S}_x^{F(j\omega,x)} = \operatorname{Re}\mathbf{S}_x^{F(j\omega,x)} + j\operatorname{Im}\mathbf{S}_x^{F(j\omega,x)} = \mathbf{S}_x^{|F(j\omega,x)|} + j\frac{\partial \arg F(j\omega, x)}{\partial x/x} \tag{68.7}$$

[1]This relationship assumes that $\mathbf{S}_x^y$ is different from zero. If $\mathbf{S}_x^y = 0$, the relative changes δy and δx may be independent. This happens, for example, in the passband of doubly terminated *LC* networks (see below), where the passband attenuation always increases independently on the sign in the variation of a reactance element.

as follows from property 12 of Table 68.1. Thus the real part of the function sensitivity gives the relative change in the magnitude response, and the imaginary part gives the change in the phase response, both with respect to a normalized element change. If one determines $\delta F = [F(j\omega, x) - F(j\omega, x_0)]/F(j\omega, x_0)$ and $\delta x = (x - x_0)/x_0$, where x_0 is the initial value of the element and the deflection $\Delta x = x - x_0$ is small, then (68.5) is used to write

$$\delta F \approx \mathbf{S}_x^{F(s,x)}\,\delta x \tag{68.8}$$

And if one determines $\delta|F| = [|F(j\omega, x)| - |F(j\omega, x_0)|]/|F(j\omega, x_0)|$, then using (68.7), one obtains

$$\delta|F| \approx \mathrm{Re}\,\mathbf{S}_x^{F(j\omega,x)}\,\delta x \tag{68.9}$$

These calculations assume that the sensitivity is also calculated at $x = x_0$.

A frequently used alternate form of (68.7) is obtained by using an attenuation function $\alpha(\omega, x) = \ln(1/|F(j\omega, x)|) = -\ln|F(j\omega, x)|$ and a phase function $\beta(\omega, x) = \arg F(j\omega, x)$ defined by the right-hand part of (68.6) [this interpretation is usually used when $F(s)$ is the filter transfer function $T(s)$]. In terms of these, (68.7) may be rewritten as

$$\mathbf{S}_x^{F(j\omega,x)} = -\frac{\partial\alpha(\omega, x)}{\partial x/x} + j\frac{\partial\beta(\omega, x)}{\partial x/x} = -\,\mathbf{S}_x[\alpha(\omega), x] + j\,\mathbf{S}_x[\beta(\omega), x] \tag{68.10}$$

From (68.7) and (68.10) one concludes that $\mathbf{S}_x^{|F(j\omega,x)|} = \mathrm{Re}\,\mathbf{S}_x^{F(j\omega,x)} = -\,\mathbf{S}_x[\alpha(\omega, x)]$. Besides, using (68.7) and (68.10), one can write that

$$\Delta \arg F(j\omega, x) = \Delta\beta(\omega, x) \approx \mathrm{Im}\,\mathbf{S}_x^{F(j\omega,x)}\,\delta x = \mathbf{S}_x[\beta(\omega, x)]\,\delta x \tag{68.11}$$

where $\Delta \arg F(j\omega, x) = \arg F(j\omega, x) - \arg F(j\omega, x_0)$.

TABLE 68.1 Properties of the Relative Sensitivity

Property Number	Relation	Property Number	Relation
1	$\mathbf{S}_x^{ky} = \mathbf{S}_{kx}^{y} = \mathbf{S}_x^{y}$	10	$\mathbf{S}_x^{y_1/y_2} = \mathbf{S}_x^{y_1} - \mathbf{S}_x^{y_2}$
2	$\mathbf{S}_x^{x} = \mathbf{S}_x^{kx} = \mathbf{S}_{kx}^{x} = 1$	11	$\mathbf{S}_{x_1}^{y} = \mathbf{S}_{x_2}^{y}\,\mathbf{S}_{x_1}^{x_2}$
3	$\mathbf{S}_{1/x}^{y} = \mathbf{S}_x^{1/y} = -\,\mathbf{S}_x^{y}$	12[a]	$\mathbf{S}_x^{y} = \mathbf{S}_x^{\lvert y\rvert} + j \arg y\,\mathbf{S}_x^{\arg y}$
4	$\mathbf{S}_x^{y_1 y_2} = \mathbf{S}_x^{y_1} + \mathbf{S}_x^{y_2}$	13[a]	$\mathbf{S}_x^{\arg y} = \frac{1}{\arg y}\mathrm{Im}\,\mathbf{S}_x^{y}$
5	$\mathbf{S}_x^{\Pi_{i=1}^{n} y_i} = \sum_{i=1}^{n} \mathbf{S}_x^{y_i}$	14[a]	$\mathbf{S}_x^{\lvert y\rvert} = \mathrm{Re}\,\mathbf{S}_x^{y}$
6	$\mathbf{S}_x^{y^n} = n\,\mathbf{S}_x^{y}$	15	$\mathbf{S}_x^{y+z} = \frac{1}{y+z}(y\,\mathbf{S}_x^{y} + z\,\mathbf{S}_x^{z})$
7	$\mathbf{S}_x^{x^n} = n\,\mathbf{S}_x^{kx^n} = n$	16	$\mathbf{S}_x^{\Sigma_{i=1}^{n} y_i} = \frac{\sum_{i=1}^{n} y_i\,\mathbf{S}_x^{y_i}}{\sum_{i=1}^{n} y_i}$
8	$\mathbf{S}_{x^n}^{y} = \frac{1}{n}\,\mathbf{S}_x^{y}$	17	$\mathbf{S}_x^{\ln y} = \frac{1}{\ln y}\,\mathbf{S}_x^{y}$
9	$\mathbf{S}_{x^n}^{x} = \mathbf{S}_{kx^n}^{x} = \frac{1}{n}$		

[a] In this relation y is a complex quantity and x is a real quantity.

When the filter function is a ratio of two polynomials $N(s)$ and $D(s)$, i.e.,

$$F(s) = \frac{N(s)}{D(s)} \tag{68.12}$$

then, assuming that the coefficients of $N(s)$ and $D(s)$ depend on the element x, and using (68.1), one derives the following form of (68.5):

$$\mathbf{S}_x^{F(s)} = x\left[\frac{\partial N(s)/\partial x}{N(s)} - \frac{\partial D(s)/\partial x}{D(s)}\right] \tag{68.13}$$

which is sometimes more convenient.

68.4 Coefficient Sensitivity

In general, a network function $F(s)$ for any active or passive lumped network is a ratio of polynomials having the form

$$F(s) = \frac{N(s)}{D(s)} = \frac{a_0 + a_1 s + a_2 s^2 + \cdots + a_m s^m}{d_0 + d_1 s + d_2 s^2 + \cdots + d_n s^n} \tag{68.14}$$

in which the coefficients a_i and d_i are real and can be functions of an arbitrary filter element x. For such an element x one may define the *relative coefficient sensitivities* as follows:

$$\mathbf{S}_x^{a_i} = \frac{\partial a_i}{\partial x}\frac{x}{a_i} \qquad \mathbf{S}_x^{d_i} = \frac{\partial d_i}{\partial x}\frac{x}{d_i} \tag{68.15}$$

or the *semirelative coefficient sensitivities* (they are even more useful):

$$\mathbf{S}_x(a_i) = x\frac{\partial a_i}{\partial x} \qquad \mathbf{S}_x(d_i) = x\frac{\partial d_i}{\partial x} \tag{68.16}$$

The coefficient sensitivities defined in this way are related to the function sensitivity introduced in the previous section. Indeed, using (68.13) and (68.16) one easily obtains that

$$\mathbf{S}_x^{F(s)} = \frac{\sum_{i=0}^{m} \mathbf{S}_x(a_i)s^i}{N(s)} - \frac{\sum_{i=0}^{n} \mathbf{S}_x(d_i)s^i}{D(s)} \tag{68.17}$$

or, in terms of relative sensitivities, that

$$\mathbf{S}_x^{F(s)} = \frac{\sum_{i=0}^{m} \mathbf{S}_x^{a_i} a_i s^i}{N(s)} - \frac{\sum_{i=0}^{n} \mathbf{S}_x^{d_i} d_i s^i}{D(s)} \tag{68.18}$$

The manner in which the filter function depends on any element x is a bilinear dependence [2]. Thus, $F(s)$ of (68.14) may also be written in the form

$$F(s) = \frac{N(s)}{D(s)} = \frac{N_1(s) + xN_2(s)}{D_1(s) + xD_2(s)} \tag{68.19}$$

where $N_1(s)$, $N_2(s)$, $D_1(s)$, and $D_2(s)$ are polynomials with real coefficients that are not functions of the filter element x. This is true whether x is chosen to be the value of a passive resistor or capacitor, the gain of some amplifier or controlled source, etc. Only for filters with ideal transformers, ideal gyrators and ideal negative impedance converters the filter functions are the biquadratic functions of the ideal element parameters [2].

Because of the bilinear dependence property there are only two ways in which a coefficient, say, a_i may depend on a network element. The first of these has the form $a_i = kx$, in which case $\mathbf{S}_x^{a_i} = 1$, and $\mathbf{S}_x(a_i) = kx$; the second possible dependence for a coefficient a_i (or d_i) is $a_i = k_0 + k_1 x$, in which case $\mathbf{S}_x^{a_i} = k_1 x/(k_0 + k_1 x)$ and $\mathbf{S}_x(a_i) = k_1 x$. In this latter situation one has two cases: 1) the parities of the terms are the same, and thus the magnitude of $\mathbf{S}_x^{a_i}$ is less than one; and 2) the terms have opposite parities, in which case the magnitude of $\mathbf{S}_x^{a_i}$ is greater than one. In the last case the relative sensitivity $\mathbf{S}_x^{a_i}$ can have an infinite value, as the result of dividing by zero. In this case a more meaningful measure of the change would be to use the semirelative coefficient sensitivity $\mathbf{S}_x(a_i)$.

8.5 Root Sensitivities

A filter function can also be represented as

$$F(s) = \frac{a_m \prod_{i=1}^{m}(s - z_i)}{d_n \prod_{i=1}^{n}(s - p_i)} \tag{68.20}$$

where z_i are zeros and p_i are poles. If $F(s)$ is also a function of the filter element x, the location of these poles and zeros will depend on this element. This dependence is described by the *semirelative root sensitivities*

$$\mathbf{S}_x(z_i) = x\frac{\partial z_i}{\partial x} \qquad \mathbf{S}_x(p_i) = x\frac{\partial p_i}{\partial x} \tag{68.21}$$

We will give calculation of the pole sensitivities only (they are used more frequently, to verify the stability calculating the absolute change Δp_i for a given δx), the calculation of the zero sensitivities follows the same pattern. Assume that p_i is a simple pole of $F(s)$ then

$$D(p_i) = D_1(p_i) + xD_2(p_i) = 0 \tag{68.22}$$

When the parameter x becomes $x + \Delta x$, the pole p_i moves to the point $p_i + \Delta p_i$. Substituting these values in (68.22) one obtains that

$$D_1(p_i + \Delta p_i) + (x + \Delta x)D_2(p_i + \Delta p_i) = 0 \tag{68.23}$$

If one uses Taylor's expansions $D_1(p_i + \Delta p_i) = D_1(p_i) + [\partial D_1(s)/\partial s]|_{s=p_i}\Delta p_i + \cdots$ and $D_2(p_i + \Delta p_i) = D_2(p_i) + [\partial D_2(s)/\partial s]|_{s=p_i}\Delta p_i + \cdots$ and substitutes them in (68.23), keeping the terms of the first order of smallness one obtains that

$$\frac{\Delta p_i}{\Delta x} = -\frac{D_2(p_i)}{D'(p_i)} \tag{68.24}$$

where $D'(p_i) = [\partial D(s)/\partial s]|_{s=p_i}$. This result allows calculation of the pole sensitivity, which becomes

$$\mathbf{S}_x(p_i) = x\frac{\partial p_i}{\partial x} = -x\frac{D_2(p_i)}{D'(p_i)} \tag{68.25}$$

One can write that

$$D_1(s) = b_0 + b_1 s + b_2 s^2 + \cdots \quad \text{and} \quad D_2(s) = c_0 + c_1 s + c_2 s^2 + \cdots.$$

Then, taking into consideration (68.19) one can write that

$$D(s) = d_0 + d_1 s + d_2 s^2 + \cdots = (b_0 + xc_0) + (b_1 + xc_1)s + (b_2 + xc_2)s^2 + \cdots$$

Differentiating this result one obtains that

$$\begin{aligned} D_2(s) &= c_0 + c_1 s + c_2 s^2 + \cdots = \partial d_0/\partial x + (\partial d_1/\partial x)s + (\partial d_2/\partial x)s^2 + \cdots \\ &= (1/x)[\mathbf{S}_x(d_0) + \mathbf{S}_x(d_1)s + \mathbf{S}_x(d_2)s^2 + \cdots] \end{aligned}$$

From the other side, $D'(s) = \partial D(s)/\partial s = d_1 + 2d_2 s + 3d_3 s^2 + \cdots$. Calculating the two last expressions at $s = p_i$ and substituting them in (68.25), one obtains that

$$\mathbf{S}_x(p_i) = -\frac{\sum_{j=0}^{n} p_i^j \mathbf{S}_x(d_j)}{\sum_{j=0}^{n-1} (j+1)d_{j+1}p_i^j} = -\frac{\sum_{j=0}^{n} d_j p_i^j \mathbf{S}_x^{d_j}}{\sum_{j=0}^{n-1} (j+1)d_{j+1}p_i^j} \tag{68.26}$$

The result (68.26) produces the pole sensitivity without representation of poles via coefficients of $D(s)$ (which is not possible if $n > 4$). It is convenient even for polynomials of low degree [3]. If p_i is a multiple root then the derivative $D'(p_i) = 0$ and $\mathbf{S}_x(p_i) = \infty$. But this does not mean that the variation δx causes infinitely large change of the pole location. This variation splits this multiple root into a group of simple roots. The location of these roots can be calculated in the following way. The roots are always satisfying the equation $D_1(s) + (x + \Delta x)D_2(s) = 0$, i.e., the equation $D(s) + \Delta x D_2(s) = 0$. One can rewrite the last equation as

$$1 + \delta x\left[\frac{xD_2(s)}{D(s)}\right] = 0 \tag{68.27}$$

where $\delta x = \Delta x/x$ as usual. The function $G(s) = [xD_2(s)/D(s)]$ can be represented as a sum of simple ratios. If the roots of $D(s)$ [i.e., poles of $F(s)$] are simple, then

$$G(s) = \sum_{i=1}^{n} \frac{K_{ip}}{s - p_i} + K_{0p} \tag{68.28}$$

where $K_{0p} = G(\infty)$ and $K_{ip} = (s - p_i)\, G(s)|_{s=p_i}$. In the vicinity of $s = p_i$, (68.27) can be substituted by

$$\lim_{s \to p_i} [1 + \delta x G(s)] = 1 + \delta x \frac{K_{ip}}{s - p_i} \tag{68.29}$$

Equating the right-hand side of (68.29) to zero and substituting $s = p_i + \Delta p_i$ in this equation, one obtains that when $\Delta x \to 0$, the pole sensitivity can be calculated as

$$\mathbf{S}_x(p_i) = \frac{\partial p_i}{\partial x/x} = -K_{ip} \tag{68.30}$$

If a pole of $G(s)$ is not simple, but multiple, with the multiplicity of k, then the limit form of (68.28) will be

$$1 + \delta x \left[\frac{K_{ip}^{(1)}}{s - p_i} + \frac{K_{ip}^{(2)}}{(s - p_i)^2} + \cdots + \frac{K_{ip}^{(k)}}{(s - p_i)^k} \right] = 0 \tag{68.31}$$

If now $s = p_i + \Delta p_i$ is substituted in (68.31) and only the largest term is kept, one finds that

$$\Delta p_i = [-\delta x K_{ip}^{(k)}]^{1/k} \tag{68.32}$$

Hence these new simple roots of $D(s)$ are equiangularly spaced on a circle around p_i.

Similar calculations with similar results can be obtained for the zeros z_i of the function $F(s)$. They, together with the calculation for poles, allow one to establish the relationship between relative sensitivity of $F(s)$ and the sensitivities of its zeros and poles. Indeed, taking into consideration (68.19), the result (68.13) can be rewritten as

$$\mathbf{S}_x^{F(s)} = \frac{xN_2(s)}{N(s)} - \frac{xD_2(s)}{D(s)} = H(s) - G(s) \tag{68.33}$$

Expanding both $H(s)$ and $G(s)$ into sums of simple ratios, one obtains

$$\mathbf{S}_x^{F(s)} = \sum_{i=1}^{n} \frac{\mathbf{S}_x(p_i)}{s - p_i} - \sum_{i=1}^{m} \frac{\mathbf{S}_x(z_i)}{s - z_i} + K_{0z} - K_{0p} \tag{68.34}$$

Here $K_{0z} = H(\infty)$, $K_{0p} = G(\infty)$, and it is assumed that both zeros and poles of $F(s)$ are simple. Finally, a useful modification is obtained for the case when a coefficient d_k in the polynomial $D(s) = d_0 + d_1 s + d_2 s^2 + \cdots$ is considered as a variable parameter. If d_k is substituted by $d_k + \Delta d_k$, the polynomial $D(s)$ becomes $D(s) + \Delta d_k s^k$. For this case the function $G(s) = (\Delta d_k s^k)/D(s)$ and one can write that

$$1 + \frac{\Delta d_k s^k}{D(s)} = 1 + \frac{\Delta d_k s^k}{(s - p_i)D'(s)} \tag{68.35}$$

Here $D'(s) = \partial D(s)/\partial s$ and it is assumed that p_i is a simple root of $D(s)$. Hence, in the vicinity of $s = p_i + \Delta p_i$, the value of Δp_i can be obtained from the equation

$$1 + \frac{\Delta d_k p_i^k}{\Delta p_i D'(p_i)} = 0 \tag{68.36}$$

From (68.36), one finds that

$$\frac{\Delta p_i}{\Delta d_k} = -\frac{p_i^k}{D'(p_k)} \tag{68.37}$$

and, if necessary, the *pole-coefficient sensitivity*

$$\mathbf{S}_{d_k}(p_i) = \frac{\partial p_i}{\partial d_k/d_k} = -d_k \frac{p_i^k}{D'(p_i)} \tag{68.38}$$

68.6 Statistical Model for One Variable

Assume that x is the value of a filter element. This x differs from the average value $\underline{x}$ in a way that cannot be controlled by the filter designer. This situation can be modeled by considering x as a random variable. Its statistical distribution depends on the manufacturing process. An approximate calculation is sufficient in most practical cases. If $F(s,x)$ is a filter function that depends on x, then the variation of $F(s,x)$ around the average value $\underline{x}$ can be approximated by

$$F(s,x) \approx F(s,\underline{x}) + (x-\underline{x})\left.\frac{\partial F(s,x)}{\partial x}\right|_{x=\underline{x}} \tag{68.39}$$

The statistical interpretation of (68.39) follows. Due to its dependence on the random variable x, $F(s,x)$ becomes a random variable as well. The values $\underline{x}$, $F(s,\underline{x}) = \underline{F}(s)$ and $\partial F(s,x)/\partial x$ calculated at $x = \underline{x}$ are constants. The last constant can be denoted as $\partial F(s,\underline{x})/\partial x$. Instead of x and $F(s,x)$ it is preferable to use their relative deviations from the average values, namely $\delta x = (x-\underline{x})/\underline{x}$ and $\delta F(s) = (F-\underline{F})/\underline{F}$. Then one obtains from (68.39) that

$$\delta F(s) \approx \left[\frac{\underline{x}}{F(s,\underline{x})}\frac{\partial F(s,\underline{x})}{\partial x}\right]\delta x = \mathbf{S}_x^{F(s,\underline{x})}\,\delta x \tag{68.40}$$

Hence, in the first-order approximation the random variables $\delta F(s)$ and δx are proportional, and the proportionality factor is the same sensitivity of $F(s,x)$ with respect to x calculated at the average point $\underline{x}$. Thus, on the $j\omega$ axis the average and the variance of $\delta F(j\omega)$ and δx are related by

$$\mu_{\delta F} \approx \mathbf{S}_x^{F(j\omega,\underline{x})}\,\mu_{\delta x} \tag{68.41}$$

and

$$\sigma_{\delta F}^2 = E\left\{\left|\frac{F(j\omega,x)-F(j\omega,\underline{x})}{F(j\omega,\underline{x})}\right|^2\right\} \approx \left|\mathbf{S}_x^{F(j\omega,\underline{x})}\right|^2\sigma_{\delta x}^2 \tag{68.42}$$

where $E\{\ \}$ means the expected value. Here, $\mu_{\delta x}$ is the average value of δx, $\mu_{\delta F}$ is the average value of δF, and $\sigma_{\delta x}^2$ and $\sigma_{\delta F}^2$ are the dispersions of these values. If the deviation δx is bound by the modulus $M_{\delta x}$, i.e., the probability distribution is concentrated in the interval $[-M_{\delta x}, M_{\delta x}]$, then the deviation δF is bound in the first approximation by

$$|\delta F| \leq M_{\delta F} \approx \left|\mathbf{S}_x^{F(j\omega,\underline{x})}\right| M_{\delta x} \tag{68.43}$$

Normally the probability distribution of x should be centered around the average value $\underline{x}$, so that it can be assumed $\mu_{\delta x} = 0$. This implies that $\mu_{\delta F} = 0$ as well.

It is not difficult to see that (68.8) and (68.40) are different by interpretation of δx and δF (since these were deflections from the nominal point, it was tacitly assumed that we were dealing with one sample of the filter, here they are random) and the point of calculation of sensitivity. The interpretation with a random variable is possible for (68.9) as well. One has to determine $\delta|F| = [|F(j\omega,x)| - |F(j\omega,\underline{x})|]/|F(j\omega,\underline{x})|$, then using (68.9) one can write

$$\mu_{\delta|F|} \approx \mathrm{Re}\,\mathbf{S}_x^{F(j\omega,\underline{x})}\,\mu_{\delta x} \tag{68.44}$$

and

$$\sigma^2_{\delta|F|} \approx \left(\text{Re}\,\mathbf{S}_x^{F(j\omega,\underline{x})}\right)^2 \sigma^2_{\delta x} \tag{68.45}$$

The result (68.11) can also be interpreted for random variables and allow one to calculate the average and the variance of the change in the filter function argument (which is hardly ever done in filter design).

68.7 Multiparameter Sensitivities and Sensitivity Measures

The multiparameter sensitivities (sometimes [4] they are called sensitivity indices) appear as an effort to introduce generalized functions that represent the influence of all filter elements. They can be used for comparison of different designs and should be minimized in the design process. The sensitivity measures appear as numbers (they are functionals of the multiparameter sensitivities) that should be minimized in the design.

First of all, the definition of function sensitivity given in (68.5) is readily extended to determine the effect on the filter function of variations of more than one component. In this case $F(s, x_1, x_2, \cdots, x_n) = F(s, \mathbf{x})$ and one may write that

$$\frac{dF(s,\mathbf{x})}{F(s,\mathbf{x})} = d[\ln F(s,\mathbf{x})] = \sum_{i=1}^{n} \mathbf{S}_{x_i}^{F(s,\mathbf{x})} \frac{dx_i}{x_i} \tag{68.46}$$

where n is the number of components being considered. Here, $\mathbf{S}_{x_i}^{F(s,\mathbf{x})} = [x_i\,\partial F(s,\mathbf{x})]/[F(s,\mathbf{x})\,\partial x_i]$. From this result one directly (substituting $s = j\omega$ and separating real and imaginary parts) obtains that

$$\frac{d\left|F(j\omega,\mathbf{x})\right|}{\left|F(j\omega,\mathbf{x})\right|} = \sum_{i=1}^{n} \text{Re}\,\mathbf{S}_{x_i}^{F(j\omega,\mathbf{x})} \frac{dx_i}{x_i} \tag{68.47}$$

and

$$d \arg F(j\omega,\mathbf{x}) = \sum_{i=1}^{n} \text{Im}\,\mathbf{S}_{x_i}^{F(j\omega,\mathbf{x})} \frac{dx_i}{x_i} \tag{68.48}$$

The results (68.47) and (68.48) are used to evaluate the deviations of the magnitude and phase values of a given filter realization from their nominal values when the circuit elements have prescribed normalized deviations $\delta x_i = (x_i - x_{i0})/x_{i0}$ $(i = 1, 2, \cdots, n)$. One can introduce a column vector of normalized deviation $\delta\mathbf{x} = [\delta x_1\,\delta x_2 \cdots \delta x_n]^t$, where t means transpose and a sensitivity row vector

$$\mathbf{S}_{\mathbf{x}}^{F(s,\mathbf{x})} = \left[\mathbf{S}_{x_1}^{F(s,\mathbf{x})} \quad \mathbf{S}_{x_2}^{F(s,\mathbf{x})} \quad \cdots \quad \mathbf{S}_{x_n}^{F(s,\mathbf{x})}\right] \tag{68.49}$$

Then defining $\delta F(s,\mathbf{x}) = [F(s,\mathbf{x}) - F(s,\mathbf{x}_0)]/F(s,\mathbf{x}_0)$ one can use (68.46) to write

$$\delta F(s,\mathbf{x}) \approx \mathbf{S}_{\mathbf{x}}^{F(s,\mathbf{x})}\,\delta\mathbf{x} \tag{68.50}$$

which is analogous to (68.8). As was mentioned before, the calculation of the filter function magnitude change is traditionally considered of primary importance in filter design. Introducing $\delta|F(j\omega,\mathbf{x})| = [|F(j\omega,\mathbf{x})| - |F(j\omega,\mathbf{x}_0)|]/|F(j\omega,\mathbf{x}_0)|$ and using (68.47), one writes

$$\delta\left|F(j\omega,\mathbf{x})\right| \approx \left[\text{Re}\,\mathbf{S}_{\mathbf{x}}^{F(j\omega,\mathbf{x})}\right]\delta\mathbf{x} \tag{68.51}$$

where the row vector

$$\left[\text{Re}\,\mathbf{S}_{\mathbf{x}}^{F(j\omega,\mathbf{x})}\right] = \left[\text{Re}\,\mathbf{S}_{x_1}^{F(j\omega,\mathbf{x})}\ \text{Re}\,\mathbf{S}_{x_2}^{F(j\omega,\mathbf{x})} \cdots \text{Re}\,\mathbf{S}_{x_n}^{F(j\omega,\mathbf{x})}\right] \tag{68.52}$$

is used. This vector is determined by the function $F(s, \mathbf{x})$ and its derivatives calculated at $\mathbf{x} = \mathbf{x}_0$. To characterize and compare the vectors of this type one can introduce different vector measures that are called *sensitivity indices.* The most frequently used ones are the *average sensitivity index*

$$\psi(F) = \sum_{i=1}^{n} \text{Re}\,\mathbf{S}_{x_i}^{F(j\omega,\mathbf{x})} \tag{68.53}$$

then the *worst-case sensitivity index*

$$\nu(F) = \sum_{i=1}^{n} |\text{Re}\,\mathbf{S}_{x_i}^{F(j\omega,\mathbf{x})}| \tag{68.54}$$

(sometimes it is called worst-case magnitude sensitivity), and, finally, the *quadratic sensitivity index*

$$\rho(F) = \left[\sum_{i=1}^{n} \left(\text{Re}\,\mathbf{S}_{x_i}^{F(j\omega,\mathbf{x})}\right)^2\right]^{1/2} \tag{68.55}$$

These sensitivity indices can be considered as multiparameter sensitivities.

If we let the individual nominal values of n elements be given as x_{i0}, then we may define a tolerance constant ε_i (positive number) by the requirement that

$$x_{i0}(1 - \varepsilon_i) \le x_i \le x_{i0}(1 + \varepsilon_i) \tag{68.56}$$

Then we may define a *worst-case measure of sensitivity*

$$M_W = \int_{\omega_1}^{\omega_2} \left(\sum_{i=1}^{n} |\text{Re}\,\mathbf{S}_{x_i}^{F(j\omega,\mathbf{x})}|\varepsilon_i\right) d\omega \tag{68.57}$$

The goal of the filter design should be the search for the set of tolerance constants yielding the least expensive in the production filter. This is a difficult problem, and at the design stage it can be modeled by the minimization of the chosen sensitivity measure. In the design based on the worst-case measure of sensitivity the usual approach [2] is to choose the tolerance constants in such a way that the contributions $|\text{Re}\,\mathbf{S}_{x_i}^{F(s,\mathbf{x})}|\varepsilon_i$ are approximately equal, i.e., the elements with lower sensitivities get wider tolerance constants.

For any values of the filter elements x_i satisfying (68.56), the magnitude characteristic will lie within the definite bounds that are apart from the nominal characteristic by the distance less than $\varepsilon_{i\,\max}\nu(F)$. If the tolerance constants are all equal to ε, then the maximum deviation from the nominal characteristic (when $\mathbf{x} = \mathbf{x}_0$) is thus given as $\varepsilon\nu(F)$. And the worst-case measure of sensitivity becomes, for this case

$$M_W = \varepsilon \int_{\omega_1}^{\omega_2} \nu(F)\, d\omega \tag{68.58}$$

Considering the imaginary parts of the sensitivity row vector one can introduce corresponding sensitivity indices and similar sensitivity measures for the filter function phase.

The element tolerances obtained using the worst-case sensitivity index and measure are extremely tight and this set of elements is frequently unfeasible. Besides, with given tolerances, the set of elements producing the worst-case sensitivity is never obtained in practice. A more feasible set of tolerances is obtained when one uses the sum of the squares of the individual functions. One may define a *quadratic measure of sensitivity* as

$$M_Q = \int_{\omega_1}^{\omega_2} \left[\sum_{i=1}^{n} \left(\text{Re}\, \mathbf{S}_{x_i}^{F(j\omega,\mathbf{x})} \right)^2 \varepsilon_i^2 \right] d\omega \qquad (68.59)$$

In the design using the sensitivity measure given by (68.59) one also tries to get the tolerances so that the contributions of each term $(\text{Re}\, \mathbf{S}_{x_i}^{F(s,\mathbf{x})})^2 \varepsilon_i^2$ are approximately equal in the considered bandwidth. Again, if the tolerances are equal then this expression is simplified to

$$M_Q = \varepsilon^2 \int_{\omega_1}^{\omega_2} \rho^2(F)\, d\omega \qquad (68.60)$$

useful for comparison of different filters.

As one can see, the multivariable sensitivities appear as a result of certain operations with the sensitivity row vector components. Additional multivariable sensitivities could be introduced, for example, the sum of magnitudes of vector components, the sum of their squares, etc. The multivariable sensitivities and the measures considered above represent the most frequently used in filter design in the context of filter characteristic variations.

The case of random variables can also be generalized so that imprecisions of the values of several elements are simultaneously considered. Around the nominal value $\underline{\mathbf{x}} = [\underline{x}_i]$ $(i = 1, 2, \cdots, n)$ the function $F(s, \underline{\mathbf{x}})$ can be approximated as

$$F(s, \mathbf{x}) \approx F(s, \underline{\mathbf{x}}) + \sum_{i=1}^{n} (x - \underline{x}_i) \frac{\partial F(s, \mathbf{x})}{\partial x_i} \qquad (68.61)$$

and from this approximation one obtains

$$\delta F(s, \mathbf{x}) \approx \sum_{i=1}^{n} \mathbf{S}_{x_i}^{F(s,\mathbf{x})}\, \delta x_i \qquad (68.62)$$

Here, $\delta F(s, \mathbf{x}) = [F(s, \mathbf{x}) - F(s, \underline{\mathbf{x}})]/F(s, \underline{\mathbf{x}})$ and $\delta x_i = (x_i - \underline{x}_i)/\underline{x}_i$ $(i = 1, 2, \cdots, n)$. This result can be rewritten as

$$\delta F(s, \mathbf{x}) = \mathbf{S}_{\mathbf{x}}^{F(s,\mathbf{x})}\, \delta \mathbf{x} \qquad (68.63)$$

and is completely analogous to (68.50). It is different in interpretation only. The components of the column vector $\delta \mathbf{x} = [\delta x_1\, \delta x_2 \cdots \delta x_n]^t$ are the random variables now and the components of the row vector $\mathbf{S}_{\mathbf{x}}^{F(s,\mathbf{x})} = [\mathbf{S}_{x_1}^{F(s,\mathbf{x})} \quad \mathbf{S}_{x_2}^{F(s,\mathbf{x})} \quad \cdots \quad \mathbf{S}_{x_n}^{F(s,\mathbf{x})}]$ are calculated at the point of $\mathbf{x} = \underline{\mathbf{x}}$.

This interpretation allows us to obtain from (68.63) that on the $j\omega$ axis

$$\mu_{\delta F} = \sum_{i=1}^{n} \mathbf{S}_{x_i}^{F(j\omega,\mathbf{x})}\, \mu_i \qquad (68.64)$$

Here, μ_i is the average of δx_i. If all μ_i are equal, i.e., $\mu_i = \mu_x$ $(i = 1, 2, \cdots, n)$, one can introduce the average sensitivity index

$$\psi(F) = \sum_{i=1}^{n} \mathbf{S}_{x_i}^{F(j\omega,\mathbf{x})} \qquad (68.65)$$

Using (68.65), the average value can be calculated as $\mu_{\delta F} = \mu_x \psi(F)$. If, in addition, the deviation of δx_i is bound by M_i, then

$$M_{\delta F} \leq \sum_{i=1}^{n} \left| \mathbf{S}_{x_i}^{F(j\omega,\mathbf{x})} \right| M_i \tag{68.66}$$

If the elements of a filter have the same precision, which means that all M_i are equal, it is reasonable to introduce the worst-case sensitivity index

$$\nu(F) = \sum_{i=1}^{n} \left| \mathbf{S}_{x_i}^{F(j\omega,\mathbf{x})} \right| \tag{68.67}$$

so that when all M_i are equal to M_x, $M_{\delta F} = M_x \nu(F)$. Finally one can calculate $\sigma_{\delta F}^2 = E\{[\delta F(j\omega, \mathbf{x})]^* \delta F(j\omega, \mathbf{x})\}$ or

$$\sigma_{\delta F}^2 = E\left\{ (\mathbf{S}_{\mathbf{x}}^{F(j\omega,\mathbf{x})}\ \delta\mathbf{x})^* (\mathbf{S}_{\mathbf{x}}^{F(j\omega,\mathbf{x})}\ \delta\mathbf{x}) \right\} \tag{68.68}$$

To take into consideration possible correlation between the components δx_i one can do the following. The value of $\mathbf{S}_{\mathbf{x}}^{F(j\omega,\mathbf{x})}\ \delta\mathbf{x}$ is a scalar. Then

$$(\mathbf{S}_{\mathbf{x}}^{F(j\omega,\mathbf{x})}\ \delta\mathbf{x}) = (\mathbf{S}_{\mathbf{x}}^{F(j\omega,\mathbf{x})}\ \delta\mathbf{x})^t = (\delta\mathbf{x})^t (\mathbf{S}_{\mathbf{x}}^{F(j\omega,\mathbf{x})})^t \tag{68.69}$$

Substituting this result into (68.68), one obtains that

$$\delta_{\delta F}^2 = E\left\{ (\mathbf{S}_{\mathbf{x}}^{F(j\omega,\mathbf{x})})^* [(\delta\mathbf{x})^* (\delta\mathbf{x})^t] (\mathbf{S}_{\mathbf{x}}^{F(j\omega,\mathbf{x})})^t \right\} \tag{68.70}$$

But the components of $\delta\mathbf{x}$ are real, i.e., $(\delta\mathbf{x})^* = \delta\mathbf{x}$ and the result of multiplication in the square brackets of (68.70) is a square $n \times n$ matrix. Then, $E\{\delta\mathbf{x}(\delta\mathbf{x})^t\}$ is also a square matrix

$$[\mathbf{P}] = \begin{bmatrix} \sigma_{x_1}^2 & \rho_{x_1 x_2} & \cdots & \rho_{x_1 x_n} \\ \rho_{x_2 x_1} & \sigma_{x_2}^2 & \cdots & \rho_{x_2 x_n} \\ & & \cdots & \\ \rho_{x_n x_1} & \cdots & & \sigma_{x_n}^2 \end{bmatrix} \tag{68.71}$$

the diagonal elements of which are variances of δx_i and off-diagonal terms are nonnormalized correlation coefficients. Then, (68.70) can be rewritten as

$$\sigma_{\delta F}^2 = (\mathbf{S}_{\mathbf{x}}^{F(j\omega,\mathbf{x})})^* [\mathbf{P}] (\mathbf{S}_{\mathbf{x}}^{F(j\omega,\mathbf{x})})^t \tag{68.72}$$

which is sometimes [4] called the propagation-of-variance formula. In the absence of correlation between the variations δx_i, the matrix $[\mathbf{P}]$ has the diagonal terms only and (68.71) becomes

$$\sigma_{\delta F}^2 = \sum_{i=1}^{n} \left| \mathbf{S}_{x_i}^{F(j\omega,\mathbf{x})} \right|^2 \sigma_{x_i}^2 \tag{68.73}$$

If all σ_{x_i} are equal to σ_x one can introduce a quadratic sensitivity index

$$\rho(F) = \left[\sum_{i=1}^{n} \left| \mathbf{S}_{x_i}^{F(j\omega,\mathbf{x})} \right|^2 \right]^{1/2} \tag{68.74}$$

and in this case $\sigma_{\delta F}^2 = \rho^2(F)\sigma_x^2$ (the value $\rho^2(F)$ is sometimes called **Schoeffler multivariable sensitivity**). One can also introduce two sensitivity measures, namely, the worst-case sensitivity measure

$$M_W = \int_{\omega_1}^{\omega_2} \left(\sum_{i=1}^{n} |\mathbf{S}_{x_i}^{F(j\omega,\mathbf{x})}| \mu_i \right) d\omega \tag{68.75}$$

and the quadratic sensitivity measure

$$M_Q = \int_{\omega_1}^{\omega_2} (\mathbf{S}_{\mathbf{x}}^{F(j\omega,\mathbf{x})})^* [\mathbf{P}] (\mathbf{S}_{\mathbf{x}}^{F(j\omega,\mathbf{x})})^t \, d\omega \tag{68.76}$$

which, when the correlation between the elements of $\delta\mathbf{x}$ is absent, becomes

$$M_Q = \int_{\omega_1}^{\omega_2} \left[\sum_{i=1}^{n} |\mathbf{S}_{x_i}^{F(j\omega,\mathbf{x})}|^2 \sigma_i^2 \right] d\omega \tag{68.77}$$

Here, for simplicity, the notation $\sigma_i = \sigma_{xi}$ is used.

The sensitivity indices and the sensitivity measures introduced for the case when $\delta\mathbf{x}$ is a random vector are cumulative; they take into consideration the variation of the amplitude and phase of the filter function. For this reason some authors prefer to use the indices as they are defined in (68.65), (68.67), and (68.74) and the measures as they are defined by (68.75) and (68.77) for the deterministic case as well (the deterministic case does not assume any correlation between the variations δx_i), with the corresponding substitution of μ_i by ε_i and σ_i^2 by ε_i^2. From the other side, one can take (68.51) and use it for the case of random vector $\delta\mathbf{x}$ considering, for example, the variation $\delta|F(j\omega, \mathbf{x})|$ as a random variable and calculating $\mu_{\delta|F|}$ and $\sigma_{\delta|F|}^2$, which will be the characteristics of this variable. In this case one can use the results (68.75), (68.77), etc., substituting $\mathbf{S}_{x_i}^{F(s,\mathbf{x})}$ by Re $\mathbf{S}_{x_i}^{F(s,\mathbf{x})}$. These possibilities are responsible for many formulations of multiparameter sensitivities that represent different measures of the vector $\mathbf{S}_{\mathbf{x}}^{F(s,\mathbf{x})}$. In the design based, for example, on (68.75) and (68.77), one determines the required $\sigma_{\delta F}^2$ using the reject probability [2] depending on the ratio of $\varepsilon_{\delta F}/\sigma_{\delta F}$. Here, $\varepsilon_{\delta F}$ is the tolerance of $|\delta F(j\omega, \mathbf{x})|$ and in many cases one takes $\varepsilon_{\delta F}/\sigma_{\delta F} = 2.5$, which gives the reject probability of 0.01. Then one determines the dispersions σ_i^2 so that the contributions of each term in (68.77) are equal. Finally, using the probability function that describes the distribution of δx_i within the tolerance borders one finds these borders (if, for example, the selected element has evenly distributed values $-\varepsilon_i \le \delta x_i \le \varepsilon_i$, then $\varepsilon_i = \sqrt{3}\sigma_i$; for Gaussian distribution one frequently accepts $\varepsilon_i = 2.5\sigma_i$).

The preliminary calculation of the coefficient sensitivities is useful for finding the sensitivity measures. If, for example, one calculates a multivariable statistical measure of sensitivity then one can consider that

$$F(j\omega, \mathbf{x}) = \frac{a_0 + a_1(j\omega) + \cdots + a_m(j\omega)^m}{d_0 + d_1(j\omega) + \cdots + d_n(j\omega)^n} = F(j\omega, \mathbf{a}, \mathbf{d}, \mathbf{x}) \tag{68.78}$$

where $\mathbf{a} = [a_0, a_1, \cdots, a_m]^t$ and $\mathbf{d} = [d_0, d_1, \cdots, d_n]^t$. Then, the component $\mathbf{S}_{x_i}^{F(j\omega,\mathbf{x})}$ defined above can be rewritten as

$$\begin{aligned} \mathbf{S}_{x_i}^{F(j\omega,\mathbf{x})} &= \frac{x_i}{F(j\omega,\mathbf{x})} \left[\frac{\partial \mathbf{a}^t}{\partial x_i} \nabla_a F(j\omega,\mathbf{x}) + \frac{\partial \mathbf{d}^t}{\partial x_i} \nabla_d F(j\omega,\mathbf{x}) \right] \\ &= \frac{1}{F(j\omega,\mathbf{x})} \left[\sum_{j=0}^{m} \frac{\partial F(j\omega,\mathbf{x})}{\partial a_j} \mathbf{S}_{x_i}(a_j) + \sum_{j=0}^{n} \frac{\partial F(j\omega,\mathbf{x})}{\partial d_j} \mathbf{S}_{x_i}(d_j) \right] \end{aligned} \tag{68.79}$$

where

$$\nabla_a F(j\omega, \mathbf{x}) = [\partial F(j\omega, \mathbf{x})/\partial a_0, \partial F(j\omega, \mathbf{x})/\partial a_1, \cdots, \partial F(j\omega, \mathbf{x})/\partial a_m]^t$$

and

$$\nabla_d F(j\omega, \mathbf{x}) = [\partial F(j\omega, \mathbf{x})/\partial d_0, \partial F(j\omega, \mathbf{x})/\partial d_1, \cdots, \partial F(j\omega, \mathbf{x})/\partial d_n]^t$$

For a given transfer function the components of the vectors $\nabla_a F(j\omega, \mathbf{x})$ and $\nabla_d F(j\omega, \mathbf{x})$ are independent of the form of the realization or the values of the elements and can be calculated in advance. If we now define a $k \times (m+1)$ matrix $\mathbf{C}_1$ as

$$\mathbf{C}_1 = \begin{bmatrix} \mathbf{S}_{x_1}(a_0) & \mathbf{S}_{x_1}(a_1) & \cdots & \mathbf{S}_{x_1}(a_m) \\ & & \cdots & \\ \mathbf{S}_{x_k}(a_0) & \mathbf{S}_{x_k}(a_1) & \cdots & \mathbf{S}_{x_k}(a_m) \end{bmatrix} \tag{68.80}$$

and $k \times (n+1)$ matrix $\mathbf{C}_2$ as

$$\mathbf{C}_2 = \begin{bmatrix} \mathbf{S}_{x_1}(d_0) & \mathbf{S}_{x_1}(d_1) & \cdots & \mathbf{S}_{x_1}(d_n) \\ & & \cdots & \\ \mathbf{S}_{x_k}(d_0) & \mathbf{S}_{x_k}(d_1) & \cdots & \mathbf{S}_{x_k}(d_n) \end{bmatrix} \tag{68.81}$$

then one can rewrite

$$\begin{aligned} \left[\mathbf{S}_{\mathbf{x}}^{F(j\omega,\mathbf{x})}\right]^t &= \left[\mathbf{S}_{x_1}^{F(j\omega,\mathbf{x})} \quad \mathbf{S}_{x_2}^{F(j\omega,\mathbf{x})} \quad \cdots \quad \mathbf{S}_{x_k}^{F(j\omega,\mathbf{x})}\right]^t \\ &= \mathbf{C}_1 \frac{\nabla_a F(j\omega, \mathbf{x})}{F(j\omega, \mathbf{x})} + \mathbf{C}_2 \frac{\nabla_d F(j\omega, \mathbf{x})}{F(j\omega, \mathbf{x})} \end{aligned} \tag{68.82}$$

Then the multiparameter statistical sensitivity measure can be rewritten as

$$\begin{aligned} M_Q = &\int_{\omega_1}^{\omega_2} \left[\left(\frac{\nabla_a F}{F}\right)^{*t} \mathbf{C}_1^t \mathbf{P} \mathbf{C}_1 \left(\frac{\nabla_a F}{F}\right) + \left(\frac{\nabla_d F}{F}\right)^{*t} \mathbf{C}_2^t \mathbf{P} \mathbf{C}_2 \left(\frac{\nabla_d F}{F}\right)\right] d\omega \\ &+ \int_{\omega_1}^{\omega_2} 2\,\mathrm{Re}\left[\left(\frac{\nabla_a F}{F}\right)^{*t} \mathbf{C}_1^t \mathbf{P} \mathbf{C}_2 \left(\frac{\nabla_d F}{F}\right)\right] d\omega \end{aligned} \tag{68.83}$$

and this definition of statistical multiparameter sensitivity measure may be directly applied to a given network realization. In a similar fashion the matrices of unnormalized coefficient sensitivities can be used with other multiparameter sensitivity measures.

68.8 Sensitivity Invariants

When one is talking about sensitivity invariants [7] it is assumed that for a filter function $F(s, \mathbf{x})$ there exists the relationship

$$\sum_{i=1}^{n} \mathbf{S}_{x_i}^{F(s,\mathbf{x})} = k \tag{68.84}$$

where $\mathbf{x} = [x_1, x_2, \cdots, x_n]^t$ as usual and k is a constant. These relationships are useful to check the sensitivity calculations. In the cases considered below, this constant can have one of three possible values, namely, 1, 0, and -1, and the sensitivity invariants are obtained from the homogeneity of some of the filter functions.

The function $F(s, \mathbf{x})$ is called homogeneous of order k with respect to the vector $\mathbf{x}$ if and only if it satisfies the relationship

$$F(s, \lambda\mathbf{x}) = \lambda^k F(s, \mathbf{x}) \tag{68.85}$$

where λ is an arbitrary scalar. For the homogeneous function $F(s, \mathbf{x})$ the sensitivities are related by (68.84). Indeed, if one takes the logarithm of both sides of (68.85) one obtains

$$\ln F(s, \lambda\mathbf{x}) = k \ln \lambda + \ln F(s, \mathbf{x}) \tag{68.86}$$

Taking the derivative of both sides of (68.86) with respect to λ one obtains that

$$\frac{1}{F(s, \lambda\mathbf{x})}\left[\sum_{i=1}^{n} \frac{\partial F(s, \lambda\mathbf{x})}{\partial(\lambda x_i)} x_i\right] = \frac{k}{\lambda} \tag{68.87}$$

Substituting in (68.87) $\lambda = 1$ gives (68.84).

Let the filter be a passive RLC circuit that includes r resistors, l inductors, and c capacitors, so that $r + l + c = n$ and $\mathbf{x} = [R_1, R_2, \cdots, R_r, L_1, L_2, \cdots, L_l, D_1, D_2, \cdots, D_c]$, where $D_i = 1/C_i$. One of the frequently used operations is the impedance scaling. If the scaling operation is applied to a port impedance or a transimpedance of the filter, i.e., $F(s, \mathbf{x}) = Z(s, \mathbf{x})$, then

$$Z(s, \lambda\mathbf{x}) = \lambda Z(s, \mathbf{x}) \tag{68.88}$$

Equation (68.88) is identical to (68.85), with $k = 1$. Then one can write that

$$\sum_{i=1}^{r} \mathbf{S}_{R_i}^{Z(s,\mathbf{x})} + \sum_{i=1}^{l} \mathbf{S}_{L_i}^{Z(s,\mathbf{x})} + \sum_{i=1}^{c} \mathbf{S}_{D_i}^{Z(s,\mathbf{x})} = 1 \tag{68.89}$$

Considering that $D_i = 1/C_i$ and $\mathbf{S}_{C_i}^{Z(s,\mathbf{x})} = -\mathbf{S}_{D_i}^{Z(s,\mathbf{x})}$ (see Table 68.1), this result can be rewritten as

$$\sum_{i=1}^{r} \mathbf{S}_{R_i}^{Z(s,\mathbf{x})} + \sum_{i=1}^{l} \mathbf{S}_{L_i}^{Z(s,\mathbf{x})} - \sum_{i=1}^{c} \mathbf{S}_{C_i}^{Z(s,\mathbf{x})} = 1 \tag{68.90}$$

If the same scaling operation is applied to a port admittance or a transadmittance of the filter, i.e., $F(s, \mathbf{x}) = Y(s, \mathbf{x})$, then

$$Y(s, \lambda\mathbf{x}) = \lambda^{-1} Y(s, \mathbf{x}) \tag{68.91}$$

But (68.91) is identical to (68.85), with $k = -1$. Then

$$\sum_{i=1}^{r} \mathbf{S}_{R_i}^{Y(s,\mathbf{x})} + \sum_{i=1}^{l} \mathbf{S}_{L_i}^{Y(s,\mathbf{x})} - \sum_{i=1}^{c} \mathbf{S}_{C_i}^{Y(s,\mathbf{x})} = -1 \tag{68.92}$$

Finally, the transfer functions (voltage or current) do not depend on the scaling operation, i.e., if $F(s,\mathbf{x}) = T(s,\mathbf{x})$, hence

$$T(s,\lambda\mathbf{x}) = T(s,\mathbf{x}) \tag{68.93}$$

which is identical to (68.85), with $k = 0$. Then

$$\sum_{i=1}^{r} \mathbf{S}_{R_i}^{T(s,\mathbf{x})} + \sum_{i=1}^{l} \mathbf{S}_{L_i}^{T(s,\mathbf{x})} - \sum_{i=1}^{c} \mathbf{S}_{C_i}^{T(s,\mathbf{x})} = 0 \tag{68.94}$$

Additional sensitivity invariants can be obtained using the relation $\mathbf{S}_{x_i}^{F(s,\mathbf{x})} = -\mathbf{S}_{1/x_i}^{F(s,\mathbf{x})}$ and using $G_i = 1/R_i$ and $\Gamma_i = 1/L_i$.

Another group of sensitivity invariants is obtained using the frequency scaling operation. The following relationship is held for a filter function:

$$F(s,R_i,\lambda L_i,\lambda C_i) = F(\lambda s,R_i,L_i,C_i) \tag{68.95}$$

Taking the logarithm of both parts, then differentiating both sides with respect to λ and substituting $\lambda = 1$ in both sides gives

$$\sum_{i=1}^{l} \mathbf{S}_{L_i}^{F(s,\mathbf{x})} + \sum_{i=1}^{c} \mathbf{S}_{C_i}^{F(s,\mathbf{x})} = \mathbf{S}_{s}^{F(s,\mathbf{x})} \tag{68.96}$$

Substituting $s = j\omega$ in (68.96) and dividing the real and imaginary parts one obtains

$$\mathrm{Re}\sum_{i=1}^{l} \mathbf{S}_{L_i}^{F(j\omega,\mathbf{x})} + \mathrm{Re}\sum_{i=1}^{c} \mathbf{S}_{C_i}^{F(j\omega,\mathbf{x})} = \omega\frac{\partial \ln|T(\omega)|}{\partial\omega} = -\frac{\partial a(\omega)}{\partial\omega} \tag{68.97}$$

and

$$\mathrm{Im}\sum_{i=1}^{l} \mathbf{S}_{L_i}^{F(j\omega,\mathbf{x})} + \mathrm{Im}\sum_{i=1}^{c} \mathbf{S}_{C_i}^{F(j\omega,\mathbf{x})} = \omega\frac{\partial \arg T(\omega)}{\partial\omega} \tag{68.98}$$

The results (68.97) and (68.98) show that in an *RLC* filter, when all inductors and capacitors (but not resistors!) are subjected to the same relative change, then the resulting change in the magnitude characteristic does not depend on the circuit realization and is determined by the slope of the magnitude characteristic at a chosen frequency. A similar statement is valid for the phase characteristic.

The sensitivity invariants for passive *RC* circuits can be obtained from the corresponding invariants for passive *RLC* circuits omitting the terms for sensitivities to the inductor variations. The results can be summarized the following way. For a passive *RC* circuit

$$\sum_{i=1}^{r} \mathbf{S}_{R_i}^{F(s,\mathbf{x})} - \sum_{i=1}^{c} \mathbf{S}_{C_i}^{F(s,\mathbf{x})} = k \tag{68.99}$$

where $k = 1$ if $F(s,\mathbf{x})$ is an input impedance or transimpedance function, then $k = 0$ if $F(s,\mathbf{x})$ is a voltage- or current-transfer function, and $k = -1$ if $F(s,\mathbf{x})$ is an input admittance or transconductance function. Application of the frequency scaling gives the result

$$\sum_{i=1}^{c} \mathbf{S}_{C_i}^{F(s,\mathbf{x})} = \mathbf{S}_{s}^{F(s,\mathbf{x})} \tag{68.100}$$

and combination of (68.99) and (68.100) gives

$$\sum_{i=1}^{r} \mathbf{S}_{R_i}^{F(s,\mathbf{x})} = \mathbf{S}_{s}^{F(s,\mathbf{x})} + k \tag{68.101}$$

and

$$\sum_{i=1}^{r} \mathbf{S}_{R_i}^{F(s,\mathbf{x})} + \sum_{i=1}^{c} \mathbf{S}_{C_i}^{F(s,\mathbf{x})} = 2\,\mathbf{S}_{s}^{F(s,\mathbf{x})} + k \tag{68.102}$$

Considering real and imaginary parts of (68.100)–(68.102), one can obtain the results that determine the limitations imposed on the sensitivity sums by the function $F(j\omega, \mathbf{x})$ when resistors and/or capacitors are subjected to the same relative change.

Finally, if the filter is not passive, then the vector of parameters

$$\mathbf{x} = [R_1, R_2, \cdots, R_r, L_1, L_2, \cdots, L_l, C_1, C_2, \cdots, C_c, R_{T1}, R_{T2}, \cdots, R_{Ta}, G_{T1}, G_{T2}, \cdots, G_{Tb}, A_{v1}, A_{v2}, \cdots, A_{vp}, A_{i1}, A_{i2}, \cdots, A_{iq}]$$

includes the components of transresistors R_{Tk}, transconductances G_{Tk}, voltage amplifiers A_{vk}, and current amplifiers A_{ik}. Applying the impedance scaling one can obtain the sensitivity invariant

$$\sum_{i=1}^{r} \mathbf{S}_{R_i}^{F(s,\mathbf{x})} + \sum_{i=1}^{l} \mathbf{S}_{L_i}^{F(s,\mathbf{x})} - \sum_{i=1}^{c} \mathbf{S}_{C_i}^{F(s,\mathbf{x})} + \sum_{i=1}^{a} \mathbf{S}_{R_{T_i}}^{F(s,\mathbf{x})} - \sum_{i=1}^{b} \mathbf{S}_{G_{T_i}}^{F(s,\mathbf{x})} = k \tag{68.103}$$

where $k = 1$ if $F(s, \mathbf{x})$ is an impedance function, then $k = 0$ if $F(s, \mathbf{x})$ is a transfer function, and $k = -1$ if $F(s, \mathbf{x})$ is an admittance function. The frequency scaling will give the same result as (68.96).

The pole (or zero) sensitivities are also related by some invariant relationships. Indeed, the impedance scaling provides the result

$$p_k\left(\lambda R_i, \lambda L_i, \frac{C_i}{\lambda}\right) = p_k(R_i, L_i, C_i) \tag{68.104}$$

Taking the derivative of both sides of (68.104) with respect to λ and substituting $\lambda = 1$ one obtains that for an arbitrary *RLC* circuit

$$\sum_{i=1}^{r} \mathbf{S}_{R_i}(p_k) + \sum_{i=1}^{l} \mathbf{S}_{L_i}(p_k) - \sum_{i=1}^{c} \mathbf{S}_{C_i}(p_k) = 0 \tag{68.105}$$

This is the relationship between semirelative sensitivities. If $p_k \neq 0$ one can divide both sides of (68.105) by p_k and obtains similar invariants for the relative sensitivities. The frequency scaling gives

$$p_k\left(R_i, \frac{L_i}{\lambda}, \frac{C_i}{\lambda}\right) = \lambda p_k(R_i, L_i, C_i) \tag{68.106}$$

and from (68.106) one obtains, for relative sensitivities only, that

$$\sum_{i=1}^{l} \mathbf{S}_{L_i}^{p_k} + \sum_{i=1}^{c} \mathbf{S}_{C_i}^{p_k} = -1 \tag{68.107}$$

The pole sensitivity invariants for passive *RC* circuits are obtained from (68.106) and (68.107), omitting the terms corresponding to the inductor sensitivities.

68.9 Sensitivity Bounds

For some classes of filters, the worst-case magnitude sensitivity index may be shown to have a lower bound [8]. Such a bound, for example, exists for filters whose passive elements are limited to resistors, capacitors, and ideal transformers, and whose active elements are limited to gyrators characterized by two gyration resistances (realized as a series connection of transresistance amplifiers and considered as different for sensitivity calculations), CCCSs, VCVSs, VCCSs, and CCVSs. Using the sensitivity invariants it is easy to show that for such a class of networks for any dimensionless transfer function $T(s)$

$$\sum_{i=1}^{n} \mathbf{S}_{x_i}^{T(s)} = 2\,\mathbf{S}_{s}^{T(s)} \tag{68.108}$$

where the x_i are taken to include only the passive elements of resistors and capacitors and the active elements of CCVS's and gyrators (if the gyrators are realized as parallel connection of transconductance amplifiers the corresponding terms should be taken with the negative sign). Substituting $s = j\omega$ in (68.108) and equating real parts one obtains that

$$\sum_{i=1}^{n} \mathbf{S}_{x_i}^{|T(j\omega)|} = 2\,\mathbf{S}_{\omega}^{|T(j\omega)|} \tag{68.109}$$

Applying (68.54) to the above one obtains for the worst-case magnitude sensitivity index

$$\nu(T) = \sum_i \left|\mathbf{S}_{x_i}^{|T(j\omega)|}\right| \geq \left|\sum_i \mathbf{S}_{x_i}^{|T(j\omega)|}\right| = \left|2\,\mathbf{S}_{\omega}^{|T(j\omega)|}\right| \tag{68.110}$$

Taking the first and last members of this expression one may define a lower bound $\mathrm{LB}\nu(T)$ of the worst-case magnitude sensitivity index as

$$\mathrm{LB}\nu(T) = \left|2\,\mathbf{S}_{\omega}^{|T(j\omega)|}\right| \tag{68.111}$$

This lower bound is a function only of the transfer function $T(s)$ and is independent on the particular synthesis technique (as soon as the above mentioned restrictions are satisfied) used to realize this transfer function. A similar lower bound may be derived [taking the imaginary parts of (68.108)] for worst-case phase sensitivity. But it is impossible to find the design path by which one will arrive at the circuit realizing this minimal bound.

68.10 Remarks on the Sensitivity Applications

The components of active *RC* filters have inaccuracies and parasitic components that distort the filter characteristics. The most important imperfections are the following.

1. The values of the resistors and capacitors and the values of transconductances (the gyrators can be considered usually as parallel connection of two transconductance amplifiers) are different from their nominal values. The evaluation of these effects is done using the worst-case or (more frequently) quadratic multiparameter sensitivity index and multiparameter sensitivity measures and the tolerances are chosen so that the contributions of the passive elements' variations in the sensitivity measure are equal.

2. The operational amplifiers have finite gain and this gain is frequency dependent. The effect of finite gain is evaluated using a semirelative sensitivity of the filter function with respect to variation of $1/A$, where A is the operational amplifier gain. This semirelative sensitivity is $-\mathbf{S}^{F(s)}(1/A) = -\partial F(s)/[F(s)\partial(1/A)] = [A^2/F(s)][\partial F(s)/\partial A] = A\,\mathbf{S}_A^{F(s)}$, which is called the gain–sensitivity product. In many cases $\mathbf{S}_A^{F(s)} \to 0$ when $A \to \infty$, whereas $\mathbf{S}_{1/A}[F(s)]$ has a limit that is different from zero. The frequency dependence is difficult to take into consideration [1]. Only in case of cascade realization, as shown below, one can evaluate the effect of this frequency dependence using the sensitivity of Q-factor.
3. Temperature dependence and aging of the passive elements and operational amplifiers can be determined. The influence of temperature, aging and other environmental factors on the value of the elements can be determined by a dependence of the probability distributions for these parameters. For example, if θ is temperature, one has to estimate $\mu_x(\theta)$ and $\sigma_x(\theta)$ before calculation of the sensitivity measures. Normally, at the nominal temperature θ_0 the average value $\mu_x(\theta_0) = 0$, and $\sigma_x(\theta_0)$ depends on the nominal precision of the elements. When the temperature changes, $\mu_x(\theta)$ increases or decreases depending on the temperature coefficients of the elements, whereas $\sigma_x(\theta)$ usually increases for any temperature variation.

8.11 Sensitivity Computations Using the Adjoint Network

The determination of the sensitivities defined in the previous sections may pose difficult computational problems. Finding the network function with the elements expressed in literal form is usually tedious and error prone, and the difficulty of such a determination increases rapidly with the number of elements. Calculating the partial derivatives, which is the most important part of the sensitivity computation, provides additional tedium and increases the possibility of error still more. Thus, in general, it is advantageous to use digital computer methods to compute sensitivities. The most obvious method for doing this is to use one of the many available computer-aided design programs (for example, SPICE) to make an analysis of the network with nominal element values, then repeat the analysis after having perturbed the value of one of the elements. This is not a desirable procedure, since it requires a large number of analyses. It can be justified if the network has some *a priori* known critical elements for which the analyses should be done.

The crucial part of sensitivity computation is, as was mentioned above, calculation of network function derivatives with respect to element variations. To simplify this part of the calculation the concept of adjoint network [9] is used. This method requires only two analyses to provide all the sensitivities of a given network immittance function.

If $\mathbf{N}$ and $\hat{\mathbf{N}}$ are linear time invariant networks, then they are said to be adjoint (to each other) if the following hold. The two networks have the same topology and ordering of branches; thus their incidence matrices [10] are equal, namely $\mathbf{A} = \hat{\mathbf{A}}$. If excitation with the unit current (unit voltage) at an arbitrary port j (port l) of network $\mathbf{N}$ yields a voltage (current) at an arbitrary port k (port m) of $\mathbf{N}$, excitation with the unit current (unit voltage) at port k (port m) of network $\hat{\mathbf{N}}$ will yield the same voltage (current) as the above in port j (port l) of $\hat{\mathbf{N}}$ (see Fig. 68.1).

Figure 68.1 also shows how the adjoint network should be constructed, as follows. (a) All resistance, capacitive and inductance branches and transformers in $\mathbf{N}$ are associated, respectively, with resistance, capacitance and inductance branches and transformers in $\hat{\mathbf{N}}$. (b) All gyrators in $\mathbf{N}$ with gyration resistance r become gyrators in $\hat{\mathbf{N}}$ with gyration resistance $-r$. (c) All VCVSs in $\mathbf{N}$ become CCCSs in $\hat{\mathbf{N}}$ with controlling and controlled branches reversing roles, and with the voltage amplification factor A_v becoming the current amplification factor $-A_i$. (d) All CCCSs in $\mathbf{N}$ become VCVSs in $\hat{\mathbf{N}}$ with controlling and controlled branches reversing roles, and with the

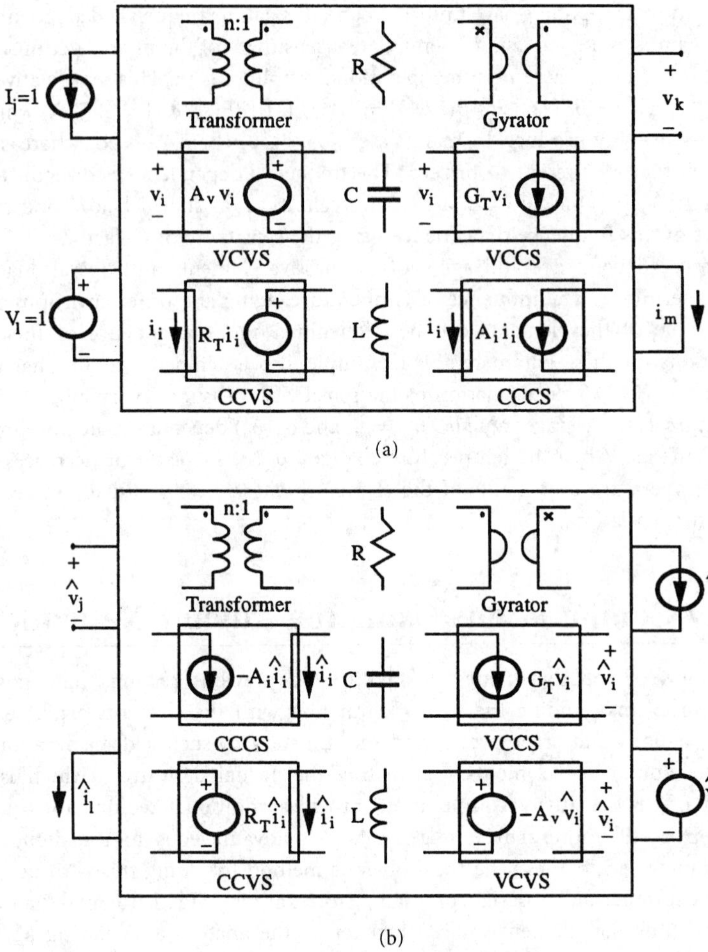

FIGURE 68.1 The components of a network (a) and its adjoint (b).

current amplification factor A_i becoming the voltage amplification factor $-A_v$. (e) All VCCSs and CCVSs have their controlling and controlled branches in **N** reversed in $\hat{\mathbf{N}}$.

Thus, the Tellegen's theorem [9] applies to the branch voltage and current variables of these two networks. If we let **V** be the vector of branch voltages and **I** be the vector of branch currents (using capital letters implies that the quantities are functions of the complex variable s) then

$$\mathbf{V}^t\hat{\mathbf{I}} = \hat{\mathbf{I}}^t\mathbf{V} = \hat{\mathbf{V}}^t\mathbf{I} = \hat{\mathbf{I}}^t\mathbf{V} = 0 \tag{68.112}$$

If in both circuits all independent sources have been removed to form n external ports (as illustrated in Fig. 68.2), then one can divide the variables in both circuits in two groups so that

$$\begin{aligned} \mathbf{I}^t &= [\mathbf{I}_p^t \mathbf{I}_b^t] & \mathbf{V}^t &= [\mathbf{V}_p^t \mathbf{V}_b^t] \\ \hat{\mathbf{I}}^t &= [\hat{\mathbf{I}}_p^t \hat{\mathbf{I}}_b^t] & \hat{\mathbf{V}}^t &= [\hat{\mathbf{V}}_p^t \hat{\mathbf{V}}_b^t] \end{aligned} \tag{68.113}$$

The first are the vectors of port variables $\mathbf{V}_p$ and $\mathbf{I}_p$ in **N** and $\hat{\mathbf{V}}_p$ and $\hat{\mathbf{I}}_p$, correspondingly, in $\hat{\mathbf{N}}$. These variables will define n-port open-circuit impedance matrices $\mathbf{Z}_{oc}$ and $\hat{\mathbf{Z}}_{oc}$ or n-port

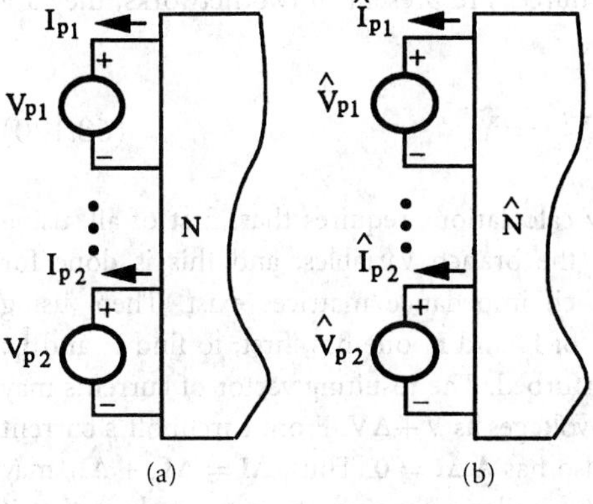

FIGURE 68.2 Separation of port variables in the network (a) and its adjoint (b).

short-circuit admittance matrices $\mathbf{Y}_{sc}$ and $\hat{\mathbf{Y}}_{sc}$ via the relationships

$$\begin{aligned} \mathbf{V}_p &= -\mathbf{Z}_{oc}\mathbf{I}_p & \hat{\mathbf{V}}_p &= -\hat{\mathbf{Z}}_{oc}\hat{\mathbf{I}}_p \\ \mathbf{I}_p &= -\mathbf{Y}_{sc}\mathbf{V}_p & \hat{\mathbf{I}}_p &= -\hat{\mathbf{Y}}_{sc}\mathbf{V}_p \end{aligned} \tag{68.114}$$

Then the rest of the variables will be nonport variables (including the variables of dependent source branches) $\mathbf{V}_b$ and $\mathbf{I}_b$ for **N** and $\hat{\mathbf{V}}_b$ and $\hat{\mathbf{I}}_b$ for $\hat{\mathbf{N}}$. These variables may define branch impedance matrices $\mathbf{Z}_b$ and $\hat{\mathbf{Z}}_b$ and branch admittance matrices $\mathbf{Y}_b$ and $\hat{\mathbf{Y}}_b$ by the relationships

$$\begin{aligned} \mathbf{V}_b &= \mathbf{Z}_b\mathbf{I}_b & \hat{\mathbf{V}}_b &= \hat{\mathbf{Z}}_b\hat{\mathbf{I}}_p \\ \mathbf{I}_b &= \mathbf{Y}_b\mathbf{V}_b & \hat{\mathbf{I}}_b &= \hat{\mathbf{Y}}_b\mathbf{V}_b \end{aligned} \tag{68.115}$$

If the branch impedance and branch admittance matrices do not exist, a hybrid matrix may be used to relate the branch variables. For **N** this may be put in the form

$$\begin{bmatrix} \mathbf{V}_{b1} \\ \mathbf{I}_{b2} \end{bmatrix} = \begin{bmatrix} \mathbf{H}_{11} & \mathbf{H}_{12} \\ \mathbf{H}_{21} & \mathbf{H}_{22} \end{bmatrix} \begin{bmatrix} \mathbf{I}_{b1} \\ \mathbf{V}_{b2} \end{bmatrix} \tag{68.116}$$

Similarly, for $\hat{\mathbf{N}}$ one may write

$$\begin{bmatrix} \hat{\mathbf{V}}_{b1} \\ \hat{\mathbf{I}}_{b2} \end{bmatrix} = \begin{bmatrix} \hat{\mathbf{H}}_{11} & \hat{\mathbf{H}}_{12} \\ \hat{\mathbf{H}}_{21} & \hat{\mathbf{H}}_{22} \end{bmatrix} \begin{bmatrix} \hat{\mathbf{I}}_{b1} \\ \hat{\mathbf{V}}_{b2} \end{bmatrix} \tag{68.117}$$

For the adjoint networks the branch impedance matrices and branch admittance matrices (if they exist) are transposed, namely

$$\mathbf{Z}_b^t = \hat{\mathbf{Z}}_b \qquad \mathbf{Y}_b^t = \hat{\mathbf{Y}}_b \tag{68.118}$$

and, if a hybrid representation is used, the matrices are connected by the relationship

$$\begin{bmatrix} \mathbf{H}_{11}^t & -\mathbf{H}_{21}^t \\ -\mathbf{H}_{12}^t & \mathbf{H}_{22}^t \end{bmatrix} = \begin{bmatrix} \hat{\mathbf{H}}_{11} & \hat{\mathbf{H}}_{12} \\ \hat{\mathbf{H}}_{21} & \hat{\mathbf{H}}_{22} \end{bmatrix} \tag{68.119}$$

As a result of these relationships, if no controlled sources are present in two networks, they are identical. In the general case it may be shown that

$$\mathbf{Z}^t_{oc} = \hat{\mathbf{Z}}_{oc} \qquad \mathbf{Y}^t_{sc} = \hat{\mathbf{Y}}_{sc} \tag{68.120}$$

The application of adjoint circuits for sensitivity calculations requires that, first of all, using the port variables as independent ones one finds the branch variables, and this is done for both circuits. Assume, for example, that the branch impedance matrices exist. Then, using unity-valued excitation currents as the components of $\mathbf{I}_p$ and $\hat{\mathbf{I}}_p$ one has, first, to find $\mathbf{I}_b$ and $\hat{\mathbf{I}}_b$. Now, in the original network let the elements be perturbed. The resulting vector of currents may thus be written as $\mathbf{I}+\Delta\mathbf{I}$ and the resulting vector of voltages as $\mathbf{V}+\Delta\mathbf{V}$. From Kirchhoff's current law we have $\mathbf{A}(\mathbf{I}+\Delta\mathbf{I}) = \mathbf{0}$ and since $\mathbf{AI} = \mathbf{0}$ one also has $\mathbf{A}\,\Delta\mathbf{I} = \mathbf{0}$. Thus, $\Delta\mathbf{I} = \Delta\mathbf{I}_p + \Delta\mathbf{I}_b$ may be substituted in any of the relations in (68.112). By similar reasoning one can conclude that it is possible as well to substitute the perturbation vector $\Delta\mathbf{V} = \Delta\mathbf{V}_p + \Delta\mathbf{V}_b$ instead of $\mathbf{V}$ in these relations. Making these substitutions one obtains

$$\begin{aligned} \hat{\mathbf{V}}^t_p\,\Delta\mathbf{I}_p + \hat{\mathbf{V}}^t_b\,\Delta\mathbf{I}_b &= 0 \\ \mathbf{I}^t_p\,\Delta\mathbf{V}_p + \hat{\mathbf{I}}^t_b\,\Delta\mathbf{V}_b &= 0 \end{aligned} \tag{68.121}$$

and subtracting these two equations one has

$$\hat{\mathbf{V}}^t_p\,\Delta\mathbf{I}_p - \hat{\mathbf{I}}^t_p\,\Delta\mathbf{V}_p + \hat{\mathbf{V}}^t_b\,\Delta\mathbf{I}_b - \hat{\mathbf{I}}^t_b\,\Delta\mathbf{V}_b = 0 \tag{68.122}$$

To a first-order approximation we have

$$\begin{aligned} \Delta\mathbf{V}_p &= -\Delta(\mathbf{Z}_{oc}\mathbf{I}_p) \approx -\Delta\mathbf{Z}_{oc}\mathbf{I}_p - \mathbf{Z}_{oc}\,\Delta\mathbf{I}_p \\ \Delta\mathbf{V}_b &= \Delta(\mathbf{Z}_b\mathbf{I}_b) \approx \Delta\mathbf{Z}_b\mathbf{I}_b + \mathbf{Z}_b\,\Delta\mathbf{I}_b \end{aligned} \tag{68.123}$$

Substituting (68.123) in (68.122) and taking into consideration that $\hat{\mathbf{V}}^t_p = -\hat{\mathbf{I}}^t_p\hat{\mathbf{Z}}^t_{oc} = -\hat{\mathbf{I}}^t_p\mathbf{Z}_{oc}$ and that $\hat{\mathbf{V}}^t_b = \hat{\mathbf{I}}^t_b\mathbf{Z}_b$, one can simplify the result (68.122) to

$$\hat{\mathbf{I}}^t_p\,\Delta\mathbf{Z}_{oc}\mathbf{I}_p = \hat{\mathbf{I}}^t_b\,\Delta\mathbf{Z}_b\mathbf{I}_b \tag{68.124}$$

Equation (68.124) clearly shows that if all currents in the original network and its adjoint are known, one can easily calculate the absolute sensitivities $\mathbf{S}(\mathbf{Z}_{ij}, \mathbf{Z}_b) \approx \Delta\mathbf{Z}_{ij}/\Delta\mathbf{Z}_b$, which can be used for calculation of the corresponding relative sensitivities. Here, $\mathbf{Z}_{ij}$ is an element of the n-port open circuit impedance matrix. Then, if necessary, these sensitivities can be used for evaluation of the transfer function sensitivity or the sensitivities of other functions derived via the n-port open circuit impedance matrix.

Usually, the transfer function calculation can be easily reduced to the calculation of a particular $\mathbf{Z}_{ij}$. In this case one can choose $I_j = 1$, $I_k = 0$ (for all $k \neq j$) as an excitation in the original network and $\hat{I}_i$, $I_k = 0$ (for all $k \neq i$) as an excitation in the adjoint network. Then, (68.124) becomes

$$\Delta\mathbf{Z}_{ij} = \hat{\mathbf{I}}^t_b\,\Delta\mathbf{Z}_b\mathbf{I}_b \tag{68.125}$$

where $\mathbf{I}_b$ and $\hat{\mathbf{I}}_b$ are the branch currents in $\mathbf{N}$ and $\hat{\mathbf{N}}$ corresponding to the indicated excitation. The relations for other types of matrices are obtained in the same manner.

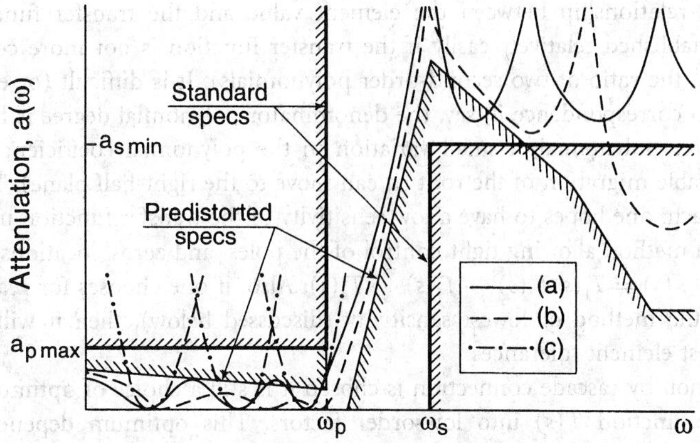

FIGURE 68.3 Attenuation requirements, their predistortion and attenuation of different realizations: (a) attenuation of a standard circuit of higher order; (b) attenuation of the nonstandard circuit; (c) passband attenuation in the stages of cascade realization.

.12 General Methods of Reducing Sensitivity

It is very desirable from the start of the realization procedure to concentrate on circuits that give lower sensitivity in comparison to other circuits. The practice of active filter realization allows formulation of some general suggestions ensuring that filter realizations will have low sensitivities to component variations.

It is possible to transfer the problem of lower sensitivity at the approximation stage, i.e., before any realization. It is obvious that a low-order transfer function $T(s)$ that just satisfies the specifications will require tighter tolerances in comparison to a higher order transfer function that easily satisfies the specifications. Figure 68.3 shows an example of such an approach for a low-pass filter. Hence, increasing the order of approximation and introducing a redundancy one achieves a set of wider element tolerances.

Usually the most critical region where it is difficult to satisfy the specifications is the edge of the passband. Two approaches can be used to find the function that will have less sensitivity in this frequency region. One way is to introduce a predistortion in the transfer function specifications. This is also shown in Fig. 68.3. The transfer function should satisfy the predistorted (tapered) specifications. It can be obtained directly if the numerical packages solving the approximation problem are available. One can also take a standard table higher order transfer function satisfying the modified specifications and then modify it to more uniformly use the tapered specifications (this allows the increase of the component tolerances even more). Another way [11] is to preserve the initial transfer function specifications and to use transfer functions with a limited value of the maximum Q of the transfer function poles. In [11] one can find such transfer functions corresponding to a Cauer approximation. The nonstandard LC circuits corresponding to these approaches cannot be tabulated and simulated; hence, neither of them is widely used. In addition, they imply the cascaded (building-block) realization that intrinsically has worse sensitivity than the realizations using simulation of doubly terminated lossless matched filters.

.13 Cascaded Realization of Active Filters

The cascaded (building-block) realization is based on the assumption that the transfer function will have low sensitivity if the realization provides tight control of the transfer function poles and zeros. The relationship between the element value and the transfer function poles and zeros can be established relatively easily if the transfer function is not more complicated than biquadratic (i.e., the ratio of two second-order polynomials). It is difficult (or even impossible) to establish such correspondence if, say, the denominator polynomial degree is higher than two. For a high degree polynomial, a small variation in the polynomial coefficient can result in a large or undesirable migration of the root (it can move to the right-half plane). This justifies the cascaded approach: one hopes to have a low sensitivity of the transfer function under realization if one chooses a method allowing tight control of the poles' and zeros' locations; hence, cascade realization with $T(s) = T_1(s)T_2(s)\cdots T_i(s)\cdots T_k(s)$. Also, if one chooses for realization of each function $T_i(s)$ the method of lowest sensitivity (discussed below), then it will be possible to obtain the largest element tolerances.

If the realization by cascade connection is chosen it is still a choice of optimum factorization of the transfer function $T(s)$ into low-order factors. This optimum depends on the filter application, the chosen method of the factor realization, and the transfer function itself. It is recommended [1] that, for realization of lower sensitivity, the poles and zeros in the partial functions $T_i(s)$ are located as far apart as possible. This statement is not always true and such a choice of poles and zeros in $T_i(s)$ is in contradiction with the requirement of high dynamic range of the stage. In general, CAD methods should be used.

The methods that are most popular for the realization of the partial transfer functions are mostly limited by the filters providing the output voltage at the output of an operational amplifier. The state–space realizations satisfy this requirement and provide the direct realizations of the polynomial coefficients. It is not occasionally that such an approach is used by a series of manufacturers. This is not the best method from the sensitivity point of view; the methods of realization using gyrators usually give better results [3]. But the cascade realization of gyrator filters require buffers between the blocks, and it is better to use this approach if the filter is not realized in cascade form. Other realization methods [1] are also occasionally used, mostly because of their simplicity.

If the transfer function is realized in a cascade form and $T(s) = T_1(s)T_2(s)\cdots T_i(s)\cdots T_k(s)$, the element x is located only in one stage. If this is the stage realizing $T_i(s)$ then

$$\mathbf{S}_x^{T(s)} = \mathbf{S}_x^{T_i(s)} \tag{68.126}$$

Assume that this $T_i(s)$ has the form

$$T_i(s) = K\frac{s^2 + (\omega_z/Q_z)s + \omega_z^2}{s^2 + (\omega_p/Q_p)s + \omega_p^2} \tag{68.127}$$

Then one can write

$$\mathbf{S}_x^{T_i(s)} = \mathbf{S}_K^{T_i(s)}\mathbf{S}_x^K + \mathbf{S}_{\omega_z}^{T_i(s)}\mathbf{S}_x^{\omega_z} + \mathbf{S}_{1/Q_z}^{T_i(s)}\mathbf{S}_x^{1/Q_z} + \mathbf{S}_{\omega_p}^{T_i(s)}\mathbf{S}_x^{\omega_p} + \mathbf{S}_{1/Q_p}^{T_i(s)}\mathbf{S}_x^{1/Q_p} \tag{68.128}$$

The second multiplier in each term of this sum depends on the stage realization method. In the first term $\mathbf{S}_K^{T_i(s)} = 1$, the first multipliers in other terms depend on the Q-factors of zeros and poles. It is enough to consider the influence of the terms related to the poles. One can notice that $\mathbf{S}_{1/Q_p}^{T_i(s)}\mathbf{S}_x^{1/Q_p} = \mathbf{S}^{T_i(s)}[T_i(s), 1/Q_p]\,\mathbf{S}_x(1/Q_p, x)$, then, calculating $\mathbf{S}^{T_i(s)}[T_i(s), 1/Q_p]$ (the calculation of the semirelative sensitivity is done for convenience of graphic representation) and $\mathbf{S}_{\omega_p}^{T_i(s)}$ one obtains

$$\mathbf{S}^{T_i(s)}[T_i(s), 1/Q_p] = \frac{1}{T_i(s)} \frac{\partial T_i(s)}{\partial(1/Q_p)} = -\frac{1}{[(s/\omega_p) + (\omega_p/s)] + (1/Q_p)} \tag{68.129}$$

and

$$\mathbf{S}_{\omega_p}^{T_i(s)} = \frac{\omega_p}{T_i(s)} \frac{\partial T_i(s)}{\partial \omega_p} = -\frac{(1/Q_p) + 2(\omega_p/s)}{[(s/\omega_p) + (\omega_p/s)] + (1/Q_p)} \tag{68.130}$$

Introducing the normalized frequency $\Omega = (\omega/\omega_p - \omega_p/\omega)$, one can find that

$$\text{Re}\; \mathbf{S}^{T_i(s)}[T_i(s), 1/Q_p] = -\frac{1/Q_p}{\Omega^2 + (1/Q_p)^2} \tag{68.131}$$

$$\text{Im}\; \mathbf{S}^{T_i(s)}[T_i(s), 1/Q_p] = \frac{\Omega}{\Omega^2 + (1/Q_p)^2} \tag{68.132}$$

$$\text{Re}\; \mathbf{S}_{\omega_p}^{T_i(s)} = \frac{\Omega\sqrt{\Omega^2 + 4} - \Omega^2 - (1/Q_p)^2}{\Omega^2 + (1/Q_p)^2} \tag{68.133}$$

$$\text{Im}\; \mathbf{S}_{\omega_p}^{T_i(s)} = \frac{(1/Q_p)\sqrt{\Omega^2 + 4}}{\Omega^2 + (1/Q_p)^2} \tag{68.134}$$

In Fig. 68.4 are shown the graphs of these four functions. They allow the following conclusions [4]. The functions reach high values in the vicinity of $\Omega = 0$, i.e., when $\omega \approx \omega_p$. This means that in the filter passband, especially if the poles and zeros are sufficiently divided (which is the condition of optimal cascading) one can neglect the contribution of zeros in modification of the transfer function $T_i(s)$. When Q_p becomes higher this vicinity of $\omega \approx \omega_p$ with a rapid change of the sensitivity functions becomes relatively smaller in the case of Re $\mathbf{S}^{T_i(s)}[T_i(s), 1/Q_p]$ [Fig. 68.4(a)] and Im $\mathbf{S}_{\omega_p}^{T_i(s)}$ [Fig. 68.4(d)] and not so small in the case of Im $\mathbf{S}^{T_i(s)}[T_i(s), 1/Q_p]$ [Fig. 68.4(b)] and Re $\mathbf{S}_{\omega_p}^{T_i(s)}$ [Fig. 68.4(c)]. Normally, the second multipliers in the terms of the sum (68.128) are all real; this means that the function Re $\mathbf{S}_{\omega_p}^{T_i(s)}$ is the most important one in estimation of the sensitivity to variations of passive elements. In many realization methods [1], [3] one obtains that $\omega_p \propto (R_1R_2C_1C_2)^{-1/2}$, which implies that $\mathbf{S}_{R_1}^{\omega_p} = \mathbf{S}_{R_2}^{\omega_p} = \mathbf{S}_{C_1}^{\omega_p} = \mathbf{S}_{C_2}^{\omega_p} = -1/2$. Thus, finally, the function Re $\mathbf{S}_{\omega_p}^{T_i(s)}$ (which can be called the main passive sensitivity term) will determine the maximum realizable Q_p for given tolerances of passive elements (or for the elements that are simulated as passive elements).

If a stage is realized using, for example a state–space approach, it includes operational amplifiers. The stage is usually designed assuming ideal operational amplifiers, then the realization errors are analyzed considering that the operational amplifiers can be described by a model

$$A(s) = \frac{A_0}{1 + (s/\omega_0)} = \frac{\text{GBW}}{s + \omega_0} \tag{68.135}$$

where A_0 is the dc gain, ω_0 is the amplifier bandwidth, and $\text{GBW} = A_0\omega_0$ is the gain–bandwidth product. If the stage transfer function is derived anew, with amplifiers described by the model (68.135), then $T_i(s)$ will no longer be a biquadratic. It will be a ratio of two higher degree polynomials, and the error analysis becomes very complicated [1]. To do an approximate analysis one can pretend that the amplifier gain is simply a real constant A. Then, the transfer function

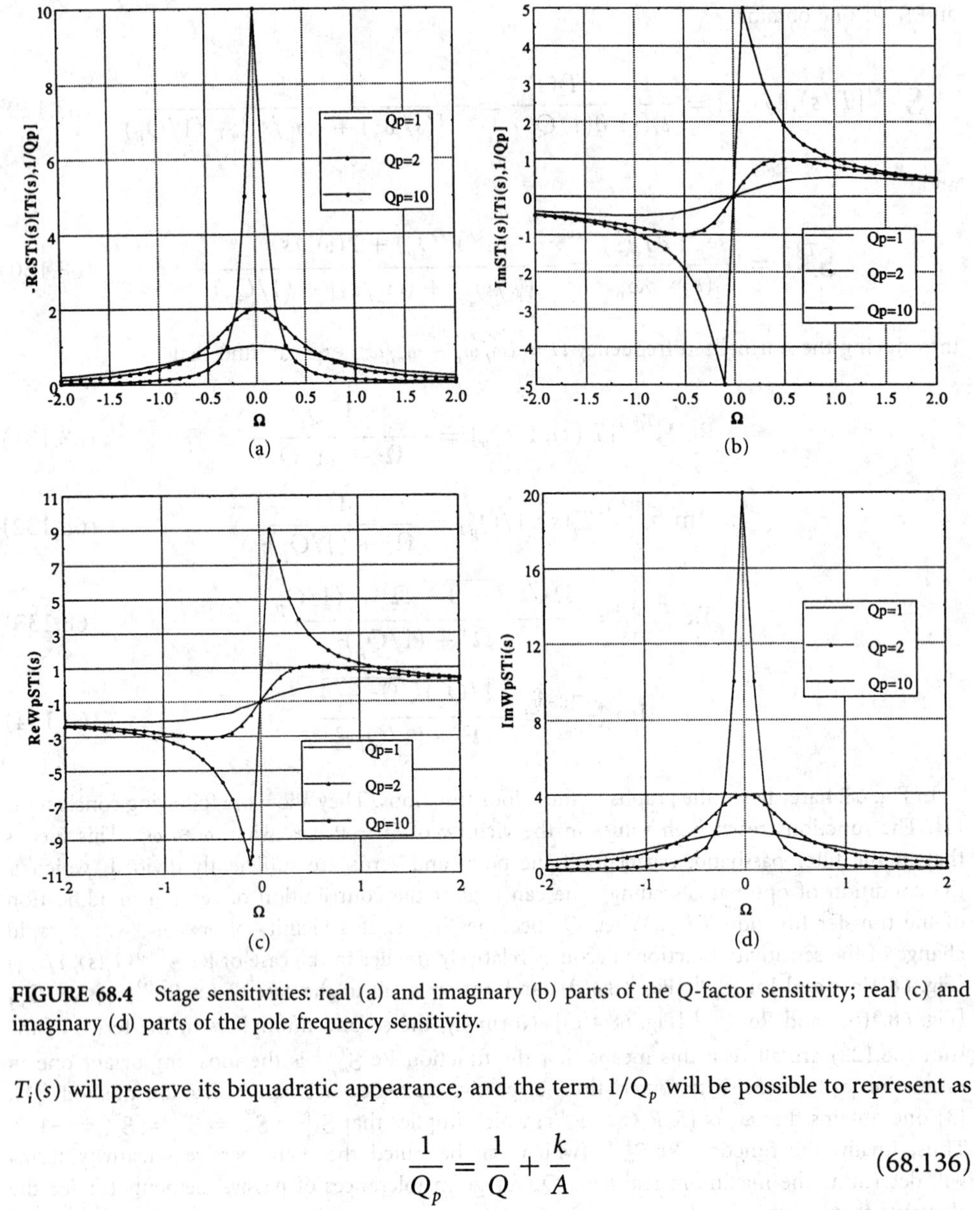

FIGURE 68.4 Stage sensitivities: real (a) and imaginary (b) parts of the Q-factor sensitivity; real (c) and imaginary (d) parts of the pole frequency sensitivity.

$T_i(s)$ will preserve its biquadratic appearance, and the term $1/Q_p$ will be possible to represent as

$$\frac{1}{Q_p} = \frac{1}{Q} + \frac{k}{A} \tag{68.136}$$

The first term in (68.136) is determined by the ratio of passive elements, the second term (k is the design constant) can be considered as an *absolute* change $\Delta(1/Q_p)$, which on the $j\omega$ axis becomes

$$\Delta\left(\frac{1}{Q_p}\right) \approx \frac{k}{A(j\omega)} = \frac{1}{A_0} + j\frac{k\omega}{\text{GBW}} \approx j\frac{k\omega}{\text{GBW}} \tag{68.137}$$

Then, in calculation of $\delta|T_i| = \text{Re}\,\mathbf{S}_K^{T_i(s)}$ the function $\text{Im}\,\mathbf{S}_{\omega_p}^{T_i(s)}$ becomes important (it can be called the main active sensitivity term) and the product $[\text{Im}\,\mathbf{S}_{\omega_p}^{T_i(s)}(k\omega/\text{GBW})]$ allows evaluation of the limitations on the Q-factor caused by the stage operational amplifiers.

The relationships given below are useful for any arbitrary realization, but pertain more to the cascade realization, where one can better control the pole's location when the circuit parameters are changing. If the filter transfer function is represented as

$$T(s) = \frac{a_m \prod_{i=1}^{m}(s - z_i)}{d_n \prod_{i=1}^{n}(s - p_i)} \tag{68.138}$$

then taking the logarithm of (68.138) and its derivative gives

$$\frac{dT(s)}{T(s)} = \frac{d(a_m/d_n)}{a_m/d_n} - \sum_{i=1}^{m} \frac{dz_i}{s - z_i} + \sum_{i=1}^{n} \frac{dp_i}{s - p_i} \tag{68.139}$$

Multiplying both sides of (68.139) by x and expressing the differentials via partial derivatives one obtains

$$\mathbf{S}_x^{T(s)} = \mathbf{S}_x^{a_m/d_n} - \sum_{i=1}^{m} \frac{\mathbf{S}_x(z_i)}{s - z_i} + \sum_{i=1}^{n} \frac{\mathbf{S}_x(p_i)}{s - p_i} \tag{68.140}$$

In the vicinity of the pole $p_i = -\sigma_i + j\omega_i$, the sensitivity is determined by the term $\mathbf{S}_x(p_i)/(s - p_i)$. Besides $\mathbf{S}_x(p_i) = -\mathbf{S}_x(\sigma_i) + j\,\mathbf{S}_x(\omega_i)$ and on the $j\omega$ axis in this region one has

$$\frac{\mathbf{S}_x(p_i)}{j\omega - p_i} = \frac{-\mathbf{S}_x(\sigma_i)\sigma_i + \mathbf{S}_x(\omega_i)(\omega - \omega_i)}{\sigma_i^2 + (\omega - \omega_i)^2} + j\,\frac{\mathbf{S}_x(\omega_i)\sigma_i + \mathbf{S}_x(\sigma_i)(\omega - \omega_i)}{\sigma_i^2 + (\omega - \omega_i)^2} \tag{68.141}$$

Hence, when $\omega \approx \omega_i$

$$\mathbf{S}_x^{|T(j\omega)|} \approx \frac{-\mathbf{S}_x(\sigma_i)\sigma_i + \mathbf{S}_x(\omega_i)(\omega - \omega_i)}{\sigma_i^2 + (\omega - \omega_i)^2} \tag{68.142}$$

and

$$\mathbf{S}_x[\arg T(j\omega)] \approx \frac{\mathbf{S}_x(\omega_i)\sigma_i + \mathbf{S}_x(\sigma_i)(\omega - \omega_i)}{\sigma_i^2 + (\omega + \omega_i)^2} \tag{68.143}$$

Usually (68.142) and (68.143) are considered at the point $\omega = \omega_i$, where $\mathbf{S}_x^{|T(j\omega)|} = -\mathbf{S}_x(\sigma_i)/\sigma_i$ and $\mathbf{S}_x[\arg T(j\omega)] = \mathbf{S}_x(\omega_i)/\sigma_i$. The frequent conclusion that follows is that the pole's movement toward the $j\omega$ axis is more dangerous (it introduces transfer function magnitude change) than the movement parallel to the $j\omega$ axis. But it is not difficult to see that in the immediate vicinity of this point, at $\omega = \omega_i \pm \sigma_i$, one has $\mathbf{S}_x^{|T(j\omega)|} = [-\mathbf{S}_x(\sigma_i) \pm \mathbf{S}_x(\omega_i)]/(2\sigma_i)$ and $\mathbf{S}_x[\arg T(j\omega)] = [\mathbf{S}_x(\omega_i) \pm \mathbf{S}_x(\sigma_i)]/(2\sigma_i)$; i.e., one has to reduce both components of the pole movement. If care is taken to get $\mathbf{S}_x(\sigma_i) = 0$, then, indeed, $\mathbf{S}_x^{|T(j\omega)|} = 0$ at $\omega = \omega_i$, but at $\omega = \omega_i + \sigma_i$ (closer to the edge of the passband) $\mathbf{S}_x^{|T(j\omega)|} = [\mathbf{S}_x(\omega_i)]/\sigma_i$ and this can result in an essential $\delta|T(j\omega)|$.

Finally, some additional relationships between different sensitivities can be obtained from the definition of $Q_p = \omega_p/(2\sigma_i) = [\sqrt{(\sigma_i^2 + \omega_i^2)}]/(2\sigma_i) \approx \omega_i/(2\sigma_i)$. One can find that

$$\mathbf{S}_x(\omega_p) = \frac{\mathbf{S}_x(\sigma_i)}{2Q_p} + \sqrt{1 - (1/4Q^2)}\;\mathbf{S}_x(\omega_i) \tag{68.144}$$

and

$$\mathbf{S}_x(Q_p) = \frac{\sqrt{4Q_p^2 - 1}\,\mathbf{S}_x(\omega_i) - (4Q_p^2 - 1)\,\mathbf{S}_x(\sigma_i)}{\omega_p} \tag{68.145}$$

for semirelative sensitivities. From this basic definition of Q-factor one can also derive the relationships

$$\mathbf{S}_x^{\omega_i} = \frac{\mathbf{S}_x^{\sigma_i}}{4Q_p^2} + \sqrt{1 - (1/4Q^2)}\,\mathbf{S}_x^{\omega_i} \tag{68.146}$$

and

$$\mathbf{S}_x^{Q_p} \approx \mathbf{S}_x^{\omega_i} - \mathbf{S}_x^{\sigma_i} \tag{68.147}$$

involving relative sensitivities. Another group of results can be obtained considering relative sensitivity of the pole $p_i = -\sigma_i + j\omega_i$. For example, one can find that

$$\mathbf{S}_x^{Q_p} = -\sqrt{4Q_p^2 - 1}\,\mathrm{Im}\,\mathbf{S}_x^{p_i} = \frac{x}{\omega_p^2}\left(\frac{\omega_i}{\sigma_i}\right)\left(\sigma_i \frac{\partial \omega_i}{\partial x} - \omega_i \frac{\partial \sigma_i}{\partial x}\right) \tag{68.148}$$

If $\partial\omega_i/\partial x = 0$ and $\partial\sigma_i/\partial x =$ constant, then

$$\mathbf{S}_x^{Q_p} \approx kQ_p \tag{68.149}$$

which shows that in this case $\mathbf{S}_x^{Q_p}$ increases proportionally to the Q-factor independently on the cause of this high sensitivity.

68.14 Simulation of Doubly Terminated Matched Lossless Filters

By cascading the first- and second-order filter sections (occasionally a third-order section is realized in odd-order filters instead of one cascade connection of a first- and second-order section) any high-order transfer function $T(s)$ can be realized. In practice, however, the resulting circuit is difficult to fabricate for high-order and/or highly selective filters. The transfer function of such filters usually contains a pair of complex-conjugate poles very close to the $j\omega$ axis. The sensitivity of this section that realizes high-Q poles is high and the element tolerances for this section can be very tight. The section can be unacceptable for fabrication.

For filters that have such high-Q transfer function poles other design techniques are often used. The most successful and widely used of these alternative strategies are based on simulating the low-sensitivity transfer function of a doubly terminated lossless (reactance) two-port.

Assume that the two-port shown in Fig. 68.5 is lossless and the transfer function $T(s) = V_2(s)/E(s)$ is realized. Considering power relations one can show [12] that for steady-state sinusoidal operation the equation

$$|\rho(j\omega)|^2 + \frac{4R_1}{R_2}\,|T(j\omega)|^2 = 1 \tag{68.150}$$

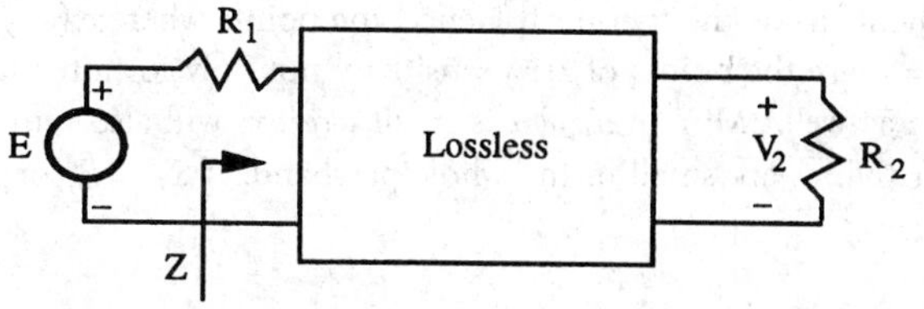

FIGURE 68.5 Lossless two-port with reactance loadings.

is valid for this circuit. Here, $\rho(j\omega) = [R_1 - Z(j\omega)]/[R_1 + Z(j\omega)]$ is the reflection coefficient and $Z(j\omega)$ is the input impedance of the loaded two-port. In many cases the filter requirements are formulated for the transducer function

$$H(s) = \sqrt{\frac{R_2}{4R_1}\frac{E(s)}{V_2(s)}} = \sqrt{\frac{R_2}{4R_1}\frac{1}{T(s)}} \tag{68.151}$$

For this function

$$\ln H(j\omega) = \alpha(\omega) + j\varphi(\omega) = -\ln|T(j\omega)| + \ln[\sqrt{(R_2/4R_1)}] - j\arg T(j\omega) \tag{68.152}$$

Here, $\alpha(\omega)$ is attenuation (it is different from the previously used only by the value of $\ln[\sqrt{(R_2/4R_1)}]$) and $\varphi(\omega) = -\beta(\omega) = -\arg T(j\omega)$ is phase. The impedance $Z(j\omega)$ satisfies the condition $\mathrm{Re}\, Z(j\omega) \geq 0$ (as for any passive circuit input impedance), which means that $|\rho(j\omega)|^2 \leq 1$. Then, as it follows from (68.150) $|H(j\omega)|^2 \geq 1$ and $\alpha(\omega) \geq 0$.

When a filter is designed using $\alpha(\omega)$ the attenuation should be small in the filter passband; moreover, the attenuation is optimized so that it is zero in one or more passband points (Fig. 68.3 shows, for example, the attenuation characteristics with three zeros in the passband). Then the attenuation partial derivative with respect to the value of each of the two-port elements is equal to zero at the attenuation zeros. Indeed, let x_i be any element of the two-port and $\partial\alpha/\partial x_i$ be the partial derivative of $\alpha(\omega, x_i)$ with respect to that element. Suppose that x_i does not have its nominal value and differs from it by a small Δx_i variation. Expanding $\alpha(\omega, x_i)$ in the Taylor series one obtains that

$$\alpha(\omega, x_i + \Delta x_i) \cong \alpha(\omega, x_i) + \Delta x_i \frac{\partial\alpha(\omega, x_i)}{\partial x_i} \tag{68.153}$$

If ω_k is an attenuation zero, then $\alpha(\omega_k, x_i) = 0$. But, as was mentioned before, $\alpha(\omega) \geq 0$ and from (68.153) one obtains that at the point $\omega = \omega_k$ one has

$$\Delta x_i \frac{\partial\alpha(\omega_k, x_i)}{\partial x_i} \geq 0 \tag{68.154}$$

Now, the variation Δx_i was of unspecified sign. Therefore, (68.154) can only be satisfied with the equality sign, which means that $\partial\alpha(\omega_k, x_i)/\partial x_i = 0$.

This result is called the **Fettweis–Orchard theorem** [4] and it explains why the preference is always given to the filter realized as a nondissipative two-port between resistive terminations or to the simulation of such a filter if the filter should be realized as an active circuit. First, considering the real parts of (68.152), one obtains that

$$x_i \frac{\partial\alpha(\omega)}{\partial x_i} = \mathbf{S}_{x_i}[\alpha(\omega)] = -\frac{x_i}{|T(j\omega)|}\frac{\partial|T(j\omega)|}{\partial x_i} = -\mathbf{S}_{x_i}^{|T(j\omega)|} \tag{68.155}$$

when x_i is any element inside the two-port. Hence, the points where $\alpha(\omega_k) = 0$ and, simultaneously, $\partial\alpha(\omega_k)/\partial x_i = 0$ are the points of zero sensitivity not only for attenuation but for transfer function magnitude as well.[2] Moreover, if α is small, $\partial\alpha/\partial x_i$ will also remain small [13], which means that the sensitivities are small in the whole passband. If $x_i = R_1$ or $x_i = R_2$, one obtains from (68.152) that

$$R_1 \frac{\partial\alpha(\omega, R_1)}{\partial R_1} = -\mathbf{S}_{R_1}^{|T(j\omega)|} - \frac{1}{2} \tag{68.156}$$

and

$$R_2 \frac{\partial\alpha(\omega, R_2)}{\partial R_2} = -\mathbf{S}_{R_2}^{|T(j\omega)|} + \frac{1}{2} \tag{68.157}$$

The derivatives $\partial\alpha/\partial R_i$ $(i = 1, 2)$ are also zero at the points where $\alpha = 0$ and they are small when α remains small [13]. This means that in the passband $\mathbf{S}_{R_1}^{|T(j\omega)|} \approx -1/2$ and $\mathbf{S}_{R_2}^{|T(j\omega)|} \approx 1/2$. Thus $|T(j\omega)|$ will share the zero sensitivity of α with respect to all the elements inside the two-port, but due to the terms $\pm 1/2$ in (68.156) and (68.157) a change either in R_1 or R_2 will produce a frequency-independent shift (which can usually be tolerated) in $|T(j\omega)|$ in addition to the small effects proportional to $\partial\alpha/\partial R_i$. This is the basis of the low sensitivity of conventional LC-ladder filters and of those active, switched-capacitor or digital filters that are based on the LC filter model. This is valid with the condition that the transfer functions of the active filter and LC prototype are the same and the parameters of the two filters enter their respective transfer functions the same way.

The Fettweis–Orchard theorem explains why the filtering characteristics sought are those with the maximum number of attenuation zeros [for a given order of the transfer function $T(s)$]. It also helps to understand why it is difficult to design a filter that simultaneously meets the requirements of α and $\varphi(\omega)$ [or to $|T(j\omega)|$ and $\beta(\omega)$]: the degrees of freedom used for optimizing $\varphi(\omega)$ will not be available to attain the maximum number of attenuation zeros. It also explains why a cascade realization is more sensitive than the realization based on the LC lossless model. Indeed, assume that, say, one of the characteristics of Fig. 68.3 is realized by three cascaded sections (with the attenuation of each section shown by the dash-and-dotted line) with each section itself realized in doubly terminated matched lossless form. Each such section of the cascaded filter will be matched at one frequency and the sensitivities to the elements that are in unmatched sections will be different from zero. In addition, the attenuation ripple in each section is usually much larger than the total ripple, and the derivative $\partial\alpha/\partial x_i$, which is, in first approximation, proportional to the attenuation ripple, will not be small. Indeed, practice shows [4] that there is, in fact, a substantial increase in sensitivity in the factored realization.

68.15 Sensitivity of Active *RC* Filters

The required component tolerances are very important factors determining the cost of filters. They are especially important with integrated realizations (where the tolerances are usually higher than in discrete technology). Also, the active filter realizations commonly require tighter tolerances than LC realizations. Yet two classes of active RC filters have tolerances comparable with those of passive LC filters. These are analog-computer and gyrator filters that simulate

[2]As a result of (68.155) and this discussion one cannot use at these points the relationship $\delta\,|T(j\omega)| \approx \mathbf{S}_{x_i}^{|T(j\omega)|}\,\delta x_i$; the relative change $\delta\,|T(j\omega)|$ is always negative and different from zero with δx_i of unspecified sign.

doubly terminated passive LC filters. The tolerance comparison [4] shows the tolerance advantages (sometimes by an order of magnitude) of the doubly terminated lossless structure as compared to any cascade realization. These are the only methods that are now used [14] for high-order high-Q sharp cutoff filters with tight tolerances. For less demanding requirements cascaded realizations could be used. The main advantages that are put forth in this case are the ease of design and simplicity of tuning. But even here the tolerance comparison [4] shows that the stages have better tolerances if they are realized using gyrators (especially C gyrators) and computer simulation methods.

.16 Errors in Sensitivity Comparisons

In conclusion, we briefly outline some common errors in sensitivity comparison. More detailed treatment can be found in [4].

1. Calculating the Wrong Sensitivities. The calculated sensitivities should have as close a relation to the filter specification as possible. In general, for a filter specified in the frequency domain the sensitivities of amplitude and phase should be calculated along the $j\omega$ axis. Sensitivities of poles, zeros, Q's, resonant frequencies, etc., should be carefully interpreted in the context of their connection with amplitude and phase sensitivities.

2. Sensitivities of Optimized Designs. The optimization should use a criterion as closely related as possible to the filter specifications. The use of a criterion that is not closely related to the filter specifications (for example, pole sensitivity) can lead to valid conclusions if the filters being compared differ by an order of magnitude in sensitivity [4]. A sensitivity comparison is valid only if all the circuits have been optimized using the criterion on which they will be compared. The optimized circuit should not be compared with a nonoptimized one. Another error is to optimize a circuit using a very narrow criterion relying on one parameter and forgetting about variations of other parameters or to optimize one part of the transfer function (usually the denominator) and forgetting about the modifying effect of the numerator.

3. Comparing the Incomparable. A frequent error occurs when comparing sensitivities with respect to different types of elements. In general, different types of elements can be realized with different tolerances, and the comparison is valid only if sensitivities are weighted proportionally. Besides, there are basic differences in variability between circuit parameters with physical dimensions and those without. The latter are often determined in the circuit as the ratio of dimensional quantities (as a result, the tolerance of the ratio will be about double the tolerances of the two-dimensional quantities determining them). In integrated technologies the dimensioned quantities usually have worse tolerances but better matching and tracking ability, especially with temperature. Hence, any conclusion involving sensitivities to different types of components, in addition, is technologically dependent.

4. Correlations Between Component Values. The correlations between components are neglected when they are essential (this is usually done for simplification of the statistical analysis). From the other side, an unwarranted correlation is introduced when it does not exist. This is a frequent case where the realization involves cancellation of terms that are equal only when the elements have their nominal values (for example, a cancellation of a pole of one section of a filter by a zero of another section, cancellation of a positive conductance by a negative conductance).

5. Incomplete Analysis. Very often only sensitivities to variations of a single component (usually an amplifier gain) are considered. This is satisfactory only if it is the most critical component which is seldom the case. Another form of incomplete analysis is to calculate only one coordinate of a complex sensitivity measure ($\mathbf{S}_x^Q$ is calculated while $\mathbf{S}_x^{\omega_0}$ is ignored). Also, frequency dependent sensitivities are calculated and compared at one discrete frequency instead of being calculated in frequency intervals.

6. First-Order Differential Sensitivities are the Most Commonly Calculated. But the fact that $\partial y/\partial x = 0$ implies that the variation of y with x is quadratic at the point considered. A consequence of this is that zero sensitivities do not imply infinitely wide tolerances for the components in question. Similarly, infinite sensitivities do not imply infinitely narrow tolerances. Infinite values arise if the nominal value of y is zero, and the finite variations of x will almost always give finite variations of y.

References

[1] L. P. Huelsman and P. E. Allen, *Introduction to the Theory and Design of Active Filters,* New York: McGraw-Hill, 1980.

[2] K. Géher, *Theory of Network Tolerances,* Budapest: Akadémiai Kiadó, 1971.

[3] W. E. Heinlein and W. H. Holmes, *Active Filters for Integrated Circuits,* London: Prentice-Hall, 1974.

[4] M. Hasler and J. Neirynck, *Electric Filters,* Dedham, MA: Artech House, 1986.

[5] *Active Inductorless Filters,* S. K. Mitra, ed., New York: IEEE, 1971.

[6] *Active RC-filters: Theory and application,* in *Benchmark Papers in Electrical Engineering and Computer Science,* vol. 15, L. P. Huelsman, ed., Stroudsburg, PA: Dowden, Hutchinson and Ross, Inc., 1976.

[7] A. F. Schwarz, *Computer-Aided Design of Microelectronic Circuits and Systems,* vol. 1. Orlando, FL: Academic, 1987.

[8] M. L. Blostein, "Some bounds on the sensitivity in *RLC* networks," in *Proc. 1st Allerton Conf. Circuits Syst. Theory,* 1963, pp. 488–501.

[9] R. K. Brayton and R. Spence, *Sensitivity and Optimization,* in *Computer-Aided Design of Electronic Circuits,* vol. 2, Amsterdam: Elsevier, 1980.

[10] C. A. Desoer and E. S. Kuh, *Basic Circuit Theory,* New York: McGraw-Hill, 1969.

[11] M. Biey and A. Premoli, *Cauer and MCPER Functions for Low-Q Filter Design,* St. Saphorin: Georgi, 1980.

[12] N. Balabanian and T. A. Bickart, *Electrical Network Theory,* New York: Wiley, 1969.

[13] H. J. Orchard, "Loss sensitivities in singly and doubly terminated filters," *IEEE Trans. Circuits Syst.,* vol. CAS-26, pp. 293–297, 1979.

[14] B. Nauta, *Analog CMOS Filters for Very High Frequencies,* Boston: Kluwer Academic, 1993.

XIV

’assive Filters

ai-Kai Chen
iversity of Illinois

69

Passive Immittances and Positive-Real Functions*

Vai-Kai Chen
niversity of Illinois, hicago

In this section on passive filters, we deal with the design of one-port networks composed exclusively of passive elements such as resistors R, inductors L, capacitors C, and coupled inductors M. The one-ports are specified by their driving-point *immittances*, *im*pedances, or *admittances*. Our basic problem is that given an immittance function, is it possible to find a one-port composed only of R, L, C, and M elements called the *RLCM one-port network* that realizes the given immittance function? This is known as the **realizability problem**, and its complete solution was first given by Brune [3].

Consider a linear *RLCM* one-port network of Fig. 69.1 excited by a voltage source $V_1(s)$. For our purposes, we assume that there are b branches and the branch corresponding to the voltage source $V_1(s)$ is numbered branch 1 and all other branches are numbered from 2 to b. The Laplace transformed Kirchhoff current law equation can be written as

$$\mathbf{A}\mathbf{I}(s) = \mathbf{0} \tag{69.1}$$

where $\mathbf{A}$ is the basis incidence matrix and $\mathbf{I}(s)$ is the branch-current vector of the network. If $\mathbf{V}_n(s)$ is the nodal voltage vector, then the branch-voltage vector $\mathbf{V}(s)$ can be expressed in terms of $\mathbf{V}_n(s)$ by

$$\mathbf{V}(s) = \mathbf{A}'\mathbf{V}_n(s) \tag{69.2}$$

where the prime denotes the matrix transpose. Taking the complex conjugate of (69.1) in conjunction with (69.2) gives

$$\mathbf{V}'(s)\bar{\mathbf{I}}(s) = \mathbf{V}_n'(s)\mathbf{A}\bar{\mathbf{I}}(s) = \mathbf{V}_n'(s)\mathbf{0} = 0 \tag{69.3}$$

or

$$\sum_{k=1}^{b} V_k(s)\bar{I}_k(s) = 0 \tag{69.4}$$

where $V_k(s)$ and $I_k(s)$ are the branch voltages and the branch currents, respectively.

*References for this chapter can be found on page 2343.

493-8341-2/95/$0.00+$.50

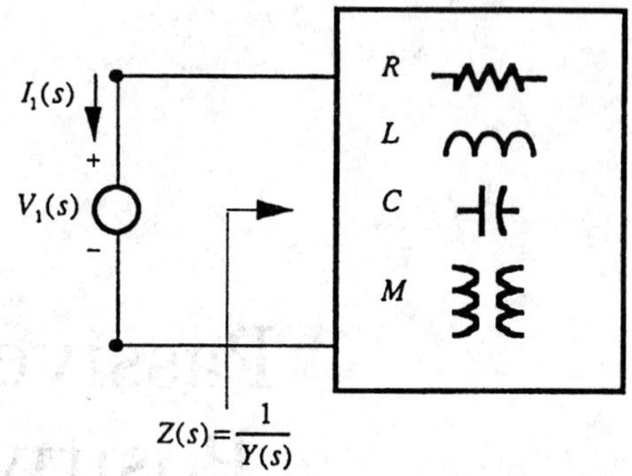

FIGURE 69.1 A general linear *RLCM* one-port network.

From Fig. 69.1 the driving-point impedance of the one-port is defined to be the ratio of $V_1(s)$ to $-I_1(s)$, or

$$Z(s) \equiv \frac{V_1(s)}{-I_1(s)} = \frac{V_1(s)\bar{I}_1(s)}{-I_1(s)\bar{I}_1(s)} = -\frac{V_1(s)\bar{I}_1(s)}{|I_1(s)|^2} \tag{69.5}$$

Eq. (69.4) can be rewritten as

$$-V_1(s)\bar{I}_1(s) = \sum_{k=2}^{b} V_k(s)\bar{I}_k(s) \tag{69.6}$$

Substituting this in (69.5) yields

$$Z(s) = \frac{1}{|I_1(s)|^2}\sum_{k=2}^{b} V_k(s)\bar{I}_k(s) \tag{69.7}$$

Likewise, the dual relation of the input admittance

$$Y(s) \equiv \frac{-I_1(s)}{V_1(s)} = \frac{1}{|V_1(s)|^2}\sum_{k=2}^{b} \bar{V}_k(s)I_k(s) \tag{69.8}$$

holds.

We now consider individual types of elements inside the one-port. For a resistive branch k of resistance R_k, we have

$$V_k(s) = R_k I_k(s) \tag{69.9}$$

For a capacitive branch of capacitance C,

$$V_k(s) = \frac{1}{sC_k} I_k(s) \tag{69.10}$$

Finally, for an inductive branch of self-inductance L_k and mutual inductances M_{kj},

$$V_k(s) = sL_k I_k(s) + \sum_{\text{all } j,\, j\neq k} sM_{kj} I_j(s) \tag{69.11}$$

Substituting these in (69.7) and grouping the summation as sums of all resistors R, all capacitors C, and all inductors LM, we obtain

$$\begin{aligned} Z(s) &= \frac{1}{|I_1(s)|^2}\left[\sum_R R_k|I_k(s)|^2 + \sum_C \frac{1}{sC_k}|I_k(s)|^2 \right. \\ &\quad \left. + \sum_{LM}\left(sL_k|I_k(s)|^2 + \sum_{\text{all } j,\, j\neq k} sM_{kj}I_j(s)\bar{I}_k(s)\right)\right] \\ &= \frac{1}{|I_1(s)|^2}\left[F_0(s) + \frac{1}{s}V_0(s) + sM_0(s)\right] \end{aligned} \tag{69.12}$$

where

$$F_0(s) \equiv \sum_R R_k|I_k(s)|^2 \geq 0 \tag{69.13a}$$

$$V_0(s) \equiv \sum_C \frac{1}{C_k}|I_k(s)|^2 \geq 0 \tag{69.13b}$$

$$M_0(s) \equiv \sum_{LM}\left(L_k|I_k(s)|^2 + \sum_{\text{all } j,\, j\neq k} M_{kj}I_j(s)\bar{I}_k(s)\right) \tag{69.13c}$$

These quantities are closely related to the average power and stored energies of the one-port under steady-state sinusoidal conditions. The average power dissipated in the resistors is

$$P_{\text{ave}} = \frac{1}{2}\sum_R R_k|I_k(j\omega)|^2 = \frac{1}{2}F_0(j\omega) \tag{69.14}$$

showing that $F_0(j\omega)$ represents twice the average power dissipated in the resistors of the one-port. The average electric energy stored in the capacitors is

$$E_C = \frac{1}{4\omega^2}\sum_C \frac{1}{C_k}|I_k(j\omega)|^2 = \frac{1}{4\omega^2}V_0(j\omega) \tag{69.15}$$

Thus, $V_0(j\omega)$ denotes $4\omega^2$ times the average electric energy stored in the capacitors. Similarly, the average magnetic energy stored in the inductors is

$$E_M = \frac{1}{4}\sum_{LM}\left[L_k|I_k(j\omega)|^2 + \sum_{\text{all } q,q\neq k} M_{kq}I_q(j\omega)\bar{I}_k(j\omega)\right] = \frac{1}{4}M_0(j\omega) \tag{69.16}$$

indicating that $M_0(j\omega)$ represents four times the average magnetic energy stored in the inductors. Therefore, all the three quantities $F_0(j\omega)$, $V_0(j\omega)$, and $M_0(j\omega)$ are real and nonnegative, and (69.12) can be rewritten as

$$Z(s) = \frac{1}{|I_1(s)|^2}\left(F_0 + \frac{1}{s}V_0 + sM_0\right) \tag{69.17}$$

Likewise, the dual result for $Y(s)$ is found to be

$$Y(s) = \frac{1}{|V_1(s)|^2}\left(F_0 + \frac{1}{\bar{s}}V_0 + \bar{s}M_0\right) \tag{69.18}$$

Now, we set $s = \sigma + j\omega$ and compute the real part and imaginary part of $Z(s)$ and obtain

$$\mathrm{Re}Z(s) = \frac{1}{|I_1(s)|^2}\left(F_0 + \frac{\sigma}{\sigma^2+\omega^2}V_0 + \sigma M_0\right) \tag{69.19}$$

$$\mathrm{Im}\,Z(s) = \frac{\omega}{|I_1(s)|^2}\left(M_0 - \frac{1}{\sigma^2+\omega^2}V_0\right) \tag{69.20}$$

where **Re** stands for "real part of" and **Im** for "imaginary part of". These equations are valid irrespective of the value of s, except at the zeros of $I_1(s)$. They are extremely important in that many analytic properties of passive impedances can be obtained from them. The following is one of such consequences:

Theorem 1: *If $Z(s)$ is the driving-point impedance of a linear, passive, lumped, reciprocal, and time-invariant one-port network N, then*

1. *Whenever* $\sigma \geq 0$, $\mathrm{Re}\,Z(s) \geq 0$.
2. *If N contains no resistors, then*

$$\sigma > 0 \text{ implies } \mathrm{Re}\,Z(s) > 0$$
$$\sigma = 0 \text{ implies } \mathrm{Re}\,Z(s) = 0$$
$$\sigma < 0 \text{ implies } \mathrm{Re}\,Z(s) < 0$$

3. *If N contains no capacitors, then*

$$\omega > 0 \text{ implies } \mathrm{Im}\,Z(s) > 0$$
$$\omega = 0 \text{ implies } \mathrm{Im}\,Z(s) = 0$$
$$\omega < 0 \text{ implies } \mathrm{Im}\,Z(s) < 0$$

4. *If N contains no self- and mutual inductors, then*

$$\omega > 0 \text{ implies } \mathrm{Im}\,Z(s) < 0$$
$$\omega = 0 \text{ implies } \mathrm{Im}\,Z(s) = 0$$
$$\omega < 0 \text{ implies } \mathrm{Im}\,Z(s) > 0$$

Similar results can be stated for the admittance function $Y(s)$ simply by replacing $Z(s)$, $\mathrm{Re}\,Z(s)$, *and* $\mathrm{Im}\,Z(s)$ *by* $Y(s)$, $\mathrm{Re}\,Y(s)$, *and* $-\mathrm{Im}\,Y(s)$, *respectively.*

The theorem states that the driving-point impedance $Z(s)$ of a passive LMC, RLM, or RC one-port network maps different regions of the complex-frequency s-plane into various regions of the Z-plane. Now, we assert that the driving-point immittance of a passive one-port is a positive-real function, and every positive-real function can be realized as the input immittance of an RLCM one-port network.

Definition 1. *Positive-real function.* A *positive-real function* $F(s)$, abbreviated as a **PR function**, is an analytic function of the complex variable $s = \sigma + j\omega$ satisfying the following three conditions:

1. $F(s)$ is analytic in the open RHS (*r*ight-*h*alf of the *s*-plane), i.e., $\sigma > 0$.
2. $F(\bar{s}) = \overline{F}(s)$ for all s in the open RHS.
3. $\mathrm{Re}\,F(s) \geq 0$ whenever $\mathrm{Re}\,s \geq 0$.

The concept of a positive-real function, as well as many of its properties, is credited to Otto Brune [3]. Our objective is to show that positive realness is a necessary and sufficient condition for a passive one-port immittance. The above definition holds for both rational and transcendental functions. A **rational function** is defined as a ratio of two polynomials. Network functions associated with any linear lumped system, with which we deal exclusively in this section, are rational. In the case of rational functions, not all three conditions in the definition are independent. For example, the analyticity requirement is implied by the other two. The second condition is equivalent to stating that $F(s)$ is real when s is real, and for a rational $F(s)$ it is always satisfied if all the coefficients of the polynomial are real.

Some important properties of a positive-real function can be stated as follows:

1. If $F_1(s)$ and $F_2(s)$ are positive real, so is $F_1[F_2(s)]$.
2. If $F(s)$ is positive real, so are $1/F(s)$ and $F(1/s)$.
3. A positive-real function is devoid of poles and zeros in the open RHS.
4. If a positive-real function has any poles or zeros on the $j\omega$-axis (0 and ∞ included), such poles and zeros must be simple. At a simple pole on the $j\omega$-axis, the residue is real positive.

Property 1 states that a positive-real function of a positive-real function is itself positive real, and property 2 shows that the reciprocal of a positive-real function is positive real. The real significance of the positive-real functions is its use in the characterization of the passive one-port immittances. This characterization is one of the most penetrating results in network theory, and is stated as

Theorem 2: *A real rational function is the driving-point immittance of a linear, passive, lumped, reciprocal, and time-invariant one-port network if and only if it is positive real.*

The necessity of the theorem follows directly from (69.19). The sufficiency was first established by Brune in 1930 by showing that any given positive-real rational function can be realized as the input immittance of a passive one-port network using only the passive elements such as resistors, capacitors, and self- and mutual inductors. A formal constructive proof will be presented in the following section.

Example 1. Consider the passive one-port of Fig. 69.2, the driving-point impedance of which is found to be

$$Z(s) = \frac{3s^2 + s + 2}{2s^2 + s + 3} \tag{69.21}$$

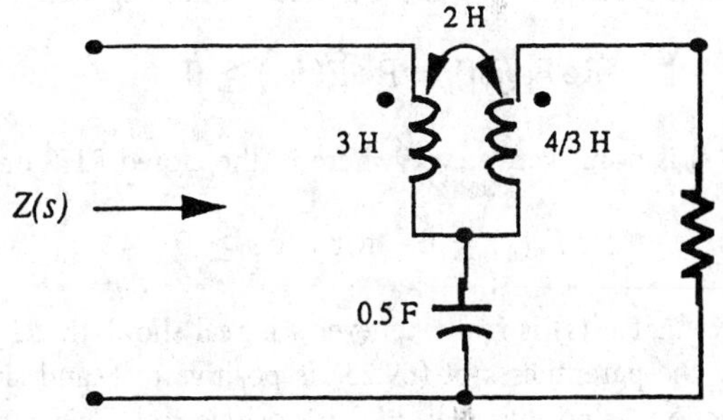

FIGURE 69.2 A passive one-port network.

To verify that the function $Z(s)$ is positive real, we compute its real part by substituting $s = \sigma + j\omega$ and obtain

$$\operatorname{Re} Z(s) = \operatorname{Re} Z(\sigma + j\omega) = \operatorname{Re} \frac{3(\sigma + j\omega)^2 + (\sigma + j\omega) + 2}{2(\sigma + j\omega)^2 + (\sigma + j\omega) + 3}$$
$$= \frac{6(\omega^2 - 1)^2 + (12\omega^2\sigma + 5\omega^2 + 6\sigma^3 + 5\sigma^2 + 14\sigma + 5)\sigma}{(2\sigma^2 - 2\omega^2 + \sigma + 3)^2 + \omega^2(4\sigma + 1)^2} \geq 0, \quad \sigma \geq 0 \tag{69.22}$$

This, in conjunction with the facts that $Z(s)$ is analytic in the open RHS and that all the coefficients of $Z(s)$ are real, shows that $Z(s)$ is positive real.

Observe that if the function $Z(s)$ is of high order, the task of ascertaining its positive realness is difficult if condition 3 of Definition 1 is employed for checking. Hence, it is desirable to have alternate but much simpler conditions for testing. For this reason, we introduce the following equivalent conditions that are relatively easy to apply:

Theorem 3: *A rational function $F(s)$ is positive real if and only if the following conditions are satisfied:*

1. *$F(s)$ is real when s is real.*
2. *$F(s)$ has no poles in the open RHS.*
3. *Poles of $F(s)$ on the $j\omega$-axis, if they exist, are simple, and residues evaluated at these poles are real and positive.*
4. $\operatorname{Re} F(j\omega) \geq 0$ *for all ω, except at the poles.*

PROOF: From the definition of a PR function, we see immediately that all the conditions are necessary. To prove sufficiency, we expand $F(s)$ in a partial fraction as

$$F(s) = \left[k_\infty s + \frac{k_0}{s} + \sum_x \left(\frac{k_x}{s + j\omega_x} + \frac{k_x}{s - j\omega_x}\right)\right] + F_1(s)$$
$$= \left(k_\infty s + \frac{k_0}{s} + \sum_x \frac{2k_x s}{s^2 + \omega_x^2}\right) + F_1(s) \tag{69.23}$$

where k_∞, k_0, and k_x are residues evaluated at the $j\omega$-axis poles $j\infty$, 0, and $j\omega_x$, respectively, and are real and positive. $F_1(s)$ is the function formed by the terms corresponding to the open LHS (*l*eft-*h*alf of the s-plane) poles of $F(s)$, and therefore is analytic in the RHS and the entire $j\omega$-axis including the point at infinity. For such a function, the minimum value of the real part throughout the region where the function is analytic lies on the boundary, namely, the $j\omega$-axis. (See, for example, Churchill [8]). This shows that the minimum value of $\operatorname{Re} F_1(s)$ for all $\operatorname{Re} s \geq 0$ occurs on the $j\omega$-axis. But according to (69.23) this value is nonnegative:

$$\operatorname{Re} F_1(j\omega) = \operatorname{Re} F(j\omega) \geq 0 \tag{69.24}$$

Thus, the real part of $F_1(s)$ is nonnegative everywhere in the closed RHS or

$$\operatorname{Re} F_1(s) \geq 0 \quad \text{for } \operatorname{Re} s \geq 0 \tag{69.25}$$

This together with the fact that $F_1(s)$ is real whenever s is real shows that $F_1(s)$ is positive real.

Since each term inside the parentheses of (69.23) is positive real, and since the sum of two or more positive-real functions is positive real, $F(s)$ is positive real. This completes the proof of the theorem.

In testing for positive realness, we may eliminate some functions from consideration by inspection because they violate certain simple necessary conditions. For example, a function cannot be PR if it has a pole or zero in the open RHS. Another simple test is that the highest powers of s in numerator and denominator not differ by more than unity, because a PR function can have at most a simple pole or zero at the origin or infinity, both of which lie on the $j\omega$-axis.

A **Hurwitz polynomial** is a polynomial devoid of zeros in the open RHS. Thus, it may have zeros on the $j\omega$-axis. To distinguish such a polynomial from the one that has zeros neither in the open RHS nor on the $j\omega$-axis, the latter is referred to as a **strictly Hurwitz polynomial.** For computational purposes, Theorem 3 can be reformulated and put in a much more convenient form.

Theorem 4: *A rational function represented in the form*

$$F(s) = \frac{P(s)}{Q(s)} = \frac{m_1(s) + n_1(s)}{m_2(s) + n_2(s)} \tag{69.26}$$

where $m_1(s)$, $m_2(s)$, and $n_1(s)$, $n_2(s)$ are the even and odd parts of the polynomials $P(s)$ and $Q(s)$, respectively, is positive real if and only if the following conditions are satisfied:

1. *$F(s)$ is real when s is real.*
2. *$P(s) + Q(s)$ is strictly Hurwitz.*
3. *$m_1(j\omega)m_2(j\omega) - n_1(j\omega)n_2(j\omega) \geq 0$ for all ω.*

A real polynomial is strictly Hurwitz if and only if the continued-fraction expansion of the ratio of the even part to the odd part or the odd part to the even part of the polynomial yields only real and positive coefficients, and does not terminate prematurely. For $P(s) + Q(s)$ to be strictly Hurwitz, it is necessary and sufficient that the continued-fraction expansion

$$\left[\frac{m_1(s) + m_2(s)}{n_1(s) + n_2(s)}\right]^{\pm 1} = \alpha_1 s + \cfrac{1}{\alpha_2 s + \cfrac{1}{\ddots + \cfrac{1}{\alpha_k s}}} \tag{69.27}$$

yields only real and positive α's, and does not terminate prematurely, i.e., k must equal the degree of $m_1(s) + m_2(s)$ or $n_1(s) + n_2(s)$, whichever is larger. It can be shown that the third condition of the theorem is satisfied if and only if its left-hand-side polynomial does not have real positive roots of odd multiplicity. This may be determined by factoring it or by the use of the Sturm's theorem, which can be found in most texts on elementary theory of equations. We illustrate the above procedure by the following examples.

Example 2. Test the following function to see if it is PR:

$$F(s) = \frac{2s^4 + 4s^3 + 5s^2 + 5s + 2}{s^3 + s^2 + s + 1} \tag{69.28}$$

For illustrative purposes, we follow the three steps outlined in the theorem, as follows:

$$F(s) = \frac{2s^4 + 4s^3 + 5s^2 + 5s + 2}{s^3 + s^2 + s + 1} = \frac{P(s)}{Q(s)} = \frac{m_1(s) + n_1(s)}{m_2(s) + n_2(s)} \tag{69.29}$$

where

$$m_1(s) = 2s^4 + 5s^2 + 2, \qquad n_1(s) = 4s^3 + 5s \tag{69.30a}$$

$$m_2(s) = s^2 + 1, \qquad n_2(s) = s^3 + s \tag{69.30b}$$

Condition 1 is clearly satisfied. To test condition 2, we perform the Hurwitz test, which gives

$$\frac{m_1(s) + m_2(s)}{n_1(s) + n_2(s)} = \frac{2s^4 + 6s^2 + 3}{5s^3 + 6s} = \frac{2}{5}s + \cfrac{1}{\frac{25}{18}s + \cfrac{1}{\frac{324}{165}s + \cfrac{1}{\frac{33}{54}s}}} \tag{69.31}$$

Since all the coefficients are real and positive and since the continued-fraction expansion does not terminate prematurely, the polynomial $P(s) + Q(s)$ is strictly Hurwitz. Thus, condition 2 is satisfied.

To test condition 3, we compute

$$\begin{aligned} m_1(j\omega)m_2(j\omega) - n_1(j\omega)n_2(j\omega) &= 2\omega^6 - 2\omega^4 - 2\omega^2 + 2 \\ &= 2(\omega^2 + 1)(\omega^2 - 1)^2 \geq 0 \end{aligned} \tag{69.32}$$

which is nonnegative for all ω, or, equivalently, which does not possess any real positive roots of odd multiplicity. Therefore, $F(s)$ is positive real.

70

Passive Cascade Synthesis*

ai-Kai Chen
iversity of Illinois,
cago

0.1 Introduction

In this part, we demonstrate that any rational positive-real function can be realized as the input immittance of a passive one-port network terminated in a resistor, thereby also proving the sufficiency of Theorem 2, Chapter 69.

Consider the even part

$$\text{Ev}\, Z(s) = r(s) = \frac{1}{2}[Z(s) + Z(-s)] \tag{70.1}$$

of a given rational positive-real impedance $Z(s)$. As in (69.26), we first separate the numerator and denominator polynomials of $Z(s)$ into even and odd parts, and write

$$Z(s) = \frac{m_1 + n_1}{m_2 + n_2} \tag{70.2}$$

Then we have

$$r(s) = \frac{m_1 m_2 - n_1 n_2}{m_2^2 - n_2^2} \tag{70.3}$$

showing that if s_0 is a zero or pole of $r(s)$, so is $-s_0$. Thus, the zeros and poles of $r(s)$ possess quadrantal symmetry, being symmetric with respect to both the real and imaginary axes. They may appear in pairs on the real axis, in pairs on the $j\omega$-axis, or in the form of sets of quadruplets in the complex-frequency plane. Furthermore, for a positive-real $Z(s)$, the $j\omega$-axis zeros are required to be of even multiplicity in order that $\text{Re}\, Z(j\omega) = r(j\omega)$ never be negative.

Suppose that we can extract from $Z(s)$ a set of open-circuit impedance parameters $z_{ij}(s)$ characterizing a component two-port network, as depicted in Fig. 70.1, which produces one

*References for this Chapter can be found on page 2333.

93-8341-2/95/$0.00+$.50

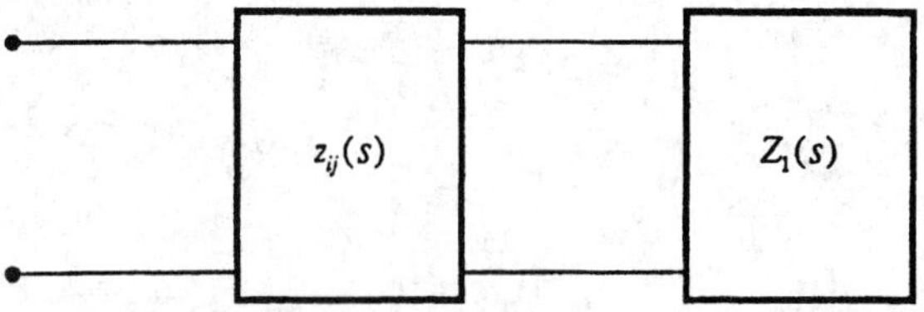

FIGURE 70.1 A two-port network terminated in $Z_1(s)$.

pair of real axis zeros, one pair of $j\omega$-axis zeros, or one set of quadruplet of zeros of $r(s)$, and leaves a rational positive-real impedance $Z_1(s)$ of lower degree, the even part of which $r_1(s)$ is devoid of these zeros but contains all other zeros of $r(s)$. After a finite q steps, we arrive at a rational positive-real impedance $Z_q(s)$, the even part $r_q(s)$ of which is devoid of zeros in the entire complex-frequency plane, meaning that its even part must be a nonnegative constant c:

$$r_q(s) = \frac{1}{2}[Z_q(s) + Z_q(-s)] = c \tag{70.4}$$

Therefore, $Z_q(s)$ is expressible as the sum of a reactance function[1] $Z_{LC}(s)$ and a resistance c:

$$Z_q(s) = Z_{LC}(s) + c \tag{70.5}$$

which can be realized as the input impedance of a lossless two-port network terminated in a c-ohm resistor, as shown in Fig. 70.2.

To motivate our discussion, we first present a theorem credited to Richards [11, 12], which is intimately tied up with the famous Bott-Duffin technique [2].

Theorem 1: *Let $Z(s)$ be a positive-real function that is neither of the form Ls nor $1/Cs$. Let k be an arbitrary positive-real constant. Then the* **Richards' function**

$$W(s) = \frac{kZ(s) - sZ(k)}{kZ(k) - sZ(s)} \tag{70.6}$$

is also positive real.

The **degree** of a rational function is defined as the sum of the degrees of its relatively prime numerator and denominator polynomials. Thus, the Richards' function $W(s)$ is also rational, the degree of which is not greater than that of $Z(s)$. It was first pointed out by Richards that if

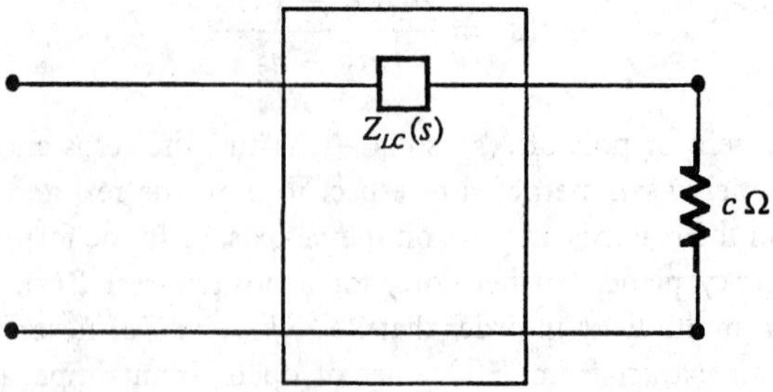

FIGURE 70.2 A two-port network terminated in a resistor.

[1]A formal definition will be given in Chapter 71.

k can be chosen so that the even part of $Z(s)$ vanishes at k, then the degree of $W(s)$ is at least two less than that of $Z(s)$. Let

$$s_0 = \sigma_0 + j\omega_0 \tag{70.7}$$

be a point in the closed RHS. Then, according to the above theorem, the function

$$\hat{W}_1(s) = \frac{s_0 Z(s) - sZ(s_0)}{s_0 Z(s_0) - sZ(s)} \tag{70.8}$$

is positive real if s_0 is positive real; and the function

$$W_1(s) = Z(s_0)\hat{W}_1(\bar{s}_0)\frac{\bar{s}_0 \hat{W}_1(s) - s\hat{W}_1(\bar{s}_0)}{\bar{s}_0 \hat{W}_1(\bar{s}_0) - s\hat{W}_1(s)} \tag{70.9}$$

is positive real if s_0 is a positive-real constant and $\hat{W}_1(s)$ is a positive-real function. Substituting (70.8) in (70.9) yields

$$W_1(s) = \frac{D_1(s)Z(s) - B_1(s)}{-C_1(s)Z(s) + A_1(s)} \tag{70.10}$$

where

$$A_1(s) = q_4 s^2 + |s_0|^2 \tag{70.11a}$$

$$B_1(s) = q_2 s \tag{70.11b}$$

$$C_1(s) = q_3 s \tag{70.11c}$$

$$D_1(s) = q_1 s^2 + |s_0|^2 \tag{70.11d}$$

$$q_1 = \frac{R_0/\sigma_0 - X_0/\omega_0}{R_0/\sigma_0 + X_0/\omega_0} \tag{70.12a}$$

$$q_2 = \frac{2|Z_0|^2}{R_0/\sigma_0 + X_0/\omega_0} \tag{70.12b}$$

$$q_3 = \frac{2}{R_0/\sigma_0 - X_0/\omega_0} \tag{70.12c}$$

$$q_4 = \frac{R_0/\sigma_0 + X_0/\omega_0}{R_0/\sigma_0 - X_0/\omega_0} = \frac{1}{q_1} \tag{70.12d}$$

in which

$$Z(s_0) = R_0 + jX_0 \equiv Z_0 \tag{70.13}$$

In the case $\omega_0 = 0$, then X_0/ω_0 must be replaced by $Z'(\sigma_0)$:

$$\frac{X_0}{\omega_0} \to Z'(\sigma_0) = \left.\frac{dZ(s)}{ds}\right|_{s=\sigma_0} \tag{70.14a}$$

For $\sigma = 0$ and $R_0 = 0$, R_0/σ_0 is replaced by $X'(\omega_0)$:

$$\frac{R_0}{\sigma_0} \to X'(\omega_0) = \left.\frac{dZ(s)}{ds}\right|_{s=j\omega_0} \tag{70.14b}$$

Definition 1. *Index set.* For a given positive-real function $Z(s)$, let s_0 be any point in the open RHS or any finite nonzero point on the $j\omega$-axis where $Z(s)$ is analytic. Then the set of four real numbers q_1, q_2, q_3, and q_4, as defined in (70.12)–(70.14), is called the **index set** assigned to the point s_0 by the positive-real function $Z(s)$.

We illustrate this concept by the following example

Example 1. Determine the index set assigned to the point $s_0 = 0.4551 + j1.099$ by the positive-real function

$$Z(s) = \frac{s^2 + s + 1}{s^2 + s + 2} \tag{70.15}$$

From definition and (70.15), we have

$$s_0 = 0.4551 + j1.099 = \sigma_0 + j\omega_0 \tag{70.16a}$$

$$\begin{aligned} Z(s_0) &= Z(0.4551 + j1.099) = 0.7770 + j0.3218 = 0.8410e^{j22.5^\circ} \\ &= R_0 + jX_0 \equiv Z_0 \end{aligned} \tag{70.16b}$$

$$|Z_0|^2 = 0.7073 \tag{70.17}$$

obtaining from (70.12)

$$q_1 = 0.707, \qquad q_2 = 0.707, \qquad q_3 = 1.414, \qquad q_4 = 1.414 \tag{70.18}$$

With these preliminaries, we now state the following theorem, which forms the cornerstone of the method of cascade synthesis of a rational positive-real impedance according to the Darlington theory [7].

Theorem 2: *Let $Z(s)$ be a positive-real function, which is neither of the form Ls nor $1/Cs$, L and C being real nonnegative constants. Let $s_0 = \sigma_0 + j\omega_0$ be a finite nonzero point in the closed RHS where $Z(s)$ is analytic, then the function*

$$W_1(s) = \frac{D_1(s)Z(s) - B_1(s)}{-C_1(s)Z(s) + A_1(s)} \tag{70.19}$$

is positive real; where A_1, B_1, C_1, and D_1 are defined in (70.11) and $\{q_1, q_2, q_3, q_4\}$ is the index set assigned to the point s_0 by $Z(s)$. Furthermore, $W_1(s)$ possesses the following attributes:

(i) If $Z(s)$ is rational, $W_1(s)$ is rational, the degree of which is not greater than that of $Z(s)$, or

$$\text{degree } W_1(s) \le \text{degree } Z(s) \tag{70.20}$$

(ii) If $Z(s)$ is rational and if s_0 is a zero of its even part $r(s)$, then

$$\text{degree } W_1(s) \le \text{degree } Z(s) - 4, \qquad \omega_0 \ne 0 \tag{70.21a}$$

$$\text{degree } W_1(s) \le \text{degree } Z(s) - 2, \qquad \omega_0 = 0 \tag{70.21b}$$

(iii) *If* $s_0 = \sigma_0 > 0$ *is a real zero of* $r(s)$ *of at least multiplicity* 2 *and if* $Z(s)$ *is rational, then*

$$\text{degree } W_1(s) \leq \text{degree } Z(s) - 4 \tag{70.22}$$

We remark that since $Z(s)$ is positive real, all the points in the open RHS are admissible. Any point on the $j\omega$-axis, exclusive of the origin and infinity, where $Z(s)$ is analytic is admissible as s_0.

We are now in a position to show that any positive-real function can be realized as the input impedance of a lossless one-port network terminated in a resistor. Our starting point is (70.19), which after solving $Z(s)$ in terms of $W_1(s)$ yields

$$Z(s) = \frac{A_1(s)W_1(s) + B_1(s)}{C_1(s)W_1(s) + D_1(s)} \tag{70.23}$$

It can be shown that $Z(s)$ can be realized as the input impedance of a two-port network N_1, which is characterized by its transmission matrix

$$\mathbf{T}_1(s) = \begin{bmatrix} A_1(s) & B_1(s) \\ C_1(s) & D_1(s) \end{bmatrix} \tag{70.24}$$

terminated in $W_1(s)$, as depicted in Fig. 70.3. To see this, we first compute the corresponding impedance matrix $\mathbf{Z}_1(s)$ of N_1 from $\mathbf{T}_1(s)$ and obtain

$$\mathbf{Z}_1(s) = \begin{bmatrix} z_{11}(s) & z_{12}(s) \\ z_{21}(s) & z_{22}(s) \end{bmatrix} = \frac{1}{C_1(s)} \begin{bmatrix} A_1(s) & A_1(s)D_1(s) - B_1(s)C_1(s) \\ 1 & D_1(s) \end{bmatrix} \tag{70.25}$$

The input impedance $Z_{11}(s)$ of N_1 with the output port terminating in $W_1(s)$ is found to be

$$Z_{11}(s) = z_{11}(s) - \frac{z_{12}(s)z_{21}(s)}{z_{22}(s) + W_1(s)} = \frac{A_1(s)W_1(s) + B_1(s)}{C_1(s)W_1(s) + D_1(s)} = Z(s) \tag{70.26}$$

The determinant of the transmission matrix $\mathbf{T}_1(s)$ is computed as

$$\det \mathbf{T}_1(s) = A_1(s)D_1(s) - B_1(s)C_1(s) = s^4 + 2(\omega_0^2 - \sigma_0^2)s^2 + |s_0|^4 \tag{70.27}$$

Observe that $\det \mathbf{T}_1(s)$ depends only upon the point s_0 and not on $Z(s)$, and that the input impedance $Z_{11}(s)$ remains unaltered if each element of $\mathbf{T}_1(s)$ is multiplied or divided by a nonzero finite quality. To complete the realization, we must now demonstrate that the two-port network N_1 is physically realizable.

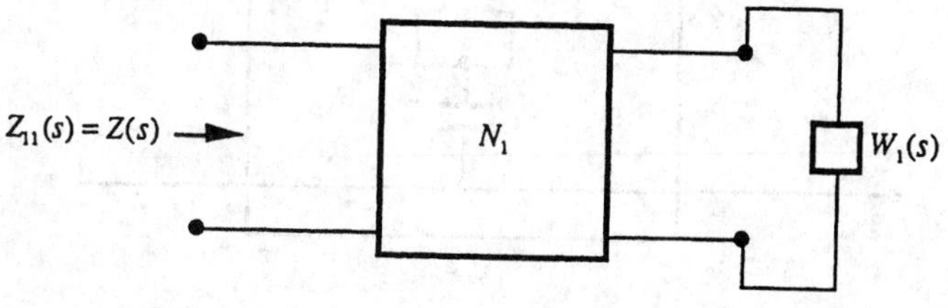

FIGURE 70.3 A two-port network N_1 terminated in the impedance $W_1(s)$.

70.2 Type-E Section

Consider the lossless nonreciprocal two-port network of Fig. 70.4 known as the **type-E section.** Our objective is to show that this two-port realizes N_1. To this end, we first compute its impedance matrix $\mathbf{Z}_E(s)$ as

$$\mathbf{Z}_E(s) = \begin{bmatrix} L_1 s + 1/Cs & Ms + 1/Cs + \zeta \\ Ms + 1/Cs - \zeta & L_2 s + 1/Cs \end{bmatrix} \tag{70.28}$$

where $M^2 = L_1 L_2$, the determinant of which is given by

$$\det \mathbf{Z}_E(s) = \frac{L_1 + L_2 - 2M + \zeta^2 C}{C} \tag{70.29}$$

a constant independent of s due to perfect coupling. From the impedance matrix $\mathbf{Z}_E(s)$, its corresponding transmission matrix $\mathbf{T}_E(s)$ is found to be

$$\mathbf{T}_E(s) = \frac{1}{MCs^2 - \zeta Cs + 1} \begin{bmatrix} L_1 Cs^2 + 1 & (L_1 + L_2 - 2M + \zeta^2 C)s \\ Cs & L_2 Cs^2 + 1 \end{bmatrix} \tag{70.30}$$

To show that the type-E section realizes N_1, we divide each element of $\mathbf{T}_1(s)$ of (70.24) by $|s_0|^2(MCs^2 - \zeta Cs + 1)$. This manipulation will not affect the input impedance $Z(s)$ but it will result in a transmission matrix having the form of $\mathbf{T}_E(s)$. Comparing this new matrix with (70.30) in conjunction with (70.11) yields the following identifications:

$$L_1 C = \frac{q_4}{|s_0|^2} \tag{70.31a}$$

$$L_1 + L_2 - 2M + \zeta^2 C = \frac{q_2}{|s_0|^2} \tag{70.31b}$$

$$C = \frac{q_3}{|s_0|^2} \tag{70.31c}$$

$$L_2 C = \frac{q_1}{|s_0|^2} \tag{70.31d}$$

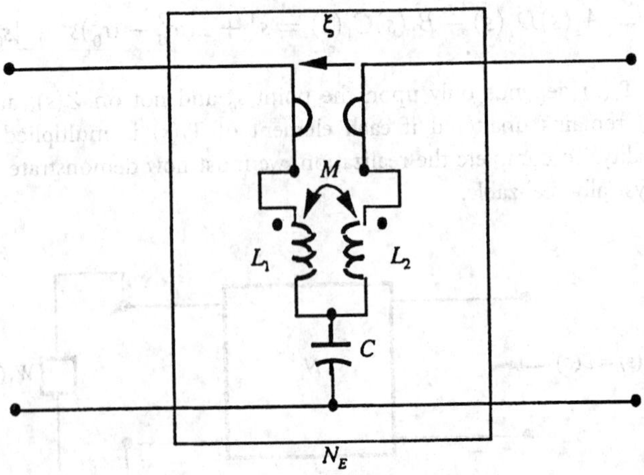

FIGURE 70.4 The type-E section.

Solving these for the element values of the type-E section, we obtain

$$L_1 = \frac{q_4}{q_3} = \frac{1}{q_1 q_3}, \qquad L_2 = \frac{q_1}{q_3} \tag{70.32a}$$

$$C = \frac{q_3}{|s_0|^2}, \qquad M = \frac{1}{q_3}, \qquad \zeta = \pm\frac{2\sigma_0}{q_3} \tag{70.32b}$$

Since the elements of the index set assigned to the point s_0 by $Z(s)$ are positive and finite for all admissible points s_0 in the closed RHS except at those admissible points $s_0 = j\omega_0$ on the $j\omega$-axis where $R_0 \neq 0$, all the elements in (70.32) are physical. But at those admissible points $s_0 = j\omega_0$ where $R_0 \neq 0$, R_0/σ_0 becomes infinity and $q_1 = q_4 = 1$ and $q_2 = q_3 = 0$. Under this situation, $W_1(s) = Z(s)$ and the corresponding two-port network N_1 degenerates into a pair of wires.

Appealing to Theorem 2 shows that if s_0 is chosen to be a complex open RHS zero of the even part $r(s)$ of $Z(s)$, the type-E section is capable of extracting a set of quadrantal zeros of $r(s)$ and leads to at least a four-degree reduction. For zeros of $r(s)$ on the $j\omega$-axis or the σ-axis, the type-E section degenerates into other types of sections, as follows:

CASE 1. $s_0 = \sigma_0 > 0$. Then we have $\zeta = 0$ and

$$L_1 = \frac{q_4}{q_3} = \frac{1}{q_1 q_3}, \qquad L_2 = \frac{q_1}{q_3}, \qquad C = \frac{q_3}{\sigma_0^2}, \qquad M = -\frac{1}{q_3} < 0 \tag{70.33}$$

The type-E section degenerates to the **Darlington type-C section,** as shown in Fig. 70.5.

CASE 2. $s_0 = j\omega_0$ *and* $R_0 = 0$. In this case, we replace R_0/σ_0 by $X'(\omega_0)$ and the gyrator in the type-E section can be avoided because $\zeta = 0$. The type-E section degenerates into the **Brune section** of Fig. 70.6, the element values of which are given by

$$L_1 = \frac{q_4}{q_3} = \frac{1}{q_1 q_3} = \frac{\omega_0 X'(\omega_0) + X_0}{2\omega_0} \tag{70.34a}$$

$$L_2 = \frac{q_1}{q_3} = \frac{[\omega_0 X'(\omega_0) - X_0]^2}{2\omega_0[\omega_0 X'(\omega_0) + X_0]} \tag{70.34b}$$

$$C = \frac{q_3}{|s_0|^2} = \frac{2}{\omega_0[\omega_0 X'(\omega_0) - X_0]} \tag{70.34c}$$

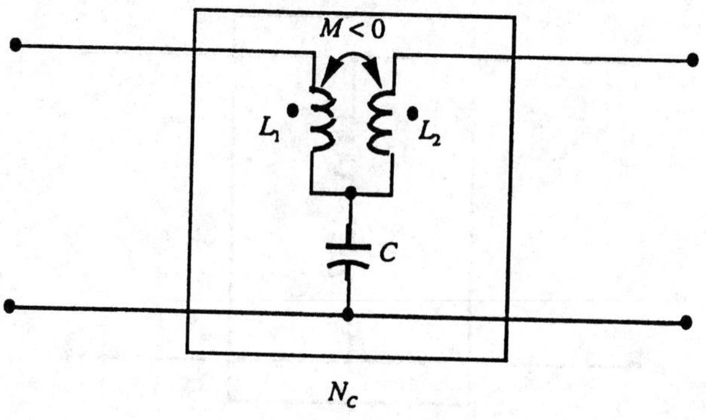

FIGURE 70.5 The Darlington type-C section.

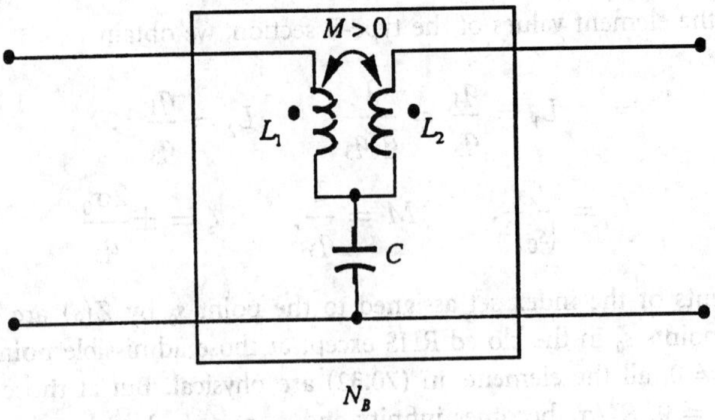

FIGURE 70.6 The Brune section.

$$M = \frac{1}{q_3} = \frac{\omega_0 X'(\omega_0) - X_0}{2\omega_0} > 0 \tag{70.34d}$$

In particular, if $X_0 = 0$ or $Z(j\omega_0) = 0$, the Brune section degenerates into the two-port network of Fig. 70.7 with element values

$$L = \frac{1}{2}X'(\omega_0), \qquad C = \frac{2}{\omega_0^2 X'(\omega_0)} \tag{70.35}$$

As ω_0 approaches zero, this degenerate Brune section goes into the **type-A section** of Fig. 70.8 with

$$L = \frac{1}{2}Z'(0) \tag{70.36}$$

When ω_0 approaches infinity, the degenerate Brune section collapses into the **type-B section** of Fig. 70.9 with

$$\frac{2}{C} = \lim_{s\to\infty} sZ(s) \tag{70.37}$$

CASE 3. $s_0 = j\omega_0$ and $R_0 \neq 0$. In this case, R_0/σ_0 is infinity and

$$q_1 = q_4 = 1, \qquad q_2 = q_3 = 0 \tag{70.38}$$

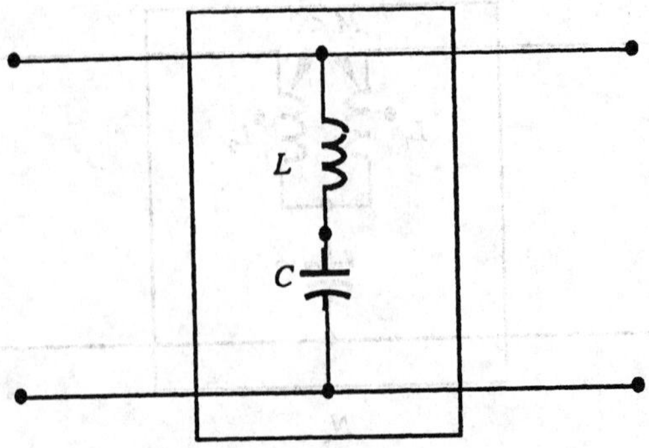

FIGURE 70.7 A degenerate Brune section.

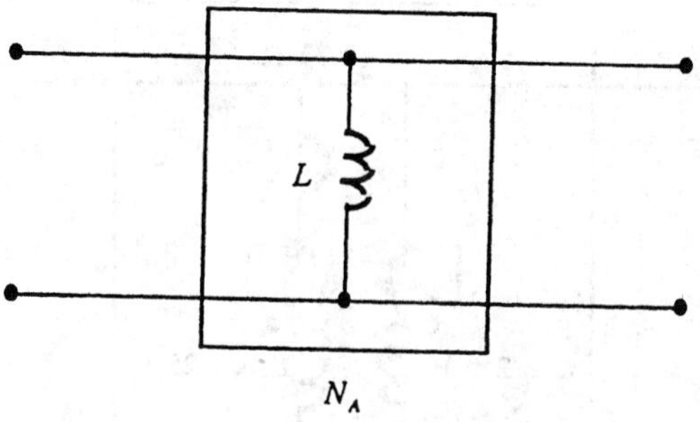

FIGURE 70.8 The type-A section.

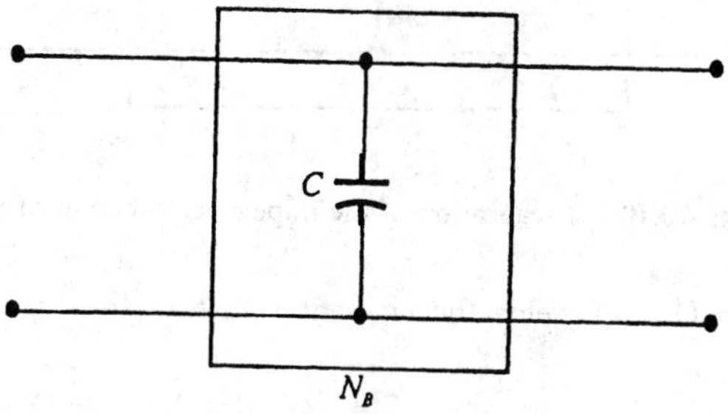

FIGURE 70.9 The type-B section.

The type-E section degenerates into a pair of wires.

Therefore, the Brune section is capable of extracting any $j\omega$-axis zero of the even part of a positive-real impedance, and leads to at least a four-degree reduction if $j\omega_0$ is nonzero and finite, a two-degree reduction otherwise. The latter corresponds to the type-A or type-B section.

Example 2. Consider the positive-real impedance

$$Z(s) = \frac{8s^2 + 9s + 10}{2s^2 + 4s + 4} \tag{70.39}$$

The zeros of its even part $r(s)$ are found from the polynomial

$$m_1 m_2 - n_1 n_2 = 16(s^4 + s^2 + 2.5) \tag{70.40}$$

obtaining

$$s_0 = \sigma_0 + j\omega_0 = 0.735 + j1.020 \tag{70.41a}$$

$$Z(s_0) \equiv R_0 + jX_0 = 2.633 + j0.4279 = 2.667e^{j9.23^\circ} \equiv Z_0 \tag{70.41b}$$

The elements of the index set assigned to s_0 by $Z(s)$ are computed as

$$q_1 = 0.7904, \qquad q_2 = 3.556, \qquad q_3 = 0.6323, \qquad q_4 = 1.265 \tag{70.42}$$

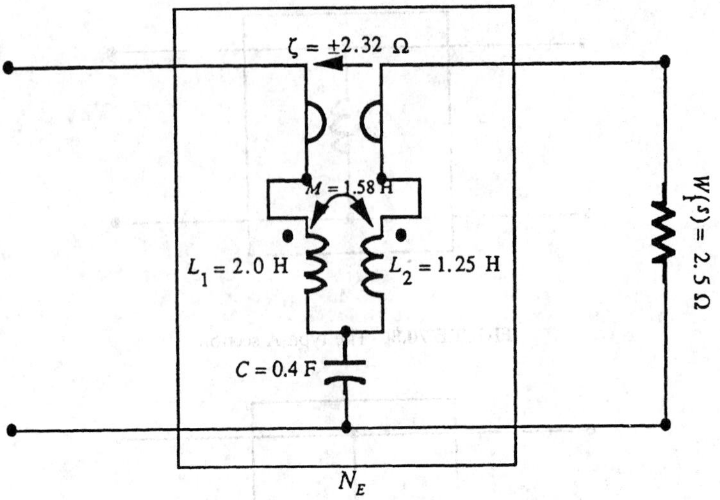

FIGURE 70.10 A realization of the impedance function of (70.39).

Substituting these in (70.32) yields the element values of the type-E section as shown in Fig. 70.10.

$$L_1 = 2\text{ H},\qquad L_2 = 1.25\text{ H},\qquad C = 0.40\text{ F} \tag{70.43a}$$

$$M = 1.58\text{ H},\qquad \zeta = \pm 2.32\ \Omega \tag{70.43b}$$

The terminating impedance $W_1(s)$ is a resistance of value

$$W_1(s) = Z(0) = 2.5\ \Omega \tag{70.44}$$

as shown in Fig. 70.10.

70.3 The Richards Section

In this part, we show that any positive real zero $s_0 = \sigma_0$ of the even part $r(s)$ of a positive-real impedance $Z(s)$, in addition to being realized by the reciprocal Darlington type-C section, can also be realized by a nonreciprocal section called the **Richards section** of Fig. 70.11.

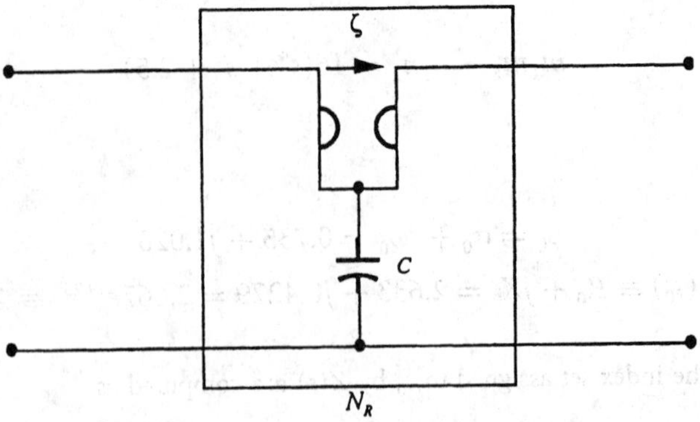

FIGURE 70.11 The Richards section.

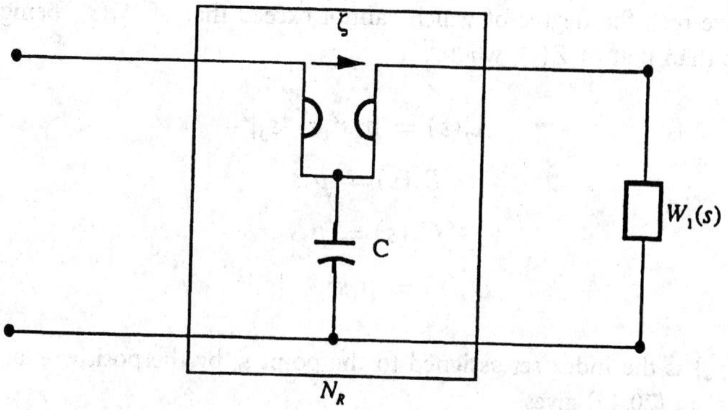

FIGURE 70.12 A realization of $Z(s)$ by Richards section.

Let $Z(s)$ be a rational positive-real function. Then according to Theorem 1, for any positive real σ_0, the function

$$W_1(s) = Z(\sigma_0)\frac{\sigma_0 Z(s) - sZ(\sigma_0)}{\sigma_0 Z(\sigma_0) - sZ(s)} \tag{70.45}$$

is also rational and positive real, the degree of which is not greater than that of $Z(s)$. As pointed out by Richards [11], if σ_0 is a zero of $r(s)$ then

$$\text{degree } W_1(s) \leq \text{degree } Z(s) - 2 \tag{70.46}$$

Inverting (70.45) for $Z(s)$ yields

$$Z(s) = \frac{\sigma_0 W_1(s) + sZ(\sigma_0)}{sW_1(s)/Z(\sigma_0) + \sigma_0} \tag{70.47}$$

This impedance can be realized by the Richards section terminated in the impedance $W_1(s)$ as indicated in Fig. 70.12 with the element values

$$C = \frac{1}{\sigma_0 Z(\sigma_0)}, \qquad \zeta = \pm Z(\sigma_0) \tag{70.48}$$

70.4 The Darlington Type-D Section

In the foregoing, we have demonstrated that the lossless two-port network N_1 can be realized by the lossless nonreciprocal type-E section, which degenerates into the classical type-A, type-B, type-C, and the Brune sections when the even part zero s_0 of the positive-real impedance is restricted to the $j\omega$-axis or the positive σ-axis. In the present section, we show that N_1 can also be realized by a lossless reciprocal two-port network by the application of Theorem 1 twice.

Let s_0 be a zero of the even part $r(s)$ of a rational positive-real impedance $Z(s)$. By Theorem 2 the function $W_1(s)$ of (70.19) is also rational positive real, and its degree is at least four or two less that that of $Z(s)$, depending on whether $\omega_0 \neq 0$ or $\omega_0 = 0$. Now apply Theorem 2 to $W_1(s)$ at the same point s_0. Then the function

$$W_2(s) = \frac{D_2(s)W_1(s) - B_2(S)}{-C_2(s)W_1(s) + A_2(s)} \tag{70.49}$$

is rational positive real, the degree of which cannot exceed that of $W_1(s)$, being at least two or four degrees less than that of $Z(s)$, where

$$A_2(s) = p_4 s^2 + |s_0|^2 \tag{70.50a}$$

$$B_2(s) = p_2 s \tag{70.50b}$$

$$C_2(s) = p_3 s \tag{70.50c}$$

$$D_2(s) = p_1 s^2 + |s_0|^2 \tag{70.50d}$$

and $\{p_1, p_2, p_3, p_4\}$ is the index set assigned to the point s_0 by the positive-real function $W_1(s)$. Solving for $W_1(s)$ in (70.49) gives

$$W_1(s) = \frac{A_2(s)W_2(s) + B_2(s)}{C_2(s)W_2(s) + D_2(s)} \tag{70.51}$$

which can be realized as the input impedance of a two-part network N_2 characterized by the transmission matrix

$$\mathbf{T}_2(s) = \begin{bmatrix} A_2(s) & B_2(s) \\ C_2(s) & D_2(s) \end{bmatrix} \tag{70.52}$$

terminated in $W_2(s)$, as depicted in Fig. 70.13.

Consider the cascade connection of the two-port N_1 of Fig. 70.3 and N_2 of Fig. 70.13 terminated in $W_2(s)$, as shown in Fig. 70.14. The transmission matrix $\mathbf{T}(s)$ of the overall two-port network N is simply the product of the transmission matrices of the individual two-ports:

$$\mathbf{T}(s) = \mathbf{T}_1(s)\mathbf{T}_2(s) \tag{70.53}$$

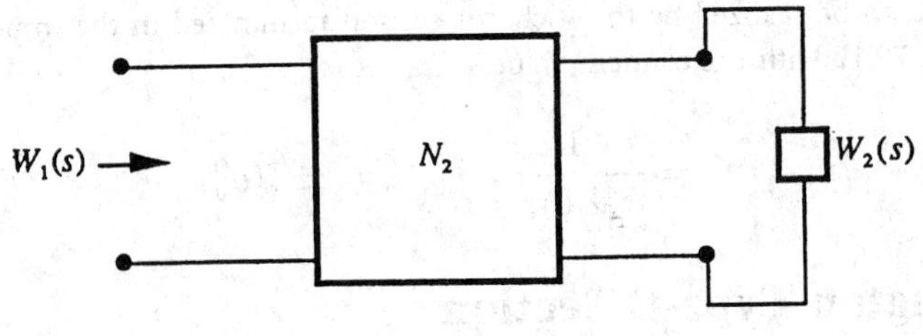

FIGURE 70.13 A realization of the impedance function $W_1(s)$.

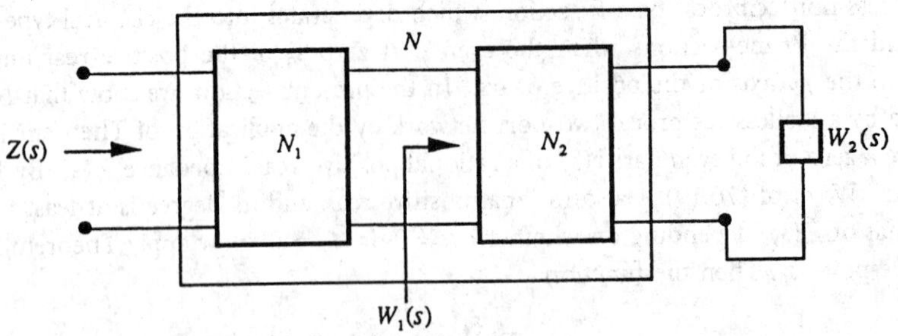

FIGURE 70.14 The cascade connection of two-port networks N_1 and N_2.

the determinant of which is found to be

$$\det \mathbf{T}(s) = [\det \mathbf{T}_1(s)][\det \mathbf{T}_2(s)] = [s^4 + 2(\omega_0^2 - \sigma_0^2)s^2 + |s_0|^4]^2 \tag{70.54}$$

Thus, when N is terminated in $W_2(s)$, the input impedance of Fig. 70.14 is $Z(s)$. This impedance remains unaltered if each element of $\mathbf{T}_1(s)$ is divided by a nonzero finite quantity. For our purposes, we stipulate that the two-port N_1 be characterized by the transmission matrix

$$\hat{\mathbf{T}}_1(s) = \frac{1}{\Delta(s)}\mathbf{T}_1(s) \tag{70.55}$$

where

$$\Delta(s) = s^4 + 2(\omega_0^2 - \sigma_0^2)s^2 + |s_0|^4 \tag{70.56}$$

Using this matrix $\hat{\mathbf{T}}_1(s)$ for N_1, the transmission matrix $\hat{\mathbf{T}}(s)$ of the overall two-port network N becomes

$$\hat{\mathbf{T}}(s) = \frac{1}{\Delta(s)}\mathbf{T}_1(s)\mathbf{T}_2(s) \equiv \frac{1}{\Delta(s)}\begin{bmatrix} A(s) & B(s) \\ C(s) & D(s) \end{bmatrix}$$

$$= \frac{1}{\Delta(s)}\begin{bmatrix} p_4q_4s^4 + |s_0|^4 + [(p_4 + q_4)|s_0|^2 + p_3q_2]s^2 & (p_2q_4 + p_1q_2)s^3 + (p_2 + q_2)|s_0|^2 s \\ (p_4q_3 + p_3q_1)s^3 + (p_3 + q_3)|s_0|^2 s & p_1q_1s^4 + |s_0|^4 + [(p_1 + q_1)|s_0|^2 + p_2q_3]s^2 \end{bmatrix} \tag{70.57}$$

The corresponding impedance matrix $\mathbf{Z}(s)$ of the overall two-port network is found to be

$$\mathbf{Z}(s) = \frac{1}{C(s)}\begin{bmatrix} A(s) & \Delta(s) \\ \Delta(s) & D(s) \end{bmatrix} \tag{70.58}$$

showing that N is reciprocal because $\mathbf{Z}(s)$ is symmetric.

Now consider the reciprocal lossless **Darlington type-D section** N_D of Fig. 70.15 with two perfectly coupled transformers

$$L_1L_2 = M_1^2, \qquad L_3L_4 = M_2^2 \tag{70.59}$$

The impedance matrix $\mathbf{Z}_D(s)$ of N_D is found to be

$$\mathbf{Z}_D(s) = \begin{bmatrix} L_1 s + \dfrac{1}{C_2 s} + \dfrac{s/C_1}{s^2 + \omega_a^2} & M_1 s + \dfrac{1}{C_2 s} + \dfrac{\omega_a^2 M_2 s}{s^2 + \omega_a^2} \\ M_1 s + \dfrac{1}{C_2 s} + \dfrac{\omega_a^2 M_2 s}{s^2 + \omega_a^2} & L_2 s + \dfrac{1}{C_2 s} + \dfrac{\omega_a^2 L_4 s}{s^2 + \omega_a^2} \end{bmatrix} \tag{70.60}$$

where $\omega_a^2 = 1/C_1L_3$.

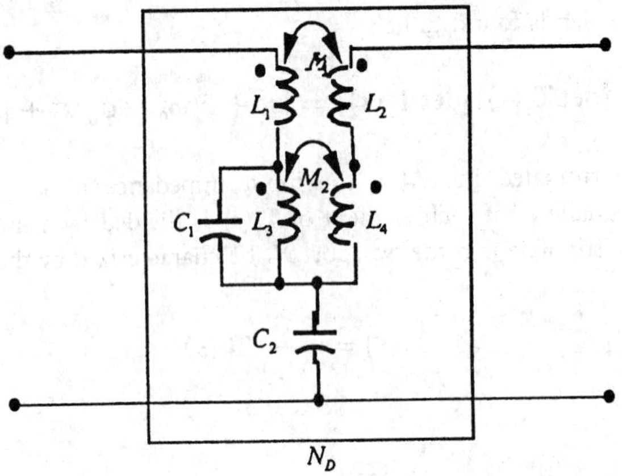

FIGURE 70.15 The Darlington type-D section N_D.

Setting $\mathbf{Z}_D(s) = \mathbf{Z}(s)$ in conjunction with (70.57) and after considerable algebraic manipulations, we can make the following identifications:

$$L_1 = \frac{p_4q_4}{p_4q_3 + p_3q_1} \tag{70.61a}$$

$$L_2 = \frac{p_1q_1}{p_4q_3 + p_3q_1} = \frac{M_1^2}{L_1} \tag{70.61b}$$

$$M_1 = \frac{1}{p_4q_3 + p_3q_1} = \sqrt{L_1L_2} \tag{70.61c}$$

$$C_2 = \frac{p_3 + q_3}{|s_0|^2} \tag{70.61d}$$

$$\omega_a^2 = \omega_1^2 = \frac{|s_0|^2(p_3 + q_3)}{p_4q_3 + p_3q_1} \tag{70.61e}$$

$$\begin{aligned} M_2 &= -\frac{\omega_1^4 - 2(\omega_0^2 - \sigma_0^2)\omega_1^2 + |s_0|^4}{\omega_1^4(p_4q_3 + p_3q_1)} \\ &= -\frac{p_3^2q_3^2|W_1(s_0) + Z(s_0)q_1|^2}{|s_0|^2(p_4q_3 + p_3q_1)(p_3 + q_3)^2} \leq 0 \end{aligned} \tag{70.61f}$$

$$L_4 = \frac{[(p_1 + q_1)|s_0|^2 + p_2q_3]\omega_1^2 - p_1q_1\omega_1^4 - |s_0|^4}{\omega_1^4(p_4q_3 + p_3q_1)} = -\frac{q_3M_2}{p_3} \tag{70.61g}$$

$$L_3 = \frac{M_2^2}{L_4} = -\frac{p_3M_2}{q_3} \tag{70.61h}$$

$$C_1 = \frac{1}{\omega_1^2L_3} = -\frac{q_3}{\omega_1^2p_3M_2} \tag{70.61i}$$

Thus, all the element values except M_2 are nonnegative, and the lossless reciprocal Darlington type-D section is equivalent to the two type-E sections in cascade.

Example 3. Consider the positive-real impedance

$$Z(s) = \frac{6s^2 + 5s + 6}{2s^2 + 4s + 4} \tag{70.62}$$

the even part of which has a zero at

$$s_0 = \sigma_0 + j\omega_0 = 0.61139 + j1.02005 \tag{70.63}$$

The elements of the index set assigned to s_0 by $Z(s)$ are given by

$$q_1 = 0.70711, \qquad q_2 = 1.76784, \qquad q_3 = 0.94283, \qquad q_4 = 1.41421 \tag{70.64}$$

The terminating impedance $W_1(s)$ is determined to be

$$W_1(s) = W_1(0) = Z(0) = 1.5\ \Omega \tag{70.65}$$

The elements of the index set assigned to the point s_0 by $W_1(s)$ are found to be

$$p_1 = 1, \qquad p_2 = 1.83417, \qquad p_3 = 0.81519, \qquad p_4 = 1 \tag{70.66}$$

Substituting these in (70.61) yields the desired element values of the type-D section, as follows:

$$L_1 = \frac{p_4 q_4}{p_4 q_3 + p_3 q_1} = 0.93086\ \text{H} \tag{70.67a}$$

$$L_2 = \frac{p_1 q_1}{p_4 q_3 + p_3 q_1} = 0.46543\ \text{H} \tag{70.67b}$$

$$M_1 = \frac{1}{p_4 q_3 + p_3 q_1} = 0.65822\ \text{H} \tag{70.67c}$$

$$C_2 = \frac{p_3 + q_3}{|s_0|^2} = 1.24303\ \text{F} \tag{70.67d}$$

$$\omega_1^2 = \frac{|s_0|^2(p_3 + q_3)}{p_4 q_3 + p_3 q_1} = 1.63656 \tag{70.67e}$$

$$M_2 = -\frac{\omega_1^4 - 2(\omega_0^2 - \sigma_0^2)\omega_1^2 + |s_0|^4}{\omega_1^4(p_4 q_3 + p_3 q_1)} = -0.61350\ \text{H} \tag{70.67f}$$

$$L_4 = -\frac{q_3 M_2}{p_3} = 0.70956\ \text{H} \tag{70.67g}$$

$$L_3 = \frac{M_2^2}{L_4} = -\frac{p_3 M_2}{q_3} = 0.53044\ \text{H} \tag{70.67h}$$

$$C_1 = \frac{1}{\omega_1^2 L_3} = -\frac{q_3}{\omega_1^2 p_3 M_2} = 1.15193\ \text{F} \tag{70.67i}$$

The complete network together with its termination is presented in Fig. 70.16.

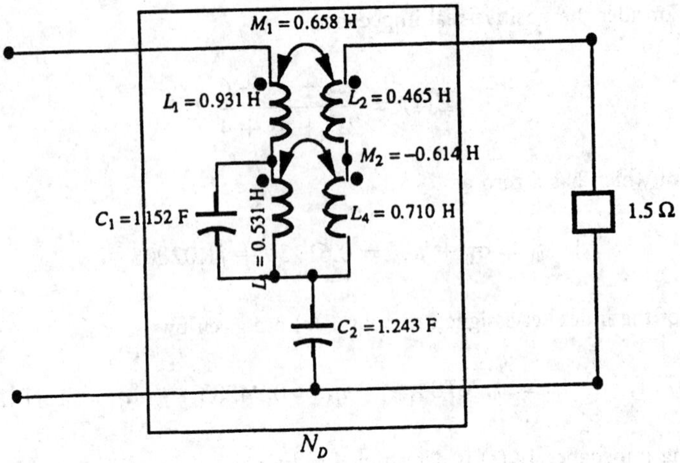

FIGURE 70.16 A Darlington type-D section terminated in a resistor.

71

Synthesis of LCM and RC One-Port Networks*

Wai-Kai Chen
University of Illinois, Chicago

71.1 Introduction

In Chapter 70 we showed that any positive-real function can be realized as the input immittance of a passive one-port network, which is describable as a lossless two-port network terminated in a resistor. Therefore, insofar as the input immittance is concerned, any passive network is equivalent to one containing at most one resistor. In this section, we consider the synthesis of a one-port network composed only of self and mutual inductors and capacitors called the **LCM one-port**, or a one-port composed only of resistors and capacitors called the **RC one-port.**

71.2 LCM One-Port Networks

Consider the input impedance $Z(s)$ of an LCM one-port network written in the form

$$Z(s) = \frac{m_1 + n_1}{m_2 + n_2} \tag{71.1}$$

the even part of which is given by

$$r(s) = \frac{m_1 m_2 - n_1 n_2}{m_2^2 - n_2^2} \tag{71.2}$$

Since the one-port is lossless, we have

$$r(j\omega) = \text{Re } Z(j\omega) = 0 \quad \text{for all } \omega \tag{71.3}$$

To make Re $Z(j\omega) = 0$, there are three nontrivial ways: (i) $m_1 = 0$ and $n_2 = 0$, (ii) $m_2 = 0$ and $n_1 = 0$, (iii) $m_1 m_2 - n_1 n_2 = 0$. The first possibility leads $Z(s)$ to n_1/m_2, the second to m_1/n_2. For the third possibility, we require that $m_1 m_2 = n_1 n_2$ or

$$(m_1 + n_1)m_2 = (m_2 + n_2)n_1 \tag{71.4}$$

*References for this Chapter can be found on page 2333.

0-8493-8341-2/95/$0.00+$.50

which is equivalent to

$$Z(s) = \frac{m_1 + n_1}{m_2 + n_2} = \frac{n_1}{m_2} \tag{71.5}$$

Therefore, the driving-point immittance of a lossless network is always the quotient of even to odd or odd to even polynomials. Its zeros and poles must occur in quadrantal symmetry, being symmetric with respect to both axes. As a result, they are simple and purely imaginary from stability considerations, or $Z(s)$ can be explicitly written as

$$Z(s) = H \frac{(s^2 + \omega_{z1}^2)(s^2 + \omega_{z2}^2)(s^2 + \omega_{z3}^2)\dots}{s(s^2 + \omega_{p1}^2)(s^2 + \omega_{p2}^2)\dots} \tag{71.6}$$

where $\omega_{z1} \geq 0$. This equation can be expanded in partial fraction as

$$Z(s) = Hs + \frac{K_0}{s} + \sum_{i=1}^{n} \frac{2K_i s}{s^2 + \omega_i^2} \tag{71.7}$$

where $\omega_{pi} = \omega_i$, and the residues H, K_0 and K_i are all real and positive.

Substituting $s = j\omega$ and writing $Z(j\omega) = \text{Re}\, Z(j\omega) + j\, \text{Im}\, Z(j\omega)$ results in an odd function known as the **reactance function** $X(\omega)$:

$$X(\omega) = \text{Im} Z(j\omega) = H\omega - \frac{K_0}{\omega} + \sum_{i=1}^{n} \frac{2K_i\omega}{-\omega^2 + \omega_i^2} \tag{71.8}$$

Taking the derivatives on both sides yields

$$\frac{dX(\omega)}{d\omega} = H + \frac{K_0}{\omega^2} + \sum_{i=1}^{n} \frac{2K_i(\omega^2 + \omega_i^2)}{(-\omega^2 + \omega_i^2)^2} \tag{71.9}$$

Since every factor in this equation is positive for all positive and negative values of ω, we conclude that

$$\frac{dX(\omega)}{d\omega} > 0 \quad \text{for} \ -\infty < \omega < \infty \tag{71.10}$$

It states that the slope of the reactance function versus frequency curve is always positive, as depicted in Fig. 71.1. Consequently, the poles and zeros of $Z(s)$ alternate along the $j\omega$-axis. This is known as the **separation property** for reactance function credited to Foster [9]. Because of this, the pole and zero frequencies of (71.6) are related by

$$0 \leq \omega_{z1} < \omega_{p1} < \omega_{z2} < \omega_{p2} < \cdots \tag{71.11}$$

We now consider the realization of $Z(s)$. If each term on the right-hand side of (71.7) can be identified as the input impedance of an LC one-port, the series connection of these one-ports would yield the desired realization. The first term is the impedance of an inductor of inductance H, and the second term corresponds to a capacitor of capacitance $1/K_0$. Each of the remaining term can be realized as a parallel combination of an inductor of inductance $2K_i/\omega_i^2$ and a capacitor of capacitance $1/2K_i$. The resulting realization is shown in Fig. 71.2 known as the **first Foster canonical form**. Likewise, if we consider the admittance function $Y(s) = 1/Z(s)$ and

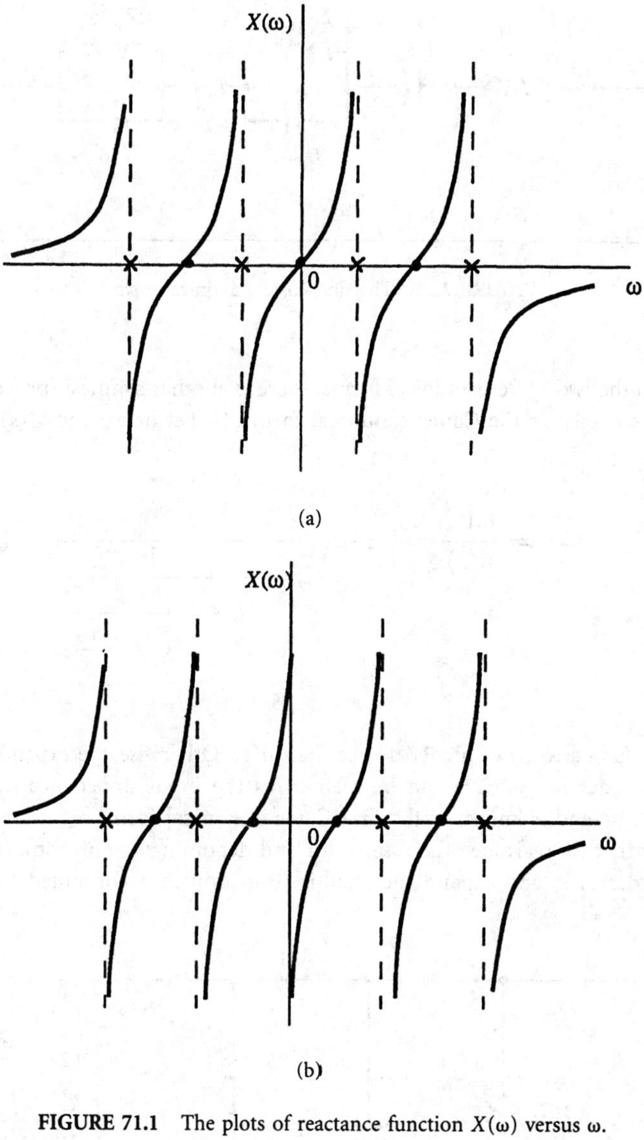

FIGURE 71.1 The plots of reactance function $X(\omega)$ versus ω.

expanded it in partial fraction, we obtain

$$Y(s) = \tilde{H}s + \frac{\tilde{K}_0}{s} + \sum_{i=1}^{n} \frac{2\tilde{K}_i s}{s^2 + \omega_i^2} \tag{71.12}$$

which can be realized by the one-port of Fig. 71.3 known as the **second Foster canonical form.** The term **canonical form** refers to a network containing the minimum number of elements to meet given specifications.

We summarize the above results by stating the following theorem.

Theorem 1: *A real rational function is the input immittance function of an LCM one-port network if and only if all of its zeros and poles are simple, lie on the jω-axis, and alternate with each other.*

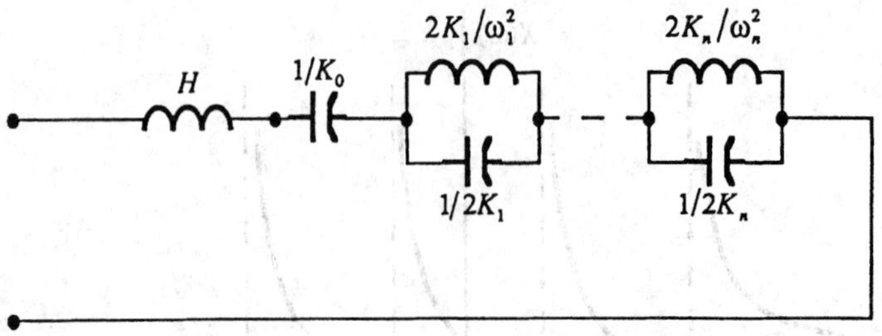

FIGURE 71.2 The first Foster canonical form.

In addition to the two Foster canonical forms, there is another synthesis procedure, that gives rise to one-ports known as the **Cauer canonical form** [4]. Let us expand $Z(s)$ in a continued fraction

$$Z(s) = \frac{m(s)}{n(s)} = L_1 s + \cfrac{1}{C_2 s + \cfrac{1}{L_3 s + \cfrac{1}{C_4 s + \cfrac{1}{\ddots}}}} \tag{71.13}$$

where $m(s)$ is assumed to be of higher degree than $n(s)$. Otherwise, we expand $Y(s) = 1/Z(s)$ instead of $Z(s)$. Equation (71.13) can be realized as the input impedance of the LC ladder network of Fig. 71.4 and is known as the **first Cauer canonical form.**

Suppose now that we rearrange the numerator and denominator polynomials $m(s)$ and $n(s)$ in ascending order of s, and expand the resulting function in a continued fraction. Such an

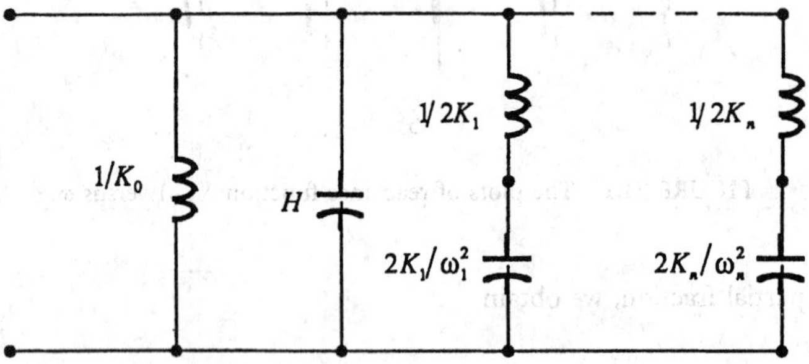

FIGURE 71.3 The second Foster canonical form.

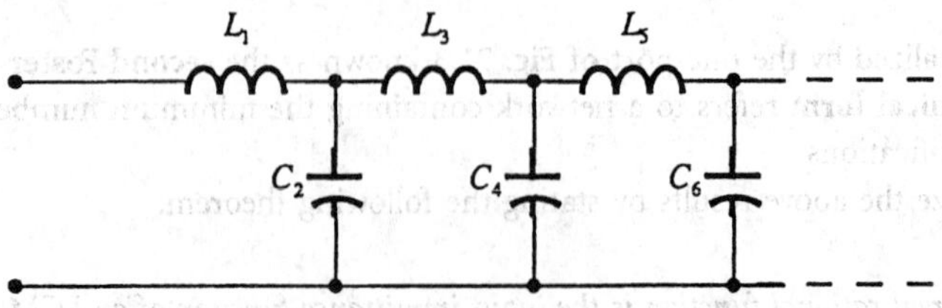

FIGURE 71.4 The first Cauer canonical form.

expansion yields

$$Z(s) = \frac{m(s)}{n(s)} = \frac{a_0 + a_2 s^2 + \cdots + a_{k-2}s^{k-2} + a_k s^k}{b_1 s + b_3 s^3 + \cdots + b_{k-1}s^{k-1}}$$
$$= \frac{1}{C_1 s} + \cfrac{1}{\cfrac{1}{L_2 s} + \cfrac{1}{\cfrac{1}{C_3 s} + \cfrac{1}{\cfrac{1}{L_4 s} + \cfrac{1}{\ddots}}}} \tag{71.14}$$

which can be realized by the LC ladder of Fig. 71.5 known as the **second Cauer canonical form.**

Example 1. Consider the reactance function

$$Z(s) = \frac{s(s^2+4)(s^2+36)}{(s^2+1)(s^2+25)(s^2+81)} \tag{71.15}$$

For the first Foster canonical form, we expand $Z(s)$ in a partial fraction

$$Z(s) = \frac{7s/128}{s^2+1} + \frac{11s/64}{s^2+25} + \frac{99s/128}{s^2+81} \tag{71.16}$$

and obtain the one-port network of Fig. 71.6.

For the second Foster canonical form, we expand $Y(s) = 1/Z(s)$ in a partial fraction

$$Y(s) = s + \frac{225/16}{s} + \frac{4851s/128}{s^2+4} + \frac{1925s/128}{s^2+36} \tag{71.17}$$

and obtain the one-port network of Fig. 71.7.

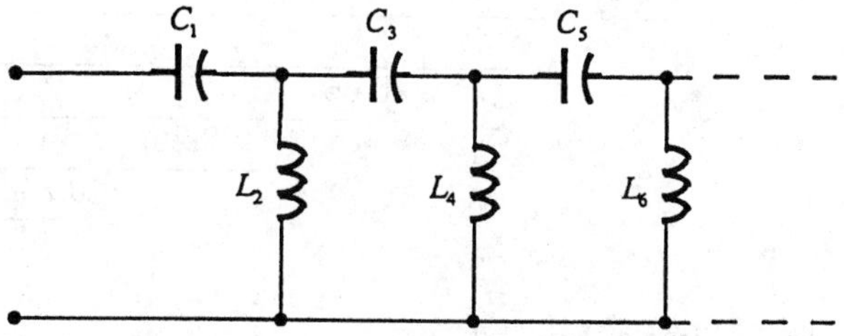

FIGURE 71.5 The second Cauer canonical form.

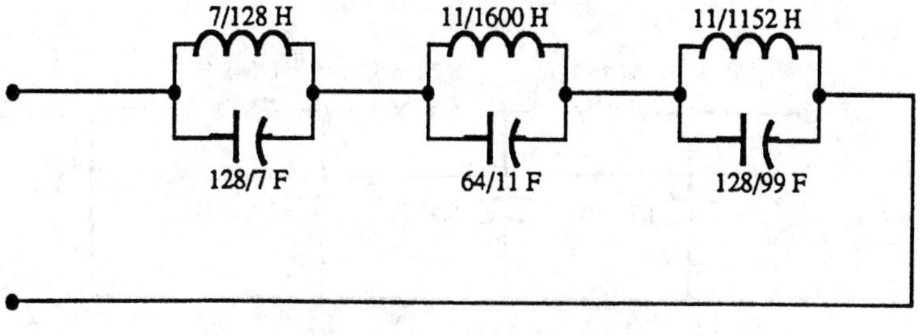

FIGURE 71.6 The first Foster canonical form.

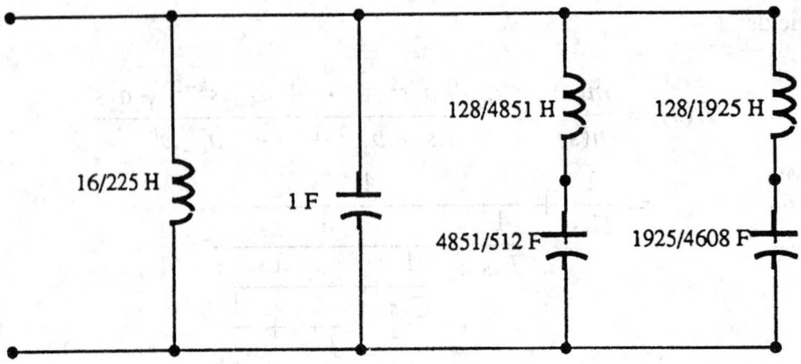

FIGURE 71.7 The second Foster canonical form.

For the first Cauer canonical form, we expand the function in a continued fraction

$$Z(s) = \cfrac{1}{s + \cfrac{1}{0.015s + \cfrac{1}{6.48s + \cfrac{1}{8.28 \times 10^{-3}s + \cfrac{1}{12.88s + \cfrac{1}{0.048s}}}}}} \tag{71.18}$$

and obtain the one-port network of Fig. 71.8.

For the second Cauer canonical form, we rearrange the polynomials in ascending order of s, then expand the resulting function in a continued fraction, and obtain

$$Z(s) = \cfrac{1}{\cfrac{14.06}{s} + \cfrac{1}{\cfrac{0.092}{s} + \cfrac{1}{\cfrac{49.84}{s} + \cfrac{1}{\cfrac{0.66}{s} + \cfrac{1}{\cfrac{192.26}{s} + \cfrac{1}{\cfrac{0.248}{s}}}}}}} \tag{71.19}$$

The desired LC ladder is shown in Fig. 71.9.

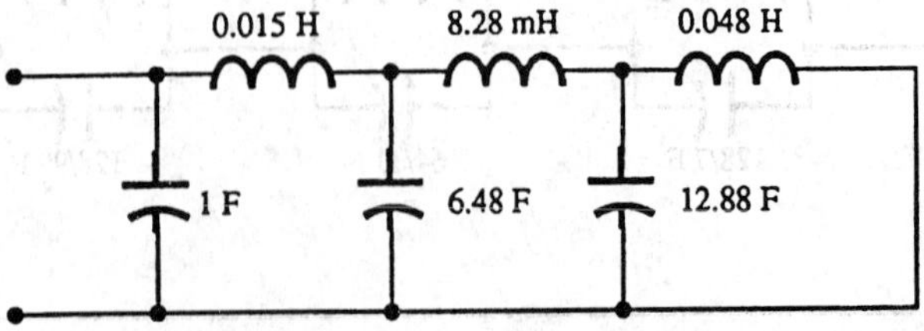

FIGURE 71.8 The first Cauer canonical form.

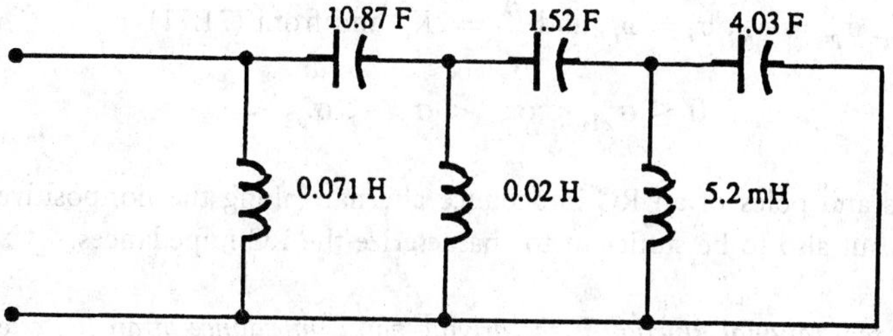

FIGURE 71.9 The second Cauer canonical form.

71.3 RC One-Port Networks

In this part, we exploit the properties of impedance functions of the RC one-ports from the known properties of the LCM one-ports of the preceding section.

From a given RC one-port N_{RC}, we construct an LC one-port N_{LC} by replacing each resistor of resistance R_i by an inductor of inductance $L_i = R_i$. Suppose that we use loop analysis for both N_{RC} and N_{LC}, and choose the same set of loop currents. In addition, assume that the voltage source at the input port is traversed only by loop current ①. Then the input impedance $Z_{LC}(s)$ of N_{LC} is determined by the equation

$$Z_{LC}(s) = \frac{\tilde{\Delta}(s)}{\tilde{\Delta}_{11}(s)} \tag{71.20}$$

where $\tilde{\Delta}$ is the loop determinant and $\tilde{\Delta}_{11}$ is the cofactor corresponding to loop current ① in N_{LC}. Similarly, the input impedance $Z_{RC}(s)$ of N_{RC} can be written as

$$Z_{RC}(s) = \frac{\Delta(s)}{\Delta_{11}(s)} \tag{71.21}$$

where Δ is the loop determinant and Δ_{11} is the cofactor corresponding to loop current ① in N_{RC}. It is not difficult to see that these loop determinants and cofactors are related by

$$\Delta(s) = \left.\frac{\tilde{\Delta}(p)}{p^r}\right|_{p^2=s} \tag{71.22a}$$

$$\Delta_{11}(s) = \left.\frac{\tilde{\Delta}_{11}(p)}{p^{r-1}}\right|_{p^2=s} \tag{71.22b}$$

where r is the order of the loop determinants $\tilde{\Delta}$ and Δ. Combining (71.20)–(71.22) yields

$$Z_{RC}(s) = \left[\frac{1}{p} Z_{LC}(p)\right]_{p^2=s} \tag{71.23}$$

This relation allows us to deduce the properties of RC networks from those of the LC networks. Substituting (71.6) and (71.7) in (71.23), we obtain the general forms of the RC impedance function as

$$Z_{RC}(s) = H\frac{(s+\sigma_{z1})(s+\sigma_{z2})(s+\sigma_{z3})\ldots}{s(s+\sigma_{p1})(s+\sigma_{p2})\ldots} = H + \frac{K_0}{s} + \sum_{i=1}^{n} \frac{\hat{K}_i}{s+\sigma_i} \tag{71.24}$$

where $\sigma_{zj} = \omega_{zj}^2$, $\sigma_{pi} = \omega_{pi}^2$, $\sigma_i = \omega_i^2$, and $\hat{K}_i = 2K_i$, and from (71.11)

$$0 \leq \sigma_{z1} < \sigma_{p1} < \sigma_{z2} < \sigma_{p2} < \cdots \tag{71.25}$$

Thus, the zeros and poles of an RC impedance alternate along the nonpositive real axis. This property turns out also to be sufficient to characterize the RC impedances.

Theorem 2: *A real rational function is the driving-point impedance of an RC one-port network if and only if all the poles and zeros are simple, lie on the nonpositive real axis, and alternate with each other, the first critical frequency (pole or zero) being a pole.*

The slope of $Z_{RC}(\sigma)$ is found from (71.24) to be

$$\frac{dZ_{RC}(\sigma)}{d\sigma} = -\frac{K_0}{\sigma^2} - \sum_{i=1}^{n} \frac{\hat{K}_i}{(\sigma + \sigma_i)^2} \tag{71.26}$$

which is negative for all values of σ, since K_0 and $\hat{K}_i$ are positive. Thus, we have

$$\frac{dZ_{RC}(\sigma)}{d\sigma} < 0 \tag{71.27}$$

A plot of $Z_{RC}(\sigma)$ as a function of σ is shown in Fig. 71.10. Since there are no poles and zeros along the positive real axis, we have

$$Z_{RC}(\infty) \leq Z_{RC}(0) \tag{71.28}$$

We now proceed to the realization of the RC one-port networks. Suppose that we are given $Z_{RC}(\sigma)$ as in (71.24). By analogy to the LC case, this impedance can be realized by the one-port network of Fig. 71.11 called the *first Foster canonical form* for the RC impedance.

To obtain the second Foster canonical form, we expand $Y_{RC}(s)/s$ in a partial fraction, where $Y_{RC}(s) = 1/Z_{RC}(s)$, and then multiply the resulting equation by s. The reason is that a direct partial-fraction expansion of $Y_{RC}(s)$ will result in negative residues. Proceeding in this way, we obtain

$$Y_{RC}(s) = K_0 + K_\infty s + \sum_{i=1}^{n} \frac{K_i s}{s + \sigma_i} \tag{71.29}$$

yielding the one-port network of Fig. 71.12.

As before, in addition to the two Foster forms, RC ladder realizations are also possible. Following the LC case, we perform a continued-fraction expansion of $Z_{RC}(s)$ and obtain

$$Z_{RC}(s) = R_1 + \cfrac{1}{C_2 s + \cfrac{1}{R_3 + \cfrac{1}{C_4 s + \cfrac{1}{\ddots}}}} \tag{71.30}$$

This expansion can be realized by the ladder network of Fig. 71.13 known as the *first Cauer canonical form* for RC impedance. If we rearrange the terms of $Z_{RC}(s)$ so that the numerator

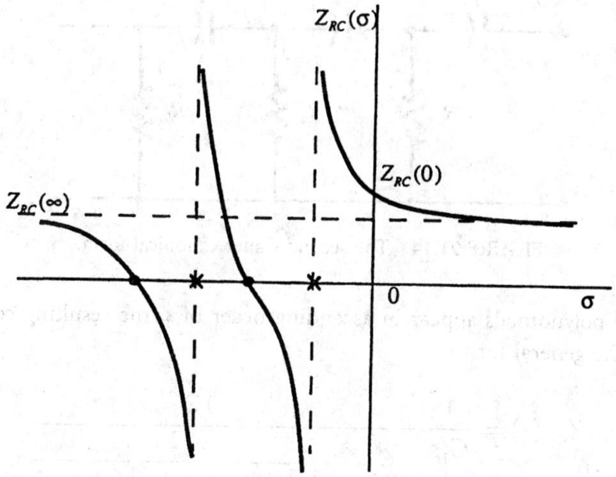

FIGURE 71.10 A plot of $Z_{RC}(\sigma)$ as a function of σ.

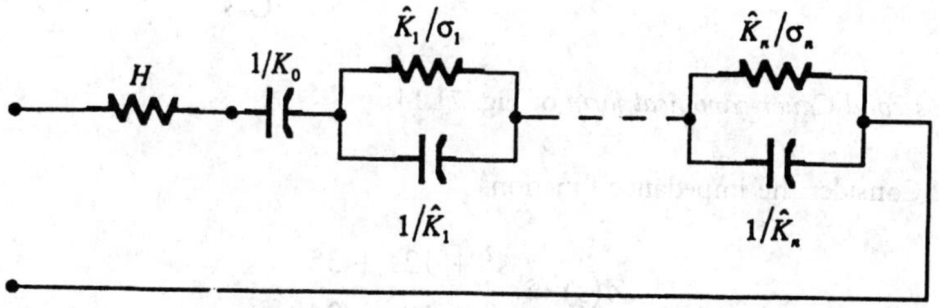

FIGURE 71.11 The first Foster canonical form.

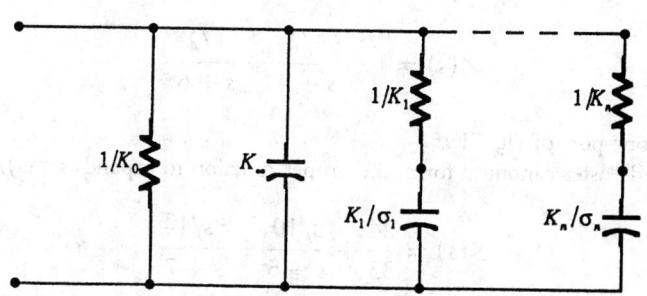

FIGURE 71.12 The second Foster canonical form.

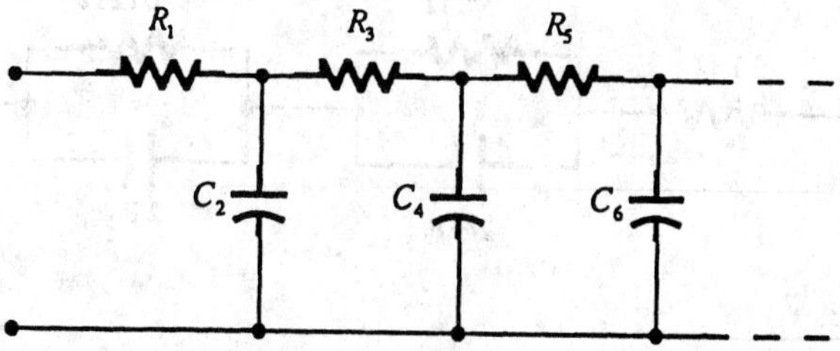

FIGURE 71.13 The first Cauer canonical form.

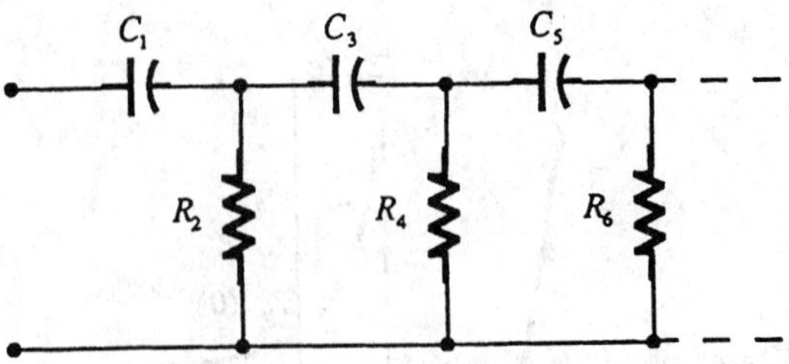

FIGURE 71.14 The second Cauer canonical form.

and denominator polynomials appear in ascending order of s, the resulting continued-fraction expansion takes the general form

$$Z_{RC}(s) = \frac{1}{C_1 s} + \cfrac{1}{\cfrac{1}{R_2} + \cfrac{1}{\cfrac{1}{C_3 s} + \cfrac{1}{\cfrac{1}{R_4} + \cfrac{1}{\cfrac{1}{C_5 s} + \cfrac{1}{\ddots}}}}} \tag{71.31}$$

yielding the *second Cauer canonical form* of Fig. 71.14.

Example 2. Consider the impedance function

$$Z(s) = \frac{s^2 + 12s + 35}{s^2 + 10s + 24} \tag{71.32}$$

To obtain the first Foster canonical form, we expand $Z(s)$ in partial fraction as

$$Z(s) = 1 + \frac{3/2}{s+4} + \frac{1/2}{s+6} \tag{71.33}$$

and obtain the one-port of Fig. 71.15.

For the second Foster canonical form, the proper function to expand is $Y(s)/s$, yielding

$$Y(s) = \frac{24}{35} + \frac{s/10}{s+5} + \frac{3s/14}{s+7} \tag{71.34}$$

The corresponding realization is shown in Fig. 71.16.

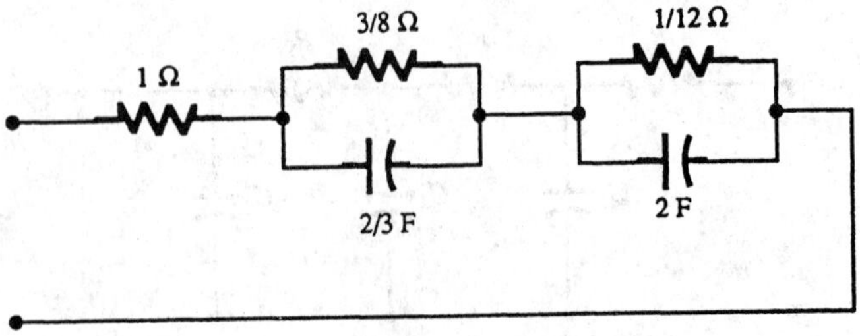

FIGURE 71.15 The first Foster canonical form.

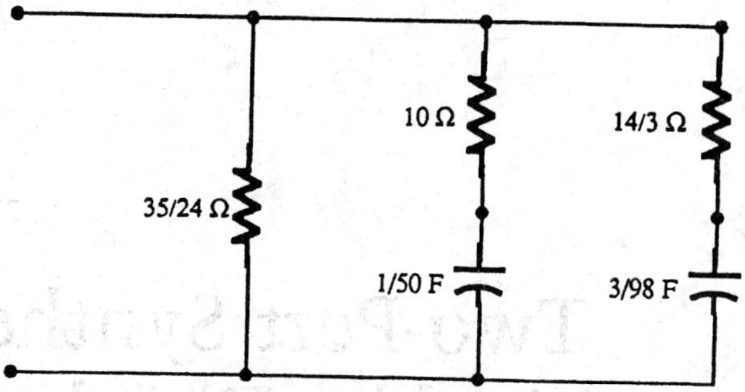

FIGURE 71.16 The second Foster canonical form.

For the first Cauer canonical form, we expand $Z(s)$ in a continued fraction as

$$Z(s) = 1 + \cfrac{1}{\cfrac{s}{2} + \cfrac{1}{\cfrac{4}{9} + \cfrac{1}{\cfrac{27s}{2} + \cfrac{1}{\cfrac{1}{72}}}}} \tag{71.35}$$

which can be realized by the ladder of Fig. 71.17. To obtain the second Cauer canonical form, we rearrange the numerator and denominator polynomials in ascending order of s, and then expand in a continued fraction

$$Z(s) = \frac{35 + 12s + s^2}{24 + 10s + s^2} = \cfrac{1}{0.69 + \cfrac{1}{\cfrac{19.76}{s} + \cfrac{1}{0.306 + \cfrac{1}{\cfrac{692.91}{s} + \cfrac{1}{8.36 \times 10^{-3}}}}}} \tag{71.36}$$

yielding the ladder of Fig. 71.18.

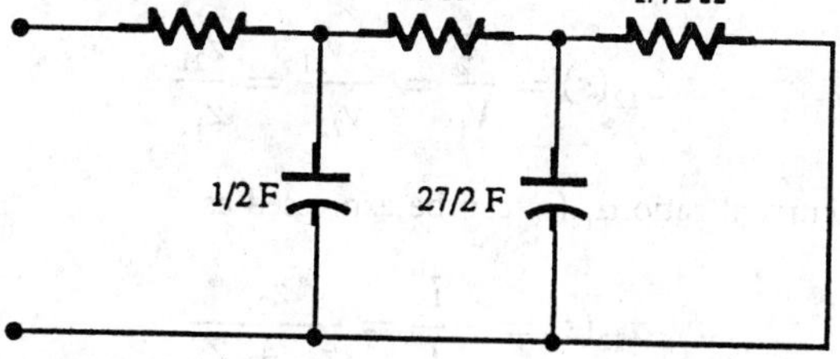

FIGURE 71.17 The first Cauer canonical form.

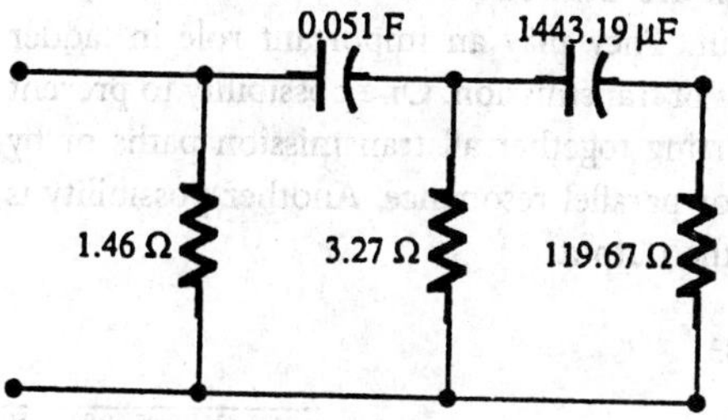

FIGURE 71.18 The second Cauer canonical form.

72

Two-Port Synthesis by Ladder Development*

Wai-Kai Chen
University of Illinois, Chicago

72.1 Introduction

In two-port synthesis, specifications are often given in terms of the transfer functions such as the transfer voltage ratio, transfer current ratio, transfer impedance, or transfer admittance. The actual realization, however, is accomplished by means of the y- or z-parameters. Figure 72.1 shows a two-port network driven by a voltage source with output terminating in an impedance $Z_2(s)$. It is straightforward to show that the transfer voltage ratio function $G_{12}(s)$ can be expressed in terms of its y-parameters $y_{ij}(s)$ or z-parameters $z_{ij}(s)$ by the equation

$$G_{12}(s) = \frac{V_2}{V_1} = \frac{-y_{21}}{y_{22} + Y_2} \tag{72.1}$$

where $Y_2(s) = 1/Z_2(s)$. When the output is open-circuited, (72.1) becomes

$$G_{12}(s) = \frac{V_2}{V_1} = \frac{-y_{21}}{y_{22}} = \frac{z_{21}}{z_{11}} \tag{72.2}$$

Likewise, the transfer current ratio $\alpha_{12}(s)$ can be expressed as

$$\alpha_{12}(s) = -\frac{I_2}{I_1} = \frac{z_{21}}{z_{22} + Z_2} \tag{72.3}$$

The **zeros of transmission** of a two-port network are defined as the frequencies at which the two-port results in zero output for a finite input. They play an important role in ladder development. There are many ways of producing zeros of transmission. One possibility to prevent the input signal from reaching the output is by shorting together all transmission paths or by opening all transmission paths by means of a series or parallel resonance. Another possibility is that signals transmitted by different paths cancel at the output.

*References for this Chapter can be found on page 2333.

0-8493-8341-2/95/$0.00+$.50

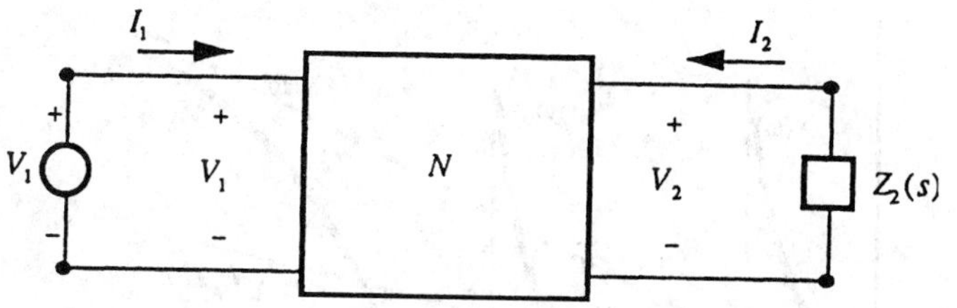

FIGURE 72.1 A terminated two-port network.

Observe from (72.1) and (72.2) that zero output, $V_2 = 0$, implies a zero for each of these functions. Therefore, zeros of transmission are zeros of $-y_{21}$ or z_{21} provided y_{21} and y_{22} or z_{21} and z_{11} have the same poles. For the ladder network, the transmission can be interrupted only by the short circuit in a shunt arm or an open circuit in a series arm. The short circuit of a shunt arm corresponds to the pole frequencies of its admittances, whereas the open circuit of a series arm corresponds to the pole frequencies of its impedances. Therefore, the zeros of transmission of a ladder network can be identified directly with the zeros of the impedances of the shunt arms and the poles of the impedances of the series arms. For the LC ladder, all zeros of transmission lie on the $j\omega$-axis, and for the RC ladder they are on the nonpositive real axis of the complex-frequency s-plane.

72.2 The LC Ladder

For LC ladders, the conditions imposed on $-y_{21}$ and y_{22} or z_{21} and z_{22} are that the driving-point functions y_{22} and z_{22} be positive real with poles and zeros interlaced on the $j\omega$-axis. The transfer functions $-y_{21}$ and z_{21}, assuming to have the same poles as y_{22} or z_{22}, must have all of its zeros on the $j\omega$-axis. However, these zeros need not be interlaced with the poles and they may not be simple. Our strategy in realization is that of carrying out the driving-point synthesis of y_{22} or z_{22}, using Cauer ladder development method, in such a way that the zeros of transmission are realized at the same time. The procedure consists of two steps: a zero-shifting step and a zero-producing step, as described below.

Zero Shifting by Partial Removal. Consider an impedance of (71.6), the partial-fraction expansion of which is given in (71.7). The first term on the right-hand side of (71.7) is due to the contribution of the pole at the infinity. If this term Hs is subtracted from $Z(s)$, the resulting function $Z(s) - Hs$ is devoid of the pole at the infinity. We say that the pole at infinity has been removed completely. Instead of complete removal of this pole, suppose that we subtract a fraction of the terms Hs from $Z(s)$ by introducing a constant k_p such that

$$Z_1(s) = Z(s) - k_p Hs, \qquad k_p < 1 \tag{72.4}$$

We say that the pole at infinity has been **partially removed** or **weakened**. The function $Z_1(s)$ that results from the partial removal of the pole at infinity still possesses the pole at infinity. Since all the zeros of $Z_1(s)$ are again located on the $j\omega$-axis, these zeros are found by substituting $s = j\omega$ in (72.4),

$$X_1(\omega) = X(\omega) - k_p H\omega \tag{72.5}$$

where $Z_1(j\omega) = jX_1(\omega)$. The zeros of $X_1(\omega)$ are values of ω satisfying the equation

$$X(\omega) = k_p H\omega \tag{72.6}$$

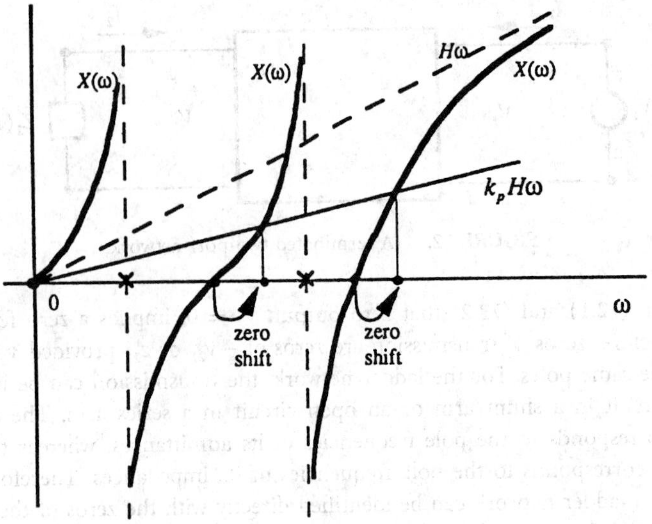

FIGURE 72.2 Zero shifting by weakening the pole at the infinity.

Solutions to this equation are found graphically from the intersections of the curves $X(\omega)$ and $k_pH\omega$, as depicted in Fig. 72.2. Observe that all the zeros in the resulting function are shifted toward the pole being weakened. The amount of shift of the zeros from their original positions depends on the value of k_p and the proximity of a zero to the pole being weakened.

We next consider the term K_0/s in (71.7) due to the pole at the origin. The partial removal of this pole is equivalent to the operation

$$Z_2(s) = Z(s) - k_p\frac{K_0}{s}, \qquad k_p < 1 \tag{72.7}$$

As before, the zeros of $Z_2(s)$ are defined by the intersections of the curves $X(\omega)$ and $-k_pK_0/\omega$ with ω,

$$X(\omega) = -k_p\frac{K_0}{\omega}, \qquad k_p < 1 \tag{72.8}$$

as illustrated in Fig. 72.3. Observe again that the zeros are shifted toward the pole being weakened, which is at the origin.

Finally, for the finite nonzero poles, the corresponding factors take the general form $2K_is/(s^2+\omega_i^2)$. The partial removal of this pair of complex conjugate poles results in the new function

$$Z_3(s) = Z(s) - k_p\frac{2K_is}{s^2+\omega_i^2}, \qquad k_p < 1 \tag{72.9}$$

The zeros of this function are defined by the intersections of the plots of $X(\omega)$ and $-k_p2K_i\omega/(\omega^2-\omega_i^2)$ with ω,

$$X(\omega) = -k_p\frac{2K_i\omega}{\omega^2-\omega_i^2}, \qquad k_p < 1 \tag{72.10}$$

as illustrated in Fig. 72.4.

Our conclusion is that the partial removal of a pole shifts the zeros toward that pole. The amount of shift depends on the value of k_p and the proximity of a zero to that pole, but in no case can a zero be shifted beyond an adjacent pole.

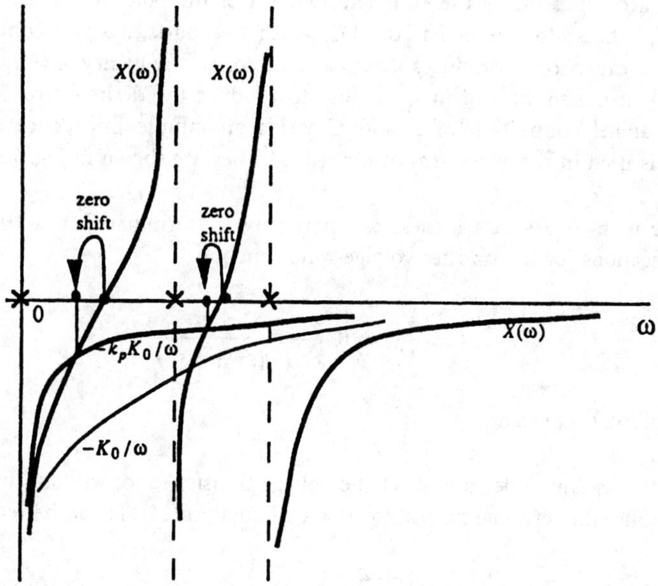

FIGURE 72.3 Zero shifting by weakening the pole at the origin.

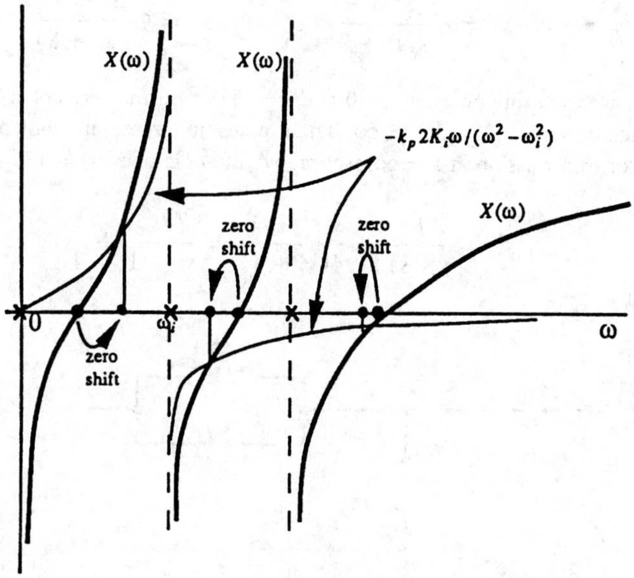

FIGURE 72.4 Zero shifting by weakening a finite nonzero pole.

Zero Producing by Complete Pole Removal. After a zero of transmission has been shifted to a desired location by the partial removal of an appropriate pole, the realization of this zero of transmission is accomplished by the complete removal of the pole of the reciprocal function corresponding to the shifted zero. For the LC ladder two-ports, a series combination of an inductor L and a capacitor C produces a zero at its resonant frequency $\omega = 1/\sqrt{LC}$, and this network is used in the shunt arm in the ladder to produce the desired zero of transmission. Likewise, the parallel connection of L and C yields an infinite impedance at its resonant frequency, and is used in the series arm of the ladder. They are shown in Fig. 72.5.

Example 1. We wish to design a lossless two-port network terminated in a 100-Ω resistor to meet the specifications for the transfer voltage-ratio function

$$G_{12}(s) = \frac{V_2}{V_1} = K\frac{s^2+4}{s^3+4s^2+9s+4} \tag{72.11}$$

within a multiplicative constant.

Since magnitude scaling does not affect the voltage transfer ratio, without loss of generality, we first assume that the terminating resistor is 1 Ω. Equation (72.11) can be rewritten as

$$G_{12}(s) = K\frac{\dfrac{s^2+4}{s(s^2+9)}}{\dfrac{4(s^2+1)}{s(s^2+9)}+1} = \frac{-y_{21}}{y_{22}+Y_2} = \frac{-y_{21}}{y_{22}+1} \tag{72.12}$$

We can make the following identifications:

$$-\,y_{21}(s) = K\frac{s^2+4}{s(s^2+9)}, \qquad y_{22}(s) = 4\frac{s^2+1}{s(s^2+9)} \tag{72.13}$$

Both functions have the same poles at $s = 0$ and $s = \pm j3$, and the zeros of transmission of the ladder are located at $s = \pm j2$ and $s = \infty$. These poles and zeros are shown in Fig. 72.6. To realize the zero of transmission at $s = \infty$, we remove the pole of $z_1(s) = 1/y_{22}(s)$ at $s = \infty$.

$$z_1(s) = \frac{1}{y_{22}(s)} = \frac{s(s^2+9)}{4(s^2+1)} = \frac{2s}{s^2+1} + \frac{s}{4} \tag{72.14}$$

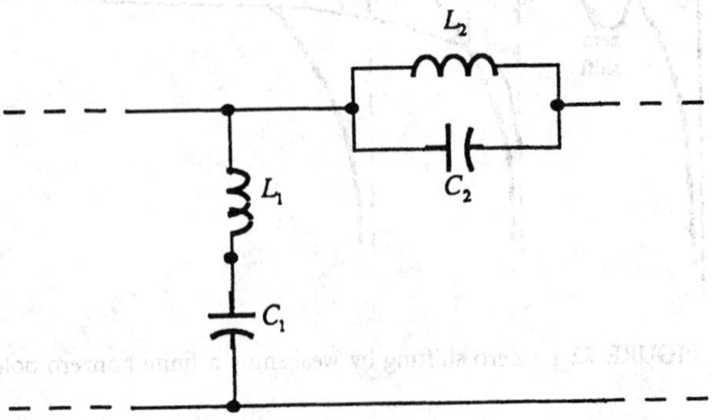

FIGURE 72.5 Zero producing sections in a ladder network.

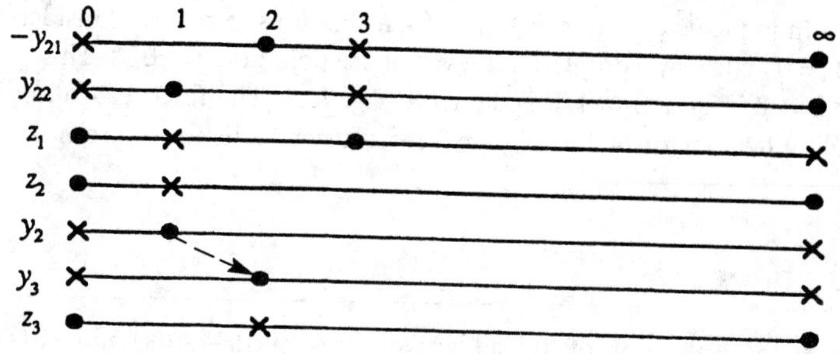

FIGURE 72.6 The poles and zeros of LC immittances.

After subtracting the term $s/4$ corresponding to an inductor of inductance 1/4 H shown in Fig. 72.7 from $z_1(s)$, the remaining impedance $z_2(s)$ is found to be

$$z_2(s) = \frac{2s}{s^2 + 1} \tag{72.15}$$

To realize the zero of transmission at $s = \pm j2$, we partially weaken the pole of the admittance

$$y_2(s) \equiv \frac{1}{z_2(s)} = \frac{s^2 + 1}{2s} \tag{72.16}$$

at $s = \infty$ in order to shift the zero at $s = \pm j$ to $s = \pm j2$, or

$$y_3(\pm j2) = y_2(s) - k_p \frac{1}{2} s \bigg|_{s=\pm j2} = \frac{-4 + 1}{\pm j2 \times 2} \mp k_p \frac{1}{2} j2 = 0 \tag{72.17}$$

yielding $k_p = 3/4$. The new function becomes

$$y_3(s) = \frac{s^2 + 1}{2s} - \frac{3}{8} s = \frac{s^2 + 4}{8s} \tag{72.18}$$

after the removal of a shunt capacitor of capacitance 3/8 F, as shown in Fig. 72.7. The factor $(s^2 + 4)$ in the numerator was anticipated because our objective was to produce a zero in the

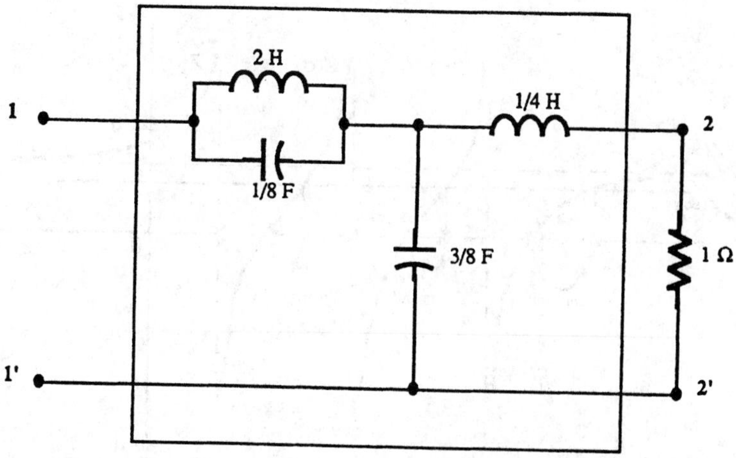

FIGURE 72.7 An LC ladder realization of the transfer voltage ratio (72.11).

driving-point admittance $y_3(s)$ at $s = \pm j2$. To realize this zero, we consider the reciprocal function $z_3(s) = 1/y_3(s)$ by complete removal of its pole at $s = \pm j2$. This yields a parallel connection of $L = 2$ H and $C = 1/8$ F shown in Fig. 72.7. The final realization is obtained by magnitude-scaling by a factor of 100. The realized constant K is found from the network to be $K = 1$.

72.3 The RC Ladder

We now consider the realization of RC ladder with prescribed $-y_{21}(s)$ and $y_{22}(s)$ or $z_{21}(s)$ and $z_{22}(s)$. Following the LC case, the zero shifting for the RC driving-point functions is accomplished by one or any combination of the following three operations:

1. The partial removal of a constant $Z(\infty)$ from $Z(s)$.
2. The partial removal of a constant $Y(0)$ from $Y(s)$.
3. The partial removal of a pole from $Z(s)$ or $Y(s)$.

The first operation permits a series resistance to be removed so that the resulting impedance is still positive real and possesses a desired zero of transmission:

$$Z_1(s) = Z(s) - k_p Z(\infty), \qquad k_p \leq 1 \tag{72.19}$$

the zeros of which occur at those values of σ satisfying

$$Z(\sigma) = k_p Z(\infty), \qquad k_p \leq 1 \tag{72.20}$$

Observe that from Fig. 72.8 all zeros are shifted toward $s = \sigma = -\infty$ by the partial removal of $Z(\infty)$ because the slope is negative for the RC impedances.

The second operation corresponds to the removal of a shunt resistance, and the remaining admittance

$$Y_1(s) = Y(s) - k_p Y(0), \qquad k_p \leq 1 \tag{72.21}$$

is still positive real, the zeros of which occur at those values of σ satisfying

$$Y(\sigma) = k_p Y(0), \qquad k_p \leq 1 \tag{72.22}$$

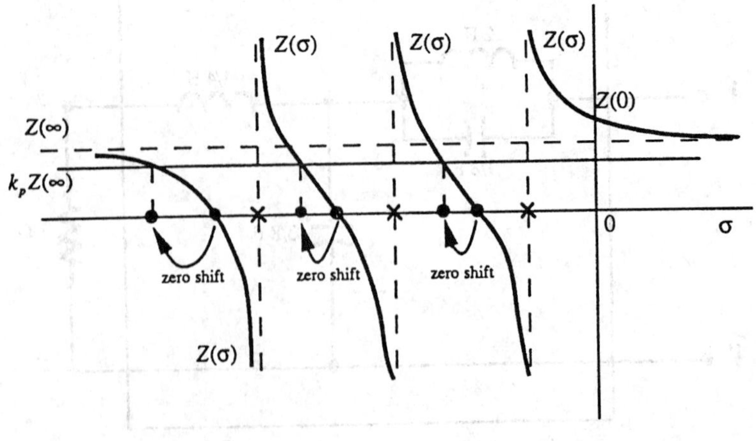

FIGURE 72.8 Zero shifting by the partial removal of $Z(\infty)$.

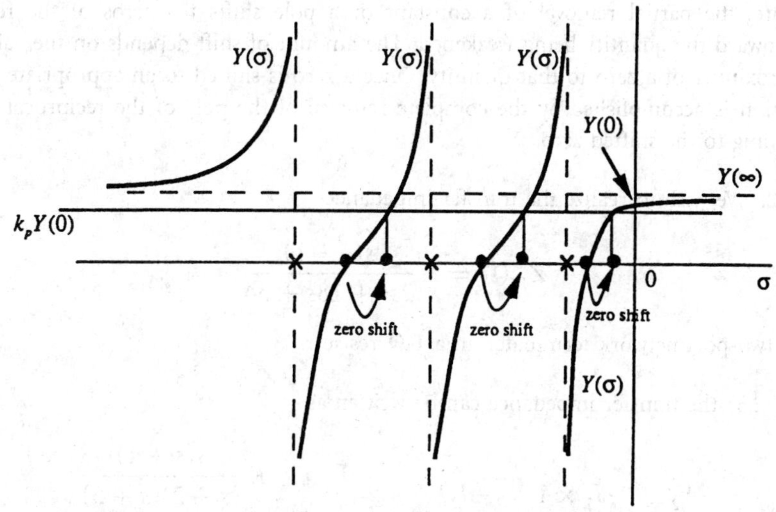

FIGURE 72.9 Zero shifting by partial removal of $Y(0)$.

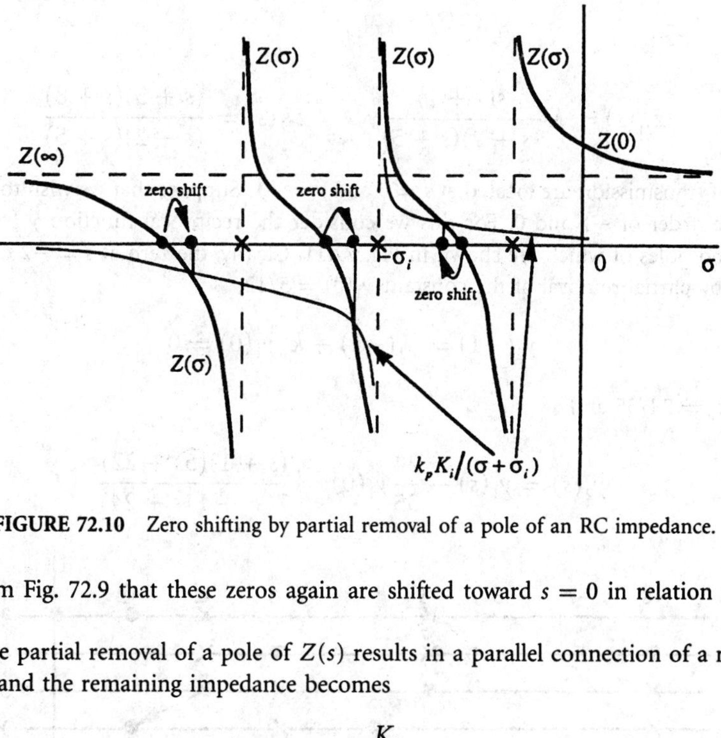

FIGURE 72.10 Zero shifting by partial removal of a pole of an RC impedance.

Observe from Fig. 72.9 that these zeros again are shifted toward $s = 0$ in relation to those of $Y(s)$.

Finally, the partial removal of a pole of $Z(s)$ results in a parallel connection of a resistor and a capacitor, and the remaining impedance becomes

$$Z_2(s) = Z(s) - k_p \frac{K_i}{s + \sigma_i}, \qquad k_p < 1 \tag{72.23}$$

the zeros of which occur at those values of σ satisfying

$$Z(\sigma) = k_p \frac{K_i}{\sigma + \sigma_i}, \qquad k_p < 1 \tag{72.24}$$

As indicated in Fig. 72.10, these zeros are again shifted toward the pole at $s = -\sigma_i$ being partially weakened.

Therefore, the partial removal of a constant or a pole shifts the zeros of the remaining function toward the quantity being weakened. The amount of shift depends on the value of k_p and the proximity of a zero to that quantity. Once a zero is shifted to an appropriate location, its realization is accomplished by the complete removal of the pole of the reciprocal function corresponding to the shifted zero.

Example 2. We wish to realize the transfer impedance

$$Z_{12}(s) = K\frac{s(s+1)}{2s^2 + 18s + 34} \tag{72.25}$$

of an RC two-port network terminated in a 1-Ω resistor.

From (72.3) the transfer impedance can be written as

$$Z_{12}(s) = \frac{V_2}{I_1} = \frac{-I_2 \times 1}{I_1} = \frac{-I_2}{I_1} = \frac{z_{21}}{z_{22}+1} = \frac{K\dfrac{s(s+1)}{(s+2)(s+5)}}{\dfrac{(s+3)(s+8)}{(s+2)(s+5)}+1} \tag{72.26}$$

identifying

$$z_{21}(s) = K\frac{s(s+1)}{(s+2)(s+5)}, \qquad z_{22}(s) = \frac{(s+3)(s+8)}{(s+2)(s+5)} \tag{72.27}$$

The zeros of transmission are located at $s = 0$ and $s = -1$. Suppose that we wish to realize these zeros in the order of -1 and 0. For this we consider the reciprocal function $y_1(s) = 1/z_{22}(s)$, the zeros and poles of which are shown in Fig. 72.11. Clearly, the zero at $s = -2$ can be shifted to $s = -1$ by partial removal of the constant $y_1(0) = 5/12$,

$$y_2(-1) = y_1(-1) - k_p y_1(0) = 0 \tag{72.28}$$

obtaining $k_p = 24/35$ and

$$y_2(s) = y_1(s) - \frac{24}{35}y_1(0) = \frac{(s+1)(5s+22)}{7(s^2+11s+24)} \tag{72.29}$$

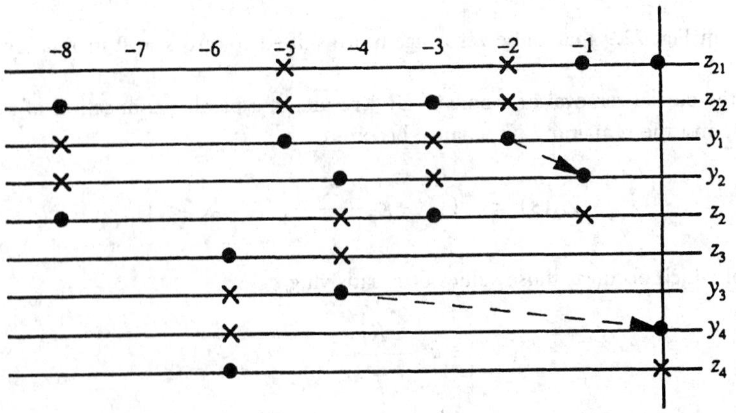

FIGURE 72.11 The zeros and poles of RC immittances.

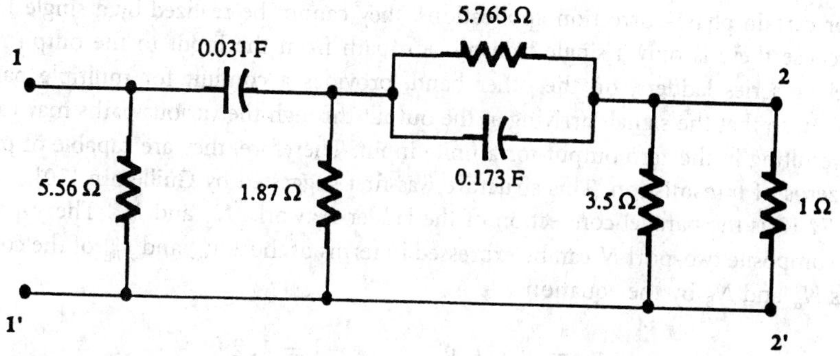

FIGURE 72.12 An RC ladder realization of the transfer impedance $Z_{12}(s)$.

This admittance has a zero at $s = -1$, as expected. The partially removed constant $k_p y_1(0)$ corresponds to a shunt resistor of resistance 3.5 Ω. To realize the zero of transmission at $s = -1$, we consider the reciprocal function $z_2(s) = 1/y_2(s)$ by the complete removal of its pole at $s = -1$. The remaining impedance $z_3(s)$ becomes

$$z_3(s) = z_2(s) - \frac{\frac{98}{17}}{s+1} = \frac{7}{5} \times \frac{s + \frac{100}{17}}{s + \frac{22}{5}} \tag{72.30}$$

The removed pole corresponds to a parallel connection of a resistor of resistance 98/17 Ω and a capacitor of capacitance 17/98 F, as illustrated in Fig. 72.12.

For the zero of transmission at $s = 0$, we consider the reciprocal function $y_3(s) = 1/z_3(s)$. The zero of $y_3(s)$ at $s = -22/5$ can be shifted to 0 by partial removal of the constant $y_3(0) = 187/350$, or

$$y_4(0) = y_3(0) - k_p y_3(0) = 0 \tag{72.31}$$

yielding $k_p = 1$. The remaining admittance $y_4(s)$ becomes

$$y_4(s) = y_3(s) - y_3(0) = \frac{\frac{9}{50}s}{s + \frac{100}{17}} \tag{72.32}$$

showing a zero at the origin, as anticipated. This zero is realized by the complete removal of the pole at the origin of its reciprocal $z_4(s) = 1/y_4(s)$, yielding

$$z_5(s) = z_4(s) - \frac{32.68}{s} = 5.56 \tag{72.33}$$

The complete realization is presented in Fig. 72.12, from which the constant K is found to be $K = 1$.

72.4 The Parallel or Series Ladders

The zeros of transmission of the LC ladders are restricted to the $j\omega$-axis, and those of the RC ladders to the negative real axis of the s-plane. For complex zeros of transmission such as those needed for certain phase-correction applications, they cannot be realized by a single LC or RC ladder because there is only a single transmission path from the input to the output. The use of parallel or series ladders, on the other hand, provides a conduit for multiple path signal transmission, so that the signals arriving at the output through the various paths may cancel one another, resulting in the zero output for a finite input. Therefore, they are capable of producing complex zeros of transmission. This structure was first suggested by Guillemin [10].

Figure 72.13 is the parallel connection of the ladder networks N_α and N_β. The y-parameters y_{ij} of the composite two-port N can be expressed in terms of those $y'_{ij\alpha}$ and $y'_{ij\beta}$ of the component two-ports N_α and N_β by the equation

$$y_{ij} = y'_{ij\alpha} + y'_{ij\beta}, \qquad i, j = 1, 2 \tag{72.34}$$

Thus, to realize $-y_{21}(s)$ and $y_{22}(s)$, we may separate them into pairs like $-y'_{21\alpha}$, $y'_{22\alpha}$ and $-y'_{21\beta}$, $y'_{22\beta}$, and realize an individual pair as an LC or RC ladder. Then connect these individual ladders in parallel to realize $-y_{21}(s)$ and $y_{22}(s)$. In order for the procedure to succeed, we must resolve the following problem. Recall that in the Cauer development of LC and RC ladders, $-y_{21}(s)$ is realized only within the multiplicative constant k. Thus, the transfer admittances realized by the component two-ports actually will be $-k_\alpha y'_{21\alpha}$ and $-k_\beta y'_{21\beta}$. The sum of these two functions will not result in the desired $-ky_{21}$ unless $k = k_\alpha = k_\beta$. To circumvent this difficulty, we introduce an additional degree of freedom by adjusting the admittance level of the α-ladder N_α by a factor b_α and the β-ladder N_β by b_β. Then the functions of the resulting realizations become

$$-y'_{21\alpha} = -b_\alpha k_\alpha y_{21\alpha}, \qquad y'_{22\alpha} = b_\alpha y_{22\alpha} \tag{72.35a}$$

$$-y'_{21\beta} = -b_\beta k_\beta y_{21\beta}, \qquad y'_{22\beta} = b_\beta y_{22\beta} \tag{72.35b}$$

where $y_{ij} = y_{ij\alpha} + y_{ij\beta}$, $i, j = 1, 2$. Substituting these in (72.34) gives

$$y_{21} = b_\alpha k_\alpha y_{21\alpha} + b_\beta k_\beta y_{21\beta} \tag{72.36a}$$

$$y_{22} = b_\alpha y_{22\alpha} + b_\beta y_{22\beta} \tag{72.36b}$$

Our objective is to choose b_α and b_β to satisfy the above equations, once k_α and k_β are known. One way to meet these requirements is to let $y'_{22\alpha}$ and $y'_{22\beta}$ have the same zeros and poles as y_{22} but different scale factors such that $y'_{22\alpha} = b_\alpha y_{22}$ and $y'_{22\beta} = b_\beta y_{22}$, obtaining from (72.36b)

$$b_\alpha + b_\beta = 1 \tag{72.37}$$

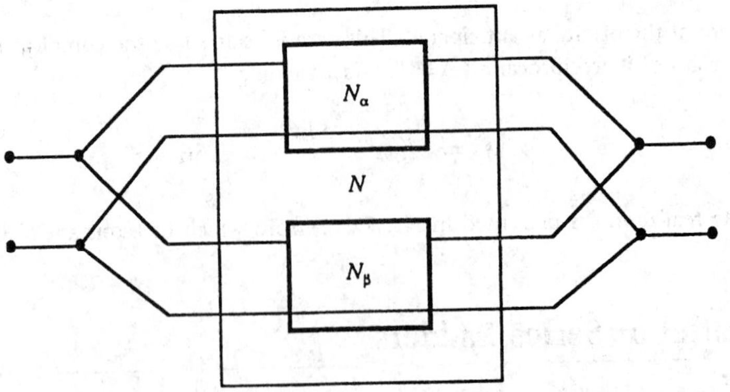

FIGURE 72.13 The parallel connection of two ladder networks.

Since we can only realize y_{21} within a multiplicative constant, we replace y_{21} by ky_{21} in (72.36a), and set

$$b_\alpha k_\alpha = b_\beta k_\beta = k \tag{72.38}$$

These two equations can be solved to yield the desired scale factors b_α and b_β.

In general, for m ladders in parallel, we require

$$b_1 + b_2 + \cdots + b_m = 1 \tag{72.39a}$$

$$b_1 k_1 = b_2 k_2 = \cdots = b_m k_m = k \tag{72.39b}$$

where b_i is the admittance scale factor for the ith ladder, and k_i is the realized multiplicative constant of the transfer admittance of the ith ladder. Once k_i are known, these m simultaneous equations can be solved for the m unknowns b_i, and the admittance level of each ladder can be scaled accordingly. The scaled ladders are connected in parallel to realize the $-y_{21}(s)$ specifications within the multiplicative constant k, and the $y_{22}(s)$ specifications exactly.

Similar results are obtained by using the z-parameters $z_{ij}(s)$ and the series connection of the component two-port networks, the details of which are omitted. However, a design example will be presented below.

Example 3. We wish to realize an RC two-port network to meet the following specifications:

$$-y_{21}(s) = k\frac{s^2+1}{(s+5)(s+9)} \tag{72.40a}$$

$$y_{22}(s) = \frac{(s+3)(s+7)}{(s+5)(s+9)} \tag{72.40b}$$

Since the zeros of transmission are located at $s = \pm j1$, they cannot be realized by a single RC ladder. For our purposes, we choose the y-parameters of the component two-ports as

$$-y_{21\alpha}(s) = k_\alpha \frac{s^2}{(s+5)(s+9)}, \qquad y_{22\alpha}(s) = y_{22}(s) \tag{72.41a}$$

$$-y_{21\beta}(s) = \frac{k_\beta}{(s+5)(s+9)}, \qquad y_{22\beta}(s) = y_{22}(s) \tag{72.41b}$$

For the α-ladder, since the zeros of transmission are all at the origin, they can be realized by the second Cauer canonical form shown in Fig. 72.14 with $k_\alpha = 0.043$. For the β-ladder, since the zeros of transmission are all at the infinity, they can be realized by the first Cauer canonical form of Fig. 72.15 with $k_\beta = 21$.

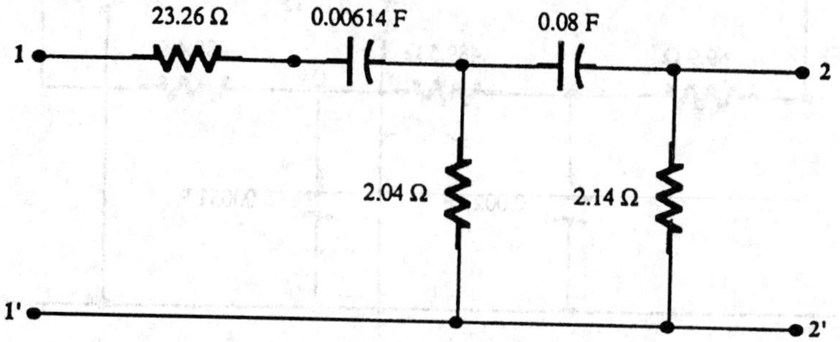

FIGURE 72.14 An RC α-ladder realization.

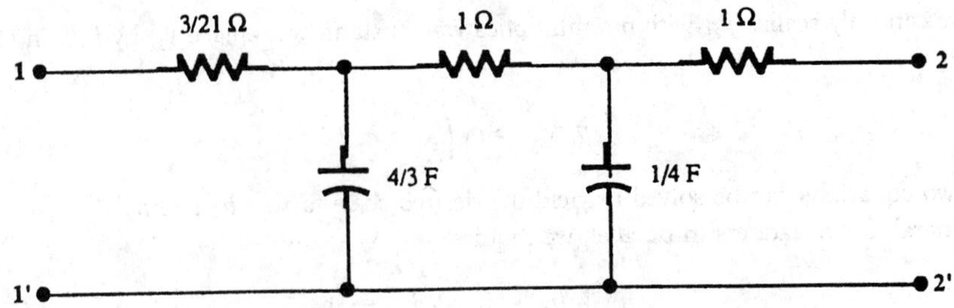

FIGURE 72.15 An RC β-ladder realization.

Our next task is to adjust the admittance level of the individual ladders so that when they are connected in parallel, the $y_{22}(s)$ specifications are realized exactly, and the $-y_{21}(s)$ specifications are realized to within a multiplicative constant k. Appealing to (72.37) and (72.38), we have

$$b_\alpha + b_\beta = 1 \tag{72.42a}$$

$$0.043 b_\alpha = 21 b_\beta = k \tag{72.42b}$$

yielding

$$b_\alpha = 0.998 \cong 1, \qquad b_\beta = 2.043 \times 10^{-3} \tag{72.43}$$

We now adjust the admittance level of the β-ladder by a factor $b_\beta = 2.043 \times 10^{-3}$, leaving the α-ladder intact. The final realization is achieved by the parallel connection of the α-ladder and the resulting β-ladder shown in Fig. 72.16. The realized multiplicative constant for the overall transfer admittance $-y_{21}(s)$ is found to be $k = 0.043$.

Example 4. We wish to realize the open-circuit transfer voltage ratio

$$G_{12}(s) = \frac{z_{21}(s)}{z_{11}(s)} = k\frac{s^2 - 2s + 10}{s^2 + 8s + 15} \tag{72.44}$$

by an RC two-port network.

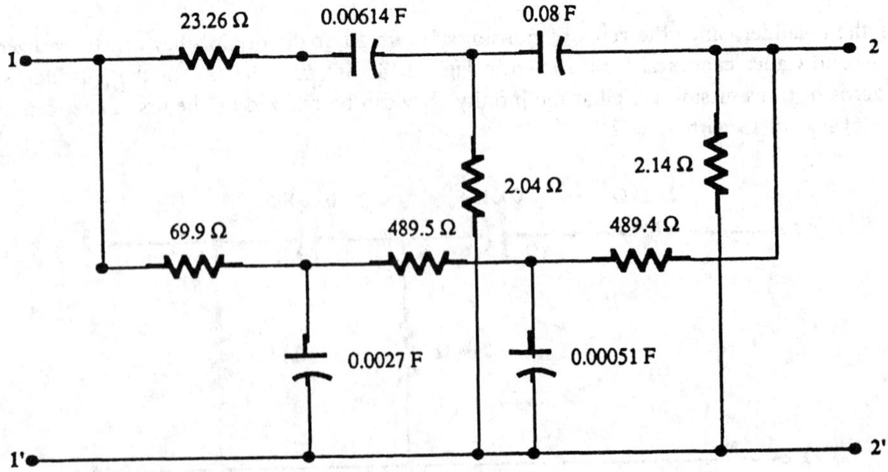

FIGURE 72.16 The parallel RC ladder realization of $-y_{21}(s)$ and $y_{22}(s)$.

First, we multiply the numerator and denominator of $G_{12}(s)$ by the factor $(s+2)$ to yield

$$G_{12}(s) = \frac{z_{21}(s)}{z_{11}(s)} = k\frac{s^3+6s+20}{(s+2)(s+3)(s+5)} \tag{72.45}$$

and then divide by $(s+1)(s+2.5)(s+4)$ in order to make the following identifications:

$$z_{21}(s) = k\frac{s^3+6s+20}{(s+1)(s+2.5)(s+4)} \tag{72.46a}$$

$$z_{11}(s) = \frac{(s+2)(s+3)(s+5)}{(s+1)(s+2.5)(s+4)} \tag{72.46b}$$

We next decompose the pairs into three pairs as follows:

$$z_{21\alpha}(s) = k_\alpha\frac{s^3}{(s+1)(s+2.5)(s+4)}, \qquad z_{11\alpha}(s) = z_{11}(s) \tag{72.47a}$$

$$z_{21\beta}(s) = k_\beta\frac{6s}{(s+1)(s+2.5)(s+4)}, \qquad z_{11\beta}(s) = z_{11}(s) \tag{72.47b}$$

$$z_{21\gamma}(s) = k_\gamma\frac{20}{(s+1)(s+2.5)(s+4)}, \qquad z_{11\gamma}(s) = z_{11}(s) \tag{72.47c}$$

1. The α-ladder. Since all the zeros of transmission are located at the origin, it can be realized by the second Cauer canonical form of Fig. 72.17 with $k_\alpha \approx 1$.

2. The β-ladder. Since one of the zeros of transmission is located at the origin, and two others at the infinity, the first half of the β-ladder can be realized as the second Cauer canonical form and the second half of the β-ladder by the first Cauer canonical form as illustrated in Fig. 72.18 with $k_\beta = 0.0117$.

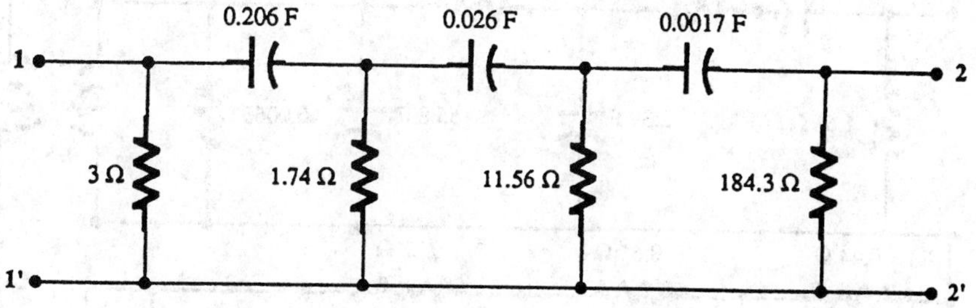

FIGURE 72.17 The α-ladder.

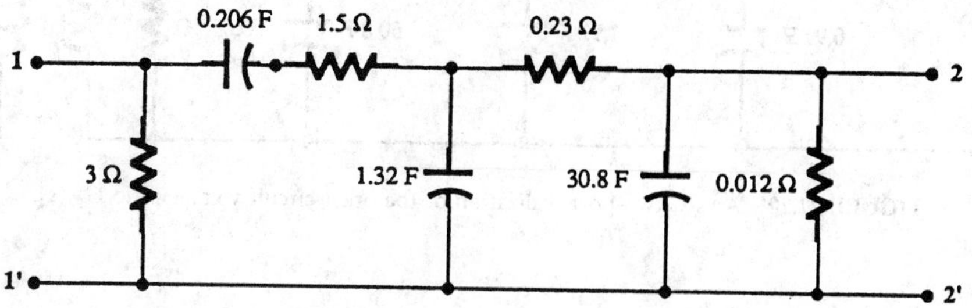

FIGURE 72.18 The β-ladder.

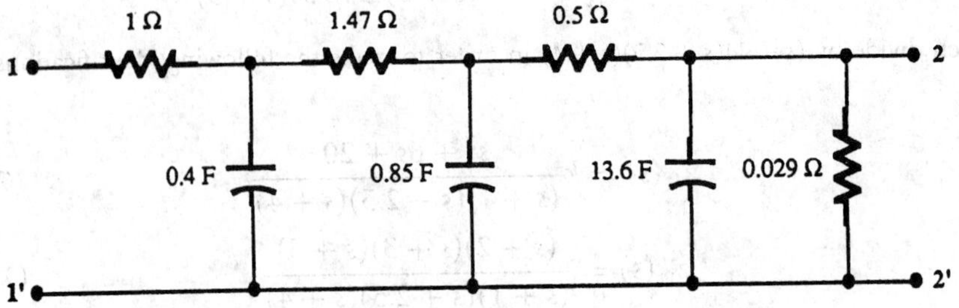

FIGURE 72.19 The γ-ladder.

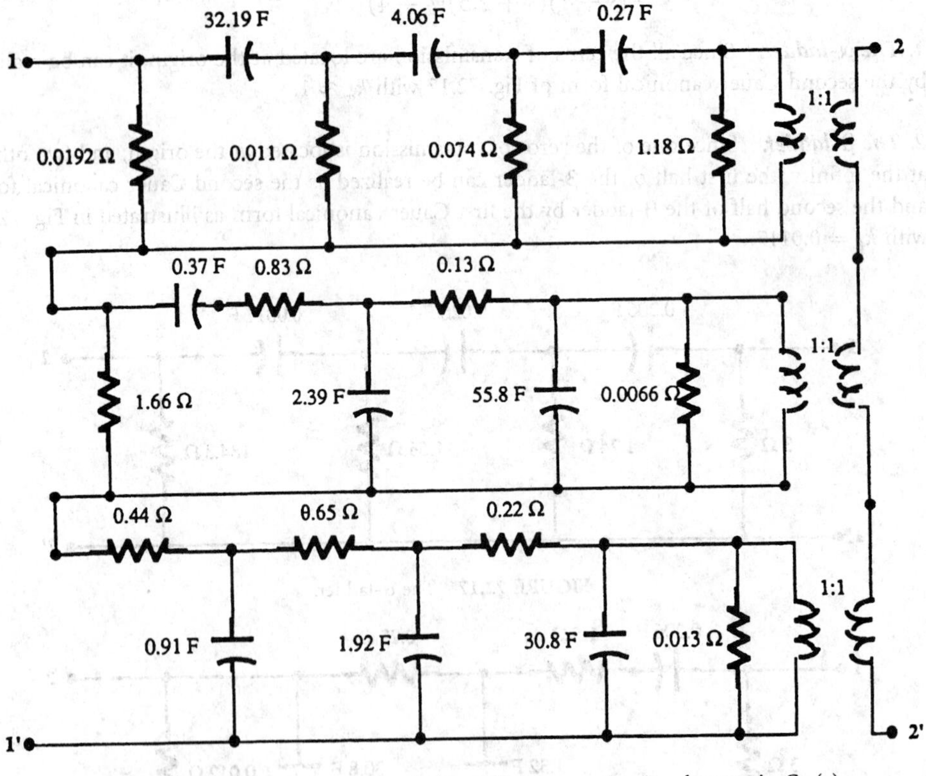

FIGURE 72.20 An RC two-port realization of the open-circuit voltage ratio $G_{12}(s)$.

3. *The γ-ladder.* Since all of its zeros of transmission are located at the infinity, it can be realized by the first Cauer canonical form of Fig. 72.19 with $k_\gamma = 0.0145$.

We next adjust the admittance level of the ladders so that when they are connected in parallel, the desired specifications are realized. From (72.39) we require that

$$b_\alpha + b_\beta + b_\gamma = 1 \tag{72.48a}$$

$$b_\alpha k_\alpha = b_\beta k_\beta = b_\gamma k_\gamma = k \tag{72.48b}$$

These equations can be solved to yield

$$b_\alpha = 0.0064, \qquad b_\beta = 0.552, \qquad b_\gamma = 0.4416 \tag{72.49}$$

with $k = 0.0064$. The final two-port network is shown in Fig. 72.20.

73

Design of Resistively Terminated Networks*

Wai-Kai Chen
University of Illinois, Chicago

73.1 Introduction

In the design of communication systems, it is frequently required to synthesize a coupling network that will transform a given frequency-dependent load impedance into another specified one. We refer to this operation as **impedance matching** or **equalization**, and the resulting coupling network as a **matching network** or **equalizer**.

Refer to the network configuration of Fig. 73.1 where the source is represented either by its Thévenin equivalent or by its Norton equivalent. Our objective here is to design a lossless two-port network or equalizer N, which when inserted between a resistive source and a resistive load will yield a preassigned transducer power-gain characteristic over the entire sinusoidal frequency spectrum. Explicit formulas for the design of Butterworth and Chebyshev LC ladder networks will be given. The more complicated situation where the load is frequency dependent will be dealt with in Chapter 74.

In the networks of Fig. 73.1, let $Z_{11}(s)$ and $Z_{22}(s)$ be the impedances looking into the input and output ports when the output and input ports are terminated in $z_2(s)$ and $z_1(s)$, respectively. The input and output **reflection coefficients** are defined by

$$\rho_{11}(s) = \frac{Z_{11}(s) - z_1(-s)}{Z_{11}(s) + z_1(s)} \tag{73.1a}$$

$$\rho_{22}(s) = \frac{Z_{22}(s) - z_2(-s)}{Z_{22}(s) + z_2(s)} \tag{73.1b}$$

respectively. We now demonstrate that the **transducer power gain** $G(\omega^2)$ defined as the ratio of average power delivered to the load to the maximum available average power at the source is given by

$$G(\omega^2) = 1 - |\rho_{11}(j\omega)|^2 \equiv |\rho_{21}(j\omega)|^2 \tag{73.2}$$

*References for this Chapter can be found on page 2333.

0-8493-8341-2/95/$0.00+$.50

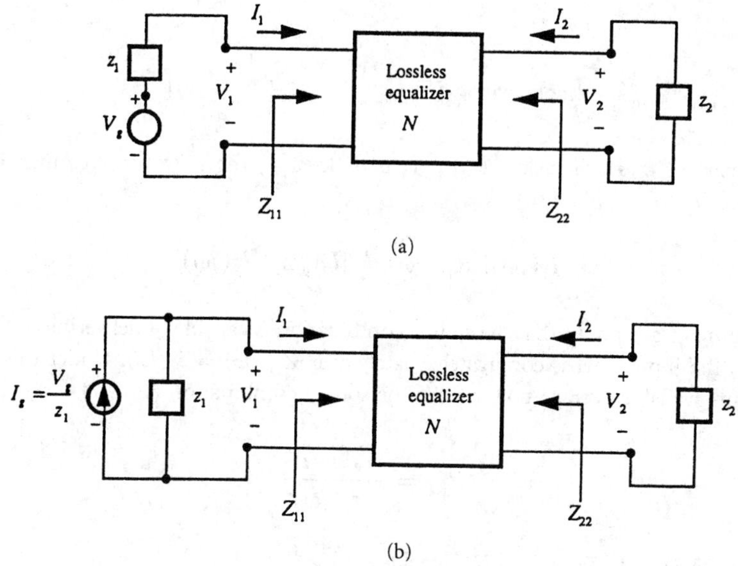

FIGURE 73.1 General broadband matching configuration.

where $\rho_{21}(s)$ is known as the **transmission coefficient**. To prove this, we first compute

$$1 - \rho_{11}(s)\rho_{11}(-s) = \frac{4r_1(s)R_{11}(s)}{[Z_{11}(s) + z_1(s)][Z_{11}(-s) + z_1(-s)]} \tag{73.3}$$

where

$$r_1(s) = \text{Ev}\, z_1(s) = \frac{1}{2}[z_1(s) + z_1(-s)] \tag{73.4a}$$

$$R_{11}(s) = \text{Ev}\, Z_{11}(s) = \frac{1}{2}[Z_{11}(s) + Z_{11}(-s)] \tag{73.4b}$$

are the even parts of $z_1(s)$ and $Z_{11}(s)$, respectively. Thus, the impedance facing the voltage source V_g is given by

$$\frac{V_g(s)}{I_1(s)} = Z_{11}(s) + z_1(s) \tag{73.5}$$

On the $j\omega$-axis, (73.3) reduces to

$$1 - |\rho_{11}(j\omega)|^2 = \frac{4r_1(j\omega)R_{11}(j\omega)}{|Z_{11}(j\omega) + z_1(j\omega)|^2} = \frac{|I_1(j\omega)|^2 R_{11}(j\omega)}{|V_g(j\omega)|^2/4r_1(j\omega)} \tag{73.6}$$

The power input to the network of Fig. 73.1(a) under sinusoidal steady state is

$$P_{\text{in}} = |I_1(j\omega)|^2 R_{11}(j\omega) \tag{73.7}$$

while the power output to the load is

$$P_{\text{out}} = |I_2(j\omega)|^2 r_2(j\omega) \tag{73.8}$$

where

$$r_2(s) = \text{Ev}\, z_2(s) = \frac{1}{2}[z_2(s) + z_2(-s)] \tag{73.9}$$

is the even part of $z_2(s)$. Since the two-port N is lossless, the power input must be equal to power output or

$$|I_1(j\omega)|^2 R_{11}(j\omega) = |I_2(j\omega)|^2 r_2(j\omega) \tag{73.10}$$

The maximum average power that the source combination is capable of delivering to the network occurs when the input port is conjugately matched or $Z_{11}(j\omega) = \bar{z}_1(j\omega)$. Under this condition, the maximum available average power from the source combination is

$$P_{\text{ava}} = \frac{|V_g(j\omega)|^2}{4r_1(j\omega)} \tag{73.11}$$

Substituting (73.8) and (73.11) in (73.6) yields

$$1 - |\rho_{11}(j\omega)|^2 = \frac{|I_2(j\omega)|^2 r_2(j\omega)}{|V_g(j\omega)|^2/4r_1(j\omega)} = \frac{P_{\text{out}}}{P_{\text{ava}}} = G(\omega^2) \tag{73.12}$$

Using (73.2) shows that

$$|\rho_{11}(j\omega)|^2 + |\rho_{21}(j\omega)|^2 = 1 \tag{73.13}$$

or

$$\begin{aligned} |\rho_{11}(j\omega)|^2 &= 1 - \frac{\text{average power to load}}{\text{average power available}} \\ &= \frac{\text{average power available} - \text{average power to load}}{\text{average power available}} \\ &= \frac{\text{“average reflected” power}}{\text{average power available}} \end{aligned} \tag{73.14}$$

$$|\rho_{21}(j\omega)|^2 = \frac{\text{average power to load}}{\text{average power available}} \tag{73.15}$$

Therefore, the magnitude squared of the reflection coefficient $|\rho_{11}(j\omega)|^2$ denotes the fraction of the maximum available average power that is reflected back to the source, and the magnitude squared of the transmission coefficient $|\rho_{21}(j\omega)|^2$ represents the fraction of the maximum available average power that is transmitted to the load from the source. In fact, their names are suggested by these interpretations. We remark that since the transducer power gain G is a function of ω^2, it is written as $G(\omega^2)$ to emphasize this.

We next express the transmission coefficient in terms of other specifications. From (73.12) we have

$$|\rho_{21}(j\omega)|^2 = 4r_1(j\omega)r_2(j\omega)\left|\frac{I_2(j\omega)}{V_g(j\omega)}\right|^2 \tag{73.16}$$

Substituting $V_g(j\omega) = z_1(j\omega)I_g(j\omega)$ in (73.16) gives

$$|\rho_{21}(j\omega)|^2 = \frac{4r_1(j\omega)r_2(j\omega)}{|z_1(j\omega)|^2}\left|\frac{I_2(j\omega)}{I_g(j\omega)}\right|^2 \tag{73.17}$$

In terms of the transfer voltage ratio and transfer impedance, we apply the relation $V_2(j\omega) = -I_2(j\omega)z_2(j\omega)$ and obtain

$$|\rho_{21}(j\omega)|^2 = \frac{4r_1(j\omega)r_2(j\omega)}{|z_2(j\omega)|^2}\left|\frac{V_2(j\omega)}{V_g(j\omega)}\right|^2 \tag{73.18}$$

$$|\rho_{21}(j\omega)|^2 = \frac{4r_1(j\omega)r_2(j\omega)}{|z_1(j\omega)z_2(j\omega)|^2}\left|\frac{V_2(j\omega)}{I_g(j\omega)}\right|^2 \tag{73.19}$$

Similarly, we can derive a relation between the output reflection coefficient $\rho_{22}(j\omega)$ and the transmission coefficient $\rho_{12}(j\omega)$ magnitude squared as

$$|\rho_{12}(j\omega)|^2 \equiv 1 - |\rho_{22}(j\omega)|^2 \tag{73.20}$$

In fact, for the lossless reciprocal two-port network N we have

$$|\rho_{21}(j\omega)|^2 = |\rho_{12}(j\omega)|^2 = 1 - |\rho_{11}(j\omega)|^2 = 1 - |\rho_{22}(j\omega)|^2 \tag{73.21}$$

73.2 Double-Terminated Butterworth Networks

In this part, we show how to design a lossless two-port network operating between a resistive generator with internal resistance R_1 and a resistive load with resistance R_2 to yield the nth-order Butterworth transducer power-gain characteristic

$$G(\omega^2) = |\rho_{21}(j\omega)|^2 = \frac{K_n}{1 + (\omega/\omega_c)^{2n}} \tag{73.22}$$

Since for a passive network $G(\omega^2)$ is bounded between 0 and 1, the DC gain K_n is restricted by

$$0 \le K_n \le 1 \tag{73.23}$$

Substituting (73.22) in (73.12) yields the squared magnitude of the input reflection coefficient as

$$|\rho_{11}(j\omega)|^2 = 1 - G(\omega^2) = 1 - |\rho_{21}(j\omega)|^2 = \frac{1 - K_n + (\omega/\omega_c)^{2n}}{1 + (\omega/\omega_c)^{2n}} \tag{73.24}$$

or

$$\rho_{11}(j\omega)\rho_{11}(-j\omega) = \alpha^{2n}\frac{1 + (\omega/\alpha\omega_c)^{2n}}{1 + (\omega/\omega_c)^{2n}} \tag{73.25}$$

where

$$\alpha = (1 - K_n)^{1/2n} \tag{73.26}$$

Appealing to analytic continuation by substituting ω by $-js$ results in

$$\rho_{11}(s)\rho_{11}(-s) = \alpha^{2n}\frac{1+(-1)^n x^{2n}}{1+(-1)^n y^{2n}} \tag{73.27}$$

where

$$y = \frac{s}{\omega_c}, \qquad x = \frac{y}{\alpha} \tag{73.28}$$

To obtain the input reflection coefficient $\rho_{11}(s)$ from $\rho_{11}(s)\rho_{11}(-s)$, we need to assign the zeros and poles of (73.27). Since $\rho_{11}(s)$ is devoid of poles in the closed RHS, we must assign all the LHS poles to $\rho_{11}(s)$. The zeros of $\rho_{11}(s)$, however, may lie in the RHS, so that in general a number of different numerators are possible. For our purposes, we choose only the LHS zeros for $\rho_{11}(s)$. Define a **minimum-phase reflection coefficient** to be one that is devoid of zeros in the open RHS. Then the minimum-phase solution of (73.27) can be written as

$$\rho_{11}(s) = \pm\alpha^n \frac{q(x)}{q(y)} \tag{73.29}$$

where $q(x)$ is the Hurwitz polynomial with unity leading coefficient formed by the LHS roots of the equation $1+(-1)^n x^{2n} = 0$. From (73.1a) the input impedance is found to be

$$Z_{11}(s) = R_1 \frac{1+\rho_{11}(s)}{1-\rho_{11}(s)} \tag{73.30}$$

Combining this with (73.29) yields

$$Z_{11}(s) = R_1 \frac{q(y) \pm \alpha^n q(x)}{q(y) \mp \alpha^n q(x)} \tag{73.31}$$

If both R_1 and R_2 are specified, then the DC gain K_n cannot be chosen independently. In fact, by substituting $s = 0$ in (73.31) and assuming that $K_n \neq 0$ we obtain

$$\frac{R_2}{R_1} = \left(\frac{1+\alpha^n}{1-\alpha^n}\right)^{\pm 1} \tag{73.32}$$

where the $\pm$ signs are determined, respectively, according to $R_2 \geq R_1$ and $R_2 \leq R_1$. Therefore, if any two of the three quantities R_1, R_2, and K_n are specified, the third one is fixed.

We now show that the input impedance $Z_{11}(s)$ can be realized by an LC ladder terminated in a resistor. In fact, explicit formulas for their element values will be given, thereby reducing the design problem to simple arithmetic. Depending upon the choice of the plus and minus signs in (73.32), two cases are distinguished.

CASE 1. $\rho_{11}(0) \geq 0$. With the choice of the plus sign, the input impedance becomes

$$Z_{11}(s) = R_1 \frac{q(y) + \alpha^n q(x)}{q(y) - \alpha^n q(x)} \tag{73.33}$$

which can be expanded in a continued fraction about infinity, as in the first Cauer canonical form, and results in an LC ladder terminated in a resistor:

$$Z_{11}(s) = L_1 s + \cfrac{1}{C_2 s + \cfrac{1}{L_3 s + \cfrac{1}{\ddots + \cfrac{1}{W}}}} \tag{73.34}$$

where W is a constant representing either a resistance or conductance. Depending upon whether n is odd or even, the LC ladder has the configuration of Fig. 73.2. The element values can be computed by the following recurrence formulas:

$$L_1 = \frac{2R_1 \sin \pi/2n}{(1-\alpha)\omega_c} \tag{73.35}$$

$$L_{2m-1}C_{2m} = \frac{4 \sin \gamma_{4m-3} \sin \gamma_{4m-1}}{\omega_c^2(1 - 2\alpha \cos \gamma_{4m-2} + \alpha^2)} \tag{73.36a}$$

$$L_{2m+1}C_{2m} = \frac{4 \sin \gamma_{4m-1} \sin \gamma_{4m+1}}{\omega_c^2(1 - 2\alpha \cos \gamma_{4m} + \alpha^2)} \tag{73.36b}$$

for $m = 1, 2, \ldots, \lceil n/2 \rceil$, the largest integer not greater than $n/2$; where

$$\gamma_m = \frac{m\pi}{2n} \tag{73.37}$$

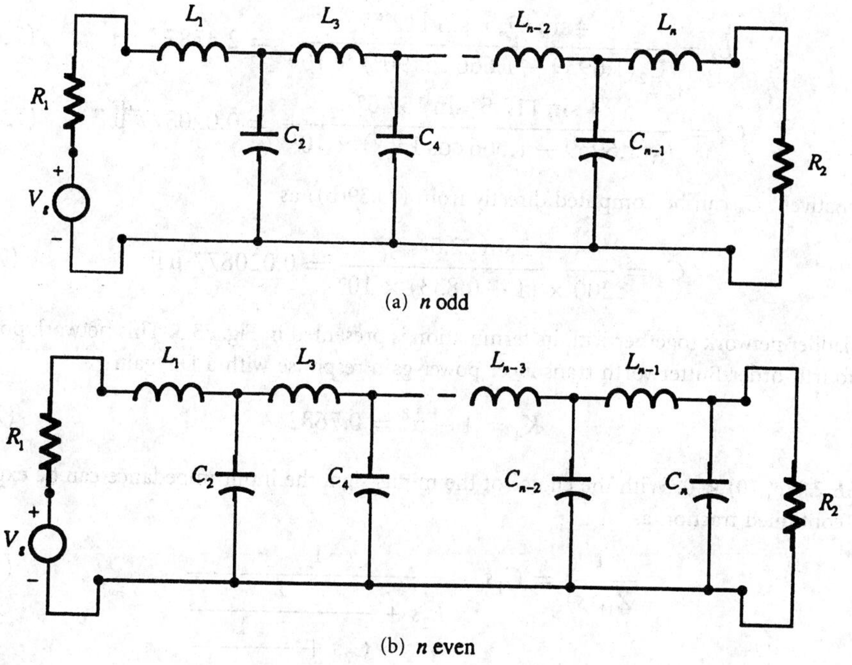

FIGURE 73.2 The Butterworth LC ladder networks for $\rho_{11}(0) \geq 0$.

The values of the final elements can also be calculated directly by

$$L_n = \frac{2R_2 \sin \pi/2n}{(1+\alpha)\omega_c}, \quad n \text{ odd} \tag{73.38a}$$

$$C_n = \frac{2 \sin \pi/2n}{R_2(1+\alpha)\omega_c}, \quad n \text{ even} \tag{73.38b}$$

A complete derivation of these formulas was first given by Bossé [1]. Hence we can calculate the element values starting from either the first or the last element. When $R_1 = R_2$, formulas (73.36) reduce to

$$L_{2m-1} = \frac{2R_1 \sin \gamma_{4m-3}}{\omega_c} \tag{73.39a}$$

$$C_{2m} = \frac{2 \sin \gamma_{4m-1}}{R_1\omega_c} \tag{73.39b}$$

Example 1. Given

$$R_1 = 70\ \Omega, \qquad R_2 = 200\ \Omega, \qquad \omega_c = 10^5 \text{ rad/s}, \qquad n = 4 \tag{73.40}$$

obtain a Butterworth LC ladder to meet these specifications.

Since $R_2 > R_1$, we choose the plus sign in (73.32) and obtain $\rho_{11}(0) \geq 0$, $\alpha = 0.833$, and $\gamma_m = 22.5m$. Thus, from (73.35) and (73.36) the element values are found to be

$$L_1 = \frac{2 \times 70 \sin 22.5°}{(1-0.833) \times 10^5} = 3.2081 \text{ mH} \tag{73.41a}$$

$$C_2 = \frac{4 \sin 22.5° \sin 67.5°}{L_1(1.6939 - 1.666 \cos 45°) \times 10^{10}} = 0.085456\ \mu\text{F} \tag{73.41b}$$

$$L_3 = \frac{4 \sin 67.5° \sin 112.5°}{C_2(1.6939 - 1.666 \cos 90°) \times 10^{10}} = 2.3587 \text{ mH} \tag{73.41c}$$

$$C_4 = \frac{4 \sin 112.5° \sin 157.5°}{L_3(1.6939 - 1.666 \cos 135°) \times 10^{10}} = 0.020877\ \mu\text{F} \tag{73.41d}$$

Alternatively, C_4 can be computed directly from (73.39(b)) as

$$C_4 = \frac{2 \sin 22.5°}{200 \times (1+0.833) \times 10^5} = 0.020877\ \mu\text{F} \tag{73.42}$$

The ladder network together with its termination is presented in Fig. 73.3. This network possesses the fourth-order Butterworth transducer power-gain response with a DC gain

$$K_4 = 1 - \alpha^8 = 0.7682 \tag{73.43}$$

CASE 2. $\rho_{11}(0) < 0$. With the choice of the minus sign, the input impedance can be expanded in a continued fraction as

$$\frac{1}{Z_{11}(s)} = C_1 s + \cfrac{1}{L_2 s + \cfrac{1}{C_3 s + \cfrac{1}{\ddots + \cfrac{1}{W}}}} \tag{73.44}$$

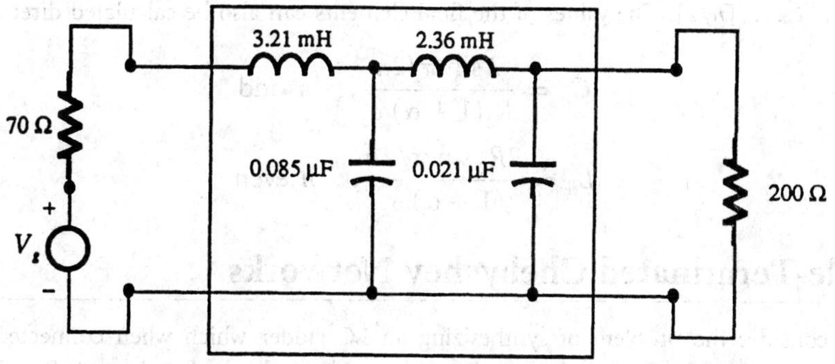

FIGURE 73.3 A fourth-order Butterworth LC ladder network.

which can be realized by the LC ladder networks of Fig. 73.4, depending on whether n is even or odd, where W is the terminating resistance or conductance. Formulas for the element values are similar to those given in (73.35)–(73.39) except that the roles of C's and L's are interchanged and R_1 and R_2 are replaced by their reciprocals:

$$C_1 = \frac{2 \sin \pi/2n}{R_1(1-\alpha)\omega_c} \tag{73.45a}$$

$$C_{2m-1}L_{2m} = \frac{4 \sin \gamma_{4m-3} \sin \gamma_{4m-1}}{\omega_c^2(1 - 2\alpha \cos \gamma_{4m-2} + \alpha^2)} \tag{73.45b}$$

$$C_{2m+1}L_{2m} = \frac{4 \sin \gamma_{4m-1} \sin \gamma_{4m+1}}{\omega_c^2(1 - 2\alpha \cos \gamma_{4m} + \alpha^2)} \tag{73.45c}$$

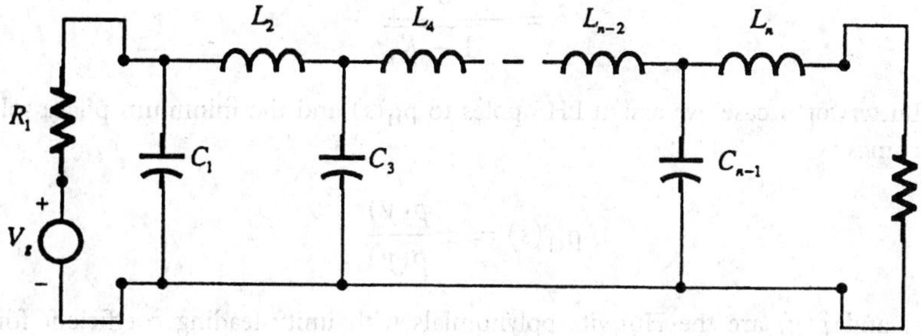

(a) *n* even

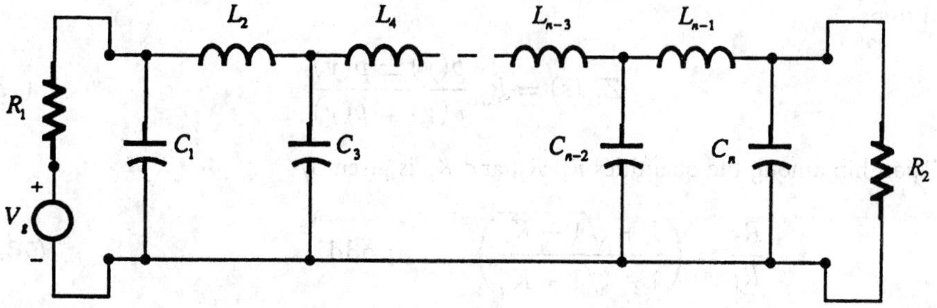

(b) *n* odd

FIGURE 73.4 The Butterworth LC ladder networks for $\rho_{11}(0) < 0$.

for $m = 1, 2, \ldots, \lceil n/2 \rceil$. The values of the final elements can also be calculated directly by

$$C_n = \frac{2\sin\pi/2n}{R_2(1+\alpha)\omega_c}, \quad n \text{ odd} \tag{73.46a}$$

$$L_n = \frac{2R_2\sin\pi/2n}{(1+\alpha)\omega_c}, \quad n \text{ even} \tag{73.46b}$$

73.3 Double-Terminated Chebyshev Networks

Now we consider the problem of synthesizing an LC ladder which when connected between a resistive source of internal resistance R_1 and a resistive load of resistance R_2 will yield a preassigned Chebyshev transducer power-gain characteristic

$$G(\omega^2) = |\rho_{21}(j\omega)|^2 = \frac{K_n}{1+\varepsilon^2 C_n^2(\omega/\omega_c)} \tag{73.47}$$

with K_n bounded between 0 and 1. Following (73.24), the squared magnitude of the input reflection coefficient can be written as

$$|\rho_{11}(j\omega)|^2 = 1 - G(\omega^2) = 1 - |\rho_{21}(j\omega)|^2 = \frac{1 - K_n + \varepsilon^2 C_n^2(\omega/\omega_c)}{1+\varepsilon^2 C_n^2(\omega/\omega_c)} \tag{73.48}$$

Appealing to analytic continuation, we obtain

$$\rho_{11}(s)\rho_{11}(-s) = (1-K_n)\frac{1+\hat{\varepsilon}^2 C_n^2(-jy)}{1+\varepsilon^2 C_n^2(-jy)} \tag{73.49}$$

where

$$\hat{\varepsilon} = \frac{\varepsilon}{\sqrt{1-K_n}} \tag{73.50}$$

As in the Butterworth case, we assign LHS poles to $\rho_{11}(s)$ and the minimum-phase solution of (73.49) becomes

$$\rho_{11}(s) = \pm\frac{\hat{p}(y)}{p(y)} \tag{73.51}$$

where $p(y)$ and $\hat{p}(y)$ are the Hurwitz polynomials with unity leading coefficient formed by the LHS roots of the equations $1 + \varepsilon^2 C_n^2(-jy) = 0$ and $1 + \hat{\varepsilon}^2 C_n^2(-jy) = 0$, respectively. From (73.1a), the input impedance of the LC ladder when the output port is terminated in R_2 is found to be

$$Z_{11}(s) = R_1\frac{p(y) \pm \hat{p}(y)}{p(y) \mp \hat{p}(y)} \tag{73.52}$$

A relationship among the quantities R_1, R_2, and K_n is given by

$$\frac{R_2}{R_1} = \left(\frac{1+\sqrt{1-K_n}}{1-\sqrt{1-K_n}}\right)^{\pm 1}, \quad n \text{ odd} \tag{73.53a}$$

$$= \left(\frac{\sqrt{1+\varepsilon^2}+\sqrt{1+\varepsilon^2-K_n}}{\sqrt{1+\varepsilon^2}-\sqrt{1+\varepsilon^2-K_n}}\right)^{\pm 1}, \quad n \text{ even} \tag{73.53b}$$

where the $\pm$ signs are determined, respectively, according to $R_2 \geq R_1$ and $R_2 \leq R_1$. Therefore, if n is odd and the DC gain is specified, the ratio of the terminating resistances is fixed by (73.53a). On the other hand, if n is even and the peak-to-peak ripple in the passband and K_n or the DC gain is specified, the ratio of the resistances is given by (73.53b).

We now show that the input impedance $Z_{11}(s)$ can be realized by an LC ladder terminated in a resistor. Again, explicit formulas for their element values will be given, thereby reducing the design problem to simple arithmetic. Depending upon the choice of the plus and minus signs in (73.51), two cases are distinguished.

CASE 1. $\rho_{11}(0) \geq 0$. With the choice of the plus sign, the input impedance becomes

$$Z_{11}(s) = R_1 \frac{p(y) + \hat{p}(y)}{p(y) - \hat{p}(y)} \tag{73.54}$$

which can be expanded in a continued fraction as in (73.34). Depending on whether n is odd or even, the corresponding LC ladder network has the configurations of Fig. 73.2. The element values can be computed by the following recurrence formulas:

$$L_1 = \frac{2R_1 \sin \pi/2n}{(\sinh a - \sinh \hat{a})\omega_c} \tag{73.55}$$

$$L_{2m-1}C_{2m} = \frac{4 \sin \gamma_{4m-3} \sin \gamma_{4m-1}}{\omega_c^2 f_{2m-1}(\sinh a,\ \sinh \hat{a})} \tag{73.56a}$$

$$L_{2m+1}C_{2m} = \frac{4 \sin \gamma_{4m-1} \sin \gamma_{4m+1}}{\omega_c^2 f_{2m}(\sinh a,\ \sinh \hat{a})} \tag{73.56b}$$

for $m = 1, 2, \ldots, \lceil n/2 \rceil$, where

$$\gamma_m = \frac{m\pi}{2n} \tag{73.57a}$$

$$\hat{a} = \frac{1}{n} \sinh^{-1}\left(\frac{\sqrt{1 - K_n}}{\varepsilon}\right) \tag{73.57b}$$

$$f_m(u, v) = u^2 + v^2 + \sin^2 \gamma_{2m} - 2uv \cos \gamma_{2m} \tag{73.57c}$$

In addition, the values of the last elements can also be computed directly by the equations

$$L_n = \frac{2R_2 \sin \pi/2n}{(\sinh a + \sinh \hat{a})\omega_c}, \quad n \text{ odd} \tag{73.58a}$$

$$C_n = \frac{2 \sin \pi/2n}{R_2(\sinh a + \sinh \hat{a})\omega_c}, \quad n \text{ even} \tag{73.58b}$$

A formal proof of these formulas was first given by Takahasi [13]. Hence, we can calculate the element values starting from either the first or the last element.

Example 2. Given

$$R_1 = 150\ \Omega, \qquad R_2 = 470\ \Omega, \qquad \omega_c = 10^8\pi \text{ rad/s}, \qquad n = 4 \tag{73.59}$$

find a Chebyshev LC ladder network to meet these specifications with peak-to-peak ripple in the passband not exceeding 1.5 dB.

Since R_1 and R_2 are both specified, the minimum passband gain $G_{\min}$ is fixed by (73.53b) as

$$G_{\min} \equiv \frac{K_n}{1+\varepsilon^2} = 1 - \left(\frac{\frac{470}{150}-1}{\frac{470}{150}+1}\right)^2 = 0.7336 \tag{73.60}$$

For the 1.5-dB ripple in the passband, the corresponding ripple factor is given by

$$\varepsilon = \sqrt{10^{0.15}-1} = 0.64229 \tag{73.61}$$

obtaining $K_4 = 1.036$, which is too large for the network to be physically realizable. Thus, let $K_4 = 1$, the maximum permissible value, and the corresponding ripple factor becomes

$$\varepsilon = \sqrt{\frac{1}{G_{\min}} - 1} = \sqrt{\frac{1}{0.7336} - 1} = 0.6026 \quad \text{or} \quad 1.345 \text{ dB} \le 1.5 \text{ dB} \tag{73.62}$$

We next compute the quantities

$$\gamma_m = \frac{m\pi}{2n} = 22.5m \tag{73.63a}$$

$$a = \frac{1}{4}\sinh^{-1}\frac{1}{0.6026} = 0.32, \qquad \hat{a} = 0 \tag{73.63b}$$

$$f_m(\sinh 0.32, 0) = \sinh^2 0.32 + \sin^2 \gamma_{2m} = 0.1059 + \sin^2 \gamma_{2m} \tag{73.63c}$$

Appealing to formulas (73.55) and (73.56), the element values are calculated as follows:

$$L_1 = \frac{2 \times 150 \sin 22.5^\circ}{(\sinh 0.32 - \sinh 0) \times 10^8\pi} = 1.123\ \mu\text{H} \tag{73.64a}$$

$$C_2 = \frac{4 \sin 22.5^\circ \sin 67.5^\circ}{L_1 10^{16}\pi^2(\sinh^2 0.32 + \sin^2 45^\circ)} = 21.062 \text{ pF} \tag{73.64b}$$

$$L_3 = \frac{4 \sin 67.5^\circ \sin 112.5^\circ}{C_2 10^{16}\pi^2(\sinh^2 0.32 + \sin^2 90^\circ)} = 1.485\ \mu\text{H} \tag{73.64c}$$

$$C_4 = \frac{4 \sin 112.5^\circ \sin 157.5^\circ}{L_3 10^{16}\pi^2(\sinh^2 0.32 + \sin^2 135^\circ)} = 15.924 \text{ pF} \tag{73.64d}$$

Alternatively, the last capacitance can also be computed directly from (73.58b) as

$$C_4 = \frac{2 \sin 22.5^\circ}{470 \times (\sinh 0.32 + \sinh 0) \times 10^8\pi} = 15.925 \text{ pF} \tag{73.65}$$

The LC ladder together with its terminations is presented in Fig. 73.5.

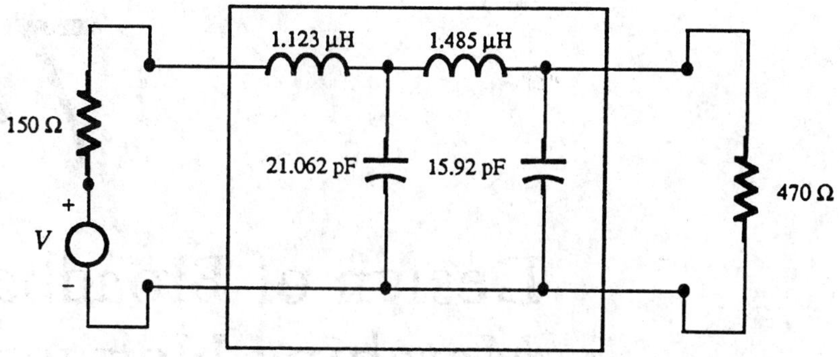

FIGURE 73.5 A fourth-order Chebyshev LC ladder network for $\rho_{11}(0) \geq 0$.

CASE 2. $\rho_{11}(0) < 0$. With the choice of the minus sign in (73.51), the input impedance, aside from the constant R_1, becomes the reciprocal of (73.54)

$$Z_{11}(s) = R_1 \frac{p(y) - \hat{p}(y)}{p(y) + \hat{p}(y)} \tag{73.66}$$

and can be expanded in a continued fraction as that shown in (73.44). Depending on whether n is even or odd, the LC ladder network has the configurations of Fig. 73.4. Formulas for the element values are similar to those given in (73.55)–(73.58) except that the roles of C's and L's are interchanged and R_1 and R_2 are replaced by their reciprocals:

$$C_1 = \frac{2 \sin \pi/2n}{R_1(\sinh a - \sinh \hat{a})\omega_c} \tag{73.67a}$$

$$C_{2m-1}L_{2m} = \frac{4 \sin \gamma_{4m-3} \sin \gamma_{4m-1}}{\omega_c^2 f_{2m-1}(\sinh a, \sinh \hat{a})} \tag{73.67b}$$

$$C_{2m+1}L_{2m} = \frac{4 \sin \gamma_{4m-1} \sin \gamma_{4m+1}}{\omega_c^2 f_{2m}(\sinh a, \sinh \hat{a})} \tag{73.67c}$$

for $m = 1, 2, \ldots, \lceil n/2 \rceil$, where γ_m and $f_m(\sinh a, \sinh \hat{a})$ are defined in (73.57). In addition, the values of the last elements can also be computed directly by the formulas

$$C_n = \frac{2 \sin \pi/2n}{R_2(\sinh a + \sinh \hat{a})\omega_c}, \quad n \text{ odd} \tag{73.68a}$$

$$L_n = \frac{2R_2 \sin \pi/2n}{(\sinh a + \sinh \hat{a})\omega_c}, \quad n \text{ even} \tag{73.68b}$$

74

Design of Broadband Matching Networks*

Wai-Kai Chen
University of Illinois, Chicago

74.1 Introduction

Refer to the network configuration of Fig. 74.1 where the source is represented either by its Thévenin equivalent or by its Norton equivalent. The load impedance $z_2(s)$ is assumed to be strictly passive over a frequency band of interest, because the matching problem cannot be meaningfully defined if the load is purely reactive. Our objective is to design an "optimum" lossless two-port network or equalizer N to match out the load impedance $z_2(s)$ to the resistive source impedance $z_1(s) = R_1$, and to achieve a preassigned transducer power-gain characteristic $G(\omega^2)$ over the entire sinusoidal frequency spectrum.

As stated in Chapter 73, the output reflection coefficient is given by

$$\rho_{22}(s) = \frac{Z_{22}(s) - z_2(-s)}{Z_{22}(s) + z_2(s)} \tag{74.1}$$

where $Z_{22}(s)$ is the impedance looking into the output port when the input port is terminated in the source resistance R_1. As shown in Chapter 73, the transducer power gain $G(\omega^2)$ is related to the transmission and reflection coefficients by the equation

$$G(\omega^2) = |\rho_{21}(j\omega)|^2 = |\rho_{12}(j\omega)|^2 = 1 - |\rho_{11}(j\omega)|^2 = 1 - |\rho_{22}(j\omega)|^2 \tag{74.2}$$

Recall that in computing $\rho_{11}(s)$ from $\rho_{11}(s)\rho_{11}(-s)$ we assign all of the LHS poles of $\rho_{11}(s)\rho_{11}(-s)$ to $\rho_{11}(s)$ because with resistive load $z_2(s) = R_2$, $\rho_{11}(s)$ is devoid of poles in the RHS. For the complex load, the poles of $\rho_{22}(s)$ include those of $z_2(-s)$, which may lie in the open RHS. As a result, the assignment of poles of $\rho_{22}(s)\rho_{22}(-s)$ is not unique. Furthermore,

*References for this Chapter can be found on page 2333.

0-8493-8341-2/95/$0.00+$.50

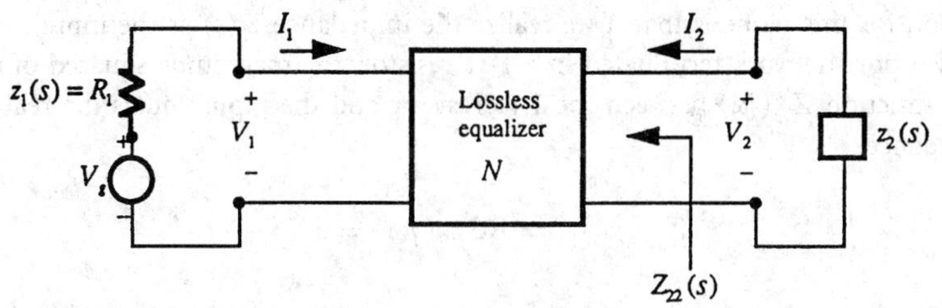

FIGURE 74.1 The schematic of broadband matching.

the nonanalyticity of $\rho_{22}(s)$ leaves much to be desired in terms of our ability to manipulate. For these reasons, we consider the normalized reflection coefficient defined by

$$\rho(s) = A(s)\rho_{22}(s) = A(s)\frac{Z_{22}(s) - z_2(-s)}{Z_{22}(s) + z_2(s)} \tag{74.3}$$

where

$$A(s) = \prod_{i=1}^{q} \frac{s - s_i}{s + s_i}, \qquad \text{Re } s_i > 0 \tag{74.4}$$

is the real all-pass function defined by the open RHS poles s_i $(i = 1, 2, \ldots, q)$ of $z_2(-s)$. An **all-pass function** is a function whose zeros are all located in the open RHS and whose poles are located at the LHS mirror image of the zeros. Therefore, it is analytic in the closed RHS and such that

$$A(s)A(-s) = 1 \tag{74.5}$$

On the $j\omega$-axis, the magnitude of $A(j\omega)$ is unity, being flat for all sinusoidal frequencies, and we have

$$|\rho(j\omega)| = |A(j\omega)\rho_{22}(j\omega)| = |\rho_{22}(j\omega)| \tag{74.6}$$

and (74.2) becomes

$$G(\omega^2) = 1 - |\rho(j\omega)|^2 \tag{74.7}$$

This equation together with the normalized reflection coefficient $\rho(s)$ of (74.3) forms the cornerstone of Youla's theory of broadband matching [14].

4.2 Basic Coefficient Constraints

In Chapter 72, we define the zeros of transmission for a terminate two-port network as the frequencies at which a zero output results for a finite input. We extend this concept by defining the zeros of transmission for a one-port impedance.

Definition 1. *Zero of transmission.* For a given impedance $z_2(s)$, a closed RHS zero of multiplicity k of the function

$$w(s) \equiv \frac{r_2(s)}{z_2(s)} \tag{74.8}$$

where $r_2(s)$ is the even part of $z_2(s)$, is called a **zero of transmission** of order k of $z_2(s)$.

The reason for this name is that if we realize the impedance $z_2(s)$ as the input impedance of a lossless two-port network terminated in a 1-Ω resistor, the magnitude squared of the transfer impedance function $Z_{12}(j\omega)$ between the 1-Ω resistor and the input equals the real part of the input impedance,

$$|Z_{12}(j\omega)|^2 = \text{Re}\, z_2(j\omega) = r_2(j\omega) \tag{74.9}$$

After appealing to analytic continuation by substituting ω by $-js$, the zeros of $r_2(s)$ are seen to be the zeros of transmission of the lossless two-port.

Consider, for example, the RC impedance $z_2(s)$ of Fig. 74.2,

$$z_2(s) = R_1 + \frac{R_2}{R_2Cs + 1} \tag{74.10}$$

the even part of which is given by

$$r_2(s) = \frac{1}{2}[z_2(s) + z_2(-s)] = \frac{R_1 + R_2 - R_1R_2^2C^2s^2}{1 - R_2^2C^2s^2} \tag{74.11}$$

obtaining

$$w(s) = \frac{r_2(s)}{z_2(s)} = \frac{R_2Cs^2 - (R_1 + R_2)/R_1R_2C}{(R_2Cs - 1)[s + (R_1 + R_2)/R_1R_2C]} \tag{74.12}$$

Thus, the impedance $z_2(s)$ has a zero of transmission of order 1 located at

$$s = \sigma_0 = \frac{1}{R_2C}\sqrt{1 + \frac{R_2}{R_1}} \tag{74.13}$$

For our purposes, the zeros of transmission are divided into four mutually exclusive classes.

Definition 2. *Classification of zeros of transmission.* Let $s_0 = \sigma_0 + j\omega_0$ be a zero of transmission of an impedance $z_2(s)$. Then s_0 belongs to one of the following four mutually exclusive classes depending on σ_0 and $z_2(s_0)$, as follows:

Class I: $\sigma_0 > 0$, which includes all the open RHS zeros of transmission.
Class II: $\sigma_0 = 0$ and $z_2(\omega_0) = 0$.
Class III: $\sigma_0 = 0$ and $0 < |z_2(j\omega_0)| < \infty$.
Class IV: $\sigma_0 = 0$ and $|z_2(j\omega_0)| = \infty$.

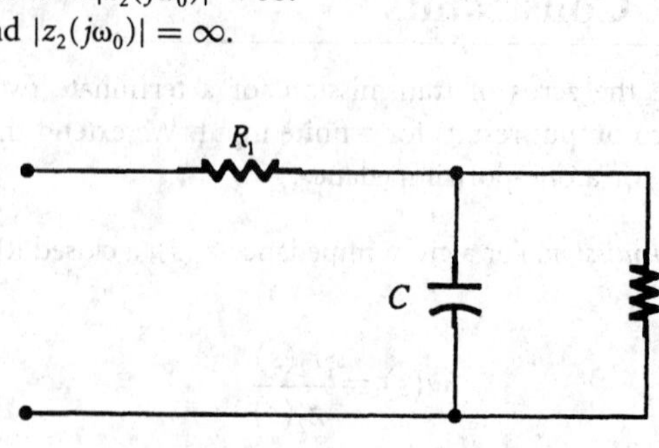

FIGURE 74.2 An RC one-port network.

For the impedance $z_2(s)$ of (74.10), its zero of transmission given in (74.13) belongs to Class I of order 1. If $z_2(s)$ is the load of the network of Fig. 74.1, it imposes the basic constraints on the normalized reflection coefficient $\rho(s)$. These constraints are important in that they are necessary and sufficient for $\rho(s)$ to be physically realizable, and are most conveniently formulated in terms of the coefficients of the Laurent series expansions of the following quantities about a zero of transmission $s_0 = \sigma_0 + j\omega_0$ of order k of $z_2(s)$:

$$\rho(s) = \rho_0 + \rho_1(s - s_0) + \rho_2(s - s_0)^2 + \cdots = \sum_{m=0}^{\infty} \rho_m (s - s_0)^m \tag{74.14a}$$

$$A(s) = A_0 + A_1(s - s_0) + A_2(s - s_0)^2 + \cdots = \sum_{m=0}^{\infty} A_m (s - s_0)^m \tag{74.14b}$$

$$F(s) \equiv 2r_2(s)A(s) = F_0 + F_1(s - s_0) + F_2(s - s_0)^2 + \cdots = \sum_{m=0}^{\infty} F_m (s - s_0)^m \tag{74.14c}$$

We remark that the expansions of the Laurent type can be found by any method because it is unique, and the resulting expansion is *the* Laurent series expansion. For the zero of transmission at infinity, the expansions take the form

$$\rho(s) = \rho_0 + \frac{\rho_1}{s} + \frac{\rho_2}{s^2} + \frac{\rho_3}{s^3} + \cdots = \sum_{m=0}^{\infty} \frac{\rho_m}{s^m} \tag{74.15a}$$

$$A(s) = A_0 + \frac{A_1}{s} + \frac{A_2}{s^2} + \frac{A_3}{s^3} + \cdots = \sum_{m=0}^{\infty} \frac{A_m}{s^m} \tag{74.15b}$$

$$F(s) = F_0 + \frac{F_1}{s} + \frac{F_2}{s^2} + \frac{F_3}{s^3} + \cdots = \sum_{m=0}^{\infty} \frac{F_m}{s^m} \tag{74.15c}$$

In fact, they can be obtained by means of the binomial expansion formula

$$(s + c)^n = s^n + ns^{n-1}c + \frac{n(n-1)}{2!}s^{n-2}c^2 + \cdots \tag{74.16}$$

which is valid for all values of n if $|s| > |c|$, and is valid only for nonnegative integers n if $|s| \leq |c|$.

Example 1. Assume that the network of Fig. 74.1 is terminated in the passive impedance

$$z_2(s) = \frac{s}{s^2 + 2s + 1} \tag{74.17}$$

and possesses the transducer power-gain characteristic

$$G(\omega^2) = \frac{K\omega^2}{\omega^4 - \omega^2 + 1}, \qquad 0 \leq K \leq 1 \tag{74.18}$$

We first compute the even part

$$r_2(s) = \frac{1}{2}[z_2(s) + z_2(-s)] = \frac{-2s^2}{(s^2 + 2s + 1)(s^2 - 2s + 1)} \tag{74.19}$$

of $z_2(s)$, and obtain the function

$$w(s) = \frac{r_2(s)}{z_2(s)} = \frac{-2s}{s^2 - 2s + 1} \tag{74.20}$$

showing that $z_2(s)$ possesses two Class II zeros of transmission at $s = 0$ and ∞. Since the pole of $z_2(-s)$ is located at $s = -1$ of order 2, the all-pass function $A(s)$ takes the form

$$A(s) = \frac{s^2 - 2s + 1}{s^2 + 2s + 1} \tag{74.21}$$

The other required functions are found to be

$$F(s) = 2A(s)r_2(s) = -\frac{4s^2}{(s^2 + 2s + 1)^2} \tag{74.22}$$

$$\rho(s)\rho(-s) = 1 - G(-s^2) = \frac{s^4 + (1 + K)s^2 + 1}{s^4 + s^2 + 1} \tag{74.23}$$

The minimum-phase solution of (74.23) is determined as

$$\pm\,\hat{\rho}(s) = \frac{s^2 + \sqrt{1 - K}s + 1}{s^2 + s + 1} \tag{74.24}$$

Now we expand the functions $A(s)$, $F(s)$ and $\hat{\rho}(s)$ in Laurent series about the zeros of transmission at the origin and at infinity, and obtain

$$A(s) = 1 - 4s + \cdots = 1 - \frac{4}{s} + \cdots \tag{74.25}$$

$$F(s) = 0 + 0 - 4s^2 + \cdots = 0 + 0 - \frac{4}{s^2} + \cdots \tag{74.26}$$

$$\pm\hat{\rho}(s) = 1 + (\sqrt{1 - K} - 1)s + \cdots = 1 + \frac{\sqrt{1 - K} - 1}{s} + \cdots \tag{74.27}$$

In both expansions, we can make the following identifications:

$$A_0 = 1, \qquad F_0 = 0, \qquad \rho_0 = 1 \tag{74.28a}$$

$$A_1 = -4, \qquad F_1 = 0, \qquad \rho_1 = \sqrt{1 - K} - 1 \tag{74.28b}$$

$$F_2 = -4 \tag{74.28c}$$

Basic Coefficient Constraints on $\rho(s)$

The basic constraints imposed on the normalized reflection coefficient $\rho(s)$ by a load impedance $z_2(s)$ are most succinctly expressed in terms of the coefficients of the Laurent series expansions (74.14) of the functions $\rho(s)$, $A(s)$, and $F(s)$ about each zero of transmission $s_0 = \sigma_0 + j\omega_0$. Depending on the classification of the zero of transmission, one of the following four sets of coefficient conditions must be satisfied:

Class I: For $x = 0, 1, 2, \ldots, k - 1$

$$A_x = \rho_x \tag{74.29a}$$

Class II: $A_x = \rho_x$ for $x = 0, 1, 2, \ldots, k-1$, and

$$\frac{A_k - \rho_k}{F_{k+1}} \geq 0 \tag{74.29b}$$

Class III: $A_x = \rho_x$ for $x = 0, 1, 2, \ldots, k-2, k \geq 2$, and

$$\frac{A_{k-1} - \rho_{k-1}}{F_k} \geq 0 \tag{74.29c}$$

Class IV: $A_x = \rho_x$ for $x = 0, 1, 2, \ldots, k-1$, and

$$\frac{F_{k-1}}{A_k - \rho_k} \geq a_{-1}, \text{ the residue of } z_2(s) \text{ evaluated at the pole } s = j\omega_0 \tag{74.29d}$$

To determine the normalized reflection coefficient $\rho(s)$ from a preassigned transducer power-gain characteristic $G(\omega^2)$, we appeal to (74.7) and analytic continuation by replacing ω by $-js$ and obtain

$$\rho(s)\rho(-s) = 1 - G(-s^2) \tag{74.30}$$

Since the zeros and poles of $\rho(s)\rho(-s)$ must appear in quadrantal symmetry, being symmetric with respect to both the real and imaginary axes of the s-plane, and since $\rho(s)$ is analytic in the closed RHS, the open LHS poles of $\rho(s)\rho(-s)$ belong to $\rho(s)$, whereas those in the open RHS belong to $\rho(-s)$. For a lumped system, $\rho(s)$ is devoid of poles on the $j\omega$-axis. For the zeros, there are no unique ways to assign them. The only requirement is that the complex-conjugate pair of zeros must be assigned together. However, if we specify that $\rho(s)$ be made a minimum-phase function, then all the open LHS zeros of $\rho(s)\rho(-s)$ are assigned to $\rho(s)$. The $j\omega$-axis zeros of $\rho(s)\rho(-s)$ are of even multiplicity, and thus they are divided equally between $\rho(s)$ and $\rho(-s)$. Therefore, $\rho(s)$ is uniquely determined by the zeros and poles of $\rho(s)\rho(-s)$ only if $\rho(s)$ is required to be minimum-phase.

Let $\hat{\rho}(s)$ be the minimum-phase solution of (74.30). Then any solution of the form

$$\rho(s) = \pm\eta(s)\hat{\rho}(s) \tag{74.31}$$

is admissible, where $\eta(s)$ is an arbitrary real all-pass function possessing the property that

$$\eta(s)\eta(-s) = 1 \tag{74.32}$$

The significance of these coefficient constraints is that they are both necessary and sufficient for the physical realizability of $\rho(s)$, and is summarized in the following theorem. The proof of this result can be found in [5].

Theorem 1: *Given a strictly passive impedance $z_2(s)$, the function defined by the equation*

$$Z_{22}(s) \equiv \frac{F(s)}{A(s) - \rho(s)} - z_2(s) \tag{74.33}$$

is positive real if and only if $|\rho(j\omega)| \leq 1$ for all ω and the coefficient conditions (74.29) are satisfied.

The function $Z_{22}(s)$ defined in (74.33) is actually the back-end impedance of a desired equalizer. To see this, we solve for $Z_{22}(s)$ in (74.3) and obtain

$$Z_{22}(s) = \frac{A(s)[z_2(s) + z_2(-s)]}{A(s) - \rho(s)} - z_2(s) = \frac{F(s)}{A(s) - \rho(s)} - z_2(s) \tag{74.34}$$

which is guaranteed to be positive real by Theorem 1. This impedance can be realized as the input impedance of a lossless two-port network terminated in a resistor. The removal of this resistor gives the desired matching network. An ideal transformer may be needed at the input port to compensate for the actual level of the generator resistance R_1.

Example 2. Design a lossless matching network to equalize the load impedance

$$z_2(s) = \frac{s}{s^2 + 2s + 1} \tag{74.35}$$

to a resistive generator of internal resistance of 0.5 Ω and to achieve the transducer power-gain characteristic

$$G(\omega^2) = \frac{K\omega^2}{\omega^4 - \omega^2 + 1}, \qquad 0 \le K \le 1 \tag{74.36}$$

From Example 1, the load possesses two Class II zeros of transmission of order 1 at $s = 0$ and $s = \infty$. The coefficients of the Laurent series expansions of the functions $A(s)$, $F(s)$, and $\rho(s)$ about the zeros of transmission at $s = 0$ and $s = \infty$ were computed in Example 1 as

$$A_0 = 1, \qquad F_0 = 0, \qquad \rho_0 = 1 \tag{74.37a}$$

$$A_1 = -4, \qquad F_1 = 0, \qquad \rho_1 = \sqrt{1 - K} - 1 \tag{74.37b}$$

$$F_2 = -4 \tag{74.37c}$$

The coefficient constraints (74.29b) for the Class II zeros of transmission of order 1 become

$$A_0 = \rho_0, \qquad \frac{A_1 - \rho_1}{F_2} \ge 0 \tag{74.38}$$

Clearly, the first condition is always satisfied. To meet the second requirement, we set

$$\frac{-4 - (\sqrt{1 - K} - 1)}{-4} \ge 0 \tag{74.39}$$

or $0 \le K \le 1$, showing that the maximum realizable K is 1. For our purposes, set $K = 1$ and choose the plus sign in (74.24). From (74.34), the equalizer back-end impedance is computed as

$$Z_{22}(s) = \frac{F(s)}{A(s) - \hat{\rho}(s)} - z_2(s) = \frac{\dfrac{-4s^2}{(s^2 + 2s + 1)^2}}{\dfrac{s^2 - 2s + 1}{s^2 + 2s + 1} - \dfrac{s^2 + 1}{s^2 + s + 1}} - \frac{s}{s^2 + 2s + 1}$$

$$= \frac{s}{3s^2 + 2s + 3} \tag{74.40}$$

This impedance can be realized as the input impedance of the parallel connection of an inductor $L = 1/3$ H, a capacitor $C = 3$ F, and a resistor $R = 0.5$ Ω. The resulting equalizer together with the load is presented in Fig. 74.3.

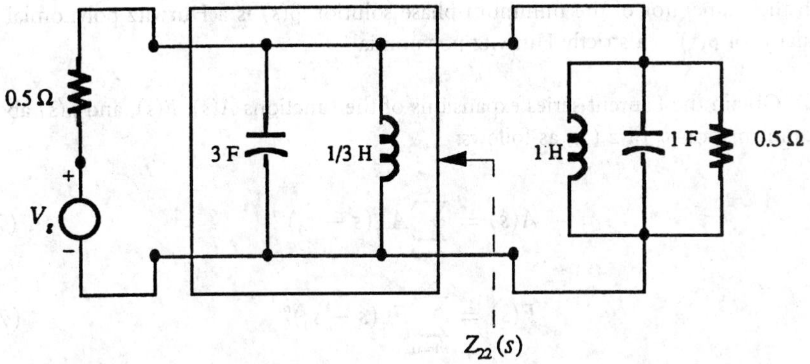

FIGURE 74.3 An equalizer having transducer power-gain characteristic (74.36).

.3 Design Procedure

We now outline an eight-step procedure for the design of an optimum lossless matching network that equalizes a frequency-dependent load impedance $z_2(s)$ to a resistive generator of internal resistance R_1 and achieves a preassigned transducer power-gain characteristic $G(\omega^2)$ over the entire sinusoidal frequency spectrum.

STEP 1. From a preassigned transducer power-gain characteristic $G(\omega^2)$, verify that $G(\omega^2)$ is an even rational real function and satisfies the inequality

$$0 \leq G(\omega^2) \leq 1 \quad \text{for all } \omega \tag{74.41}$$

The gain level is usually not specified to allow some flexibility.

STEP 2. From a prescribed strictly passive load impedance $z_2(s)$, compute

$$r_2(s) \equiv \text{Ev}\, z_2(s) = \frac{1}{2}[z_2(s) + z_2(-s)] \tag{74.42}$$

$$A(s) = \prod_{i=1}^{q} \frac{s - s_i}{s + s_i}, \qquad \text{Re}\, s_i > 0 \tag{74.43}$$

where s_i $(i = 1, 2, \ldots, q)$ are the open RHS poles of $z_2(-s)$, and

$$F(s) = 2A(s)r_2(s) \tag{74.44}$$

STEP 3. Determine the locations and the orders of the zeros of transmission of $z_2(s)$, which are defined as the closed RHS zeros of the function

$$w(s) = \frac{r_2(s)}{z_2(s)} \tag{74.45}$$

and divide them into respective classes according to Definition 2.

STEP 4. Perform the unique factorization of the function

$$\hat{\rho}(s)\hat{\rho}(-s) = 1 - G(-s^2) \tag{74.46}$$

in which the numerator of the minimum-phase solution $\hat{\rho}(s)$ is a Hurwitz polynomial and the denominator of $\hat{\rho}(s)$ is a strictly Hurwitz polynomial.

STEP 5. Obtain the Laurent series expansions of the functions $A(s)$, $F(s)$, and $\hat{\rho}(s)$ about each zero of transmission s_0 of $z_2(s)$, as follows:

$$A(s) = \sum_{m=0}^{\infty} A_m (s - s_0)^m \tag{74.47a}$$

$$F(s) = \sum_{m=0}^{\infty} F_m (s - s_0)^m \tag{74.47b}$$

$$\hat{\rho}(s) = \sum_{m=0}^{\infty} \rho_m (s - s_0)^m \tag{74.47c}$$

They may be obtained by any available methods.

STEP 6. According to the classes of zeros of transmission, list the basic constraints (74.29) imposed on the coefficients of (74.47). The gain level is ascertained from these constraints. If not all the constraints are satisfied, consider the more general solution

$$\rho(s) = \pm\eta(s)\hat{\rho}(s) \tag{74.48}$$

where $\eta(s)$ is an arbitrary real all-pass function. Then repeat Step 5 for $\rho(s)$, starting with lower-order $\eta(s)$. If the constraints still cannot be satisfied, modify the preassigned transducer power-gain characteristics $G(\omega^2)$. Otherwise, no match exists.

STEP 7. Having successfully carried out Step 6, the equalizer back-end impedance is determined by the equation

$$Z_{22}(s) \equiv \frac{F(s)}{A(s) - \rho(s)} - z_2(s) \tag{74.49}$$

where $\rho(s)$ may be $\hat{\rho}(s)$, and $Z_{22}(s)$ is guaranteed to be positive real.

STEP 8. Realize $Z_{22}(s)$ as the input impedance of a lossless two-port network terminated in a resistor. An ideal transformer may be required at the input port to compensate for the actual level of the generator resistance R_1. This completes the design of an equalizer.

Example 3. Design a lossless matching network to equalize the RLC load as shown in Fig. 74.4 to a resistive generator and to achieve the fifth-order Butterworth transducer power-gain characteristic with a maximal DC gain. The cut-off frequency is 10^8 rad/s.

To simplify the computation, we magnitude-scale the network by a factor of 10^{-2} and frequency-scale it by a factor of 10^{-8}. Thus, s denotes the normalized complex frequency and ω the normalized real frequency. The load impedance becomes

$$z_2(s) = \frac{s^2 + s + 1}{s + 1} \tag{74.50}$$

We now follow the eight steps outlined above to design a lossless equalizer to meet the desired specifications.

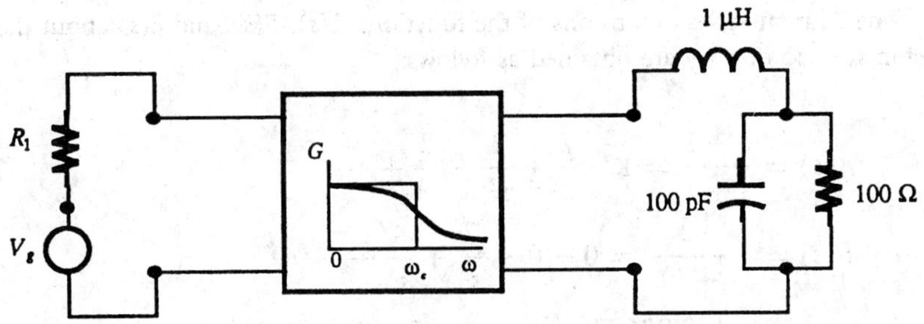

FIGURE 74.4 The broadband matching of an RLC load to a resistive generator.

STEP 1. The fifth-order Butterworth transducer power-gain characteristic is given by

$$G(\omega^2) = \frac{K_5}{1+\omega^{10}}, \qquad 0 \le K_5 \le 1 \tag{74.51}$$

Our objective is to maximize the DC gain K_5.

STEP 2. From the load impedance $z_2(s)$, we compute the functions

$$r_2(s) = \frac{1}{2}[z_2(s) + z_2(-s)] = \frac{1}{1-s^2} \tag{74.52}$$

$$A(s) = \frac{s-1}{s+1} \tag{74.53}$$

where $s = 1$ is the open RHS pole of $z_2(-s)$, and

$$F(s) = 2A(s)r_2(s) = \frac{-2}{(s+1)^2} \tag{74.54}$$

STEP 3. The zeros of transmission of $z_2(s)$ are defined by the closed RHS zero of the function.

$$w(s) = \frac{r_2(s)}{z_2(s)} = \frac{1}{(s^2+s+1)(1-s)} \tag{74.55}$$

indicating that $s = \infty$ is a Class IV zero of transmission of order 3.

STEP 4. Substituting (74.51) in (74.46) with $-js$ replacing ω gives

$$\rho(s)\rho(-s) = 1 - G(-s^2) = 1 - \frac{K_5}{1-s^{10}} = \alpha^{10}\frac{1-x^{10}}{1-s^{10}} \tag{74.56}$$

where

$$\alpha = (1-K_5)^{1/10}, \qquad x = \frac{s}{\alpha} \tag{74.57}$$

The minimum-phase solution of (74.56) is found to be

$$\hat{\rho}(s) = \frac{s^5 + 3.23607\alpha s^4 + 5.23607\alpha^2 s^3 + 5.23607\alpha^3 s^2 + 3.23607\alpha^4 s + \alpha^5}{s^5 + 3.23607 s^4 + 5.23607 s^3 + 5.23607 s^2 + 3.23607 s + 1} \tag{74.58}$$

The maximum attainable DC gain will be ascertained later from the coefficient conditions.

STEP 5. The Laurent series expansions of the functions $A(s)$, $F(s)$, and $\hat{\rho}(s)$ about the zero of transmission $s_0 = \infty$ of $z_2(s)$ are obtained as follows:

$$A(s) = \frac{s-1}{s+1} = 1 - \frac{2}{s} + \frac{2}{s^2} - \frac{2}{s^3} + \cdots \tag{74.59a}$$

$$F(s) = \frac{-2}{(s+1)^2} = 0 + 0 - \frac{2}{s^2} + \frac{4}{s^3} + \cdots \tag{74.59b}$$

$$\hat{\rho}(s) = 1 + \frac{3.23607(\alpha-1)}{s} + \frac{5.23607(\alpha-1)^2}{s^2} + \frac{5.23607(\alpha^3 - 3.23607\alpha^2 + 3.23607\alpha - 1)}{s^3} + \cdots \tag{74.59c}$$

STEP 6. For a Class IV zero of transmission of order 3, the coefficient conditions are, from (74.29d) with $k = 3$,

$$A_m = \rho_m, \qquad m = 0, 1, 2 \tag{74.60a}$$

$$\frac{F_2}{A_3 - \rho_3} \geq a_{-1}(\infty) = 1 \tag{74.60b}$$

where $a_{-1}(\infty)$ is the residue of $z_2(s)$ evaluated at the pole $s = \infty$, which is also the zero of transmission of $z_2(s)$. Substituting the coefficients in (74.59) in (74.60) yields the constraints imposed on K_5 as

$$A_0 = 1 = \rho_0 \tag{74.61a}$$

$$A_1 = -2 = \rho_1 + 3.23607(\alpha - 1) \tag{74.61b}$$

$$A_2 = 2 = \rho_2 = 5.23607(\alpha-1)^2 \tag{74.61c}$$

yielding $\alpha = 0.3819664$, and

$$\frac{F_2}{A_3 - \rho_3} = \frac{-2}{-2 - 5.23607(\alpha^3 - 3.23607\alpha^2 + 3.23607\alpha - 1)} \geq a_{-1}(\infty) = 1 \tag{74.62}$$

This inequality is satisfied for $\alpha = 0.3819664$. Hence, we choose $\alpha = 0.3819664$ and obtain from (74.57) the maximum realizable DC gain K_5 as

$$K_5 = 1 - \alpha^{10} = 0.99993 \tag{74.63}$$

With this value of K_5, the minimum-phase reflection coefficient becomes

$$\hat{\rho}(s) = \frac{s^5 + 1.23607s^4 + 0.76393s^3 + 0.2918s^2 + 0.068884s + 0.0081307}{s^5 + 3.23607s^4 + 5.23607s^3 + 5.23607s^2 + 3.23607s + 1} \tag{74.64}$$

STEP 7. The equalizer back-end impedance is determined by

$$Z_{22}(s) \equiv \frac{F(s)}{A(s)-\hat{\rho}(s)} - z_2(s) = \frac{\dfrac{-2}{(s+1)^2}}{\dfrac{s-1}{s+1}-\hat{\rho}(s)} - \frac{s^2+s+1}{s+1} \tag{74.65}$$

$$= \frac{0.94427s^4+2.1115s^3+2.6312s^2+2.1591s+0.9919}{1.0557s^3+2.3607s^2+2.3131s+1.0081}$$

STEP 8. Expanding $Z_{22}(s)$ in a continued fraction results in

$$Z_{22}(s) = 0.894s + \cfrac{1}{1.88s + \cfrac{1}{1.25s + \cfrac{1}{0.455s + \cfrac{1}{0.984}}}} \tag{74.66}$$

which can be identified as an LC ladder network terminated in a resistor. Denormalizing the element values with regard to magnitude-scaling by a factor of 100 and frequency-scaling by a factor 10^8 gives the final design of the equalizer of Fig. 74.5.

Example 4. Design a lossless equalizer to match the load

$$z_2(s) = \frac{s^2+9s+8}{s^2+2s+2} \tag{74.67}$$

to a resistive generator and to achieve the largest flat transducer power gain over the entire sinusoidal frequency spectrum.

STEP 1. For truly-flat transducer power gain, let

$$G(\omega^2) = K, \qquad 0 \le K \le 1 \tag{74.68}$$

STEP 2. The following functions are computed from $z_2(s)$:

$$r_2(s) = \frac{(s^2-4)^2}{(s^2+2s+2)(s^2-2s+2)} \tag{74.69a}$$

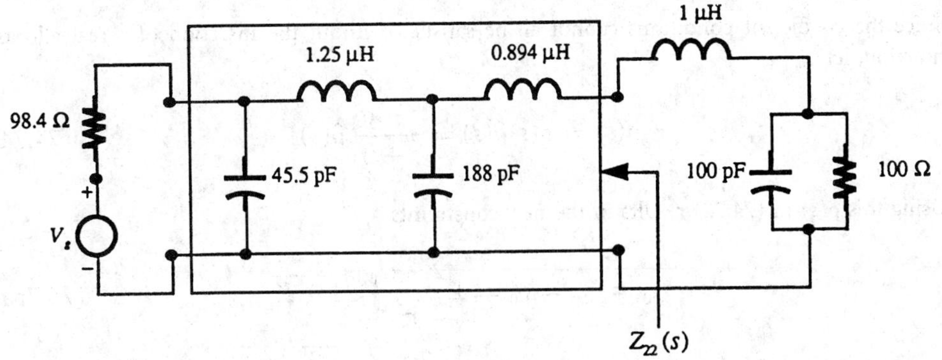

FIGURE 74.5 A fifth-order Butterworth broadband matching equalizer.

$$A(s) = \frac{s^2 - 2s + 2}{s^2 + 2s + 2} \tag{74.69b}$$

$$F(s) = \frac{2(s^4 - 8s^2 + 16)}{(s^2 + 2s + 2)^2} \tag{74.69c}$$

STEP 3. Since

$$w(s) = \frac{r_2(s)}{z_2(s)} = \frac{(s^2 - 4)^2}{(s^2 - 2s + 2)(s^2 + 9s + 8)} \tag{74.70}$$

the load impedance $z_2(s)$ possesses a Class I zero of transmission of order 2 at $s = 2$.

STEP 4. Substituting (74.68) in (74.46) yields

$$\rho(s)\rho(-s) = 1 - K \tag{74.71a}$$

the minimum-phase solution of which is found to be

$$\hat{\rho}(s) = \pm\sqrt{1 - K} \tag{74.71b}$$

STEP 5. For a Class I zero of transmission of order 2, the coefficient conditions (74.29a) become

$$A(2) = \hat{\rho}(2) \tag{74.72a}$$

$$A_1 = \left.\frac{dA(s)}{ds}\right|_{s=2} = \rho_1 = \left.\frac{d\rho(s)}{ds}\right|_{s=2} \tag{74.72b}$$

The Laurent series expansions of the functions $A(s)$, $F(s)$, and $\hat{\rho}(s)$ about the zero of transmission at $s = 2$ are not needed.

STEP 6. Substituting (74.69b) and (74.71b) in (74.72) gives

$$A_0 = A(2) = 0.2 = \rho_0 = \pm\sqrt{1 - K} \tag{74.73a}$$

$$A_1 = \left.\frac{dA(s)}{ds}\right|_{s=2} = 0.08 \neq \rho_1 = 0 \tag{74.73b}$$

Since the coefficient conditions cannot all be satisfied without the insertion of a real all-pass function, let

$$\rho(s) = \eta(s)\hat{\rho}(s) = \frac{s - \sigma_1}{s + \sigma_1}\hat{\rho}(s) \tag{74.74}$$

Using this $\rho(s)$ in (74.72) results in the new constraints

$$A_0 = 0.2 = \rho_0 = \pm\frac{2 - \sigma_1}{2 + \sigma_1}\sqrt{1 - K} \tag{74.75a}$$

$$A_1 = 0.08 = \rho_1 = \pm\frac{2\sigma_1\sqrt{1 - K}}{(2 + \sigma_1)^2} \tag{74.75b}$$

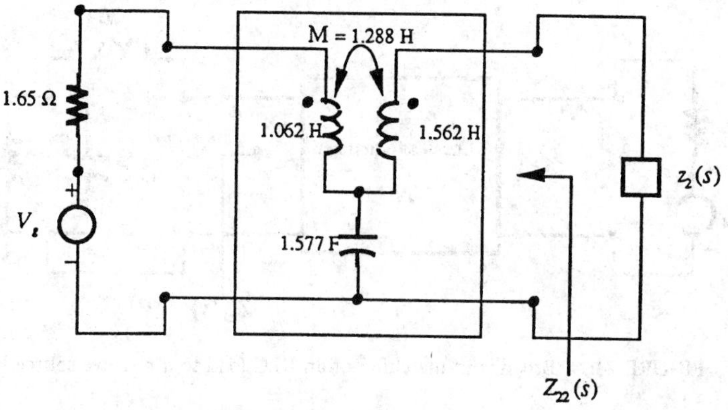

FIGURE 74.6 A lossless equalizer having truly-flat transducer power gain.

which can be combined to yield

$$\sigma_1^2 + 5\sigma_1 - 4 = 0 \tag{74.76}$$

obtaining $\sigma_1 = 0.70156$ or -5.7016. Choosing $\sigma_1 = 0.70156$ and the plus sign for $\rho(s)$, the maximum permissible flat transducer power gain is found to be

$$K_{\max} = 0.82684 \tag{74.77}$$

STEP 7. The equalizer back-end impedance is determined as

$$Z_{22}(s) \equiv \frac{F(s)}{A(s) - \rho(s)} - z_2(s) = \frac{1.4161s + 0.8192}{0.58388s + 0.49675} \tag{74.78}$$

STEP 8. The positive-real impedance $Z_{22}(s)$ can be realized as the input impedance of a lossless two-port network terminated in a resistor. The overall network is presented in Fig. 74.6.

4.4 Explicit Formulas for the RLC Load

In many practical cases, the source can usually be represented by an ideal voltage source in series with a pure resistor, which may be the Thévenin equivalent of some other network, and the load is composed of the parallel combination of a resistor and a capacitor and then in series with an inductor, as shown in Fig. 74.7, which may include the parasitic effects of a physical device. The problem is to match out this load and source over a preassigned frequency band to within a given tolerance, and to achieve a prescribed transducer power-gain characteristic $G(\omega^2)$. In the case that $G(\omega^2)$ is of Butterworth or Chebyshev type of response, explicit formulas for the design of such optimum matching networks for any RLC load of the type shown in Fig. 74.7 are available, thereby avoiding the necessity of applying the coefficient constraints and solving the nonlinear equations for selecting the optimum design parameters. As a result, we reduce the design of these equalizers to simple arithmetic.

Butterworth Networks

Refer to Fig. 74.7. We wish to match out the load impedance

$$z_2(s) = Ls + \frac{R}{RCs + 1} \tag{74.79}$$

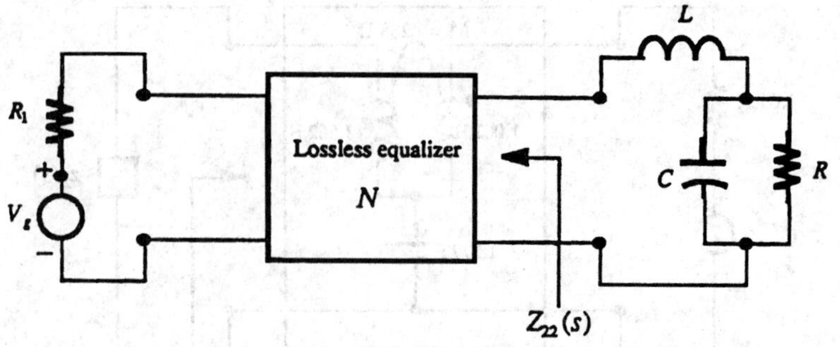

FIGURE 74.7 Broadband matching of an RLC load to a resistive source.

to a resistive generator and to achieve the nth-order Butterworth transducer power-gain characteristic

$$G(\omega^2) = \frac{K_n}{1 + (\omega/\omega_c)^{2n}}, \qquad 0 \le K_n \le 1 \tag{74.80}$$

with maximum attainable DC gain K_n, where ω_c is the 3-dB bandwidth or the radian cutoff frequency.

The even part $r_2(s)$ of $z_2(s)$ is found to be

$$r_2(s) = \frac{-R}{R^2C^2s^2 - 1} \tag{74.81}$$

Since $z_2(-s)$ has an open RHS pole at $s = 1/RC$, the all-pass real function defined by this pole is given by

$$A(s) = \frac{s - 1/RC}{s + 1/RC} = \frac{RCs - 1}{RCs + 1} \tag{74.82}$$

yielding

$$F(s) = 2A(s)r_2(s) = \frac{-2R}{(RCs + 1)^2} \tag{74.83}$$

We next replace ω by $-js$ in (74.80) and substitute the resulting equation in (74.46) to obtain

$$\rho(s)\rho(-s) = \alpha^{2n}\frac{1 + (-1)^n x^{2n}}{1 + (-1)^n y^{2n}} \tag{74.84}$$

where

$$y = \frac{s}{\omega_c}, \qquad x = \frac{y}{\alpha} \tag{74.85a}$$

$$\alpha = (1 - K_n)^{1/2n} \tag{74.85b}$$

As previously shown in (73.29), the minimum-phase solution $\hat{\rho}(s)$ of (74.84) is found to be

$$\hat{\rho}(s) = \alpha^n \frac{q(x)}{q(y)} \tag{74.86}$$

For our purposes, we consider the more general solution

$$\rho(s) = \pm\eta(s)\hat{\rho}(s) \tag{74.87}$$

where $\eta(s)$ is an arbitrary first-order real all-pass function of the form

$$\eta(s) = \frac{s - \sigma_1}{s + \sigma_1}, \qquad \sigma_1 \geq 0 \tag{74.88}$$

Since the load impedance $z_2(s)$ possesses a Class IV zero of transmission at infinity of order 3, the coefficient constraints become

$$A_m = \rho_m, \qquad m = 0, 1, 2 \tag{74.89a}$$

$$L_a \equiv \frac{F_2}{A_3 - \rho_3} \geq L \tag{74.89b}$$

After substituting the coefficients F_2, A_3, and ρ_3 from the Laurent series expansions of $F(s)$, $A(s)$, and $\rho(s)$ in (74.89), (74.89a) can all be satisfied by requiring that the DC gain be

$$K_n = 1 - \left[1 - \frac{2(1 - RC\sigma_1)\sin\gamma_1}{RC\omega_c}\right]^{2n} \tag{74.90}$$

where γ_m is defined in (73.37), and after considerable mathematical manipulations the constraint (74.89b) becomes

$$L_a = \frac{4R\sin\gamma_1\sin\gamma_3}{(1 - RC\sigma_1)\left[RC\omega_c^2(\alpha^2 - 2\alpha\cos\gamma_2 + 1) + 4\sigma_1\sin\gamma_1\sin\gamma_3\right]} \geq L \tag{74.91}$$

The details of these derivations can be found in [6]. Thus, with K_n as specified in (74.90), a match is possible if and only if the series inductance L does not exceed a critical inductance L_a. To show that any RLC load can be matched, we must demonstrate that there exists a nonnegative real σ_1 such that L_a can be made at least as large as the given inductance L and satisfies the constraint (74.90) with $0 \leq K_n \leq 1$. To this end, four cases are distinguished. Let

$$L_{a1} = \frac{R^2C\omega_c\sin\gamma_3}{[(RC\omega_c - \sin\gamma_1)^2 + \cos^2\gamma_1]\omega_c\sin\gamma_1} > 0 \tag{74.92}$$

$$L_{a2} = \frac{8R\sin^2\gamma_1\sin\gamma_3}{[(RC\omega_c - \sin\gamma_3)^2 + (1 + 4\sin^2\gamma_1)\sin\gamma_1\sin\gamma_3]\omega_c} > 0 \tag{74.93}$$

CASE 1. $RC\omega_c \geq 2\sin\gamma_1$ *and* $L_{a1} \geq L$. Under this situation, $\sigma_1 = 0$ and the maximum attainable K_n is given by (74.90). The equalizer back-end impedance $Z_{22}(s)$ can be expanded in a continued fraction as

$$Z_{22}(s) = (L_{a1} - L)s + \cfrac{1}{C_2 s + \cfrac{1}{L_3 s + \cfrac{1}{\ddots + \cfrac{1}{W}}}} \tag{74.94}$$

where W is a constant representing either a resistance or a conductance, and

$$L_1 = L_{a1} \tag{74.95a}$$

$$C_{2m}L_{2m-1} = \frac{4\sin\gamma_{4m-1}\sin\gamma_{4m+1}}{\omega_c^2(1-2\alpha\cos\gamma_{4m}+\alpha^2)}, \qquad m \le \frac{1}{2}(n-1) \tag{74.95b}$$

$$C_{2m}L_{2m+1} = \frac{4\sin\gamma_{4m+1}\sin\gamma_{4m+3}}{\omega_c^2(1-2\alpha\cos\gamma_{4m+2}+\alpha^2)}, \qquad m < \frac{1}{2}(n-1) \tag{74.95c}$$

where $m = 1, 2, \ldots, \left\lceil \frac{1}{2}(n-1) \right\rceil$, $n > 1$. In addition, the final reactive element can also be computed directly by the formulas

$$C_{n-1} = \frac{2(1+\alpha^n)\sin\gamma_1}{R(1-\alpha^n)(1+\alpha)\omega_c}, \qquad n \text{ odd} \tag{74.96a}$$

$$L_{n-1} = \frac{2R(1-\alpha^n)\sin\gamma_1}{(1+\alpha^n)(1+\alpha)\omega_c}, \qquad n \text{ even} \tag{74.96b}$$

Equation (74.94) can be identified as an LC ladder terminated in a resistor, as depicted in Fig. 74.8. The terminating resistance is determined by

$$R_{22} = R\frac{1-\alpha^n}{1+\alpha^n} \tag{74.97}$$

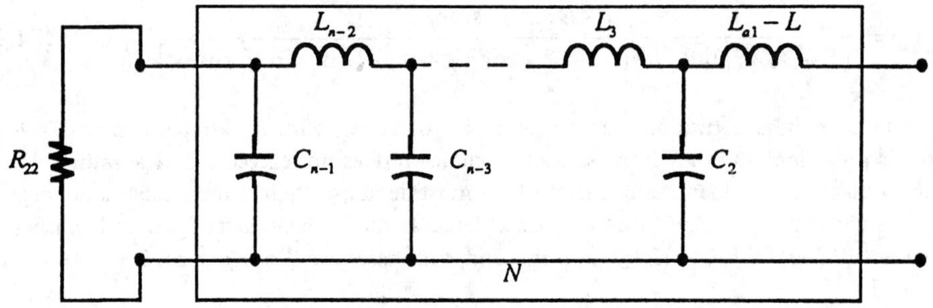

(a) *n* odd

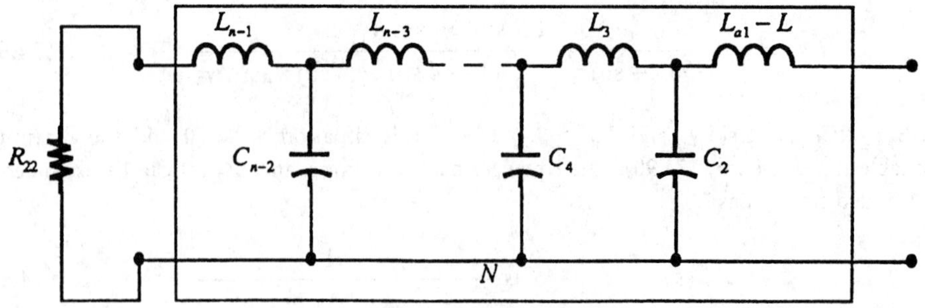

(b) *n* even

FIGURE 74.8 The nth-order Butterworth ladder network N.

CASE 2. $RC\omega_c \geq 2\sin\gamma_1$ and $L_{a1} < L$. Under this situation, σ_1 is nonzero and can be determined by the formula

$$\sigma_1 = \frac{1}{RC}\left(1 + 2\sqrt{p}\sinh\frac{\varphi}{3} - \frac{2RC\omega_c \sin^2\gamma_1 + \sin\gamma_3}{3\sin\gamma_1}\right) \tag{74.98}$$

where

$$p = \frac{(RC\omega_c - 2\sin\gamma_1)^2 \sin\gamma_3}{9\sin\gamma_1} > 0 \tag{74.99a}$$

$$w = \frac{(2RC\omega_c \sin^2\gamma_1 + \sin\gamma_3)}{54\sin^3\gamma_1}\left[3(RC\omega_c - 2\sin\gamma_1)^2 \sin\gamma_1 \sin\gamma_3 + (2RC\omega_c \sin^2\gamma_1 + \sin\gamma_3)^2\right] - \frac{R^2 C \sin\gamma_3}{2L\sin\gamma_1} \tag{74.99b}$$

$$\varphi = \sinh^{-1}\frac{w}{(\sqrt{p})^3} \tag{74.99c}$$

Using this value of σ_1, the DC gain K_n is computed by (74.90).

CASE 3. $RC\omega_c < 2\sin\gamma_1$ and $L_{a2} \geq L$. Then we have

$$K_n = 1 \tag{74.100a}$$

$$\sigma_1 = \frac{1}{RC}\left(1 - \frac{RC\omega_c}{2\sin\gamma_1}\right) > 0 \tag{74.100b}$$

CASE 4. $RC\omega_c < 2\sin\gamma_1$ and $L_{a2} < L$. Then the desired value of σ_1 can be computed by formula (74.98). Using this value of σ_1, the DC gain K_n is computed by (74.90).

Example 5. Let

$$R = 100\,\Omega, \qquad C = 100\text{ pF}, \qquad L = 0.5\ \mu\text{F} \tag{74.101a}$$

$$n = 6, \qquad \omega_c = 10^8\text{ rad/s} \tag{74.101b}$$

From (74.92), we first compute

$$L_{a1} = \frac{100\sin 45^\circ}{[(1-\sin 15^\circ)^2 + \cos^2 15^\circ] \times 10^8 \sin 15^\circ} = 1.84304\ \mu\text{H} \tag{74.102}$$

Since $L_{a1} > L$ and

$$RC\omega_c = 1 > 2\sin 15^\circ = 0.517638 \tag{74.103}$$

Case 1 applies and the matching network can be realized as an LC ladder terminating in a resistor as shown in Fig. 74.8b. With $\sigma_1 = 0$, the maximum attainable DC gain K_6 is from (74.90)

$$K_6 = 1 - \left(1 - \frac{2\sin 15^\circ}{RC\omega_c}\right)^{12} = 0.999841 \tag{74.104}$$

giving from (74.85b)

$$\alpha = (1 - K_6)^{1/12} = 0.482362 \tag{74.105}$$

Applying formulas (74.95) yields the element values of the LC ladder network, as follows:

$$L_1 = L_{a1} = 1.84304\ \mu\text{H} \tag{74.106a}$$

$$C_2 = \frac{4 \sin 45^\circ \sin 75^\circ}{1.84304 \times 10^{-6} \times 10^{16}(1 - 2 \times 0.482362 \cos 60^\circ + 0.482362^2)} = 197.566\ \text{pF} \tag{74.106b}$$

$$L_3 = \frac{4 \sin 75^\circ \sin 105^\circ}{197.566 \times 10^{-12} \times 10^{16}(1 - 2 \times 0.482362 \cos 90^\circ + 0.482362^2)} = 1.53245\ \mu\text{H} \tag{74.106c}$$

$$C_4 = \frac{4 \sin 105^\circ \sin 135^\circ}{1.53245 \times 10^{-6} \times 10^{16}(1 - 2 \times 0.482362 \cos 120^\circ + 0.482362^2)} = 103.951\ \text{pF} \tag{74.106d}$$

$$L_5 = \frac{4 \sin 135^\circ \sin 165^\circ}{103.951 \times 10^{-12} \times 10^{16}(1 - 2 \times 0.482362 \cos 150^\circ + 0.482362^2)} = 0.34051\ \mu\text{H} \tag{74.106e}$$

The last reactive elements L_5 can also be calculated directly from (74.96b) as

$$L_5 = \frac{2 \times 100(1 - 0.482362^6) \sin 15^\circ}{(1 + 0.482362^6)(1 + 0.482362)10^8} = 0.34051\ \mu\text{H} \tag{74.107}$$

Finally, the terminating resistance is determined from (74.97) as

$$R_{22} = 100 \frac{1 - 0.482362^6}{1 + 0.482362^6} = 97.512\ \Omega \tag{74.108}$$

The matching network together with its terminations is presented in Fig. 74.9. We remark that for computational accuracy we retain five significant figures in all the calculations. In practice, one or two significant digits are sufficient, as indicated in the figure.

Example 6. Let

$$R = 100\ \Omega, \qquad C = 50\ \text{pF}, \quad L = 0.5\ \mu\text{F} \tag{74.109a}$$

$$n = 5, \qquad \omega_c = 10^8\ \text{rad/s} \tag{74.109b}$$

Since

$$RC\omega_c = 0.5 < 2 \sin 18^\circ = 0.618 \tag{74.110}$$

and from (74.93)

$$L_{a2} = 1.401\ \mu\text{H} > L = 0.5\ \mu\text{H} \tag{74.111}$$

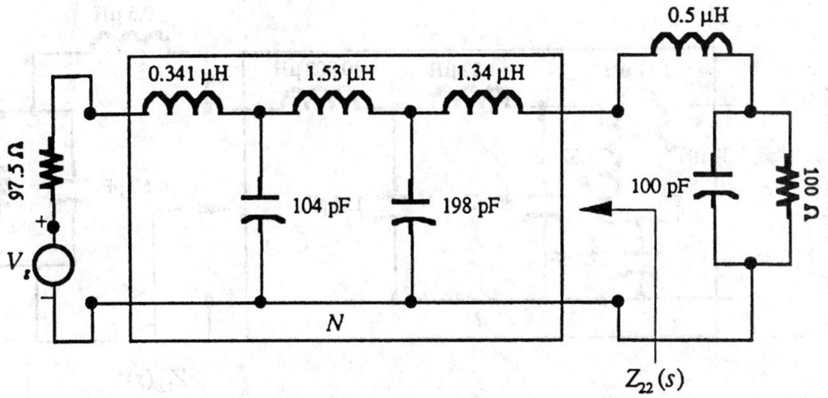

FIGURE 74.9 A sixth-order Butterworth matching network.

Case 3 applies, and we have $K_5 = 1$ and from (74.100b)

$$\sigma_1 = 0.381966 \times 10^8 \tag{74.112}$$

The normalized reflection coefficient is found to be

$$\rho(y) = \frac{(y - 0.381966)y^5}{(y + 0.381966)(y^5 + 3.23607y^4 + 5.23607y^3 + 5.23607y^2 + 3.23607y + 1)} \tag{74.113}$$

where $y = s/10^8$. Finally, we compute the equalizer back-end impedance as

$$\begin{aligned}
\frac{Z_{22}(s)}{100} &= \frac{F(y)}{A(y) - \rho(y)} - z_2(y) \\
&= \frac{\dfrac{-2}{(0.5y+1)^2}}{\dfrac{0.5y-1}{0.5y+1} - \rho(s)} - 0.5y - \frac{1}{0.5y+1} \\
&= \frac{2.573y^5 + 4.1631y^4 + 5.177y^3 + 4.2136y^2 + 2.045y + 0.38197}{2.8541y^4 + 4.618y^3 + 4.118y^2 + 2.045y + 0.38197} \\
&= 0.9015y + \cfrac{1}{1.949y + \cfrac{1}{1.821y + \cfrac{1}{0.8002y + \cfrac{1}{\dfrac{1.005y + 0.3822}{0.9944y + 0.3822}}}}}
\end{aligned} \tag{74.114}$$

The final matching network together with its terminations is presented in Fig. 74.10.

Example 7. Let

$$R = 100\ \Omega, \qquad C = 50\ \text{pF}, \qquad L = 3\ \mu\text{F} \tag{74.115a}$$

$$n = 4, \qquad \omega_c = 10^8\ \text{rad/s} \tag{74.115b}$$

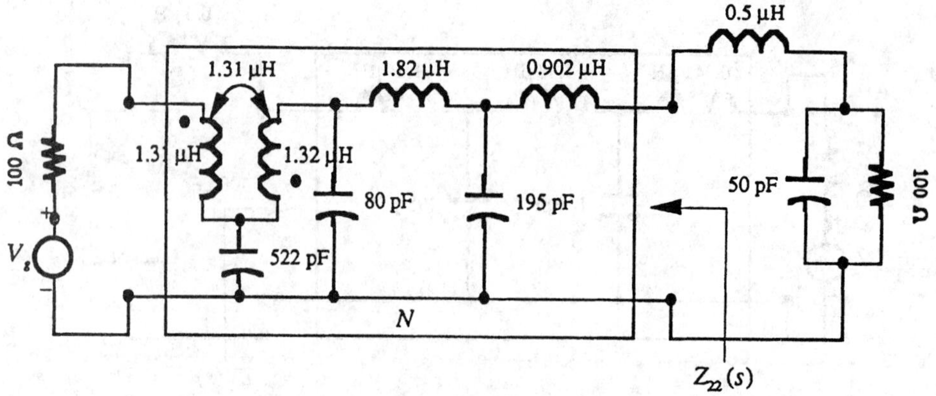

FIGURE 74.10 A fifth-order Butterworth matching network.

Since

$$RC\omega_c = 0.5 < 2\sin 22.5^\circ = 0.76537 \tag{74.116}$$

and from (74.93)

$$L_{a2} = 1.462\ \mu\text{H} < L = 3\ \mu\text{H} \tag{74.117}$$

Case 4 applies, and from (74.98)

$$\sigma_1 = 1.63129 \times 10^8 \tag{74.118}$$

where (74.99)

$$p = 0.0188898, \qquad w = 0.2304, \qquad \varphi = 5.17894 \tag{74.119}$$

From (74.90), the maximum attainable DC gain is obtained as

$$K_4 = 1 - \left[1 - \frac{2(1 - 100 \times 50 \times 10^{-12} \times 1.63129 \times 10^8)\sin 22.5^\circ}{100 \times 50 \times 10^{-12} \times 10^8}\right]^8$$
$$= 0.929525 \tag{74.120}$$

giving from (74.85b)

$$\alpha = (1 - K_4)^{1/8} = 0.717802 \tag{74.121}$$

Finally, the normalized reflection coefficient $\rho(s)$ is obtained as

$$\rho(s) = \frac{(y - 1.63129)(y^4 + 1.87571y^3 + 1.75913y^2 + 0.966439y + 0.265471)}{(y + 1.63129)(y^4 + 2.61313y^3 + 3.41421y^2 + 2.61313y + 1)} \tag{74.122}$$

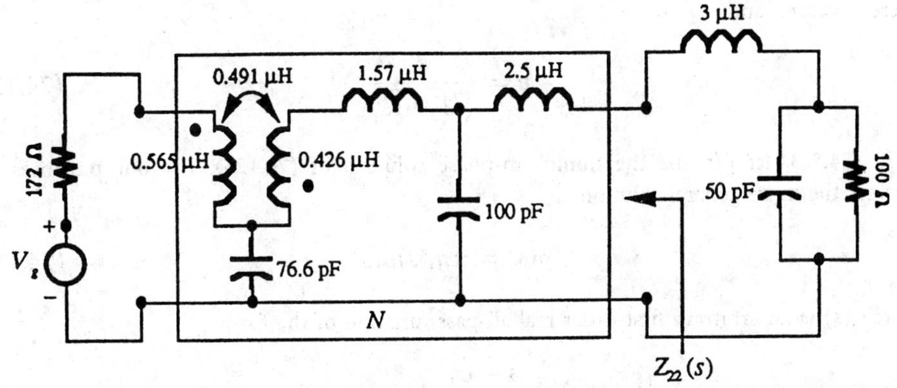

FIGURE 74.11 A fourth-order Butterworth matching network.

where $y = s/10^8$. Finally, we compute the equalizer back-end impedance as

$$\frac{Z_{22}(s)}{100} = \frac{F(y)}{A(y) - \rho(y)} - z_2(y) = \frac{\dfrac{-2}{(0.5y+1)^2}}{\dfrac{0.5y-1}{0.5y+1} - \rho(s)} - 0.5y - \frac{1}{0.5y+1}$$

$$= \frac{3.3333y^4 + 7.4814y^3 + 8.6261y^2 + 5.9748y + 2.0642}{1.3333y^3 + 2.9926y^2 + 2.9195y + 1.1982} \tag{74.123}$$

$$= 2.5y + \cfrac{1}{1.0046y + \cfrac{1}{1.5691y + \cfrac{1}{\cfrac{1.0991y + 2.0642}{0.84591y + 1.1982}}}}$$

The final matching network together with its terminations is presented in Fig. 74.11.

Chebyshev Networks

Refer again to Fig. 74.7. We wish to match out the load impedance

$$z_2(s) = Ls + \frac{R}{RCs + 1} \tag{74.124}$$

to a resistive generator and to achieve the nth-order Chebyshev transducer power-gain characteristic

$$G(\omega^2) = \frac{K_n}{1 + \varepsilon^2 C_n^2(\omega/\omega_c)}, \qquad 0 \le K_n \le 1 \tag{74.125}$$

with maximum attainable constant K_n. Following (74.84), we obtain

$$\rho(s)\rho(-s) = (1 - K_n)\frac{1 + \hat{\varepsilon}^2 C_n^2(-jy)}{1 + \varepsilon^2 C_n^2(-jy)} \tag{74.126}$$

where $y = s/\omega_c$ and

$$\hat{\varepsilon} = \frac{\varepsilon}{\sqrt{1 - K_n}} \tag{74.127}$$

As in (74.86), let $\hat{\rho}(s)$ be the minimum-phase solution of (74.126). For our purposes, we consider the more general solution

$$\rho(s) = \pm\eta(s)\hat{\rho}(s) \tag{74.128}$$

where $\eta(s)$ is an arbitrary first-order real all-pass function of the form

$$\eta(s) = \frac{s - \sigma_1}{s + \sigma_1}, \qquad \sigma_1 \geq 0 \tag{74.129}$$

Since the load impedance $z_2(s)$ possesses a Class IV zero of transmission at infinity of order 3, the coefficient constraints become

$$A_m = \rho_m, \qquad m = 0, 1, 2 \tag{74.130a}$$

$$\frac{F_2}{A_3 - \rho_3} \geq L \tag{74.130b}$$

After substituting the coefficients F_2, A_3, and ρ_3 from the Laurent series expansions of $F(s)$, $A(s)$, and $\rho(s)$ in (74.130a), they lead to the constraints on the constant K_n as

$$K_n = 1 - \varepsilon^2 \sinh^2\left\{n \sinh^{-1}\left[\sinh a - \frac{2(1 - RC\sigma_1)\sin\gamma_1}{RC\omega_c}\right]\right\} \tag{74.131}$$

where γ_m is defined in (73.37), and

$$a = \frac{1}{n}\sinh^{-1}\frac{1}{\varepsilon} \tag{74.132}$$

To apply the constraint (74.130b), we rewrite it as

$$L_b \equiv \frac{F_2}{A_3 - \rho_3} \geq L \tag{74.133}$$

After substituting the coefficients F_2, A_3, and ρ_3 from the Laurent series expansions of $F(s)$, $A(s)$, and $\rho(s)$ in (74.133) and after considerable mathematical manipulations, we obtain

$$L_b = \frac{4R\sin\gamma_1\sin\gamma_3}{(1 - RC\sigma_1)[RC\omega_c^2 f_1(\sinh a, \sinh\hat{a}) + 4\sigma_1\sin\gamma_1\sin\gamma_3]} \geq L \tag{74.134}$$

where

$$\hat{a} = \frac{1}{n}\sinh^{-1}\frac{1}{\hat{\varepsilon}} = \frac{1}{n}\sinh^{-1}\frac{\sqrt{1 - K_n}}{\varepsilon} \tag{74.135}$$

$$f_m(x, y) = x^2 + y^2 - 2xy\cos\gamma_{2m} + \sin^2\gamma_{2m}, \quad m = 1, 2, \ldots, \left\lceil\frac{1}{2}n\right\rceil \tag{74.136}$$

The details of these derivations can be found in [6]. Thus, with K_n as specified in (74.131), a match is possible if and only if the series inductance L does not exceed a critical inductance L_b. To show that any RLC load can be matched, we must demonstrate that there exists a nonnegative real σ_1 such that L_b can be made at least as large as the given inductance L and satisfies the constraint (74.131) with $0 \leq K_n \leq 1$. To this end, four cases are distinguished. Let

$$L_{b1} = \frac{R^2 C\omega_c \sin\gamma_3}{[(1 - RC\omega_c \sinh a \sin\gamma_1)^2 + R^2C^2\omega_c^2 \cosh^2 a \cos^2\gamma_1]\omega_c \sin\gamma_1} > 0 \tag{74.137}$$

$$L_{b2} = \frac{8R\sin^2\gamma_1 \sin\gamma_3}{[(RC\omega_c \sinh a - \sin\gamma_3)^2 + (1+4\sin^2\gamma_1)\sin\gamma_1 \sin\gamma_3 + R^2C^2\omega_c^2 \sin^2\gamma_2]\omega_c \sinh a} > 0 \tag{74.138}$$

Observe that both L_{b1} and L_{b2} are positive.

CASE 1. $RC\omega_c \sinh a \geq 2\sin\gamma_1$ *and* $L_{b1} \geq L$. Under this situation, $\sigma_1 = 0$ and the maximum attainable K_n is given by

$$K_n = 1 - \varepsilon^2 \sinh^2\left\{n \sinh^{-1}\left[\sinh a - \frac{2\sin\gamma_1}{RC\omega_c}\right]\right\} \tag{74.139}$$

The equalizer back-end impedance $Z_{22}(s)$ can be expanded in a continued fraction as in (74.94) with L_{b1} replacing L_{a1} and realized by the LC ladders of Fig. 74.8 with

$$L_1 = L_{b1} \tag{74.140a}$$

$$C_{2m}L_{2m-1} = \frac{4\sin\gamma_{4m-1}\sin\gamma_{4m+1}}{\omega_c^2 f_{2m}(\sinh a, \sinh\hat{a})}, \qquad m \leq \frac{1}{2}(n-1) \tag{74.140b}$$

$$C_{2m}L_{2m+1} = \frac{4\sin\gamma_{4m+1}\sin\gamma_{4m+3}}{\omega_c^2 f_{2m+1}(\sinh a, \sinh\hat{a})}, \qquad m < \frac{1}{2}(n-1) \tag{74.140c}$$

where $m = 1, 2, \ldots, \left\lceil \frac{1}{2}(n-1) \right\rceil$, $n > 1$. In addition, the final reactive element can also be computed directly by the formulas

$$C_{n-1} = \frac{2(\sinh na + \sinh n\hat{a})\sin\gamma_1}{R\omega_c(\sinh a + \sinh\hat{a})(\sinh na - \sinh n\hat{a})}, \qquad n \text{ odd} \tag{74.141a}$$

$$L_{n-1} = \frac{2R(\cosh na - \cosh n\hat{a})\sin\gamma_1}{\omega_c(\sinh a + \sinh\hat{a})(\cosh na + \cosh n\hat{a})}, \qquad n \text{ even} \tag{74.141b}$$

The terminating resistance is determined by

$$R_{22} = R\frac{\sinh na - \sinh n\hat{a}}{\sinh na + \sinh n\hat{a}}, \qquad n \text{ odd} \tag{74.142a}$$

$$R_{22} = R\frac{\cosh na - \cosh n\hat{a}}{\cosh na + \cosh n\hat{a}}, \qquad n \text{ even} \tag{74.142b}$$

CASE 2. $RC\omega_c \sinh a \geq 2 \sin \gamma_1$ and $L_{b1} < L$. Under this situation, σ_1 is nonzero and can be determined by the formula

$$\sigma_1 = \frac{1}{RC}\left(1 + 2\sqrt{q}\sinh\frac{\varphi}{3} - \frac{2RC\omega_c \sin^2\gamma_1 \sinh a + \sin\gamma_3}{3\sin\gamma_1}\right) \tag{74.143}$$

where

$$q = \frac{(RC\omega_c \sinh a - 2\sin\gamma_1)^2 \sin\gamma_3 + 3R^2C^2\omega_c^2 \sin\gamma_1 \cos^2\gamma_1}{9\sin\gamma_1} > 0 \tag{74.144a}$$

$$\begin{aligned}\zeta = {} & \frac{(2RC\omega_c \sin^2\gamma_1 \sinh a + \sin\gamma_3)}{54\sin^3\gamma_1} \\ & \times [3(RC\omega_c \sinh a - 2\sin\gamma_1)^2 \sin\gamma_1 \sin\gamma_3 \\ & + 2.25R^2C^2\omega_c^2 \sin^2\gamma_2 + (2RC\omega_c \sin^2\gamma_1 \sinh a + \sin\gamma_3)^2] \\ & - \frac{R^2C\sin\gamma_3}{2L\sin\gamma_1}\end{aligned} \tag{74.144b}$$

$$\varphi = \sinh^{-1}\frac{\zeta}{(\sqrt{q})^3} \tag{74.144c}$$

Using this value of σ_1, the constant K_n is computed by (74.131).

CASE 3. $RC\omega_c \sinh a < 2\sin\gamma_1$ and $L_{b2} \geq L$. Then we have

$$K_n = 1 \tag{74.145a}$$

$$\sigma_1 = \frac{1}{RC}\left(1 - \frac{RC\omega_c \sinh a}{2\sin\gamma_1}\right) > 0 \tag{74.145b}$$

CASE 4. $RC\omega_c \sinh a < 2\sin\gamma_1$ and $L_{b2} < L$. Then the desired value of σ_1 can be computed by formula (74.143). Using this value of σ_1, the constant K_n is computed by (74.131).

Example 8. Let

$$R = 100\ \Omega, \qquad C = 500\ \text{pF}, \qquad L = 0.5\ \mu\text{F} \tag{74.146a}$$

$$n = 6, \qquad \varepsilon = 0.50885\ \text{(1-dB ripple)}, \qquad \omega_c = 10^8\ \text{rad/s} \tag{74.146b}$$

From (74.137), we first compute

$$\begin{aligned} L_{b1} &= \frac{500\sin 45^\circ}{[(1 - 5\sinh 0.237996\sin 15^\circ)^2 + 25\cosh^2 0.237996\cos^2 15^\circ]10^8 \sin 15^\circ} \\ &= 0.54323\ \mu\text{H}\end{aligned} \tag{74.147}$$

Since $L_{b1} > L$ and

$$RC\omega_c \sinh a = 5\sinh 0.237996 = 1.20125 > 2\sin 15^\circ = 0.517638 \tag{74.148}$$

Case 1 applies and the matching network can be realized as an LC ladder terminating in a resistor as shown in Fig. 74.8b. With $\sigma_1 = 0$, the maximum attainable DC gain K_6 is from (74.131)

$$K_6 = 1 - 0.50885^2 \sinh^2\left\{6 \sinh^{-1}\left[\sinh 0.237996 - \frac{2\sin 15^\circ}{5}\right]\right\} = 0.78462 \quad (74.149)$$

giving from (74.127) and (74.135)

$$\hat{\varepsilon} = \frac{0.50885}{\sqrt{1 - 0.78462}} = 1.0964 \quad (74.150a)$$

$$\hat{\alpha} = \frac{1}{6}\sinh^{-1}\frac{1}{1.0964} = 0.13630 \quad (74.150b)$$

Applying formulas (74.140) yields the element values of the LC ladder network, as follows:

$$L_1 = L_{b1} = 0.54323\ \mu\text{H}, \qquad C_2 = 634\ \text{pF}, \qquad L_3 = 0.547\ \mu\text{H} \quad (74.151a)$$

$$C_4 = 581\ \text{pF}, \qquad L_5 = 0.329\ \mu\text{H} \quad (74.151b)$$

The last reactive element L_5 can also be calculated directly from (74.141b) as

$$L_5 = \frac{2 \times 100[\cosh(6 \times 0.237996) - \cosh(6 \times 0.13630)]\sin 15^\circ}{10^8(\sinh 0.237996 + \sinh 0.13630)[\cosh(6 \times 0.237996) + \cosh(6 \times 0.13630)]}$$
$$= 0.328603\ \mu\text{H} \quad (74.152)$$

Finally, the terminating resistance is determined from (74.142b) as

$$R_{22} = 100\frac{\cosh(6 \times 0.237996) - \cosh(6 \times 0.13630)}{\cosh(6 \times 0.237996) + \cosh(6 \times 0.13630)} = 23.93062\ \Omega \quad (74.153)$$

The matching network together with its terminations is presented in Fig. 74.12. We remark that for computational accuracy we retain five significant figures in all the calculations. In practice, one or two significant digits are sufficient, as indicated in the figure.

Example 9. Let

$$R = 100\ \Omega, \qquad C = 500\ \text{pF}, \qquad L = 1\ \mu\text{H} \quad (74.154a)$$

$$n = 5, \qquad \varepsilon = 0.50885\ (1\text{-dB ripple}), \qquad \omega_c = 10^8\ \text{rad/s} \quad (74.154b)$$

From (74.137), we first compute

$$L_{b1} = 0.52755\ \mu\text{H} \quad (74.155)$$

Since $L_{b1} < L$ and

$$RC\omega_c \sinh a = 5 \sinh 0.28560 = 1.44747 > 2 \sin 18^\circ = 0.61803 \quad (74.156)$$

Case 2 applies. From (74.144), we obtain

$$q = 7.73769, \qquad \zeta = 7.84729, \qquad \varphi = 0.356959 \quad (74.157)$$

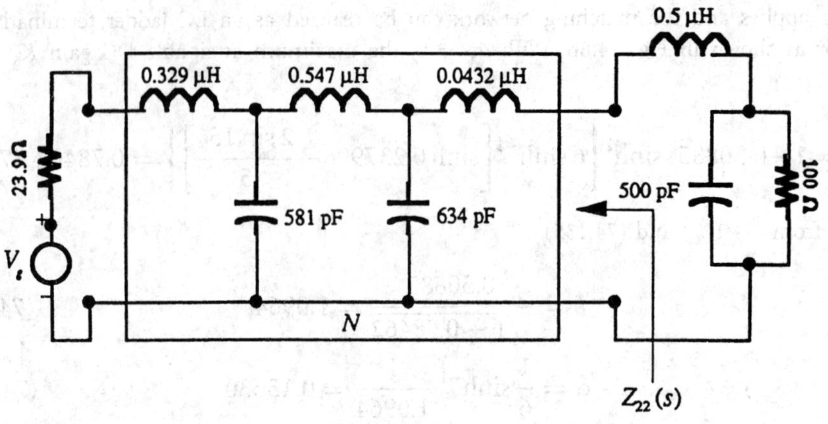

FIGURE 74.12 A sixth-order Chebyshev matching network.

Substituting these in (74.143) gives

$$\sigma_1 = 0.0985305 \times 10^8 \tag{74.158}$$

From (74.131), the maximum attainable constant K_5 is found to be

$$K_5 = 0.509206 \tag{74.159}$$

The rest of the calculations proceed exactly as in the previous example, and the details are omitted.

Example 10. Let

$$R = 100\ \Omega, \qquad C = 100\ \text{pF}, \qquad L = 0.5\ \mu\text{F} \tag{74.160a}$$

$$n = 5, \qquad \varepsilon = 0.76478\ (\text{2-dB ripple}), \qquad \omega_c = 10^8\ \text{rad/s} \tag{74.160b}$$

We first compute

$$a = \frac{1}{5}\sinh^{-1}\frac{1}{\varepsilon} = 0.2166104 \tag{74.161}$$

and from (74.138)

$$L_{b2} = 2.72234\ \mu\text{H} \tag{74.162}$$

Since $L_{b2} \geq L$ and

$$RC\omega_c \sinh a = \sinh 0.2166104 = 0.218308 < 2\sin 18^\circ = 0.618034 \tag{74.163}$$

Case 3 applies. Then $K_5 = 1$ and from (74.145b)

$$\sigma_1 = \frac{1}{10^{-8}}\left(1 - \frac{\sinh 0.2166104}{2\sin 18^\circ}\right) = 0.64677 \times 10^8 \tag{74.164}$$

For $K_5 = 1$, (74.126) degenerates into

$$\rho(y)\rho(-y) = \frac{\varepsilon^2 C_5^2(-jy)}{1 + \varepsilon^2 C_5^2(-jy)} \tag{74.165}$$

where $y = s/10^8$, the minimum-phase solution of which is found to be

$$\hat{\rho}(y) = \frac{y^5 + 1.25y^3 + 0.3125y}{y^5 + 0.706461y^4 + 1.49954y^3 + 0.693477y^2 + 0.459349y + 0.0817225} \tag{74.166}$$

A more general solution is given by

$$\rho(y) = \frac{y - 0.64677 \times 10^8}{y + 0.64677 \times 10^8}\hat{\rho}(y) \tag{74.167}$$

Finally, we compute the equalizer back-end impedance as

$$\begin{aligned}
\frac{Z_{22}(y)}{100} &= \frac{F(y)}{A(y) - \rho(y)} - z_2(y) \\
&= \frac{\dfrac{-2}{(y+1)^2}}{\dfrac{y-1}{y+1} - \rho(x)} - 0.5y - \frac{1}{y+1} \\
&= \frac{1.63267y^5 + 0.576709y^4 + 2.15208y^3 + 0.533447y^2 + 0.554503y + 0.0528557}{0.734663y^4 + 0.259505y^3 + 0.639438y^2 + 0.123844y + 0.0528557} \\
&= 2.222y + \cfrac{1}{1.005y + \cfrac{1}{3.651y + \cfrac{1}{0.8204y + \cfrac{1}{\cfrac{0.2441y + 0.05286}{0.02736y + 0.05286}}}}}
\end{aligned} \tag{74.168}$$

The final matching network together with its terminations is presented in Fig. 74.13.

Example 11. Let

$$R = 100\ \Omega, \qquad C = 100\ \text{pF}, \qquad L = 3\ \mu\text{F} \tag{74.169a}$$

$$n = 4, \qquad \varepsilon = 0.76478\ (2\text{-dB ripple}), \qquad \omega_c = 10^8\ \text{rad/s} \tag{74.169b}$$

We first compute

$$a = \frac{1}{4}\sinh^{-1}\frac{1}{0.76478} = 0.27076 \tag{74.170}$$

and from (74.138)

$$L_{b2} = 2.66312\ \mu\text{H} \tag{74.171}$$

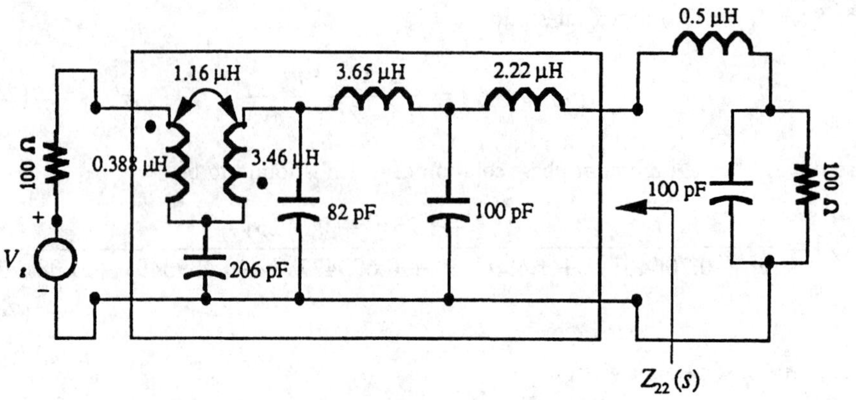

FIGURE 74.13 A fifth-order Chebyshev matching network.

Since $L_{b2} < L$ and

$$RC\omega_c \sinh a = \sinh 0.27076 = 0.27408 < 2\sin 22.5^\circ = 0.765367 \tag{74.172}$$

Case 4 applies. From (74.144), we obtain

$$q = 0.349261, \qquad \zeta = 0.390434, \qquad \varphi = 1.39406 \tag{74.173}$$

Substituting these in (74.143) gives

$$\sigma_1 = 0.694564 \times 10^8 \tag{74.174}$$

obtaining from (74.127) and (74.131)

$$K_4 = 0.984668, \qquad \hat{\varepsilon} = 6.17635 \tag{74.175}$$

From (74.126)

$$\rho(y)\rho(-y) = (1 - K_4)\frac{1 + \hat{\varepsilon}^2 C_4^2(-jy)}{1 + \varepsilon^2 C_4^2(-jy)} \tag{74.176}$$

where $y = s/10^8$, the minimum-phase solution of which is found to be

$$\hat{\rho}(y) = \frac{y^4 + 0.105343y^3 + 1.00555y^2 + 0.0682693y + 0.126628}{y^4 + 0.716215y^3 + 1.25648y^2 + 0.516798y + 0.205765} \tag{74.177}$$

A more general solution is given by

$$\rho(y) = \frac{y - 0.694564 \times 10^8}{y + 0.694564 \times 10^8}\hat{\rho}(s) \tag{74.178}$$

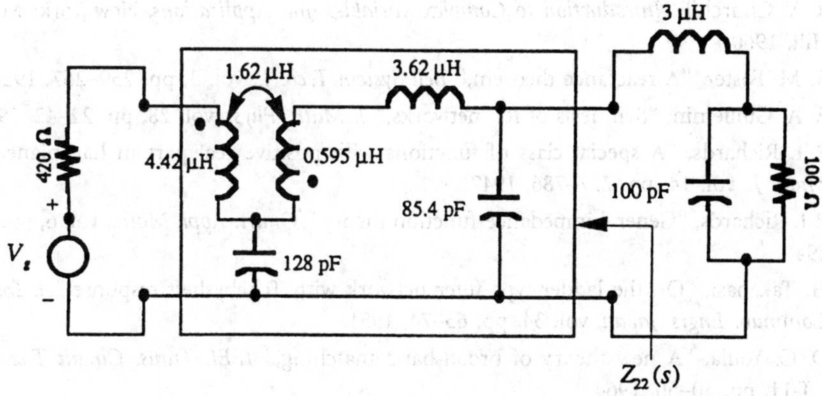

FIGURE 74.14 A fourth-order Chebyshev matching network.

Finally, we compute the equalizer back-end impedance as

$$\begin{aligned}\frac{Z_{22}(y)}{100} &= \frac{F(y)}{A(y) - \rho(y)} - z_2(y) = \frac{\dfrac{-2}{(y+1)^2}}{\dfrac{y-1}{y+1} - \rho(s)} - \frac{3y^2 + 3y + 1}{y+1} \\ &= \frac{0.780486y^2 + 0.320604y + 0.23087}{0.666665y^3 + 0.273851y^2 + 0.413057y + 0.0549659} \\ &= \cfrac{1}{0.8542y + \cfrac{1}{3.616y + \cfrac{1}{\cfrac{0.1219y + 0.2309}{0.2159y + 0.05497}}}}\end{aligned} \tag{74.179}$$

The final matching network together with its terminations is presented in Fig. 74.14.

erences

[1] G. Bossé, "Siebketten ohne Dämpfungsschwankungen im Durchlassbereich (Potenzketten)," *Frequenz*, vol. 5, pp. 279–284, 1951.

[2] R. Bott and R. J. Duffin, "Impedance synthesis without the use of transformers," *J. Appl. Phys.*, vol. 20, p. 816, 1949.

[3] O. Brune, "Synthesis of a finite two-terminal network whose driving-point impedance is a prescribed function of frequency," *J. Math. Phys.*, vol. 10, pp. 191–236, 1931.

[4] W. Cauer, *Synthesis of Linear Communication Networks*, New York: McGraw-Hill, chapters 5 and 6, 1958.

[5] W. K. Chen, *Theory and Design of Broadband Matching Networks*, Cambridge, UK: Pergamon Press, 1976.

[6] W. K. Chen, "Explicit formulas for the synthesis of optimum broadband impedance-matching networks," *IEEE Trans. Circuits Syst.*, vol. CAS-24, pp. 157–169, 1977.

[7] W. K. Chen, *Passive and Active Filters: Theory and Implementations*, New York: John Wiley & Sons, 1986.

[8] R. V. Churchill, *Introduction to Complex Variables and Applications*, New York: McGraw-Hill, 1960.

[9] R. M. Foster, "A reactance theorem," *Bell System Tech. J.*, vol. 3, pp. 259–267, 1924.

[10] E. A. Guillemin, "Synthesis of RC networks," *J. Math. Phys.*, vol. 28, pp. 22–42, 1949.

[11] P. I. Richards, "A special class of functions with positive real part in half-plane," *Duke Math. J.*, vol. 14, pp. 777–786, 1947.

[12] P. I. Richards, "General impedance-function theory," *Quart. Appl. Math.*, vol. 6, pp. 21–29, 1948.

[13] H. Takahasi, "On the ladder-type filter network with Tchebysheff response," *J. Inst. Elec. Commun. Engrs. Japan*, vol. 34, pp. 65–74, 1951.

[14] D. C. Youla, "A new theory of broad-band matching," *IEEE Trans. Circuit Theory*, vol. CT-11, pp. 30–50, 1964.

XV

ctive Filters

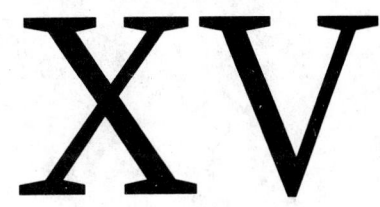

wrence P. Huelsman
versity of Arizona

75

Low-Gain Active Filters

lip E. Allen
ia Institute of ology

amin J. Blalock
ia Institute of ology

hen W. Milam
ia Institute of ology

.1 Introduction

Active filters consist of only amplifiers, resistors, and capacitors. Complex roots are achieved by the use of feedback eliminating the need for inductors. The gain of the amplifier can be finite or infinite (an op-amp). This section describes active filters using low-gain or finite-gain amplifiers. Filter design equations and examples will be given along with the performance limits of low-gain amplifier filters.

.2 First- and Second-Order Transfer Functions

Before discussing the characteristics and the synthesis of filters it is important to understand the transfer functions of first- and second-order filters. Later we will explain the implementations of these filters and show how to construct higher order filters from first- and second-order sections. The transfer functions of most first- and second-order filters are examined in the following.

First-Order Transfer Functions

A standard form of the transfer function of a first-order low-pass filter is

$$T_{LP}(s) = \frac{T_{LP}(j0)\omega_o}{s + \omega_o} \tag{75.1}$$

3-8341-2/95/$0.00 + $.50

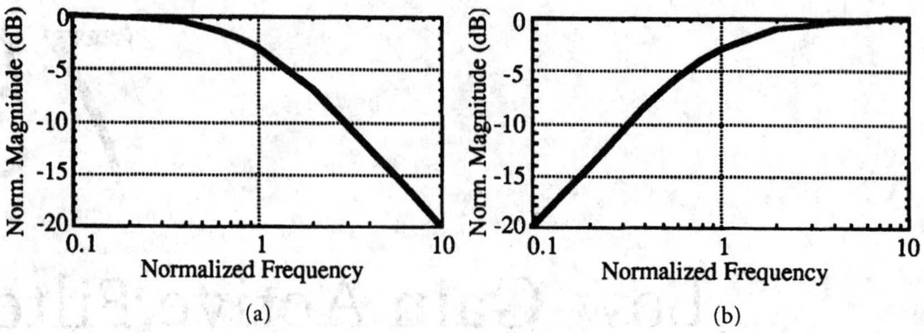

FIGURE 75.1 Normalized magnitude response of (a) first-order low-pass and (b) first-order high-pass filters.

where $T_{LP}(j0)$ is the value of $T_{LP}(s)$ at dc and ω_o is the **pole frequency**. It is common practice to normalize both the magnitude and frequency. Normalizing (75.1) yields

$$T_{LP\,n}(s_n) = \frac{T_{LP}(s_n\omega_o)}{|T_{LP}(j0)|} = \frac{1}{s_n + 1} \tag{75.2}$$

where $s_n = s/\omega_o$ and amplitude has been normalized as

$$T_{LP\,n}(s) = \frac{T_{LP}(s)}{|T_{LP}(j0)|} \tag{75.3}$$

The equivalent normalized forms of a first-order high-pass filter are

$$T_{HP}(s) = \frac{T_{HP}(j\infty)s}{s + \omega_o} \tag{75.4}$$

and

$$T_{HP\,n}(s_n) = \frac{T_{HP}(s_n\omega_o)}{|T_{HP}(j\infty)|} = \frac{s_n}{s_n + 1} \tag{75.5}$$

where $T_{HP}(j\infty) = T_{HP}(s) \mid$ at $\omega = \infty$. The normalized magnitude responses of these functions are shown in Fig. 75.1.

Second-Order Transfer Functions

The standard form of a second-order low-pass filter is given as

$$T_{LP}(s) = \frac{T_{LP}(j0)\omega_o^2}{s^2 + \left(\frac{\omega_o}{Q}\right)s + \omega_o^2} \tag{75.6}$$

where $T_{LP}(j0)$ is the value of $T_{LP}(s)$ at dc, ω_o is the pole frequency, and Q is the **pole Q** or the **pole quality factor**. The **damping factor** ζ, which may be better known to the reader, is given as

$$\zeta = \frac{1}{2Q} \tag{75.7}$$

The poles of the transfer function of (75.7) are

$$p_1, p_2 = \frac{-\omega_o}{2Q} \pm j\left(\frac{\omega_o}{2Q}\right)\sqrt{4Q^2 - 1} \tag{75.8}$$

Normalization of (75.6) in both amplitude and frequency gives

$$T_{\text{LP}\,n}(s_n) = \frac{T_{\text{LP}}(s_n\omega_o)}{|T_{\text{LP}}(j0)|} = \frac{1}{s_n^2 + \dfrac{s_n}{Q} + 1} \tag{75.9}$$

where $s_n = s/\omega_o$. The standard second-order, high-pass and bandpass transfer functions are

$$T_{\text{HP}}(s) = \frac{T_{\text{HP}}(j\infty)s^2}{s^2 + \left(\dfrac{\omega_o}{Q}\right)s + \omega_o^2} \tag{75.10}$$

and

$$T_{\text{BP}}(s) = \frac{T_{\text{BP}}(j\omega_o)\left(\dfrac{\omega_o}{Q}\right)s}{s^2 + \left(\dfrac{\omega_o}{Q}\right)s + \omega_o^2} \tag{75.11}$$

where $T_{\text{BP}}(j\omega_o) = T_{\text{BP}}(s)$ at $s = j\omega = j\omega_o$. The poles of the second-order high-pass and bandpass transfer functions are given by (75.8).

We can normalize these equations as we did for $T_{\text{LP}}(s)$ to get

$$T_{\text{HP}\,n}(s_n) = \frac{T_{\text{HP}}(s_n\omega_o)}{|T_{\text{HP}}(j\infty)|} = \frac{s_n^2}{s_n^2 + \dfrac{s_n}{Q} + 1} \tag{75.12}$$

$$T_{\text{BP}\,n}(s_n) = \frac{T_{\text{BP}}(s_n\omega_o)}{|T_{\text{BP}}(j\omega_o)|} = \frac{\dfrac{s_n}{Q}}{s_n^2 + \dfrac{s_n}{Q} + 1} \tag{75.13}$$

where

$$T_{\text{HP}\,n}(s) = \frac{T_{\text{HP}}(s)}{|T_{\text{HP}}(j\infty)|} \tag{75.14}$$

and

$$T_{\text{BP}\,n}(s) = \frac{T_{\text{BP}}(s)}{|T_{\text{BP}}(j\omega_o)|} \tag{75.15}$$

Two other types of second-order transfer function filters that we have not covered here are the bandstop and the all-pass. These transfer functions have the same poles as the previous ones. However, the zeros of the bandstop transfer function are on the $j\omega$ axis while the zeros of the all-pass transfer function are quadratically symmetrical to the poles (they are mirror images of the poles in the right-half plane). Both of these transfer functions can be implemented by a second-order biquadratic transfer function whose transfer function is given as

$$T_{BQ}(s) = \frac{K\left(s^2 \pm \left(\frac{\omega_z}{Q_z}\right)s + \omega_z^2\right)}{s^2 + \left(\frac{\omega_p}{Q_p}\right)s + \omega_p^2} \tag{75.16}$$

where K is a constant, ω_z is the zero frequency, Q_z the zero Q, ω_p is the pole frequency, and Q_p the pole Q.

Frequency Response (Magnitude and Phase)

The magnitude and phase response of the normalized second-order low-pass transfer function is shown in Fig. 75.2, where Q is a parameter. In this figure we see that Q influences the frequency response near ω_o. If Q is greater than 0.707, then the normalized magnitude response has a peak value of

$$|T_n[\omega_n(\max)]| = \frac{Q}{\sqrt{1 - (1/4Q^2)}} \tag{75.17}$$

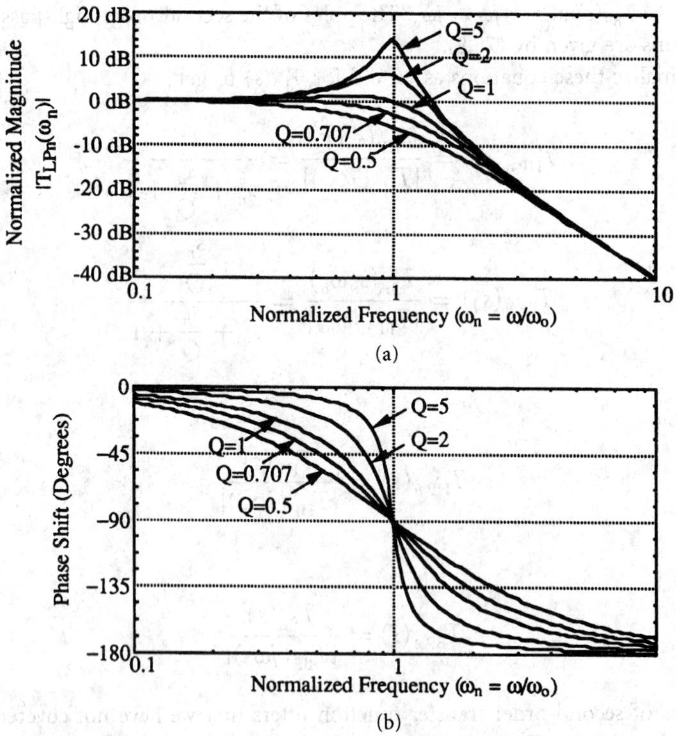

FIGURE 75.2 (a) Normalized magnitude and (b) phase response of the standard second-order low-pass transfer function with Q as a parameter.

at a frequency of

$$\omega_n(\max) = \sqrt{1 - \frac{1}{2Q^2}} \tag{75.18}$$

If the transfer function is multiplied by -1, the phase shift is shifted vertically by $\pm 180°$.

The magnitude and phase response of the normalized second-order high-pass transfer function is shown in Fig. 75.3. For Q greater than 0.707 the peak value of the normalized magnitude response is as

$$\omega_n(\max) = \frac{1}{\sqrt{1 - \frac{1}{2Q^2}}} \tag{75.19}$$

The normalized frequency response of the standard second-order bandpass transfer function is shown in Fig. 75.4. The slopes of the normalized magnitude curves at frequencies much greater or much less than ω_o are ± 20 dB/decade rather than ± 40 dB/decade of the second-order high- and low-pass transfer functions. This difference is because one pole is causing the high-frequency roll-off while the other pole is causing the low-frequency roll-off. The peak of the magnitude occurs at $\omega = \omega_o$ or $\omega_n = 1$.

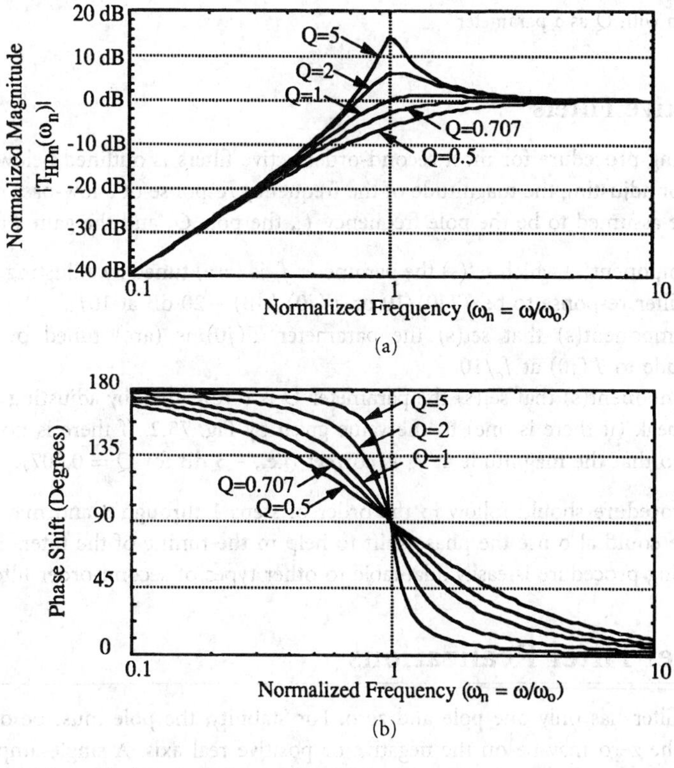

FIGURE 75.3 (a) Normalized magnitude and (b) phase response of the standard second-order high-pass transfer function with Q as a parameter.

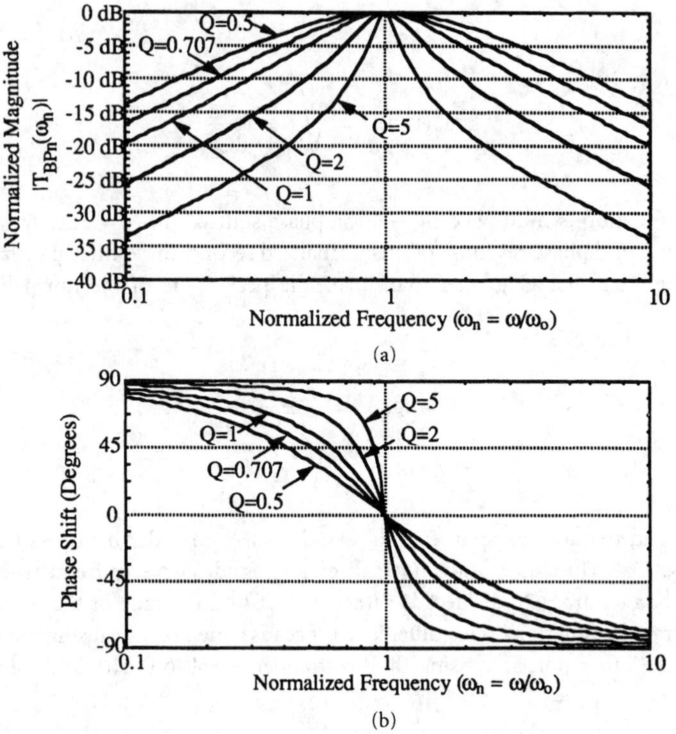

FIGURE 75.4 (a) Normalized magnitude and (b) phase response of the standard second-order bandpass transfer function with Q as a parameter.

Tuning Active Filters

A general tuning procedure for most second-order active filters is outlined below. This method is illustrated for adjusting the magnitude of the frequency response of a low-pass filter. The filter parameters are assumed to be the pole frequency f_o, the pole Q, and the gain $T(j0)$.

1. The component(s) which set(s) the parameter f_o is (are) tuned by adjusting the magnitude of the filter response to be $T(j0)/10$ or $T(j0)$ (dB) -20 dB at $10f_o$.
2. The component(s) that set(s) the parameter $T(j0)$ is (are) tuned by adjusting the magnitude to $T(j0)$ at $f_o/10$.
3. The component(s) that set(s) the parameter Q is (are) tuned by adjusting the magnitude of the peak (if there is one) to the value given by Fig. 75.2. If there is no peaking, then adjust so that the magnitude at f_o is correct (i.e., -3 dB for $Q = 0.707$).

The tuning procedure should follow in the order of steps 1 through 3 and may be repeated if necessary. One could also use the phase shift to help in the tuning of the filter. The concept of the above tuning procedure is easily adaptable to other types of second-order filters.

75.3 First-Order Filter Realizations

A first-order filter has only one pole and zero. For stability, the pole must be on the negative real axis but the zero may be on the negative or positive real axis. A single-amplifier low-gain realization of a first-order low-pass filter is shown in Fig. 75.5(a). The transfer function for this filter is

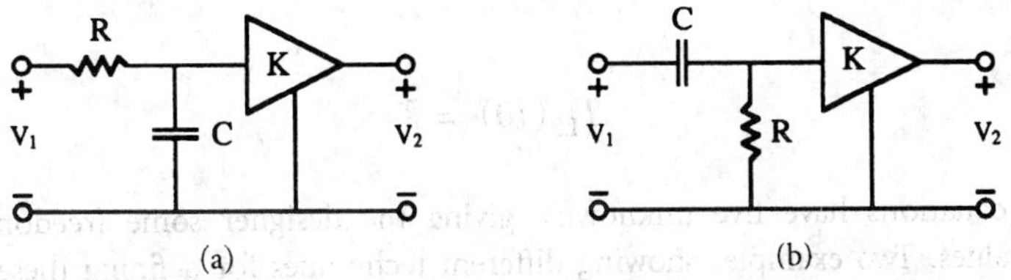

FIGURE 75.5 (a) Low-pass and (b) high-pass first-order filters.

$$T(s) = \frac{V_2(s)}{V_1(s)} = \frac{K/RC}{s + 1/RC} = \frac{K\omega_o}{s + \omega_o} = \frac{T_{LP}(0)\omega_o}{s + \omega_o} \tag{75.20}$$

The low-frequency ($\omega \ll \omega_o$) gain magnitude and polarity are set by K.

First-order high-pass filters can also be realized in a single-amplifier low-gain circuit [see Fig. 75.5(b)]. The transfer function for this filter is given in 75.4, and in this case $T_{HP}(j\infty) = K$. Both low-pass and high-pass magnitude responses are shown in Fig. 75.1. Note that the first-order filters exhibit no peaking and the frequency dependent part of the magnitude response approaches ± 20 dB/decade.

.4 Second-Order Positive-Gain Filters

Practical realizations for second-order filters using positive gain amplifiers are presented here. These filters are easy to design and have been extensively used in many applications. The first realization is a low-pass Sallen and Key filter and is shown in Fig. 75.6 [8].

The transfer function of Fig. 75.6 is

$$\frac{V_2(s)}{V_1(s)} = \frac{\dfrac{K}{R_1R_3C_2C_4}}{s^2 + s\left(\dfrac{1}{R_3C_4} + \dfrac{1}{R_1C_2} + \dfrac{1}{R_3C_2} - \dfrac{K}{R_3C_4}\right) + \dfrac{1}{R_1R_3C_2C_4}} \tag{75.21}$$

Equating this transfer function with the standard form of a second-order low-pass filter given in (75.6) gives

$$\omega_o = \frac{1}{\sqrt{R_1R_3C_2C_4}} \tag{75.22}$$

$$\frac{1}{Q} = \sqrt{\frac{R_3C_4}{R_1C_2}} + \sqrt{\frac{R_1C_4}{R_3C_2}} + (1-K)\sqrt{\frac{R_1C_2}{R_3C_4}} \tag{75.23}$$

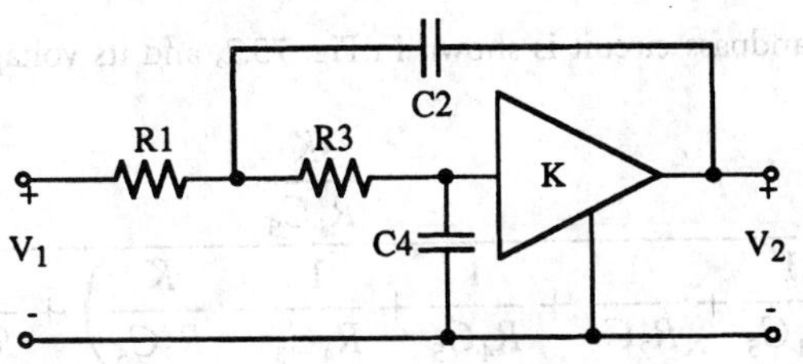

FIGURE 75.6 Low-pass Sallen and Key filter.

and

$$T_{\text{LP}}(j0) = K \tag{75.24}$$

These three equations have five unknowns, giving the designer some freedom in selecting component values. Two examples showing different techniques for defining these values of the circuit in Fig. 75.6 are given below.

Example 1.

An Equal-Resistance Equal-Capacitance Low-Pass Filter. For this example, $R = R_1 = R_3$ and $C = C_2 = C_4$. A Butterworth low-pass filter characteristic is needed with $\omega_o = 6283$ rad/s (1 kHz) and $Q = 0.7071$. With these constraints,

$$\omega_o = \frac{1}{RC} \qquad \frac{1}{Q} = 3 - K \tag{75.25}$$

and $RC = 159$ μs, $T_{\text{LP}}(j0) = K = 1.586$. Selecting $C = 0.1$ μF yields $R = 1.59$ kΩ.

Example 2.

A Unity-Gain Low-Pass Filter. For this example let $K = 1$. Therefore, (75.23) becomes

$$\frac{1}{Q} = \sqrt{\frac{R_3 C_4}{R_1 C_2}} + \sqrt{\frac{R_1 C_4}{R_3 C_2}} = \sqrt{\frac{C_4}{C_2}}\left(\sqrt{\frac{R_3}{R_1}} + \sqrt{\frac{R_1}{R_3}}\right) \tag{75.26}$$

The desired transfer function, with the complex frequency normalized by a factor of 10^4 rad/s, is

$$\frac{V_2(s)}{V_1(s)} = \frac{0.988}{s^2 + 0.179s + 0.988} = \frac{T_{\text{LP}}(j0)\omega_o^2}{s^2 + \dfrac{\omega_o}{Q}s + \omega_o^2} \tag{75.27}$$

and $\omega_o = 0.994$ rad/s, $Q = 5.553$. To obtain a real-valued resistor ratio, pick

$$\frac{C_4}{C_2} \leq \frac{1}{4Q^2} = 0.00811 \tag{75.28}$$

or $C_4/C_2 = 0.001$ in this case. Equation (75.26) yields two solutions for the ratio R_3/R_1, 30.397 and 0.0329. From (75.22) with $\omega_o = 9.94$ krad/s, $R_1C_2 = 577$ μs. If C_2 is selected as 0.1 μF, then $C_4 = 100$ pF, $R_1 = 5.77$ kΩ, and $R_3 = 175.4$ kΩ.

A Sallen and Key bandpass circuit is shown in Fig. 75.7, and its voltage transfer function is

$$\frac{V_2(s)}{V_1(s)} = \frac{\dfrac{sK}{R_1C_5}}{s^2 + s\left(\dfrac{1}{R_1C_5} + \dfrac{1}{R_2C_5} + \dfrac{1}{R_4C_5} + \dfrac{1}{R_4C_3} - \dfrac{K}{R_2C_5}\right) + \dfrac{1}{R_4C_3C_5}\left(\dfrac{1}{R_1} + \dfrac{1}{R_2}\right)} \tag{75.29}$$

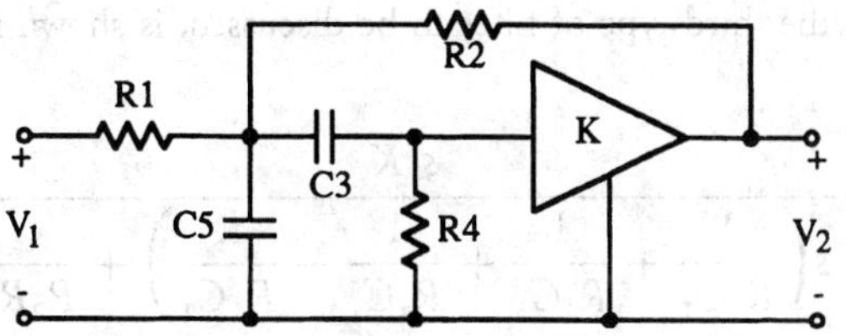

FIGURE 75.7 Bandpass Sallen and Key filter.

Equating this transfer function to that of (75.11) results in

$$\omega_o = \sqrt{\frac{1+\dfrac{R_1}{R_2}}{R_1R_4C_3C_5}} \tag{75.30}$$

$$\frac{1}{Q} = \frac{\left(1+\dfrac{R_1}{R_2}(1-K)\right)\sqrt{\dfrac{R_4C_3}{R_1C_5}} + \sqrt{\dfrac{R_1C_3}{R_4C_5}} + \sqrt{\dfrac{R_1C_5}{R_4C_3}}}{\sqrt{1+\dfrac{R_1}{R_2}}} \tag{75.31}$$

and

$$T_{\text{BP}}(j0) = \frac{\dfrac{K}{R_1C_5}}{\dfrac{1}{R_1C_5}+\dfrac{1}{R_2C_5}+\dfrac{1}{R_4C_5}+\dfrac{1}{R_4C_3}-\dfrac{K}{R_2C_5}} \tag{75.32}$$

These three equations contain six unknowns. A common constraint is to set $K = 2$, requiring the gain block to have two equal-valued resistors connected around an op-amp.

Example 3.

An Equal-Capacitance Gain-of-2 Bandpass Filter. Design a bandpass filter with $\omega_n = 1$ and $Q = 2$. Arbitrarily select $C_3 = C_5 = 1$ F and $K = 2$. If R_1 is selected to be 1 Ω, then (75.11), (75.29), and (75.30) give

$$\omega_o^2 = \left(1+\frac{R_1}{R_2}\right)\frac{1}{R_4} = 1 \tag{75.33}$$

and

$$\frac{\omega_o}{Q} = 1 - \frac{1}{R_2} + \frac{2}{R_4} = \frac{1}{2} \tag{75.34}$$

Solving these equations gives $R_2 = 0.7403\ \Omega$ and $R_4 = 2.3508\ \Omega$. Practical values are achieved after frequency and impedance denormalizations are made.

The high-pass filter, the third type of filter to be discussed, is shown in Fig. 75.8. Its transfer function is

$$\frac{V_2(s)}{V_1(s)} = \frac{s^2K}{s^2 + s\left(\frac{1}{R_2C_1} + \frac{1}{R_4C_3} + \frac{1}{R_4C_1} - \frac{K}{R_2C_1}\right) + \frac{1}{R_2R_4C_1C_3}} \tag{75.35}$$

Equating (75.35) to (75.10) results in

$$\omega_o = \frac{1}{\sqrt{R_2R_4C_1C_3}} \tag{75.36}$$

$$\frac{1}{Q} = \sqrt{\frac{R_4C_3}{R_2C_1}} + \sqrt{\frac{R_2C_1}{R_4C_3}} + \sqrt{\frac{R_2C_3}{R_4C_1}} - K\sqrt{\frac{R_4C_3}{R_2C_1}} \tag{75.37}$$

and

$$T_{\mathrm{HP}}(j\infty) = K \tag{75.38}$$

The design procedure using these equations is similar to that for the low-pass and bandpass filters.

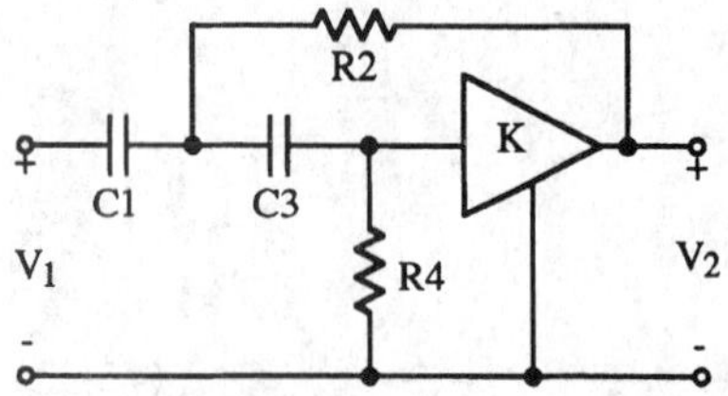

FIGURE 75.8 High-pass Sallen and Key filter.

The last type of second-order filter in this section is the notch filter, shown in Fig. 75.9. The general transfer function for the notch filter is

$$\frac{V_2(s)}{V_1(s)} = T_{\mathrm{N}}(j0)\frac{s^2 + \omega_z^2}{s^2 + \frac{\omega_p}{Q_p}s + \omega_p^2} \tag{75.39}$$

where ω_p and ω_z are the pole and zero frequencies (ω_o), respectively, and Q_p is the pole Q. For the circuit of Fig. 75.9(a),

$$T_{\mathrm{N}}(j0) = \frac{\omega_z^2}{\omega_p^2} \qquad R_2C_4 = \frac{\sqrt{1+a}}{\omega_p} \tag{75.40}$$

and

$$R_2 = (1+a)R_1 = 2R_3 = aR_4 \qquad C_4 = \frac{\omega_z^2}{\omega_p^2}C_1 = C_2 + C_1 = \frac{C_5}{2} \tag{75.41}$$

For the circuit of Fig. 75.9(b),

$$T_{\mathrm{N}}(j0) = 1 \qquad R_2C_4 = \frac{\sqrt{1+a}}{\omega_p} \tag{75.42}$$

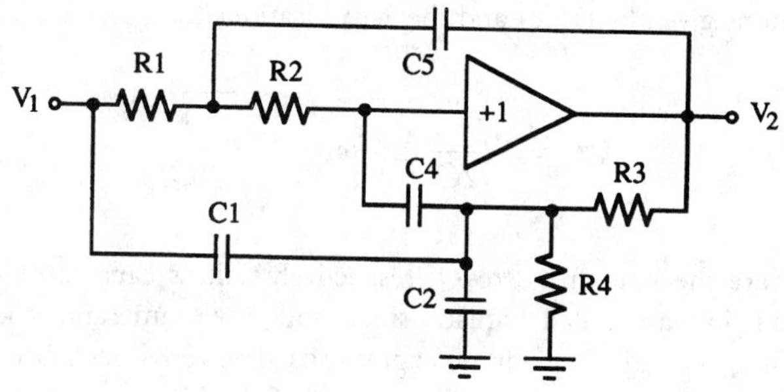

(a)

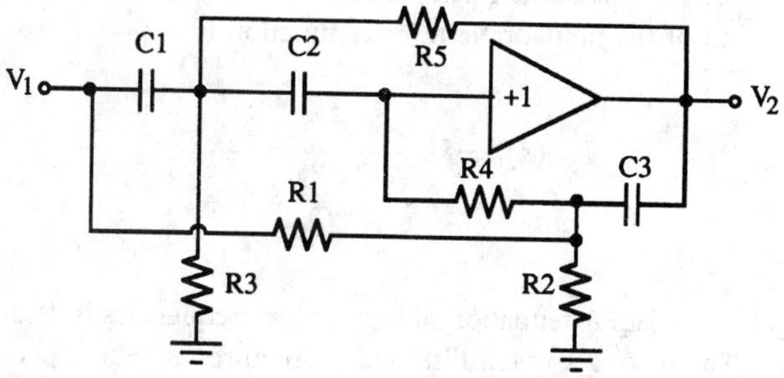

(b)

FIGURE 75.9 Notch filters derived from center-loaded twin-T networks: (a) $\omega_z \geq \omega_p$, (b) $\omega_z \leq \omega_p$ [10].

and

$$R_4 = (1+a)\frac{\omega_z^2}{\omega_p^2} R_1 = (1+a)(R_1 + R_2) = aR_3 = 2R_5, \qquad C_1 = C_2 = \frac{C_3}{2} \tag{75.43}$$

The Q for both notch circuits is

$$Q = \frac{\sqrt{1+a}}{a} \tag{75.44}$$

so a is a helpful design parameter for these circuits. For typical values of Q, say 0.5, 1, and 5, the corresponding values of a are 4.828, 1.618, and 0.2210, respectively.

75.5 Second-Order Biquadratic Filters

Next, second-order biquadratic functions will be described. These functions are general realizations of the second-order transfer function. Filters implementing biquadratic functions, often referred to as biquads, are found in many signal processing applications.

The Biquadratic Transfer Function

The general form of the second-order biquadratic transfer function is

$$T(s) = H\frac{s^2 + b_1 s + b_0}{s^2 + a_1 s + a_0} = H\frac{s^2 + \dfrac{\omega_z}{Q_z}s + \omega_z^2}{s^2 + \dfrac{\omega_p}{Q_p}s + \omega_p^2} \tag{75.45}$$

with the pole locations given by (75.8) and the zero locations by

$$z_1, z_2 = -\frac{\omega_z}{2Q_z} \pm j\omega_z \sqrt{1 - \frac{1}{4Q_z^2}} \tag{75.46}$$

where Q_p and Q_z are the pole and zero Q, respectively. Filters capable of implementing this voltage transfer function are called biquads since both the numerator and denominator of their transfer functions contain biquadratic expressions. The zeros described by the numerator of (75.45) strongly influence the magnitude response of the biquadratic transfer function and determine the filter type (low-pass, high-pass, etc.).

The notch filter form of the biquadratic transfer function is

$$T_{\mathrm{NF}}(s) = H \frac{s^2 + \omega_z^2}{s^2 + \dfrac{\omega_p}{Q_p} s + \omega_p^2} \tag{75.47}$$

with zeros located at $s = \pm j\omega_z$. Attenuation of high or low frequencies is determined by selection of ω_z relative to ω_p. The low-pass notch filter (LPN) requires $\omega_z > \omega_p$, shown in Fig. 75.10(a) and the high-pass notch filter (HPN) requires $\omega_p > \omega_z$, shown in Fig. 75.10(b).

An all-pass filter implemented using the biquadratic transfer function has the general form

$$T_{\mathrm{AP}}(s) = H \frac{s^2 - \dfrac{\omega_z}{Q_z} s + \omega_z^2}{s^2 + \dfrac{\omega_p}{Q_p} s + \omega_p^2} \tag{75.48}$$

The all-pass magnitude response is independent of frequency; i.e., its magnitude is constant. The all-pass filter finds use in shaping the phase response of a system. To accomplish this, the all-pass has right-half plane zeros that are mirror images around the imaginary axis of left-half plane poles.

Biquad Implementations

A single-amplifier low-gain realization [4] of the biquadratic transfer function with transmission zeros on the $j\omega$ axis is shown in Fig. 75.11. Zeros are generated by the circuit's input "twin-T"

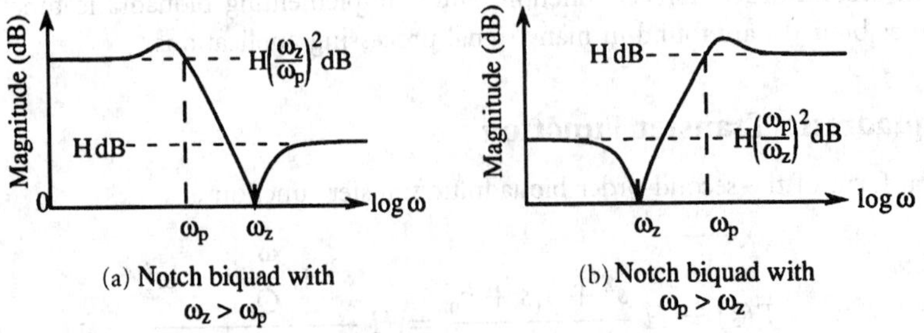

FIGURE 75.10 Frequency response of second-order notch filters. (a) $\omega_z > \omega_p$. (b) $\omega_p > \omega_z$.

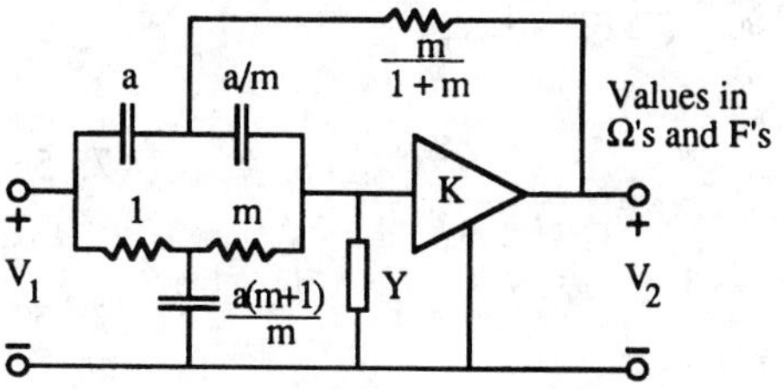

FIGURE 75.11 A low-gain realization for biquadratic network functions [5].

RC network. This filter is well suited for implementing elliptic functions. Its transfer function has the form

$$T(s) = H\frac{s^2+\omega_z^2}{s^2+\dfrac{\omega_p}{Q_p}s+\omega_p^2} = H\frac{s^2+b_0^2}{s^2+a_1s+a_0^2} \tag{75.49}$$

The selection of element $Y(s)$ in the filter circuit can be a resistor or capacitor. If the condition $\omega_p > \omega_z$ is desired, $Y = 1/R$ is chosen. For this choice the resulting transfer function has the form

$$T(s) = \frac{V_2(s)}{V_1(s)} = \frac{K(s^2+1/a^2)}{s^2+(m+1/a)[1/R+(2-K)/m]s+[1+(m+1)/R]/a^2} \tag{75.50}$$

where K, m, and a are defined in Fig. 75.11. The design parameter m is used to control the spread of circuit element values. The design equations for this filter result from equating (75.49) and (75.50) are

$$a = \frac{1}{\sqrt{b_o}} \tag{75.51}$$

$$R = \frac{m+1}{a_o/b_o-1} \tag{75.52}$$

$$K = 2+\frac{m}{m+1}\left(\frac{a_o}{b_o}-1-\frac{a_1}{\sqrt{b_o}}\right) \tag{75.53}$$

and

$$H = K \tag{75.54}$$

If the condition $\omega_z > \omega_p$ is desired, then $Y = s(aC)$ is chosen. This choice has the following transfer function:

$$\frac{V_2(s)}{V_1(s)} = \frac{\dfrac{K(s^2+1/a^2)}{(m+1)C+1}}{s^2+\left\{\dfrac{(m+1)[C+(2-K)/m]}{a[(m+1)C+1]}\right\}s+\dfrac{1}{a^2[(m+1)C+1]}} \tag{75.55}$$

The design equations follow from equating (75.49) and (75.55) to get

$$a = \frac{1}{\sqrt{b_o}} \tag{75.56}$$

$$C = \frac{b_o/a_o - 1}{m+1} \tag{75.57}$$

$$K = 2 + \frac{m}{m+1}\left(\frac{b_o}{a_o} - 1 - \frac{a_1\sqrt{b_o}}{a_o}\right) \tag{75.58}$$

and

$$H = \frac{a_o}{b_o}K \tag{75.59}$$

As before, the factor m is chosen arbitrarily to control the spread of element values.

Example 4.

A Low-Pass Elliptic Filter. It is desired to realize the elliptic voltage transfer function given as

$$\frac{V_2(s)}{V_1(s)} = \frac{H(s^2 + 2.235\,990)}{s^2 + 0.641\,131s + 1.235\,820}$$

This transfer function will have a 1-dB ripple in the passband, and at least 6 dB of attenuation for all frequencies greater than 1.2 rad/s [12]. Obviously, the second choice described above applies. From (75.56)–(75.59), with $m = 0.2$, we find that $a = 0.66875$, $C = 0.67443$, $K = 2.0056$, and $H = 1.1085$.

Biquadratic transfer functions can also be implemented using the two-amplifier low-gain configuration in Fig. 75.12 [5]. When zeros located off the $j\omega$ axis are desired, the design equations for the two-amplifier low-gain configuration are simpler than those required by the single-amplifier low-gain configuration. Also note that the required gain blocks of -1 and $+2$ are readily implemented using operational amplifiers. The transfer function for this configuration is

$$\frac{V_2(s)}{V_1(s)} = \frac{2(Y_1 - Y_2)}{Y_3 - Y_4} \tag{75.60}$$

The values of the admittances are determined by separately dividing the numerator and denominator by $s + c$, where c is a convenient value greater than zero. Partial fraction expansion

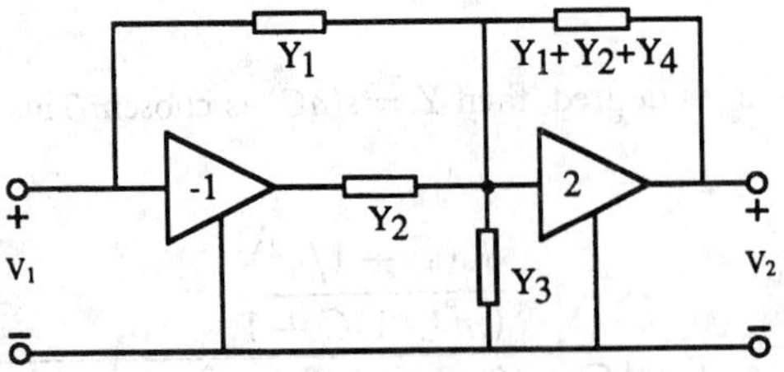

FIGURE 75.12 Two-amplifier low-gain biquad configuration.

results in expressions that can be implemented using *RC* networks. From the partial fraction expansion of the numerator divided by $s + c$, the pole residue at $s = -c$ is

$$k_b = \frac{H(c^2 - cb_1 + b_o)}{-c} \tag{75.61}$$

Depending on c, b_1, and b_o, the quality k_b can be positive or negative. If positive, Y_1 is

$$Y_1 = \frac{Hs}{2} + \frac{Hb_o}{2c} + \frac{1}{2/k_b + 2c/k_b s} \tag{75.62}$$

with $Y_2 = 0$, removing the inverting gain amplifier from the circuit. The *RC* network used to realize Y_1 when k_b is positive is shown in Fig. 75.13(a). If k_b is negative, Y_1 and Y_2 become

$$Y_1 = \frac{Hs}{2} + \frac{Hb_o}{2c} \quad \text{and} \quad Y_2 = \frac{1}{2/|k_b| + 2c/|k_b|s} \tag{75.63}$$

Realizations for Y_1 and Y_2 for this case are shown in Fig. 75.13(b). In determining Y_3 and Y_4, the partial fraction expansion of the denominator divided by $s + c$ yields a pole residue at $s = -c$,

$$k_a = \frac{c^2 - ca_1 + a_o}{-c} \tag{75.64}$$

If k_a is positive, then $Y_4 = 0$ and Y_3 is

$$Y_3 = s + \frac{a_o}{c} + \frac{1}{1/k_a + c/k_a s} \tag{75.65}$$

The realization of Y_3 is shown in Fig. 75.14(a). For a negative k_a, Y_3 and Y_4 become

$$Y_3 = s + \frac{a_o}{c} \tag{75.66}$$

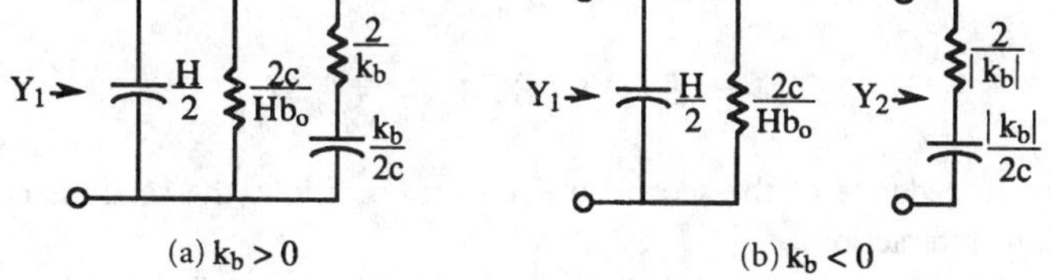

FIGURE 75.13 Realizations for $Y_1(s)$ and $Y_2(s)$ in Fig. 75.12.

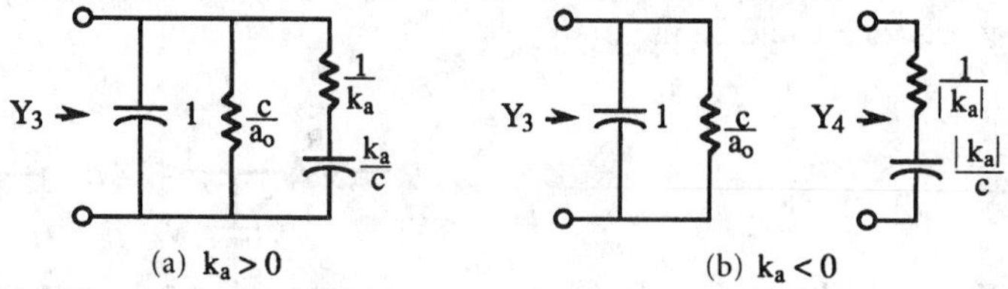

FIGURE 75.14 Realizations for $Y_3(s)$ and $Y_4(s)$ in Fig. 75.12.

and

$$Y_4 = \frac{1}{1/|k_a| + c/|k_a|s} \tag{75.67}$$

RC network realizations for Y_3 and Y_4 are shown in Fig. 75.14(b).

Example 5.

An All-Pass Function [5]. It is desired to use the configuration of Fig. 75.12 to realize the following all-pass function:

$$\frac{V_2(s)}{V_1(s)} = \frac{s^2 - 4s + 4}{s^2 + 4s + 4}$$

The choice of $c = +2$ will simplify the network. For this choice the numerator partial-fraction expansion is

$$\frac{s^2 - 4s + 4}{s + 2} = s + 2 - \frac{8s}{s + 2}$$

Hence $Y_1(s) = (s + 2)/2$ and $Y_2(s) = 4s/(s + 2)$. The denominator partial-fraction expansion is

$$\frac{s^2 + 4s + 4}{s + 2} = s + 2$$

Thus, $Y_3(s) = s + 2 + [4s/(s + 2)]$ and $Y_4(s) = 0$.

75.6 Higher Order Filters

Many applications require filters of order greater than two. One way to realize these filters is simply to cascade second-order filters to implement the higher order filter. If the order is odd, then one first-order (or third-order) section will be required. The advantage of the cascade approach is that it builds on the precious techniques described in this section. The desired high-order transfer function of the filter $T(s)$ will be broken into second-order functions $T_k(s)$ so that

$$T(s) = T_1(s)T_2(s)T_3(s)\cdots T_{n/2}(s) \tag{75.68}$$

Since the output impedance of the second-order sections is low, the sections can be cascaded without significant interaction.

If $T(s)$ is an odd-order function then a first-order passive network (like those shown in Fig. 75.15) can be added to the cascade of second-order functions. Both of the sections shown have nonzero output impedances, so it is advisable to use them only in the last stage of a cascade filter. Alternately, Fig. 75.5 could be used.

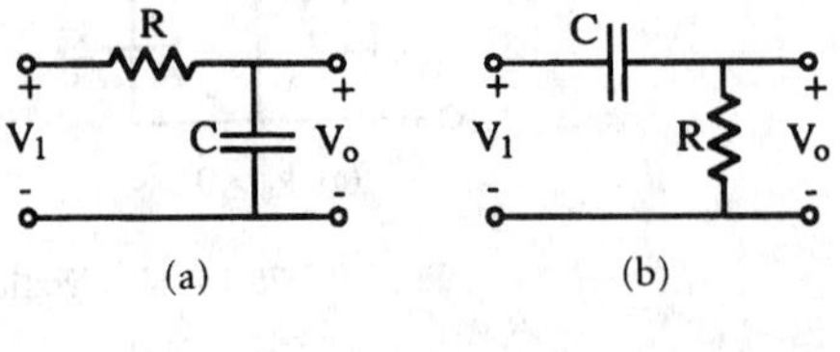

FIGURE 75.15 First-order filter sections: (a) low-pass, (b) high-pass.

Several considerations should be taken into account when cascading second-order sections together. Dynamic range is one of these considerations. To maximize the dynamic range of the filter the peak gain of each of the individual transfer functions should be equal to the maximum gain of the overall transfer function [10].

To maximize the signal-to-noise ratio of the cascade filter, the magnitude curve in the passband of each of the individual transfer functions should be flat when possible. Otherwise, signals with frequencies in a minimum gain region of an individual transfer function will have a lower signal-to-noise ratio than other signals [10]. Another consideration is minimum noise which is achieved by placing high gain stages first.

Designing a higher order filter as a cascade of second-order sections allows the designer many options. For instance, a fourth-order bandpass filter could result from the cascade of two second-order bandpass sections or one second-order low-pass and a second-order high-pass. Since the bandpass section is the easiest of the three types to tune [10] it might be selected instead of the low-pass-high-pass combination. Other guidelines for the cascade of second-order sections include putting the low-pass or bandpass sections at the input of the cascade in order to decrease the high-frequency content of the signal, which could avoid slewing in following sections. Using a bandpass or high-pass section at the output of the cascade can reduce dc offsets or low-frequency ripple.

One option that is now available is to use CAD tools in the design of filters. When high-order filters are needed, CAD tools can save the designer time. The examples that are worked later in this section require the use of filter tables interspersed with calculations. A CAD tool such as MicroSim Filter Designer [6] can hasten this process since it eliminates the need to refer to tables. The designer can also compare filters of different orders and different approximation methods (Butterworth, Chebyshev, etc.) to determine which one is the most practical for a specific application. The Filter Designer is able to design using the cascade method as discussed in this section, and after proposing a circuit it can generate a netlist so that a program such as SPICE can simulate the filter along with additional circuitry, if necessary.

The Filter Designer allows the user to specify a filter with either a passband/stopband description, center frequency/bandwidth description. or a combination of filter order and type (low-pass, high-pass, etc.). Then an approximation type is chosen, with Butterworth, Chebyshev, inverse Chebyshev, Bessel, and elliptic types available. A Bode plot, pole-zero plot, step or impulse response plot of the transfer function can be inspected. The designer can then select a specific circuit that will be used to realize the filter. The Sallen and Key low-pass and high-pass stages and the biquad multiloop feedback (Fig. 75.16) stages are included. The *s*-coefficients of the filter can also be displayed.

The Filter Designer allows modifications to the specifications, *s*-coefficients, or actual components of the circuit in order to examine their effect on the transfer function. There is also

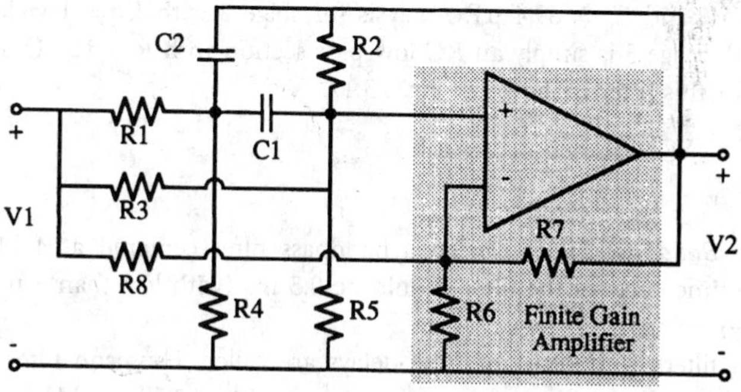

FIGURE 75.16 Biquad multiloop feedback stage.

a rounding function that will round components to 1, 5, or 10 percent values for resistors and 5 percent for capacitors and then recompute the transfer function based on these new component values. Finally, the package offers a resize function that will rescale components without changing the transfer function.

Example 6.

A Fifth-Order Low-Pass Filter. Design a low-pass filter using the Chebyshev approximation with a 1-dB equiripple passband, a 500 Hz passband, 45 dB attenuation at 1 kHz, and a gain of 10. Using a nomograph for Chebyshev magnitude functions, the order of this function is found to be five [5]. Using this information and a chart of quadratic factors of 1.0-dB equal-ripple magnitude low-pass functions, the transfer function is found to be

$$T(s_n) = \frac{0.9883K_1}{s_n^2 + 0.1789s_n + 0.9883} \cdot \frac{0.4293K_2}{s_n^2 + 0.4684s_n + 0.4293} \cdot \frac{0.2895K_3}{s_n + 0.2895}$$

Denormalizing the frequency with $s_n = s/500(2\pi)$ gives

$$T_1(s) = \frac{9{,}751{,}000K_1}{(s^2 + 562s + 9{,}751{,}000)}$$

$$T_2(s) = \frac{4{,}240{,}000K_2}{(s^2 + 1470s + 4{,}240{,}000)}$$

and

$$T_3(s) = \frac{909K_3}{s + 909}$$

Stages 1 and 2 will be Sallen and Key low-pass filters with $R = 10\ \text{k}\Omega$. Following the guidelines given above, stage 1 will have the gain of 10 and successive stages will have unity gain. Using (75.23) for stage 1 ($R_2 = R_4 = R$),

$$\frac{1}{Q} = 2\sqrt{\frac{C_4}{C_2}} + (1 - K)\sqrt{\frac{C_2}{C_4}} = 0.1800$$

so $C_4 = 4.696C_2$. Substituting this into (75.22),

$$\omega_n = \frac{1}{2.167RC_2} = 3123\ \text{rad/s}$$

Thus $C_2 = 14.8$ nF and $C_4 = 69.4\ \mu$F. Analysis for stage 2 with $K_2 = 1$ yields $C_2 = 0.135\ \mu$F and $C_4 = 17.4$ nF. Stage 3 is simply an RC low-pass section with $R = 10\ \text{k}\Omega$ and $C = 0.110\ \mu$F. The schematic for this filter is shown in Fig. 75.17.

Example 7.

A Fourth-Order Bandpass Filter. Design a bandpass filter centered at 4 kHz with a 1 kHz bandwidth. The time delay of the filter should be 0.5 ms (with less than 5 percent error at the center frequency).

The group of filters with constant time delays are called Thomson filters. For this design, begin with a low-pass filter with the normalized bandwidth of 1 kHz/4 kHz, or 0.25. Consulting a table for delay error in a Thomson filter [5] shows that a fourth-order filter is needed. From

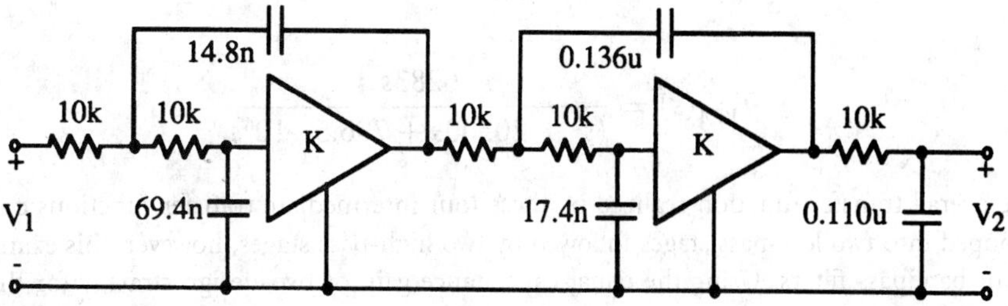

FIGURE 75.17 A fifth-order low-pass filter.

tables for Thomson filters, the quadratic factors for the fourth-order low-pass filter give the overall transfer function

$$T_{LP}(s_n) = \frac{1}{(s_n^2 + 4.208s_n + 11.488)(s_n^2 + 5.792s_n + 9.140)}$$

To convert $T_{LP}(s_n)$ to a bandpass function $T_{BP}(s_n)$, the transformation

$$s_n = \frac{\omega_R}{BW}\frac{s_{bn}^2 + 1}{s_{bn}} = 4\frac{s_{bn}^2 + 1}{s_{bn}}$$

is used, giving

$$T_{BP\,n}(s) = \frac{s^4/256}{(s^4 + 1.05s^3 + 2.72s^2 + 1.05s + 1)(s^4 + 1.45s^3 + 2.57s^2 + 1.45s + 1)}$$

ω_R and BW are the geometric center frequency and bandwidth of the bandpass filter. The polynomials in the denominator of this expression are difficult to factor, so an alternate method can be used to find the poles of the normalized bandpass function. Each pole p_{1n} of the normalized low-pass transfer function generates two poles in the normalized bandpass transfer function [5], [10] given by

$$p_{bn} = \frac{BW\ p_{1n}}{2\omega_R} \pm \sqrt{\left(\frac{BW\ p_{1n}}{2\omega_R}\right)^2 - 1} \qquad (75.69)$$

Using (75.69) with each of the poles of $T_{LP}(s_n)$ gives the following pairs of poles (which are also given in tables for Thomson filters):

$$p_{b1n} = -0.1777 \pm j0.6919 \qquad p_{b2n} = -0.3483 \pm j1.356$$
$$p_{b3n} = -0.3202 \pm j0.8310 \qquad p_{b4n} = -0.4038 \pm j1.0478$$

Note that each low-pass pole does not generate a conjugate pair, but when taken together, the bandpass poles generated by the low-pass poles will be conjugates. The unnormalized transfer functions generated by the normalized bandpass poles are

$$T_{BP\,1}(s) = \frac{6283s}{(s^2 + 8932s + 322.3 \cdot 10^6)}$$

$$T_{BP\,2}(s) = \frac{6283s}{(s^2 + 1751s + 1238 \cdot 10^6)}$$

$$T_{BP\,3}(s) = \frac{6283s}{(s^2 + 16100s + 501.0 \cdot 10^6)}$$

and

$$T_{BP4}(s) = \frac{6283s}{(s^2 + 20300s + 796.5 \cdot 10^6)}$$

The overall transfer function realized by these four intermediate transfer functions could also be grouped into two low-pass stages followed by two high-pass stages, however, this example will use four bandpass filters. Using the equal-capacitance gain of two-design strategy for the Sallen and Key bandpass stage,

$$\omega_n^2 = \left(1 + \frac{R_1}{R_2}\right)\frac{1}{R_1 R_4 C^2} \tag{75.70}$$

$$\frac{1}{Q} = \left(\left(1 - \frac{R_1}{R_2}\right)\sqrt{\frac{R_4}{R_1}} + 2\sqrt{\frac{R_1}{R_4}}\right)\sqrt{\frac{R_2}{R_1 + R_2}} \tag{75.71}$$

and

$$H_o = \left(\frac{\frac{K}{R_1}}{\frac{1}{R_1} + \frac{1}{R_2} + \frac{2}{R_4} - \frac{K}{R_2}}\right) \tag{75.72}$$

Making the further assignment $R_2 = R_1$ simplifies the equations to

$$\omega_n^2 = \frac{2}{R_1 R_4 C^2} \qquad \frac{1}{Q} = 2\sqrt{\frac{R_1}{R_4}} \qquad \text{and } H_o = 2Q^2$$

Selecting $C = 1$ nF gives the component values shown in Fig. 75.18, the schematic diagram for the fourth-order bandpass filter. While the gain of the filter is greater than unity, the output can be attenuated if unity gain is desired.

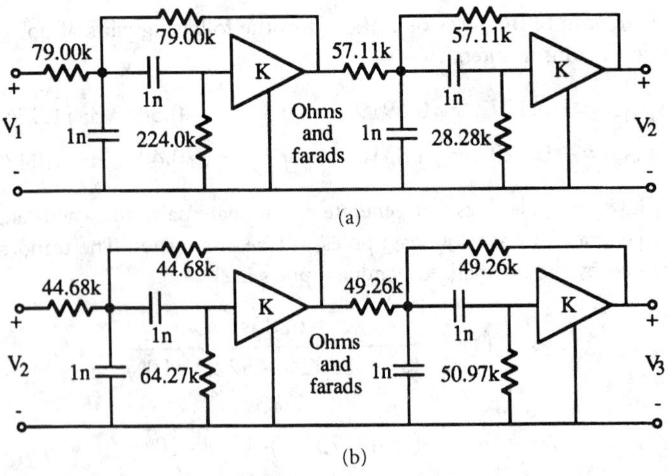

FIGURE 75.18 Fourth-order bandpass filter. (a) Stages 1 and 2. (b) Stages 3 and 4.

Example 8.

A Sixth-Order High-Pass Filter. Design a Butterworth high-pass filter with a −3-dB frequency of 100 kHz and a stopband attenuation of 30 dB at 55 kHz. A nomograph for Butterworth (maximally flat) functions indicates that this filter must have an order of at least six. A table for quadratic factors of Butterworth functions [5] gives

$$T(s_n) = \frac{K_1}{(s_n^2 + 0.518s_n + 1)} \frac{K_2}{(s_n^2 + 1.414s_n + 1)} \frac{K_3}{(s_n^2 + 1.932s_n + 1)}$$

Using $s_{nH} = 1/s_n$ for low-pass to high-pass transformation and $s_n = s/2\pi \cdot 10^5$ to denormalize the frequency,

$$T_1(s) = \frac{s^2 K_1}{s^2 + 325{,}500s + 394.8 \cdot 10^9}$$

$$T_2(s) = \frac{s^2 K_2}{s^2 + 888{,}600s + 394.8 \cdot 10^9}$$

and

$$T_3(s) = \frac{s^2 K_3}{s^2 + 1{,}214{,}000s + 394.8 \cdot 10^9}$$

The Sallen and Key high-pass filter can be used to implement all three of these stages. Picking a constant R, constant C design style with $\omega_n = 10^5$ rad/s gives $R = 10$ kΩ and $C = 1$ nF for each stage. Equation (75.23) then simplifies to

$$\frac{1}{Q} = 3 - K$$

for each stage. Thus $K_1 = 2.482$, $K_2 = 1.586$, and $K_3 = 1.068$, and the resulting high-pass filter is shown in Fig. 75.19.

Example 9.

A Fifth-Order Low-Pass Filter. Design a low-pass filter with an elliptic characteristic such that it has a 1-dB ripple in the passband (which is 500 Hz) and 65 dB attenuation at 1 kHz. From a nomograph for elliptic magnitude functions, the required order of this filter is five. A chart for a fifth-order elliptic filter with 1-dB passband ripple gives

$$T(s_n) = \frac{H(s_n^2 + 4.365)(s_n^2 + 10.568)}{(s_n + 0.3126)(s_n^2 + 0.4647s_n + 0.4719)(s_n^2 + 0.1552s_n + 0.9919)}$$

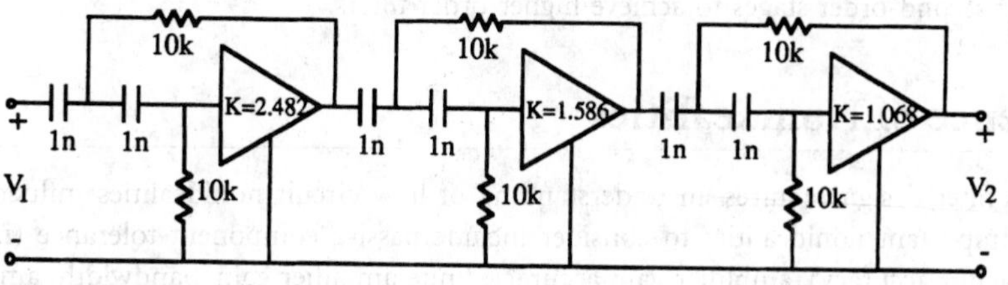

FIGURE 75.19 A sixth-order high-pass filter (all component values are Ω or F).

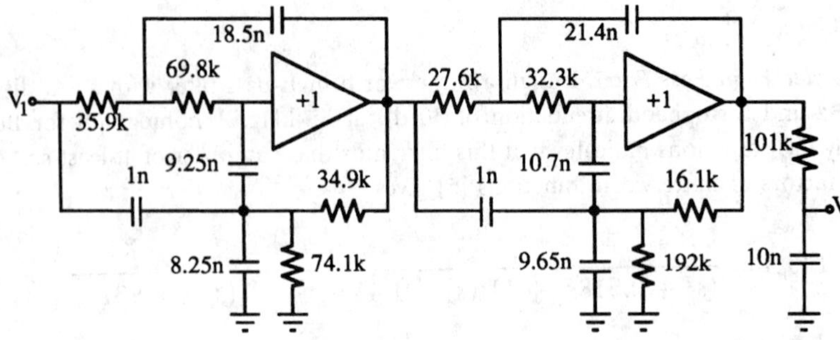

FIGURE 75.20 Fifth-order low-pass filter (all component values are Ω or F).

This transfer function can be realized with two notch filters like Fig. 75.9(a) and a first-order low-pass section. Denormalizing the frequency with $s_n = s/1000\pi$ gives

$$T_1(s) = \frac{H_1(s^2 + 43{,}080{,}000)}{s^2 + 1{,}460s + 4{,}657{,}000}$$

$$T_2(s) = \frac{H_2(s^2 + 104{,}300{,}000)}{s^2 + 487.6s + 9{,}790{,}000}$$

and

$$T_3(s) = \frac{982.1}{s + 982.1}$$

For the first stage, use (74.40) and (75.44), $H_1 = 0.1081$ and $a = 0.9429$. Setting $C_1 = 1$ nF and applying (75.40) and (75.41),

$$C_1 = 1 \text{ nF} \qquad C_4 = 9.25 \text{ nF} \qquad C_2 = 8.25 \text{ nF} \qquad C_5 = 18.5 \text{ nF}$$
$$R_2 = 69.8 \text{ k}\Omega \qquad R_1 = 35.9 \text{ k}\Omega \qquad R_3 = 34.9 \text{ k}\Omega \qquad R_4 = 74.1 \text{ k}\Omega$$

Similarly, in the second stage $H_2 = 0.0939$, $a = 0.1684$, and

$$C_1 = 1 \text{ nF} \qquad C_4 = 10.7 \text{ nF} \qquad C_2 = 9.65 \text{ nF} \qquad C_5 = 21.4 \text{ nF}$$
$$R_2 = 32.3 \text{ k}\Omega \qquad R_1 = 27.6 \text{ k}\Omega \qquad R_3 = 16.1 \text{ k}\Omega \qquad R_4 = 192 \text{ k}\Omega$$

The third stage can be a simple RC low-pass with $C = 10$ nF and $R = 101$ kΩ. The resulting cascade filter is shown in Fig. 75.20.

These examples illustrate both the use of the Filter Designer and the method of cascading first- and second-order stages to achieve higher order filters.

75.7 Influence of Nonidealities

Effective filter design requires an understanding of how circuit nonidealities influence performance. Important nonidealities to consider include passive component tolerance (i.e., resistor and capacitor accuracy), amplifier gain accuracy, finite amplifier gain–bandwidth, amplifier slew rate and noise.

Sensitivity Analysis

Classical sensitivity functions [5] are valuable tools for analyzing the influence of nonidealities on filter performance. The sensitivity function, S_x^y [2], by definition, describes the change in a performance characteristic of interest, say y, due to the change in nominal value of some element x,

$$S_x^y = \left(\frac{\partial y}{\partial x}\right)\frac{x}{y} = \frac{\partial y/y}{\partial x/x} = \frac{\partial(\ln y)}{\partial(\ln x)} \tag{75.73}$$

The mathematical properties of the sensitivity function are given in Table 75.1 [5] for convenient reference.

A valuable result of the sensitivity function is that it enables us to estimate the percentage change in a performance characteristic due to variation in passive circuit elements from their nominal values. Consider for example a low-pass filter's cutoff frequency ω_o, which is a function of R_1, R_2, C_1, and C_2. Using sensitivity functions, the percentage change in ω_o is estimated as [3]

$$\frac{\Delta\omega_o}{\omega_o} \approx S_{R_1}^{\omega_o}\left(\frac{\Delta R_1}{R_1}\right) + S_{R_2}^{\omega_o}\left(\frac{\Delta R_2}{R_2}\right) + S_{C_1}^{\omega_o}\left(\frac{\Delta C_1}{C_1}\right) + S_{C_2}^{\omega_o}\left(\frac{\Delta C_2}{C_2}\right) \tag{75.74}$$

Note that the quantity $\Delta\omega_o/\omega_o$ represents a small differential change in the cutoff frequency ω_o. When considering the root locus of a filter design, the change in pole location due to a change in gain K, for example, could be described by [3]

$$S_K^{p_1} = \left(\frac{\partial\sigma_o}{\partial K}\right) + j\frac{\partial\omega_o}{\partial K}\left(\frac{K}{\omega_o}\right) \tag{75.75}$$

TABLE 75.1 Properties of Sensitivity Functions [5].

$S_x^{ky} = S_{kx}^y = S_x^y$	$S_x^{y_1/y_2} = S_x^{y_1} - S_x^{y_2}$
$S_x^x = S_x^{kx} = S_{kx}^{kx} = 1$	$S_{x_1}^y = S_{x_2}^y S_{x_1}^{x_2}$
$S_{1/x}^y = S_x^{1/y} = -S_x^y$	$S_x^y = S_x^{\lvert y\rvert} + j\arg y\, S_x^{\arg y}$
$S_x^{y_1 y_2} = S_x^{y_1} + S_x^{y_2}$	$S_x^{\arg y} = \frac{1}{\arg y}\operatorname{Im} S_x^y$ (*)
$S_x^{\prod_{i=1}^n y_i} = \sum_{i=1}^n S_x^{y_i}$	$S_x^{\lvert y\rvert} = \operatorname{Re} S_x^y$ (*)
$S_x^{y^n} = nS_x^y$	$S_x^{y+z} = \frac{1}{y+z}(yS_x^y + zS_x^z)$
$S_x^{x^n} = S_x^{kx^n} = n$	$S_x^{\sum_{i=1}^n y_i} = \frac{\sum_{i=1}^n y_i S_x^{y_i}}{\sum_{i=1}^n y_i}$
$S_{x^n}^y = \frac{1}{n}S_x^y$	$S_x^{\ln y} = \frac{1}{\ln y}S_x^y$
$S_{x^n}^x = S_{kx^n}^x = \frac{1}{n}$	

Relations denoted by "(*)" use y to indicate a complex quantity and x to indicate a real quantity.

where the pole $p_1 = \sigma_o + j\omega_o$. Furthermore, the filter transfer function's magnitude and phase sensitivity is defined by [3]

$$S_x^{T(j\omega)} = S_x^{|T(j\omega)|} + S_x^{\theta(\omega)} = \text{Re } S_x^{T(j\omega)} + \frac{1}{\theta(\omega)} \text{Im } S_x^{T(j\omega)} \tag{75.76}$$

and

$$S_x^{T(j\omega)} = \frac{x}{|T(j\omega)|}\frac{\partial}{\partial x}|T(j\omega)| + jx\frac{\partial\theta(\omega)}{\partial x} \tag{75.77}$$

where $T(j\omega) = |T(j\omega)| \exp[j\theta(\omega)]$. The interested reader is directed to [5, ch. 3] for a detailed discussion of sensitivity.

Gain–Bandwidth

Now let us consider some of the amplifier nonidealities that influence the performance of our low-gain amplifier filter realizations. There are two amplifier implementations of interest each using an operational amplifier. They are the noninverting configuration and the inverting configuration shown in Fig. 75.21.

The noninverting amplifier is described by

$$K = \frac{V_2}{V_1} = \frac{A_d(s)}{1 + A_d(s)/K_o} \tag{75.78}$$

where $K_o = 1 + R_B/R_A$ is the ideal gain, and $A_d(s)$ is the operational amplifier's differential gain. Using the dominant-pole model for the op-amp,

$$A_d(s) = \frac{\text{GB}}{s + \omega_a} = \frac{A_o\omega_a}{s + \omega_a} \tag{75.79}$$

where A_o is the dc gain, ω_a is the dominant pole, and GB is the gain–bandwidth product of the op-amp. Inserting (75.79) into (75.78) gives

$$K = \frac{\text{GB}}{s + \omega_a(1 + A_o/K_o)} \approx \frac{\text{GB}}{s + \text{GB}/K_o} \tag{75.80}$$

The approximation of the K expression is valid provided $A_o \gg K_o$. The magnitude and phase of (75.80) are

$$|K(j\omega)| = \frac{\text{GB}}{\sqrt{\omega^2 + (\text{GB}/K_o)^2}} \tag{75.81}$$

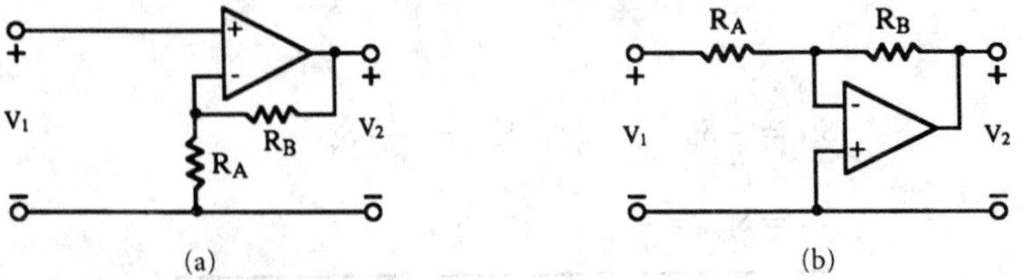

FIGURE 75.21 (a) Noninverting amplifier and (b) inverting amplifier.

and

$$\arg[K(j\omega)] = -\tan^{-1}\left(\frac{\omega}{\mathrm{GB}}K_o\right) \tag{75.82}$$

Note that for $\omega > \mathrm{GB}/K_o$, (75.81) and (75.82) may be approximated by

$$|K(j\omega)| \approx K_o \tag{75.83}$$

and

$$\arg[K(j\omega)] \approx -\frac{\omega}{\mathrm{GB}}K_o \tag{75.84}$$

The phase expression (75.84) describes the phase lag introduced to an active filter realization by the op-amp's finite gain–bandwidth product. To illustrate the importance of taking this into account, consider a positive-gain filter with $K_o = 3$ and operation frequency $\omega = \mathrm{GB}/30$ [5]. According to (75.84), a phase lag of 5.73° is introduced to the filter by the op-amp. Such phase lag might significantly impair the performance of some filter realizations.

The inverting amplifier of Fig. 75.21(b) is described by

$$K = \frac{V_2}{V_1} = \frac{-A_d(s)[R_B/(R_A + R_B)]}{1 + A_d(s)[R_A/(R_A + R_B)]} \tag{75.85}$$

Inserting (75.79) into (75.85) gives

$$K = \frac{-[R_B/(R_A + R_B)]\,\mathrm{GB}}{s + \omega_a[1 + A_o R_A/(R_A + R_B)]} \approx \frac{-[R_B/(R_A + R_B)]\,\mathrm{GB}}{s + \mathrm{GB}[R_A/(R_A + R_B)]} \tag{75.86}$$

where as before the approximation is valid if $A_o \gg (R_A + R_B)/R_A$. The magnitude and phase expressions for (75.86) are

$$|K(j\omega)| = \frac{\mathrm{GB}\,R_B/(R_A + R_B)}{\sqrt{\omega^2 + [\mathrm{GB}\,R_A/(R_A + R_B)]^2}} \tag{75.87}$$

and

$$\arg[K(j\omega)] = \pi - \tan^{-1}\left[\frac{\omega}{\mathrm{GB}}\left(1 + \frac{R_B}{R_A}\right)\right] \tag{75.88}$$

If $\omega < \mathrm{GB}\,R_A/(R_A + R_B)$, then (75.87) and (75.88) are approximated by

$$|K(j\omega)| \approx \frac{R_B}{R_A} \tag{75.89}$$

and

$$\arg[K(j\omega)] \approx \pi - \left[\frac{\omega}{\mathrm{GB}}\left(1 + \frac{R_B}{R_A}\right)\right] \tag{75.90}$$

An important limitation of the inverting configuration is its bandwidth [5], compared to that of the noninverting configuration, when small values of gain K are desired. Take for example the case where a unity-gain amplifier is needed. Using the amplifier of Fig. 75.21(a), the resistor values would be $R_B = 0$ and $R_A = \infty$. From (75.80) we see that this places a pole at $s = -\text{GB}$. Using the amplifier of Fig. 75.21(b), the resistors would be selected such $R_B = R_A$. From (75.86) we see that a pole located at $s = -\text{GB}/2$ results. Hence, for unity-gain realizations, the inverting configuration has half the bandwidth of the noninverting case. However, for higher gains, as the ratio R_B/R_A becomes large, both configurations yield a pole location that approaches $s = -(\text{GB})R_A/R_B$.

When using the inverting configuration described above to implement a negative-gain realization, the input impedance to the amplifier is approximately R_A. R_A must then be carefully selected to avoid loading the RC network. Another problem is related to maintaining low phase lag when a high Q is desired [5]. A Q of 10 requires that the ratio R_B/R_A be approximately 900. However, to avoid a phase lag of 6° (75.90) indicates that for $Q = 10$ the maximum filter design frequency cannot exceed approximately GB/8500. If GB = 1 MHz, say if the 741 op-amp is being used, then the filter cannot function properly above a frequency of about 100 Hz.

We have seen that a practical amplifier implementation contributes a pole to the gain K of a filter due to the op-amp's finite gain–bandwidth product GB. This means that an active filter's transfer function will have an additional pole that is attributed to the amplifier. The poles then of an active filter will be influenced by this additional pole. Consider the positive-gain amplifier used in the Sallen–Key low-pass filter (Fig. 75.6). The filter's ideal transfer function is expressed in (75.21). For the equal R, equal C case ($R_1 = R_3$, $C_2 = C_4$) we have $R_B = R_A[2 - (1/Q)]$. By inserting (75.80) into (75.21), the transfer function becomes [3]

$$T(s_n) = \frac{V_2}{V_1} = \frac{\text{GB}_n}{s_n^3 + s_n^2\left(3 + \dfrac{\text{GB}_n}{1 - \dfrac{1}{Q}}\right) + s_n\left(1 + \dfrac{\text{GB}_n}{3Q - 1}\right) + \dfrac{\text{GB}_n}{3 - \dfrac{1}{Q}}} \tag{75.91}$$

where the normalized terms are defined by $s_n = s/\omega_o$ and $\text{GB}_n = \text{GB}/\omega_o$. The manner in which the poles of (75.91) are influenced by the amplifier's finite gain–bandwidth product is illustrated in Fig. 75.22(a). This plot is the upper left-half s_n plane. The solid lines correspond to constant values of Q and the circles correspond to discrete values of GB_n, which were the same for each Q. For $Q = 0.5$ the GB_n values are labeled. Note that the $\text{GB}_n = \infty$ case shown is the ideal case where amplifier GB is infinite. The poles shift towards the origin as ω_o approaches a value near GB. Furthermore, for large design values of Q, i.e., greater than 3, the actual Q value will decrease as ω_o approaches GB.

As another example, consider the Sallen–Key low-pass again but for the case where positive unity-gain and equal resistance ($R_1 = R_3$) are desired. When the amplifier's gain–bandwidth product is taken into account the filter's transfer function becomes [3]

$$T(s) = \frac{V_2}{V_1} = \frac{\text{GB}_n}{s_n^3 + s_n^2\left(2Q + \dfrac{1}{Q} + \text{GB}_n\right) + s_n\left(1 + \dfrac{\text{GB}_n}{Q}\right) + \text{GB}_n} \tag{75.92}$$

where again $s_n = s/\omega_o$ and $\text{GB}_n = \text{GB}/\omega_0$. The poles of (75.92) are shown in Fig. 75.22(b). The plots are similar but the (b) plot constant GB_n values for the range of Q shown are closer together than the (a) plot. This means that the poles of the unity-gain, equal R design are less sensitive to amplifier gain–bandwidth product than the equal R, equal C case. The influence GB has on Sallen–Key low-pass filter parameters is summarized in Table 75.2.

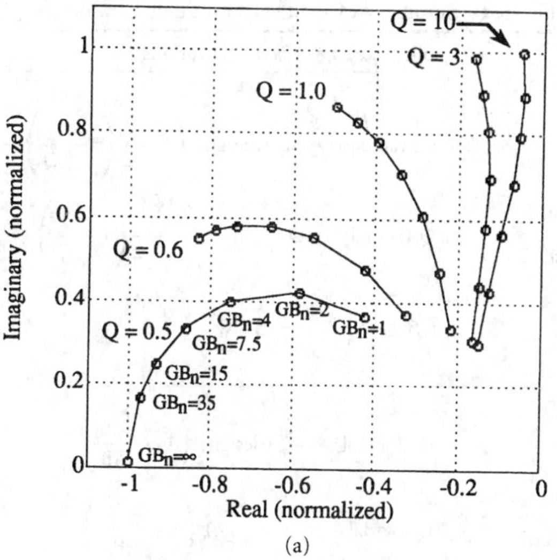

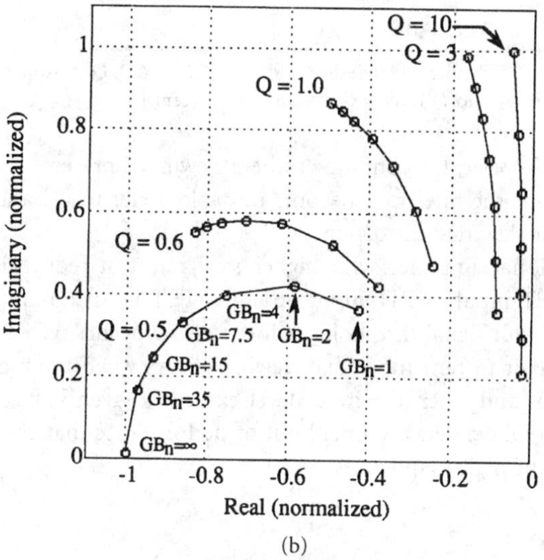

FIGURE 75.22 Effect of GB on the poles of the Sallen–Key second-order low-pass case. (a) Equal R, equal C case, (b) unity gain, equal R case.

Slew Rate

Slew rate is another operational amplifier parameter that warrants consideration in active filter design. When a filter signal level is large and/or the frequency is high, op-amp slew rate (SR) limitations can affect the small-signal behavior of amplifiers used in low-gain filter realizations. The amplifier's amplitude and phase response can be distorted by large-signal amplitudes. An analysis of slew rate induced distortion is presented in [1]. For the inverting amplifier [Fig. 75.21(b)], slewing increases the small-signal phase lag. This distortion is described by [5]

$$\arg[K(j\omega)] = \pi - \tan^{-1}\left(\frac{(\omega/\,\mathrm{GB})(1 + R_B/R_A)}{N(A)}\right) \tag{75.93}$$

where $N(A)$ is a describing function and A is the peak amplitude of the sinusoidal voltage input to the amplifier. $N(A) = 1$ while the op-amp is not slewing, yielding the same phase lag

TABLE 75.2 Summary of Finite GB on Filter Parameters.

Sallen–Key Low-Pass Realizations	
Equal R, equal C	$\omega_0 \text{ (actual)} \approx \omega_o \text{ (design)} \left(1 - \frac{1}{2}\left(3 - \frac{1}{Q}\right)^2 \frac{\omega_o}{\text{GB}}\right)$
	$Q \text{ (actual)} \approx Q \text{ (design)} \left(1 - \frac{1}{2}\left(3 - \frac{1}{Q}\right)^2 \frac{\omega_o}{\text{GB}}\right)$
	$\text{GB} \geq 45\left(1 - \frac{1}{3Q}\right)^2 \omega_0^*$
Unity-gain, equal R	$\omega_o \text{ (actual)} \approx \omega_o \text{ (design)} \left(1 - \frac{\omega_o Q}{\text{GB}}\right)$
	$Q \text{ (actual)} \approx Q \text{ (design)} \left(1 + \frac{\omega_o Q}{\text{GB}}\right)$
	$\text{GB} \geq 10\, Q\omega_0^*$

The "*" denotes the condition for $\Delta\omega_o/\omega_o = \Delta Q/Q \leq 10$ percent, i.e., for ω_o and Q to change less than 10 percent [3].

described by (75.88). However, for conditions where slewing is present, $N(A)$ becomes less than 1. Under such conditions the filter may become unstable. Refer to [1] and [5, ch. 4, sect. 6] for a detailed discussion of slew rate distortion.

Selecting an operational amplifier with higher slew rate can reduce large-signal distortion. Given the choice of FET input or BJT input op-amps, FET input op-amps tend to have higher slew rates and higher input signal thresholds before slewing occurs. When selecting an amplifier for a filter it is important to remember that slewing occurs when the magnitude of the output signal slope exceeds SR and when the input level exceeds a given voltage threshold. Often the slew rate of a filter's amplifier sets the final limit of performance that can be attained in a given active filter realization at high frequencies.

Noise [7]

The noise performance of a given filter design will ultimately determine the circuit's signal-to-noise ratio and dynamic range. For these reasons a strong understanding of noise analysis is required for effective filter design. In this section noise analysis of the noninverting and inverting op-amp configurations are discussed for the implementation of the gain block K. An example noise analysis is illustrated for a simple filter design.

Proceeding with noise analysis, consider the four noise sources associated with operational amplifier. Shown in Fig. 75.23, E_{n1} and E_{n2} represent the op-amp's voltage noise referred to the negative and positive inputs, respectively, and I_{n1} and I_{n2} represent the op-amp's current noise referred to the negative and positive inputs, respectively. E_{n1} and E_{n2} are related to E_n, from an op-amp manufacturer's data sheet, through the relationship

$$E_n = \sqrt{E_{n1}^2 + E_{n2}^2} \quad [\text{V}/\sqrt{\text{Hz}}]. \tag{75.94}$$

I_n is also usually reported in an op-amp manufacturer's data sheet. The value given for I_n is used for I_{n1} and I_{n2}.

$$I_n = I_{n1} = I_{n2} \quad [\text{A}/\sqrt{\text{Hz}}] \tag{75.95}$$

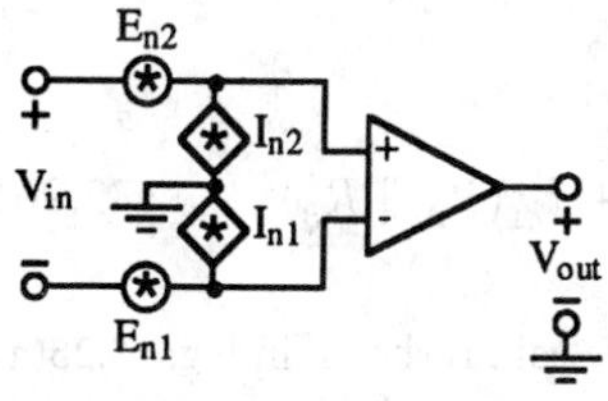

FIGURE 75.23 Operational amplifier noise model with four noise sources.

The analysis of the filter noise performance begins with the identification of all noise sources, which include the op-amp noise sources and all resistor thermal noise ($E_t = \sqrt{4\,\mathrm{kT}\,R\Delta f}$ where 4 kT is approximately $1.65 \times 10^{-20}\,\mathrm{V}^2/\Omega\cdot$ Hz at room temperature and Δf is the noise bandwidth). To begin the analysis, uncorrelated independent signal generators can be substituted for each noise source, voltage sources for noise voltage sources and current sources for current noise sources. By applying superposition and using the rms value of the result, the noise contribution of each element to the total output noise E_{no} can be determined. Only squared rms values can be summed to determine E_{no}^2. The resultant expression then for E_{no}^2 should never involve subtraction of noise sources, only addition. Sources of noise in a circuit always add to the total noise at the circuit's output. The total noise referred to the input E_{ni} is simply E_{no} divided by the circuit's transfer function gain. Capacitors ideally only influence noise performance through their effect on the circuit's transfer function. That is, the transfer gain seen by a noise source to the circuit's output can be influenced by capacitors depending on the circuit. As a result, of course, a circuit's noise performance can be a strong function of frequency.

Consider again the noninverting op-amp configuration, which can be used to implement positive K. The circuit is shown in Fig. 75.24(a) with appropriate noise sources included. Resistor R_s has been included to represent the driving source's (V_1) output resistance. The simplified circuit lumps together the amplifier's noise source's into E_{na}, shown in Fig. 75.24(b). E_{na} is described by (75.96). The circuit in Fig. 75.24(b) can be used in place of the positive gain block K to simplify the noise analysis of a larger circuit. R_s obviously will be determined by the

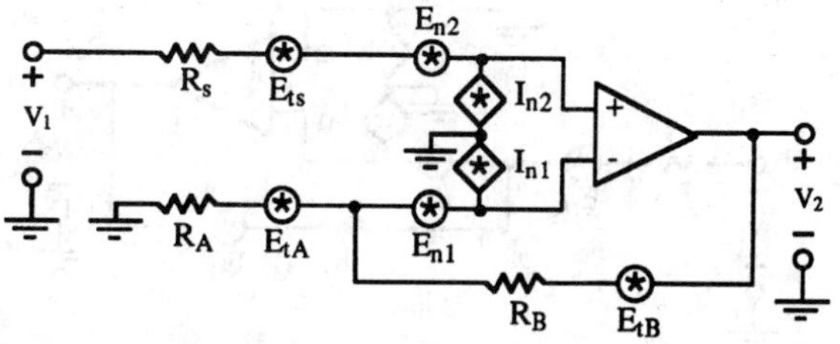

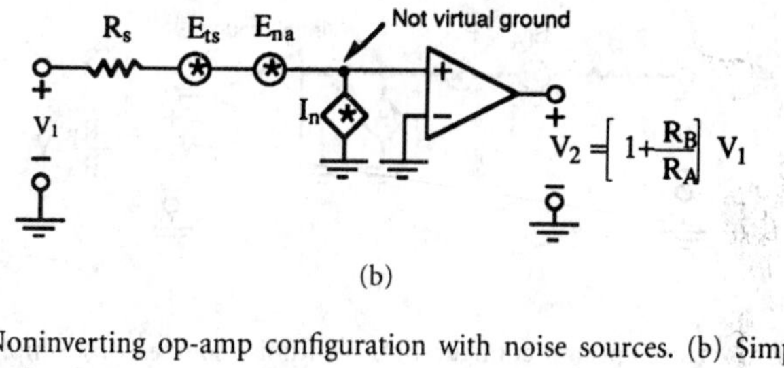

FIGURE 75.24 (a) Noninverting op-amp configuration with noise sources. (b) Simplified noninverting amplifier circuit with noise sources.

circuit seen looking back from the amplifier's input:

$$E_{na}^2 = E_n^2 + \left(\frac{R_A}{R_A + R_B}\right)^2 (E_{tB}^2) + \left(\frac{R_B}{R_A + R_B}\right)^2 (E_{tA}^2) + I_{n1}^2 (R_A \parallel R_B)^2 \qquad (75.96)$$

The inverting op amp configuration for implementing a negative gain is shown in Fig. 75.25(a) with noise sources included. Here again, the op-amp's noise sources can be lumped together and referred to the input as noise source E_{nb} shown in Fig. 75.25(b). For simplicity, R_s has been lumped together with R_A.

E_{nb} is described by (75.97). The simplified circuit in Fig. 75.25(b) is convenient for simplifying the noise analysis of larger circuit that uses a negative gain block.

$$E_{nb}^2 = \left(1 + \frac{R_A}{R_B}\right)^2 (E_n^2) + \left(\frac{R_A}{R_B}\right)^2 (E_{tB}^2) \qquad (75.97)$$

Example 10.

First-Order Low-Pass Filter Noise Analysis. Noise analysis of a low-pass filter with noninverting midband gain of 6 dB and −20 dB/decade gain roll-off after a 20 kHz cutoff frequency is desired. Recall from Section 75.3 that $\omega_o = RC$. Let us designate R_1 for R and C_2 for C. To obtain $\omega_o = 2\pi(20\text{ kHz})$, commercially available component values $R_1 = 16.9\text{ k}\Omega$ and $C_2 = 470$ pF (such as a 1 percent metal-film resistor for R_1 and a ceramic capacitor for C_2) can be used. Implement K with the noninverting op-amp configuration and use Fig. 75.24(b) as a starting point of the noise analysis. The midband gain of 6 dB can be achieved by selecting $R_A = R_B$. For this example let $R_A = R_B = 10.0\text{ k}\Omega$, also available in 1 percent metal film. Including the thermal noise source for R_1, the circuit shown in Fig. 75.26 results.

From this circuit we can easily derive the output noise of the low-pass filter.

$$E_{no}^2 = K^2 \left[E_{na}^2 + \left| \frac{Z_2}{R_1 + Z_2} \right|^2 (E_{t1}^2) + |R_1 \| Z_2|^2 (I_n^2) \right] \qquad (75.98)$$

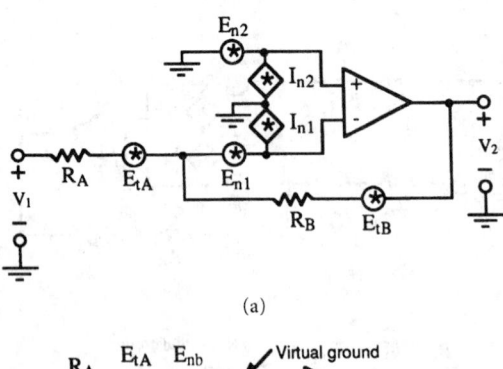

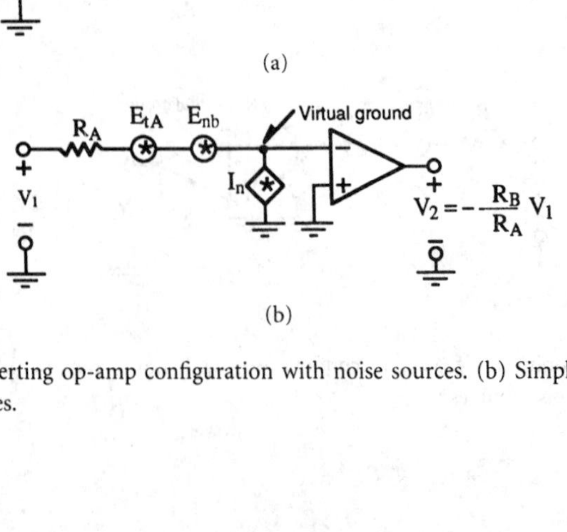

FIGURE 75.25 (a) Inverting op-amp configuration with noise sources. (b) Simplified inverting amplifier circuit with noise sources.

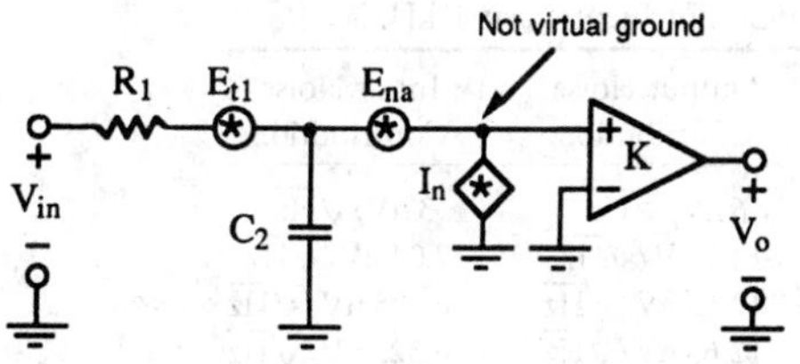

FIGURE 75.26 Positive gain first-order low-pass filter with noise sources.

where $Z_2 = 1/sC_2$, the impedance of the capacitor and E_{na} is defined by (75.96). The commercial op-amp OP-27 can be used to implement K and thus determines the values of E_n and I_n. For OP-27, $E_n = 3\ \text{nV}/\sqrt{\text{Hz}}$ and $I_n = 0.4\ \text{pA}/\sqrt{\text{Hz}}$. A comparison of the hand analysis to a PSpice simulation of this low-pass filter is shown in Fig. 75.27. The frequency dependence of K was not included in the hand analysis since the filter required a low midband gain of two and low cutoff frequency relative to the OP-27's 8 MHz gain–bandwidth product. The simulated frequency response is shown in Fig. 75.28. The computer simulation did include a complete model for the OP-27.

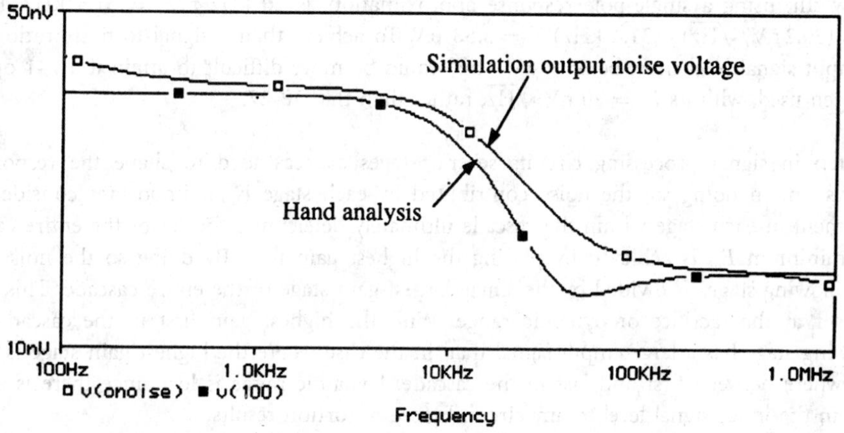

FIGURE 75.27 Comparison of calculated output noise with simulation.

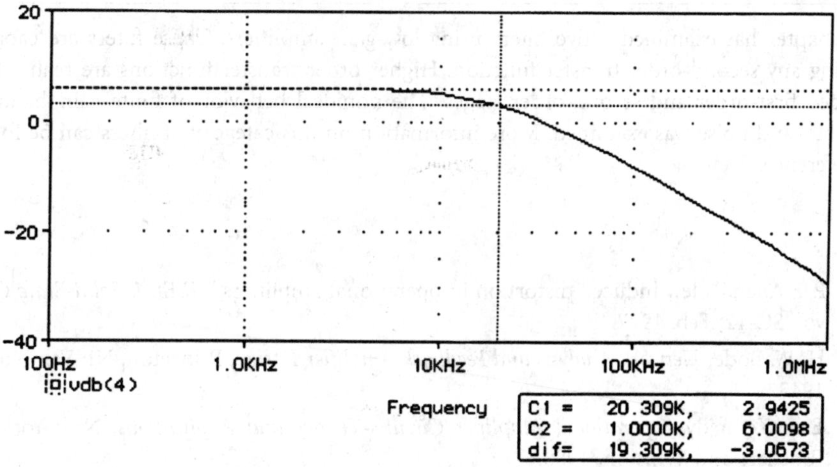

FIGURE 75.28 Simulated low-pass filter frequency response. The cursors indicate a −3-dB frequency of 20 kHz.

TABLE 75.3 Noise Contributions of Each Circuit Element at 1 kHz.

Noise Source	Noise Value	Gain Multiplier	Output Noise Contribution	Input Noise Contribution
E_n	$3\,\text{nV}/\sqrt{\text{Hz}}$	2	$6\,\text{nV}/\sqrt{\text{Hz}}$	$3\,\text{nV}/\sqrt{\text{Hz}}$
I_n	$0.4\,\text{pA}/\sqrt{\text{Hz}}$	35.2 k	$14.1\,\text{nV}/\sqrt{\text{Hz}}$	$7.04\,\text{nV}/\sqrt{\text{Hz}}$
R_A	$12.65\,\text{nV}/\sqrt{\text{Hz}}$	1	$12.65\,\text{nV}/\sqrt{\text{Hz}}$	$6.325\,\text{nV}/\sqrt{\text{Hz}}$
R_B	$12.65\,\text{nV}/\sqrt{\text{Hz}}$	1	$12.65\,\text{nV}/\sqrt{\text{Hz}}$	$6.325\,\text{nV}/\sqrt{\text{Hz}}$
R_1	$16.4\,\text{nV}/\sqrt{\text{Hz}}$	1.998	$32.8\,\text{nV}/\sqrt{\text{Hz}}$	$16.4\,\text{nV}/\sqrt{\text{Hz}}$
Total Noise Contributions			$40.4\,\text{nV}/\sqrt{\text{Hz}}$	$20.2\,\text{nV}/\sqrt{\text{Hz}}$

As indicated by Fig. 75.27, the output noise predicted by (75.98) agrees well with the simulation. Analyzing the noise contribution of each circuit element to the filter's total output noise can provide insight for improving the design. To this end, Table 75.3 was generated for this filter.

Note that at 1 kHz R_1 is the dominant source of noise. A smaller value of R_1 could be chosen at the expense of increasing C_2. Care must be taken in doing so because changing C_2 also changes the gain multiplier for I_n and R_1's noise source, as described in (75.98). The noise bandwidth, using a single-pole response approximation, is 20 kHz($\pi/2$) = 31.4 kHz. Hence, $E_{ni} = (20.2\,\text{nV}/\sqrt{\text{Hz}}) \cdot (31.4\text{ kHz})^{1/2} = 3.58\ \mu\text{V}$. To achieve then a signal-to-noise ratio of 10, the input signal level must be 35.8 μV. This would be more difficult to attain if a 741 op-amp has been used, with its $E_n = 20\text{ nV}/\sqrt{\text{Hz}}$, rather than the OP-27.

Often in signal processing circuits several stages are cascaded to shape the response of the system. In doing so, the noise contributed by each stage is an important consideration. Placement of each stage within the cascade ultimately determines the E_{ni} of the entire cascade. The minimum E_{ni} is achieved by placing the highest gain first. By doing so the noise of all the following stages is divided by the single largest gain stage of the entire cascade. This comes however at the sacrifice of dynamic range. With the highest gain first in the cascade, each following stage has a larger input signal than in the case where the highest gain stage is last or somewhere between first and last in the cascade. Dynamic range is lost since there is a finite limit in the input signal level to any circuit before distortion results.

75.8 Summary

This chapter has examined active filters using low-gain amplifiers. These filters are capable of realizing any second-order transfer function. Higher order transfer functions are realized using cascaded first-order and second-order stages. The nonideal behavior of finite gain–bandwidth, slew rate, and noise was examined. More information on this category of filters can be found in the references.

References

[1] P. E Allen, "Slew induced distortion in operational amplifiers," *IEEE J. Solid-State Circuits*, vol. SC-12, Feb. 1977.

[2] H. W. Bode, *Network Analysis and Feedback Amplifier Design*, Princeton, NJ: Van Nostrand, 1945.

[3] E. J. Kennedy, *Operational Amplifier Circuits-Theory and Applications*, New York: Holt, Rinehart and Winston, 1988.

[4] W. J. Kerwin and L. P. Huelsman, "The design of high-performance active *RC* bandpass filters," In *Proc. IEEE Int. Conv. Rec.* pt. 10, Mar. 1966, pp. 74–80.

[5] L. P. Huelsman and P. E. Allen, *Introduction to the Theory and Design of Active Filters,* New York: McGraw-Hill, 1980.

[6] MicroSim Corporation, *PSpice Circuit Synthesis, version 4.05*, Irvine, CA: MicroSim, 1991.

[7] C. D. Motchenbacher and J. A. Connelly, *Low Noise Electronic System Design*, New York: Wiley, 1993.

[8] R. P. Sallen and E. L. Key "A practical method of designing *RC* active filters," *IRE Trans. Circuit Theory*, vol. CT-2, 1955.

[9] R. Schaumann, M. S. Ghausi, and K. R. Laker, *Design of Analog Filters: Passive, Active RC, and Switched Capacitor*, Englewood Cliffs, NJ: Prentice-Hall, 1990.

[10] A. S. Sedra and P. O. Brackett, *Filter Theory and Design: Active and Passive*, Portland, OR: Matrix, 1978.

[11] A. S. Sedra and K. C. Smith, *Microelectronic Circuits*, New York: Holt, Rinehart and Wilson, 1987.

[12] A. Zverev, *Handbook of Filter Synthesis*, New York: Wiley, 1967.

76

Single-Amplifier Multiple-Feedback Filters

F. William Stephenson
Virginia Polytechnic Institute and State University

76.1 Introduction

In this section we will consider the design of second-order sections that incorporate a single operational amplifier. Such designs are based upon one of the earliest approaches to RC active filter synthesis, which has proven to be a fundamentally sound technique for over 30 years. Furthermore, this basic topology has formed the basis for designs as technology has evolved from discrete component assemblies to monolithic realizations. Hence, the circuits presented here truly represent reliable well-tested building blocks for sections of modest selectivity.

76.2 General Structure for Single-Amplifier Filters

The general structure of Fig. 76.1 forms the basis for the development of infinite-gain single-amplifier configurations. Simple circuit analysis may be invoked to obtain the open-circuit voltage transfer function. For the passive network:

$$I_1 = y_{11}V_1 + y_{12}V_2 + y_{13}V_3 \tag{76.1}$$

$$I_2 = y_{21}V_1 + y_{22}V_2 + y_{23}V_3 \tag{76.2}$$

$$I_3 = y_{31}V_1 + y_{32}V_2 + y_{33}V_3 \tag{76.3}$$

For the amplifier, ideal except for finite gain A:

$$V_3 = -AV_2 \tag{76.4}$$

0-8493-8341-2/95/$0.00+$.50

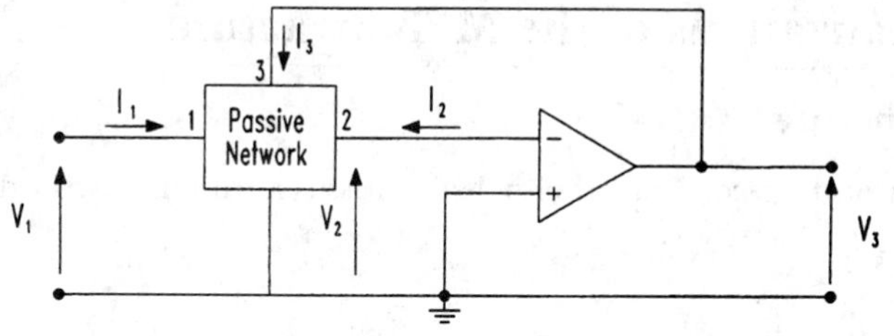

FIGURE 76.1 General infinite-gain single-amplifier structure.

Noting that $I_2 = 0$, the above equations reduce to the following expression for the voltage transfer function:

$$\frac{V_3}{V_1} = \frac{-y_{31}}{y_{32} + \dfrac{y_{33}}{A}} \tag{76.5}$$

As $A \rightarrow \infty$, which we can expect at low frequencies, the above expression reduces to the more familiar

$$\frac{V_3}{V_1} = -\frac{y_{31}}{y_{32}} \tag{76.6}$$

Theoretically, a wide range of transfer characteristics can be realized by appropriate synthesis of the passive network [1]. However, it is not advisable to extend synthesis beyond second-order functions for structures containing only one operational amplifier due to the ensuing problems of sensitivity and tuning. Furthermore, notch functions require double-element replacements [2] or parallel ladder arrangements [3], which are nontrivial to design, and whose performance is inferior to that resulting from other topologies such as those discussed in Chapters 77 and 78.

While formal synthesis techniques could be used to meet particular requirements, the most common approach is to use a double-ladder realization of the passive network, as shown in Fig. 76.2. This arrangement, commonly referred to as the multiple-loop feedback (MFB) structure, is described by the following voltage transfer ratio:

$$\frac{V_3}{V_1} = \frac{-Y_1 Y_3}{Y_5(Y_1 + Y_2 + Y_3 + Y_4) + Y_3 Y_4} \tag{76.7}$$

This negative feedback arrangement yields highly stable realizations. The basic all-pole (low-pass, bandpass, high-pass) functions can be realized by single-element replacements for the admittances $Y_1, \cdots, Y_5$, as described in the following subsection.

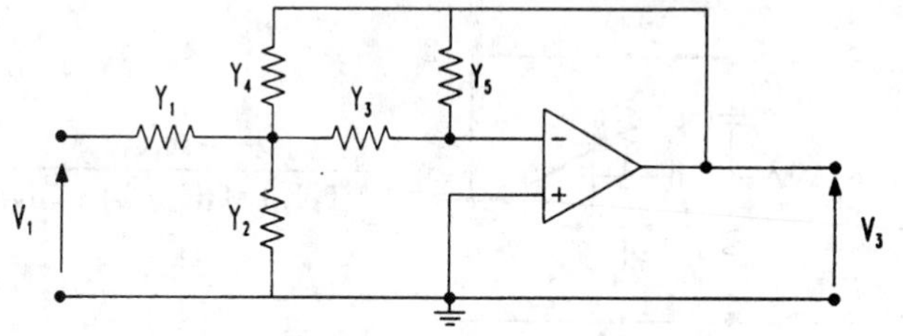

FIGURE 76.2 General double ladder multiple-feedback network.

76.3 All-Pole Realizations of the MFB Structure

Low-Pass Structure

The general form of the second-order all-pole low-pass structure is described by the following transfer ratio:

$$\frac{V_3}{V_1} = \frac{H}{s^2 + \dfrac{\omega_p s}{Q_p} + \omega_p^2} \tag{76.8}$$

By comparing the above requirement with (76.7) it is clear that both Y_1 *and* Y_3 must represent conductances. Furthermore, by reviewing the requirements for the denominator, Y_5 *and* Y_2 must be capacitors, while Y_4 is a conductance.

High-Pass Structure

The general form of the second-order all-pole high-pass transfer function is

$$\frac{V_3}{V_1} = \frac{Hs^2}{s^2 + \dfrac{\omega_p s}{Q_p} + \omega_p^2} \tag{76.9}$$

TABLE 76.1 MFB All-Pole Realizations.

Filter Type	Network	Voltage Transfer Function
Lowpass	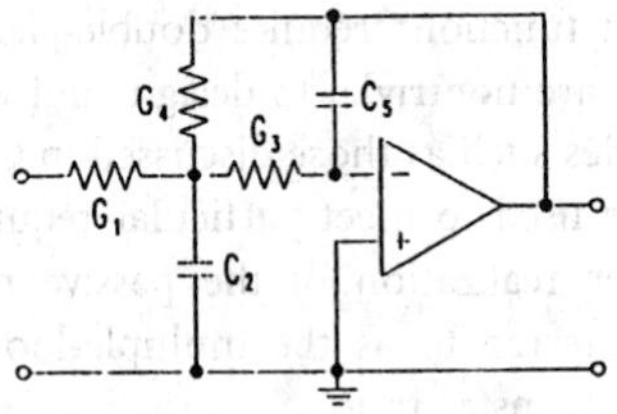	$\dfrac{-G_1G_3}{s^2C_2C_5 + sC_5(G_1 + G_3 + G_4) + G_3G_4}$
Highpass	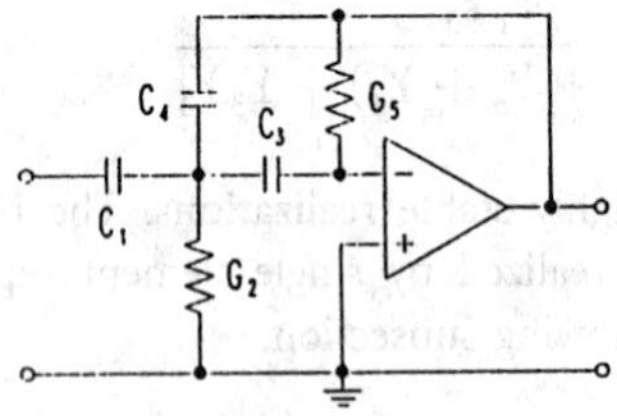	$\dfrac{-s^2C_1C_3}{s^2C_3C_4 + sG_5(C_1 + C_3 + C_4) + G_2G_5}$
Bandpass	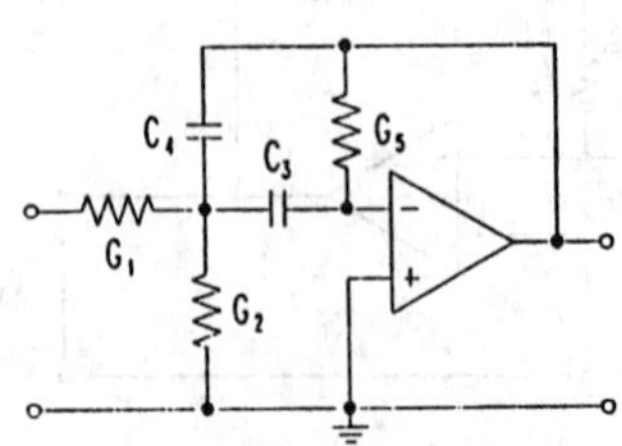	$\dfrac{-sG_1C_3}{s^2C_3C_4 + sG_5(C_3 + C_4) + G_5(G_1 + G_2)}$

With reference to Eq. (76.7), it is seen that both Y_1 *and* Y_3 must represent capacitors. There is a need for a third capacitor ($Y_4 = sC_4$) to yield the s^2 term in the denominator function. The remaining two elements Y_2 *and* Y_5, represent conductances.

Bandpass Structure

The general form of the second-order all-pole bandpass transfer function is

$$\frac{V_3}{V_1} = \frac{Hs}{s^2 + \dfrac{\omega_p s}{Q_p} + \omega_p^2} \tag{76.10}$$

Two solutions exist since Y_1 *and* Y_3 can be either capacitive or conductive. Choosing $Y_1 = G_1$ and $Y_3 = sC_3$ yields $Y_4 = sC_4$ and Y_2, Y_5 are both conductances.

The general forms of the above realizations are summarized in Table 76.1 [4].

76.4 MFB All-Pole Designs

MFB designs are typically reserved for sections having a pole-Q of 10 or less. One of the reasons for this constraint is the reliance upon component ratios for achieving Q. This can be illustrated by consideration of the low-pass structure for which

$$\frac{V_3}{V_1} = \frac{-G_1 G_3}{s^2 C_2 C_5 + sC_5(G_1 + G_3 + G_4) + G_3 G_4} \tag{76.11}$$

By comparison with (76.8),

$$Q_p = \frac{\sqrt{C_2 C_5 G_3 G_4}}{C_5(G_1 + G_3 + G_4)}$$

or, in terms of component ratios:

$$Q_p = \frac{\sqrt{C_2}}{\sqrt{C_5}} \left\{ \frac{1}{\dfrac{G_1}{\sqrt{G_3 G_4}} + \sqrt{\dfrac{G_3}{G_4}} + \sqrt{\dfrac{G_4}{G_3}}} \right\} \tag{76.12}$$

Hence, high Q_p can only be achieved by means of high component spreads. In general terms, a Q_p of value n requires a component spread proportional to n^2.

Filter design is effected by means of coefficient matching. Thus, for the low-pass case, comparison of like coefficients in (76.8) and the transfer ratio in Table 76.1 yields

$$G_1 G_3 = H \tag{76.13}$$

$$C_2 C_5 = 1 \tag{76.14}$$

$$C_5(G_1 + G_3 + G_4) = \frac{\omega_p}{Q_p} \tag{76.15}$$

$$G_3 G_4 = \omega_p^2 \tag{76.16}$$

These equations do not yield an equal-capacitor solution but can be solved for equal-resistor pairs. Hence, if $G_1 = G_3$,

$$G_1 = G_3 = \sqrt{H} \qquad \text{[from (76.13)]}$$

$$G_4 = \frac{\omega_p^2}{\sqrt{H}} \qquad \text{[from (76.16)]}$$

Then,

$$C_5 = \frac{\omega_p \sqrt{H}}{Q_p(2H + \omega_p^2)} = \frac{1}{C_2}$$

An alternative solution for which $G_3 = G_4$ is shown in Table 76.2, together with equal-capacitor designs for the bandpass and high-pass cases.

The conditions [4] for maximum Q_p in the bandpass realization require $C_3 = C_4$ and $G_1 = G_2 = nG_5$, where n is a real number. This yields a maximum Q_p of $\sqrt{n/2}$, and requires that $H = \omega_p Q_p$.

TABLE 76.2 Element Values for the MFB All-Pole Realizations

Element (Table 76.1)	Low-Pass	Bandpass	High-Pass
Y_1	$G_1 = \dfrac{H}{\omega_p}$	$G_1 = H$	$C_1 = H$
Y_2	$C_2 = \dfrac{Q_p(2\omega_p^2 + H)}{\omega_p^2}$	$G_2 = 2\omega_p Q_p - H$	$G_2 = \omega_p(2 + H)Q_p$
Y_3	$G_3 = \omega_p$	$C_3 = 1$	$C_3 = 1$
Y_4	$G_4 = G_3$	$C_4 = C_3$	$C_4 = C_3$
Y_5	$C_5 = \dfrac{\omega_p^2}{Q_p(2\omega_p^2 + H)}$	$G_5 = \dfrac{\omega_p}{2Q_p}$	$G_5 = \dfrac{\omega_p}{Q_p(2 + H)}$

Example 1. Using the cascade approach, design a four-pole Butterworth bandpass filter having a Q of 5, a center frequency of (1.5) kHz, and midband gain of 20 dB. Assume that only 6800 pF capacitors are available.

Solution. The lowpass prototype is the second-order Butterworth characteristic having a dc gain of 10 (i.e., 20 dB). Thus,

$$T(s) = \frac{10}{s^2 + \sqrt{2}s + 1} \qquad \text{(i)}$$

The low-pass-to-bandpass frequency transformation for a Q of 5 entails replacing s in (i) by $5(s + 1/s)$. This yields the following bandpass function for realization:

$$\frac{V_o}{V_i} = \frac{0.4s^2}{s^4 + 0.28284s^3 + 2.04s^2 + 0.28284s + 1}$$

$$= \underbrace{\frac{-sH_1}{(s^2 + .15142s + 1.15218)}}_{\text{(section 1)}} \cdot \underbrace{\frac{-sH_2}{(s^2 + .13142s + .86792)}}_{\text{(section 2)}} \qquad \text{(ii)}$$

$$Q_1 = Q_2 = 7.089$$

$$\omega_{p1} = 1.0734 \qquad \omega_{p2} = 0.9316$$

As expected, the Q-factors of the cascaded sections are equal in the transformed bandpass characteristic. However, the order of cascade is still important. So as to reduce the noise output of the filter, it is necessary to apportion most of the gain to section 1 of the cascade. Section 2 then filters out the noise without introducing excessive passband gain. In the calculation that follows, it is important to note that the peak gain of a bandpass section is given by HQ/ω_p.

Since the overall peak gain of the cascade is to be 10, let this also be the peak gain of section 1. Hence,

$$\frac{H_1 Q_1}{\omega_{p1}} = 6.6041 H_1 = 10$$

giving $H_1 = 1.514$.

Furthermore, from (ii):

$$H_1 H_2 = 0.4$$

so that $H_2 = 0.264$.

The design of each bandpass section proceeds by coefficient matching, conveniently simplified by Table 76.2. Setting $C_3 = C_4 = 1$ F, the normalized resistor values for section 1 may be determined as

$$R_1 = 0.661\ \Omega; \qquad R_2 = 0.073\ \Omega; \qquad R_5 = 13.208\ \Omega$$

The impedance denormalization factor is determined as

$$z_n = \frac{10^{12}}{2\pi \times 1500 \times 6800} = 15{,}603$$

Thus, the final component values for section 1 are

$$C_1 = C_2 = 6800\ \text{pF}$$

$$\left.\begin{aligned} R_1 &= 10.2\ \text{k}\Omega \\ R_2 &= 1.13\ \text{k}\Omega \\ R_5 &= 205\ \text{k}\Omega \end{aligned}\right\} \quad \text{Standard 1 percent values}$$

Note the large spread in resistance values ($R_5/R_2 \simeq 4Q^2$) and the fact that this circuit is only suitable for low-Q realizations. It should also be noted that the amplifier open-loop gain at ω_p must be much greater than $4Q^2$ if it is not to cause significant differences between the design and measured values of Q.

The component values for section 2 are determined in an identical fashion.

Example 2. Design the MFB bandpass filter characterized in Example 1 as a high-pass/low-pass cascade of second-order sections. Use the design equations of Table 76.2 and, where possible, set capacitors equal to 5600 pF. It is suggested that you use the same impedance denormalization factor in each stage. Select the nearest preferred 1 percent resistor values.

Solution. Since the peak gain of the overall cascade is to be 10, let this also be the gain of stage 1 (this solution yields the best noise performance). The peak gain of the low-pass section is given by

$$\frac{H_1 Q}{\sqrt{1-1/2Q^2}} = 7.16H_1 = 10 \qquad \therefore\ \underline{\underline{H_1 = 1.397}}$$

The overall transfer function (from Example 1) is

$$\frac{V_o}{V_i} = \frac{0.4s^2}{s^4 + 0.2824s^3 + 2.04s^2 + 0.2824s + 1}$$

$$\therefore\ H_1 H_2 = 0.4 \quad \text{so that} \quad \underline{\underline{H_2 = 0.286}}$$

Thus, assuming a low-pass/high-pass cascade, we have

$$\underbrace{\frac{1.397}{s^2 + 0.15145 + 1.1522}}_{\text{section 1}} \cdot \underbrace{\frac{0.286s^2}{s^2 + 0.1314s + 0.8679}}_{\text{section 2}}$$

Design the low-pass section (section 1) using Table 76.2 to yield

$$G_1 = \frac{H_1}{\sqrt{1.15218}} = 1.301, \text{ so that } \ R_1 = 0.7684\ \Omega$$

$$C_2 = \frac{(2\omega_p^2 + H_1)}{\omega_p^2} Q_p = 22.77\ \text{F}$$

$$G_3 = \omega_p = 1.0733, \quad \text{so that} \ \ R_3 = 0.9316\ \Omega = R_4$$

$$C_5 = 1/C_2 = 0.0439\ \text{F}$$

Now, design the high-pass section (section 2)

$$C_1' = 0.286\ \text{F}$$

$$G_2' = \omega_p(2 + H_2)Q = 15.099, \quad \text{so that} \ \ R_2' = 0.0662\ \Omega$$

$$C_3' = C_4' = 1\ \text{F}$$

$$G_5' = \frac{\omega_p}{Q_p(2+H)} = 0.0574, \quad \text{so that} \ \ R_5' = 17.397\ \Omega$$

To obtain as many 5600 pF capacitors as possible, the two sections should be denormalized separately. However, in this example, a single impedance denormalization will be used. Setting $C_3' = C_4' = 5600$ pF yields $z_n = 18\,947$.

This leads to the following component values:

Low-pass stage	High-pass stage
$R_1 = 2.204$ kΩ (2.21 kΩ)	$C_1' = 1602$ pF
$C_2 = 0.128$ μF	$R_2' = 1.254$ kΩ (1.24 kΩ)
$R_3 = 17.651$ kΩ (17.8 kΩ) $= R_4$	$C_3' = C_4' = 5600$ pF
$C_5 = 246$ pF	$R_5' = 329.62$ kΩ (332 kΩ)

1 percent values in parentheses.

76.5 Practical Considerations in the Design of MFB Filters

Sensitivity, the effects of finite amplifier gain, and tuning are all of importance in practical designs. The following discussion is based upon the bandpass case.

Sensitivity

Taking account of finite amplifier gain **A**, but assuming $R_{in} = \infty$ and $R_o = 0$ for the amplifier, the bandpass transfer function becomes

$$\frac{V_3}{V_1} = \frac{-sG_1/C_4\left(1+\dfrac{1}{A}\right)}{s^2 + s\left\{\dfrac{G_5(C_3+C_4)}{C_3C_4} + \dfrac{G_1+G_2}{C_4(1+A)}\right\} + \left\{\dfrac{G_5(G_1+G_2)}{C_3C_4}\right\}} \tag{76.17}$$

which is identical to the expression in Table 76.1 if $A = \infty$.

Assuming a maximum Q design,

$$Q = \frac{Q_p}{1 + \dfrac{2Q_p^2}{(1+A)}} \tag{76.18}$$

where Q_p is the desired selectivity and Q is the actual Q-factor in the presence of finite amplifier gain.

If $A \gg 2Q - 1$, the classical Q-sensitivity may be derived as

$$S_A^Q = \frac{2Q^2}{A} \tag{76.19}$$

which is uncomfortably high. By contrast, the passive sensitivities are relatively low:

$$S_{C_3,C_4}^Q = 0 \qquad S_{G_5}^Q = -0.5 \qquad S_{G_1,G_2}^Q = 0.25$$

while the ω_p sensitivities are all ± 0.5.

Effect of Finite Amplifier Gain

The effect of finite amplifier gain can be further illustrated by plotting (76.18) for various Q-factors, and for two commercial operational amplifiers. Assuming a single-pole roll-off model, the frequency dependence of open-loop gain for $\mu A741$ and $LF351$ amplifiers is as follows.

Frequency (Hz)	Gain	
	$\mu A741$	LF351
10	1.3×10^5	3.16×10^5
100	10^4	3.16×10^4
1000	10^3	3.16×10^3
10 000	10^2	3.16×10^2

Figure 76.3 shows the rather dramatic fall-off in actual Q as frequency increases (and hence gain decreases). Thus, for designs of modest Q (note that $A \gg 2Q - 1$), a very high quality amplifier is needed if the center frequency is more than a few kilohertz. For example, the LF351 with a unity gain frequency of 4 MHz will yield 6 percent error in Q at a frequency of only 1 kHz.

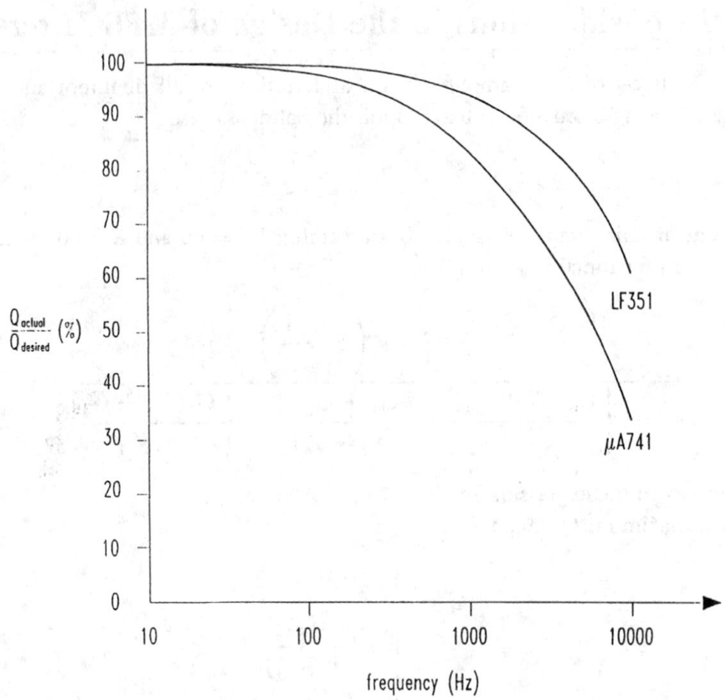

FIGURE 76.3 Effect of finite amplifier gain on the design Q for MFB bandpass realizations using two commercial operational amplifiers.

Tuning

Limited functional tuning of the bandpass section is possible. For example, the midband (peak) gain

$$K_o = \frac{G_1 C_3}{G_5(C_3 + G_4)} \tag{76.20}$$

may be adjusted by means of either G_1 or G_5.

Subsequent adjustment of either Q_p or ω_p is possible via G_2. In view of the discussion above, it is most likely that adjustment of Q_p will be desired. However, since the expressions for Q_p and ω_p are so similar, any adjustment of Q_p is likely to require an iterative procedure to ensure that ω_p does not change undesirably.

A more desirable functional tuning result is obtained in circumstances in which it is necessary to preserve a constant bandwidth, i.e., in a spectrum analyzer. Since

$$\omega_p = \sqrt{G_5(G_1 + G_2)/C_3 C_4} \tag{76.21}$$

and

$$B = \frac{\omega_p}{Q_p} = G_5(C_3 + C_4) \tag{76.22}$$

adjustment of G_2 will allow for a frequency sweep without affecting K_o or B.

An alternative to functional tuning may be found by adopting deterministic [5] or automatic [6] tuning procedures. These are particularly applicable to hybrid microelectronic or monolithic realizations.

76.6 Modified Multiple-Loop Feedback (MMFB) Structure

In negative feedback topologies such as the MFB, "high" values of Q_p are obtained at the expense of large spreads in element values. By contrast, in positive feedback topologies such as those attributed to Sallen and Key, Q_p is enhanced by subtracting a term from the s^1 (damping) coefficient in the denominator. The two techniques are combined in the MMFB (Deliyannis) arrangement [7] of Fig. 76.4.

Analysis of the circuit yields the bandpass transfer function given by

$$\frac{V_o}{V_i} = \frac{-sC_3G_1(1+k)}{s^2C_3C_4 + s\{G_5(C_3+C_4) - kC_3G_1\} + G_1G_5} \tag{76.23}$$

where $k = G_b/G_a$, and the Q-enhancement term "$-kC_3G_1$" signifies the presence of positive feedback. This latter term is also evident in the expression for Q_p:

$$Q_p = \frac{\sqrt{G_1/G_5}}{\left\{\sqrt{\dfrac{C_4}{C_3}} + \sqrt{\dfrac{C_3}{C_4}} - k\dfrac{G_1}{G_5}\sqrt{\dfrac{C_1}{C_2}}\right\}} \tag{76.24}$$

The design process consists of matching coefficients in (76.23) with those of the standard bandpass expression of (76.10). The design steps have been conveniently summarized by Huelsman [8] for the equal-capacitor solution and the following procedure is essentially the same as that described by him.

Example 3. Design a second-order bandpass filter with a center frequency of 1 kHz, a pole-Q of 8, and a maximum resistance spread of 50. Assume that the only available capacitors are of value 6800 pF.

1. The above constraint suggests an equal-valued capacitor solution. Thus, set $C_3 = C_4 = C$.
2. Determine the resistance ratio parameter n_o that would be required if there were no positive feedback. From Section 76.4, $n_o = 4Q_p^2 = 256$.

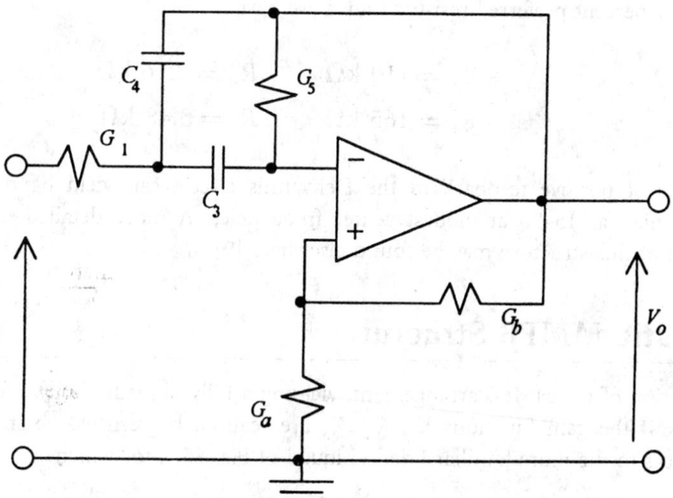

FIGURE 76.4 Modified multiple-loop feedback (MMFB) structure.

3. Select the desired ratio n (where n is greater than 1 but less than 256) and use it to determine the amount of positive feedback k. From (76.24),

$$Q_p = \frac{\sqrt{n}}{2 - kn}$$

so that

$$k = \frac{1}{\sqrt{n}}\left\{\frac{2}{\sqrt{n}} - \frac{1}{Q_p}\right\}$$

Since $n = 50$ and $Q_p = 8$, $k = 0.0316$.

4. A convenient value may now be selected for R_B. If $R_B = 110\ \text{k}\Omega$, then $R_A = R_B\,(0.0316) = 3.48\ \text{k}\Omega$.
5. Since, from (76.23),

$$\omega_p = \sqrt{\frac{G_1 G_5}{C_3 C_4}}$$

and $G_1/G_5 = n$ we may determine G_5 as

$$G_5 = \frac{\omega_p C}{\sqrt{n}}$$

Since $C = 6800$ pF, $n = 50$, and $G_5 = 1/R_5$:

$$R_5 = \frac{\sqrt{50}}{2\pi 10^3 \times 6.8 \times 10^{-9}} = 165.5\ \text{k}\Omega$$

Hence $R_1 = R_5/n = 3.31\ \text{k}\Omega$.

6. Using 1 percent preferred resistor values, we have

$$R_B = 110\ \text{k}\Omega \qquad R_A = 3.16\ \text{k}\Omega$$
$$R_5 = 165\ \text{k}\Omega \qquad R_1 = 3.48\ \text{k}\Omega$$

Judicious use of positive feedback in the Deliyannis circuit can yield bandpass filters with Q values as high as 15–20 at modest center frequencies. A more detailed discussion of the optimization of this structure may be found elsewhere [9].

76.7 Biquadratic MMFB Structure

A generalization of the MMFB arrangement, yielding a fully biquadratic transfer ratio is shown in Fig. 76.5. If the gain functions K_1, K_2, K_3 are realized by resistive potential dividers, the circuit reduces to the more familiar Friend biquad of Fig. 76.6, for which

$$\frac{V_o}{V_i} = \frac{cs^2 + ds + e}{s^2 + as + b} \tag{76.25}$$

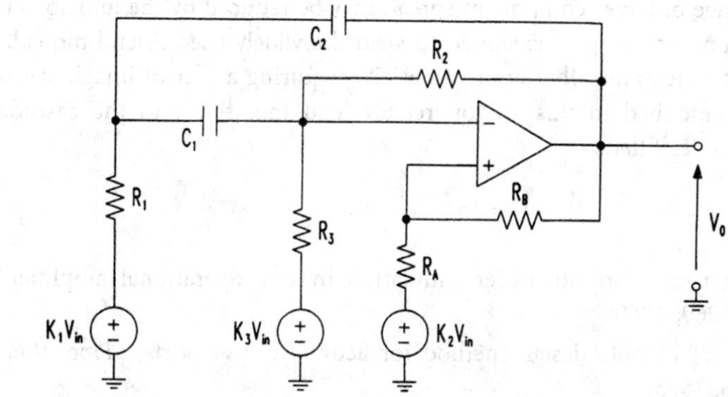

FIGURE 76.5 A generalization of the MMFB circuit.

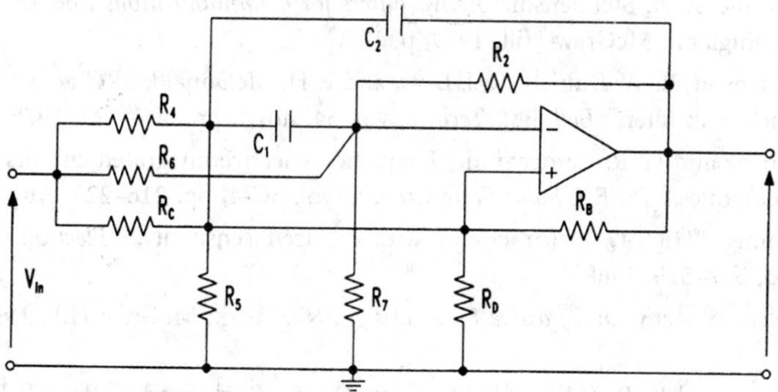

FIGURE 76.6 The Friend biquad.

where

$$K_1 = \frac{R_5}{R_4 + R_5}; \quad K_2 = \frac{R_D}{R_c + R_D}; \quad K_3 = \frac{R_7}{R_6 + R_7}$$
$$R_1 = \frac{R_4 R_5}{R_4 + R_5}; \quad R_A = \frac{R_c R_D}{R_c + R_D}; \quad R_3 = \frac{R_6 R_7}{R_6 + R_7} \tag{76.26}$$

This structure is capable of yielding a full range of biquads of modest pole Q, including notch functions derived as elliptic characteristics of low modular angle. It was used extensively in the Bell System, where the benefits of large-scale manufacture were possible. Using the standard tantalum thin film process, and deterministic tuning by means of laser trimming, quite exacting realizations were possible [10]. The structure is less suited to discrete component realizations. Although design is possible by coefficient matching, the reader is referred to an excellent step-by-step procedure developed by Huelsman [11].

76.8 Conclusions

The multiple-feedback structure is one of the most basic active filter building blocks. It is extremely reliable when used to realize low-Q (< 10), low frequency (up to 15 kHz) second-order sections of the low-pass, bandpass, and high-pass forms. Stability is ensured by the negative feedback topology, though component spreads are proportional to Q^2.

The disadvantage of larger component spread may be reduced by the judicious use of positive feedback. This approach may be extended to yield the widely used Friend biquad, which allows the realization of notch and other approximations requiring a pair of imaginary zeros.

All networks described in this section readily lend themselves to the cascade method for realizing higher-order filters.

References

[1] G. K. Aggarwal, "On *n*th order simulation by one operational amplifier," *Proc. IEEE*, vol. 52, p. 969, 1969.

[2] P. L. Taylor, "Flexible design method for active *RC* two-ports," *Proc. IEE*, vol. 110, pp. 1607–1616, 1963.

[3] A. G. J. Holt and J. I. Sewell, "Active *RC* filters employing a single operational amplifier to obtain biquadratic responses," *Proc. IEE*, vol. 112, pp. 2227–2234, 1965.

[4] P. Bowron and F. W. Stephenson, *Active Filters for Communications and Instrumentation*, Berkshire, England: McGraw-Hill, 1979, p. 170.

[5] R. A. Friedenson, R. W. Daniels, R. J. Dow, and P. H. McDonald, "*RC* active filters for the *D*3 channel bank filter," *Bell Syst. Tech. J.*, vol. 54, no. 3, pp. 507–529, 1975.

[6] A. B. Grebene and H. R. Camenzind, "Frequency-selective integrated circuits using phase-locked techniques," *IEEE J. Solid-State Circuits*, vol. SC-4, pp. 216–225, Aug. 1969.

[7] T. Deliyannis, "High-Q factor circuit with reduced sensitivity," *Electron. Lett.*, vol. 4, no. 26, pp. 577–579, 1968.

[8] L. P. Huelsman, *Active and Passive Filter Design*, New York: McGraw-Hill, 1993, ch. 5, pp. 277–278.

[9] M. S. Ghausi and K. R. Laker, *Modern Filter Design*, Englewood Cliffs, NJ: Prentice Hall, 1981, ch. 4, pp. 197–201.

[10] J. J. Friend, C. A. Harris, and D. Hilberman, "STAR: An active biquadratic filter section," *IEEE Trans. Circuits Syst.*, vol. CAS-22, pp. 115–121, Feb. 1975.

[11] L. P. Huelsman, "Multiple-loop feedback filters," in *RC Active Filter Design Handbook* (F. W. Stephenson, Ed.), New York: Wiley, 1985, ch. 7, pp. 201–203.

77

Multiple-Amplifier Biquads

Norbert J. Fliege
Technische Universität Hamburg, Germany

77.1 Introduction

The step from single-amplifier to multiple-amplifier biquadratic filter sections provides several benefits. The most important benefits are as follows.

- Reduced passive element spread, i.e., the ratio between the largest and the smallest values of resistors and/or capacitors can be reduced compared to the single-amplifier case.
- The required amplifier gains in some circuits grow linearly or less with the *Q*-factor of the complex pole pairs.
- Multiple-amplifier biquads often provide lower sensitivities to both passive and active components.
- Most of the multiple-amplifier biquads are more universal filter structures realizing the general biquadratic transfer function.
- Most of the filter parameters such as pole and zero *Q*-factors, pole and zero frequencies, and the gain factor of the transfer function can be tuned independently.
- Designing multiple-amplifier filters, often the values of the capacitors can be chosen freely. The filter parameters are then determined by resistors, which is less costly than by capacitors.

On the other hand, these benefits must be paid for by increased space requirements and increased power dissipation. However, today there are low-cost and low-power integrated-circuit op-amps available with up to four op-amps on one chip. Therefore, size and power dissipation are often no longer the main problem.

In the following, we will first consider biquadratic filter sections and dual-amplifier twin-*T* biquads. Both circuit families are directly derived from single-amplifier circuits. Next we will derive filter circuits having a quite different origin: filters that are derived from the generalized

8493-8341-2/95/$0.00 + $.50

impedance converter (GIC) and filters derived from the state-variable representation of linear systems on the analog computer. Finally, we will briefly consider filter circuits based on first-order all-pass sections.

77.2 Biquads with Decoupled Time Constants

One of the simplest methods for improving the performance of a biquad is demonstrated in Fig. 77.1.

Figure 77.1(a) shows a well-known Sallen and Key bandpass circuit [12], which offers moderate pole sensitivities with respect to the passive elements. It can easily be shown that this circuit has the following transfer voltage ratio:

$$H(s) = \frac{V_o}{V_i} = \frac{-K \cdot sT_2}{s^2 T_1 T_2 (1+K) + s(T_{12} + T_1 + T_2) + 1} \tag{77.1}$$

with the amplifier gain K and the time constants $T_1 = R_1C_1$, $T_2 = R_2C_2$, and $T_{12} = R_1C_2$. The gain requirement is more than $4Q_p^2 - 1$ with Q_p being the Q-factor of the pole pair. This relationship limits the circuit to low- or medium-Q applications.

If we insert another amplifier between the two RC networks R_1C_1 and R_2C_2, we obtain the biquad in Fig. 77.1(b) [5], [14], which possesses a transfer voltage ratio

$$H(s) = \frac{V_o}{V_i} = \frac{K_1K_2 \cdot sT_2}{s^2 T_1 T_2 (1 - K_1K_2) + s(T_1 + T_2) + 1} \tag{77.2}$$

Here the product K_1K_2 of the gain factors plays the role of the gain $-K$ in Fig. 77.1(a). One of the gain factors must be positive, the other negative. A comparison of (77.1) and (77.2) shows that after isolating both RC networks the "cross time constant" T_{12} disappears. The two time constants T_1 and T_2 are decoupled. From this change we can derive two benefits. Both factors K_1 and K_2 require only a gain of approximately $2Q_p$ and, as will be shown later, this circuit can be designed with zero Q-sensitivity.

Next, we generalize the structure in Fig. 77.1(b) by replacing the passive elements by general admittances; see Fig. 77.2. In this circuit, the two subnetworks Y_{1a}, Y_{1b} and Y_{2a}, Y_{2b} are decoupled. By a simple analysis we obtain the transfer voltage ratio of the general circuit in Fig. 77.2:

$$H(s) = \frac{V_o}{V_i} = \frac{K_1K_2Y_{1a}Y_{2a}}{(Y_{2a} + Y_{2b})(Y_{1a} + Y_{1b}) - K_1K_2Y_{1b}Y_{2a}} \tag{77.3}$$

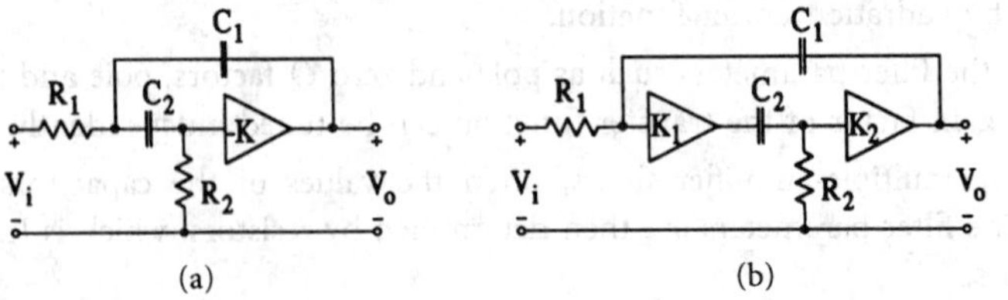

FIGURE 77.1 (a) Sallen and Key bandpass filter and (b) dual-amplifier bandpass filter with decoupled time constants.

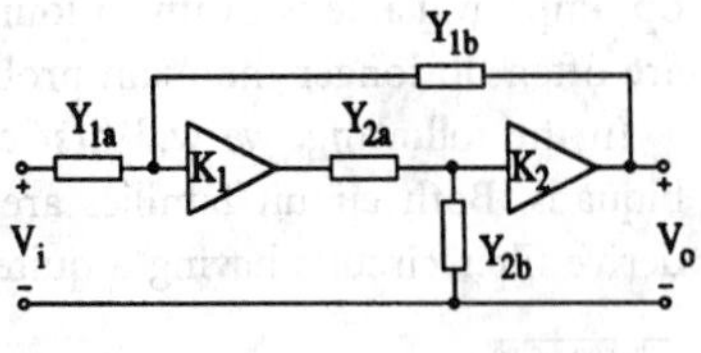

FIGURE 77.2 General biquad with decoupled networks.

By prespecifying the types of passive elements in Fig. 77.2 and (77.3), respectively, we next will derive a low-pass, a bandpass, and a high-pass filter section.

With the prespecified elements $Y_{1a} = G_1$, $Y_{1b} = sC_1$, $Y_{2a} = G_2$, and $Y_{2b} = sC_2$ we obtain from (77.3) the low-pass transfer function

$$H_{LP}(s) = \frac{V_o}{V_i} = \frac{K_1K_2G_1G_2}{s^2C_1C_2 + s[G_1C_2 + G_2C_1(1 - K_1K_2)] + G_1G_2} = \frac{K_1K_2}{s^2T_1T_2 + s[T_2 + T_1(1 - K_1K_2)] + 1} \tag{77.4}$$

If we predefine $K_1 = K_2 = 1$ we simply get voltage followers in the filter circuit and additionally a simple design procedure resulting in a low-sensitivity filter section. Figure 77.3 shows the low-pass filter section with two op-amps and four passive elements.

For $K_1 = K_2 = 1$ we will have a dc gain $H_0 = 1$, a pole frequency

$$\omega_p = 1/\sqrt{T_1T_2} \tag{77.5}$$

and a Q-factor

$$Q_p = \sqrt{T_2/T_1} \tag{77.6}$$

Thus, designing the filter section, from the predefined parameters ω_p and Q_p we determine the time constants

$$T_2 = \frac{Q_p}{\omega_p} \qquad T_1 = \frac{1}{Q_p\omega_p} \tag{77.7}$$

Finally, we choose the values of the capacitors and calculate the resistors from the time constants.

From (77.5) and (77.6) we can immediately read the pole sensitivities:

$$S_{R_1}^{\omega_p} = S_{C_1}^{\omega_p} = S_{R_2}^{\omega_p} = S_{C_2}^{\omega_p} = -\frac{1}{2} \tag{77.8}$$

$$S_{R_1}^{Q_p} = S_{C_1}^{Q_p} = -S_{R_2}^{Q_p} = -S_{C_2}^{Q_p} = -\frac{1}{2} \tag{77.9}$$

This result is comparable with the pole sensitivities of passive second-order RLC networks.

Next we will develop a high-pass filter section by prespecifying $Y_{1a} = sC_1$, $Y_{1b} = G_1$, $Y_{2a} = sC_2$, and $Y_{2b} = G_2$ resulting in a transfer voltage ratio

$$H_{HP}(s) = \frac{V_o}{V_i} = \frac{s^2K_1K_2C_1C_2}{s^2C_1C_2 + s[G_2C_1 + G_1C_2(1 - K_1K_2)] + G_1G_2} = \frac{s^2K_1K_2T_1T_2}{s^2T_1T_2 + s[T_1 + T_2(1 - K_1K_2)] + 1} \tag{77.10}$$

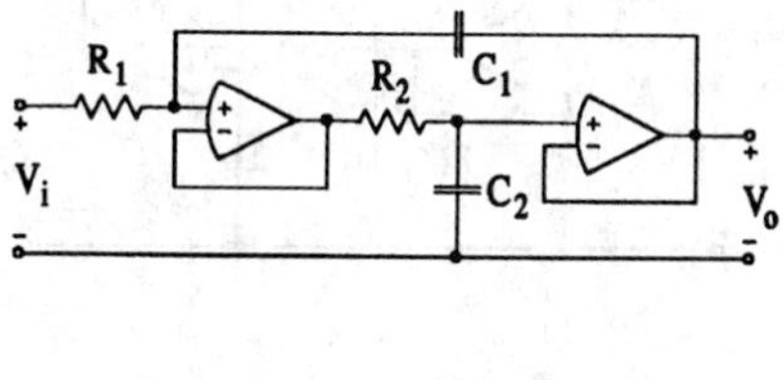

FIGURE 77.3 Dual op-amp low-pass filter section.

The only difference of this transfer function compared with the low-pass transfer function in (77.4) is the high-pass term (with s^2) in the numerator and the fact that the time constants T_1 and T_2 are interchanged in the denominator polynomial. Thus we can transfer the results in (77.5)–(77.9) to the high-pass filter by only replacing the index 1 by 2 and vice versa. Figure 77.4 shows the high-pass filter section.

The bandpass filter circuit mentioned at the beginning of this section is defined by the following types of elements: $Y_{1a} = G_1$, $Y_{1b} = sC_1$, $Y_{2a} = sC_2$, and $Y_{2b} = G_2$. Its transfer function is written in (77.2). The bandpass design is somewhat different from that of the low-pass and the high-pass filter described above. If we set the design values

$$G_1 = G_2 = 1 \qquad C_1 = C_2 = \frac{1}{2Q_p} \tag{77.11}$$

and

$$-K_1K_2 = 4Q_p^2 - 1 \tag{77.12}$$

we obtain a special design where all passive Q-sensitivities are zero and all other sensitivities are very low [5], [14]:

$$S_{R_1}^{Q_p} = S_{C_1}^{Q_p} = S_{R_2}^{Q_p} = S_{C_2}^{Q_p} = 0 \tag{77.13}$$

$$S_{R_1}^{\omega_p} = S_{C_1}^{\omega_p} = S_{R_2}^{\omega_p} = S_{C_2}^{\omega_p} = -\frac{1}{2} \tag{77.14}$$

$$S_{K_1}^{Q_p} = S_{K_2}^{Q_p} = \frac{-K_1K_2}{2(1-K_1K_2)} < \frac{1}{2} \tag{77.15}$$

$$S_{K_1}^{\omega_p} = S_{K_2}^{\omega_p} = -\frac{1}{2}\left(1 - \frac{1}{4Q_p^2}\right) \tag{77.16}$$

Figure 77.5 shows the bandpass filter section realized with two op-amps. In this circuit K_1 is a noninverting and K_2 an inverting amplifier. The resistor R_2 of the passive network R_2, C_2 is realized by the input resistor of the inverting amplifier.

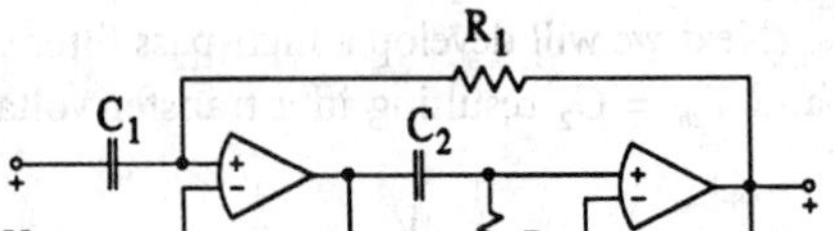

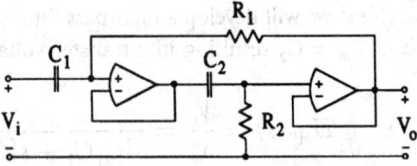

FIGURE 77.4 Dual op-amp high-pass filter section.

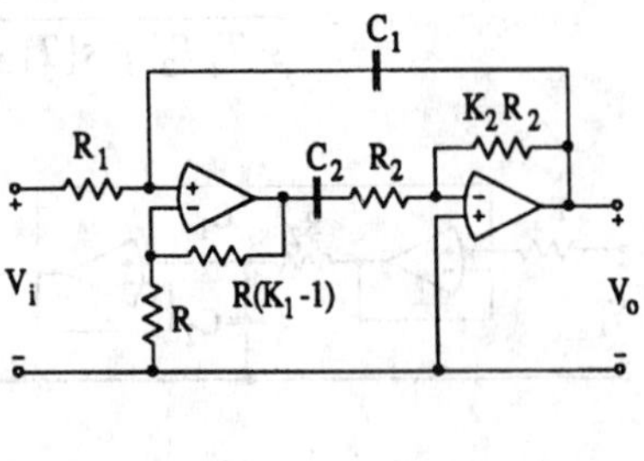

FIGURE 77.5 Dual op-amp bandpass filter section.

77.3 Dual-Amplifier Twin-*T* Biquads

Twin-*T* feedback networks are easily tunable and provide relatively favorable sensitivity properties. In active filters, a twin-*T* network is connected between the input and the output of an inverting amplifier. It has been shown [6], that the sensitivity to the active element can be substantially reduced when a symmetrical feedback network is used, i.e., when the dc gain of the feedback network is equal to the high-frequency gain.

In a single-amplifier twin-*T* biquad, the output port of the feedback network is loaded by the input resistance of the inverting amplifier. As a consequence, the feedback network is no longer symmetrical. Here again we can significantly improve the behavior of the twin-*T* biquad by introducing a second amplifier: inserting a voltage follower between the twin-*T* network and the inverting amplifier maintains the symmetry of the feedback network. Figure 77.6 shows the corresponding dual-amp twin-*T* resonator[1] with the usually chosen passive element relations.

From the resonator in Fig. 77.6 several filter circuits can be derived by inserting the input voltage source in one of the grounded branches and by taking one of the two amplifier outputs as output terminal of the filter section. The voltage transfer ratios of all these circuits will have different numerator polynomials but the same denominator polynomial. As an example, Fig. 77.7 shows a band-rejection filter derived from the dual-amp twin-*T* resonator in Fig. 77.6.

The voltage transfer ratio of the band-rejection filter in Fig. 77.7 can be calculated to be

$$H_{\mathrm{BR}}(s) = \frac{V_o}{V_i} = \frac{s^2T^2 + 1}{s^2T^2 + s4T/K + 1} \tag{77.17}$$

with the time constant $T = RC$ and the gain $K = (R_0 + R_1)/R_1$ of the series connection of the two op-amp circuits. Designing this filter, we determine the time constant T from the notch frequency ω_z or pole frequency ω_p, respectively,

$$T = 1/\omega_z = 1/\omega_p \tag{77.18}$$

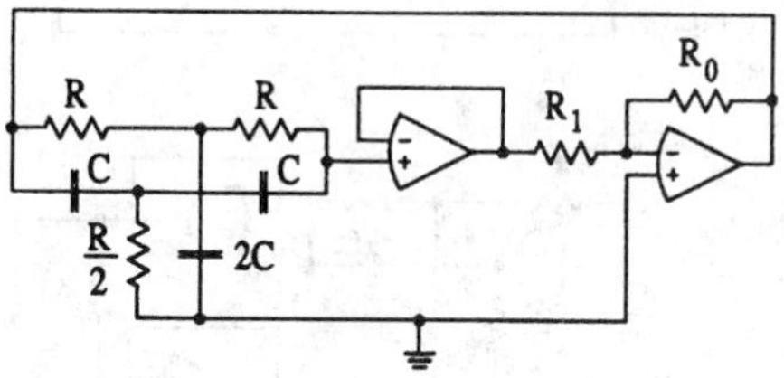

FIGURE 77.6 Dual-amp twin-*T* resonator.

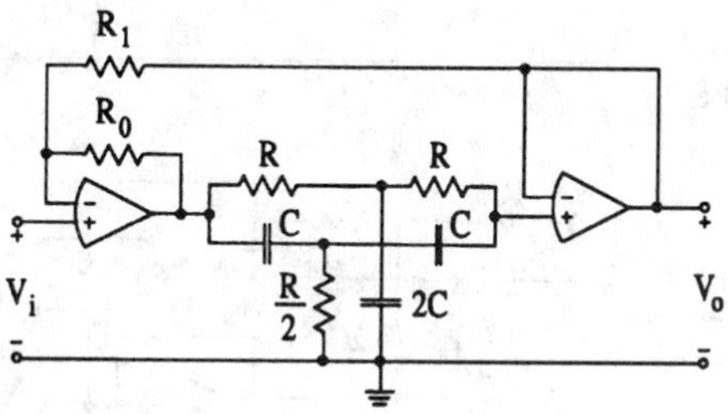

FIGURE 77.7 Dual-amp twin-*T* band-rejection filter.

[1]A resonator is a circuit without input and output terminals that only serves to show the feedback mechanism.

and the amplifier gain from the Q-factor:

$$K = 4 \cdot Q_p \tag{77.19}$$

In [6], it is shown that the filter has favorably low gain–sensitivity products.

In order to obtain further filter variants, in [6] a complementary circuit is derived from the resonator in Fig. 77.6, see Fig. 77.8.

This resonator is especially useful for deriving filter sections with frequently applied transfer functions. In Figs. 77.9–77.13 some of these filter sections are depicted. Their voltage transfer ratios are given by

$$H(s) = \frac{V_o}{V_i} = \frac{N(s)}{s^2T^2 + s4T/K + 1} \tag{77.20}$$

The band-rejection filter in Fig. 77.9 has a numerator polynomial $N(s) = s^2T^2 + 1$. The filter section in Fig. 77.10 is low-pass with a numerator $N(s) = 1$, the filter section in Fig. 77.11 is high-pass with $N(s) = s^2T^2$. The design of these sections is exactly the same as that of the filter in Fig. 77.7. It also provides the same pole sensitivities.

The two structures in Fig. 77.12 and 77.13 are particularly suitable for realizing elliptical filter sections (also called Cauer filters). The Cauer low-pass filter section in Fig. 77.12 has a numerator polynomial

$$N(s) = s^2R^2CC_1 + 1, \qquad C_1 + C_2 = C \tag{77.21}$$

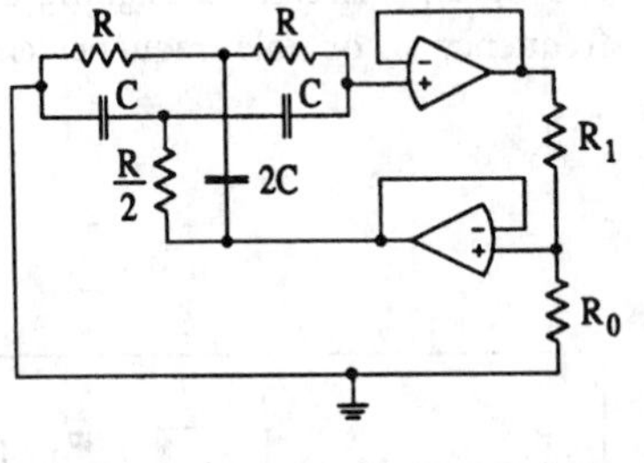

FIGURE 77.8 Complementary circuit of the resonator in Fig. 77.6.

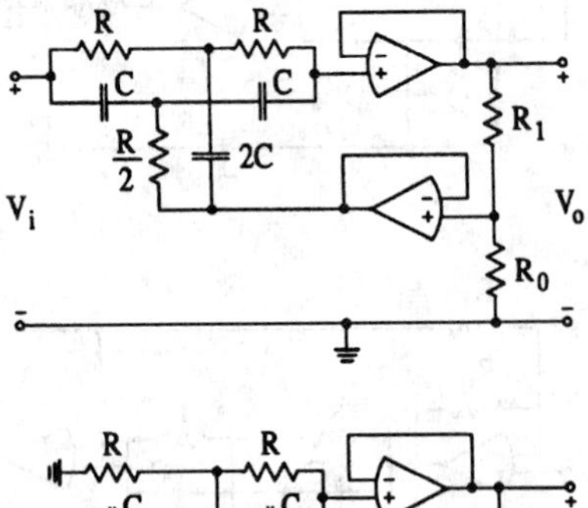

FIGURE 77.9 Twin-T band-rejection filter.

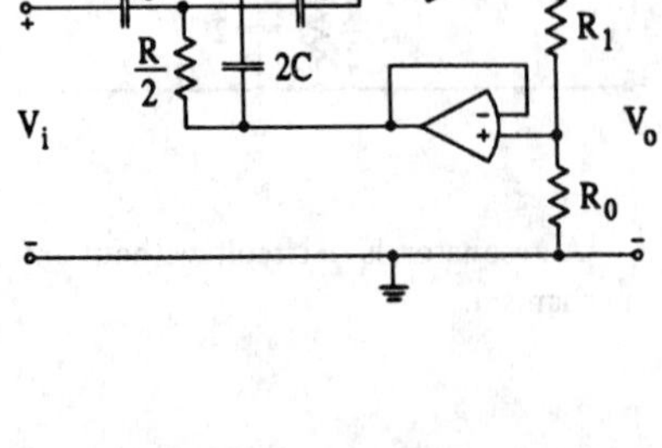

FIGURE 77.10 Twin-T low-pass filter.

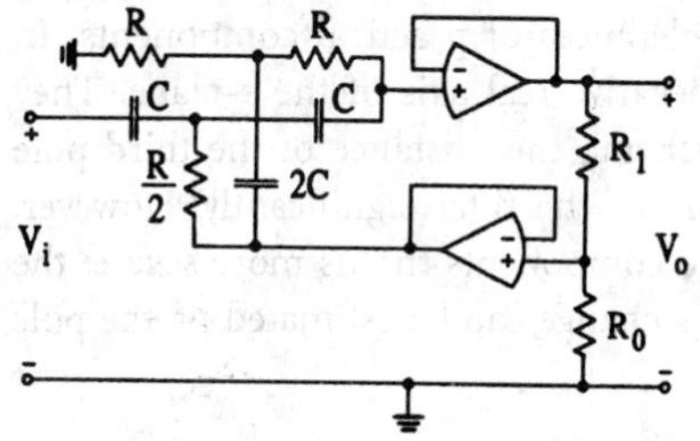

FIGURE 77.11 Twin-*T* high-pass filter.

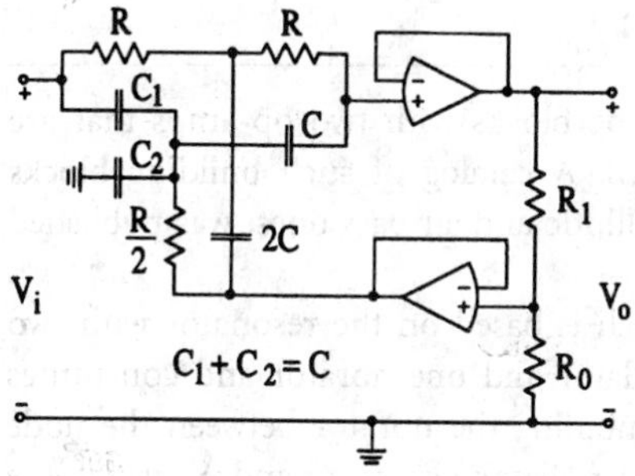

FIGURE 77.12 Twin-*T* Cauer low-pass filter section.

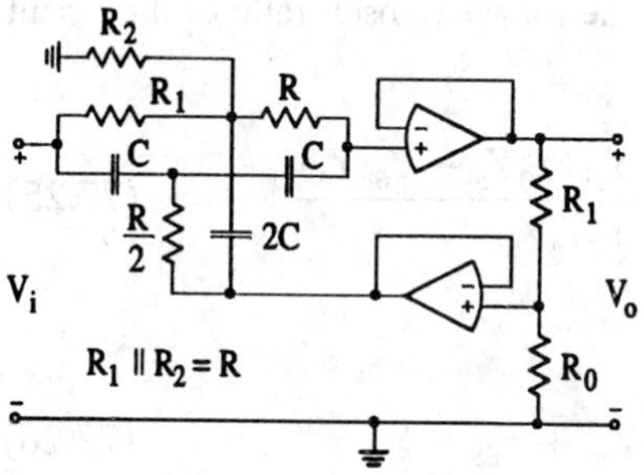

FIGURE 77.13 Twin-*T* Cauer high-pass filter section.

In this circuit, the first capacitor C is split into a parallel connection of two capacitors C_1 and C_2, where C_2 is grounded. The zeros are on the $j\omega$ axis at positions $\pm j\omega_z$ with $\omega_z = 1/\sqrt{R^2CC_1}$. The notch frequency ω_z is greater than the pole frequency ω_p. Designing the filter section, we determine the parameters T and K as described above. The splitting ratio of the input capacitor is determined by the poles and zeros:

$$\frac{C_2}{C_1} = \frac{\omega_z^2}{\omega_p^2} - 1 \tag{77.22}$$

Finally, in Fig. 77.13 we have a Cauer high-pass filter section with a numerator polynomial

$$N(s) = s^2RR_1C^2 + 1 \qquad R_1\|R_2 = R \tag{77.23}$$

Here the zero frequency ω_z is less than the pole frequency ω_p. As in the low-pass case, the splitting ratio of the input resistors is determined by these two frequencies:

$$\frac{R_1}{R_2} = \frac{\omega_p^2}{\omega_z^2} - 1 \tag{77.24}$$

The circuit in Fig. 77.6 is actually of third order. By matching the passive components, as shown in Fig. 77.6, we obtain a second-order transfer function. The question may arise as to what happens if we have a small mismatch due to the tolerances of practical components. In this case, a third pole and a third zero appears on the negative real axis of the s-plane. They do not cancel each other exactly, but approximately. In general, the existance of the third pole and the third zero does not affect the frequency response of the filter significantly. However, there is a second effect due to a small mismatch of passive components that is more severe: the position of the pole pair desired by design is changed. This change can be estimated by the pole sensitivities with respect to the passive elements [6].

77.4 GIC-Derived Dual-Amplifier Biquads

In this section, we consider a class of biquadratic building blocks with two op-amps that are derived from the generalized impedance converter (GIC). A catalog of such building blocks realizing a wide variety of network functions, including elliptic and all-pass ones, was published by Fliege [4].

Figure 77.14(b) shows the general filter structure, which is based on the resonator with two nullors in Fig. 77.14(a). Each nullor consists of one nullator and one norator and constitutes a model for the ideal op-amp, see also Chapter 22. Combining the norator between the node between the admittances Y_2 and Y_3 and ground and the nullator across Y_4 and Y_6 yields the op-amp μ_1 in Fig. 77.14(b). It can readily be verified that the voltage transfer ratio of the circuit in Fig. 77.14(b) with $\mu_i = \infty$, $i = 1, 2$, is given by

$$H(s) = \frac{V_o}{V_i} = \frac{Y_{6b}(Y_2Y_4 + Y_{1a}Y_4) + Y_{1b}(Y_3Y_5 - Y_{6a}Y_4)}{Y_1Y_3Y_5 + Y_2Y_4Y_6} \tag{77.25}$$

with

$$Y_1 = Y_{1a} + Y_{1b} \qquad Y_6 = Y_{6a} + Y_{6b} \tag{77.26}$$

If we choose the node between Y_4 and Y_5 as the output of the building block we will obtain a similar transfer function. We only have to interchange the admittances Y_1 by Y_6, Y_2 by Y_5, and Y_3 by Y_4.

First, we will derive a second-order low-pass building block from the general structure and we will take this low-pass filter as a prototype for the whole circuit family to explain their advantageous characteristics. If we choose the passive elements of the building block as $Y_{1a} = G_1$,

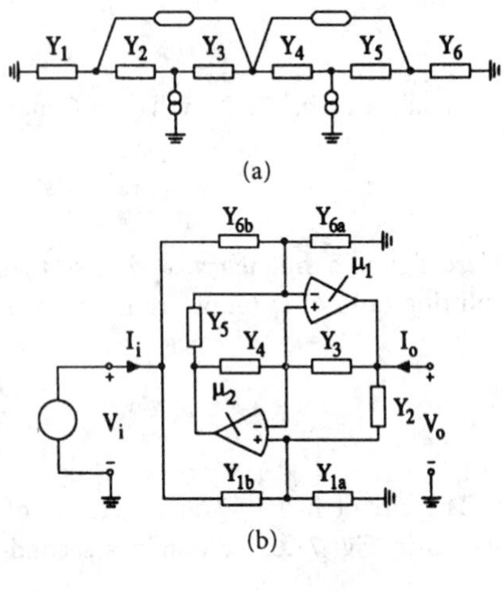

FIGURE 77.14 Resonator with (a) two nullors and (b) the corresponding general dual-amplifier filter structure.

$Y_{1b} = 0$, $Y_2 = G_2$, $Y_3 = sC_3$, $Y_4 = G_4$, $Y_5 = G_5 + sC_5$, $Y_{6a} = 0$, and $Y_{6b} = G_6$ we obtain the low-pass building block in Fig. 77.15. Its voltage transfer ratio reads

$$H_{LP}(s) = \frac{V_o}{V_i} = \frac{G_6 G_4 (G_2 + G_1)}{s^2 C_3 C_5 G_1 + s C_3 G_5 G_1 + G_2 G_4 G_6}$$
$$= H_0 \cdot \frac{\omega_p^2}{s^2 + s\omega_p/Q_p + \omega_p^2} \tag{77.27}$$

with dc gain

$$H_0 = (1 + \alpha_{21})/\alpha_{21} \tag{77.28}$$

pole frequency

$$\omega_p = \sqrt{\alpha_{21}/(T_{34} T_{56})} \tag{77.29}$$

and Q-factor

$$Q_p = \omega_p \cdot T_{55} \tag{77.30}$$

Here, the time constants are defined as

$$T_{\mu\nu} = C_\mu / G_\nu \tag{77.31}$$

and the conductance ratios as

$$\alpha_{\mu\nu} = G_\mu / G_\nu \tag{77.32}$$

Figure 77.15(b) shows the location of the finite poles and zeros in the s-plane.

In the following, we predefine $G_1 = G_2$, i.e., $\alpha_{21} = 1$, and $T_{34} = T_{56} = T$. This provides optimum pole sensitivities with respect to the passive elements and with respect to the gain-bandwidth products of the operational amplifiers. Hence, the gain factor H_0 always has the value 2.

Based on these parameter restrictions, we can establish an easy design procedure. First, we can choose the capacitors $C_3 = C_5 = C$. Then, given the pole parameters ω_p and Q_p, we determine the two resistors

$$R_4 = R_6 = R = \frac{1}{\omega_p \cdot C} \tag{77.33}$$

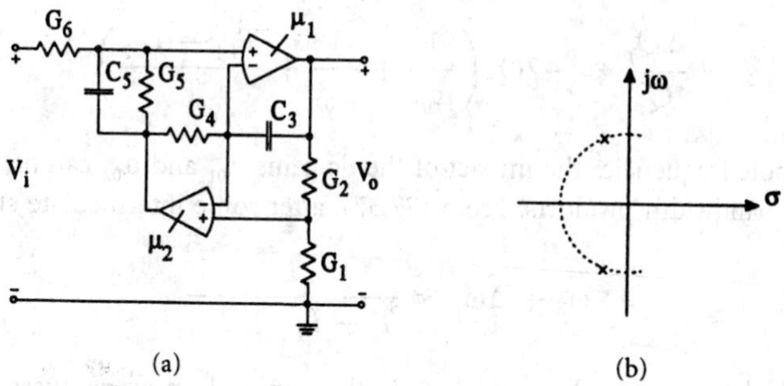

FIGURE 77.15 Second-order low-pass building block.

and from (77.30) the resistor

$$R_5 = Q_p \cdot R \tag{77.34}$$

There are three key features that make these dual-amplifier biquads particularly favorable for practical applications.

- The biquad building blocks have a low spread of elements. The resistors $R_1 = R_2$ can be chosen freely. The same holds for the two capacitors $C_3 = C_5$. The frequency-determining resistors $R_4 = R_6$ have equal values, too. There is only the Q-determining resistor R_5, which differs by a factor of Q_p.
- The relative sensitivities of the gain and the pole parameters with respect to the passive elements are of the same order of magnitude as in case of a second-order passive *RLC* network, see (77.35).
- The impact of the op amp gain and gain-bandwidth product on the pole parameters is extremely low.

A sensitivity analysis of the biquad circuit in Fig. 77.15 gives the following results;

$$\begin{aligned} S_{R_1}^{H_0} &= -S_{R_2}^{H_0} = -\frac{1}{2} \\ S_{R_1}^{\omega_p} &= -S_{R_2}^{\omega_p} = -S_{C_3}^{\omega_p} = -S_{R_4}^{\omega_p} = -S_{C_5}^{\omega_p} = -S_{R_6}^{\omega_p} = \frac{1}{2} \\ S_{R_1}^{Q_p} &= -S_{R_2}^{Q_p} = -S_{C_3}^{Q_p} = -S_{R_4}^{Q_p} = +S_{C_5}^{Q_p} = -S_{R_6}^{Q_p} = \frac{1}{2} \\ S_{R_5}^{Q_p} &= 1. \end{aligned} \tag{77.35}$$

In most practical applications, we can describe the op-amp dynamics by a one-pole model:

$$\mu(s) = \frac{\mu_0}{1 + (s/\omega_c)} = \frac{1}{(1/\mu_0) + (s/\omega_T)} \tag{77.36}$$

with the dc gain μ_0, the 3-dB cutoff frequency ω_c and the gain–bandwidth product $\omega_T = \mu_0\omega_c = 1/T_T$.

If the circuit elements are chosen properly, e.g., choosing network parameters $\alpha_{21} = 1$ and $T_{34} = T_{56} = T = 1/\omega_p$, an analysis of the impact of the parameters μ_{0i} and $T_{Ti} = 1/\omega_{Ti}$ of the two op-amps μ_i, $i = 1, 2$, on the pole parameters of the transfer function yields

$$\frac{\Delta\omega_p}{\omega_p} \approx \frac{1}{\mu_{01}} - \frac{1}{\mu_{02}} - \frac{T_{T1} + T_{T2}}{T} \tag{77.37}$$

$$\frac{\Delta Q_p}{Q_p} \approx -2Q_p\left(\frac{1}{\mu_{01}} + \frac{1}{\mu_{02}} + \frac{T_{T1} - T_{T2}}{T}\right) \tag{77.38}$$

In case of high pole frequencies the impact of the dc gains μ_{01} and μ_{02} can be neglected against that of the gain–bandwidth products. From (77.37), after some intermediate steps, we obtain

$$\omega_p + \Delta\omega_p \approx \frac{1}{T + T_{T1} + T_{T2}} \tag{77.39}$$

In case of an ideal op-amp we have $\omega_p = 1/T$. the nonideal op-amp causes a pole frequency change $\Delta\omega_p$ due to the time constants T_{Ti}. The frequency determining time constant T of the passive network is increased by the two time constants T_{Ti} of the op-amps, which are reciprocals of the gain–bandwidth products ω_{Ti}.

The most interesting result is the Q-factor change in (77.38). If the frequency responses of the two op-amps are matched, as is usually the case with dual packages, the impact of the two gain–bandwidth products cancel out. Thus, we have nearly no Q enhancement at higher pole frequencies.

To make the GIC-derived biquad almost independent of op-amp parameters, the above-mentioned conditions ($\alpha_{21} = 1$, $T_{34} = T_{56}$) must be met by the passive elements. In a more general view, this result holds for the whole family of GIC-derived building blocks. Each of these building blocks has two frequency-determining time constants and one resistive voltage divider (G_1 and G_2 or G_3 and G_4). In any case, to obtain independance of op-amp parameters we have to choose equal time constants and a voltage divider with equal resistors. If the time constants or the resistors, respectively, do not match exactly, the independance of op-amp parameters remains nearly unchanged.

Next we will derive a bandpass building block from the general biquad in Fig. 77.14(b). If we choose the passive elements as $Y_{1a} = G_1$, $Y_{1b} = 0$, $Y_2 = G_2$, $Y_3 = G_3$, $Y_4 = sC_4$, $Y_5 = G_5$, $Y_{6a} = sC_6$, and $Y_{6b} = G_6$, we obtain the bandpass building block in Fig. 77.16, which has a voltage transfer ratio

$$\begin{aligned} H_{\text{BP}}(s) = \frac{V_o}{V_i} &= \frac{G_6(sC_4G_2 + sC_4G_1)}{s^2C_4C_6G_2 + sC_4G_2G_6 + G_1G_3G_5} \\ &= H_0 \cdot \frac{s}{s^2 + s\omega_p/Q_p + \omega_p^2} \end{aligned} \tag{77.40}$$

with pole frequency

$$\omega_p = \sqrt{\alpha_{12}/(T_{43}T_{65})} \tag{77.41}$$

Q-factor

$$Q_p = \omega_p \cdot T_{66} \tag{77.42}$$

and midband gain

$$H_0 Q_p = (1 + \alpha_{12}) \tag{77.43}$$

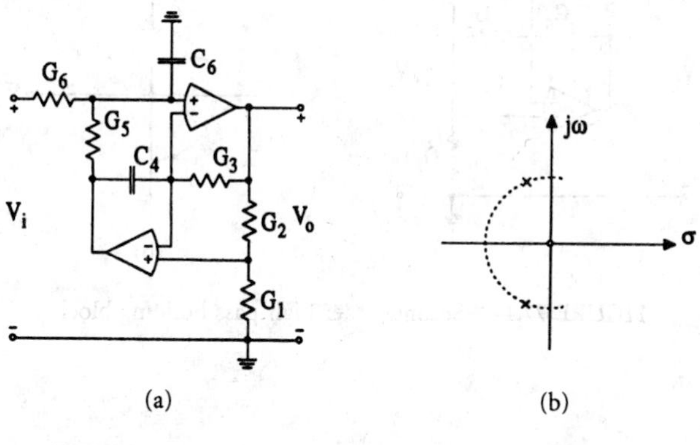

FIGURE 77.16 Second-order bandpass building block.

The filter design is similar to that of the low-pass building block. First we choose the two resistors $R_1 = R_2$ and the two capacitors $C_4 = C_6 = C$. Then, from the pole frequency ω_p we determine the two resistors

$$R_3 = R_5 = R = \frac{1}{\omega_p \cdot C} \tag{77.44}$$

and from the Q-factor Q_p the resistor

$$R_6 = Q_p \cdot R \tag{77.45}$$

The pole sensitivities with respect to the passive elements and the relationship between the gain and the gain–bandwidth product of the op-amps are the same as in case of the low-pass filters. It should be mentioned that the maximum of the magnitude response $|H(j\omega)|$ occurs at $\omega = \omega_p$ and has exactly the value 2, independent of the Q-factor and the time constants of the circuit,[2] see (77.43).

We can derive a high-pass building block by choosing the following elements: $Y_{1a} = G_1$, $Y_{1b} = 0$, $Y_2 = G_2$, $Y_3 = G_3$, $Y_4 = sC_4$, $Y_5 = G_5$, $Y_{6a} = G_6$, and $Y_{6b} = sC_6$; see Fig. 77.17.

This filter circuit has a voltage transfer ratio

$$\begin{aligned} H_{\mathrm{HP}}(s) = \frac{V_o}{V_i} &= \frac{sC_6(sC_4G_2 + sC_4G_1)}{s^2C_4C_6G_2 + sC_4G_2G_6 + G_1G_3G_5} \\ &= (1+\alpha_{12}) \cdot \frac{s^2}{s^2 + s\omega_p/Q_p + \omega_p^2} \end{aligned} \tag{77.46}$$

with ω_p as in (77.41) and Q_p as in (77.42).

The design of the high-pass building block in Fig. 77.17 is identical to that of the bandpass described above. Both building blocks have also the same pole sensitivities and the same impact of the op-amps on the pole parameters.

If we feed the input signal not only through the capacitor C_6, but additionally through the element G_1, and if we set again $\alpha_{12} = 1$ we obtain the all-pass building block in Fig. 77.18.

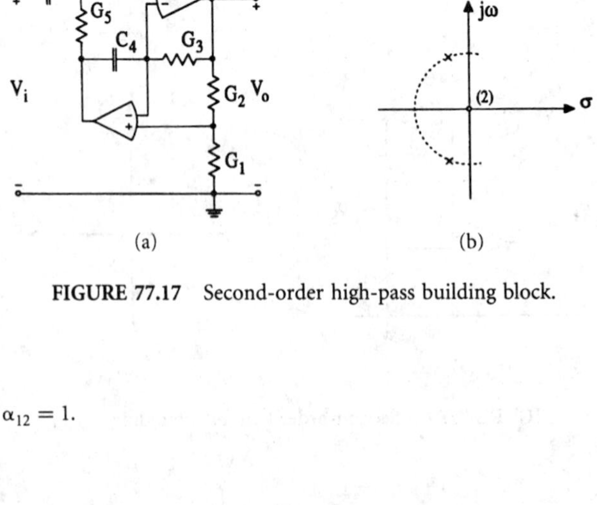

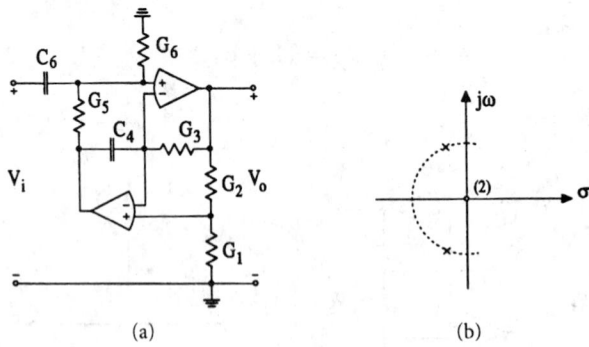

FIGURE 77.17 Second-order high-pass building block.

[2]Only assuming $\alpha_{12} = 1$.

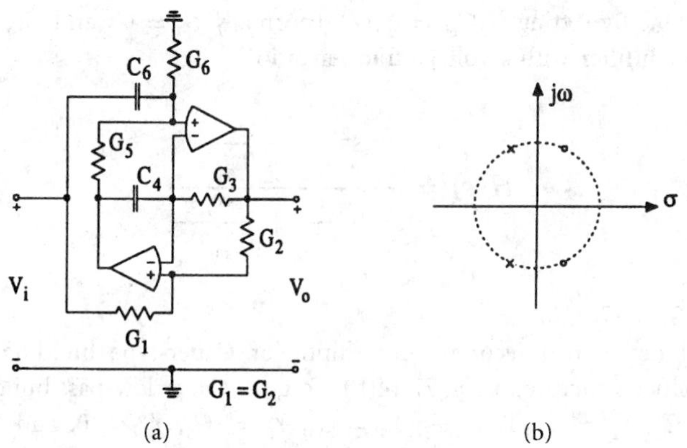

FIGURE 77.18 Second-order all-pass building block.

Its voltage transfer ratio reads

$$H_{\mathrm{AP}}(s) = \frac{V_o}{V_i} = \frac{s^2 C_6 C_4 G_2 - s C_4 G_1 G_6 + G_1 G_3 G_5}{s^2 C_4 C_6 G_2 + s C_4 G_2 G_6 + G_1 G_3 G_5}$$

$$= \frac{s^2 - \dfrac{s}{T_{66}} + \dfrac{1}{T_{43} T_{65}}}{s^2 + \dfrac{s}{T_{66}} + \dfrac{1}{T_{43} T_{65}}} \tag{77.47}$$

The all-pass building block is designed exactly as the high-pass or bandpass circuit.

From the all-pass building block, we can derive a second-order notch filter by adding a conductor G_{6b} in parallel with the capacitor G_6; see Fig. 77.19.

The circuit in Fig. 77.19 has the general voltage transfer ratio

$$H(s) = \frac{V_o}{V_i} = \frac{s C_6 C_4 G_2 + s C_4 (G_2 G_{6b} - G_1 G_{6a}) + G_1 G_3 G_5}{s^2 C_4 C_6 G_2 + s C_4 G_2 G_6 + G_1 G_3 G_5} \tag{77.48}$$

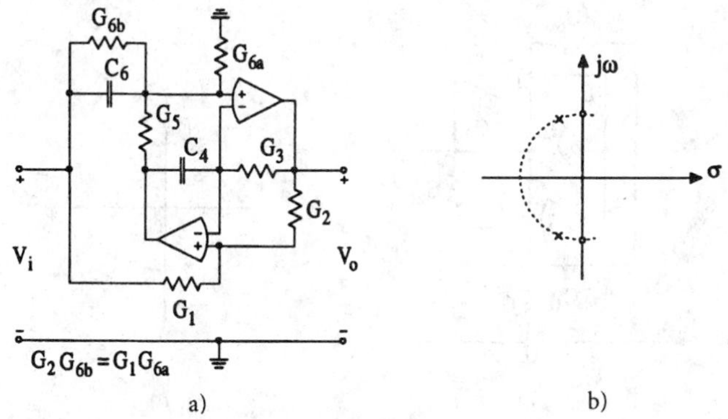

FIGURE 77.19 Notch filter building block.

with $G_6 = G_{6a} + G_{6b}$. By setting $G_2G_{6b} = G_1G_{6a}$ (normally $G_1 = G_2$ and $G_{6a} = G_{6b}$) we obtain a second-order notch filter with a voltage transfer ratio

$$H(s) = \frac{s^2 + \dfrac{\alpha_{12}}{T_{43}T_{65}}}{s^2 + \dfrac{s}{T_{66}} + \dfrac{\alpha_{12}}{T_{43}T_{65}}} \tag{77.49}$$

where $\alpha_{12} = 1$ if $G_1 = G_2$.

Finally we will derive two second-order elliptic or Cauer-type building blocks from the general dual-amplifier structure in Fig. 77.14(b). To get a Cauer low-pass building block we take $Y_{1a} = G_1$, $Y_{1b} = C_1$, $Y_2 = G_2$, $Y_3 = G_3$, $Y_4 = G_4$, $Y_5 = sC_5$, $Y_{6a} = 0$, and $Y_{6b} = G_6$; see Fig. 77.20.

This filter circuit has a voltage transfer ratio

$$\begin{aligned} H(s) = \frac{V_o}{V_i} &= \frac{s^2C_1G_3C_5 + G_6(G_2G_4 + G_1G_4)}{s^2C_1G_3C_5 + sG_1G_3C_5 + G_2G_4G_6} \\ &= \frac{s^2 + \dfrac{\alpha_{43}(1+\alpha_{12})}{T_{12}T_{56}}}{s^2 + \dfrac{s}{T_{11}} + \dfrac{\alpha_{43}}{T_{12}T_{56}}} \\ &= \frac{s^2 + \omega_z^2}{s^2 + s\omega_p/Q_p + \omega_p^2} \end{aligned} \tag{77.50}$$

The magnitude ω_z of the zeros is always greater than the magnitude ω_p of the poles. Additionally, without any matching of elements, the real part of the zeros is always zero.

The design can proceed in the following steps. First we predefine $\alpha_{43} = 1$ and choose due to practical considerations the values of the resistors $R_3 = R_4$. We also select the two capacitors C_1 and C_5. Then, comparing the first coefficient in the denominator of $H(s)$ in (77.50), namely $T_{11} = Q_p/\omega_p$ with $T_{11} = C_1R_1$ we determine the resistor

$$R_1 = \frac{Q_p}{\omega_p C_1} \tag{77.51}$$

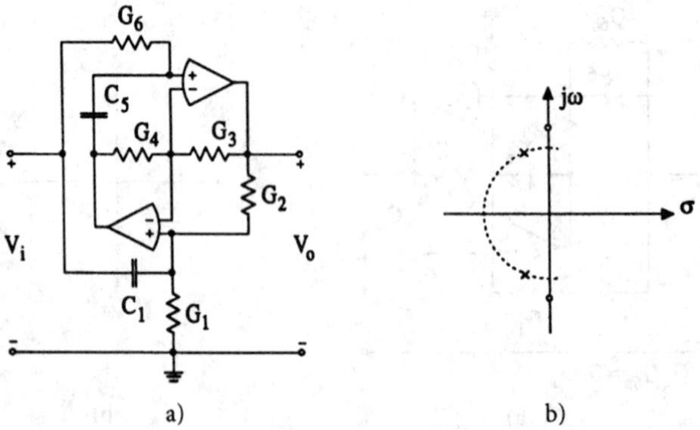

FIGURE 77.20 Second-order Cauer low-pass building block.

Further, comparing the last coefficients in the numerator and denominator yields

$$\frac{\omega_z^2}{\omega_p^2} = 1 + \alpha_{12} \tag{77.52}$$

With $\alpha_{12} = R_2/R_1$ we can solve (77.52) for

$$R_2 = R_1 \cdot \left(\frac{\omega_z^2}{\omega_p^2} - 1\right) \tag{77.53}$$

Finally, from $\omega_p^2 = 1/(T_{12}T_{56})$ we can determine the resistor

$$R_6 = \frac{1}{\omega_p^2 C_1 C_5 R_2} \tag{77.54}$$

In order to get the three resistors R_1, R_2, and R_6 in the same order of magnitude it might be advisable to predefine different values for the capacitors C_1 and C_5.

A Cauer high-pass building block is depicted in Fig. 77.21. Its voltage transfer ratio reads

$$\begin{aligned} H(s) = \frac{V_o}{V_i} &= \frac{s^2C_2G_4C_6 + G_1(G_3G_5 - G_6G_4)}{s^2C_2G_4C_6 + sG_6G_4C_2 + G_1G_3G_5} \\ &= \frac{s^2 + \dfrac{\alpha_{34} - \alpha_{65}}{T_{21}T_{65}}}{s^2 + s\dfrac{1}{T_{11}} + \dfrac{\alpha_{43}}{T_{12}T_{56}}} \\ &= \frac{s^2 + \omega_z^2}{s^2 + s\omega_p/Q_p + \omega_p^2} \end{aligned} \tag{77.55}$$

The design of this building block is similar to that of the low-pass circuit. First we choose the resistors $R_3 = R_4$ and the capacitors C_2 and C_6 and then, with predefined parameters Q_p, ω_p,

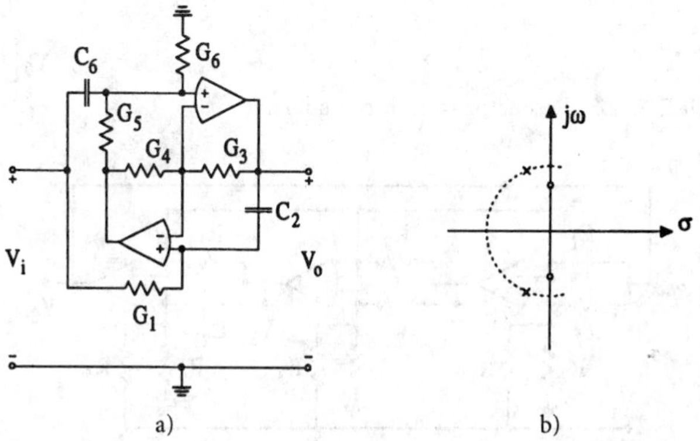

FIGURE 77.21 Second-order Cauer high-pass building block.

and ω_z, we determine the remaining three resistors:

$$R_6 = \frac{Q_p}{\omega_p C_6} \tag{77.56}$$

$$R_5 = R_6 \cdot \left(1 - \frac{\omega_z^2}{\omega_p^2}\right) \tag{77.57}$$

$$R_1 = \frac{1}{\omega_p^2 C_6 C_2 R_5} \tag{77.58}$$

77.5 GIC-Derived Three-Amplifier Biquads

In order to get more flexibility for realizing arbitrary second-order transfer functions and to obtain less resistor spread for realizing a given pole-Q, we can extend the resonator in Fig. 77.14(a) to a resonator with three nullors; see Fig. 77.22.

There are two different biquads known from the literature that are based on the resonator in Fig. 77.22. The first one is proposed by Mikhael and Bhattacharyya [9] and is shown in Fig. 77.23.

This filter circuit requires a small resistor spread to realize high pole-Q. The zeros of the transfer function are formed with a resistive feed-forward network providing a flexible design with arbitrary numerator coefficients. The voltage transfer ratio V_3/V_i is

$$H_3(s) = \frac{V_3}{V_i} = \frac{N_3(s)}{D(s)} \tag{77.59}$$

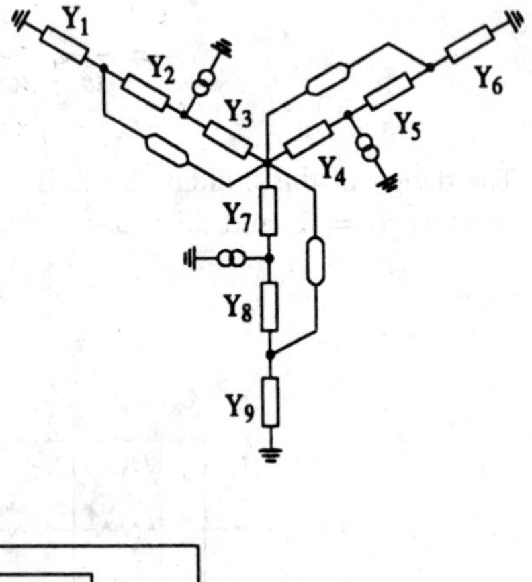

FIGURE 77.22 Resonator with three nullors.

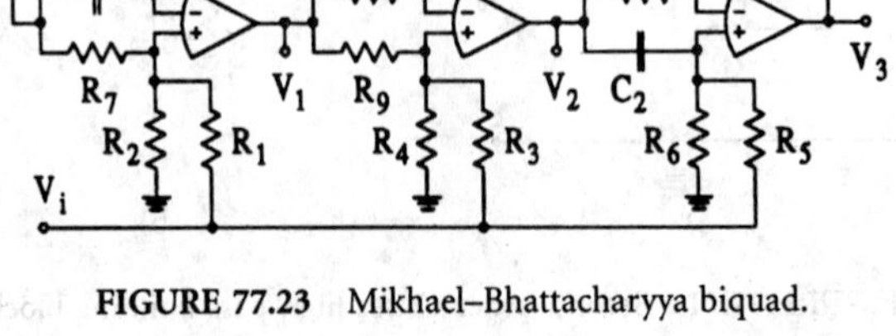

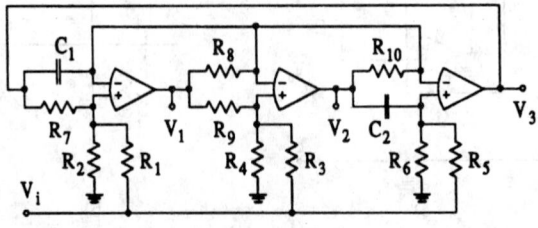

FIGURE 77.23 Mikhael–Bhattacharyya biquad.

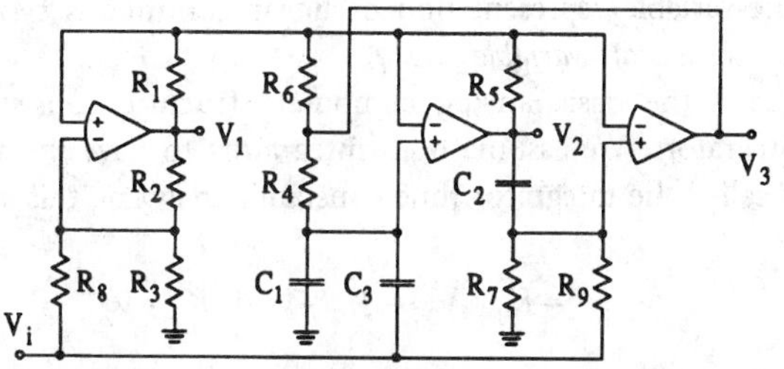

FIGURE 77.24 Padukone–Mulawka–Ghausi biquad.

with

$$N_3(s) = s^2C_1C_2G_1G_9 + sC_2(G_7G_8G_3 + G_8G_2G_3 - G_8G_1G_4) + (G_9G_7G_{10}G_5 + G_9G_{10}G_2G_5 - G_9G_{10}G_4G_6) \tag{77.60}$$

and

$$D(s) = s^2C_1C_2(G_1G_9 + G_2G_9) + sC_2(G_7G_8G_3 + G_7G_8G_4) + (G_9G_7G_{10}G_5 + G_9G_{10}G_7G_6) \tag{77.61}$$

The two other output nodes lead to similar transfer expressions.

The second biquad, which is based on the resonator in Fig. 77.22, was proposed by Padukone, Mulawka, and Ghausi [11] and is depicted in Fig. 77.24. Assuming ideal op-amps and choosing V_3 as output voltage, we obtain the following transfer function:

$$H_3(s) = \frac{V_3}{V_i} = \frac{N_3(s)}{D(s)} \tag{77.62}$$

with

$$N_3(s) = s^2[C_2C_3(G_2G_6 - G_1G_3) + C_1C_2G_1G_8] + s[C_1G_2G_5G_9 - C_3G_2G_5G_7 + C_2G_1G_4G_8] + (G_2G_4G_5G_9) \tag{77.63}$$

and

$$D(s) = s^2(C_1 + C_3)C_2G_2G_6 + sC_2G_1G_4(G_3 + G_8) + G_2G_4G_5(G_7 + G_9) \tag{77.64}$$

It has been shown [11] that the pole sensitivities to all passive components are not greater than unity. The filter section has the main advantage of being particularly insensitive to gain–bandwidth variations even when the op-amps are mismatched.

77.6 State-Variable-Based Biquads

A frequency used multiple-amplifier biquad is the circuit proposed by Kerwin, Huelsman and Newcomb [8]. This filter circuit has extreme flexibility, good performance, and low sensitivities to the passive components. The filter is based on analog computer structures [13], which are derived from the state-variable representation of linear continuous systems. Therefore, these filters are also referred to as *state-variable filters.*

Figure 77.25(a) shows the basic analog computer structure consisting of one summing amplifier and two integrators. We assume both integrators to have the same transfer function $-1/(sT)$, where T is called the integrator time constant. Analyzing this structure yields

$$V_1 = -K_1 \cdot V_3 + K_2 \cdot V_i + K_3 \cdot V_2 \tag{77.65}$$

$$V_2 = -\frac{1}{sT} \cdot V_1 \tag{77.66}$$

$$V_3 = -\frac{1}{sT} \cdot V_2 \tag{77.67}$$

which results in

$$H_{\mathrm{HP}}(s) = \frac{V_1}{V_i} = \frac{s^2T^2K_2}{s^2T^2 + sTK_3 + K_1} \tag{77.68}$$

Using (77.66), we can immediately derive the voltage transfer ratio V_2/V_i from (77.68):

$$H_{\mathrm{BP}}(s) = \frac{V_2}{V_i} = \frac{-sTK_2}{s^2T^2 + sTK_3 + K_1} \tag{77.69}$$

Finally, with (77.67) we obtain from (77.69)

$$H_{\mathrm{LP}}(s) = \frac{V_3}{V_i} = \frac{K_2}{s^2T^2 + sTK_3 + K_1} \tag{77.70}$$

Thus, the structure in Fig. 77.25(a) simultaneously realizes a high-pass filter, a bandpass filter, and a low-pass filter. The corresponding filter circuit, proposed by Kerwin, Huelsman, and Newcomb is depicted in Fig. 77.25(b). The integrators consist of one op-amp, one resistor R, and one capacitor C. The time constant is given by $T = RC$. The three gain factors of the

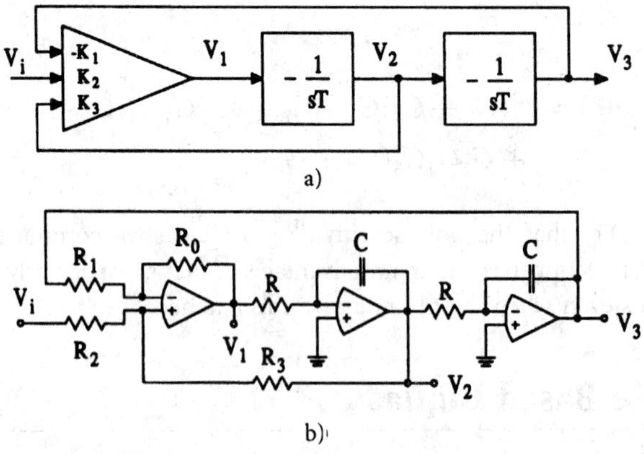

FIGURE 77.25 (a) Second-order analog computer structure and (b) state-variable filter section proposed by Kerwin, Huelsman, and Newcomb.

summing amplifier are determined by the four resistors R_0–R_3:

$$K_1 = \frac{R_0}{R_1} \tag{77.71}$$

$$K_2 = \frac{R_3}{R_2 + R_3}\left(1 + \frac{R_0}{R_1}\right) \tag{77.72}$$

$$K_3 = \frac{R_2}{R_2 + R_3}\left(1 + \frac{R_0}{R_1}\right) \tag{77.73}$$

Obviously, the integrator time constant T plays the role of a reciprocal normalization frequency. Thus, if we refer the frequency variable s to $1/T$, we obtain from (77.70) the normalized low-pass transfer function

$$H_{\text{LP}}(s) = \frac{V_3}{V_i} = \frac{K_2}{s^2 + sK_3 + K_1} \tag{77.74}$$

Given the normalized pole frequency ω_p and the pole Q-factor Q_p, we can design the filter section by equating

$$K_1 = \omega_p^2 \tag{77.75}$$

and

$$K_3 = \omega_p / Q_p \tag{77.76}$$

Then K_2 is fixed by the dc gain of the transfer function:

$$K_2 = H_0 \tag{77.77}$$

The state-variable filter circuit can be extended to a general biquad by adding an output amplifier that sums the three voltages V_1, V_2, and V_3. Figure 77.26(a) shows a state-variable filter with an output amplifier summing the voltages V_1, V_2, and V_3 of the circuit in Fig. 77.25(b). This biquad has been proposed also by Kerwin, Huelsman and Newcomb [8]. As alternative circuits, the amplifiers in Fig. 77.26(b) and (c) can be used. Figure 77.26(b) shows an output amplifier that realizes the following sum:

$$V_o = \alpha_1 V_1 + \alpha_2 V_2 + \alpha_3 V_3 \tag{77.78}$$

with

$$\alpha_1 = -\frac{R_{10}}{R_{11}} \qquad \alpha_2 = \frac{R_{14}}{R_{12} + R_{14}}\left(1 + \frac{R_{10}}{R_{11} \| R_{13}}\right) \qquad \alpha_3 = -\frac{R_{10}}{R_{13}} \tag{77.79}$$

If we solve (77.68)–(77.70) for the three output voltages and substitute them in (77.79) with normalized variables s we obtain

$$H(s) = \frac{V_o}{V_i} = -K_2 \frac{|\alpha_1| s^2 + |\alpha_2| s + |\alpha_3|}{D(s)} \tag{77.80}$$

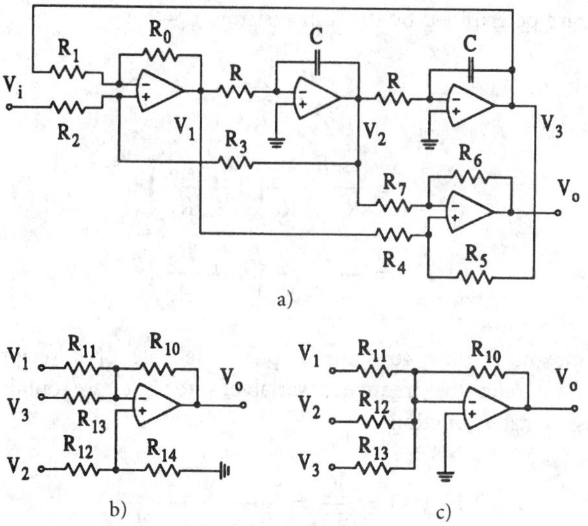

FIGURE 77.26 (a) State-variable filter with output amplifier, (b) output amplifier with inverting and noninverting inputs, and (c) output amplifier with three inverting inputs.

All numerator coefficients have the same sign. Therefore, the zeros of the transfer function are in the left-*s*-half plane. If we set $\alpha_2 = 0$, i.e., if we delete the voltage divider R_{12}, R_{14} and ground the noninverting input terminal of the op-amp, we obtain zeros on the $j\omega$ axis.

When designing the biquad, R_{14} and R_{10} may be used to scale the impedance level of the two resistive subnetworks. Then from the three numerator coefficients or from the overall gain constant of the transfer function, the zero frequency ω_z and the zero Q-factor Q_z we can easily determine the remaining resistors R_{11}, R_{12}, and R_{13}.

The output amplifier in Fig. 77.26(c) has three inverting inputs. Summing the voltages V_1, V_2, and V_3 leads to a numerator polynomial where the sign of the middle coefficient is different from the sign of the other two. Thus, the zeros are in the right-*s*-half plane. Again, we can delete the resistor R_{12} to realize zeros on the $j\omega$ axis.

A second state-variable biquad circuit proposed by Tow and Thomas [3], [16], [17] yields similar performance to that of the Kerwin–Huelsman–Newcomb circuit. It uses a feedback loop with one damped integrator, one integrator, and one inverting amplifier; see Fig. 77.27(a). Figure 77.27(b) shows the Tow–Thomas circuit with three op-amps.

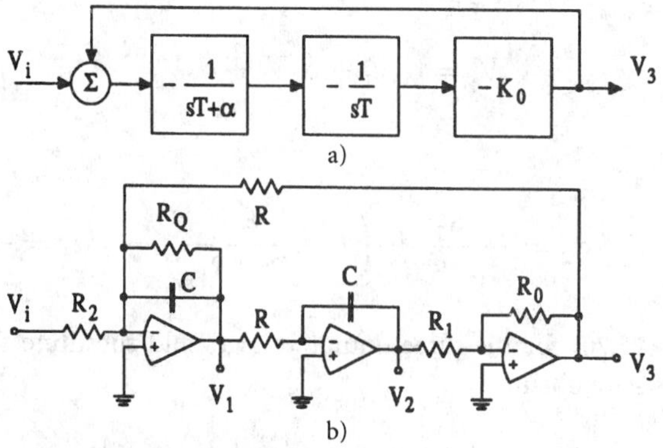

FIGURE 77.27 (a) Principle and (b) three-op-amp realization of the Tow–Thomas filter.

The damped integrator has a transfer function

$$\left.\frac{V_1}{V_3} = \frac{-1}{sT + \alpha}\right|_{V_i=0} \tag{77.81}$$

with $T = RC$ and $\alpha = R/R_Q$. The inverting amplifier has a gain

$$\frac{V_3}{V_2} = -\frac{R_0}{R_1} = -K_0 \tag{77.82}$$

An analysis of the circuit in Fig. 77.27(b) with V_i being the input voltage and V_3 the output voltage yields a transfer function

$$H_{\text{LP}}(s) = \frac{V_3}{V_i} = \frac{-K_0\alpha_2}{s^2T^2 + sTK_Q + K_0} \tag{77.83}$$

with

$$K_0 = \frac{R_0}{R_1} \qquad \alpha_2 = \frac{R}{R_2} \qquad K_Q = \frac{R}{R_Q} \tag{77.84}$$

For the design of the filter, the integrator time constant T serves as a reciprocal normalization frequency. Then, from the predefined normalized pole frequency we can determine the resistor ratio K_0, from the pole Q-factor the ratio K_Q, and from the dc gain of the filter section the ratio α_2. Choosing convenient values for C and R_0, we finally determine the resistors R, R_1, R_Q, and R_2 from the parameters T, K_0, K_Q, and α_2, respectively.

The filter circuit in Fig. 77.26 requires an additional op-amp to realize a transfer function with a general second-degree numerator polynomial. An alternative method is to feed fractions of the input signal forward into the input of each op-amp. This is realized in the multiple-input Tow–Thomas biquad [3]; see Fig. 77.28. The transfer function of this circuit can be calculated to be

$$H(s) = \frac{V_o}{V_i} = -\frac{s^2T^2\alpha_4 + sT[K_Q\alpha_4 - K_0\alpha_3] + [K_0(\alpha_2 - K_Q\alpha_3)]}{s^2T^2 + sTK_Q + K_0} \tag{77.85}$$

with K_0, α_2, and K_Q as defined in (77.84) and

$$\alpha_3 = \frac{R}{R_3} \qquad \alpha_4 = \frac{R_0}{R_4} \tag{77.86}$$

Thus, arbitrary numerator coefficients can be predescribed. In particular, if we choose $\alpha_3 = \alpha_4 = 0$ we obtain the low-pass filter circuit in Fig. 77.27(b) and the transfer function in (77.83).

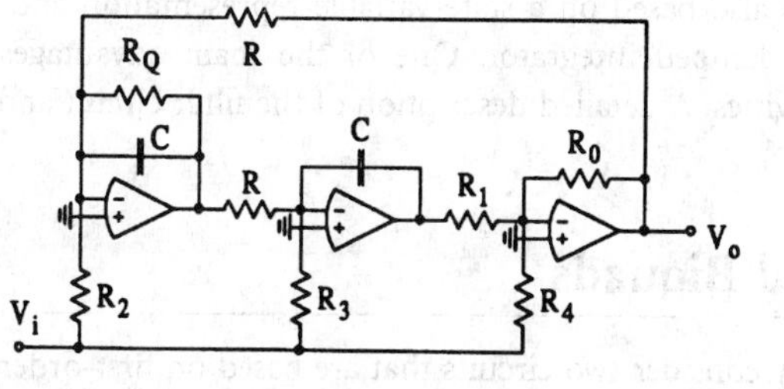

FIGURE 77.28 Generalized Tow–Thomas biquad.

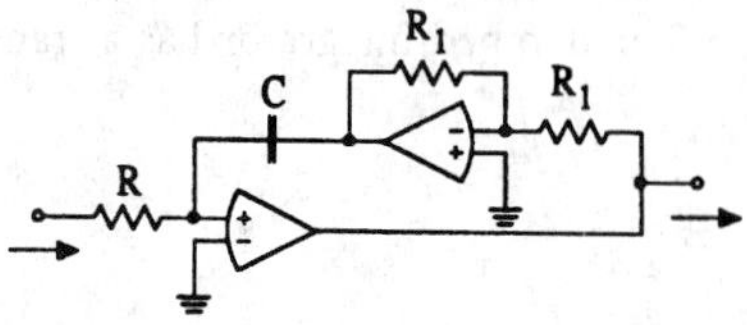

FIGURE 77.29 Noninverting integrator using an additional op-amp to compensate for phase lag.

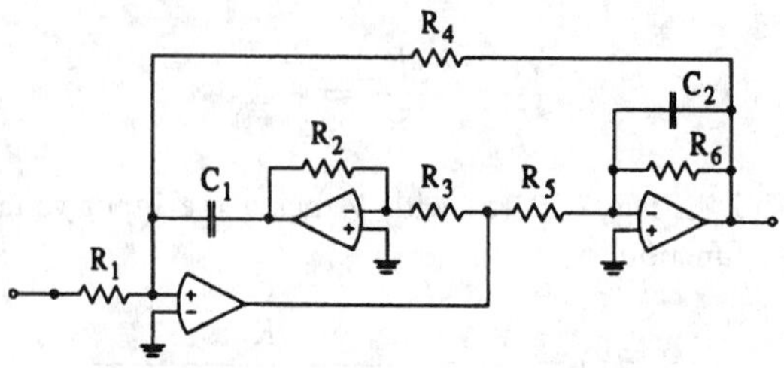

FIGURE 77.30 Åkerberg–Mossberg biquad.

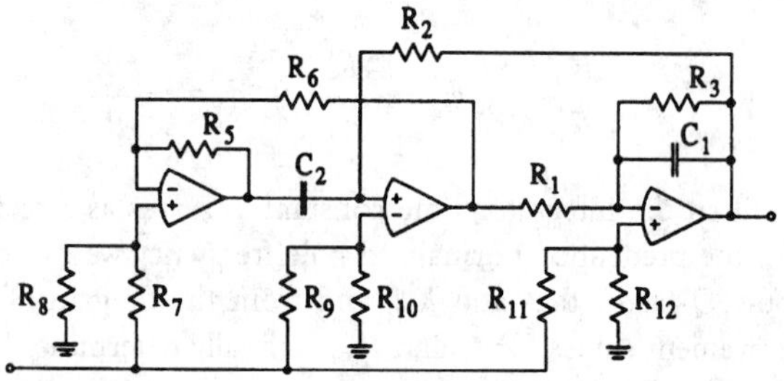

FIGURE 77.31 Berka–Herpy biquad.

When the state-variable filters described above are used to realize high-Q filter functions, the Q practically obtained is usually higher than that desired in the design. This effect is called Q enhancement and is caused by the phase lag introduced by the nonideal op-amps. One way to solve this problem is to use integrators with phase compensation.

Figure 77.29 shows a noninverting integrator with an additional op-amp for phase lag compensation. A detailed description of this circuit can be found in [7]. Putting this noninverting integrator together with an inverting integrator in a feedback loop results in a resonator with a Q-factor that is almost independent of the gain–bandwidth product of the op amps. Thus, nearly no Q enhancement occurs.

Exactly this feedback loop is used in the Åkerberg–Mossberg biquad [1]; see Fig. 77.30. In this circuit, a noninverting integrator with phase lag compensation together with an inverting damped integrator is connected as a feedback loop. More details about this filter section can be found in [1], [6], [7].

Finally, let us consider the general biquad proposed by Berka and Herpy [2]; see Fig. 77.31. This biquad is also based on a state-variable representation and requires a second-order differentiator and a damped integrator. One of the main advantages of this circuit are the extremely low sensitivities. A detailed description of the filter circuit and its design can be found in [6].

77.7 All-Pass-Based Biquads

Finally, we will briefly consider two circuits that are based on first-order all-pass sections. Figure 77.32 shows the classical filter circuit introduced by Tarmy and Ghausi [15].

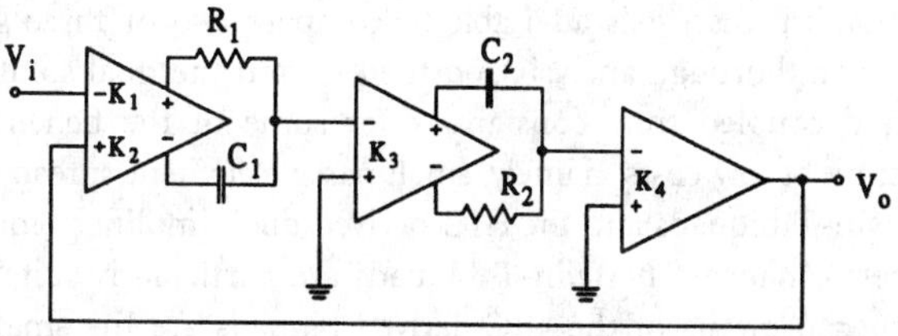

FIGURE 77.32 Tarmy–Ghausi circuit.

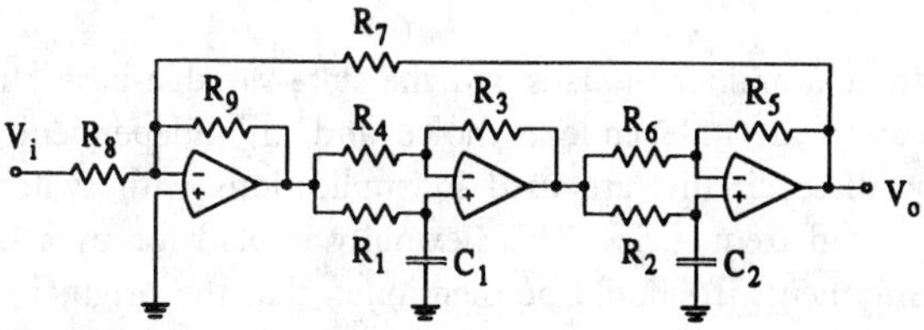

FIGURE 77.33 Tarmy–Ghausi circuit modified by Moschytz.

In this circuit amplifiers with differential inputs and differential outputs are used. In the feedback loop, two all-pass sections with voltage transfer ratios

$$H_i(s) = \frac{sT_i - 1}{sT_i + 1} \tag{77.87}$$

$T_i = R_iC_i$, $i = 1, 2$, and one inverting amplifier are cascaded. The overall transfer function is

$$H(s) = \frac{V_o}{V_i} = \alpha_1 \frac{s^2 T_1 T_2 - s(T_1 + T_2) + 1}{s^2 T_1 T_2 + s\alpha_2(T_1 + T_2) + 1} \tag{77.88}$$

with

$$\alpha_1 = \frac{K_1K_3K_4}{1 + K_2K_3K_4} \qquad \alpha_2 = \frac{1 - K_2K_3K_4}{1 + K_2K_3K_4} \tag{77.89}$$

The main advantages of the Tarmy–Ghausi circuit are its low sensitivities to the gain-bandwidth products of the op amps and thus its favorable performance at high pole frequencies and high pole *Q*-factors.

Most of the popular low-cost op-amps are not configured for differential output operations. This problem is bypassed by a circuit proposed by Moschytz [10]; see Fig. 77.33.

The circuit in Fig. 77.33 is also based on first-order all-pass sections. But the all-pass circuits are realized by means of only single-ended op-amps. Therefore, this circuit is more convenient for practical realizations.

77.8 Summary

In this chapter, we have shown that a lot of different multiple-amplifier biquads are known from the literature. Hence, a design engineer who aims to realize a high performance active filter is in the favorable situation to find a rich variety of alternative circuits and design methods. On the other hand, this variety is also confusing and the design engineer may need assistance in deciding among all these circuits and methods. Unfortunately, the multiple-amplifier biquads mentioned in this chapter cannot be classified in a simple way. Therefore, only a rough guidance can be given. In a practical case, it is advisable to compare two or three solutions next to the predefined demands by a thorough analysis and then to find the final solution.

The biquads with decoupled time constants offer some of the benefits mentioned at the beginning of this section at low costs, namely, small passive element spread and low sensitivities. The dual-amplifier twin-T biquads and the GIC-derived dual-amplifier biquads offer a trade-off between costs and performance. The twin-T biquads are particularly suitable for the design of elliptic filters. The special merits of the GIC-derived biquads are the small number of passive components and the independence of the op-amp parameters. Thus, this biquad represents a robust filter solution for many different applications. Additionally, these filters can be easily designed.

The GIC-derived three-amplifier biquads and the state-variable-based filters offer additional flexibility with respect to an independent choice and an independent tuning of the filter parameters. Typically, these circuits are used in applications with switched parameters, e.g., filters with switched cutoff frequencies. This flexibility is paid for by a higher number of op amps and passive components. It should be mentioned that the original state-variable circuits, i.e., two integrators in a loop, are rather sensitive to the phase lag introduced by the nonideal op-amps. This is always a problem when simultaneously high-frequency and high-Q performance shall be achieved. In such applications, biquads with compensated phase lag should be applied, e.g., the biquads proposed by Åkerberg and Mossberg, Berka and Herpy, or Tarmy and Ghausi.

References

[1] D. Åkerberg and K. Mossberg, "A versatile active *RC* building block with inherent compensation for the finite bandwidth of the amplifier," *IEEE Trans. Circuits Syst.*, vol. CAS-21, pp. 75–78, 1974.

[2] J. C. Berka and M. Herpy, "Novel active *RC* building block with optimal sensitivity," *Electron. Lett.*, vol. 17, pp. 887–888, 1981.

[3] P. E. Fleischer and J. Tow, "Design formulas for biquad active filters using three operational amplifiers," *Proc. IEEE*, vol. 61, pp. 662–663, 1973.

[4] N. J. Fliege, "A new class of second-order *RC*-active filters with two operational amplifiers," *Nachrichtentech. Z.*, vol. 26, pp. 279–282, 1973.

[5] P. R. Geffe, "A *Q*-invariant active resonator," *Proc. IEEE* (Lett.), vol, 57, p. 1442, 1969.

[6] M. Herpy and J.-C. Berka, *Active RC Filter Design*, Amsterdam: Elsevier, 1986.

[7] L. P. Huelsman and P. E. Allen, *Introduction to the Theory and Design of Active Filters*, New York: McGraw-Hill, 1980.

[8] W. J. Kerwin, L. P. Huelsman, and R. W. Newcomb, "State-variable synthesis for insensitive integrated circuit transfer functions," *IEEE J. Solid-State Circuits*, vol. SC-2, pp. 87–92, 1967.

[9] W. B. Mikhael and B. B. Bhattacharyya, "A practical design for insensitive *RC*-active filters," *IEEE Trans. Circuits Syst.*, vol. CAS-22, pp. 407–415, 1975.

[10] G. S. Moschytz, "High-*Q* factor insensitive active *RC* networks, similar to the Tarmy–Ghausi circuit but using single-ended operational amplifiers," *Electron. Lett.*, vol. 8, pp. 458–459, 1972.

[11] P. Padukone, J. Mulawka, and M. S. Ghausi, "An active biquadratic section with reduced sensitivity to operational amplifier imperfections," *J. Franklin Inst.*, vol. 30, no. 1, pp. 27–40, 1980.

[12] R. P. Sallen and E. L. Key, "A practical method of designing *RC* active filters," *IRE Trans. Circ. Theory*, vol. CT-2, pp. 75–85, 1955.

[13] W. Schüßler, "Schaltung und Messung von Übertragungsfunktionen an einem Analogrechner," *Archiv der Elektrischen Übertragung AEÜ*, vol. 13, pp. 405–419, 1959.

[14] M. A. Soderstrand and S. K. Mitra, "Extremely low sensitivity active *RC* filter," *Proc. IEEE* (Lett.), vol. 57, p. 2175, 1969.

[15] R. Tarmy and M. S. Ghausi, "Very high-*Q* insensitive *RC* networks," *IEEE Trans. Circ. Theory*, vol. CT-17, pp. 358–366, 1970.

[16] L. C. Thomas, "The biquad: Part I—Some practical design considerations," *IEEE Trans. Circ. Theory*, vol. CT-18, pp. 350–357; ——, "The biquad: Part II—A multipurpose active filtering system," *IEEE Trans. Circ. Theory*, vol. CT-18, pp. 358–361, 1971.

[17] J. Tow, "Design formulas for active *RC* filters using operational-amplifier biquad," *Electron. Lett.*, vol. 5, pp. 339–341, 1969.

78

The Current Generalized Immittance Converter (CGIC) Biquads

Wasfy B. Mikhael
University of Central Florida

78.1 Introduction

CGIC's have been used to realize high-performance active biquads with 2 OA's, 3 OA's, or *n* OA's per section [1] and [5].

In this chapter two and three-amplifier biquads are presented that are based on CGIC's. Although several biquads have been reported, the ones presented here have proved to be clearly superior to single-amplifier biquads. This is because the design carried out [1] was constrained by stringent performance criteria satisfying important properties and features such as stability,

Wasfy B. Mikhael, "Chapter 9: Biquad II" in *RC Active Filter Design Handbook*, Stephenson, New York: Wiley, 1985.

0-8493-8341-2/95/$0.00 + $.50

versatility, insensitivity to component tolerances and drift, low dependence on the op-amp (OA) frequency limitations, finite gain, tunability, small spread in component values, and minimum total capacitance.

In addition, the performance of the CGIC-based biquads presented here are comparable to multiple OA biquads. On the other hand, the 3-OA CGIC biquad is shown to yield additional performance improvements over the 2-OA CGIC biquad. Also, the CGIC biquads use the OA's in the differential mode.

In the following discussion the generalized structure of the 2-OA and 3-OA CGIC biquads are presented. Illustrative examples of the element identification to realize the most commonly used biquads are tabulated. Stability and sensitivity properties are discussed. A design procedure for each biquad is described that minimizes the active sensitivities while maintaining the filter's stability. Several second-order design examples are given. A sixth-order Chebyshev LPF and a sixth-order elliptic BPF are designed using the design values and tuning procedure suggested. The excellent performance of the resulting realizations is experimentally verified. A universal 2-OA GIC hybrid implementation using thick film is also described. 2-OA CGIC biquadratic active filter realizations for extended high-frequency applications, employing the composite operational amplifiers technique [6], [7] are given.

78.2 Biquadratic Structure Using the Antoniou CGIC (2OA-CGIC Biquad)

Consider the network of Fig. 78.1, which is simply Antoniou's CGIC [2], with two new ports created across 3-G and 4-G. This is represented symbolically in Fig. 78.2.

A new configuration is now obtained, as shown in Fig. 78.3. A synthesis procedure is now described that uses this configuration. The transfer functions between the input and output

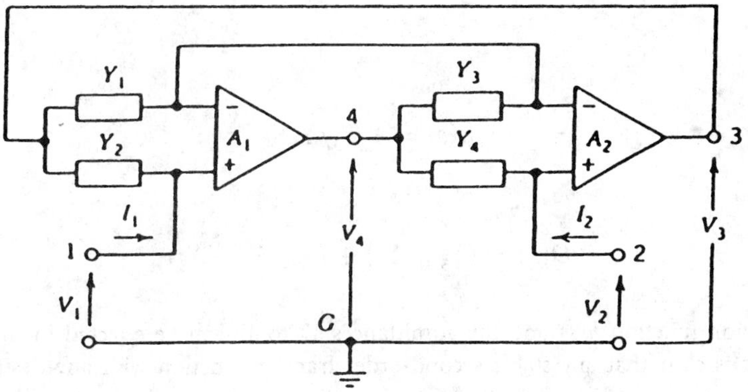

FIGURE 78.1 Antoniou's CGIC with additional ports 3G and 4G.

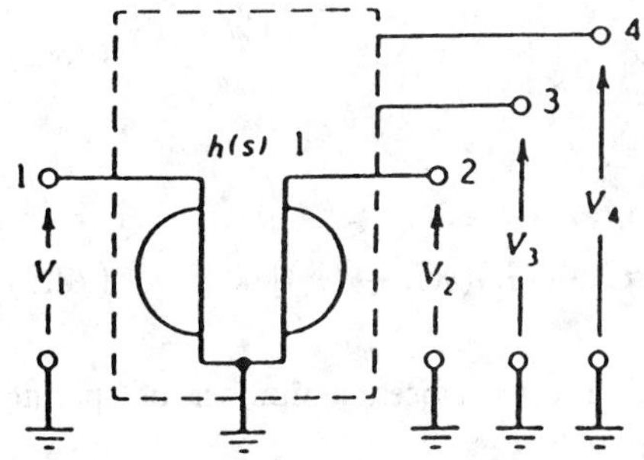

FIGURE 78.2 Symbolic representation of the CGIC in Fig. 78.1.

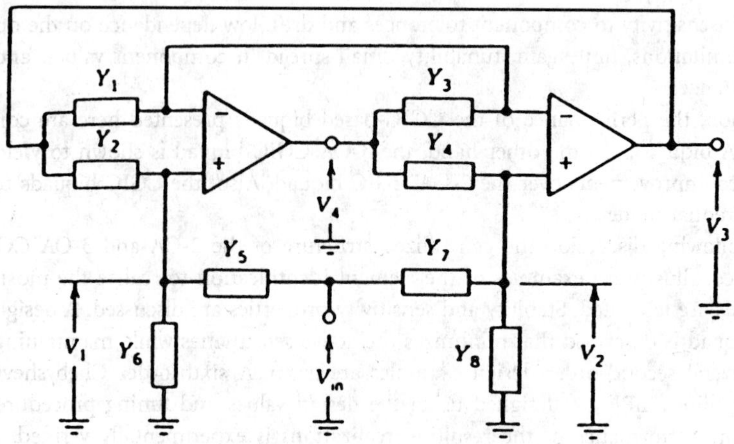

FIGURE 78.3 The basic configuration.

terminals 2, 3, and 4, assuming ideal OA's are readily obtained as

$$\frac{V_3}{V_1} = T_1 = [Y_5 + h(s)[Y_7(1 + Y_6/Y_2) - Y_5Y_8/Y_2]]/D(s) \tag{78.1a}$$

$$\frac{V_4}{V_i} = T_2 = [Y_5(1 + Y_8/Y_4) - Y_6Y_7/Y_4 + h(s)Y_7]/D(s) \tag{78.1b}$$

$$\frac{V_2}{V_i} = T_3 = [Y_5 + h(s)Y_7]/D(s) \tag{78.1c}$$

where

$$h(s) = Y_2Y_3/Y_1Y_4$$

$$D(s) = (Y_5 + Y_6) + h(s)(Y_7 + Y_8) \tag{78.1d}$$

The conversion function $h(s)$ and the admittances Y_5 to Y_8 can be selected in many different ways, and it is clear that any stable second-order transfer function with any desired zero and pole locations can be realized.

Letting

$$Y_i = sC_i + G_i \tag{78.2}$$

where $i = 1$ to 4, we have from (78.1)

$$h(s) = (sC_2 + G_2)(sC_3 + G_3)/[(sC_1 + G_1)(sC_4 + G_4)] \tag{78.3}$$

Clearly, by omitting one or more conductances and/or capacitances, a number of specific conversion functions can be generated.

78.3 Realization of the Most Commonly Used Biquad Functions Using the 2-OA CGIC

The most frequently used second-order transfer functions have already been described in Chapter 75. Realizations of these functions are given in Table 78.1. The LP, HP, BP, and BS sections are produced by choosing $h(s)$ in a simple manner such as $K_1 s$, $K_2 s^2$, $K_3 s + K_4 s^2$ or their reciprocals. Circuits 3, 4, and 7 can be regarded as realizations of simple RLC networks [2], [3]. Second-order AP sections can be obtained from circuits 11 and 12 of Table 78.1.

Figures 78.1 and 78.2 and Table 78.1 show that, with the exception of circuit 10, the response is obtained from the output of an OA. Owing to the low output resistance of the amplifier, any number of sections can be cascaded without using isolation amplifiers.

78.4 Stability During Activation and Sensitivity Using the 2-OA CGIC Biquad

Stability Properties

It has been shown [4] that some networks using CGIC's can be conditionally stable, that is, a circuit can lock in an unstable mode during activation (just after switching on the power supply). For amplifiers with a finite open loop gain A, the circuit of Fig. 78.3 gives

$$\frac{V_k}{V_i} = \frac{N_k(s)}{D(s)}$$

where $k = 2, 3, 4$ and

$$\begin{aligned} D(s) &= M_1 Y_1 + M_2 Y_3 \\ &\quad + (1 + M_1)(1 + M_2)[Y_1/A_1 + Y_3/A_2 + Y_1/A_1A_2 + Y_3/A_1A_2] \end{aligned} \tag{78.4}$$

$$M_1 = \frac{Y_5 + Y_6}{Y_2}$$

$$M_2 = \frac{Y_7 + Y_8}{Y_4}$$

It can be easily shown that for the circuit in Fig. 78.3 low-frequency unstable modes cannot arise during activation. This is due to the absence of differences and changes of sign in the denominator coefficients. Thus any combination of OA gains that occur during transients and power supply switching does not result in saturation or low-frequency instability before reaching the steady state [5].

Sensitivity Analysis

The pole Q-factor Q_p, the undamped frequency of oscillation ω_p, the notch frequency ω_n, and the multiplier constants H_{BS}, H_{LP}, H_{HP}, and H_{BP}, as well as Q_z, ω_z, and H_{AP}, have been previously defined in Chapter 75. The sensitivity of a quantity x with respect to variations in an element e is denoted by S_e^x. For ideal amplifiers, the use of Table 78.1 leads to

$$0 \leq S_e^x \leq 1 \tag{78.5}$$

where x represents any one of these H, ω, or Q qualities and e represents any capacitance or conductance.

TABLE 78.1 Element Identification for Realizing the Most Commonly Used Transfer Functions

Circuit Number	$h(s)$	Y_1	Y_2	Y_3	Y_4	Y_5	Y_6	Y_7	Y_8	Transfer Function	Remarks
1	$\frac{sC_2(sC_3+G_3)}{G_1G_4}$	G_1	sC_2	sC_3+G_3	G_4	G_5	0	0	G_8	$T_2=\frac{G_1G_5(G_4+G_8)}{G_1G_5G_4+sC_2G_3G_8+s^2C_2C_3G_8}$	LP
2	$\frac{G_2G_3}{sC_1(sC_4+G_4)}$	sC_1	G_2	G_3	$sC_4+G_4^a$	0	G_6	G_7	$sC_8^a+G_8^a$	$T_1=\frac{G_2G_3G_7\left(1+\frac{G_6}{G_2}\right)}{G_2G_3(G_7+G_8)+s(C_2G_4G_6+C_8G_2G_3)+s^2C_1C_4G_6}$	LP
3	$\frac{sC_3G_2}{G_1G_4}$	G_1	G_2	sC_3	G_4	0	G_6	sC_7	$sC_8^a+G_8$	$T_1=\frac{s^2C_3C_7(G_2+G_6)}{G_6G_1G_4+sC_3G_2G_8+s^2(C_7+C_8)G_2C_3}$	HP
4	$\frac{sC_3G_2}{G_1G_4}$	G_1	G_2	sC_3	G_4	0	G_6	$\frac{sC_7}{1+sC_7R_7}$	0	$T_1=\frac{s^2C_3C_7(G_2+G_6)}{G_6G_1G_4+sC_7G_1G_4G_6R_7+s^2C_3C_7G_2}$	HP
5	$\frac{s^2C_2C_3}{G_1G_4}$	G_1	sC_2	sC_3	G_4	sC_5	G_6	0	G_8	$T_2=\frac{sC_5\left(1+\frac{G_8}{G_4}\right)G_1G_4}{G_1G_4G_6+sC_5G_1G_4+s^2C_2C_3G_8}$	BP
6	$\frac{G_2G_3}{sC_1(sC_4+G_4)}$	sC_1	G_2	G_3	$sC_4+G_4^a$	0	G_6	sC_7	G_8	$T_1=\frac{sC_7G_2G_3\left(1+\frac{G_6}{G_2}\right)}{G_2G_3G_8+s(C_7G_2G_3+C_1G_4G_8)+s^2C_1C_4G_6}$	BP
7	$\frac{sC_3G_2}{G_1G_4}$	G_1	G_2	sC_3	G_4	0	G_6	G_7	$sC_8+G_8^a$	$T_1=\frac{sC_3G_7(G_2+G_6)}{G_1G_4G_6+sC_3G_2(G_7+G_8)+s^2C_3C_8G_2}$	BP
8	$\frac{G_2G_3}{s^2C_1C_4}$	sC_1	G_2	G_3	sC_4	G_5	0	G_7	sC_8	$T_2=\frac{s^2C_1G_5(C_4+C_8)+G_2G_3G_7}{s^2C_1G_5C_4+sC_8G_2G_3+G_2G_3G_7}$	N
9	$\frac{sG_2C_3}{G_1G_4}$	G_1	G_2	sC_3	G_4	G_5	0	sC_7	G_8	$T_2=\frac{s^2C_3C_7G_2+G_1G_5(G_4+G_8)}{s^2C_3C_7G_2+sC_3G_2G_8+G_1G_4G_5}$	N
10	$\frac{sG_2C_3}{G_1G_4}$	G_1	G_2	sC_3	G_4	G_5	G_6^a	sC_7	$sC_8^a+G_8$	$T_1=\frac{\left(s^2+\frac{G_1G_4G_5}{C_3C_7G_2}\right)\left(\frac{C_7}{C_7+C_8}\right)}{s^2+\frac{G_8}{C_7+C_8}+\frac{G_1G_4(G_5+G_6)}{C_3G_2(C_7+C_8)}}$	N
11	$\frac{G_2G_3}{s^2C_1C_4}$	sC_1	G_2	G_3	sC_4	G_5	G_6^a	G_7	sC_8	$T_1=\frac{s^2C_1G_4G_5-sG_5C_8G_3+G_3G_7(G_2+G_6)}{s^2C_1C_4(G_5+G_6)+sC_8G_2G_3+G_2G_3G_7}$	Nonminimum phase For all pass: $G_6=0$ $G_5=G_2$
12[b]	$\frac{sG_2G_3}{G_1G_4}$	G_1	G_2	sC_3	G_4	G_5	G_6	sC_7	G_8	$T_1=\frac{s^2C_3C_7(G_2+G_6)-sC_3G_5G_8+G_1G_4G_5}{s^2C_3C_7G_2+sC_3G_2G_8+(G_5+G_6)G_1G_4}$	Nonminimum phase For all pass: $G_6=0$ $G_2=G_5$

[a]These elements can be set equal to zero.
[b]A special case of Circuit 12—namely, the all-pass case—has also been independently proposed by J. T. Lim, Bell Northern Research, Canada, as communicated privately by him to Dr. A. Antoniou.

78.5 Design and Tuning Procedure of the 2-OA CGIC Biquad

A design procedure is now described. Table 78.1 shows that there are several degrees of freedom in the choice of element values. These may be used to minimize $S_A^{Q_p}$ or the spread of element values. By using minimum sensitivity (to the OA parameters) constraints in circuits 1, 3, 7, 10, and 12, possible sets of element values for LP, HP, BP, BS, and AP sections have been obtained, as shown in Table 78.2. In addition, this choice of elements guarantees stable operation with real OA's. It is seen that the notch frequency ω_n, the undamped frequency of oscillation ω_p and Q-factor Q_p can be easily adjusted by sequentially trimming three distinct resistors. A tuning sequence is also given in Table 78.2.

It is to be noted from Table 78.2 for the design of the BS sections, that

$$\omega_n^2 = \frac{\omega_p G_s}{C_s} \tag{78.6}$$

where

$$G_5 + G_6 = G_1 \qquad C_7 + C_8 + C \quad \text{and} \quad \omega_p = \frac{G}{C}$$

Hence

$$\frac{\omega_n^2}{\omega_p^2} = \frac{G_5}{G} \cdot \frac{C}{C_7} \tag{78.7}$$

where

$$\frac{G_5}{G} \text{ is always} \leq 1 \quad \text{and} \quad \frac{C}{C_7} \text{ is always} \geq 1$$

It is seen from (78.7) that for $\omega_n \leq \omega_p$, C_8 can be set to zero ($C_7 = C$), and only two capacitors per section are used. For $\omega_n > \omega_p$, three capacitors are required. In all cases, regardless of the choice of C_7 and C_8, the total capacitance per section is equal to $2C$.

In most applications where notch sections are used, ω_n/ω_p is close to unity and care should be taken when choosing G_5 and C_7 in (78.7). A suitable choice of G_5, G_6, G_7, and C_8 values may be obtained by letting

$$\frac{G_5}{G} = \frac{1}{K} \quad \text{and} \quad \frac{C}{C_7} = 2$$

Thus K is approximately equal to 2 ($K < 2$ for $\omega_n > \omega_p$ and $K > 2$ for $\omega_n \leq \omega_p$). The suggested choice yields a capacitor spread of 2 and both R_5 and R_6 are approximately equal to $2R$.

78.6 Design Examples Using the 2-OA CGIC Biquad

Example 1. Design a second-order Butterworth ($Q_p = 0.707$) LP filter having a cutoff frequency $f_p = 20{,}000/2\pi$ Hz.

TABLE 78.2 Design Values and Tuning Procedure

Circuit number (from Table 9.1)	Design Values	Transfer Function Realized	Tuning Sequence ω_o	ω_p	Q_p
1	$G_1 = G_4 = G_5 = G_8 = G,\ G_3 = G/Q_p,\ C_2 = C_3 = C$, where $C = \dfrac{G}{\omega_p}$	$T_2 = \dfrac{2\omega_p^2}{D(s)}$	—	G_8	G_3
3	$G_1 = G_2 = G_4 = G_6 = G,\ G_8 = G/Q_p,\ C_8 = 0,\ C_3 = C_7 = C$, where $C = \dfrac{G}{\omega_p}$	$T_1 = \dfrac{2s^2}{D(s)}$	—	G_4	G_8
7	$G_1 = G_2 = G_4 = G_6 = G,\ G_7 = G/Q_p,\ G_8 = 0,\ C_3 = C_8 = C$, where $C = \dfrac{G}{\omega_p}$	$T_1 = \dfrac{\left(\dfrac{2\omega_p}{Q_p}\right)s}{D(s)}$	—	G_2	G_7
10	$G_1 = G_2 = G_4 = G_5 + G_6 = G,\ G_8 = G/Q_p,\ C_3 = C_7 + C_8 = C$, where $\omega_p = \dfrac{G}{C}$ and $\omega_n^2 = \omega_p \dfrac{G_5}{C_7}$	$T_3 = \dfrac{C_7}{C}\dfrac{(s^2 + \omega_n^2)}{D(s)}$	G_2	G_6	G_8
12	$G_1 = G_2 = G_4 = G_5 = G,\ G_6 = 0,\ C_3 = C_7 = C,\ G_8 = G/Q_p$, where $\omega_p = \dfrac{G}{C}$	$T_1 = \dfrac{D(-s)}{D(s)}$		G_4	G_8

Note: $D(s) = s^2 + (\omega_p/Q_p)s + \omega_p^2$.

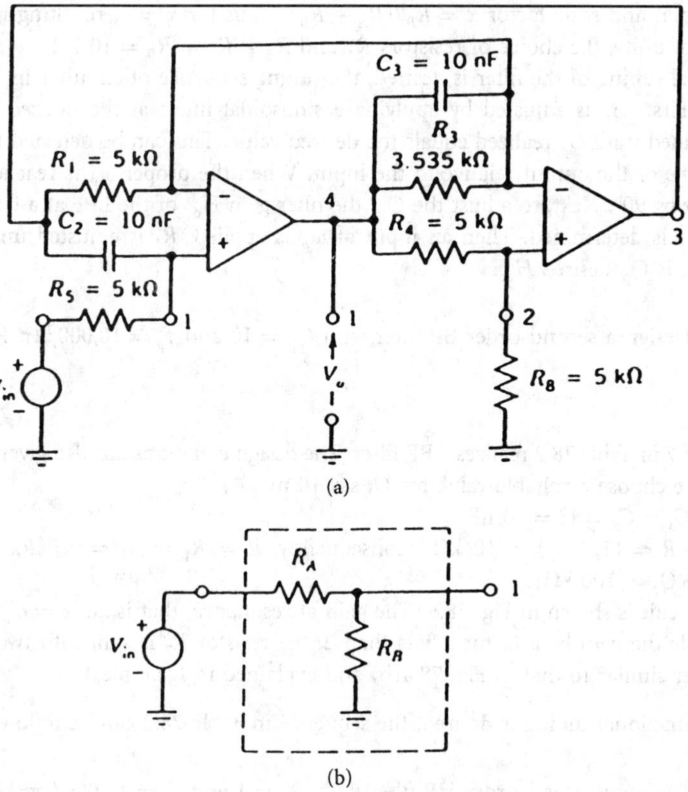

FIGURE 78.4 (a) Second-order Butterworth LPF design example; (b) controlling the gain factor of the LPF.

Procedure

1. Circuit 1 in Table 78.2 realizes an LPF. The design equations are also given in Table 78.2.
2. First we choose an appropriate value for C, say 10 nF. Thus $C = C_2 = C_3 = 10$ nF.
3. Now,

$$R = \frac{1}{C\omega_p} = \frac{1}{20{,}000 \times 10^{-8}}$$

   Therefore, $R = 5$ kΩ
4. Consequently, $R = R_1 = R_4 = R_5 = R_8 = 5$ kΩ and $R_3 = RQ_p = 3.535$ kΩ.
5. The circuit is shown in Fig. 78.4(a). It is noted that the low-frequency gain of the LP filter H_{LP}, is 2.

A simple procedure can be followed to scale H_{LP} by a factor x less than unity, that is, effectively multiplying the transfer function realized by x. This is done by replacing the resistance R_5 by two resistors R_A, and R_B (in series with the input V_{in}) in the manner shown in Fig. 78.4(b), where

$$\begin{aligned} R_5 &= R = 5\ \text{k}\Omega \\ &= R_A \parallel R_B \end{aligned}$$

The desired gain and scale factor $x = R_B/(R_A + R_B)$. Thus for $x = \frac{1}{2}$, resulting in a dc gain of the LP filter of unity, the choice of resistors R_A and R_B is $R_A = R_B = 10$ kΩ.

If functional tuning of the filter is desired, the tuning sequence of circuit 1 in Table 78.2 can be followed. First, ω_p is adjusted by applying a sinusoidal input at the desired ω_p frequency. Then R_B is tuned until ω_p realized equals the desired value. This can be detected by monitoring the phase angle of the output relative to the input. When the proper ω_p is reached, the output lags the input by 90°. Next, to adjust the Q_p, the filter gain H_{dc} of the LPF at a frequency much lower than ω_p is determined. Then an input at ω_p is applied. R_3 is adjusted until the gain of the LPF at ω_p is Q_p desired H_{dc}.

Example 2. Design a second-order BP filter with $Q_p = 10$ and $f_p = 10{,}000/2\pi$ Hz.

Procedure

1. Circuit 7 in Table 78.2 realizes a BP filter. The design equations are also given in Table 78.2.
2. First we choose a suitable value for C, say 10 nF.
3. Thus $C_3 = C_8 = C = 10$ nF.
4. Hence $R = (1/C\omega_p) = 10$ kΩ. Consequently, $R = R_1 = R_2 = R_4$, $R_6 = 10$ kΩ and $R_7 = RQ_p = 100$ kΩ.
5. The circuit is shown in Fig. 78.5. The gain at resonance, that is, at $\omega = \omega_p$, is equal to 2. To scale the gain by a factor x less than 2, the resistor R_7 is split into two resistors in a manner similar to that in Fig. 78.4(b) and explained in Example 1.

Again, if functional tuning is desired, the sequence in Table 78.2 can be followed.

Example 3. Design a second-order HP filter with $Q_p = 1$ and $f_p = 10{,}000/2\pi$ Hz.

Procedure

1. Circuit 3 in Table 78.2 realizes an HP filter.
2. Let us choose $C_3 = C_7 = C = 5$ nF. Hence R can be computed as

$$R = \frac{1}{\omega_p C} = \frac{1}{10\,000 \times 5 \times 10^{-9}}$$

Therefore, $R = 20$ kΩ.

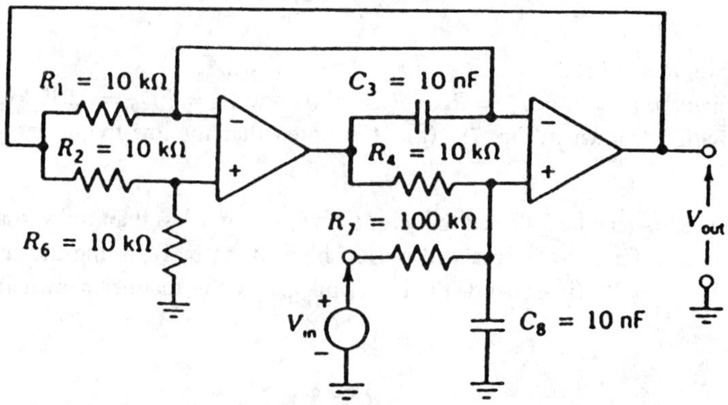

FIGURE 78.5 Design of a second-order BPF.

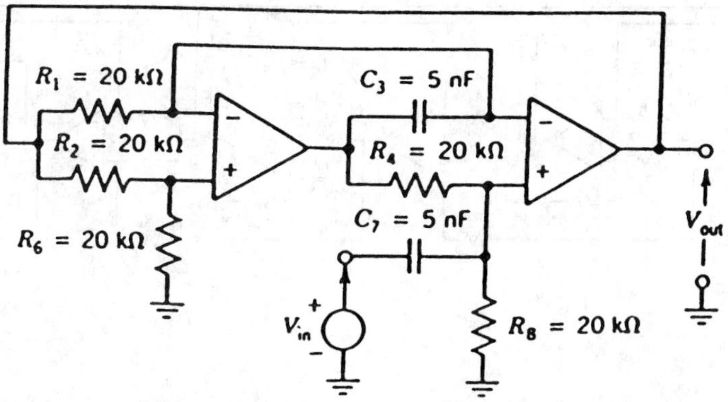

FIGURE 78.6 Design of a second-order HPF.

3. Consequently, $R_1 = R_2 = R_4 = R_6 = R = 20$ kΩ and $R_8 = RQ = 20$ kΩ.
4. The realization is shown in Fig. 78.6. The gain at high-frequency H_{HP} is equal to 2.

78.7 Practical High-Order Design Examples Using the 2-OA CGIC Biquad

Using Table 78.2, 1 percent metal-film resistors, 2 percent polystyrene capacitors and μA741 OA's, a sixth-order Chebyshev LP filter and a sixth-order elliptic BP filter were designed and constructed. The LP filter has a maximum passband attenuation of 1.0 dB; bandwidth = 3979 Hz. The BP filter has the following specifications:

Center frequency = 1500 Hz.
Passband = 60 Hz.
Maximum passband attenuation = 0.3 dB.
Minimum stopband attenuation outside the frequency range 1408 → 1595 Hz = 38 dB.

Low-Pass Filter

The realization uses cascaded section of type 1, in Table 78.2, as shown in Fig. 78.7(a). The measured frequency response (input level = 50 mV), shown in Fig. 78.7(b) and (c), agrees with the theoretical response. The effect of dc-supply variations is illustrated in Fig. 78.7(d). The deviation in the passband ripple is about 0.1 dB for supply voltage in the range 5 to 15 V. The effect of temperature variations is illustrated in Fig. 78.7(e), which shows the frequency response at −10°C (right-hand curve), 20°C, and 70°C (left-hand curve). The last peak has been displaced horizontally by 42 Hz, which corresponds to a change of 133 ppm/°C. The frequency displacement is due to passive element variations and is within the predicted value.

Bandpass Filter

The realization uses cascaded sections of the types 7 and 10, in Table 78.2, and is shown in Fig. 78.8(a). The measured frequency response is shown in Fig. 78.8(b) and (c), and it is in agreement with the theoretical response. Fig. 78.8(d) shows the frequency response for supply voltages of 7.5 V (lower curve) and 15 V; the input is 0.3 V. The passband ripple remains less than 0.39 dB and the deviation in the stopband is negligible. Fig. 78.8(e) and (f) illustrate the effect of temperature variations. The passband ripple remains less than 0.35 dB in the

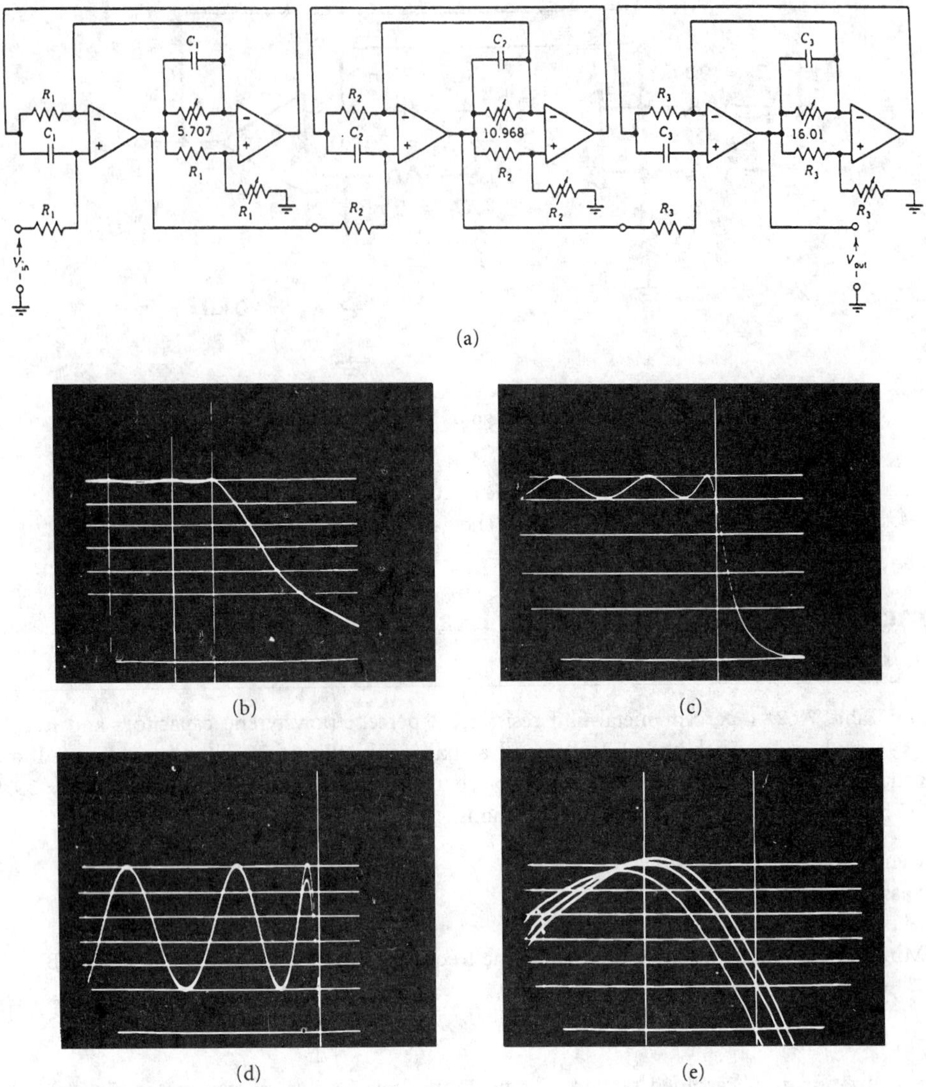

FIGURE 78.7 Realization of the sixth-order Chebyshev low-pass filter. Frequency responses: (b) Logarithmic gain scale and linear frequency scale; (c) linear gain and frequency scales; (d) for supply voltages ±5 V (lower curve) and ±15 V, input level = 0.05 V; (e) at temperatures −10°C (right-hand curve), 20°C and 70°C (left-hand curve).

temperature range −10 to 70°C. A center frequency displacement of 15 Hz has been measured that corresponds to a change of 125 ppm/°C.

78.8 Universal 2-OA CGIC Biquad

Study of Table 78.2 suggests that several circuits may be combined to form a universal biquad. This can be achieved on a single substrate using thick-film technology.

Upon examining the element identification and design values in Table 78.2 it is easy to see that one common thick-film substrate can be made to realize circuits 1, 3, 7, and 10 in Table 78.2 (other circuits from Table 78.1 can be included if desired) with no duplication in OA's and

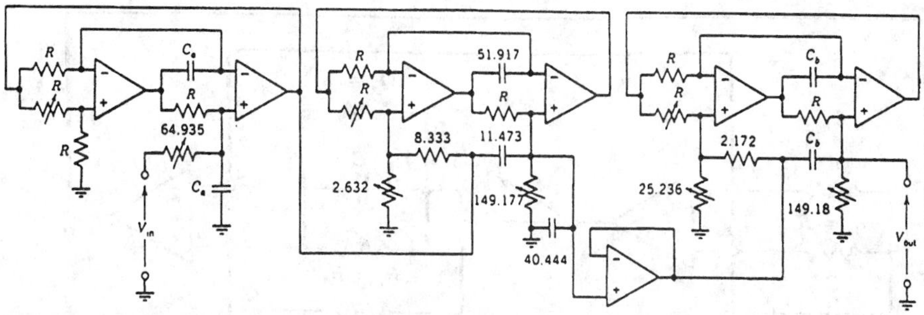

(a)

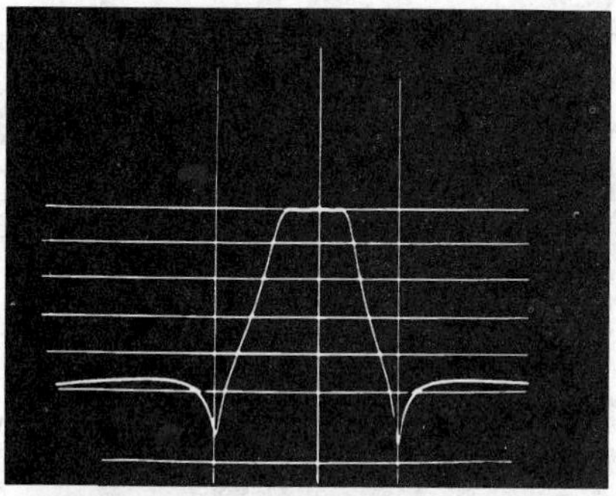

(b)

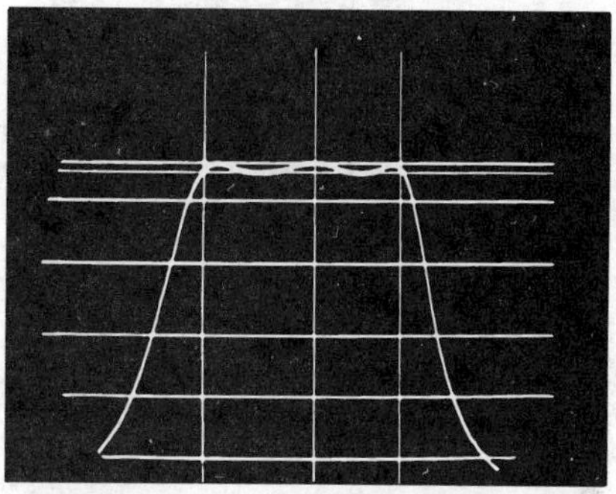

(c)

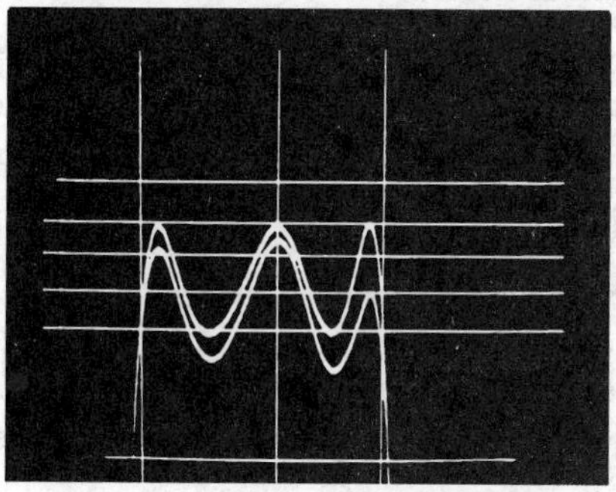

(d)

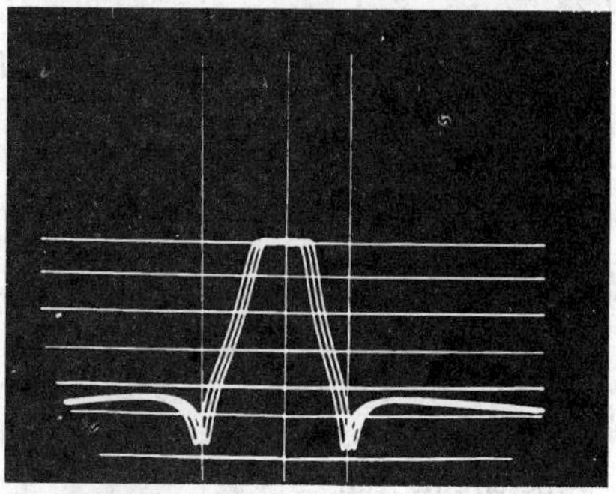

(e)

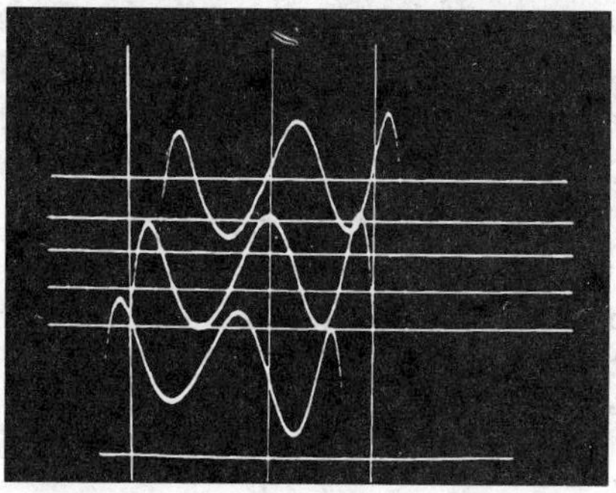

(f)

FIGURE 78.8 Realization of the sixth-order elliptic bandpass filter. Frequency responses of bandpass filter: (b) logarithmic gain and linear frequency scales; (c) linear gain and frequency scales; (d) for supply voltages of ±7.5 V (lower curve) and ±15 V, input level of 0.3 V; (e) at temperatures of 10°C (right-hand curve), 20°C, and 70°C (left-hand curve); and (f) expanded passband of Fig. 78.8(e).

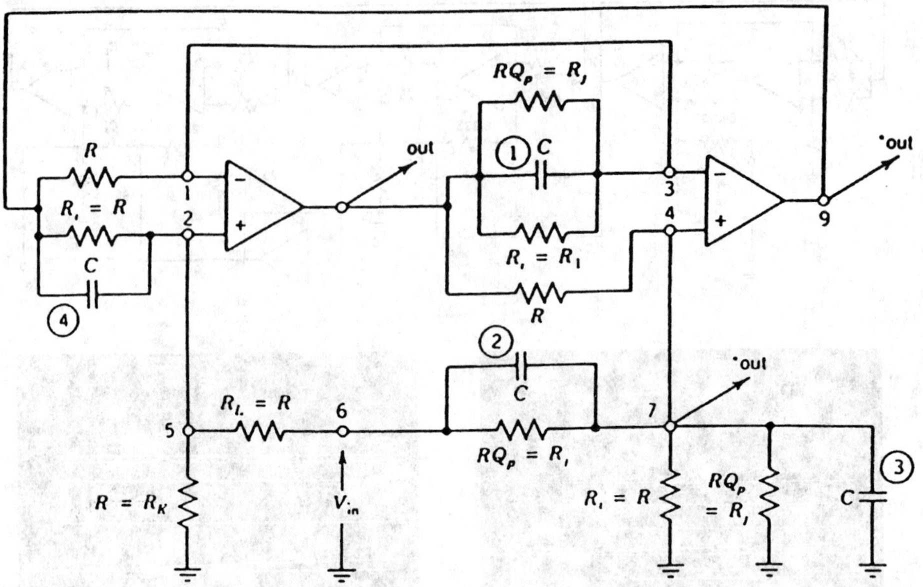

FIGURE 78.9 Superposition of circuits 1, 3, 7, and 10 from Table 78.2. Note: * = output terminals.

chip capacitors and minimal duplication in resistors. The superposition of circuits 1, 3, 7, and 10 is shown in Fig. 78.9. The following items should be noted.

1. Each resistor having the same subscript represents one resistor only and needs to appear once in a given biquad realization and thus once on the substrate. As an example, for $R_J = RQ_P$, only one R_J is needed with connection to several nodes. The unwanted connections may be opened during the trimming process according to the type of circuit required.
2. Three capacitor pads are needed; they are marked 1, 2, and 3 in Fig. 78.9. To obtain capacitor 4, either capacitor 2 or 3 connections are made common with capacitor 4 terminals. The capacitor pad terminal are available on the external terminals of the substrate.[1] The chip capacitors are reflow-soldered in the appropriate locations based on the circuit realized.

A dual CGIC universal biquad implemented using thick-film resistors, chip NPO capacitors, and one quad OA is shown in Fig. 78.10. Note that this hybrid array is capable of realizing gyrators and simulating inductors and super capacitors. Sample results are given in Fig. 78.11 for realizing different biquadratic functions using this implementation.

78.9 The 3-OA CGIC Biquadratic Structure

Consider the circuit shown in Fig. 78.12 where the output can be taken to be either V_1, V_2, or V_3. Assuming ideal OA's, the transfer functions between the input and output terminals 1, 2, and 3 can be readily obtained as

$$\begin{aligned} V_1/V_i &= T_1 \\ &= \{Y_{11}Y_{12}[Y_1(1 + Y_4/Y_9) - Y_2Y_3/Y_9] \\ &\quad +(Y_{12}Y_3Y_7Y_8)/Y_9 + [(1 + Y_4/Y_9)Y_5 - Y_3Y_6/Y_9]Y_7Y_{10}\}/D(s) \end{aligned} \tag{78.8a}$$

[1]This can readily be understood by examining the different realizations to be obtained from this layout.

FIGURE 78.10 Dual CGIC universal biquad implemented using thick-film resistors; chip NPO capacitors and one quad OA.

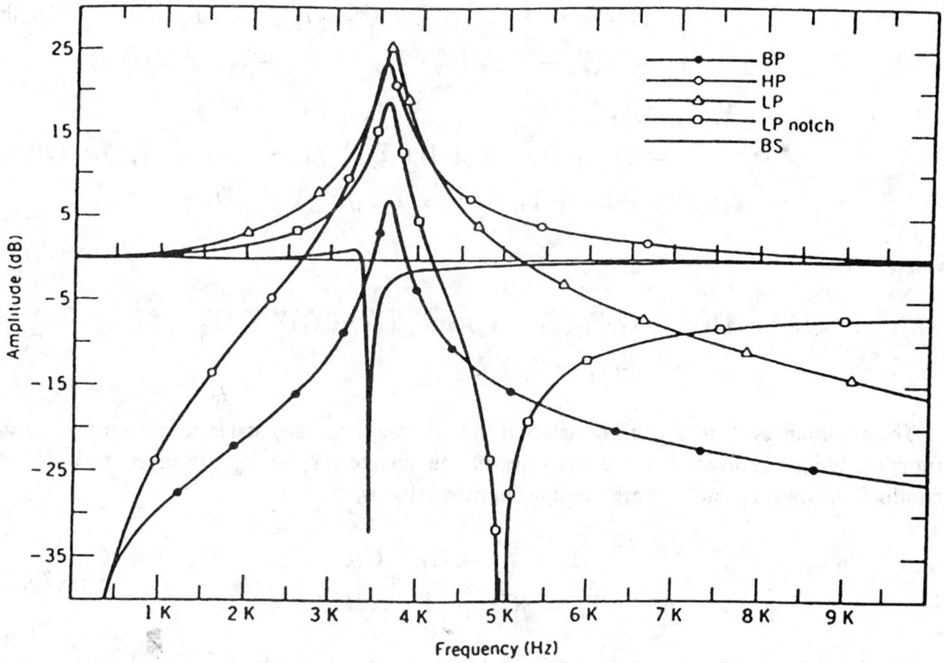

FIGURE 78.11 Different second-order realizations obtained from the universal CGIC biquad.

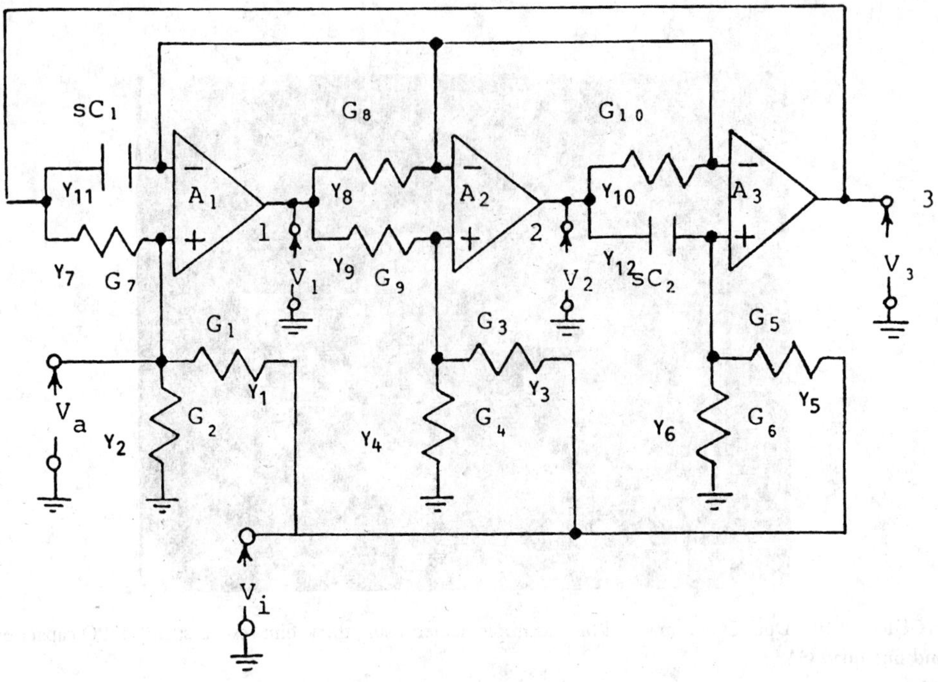

FIGURE 78.12 30A-CGIC Biquad.

$$
\begin{aligned}
V_2/V_i &= T_2 \\
&= \{Y_{11}Y_{12}Y_1 + Y_{12}[Y_1Y_6 + (Y_3Y_7Y_8)/Y_9 - Y_5Y_6] \\
&\quad + [Y_3Y_6 - Y_4Y_5](Y_7Y_8Y_{12})/Y_9Y_{11} + Y_5Y_7Y_{10}\}/D(s)
\end{aligned} \tag{78.8b}
$$

$$
\begin{aligned}
V_3/V_i &= T_3 \\
&= \{Y_1Y_{11}Y_{12} + Y_{12}[(1 + Y_2/Y_7)Y_3 - Y_1Y_4/Y_7]Y_8/Y_9 \\
&\quad + [(1 + Y_2/Y_7)Y_5 - Y_1Y_6/Y_7]Y_7Y_{10}\}D(s)
\end{aligned} \tag{78.8c}
$$

where

$$
\begin{aligned}
D(s) = {} & Y_{11}Y_{12}(Y_1 + Y_2) + Y_{12}(Y_3 + Y_4)Y_7Y_8/Y_9 \\
& + (Y_5 + Y_6)Y_7Y_{10}
\end{aligned} \tag{78.9}
$$

The admittances Y_1 to Y_{12} can be selected in many ways and any stable second-order transfer function can be realized. For the purposes of this chapter, Y_1 to Y_{10} are taken to be purely conductive, while Y_{11} and Y_{12} are purely capacitive, that is,

$$
\begin{gathered}
Y_1 - Y_{10} = G_1 - G_{10} \\
Y_{11} = sC_1 \qquad Y_{12} = sC_2
\end{gathered} \tag{78.10}
$$

Any rational and stable transfer function can be expressed as a product of second-order transfer functions of the form

$$
T(s) = (a_2s^2 + a_1s + a_0)/(b_2s^2 + b_1s + b_0) \tag{78.11}
$$

where $a_1 = a_2 = 0$, $a_0 = a_1 = 0$, $a_0 = a_2 = 0$, or $a_1 = 0$ for an LP, HP, BP, or N section, respectively. These section can be realized by choosing the G_i's ($i = 1$–10) properly in (78.8). By comparing (78.8) to (78.11), circuits 1–4 in Table 78.3 can be obtained.

All-pass transfer functions can be realized by setting $a_2 = b_2$, $a_1 = -b_1$, $a_0 = b_0$; these can be obtained from circuit 5 of Table 78.3.

It can be easily shown that this biquad possesses similar excellent low sensitivity properties and stability during activation as those of the 2-OA CGIC biquad, given in Section 78.4.

.10 Design and Tuning Procedure of the 3-OA CGIC Biquad

Several degrees of freedom exist in choosing element values, as shown in Table 78.3. These are used to satisfy the constraints of the given design, namely, those of low sensitivity, reduced dependence on the OA finite gain and bandwidth, and low element-spread design, in circuits 1–5 in Table 78.3. Using these constraints, a possible design for LP, HP, BP, AP, and N sections is obtained, as indicated in Table 78.4. It is seen that Q_p, ω_p, ω_n, Q_z, and ω_z, can be independently adjusted by trimming at most three resistors. A trimming sequence is also given in Table 78.4.

.11 A Practical Sixth-Order Elliptic BP Filter Design Using the 3-OA CGIC Biquad

The sixth-order elliptic bandpass filter specified in Chapter 78.7 was designed using the 3-OA CGIC Biquad and similar components to those in Chapter 78.7.

The realization shown in Fig. 78.13(a) uses cascaded sections of the types 3 and 4 in Table 78.4. The element design values are also given in Fig. 78.13(a). The measured frequency response is shown in Fig. 78.13(b) and (c); it is in agreement with the theoretical response. Fig. 78.13(d) shows the frequency response for supply voltages of ±7.5 V (lower curve), and ±15 V (upper curve); the input voltage is 0.3 V. The passband ripple remains less than 0.34 dB and the deviation in the stopband is negligible. Fig. 78.13(e) and (f) illustrate the effect of temperature variations. The passband ripple remains less than 0.5 dB in the temperature range −10°C (right-hand curve) to 70°C (left-hand curve). A center frequency displacement of 9 Hz has been measured, which corresponds to a change of 75 ppm/°C.

These results illustrate the additional performance improvements compared with the results in Section 78.7 using the 2-OA CGIC biquad.

.12 The Composite GIC Biquad: 2 C 2-OA CGIC

To obtain the composite CGIC biquad [7], each single OA in the original 2-OA CGIC biquad is replaced by a composite amplifier each constructed using two regular OA's [6] and denoted C2OA. All possible combinations of the four C2OA structures in [7] were used to replace the two OA's in the CIC network. Although several useful combinations were obtained, it was found that the best combination is shown in Fig. 78.14, where A_1 is replaced by C2OA-4 and A_2 is replaced by C2OA-3 in the CGIC of Fig. 78.1.

Computer simulation plots and experimental results of Fig. 78.15 show clearly the considerable improvements of the new CGIC filter responses over those of a 2-OA CGIC implemented using regular OA's.

Similar improvements in inductance simulation applications, employing the 2C2OA CGIC, have been reported.

TABLE 78.3 Element Identification for Realizing the Most Commonly Used Transfer Functions

Circuits Number	G_1	G_2	G_3	G_4	G_5	G_6	Transfer Function	Remarks
1	0		0			*	$T_3 = \dfrac{\left(1+\dfrac{G_2}{G_7}\right)\dfrac{G_5G_7G_{10}}{C_1C_2}}{s^2G_2 + s\dfrac{G_4G_7G_8}{C_1G_9} + \dfrac{(G_5+G_6)G_7G_{10}}{C_1C_2}}$	LP
2		*	0		0		$T_1 = \dfrac{s^2G_1\left(1+\dfrac{G_4}{G_9}\right)}{s^2(G_1+G_2) + s\dfrac{G_4G_7G_8}{C_1G_9} + \dfrac{G_6G_7G_{10}}{C_1C_2}}$	HP
3	0			*	0		$T_3 = \dfrac{s\left(1+\dfrac{G_2}{G_7}\right)\dfrac{G_3G_7G_8}{C_1G_9}}{s^2G_2 + s\dfrac{(G_3+G_4)G_7G_8}{C_1G_9} + \dfrac{G_6G_7G_{10}}{C_1C_2}}$	BP
4		*	0			*	$T_1 = G_1\left(1+\dfrac{G_4}{G_9}\right)\dfrac{s^2 + \dfrac{G_5G_7G_{10}}{C_1C_2G_1}}{s^2(G_1+G_2) + s\dfrac{G_4G_7G_8}{C_1G_9} + \dfrac{(G_5+G_6)G_7G_{10}}{C_1C_2}}$	N
5		*	0			0	$T_3 = \dfrac{s^2G_1 - s\dfrac{G_1G_4G_8}{C_1G_9} + \left(1+\dfrac{G_2}{G_7}\right)\dfrac{G_5G_7G_{10}}{C_1C_2}}{s^2(G_1+G_2) + s\dfrac{G_4G_7G_8}{C_1G_9} + \dfrac{G_5G_7G_{10}}{C_1C_2}}$	nonminimum phase†

$Y_7 - Y_{10} = G_7 - G_{10}$ always.
$Y_{11} - sC_1$, $Y_{12} = sC_2$.
*These elements can be set to zero.
†For all-pass $G_2 = 0$, $G_7 = G_1$.

TABLE 78.4 Design Values and Tuning Procedures

Circuit Number	Design Values	Transfer Function Realized	Tuning Sequence ω_n	ω_p	Q_p	Q_z
1	$R_2 = R$, $R_4 = RQ_p^{1/2}$, $R_5 = 2R/(\alpha H_{LP})$, $R_6 = R/[\alpha(1 - H_{LP}/2)]$, $H_{LP} \leq 2$	$T_3 = H_{LP}\dfrac{\omega_p^2}{D(s)}$	—	R_5	R_4	—
2	$R_1 = R(1 + Q_p^{-1/2})/H_{HP}$, $R_2 = R(1 + Q_p^{-1/2})/[1 + Q_p^{-1/2} - H_{HP}]$, $R_4 = RQ_p^{1/2}$, $R_6 = R/\alpha$, $H_{HP} \leq 1 + Q_p^{-1/2}$	$T_1 = H_{HP}\dfrac{s^2}{D(s)}$	—	R_6	R_4	—
3	$R_2 = R$, $R_3 = 2RQ_p^{1/2}H_{BP}$, $R_4 = RQ_p^{1/2}/(1 - H_{BP}/2)$, $R_6 = R/\alpha$, $H_{BP} \leq 2$	$T_3 = H_{BP}\dfrac{\frac{s\omega_p}{Q_p}}{D(s)}$	—	R_6	R_3	—
4	$R_1 = R(1 + Q_p^{-1/2})/H_N$, $R_2 = 1/(G - G_1)$, $R_4 = RQ_p^{1/2}$, $R_6 = 1/(\alpha G - G_5)$, $R_5 = R\omega_p^2(1 + Q_p^{-1/2})/(\alpha H_N \omega_n^2)$, $H_N \leq (1 + Q_p^{-1/2})$, $H_N \leq \omega_p^2(1 + Q_p^{-1/2})/\omega_n^2$ for $\omega_n > \omega_p$	$T_1 = H_N\dfrac{(s^2 + \omega_n^2)}{D(s)}$	R_5	R_6	R_4	—
5	For all-pass: $R_1 = R$, $R_4 = RQ_p^{1/2}$, $R_5 = R/\alpha$, $R_1 = R_7$, $R_2 = \infty$, $H_{AP} = 1$	$T_1 = H_{AP}\dfrac{s^2 - \frac{\omega_z}{Q_z}s + \omega_z^2}{D(s)}$		R_5	R_4	

$D(s) = s^2 + (\omega_p/Q_p)s + \omega_p^2$, $\alpha = 2Q_p^{1/2}/(1 + Q_p^{1/2})$.
$C_1 = C_2 = C = 1/(\omega_p R)$, $R_{10} = \alpha R$, $R_8 = RQ_p^{1/2}$, $R_7 = R_8 = R$.

SECTION NO.	R_1	R_2	R_3	R_4	R_5	R_6	R_7	R_8	R_9	R_{10}	C_1	C_2
1	-	2	11.396	-	-	1.175	2	11.396	2	3.403	53.052	53.052
2	8.939	2.576	-	17.273	4.592	1.426	2	17.273	2	3.585	51.917	51.917
3	4.309	3.732	-	17.273	2.611	1.948	2	17.273	2	3.585	54.211	54.211

Resistors in Kilo ohms and capacitors in Nanofarads

(a)

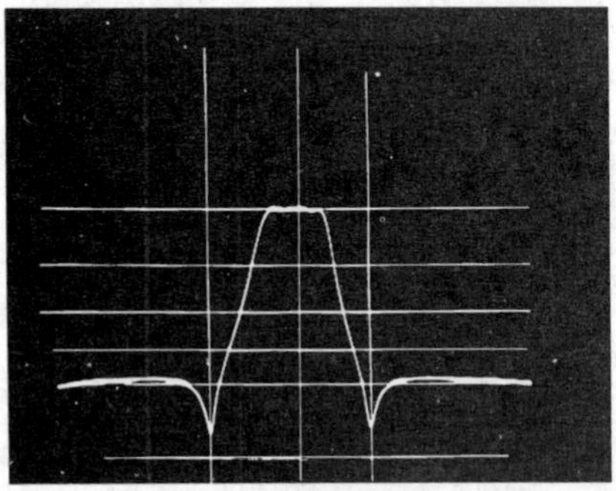

(b)

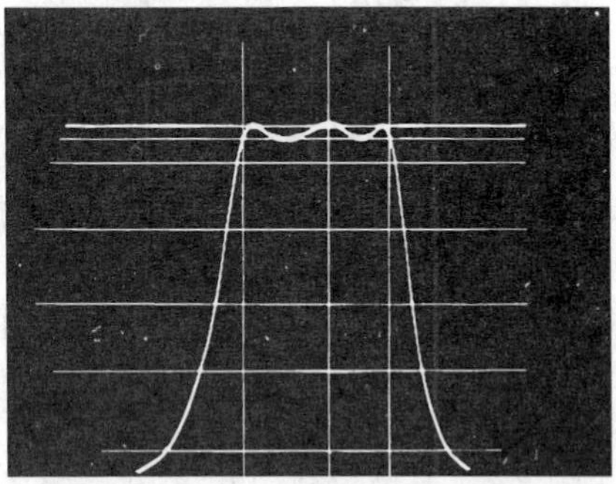

(c)

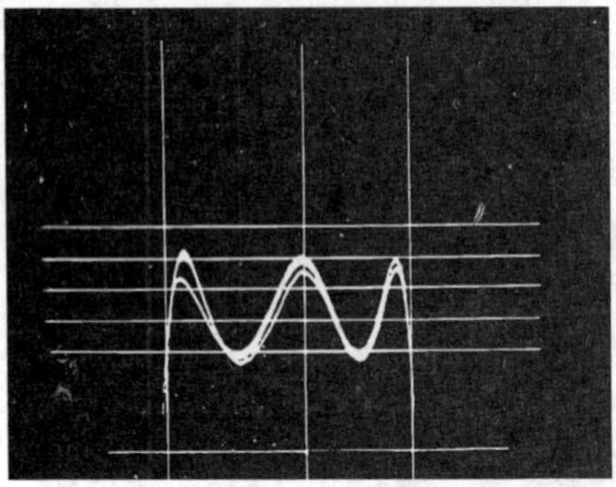

(d)

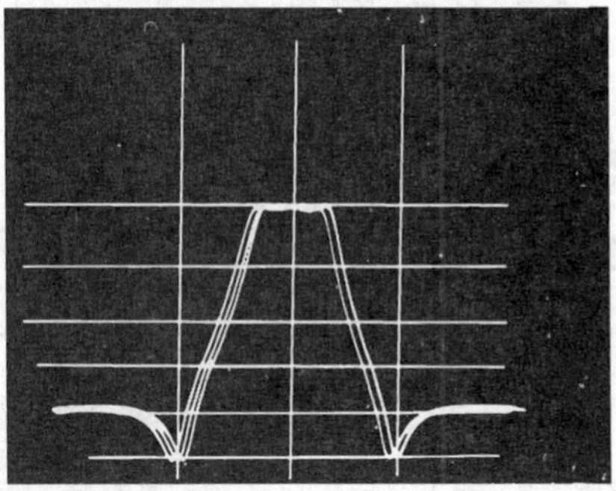

(e)

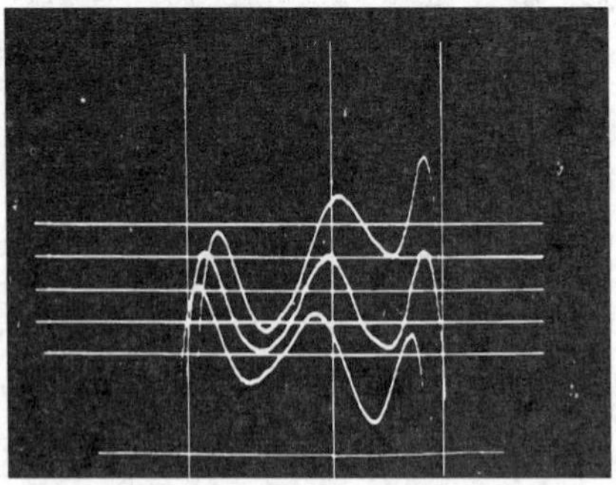

(f)

FIGURE 78.13 Sixth-order elliptic bandpass filter. (a) Resistors in kΩ and capacitors in nF. (b)–(f) Frequency response using 30A-CGIC. (b) Logarithmic gain and linear frequency scales. (c) Linear gain and frequency scales. (d) Frequency response for supply voltages of ±7.5 V (lower curve) and ±15 V, input level of 0.3 V. (e) Frequency response. (f) At temperatures of −10°C (right-hand curve), 20°C and 70°C (left-hand curve).

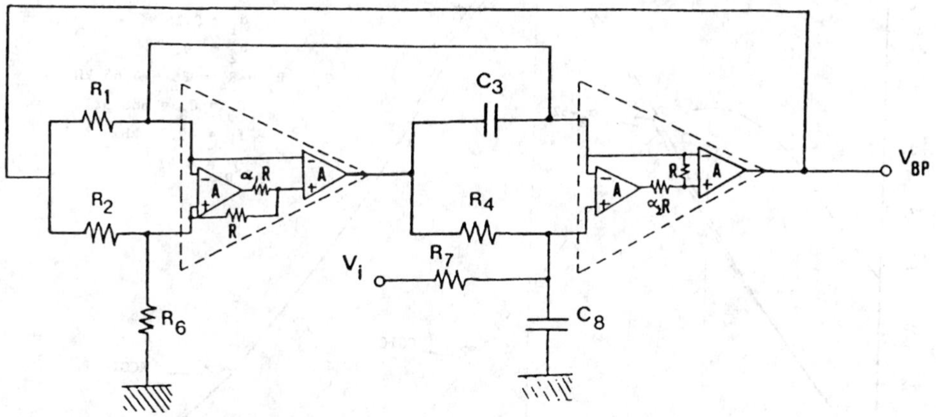

FIGURE 78.14 Practical BP filter realization of the composite GIC using C2OA-4 and C2OA-3.

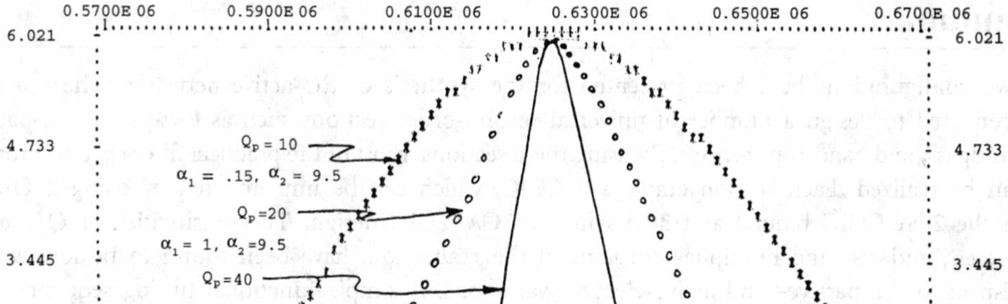

(a)

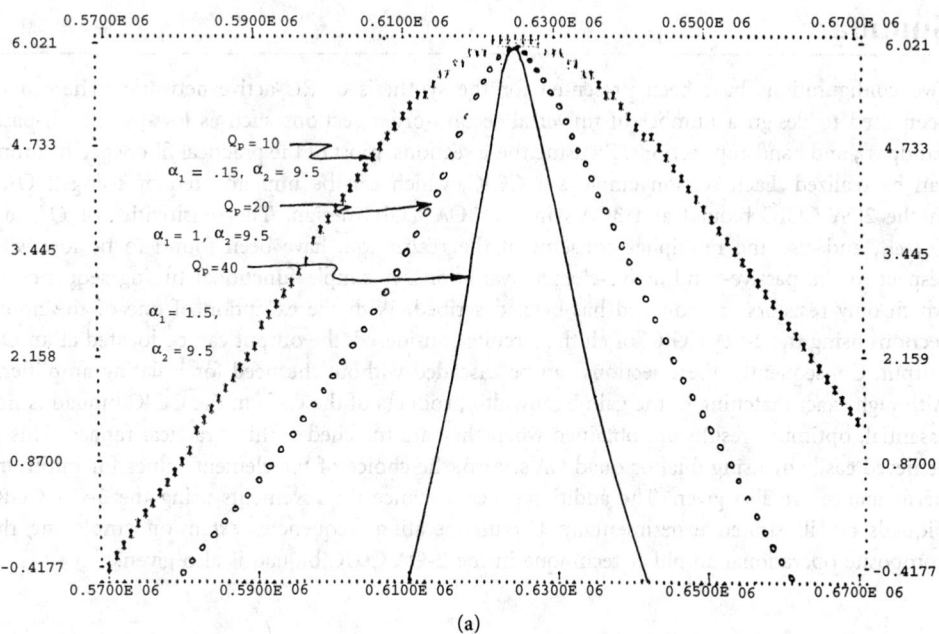

FIGURE 78.15 (a) Computer plots of the composite GIC BP filter frequency responses for $f_o = 100$ kHz and $Q_p = 10, 20, 40$. (Continued).

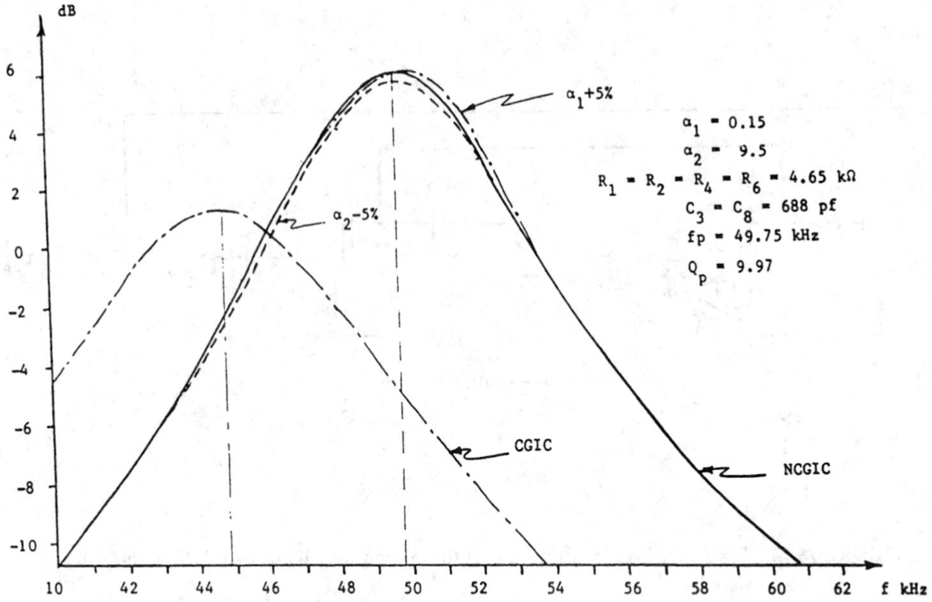

(b)

FIGURE 78.15 (Continued). (b) Experimental results of single and composite GIC BP filters of Fig. 6.3, and the response sensitivity for a_1 and a_2 variations ($f_o = 50$ kHz, $Q_p = 10$).

78.13 Summary

Two configurations have been presented for the synthesis of RC-active networks. These have been used to design a number of universal second-order sections such as low-pass, high-pass, bandpass, and bandstop sections. By using these sections, most of the practical filter specifications can be realized. Each section employs a CGIC, which can be implemented by using 2 OA's in the 2OA CGIC biquad and 3OA's in the 3OA CGIC design. The sensitivities of Q_p, ω_p, Q_z, ω_z, and also the multiplier constant of the realization, have been found to be low with respect to the passive- and active-element variations. A simple functional tuning sequence in which only resistors are adjusted has been described. With the exception of one of the notch sections using the 2-OA CGIC, of all the circuits considered, the output can be located at an OA output. Consequently, these sections can be cascaded without the need for isolating amplifiers. Although exact matching of the gain bandwidth products of the OA's in the CGIC biquad is not essential, optimum results are obtained when they are matched within practical ranges. This is achieved easily by using dual or quad OA's. Also, the choice of the element values for optimum performance are also given. The additional performance improvements using the 3-OA CGIC biquads are illustrated experimentally. Useful operating frequencies extension employing the composite operational amplifier technique in the 2-OA CGIC biquad is also given.

References

[1] W. B. Mikhael and B. B. Bhattacharyya, "A practical design for insensitive RC-active filters," *IEEE Trans. Circuits Syst.*, vol. CAS-21, pp. 75–78, Jan. 1974.

[2] A. Antoniou, "Realization of gyrators using operational amplifiers and their use in RC active network synthesis," *Proc. IEEE.* vol. 116, pp. 1838–1850, Nov. 1969.

[3] J. Valihora, "Modern technology applied to network implementation," in *Proc. IEEE Int. Symp. Circuit Theory*, Apr. 1972, pp. 169–173.

[4] A. Antoniou, "Stability properties of some gyrator circuits," *Electron Lett.*, vol. 4, pp. 510–512, 1968.

[5] B. B. Bhattacharyya, W. B. Mikhael, and A. Antoniou, "Design of *RC*-active networks by using generalized immittance converters," in *Proc. IEEE Int. Symp. Circuit Theory*, Apr. 1973, pp. 290–294.

[6] W. B. Mikhael and S. Michael, "Composite operational amplifiers: Generations and finite gain applications," *IEEE Trans. Circuit Syst.*, vol. CAS-34, pp. 449–460, May 1987.

[7] S. Micheal and W. B. Mikhael, "High frequency filtering and inductance simulation using new composite generalized immittance converters," in *Proc. IEEE ISCAS*, Kyoto, Japan, June 1985, pp. 299–300.

79

High-Order Filters

Rolf Schaumann
Portland State University, Oregon

79.1 Introduction

With the realization of second-order filters discussed in the previous sections of this chapter, we shall now treat methods for practical filter implementations of order higher than two. Specifically, we shall investigate how to realize efficiently, with low sensitivities to component tolerances, the input-to-output voltage transfer function

$$H(s) = \frac{V_{out}}{V_{in}} = \frac{N(s)}{D(s)} = \frac{a_m s^m + a_{m-1}s^{m-1} + \cdots + a_1 s + a_0}{s^n + b_{n-1}s^{n-1} + \cdots + b_1 s + b_0} \tag{79.1}$$

where $n \geq m$ and $n > 2$. The sensitivity behavior of high-order filter realizations shows that, in general, it is not advisable to realize the transfer function $H(s)$ in the so-called **direct form** [5, ch. 3] (see also Chapter 68 in this book). By direct form we mean an implementation of (79.1) that uses only one or maybe two active devices, such as operational amplifiers (op-amps) or operational transconductance amplifiers (OTAs), embedded in a high-order passive RC network. Although it is possible in principle to realize (79.1) in direct form, the resulting circuits are normally so sensitive to component tolerances as to be impractical. Since the direct form for the realization of higher-order functions is ruled out, we shall in this section present those methods that result in designs of practical manufacturable active filters with acceptably low sensitivity, the *cascade* approach, the *multiple-loop feedback topology*, and *ladder simulations*. Both cascade and multiple-loop feedback techniques are modular, with active biquads used as the fundamental building blocks. The ladder simulation method seeks active realizations that inherit the low passband sensitivity properties of passive doubly-terminated LC ladder filters.

In the *cascade* approach to be discussed first, a high-order function $H(s)$ is factored into low- (first- or second-) order subnetworks that are realized as discussed in the previous chapters of this section and connected in cascade such that their product implements the prescribed function $H(s)$. The method is widely employed in industry; it is well understood, very easy to design, and efficient in its use of active devices. It uses a modular approach and results in filters that for the most part show satisfactory performance in practice. The main advantages

0-8493-8341-2/95/$0.00 + $.50

of cascade filters are their generality, i.e., any arbitrary stable transfer function can be realized as a cascade circuit, and very easy tuning because each biquad is responsible for the realization of only *one* pole pair (and zero pair): the realizations of the individual critical frequencies of the filter are *decoupled* from each other. The disadvantage of this decoupling is that for filters of high order, say $n > 8$, with stringent requirements and tight tolerances, cascade designs are often found to remain still too sensitive to component variations in the passband. In these cases, the following approaches lead to more reliable circuits.

The *multiple-loop feedback* or *coupled-biquad methods* also split the high-order transfer function into second-order subnetworks. These are interconnected in some type of feedback configuration that introduces *coupling* chosen to reduce the transfer function sensitivities. The multiple-loop feedback approach retains the modularity of cascade designs but at the same time yields high-order filter realizations with noticeably better passband sensitivities. Of the numerous topologies that have been proposed in the literature, see e.g., [5, ch. 6], we shall discuss only the FLF (follow-the-leader feedback) and the LF (leapfrog) methods. Both are particularly well suited for all-pole characteristics but can be extended to realizations of general high-order transfer functions. Although based on coupling of biquads in a feedback configuration, the LF procedure is actually derived from an *LC* ladder simulation and will, therefore, be treated as part of the following method.

As the name implies, the *ladder simulation approach* uses an active circuit to simulate the behavior of doubly terminated *LC* ladders in an attempt to inherit their excellent low passband sensitivity properties. The methods fall into two groups. One is based on *element substitution*, where the inductors are simulated via electronic circuits, and the resulting active "components" are inserted into the *LC* filter topology. The second group may be labeled *operational simulation* of the *LC* ladder where the active circuit is configured to realize the *internal operation*, i.e., the equations, of the *LC* prototype. Active filters simulating the structure or the behavior of *LC* ladders have been found to have the lowest sensitivities among active filters and, consequently, to be the most appropriate if filter requirements are stringent. They can draw on the wealth of information available for lossless filters, e.g., [5, ch. 2], [7], that can be used directly in the design of active ladder simulations. A disadvantage of this design method is that a passive *LC* prototype must, of course, exist[1] before an active simulation can be attempted.

79.2 Cascade Realizations

Without loss of generality we may assume in our discussion of *active* filters that the polynomials $N(s)$ and $D(s)$ of (79.1) are *even*, i.e., both n and m are even. An odd function can always be factored into the product of an even function and a first-order function, where the latter can easily be realized by a *passive RC* network and can be appended to the high-order active filter as an additional section. Thus, we can factor (79.1) into the product of second-order pole–zero pairs, so that the high-order transfer function $H(s)$ is factored into the product of the second-order functions

$$T_i(s) = k_i \frac{\alpha_{2i}s^2 + \alpha_{1i}s + \alpha_{0i}}{s^2 + s\omega_{0i}/Q_i + \omega_{0i}^2} = k_i t_i(s) \tag{79.2}$$

[1]The realizability conditions for passive *LC* filters are more restrictive than those for active *RC* filters; specifically, the numerator $N(s)$ of (79.1) must be strictly even or odd so that only $j\omega$-axis transmission zeros can be realized.

such that

$$H(s) = \prod_{i=1}^{n/2} T_i(s) = \prod_{i=1}^{n/2} k_i \frac{\alpha_{2i}s^2 + \alpha_{1i}s + \alpha_{0i}}{s^2 + s\omega_{0i}/Q_i + \omega_{0i}^2} = \prod_{i=1}^{n/2} k_i t_i(s) \qquad (79.3)$$

In (79.2) ω_{0i} is the pole frequency and Q_i the pole quality factor; the coefficients α_{2i}, α_{1i}, and α_{0i} determine the type of second-order function $T_i(s)$, which can be realized by an appropriate choice of biquad from the literature, e.g., [5, ch. 5] (see also Chapters 76.3 and 76.4). The transfer functions $T_i(s)$ of the individual biquads are scaled by a suitably defined *gain constant* k_i, e.g., such that the leading coefficient in the numerator of the *gain-scaled* transfer function $t_i(s)$ is unity or such that $|t_i(j\omega)| = 1$ at some desired frequency. If we assume now that the output impedances of the biquads are small compared to their input impedances, all second-order blocks can be connected in cascade as in Fig. 79.1 without causing mutual interactions due to loading. The product of the biquadratic functions is then realized as required by (79.3).

The process is straightforward and leads to a possible cascade design, but several questions must still be answered.

1. Which pair of zeros of (79.1) should be assigned to which pole-pair when the biquadratic functions $T_i(s)$ are formed? Since we have $n/2$ pole pairs and $n/2$ zero pairs (counting zeros at 0 and at ∞) we can select from $(n/2)!$ possible *pole–zero pairings.*
2. In which order should the biquads be cascaded? For $n/2$ biquads, we have $(n/2)!$ possible ways of *section ordering.*
3. How should the gain constants k_i in (79.2) be chosen to determine the signal level for each biquad? In other words, what is the optimum *gain assignment*?

Because the total prescribed transfer function $H(s)$ is simply the product of $T_i(s)$, the choices in 1, 2, and 3 are quite arbitrary as far as $H(s)$ is concerned. However, they determine significantly the dynamic range,[2] i.e., the distance between the maximum possible undistorted signal (limited by the active devices) and the noise floor, because the maximum and minimum signal levels through the cascade filter depend on the pole–zero pairings, cascading sequence, and gain assignment.

There exist well-developed techniques and algorithms for answering the questions of pole–zero pairing, section ordering, and gain assignment exactly [3, ch. 1], [5, ch. 6]; they rely heavily on computer routines and are too lengthy to be treated in this text. We shall only give a few rules-of-thumb that are based on the intuitive observation that in the passband the magnitude of each biquad should vary as little as possible. This keeps signal levels as equal as possible versus frequency and avoids in-band attenuation and the need for subsequent amplification that will at the same time raise the noise floor and thereby further limit the dynamic range. Note also, that in-band signal amplification may overdrive the amplifier stages and cause distortion. The simple rules below provide generally adequate designs.

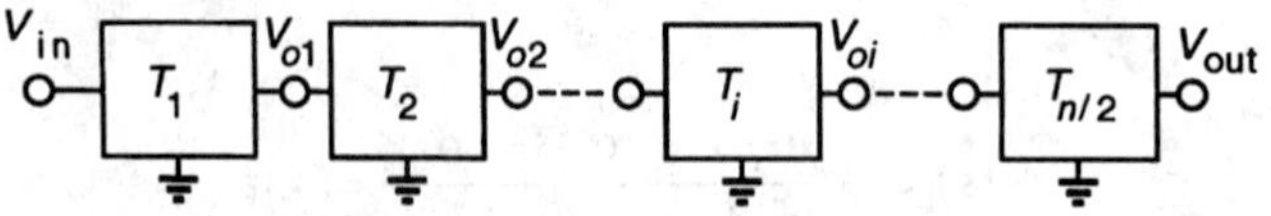

FIGURE 79.1 Cascade realization of an nth-order transfer function; n is assumed even.

[2]Pole–zero pairing also affects to some extent the sensitivity performance, but the effect usually is not very strong and will not be discussed in this text. For a detailed treatment see [3].

1. Assign each zero or zero pair to the closest pole pair.
2. Sequence the sections in the order of increasing values of Q_i, i.e., $Q_1 < Q_2 < \cdots < Q_{n/2}$, so that the section with the flattest transfer function magnitude comes first, the next flattest one follows, and so on. If the position of any section is predetermined,[3] use the sequencing rule for the remaining sections.

After performing steps 1 and 2, assign the gain-scaling constants k_i such that the maximum output signals of all sections in the cascade are the same, i.e.,

$$\max |V_{oi}(j\omega)| = \max |V_{o,n/2}(j\omega)| = \max |V_{\text{out}}(j\omega)| \qquad i = 1, \cdots, n/2 - 1 \quad (79.4)$$

This can be done readily with the help of a network analysis routine or simulator, such as SPICE, by computing the output signals at each biquad: using the notation, from (79.3),

$$H(s) = \prod_{i=1}^{n/2} T_i(s) = \prod_{i=1}^{n/2} k_i t_i(s) = \prod_{i=1}^{n/2} k_i \prod_{i=1}^{n/2} t_i(s) \quad (79.5)$$

and

$$H_i(s) = \prod_{j=1}^{i} T_j(s) = \prod_{j=1}^{i} k_j t_j(s) = \prod_{j=1}^{i} k_j \prod_{j=1}^{i} t_j(s) \quad (79.6)$$

we label the total prescribed gain constant

$$K = \prod_{i=1}^{n/2} k_i \quad (79.7)$$

such that

$$\max \left| \prod_{i=1}^{n/2} t_i(j\omega) \right| = M_{n/2} \quad (79.8a)$$

is some given value. Further, let us denote the maxima of the intermediate gain-scaled transfer functions by M_i, i.e.,

$$\max \left| \prod_{k=1}^{i} t_k(j\omega) \right| = M_i \qquad i = 1, \cdots, n/2 - 1 \quad (79.8b)$$

then we obtain [5, ch. 6]

$$k_1 = K \frac{M_{n/2}}{M_1} \quad \text{and} \quad k_j = \frac{M_{j-1}}{M_j} \qquad j = 2, \cdots, n/2 \quad (79.9)$$

Choosing the gain constants as in (79.9) guarantees that the same maximum voltage appears at all biquad outputs to assure that the largest possible signal can be processed without distortion.

[3]Such as, e.g., placing a lowpass at the input may be preferred to avoid unnecessary high-frequency signals from entering the filter.

To illustrate the process, let us realize the sixth-order transfer function

$$H(s) = \frac{0.7560s^3}{(s^2 + 0.5704s + 1)(s^2 + 0.4216s + 2.9224)(s^2 + 0.1443s + 0.3422)} \tag{79.10}$$

where the frequency parameter s is normalized with respect to $\omega_n = 130.59 \cdot 10^3\ \text{s}^{-1}$. It defines a sixth-order bandpass filter with a 1-dB equiripple passband in 12 kHz $\leq f \leq$ 36 kHz and at least 25 dB attenuation in $f \leq$ 4.8 kHz and $f \geq$ 72 kHz [5, ch. 1]. $H(s)$ can be factored into the product of

$$T_1(s) = k_1 t_1(s) = \frac{k_1 s}{s^2 + 0.5704s + 1} \quad \text{with} \quad Q_1 = \frac{\sqrt{1}}{0.5704} \approx 1.75 \tag{79.11a}$$

$$T_2(s) = k_2 t_2(s) = \frac{k_2 s}{s^2 + 0.4216s + 2.9224} \quad \text{with} \quad Q_2 = \frac{\sqrt{2.9224}}{0.4216} \approx 4.06 \tag{79.11b}$$

$$T_3(s) = k_3 t_3(s) = \frac{k_3 s}{s^2 + 0.1443s + 0.3422} \quad \text{with} \quad Q_3 = \frac{\sqrt{0.3422}}{0.1443} \approx 4.06 \tag{79.11c}$$

which are cascaded in the order of increasing values of Q, i.e., $T_1 T_2 T_3$. We can compute readily the maximum values at the output of the sections in the cascade as $M_1 = |t_1|_{\max} \approx 1.75$, $M_2 = |t_1 t_2|_{\max} \approx 1.92$ and $M_3 = |t_1 t_2 t_3|_{\max} \approx 1.32$ to yield by (79.9)

$$k_1 = 0.7560 \frac{1.32}{1.75} \approx 0.57 \quad k_2 = \frac{1.75}{1.92} \approx 0.91 \quad k_3 = \frac{1.92}{1.32} \approx 1.45 \tag{79.12}$$

Let us build the sections with the Åckerberg–Mossberg circuit shown in Fig. 79.2, which realizes

$$\frac{V_o}{V_i} = -\frac{k\omega_0^2}{s^2 + s\omega_0/Q + \omega_0^2} \quad \text{with} \quad \omega_0 = \frac{1}{RC} \tag{79.13}$$

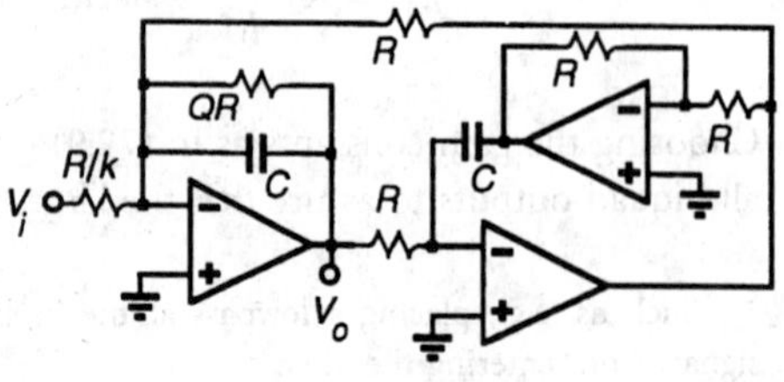

FIGURE 79.2 Åckerberg–Mossberg bandpass circuit.

We obtain readily from (79.11)–(79.13) the following component values:

$$\text{for } T_1: R = 1.53\ \text{k}\Omega,\quad QR = 2.68\ \text{k}\Omega,\ R/k = 2.68\ \text{k}\Omega$$
$$\text{for } T_2: R = 0.896\ \text{k}\Omega,\ QR = 3.63\ \text{k}\Omega,\ R/k = 1.68\ \text{k}\Omega$$
$$\text{for } T_3: R = 2.62\ \text{k}\Omega,\quad QR = 10.6\ \text{k}\Omega,\ R/k = 1.06\ \text{k}\Omega$$

The three sections are then interconnected in cascade in the order $T_1T_2T_3$.

79.3 Multiple-Loop Feedback Realizations

These topologies are also based on biquad building blocks that are then embedded, as the name implies, into multiple-loop resistive feedback configurations. The resulting coupling between sections is selected such that transfer function sensitivities are reduced below those of cascade circuits. It has been shown that the sensitivity behavior of the different available configurations is comparable; we shall, therefore, concentrate our discussion only on the follow-the-leader feedback (FLF) and, as part of the ladder simulation techniques, on the leapfrog (LF) topologies, which have the advantage of being relatively easy to derive without any sacrifice in performance. Our derivation will reflect the fact that both configurations[4] are particularly convenient for geometrically symmetrical bandpass functions and that the LF topology is obtained from a direct simulation of an *LC* low-pass ladder.

The Follow-the-Leader Feedback Topology

The follow-the-leader feedback (FLF) topology consists of a cascade of biquads whose outputs are fed back into a summer at the filter's input. At the same time the biquad outputs may be fed forward into a second summer at the filter's output to permit an easy realization of arbitrary transmission zeros. The actual implementation of the summers and the feedback factors is shown in Fig. 79.3; if there are n noninteracting biquads, the order of the realized transfer

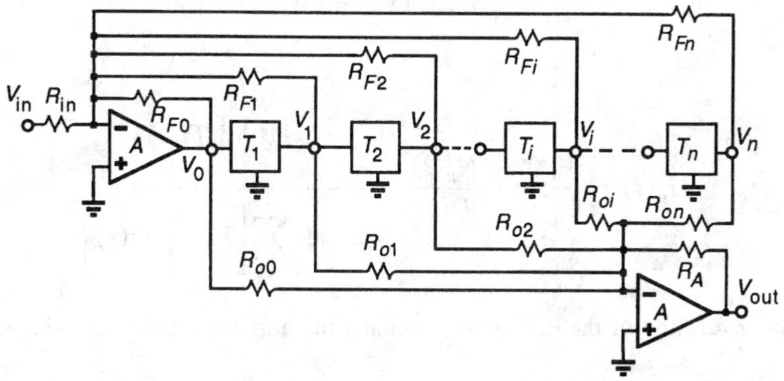

FIGURE 79.3 FLF circuit built from second-order sections $T_i(s)$ and a feedback network consisting of an op-amp summer with resistors R_{Fi}. Also shown is an output summer with resistors R_{oi} to facilitate the realization of arbitrary transmission zeros.

[4]As are all other multiple-loop feedback circuits.

function $H(s)$ is $2n$. Assuming that the two summer op-amps are ideal, routine analysis yields

$$-V_0 = \frac{R_{F0}}{R_{\text{in}}} V_{\text{in}} + \sum_{i=1}^{n} \frac{R_{F0}}{R_{Fi}} V_i = \alpha V_{\text{in}} + \sum_{i=1}^{n} F_i V_i \tag{79.14}$$

where we defined $\alpha = R_{F0}/R_{\text{in}}$ and the feedback factors $F_i = R_{F0}/R_{Fi}$. Similarly, we find for the output summer

$$V_{\text{out}} = -\sum_{i=0}^{n} K_i V_i = -\sum_{i=0}^{n} \frac{R_A}{R_{oi}} V_i \tag{79.15}$$

from which the definition of the resistor ratios K_i is apparent. Any of the parameters F_i and K_i may, of course, be reduced to zero by replacing the corresponding resistor, R_{Fi} or R_{oi}, respectively, by an open circuit. Finally, the internal voltages V_i can be computed from

$$V_i = V_0 \prod_{j=1}^{i} T_j(s) \qquad i = 1, \cdots, n \tag{79.16}$$

so that with (79.14)

$$H_0(s) = \frac{V_0}{V_{\text{in}}} = -\frac{\alpha}{1 + \sum_{k=1}^{n} \left[F_k \prod_{j=1}^{k} Tj(s) \right]} \tag{79.17}$$

which with (79.16) yields

$$H_i(s) = \frac{V_i}{V_{\text{in}}} = -\frac{N_i(s)}{D(s)} = -\frac{\alpha \prod_{j=1}^{i} Tj(s)}{1 + \sum_{k=1}^{n} \left[F_k \prod_{j=1}^{k} Tj(s) \right]} \qquad i = 1, \cdots, n \tag{79.18}$$

Note that

$$H_n(s) = \frac{V_n}{V_{\text{in}}} = -\frac{N_n(s)}{D(s)} = -\frac{\alpha \prod_{j=1}^{n} Tj(s)}{1 + \sum_{k=1}^{n} \left[F_k \prod_{j=1}^{k} Tj(s) \right]} \tag{79.19}$$

is the transfer function of the FLF network *without* the output summer, i.e., with $R_{oi} = \infty$ for all i.

By (79.19), the transmission zeros of $H_n(s)$ are set by the zeros of $T_j(s)$, i.e., by the *feed-forward* path, whereas the poles of $H_n(s)$ are determined by the *feedback* network and involve both the poles and zeros of the biquads $T_j(s)$ and the feedback factors F_k. Typically, an FLF network is designed with second-order bandpass biquads

$$T_i(s) = A_i \frac{s/Q_i}{s^2 + s/Q_i + 1} = A_i t_i(s) \tag{79.20}$$

so that $H_n(s)$ has all transmission zeros at the origin, i.e., it is an all-pole bandpass function. Note that in (79.20) the frequency parameter s is normalized to the pole frequency ω_0 ($\omega_{0i} = \omega_0$ for all i is assumed), and Q_i and A_i are the section's pole quality factor and midband gain, respectively. Designing an FLF network with arbitrary zeros requires second-order sections with finite transmission zeros, which leads to quite difficult design procedures. It is much simpler to use (79.15) with (79.18) to yield

$$H(s) = \frac{V_{\text{out}}}{V_{\text{in}}} = \frac{N_{2n}(s)}{D_{2n}(s)} = \alpha \frac{K_0 + \sum_{k=1}^{n}\left[K_k \prod_{j=1}^{k} Tj(s)\right]}{1 + \sum_{k=1}^{n}\left[F_k \prod_{j=1}^{k} Tj(s)\right]} \tag{79.21}$$

which realizes the transfer function of the complete circuit in Fig. 79.3 with an arbitrary numerator polynomial $N_{2n}(s)$, even for second-order bandpass functions $T_i(s)$ as in (79.20). It is a ratio of two polynomials whose roots can be set by an appropriate choice of the functions $T_i(s)$, the parameters K_i for the transmission zeros, and the feedback factors F_i for the poles.

We shall illustrate the design procedure by considering a specific case. For $n = 3$, (79.21) becomes

$$H(s) = \alpha \frac{K_0 + K_1T_1 + K_2T_1T_2 + K_3T_1T_2T_3}{1 + F_1T_1 + F_2T_1T_2 + F_3T_1T_2T_3} \tag{79.22}$$

Next we transform the bandpass functions $T_i(s)$ in (79.20) into low-pass functions by the low-pass-to-bandpass transformation (see Chapter 67)

$$p = Q\frac{s^2 + 1}{s} \tag{79.23}$$

where $Q = \omega_0/B$ is the "quality factor" of the high-order bandpass with bandcenter ω_0 and bandwidth B, and p is the normalized lowpass frequency. This step transforms the bandpass functions (79.20) with *all identical* pole frequencies ω_0 into the first-order low-pass functions

$$T_{i\text{LP}}(p) = \frac{A_iQ/Q_i}{p + Q/Q_i} = A_i\frac{q_i}{p + q_i} \tag{79.24}$$

where $q_i = Q/Q_i$ and A_i is the dc gain of the lowpass section. Applying (79.23) also to the prescribed function $H(s)$ of order $2n$ in (79.22) converts it into a prototype low-pass function $H_{\text{LP}}(p)$ of order n. Substituting (79.24) into the numerator and denominator expressions of that function, of order $n = 3$ in our case, shows that the zeros and poles, respectively, are determined by

$$N_3(p) = \alpha K_0 \prod_{j=1}^{3}(p + q_j) + \sum_{j=1}^{2}\left[k_j \prod_{i=j+1}^{3}(p + q_i)\right] + k_3 \tag{79.25a}$$

and

$$D_3(p) = \prod_{j=1}^{3}(p + q_j) + \sum_{k=1}^{2}\left[f_k \prod_{i=k+1}^{3}(p + q_i)\right] + f_3 \tag{79.25b}$$

where we introduced the abbreviations

$$f_i = F_i \prod_{j=1}^{i} A_j q_j \quad \text{and} \quad k_i = \alpha K_i \prod_{j=1}^{i} A_j q_j \tag{79.26}$$

To realize the prescribed third-order function

$$H_{\text{LP}}(p) = \frac{V_o}{V_i} = \frac{a_3 p^3 + a_2 p^2 + a_1 p + a_0}{p^3 + b_2 p^2 + b_1 p + b_0} \tag{79.27}$$

we compare coefficients between (79.25) and (79.27). For the denominator terms we obtain

$$\begin{aligned} b_2 &= q_1 + q_2 + q_3 + f_1 \\ b_1 &= q_1 q_2 + q_1 q_3 + q_2 q_3 + f_1(q_2 + q_3) + f_2 \\ b_0 &= q_1 q_2 q_3 + f_1 q_2 q_3 + f_2 q_3 + f_3 \end{aligned}$$

These are three equations in six unknowns, f_i and q_i, $i = 1, \cdots, 3$, which can be written more conveniently in matrix form:

$$\begin{pmatrix} 1 & 0 & 0 \\ q_2 + q_3 & 1 & 0 \\ q_2 q_3 & q_3 & 1 \end{pmatrix} \begin{pmatrix} f_1 \\ f_2 \\ f_3 \end{pmatrix} = \begin{pmatrix} b_2 - (q_1 + q_2 + q_3) \\ b_1 - (q_1 q_2 + q_1 q_3 + q_2 q_3) \\ b_0 - q_1 q_2 q_3 \end{pmatrix} \tag{79.28}$$

The transmission zeros are found via an identical process: the unknown parameters k_i are computed from an equation of the form (79.28) with f_i replaced by $k_i/(\alpha K_0)$ and b_i replaced by a_i/a_3, $i = 1, \cdots, 3$. Also, $K_0 = \alpha_3/\alpha$.

The unknown parameters f_i can be solved from the matrix expression (79.28). It is a set of *linear* equations whose coefficients are functions of the prescribed coefficients b_i and of the numbers q_i, which for given Q are determined by the quality factors Q_i of the second-order sections $T_i(s)$. Thus, the Q_i are *free* parameters that may be selected to satisfy any criteria, which may lead to a better-working circuit. The free design parameters may be chosen for example to reduce a circuit's sensitivity to element variations. This leads to a multiparameter (the n Q_i-values) optimization problem whose solution requires the availability of the appropriate computer algorithms. If such software is not available, specific values of Q_i can be chosen. The design becomes particularly simple if all the Q_i-factors are equal, a choice that has the additional practical advantage of resulting in *all identical* second-order building blocks, $T_i(s) = T(s)$. For this reason, this approach has been referred to as the "primary resonator block" (PRB) technique. The passband sensitivity performance of PRB circuits is almost as good as that of fully optimized FLF structures. The relevant equations are derived in the following.

With $q_i = q$ for all i we find from (79.28)

$$\begin{pmatrix} 1 & 0 & 0 \\ 2q & 1 & 0 \\ q^2 & q & 1 \end{pmatrix} \begin{pmatrix} f_1 \\ f_2 \\ f_3 \end{pmatrix} = \begin{pmatrix} b_2 - 3q \\ b_1 - 3q^2 \\ b_0 - q^3 \end{pmatrix} \tag{79.29}$$

which shows that

$$\begin{aligned} F_1 A_1 q &= f_1 = b_2 - 3q \\ F_2 A_1 A_2 q^2 &= f_2 = b_1 - 3q^2 - 2q f_1 \\ F_3 A_1 A_2 A_3 q^3 &= f_3 = b_0 - q^3 - q^2 f_1 - q f_2 \end{aligned} \tag{79.30}$$

The system (79.30) represents three equations for the four unknowns q, f_i, $i = 1, \cdots, 3$ (in general, one obtains n equations for $n + 1$ unknowns) so that one parameter, q, can still be used for optimization purposes. This single degree of freedom is often eliminated by choosing

$$q = \frac{b_{n-1}}{nb_n} \tag{79.31a}$$

i.e., $q = b_2/3$ in our example, which means $f_1 = 0$. The remaining feedback factors can then be computed recursively from (79.30). The systematic nature of the equations makes it apparent how to proceed for $n > 3$. As a matter of fact, it is not difficult to show that, with $f_1 = 0$, in general

$$f_2 = b_{n-2} - \frac{n(n-1)}{2!} q^2 b_n \tag{79.31b}$$

$$f_i = b_{n-i} - \frac{q^i}{(n-i)!}\left[\frac{n!}{i!} b_n + \sum_{j=1}^{i-1} \frac{f_j}{q^j} \frac{(n-j)!}{(i-j)!}\right] \qquad i = 3, \cdots, n \tag{79.31c}$$

Note that b_n usually equals unity. As mentioned earlier, equations of identical form, with f_i replaced by $k_i/(\alpha K_0)$ and b_i by a_i/a_n with $K_0 = a_n/\alpha$, are used to determine the summing coefficients K_i of the output summer, which establishes the transmission zeros. Thus, given a geometrically symmetrical bandpass function with center frequency ω_0, quality factor Q, and bandwidth B, (79.31) can be used to calculate the parameter q and all feedback and summing coefficients required for a PRB design. All second-order bandpass sections, (79.20), are tuned to the same pole frequency $\omega = \omega_0$ and have the same pole quality factor $Q_p = Q/q$, where $Q = \omega_0/B$.

As discussed, the design procedure computes only the *products* f_i and k_i; the actual values of F_i, K_i, and A_i are not uniquely determined. As a matter of fact, the gain constants A_i are *free* parameters that are selected to maximize the circuit's dynamic range in much the same way as for cascade designs.[5] With a few simple approximations it can be shown [5, ch. 6] that the appropriate choice of gain constants in the FLF circuit is

$$A_i \approx \sqrt{1 + (Q_i/Q)^2} = \sqrt{1 + q_i^{-2}} \qquad i = 1, \cdots, n \tag{79.32a}$$

The same equation holds for the PRB case, where $Q_i = Q_p$ for all i so that

$$A_i = A \approx \sqrt{1 + (Q_p/Q)^2} = \sqrt{1 + q^{-2}} \tag{79.32b}$$

The following example illustrates the complete FLF (PRB) design process.

Let us illustrate the multiple-loop feedback procedure by realizing again the bandpass function (79.10), but now as an FLF (PRB) circuit. The previous data specify that the bandpass function should be converted into a prototype low-pass function with bandcenter $\omega_0/(2\pi) = \sqrt{12 \cdot 36}$ kHz and bandwidth $B/(2\pi) = (36\text{–}12)$ kHz by the transformation (79.23),

$$p = Q\frac{s^2 + 1}{s} = \frac{\omega_0}{B}\frac{s^2 + 1}{s} = \frac{\sqrt{12 \cdot 36}}{36 - 12}\frac{s^2 + 1}{s} = 0.866\frac{s^2 + 1}{s} \tag{79.33}$$

[5]For practical FLF (PRB) designs, this step of scaling the signal levels is very important because an inadvertently poor choice of gain constants can result in *very* large internal signals and, consequently, very poor dynamic range.

where s is normalized by ω_0 as before. Substituting (79.33) into (79.10) results in

$$H_{\mathrm{LP}}(s) = \frac{0.491}{p^3 + 0.984p^2 + 1.236p + 0.491} \tag{79.34}$$

which corresponds to (79.27) with $a_i = 0$, $i = 1, 2, 3$. To realize this function, we need three first-order low-pass sections of the form (79.24), i.e.,

$$T_{\mathrm{LP}}(p) = A\frac{q}{p+q} \tag{79.35}$$

where $A = \sqrt{1 + 1/q^2}$ from [79.32(b)]. We proceed with [79.31(a)] to choose $q = 0.984/3 = 0.328$, so that $A = 3.209$. With these numbers, we can apply (79.33) to (79.35) to obtain the second-order bandpass that must be realized:

$$T_{\mathrm{BP}}(p) = 3.209\frac{0.328}{0.866\dfrac{s^2+1}{s} + 0.328} = \frac{1.2153s}{s^2 + 0.379s + 1} \tag{79.36}$$

The feedback resistors we obtain by (79.30):

$$\begin{aligned} 1.052\,F_1 &= f_1 = 0 \rightarrow F_1 = 0 \\ 1.108\,F_2 &= f_2 = b_1 - 3q^2 = 0.913 \rightarrow F_2 = 0.777 \\ 1.166\,F_3 &= f_3 = b_0 - q^3 - qf_2 = 0.371 \rightarrow F_3 = 0.318 \end{aligned} \tag{79.37}$$

Choosing, e.g., $R_{F0} = 10\ \mathrm{k\Omega}$ results in $R_{Fi} = R_{F0}/F_i$, that is $R_{F1} = \infty$, $R_{F2} = 12.9\ \mathrm{k\Omega}$, and $R_{F3} = 31.5\ \mathrm{k\Omega}$. Also, $R_{F0}/R_{\mathrm{in}} = \alpha = 0.491(Aq)^{-3} = 0.421 \rightarrow R_{\mathrm{in}} = 23.7\ \mathrm{k\Omega}$.

There remains the choice of second-order bandpass sections. Notice from (79.19) that the sections must have a positive, i.e., noninverting, gain to keep the feedback loops stable. We choose GIC (general impedance converter) sections [5, ch. 4], one of which is shown explicitly in Fig. 79.4. They realize,

$$T(s) = \frac{V_o}{V_i} = \frac{\left(1 + \dfrac{G_3}{G_2}\right)\dfrac{G_1}{C}s}{s^2 + \dfrac{G_1}{C}s + \left(\dfrac{G}{C}\right)^2\dfrac{G_3}{G_2}} = \frac{K\dfrac{\omega_0}{Q}s}{s^2 + \dfrac{\omega_0}{Q}s + \omega_0^2} \tag{79.38}$$

Comparing (79.38) to (79.36), choosing $R_3 = 5\ \mathrm{k\Omega}$, $C = 2\ \mathrm{nF}$, and remembering that s is normalized with respect to $\omega_0 = 130.59 \cdot 10^3\ \mathrm{s}^{-1}$ results in

$$R = 5.45\ \mathrm{k\Omega}, \qquad R_1 = 10.17\ \mathrm{k\Omega}, \qquad R_2 = 10.14\ \mathrm{k\Omega}$$

The Leapfrog Topology

The leapfrog (LF) configuration is pictured in Fig. 79.5. Each of the boxes labeled T_i realizes a second-order transfer function. The feedback loops comprise always two sections; thus, inverting and noninverting sections must alternate to keep the loop gains negative and the loops stable. If the circuit is derived from a double resistively terminated lossless ladder filter as is normally the case, T_1 and T_n are lossy and all the internal sections are lossless. A lossless block implies

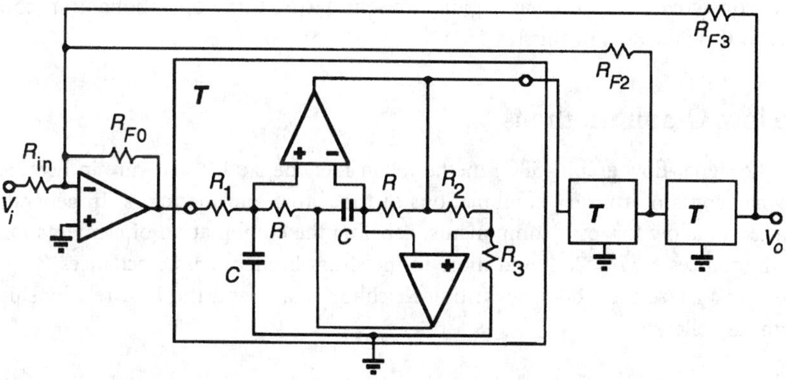

FIGURE 79.4 PRB topology with three identical GIC bandpass sections. The circuit for one section is shown explicitly, as is the input summer with feedback resistors.

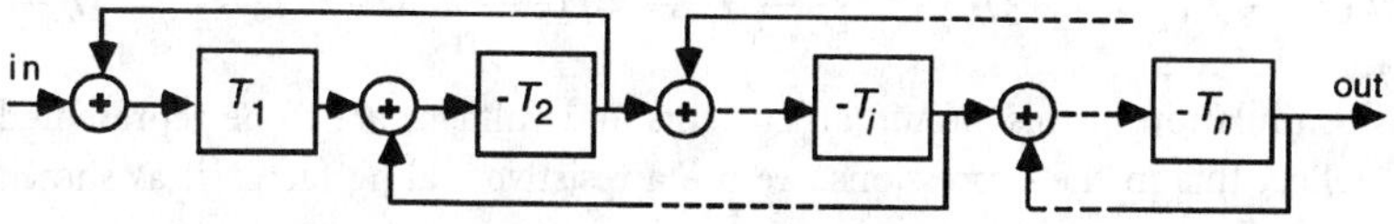

FIGURE 79.5 The leapfrog topology.

a function T_i within infinite Q, which may not be stable by itself, but the overall feedback connection guarantees stability.

An LF circuit can be derived from the configuration in Fig. 79.5 by direct analysis with, e.g., bandpass transfer functions as in (79.20) assumed for the blocks T_i. Comparing the resulting equation with that of a prescribed filter yields expressions that permit determining all circuit parameters in a similar way as for FLF filters [1]. Because the topology is identical to that derived from a signal-flow graph representation of an *LC* ladder filter, we shall not consider the details of the LF approach here, but instead proceed directly to the ladder simulation techniques.

79.4 Ladder Simulation

Although transfer functions of LC ladders are more restrictive than those realizable by cascade circuits,[1] lossless ladder filters designed for maximum power transfer have received considerable attention in the active filter literature because of their significant advantage of having the lowest possible sensitivities to component tolerances in the passband. Indeed, the majority of high-order active filters with demanding specifications are being designed as simulated *LC* ladders. Many active circuit structures have been developed that simulate the performance of passive *LC* ladders and inherit their good sensitivity performance. The ladder simulations can be classified into two main groups: *operational simulation* and *element substitution.* Starting from an existing *LC* prototype ladder, the operational simulation models the internal operation of the ladder by simulating the circuit equations, i.e., Kirchhoff's voltage and current laws and the *I–V* relationships of the ladder arms. Fundamentally, this procedure simulates the *signal-flow graph* (SFG) of the ladder, where all voltages and all currents are considered signals that are integrated on the inductors and capacitors, respectively. The SFG method will be developed in the next subsection. The element substitution procedure replaces all inductors or inductive branches by active networks whose input impedance is inductive over the frequency range of interest. A practical approach to this method will be presented in the section on element substitution. For more detailed discussions of these important and practical procedures, we refer the reader to a modern text on active filters, such as [5] and, for approaches using operational transconductance amplifiers, to Chapter 80.3 in this text.

Signal-Flow Graph Methods

To derive the signal-flow graph (SFG) method, consider the ladder structure in Fig. 79.6, whose branches may contain arbitrary combinations of capacitors and inductors. In general, resistors are permitted to allow for lossy components. Labeling the combination of elements in the series arms as admittances Y_i, $i = 2$, 4, and those in the shunt branches as impedances Z_j, $j = 1, 3, 5$, we can readily analyze the ladder by writing Kirchhoff's laws and the I–V relationships for the ladder arms as follows:

$$\begin{aligned} &I_i = G_i(V_i - V_1), \qquad V_1 = Z_1 I_1 = Z_1(I_i - I_2) \\ &I_2 = Y_2 V_2 = Y_2(V_1 - V_3), \quad V_3 = Z_3 I_3 = Z_3(I_2 - I_4) \\ &I_4 = Y_4 V_4 = Y_4(V_3 - V_5), \qquad V_5 = V_o = Z_5 I_5 = Z_5(I_4 - I_6), \qquad I_6 = G_0 V_o \end{aligned} \tag{79.39}$$

In the active simulation of this circuit, all currents and voltages are to be represented as *voltage* signals. To reflect this in the expressions, we use a resistive scaling factor R as shown in one of these equations as an example,

$$V_3 = \frac{Z_3}{R} I_3 R = \frac{Z_3}{R}(I_2 R - I_4 R) \tag{79.40}$$

and introduce the notation

$$I_k R = i_k \qquad V_k = v_k \qquad G_i R = g_i \qquad \frac{Z_k}{R} = z_k \qquad Y_k R = y_k \tag{79.41}$$

The lowercase symbols are used to represent the *scaled* quantities; notice that z_k and y_k are dimensionless voltage transfer functions, also called *transmittances*, and that both i_k and v_k are voltages. We shall retain the symbol i_k in order to remind ourselves of the origin of that signal as a current in the original ladder. Equations (79.39) then take on the form

$$\begin{aligned} &i_i = g_i[v_i + (-v_1)] \qquad -v_1 = -z_1 i_1 = -z_1[i_i + (-i_2)] \\ &-i_2 = y_2(-v_2) = y_2[(-v_1) + v_3] \quad v_3 = -z_3(-i_3) = -z_3[(-i_2) + i_4] \\ &i_4 = y_4 v_4 = y_4[v_3 + (-v_5)] \qquad -v_5 = -v_o = -z_5 i_5 = -z_5[i_4 + (-i_6)] \\ &-i_6 = g_0(-v_0) \end{aligned} \tag{79.42}$$

where we have made all the transmittances z_i inverting and assigned signs to the signals in a consistent fashion such that only *signal addition* is required in the circuit to be derived from

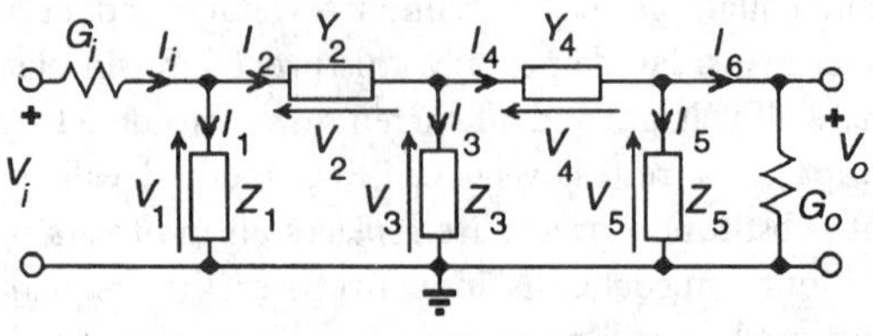

FIGURE 79.6 Ladder network.

these equations. We made this choice because addition can be performed with no additional circuitry at op-amp summing nodes (see below), whereas subtraction would require either differential amplifiers or inverters. The price to be paid for this convenience is that in some cases the overall transfer function suffers a sign inversion (a 180° phase shift), which is of no consequence in most cases. The signal-flow block diagram implementing (79.42) is shown in Fig. 79.7. As is customary, all voltage signals are drawn at the bottom of the diagram, and those derived from currents at the top. We observe that the circuit consists of a number of interconnected loops of two transmittances each and that all loop gains are negative as required for stability. Notice that redrawing this figure in the form of Fig. 79.5 results in an identical configuration, i.e., as mentioned earlier, the leapfrog method is derived from a ladder simulation technique.

To determine how the transmittances are to be implemented we need to know which elements are in the ladder arms. Consider first the simple case of an all-pole lowpass ladder, where $Z_i = 1/(sC_i)$ and $Y_j = 1/(sL_j)$, i.e., $z_i = 1/(sC_iR)$ and $y_i = 1/(sL_i/R)$ (see Fig. 79.11). Evidently then, for this case all transmittances are integrators. Suitable circuits are shown in Fig. 79.8, where for each integrator we have used two inputs in anticipation of the final realization which has to sum two signals as indicated in Fig 79.7. The circuits realize

$$V_0 = \pm\frac{G_1V_1 + G_2V_2}{sC + G_3} \tag{79.43}$$

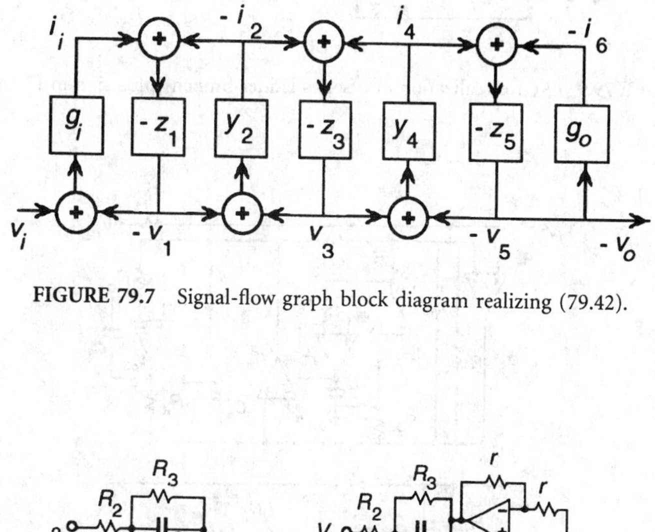

FIGURE 79.7 Signal-flow graph block diagram realizing (79.42).

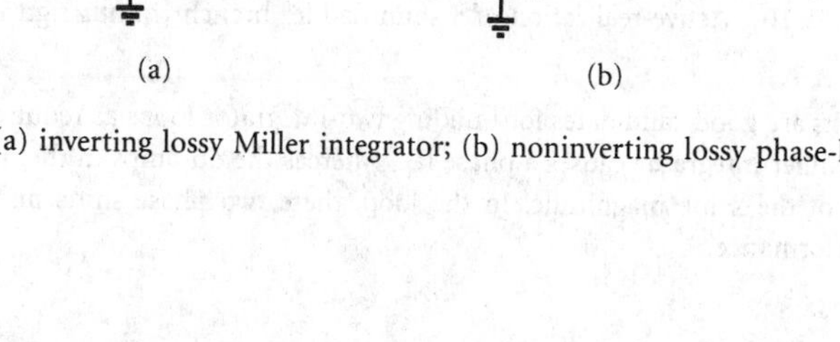

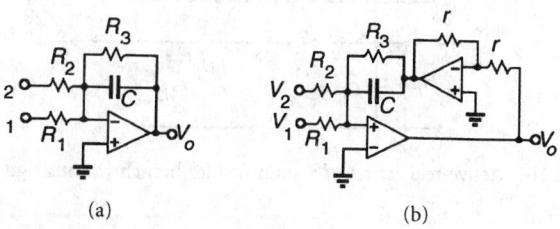

FIGURE 79.8 (a) inverting lossy Miller integrator; (b) noninverting lossy phase-lead integrator.

where the plus sign is valid for the series transmittances $y_i(s)$ in Fig. 79.8(b) and the minus sign for the shunt transmittances $z_i(s)$[6] in Fig. 79.8(a). G_3 is zero if the integration is lossless as required in an internal branch of the ladder; in the two end branches, G_3 is used to implement the source and load resistors of the ladder.

The next question to be answered is what circuitry will realize more general ladder arms that may contain both series and parallel *LC* elements, and resistors to handle the source or load terminations (or losses if required). Such general series and shunt branches are shown at the bottom of Figs. 79.9 and 79.10, respectively, where we have labeled the capacitors in the *passive* network as "$\hat{C}$" to be able to distinguish them from the capacitors in the active circuit (labeled C without the circumflex). We have chosen the signs of the relevant voltages and currents in Figs. 79.9 and 79.10 appropriately to obtain *noninverting transmittances* $y(s)$ for the series arms and *inverting transmittances* $z(s)$ for the shunt arms as requested in Fig. 79.7. Recall that this choice permits the flowgraph to be realized with only *summing* functions. For easy reference, the active *RC* circuits realizing these branches are shown directly above the passive *LC* arms [5, ch. 6].

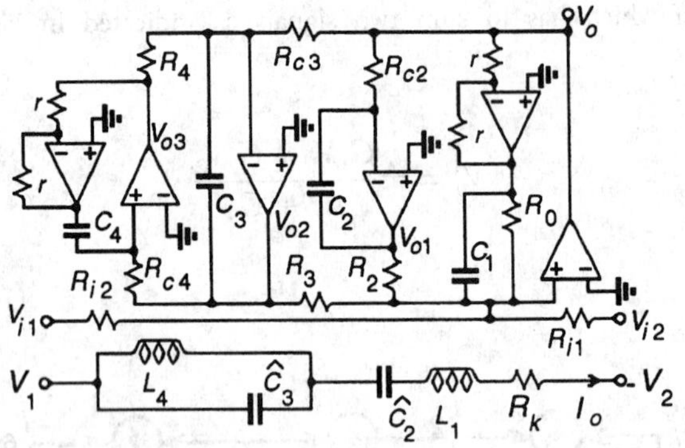

FIGURE 79.9 Active realization of a series ladder branch [plus sign in (79.46)].

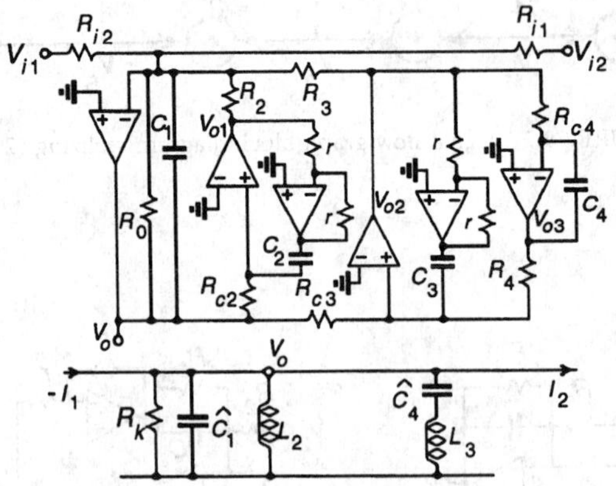

FIGURE 79.10 Active realization of a shunt ladder branch [minus sign in (79.46)].

[6]These two circuits are good candidates for building two-integrator loops as required in Fig. 79.7 because the op-amp in the Miller integrator causes a phase lag, whereas the op-amps in the noninverting integrator cause a phase lead of the same magnitude. In the loop, these two phase shifts just cancel and cause no errors in circuit performance.

The passive series branch in Fig. 79.9 realizes the current I_o,

$$I_o R_p = Y(s) R_p (aV_1 + bV_2)$$
$$= \frac{1}{R_k/R_p + sL_1/R_p + \dfrac{1}{s\hat{C}_2 R_p} + \dfrac{1}{s\hat{C}_3 R_p + \dfrac{1}{sL_4/R_p}}}(aV_1 + bV_2) \quad (79.44a)$$

which was converted into a voltage through multiplication with a scaling resistor R_p. (p stands for *passive*; R_p is the resistor used to scale the *passive* circuit.) Also, we have multiplied the input signals by two constants a and b in anticipation of future scaling possibilities in the *active* circuit. Using, as before, *lowercase* symbols for the *normalized* variables, we obtain for the series branch

$$i_o = y(s)(av_1 + bv_2) = \frac{1}{r_k + sl_1 + \dfrac{1}{sc_2} + \dfrac{1}{sc_3 + \dfrac{1}{sl_4}}}(av_1 + bv_2) \quad (79.44b)$$

In an analogous fashion we find for the voltage V_o in the passive shunt branch in Fig. 79.10, after impedance-level scaling with R_p and signal-level scaling with a and b, the expression

$$V_o = -\frac{Z(s)}{R_p}(aI_1 R_p + bI_2 R_p)$$
$$= -\frac{1}{G_k R_p + s\hat{C}_1 R_p + \dfrac{1}{sL_2/R_p} + \dfrac{1}{sL_3/R_p + \dfrac{1}{s\hat{C}_4 R_p}}}(aI_1 R_p + bI_2 R_p) \quad (79.45a)$$

which with lowercase notation gives

$$v_o = -z(s)(ai_1 + bi_2) = -\frac{1}{g_k + sc_1 + \dfrac{1}{sl_2} + \dfrac{1}{sl_3 + \dfrac{1}{sc_4}}}(ai_1 + bi_2) \quad (79.45b)$$

Turning now to the active *RC* branches in Figs. 79.9 and 79.10, elementary analysis of the two circuits, assuming ideal op-amps,[7] results in

$$V_o = \pm \frac{R_a G_{i1} V_{i1} + R_a G_{i2} V_{i2}}{R_a G_0 + sC_1 R_a + \dfrac{R_a G_2}{sC_2 R_{c2}} + \dfrac{R_a G_3}{sC_3 R_{c3} + \dfrac{G_4 R_{c3}}{sC_4 R_{c4}}}} \quad (79.46)$$

[7]Using a more realistic op amp model, $A(s) \approx \omega_t/s$, one can show [5, ch. 6] that (79.46) to a first-order approximation is multiplied by $(1 + j\omega/\omega_t) \approx \exp(j\omega/\omega_t)$ for the series arm [plus sign in (79.46)] and by $(1 - j\omega/\omega_t) \approx \exp(-j\omega/\omega_t)$ for the shunt arm [minus sign in (79.46)]. Thus, in the loop gains determined by the product of an inverting and a noninverting branch, op-amp effects cancel to a first order, justifying the assumption of ideal op-amps. See also footnote 6.

where we used a normalizing resistor R_a. (*a* stands for *active*; R_a is the resistor used to scale the *active* circuit.) In (79.46) the plus sign is valid for the series arm, Fig. 79.9, and the minus sign for the shunt arm, Fig. 79.10.

Notice that (79.46) is of the same form as (79.44) and (79.45) so that we can compare the expressions term-by-term to obtain the component values required for the active circuit(s) to realize the prescribed passive ladder arms. Thus, we find for the *series* branch, Fig. 79.9, by comparing coefficients between (79.46) and (79.44a), and assuming all equal capacitors C in the active circuit:

$$R_{i1} = \frac{R_a}{a} \quad R_{i2} = \frac{R_a}{b} \quad R_0 = \frac{R_a R_p}{R_k} \quad C = \frac{L_1}{R_a R_p}$$
$$R_{c2}R_2 = \frac{\hat{C}_2}{C} R_a R_p \quad R_{c3}R_3 = \frac{\hat{C}_3}{C} R_a R_p \quad R_{c4}R_4 = \frac{L_4 \hat{C}_3}{C^2} \tag{79.47a}$$

In an identical fashion we obtain from (79.46) and [79.45(a)] for the components of the *shunt* arm

$$R_{i1} = \frac{R_a}{a} \quad R_{i2} = \frac{R_a}{b} \quad R_0 = R_k \frac{R_a}{R_p}$$
$$C = \hat{C}_1 \frac{R_p}{R_a} \quad R_{c2}R_2 = \frac{L_2}{C}\frac{R_a}{R_p} \tag{79.47b}$$
$$R_{c3}R_3 = \frac{L_3}{C}\frac{R_a}{R_p} \quad R_{c4}R_4 = \frac{L_3 \hat{C}_4}{C^2}$$

The scaling resistors R_a and R_p are chosen to obtain convenient element values. Note that each of the last three equations for both circuits determines only the *product* of two resistors. This leaves three degrees of freedom, which are normally chosen to maximize dynamic range by equalizing the maximum signal levels at all op-amp outputs. We shall provide some discussion of these matters in the next subsection.

We have displayed the active and passive branches in Figs. 79.9 and 79.10 together to illustrate the one-to-one correspondence of the circuits. For example, if we wish to design an all-pole low-pass filter, such as the one in Fig. 79.11, where the internal series arms consist of a single inductor and each internal shunt arm of a single capacitor, the corresponding active realizations reduce to those of Fig. 79.8(b) and (a), respectively, with $R_3 = \infty$. For the end branches, R_s in series with L_1 and $\hat{C}_4$ in parallel with R_l, we obtain the circuits in Fig. 79.8(b) and (a), respectively, with R_3 *finite* to account for the source and load resistors. The remaining branches in the active circuits are absent. Assembling the resulting circuits as prescribed in Fig. 79.7 leads to the active SFG filter in Fig. 79.12, where we have for convenience chosen all identical capacitors C and have multiplied the input by an arbitrary constant K because the active circuit may realize a gain scaling factor. The component values are computed from a set of equations similar to (79.47). To show in some detail how they are arrived at, we derive the equations for each integrator in Fig. 79.12 and compare them to the corresponding arm in the passive ladder. Recalling that signals with lowercase symbols in the active circuit are voltages, we obtain

$$\frac{G_1 R_a V_{\text{in}} + G_2 R_a(-v_2)}{sCR_a + G_3 R_a} \rightarrow \frac{KV_{\text{in}} - V_2}{sL_1/R_p + R_s/R_p} \tag{79.48a}$$

$$\frac{G_4 R_a i_1 + G_5 R_a(-i_3)}{sCR_a} \rightarrow \frac{I_1 - I_3}{s\hat{C}_2 R_p} \tag{79.48b}$$

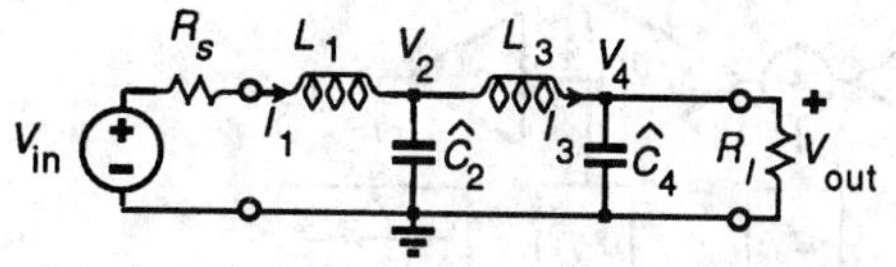

FIGURE 79.11 Fourth-order all-pole low-pass ladder.

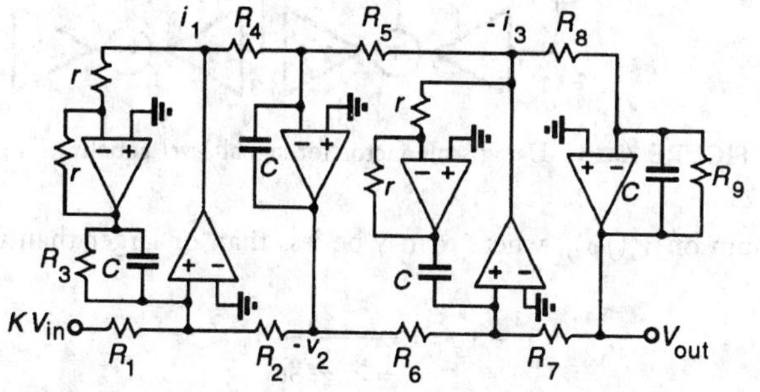

FIGURE 79.12 Active realization of the ladder in Fig. 79.11.

$$\frac{G_6R_a(-v_2)+G_7R_aV_{\text{out}}}{sCR_a} \rightarrow \frac{-V_2+V_4}{sL_3/R_p} \tag{79.48c}$$

$$\frac{G_8R_a(-i_3)}{sCR_a+G_9R_a} \rightarrow \frac{-i_3}{s\hat{C}_4R_p+R_p/R_l} \tag{79.48d}$$

where we used scaling resistors for the active (R_a) and the passive (R_p) circuits as before. Choosing a convenient value for C in the active circuit and equating the time constants and the dc gains in (79.48a) results in the following expressions:

$$R_3=\frac{L_1}{C}\frac{1}{R_s} \qquad R_1=R_3\frac{R_s}{KR_p} \qquad R_2=R_3\frac{R_s}{R_p} \tag{79.49a}$$

Similarly,

$$R_4=R_5=\frac{\hat{C}_2}{C}R_p \quad R_6=R_7=\frac{L_3}{C}\frac{1}{R_p} \quad R_8=\frac{\hat{C}_4}{C}R_p \quad R_9=\frac{\hat{C}_4}{C}R_l=\frac{R_l}{R_p}R_8 \tag{79.49b}$$

Maximization of the Dynamic Range

The remaining task in a signal-flow graph simulation of an *LC* ladder is that of voltage-level scaling for dynamic range maximization. As in cascade design, we need to achieve that all op-amps in $0 \le \omega \le \infty$ see the same maximum signal level so that no op-amp becomes overdriven sooner than any other one. It may be accomplished by noting that a scale factor can be inserted into any signal line in a signal-flow graph *as long as the loop gains are not changed.* Such signal level scaling will not affect the transfer function except for an overall gain factor. The procedure can be illustrated in flow diagram in Fig. 79.7. If we simplify the circuit by combining the self-loops at input and output and employ the scale factors α, β, γ, δ, and their inverses to the loops in Fig. 79.7, we obtain the modified flow diagram in Fig. 79.13. Simple analysis shows that the transfer function has not changed except for a multiplication by the factor αβγδ, which is canceled by the multiplier $(\alpha\beta\gamma\delta)^{-1}$ at the input. To understand how the scale factors are computed, assume that in Fig. 79.7, i.e., before scaling, the maximum of $|i_4(j\omega)|$ is α times as

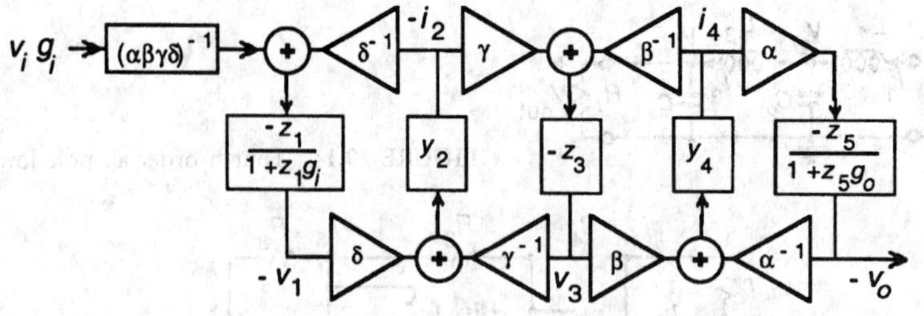

FIGURE 79.13 Using scale factor for signal-level equalization.

large as the maximum of $|v_o(j\omega)|$, where α may be less than or larger than unity. Since

$$-v_o = \frac{-z_5}{1 + z_5 g_0} i_4$$

the maxima can be equalized, i.e., the level of v_o can be increased by α, if we apply a gain scale factor α as indicated in Fig. 79.13. To keep the loop gain constant, a second scale factor α^{-1} is inserted as shown. Continuing, if in Fig. 79.7 max $|v_3(j\omega)|$ is β times as large as max $|i_4(j\omega)|$, we raise the level of $|i_4|$ by a gain factor β and correct the loop gain by a second factor β^{-1}, as shown in Fig. 79.13. In a similar fashion we obtain the maxima of the remaining voltages and apply scale factors γ, δ, γ^{-1}, and δ^{-1}, as appropriate, in Fig. 79.13.

It is easy to determine the relevant maxima needed for calculating the gain factors. Recall that the node voltages v_i, v_1, i_2, v_3, i_4, and $v_5 = v_o$ in the signal-flow graph of Fig. 79.13 correspond directly to the actual currents and voltages in the original ladder, Fig. 79.6, and their maxima in $0 \leq \omega \leq \infty$ can readily be evaluated with any network analysis program. For the circuit in Fig. 79.13, the scale factors that ensure that the currents (normalized by R_p) in all series ladder arms and the voltages in all shunt ladder arms have the same maxima are then obtained as

$$\alpha = \frac{\max|v_o|}{\max|i_4|} \qquad \beta = \frac{\max|i_4|}{\max|v_3|} \qquad \gamma = \frac{\max|v_3|}{\max|i_2|} \qquad \delta = \frac{\max|i_2|}{\max|v_i|} \tag{79.50}$$

The procedure so far takes care of ladders with only one element in each branch, such as all-pole low-pass filters (Fig. 79.11). In the more general case, we must also equalize the maxima of the magnitudes of V_{o1}, V_{o2}, and V_{o3} at the *internal* op-amp outputs in Figs. 79.9 and 79.10. This is achieved easily if we remember that the "external" voltage maxima of $|V_{i1}|$, $|V_{i2}|$, and $|V_o|$ are already equalized by the previous steps leading to (79.50) and that in the passive *series* branch V_{o1} represents the voltage on $\hat{C}_2$, V_{o2} stands for the voltage on $\hat{C}_3$, and V_{o3} corresponds to the current (times R_p) through the inductor L_4. After finding the maxima of these signals with the help of a network analysis program and computing the scale factors

$$m_1 = \frac{\max|V_{o1}|}{\max|V_o|} \qquad m_2 = \frac{\max|V_{o2}|}{\max|V_o|} \qquad m_3 = \frac{\max|V_{o3}|}{\max|V_o|} \tag{79.51}$$

we can equalize all internal branch voltages of the *series* arm (Fig. 79.9) by modifying the design equations [79.47(a)] as follows [5, ch. 6]:

$$\begin{gathered} R_{i1} = \frac{R_a}{a} \quad R_{i2} = \frac{R_a}{b} \quad R_0 = \frac{R_a R_p}{R_k} \quad C = \frac{L_1}{R_a R_p} \\ m_1 R_{c2} \frac{R_2}{m_1} = \frac{\hat{C}_2}{C} R_a R_p \quad m_2 R_{c3} \frac{R_3}{m_2} = \frac{\hat{C}_3}{C} R_a R_p \quad \frac{m_3}{m_2} R_{c4} \frac{R_4}{m_3/m_2} = \frac{L_4 \hat{C}_3}{C^2} \end{gathered} \tag{79.52}$$

A possible choice for the element values given by products is

$$m_1 R_{c2} = \frac{R_2}{m_1} = \sqrt{\frac{\hat{C}_2}{C} R_a R_p} \qquad m_2 R_{c3} = \frac{R_3}{m_2} = \sqrt{\frac{\hat{C}_3}{C} R_a R_p}$$
$$\frac{m_3}{m_2} R_{c4} = \frac{m_2}{m_3} R_4 = \frac{\sqrt{L_4 \hat{C}_3}}{C} \qquad (79.53a)$$

In the passive *shunt* branch, V_{o1} and V_{o2} represent the currents (times R_p) through the inductors L_2 and L_3, respectively, and V_{o3} stands for the voltage across $\hat{C}_4$. In an identical fashion we obtain then with (79.50) from (79.47b) for the components of the *shunt* arm

$$R_{i1} = \frac{R_a}{a} \qquad R_{i2} = \frac{R_a}{b} \qquad R_0 = R_k \frac{R_a}{R_p} \qquad C = \hat{C}_1 \frac{R_p}{R_a}$$
$$m_1 R_{c2} = \frac{R_2}{m_1} = \sqrt{\frac{L_2}{C} \frac{R_a}{R_p}} \qquad m_2 R_{c3} = \frac{R_3}{m_2} = \sqrt{\frac{L_3}{C} \frac{R_a}{R_p}} \qquad (79.53b)$$
$$\frac{m_3}{m_2} R_{c4} = \frac{m_2}{m_3} R_4 = \frac{\sqrt{L_3 \hat{C}_4}}{C}$$

We shall next present an example [2], [5, ch. 6] in which the reader may follow the different steps discussed.

To simulate the fourth-order elliptic lowpass ladder filter in Fig. 79.14(a) by the signal-flow graph technique, we first reduce the loop count by a source transformation. The resulting circuit, after impedance scaling, is shown in Fig. 79.14(b). To use signal-level scaling for optimizing the dynamic range, the relevant maxima of the *LC* ladder currents and voltages are needed. Table 79.1 lists the relevant data obtained with the help of network analysis software. From these

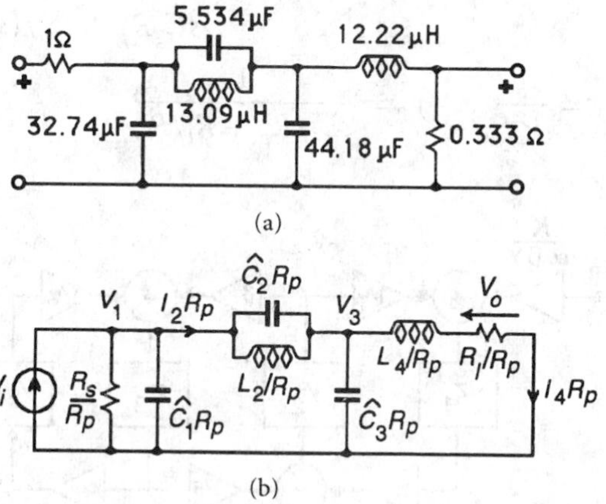

FIGURE 79.14 (a) Fourth-order elliptic *LC* low-pass ladder; (b) the circuit after source transformation and impedance scaling.

TABLE 79.1 Voltage and Current Maxima for Fig. 79.14(a).

Voltage or Current	Maximum of Voltage or Current
voltage across $\hat{C}_1$	0.699 V
current through L_2	1.550 A
voltage across $\hat{C}_2$	1.262 V
current through $(L_2 \| \hat{C}_2)$	1.125 A
voltage across $\hat{C}_3$	0.690 V
current through L_4	0.866 A

numbers we find

$$\alpha = \frac{\max|v_3|}{\max|i_4 R_p|} = \frac{0.69}{0.866} = 0.797 \qquad \beta = \frac{\max|i_2 R_p|}{\max|v_3|} = \frac{1.125}{0.690} = 1.630$$
$$\gamma = \frac{\max|v_1|}{\max|i_2 R_p|} = \frac{0.699}{1.125} = 0.621 \tag{79.54}$$

For ease of reference, the signal-flow graph with scaling factors is shown in Fig. 79.15. Let $R_p = 1\Omega$. Note that the input signal voltage $i_{\text{in}} = R_p V_i / R_s$ has been multiplied by a factor K to permit realizing an arbitrary signal gain. If the desired gain is unity, and since the dc gain in the passive LC ladder equals

$$\frac{V_o}{V_i} = \frac{R_l}{R_s + R_l} = \frac{0.333}{1.333} = 0.25$$

we need to choose $K = 1/0.25 = 4$. Thus, the input signal i_i is multiplied by

$$\frac{K}{\alpha\beta\gamma} = \frac{4}{0.797 \cdot 1.630 \cdot 0.621} = 4.958 \tag{79.55}$$

The transmittances are defined as

$$z_1 = \frac{1}{s\hat{C}_1 R_p + G_s R_p} \qquad y_2 = \frac{1}{\dfrac{1}{s\hat{C}_2 R_p + R_p/(sL_p)}}$$
$$z_3 = \frac{1}{s\hat{C}_3} \qquad y_4 = \frac{1}{sL_4/R_p + R_l/R_p} \tag{79.56}$$

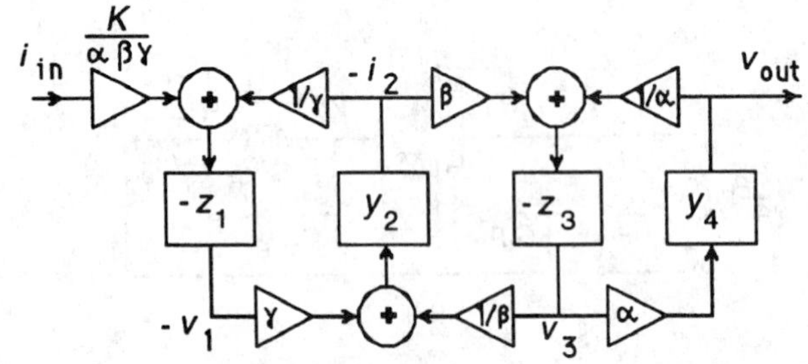

FIGURE 79.15 Signal-flow graph for the filter in Fig. 79.14.

With the numbers in (79.54) and (79.55) we obtain from the signal-flow graph

$$-v_1 = -z_1\left(4.958 i_{in} - \frac{1}{0.621} i_2\right) \qquad -i_2 = y_2\left(-0.621 v_1 + \frac{1}{1.630} v_3\right)$$
$$v_3 = -z_3\left(-1.630 i_2 + \frac{1}{0.797} v_{out}\right) \qquad v_{out} = y_4\, 0.792 v_3 \tag{79.57}$$

The reader is encouraged to verify from (79.56) and (79.57) and Figs. 79.9 and 79.10 that the active SFG realization of the ladder is as shown in Fig. 79.16. For instance, the circuitry between the nodes v_1, v_3, and i_2 implements the parallel LC series branch to realize the finite transmission zero, obtained from Fig. 79.9 by setting the components G_0, C_1, G_2, G_{c2}, and C_2 to zero. The element values in the active circuit are now determined readily by comparing the circuit equations for the active circuit,

$$-v_1 = -\frac{G_1 R_{a1} i_{in} + G_2 R_{a1}(-i_2)}{sCR_{a1} + G_8 R_{a1}} \qquad -i_2 = \frac{\dfrac{G_5 R_{a2}(-v_1) + G_6 R_{a2} v_3}{G_9 R_{a2}}}{sCR_{10} + \dfrac{G_{12} R_{10}}{sCR_{11}}}$$
$$v_3 = -\frac{G_3 R_{a3}(-i_2) + G_4 R_{a3} v_o}{sCR_{a3}} \qquad v_o = \frac{G_7 R_{a4} v_3}{sCR_{a4} + G_{13} R_{a4}} \tag{79.58}$$

to (79.57) and using (79.56).[8] Comparing the coefficients results in

$$G_1 R_{a1} = 4.958 \qquad G_2 R_{a1} = \frac{1}{0.621} \qquad G_5 R_{a2} = 0.621 \qquad G_6 R_{a2} = \frac{1}{1.630}$$
$$G_3 R_{a3} = 1.630 \qquad G_4 R_{a3} = \frac{1}{0.797} \qquad G_7 R_{a4} = 0.797$$

Further,

$$\hat{C}_1 R_p = CR_{a1} \qquad \hat{C}_2 R_p = C\frac{R_9 R_{10}}{R_{a2}} \qquad \hat{C}_3 R_p = CR_{a3} \qquad \frac{L_4}{R_p} = CR_{a4}$$

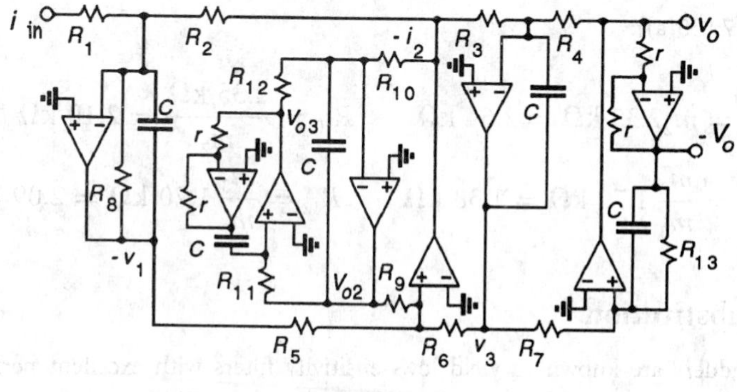

FIGURE 79.16 Active realization of the circuit in Fig. 79.14.

[8] Observe that for greater flexibility we permitted a different scaling resistor R_{ai}, $i = 1, \cdots, 4$, in each branch.

Thus, with $R_p = 1\ \Omega$ and choosing $C = 5$ nF

$$R_{a1} = \frac{\hat{C}_1}{C} R_p = 6.55\ \text{k}\Omega \qquad R_{a3} = \frac{\hat{C}_3}{C} R_p = 8.84\ \text{k}\Omega \qquad R_{a4} = \frac{L_4}{CR_p} = 2.44\ \text{k}\Omega$$

R_{a2} is undetermined; choosing $R_{a2} = 5$ kΩ leads to the feed-in resistors for each branch

$$R_1 = \frac{R_{a1}}{4.958} = 0.800\ \text{k}\Omega \qquad R_2 = 0.621 R_{a1} = 4.07\ \text{k}\Omega$$

$$R_3 = \frac{R_{a3}}{1.630} = 5.42\ \text{k}\Omega$$

$$R_4 = 0.797 R_{a3} = 7.05\ \text{k}\Omega \qquad R_5 = \frac{R_{a2}}{0.621} = 8.05\ \text{k}\Omega$$

$$R_6 = 1.630 R_{a3} = 8.15\ \text{k}\Omega \qquad R_7 = \frac{R_{a4}}{0.797} = 3.06\ \text{k}\Omega$$

The remaining components are determined from the equations

$$R_8 = R_s \frac{R_{a1}}{R_p} = 6.55\ \text{k}\Omega \qquad R_{13} = \frac{R_{a4} R_p}{R_l} = 7.33\ \text{k}\Omega$$

and

$$R_9 R_{10} = R_p R_{a2} \frac{\hat{C}_2}{C} = [2.35\ \text{k}\Omega]^2 \qquad R_{11} R_{12} = \frac{L_2}{C} \frac{R_9 R_{10}}{R_{a2} R_p} = [1.70\ \text{k}\Omega]^2$$

Since only the products of these resistors are given, we select their values uniquely for dynamic range maximization. According to (79.51) we compute from Table 79.1

$$m_2 = \frac{\max |v_{c2}|}{\max |i_2|} = \frac{1.262}{1.125} = 1.12 \quad \text{and} \quad m_3 = \frac{\max |i_{L2}|}{\max |i_2|} = \frac{1.550}{1.125} = 1.38$$

to yield from [79.53(a)]:

$$R_9 = m_2 2.35\ \text{k}\Omega = 2.64\ \text{k}\Omega \qquad R_{10} = \frac{2.35\ \text{k}\Omega}{m_2} = 2.10\ \text{k}\Omega$$

$$R_{11} = \frac{m_2}{m_3} 1.70\ \text{k}\Omega = 1.38\ \text{k}\Omega \qquad R_{12} = \frac{m_3}{m_2} 1.70\ \text{k}\Omega = 2.09\ \text{k}\Omega$$

Element Substitution

Because *LC* ladders are known to yield low-sensitivity filters with excellent performance, but high-quality inductors cannot be implemented in microelectronic form, an appealing solution to the filter design problem is to retain the ladder structure and to simulate the behavior of the inductors by circuits consisting of resistors, capacitors, and op-amps. A proven technique uses *impedance converters,* electronic circuits whose input impedance is proportional to frequency when loaded by the appropriate element at the output. The best-known impedance converter is the gyrator, a two-port circuit whose input impedance is inversely proportional to the load impedance, i.e.,

$$Z_{\text{in}}(s) = \frac{r^2}{Z_L(s)} \tag{79.59}$$

The parameter r is called the **gyration resistance.** Clearly, when the load is a capacitor, $Z_L(s) = 1/(sC)$, $Z_i(s) = sr^2C$ is the impedance of an inductor of value $L = r^2C$. Gyrators are very easy to realize with transconductors (voltage-to-current converters) and are widely used in transconductance-C filters; see Chapter 80.3. However, no high-quality gyrators with good performance beyond the audio range have been designed to date with op-amps. If op-amps are to be used, a different kind of impedance converter is employed, one that converts a load resistor R_L into an inductive impedance, such that

$$Z_{\text{in}}(s) = (sk)R_L \tag{79.60}$$

A good circuit that performs this function is Antoniou's general impedance converter (GIC) shown in Fig. 79.17(a). The circuit, with elements slightly rearranged was encountered in Fig. 79.4, where we used the GIC to realize a second-order bandpass function. The circuit is readily analyzed if we recall that the voltage measured between the op-amp input terminals and the currents flowing into the op-amp input terminals are zero. Thus, we obtain from Fig. 79.17(a) the set of equations

$$V_o = V_i \qquad \frac{I_4}{sC} = I_3R \qquad I_2R = I_1R_1 \qquad I_2 = I_3 \qquad I_i = I_1 \qquad I_4 = I_o \tag{79.61}$$

These equations indicate that the terminal behivior of the general impedance converter is described by

$$V_o = V_i \qquad I_o = sCR_1I_i = skI_i \tag{79.62}$$

that is

$$\frac{V_i}{I_i} = Z_{\text{in}}(s) = sk\frac{V_o}{I_o} = skZ_L(s) \tag{79.63}$$

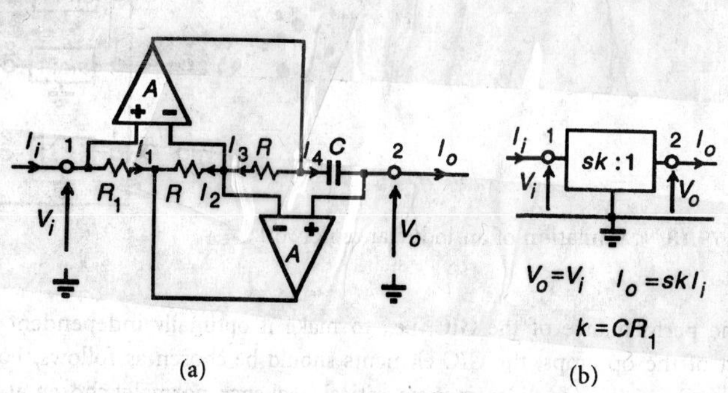

FIGURE 79.17 General impedance converter: (a) circuit; (b) symbolic representation.

Notice that the input impedance is inductive as prescribed by (79.60) if the load is a resistor.[9] Figure 79.17(b) also shows the circuit symbol we shall use for the GIC in the following to keep the circuit diagrams simple. This impedance converter and its function of converting a resistive load into an inductive input impedance is the basis for Gorski–Popiel's embedding technique [5, ch. 6], which permits replacing the inductors in an LC filter by resistors.

Inductor Replacement—Gorski–Popiel's Method

To understand the behavior of the circuit in Fig. 79.18, recall from (79.62) that the voltages at terminals 1 and 2 of the GICs are the same. Then, by superposition the typical input current I_{oi} into the resistive network $\boldsymbol{R}$ can be written as

$$I_{oi} = \frac{1}{R_{i1}}V_1 + \frac{1}{R_{i2}}V_2 + \cdots + \frac{1}{R_{in}}V_n = \sum_{j=1}^{n}\frac{1}{R_{ij}}V_j, \quad i = 1, \ldots, n \tag{79.64}$$

where the parameters R_{ij} are given by the resistors and the configuration of $\boldsymbol{R}$. Using the current relationship $I_{oi} = skI_i$ of the impedance converters in (79.64) results in

$$I_i = \sum_{j=1}^{n}\frac{1}{skR_{ij}}V_j \quad i = 1, \ldots, n \tag{79.65}$$

which makes the combined network consisting of the n GICs and the network $\boldsymbol{R}$ look purely inductive, i.e., each resistor R_r in the network $\boldsymbol{R}$ appears replaced by an inductor of value

$$L_r = skR_r = sCR_1R_r \tag{79.66}$$

Examples of this process are contained in Fig. 79.19. Figures 79.19(a) and (b) illustrate the conventional realizations of a grounded and a floating inductor requiring one and two, respectively, converters. This is completely anılogous to the use of gyrators, where a grounded inductor requires one gyrator, but a floating inductor is realized with two gyrators connected back-to-back. Figures 79.19(c) and (d) show ıow the process is extended to complete inductive subnetworks: the GICs are used to isolate tle subnetworks from the remainder of the circuit; there is no need to convert each inductor separately. The method is further illustrated with the filter in Fig. 79.20. The LC ladder is designed to specifications by use of the appropriate design tables [4], [7] or filter design software [8]; see also Section XIII in this book. In the resulting ladder the inductive subnetworks are separated as shown in Fig. 79.20(a) by the five dashed cuts.

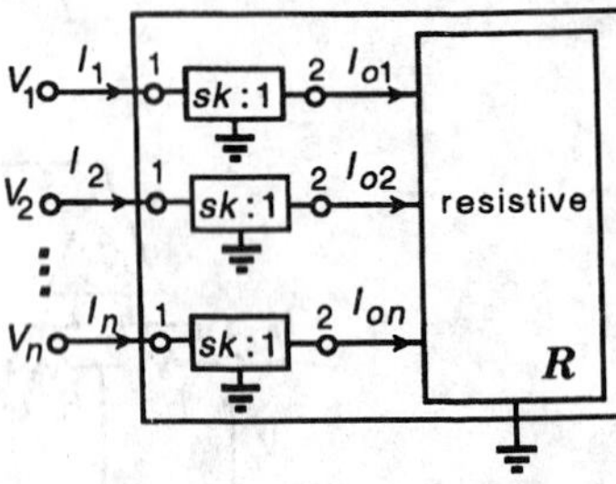

FIGURE 79.18 Simulation of an inductance network.

[9]To optimize the performance of the GIC, i.e., to make it optimally independent of the finite gain–bandwidth product of the op-amps, the GIC elements should be chosen as follows. For an arbitrary load $Z_L(s)$ one chooses $\omega_c C = 1/|Z_L(j\omega_c)|$. ω_c is some critical frequency, normally chosen at the upper passband corner. If the load is resistive, $Z_L = R_L$, select $C = 1/(\omega_c R_L)$ [6].

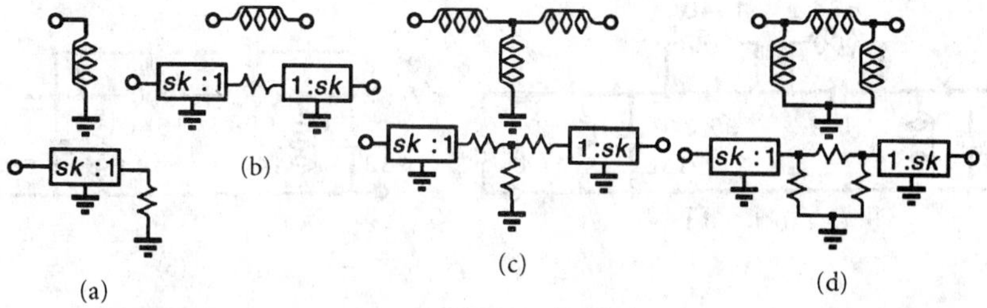

FIGURE 79.19 Elementary inductance networks and their GIC-R equivalents: (a) grounded inductor, (b) floating inductor, (c) inductive T, (d) inductive Π.

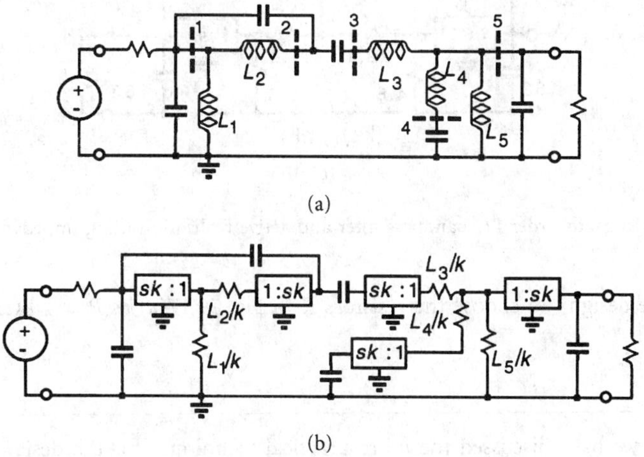

FIGURE 79.20 *LC* ladder realization with GICs.

The cuts are repaired by inserting GICs with conversion factor k and the correct orientation (output, terminal 2 in Fig. 79.17, toward the inductors). Note that all GICs must have the same conversion factor k. Finally, the inductors L_i are replaced by resistors of value L_i/k as shown in Fig. 79.20(b). A numerical example will illustrate the design steps.

Assume we wish to implement a sixth-order bandpass filter with the following specifications:

Maximally flat passband with ≤ 3 dB attenuation in 900 Hz $\leq f \leq$ 1200 Hz.
Transmission zero at $f_z = 1582.5$ Hz; source and load resistors $R = 3$ kΩ.

Using the appropriate software or tables, the *LC* filter in Fig. 79.21(a) is found with element values in kΩ, mH, and nF. Figure 79.21(b) shows the circuit redrawn to help identify the inductive subnetwork and the locations of the cuts. Note that only three impedance converters are used rather than six (two each for the two floating inductors and one each for the grounded ones) if a direct conversion of the individual inductors had been attempted.

For the final active realization, we assumed a conversion parameter $k = CR_1 = 30$ μs. Finally, to design the GICs we compute the impedances seen into the nodes where the GICs see the (now) resistive subnetworks. Using analysis software to compute $|V_i/I|$ at these nodes, we find

$$\text{for GIC}_a, \quad |Z_a| \approx 18 \text{ k}\Omega, \quad \text{and for GIC}_b \text{ and GIC}_c, \quad |Z_b| \approx |Z_c| \approx 4.7 \text{ k}\Omega$$

so that with $\omega_c = 2\pi \cdot 1.2$ kHz and $k = 30$ μs the design elements are

$$C_a = \frac{1}{\omega_c |Z_a|} \approx 7.4 \text{ nF} \quad R_{1a} = \frac{k}{C_a} \approx 4 \text{ k}\Omega \quad C_b = C_c \approx 28 \text{ nF} \quad R_b = R_c \approx 1 \text{ k}\Omega$$

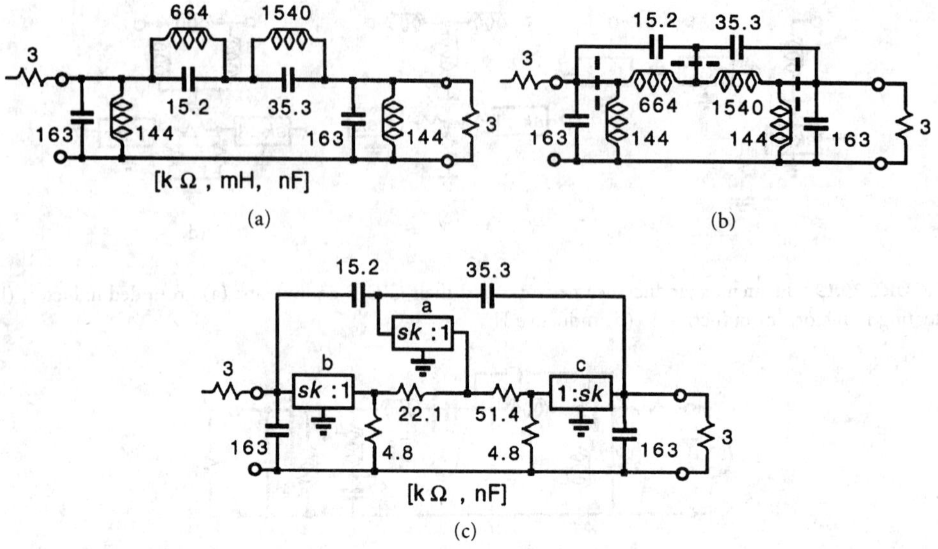

FIGURE 79.21 Sixth-order *LC* bandpass filter and active realization using impedance converters.

To complete the design, we choose the resistors R in Fig. 79.17(a) as $R = 1$ kΩ.

79.5 Summary

In this chapter, we have discussed the more practical techniques for the design of active filters of order higher than two: *cascade* design, *multiple-loop feedback* approaches, and methods that *simulate the behavior of LC ladder* filters. We have pointed out that direct realization methods are impractical because they result in high sensitivities to component values. In many applications, a cascade design leads to satisfactory results. The practical advantages of cascade circuits are modularity, ease of design, flexibility, very simple tuning procedures, and economical use of op-amps, with as few as one op-amp per pole pair. Also, we point out again that an arbitrary transfer function can be realized with the cascade design method, i.e., no restrictions are placed on the permitted locations of poles and zeros.

For challenging filter requirements the cascade topologies may still be too sensitive to parameter changes. In those cases the designer can use the FLF configuration or, for best performance, provided that a passive prototype ladder exists, a ladder simulation. FLF circuits retain the advantage of modularity, but if optimal performance is desired, computer-aided optimization routines must generally be used to adjust the available free design parameters. However, excellent performance with very simple design procedures, no optimization and high modularity can be obtained by use of the *primary-resonator-block* (PRB) method, where all biquad building blocks are identical.

From the point of view of minimum passband sensitivity to component tolerances, the best active filters are obtained by simulating *LC ladders.* If the prescribed transfer characteristic can at all be realized as a passive *LC* ladder, then the designer can make use of the wealth of available information about the design of such circuits, and simple procedures are available for "translating" the passive *LC* circuit into its active counterpart. We may either take the passive circuit and replace the inductors by active networks, or we imitate the mathematical behavior of the whole *LC* circuit by realizing the integrating action of inductors and capacitors via active *RC* integrators. Both approaches result in active circuits of high quality; a disadvantage is that more op amps may be required than in cascade or FLF methods. This drawback is offset, however, by the fact that the sensitivities to component tolerances are 'almost' as low as those of the originating ladder.

An alternative method of an active realization of a passive *LC* ladder is obtained by scaling each impedance of the passive circuit by $1/(ks)$ (Bruton's transformation) [5, ch. 6]. This impedance transformation converts resistors into capacitors, inductors into resistors, and capacitors into "*frequency-dependent negative resistors* (FDNRs)," which can readily be realized with Antoniou's GIC circuit. The procedure is especially useful for passive prototype circuits that have only *grounded* capacitors because capacitors are converted to FDNRs and floating FDNRs are very difficult to implement in practice. The method results in biasing difficulties for active elements, and because the *entire* ladder must be transformed, the active circuit no longer contains source and load resistors. If these two components are prescribed and must be maintained as in the original passive circuit, additional buffers are required. The method was not discussed in this test because it shows no advantages over the Gorski–Popiel procedure.

An important practical aspect is the *limited dynamic range* of active filters, restricted by noise and by the finite linear signal swing of op-amps. Dynamic range maximization should be addressed whenever possible. Apart from designing low-noise circuits, the procedures always proceed to equalize the op-amp output voltages by exploiting available free gain constants or impedance scaling factors. It is a disadvantage of the element substitution method that no general dynamic range scaling method appears to be available.

The methods discussed in this chapter have dealt with the design of filters in *discrete* form, that is, separate passive components and operational or transconductance amplifiers are assembled on, e.g., a printed circuit board to make up the desired filter. Chapter 80 addresses modifications in the design approach that lead to filters realizable in fully integrated form.

References

[1] K. R. Laker and M. S. Ghausi, "A comparison of active multiple-loop feedback techniques for realizing high-order bandpass filters," *IEEE Trans. Circuits Syst.*, vol. CAS-21, pp. 774–783, Nov. 1974.

[2] K. Martin and A. S. Sedra, "Design of Signal-flow Graph (SFG) Active Filters," *IEEE Trans. Circuits Syst.*, vol. CAS-25, pp. 185–195, Sept. 1978.

[3] G. S. Moschytz, *Linear Integrated Networks—Design*, New York, Van Nostrand Reinhold, 1975.

[4] R. Saal, *Handbook of Filter Design*, Berlin: AEG-Telefunken, 1979.

[5] R. Schaumann, M. S. Ghausi, and K. R. Laker, *Design of Analog Filters: Passive, Active RC and Switched Capacitor*, Englewood Cliffs, NJ: Prentice-Hall, 1990.

[6] A. S. Sedra and P. O. Brackett, *Filter Theory and Design: Active and Passive*, Portland, OR: Matrix, 1978.

[7] A. I. Zverev, *Handbook of Filter Synthesis*, New York: Wiley, 1967.

[8] B. M. Silveira, C. Ouslis, and A. S. Sedra, "Passive Ladder Synthesis in FiltorX," in *Proc. IEEE Int. Symp. Circuits Syst.*, 1993, pp. 1431–1434.

80

Continuous-Time Integrated Filters

Rolf Schaumann
Portland State University, Oregon

80.1 Introduction

All modern signal-processing systems include various types of electrical filters that the designer has to realize in an appropriate technology. The literature contains many well-defined filter design techniques, and computer programs are available that help the designer find the appropriate transfer function that describes the required filter characteristics mathematically [4]. We also refer the reader to Sections XIII, XIV, and the remaining chapters of Section XV in this text.

Once the filter's transfer function is obtained, implementation methods must be found that are compatible with the technology selected for the design of the total system. In some situations, considerations of power consumption, frequency range, signal level, or production numbers may dictate *discrete* (*passive* or *active*) filter realizations. Often, however, as much as possible of the total system must be *fully integrated* in microelectronic form, so that the filters should also be implemented in the same technology.

Often digital (Section XVI) or sampled-data (Chapter 81) implementations are suitable for realizing the filter requirements, but in many signal-processing situations filters must interface with the "real world," where the input and output signals take on continuous values as functions of the continuous variable time, i.e., they are *continuous-time* (*c-t*) signals. In these situations *c-t* antialiasing and reconstruction filters are often required. Because the performance of the total filter system is of relevance and not just the performance of the intrinsic filter, the designer may have to consider whether it might not be preferable to implement the entire system in the *c-t* domain rather than as a digital or sampled-data system. At least at low-frequencies the latter methods have the advantages of very high accuracy, better signal-to-noise ratio, and little or no parameter drifts, but they entail a number of problems connected with analog-to-digital (A/D) and digital-to-analog (D/A) conversion (Chapter 64), sample-and-hold, switching, antialiasing, and reconstruction circuitry.

0-8493-8341-2/95/$0.00 + $.50

Traditionally, *c-t* filters were implemented as discrete designs. Well-understood procedures exist for deriving passive *LC* filters (Section XIV) from a given transfer function with prescribed complex natural frequencies, e.g., [4, ch. 2]. Since to date no practical methods exist for building high-quality, i.e., low-loss inductors on an integrated circuit (IC) chip, the required complex natural frequencies must be realized by using *gain*, i.e., as we saw earlier in this section, by embedding an operational amplifier (op-amp; see Chapter 11) in an *RC* feedback network [4, ch. 4, 5]. Since op-amps, resistors, and capacitors can be implemented on an integrated circuit, with active *RC* networks it appears that the problem of monolithic filter design is solved in principle: all active devices and any necessary capacitors and resistors can be integrated together on one silicon chip. Although this conclusion is correct, the designer needs to consider four other factors that are important in integrated *c-t* filter design and perhaps are not immediately obvious.

The first item concerns the most important design task for achieving commercially practical designs: integrated filters must be *electronically tunable*, preferably by an *automatic* tuning scheme. Because of its importance, we shall devote Section 80.4 to this topic. The second item deals with the economics of practical implementations of active filters: in *discrete* designs, the cost of components and stocking them usually necessitate designing the filter with a minimum number of active devices, such as one or possibly two op-amps per pole pair, and the smallest number of different (if possible, all identical) capacitors. In *integrated* realizations, capacitors are determined by processing mask dimensions and the number of different capacitor values is unimportant, as long as the *element spread* is not excessive. Further, active devices frequently occupy less chip area than passive elements so that it is often preferable to use active elements instead of passive ones.[1] Third, the designer should remember that IC technology cannot easily generate accurate absolute component values, but that *ratios* of like components, such as capacitor ratios, can be realized very precisely. Finally, the last observation pertains to the fact that filters usually have to share an integrated circuit with other, possibly switched or digital, systems so that the ac ground lines (power supply and ground wires) are likely to contain switching transients and generally are noisy. Measuring the analog signals relative to ac ground, therefore, may result in designs with poor signal-to-noise ratio and low power-supply rejection. The situation is remedied in practice by building continuous-time filters in fully differential, balanced form, where the signals are referred to each other as $V = V^{+} - V^{-}$, as shown in Fig. 80.4(b), (c) and (d). An additional advantage of this arrangement is that the signal range is doubled (for an added 6 dB of signal-to-noise ratio) and that the even-order harmonics in the nonlinear operation of the active devices cancel. All filters in this chapter will, therefore, be designed in fully differential form.

Today, *c-t* filters integrated in bipolar, CMOS, or BiCMOS technology are no longer academic curiosities but a commercial reality. In the following we shall discuss the main methods that have proven to be reliable. First, we shall present *MOSFET-C filters*, whose design methods resemble most closely the standard active *RC* procedures discussed in the earlier chapters of this section; they can, therefore, be most readily understood by the reader without requiring further background. Next, we shall introduce the *transconductance-C* (also referred to as g_m-*C*) technique, which is currently the predominant method for *c-t* integrated filters. Designs based on transconductors lead to filters for higher operating frequencies, important for modern communication systems.

80.2 MOSFET-*C* Filters

As mentioned in the introduction, the MOSFET-*C* method follows standard op-amp-based active filter techniques [6], which, as we saw in previous chapters, rely heavily on the availability

[1]However, keeping the number of active devices small remains important because active devices consume power and generate noise.

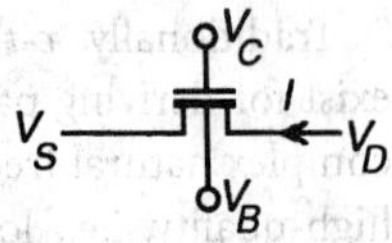

FIGURE 80.1 MOS transistor biased in the triode region.

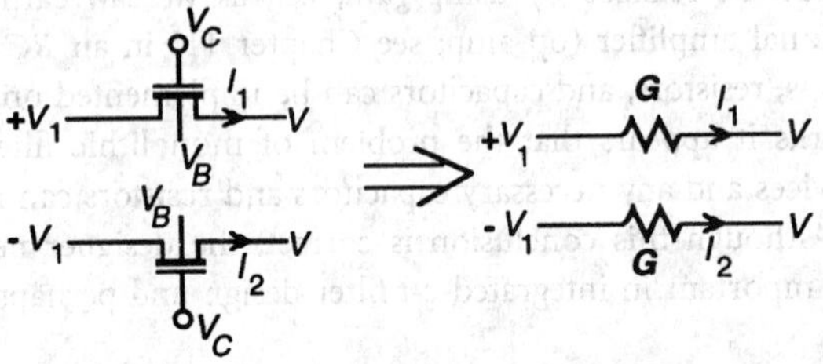

(a) (b)

FIGURE 80.2 (a) Balanced linear MOSFET conductance; (b) equivalent resistor circuit.

of integrators and summers. The only difference is that the method replaces the resistors used in the conventional active *RC* integrating and summing circuitry by MOSFET devices biased in the triode (ohmic) region ($V_C - V_S - V_T > V_D - V_S$, see Chapter 56.2). Defining the source (V_S), drain (V_D), gate (V_C), and substrate (V_B) voltages as shown in Fig. 80.1, the resulting nonlinear drain current I becomes

$$\begin{aligned} I &= \frac{W}{L}\mu C_{ox}\{(V_C - V_T)(V_D - V_S) + a_2(V_D^2 - V_S^2) + a_3(V_D^3 - V_S^3) + \cdots\} \\ &= \left[\frac{W}{L}\mu C_{ox}(V_C - V_T)\right](V_D - V_S) + b_2(V_D^2 - V_S^2) + b_3(V_D^3 - V_S^3) + \cdots \end{aligned} \tag{80.1}$$

where W and L are the channel width and length, respectively, μ is the effective mobility, C_{ox} is the gate capacitance per unit area, and V_T is the threshold voltage of the device. The term in brackets is a tunable conductor,

$$G(V_C) = \frac{W}{L}\mu C_{ox}(V_C - V_T) \tag{80.2}$$

where the gate voltage V_C is used for control or *tuning*. Thus, the drain current of a MOSFET in the triode region is proportional to the drain-to-source voltage, but contains nonlinear second- and higher order terms. Third- and higher order terms can be shown to be small and will be neglected in the following. Now consider placing two of these MOS conductances in parallel as shown in Fig. 80.2(a). Note that the two devices are driven in balanced form by $+V_1$ and $-V_1$ at one pair of terminals and that the other terminals are at the same voltage V. Applying these conditions to (80.1) results in

$$I_1 = G(V_C)(+V_1 - V) + b_2[(+V_1)^2 - V^2] \tag{80.3a}$$
$$I_2 = G(V_C)(-V_1 - V) + b_2[(-V_1)^2 - V^2] \tag{80.3b}$$

Consequently, apart from the neglected high-order odd terms, the difference current $I_1 - I_2$ is perfectly linear in the applied signal voltage V_1:

$$I_1 - I_2 = 2G(V_C)V_1 \tag{80.4}$$

Thus, in the MOSFET-C method, the even-order nonlinearities can be shown to be eliminated by carefully balanced circuit design where all signals are measured strictly differentially [6]. Note that the expression for $I_1 - I_2$ in (80.4) is the same as that for the linear resistive equivalent circuit in Fig. 80.2(b). This means that the MOSFET circuit in Fig. 80.2(a) can be substituted for a pair of resistors in any *appropriate* active *RC* circuit, where appropriate means that the resistor pair is driven by balanced signals at one end and that the voltages at the other two terminals are the same (in practice usually virtual ground). In addition to these conditions, the substitution is valid if the MOSFETs are operated in the triode region and if $v_1(t)$ and $v(t)$ are safely within the range $|V_C - V_B|$. Of course, in practice there remains some small distortion due to mismatches and the neglected odd-order terms. Odd-order nonlinearities are usually small enough to be negligible. Typically, the remaining nonlinearities, arising from odd harmonics, device mismatch, and body effects, are found to be of the order of 0.1 percent for 1 V signals.

In the next section, we shall discuss the MOSFET-*C* integrator that is the fundamental building block used in *all* MOSFET-*C* active filter designs. The integrator will initially be used to construct first- and second-order MOSFET-*C* sections from which higher order filters can be assembled by the cascade method. In the section on ladder simulations we shall then show how simulated *LC* ladder filters are designed by use of integrators.

Basic Building Blocks

The method of MOSFET substitution for resistors in an active *RC* prototype works correctly, if the active *RC* circuits are of a form where all resistors come in balanced pairs, with one end of the resistor pair voltage-driven and the two other terminals seeing the same voltage [see Fig. 80.2(a)]. Active *RC* circuits do not normally satisfy these conditions, but many can readily be converted into that form if a balanced symmetrical op-amp as pictured in Fig. 80.3 is available. It is important to note that this structure is not simply a differential op-amp, but that the input and output voltages are symmetrical with respect to a ground reference. Using a balanced op-amp, the conversion of many active *RC* prototypes into balanced form proceeds simply by taking the single-ended circuits and mirroring them at ground as shown below.

Integrators

Mirroring the active *RC* integrator structure in Fig. 80.4(a) at ground and using the balanced op-amp in Fig. 80.3 leads to the balanced integrator in Fig. 80.4(b). Note that the two resistors are connected in the configuration prescribed in Fig. 80.2 with $V = 0$ (virtual ground) so that they may be replaced by the MOSFET equivalent in Fig. 80.4(c). Analysis of the circuit in the time domain results in

$$-v_{out} = v^{+} - \frac{1}{C}\int_{-\infty}^{t} i_2(t)\,dt \qquad +v_{out} = v^{-} - \frac{1}{C}\int_{-\infty}^{t} i_1(t)\,dt \tag{80.5a}$$

$$v_{out} - (-v_{out}) = v^{-} - v^{+} - \frac{1}{C}\int_{-\infty}^{t} [i_1(t) - i_2(t)]\,dt \tag{80.5b}$$

$$2v_{out}(t) = 0 - \frac{1}{C}\int_{-\infty}^{t} 2G(V_C)v_1(t)\,dt = -\frac{1}{C}\int_{-\infty}^{t} \frac{2v_1(t)}{R(V_C)}\,dt \tag{80.5c}$$

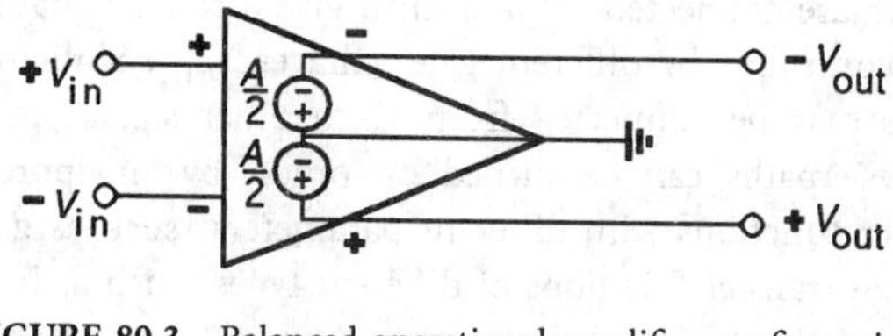

FIGURE 80.3 Balanced operational amplifier configuration.

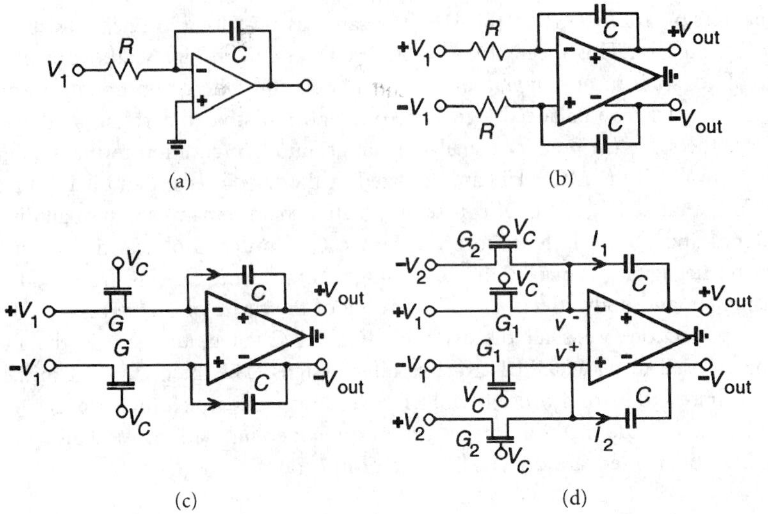

FIGURE 80.4 (a) Active *RC* integrator; (b) fully balanced equivalent; (c) MOSFET-*C* equivalent, (d) differential MOSFET-*C* integrator.

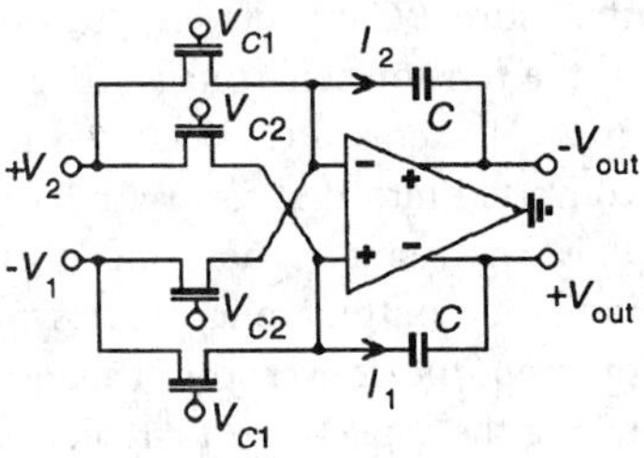

FIGURE 80.5 Integrator using a four-MOSFET configuration for greater linearity.

or, after simplifying and a Laplace transform, in the frequency domain

$$v_{out} = -\frac{1}{CR(V_C)}\int_{-\infty}^{t} v_1(t)\,dt \rightarrow V_{out}(s) = -\frac{1}{sCR(V_C)}V_1(s) \tag{80.5d}$$

We see that the MOSFET-*C* integrator realizes exactly the same transfer function as the active *RC* prototype and that the integration time constant $CR(V_C)$ is tunable by the control voltage V_C that is applied to all gates. Note that the circuit in Fig. 80.4(c) is *not* a differential integrator. To build a differential integrator, one must connect a second pair of balanced resistors (MOSFETs) with inputs $-V_2$ and $+V_2$ as in Fig. 80.5(d) to yield

$$v_{out}(t) = -\frac{1}{C}\int_{-\infty}^{t}\left[\frac{v_1(t)}{R_1} - \frac{v_2(t)}{R_2}\right]dt \rightarrow V_{out}(s) = -\frac{1}{sC}\left[\frac{V_1(s)}{R_1} - \frac{V_2(s)}{R_2}\right] \tag{80.6}$$

The same principle can also be used to build *programmable* integrators, and, therefore, programmable filters: consider in Fig. 80.4(d) the terminals for V_1 and $-V_2$ connected, i.e., $V_1 = -V_2$, so that the two resistors are connected in parallel to give a resistive path $G_1 + G_2$, and let the two resistor values be controlled by different gate voltages $V_{C1,2}$. Similarly, additional *balanced* MOSFET-resistor paths may be connected from V_1 to other signals to the integrator inputs (summing nodes). These paths can be turned on or off by an appropriate choice of gate voltages, so that transfer functions with different parameters (such as gains, quality factors or pole frequencies) or even transfer functions of different types (such as bandpass, low-pass, etc.) are obtainable.

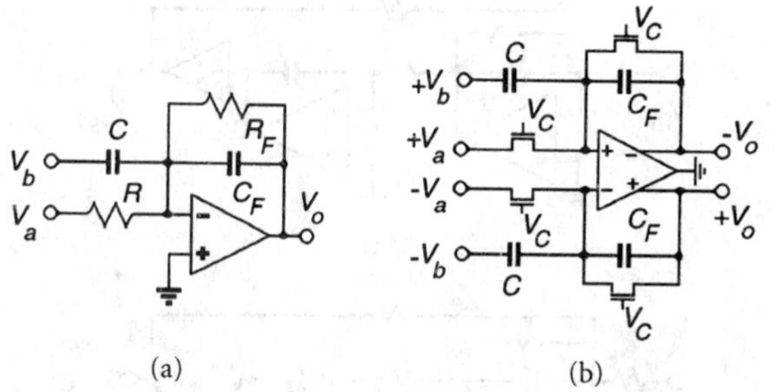

FIGURE 80.6 (a) Lossy integrator with capacitive feed-in branch; (b) MOSFET-*C* equivalent.

If better linearity is required than is obtainable with the two-transistor MOSFET circuit in Fig. 80.4(c) that replaces each resistor in the balanced active *RC* structure by one MOSFET, a four-MOSFET cross-coupled modification for each resistor pair can be used instead [5, paper 2-B.7]. The configuration is illustrated in Fig. 80.5. The current difference can be shown to equal

$$\Delta I = I_1 - I_2 = K(V_{C1} - V_{C2})[V_1 - (-V_1)] \tag{80.7}$$

that is, ΔI is proportional to the product of the difference of the input signals $\pm V_1$ and the difference of the applied gate voltages V_{Ci}, $i = 1, 2$. This indicates that one may interchange the input and gate control voltages and reduce the drive requirements of the previous op-amp stage because no resistive current but only the small gate-capacitor current flows with the input applied at the control gates. The price paid for this advantage is that the requirements on the control voltage source become more difficult and may necessitate an appropriate common-mode voltage on the signal lines.

Additional resistive or capacitive inputs may, of course, be used to design more general lossy integrators as illustrated in Fig. 80.6. By writing the node equation at the summing node, this circuit may be analyzed to realize

$$V_o = -\frac{1}{sC_F + G_F}(GV_a + sCV_b) \tag{80.8}$$

where the conductors are implemented in Fig. 80.6(b) as

$$R(V_C) = \frac{1}{\mu C_{ox}(W/L)(V_C - V_T)} \qquad R_F(V_C) = \frac{1}{\mu C_{ox}(W_F/L_F)(V_C - V_T)} \tag{80.9}$$

as was derived in (80.2). The controlling gate voltages V_C for the two resistors R and R_F may in general be different so that the resistor values may be tuned (or turned on or off) independently.

With lossy and lossless MOSFET-*C* integrators available we can obtain not only first- and second-order filters as shown in the next subsection, but also simulations of *LC* ladders in much the same way as was done in Chapter 79 for active *RC* circuits. We shall discuss this procedure in some detail in the section on ladder simulations.

First- and Second-Order Sections

Based on the principle of balancing a single-ended structure, appropriate[2] standard classical active *RC* filters from the literature can be converted into balanced form and resistors be replaced by MOSFETs. As an illustration, consider the single-ended prototype (Tow–Thomas) biquad in Fig. 80.7(a). As is typical for second-order active *RC* filter sections (see Chapter 77), the circuit

[2]All resistors must be voltage-driven from one side and see the same voltage, usually virtual ground, on the other.

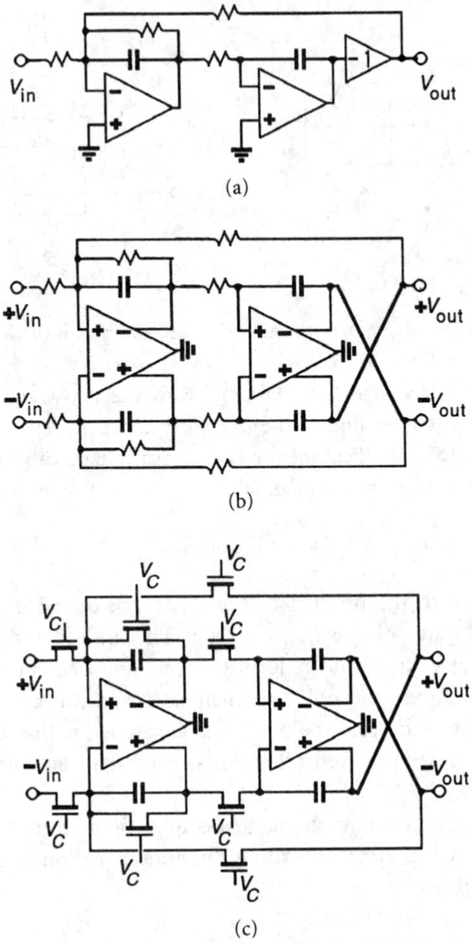

FIGURE 80.7 (a) Active *RC* prototype (Tow–Thomas biquad) for conversion into a MOSFET-*C* structure; (b) fully balanced version of the biquad with resistors; (c) fully balanced version of the biquad with MOSFETs.

is a two-integrator loop consisting of inverting lossy and lossless integrators (in addition to an inverter to keep the loop gain negative for stability reasons). The realized transfer function is not important to our discussion, but we point out that all resistors are voltage-driven (by the signal source or by an op-amp) and at their other ends all resistors are connected to an op-amp input, i.e., they are at virtual ground. Thus, this circuit satisfies our earlier condition for conversion to a MOSFET-*C* structure. Fig. 80.7(b) shows the balanced active *RC* equivalent necessary to eliminate the nonlinear performance. Replacing the resistors by MOSFETs biased in the triode region, with aspect ratios W/L appropriately chosen to realize the prescribed resistor values

$$G_i = \frac{W_i}{L_i}\,\mu C_{ox}(V_C - V_T) \tag{80.10}$$

with excellent resistor matching given by aspect ratios

$$\frac{G_i}{G_k} = \frac{(W/L)_i}{(W/L)_k} \tag{80.11}$$

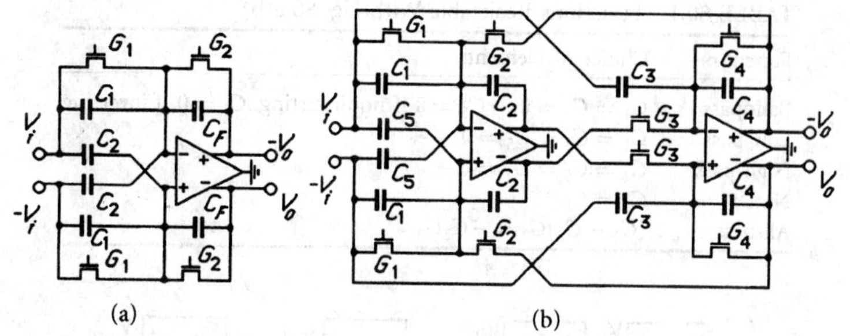

FIGURE 80.8 General (a) first- and (b) second-order MOSFET-*C* sections. The MOSFETs are labeled by the conductance values they are to implement.

leads to the final MOSFET-*C* version in Fig. 80.7(c). The voltage-variable resistors given by (80.10) permit loss and time constants to be electronically tuned. Note that the inverter in the original circuit is not needed in Fig. 80.7(b) and (c) because inversion in the differential topology is obtained by crossing wires.

The circuit may be made *programmable* by connecting additional MOSFET resistors with appropriate W/L ratios in parallel with the fundamental ones shown in Fig. 80.7(c) and then switching them on or off as required by the desired values of the filter coefficients.

The above method indicates how first- or second-order sections can be obtained by choosing any suitable configuration from the active *RC* filter literature [4, ch. 5] and converting it to balanced MOSFET-*C* form. We shall next show a couple of entirely general first- and second-order sections that can be developed from the integrator in Fig. 80.6(b). The resulting circuits are shown in Fig. 80.8 [6, paper 2-A.2]. If we combine the inputs $V_a = V_b = V_i$ of the integrator in Fig. 80.6(b) and add two further cross-coupled feed-in capacitors, we obtain the first-order circuit in Fig. 80.8(a). Writing the node equation at the (inverting or non-inverting)[3] op-amp input results in the transfer function

$$T_1(s) = \frac{V_o}{V_i} = \frac{s(C_1 - C_2) + G_1}{sC_F + G_2} \tag{80.12}$$

Similarly, if we combine two such integrators in a loop, with individual signals paths and signs selected to assure negative feedback, we obtain the general second-order section in Fig. 80.8(b). Writing again the node equations at the op-amp input nodes leads to the transfer function

$$T_2(s) = \frac{V_o}{V_i} = \frac{s^2C_2C_3 + s(C_1 - C_5)G_3 + G_1G_3}{s^2C_2C_4 + sC_2G_4 + G_2G_3} \tag{80.13}$$

Note that the circuits can realize zeros anywhere on the real axis or in the s-plane, depending on the choice of element values. Consequently, these sections can be used to implement arbitrary high-order transfer functions in a cascade topology. Specifically, for the indicated choice of elements the general biquad function in (80.13) realizes the different transfer functions in Table 80.1.

Cascade Realizations

The realization of high-order transfer functions as a connection of low-order sections, including the *cascade, multiple-loop feedback,* and *coupled-biquad* approaches, is identical to that discussed for discrete active *RC* filters. The difference lies only in the implementation of the sections in

[3] Because the circuit is completely symmetrical, it is only necessary to derive the equations for one side, e.g., for V_o^+; the expressions for the other side, i.e., V_o^-, are the same.

TABLE 80.1 Functions Realizable With Fig. 80.8(b)

Function	Choice of Elements
Bandpass	$G_1 = C_3 = 0$; $C_5 = 0$ if noninverting, $C_1 = 0$ if inverting
Low-pass	$C_1 = C_3 = C_5 = 0$
High-pass	$C_1 = C_5 = G_1 = 0$
Notch	$C_1 = C_5$
All-pass	$(C_1 - C_5)G_3 = -C_2G_4$

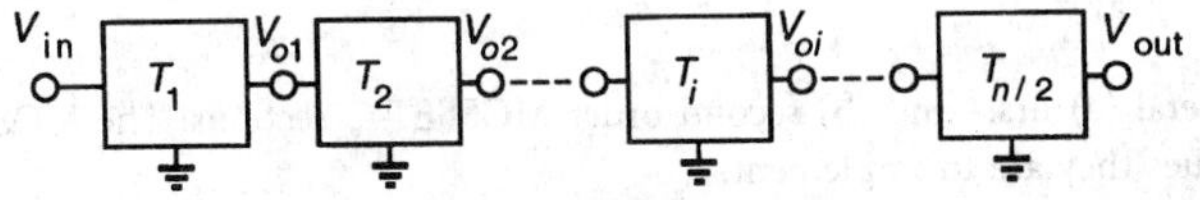

FIGURE 80.9 Cascade realization of an nth-order transfer function.

fully integrated form. We shall not repeat the process here but only discuss the cascade design, the most prevalent method, in terms of an example and encourage the reader to refer to the earlier discussion in Chapter 79 for details. To repeat briefly: if a high-order function

$$H(s) = \frac{V_{out}}{V_{in}} = \frac{N(s)}{D(s)} = \frac{a_m s^m + a_{m-1}s^{m-1} + \cdots + a_1 s + a_0}{s^n + b_{n-1}s^{n-1} + \cdots + b_1 s + b_0} \tag{80.14}$$

with $m \le n$ and $n > 2$ is given, it is factored into second-order (and one first-order if n is odd) sections,

$$H(s) = \prod_{i=1}^{n/2} T_i(s) = \prod_{i=1}^{n/2} k_i \frac{\alpha_{2i}s^2 + \alpha_{1i}s + \alpha_{0i}}{s^2 + s\omega_{0i}/Q_i + \omega_{0i}^2} = \prod_{i=1}^{n/2} k_i t_i(s) \tag{80.15}$$

where each of the second-order functions

$$T_i(s) = \frac{V_{oi}}{V_{o\,i-1}} = k_i \frac{\alpha_{2i}s^2 + \alpha_{1i}s + \alpha_{0i}}{s^2 + s\omega_{0i}/Q_i + \omega_{0i}^2} = k_i t_i(s) \tag{80.16}$$

is implemented as a suitable biquad with specified pole quality factor Q_i and pole frequency ω_{0i}, such as the one in Fig. 80.8(b) realizing (80.13). As was explained in Chapter 79, k_i is a suitable gain constant, chosen to equalize the signal levels in order to optimize the dynamic range, and $t_i(s)$ is a gain-scaled transfer function. Note that in writing (80.15) we assumed that n is even. Fig. 80.9 shows the general structure of the filter.

To provide an example, assume we wish to realize a delay of $\tau_D = 0.187$ μs via a fifth-order Bessel approximation, but with a transmission zero at $f/f_n = 4.67$ to improve the attenuation in the stopband.[4] The transfer function is found to be

$$H_5(s) = \frac{43.3315(s^2 + 21.809)}{(s + 3.6467)(s^2 + 6.7040s + 14.2729)(s^2 + 4.6494s + 18.1563)} \tag{80.17}$$

The normalizing frequency is $f_n = 1/(2\pi\tau_D) = 850$ kHz [4, ch.1]. Let us choose a cascade implementation with the circuits in Fig. 80.8. Factoring (80.17) leads to the first-order and the

[4]Note that the transmission zeros on the $j\omega$ axis can be added to a linear-phase network, such as a Bessel filter, without changing the phase because on the $j\omega$ axis a zero factor $(\omega_z^2 - \omega^2)$ is a purely real number.

two second-order functions on the left-hand side of (80.18) to be realized by the functions on the right, which are obtained from (80.12) and (80.13):

$$T_1(s) = \frac{3.6467}{s + 3.6467} \rightarrow T_1(s) = \frac{V_{o1}}{V_{in}} = \frac{G_1}{sC_F + G_2} \tag{80.18a}$$

$$T_2(s) = \frac{14.2729}{s^2 + 6.7040s + 14.2729} \rightarrow T_2(s) = \frac{V_{o2}}{V_{o1}} = \frac{G_1 G_3}{s^2 C_2 C_4 + sC_2 G_4 + G_2 G_3} \tag{80.18b}$$

$$T_3(s) = \frac{0.8325(s^2 + 21.809)}{s^2 + 4.6494s + 18.1563} \rightarrow T_3(s) = \frac{V_{o3}}{V_{o2}} = \frac{s^2 C_2 C_3 + G_1 G_3}{s^2 C_2 C_4 + sC_2 G_4 + G_2 G_3} \tag{80.18c}$$

The components of the first- and second-order filters in Fig. 80.8 are to be determined from these equations. The gain constants in the function were chosen to result in unity gain at dc. Comparing coefficients, we find with $\omega_n = 1/\tau_D \approx 2\pi \cdot 850$ krad/s$^{-1} \approx 5.341$ Mrad/s^{-1}:

From (80.18a)
$$G_1 = G_2 = 3.6467\omega_n C_F \tag{80.19a}$$

Choosing $C_F = 2$ pF gives $G_1 = G_2 = 38.952$ μS; also $C_1 = C_2 = 0$.

From (80.18b)
$$G_1 G_3 = G_2 G_3 = 14.2729\omega_n^2 C_2 C_4; \quad G_4 = 6.7040\omega_n C_4 \tag{80.19b}$$

Choosing $C_2 = C_4 = 2$ pF results in $G_1 = G_2 = G_3 = 40.354$ μS and $G_4 = 71.6$ μS; also, $C_1 = C_3 = C_5 = 0$.

From (80.18c)
$$G_1 G_3 = 21.908\omega_n^2 C_2 C_3; \quad G_2 G_3 = 18.1563\omega_n^2 C_2 C_4$$
$$G_4 = 4.6494\omega_n C_4; \quad C_3 = 0.8325 C_4 \tag{80.19c}$$

Choosing $C_2 = 2$ pF and $C_4 = 10$ pF yields $C_3 = 8.325$ pF, $G_1 = G_2 = G_3 = 101.77$ μS and $G_4 = 248.32$ μS; also, $C_1 = C_5 = 0$. The remaining task is to convert the resistors into MOSFET devices. Assume the process provides transistors with $\mu C_{ox} = 120$ μA/V^2 and $V_T = 0.9$ V. For the choice of $V_C = 2$ V, the aspect ratios are then calculated from the above conductance values and from (80.10) via

$$\frac{W_i}{L_i} = \frac{G_i(V_C)}{\mu C_{ox}(V_C - V_T)} = \frac{G_i(V_C)}{120\mu\text{A/V}^2 \cdot 1.1\text{ V}} = \frac{G_i/\mu\text{S}}{132} \tag{80.20}$$

For instance, in the first-order section we find $W_1/L_1 = W_2/L_2 = 38.952/132 = 1/3.389$. The resulting circuit is shown in Fig. 80.10.

Ladder Simulations

Using MOSFET-*C* integrators, the ladder simulation method for the MOSFET-*C* approach is entirely analogous to the active *RC* procedures discussed earlier in this book. The process is illustrated by a step-by-step generic example, which should guide the reader when implementing a specific design.

Assume a fifth-order elliptic low-pass filter is prescribed with the transfer function

$$H_{ell}(s) = \frac{(s^2 + \omega_1^2)(s^2 + \omega_2^2)}{(s + a)(s^2 + bs + c)(s^2 + ds + e)} \tag{80.21}$$

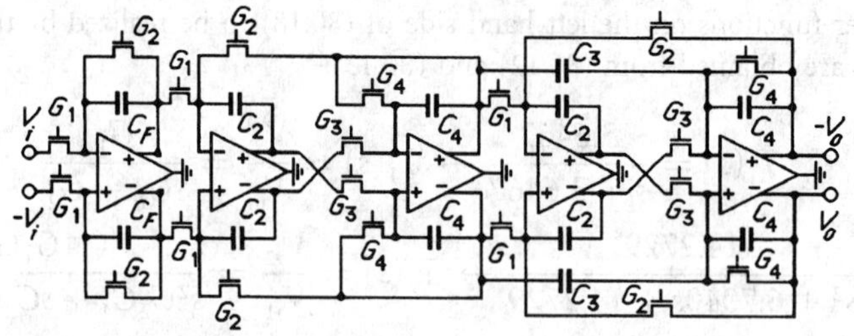

FIGURE 80.10 Circuit to realize the filter described by (80.17). Note that for easy reference we have kept the subscripts on the elements in each section the same as in Fig. 80.8.

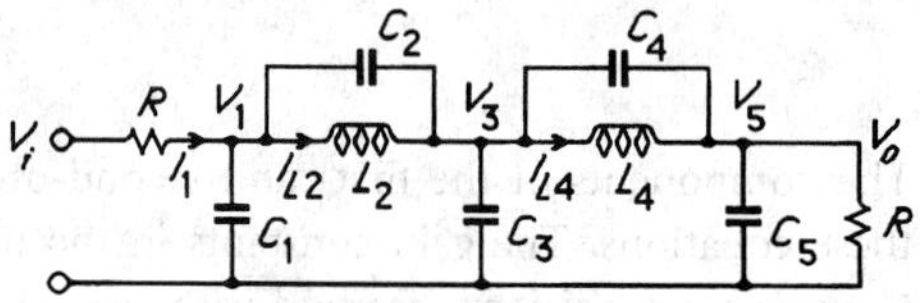

FIGURE 80.11 Fifth-order elliptic *LC* low-pass filter.

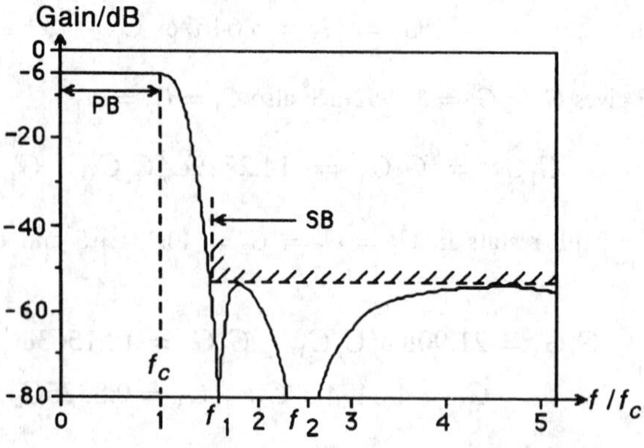

FIGURE 80.12 Transfer function magnitude of a fifth-order elliptic low-pass function.

which with the help of readily available ladder synthesis software is realized by the *LC* ladder structure in Fig. 80.11, with known component values. A plot of such a function is shown in Fig. 80.12. The two transmission zeros f_1 and f_2 in the figure are obtained when L_2, C_2, and L_4, C_4, respectively, resonate. The *LC* active simulation proceeds by deriving the signal-flow graph [4, ch. 6] equations or by writing the loop and node equations of the ladder along with the *V–I* relationships describing the functions of the elements

$$I_1 = \frac{V_i - V_1}{R} \qquad I_{L2} = \frac{V_1 - V_3}{sL_2} \qquad I_{L4} = \frac{V_3 - V_5}{sL_4} \tag{80.22a}$$

$$V_1 = \frac{I_1 - [I_{L2} + sC_2(V_1 - V_3)]}{sC_1} \tag{80.22b}$$

$$V_3 = \frac{I_{L2} - I_{L4} + sC_2(V_1 - V_3) - sC_4(V_3 - V_5)]}{sC_3} \tag{80.22c}$$

$$V_5 = V_o = \frac{I_{L4} + sC_4(V_3 - V_5)}{sC_5 + 1/R} \tag{80.22d}$$

We recognize that these equations represent *integrations* of voltages into currents and currents into voltages. We also note that the currents through the capacitors C_2 and C_4 can be taken care of efficiently without resorting to integration. By connecting C_2 and C_4 directly to the voltage nodes V_1, V_3, and V_3, V_5, respectively, they conduct the currents as prescribed in (80.22b)–(80.22d). Next we reformat (80.22a)–(80.22d) by eliminating I_1 from (80.22a) and (80.22b) and rewriting (80.22b), (80.22c), and (80.22d) such that V_1, V_3, and V_5, respectively, appear only on the left-hand side. The result is the new set of equations

$$I_{L2} = \frac{V_1 - V_3}{sL_2}, \quad I_{L4} = \frac{V_3 - V_5}{sL_4} \tag{80.23a}$$

$$V_1 = \frac{V_i G - I_{L2} + sC_2 V_3}{s(C_1 + C_2) + G} \tag{80.23b}$$

$$V_3 = \frac{I_{L2} - I_{L4} + sC_2 V_1 + sC_4 V_5}{s(C_2 + C_3 + C_4)} \tag{80.23c}$$

$$V_5 = V_o = \frac{I_{L4} + sC_4 V_3}{s(C_4 + C_5) + G} \tag{80.23d}$$

The fact that all signals in the MOSFET-*C* circuit are voltages that in turn produce currents summed at the op-amp inputs, and that the integration constant must be *time* rather than *capacitance*, is handled by scaling the equations by a resistor *R*. We illustrate the process on (80.23a) and (80.23b):

$$I_{L4} = \frac{V_3 - V_5}{sL_4} \rightarrow \frac{I_{L4}}{G} = \frac{GV_3 - GV_5}{s(L_4 G)G} \tag{80.24a}$$

$$V_1 = \frac{V_i G - I_{L2} + sC_2 V_3}{s(C_1 + C_2) + G} \rightarrow V_1 = \frac{V_i G - (I_{L2}/G)G + sC_2 V_3}{s(C_1 + C_2) + G} \tag{80.24b}$$

Notice that $L_4 G^2$ has the unit of *farad*, i.e., it is a capacitor which, however, in Figs. 80.13 and 80.14 will be labeled L_4 and L_2, respectively, to help keep track of its origin. Similarly, I_{L2}/G and I_{L4}/G will be labeled V_{I2} and V_{I4}, respectively. The integrators can now be realized as in Fig. 80.6(b) and then interconnected as the equations prescribe. Fig. 80.13 illustrates the process for (80.24a) and (80.24b). The MOSFET-*C* implementation of all appropriately scaled equations (80.23a)–(80.23d) leads to the fifth-order elliptic filter shown in Fig. 80.14. As given in (80.10), the aspect ratio of each MOSFET and the control voltage V_C is adjusted to realize the corresponding resistor values of the standard active *RC* implementation and all MOSFET gates are controlled by the same V_C for tuning purposes. Arrays of MOSFETs controlled by different values of V_C can be used to achieve programmable filter coefficients.

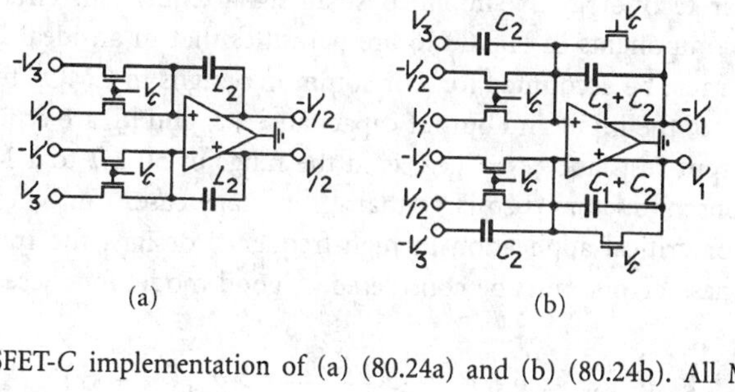

FIGURE 80.13 MOSFET-*C* implementation of (a) (80.24a) and (b) (80.24b). All MOSFETs realize the value *G*.

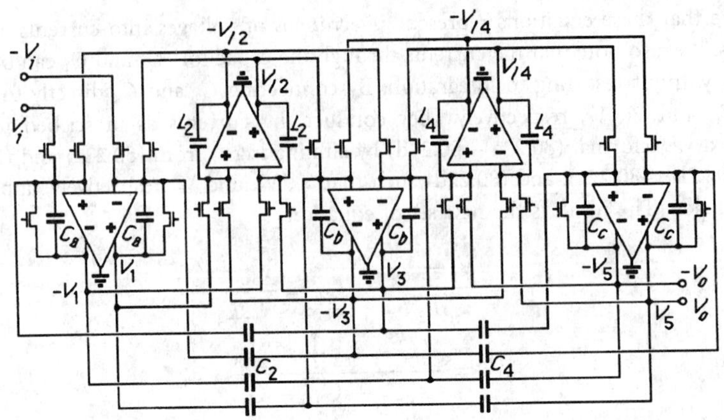

FIGURE 80.14 Operational simulation of the *LC* ladder of Fig. 80.11 with MOSFET-*C* integrators; $C_a = C_1 + C_2$, $C_b = C_2 + C_3 + C_4$, $C_c = C_4 + C_5$.

An often-cited advantage of the MOSFET-*C* technique is the reduced sensitivity to parasitic capacitors, whereas the g_m-*C* approach discussed next must carefully account for parasitics by predistortion. Note from Figs. 80.6(b), 80.8, 80.10, and 80.14 that all capacitors and the MOSFET resistors are connected to voltage-driven or to virtual ground nodes so that parasitic capacitors to ground are of no consequence as long as amplifiers with sufficiently high gain and wide bandwidth are used. Fortunately, such amplifiers are being developed [5, paper 2-B.5] so that MOSFET-*C* circuits promise to become increasingly attractive in the future.

80.3 g_m-C Filters

At the time of this writing, the dominant active device used in the design of integrated continuous-time filters is the *transconductor* (g_m) or the *operational transconductance amplifier* (OTA) [7, ch. 5]. Both names, g_m-*C* filters and OTA-*C* filters, are used in the literature; we shall use the term g_m-*C* filter in this text. The main reasons for this prevalence appear to be the simple systematic design methods for g_m-*C* filters and, especially, the higher frequency range in which g_m-based filters can operate. Also, OTAs often have simpler circuitry (fewer elements) than op-amps. A transconductor is a voltage-to-current converter described by the equation

$$I_{\text{out}} = g_m(s)V_{\text{in}} \tag{80.25}$$

where $g_m(s)$ is the frequency-dependent transconductance parameter with units of [Ampere/Volt] or [Siemens], abbreviated [S]. Typical values for g_m are tens to hundreds of μS in CMOS and up to mS in bipolar technology. A simplified small-signal equivalent circuit is shown in Fig. 80.15. The dashed components in Fig. 80.15 are parasitics that in an ideal OTA should be zero but that in practice must be accounted for. For common designs in CMOS technology, the input conductance g_i is zero, the input and output capacitances, c_i and c_o, are typically of the order of 0.05 pF and the output resistance $r_o = 1/g_o$ is in the range of 50 kΩ to 1 MΩ. The bandwidth of well-designed transconductors is so large that g_m in many cases can be regarded as constant, $g_m(s) = g_{m0}$, but for critical application in high-frequency designs the transconductance pole and the resulting phase errors must be considered. A good model for these situations is

$$g_m(s) \approx g_{m0}e^{-s\tau} \approx \frac{g_{m0}}{1 + s\tau} \approx g_{m0}(1 - s\tau) \tag{80.26}$$

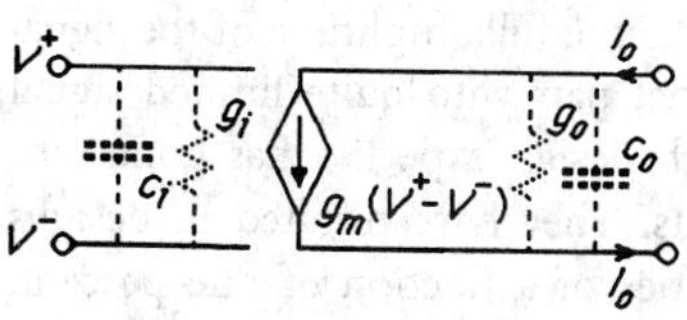

FIGURE 80.15 Small-signal equivalent circuit.

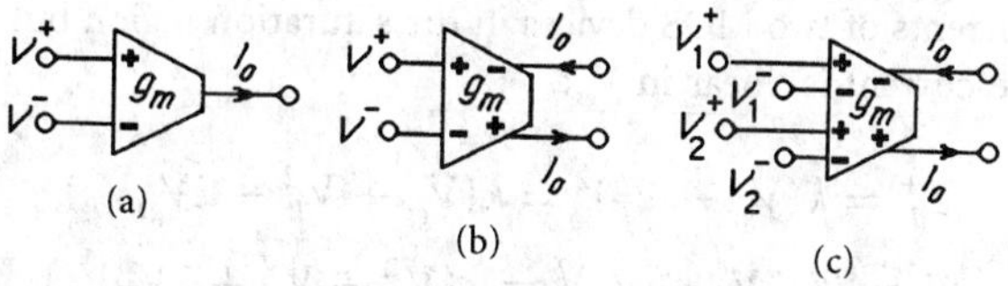

FIGURE 80.16 OTA symbols (a) differential input—single-ended output: $I_o = g_m(V^+ - V^-)$; (b) fully differential; (c) with multiple inputs: $I_o = g_m[(V_1^+ - V_1^-) + (V_2^+ - V_2^-)]$.

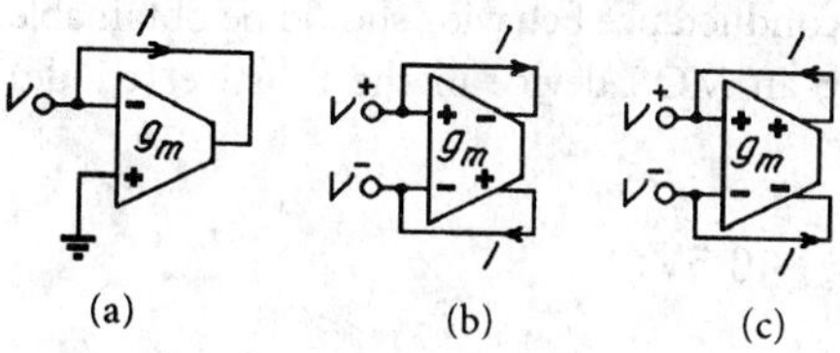

FIGURE 80.17 Simulated resistors: (a) positive single-ended of value $V/I = 1/g_m$, (b) positive differential of value $(V^+ - V^-)/I = 1/g_m$, (c) negative differential.

where $f = 1/(2\pi\tau)$ is the pole location (typically at 50 MHz to several 100 MHz) and the phase error $\Delta\phi = -\omega\tau$ is considered small, i.e., $\omega\tau \ll 1$. The three different approximations for representing the frequency dependence are equivalent; the pole is used most often, the zero frequently results in simpler equations and algebra, and the phase may give better insight into the behavior of feedback loops.

The most commonly used circuit symbols are shown in Fig. 80.16. Note that OTA designs with multiple differential inputs as in Fig. 80.16(c) are readily available and often lead to simpler filter designs with less silicon area and power consumption because only the OTA input stages must be duplicated whereas the remaining parts of the OTA, such as output and common-mode feedback circuitry, can be shared. Essentially, if two OTAs with the same g_m value in a filter have a common output node (a frequent situation), the two OTAs can be merged into the circuit of Fig. 80.16(c), thus saving circuitry and power.

Filter design methods discussed in this section use only OTAs and capacitors: OTAs to provide gain and capacitors to provide integration. To establish time constants, resistors may also be required, but their function can be obtained from OTAs: "resistors" of value $1/g_m$ can be simulated by connecting the OTA output to its input with the polarities shown in Fig. 80.17(a) and (b). Inverting one pair of terminals results in a negative resistor as in Fig. 80.17(c). Since transconductors and capacitors can be used to build all components necessary for designing filters, they are called transconductance-C or g_m-C filters. We shall discuss the remaining "composite" building blocks, integrators and gyrators,[5] in the section on cascade realizations.

In this subsection we shall not go into the electronic circuit design methods for OTAs, but refer the reader to the literature, which contains a great number of useful transconductance designs in all current technologies. In [5] are numerous papers that contain practical transconductance circuits. The most popular designs currently use CMOS, but bipolar and BiCMOS are also widely employed, and GaAs has been proposed for applications at the highest frequencies or under unusually severe environmental conditions. Since transconductors are almost always used in

[5]A gyrator is a two-port circuit whose input impedance is inversely proportional to the load impedance: $Z_{in}(s) = r^2/Z_{load}(s)$. If $Z_{load} = 1/(sC)$, the input is inductive, $Z_{in}(s) = sr^2C = sL$. r is called the gyration resistance.

open loop without local feedback, their input stages must handle the full amplitude of the signal to be processed. Typically, the OTA input stage is a differential pair with quite limited signal swing before nonlinearities become unacceptable. Thus, much design expertise has gone into developing linearization schemes for transconductance circuits. They have resulted in designs that can handle signals of the order of volts with nonlinearities of a fraction of one percent. The most commonly employed approach uses variations of the principle of taking the difference between the drain currents of two MOS devices in the saturation region but driven differentially, so that the difference current is linear in V_{gs}:

$$\begin{aligned} I_d^+ &= k(V_{gs} - V_T)^2 = k(V_{gs}^2 + V_T^2 - 2V_{gs}V_T) \\ I_d^- &= k(-V_{gs} - V_T)^2 = k(V_{gs}^2 + V_T^2 + 2V_{gs}V_T) \\ \Delta I_d &= I_d^+ - I_d^- = -4V_{gs}V_T \end{aligned} \tag{80.27}$$

Another approach reasons that the most linear (trans)conductance behavior should be obtainable from the current through a resistor. Thus, operating an MOS device in the resistive (triode) region,

$$I_d = k[(V_{gs} - V_T)V_{ds} - 0.5V_{ds}^2]$$

and taking the derivative with respect to V_{gs} for constant $V_{ds} = V_{DS}$ results in a perfectly linear transconductance,

$$g_m = \frac{dI_d}{dV_{gs}} = kV_{DS} \tag{80.28}$$

that furthermore can be adjusted (tuned) by varying a dc bias voltage (V_{DS}) as long as V_{DS} stays small enough for the transistor to remain in the triode region. The circuitry surrounding the triode-region MOS device must assure that V_{DS} remains constant and independent of the signal.

As mentioned, the literature contains numerous practical CMOS or bipolar transconductance designs that require low power-supply voltages (±2.5 V, or 0 to 5 V, or even less for low-power applications), and have acceptable signal swing (of the order of volts), excellent linearity (as low as a small fraction of one percent), and wide bandwidth (up to several hundred megahertz and even into the gigahertz range). Two further aspects of OTA design should be stressed at this point. First, since g_m-C filters often contain many transconductors, the designer ought to strive for simple OTA circuitry. It saves silicon real estate and at the same time often results in better frequency performance because of reduced parasitics at internal nodes. We point out though that there exists a trade-off between simple circuitry and large voltage swing: a wide linear signal range requires special linearizing circuit techniques and, therefore, additional components. The second issue pertains to tuning. We mentioned earlier that continuous-time filters always require tuning steps to eliminate the effects of fabrication tolerances and component drifts. In IC technologies this implies that the circuit components must be electronically adjustable. Since (MOS) capacitors are generally fixed, all tuning must be handled via the transconductance cells by changing one or more bias points. Usually two adjustments are needed: the magnitude of g_m must be varied to permit tuning the frequency parameters set by g_m/C-ratios and, as explained later, the phase $\Delta\phi$ must be varied to permit tuning the quality factors. We shall discuss tuning methods in some detail in Section 80.4.

In the next subsection we introduce the central building blocks from which g_m-C filters are constructed, the integrator and the gyrator. Just as we saw in the discussion of MOSFET-C circuits and in the earlier treatment of active RC filters, integrators are fundamental to the development of active filter structures, both for second-order sections and cascade designs, as well as for higher order *LC* ladder simulations. Gyrators along with capacitors are used to replace inductors in passive *RLC* filters so that many passive filter structures, such as *LC* ladders, can be directly translated into g_m-C form.

Basic Building Blocks

Integrators

Integrators are obtained readily by loading an OTA with a floating or a grounded capacitor as shown in Fig. 80.18. Observe that the simpler technology of grounded capacitors requires four times the capacitor value and silicon area. Ideally, the integrator realizes the transfer function

$$\frac{V_o}{V_i} = -\frac{g_m}{sC} \tag{80.29a}$$

but notice that the function is sensitive to unavoidable parasitic capacitors as well as to the OTA output conductance g_o. Observe from Fig. 80.19 that the output conductance is in parallel with the integrating capacitor C, and the output capacitances C_o from the positive and negative output nodes of the OTA circuitry to ground add to the value of C. Furthermore, the designer should bear in mind that in IC technology floating capacitors have a substantial parasitic capacitance C_s (about 10 percent of the value of C) from the bottom plate to the substrate, i.e., to ac ground. To maintain symmetry, the designer may wish to split the integrating capacitor into two halves connected such that the parasitic bottom plate capacitors 0.5 C_s appear at the two OTA outputs. The situation is illustrated in Fig. 80.19. Taking the parasitics into consideration, evidently, the integrator realizes

$$\frac{V_o}{V_i} = -\left.\frac{g_m}{sC_{\text{int}} + g_o}\right|_{s=j\omega} = -\frac{g_m}{j\omega C_{\text{int}}\left(1 - j\dfrac{1}{Q_{\text{int}}}\right)} \tag{80.29b}$$

that is, it becomes lossy with a finite integrator quality factor

$$Q_{\text{int}} = \frac{\omega C_{\text{int}}}{g_o} \tag{80.30a}$$

and an effective integrating capacitor equal to

$$C_{\text{int}} = C + \frac{1}{2}\left(\frac{C_s}{2} + C_o\right) \tag{80.30b}$$

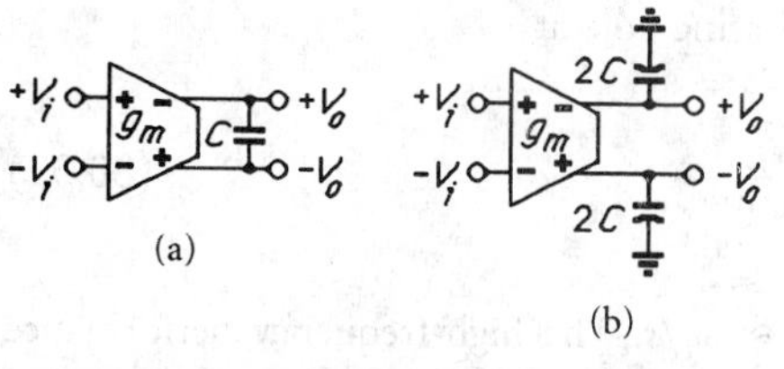

FIGURE 80.18 Integrator. The integrator capacitor may be floating (a) or grounded (b). Note that the grounded-capacitor realization requires four times the area.

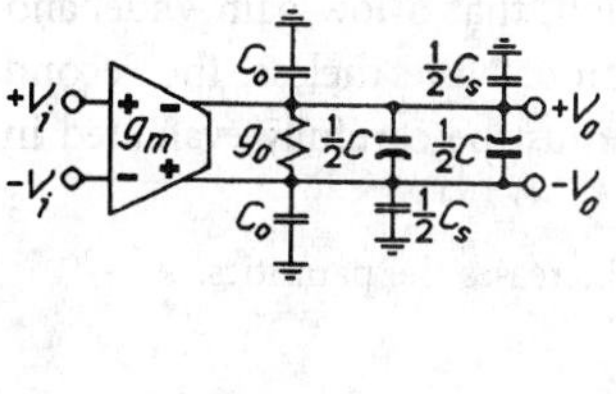

FIGURE 80.19 Parasitic capacitors associated with a g_m-C integrator.

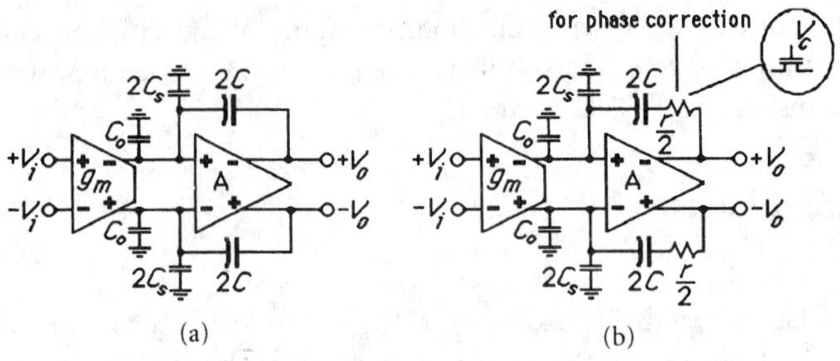

FIGURE 80.20 (a) g_m-C–op-amp integrator; (b) nominal correction for phase errors.

To maintain the correct nominal designed integration constant, the circuit capacitor C should be predistorted appropriately to reflect the parasitics appearing at the integration nodes, i.e., the parasitics should be estimated as best as possible, for example from a layout process file, and their values subtracted from the nominal value of C in the final layout. If grounded capacitors are used as in Fig. 80.18(b), the bottom plate should, of course, be connected to ground so that the substrate capacitances are connected between ground and the power supply. Thus, they are shorted out for signal currents and play no role. Observe that the presence of parasitic capacitors tends to limit the high-frequency performance of these filters because high-frequency filters require large time constants g_m/C, i.e., small capacitors.[6] The smallest capacitor C, however, must obviously be larger than the sum of all parasitics connected at the integrator output nodes to be able to absorb these parasitics. Because the values of the parasitic capacitors can generally only be estimated, one chooses typically C to be three to five times larger than the expected parasitics to maintain some predictability in the design. The reader will notice that integrators with grounded capacitors, Fig. 80.18(b), have a small advantage in high-frequency circuits where parasitic capacitors become large relative to C. Because of the problem with parasitic capacitors, an alternative "g_m-C–op-amp" integrator is also used that employs an op-amp to minimize their effects for the price of a second active device, increased noise, silicon area, and power consumption. Fig. 80.20(a) shows the configuration. Notice that now the parasitic capacitors play no role because they are connected between ground and virtual ground (the op-amp inputs) so that they are never charged. A more careful analysis shows that the integrator realizes

$$\frac{V_o}{V_i} = \frac{g_m}{sC}\frac{1}{1+\dfrac{1}{A(s)}\left(1+\dfrac{C_p}{C}\right)} \approx \frac{g_m}{sC}\frac{1}{1+\dfrac{s}{\omega_t}\left(1+\dfrac{C_p}{C}\right)} \tag{80.31}$$

where $C_p = 0.5(C_o + 2C_s)$ represents the total parasitic capacitance at each of the op-amp input terminals. Evidently, the integrator has acquired a parasitic pole at

$$s = -\omega_t \frac{C}{C + C_p} \tag{80.32}$$

where we have modeled the amplifier gain as $A(s) \approx \omega_t/s$. The high-frequency performance is now limited by the op-amp behavior, It has been shown, though, that a low-gain wideband amplifier, essentially a second OTA, can be used for this application. Nevertheless, the second active device introduces parasitic poles (and zeros) whose effects must be carefully evaluated in

[6]Increasing g_m is generally not a satisfactory solution because it also increases the parasitics.

practice. The dominant pole introduced by the op-amp may be canceled *nominally* by an rC phase lead as shown in Fig. 80.20(b). The circuit realizes

$$\frac{V_o}{V_i} = \frac{g_m}{sC} \frac{1 + sCr}{1 - \frac{\omega}{\omega_t}\omega C_p r + \frac{s}{\omega_t}\left(1 + \frac{C_p}{C}\right)} \approx \frac{g_m}{sC} \frac{1 + sCr}{1 + \frac{s}{\omega_t}\left(1 + \frac{C_p}{C}\right)} \tag{80.33}$$

so that r should be chosen as

$$r = \frac{1}{\omega_t C}\left(1 + \frac{C_p}{C}\right) \tag{80.34}$$

to cancel the pole. The small resistor r may be a MOSFET in the triode region as indicated in Fig. 80.20(b) so that $r(V_C)$ becomes variable for any necessary fine adjustments. Notice that the cancellation is only *nominal* since (80.34) can never be satisfied exactly because of the uncertain values of ω_t and C_p.

Gyrators

A gyrator is defined by the equations

$$V_i = -rI_o \qquad V_o = rI_i \tag{80.35a}$$

that is, the input impedance $Z_{\text{in}}(s)$ is inversely proportional to the load impedance $Z_{\text{load}}(s)$:

$$Z_{\text{in}}(s) = \frac{V_i}{I_i} = -r^2\frac{I_o}{V_o} = r^2\frac{1}{Z_{\text{load}}(s)} \tag{80.35b}$$

r is the so-called gyration resistance. If a gyrator is loaded by a capacitor $Z_{\text{load}}(s) = 1/(sC)$, the input impedance is proportional to frequency, i.e., it behaves like an inductor of value sr^2C:

$$Z_{\text{in}}(s) = sr^2C = sL \tag{80.36}$$

Equations (80.35a) indicate that a gyrator can be interpreted as a connection of an inverting and a noninverting transconductor of value $g_m = 1/r$ as shown in Fig. 80.21(a). This fact makes it very easy to build excellent gyrators with OTAs (see Fig. 80.21(b), and (c)), whereas it has been found quite difficult to obtain good gyrators with op-amps. An exception is Antoniou's general impedance converter circuit [4, ch. 5], but it is useful only at relatively moderate frequencies (up to about 5 to 10 percent of the op-amp's gain–bandwidth product); also, the circuit contains resistors that are not voltage-driven and, therefore, cannot readily be translated into a MOSFET-C equivalent, as was discussed previously. The availability of good gyrators provides us with a convenient method for g_m-C high-frequency integrated ladder filters that is based on inductor-replacement to be discussed in the section on element replacement methods.

Notice that the comments made earlier about the effects of parasitic capacitors apply also to inductor simulation: the parasitic input and output capacitors of the OTAs and the bottom-plate-to-substrate capacitances (Fig. 80.19) add to the capacitor used to set the value of the simulated inductor. For instance, using the same notation as in Fig. 80.19, the effective capacitor in Fig. 80.21(b) equals

$$C_{\text{eff}} = C + \frac{1}{2}\left(\frac{Cs}{2} + C_i + C_o\right) \tag{80.37}$$

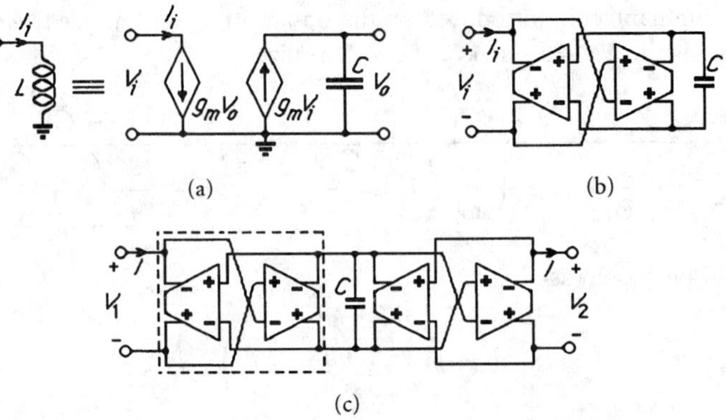

FIGURE 80.21 (a) Controlled-source realization of capacitively-loaded gyrator to realize a grounded inductor; (b) differential g_m-C implementation; (c) a floating inductor requires two gyrators.

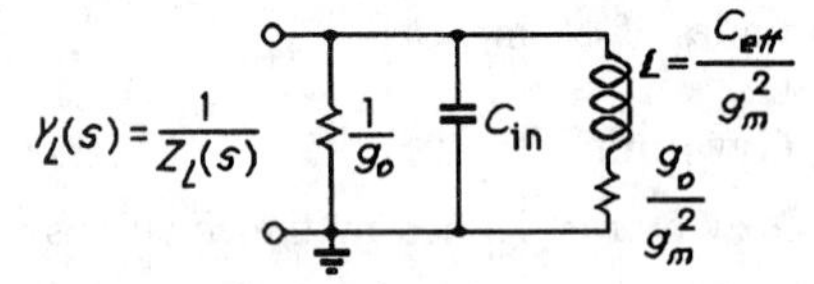

FIGURE 80.22 Passive equivalent circuit for the grounded inductor simulated in Fig. 80.21(b).

where C_i is the parasitic capacitance at the input terminals of the OTA, see Fig. 80.15, and C is assumed to be split as in Fig. 80.19. Note also that a parasitic capacitor of value $C_{\text{in}} = C_i + C_o$ is measured across the inductor of Fig. 80.20(b) so that the inductor

$$L_{\text{eff}} = \frac{C_{\text{eff}}}{g_m^2} \tag{80.38}$$

has a self-resonance frequency

$$\omega_0 = \frac{1}{\sqrt{L_{\text{eff}} C_{\text{in}}}} \tag{80.39}$$

Finally, to complete the inductor model, recall that the OTAs have finite differential output conductances[7] g_o (see Fig. 80.15), which appear across the input and load terminals of the gyrator in Fig. 80.21(b). Consequently, the full inductive admittance realized by Fig. 80.21(b) equals

$$Y_L(s) = g_o + sC_{\text{in}} + \frac{g_m^2}{sC_{\text{eff}} + g_o} \tag{80.40}$$

to yield the equivalent circuit in Fig. 80.22 for the inductor L. The designer should keep this circuit in mind when using this method to develop a filter.

Evidently, the realized quality factor of the inductor equals $Q_L = \omega C_{\text{eff}}/g_o$, which means OTAs with large output resistance $r_o = 1/g_o$ are needed for high-quality inductor simulations.

We shall discuss next first- and second-order g_m-C sections used as building blocks for the cascade approach to high-order filter design. As was the case for the MOSFET-C method, we shall see that g_m-C sections are constructed by interconnecting integrators.

[7]As was mentioned, the input conductances can normally be neglected.

First- and Second-Order Sections

Consider the integrator in Fig. 80.18(a) loaded by the resistor in Fig. 80.17(b) to make it lossy. Let the input signals also be fed through capacitors into the integrator output nodes as shown in Fig. 80.23. The circuit is readily analyzed by writing Kirchhoff's current law at the output node to yield

$$\frac{V_o}{V_i} = -\frac{sC_1 + g_{m1}}{s(C_1 + C) + g_{m2}} \tag{80.41}$$

The circuit may realize any desired first-order function by choosing the appropriate values for the transconductances and the capacitors. For example, a low-pass can be obtained by setting $C_1 = 0$; a high-pass results from $g_{m1} = 0$ and possibly $C = 0$; and $C = 0$ and $g_{m1} = -g_{m2}$ results in an all-pass function.

A second-order block can be designed by interconnecting two integrators in a variety of feedback configurations. To keep the method more transparent we show one such possibility in Fig. 80.24 with single-ended outputs. A differential structure is obtained by mirroring the circuit at ground and duplicating the appropriate components, as was demonstrated in connection with Fig. 80.4. Let us disregard for now the dashed OTA with inputs V_a and V_b and apply an input signal V_i; writing the node equations for Fig. 80.24 we obtain

$$g_{m3}(V_i - V_2) + g_{m4}(V_1 - V_3) = 0 \quad sC_1V_1 = -g_{m1}V_3 \quad sC_2V_2 = -g_{m2}V_1 \tag{80.42}$$

By eliminating two of the three voltages V_1, V_2, or V_3 from these equations we obtain the bandpass, low-pass and high-pass functions, respectively,

$$\frac{V_1}{V_i} = H_{\rm BP}(s) = \frac{-sC_2g_{m1}g_{m3}}{s^2C_1C_2g_{m4} + sC_2g_{m1}g_{m4} + g_{m1}g_{m2}g_{m3}} \tag{80.43a}$$

$$\frac{V_2}{V_i} = H_{\rm LP}(s) = \frac{g_{m1}g_{m2}g_{m3}}{s^2C_1C_2g_{m4} + sC_2g_{m1}g_{m4} + g_{m1}g_{m2}g_{m3}} \tag{80.43b}$$

$$\frac{V_3}{V_i} = H_{\rm HP}(s) = \frac{s^2C_1C_2g_{m3}}{s^2C_1C_2g_{m4} + sC_2g_{m1}g_{m4} + g_{m1}g_{m2}g_{m3}} \tag{80.43c}$$

In *any* electrical network, one can generate different numerator polynomials, i.e., different transmission zeros, without disturbing the poles, i.e., the system polynomial, by applying an input *voltage* to any node that is lifted off *ground*, or by sending an input *current* into any

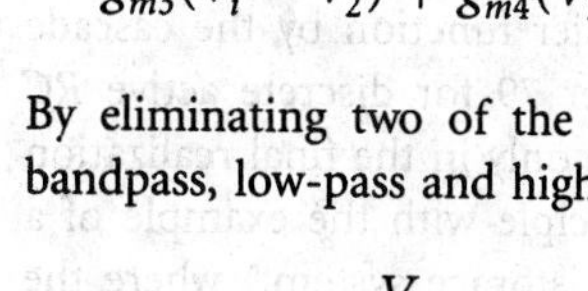

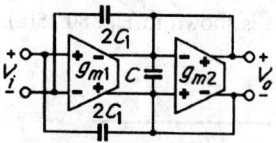

FIGURE 80.23 First-order g_m-C section.

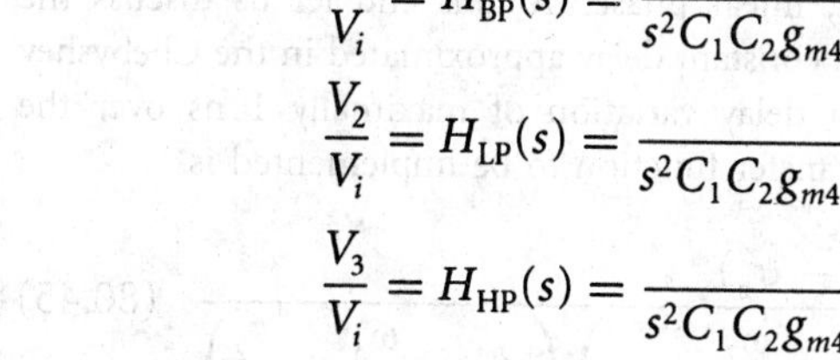

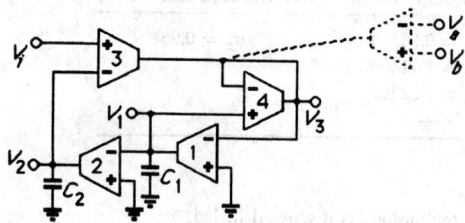

FIGURE 80.24 A general g_m-C biquadratic section.

floating node. The second of these possibilities is illustrated in Fig. 80.24 in dashed form where a current $g_m(V_b - V_a)$ is sent into node 3. We leave the analysis to the reader. We shall demonstrate the first possibility by applying V_i to the noninverting terminals of OTA_1 and OTA_2, which are grounded in Fig. 80.24. The reader may show by routine analysis that lifting these two nodes off ground and then applying V_i to both of them (in addition to the original input) results in the complete biquadratic transfer function

$$\frac{V_3}{V_i} = \frac{s^2 C_1 C_2 g_{m3} + s g_{m1}(C_2 g_{m4} - C_1 g_{m3}) + g_{m1} g_{m2} g_{m3}}{s^2 C_1 C_2 g_{m4} + s C_2 g_{m1} g_{m4} + g_{m1} g_{m2} g_{m3}} \tag{80.44}$$

with which, for example, a notch filter may be realized by setting $C_2 g_{m4} = C_1 g_{m3}$.

A great variety of second-order sections can be found in the literature. Many are designed for specific transfer functions rather than the general circuit in Fig. 80.24, and are often simpler and contain fewer OTAs. Readers are well advised to scan the available literature for the best circuit for their applications.

Cascade Realizations

Apart from modularity and simple design methods, the main advantage of the cascade approach is its generality: a cascade structure can realize a transfer function with arbitrary zero locations, whereas simulations of lossless *LC ladders* discussed below are restricted to $j\omega$ axis transmission zeros. For g_m-C integrated filters, implementing a prescribed transfer function by the cascade method follows the same principles that were discussed in Chapter 79 for discrete active *RC* filters and leads to the filter structure in Fig. 80.9; the difference lies only in the final realization of the sections in monolithic form. We shall demonstrate the principle with the example of a high-frequency filter for the read/write channel of a magnetic disk storage system,[8] where the most critical specification is constant delay, i.e., linear phase. To this end let us discuss the design of a seventh-order cascade low-pass with a constant delay approximated in the Chebyshev (equiripple) sense. The specifications call for a delay variation of maximally 1 ns over the passband and a bandwidth $f_c = 10$ MHz. The transfer function to be implemented is

$$H_7(s) = \frac{K_0(s^2 - \sigma_z^2)}{(s+\sigma)\left(s^2 + s\dfrac{\omega_1}{Q_1} + \omega_1^2\right)\left(s^2 + s\dfrac{\omega_2}{Q_2} + \omega_2^2\right)\left(s^2 + s\dfrac{\omega_3}{Q_3} + \omega_3^2\right)} \tag{80.45}$$

with the required parameters given in Table 80.2. The purpose of the symmetrical pair of zeros at $\pm\sigma_z$ is magnitude equalization to effect a gain boost for pulse slimming. Note that these zeros do not affect the phase or the delay because the factor $(\omega^2 + \sigma_z^2)$ is real on the $j\omega$ axis. The fully differential second-order circuit chosen for the low-pass sections is shown in Fig. 80.25(a).

TABLE 80.2 Filter Parameters for (80.45)

	Pole Frequency (Normalized to 10^7 s^{-1})	Pole Quality Factor	Zero (Normalized to 10^7 s^{-1})
Biquad 1	$\omega_1 = 1.14762$	$Q_1 = 0.68110$	$\sigma_z = 0.95$
Biquad 2	$\omega_2 = 1.71796$	$Q_2 = 1.11409$	
Biquad 3	$\omega_3 = 2.31740$	$Q_3 = 2.02290$	
Section 4	$\sigma = 0.86133$	—	

[8] A similar commercially successful design in bipolar technology is discussed in [1].

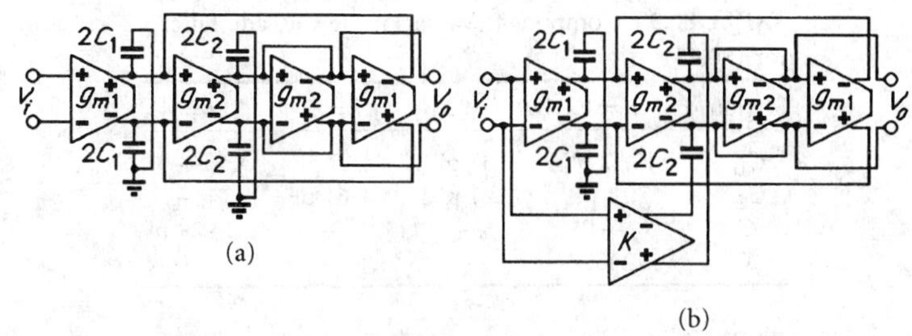

FIGURE 80.25 (a) Second-order low-pass g_m-C section; (b) an equalizer section.

Simple analysis shows that it realizes the function

$$H_{\mathrm{LP}}(s) = \frac{V_{\mathrm{LP}}}{V_i} = \frac{g_{m1}g_{m2}}{s^2 C_1 C_2 + sC_1 g_{m2} + g_{m1}g_{m2}} = \frac{\omega_0^2}{s^2 + s\omega_0/Q_0 + \omega_0^2} \tag{80.46}$$

where

$$\omega_0 = \sqrt{\frac{g_{m1}g_{m2}}{C_1 C_2}} \tag{80.47a}$$

and

$$Q_0 = \sqrt{\frac{g_{m1}C_2}{g_{m2}C_1}} \tag{80.47b}$$

are the pole frequency and pole quality factor, respectively. There remains the question of how to obtain the two real transmission zeros at $\pm\sigma_z$. For this purpose we recall that the zeros of a transfer function can be changed without destroying the poles by feeding the input signal or a fraction thereof into any of the ground nodes lifted off ground. For the situation at hand this can be accomplished by lifting the capacitors $2C_2$ in the low-pass off ground and feeding KV_i with the appropriate polarity into the terminals so generated. Fig. 80.25(b) shows the resulting circuit, which may be analyzed to yield the transfer function

$$\frac{V_o}{V_i} = \frac{-Ks^2 C_1 C_2 + g_{m1}g_{m2}}{s^2 C_1 C_2 + sC_1 g_{m2} + g_{m1}g_{m2}} \tag{80.48}$$

with zeros at $\sigma_z = \pm\omega_0 K^{-1/2}$. Observe that the two transconductors labeled g_{m1} and g_{m2}, respectively, have common output terminals; they can, therefore, be merged into one double-input transconductor each as discussed in connection with Fig. 80.16(c). Further, we need the circuit in Fig. 80.23 with $C_1 = 0$ to realize a first-order section, a lossy integrator with the function

$$\frac{V_o}{V_i} = -\frac{g_{m1}}{sC + g_{m2}} \tag{80.49}$$

TABLE 80.3 Component Values of the Cascade Filter for (80.45)

Section	$i = 1$	$i = 2$	$i = 3$	$i = 4$
C_{1i}	15.77 pF	6.31 pF	2.45 pF	—
C_{2i}	7.12 pF	7.81 pF	10.61 pF	—
C	—	—	—	14.29 pF
K	1.459	—	—	—

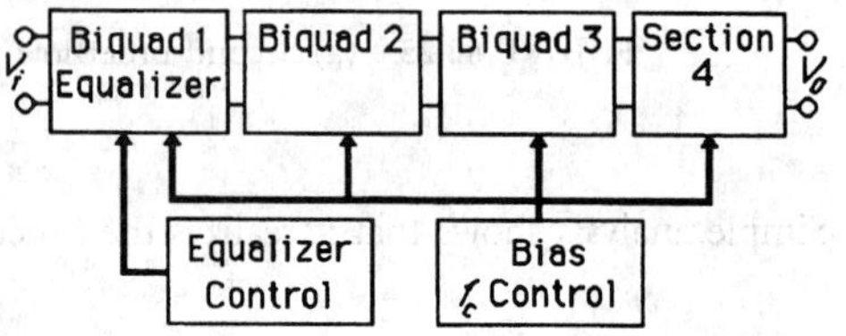

FIGURE 80.26 Structure of the seventh-order low-pass filter including control circuitry.

Finally, we must determine the component values. To this end we compute from (80.47) and (80.49) the relationships

$$C_1 = \frac{g_m}{\omega_i Q_i} = \frac{12.5}{\omega_i Q_i}\ \text{pF} \quad C_2 = g_m \frac{Q_i}{\omega_i} = 12.5\frac{Q_i}{\omega_i}\ \text{pF}$$
$$C = \frac{g_m}{\sigma} = \frac{12.5}{\sigma}\ \text{pF} \quad K = \left(\frac{\omega_1}{\sigma_z}\right)^2 \tag{80.50}$$

where we used that ω_i in Table 80.2 was normalized by a factor $10^7\ \text{s}^{-1}$ and we assumed for simplicity that all OTAs have the same transconductance values,[9] $g_{m1} = g_{m2} = g_m = 125\ \mu\text{S}$. Lastly, to illustrate the need to account for parasitics, we observe that in Fig. 80.25 the capacitor C_1 is paralleled by $2C_o$ and $1C_i$, and the capacitor C_2 by $2C_o$ and $3C_i$. The third input capacitor in parallel with C_2 is arrived at by the fact that each biquad must drive the next biquad in the cascade connection and there sees an OTA input. The capacitor C in Fig. 80.23 is paralleled by $2C_o$ and $1C_i$. To arrive at numerical values we shall assume $C_o = 0.09$ pF and $C_i = 0.04$ pF. Consequently, the capacitor values are obtained from (80.50) as

$$C_1 = \frac{12.5}{\omega_i Q_i}\ \text{pF} - 0.22\ \text{pF} \quad C_2 = 12.5\frac{Q_i}{\omega_i}\ \text{pF} - 0.3\ \text{pF} \quad C = \frac{12.5}{\sigma}\ \text{pF} - 0.22\ \text{pF} \tag{80.51}$$

Table 80.3 contains the computed capacitor values and Fig. 80.26 shows the final cascade block diagram with the equalizer section leading and the first-order low-pass at the end. The two control blocks are necessary to be able to tune the filter electronically: the gain K is varied via the *Equalizer Control* block to set the position of the zeros and thereby the amount of gain boost, i.e., the pulse slimming. Electronic tuning of the frequency parameter g_m/C is accomplished via adjusting the bias currents of the OTAs by the block labeled *Bias* f_c *Control.* Thereby uncontrollable changes in the value g_m/C due to process variations or temperature can be accounted for. Details of such a control scheme are discussed in Chapter 80.4.

[9]This assumption is for convenience of design and layout because it permits a given transconductance cell to be used throughout the circuit. An advantage of different transconductance values in different sections is that capacitor values may be equalized, since by (80.50) the capacitors are proportional to g_m.

Ladder Simulations

As mentioned earlier, the main reason for using the popular *LC* ladder simulation method for filter design is the generally lower passband sensitivity of this topology to component tolerances [4, ch. 3], see also Chapter 68. As before, the procedures are in principle identical to those discussed in connection with discrete circuits and are best illustrated with the help of a generic example. Let us consider again the classical ladder structure in Fig. 80.11, which realizes the fifth-order elliptic low-pass characteristic (80.21) and is described by (80.22). Two methods are available to simulate the ladder. The first and most intuitive method to be discussed replaces the inductors L_2 and L_4 by capacitively loaded gyrators (Fig. 80.21). The second method recognizes that the inductors and the grounded capacitors in Fig. 80.11 perform the function of integration. This signal-flow graph method is completely analogous to the process discussed earlier for MOSFET-*C* filters and will be presented in the section on signal-flow graph methods.

Element Replacement Methods

Replacing the inductors L_2 and L_4 by capacitively-loaded *gyrators* leads to the circuit in Fig. 80.27(a). It is obtained by first converting the voltage source to a current source (Norton transformation), which also converts the series source resistor R into a shunt resistor. The first OTA in Fig. 80.27(a) performs the source transformation, the second OTA is the grounded resistor. Since the two inductors are floating, two gyrators, i.e., four OTAs [see Fig. 80.21(c)] are used for the implementation. Note that all OTAs are identical, and that all capacitors except C_2 and C_4 could be grounded; for example, instead of connecting C_1 *between* nodes A and B in Fig. 80.27(a), capacitors of value $2C_1$ could be connected from *both* nodes A and B to ground. Comparing the active circuit with the *LC* prototype identifies readily both structure[10] and components. The element values are obtained directly from the prototype *LC* ladder, i.e., from published tables, e.g., [8], or from appropriate synthesis software. Labeling for the moment the normalized components in the prototype *LC* ladder by the subscript n, i.e., R_n, $C_{i,n}$, and $L_{i,n}$, the transformation into the real component values with units of [F] and [H] is achieved by the following equations:

$$R = \frac{R_n}{g_m}, \quad C_i = C_{i,n}\frac{g_m}{\omega_c}, \quad L_i = L_{i,n}\frac{1}{g_m\omega_c} = \frac{C_{Li}}{g_m^2} \rightarrow C_{Li} = L_{i,n}\frac{g_m}{\omega_c} \tag{80.52}$$

g_m is the transconductance value chosen by the designer, ω_c is the normalizing frequency (usually the specified passband corner frequency), and R_n is in most prototype designs normalized to $R_n = 1$. Naturally, as discussed earlier, all capacitor values must be predistorted to account for the parasitic capacitors appearing at the capacitor nodes. For example, note that C_{L2} is paralleled by two OTA input and two OTA output capacitors, and for symmetrical layout (Fig. 80.19), by $0.25C_s$. We see that the element replacement method is a very straightforward design process; it has been found to work very well in practice. Fig. 80.27(b) shows the same circuit but realized with dual-input OTAs [Fig. 80.16(c)]. As was pointed out a number of times, this merging of OTAs can always be used when two (or more) OTAs share a common output node. It results in simplified circuitry, and possibly reduced power consumption and chip area. In such cases only the linearized input stages of the OTA must be duplicated, but bias, output, and common-mode rejection circuitry can be shared. Observe also that the input voltage is applied to both inputs of the first dual-input OTA, thereby doubling its value of g_m. This multiplies the transfer function by a factor of two and eliminates the 6-dB loss inherent in the *LC* ladder (see Fig. 80.12).

[10]The number of floating capacitors is, of course, doubled because of the balanced differential structure of the active implementation.

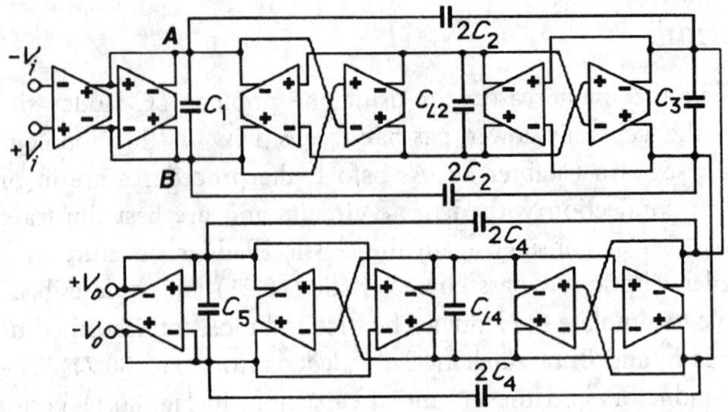

(a)

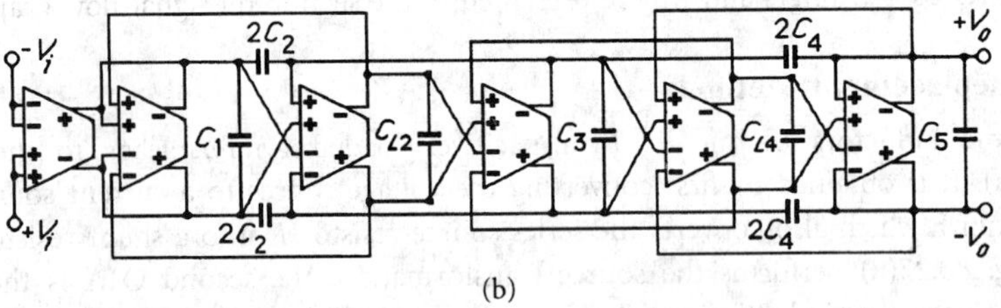

(b)

FIGURE 80.27 (a) Transconductor-C simulation by the element replacement method of the fifth-order elliptic low-pass ladder in Fig. 80.11, including source and load resistors. All g_m cells are identical. Note that the floating inductors require two gyrators for implementation. (b) The circuit with dual-input OTAs.

A small example will show the design process. For an antialiasing application we need to realize a third-order elliptic low-pass filter with $f_c = 10$ MHz bandwidth, 0.5-dB passband ripple, 17.5-MHz stopband edge and 23-dB stopband attenuation. It leads to the transfer function

$$H(s) = \frac{0.28163(s^2 + 3.2236)}{(s + 0.7732)(s^2 + 0.5016s + 1.1742)} \tag{80.53}$$

and the normalized element values $R_n = 1$, $C_{1n} = C_{3n} = 1.293$, $C_{2n} = 0.3705$ and $L_{2n} = 0.8373$. The topology is as in Fig. 80.11 with $C_{4n} = L_{4n} = C_{5n} = 0$. Fig. 80.28 shows the active circuit. Observe that we have realized each of the floating capacitors C_i as $0.5C_i + 0.5C_i$ with inverted bottom-plate connections: this design choice preserves symmetry in the balanced differential layout by placing the unavoidable substrate capacitors of value $\approx 0.1 \times (0.5C_{L2})$ *at each* of the upper and lower nodes of C_{L2} and $\approx 0.1 \times (0.5C_i + C_2)$ at the upper and lower nodes of C_i, $i = 1, 3$. Choosing the value of transconductance as $g_m \approx 180$ μS, using (80.51), $\omega_c = 2\pi \cdot 10 \cdot 10^6$ s^{-1}, and observing the necessary predistortion for the *differential* parasitic OTA input and output capacitors $C_i = 0.03$ pF and $C_o = 0.08$ pF, respectively, and for the bottom-plate capacitors C_s assumed to be 10 percent of the corresponding circuit capacitor value, results in

$$\begin{aligned}
C_1 &= C_{1n}\frac{g_m}{\omega_c} - 3C_o - 2C_i - 0.1\frac{1}{2}\left(C_2 + \frac{C_1}{2}\right) = 3.191 \text{ pF} \\
C_2 &= C_{2n}\frac{g_m}{\omega_c} = 1.061 \text{ pF} \\
C_3 &= C_{3n}\frac{g_m}{\omega_c} - 2(C_o + C_i) - 0.1\left(\frac{C_2}{2} + \frac{C_3}{4}\right) = 3.268 \text{ pF} \\
C_{L2} &= L_{2n}\frac{g_m}{\omega_c} - 2(C_o + C_i) - 0.1\frac{C_L}{4} = 2.120 \text{ pF}
\end{aligned} \tag{80.54}$$

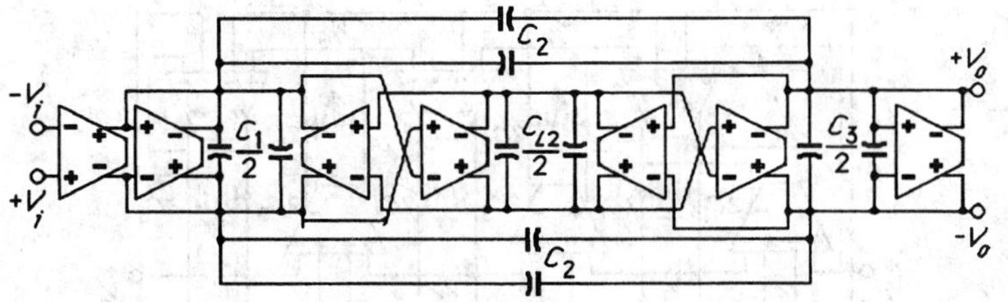

FIGURE 80.28 (a) Realization of the third-order elliptic low-pass ladder of (80.51). The capacitor values indicated refer to *each* of the pair of capacitors.

Notice that in a ladder structure all parasitic capacitors can be absorbed in the circuit capacitors. Consequently, no new parasitic poles or zeros are created that might destroy the transfer function shape. This is a further important advantage of the ladder-simulation method for g_m-C filters.

Signal-Flow Graph Methods

As we have seen earlier, the signal-flow graph (SFG) or operational-simulation method takes the circuit equations describing the ladder (Kirchhoff's laws and the I–V relationships for the elements) and realizes them directly via summers (for Kirchhoff's laws) and integrators (for inductors and grounded capacitors). The procedure was detailed in connection with (80.22)–(80.24). Recall that (80.23) represents *integrations* of voltages into currents and currents into voltages. As was the case for the MOSFET-C design, all signals in the SFG g_m-C circuit are voltages that here are summed at the OTA inputs, then multiplied by g_m to produce an output current, that is integrated by a capacitor to produce a voltage as input for the OTA of the next stage. To reflect these facts in the relevant equations, we scale (80.23) analogously to (80.24) to obtain

$$\frac{I_{L2}}{g_m} = \frac{g_m(V_1 - V_3)}{s(L_{2n}g_m)g_m} \quad \frac{I_{L4}}{g_m} = \frac{g_m(V_3 - V_5)}{s(L_{4n}g_m)g_m} \tag{80.55a}$$

$$V_1 = \frac{g_m V_i - (I_{L2}/g_m)g_m + sC_{2n}V_3}{s(C_{1n} + C_{2n}) + g_m} \tag{80.55b}$$

$$V_3 = \frac{(I_{L2}/g_m)g_m - (I_{L4}/g_m)g_m + sC_{2n}V_1 + sC_{4n}V_5}{s(C_{2n} + C_{3n} + C_{4n})} \tag{80.55c}$$

$$V_5 = V_o = \frac{(I_{L4}/g_m)g_m + sC_{4n}V_3}{s(C_{4n} + C_{5n}) + g_m} \tag{80.55d}$$

The scaling factor in this case is the design transconductance g_m. Note that $L_i g_m^2$ in (80.55a) has units of capacitance and that source and load resistors in (80.55b) and (80.55d) have the value $1/g_m$. Implementing these equations with lossless or lossy, as appropriate, integrators in fully differential form results in the circuit in Fig. 80.29, where we used a signal notation similar to that in Fig. 80.14 and chose all integrating capacitors grounded. Starting from a normalized LC prototype, the actual component values are obtained again via (80.52). Note that the OTA at the input performs the voltage-to-current conversion (V_i to I_1) and that the last OTA both here and in Fig. 80.27 implements the load resistor. The second OTA in Fig. 80.27, realizing the source resistor, is saved in Fig. 80.29 by sending the current I_1 directly into the integrating node (the capacitor C_1), as suggested by (80.55b).[11] Also observe that circuit complexity in Fig. 80.29

[11]Had we used this "resistor," the current into the integration node would have been realized as $g_m V_i = g_m V_i \times (1/g_m) \times g_m$, clearly a redundant method.

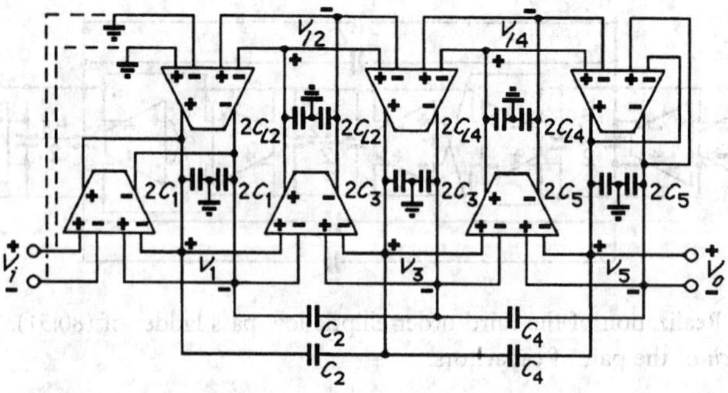

FIGURE 80.29 Signal-flow graph g_m-C realization of a fifth-order elliptic low-pass function. The circuit is an active simulation of the *LC* ladder in Fig. 80.11, including source and load resistors. All g_m cells are identical.

was kept low by resorting to OTAs with dual inputs. We note again that all transconductors in Fig. 80.27–80.29 are identical[12] so that a single optimized g_m cell (an *analog gate*) can be used throughout the filter chip for an especially simple IC design process, analogous to that of a gate array. The inherent 6-dB loss of the *LC* ladder can also be eliminated, if desired by lifting the two grounded inputs of the second OTA filter in Fig. 80.29 off ground connecting them to the input voltage $\pm V_i/2$ as indicated by the dashed connections.

As a simple design example consider again the realization of the third-order elliptic low-pass filter described by (80.53). The nominal *LC* elements were given earlier; the circuit is described by (80.55) with $C_{4n} = C_{5n} = L_{4n} = 0$, i.e.,

$$\frac{I_{L2}}{g_m} = \frac{g_m(V_1 - V_3)}{s(L_{2n}g_m)g_m} \tag{80.56a}$$

$$V_1 = \frac{2g_m V_i - (I_{L2}/g_m)g_m + sC_{2n}V_3}{s(C_{1n} + C_{2n}) + g_m} \tag{80.56b}$$

$$V_3 = V_o = \frac{(I_{L2}/g_m)g_m + sC_{2n}V_1}{s(C_{2n} + C_{3n}) + g_m} \tag{80.56c}$$

which leads to the circuit in Fig. 80.30(a), as the reader may readily verify. Note the connection of the input to realize the term $2g_m V_i$ in (80.56b) and eliminate the 6-dB loss as mentioned earlier. We observe again that the capacitors must be predistorted to account for parasitics as was discussed in connection with (80.54); specifically, labeling as before the *differential* OTA input and output capacitors as C_i and C_o, respectively, we find

$$C_1 \approx C_{1,\text{nominal}} - 2C_o - 2C_i; \quad C_{L2} \approx C_{L2,\text{nominal}} - C_o - 2C_i; \quad C_3 \approx C_{3,\text{nominal}} - C_o - 2C_i$$

Fig. 80.30(b) shows the experimental performance of the circuit fabricated in 2-μm CMOS technology with design-automation software that uses the OTAs from a design library as "analog gates," and *automatically* lays out the chip and predistorts the capacitors according to the process file. The transconductance value used is $g_m \approx 180$ μS, and the capacitors are approximately the same as the ones for the example in Fig. 80.28. We shall see in Section 80.5 that apart from differences due to layout parasitics it is no coincidence that the element values are essentially

[12]This can generally be achieved in g_m-C ladder simulations; the only exception occurs for those *LC* ladders that require unequal terminating resistors, such as even-order Chebyshev filters; in that case, one of the transconductors will also be different.

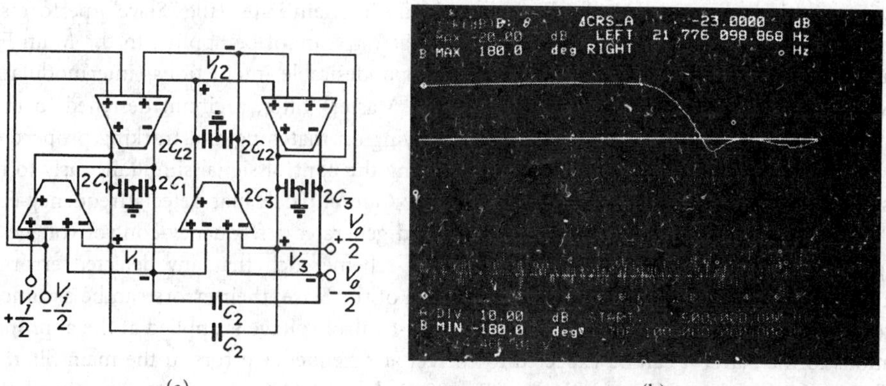

(a) (b)

FIGURE 80.30 (a) SFG g_m-C third-order elliptic low-pass filter. (b) Experimental performance in 2-μm CMOS technology.

the same as before. Notice that the filter meets all design specifications. The lower trace is the thermal noise at ≈ 70 dB below the signal; the measured total harmonic distortion was THD< 1 percent for a 2-V_{p-p} input signal at 3 MHz.

80.4 Tuning

To obtain accurate filter performance with frequency-parameters set by *RC* products or C/g_m ratios, accurate *absolute* values of components are required. These must be realized by the IC process *and maintained during operation.* Although IC processing is very reliable in realizing accurate *ratios* of like components on a chip, the processing tolerances of *absolute values* must be expected to be of the order of 20–50 percent or more. Tolerances of this magnitude are generally far too large for an untuned filter to perform within specifications. Consequently, filters *must be tuned* to their desired performance by adjusting element values. Clearly, in fully integrated filters, where all components are on a silicon chip, tuning must be performed electronically by some suitably designed automatic control circuitry that is part of the total continuous-time filter system. Tuning implies *measuring* filter performance, *comparing* it with a standard, *calculating* the error, and *applying a correction* to the system to reduce the error. An accurate *reference frequency*, e.g., a system clock (V_{REF} in Fig. 80.31), is usually used as a convenient standard. From the filter's response to the signal at this known frequency, errors are detected and the appropriate corrections are applied via the control circuitry [4, ch. 7].

Figure 80.31 shows a block diagram of this "Master–Slave" scheme that is followed by most currently proposed designs. They are based on the premise that the filter must operate continuously, i.e., the signal cannot be switched off occasionally to permit tuning in a signal-free environment, thereby alleviating serious matching, noise, and crosstalk problems. The figure should help understand the principle of the operation without having to consider the actual circuitry, which is very implementation-specific. Reference [6, pt. 6] contains many papers

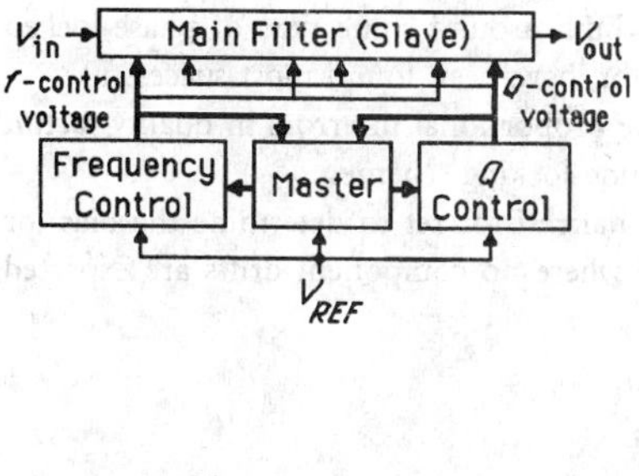

FIGURE 80.31 Block diagram of Master–Slave control system for integrated c-t filters.

showing practical approaches to the tuning problem. The Main Filter (the "Slave") performs the required signal processing. Since the reference signal V_{REF} cannot be applied to the Main Filter simultaneously with the main signal V_{in} because of undesirable interactions (intermodulation) between V_{in} and V_{REF}, it is instead applied to the "Master" filter, a circuit designed to model the Slave's behavior that is relevant for tuning. Relying on matching and tracking properties of Master and Slave, tuning is accomplished by applying the control signal simultaneously to both Master and Slave. The system contains a *Frequency-Control* block that detects frequency-errors in the Master's response to the reference signal and generates a frequency-control voltage that is applied to the Master in a closed-loop control scheme such that any detected errors are minimized.[13] Since the Master is an accurate model of the Slave, their errors can be assumed to match and track. Consequently, when the frequency-control voltage is applied at the appropriate locations to the Slave, it can be expected to correct any frequency errors in the main filter.

The purpose of the additional block in Fig. 80.31, labeled *Q-Control*, is to tune the *shape* of the transfer characteristic. Once the frequency parameters are correct as tuned by the f-control loop, the transfer function shape is determined only by the quality factors Q_i of the poles and zeros. Q, as a ratio of two frequencies, is a *dimensionless* number and as such is determined by a *ratio* of like components [resistor, capacitor, and/or g_m ratio; see e.g., (80.47b)]. At *fairly low frequencies and moderate values of* Q, the quality factor is realizable quite accurately in an IC design; for high-frequency, high-Q designs, however, Q is found to be a very sensitive function of parasitic elements and phase shifts so that it is unreasonable to expect Q to turn out correctly without tuning. Therefore, including for generality a Q-Control block[14] permits automatic tuning of the transfer function shape by a scheme that is completely analogous to the f-control method, as illustrated in Fig. 80.31.

With few exceptions, all currently proposed automatic tuning schemes follow the Master-Slave approach. A different concept proposed in [2] uses adaptive techniques to tune all poles *and* zeros of a filter function for improved tuning accuracy. Although the Master–Slave and the adaptive techniques work well, generally, the necessary circuitry has been found too large, noisy, and power hungry for many practical applications. Thus, alternative choices use a simple post-fabrication trim step[15] to eliminate the large fabrication tolerances, possibly together with careful design to make the electronic devices (OTAs) independent of temperature variations [1], [3].

80.5 Discussion and Conclusions

We have discussed in this section one of the fastest growing research and development areas in the topic of continuous-time filters, *viz.*, the field of fully integrated filters. Growing pressures to reduce costs and size, and improve reliability lead to increasingly larger parts of electronic systems being placed onto integrated circuits, a trend that *c-t* filters need to follow. We have seen that in most respects the methods for IC *c-t* filter design are identical to well-known active *RC* techniques, and follow directly the well-understood and proven standard active *RC* biquad-cascade or ladder-simulation methodologies. The difference lies in the final implementation, which the designer may adapt to any IC process and power supply level appropriate for the implementation of the system. Signal-to-noise ratios of the order of 70 to 80 dB and better, and

[13]In the designs reported to date, Frequency-Control blocks built around some type of phase-locked loop scheme, using a multiplier or an EXOR gate as phase detector, have been found most successful.

[14]Because magnitude errors can be shown in most cases to be proportional to errors in quality factor, the Q-Control block is normally implemented around an amplitude-locking scheme.

[15]For example, an on-chip laser-trimmed resistor or an external resistor set to determine the bias for OTAs. It works for applications with very low quality factors or where no component drifts are expected during operation.

distortion levels of less than 0.5 percent at signal levels of 1 V are obtainable. The two most prominent design approaches are the MOSFET-*C* and the g_m-*C* methods, both of which are easy and systematic to use and lead to filters that have proven themselves in practice. At the time of this writing, g_m-*C* filters appear to have the edge in high-frequency performance: with OTA bandwidths reaching gigahertz frequencies even for CMOS technology, *c-t* filters can be designed readily for applications in the 100-MHz range, i.e., to about 10 to 20 percent of the active-device bandwidth. A second important difference that requires the designer's attention is that the IC filter must be automatically tunable. This implies electronically variable components, OTAs or MOSFET "resistors," in the filter and some automatic control scheme for detecting errors and providing adjustments. Only simple and efficient designs for such tuning schemes will guarantee the ultimate commercial acceptance of integrated filters. At the present time, many approaches have been proposed that the reader may modify or adapt to his/her system [5].

We point out again that the cascade design is the more general procedure because it permits the realization of arbitrary transmission zeros anywhere in the complex *s*-plane, as required for example in gain or phase equalizers. Lossless ladders on the other hand are more restrictive because their transmission zeros are constrained to lie on the $j\omega$ axis, but they have lower passband sensitivities to component tolerances than cascade filters. Since good gyrators are readily available for g_m-*C* filters—in contrast to MOSFET-*C* designs—two competing implementation methods appear to suggest themselves in the g_m-*C* approach: the signal-flow graph and the element-substitution methods. In fact, there is no discernible difference between the two methods; indeed, frequently they lead to the same structures. For instance, the reader may readily verify that apart from a minor wiring change at the inputs, the circuits in Figs. 80.27(b) and 80.29 are identical, as are the ones in Figs. 80.28 (after redrawing it for dual-input OTAs) and Fig. 80.30. The wiring change is illustrated in Fig. 80.32. Figure 80.32(a) shows the input section of Fig. 80.29 (wired for 0-dB dc gain) and the wiring in Fig. 80.32(b) is that of Fig. 80.27(b). Notice that both circuits realize

$$(sC_1 + g_m)V_1 = 2g_m V_i - g_m V_{I2} \tag{80.57}$$

The designer, therefore, may choose a ladder method based on his/her familiarity with the procedure, available tables, or prototype designs; the final circuits are usually the same.

A further item merits reemphasizing at this point: As the reader may verify from the examples presented, g_m-*C* ladder structures generally have at all circuit nodes a design capacitor that can be used to absorb parasitics by predistortion as was discussed earlier. It may be preferable, therefore, to avoid the g_m-*C*–op-amp integrator on Fig. 80.20 with its increased noise level, power consumption, and with its associated parasitic poles (and zeros), and use parasitics absorption, as we have done in our examples. This method will *not* introduce any new parasitic critical frequencies into the filter and result in less distortion of the transfer function shape and

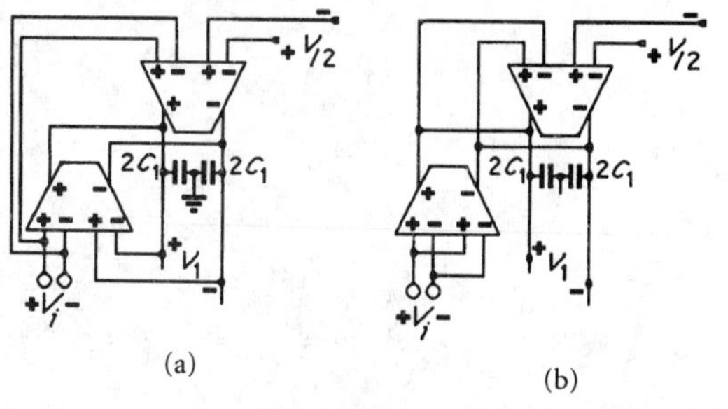

FIGURE 80.32 Excerpt from Fig. 80.28 (a) to illustrate alternative wiring (b) at the filter input.

easier tuning. As we mentioned earlier, this feature is a substantial advantage of g_m-C ladder filters.

References

[1] G. A. De Veirman and R. G. Yamasaki, "Design of a bipolar 10-MHz continuous-time 0.05° equiripple linear phase filter," *IEEE J. Solid-State Circuits*, vol. SC-27, pp. 324–331, Mar. 1992.

[2] K. A. Kozma, D. A. Johns, and A. S. Sedra, "Automatic tuning of continuous-time filters using an adaptive tuning technique," *IEEE Trans. Circuit Syst.*, vol. CAS-38, pp. 1241–1248, 1991.

[3] C. A. Laber and P. R. Gray, "A 20 MHz BiCMOS parasitic-insensitive continuous-time filter and second-order equalizer optimized for disk-drive read channels," *IEEE J. Solid-State Circuits*, vol. SC-28, pp. 462–470, Apr. 1993.

[4] R. Schaumann, M. S. Ghausi, and K. R. Laker, *Design of Analog Filters: Passive, Active RC and Switched Capacitor*, Englewood Cliffs, NJ: Prentice-Hall, 1990.

[5] Y. Tsividis and J. A. Voormann, Eds., *Integrated Continuous-Time Filters: Principles, Design, and Implementations*, New York: IEEE, 1993.

[6] Y. Tsividis, M. Banu, and J. Khoury, "Continuous-time MOSFET-C filters in VLSI," *IEEE Trans Circuits Syst.*, vol. CAS-33, Special Issue on VLSI Analog and Digital Signal Processing, pp. 125–140, 1986; see also *IEEE J. Solid-State Circuits*, vol. SC-21, pp. 15–30, 1986.

[7] C. Toumazou, F. J. Lidgey, and D. G. Haigh, Eds., *Analogue IC Design: The Current-Mode Approach*, London: IEE—Peter Peregrinus, 1990.

[8] A. I. Zverev, *Handbook of Filter Synthesis*, New York: Wiley, 1967.

81

Switched Capacitor Filters

**dgar
ánchez-Sinencio**
xas A&M University

sé Silva-Martínez
*tional Institute for
search in Astrophysics,
tics & Electronics, Mexico*

1.1 Introduction

The need to have monolithic analog filters motivated circuit designers in the late 1970s to investigate alternatives to conventional active-*RC* filters. A practical alternative appeared: switched-capacitors (SC) filters [1]–[3]. The original idea was to replace a resistor by a switched capacitor simulating the resistor. Thus, this equivalent resistor could be implemented with a capacitor, and two switches operating with a two-clock phases. SC filters consist of switches, capacitors, and op-amps. They are characterized by difference equations in contrast to differential equations for continuous-time filters. Simultaneously, the mathematical operator to handle sample-data systems such as switched-capacitor circuits is the z-transform, and the Laplace transform for continuous-time circuits. Several key properties of SC circuits have made them very popular in industrial environments:

1. Reduced silicon area, since the equivalent of large resistors can be simulated using small-size capacitors. Furthermore, positive and/or negative equivalent resistors can be easily implemented with SC techniques.
2. The time constants (*RC* products) from active-*RC* filters become capacitor ratios

93-8341-2/95/$0.00+$.50

multiplied by the clock period T. That is

$$RC \Rightarrow \frac{C}{C_R}T = \frac{C}{C_R f_C}$$

where f_c is the clock frequency.

The above expression can be realized in real applications with a good accuracy of nearly 0.1 percent.

3. Typically, the load of an SC circuit is mainly capacitive, therefore the required op-amps do not require a low-impedance output stage. This allows the use of a single-stage operational transconductance amplifier, which is especially useful in high-speed applications.
4. SC filters can be implemented in a digital circuit process technology. Thus, useful mixed-mode signal circuits can be economically realized in conventional MOS technology.
5. The SC design technique has matured. In the audio range, SC design techniques have became the dominant design approach. Furthermore, many circuits in communication applications use SC implementations.

In what follows, we will discuss basic building blocks involved in low- and high-order filters. Limitations and practical considerations will be presented. Furthermore, due to the industrial push from 5-V to 3.3-V power supplys, which are common is portable equipment, a brief discussion on low-voltage circuit design is included.

81.2 Basic Building Blocks

The basic building blocks involved in SC circuits are voltage gain amplifiers, integrators, and second-order filters. A key building block is the integrator. In fact, by means of a two-integrator loop a second-order (biquadratic) filter can be realized. Furthermore, a cascade connection of biquadratic filters yields higher order filters.

Gain Amplifiers

The gain amplifier is a basic building block in switched-capacitor circuits. Both the peak gain of a second-order filter and the link between the resonators in a ladder filter are controlled by voltage amplifier stages rather than by integrators. Many other applications require voltage gain stages. A voltage amplifier can be implemented by using two capacitors and an operational amplifier, as shown in Fig. 81.1(a). Ideally, the gain of this amplifier is given by $-C_S/C_I$. This topology is compact, versatile, and time continuous. Although this gain amplifier is quite simple, several second-order effects present in the op-amp make their design more complex. For example, a drawback of this topology is the lack of dc feedback.

For dc, the capacitors behave as open circuits, hence the operating point of the operational amplifier is not stabilized by the integrating capacitor C_I. In addition, the leakage current present at the input of the op-amp is integrated by the integrating capacitor, whose voltage eventually saturates the circuit. The leakage current I_{leak} in switched-capacitors circuits is a result of the diodes associated with the bottom plate of the capacitors and the switches (drain and source junctions). This leakage current is about 1 nA/cm^2. Analysis of Fig. 81.1(a) yields

$$v_0(t) = v_0(t_0) - \frac{C_S}{C_I}V_i(t) - \frac{I_{\text{leak}}}{C_I}(t - t_0) \tag{81.1}$$

Note that the dc output voltage is not defined for $t = t_0$. In addition, the leakage current present at the input of the op-amp is integrated by C_I and eventually saturates the circuit. To

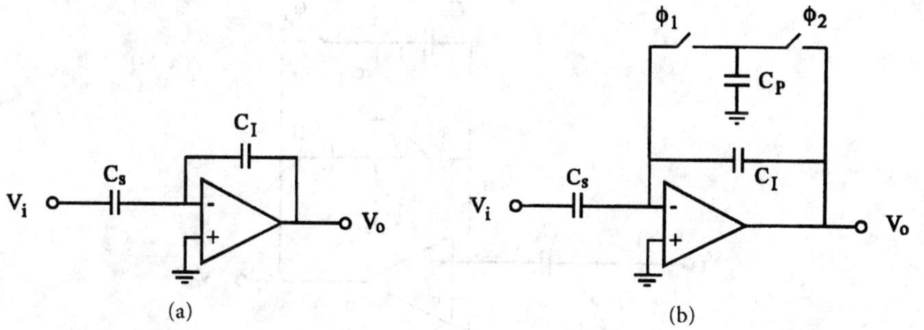

FIGURE 81.1 Gain amplifiers. (a) Without dc feedback and (b) with dc feedback.

overcome this drawback, a switched-capacitor resistor can be added as shown in Fig. 81.1(b). The switched-capacitor resistor gives a dc path for the leakage current but reduces further the low-frequency gain. A detailed analysis of this topology shows that the dc output voltage is equal to $-I_{\text{leak}}T/C_P$, with C_P the parasitic capacitor associated with the feedback path. However, using typical analysis methods for switched-capacitor networks, it can also be shown that the z domain transfer function of this topology is

$$H(z) = \frac{V_0(z)}{V_i(z)} \cong -\frac{C_S}{C_I}\frac{1 - z^{-1}}{1 - \left(1 - \dfrac{C_P}{C_I}\right)z^{-1}} \tag{81.2}$$

with $z = e^{j2\pi fT}$. For low frequencies, $z \approx 1$, the transfer function is very small, and only for high frequencies, $z \approx -1$, the circuit behaves as a voltage amplifier.

An offset free voltage amplifier is shown in Fig. 81.2. During the clock phase ϕ_2, the output voltage is equal to the op-amp offset voltage and it is sampled by both the integrating and sampling capacitors.

Because of the sampling of the offset voltage during the previous clock phase, during the integrating clock phase the charge injected by the sampling capacitor is equal to $C_S V_i$, and the charge extracted from the integrating capacitor becomes equal to $C_I V_0$. In this case, the offset voltage does not affect the charge recombination and the z domain transfer function, during the clock phase ϕ_1, becomes

$$H(z) = -\frac{C_S}{C_I} \tag{81.3}$$

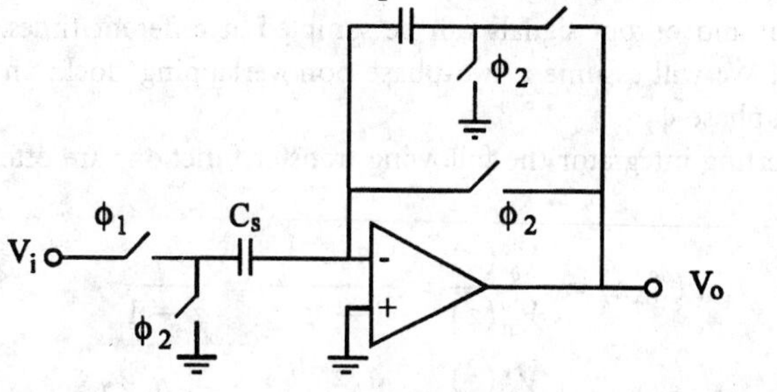

FIGURE 81.2 Gain amplifier available during the ϕ_1 clock phase.

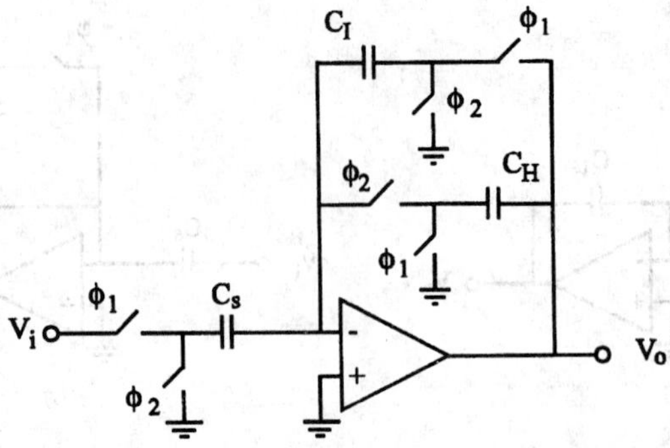

FIGURE 81.3 Gain amplifier available during both clock phases.

Equation (81.3) shows that ideally this topology is insensitive to the op-amp offset voltage. The topology behaves as an inverting amplifier if the clock phases shown in Fig. 81.2 are used. A noninverting amplifier is obtained if the clock phases associated with the sampling capacitor are interchanged. Because during ϕ_2 the op-amp output is short circuited with the inverting input, this topology presents two drawbacks. First, the amplifier output is only available during the clock phase ϕ_1. This limitation could be important in complex applications wherein the output of the amplifier has to be used during both clock phases. The second disadvantage of this topology is the large excursion of the op-amp output voltage. During the first clock phase, the op-amp output is equal to the op-amp offset voltage and in the next clock phase this voltage is equal to $-(C_S/C_I)V_i$. Hence an op-amp with large slew rate may be required.

Another interesting topology is shown in Fig. 81.3. During ϕ_2, the op-amp output voltage is equal to the previous voltage plus the op-amp offset voltage plus V_o/A_V, where A_V is the open-loop dc gain of the op-amp. In this clock phase both capacitors C_I and C_S are charged to the voltage at the inverting terminal of the op-amp. This voltage is approximately equal to the op-amp offset voltage plus V_o/A_V. During the next clock phase, the sampling capacitor is charged to $C_S(V_i - V_-)$, but because it was precharged to $-C_S V_-$, the injected charge to C_I is equal to $C_S V_i$. As a result of this charge cancellation, the op-amp output voltage is equal to $-(C_S/C_I)V_i$. Therefore, this topology has low sensitivity to the op-amp offset voltage and to the op-amp finite dc gain. A minor drawback of this topology is that the op-amp stays in open loop during the nonoverlapping clock phases transition times. This fact produces spikes during these time intervals. A solution for this is to connect a small capacitor between the op-amp output and the left-hand plate of C_S.

First-Order Blocks

The standard stray-insensitive integrators are shown in Fig. 81.4. Note that in sampled data systems both input and output signals can be sampled at different times. This yields different transfer functions. We will assume a two-phase nonoverlapping clock: an odd clock phase ϕ_1, and an even clock phase ϕ_2.

For the noninverting integrator the following transfer functions are often used:

$$H^{oo}(z) = \frac{V_o^o(z)}{V_{in}^o(z)} = \frac{a_p z^{-1}}{1 - z^{-1}} = \frac{a_p}{z - 1} \tag{81.4a}$$

$$H^{oe}(z) = \frac{V_o^e(z)}{V_{in}^o(z)} = \frac{a_p z^{-1/2}}{1 - z^{-1}} = \frac{a_p}{z^{1/2} - z^{-1/2}} \tag{81.4b}$$

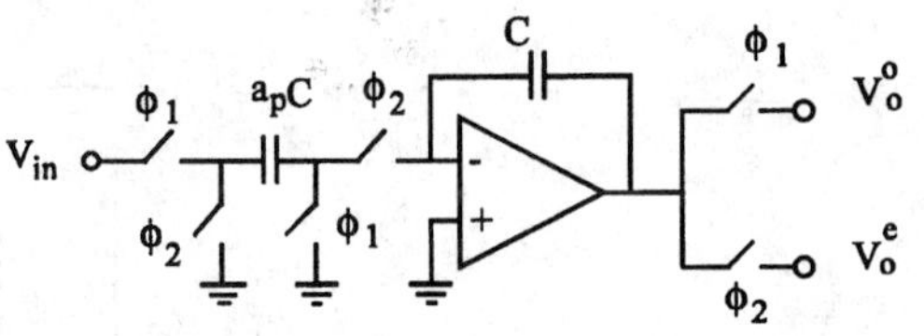

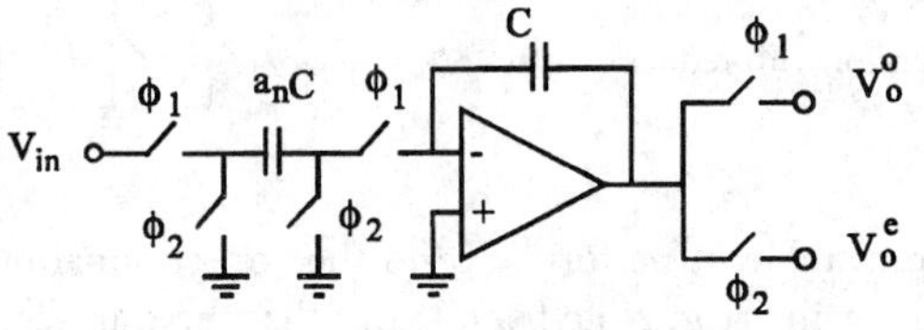

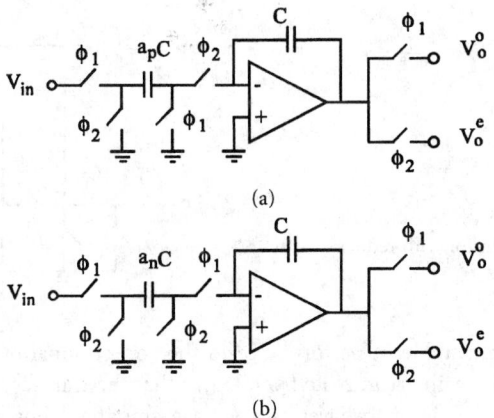

FIGURE 81.4 Conventional stray-insensitive SC integrators: (a) noninverting, (b) inverting.

For the inverting integrator

$$H^{oo}(z) = \frac{V_o^o(z)}{V_{in}^o(z)} = -\frac{a_n}{1 - z^{-1}} = -\frac{a_n z}{z - 1} \tag{81.5a}$$

$$H^{oe}(z) = \frac{V_{in}^e(z)}{V_{in}^o(z)} = -\frac{a_n z^{-1/2}}{1 - z^{-1}} = -\frac{a_n}{z^{1/2} - z^{-1/2}} \tag{81.5b}$$

where z^{-1} represents a unit delay. A crude demonstration showing the integration nature of these SC integrators [1], [8] in the s-domain is to consider high-sampling rate, that is, a clock frequency ($f_c = 1/T$) much higher than the operating signal frequencies. Thus, let us consider (81.4a), and assuming high sampling rate we can write a mapping from the z- to the s-domain:

$$z \cong 1 + sT \tag{81.6}$$

then

$$H(s) = \left.\frac{a_p}{z - 1}\right|_{z \cong 1+sT} \cong \frac{1}{(T/a_p)s} \tag{81.7}$$

This last expression corresponds to a continuous-time noninverting integrator with a time constant of $T/a_p = 1/f_c a_p$, that is, a capacitance ratio times the clock period.

In many applications the capacitor ratios associated with the integrators are very large, thus the total capacitance becomes excessive. This is particularly critical for biquadratic filters with high Q, where the ratio between the largest and smallest capacitance is proportional to the quality factor Q. A suitable inverting SC integrator for high-Q applications [10] is shown in Fig. 81.5; the corresponding transfer function is:

$$H^{oe}(z) = \frac{V_o^e(z)}{V_{in}^o(z)} = -\frac{C_1 C_3}{C_2 C_4'} \frac{1}{1 - z^{-1}} \tag{81.8}$$

where

$$C_4' = C_4 + C_3.$$

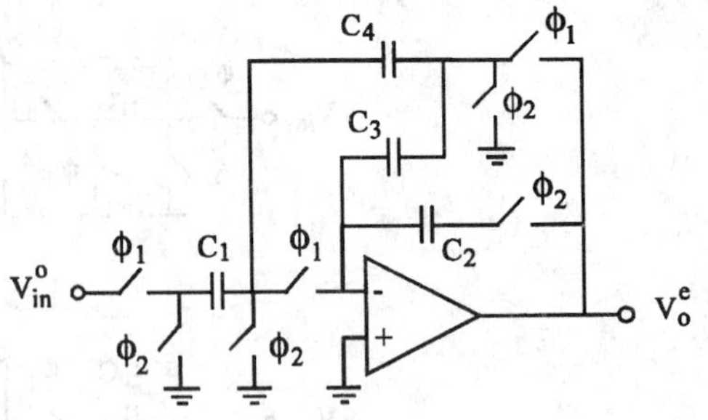

FIGURE 81.5 A SC integrator with reduced capacitance spread.

This integrator is comparable in performance to the conventional of Fig. 81.3, in terms of stray sensitivity and finite gain error. Note from (81.8) that the transfer function is only defined during ϕ_2. During ϕ_1, the circuit behaves as a voltage amplifier, thus high-slew-rate op-amps could be required. A serious drawback is the increased offset in comparison with the standard SC integrators. However, in typical two integrator loop filters, the other integrator can be chosen to be offset and low dc gain compensated as shown in Fig. 81.6.

The SC integrator performs the integration by means of C_1 and C_F and the hold capacitor C_h stores the offset voltage. The voltage across C_h compensates the offset voltage and the dc gain error of the op-amp. Note that the SC integrator of Fig. 81.6 can operate as a noninverting integrator if the clocking in parenthesis is employed. C_M provides a time-continuous feedback around the op-amp. The transfer function for infinite op-amp gain is

$$H^{oo}(z) = \frac{V_o^o(z)}{V_{\text{in}}^o(z)} = -\frac{C_1}{C_F}\frac{1}{1 - z^{-1}} \tag{81.9}$$

Furthermore, if the dc offset cannot be tolerated in a certain application, an autozeroing method can be used to compensate for the dc offset. Next we discuss a general form of a first-order building block; see Fig. 81.7.

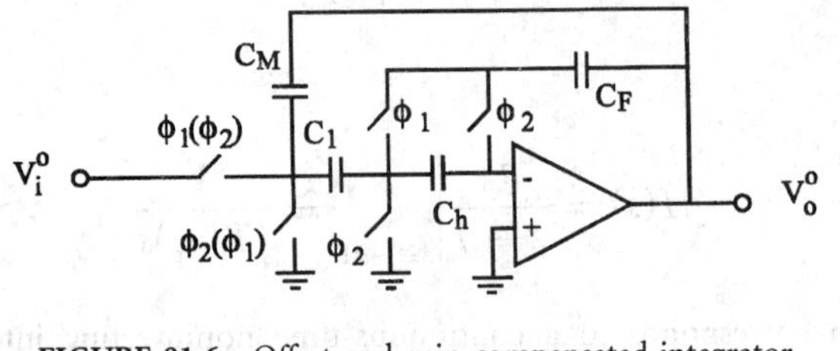

FIGURE 81.6 Offset and gain compensated integrator.

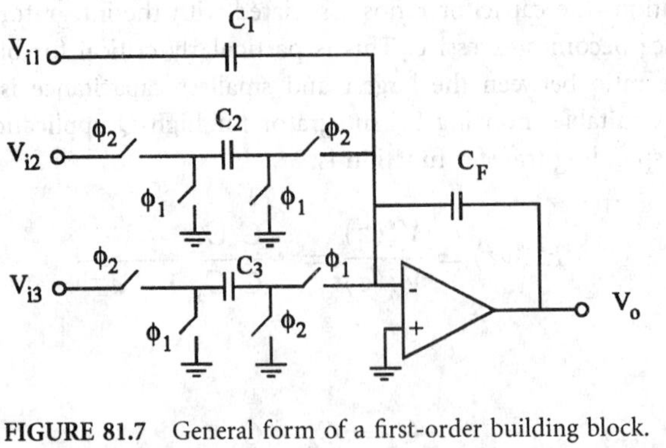

FIGURE 81.7 General form of a first-order building block.

The output voltage can be expressed as

$$V_o^e = -\frac{C_1}{C_F} V_{i_1}^e - \frac{C_2}{C_F} \frac{1}{1 - z^{-1}} V_{i_2}^e + \frac{C_3}{C_F} \frac{z^{-1}}{1 - z^{-1}} V_{i_3}^e \tag{81.10}$$

Observe that the capacitor C_3 and switches can be considered as the implementation of a negative resistor. Also note that $V_{i_2}^e$ could be V_0^e, this connection would make the integrator a lossy one. In that case (81.10) can be written as

$$V_o^e \frac{\left(1 + \frac{C_2}{C_F}\right) z - 1}{z - 1} = -\frac{C_1}{C_F} V_{i_1}^e + \frac{C_3}{C_F} \frac{1}{z - 1} V_{i_3}^e \quad \text{for } V_{i_2}^e = V_0^e \tag{81.11}$$

The building block of Fig. 81.7 is the basis of higher order filters. This will be shown next.

SC Biquadratic Sections

The circuit shown in Fig. 81.8 can implement any pair of poles and zeros in the z-domain. For $C_A = C_B = 1$

$$H^{ee}(z) = \frac{V_o^e(z)}{V_{\text{in}}^e(z)} = -\frac{(C_5 + C_6)z^2 + (C_1 C_2 - C_5 - 2C_6)z + C_6}{z^2 + (C_2 C_3 + C_2 C_4 - 2)z + (1 - C_2 C_4)} \tag{81.12}$$

Simple design equations follows:

$$\text{Low-pass case:} \quad C_5 = C_6 = 0 \tag{81.13a}$$

$$\text{High-pass case:} \quad C_1 = C_5 = 0 \tag{81.13b}$$

$$\text{Band-pass case:} \quad C_1 = C_6 = 0 \tag{81.13c}$$

Comparing the coefficients of the denominator of (81.12) with the general expression $z^2 - (2r \cos\theta)z + r^2$, we can obtain the following expressions:

$$C_2 C_4 = 1 - r^2 \tag{81.14a}$$

$$C_2 C_3 = 1 - 2r \cos\theta + r^2 \tag{81.14b}$$

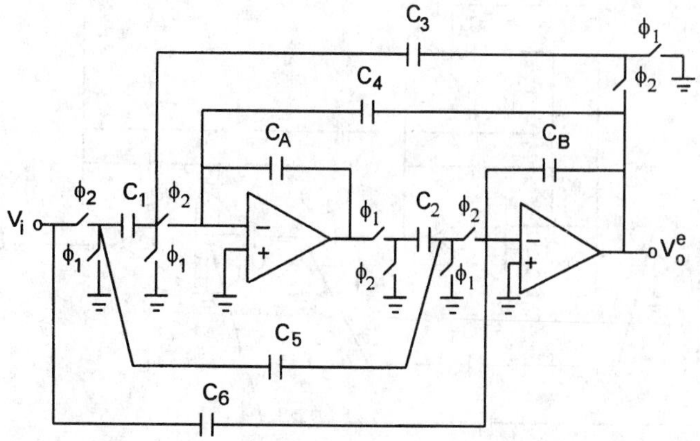

FIGURE 81.8 A SC biquadratic section [Martin–Sedra].

For equal voltage levels at the two integrator outputs, and assuming Q greater than 3, and high-sampling rate ($\theta = \omega_{od}T \ll 1$) we can write

$$C_2 = C_3 = \sqrt{1 + r^2 - 2r\cos\theta} \cong \omega_{od}T \tag{81.14c}$$

$$C_4 = \frac{1 - r^2}{C_2} \cong \frac{1}{Q} \tag{81.14d}$$

The capacitance spread for high sampling rate and high Q can be expressed as

$$\frac{C_{\max}}{C_{\min}} = \max\left[\frac{C_1}{C_2}, \frac{C_1}{C_4}\right] = \max\left\{\frac{1}{\omega_{od}T}, Q\right\} \tag{81.15}$$

In some particular cases this capacitance spread becomes prohibited. For such cases the SC integrators shown in Figs. 81.5 and 81.6 can be used to replace the conventional building blocks. This combination yields the SC biquadratic section shown in Fig. 81.9. This structure, besides offering a reduction of the total capacitance, also yields a reduction of the offset voltage of the op-amps. Note that the capacitors C_h does not play an important role in the design, and can be chosen to have a small value. The design equations for the zeros are similar to (8.13a–81.13c). For the poles, comparing with $z^2 - (2r\cos\theta)z + r^2$ and the analysis of Fig. 81.9, we can write

$$\frac{C_2C_3}{C_A + C_B} = 1 + r^2 - 2r\cos\theta \tag{81.16a}$$

$$\frac{C'_AC_2C_4}{C_{A_1}C_AC_B} = 1 - r^2 \tag{81.16b}$$

where $C_{A_1} = C'_A + C''_A$.

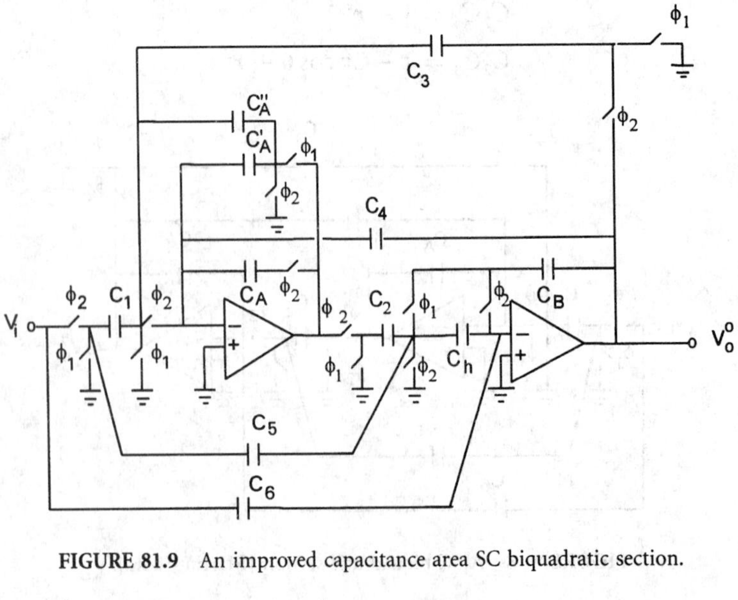

FIGURE 81.9 An improved capacitance area SC biquadratic section.

Simple design equations can be obtained assuming high-sampling rate, large Q, and $C_2 = C_3 = C_4 = C'_A = C_h = 1$; then

$$C_A + C_B \cong \frac{1}{\omega_{od} T} \tag{81.17a}$$

$$C''_A \cong Q\omega_{od} T - 1 \tag{81.17b}$$

Another common situation is the use of SC filters in high frequencies; in such cases a structure with a minimum gain–bandwidth product ($GB = \omega_u$) effect is desirable. This structure is shown in Fig. 81.10 and is often referred to as a decoupled structure. It is worthwhile to mention that two SC architectures can have ideally the same transfer function; however, with real op-amps, their frequency (and time) response differs significantly. A rule of thumb to reduce the GB effects in SC filters is to avoid a direct connection between the output of one op-amp to the input of another op-amp. It is desirable to transfer the output of an op-amp to a grounded capacitor, and in the next clock phase transfer the capacitor charge into the op-amp input. More discussion on ω_u effects is given in the next section. Analysis of Fig. 81.10 yields the following expressions:

$$r^2 = \frac{1 + a_9}{1 + a_8} \tag{81.18a}$$

$$2r \cos\theta = \frac{2 + a_8 + a_9 - a_2 a_7}{1 + a_8} \tag{81.18b}$$

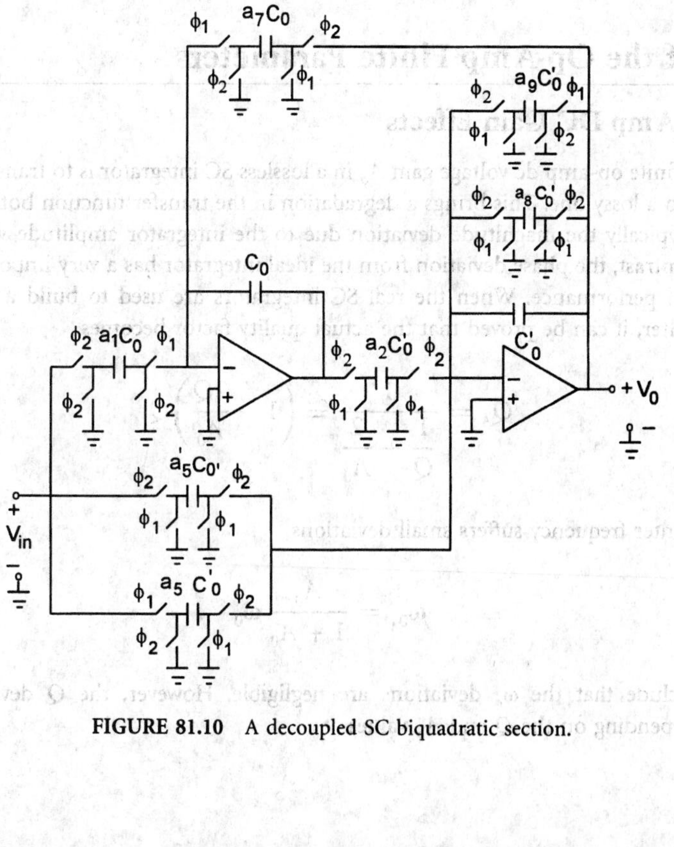

FIGURE 81.10 A decoupled SC biquadratic section.

If the input is sampled during ϕ_2 and held during ϕ_1, the ideal transfer function is given by

$$H^e(z) = \frac{V_o^e(z)}{V_i^e(z)} = \frac{-a_5'}{1+a_8} \frac{z^2 - z(a_5' + a_5 - a_1 a_2)/a_5' + a_5/a_5'}{z^2 - z\dfrac{2 + a_8 + a_9 - a_2 a_7}{1+a_8} + \dfrac{1+a_9}{1+a_8}} \tag{81.19}$$

The capacitor $a_9 C_0'$ can be used as a design trade off parameter to optimize the biquad performance. A simple set of design equations follows:

$$a_8 = \frac{(1+a_9) - r^2}{r^2} \tag{81.20a}$$

$$a_2 = a_7 = \sqrt{\frac{1 + r^2 - 2r\cos\theta}{r^2}(1+a_9)} \tag{81.20b}$$

Under high-sampling rate and high Q, the following expressions can be obtained:

$$\omega_0 \cong f_c \sqrt{\frac{a_2 a_7}{1+a_8}} \tag{81.21a}$$

and

$$Q \cong \sqrt{\frac{a_2 a_7 (1+a_8)}{a_8 - a_9}} \tag{81.21b}$$

A trade-off between Q-sensitivity and total capacitance is given by a_8 and a_9.

81.3 Effects of the Op-Amp Finite Parameters

Finite Op-Amp DC Gain Effects

The effect of finite op-amp dc voltage gain A_o in a lossless SC integrator is to transform a lossless integrator into a lossy one. This brings a degradation in the transfer function both in amplitude and phase. Typically the magnitude deviation due to the integrator amplitude variation is not critical. By contrast, the phase deviation from the ideal integrator has a very important influence on the overall performance. When the real SC integrators are used to build a two-integrator biquadratic filter, it can be proved that the actual quality factor becomes

$$Q_A = \frac{1}{\dfrac{1}{Q} + \dfrac{2}{A_0}} \cong \left(1 - \frac{2Q}{A_0}\right) Q \tag{81.22}$$

The actual center frequency suffers small deviations.

$$\omega_{0_A} = \frac{A_0}{1 + A_0} \omega_0 \tag{81.23}$$

We can conclude that the ω_o deviations are negligible. However, the Q deviations can be significant depending on the Q and A_o values.

Finite Op-Amp Gain–Bandwidth Product Effects

The op-amp bandwidth is very critical for high-frequency applications. The analysis is carried out when the op-amp voltage gain is modeled with one dominant pole, i.e.,

$$A_v(s) = \frac{A_o}{1 + s/\omega_3} = \frac{A_o\omega_3}{s + \omega_3} = \frac{\omega_u}{s + \omega_3} \cong \frac{\omega_u}{s} \tag{81.24}$$

where A_o is the dc gain, ω_u is approximately the unity-gain bandwidth (GB), and ω_3 is the op-amp bandwidth. The analysis taking into account $A_v(s)$ is rather cumbersome since the op-amp input–output characterization is a continuous-time system modeled by a first-order differential equation, and the rest of the SC circuit is characterized by discrete-time systems modeled by difference equations. It can be shown that the step response of a single op-amp SC circuit due to step inputs applied at $t = t_1$ is

$$V_0(t) = V_0(t_1)e^{-(t-t_1)\alpha\omega_u} + V_{od}\{1 - e^{-(t-t_1)\alpha\omega_u}\} \tag{81.25}$$

where v_{od} is the final output voltage, which is a function of the initial conditions, inputs, and filter architecture, and α is a topology dependent voltage divider $0 < \alpha \leq 1$.

$$\alpha = \frac{\sum C_f}{\sum_i C_i} \tag{81.26}$$

where the C_f sum consists of all feedback capacitors connected directly between the op-amp output and the negative input terminal, and the C_i sum comprises overall capacitors connected to the negative op-amp terminal. Note that the $\alpha\omega_u$ product determines the rise time of the response, therefore both α and ω_u should be maximized. For the multiple op-amp case the basic concept prevails. Observe that for a two-clock phases SC filters α becomes an $\mathbf{M} \times \mathbf{M}$ matrix **A** for each clock-phase, where **M** is the number of op-amps in the filter architecture. For the common case that $t - t_1 = T/2$ at the end of any clock phase, the figure of merit to be maximized becomes $\alpha T\omega_u/2$. This means that a rule of thumb yielding reduced gain–bandwidth product effects requires that

$$\alpha T\omega_u/2 \geq 5 \tag{81.27}$$

This rule is based on the fact that five time constants are required to obtain the steady-state response with magnitude error of less than 1 percent. This rule of thumb also applies for each entry of **A** for the case of multiple op-amp case. For a fixed clock frequency, $f_c = 1/T$, there are two components to be optimized: ω_u, which is a function of the op-amp design and process technology, and **A**, which is matrix filter topology dependent. The minimum ω_u effects biquadratic SC filter presented in the previous section (Fig. 81.9) has the property that **A** has been optimized. For illustration on how to determine the **A** matrix of an SC filter, the **A** matrix for the minimum ω_u effects biquadratic is given by

$$\mathbf{A}\big|_{\phi_1} = \begin{bmatrix} \dfrac{1}{1 + a_1 + a_7} & 0 \\ 0 & 1 \end{bmatrix} \tag{81.28}$$

$$\mathbf{A}\big|_{\phi_2} = \begin{bmatrix} 1 & \dfrac{a_2}{1 + a_2 + a_5 + a_5' + a_8 + a_9} \\ 0 & \dfrac{1 + a_8}{1 + a_2 + a_5 + a_5' + a_8 + a_9} \end{bmatrix} \tag{81.29}$$

Thus the designer should use the extra degrees of freedom to maximize the entries of **A** during all phases. Another design consideration could be to maximize ω_u. This last consideration should be carefully addressed since very large ω_u values, as it will be shown, can cause excessive noise. Therefore, a judiciously trade-off must be used in choosing the ω_u of each op-amp in the filter topology.

81.4 Noise and Clock Feedthrough

The lower range of signals that can be processed by the electronic devices is limited by several unwanted signals that appear at the output of the circuit. The rms values of these electrical signals determine the noise level of the system, and it represents the lowest limit for the incoming signals to be processed. Input signals smaller than the noise level, in most of the cases, cannot be driven by the circuit.

The most critical noise sources are those due to: i) the used elements (transistors, diodes, resistors, etc.); ii) the noise induced by the clocks; iii) the harmonic distortion components generated due to the intrinsic nonlinear characteristics of the devices; and iv) the noise induced by the surrounding circuitry. In this section, types i), ii), and iii) are considered. The noise generated by the surrounding circuitry and coupled to the output of the switched-capacitor circuit can be further reduced by using fully differential structures. These structures are not treated here but there are excellent references in the literature [2], [4], [6], [9].

Noise Due to the Op-Amp

In an MOS transistor, the noise is generated by different mechanisms but there are two dominant noise sources: channel thermal noise and $1/f$ or flicker noise. A discussion of the nature of these noise sources follows.

Thermal Noise

The flow of the carriers due to the drain-source voltage takes place on the source-drain channel, mostly like in a typical resistor. Therefore, due to the random flow of the carriers, thermal noise is generated. For an MOS transistor biased in the linear region the spectral density of the input referred thermal noise is approximately given by [1]–[4]

$$V_{\text{eqth}}^2 = 4kTR_{\text{on}} \tag{81.30}$$

where R_{on}, k, and T are the drain–source resistance of the transistor, the Boltzmann constant, and the temperature (in Kelvin degrees), respectively. In saturation, the spectral noise density can be calculated by the same expression but with R_{on} equal to $2/3g_m$, being g_m the small-signal transconductance of the transistor.

1/*f* Noise

This type of noise if mainly due to the imperfections in the silicon–silicon oxide interface. The surface states and the traps in this interface randomly interfere with the charges flowing through the channel, hence the generated noise is strongly dependent of the technology. The $1/f$ noise (or flicker noise) is also inversely proportional to the gate area because at larger areas more traps and surface states are present and some averaging occurs. The spectral density of the input referred $1/f$ noise is commonly characterized as

$$V_{\text{eq}1/f}^2 = \frac{k_F}{WLf} \tag{81.31}$$

where the product WL, f, and k_F are the gate area of the transistor, the frequency in hertz, and the flicker constant, respectively. The spectral noise density of the MOS transistor is composed by both components, therefore the input referred spectral noise density of a transistor operated in its saturation region becomes

$$V_{eq}^2 = \frac{8}{3}\frac{kT}{g_m} + \frac{k_F}{WLf} \tag{81.32}$$

Op-Amp Noise Contributions

In an op-amp, the output referred noise density is composed by the noise contribution of all transistors, hence the noise level is function of the op-amp architecture. A typical unbuffered folded cascode op-amp (folded cascode OTA) is shown in Fig. 81.11. For the computation of the noise level, the contribution of each transistor has to be evaluated. This can be done by obtaining the OTA output current generated by the gate-referred noise of all the transistors. For instance, the spectral density of the output referred noise current due to M_1 is straightforwardly determined because the gate referred noise is at the input of the OTA, leading to

$$i_{01}^2 = G_m^2 V_{eq1}^2 \tag{81.33}$$

where G_m (equal to g_{m1} at low frequencies) is the OTA transconductance and V_{eq1}^2 is the input referred noise density of M_1. Similarly, the contributions of M_2 and M_5 to the spectral density of the output referred noise current are given by

$$\begin{aligned} i_{o2}^2 &= g_{m2}^2 v_{eq2}^2 \\ i_{o5}^2 &= g_{m5}^2 v_{eq5}^2 \end{aligned} \tag{81.34}$$

The noise contributions of transistors M_3 and M_4 are very small in comparison to the other components. This is because their noise drain currents, due to the source degeneration implicit in these transistors, are determined by the equivalent conductance associated with their sources, instead of those by their transconductance. Since in a saturated MOS transistor the equivalent conductance is much smaller than the transistor transconductance, this noise drain current contribution can be neglected. The noise contribution of M_6 is mainly common-mode noise,

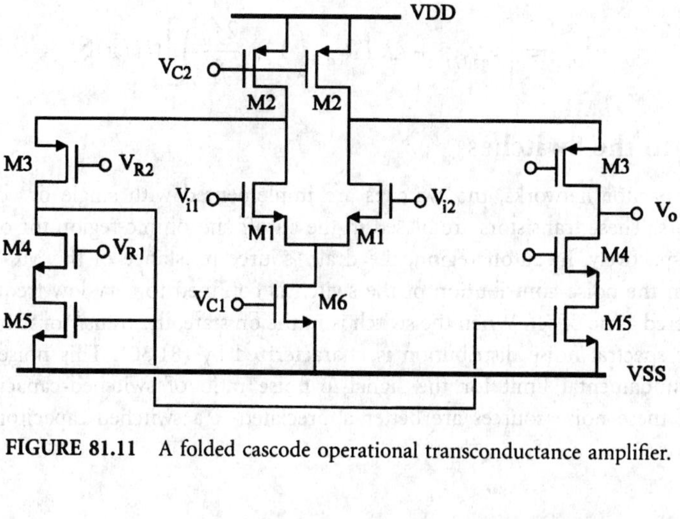

FIGURE 81.11 A folded cascode operational transconductance amplifier.

therefore it is almost canceled at the OTA output due to the current subtraction. The spectral density of the total output referred noise current can be approximately calculated as

$$i_0^2 = 2[G_m^2 v_{\text{eqI}}^2 + g_{m2}^2 v_{\text{eq2}}^2 + g_{m5}^2 v_{\text{eq5}}^2] \tag{81.35}$$

The factor 2 is the result of the pair of transistors M_1, M_2, and M_5. From this equation, the OTA input referred noise density becomes

$$v_{\text{OTA in}}^2 = 2v_{\text{eqI}}^2 \left[1 + \frac{g_{m2}^2 v_{\text{eq2}}^2 + g_{m5}^2 v_{\text{eq5}}^2}{G_m^2 v_{\text{eqI}}^2} \right] \tag{81.36a}$$

According with this result, if G_m is larger than g_{m2} and g_{m5}, the OTA input referred noise density is mainly determined by the OTA input stage. In that case and using (81.32), (81.36) yields

$$v_{\text{OTAin}}^2 \cong 2v_{\text{eqI}}^2 = v_{\text{eq}1/f}^2 + 4kTR_{\text{eqth}} \tag{81.36b}$$

where the factor 2 has been included in $v_{\text{eq}1/f}$ and R_{eqth}. In (81.36b), $v_{\text{eq}1/f}^2$ is the equivalent $1/f$ noise density and R_{eqth} is the equivalent resistance for noise, equal to $4/3g_m$.

Noise in a Switched-Capacitor Integrator

In a switched-capacitor lossless integrator, the output referred noise density component due to the OTA is frequency limited by the gain–bandwidth product of the OTA. In order to avoid misunderstandings, in this section f_u (the unity-gain frequency of the OTA in hertz) instead that ω_u (in radians per second) is used. Since f_u must be higher than the clock frequency f_c and due to the sampled nature of the SC integrator, the OTA high-frequency noise is folded back into the integrator baseband. In the case of the SC integrator and assuming that the flicker noise is not folded back, the output referred spectral noise density becomes

$$V_{o\,\text{eq}\,\text{I}}^2 = \left[v_{\text{eq}\,1/f}^2 + 4kTR_{\text{eqth}} \left(1 + \frac{2f_u}{f_c} \right) \right] |1 + H(z)|^2 \tag{81.37a}$$

where the folding factor is equal to f_u/f_c, and $H(z)$ is the z-domain transfer function of the integrator. The factor $2f_u/f_c$ is the result of both positive and negative foldings. Typically, the frequency range of the signal to be processed is around and below the unity-gain frequency of the integrator, therefore $|H(z) > 1|$ and (81.37a) can be approximated as

$$v_{o\,\text{eq}\,\text{I}}^2 = \left[v_{\text{eq}\,1/f}^2 + 4kTR_{\text{eqth}} \left(1 + \frac{2f_u}{f_c} \right) \right] |H(z)|^2 \tag{81.37b}$$

Noise Due to the Switches

In switched-capacitor networks, the switches are implemented with single or complementary MOS transistors. These transistors are biased in the cutoff and ohmic region for open and close operations, respectively. In cutoff region, the drain–source resistance of the MOS transistor is very high, then the noise contribution of the switch is confined to very low frequencies and it can be considered as dc offset. When the switch is in the on state, the transistor is biased in linear region and its spectral noise distribution is characterized by (81.30). This noise contribution is the most fundamental limit for the signal-to-noise ratio of switched-capacitor networks. The effects of these noise sources are better appreciated if a switched-capacitor integrator is considered.

Let us consider the SC integrator of Fig. 81.12 and assume that $\phi_1 = \phi_1'$ and $\phi_2 = \phi_2'$. The spectral noise density of the ϕ_1 driven switches are low-pass filtered by the on resistance of the switches R_{on} and the sampling capacitor C_S. The cutoff frequency of this continuous-time filter is given by $f_{on} = 1/(2\pi R_{on} C_S)$. Typically f_{on} is higher than the clock frequency, therefore, the high-frequency noise is folded back into the baseband of the integrator when it is sampled by the switched-capacitor integrator. Taking into account the folding effects, the output referred spectral noise density component becomes

$$v_{o\,\mathrm{eq}\,2}^2 = 4kTR_{on}\left[1 + \frac{2f_{on}}{f_c}\right]|H(z)|^2 \tag{81.38a}$$

with R_{on} the switch resistance. After several mathematical manipulations and using $2\pi f_{on} = 1/R_{on}C_S$, (81.38a) yields

$$v_{o\,\mathrm{eq}\,2}^2 = 4kTR_{int}\left[\frac{f_c + 2f_{on}}{2\pi f_{on}}\right]|H(z)|^2 = 4kT\frac{1}{f_c C_S}\left[\frac{f_c + 2f_{on}}{2\pi f_{on}}\right]|H(z)|^2 \tag{81.38b}$$

where R_{int} is the equivalent switched-capacitor resistance, $R_{int} = 1/f_c C_S$. For the noise induced by the ϕ_2 driven switches, the situation is slightly different. While for low-frequency signals the inverting input of the OTA can be considered as a virtual ground; for high frequencies the finite f_u of the OTA limits its frequency response. Hence, the folding factor is slightly lower than that considered in (81.38a). For infinite f_u and considering the contribution of all switches, the output referred spectral noise density of the switched capacitor integrator becomes

$$v_{o\,\mathrm{eq}\,2T}^2 = 4kTR_{int}\left[\frac{f_c + 2f_{on}}{\pi f_{on}}\right]|H(z)|^2 \tag{81.39}$$

More detailed computations show that this noise contribution is almost equal to

$$v_{o\,\mathrm{eq}\,2T}^2 \cong 4kTR_{int}|H(z)|^2 = 4kT\frac{1}{f_C C_S}|H(z)|^2 \tag{81.40}$$

which represents the noise spectrum of a continuous-time resistor of value R_{int}. Adding the noise contributions of the op-amp and the switches and referring this noise to the integrator input the spectral noise density becomes

$$v_{\mathrm{ineq}}^2 \cong 4kT\frac{1}{f_C C_S}[1 + A] + v_{\mathrm{eq}\,1/f} \tag{81.41}$$

where

$$A = \frac{R_{eqth}}{R_{int}}\left(1 + \frac{2f_u}{f_c}\right)$$

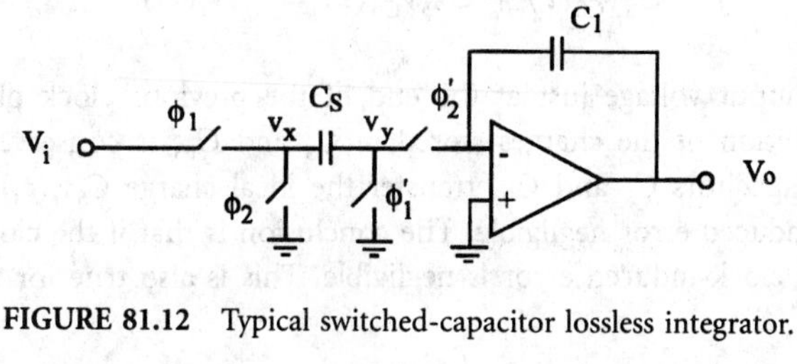

FIGURE 81.12 Typical switched-capacitor lossless integrator.

In A, R_{eqth} is related to the equivalent OTA noise resistance and, in most of the cases, is determined by the OTA input stage. Increasing the transconductance of the differential pair [see (81.32)] R_{eqth} decreases. Typically $A \ll 1$, hence the most important noise contribution comes from the switches.

According to (81.41), the switched-capacitor integrator noise density is proportional to R_{int} ($= 1/f_c C_S$). Therefore, for *low-noise applications* it is desirable to design the integrators with *large capacitors.* However, the costs for the noise reduction are larger silicon area and higher power consumption. This last result is because the slew rate of the OTA is inversely proportional to the load capacitance and in order to maintain the specifications it is necessary to increase the current drive capability of the OTA. Clearly, there is a trade-off between noise level, silicon area, and power consumption.

Clock Feedthrough

Another factor that limits the accuracy of switched-capacitor networks is the charge induced by the clocking of the switches. These charges are induced by the gate–source capacitance, the gate–drain capacitance, and the charge stored in the channel. Furthermore, some of these charges are input signal dependent and introduce distortion in the circuit. Although these errors cannot be canceled, there are some techniques that reduce these effects.

The analysis of the clock feedthrough is very difficult because it depends on the order of the clock phases, the relative delay of the clock phases, as well as the speed of the clock transitions. For instance, let's consider in Fig. 81.12 the case when ϕ_1 goes down before ϕ_1'. This situation is shown in Fig. 81.13(a).

While C_{P1} is connected between two low-impedance nodes, C_{P2} is connected between ϕ_1, a low-impedance node, and v_x. For $\phi_1 > v_i + V_T$ the transistor is on and the current injected by C_{P2} is absorbed by the drain–source resistance; then v_x remains at a voltage equal to v_i. When the transistor is turned off, $\phi_1 < v_i + V_T$, charge conservation at node v_x leads to

$$v_x = v_i + \frac{C_{P2}}{C_S + C_{P2}}(\mathrm{VSS} - v_i - V_T) \tag{81.42}$$

where VSS is the low level of ϕ_1 and ϕ_2. During the next clock phase, the charge of C_S is injected to C_I. Thus, a charge error proportional to $C_{P2}/(C_S + C_{P2})$ is induced by C_{P2}. In addition, an offset voltage proportional to VSS$-V_T$ is also generated. Because the threshold voltage V_T is a nonlinear function of v_i, an additional error in the transfer function and harmonic distortion components appear at the output of the integrator. The same effect occurs when the clock phases ϕ_2 and ϕ_2' have a similar sequence.

Let us consider the case when ϕ_1' is opened before ϕ_1; the situation is shown in the Fig. 81.13(b). Before the transistor turns off, $v_x = v_i$ and $v_y = 0$. When the transistor is turned off, VSS $< \phi_1 < v_i + V_T$, the charge is recombined between C_S, C_{P2}, and C_{P3}. After the charge redistribution, the charge conservation at node Y leads to

$$C_S v_{CS}(t) + C_{P3} v_{C3}(t) = C_S v_i(t_0) \tag{81.43}$$

where $v_i(t_0)$ is the input voltage just at the end of the previous clock phase. Observe from (81.43) that the addition of the charges stored in C_S and C_{P3} is conserved. During the next clock phase, both capacitors C_S and C_{P3} transfer the ideal charge $C_S v_i(t_0)$ to C_I, making the clock feedthrough induced error negligible. The conclusion is that if the clock phase ϕ_1' is a bit delayed than ϕ_1 the clock-induced error is negligible. This is also true for the clock phases ϕ_2 and ϕ_2'.

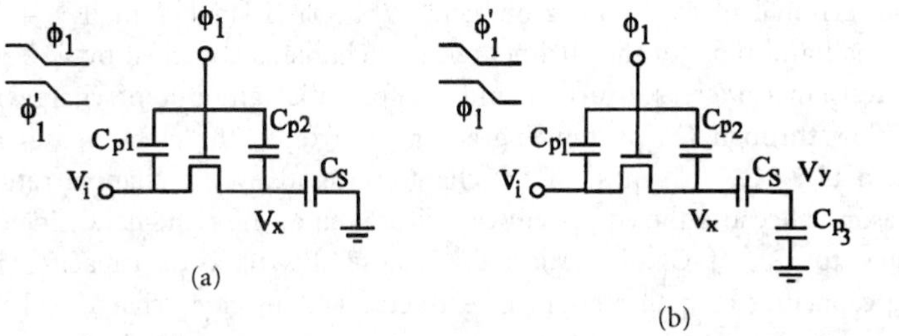

FIGURE 81.13 Charge induced due to the clocks. (a) If ϕ_1 goes down before ϕ_1' and (b) if ϕ_1' goes down before ϕ_1.

In Fig. 81.12, the right-hand switches also introduce clock feedthrough, but unlike the clock feedthrough previously analyzed, this is input signal independent. When the clock phase ϕ_2' goes down, the gate–source overlap capacitor extracts from the summing node the following charge:

$$\Delta Q = C_{GS}(V_{SS} - V_T) \tag{81.44}$$

In this case, V_T does not introduce distortion because v_y is almost at zero voltage for both clock phases. The main effect of C_{GS} is to introduce an offset voltage. The same analysis reveals that the bottom right-hand switch introduces a similar offset voltage.

From the previous analysis it can be seen that the clock feed-through can be reduced by using minimum dimension transistors. This implies minimum parasitic capacitors and minimum induced charge from the channel. If possible, the clock phases should be arranged for minimum clock feedthrough.

Channel Mobile Charge

Charge injection also occurs due to the mobile charge in the channel. If the transistor is biased in linear region, the channel mobile charge can be approximated by

$$Q_{ch} = C_{GC}(v_{GS} - V_T) \tag{81.45}$$

where C_{GC} is the gate–channel capacitor; see Fig. 81.14.

When the switch is turned off this charge is released and part of it goes to the sampling capacitor. Fortunately, the previously discussed technique for the reduction of clock feedthrough also reduces the effects of the channel mobile charge injection. The mobile charge injected to the sampling capacitor is a function of several parameters; e.g., input signal, falling rate of the clock, overlap capacitors, gate capacitor, threshold voltage, integrating capacitor, and the supply voltages.

The effects of the channel mobile charges are severe when the ϕ_1-driven transistor is opened before the ϕ_1'-driven transistor. In this case, the situation is similar to that shown in Fig. 81.13(a),

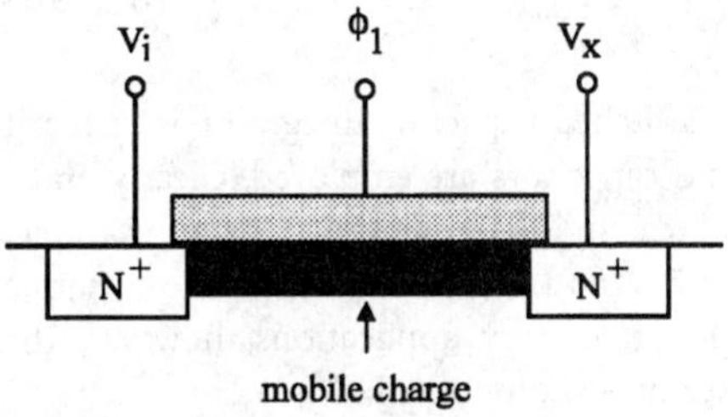

FIGURE 81.14 Cross section of the MOS transistor.

in which one terminal of C_S is still grounded. While ϕ_1 is higher than $v_i + V_T$ the channel resistance is small and most of the channel released charge is absorbed by v_i. For $\phi_1 < v_i - V_T$, the channel resistance increases further and a substantial amount of charge released by the channel will flow through C_S, introducing a charge error. If the clock phases are arranged as these shown in Fig. 81.13(b), most of the charge released by the channel returns back to v_i. The main reason is because the equivalent capacitor seen at the right-hand side of the transistor is nearly equal to C_{P3}, if C_{P2} is neglected. Because this parasitic capacitor is smaller than the sampling capacitor, a small amount of extracted (or injected) charge will produce a huge variation on v_y [see Fig. 81.13(b)], pushing back most of the mobile charges.

Dynamic Range

The dynamic range is defined as the ratio of the maximum signal that the circuit can drive without significant distortion to the noise level. The maximum distortion tolerated by the circuit depends on the application, but −60 dB is commonly used. Since the linearity of the capacitors is good enough and if the harmonic distortion components introduced by the OTA input stage are small, the major limitation for the distortion is determined by the output stage of the OTA. For the folded cascode OTA of Fig. 81.11 this limit is

$$v_{o\,\max} \cong V_{R2} + V_{TP3} \tag{81.46}$$

If the reference voltage V_{R2} is maximized, (81.46) yields

$$v_{o\,\max} \cong V_{DD} - 2V_{DSATP} \tag{81.47}$$

A similar expression can be obtained for the lowest limit. Assuming a symmetrical single-side output stage, from (81.47), the maximum rms value of the OTA output voltage is given by

$$v_{oRMS} \cong (V_{DD} - 2V_{DSATP})/\sqrt{2} \tag{81.48}$$

If the in-band noise, integrated up to $\omega = 1/R_{int}C_I$, is considered and if the most important term of (81.41) is retained, the dynamic range of the single ended switched-capacitor integrator becomes

$$DR \cong \frac{(V_{DD} - 2V_{DSATP})}{\sqrt{2}\sqrt{4kT/C_I}} \tag{81.49a}$$

At room temperature, this equation is reduced to the following expression:

$$DR \cong 5.6 \times 10^9 \sqrt{C_I}(V_{DD} - 2V_{DSATP}) \tag{81.49b}$$

According with this result, the dynamic range of the switched-capacitor integrator is reduced when the power supplies are scaled down and minimum capacitors are employed. Clearly, there is a compromise between power consumption, silicon area, and dynamic range. As an example, for the case of $C_I = 1.0$ pF and supply voltages of ±1.5 V and neglecting V_{DSATP}, the dynamic range of a single integrator is around 78 dB. For low-frequency applications, however, the dynamic range is lower due to the low-frequency flicker noise component.

81.5 Design Considerations for Low-Voltage Switched-Capacitor Circuits

fOR THE TYPICAL DIGITAL SUPPLY VOLTAGES, 0–5 v, SWITCHED-CAPACITOR NETWORKS achieve dynamic ranges of the order of 80–100 dB. As long as the power supplies are reduced, the swing of the signal decreases and the resistance of the switches increases further. Both effects reduce the dynamic range of the switched-capacitor networks. A discussion of these topics follows.

Low-Voltage Operational Amplifiers

The design techniques for low-voltage low-power amplifiers for switched-capacitor circuits have been addressed by several authors [4], [7]–[9]. The implementation of op-amps for low-voltage applications does not seem to be a fundamental limitation as long as the transistor threshold voltage is smaller than (VDD – VSS)/2. This limitation will become clear in the design example presented in this section. The design of the operational amplifier is strongly dependent on the application. For high-frequency circuits, the folded-cascode op-amp is suitable, but the swing of the signals at the output stage is limited by the cascode transistors. If large output voltage swing is needed, complementary stages are desirable. To illustrate the design trade-offs involved in the design of a low-voltage OTA, let us consider the folded cascode OTA of Fig. 81.11. For low-voltage applications and small signals, the transistors have to be biased with very low $V_{GS} - V_T$. For ± 0.75 V applications and $V_T = 0.5$ V, $V_{DSAT1} + V_{DSAT6}$ must be lower than 0.25 V, otherwise the transistor M_6 goes to the triode region. For large signals, however, the variations of the input signal produce variations at the source voltage of M_1. If an input of the OTA is grounded, it can be shown that these voltage variations are of the order of $\pm 1.44\ V_{DSAT1}$. Hence, for a proper operation of the OTA input stage it is desirable to satisfy

$$0.25 > 2.44V_{DSAT\,1} + V_{DSAT\,6} \tag{81.50}$$

It has to be taken into account that the threshold voltage of M_1 increases if an N-well process is used, due to the body effects. In critical applications, PMOS transistors fabricated in a different well with their source tied to their own well can be used. The dimensioning of the transistors and the bias conditions are directly related to the application. For instance, if the switched-capacitor integrator must slew 1 V into 4 μs and the load capacitor is of the order of 10 pF, the OTA output current must be equal to or higher than 2.5 μA. Typically, for the folded cascode OTA the dc current of both the output and the input stages are the same. Therefore, the bias current for M_1, M_3, M_4, and M_5 can be equal to 2.5 μA. The bias current for M_2 and M_6 is 5 μA. If $V_{GS1} - V_{T1}$ is equal to 0.06 V the dimensions of M_1 can be computed. Similarly, the dimensions of the transistors can be calculated, most of them designed to maximize the output range of the OTA. The dimensions and the bias condition for the transistors are given in Table 81.1.

TABLE 81.1 Dimension and Bias Current for the Transistors

Transistor	$W(\mu m)/L(\mu m)$	I_{BIAS}
M_1	48/2.4	2.5 μA
M_2	120/2.4	5.0 μA
M_3	60/2.4	2.5 μA
M_4	60/4.2	2.5 μA
M_5	60/4.2	2.5 μA
M_6	60/4.2	5.0 μA

A very important issue in the design of low-voltage amplifiers is the reference voltage. In the folded-cascode of Fig. 81.11, the values of the reference voltages V_{R1} and V_{R2} must be optimized for maximum swing of the output signal.

Analog Switches

For low-voltage applications, the highest voltage that can be processed is limited by the analog switches rather than by the op-amps. For a single NMOS transistor, the switch resistance is approximately given by

$$R_{DS} = \frac{1}{\mu_n C_{OX} \frac{W}{L}(V_{GS} - V_T)} \tag{81.51}$$

where μ_n and C_{OX} are technological parameters. According to (81.51), the switch resistance increases further when V_{GS} approaches to V_T. This effect is shown in Fig. 81.15 for the case VDD = –VSS = 0.75 V and V_T = 0.5 V. From this figure, the switch resistance is higher than 300 kΩ for input signals of 0.2 V. However, for a drain–source voltage higher than $V_{GS} - V_T$ the transistor saturates and does not behave as a switch anymore. This limitation clearly reduces further the dynamic range of the switched-capacitor circuits.

A solution for this drawback is to generate the clocks from higher voltage supplies. A simplified diagram of a voltage doubler is depicted in Fig. 81.16(a). During the clock phase ϕ_1, the capacitor C_1 is charged to VDD and during the next clock phase its negative plate is connected to VDD. Hence, at the beginning of ϕ_2, the voltage at the top plate of C_1 is equal to 2VDD – VSS. After several clock cycles, if C_{LOAD} is not further discharged, the charge is recombined leading to an output voltage equal to 2 VDD – VSS. A practical implementation for an N-well process is shown in Fig. 81.16(b).

In this circuit, the transistors M_1, M_2, M_3, and M_4 behave as the switches S_1, S_2, S_3, and S_4 respectively, of Fig. 81.16(a). While for M_1 and M_2 the normal clocks are used, special clock phases are generated for M_3 and M_4 because they drive higher voltages. The circuit operates as follows.

During ϕ_1, M_8 is opened because ϕ_2' is high. The voltage at node v_y is higher than VDD because the capacitors C_3 and C_P were charged to VDD during the previous clock phase ϕ_2 and,

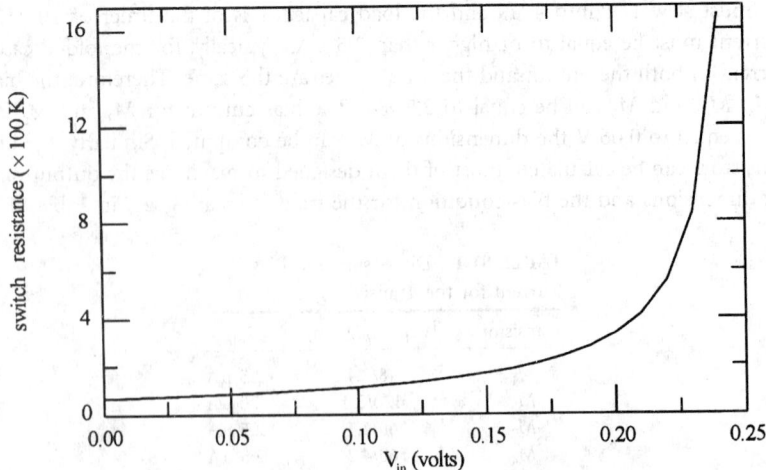

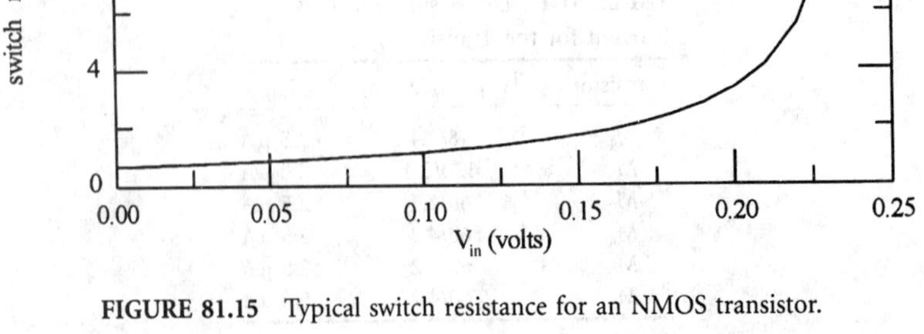

FIGURE 81.15 Typical switch resistance for an NMOS transistor.

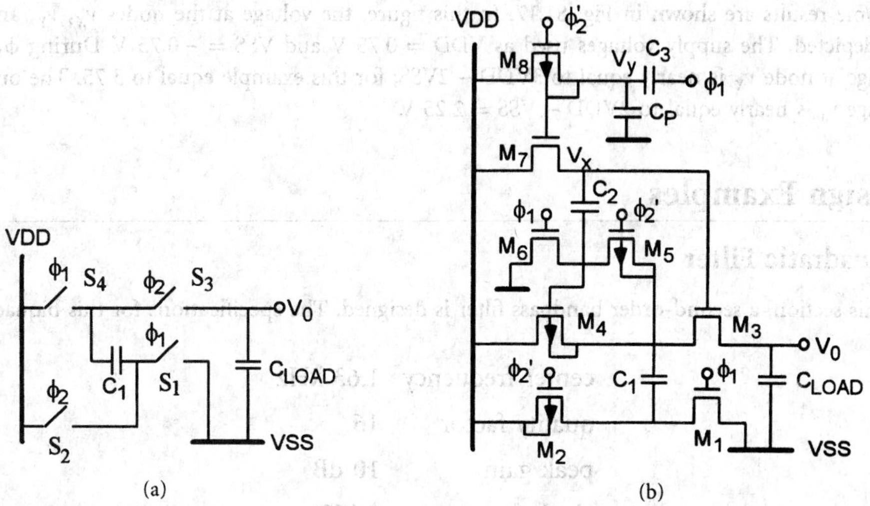

FIGURE 81.16 Voltage doubler. (a) Simplified diagram and (b) transistor level diagram.

when ϕ_1 goes up, charge is injected to the node through the capacitor C_3. Since v_y is higher than VDD, M_7 is turned on. The bottom plate of C_2 is connected to VSS by M_6 and C_2 is charged to VDD – VSS.

During ϕ_2, the refresh clock phase, the bottom plate of C_1 is connected to VDD by the PMOS transistor M_2. Note that if an NMOS transistor is employed, the voltage at the bottom plate of C_1 is equal to VDD – V_T, resulting in lower output voltage. During ϕ_2 if C_1 is not discharged, the voltage at its top plate is 2VDD – VSS. The voltage at node v_x becomes close to 3VDD – 2VSS volts, turning M_3 on and enabling the charge recombination of C_1 and C_{LOAD}. As a result, after several clock periods, the output voltage v_0 becomes equal to 2VDD – VSS. It has to be noted that v_y is precharged to VDD during this clock phase and that M_7 is turned off, keeping the voltage v_x high. Also, in order to avoid discharges the gate of M_4 is connected to the bottom plate of C_2. Thus, M_4 is turned off during the refresh phase.

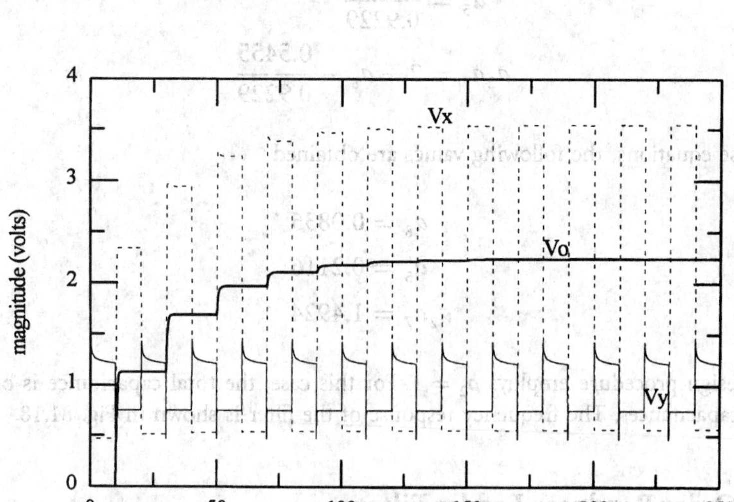

FIGURE 81.17 Time response of the voltage doubler at start up.

Some results are shown in Fig. 81.17. In this figure, the voltage at the nodes v_X, v_Y, and v_0 are depicted. The supply voltages used as VDD = 0.75 V and VSS = −0.75 V. During ϕ_1 the voltage at node v_X is nearly equal to 3VDD − 2VSS; for this example equal to 3.75. The output voltage v_0 is nearly equal to 2VDD − VSS = 2.25 V.

81.6 Design Examples

Biquadratic Filter

In this section, a second-order bandpass filter is designed. The specifications for this biquad are

center frequency	1.63 kHz
quality factor	16
peak gain	10 dB
clock frequency	8 kHz

A transfer function that realizes this filter is given by the following expression:

$$H(z) = \frac{0.1953(z-1)z}{z^2 + 0.5455z + 0.9229} \tag{81.52}$$

This transfer function can be implemented by using the biquads presented in Section 81.2. For the biquad of Fig. 81.10, and employing $a_1 = a_5 = a_9 = 0$, the circuit behaves as a bandpass filter. Equating the terms of (81.52) with the terms of (81.19), the following equations are obtained:

$$\begin{aligned} a_8 &= \frac{1}{0.9229} - 1 \\ a_5' &= \frac{0.1953}{0.9229} \\ a_2 a_7 &= 2 + a_8 - \frac{0.5455}{0.9229} \end{aligned} \tag{81.53}$$

Solving these equations, the following values are obtained

$$\begin{aligned} a_8 &= 0.0835 \\ a_5' &= 0.2116 \\ a_2 a_7 &= 1.4924 \end{aligned}$$

A typical design procedure employs $a_2 = 1$. For this case, the total capacitance is of the order of 32 unit capacitances. The frequency response of the filter is shown in Fig. 81.18.

A Sixth-Order Bandpass Ladder Filter

In this section, the design procedure for a sixth-order bandpass filter based on an *RLC* prototype is considered. The ladder filters are very attractive because of their low passband sensitivity to

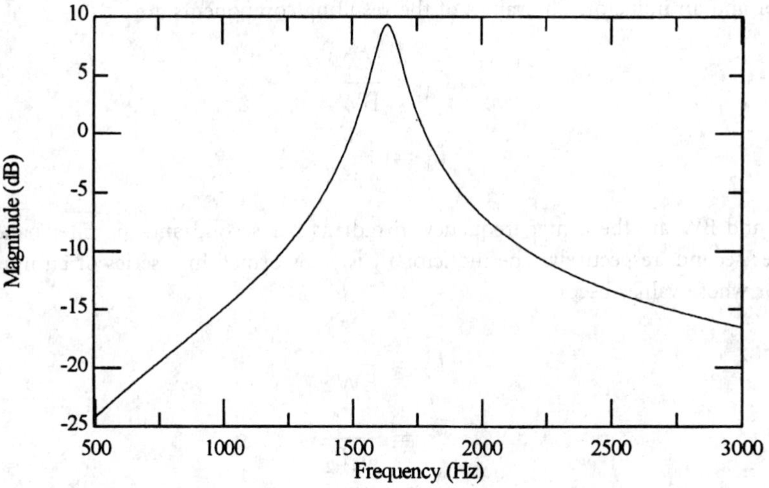

FIGURE 81.18 Frequency response of the second-order bandpass filter.

tolerances in the filter components. Let us consider the following specifications:

center frequency	100 kHz
bandwidth	2.5 kHz
passband ripple	< 0.5 dB
clock frequency	2 MHz

The design starts with a *RLC* low-pass prototype; see Fig. 81.19.

The values of the components for a prototype with 1 rad/s passband frequency can be obtained from tables or computed from well-known expressions. For this example, the passive components are

$$R_1 = 1\ \Omega$$
$$C_1 = 1.5963\ \text{F}$$
$$L_2 = 1.0967\ \text{H}$$

Using the low-pass to bandpass transformation, the capacitor C_1 is transformed in a parallel of

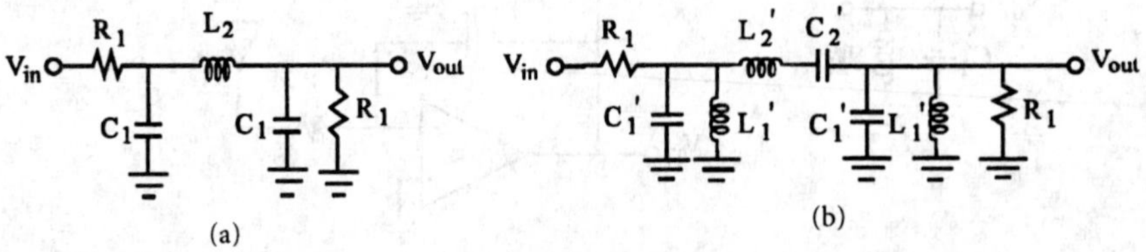

FIGURE 81.19 *RLC* prototypes. (a) Third-order low-pass filter and (b) sixth-order bandpass filter.

a capacitor and an inductor. The values of the resulting components are

$$C_1' = \frac{C_1}{\mathrm{BW}}$$
$$L_1' = \frac{1}{\omega_0^2 C_1'} \tag{81.54}$$

where ω_0 and BW are the center frequency in radians per second and the filter bandwidth in radians per second, respectively. The inductor L_2 is transformed in a series of an inductor and a capacitor whose values become

$$L_2' = \frac{L_2}{\mathrm{BW}}$$
$$C_2' = \frac{1}{\omega_0^2 L_2'} \tag{81.55}$$

The bandpass prototype is shown in Fig. 81.19(b). Before making the denormalizations it is desirable to transform the passive prototype to an active implementation. The grounded *LC* tank circuit can be simulated by the *RC* active implementation shown in Fig. 81.20.

The value of the simulated inductance is

$$L_1' = R_1^2 C_1' \tag{81.56}$$

It is important to note that the output of the active *RC* circuit is a low-impedance node. Therefore, the inverting input of the op-amps must be used for current injection. Similarly, a floating resonator can be simulated by the circuit shown in Fig. 81.21. Obviously, this is an expensive implementation but typically several active elements can be shared in the final design. Using typical circuit analysis techniques it can be shown that i_{12} for the passive *LC* tank is related to the components by the following expression:

$$i_{12} = \frac{sC_2'}{1 + s^2 L_2' C_2'}(v_{01} - v_{02}) \tag{81.57}$$

and for the active implementation

$$i_{12} = \frac{\left(\frac{R_2}{R_Q} C_Q\right)s}{1 + s^2 R_2^2 C_2^2}(v_{01} - v_{02}) \tag{81.58}$$

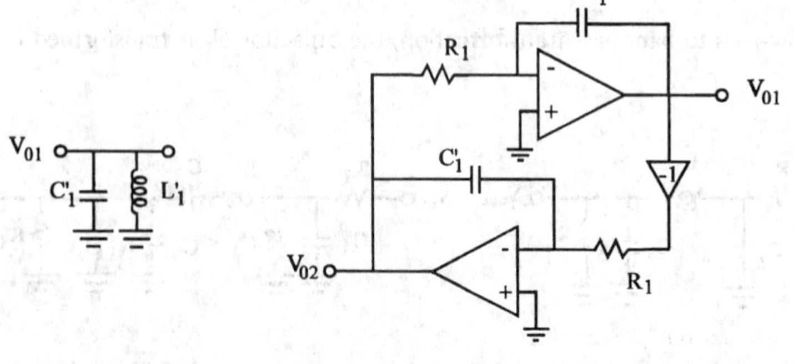

FIGURE 81.20 *RC* active implementation of a grounded *LC* tank circuit.

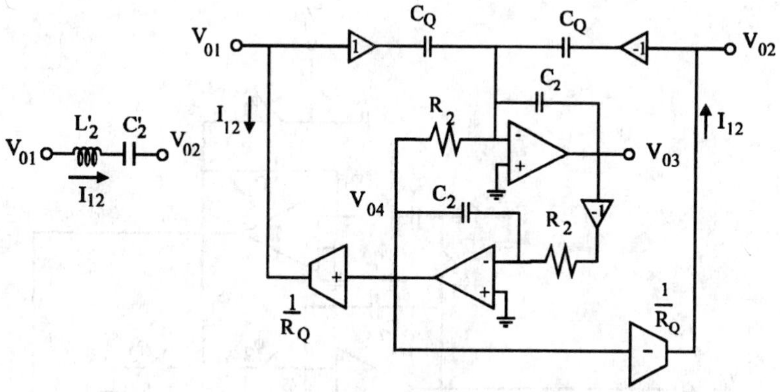

FIGURE 81.21 *RC* active implementation of a floating *LC* tank circuit.

Comparing (81.57) and (81.58), the active and passive implementations are equivalents if the following constraints are satisfied:

$$C_2' = \frac{R_2}{R_Q} C_Q \tag{81.59}$$

$$L_2'C_2' = R_2^2 C_2^2 \tag{81.60}$$

Obviously, the resistors R_2 can be replaced by switch capacitor equivalents. Using these building blocks, the implementation of the sixth-order bandpass filter is straightforward. The sixth-order switched-capacitor bandpass filter is shown in Fig. 81.22. The implementation of the $L_1'C_1'$ tank circuit is straightforward from Fig. 81.20. The resonant frequency of the switched-capacitor resonator is determined by the capacitors θC_1. The node voltage v_{01} is taken at the output of the first op-amp. The input resistor is implemented by the capacitor C_R and its associated switches. In Fig. 81.22, the resonator associated with v_{03} and v_{04} and the capacitors C_Q implement the $L_2'C_2'$ floating tank circuit. The final step of the design is to compute the values of the components. A simplified design follows, but more detailed procedures are presented in [1], [2], [6].

The loops of the switched-capacitor filter implementation are of the lossless discrete integrator (LDI) type and the resistors are of the serial type. Hence, if an LDI prewarping scheme is used the analog center frequency of the filter should be mapped to the desired value, mainly for high-*Q* filters wherein the interaction between resonators is very small. Since the resistors are not LDI, a distortion in the quality factor of the filter sections occurs, and thus increasing the passband ripple. The LDI transformation relates the analog and discrete frequencies by the following expression [1], [6]:

$$\omega_{\text{analog}} = 2f_c \sin\left(\frac{\omega_{\text{discrete}}}{2f_c}\right) \tag{81.61}$$

Applying the LDI transformation to the center frequency of the filter, the predistorted center frequency is equal to 99.56 kHz. Using this prewarped frequency and (81.54)–(81.60), the

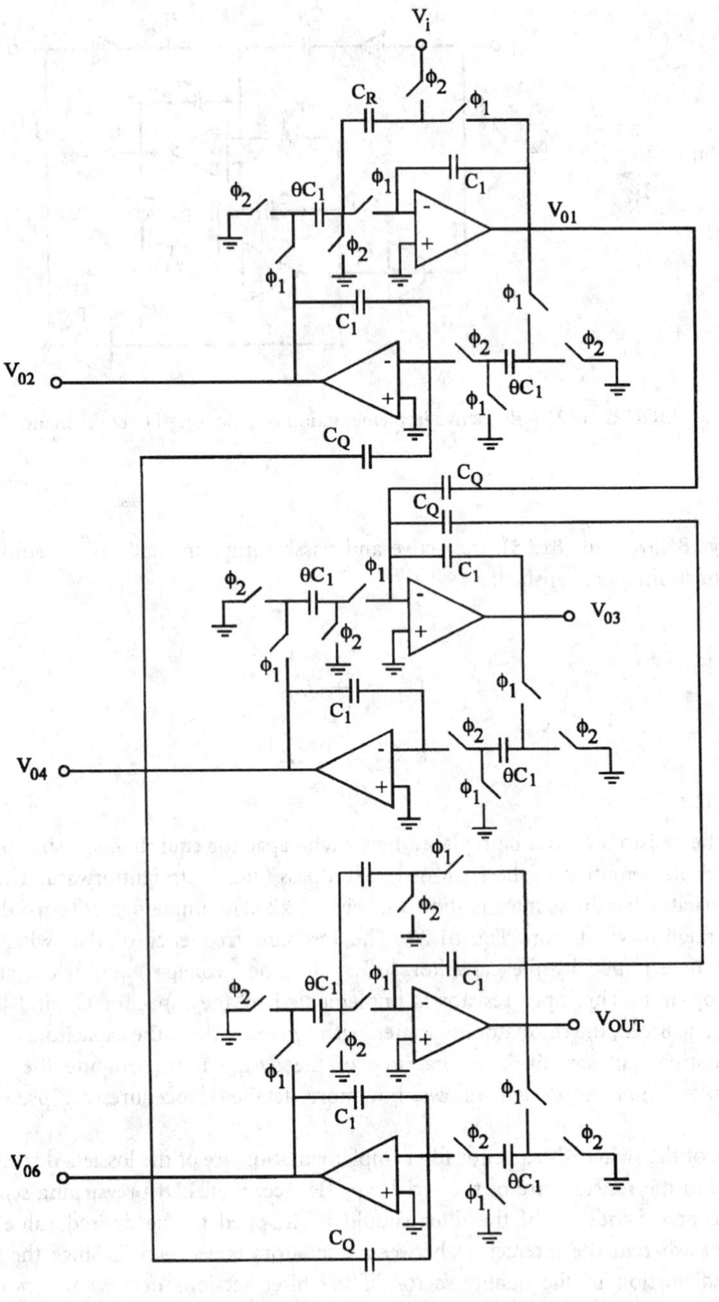

FIGURE 81.22 Sixth-order bandpass ladder filter.

following component values are obtained:

$$R_1 = 1\ \Omega$$

$$C_1' = 1.01623 \times 10^{-4}\ \text{F}$$

$$L_1' = 2.51316 \times 10^{-8}\ \text{H}$$

$$C_2' = 3.65803 \times 10^{-8}\ \text{F}$$

$$L_2' = 6.98181 \times 10^{-5}\ \text{H}$$

An impedance denormalization such that C_1' is scaled to 15 pF leads to the following passive component values:

$$R_1 = 6.77487\ \text{M}\Omega$$
$$C_1' = 15.000\ \text{pF}$$
$$L_1' = 0.17026\ \text{H}$$
$$C_2' = 5.3994\ \text{fF}$$
$$L_2' = 473.082\ \text{H}$$

In addition, the continuous-time and switched-capacitor resistors are related by the following expression:

$$C_{eq} = \frac{1}{f_c R_{cont}}$$

where R_{cont} is the time continuous resistance and C_{eq} the capacitor of the switched-capacitor resistor. Using $f_c = 2$ MHz, the capacitors needed for the simulation of R_1 can be computed from the previous expression. From this value, C_R $(= 1/f_c R_1)$ can be obtained. For the computation of the other elements, it can be noted that $L_1'C_1'$ is equal to $L_2'C_2'$. Therefore the implementation of the floating tank can be carried out with the resonator used in the first tank. For minimum capacitor spread C_Q/C_1' is commonly designed equal to R_Q/R_2, leading (81.59) to

$$\frac{C_Q}{C_1'} = \frac{R_Q}{R_2}\frac{C_2'}{C_1'} = \sqrt{\frac{C_2'}{C_1'}} \tag{81.62}$$

In the switched-capacitor filter, the transconductor $1/R_Q$ is implemented by C_Q/C_1R. Hence, the final values of the capacitors are

$$C_R = 0.0738\ \text{pF}$$
$$C_1 = 15.000\ \text{pF}$$
$$\theta C_1 = 4.6900\ \text{pF}$$
$$C_Q = 0.2833\ \text{pF}$$

While the filter center frequency of this design is accurate, the passband ripple is increased to around 0.8 dB instead of 0.5 dB. This is because the resistors used in the terminals of the filter are not implemented as LDI resistors. However, this effect can be partially corrected adjusting the resonant frequency of the second loop. If θC_1 is equal to 4.685 pF for the second resonator, the ripple is decreased to around 0.55 dB. The results for this case are shown in Fig. 81.23.

A Programmable Switched-Capacitor Filter

In most of the programmable filters, the important parameters are the resonant frequencies, the pole-quality factor, and sometimes, the peak gain. Thus, programmable low-pass, bandpass, high-pass, and bump equalizers are typically designed. Nevertheless, in some applications it is more important to control the gain at the frequency bands instead of that at the resonant frequency, the quality factor, or the peak gain. A typical approach for the implementation of these systems is to employ a parallel of a low-pass, bandpass, and a high-pass filter with programmable peak gain. For a second-order system, the implementation of this approach needs

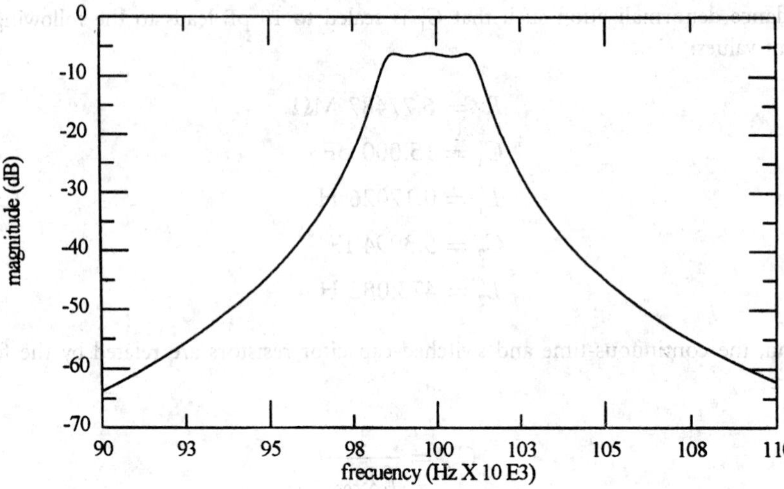

FIGURE 81.23 Frequency response for the sixth-order bandpass filter.

at least six operational amplifiers, three capacitor banks, and the implementation of six poles. Therefore, the number of switches, the power dissipation, and the silicon area needed for these structures are considerable. Another approach follows in this section.

In order to independently control the low-, medium-, and high-frequency bands it is required to realize the following transfer function:

$$H(s) = \frac{K_1 s^2 + K_2 \text{BW}s + K_3 \omega_0^2}{s^2 + \text{BW}s + \omega_0^2} \tag{81.63}$$

where BW and ω_0 are the bandwidth and the frequency of the poles, respectively. The control of the frequency bands is carried out by the parameters K_1, K_2, and K_3. From (81.63) it is clear that the dc gain and the high-frequency gain are equal to K_3 and K_1, respectively. A disadvantage of this expression is the lack of good control in the medium frequencies. If K_1 is equal to K_3, the filter gain at the resonant frequency depends on K_2. However, for the general case it is affected by the parameters K_1 and K_3. In addition, the shape of the transfer function is not well behaved for moderate- and high-Q applications. The behavior of the transfer function is improved if the following equation is employed:

$$H(s) = K_1 \frac{s^2 + K_2 K_3 \text{BW}s + K_3 \omega_0^2}{s^2 + K_1 \text{BW}s + K_1 \omega_0^2} \tag{81.64}$$

For low and high frequencies the behavior of this equation is similar to that of (81.63). At medium frequencies the behavior of the transfer function is better controlled by the parameter K_2 than for the case of (81.63). For the medium-frequency band, the gain is related to the product $K_2 K_3$ instead of that of the absolute value of K_2. Therefore, the effect of parameter K_2 on the transfer function is related to K_3.

A block diagram representation of (81.64) is presented in Fig. 81.24. In this figure, the filter bandwidth BW is equal to ω_0/Q, with Q equal to the quality factor of the filter. The control parameters are the gain factors K_1, K_2, and K_3. The implementation of this block diagram can be carried out by using various techniques, e.g., OTA-C, MOSFET-C, or switched-capacitor. A switched-capacitor implementation is shown in Fig. 81.25. For high sampling rate $2\pi f_c \gg \omega_0$, the capacitors θC_i are related with the pole frequency by the equation $\theta = \omega_0 T$. The capacitor banks control the gain of the three frequency bands. For the computation of the total capacitance,

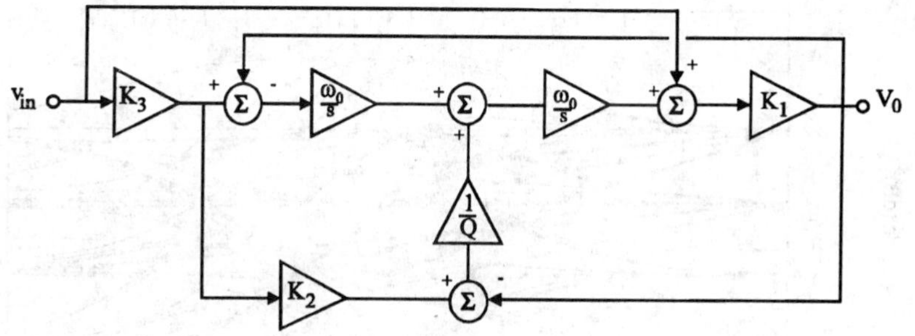

FIGURE 81.24 A flow diagram representation of (81.64).

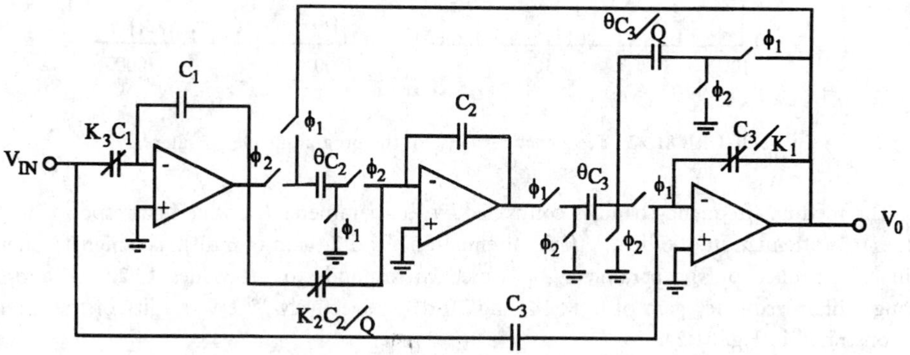

FIGURE 81.25 Switched-capacitor realization for the programmable filter.

the capacitors can be associated in the following groups: first group (K_3C_1, C_1), second group ($\theta C_2, K_2C_2/Q, C_2$), and third group ($\theta C_3/Q, \theta C_3, C_3/K_1, C_3$).

The versatility of the topology is shown in Figs. 81.26 and 81.27. For these results, the following design parameters have been employed: clock frequency = 128 kHz, $\omega_0 = 6.2832 * 350$ rad/s and $Q = 0.5$. For Fig. 81.26, the parameter K_3 is equal to 10, fixing the low-frequency gain at 20

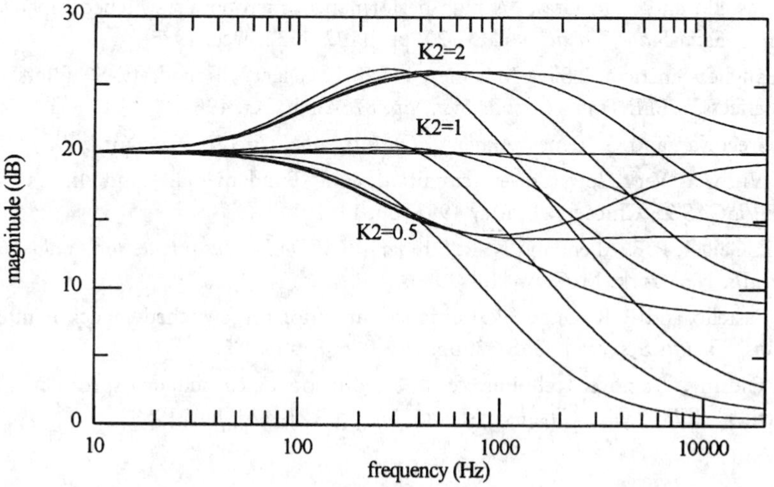

FIGURE 81.26 Effect of the parameter K_2 on the filter transfer function. For this plot $K_3 = 10$ and K_1 has been varied (1, 2.5, 5, and 10).

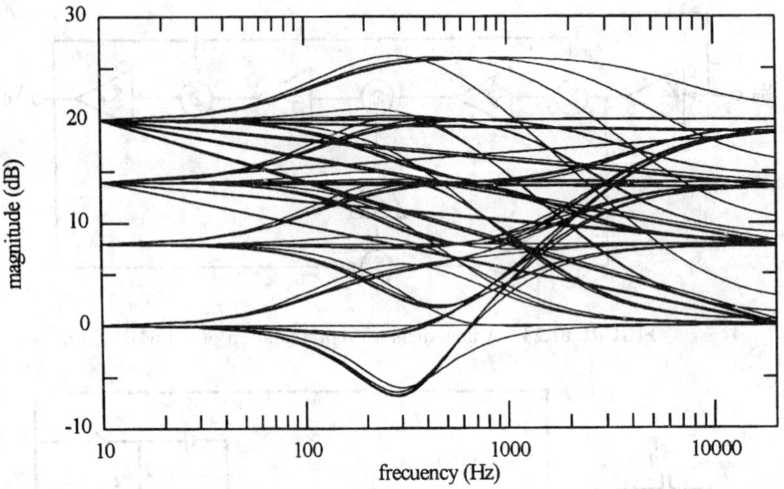

FIGURE 81.27 Frequency response of the programmable SC filter.

dB. The medium-frequency band is controlled by the parameter K_2, which corresponds to the values 0.5 (attenuation of 6 dB), 1 (0 dB attenuation), and 2 (gain of 6 dB). The high-frequency gain is controlled by the parameter K_1, which corresponds to the values 1, 2.5, 5, and 10, giving a high-frequency gain of 0, 8, 14, and 20 dB, respectively. The versatility of the filter is demonstrated in Fig. 81.27.

References

[1] P. E. Allen and E. Sánchez-Sinencio, *Switched-Capacitor Circuits.* New York: Van Nostrand, 1984.

[2] R. Unbehauen and A. Cichocki, *MOS Switched-Capacitor and Continuous-Time Integrated Circuits and Systems.* Berlin: Springer-Verlag, 1989.

[3] R. W. Brodersen, P. R. Gray, and D. A. Hodges, "MOS switched-capacitor filters," *Proc. IEEE,* vol. 67, pp. 61–75, Jan. 1979.

[4] R. Castello and P. R. Gray, "A high-performance micropower switched-capacitor filter," *IEEE J. Solid-State Circuits,* vol. SC-20, pp. 1122–1132, Dec. 1985.

[5] E. Sánchez-Sinencio, J. Silva-Martínez, and R. L. Geiger, "Biquadratic SC filters with small GB effects," *IEEE Trans. Circuits Syst.*, pp. 876–884, Oct. 1984.

[6] R. Gregorian and G. Temes, *Analog MOS Integrated Circuits,* New York: Wiley, 1986.

[7] E. Vittoz, "Very low power circuit design: Fundamentals and limits," in *Proc. IEEE/ISCAS'93,* Chicago, IL, May 1993, pp. 1451–1453.

[8] R. L. Geiger, P. E. Allen, and N. R. Strader, *VLSI Design Techniques for Analog and Digital Circuits,* New York: McGraw-Hill, 1990.

[9] R. Castello and P. R. Gray, "Performance limitations in switched-capacitor filters," *IEEE Trans. Circuits Syst.*, vol. CAS-32, pp. 865–876, Sept. 1985.

[10] H. Qiuting, "A novel technique for the reduction of capacitance spread in high-*Q* SC circuits," *IEEE Trans. Circuits Syst.*, vol. 36, pp. 121–126, Jan. 1989.

XVI

Digital Filters

. C. Lim
ational University of Singapore

H. Er
vang Technological
versity

dreas Antoniou
versity of Victoria

Montgomery Smith
University of Tennessee
e Institute

ce W. Bomar
University of Tennessee
e Institute

C. Lim
onal University of
apore

pio Saramäki
pere University of
nology

82

FIR Filters

2.1 Properties of FIR Filters

M. H. Er

Linear Phase Property

The finite-impulse response (FIR) filter is characterized by a unit-sample response that has a finite duration. One of the advantages of FIR filters compared to their infinite-impulse response (IIR) counterparts is that FIR filters can be designed with exactly linear phase. Linear phase response is important for applications where phase distortion due to nonlinear phase can degrade performance, such as in data transmission and television applications.

A FIR causal filter can be characterized by the transfer function [1]

$$H(z) = \sum_{n=0}^{N-1} h(nT)z^{-n} \tag{82.1}$$

where $h(nT)$ is the impulse response of the filter, N is the filter length, and T is the sampling interval. Using the relationship that

$$H(z) = \frac{Y(z)}{X(z)} \tag{82.2}$$

the difference equation of a FIR filter can be obtained by taking the inverse Z-transform of (82.1), that is

$$y(iT) = \sum_{n=0}^{N-1} h(nT)X(iT - nT) \tag{82.3}$$

which says that the current output of a FIR causal filter is the weighted sum of the current and past inputs. The weighting coefficients are given by the impulse response of the filter.

3-8341-2/95/$0.00 + $.50

From (82.1), the frequency response can be obtained by replacing $z = e^{j\omega T}$ as

$$H(e^{j\omega T}) = \sum_{n=0}^{N-1} h(nT)e^{-j\omega nT} = M(\omega)e^{j\phi(\omega)} \tag{82.4}$$

where $M(\omega)$ and $\phi(\omega)$ are the magnitude and phase responses respectively defined as

$$M(\omega) = |H(e^{j\omega T})| \tag{82.5a}$$

$$\phi(\omega) = \arg H(e^{j\omega T}) \tag{82.5b}$$

The phase delay and group (time) delay functions of a filter are defined as

$$\tau_p = \frac{\phi(\omega)}{\omega} \tag{82.6}$$

and

$$\tau_g = -\frac{d\phi(\omega)}{d\omega} \tag{82.7}$$

respectively. Filters for which τ_p and τ_g are independent of frequency are referred to as constant time delay or linear phase filters. Hence, the phase response of a linear phase filter is given by

$$\phi(\omega) = -\tau\omega, \qquad -\pi < \omega < \pi \tag{82.8}$$

where τ is a constant phase delay in samples.

From (82.4), (82.5b), and (82.8), the phase response can be expressed as

$$\phi(\omega) = -\tau\omega = \tan^{-1} \frac{-\sum_{n=0}^{N-1} h(nT)\sin(\omega nT)}{\sum_{n=0}^{N-1} h(nT)\cos(\omega nT)} \tag{82.9}$$

Consequently,

$$\tan(\omega\tau) = \frac{\sum_{n=0}^{N-1} h(nT)\sin(\omega nT)}{\sum_{n=0}^{N-1} h(nT)\cos(\omega nT)} \tag{82.10}$$

Using the definition $\tan(\omega\tau) = \sin(\omega\tau)/\cos(\omega\tau)$, (82.10) can be reexpressed as

$$\sum_{n=0}^{N-1} h(nT)\sin(\omega\tau - \omega nT) = 0 \tag{82.11}$$

It can be shown [1] that a solution to (82.11) is given by

$$\tau = \frac{(N-1)T}{2} \tag{82.12}$$

and

$$h(nT) = h[(N-1-n)T] \qquad 0 \le n \le N-1 \tag{82.13}$$

Hence, FIR filters can be designed to have constant phase and group delays if the conditions of (82.12) and (82.13) are satisfied. The symmetry property of (82.13) can also lead to efficient filter realizations.

In applications where only constant group delay is needed, the phase response can have the form

$$\phi(\omega) = \phi_0 - \tau\omega \tag{82.14}$$

where ϕ_0 is a constant. With $\phi_0 = \pm\pi/2$, it can be shown [1] that the impulse response is of the form

$$h(nT) = -h[(N-1-n)T] \qquad 0 \le n \le N-1 \tag{82.15}$$

In this case, the impulse response exhibits antisymmetrical property.

Frequency Response of Linear Phase FIR Filters

The frequency response of a causal linear phase FIR filter can be simplified to some simple forms using (82.13) and (82.15) and the values of N as follows:

(1) Symmetric Impulse Response and N = Odd. In this case,

$$\begin{aligned} H(e^{j\omega T}) = &\sum_{n=0}^{(N-3)/2} h(nT)e^{-j\omega nT} + h\left[\frac{(N-1)T}{2}\right]e^{-j\omega(N-1)T/2} \\ &+ \sum_{n=(N+1)/2}^{N-1} h(nT)e^{-j\omega nT} \end{aligned} \tag{82.16}$$

Using (82.13), letting $m = N - 1 - n$ and changing the limits of summation and finally letting $m = n$, the last summation in (82.16) can be reexpressed as

$$\sum_{n=(N+1)/2}^{N-1} h(nT)e^{-j\omega nT} = \sum_{n=0}^{(N-3)/2} h(nT)e^{-j\omega(N-1-n)T} \tag{82.17}$$

Substituting (82.17) into (82.16), one obtains

$$\begin{aligned} H(e^{j\omega T}) = &\sum_{n=0}^{(N-3)/2} h(nT)[e^{-j\omega nT} + e^{-j\omega(N-1-n)T}] \\ &+h\left[\frac{(N-1)T}{2}\right]e^{-j\omega(N-1)T/2} \end{aligned} \tag{82.18}$$

Factoring $e^{-j\omega(N-1)T/2}$ in (82.18) and letting $k = (N-1)/2 - n$, (82.18) can be reexpressed as

$$H(e^{j\omega T}) = e^{-j\omega(N-1)T/2}\left\{\sum_{k=1}^{(N-1)/2} h\left[\left(\frac{N-1}{2} - k\right)T\right]\left(e^{j\omega kT} + e^{-j\omega kT}\right) + h\left[\frac{(N-1)T}{2}\right]\right\} \quad (82.19)$$

Using the property that $e^{j\theta} + e^{-j\theta} = 2\cos\theta$, (82.19) can be simplified to the form

$$H(e^{j\omega T}) = e^{-j\omega(N-1)T/2}\left\{\sum_{k=1}^{(N-1)/2} 2h\left[\left(\frac{N-1}{2} - k\right)T\right]\cos(\omega kT) + h\left[\frac{(N-1)T}{2}\right]\right\} \quad (82.20)$$

Letting $a(o) = h[(N-1)T/2]$ and $a(k) = 2h[((N-1)/2 - k)T]$, (82.20) can be simplified further to

$$H(e^{j\omega T}) = e^{-j\omega(N-1)T/2}\left[\sum_{k=0}^{(N-1)/2} a(k)\cos(\omega kT)\right] \quad (82.21)$$

(2) Symmetric Impulse Response and N = Even. For this case, the frequency response takes the form

$$H(e^{j\omega T}) = e^{-j\omega(N-1)T/2}\left\{\sum_{k=0}^{N/2-1} 2h(kT)\cos\left[\omega\left(\frac{N}{2} - k - \frac{1}{2}\right)T\right]\right\} \quad (82.22)$$

Letting $b(k) = 2h[(N/2 - k)T]$, $k = 1, 2, \cdots, N/2$, (82.22) can be expressed as

$$H(e^{j\omega T}) = e^{-j\omega(N-1)T/2}\left\{\sum_{k=1}^{N/2} b(k)\cos\left[\omega\left(k - \frac{1}{2}\right)T\right]\right\} \quad (82.23)$$

An interesting feature of this frequency response is that $H(e^{j\omega T})$ is always equal to zero for $\omega = \pi$, independent of $b(k)$. This implies that high-pass filter characteristics cannot be realized with this type of filter.

(3) Antisymmetric Impulse Response and N = Odd. For this case, the derivation of the frequency response is the same as that in (1) except that the cosine summations are replaced by the sine summations multiplied by j because of (82.15). Hence, the frequency response is given by

$$H(e^{j\omega T}) = je^{-j\omega(N-1)T/2}\left\{\sum_{k=1}^{(N-1)/2} 2h\left[\left(\frac{N-1}{2} - k\right)T\right]\sin(\omega kT) + h\left[\frac{(N-1)T}{2}\right]\right\} \quad (82.24)$$

TABLE 82.1 Frequency Response of Linear Phase FIR Filters

$h(nT)$	N	$H(e^{j\omega T})$
Symmetrical	Odd	$e^{-j\omega(N-1)T/2}\sum_{k=0}^{(N-1)/2} a(k)\cos(\omega kT)$
	Even	$e^{-j\omega(N-1)T/2}\sum_{k=1}^{N/2} b(k)\cos\left[\omega\left(k-\frac{1}{2}\right)T\right]$
Antisymmetrical	Odd	$je^{-j\omega(N-1)T/2}\sum_{k=1}^{(N-1)/2} c(k)\sin(\omega kT)$
	Even	$je^{-j\omega(N-1)T/2}\sum_{k=1}^{N/2} d(k)\sin\left[\omega\left(k-\frac{1}{2}\right)T\right]$

where $a(o) = h\left[\frac{(N-1)T}{2}\right]$, $a(k) = c(k) = 2h\left[\left(\frac{N-1}{2} - k\right)T\right]$, $b(k) = d(k) = 2h\left[\left(\frac{N}{2} - k\right)T\right]$

It should be noted that for odd values of N, (82.15) requires that $h[(N-1)T/2] = 0$. Letting $c(k) = 2h[((N-1)/2 - k)T]$, $k = 1, 2, \cdots, (N-1)/2$, (82.24) becomes

$$H(e^{j\omega T}) = je^{-j\omega(N-1)T/2}\left\{\sum_{k=1}^{(N-1)/2} c(k)\sin(\omega kT)\right\} \tag{82.25}$$

A notable feature of this frequency response is that at frequencies $\omega = 0$ and $\omega = \pi$, the frequency response is always zero, independent of $c(k)$.

(4) Antisymmetric Impulse Response and N = Even. For this case, the frequency response is the same as that in (2) except the cosine summations become sine summations multiplied by j as follows:

$$H(e^{j\omega T}) = je^{-j\omega(N-1)T/2}\left\{\sum_{k=0}^{N/2-1} 2h(kT)\sin\left[\omega\left(\frac{N}{2} - k - \frac{1}{2}\right)T\right]\right\} \tag{82.26}$$

Letting $d(k) = 2h[(N/2 - k)T]$, $k = 1, 2, \cdots, N/2$, (82.26) becomes

$$H(e^{j\omega T}) = je^{-j\omega(N-1)T/2}\left\{\sum_{k=1}^{N/2} d(k)\sin\left[\omega\left(k - \frac{1}{2}\right)T\right]\right\} \tag{82.27}$$

In this case, the frequency response is zero at $\omega = 0$, independent of $d(k)$.

In summary, the frequency responses of the four possible types of FIR filters with linear phase are given in Table 82.1.

Locations of Zeros of Linear Phase FIR Filters

The symmetric and antisymmetric conditions of the impulse response given by (82.13) and (82.15) impose certain constraints on the zeros of the transfer function $H(z)$ [2]. For the case where N is an odd value, $H(z)$ can be written as

$$H(z) = z^{-(N-1)/2} \sum_{k=0}^{(N-1)/2} \frac{a(k)}{2}(z^k \pm z^{-k}) \tag{82.28}$$

where the $\pm$ sign corresponds to symmetry and antisymmetry in the impulse response respectively, and $a(o)$ and $a(k)$ are defined in Table 82.1.

Substituting z^{-1} for z in (82.28), one obtains

$$H(z^{-1}) = z^{(N-1)/2} \sum_{k=0}^{(N-1)/2} \frac{a(k)}{2}(z^{-k} \pm z^{k}) \tag{82.29}$$

It follows from (82.28) and (82.29) that

$$H(z^{-1}) = \pm z^{(N-1)} H(z) \tag{82.30}$$

Equation 82.30 shows that $H(z)$ and $H(z^{-1})$ are identical to within a delay of $(N-1)$ samples and a multiplier of ± 1. Thus, the zeros of $H(z^{-1})$ are identical to the zeros of $H(z)$. Therefore, if $z_i = r_i e^{j\phi_i}$ is a zero of $H(z)$, then $z_i^{-1} = (1/r_i)e^{-j\phi_i}$ must also be a zero of $H(z)$. This has the following implications on the zero locations:

1. If $r_i = 1$ and $\phi_i = 0$ or π, then the zeros lie at either $z = +1$ or $z = -1$. In these cases, the zero is its own complex conjugate.
2. If $r_i = 1$ and $\phi_i \neq 0$ or π, then the zeros of $H(z)$ that are on the unit circle are also zeros of $H(z^{-1})$ that are on the unit circle. Hence, the zeros occur in complex conjugate pairs on the unit circle.
3. If $r_i \neq 1$ and $\phi_i = 0$ or π, then the zeros are real and occur in reciprocal pairs on the unit circle.
4. If $r_i \neq 1$ and $\phi_i \neq 0$ or π, then the zeros occur in quadruplets with complex conjugate reciprocal pairs off the unit circle.

Figure 82.1 shows the possible types of zeros for linear phase FIR filters.

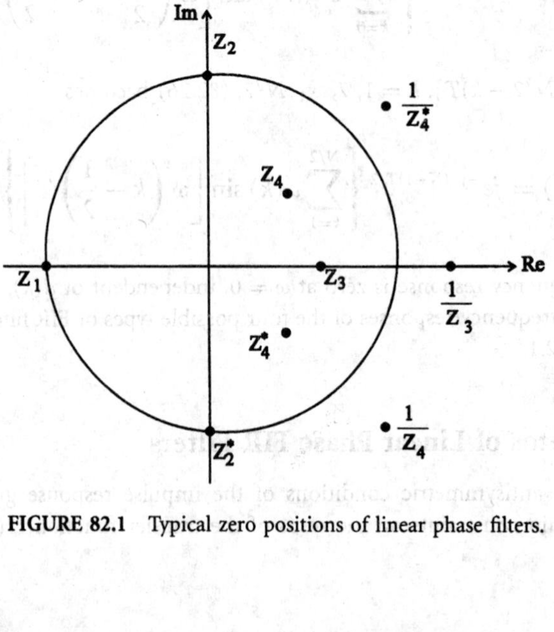

FIGURE 82.1 Typical zero positions of linear phase filters.

82.2 Design of FIR Filters

M. H. Er

Windowing Techniques

Windowing is one of the earliest techniques for designing FIR filters [3], [4]. The technique is simple because the filter coefficients can be obtained in closed form without the need for solving complex optimization problems as in some other sophisticated FIR design techniques. Hence, the design time is very short and the technique remains an attractive tool for FIR filter design.

To understand the windowing technique, first consider the process of obtaining a finite-length impulse response by truncating an infinite-duration impulse response sequence. Suppose $H_d(e^{j\omega T})$ is an ideal desired low-pass response with cutoff frequency ω_c. As the frequency response of an FIR filter is a periodic function, it can be expressed as a Fourier series as

$$H_d(e^{j\omega T}) = \sum_{n=-\infty}^{\infty} h_d(nT)e^{-j\omega nT} \tag{82.31}$$

where

$$h_d(nT) = \frac{T}{2\pi}\int_{-\pi/T}^{\pi/T} H_d(e^{j\omega T})e^{j\omega nT}\, d\omega \tag{82.32}$$

In general, $H_d(e^{j\omega T})$ is piecewise constant with a certain passband and stopband and with discontinuities at the boundaries between bands. Hence, the impulse sequence $h_d(nT)$ is of infinite duration. For example, for the ideal low-pass response shown in Fig. 82.2, the corresponding impulse response sequence is given by

$$h_d(nT) = \frac{\omega_c T}{\pi}\left(\frac{\sin \omega_c nT}{\omega_c nT}\right) \qquad -\infty \le n \le \infty \tag{82.33}$$

It is clear that (82.33) is a noncausal IIR filter. Also it is unstable and therefore unrealizable.

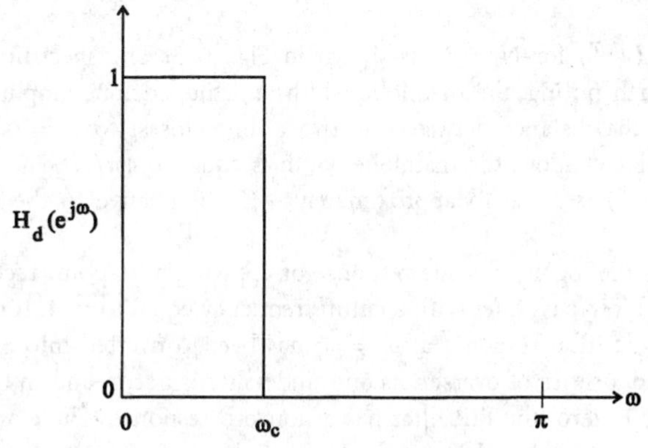

FIGURE 82.2 An ideal low-pass filter specification.

The Rectangular Window. One way to obtain a finite-duration causal impulse response is to simply truncate $h_d(nT)$ and introduce sufficient delay to obtain a causal impulse response, i.e., define

$$h(nT) = \begin{cases} h'_d(nT) & 0 \leq n \leq N-1 \\ 0 & \text{elsewhere} \end{cases} \tag{82.34}$$

where $h'_d(nT)$ is a delay version of $h_d(nT)$.

This can be represented as the product of the desired impulse response and a finite-duration window $w_r(nT)$, i.e.,

$$h(nT) = h'_d(nT)w_r(nT) \tag{82.35}$$

where $w_r(nT)$ is the rectangular window function defined as

$$w_r(nT) = \begin{cases} 1 & 0 \leq n \leq N-1 \\ 0 & \text{elsewhere} \end{cases} \tag{82.36}$$

Let $\theta = \omega T$ and using the fact that multiplication of two discrete-time sequences corresponds to a convolution of their Fourier transforms. Hence,

$$H(e^{j\omega}) = \frac{1}{2\pi}\int_{-\pi}^{\pi} H'_d(e^{j\theta})W_r(e^{j(\omega-\theta)})\,d\theta \tag{82.37}$$

where $W_r(e^{j\theta})$ is the spectrum of the rectangular window.

Since the two functions in the integral are periodic, a circular convolution results and the limits of integration are taken over one period. Thus the frequency response $H(e^{j\omega})$ will be a "smeared" version of the desired response $H'_d(e^{j\omega})$ and the discontinuities in the desired frequency response become transition bands of $H(e^{j\omega})$. To understand this, it is instructive to examine the frequency response for the causal rectangular window, that is,

$$\begin{aligned} W_r(e^{j\omega T}) &= \sum_{n=0}^{N-1} e^{-j\omega nT} \\ &= e^{-j\omega(N-1)T/2}\frac{\sin(\omega NT/2)}{\sin(\omega T/2)} \end{aligned} \tag{82.38}$$

The spectrum $W_r(e^{j\omega T})$ for $N = 31$ is shown in Fig. 82.3. The spectrum $W_r(e^{j\omega T})$ has two features that are worth noting, the mainlobe width and the sidelobe amplitude. The mainlobe width is defined as the distance between the two points closest to $\omega = 0$, where $W_r(e^{j\omega T})$ is zero. For a rectangular window, the mainlobe width is equal to $4\pi/N$. The maximum sidelobe amplitude for $W_r(e^{j\omega T})$ is equal to approximately -13 dB relative to the maximum value at $\omega = 0$.

Figure 82.4 shows the log–magnitude response of applying a 31-point rectangular window to approximate an ideal low-pass filter with a cutoff frequency equal to $\pi/4$. It can be seen that the sharp transition in the ideal response at $\omega = \omega_c$ has been converted into a gradual transition. Also, in the passband a series of overshoots and undershoots occur, and in the stopband, where the desired response is zero, the FIR filter has a nonzero response. These are the results of the convolution between $W_r(e^{j\omega T})$ and $H_d(e^{j\omega T})$. The mainlobe of $W_r(e^{j\omega T})$ causes the smearing of the desired response and the sidelobes of $W_r(e^{j\omega T})$ appear as overshoots and undershoots to the

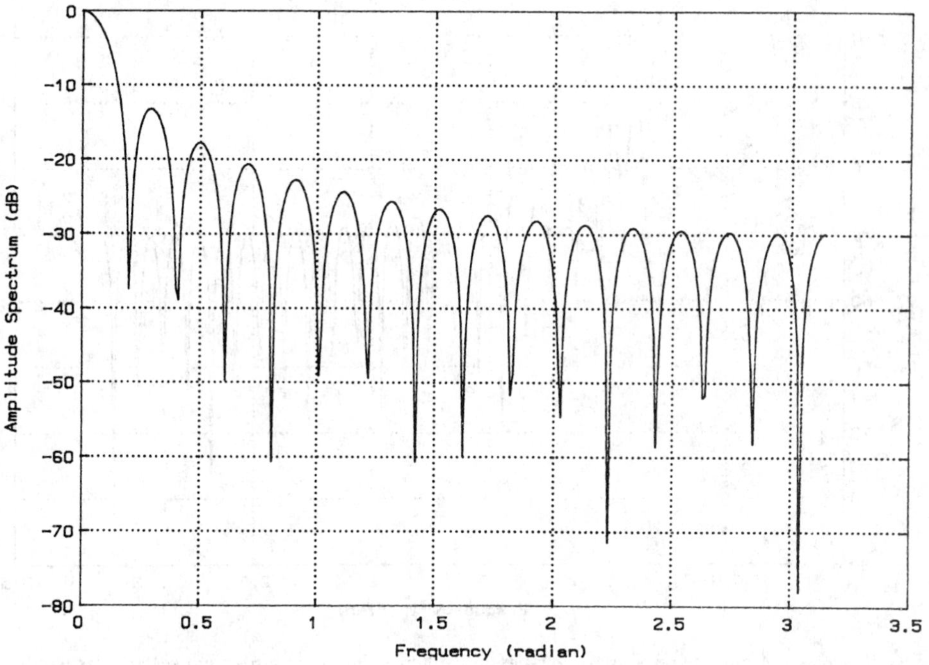

FIGURE 82.3 Fourier transform of the rectangular window.

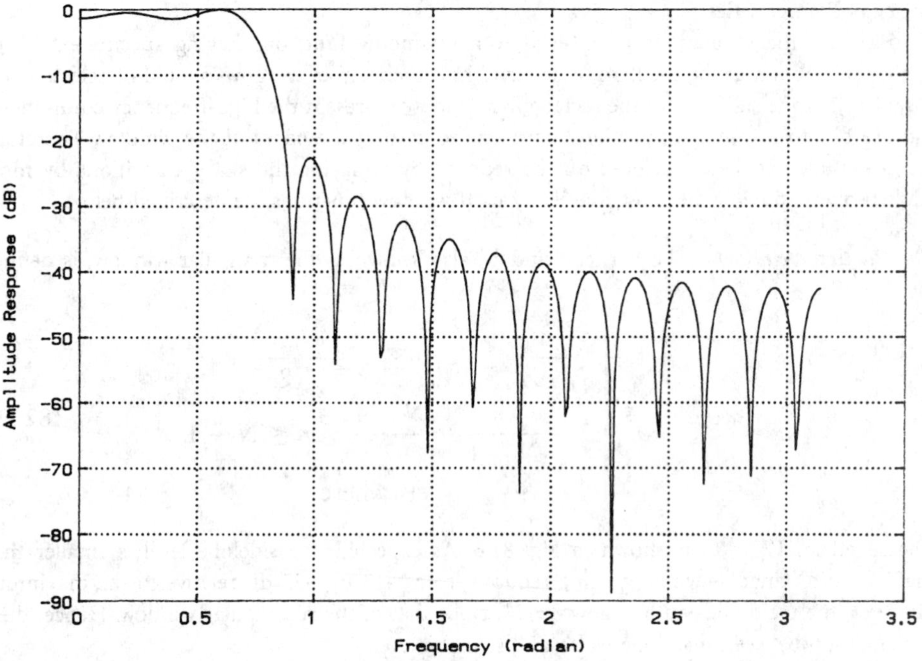

FIGURE 82.4 Magnitude response of low-pass FIR filter design using a 31-point rectangular window.

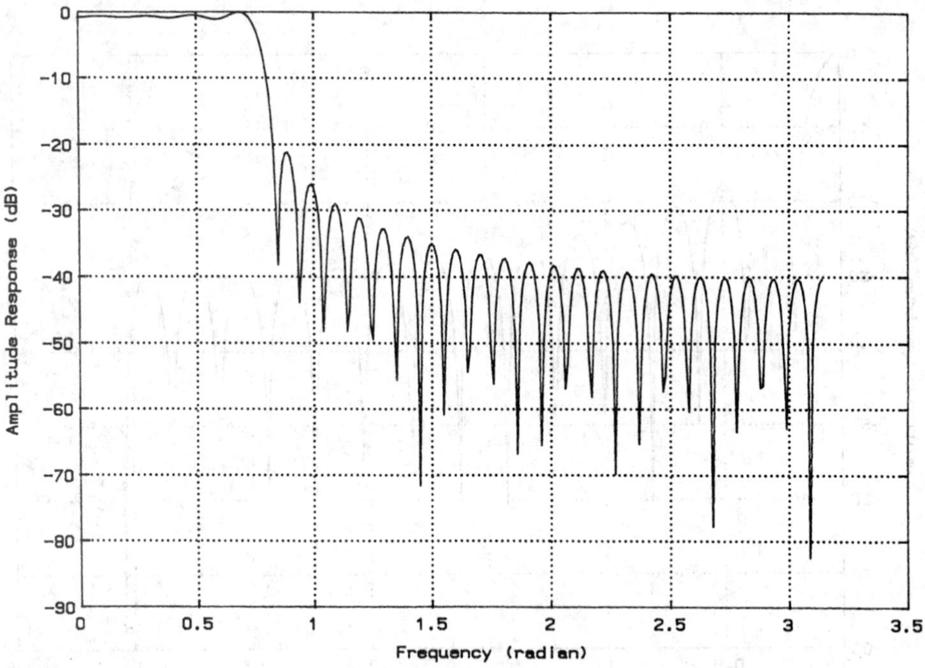

FIGURE 82.5 Magnitude response of low-pass FIR filter design using a 61-point rectangular window.

desired response. It is interesting to note that there will always be oscillations in the function $H(e^{j\omega T})$ in the vicinity of the steep transitions in $H_d'(e^{j\omega T})$, no matter how large the value of N as shown in Fig. 82.5 for $N = 61$. This result is known as the Gibbs phenomenon [2] in the theory of Fourier series.

To reduce the oscillations in $H(e^{j\omega T})$, other window functions having spectra exhibiting smaller sidelobes must be used. To understand how the form of windows should be selected, it is observed that the sidelobes of the rectangular window represent the high-frequency components and are due to the sharp transitions from one to zero at the edges of the window. Therefore, the amplitudes of these sidelobes can be reduced by replacing the sharp transitions by more gradual ones. Some of the most frequently used window functions are described below.

The Bartlett Window. The Bartlett window, also known as the triangular window, is defined as

$$w_t(nT) = \begin{cases} \dfrac{2n}{N-1} & 0 \le n \le \dfrac{N-1}{2} \\ 2 - \dfrac{2n}{N-1} & \dfrac{N-1}{2} \le n \le N-1 \\ 0 & \text{elsewhere} \end{cases} \tag{82.39}$$

The spectrum $W_t(e^{j\omega T})$ is shown in Fig. 82.6. As expected, the sidelobe level is smaller than that of the rectangular window, being reduced from -13 to -25 dB relative to the maximum. However, the mainlobe width is now $8\pi/N$, twice that of the rectangular window. Hence, there is a trade-off between mainlobe width and sidelobe level.

Figure 82.7 illustrates the FIR low-pass magnitude response obtained by using the Bartlett window. Comparing Fig. 82.7 to Fig. 82.4, it is observed that the Bartlett window produces a smoother magnitude response.

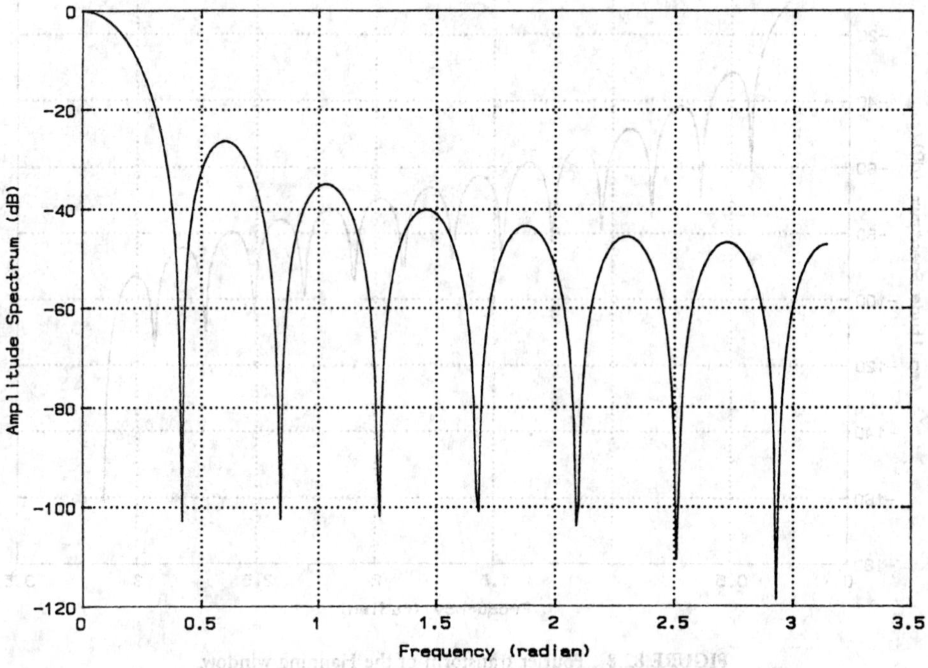

FIGURE 82.6 Fourier transform of the Bartlett window.

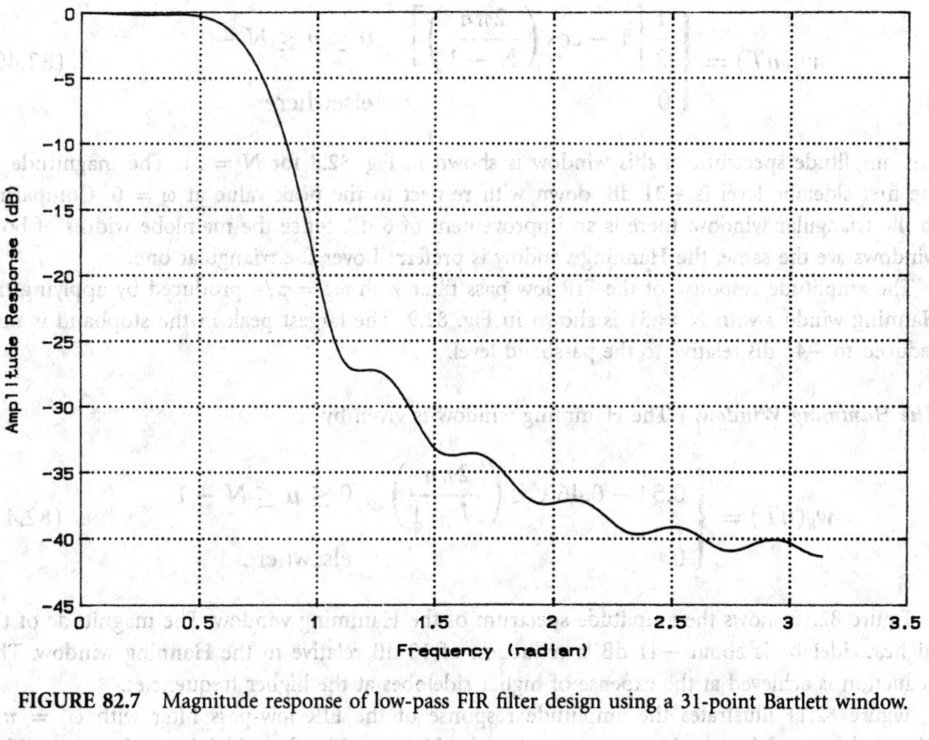

FIGURE 82.7 Magnitude response of low-pass FIR filter design using a 31-point Bartlett window.

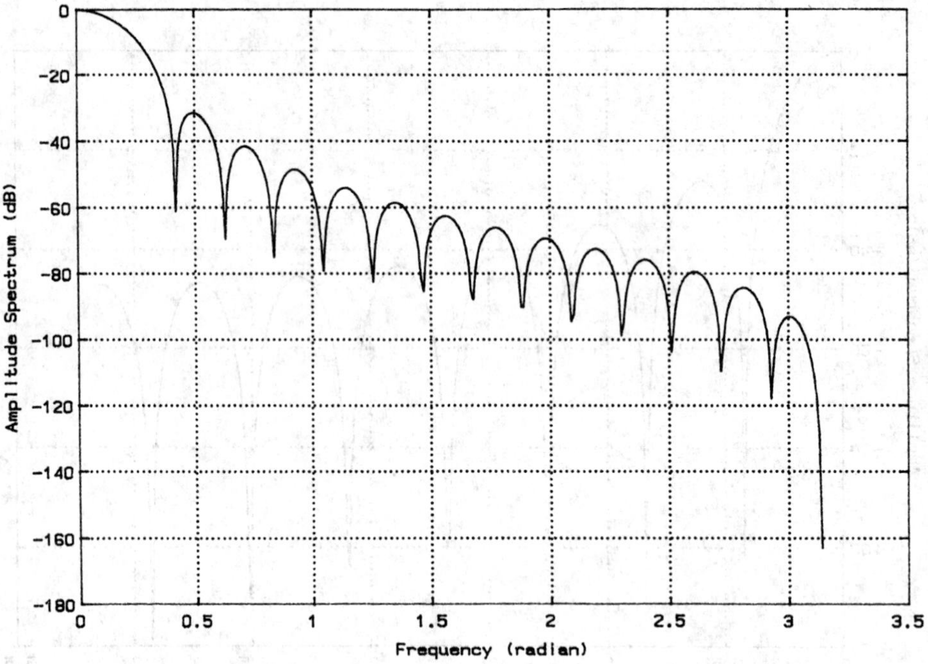

FIGURE 82.8 Fourier transform of the Hanning window.

The Hanning Window. The Hanning window, also known as the raised-cosine window, is given by

$$w_c(nT) = \begin{cases} \dfrac{1}{2}\left[1 - \cos\left(\dfrac{2\pi n}{N-1}\right)\right] & 0 \le n \le N-1 \\ 0 & \text{elsewhere} \end{cases} \tag{82.40}$$

The amplitude spectrum of this window is shown in Fig. 82.8 for $N = 31$. The magnitude of the first sidelobe level is -31 dB, down with respect to the peak value at $\omega = 0$. Comparing to the triangular window, there is an improvement of 6 dB. Since the mainlobe widths of both windows are the same, the Hanning window is preferred over the triangular one.

The amplitude response of the FIR low-pass filter with $\omega_c = \pi/4$ produced by applying the Hanning window with $N = 31$ is shown in Fig. 82.9. The largest peak in the stopband is now reduced to -44 dB relative to the passband level.

The Hamming Window. The Hamming window is given by

$$w_h(nT) = \begin{cases} 0.54 - 0.46\cos\left(\dfrac{2\pi n}{N-1}\right) & 0 \le n \le N-1 \\ 0 & \text{elsewhere} \end{cases} \tag{82.41}$$

Figure 82.10 shows the amplitude spectrum of the Hamming window. The magnitude of the highest sidelobe is about -41 dB, a reduction of 10 dB relative to the Hanning window. This reduction is achieved at the expense of higher sidelobes at the higher frequencies.

Figure 82.11 illustrates the amplitude response of the FIR low-pass filter with $\omega_c = \pi/4$ obtained by applying the Hamming window for $N = 31$. The first sidelobe peak is -51 dB, a

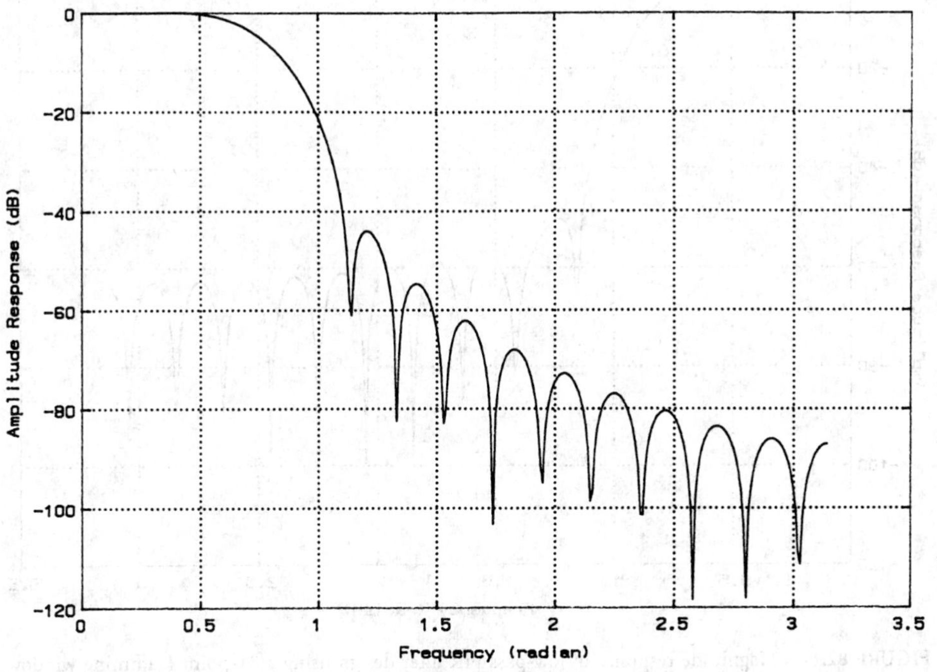

FIGURE 82.9 Magnitude response of low-pass FIR filter design using a 31-point Hanning window.

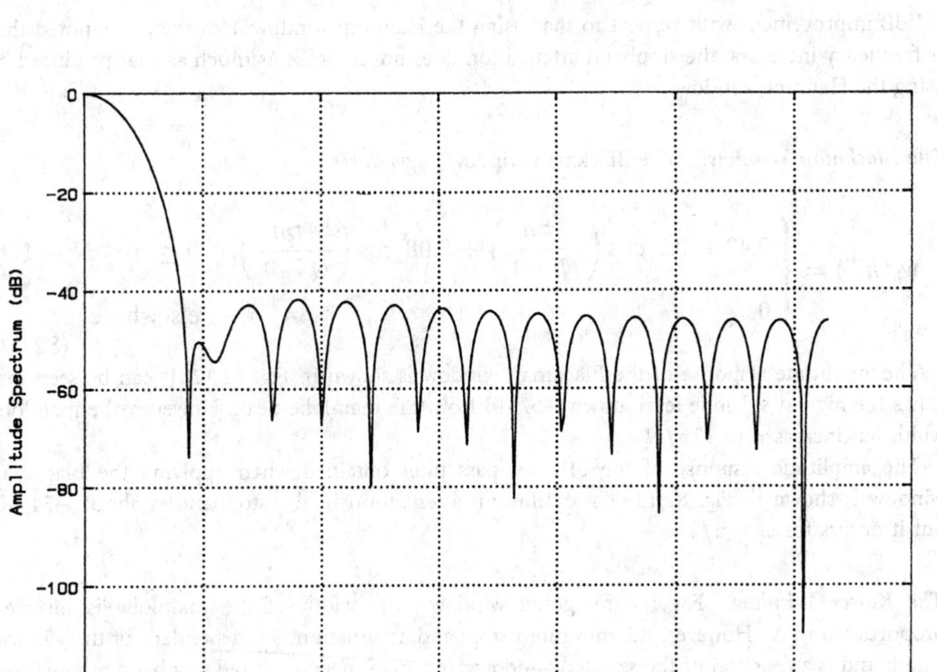

FIGURE 82.10 Fourier transform of the Hamming window.

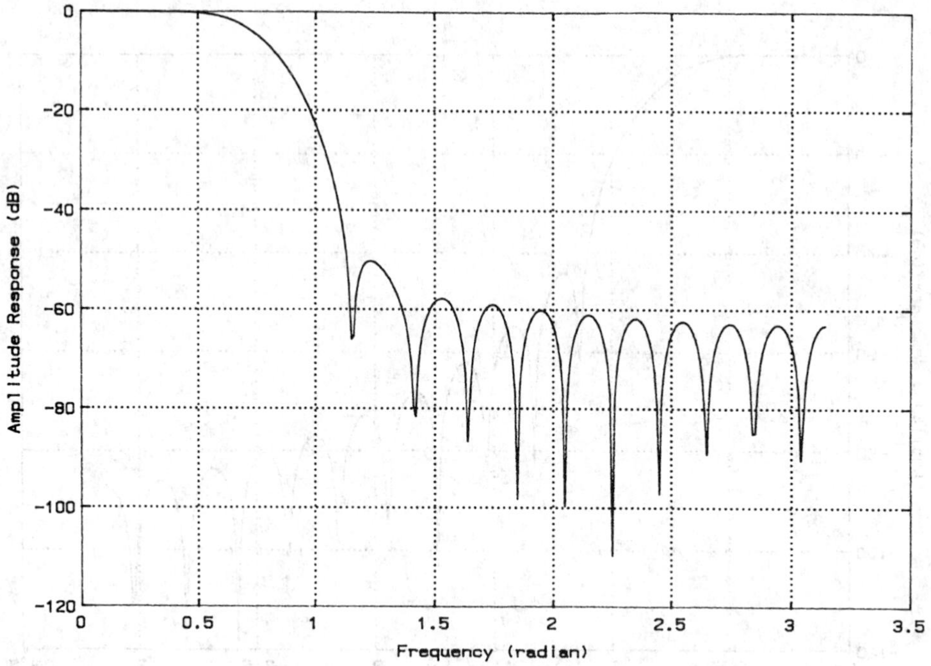

FIGURE 82.11 Magnitude response of low-pass FIR filter design using a 31-point Hamming window.

−7 dB improvement with respect to that using the Hanning window. However, it is noted that as frequency increases, the stopband attenuation does not increase as much as that produced by using the Hanning window.

The Blackman Window. The Blackman window is given by

$$w_b(nT) = \begin{cases} 0.42 - 0.5\cos\left(\dfrac{2\pi n}{N-1}\right) + 0.08\cos\left(\dfrac{4\pi n}{N-1}\right) & 0 \le n \le N-1 \\ 0 & \text{elsewhere} \end{cases} \tag{82.42}$$

The amplitude response of the Blackman window is shown in Fig. 82.12. It can be seen that it has the highest sidelobe level, down −57 dB from the mainlobe peak. However, the mainlobe width has increased to $12\pi/N$.

The amplitude response of the FIR low-pass filter obtained when applying the Blackman window is shown in Fig. 82.13. The minimum attenuation in the stopband is about −74 dB, but it occurs for $\omega > \pi/2$.

The Kaiser Window. For the foregoing windows, the width of the mainlobe is inversely proportional to N. However, the minimum stopband attenuation is independent of the window length and is a function of the selected window. Hence, to meet a desired stopband attenuation, the designer is forced to select a window that meets the design specifications. It is worth noting that windows with low sidelobe levels have broader mainlobe widths, hence requiring an increase in the order of the filter N to achieve the desired transition width.

In 1974, Kaiser [4] introduced a new window, now known as the Kaiser window, based on discrete-time approximations of the prolate spheroidal wave functions. This window has

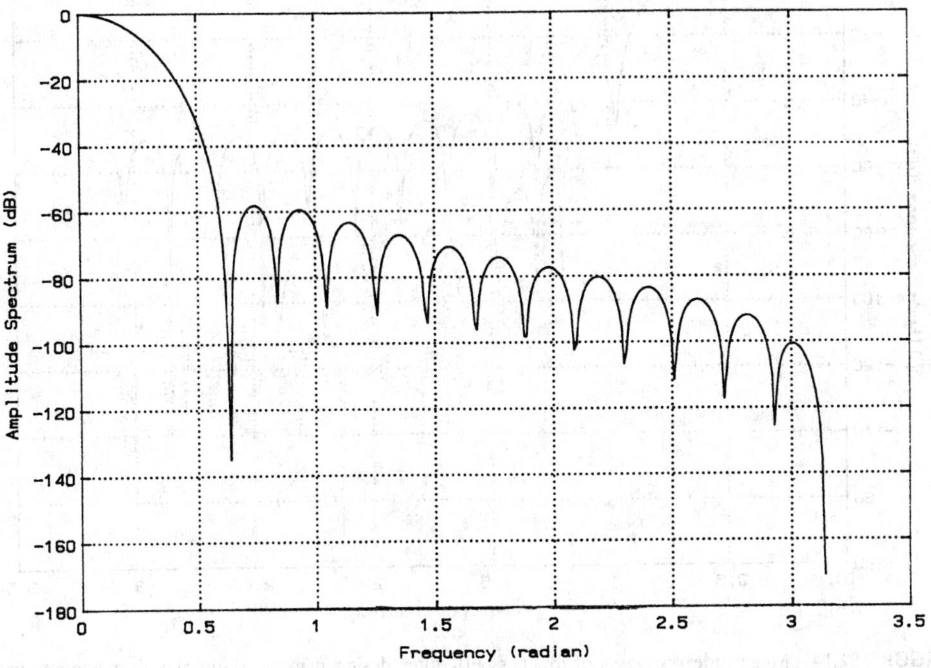

FIGURE 82.12 Fourier transform of the Blackman window.

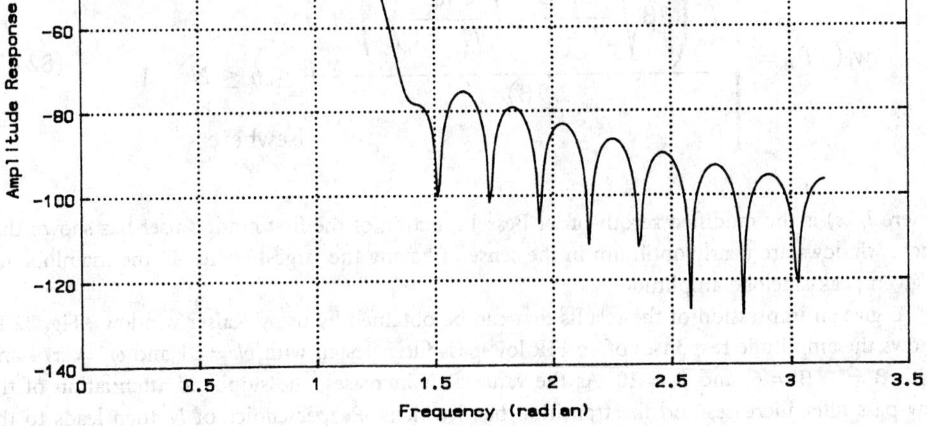

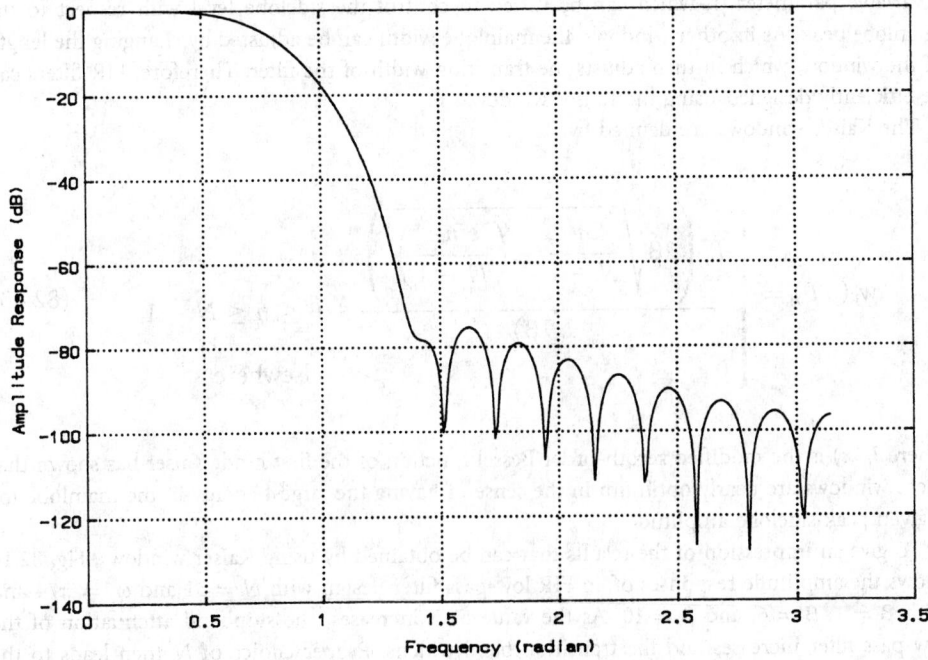

FIGURE 82.13 Magnitude response of low-pass FIR filter design using a 31-point Blackman window.

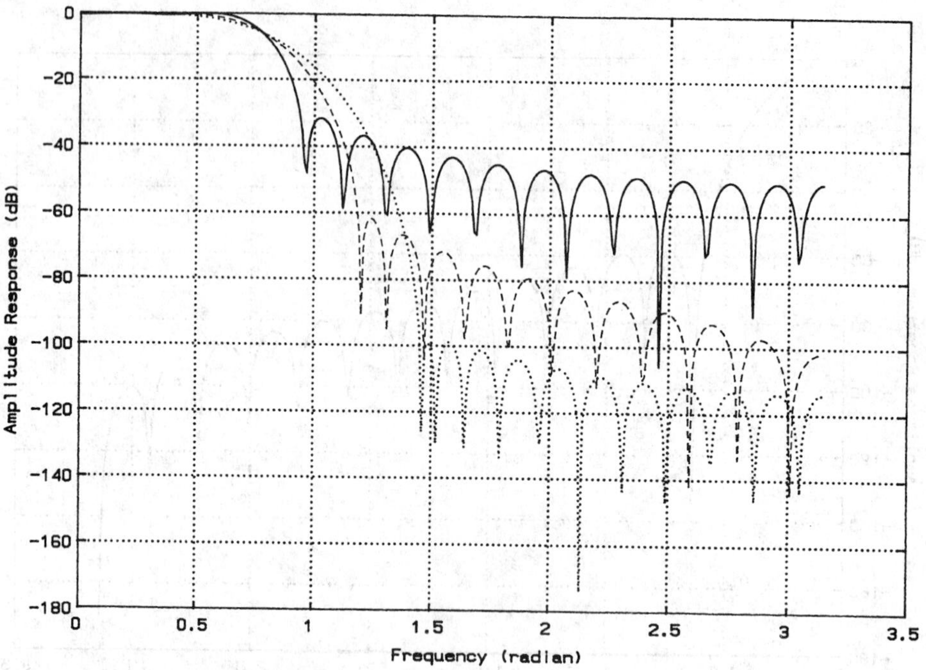

FIGURE 82.14 Magnitude responses of low-pass FIR filter design using a 31-point Kaiser window with $\beta = 1$ (solid line), $\beta = 6$ (dash line), and $\beta = 10$ (dotted line).

a variable parameter β, which can be varied to control the sidelobe level with respect to the mainlobe peak. As in other windows, the mainlobe width can be adjusted by changing the length of the window, which in turn adjusts the transition width of the filter. Therefore, FIR filters can be efficiently designed using the Kaiser window.

The Kaiser windows are defined by

$$w_k(nT) = \begin{cases} \dfrac{I_0\left(2\beta\sqrt{\dfrac{n}{N-1} - \left(\dfrac{n}{N-1}\right)^2}\right)}{I_0(\beta)} & 0 \le n \le N-1 \\ 0 & \text{elsewhere} \end{cases} \tag{82.43}$$

where $I_0(x)$ is the modified zeroth-order Bessel function of the first kind. Kaiser has shown that these windows are nearly optimum in the sense of having the largest energy in the mainlobe for a given peak sidelobe amplitude.

To give an impression of the results that can be obtained by using Kaiser windows, Fig. 82.14 shows the amplitude responses of an FIR low-pass filter design with $N = 31$ and $\omega_c = \pi/4$ and with $\beta = 1$, $\beta = 6$, and $\beta = 10$. As the value of β increases, the stopband attenuation of the low-pass filter increases and the transition band widens. Proper choice of N then leads to the final design.

The windows discussed above are compared in terms of their mainlobe width and the maximum sidelobe level in Table 82.2.

TABLE 82.2 Spectral Properties of N-Point Windows

Window	Mainlobe Width	Peak Amplitude of Sidelobe (dB)
Rectangular	$4\pi/N$	−13
Bartlett	$8\pi/N$	−25
Hanning	$8\pi/N$	−31
Hamming	$8\pi/N$	−41
Blackman	$12\pi/N$	−57

References

[1] L. R. Rabiner and B. Gold, *Theory and Application of Digital Signal Processing*, Englewood Cliffs, NJ: Prentice-Hall, 1975.

[2] A. V. Oppenheim and R. W. Schafer, *Digital Signal Processing*, Englewood Cliffs, NJ: Prentice-Hall, 1975.

[3] L. R. Rabiner, "Techniques for designing finite duration impulse response digital filters," *IEEE Trans. Commun. Technol.*, vol. COM-19, pp. 188–195, Apr. 1971.

[4] J. F. Kaiser, "Nonrecursive digital filter design using the I_0–sinh window function," in *Selected Papers in Digital Signal Processing, II*, New York: IEEE, 1976.

Equiripple FIR Filters

Andreas Antoniou

Introduction

The design of FIR filters can be accomplished either through noniterative or iterative methods. Noniterative methods entail the use of a small set of closed-form formulas and are, as a consequence, simple to apply. A frequently used method of this class is through the use of the Fourier series in conjunction with window functions. Iterative methods are based on the application of optimization techniques. These are characterized by a substantial increase in the computational complexity, but often lead to designs that are optimal in some respect.

This subsection deals with an iterative method for the design of FIR filters known as the *weighted-Chebyshev* method. In this approach, an error function is formulated for the desired filter in terms of a linear combination of cosine functions and is then minimized by using a very efficient multivariable optimization algorithm known as the *Remez exchange algorithm*. When convergence is achieved, the error function becomes equiripple, as in other Chebyshev solutions. The amplitude of the error in different frequency bands of interest is controlled by applying weighting to the error function.

The weighted-Chebyshev method is very flexible and can be used to obtain optimal solutions for most types of FIR filters, e.g., digital differentiators, Hilbert transformers, and low-pass, high-pass, bandpass, bandstop, and multiband filters with piecewise-constant amplitude responses. Furthermore, it can be used to design filters with arbitrary amplitude responses. Consequently, it is widely used. In common with other optimization methods, the weighted-Chebyshev method requires a large amount of computation; however, as the cost of computation is becoming progressively cheaper and cheaper with time, this disadvantage is not a serious one.

The underlying principles of the weighted-Chebyshev method were proposed during the early 1970s [1]–[3] and a series of developments soon after [4]–[8] led to the well-known computer program of McClellan, Parks, and Rabiner [9]. Some more recent enhancements to the method are reported in [10] and [11]. A detailed treatment of the subject can be found in [12].

Problem Formulation

An FIR filter with a symmetrical impulse response and odd length N can be represented by the transfer function

$$H(z) = \sum_{n=0}^{N-1} h(nT)z^{-n}$$

If we assume a sampling rate $\omega_s = 2\pi$, we have $T = 2\pi/\omega_s = 1$ s, and hence the frequency response of the filter can be expressed as

$$H(e^{j\omega T}) = e^{-jc\omega}P_c(\omega)$$

where

$$P_c(\omega) = \sum_{k=0}^{c} a_k \cos k\omega \tag{82.44}$$

with

$$a_0 = h(c)$$
$$a_k = 2h(c-k) \quad \text{for } k = 1, 2, \cdots, c$$
$$c = (N-1)/2$$

For a desired frequency response $e^{-jc\omega}D(\omega)$ and a specified weighting function $W(\omega)$, an error function $E(\omega)$ can be constructed as

$$E(\omega) = W(\omega)[D(\omega) - P_c(\omega)] \tag{82.45}$$

If it were possible to minimize the magnitude of the above error such that

$$|E(\omega)| \le \delta_p$$

with respect to some compact subset of the frequency interval $[0, \pi]$, say Ω, a filter would be obtained in which

$$|E_0(\omega)| = |D(\omega) - P_c(\omega)| \le \frac{\delta_p}{|W(\omega)|} \quad \text{for } \omega \in \Omega \tag{82.46}$$

In an equiripple filter, the magnitude of the error oscillates uniformly between zero and some maximum in each passband and stopband. In a low-pass equiripple filter, the amplitude response assumes the form depicted in Fig. 82.15, where δ_p and δ_a are the amplitudes of the passband and stopband ripples, and ω_p and ω_a are the passband and stopband edges, respectively. Hence, we require

$$D(\omega) = \begin{cases} 1 & \text{for } 0 \le \omega \le \omega_p \\ 0 & \text{for } \omega_a \le \omega \le \pi \end{cases}$$

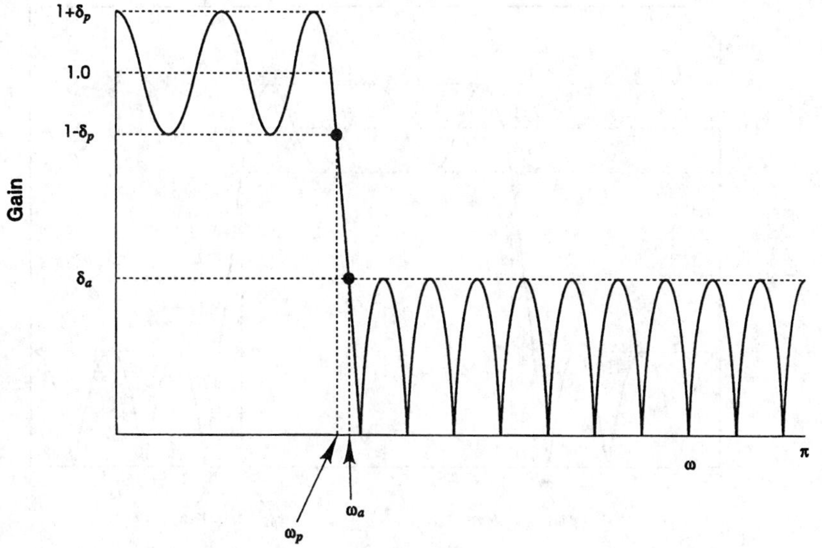

FIGURE 82.15 Amplitude response of equiripple low-pass filter. (Reproduced from A. Antoniou, *Digital Filters: Analysis, Design, and Applications.* New York: McGraw-Hill, 1993, by permission of McGraw-Hill, Inc.)

with

$$|E_0(\omega)| \leq \begin{cases} \delta_p & \text{for } 0 \leq \omega \leq \omega_p \\ \delta_a & \text{for } \omega_a \leq \omega \leq \pi \end{cases} \tag{82.47}$$

Therefore, from (82.46) and (82.47) we deduce

$$W(\omega) = \begin{cases} 1 & \text{for } 0 \leq \omega \leq \omega_p \\ \delta_p/\delta_a & \text{for } \omega_a \leq \omega \leq \pi \end{cases}$$

Similarly, for high-pass filters, we obtain

$$D(\omega) = \begin{cases} 0 & \text{for } 0 \leq \omega \leq \omega_a \\ 1 & \text{for } \omega_p \leq \omega \leq \pi \end{cases}$$

and

$$W(\omega) = \begin{cases} \delta_p/\delta_a & \text{for } 0 \leq \omega \leq \omega_a \\ 1 & \text{for } \omega_p \leq \omega \leq \pi \end{cases}$$

Bandpass and Bandstop Filters. The above formulation can be easily extended to other types of filters. For bandpass filters, we have

$$D(\omega) = \begin{cases} 0 & \text{for } 0 \leq \omega \leq \omega_{a1} \\ 1 & \text{for } \omega_{p1} \leq \omega \leq \omega_{p2} \\ 0 & \text{for } \omega_{a2} \leq \omega \leq \pi \end{cases}$$

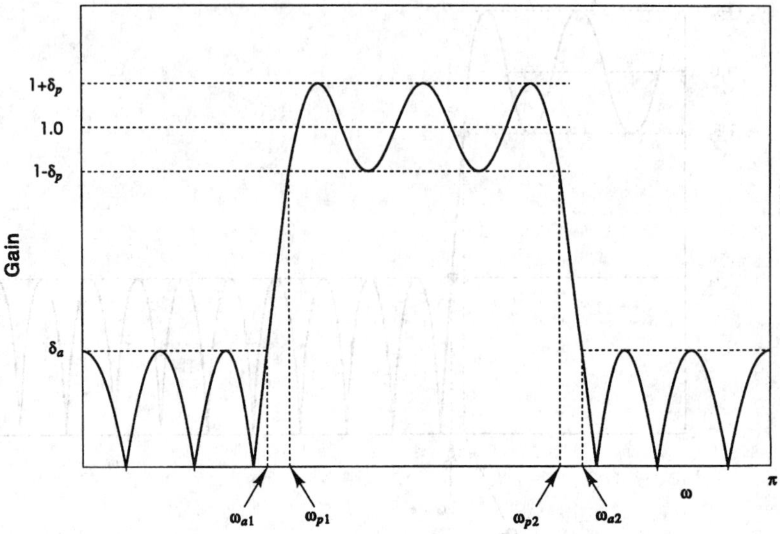

FIGURE 82.16 Amplitude response of equiripple bandpass filter. (Reproduced from A. Antoniou, *Digital Filters: Analysis, Design, and Applications.* New York: McGraw-Hill, 1993, by permission of McGraw-Hill, Inc.)

and

$$W(\omega) = \begin{cases} \delta_p/\delta_a & \text{for } 0 \le \omega \le \omega_{a1} \\ 1 & \text{for } \omega_{p1} \le \omega \le \omega_{p2} \\ \delta_p/\delta_a & \text{for } \omega_{a2} \le \omega \le \pi \end{cases}$$

where δ_p and δ_a are the passband and stopband ripples, respectively, ω_{p1} and ω_{p2} are passband edges, and ω_{a1} and ω_{a2} are stopband edges, as depicted in Fig. 82.16. On the other hand, for bandstop filters

$$D(\omega) = \begin{cases} 1 & \text{for } 0 \le \omega \le \omega_{p1} \\ 0 & \text{for } \omega_{a1} \le \omega \le \omega_{a2} \\ 1 & \text{for } \omega_{p2} \le \omega \le \pi \end{cases}$$

and

$$W(\omega) = \begin{cases} 1 & \text{for } 0 \le \omega \le \omega_{p1} \\ \delta_p/\delta_a & \text{for } \omega_{a1} \le \omega \le \omega_{a2} \\ 1 & \text{for } \omega_{p2} \le \omega \le \pi \end{cases}$$

The Alternation Theorem. An effective approach for the design of equiripple filters is to solve the minimax problem

$$\underset{\mathbf{x}}{\text{minimize}} \left\{ \max_{\omega} |E(\omega)| \right\} \tag{82.48}$$

where

$$\mathbf{x} = [a_0 \quad a_1 \quad \cdots \quad a_c]^T$$

is a column vector whose elements are the coefficients of the transfer function of the filter, which happen to be the values of the impulse response. The solution of this problem exists by virtue of the so-called *alternation theorem* [13], which is as follows.

Theorem: *If $P_c(\omega)$ is a linear combination of $r = c + 1$ cosine functions of the form*

$$P_c(\omega) = \sum_{k=0}^{c} a_k \cos k\omega$$

then a necessary and sufficient condition that $P_c(\omega)$ be the unique, best, weighted-Chebyshev approximation to a continuous function $D(\omega)$ on Ω, where Ω is a compact subset of the frequency interval $[0, \pi]$, is that the weighted error function $E(\omega)$ exhibit at least $r + 1$ extremal frequencies in Ω, i.e., there must exist at least $r + 1$ points $\hat{\omega}_i$ in Ω such that

$$\hat{\omega}_0 < \hat{\omega}_1 < \cdots < \hat{\omega}_r$$

$$E(\hat{\omega}_i) = -E(\hat{\omega}_{i+1}) \quad \text{for } i = 0, 1, \cdots, r - 1$$

and

$$|E(\hat{\omega}_i)| = \max_{\omega\in\Omega} |E(\omega)| \quad \text{for } i = 0, 1, \cdots, r$$

From the alternation theorem and (82.45) we can write

$$E(\hat{\omega}_i) = W(\hat{\omega}_i)[D(\hat{\omega}_i) - P_c(\hat{\omega}_i)] = (-1)^i\delta \tag{82.49}$$

for $i = 0, 1, \cdots, r$, where δ is a constant. This system of equations can be put in matrix form as

$$\begin{bmatrix} 1 & \cos\hat{\omega}_0 & \cos 2\hat{\omega}_0 & \cdots & \cos c\hat{\omega}_0 & \dfrac{1}{W(\hat{\omega}_0)} \\ \vdots & \vdots & \vdots & & \vdots & \vdots \\ 1 & \cos\hat{\omega}_r & \cos 2\hat{\omega}_r & \cdots & \cos c\hat{\omega}_r & \dfrac{(-1)^r}{W(\hat{\omega}_r)} \end{bmatrix} \begin{bmatrix} a_0 \\ a_1 \\ \vdots \\ a_c \\ \delta \end{bmatrix} = \begin{bmatrix} D(\hat{\omega}_0) \\ D(\hat{\omega}_1) \\ \vdots \\ D(\hat{\omega}_r) \end{bmatrix} \tag{82.50}$$

If the extremal frequencies (or extremals for short) were known, coefficients a_k and, in turn, the frequency response of the filter could be computed using (82.44). The solution of this system exists since the above $(r + 1) \times (r + 1)$ matrix is nonsingular [13].

Remez Exchange Algorithm

The Remez exchange algorithm is an *iterative multivariable* algorithm, which is naturally suited for the solution of the minimax problem stated in (82.48). It is based on the second optimization method of Remez [14] and involves the following basic steps.

Algorithm 1: Basic Remez Exchange Algorithm

1. Initialize extremals $\hat{\omega}_0, \hat{\omega}_1, \cdots, \hat{\omega}_r$ ensuring that an extremal is assigned at each band edge.
2. Locate the frequencies $\hat{\hat{\omega}}_0, \hat{\hat{\omega}}_1, \cdots, \hat{\hat{\omega}}_\rho$ at which $|E(\omega)|$ is maximum and $|E(\hat{\hat{\omega}}_i)| \geq \delta$. These frequencies are *potential* extremals for the next iteration.

3. Compute the convergence parameter

$$Q = \frac{\max\left|E(\hat{\omega}_i)\right| - \min\left|E(\hat{\omega}_i)\right|}{\max\left|E(\hat{\omega}_i)\right|}$$

where $i = 0, 1, \cdots, \rho$.

4. Reject $\rho - r$ *superfluous* potential extremals $\hat{\omega}_i$ according to an appropriate rejection criterion and renumber the remaining $\hat{\omega}_i$ sequentially; then set $\hat{\omega}_i = \hat{\omega}_i$ for $i = 0, 1, \cdots, r$.
5. If $Q > \varepsilon$, where ε is a convergence tolerance (say $\varepsilon = 0.01$), repeat from step 2; otherwise continue to step 6.
6. Compute $P_c(\omega)$ using the last set of extremals; then deduce $h(n)$, the impulse response of the required filter, and stop.

The amount of computation required by the algorithm tends to depend quite heavily on the initialization scheme used in step 1, on the search method used for the location of the maxima of the error function in step 2, and on the criterion used to reject superfluous frequencies $\hat{\omega}_i$ in step 4.

Initialization of Extremals. The simplest scheme for the initialization of extremals $\hat{\omega}_i$ for $i = 0, 1, \cdots, r$ is to assume that they are uniformly spaced in the frequency bands of interest. If there are J distinct bands in the required filter of widths $B_1, B_2, \cdots, B_J$ and extremals are to be located at the left-hand and right-hand band edges of each band, the sum of these bandwidths should be divided into $r+1-J$ intervals. Thus the average interval between adjacent extremals is

$$W_0 = \frac{1}{r+1-J}\sum_{j=1}^{J} B_j$$

Since the quantities B_j/W_0 need not be integers, the use of W_0 for the generation of the extremals will almost always result in a fractional interval in each band. This problem can be avoided by rounding the number of intervals B_j/W_0 to the nearest integer and then readjusting the frequency interval for the corresponding band accordingly. This can be achieved by letting the number of intervals in bands j and J be

$$m_j = \text{Int}\left(\frac{B_j}{W_0} + 0.5\right) \quad \text{for } j = 1, 2, \cdots, J-1$$

and

$$m_J = r - \sum_{j=1}^{J-1}(m_j + 1)$$

respectively, and then recalculating the frequency intervals for the various bands as

$$W_j = \frac{B_j}{m_j} \quad \text{for } j = 1, 2, \cdots, J$$

A more sophisticated initialization scheme, which was found to give good results, is described in [15].

Location of Maxima of the Error Function. The frequencies $\hat{\omega}_i$, which *must include maxima at band edges* if $|E(\hat{\omega}_i)| \geq |\delta|$, can be located by simply evaluating $|E(\omega)|$ over a dense set of frequencies. A reasonable number of frequency points that yields sufficient accuracy in the determination of the frequencies $\hat{\omega}_i$ is $8(N+1)$. This corresponds to about 16 frequency points per ripple of $|E(\omega)|$. A suitable frequency interval for the jth band is $w_j = W_j/S$ with $S = 16$.

The above *exhaustive* search can be implemented in terms of Algorithm 2 below, where ω_{Lj} and ω_{Rj} are the left-hand and right-hand edges in band j; W_j is the interval between adjacent extremals and m_j is the number of intervals W_j in band j; w_j is the interval between successive samples of $|E(\omega)|$ in interval W_j and S is the number of intervals w_j in each interval W_j; N_j is the total number of intervals w_j in band j; and J is the number of bands.

Algorithm 2: Exhaustive Step-by-Step Search

1. Set $N_j = m_j S$, $w_j = B_j/N_j$ and $e = 0$.
2. For each of bands $1, 2, \cdots, j, \cdots, J$ do the following. For each of frequencies $\omega_{1j} = \omega_{Lj}$, $\omega_{2j} = \omega_{Lj} + \omega_j, \cdots, \omega_{ij} = \omega_{Lj} + (i-1)\omega_j, \cdots, \omega_{N_j+1} = \omega_{Rj}$, set $\hat{\omega}_e = \omega_{ij}$ and $e = e + 1$ provided that $|E(\omega_{ij})| \geq |\delta|$ and one of the following conditions holds:
   i. Case $\omega_{ij} = \omega_{Lj}$: if $|E(\omega_{ij})|$ is maximum at $\omega_{ij} = \omega_{Lj}$ (i.e., $|E(\omega_{Lj})| > |E(\omega_{Lj} + \varepsilon)|$);
   ii. Case $\omega_{Lj} < \omega_{ij} < \omega_{Rj}$: if $|E(\omega)|$ is maximum at $\omega = \omega_{ij}$ (i.e., $|E(\omega_{ij} - \omega_j)| < |E(\omega_{ij})| > |E(\omega_{ij} + w_j)|$);
   iii. Case $\omega_{ij} = \omega_{Rj}$: if $|E(\omega_{ij})|$ is maximum at $\omega_{ij} = \omega_{Rj}$ (i.e., $|E(\omega_{Rj})| > |E(\omega_{Rj} - \varepsilon)|$.

The parameter ε in steps i and iii is a small positive constant and a value $10^{-2} w_j$ yields satisfactory results.

In practice, $|E(\omega)|$ is maximum at an interior left-hand band edge[1] if its first derivative at the band edge is negative, and a mirror-image situation applies at an interior right-hand band edge. In such cases, $|E(\omega)|$ has a zero immediately to the right or left of the band edge and the inequality in step i or iii may sometimes fail to identify a maximum. However, the problem can be avoided by using the inequality $|E(\omega_{Lj} - \varepsilon)| > |E(\omega_{Lj})|$ in step i and $|E(\omega_{Rj})| < |E(\omega_{Rj} + \varepsilon)|$ in step iii for interior band edges.

In rare circumstances, a maximum of $|E(\omega)|$ may occur between a band edge and the first sample point. Such a maximum may be missed by Algorithm 2, but the problem can be easily identified since the number of potential extremals will then be less than the minimum. The remedy is to check the number of potential extremals at the end of each iteration and if it is found to be less than $r + 1$, the density of sample points, i.e., S, is doubled and the iteration is repeated. If the problem persists, the process is repeated until the required number of potential extremals is obtained. If a value of S equal to or less than 256 does not resolve the problem, the loss of potential extremals is most likely due to some other reason.

An important precaution in the implementation of the preceding search method is to ensure that extremals belong to the dense set of frequency points to avoid numerical ill-conditioning in the computation of $E(\omega)$ [see (82.49) and (82.51)]. In addition, the condition $|E(\omega_{ij})| \geq |\delta|$ should be replaced by $|E(\omega_{ij})| > |\delta| - \varepsilon_1$, where ε_1 is a small positive constant, say 10^{-6}, to ensure that no maxima are missed owing to roundoff errors.

The search method is very reliable and its use in Algorithm 1 leads to a *robust* algorithm since the entire frequency axis is searched using a dense set of frequency points. Its disadvantage is that it requires a considerable amount of computation and is, therefore, inefficient.

A more efficient version of Algorithm 2 is obtained by maintaining all the interior band edges as extremals throughout the optimization independently of the behavior of the error function at the band edges. However, the algorithm obtained tends to be somewhat less robust, i.e., it tends to fail more frequently than Algorithm 2.

[1]An interior band edge is one in the range $0 < \omega < \pi$, i.e., not at $\omega = 0$ or π.

***Computation of* $|E(\omega)|$ *and* $P_c(\omega)$.** In steps 2 and 6 of the basic Re nez algorithm (Algorithm 1), $|E(\omega)|$ and $P_c(\omega)$ need to be evaluated. This can be done by determining coefficients a_k by inverting the matrix in Eq. (82.50). This approach is inefficient and may be subject to numerical ill-conditioning, in particular, if δ is small and N is large. An alternative and more efficient approach is to deduce δ analytically and then interpolate $P_c(\omega)$ on the r frequency points using the *barycentric* form of the *Lagrange interpolation* formula. The necessary formulation is as follows.

Parameter δ can be deduced as

$$\delta = \frac{\sum_{k=0}^{r} \alpha_k D(\hat{\omega}_k)}{\sum_{k=0}^{r} \frac{(-1)^k \alpha_k}{W(\hat{\omega}_k)}}$$

and $P_c(\omega)$ is given by

$$P_c(\omega) = \begin{cases} C_k & \text{for } \omega = \hat{\omega}_0, \hat{\omega}_1, \cdots, \hat{\omega}_{r-1} \\ \dfrac{\sum_{k=0}^{r-1} \frac{\beta_k C_k}{x - x_k}}{\sum_{k=0}^{r-1} \frac{\beta_k}{x - x_k}} & \text{otherwise} \end{cases} \tag{82.51}$$

where

$$\alpha_k = \prod_{i=0,\, i\neq k}^{r} \frac{1}{x_k - x_i}$$

$$C_k = D(\hat{\omega}_k) - (-1)^k \frac{\delta}{W(\hat{\omega}_k)}$$

$$\beta_k = \prod_{i=0,\, i\neq k}^{r-1} \frac{1}{x_k - x_i}$$

with

$$x = \cos\omega \quad \text{and} \quad x_i = \cos\hat{\omega}_i \quad \text{for } i = 0, 1, 2, \cdots, r$$

In step 2 of the Remez algorithm, $|E(\omega)|$ often needs to be evaluated at a frequency that was an extremal during the previous iteration. For these cases, the magnitude of the error function is simply $|\delta|$, according to (82.49), and need not be evaluated. This would reduce the amount of computation to some extent.

An alternative formulation that simplifies the implementation of the Remez exchange algorithm can be found in [12].

Rejection of Superfluous Potential Extremals. The solution of (82.50) can be obtained only if *precisely* $r+1$ extremals are available. By differentiating $E(\omega)$, one can show that in a filter with one frequency band of interest (e.g., a digital differentiator) the number of maxima in $|E(\omega)|$ (potential extremals in step 2 of Algorithm 1) can be as high as $r+1$. In the weighted-Chebyshev method, band edges at which $|E(\omega)|$ is maximum or $|E(\omega)| \geq |\delta|$ are treated as potential extremals (see Algorithm 2). Therefore, whenever the number of frequency bands is increased by one, the number of potential extremals is increased by 2, i.e., for a filter with J bands there can be as many as $r + 2J - 1$ frequencies $\hat{\omega}_i$ and a maximum of $2J - 2$ superfluous $\hat{\omega}_i$ may occur. This problem is overcome by rejecting $\rho - r$ of the potential extremals $\hat{\omega}_i$, if $\rho > r$, in step 4 of the algorithm.

A simple rejection scheme is to reject the $\rho - r$ frequencies $\hat{\omega}_i$ that yield the lowest $|E(\hat{\omega}_i)|$ and then renumber the remaining $\hat{\omega}_i$ from 0 to r [8]. This strategy is based on the well-known fact that the magnitude of the error in a given band is inversely related to the density of extremals in that band, i.e., a low density of extremals results in a large error and a high density results in a small error. Conversely, a low band error is indicative of a high density of extremals, and rejecting superfluous $\hat{\omega}_i$ in such a band is the appropriate course of action.

A problem with the scheme just described is that whenever a frequency remains an extremal in two successive iterations, $|E(\omega)|$ assumes the value of $|\delta|$ in the second iteration by virtue of (82.49). In practice, there are almost always several frequencies that remain extremals from one iteration to the next, and the value of $|E(\omega)|$ at these frequencies will be the same. Consequently, the rejection of potential extremals on the basis of the magnitude of the error can become arbitrary and may lead to the rejection of potential extremals in bands where the density of extremals is low. This tends to increase the number of iterations, and it may even prevent the algorithm from converging on occasion. This problem can to some extent be alleviated by rejecting only potential extremals that are not band edges.

An alternative rejection scheme based on the aforementioned strategy, which gives excellent results for two-band and three-band filters, involves ranking the frequency bands in the order of lowest average band error, dropping the band with the highest average error from the list, and then rejecting potential extremals, one per band, in a cyclic manner starting with the band with the lowest average error [11]. The steps involved are as follows.

Algorithm 3: Rejection of Superfluous Potential Extremals

1. Compute the average band errors

$$E_j = \frac{1}{\nu_j} \sum_{\hat{\omega}_i \in \Omega_j} \left| E\left(\hat{\omega}_i\right) \right| \quad \text{for } j = 1, 2, \cdots, J$$

where Ω_j is the set of potential extremals in band j given by

$$\Omega_j = \left\{ \hat{\omega}_i \colon \omega_{Lj} \le \hat{\omega}_i \le \omega_{Rj} \right\}$$

ν_j is the number of potential extremals in band j, and J is the number of bands.

2. Rank the J bands in the order of lowest average error and let $l_1, l_2, \cdots, l_J$ be the ranked list obtained, i.e., l_1 and l_J are the bands with the lowest and highest average errors, respectively.
3. Reject one $\hat{\omega}_i$ in each of bands $l_1, l_2, \cdots, l_{J-1}, l_1, l_2, \cdots$ until $\rho - r$ superfluous $\hat{\omega}_i$ are rejected. In each case, reject the $\hat{\omega}_i$, other than a band edge, that yields the lowest $|E(\hat{\omega}_i)|$ in the band.

For example, if $J = 3$, $\rho - r = 3$, and the average errors for bands 1, 2, and 3 are respectively, 0.05, 0.08, and 0.02, then $\hat{\omega}_i$ are rejected in bands 3, 1, and 3. Note that potential extremals are not rejected in band 2, which is the band of highest average error.

Computation of Impulse Response. The impulse response in step 6 of Algorithm 1 can be determined by noting that function $P_c(\omega)$ is the frequency response of a noncausal version of the required filter. The impulse response of this filter, represented by $h_0(n)$ for $-c \le n \le c$, can be determined by computing $P_c(k\Omega)$ for $k = 0, 1, 2, \cdots, c$, where $\Omega = 2\pi/N$, and then using the *inverse discrete Fourier transform*. It can be shown that

$$h_0(n) = h_0(-n) = \frac{1}{N}\left[P_c(0) = \sum_{k=1}^{c} 2P_c(k\Omega)\cos\left(\frac{2\pi kn}{N}\right)\right]$$

for $n = 0, 1, 2, \cdots, c$. Therefore, the impulse response of the required causal filter is given by

$$h(n) = h_0(n - c)$$

for $n = 0, 1, 2, \cdots, N - 1$.

Improved Search Methods

For a filter of length N, with the number of intervals w_j in each interval W_j equal to S, the exhaustive step-by-step search described (Algorithm 2) requires about $S \times (N + 1)/2$ function evaluations, where each function evaluation entails $N - 1$ additions, $(N + 1)/2$ multiplications, and $(N + 1)/2$ divisions [see (82.51)].

A Remez optimization usually requires four to eight iterations for low-pass or high-pass filters, four to ten iterations for bandpass filters, and four to 12 iterations for bandstop filters. Further, if prescribed specifications are to be achieved and the appropriate value of N is unknown, typically two to four Remez optimizations have to be performed.[2] Thus, if $N = 101$, $S = 16$, number of Remez optimizations = 4, iterations per optimization = 6, the design would entail 24 iterations, 19 200 function evaluations, 1.92×10^6 additions, 0.979×10^6 multiplications, and 0.979×10^6 divisions. This is in addition to the computation required for the evaluation of δ and coefficients α_k, C_k, and β_k once per iteration. In effect, the amount of computation required to complete a design is quite substantial.

The bulk of the computation in Algorithm 2 is carried out to locate the maxima of $|E(\omega)|$, and the large amount of computation is a consequence of the exhaustive character of the search. Therefore, any attempt to reduce the computational complexity of the Remez exchange algorithm must of necessity involve a more efficient search for the maxima of $|E(\omega)|$.

The error function in the weighted-Chebyshev method is well behaved in practice, and is normally unimodal between successive zeros, as can be seen in Figs. 82.15 and 82.16. Hence, the maxima of $|E(\omega)|$ can be located through more sophisticated search methods that utilize gradient information. Two such methods are the so-called *selective step-by-step search* and *cubic-interpolation search* reported in [10] and [11]. Collectively, the two search methods can reduce the amount of computation to about one-fifth the amount required by the exhaustive search.

Selective Step-by-Step Search. The underlying principle in the development of the selective step-by-step search is that normally there is strict alternation between the maxima and the zeros of $|E(\omega)|$. In a given iteration, the maxima of $|E(\omega)|$ are either old maxima from the previous iteration that have moved or new maxima introduced at band edges. New interior maxima may also arise, in theory, but such occurrences are quite rare in practice.

The selective step-by-step search involves three distinct parts as follows.

1. Maxima that correspond to previous maxima are located by searching in the neighborhoods of the most recent set of extremals in a step-by-step fashion using the first derivative of $|E(\omega)|$. If the first derivative is positive at an extremal, the search is carried out to the right of the extremal; otherwise, the search is carried out to the left of the extremal.

[2]See subsection on prescribed specifications.

2. New maxima at band edges can be located by noting the circumstances under which new maxima can arise. These are as follows:
   - To the right of $\omega = 0$ (first band), if there is an extremal and $|E(\omega)|$ has a minimum at $\omega = 0$.
   - To the left of $\omega = \pi$ (last band), if there is an extremal and $|E(\omega)|$ has a minimum at $\omega = \pi$.
   - At $\omega = 0$, if there is no extremal at $\omega = 0$.
   - At $\omega = \pi$, if there is no extremal at $\omega = \pi$.
   - To the right of an interior left-hand edge.
   - To the left of an interior right-hand edge.
   - At $\omega = \omega_{Lj}$, if there is no extremal at $\omega = \omega_{Lj}$.
   - At $\omega = \omega_{Rj}$, if there is no extremal at $\omega = \omega_{Rj}$.
3. New interior maxima, which cannot be located by the checks in (1) and (2), can be found by noting the presence of large gaps in the set of potential extremals identified in (1) and (2). If the difference between two consecutive potential extremals exceeds 1.5 to 2 times the initial interval between extremals (i.e., W_j), then the interval is checked for additional maxima.

If a selective step-by-step search based on the above principles is used in Algorithm 1, then at the start of the optimization the distance between a typical extremal $\hat{\omega}_i$ and the nearby maximum point $\hat{\hat{\omega}}_i$ will be less than half the period of the corresponding ripple of $|E(\omega)|$, owing to the relative symmetry of the ripples of the error function. In effect, during the first iteration less than half of the combined width of the different bands needs to be searched. Thus the number of function evaluations required would be reduced from about 16 to less than 8 per extremal in practice. This will reduce the number of function evaluations by more than 50 percent relative to that required by the exhaustive search of Algorithm 1 without degrading the accuracy of the optimization in any way. As the optimization progresses and the solution is approached, extremal $\hat{\omega}_i$ and maximum point $\hat{\hat{\omega}}_i$ tend to coincide and, therefore, the cumulative length of the frequency range that has to be searched is progressively reduced, thereby resulting in further economies in the number of function evaluations. In the last iteration, only two or three function evaluations are needed (including derivatives) per ripple. As a result, the total number of function evaluations can be reduced by 65 to 70 percent relative to that required by the exhaustive search [10], [11].

Cubic-Interpolation Search. The maxima in item (1) of the above method can also be found through the use of one stage of polynomial interpolation. Either quadratic or cubic interpolation can be used. In these methods, a polynomial approximation is obtained for the magnitude of the error function in the neighborhood of a given extremal and the location of the maximum is determined by finding the point at which the first derivative is zero. Although cubic interpolation entails a more complicated formulation than quadratic interpolation, it leads to improved accuracy, which tends to translate into improved efficiency.

Several choices are possible in setting up a cubic-interpolation search for the problem at hand. One that was found to work well in practice entails evaluating $|E(\omega)|$ at three frequency points and its derivative at one point. Choosing the extremal itself as one of the points reduces the computation further since the value of $|E(\omega)|$ is known to be $|\delta|$ from the previous iteration. Thus this scheme entails three function evaluations per extremal.

The computational complexity of the cubic-interpolation search described remains constant from iteration to iteration since the number of function evaluations required to perform an interpolation is constant. At the start of the optimization, the cubic-interpolation search is more efficient than the selective step-by-step search. However, as the solution is approached the number of function evaluations required by the selective search is progressively reduced, as was stated earlier, and at some point the selective search becomes more efficient. A prudent strategy under these circumstances is to use the cubic-interpolation search at the start of the optimization and switch over to the selective step-by-step search when some suitable criterion is satisfied. Extensive experimentation has shown that computational advantage can be gained by using the cubic-interpolation search if parameter Q (see Algorithm 1) is greater than about 0.65 and the selective search otherwise. The use of the cubic-interpolation search along with the selective step-by-step search of the preceding section can reduce the number of function evaluations by 70 to 85 percent relative to that required by the exhaustive search [10], [11].

More information, including the necessary formulation as well as a practical and efficient implementation of the Remez exchange algorithm in terms of the above search methods, can be found in [12].

Example 1. The Remez algorithm was used with 1) the exhaustive search, 2) the selective step-by-step search, and 3) the selective search in conjunction with the cubic-interpolation search to design an FIR equiripple high-pass filter satisfying the following specifications:

Filter length N: 23;
Passband edge ω_p: 2.0 rad/s;
Stopband edge ω_a: 1.0 rad/s;
Ratio δ_p/δ_a: 15.0;
Sampling frequency ω_s: 2π rad/s.

The progress of the design is illustrated in Table 82.3. As can be seen, the exhaustive and selective search methods required six iterations each, whereas the selective search in conjunction with cubic interpolation required seven iterations. However, the number of function evaluations (evaluations of $P_c(\omega)$ using Eq. (82.51) plus evaluations of its first or second derivative) decreased from 1013 in the first method to 350 in the second method to 259 in the third method. In the Remez algorithm, approximately 80 to 90 percent of the computational effort involves function evaluations. In effect, relative to that required by the exhaustive search, the use of the selective step-by-step search reduced the amount of computation by about 65.4 percent, and the use of the selective step-by-step search in conjunction with the cubic-interpolation search reduced the amount of computation by about 74.4 percent.

The three methods resulted in approximately the same impulse responses, as can be seen in Table 82.4, and the passband ripple and minimum stopband attenuation obtained in each case were 0.043 and 75.7 dB, respectively. The amplitude response of the filter is illustrated in Fig. 82.17.

TABLE 82.3 Progress in Design of High-Pass Filter (Example 1)

	Exhaustive Search		Selective Search		Selective Search with Cubic Interpolation	
Iteration Number	Q	FE's	Q	FE's	Q	FE's
1	0.9912	169	0.9912	93	0.9912	66
2	0.9207	168	0.9207	86	0.9406	44
3	0.9480	169	0.9480 $\hat{\omega}_{32}$ rejected	55	0.8830	42
4	0.7249	169	0.7249	62	0.6952	31
5	0.0923	169	0.0923	31	0.1417	31
6	0.0017	169	0.0017	23	0.0102	23
7	—	—	—	—	0.0000	22
Total FE's		1013		350		259

TABLE 82.4 Impulse Response of High-Pass Filter (Example 1)

n	$h_0(n) = h_0(-n)$ Exhaustive or Selective Search	Selective Search with Cubic Interpolation
0	$5.034\,954 \times 10^{-1}$	$5.035\,077 \times 10^{-1}$
1	$-3.123\,538 \times 10^{-1}$	$-3.123\,535 \times 10^{-1}$
2	$-3.085\,731 \times 10^{-3}$	$-3.097\,829 \times 10^{-3}$
3	$8.932\,914 \times 10^{-2}$	$8.932\,911 \times 10^{-2}$
4	$2.053\,235 \times 10^{-3}$	$2.063\,564 \times 10^{-3}$
5	$-3.898\,118 \times 10^{-2}$	$-3.898\,177 \times 10^{-2}$
6	$-8.467\,375 \times 10^{-4}$	$-8.540\,079 \times 10^{-4}$
7	$1.660\,800 \times 10^{-2}$	$1.660\,858 \times 10^{-2}$
8	$5.401\,585 \times 10^{-5}$	$5.892\,008 \times 10^{-5}$
9	$-6.100\,465 \times 10^{-3}$	$-6.101\,979 \times 10^{-3}$
10	$7.298\,411 \times 10^{-4}$	$7.281\,192 \times 10^{-4}$
11	$9.275\,654 \times 10^{-4}$	$9.285\,482 \times 10^{-4}$

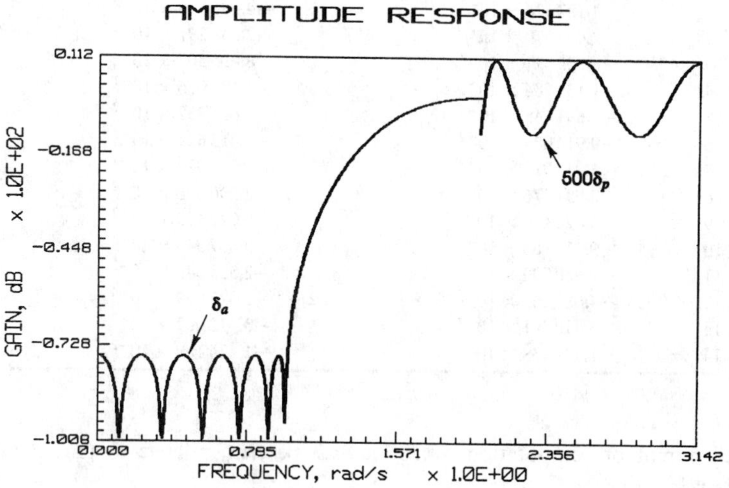

FIGURE 82.17 Amplitude response of equiripple high-pass filter (Example 1).

Example 2. In this example, the Remez algorithm was used with 1) the exhaustive search, 2) the selective step-by-step search, and 3) the selective step-by-step search in conjunction with the cubic-interpolation search to design an FIR equiripple bandstop filter satisfying the following specifications:

Filter length N: 29;
Lower passband edge ω_{p1}: 0.8 rad/s;
Upper passband edge ω_{p2}: 2.1 rad/s;
Lower stopband edge ω_{a1}: 1.1 rad/s;
Upper stopband edge ω_{a2}: 1.8 rad/s;
Ratio δ_{p1}/δ_a: 5.0;
Ratio δ_{p1}/δ_{p2}: 2.0;
Sampling frequency ω_s: 2π rad/s.

The progress of the design is illustrated in Table 82.5. In this example, each of the three methods required four iterations. The number of function evaluations decreased from 804 in the first method to 190 in the second method to 131 in the third method. In effect, the use of the selective step-by-step search reduced the amount of computation by about 76.4 percent, and the use of the selective step-by-step search in conjunction with the cubic-interpolation search

TABLE 82.5 Progress in Design of Bandstop Filter (Example 2)

Iteration Number	Exhaustive Search		Selective Search		Selective Search with Cubic Interpolation	
	Q	FE's	Q	FE's	Q	FE's
1	0.6836	201	0.6836	79	0.6940	36
2	0.3138	201	0.3138	51	0.2378	36
3	0.0804	201	0.0804	34	0.0675	32
4	0.0000	201	0.0000	26	0.0007	27
Total FE's		804		190		131

TABLE 82.6 Impulse Response of Bandstop Filter (Example 2)

n	$h_0(n) = h_0(-n)$ Exhaustive or Selective Search	$h_0(n) = h_0(-n)$ Selective Search with Cubic Interpolation
0	$6.656\,629 \times 10^{-1}$	$6.656\,478 \times 10^{-1}$
1	$-4.187\,326 \times 10^{-2}$	$-4.186\,510 \times 10^{-2}$
2	$2.635\,370 \times 10^{-1}$	$2.635\,297 \times 10^{-1}$
3	$8.005\,521 \times 10^{-2}$	$8.005\,307 \times 10^{-2}$
4	$-1.131\,284 \times 10^{-1}$	$-1.131\,056 \times 10^{-1}$
5	$-3.691\,645 \times 10^{-2}$	$-3.691\,932 \times 10^{-2}$
6	$-9.914\,085 \times 10^{-4}$	$-1.013\,621 \times 10^{-3}$
7	$-3.018\,917 \times 10^{-2}$	$-3.017\,017 \times 10^{-2}$
8	$2.931\,776 \times 10^{-2}$	$2.930\,006 \times 10^{-2}$
9	$5.022\,490 \times 10^{-2}$	$5.022\,450 \times 10^{-2}$
10	$-9.715\,988 \times 10^{-3}$	$-9.687\,345 \times 10^{-3}$
11	$-2.550\,790 \times 10^{-2}$	$-2.553\,543 \times 10^{-2}$
12	$-4.023\,265 \times 10^{-4}$	$-3.827\,029 \times 10^{-4}$
13	$-3.410\,741 \times 10^{-2}$	$-3.412\,007 \times 10^{-2}$
14	$-1.421\,939 \times 10^{-2}$	$-1.424\,189 \times 10^{-2}$

reduced the amount of computation by about 83.7 percent, relative to that required by the exhaustive search.

The three methods resulted in approximately the same impulse responses, as can be seen in Table 82.6. The amplitude response of the filter is illustrated in Fig. 82.18; the passband ripples obtained for the two passbands were 1.78 and 0.89 dB, respectively, and the minimum stopband attenuation was 33.79 dB.

Prescribed Specifications

Given a filter length N, a set of passband and stopband edges, and a ratio δ_p/δ_a, an FIR filter with approximately piecewise-constant amplitude-response specifications can be readily designed. While the filter obtained will have passband and stopband edges at the correct locations and the ratio δ_p/δ_a will be as required, the amplitudes of the passband and stopband ripples are highly unlikely to be precisely as specified. An acceptable design can be obtained by predicting the value of N on the basis of the required specifications and then designing filters for increasing or decreasing values of N until the lowest value of N that satisfies the specifications is found.

A reasonably accurate *empirical* formula for the prediction of N for the case of low-pass and high-pass filters, due to Herrmann, Rabiner, and Chan [16], is

$$N = \text{Int}\left[\frac{(D - FB^2)}{B} + 1.5\right] \tag{82.52}$$

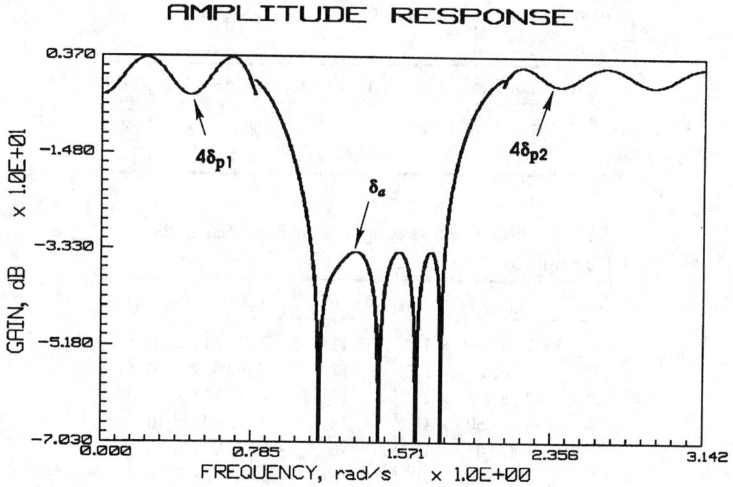

FIGURE 82.18 Amplitude response of equiripple bandstop filter (Example 2).

where

$$B = |\omega_a - \omega_p|/2\pi$$

$$D = [0.005309(\log \delta_p)^2 + 0.07114 \log \delta_p - 0.4761] \log \delta_a$$
$$-[0.00266(\log \delta_p)^2 + 0.5941 \log \delta_p + 0.4278]$$

$$F = 0.51244(\log \delta_p - \log \delta_a) + 11.012$$

This formula can also be used to predict the filter length in the design of bandpass, bandstop, and multiband filters in general. In these filters, a value of N is computed for each transition band between a passband and stopband or a stopband and passband using (82.52) and the largest value of N so obtained is taken to be the predicted filter length. *Prescribed specifications* can be achieved by using the following design algorithm.

Algorithm 4: Design of Filters Satisfying Prescribed Specifications

1. Compute N using (82.52); if N is even, set $N = N + 1$.
2. Design a filter of length N and determine the minimum value of δ, say $\breve{\delta}$.
   i. If $\breve{\delta} > \delta_p$, then do the following:
      a. Set $N = N + 2$, design a filter of length N, and find $\breve{\delta}$;
      b. If $\breve{\delta} \leq \delta_p$, then go to step 3; else, go to step 2ia.
   ii. If $\breve{\delta} < \delta_p$, then do the following:
      a. Set $N = N - 2$, design a filter of length N, and find $\breve{\delta}$;
      b. If $\breve{\delta} > \delta_p$, then go to step 4; else, go to step 2iia.
3. Use the last set of extremals and the corresponding value of N to obtain the impulse response of the required filter and stop.
4. Use the last but one set of extremals and the corresponding value of N to obtain the impulse response of the required filter and stop.

Example 3. Algorithm 4 was used to design an FIR equiripple bandpass filter that would satisfy the following specifications:

Odd filter length;
Maximum passband ripple A_p: 0.5 dB;

TABLE 82.7 Progress in Design of Bandpass Filter (Example 3)

N	Iterations	FE's	A_p, dB	A_{a1}, dB	A_{a2}, dB
41	8	550	0.47	50.4	30.4
39	7	527	0.67	47.5	27.5

TABLE 82.8 Impulse Response of Bandpass Filter (Example 3)

n	$h_0(n) = h_0(-n)$	n	$h_0(n) = h_0(-n)$
0	$2.761\,666 \times 10^{-1}$	11	$2.726\,816 \times 10^{-2}$
1	$1.660\,224 \times 10^{-2}$	12	$-2.663\,859 \times 10^{-2}$
2	$-2.389\,235 \times 10^{-1}$	13	$-1.318\,252 \times 10^{-2}$
3	$-3.689\,501 \times 10^{-2}$	14	$6.312\,944 \times 10^{-3}$
4	$1.473\,038 \times 10^{-1}$	15	$-5.820\,976 \times 10^{-3}$
5	$2.928\,852 \times 10^{-2}$	16	$5.827\,957 \times 10^{-3}$
6	$-4.770\,552 \times 10^{-2}$	17	$1.528\,658 \times 10^{-2}$
7	$-2.008\,131 \times 10^{-3}$	18	$-8.288\,708 \times 10^{-3}$
8	$-1.875\,082 \times 10^{-2}$	19	$-1.616\,904 \times 10^{-2}$
9	$-2.262\,965 \times 10^{-2}$	20	$1.092\,728 \times 10^{-2}$
10	$3.860\,990 \times 10^{-2}$	—	—

Minimum stopband attenuation A_{a1}: 50.0 dB;
Minimum stopband attenuation A_{a2}: 30.0 dB;
Lower passband edge ω_{p1}: 1.2 rad/s;
Upper passband edge ω_{p2}: 1.8 rad/s;
Lower stopband edge ω_{a1}: 0.9 rad/s;
Upper stopband edge ω_{a2}: 2.1 rad/s;
Sampling frequency ω_s: 2π rad/s.

The progress of the design is illustrated in Table 82.7. As can be seen, a filter of length 41 was initially predicted, which was found to have a passband ripple of 0.47 dB, a minimum stopband attenuation of 50.4 in the lower stopband, and 30.4 dB in the upper stopband, i.e., the required specifications were satisfied. Then a filter length of 39 was tried and found to violate the specifications. Hence the first design is the required filter. The impulse response is given in Tab le 82.8. The corresponding amplitude response is depicted in Fig. 82.19.

Generalization

There are four types of constant-delay FIR filters. The impulse response can be *symmetrical* or *antisymmetrical*, and the filter length can be *odd* or *even*. In the preceding subsections, we considered the design of filters with symmetrical impulse response and odd length. In this subsection, we show that the Remez algorithm can also be applied for the design of other types of filters.

Antisymmetrical Impulse Response and Odd Filter Length. Assuming that $\omega_s = 2\pi$, the frequency response of an FIR filter with *antisymmetrical* impulse response and *odd* length can be expressed as

$$H(e^{j\omega T}) = e^{-jc\omega} jP'_c(\omega)$$

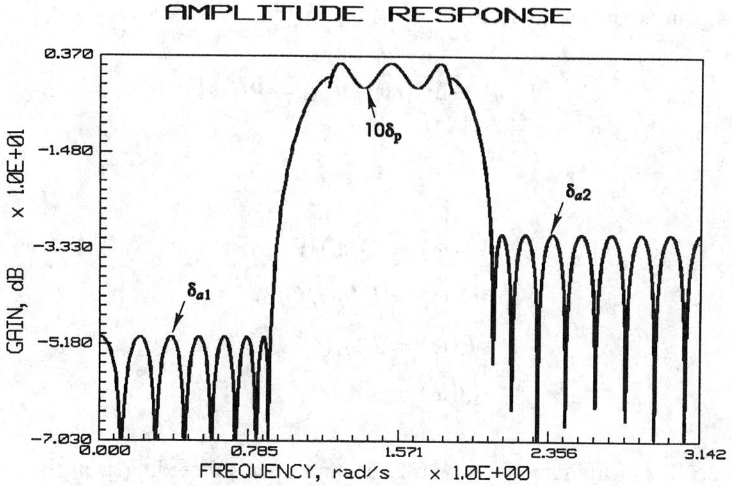

FIGURE 82.19 Amplitude response of equiripple bandpass filter (Example 3).

where

$$P_c'(\omega) = \sum_{k=1}^{c} a_k \sin k\omega \tag{82.53}$$

$$a_k = 2h(c-k) \quad \text{for } k = 1, 2, \cdots, c$$
$$c = (N-1)/2$$

A filter with a desired frequency response $e^{-jc\omega} jD(\omega)$ can be designed by constructing the error function

$$E(\omega) = W(\omega)[D(\omega) - P_c'(\omega)] \tag{82.54}$$

and then minimizing $|E(\omega)|$ with respect to some compact subset of the frequency interval $[0, \pi]$. From (82.53), $P_c'(\omega)$ can be expressed as [6]

$$P_c'(\omega) = \sin \omega \; P_{c-1}(\omega) \tag{82.55}$$

where

$$P_{c-1}(\omega) = \sum_{k=0}^{c-1} \tilde{c}_k \cos k\omega$$

and

$$a_1 = \tilde{c}_0 - \tfrac{1}{2}\tilde{c}_2$$
$$a_k = \tfrac{1}{2}(\tilde{c}_{k-1} - \tilde{c}_{k+1}) \quad \text{for } k = 2, 3, \cdots, c-2$$
$$a_{c-1} = \tfrac{1}{2}\tilde{c}_{c-2}$$
$$a_c = \tfrac{1}{2}\tilde{c}_{c-1}$$

Hence (82.54) can be put in the form

$$E(\omega) = \tilde{W}(\omega)[\tilde{D}(\omega) - \tilde{P}(\omega)] \tag{82.56}$$

where

$$\begin{aligned}\tilde{W}(\omega) &= Q(\omega)W(\omega)\\ \tilde{D}(\omega) &= D(\omega)/Q(\omega)\\ \tilde{P}(\omega) &= P_{c-1}(\omega)\\ Q(\omega) &= \sin\omega\end{aligned}$$

Evidently, (82.56) is of the same form as (82.45), and upon proceeding as in the subsection on problem formulation, one can obtain the system of equations

$$\begin{bmatrix} 1 & \cos\hat{\omega}_0 & \cos 2\hat{\omega}_0 & \cdots & \cos(c-1)\hat{\omega}_0 & \dfrac{1}{\tilde{W}(\hat{\omega}_0)} \\ 1 & \cos\hat{\omega}_1 & \cos 2\hat{\omega}_1 & \cdots & \cos(c-1)\hat{\omega}_1 & \dfrac{-1}{\tilde{W}(\hat{\omega}_1)} \\ \vdots & \vdots & \vdots & & \vdots & \vdots \\ 1 & \cos\hat{\omega}_r & \cos 2\hat{\omega}_r & \cdots & \cos(c-1)\hat{\omega}_r & \dfrac{(-1)^r}{\tilde{W}(\hat{\omega}_r)} \end{bmatrix} \begin{bmatrix} a_0 \\ a_1 \\ \vdots \\ a_{c-1} \\ \delta \end{bmatrix} = \begin{bmatrix} \tilde{D}(\hat{\omega}_0) \\ \tilde{D}(\hat{\omega}_1) \\ \vdots \\ \tilde{D}(\hat{\omega}_r) \end{bmatrix}$$

where $r = c$ is the number of cosine functions in $P_{c-1}(\omega)$. The above system is the same as that in (82.50) except that the number of extremals has been reduced from $c + 2$ to $c + 1$; therefore, the application of the Remez algorithm follows the methodology detailed in the subsections on problem formulation and the Remez exchange algorithm.

The use of Algorithm 1 yields the optimum $P_{c-1}(\omega)$ and from (82.55), the cosine function $P'_c(\omega)$ can be formed. Now $jP'_c(\omega)$ is the frequency response of a noncausal version of the required filter. The impulse response of this filter can be obtained as

$$h_0(n) = -h_0(-n) = -\frac{1}{N}\left[\sum_{k=1}^{c} 2P'_c(k\Omega)\sin\left(\frac{2\pi kn}{N}\right)\right]$$

for $n = 0, 1, 2, \cdots, c$, where $\Omega = 2\pi/N$, by using the inverse discrete Fourier transform. The impulse response of the corresponding causal filter is given by

$$h(n) = h_0(n - c)$$

for $n = 0, 1, 2, \cdots, N - 1$.

The Remez algorithm can also be applied for the design of filters with symmetrical or antisymmetrical impulse response and even N. However, these filters are used less frequently. The reader is referred to [6] and [12] for more details.

Digital Differentiators

The Remez algorithm can be easily applied for the design of *equiripple digital differentiators.* The ideal frequency response of a causal differentiator is of the form $e^{-jc\omega} jD(\omega)$, where

$$D(\omega) = \omega \quad \text{for } 0 < |\omega| < \pi \tag{82.57}$$

and $c = (N - 1)/2$. Since the frequency response is antisymmetrical, differentiators can be designed in terms of filters with antisymmetrical impulse response of either odd or even length.

Problem Formulation. Assuming odd filter length, (82.54) and (82.57) give the error function

$$E(\omega) = W(\omega)[\omega - P'_c(\omega)] \quad \text{for } 0 < \omega \le \omega_p$$

where ω_p is the required bandwidth. Equiripple absolute or relative error may be required, depending on the application at hand. Hence, $W(\omega)$ can be chosen to be either unity or $1/\omega$. In the latter case, which is the more meaningful of the two in practice, $E(\omega)$ can be expressed as

$$E(\omega) = 1 - \frac{1}{\omega} P'_c(\omega) \quad \text{for } 0 < \omega \le \omega_p$$

and from (82.55)

$$E(\omega) = 1 - \frac{\sin \omega}{\omega} P_{c-1}(\omega) \quad \text{for } 0 < \omega \le \omega_p \tag{82.58}$$

Therefore, the error function can be expressed as in (82.56) with

$$\tilde{W}(\omega) = \frac{1}{\tilde{D}(\omega)} = \frac{\sin \omega}{\omega}$$

$$\tilde{P}(\omega) = P_{c-1}(\omega)$$

Prescribed Specifications. A digital differentiator is fully specified by the constraint

$$|E(\omega)| \le \delta_p \quad \text{for } 0 < \omega \le \omega_p$$

where δ_p is the maximum passband error and ω_p is the bandwidth of the differentiator.

The differentiator length N that will just satisfy the required specifications is not normally known *a priori* and, although it may be determined on a hit-or-miss basis, a large number of designs may need to be carried out. In filters with approximately piecewise-constant amplitude responses, N can be predicted using the empirical formula of (82.52). In the case of differentiators, N can be predicted by noting a useful property of digital differentiators. If δ and δ_1 are the maximum passband errors in differentiators of lengths N and N_1, respectively, then the quantity $\ln(\delta/\delta_1)$ is *approximately linear* with respect to $N - N_1$ for a wide range of values of N_1 and ω_p. Assuming linearity, we can show that [17]

$$N = N_1 + \frac{\ln(\delta/\delta_1)}{\ln(\delta_2/\delta_1)}(N_2 - N_1) \tag{82.59}$$

where δ_2 is the maximum passband error in a differentiator of length N_2.

By designing two low-order differentiators, a fairly accurate prediction of the required value of N can be obtained by using (82.59). Once a filter order is predicted a series of differentiators can be designed with increasing or decreasing N until a design that just satisfies the specifications is obtained.

TABLE 82.9 Progress in Design of Digital Differentiator (Example 4)

N	Iterations	FE's	δ_p
21	4	145	1.075×10^{-2}
23	4	162	6.950×10^{-3}
55	7	815	8.309×10^{-6}
53	7	757	1.250×10^{-5}

TABLE 82.10 Impulse Response of Digital Differentiator (Example 4)

n	$h(n) = -h_0(-n)$	n	$h_0(n) = -h_0(-n)$
0	0.0	14	$1.762\,268 \times 10^{-2}$
1	$-9.933\,416 \times 10^{-1}$	15	$-1.313\,097 \times 10^{-2}$
2	$4.868\,036 \times 10^{-1}$	16	$9.615\,295 \times 10^{-3}$
3	$-3.138\,353 \times 10^{-1}$	17	$-6.902\,518 \times 10^{-3}$
4	$2.245\,441 \times 10^{-1}$	18	$4.844\,090 \times 10^{-3}$
5	$-1.690\,252 \times 10^{-1}$	19	$-3.312\,235 \times 10^{-3}$
6	$1.306\,918 \times 10^{-1}$	20	$2.197\,502 \times 10^{-3}$
7	$-1.024\,631 \times 10^{-1}$	21	$-1.407\,064 \times 10^{-3}$
8	$8.081\,083 \times 10^{-2}$	22	$8.632\,670 \times 10^{-4}$
9	$-6.377\,426 \times 10^{-2}$	23	$-5.023\,168 \times 10^{-4}$
10	$5.016\,708 \times 10^{-2}$	24	$2.729\,367 \times 10^{-4}$
11	$-3.921\,782 \times 10^{-2}$	25	$-1.349\,790 \times 10^{-4}$
12	$3.039\,137 \times 10^{-2}$	26	$5.859\,128 \times 10^{-5}$
13	$-2.329\,439 \times 10^{-2}$	27	$-1.634\,535 \times 10^{-5}$

Example 4. The selective step-by-step search with cubic interpolation was used in Algorithm 4 to design a digital differentiator that should satisfy the following specifications:

Odd differentiator length;
Bandwidth ω_p: 2.75 rad/s;
Maximum passband ripple δ_p: 1.0×10^{-5};
Sampling frequency ω_s: 2π rad/s.

The progress of the design is illustrated in Table 82.9. First, differentiators of lengths 21 and 23 were designed and the required N to satisfy the specifications was predicted to be 55 using (82.59). This differentiator length was found to satisfy the specifications, and a design for length 53 was then carried out. The second design was found to violate the specifications and hence the first design is the required differentiator. The impulse response of this differentiator is given in Table 82.10. The amplitude response and passband relative error of the differentiator are plotted in Fig. 82.20(a) and (b).

Arbitrary Amplitude Responses

Very frequently FIR filters are required whose amplitude responses cannot be described by analytical functions. For example, in the design of two-dimensional filters through the singular-value decomposition [18], [19], the required two-dimensional filter is obtained by designing a set of one-dimensional digital filters whose amplitude responses turn out to have arbitrary shapes. In these applications, the desired amplitude response $D(\omega)$ is specified in terms of a table that lists a prescribed set of frequencies and the corresponding values of the required filter gain. Filters of this class can be readily designed by employing some interpolation scheme that can be used to evaluate $D(\omega)$ and its first derivative with respect to ω at any ω. A suitable scheme is to fit a set of third-order polynomials to the prescribed amplitude response.

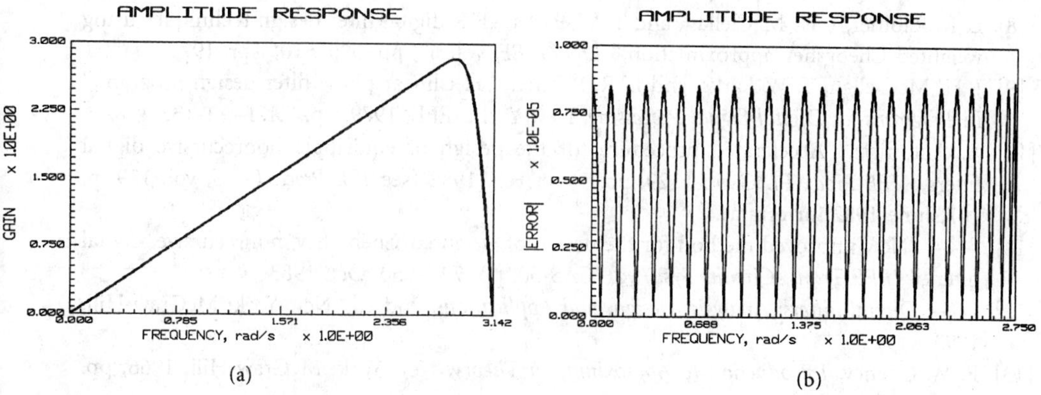

FIGURE 82.20 Design of digital differentiator (Example 4): (a) amplitude response; (b) passband relative error.

Multiband Filters

The algorithms presented in the previous subsections can also be used to design *multiband* filters. While there is no theoretical upper limit on the number of bands, in practice, the design tends to become more and more difficult as the number of bands is increased. The reason is that the difference between the number of possible maxima in the error function and the number of extremals increases linearly with the number of bands, e.g., if the number of bands is 8, then the difference is 14. As a consequence, the number of potential extremals that need to be rejected is large and the available rejection techniques become inefficient. The end result is that the number of iterations is increased quite significantly, and convergence is slow and sometimes impossible.

In mathematical terms, the above difficulty is attributed to the fact that, in the weighted-Chebyshev methods considered here, the approximating polynomial becomes seriously *under-determined* if the number of bands exceeds three. The problem can be overcome by using the generalized Remez method described in [15]. This approach was found to yield better results for filters with more than five bands.

References

[1] O. Herrmann, "Design of nonrecursive digital filters with linear phase," *Electron. Lett.*, vol. 6, pp. 182–184, May 1970.

[2] E. Hofstetter, A. Oppenheim, and J. Siegel, "A new technique for the design of non-recursive digital filters," *5th Annu. Princeton Conf. Informat. Sci., Syst.*, Mar. 1971, pp. 64–72.

[3] T. W. Parks and J. H. McClellan, "Chebyshev approximation for nonrecursive digital filters with linear phase," *IEEE Trans. Circuit Theory*, vol. CT-19, pp. 189–194, Mar. 1972.

[4] ——, "A program for the design of linear phase finite impulse response digital filters," *IEEE Trans. Audio Electroacoust.*, vol. AU-20, pp. 195–199, Aug. 1972.

[5] L. R. Rabiner and O. Herrmann, "On the design of optimum FIR low-pass filters with even impulse response duration," *IEEE Trans. Audio Electroacoust.*, vol. AU-21, pp. 329–336, Aug. 1973.

[6] J. H. McClellan and T. W. Parks, "A unified approach to the design of optimum FIR linear-phase digital filters," *IEEE Trans. Circuit Theory*, vol. CT-20, pp. 697–701, Nov. 1973.

[7] J. H. McClellan, T. W. Parks, and L. R. Rabiner, "A computer program for designing optimum FIR linear phase digital filters," *IEEE Trans. Audio Electroacoust.*, vol. AU-21, pp. 506–526, Dec. 1973.

[8] L. R. Rabiner, J. H. McClellan, and T. W. Parks, "FIR digital filter design techniques using weighted Chebyshev approximation," *Proc. IEEE*, vol. 63, pp. 595–610, Apr. 1975.
[9] J. H. McClellan, T. W. Parks, and L. R. Rabiner, "FIR linear phase filter design program," in *Programs for Digital Signal Processing*, New York: IEEE, 1979, pp. 5.1-1–5.1-13.
[10] A. Antoniou, "Accelerated procedure for the design of equiripple nonrecursive digital filters," *IEE Proc., Pt. G*, vol. 129, pp. 1–10, Feb. 1982 (see *IEE Proc., Pt. G*, vol. 129, p. 107, June 1982 for errata).
[11] ——, "New improved method for the design of weighted-Chebyshev, nonrecursive, digital filters," *IEEE Trans. Circuits Syst.*, vol. CAS-30, pp. 740–750, Oct. 1983.
[12] ——, *Digital Filters: Analysis, Design, and Applications*, 2nd ed., New York: McGraw-Hill, 1993.
[13] E. W. Cheney, *Introduction to Approximation Theory*, New York: McGraw-Hill, 1966, pp. 72–100.
[14] E. Ya. Remes, *General Computational Methods for Tchebycheff Approximation*, Kiev: Atomic Energy Comm., 1957, Translation 4491, pp. 1–85, 1957.
[15] D. J. Shpak and A. Antoniou, "A generalized Reméz method for the design of FIR digital filters," *IEEE Trans. Circuits Syst.*, vol. 37. pp. 161–174, Feb. 1990.
[16] O. Herrmann, L. R. Rabiner, and D. S. K. Chan, "Practical design rules for optimum finite impulse response low-pass digital filters," *Bell Syst. Tech. J.*, vol. CAS-52, pp. 769–799, July–Aug. 1973.
[17] A. Antoniou and C. Charalambous, "Improved design method for Kaiser differentiators and comparison with equiripple method," *IEE Proc., Pt. E*, vol. 128, pp. 190–196, Sept. 1981.
[18] A. Antoniou and W.-S. Lu, "Design of two-dimensional digital filters by using the singular value decomposition," *IEEE Trans. Circuits Syst.*, vol. CAS-34, pp. 1191–1198, Oct. 1987.
[19] W.-S. Lu, H.-P. Wang, and A. Antoniou, "Design of two-dimensional FIR digital filters using the singular-value decomposition," *IEEE Trans. Circuits Syst.*, vol. CAS37, pp. 35–46, Jan. 1990.

Least Squares and Related Techniques

L. Montgomery Smith and Bruce W. Bomar

Introduction and Statement of the Problem

One optimal method of FIR filter design is based upon the approach of minimizing—in some specified manner—the mean square error between the transfer function of the filter and the frequency characteristics of a desired filter response. In a manner similar to the methods discussed in previous subsections, the mathematical objective of least-squares filter design is to determine the values for $N+1$ impulse response samples

$$h(0),\ h(1),\ h(2), \cdots, h(N) \tag{82.60}$$

so that the transfer function of the filter, which is given by the Fourier transform of the impulse response samples,

$$H(e^{j\omega}) = \sum_{n=0}^{N} h(n)e^{-j\omega n} \tag{82.61}$$

approximates a desired response $H_d(e^{j\omega})$ in a least-squares sense. The impulse response sequence values can be found from $H(e^{j\omega})$ by means of the inverse Fourier transform given by

$$h(n) = \frac{1}{2\pi}\int_{-\pi}^{\pi} H(e^{j\omega})e^{j\omega n}\,d\omega \tag{82.62}$$

In this presentation, attention is restricted to causal FIR filters, so that impulse response samples outside the range $(0, N)$ are assumed to be identically zero. This specified region of support is not a limitation to the methods discussed here, but is used since it is the more common application encountered in practice.

In the subsections that follow, four approaches that fall into the category of least-squares FIR filter design are presented: unweighted least squares, frequency sampling, weighted least squares, and eigenfilter methods. These techniques differ in the specific optimization criteria chosen, as well as sometimes in the overall design approach. All methods have been tested effective, and for the most part can claim computational efficiency as contrasted with minimax designs, which are by their nature iterative.

Unweighted Least-Squares Designs

In unweighted least-square filter designs, the error in no particular frequency band is emphasized over that in any other [15]. Such filters minimize the output error variance for a white noise input signal. Interestingly, an optimal unweighted-mean-square design is equivalent to a rectangular window design. To show this, define the mean-square error between the realized and desired frequency responses by

$$\epsilon = \frac{1}{2\pi}\int_{-\pi}^{\pi} |H(e^{j\omega}) - H_d(e^{j\omega})|^2\,d\omega \tag{82.63}$$

By Parseval's relation, this is equal to

$$\begin{aligned}\epsilon &= \sum_{n=-\infty}^{\infty} |h(n) - h_d(n)|^2 \\ &= \sum_{n=-\infty}^{-1} |h_d(n)|^2 + \sum_{n=0}^{N} |h(n) - h_d(n)|^2 + \sum_{n=N+1}^{\infty} |h_d(n)|^2\end{aligned} \tag{82.64}$$

where $h_d(n)$ is the impulse response corresponding to the desired frequency response $H_d(e^{j\omega})$ found by means of the inverse Fourier transform given in (82.62). Since $h(n)$ is restricted by design to be zero for $n \leq -1$ and $n \geq N+1$, summations over those indexes involve only the squared magnitude of the desired impulse response samples, and thus cannot be reduced by any choice of $h(n)$. In the region $0 \leq n \leq N$, the sum of squares can be made zero by choosing $h(n)$ equal to $h_d(n)$. Thus, the unweighted least-squares filter is given by

$$h(n) = \begin{cases} h_d(n) & \text{for } 0 \leq n \leq N \\ 0 & \text{otherwise} \end{cases} \tag{82.65}$$

with $h_d(n)$ as determined from $H_d(e^{j\omega})$ using (82.62). In some cases, the desired impulse response coefficients can be calculated in closed form from the inverse Fourier transform. In other cases, a numerical approximation to the integral must be performed, perhaps utilizing the discrete Fourier transform (DFT).

This is, of course, the same result as that obtained by the windowing method using the rectangular window function. While it does optimize the specified criterion, it suffers from the same disadvantages discussed earlier. Specifically, for frequency selective filters (e.g., low-pass, high-pass, bandpass, and bandstop), the frequency response near band edges deviates greatly from that desired, although in other regions, it approximates it with high fidelity. Since this property introduces the largest errors near the critical design frequencies, filters designed in this manner are usually undesirable. This is not always the case, however. For example, deconvolution or D/A conversion compensation filters often have smooth responses with no particular weighting associated with any frequency band, and so this method may be applicable there.

Example. Consider the design of a Nth-order (N even) linear phase low-pass filter with the following desired frequency response:

$$H_d(e^{j\omega}) = \begin{cases} e^{-j\omega N/2} & \text{for } |\omega| \leq \omega_c \\ 0 & \text{otherwise} \end{cases}$$

The impulse response for this transfer function is given by

$$\begin{aligned} h_d(n) &= \frac{1}{2\pi} \int_{-\omega_c}^{\omega_c} e^{j\omega(n-N/2)}\, d\omega \\ &= \frac{\sin[\omega_c(n - N/2)]}{\pi(n - N/2)} \end{aligned}$$

The impulse response coefficients for the actual Nth-order FIR filter are those for the filter above in the range $0 \leq n \leq N$.

As a specific numerical example, let $N = 12$ and $\omega_c = 0.35\pi$. The impulse response coefficients are given in the UWLS column of Table 82.11. The magnitude frequency response for this filter is shown in Fig. 82.21. Note the large deviations from the ideal in the region near the cutoff frequency.

Frequency Sampling Methods

Another method for FIR filter design utilizes linear algebraic methods. In this approach, the desired frequency response $H_d(e^{j\omega})$ is specified at $N+1$ discrete frequencies $\{\omega_0, \omega_1, \omega_2, \cdots, \omega_N\}$. Then, the filter impulse response coefficients are chosen so that the transfer function $H(e^{j\omega})$ is equal to the desired response at those particular frequencies. Since the transfer function at any given frequency as given by (82.60) is a sum of products involving the impulse response

TABLE 82.11 Filter Coefficients

	UWLS	FSLS	WLS	EF
$h(0) = h(12)$	0.016 394	0.019 793	0.012 229	0.016 709
$h(1) = h(11)$	−0.045 016	−0.045 945	−0.036 234	−0.031 425
$h(2) = h(10)$	−0.075 683	−0.080 207	−0.066 372	−0.063 205
$h(3) = h(9)$	−0.016 598	−0.013 173	−0.015 708	−0.014 816
$h(4) = h(8)$	0.128 759	0.131 596	0.124 500	0.124 093
$h(5) = h(7)$	0.283 616	0.288 841	0.281 488	0.281 006
$h(6)$	0.350 000	0.350 694	0.350 297	0.350 000

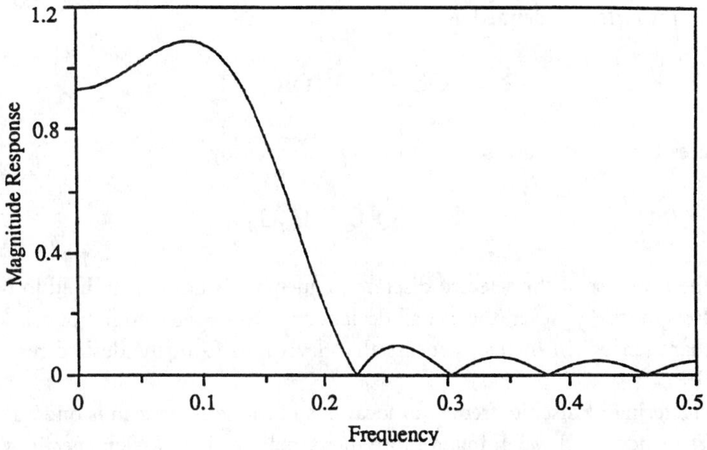

FIGURE 82.21 Magnitude frequency response for a 12th-order lowpass filter designed using the unweighted least-squares method plotted versus frequency f where $\omega = 2\pi f$.

coefficients, this procedure can be written in matrix-vector form as

$$\begin{bmatrix} 1 & e^{-j\omega_0} & e^{-j2\omega_0} & \cdots & e^{-jN\omega_0} \\ 1 & e^{-j\omega_1} & e^{-j2\omega_1} & \cdots & e^{-jN\omega_1} \\ \vdots & \vdots & \vdots & \ddots & \vdots \\ 1 & e^{-j\omega_N} & e^{-j2\omega_N} & \cdots & e^{-jN\omega_N} \end{bmatrix} \begin{bmatrix} h(0) \\ h(1) \\ \vdots \\ h(N) \end{bmatrix} = \begin{bmatrix} H_d(e^{j\omega_0}) \\ H_d(e^{j\omega_1}) \\ \vdots \\ H_d(e^{j\omega_N}) \end{bmatrix} \quad (82.66)$$

In more compact form, this can be written

$$\mathbf{Qh} = \mathbf{H}_d \quad (82.67)$$

where the definitions for the matrix $\mathbf{Q}$ and vectors $\mathbf{h}$ and $\mathbf{H}_d$ should be apparent from (82.66). The impulse response coefficients are found simply by solving

$$\mathbf{h} = \mathbf{Q}^{-1}\mathbf{H}_d \quad (82.68)$$

This method guarantees that the frequency response of the filter exactly matches that of the desired response at the selected discrete frequencies. As such, it implicitly allows for frequency band weighting, since the specified frequencies do not have to be uniformly spaced over normalized frequency space. By clustering more of the selected frequencies in a given band, the response in that band is usually closer to the desired response, while the response in other regions where the frequency grid is more sparse will have larger deviations.

However, a problem may arise at frequencies other than the selected discrete set. The response at these other frequencies is not controlled in the design algorithm and often deviates greatly from the desired response. The result is often a filter design with unacceptable ripple in both the passbands and stopbands. In some cases, this ripple behavior can be quite severe.

To alleviate this problem, a least-squares solution [2] can be obtained by specifying the desired response at $M + 1$ discrete frequencies, where $M > N$ (often $M \gg N$). In this case, the matrix-vector formulation of (82.66) and (82.67) is still applicable, however, the matrix $\mathbf{Q}$ is $M + 1$ rows by $N + 1$ columns and the $\mathbf{H}_d$ vector consists of $M + 1$ elements. With the mean-square error criterion defined as

$$\epsilon = (\mathbf{Qh} - \mathbf{H}_d)^{\dagger}(\mathbf{Qh} - \mathbf{H}_d) \quad (82.69)$$

the least-squares (or pseudo-inverse) solution is then given by

$$\mathbf{h} = (\mathbf{Q}^{\dagger}\mathbf{Q})^{-1}\mathbf{Q}^{\dagger}\mathbf{H}_d \quad (82.70)$$

In this case, the response at the selected discrete frequencies is not constrained to be exactly that of the desired response. However, the overall deviations at those selected frequencies is minimized in a mean-square sense. This helps to ensure that deviations from the desired response are kept under control.

Although clustering of discrete frequency locations in the selected grid is one way of effectively performing frequency band weighting, a more direct manner is to assign specific weights to the desired response. In this case, the error criterion is defined as

$$\epsilon = (\mathbf{Qh} - \mathbf{H}_d)^{\dagger}\mathbf{W}(\mathbf{Qh} - \mathbf{H}_d) \quad (82.71)$$

where $\mathbf{W}$ is a positive definite diagonal weighting matrix whose mth element represents the weighting assigned to the desired frequency response sample $H_d(e^{j\omega_m})$. The least-squares solution in this case is given by

$$\mathbf{h} = (\mathbf{Q}^{\dagger}\mathbf{W}\mathbf{Q})^{-1}\mathbf{Q}^{\dagger}\mathbf{W}\mathbf{H}_d \quad (82.72)$$

By means of the weighting assigned to each frequency sample, the achieved response can be made to approach the desired response as closely as the designer wishes.

Computational complexity in all the previously discussed frequency sampling design algorithms can often be reduced by exploiting symmetry constraints in the impulse response. Usually, FIR filters are desired to be linear phase. This property can be ensured in the design by constraining $h(n) = h(N - n)$ or $h(n) = -h(N - n)$, so that $H(e^{j\omega}) = A(e^{j\omega})e^{-j\omega N/2}$ or $H(e^{j\omega}) = jA(e^{j\omega})e^{-j\omega N/2}$, respectively, where $A(e^{j\omega})$ is a purely real function. (Other symmetry constraints, such as those for Hilbert transformers, are also possible. However, the basic principle is illustrated by these two more common cases.) Under these constraints, only $N/2 + 1$ (for N even) or $(N + 1)/2$ (for N odd) distinct impulse response samples exist, so that the order of the system of linear design equations is approximately halved. In addition, the complex phase factor $e^{-j\omega N/2}$ can be eliminated in the design process, so that the defining equations become purely real, and thus the elements in the design matrix and vectors are also real. This further reduces computational loading over that that is implicit in the previously presented general design equations.

Frequency sampling design methods have the advantages of being applicable to a wide range of filter classes and easily programmed for numerical implementation. In some cases, where the response at certain specific frequencies is of extreme importance, they can be the methods of choice. However, some ambiguity exists in the choice of the discrete frequency grid, and different grids will produce different results. Selecting the grid thus becomes something of an ad hoc process. Another disadvantage is associated with the lack of design constraints to frequencies between those selected. As mentioned earlier, the response between the specified frequencies can sometimes deviate significantly from the desired response, yielding unsatisfactory results.

Example. Consider the design of a Nth-order (N even) linear phase low-pass filter with the same desired frequency characteristics as in the previous example. The linear phase property can

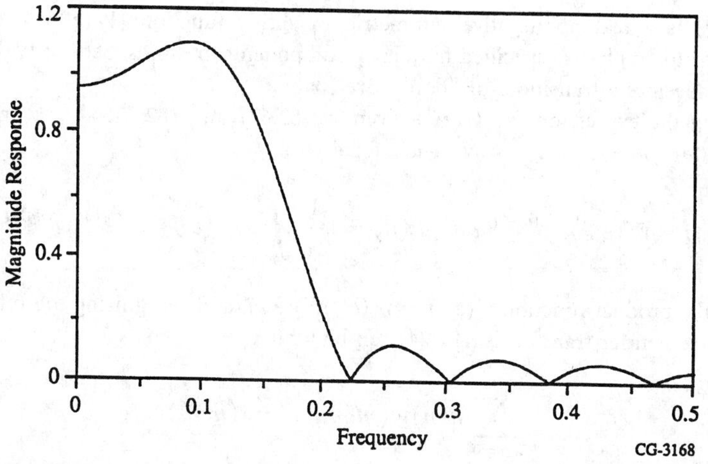

FIGURE 82.22 Magnitude frequency response for a 12th-order lowpass filter designed using the unweighted least-squares method plotted versus frequency f where $\omega = 2\pi f$.

be guaranteed by constraining $h(n) = h(N - n)$ and for N even, the resulting transfer function written

$$H(e^{j\omega}) = \left[h(N/2) + 2\sum_{n=1}^{N/2} h(n + N/2)\cos(\omega n) \right] e^{-j\omega N/2}$$

The design problem is thus that of finding coefficients $\{h(N/2), h(N/2+1), \cdots, h(N)\}$ so that

$$h(N/2) + 2\sum_{n=1}^{N/2} h(n + N/2)\cos(\omega n) \cong \begin{cases} 1 & \text{for } |\omega| \le \omega_c \\ 0 & \text{otherwise} \end{cases}$$

If the desired response is specified on a $M + 1$-point frequency grid where $M > N$ and appropriate weighting is given to each frequency sample, the least-squares (pseudo-inverse) solution of linear equations such as that in (82.70) or (82.72) must be used.

As a specific numerical example, let $N = 12$ and $\omega_c = 0.35\pi$ as in the previous case. For illustration, let the desired response be specified on a $M + 1 = 11$-point grid defined by $\{0, 0.1\pi, 0.2\pi, \cdots, \pi\}$. (Note that the response is not specified in the range $0.3\pi < |\omega| < 0.4\pi$ around the cutoff frequency.) Solving the least-squares (pseudo-inverse) solution of linear equations yields the impulse response coefficients in the FSLS column of Table 82.11. The magnitude frequency response for this filter is shown in Fig. 82.22. Note that although the result is quite similar to that obtained in the previous example, deviations from the desired response at the specified grid points are lower.

Weighted Least-Squares Methods

For cases where filter approximation over a continuous range of frequencies is desired inclusive of weighting, an approach resulting in a linear algebraic design method can be employed [1], [3], [6], [7], [14]. In this technique, the error criterion is a weighted mean-square error defined by

$$\epsilon = \frac{1}{2\pi}\int_{-\pi}^{\pi} W(e^{j\omega})|H(e^{j\omega}) - H_d(e^{j\omega})|^2\, d\omega \qquad (82.73)$$

where $W(e^{j\omega})$ is a real nonnegative symmetric weighting function $[W(e^{j\omega}) \geq 0,\ W(e^{j\omega}) = W(e^{-j\omega})]$ used to emphasize specified frequency components or bands. Setting $W(e^{j\omega}) = 0$ in a given region specifies a transition, or "don't care" band.

Substituting the expression for $H(e^{j\omega})$ given in (82.67) into (82.73) above, and evaluating $\partial\epsilon/\partial h(m) = 0$ for $m = 0, 1, 2, \cdots, N$ yields, for each m,

$$\sum_{n=0}^{N}\left[\frac{1}{2\pi}\int_{-\pi}^{\pi} W(e^{j\omega})e^{j\omega(m-n)}\,d\omega\right]h(n) = \frac{1}{2\pi}\int_{-\pi}^{\pi} W(e^{j\omega})H_d(e^{j\omega})e^{j\omega m}\,d\omega \qquad (82.74)$$

By defining the product function $G(e^{j\omega}) = W(e^{j\omega})H_d(e^{j\omega})$, and recognizing the integral expressions as inverse Fourier transforms, (82.74) can be written

$$\sum_{n=0}^{N} h(n)w(m-n) = g(m) \qquad (82.75)$$

for $m = 0, 1, 2, \cdots, N$, where $w(n)$ and $g(n)$ are the inverse Fourier transforms of the functions $W(e^{j\omega})$ and $G(e^{j\omega})$, respectively. The real and symmetric constraints on the frequency domain weighting function $W(e^{j\omega})$ force the sequence $w(n)$ also to be real and symmetric, so the design equations can be written in matrix-vector form as

$$\begin{bmatrix} w(0) & w(1) & \cdots & w(N) \\ w(1) & w(0) & \cdots & w(N-1) \\ \vdots & \vdots & \ddots & \vdots \\ w(N) & w(N-1) & \cdots & w(0) \end{bmatrix}\begin{bmatrix} h(0) \\ h(1) \\ \vdots \\ h(N) \end{bmatrix} = \begin{bmatrix} g(0) \\ g(1) \\ \vdots \\ g(N) \end{bmatrix} \qquad (82.76)$$

or more compactly,

$$\mathbf{Wh} = \mathbf{g}, \qquad \mathbf{h} = \mathbf{W}^{-1}\mathbf{g} \qquad (82.77)$$

In this way, it can be seen that the matrix **W** is Toeplitz, a property that can be exploited to reduce computation in the actual design procedure. (The **W** matrix in (82.76) should not be confused with the diagonal weighting matrix used in frequency sampling design methods.) If, in addition, symmetry constraints are also imposed on the impulse response coefficients, computation can be further reduced by the method described in [8].

Provided that the sequences $w(n)$ and $g(n)$ can be evaluated in closed form, this approach has the advantage over frequency sampling methods of minimizing the weighted mean-square error over every point in frequency space. For frequency selective multiband filters, closed-form expressions involving the band weights and the sinc function ($\mathrm{sinc}(x) = \sin(x)/x$) can be found. In cases where closed-form expressions cannot be obtained, numerical approximations can be used. However, if a DFT or rectangular rule integration is employed to estimate $w(n)$ and $g(n)$, it can be shown that the results are identical to those obtained by frequency sampling methods with the same frequency grid.

Example. Again, consider the design of a Nth-order (N even) linear phase low-pass filter with the same desired frequency characteristics as in the previous examples. Let the weighting function be specified as

$$W(e^{j\omega}) = \begin{cases} 1 & \text{for } |\omega| \leq \omega_p \\ 0 & \text{for } \omega_p < |\omega| < \omega_s \\ 1 & \text{for } \omega_s \leq |\omega| \leq \pi \end{cases}$$

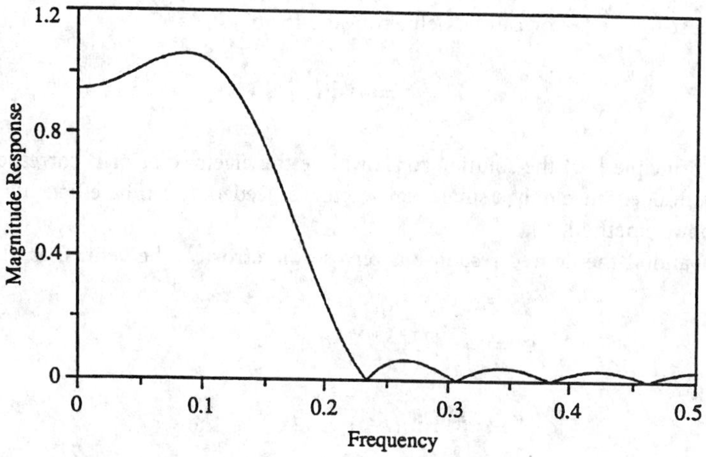

FIGURE 82.23 Magnitude frequency response for a 12th-order lowpass filter designed using the unweighted least-squares method plotted versus frequency f where $\omega = 2\pi f$.

where ω_p and ω_s are the passband and stopband edges, respectively, chosen so that $\omega_p < \omega_c < \omega_s$. Note that the response is given zero weighting in the range $\omega_p < |\omega| < \omega_s$ around the cutoff frequency. However, unlike the previous example, all points in the stopband and passband are taken into consideration, rather than discrete points on a finite grid. In this example, the sequences $w(n)$ and $g(n)$ as defined above can be found in closed form and are given by

$$w(n) = \delta(n) + \frac{\sin(\omega_p n)}{\pi n} - \frac{\sin(\omega_s n)}{\pi n}$$
$$g(n) = \frac{\sin[\omega_p(n - N/2)]}{\pi(n - N/2)}$$

These expressions can be used to form the **W** matrix and **g** vector for use in the design equation (82.77).

As a specific numerical case, let $N = 12$ as in the previous examples with $\omega_p = 0.3\pi$ and $\omega_s = 0.4\pi$. Solving the solution of linear equations yields the impulse response coefficients in the WLS column of Table 82.11. The magnitude frequency response for this filter is shown in Fig. 82.23. Note that both the passband and stopband ripple have been reduced from the previous examples by utilizing a "don't care" transition band, and placing equal weighting on all other points in frequency space.

Eigenfilter Method

With this method the least-squares design problem is reformulated such that the impulse response samples can be obtained by finding the eigenvector of a matrix [9], [12], [13], [16]. This can be done by defining the vectors

$$\mathbf{h} = [h(0) \quad h(1) \quad h(2) \cdots h(N)]^t$$
$$\mathbf{c}(e^{j\omega}) = [1 \quad e^{-j\omega} \quad e^{-j2\omega} \ldots e^{-jN\omega}]^t \qquad (82.78)$$

so that

$$H(e^{j\omega}) = \mathbf{h}^t \mathbf{c}(e^{j\omega}) \qquad (82.79)$$

Then, if the least-squares error can be defined in the form

$$\epsilon = \mathbf{h}^t \mathbf{P} \mathbf{h} \tag{82.80}$$

by Rayleigh's principle [10] the solution for **h** will be the eigenvector of **P** corresponding to its minimum eigenvalue. Since only a single eigenvector is needed, it can be efficiently found using the iterative power method [5].

In the stopband(s) the desired response is zero so the error can be defined as

$$\begin{aligned}\epsilon_s &= \int_{R_s} |H(e^{j\omega})|^2 \, d\omega \\ &= \int_{R_s} \mathbf{h}^t \mathbf{c}(e^{j\omega})[\mathbf{h}^t \mathbf{c}(e^{j\omega})]^\dagger \, d\omega \\ &= \mathbf{h}^t \left[\int_{R_s} \mathbf{c}(e^{j\omega})\mathbf{c}^\dagger(e^{j\omega}) \, d\omega \right] \mathbf{h} \\ &= \mathbf{h}^t \mathbf{P}_s \mathbf{h}\end{aligned} \tag{82.81}$$

where R_s is the portion of $-\pi \leq \omega \leq \pi$ containing stopbands. Likewise, in the passband(s) the error can be defined as

$$\begin{aligned}\epsilon_p &= \int_{R_p} \left| \frac{H(e^{j\omega_0})}{H_d(e^{j\omega_0})} H_d(e^{j\omega}) - H(e^{j\omega}) \right|^2 d\omega \\ &= \mathbf{h}^t \left\{ \int_{R_p} \left[\frac{H_d(e^{j\omega})}{H_d(e^{j\omega_0})} \mathbf{c}(e^{j\omega_0}) - \mathbf{c}(e^{j\omega}) \right] \left[\frac{H_d^*(e^{j\omega})}{H_d^*(e^{j\omega_0})} \mathbf{c}^\dagger(e^{j\omega_0}) - \mathbf{c}^\dagger(e^{j\omega}) \right] d\omega \right\} \mathbf{h} \\ &= \mathbf{h}^t \mathbf{P}_p \mathbf{h}\end{aligned} \tag{82.82}$$

where ω_0 is a convenient reference frequency in the passband for which $H_d(e^{j\omega_0}) \neq 0$ and R_p is the portion of $-\pi \leq \omega \leq \pi$ containing passbands. The reason for including the reference frequency ratio $H(e^{j\omega_0})/H_d(e^{j\omega_0})$ is to bring **h** into the equation leading to the eigenformulation.

Given $\mathbf{P}_s$ and $\mathbf{P}_p$, the total error can be defined as

$$\mathbf{P} = (1 - \alpha)\mathbf{P}_p + \alpha \mathbf{P}_s \tag{82.83}$$

where α controls the relative accuracy of approximation in the passband and stopband. If desired, a continuous weighting function like that used with the weighted least-squares method can be incorporated into the eigenformulation as discussed in [9]. Equation (82.83) is easily extended to the case of filters with multiple passbands and stopbands where individual band weighting is desired.

Example. As an example of this approach, consider again designing a Nth-order (N even) linear phase low-pass filter with the same frequency specifications as in the previous examples. Then, in the stopband ($\omega_c < \omega_s \leq |\omega| \leq \pi$),

$$\mathbf{P}_s = \int_{-\pi}^{-\omega_s} \mathbf{c}(e^{j\omega})\mathbf{c}^\dagger(e^{j\omega}) \, d\omega + \int_{\omega_s}^{\pi} \mathbf{c}(e^{j\omega})\mathbf{c}^\dagger(e^{j\omega}) \, d\omega$$

so that the elements of $\mathbf{P}_s$ are then given by

$$[\mathbf{P}_s]_{mn} = 2\int_{\omega_s}^{\pi} \cos[(m-n)\omega]\, d\omega$$

$$= 2\frac{\sin[(m-n)\pi]}{(m-n)} - 2\frac{\sin[(m-n)\omega_s]}{(m-n)}$$

where $m, n = 0, 1, \cdots, N$.

In the passband ($|\omega| \le \omega_p < \omega_c$), it is convenient to choose a reference frequency of $\omega_0 = 0$ giving

$$\mathbf{c}(e^{j\omega_0}) = [1 \quad 1 \cdots 1] = \mathbf{1}$$

so that

$$\mathbf{P}_p = \int_{-\omega_p}^{\omega_p} [e^{-j\omega N/2}\mathbf{1} - \mathbf{c}(e^{j\omega})][e^{j\omega N/2}\mathbf{1}^t - \mathbf{c}^\dagger(e^{j\omega})]\, d\omega$$

The elements of $\mathbf{P}_p$ are thus

$$[\mathbf{P}_p]_{mn} = 2\omega_p + 2\frac{\sin[(m-n)\omega_p]}{m-n} - 2\frac{\sin[(m-N/2)\omega_p]}{n-N/2} - 2\frac{\sin[(n-N/2)\omega_p]}{n-N/2}$$

Choosing $N = 12$ and $\alpha = 0.5$ gives a filter comparable to that of the previous examples. Since most computer programs for finding eigenvectors return a normalized eigenvector with $\mathbf{h}^t\mathbf{h} = 1$, it is necessary to scale the eigenvector for the desired passband amplitude. In this example, the coefficients were scaled so that $h(6) = 0.35$ to simplify comparison of the results with the previous examples. The resulting coefficients are given in the EF column of Table 82.11. The magnitude frequency response of this filter is given in Fig. 82.24. Notice the similarity to the response obtained by the weighted least-squares method shown in Fig. 82.23.

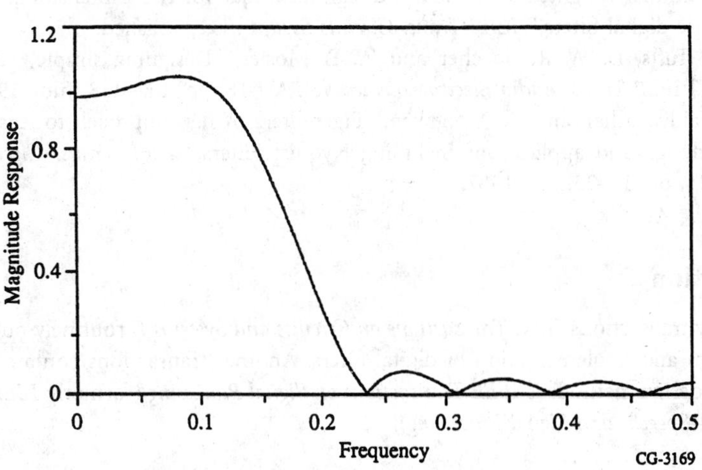

FIGURE 82.24 Magnitude frequency response for a 12th-order lowpass filter designed using the unweighted least-squares method plotted versus frequency f where $\omega = 2\pi f$.

References

[1] M. O. Ahmad and J-D. Wang, "An analytical least-square solution to the design problem of two-dimensional FIR filters with quadrantally symmetric or antisymmetric frequency response," *IEEE Trans. Circuits Syst.*, vol. 36, pp. 968–979, July 1989.

[2] V. R. Algazi and M. Suk, "On the frequency weighted least-square design of finite duration filters," *IEEE Trans. Circuits Syst.*, vol. CAS-22, pp. 943–953, Dec. 1975.

[3] V. R. Algazi, M. Suk, and C-S. Rim, "Design of almost minimax FIR filters in one and two dimensions by WLS techniques," *IEEE Trans. Circuits Syst.*, vol. CAS-33, pp. 590–596, June 1986.

[4] C. S. Burrus, A. W. Soewito, and R. A. Gopinath, "Least squared error FIR filter design with transition bands," *IEEE Trans. Signal Processing*, vol. 40, pp. 1327–1340, June 1992.

[5] J. N. Franklin, *Matrix Theory*, Englewood Cliffs, NJ: Prentice-Hall, 1968.

[6] Y. C. Lim, J.-H. Lee, C. K. Chen, and R.-H. Yang, "A weighted least-squares algorithm for quasi-equiripple FIR and IIR digital filter design," *IEEE Trans. Signal Processing*, vol. 40, pp. 551–558, Mar. 1992.

[7] Y. C. Lim, R.-H. Yang, and S. N. Koh, "The design of weighted minimax quadrature mirror filters," *IEEE Trans. Signal Processing*, vol. 41, pp. 1780–1789, May 1993.

[8] G. A. Merchant and T. W. Parks, "Efficient solution of a Toeplitz-plus-Hankel coefficient matrix system of equations," *IEEE Trans. Acoust., Speech, Signal Processing*, vol. ASSP-30, pp. 40–44, Feb. 1982.

[9] T. Q. Nguyan, "The design of arbitrary FIR digital filters using the eigenfilter method," *IEEE Trans. Signal Processing*, vol. 41, pp. 1128–1139, Mar. 1993.

[10] B. Nobel and J. W. Daniel, *Applied Linear Algebra*, Englewood Cliffs, NJ: Prentice-Hall, 1977.

[11] G. Oetken, T. W. Parks, and H. W. Schüssler, "New results in the design of digital interpolators," *IEEE Trans. Acoust., Speech, Signal Processing*, vol. ASSP-23, pp. 301–309, June 1975.

[12] S-C. Pei and J-J. Shyu, "Eigen-approach for designing FIR filters and all-pass phase equalizers with prescribed magnitude and phase response," *IEEE Trans. Circuits Syst. II*, vol. 39, pp. 137–146, Mar. 1992.

[13] ——, "Complex eigenfilter design of arbitrary complex coefficient FIR digital filters," *IEEE Trans. Circuits Syst. II*, vol. 40, pp. 32–40, Jan. 1993.

[14] L. M. Smith, "Weighted least-squares design technique for two-dimensional finite impulse response digital filters," *IEEE Trans. Circuits Syst.*, to be published.

[15] D. W. Tufts, D. W. Rorabacher and W. E. Mosier, "Designing simple, effective digital filters," *IEEE Trans. Audio Electroacoustics*, vol. AU-18, pp. 142–158, June 1970.

[16] P. P. Vaidyanathan and T. Q. Nguyen, "Eigenfilters: A new approach to least-squares FIR filter design and applications including Nyquist Filters," *IEEE Trans. Circuits Syst.*, vol. CAS-34, pp. 11–23, Jan. 1987.

Further Information

The monthly transactions *IEEE Transactions on Circuits and Systems II* routinely publishes papers on the design and implementation of digital filters. Another transactions containing papers on digital filter design methods is *IEEE Transactions on Signal Processing* (formerly *IEEE Transactions on Acoustics, Speech, and Signal Processing*).

Linear Programming (LP) and Mixed Integer Linear Programming (MILP) Design of FIR Filters

Y. C. Lim

Linear Programming [1]

Linear programming is an optimization problem that deals with the maximization or minimization of a linear function called the objective function subject to linear constraints. The problem expressed in (82.84a)–(82.84f) is a linear programming problem with J variables and K constraints. The objective function is

$$f = \mathbf{c}^T\mathbf{X} \tag{82.84a}$$

and the constraints are

$$\mathbf{AX} \leq \mathbf{b} \tag{82.84b}$$

where

$$\mathbf{A} = \begin{bmatrix} a_{11} & a_{12} & \cdots & a_{1J} \\ a_{21} & \ddots & & \vdots \\ \vdots & & \ddots & \vdots \\ a_{K1} & \cdots & \cdots & a_{KJ} \end{bmatrix} \tag{82.84c}$$

$$\mathbf{X} = [x_1 \quad x_2 \quad \cdots \quad x_J]^T \tag{82.84d}$$

$$\mathbf{b} = [b_1 \quad b_2 \quad \cdots \quad b_K]^T \tag{82.84e}$$

$$\mathbf{c} = [c_1 \quad c_2 \quad \cdots \quad c_J]^T \tag{82.84f}$$

Formulation of the Linear Phase FIR Filter Design Problem as an LP Program [2], [3]

We shall choose the design of a low-pass filter as shown in Fig. 82.25 as an example to illustrate the formulation of a filter design problem as an LP problem. Let the peak stopband ripple be δ_s, the peak passband ripple be δ_p and the passband gain be d. Let ω_1 and ω_2 be the passband edge and stopband edge, respectively. Let

$$0 \leq \omega_p \leq \omega_1 \tag{82.85a}$$

$$\omega_2 \leq \omega_s \leq \pi \tag{82.85b}$$

Let the frequency response of the filter be given by

$$H(e^{j\omega T}) = e^{-j\alpha\omega}P(\omega)$$

where α is a constant and $P(\omega)$ is a trigonometrical function (see Table 82.1). From 82.25, we have

$$P(\omega_p) \leq d + \delta_p \tag{82.86a}$$

$$P(\omega_p) \geq d - \delta_p \tag{82.86b}$$

$$P(\omega_s) \leq \delta_s \tag{82.86c}$$

$$P(\omega_s) \geq -\delta_s \tag{82.86d}$$

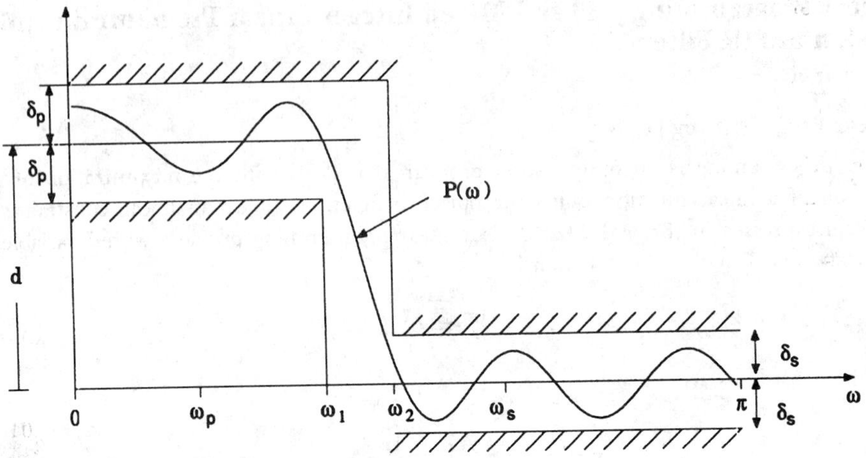

FIGURE 82.25 The frequency response of a low-pass filter.

The set of constraints in (82.86a)–(82.86d) can be generalized to (82.87a)–(82.87f).

$$P(\omega) \leq G(\omega) \times d + k(\omega) \times \delta \quad (82.87a)$$

$$P(\omega) \geq G(\omega) \times d - k(\omega) \times \delta \quad (82.87b)$$

where

$$G(\omega_p) = 1 \quad (82.87c)$$

$$G(\omega_s) = 0 \quad (82.87d)$$

$$k(\omega_p) \times \delta = \delta_p \quad (82.87e)$$

$$k(\omega_s) \times \delta = \delta_s \quad (82.87f)$$

$G(\omega)$ is the gain weighting function and $k(\omega)$ is the ripple weighting function. By using suitable function values for $G(\omega)$ and $k(\omega)$, (82.87a) and (82.87b) may be used to describe arbitrary frequency response requirements for a filter.

A set of inequalities can be obtained by evaluating (82.87a) and (82.87b) on a dense grid of frequencies along the ω axis. Minimizing δ (or maximizing $-\delta$) subject to this set of constraints is a linear programming problem. The method can be easily extended to the design of 2-D FIR filters by taking $P(\omega)$ to be the frequency response of a 2-D filter.

It should be noted that the density of the frequency grid points where the inequalities of (82.87a) and (82.87b) are evaluated must be sufficiently high so that the frequency response of the filter does not violate implied specifications at frequencies between the frequency grid points. If $\hat{\delta}$ is the actual peak ripple magnitude, a misleading factor J given by

$$J = (\hat{\delta} - \delta)/\hat{\delta} \quad (82.88)$$

may be defined. The higher the grid density used, the smaller the value of J. In [4] a grid spacing of λ given by

$$\lambda = \frac{0.5 - \sum_i \Delta_i}{kN} \quad (82.89)$$

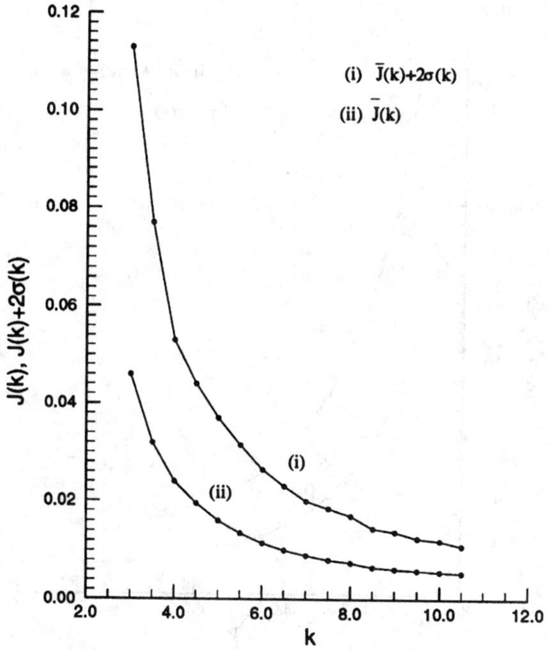

FIGURE 82.26 $\bar{J}(k)$ versus k plots for 1-D linear phase FIR filter. σ is the standard deviation [4].

was proposed where Δ_i is the ith normalized transition width, N is the filter length, and k is a constant. $\sum_i \Delta_i$ is the sum of all the transition widths. The relationship between k and the mean value of J denoted by $\bar{J}$ for 1-D and 2-D FIR linear phase filter are shown in Figs. 82.26 and 82.27, respectively.

Linear Programming Design of Nonlinear Phase FIR Filters [5]

Let the desired frequency response at ω be $d(\omega)e^{j\Phi(\omega)}$, where $d(\omega)$ and $e^{j\Phi(\omega)}$ are the desired magnitude and phase responses, respectively. The frequency response ripple $E(\omega)e^{j\psi(\omega)}$ is thus given by

$$E(\omega)e^{j\psi(\omega)} = H(e^{j\omega}) - d(\omega)e^{j\Phi(\omega)} \tag{82.90}$$

where $E(\omega)$ and $\psi(\omega)$ are the magnitude and phase of the frequency response ripple. It is desired to minimize a weighted version of $E(\omega)$. This can be achieved by minimizing the weighted version of $\mathrm{Re}\{E(\omega)e^{j\psi(\omega)}e^{j\theta}\}$ on a dense grid of θ ranging from $\theta = 0$ through $\theta = \pi$, where $\mathrm{Re}\{x\}$ denotes the real part of x. If θ is uniformly distributed with spacing $\Delta\theta$, then

$$E(\omega) \leq \max_{\theta} |\mathrm{Re}\{E(\omega)e^{j\psi(\omega)}e^{j\theta}\}| \tag{82.91}$$

where $\max_\theta |x(\theta)|$ denotes the maximum value of $|x(\theta)|$ for all values of θ. If $h(n)$ is the nth impulse response of the FIR filter under consideration, then

$$H(e^{j\omega}) = \sum_n h(n)e^{-j\omega n} \tag{82.92}$$

The real part of $E(\omega)e^{j\psi(\omega)}e^{j\theta}$ is given by

$$\begin{aligned}\mathrm{Re}\{E(\omega)e^{j\psi(\omega)}e^{j\theta}\} = {} & \textstyle\sum_n h(n)\{\cos(\omega n)\cos(\theta) + \sin(\omega n)\sin(\theta)\} \\ & -d(\omega)\{\cos[\Phi(\omega)]\cos(\theta) - \sin[\Phi(\omega)]\sin(\theta)\}\end{aligned} \tag{82.93}$$

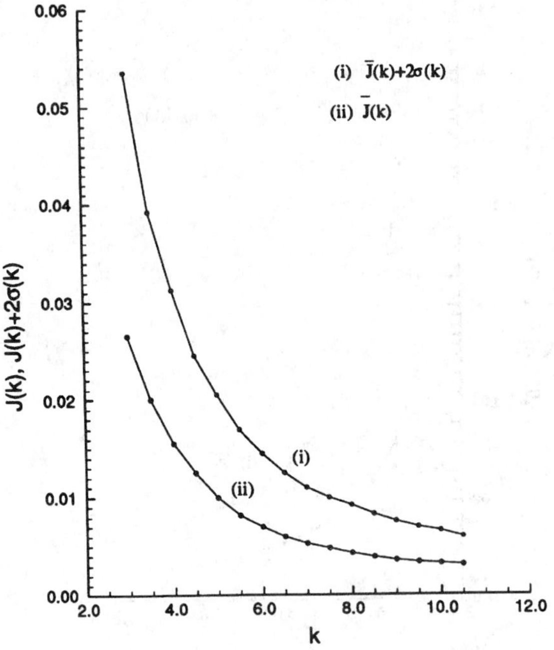

FIGURE 82.27 $\bar{J}(k)$ versus k plots for 2-D linear phase FIR filter. σ is the standard deviation [4].

The linear programming problem for the design of nonlinear phase FIR filters can now be formulated as follows:

$$\text{minimize } \delta \tag{82.94a}$$

subject to

$$\text{Re}\{E(\omega)e^{j\psi(\omega)}e^{j\theta}\} \leq k(\omega) \times \delta \tag{82.94b}$$

$$\text{Re}\{E(\omega)e^{j\psi(\omega)}e^{j\theta}\} \geq -k(\omega) \times \delta \tag{82.94c}$$

Design of FIR Filters with Discrete-Value Coefficients [6]–[10]

In a full custom hardware implementation, the coefficient multiplier is the most expensive and the most important speed determining component. Consequently, the primary concern in the design of filters for hardware implementation is to reduce the multiplier complexity. If a fixed point hardware multiplier is employed, all the coefficient values will have the same quantization step size. In this case, if the gain of the filter d is scaled by a factor of 1/(coefficient quantization step size), all the coefficient values are integers. Such a filter with integer coefficient values can be optimized by using a mixed integer linear programming (MILP) algorithm. MILP is LP with integer constraints imposed on those LP variables corresponding to the filter's coefficient values.

In binary arithmetic, multiplying a number by an integer power of two is a very simple process; it is only a matter of shifting the bit position of the number. A full fletch multiplier is not needed. Hence, if each coefficient value of the filter is a sum of a limited number of signed power-of-two (SPT) terms, the filter can be implemented without using multipliers. MILP may also be used to optimize such filters.

The passband gain of an infinite-precision-coefficient-value filter may be arbitrarily scaled by scaling its coefficient values. Scaling the coefficient values corresponds to scaling the passband gain and peak ripple magnitude of the filter by the same scaling constant. Hence, the normalized peak ripple magnitude, i.e., peak ripple magnitude divided by passband gain, remains unchanged. However, this is not the case in the design of discrete-coefficient-value filters; the coefficient values cannot be arbitrarily scaled and still maintain their discrete nature. Thus, it is necessary to reoptimize the coefficient values when the passband gain is scaled. As a consequence, the normalized peak ripple magnitude, (weighted peak ripple magnitude/passband gain) changes as the passband gain changes.

If d is the passband gain of the filter, it may be allowed to vary within predetermined bounds in such a way that δ/d is minimized. This can be achieved by minimizing the objective function $\delta - \beta d$, where β is a constant to be updated iteratively using the following algorithm [10].

1. Round the coefficient values of the optimum infinite-precision-coefficient-value filter to the nearest discrete value. Let β be the value of δ/d of this rounded coefficient value filter. Set upper and lower bounds for d.
2. Perform an MILP run minimizing $\delta - \beta d$. Let β_2 be equal to δ/d of the optimum solution of the MILP run.
3. If $\beta = \beta_2$, this is the desired solution; stop. If $\beta \neq \beta_2$, let $\beta = \beta_2$ and go to step 2.

The design of a one-dimensional low-pass filter satisfying the following set of specifications is used to illustrate the above method: filter length = 20; passband edge = 0.1× sampling frequency; stopband edge = 0.2× sampling frequency; each coefficient value is a sum or difference of not more than two power-of-two terms; the peak ripple magnitude is to be minimized. Without imposing discrete constraints on the coefficient values, the peak ripple magnitude is 0.01245. Rounding each coefficient value to the nearest sum or difference of two power-of-two terms yields a design with normalized peak ripple magnitude equal to 0.05134. Letting $\beta = 0.05134$ produces a design with normalized peak ripple magnitude equal to 0.01299. Letting $\beta = 0.01299$ produces the same solution. Hence, the solution with $\beta = 0.01299$ is the optimum solution.

References

[1] G. Hadley, *Linear Programming*, Reading, MA: Addison-Wesley, new edition 1975.

[2] L. R. Rabiner, "Linear Program design of finite impulse response (FIR) digital filters," *IEEE Trans. Audio Electroacoust*, vol. AU-20, pp. 280–288, Oct. 1972.

[3] Y. C. Lim, "Efficient special purpose linear programming for FIR filter design," *IEEE Trans. Acoust., Speech, Signal Processing*, vol. ASSP-31, pp. 963–968, Aug. 1983.

[4] R. H. Yang and Y. C. Lim, "Grid density for design of one- and two-dimensional FIR filters," *Electron. Lett.*, vol. 22, pp. 2053–2055, Oct. 1991.

[5] X. Chen and T. W. Park, "Design of FIR filters in the complex domain," *IEEE Trans. Acoust., Speech, Signal Processing*, vol. ASSP-35, pp. 144–153, Feb. 1987.

[6] Y. C. Lim and A. G. Constantinides, "New integer programming scheme for nonrecursive digital filter design," *Electron. Lett.*, vol. 15, pp. 812, 813, Dec. 1979.

[7] D. M. Kodek, "Design of optimal finite wordlength FIR digital filters using integer programming techniques," *IEEE Trans. Acoust., Speech, Signal Processing*, vol. ASSP-28, pp. 304–308, June 1980.

[8] Y. C. Lim, S. R. Parker, and A. G. Constantinides, "Finite wordlength FIR filter design using integer programming over a discrete coefficient space," *IEEE Trans. Acoust., Speech, Signal Processing*, vol. ASSP-30, pp. 661–664, Aug. 1982.

[9] P. Siohan and A. Benslimane, "Finite precision design of optimal linear phase 2-D FIR digital filters," *IEEE Trans. Circuits Syst.*, vol. 36, pp. 11–22, Jan. 1989.

[10] Y. C. Lim, "Design of discrete-coefficient-value linear phase FIR filters with optimum normalized peak ripple magnitude," *IEEE Trans. Circuits Syst.*, vol. CAS-37, pp. 1480–1486, Dec. 1990.

Design of Computationally Efficient FIR Filters Using Periodic Subfilters as Building Blocks

Tapio Saramäki

Introduction

For many digital signal processing applications, FIR filters are preferred over their IIR counterparts as the former can be designed with exactly linear phase and they are free of stability problems and limit cycle oscillations. The major drawback of FIR filters is that they require, especially in applications demanding narrow transition bands, considerably more arithmetic operations and hardware components than do comparable IIR filters. Ignoring the correction term for very low-order filters, the minimum order of an optimum linear-phase low-pass FIR filter can be approximated [1] by

$$N \approx \Phi(\delta_p, \delta_s)/(\omega_s - \omega_p) \tag{82.95a}$$

where

$$\begin{aligned}\Phi(\delta_p, \delta_s) = {} & 2\pi[0.005\,309(\log_{10}\delta_p)^2 + 0.071\,14\log_{10}\delta_p - 0.4761]\log_{10}\delta_s \\ & -2\pi[0.002\,66(\log_{10}\delta_p)^2 + 0.5941\log_{10}\delta_p + 0.4278]\end{aligned} \tag{82.95b}$$

Here, ω_p and ω_s are the passband and stopband edge angles, whereas δ_p and δ_s are the passband and stopband ripple magnitudes. From the above estimate, it is seen that as the transition bandwidth $\omega_s - \omega_p$ is made smaller, the required filter order increases inversely proportionally to it. Since the direct-form implementation exploiting the coefficient symmetry requires approximately $N/2$ multipliers, this kind of implementation becomes very costly if the transition bandwidth is small.

The cost of implementation of a narrow transition-band FIR filter can be significantly reduced by using multiplier-efficient realizations, fast convolution algorithms, or multirate filtering. This section considers those multiplier-efficient realizations that use as basic building blocks the transfer functions obtained by replacing each unit delay in a conventional transfer function by multiple delays. We concentrate on the synthesis techniques described in [2]–[4], [6], and [8]–[10].

Frequency-Response Masking Approach

A very elegant approach to significantly reducing the implementation cost of an FIR filter has been proposed by Lim [3]. In this approach, the overall transfer function is constructed as

$$H(z) = F(z^L)G_1(z) + [z^{-LN_F/2} - F(z^L)]G_2(z) \tag{82.96a}$$

where

$$F(z^L) = \sum_{n=0}^{N_F} f(n)z^{-nL} \qquad f(N_F - n) = f(n) \tag{82.96b}$$

$$G_1(z) = z^{-M_1}\sum_{n=0}^{N_1} g_1(n)z^{-n} \qquad g_1(N_1 - n) = g_1(n) \tag{82.96c}$$

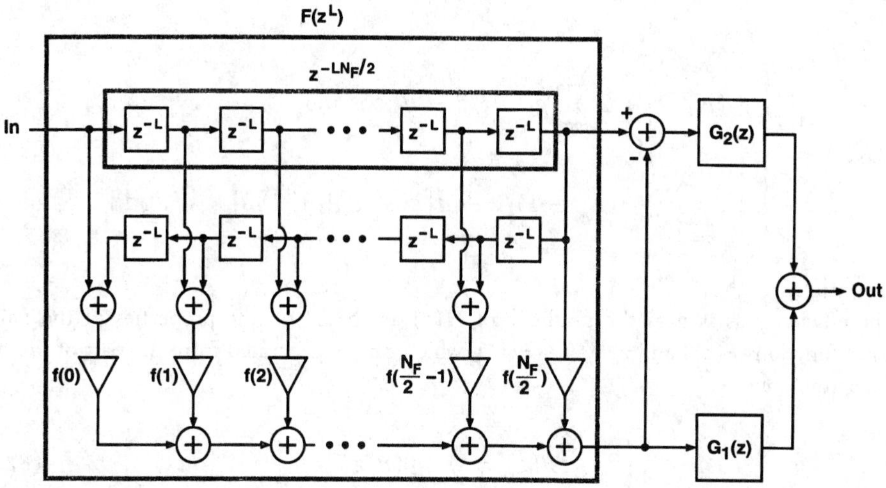

FIGURE 82.28 An efficient implementation for a filter synthesized using the frequency-response masking approach.

and

$$G_2(z) = z^{-M_2} \sum_{n=0}^{N_2} g_2(n) z^{-n} \qquad g_2(N_2 - n) = g_2(n) \tag{82.96d}$$

Here, N_F is even, whereas both N_1 and N_2 are either even or odd. For $N_1 \geq N_2$, $M_1 = 0$ and $M_2 = (N_1 - N_2)/2$, whereas for $N_1 < N_2$, $M_1 = (N_2 - N_1)/2$ and $M_2 = 0$. These selections guarantee that the delays of both of the terms of $H(z)$ are equal. An efficient implementation for the overall filter is depicted in Fig. 82.28, where the delay term $z^{-LN_F/2}$ is shared with $F(z^L)$. Also, $G_1(z)$ and $G_2(z)$ can share their delays if a transposed direct-form implementation (exploiting the coefficient symmetry) is used.

The frequency response of the overall filter can be written as

$$H(e^{j\omega}) = H(\omega) e^{-j(LN_F + \max\{N_1, N_2\})\omega/2} \tag{82.97}$$

where $H(\omega)$ denotes the *zero-phase frequency response* of $H(z)$ and can be expressed as

$$H(\omega) = H_1(\omega) + H_2(\omega) \tag{82.98a}$$

where

$$H_1(\omega) = F(L\omega) G_1(\omega) \tag{82.98b}$$

and

$$H_2(\omega) = [1 - F(L\omega)] G_2(\omega) \tag{82.98c}$$

with

$$F(\omega) = f(N_F/2) + 2 \sum_{n=1}^{N_F/2} f(N_F/2 - n) \cos n\omega \tag{82.98d}$$

and

$$G_k(\omega) = \begin{cases} g_k(N_k/2) + 2\sum_{n=1}^{N_k/2} g_k(N_k/2 - n)\cos n\omega & N_k \text{ even} \\ 2\sum_{n=0}^{(N_k-1)/2} g_k[(N_k-1)/2 - n]\cos[(n+1/2)\omega] & N_k \text{ odd} \end{cases} \quad (82.98e)$$

for $k = 1, 2$.

The efficiency as well as the synthesis of $H(z)$ are based on the properties of the pair of transfer functions $F(z^L)$ and $z^{-LN_F/2} - F(z^L)$, which can be generated from the pair of *prototype* transfer functions

$$F(z) = \sum_{n=0}^{N_F} f(n)z^{-n} \quad (82.99)$$

and $z^{-N_F/2} - F(z)$ by replacing z^{-1} by z^{-L}, that is, by substituting for each unit delay L unit delays. The order of the resulting filters is increased to LN_F, but since only every Lth impulse response value is nonzero, the filter complexity (number of adders and multipliers) remains the same. The above prototype pair forms a *complementary* filter pair since their zero-phase frequency responses, $F(\omega)$ and $1 - F(\omega)$ with $F(\omega)$ given by (82.98d), add up to unity. Figure 82.29(a) illustrates the relations between these responses in the case where $F(z)$ and $z^{-N_F/2} - F(z)$ is a low-pass–high-pass filter pair with edges at θ and ϕ.

The substitution $z^{-L} \mapsto z^{-1}$ preserves the complementary property resulting in the *periodic* responses $F(L\omega)$ and $1 - F(L\omega)$, which are frequency-axis compressed versions of the prototype responses such that the interval $[0, L\pi]$ is shrunk onto $[0, \pi]$ [see Fig. 82.29(b)]. Since the periodicity of the prototype responses is 2π, the periodicity of the resulting responses is $2\pi/L$ and they contain several passband and stopband regions in the interval $[0, \pi]$.

For a low-pass filter $H(z)$, one of the transition bands provided by $F(z^L)$ or $z^{-LN_F/2} - F(z^L)$ is used as that of the overall filter. In the first case, denoted by Case A, the edges are given by

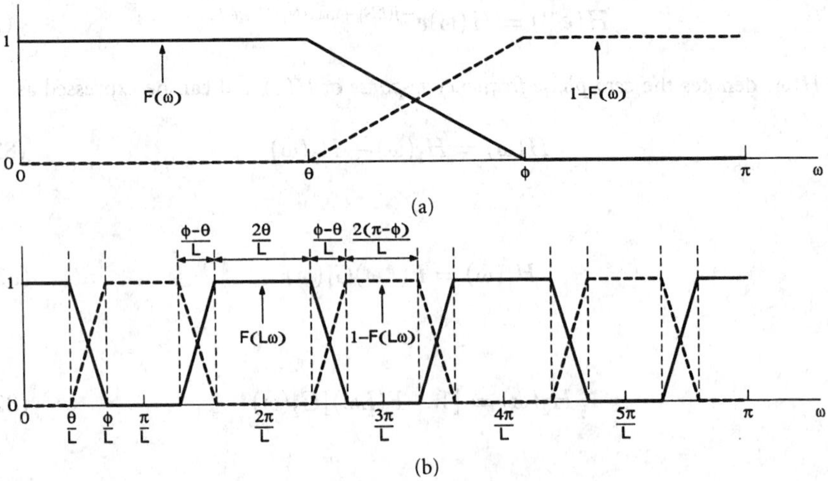

FIGURE 82.29 Generation of a complementary periodic filter pair by starting with a low-pass–high-pass complementary pair. (a) Prototype filter responses $F(\omega)$ and $1 - F(\omega)$. (b) Periodic responses $F(L\omega)$ and $1 - F(L\omega)$ for $L = 6$.

(see Fig. 82.30)

$$\omega_p = (2l\pi + \theta)/L \qquad \omega_s = (2l\pi + \phi)/L \tag{82.100}$$

where l is a fixed integer, and in the second case, referred to as Case B, by (see Fig. 82.31)

$$\omega_p = (2l\pi - \phi)/L \qquad \omega_s = (2l\pi - \theta)/L \tag{82.101}$$

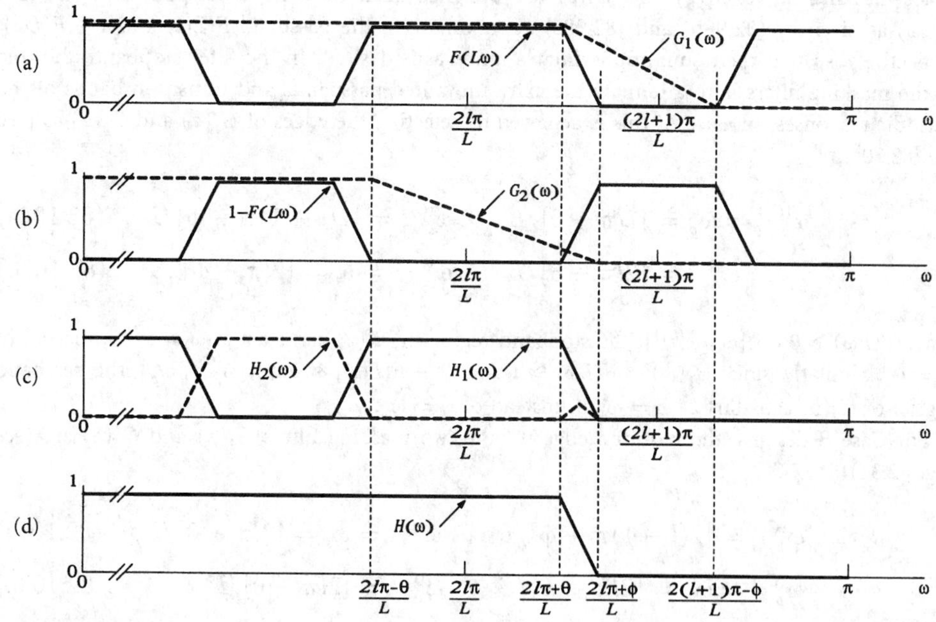

FIGURE 82.30 Case A design of a low-pass filter using the frequency-response masking technique.

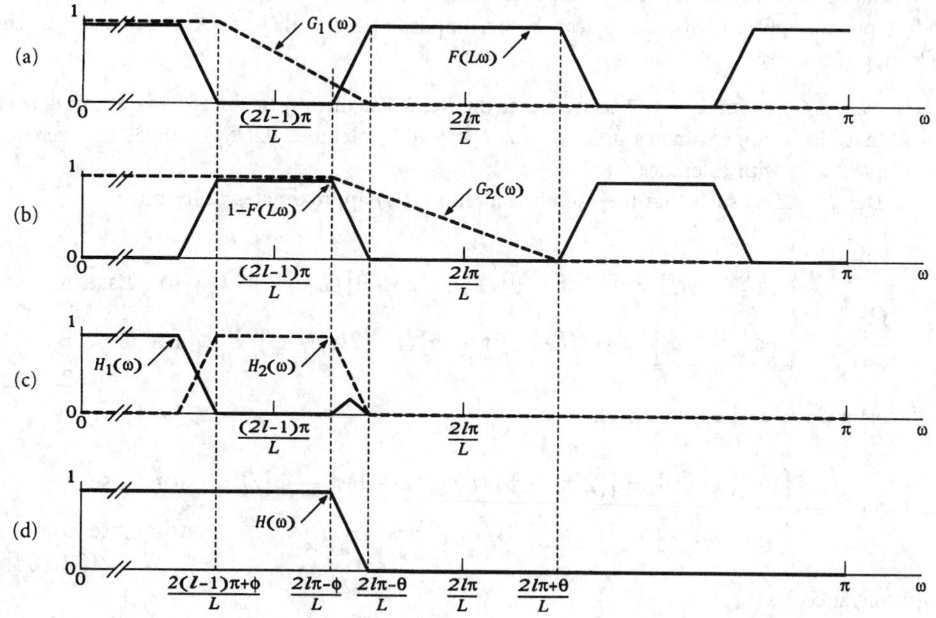

FIGURE 82.31 Case B design of a low-pass filter using the frequency-response masking technique.

The widths of these transition bands are $(\phi - \theta)/L$, which is only $1/L$th of that of the prototype filters. Since the filter order is roughly inversely proportional to the transition bandwidth, this means that the arithmetic complexity of the periodic transfer functions to provide one of the transition bands is only $1/L$th of that of a conventional nonperiodic filter. Note that the orders of both the periodic filters and the corresponding nonperiodic filters are approximately the same, but the conventional filter does not contain zero-valued impulse response samples.

In order to exploit the attractive properties of the periodic transfer functions, the two low-order masking filters $G_1(z)$ and $G_2(z)$ are designed such that the subresponses $H_1(\omega)$ and $H_2(\omega)$ as given by (82.98b) and (82.98c) approximate in the passband $F(L\omega)$ and $1 - F(L\omega)$, respectively, so that their sum approximates unity, as is desired. In the filter stopband, the role of the masking filters is to attenuate the extra unwanted passbands and transition bands of the periodic responses. In Case A, this is achieved by selecting the edges of $G_1(z)$ and $G_2(z)$ as (see Fig. 82.30)

$$\omega_p^{(G_1)} = \omega_p = [2l\pi + \theta]/L \qquad \omega_s^{(G_1)} = [2(l+1)\pi - \phi]/L \tag{82.102a}$$

$$\omega_p^{(G_2)} = [2l\pi - \theta]/L \qquad \omega_s^{(G_2)} = \omega_s = [2l\pi + \phi]/L \tag{82.102b}$$

Since $F(L\omega) \approx 0$ on $[\omega_s, \omega_s^{(G_1)}]$, the stopband region of $G_1(z)$ can start at $\omega = \omega_s^{(G_1)}$, instead of $\omega = \omega_s$. Similarly, since $H_1(\omega) \approx F(L\omega) \approx 1$ and $[1 - F(L\omega)] \approx 0$ on $[\omega_p^{(G_2)}, \omega_p]$, the passband region of $G_2(z)$ can start at $\omega = \omega_p^{(G_2)}$, instead of $\omega = \omega_p$.

For Case B designs, the required edges of the two masking filters $G_1(z)$ and $G_2(z)$ are (see Fig. 82.31)

$$\omega_p^{(G_1)} = [2(l-1)\pi + \phi]/L \qquad \omega_s^{(G_1)} = \omega_s = [2l\pi - \theta]/L \tag{82.103a}$$

$$\omega_p^{(G_2)} = \omega_p = [2l\pi - \phi]/L \qquad \omega_s^{(G_2)} = [2l\pi + \theta]/L \tag{82.103b}$$

The effects of the ripples of the subresponses on the ripples of the overall response $H(\omega)$ have been studied carefully in [3]. Based on these observations, the design of $H(z)$ with passband and stopband ripples of δ_p and δ_s can be accomplished for both Case A and Case B in the following two steps:

1. Design $G_k(z)$ for $k = 1, 2$ using either the Remez algorithm or linear programming such that $G_k(\omega)$ approximates unity on $[0, \omega_p^{(G_k)}]$ with tolerance $0.85\delta_p \cdots 0.9\delta_p$ and zero on $[\omega_s^{(G_k)}, \pi]$ with tolerance $0.85\delta_s \cdots 0.9\delta_s$.
2. Design $F(L\omega)$ such that the overall response $H(\omega)$ approximates unity on

$$\Omega_p^{(F)} = \begin{cases} [\omega_p^{(G_2)}, \omega_p] = [[2l\pi - \theta]/L, [2l\pi + \theta]/L] & \text{for Case A} \\ \left[\omega_p^{(G_1)}, \omega_p\right] = [[2(l-1)\pi + \phi]/L, [2l\pi - \phi]/L] & \text{for Case B} \end{cases} \tag{82.104a}$$

with tolerance δ_p and approximates zero on

$$\Omega_s^{(F)} = \begin{cases} [\omega_s, \omega_s^{(G_1)}] = [[2l\pi + \phi]/L, [2(l+1)\pi - \phi]/L] & \text{for Case A} \\ \left[\omega_s, \omega_s^{(G_2)}\right] = [[2l\pi - \theta]/L, [2l\pi + \theta]/L] & \text{for Case B} \end{cases} \tag{82.104b}$$

with tolerance δ_s.

The design of $F(L\omega)$ can be performed conveniently using linear programming [3]. Another, computationally more efficient, alternative is to use the Remez algorithm [10]. Its use is based

TABLE 82.12 Error Function for Designing $F(\omega)$ Using the Remez Algorithm

$$E_F(\omega) = W_F(\omega)[F(\omega) - D_F(\omega)],$$

where

$$D_F(\omega) = [u(\omega) + l(\omega)]/2, \quad W_F(\omega) = 2/[u(\omega) - l(\omega)]$$

with

$$u(\omega) = \min(\Psi_1(\omega) + \psi_1(\omega), \Psi_2(\omega) + \psi_2(\omega))$$

$$l(\omega) = \max(\Psi_1(\omega) - \psi_1(\omega), \Psi_2(\omega) - \psi_2(\omega))$$

$$\Psi_k(\omega) = \frac{D_H[h_k(\omega)] - G_2[h_k(\omega)]}{G_1[h_k(\omega)] - G_2[h_k(\omega)]}, \quad k = 1, 2$$

$$\psi_k(\omega) = \frac{1/W_H[h_k(\omega)]}{|G_1[h_k(\omega)] - G_2[h_k(\omega)]|}, \quad k = 1, 2$$

and

$$h_1(\omega) = (2l\pi + \omega)/L, \quad h_2(\omega) = \begin{cases} (2l\pi - \omega)/L & \text{for } \omega \in [0, \theta] \\ [2(l+1)\pi - \omega]/L & \text{for } \omega \in [\phi, \pi] \end{cases}$$

for Case A and

$$h_1(\omega) = (2l\pi - \omega)/L, \quad h_2(\omega) = \begin{cases} (2l\pi + \omega)/L & \text{for } \omega \in [0, \theta] \\ [2(l-1)\pi + \omega]/L & \text{for } \omega \in [\phi, \pi] \end{cases}$$

for Case B.

on the fact that

$$|E_H(\omega)| \le 1 \quad \text{for } \omega \in \Omega_p^{(F)} \cup \Omega_s^{(F)} \tag{82.105a}$$

where

$$E_H(\omega) = W_H(\omega)[H(\omega) - D_H(\omega)] \tag{82.105b}$$

is satisfied when $F(\omega)$ is designed such that the maximum absolute value of the error function given in Table 82.12 becomes less than or equal to unity on $[0, \theta] \cup [\phi, \pi]$.

For Step 2 of the above algorithm, $D_H(\omega) = 1$ and $W_H(\omega) = 1/\delta_p$ on $\Omega_p^{(F)}$, whereas $D_H(\omega) = 0$ and $W_H(\omega) = 1/\delta_s$ on $\Omega_s^{(F)}$, giving for $k = 1, 2$

$$D_H[h_k(\omega)] = \begin{cases} 1 & \text{for } \omega \in [0, \theta] \\ 0 & \text{for } \omega \in [\phi, \pi] \end{cases} \qquad W_H[h_k(\omega)] = \begin{cases} 1/\delta_p & \text{for } \omega \in [0, \theta] \\ 1/\delta_s & \text{for } \omega \in [\phi, \pi] \end{cases} \tag{82.106a}$$

for Case A and

$$D_H[h_k(\omega)] = \begin{cases} 1 & \text{for } \omega \in [0, \theta] \\ 0 & \text{for } \omega \in [\phi, \pi] \end{cases} \qquad W_H[h_k(\omega)] = \begin{cases} 1/\delta_s & \text{for } \omega \in [0, \theta] \\ 1/\delta_p & \text{for } \omega \in [\phi, \pi] \end{cases} \tag{82.106b}$$

for Case B. Even though the resulting error function looks very complicated, it is straightforward to use the subroutines EFF and WATE in the Remez algorithm described in [5] for optimally designing $F(z)$.

The order of $G_1(z)$ can be considerably reduced by allowing larger ripples on those regions of $G_1(z)$ where $F(L\omega)$ has one of its stopbands. As a rule of thumb, the ripples on these regions can be selected to be 10 times larger [3]. Similarly, the order of $G_2(z)$ can be decreased by allowing (ten times) larger ripples on those regions where $F(L\omega)$ has one of its passbands.

In practical filter synthesis problems, ω_p and ω_s are given and l, L, θ, and ϕ must be determined. To ensure that (82.100) yields a desired solution with $0 \le \theta < \phi \le \pi$, it is required that (see Fig. 82.30)

$$\frac{2l\pi}{L} \le \omega_p \qquad \omega_s \le \frac{(2l+1)\pi}{L} \tag{82.107a}$$

for some positive integer l, giving

$$l = \lfloor L\omega_p/(2\pi) \rfloor, \qquad \theta = L\omega_p - 2l\pi, \qquad \phi = L\omega_s - 2l\pi \tag{82.107b}$$

where $\lfloor x \rfloor$ stands for the largest integer that is smaller than or equal to x. Similarly, to ensure that (82.101) yields a desired solution with $0 \le \theta < \phi \le \pi$, it is required that (see Fig. 82.31)

$$\frac{(2l-1)\pi}{L} \le \omega_p \qquad \omega_s \le \frac{2l\pi}{L} \tag{82.108a}$$

for some positive integer l, giving

$$l = \lceil L\omega_s/(2\pi) \rceil, \qquad \theta = 2l\pi - L\omega_s, \qquad \phi = 2l\pi - L\omega_p \tag{82.108b}$$

where $\lceil x \rceil$ stands for the smallest integer that is larger than or equal to x. For any set of ω_p, ω_s, and L, either (82.107b) or (82.108b) (not both) will yield the desired θ and ϕ, provided that L is not too large. If $\theta = 0$ or $\phi = \pi$, then the resulting specifications for $F(\omega)$ are meaningless and the corresponding value of L cannot be used.

The remaining problem is to determine L to minimize the number of multipliers, which is $N_F/2 + 1 + \lfloor (N_1+2)/2 \rfloor + \lfloor (N_2+2)/2 \rfloor$ or $N_F + N_1 + N_2 + 3$ depending on whether the symmetries in the filter coefficients are exploited or not. Hence, in both cases, a good measure of the filter complexity is the sum of the orders of the subfilters. Instead of determining the actual minimum filter orders for various values of L, the computational workload can be significantly reduced based on the use of the estimation formula given by (82.95a) and (82.95b). Since the widths of transition bands of $F(z)$, $G_1(z)$, and $G_2(z)$ are $\phi - \theta$, $(2\pi - \phi - \theta)/L$ and $(\phi + \theta)/L$, respectively, good estimates for the corresponding filter orders are

$$N_F \approx \frac{\Phi(\delta_p, \delta_s)}{\phi - \theta} \qquad N_1 \approx \frac{L\Phi(\delta_p, \delta_s)}{2\pi - \phi - \theta} \qquad N_2 \approx \frac{L\Phi(\delta_p, \delta_s)}{\phi + \theta} \tag{82.109}$$

For the optimum nonperiodic direct-form design, the transition bandwidth is $\omega_s - \omega_p = (\phi - \theta)/L$, giving

$$N_{\text{opt}} \approx \frac{L\Phi(\delta_p, \delta_s)}{\phi - \theta} \tag{82.110}$$

The sum of the subfilter orders can be expressed in terms of N_{opt} as follows:

$$N_{ove} = N_{opt}\left[\frac{1}{L} + \frac{\phi - \theta}{2\pi - \phi - \theta} + \frac{\phi - \theta}{\phi + \theta}\right] \tag{82.111}$$

The smallest values of N_{ove} are typically obtained at those values of L for which $\theta + \phi \approx \pi$ and, correspondingly, $2\pi - \theta - \phi \approx \pi$. In this case, $N_1 \approx N_2$ and (82.111) reduces, after substituting $\phi - \theta = L(\omega_s - \omega_p)$, to

$$N_{ove} = N_{opt}\left[\frac{1}{L} + 2L(\omega_s - \omega_p)/\pi\right] \tag{82.112}$$

At these values of L, N_F decreases and $N_1 \approx N_2$ increases inversely proportionally to L with the minimum of N_{ove},

$$N_{ove} = 2N_{opt}\sqrt{\frac{2(\omega_s - \omega_p)}{\pi}} \tag{82.113}$$

taking place at

$$L_{opt} = 1\Big/\sqrt{\frac{2(\omega_s - \omega_p)}{\pi}} \tag{82.114}$$

If for $L = L_{opt}$, $\theta + \phi$ is not approximately equal to π, then L minimizing the filter complexity can be found in the near vicinity of L_{opt}. The following example illustrates the use of the above estimation formulas.

Example 1. Consider the specifications: $\omega_p = 0.4\pi$, $\omega_s = 0.402\pi$, $\delta_p = 0.01$, and $\delta_s = 0.001$. For the optimum conventional direct-form design, $N_{opt} = 2541$, requiring 1271 multipliers when the coefficient symmetry is exploited. Equation (82.114) gives $L_{opt} = 16$. Table 82.13 shows, for the admissible values of L in the vicinity of this value, l, θ, ϕ, the estimated orders for the subfilters, and the sum of the subfilter orders as well as whether the overall filter is a Case A or Case B design. For N_F, the minimum even order larger than or equal to the estimated order is used, whereas N_2 is forced to be even (odd) if N_1 is even (odd).

TABLE 82.13 Estimated Filter Orders for the Admissible Values of L in the Vicinity of $L_{opt} = 16$

L	Case	l	θ	ϕ	N_F	N_1	N_2	$N_F + N_1 + N_2$
8	B	2	0.784π	0.8π	318	98	26	442
9	B	2	0.382π	0.4π	282	38	58	378
11	A	2	0.4π	0.422π	232	47	69	348
12	A	2	0.8π	0.824π	212	162	38	412
13	B	3	0.774π	0.8π	196	155	43	394
14	B	3	0.372π	0.4π	182	58	92	332
16	A	3	0.4π	0.432π	160	70	98	328
17	A	3	0.8π	0.834π	150	236	54	440
18	B	4	0.764π	0.8π	142	210	58	410
19	B	4	0.362π	0.4π	134	78	128	340
21	A	4	0.4π	0.442π	122	92	128	342
22	A	4	0.8π	0.844π	116	314	68	498

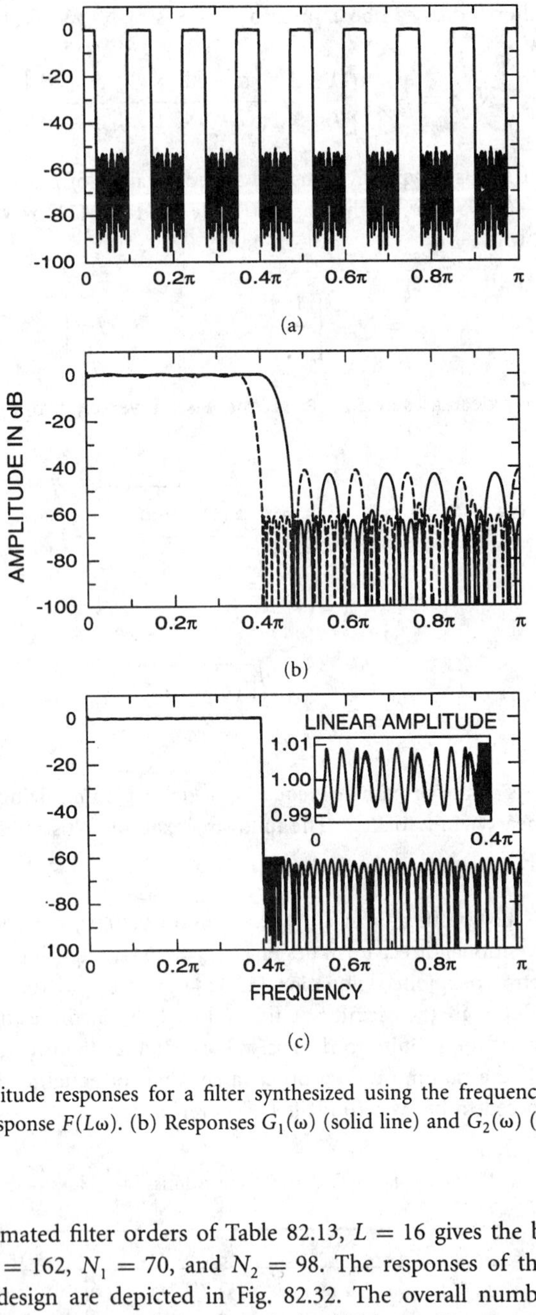

FIGURE 82.32 Amplitude responses for a filter synthesized using the frequency-response masking approach. (a) Periodic response $F(L\omega)$. (b) Responses $G_1(\omega)$ (solid line) and $G_2(\omega)$ (dashed line). (c) Overall response.

Also with the estimated filter orders of Table 82.13, $L = 16$ gives the best result. The actual filter orders are $N_F = 162$, $N_1 = 70$, and $N_2 = 98$. The responses of the subfilters as well as that of the overall design are depicted in Fig. 82.32. The overall number of multipliers and adders for this design are 168 and 330, respectively, which are 13 percent of those required by an equivalent conventional direct-form design (1271 and 2541). The overall filter order is 2690, which is only 6 percent higher than that of the direct-form design (2541).

Multistage Frequency-Response Masking Approach

If the order of $F(z)$ is too high, its complexity can be reduced by implementing it using the frequency-response masking technique. Extending this to an arbitrary number of stages results in the multistage frequency-response masking approach [3], [4], where $H(z)$ is generated iteratively as

$$H(z) \equiv F^{(0)}(z) = F^{(1)}(z^{L_1})G_1^{(1)}(z) + [z^{-L_1N_F^{(1)}/2} - F^{(1)}(z^{L_1})]G_2^{(1)}(z) \tag{82.115a}$$

$$F^{(1)}(z) = F^{(2)}(z^{L_2})G_1^{(2)}(z) + [z^{-L_2N_F^{(2)}/2} - F^{(2)}(z^{L_2})]G_2^{(2)}(z) \tag{82.115b}$$

$$\vdots \quad \vdots \quad \vdots$$

$$F^{(R-1)}(z) = F^{(R)}(z^{L_R})G_1^{(R)}(z) + [z^{-L_RN_F^{(R)}/2} - F^{(R)}(z^{L_R})]G_2^{(R)}(z) \tag{82.115c}$$

Here, the $G_1^{(r)}(z)$'s and $G_2^{(r)}(z)$'s for $r = 1, 2, \cdots, R$ as well as $F^{(R)}(z)$ are the filters to be designed. For implementation purposes, $H(z)$ can be expressed in the form shown in Table 82.14. Fig. 82.33 shows an efficient implementation for a three-stage filter, where the delay terms z^{-M_3}, z^{-m_2}, and z^{-m_1} can be shared with $F^{(3)}(z^{\hat{L}_3})$. In order to obtain a desired overall solution, the orders of the $G_1^{(r)}(z)$'s and $G_2^{(r)}(z)$'s for $r = 2, 3, \cdots, R$, denoted by $N_1^{(r)}$ and $N_2^{(r)}$ in Table 82.14, have to be even.

Given the filter specifications and the L_r's for $r = 1, 2, \cdots, R$, the $G_1^{(r)}(z)$'s and $G_2^{(r)}(z)$'s as well as $F^{(R)}(z)$ can be synthesized in the following steps:

1. Set $r = 1$, $L = L_1$, and

$$D_H(\omega) = \begin{cases} 1 & \text{for } \omega \in [0, \omega_p] \\ 0 & \text{for } \omega \in [\omega_s, \pi] \end{cases} \qquad W_H(\omega) = \begin{cases} 1/\delta_p & \text{for } \omega \in [0, \omega_p] \\ 1/\delta_s & \text{for } \omega \in [\omega_s, \pi] \end{cases} \tag{82.116}$$

2. Determine whether $F^{(r-1)}(z)$ is a Case A or Case B design as well as θ, ϕ, and l for $F^{(r)}(z)$ according to (82.107b) or (82.108b). Also, determine $\omega_p^{(G_k)}$ and $\omega_s^{(G_k)}$ for $k = 1, 2$ from (82.102a) and (82.102b) or (82.103a) and (82.103b).

TABLE 82.14 Implementation Form for the Transfer Function in the Multistage Frequency-Response Masking Approach

$$H(z) \equiv F^{(0)}(z^{\hat{L}_0}) = F^{(1)}(z^{\hat{L}_1})G_1^{(1)}(z^{\hat{L}_0}) + [z^{-M_1} - F^{(1)}(z^{\hat{L}_1})]G_2^{(1)}(z^{\hat{L}_0})$$

$$F^{(1)}(z^{\hat{L}_1}) = F^{(2)}(z^{\hat{L}_2})G_1^{(2)}(z^{\hat{L}_1}) + [z^{-M_2} - F^{(2)}(z^{\hat{L}_2})]G_2^{(2)}(z^{\hat{L}_1})$$

$$\vdots \quad \vdots \quad \vdots$$

$$F^{(R-1)}(z^{\hat{L}_{R-1}}) = F^{(R)}(z^{\hat{L}_R})G_1^{(R)}(z^{\hat{L}_{R-1}}) + [z^{-M_R} - F^{(R)}(z^{\hat{L}_R})]G_2^{(R)}(z^{\hat{L}_{R-1}}),$$

where

$$\hat{L}_0 = 1, \quad \hat{L}_r = \prod_{k=1}^{r} L_k, \quad r = 1, 2, \cdots, R$$

$$M_R = \hat{L}_R N_F^{(R)}/2, \quad M_{R-r} = M_{R-r+1} + m_{R-r}, \quad r = 1, 2, \cdots, R-1$$

$$m_{R-r} = \hat{L}_{R-r} \max\{N_1^{(R-r+1)}, N_2^{(R-r+1)}\}/2, \quad r = 1, 2, \cdots, R-1$$

$N_F^{(R)}$ is the order of $F^{(R)}(z)$.

$N_1^{(r)}$ and $N_2^{(r)}$ are the orders of $G_1^{(r)}(z)$ and $G_2^{(r)}(z)$, respectively.

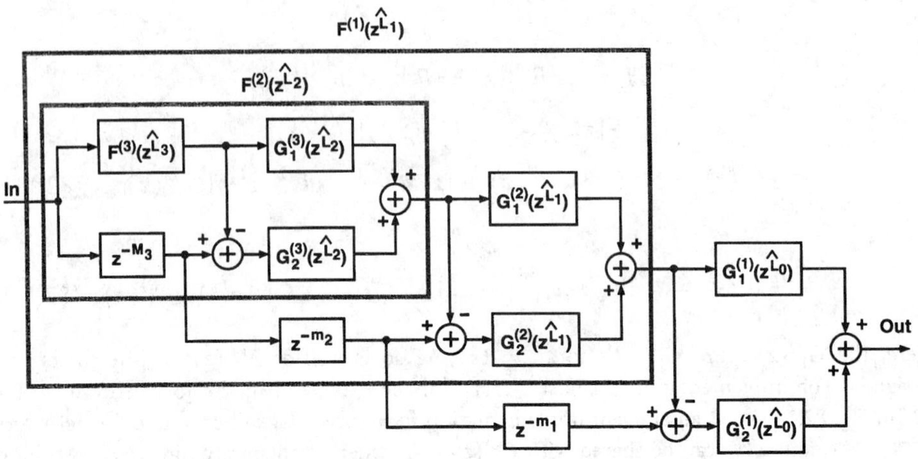

FIGURE 82.33 An implementation for a filter synthesized using the three-stage frequency-response masking approach.

3. Design $G_k^{(r)}(z)$ for $k = 1, 2$ using either the Remez algorithm or linear programming, in such a way that

$$\max_{\omega\in[0,\omega_p^{(G_k)}]\cup[\omega_s^{(G_k)},\pi]} |W_H(\omega)[G_k^{(r)}(\omega) - D_H(\omega)]| \leq 0.9 \tag{82.117}$$

4. Determine $W_F(\omega)$ and $D_F(\omega)$ from Table 82.12.
5. If $r = R$, then go to the next step. Otherwise, set $r = r + 1$, $L = L_r$, $W_H(\omega) = W_F(\omega)$, $D_H(\omega) = D_F(\omega)$, $\omega_p = \theta$, $\omega_s = \phi$, and go to Step 2.
6. Design $F^{(R)}(z)$, using either the Remez algorithm or linear programming, in such a way that

$$\max_{\omega\in[0,\theta]\cup[\phi,\pi]} |W_F(\omega)[F^{(R)}(\omega) - D_F(\omega)]| \leq 1 \tag{82.118}$$

In the above algorithm, $G_1^{(1)}(z)$ and $G_2^{(1)}(z)$ are determined like in the one-stage frequency-response masking technique. The remaining filter part as given by (82.115b) has then to be designed such that the maximum absolute value of the error function given in Table 82.12 becomes less than or equal to unity on $[0, \theta] \cup [\phi, \pi]$. Using the substitutions $\omega_p = \theta$ and $\omega_s = \phi$, the synthesis problem for $F^{(1)}(z)$ becomes the same as for the overall filter with the only exception that the desired function $D_F(\omega)$ and the weighting function $W_F(\omega)$ are not constants in the passband and stopband regions. Therefore, the following $G_1^{(r)}(z)$'s and $G_2^{(r)}(z)$'s can be designed in the same manner. Finally, $F^{(R)}(z)$ is determined at Step 6 like $F(z)$ in one-stage designs.

Given the filter specifications, the remaining problem is to select R as well as the L_r's to minimize the filter complexity. This problem has been considered in [4]. Assuming that for all the selected L_r's, $\theta + \phi \approx \pi$, the sum of the estimated orders of $F^{(R)}(z)$ and the $G_1^{(r)}(z)$'s and $G_2^{(r)}(z)$'s becomes

$$N_{\text{ove}}(R) = \left[1\bigg/\prod_{r=1}^{R} L_r + [2(\omega_s - \omega_p)/\pi]\sum_{r=1}^{R} L_r\right] N_{\text{opt}} \tag{82.119}$$

The minimum of $N_{\text{ove}}(R)$ taking place at

$$L_1 = L_2 = \cdots = L_R = L_{\text{opt}}(R) = \left[\frac{2(\omega_s - \omega_p)}{\pi}\right]^{-1/(R+1)} \tag{82.120}$$

is

$$N_{\text{ove}}(R) = (R+1)\left[\frac{(2\omega_s - \omega_p)}{\pi}\right]^{R/(R+1)} N_{\text{opt}} \tag{82.121}$$

The derivation of the above formula is based on the assumption that the orders of all the $G_1^{(r)}(z)$'s and $G_2^{(r)}(z)$'s for $r = 1, 2, \cdots, R$ are equal, which is seldom true. Therefore, in order to minimize the overall filter complexity, the values of the L_r's should be varied in the vicinity of $L_{\text{opt}}(R)$. Given ω_p, ω_s, and R, good values for the L_r's can be obtained by the following procedure:

1. Set $r = 1$.
2. Determine $L = L_{\text{opt}}(R + 1 - r)$ from (82.120).
3. For values of $\tilde{L}_r$ in the vicinity of L determine $\theta(\tilde{L}_r)$ and $\phi(\tilde{L}_r)$.
4. If $r = R$, then go to Step 7. Otherwise, go to the next step.
5. Determine $L_r = \tilde{L}_r$ minimizing

$$(R+1-r)\left[\frac{2[\phi(\tilde{L}_r) - \theta(\tilde{L}_r)]}{\pi}\right]^{(R-r)/(R+1-r)} + \frac{\phi(\tilde{L}_r) - \theta(\tilde{L}_r)}{\theta(\tilde{L}_r) + \phi(\tilde{L}_r)} + \frac{\phi(\tilde{L}_r) - \theta(\tilde{L}_r)}{2\pi - \theta(\tilde{L}_r) - \phi(\tilde{L}_r)} \tag{82.122}$$

6. Set $r = r + 1$, $\omega_p = \theta(L_r)$, $\omega_s = \phi(L_r)$, and go to Step 2.
7. Determine $L_R = \tilde{L}_R$ minimizing

$$\frac{1}{\tilde{L}_R} + \frac{\phi(\tilde{L}_R) - \theta(\tilde{L}_R)}{\theta(\tilde{L}_R) + \phi(\tilde{L}_R)} + \frac{\phi(\tilde{L}_R) - \theta(\tilde{L}_R)}{2\pi - \theta(\tilde{L}_R) - \phi(\tilde{L}_R)} \tag{82.123}$$

At the first step in this procedure, L_1 is determined to minimize the estimated overall complexity of $G_1^{(1)}(z)$, $G_2^{(1)}(z)$, and the remaining $F^{(1)}(z)$, which is given by (82.122) as a fraction of N_{opt}. Compared to (82.111) for the one-stage design, $1/\tilde{L}_r$ is replaced in (82.122) by the first term. This term estimates the complexity of $F^{(1)}(z)$ based on the use of (82.121) with $\omega_p = \theta(\tilde{L}_r)$ and $\omega_s = \phi(\tilde{L}_r)$ and the fact that it is an $R-1$ stage design. Also, L_2 is redetermined based on the same assumptions and the process is continued in the same manner. Finally, L_R is determined to minimize the sum of the estimated orders of $G_1^{(R)}(z)$, $G_2^{(R)}(z)$, and $F^{(R)}(z)$ like in the one-stage design [cf. (82.111)].

Example 2. Consider the specifications of Example 1. For a two-stage design, the above procedure gives $L_1 = L_2 = 6$. For these values, $F^{(0)}(z) \equiv H(z)$ and $F^{(1)}(z)$ are Case A designs ($l = 1$) with $\theta = 0.4\pi$ and $\phi = 0.412\pi$; and $\theta = 0.4\pi$ and $\phi = 0.472\pi$, respectively. The minimum orders of $G_1^{(1)}(z)$, $G_2^{(1)}(z)$, $G_1^{(2)}(z)$, $G_2^{(2)}(z)$, and $F^{(2)}(z)$ are 26, 40, 28, 36, and 74, respectively. Compared with the conventional direct-form FIR filter of order 2541, the number of multipliers and adders required by this design (107 and 204) are only 8 percent at the expense of a 15 percent increase in the overall filter order (to 2920). For a three-stage design,

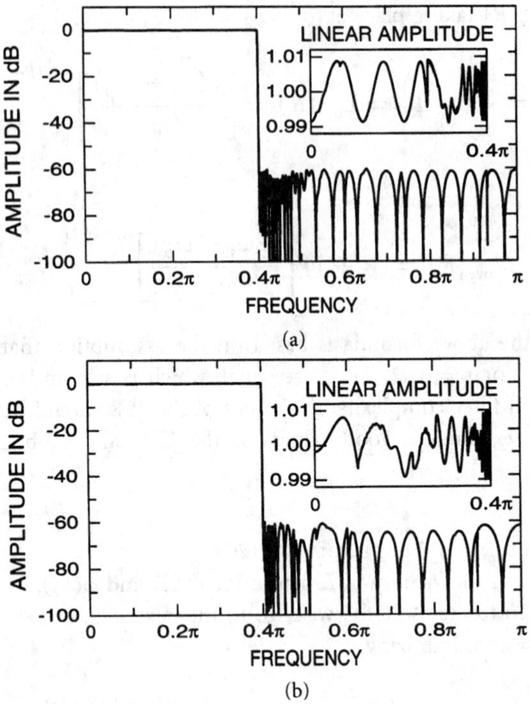

FIGURE 82.34 Amplitude responses for filters synthesized using the multistage frequency-response masking approach. (a) Two-stage filter. (b) Three-stage filter.

we get $L_1 = L_2 = L_3 = 4$. In this case, $F^{(0)}(z)$, $F^{(1)}(z)$, and $F^{(2)}(z)$ are Case B designs ($l = 1$) with $\theta = 0.392\pi$ and $\phi = 0.4\pi$; $\theta = 0.4\pi$ and $\phi = 0.432\pi$; and $\theta = 0.272\pi$ and $\phi = 0.4\pi$, respectively. The minimum orders of $G_1^{(1)}(z)$, $G_2^{(1)}(z)$, $G_1^{(2)}(z)$, $G_2^{(2)}(z)$, $G_1^{(3)}(z)$, $G_2^{(3)}(z)$, and $F^{(3)}(z)$ are 16, 28, 18, 24, 16, 32, and 40, respectively. The number of multipliers and adders (94 and 174) are only 7 percent of those required by the direct-form equivalent at the expense of a 26 percent increase in the overall filter order (to 3196). The amplitude responses of the resulting two-stage and three-stage designs are depicted in Fig. 82.34.

Design of Narrowband Filters

Another general approach for designing multiplier-efficient FIR filters has been proposed by Jing and Fam [2]. This design technique is based on iteratively using the fact that there exist efficient implementation forms for filters with $\omega_s < \pi/2$ and for filters with $\omega_p > \pi/2$. A filter with $\omega_s < \pi/2$ is called a *narrowband* filter while that with $\omega_p > \pi/2$ is called a *wideband* filter. This subsection considers the design of narrowband filters, whereas the next subsection is devoted to the design of wideband filters. Finally, these techniques are combined, resulting in the Jing–Fam approach.

When the stopband edge of $H(z)$ is less than $\pi/2$, the first transition band of $F(z^L)$ can be used as that of $H(z)$ (see Fig. 82.35), that is,

$$\omega_p = \theta/L \qquad \omega_s = \phi/L \tag{82.124}$$

In this case, the overall transfer function can be written in the following, simplified form [6] and [8]:

$$H(z) = F(z^L)G(z) \tag{82.125}$$

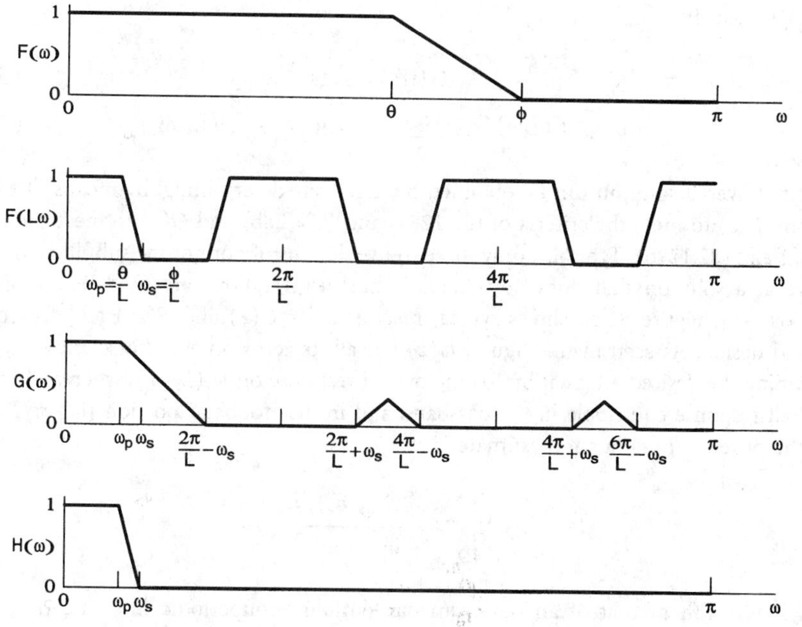

FIGURE 82.35 Synthesis of a narrowband filter as a cascade of a periodic and a nonperiodic filter.

where the orders of both $F(z)$ and $G(z)$ can be freely selected to be either even or odd. As shown in Fig. 82.35, the role of $G(z)$ is to provide the desired attenuation on those regions where $F(z^L)$ has extra unwanted passband and transition band regions, that is, on

$$\Omega_s(L, \omega_s) = \bigcup_{k=1}^{\lfloor L/2 \rfloor} \left[k\frac{2\pi}{L} - \omega_s,\ \min\left(k\frac{2\pi}{L} + \omega_s,\ \pi\right)\right] \tag{82.126}$$

There exist two ways of designing the subfilters $F(z^L)$ and $G(z)$. In the first case, they are determined, by means of the Remez algorithm, to satisfy

$$1 - \delta_p^{(F)} \le F(\omega) \le 1 + \delta_p^{(F)} \quad \text{for } \omega \in [0, L\omega_p] \tag{82.127a}$$

$$-\delta_s \le F(\omega) \le \delta_s \quad \text{for } \omega \in [L\omega_s, \pi] \tag{82.127b}$$

$$1 - \delta_p^{(G)} \le G(\omega) \le 1 + \delta_p^{(G)} \quad \text{for } \omega \in [0, \omega_p] \tag{82.127c}$$

$$-\delta_s \le G(\omega) \le \delta_s \quad \text{for } \omega \in \Omega_s(L, \omega_s) \tag{82.127d}$$

where

$$\delta_p^{(G)} + \delta_p^{(F)} = \delta_p \tag{82.127e}$$

The ripples $\delta_p^{(F)}$ and $\delta_p^{(G)}$ can be selected, e.g., to be half the overall ripple δ_p. In the above specifications, both $F(z^L)$ and $G(z)$ have $[0, \omega_p]$ as a passband region.

Another approach, leading to a considerable reduction in the order of $G(z)$, is to design simultaneously $F(\omega)$ to satisfy

$$1 - \delta_p \le F(\omega)G(\omega/L) \le 1 + \delta_p \quad \text{for } \omega \in [0, L\omega_p] \tag{82.128a}$$

$$-\delta_s \le F(\omega)G(\omega/L) \le \delta_s \quad \text{for } \omega \in [L\omega_s, \pi] \tag{82.128b}$$

and $G(\omega)$ to satisfy

$$G(0) = 1 \tag{82.129a}$$

$$-\delta_s \le F(L\omega)G(\omega) \le \delta_s \quad \text{for } \omega \in \Omega_s(L, \omega_s) \tag{82.129b}$$

The desired overall solution can be obtained by iteratively determining, by means of the Remez algorithm, $F(z)$ to meet the criteria of (82.128a) and (82.128b) and $G(z)$ to meet the criteria of (82.129a) and (82.129b). Typically, only three to five designs of both of the subfilters are required to arrive at a solution that does not change if further iterations are used. For more details, see [8] or [10]. Figure 82.36 shows typical responses for $G(z)$ and $F(z^L)$ and for the overall optimized design. As seen in this figure, $G(z)$ has all its zeros on the unit circle concentrating on providing the desired attenuation for the overall response on $\Omega_s(L, \omega_s)$, whereas $F(z^L)$ makes the overall response equiripple in the passband and in the stopband portion $[\omega_s, \pi/L]$.

For the order of $F(z)$, a good estimate is

$$N_F \approx \frac{\Phi(\delta_p, \delta_s)/L}{\omega_s - \omega_p} \tag{82.130}$$

so that it is $1/L$th of that of an optimum conventional nonperiodic filter meeting the given overall criteria. The order of $G(z)$, in turn, can be estimated accurately [10] by

$$N_G = \cosh^{-1}(1/\delta_s)\left[\frac{1}{X\left(\omega_p, \dfrac{2\pi}{L} - \dfrac{\omega_p + 2\omega_s}{3}\right)} + \frac{L/2}{X\left(\dfrac{L\omega_p}{2}, \pi - \dfrac{L(\omega_p + 2\omega_s)}{6}\right)}\right] \tag{82.131a}$$

where

$$X(\omega_1, \omega_2) = \cosh^{-1}[(2\cos\omega_1 - \cos\omega_2 + 1)/(1 + \cos\omega_2)] \tag{82.131b}$$

The minimization of the number of multipliers, $\lfloor (N_F + 2)/2 \rfloor + \lfloor (N_F + 2)/2 \rfloor$, with respect to L can be performed conveniently by evaluating the sum of the above estimated orders for admissible values of L, $2 \le L < \pi/\omega_s$. The upper limit is a consequence of the fact that the stopband edge angle of $F(z)$, $\phi = L\omega_s$, must be less than π. The following example illustrates the minimization of the filter complexity.

Example 3. The narrowband specifications are $\omega_p = 0.025\pi$, $\omega_s = 0.05\pi$, $\delta_p = 0.01$, and $\delta_s = 0.001$. Figure 82.37(a) shows the estimated N_F, N_G, and $N_F + N_G$ as functions of L, whereas Fig. 82.37(b) shows the corresponding actual minimum orders. It is seen that the estimated orders are so close to the actual ones that the minimization of the filter complexity can be accomplished based on the use of the above estimation formulas. It is also observed that $N_F + N_G$ is a unimodal function of L. With the estimates, $L = 8$ gives the best result. The estimated orders are $N_F = 25$ and $N_G = 19$, whereas the actual orders are $N_F = 26$ and $N_G = 19$. The amplitude responses for the subfilters and for the overall filter are depicted in Fig. 82.36. This design requires 24 multipliers and 45 adders. The minimum order of a conventional direct-form design is 216, requiring 109 multipliers and 216 adders. The price paid for these 80 percent reductions in the filter complexity is a 5 percent increase in the overall filter order (from 216 to 227).

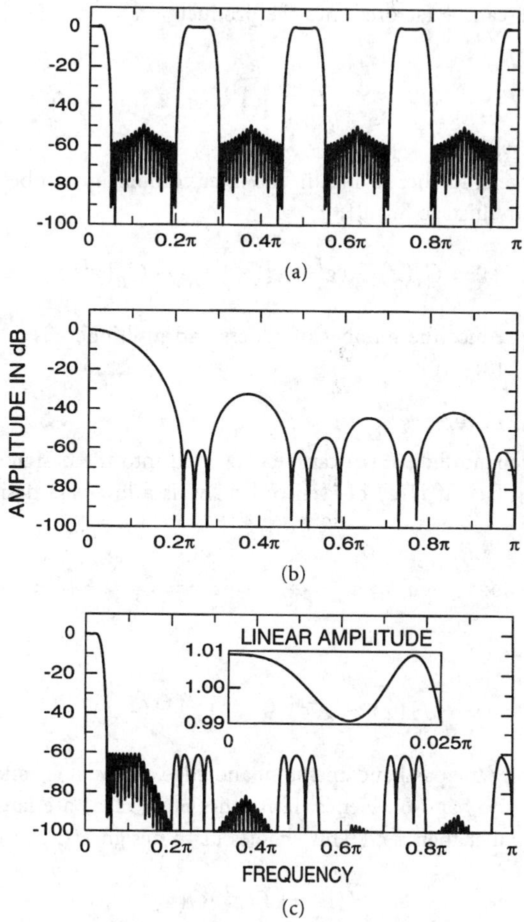

FIGURE 82.36 Typical amplitude responses for a filter of the form $H(z) = F(z^L)G(z)$. $L = 8$, $\omega_p = 0.025\pi$, $\omega_s = 0.05\pi$, $\delta_p = 0.01$, and $\delta_s = 0.001$. (a) $F(z^L)$ of order 26 in z^L. (b) $G(z)$ of order 19. (c) Overall filter.

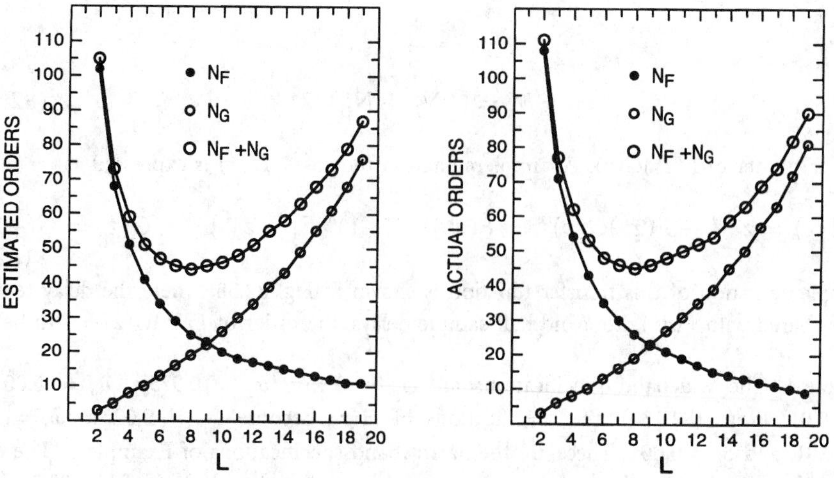

FIGURE 82.37 Estimated and actual subfilter orders as well as the sum of the subfilter orders versus L in a typical narrowband case.

In the cases where L can be factored into the product

$$L = \prod_{r=1}^{R} L_r \tag{82.132}$$

where the L_r's are integers, further savings in the filter complexity can be achieved by designing $G(z)$ in the following multistage form [8]:

$$G(z) = G_1(z)G_2(z^{L_1})G_3(z^{L_1L_2}) \cdots G_R(z^{L_1L_2 \cdots L_{R-1}}) \tag{82.133}$$

Another alternative to reduce the number of adders and multipliers is to use special structures for implementing $G(z)$ [8]–[10].

Design of Wideband Filters

The synthesis of a wideband filter $H(z)$ can be converted into the design of a narrowband filter based on the following fact. If $\hat{H}(z)$ of even order $2M$ is a low-pass design with the following edges and ripples:

$$\hat{\omega}_p = \pi - \omega_s \qquad \hat{\omega}_s = \pi - \omega_p \qquad \hat{\delta}_p = \delta_s \qquad \hat{\delta}_s = \delta_p \tag{82.134}$$

then

$$H(z) = z^{-M} - (-1)^M \hat{H}(-z) \tag{82.135}$$

is a low-pass filter having the passband and stopband edge angles at ω_p and ω_s and the passband and stopband ripples of δ_p and δ_s. Hence, if ω_p and ω_s of $H(z)$ are larger than $\pi/2$, then $\hat{\omega}_p$ and $\hat{\omega}_s$ of $\hat{H}(z)$ are smaller than $\pi/2$. This enables us to design $\hat{H}(z)$ in the form

$$\hat{H}(z) = F(z^L)G(z) \tag{82.136}$$

using the techniques of the previous subsection, yielding

$$H(z) = z^{-M} - (-1)^M F[(-z)^L]G(-z) \tag{82.137a}$$

where

$$M = (LN_F + N_G)/2 \tag{82.137b}$$

is half the order of $F(z^L)G(z)$. For implementation purposes, $H(z)$ is expressed as

$$H(z) = z^{-M} - \hat{F}(z^L)\hat{G}(z) \qquad \hat{F}(z^L) = (-1)^M F[(-z)^L] \qquad \hat{G}(z) = G(-z) \tag{82.138}$$

An implementation of this transfer function is shown in Fig. 82.38, where the delay term z^{-M} can be shared with $\hat{F}(z^L)$. To avoid half-sample delays, the order of $\hat{F}(z^L)\hat{G}(z)$ has to be even.

Example 4. The wideband specifications are $\omega_p = 0.95\pi$, $\omega_s = 0.975\pi$, $\delta_p = 0.001$, and $\delta_s = 0.01$. From (82.134), the specifications of $\hat{H}(z)$ become $\hat{\omega}_p = 0.025\pi$, $\hat{\omega}_s = 0.05\pi$, $\hat{\delta}_p = 0.01$, and $\hat{\delta}_s = 0.001$. These are the narrowband specifications of Example 3. The desired wideband design is thus obtained by using the subfilters $F(z^L)$ and $G(z)$ of Fig. 82.36 ($L = 8$, $N_F = 26$, and $N_G = 19$). However, the overall order is odd (227). A solution with even order is achieved by increasing the order of $G(z)$ by one ($N_G = 20$). Figure 82.39 shows the amplitude

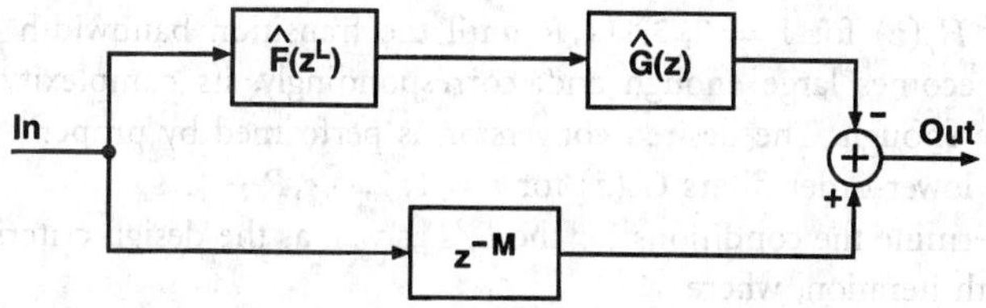

FIGURE 82.38 Implementation for a wideband filter in the form $H(z) = z^{-M} - (-1)^M F[(-z)^L]G(-z)$. $\hat{F}(z^L) = (-1)^M F[(-z)^L]$ and $\hat{G}(z) = G(-z)$.

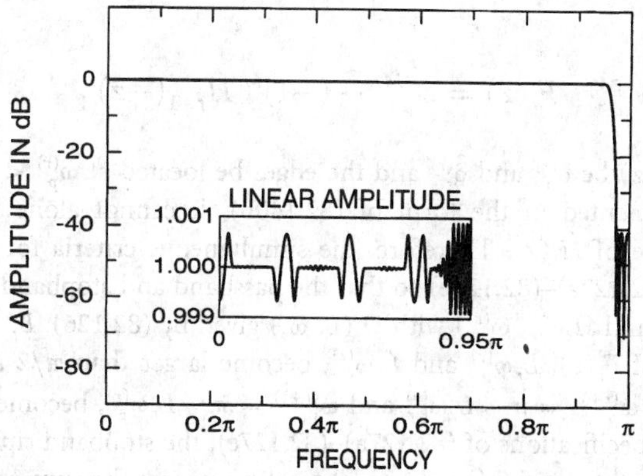

FIGURE 82.39 Amplitude response for a wideband filter implemented as shown in Fig. 82.38.

response of the resulting filter, requiring 25 multipliers, 46 adders, and 228 delay elements. The corresponding numbers for a conventional direct-form equivalent of order 216 are 109, 216, and 216, respectively.

Generalized Designs

The Jing–Fam approach [2] is based on iteratively using the facts that a narrowband filter can be implemented effectively as $H(z) = F(z^L)G(z)$ and a wideband filter in the form of (82.137a) and (82.137b). In this approach, a narrowband filter is generated [9] as

$$H(z) \equiv \hat{H}_1(z) = G_1(z)F_1(z^{L_1}) \tag{82.139a}$$

where

$$F_1(z) = z^{-M_1} - (-1)^{M_1}\hat{H}_2(-z) \qquad \hat{H}_2(z) = G_2(z)F_2(z^{L_2}) \tag{82.139b}$$

$$F_2(z) = z^{-M_2} - (-1)^{M_2}\hat{H}_3(-z) \qquad \hat{H}_3(z) = G_3(z)F_3(z^{L_3}) \tag{82.139c}$$

$$\vdots \qquad \vdots \qquad \vdots$$

$$F_{R-2}(z) = z^{-M_{R-2}} - (-1)^{M_{R-2}}\hat{H}_{R-1}(-z)$$

$$\hat{H}_{R-1}(z) = G_{R-1}(z)F_{R-1}(z^{L_{R-1}}) \tag{82.139d}$$

$$F_{R-1}(z) = z^{-M_{R-1}} - (-1)^{M_{R-1}}\hat{H}_R(-z) \qquad \hat{H}_R(z) = G_R(z) \tag{82.139e}$$

with M_r for $r = 1, 2, \ldots, R-1$ being half the order of $\hat{H}_{r+1}(z)$. Here, the basic idea is to convert iteratively the design of the narrowband overall filter into the designs of narrowband transfer functions $\hat{H}_r(z)$ for $r = 2, 3, \cdots, R$ until the transition bandwidth of the remaining $\hat{H}_R(z) = G_R(z)$ becomes large enough and, correspondingly, its complexity (the number of multipliers) is low enough. The desired conversion is performed by properly selecting the L_r's and designing the lower-order filters $G_r(z)$ for $r = 1, 2, \cdots, R-1$.

In order to determine the conditions for the L_r's as well as the design criteria for the $G_r(z)$'s, we consider the rth iteration, where

$$\hat{H}_r(z) = G_r(z)F_r(z^{L_r}) \tag{82.140a}$$

with

$$F_r(z) = z^{-M_r} - (-1)^{M_r}\hat{H}_{r+1}(-z) \tag{82.140b}$$

Let the ripples of $\hat{H}_r(z)$ be $\hat{\delta}_p^{(r)}$ and $\hat{\delta}_s^{(r)}$ and the edges be located at $\omega_p^{(r)} < \pi/2$ and $\omega_s^{(r)} < \pi/2$. Since $F_r(z)$ is implemented in the form of (82.140b), it cannot alone take care of shaping the passband response of $\hat{H}_r(z)$. Therefore, the simultaneous criteria for $G_r(z)$ and $F_r(z)$ are stated according to (82.127a)–(82.127e) so that the passband and stopband regions of $G_r(z)$ are, respectively, $[0, \omega_p^{(r)}]$ and $\Omega_s(L_r, \omega_s^{(r)})$ with $\Omega_s(L, \omega_s)$ given by (82.126). L_r has to be determined such that the edges of $F_r(z)$, $L_r\omega_p^{(r)}$ and $L_r\omega_s^{(r)}$, become larger than $\pi/2$ and, correspondingly, the edges of $\hat{H}_{r+1}(z)$, $\omega_p^{(r+1)} = \pi - L_r\omega_s^{(r)}$ and $\omega_s^{(r+1)} = \pi - L_r\omega_p^{(r)}$, become less than $\pi/2$.

In the case of the specifications of (82.127a)–(82.127e), the stopband ripple of $G_r(z)$, denoted for later use by $\delta_s^{(r)}$, and that of $F_r(z)$ are equal to $\hat{\delta}_s^{(r)}$, whereas the sum of the passband ripples is equal to $\hat{\delta}_p^{(r)}$. Denoting by $\delta_p^{(r)}$ the passband ripple selected for $G_r(z)$, the corresponding ripple of $F_r(z)$ is $\hat{\delta}_p^{(r)} - \delta_p^{(r)}$. Since $F_r(z)$ and $\hat{H}_{r+1}(z)$ interchange the ripples, the ripple requirements for $\tilde{H}_{r+1}(z)$ are $\hat{\delta}_p^{(r+1)} = \hat{\delta}_s^{(r)}$ and $\hat{\delta}_s^{(r+1)} = \hat{\delta}_p^{(r)} - \delta_p^{(r)}$.

The criteria for the $G_r(z)$'s for $r = 1, 2, \cdots, R$ can thus be stated as

$$1 - \delta_p^{(r)} \le G_r(\omega) \le 1 + \delta_p^{(r)} \quad \text{for} \quad \omega \in [0, \omega_p^{(r)}] \tag{82.141a}$$

$$-\delta_s^{(r)} \le G_r(\omega) \le \delta_s^{(r)} \quad \text{for} \quad \omega \in \Omega_s^{(r)} \tag{82.141b}$$

where

$$\Omega_s^{(r)} = \begin{cases} \bigcup_{k=1}^{\lfloor L_r/2 \rfloor} \left[k\frac{2\pi}{L_r} - \omega_s^{(r)}, \min\left(k\frac{2\pi}{L_r} + \omega_s^{(r)}, \pi\right)\right] & \text{for } r < R \\ \left[\omega_s^{(R)}, \pi\right] & \text{for } r = R \end{cases} \tag{82.141c}$$

Here, the $\omega_p^{(r)}$'s and $\omega_s^{(r)}$'s for $r = 2, 3, \cdots, R$ are determined iteratively as

$$\omega_p^{(r)} = \pi - L_{r-1}\omega_s^{(r-1)} \qquad \omega_s^{(r)} = \pi - L_{r-1}\omega_p^{(r-1)} \tag{82.141d}$$

where $\omega_p^{(1)} = \omega_p$ and $\omega_s^{(1)} = \omega_s$ are the edges of the overall design, and the $\delta_s^{(r)}$'s as

$$\delta_s^{(r)} = \begin{cases} \delta_p - \sum_{\substack{k=1 \\ k \text{ odd}}}^{r-1} \delta_p^{(k)} & \text{for } r \text{ even} \\ \delta_s - \sum_{\substack{k=2 \\ k \text{ even}}}^{r-1} \delta_p^{(k)} & \text{for } r \text{ odd} \end{cases} \tag{82.141e}$$

where δ_p and δ_s are the ripple values of the overall filter and $\delta_p^{(r)}$ is the passband ripple selected for $G_r(z)$. In order for the overall filter to meet the given ripple requirements, $\delta_s^{(R)}$ and the $\delta_p^{(r)}$'s have to satisfy for R even

$$\sum_{\substack{k=2\\k\text{ even}}}^{R} \delta_p^{(k)} = \delta_s, \qquad \delta_s^{(R)} + \sum_{\substack{k=1\\k\text{ odd}}}^{R-1} \delta_p^{(k)} = \delta_p \tag{82.142a}$$

or for R odd

$$\sum_{\substack{k=1\\k\text{ odd}}}^{R} \delta_p^{(k)} = \delta_p, \qquad \delta_s^{(R)} + \sum_{\substack{k=2\\k\text{ even}}}^{R-1} \delta_p^{(k)} = \delta_s \tag{82.142b}$$

In the above, the L_r's have to be determined such that the $\omega_s^{(r)}$'s for $r < R$ become smaller than $\pi/2$. It is also desired that for the last filter stage $G_R(z)$, $\omega_s^{(R)}$ is smaller than $\pi/2$.

If $2\pi/L_r - \omega_s^{(r)} < \pi/2$ for $r < R$ or $\omega_s^{(R)} < \pi/2$, then the arithmetic complexity of $G_r(z)$ can be reduced by designing it, using the techniques of previous subsections, in the form

$$G_r(z) = G_r^{(1)}(z^{K_r})G_r^{(2)}(z) \tag{82.143}$$

It is preferred to design the subfilters of $G_r(z)$ in such a way that the passband shaping is done entirely by $G_r^{(1)}(z^{K_r})$. The number of multipliers in the $G_r(z)$'s for $r = 1, 2, \cdots, R-1$ can be reduced by the experimentally observed fact that the overall filter still meets the given criteria when the stopband regions of these filters are decreased by using in (82.141c) the substitution

$$(2\omega_s^{(r)} + \omega_p^{(r)})/3 \mapsto \omega_s^{(r)} \tag{82.144}$$

After some manipulations, $H(z)$ as given by (82.139a)–(82.139e) and (82.143) can be rewritten in the explicit form shown in Table 82.15. If $G_r(z)$ is a single-stage design, then $G_r^{(1)}(z^{K_r}) \equiv 1$. In order to obtain the desired overall solution, the overall order of $G_r(z)$ for $r \geq 2$, denoted by N_r in Table 82.15, has to be even. Realizations for the overall transfer function are given in Fig. 82.40, where

$$m_r = \hat{M}_r - \hat{M}_{r+1} = \tfrac{1}{2}\hat{L}_r N_r \qquad r = 2, 3, \cdots, R-1 \qquad m_R = \hat{M}_R \tag{82.145}$$

The structure of Fig. 82.40(b) is preferred since the delay terms z^{-m_r} can be shared with $H_R^{(1)}(z^{K_R\hat{L}_R})$ or, if this filter stage is not present, with $H_R^{(2)}(z^{L_R})$. This is because the overall order of this filter stage is usually larger than the sum of the m_r's.

If the edges ω_p and ω_s of the overall filter are larger than $\pi/2$, then we set $H(z) \equiv F_1(z)$. In this case, $\delta_p^{(1)} \equiv 0$, $L_1 \equiv 1$, and $G_1(z)$, $\omega_p^{(1)}$, and $\omega_s^{(1)}$ are absent. Furthermore, $\omega_p^{(2)} = \pi - \omega_s$ and $\omega_s^{(2)} = \pi - \omega_p$, and $H_1(z)$ is absent in Fig. 82.40 and in Table 82.15.

The remaining problem is to select R, the L_r's, the K_r's, and the ripple values such that the filter complexity is minimized. The following example illustrates this.

Example 5. Consider the specifications of Example 1, that is, $\omega_p = 0.4\pi$, $\omega_s = 0.402\pi$, $\delta_p = 0.01$, and $\delta_s = 0.001$. In this case, the only alternative is to select $L_1 = 2$. The resulting passband and stopband regions for $G_1(z)$ are (the substitution of (82.144) is used)

$$\Omega_p^{(1)} = [0, 0.4\pi] \qquad \Omega_s^{(1)} = [0.5987\pi, \pi]$$

TABLE 82.15 Explicit Form for the Transfer Function in the Jing–Fam Approach

$$H(z) = H_1(z^{\hat{L}_1})\{I_2 z^{-\hat{M}_2} + H_2(z^{\hat{L}_2})[I_3 z^{-\hat{M}_3} + H_3(z^{\hat{L}_3})(\cdots$$

$$\{I_{R-1} z^{-\hat{M}_{R-1}} + H_{R-1}(z^{\hat{L}_{R-1}})[I_R z^{-\hat{M}_R} + H_R(z^{\hat{L}_R})]\}\cdots)]\},$$

where

$$H_r(z^{\hat{L}_r}) = H_r^{(1)}(z^{K_r \hat{L}_r}) H_r^{(2)}(z^{\hat{L}_r})$$

$$H_r^{(1)}(z) = G_r^{(1)}(J_r^{(1)} z), \quad H_r^{(2)}(z) = S_r G_r^{(2)}(J_r^{(2)} z)$$

$$S_1 = 1, \quad S_r = -(-1)^{\hat{M}_r/\hat{L}_r}, \quad r = 2, 3, \cdots, R$$

$$J_1^{(2)} = 1, \quad J_2^{(2)} = -1, \quad J_r^{(2)} = -[J_{r-1}^{(2)}]^{L_{r-1}}, \quad r = 3, 4, \cdots, R$$

$$J_r^{(1)} = [J_r^{(2)}]^{K_r}$$

$$\hat{L}_1 = 1, \quad \hat{L}_r = \prod_{k=1}^{r-1} L_k, \quad r = 2, 3, \cdots, R$$

$$\hat{M}_R = \tfrac{1}{2}\hat{L}_R N_R, \quad \hat{M}_{R-r} = \hat{M}_{R-r+1} + \tfrac{1}{2}\hat{L}_{R-r} N_{R-r}, \quad r = 1, 2, \cdots, R-2$$

$$I_2 = 1, \quad I_r = [J_{r-1}^{(2)}]^{\hat{M}_r/\hat{L}_{r-1}}, \quad r = 3, 4, \ldots, R$$

$$N_r = K_r N_r^{(1)} + N_r^{(2)}$$

$N_r^{(1)}$ and $N_r^{(2)}$ are the orders of $G_r^{(1)}(z)$ and $G_r^{(2)}(z)$, respectively.

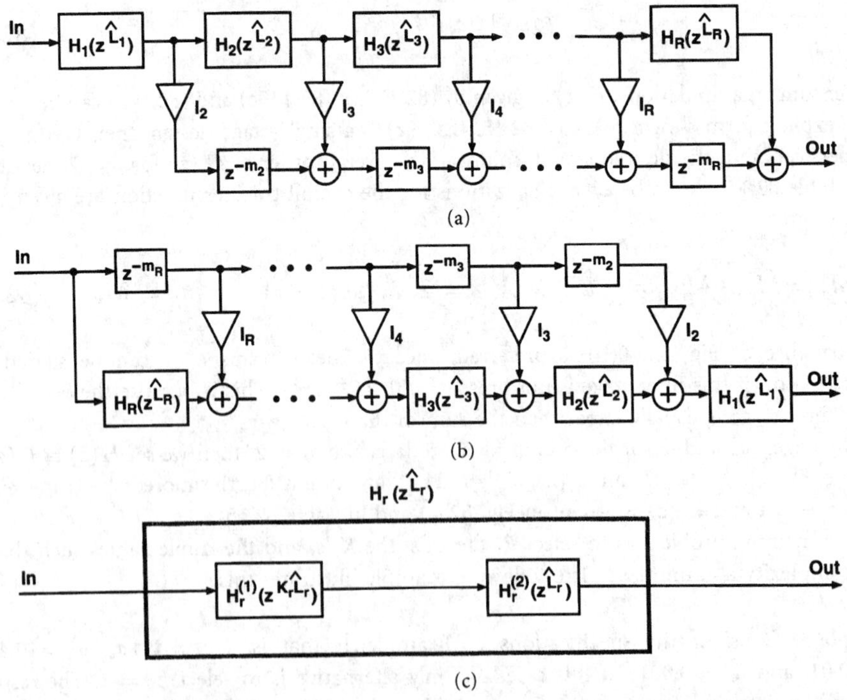

FIGURE 82.40 Implementations for a filter synthesized using the Jing–Fam approach. (a) Basic structure. (b) Transposed structure. (c) Structure for the subfilter $H_r(z^{\hat{L}_r})$.

For $\hat{H}_2(z)$, the edges become $\omega_p^{(2)} = \pi - L_1\omega_s = 0.196\pi$ and $\omega_s^{(2)} = \pi - L_1\omega_p = 0.2\pi$. For L_2, there are two alternatives to make the edges of $\hat{H}_3(z)$, $\omega_p^{(3)} = \pi - L_2\omega_s^{(2)}$ and $\omega_s^{(3)} = \pi - L_2\omega_p^{(2)}$, less than $\pi/2$. These are $L_2 = 3$ and $L_2 = 4$. For $R = 5$ stages, there are the following four alternatives to make all the $\omega_s^{(r)}$'s smaller than $\pi/2$:

$$\begin{array}{llll} L_1 = 2 & L_2 = 4 & L_3 = 3 & L_4 = 2 \\ L_1 = 2 & L_2 = 4 & L_3 = 4 & L_4 = 4 \\ L_1 = 2 & L_2 = 3 & L_3 = 2 & L_4 = 4 \\ L_1 = 2 & L_2 = 3 & L_3 = 2 & L_4 = 3 \end{array}$$

Among these alternatives, the first one results in an overall filter with minimum complexity. In this case, the edges of $\hat{H}_3(z)$, $\hat{H}_4(z)$, and $\hat{H}_5(z) \equiv G_5(z)$ become as shown in Table 82.16. The corresponding passband and stopband regions for $G_2(z)$, $G_3(z)$, $G_4(z)$, and $G_5(z)$ are

$$\Omega_p^{(2)} = [0, 0.196\pi] \qquad \Omega_s^{(2)} = [0.3013\pi, 0.6987\pi] \cup [0.8013\pi, \pi]$$

$$\Omega_p^{(3)} = [0, 0.2\pi] \qquad \Omega_s^{(3)} = [0.4560\pi, 0.8773\pi]$$

$$\Omega_p^{(4)} = [0, 0.352\pi] \qquad \Omega_s^{(4)} = [0.616\pi, \pi]$$

$$\Omega_p^{(5)} = [0, 0.2\pi] \qquad \Omega_s^{(5)} = [0.296\pi, \pi]$$

What remains is to determine the ripple requirements. From (82.142b), it follows for $R = 5$, $\delta_p^{(1)} + \delta_p^{(3)} + \delta_p^{(5)} = \delta_p$ and $\delta_p^{(2)} + \delta_p^{(4)} + \delta_s^{(5)} = \delta_s$. By simply selecting the ripple values in these summations to be equal, the required ripples for the $G_r(z)$'s become as shown in Table 82.16.

TABLE 82.16 Data for a Filter Designed Using the Jing–Fam Approach

	$r = 1$	$r = 2$	$r = 3$	$r = 4$	$r = 5$
$\omega_p^{(r)}$	0.4π	0.196π	0.2π	0.352π	0.2π
$\omega_s^{(r)}$	0.402π	0.2π	0.216π	0.4π	0.296π
$\delta_p^{(r)}$	$\frac{1}{3}\times 10^{-2}$	$\frac{1}{3}\times 10^{-3}$	$\frac{1}{3}\times 10^{-2}$	$\frac{1}{3}\times 10^{-3}$	$\frac{1}{3}\times 10^{-2}$
$\delta_s^{(r)}$	10^{-3}	$\frac{2}{3}\times 10^{-2}$	$\frac{2}{3}\times 10^{-3}$	$\frac{1}{3}\times 10^{-2}$	$\frac{1}{3}\times 10^{-3}$
L_r	2	4	3	2	—
K_r	—	3	2	—	3
$N_r^{(1)}$	—	20	11	—	22
$N_r^{(2)}$	31	10	8	26	14
N_r	31	70	30	26	80
$\hat{L}_r$	1	2	8	24	48
$J_r^{(1)}$	—	−1	1	—	−1
$J_r^{(2)}$	1	−1	−1	1	−1
$\hat{M}_r$	—	2422	2352	2232	1920
I_r	—	1	1	−1	1
S_r	1	1	−1	1	−1
m_r	—	70	120	312	1920

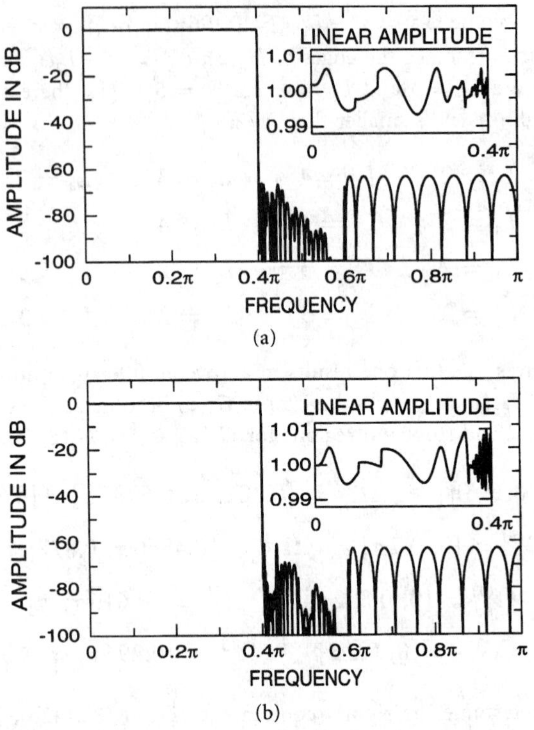

FIGURE 82.41 Amplitude responses for filters synthesized using the Jing–Fam approach.

The first and fourth subfilters are single-stage filters since their stopband edges are larger than $\pi/2$, whereas the remaining three filters are two-stage designs. The parameters describing the overall filter are shown in Table 82.16, whereas Fig. 82.41(a) depicts the response of this filter. The number of multipliers and the order of this design are 78 and 4875, whereas the corresponding numbers for the direct-form equivalent are 1271 and 2541. The number of multipliers required by the proposed design is thus only 6 percent of that of the direct-form filter. Since the complexity of $H_5(z^{\hat{L}_5})$ is similar to those of the earlier filter stages, $R = 5$ is a good selection in this example.

The overall filter order as well as the number of multipliers can be decreased by selecting smaller ripple values for the first stages, thereby allowing larger ripples for the last stages. Proper selections for the ripple requirements and filter orders are shown in Table 82.17. The first four filters have been optimized such that their passband variations are minimized. The first criteria are met by a half-band filter of order 34, having the passband and stopband edges at 0.4013π and 0.5987π. Since every second impulse response coefficient of this filter is zero-valued except for the central coefficient with an easily implementable value of 1/2, this filter requires only 9 multipliers. For the last stage, K_5 is reduced to 2 to decrease the overall filter order. The order of the resulting overall filter [see Fig. 82.41(b)] is 3914, which is 54 percent higher than that of the direct-form equivalent. The number of multipliers is reduced to 70.

The Jing–Fam approach cannot be applied directly for synthesizing filters whose edges are very close to $\pi/2$. This problem can, however, be overcome by slightly changing the sampling rate or, if this is not possible, by shifting the edges by a factor of 3/2 by using decimation by this factor at the filter input and interpolation by the same factor at the filter output [2]. One attractive feature of the Jing–Fam approach is that it can be combined with multirate filtering to reduce the filter complexity even further [7].

TABLE 82.17 Data for Another Filter Designed Using the Jing–Fam Approach

$\delta_p^{(r)}$	7.3×10^{-4}	7.1×10^{-5}	3.5×10^{-4}	12.1×10^{-5}	89.2×10^{-4}
$\delta_s^{(r)}$	10^{-3}	92.7×10^{-4}	92.9×10^{-5}	89.2×10^{-4}	80.8×10^{-5}
K_r	—	3	2	—	2
$N_r^{(1)}$	—	22	13	—	27
$N_r^{(2)}$	34	10	8	24	6

When comparing the above designs with the filters synthesized using the multistage frequency-response masking technique (Example 2), it is observed that the above designs require slightly fewer multipliers at the expense of an increased overall filter order. Both of these general approaches are applicable to those specifications that are not very narrowband or very wideband. For most very narrowband and wideband cases, filters synthesized in the simplified forms $H(z) = F(z^L)G(z)$ and $H(z) = z^{-M} - (-1)^M F[(-z)^L]G(-z)$, respectively, give the best results (see Examples 3 and 4).

References

[1] O. Herrmann, L. R. Rabiner, and D. S. K. Chan, "Practical design rules for optimum finite impulse response lowpass digital filters," *Bell System Tech. J.*, vol. 52, pp. 769–799, July–Aug. 1973.

[2] Z. Jing and A. T. Fam, "A new structure for narrow transition band, lowpass digital filter design," *IEEE Trans. Acoust., Speech, Signal Processing*, vol. ASSP-32, pp. 362–370, Apr. 1984.

[3] Y. C. Lim, "Frequency-response masking approach for the synthesis of sharp linear phase digital filters," *IEEE Trans. Circuits Syst.*, vol. CAS-33, pp. 357–364, Apr. 1986.

[4] Y. C. Lim and Y. Lian "The optimum design of one- and two-dimensional FIR filters using the frequency response masking technique," *IEEE Trans. Circuits Syst.—II: Analog and Digital Signal Processing*, vol. 40, pp. 88–95, Feb. 1993.

[5] J. H. McClellan, T. W. Parks, and L. R. Rabiner, "A computer program for designing optimum FIR linear phase digital filters," *IEEE Trans. Audio Electroacoust.*, vol. AU-21, pp. 506–526, Dec. 1973; also reprinted in *Selected Papers in Digital Signal Processing, II*, IEEE Digital Signal Processing Comm. and IEEE-ASSP, Eds., New York: IEEE Press, 1976, pp. 97–117.

[6] Y. Neuvo, C.-Y. Dong, and S. K. Mitra, "Interpolated finite impulse response filters," *IEEE Trans. Acoust., Speech, Signal Processing*, vol. ASSP-32, pp. 563–570, June 1984.

[7] T. Ramstad and T. Saramäki, "Multistage, multirate FIR filter structures for narrow transition-band filters," in *Proc. 1990 IEEE Int. Symp. Circuits Syst.*, New Orleans, LA, 1990, pp. 2017–2021.

[8] T. Saramäki, Y. Neuvo, and S. K. Mitra, "Design of computationally efficient interpolated FIR filters," *IEEE Trans. Circuits Syst.*, vol. CAS-35, pp. 70–88, Jan. 1988.

[9] T. Saramäki and A. T. Fam, "Subfilter approach for designing efficient FIR filters," in *Proc. 1988 IEEE Int. Symp. Circuits Syst.*, Espoo, Finland, pp. 2903–2915, 1988.

[10] T. Saramäki, "Finite impulse response filter design," in *Handbook for Digital Signal Processing*, S. K. Mitra and J. F. Kaiser, Eds., New York: Wiley, 1993, ch. 4, pp. 155–277.

83

IIR Filters

Sawasd Tantaratana
University of Massachusetts at Amherst

Stuart S. Lawson
University of Warwick, United Kingdom

Y. C. Lim
National University of Singapore

83.1 Properties of IIR Filters

Sawasd Tantaratana

System Function and Impulse Response

A digital filter with impulse response having infinite length (i.e., its values outside a finite interval cannot all be zero) is termed infinite impulse response (IIR) filter. The most important class of IIR filters can be described by the difference equation

$$\begin{aligned} y(n) = {} & b_0x(n) + b_1x(n-1) + \cdots + b_Mx(n-M) \\ & - a_1y(n-1) - a_2y(n-2) - \cdots - a_Ny(n-N) \end{aligned} \tag{83.1}$$

where $x(n)$ is the input, $y(n)$ is the output of the filter, $\{a_1, a_2, \cdots, a_N\}$ and $\{b_0, b_1, \cdots, b_M\}$ are constant coefficients. We assume that $a_N \neq 0$. The impulse response is the output of the system when it is driven by a unit impulse at $n = 0$, with the system being initially at rest, i.e., the output being zero prior to applying the input. We denote the impulse response by $h(n)$. With $x(0) = 1$, $x(n) = 0$ for $n \neq 0$, and $y(n) = 0$ for $n < 0$, we can compute $h(n)$, $n \geq 0$, from (83.1) in a recursive manner. Taking the z-transform of (83.1), we obtain the system function

$$H(z) = \frac{Y(z)}{X(z)} = \frac{b_0 + b_1z^{-1} + \cdots + b_Mz^{-M}}{1 + a_1z^{-1} + a_2z^{-2} + \cdots + a_Nz^{-N}} \tag{83.2}$$

0-8493-8341-2/95/$0.00+$.50

N is the order of the filter. The system function and the impulse response are related through the z-transform and its inverse, i.e.,

$$H(z) = \sum_{n=-\infty}^{\infty} h(n)z^{-n} \qquad h(n) = \frac{1}{2\pi j}\oint_C H(z)z^{n-1}\,dz \tag{83.3}$$

where C is a closed counterclockwise contour in the region of convergence. See Chapter 3 for z-transform.

We assume that $M \leq N$. Otherwise, the system function can be written as

$$\begin{aligned} H(z) &= [c_0 + c_1 z^{-1} + \cdots + c_{M-N} z^{-(M-N)}] \\ &+ \left[\frac{b_0' + b_1' z^{-1} + \cdots + b_N' z^{-N}}{1 + a_1 z^{-1} + a_2 z^{-2} + \cdots + a_N z^{-N}}\right] \end{aligned} \tag{83.4}$$

which is an FIR filter in parallel with an IIR filter, or as

$$H(z) = [c_0' + c_1' z^{-1} + \cdots + c_{M-N}' z^{-(M-N)}]\left[\frac{b_0'' + b_1'' z^{-1} + \cdots + b_N'' z^{-N}}{1 + a_1 z^{-1} + a_2 z^{-2} + \cdots + a_N z^{-N}}\right] \tag{83.5}$$

which is an FIR filter in cascade with an IIR filter. FIR filters are covered in Chapter 82. This chapter (Chapter 83) covers IIR filters.

For ease of implementation, it is desirable that the coefficients $\{a_1, a_2, \cdots, a_N\}$ and $\{b_0, b_1, \cdots, b_M\}$ be real numbers (as opposed to complex numbers), which is another assumption that we make, unless it is specified otherwise.

Causality (Physical Realizability)

A causal (physically realizable) filter is one whose output value does not depend on the future input values. A noncausal filter cannot be realized in real time since some future input values are needed in computing the current output value. The difference equation in (83.1) can represent a causal system or a noncausal system. If the output $y(n)$ is calculated, for an increasing value of n, from $x(n), x(n-1), \cdots, x(n-M), y(n-1), \cdots, y(n-N)$, according to the right-hand side of (83.1), then the difference equation represents a causal system. On the other hand, we can rewrite (83.1) as

$$\begin{aligned} y(n-N) = \frac{1}{a_N}[&b_0 x(n) + b_1 x(n-1) + \cdots + b_M x(n-M) \\ &- y(n) - a_1 y(n-1) - \cdots - a_{N-1} y(n-N+1)] \end{aligned} \tag{83.6}$$

If the system calculates $y(n-N)$, for a decreasing value of n, using the right-hand side of (83.6), then the system is noncausal since $y(n-N)$ depends on $x(n), \cdots, x(n-M)$, which are future input values. We shall assume that the IIR filter represented by (83.1) is causal. It follows from this assumption the $h(n) = 0$ for $n < 0$ and that the convergence region of $H(z)$ is of the form: $|z| > r_0$, where r_0 is a nonnegative constant.

Noncausal filters are useful in practical applications where the output need not be calculated in real time or where the variable n does not represent time, such as in image processing where n is a spatial variable. Generally, a noncausal filter can be modified to be causal by adding sufficient delay at the output.

Poles and Zeros

Rewriting (83.2) we have

$$H(z) = z^{N-M} \frac{b_0 z^M + b_1 z^{M-1} + \cdots + b_{M-1} z + b_M}{z^N + a_1 z^{N-1} + \cdots + a_{N-1} z + a_N} \tag{83.7}$$

Assuming b_0, $b_M \neq 0$, then there are N poles given by the roots of the denominator polynomial and M zeros given by the roots of the numerator polynomial. In addition, there are $N - M$ zeros at the origin on the complex plane. The locations of the poles and zeros can be plotted on the complex z plane. Denoting the poles by $p_1, p_2, \cdots, p_N$ and the nonzero zeros by $q_1, q_2, \cdots, q_M$, we can write

$$H(z) = b_0 z^{N-M} \frac{(z - q_1)(z - q_2) \cdots (z - q_M)}{(z - p_1)(z - p_2) \cdots (z - p_N)} \tag{83.8}$$

Since we assume that the coefficients $\{a_1, a_2, \cdots, a_N\}$ and $\{b_0, b_1, \cdots, b_M\}$ are real-valued, for each complex-valued pole (i.e., pole off the real axis on the z plane) there must be another pole that is the complex conjugate of the first. Similarly, complex-valued zeros must exist in complex-conjugate pairs. The combination of a complex-conjugate pole pair (or zero pair) yields a second-order polynomial with real coefficients. Real-valued pole (or zero) can appear single in (83.8).

It is clear from (83.8) that knowing all the pole and zero locations, we can write the system function to within a constant factor. Since the constant factor is only a gain, which can be adjusted as desired, specifying the locations of the poles and zeros essentially specifies the system function of the IIR filter.

Stability

A causal IIR filter is stable (in the sense that a bounded input gives rise to a bounded output) if all the poles lie inside the unit circle. If there are one or more simple poles on the unit circle (and all the others lie inside the unit circle), then the filter is marginally stable, giving a sustained oscillation. If there are multiple poles (more than one pole at the same location) on the unit circle or if there is at least one pole outside the unit circle, a slight input will give rise to an output with increasing magnitude. For most practical filters, all the poles are designed to lie inside the unit circle. In some special IIR systems (such as oscillators), poles are placed on the unit circle to obtain the desired result.

Given the system function in the form of (83.2) or (83.7), the stability can be verified by finding all the poles of the filters and checking to see if all of them are inside the unit circle. Equivalently, stability can be verified directly from the coefficients $\{a_i\}$, using the Schür–Cohn algorithm [4]. For a second-order system, if the coefficients a_1 and a_2 lie inside the triangle in Fig. 83.1, then the system is stable.

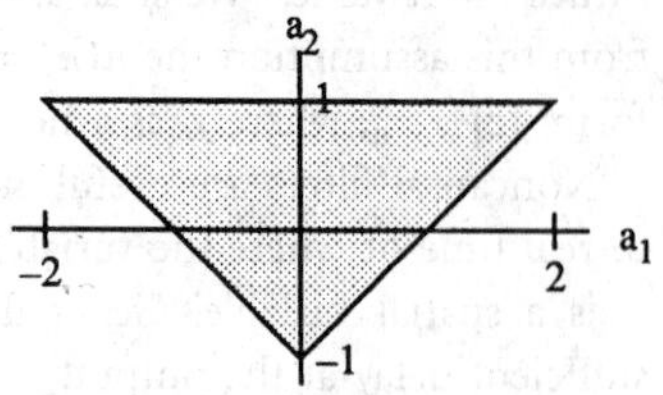

FIGURE 83.1 Region for the coefficients a_1 and a_2 that yield a stable second-order IIR filter.

Frequency Response

The frequency response of the IIR filter is the value of the system function evaluated on the unit circle on the complex plane, i.e., with $z = e^{j2\pi f}$, where f varies from 0 to 1, or from $-1/2$ to 1/2. The variable f represents the digital frequency. For simplicity, we write $H(f)$ for $H(z)|_{z=\exp(j2\pi f)}$. Therefore,

$$H(f) = b_0 e^{j2\pi(N-M)f} \frac{(e^{j2\pi f} - q_1)(e^{j2\pi f} - q_2)\cdots(e^{j2\pi f} - q_M)}{(e^{j2\pi f} - p_1)(e^{j2\pi f} - p_2)\cdots(e^{j2\pi f} - p_N)} \tag{83.9}$$

$H(f)$ is generally a complex function of f, consisting of the real part $H_R(f)$ and the imaginary part $H_I(f)$. It can also be expressed in terms of the magnitude $|H(f)|$ and the phase $\theta(f)$

$$H(f) = H_R(f) + jH_I(f) = |H(f)|e^{j\theta(f)} \tag{83.10}$$

From (83.9) we see that the magnitude response $|H(f)|$ equals the product of the magnitudes of the individual factors in the numerator, divided by the product of the magnitudes of the individual factors in the denominator. The magnitude square can be written as

$$|H(f)|^2 = H(f)H^*(f) = [H_R(f)]^2 + [H_I(f)]^2 \tag{83.11}$$

Since $H^*(f) = H^*(1/z^*)|_{z=\exp(j2\pi f)}$ and $H^*(1/z^*) = H(z^{-1})$ when all the coefficients of $H(z)$ are real, we have

$$|H(f)|^2 = H(z)H(z^{-1})\Big|_{z=\exp(j2\pi f)} \tag{83.12}$$

Using (83.12), the magnitude square can be put in the form

$$|H(f)|^2 = \frac{\sum_{k=0}^{M} \tilde{b}_k \cos(2\pi k f)}{\sum_{k=0}^{N} \tilde{a}_k \cos(2\pi k f)} \tag{83.13}$$

where the coefficients are given by

$$\begin{aligned} \tilde{b}_0 &= \sum_{j=0}^{M} b_j^2 \quad \tilde{b}_k = 2\sum_{j=k}^{M} b_j b_{j-k} \quad k = 1, \cdots, M \\ \tilde{a}_0 &= \sum_{j=0}^{N} a_j^2 \quad \tilde{a}_k = 2\sum_{j=k}^{N} a_j a_{j-k} \quad k = 1, \cdots, N \end{aligned} \tag{83.14}$$

with the understanding that $a_0 = 1$. Given $\{\tilde{b}_0, \tilde{b}_1, \cdots, \tilde{b}_M\}$ we can find $\{b_0, b_1, \cdots, b_M\}$ and vice versa. Similarly, $\{\tilde{a}_0, \tilde{a}_1, \cdots, \tilde{a}_N\}$ and $\{a_1, a_2, \cdots, a_N\}$ can be computed from each other. The form in (83.13) is useful in computer-aided design of IIR filters using linear programming [1].

We see from (83.9) that the phase response $\theta(f)$ equals the sum of the phases of the individual factors in the numerator, minus the sum of the phases of the individual factors in the denominator. The phase can be written in terms of the real and imaginary parts of $H(f)$ as

$$\theta(f) = \arctan\left[\frac{H_I(f)}{H_R(f)}\right] \tag{83.15}$$

A filter having linear phase in a frequency band (e.g., in the passband) means that there is no phase distortion in that band.

The group delay is defined as

$$\tau(f) = -\frac{1}{2\pi}\frac{d}{df}\theta(f) \tag{83.16}$$

The group delay corresponds to the delay, from the input to the output, of the envelope of a narrowband signal [3]. A linear phase gives rise to a constant group delay. Nonlinearity in the phase appears as deviation of the group delay from a constant value.

The magnitude response of an IIR filter does not change, except for a constant factor, if a zero is replaced by the reciprocal of its complex conjugate, i.e., if $(z - q)$ is replaced with $(z - 1/q^*)$. This can be seen as follows. Letting $\tilde{H}(z)$ be the system function without the factor $(z - q)$, we have

$$H(z) = \tilde{H}(z)(z-q) = \tilde{H}(z)(z-1/q^*)\frac{(z-q)}{(z-1/q^*)} = \hat{H}(z)\frac{(z-q)}{(z-1/q^*)} \tag{83.17}$$

where $\hat{H}(z)$ is $H(z)$ with the zero at q being replaced with a zero at $1/q^*$. It follows from (83.17) that

$$\begin{aligned}|H(f)|^2 &= H(z)H^*(1/z^*)\big|_{z=\exp(j2\pi f)}\\ &= \hat{H}(z)\hat{H}^*(1/z^*)\frac{(z-q)(z^{-1}-q^*)}{(z-1/q^*)(z^{-1}-1/q)}\bigg|_{z=\exp(j2\pi f)}\\ &= |q|^2|\hat{H}(f)|^2\end{aligned}$$

Similarly, replacing the pole at p with a pole at $1/p^*$ will not alter the magnitude of the response except for a constant factor. This property is useful in changing an unstable IIR filter to a stable one without altering the magnitude response.

Compared to an FIR filter, an IIR filter requires a much lower order than a FIR filter to achieve the same requirement of the magnitude response. However, the phase of a stable causal IIR filter cannot be made linear. This is the major reason not to use an IIR filter in applications where linear phase is essential. Nevertheless, using phase compensation such as allpass filters (see the subsection on allpass filters), the phase of an IIR filter can be adjusted close to linear. This process increases the order of the overall system, however. Note that if causality is not required, then a linear-phase IIR filter can be obtained using a time-reversal filter [1].

Realizations

A realization of an IIR filter according to (83.1) is shown in Fig. 83.2(a), which is called Direct Form I. By rearranging the structure, we can obtain Direct Form II, as shown in Fig. 83.2(b). Through transposition, we can obtain Transposed Direct Form I and Transposed Direct Form II as shown in Fig. 83.2(c) and (d).

The system function can be put in the form

$$H(z) = \prod_{i=1}^{K}\frac{b_{i0} + b_{i1}z^{-1} + b_{i2}z^{-2}}{1 + a_{i1}z^{-1} + a_{i2}z^{-2}} \tag{83.18}$$

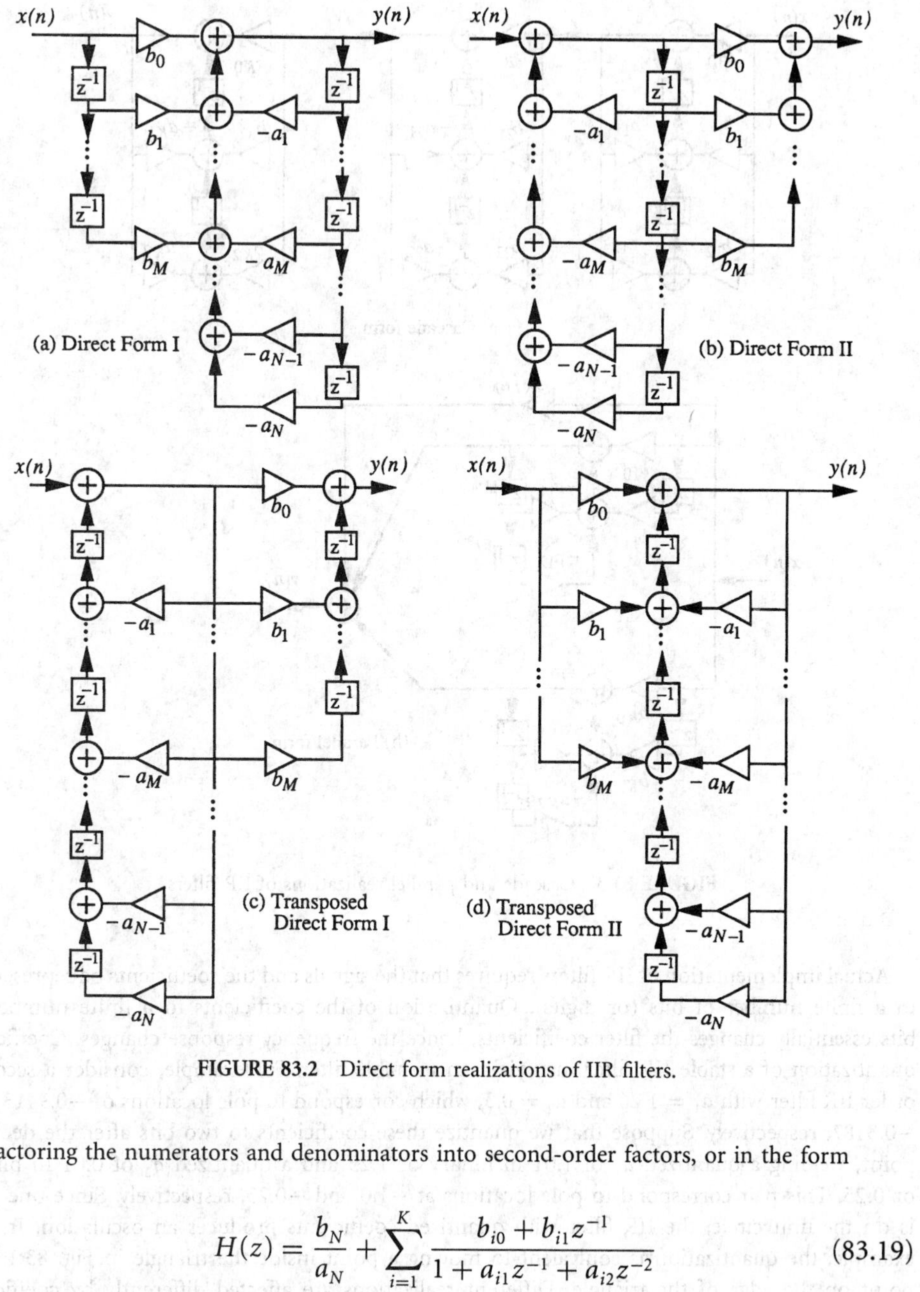

FIGURE 83.2 Direct form realizations of IIR filters.

by factoring the numerators and denominators into second-order factors, or in the form

$$H(z) = \frac{b_N}{a_N} + \sum_{i=1}^{K} \frac{b_{i0} + b_{i1}z^{-1}}{1 + a_{i1}z^{-1} + a_{i2}z^{-2}} \tag{83.19}$$

by partial fraction expansion. The value of K is $N/2$ when N is even and it is $(N+1)/2$ when N is odd. When N is odd, one of a_{i2} must be zero, as well as one of b_{i2} in (83.18) and one of b_{i1} in (83.19). All the coefficients in (83.18) and (83.19) are real numbers. According to (83.18), the IIR filter can be realized by K second-order IIR filters in cascade, as shown in Fig. 83.3(a). According to (83.19), the IIR filter realized by K second-order IIR filters and one scaler (i.e., b_N/a_N) in parallel, as depicted in Fig. 83.3(b). Each second-order subsystem can use any of the structures given in Fig. 83.2.

There are many other realizations for IIR filters, such as state-space structures [2], wave structures (Section 83.3), and lattice structures (Section 83.4).

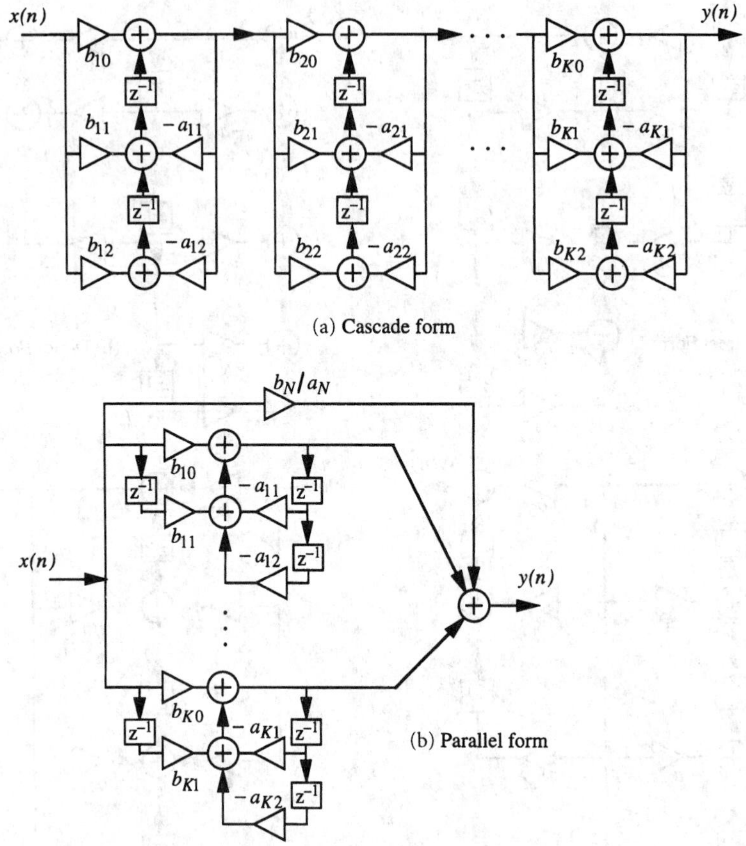

FIGURE 83.3 Cascade and parallel realizations of IIR filters.

Actual implementation of IIR filters requires that the signals and the coefficients be represented in a finite number of bits (or digits). Quantization of the coefficients to a finite number of bits essentially changes the filter coefficients, hence the frequency response changes. Coefficient quantization of a stable IIR filter may yield an unstable filter. For example, consider a second-order IIR filter with $a_1 = 1.26$ and $a_2 = 0.3$, which correspond to pole locations of -0.9413 and -0.3187, respectively. Suppose that we quantize these coefficients to two bits after the decimal point, yielding a quantized a_1 of 1.01 in binary or 1.25 and a quantized a_2 of 0.01 in binary or 0.25. This pair correspond to pole locations at -1.0 and -0.25, respectively. Since one pole is on the unit circle, the IIR filter with quantized coefficients produces an oscillation. In this example, the quantization is equivalent to moving a point inside the triangle in Fig. 83.1 to a point on the edge of the triangle. Different realizations are affected differently by coefficient quantization. Chapter 84 investigates coefficient quantization and roundoff noise in detail.

Minimum Phase

An IIR filter is a minimum-phase filter if all the zeros and poles are inside the unit circle. A minimum-phase filter introduces the smallest group delay among all filters that have the same magnitude response. A minimum-phase IIR filter can be constructed from a nonminimum-phase filter by replacing each zero (or pole) outside the unit circle with a zero (or pole) that is the reciprocal of its complex conjugate, as illustrated in Fig. 83.4. This process moves all zeros

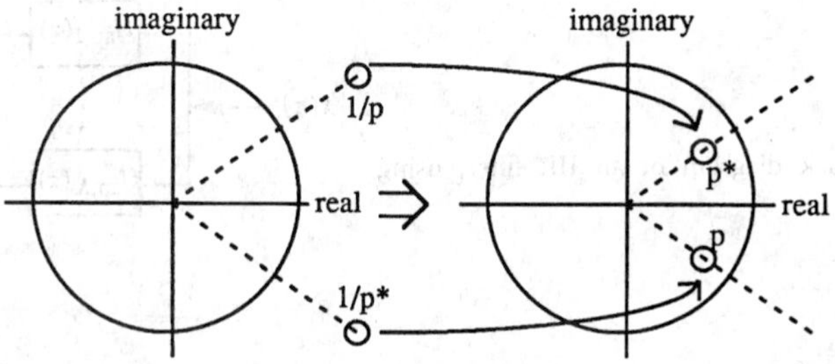

FIGURE 83.4 Changing a zero location to obtain a minimum-phase filter.

and poles outside the unit circle to the inside. The magnitude response does not change, except for a constant factor, which is easily adjusted.

Given an IIR filter $H(z)$ with input $x(n)$ and output $y(n)$, the inverse filter $1/H(z)$ can reconstruct $x(n)$ from $y(n)$ by feeding $y(n)$ to the input of $1/H(z)$. Assuming that both the filter and the inverse filter are causal, both of them can be stable only if $H(z)$ is a minimum-phase filter.

Allpass Filters

An allpass filter has a magnitude response of unity (or constant). An Nth order IIR allpass filter with real coefficients has a system function given by

$$H(z) = z^{-N}\frac{D(z)}{D(z^{-1})} = z^{-N}\frac{a_N z^N + \cdots + a_2 z^2 + a_1 z + 1}{1 + a_1 z^{-1} + a_2 z^{-2} + \cdots + a_N z^{-N}} \tag{83.20}$$

$$= z^{-N}\frac{(1 - p_1 z)(1 - p_2 z)\cdots(1 - p_N z)}{(1 - p_1 z^{-1})(1 - p_2 z^{-1})\cdots(1 - p_N z^{-1})} \tag{83.21}$$

Since $H(z)H(z^{-1}) = 1$, it follows that $|H(f)|^2 = 1$. The factor z^{-N} is included so that the filter is causal. Equation (83.21) implies that zeros and poles come in reciprocal pairs: if there is a pole at $z = p$, then there is a zero at $z = 1/p$, as illustrated in Fig. 83.5.

Since the coefficients are real, poles and zeros off the real axis must exist in quadruplets: poles at p and p^* and zeros at $1/p$ and $1/p^*$, where $|p| < 1$ for stability. For poles and zeros on the real axis, they exist in reciprocal pairs: pole at p and zero at $1/p$, where p is real and $|p| < 1$ for stability. Since the numerator and the denominator in (83.20) share the same set of coefficients, we need only N multiplications in realizing an Nth order allpass filter.

The system function in (83.20) can be written as the product (or sum) of first and second-order allpass filters. The system function and the phase response of a first-order allpass filter is

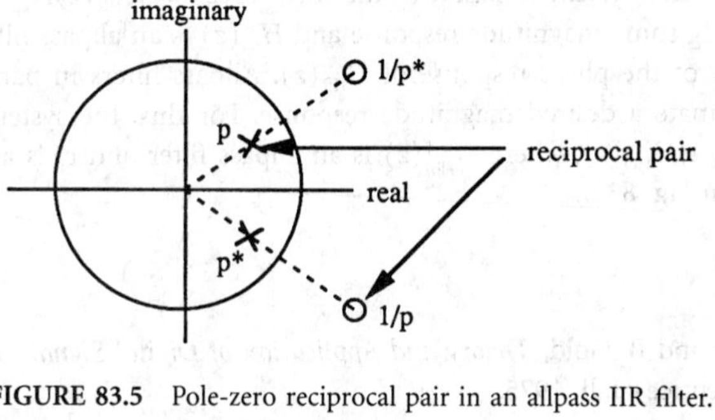

FIGURE 83.5 Pole-zero reciprocal pair in an allpass IIR filter.

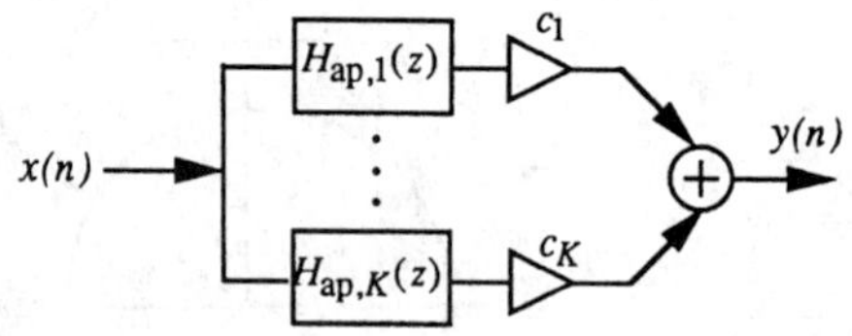

FIGURE 83.6 Block diagram of an IIR filter, using allpass filters.

given by

$$H(z) = \frac{a_1 z + 1}{z + a_1} \tag{83.22}$$

$$\theta(f) = \arctan\left[\frac{(a_1^2 - 1)\sin(\omega)}{2a_1 + (a_1^2 + 1)\cos(\omega)}\right] \tag{83.23}$$

where $\omega = 2\pi f$. For a second-order allpass filter, they are

$$H(z) = \frac{a_2 z^2 + a_1 z + 1}{z^2 + a_1 z + a_2} \tag{83.24}$$

$$\theta(f) = \arctan\left[\frac{2a_1(a_2 - 1)\sin(\omega) + (a_2^2 - 1)\sin(2\omega)}{2a_2 + a_1^2 + 2a_1(a_2 + 1)\cos(\omega) + (a_2^2 + 1)\cos(2\omega)}\right] \tag{83.25}$$

The group delay $\tau(f)$ of an allpass filter is always ≥ 0. The output signal energy of an allpass filter is the same as the input signal energy, i.e., $\sum_{n=-\infty}^{\infty} |y(n)|^2 = \sum_{n=-\infty}^{\infty} |x(n)|^2$, which means that the allpass filter is a lossless system. Note that if we attempt to find a minimum-phase filter from a stable allpass filter, by moving all the zeros inside the unit circle, all poles and zeros would cancel out, yielding the trivial filter with a system function of unity.

A more general form of (83.20), allowing the coefficients to be complex, is Nth order allpass filter with system function

$$H(z) = z^{-N}\frac{D^*(z^*)}{D(z^{-1})} = z^{-N}\frac{a_N^* z^N + \cdots + a_2^* z^2 + a_1^* z + 1}{1 + a_1 z^{-1} + a_2 z^{-2} + \cdots + a_N z^{-N}} \tag{83.26}$$

$$= z^{-N}\frac{(1 - p_1^* z)(1 - p_2^* z)\cdots(1 - p_N^* z)}{(1 - p_1 z^{-1})(1 - p_2 z^{-1})\cdots(1 - p_N z^{-1})} \tag{83.27}$$

Therefore, for a pole at $z = p$ there is a zero at $z = 1/p^*$, i.e., poles and zeros exist in reciprocal–conjugate pairs.

Allpass filters have been used as building blocks for various applications [5]. Particularly, an allpass filter can be designed to approximate a desired phase response. Therefore, an allpass filter in cascade with an IIR filter can be used to compensate the nonlinear phase of the IIR filter. Such a cascade filter has system function of the form $H(z) = H_{IIR}(z)H_{ap}(z)$, where $H_{IIR}(z)$ is an IIR filter satisfying some magnitude response and $H_{ap}(z)$ is an allpass filter that compensates for the nonlinearity of the phase response of $H_{IIR}(z)$. Allpass filters in parallel connection can be used to approximate a desired magnitude response. For this, the system function is in the form $H(z) = \sum_{k=1}^{K} c_i H_{ap,i}(z)$, where $H_{ap,i}(z)$ is an allpass filter and c_i is a coefficient. A block diagram is shown in Fig. 83.6.

References

[1] L. R. Rabiner and B. Gold, *Theory and Application of Digital Signal Processing*, Englewood Cliffs, NJ: Prentice-Hall, 1975.

[2] R. A. Roberts and C. T. Mullis, *Digital Signal Processing*, Reading, MA: Addison-Wesley, 1987.

[3] A. V. Oppenheim and R. W. Schafer, *Discrete-Time Signal Processing*, Englewood Cliffs, NJ: Prentice-Hall, 1989.

[4] J. G. Proakis and D. G. Manolakis, *Digital Signal Processing Principles, Algorithms, and Applications*, 2nd ed. New York: Macmillan, 1992.

[5] P. A. Regalia, S. K. Mitra, and P. P. Vaidyanathan, "The digital all-pass filter: A versatile signal processing building block," *Proc. IEEE*, vol. 76, pp. 19–37, Jan. 1988.

83.2 Design of IIR Filters

Sawasd Tantaratana

Introduction

A filter is generally designed to satisfy a frequency response specification. IIR filter design normally focuses on satisfying a magnitude response specification. If the phase response is essential, it is usually satisfied by a phase compensation filter, such as an allpass filter (see Section 83.1). We will adopt a magnitude specification that is normalized so that the maximum magnitude is 1. The magnitude square in the passband must be at least $1/(1+\varepsilon^2)$ and at most 1; while it must be no larger than δ^2 in the stopband, where ε and δ are normally small. The passband edge is denoted by f_p and the stopband edge by f_s. Figure 83.7(a) shows such a specification for a low-pass filter. The region between the passband and the stopband is the transition band. There is no constraint on the response in the transition band. Another specification that is often used is shown in Fig. 83.7(b), using δ_1 and δ_2 to specify the acceptable magnitude. Given δ_1 and δ_2, they can be converted to ε and δ using $\varepsilon = 2\delta_1^{0.5}/(1-\delta_1)$ and $\delta = \delta_2/(1+\delta_1)$. The magnitude is often specified in dB, which is $20\log_{10}|H(f)|$. Specifications for other types of filters (high-pass, bandpass, and bandstop) are similar.

We can classify various IIR filter design methods into three categories: the design using analog prototype filter, the design using digital frequency transformation, and computer-aided design. In the first category, an analog filter is designed to the (analog) specification and the analog filter transfer function is transformed to digital system function using some kind of transformation. The second category assumes that a digital low-pass filter can be designed. The desired digital filter is obtained from the digital low-pass filter by a digital frequency transformation. The last category uses some algorithm to choose the coefficients so that the response is as close (in some

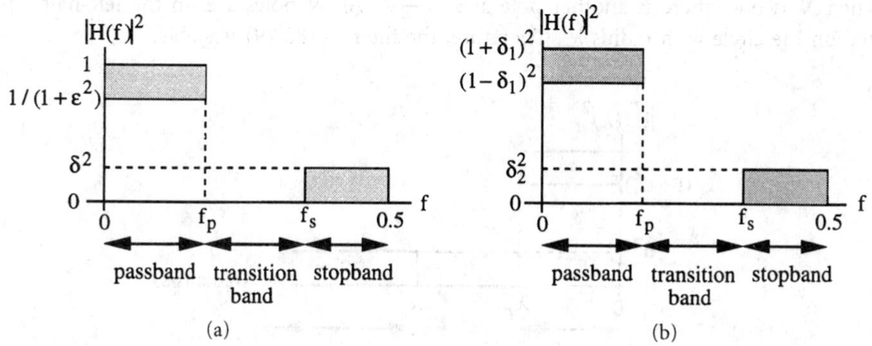

FIGURE 83.7 Specifications for a digital low-pass filter. (a) Specification using ε and δ. (b) Specification using δ_1 and δ_2.

sense) as possible to the desired filter. Design methods in the first two categories are simple to do, requiring only a handheld calculator. Computer-aided design requires some computer programming, but it can be used to design nonstandard filters.

Analog Filters

Here, we describe four basic types of analog low-pass filters that can be used as prototype for designing IIR filters. For each type, we give the transfer function, its magnitude response, and the order N needed to satisfy the (analog) specification. We will use $H_a(s)$ for the transfer function of an analog filter, where s is the variable in the Laplace transform. Each of these filters have all its poles on the left-half s plane, so that it is stable. We will use the variable λ to represent the analog frequency in radians/second. The frequency response $H_a(\lambda)$ is the transfer function evaluated at $s = j\lambda$. The analog low-pass filter specification is given by

$$\begin{aligned}(1+\varepsilon^2)^{-1} \le |H_a(\lambda)|^2 \le 1 \quad &\text{for } 0 \le (\lambda/2\pi) \le (\lambda_p/2\pi)\ \text{Hz} \\ 0 \le |H_a(\lambda)|^2 \le \delta^2 \quad &\text{for } (\lambda_s/2\pi) \le (\lambda/2\pi) < \infty\ \text{Hz}\end{aligned} \tag{83.28}$$

where λ_p and λ_s are the passband edge and stopband edge, respectively. The specification is sketched in Fig. 83.8.

Butterworth Filters. The transfer function of an Nth order Butterworth filter is given by

$$H_a(s) = \begin{cases} \displaystyle\prod_{i=1}^{N/2} \frac{1}{(s/\lambda_c)^2 - 2\,\mathrm{Re}(s_i)(s/\lambda_c) + 1} & N = \text{even} \\ \displaystyle\frac{1}{(s/\lambda_c)+1} \prod_{i=1}^{(N-1)/2} \frac{1}{(s/\lambda_c)^2 - 2\,\mathrm{Re}(s_i)(s/\lambda_c) + 1} & N = \text{odd} \end{cases} \tag{83.29}$$

where $\lambda_p \le \lambda_c \le \lambda_s$ and $s_i = \exp\{j(1 + (2i-1)/N)\pi/2\}$. The magnitude response square is

$$|H_a(\lambda)|^2 = \frac{1}{1 + (\lambda/\lambda_c)^{2N}} \tag{83.30}$$

Figure 83.9 shows the magnitude response $|H_a(\lambda)|$, with $\lambda_c = 1$. Note that a Butterworth filter is an all-pole (no zero) filter, with the poles being at $s = \lambda_c s_i$ and $s = \lambda_c s_i^*$, $i = 1, \cdots, N/2$ if N is even or $i = 1, \cdots, (N-1)/2$ if N is odd, where x^* denotes the complex conjugate of x. When N is odd, there is another pole at $s = -\lambda_c$. All N poles are on the left-half s plane, located on the circle with radius λ_c. Therefore, the filter in (83.29) is stable.

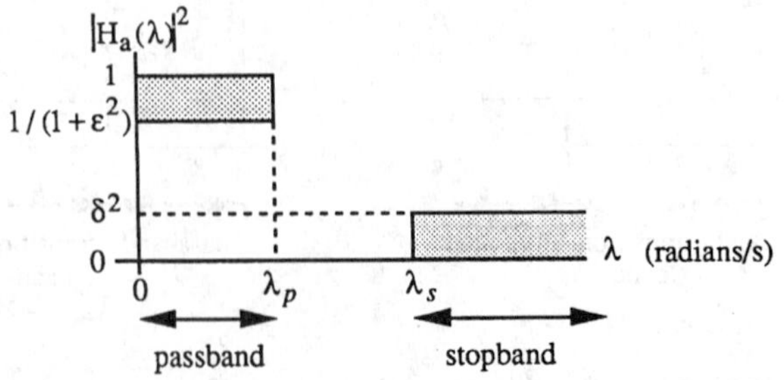

FIGURE 83.8 Specification for an analog low-pass filter.

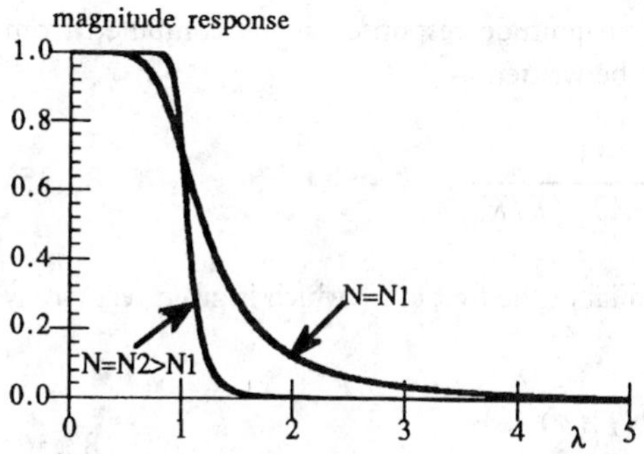

FIGURE 83.9 Magnitude responses of Butterworth filters.

To satisfy the specification in (83.28), the filter order can be calculated from

$$N = \text{integer} \geq \frac{\log[\varepsilon/(\delta^{-2}-1)^{1/2}]}{\log[\lambda_p/\lambda_s]} \tag{83.31}$$

The value of λ_c can be chosen as any value in the following range:

$$\lambda_p \varepsilon^{-1/N} \leq \lambda_c \leq \lambda_s(\delta^{-2}-1)^{-1/(2N)} \tag{83.32}$$

If we choose $\lambda_c = \lambda_p \varepsilon^{-1/N}$, then the magnitude response square passes through $1/(1+\varepsilon^2)$ at $\lambda = \lambda_p$. If we choose $\lambda_c = \lambda_s(\delta^{-2}-1)^{-1/(2N)}$, then the magnitude response square passes through δ^2 at $\lambda = \lambda_s$. If λ_c is between these two values, then the magnitude square will be $\geq 1/(1+\varepsilon^2)$ at $\lambda = \lambda_p$, and $\leq \delta^2$ at $\lambda = \lambda_s$.

Chebyshev Filters (Type-I Chebyshev Filters). A Chebyshev filter is also an all-pole filter. The Nth order Chebyshev filter has a transfer function given by

$$H_a(s) = C \prod_{i=1}^{N} \frac{1}{(s-p_i)} \tag{83.33}$$

where

$$C = \begin{cases} -\prod_{i=1}^{N} p_i & N \text{ is odd} \\ (1+\varepsilon^2)^{-1/2} \prod_{i=1}^{N} p_i & N \text{ is even} \end{cases} \tag{83.34a}$$

$$p_i = -\lambda_p \sinh(\phi) \sin\left(\frac{2i-1}{2N}\pi\right) + j\lambda_p \cosh(\phi) \cos\left(\frac{2i-1}{2N}\pi\right) \tag{83.34b}$$

$$\phi = \frac{1}{N} \ln\left[\frac{1+(1+\varepsilon^2)^{1/2}}{\varepsilon}\right] \tag{83.34c}$$

The value of C normalizes the magnitude so that the maximum magnitude is 1. Note that C is always a positive constant. The poles are on the left-half s plane, lying on an ellipse centered at the origin with a minor radius of $\lambda_p \sinh(\phi)$ and major radius of $\lambda_p \cosh(\phi)$. Except for one pole when N is odd, all the poles have a complex conjugate pair. Specifically, $p_i = p^*_{N-i+1}$, $i = 1, 2, \cdots, N/2$ or $(N-1)/2$. Combining each complex conjugate pair in (83.33) yields a second-order factor with real coefficients. The magnitude response can be computed from (83.34a)–(83.34c) with $s = j\lambda$. Its square can also be written as

$$|H_a(\lambda)|^2 = \frac{1}{1 + \varepsilon^2 T_N^2(\lambda/\lambda_p)} \tag{83.35}$$

where $T_N(x)$ is the Nth degree Chebyshev polynomial of the first kind, which is given recursively by

$$\begin{aligned} T_0(x) &= 1 \qquad T_1(x) = x \\ T_{n+1}(x) &= 2xT_n(x) - T_{n-1}(x) \qquad n \geq 1 \end{aligned} \tag{83.36}$$

Notice that $T_N^2(\pm 1) = 1$. Therefore, we have from (83.35) that the magnitude square passes through $1/(1+\varepsilon^2)$ at $\lambda = \lambda_p$, i.e., $|H_a(\lambda_p)|^2 = 1/(1+\varepsilon^2)$. Note also that $T_N(0) = (-1)^{N/2}$ for even N and it is 0 for odd N. Therefore, $|H_a(0)|^2$ equals $1/(1+\varepsilon^2)$ for even N and it equals 1 for odd N. Figure 83.10 shows some examples of magnitude response square.

The filter order required to satisfy the specification in (83.28) is

$$\begin{aligned} N &\geq \frac{\cosh^{-1}[(\delta^{-2} - 1)^{1/2}/\varepsilon]}{\cosh^{-1}(\lambda_s/\lambda_p)} \\ &= \frac{\log\{[(\delta^{-2} - 1)^{1/2}/\varepsilon] + [(\delta^{-2} - 1)/\varepsilon^2 - 1]^{1/2}\}}{\log\{(\lambda_s/\lambda_p) + [(\lambda_s/\lambda_p)^2 - 1]^{1/2}\}} \end{aligned} \tag{83.37}$$

which can be computed knowing ε, δ, λ_p, and λ_s.

Inverse Chebyshev Filters (Type-II Chebyshev Filters). Notice from Fig. 83.10 that the Chebyshev filter has magnitude response containing equiripples in the passband. The equiripples can be arranged to go inside the stopband, for which case we obtain inverse Chebyshev filters. The magnitude response square of the inverse Chebyshev filter is

$$|H_a(\lambda)|^2 = \frac{1}{1 + \dfrac{(\delta^{-2} - 1)}{T_N^2(\lambda_s/\lambda)}} \tag{83.38}$$

Since $T_N^2(\pm 1) = 1$, (83.38) gives $|H_a(\lambda_s)|^2 = \delta^2$. Figure 83.11 depicts some examples of (83.38). Note that $|H_a(\infty)|$ equals 0 if N is odd and it equals δ if N is even.

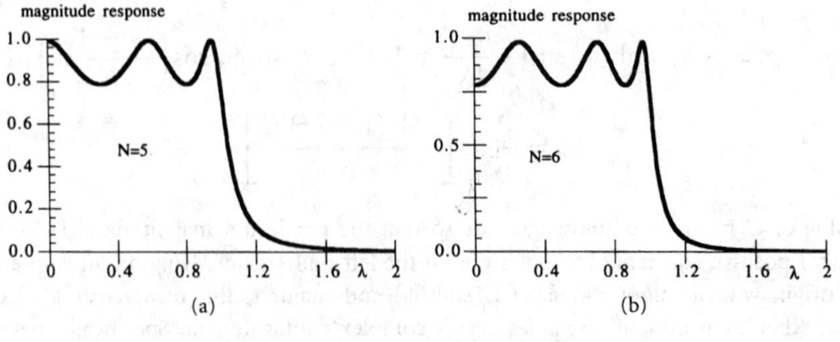

FIGURE 83.10 Magnitude responses of Chebyshev filters.

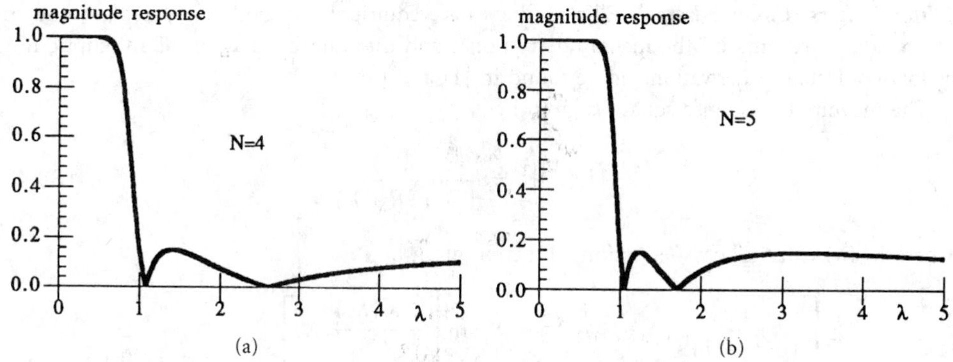

FIGURE 83.11 Magnitude responses of inverse Chebyshev filters.

The transfer function giving rise to (83.38) is given by

$$H_a(s) = \begin{cases} C\displaystyle\prod_{i=1}^{N}\frac{(s-q_i)}{(s-p_i)} & N \text{ is even} \\ \dfrac{C}{(s-p_{(N+1)/2})}\displaystyle\prod_{\substack{i=1\\ i\neq(N+1)/2}}^{N}\frac{(s-q_i)}{(s-p_i)} & N \text{ is odd} \end{cases} \tag{83.39}$$

where

$$C = \begin{cases} \displaystyle\prod_{i=1}^{N}\frac{p_i}{q_i} & N \text{ is even} \\ -p_{(N+1)/2}\displaystyle\prod_{\substack{i=1\\ i\neq(N+1)/2}}^{N}\frac{p_i}{q_i} & N \text{ is odd} \end{cases} \tag{83.40a}$$

$$p_i = \frac{\lambda_s}{\alpha_i^2+\beta_i^2}(\alpha_i - j\beta_i) \quad q_i = j\frac{\lambda_s}{\cos\left(\dfrac{2i-1}{2N}\pi\right)} \tag{83.40b}$$

$$\alpha_i = -\sinh(\phi)\sin\left(\frac{2i-1}{2N}\pi\right) \quad \beta_i = \cosh(\phi)\cos\left(\frac{2i-1}{2N}\pi\right) \tag{83.40c}$$

$$\phi = \frac{1}{N}\cosh^{-1}(\delta^{-1}) = \frac{1}{N}\ln[\delta^{-1} + (\delta^{-2}-1)^{1/2}] \tag{83.40d}$$

Note that the zeros are on the imaginary axis on the s plane. The filter order N required to satisfy the specification in (83.28) is the same as the order for the Chebyshev filter, given by (83.37).

Another form for the inverse Chebyshev filter has a magnitude response square given by

$$|H_a(\lambda)|^2 = \frac{1}{1+\varepsilon^2\dfrac{T_N^2(\lambda_s/\lambda_p)}{T_N^2(\lambda_s/\lambda)}} \tag{83.41}$$

which passes through $1/(1+\varepsilon^2)$ at $\lambda = \lambda_p$. For further details of this form see [2].

Elliptic Filters (Cauer Filters). Elliptic filters have equiripples in both the passband and the stopband. We summarize the magnitude response and the transfer function of an elliptic filter as follows. Detail of derivation can be found in [1] and [6].

The magnitude response square is given by

$$|H_a(\lambda)|^2 = \frac{1}{1+\varepsilon^2 R_N^2(\lambda)} \tag{83.42}$$

where $R_N(\lambda)$ is the Chebyshev rational function given by

$$R_N(\lambda) = \begin{cases} \dfrac{(\delta^{-2}-1)^{1/4}}{\varepsilon^{1/2}}\lambda \displaystyle\prod_{i=1}^{(N-1)/2} \dfrac{\lambda^2 - \lambda_r \mathrm{sn}^2\left[\dfrac{2iK(\lambda_r)}{N}, \lambda_r\right]}{\lambda^2 \lambda_r \mathrm{sn}^2\left[\dfrac{2iK(\lambda_r)}{N}, \lambda_r\right] - 1} & N = \text{odd} \\ \dfrac{(\delta^{-2}-1)^{1/4}}{\varepsilon^{1/2}} \displaystyle\prod_{i=1}^{N/2} \dfrac{\lambda^2 - \lambda_r \mathrm{sn}^2\left[\dfrac{(2i-1)K(\lambda_r)}{N}, \lambda_r\right]}{\lambda^2 \lambda_r \mathrm{sn}^2\left[\dfrac{(2i-1)K(\lambda_r)}{N}, \lambda_r\right] - 1} & N = \text{even} \end{cases} \tag{83.43}$$

Here, $\lambda_r = \lambda_p/\lambda_s$, $K(t)$ is the complete elliptic integral of the first kind given by

$$K(t) = \int_0^{\pi/2} \frac{d\theta}{[1-t^2\sin^2\theta]^{1/2}} = \int_0^1 \frac{dx}{[(1-x^2)(1-t^2x^2)]^{1/2}} \tag{83.44}$$

The Jacobian elliptic sine function $\mathrm{sn}[u,t]$ is defined as

$$\mathrm{sn}[u,t] = \sin\phi \quad \text{if } u = \int_0^{\phi} \frac{d\theta}{[1-t^2\sin^2\theta]^{1/2}} \tag{83.45}$$

The integral

$$F(\phi,t) = \int_0^{\phi} \frac{d\theta}{[1-t^2\sin^2\theta]^{1/2}} = \int_0^{\sin\phi} \frac{dx}{[(1-x^2)(1-t^2x^2)]^{1/2}} \tag{83.46}$$

is called the elliptic integral of the first kind. Note that $K(t) = F(\pi/2, t)$.

The transfer function corresponding to the magnitude response in (83.42) is

$$H_a(s) = \begin{cases} \dfrac{C}{(s+p_0)} \displaystyle\prod_{i=1}^{(N-1)/2} \dfrac{(s^2+B_i)}{(s^2+A_{i1}s+A_{i2})} & N \text{ odd} \\ C\displaystyle\prod_{i=1}^{N/2} \dfrac{(s^2+B_i)}{(s^2+A_{i1}s+A_{i2})} & N \text{ even} \end{cases} \tag{83.47}$$

where

$$C = \begin{cases} p_0 \displaystyle\prod_{i=1}^{(N-1)/2} \dfrac{A_{i2}}{B_i} & N \text{ odd} \\ \dfrac{1}{(1+\varepsilon^2)^{1/2}} \displaystyle\prod_{i=1}^{N/2} \dfrac{A_{i2}}{B_i} & N \text{ even} \end{cases} \tag{83.48}$$

The pole p_0, and the coefficients B_i, A_{i1}, A_{i2} are calculated as follows:

$$\lambda_r = \frac{\lambda_p}{\lambda_s} \quad \lambda_c = \sqrt{\lambda_p \lambda_s} \quad \alpha = 0.5\frac{1-(1-\lambda_r^2)^{1/4}}{1+(1-\lambda_r^2)^{1/4}} \tag{83.49a}$$

$$\beta = e^{-\pi K[(1-\lambda_r^2)^{1/2}]/K(\lambda_r)} \approx \alpha + 2\alpha^5 + 15\alpha^9 + 150\alpha^{13} \tag{83.49b}$$

$$\gamma = \frac{1}{2N}\ln\left[\frac{(1+\varepsilon^2)^{1/2}+1}{(1+\varepsilon^2)^{1/2}-1}\right] \tag{83.49c}$$

$$\sigma = \left|\frac{2\beta^{1/4}\sum_{k=0}^{\infty}(-1)^k\beta^{k(k+1)}\sinh[(2k+1)\gamma]}{1+2\sum_{k=1}^{\infty}(-1)^k\beta^{k^2}\cosh[2k\gamma]}\right| \tag{83.49d}$$

$$\zeta = (1+\lambda_r\sigma^2)\left(1+\frac{\sigma^2}{\lambda_r}\right) \quad \eta = \begin{cases} i & N \text{ odd} \\ i-0.5 & N \text{ even} \end{cases} \tag{83.49e}$$

$$\psi_i = \frac{2\beta^{1/4}\sum_{k=0}^{\infty}(-1)^k\beta^{k(k+1)}\sin[(2k+1)\pi\eta/N]}{1+2\sum_{k=1}^{\infty}(-1)^k\beta^{k^2}\cos[2k\pi\eta/N]} \tag{83.49f}$$

$$\mu_i = \left[(1-\lambda_r\psi_i^2)\left(1-\frac{\psi_i^2}{\lambda_r}\right)\right]^{1/2} \tag{83.49g}$$

$$p_0 = \lambda_c\sigma \quad B_i = \frac{\lambda_c^2}{\psi_i^2} \quad A_{i1} = \frac{2\lambda_c\sigma\mu_i}{1+\sigma^2\psi_i^2} \quad A_{i2} = \lambda_c^2\frac{\sigma^2\mu_i^2+\zeta\psi_i^2}{[1+\sigma^2\psi_i^2]^2} \tag{83.49h}$$

The infinite summations above converge very quickly, so that only a few terms are needed in actual calculation. A simple program can be written to compute the values in (83.49a)–(83.49h).

The filter order required to satisfy (83.28) is calculated from

$$N \geq \frac{1}{\log(\beta)}\log\left[\frac{\varepsilon^2}{16(\delta^{-2}-1)}\right] \tag{83.50}$$

where β is given by (83.49b). An example of the magnitude response is plotted in Fig. 83.12. We see that there are ripples in both the passband and the stopband.

Comparison. In comparing the filters given above, the Butterworth filter requires the highest order and the elliptic filter requires the smallest order to satisfy the same passband and stopband specifications. The Butterworth filter and the inverse Chebyshev filter have nicer (closer to linear) phase characteristics in the passband than Chebyshev and elliptic filters. The magnitude responses of the Butterworth and Chebyshev filters decrease monotonically in the stopband to zero, which reduces the aliasing caused by some analog-to-digital transformations.

Design Using Analog Prototype Filters

In this subsection, we consider designing IIR filters using analog prototype filters. This method is suitable for designing the standard types of filters: low-pass filter (LPF), high-pass filter (HPF),

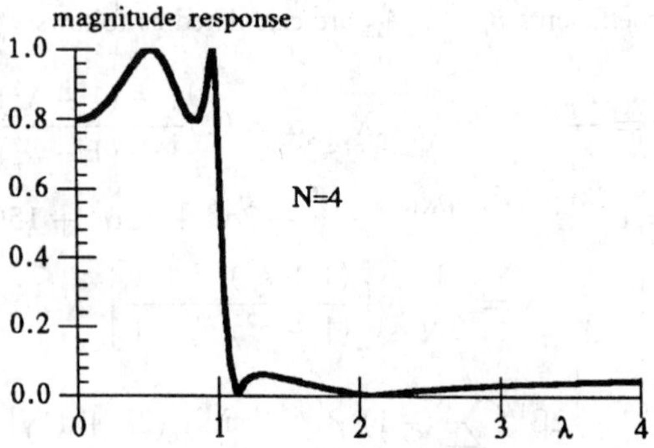

FIGURE 83.12 Magnitude response of elliptic filter.

bandpass filter (BPF), and bandstop filter (BSF). The basic idea is to transform the digital specification to analog specification, design an analog filter, and then transform the analog filter transfer function to digital filter system function. Several types of transformation have been studied.

The design steps are outlined by Fig. 83.13. Given the desired magnitude response $|H^x(f)|$ of digital LPF, HPF, BPF, or BSF, it is transformed to analog magnitude specification (of the corresponding type: LPF, HPF, BPF, or BSF) $|H_a^x(\lambda)|$. The analog magnitude specification is then transformed to analog LPF magnitude specification $|H_a(\lambda)|$. We then design an analog prototype filter as discussed in the subsection on analog filters, obtaining analog LPF transfer function $H_a(s)$. Next, the analog LPF transfer function is transformed to analog transfer function $H_a^x(s)$ of the desired type (LPF, HPF, BPF, or BSF), followed by a transformation to digital filter system function $H^x(z)$. By combining the appropriate steps, we can obtain transformations to go directly from $|H^x(f)|$ to $|H_a(\lambda)|$ and directly from $H_a(s)$ to $H(z)$, as indicated by the dotted lines in Fig. 83.13. Note that for designing digital LPF, the middle steps involving $|H_a^x(\lambda)|$ and $H_a^x(s)$ are not applicable.

Transformations

There are several types of transformations. They arise from approximating continuous-time signals and systems by discrete-time signals and systems. Table 83.1 shows several transformations, with their advantages and disadvantages. The constant T is the sampling interval. The resulting mapping is used for transforming $H_a(s)$ to $H(z)$. For example, in the backward difference approximation we obtain $H(z)$ by replacing the variable s with $(1-z^{-1})/T$ in $H_a(s)$, i.e.,

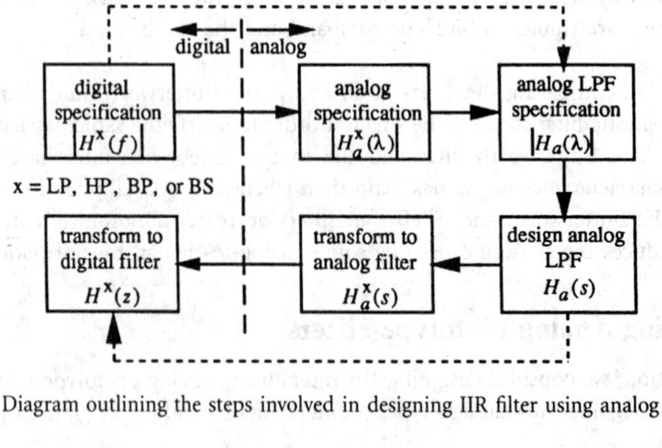

FIGURE 83.13 Diagram outlining the steps involved in designing IIR filter using analog prototype filter.

$H(z) = H_a(s)\big|_{s=(1-z^{-1})/T}$. The bilinear transformation is the best all-around method, followed by the impulse invariant method. Therefore, we describe these two transformations in more detail.

Bilinear Transformation. Using this transformation, the analog filter is converted to digital filter by replacing s in the analog filter transfer function with $(2/T)(1-z^{-1})/(1+z^{-1})$, i.e.,

$$H(z) = H_a(s)\big|_{s=(2/T)(1-z^{-1})/(1+z^{-1})} \tag{83.51}$$

From the mapping, we can show as follows that the imaginary axis on the s plane is mapped to the unit circle on the z plane. Letting $s = j\lambda$, we have

$$\begin{aligned} z &= \frac{(2/T)+s}{(2/T)-s} = \frac{(2/T)+j\lambda}{(2/T)-j\lambda} \\ &= \frac{\sqrt{(2/T)^2+(\lambda)^2}e^{j\arctan(\lambda/(2/T))}}{\sqrt{(2/T)^2+(\lambda)^2}e^{j\arctan(-\lambda/(2/T))}} = e^{j2\arctan(\lambda T/2)} \end{aligned} \tag{83.52}$$

which is the unit circle on the z plane as λ goes from $-\infty$ to ∞. Writing $z = e^{j2\pi f}$ in (83.52), we obtain the relation between the analog frequency λ and the digital frequency f:

$$f = \frac{1}{\pi}\arctan\left(\frac{\lambda T}{2}\right) \qquad \lambda = \frac{2}{T}\tan(\pi f) \tag{83.53}$$

which is plotted in Fig. 83.14. Equation (83.53) is used for converting digital specification to analog specification, i.e., $\lambda_s = (2/T)\tan(\pi f_s)$ and $\lambda_p = (2/T)\tan(\pi f_p)$. In a complete design process, starting from the digital specification and ending at the digital filter system function, as outlined in Fig. 83.13, the sampling interval T is canceled out in the process. Hence, it has no effect and any convenient value (such as 1 or 2) can be used.

Impulse Invariant Method. This method approximates the analog filter impulse response $h_a(t)$ by its samples separated by T seconds. The result is the impulse response $h(n)$ of the digital filter, i.e., $h(n) = h_a(nT)$. From this relation, it can be shown that

$$H(f) = \frac{1}{T}\sum_{k=-\infty}^{\infty} H_a(\lambda)\bigg|_{\lambda=2\pi(f+k)/T} = \frac{1}{T}\sum_{k=-\infty}^{\infty} H_a\left(2\pi\frac{f+k}{T}\right) \tag{83.54}$$

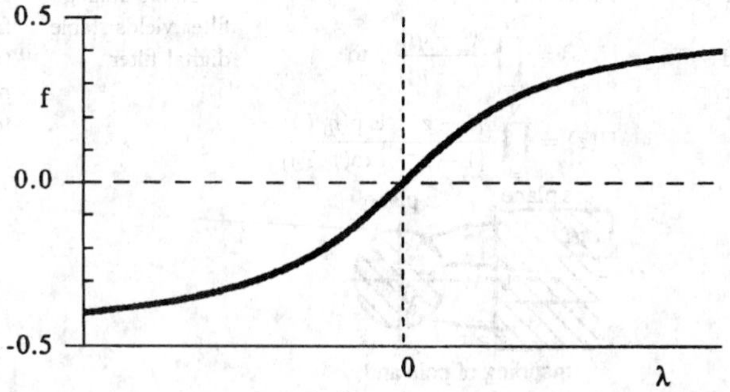

FIGURE 83.14 Relation between λ and f for bilinear transformation.

TABLE 83.1 Various Types of Analog-to-Digital Transformation

Type of Transformation	Principle	Resulting mapping	Advantages	Disadvantages
Backward difference approximation	$\frac{dy}{dt} \approx \frac{y(n) - y(n-1)}{T}$	$s = \frac{1 - z^{-1}}{T}, \quad z = \frac{1}{1 - sT}$ s plane z plane 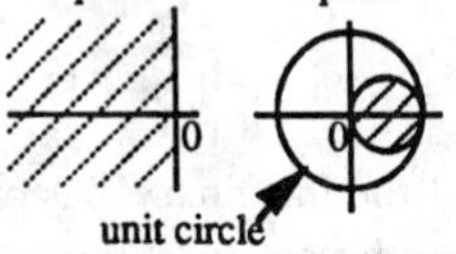 	1. Stable analog filter yields stable digital filter.	1. Left-half s plane is not mapped onto unit circle on z plane. Pole locations will be in the circle centered at 0.5 with radius 0.5 on the z plane.
Forward difference approximation	$\frac{dy}{dt} \approx \frac{y(n+1) - y(n)}{T}$	$s = \frac{z - 1}{T}, \quad z = 1 + sT$ s plane z plane 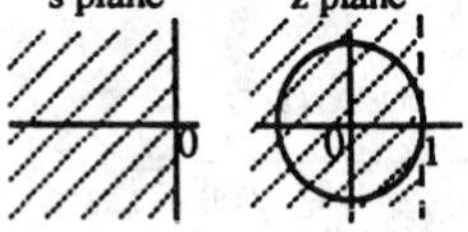 		1. Stable analog filter does not yield stable digital filter. 2. Left-half s plane is not mapped onto unit circle on z plane.
Impulse invariant method	Sample the analog impulse response: $h(n) = h_a(nT)$	Transform $H_a(s) = \sum_{i=1}^{N} \frac{b_i}{(s - p_i)}$ to $H(z) = \sum_{i=1}^{N} \frac{b_i}{[1 - z^{-1} \exp(p_i T)]}$ s plane z plane 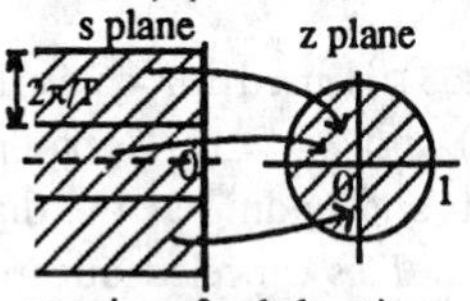 mapping of pole locations	1. Preserve shape of impulse response. 2. Stable analog filter yields stable digital filter. 3. The analog frequency and digital frequency are linearly related, $\lambda T = 2\pi f$, $-0.5 \leq f \leq 0.5$.	1. Aliasing in the frequency domain (due to many-to-one mapping from s plane to z plane). $T \cdot H(f) = \sum_{k=-\infty}^{\infty} H_a\left(2\pi \frac{f+k}{T}\right)$
Bilinear transformation	Approximate $y(t) = \int_{t-T}^{t} y'(\tau)\, d\tau + y(t-T)$ by $y(n) = \frac{T}{2}[y'(n) + y'(n-1)] + y(n-1)$	$s = \frac{2}{T}\frac{1 - z^{-1}}{1 + z^{-1}}, \quad z = \frac{(2/T) + s}{(2/T) - s}$ s plane z plane 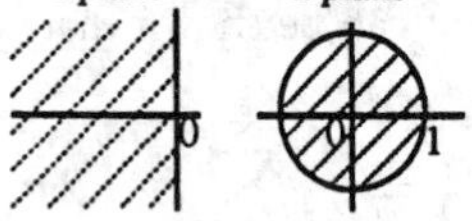 	1. Stable analog filter yields stable digital filter. 2. Left-half s plane is mapped onto unit circle on z plane, a one-to-one mapping.	1. Frequency warping—nonlinear relation between analog frequency and digital frequency.
Matched z transformation	Map each pole and zero on s plane directly to pole and zero on the z plane.	Transform $H_a(s) = \prod_{i=1}^{N} \frac{(s - q_i)}{(s - p_i)}$ to $H(z) = \prod_{i=1}^{N} \frac{[1 - z^{-1} \exp(q_i T)]}{[1 - z^{-1} \exp(p_i T)]}$ s plane z plane 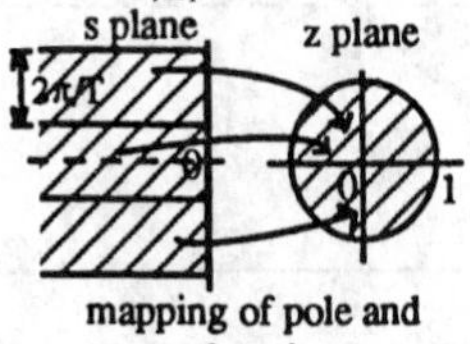 mapping of pole and zero locations	1. Stable analog filter yields stable digital filter.	1. Aliasing in the frequency domain (due to many-to-one mapping from s plane to z plane).

The analog and digital frequencies are related by

$$f = \frac{\lambda T}{2\pi} \qquad |f| \leq 0.5 \tag{83.55}$$

From (83.54), the digital filter frequency response is the sum of shifted versions of the analog filter frequency response. There is aliasing if $H_a(\lambda)$ is not zero for $|\lambda/2\pi| > 1/(2T)$. Therefore, the analog filter used in this method should have a frequency response that goes to zero quickly as λ goes to ∞. Because of the aliasing, this method cannot be used for designing a high-pass filter. Writing the analog filter transfer function in the form

$$H_a(s) = \sum_{i=1}^{N} \frac{b_i}{(s - p_i)} \tag{83.56}$$

It follows that the analog impulse response is given by $h_a(t) = \sum_{i=1}^{N} b_i e^{p_i t}$ and the digital filter system function can be obtained as

$$\begin{aligned} H(z) &= \sum_{n=0}^{\infty} h(n) z^{-n} = \sum_{n=0}^{\infty} h_a(nT) z^{-n} \\ &= \sum_{i=1}^{N} b_i \sum_{n=0}^{\infty} (e^{p_i T} z^{-1})^n = \sum_{i=1}^{N} \frac{b_i}{(1 - e^{p_i T} z^{-1})} \end{aligned} \tag{83.57}$$

Therefore, an analog filter transfer function $H_a(s) = \sum_{i=1}^{N} b_i/(s - p_i)$ gets transformed to a digital filter system function $H(z) = \sum_{i=1}^{N} b_i/(1 - e^{p_i T} z^{-1})$, as shown in Table 83.1. Similar to the bilinear transformation, in a complete design process the choice of T has no effect (except for the final magnitude scaling factor).

Low-Pass Filters (LPF's)

We give one example in designing an LPF using the impulse invariant method and one example using the bilinear transformation.

In this example, suppose that we wish to design a digital filter using an analog Butterworth prototype filter. The digital filter specification is

$$\begin{aligned} &20 \log |H(f)| \geq -2\text{dB} \quad \text{for } 0 \leq f \leq 0.11 \\ &20 \log |H(f)| \leq -10\,\text{dB} \quad \text{for } 0.2 \leq f \leq 0.5 \end{aligned}$$

where the log is of base 10. Therefore, we have $\varepsilon = 0.7648$, $\delta = 0.3162$, $f_p = 0.11$, and $f_s = 0.2$. Let us use the impulse invariant method. Therefore, the analog passband edge and stopband edge are $\lambda_p = 0.22\pi/T$ and $\lambda_s = 0.4\pi/T$, respectively. We use the same ripple requirements: $\varepsilon = 0.7648$ and $\delta = 0.3162$. Using these values, a Butterworth filter order is calculated from (83.31), yielding $N \geq 2.3$. So, we choose $N = 3$. With $\lambda_c = \lambda_p \varepsilon^{-1/N} = 0.2406\pi/T$, we find the analog filter transfer function to be

$$H_a(s) = \frac{\lambda_c^3}{(s+\lambda_c)(s^2+\lambda_c s+\lambda_c^2)}$$

$$= \lambda_c\left[\frac{1}{s+\lambda_c} + \frac{-0.5-j0.5/\sqrt{3}}{s+0.5(1-j\sqrt{3})\lambda_c} + \frac{-0.5+j0.5/\sqrt{3}}{s+0.5(1+j\sqrt{3})\lambda_c}\right]$$

$$= \frac{0.7559}{T}\left[\frac{1}{s+0.7559/T} + \frac{-0.5-j0.5/\sqrt{3}}{s+0.3779(1-j\sqrt{3})/T} + \frac{-0.5+j0.5/\sqrt{3}}{s+0.3779(1+j\sqrt{3})/T}\right]$$

Using (83.56) and (83.57) we obtain the digital filter system function:

$$H(z) = \frac{0.7559}{T}\left[\frac{1}{1-e^{-0.7559}z^{-1}} + \frac{-0.5-j0.5/\sqrt{3}}{1-e^{-0.3779(1-j\sqrt{3})}z^{-1}} + \frac{-0.5+j0.5/\sqrt{3}}{1-e^{-0.3779(1+j\sqrt{3})}z^{-1}}\right]$$

$$= \frac{0.7559}{T}\left[\frac{1}{1-0.4696z^{-1}} - \frac{1-0.7846z^{-1}}{1-1.0873z^{-1}+0.4696z^{-2}}\right]$$

Due to aliasing, the maximum value of the resulting magnitude response (which is at $f=0$ or $z=1$) is no longer equal to 1, although the analog filter has maximum magnitude (at $\lambda=0$ or $s=0$) of 1. Note that the choice of T affects only the scaling factor, which is only a constant gain factor. If we adjust the system function so that the maximum magnitude is 1, i.e., $|H(f)|=1$, we have

$$H(z) = 0.7565\left[\frac{1}{1-0.4696z^{-1}} - \frac{1-0.7846z^{-1}}{1-1.0873z^{-1}+0.4696z^{-2}}\right]$$

The magnitude response in dB and the phase response are plotted in Fig. 83.15. From the result, $|H(f)| = -1.97$ dB at $f=0.11$ and $|H(f)| = -13.42$ dB at $f=0.2$; both satisfy the desired specification. The aliasing in this example is small enough that the resulting response still meets the specification. It is possible that the aliasing is large enough that the designed filter does not meet the specification. To compensate for the unknown aliasing, we may want to use smaller ε and δ in designing the analog prototype filter.

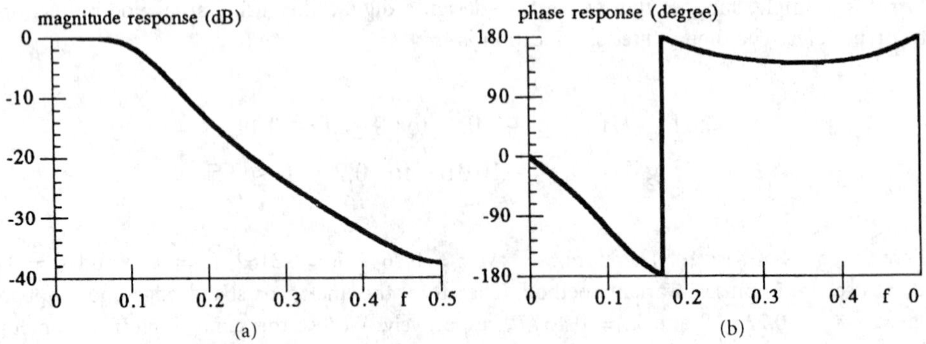

FIGURE 83.15 Frequency response of the LP filter designed using impulse invariant method.

In this next example, we demonstrate the design method using bilinear transformation, with an analog elliptic prototype filter. Let the desired filter specification be

$$|H(f)|^2 \geq 0.8 \quad (\text{or } -0.97 \text{ dB}) \quad \text{for } 0 \leq f \leq 0.1125$$
$$|H(f)|^2 \leq 2.5 \times 10^{-5} \quad (\text{or } -46.02 \text{ dB}) \quad \text{for } 0.15 \leq f \leq 0.5$$

which means $\varepsilon = 0.5$, $\delta = 0.005$, $f_p = 0.1125$, and $f_s = 0.15$. For bilinear transformation, we calculate the analog passband and stopband edges as $\lambda_p = (2/T)\tan(\pi f_p) = 0.7378/T$ and $\lambda_s = (2/T)\tan(\pi f_s) = 1.0190/T$, respectively. Therefore, $\lambda_p/\lambda_s = 0.7240$. From (83.50) we obtain the order $N \geq 4.8$. So, we use $N = 5$. The analog elliptic filter transfer function is calculated from (83.47)–(83.49) to be

$$H_a(s) = \frac{7.8726 \times 10^{-3}\left[\left(\frac{sT}{2}\right)^2 + 0.6006\right]\left[\left(\frac{sT}{2}\right)^2 + 0.2782\right]}{\left[\left(\frac{sT}{2}\right) + 0.1311\right]\left[\left(\frac{sT}{2}\right)^2 + 0.1689\left(\frac{sT}{2}\right) + 0.0739\right]\left[\left(\frac{sT}{2}\right)^2 - 0.0457\left(\frac{sT}{2}\right) + 0.1358\right]}$$

To convert to digital filter system function, we replace s with $(2/T)(1 - z^{-1})/(1 + z^{-1})$. Equivalently, we replace $sT/2$ with $(1 - z^{-1})/(1 + z^{-1})$, yielding

$$H(z) = \frac{1.0511 \times 10^{-2}(1 + z^{-1})(1 - 0.4991z^{-1} + z^{-2})(1 - 1.1294z^{-1} + z^{-2})}{(1 - 0.7682z^{-1})(1 - 1.4903z^{-1} + 0.7282z^{-2})(1 - 1.5855z^{-1} + 1.0838z^{-2})}$$

Note that the choice of T has no effect on the resulting system function. The magnitude response in dB and the phase response are plotted in Fig. 83.16, which satisfies the desired magnitude specification. Note the equiripples in both the passband and the stopband.

High-Pass Filters (HPF's)

As mentioned above, the impulse invariant method is not suitable for high-pass filters, due to aliasing. Therefore, we will only discuss the bilinear transformation. In addition to the procedure used with designing an LPF, we need to transform the analog high-pass specification to analog low-pass specification and transform the resulting analog LPF to analog HPF. There is a simple transformation for this job: replacing s in the analog LPF transfer function with $1/s$. In terms of the frequency, $j\lambda$ becomes $1/j\lambda = j(-1/\lambda)$, i.e., a low frequency is changed to a (negative) high frequency. Therefore, an analog LPF becomes an analog HPF. When combined with the bilinear transformation, this process gives the transformation

$$s = \frac{T}{2}\frac{(1 + z^{-1})}{(1 - z^{-1})} \quad \text{or} \quad z = \frac{s + (T/2)}{s - (T/2)} \tag{83.58}$$

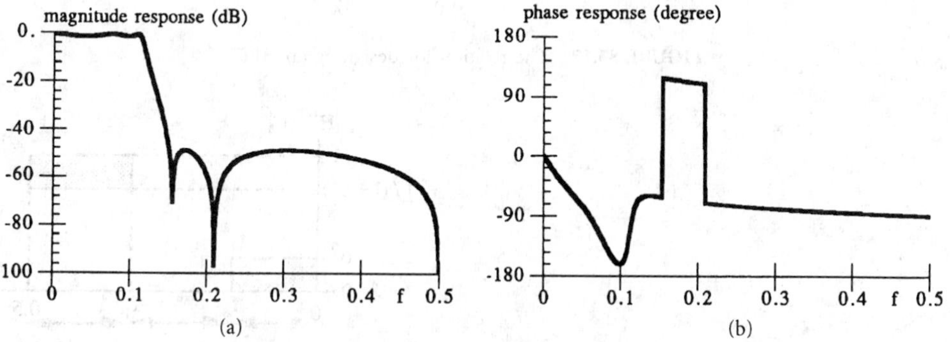

FIGURE 83.16 Frequency response of the LPF designed using bilinear transformation.

Writing $s = j\lambda$, we can show that $z = \exp\{j[2\arctan(2\lambda/T) - \pi]\}$. With $z = \exp(j2\pi f)$, we have

$$\lambda = \frac{T}{2}\tan[\pi(f + 0.5)] \tag{83.59}$$

To write f in terms of λ, we can show that, after adjusting the range of f to $[-1/2, 1/2]$,

$$f = \begin{cases} -\frac{1}{2} + \frac{1}{\pi}\arctan(2\lambda/T) & \lambda > 0 \\ \frac{1}{2} + \frac{1}{\pi}\arctan(2\lambda/T) & \lambda < 0 \end{cases} \tag{83.60}$$

Equations (83.59) and (83.60) give the relation between the digital frequency and the analog frequency, corresponding to the transformation in (83.58). This relation is plotted in Fig. 83.17, from which we see that a low digital frequency corresponds to a high analog frequency and vice versa.

We can summarize the design steps as follows. Given a digital HPF specification as in Fig. 83.18, it is converted to an analog LPF specification using (83.60) to obtain the passband and stopband edges λ_p and λ_s from f_p and f_s, respectively. With λ_p, λ_s, ε, and δ, we design the low-pass analog prototype filter. Let the transfer function be $H_a(s)$. This transfer function is then converted to digital HPF system function by replacing s with $(T/2)(1+z^{-1})/(1-z^{-1})$. Note that this corresponds to the procedure in Fig. 83.13, with the bypass of the "analog specification" block and the "transform to analog filter" block, as indicated by the dotted lines in Fig. 83.13.

As an example, consider designing a digital HPF with the following specification:

$$|H^{HP}(f)|^2 \geq 0.8 \quad (\text{or } -0.97 \text{ dB}) \quad \text{for } 0.4 \leq f \leq 0.5$$
$$|H^{HP}(f)|^2 \leq 2.5 \times 10^{-5} \quad (\text{or } -46.02\text{dB}) \quad \text{for } 0 \leq f \leq 0.3$$

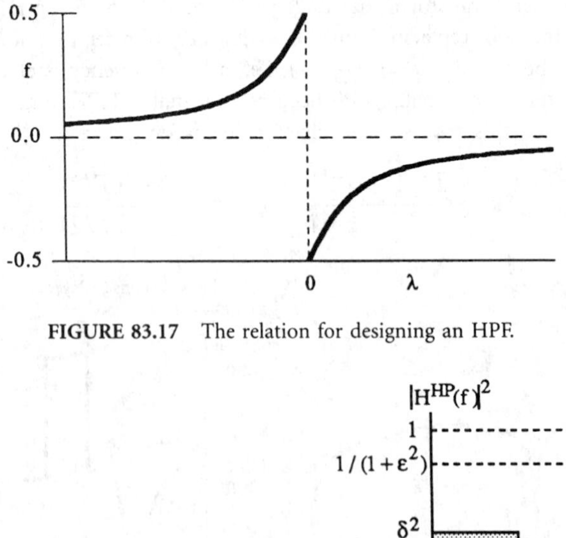

FIGURE 83.17 The relation for designing an HPF.

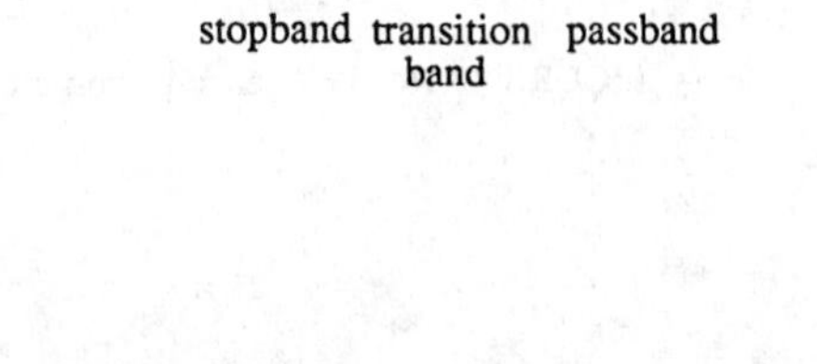

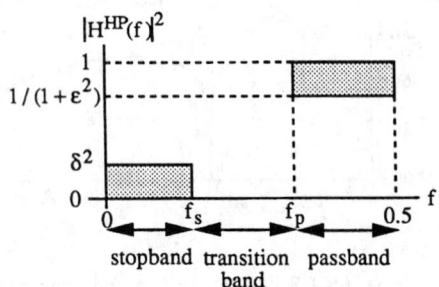

FIGURE 83.18 Digital HPF specification.

which means $\varepsilon = 0.5$, $\delta = 0.005$, $f_p = 0.4$, and $f_s = 0.3$. Since T does not affect the result, we let $T = 2$ for convenience. We calculate the analog LPF passband and stopband edges as $\lambda_p = \tan[\pi(0.5 + f_p)] = -0.3249$ rad/s and $\lambda_s = \tan[\pi(0.5 + f_s)] = -0.7265$ rad/s. Since the magnitude response is symmetry with respect to $\lambda = 0$, we use $\lambda_p = 0.3249$ rad/s and $\lambda_s = 0.7265$ rad/s. Therefore, $\lambda_s/\lambda_p = 2.2361$. Suppose that we choose the inverse Chebyshev filter as the analog prototype filter. From (83.37), we obtain the order $N \geq 4.6$. So, we use $N = 5$. From (83.39), the low-pass analog inverse Chebyshev filter transfer function is

$$H_a(s) = \frac{1.8160 \times 10^{-2}(s^2 + 0.5835)(s^2 + 1.5276)}{(s + 0.4822)(s^2 + 0.6772s + 0.2018)(s^2 - 0.2131s + 0.1663)}$$

To convert to digital filter system function, we replace s with $(1 + z^{-1})/(1 - z^{-1})$, yielding

$$H^{\mathrm{HP}}(z) = \frac{1.8920 \times 10^{-2}(1 - z^{-1})(1 + 0.5261z^{-1} + z^{-2})(1 - 0.4175z^{-1} + z^{-2})}{(1+0.3493z^{-1})(1+0.8496z^{-1} + 0.2792z^{-2})(1+1.2088z^{-1} + 0.6910z^{-2})}$$

The magnitude response and the phase response are plotted in Fig. 83.19.

Bandpass Filters (BPF's)

A magnitude response specification for a digital BPF is depicted in Fig. 83.20(a). Note that there are two passband edges (f_{p1} and f_{p2}) and two stopband edges (f_{s1} and f_{s2}). For the bilinear transformation $s = (2/T)(1 - z^{-1})/(1 + z^{-1})$ we can transform the digital BPF specification to an analog BPF specification by letting

$$\lambda_{p1} = \frac{2}{T}\tan(\pi f_{p1}) \quad \lambda_{p2} = \frac{2}{T}\tan(\pi f_{p2})$$
$$\lambda_{s1} = \frac{2}{T}\tan(\pi f_{s1}) \quad \lambda_{s2} = \frac{2}{T}\tan(\pi f_{s2}) \tag{83.61}$$

and keeping the same ε and δ.

Now, we need a transformation between an analog BPF and an analog LPF. To distinguish between the variable s and λ for the two filters, let us use s' and λ' for the analog LPF and s and λ for the analog BPF, respectively. A transformation for converting an analog LPF to an analog BPF is given by

$$s' = \frac{s^2 + \lambda_0^2}{Ws} \quad \text{or} \quad \lambda' = \frac{\lambda^2 - \lambda_0^2}{W\lambda} \tag{83.62}$$

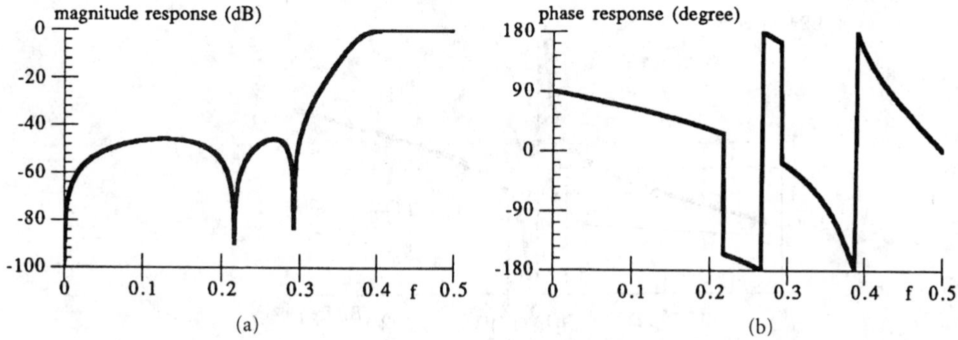

FIGURE 83.19 Frequency response of the HPF designed using bilinear transformation.

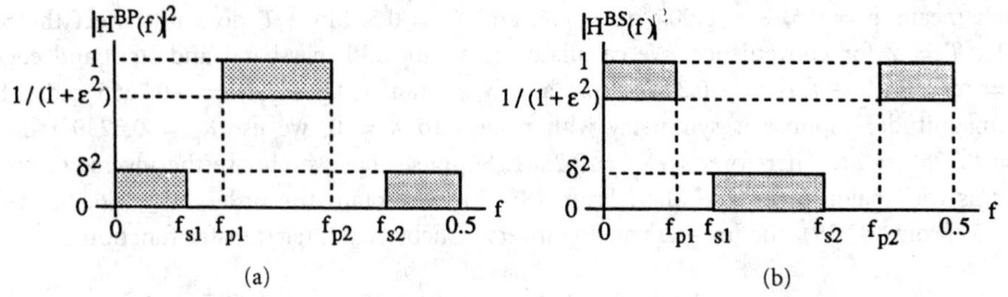

FIGURE 83.20 Magnitude specifications for digital BPF and BSF. (a) Digital BPF specification. (b) Digital BSF specification.

where

$$W = \lambda_{p2} - \lambda_{p1} \quad \text{and} \quad \lambda_0^2 = \lambda_{p1}\lambda_{p2} \tag{83.63}$$

Figure 83.21 depicts an example of the relation between λ and λ'. Note that λ_{p1} and λ_{p2} get mapped to $\lambda' = -1$ and $+1$, respectively. Therefore, the analog LPF has a passband edge of 1. The values of λ_{s1} and λ_{s2} get mapped to $\lambda'_{s1} = -|(\lambda_{s1}^2 - \lambda_0^2)/(W\lambda_{s1})|$ and $\lambda'_{s2} = |(\lambda_{s2}^2 - \lambda_0^2)/(W\lambda_{s2})|$. However, these two values may not be negative of each other. Since the analog LPF must have a symmetric magnitude response, we must use the more stringent of the two stopband edges, i.e., the smaller of $|\lambda'_{s1}|$ and $|\lambda'_{s2}|$. Letting

$$\lambda'_s = \min\{|\lambda'_{s1}|, |\lambda'_{s2}|\} = \min\left\{\left|\frac{\lambda_{s1}^2 - \lambda_0^2}{W\lambda_{s1}}\right|, \left|\frac{\lambda_{s2}^2 - \lambda_0^2}{W\lambda_{s2}}\right|\right\} \tag{83.64}$$

we now have the analog LPF specification. Therefore, a prototype analog LPF can be designed.

The design process can be summarized as follows. First, the desired digital BPF magnitude specification is converted to an analog BPF magnitude specification using (83.61). Then the analog BPF specification is converted to an analog LPF specification using λ'_s calculated from (83.64) and $\lambda'_p = 1$. Next, a prototype analog LPF is designed with the values of ε, δ, $\lambda'_p = 1$, and λ'_s, yielding an analog LPF transfer function $H_a(s')$. The LPF transfer function is converted to an analog BPF transfer function $H_a^{\mathrm{HP}}(s)$, using the transformation (from s' to s) given in (83.62). Finally, the analog BPF transfer function is converted to a digital BPF transfer function $H^{\mathrm{BP}}(z)$ using the bilinear transformation $s = (2/T)(1 - z^{-1})/(1 + z^{-1})$. As before, the value of T does not affect the result.

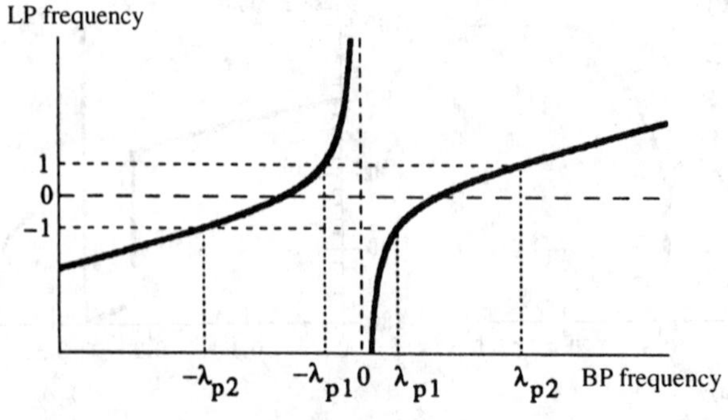

FIGURE 83.21 Relation between λ and λ' for bandpass-to-low-pass conversion.

For example, let the desired digital BPF have the following specification:

$$|H^{BP}(f)|^2 \begin{cases} \geq 0.8 \quad (-0.97\text{ dB}) & \text{for } 0.25 \leq f \leq 0.3 \\ \leq 2.5 \times 10^{-5} \quad (-46.02\text{ dB}) & \text{for } 0 \leq f \leq 0.2 \quad \text{and } 0.35 \leq f \leq 0.5 \end{cases}$$

which means $\varepsilon = 0.5$, $\delta = 0.005$, $f_{p1} = 0.25$, $f_{p2} = 0.3$, $f_{s1} = 0.2$, and $f_{s2} = 0.35$. Let $T = 2$ for convenience. Using $\lambda = \tan(\pi f)$, we obtain the analog BPF passband and stopband edges as $\lambda_{p1} = \tan(\pi f_{p1}) = 1.0$ rad/s, $\lambda_{p2} = \tan(\pi f_{p2}) = 1.3764$ rad/s, $\lambda_{s1} = \tan(\pi f_{s1}) = 0.7265$ rad/s, and $\lambda_{s2} = \tan(\pi f_{s2}) = 1.9626$ rad/s. Therefore, $\lambda_0^2 = 1.3764$ and $W = 0.3764$. So, we have $\lambda'_s = \min\{3.1030, 3.3509\} = 3.1030$ rad/s. Suppose that we use the elliptic LPF as an analog prototype filter. With $\varepsilon = 0.5$, $\delta = 0.005$, $\lambda'_p = 1$, and $\lambda'_s = 3.1030$ rad/s, we need an elliptic filter of order $N = 3$. The low-pass analog elliptic filter transfer function is

$$H_a(s') = \frac{4.1129 \times 10^{-2}(s'^2 + 12.6640)}{(s' + 0.5174)(s'^2 + 0.4763s' + 1.0067)}$$

Replacing s' with $(s^2 + 1.3764)/(0.3764s)$ yields the analog BPF transfer function

$$H_a^{BP}(s) = \frac{1.5480 \times 10^{-2}s(s^4 + 4.5467s^2 + 1.8944)}{(s^2 + 0.1947s + 1.3764)(s^4 + 0.1793s^3 + 2.8953s^2 + 0.2467s + 1.8944)}$$

Note that an Nth-order LPF becomes a $2N$th-order BPF. To convert to digital filter system function, we replace s with $(1 - z^{-1})/(1 + z^{-1})$, yielding

$$H^{BP}(z) = \frac{7.2077 \times 10^{-3}(1 - z^{-2}) \times (1 + 0.4807z^{-1} + 1.1117z^{-2} + 0.4807z^{-3} + z^{-4})}{(1 + 0.2928z^{-1} + 0.8485z^{-2}) \times (1 + 0.5973z^{-1} + 1.8623z^{-2} + 0.5539z^{-3} + 0.8629z^{-4})}$$

The magnitude and phase responses are plotted in Fig. 83.22.

Note that for the transformation in (83.62), we can also let $W = \lambda_{s2} - \lambda_{s1}$ and $\lambda_0^2 = \lambda_{s1}\lambda_{s2}$, instead of (83.63). Such a choice will give $\lambda'_s = 1$. The passband edge for the prototype LPF is now calculated from $\lambda'_p = \min\{|(\lambda_{p1}^2 - \lambda_0^2)/(W\lambda_{p1})|, |(\lambda_{p2}^2 - \lambda_0^2)/(W\lambda_{p2})|\}$.

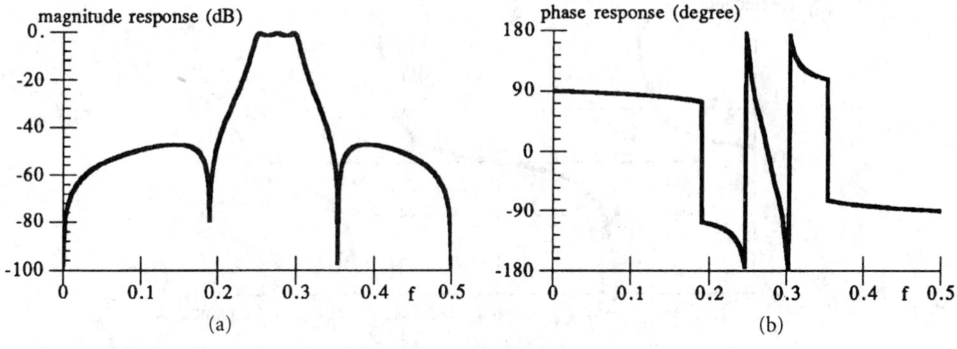

FIGURE 83.22 Frequency response of the designed digital BPF.

Bandstop Filters (BSF's)

A digital BSF specification is depicted in Fig. 83.20(b). As in the case of the BPF there are two passband edges (f_{p1} and f_{p2}) and two stopband edges (f_{s1} and f_{s2}). A transformation from analog BSF to analog LPF is given by

$$s' = \frac{Ws}{s^2 + \lambda_0^2} \quad \text{or} \quad \lambda' = \frac{W\lambda}{\lambda_0^2 - \lambda^2} \tag{83.65}$$

where W and λ_0^2 are given by (83.63). Note that the expression for s in (83.65) is the reciprocal of that in (83.62). The relation between the LPF frequency λ' and the BSF frequency λ is depicted in Fig. 83.23. The passband edges λ_{p1}, and λ_{p2} get mapped to $\lambda' = 1$ and -1, respectively. The values of λ_{s1} and λ_{s2} get mapped to $\lambda'_{s1} = |W\lambda_{s1}/(\lambda_0^2 - \lambda_{s1}^2)|$ and $\lambda'_{s2} = -|W\lambda_{s2}/(\lambda_0^2 - \lambda_{s2}^2)|$. Therefore, the passband edge and stopband edge of the prototype analog LPF are 1 and λ'_s, respectively, where

$$\lambda'_s = \min\{|\lambda'_{s1}|, |\lambda'_{s2}|\} = \min\left\{\left|\frac{W\lambda_{s1}}{\lambda_0^2 - \lambda_{s1}^2}\right|, \left|\frac{W\lambda_{s2}}{\lambda_0^2 - \lambda_{s2}^2}\right|\right\} \tag{83.66}$$

The design process for the BSF can follow the same process as the design for the BPF, except that we use (83.65) and (83.66) instead of (83.62) and (83.64).

Similar to the case of the BPF, we can also let $W = \lambda_{s2} - \lambda_{s1}$ and $\lambda_0^2 = \lambda_{s1}\lambda_{s2}$, instead of (83.63), for the transformation in (83.65). The stopband edge and the passband edge for the prototype LPF are now $\lambda'_s = 1$ and $\lambda'_p = \min\{|W\lambda_{p1}/(\lambda_{p1}^2 - \lambda_0^2)|, |W\lambda_{p2}/(\lambda_{p2}^2 - \lambda_0^2)|\}$.

Design Using Digital Frequency Transformations

This method assumes that we can design a digital LPF. The desired filter is then obtained from the digital LPF by transforming the digital LPF in the z domain. Let us denote the z variable for the digital LPF by z' and that for the desired digital filter by z. Similarly, we use f' for the digital frequency of the digital LPF and f for the frequency of the desired digital filter. Suppose that the digital LPF has system function $H(z')$ and the desired digital filter has system function $H^x(z)$, where x stands for LP, HP, BP, or BS. The system function $H^x(z)$ is obtained from $H(z')$ by replacing z' with an appropriate function of z. The LPF $H(z')$ can be designed using the method discussed in the subsection on design using analog prototype filters, or by some other means. The specification for the digital LPF is obtained from the specification of the desired digital filter through the relation between f' and f. The relation depends on the specific transformation. Note that the difference between the method in this subsection and the

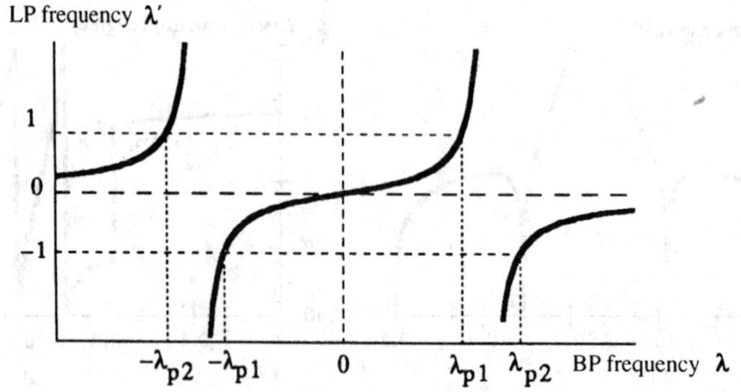

FIGURE 83.23 Relation between λ and λ' for bandstop-to-low-pass conversion.

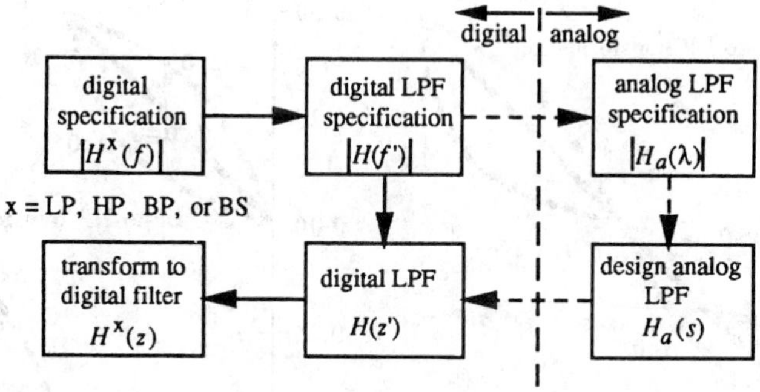

FIGURE 83.24 Design process using digital frequency transformation.

method described above is that the transformation between the desired type of filter and the LPF is in the digital domain (the z domain) for the current method, whereas it is in the analog domain (the s domain) in the previous method. Figure 83.24 shows the design process using digital frequency transformation. The advantage of the current method is that in designing a desired digital HPF, BPF, or BSF, we design a digital LPF, which can make use of the impulse invariant method, in addition to the bilinear transformation. This is not the case for the method discussed previously, due to excessive aliasing.

Low-Pass Filters

We can transform a digital LPF to a digital LPF using the transformation

$$z' = \frac{z+\alpha}{1+\alpha z} \qquad |\alpha| < 1 \tag{83.67}$$

With $z = \exp(j2\pi f)$ and $z' = \exp(j2\pi f')$, we can show that the two digital frequencies are related by

$$f' = \frac{1}{2\pi} \arctan\left[\frac{(1-\alpha^2)\sin 2\pi f}{2\alpha + (1+\alpha^2)\cos 2\pi f}\right] \tag{83.68}$$

The relation (83.68) is plotted in Fig. 83.25(a). If $\alpha = 0$, then $z' = z$ and $f' = f$, which is the trivial case. When $\sigma \neq 0$, there is frequency warping introduced by the transformation. After choosing α, we can transform the desired digital LPF specification to another digital LPF specification, i.e., calculate f_p' and f_s' from f_p and f_s. With f_p, f_s, ε and δ, a digital LPF can then be designed to satisfy the specification. The resulting system function is then transformed to the desired LPF using (83.67). This method may yield a filter of lower or higher order (due to the frequency warping) compared to the case that there is no digital frequency transformation ($\alpha = 0$).

As an alternative to specifying α, we can specify f_p', which, together with f_p, specifies the value of α, according to (83.68). With the value of α, we can calculate f_s' from f_s. We can also exchange the role of the passband edge with the stopband edge, i.e., we specify f_s' and compute α from the values of f_s' and f_s. With α, f_p' can be determined from f_p.

High-Pass Filters

To transform a digital LPF to a digital HPF, we can use the transformation

$$z' = -\frac{z+\alpha}{1+\alpha z} \qquad |\alpha| < 1 \tag{83.69}$$

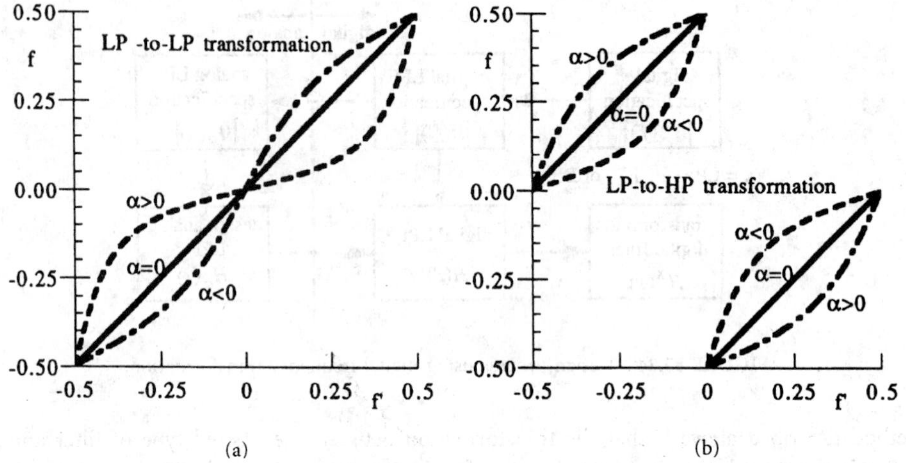

FIGURE 83.25 Frequency relation for digital frequency transformations: (a) low-pass to low-pass; (b) low-pass to high-pass.

Substituting $z = \exp(j2\pi f)$ and $z' = \exp(j2\pi f')$, we obtain the relation between the two digital frequencies:

$$f' = \frac{1}{2\pi} \arctan\left[\frac{-(1-\alpha^2)\sin 2\pi f}{-2\alpha - (1+\alpha^2)\cos 2\pi f}\right] \tag{83.70}$$

This relation is plotted in Fig. 83.25(b). If $\alpha = 0$, then $z' = -z$ and $f' = f + \pi$. The design process proceeds as follows. After choosing the value of α, the desired digital HPF specification is transformed to the digital LPF specification, using the relation in (83.70). Using the resulting values of f_s', f_p', together with the ripple specifications (ε and δ), a digital LPF is designed to satisfy the specification. The resulting LPF system function $H(z')$ is then transformed to the desired HPF by substituting z' with $-(z+\alpha)/(1+\alpha z)$, given by (83.69). Similar to the case of the LPF, we can specify f_p' (instead of α) and calculate the required value of α from f_p' and f_p.

Band-Pass Filters

To transform a digital LPF to a digital HPF, we can use the transformation

$$z' = -\frac{1 + \dfrac{2\alpha k}{k+1} z + \dfrac{k-1}{k+1} z^2}{\dfrac{k-1}{k+1} + \dfrac{2\alpha k}{k+1} z + z^2} \qquad |\alpha| < 1, \quad k > 0 \tag{83.71}$$

This implies the following relation between the two digital frequencies:

$$f' = \frac{1}{2\pi} \arctan\left[\frac{(1-b)\{2a \sin 2\pi f + (1+b)\sin 4\pi f\}}{-a^2 - 2b - 2a(1+b)\cos 2\pi f - (b^2+1)\cos 4\pi f}\right] \tag{83.72}$$

where $a = 2\alpha k/(k+1)$ and $b = (k-1)/(k+1)$. An example of (83.72) is plotted in Fig. 83.26. If $\alpha = 0$, the curve would be odd symmetric with respect to $f = 0.25$. The design process is similar to the case of the HPF, except that there are now two passband edges, f_{p1}', and $|f_{p2}'|$, and two stopband edges, f_{s1}' and $|f_{s2}'|$, for the digital LPF to satisfy (see Fig. 83.26). To satisfy both sets, we let $f_p' = \max\{f_{p1}', |f_{p2}'|\}$ and $f_s' = \min\{f_{s1}', |f_{s2}'|\}$ be the passband and stopband edges for the digital LPF filter.

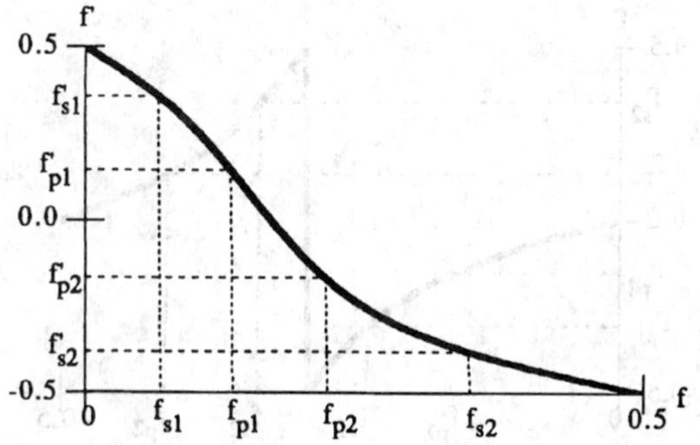

FIGURE 83.26 Frequency relation for BP to LP digital frequency transformation.

If we specify f_p', then together with f_{p1} and f_{p2} they determine the values of α and k:

$$\alpha = \frac{\cos[\pi(f_{p2}+f_{p1})]}{\cos[\pi(f_{p2}-f_{p1})]} \qquad k = \cot[\pi(f_{p2}-f_{p1})]\tan(\pi f_p') \tag{83.73}$$

With the value of α and k, we calculate the values of f_{s1}' and f_{s2}' from (83.72) and let $f_s' = \min\{f_{s1}', |f_{s2}'|\}$. Thus we have f_p', f_s', ε, and δ as the digital LPF specification. After a digital LPF is designed to satisfy this specification, it is converted to digital BPF by the transformation in (83.71)

Bandstop Filters

To transform a digital LPF to a digital BSF, we can use

$$z' = \frac{1 + \dfrac{2\alpha k}{k+1} z + \dfrac{k-1}{k+1} z^2}{\dfrac{k-1}{k+1} + \dfrac{2\alpha k}{k+1} z + z^2} \qquad |\alpha| < 1 \quad k > 0 \tag{83.74}$$

The corresponding relation between the two digital frequencies is

$$f' = \frac{1}{2\pi}\arctan\left[\frac{-(1-b)\{2a\sin 2\pi f + (1+b)\sin 4\pi f\}}{a^2 + 2b + 2a(1+b)\cos 2\pi f + (b^2+1)\cos 4\pi f}\right] \tag{83.75}$$

where $a = 2\alpha k/(k+1)$ and $b = (k-1)/(k+1)$. An example is plotted in Fig. 83.27. The design process is the same as described in the subsection on bandpass filters.

When f_p' is specified, together with f_{p1} and f_{p2}, the values of α and k can be calculated from

$$\alpha = \frac{\cos[\pi(f_{p2}+f_{p1})]}{\cos[\pi(f_{p2}-f_{p1})]} \qquad k = \tan[\pi(f_{p2}-f_{p1})]\tan(\pi f_p') \tag{83.76}$$

With these values, we can calculate the values of f_{s1}' and f_{s2}' from (83.75). Letting $f_s' = \min\{f_{s1}', |f_{s2}'|\}$, we now have f_p', f_s', ε, and δ, which constitute the digital LPF specification. A digital LPF is then designed and converted to digital BSF by the transformation in (83.74).

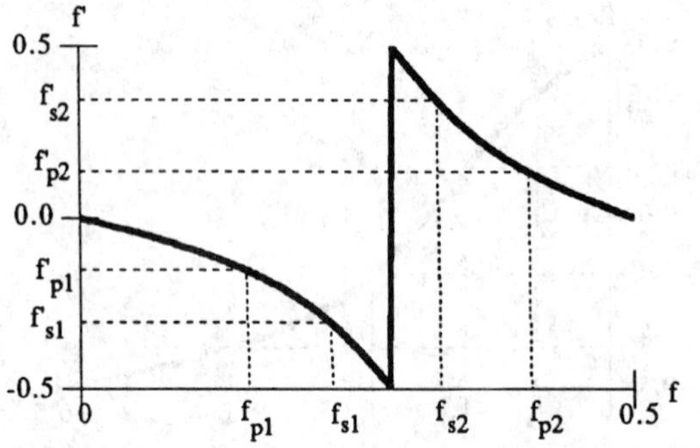

FIGURE 83.27 Frequency relation for BS to LP digital frequency transformation.

Computer-Aided Designs

The general idea is to use an algorithm to search for the set of coefficients such that the resulting response (magnitude and/or phase) is "close" to the desired response. The "closeness" is in some well-defined sense. The advantage of such method is that it can be used to design nonstandard filters, such as multiband filters, phase equalizers, differentiators, etc. However, it requires a computer program to execute the algorithm. In addition, it usually cannot directly determine the order N such that the passband and stopband ripples are within the desired ranges. The order is usually determined through several trials.

Let us put the IIR system function in the form

$$H(z) = b_0 \prod_{i=1}^{K} \frac{1 + b_{i1} z^{-1} + b_{i2} z^{-2}}{1 + a_{i1} z^{-1} + a_{i2} z^{-2}} = \frac{N(z)}{D(z)} \tag{83.77}$$

The constant b_0 is included in the numerator $N(z)$. The design problem involves choosing $4K + 1$ coefficients: $(b_0, b_{11}, b_{12}, a_{11}, a_{12}, \cdots, b_{K1}, b_{K2}, a_{K1}, a_{K2})$, which will be denoted by the vector $\underline{\gamma}$.

The frequency response is written in terms of the magnitude response and phase response as

$$H(f) = \frac{N(f)}{D(f)} = A(f) e^{j\theta(f)} \tag{83.78}$$

where $A(f) = |H(f)|$ is the magnitude response and $\theta(f) = \arctan\{\text{Imag}[H(f)]/\text{Real}[H(f)]\}$ is the phase response. The group delay is

$$\tau(f) = -\frac{1}{2\pi} \frac{d}{df} \theta(f) \tag{83.79}$$

Suppose that the desired frequency response is given by

$$H_d(f) = \frac{N_d(f)}{D_d(f)} = A_d(f) e^{j\theta_d(f)} \tag{83.80}$$

with $\tau_d(f)$ being the group delay. As mentioned above, $H_d(f)$ does not have to be one of the standard filters.

Our objective is to choose the coefficients $\underline{\gamma}$ so that the resulting filter response is close to the desired response. The closeness may be defined over the whole frequency band $0 \le f \le 0.5$, or only over certain frequency bands. For example, if some frequency bands are designated as transition bands, then they are not included in the computation of the closeness. To use computer algorithms, we sample the response at discrete frequencies, say at L frequencies: $0 \le f_i, \cdots, f_L \le 0.5$. These frequencies must be sufficiently dense over the frequency bands that the closeness is computed. To accommodate different levels of ripples in various frequency bands, we also include a weighting function $W(f_i)$, $1 \le i \le L$, in the computation of the closeness.

A suitable measure for the closeness is through the pth error, defined as

$$E(\underline{f}, \underline{\gamma}) = \left[\sum_{i=1}^{L} W(f_i) |H_d(f_i) - H(f_i)|^p \right]^{1/p} \tag{83.81}$$

where $\underline{f}$ is the vector representing $(f_1, \cdots, f_L)$. If only the magnitude is of interest, then the error is

$$E_A(\underline{f}, \underline{\gamma}) = \left[\sum_{i=1}^{L} W(f_i) |A_d(f_i) - A(f_i)|^p \right]^{1/p} \tag{83.82}$$

and if only the group delay is of interest the error is

$$E_\tau(\underline{f}, \underline{\gamma}) = \left[\sum_{i=1}^{L} W(f_i) |\tau_d(f_i) - \tau(f_i)|^p \right]^{1/p} \tag{83.83}$$

The magnitude error and the group delay error can also be linearly combined as

$$E_c(\underline{f}, \underline{\gamma}) = \beta E_A(\underline{f}, \underline{\gamma}) + (1 - \beta) E_\tau(\underline{f}, \underline{\gamma}) \quad 0 \le \beta \le 1 \tag{83.84}$$

Given the weighting function, we seek the set of coefficients $\underline{\gamma}$ that minimizes one of the errors in (83.81)–(83.84). When $p = 1$, the error is the weighted absolute error and the minimization yields the minimum absolute error solution. When $p = 2$, the error is the weighted root-mean-square error and the solution is the minimum weighted root-mean-square error. When $p \to \infty$, the largest error dominates. Consequently, the pth error is the maximum over $1 \le i \le L$ of the weighted error. For example, (83.81) becomes

$$E(\underline{f}, \underline{\gamma}) = \max_{1 \le i \le L} W(f_i) |H_d(f_i) - H(f_i)| \tag{83.85}$$

Minimizing $E(\underline{f}, \underline{\gamma})$ yields the minimax solution. A minimax solution gives equiripples in each of the bands having equal weighting function inside the band.

There are algorithms for obtaining the solutions that minimize the above errors; we do not present them here. For detail see the references. Several programs have been published previously in [3].

eferences

[1] R. W. Daniels, *Approximation Methods for the Design of Passive, Active, and Digital Filters*, New York: McGraw-Hill, 1974.

[2] L. R. Rabiner and B. Gold, *Theory and Application of Digital Signal Processing*, Englewood Cliffs, NJ: Prentice-Hall, 1975.

[3] IEEE DSP Comm., IEEE, New York: *Programs for Digital Signal Processing*. 1979.

[4] A. V. Oppenheim and R. W. Schafer, *Discrete-Time Signal Processing*, Englewood Cliffs, NJ: Prentice-Hall, 1989.

[5] J. G. Proakis and D. G. Manolakis, *Digital Signal Processing Principles, Algorithms, and Applications*, 2nd ed., New York: Macmillan, 1992.
[6] A. Antoniou, *Digital Filters Analysis, Design, and Applications*, 2nd ed., New York: McGraw-Hill, 1993.
[7] A. G. Constantinides, "Spectral transformations for digital filters," *Proc. IEE*, vol. 117, no. 8, pp. 1585–1590, Aug. 1970.
[8] K. Steiglitz, "Computer-aided design of recursive digital filters," *IEEE Trans. Audio Electroacoust.*, vol. AU-18, 123–129, June 1970.
[9] A. G. Deczky, "Synthesis of recursive digital filters using the minimum p-error criterion," *IEEE Trans. Audio Electroacoust.*, vol. AU-20, pp. 257–263, Oct. 1972.
[10] L. R. Rabiner, N. Y. Graham, and H. D. Helms, "Linear programming design of IIR digital filters with arbitrary magnitude function," *IEEE Trans. Acoust., Speech, Signal Processing*, vol. ASSP-22, pp. 117–123, Apr. 1974.
[11] J. W. Bandler and B. J. Bardakjian, "Least *p*th optimization of recursive digital filters," *IEEE Trans. Audio Electroacoust.*, vol. AU-21, pp. 460–470, Oct. 1973.

83.3 Wave Digital Filters

Stuart S. Lawson

Introduction

The wave digital filter (WDF) belongs to the subclass of digital filters that are derived from suitable analog networks by the application of the bilinear transformation. However, they differ from most other methods that use the bilinear transformation because the design technique aims to preserve desirable attributes in the original analog reference network. This is achieved by using a scattering parameter formulation that can be viewed as a linear transformation on the voltages and currents to yield wave variables that will be used in the final wave digital filter. In this subsection we will see how the wave digital equivalents of well-known analog components such as resistors, inductors, capacitors, and unit elements are derived. More importantly, the design process will be explained in detail with many examples to illustrate its use.

The desirable attributes we mentioned above relate to the low sensitivity to element variations in the case of double-terminated lossless analog filters. Sensitivity in digital filters is not quite the same as it is in analog filters, but normally we can say that low sensitivity implies low coefficient quantization error and in certain circumstances low roundoff noise. These analog filters can be designed so that, at frequencies of minimum loss, the source will deliver maximum power into the load. At these points of maximum available power transfer (MAP), the derivative of the loss with respect to any reactive component is zero. Although the number of MAP points in the filter's passband is finite, we can be fairly certain that at other passband frequencies, the attenuation sensitivity will be small. This can be achieved for the classical filter approximations, e.g., Chebyshev, Butterworth, elliptic, etc. This property has led to the replacement of bulky inductors by gyrators and to the concept of the frequency dependent negative resistance. Effectively, the inductor is being simulated by active circuit elements such as operational amplifiers. The technique has been extended to the design of switched capacitor filters as well.

Similarly, we can design digital filters that are modeled on the behavior of lossless analog networks. The approach we will take consists of representing the behavior of the analog components in the filter by wave variables. The analog filters that we shall call henceforth **reference** filters, that will be of use to us are as follows:

1. Double-terminated lossless ladder network.
2. Double-terminated lossless lattice network.
3. Double-terminated lossless transmission-line network.

To derive a WDF there are two main design approaches that both lead to the same type of structure. We will call the first approach the **one-port approach** as it treats resistors, inductors, and capacitors as one-ports and derives their digital equivalents. Interconnection is provided by **adaptors**, which are digital representations of series and parallel junctions. In the **two-port approach**, all components are treated as two-ports and interconnection can be performed directly without the use of adaptors. However, there are special arrangements for the terminations.

Wave Digital Filter Principles

The first stage in transforming an analog network into wave digital form is to represent the circuit equations in terms of wave variables. For an n-port circuit element, the transformation from voltage and current V_k and I_k, to wave variables A_k and B_k, $k = 1, \cdots, n$ is as follows:

$$\begin{bmatrix} A_k \\ B_k \end{bmatrix} = \begin{bmatrix} 1 & R_k \\ 1 & -R_k \end{bmatrix} \begin{bmatrix} V_k \\ I_k \end{bmatrix} \tag{83.86}$$

where R_k is the **port resistance** at port k and is a free parameter whose value will be set to avoid delay-free loops. A_k and B_k are known, respectively, as the incident and reflected wave at port k from the analogy with scattering parameter theory. This transformation is known as the voltage wave formulation since A_k and B_k have the dimensions of volts. It is clear that the transformation from voltage and current to voltage waves is linear.

For a given one-port circuit element ($n = 1$), the voltage–current relationship is known and is of the form $V = IZ$ for an impedance Z. Thus if we eliminate V and I using (83.86) we find that

$$\frac{B}{A} = \frac{Z - R}{Z + R} \tag{83.87}$$

Equation (83.87) is the reflection coefficient S for the impedance Z. Thus $B = SA$ and we can think of A as input. B as output, and S as the transfer function.

The behavior of a passive two-port (Fig. 83.28) can be described by *ABCD* parameters in the form

$$\begin{bmatrix} V_1 \\ I_1 \end{bmatrix} = \begin{bmatrix} A & B \\ C & D \end{bmatrix} \begin{bmatrix} V_2 \\ I_2 \end{bmatrix} \tag{83.88}$$

where both currents I_1 and I_2 flow into the network B and D will subsequently have different signs than usual. Our convention makes the algebra a little easier when cascading two ports. We shall refer to the 2×2 matrix of (83.88) as a *modified ABCD* matrix.

The voltages and currents in (83.86) and (83.88) can be eliminated to obtain the following:

$$\begin{bmatrix} A_1 \\ B_1 \end{bmatrix} = \begin{bmatrix} \alpha & \beta \\ \gamma & \delta \end{bmatrix} \begin{bmatrix} A_2 \\ B_2 \end{bmatrix} \tag{83.89}$$

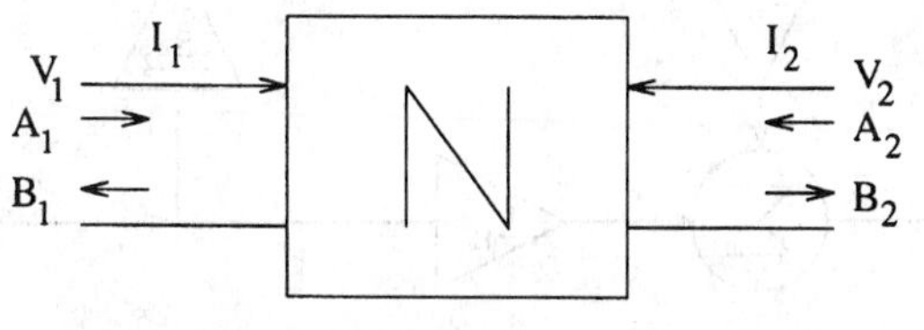

FIGURE 83.28 Passive two-port network.

where

$$\alpha = \frac{1}{2}(A + CR_1 + BG_2 + DR_1G_2)$$

$$\beta = \frac{1}{2}(A + CR_1 - BG_2 - DR_1G_2)$$

$$\gamma = \frac{1}{2}(A - CR_1 + BG_2 - DR_1G_2)$$

$$\delta = \frac{1}{2}(A - CR_1 - BG_2 + DR_1G_2)$$

and $G_2 = 1/R_2$.

In a similar way to the one-port case we are interested in obtaining the relationship

$$\underline{B} = \mathbf{S}\underline{A} \tag{83.90}$$

but now **S** is a 2 × 2 scattering matrix and $\underline{B}$ and $\underline{A}$ are the 2-element output and input vectors, respectively. The elements of **S** can be derived from (83.89) and are as follows:

$$\left.\begin{aligned} S_{11} &= \delta/\beta \\ S_{12} &= -\Delta/\beta \\ S_{21} &= 1/\beta \\ S_{22} &= -\alpha/\beta \end{aligned}\right\} \tag{83.91}$$

where

$$\Delta = \begin{vmatrix} \alpha & \beta \\ \gamma & \delta \end{vmatrix} = -R_1G_2$$

The signal-flow graph (SFG) of a general WDF two port is shown in Fig. 83.29 using the scattering parameter representation of (83.90). Cascading two ports together, it is clear that loops will be formed due to the terms S_{11} and S_{22}. These loops cause problems only if they are delay free and so it is necessary to ensure that either the transfer function for S_{11} or S_{22} has a factor of z^{-1}. This leads to two possibilities when deriving SFG's for two ports.

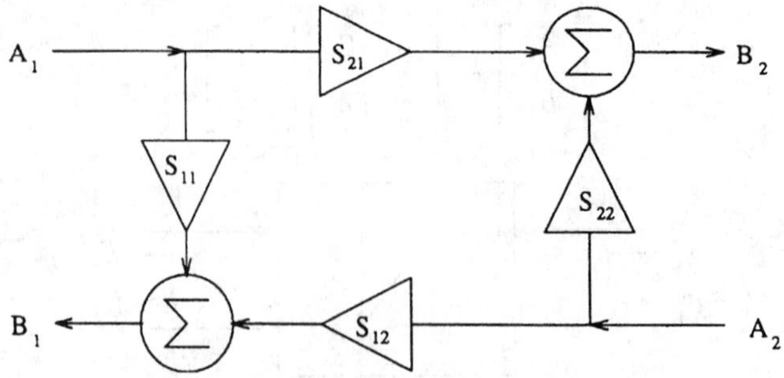

FIGURE 83.29 Signal-flow graph of general WDF two port.

An important and useful side effect of using wave variables is that the resulting WDF network has two inputs and two outputs. With reference to Fig. 83.29, normally A_2 is set to zero, the two outputs are then related by the following:

$$|B_1|^2 + |B_2|^2 = 1$$

so that if B_1 is the low-pass filtered version of A_1, then B_2 is the high-pass filtered version, etc.

Building Blocks

One Ports

For an inductor $V = (sL)I$, and on using (83.87) with $R = L$, we find that

$$\frac{B}{A} = \frac{s-1}{s+1} \tag{83.92}$$

and, on applying the bilinear transformation, we get simply that

$$B = -z^{-1}A \tag{83.93}$$

The SFG or *wave-flow diagram (WFD)* of (83.93) is shown in Fig. 83.30. Thus for an inductor, the WFD is nothing more than a delay and sign inversion. However, the consequences of setting $R = L$ will be felt later when we wish to interconnect components.

Using a similar technique, the WFD's for other two-terminal devices can be easily derived. The most useful ones are summarized in Fig. 83.30.

Two Ports

Introduction. In this subsection we look at the WFD's of series inductors and shunt capacitances treated as two ports. With these two, we are able to derive straightforwardly the WFD's of other two ports required in modeling double-terminated lossless ladder networks. It turns out that for each case, series L or shunt C, there are two WFD's corresponding to whether S_{11} or S_{22} has a delay-free path.

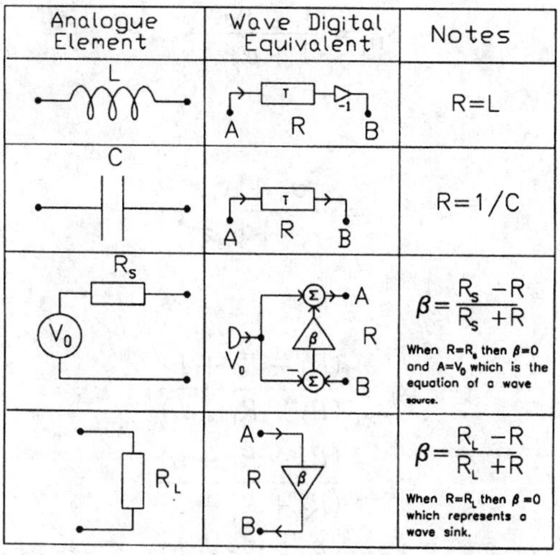

FIGURE 83.30 Wave-flow diagrams (WFD) of basic one-port circuit elements.

Series Inductor. The modified *ABCD* matrix of a series inductor is given by

$$\begin{bmatrix} 1 & -sL \\ 0 & -1 \end{bmatrix} \tag{83.94}$$

The elements of the scattering matrix **S** are

$$\left.\begin{aligned} S_{11} &= \frac{(R_2 - R_1 + sL)}{(R_2 + R_1 + sL)} \\ S_{12} &= \frac{2R_1}{(R_2 + R_1 + sL)} \\ S_{21} &= \frac{2R_2}{(R_2 + R_1 + sL)} \\ S_{22} &= \frac{(R_1 - R_2 + sL)}{(R_2 + R_1 + sL)} \end{aligned}\right\} \tag{83.95}$$

subject to the following constraints:

$$\left.\begin{aligned} S_{11} + S_{12} &= 1 \\ S_{21} + S_{22} &= 1 \end{aligned}\right\} \tag{83.96}$$

Combining (83.90) and (83.95) gives

$$\left.\begin{aligned} B_1 &= S_{11}(A_1 - A_2) + A_2 \\ B_2 &= S_{22}(A_2 - A_1) + A_1 \end{aligned}\right\} \tag{83.97}$$

Thus we need only realize S_{11} and S_{22} to define the series inductance. Finally, if we now apply the bilinear transformation to the equations for S_{11} and S_{22} we find that, after simplification,

$$S_{11} = \frac{\alpha_1 + \alpha_3 z^{-1}}{1 + \alpha_2 z^{-1}} \tag{83.98}$$

and

$$S_{22} = -\frac{\alpha_3 + \alpha_1 z^{-1}}{1 + \alpha_2 z^{-1}} \tag{83.99}$$

where

$$\begin{aligned} \alpha_1 &= \frac{(R_2 - R_1 + L)}{(R_2 + R_1 + L)} \\ \alpha_2 &= \frac{(R_2 + R_1 - L)}{(R_2 + R_1 + L)} \\ \alpha_3 &= \frac{(R_2 - R_1 - L)}{(R_2 + R_1 + L)} \end{aligned}$$

In addition, a linear constraint exists between these multiplier constants,

$$\alpha_1 + \alpha_2 = 1 + \alpha_3$$

which will enable us to reduce their number to two. To avoid delay-free loops on interconnection, using the arguments of the previous subsection, either α_1 or α_3 must be zero. This gives rise to two possible SFG's.

For the first case, let $\alpha_1 = 0$ then $R_1 = R_2 + L$ and

$$S_{11} = \frac{\alpha_3 z^{-1}}{1 + \alpha_2 z^{-1}} \quad (83.100)$$

and

$$S_{22} = -\frac{\alpha_3}{1 + \alpha_2 z^{-1}} \quad (83.101)$$

and

$$\alpha_2 = 1 + \alpha_3 = R_2/R_1 \quad (83.102)$$

It is important to note that (i) there is now only one independent multiplier constant, (ii) port 1 resistance is dependent on the value of the inductor and port 2 resistance, and (iii) $S_{11} = -S_{22}z^{-1}$.

Finally, the WFD of the series inductor can be found by combining (83.97) and (83.100)–(83.102). It is canonic in delay and multipliers.

For the second case, let $\alpha_3 = 0$ then $R_2 = R_1 + L$ and

$$S_{11} = \frac{\alpha_1}{1 + \alpha_2 z^{-1}} \quad (83.103)$$

and

$$S_{22} = -\frac{\alpha_1 z^{-1}}{1 + \alpha_2 z^{-1}} \quad (83.104)$$

and

$$\alpha_2 = 1 - \alpha_1 = R_1/R_2 \quad (83.105)$$

Again, it is important to note that (i) there is now only one independent multiplier constant, (ii) port 2 resistance is dependent on the value of the inductor and port 1 resistance, and (iii) $S_{22} = -S_{11}z^{-1}$.

Finally, the WFD of the series inductor can be found by combining (83.97) and (83.103)–(83.105). It is also canonic in delay and multipliers.

Shunt Capacitor. For a shunt capacitor of value C farads, the WFD equations are as follows:

$$\left.\begin{aligned} B_1 &= (S_{11}A_1 + S_{22}A_2) + A_2 \\ B_2 &= (S_{11}A_1 + S_{22}A_2) + A_1 \end{aligned}\right\} \qquad (83.106)$$

As in the series inductor case, there are two cases to consider, either $\alpha_1 = 0$ or $\alpha_3 = 0$. If $\alpha_1 = 0$ then

$$G_1 = G_2 + C \qquad (83.107)$$

$$S_{22} = \frac{-\alpha_3}{1 + \alpha_2 z^{-1}} \qquad (83.108)$$

$$S_{11} = -S_{22} z^{-1} \qquad (83.109)$$

$$\alpha_2 = 1 - \alpha_3 = G_2/G_1 \qquad (83.110)$$

and if $\alpha_3 = 0$ then,

$$G_2 = G_1 + C \qquad (83.111)$$

$$S_{11} = \frac{\alpha_1}{1 + \alpha_2 z^{-1}} \qquad (83.112)$$

$$S_{22} = -S_{11} z^{-1} \qquad (83.113)$$

$$\alpha_2 = 1 + \alpha_1 = G_1/G_2 \qquad (83.114)$$

The signal flow graphs of the series inductor and shunt capacitance for each of the two cases are to be found in Figs. 83.31 and 83.32, respectively.

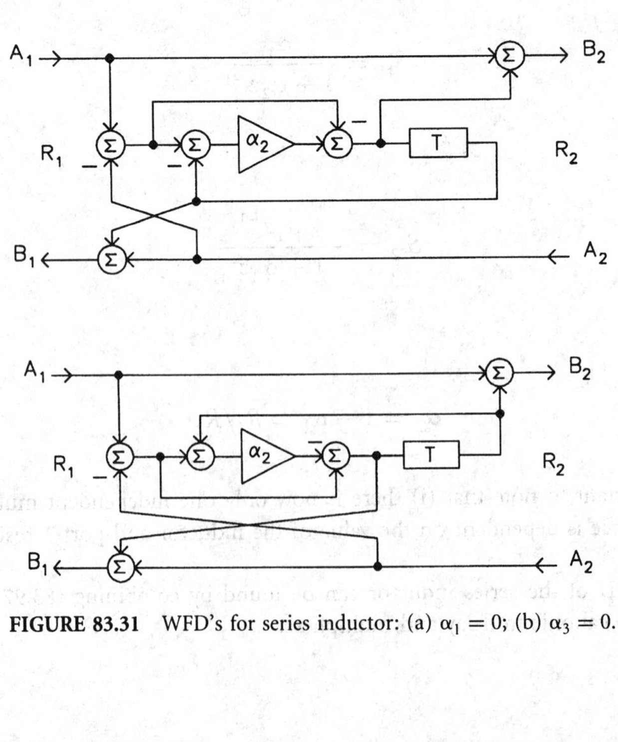

FIGURE 83.31 WFD's for series inductor: (a) $\alpha_1 = 0$; (b) $\alpha_3 = 0$.

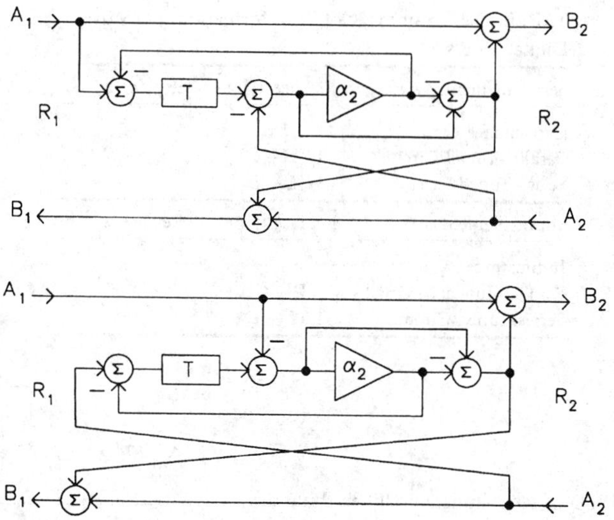

FIGURE 83.32 WFD's for shunt capacitor: (a) $\alpha_1 = 0$; (b) $\alpha_3 = 0$.

The Unit Element. The modified *ABCD* matrix of a lossless transmission line (unit element) is

$$\begin{bmatrix} \cos\theta & -jZ_0 \sin\theta \\ jY_0 \sin\theta & -\cos\theta \end{bmatrix} \tag{83.115}$$

where Z_0 is the characteristic impedance of the line and $Y_0 = 1/Z_0$. Also, $\theta = k\Omega$, where k is the line constant and Ω is the angular frequency.

On substituting for A, B, C, and D from (83.115) into (83.89) and (83.91), setting $R_1 = R_2 = Z_0$ and applying the bilinear transform, we find that

$$\left.\begin{aligned} S_{11} &= 0 \\ S_{12} &= z^{-1/2} \\ S_{21} &= z^{-1/2} \\ S_{22} &= 0 \end{aligned}\right\} \tag{83.116}$$

Thus the WFD of a unit element is a half unit delay path from port 1 to port 2 and the same from port 2 to port 1. The delays may be combined into either path to form a unit delay. The effect on the overall transfer function being a linear phase shift.

Building Blocks for Other Series and Shunt Elements. The WFD's of other network elements, e.g., a series-tuned circuit in the shunt-arm or a series capacitance can be obtained by simple transformations on the WFD's of a series inductor or shunt capacitor.

As an example consider the series-tuned circuit in the shunt arm, the impedance Z of which is given by

$$Z = sL + 1/sC \tag{83.117}$$

Applying the bilinear transformation we find that

$$Z = (D + L)\frac{(1 + \mathscr{T})}{(1 - \mathscr{T})} \tag{83.118}$$

TABLE 83.2 Summary of Transformations for Wave Digital Filters

Series Elements	Replace L by	Replace z^{-1} by
Capacitance C	D	$-z^{-1}$
Parallel-Tuned Circuit	$1/(\Gamma + C)$	$-\mathcal{T}$
Series-Tuned Circuit	$1/(L + D)$	$\mathcal{T}$
Shunt Elements	**Replace C by**	**Replace z^{-1} by**
Inductance L	Γ	$-z^{-1}$
Parallel-Tuned Circuit	$\Gamma + C$	$-\mathcal{T}$
Series-Tuned Circuit	$1/(L + D)$	$\mathcal{T}$

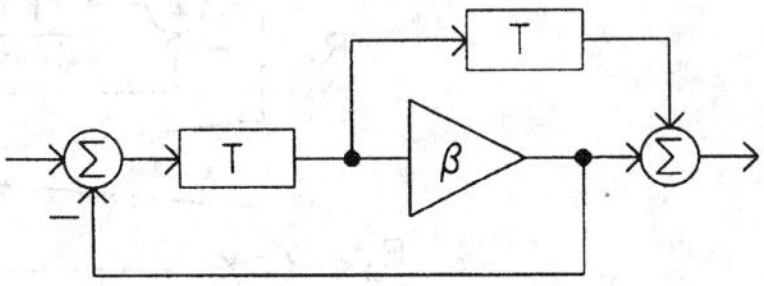

FIGURE 83.33 WFD of series-tuned circuit in shunt arm.

where

$$\mathcal{T} = \frac{z^{-1}(\beta + z^{-1})}{(1 + \beta z^{-1})}$$

$$D = 1/C$$

and

$$\beta = \frac{(D - L)}{(D + L)}$$

By comparing this result with that for a shunt capacitance, i.e.,

$$Z = 1/sC = \frac{(1 + z^{-1})}{(1 - z^{-1})C} \tag{83.119}$$

then, to obtain the two-port WFD of a series-tuned circuit in the shunt arm from a shunt capacitance, z^{-1} must be replaced by $\mathcal{T}$ and C by $1/(D + L)$. The transformation just discussed together with those for other useful circuit elements are to be found in Table 83.2. Note that $\mathcal{T}$ is an allpass function, its WFD shown in Fig. 83.33.

Interconnections

Interconnecting One Ports and Two Ports. In the one-port approach, the port resistance of each inductor and capacitor in an *LC* ladder had been set equal to the corresponding component value. This led to very simple WFD's for these components. To interconnect components with different port resistances, a building block known as an **adaptor** is required. An adaptor is an n-port device in general and its equations are derived by applying Kirchoff's current and voltage laws at an electrical junction, yielding serial or parallel versions. For most filters of interest, only two- and three-port adaptors are required.

The equations of a three-port parallel adaptor, which has 3 inputs and 3 outputs, are as follows:

$$B_k = A_0 - A_k \tag{83.120}$$

where

$$A_0 = \sum_{k=1}^{3} \beta_k A_k$$

$$\beta_k = \frac{2G_k}{(G_1 + G_2 + G_3)}$$

$$G_k = \frac{1}{R_k}$$

Also, $\sum_{k=1}^{3} \beta_k = 2$, so that one multiplier may be eliminated. For interconnection purposes, it is necessary to ensure that at least one port is **reflection-free**, i.e., there is no path from input to output of that port so as to avoid possible delay-free loops. For example, if we wished port 2 to be reflection-free then we would set $\beta_2 = 1$. In this case the number of independent multipliers drops to one. Furthermore, we have $G_2 = G_1 + G_3$. The WFD of the 3-port parallel adaptor with port 2 reflection-free and its symbol are shown in Fig. 83.34.

The corresponding equations for a three-port series adaptor are

$$B_k = A_k - \beta_k A_0 \tag{83.121}$$

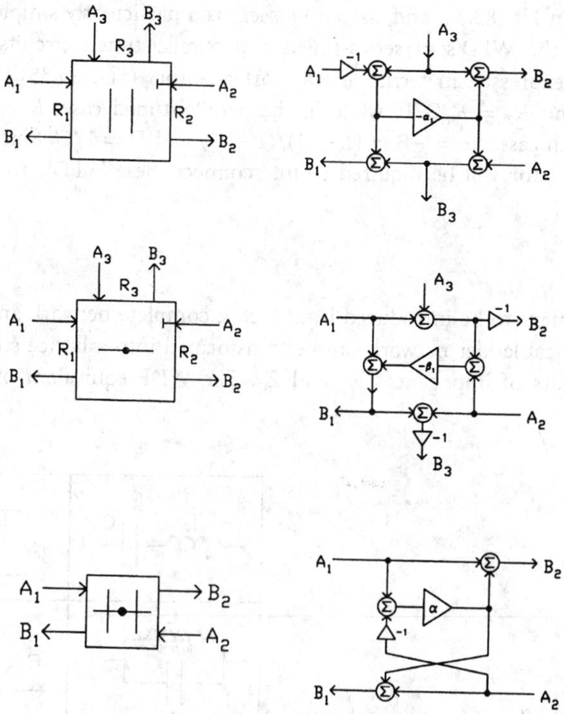

FIGURE 83.34 WFD's of two- and three-port adaptors.

where

$$A_0 = \sum_{k=1}^{3} A_k$$

$$\beta_k = \frac{2R_k}{(R_1 + R_2 + R_3)}$$

Again we have $\sum_{k=1}^{3} \beta_k = 2$ so that one multiplier can be eliminated. An additional multiplier can be eliminated by forcing one port to be reflection-free, port 2 here as in the parallel case. The resulting WFD and its symbol are shown in Fig. 83.34.

The WFD's of two-port serial and parallel adaptors are essentially the same and we will refer to them just as two-port adaptors. There are several alternative realizations, each with their particular uses. The input/output equations of one such realization are as follows:

$$\left.\begin{aligned} B_1 &= A_2 + \alpha(A_2 - A_1) \\ B_2 &= A_1 + \alpha(A_2 - A_1) \end{aligned}\right\} \tag{83.122}$$

where

$$\alpha = \frac{(R_1 - R_2)}{(R_1 + R_2)}$$

The WFD is shown in Fig. 83.34, and, as can be seen, is a particularly simple structure.

It turns out that the WFD's of series-tuned and parallel-tuned circuits, discussed earlier, can alternatively be expressed in terms of two-port adaptors (Fig. 83.35). In the series-tuned case, $R_1 = L + D$ and $R_2 = R_1 D/L$, while in the parallel-tuned case, $R_1 = LD/(L + D)$ and $R_2 = R_1 L/D$. For both cases, $\alpha = -\beta = (L - D)/(L + D)$ and $D = 1/C$. Since R_1 is determined, another two-port adaptor will be required to interconnect these WFD's to the rest of a WDF network.

Lattice Adaptor

The final circuit element to be introduced is, in fact, a complete network and this is the **lattice adaptor.** A symmetrical ladder network can be transformed into a lattice structure (Fig. 83.36) consisting of two pairs of impedances Z_1 and Z_2. The WDF equivalent of the lattice can be

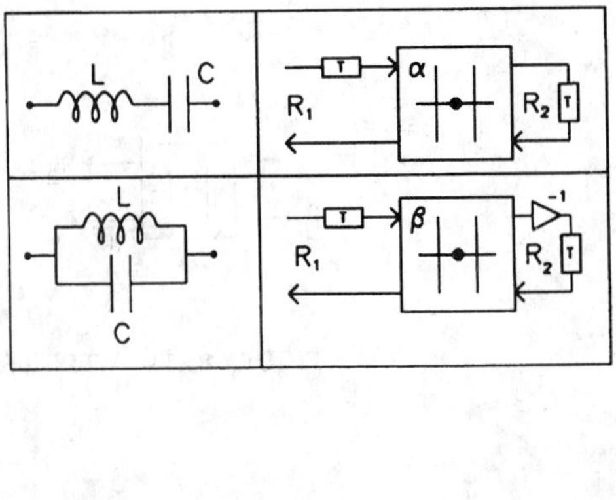

FIGURE 83.35 WFD's of series and parallel tuned circuits using adaptors.

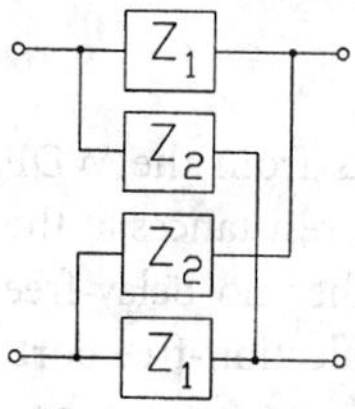

FIGURE 83.36 Double-terminated analog lattice network.

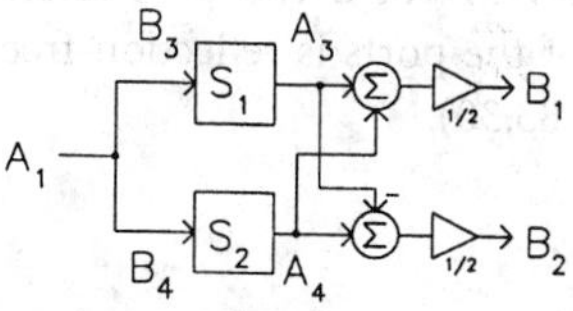

FIGURE 83.37 WFD of lattice WDF structure.

represented by the following set of equations:

$$\left.\begin{aligned} B_1 &= (A_3 + A_4)/2 \\ B_2 &= (A_4 - A_3)/2 \\ B_3 &= A_1 - A_2 \\ B_4 &= A_1 + A_2 \end{aligned}\right\} \tag{83.123}$$

where, in addition, $A_3 = S_1 B_3$, $A_4 = S_2 B_4$ and

$$S_k = \frac{Z_k - R}{Z_k + R}$$

is the reflectance of the impedance Z_k, $k = 1, 2$ and $R = R_1 = R_2$.

Equation (83.123) defines a four-port Lattice adaptor. In practice, A_2 is set to zero and the adaptor simplifies to the WFD shown in Fig. 83.37. The main advantage of using the lattice equivalent of a symmetrical ladder network is in the reduction of the number of multipliers and adders in the digital structure. In addition, the lattice WDF does not suffer from the high stopband sensitivity of its analog counterpart. The main reasons are (i) the use of wave variables and (ii) identical impedances can be realized with arbitrary accuracy using a digital representation.

The reflectances can be realized as cascades of first- and second-order allpass sections and design techniques exist that allow the coefficients to be determined easily for the classical filter approximations such as Butterworth, Chebyshev, and elliptic. The WFD of a second-order allpass section using two-port adaptors is shown in Fig. 83.38.

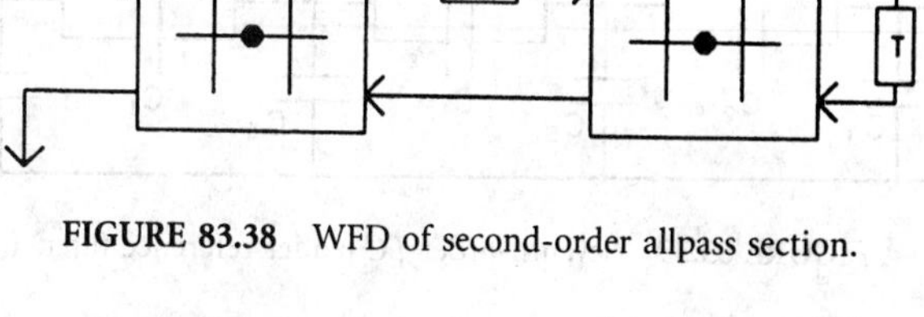

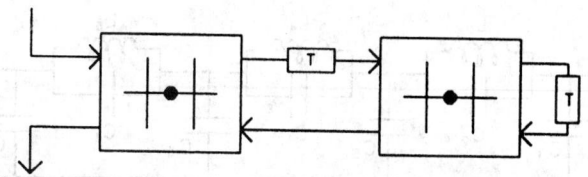

FIGURE 83.38 WFD of second-order allpass section.

WDF Networks

A few simple rules must be adhered to when constructing complete networks from the WDF building blocks described here. First, blocks can be cascaded so long as the port resistances at the interconnections are equal. Second, on interconnecting building blocks together no delay-free loops must be created. This is assured if the building blocks chosen have one reflection-free port. Although there are two approaches to WDF synthesis, i.e., the one- or two-port approaches, essentially they yield the same filter structures. For example, the WFD of a two-port series inductor is equivalent to a three-port series adaptor in which one of the ports is reflection-free and another is connected to the WFD of a one-port inductor (Fig. 83.30).

Filter Design Methods

Design Choices

The reference filters that are going to be used here are double-terminated *LC* ladders and lattices together with filters constructed from a cascade of unit elements of differing characteristic impedances. Each of these reference filters is lossless and potentially low roundoff noise WDF's can be derived.

Various design choices are open to us. First, analog filter tables or synthesis can be used to provide component values that can then be transformed using simple relationships into multiplier coefficient values. In two examples, we will look at the design of a low-pass ladder and unit element based WDF. In a further example we will use a method due to Gazsi that gives explicit formulas for the multiplier coefficients of lattice-based WDF for Butterworth, Chebyshev, inverse Chebyshev, and Cauer approximations. For nonstandard specifications, e.g., for simultaneous magnitude and phase or delay requirements, some form of mathematical optimization is necessary. This will be explored in the last example. First, the methods will be introduced.

Design Using Filter Tables

To design a WDF from the seventh-order *LC* ladder reference filter shown in Fig. 83.39 we can start from the source or load end. If we begin the WDF synthesis from the source, then the resistive voltage source is replaced by its WDF equivalent from Fig. 83.30 with the port resistance $R = R_s$. The next block is the WFD of a shunt capacitor taken from Fig. 83.32, where $G_1 = G_s + C_1$ and $\alpha_1 = G_s/G_1$. As we add additional blocks, each output port resistance/admittance is defined in terms of that input port resistance/admittance and the

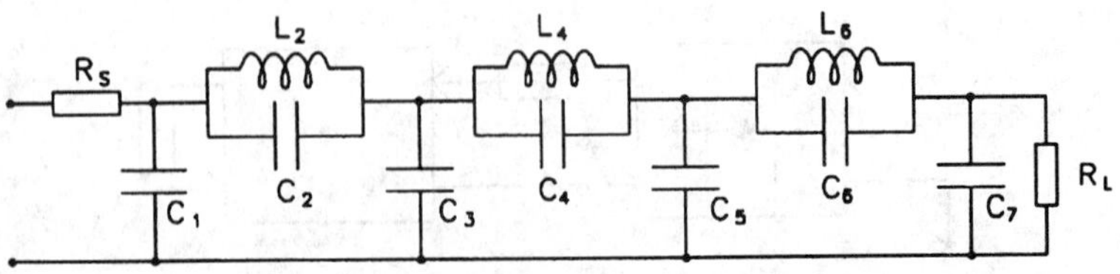

FIGURE 83.39 Seventh-order *LC* ladder reference filter.

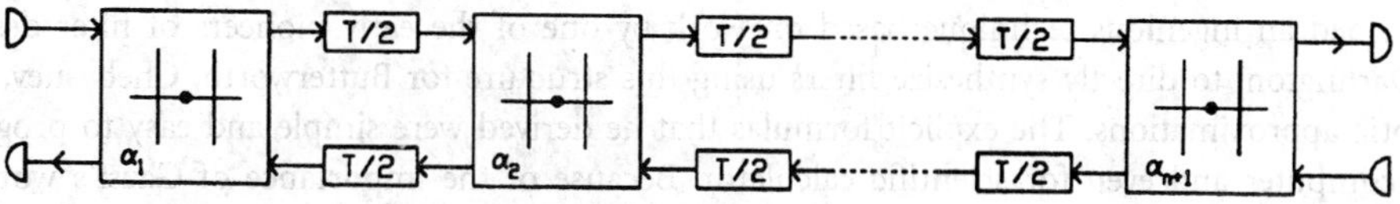

FIGURE 83.40 UE wave digital filter structure.

component value(s). The remaining design equations are as follows:

$$\left.\begin{aligned}
&R_2 = R_1 + \frac{L_2}{1+L_2C_2} \quad G_3 = G_2 + C_3 \quad R_4 = R_3 + \frac{L_4}{1+L_4C_4}\\
&G_5 = G_4 + C_5 \quad R_6 = R_5 + \frac{L_6}{1+L_6C_6} \quad G_7 = G_6 + C_7\\
&\alpha_2 = R_1/R_2 \quad \alpha_3 = \frac{(1-L_2C_2)}{(1+L_2C_2)} \quad \alpha_4 = G_2/G_3\\
&\alpha_5 = R_3/R_4 \quad \alpha_6 = \frac{(1-L_4C_4)}{(1+L_4C_4)} \quad \alpha_7 = G_4/G_5\\
&\alpha_8 = R_5/R_6 \quad \alpha_9 = \frac{(1-L_6C_6)}{(1+L_6C_6)} \quad \alpha_{10} = G_6/G_7\\
&\alpha_{11} = \frac{(G_7 - G_L)}{(G_7 + G_L)}
\end{aligned}\right\} \tag{83.124}$$

Note that $G_k = 1/R_k$ and the number of multiplier coefficients is equal to 1 less than the number of passive components. The analog component values can be found from tables or by synthesis. The overall transfer function of the WDF, $G(z)$, is given by

$$G(z) = (1 + R_7/R_L)H(z) \tag{83.125}$$

where $H(z)$ is the transfer function obtained by directly bilinearly transforming the analog transfer function $H(s)$. If we had designed the WDF from the load-end then the overall transfer function would have been equal to $2H(z)$.

To design a WDF based on a cascade of commensurate transmission line sections or unit elements (UE), we replace each UE by its WDF equivalent, derived in an earlier section. Because their characteristic impedances will, in general, be different, UE's must be interconnected using two-port adaptors. The structure that is obtained is illustrated in Fig. 83.40, where the multiplier (adaptor) values $\{\alpha_k\}$ are given by the following:

$$\alpha_k = \frac{(Z_{k-1} - Z_k)}{(Z_{k-1} + Z_k)} \tag{83.126}$$

for $k = 1, 2, \cdots, (n+1)$. Z_k is the characteristic impedance of the kth unit element with $Z_0 = R_s$ and $Z_{n+1} = R_L$.

Direct Synthesis Method

Introduction. The lattice WDF of Fig. 83.37 is arguably the most useful WDF structure because (i) it is economical in its use of multipliers and adders for a given order, (ii) it can realize both minimum and nonminimum transfer functions, and (iii) specifications can be met with very low coefficient wordlengths. This is in addition to the general advantages of WDF's. In 1985, Gazsi described an ingenious technique based on work by one of the early pioneers of filter design, S. Darlington, to directly synthesize filters using this structure for Butterworth, Chebyshev, and elliptic approximations. The explicit formulas that he derived were simple and easy to program for computer and even for scientific calculator. Because of the importance of Gazsi's work to practical filter design, his method is shown here for the more useful elliptic low-pass filter case.

Determination of Filter Order. To determine the minimum filter order n required for a given specification we use the following expression:

$$n = \frac{8\ \ln(4\epsilon_s/\epsilon_p)}{\ln(2k_5)} \tag{83.127}$$

where the passband and stopband ripple factors ϵ_p and ϵ_s are related, respectively, to the passband ripple a_p, and the minimum stopband attenuation a_s, by the following expressions:

$$\left.\begin{aligned} a_p &= 10\ \log(1+\epsilon_p^2) \\ a_s &= 10\ \log(1+\epsilon_s^2) \end{aligned}\right\} \tag{83.128}$$

a_p and a_s are measured in dB's. To determine k_4 we use the recurrence relationship

$$k_i = k_{i-1}^2 + \sqrt{(k_{i-1}^4 - 1)}, \quad i = 2, \cdots, 5 \tag{83.129}$$

The initial value k_1 is calculated from the following set of equations:

$$\left.\begin{aligned} \omega_p &= 2\pi f_p \\ \omega_s &= 2\pi f_s \\ \Omega_p &= \tan(\omega_p T/2) \\ \Omega_s &= \tan(\omega_s T/2) \\ k_1 &= \sqrt{\Omega_s/\Omega_p} \end{aligned}\right\} \tag{83.130}$$

where T is the sampling interval. For low-pass designs the order will be odd so that after using (83.127), we need to round up to the nearest odd number N, say. Because of this a **design margin** will be created that can be exploited when determining the actual values of the passband and stopband parameters.

Calculation of Filter Parameters. We can calculate the bounds of the closed interval $[f_{s1}, f_s]$ for the stopband edge frequency using the following set of equations:

$$\left.\begin{aligned} r_1 &= \sqrt{\epsilon_s/\epsilon_p} \\ r_i &= r_{i-1}^2 + \sqrt{r_{i-1}^4 - 1}, \quad i = 2, 3 \\ s_5 &= \frac{1}{2}(\sqrt[N]{2r_3})^4 \\ s_i &= \sqrt{\frac{1}{2}(s_{i+1} + s_{i+1}^{-1})}, \quad i = 4, \cdots, 1 \\ f_{s1} &= \tan^{-1}(\Omega_p s_1^2)/(\pi T) \end{aligned}\right\} \tag{83.131}$$

After choosing a final stopband edge frequency $\hat{f}_s$, such that $f_{s1} \leq \hat{f}_s \leq f_s$, we can further calculate the bounds of the closed interval $[\epsilon_{p1}, \epsilon_p]$ for the passband ripple factor using the following set of equations:

$$\left.\begin{aligned}
p_1 &= \sqrt{\hat{\Omega}_s/\Omega_p} \\
p_i &= p_{i-1}^2 + \sqrt{p_{i-1}^4 - 1}, \quad i = 2, \cdots, 5 \\
q_4 &= \frac{1}{2}(\sqrt{2p_5})^N \\
q_i &= \sqrt{\frac{1}{2}(q_{i+1} + q_{i+1}^{-1})}, \quad i = 3, \cdots, 1 \\
\epsilon_{p1} &= \epsilon s/q_1^2
\end{aligned}\right\} \tag{83.132}$$

Once a value for passband ripple factor has been chosen, the revised stopband ripple factor can be determined using $\epsilon_s = \epsilon_p q_1^2$.

Calculation of Filter (Adaptor) Coefficients. Having determined the final values of all the filter parameters (denoted by carets on the variables), the coefficients $\{\beta_i\}$ can now be computed. First, Gazsi defines some auxiliary variables:

$$\left.\begin{aligned}
v_1 &= \hat{\epsilon}_p^{-1} + \sqrt{\hat{\epsilon}_p^{-2} + 1} \\
v_i &= q_i v_{i-1} + \sqrt{1 + (q_i v_{i-1})^2}, \quad i = 2, 3 \\
w_6 &= \sqrt[N]{\frac{q_4}{v_3} + \sqrt{\left(\frac{q_4}{v_3}\right)^2 + 1}} \\
w_i &= \frac{1}{2p_i}(w_{i+1} - w_{i+1}^{-1}), \quad i = 5, \cdots, 1
\end{aligned}\right\} \tag{83.133}$$

from which the coefficient of the first-order section can be determined thus

$$\beta_0 = \frac{1 + w_1 p_1 \Omega_p}{1 - w_1 p_1 \Omega_p} \tag{83.134}$$

To calculate the coefficients of the second-order sections, further auxiliary variables are required as follows:

$$\left.\begin{aligned}
c_{4,i} &= \frac{p_5}{\sin(i\pi/N)} \\
c_{j-1,i} &= \frac{1}{2p_j}(c_{j,i} + c_{j,i}^{-1}), \quad j = 4, \cdots, 1 \\
y_i &= 1/c_{0,i} \\
B_i &= \frac{w_1^2 + y_i^2}{1 + (w_1 y_i)^2}(p_1 \Omega_p)^2 \\
A_i &= \frac{-2w_1 p_1 \Omega_p}{1 + (w_1 y_i)^2}\sqrt{1 - (p_1^2 + p_1^{-2} - y_i^2)y_i^2}
\end{aligned}\right\} \tag{83.135}$$

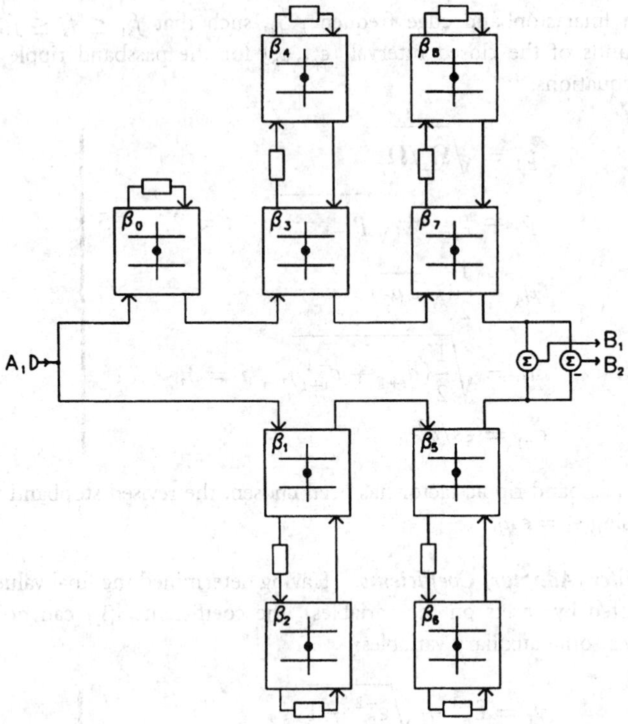

FIGURE 83.41 Lattice WDF structure showing first- and second-order sections and coefficient labeling scheme.

for $i = 1, 2, \cdots, (N-1)/2$. We can then finally calculate the coefficients for each second-order section in turn from the following formulas:

$$\left.\begin{aligned} \beta_{2i-1} &= \frac{A_i - B_i - 1}{A_i + B_i + 1} \\ \beta_{2i} &= \frac{1 - B_i}{1 + B_i} \end{aligned}\right\} \qquad (83.136)$$

Figure 83.41 shows a ninth-order lattice WDF structure and the coefficient numbering scheme. The difference in order between the upper and lower arms should never exceed one.

Optimization

In some design problems, the specifications cannot be met by the classical filter approximations. For example, it may be necessary to have an approximately linear phase (ALP) passband characteristic. FIR filters can, of course, be used to give exact linear phase but generally, the filter order is high. Recursive filters and WDF's in particular, offer an attractive alternative. The lattice WDF structure introduced earlier can be used for the design of ALP filters. However the design problem must be formulated as the minimization of some error function subject to constraints. We can then use existing optimization methods such as those available from the NAG or LINPACK library. An L_2 error norm defined as follows has been found to be useful:

$$\epsilon(\underline{\alpha}) = \sum_{i=1}^{n} W_i^g \xi_i^2 + \sum_{i=1}^{m} W_i^d \psi_i^2 \qquad (83.137)$$

where $\underline{\alpha}$ is the vector of filter coefficients, ξ_i and ψ_i are, respectively, the gain and delay errors at the ith frequency point, and W_i^g and W_i^d are the weights, respectively, for the gain and delay errors at i. The gain error is normally defined as the difference between the computed and desired gain, normally unity in the passband and zero in the stopband. Similarly the delay error is the difference between the computed and desired delay. The desired delay is not fixed but is used as an additional optimization parameter. However, there may be cases in which the delay is specified to be less than some prespecified value. This can be accommodated by an additional constraint. If we had used phase instead of delay then the additional optimization parameter would have been the phase gradient. Finally, n and m are the number of frequency points used in gain and delay error calculations, respectively.

The design objective is to find the value of the coefficient vector $\underline{\alpha}$ that minimizes the error function, $\epsilon(\underline{\alpha})$ subject to a stability constraint. To satisfy this constraint, it is necessary that $|\alpha_i| < 1$ for every i. Many optimization techniques can be applied to this problem, such as quasi-Newton, linear and quadratic programming, simulated annealing, as well as genetic algorithms. We will look at a design example using optimization in the next subsection.

Design Examples

Example I: Low-Pass Ladder WDF

An elliptic low-pass filter is to be designed with the following specification: $f_p = 0.1$, $f_s = 0.13$, $a_p = 0.1$ dB and $a_s = 60$ dB. The sampling frequency is normalized to unity. To calculate the minimum order we need to find the prewarped edge frequencies using the bilinear transform of (83.130). On using (83.128), we find that a seventh-order filter will meet the specification. The next stage is to determine the component values of the reference filter using analog filter design tables such as those of Saal or Zverev. These tables are tabulated according to reflection coefficient ρ, which is related to a_p by the relationship $a_p = -20\log(\sqrt{1-\rho^2})$. The stopband edge frequency Ω_s relative to the passband edge frequency is 1.3319. The tables yield a design C071549, meaning seventh-order, $\rho = 0.15$, and $\theta = \arcsin(1/\Omega_s) = 49°$. The minimum stopband attenuation will be 61.17 dB. The analog component values are given in Table 83.3.

The ladder WDF coefficient values are computed using the design equations of (83.124) and are shown in Table 83.4. The final structure in block diagram form is shown in Fig. 84.42.

TABLE 83.3 Reference Filter Component values for Example I

$R_s = 1$	$C_1 = 1.085\,11$
$C_2 = 0.113\,38$	$L_2 = 1.298\,68$
$C_3 = 1.630\,31$	
$C_4 = 0.549\,52$	$L_4 = 1.002\,91$
$C_5 = 1.473\,76$	
$C_6 = 0.397\,71$	$L_6 = 1.012\,01$
$C_7 = 0.863\,96$	$R_L = 1.0$

TABLE 83.4 Ladder Wave Digital Filter coefficient values for Example I

$\alpha_1 = 0.4796$	$\alpha_2 = 0.2976$
$\alpha_3 = 0.7433$	$\alpha_4 = 0.2757$
$\alpha_5 = 0.4073$	$\alpha_6 = 0.2894$
$\alpha_7 = 0.3835$	$\alpha_8 = 0.3670$
$\alpha_9 = 0.4260$	$\alpha_{10} = 0.5038$
$\alpha_{11} = 0.2704$	

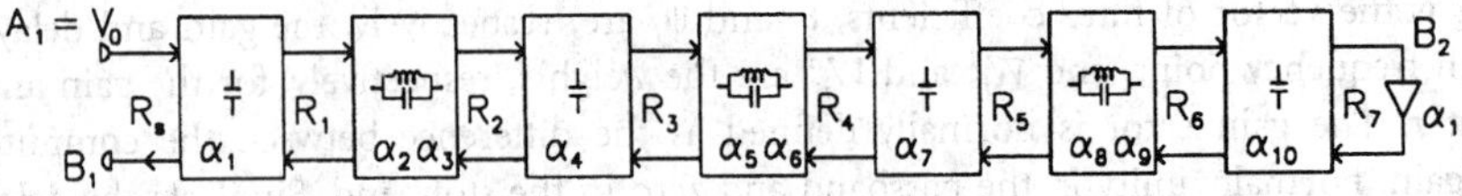

FIGURE 83.42 Block diagram of seventh-order ladder WDF designed from the source end.

Example II: Low-Pass Unit Element WDF

The next example is the design of a WDF based on a cascade of unit elements, the structure for which was shown in Fig. 83.40. Although design can be performed either from the source or the load end, the final structure is the same.

The transfer function that can be achieved with this structure will be all-pole and the frequency response will be periodic. Levy has produced tables of characteristic impedances for low-pass odd-order double-terminated filters based on the Chebyshev approximation, the magnitude-squared function for which is

$$|G(\theta)|^2 = \frac{1}{1 + h^2 T_n^2(\sin\theta / \sin\theta_p)} \tag{83.138}$$

where T_n denotes the nth-order Chebyshev polynomial and θ_p is the cutoff parameter. Note that $|G(\theta)|^2$ does not fall to zero in the stopband as it would for lumped filters.

The relationship between θ and the discrete-time frequency variable ω is $\theta = \omega T/2$, which is linear. Thus, no predistortion of the edge frequencies is required. To estimate the filter order N we use the following expression:

$$N = \frac{\log(2\epsilon_s) - \log(\epsilon_p)}{\log[2\sin(\theta_s)] - \log[\sin(\theta_p)]} \tag{83.139}$$

where $\theta_p = \pi f_p/F$ is the passband edge parameter and $\theta_s = \pi f_s/F$ is the stopband edge parameter, F is the sampling frequency, and other terms are as defined in a previous section.

To use Levy's tables, two additional parameters need to be evaluated. The first is bandwidth BW, defined as $\text{BW} = 4f_p/F$ and the second is VSWR S, which is related to passband ripple by $S = 2\epsilon - 1 + 2\sqrt{\epsilon^2 - \epsilon}$, where $a_p = 10\log(\epsilon)$. The characteristic impedances have symmetry according to the following relationship: $Z_k = Z_{N-k+1}$ for $k = 1, \cdots, (N-1)/2$. In addition, $Z_0 = R_S = 1$ and $Z_{N+1} = R_L = 1$.

Having determined the characteristic impedances, the adaptor multiplier coefficients can be computed using the following expression, for $k = 1, \cdots, (N+1)$:

$$\alpha_k = \frac{Z_{k-1} - Z_k}{Z_{k-1} + Z_k} \tag{83.140}$$

The specification for a low-pass UE WDF is as follows: $f_p = 0.1$, $f_s = 0.2$, $a_p = 0.5$ dB and $a_s = 60$ dB. Using (83.139), the minimum filter order is 7. BW is 0.4 and $S = 1.98$. The closest tabulated design is for $S = 2$. The characteristic impedances are as follows:

$$Z_0 = Z_8 = 1$$

$$Z_1 = Z_7 = 5.622$$

$$Z_2 = Z_6 = 0.2557$$

$$Z_3 = Z_5 = 8.329$$

$$Z_4 = 0.2373$$

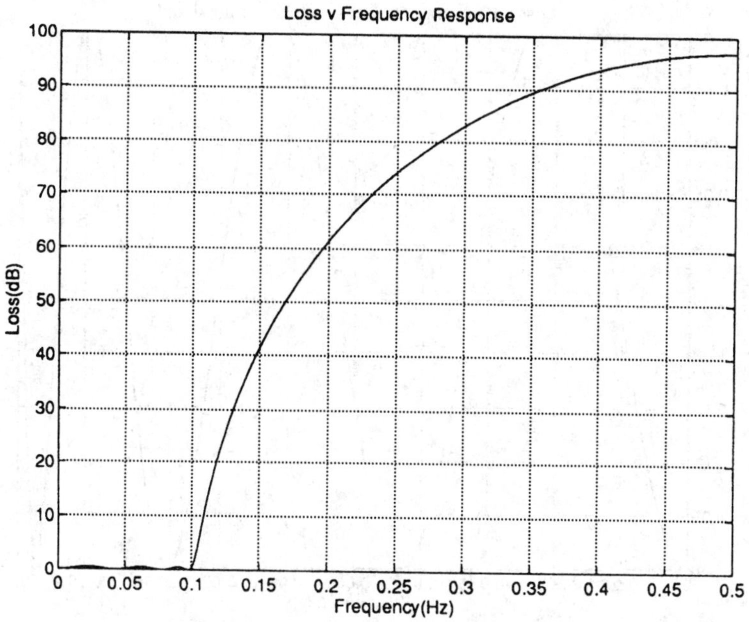

FIGURE 83.43 Loss versus frequency response for UEWDF of Example II.

Using (83.140), we find the multiplier coefficients to be as follows:

$$\alpha_1 = -\alpha_8 = -0.697\,98$$
$$\alpha_2 = -\alpha_7 = 0.912\,99$$
$$\alpha_3 = -\alpha_6 = -0.940\,43$$
$$\alpha_4 = -\alpha_5 = 0.944\,60$$

The resulting UEWDF network was analyzed in the frequency domain and its loss versus frequency response is shown in Fig. 83.43. The passband is shown in more detail in Fig. 83.44.

Example III: Low-Pass Lattice WDF

Using the same specification as in Example I, we will design a lattice WDF using Gazsi's method. The minimum order will be 7 and applying the various formulas given in an earlier section, the coefficients are found to be as follows:

$$\beta_0 = 0.726\,66 \qquad \beta_1 = -0.624\,55$$
$$\beta_2 = 0.903\,58 \qquad \beta_3 = -0.801\,28$$
$$\beta_4 = 0.833\,18 \qquad \beta_5 = -0.941\,51$$
$$\beta_6 = 0.797\,89$$

In the design process a design margin is created for the stopband frequency between 0.1279 and 0.13; a value of 0.13 was chosen. In addition, a design margin for the passband ripple between 0.081 654 and 0.1 was created and a value of 0.09 was chosen. The loss versus frequency response is shown in Fig. 83.45 and the passband is shown in more detail in Fig. 83.46.

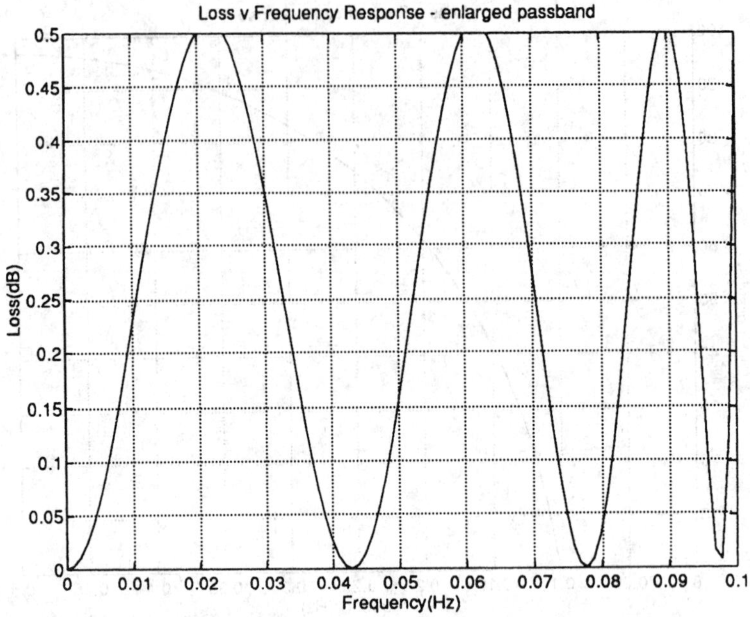

FIGURE 83.44 Passband loss versus frequency response for Example II.

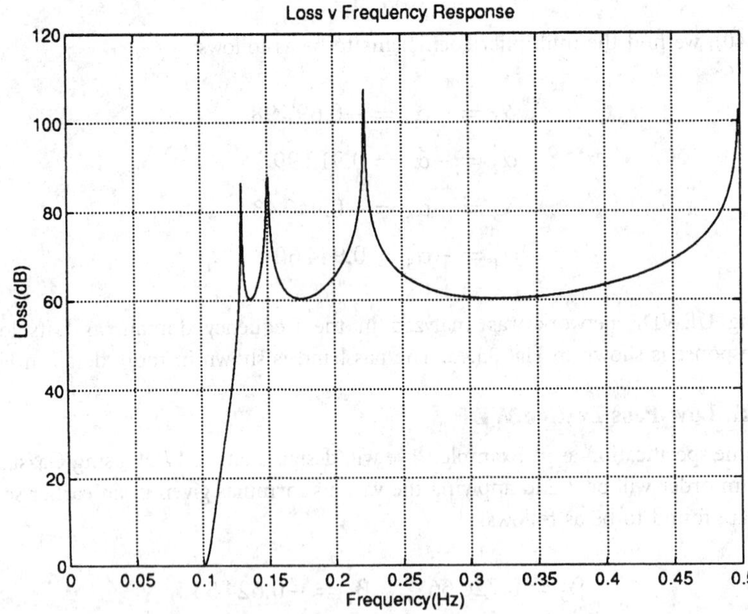

FIGURE 83.45 Loss versus frequency response for LTWDF of Example III.

Example IV: Approximately Linear Phase (ALP) Bandpass Lattice WDF

The next design is that of an ALP bandpass lattice WDF. As no explicit equations for the coefficients exist in this case, some form of optimization is required. The specifications are as follows: $f_{p1} = 0.2$, $f_{p2} = 0.3$, $f_{s1} = 0.15$, and $f_{s2} = 0.35$. The group delay is required to be constant in the interval [0.175, 0.325]. The passband ripple should be less than 0.1 dB and the minimum stopband attenuation greater than 50 dB.

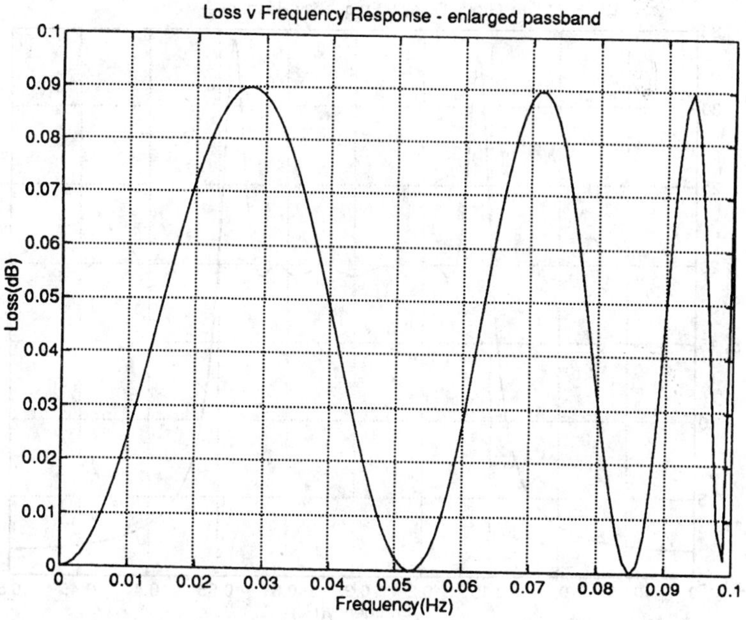

FIGURE 83.46 Passband loss versus frequency response for Example III.

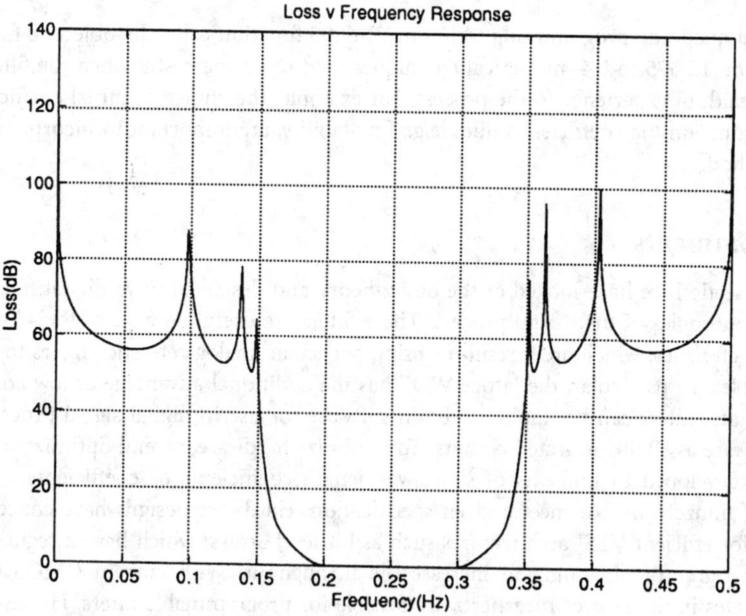

FIGURE 83.47 Loss versus frequency response for Example IV.

No expression exists for the order of ALP filters but a lower bound would be the order of a minimum phase design which, in this case, would be 10. To meet the magnitude specifications as well as maintaining a flat delay requires considerably greater order. In fact, in this case an order of 22 yields a fairly good result. The loss versus frequency response is shown in Fig. 83.47, while the delay characteristic is to be found in Fig. 83.48. The delay error, defined as $200(1-\lambda)/(1+\lambda)$, where $\lambda = \tau_{min}/\tau_{max}$ is 8.9 percent. The optimization technique was based on

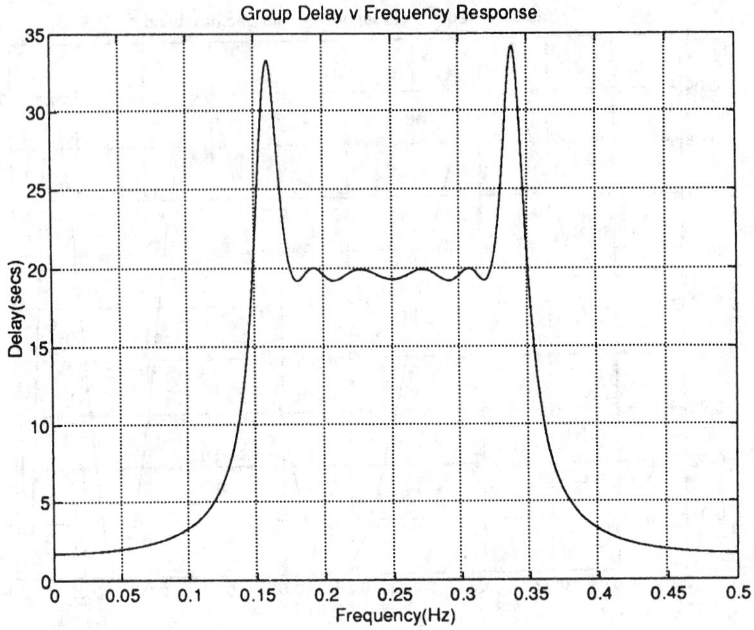

FIGURE 83.48 Group delay versus frequency response for Example IV.

a sequential quadratic programming technique. In 495 function calls, the objective function was reduced from 12 375 to 0.4. numerical techniques tend to be successful when the filter designer adds his wealth of experience to the process. For example, the choice of initial coefficient values and constraints on the coefficient values, e.g., for stability, are important to incorporate into the design method.

Final Comments

In this subsection we have looked at the basic theory and design of wave digital filters together with some examples of the design process. These filters are useful because of their low roundoff noise characteristics, which are a result of using particular analog reference filters together with wave variables. In particular, the lattice WDF has the additional advantage of low complexity.

Wave digital filters can be implemented in software for use in digital signal processing chips or in hardware as VLSI integrated circuits. To minimize hardware, several optimization methods have been developed to find sets of short wordlength coefficients or coefficients with a small number of nonzero bits to meet a given specification. Hardware designs have concentrated on searching for efficient VLSI architectures such as bit-level arrays, which have a regular structure and some degree of pipelining to increase the throughput. With current CMOS technology, sampling rates in the tens of megahertz are feasible for programmable filters. However for fixed coefficient filters, more can be done to minimize the arithmetic required and so rates in the low hundreds of megahertz can be expected.

In conclusion, the wave digital filter concept has been generalized in the 1-D case, used also for multirate filtering and extended successfully to 2-D and higher dimensions. A fruitful and relatively new area is their use in modeling physical systems described by sets of partial differential equations.

Readers who are interested in obtaining various software packages for the design, analysis, and simulation of WDF's are invited to contact the author, whose email address is s.lawson@eng.warwick.ac.uk.

References

[1] A. Antoniou, *Digital Filters: Analysis, Design and Applications*, 2nd ed., New York: McGraw-Hill, 1993.
[2] A. Fettweis, "Wave digital filters: Theory and practice," *Proc. IEEE*, vol. 74, pp. 270–327, 1986.
[3] L. Gazsi, "Explicit formulas for lattice wave digital filters," *IEEE Trans. Circuits Syst.*, vol. CAS-32, pp. 68–88, 1985.
[4] S. S. Lawson and A. R. Mirzai, *Wave Digital Filters*, New York: Ellis-Horwood, 1990.
[5] S. S. Lawson and A. Wicks, "Improved design of digital filters satisfying combined loss and delay specification," *IEE Proc., Pt. G*, vol. 140, pp. 223–229, 1993.
[6] R. Levy, "Tables of element values for the distributed low-pass prototype filter," *IEEE Trans. Microwave Theory Tech.*, vol. MTT-13, pp. 514–536, 1965.
[7] P. A. Regalia, S. K. Mitra, and P. P. Vaidyanathan, "The digital all-pass filter: A versatile signal processing building block," *Proc. IEEE*, vol. 76, pp. 19–37, 1988.
[8] A. I. Zverev, *Handbook of Filter Synthesis*, New York: Wiley, 1967.

83.4 Lattice Filters [1]–[3]

Y. C. Lim

Lattice Filters

There are several families of lattice structures for the implementation of IIR filters. Two of the most commonly encountered families are the tapped numerator structure shown in Fig. 83.49 [3] and the injected numerator structure shown in Fig. 83.50 [3]. It should be noted that not all the taps and injectors of the filters are nontrivial. For example, if $\lambda_i = 0$ for all i, the structure of Fig. 83.49 simplifies to that of Fig. 83.51 [1]. If $\phi_i = 0$ for $i \geq 1$, the structure of Fig. 83.50 reduces to that of Fig. 83.52. For both families, the denominator of the filter's transfer function is synthesized using a lattice network. The transfer function's numerator of the tapped numerator structure is realized by a weighted sum of the signals tapped from $N+1$ appropriate points of the lattice. For the injected numerator structure, the transfer function's numerator is realized by weighting and injecting the input into $N+1$ appropriate points on the lattice. The lattice itself may appear in several forms as shown in Fig. 83.53 [1]. Figure 83.54 shows the structure of a third-order injected numerator filter synthesized using the one-multiplier lattice.

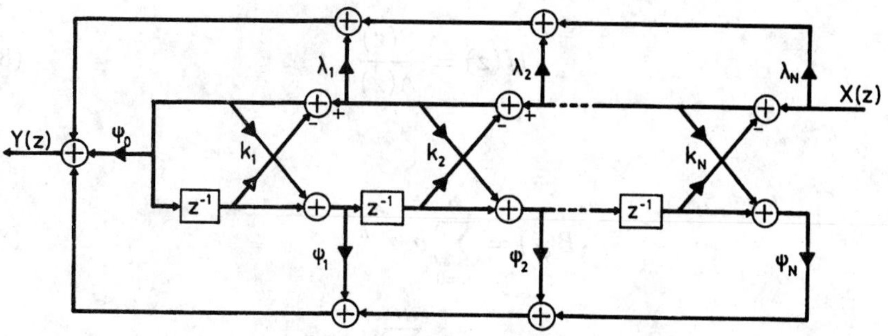

FIGURE 83.49 The general structure of a topped numerator filter.

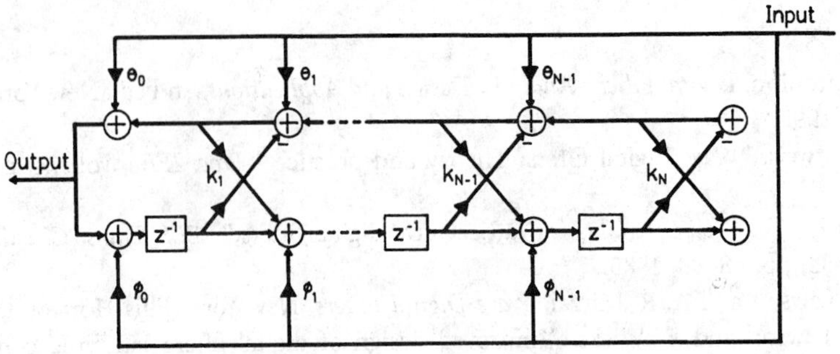

FIGURE 83.50 The general structure of an injected numerator filter.

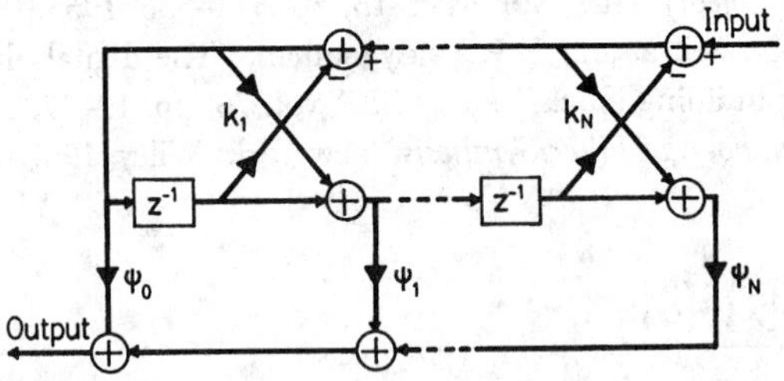

FIGURE 83.51 The structure of a topped numerator filter with $\lambda_i = 0$ for all i.

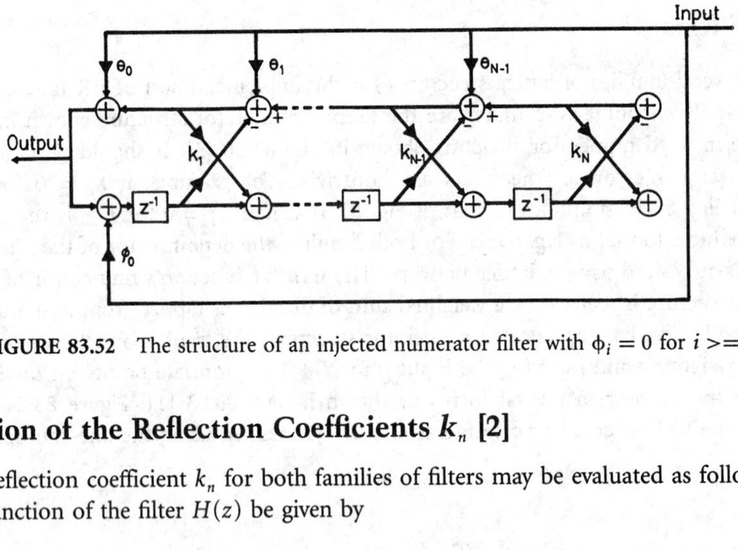

FIGURE 83.52 The structure of an injected numerator filter with $\phi_i = 0$ for $i >= 1$.

Evaluation of the Reflection Coefficients k_n [2]

The nth reflection coefficient k_n for both families of filters may be evaluated as follows. Let the transfer function of the filter $H(z)$ be given by

$$H(z) = \frac{B(z)}{A(z)} \tag{83.141}$$

where

$$B(z) = \sum_{n=0}^{N} b_n z^{-n} \tag{83.142}$$

$$A(z) = 1 + \sum_{n=1}^{N} a_n z^{-n} \tag{83.143}$$

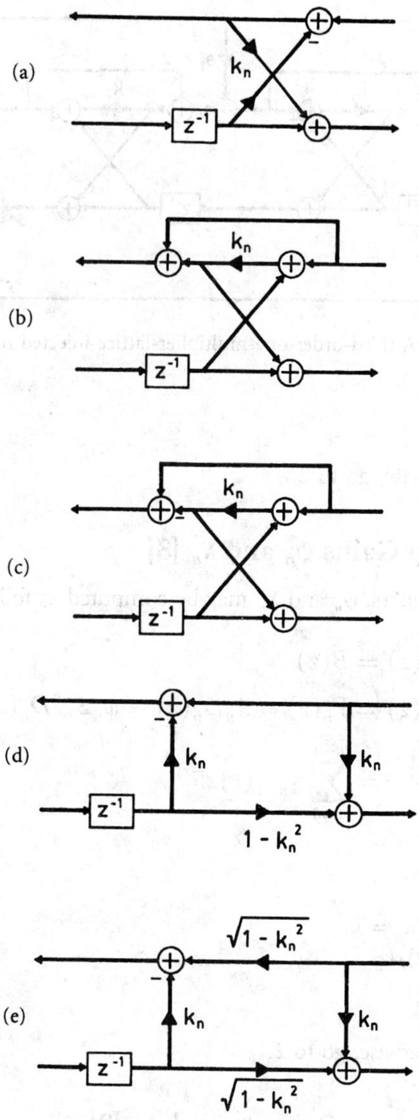

FIGURE 83.53 (a) two-multiplier lattice. (b) and (c) one-multiplier lattice. (d) three-multiplier lattice. (e) four-multiplier lattice.

Define

$$D_N(z) = A(z) \tag{83.144}$$

$$D_{n-1}(z) = \frac{D_n(z) - k_n z^{-n} D_n(z^{-1})}{1 - k_n^2} \tag{83.145}$$

$$= 1 + \sum_{r=1}^{n-1} d_{n-1}(r) z^{-r} \tag{83.146}$$

1. Set $n = N$.
2. Compute $D_n(z)$.

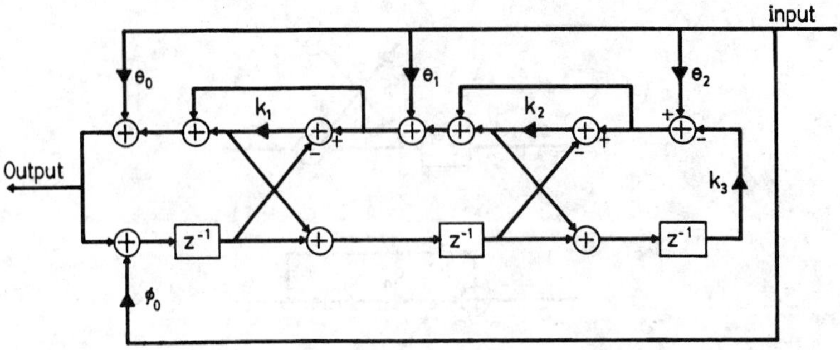

FIGURE 83.54 A third-order one-multiplier-lattice injected numerator filter.

3. $k_n = d_n(n)$
4. Decrement n.
5. If $n = 0$, stop; otherwise, go to 2.

Evaluation of the Tap Gains ψ_n and λ_n [3]

For the tapped numerator filters, ψ_n and λ_n may be computed as follows. Define

$$\Gamma_N(z) = B(z) \tag{83.147}$$

$$\Gamma_{n-1}(z) = \Gamma_n(z) - \lambda_n D_n(z) - \psi_n z^{-n} D_n(z^{-1}) \tag{83.148}$$

$$= \sum_{r=0}^{n-1} \gamma_{n-1}(r) z^{-r} \tag{83.149}$$

1. Set $n = N$.
2. Compute $\Gamma_n(z)$.
   Set either $\psi_n = 0$ or $\lambda_n = 0$.
   If $\psi_n = 0$, $\lambda_n = \gamma_n(n)/k_n$.
   If $\lambda_n = 0$, $\psi_n = \gamma_n(n)$.
3. Decrement n.
4. If $n = -1$, stop; otherwise, go to 2.

Evaluation of the Injector Gains θ_n and ϕ_n [3]

For the injected numerator filters, θ_n and ϕ_n may be computed as follows. Define

$$L_0^0(z) = \begin{bmatrix} 1 & 0 \\ 0 & 1 \end{bmatrix} \tag{83.150}$$

$$L_m^n(z) = \begin{bmatrix} 1 & k_n z^{-1} \\ k_n & z^{-1} \end{bmatrix} L_m^{n-1}(z), \qquad n > m \tag{83.151}$$

$$= \begin{bmatrix} P_m^n(z) & Q_m^n(z) \\ R_m^n(z) & S_m^n(z) \end{bmatrix} \tag{83.152}$$

$$P_m^n(z) = 1 + \sum_{r=1}^{n-m-1} p_m^n(r) z^{-r} \tag{83.153}$$

$$Q_m^n(z) = \sum_{r=1}^{n-m} q_m^n(r) z^{-r} \tag{83.154}$$

$$\Xi_{N-1} = B(z) + \phi_0 Q_0^N(z) \tag{83.155}$$

$$\Xi_{n-1}(z) = \Xi_n(z) + \phi_{N-n} Q_{N-n}^N(z) - \theta_{N-n-1} P_{N-n-1}^N(z) \tag{83.156}$$

$$= \sum_{r=0}^{n-1} \xi_{n-1}(r) z^{-r} \tag{83.157}$$

1. $\phi_0 = -b_N / q_0^N(N)$

   Set $n = 0$.

2. Increment n.
   Compute $\Xi_{N-n}(z)$.
   Set either $\phi_n = 0$ or $\theta_{n-1} = 0$.
   If $\phi_n = 0$, $\theta_{n-1} = \xi_{N-n}(N-n)/p_{n-1}^N(N-n)$.
   If $\theta_{n-1} = 0$, $\phi_n = -\xi_{N-n}(N-n)/q_n^N(N-n)$.
3. If $n = N - 1$ go to 4; otherwise, go to 2.
4. $\theta_{N-1} = \xi_0(0)$. Stop.

eferences

[1] A. H. Gray, Jr. and J. D. Markel, "Digital lattice and ladder filter synthesis," *IEEE Trans. Audio Electroacoust.*, vol. AU-21, pp. 491–500, Dec. 1973.

[2] A. H. Gray, Jr. and J. D. Markel, "A normalized digital filter structure," *IEEE Trans. Acoustics, Speech, Signal Processing*, vol. ASSP-23, pp. 268–277, June 1975.

[3] Y. C. Lim, "On the synthesis of the IIR digital filters derived from single channel AR lattice network," *IEEE Trans. Acoustics, Speech, Signal Processing*, vol. ASSP-32, pp. 741–749, Aug. 1984.

84

Finite Wordlength Effects

Bruce W. Bomar
University of Tennessee Space Institute

Practical digital filters must be implemented with finite precision numbers and arithmetic. As a result, both the filter coefficients and the filter input and output signals are in discrete form. This leads to four types of finite wordlength effects.

Discretization (quantization) of the filter coefficients has the effect of perturbing the location of the filter poles and zeroes. As a result, the actual filter response differs slightly from the ideal response. This *deterministic* frequency response error is referred to as **coefficient quantization error**.

The use of finite precision arithmetic makes it necessary to quantize filter calculations by rounding or truncation. **Roundoff noise** is that error in the filter output that results from rounding or truncating calculations within the filter. As the name implies, this error looks like low-level noise at the filter output.

Quantization of the filter calculations also renders the filter slightly nonlinear. For large signals this nonlinearity is negligible and roundoff noise is the major concern. However, for recursive filters with a zero or constant input, this nonlinearity can cause spurious oscillations called **limit cycles**.

With fixed-point arithmetic it is possible for filter calculations to overflow. The term **overflow oscillation**, sometimes also called **adder overflow limit cycle**, refers to a high-level oscillation that can exist in an otherwise stable filter due to the nonlinearity associated with the overflow of internal filter calculations.

In this chapter, we examine each of these finite wordlength effects. Both fixed-point and floating-point number representations are considered.

0-8493-8341-2/95/$0.00+$.50

84.1 Number Representation

In digital signal processing, $(B+1)$-bit fixed-point numbers are usually represented as two's-complement signed fractions in the format

$$b_0 \cdot b_{-1}b_{-2}\cdots b_{-B}$$

The number represented is then

$$X = -b_0 + b_{-1}2^{-1} + b_{-2}2^{-2} + \cdots + b_{-B}2^{-B} \quad (84.1)$$

where b_0 is the sign bit and the number range is $-1 \leq X < 1$. The advantage of this representation is that the product of two numbers in the range from -1 to 1 is another number in the same range.

Floating-point numbers are represented as

$$X = (-1)^s m 2^c \quad (84.2)$$

where s is the *sign bit*, m is the **mantissa**, and c is the *characteristic* or *exponent*. To make the representation of a number unique, the mantissa is *normalized* so that $0.5 \leq m < 1$.

Although floating-point numbers are always represented in the form of (84.2), the way in which this representation is actually *stored* in a machine may differ. Since $m \geq 0.5$, it is not necessary to store the 2^{-1}-weight bit of m, which is always set. Therefore, in practice numbers are usually stored as

$$X = (-1)^s (0.5 + f) 2^c \quad (84.3)$$

where f is an unsigned fraction, $0 \leq f < 0.5$.

Most floating-point processors now use the IEEE Standard 754 32-bit floating-point format for storing numbers. According to this standard the exponent is stored as an unsigned integer p where

$$p = c + 126 \quad (84.4)$$

Therefore, a number is stored as

$$X = (-1)^s (0.5 + f) 2^{p-126} \quad (84.5)$$

where s is the sign bit, f is a 23-b unsigned fraction in the range $0 \leq f < 0.5$, and p is an 8-b unsigned integer in the range $0 \leq p \leq 255$. The total number of bits is $1 + 23 + 8 = 32$. For example, in IEEE format 3/4 is written $(-1)^0(0.5 + 0.25)2^0$ so $s = 0$, $p = 126$, and $f = 0.25$. The value $X = 0$ is a unique case and is represented by all bits zero (i.e., $s = 0$, $f = 0$, and $p = 0$). Although the 2^{-1}-weight mantissa bit is not actually stored, it does exist so the mantissa has 24 b plus a sign bit.

84.2 Fixed-Point Quantization Errors

In fixed-point arithmetic, a multiply doubles the number of significant bits. For example, the product of the two 5-b numbers 0.0011 and 0.1001 is the 10-b number 00.000 110 11. The extra bit to the left of the decimal point can be discarded without introducing any error. However, the least significant four of the remaining bits must ultimately be discarded by some form of quantization so that the result can be stored to 5 b for use in other calculations. In the example above this results in 0.0010 (quantization by rounding) or 0.0001 (quantization by truncating). When a sum of products calculation is performed, the quantization can be performed either after each multiply or after all products have been summed with double-length precision.

We will examine three types of fixed-point quantization—rounding, truncation, and magnitude truncation. If X is an exact value then the rounded value will be denoted $Q_r(X)$, the truncated value $Q_t(X)$, and the magnitude truncated value $Q_{mt}(X)$. If the quantized value has B bits to the right of the decimal point, the quantization step size is

$$\Delta = 2^{-B} \tag{84.6}$$

Since rounding selects the quantized value nearest the unquantized value, it gives a value which is never more than $\pm\Delta/2$ away from the exact value. If we denote the rounding error by

$$\epsilon_r = Q_r(X) - X \tag{84.7}$$

then

$$-\frac{\Delta}{2} \leq \epsilon_r \leq \frac{\Delta}{2} \tag{84.8}$$

Truncation simply discards the low-order bits giving a quantized value that is always less than or equal to the exact value so

$$-\Delta < \epsilon_t \leq 0 \tag{84.9}$$

Magnitude truncation chooses the nearest quantized value that has a magnitude less than or equal to the exact value so

$$-\Delta < \epsilon_{mt} < \Delta \tag{84.10}$$

The error resulting from quantization can be modeled as a random variable uniformly distributed over the appropriate error range. Therefore, calculations with roundoff error can be considered error-free calculations that have been corrupted by additive white noise. The mean of this noise for rounding is

$$m_{\epsilon_r} = E\{\epsilon_r\} = \frac{1}{\Delta}\int_{-\Delta/2}^{\Delta/2} \epsilon_r \, d\epsilon_r = 0 \tag{84.11}$$

where $E\{\}$ represents the operation of taking the expected value of a random variable. Similarly, the variance of the noise for rounding is

$$\sigma_{\epsilon_r}^2 = E\{(\epsilon_r - m_{\epsilon_r})^2\} = \frac{1}{\Delta}\int_{-\Delta/2}^{\Delta/2} (\epsilon_r - m_{\epsilon_r})^2 \, d\epsilon_r = \frac{\Delta^2}{12} \tag{84.12}$$

Likewise, for truncation,

$$\begin{aligned} m_{\epsilon_t} &= E\{\epsilon_t\} = -\frac{\Delta}{2} \\ \sigma^2_{\epsilon_t} &= E\{(\epsilon_t - m_{\epsilon_t})^2\} = \frac{\Delta^2}{12} \end{aligned} \tag{84.13}$$

and, for magnitude truncation

$$\begin{aligned} m_{\epsilon_{mt}} &= E\{\epsilon_{mt}\} = 0 \\ \sigma^2_{\epsilon_{mt}} &= E\{(\epsilon_{mt} - m_{\epsilon_{mt}})^2\} = \frac{\Delta^2}{3} \end{aligned} \tag{84.14}$$

84.3 Floating-Point Quantization Errors

With floating-point arithmetic it is necessary to quantize after both multiplications and additions. The addition quantization arises because, prior to addition, the mantissa of the smaller number in the sum is shifted right until the exponent of both numbers is the same. In general, this gives a sum mantissa that is too long and so must be quantized.

We will assume that quantization in floating-point arithmetic is performed by rounding. Because of the exponent in floating-point arithmetic, it is the relative error that is important. The relative error is defined as

$$\varepsilon_r = \frac{Q_r(X) - X}{X} = \frac{\epsilon_r}{X} \tag{84.15}$$

Since $X = (-1)^s m 2^c$, $Q_r(X) = (-1)^s Q_r(m) 2^c$ and

$$\varepsilon_r = \frac{Q_r(m) - m}{m} = \frac{\epsilon}{m} \tag{84.16}$$

If the quantized mantissa has B bits to the right of the decimal point, $|\epsilon| < \Delta/2$ where, as before, $\Delta = 2^{-B}$. Therefore, since $0.5 \le m < 1$,

$$|\varepsilon_r| < \Delta \tag{84.17}$$

If we assume that ϵ is uniformly distributed over the range from $-\Delta/2$ to $\Delta/2$ and m is uniformly distributed over 0.5 to 1,

$$\begin{aligned} m_{\varepsilon_r} &= E\left\{\frac{\epsilon}{m}\right\} = 0 \\ \sigma^2_{\varepsilon_r} &= E\left\{\left(\frac{\epsilon}{m}\right)^2\right\} = \frac{2}{\Delta}\int_{1/2}^{1}\int_{-\Delta/2}^{\Delta/2} \frac{\epsilon^2}{m^2}\, d\epsilon\, dm \\ &= \frac{\Delta^2}{6} = (0.167)2^{-2B} \end{aligned} \tag{84.18}$$

In practice, the distribution of m is not exactly uniform. Actual measurements of roundoff noise in [1] suggested that

$$\sigma^2_{\varepsilon_r} \approx 0.23\Delta^2 \tag{84.19}$$

while a detailed theoretical and experimental analysis in [2] determined

$$\sigma^2_{\varepsilon_r} \approx 0.18\Delta^2 \tag{84.20}$$

From (84.15) we can represent a quantized floating-point value in terms of the unquantized value and the random variable ε_r using

$$Q_r(X) = X(1 + \varepsilon_r) \tag{84.21}$$

Therefore, the finite-precision product X_1X_2 and the sum $X_1 + X_2$ can be written

$$fl(X_1X_2) = X_1X_2(1 + \varepsilon_r) \tag{84.22}$$

and

$$fl(X_1 + X_2) = (X_1 + X_2)(1 + \varepsilon_r) \tag{84.23}$$

where ε_r is zero-mean with the variance of (84.20).

84.4 Roundoff Noise

To determine the roundoff noise at the output of a digital filter we will assume that the noise due to a quantization is stationary, white, and uncorrelated with the filter input, output, and internal variables. This assumption is good if the filter input changes from sample to sample in a sufficiently complex manner. It is not valid for zero or constant inputs for which the effects of rounding are analyzed from a limit cycle perspective.

To satisfy the assumption of a sufficiently complex input, roundoff noise in digital filters is often calculated for the case of a zero-mean white noise filter input signal $x(n)$ of variance σ_x^2. This simplifies calculation of the output roundoff noise because expected values of the form $E\{x(n)x(n-k)\}$ are zero for $k \neq 0$ and give σ_x^2 when $k = 0$. This approach to analysis has been found to give estimates of the output roundoff noise that are close to the noise actually observed for other input signals.

Another assumption that will be made in calculating roundoff noise is that the product of two quantization errors is zero. To justify this assumption, consider the case of a 16-b fixed-point processor. In this case a quantization error is of the order 2^{-15}, while the product of two quantization errors is of the order 2^{-30}, which is negligible by comparison.

If a linear system with impulse response $g(n)$ is excited by white noise with mean m_x and variance σ_x^2, the output is noise of mean [3, pp. 788–790]

$$m_y = m_x \sum_{n=-\infty}^{\infty} g(n) \tag{84.24}$$

and variance

$$\sigma_y^2 = \sigma_x^2 \sum_{n=-\infty}^{\infty} g^2(n) \tag{84.25}$$

Therefore, if $g(n)$ is the impulse response from the point where a roundoff takes place to the filter output, the contribution of that roundoff to the variance (mean-square value) of the output roundoff noise is given by (84.25) with σ_x^2 replaced with the variance of the roundoff. If there is more than one source of roundoff error in the filter, it is assumed that the errors are uncorrelated so the output noise variance is simply the sum of the contributions from each source.

Roundoff Noise in FIR Filters

The simplest case to analyze is a finite impulse response (FIR) filter realized via the convolution summation

$$y(n) = \sum_{k=0}^{N-1} h(k)x(n-k) \qquad (84.26)$$

When fixed-point arithmetic is used and quantization is performed after each multiply, the result of the N multiplies is N-times the quantization noise of a single multiply. For example, rounding after each multiply gives, from (84.6) and (84.12), an output noise variance of

$$\sigma_o^2 = N\frac{2^{-2B}}{12} \qquad (84.27)$$

Virtually all digital signal processor integrated circuits contain one or more double-length accumulator registers which permit the sum-of-products in (84.26) to be accumulated without quantization. In this case only a single quantization is necessary following the summation and

$$\sigma_o^2 = \frac{2^{-2B}}{12} \qquad (84.28)$$

For the floating-point roundoff noise case we will consider (84.26) for $N = 4$ and then generalize the result to other values of N. The finite-precision output can be written as the exact output plus an error term $e(n)$. Thus,

$$\begin{aligned} y(n) + e(n) = (\{[h(0)x(n)[1+\varepsilon_1(n)] \\ +h(1)x(n-1)[1+\varepsilon_2(n)]][1+\varepsilon_3(n)] \\ +h(2)x(n-2)[1+\varepsilon_4(n)]\}\{1+\varepsilon_5(n)\} \\ +h(3)x(n-3)[1+\varepsilon_6(n)])[1+\varepsilon_7(n)] \end{aligned} \qquad (84.29)$$

In (84.29), $\varepsilon_1(n)$ represents the error in the first product, $\varepsilon_2(n)$ the error in the second product, $\varepsilon_3(n)$ the error in the first addition, etc. Notice that it has been assumed that the products are summed in the order implied by the summation of (84.26).

Expanding (84.29), ignoring products of error terms, and recognizing $y(n)$ gives

$$\begin{aligned} e(n) = h(0)x(n)[\varepsilon_1(n)+\varepsilon_3(n)+\varepsilon_5(n)+\varepsilon_7(n)] \\ +h(1)x(n-1)[\varepsilon_2(n)+\varepsilon_3(n)+\varepsilon_5(n)+\varepsilon_7(n)] \\ +h(2)x(n-2)[\varepsilon_4(n)+\varepsilon_5(n)+\varepsilon_7(n)] \\ +h(3)x(n-3)[\varepsilon_6(n)+\varepsilon_7(n)] \end{aligned} \qquad (84.30)$$

Assuming that the input is white noise of variance σ_x^2 so that $E\{x(n)x(n-k)\}$ is zero for $k \neq 0$, and assuming that the errors are uncorrelated,

$$E\{e^2(n)\} = [4h^2(0) + 4h^2(1) + 3h^2(2) + 2h^2(3)]\sigma_x^2\sigma_{\varepsilon_r}^2 \qquad (84.31)$$

In general, for any N,

$$\sigma_o^2 = E\{e^2(n)\} = \left[Nh^2(0) + \sum_{k=1}^{N-1}(N+1-k)h^2(k)\right]\sigma_x^2\sigma_{\varepsilon_r}^2 \qquad (84.32)$$

Notice that if the order of summation of the product terms in the convolution summation is changed, then the order in which the $h(k)$'s appear in (84.32) changes. If the order is changed so that the $h(k)$ with smallest magnitude is first, followed by the next smallest, etc., then the roundoff noise variance is minimized. However, performing the convolution summation in nonsequential order greatly complicates data indexing and so may not be worth the reduction obtained in roundoff noise.

Roundoff Noise in Fixed-Point IIR Filters

To determine the roundoff noise of a fixed-point infinite impulse response (IIR) filter realization, consider a causal first-order filter with impulse response

$$h(n) = a^n u(n) \tag{84.33}$$

realized by the difference equation

$$y(n) = ay(n-1) + x(n) \tag{84.34}$$

Due to roundoff error, the output actually obtained is

$$\hat{y}(n) = Q\{ay(n-1) + x(n)\} = ay(n-1) + x(n) + e(n) \tag{84.35}$$

where $e(n)$ is a random roundoff noise sequence. Since $e(n)$ is injected at the same point as the input, it propagates through a system with impulse response $h(n)$. Therefore, for fixed-point arithmetic with rounding, the output roundoff noise variance from (84.6), (84.12), (84.25), and (84.33) is

$$\sigma_o^2 = \frac{\Delta^2}{12} \sum_{n=-\infty}^{\infty} h^2(n) = \frac{\Delta^2}{12} \sum_{n=0}^{\infty} a^{2n} = \frac{2^{-2B}}{12} \frac{1}{1-a^2} \tag{84.36}$$

With fixed-point arithmetic there is the possibility of overflow following addition. To avoid overflow it is necessary to restrict the input signal amplitude. This can be accomplished by either placing a *scaling* multiplier at the filter input or by simply limiting the maximum input signal amplitude. Consider the case of the first-order filter of (84.34). The transfer function of this filter is

$$H(e^{j\omega}) = \frac{Y(e^{j\omega})}{X(e^{j\omega})} = \frac{1}{e^{j\omega} - a} \tag{84.37}$$

so

$$|H(e^{j\omega})|^2 = \frac{1}{1 + a^2 - 2a\cos(\omega)} \tag{84.38}$$

and

$$|H(e^{j\omega})|_{\max} = \frac{1}{1-|a|} \tag{84.39}$$

The peak gain of the filter is $1/(1-|a|)$ so limiting input signal amplitudes to $|x(n)| \leq 1 - |a|$ will make overflows unlikely.

An expression for the output roundoff noise-to-signal ratio can easily be obtained for the case where the filter input is white noise, uniformly distributed over the interval from $-(1-|a|)$ to $(1-|a|)$ [4], [5]. In this case

$$\sigma_x^2 = \frac{1}{2(1-|a|)}\int_{-(1-|a|)}^{1-|a|} x^2\,dx = \frac{1}{3}(1-|a|)^2 \tag{84.40}$$

so, from (84.25),

$$\sigma_y^2 = \frac{1}{3}\frac{(1-|a|)^2}{1-a^2} \tag{84.41}$$

Combining (84.36) and (84.41) then gives

$$\frac{\sigma_o^2}{\sigma_y^2} = \left(\frac{2^{-2B}}{12}\frac{1}{1-a^2}\right)\left(3\frac{1-a^2}{(1-|a|)^2}\right) = \frac{2^{-2B}}{12}\frac{3}{(1-|a|)^2} \tag{84.42}$$

Notice that the noise-to-signal ratio increases without bound as $|a| \to 1$.

Similar results can be obtained for the case of the causal second-order filter realized by the difference equation

$$y(n) = 2r\cos(\theta)y(n-1) - r^2 y(n-2) + x(n) \tag{84.43}$$

This filter has complex-conjugate poles at $re^{\pm j\theta}$ and impulse response

$$h(n) = \frac{1}{\sin(\theta)} r^n \sin[(n+1)\theta]u(n) \tag{84.44}$$

Due to roundoff error, the output actually obtained is

$$\hat{y}(n) = 2r\cos(\theta)y(n-1) - r^2 y(n-2) + x(n) + e(n) \tag{84.45}$$

There are two noise sources contributing to $e(n)$ if quantization is performed after each multiply, and there is one noise source if quantization is performed after summation. Since

$$\sum_{n=-\infty}^{\infty} h^2(n) = \frac{1+r^2}{1-r^2}\frac{1}{(1+r^2)^2 - 4r^2\cos^2(\theta)} \tag{84.46}$$

the output roundoff noise is

$$\sigma_o^2 = \nu\frac{2^{-2B}}{12}\frac{1+r^2}{1-r^2}\frac{1}{(1+r^2)^2 - 4r^2\cos^2(\theta)} \tag{84.47}$$

where $\nu = 1$ for quantization after summation, and $\nu = 2$ for quantization after each multiply.

To obtain an output noise-to-signal ratio we note that

$$H(e^{j\omega}) = \frac{1}{1 - 2r\cos(\theta)e^{-j\omega} + r^2 e^{-j2\omega}} \tag{84.48}$$

and, using the approach of [6],

$$|H(e^{j\omega})|^2_{\max} = \frac{1}{4r^2\left\{\left[\mathrm{sat}\left(\frac{1+r^2}{2r}\cos(\theta)\right) - \frac{1+r^2}{2r}\cos(\theta)\right]^2 + \left[\frac{1-r^2}{2r}\sin(\theta)\right]^2\right\}} \tag{84.49}$$

where

$$\mathrm{sat}(\mu) = \begin{cases} 1 & \mu > 1 \\ \mu & -1 \le \mu \le 1 \\ -1 & \mu < -1 \end{cases} \tag{84.50}$$

Following the same approach as for the first-order case then gives

$$\frac{\sigma_o^2}{\sigma_y^2} = \nu \frac{2^{-2B}}{12}\frac{1+r^2}{1-r^2}\frac{3}{(1+r^2)^2 - 4r^2\cos^2(\theta)} \times \frac{1}{4r^2\left\{\left[\mathrm{sat}\left(\frac{1+r^2}{2r}\cos(\theta)\right) - \frac{1+r^2}{2r}\cos(\theta)\right]^2 + \left[\frac{1-r^2}{2r}\sin(\theta)\right]^2\right\}} \tag{84.51}$$

Figure 84.1 is a contour plot showing the noise-to-signal ratio of (84.51) for $\nu = 1$ in units of the noise variance of a single quantization, $2^{-2B}/12$. The plot is symmetrical about $\theta = 90°$, so only the range from 0° to 90° is shown. Notice that as $r \to 1$, the roundoff noise increases without bound. Also notice that the noise increases as $\theta \to 0°$.

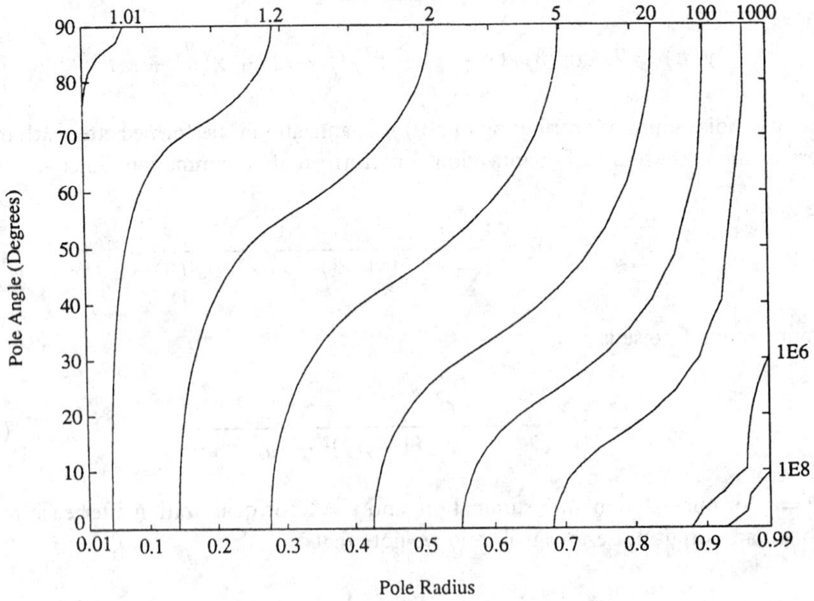

FIGURE 84.1 Normalized fixed-point roundoff noise variance.

It is possible to design state-space filter realizations that minimize fixed-point roundoff noise [7]–[10]. Depending on the transfer function being realized, these structures may provide a roundoff noise level that is orders-of-magnitude lower than for a nonoptimal realization. The price paid for this reduction in roundoff noise is an increase in the number of computations required to implement the filter. For an Nth-order filter the increase is from roughly $2N$ multiplies for a direct form realization to roughly $(N+1)^2$ for an optimal realization. However, if the filter is realized by the parallel or cascade connection of first- and second-order optimal subfilters, the increase is only to about $4N$ multiplies. Furthermore, near-optimal realizations exist that increase the number of multiplies to only about $3N$ [10].

Roundoff Noise in Floating-Point IIR Filters

For floating-point arithmetic it is first necessary to determine the injected noise variance of each quantization. For the first-order filter this is done by writing the computed output as

$$y(n) + e(n) = [ay(n-1)(1+\varepsilon_1(n)) + x(n)](1+\varepsilon_2(n)) \tag{84.52}$$

where $\varepsilon_1(n)$ represents the error due to the multiplication and $\varepsilon_2(n)$ represents the error due to the addition. Neglecting the product of errors, (84.52) becomes

$$\begin{aligned} y(n) + e(n) \approx\ & ay(n-1) + x(n) + ay(n-1)\varepsilon_1(n) \\ & + ay(n-1)\varepsilon_2(n) + x(n)\varepsilon_2(n) \end{aligned} \tag{84.53}$$

Comparing (84.34) and (84.53), it is clear that

$$e(n) = ay(n-1)\varepsilon_1(n) + ay(n-1)\varepsilon_2(n) + x(n)\varepsilon_2(n) \tag{84.54}$$

Taking the expected value of $e^2(n)$ to obtain the injected noise variance then gives

$$\begin{aligned} E\{e^2(n)\} =\ & a^2E\{y^2(n-1)\}E\{\varepsilon_1^2(n)\} + a^2E\{y^2(n-1)\}E\{\varepsilon_2^2(n)\} \\ & + E\{x^2(n)\}E\{\varepsilon_2^2(n)\} + E\{x(n)y(n-1)\}E\{\varepsilon_2^2(n)\} \end{aligned} \tag{84.55}$$

To carry this further it is necessary to know something about the input. If we assume the input is zero-mean white noise with variance σ_x^2, then $E\{x^2(n)\} = \sigma_x^2$ and the input is uncorrelated with past values of the output so $E\{x(n)y(n-1)\} = 0$ giving

$$E\{e^2(n)\} = 2a^2\sigma_y^2\sigma_{\varepsilon_r}^2 + \sigma_x^2\sigma_{\varepsilon_r}^2 \tag{84.56}$$

and

$$\begin{aligned} \sigma_o^2 &= (2a^2\sigma_y^2\sigma_{\varepsilon_r}^2 + \sigma_x^2\sigma_{\varepsilon_r}^2)\sum_{n=-\infty}^{\infty} h^2(n) \\ &= \frac{2a^2\sigma_y^2 + \sigma_x^2}{1-a^2}\sigma_{\varepsilon_r}^2 \end{aligned} \tag{84.57}$$

However,

$$\sigma_y^2 = \sigma_x^2 \sum_{n=-\infty}^{\infty} h^2(n) = \frac{\sigma_x^2}{1-a^2} \tag{84.58}$$

so

$$\sigma_o^2 = \frac{1+a^2}{(1-a^2)^2}\sigma_{\varepsilon_r}^2\sigma_x^2 = \frac{1+a^2}{1-a^2}\sigma_{\varepsilon_r}^2\sigma_y^2 \tag{84.59}$$

and the output roundoff noise-to-signal ratio is

$$\frac{\sigma_o^2}{\sigma_y^2} = \frac{1+a^2}{1-a^2}\sigma_{\varepsilon_r}^2 \tag{84.60}$$

Similar results can be obtained for the second-order filter of (84.43) by writing

$$\begin{aligned} y(n)+e(n) = ([2r\cos(\theta)y(n-1)(1+\varepsilon_1(n)) - r^2y(n-2)(1+\varepsilon_2(n))] \\ \times[1+\varepsilon_3(n)] + x(n))(1+\varepsilon_4(n)) \end{aligned} \tag{84.61}$$

Expanding with the same assumptions as before gives

$$\begin{aligned} e(n) \approx\ & 2r\cos(\theta)y(n-1)[\varepsilon_1(n)+\varepsilon_3(n)+\varepsilon_4(n)] \\ & -r^2y(n-2)[\varepsilon_2(n)+\varepsilon_3(n)+\varepsilon_4(n)] + x(n)\varepsilon_4(n) \end{aligned} \tag{84.62}$$

and

$$\begin{aligned} E\{e^2(n)\} =\ & 4r^2\cos^2(\theta)\sigma_y^2 3\sigma_{\varepsilon_r}^2 + r^2\sigma_y^2 3\sigma_{\varepsilon_r}^2 \\ & +\sigma_x^2\sigma_{\varepsilon_r}^2 - 8r^3\cos(\theta)\sigma_{\varepsilon_r}^2 E\{y(n-1)y(n-2)\} \end{aligned} \tag{84.63}$$

However,

$$\begin{aligned} & E\{y(n-1)y(n-2)\} \\ & = E\{[2r\cos(\theta)y(n-2) - r^2y(n-3) + x(n-1)]y(n-2)\} \\ & = 2r\cos(\theta)E\{y^2(n-2)\} - r^2E\{y(n-2)y(n-3)\} \\ & = 2r\cos(\theta)E\{y^2(n-2)\} - r^2E\{y(n-1)y(n-2)\} \\ & = \frac{2r\cos(\theta)}{1+r^2}\sigma_y^2 \end{aligned} \tag{84.64}$$

so

$$E\{e^2(n)\} = \sigma_{\varepsilon_r}^2\sigma_x^2 + \left[3r^4 + 12r^2\cos^2(\theta) - \frac{16r^4\cos^2(\theta)}{1+r^2}\right]\sigma_{\varepsilon_r}^2\sigma_y^2 \tag{84.65}$$

and

$$\begin{aligned} \sigma_o^2 & = E\{e^2(n)\}\sum_{n=-\infty}^{\infty} h^2(n) \\ & = \xi\left[\sigma_{\varepsilon_r}^2\sigma_x^2 + \left[3r^4 + 12r^2\cos^2(\theta) - \frac{16r^4\cos^2(\theta)}{1+r^2}\right]\sigma_{\varepsilon_r}^2\sigma_y^2\right] \end{aligned} \tag{84.66}$$

where from (84.46),

$$\xi = \sum_{n=-\infty}^{\infty} h^2(n) = \frac{1+r^2}{1-r^2}\frac{1}{(1+r^2)^2 - 4r^2\cos^2(\theta)} \tag{84.67}$$

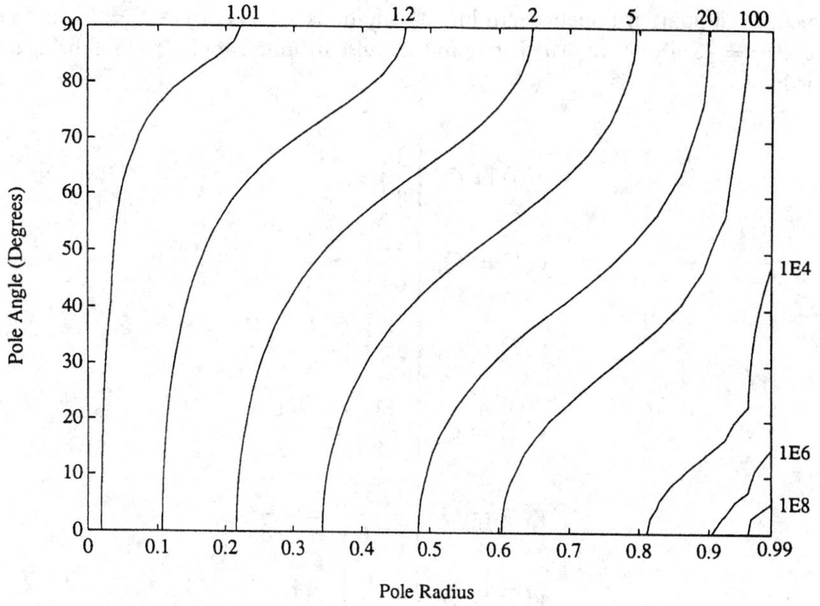

FIGURE 84.2 Normalized floating-point roundoff noise variance.

Since $\sigma_y^2 = \xi\sigma_x^2$, the output roundoff noise-to-signal ratio is then

$$\frac{\sigma_o^2}{\sigma_y^2} = \xi\left[1 + \xi\left[3r^4 + 12r^2\cos^2(\theta) - \frac{16r^4\cos^2(\theta)}{1+r^2}\right]\right]\sigma_{\varepsilon_r}^2 \tag{84.68}$$

Figure 84.2 is a contour plot showing the noise-to-signal ratio of (84.68) in units of the noise variance of a single quantization $\sigma_{\varepsilon_r}^2$. The plot is symmetrical about $\theta = 90°$, so only the range from 0° to 90° is shown. Notice the similarity of this plot to that of Fig. 84.1 for the fixed-point case. It has been observed that filter structures generally have very similar fixed-point and floating-point roundoff characteristics [2]. Therefore, the techniques of [7]–[10], which were developed for the fixed-point case, can also be used to design low-noise floating-point filter realizations. Furthermore, since it is not necessary to scale the floating-point realization, the low-noise realizations need not require significantly more computation than the direct form realization.

84.5 Limit Cycles

A limit cycle, sometimes referred to as a **multiplier roundoff limit cycle**, is a low-level oscillation that can exist in an otherwise stable filter as a result of the nonlinearity associated with rounding (or truncating) internal filter calculations [11]. Limit cycles require recursion to exist and do not occur in nonrecursive FIR filters.

As an example of a limit cycle, consider the second-order filter realized by

$$y(n) = Q_r\left\{\frac{7}{8}y(n-1) - \frac{5}{8}y(n-2) + x(n)\right\} \tag{84.69}$$

where $Q_r\{\}$ represents quantization by rounding. This is a stable filter with poles at $0.4375 \pm j0.6585$. Consider the implementation of this filter with 4-b (3 b and a sign bit) two's complement fixed-point arithmetic, zero initial conditions ($y(-1) = y(-2) = 0$), and an input sequence $x(n) = \frac{3}{8}\delta(n)$, where $\delta(n)$ is the unit impulse or unit sample. The following sequence is obtained;

$$
\begin{aligned}
y(0) &= Q_r\left\{\frac{3}{8}\right\} = \frac{3}{8}\\
y(1) &= Q_r\left\{\frac{21}{64}\right\} = \frac{3}{8}\\
y(2) &= Q_r\left\{\frac{3}{32}\right\} = \frac{1}{8}\\
y(3) &= Q_r\left\{-\frac{1}{8}\right\} = -\frac{1}{8}\\
y(4) &= Q_r\left\{-\frac{3}{16}\right\} = -\frac{1}{8}\\
y(5) &= Q_r\left\{-\frac{1}{32}\right\} = 0\\
y(6) &= Q_r\left\{\frac{5}{64}\right\} = \frac{1}{8}\\
y(7) &= Q_r\left\{\frac{7}{64}\right\} = \frac{1}{8}\\
y(8) &= Q_r\left\{\frac{1}{32}\right\} = 0\\
y(9) &= Q_r\left\{-\frac{5}{64}\right\} = -\frac{1}{8}\\
y(10) &= Q_r\left\{-\frac{7}{64}\right\} = -\frac{1}{8}\\
y(11) &= Q_r\left\{-\frac{1}{32}\right\} = 0\\
y(12) &= Q_r\left\{\frac{5}{64}\right\} = \frac{1}{8}\\
&\vdots
\end{aligned}
\tag{84.70}
$$

Notice that while the input is zero except for the first sample, the output oscillates with amplitude 1/8 and period 6.

Limit cycles are primarily of concern in fixed-point recursive filters. As long as floating-point filters are realized as the parallel or cascade connection of first- and second-order subfilters, limit cycles will generally not be a problem since limit cycles are practically not observable in first- and second-order systems implemented with 32-b floating-point arithmetic [12]. It has been shown that such systems must have an extremely small margin of stability for limit cycles to exist at anything other than underflow levels, which are at an amplitude of less than 10^{-38} [12].

There are at least three ways of dealing with limit cycles when fixed-point arithmetic is used. One is to determine a bound on the maximum limit cycle amplitude, expressed as an integral number of quantization steps [13]. It is then possible to choose a word length that makes the limit cycle amplitude acceptably low. Alternately, limit cycles can be prevented by randomly rounding calculations up or down [14]. However, this approach is complicated to implement. The third approach is to properly choose the filter realization structure and then quantize the filter calculations using magnitude truncation [15], [16]. This approach has the disadvantage of producing more roundoff noise than truncation or rounding [see (84.12)–(84.14)].

84.6 Overflow Oscillations

With fixed-point arithmetic it is possible for filter calculations to overflow. This happens when two numbers of the same sign add to give a value having magnitude greater than one. Since numbers with magnitude greater than one are not representable, the result overflows. For example, the two's complement numbers 0.101 (5/8) and 0.100 (4/8) add to give 1.001 which is the two's complement representation of $-7/8$.

The overflow characteristic of two's complement arithmetic can be represented as $R\{\}$ where

$$R\{X\} = \begin{cases} X - 2 & X \geq 1 \\ X & -1 \leq X < 1 \\ X + 2 & X < -1 \end{cases} \tag{84.71}$$

For the example just considered, $R\{9/8\} = -7/8$.

An overflow oscillation, sometimes also referred to as an *adder overflow limit cycle*, is a high-level oscillation that can exist in an otherwise stable fixed-point filter due to the gross nonlinearity associated with the overflow of internal filter calculations [17]. Like limit cycles, overflow oscillations require recursion to exist and do not occur in nonrecursive FIR filters. Overflow oscillations also do not occur with floating-point arithmetic due to the virtual impossibility of overflow.

As an example of an overflow oscillation, once again consider the filter of (84.69) with 4-b fixed-point two's complement arithmetic and with the two's complement overflow characteristic of (84.71):

$$y(n) = Q_r\left\{R\left[\frac{7}{8}y(n-1) - \frac{5}{8}y(n-2) + x(n)\right]\right\} \tag{84.72}$$

In this case we apply the input

$$\begin{aligned} x(n) &= -\frac{3}{4}\delta(n) - \frac{5}{8}\delta(n-1) \\ &= \left\{-\frac{3}{4}, -\frac{5}{8}, 0, 0, \ldots\right\}, \end{aligned} \tag{84.73}$$

giving the output sequence

$$
\begin{aligned}
y(0) &= Q_r\left\{R\left[-\frac{3}{4}\right]\right\} = Q_r\left\{-\frac{3}{4}\right\} = -\frac{3}{4} \\
y(1) &= Q_r\left\{R\left[-\frac{41}{32}\right]\right\} = Q_r\left\{\frac{23}{32}\right\} = \frac{3}{4} \\
y(2) &= Q_r\left\{R\left[\frac{9}{8}\right]\right\} = Q_r\left\{-\frac{7}{8}\right\} = -\frac{7}{8} \\
y(3) &= Q_r\left\{R\left[-\frac{79}{64}\right]\right\} = Q_r\left\{\frac{49}{64}\right\} = \frac{3}{4} \\
y(4) &= Q_r\left\{R\left[\frac{77}{64}\right]\right\} = Q_r\left\{-\frac{51}{64}\right\} = -\frac{3}{4} \\
y(5) &= Q_r\left\{R\left[-\frac{9}{8}\right]\right\} = Q_r\left\{\frac{7}{8}\right\} = \frac{7}{8} \\
y(6) &= Q_r\left\{R\left[\frac{79}{64}\right]\right\} = Q_r\left\{-\frac{49}{64}\right\} = -\frac{3}{4} \\
y(7) &= Q_r\left\{R\left[-\frac{77}{64}\right]\right\} = Q_r\left\{\frac{51}{64}\right\} = \frac{3}{4} \\
y(8) &= Q_r\left\{R\left[\frac{9}{8}\right]\right\} = Q_r\left\{-\frac{7}{8}\right\} = -\frac{7}{8} \\
&\vdots
\end{aligned}
\qquad (84.74)
$$

This is a large-scale oscillation with nearly full-scale amplitude.

There are several ways to prevent overflow oscillations in fixed-point filter realizations. The most obvious is to scale the filter calculations so as to render overflow impossible. However, this may unacceptably restrict the filter dynamic range. Another method is to force completed sums-of-products to saturate at ± 1, rather than overflowing [18], [19]. It is important to saturate only the completed sum, since intermediate overflows in two's complement arithmetic do not affect the accuracy of the final result. Most fixed-point digital signal processors provide for automatic saturation of completed sums if their *saturation arithmetic* feature is enabled. Yet another way to avoid overflow oscillations is to use a filter structure for which any internal filter transient is guaranteed to decay to zero [20]. Such structures are desirable anyway, since they tend to have low roundoff noise and be insensitive to coefficient quantization [21].

84.7 Coefficient Quantization Error

Each filter structure has its own finite, generally nonuniform grids of realizable pole and zero locations when the filter coefficients are quantized to a finite word length. In general the pole and zero locations desired in a filter do not correspond exactly to the realizable locations. The error in filter performance (usually measured in terms of a frequency response error) resulting from the placement of the poles and zeroes at the nonideal but realizable locations is referred to as coefficient quantization error.

Consider the second-order filter with complex-conjugate poles

$$\begin{aligned}\lambda &= re^{\pm j\theta} \\ &= \lambda_r \pm j\lambda_i \\ &= r\cos(\theta) \pm jr\sin(\theta)\end{aligned} \tag{84.75}$$

and transfer function

$$H(z) = \frac{1}{1 - 2r\cos(\theta)z^{-1} + r^2 z^{-2}} \tag{84.76}$$

realized by the difference equation

$$y(n) = 2r\cos(\theta)y(n-1) - r^2 y(n-2) + x(n) \tag{84.77}$$

Figure 84.3 from [5] shows that quantizing the difference equation coefficients results in a nonuniform grid of realizable pole locations in the z plane. The grid is defined by the intersection of vertical lines corresponding to quantization of $2\lambda_r$ and concentric circles corresponding to quantization of $-r^2$. The sparseness of realizable pole locations near $z = \pm 1$ will result in a large coefficient quantization error for poles in this region.

Figure 84.4 gives an alternative structure to (84.77) for realizing the transfer function of (84.76). Notice that quantizing the coefficients of this structure corresponds to quantizing λ_r and λ_i. As shown in Fig. 84.5 from [5], this results in a uniform grid of realizable pole locations. Therefore, large coefficient quantization errors are avoided for all pole locations.

It is well established that filter structures with low roundoff noise tend to be robust to coefficient quantization, and visa versa [22]–[24]. For this reason, the uniform grid structure of

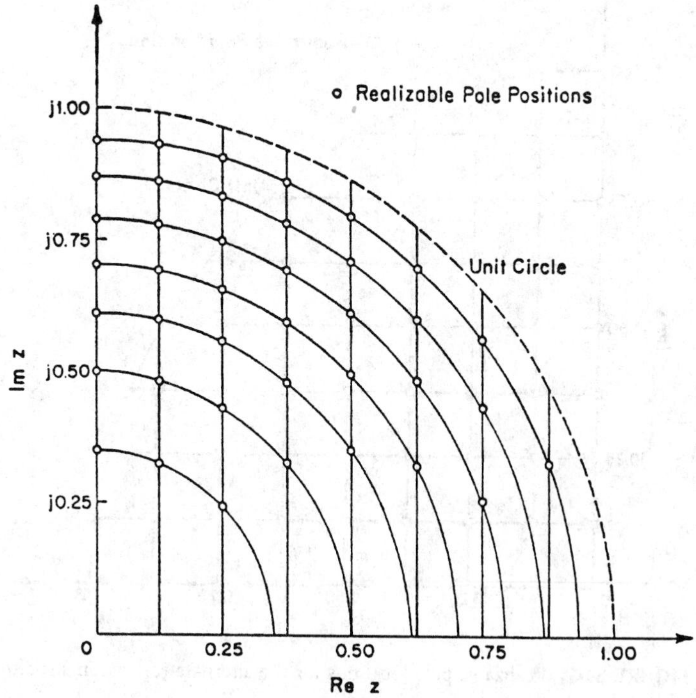

FIGURE 84.3 Realizable pole locations for the difference equation of (84.76).

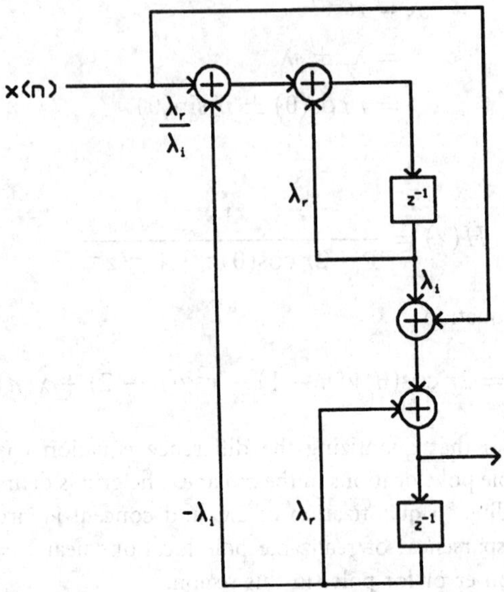

FIGURE 84.4 Alternate realization structure.

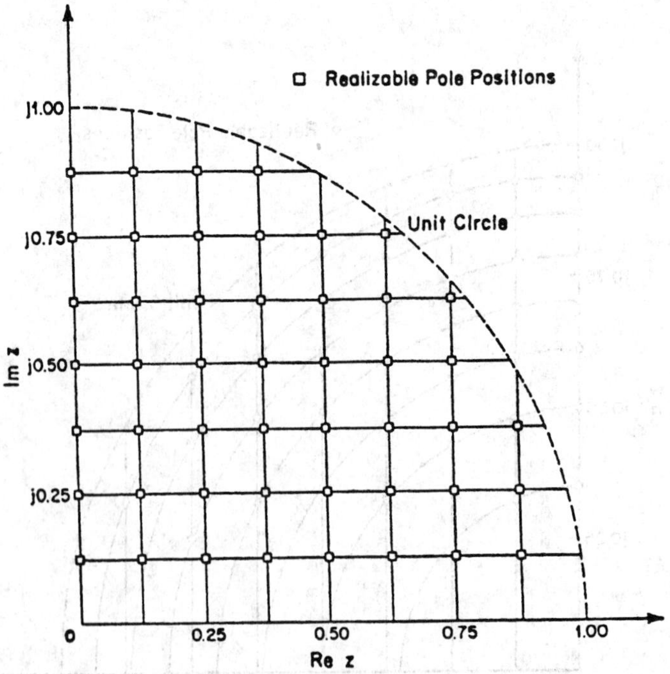

FIGURE 84.5 Realizable pole locations for the alternate realization structure.

Fig. 84.4 is also popular because of its low roundoff noise. Likewise, the low-noise realizations of [7]–[10] can be expected to be relatively insensitive to coefficient quantization, and digital wave filters and lattice filters that are derived from low-sensitivity analog structures tend to have not only low coefficient sensitivity, but also low roundoff noise [25], [26].

It is well known that in a high-order polynomial with clustered roots, the root location is a very sensitive function of the polynomial coefficients. Therefore, filter poles and zeros can be much more accurately controlled if higher order filters are realized by breaking them up into the parallel or cascade connection of first- and second-order subfilters. One exception to this rule is the case of linear-phase FIR filters in which the symmetry of the polynomial coefficients and the spacing of the filter zeros around the unit circle usually permits an acceptable direct realization using the convolution summation.

Given a filter structure it is necessary to assign the ideal pole and zero locations to the realizable locations. This is generally done by simply rounding or truncating the filter coefficients to the available number of bits, or by assigning the ideal pole and zero locations to the nearest realizable locations. A more complicated alternative is to consider the original filter design problem as a problem in discrete optimization, and choose the realizable pole and zero locations that give the best approximation to the desired filter response [27]–[30].

84.8 Realization Considerations

Linear-phase FIR digital filters can generally be implemented with acceptable coefficient quantization sensitivity using the direct convolution sum method. When implemented in this way on a digital signal processor, fixed-point arithmetic is not only acceptable but may actually be preferable to floating-point arithmetic. Virtually all fixed-point digital signal processors accumulate a sum of products in a double-length accumulator. This means that only a single quantization is necessary to compute an output. Floating-point arithmetic, on the other hand, requires a quantization after every multiply and after every add in the convolution summation. With 32-b floating-point arithmetic these quantizations introduce a small enough error to be insignificant for many applications.

When realizing IIR filters, either a parallel or cascade connection of first- and second-order subfilters is almost always preferable to a high-order direct-form realization. With the availability of very low-cost floating-point digital signal processors, like the Texas Instruments TMS320C32, it is highly recommended that floating-point arithmetic be used for IIR filters. Floating-point arithmetic simultaneously eliminates most concerns regarding scaling, limit cycles, and overflow oscillations. Regardless of the arithmetic employed, a low roundoff noise structure should be used for the second-order sections. Good choices are given in [2] and [10]. Recall that realizations with low fixed-point roundoff noise also have low floating-point roundoff noise. The use of a low roundoff noise structure for the second-order sections also tends to give a realization with low coefficient quantization sensitivity. First-order sections are not as critical in determining the roundoff noise and coefficient sensitivity of a realization, and so can generally be implemented with a simple direct form structure.

eferences

[1] C. Weinstein and A. V. Oppenheim, "A comparison of roundoff noise in floating-point and fixed-point digital filter realizations," *Proc. IEEE*, vol. 57, pp. 1181–1183, June 1969.

[2] L. M. Smith, B. W. Bomar, R. D. Joseph, and G. C. Yang, "Floating-point roundoff noise analysis of second-order state-space digital filter structures," *IEEE Trans. Circuits Syst. II*, vol. 39, pp. 90–98, Feb. 1992.

[3] J. G. Proakis and D. G. Manolakis, *Introduction to Digital Signal Processing*, 1st ed., New York: Macmillan, 1988.

[4] A. V. Oppenheim and R. W. Schafer, *Digital Signal Processing*, Englewood Cliffs, NJ: Prentice-Hall, 1975.

[5] A. V. Oppenheim and C. J. Weinstein, "Effects of finite register length in digital filtering and the fast Fourier transform," *Proc. IEEE* vol. 60, pp. 957–976, Aug. 1972.

[6] B. W. Bomar and R. D. Joseph, "Calculation of L_∞ norms for scaling second-order state-space digital filter sections," *IEEE Trans. Circuits Syst.*, vol. CAS-34, pp. 983–984, Aug. 1987.

[7] C. T. Mullis and R. A. Roberts, "Synthesis of minimum roundoff noise fixed-point digital filters," *IEEE Trans. Circuits Syst.*, vol. CAS-23, pp. 551–562, Sept. 1976.

[8] L. B. Jackson, A. G. Lindgren, and Y. Kim, "Optimal synthesis of second-order state-space structures for digital filters," *IEEE Trans. Circuits Syst.*, vol. CAS-26, pp. 149–153, Mar. 1979.

[9] C. W. Barnes, " On the design of optimal state-space realizations of second-order digital filters," *IEEE Trans. Circuits Syst.*, vol. CAS-31, pp. 602–608, July 1984.

[10] B. W. Bomar, "New second-order state-space structures for realizing low roundoff noise digital filters," *IEEE Trans. Acoust., Speech, Signal Processing*, vol. ASSP-33, pp. 106–110, Feb. 1985.

[11] S. R. Parker and S. F. Hess, "Limit-cycle oscillations in digital filters," *IEEE Trans. Circuit Theory*, vol. CT-18, pp. 687–697, Nov. 1971.

[12] P. H. Bauer, "Limit cycle bounds for floating-point implementations of second-order recursive digital filters," *IEEE Trans. Circuits Syst. II*, vol. 40, pp. 493–501, Aug. 1993.

[13] B. D. Green and L. E. Turner, "New limit cycle bounds for digital filters," *IEEE Trans. Circuits Syst.*, vol. 35, pp. 365–374, Apr. 1988.

[14] M. Buttner, "A novel approach to eliminate limit cycles in digital filters with a minimum increase in the quantization noise," in *Proc. 1976 IEEE Int. Symp. Circuits Syst.*, Apr. 1976, pp. 291–294.

[15] P. S. R. Diniz and A. Antoniou, "More economical state-space digital filter structures which are free of constant-input limit cycles," *IEEE Trans. Acoust., Speech, Signal Processing*, vol. ASSP-34, pp. 807–815, Aug. 1986.

[16] B. W. Bomar, "Low-roundoff-noise limit-cycle-free implementation of recursive transfer functions on a fixed-point digital signal processor," *IEEE Trans. Industr. Electron.*, vol. 41, pp. 70–78, Feb. 1994.

[17] P. M. Ebert, J. E. Mazo, and M. G. Taylor, "Overflow oscillations in digital filters," *Bell Syst. Tech. J.*, vol. 48. pp. 2999–3020, Nov. 1969.

[18] A. N. Willson Jr., "Limit cycles due to adder overflow in digital filters," *IEEE Trans. Circuit Theory*, vol. CT-19, pp. 342–346, July 1972.

[19] J. H. F. Ritzerfield, "A condition for the overflow stability of second-order digital filters that is satisfied by all scaled state-space structures using saturation," *IEEE Trans. Circuits Syst.*, vol. 36, pp. 1049–1057, Aug. 1989.

[20] W. T. Mills, C. T. Mullis, and R. A. Roberts, "Digital filter realizations without overflow oscillations," *IEEE Trans. Acoust., Speech, Signal Processing*. vol. ASSP-26, pp. 334–338, Aug. 1978.

[21] B. W. Bomar, "On the design of second-order state-space digital filter sections," *IEEE Trans. Circuits Syst.*, vol. 36, pp. 542–552, Apr. 1989.

[22] L. B. Jackson, "Roundoff noise bounds derived from coefficient sensitivities for digital filters," *IEEE Trans. Circuits Syst.*, vol. CAS-23, pp. 481–485, Aug. 1976.

[23] D. B. V. Rao, "Analysis of coefficient quantization errors in state-space digital filters," *IEEE Trans. Acoust., Speech, Signal Processing*, vol. ASSP-34, pp. 131–139, Feb. 1986.

[24] L. Thiele, "On the sensitivity of linear state-space systems," *IEEE Trans. Circuits Syst.*, vol. CAS-33, pp. 502–510, May 1986.

[25] A. Antoniou, *Digital Filters: Analysis and Design*, New York: McGraw-Hill, 1979.

[26] Y. C. Lim, "On the synthesis of IIR digital filters derived from single channel AR lattice network," *IEEE Trans. Acoust., Speech, Signal Processing*, vol. ASSP-32, pp. 741–749, Aug. 1984.

[27] E. Avenhaus, "On the design of digital filters with coefficients of limited wordlength," *IEEE Trans. Audio Electroacoust.*, vol. AU-20, pp. 206–212, Aug. 1972.

[28] M. Suk and S. K. Mitra, "Computer-aided design of digital filters with finite wordlengths," *IEEE Trans. Audio Electroacoust.*, vol. AU-20, pp. 356–363, Dec. 1972.

[29] C. Charalambous and M. J. Best, "Optimization of recursive digital filters with finite wordlengths," *IEEE Trans. Acoust., Speech, Signal Processing*, vol. ASSP-22, pp. 424–431, Dec. 1979.

[30] Y. C. Lim, "Design of discrete-coefficient-value linear-phase FIR filters with optimum normalized peak ripple magnitude," *IEEE Trans. Circuits Syst.*, vol. 37, pp. 1480–1486, Dec. 1990.

85

Aliasing-Free Reconstruction Filter Bank

Truong Q. Nguyen
University of Wisconsin

85.1 Introduction

In this chapter, we present the theory and design of *aliasing-free* and *perfect-reconstruction* filter banks. Section 85.2 covers the fundamentals of multirate systems including downsampling/upsampling, delay chains, serial-parallel/parallel-serial converters, polyphase representations, noble identities, decimation, and interpolation filters. These multirate components are essential in the theory and implementation of filter banks.

The basic operations of and the reconstruction errors in a maximally-decimated uniform filter bank are discussed in Section 85.3. By using the polyphase representation, the necessary and sufficient conditions for aliasing-free filter banks are given in Section 85.3. Moreover, several examples of alias-free filter banks such as the two-channel quadrature-mirror-filter (QMF) bank, the two-channel allpass-based IIR filter bank, and the M-channel DFT filter bank are elaborated in details.

Section 85.4 presents the theory and lattice structures for perfect-reconstruction (PR) filter banks. The section is divided into three subsections: paraunitary filter bank, linear-phase filter bank, and cosine-modulated filter bank.

0-8493-8341-2/95/$0.00 + $.50

The design methods for filter banks are discussed in Section 85.5. The first subsection considers the design method based on lattice structure realization and the next subsection presents an alternative method based on filter coefficients in the form of quadratic-constrained least-squares (QCLS) optimization.

Compactly supported wavelet is closely related to the two-channel filter bank. It can be obtained by iterating on a two-channel digital filter bank. Chapter 4 discusses this relation in detail. Other issues on wavelet such as regularity, maximally flat filter, tight frames, M-band wavelets, wavelet packets, etc. are also presented in Chapter 4.

Notations. The variable ω is used as the frequency variable, whereas the term "normalized frequency" is used to denote $f = \omega/2\pi$. Boldfaced quantities denote matrices and column vectors, with upper case used for the former and lower case for the latter, as in $\mathbf{A}$, $\mathbf{h}(z)$, etc. The superscripts (T) and $(^\dagger)$ stand for matrix transposition and transposition with coefficient conjugation, respectively. $\tilde{\mathbf{E}}(z) = \mathbf{E}^\dagger(z^{-1})$. The k by k identity matrix is denoted as $\mathbf{I}_k$ and the exchange matrix $\mathbf{J}$ is defined to be

$$\mathbf{J} = \begin{pmatrix} 0 & \cdots & 0 & 1 \\ 0 & \cdots & 1 & 0 \\ \vdots & ⋰ & \vdots & \vdots \\ 1 & \cdots & 0 & 0 \end{pmatrix}$$

85.2 Fundamentals of a Multirate System

Downsampling [26], [29]

Figure 85.1 shows the block diagram of a downsampler by a factor of M. The output signal $y_D(n)$ in terms of $x(n)$ is

$$\begin{cases} y_D(n) = x(nM) \\ Y_D(z) = \dfrac{1}{M}\displaystyle\sum_{k=0}^{M-1} X(z^{1/M}W_M^k) \qquad W_M = e^{-j2\pi/M} \\ Y_D(e^{j\omega}) = \dfrac{1}{M}\displaystyle\sum_{k=0}^{M-1} X(e^{(j/M)(\omega-2\pi k)}) \end{cases} \tag{85.1}$$

It is clear from the above expression that the output of a downsampler has M copies of the *stretched* input spectrum. The first term in the summation is the input spectrum $X(e^{j(\omega/M)})$ $(k = 0)$ and the remaining $(M-1)$ terms are the aliased versions of $X(e^{j(\omega/M)})$, $1 \le k \le M-1$. If the input spectrum $X(e^{j\omega})$ is bandlimited to $-\pi/M < \omega \le \pi/M$, then there is no contribution from the aliased versions in the frequency range $-\pi < \omega \le \pi$. On the other hand, if $X(e^{j\omega})$ is not bandlimited to the above frequency range, then the downsampled output is aliased. This is the main reason for low-pass filtering the input signal before downsampling. An example for $M = 2$, where $X(e^{j\omega})$ is not bandlimited, is shown in Fig. 85.1(b) and (c).

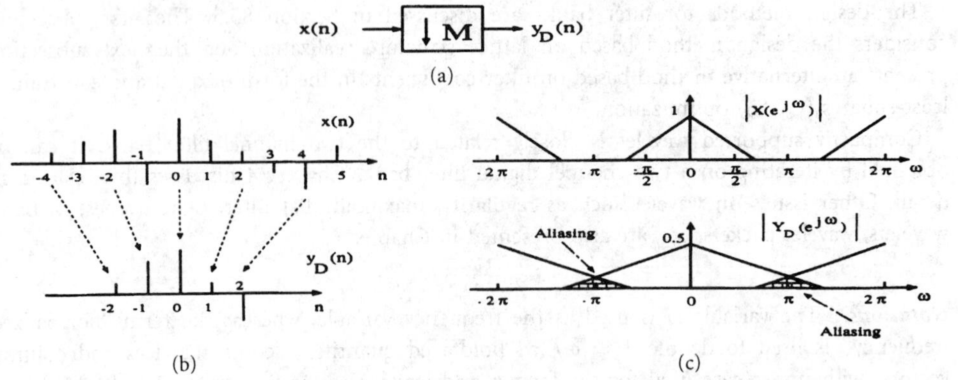

FIGURE 85.1 Downsampling: (a) Block diagram; (b), (c) time and frequency-domain example for $M = 2$.

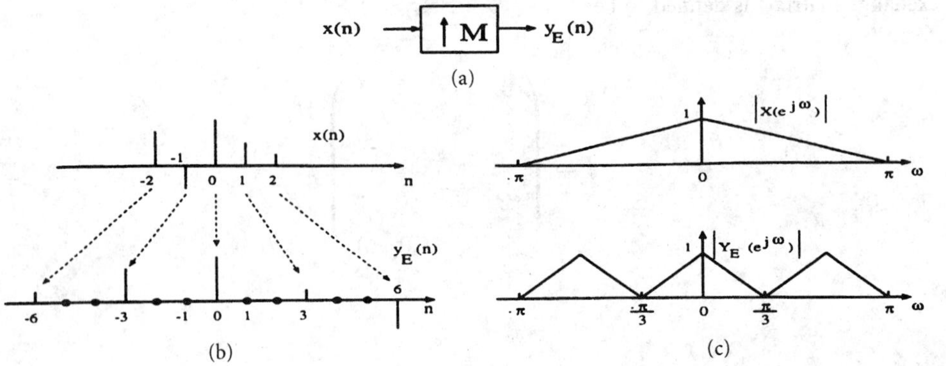

FIGURE 85.2 Upsampling: (a) Block diagram; (b), (c) time and frequency-domain example for $M = 3$.

Upsampling [26], [29]

Figure 85.2(a) shows the block diagram of an upsampler by a factor of M. The output signal $y_E(n)$ in terms of $x(n)$ is

$$y_E(n) = \begin{cases} x(n/M) & n \text{ is a multiple of } M \\ 0 & \text{otherwise} \end{cases} \qquad Y_E(z) = X(z^M) \quad Y_E(e^{j\omega}) = X(e^{jM\omega}) \tag{85.2}$$

The output $y_E(n)$ is a *compressed* copy of the output. Besides the original input spectrum, there are images repeated at $\omega_k = 2\pi k/M$. Consequently, it is necessary to low-pass the output to suppress the images. An example for $M = 3$ is shown in Fig. 85.2(b) and (c).

Delay Chain, Serial–Parallel Converter, Parallel–Serial Converter [29]

Figure 85.3(a) shows the block diagram of a delay chain where the transfer function from the input to the kth output is z^{-k}. By itself, the delay chain is not very interesting. However, using it in cascade with a set of decimators (or expanders), a serial-to-parallel (S/P) [or parallel-to-serial (P/S)] converter can be implemented.

Figure 85.3(b) shows such implementation for a S/P converter. The output at the kth branch is $x(nM - k)$, which implies that the input sequence is selected in a *counterclockwise* fashion. In

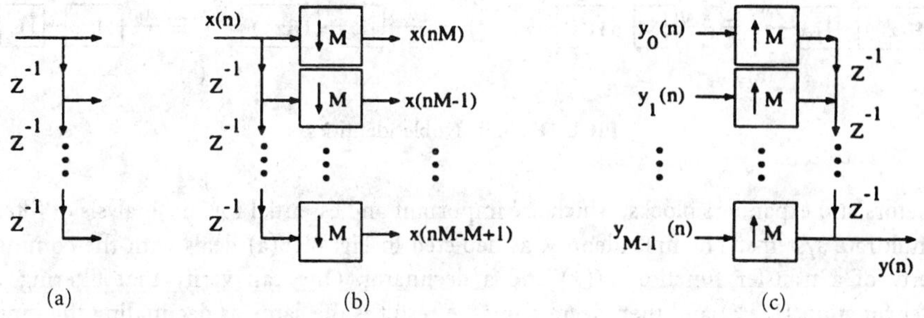

FIGURE 85.3 (a) Delay chain, (b) serial-parallel converter, (c) parallel-serial converter.

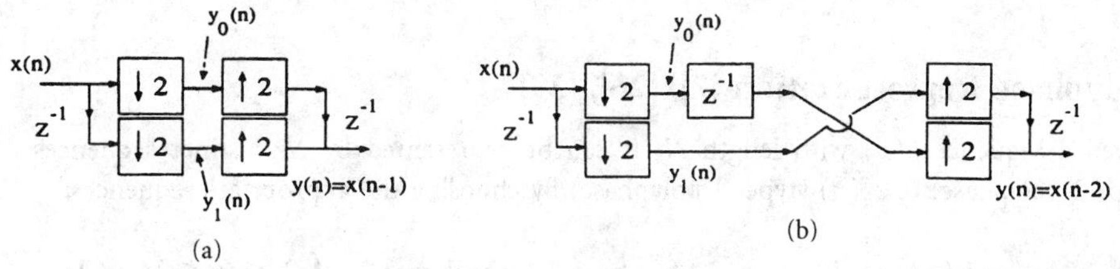

FIGURE 85.4 Simple perfect-reconstruction filter bank.

other words, the order for selecting the signals at the kth branch is

$$k = 0, M-1, M-2, \cdots, 2, 1, 0, M-1, M-2, \cdots \tag{85.3}$$

The output rate of a S/P converter is $1/M$ of the input rate.

On the other hand, a P/S converter can be obtained by cascading a set of expanders with a reverse-ordered delay chain, as depicted in Fig. 85.3(c). The output $y(n)$ is an interleaved combination of the signals $y_k(n)$, thus, its rate is M times the rate of $y_k(n)$. Assuming that $y_k(n)$ are causal, $y(n)$ is

$$y_{M-1}(0)y_{M-2}(0)\cdots y_0(0)y_{M-1}(1)y_{M-2}(1)\cdots y_0(1)\cdots \tag{85.4}$$

which implies $y_k(n)$ are selected in a *clockwise* fashion.

Simple Perfect-Reconstruction (PR) Multirate Systems

Figure 85.4 shows two simple multirate systems where the output signals are delayed versions of the input signal. The first system [Fig. 85.4(a)] blocks the input signal into the even-indexed and odd-indexed sequences $y_k(n)$ using a S/P converter and combines them together using a P/S converter. The overall delay for this system is 1 sample. On the other hand, if the orders of $y_k(n)$ are switched (after delaying $y_0(n)$ by a sample), as shown in Fig. 85.4(b), then it can be verified that the overall delay of the system is 2 samples.

The above simple PR filter banks do not process the subband signals $y_k(n)$, and thus are not useful filter banks in practice. They are, however, important in demonstrating the existence of PR filter banks.

Noble Identities [7], [29]

The components of a multirate system are filtering, decimators, expanders, S/P, and P/S converters. The two *noble identities* in this subsection allow interchangeability of the filtering,

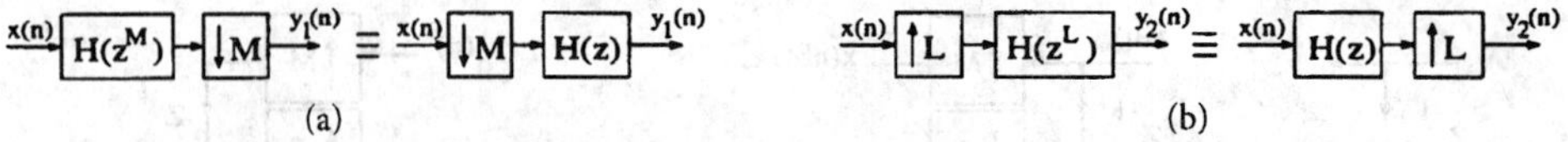

FIGURE 85.5 Noble identities.

decimators and expanders blocks, which are important and essential for the analysis of filter bank and multirate systems. The first identity, as depicted in Fig. 85.5(a) deals with the commutative property of a transfer function $H(z)$ and a decimator. One can verify that filtering with a transfer function $H(z^M)$ and then decimating the result is the same as decimating the input first and then filtering with the transfer function $H(z)$. The second identity [Fig. 85.5(b)] shows the equivalent operation between a transfer function $H(z)$ and an expander.

Polyphase Representation [7], [26], [29]

Given a sequence $h(n)$ with length N, it can be represented by M distinct sequences $e_k(n)$ (type-I polyphase) or $r_k(n)$ (type-II polyphase) by choosing the appropriate sequences:

$$\begin{cases} e_k(n) = h(nM + k), & \text{type-I polyphase} \\ r_k(n) = h(nM + M - 1 - k) & \text{type-II polyphase} \end{cases} \quad \begin{cases} 0 \le k \le M - 1 \\ 0 \le n \le \lceil \frac{N}{M} \rceil \end{cases} \tag{85.5}$$

The equivalent z-domain representation is (as shown in Fig. 85.6):

$$\underset{\text{type-I polyphase}}{H(z) = \sum_{k=0}^{M-1} z^{-k} E_k(z^M)} \qquad \underset{\text{type-II polyphase}}{H(z) = \sum_{k=0}^{M-1} z^{-(M-1-k)} R_k(z^M)} \qquad R_k(z) = E_{M-1-k}(z) \tag{85.6}$$

Example. Suppose that $h(n) = n$, $0 \le n \le 13$, then the polyphase representation $e_k(n)$ and $r_k(n)$ for $M = 3$ are

$$\underset{\text{type-I polyphase}}{\begin{cases} e_0 = 0, 3, 6, 9, 12 \\ e_1 = 1, 4, 7, 10, 13 \\ e_2 = 2, 5, 8, 11 \end{cases}} \qquad \underset{\text{type-II polyphase}}{\begin{cases} r_0 = 2, 5, 8, 11 \\ r_1 = 1, 4, 7, 10, 13 \\ r_2 = 0, 3, 6, 9, 12 \end{cases}} \tag{85.7}$$

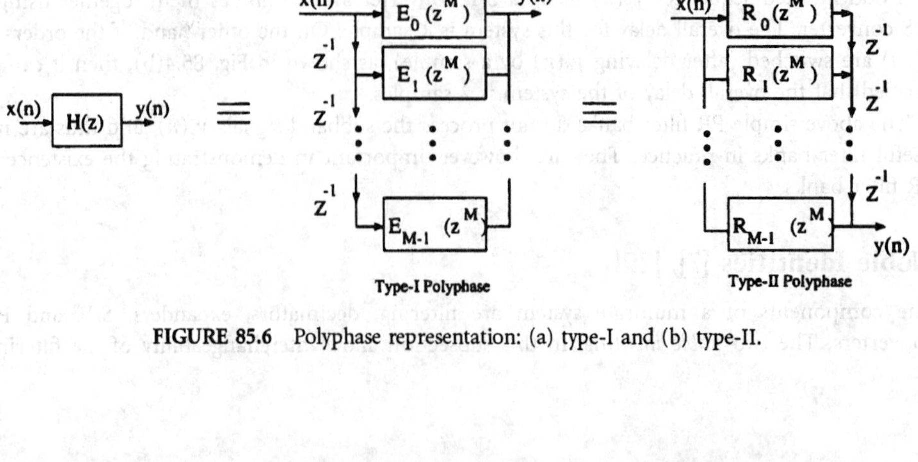

FIGURE 85.6 Polyphase representation: (a) type-I and (b) type-II.

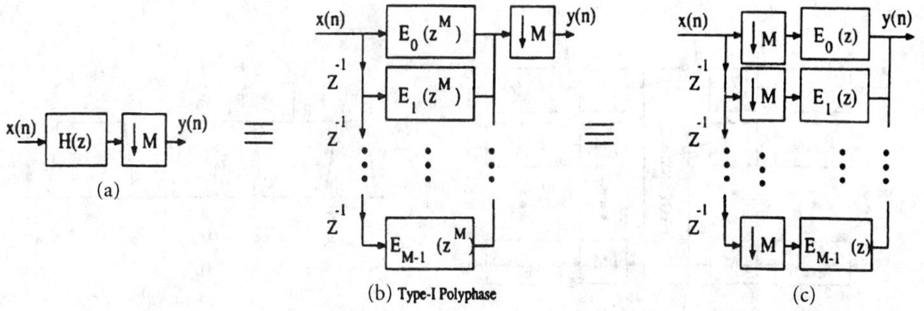

FIGURE 85.7 Decimation filtering and the efficient polyphase implementation.

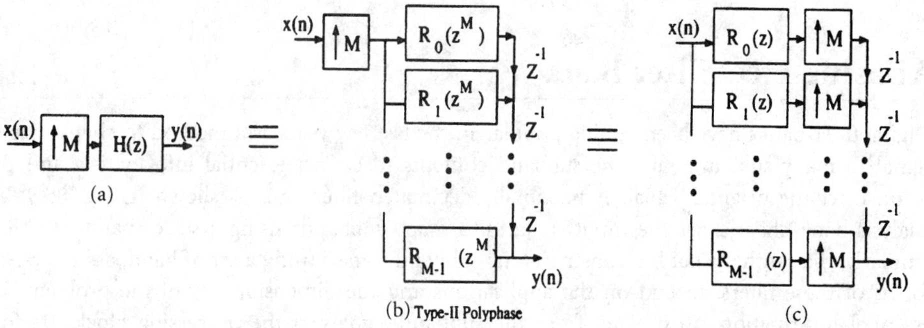

FIGURE 85.8 Interpolation filtering and the efficient polyphase implementation.

In summary, filtering by $H(z)$ is the same as filtering with the corresponding polyphase components $E_k(z)$ [or $R_k(z)$] of $H(z)$. Since the number of nonzero coefficients in $E_k(z)$ [or $R_k(z)$] is the same as that of $H(z)$, the above polyphase implementations do not offer any advantages. However, if $H(z)$ is proceeded (or preceded) with decimators (or expanders), as in a decimation (or interpolation) filter, then the polyphase implementations offer significant improvement on complexity, as elaborated in the next two subsections.

Decimation Filter [26], [29]

Figure 85.7 shows the decimation filter, its polyphase implementations, before and after using the noble identity. The equivalent operations for a decimation filter are thus, S/P converter, polyphase filtering (at the lower rate) and combining at the output. Comparing with the decimation filter in Fig. 85.7(a) which operates at the input signal rate, the equivalent polyphase implementation in Fig. 85.7(c) operates at $1/M$ the input signal rate.

Interpolation Filter [26], [29]

Figure 85.8 shows the interpolation filter, its polyphase implementations, before and after using the noble identity. The equivalent operations for an interpolation filter are thus, polyphase filtering (at the lower rate), S/P converter, and combining at the output. Comparing to the interpolation filter in Fig. 85.8(a), which operates at M times the input signal rate, the equivalent polyphase implementation in Fig. 85.8(c) operates at the input signal rate.

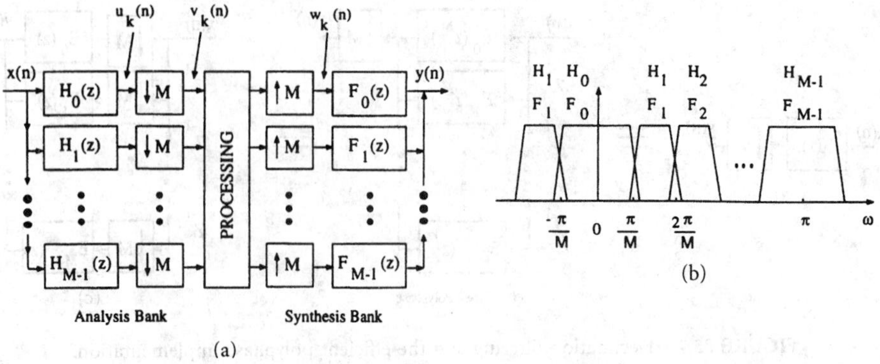

FIGURE 85.9 M-channel maximally-decimated uniform filter bank.

85.3 Aliasing-Free Filter Bank

Digital filter banks have been getting popular in the last few years as a method to channelize the signal to many subbands, use the subband contents to extract essential information and then reconstruct the original signal. A maximally decimated filter bank is shown in Fig. 85.9. The analysis bank channelizes the input signal into many subbands using a set of bandpass filters. Similarly, the synthesis bank reconstructs the subband signals using a set of bandpass filters. The forms of these filters depend on the applications and the dimensionality of the problem. The essential information is extracted from the subband signals in the processing block. Its form varies and depends on the applications. For example, in the audio/video compression system, the spectral contents of the subband signals are coded depending on their energies whereas in a radar system, the subband signals might be used to null out a narrowband interference adaptively. Filter banks found applications in audio/image compression, image analysis and enhancement, robotics, computer vision, echo-cancellation, radar, voice privacy, communications, etc.

Figure 85.9(a) illustrates a typical M-channel maximally decimated filter bank where $H_k(z)$ and $F_k(z)$, $0 \leq k \leq M-1$, are analysis and synthesis filters, respectively. Figure 85.9(b) shows the frequency responses of $H_k(e^{j\omega})$ and $F_k(e^{j\omega})$. The analysis filters $H_k(z)$ channelize the input signal $x(n)$ into M subband signals, which are downsampled (decimated) by a factor M. At the receiving end, the M subband signals are decoded, interpolated, and recombined using a set of synthesis filters $F_k(z)$. The decimator, which decreases the sampling rate of the signal, and the expander, which increases the sampling rate of the signals, are denoted by the boxes with down-arrows and up-arrows, respectively, as shown in the figure.

Reconstructed Error [6], [7], [29]

Since the analysis filters $H_k(z)$ are not ideal bandpass filters, the signals $u_k(n)$ in Fig. 85.9(a) are not bandlimited to π/M, which implies that the signals $v_k(n)$ have aliased components of $u_k(n)$ (see the subsection on downsampling). The aliased levels at $v_k(n)$ depend on the stopband attenuation of $H_k(e^{j\omega})$ and their transition bands. The interpolated signals $w_k(n)$ have M "images" of the compressed spectrum $V_k(e^{j\omega})$ [assuming that no processing has been done on $v_k(n)$]. These images are filtered by the synthesis filters $F_k(z)$. There are two types of errors at the reconstructed output signal $y(n)$: *distortions* (magnitude and phase) and *aliasing*. The nonideal filtering characteristics of $H_k(z)$ and $F_k(z)$ contribute to both the distortions and aliasing errors, whereas the changes in sampling rates (decimation and interpolation) contribute to the aliasing error. A system with no aliasing error (i.e., aliasing cancellation) is called an alias-free filter bank and a system with no distortion and aliasing errors is called a perfect-reconstruction filter bank.

In terms of the input signal and the filters $H_k(z)$ and $F_k(z)$, $Y(z)$ is

$$Y(z) = \sum_{k=0}^{M-1} T_k(z) X(zW^k) \qquad W = e^{-j2\pi/M} \tag{85.8}$$

where $T_k(z)$ are

$$T_k(z) = \frac{1}{M} \sum_{l=0}^{M-1} F_l(z) H_l(zW^k) \tag{85.9}$$

The transfer functions $T_0(z)$ and $T_k(z)$, $1 \le k \le M-1$ are the distortion and alias transfer functions, respectively. The objective in designing an alias-free filter bank or a PR filter bank is to find a set of filters $H_k(z)$ and $F_k(z)$ such that

$$\underbrace{T_k(z) = 0 \qquad 1 \le k \le M-1}_{\text{alias-free}} \qquad \text{or} \qquad \underbrace{\begin{cases} T_0(z) = z^{-n_0} \\ T_k(z) = 0 \qquad 1 \le k \le M-1 \end{cases}}_{\text{perfect-reconstruction}} \tag{85.10}$$

Polyphase Representation of Filter Bank [7], [29]

Using type-I and type-II polyphase representations for the analysis and synthesis bank yields

$$\begin{bmatrix} H_0(z) \\ H_1(z) \\ \vdots \\ H_{M-1}(z) \end{bmatrix} = \begin{bmatrix} E_{00}(z^M) & E_{01}(z^M) & \cdots & E_{0,M-1}(z^M) \\ E_{10}(z^M) & E_{11}(z^M) & \cdots & E_{1,M-1}(z^M) \\ \vdots & \vdots & \ddots & \vdots \\ E_{M-1,0}(z^M) & E_{M-1,1}(z^M) & \cdots & E_{M-1,M-1}(z^M) \end{bmatrix} \begin{bmatrix} 1 \\ z^{-1} \\ \vdots \\ z^{-(M-1)} \end{bmatrix}$$

$$= \mathbf{E}(z^M) \begin{bmatrix} 1 \\ z^{-1} \\ \vdots \\ z^{-(M-1)} \end{bmatrix} \tag{85.11}$$

$$[F_0(z) \; F_1(z) \; \cdots \; F_{M-1}(z)] = (z^{-(M-1)} \; \cdots \; z^{-1} \; 1)\mathbf{R}(z^M)$$

Here, $\mathbf{E}(z)$ and $\mathbf{R}(z)$ are the polyphase transfer matrices of the analysis and synthesis filter banks, respectively. Given any sets of analysis and synthesis filters, one can always find the corresponding transfer matrices $\mathbf{E}(z)$ (type-I polyphase) and $\mathbf{R}(z)$ (type-II polyphase).

Figure 85.10(a) is the polyphase representation of the structure in Fig. 85.9(a). Using the noble identities to move the decimators to the left of $\mathbf{E}(z^M)$ and the expanders to the right of $\mathbf{R}(z^M)$, Fig. 85.10(a) becomes the structure in Fig. 85.10(b). A few words on the implementation efficiency of this representation are needed here. As indicated in the figure, the input is blocked into a vector using a serial-parallel converter (implemented as cascade of delay chain and decimators), filtered by $\mathbf{R}(z)\mathbf{E}(z)$ and then recombined using a parallel-serial converter (implemented as a cascade of expanders and reversed-ordered delay chain). Comparing to the system in Fig. 85.9(a), the total number of nonzero coefficients in $\mathbf{E}(z)$ and $\mathbf{R}(z)$ are the same

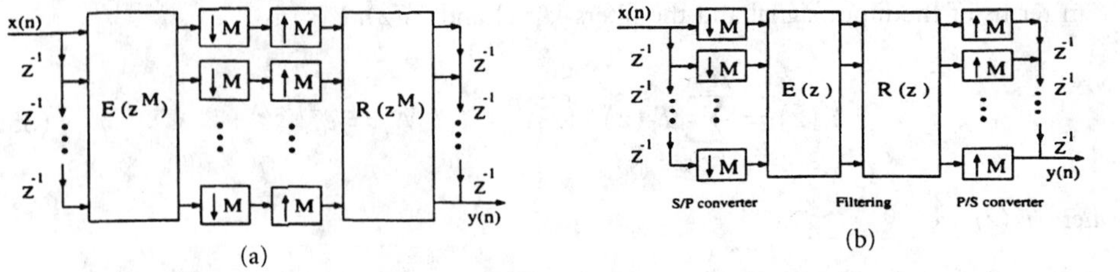

FIGURE 85.10 Polyphase representation of M-channel uniform filter bank.

as that in $H_k(z)$ and $F_k(z)$, respectively. The main difference is the rate of operation. In other words, the filtering operations in the original filter bank [Fig. 85.9(a)] are done at the input rate, whereas the filtering operations in the polyphase representation [Fig. 85.10(b)] are done at $1/M$ the input rate.

Necessary and Sufficient Condition for Alias-Free Filter Bank [7], [29]

Defining $\mathbf{P}(z) = \mathbf{R}(z)\mathbf{E}(z)$, then the necessary and sufficient condition on $\mathbf{P}(z)$ such that the output signal $y(n)$ has no aliased component of $x(n)$, i.e., $T_k(z) = 0$ for $1 \le k \le M-1$, is

$$\mathbf{P}(z) = \text{pseudo-circulant matrix}$$

$$= \begin{pmatrix} P_0(z) & P_1(z) & \cdots & P_{M-2}(z) & P_{M-1}(z) \\ z^{-1}P_{M-1}(z) & P_0(z) & \cdots & P_{M-3}(z) & P_{M-2}(z) \\ \vdots & \vdots & \ddots & \vdots & \vdots \\ z^{-1}P_1(z) & z^{-1}P_2(z) & \cdots & z^{-1}P_{M-1}(z) & P_0(z) \end{pmatrix} \tag{85.12}$$

In other words, as long as the product of the polyphase matrices $\mathbf{R}(z)$ and $\mathbf{E}(z)$ is a pseudo-circulant matrix, then the only distortions at the output are the amplitude and phase distortions [represented in $T_0(z)$]. Examples of alias-free (non-PR) filter banks are the two-channel quadrature mirror filter (QMF) bank (known as the Johnston filter bank), the two-channel allpass-based IIR filter bank, and the M-channel DFT filter bank. These will all be considered in the next three subsections.

Two-Channel QMF (Johnston) Filter Bank [20]

Figure 85.11 shows a two-channel filter bank, where $H_0(z)$ and $H_1(z)$ are analysis filters and $F_0(z)$ and $F_1(z)$ are synthesis filters, respectively. From (85.8), it can be verified that ($M = 2$)

$$Y(z) = T_0(z)X(z) + T_1(z)X(-z) \tag{85.13}$$

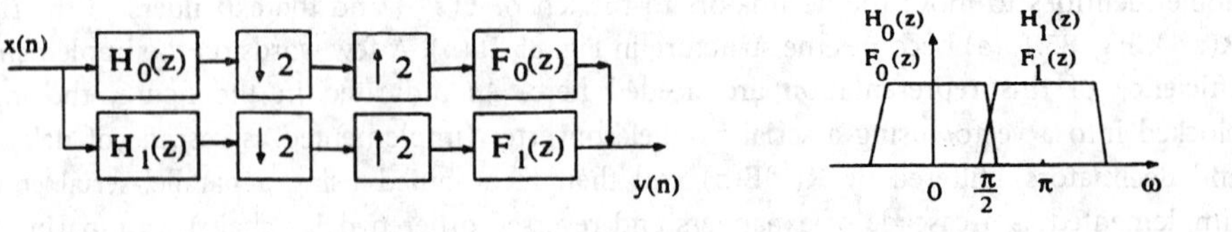

FIGURE 85.11 Two-channel filter bank.

where $T_0(z)$ and $T_1(z)$ are the distortion and aliasing functions, respectively. $T_k(z)$ are

$$\begin{cases} T_0(z) = F_0(z)H_0(z) + F_1(z)H_1(z) \\ T_1(z) = F_0(z)H_0(-z) + F_1(z)H_1(-z) \end{cases} \tag{85.14}$$

For arbitrary choices of $H_k(z)$, the only choices of $F_k(z)$ such that aliasing is canceled are [1]–[4] (i.e., $T_1(z) = 0$): $F_1(z) = -H_0(-z)$ and $F_0(z) = H_1(-z)$. With the above choices of $F_k(z)$, the distortion function $T_0(z)$ becomes

$$T_0(z) = H_0(z)H_1(-z) - H_0(-z)H_1(z) \tag{85.15}$$

The objective in an alias-free filter bank is to find $H_k(z)$ such that the distortion function $T_0(z)$ approximates a delay. The Johnston filter bank assumes that

$$H_0(z) \text{ is an even-length linear-phase (symmetric) filter and } H_1(z) = H_0(-z) \tag{85.16}$$

Note that this is a two-channel DFT filter bank where $H_1(z)$ is a shifted version of $H_0(z)$. The analysis filter bank is

$$\begin{aligned} \begin{pmatrix} H_0(z) \\ H_1(z) \end{pmatrix} &= \begin{pmatrix} E_0(z^2) & E_1(z^2) \\ E_0(z^2) & -E_1(z^2) \end{pmatrix} \begin{pmatrix} 1 \\ z^{-1} \end{pmatrix} \\ &= \begin{pmatrix} 1 & 1 \\ 1 & -1 \end{pmatrix} \begin{pmatrix} E_0(z^2) & 0 \\ 0 & E_1(z^2) \end{pmatrix} \begin{pmatrix} 1 \\ z^{-1} \end{pmatrix} \end{aligned} \tag{85.17}$$

where $E_k(z)$ are the polyphase components of the low-pass filter $H_0(z)$. The corresponding synthesis filters are

$$\begin{aligned} [F_0(z) \quad F_1(z)] &= [H_1(-z) \quad -H_0(-z)] \\ &= (z^{-1} \quad 1) \begin{pmatrix} E_1(z^2) & 0 \\ 0 & E_0(z^2) \end{pmatrix} \begin{pmatrix} 1 & 1 \\ 1 & -1 \end{pmatrix} \end{aligned} \tag{85.18}$$

and the overall distortion function $T_0(z)$ is $T_0(z) = z^{-1}E_0(z^2)E_1(z^2)$, which can be equalized using either IIR or FIR filters. Figure 85.12 shows the equivalent system for the two-channel Johnston filter bank. Several designs have been tabulated in [20].

Let N be the length of $H_0(z)$, then the number of coefficients in $E_k(z)$ are $N/2$. From Fig. 85.12, the total number of multiplications and additions in the analysis bank are N and N, respectively. Since the filtering is computed at the lower rate, then the effective complexity is $N/2$ multiplications and $N/2$ additions per unit time, respectively. The computation complexity of the synthesis bank is the same as that of the analysis bank.

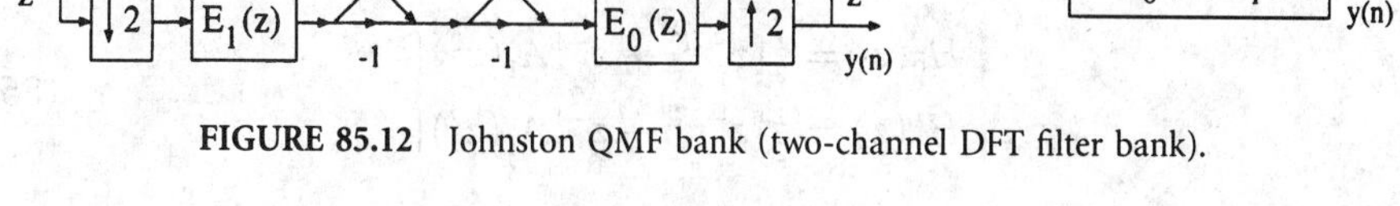

FIGURE 85.12 Johnston QMF bank (two-channel DFT filter bank).

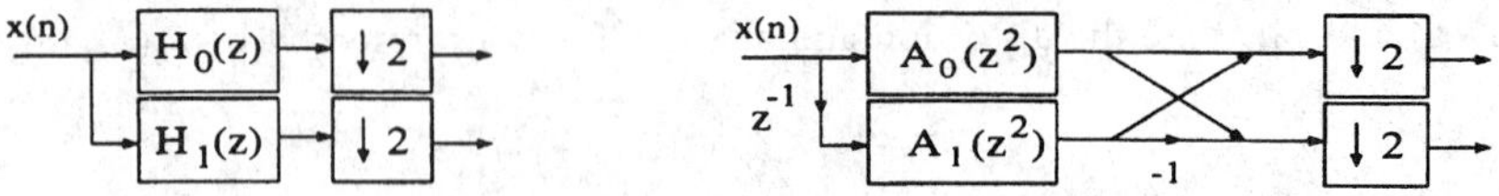

FIGURE 85.13 Two-channel allpass-based IIR filter bank.

Two-Channel Allpass-Based IIR Filter Bank [22], [23], [29]

It is clear from the above derivation that the DFT filter bank is not a PR system and the distortion function $T_0(z)$ is a product of the two polyphase components of $H_0(z)$, i.e., $E_0(z)$ and $E_1(z)$. If these polyphase components have unit magnitudes, then there will be no amplitude distortion at the output of the filter bank. In other words, if the polyphase components are allpass functions, then $|Y(e^{j\omega})| = |X(e^{j\omega})|$. For example, if both polyphase components are delays (which is a special case of allpass function), then $H_0(z)$ and $H_1(z)$ are sum and difference of delays (which is not very interesting). However, if both polyphase components are IIR allpass functions, i.e.,

$$E_0(z) = A_0(z) \qquad E_1(z) = A_1(z) \tag{85.19}$$

then $H_0(z)$ and $H_1(z)$ are sum and differences of two allpass filters $A_0(z)$ and $A_1(z)$, as depicted in Fig. 85.13, i.e.,

$$\begin{cases} H_0(z) = \frac{1}{2}[A_0(z^2) + z^{-1}A_1(z^2)] \\ H_1(z) = \frac{1}{2}[A_0(z^2) - z^{-1}A_1(z^2)] \end{cases} \tag{85.20}$$

With the above choices of $H_k(z)$, the only distortion at the output is phase distortion (being the sum of the phase responses of the allpass filters), which could be equalized using an allpass filter. Butterworth, Chebyshev, and elliptic filters (with appropriate frequency specifications) can be decomposed into sum and difference of two allpass filters [22], as in (85.20). Given the filter, the synthesis procedure to find $A_k(z)$ is given in [22].

Choice of $A_k(z)$. For any choice of $A_k(z)$, it is not clear that the resulting analysis filters $H_k(z)$ approximate ideal low-pass and high-pass responses. Thus, it is important to choose $A_k(z)$ carefully to obtain desirable frequency characteristics. Let $\phi_0(\omega)$ and $\phi_1(\omega)$ be the phase responses of $A_0(e^{j\omega})$ and $A_1(e^{j\omega})$, respectively, then

$$|H_0(e^{j\omega})| = \frac{1}{2}\left|e^{j\phi_0(2\omega)} + e^{j(\phi_1(2\omega)-\omega)}\right| = \frac{1}{2}\left|1 + e^{j(\phi_1(2\omega)-\phi_0(2\omega)-\omega)}\right| \tag{85.21}$$

To obtain a good low-pass filter characteristic for $\mathbf{H}_0(z)$, (85.21) implies that the phases of the allpass functions should be in-phase in the passband region and out-of-phase in the stopband region. The choice of $\phi_0(2\omega)$ is irrelevant here and should be chosen such that the filters have approximately linear phase in their passband regions. Let $A_0(z) = z^{-K}$ (i.e., $\phi_0(\omega) = -K\omega$), then $A_1(z)$ should have the following phase response:

$$\phi_1(2\omega) - 2K\omega - \omega = \begin{cases} 0 & \text{in the passband region} \\ \pi & \text{in the stopband region} \end{cases} \tag{85.22}$$

and the filters are

$$\begin{cases} H_0(z) = \frac{1}{2}[z^{-2K} + z^{-1}A_1(z^2)] \\ H_1(z) = \frac{1}{2}[z^{-2K} - z^{-1}A_1(z^2)] \end{cases} \tag{85.23}$$

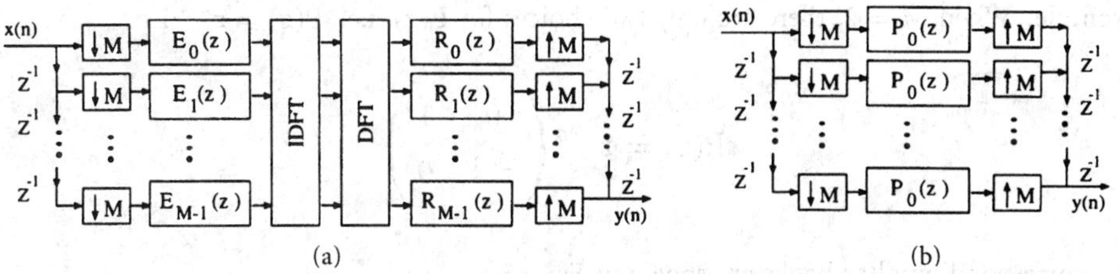

FIGURE 85.14 M-channel uniform DFT filter bank.

Methods for designing an allpass transfer function approximating a given phase response are elaborated in [23].

For a given K, the polyphase component $A_1(z)$ is a K-order allpass transfer function, which can be decomposed into a cascade of K first-order allpasses. Each first-order allpass transfer function is implemented by one multiplication and two additions. For the K allpass sections in the analysis bank, K multiplications and $2K$ additions are needed. The effective (computed at the lower rate) computation complexity for the analysis bank is $K/2$ multiplications and $K+1$ additions, respectively (with the additional cost of a two-point DFT).

M-Channel DFT Filter Bank [21], [29]

The analysis and synthesis filters $H_k(z)$ and $F_k(z)$ are uniformly shifted versions of the lowpass filters $H_0(z)$ and $F_0(z)$, respectively, i.e.,

$$\begin{cases} H_0(z) = \sum_{k=0}^{M-1} z^{-k} E_k(z^M) & H_k(z) = H_0(zW^k) \quad W = e^{-j2\pi/M} \\ F_0(z) = \sum_{k=0}^{M-1} z^{-(M-1-k)} R_k(z^M) & F_k(z) = F_0(zW^k) \quad W = e^{-j2\pi/M} \end{cases} \tag{85.24}$$

Figure 85.14(a) shows the polyphase representation of the M-channel DFT filter bank. It is clear that PR cannot be achieved. By choosing $R_k(z) = \prod_{l \neq k} E_l(z)$, then Fig. 85.14(a) simplifies to Fig. 85.14(b), which implies that the overall distortion function is

$$T_0(z) = z^{-(M-1)} P_0(z^M) \quad \text{where} \quad P_0(z) = \prod_{k=0}^{M-1} E_k(z) \tag{85.25}$$

Similarly, as in the two-channel DFT filter bank and the two-channel allpass-based IIR filter bank, the output $y(n)$ should be equalized to minimize the distortion function $T_0(z)$.

85.4 Perfect-Reconstruction Filter Bank

Necessary and Sufficient Condition. Equation (85.12) gives the necessary and sufficient condition for an alias-free filter bank. If there exists a transfer function $P_k(z) = z^{-m_0}$ and all other functions $P_l(z) = 0$, $l \neq k$, then the filter bank is a PR one, i.e., [7], [29]

$$\mathbf{P}(z) = \mathbf{R}(z)\mathbf{E}(z) = z^{-m_0} \begin{pmatrix} \mathbf{0} & \mathbf{I}_{M-r} \\ z^{-1}\mathbf{I}_r & \mathbf{0} \end{pmatrix} \tag{85.26}$$

The resulting overall delay is $r + M - 1 + m_0 M$, i.e., $T_0(z) = cz^{-(r+M-1+m_0M)}$.

Example. When $M = 2$, there are only two choices for $\mathbf{P}(z)$, i.e., $\mathbf{P}(z) = z^{-m_0}\mathbf{I}$ or

$$\mathbf{P}(z) = z^{-m_0}\begin{pmatrix} 0 & 1 \\ z^{-1} & 0 \end{pmatrix}$$

The corresponding filter banks are shown in Fig. 85.4.

The PR filter bank where the polyphase transfer matrices $\mathbf{R}(z)$ and $\mathbf{E}(z)$ satisfy (85.26) is called a biorthogonal filter bank. Example of biorthogonal filter bank is the two-channel linear-phase PR filter bank [4] (see the subsection) and the three-channel linear-phase PR filter bank [10].

Paraunitary Filter Bank [6]–[9], [29]

The paraunitary filter bank is a special case of biorthogonal filter bank where $\mathbf{E}(z)$ and $\mathbf{R}(z)$ are related as

$$\mathbf{R}(z) = \mathbf{E}^{-1}(z) = \tilde{\mathbf{E}}(z) = \mathbf{E}^T(z^{-1}) \tag{85.27}$$

The synthesis filters $F_k(z)$ are related to analysis filters $H_k(z)$ (of length N) as

$$F_k(z) = z^{-(N-1)}H_k(z^{-1}) \qquad f_k(n) = h_k(N-1-n) \tag{85.28}$$

Factorization. Any causal degree-J FIR paraunitary $\mathbf{E}(z)$ can be expressed as [12], [61]

$$\mathbf{E}(z) = \mathbf{V}_J(z)\mathbf{V}_{J-1}(z)\cdots\mathbf{V}_1(z)\mathbf{U} \quad \text{where} \quad \begin{cases} \mathbf{V}_k(z) = \mathbf{I} - \mathbf{v}_k\mathbf{v}_k^\dagger + z^{-1}\mathbf{v}_k\mathbf{v}_k^\dagger \\ \mathbf{U}^\dagger\mathbf{U} = \mathbf{I} \end{cases} \tag{85.29}$$

Here, $\mathbf{v}_k$ is a unit-norm column vector. The constant unitary matrix $\mathbf{U}$ can be factorized as $\mathbf{U} = \mathbf{U}_1\mathbf{U}_2\cdots\mathbf{U}_{M-1}\mathbf{D}$, where $\mathbf{D}$ is a diagonal matrix with diagonal elements $D_{kk} = e^{j\theta_k}$ and $\mathbf{U}_k = \mathbf{I} - 2\mathbf{u}_k\mathbf{u}_k^\dagger$ for unit-norm column vector $\mathbf{u}_k$.

The above factorization is complete in the sense that it covers all paraunitary filter banks, i.e., given any paraunitary filter bank, there exists a set of vectors $\mathbf{v}_k$ and $\mathbf{u}_k$ that implement it. Synthesis procedure is discussed in [9], [29]. Alternative factorizations based on Givens rotation are discussed in [8].

Power Complementary. A set of filters $H_k(z)$ is said to be power-complementary if their frequency responses $H_k(e^{j\omega})$ satisfy the property

$$\sum_k |H_k(e^{j\omega})|^2 = c$$

where c is a positive constant. The analysis (and synthesis) filters of a paraunitary filter bank satisfy the power complementary property.

Two-Channel Paraunitary Filter Bank [1]–[3]

Let

$$H_1(z) = -z^{-(N-1)}H_0(-z^{-1}) \qquad F_k(z) = z^{-(N-1)}H_k(z^{-1}) \tag{85.30}$$

where N is the filter's length (even N). Substituting the above relations into (85.14) yields $T_1(z) = 0$ and

$$\begin{aligned} T_0(z) &= [H_0(z)H_0(z^{-1}) + H_1(z)H_1(z^{-1})]z^{-(N-1)} \\ &= [H_0(z)H_0(z^{-1}) + H_0(-z)H_0(-z^{-1})]z^{-(N-1)} \end{aligned}$$

For PR system, $T_0(z)$ should be a delay, i.e., $z^{-(N-1)}$, which implies that

$$H_0(z)H_0(z^{-1}) + H_0(-z)H_0(-z^{-1}) = 1 \tag{85.31}$$

Let $G(z) = H_0(z)H_0(z^{-1})$, then (85.31) implies that $G(z)$ is a halfband filter and $H_0(z)$ is a spectral factor of a halfband filter [29], [40]. In summary, the two-channel paraunitary filter bank has the following properties:

$$\begin{cases} H_0(z) \text{ is a spectral factor of a halfband filter} \\ H_1(z) = -z^{-(N-1)}H_0(-z^{-1}) \\ F_k(z) = z^{-(N-1)}H_k(z^{-1}) \end{cases} \tag{85.32}$$

A design procedure for a two-channel paraunitary filter bank [2] would be (a). Design an equiripple halfband filter $\hat{G}(z)$ using Remez algorithm of length $2N - 1$ [37] (b). Measure the stopband attenuation (δ) and form a nonnegative function $G(z) = \hat{G}(z) + \delta z^{-(N-1)}$ (c). Find the spectral factorization $H_0(z)$ of $\hat{G}(z)$ and (d). $H_1(z)$ and $F_k(z)$ are computed using (85.32).

The above two-channel paraunitary filter bank can also be realized in a lattice structure as shown in Fig. 85.15(a), where

$$\mathbf{U}_k = \begin{pmatrix} \cos\theta_k & \sin\theta_k \\ -\sin\theta_k & \cos\theta_k \end{pmatrix},$$

and Fig. 85.15(b), where

$$\mathbf{V}_k = \begin{pmatrix} 1 & \alpha_k \\ -\alpha_k & 1 \end{pmatrix},$$

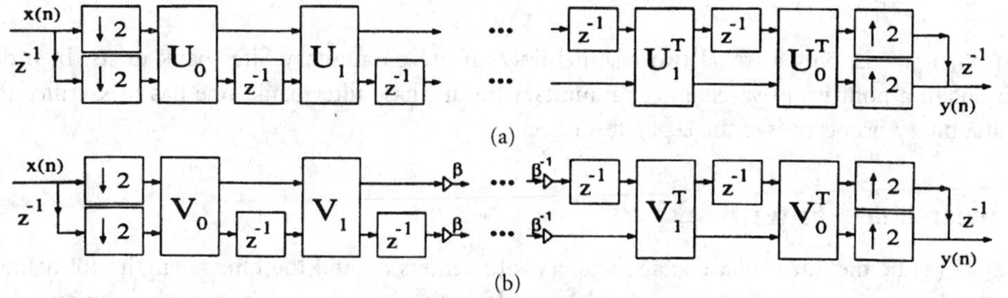

FIGURE 85.15 Factorization of two-channel paraunitary filter bank.

and

$$\beta = 1/\sqrt{\prod_k 1 + \alpha_k^2}.$$

The lattice structure is complete in the sense that for any choices of θ_k (or α_k), the resulting filter satisfies (85.32) (and it thus paraunitary) and given any paraunitary filter bank, then there exists a unique set of angle θ_k (or α_k) that implements it. Given $H_0(z)$, the procedure to synthesize θ_k (or α_k), is given in [3]. The orthogonal Daubechies wavelet [32] is a paraunitary filter bank and can be implemented using the lattice structure below. For example, the D_6 wavelet has lattice angles $\theta_0 = \pi/3$ and $\theta_1 = -\pi/12$.

We only consider the implementation complexity of the structure in Fig. 85.15(b) here. Let N be the length of $H_0(z)$, $(N = 2J)$, then the number of lattice sections V_k in the analysis bank is J. Each lattice section requires 2 multiplications and 2 additions, which implies that the number of multiplications and additions in the analysis bank is $2J + 2$ (for the extra multiplications β) and $2J$, respectively. The effective computation complexity (at the input rate) is $J + 1$ multiplications and J additions.

For a given design specification (stopband attenuation, transition band, etc.), [29, p. 311] compares the complexity between the Johnston's filter and the paraunitary filter bank.

Two-Channel Paraunitary Linear-Phase Filter Bank (FIR) [29]

$$\begin{cases} H_0(z) \text{ is an even-length} & \\ \quad \text{symmetric filter} & H_0(z) = z^{-(N-1)}H_0(z^{-1}) \\ H_1(z) \text{ is an even-length} & \\ \quad \text{antisymmetric filter} & H_1(z) = -z^{-(N-1)}H_1(z^{-1}) \\ F_0(z) = z^{-(N-1)}H_0(z^{-1}) = H_0(z) & F_1(z) = z^{-(N-1)}H_1(z^{-1}) = -H_1(z) \end{cases} \tag{85.33}$$

With the above choices of filters, aliasing is canceled and the distortion function becomes

$$T_0(z) = H_0^2(z) - H_1^2(z) = [H_0(z) + H_1(z)][H_0(z) - H_1(z)] \tag{85.34}$$

which should be a delay (z^{-n_0}) for a PR filter bank. From (85.34), it is clear that both $[H_0(z) + H_1(z)]$ and $[H_0(z) - H_1(z)]$ should be delays, which implies that

$$\begin{cases} H_0(z) = \frac{1}{2}(z^{-n_1} + z^{-n_2}) \\ H_1(z) = \frac{1}{2}(z^{-n_1} - z^{-n_2}) \end{cases} \quad n_0 = n_1 + n_2 \tag{85.35}$$

In other words, only a trivial two-channel linear-phase paraunitary filter bank exists. In order to obtain a nontrivial two-channel paraunitary linear-phase filter bank, one has to sacrifice the paraunitary property (see the separate subsection).

Linear-Phase Filter Bank

Let $H_k(z)$ be the linear-phase analysis filters with centers c_k, and the filters' lengths are defined to be $N_k = 2c_k + 1$. For a given number of channels (M), N_k can be uniquely expressed as $N_k = m_k M + i_k$, where m_k and i_k are the modulo and the remainder parts of N_k with respect to M. The polyphase representation of $H_k(z)$ is $H_k(z) = \sum_{l=0}^{M-1} z^{-l} E_{k,l}(z^M)$, where $E_{k,l}(z)$ are type-I polyphase components of $H_k(z)$ [7]. $E_{k,l}(z)$ are related [10], [29] as

$$E_{k,l}(z^{-1}) = z^{m_k} J_k \times \begin{cases} E_{k,(i_k-l)}(z), & l \le i_k \\ z^{-1} E_{k,(M+i_k-l)}(z), & l > i_k \end{cases}$$

where

$$J_k = \begin{cases} +1 & h_k(n) \text{ is symmetric} \\ -1 & h_k(n) \text{ is antisymmetric} \end{cases}$$

For a special case where all i_k are equal to a constant, then the total number of linear-phase PR filter bank [10] is

$$\text{total} = \psi(\Delta_M - \Theta_M) + M\Theta_M$$

where

$$\psi = \begin{cases} M/2 & M \text{ is even} \\ M & M = 4r + 1 \\ 0 & \text{otherwise} \end{cases} \qquad \Delta_M = \begin{cases} M/2 & \text{even } M \\ (M+1)/2 & \text{odd } M \end{cases}$$

$$\Theta_M = \begin{cases} M/2 & \text{even } M \\ (M-1)/2 & \text{odd } M \end{cases}$$

In the following subsections, the theory and analysis of the two-channel linear-phase biorthogonal filter bank and the M-channel linear-phase paraunitary filter bank are covered in detail.

Two-Channel Biorthogonal Linear-Phase Filter Bank [4], [5]

There are two types of biorthogonal filter banks [4]:

- Type A (SAOO) [4]: Both filters have odd orders (even length). $H_0(z)$ is symmetric and $H_1(z)$ is antisymmetric.
- Type B (SSEE) [4]: Both filters have even orders (odd length). Both filters have symmetric impulse responses.

Type A: (SAOO). Considering a pair of even-length (N) FIR transfer functions where the first function is symmetric and the other is antisymmetric, (we only consider the case of equal length here, the general case is discussed in [4]) and defining

$$\begin{pmatrix} P(z) \\ Q(z) \end{pmatrix} = \frac{1}{2} \begin{pmatrix} 1 & 1 \\ 1 & -1 \end{pmatrix} \begin{pmatrix} H_0(z) \\ H_1(z) \end{pmatrix} \tag{85.36}$$

then it can be proved that $Q(z) = z^{-(N-1)} P(z^{-1}) = \tilde{P}(z)$, $[q(n) = p(N-1-n)]$. Thus, as long as we can find a one-input/two-output system [as depicted in Fig. 85.16(a)] that gives us the transfer function pair $[P(z), \tilde{P}(z)]$, then the analysis filters (with the above symmetric properties) can be found, as depicted in Fig. 85.16(b).

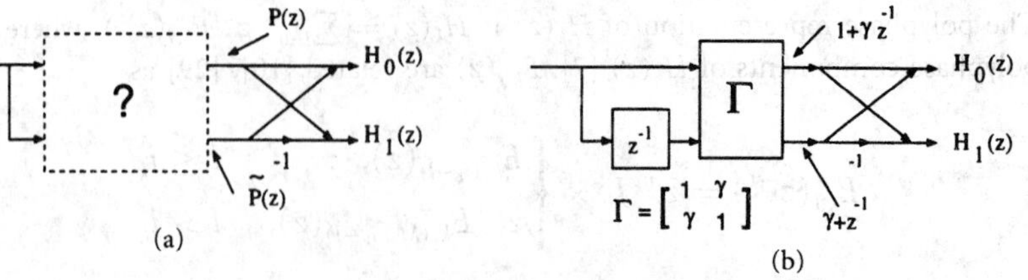

FIGURE 85.16 Pertaining to the two-channel biorthogonal linear-phase filter bank (type A).

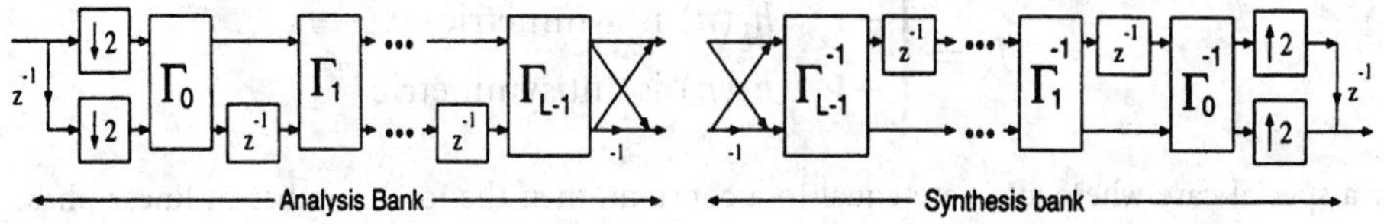

FIGURE 85.17 Lattice structure for the two-channel biorthogonal linear-phase filter bank (type A).

The simplest choice for the black box that gives the pair $[P(z), \tilde{P}(z)]$ is $\Gamma\Lambda(z)$ where [see Fig. 85.16(b)]

$$\Gamma = \begin{pmatrix} 1 & \gamma \\ \gamma & 1 \end{pmatrix} \qquad \Lambda(z) = \begin{pmatrix} 1 & 0 \\ 0 & z^{-1} \end{pmatrix} \tag{85.37}$$

which yields $P(z) = 1 + \gamma z^{-1}$, $H_0(z) = 0.5[(1+\gamma) + (1+\gamma)z^{-1}]$ and $H_1(z) = 0.5[(1-\gamma) - (1-\gamma)z^{-1}]$. The resulting analysis filters have length 2. In order to increase their lengths (and keep them even), the additional block must have only z^{-2} powers (not z^{-1}). Figure 85.17 shows the general lattice structure for a biorthogonal linear-phase type A (SAOO) filter bank. Since Γ_k is nonsingular (except for the case where $\gamma_k = 1$, which is discussed in [4]), one can reconstruct the input by inverting the operations in the analysis bank (Fig. 85.17). The above lattice structure is the complete structure for a two-channel biorthogonal linear-phase filter bank (type A). In other words, given any type A pair of filters $[H_0(z), H_1(z)]$, its polyphase transfer matrix $\mathbf{E}(z)$ can be factorized as

$$\mathbf{E}(z) = \Gamma_{L-1}\Lambda(z)\Gamma_{L-2}\Lambda(z)\cdots\Lambda(z)\Gamma_0 \tag{85.38}$$

where Γ_k and $\Lambda(z)$ are shown in (85.37), respectively.

Let N ($N = 2J$) be the length of $H_0(z)$, then the number of lattice sections Γ_k in the analysis bank is J. Each lattice section requires 2 multiplications and 2 additions, which implies that the number of multiplications and additions at the analysis bank is $2J$ and $2J$, respectively. The effective computation complexity (at the input rate) is J multiplications and J additions. An alternative implementation that uses one multiplication and three additions per lattice section is discussed in [29, p. 343].

Type B: (SSEE) [4]. Consider a pair of odd-length (even-order) symmetric FIR transfer functions. Let N_k be the lengths of $H_k(z)$, then their sum should satisfy the condition $N_1 = N_0 + 4\alpha + 2$, where α is any nonnegative number [4]. It can be shown that the polyphase transfer matrix $\mathbf{E}(z)$ can be factorized as [4]

$$\mathbf{E}(z) = \prod_{k=0}^{L-1} \mathbf{A}_{L-1-k}(z) \quad \text{where} \quad \mathbf{A}_k(z) = \mathbf{B}_k(z)\begin{pmatrix} \alpha_k z^{-2K_k} & 0 \\ 0 & 1 \end{pmatrix} \tag{85.39}$$

The above lattice structure implementation in (85.39) is complete. In other words, given a pair of type B biorthogonal linear-phase filter banks, one can find the corresponding lattice sections $\mathbf{A}_k(z)$. The biorthogonal Daubechies wavelets [32] are a type B linear-phase filter bank and they can be implemented using the above lattice structure. The lattice structure offers efficient implemention (see [4]).

M-Channel Linear-Phase Paraunitary Filter Bank ($M > 2$) [11]

Although linear-phase and paraunitary properties cannot be simultaneously imposed in the case of two-channel, they can coexist for filter bank with more channels [11]. For instance, DCT (discrete-cosine-transform) and LOT (lapped orthogonal transform) are two examples where both the analysis and synthesis filters $H_k(z)$ and $F_k(z)$ are linear-phase FIR filters and the corresponding filter banks are paraunitary filter bank. Assuming that

$$\begin{cases} M \text{ is even} \quad L = NM \\ f_k(n) = h_k(N-1-n) \end{cases} \tag{85.40}$$

it is shown that $M/2$ filters (in analysis or synthesis) have symmetric impulse responses and the other $M/2$ filters have antisymmetric impulse responses. Under the assumptions on N, M, and on the filter symmetry, the polyphase transfer matrix $\mathbf{E}(z)$ of the LPPUFB (linear-phase paraunitary filter bank) of degree $K-1$ can be decomposed as a product of orthogonal factors and delays [11], i.e.,

$$\mathbf{E}(z) = \mathbf{S}\mathbf{Q}\mathbf{T}_{N-1}\Lambda(z)\mathbf{T}_{N-2}\Lambda(z)\cdots\Lambda(z)\mathbf{T}_0\mathbf{Q} \tag{85.41}$$

where

$$\mathbf{Q} = \begin{pmatrix} \mathbf{I}_{M/2} & \mathbf{0}_{M/2} \\ \mathbf{0}_{M/2} & \mathbf{J}_{M/2} \end{pmatrix} \qquad \Lambda(z) = \begin{pmatrix} \mathbf{I}_{M/2} & \mathbf{0}_{M/2} \\ \mathbf{0}_{M/2} & z^{-1}\mathbf{I}_{M/2} \end{pmatrix}$$
$$\mathbf{S} = \frac{1}{\sqrt{2}}\begin{pmatrix} \mathbf{S}_0 & \mathbf{0}_{M/2} \\ \mathbf{0}_{M/2} & \mathbf{S}_1 \end{pmatrix}\begin{pmatrix} \mathbf{I}_{M/2} & \mathbf{J}_{M/2} \\ \mathbf{I}_{M/2} & -\mathbf{J}_{M/2} \end{pmatrix} \tag{85.42}$$

where $\mathbf{S}_0$ and $\mathbf{S}_1$ can be any $M/2 \times M/2$ orthogonal matrices. $\mathbf{T}_i$ are $M \times M$ orthogonal matrices

$$\mathbf{T}_i = \begin{pmatrix} \mathbf{I}_{M/2} & \mathbf{I}_{M/2} \\ \mathbf{I}_{M/2} & -\mathbf{I}_{M/2} \end{pmatrix}\begin{pmatrix} \mathbf{U}_i & \mathbf{0} \\ \mathbf{0} & \mathbf{V}_i \end{pmatrix}\begin{pmatrix} \mathbf{I}_{M/2} & \mathbf{I}_{M/2} \\ \mathbf{I}_{M/2} & -\mathbf{I}_{M/2} \end{pmatrix} \tag{85.43}$$

where $\mathbf{U}_i$ and $\mathbf{V}_i$ are arbitrary orthogonal matrices. The above factorization in (85.41) covers all LPPUFB for even number of channels. In other words, given any set of filters $H_k(z)$ that belong to a LPPUFB, one can obtain the corresponding matrices **S**, **Q**, and $\mathbf{T}_k(z)$. The synthesis procedure is given in [11].

GenLOT (Generalized Lapped Orthogonal Transform) [12]

The basis functions (analysis filters) of the DCT and of the LOT are linear phase and orthonormal. They must belong to the class of LPPUFB, discussed in the previous subsection. In other words, the PTM (polyphase transfer matrix) of the DCT and LOT must be in the form of (85.41). It is shown in [14] that the PTM of the LOT can be represented as in Fig. 85.18. Since the factorization in (85.41) covers all M-channel linear-phase paraunitary filter banks, the PTM in Fig. 85.18 should have the form as in (85.41). It does not, however. Our objective below is to derive another lattice form for (85.41) in which both the DCT and the LOT are special cases.

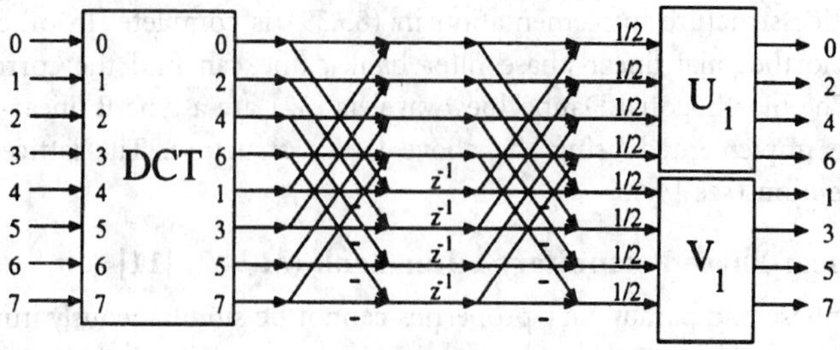

FIGURE 85.18 Polyphase transfer matrix of the lapped orthogonal transform.

We will abbreviate the notation for (85.41) as

$$\mathbf{E}(z) = \mathbf{SQT}_{N-1}\left(\prod_{i=N-2}^{0} \mathbf{\Lambda}(z)\mathbf{T}_i\right)\mathbf{Q} \tag{85.44}$$

Let

$$\mathbf{W} = \frac{1}{\sqrt{2}}\begin{bmatrix} \mathbf{I}_{M/2} & \mathbf{I}_{M/2} \\ \mathbf{I}_{M/2} & -\mathbf{I}_{M/2} \end{bmatrix} \quad \text{and} \quad \mathbf{\Phi}_i = \begin{bmatrix} \mathbf{U}_i & \mathbf{0}_{M/2} \\ \mathbf{0}_{M/2} & \mathbf{V}_i \end{bmatrix} \tag{85.45}$$

where $\mathbf{U}_i$ and $\mathbf{V}_i$ can be any $M/2 \times M/2$ orthogonal matrices, then $\mathbf{T}_i$ can be expressed as $\mathbf{T}_i = \mathbf{W}\mathbf{\Phi}_i\mathbf{W}$ and $\mathbf{SQT}_{N-1}$ can be simplified to

$$\mathbf{SQT}_{N-1} = \begin{bmatrix} \mathbf{S}_0\mathbf{U}_{N-1} & \mathbf{0}_{M/2} \\ \mathbf{0}_{M/2} & \mathbf{S}_1\mathbf{V}_{N-1} \end{bmatrix}\mathbf{W}.$$

As $\mathbf{U}_{N-1}$ and $\mathbf{S}_0$ are generic orthogonal matrices, and the product $\mathbf{S}_0\mathbf{U}_{N-1}$ is also a generic orthogonal matrix, we can discard the term $\mathbf{S}_0$ without any loss of generality. The same is valid for $\mathbf{S}_1$ with regard to $\mathbf{V}_{N-1}$. Therefore, we get $\mathbf{SQT}_{N-1} = \mathbf{\Phi}_{N-1}\mathbf{W}$ and (85.44) reduces to

$$\mathbf{E}(z) = \left(\prod_{i=N-1}^{1} \mathbf{\Phi}_i\mathbf{W}\mathbf{\Lambda}(z)\mathbf{W}\right)\mathbf{E}_0 \tag{85.46}$$

where $\mathbf{E}_0 = \mathbf{\Phi}_0\mathbf{W}\mathbf{Q}$ is a general $M \times M$ orthogonal matrix with symmetric basis functions, i.e., the PTM of order 0 of a LPPUFB. Since an order-n PTM leads to filters of length $(n+1)M$, a LPPUFB with filter length $nM + M$ can be obtained from one with filter length nM by adding a stage to the PTM of the latter. For any $N > 1$, any PTM of a LPPUFB can be expressed as

$$\mathbf{E}(z) = \mathbf{K}_{N-1}(z)\mathbf{K}_{N-2}(z)\cdots\mathbf{K}_1(z)\mathbf{E}_0 \quad \text{where} \quad \mathbf{K}_i(z) = \mathbf{\Phi}_i\mathbf{W}\mathbf{\Lambda}(z)\mathbf{W} \tag{85.47}$$

The GenLOT is defined as a LPPUFB obeying (85.47) where $\mathbf{E}_0$ is chosen to be the DCT matrix, which we denote as $\mathbf{D}$ [Fig. 85.19(a)]. The output of the DCT is, then, separated into groups of even and odd indexed coefficients [Fig. 85.19(b)]. The GenLOT with $N-1$ stages after the DCT has basis functions (filters) with length $L = NM$ and has its PTM defined as

$$\mathbf{E}(z) = \mathbf{K}_{N-1}(z)\mathbf{K}_{N-2}(z)\cdots\mathbf{K}_1(z)\mathbf{D} \tag{85.48}$$

The implementation flow-graphs for the analysis and synthesis sections are shown in Fig. 85.19(b). In this figure, each branch carries $M/2$ samples.

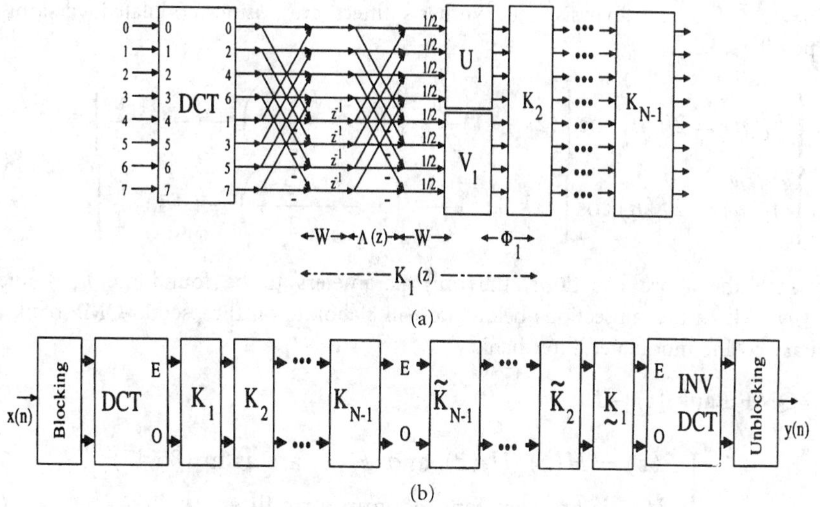

FIGURE 85.19 GenLOT—Generalized lapped orthogonal transform.

The class of GenLOT's, defined in this way, allows us to view the DCT and LOT as special cases, respectively, for $N = 1$ and $N = 2$. The degrees of freedom reside on the matrices $\mathbf{U}_i$ and $\mathbf{V}_i$ that are only restricted to be real $M/2 \times M/2$ orthogonal matrices. Thus, each one can be parameterized into a set of $M(M-2)/8$ plane rotations, (or less, for constrained solutions). Similarly, to the lattice factorization in (85.41), the above factorization in (85.48) is a general factorization that covers all LPPUFB (even M).

Cosine-Modulated Filter Bank [13]–[16]

Let $H(z)$ and $F(z)$ be the prototype filters of the analysis and synthesis banks, respectively [as shown in Fig. 85.20(a)]. The cosine-modulated analysis and synthesis filters $H_k(z)$ and $F_k(z)$ are defined as

$$\begin{cases} H_k(z) = a_k b_k U_k(z) + a_k^* b_k^* U_k^*(z) \\ F_k(z) = a_k^* b_k V_k(z) + a_k b_k^* V_k^*(z) \end{cases} \quad \text{where} \quad \begin{cases} U_k(z) = H(zW^{k+\frac{1}{2}}) \\ V_k(z) = F(zW^{k+\frac{1}{2}}) \end{cases}$$
$$\begin{cases} a_k = e^{j(-1)^k \frac{\pi}{4}} \\ b_k = W^{-\frac{1}{2}(k+\frac{1}{2})} \end{cases} \tag{85.49}$$

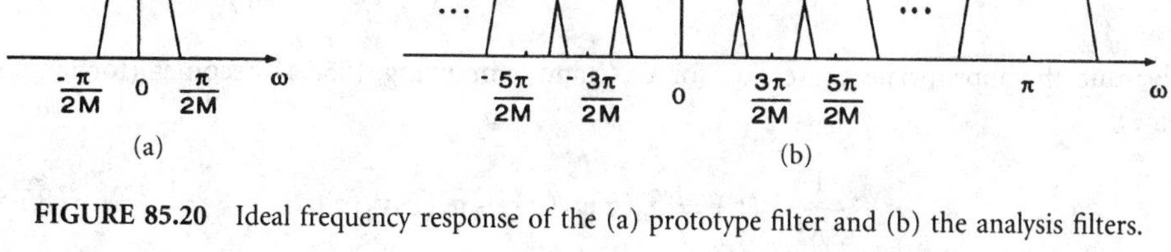

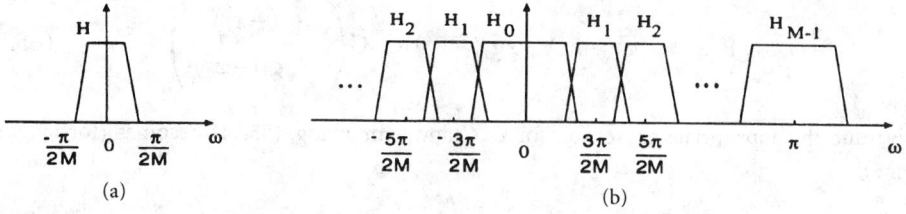

FIGURE 85.20 Ideal frequency response of the (a) prototype filter and (b) the analysis filters.

and $W = e^{-jk\pi/M}$. The analysis and synthesis filters are cosine-modulated versions of the prototype filters, i.e.,

$$\begin{cases} h_k(n) = 2h(n)\cos\left[(2k+1)\dfrac{\pi}{2M}\left(n - \dfrac{N-1}{2}\right) + (-1)^k\dfrac{\pi}{4}\right] \\ f_k(n) = 2f(n)\cos\left[(2k+1)\dfrac{\pi}{2M}\left(n - \dfrac{N-1}{2}\right) - (-1)^k\dfrac{\pi}{4}\right] \end{cases} \tag{85.50}$$

Clearly from the above equations, the only parameters to be found are $h(n)$ and $f(n)$, $0 \le n \le N-1$. In the subsections below, we will elaborate on the pseudo-QMF bank and the paraunitary cosine-modulated filter bank.

Pseudo-QMF Bank [17]–[19]

$$\begin{cases} F(z) = H(z),\ H_k(z) \text{ and } F_k(z) \text{ are as in (85.47)} \\ H(z) \text{ is an even-length symmetric filter} \\ H(z) \text{ is a spectral factor of an } M\text{th band filter;} \end{cases} \tag{85.51}$$

The above choices of $H_k(z)$ and $F_k(z)$ ensure that the aliasing from an adjacent band is canceled (i.e., $T_1(z) \approx T_{M-1}(z) \approx 0$). The aliasing levels from $T_2(z)$ to $T_{M-2}(z)$ are comparable to the stopband attenuation of $H(e^{j\omega})$. The distortion function $T_0(z)$ is a delay. In summary, as long as one can design a low-pass filter with high stopband attenuation and satisfies (85.51), then the only reconstruction error in the pseudo-QMF bank is the aliasing components in $T_k(z)$, $2 \le k \le M-2$.

Paraunitary Cosine-Modulated Filter Bank [13]

Let $H(z) = \sum_{l=0}^{2M-1} z^{-l} G_l(z^{2M})$ be the linear-phase prototype filters, where $G_l(z)$ is the type-I polyphase components of $H(z)$. $\mathbf{E}(z)$ can be expressed in terms of $G_l(z)$ and $\hat{\mathbf{C}}$ as [13]

$$\mathbf{E}(z) = \hat{\mathbf{C}}\begin{pmatrix} \mathbf{g}_0(-z^2) \\ z^{-1}\mathbf{g}_1(-z^2) \end{pmatrix} \tag{85.52}$$

where

$$\begin{cases} \mathbf{g}_0(z) = \text{diag}[G_0(z) \quad G_1(z) \quad \cdots \quad G_{M-1}(z)] \\ \mathbf{g}_1(z) = \text{diag}[G_M(z) \quad G_{M+1}(z) \quad \cdots \quad G_{2M-1}(z)] \\ \left[\hat{\mathbf{C}}\right]_{k,l} = 2\cos\left((2k+1)\dfrac{\pi}{2M}\left(l - \dfrac{N-1}{2}\right) + (-1)^k\dfrac{\pi}{4}\right) \end{cases} \tag{85.53}$$

Using the above $\mathbf{E}(z)$, one obtains the expression for $\mathbf{P}(z) \triangleq \tilde{\mathbf{E}}\mathbf{E}(z)$ as follows:

$$\mathbf{P}(z) = (\tilde{\mathbf{g}}_0(-z^2) \quad z\tilde{\mathbf{g}}_1(-z^2))\hat{\mathbf{C}}^T\hat{\mathbf{C}}\begin{pmatrix} \mathbf{g}_0(-z^2) \\ z^{-1}\mathbf{g}_1(-z^2) \end{pmatrix} \tag{85.54}$$

Substitute the appropriate expression for $\hat{\mathbf{C}}^T\hat{\mathbf{C}}$ and simplifying, (85.54) becomes (for length $= 2mM$)

$$\frac{1}{2M}\mathbf{P}(z) = \tilde{\mathbf{g}}_0(z)\mathbf{g}_0(z) + \tilde{\mathbf{g}}_1(z)\mathbf{g}_1(z) \tag{85.55}$$

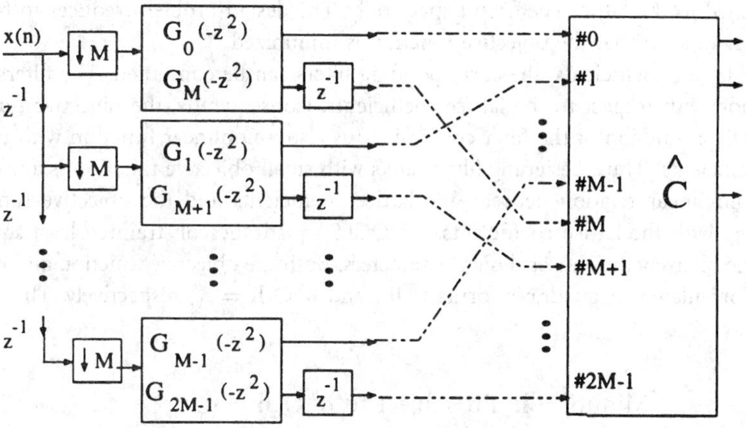

FIGURE 85.21 Lattice structure implementation of paraunitary cosine-modulated filter bank.

The above equations are the necessary and sufficient conditions on the polyphase filters $G_l(z)$ for a paraunitary cosine-modulated filter bank. Writing (85.52) in terms of the polyphase filters $G_l(z)$, we have

$$\tilde{G}_l(z)G_l(z) + \tilde{G}_{M+l}(z)G_{M+l}(z) = \frac{1}{2M}(-z)^{-s} \tag{85.56}$$

In summary, as long as the polyphase components $G_l(z)$ of the linear-phase prototype filter $H(z)$ satisfy the conditions in (85.56), then the cosine-modulated filter bank is paraunitary.

Lattice Structure [13], [29]. The above condition (85.56) is the same as the condition in the paraunitary two-channel filter bank. Thus, the cosine-modulated filter bank can be implemented as a parallel bank of two-channel paraunitary filter banks, as depicted in Fig. 85.21.

The theory for the paraunitary cosine-modulated filter bank where the length of the analysis filters are arbitrary is discussed in [16]. It turns out that the same PR conditions on the polyphase components as in (85.56) hold for the arbitrary-length case [16]. The above paraunitary cosine-modulated filter bank is general and covers all paraunitary cosine-modulated filter banks. The implementation is efficient because of the lattice structures associated with the pair $[G_k(z), G_{k+M}(z)]$ and the matrix $\hat{\mathbf{C}}$. See [13], [29] for the details.

Biorthogonal cosine-modulated filter banks allow additional properties such as allpass polyphase components and variable overall delay [15]. By trading-off between the filter attenuation and the overall system delay, it is possible to obtain a PR cosine-modulated filter bank where the overall delay is a variable.

85.5 Design of Filter Bank

The problem of designing a PR filter bank can be formulated as finding **h** such that it

$$\text{Minimize } \Phi \text{ subject to PR conditions} \tag{85.57}$$

where Φ is the objective function. The specific form of Φ depends on the application of the filter bank. Some popular objective functions are the passband and stopband attenuation and/or coding gain, etc. Since all of the above PR filter banks are characterized by the appropriate set of lattice coefficients, therefore, the PR conditions are satisfied automatically as long as the search space is constrained to the lattice coefficient space only. The design problem reduces to finding the lattice coefficient such that the objective function is minimized.

For any set of lattice coefficients, the corresponding filters can be computed. The filters are a nonlinear function with respect to the lattice coefficients. Consequently, the objective function (which is normally a function of the filter coefficients) is also a nonlinear function with respect to the lattice coefficients. Thus, designing filter banks with small objective function is a problem because of the nonlinear relation between the lattice coefficients and the objective function. Instead of dealing with the lattice coefficients, the QCLS (quadratic constrainted least squares) approach uses the filter coefficients as design parameters. Both the objective function and the PR conditions are formulated in quadratic forms $\mathbf{h}^T\mathbf{P}\mathbf{h}$ and $\mathbf{h}^T\mathbf{Q}_k\mathbf{h} = c_k$, respectively. The design problem becomes

$$\text{Minimize } \mathbf{h}^T\mathbf{P}\mathbf{h} \text{ subject to } \mathbf{h}^T\mathbf{Q}_k\mathbf{h} = c_k \tag{85.58}$$

The advantage here is the derivatives and the Hessian of both the objective function and the PR conditions can be computed exactly, which helps the minimization algorithm.

Lattice-Structure Approach

Most of the PR filter banks such as the

- two-channel paraunitary filter bank,
- two-channel biorthogonal linear-phase filter bank,
- M-channel linear-phase paraunitary filter bank,
- GenLOT,
- paraunitary cosine-modulated filter bank,

can be implemented based on lattice structure (see the separate subsections). For any set of lattice coefficients, the properties associated with the filter banks (such as paraunitary, linear-phase, cosine-modulation, etc.) are structurally imposed. Consequently, the only question to be addressed in the design is to find the appropriate set of lattice coefficients such that the resulting filters have desirable frequency responses. The design problem can be formulated as

$$\min \Phi$$

where the parameter space is the lattice coefficients k_i and Φ is the objective function. The objective function could take many different forms depending on the applications. Examples are stopband attenuation, coding gain (the optimization becomes maximization problem in this case), interference energy, etc. Design examples of lattice-based design method can be found in the [3], [4], [7]–[13].

Consider the lattice structure for the two-channel paraunitary filter bank in Fig. 85.15. The analysis filters $H_k(z)$ are nonlinear functions with respect to the lattice coefficients (in blocks $\mathbf{U}_k$ and $\mathbf{V}_k$). In other words, a small deviation in the lattice coefficients at the first few stages affects the frequency responses of $H_k(z)$ greatly. On the other hand, the effect on $H_k(z)$ by a small deviation in the lattice coefficients at the last few stages is minimal. Since the relations between the lattice coefficients and the filters are nonlinear, designing filter banks with high attenuation is a challenging task. Alternative methods on designing filter banks where the parameters are the filter coefficients are presented below.

Quadratic-Constrained Least-Squares (QCLS) Approach

For a paraunitary filter bank, the PR condition in (85.10) becomes

$$\begin{cases} T_0(z) = z^{-n_0} \\ T_k(z) = 0 \quad 1 \le k \le M-1 \end{cases} \qquad T_k(z) = \frac{1}{M}\sum_{l=0}^{M-1} z^{-(N-1)} H_l(z^{-1}) H_l(zW^k) \tag{85.59}$$

where the parameters to be found are the coefficients of $H_k(z)$. Let $\mathbf{h}$ be a vector consisting of all the filter coefficients $h_k(n)$, $0 \le k \le M-1$, and $0 \le n \le N-1$, i.e.,

$$\mathbf{h} = [h_0(0) \quad \cdots \quad h_0(N-1) \quad \cdots \quad h_{M-1}(0) \quad \cdots \quad h_{M-1}(N-1)]^T \tag{85.60}$$

then (85.59) can be written in the following form:

$$\begin{cases} \mathbf{h}^T \mathbf{Q}_k \mathbf{h} = 0 \\ \mathbf{h}^T \mathbf{S}_k \mathbf{h} = 1 \end{cases} \tag{85.61}$$

where $\mathbf{Q}_k$ and $\mathbf{S}_k$ depend on the filter bank parameters. Furthermore, it is possible to express the objective function as a quadratic form in $\mathbf{h}$ [38], i.e.,

$$\Phi = \mathbf{h}^T \mathbf{P} \mathbf{h} \tag{85.62}$$

where $\mathbf{P}$ is a real, symmetric and positive-definite matrix. Combining (85.61) and (85.62), the optimized filter $H_k(z)$ is precisely $\mathbf{h}_{\text{opt}}$ such that

$$\mathbf{h}_{\text{opt}} = \min \mathbf{h}^T \mathbf{P} \mathbf{h} \text{ subject to } \begin{cases} \mathbf{h}^T \mathbf{Q}_k \mathbf{h} = 0 \\ \mathbf{h}^T \mathbf{S}_k \mathbf{h} = 1 \end{cases} \tag{85.63}$$

In summary, we would like to formulate the design problem into a least-squares optimization problem with quadratic constraints as in (85.63). Since $\mathbf{Q}_k$ is normally not positive definite, it is difficult to solve the above minimization problem. However, there are optimization procedures that approximately solve (85.63) by linearizing the quadratic constraints [35]. Using these procedures will yield an approximate solution (i.e., the constraints are not satisfied exactly). However, the errors are very small and can be ignored in most practical cases, as we will demonstrate in the examples below.

In the following subsections, we will express the appropriate PR conditions in quadratic form as in (85.63) for the two-channel PR linear-phase filter bank, The NPR pseudo-QMF bank, PR cosine-modulated filter bank, and the M-channel orthonormal filter bank. The forms of $\mathbf{Q}_k$ and $\mathbf{S}_k$ will be derived in detail.

Two-Channel Linear-Phase Biorthogonal (SAOO) Filter Bank

In this subsection, the PR condition for the two-channel linear-phase (SAOO) biorthogonal filter bank is formulated as a QCLS problem. Let $H_0(z) = \sum_{n=0}^{2m-1} h_0(n)z^{-n}$ and $H_1(z) = \sum_{n=0}^{2m-1} h_1(n)z^{-n}$ be the symmetric and antisymmetric linear-phase filters, respectively. Because of the symmetries in both $h_0(n)$ and $h_1(n)$, the only parameters in this filter bank are $h_0(n)$ and $h_1(n)$ for $0 \le n \le (m-1)$. In order to cancel aliasing at the output, the synthesis filters must be $F_0(z) = -H_1(-z)$ and $F_1(z) = H_0(-z)$. Using the above choices for $F_i(z)$, the overall distortion function $T(z)$ for a PR system (which should be a delay) is

$$T(z) \triangleq \sum_{n=0}^{4m-2} t(n)z^{-n} = -H_0(z)H_1(-z) + H_0(-z)H_1(z) = z^{-n_0} \tag{85.64}$$

where n_0 is a positive integer. One can verify that $T(z)$ is a symmetric transfer function and consequently, $n_0 = 2m - 1$. Thus, $t(n)$ must satisfy the following conditions for a PR system

$$t(n) = \begin{cases} 0 & 0 \le n \le 2m-2 \\ 1 & n = 2m-1 \end{cases} \tag{85.65}$$

Substituting $(-z)$ for z in (85.64), one obtains $T(z) = -T(-z) = z^{-1}\hat{T}(z^2)$, where $\hat{T}(z)$ is an arbitrary polynomial. In other words, $t(n) = 0$ for even n. Consequently, the $2m$ conditions in (85.65) reduce to the following m conditions:

$$t(2k+1) = \begin{cases} 0 & 0 \le k \le m-2 \\ 1 & k = m-1 \end{cases} \tag{85.66}$$

Let

$$\begin{cases} \mathbf{h} = [h_0(0) \quad \cdots \quad h_0(m-1) \quad h_1(0) \quad \cdots \quad h_1(m-1)]^T \\ \mathbf{e}(z) = (1 \quad z^{-1} \quad \cdots \quad z^{-(m-1)})^T \end{cases} \tag{85.67}$$

It is our objective to express the above m conditions in (85.66) in terms of the unknown variable $\mathbf{h}$. The polynomials $H_0(z)$, $H_1(z)$, $H_0(-z)$, and $H_1(-z)$ can be written in the following forms:

$$\begin{cases} H_0(z) = \mathbf{h}^T \begin{pmatrix} \mathbf{e}(z) + z^{-m}\mathbf{J}\mathbf{e}(z) \\ \mathbf{0} \end{pmatrix} \\ H_0(-z) = \mathbf{h}^T \begin{pmatrix} \mathbf{U}\mathbf{e}(z) + (-1)^m z^{-m}\mathbf{J}\mathbf{U}\mathbf{e}(z) \\ \mathbf{0} \end{pmatrix} \\ H_1(z) = \mathbf{h}^T \begin{pmatrix} \mathbf{0} \\ \mathbf{e}(z) - z^{-m}\mathbf{J}\mathbf{e}(z) \end{pmatrix} \\ H_1(-z) = \mathbf{h}^T \begin{pmatrix} \mathbf{0} \\ \mathbf{U}\mathbf{e}(z) - (-1)^m z^{-m}\mathbf{J}\mathbf{U}\mathbf{e}(z) \end{pmatrix} \end{cases} \tag{85.68}$$

where $\mathbf{J}$ is the exchange matrix and $\mathbf{U}$ is a diagonal matrix with elements $U_{k,k} = (-1)^k$. Substituting the above relations into (85.64), $T(z)$ is simplified to

$$T(z) = \mathbf{h}^T \begin{pmatrix} \mathbf{0} & \Gamma(z) \\ \mathbf{0} & \mathbf{0} \end{pmatrix} \mathbf{h} \tag{85.69}$$

where

$$\begin{aligned} \Gamma(z) = {} & [\mathbf{U}\mathbf{e}(z)\mathbf{e}^T(z) - \mathbf{e}(z)\mathbf{e}^T(z)\mathbf{U}] + z^{-m}[(-1)^m(\mathbf{J}\mathbf{U}\mathbf{e}(z)\mathbf{e}^T(z) + \mathbf{e}(z)\mathbf{e}^T(z)\mathbf{U}\mathbf{J}) \\ & -(\mathbf{U}\mathbf{e}(z)\mathbf{e}^T(z)\mathbf{J} + \mathbf{J}\mathbf{e}(z)\mathbf{e}^T(z)\mathbf{U})] \\ & + z^{-2m}(-1)^m[\mathbf{J}\mathbf{e}(z)\mathbf{e}^T(z)\mathbf{U}\mathbf{J} - \mathbf{J}\mathbf{U}\mathbf{e}(z)\mathbf{e}^T(z)\mathbf{J}] \end{aligned} \tag{85.70}$$

The only matrix in $\Gamma(z)$ that is a function of z is $\mathbf{e}(z)\mathbf{e}^T(z)$. Substituting

$$\mathbf{e}(z)\mathbf{e}^T(z) = \sum_{n=0}^{2m-2} z^{-n}\mathbf{D}_n \quad \text{where} \quad [\mathbf{D}_n]_{i,j} = \begin{cases} 1 & i+j=n \\ 0 & \text{otherwise} \end{cases}$$

in $\Gamma(z)$, then the right-hand side of (85.69) is a polynomial of the form

$$\sum_{k=0}^{4m-2} z^{-k}\mathbf{h}^T \begin{pmatrix} \mathbf{0} & \Gamma_n \\ \mathbf{0} & \mathbf{0} \end{pmatrix} \mathbf{h}$$

where Γ_k are constant matrices depending on $\mathbf{D}_k$, $\mathbf{J}$, and $\mathbf{U}$. Comparing term-by-term in (85.69), (85.66) becomes

$$\begin{cases} \mathbf{h}^T\mathbf{Q}_{2k+1}\mathbf{h} = 0 & 0 \le k \le m-2 \\ \mathbf{h}^T\mathbf{Q}_{2m-1}\mathbf{h} = 1 \end{cases} \tag{85.71}$$

where

$$\mathbf{Q}_n = \begin{pmatrix} \mathbf{0} & \Gamma_n \\ \mathbf{0} & \mathbf{0} \end{pmatrix}$$

and

$$\Gamma_n = \begin{cases} \mathbf{U}\mathbf{D}_n - \mathbf{D}_n\mathbf{U} & 0 \le n \le m-1 \\ \mathbf{U}\mathbf{D}_n - \mathbf{D}_n\mathbf{U} + (-1)^m(\mathbf{J}\mathbf{U}\mathbf{D}_{n-m} + \mathbf{D}_{n-m}\mathbf{U}\mathbf{J}) & \\ \quad -(\mathbf{U}\mathbf{D}_{n-m}\mathbf{J} + \mathbf{J}\mathbf{D}_{n-m}\mathbf{U}) & m \le n \le 2m-2 \\ (-1)^m(\mathbf{J}\mathbf{U}\mathbf{D}_{m-1} + \mathbf{D}_{M-1}\mathbf{U}\mathbf{J}) & \\ \quad -(\mathbf{U}\mathbf{D}_{m-1}\mathbf{J} + \mathbf{J}\mathbf{D}_{m-1}\mathbf{U}) & n = 2m-1 \end{cases} \tag{85.72}$$

In summary, the PR condition in (85.64) is rewritten as m quadratic constraints on $\mathbf{h}$ as in (85.71).

Design Procedure.

- Given m and the passband and stopband edges of $H_k(z)$, compute $\mathbf{P}$ using the eigenfilter technique [38]. Since the polyphase matrices are not lossless, the frequency error Φ must include both passband and stopband errors in $H_k(z)$.
- Compute $\mathbf{Q}_{2k+1}$ from (85.72).
- Design a low-pass and a high-pass linear-phase filter with the same specifications as in $H_0(z)$ and $H_1(z)$, using any filter design method. Use their coefficients as an initialized value for $\mathbf{h}$ in the quadratic-constrained minimization problem as in (85.63). Use any nonlinear optimization algorithm (such as IMSL [36]) to solve the above minimization problem.

Example. Let $m = 31$ and the passband and stopband edges of $H_0(z)$ and $H_1(z)$ be 0.414π and 0.586π, respectively. The magnitude responses of $H_k(z)$ are plotted in Fig. 85.22. The stopband attenuations for $H_0(z)$ and $H_1(z)$ are -57.4 dB and -57.9 dB, respectively, which are about 14.9 dB and 15.4 dB better than those designed using the lattice approach [4]. The above filter bank

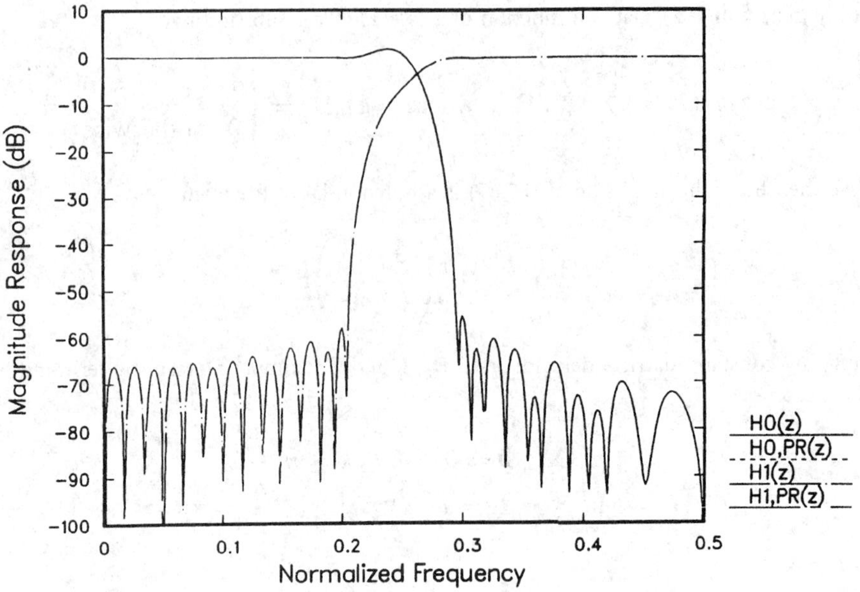

FIGURE 85.22 Magnitude response plots for the two-channel linear-phase filter bank.

is only an approximate PR filter bank since the quadratic constraints are approximately satisfied (the error is about 1×10^{-15}). To obtain a PR filter bank, we synthesize the lattice coefficients using the procedure in [4]. With these lattice coefficients, the corresponding PR analysis filters $H_{0,\mathrm{PR}}(z)$ and $H_{1,\mathrm{PR}}(z)$ are computed and their magnitude responses, together with those of $H_k(z)$, are plotted in Fig. 85.22. We observe practically no difference between the PR analysis filters $H_{k,\mathrm{PR}}(z)$ and the analysis filters $H_k(z)$ designed using the proposed method.

M-Channel Near-Perfect-Reconstruction (NPR) Pseudo-QMF Bank

The pseudo-QMF banks belong to the family of modulated filter banks. Pseudo-QMF theory is well known and is widely used. The analysis and synthesis filters are cosine-modulated versions of a prototype filter. Since the desired analysis and synthesis filters have narrow transition bands and high stopband attenuation, the overlap between nonadjacent filters is negligible. Moreover, the significant aliasing terms from the overlap of the adjacent filters are canceled by the filter designs. The prototype filter $H(z)$ is found by minimizing an objective function consisting of the stopband attenuation and the overall distortion. Although it is possible to obtain a pseudo-QMF bank with high attenuation, the overall distortion level might be high (on the order of -40 dB). In summary, the overall distortion of the pseudo-QMF bank is not sufficiently small for applications where a -100 dB error level is required.

The NPR pseudo-QMF bank is a special case of the pseudo-QMF bank where the prototype filter $H(z)$ is a *linear-phase spectral factor of a 2Mth band filter*. The analysis and synthesis filters $h_k(n)$ and $f_k(n)$ are cosine-modulated versions of the prototype filter $h(n)$. With the above properties, the output of an NPR filter bank does not have any amplitude or phase distortions. The only reconstruction error is the aliasing error, which is comparable to the stopband attenuation. Even though $H(z)$ is a spectral factor of a $2M$th band filter, no spectral factorization is needed in the new approach. In other words, the $2M$th band constraints are imposed approximately. This approach yields NPR solutions where there is some aliasing at the reconstructed output (the level is comparable to the stopband attenuation). In order to obtain total aliasing cancellation (and thus, PR), not only the prototype filter $h(n)$ should be a spectral factor of a $2M$th band filter, but each polyphase component (in an M-phase decomposition) of $h(n)$ should be a spectral factor of a halfband filter.

QCLS Formulation (Even Length) [19]. Let $H(z) = \sum_{n=0}^{N-1} h(n)z^{-n}$ be the real-coefficient, linear-phase, even length prototype filter of length N, where $N = 2(mM + m_1)$ and $0 \leq m_1 \leq M - 1$. Assume that $H(z)$ is a linear-phase spectral factor of a $2M$th band filter $G(z)$, i.e., $G(z) = z^{-(N-1)}H(z)\tilde{H}(z) = H^2(z)$ in lieu of linear-phase property of $H(z)$. The analysis and synthesis filters $h_k(n)$ and $f_k(n)$ are cosine-modulated versions of $h(n)$. Defining $\mathbf{h}$ and $\mathbf{e}(z)$ as

$$\begin{cases} \mathbf{h} = [h(0) \quad h(1) \quad \cdots \quad h(mM + m_1 - 1)]^T \\ \mathbf{e}(z) = [1 \quad z^{-1} \quad \cdots \quad z^{-(mM+m_1-1)}]^T \end{cases} \tag{85.73}$$

then the prototype filter $H(z)$ can be represented as

$$H(z) = \mathbf{h}^T(\mathbf{I} \quad \mathbf{J})\begin{pmatrix} \mathbf{e}(z) \\ z^{-(mM+m_1)}\mathbf{e}(z) \end{pmatrix},$$

where the dimensions of both $\mathbf{I}$ and $\mathbf{J}$ are $(mM + m_1) \times (mM + m_1)$. Using the above notation, the $2M$th band filter $G(z)$ is

$$\begin{aligned} G(z) &= \sum_{n=0}^{4mM+4m_1-2} g(n)z^{-n} \\ &= H^2(z) = \mathbf{h}^T(\mathbf{I} \quad \mathbf{J})\begin{pmatrix} \mathbf{e}(z) \\ z^{-(mM+m_1)}\mathbf{e}(z) \end{pmatrix}[\mathbf{e}^T(z) \quad z^{-(mM+m_1)}\mathbf{e}^T(z)]\begin{pmatrix} \mathbf{I} \\ \mathbf{J} \end{pmatrix}\mathbf{h} \\ &= \mathbf{h}^T[\mathbf{V}(z) + z^{-(mM+m_1)}(\mathbf{J}\mathbf{V}(z) + \mathbf{V}(z)\mathbf{J}) + z^{-2(mM+m_1)}\mathbf{J}\mathbf{V}(z)\mathbf{J}]\mathbf{h} \end{aligned} \tag{85.74}$$

where

$$\mathbf{V}(z) = \mathbf{e}(z)\mathbf{e}^T(z) = \begin{pmatrix} 1 \\ z^{-1} \\ \vdots \\ z^{-(mM+m_1-1)} \end{pmatrix}(1 \;\; z^{-1} \;\; \cdots \;\; z^{-(mM+m_1-1)}) = \sum_{n=0}^{2mM+2m_1-2} z^{-n}\mathbf{D}_n \tag{85.75}$$

Here, $\mathbf{D}_n$ is defined as

$$[\mathbf{D}_n]_{i,j} = \begin{cases} 1 & i + j = n \\ 0 & \text{otherwise} \end{cases}$$

Substituting (85.75) into (85.74) and simplifying, we have

$$G(z) = \sum_{n=0}^{4mM+4m_1-2} g(n)z^{-n} = \mathbf{h}^T\left(\sum_{n=0}^{4mM+4m_1-2} z^{-n}\mathbf{S}_n\right)\mathbf{h} \tag{85.76}$$

where $\mathbf{S}_n$ depends on $\mathbf{D}_n$ and $\mathbf{J}$ as follows:

$$\mathbf{S}_n = \begin{cases} \mathbf{D}_n & 0 \le n \le mM + m_1 - 1 \\ \mathbf{D}_n + \mathbf{J}\mathbf{D}_{n-mM-m_1} \\ \quad +\mathbf{D}_{n-mM-m_1}\mathbf{J} & mM + m_1 \le n \le 2(mM + m_1 - 1) \\ \mathbf{J}\mathbf{D}_{mM+m_1-1} + \mathbf{D}_{mM+m_1-1}\mathbf{J} & n = 2(mM + m_1) - 1 \\ \mathbf{J}\mathbf{D}_{n-mM-m_1} + \mathbf{D}_{n-mM-m_1}\mathbf{J} \\ \quad +\mathbf{J}\mathbf{D}_{n-2mM-2m_1}\mathbf{J} & 2(mM + m_1) \le n \le 3(mM + m_1) - 2 \\ \mathbf{J}\mathbf{D}_{n-2mM-2m_1}\mathbf{J} & 3(mM + m_1) - 1 \le n \le 4(mM + m_1) - 2. \end{cases} \tag{85.77}$$

The objective is to find $\mathbf{h}$ such that $G(z)$ is a $2M$th band filter, i.e.

$$g_n = \begin{cases} 0 & n = 2(mM + m_1) - 1 - 2lM \quad \begin{cases} 1 \le l \le m - 1 & m_1 = 0 \\ 1 \le l \le m & m_1 \ne 0 \end{cases} \\ \dfrac{1}{2M} & n = 2(mM + m_1) - 1 \end{cases} \tag{85.78}$$

Equating the terms with the same power of z^{-1} in (85.76) and using (8 5.77) and (85.78), the following m constraints on $\mathbf{h}$ are obtained:

$$\begin{cases} \mathbf{h}^T\mathbf{D}_n\mathbf{h} = 0 & \begin{cases} \left\lfloor \dfrac{(m+1)}{2} \right\rfloor \le l \le (m-1) & m_1 = 0 \\ \left\lfloor \dfrac{(m+1)}{2} \right\rfloor \le l \le m & m_1 \ne 0 \end{cases} \\ \mathbf{h}^T(\mathbf{D}_n + \mathbf{J}\mathbf{D}_{n-mM-m_1} + \mathbf{D}_{n-mM-m_1}\mathbf{J})\mathbf{h} = 0 & 1 \le l \le \left\lfloor \dfrac{(m+1)}{2} \right\rfloor - 1 \\ \mathbf{h}^T(\mathbf{J}\mathbf{D}_{mM+m_1-1} + \mathbf{D}_{mM+m_1-1}\mathbf{J})\mathbf{h} = \dfrac{1}{2M} \end{cases} \tag{85.79}$$

for $n = 2M(m-l)+2m_1-1$. In summary, as long as $\mathbf{h}$ satisfies the m conditions in (85.79), the resulting pseudo-QMF bank has not amplitude or phase distortions. The only reconstruction error is aliasing which can be minimized by finding solutions with high stopband attenuation. The optimized filter $H(z)$ of the NPR pseudo-QMF bank is the solution of

$$\mathbf{h}_{\text{opt}} = \text{Min}\, \mathbf{h}^T\mathbf{P}\mathbf{h} \quad \text{subject to} \quad (85.79) \tag{85.80}$$

Example. In this example, a 32-channel pseudo-QMF bank is designed using the above method. Let $m = 8$, $m_1 = 0$, $M = 32$ ($N = 512$). The magnitude responses of the optimized analysis filters $H_k(z)$ are plotted in Fig. 85.23. The stopband attenuation and the aliasing level are about 96 dB and −96 dB, respectively. The amplitude distortion in this example is about 1×10^{-12}.

M-Channel Paraunitary Cosine-Modulated Filter Bank

Recently, the paraunitary cosine-modulated filter bank has emerged as an optimal filter bank with respect to implementation cost and design ease. The impulse responses of the analysis

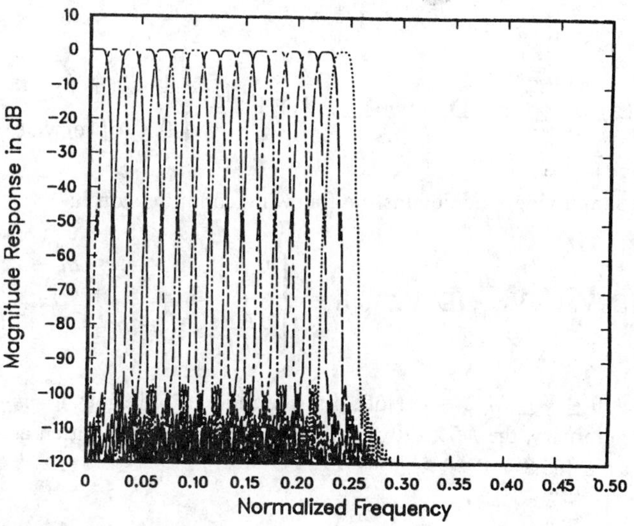

FIGURE 85.23 Magnitude response plots for the NPR pseudo-QMF bank.

filters $h_k(n)$ and $f_k(n)$ are cosine-modulated versions of the prototype filter $h(n)$, i.e.,

$$\begin{cases} h_k(n) = 2h(n)\cos\left((2k+1)\dfrac{\pi}{2M}\left(n - \dfrac{N-1}{2}\right) + (-1)^k\dfrac{\pi}{4}\right) \\ f_k(n) = 2h(n)\cos\left((2k+1)\dfrac{\pi}{2M}\left(n - \dfrac{N-1}{2}\right) - (-1)^k\dfrac{\pi}{4}\right) \end{cases} \tag{85.81}$$

where n and k are in the range of $0 \le n \le N-1$ and $0 \le k \le M-1$, respectively. Here, the lengths of $H_k(z)$ and $F_k(z)$ are assumed to be multiples of $2M$, i.e., $N = 2mM$. The PR condition is shown in (85.56) for the case of even M. It is our objective to rewrite these PR conditions in quadratic form (85.61). Here we only consider the even M case since the odd M case is very similar. In terms of the variables $h(n)$, the polyphase filter $G_k(z)$ is

$$G_k(z) = \mathbf{h}^T\mathbf{V}_k\mathbf{e}(z) \quad \text{and} \quad G_k(z^{-1}) = z^{m-1}\mathbf{h}^T\mathbf{V}_k\mathbf{J}\mathbf{e}(z) \tag{85.82}$$

where

$$\begin{cases} \mathbf{h} = [h(0) \quad h(1) \quad \cdots \quad h(mM-1)]^T \\ \mathbf{e}(z) = (1 \quad z^{-1} \quad \cdots \quad z^{-(m-1)})^T \end{cases}$$

$$[\mathbf{V}_k]_{i,j} = \begin{cases} 1 & \begin{cases} i = k + 2jm & k + 2jM < mM \\ i = 2M(m-j) - 1 - k & k + 2jM \ge mM \end{cases} \\ 0 & \text{otherwise} \end{cases} \tag{85.83}$$

Note that the dimensions of $\mathbf{h}$, $\mathbf{e}(z)$ and $\mathbf{V}_k$ are $(mM \times 1)$, $(m \times 1)$, and $(mM \times m)$, respectively. Equation (85.56) is simplified to

$$\mathbf{h}^T[\mathbf{V}_k\mathbf{J}\mathbf{e}(z)\mathbf{e}^T(z)\mathbf{V}_k^T + \mathbf{V}_{M+k}\mathbf{J}\mathbf{e}(z)\mathbf{e}^T(z)\mathbf{V}_{M+k}^T]\mathbf{h} = \frac{1}{2M}z^{-(m-1)} \tag{85.84}$$

Substituting

$$\mathbf{e}(z)\mathbf{e}^T(z) = \sum_{n=0}^{2m-2} z^{-n}\mathbf{D}_n \quad \text{where} \quad [\mathbf{D}_n]_{i,j} = \begin{cases} 1 & i+j=n \\ 0 & \text{otherwise} \end{cases} \tag{85.85}$$

into (85.84) and simplifying, one obtains the following conditions on **h**:

$$\mathbf{h}^T[\mathbf{V}_k\mathbf{J}\mathbf{D}_n\mathbf{V}_k^T + \mathbf{V}_{M+k}\mathbf{J}\mathbf{D}_n\mathbf{V}_{M+k}^T]\mathbf{h} = \begin{cases} 0 & 0 \le n \le m-2 \\ \dfrac{1}{2M} & n = m-1 \end{cases} \tag{85.86}$$

for k in the range $0 \le k \le M/2 - 1$. Note that the index n only goes to $m-1$ since (85.56) is symmetric. In summary, the $M/2$ PR conditions in (85.56) are rewritten as $mM/2$ quadratic constraints in **h** as in (85.86).

Design Procedure.

- Given M, m and the stopband edge of $H(z)$, compute **P** using the eigenfilter technique [38] (only the stopband of $|H(\mathbf{e}^{j\omega})|$ is needed because of the power-complementary property).
- For each k in the range $0 \le k \le M/2 - 1$, compute the m conditions using (85.86). The total number of conditions is $mM/2$.
- Design a low-pass filter with the same specifications as in $H(z)$ and use its coefficients as an initialized value for **h** in the quadratic-constrainted minimization problem (85.63). Use any nonlinear minimization algorithm (such as IMSL [36]) to solve the above minimization problem.

Example. Let $M = 16$ and $m = 8$; thus the filter length is $N = 256$. The magnitude responses of the optimized analysis filters $H_k(z)$ are shown in Fig. 85.24. The stopband attenuation of the optimized analysis filters is about -82 dB, which is much higher than those designed using conventional approaches. Keep in mind that the above filter bank is only an approximate PR filter bank since the involved quadratic constraints are approximately satisfied (the error is about 1×10^{-9}). This error is very small and can be considered to be negligible for all practical purposes.

M-Channel Linear-Phase Paraunitary Filter Bank

Let $H_k(z)$ be the analysis linear-phase filters (of lengths N) of an M-channel paraunitary linear-phase filter bank. The synthesis filters $F_k(z)$ are $F_k(z) = z^{-(N_k-1)}H_k(z^{-1}) = J_kH_k(z)$, where

$$J_k = \begin{cases} 1 & H_k(z) \text{ is symmetric} \\ -1 & H_k(z) \text{ is antisymmetric} \end{cases}$$

It is shown that there are

$$\begin{cases} M/2 \text{ symmetric and } M/2 \text{ antisymmetric filters} & \text{even } M \\ (M+1)/2 \text{ symmetric and } (M-1)/2 \text{ antisymmetric filters} & \text{odd } M \end{cases}$$

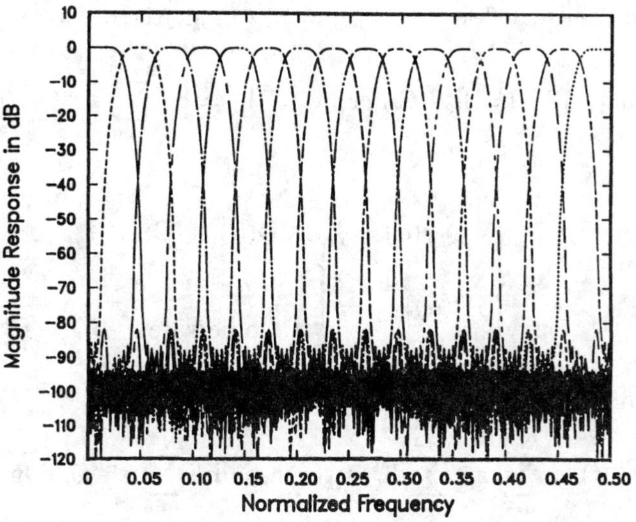

FIGURE 85.24 Magnitude repsonse plots for the cosine-modulated filter bank.

Moreover, in a linear-phase orthonormal filter bank with high attenuation, the even-indexed and the odd-indexed analysis filters $H_k(z)$ should be symmetric and antisymmetric, respectively. The reconstructed output $\hat{X}(z)$ is

$$\hat{X}(z) = \frac{1}{M}\sum_{l=0}^{M-1} T_l(z)X(zW^l) \quad \text{where} \quad T_l(z) = \sum_{k=0}^{M-1} J_k H_k(z) H_k(zW^l) \tag{85.87}$$

Since the filter bank is a perfect-reconstruction one, then

$$\begin{cases} T_0(z) = z^{-(N-1)} \\ T_l(z) = 0, \qquad l \neq 0. \end{cases}$$

We will formulate the above PR conditions in quadratic form (QCLS) as in (85.63). Instead of analyzing the complex-coefficient $T_l(z)$, we will work with the real-coefficient function $T_l(zW^{-l/2})$ for simplicity, i.e.,

$$T_l(zW^{-l/2}) = \sum_{k=0}^{M-1} J_k H_k(zW^{-l/2}) H_k(zW^{l/2}) \tag{85.88}$$

Here, only the even N case is considered, since the odd N case can be derived similarly. Let $\mathbf{h}_k$ be the vector of the first $N/2$ elements of $h_k(n)$ and $\mathbf{e}$ is the vector of the delay chain, i.e.,

$$\begin{cases} \mathbf{h}_k = [h_k(0) \quad h_k(1) \quad \cdots \quad h_k(L-1)]^T \\ \mathbf{e} = (1 \quad z^{-1} \quad \cdots \quad z^{-(N-1)})^T \end{cases} \tag{85.89}$$

where $L = N/2$, then the function $H_k(zW^{-l/2})$ is $H_k(zW^{-l/2}) = \mathbf{h}_k^T \mathbf{C}\Lambda_l \mathbf{e}$ where

$$\mathbf{C} = \begin{cases} (\mathbf{I} \quad \mathbf{J}) & H_k(z) \text{ is symmetric} \\ (\mathbf{I} \quad -\mathbf{J}) & H_k(z) \text{ is antisymmetric} \end{cases}$$

and Λ_l is a diagonal matrix of element $[\Lambda_l]_{m,m} = W^{-ml/2}$. Moreover,

$$H_k(zW^{-l/2})H_k(zW^{l/2}) = \mathbf{h}_k^T J_k \mathbf{C}\Lambda_l \mathbf{e}\mathbf{e}^T \Lambda_l^{-1}\mathbf{C}^T \mathbf{h}_k = \mathbf{h}_k^T \sum_{\alpha=0}^{2N-1} z^{-\alpha}\mathbf{Q}_{l,k}(\alpha)\mathbf{h}_k \tag{85.90}$$

where

$$\begin{cases} \mathbf{Q}_{l,k}(\alpha) = J_k \mathbf{C}\Lambda_l \mathbf{D}(\alpha)\Lambda_l^{-1}\mathbf{C}^T \\ [\mathbf{D}(\alpha)]_{m,n} = \begin{cases} 1 & m+n=\alpha \\ 0 & \text{otherwise} \end{cases} \end{cases} \tag{85.91}$$

Consequently, $T_l(zW^{-l/2})$ is

$$T_l(zW^{-l/2}) = \sum_{\alpha=0}^{2N-1} z^{-\alpha} \sum_{k=0}^{M-1} \mathbf{h}_k^T \mathbf{Q}_{l,k}(\alpha)\mathbf{h}_k = \mathbf{h}^T \sum_{\alpha=0}^{2N-1} z^{-\alpha}\mathbf{Q}_l(\alpha)\mathbf{h} \tag{85.92}$$

where

$$\begin{cases} \mathbf{h} = (\mathbf{h}_0^T \quad \mathbf{h}_1^T \quad \cdots \quad \mathbf{h}_{M-1}^T)^T \\ \mathbf{Q}_l(\alpha) = \text{diag}[\mathbf{Q}_{l,0}(\alpha) \quad \mathbf{Q}_{l,1}(\alpha) \quad \cdots \quad \mathbf{Q}_{l,M-1}(\alpha)] \end{cases} \tag{85.93}$$

Thus, the PR conditions become (note that only the first N coefficients of $T_l(zW^{-l/2})$ are considered since it is an even-order linear-phase function),

$$\mathbf{h}^T\mathbf{Q}_l(\alpha)\mathbf{h} = \begin{cases} \text{constant} & l=0, \quad \alpha = N-1 \\ 0 & \text{otherwise} \end{cases} \tag{85.94}$$

for $0 \le l \le M/2$ and $0 \le \alpha \le N-1$. The number of conditions here is $(M/2+1)N$, which can be large for filter banks with many channels and large lengths. The number of conditions can be reduced (approximately by half) by imposing relations on the analysis filters, such as the pairwise-mirror-image property. In other words, the filters are related as $H_{M-1-k}(z) = H_k(-z)$. The design problem is reduced to finding the vector $\mathbf{h}_{\text{opt}}$ such that

$$\mathbf{h}_{\text{opt}} = \text{Min}\, \mathbf{h}^T\mathbf{P}\mathbf{h} \quad \text{subject to} \quad (85.94) \tag{85.95}$$

The objective function $\mathbf{h}^T\mathbf{P}\mathbf{h}$ consists of all the stopband errors of $H_k(e^{j\omega})$. The design procedure is as follows.

Design Procedure.

- Given M, N, and the cutoff frequencies of $H_k(z)$, compute $\mathbf{P}$ (note that only the stopbands of $|H(e^{j\omega})|$ are included in P because of the power-complementary property).
- Compute the matrices $\mathbf{Q}_l(\alpha)$ for $0 \le l \le M/2$ and $0 \le \alpha \le N-1$.
- Design the initialize filters $H_k(z)$ and use their coefficients for initialization.
- Find $\mathbf{h}_{\text{opt}}$ by solving the minimization problem in (85.95).

Example. Let $M = 8$, $N = 40$, and the filters $H_k(z)$ satisfy the pairwise-mirror-image property. We design the paraunitary linear-phase filter bank using the above quadratic-constrained formulation. The magnitude response plots of the analysis filters are plotted in Fig. 85.25.

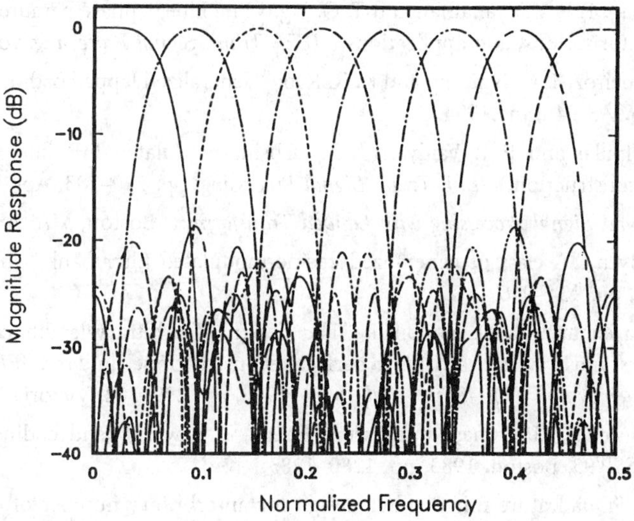

FIGURE 85.25 Magnitude response plots for the paraunitary linear-phase filter bank.

eferences

[1] F. Mintzer, "Filters for distortion-free two-band multirate filter banks," *IEEE Trans. Acoust., Speech, Signal Processing*, pp. 626–630, June 1985.

[2] M. J. Smith and T. P. Barnwell, III, "Extract reconstruction techniques for tree-structured subband coders," *IEEE Trans. Acoust., Speech, Signal Processing*, vol. ASSP-34, pp. 434–441, June 1986.

[3] P. P Vaidyanathan and P. Q. Hoang, "Lattice structures for optimal design and robust implementation of two-channel perfect-reconstruction QMF banks," *IEEE Trans. Acoust., Speech, Signal Processing*, pp. 81–94, Jan. 1988.

[4] T. Q. Nguyen and P. P. Vaidyanathan, "Two channel perfect reconstruction FIR QMF structures which yield linear phase analysis and synthesis filters," *IEEE Trans. Acoust., Speech, Signal Processing*, pp. 676–690, May 1989.

[5] M. Vetterli and D. Le Gall, "Perfect reconstruction filter banks: Some properties and factorizations," *IEEE Trans. Acoust., Speech, Signal Processing*, pp. 1057–1071, July 1989.

[6] M. Vetterli, "A theory of multirate filter banks," *IEEE Trans. Acoust., Speech, Signal Processing*, pp. 356–372, Mar. 1987.

[7] P. P. Vaidyanathan, "Theory and design of M-channel maximally decimated quadrature mirror filters with arbitrary M, having perfect reconstruction property," *IEEE Trans. Acoust., Speech, Signal Processing*, vol. ASSP-35, pp. 476–492, Apr. 1987.

[8] Z. Doganata, P. P. Vaidyanathan, and T. Q. Nguyen, "General synthesis procedures for FIR lossless transfer matrices for perfect reconstruction multirate filter bank application," *IEEE Trans. Acoust., Speech, Signal Processing*, pp. 1561–1574, Oct. 1988.

[9] P. P. Vaidyanathan, T. Q. Nguyen, Z. Doganata, and T. Saramaki, "Improved technique for design of perfect reconstruction FIR QMF banks with lossless polyphase matrices," *IEEE Trans. Acoust., Speech, Signal Processing*, pp. 1042–1056, July 1989.

[10] T. Q. Nguyen and P. P. Vaidyanathan, "Structures for M-channel perfect-reconstruction FIR QMF banks which yield linear-phase analysis filters," *IEEE Trans. Acoust., Speech, Signal Processing*, pp. 433–446, Mar. 1990.

[11] A. K. Soman, P. P. Vaidyanathan and T. Q. Nguyen, "Linear-phase paraunitary filter banks: theory, factorizations, and applications," *IEEE Trans. Signal Processing*, vol. 41, Dec. 1993.

[12] R. L. de Queiroz, T. Q. Nguyen, and K. R. Rao, "Generalized lapped orthogonal transform," in *Proc. ISCAS 94*, June 1994.

[13] R. D. Koilpillai and P. P. Vaidyanathan, "Cosine-modulated FIR filter banks satisfying perfect reconstruction," *IEEE Trans. Signal Processing*, pp. 770–783, Apr. 1992.

[14] H. S. Malvar, *Signal Processing with Lapped Transforms.* Boston, MA: Artech, 1992.

[15] T. Q. Nguyen, "A class of generalized cosine-modulated filter bank," in *Proc. ISCAS'92*, May 1992, pp. 934–946.

[16] T. Q. Nguyen and R. D. Koilpillai, "The design of arbitrary-length cosine-modulated filter banks and wavelets, satisfying perfect reconstruction," in *Proc. IEEE-SP Int. Symp. Time-Frequency and Time-Scale Analy.*, Oct. 1992, pp. 299–302, Victoria, B.C., Canada.

[17] J. H. Rothweiler, "Polyphase quadrature filters—A new subband coding technique," in *Proc. ICASSP'93*, Boston, 1983, pp. 1280–1283.

[18] P. L. Chu, "Quadrature mirror filter design for an arbitrary number of equal bandwidth channels," *IEEE Trans. Acoust., Speech, Signal Processing*, 203–218, Feb. 1985.

[19] T. Q. Nguyen, "Near-perfect-reconstruction pseudo-QMF banks," *IEEE Trans. Signal Processing*, pp. 65–76, Jan. 94.

[20] J. D. Johnston, "A filter family designed for use in quadrature mirror filter banks," in *Proc. ICASSP'80*, Apr. 1980, pp. 291–294.

[21] K. Swaminathan and P. P. Vaidyanathan, "Theory and design of uniform DFT, parallel quadrature mirror filter banks," *IEEE Trans. Circuits Syst.*, pp. 1170–1191, Dec. 1986.

[22] P. A. Regalia, S. K. Mitra, and P. P. Vaidyanathan, "The digital all-pass filter: A versatile signal processing building block," *Proc. IEEE*, pp. 19–37, Jan. 1988.

[23] T. I. Laakso, T. Q. Nguyen, and R. D. Koilpillai, "Designing allpass filters using the eigenfilter method," in *Proc. ICASSP'93.* pp. III-77–III-80, Minneapolis, Apr. 1993.

[24] T. Q. Nguyen, "A quadratic constrained least-squares approach to the design of digital filter banks," in *Proc. ISCAS'92*, San Diego, May 1992, pp. 1344–1347.

[25] K. Nayebi, T. P. Barnwell, III, and M. J. T. Smith, "Time-domain filter bank analysis: A new design theory," *IEEE Trans. Signal Processing*, vol. 40, June 1992.

[26] R. E. Crochiere and L. R. Rabiner, *Multirate Signal Processing*, Englewood Cliffs, NJ: Prentice-Hall, 1983.

[27] R. Gopinath and C. S. Burrus, "Wavelet transforms and filter banks," in *Wavelets and Applications*, C. H. Chui, ed., New York: Academic, 1991.

[28] A. N. Akansu and R. A. Haddad, *Multiresolution Signal Decomposition: Transforms, Subbands and Wavelets*, New York: Academic, 1992.

[29] P. P. Vaidyanathan, *Multirate Systems and Filters Banks*, Englewood Cliffs, NJ: Prentice-Hall, 1993.

[30] I. Daubechies, "Orthonormal bases of compactly supported wavelets," *Commun. Pure Appl. Mathemat.*, vol. XLI, pp. 909–996, 1988.

[31] G. Strang, "Wavelets and dilation equations: A brief introduction," *SIAM Rev.*, vol. 31, pp. 614–627, Dec. 1989.

[32] I. Daubechies, *Ten Lectures on Wavelets*, CBMS-NSF Series on Applied Mathematics, SIAM, 1992.

[33] M. Vetterli and C. Herley, "Wavelets and filter banks," *IEEE Trans. Signal Processing*, vol. SP-40, 1992.

[34] H. Zou and A. H. Tewfik, "Discrete orthogonal M-band wavelet decompositions," in *Proc. ICASSP'92.*

[35] K. Schittkowski, "NLPQL: A FORTRAN subroutine solving constrained nonlinear programming problems, Clyde L. Monma, Ed. *Ann. Operat. Res.*, vol. 5, pp. 485–500, 1986.

[36] IMSL: A FORTRAN Mathematical Package.

[37] J. H. McClellan, T. W. Parks, and L. R. Rabiner, "A computer program for designing optimum FIR linear phase digital filters," *IEEE Trans. Audio Electroacoust.*, vol. AU-21, pp. 506–526, Dec. 1973.

[38] P. P. Vaidyanathan and T. Q. Nguyen, "Eigenfilters: A new approach to least squares FIR filter design and applications including Nyquist filters," *IEEE Trans. Circuits Syst.*, pp. 11–23, Jan. 1987.

[39] A. V. Oppenheim and R. W. Schafer, *Discrete-time Signal Processing*, Englewood Cliffs, NJ: Prentice-Hall, 1989.

86

VLSI Implementation of Digital Filters

Joseph B. Evans
University of Kansas

86.1 Introduction

Digital implementations of filters are preferred over analog realizations for many reasons. Improvements in VLSI technology have enabled digital filters to be used in an increasing number of application domains.

There are a variety of methods that can be used to implement digital filters. In this discussion, we will focus on the use of traditional VLSI digital logic families such as CMOS, rather than more exotic approaches. The vast majority of implementations encountered in practice make use of traditional technologies because the performance and cost characteristics of these approaches are so favorable.

Digital filter implementations can be classified into several categories based on the architectural approach used; these are general purpose, special purpose, and programmable logic implementations. The choice of a particular approach should be based upon the flexibility and performance required by a particular application. General purpose architectures possess a great deal of flexibility, but are somewhat limited in performance, being best suited for relatively low sampling frequencies, usually under 1 MHz. Special purpose architectures are capable of much higher performance, with sampling frequencies as high as 180 MHz, but are often only configurable for one application domain. Programmable logic implementations lie somewhere between these extremes, providing both flexibility and reasonably high performance, with sampling rates as high as 40 MHz.

Digital filtering implementations have been strongly influenced by the evolution of VLSI technology. The regular computational structures encountered in filters are well suited for VLSI implementation. This regularity often translates into efficient parallelism and pipelining. Further, the small set of computational structures required make automatic synthesis of special purpose and programmable logic designs feasible. The design automation of digital filter implementation is relatively simple compared to the general design synthesis problem. For this reason, digital filters are often the test case for evaluating new device and computer-aided design technologies.

0-8493-8341-2/95/$0.00 + $.50

6.2 General Purpose Processors

General purpose digital signal processors are by far the most commonly used method for digital filter implementation, particularly at audio bandwidths. These systems possess architectures well suited to digital filtering, as well as other digital signal processing algorithms.

Historical Perspective

General purpose digital signal processors trace their lineage back to the microprocessors of the early 1980's. The generic microprocessors of that period were ill suited for the implementation of DSP algorithms, due to the lack of hardware support for numerical algorithms of significant complexity in those architectures. The primary requirement for digital signal processing implementation was identified to be hardware support for multiplication, due to the large number of multiply–accumulate operations in digital signal processing algorithms and their large contribution to computational delays. The earliest widely available single chip general purpose DSP implementation was from AT & T, which evolved into the AT & T DSP20 family. This was soon followed by products such as Texas Instruments TMS32010 and NEC 7720. The early DSP chips exhibited several shortcomings, such as difficult programming paradigms, awkward architectures for many applications, and limited numerical precision. Many of these difficulties were imposed by the limits of the VLSI technology of the time, some by inexperience in this particular application area. Despite these shortcomings, however, the early processors were well suited to the implementation of digital filter algorithms, because digital filtering was identified as one of the target areas for these architectures. This match between architecture and algorithms continues to be exhibited in current general purpose DSP chips.

Current Processors

There are a variety of general purpose digital signal processors currently commercially available. We will look at several of these architectures in detail, although this discussion will not be comprehensive by any means. The processors are best classified in two categories, fixed point processors and floating point processors. In both cases, these architectures are commonly based on a single arithmetic unit shared amongst all computations, which leads to constraints on the sampling rates that may be attained.

Fixed point processors exhibit extremely high performance in terms of maximum throughput as compared to their floating point counterparts. In addition, fixed point processors are typically inexpensive as compared to floating point options, due to the smaller integrated circuit die area occupied by fixed point processing blocks. A major difficulty encountered in implementing filters on fixed point processors is that overflow and underflow need to be prevented by careful attention to scaling, and roundoff effects may be significant.

Floating point processors, on the other hand, are significantly easier to program, particularly in the case of complex algorithms, at the cost of lower performance and larger die area. Given the regular structure of most digital filtering algorithms and computer-aided design support for filters based on limited precision arithmetic, fixed point implementations may be the more cost effective option for this type of algorithm. Because of the prevalence of both types of general purpose processors, examples of each will be examined in detail.

Two widely used floating point processor families will be studied, although there are many contenders in this field. These families are the Texas Instruments family of floating point DSP's, in particular the TI TMS320C3x [14] and the Analog Devices ADSP-210200 family [12].

The architecture of the TI TMS320C30 is illustrated in Fig. 86.1. The floating point word size used by this processor is 32b. The most prominent feature of this chip is the floating point arithmetic unit, which contains a floating point multiplier and adder. This unit is highly pipelined to support high throughput, at the cost of latency; when data is input to the multiplier, for example, the results will not appear on the output from that unit until several clock cycles later. Other features include a separate integer unit for control calculations, and significant amounts (2k words) of SRAM for data and on-chip instruction memory. On-chip ROM (4k words) is also optionally provided in order to eliminate the need for an external boot ROM in some applications. This chip also includes a 64-word instruction cache to allow its use with lower speed memories. The modified Harvard architecture, that is, the separate data and instruction buses, provide for concurrent instruction and data word transfers within one cycle time. The TMS320C30 is currently available with instruction cycle times as low as 60 ns. A code segment that implements portions of an FIR filter is

```
      RPTS    RC
      MPYF3   *AR0++(1),  *AR1++(1)%, R0
||    ADDF3   R0,R2,R2              ...
      ADDF    R0,R2,R0
```

where the MPYF3 instruction performs a pipelined multiply operation in parallel with data and coefficient pointer increments. The ADDF3 instruction is performed in parallel with the MPYF3 instruction, as denoted by the "||" symbol. Because these operations are in parallel, only one instruction cycle per tap is required. An FIR filter tap has been benchmarked at 60 ns on this chip. Similarly, a typical biquad IIR filter code segment is

```
      MPYF3   *AR0,*AR1,R0
      MPYF3   *++AR0(1), *AR1−(1)%,R1
      MPYF3   *++AR0(1), *AR1,R0
||    ADDF3   R0,R2,R2
      MPYF3   *++AR0(1), *AR1−(1)%,R0
||    ADDF3   R0,R2,R2
      MPYF3   *++AR0(1),R2,R2
||    STF     R2, *AR1++(1)%
      ADDF    R0,R2
      ADDF    R1,R2,R0
```

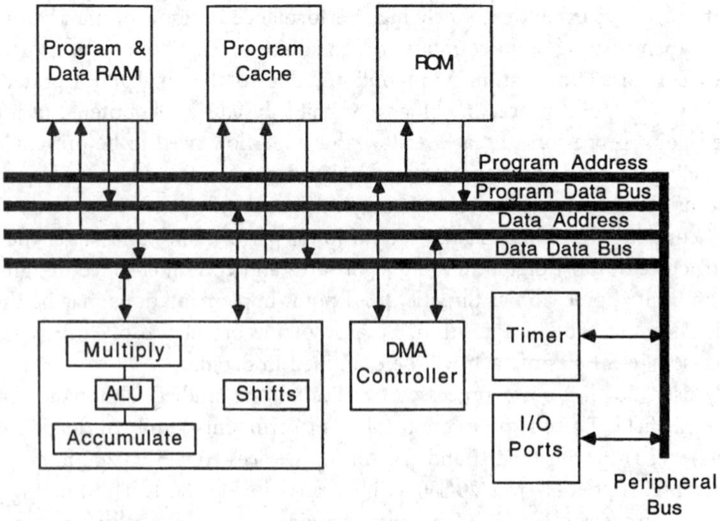

FIGURE 86.1 The Texas Instruments TMS320C30 architecture.

where the MPYF3 and ADDF3 instructions implement the primary filter arithmetic and memory pointer modification operations in parallel, as in the previous example. The biquad IIR benchmark on this processor is 300 ns.

Another floating point chip worthy of note is the Analog Devices ADSP-21020 series. The architecture of the ADSP-21020 chip is shown in Fig. 86.2. This chip can be seen to share a number of features with the TMS320C3x family, that is a 32 × 32-b floating-point multiply-accumulate unit (not pipelined), modified Harvard architecture, and 16 words of on-chip memory. In this case, the scratchpad memory is organized into register files, much like a general purpose RISC architecture register set. The memory capacity of this device is significantly smaller than that of its competitors. As in the case of the TMS320C3x, on the other hand, an instruction cache (32 words) is also provided. The cycle time for the ADSP-21020 is 40 ns. An N tap FIR filter code segment illustrates the operation of this device,

```
         i0=coef;
         f9=0.0;
         f1=0; f4=dm(i0,m0); f5=pm(i8,m8);
         lcntr=N, DO bottom until lce;
bottom:        f1=f1+f9; f9=f4*f5; f4=dm(i0,m0); f5=pm(i8,m8);
         f1=f1+f9;
```

where the "*" and "+" instructions perform the multiply–accumulate operations and the dm( ) and pm( ) instructions perform the memory address update operations in parallel. An FIR filter tap thus executes in one instruction per tap on the ADSP-21020, or 40 ns. An IIR filter biquad section requires 200 ns on this chip. Note that while the assembly language for the Analog Devices chip is significantly different from that of the Texas Instruments chip, the architectural similarities are striking.

Two families of fixed point digital signal processors will also be examined and compared. These are the Texas Instruments TMS320C5x family [15], and the Motorola 56000 series of devices [10].

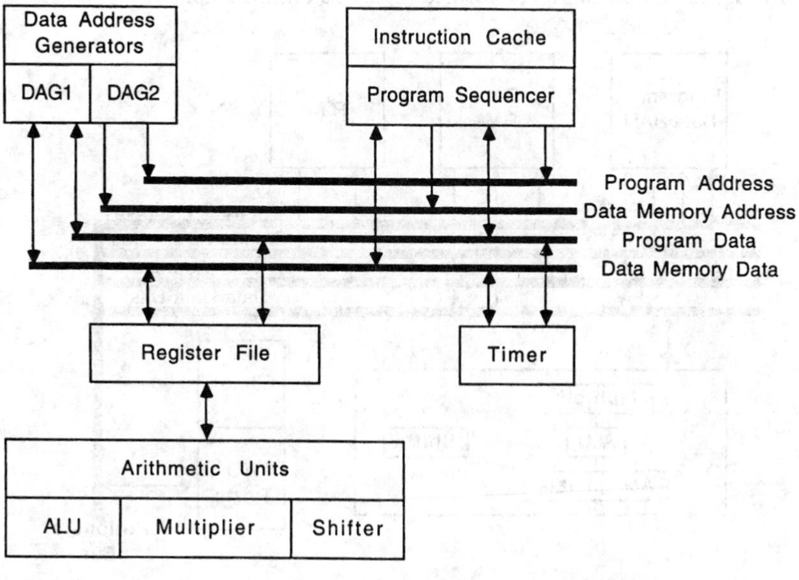

FIGURE 86.2 The Analog Devices ADSP-21020 architecture.

The Texas Instruments TMS320C5x series devices are high performance digital signal processors derived from the original TI DSP chip, the TMS32010, and its successor, the TMS320C2x. The architecture of the TMS320C50 is shown in Fig. 86.3. The chip is based on the Harvard architecture, that is, separate data and instruction buses. This additional bandwidth between processing elements supports rapid concurrent transfers of data and instructions. This chip uses a 16 × 16-b fixed point multiplier and a 32 b accumulator, and up to 10 k words on-chip scratchpad RAM. This architecture supports instruction rates of 50 ns. An FIR filter code segment is shown below, where the primary filter tap operations are performed by the MACD instruction

```
RPTK N
MACD *-, COEFFP
```

This exhibits a general similarity with that for the TI floating point chips, in particular a single instruction cycle per tap, although in this case a single instruction is executed as opposed to two parallel instruction of the TMS320C3x. The memory addressing scheme is also significantly different. An FIR filter on the TMS320C5x can thus be implemented in 25 ns per tap. An *N*th-order IIR filter code segment is shown below, where the MACD and AC instructions perform the primary multiplication operations,

```
ZPR
LACC    *,15,AR1
RPT     #(N-2)
AC      COEFFB, *-
APAC
SACH    *,1
ADRK    N-1
RPTZ    #(N-1)
MACD    COEFFA, *-
LTA     *, AR2
SACH    *,1
```

A single IIR biquad section can be performed in 250 ns on this chip.

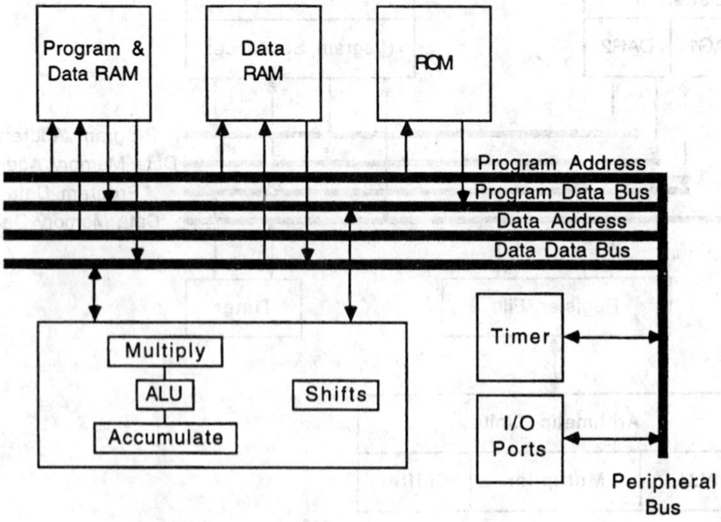

FIGURE 86.3 The Texas Instruments TMS320C50 architecture.

The Motorola 56001 series is a fixed point architecture with 24-b word size, as opposed to the smaller word sizes in most fixed point DSP chips. The architecture of the 56001 is depicted in Fig. 86.4. This chip shares many of the same features as other DSP chips, that is, Harvard architecture, on-chip scratchpad memory (512 words), and hardware multiply–accumulate support, in this case 24 × 24-b operators that form a 56-b result. The instruction cycle time of the Motorola 56001 is 97.5 ns. An FIR filter implemented on the 56001 might use the code segment shown below,

```
MOVE   #AADDR,R0
MOVE   #BADDR+n,R4
NOP
CLR    A              X:(R0)+,X0    Y:(R4)−,Y0
REP    #N
MAC    X0,Y0,A        X:(R0)+,X0    Y:(R4)−,Y0
RND    A
```

where the MAC instruction retrieves data from the appropriate registers, loads it into the multiplier, and leaves the result in the accumulator. The 56001 can perform FIR filtering at a rate of one instruction per tap, or 97.5 ns per tap. An IIR filter code segment uses the MAC instruction, as well as several others to set up the registers for the arithmetic unit, as shown below.

```
OR     #$08,MR
RND    A·          X:(R0)−,X0    Y:(R4)+,Y0
MAC    −Y0,X0,A    X:(R0)−,X1    Y:(R4)+,Y0
MAC    −Y0,X1,A    X1,X:(R0)+    Y:(R4)+,Y0
MAC    Y0,X0,A     A,X:(R0)      Y:(R4), Y0
MAC    Y0,X1,A                   1
MOVE               A,X:OUTPUT
```

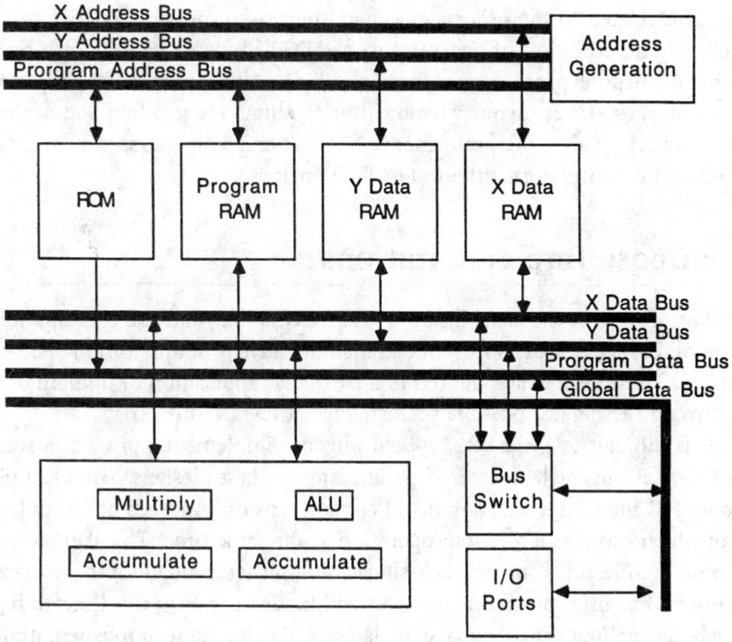

FIGURE 86.4 The Motorola 56001 architecture.

The 56001 can compute a second order IIR biquad in seven instruction cycles, or 682.5 ns.

From these examples, it can be seen that the current general purpose DSP processors possess many common features which make them well suited for digital filtering. The hardware multiply–accumulate unit, Harvard architecture, and on-chip memory are consistent characteristics of these devices. The major shortcoming of such architectures for digital filtering is the necessity to multiplex a single arithmetic unit, which implies that sampling rates above $1/NT$ are not possible, where N is the number of atomic operations (e.g., FIR filter taps) and T is the time to complete those operations.

Future Directions

Several trends have become apparent as VLSI technology has improved. One trend of note is the increasing use of parallelism, both on-chip and between chips. The support for multiprocessor communications in the TI TMS320C40 provides an avenue for direct parallel implementation of algorithms. Commercially available products with multiple fixed point DSP processors on a single chip have also appeared, such as the SPROC-1000 from STAR Semiconductor, with four DSP core processors.

Another trend has been the development of better programming interfaces for the general purpose chips. In particular, high-level language compilers have improved to the point where they provide for reasonably good performance for complex algorithms, although still not superior to that obtained by manual assembly language programming.

Another trend that is worthy of note is the development of low-power DSP implementations. These devices are targeted at the wireless Personal Communications System (PCS) marketplace, where minimum power usage is critical. The developments in this area have been particularly striking, given the strong dependence of power consumption on clock frequency, which is usually high in DSP implementations. Through a combination of careful circuit design, power supply voltage reductions, and architectural innovations, extremely low-power implementations have been realized.

A final trend is related to the process of general purpose processors relative to digital signal processing chips. The evolution of general purpose DSP implementations may have come full circle, as general purpose processors such as the DEC Alpha possess on-chip floating point multiplication units, as well as memory bandwidths equaling or exceeding that of the DSP chips. These features are reflected in the performance of these chips on standard benchmarks [13], in which the DEC Alpha outperforms the fastest DSP engines.

86.3 Special Purpose Implementations

The tremendous growth in the capabilities of VLSI technology and the corresponding decrease in the fabrication costs has lead to the wide availability advent of application-specific integrated circuits (ASIC's). These devices are tailored to a particular application or domain of applications in order to provide the highest possible performance at low per-unit costs.

Although it is difficult to generalize, special purpose implementations share some common features. The first is the high degree of parallelism in these designs. For example, a typical special purpose FIR filter implementation will contain tens or hundreds of multiply–accumulate units, each of which executes a filter tap operation at the same time. This is in contrast to most general purpose architectures, in which a single multiply–accumulate unit is shared. Another common feature is extensive pipelining between arithmetic operators; this leads to high sampling rates and high throughput, at some cost in latency. Finally, these implementations are often lacking in flexibility, being designed for specific application domains. The number of filter taps may be fixed, or the filter coefficients themselves may be fixed. In almost all instances, these implementations are based on fixed point arithmetic.

Because the implementation cost of multiplication operations is so large compared to other operations, significant research effort has been expended on developing fast and efficient multiplier architectures, as well as digital filter design techniques that can be used to reduce the number of multiplications. A large number of multiplier architectures have been developed, ranging from bit-serial structures to bit and word level pipelined array designs [9]. The most appropriate architecture is a function of the application requirements, as various area versus speed options are available. The other major research direction is the minimization of multiplication operations. In this case, multiplications are eliminated by conscientious structuring of the realization, as in linear phase filters, circumvented by use of alternate number systems such as the residue number system (RNS), or simplified to a limited number of shift-and-add operations. The later option has been used successfully in a large number of both FIR and IIR realizations, some of which will be discussed below.

Historically, bit-serial implementations of digital filters have been of some interest to researchers and practitioners in the early days of VLSI because of the relatively high cost of silicon area devoted to both devices and routing [2]. Even in primitive technologies, bit-serial implementations could exploit the natural parallelism in digital filtering algorithms.

As clock rates have risen, the silicon area has be come more economical, parallel implementations have become the most effective way of implementing high performance digital filters. The concept of the systolic array has strongly influenced the implementation of both FIR and IIR filters [8]. Systolic arrays are characterized by spatial and temporal locality, that is, algorithms and processing elements should be structured to minimize interconnection distances between nodes and to provide at least a single delay element between nodes. Interconnection distances need to be kept to a minimum to reduce delays associated with signal routing, which is becoming the dominant limiting factor in VLSI systems. Imposing pipeline delays between nodes minimizes computational delay paths and leads to high throughput.

These characteristic features of special purpose digital filter designs will be illustrated by examples of FIR and IIR filter implementations.

FIR Filter Examples

FIR filters may be implemented in a number of ways, depending on application requirements. The primary factors that must be considered are the filter length, sampling rate, and area, which determine the amount of parallelism that can be applied. Once the degree of parallelism and pipelining are determined, the appropriate general filter structure can be determined.

A typical high-performance FIR filter implementation [7] provides sampling rates of 180 MHz for 32 linear phase taps. This chip uses canonical signed digit (CSD) coefficients. This representation is based on a number system which the digits take the values $(-1, 0, 1)$. A filter tap can be implemented with a small number of these digits, and hence that tap requires a small number of shift-and-add operations. Each coefficient is implemented based on 2-b shift-and-add units, as depicted in Fig. 86.5. Delay elements are bypassed during configuration to allow realization of coefficients with additional bits. This chip also makes use of extensive pipelining, carry-save addition, and advanced single-phase clocking techniques to provide high throughput.

In part due to the highly structured nature of FIR filtering algorithms, automatic design tools have been used to successfully implement high-performance FIR filters similar to that just presented. These methods often integrate the filter and architectural design into a unified process that can effectively utilize silicon area to provide the desired performance.

At the other extreme of performance is the Motorola 56200 FIR filter chip [10]. This chip, while currently several years old, represents an approach to the custom implementation of long (several hundred taps) FIR filters. In this case, a single processing element is multiplexed among

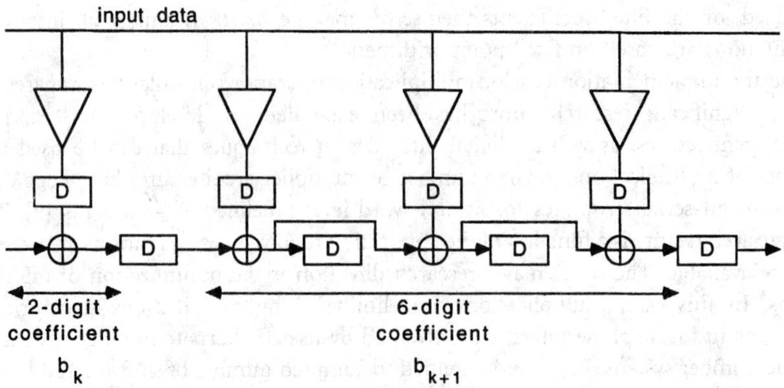

FIGURE 86.5 Custom FIR filter architecture for 180 MHz sampling rates.

all of the filter taps, similar in concept to the approach used in general purpose DSP processors as shown in Fig. 86.6. Due to the regularity of the filter structure, extensive pipelining in the arithmetic unit can be used to support a large number of taps at audio rates. This chip can be used to realize a 256 tap FIR filter at sampling rates up to 19 kHz, with higher performance for shorter filters. Longer filters can be implemented using cascaded processors.

A comparison of implementations [5]–[7], [11], [19]–[21] illustrates the range of design and performance options. This is illustrated in Table 86.1, where the "score" is calculated according to the sampling rate multiplied by the number of taps per unit area, with normalization for the particular technology used. This simplistic comparison does not consider differences in word length or coefficient codings, but it does provide some insight into the results of the various design approaches. A significant number of other digital FIR filtering chips exist, both research prototypes and commercial products; this exposition only outlines some of the architectural options.

IIR Filter Examples

Custom IIR filter implementations are also most commonly based on parallel architectures, although there are somewhat fewer custom realizations of IIR filters that FIR filters. A significant difficulty in the implementation of high-performance IIR filters in the need for feedback in the computation of an IIR filter section. This limits the throughput that can be attained to at least one multiply–accumulate cycle in a straightforward realization. Another difficulty is the

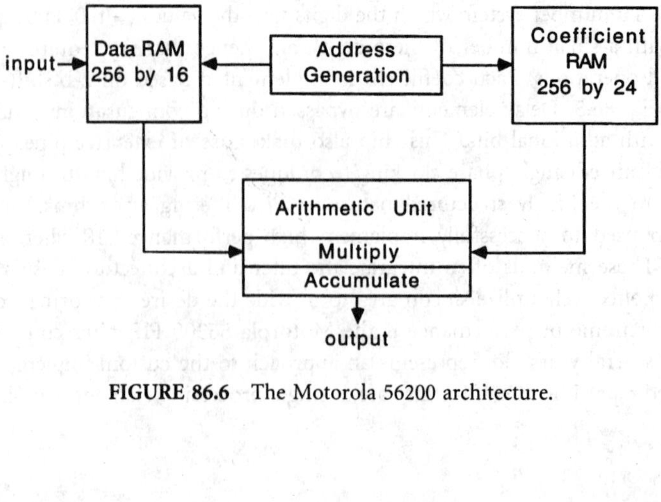

FIGURE 86.6 The Motorola 56200 architecture.

TABLE 86.1 FIR Filter ASIC Comparison

Design	Taps	Area (mm^2)	Rate (MHz)	Technology (μm)	Score
[7]	32	20.1	180.0	1.2	495.2
[21]	43	40.95	150.0	1.2	139.3
[20]	64	48.65	100.0	0.8	33.68
[11]	64	225	22.0	1.5	21.12
[19]	16	17.65	30.0 (est.)	1.25	53.12
[5]	4	25.8 (est.)	37.0	1.25	11.20
[6]	40	22	100.0	0.9	132.5

numerical stability of IIR filters with short coefficients, which makes aggressive quantization of coefficients less promising.

In order to address the difficulties with throughput limitation due to feedback, structures based on systolic concepts have been developed. Although the feedback problem imposes a severe constraint on the implementation, use of bit and word level systolic structures that pipeline data most significant digit first can minimize the impact of this restriction [17]. Using these techniques, and a Signed Binary Number Representation (SBNR) similar to a CSD code, first-order sections with sampling rates of 15 MHz have been demonstrated in a 1.5 μm standard cell process in an area of 21.8 mm^2. This particular design uses fairly large coefficient and data words, however, at 12 b and 11 b, respectively.

The numerical stability problem has been addressed by a variety of techniques. One of these is based on minimizing limited precision effects by manipulation of traditional canonical filter structures and clever partitioning of arithmetic operations. A more recent and general approach is based on modeling the digital implementations of filters after their analog counterparts; these classes of filters are known as wave digital filters (WDF's) [4]. WDF's exhibit good passband ripple and stopband attenuation, with high tolerance to limited wordlength effects. Because of the later property, efficient implementations based on short word sizes are feasible. A WDF design for a second-order section in custom 1.5-μm CMOS based on a restricted coefficient set akin to CSD supports to 10 MHz sampling rates in an area of 12.9 mm^2 [16].

Future Trends

The future trends in digital filter implementation appear to be a fairly straightforward function of the increasing capability of VLSI devices. In particular, more taps and filter sections per chip and higher sampling rates are becoming achievable. Related to these trends are higher degrees of on-chip parallelism. Further programmability is more reasonable as density and speed margins increase, although there is still a high cost in area and performance. Finally, special purpose implementations show extraordinary promise in the area of low-power systems, where custom circuit design techniques and application-specific architectural features can be combined to best advantage.

86.4 Programmable Logic Implementations

The rapid evolution of VLSI technology has enabled the development of several high-density programmable logic architectures. There are several novel features that make these devices of interest beyond their traditional field of state machine implementation. In particular, the density of the largest of these devices is currently approximately 20 000 gates, which encompasses the level of complexity found in the majority of ASIC's (although many ASIC's are over an order of magnitude more complex). This level of complexity is sufficient to support many designs that would traditionally need to be implemented as ASIC's. The speed of the new programmable logic devices (PLD's) and field programmable gate arrays (FPGA's) is quite reasonable, with toggle rates on the order of 150 MHz. While this is not as great as custom implementations, it does allow many applications to be realized in this new technology.

One of the most significant features of FPGA implementations is the capability for in-system reprogrammability in many FPGA families. Unlike traditional field programmable parts based on antifuse technology and that can only be programmed once, many of the new architectures are based on memory technology. This means the entirely new computational architectures can be implemented simply by reprogramming the logic functions and interconnection routing on the chip. Ongoing research efforts have been directed toward using FPGA's as generalized coprocessors for supercomputing applications.

The implications of the PLD technology for filter implementation are significant. These new devices provide an enormous amount of flexibility, which can be used in the implementation of a variety of novel architectures on a single chip. This is particularly useful for rapid prototyping of digital filtering algorithms, where several high-performance designs can be evaluated in a target environment on the same hardware platform.

Because many of the new PLD's are based on SRAM technology, the density of these devices can be expected to grow in parallel with the RAM growth curve, that is, at approximately 60 percent per year. Further, since these devices may be used for a large variety of applications, they will most likely become high-volume commodity parts, and hence prices will be low compared to more specialized DSP chips. This implies that new digital signal processing systems that were not previously technically and economically feasible to implement in this technology may be implemented using PLD's in coming years.

One of the extra costs of this approach, as opposed to the full custom strategy, is the need for support chips. Several chips are needed, include memory to store the PLD configurations, as well as logic to control the downloading of the program.

We will next examine the implementation of several FIR and IIR digital filtering architectures based on FPGA's.

FIR Filter Implementations

Several approaches to the FPGA implementation of FIR filters can be taken. Due to the flexibility of these parts, switching from one architecture to the next only requires reprogramming the device, subject to constraints on I/O pin locations. Two fundamental strategies for realizing FIR filters will be illustrated here, one which is suited to relatively short filters (or longer filters cascaded across several chips) operating at high rates, and another which is suited for longer filters at lower rates.

A high-performance FIR filter example [3], illustrated in Fig. 86.7, is based on the observation that since the entire device is reprogrammable, architectures in which filter coefficient multiplications are implemented as "hardwired" shifts can be easily reconfigured depending on the desired filter response. In this example, each of the coefficients is represented in a canonical signed digit (CSD) code with a limited number of nontrivial (e.g., nonzero) bits, which allows each tap to be implemented as a small number of shift and add operations. A filter tap may be implemented in two columns of logic blocks on a Xilinx 3100-series FPGA [18], where the two columns of full adders and associated delays implement a tap based on CSD coefficients with two nontrivial bits. With this architecture, up to 11 taps can be implemented on a single Xilinx XC3195 FPGA at sampling rates of above 40 MHz. Longer filters may be implemented by a cascade of FPGA devices.

An FIR filter architecture for longer filters is based upon implementation of several traditional multiply–accumulate (MAC) units on one chip, as shown in Fig. 86.8. Each of these MAC units can then be shared among a large number of filter tap computations, much as the single MAC unit in the Motorola 56200 is multiplexed. Since four multipliers can be implemented in current

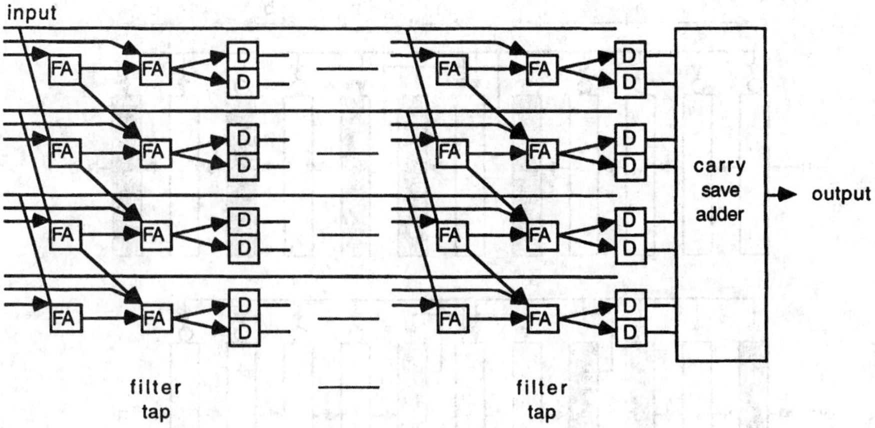

FIGURE 86.7 High-performance FIR architecture on FPGA's.

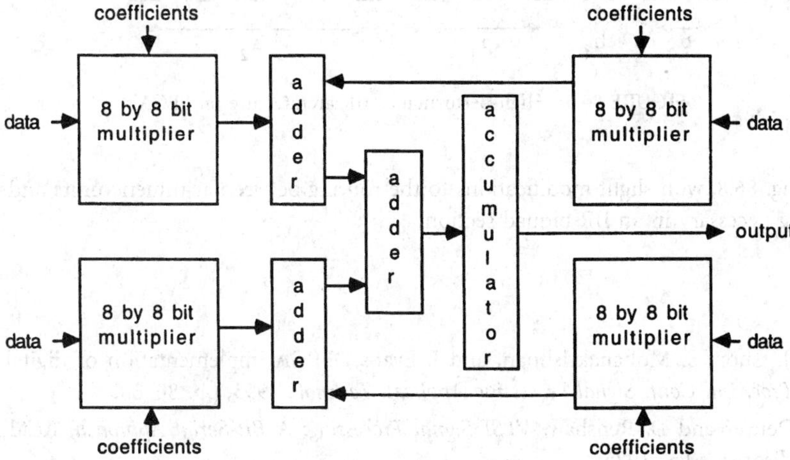

FIGURE 86.8 FIR architecture on FPGA's for large number of taps.

technology, the inherent parallelism of FIR filters can be exploited to support sampling rates of up to 1.25 MHz for 32 taps.

IIR Filter Implementations

As in the case of FIR filters, IIR filters can be implemented using a "hardwired" architecture suited to high performance, or a more traditional approach based on general multiply–accumulate units. In the case of IIR filters, however, the hardwired implementation is significantly more desirable than the alternate approach due to the difficulty in rescheduling multiplexed processing elements in a system with feedback.

An architecture that is reconfigured to implement different filters will generally provide both high performance and good area efficiency. An example of such a system is shown in Fig. 86.9, in which two IIR biquad sections are implemented on a single FPGA using a traditional canonical filter structure [1]. Each of the columns realizes a shift-and-add for one nontrivial bit of a coefficient, where the shaded blocks also contain delay registers. This implementation yields sampling rates of better than 10 MHz for typical coefficients.

A more traditional approach to the realization of IIR filters using multiply–accumulate units is also possible, but may be less efficient. The general architecture is similar to that of the FIR

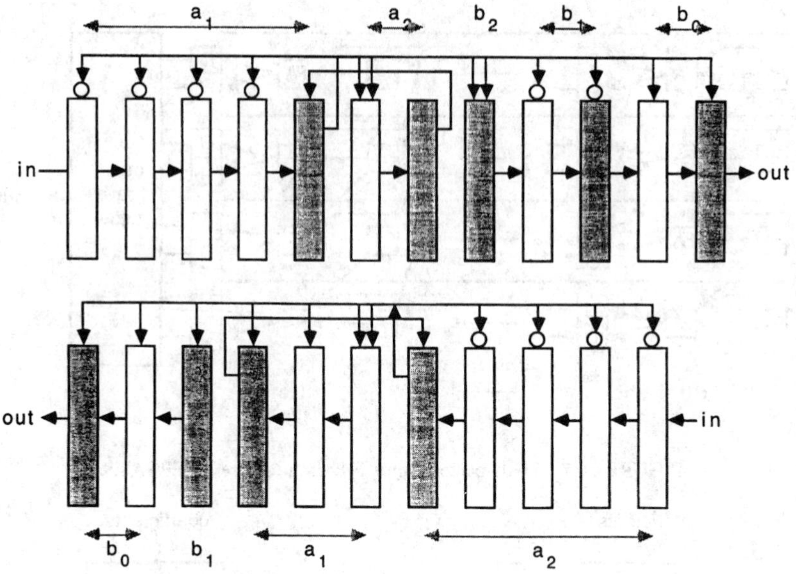

FIGURE 86.9 High-performance IIR architecture on FPGA's.

filter in Fig. 86.8, with slight modifications to the routing between arithmetic units and support for scaling necessary in an IIR biquad section.

References

[1] C.-J. Chou, S. Mohanakrishnan, and J. Evans, "FPGA implementation of digital filters," in *Proc. Int. Conf. Signal Processing Applicat. Technol.*, 1993, pp. 80–88.

[2] P. Denyer and D. Renshaw, *VLSI Signal Processing: A Bit-Serial Approach,* Reading MA: Addison-Wesley, 1985.

[3] J. Evans, "An efficient FIR filter architecture," *IEEE Int. Symp. Circuits Syst.*, 1993, pp. 627–630.

[4] A. Fettweis, "Wave digital filters: Theory and practice," *Proc. IEEE,* vol. 74, pp. 270–327, Jan. 1986.

[5] P. Hartley, P. Corbett, P. Jacob, S. Karr, "A high speed FIR filter designed by compiler," in *Proc. IEEE Cust. IC Conf.*, 1989, pp. 20.2.1–20.2.4.

[6] M. Hatamian and S. Rao, "A 100 MHz 40-tap programmable FIR filter chip," in *Proc. IEEE Int. Symp. Circuits Syst.*, 1990, pp. 3053–3056.

[7] K.-Y. Khoo, A. Kwentus, A. Willson, Jr., "An efficient 175 MHz programmable FIR digital filter," in *Proc. IEEE Int. Symp. Circuits Syst.*, 1993, pp. 72–75.

[8] S. Y. Kung, *VLSI Array Processors,* Englewood Cliffs, NJ: Prentice-Hall, 1988.

[9] G.-K. Ma and F. Taylor, "Multiplier policies for digital signal processing," *IEEE ASSP Mag.*, pp. 6–20, Jan. 1990.

[10] Motorola, *DSP56001 Digital Signal Processor User's Manual,* Phoenix, AZ: Motorola, 1989.

[11] P. Ruetz, "The architectures and design of a 20-MHz real-time DSP chip set," *IEEE J. Solid-State Circuits,* vol. 24, pp. 338–348, Feb. 1989.

[12] W. Schweber, "Floating-point DSP for high-speed signal processing," *Analog Dialogue,* vol. 25, no. 4, Analog Devices, pp. 3–5, 1993.

[13] L. Stewart, A. Payne, and T. Levergood, "Are DSP chips obsolete?", in *Proc. Int. Conf. Signal Processing Applicat. Technol.*, Nov, 1992, pp. 178–187.

[14] Texas Instruments, *TMS320C3x User's Manual,* Dallas, TX: Texas Instruments, 1992.

[15] Texas Instruments, *TMS320C5x User's Manual,* Dallas, TX: Texas Instuments, 1993.

[16] A. Wicks and S. Summerfield, "VLSI implementation of high speed wave digital filters based on a restricted coefficient set," in *Proc. IEEE Int. Symp. Circuits Syst.*, 1993, pp. 603–606.

[17] R. Woods, J. McCanny, S. Knowles, and O. McNally, "A high performance IIR digital filter chip," in *Proc. IEEE Int. Symp. Circuits Syst.*, 1990, pp. 1410–1413.

[18] Xilinx, *The Programmable Logic Data Book,* San Jose, CA: Xilinx, 1993.

[19] F. Yassa, J. Jasica, R. Hartley, S. Novjam, "A silicon compiler for digital signal processing; Methodology, implementation, and applications," *Proc. IEEE*, vol. 75, pp. 1272–1282, Sept. 1987.

[20] T. Yoshino, R. Jain, P. Yang, H. Davis, W. Gass, A. Shah, "A 100-MHz 64-tap FIR digital filter in 0.8 μm BiCMOS gate array," *IEEE J. Solid-State Circuits*, vol. 25, pp. 1494–1501, June 1990.

[21] J. Laskowski, H. Samuel, "A 150-MHz 43-Tap Half-Band FIR digital filter in 1.2 μm CMOS Generated by Compiler," *Proc. IEEE Cust. IC Conf.*, 1992, pp. 11.4.1–11.4.4.

...rther Information

The *IEEE Transactions on Circuits and Systems II: Analog and Digital Signal Processing* frequently contains papers on the VLSI implementations of digital filters as well as design methods for efficient implementation. The *IEEE Transactions on Signal Processing* often includes articles in these areas as well. Papers in the *IEEE Journal on Solid-State Circuits*, the *IEEE Transactions on VLSI Systems*, and the *IEE Electronics Letters* regularly cover particular implementations of digital filters.

The conference proceedings for the *IEEE International Symposium on Circuits and Systems* and the *IEEE International Conference on Acoustics, Speech, and Signal Processing* also contain a wealth of information on digital filter implementation.

The textbook *VLSI Array Processors* by S. Y. Kung discusses the concept of systolic arrays at length.

87

Two-Dimensional FIR Filters

R. Ansari
Bell Communications Research

A. E. Cetin
Bilkent University and Koc University

87.1 Introduction

In this chapter, methods of designing two-dimensional (2-D) finite-extent impulse response (FIR) discrete-time filters are described. Two-dimensional FIR filters offer the advantages of phase linearity and guaranteed stability, which make them attractive in applications. Over the years an extensive array of techniques for designing 2-D FIR filters has been accumulated [13], [21], [28]. These techniques can be conveniently classified into the two categories of general and specialized designs. Techniques in the category of general design are intended for approximation of *arbitrary* desired frequency responses, usually with no structural constraints on the filter. These techniques include approaches such as windowing of the ideal impulse response [20] or the use of suitable optimality criteria possibly implemented with iterative algorithms. On the other hand, techniques in the category of special design are applicable to restricted classes of filters, either due to the nature of the response being approximated or due to the imposition of structural constraints on the filter used in the design. The specialized designs are a consequence of the observation that commonly used filters have characteristic underlying features that can be exploited to simplify the problem of design and implementation. The stopbands and passbands of filters encountered in practice are often defined by straight-line, circular, or elliptical boundaries. Specialized design methodologies have been developed for handling these cases and they are typically based on techniques such as the transformation of one-dimensional (1-D) filters or the rotation and translation of separable filter responses. If the desired response possesses symmetries, then the symmetries imply relationships among the filter coefficients that are exploited in both the design and the implementation of the filters. In some

0-8493-8341-2/95/$0.00 + $.50

design problems it may be advantageous to impose structural constraints in the form of parallel and cascade connections.

The material in this chapter is organized as follows. A preliminary discussion of characteristics of 2-D FIR filters and of issues relevant to the design methods appears in Section 87.2. Following this, methods of general and special FIR filter design are described in Sections 87.3 and 87.4, respectively. Several examples of design illustrating the procedure are also presented. Issues in 2-D FIR filter implementation are briefly discussed in Section 87.5. Finally, additional topics are outlined and a list of sources for further information is provided.

87.2 Preliminary Design Considerations

In any 2-D filter design there is a choice between FIR and IIR filters, and their relative merits are briefly examined next. Two-dimensional FIR filters offer certain advantages over 2-D IIR filters and as a result, FIR filters have found widespread use in applications such as image and video processing. One key attribute of an FIR filter is that it can be designed with a strictly linear passband phase, and it can be implemented with small delays without the need to reverse the signal array during processing. A 2-D FIR filter impulse response has only a finite number of nonzero samples, which guarantees stability. On the other hand, stability is difficult to test in the case of 2-D IIR filters due to the absence of a 2-D counterpart of the fundamental theorem of algebra, and a 2-D polynomial is almost never factorizable. If a 2-D FIR filter is implemented nonrecursively with finite precision, then it does not exhibit limit cycle oscillations. Arithmetic quantization noise and coefficient quantization effects in FIR filter implementation are usually very low. A key disadvantage of FIR filters is that they typically have higher computational complexity than IIR filters for meeting the same specifications, especially in cases where the specifications are stringent.

The term 2-D FIR filter refers to a linear shift-invariant system whose input–output relation is represented by a convolution [13]

$$y(n_1, n_2) = \sum\sum_{(k_1,k_2)\in I} h(k_1, k_2)x(n_1 - k_1, n_2 - k_2) \tag{87.1}$$

where $x(n_1, n_2)$ and $y(n_1, n_2)$ are the input and the output sequences, respectively, $h(n_1, n_2)$ is the impulse response sequence, and I is the support of the impulse response sequence. FIR filters have compact support, meaning that only a finite number of coefficients are nonzero. This makes the impulse response sequence of FIR filters absolutely summable, thereby ensuring filter stability. Usually the filter support I is chosen to be a rectangular region centered at the origin, e.g., $I = \{(n_1, n_2): -N_1 \leq n_1 \leq N_1, -N_2 \leq n_2 \leq N_2\}$. However, there are some important cases where it is more advantageous to select a nonrectangular region as the filter support [30].

Once the extent of the impulse response support is determined, the sequence $h(n_1, n_2)$ should be chosen in order to meet given filter specifications under suitable approximation criteria. These aspects are elaborated on in the next subsection. This is followed by a discussion of phase linearity and filter response symmetry considerations and then, some guidelines on using the design methods are provided.

Filter Specifications and Approximation Criteria

The problem of designing a 2-D FIR filter consists of determining the impulse response sequence $h(n_1, n_2)$ or its system function $H(z_1, z_2)$ in order to satisfy given requirements on the filter response. The filter requirements are usually specified in the frequency domain, and only this case is considered here. The frequency response,[1] $H(\omega_1, \omega_2)$, corresponding to the impulse

[1]Here, $\omega_1 = 2\pi f_1$, and $\omega_2 = 2\pi f_2$ are the horizontal and vertical angular frequencies, respectively.

response $h(n_1, n_2)$ with a support I is expressed as

$$H(\omega_1, \omega_2) = \sum_{(n_1,n_2)\in I}\sum h(n_1, n_2)e^{-j(\omega_1 n_1 + \omega_2 n_2)} \tag{87.2}$$

Note that $H(\omega_1, \omega_2) = H(\omega_1 + 2\pi, \omega_2) = H(\omega_1, \omega_2 + 2\pi)$ for all (ω_1, ω_2). In other words, $H(\omega_1, \omega_2)$ is a periodic function with a period 2π in both ω_1 and ω_2. This implies that by defining $H(\omega_1, \omega_2)$ in the region $\{-\pi < \omega_1 \le \pi, -\pi < \omega_2 \le \pi\}$ the frequency response of the filter for all (ω_1, ω_2) is determined.

For 2-D FIR filters the specifications are usually given in terms of the magnitude response $|H(\omega_1, \omega_2)|$. Attention in this subsection is confined to the case of a two-level magnitude design, where the desired magnitude levels are either 1.0 (in the passband) or 0.0 (in the stopband). Some of the procedures can be easily modified to accommodate multilevel magnitude specifications, as, for instance, in a case that requires the magnitude to increase linearly with distance from the origin in the frequency domain.

Consider the design of a 2-D FIR low-pass filter whose specifications are shown in Fig. 87.1. The magnitude of the low-pass filter ideally takes the value 1.0 in the passband region F_p, which is centered around the origin $(\omega_1, \omega_2) = (0, 0)$ and 0.0 in the stopband region F_s. As a magnitude discontinuity is not possible with a finite filter support I, it is necessary to interpose a transition region F_t between F_p and F_s. Also, magnitude bounds $|H(\omega_1, \omega_2) - 1| \le \delta_p$ in the passband and $|H(\omega_1, \omega_2)| \le \delta_s$ in the stopband are specified, where the parameters δ_p and δ_s are positive real numbers, typically much less than 1.0. The frequency response $H(\omega_1, \omega_2)$ is assumed to be real. Consequently, the low-pass filter is specified in the frequency domain by the regions F_p, F_s, and the tolerance parameters δ_p and δ_s. A variety of stopband and passband shapes can be specified in a similar manner.

In order to meet given specifications, an adequate filter order (the number of nonzero impulse response samples) needs to be determined. If the specifications are stringent, with tight tolerance parameters and small transition regions, then the filter support region I must be large. In other words, there is a trade-off between the filter support region I and the frequency domain specifications. In the general case the filter order is not known *a priori*, and may be determined either through an iterative process or using estimation rules if available. If the filter order is given, then in order to determine an optimum solution to the design problem, an appropriate optimality criterion is needed. Commonly used criteria in 2-D filter design are minimization of the L_p norm (p finite) of the approximation error, or the L_∞ norm. If desired, a maximal flatness requirement at desired frequencies can be imposed [22]. It should be noted that if the specifications are given in terms of the tolerance bounds on magnitude, as described above, then

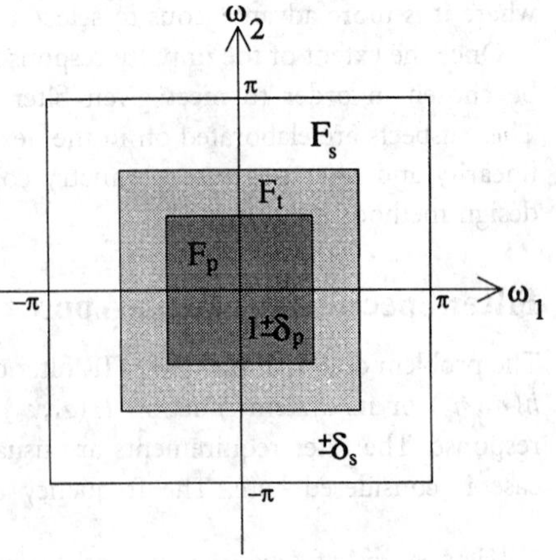

FIGURE 87.1 Frequency response specifications for a 2-D low-pass filter ($|H(\omega_1, \omega_2) - 1| \le \delta_p$ for $(\omega_1, \omega_2) \in F_p$ and $|H(\omega_1, \omega_2)| \le \delta_s$ for $(\omega_1, \omega_2) \in F_s$).

the use of the L_∞ criterion is appropriate. However, the use of other criteria such as a weighted L_2 norm can serve to arrive at an almost minimax solution [2].

Zero-Phase FIR Filters and Symmetry Considerations

Phase linearity is important in many filtering applications. As in the 1-D case a number of conditions for phase linearity can be obtained depending on the nature of symmetry. But the discussion here is limited to the case of "zero phase" design, with a purely real frequency response. A salient feature of 2-D FIR filters is that realizable FIR filters that have purely real frequency responses are easily designed. The term "zero phase" is somewhat misleading in the sense that the frequency response may be negative at some frequencies. The term should be understood in the sense of "zero phase in passband", because the passband frequency response is within a small deviation of the value 1.0. The frequency response may assume negative values in the stopband region, where phase linearity is immaterial. In frequency domain the zero-phase or real frequency response condition corresponds to

$$H(\omega_1, \omega_2) = H^*(\omega_1, \omega_2) \tag{87.3}$$

where $H^*(\omega_1, \omega_2)$ denotes the complex conjugate of $H(\omega_1, \omega_2)$. The condition (87.3) is equivalent to

$$h(n_1, n_2) = h^*(-n_1, -n_2) \tag{87.4}$$

in the spatial domain. Making a common practical assumption that $h(n_1, n_2)$ is real, the above condition reduces to

$$h(n_1, n_2) = h(-n_1, -n_2) \tag{87.5}$$

implying a region of support with the above symmetry about the origin.

Henceforth only the design of zero-phase FIR filters is considered. With $h(n_1, n_2)$ real, and satisfying (87.5), the frequency response $H(\omega_1, \omega_2)$ is expressed as

$$\begin{aligned} H(\omega_1, \omega_2) &= h(0,0) + \sum_{(n_1,n_2)\in I_1} h(n_1, n_2) e^{-j(\omega_1 n_1 + \omega_2 n_2)} \\ &\quad + \sum_{(n_1,n_2)\in I_2} h(n_1, n_2) e^{-j(\omega_1 n_1 + \omega_2 n_2)} \\ &= h(0,0) + \sum_{(n_1,n_2)\in I_1} 2h(n_1, n_2)\cos(\omega_1 n_1 + \omega_2 n_2) \end{aligned} \tag{87.6}$$

where I_1 and I_2 are disjoint regions such that $I_1 \cup I_2 \cup \{(0,0)\} = I$, and if $(n_1, n_2) \in I_1$, then $(-n_1, -n_2) \in I_2$.

In order to understand the importance of phase linearity in image processing, consider an example that illustrates the effect of nonlinear-phase filters on images. In Fig. 87.2(a) an image that is corrupted by white Gaussian noise, $\sim N(0, 6.5)$, is shown. This image is filtered with a nonlinear-phase low-pass filter and the resultant image is shown in Fig. 87.2(b). It is observed that edges and textured regions are severely distorted in Fig. 87.2(b). This is due to the fact that the spatial alignment of frequency components that define an edge in the original is altered by the phase nonlinearity. The same image is also filtered with a zero-phase low-pass filter $H(\omega_1, \omega_2)$, which has the same magnitude characteristics as the nonlinear phase filter. The resulting image is shown in Fig. 87.2(c). It is seen that the edges are perceptually

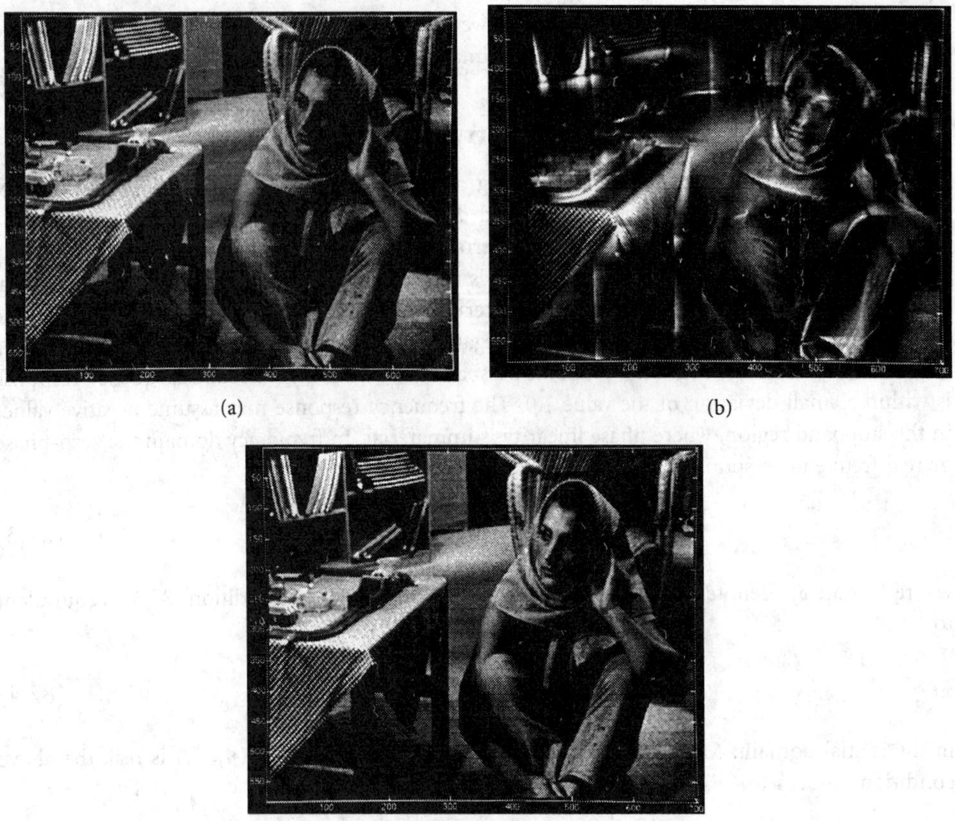

(a) (b) (c)

FIGURE 87.2 (a) Original image of 696 × 576 pixels corrupted with Gaussian noise; (b) nonlinear-phase low-pass filtered image; (c) zero-phase low-pass filtered image.

preserved in Fig. 87.2(c), although blurred due to the low-pass nature of the filter. In this example a separable zero-phase low-pass filter $H(\omega_1, \omega_2) = H_1(\omega_1)H_1(\omega_2)$ is used, where $H_1(\omega)$ is a 1-D Lagrange filter with a cutoff $\pi/2$. In spatial domain $h(n_1, n_2) = h_1(n_1)h_1(n_2)$, where $h_1(n) = \{\cdots, 0, -1/32, 0, 9/32, 1/2, 9/32, 0, -1/32, 0, \cdots, \}$ is the impulse response of the seventh-order symmetric (zero-phase) 1-D Lagrange filter. The nonlinear phase filter is a cascade of the above zero-phase filter with an allpass filter.

In some filter design problems, symmetries in the frequency domain specifications can be exploited by imposing restrictions on the filter coefficients and the shape of the support region for the impulse response. Several symmetries that can be exploited are extensively studied in [30], [42], [43]. For example, a condition often encountered in practice is the symmetry with respect to each of the two frequency axes. In this case the frequency response of a zero-phase filter satisfies

$$H(\omega_1, \omega_2) = H(-\omega_1, \omega_2) = H(\omega_1, -\omega_2) \tag{87.7}$$

This yields an impulse response that is symmetric with respect to the n_1 and n_2 axes, i.e.,

$$h(n_1, n_2) = h(-n_1, n_2) = h(n_1, -n_2) \tag{87.8}$$

By imposing symmetry conditions one reduces the number of independently varying filter coefficients that must be determined in the design. This can be exploited in reducing both the computational complexity of the filter design and the number of arithmetic operations required in the implementation.

Guidelines on the Use of the Design Techniques

The design techniques described in this subsection are classified into the two categories of general and specialized designs. The user should use the techniques of general design in cases requiring approximation of arbitrary desired frequency responses, usually with no structural constraints on the filter. The specialized designs are recommended in cases where filters exhibit certain underlying features that can be exploited to simplifying the problem of design and implementation.

In the category of general design, four methods are described. Of these, the windowing procedure is quick and simple. It is useful in situations where implementation efficiency is not critical, especially in single-use applications. The second procedure is based on linear programming, and is suitable for design problems where equiripple solutions are desired to meet frequency domain specifications. The remaining two procedures may also be used for meeting frequency domain specifications, and lead to nearly equiripple solutions. The third procedure provides solutions for L_p approximations. The fourth procedure is an iterative method that is easy to implement, and is convenient in situations where additional constraints are to be placed on the filter.

In the category of specialized design described here, the solutions are derived from 1-D filters. These often lead to computationally efficient implementation, and are recommended in situations where low implementation complexity is critical, and the filter characteristics possess features that can be exploited in the design. An important practical class of filters is one where specifications can be decomposed into a set of separable filter designs requiring essentially the design of suitable 1-D filters. Here the separable design procedure should be used. Another class of filters is one where the passbands and stopbands are characterized by circular, elliptical, or special straight-line boundaries. In this case a frequency transformation method, called the McClellan transformation procedure, is convenient to use. The desired 2-D filter constant-magnitude contours are defined by a proper choice of parameters in a transformation of variables applied to a 1-D zero-phase filter. Finally, in some cases, filter specifications are characterized by ideal frequency responses in which passbands and stopbands are separated by straight-line boundaries that are not suitable for applying the McClellan transformation procedure. In this case, the design may usually be carried out by nonrectangular transformations and sampling grid conversions. The importance of this design method stems from the implementation efficiency that results from a generalized notion of separable processing.

87.3 General Design Methods for Arbitrary Specifications

Some general methods of meeting arbitrary specifications are now described. These are typically based on extending techniques of 1-D design. However there are important differences. The Parks–McClellan procedure for minimax approximation based on the alternation theorem does not find a direct extension. This is because the set of cosine functions used in the 2-D approximation does not satisfy the Haar condition in the domain of interest [23], and the Chebyshev approximation does not have a unique solution. However, techniques that employ exchange algorithms have been developed for the 2-D case [18], [23], [34].

Here we consider four procedures in some detail. The first technique is based on windowing. It is a simple technique, but is not optimum for Chebyshev approximation. The second technique is based on frequency sampling, and this can be used to arrive at equiripple solutions using linear programming. Finally, two techniques for arriving iteratively at a nearly equiripple solution are described. The first of these is based on L_p approximations using nonlinear optimization. The second is based on the use of alternating projections in the sample and the frequency domains.

Design of 2-D FIR Filters by Windowing

This design method is basically an extension of the window-based 1-D FIR filter design to the case of 2-D filters. An ideal impulse response sequence, which is usually an infinite-extent sequence, is suitably windowed to make the support finite. One-dimensional FIR filter design by windowing, and some examples of 1-D windows are described in detail in the first subsection in Section 87.2.

Let $h_{id}(n_1, n_2)$ and $H_{id}(\omega_1, \omega_2)$ be the impulse and frequency responses of the ideal filter, respectively. The impulse response of the required 2-D filter $h(n_1, n_2)$ is obtained as a product of the ideal impulse response sequence and a suitable 2-D window sequence, which has a finite extent support I, that is,

$$h(n_1, n_2) = \begin{cases} h_{id}(n_1, n_2)w(n_1, n_2) & (n_1, n_2) \in I \\ 0 & \text{otherwise} \end{cases} \tag{87.9}$$

where $w(n_1, n_2)$ is the window sequence. The resultant frequency response $H(\omega_1, \omega_2)$, is a smoothed version of the ideal frequency response as $H(\omega_1, \omega_2)$ is related to the $H_{id}(\omega_1, \omega_2)$ via the periodic convolution, that is,

$$H(\omega_1, \omega_2) = \frac{1}{4\pi^2} \int_{-\pi}^{\pi} \int_{-\pi}^{\pi} H_{id}(\Omega_1, \Omega_2) W(\omega_1 - \Omega_1, \omega_2 - \Omega_2)\, d\Omega_1\, d\Omega_2 \tag{87.10}$$

where $W(\omega_1, \omega_2)$ is the Fourier Transform of the window sequence $w(n_1, n_2)$.

As in the 1-D case, a 2-D window sequence $w(n_1, n_2)$ should satisfy three requirements:

- It must have a finite-extent support I,
- its discrete-space Fourier transform should in some sense approximate the 2-D impulse function $\delta(\omega_1, \omega_2)$, and
- it should be real, with a zero-phase discrete-space Fourier transform.

Usually, 2-D windows are derived from 1-D windows. Three methods of constructing windows are briefly examined. One method is to obtain a separable window from two 1-D windows, that is

$$w_r(n_1, n_2) = w_1(n_1)w_2(n_2) \tag{87.11}$$

where $w_1(n)$ and $w_2(n)$ are the 1-D windows. Thus, the support of the resultant 2-D window $w_r(n_1, n_2)$, is a rectangular region. The frequency response of the 2-D window is also separable, i.e., $W_r(\omega_1, \omega_2) = W_1(\omega_1)W_2(\omega_2)$.

The second method of constructing a window, due to Huang [20], consists of sampling the surface generated by rotating a 1-D continuous-time window $w(t)$ as follows:

$$w_c(n_1, n_2) = w\left(\sqrt{n_1^2 + n_2^2}\right) \tag{87.12}$$

where $w(t) = 0$, $t \geq N$. The impulse response support is $I = \{n_1, n_2\colon \sqrt{n_1^2 + n_2^2} < N\}$. Note that the 2-D Fourier transform of the $w_c(n_1, n_2)$ is not equal to the circularly rotated version of the Fourier transform of $w(t)$.

Finally, in the third method, proposed by Yu and Mitra [51], the window is constructed by using a 1-D to 2-D transformation belonging to a class called the McClellan transformations [31]. These transformations are discussed in greater detail in Section 87.4. Here we consider a special case of the transform that produces approximately circular contours in the 2-D frequency domain. Briefly, the discrete-space frequency transform of the 2-D window sequence obtained with a McClellan transformation applied to a 1-D window is given by

$$\begin{aligned} T(\omega_1,\omega_2) &= \sum_{n=-N}^{N} w(n)e^{-j\omega n}\Big|_{\cos(\omega)=0.5\cos(\omega_1)+0.5\cos(\omega_2)+0.5\cos(\omega_1)\cos(\omega_2)-0.5} \\ &= w(0)+\sum_{n=1}^{N} w(n)\cos(n\omega)\Big|_{\cos(\omega)=0.5\cos(\omega_1)+0.5\cos(\omega_2)+0.5\cos(\omega_1)\cos(\omega_2)-0.5} \\ &= \sum_{n=0}^{N} b(n)\cos^n(\omega)\Big|_{\cos(\omega)=0.5\cos(\omega_1)+0.5\cos(\omega_2)+0.5\cos(\omega_1)\cos(\omega_2)-0.5} \end{aligned} \tag{87.13}$$

where $w(n)$ is an arbitrary symmetric 1-D window of duration $2N+1$ centered at the origin, and the coefficients $b(n)$ are obtained from $w(n)$ via Chebyshev polynomials [31]. After some algebraic manipulations it can be shown that

$$T(\omega_1,\omega_2) = \sum_{n_1=-N}^{N}\sum_{n_2=-N}^{N} w_t(n_1,n_2)e^{-j(n_1\omega_1+n_2\omega_2)} \tag{87.14}$$

where $w_t(n_1, n_2)$ is a zero-phase 2-D window of size $(2N+1)\times(2N+1)$ obtained by using the McClellan transformation.

The construction of 2-D windows using the above three methods is now examined. In the case of windows obtained by the separable and the McClellan transformation approaches, the 1-D prototype is a Hamming window,

$$w_h(n) = \begin{cases} 0.54+0.46\cos(\pi n/N) & |n|<N \\ 0 & \text{otherwise} \end{cases} \tag{87.15}$$

In the second case $w_c(n_1, n_2) = 0.54 + 0.46\cos(\pi\sqrt{n_1^2+n_2^2}/N)$. By selecting $w_1(n) = w_2(n) = w_h(n)$ in (87.11), we get a 2-D window $w_r(n_1, n_2)$ of support $I = \{|n_1| < N, |n_2| < N\}$, which is a square-shaped symmetric region centered at the origin. For $N = 6$ the region of support, I contains $11 \times 11 = 121$ points. Figure 87.3(a) shows the frequency response of this window. A second window is designed by using (87.12), i.e., $w_c(n_1, n_2) = w_h(\sqrt{n_1^2+n_2^2})$. For $N = 6$, the frequency response of this filter is shown in Fig. 87.3(b). The region of support is almost circular and it contains 113 points. As can be seen from these examples, the 2-D windows may not behave as well as 1-D windows. Speake and Mersereau [44] compared these two methods and observed that the mainlobe width and the highest attenuation level of the sidelobes of the 2-D windows differ from their 1-D prototypes.

Let us construct a 2-D window by the McClellan transformation with a 1-D Hamming window of order 13 ($n = 6$) as the prototype. The frequency response of the 2-D window

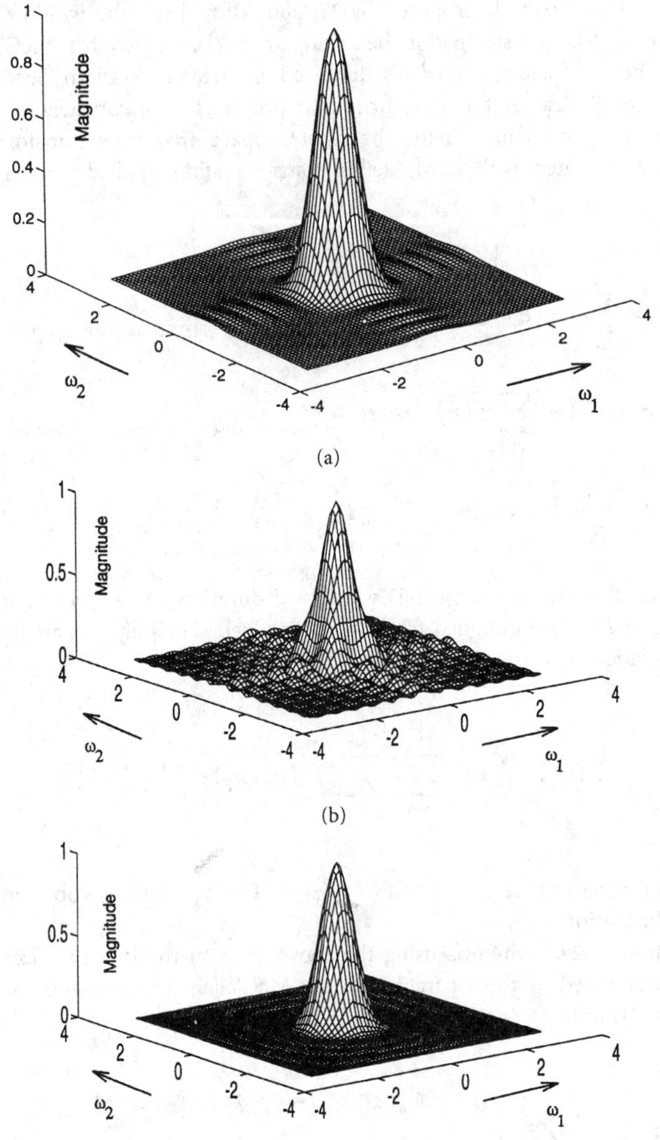

FIGURE 87.3 Frequency responses of the (a) separable, (b) Huang, and (c) McClellan 2-D windows generated from a Hamming window of order 13 ($N = 6$).

$w_t(n_1, n_2)$ is shown in Fig. 87.3(c). The frequency response of this window is almost circularly symmetric and it preserves the features of its 1-D prototype.

Consider the design of a circularly symmetric low-pass filter. The ideal frequency response for $(\omega_1, \omega_2) \in [-\pi, \pi] \times [-\pi, \pi]$ is given by

$$H_{id}(\omega_1, \omega_2) = \begin{cases} 1 & \sqrt{\omega_1^2 + \omega_2^2} \le \omega_c \\ 0 & \text{otherwise} \end{cases} \tag{87.16}$$

whose impulse response is given by

$$h_{id}(n_1, n_2) = \frac{\omega_c J_1\left(\omega_c\sqrt{n_1^2 + n_2^2}\right)}{2\pi\sqrt{n_1^2 + n_2^2}} \tag{87.17}$$

where $J_1(\cdot)$ is the first-order Bessel function of the first kind and ω_c is the cutoff frequency. The frequency response of the 2-D FIR filter obtained with a rectangular window of size $(2 \times 5 + 1)$ by $(2 \times 5 + 1)$ is shown in Fig. 87.4(a). Note the Gibbs-phenomenon-type ripples at the passband edges. In Fig. 87.4(b) the separable window of Fig. 87.3(a), derived from a Hamming window, is used to design the 2-D filter. Note that this 2-D filter has smaller ripples at the passband edges.

In windowing methods, it is often assumed that $H_{id}(\omega_1, \omega_2)$ is given. However, if the specifications are given as described in the first subsection in Section 87.2, then a proper $H_{id}(\omega_1, \omega_2)$ should be constructed.

The ideal magnitudes are either 1.0 (in passband) or 0.0 (in stopband). However, there is a need to define a *cutoff* boundary that lies within the transition band. This can be accomplished by using a suitable notion of "midway" cutoff between the transition boundaries. In practical cases where transition boundaries are given in terms of straight-line segments or smooth curves such as circles and ellipses, the construction of "midway" cutoff boundary is relatively straightforward. The ideal impulse response $h_{id}(n_1, n_2)$ is computed from the desired frequency response $H_{id}(\omega_1, \omega_2)$, either analytically (if possible), or by using the discrete Fourier transform (DFT). In the latter case the desired response $h_{id}(\omega_1, \omega_2)$ is first sampled on a rectangular grid in the Fourier domain, then an inverse DFT computation is carried out via a 2-D fast Fourier transform (FFT) algorithm to obtain an approximation to the sequence $h_{id}(n_1, n_2)$. The resulting sequence is an aliased version of the ideal impulse response. Therefore a sufficiently dense grid should be used in order to reduce the effects of aliasing.

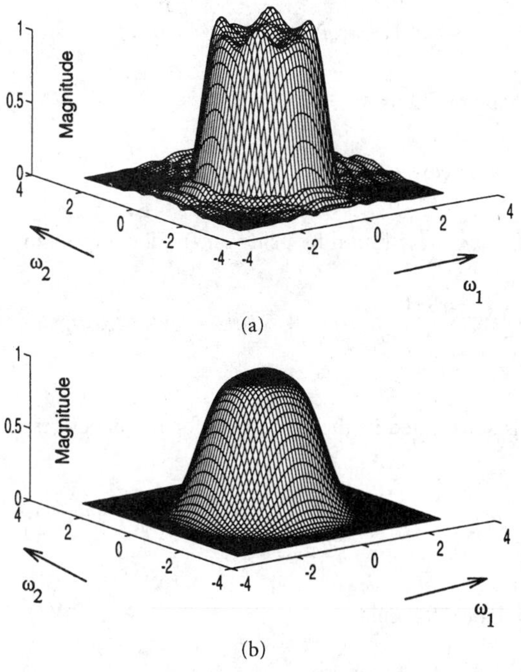

FIGURE 87.4 Frequency responses of the 2-D filters designed with (a) a rectangular window and (b) the separable window of Fig. 87.3(a).

In practice, several trials may be needed to design the final filter satisfying bounds both in the passbands and stopbands. The filter support is adjusted to obtain the smallest order to meet given requirements.

Filter design with windowing is a simple approach that is suitable for applications where a quick and suboptimal design is needed. Additional information on windowing can be found in [24], [44].

Frequency Sampling and Linear Programming-Based Method

This method is based on the application of the sampling theorem in the frequency domain. Consider the design of a 2-D filter with impulse response support of $N_1 \times N_2$ samples. The frequency response of the filter can be obtained from a conveniently chosen set of its samples on an $N_1 \times N_2$ grid. For example, the DFT of the impulse response can be used to interpolate the response for the entire region $[0, 2\pi] \times [0, 2\pi]$. The filter design then becomes a problem of choosing an appropriate set of DFT coefficients [19].

One choice of DFT coefficients consists of the ideal frequency response values, assuming a suitable cutoff. However, the resultant filters usually exhibit large magnitude deviations away from the DFT sample locations in the filter passbands and stopbands. The approximation error can be reduced by allowing the DFT values in the transition band to vary, and choosing them to minimize the deviation of the magnitude from the desired values. Another option is to allow all the DFT values to vary, and pick the optimal set of values for minimum error. The use of DFT-based interpolation allows for a computationally efficient implementation. The implementation cost of the method basically consists of a 2-D array product and inverse discrete Fourier transform (IDFT) computation, with appropriate addition.

Let us consider the set $S \subset Z^2$ that defines the equispaced frequency locations $(2k_1\pi/N_1, 2k_2\pi/N_2)$:

$$S = \{k_1 = 0, 1, \cdots, N_1 - 1;\ k_2 = 0, 1, \cdots, N_2 - 1\} \tag{87.18}$$

The DFT values can be expressed as

$$H_{\mathrm{DFT}}[k_1, k_2] = H(\omega_1, \omega_2)\big|_{(\omega_1,\omega_2)=(2k_1\pi/N_1, 2k_2\pi/N_2)}, \quad (k_1, k_2) \in S \tag{87.19}$$

The filter coefficients $h(n_1, n_2)$ are found by using an IDFT computation

$$h(n_1, n_2) = \frac{1}{N_1 N_2} \sum_{k_1=0}^{N_1-1} \sum_{k_2=0}^{N_2-1} H_{\mathrm{DFT}}[k_1, k_2] e^{j((2\pi/N_1)k_1 n_1 + (2\pi/N_2)k_2 n_2)} \qquad (n_1, n_2) \in S \tag{87.20}$$

If the above equation is substituted in the expression for frequency response

$$H(\omega_1, \omega_2) = \sum_{n_1=0}^{N_1-1} \sum_{n_2=0}^{N_2-1} h(n_1, n_2) e^{-j(\omega_1 n_1 + \omega_2 n_2)} \tag{87.21}$$

we arrive at the interpolation formula

$$H(\omega_1, \omega_2) = \sum_{k_1=0}^{N_1-1} \sum_{k_2=0}^{N_2-1} H_{\mathrm{DFT}}[k_1, k_2] A_{k_1 k_2}(\omega_1, \omega_2) \tag{87.22}$$

where

$$A_{k_1k_2}(\omega_1, \omega_2) = \frac{1}{N_1N_2}\left(\frac{1 - e^{-jN_1\omega_1}}{1 - e^{-j(\omega_1 - 2\pi k_1/N_1)}}\right)\left(\frac{1 - e^{-jN_2\omega_2}}{1 - e^{-j(\omega_2 - 2\pi k_1/N_2)}}\right) \tag{87.23}$$

Equation (87.22) serves as the basis of the frequency sampling design. As mentioned before, if the H_{DFT} are chosen directly according to the ideal response, then the magnitude deviations are usually large. To reduce the ripples, one option is to express the set S as the disjoint union of two sets S_t and S_c, where S_t contains indexes corresponding to the transition band F_t and S_c contains indexes corresponding to the "care"-bands, i.e., the union of the passbands and stopbands $F_p \cup F_s$. The expression for frequency response in (87.22) can be split into two summations, one over S_t and the other over S_c:

$$H(\omega_1, \omega_2) = \sum_{S_t} H_{\text{DFT}}[k_1, k_2]A_{k_1k_2}(\omega_1, \omega_2) + \sum_{S_c} H_{\text{DFT}}[k_1, k_2]A_{k_1k_2}(\omega_1, \omega_2) \tag{87.24}$$

where the first term on the right-hand side is optimized. The design equations can be put in the form

$$1 - \alpha\delta \leq H(\omega_1, \omega_2) \leq 1 + \alpha\delta \qquad (\omega_1, \omega_2) \in F_p \tag{87.25}$$

and

$$-\delta \leq H(\omega_1, \omega_2) \leq \delta \qquad (\omega_1, \omega_2) \in F_s \tag{87.26}$$

where δ is the peak approximation error in the stopband and $\alpha\delta$ is the peak approximation error in the passband, and where α is any positive constant defining the relative weights of the deviations. The problem is readily cast as a linear programming problem with a sufficiently dense grid of points.

For equiripple design, all the DFT values H_{DFT} over S_t and S_c are allowed to vary. An example of this design follows.

Example 1. The magnitude response for the approximation of a circularly symmetric response is shown in Fig. 87.5. Here the passband is the interior of the circle $R_1 = \pi/3$ and the stopband is the exterior of the circle $R_2 = 2\pi/3$. With $N_1 = N_2 = 9$, the passband ripple is 0.08 dB and the minimum stopband attenuation is 32.5 dB [19].

FIR Filters Optimal in L_p Norm

A criterion different from the minimax criterion is briefly examined. Let us define the error at the frequency pair (ω_1, ω_2) as follows:

$$E(\omega_1, \omega_2) = H(\omega_1, \omega_2) - H_{id}(\omega_1, \omega_2) \tag{87.27}$$

One design approach is to minimize the L_p norm of the error

$$\mathcal{E}_p = \left(\frac{1}{4\pi^2}\int_{-\pi}^{\pi}\int_{-\pi}^{\pi} |E(\omega_1, \omega_2)|^p \, d\omega_1 \, d\omega_2\right)^{1/p} \tag{87.28}$$

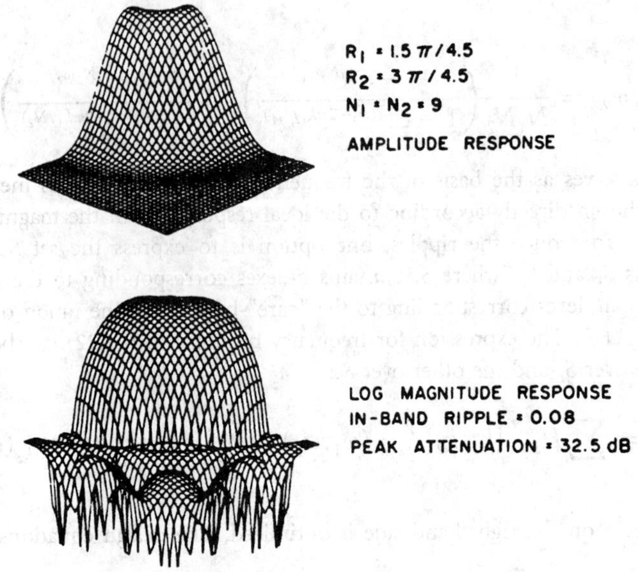

FIGURE 87.5 Frequency response of the circularly symmetric filter obtained by using the frequency sampling method. Figure adapted from [19] with IEEE permission.

Filter coefficients are selected by a suitable algorithm. For $p = 2$ Parseval's relation implies that

$$\mathscr{E}_2^2 = \sum_{n_1=-\infty}^{\infty} \sum_{n_2=-\infty}^{\infty} [h(n_1, n_2) - h_{id}(n_1, n_2)]^2 \tag{87.29}$$

By minimizing (87.29) with respect to the filter coefficients $h(n_1, n_2)$, which are nonzero only in a finite-extent region I, one gets

$$h(n_1, n_2) = \begin{cases} h_{id}(n_1, n_2) & (n_1, n_2) \in I \\ 0 & \text{otherwise} \end{cases} \tag{87.30}$$

which is the filter designed by using a straightforward rectangular window. Due to the Gibbs phenomenon it may have larger variations at the edges of passband and stopband regions. A suitable weighting function can be used to reduce the ripple [2], and an approximately equiripple solution can be obtained.

For the general case of $p \neq 2$ [30], the minimization of (87.28) is a nonlinear optimization problem. The integral in (87.28) is discretized and minimized by using an iterative nonlinear optimization technique. The solution for $p = 2$ is easy to obtain using linear equations. This serves as an excellent initial estimate for the coefficients in the case of larger values of p. As p increases, the solution becomes approximately equiripple. The error term $E(\omega_1, \omega_2)$ in (87.28) is nonuniformly weighted in passbands and stopbands, with larger weight given close to band edges, where deviations are typically larger.

Iterative Method for Approximate Minimax Design

We now consider a simple procedure based on alternating projections in the sample and frequency domains, which leads to an approximately equiripple response. In this method the zero-phase FIR filter design problem is formulated to alternately satisfy the frequency domain constraints on the magnitude response bounds and spatial domain constraints on the impulse response support [10], [11]. The algorithm is iterative and each iteration requires two 2-D FFT computations.

As pointed out in Section 87.2, 2-D FIR filter specifications are given as requirements on the magnitude response of the filter. It is desirable that the frequency response $H(\omega_1, \omega_2)$ of the zero-phase FIR filter be within prescribed upper and lower bounds in its passbands and stopbands. Let us specify bounds on the frequency response $H(\omega_1, \omega_2)$ of the minimax FIR filter $h(n_1, n_2)$ as follows:

$$H_{id}(\omega_1, \omega_2) - E_d(\omega_1, \omega_2) \leq H(\omega_1, \omega_2) \leq H_{id}(\omega_1, \omega_2) + E_d(\omega_1, \omega_2) \quad \omega_1, \omega_2 \in R \tag{87.31}$$

where $H_{id}(\omega_1, \omega_2)$ is the ideal filter response, $E_d(\omega_1, \omega_2)$ is a positive function of (ω_1, ω_2), which may take different values in different passbands and stopbands, and R is a region defined in (87.28) consisting of passbands and stopbands of the filter (note that $H(\omega_1, \omega_2)$ is real for a zero-phase filter). Usually, $E_d(\omega_1, \omega_2)$ is chosen constant in a passband or a stopband. Inequality (87.31) is the frequency domain constraint of the iterative filter design method.

In spatial domain the filter must have a finite-extent support I, which is a symmetric region around the origin. The spatial domain constraint requires that the filter coefficients must be equal to zero outside the region I.

The iterative method begins with an arbitrary finite-extent real sequence $h_0(n_1, n_2)$ that is symmetric $[h_0(n_1, n_2) = h_0(-n_1, -n_2)]$. Each iteration consists of making successive imposition of spatial and frequency domain constraints onto the current iterate. The kth iteration consists of the following steps:

- Compute the Fourier transform of the kth iterate $h_k(n_1, n_2)$ on a suitable grid of frequencies by using a 2-D FFT algorithm.
- Impose the frequency domain constraint as follows:

$$G_k(\omega_1, \omega_2) = \begin{cases} H_{id}(\omega_1, \omega_2) + E_d(\omega_1, \omega_2) & \text{if } H_k(\omega_1, \omega_2) > H_{id}(\omega_1, \omega_2) + E_d(\omega_1, \omega_2) \\ H_{id}(\omega_1, \omega_2) - E_d(\omega_1, \omega_2) & \text{if } H_k(\omega_1, \omega_2) < H_{id}(\omega_1, \omega_2) - E_d(\omega_1, \omega_2) \\ H_k(\omega_1, \omega_2) & \text{otherwise} \end{cases} \tag{87.32}$$

- Compute the inverse Fourier Transform of $G_k(\omega_1, \omega_2)$, and
- zero out $g_k(n_1, n_2)$ outside the region I to obtain h_{k+1}.

The flow diagram of this method is shown in Fig. 87.6. It can be proven that the algorithm converges for all symmetric input sequences. This method requires the specification of the bounds or equivalently, $E_d(\omega_1, \omega_2)$, and the filter support I. In 2-D filter design, filter-order estimates for prescribed frequency domain specifications are not available. Therefore successive reduction of bounds is used. If the specifications are too tight then the algorithm does not converge. In such cases one can either progressively enlarge the filter support region, or relax the bounds on the ideal frequency response.

The size of the 2-D FFT must be chosen sufficiently large. The passband and stopband edges are very important for the convergence of the algorithm. These edges must be represented accurately on the frequency grid of the FFT algorithm.

The shape of the filter support is very important in any 2-D filter design method. The support should be chosen to exploit the symmetries in the desired frequency response. For example, diamond-shaped supports show a clear advantage over the commonly assumed rectangular regions in designing diamond filters or 90° fan filters [4], [5].

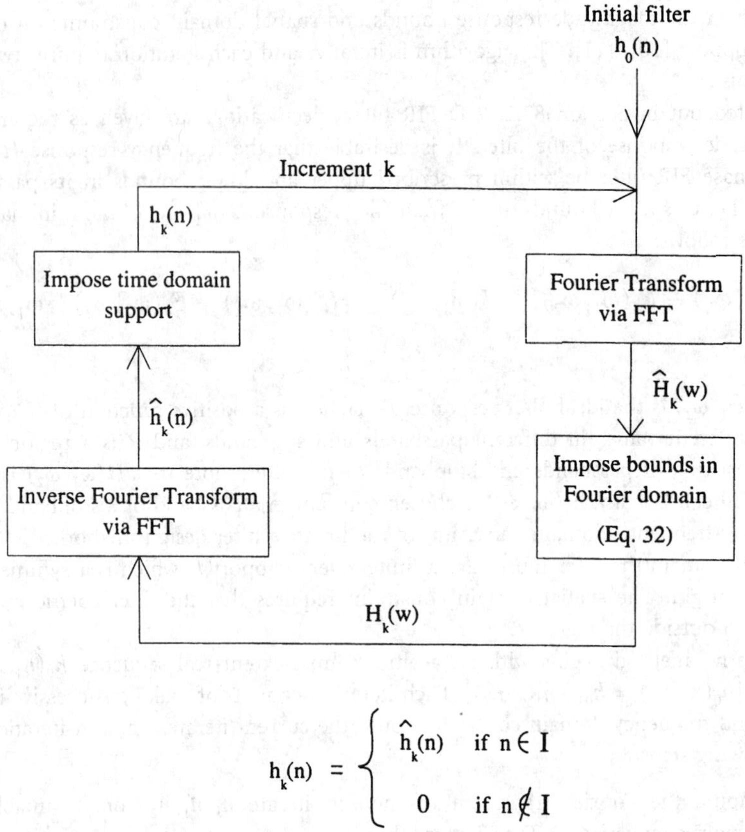

FIGURE 87.6 Flow diagram of the iterative filter design algorithm.

Since there are efficient FFT routines, 2-D FIR filters with large orders can be designed by using this procedure.

Example 2. Let us consider the design of a circularly symmetric low-pass filter. Maximum allowable deviation is $\delta_p = \delta_s = 0.05$ in both the passband and the stopband. The passband and stopband cutoff boundaries have a radius of 0.43π and 0.63π, respectively. This means that the functions $E_d(\omega_1, \omega_2) = 0.05$ in the passband and the stopband. In the transition band the frequency response is conveniently bounded by the lower bound of the stopband and the upper bound of the passband. The filter support is a square-shaped 17×17 region. The frequency response of this filter is shown in Fig. 87.7.

Example 3. Let us now consider an example in which we observe the importance of filter support. We design a fan filter whose specifications are shown in Fig. 87.8. Maximum allowable deviation is $\delta_p = \delta_s = 0.1$ in both the passband and the stopband. If one uses a 7×7 square-shaped support that has 49 points, then it cannot meet the design specifications. However, a diamond-shaped support

$$I_d = \{-5 \le n_1 + n_2 \le 5\} \cap \{-5 \le n_1 - n_2 \le 5\} \tag{87.33}$$

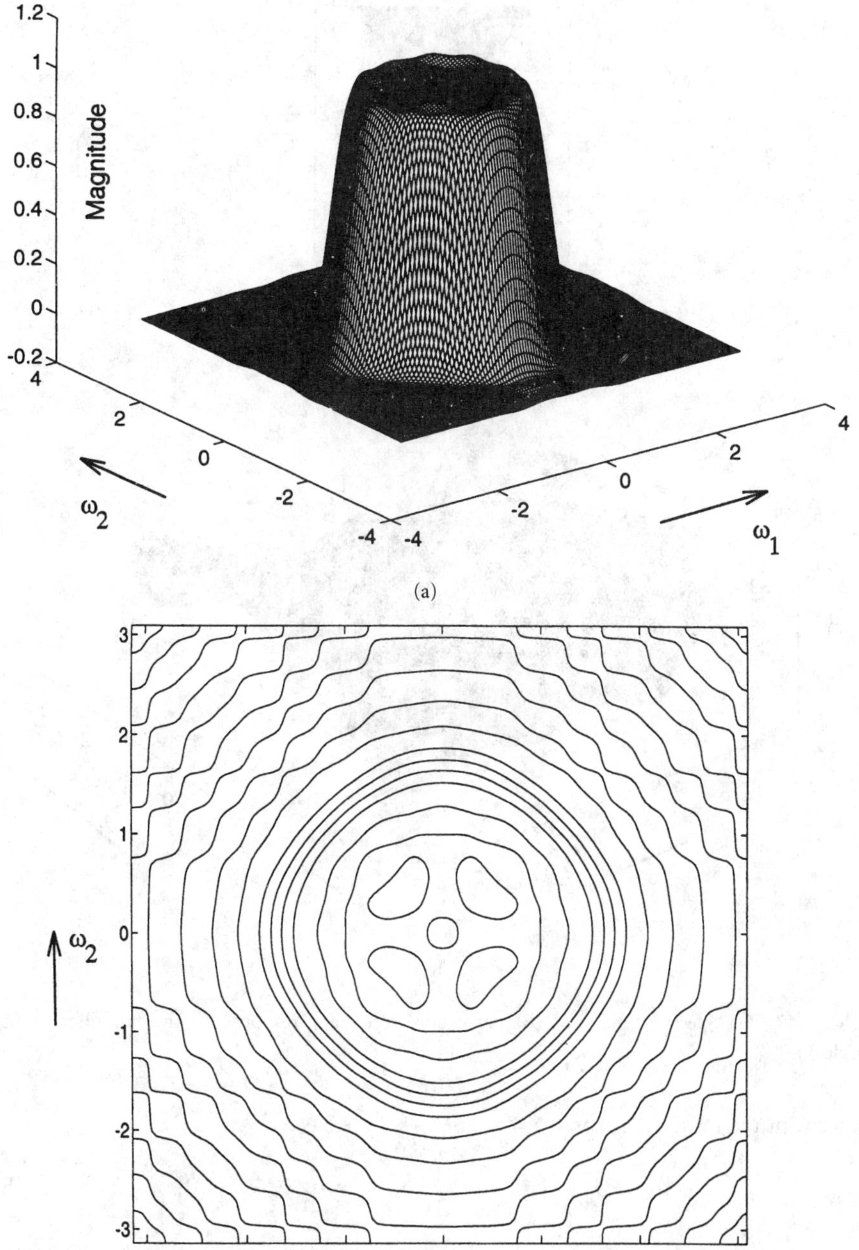

FIGURE 87.7 (a) Frequency response and (b) contour plot of the low-pass filter of Example 2.

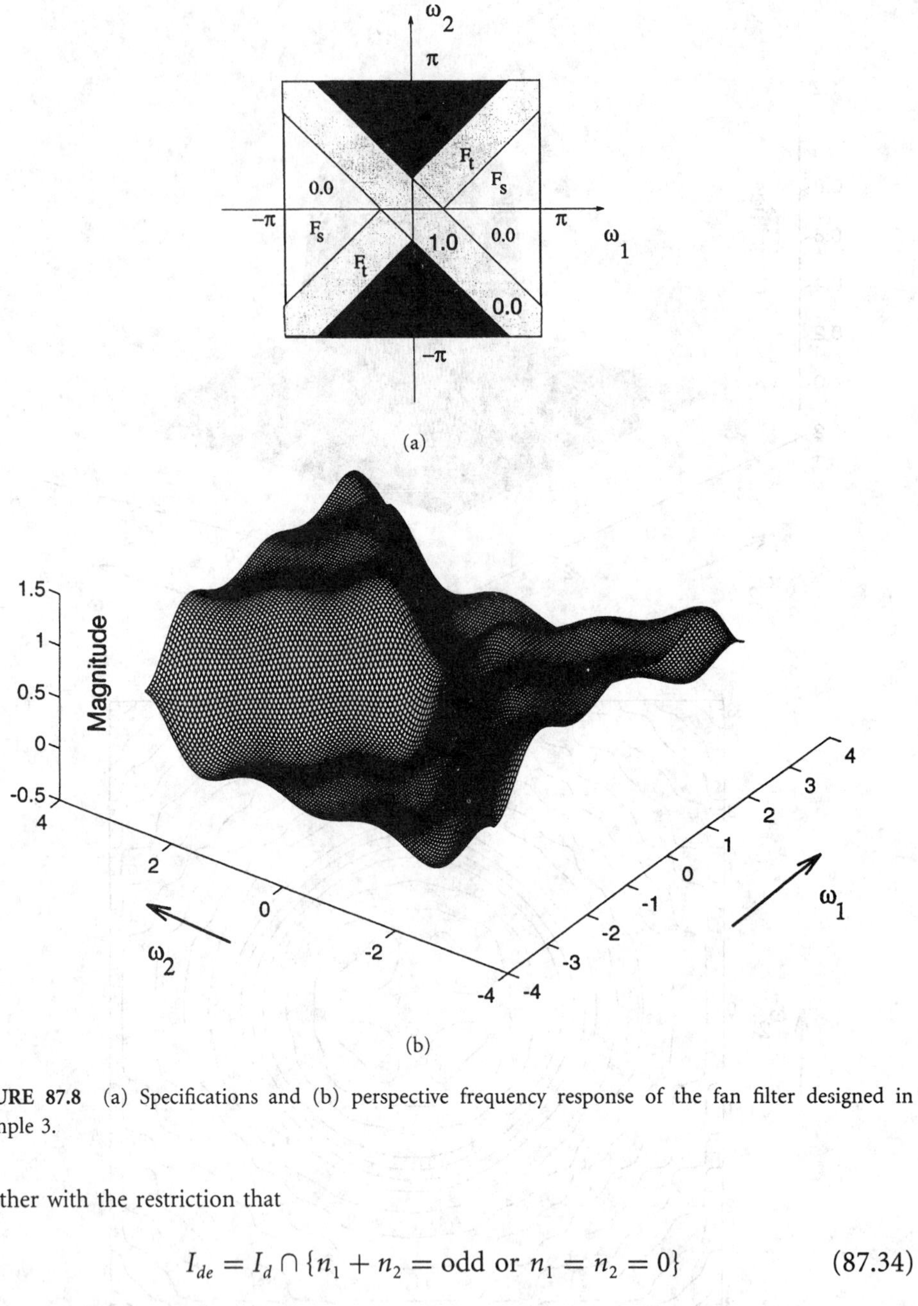

FIGURE 87.8 (a) Specifications and (b) perspective frequency response of the fan filter designed in Example 3.

together with the restriction that

$$I_{de} = I_d \cap \{n_1 + n_2 = \text{odd or } n_1 = n_2 = 0\} \tag{87.34}$$

produces a filter satisfying the bounds. The filter support region I_{de} contains 37 points. The resultant frequency response is shown in Fig. 87.8.

87.4 Special Design Procedure for Restricted Classes

Many cases of practical importance typically require filters belonging to restricted classes. The stopbands and passbands of these filters are often defined by straight-line, circular, or elliptical boundaries. In these cases, specialized procedures lead to efficient design and low-cost implementation. The filters in these cases are derived from 1-D prototypes.

Separable 2-D FIR Filter Design

The design of 2-D FIR filters composed of 1-D building blocks is briefly discussed. In cases where the specifications are given in terms of multiple passbands in the shapes of rectangles with sides parallel to the frequency axes, the design problem can be decomposed into multiple designs. The resulting filter is a parallel connection of component filters that are themselves separable filters. The separable structure was encountered earlier in the construction of 2-D windows from 1-D windows in the first subsection in Section 87.3. The design approach is essentially the same. We will confine the discussion to the case of cascade structures, which is a simple and very important practical case.

The frequency response of the 2-D separable FIR filter is expressed as

$$H(\omega_1, \omega_2) = H_1(\omega_1)H_2(\omega_2) \tag{87.35}$$

where $H_1(\omega)$ and $H_2(\omega)$ are frequency responses of two 1-D zero-phase FIR filters of durations N_1 and N_2, respectively. The corresponding 2-D filter is also a zero-phase FIR filter with $N_1 \times N_2$ coefficients, and its impulse response is given by

$$h(n_1, n_2) = h_1(n_1)h_2(n_2) \tag{87.36}$$

where $h_1(n)$ and the $h_2(n)$ are the impulse responses of the 1-D FIR filters.

If the ideal frequency response can be expressed in a separable cascade form as in (87.35), then the design problem is reduced to the case of appropriate 1-D filter designs. A simple but important example is the design of a 2-D low-pass filter with a symmetric square-shaped passband PB $= \{(\omega_1, \omega_2)\colon\ |\omega_1| < \omega_c, |\omega_2| < \omega_c\}$. Such a low-pass filter can be designed from a single 1-D FIR filter with a cutoff frequency of ω_c by using (87.9). A low-pass filter constructed in this way is shown in Fig. 87.2(c). The frequency response of this 2-D filter, whose 1-D prototypes are seventh-order Lagrange filters is shown in Fig. 87.9.

This design method is also used in designing 2-D filter banks, which are utilized in subband coding of images and video signals [47], [49], [50]. The four filters $H_{i,j}(\omega_1, \omega_2)$, $i, j = 0, 1$, of a

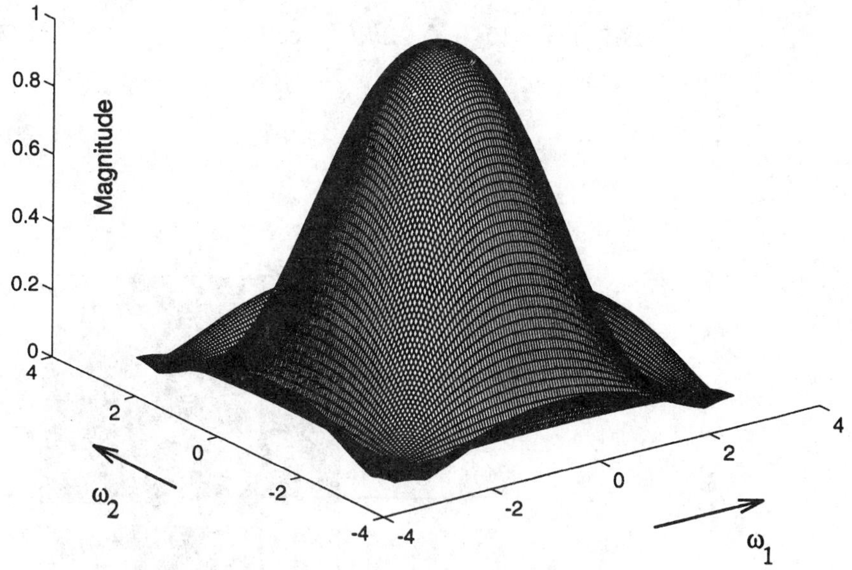

FIGURE 87.9 Frequency response of the separable low-pass filter $H(\omega_1, \omega_2) = H_1(\omega_1)H_1(\omega_2)$, where $H_1(\omega)$ is a seventh-order Lagrange filter.

separable 2-D filter bank are constructed from two 1-D filters as follows:

$$H_{i,j}(\omega_1, \omega_2) = H_i(\omega_1)H_j(\omega_2), \quad i, j = 0, 1 \tag{87.37}$$

where the 1-D filters $H_0(\omega_1)$ and $H_1(\omega_2)$ are low-pass (with a cut-off frequency at $\pi/2$) and high-pass filters of a 1-D subband filter bank, respectively [49]. Any 1-D filter bank described in Chapter 85 can be used in (87.37) to design 2-D filter banks. In this way the 2-D frequency domain is divided into four regions of equal area in a rectangular way. The ideal passband regions of the filters $H_{i,j}(\omega_1, \omega_2)$, are shown in Fig. 87.10.

Frequency Transformation Method

In this method a 2-D zero-phase FIR filter is designed from a 1-D zero-phase filter by a clever substitution of variables. The design procedure was first proposed by McClellan [31] and the frequency transformation is usually called the McClellan transformation [13], [33], [35], [36].

Let $H_1(\omega)$ be the frequency response of a 1-D zero-phase filter with $2N + 1$ coefficients. The key idea of this method is to find a suitable transformation $\omega = G(\omega_1, \omega_2)$ such that the 2-D frequency response $H(\omega_1, \omega_2)$, which is given by

$$H(\omega_1, \omega_2) = H_1(\omega)\big|_{\omega=G(\omega_1,\omega_2)} \tag{87.38}$$

approximates the desired frequency response $H_{id}(\omega_1, \omega_2)$.

Since the 1-D filter is a zero-phase filter, its frequency response is real, and it can be written as follows:

$$H_1(\omega) = h_1(0) + \sum_{n=1}^{N} 2h_1(n)\cos(\omega n) \tag{87.39}$$

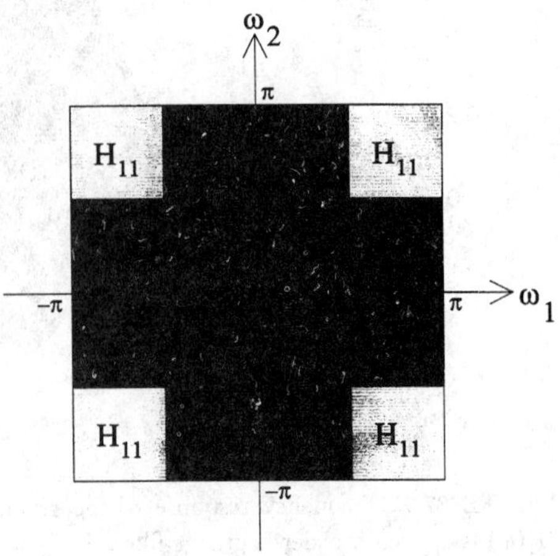

FIGURE 87.10 Ideal passband regions of the separable filters of a rectangular filter bank.

where the term $\cos(\omega n)$ can be expressed as a function of $\cos(\omega)$ by using the nth Chebyshev polynomial T_n,[2] i.e.,

$$\cos(\omega n) = T_n[\cos(\omega)]. \tag{87.40}$$

Using (87.40), the 1-D frequency response can be written as

$$H_1(\omega) = \sum_{n=0}^{N} 2b(n)[\cos(\omega)]^n \tag{87.41}$$

where the coefficients $b(n)$ are related to the filter coefficients $h(n)$.

In this design method the key step is to substitute a transformation function $F(\omega_1, \omega_2)$ for $\cos(\omega)$ in (87.41). In other words the 2-D frequency response $H(\omega_1, \omega_2)$ is obtained as follows:

$$\begin{aligned} H(\omega_1, \omega_2) &= H_1(\omega)\big|_{\cos(\omega)=F(\omega_1,\omega_2)} \\ &= \sum_{n=0}^{N} 2b(n)[F(\omega_1, \omega_2)]^n \end{aligned} \tag{87.42}$$

The function $F(\omega_1, \omega_2)$ is called the McClellan transformation.

The frequency response $H(\omega_1, \omega_2)$ of the 2-D FIR filter is determined by two free functions, the 1-D prototype frequency response $H_1(\omega)$ and the transformation $F(\omega_1, \omega_2)$. In order to have $H(\omega_1, \omega_2)$ be the frequency response of an FIR filter, the transformation $F(\omega_1, \omega_2)$, must itself be the response of a 2-D FIR filter. McClellan proposed $F(\omega_1, \omega_2)$ to be the frequency response of a 3×3 zero-phase filter in [31]. In this case the transformation $F(\omega_1, \omega_2)$, can be written as follows:

$$F(\omega_1, \omega_2) = A + B\cos(\omega_1) + C\cos(\omega_2) + D\cos(\omega_1 - \omega_2) + E\cos(\omega_1 + \omega_2) \tag{87.43}$$

where the real parameters A, B, C, D, and E, are related to the coefficients of the 3×3 zero-phase FIR filter. For $A = -1/2$, $B = C = 1/2$, $D = E = 1/4$, the contour plot of the transformation $F(\omega_1, \omega_2)$ is shown in Fig. 87.11. Note that in this case the contours are approximately circularly symmetric around the origin. It can be seen that the deviation from the circularity, expressed as a fraction of the radius, decreases with the radius. In other words, the distortion from a circular response is larger for large radii. It is observed from Fig. 87.11 that, with the above choice of parameters, A, B, C, D, and E, the transformation is bounded ($|F(\omega_1, \omega_2)| \leq 1$), which implies that $H(\omega_1, \omega_2)$ can take only the values that are taken by the 1-D prototype filter $H_1(\omega)$. Since $|\cos(\omega)| \leq 1$, the transformation $F(\omega_1, \omega_2)$, which replaces $\cos(\omega)$ in (87.42) must also take values between 1 and -1. If a particular transformation does not obey these bounds, then it can be scaled such that the scaled transformation satisfies the bounds.

If the transformation $F(\omega_1, \omega_2)$ is real [it is real in (87.43)] then the 2-D filter $H(\omega_1, \omega_2)$ will also be real or in other words it will be a zero-phase filter. Furthermore, it can be shown that the 2-D filter $H(\omega_1, \omega_2)$, is an FIR filter with a support containing $(2M_1N + 1) \times (2M_2N + 1)$ coefficients, if the transformation $F(\omega_1, \omega_2)$ is an FIR filter with $(2M_1+1)\times(2M_2+1)$ coefficients, and the order of the 1-D prototype filter is $2N + 1$. In (87.43) $M_1 = M_2 = 1$. As it can be intuitively guessed, one can design a 2-D approximately circularly symmetric low-pass (high-pass) [bandpass] filter with the above McClellan transformation by choosing the 1-D prototype filter $H_1(\omega)$, a low-pass (high-pass) [bandpass] filter.

[2]Chebyshev polynomials are recursively defined as follows: $T_0(x) = 1$, $T_1(x) = x$ and $T_n(x) = 2xT_{n-1}(x) - T_{n-2}(x)$.

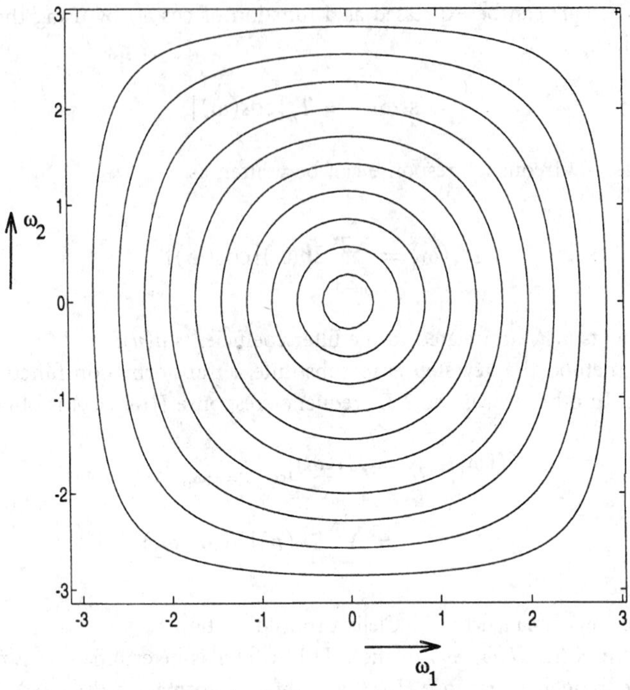

FIGURE 87.11 Contour plot of the McClellan transformation $F(\omega_1, \omega_2) = 0.5\cos(\omega_1) + 0.5\cos(\omega_2) + 0.5\cos(\omega_1)\cos(\omega_2) - 0.5$.

We will present some examples to demonstrate the effectiveness of the McClellan transformation.

Example 4. 2-D Window Design by Transformations [51]: In this example we design 2-D windows by using the McClellan transformation. Actually, we briefly mentioned this technique in the first subsection in Section 87.3. The 1-D prototype filter is chosen as an arbitrary 1-D symmetric window centered at the origin. Let $w(n)$ be the 1-D window of size $2N + 1$, and $W(\omega) = \sum_{n=-N}^{N} w(n)\exp(-j\omega n)$ be its frequency response. The transformation $F(\omega_1, \omega_2)$ is chosen as in (87.43) with the parameters $A = -1/2$, $B = C = 1/2$, $D = E = 1/4$, of Fig. 87.11. This transformation $F(\omega_1, \omega_2)$ can be shown to be equal to

$$F(\omega_1, \omega_2) = 0.5\,\cos(\omega_1) + 0.5\,\cos(\omega_2) + 0.5\,\cos(\omega_1)\cos(\omega_2) - 0.5 \tag{87.44}$$

The frequency response of the McClellan window $H_t(\omega_1, \omega_2)$ is given by

$$H_t(\omega_1, \omega_2) = W(\omega)\big|_{\cos(\omega)=F(\omega_1,\omega_2)} \tag{87.45}$$

The resultant 2-D zero-phase window $w_t(n_1, n_2)$ is centered at the origin and of size $(2N + 1) \times (2N + 1)$ because $M_1 = M_2 = 1$. The window coefficients can be either computed by using the inverse Chebyshev relation,[3] or by computing the inverse Fourier Transform of (87.45). The frequency response of a 2-D window constructed from a 1-D Hamming window of order 13 is shown in Fig. 87.3(c). The size of the window is 13×13.

[3] $1 = T_0, (x)$, $x = T_1(x) - T_0(x)$, $x^2 = 1/2[T_0(x) + T_2(x)]$, $x^3 = 1/4[3T_1(x) + T_3(x)]$, etc.

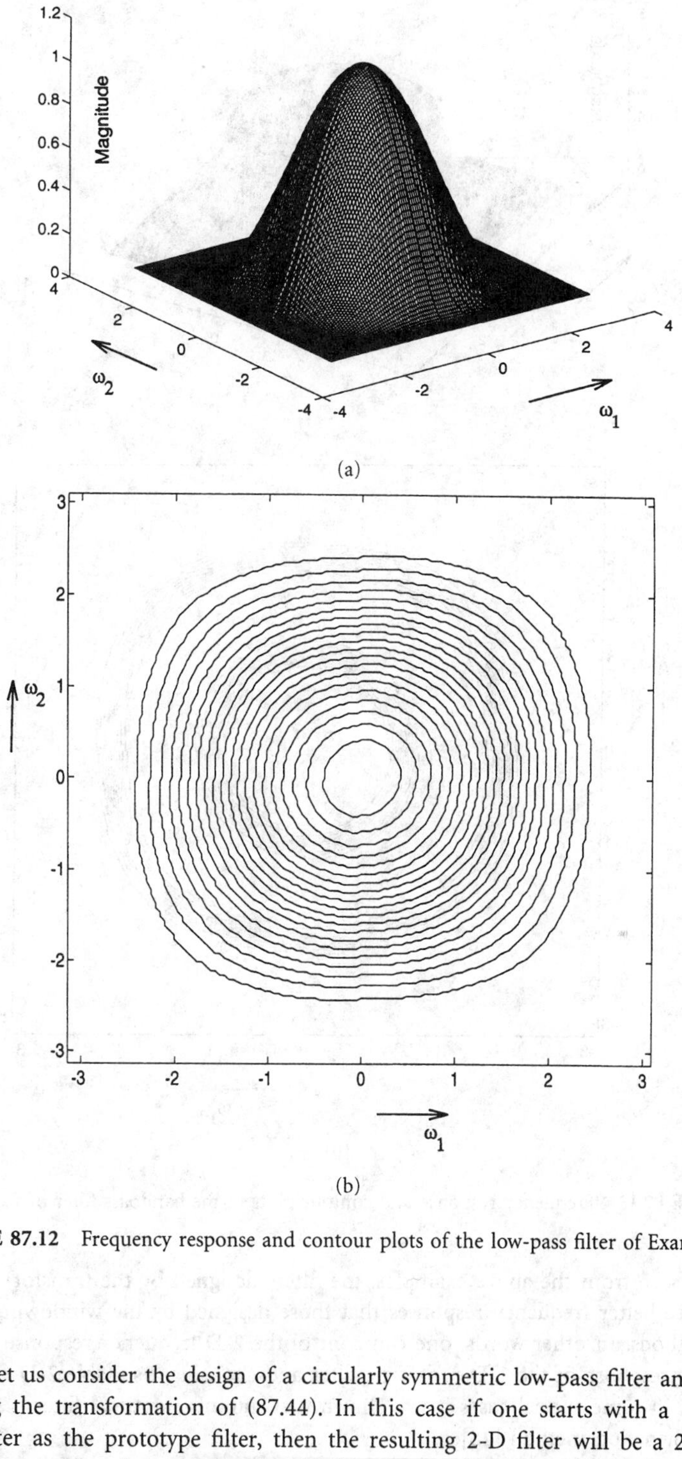

FIGURE 87.12 Frequency response and contour plots of the low-pass filter of Example 5.

Example 5. Let us consider the design of a circularly symmetric low-pass filter and a bandpass filter by using the transformation of (87.44). In this case if one starts with a 1-D low-pass (bandpass) filter as the prototype filter, then the resulting 2-D filter will be a 2-D circularly symmetric low-pass (bandpass) filter due to the almost circularly symmetric nature of the transformation. In this example the Lagrange filter of the seventh-order used in Section 87.2 is used as the prototype. The prototype 1-D bandpass filter of the fifteenth-order is designed by using the Parks–McClellan algorithm [39]. The frequency response and contour plots of the low-pass and bandpass filter are shown in Figs. 87.12 and 87.13, respectively.

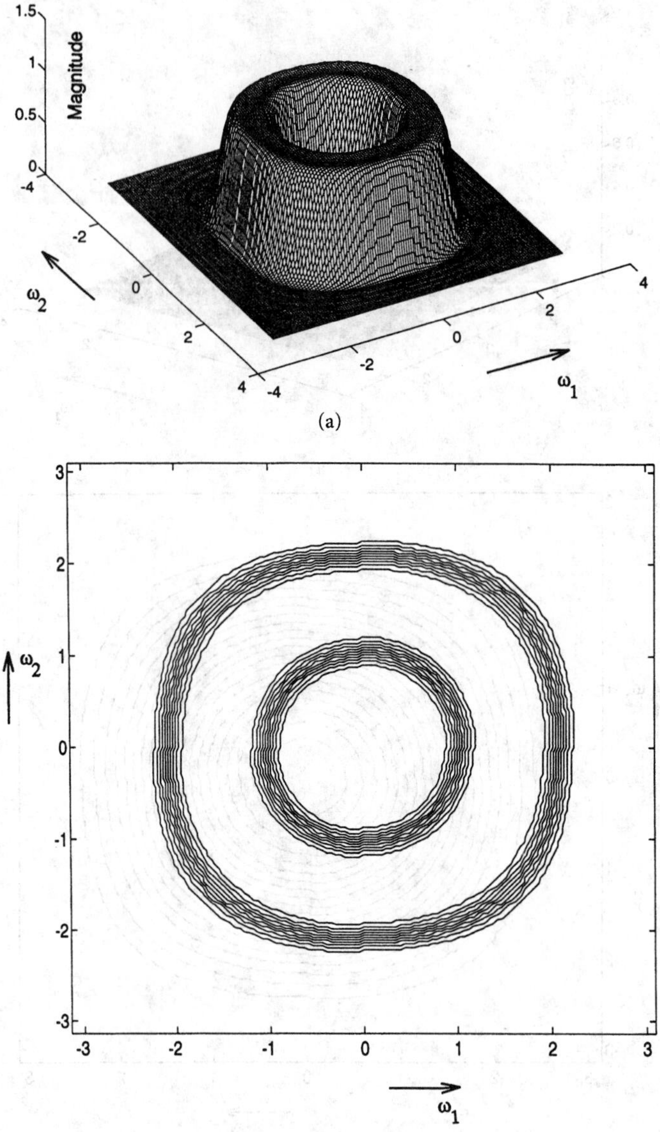

FIGURE 87.13 Frequency response and contour plots of the bandpass filter of Example 5.

As can be seen from the above examples, the filters designed by the transformation method appear to have better frequency responses that those designed by the windowing or frequency sampling methods. In other words, one can control the 2-D frequency response by controlling the frequency response of the 1-D prototype filter and choosing a suitable 2-D transformation. Furthermore, in some special cases it was shown that minimax optimal filters can be designed by the transformation method [18].

We have considered specific cases of the special transformations given by (87.43). By varying the parameters in (87.43) or expanding the transformation to include additional terms, a wider class of contours can be approximated. Ideally the frequency transformation approach requires the simultaneous optimal selection of the transformation $F(\omega_1, \omega_2)$ and the 1-D prototype filter $H_1(\omega)$ to approximate a desired 2-D frequency response. This can be posed as a nonlinear optimization problem. However, a suboptimal two-stage design by separately choosing $F(\omega_1, \omega_2)$ and $H_1(\omega)$ works well in practice. The transformation $F(\omega_1, \omega_2)$ should approximate 1 (−1) in the passband (stopband) of the desired filter. The contour produced by the transformation corresponding to the 1-D passband (stopband) edge frequency ω_p (ω_s), should ideally map to the given passband (stopband) boundary in the 2-D specifications. However, this cannot be achieved in general given the small number of variable parameters in the transformation. The parameters are therefore selected to minimize a suitable norm of the error between actual and ideal (constant) values of the transformation over the boundaries.

Various transformations and design considerations are described in [35], [36], [38], and [41]. The use of this transformation in exact reconstruction filter bank design was proposed in [6].

Filters designed by the transformation method can be implemented in a computationally efficient manner [13], [28]. The key idea is to implement (87.42) instead of implementing the filter by using the direct convolution sum. Implementing the transformation $F(\omega_1, \omega_2)$, which is an FIR filter of low-order, in a modular structure realizing (87.42) is more advantageous than ordinary convolution sum [13], [32].

In the case of circular passband design, it was observed that for a low-order transformation, the transformation contours exhibit large deviations from circularity. A simple artifice to overcome this problem in approximating wideband responses is to use decimation of a 2-D narrowband filter impulse response [17]. The solution consists of transforming the specifications to an appropriate narrowband design, where the deviation from circularity is smaller. The narrow passband can be expanded by decimation while essentially preserving the circularity of the passband.

Design Using Nonrectangular Transformations and Sampling Rate Conversions

In some filter specifications the desired responses are characterized by ideal frequency responses in which passbands and stopbands are separated by straight-line boundaries that are not necessarily parallel to the frequency axes. Examples of these are the various kinds of fan filters [4], [14], [16], [25] and diamond-shaped filters [5], [46]. Other shapes with straight-line boundaries are also approximated [7], [8], [12], [26], [27], [48]. Several design methods applicable in such cases have been developed and these methods are usually based on transformations related to concepts of sampling rate conversions. Often, alternate frequency domain interpretations are used to explain the design manipulations. A detailed treatment of these methods is beyond the scope of this chapter. However, some key ideas are described, and a specific case of a diamond filter is used to illustrate the methods. The importance of these design methods stems from the implementation efficiency that results from a generalized notion of separable processing.

In the family of methods considered here, manipulations of a separable 2-D response using a combination of several steps is carried out. In the general case of designing filters with straight-line boundaries, it is difficult to describe a systematic procedure. However, in a given design problem, an appropriate set of steps in the design is suggested by the nature of the desired response.

Some underlying ideas can be understood by examining the problem of obtaining a filter with parallelogram-shaped passband region. The sides of the parallelogram are assumed to be tilted with respect to the frequency axes. One approach to solving this problem is to perform the following series of manipulations on a separable prototype filter with a rectangular passband. The prototype filter impulse response is upsampled on a *nonrectangular* grid. The upsampling is done by an integer factor greater than one and it is defined by a nondiagonal nonsingular integer matrix [37]. The upsampling produces a parallelogram by a rotation and compression of the frequency response of the prototype filter together with a change in the periodicity. The matrix elements are chosen to produce the desired orientation in the resultant response. Depending on the desired response, cascading to eliminate unwanted portions of the passband in the frequency response, along with possible shifts and additions, may be used. The nonrectangular upsampling is then followed by a rectangular decimation of the sequence to expand the passband out to the desired size. In some cases the operations of the upsampling transformation and decimation can be combined by the use of nonrectangular decimation of impulse response samples. Results using such procedures produce efficient filter structures that are implemented with essentially 1-D techniques but where the orientations of processing are not parallel to the sample coordinates.

Consider the case of a diamond filter design shown in Fig. 87.14. Note that the filter in Fig. 87.14 can be obtained from the filter in Fig. 87.15 by a transformation of variables. If $F_a(z_1, z_2)$ is the transfer function of the filter approximating the response in Fig. 87.15(a), then the diamond filter transfer function $D(z_1, z_2)$ given by

$$D(z_1, z_2) = F_a(z_1^{1/2} z_2^{1/2}, z_1^{-1/2} z_2^{1/2}) \qquad (87.46)$$

will approximate the response in Fig. 87.14(a). The response in Fig. 87.15(a) can be expressed as the sum of the two responses shown in Fig. 87.15(b) and (c). We observe that if $F_b(z_1, z_2)$ is the transfer function of the filter approximating the response in Fig. 87.15(b) then

$$F_c(z_1, z_2) = F_b(-z_1, -z_2) \qquad (87.47)$$

will approximate the response in Fig. 87.15(c). This is due to the fact that negating the arguments shifts the (periodic) frequency response of F_b by (π, π). The response in Fig. 87.15(b) can be expressed as the product of two ideal 1-D low-pass filters, one horizontal and one vertical, which have the response shown in Fig. 87.15(d). This 1-D frequency response can be approximated by a halfband filter. Such an approximation will produce a response in which the transition band straddles both sides of the cutoff frequency boundaries in Fig. 87.15(a). If we wish to constrain the transition band to lie within the boundaries of the diamond-shaped region in Fig. 87.14(a), then we should choose a 1-D filter whose stopband interval is $(\pi/2, \pi)$. Let $H(z)$ be the transfer function of the prototype 1-D low-pass filter approximating the response in Fig. 87.15(d) with a suitably chosen transition boundary. The transfer function $H(z)$ can be expressed as

$$H(z) = T_1(z^2) + zT_2(z^2) \qquad (87.48)$$

The transfer function F_a is given by

$$F_a(z_1, z_2) = H(z_1)H(z_2) + H(-z_1)H(-z_2) \qquad (87.49)$$

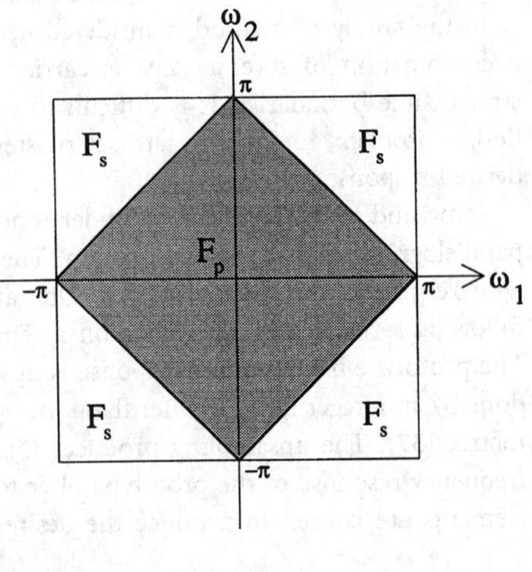

FIGURE 87.14 Ideal frequency response of a diamond filter.

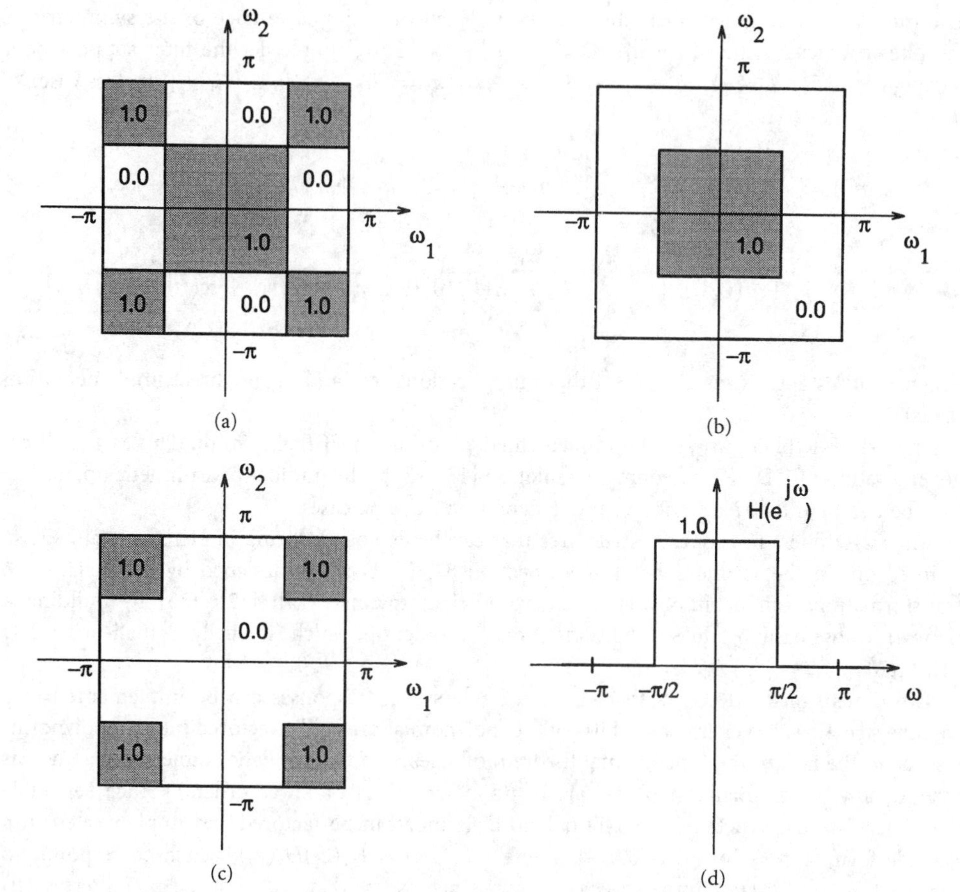

FIGURE 87.15 Ideal frequency responses of the filters: (a) $F_a(z_1, z_2)$, (b) $F_b(-z_1, -z_2)$, (c) $F_c(z_1, z_2)$, and (d) $H(z)$ in obtaining a diamond filter.

Combining (87.46), (87.48), and (87.49), we get

$$D(z_1, z_2) = 2T_1(z_1 z_2)T_1(z_1^{-1}z_2) + 2z_2 T_2(z_1 z_2)T_2(z_1^{-1}z_2) \tag{87.50}$$

As mentioned before, $H(z)$ can be chosen to be a halfband filter with

$$T_1(z^2) = 0.5 \tag{87.51}$$

The filter T_2 can be either FIR or IIR. It should be noted that the result can also be obtained as a nonrectangular downsampling, by a factor of 2, of the impulse response of the filter $F_b(-z_1, -z_2)$.

Another approach that utilizes multirate concepts is based on the novel idea of applying frequency masking in the two-dimensional case [29].

87.5 2-D FIR Filter Implementation

The straightforward way to implement 2-D FIR filters is to evaluate the convolution sum given in (87.1). Let us assume that the FIR filter has L nonzero coefficients in its region of support I. In order to get an output sample, L multiplications and L additions need to be performed.

The number of arithmetic operations can be reduced by taking advantage of the symmetry of the filter coefficients, that is, $h(n_1, n_2) = h(-n_1, -n_2)$. For example, let the filter support be a rectangular region $I = \{n_1 = -N_1, \cdots, 0, 1, \cdots, N_1, n_2 = -N_2, \cdots, 0, 1, \cdots, N_2\}$. In this case,

$$y(n_1, n_2) = \sum_{k_1=-N_1}^{N_1} \sum_{k_2=1}^{N_2} h(k_1, k_2)[x(n_1 - k_1, n_2 - k_2) + x(n_1 + k_1, n_2 + k_2)]$$
$$+ h(0,0)x(n_1, n_2) + \sum_{k_1=1}^{N_1} h(k_1, 0)[x(n_1 - k_1, n_2) + x(n_1 + k_1, n_2)], \tag{87.52}$$

which requires approximately half of the multiplications required in the direct implementation (87.1).

Any 2-D FIR filter can also be implemented by using an FFT algorithm. This is the direct generalization of 1-D FFT based implementation [13], [28]. The number of arithmetic operations may be less than the space domain implementation in some cases.

Some 2-D filters have special structures that can be exploited during implementation. As we pointed out in the second subsection in Section 87.4, 2-D filters designed by McClellan-type transformations can be implemented in an efficient manner [13], [32], [33] by building a network whose basic module is the transformation function, which is usually a small-order 2-D FIR filter.

Two-dimensional FIR filters that have separable system responses can be implemented in a cascade structure. In general, an arbitrary 2-D polynomial cannot be factored into subpolynomials due to the lack of the fundamental theorem of algebra in two or higher dimensions (whereas in 1-D, any polynomial can be factored into polynomials of lower orders). Since separable 2-D filters are constructed from 1-D polynomials they can be factored and implemented in a cascade form. Let us consider (87.35), where $H(z_1, z_2) = H_1(z_1)H_2(z_2)$, which corresponds to $h(n_1, n_2) = h_1(n_1)h_2(n_2)$ in the space domain. Let us assume that orders of the 1-D filters $h_1(n)$ and $h_2(n)$ are $2N_1 + 1$ and $2N_2 + 1$, respectively. In this case the 2-D filter $h(n_1, n_2)$ has the same rectangular support I as in (87.52). In this case

$$y(n_1, n_2) = \sum_{k_2=-N_2}^{N_2} h_2(k_2) \sum_{k_1=-N_1}^{N_1} h(k_1)x(n_1 - k_1, n_2 - k_2) \tag{87.53}$$

The 2-D filtering operation in (87.53) is equivalent to a two stage 1-D filtering in which the input image $x(n_1, n_2)$ is first filtered horizontally line by line by $h_1(n)$, then the resulting output is filtered vertically column by column by $h_2(n)$. In order to produce an output sample, the direct implementation requires $(2N_1 + 1) \times (2N_2 + 1)$ multiplications, whereas the separable implementation requires $(2N_1 + 1) + (2N_2 + 1)$ multiplications, which is computationally much more efficient than the direct form realization. This is achieved at the expense of additional memory space (separable implementation requires a buffer to store the results of first stage during the implementation). By taking advantage of the symmetric nature of h_1 and h_2, the number of multiplications can be further reduced.

Filter design methods that impose structural constraints like cascade, parallel, and other forms are proposed by several researchers [15], [45]. These filters can be efficiently implemented because of their special structures. Unfortunately, the design procedure requires nonlinear optimization techniques that may be very complicated.

With advances in VLSI technology, the implementation of 2-D FIR filters using high-speed digital signal processors is becoming increasingly common in complex image processing systems.

Acknowledgement

The authors would like to express their sincere thanks to Omer Nezih Gerek for the help he provided in preparing the figures for the chapter.

References

[1] A. Abo-Taleb and M. M. Fahmy, "Design of FIR two-dimensional digital filters by successive projections," *IEEE Trans. Circuits Syst.*, vol. CAS-31, pp. 801–805, 1984.

[2] V. R. Algazi, M. Suk, C.-S. Rim, "Design of almost minimax FIR filter in one and two dimensions by WLS techniques," *IEEE Trans. Circuits and Systems*, pp. 590–596, June 1986.

[3] S. A. H. Aly and M. M. Fahmy, "Symmetry in two-dimensional rectangularly sampled digital filters," *IEEE Trans. Acoust., Speech, Signal Processing*, vol. ASSP-29, pp. 794–805, 1981.

[4] R. Ansari, "Efficient IIR and FIR fan filters," *IEEE Trans. Circuits Syst.*, vol. CAS-34, pp. 941–945, Aug. 1987.

[5] R. Ansari, H. P. Gaggioni, and D. J. Le Gall, "HDTV coding using a nonrectangular subband decomposition," *Proc. SPIE*, vol. 1001, Conf. Visual Commun. Image Processing, Boston, MA, pp. 821–824, Nov. 1988.

[6] R. Ansari and C. Guillemot, "Exact reconstruction filter banks using diamond FIR filters," in *Proc. 1990 Bilkent Int. Conf. New Trends Commun. Control, Signal Processing*, E. Arikan, Ed. Holland: Elsevier, 1990, pp. 1412–1421.

[7] R. H. Bamberger and M. J. T. Smith, "Efficient 2-D analysis/synthesis filter banks for directional image component representation," in *Proc. IEEE Int. Symp. Circuits Syst.*, May 1990, pp. 2009–2012.

[8] ——, "A filter bank for the directional decomposition of images: Theory and design," *IEEE Trans. Acoust., Speech, Signal Processing*, Apr. 1992.

[9] C. Charalambous, "The performance of an algorithm for minimax design of two-dimensional linear-phase FIR digital filters," *IEEE Trans. Circuits Syst.*, vol. CAS-32, pp. 1016–1028, 1985.

[10] A. E. Cetin and R. Ansari, "An iterative procedure for designing two dimensional FIR filters," in *Proc. IEEE Int. Symp. Circuits Syst.* (ISCAS'87), 1987, pp. 1044–1047.

[11] ——, "Iterative procedure for designing two-dimensional FIR filters," *Electron. Lett.*, vol. 23, pp. 131–133, Jan. 1987.

[12] T. Chen and P. P. Vaidyanathan, "Multidimensional multirate filters and filter banks derived from one-dimensional filters," *IEEE Trans. Signal Processing*, vol. 41, pp. 1035–1047, Mar. 1993.

[13] D. Dudgeon and R. M. Mersereau, *Multidimensional Digital Signal Processing*, Englewood Cliffs, NJ: Prentice-Hall, 1984.

[14] P. Embree, J. P. Burg, and M. M. Backus, "Wideband velocity filtering—The pie slice process," *Geophys.*, vol. 28, pp. 948–974, 1963.

[15] O. D. Faugeras and J. F. Abramatic, "2-D FIR filter design from independent 'small' generating kernels using a mean square and Tchebyshev error criterion," in *Proc. IEEE Int. Conf. Acoust., Speech, Signal Processing*, 1979, pp. 1–4.

[16] A. P. Gerheim, "Synthesis procedure for 90° fan filters," *IEEE Trans. Circuits Syst.*, vol. CAS-30, pp. 858–864, Dec. 1983.

[17] C. Guillemot and R. Ansari, "Two-dimensional filters with wideband circularly symmetric frequency response," *IEEE Trans. Circuits and Systems II: Analog and Digital Signal Processing*, pp. 703–707, Oct. 1994.

[18] D. B. Harris and R. M. Mersereau, "A comparison of algorithms for minimax design of two-dimensional linear phase FIR digital filters," *IEEE Trans. Acoust., Speech, Signal Processing*, vol. ASSP-25, pp. 492–500, 1977.

[19] J. V. Hu and L. R. Rabiner, "Design techniques for two-dimensional digital filters," *IEEE Trans. Audio Electroacoust.*, vol. AU-20, 249–257, 1972.

[20] T. S. Huang, "Two-dimensional windows," *IEEE Trans. Audio Electroacoust.*, vol. AU-20, pp. 88–90, Jan. 1972.

[21] T. S. Huang, Ed., *Two-Dimensional Digital Signal Processing I: Linear Filters*, New York: Springer-Verlag, 1981.

[22] Y. Kamp and J. P. Thiran, "Maximally flat nonrecursive two-dimensional digital filters," *IEEE Trans. Circuits Syst.*, vol. CAS-21, pp. 437–449, May 1974.

[23] ——, "Chebyshev approximation for two-dimensional nonrecursive digital filters," *IEEE Trans. Circuits Syst.*, vol. CAS-22, pp. 208–218, 1975.

[24] H. Kato and T. Furukawa, "Two-dimensional type-preserving circular windows," *IEEE Trans. Acoust., Speech, Signal Processing*, vol. ASSP-29, pp. 926–928, 1981.

[25] A. H. Kayran and R. A. King, "Design of recursive and nonrecursive fan filters with complex transformations," *IEEE Trans. Circuits Syst.*, vol. CAS-30, pp. 849–857, Dec. 1983.

[26] C.-L. Lau and R. Ansari, "Two-dimensional digital filter and implementation based on generalized decimation," in *Proc. Princeton Conf.*, Princeton, NJ, Mar. 1986.

[27] ——, "Design of two-dimensional filters using sampling rate alteration," in *Proc. IEEE Int. Symp. Circuits Syst.*, pp. 474–477, 1984.

[28] J. S. Lim, *Two-Dimensional Signal and Image Processing*, Englewood Cliffs, NJ: Prentice-Hall, 1990.

[29] Y. C. Lim and Y. Lian, "The optimum design of one- and two-dimensional FIR filters using the frequency response masking technique," *IEEE Trans. Circuits Syst. II: Analog and Digital Signal Processing*, vol. 40, pp. 88–95, 1993.

[30] J. H. Lodge and M. M. Fahmy, "An efficient l_p optimization technique for the design of two-dimensional linear-phase FIR digital filters," *IEEE Trans. Acoust., Speech, Signal Processing*, vol. ASSP-28, pp. 308–313, 1980.

[31] J. H. McClellan, "The design of two-dimensional filters by transformations," in *Proc. 7th Annu. Princeton Conf. Informat. Sci. Syst.*, pp. 247–251, 1973.

[32] J. H. McClellan and D. S. K. Chan, "A 2-D FIR filter structure derived from the Chebyshev recursion," *IEEE Trans. Circuits Syst.*, vol. 24, pp. 372–378, 1977.

[33] W. F. G. Mecklenbrauker and R. M. Mersereau, "McClellan transformation for 2-D digital filtering: II—Implementation," *IEEE Trans. Circuits Syst.*, vol. CAS-23, pp. 414–422, 1976.

[34] R. M. Mersereau, D. B. Harris, and H. S. Hersey, "An efficient algorithm for the design of two-dimensional digital filters," in *Proc. 1974 Int. Symp. Circuits Syst.*, 1975, pp. 443–446.

[35] R. M. Mersereau, W. F. G. Mecklenbrauker, and T. F. Quatieri, Jr., "McClellan transformation for 2-D digital filtering: I—Design," *IEEE Trans. Circuits Syst.*, vol. CAS-23, pp. 405–414, 1976.

[36] R. M. Mersereau, "The design of arbitrary 2-D zero-phase FIR filters using transformations," *IEEE Trans. Circuits Syst.*, vol. 27, pp. 372–378, 1980.

[37] R. M. Mersereau and T. C. Speake, "The processing of periodically sampled multidimensional signals," *IEEE Trans. Acoust., Speech, Signal Processing*, vol. ASSP-31, pp. 188–194, Feb. 1983.

[38] D. T. Nguyen and M. N. S. Swamy, "Scaling free McClellan transfer for 2-D digital filters," *IEEE Trans. Circuits Syst.*, vol. CAS-33, pp. 108, 109, Jan. 1986.

[39] T. W. Parks and J. H. McClellan, "Chebyshev approximation for nonrecursive digital filters with linear phase," *IEEE Trans. Circuit Theory*, vol. 19, pp. 189–194, 1972.

[40] S.-C. Pei and J.-J. Shyu, "Design of 2-D FIR digital filters by McClellan transformation and least squares eigencontour mapping," *IEEE Trans. Circuits Syst. II: Analog and Digital Signal Processing*, vol. 40, pp. 546–555, 1993.

[41] E. Z. Psarakis and G. V. Moustakides, "Design of two-dimensional zero-phase FIR filters via the generalized McClellan transform," *IEEE Trans. Circuits Syst.*, pp. 1355–1363, vol. CAS-38, Nov. 1991.

[42] P. K. Rajan, H. C. Reddy, and M. N. S. Swamy, "Fourfold rational symmetry in two-dimensional FIR digital filters employing transformations with variable parameters," *IEEE Trans. Acoust., Speech, Signal Processing*, vol. ASSP-31, pp. 488–499, 1982.

[43] V. Rajaravivarma, P. K. Rajan, and H. C. Reddy, "Design of multidimensional FIR digital filters using the symmetrical decomposition technique," *IEEE Trans. Signal Processing*, vol. 42, pp. 164–174, Jan. 1994.

[44] T. Speake and R. M. Mersereau, "A comparison of different window formulas for two-dimensional FIR filter design," in *Proc. IEEE Int. Conf. Acoust., Speech, Signal Processing*, 1979, pp. 5–8.

[45] S. Treitel and J. L. Shanks, "The design of multistage separable planar filters," *IEEE Trans. Geosci. Electron.*, vol. GE-9, pp. 10–27, 1971.

[46] G. J. Tonge, "The sampling of television images," Experiment. Development Rep. 112/81, Independent Broadcasting Authority, 1981.

[47] M. Vetterli, "A theory of multirate filter banks," *Signal Processing*, vol. 6, pp. 97–112, 1984.

[48] E. Viscito and J. P. Allebach, "The analysis and design of multidimensional FIR perfect reconstruction filter banks for arbitrary sampling lattices," *IEEE Trans. Circuits Syst.*, vol. CAS-38, pp. 29–41, Jan. 1991.

[49] J. W. Woods and S. D. O'Neill, "Subband coding of images," *IEEE Trans. Acoust., Speech, Signal Processing*, vol. ASSP-34, pp. 1278–1288, 1986.

[50] J. W. Woods, Ed., *Subband Image Coding*, Kluwer Academic, Norwell, MA, 1990.

[51] T.-H. Yu and S. K. Mitra, "A new two-dimensional window," *IEEE Trans. Acoust., Speech, Signal Processing*, vol. ASSP-33, pp. 1058–1061, 1985.

Further Information

Most of the research papers describing the advances of 2-D FIR filter design methods appear in *IEEE Transactions on Signal Processing, IEEE Transactions on Image Processing, IEEE Transactions on Circuits and Systems, Electronics Letters*, and the *Proceedings of the International Conference on Acoustics, Speech, and Signal Processing* (ICASSP) and the *Proceedings of the International Symposium on Circuits and Systems.*

88

Two-Dimensional IIR Filters

A. G. Constantinides
Imperial College of Science, Technology, and Medicine

X. J. Xu
Imperial College of Science, Technology, and Medicine

88.1 Introduction

A linear 2-D IIR digital filter can be characterized by its transfer function

$$H(z_1, z_2) = \frac{N(z_1, z_2)}{D(z_1, z_2)} = \frac{\sum_{i=0}^{N_2}\sum_{j=0}^{M_2} a_{ij} z_1^{-i} z_2^{-j}}{\sum_{i=0}^{N_1}\sum_{j=0}^{M_1} b_{ij} z_1^{-i} z_2^{-j}} \tag{88.1}$$

where the sampling period $T_i = 2\pi/\omega_{si}$ for $i = 1, 2$ with ω_{si} and the sampling frequencies a_{ij} and b_{ij} are real numbers known as the coefficients of the filter. Without loss of generality we can assume $M_1 = M_2 = N_1 = N_2 = M$ and $T_1 = T_2 = T$. Designing a 2-D filter is to calculate the filter coefficients a_{ij} and b_{ij} in such a way that the amplitude response and/or the phase response (group delay) of the designed filter approximates to some ideal responses while maintaining the stability of the designed filter. The latter requires that

$$D(z_1, z_2) \neq 0 \quad \text{for } |z_i| \geq 1 \qquad i = 1, 2 \tag{88.2}$$

The amplitude response of the 2-D filter is expressed as

$$M(\omega_1, \omega_2) = |H(e^{j\omega_1 T}, e^{j\omega_2 T})| \tag{88.3}$$

0-8493-8341-2/95/$0.00+$.50

the phase response as

$$\phi(\omega_1, \omega_2) = \arg H(e^{j\omega_1 T}, e^{j\omega_2 T}) \tag{88.4}$$

and the two group delay functions as

$$\tau_i(\omega_1, \omega_2) = \frac{d\phi(\omega_1, \omega_2)}{d\omega_i} \qquad i = 1, 2 \tag{88.5}$$

Equation (88.1) is the general form of transfer functions of the nonseparable numerator and denominator 2-D IIR filters. It can involve two subclasses, namely, the separable product transfer function

$$\begin{aligned} H(z_1, z_2) &= H_1(z_1)H_2(z_2) \\ &= \frac{\sum_{i=0}^{N_2} a_{1i} z_1^{-i} \sum_{j=0}^{M_2} a_{2i} z_2^{-j}}{\sum_{i=0}^{N_1} b_{1i} z_1^{-i} \sum_{j=0}^{M_1} b_{2i} z_2^{-j}} \end{aligned} \tag{88.6}$$

and the separable denominator, nonseparable numerator transfer function given by

$$H(z_1, z_2) = \frac{\sum_{i=0}^{N_2} \sum_{j=0}^{M_2} a_{ij} z_1^{-i} z_2^{-j}}{\sum_{i=0}^{N_1} b_{1i} z_1^{-i} \sum_{j=0}^{M_1} b_{2j} z_2^{-j}} \tag{88.7}$$

The stability constraints for the above two transfer functions are the same as those for the individual two 1-D cases. These are easy to check and correspondingly the transfer function is easy to stabilize if the designed filter is found to be unstable. Therefore, in the design of the above two classes, in order to reduce the stability problem to that of the 1-D case, the denominator of the 2-D transfer function is chosen to have two 1-D polynomials in z_1 and z_2 variables in cascade. However, in the general formulation of nonseparable numerator and denominator filters, this oversimplification is removed. The filters of this type are generally designed either through transformation of 1-D filters, or through optimization approaches, as is discussed in the following.

88.2 Transformation Techniques

Analog Filter Transformations

In the design of 1-D analog filters, a group of analog filter transformations of the form $s = g(s')$ is usually applied to normalized continuous low-pass transfer functions like those obtained by using the Bessel, Butterworth, Chebyshev, and elliptic approximations. These transformations can be used to design low-pass, high-pass, bandpass, or bandstop filters satisfying piecewise-constant amplitude response specifications. Through the application of the bilinear transformation, corresponding 1-D digital filters can be designed, and since 2-D digital filters can be designed in terms of 1-D filters, these transformations are of considerable importance in the design of 2-D digital filters as well. In the 2-D cases, the transformations have a form of

$$s = g(s_1, s_2) \tag{88.8}$$

As a preamble, in this subsection two groups of transformations of interest in the design of 2-D digital filters are introduced, which essentially produce 2-D continuous transfer functions from 1-D ones.

Rotated Filter

The first group of transformations, suggested by Shanks, Treitel, and Justice [1], are of the form

$$g_1(s_1, s_2) = -s_1 \sin\beta + s_2 \cos\beta \tag{88.9a}$$

$$g_2(s_1, s_2) = s_1 \cos\beta + s_2 \sin\beta \tag{88.9b}$$

They map 1-D into 2-D filters with arbitrary directionality in a 2-D frequency response plane. These filters are called rotated filters because they are obtained by rotating 1-D filters.

If $H(s)$ is a 1-D continuous transfer function, then a corresponding 2-D continuous transfer function can be generated as

$$H_{D1}(s_1, s_2) = H(s)\big|_{s=g_1(s_1,s_2)} \tag{88.10a}$$

$$H_{D2}(s_1, s_2) = H(s)\big|_{s=g_2(s_1,s_2)} \tag{88.10b}$$

by replacing the s in $H(s)$ with $g_1(s_1, s_2)$ and $g_2(s_1, s_2)$, respectively.

It is easy to show [2] that a transformation of $g_1(s_1, s_2)$ or $g_2(s_1, s_2)$ will give rise to a contour in the amplitude response of the 2-D analog filter that is a straight line rotated by an angle β with respect to the s_1 or s_2 axis, respectively. Figure 88.1 illustrates an example of 1-D to 2-D analog transformation by (88.10a) for $\beta = 0°$ and $\beta = 45°$.

The rotated filters are of special importance in the design of circularly symmetric filters, as will be discussed in Section 88.4.

Transformation Using a Two-Variable Reactance Function

The second group of transformations is based on the use of a two-variable reactance function. One of the transformations was suggested by Ahmadi, Constantinides, and King [3], [4]. This is given by

$$g_3(s_1, s_2) = \frac{a_1 s_1 + a_2 s_2}{1 + b s_1 s_2} \tag{88.11}$$

where a_1, a_2, and b are all positive constants.

Let us consider a 2-D filter designed by using a 1-D analog low-pass filter with cutoff frequency Ω_c. Equation (88.11) results in

$$\Omega_2 = \frac{\Omega_c - a_1\Omega_1}{a_2 - b\Omega_c\Omega_1} \tag{88.12}$$

The mapping of $\Omega = \Omega_c$ onto the (Ω_1, Ω_2) plane for various values of b is depicted in Fig. 88.2 [5]. The cutoff frequencies along the Ω_1 and Ω_2 axes can be adjusted by simply varying a_1 and a_2. On the other hand, the convexity of the boundary can be adjusted by varying b. We note that b must be greater than zero to preserve stability. Also, it should be noted that $g_3(s_1, s_2)$

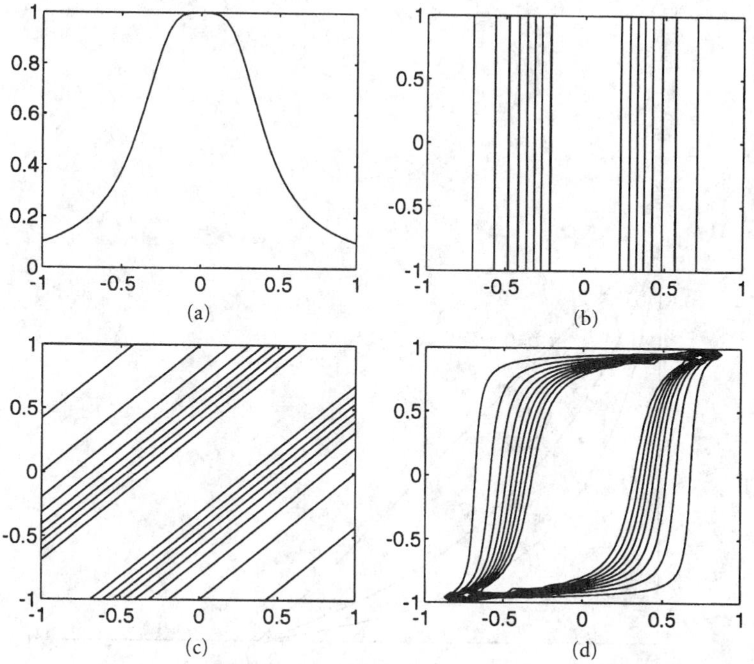

FIGURE 88.1 1-D to 2-D analog filter transformation. (a) Amplitude response of 1-D analog filter. (b) Contour plot of 2-D filter, $\beta = 0°$. (c) Contour plot of 2-D filter, $\beta = 45°$. (d) Contour plot of 2-D filter, $\beta = 45°$, after applying double bilinear transformation.

becomes a low-pass to bandpass transformation along $s_1 = s_2$, and therefore the designed filter will behave like a bandstop filter along $\Omega_1 = \Omega_2$. This problem can be overcome by using a guard filter of any order.

King and Kayran [5] have extended the above technique by using a higher order reactance function of the form

$$g_4(s_1, s_2) = \frac{a_1 s_1 + a_2 s_2}{1 + b_1(s_1^2 + s_2^2) + b_2 s_1 s_2} \tag{88.13}$$

and they proved that the stability of $g_4(s_1, s_2)$ is ensured if

$$b_1 > 0 \tag{88.14}$$

and

$$b_1 > \frac{b_2^2}{4} - b_1^2 > 0 \tag{88.15}$$

However, it is necessary as earlier to include a guard filter, which may have the simple form of

$$G(z_1, z_2) = \frac{(1 + z_1)(1 + z_2)}{(d_1 + z_1)(d_2 + z_2)} \tag{88.16}$$

in order to remove the high-pass regions along all radii except the coordinate axes.

Then, through an optimization procedure, the coefficients of $g_4(s_1, s_2)$ and $G(z_1, z_2)$ are calculated subject to the constraints of (88.14) and (88.15), so that the cutoff frequency of the 1-D filter is mapped into a desired cutoff boundary in the (Ω_1, Ω_2) plane.

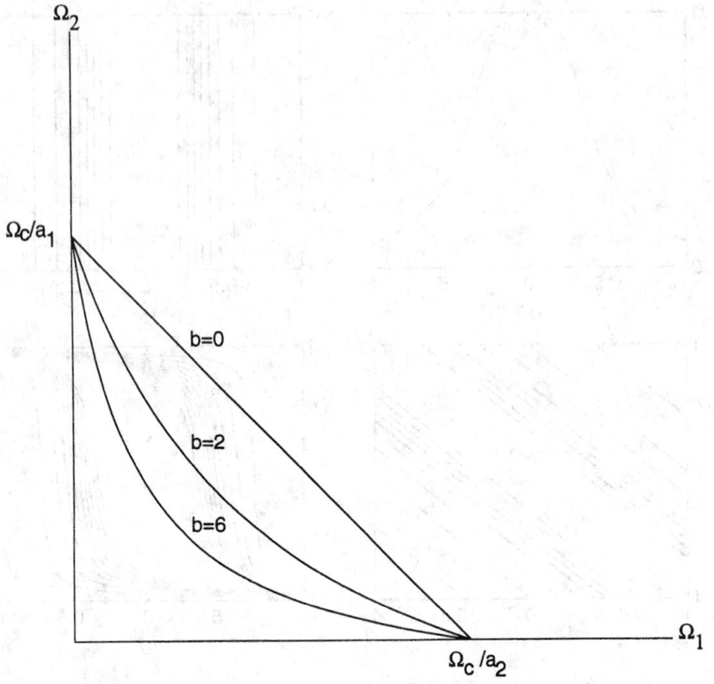

FIGURE 88.2 Plot of (88.12) for different values of b.

Spectral Transformations

Spectral transformation is another kind of important transformation in the design of both 1-D and 2-D IIR filters. In this subsection, three groups of spectral transformations are discussed. Among them, the linear transformations map frequency axes onto frequency axes in the (Ω_1, Ω_2) plane, the complex transformation is of wide applications to the design of fan filters, and the Constantinides transformations transform a discrete function into another discrete function and through which any transformation of a low-pass filter to another low-pass, high-pass, bandpass or bandstop filter becomes possible.

Linear Transformations

Consider a group of linear transformations that map frequency axes onto themselves in the (Ω_1, Ω_2) plane. There are eight possible such transformations [7], [8] and they have the algebraic structure of a finite group under the operation of multiplication [2], with each transformation can be expressed as

$$\begin{bmatrix} \omega_1 \\ \omega_2 \end{bmatrix} := D(T) \begin{bmatrix} \omega_1 \\ \omega_2 \end{bmatrix} \tag{88.17}$$

where $D(T)$ is a 2×2 unitary matrix representing transformation T_8. The eight transformations and their effect on the frequency response of the digital filter are as follows with a multiplication table being illustrated in Table 88.1 [2].

1. Identity (I):

$$D(I) = \begin{bmatrix} 1 & 0 \\ 0 & 1 \end{bmatrix}.$$

2. Reflection about the ω_1 axis ($\rho_{\omega 1}$):

$$D(\rho_{\omega 1}) = \begin{bmatrix} 1 & 0 \\ 0 & -1 \end{bmatrix}$$

3. Reflection about the ω_2 axis ($\rho_{\omega 2}$):

$$D(\rho_{\omega 2}) = \begin{bmatrix} -1 & 0 \\ 0 & 1 \end{bmatrix}$$

4. Reflection about the ψ_1 axis ($\rho_{\psi 1}$):

$$D(\rho_{\psi 1}) = \begin{bmatrix} 0 & 1 \\ 1 & 0 \end{bmatrix}$$

5. Reflection about the ψ_2 axis ($\rho_{\psi 2}$):

$$D(\rho_{\psi 2}) = \begin{bmatrix} 0 & -1 \\ -1 & 0 \end{bmatrix}$$

6. Counterclockwise rotation by 90°(R_4):

$$D(R_4) = \begin{bmatrix} 0 & -1 \\ 1 & 0 \end{bmatrix}$$

7. Counterclockwise rotation by 180°(R_4^2):

$$D(R_4^2) = \begin{bmatrix} -1 & 0 \\ 0 & -1 \end{bmatrix}$$

8. Counterclockwise rotation by 270°(R_4^3):

$$D(R_4^3) = \begin{bmatrix} 0 & 1 \\ -1 & 0 \end{bmatrix}$$

TABLE 88.1 Multiplication Table of Group

	I	$\rho_{\omega 1}$	$\rho_{\omega 2}$	$\rho_{\psi 1}$	$\rho_{\psi 2}$	R_4	R_4^2	R_4^3
I	I	$\rho_{\omega 1}$	$\rho_{\omega 2}$	$\rho_{\psi 1}$	$\rho_{\psi 2}$	R_4	R_4^2	R_4^3
$\rho_{\omega 1}$	$\rho_{\omega 1}$	I	R_4^2	R_4^3	R_4	$\rho_{\psi 2}$	$\rho_{\omega 2}$	$\rho_{\psi 1}$
$\rho_{\omega 2}$	$\rho_{\omega 2}$	R_4^2	I	R_4	R_4^3	$\rho_{\psi 1}$	$\rho_{\omega 1}$	$\rho_{\psi 2}$
$\rho_{\psi 1}$	$\rho_{\psi 1}$	R_4	R_4^3	I	R_4^2	$\rho_{\omega 1}$	$\rho_{\psi 2}$	$\rho_{\omega 2}$
$\rho_{\psi 2}$	$\rho_{\psi 2}$	R_4^3	R_4	R_4^2	I	$\rho_{\omega 2}$	$\rho_{\psi 1}$	$\rho_{\omega 1}$
R_4	R_4	$\rho_{\psi 1}$	$\rho_{\psi 2}$	$\rho_{\omega 2}$	$\rho_{\omega 1}$	R_4^2	R_4^3	I
R_4^2	R_4^2	$\rho_{\omega 2}$	$\rho_{\omega 1}$	$\rho_{\psi 2}$	$\rho_{\psi 1}$	R_4^3	I	R_4
R_4^3	R_4^3	$\rho_{\psi 2}$	$\rho_{\psi 1}$	$\rho_{\omega 1}$	$\rho_{\omega 2}$	I	R_4	R_4^2

In the above symbolic representation, ψ_1 and ψ_2 represent axes that are rotated by 45° in the counterclockwise sense relative to the ω_1 and ω_2 axes, respectively, and R_k denotes rotation by $360°/k$ in the counterclockwise sense. These transformations could equivalently be defined in the (z_1, z_2) domain by complex conjugating and/or interchanging the complex variables z_1 and z_2 in the filter transfer function.

An important property of the group is that each transformation distributes over a product of functions of ω_1 and ω_2, that is,

$$T_\delta\left[\prod_{i=1}^{K} F_i(\omega_1, \omega_2)\right] = \prod_{i=1}^{K} T_\delta[F_i(\omega_1, \omega_2)] \tag{88.18}$$

where T_δ represents any of the eight transformation operators. The validity of this property follows the definition of the transformations [8].

In the (z_1, z_2) domain, if $H(z_1, z_2)$ represents a causal filter with impulse response $h(n_1, n_2)$, then the filter represented by $H(z_1^{-1}, z_2^{-1})$ will have an impulse response $h(-n_1, -n_2)$, and is therefore, noncausal, since $h(-n_1, -n_2) \neq 0$ for $n_1 < 0$, $n_2 < 0$. Such a filter can be implemented in terms of causal transfer function $H(z_1, z_2)$, i.e., by rotating the n_1 and n_2 axes of the input signal by 180°, processing the rotated signal by the causal filter, and then rotating the axes of the output signal by 180°, as illustrated in Fig. 88.3(b).

Noncausal filters can be used for the realization of zero-phase filters by cascading K pairs of filters whose transfer functions are $H_i(z_1, z_2)$ and $H_i(z_1^{-1}, z_2^{-1})$ for $i = 1, 2, \cdots, K$, as depicted in Fig. 88.3(c).

Complex Transformation and 2-D Fan Filters

Complex Transformation. A complex transformation is of the form [9]

$$z = e^{\phi} z_1^{\alpha_1/\beta_1} z_2^{\alpha_2/\beta_2} \tag{88.19}$$

by which a 2-D filter $H(z_1, z_2)$ can be derived from 1-D filter $H_1(z)$. The corresponding frequency transformation of (88.19) is

$$\exp(j\omega) \rightarrow \exp\left\{j\left(\phi + \omega_1\frac{\alpha_1}{\beta_1} + \omega_2\frac{\alpha_2}{\beta_2}\right)\right\} \tag{88.20}$$

or

$$\omega \rightarrow \phi + \omega_1\frac{\alpha_1}{\beta_1} + \omega_2\frac{\alpha_2}{\beta_2} \tag{88.21}$$

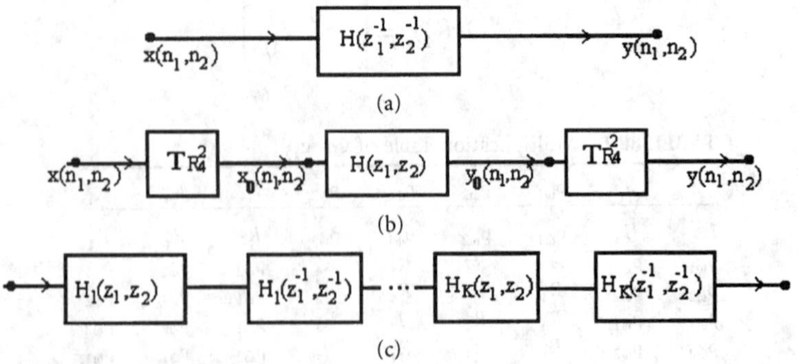

FIGURE 88.3 Design of zero-phase filters: (a) noncausal filter, (b) equivalent causal implementation, (c) cascade zero-phase filter.

There are three major effects of transformation (88.20) on the resulting filter:

1. Frequency shifting along the ω_1 axis. The frequency response of the resulting filter will be shifted by ϕ along the ω_1 axis.
2. Rotation of the frequency response. The angle of rotation is

$$\theta = \arctan\left(\frac{\alpha_2}{\beta_2}\right) \tag{88.22}$$

Since the original filter is 1-D and a function of z_1, the angle of rotation will be defined by the fractional power of z_2.
3. Scaling the frequency response along the ω_1 axis. The fractional power of z_1 will scale the frequency response by a factor β_1/α_1. However, the periodicity of the frequency response will be $(\alpha_1/\beta_1)2\pi$ instead of 2π. Other effects may also be specified [9].

The complex transformation is of importance in the design of fan filters. By using a prototype lowpass filter with a cutoff frequency at $\omega_c = \pi/2$, and the transformation (88.19), one obtains the shifted, scaled, and rotated characteristics in the frequency domain. We denote the transformed filter by

$$H\left(z_1, z_2; \frac{\alpha_1}{\beta_1}, \frac{\alpha_2}{\beta_2}\right) = H_1(z)\big|_{z=e^{\phi}z_1^{\alpha_1/\beta_1}z_2^{\alpha_2/\beta_2}} \tag{88.23}$$

In general, the filter coefficients in function H will be complex and the variables z_1 and z_2 will have rational noninteger powers. However, appropriate combinations of transformed filters will remove both of these difficulties, as will be shown in the following.

Symmetric Fan Filters. An ideal symmetric fan filter has the specification of

$$H_{f1}(e^{j\omega_1 T}, e^{j\omega_2 T}) = \begin{cases} 1 & \text{for } |\omega_1| \geq |\omega_2| \\ 0 & \text{otherwise} \end{cases} \tag{88.24}$$

We introduce transfer function $\hat{H}_1(z_1, z_2)$, $\hat{H}_2(z_1, z_2)$, $\hat{H}_3(z_1, z_2)$, and $\hat{H}_4(z_1, z_2)$ of four filters generated by (88.23) with $(\alpha_1/\beta_1, \alpha_2/\beta_2) = (1/2, 1/2)\ (-1/2, 1/2)$, $(1/2, -1/2)$, and $(-1/2, -1/2)$, respectively, and $\phi = \pi/2$. The responses of the transformed filters, $\hat{H}_i(z_1, z_2)$, $i = 1, 2, 3, 4$, can be found in Fig. 88.4 together with the prototype filter.

In this design procedure, the filters in Fig. 88.4 will be used as the basic building blocks for a fan filter specified by (88.24). One can construct the following filter characteristics:

$$\begin{aligned} G_{11}(z_1, z_2) = {} & \hat{H}_1(z_1, z_2)\hat{H}_1^*(z_1^{-1}, z_2^{-1})\hat{H}_2(z_1, z_2)\hat{H}_2^*(z_1^{-1}, z_2^{-1}) \\ & + \hat{H}_3(z_1, z_2)\hat{H}_3^*(z_1^{-1}, z_2^{-1})\hat{H}_4(z_1, z_2)\hat{H}_4^*(z_1^{-1}, z_2^{-1}) \end{aligned} \tag{88.25a}$$

$$\begin{aligned} G_{22}(z_1, z_2) = {} & \hat{H}_1(z_1, z_2)\hat{H}_1^*(z_1^{-1}, z_2^{-1})\hat{H}_3(z_1, z_2)\hat{H}_3^*(z_1^{-1}, z_2^{-1}) \\ & + \hat{H}_2(z_1, z_2)\hat{H}_2^*(z_1^{-1}, z_2^{-1})\hat{H}_4(z_1, z_2)\hat{H}_4^*(z_1^{-1}, z_2^{-1}) \end{aligned} \tag{88.25b}$$

which are shown in Fig. 88.5.

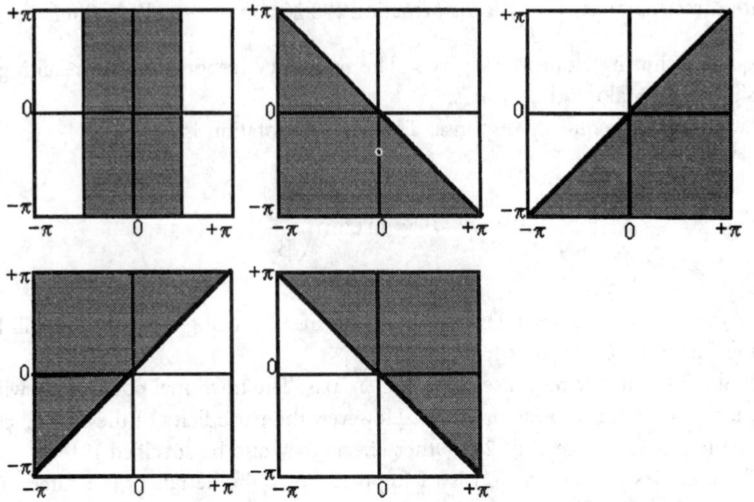

FIGURE 88.4 Basic building blocks of symmetric fan filters.

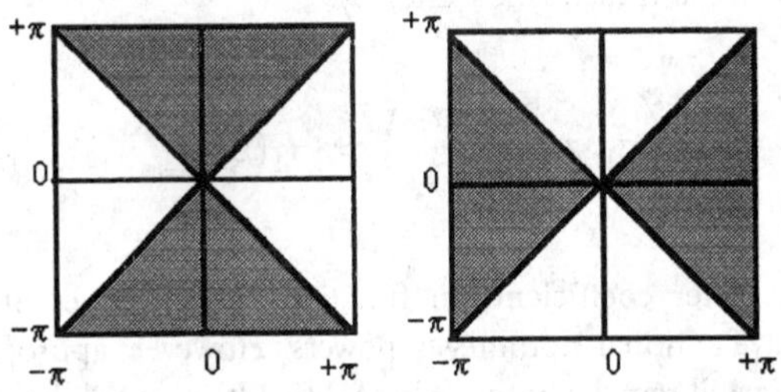

FIGURE 88.5 Amplitude characteristics of $G_{11}(z_1, z_2)$ and $G_{22}(z_1, z_2)$.

Quadrant Fan Filters. The frequency characteristic of a quadrant fan filter is specified as

$$H_{f2}(e^{j\omega_1 T}, e^{j\omega_2 T}) = \begin{cases} 1 & \text{for } \omega_1\omega_2 \geq 0 \\ 0 & \text{otherwise} \end{cases} \tag{88.26}$$

We consider the same ideal prototype filter. Then the transformed filters $\hat{H}_{14}(z_1, z_2)$, $\hat{H}_{12}(z_1, z_2)$, $\hat{H}_{23}(z_1, z_2)$ and $\hat{H}_{34}(z_1, z_2)$ are obtained via (88.23) with $(\alpha_1/\beta_1, \alpha_2/\beta_2)$ equal to $(1, 0)$, $(0, 1)$, $(-1, 0)$, and $(0, -1)$, respectively, and $\phi = \pi/2$. The subscripts on $\hat{H}$ refer to quadrants to which the low-pass filter characteristics have been shifted. Figure 88.6 illustrates the amplitude responses of these transformed filters together with the prototype.

The filters in Fig. 88.6 will be used as the basic building blocks for fan filters specified by (88.26). In a similar manner to (88.25a) and (88.25b), the filter characteristics $G_{13}(z_1, z_2)$ and $G_{23}(z_1, z_2)$ can be constructed as follows:

$$\begin{aligned} G_{13}(z_1, z_2) = {} & \hat{H}_{12}(z_1, z_2)\hat{H}^*_{12}(z_1^{-1}, z_2^{-1})\hat{H}_{14}(z_1, z_2)\hat{H}^*_{14}(z_1^{-1}, z_2^{-1}) \\ & + \hat{H}_{23}(z_1, z_2)\hat{H}^*_{23}(z_1^{-1}, z_2^{-1})\hat{H}_{34}(z_1, z_2)\hat{H}^*_{34}(z_1^{-1}, z_2^{-1}) \end{aligned} \tag{88.27a}$$

$$\begin{aligned} G_{24}(z_1, z_2) = {} & \hat{H}_{12}(z_1, z_2)\hat{H}^*_{12}(z_1^{-1}, z_2^{-1})\hat{H}_{23}(z_1, z_2)\hat{H}^*_{23}(z_1^{-1}, z_2^{-1}) \\ & + \hat{H}_{14}(z_1, z_2)\hat{H}^*_{14}(z_1^{-1}, z_2^{-1})\hat{H}_{34}(z_1, z_2)\hat{H}^*_{34}(z_1^{-1}, z_2^{-1}) \end{aligned} \tag{88.27b}$$

whose amplitude responses are depicted in Fig. 88.7.

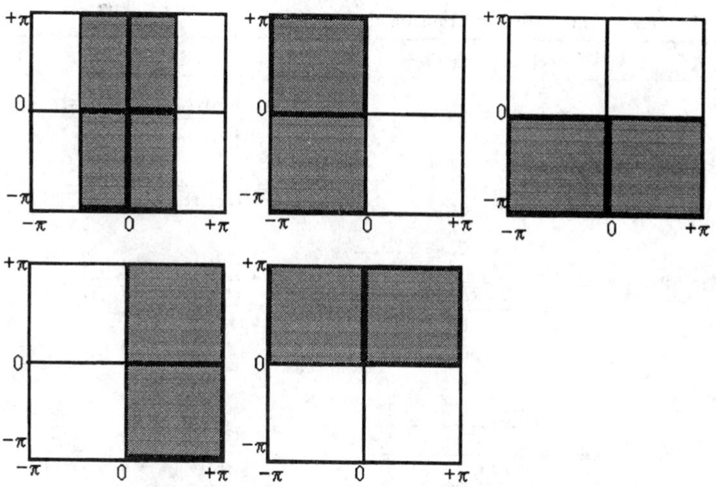

FIGURE 88.6 Basic building blocks for quadrant fan filter design.

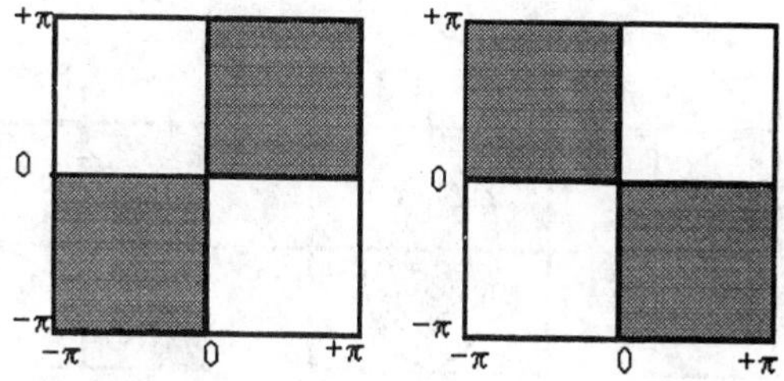

FIGURE 88.7 Two quadrant pass filters.

Constantinides Transformations

The so-called Constantinides transformations [10], [19] are of importance in the design of 1-D digital filters, and are of the form

$$z = f(\overline{z}) = e^{jl\pi} \prod_{t=1}^{m} \frac{\overline{z} - a_t^*}{1 - a_t \overline{z}} \tag{88.28}$$

where l and m are integers and a_t^* is the complex conjugate of a_t.

Pendergrass, Mitra, and Jury [11] showed that, in the decision of 2-D IIR filters, this group of transformations is as useful as in 1-D cases. By choosing the parameters l, m, and a_t in (88.28) properly, a set of four specific transformations can be obtained that can be used to transform a low-pass transfer function into a corresponding low-pass, high-pass, bandpass, or a bandstop transfer function. These transformations are summarized in Table 88.2, where subscript i is included to facilitate the application of the transformations to 2-D discrete transfer functions.

Let Ω_i and ω_i for $i = 1, 2$ be the frequency variables in the original and transformed transfer function, respectively. Suppose $H_L(z_1, z_2)$ is a low-pass transfer function with respect to z_i, if each of z_1 and z_2 is transformed to be low-pass, bandpass, or high-pass, then a number of different 2-D filter combinations can be achieved. As an example, some of the possible amplitude responses are illustrated in Fig. 88.8(a)–(f) [2].

TABLE 88.2 Constantinides Transformations

Type	Transformation	Parameters
LP to LP	$z_i = \dfrac{\bar{z}_i - \alpha_i}{1 - \alpha_i \bar{z}_i}$	$\alpha_i = \dfrac{\sin[(\Omega_{pi} - \omega_{pi})T_i/2]}{\sin[(\Omega_{pi} + \omega_{pi})T_i/2]}$
LP to HP	$z_i = \dfrac{\bar{z}_i - \alpha_i}{1 - \alpha_i \bar{z}_i}$	$\alpha_i = \dfrac{\cos[(\Omega_{pi} - \omega_{pi})T_i/2]}{\cos[(\Omega_{pi} + \omega_{pi})T_i/2]}$
LP to BP	$z_i = -\dfrac{\bar{z}_i^2 - \dfrac{2\alpha_i k_i}{k_i+1}\bar{z}_i + \dfrac{k_i - 1}{k_i + 1}}{1 - \dfrac{2\alpha_i k_i}{k_i+1}\bar{z}_i + \dfrac{k_i - 1}{k_i + 1}\bar{z}_i^2}$	$\alpha_i = \dfrac{\cos[(\omega_{p2i} + \omega_{p1i})T_i/2]}{\cos[(\omega_{p2i} + \omega_{p1i})T_i/2]}$ $k_i = \tan\dfrac{\Omega_{pi}T_i}{2} \cot\dfrac{(\omega_{p2i} - \omega_{p1i})T_i}{2}$
LP to BS	$z_i = -\dfrac{\bar{z}_i^2 - \dfrac{2\alpha_i}{1+k_i}\bar{z}_i + \dfrac{1 - k_i}{1 + k_i}}{1 - \dfrac{2\alpha_i}{1+k_i}\bar{z}_i + \dfrac{1 - k_i}{1 + k_i}\bar{z}_i^2}$	$\alpha_i = \dfrac{\cos[(\omega_{p2i} - \omega_{p1i})T_i/2]}{\cos[(\omega_{p2i} - \omega_{p1i})T_i/2]}$ $k_i = \tan\dfrac{\Omega_{pi}T_i}{2} \tan\dfrac{(\omega_{p2i} - \omega_{p1i})T_i}{2}$

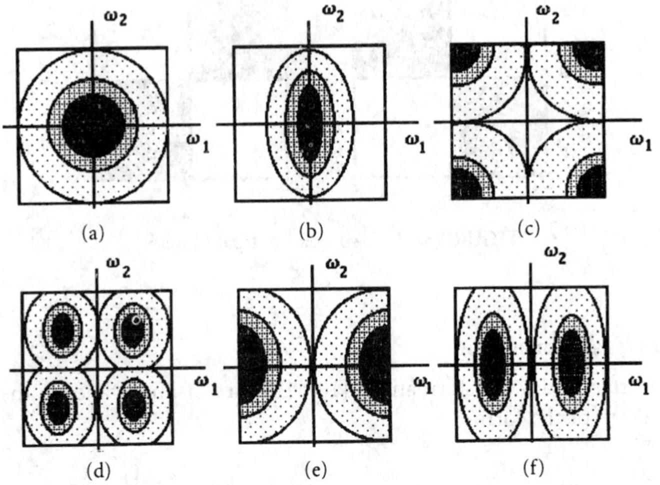

FIGURE 88.8 Application of Constantinides transformations to 2-D IIR filters. (a) Circularly symmetric low-pass filter. (b) LP to LP for z_1 and z_2. (c) LP to HP for z_1 and z_2. (d) LP to BP for z_1 and z_2. (e) LP to HP for z_1 and LP to LP for z_2. (f) LP to BP for z_1 and LP to LP for z_2.

88.3 Design of Separable Product Filters

A 2-D IIR filter is characterized by a separable product transfer function of the form

$$H(z_1, z_2) = H_1(z_1)H_2(z_2) \tag{88.29}$$

if its passband or stopband is of the shape of a rectangular domain. The design of such a class of filters can be accomplished by using the method proposed by Hirano and Aggarwal [12]. The method can be used to design filters with quadrantal or half-plane symmetry.

Design of Quadrantally Symmetric Filters

A 2-D filter is said to be quadrantally symmetric if its amplitude response satisfies the equality

$$|H(z_1, z_2)| = |H(z_1^*, z_2^*)| = |H(z_1^*, z_2)| = |H(z_1, z_2^*)| \tag{88.30}$$

Consider two 1-D bandpass filters specified by two transfer functions $H_1(z)$ and $H_2(z)$, respectively, and let $z = z_1$ in the first one and $z = z_2$ in the second. If their frequency responses can be expressed by

$$|H_1(e^{j\omega_1 T_1})| = \begin{cases} 1 & \omega_{12} \le \omega_1 \le \omega_{13} \\ 0 & 0 \le \omega_1 \le \omega_{11} \text{ or } \omega_{14} \le \omega_1 \le \infty \end{cases} \tag{88.31a}$$

and

$$|H_2(e^{j\omega_2 T_2})| = \begin{cases} 1 & \omega_{22} \le \omega_2 \le \omega_{23} \\ 0 & 0 \le \omega_2 \le \omega_{21} \text{ or } \omega_{24} \le \omega_2 \le \infty \end{cases} \tag{88.31b}$$

respectively, and since

$$|H(e^{j\omega_1 T_1}, e^{j\omega_2 T_2})| = |H_1(e^{j\omega_1 T_1})|\,|H_2(e^{j\omega_2 T_2})|$$

Equations (88.31a) and (88.31b) give

$$|H(e^{j\omega_1 T_1}, e^{j\omega_2 T_2})| = \begin{cases} 1 & \omega_{12} \le \omega_1 \le \omega_{13} \text{ and } \omega_{22} \le \omega_2 \le \omega_{23} \\ 0 & \text{otherwise} \end{cases}$$

Evidently, the 2-D filter obtained will pass frequency components that are in both the passband of $H_1(z_1)$ and $H_2(z_2)$; that is, the passband of $H(z_1, z_2)$ will be a rectangle with sides $\omega_{13} - \omega_{12}$ and $\omega_{23} - \omega_{22}$. On the other hand, frequency components that are in the stopband of either the first filter or the second filter will be rejected. Hence, the stopband of $H(z_1, z_2)$ consists of the domain obtained by combining the stopbands of the two filters.

By a similar method, if each of the two filters is allowed to be a low-pass, bandpass, or high-pass 1-D filter, then nine different rectangular passbands can be achieved, as illustrated in Fig. 88.9. The cascade arrangement of any two of those filters may be referred to as a generalized bandpass filter [12].

Another typical simple characteristic is specified by giving the rectangular stopband region, which is referred to as the rectangular stop filter. Also, there are nine possible types of such a filter which are complementary to that of the generalized bandpass filter shown in Fig. 88.9 (i.e., considering the shadowed region as the stopband). This kind of bandstop filter can be realized in the form of

$$H(z_1, z_2) = H_A(z_1)H_A(z_2) - e^{jk\pi}[H_1(z_1)H_2(z_2)]^2 \tag{88.32}$$

where $H_1(z_1)H_2(z_2)$ is a generalized bandpass filter described above, and $H_A(z_1)$ and $H_A(z_2)$ are allpass 1-D filters [13] whose poles are the poles of $H_1(z_1)$ and $H_2(z_2)$, respectively. Equation (88.32) can be referred to as a generalized bandstop filter.

Extending the principles discussed above, if a 2-D filter is constructed by cascading K 2-D filters with passbands P_i, stopbands S_i, and transfer functions $H_i(z_1, z_2)$, the overall transfer function is obtained as

$$H(z_1, z_2) = \prod_{i=1}^{K} H_i(z_1, z_2)$$

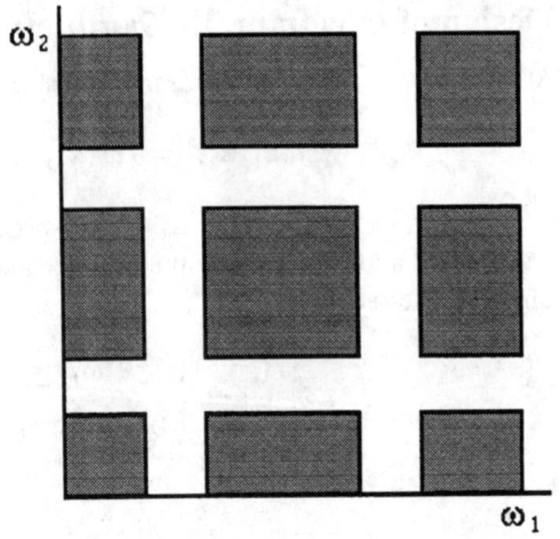

FIGURE 88.9 Idealized amplitude response of generalized bandpass filter.

and the passband P and stopband S of the cascaded 2-D filter are defined by

$$P = \bigcap_{i=1}^{K} P_i \quad \text{and} \quad S = \bigcup_{i=1}^{K} S_i$$

that is, the only frequency components not to be rejected will be those that will be passed by each and every filter in the cascade arrangement.

On the other hand, if a 2-D filter is constructed by connecting K 2-D filters in parallel, then

$$H(z_1, z_2) = \sum_{i=1}^{K} H_i(z_1, z_2)$$

Assuming that all the parallel filters have the same phase shift, and the passband $P_i = 1$ of the various filters are not overlapping, then the passband and stopband of the parallel arrangement is given by

$$P = \bigcup_{i=1}^{K} P_i \quad \text{and} \quad S = \bigcap_{i=1}^{K} S_i$$

Parallel IIR filters are more difficult to design than cascade ones, due to the requirement that the phase shifts of the various parallel filters be equal. However, if all the data to be filtered are available at the start of the processing, zero-phase filters can be used.

By combining parallel and cascade subfilters, 2-D IIR filters whose passbands or stopbands are combinations of rectangular regions can be designed, as illustrated by the following example.

Example 1. Design a 2-D IIR filter whose passband is the area between two overlapping rectangles, as depicted in Fig. 88.10(a).

Solution.

1. Construct a first 2-D low-pass filter with rectangular passband $(\omega_{12}, \omega_{22})$ by using 1-D low-pass filters, as shown in Fig. 88.10(b).
2. Construct a second 2-D low-pass filter with rectangular passband $(\omega_{11}, \omega_{21})$ by 1-D low-pass filters. Then using (88.32) to construct a 2-D high-pass filter with rectangular stopband $(\omega_{11}, \omega_{21})$, as shown in Fig. 88.10(c).

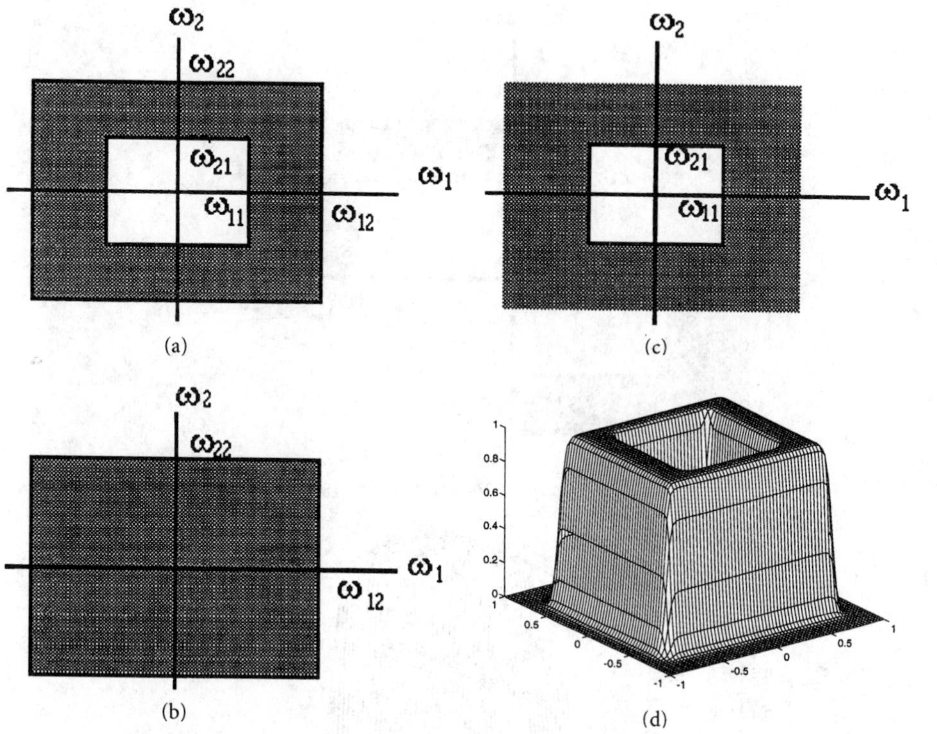

FIGURE 88.10 Amplitude response for the filter of Example 1: (a) specified response; (b) response of the 2-D low-pass filter; (c) response of the 2-D high-pass filter; (d) 3-D plot of the amplitude response of a real 2-D IIR filter.

3. Cascade the first 2-D low-pass filter with the 2-D high-pass filter to obtain the required filter. The amplitude response of a practically designed 2-D filter is in Fig. 88.10(d).

Design of Half-Plane Symmetric Filters

A 2-D filter is said to be half-plane symmetric if its amplitude response satisfies

$$|H(z_1, z_2)| = |H(z_1^*, z_2^*)| \tag{88.33}$$

but

$$|H(z_1, z_2)| \neq |H(z_1^*, z_2)| \neq |H(z_1, z_2^*)| \tag{88.34}$$

that is, half-plane symmetry does not imply quadrantal symmetry.

The design of those filters can be accomplished by cascading two quadrant pass filters derived previously in the subsection on complex transformations and 2-D fan filters with quadrantally symmetric filters, as is demonstrated in Example 2.

Example 2. Design a 2-D half-plane symmetric filter whose passband is defined in Fig. 88.11(a).

Solution.

1. Follow the steps 1 to 3 in Example 1 to construct a 2-D bandpass filter with the passband being the area between two overlapping rectangles, as shown in Fig. 88.10(a).

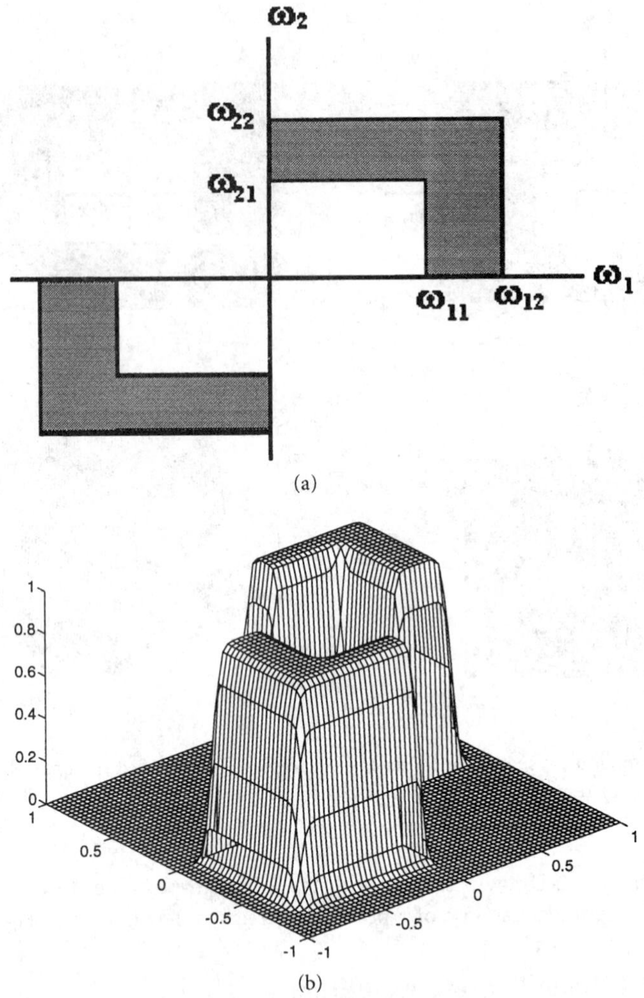

FIGURE 88.11 Amplitude response for the filter of Example 2: (a) specified response; (b) 3-D plot of the amplitude response of a real 2-D IIR filter.

2. Construct a two quadrant (first and third quadrants) pass filter as depicted in Fig. 88.7(a).
3. Cascade the 2-D bandpass filter with the two quadrant pass filters to obtain the required filter.

The amplitude response of a practically designed filter is shown in Fig. 88.11(b).

Design of Filters Satisfying Prescribed Specifications

The problem of designing filters that satisfy prescribed specifications to a large extent has been solved for the case of 1-D filters, and by extending the available methods, 2-D IIR filters satisfying prescribed specifications can also be designed.

Assume that we are to design a 2-D low-pass filter with the following specification:

$$|H(\omega_1, \omega_2)| = \begin{cases} 1 \pm \Delta p & \text{for } 0 \leq |\omega_i| \leq \omega_{pi} \\ \Delta a & \text{for } \omega_{ai} \leq |\omega_i| \leq \omega_{si}/2, \qquad i = 1, 2 \end{cases} \tag{88.35}$$

where ω_{pi} and ω_{ai}, $i = 1, 2$ are passband and stopband edges along the ω_1 and ω_2 axes, respectively. The two 1-D filters that are cascaded to form a 2-D one are specified by ω_{pi}, ω_{ai}, δ_{pi} and δ_{ai} $(i = 1, 2)$ as the passband edge, stopband edge, passband ripple, and stopband loss, respectively. From (88.29) we have

$$\max\{M(\omega_1, \omega_2)\} = \max\{M_1(\omega_1)\} \max\{M_2(\omega_2)\}$$

and

$$\min\{M(\omega_1, \omega_2)\} = \min\{M_1(\omega_1)\} \min\{M_2(\omega_2)\}$$

Hence the derived 2-D filter will satisfy the specifications of (88.35) if the following constraints are satisfied

$$(1 + \delta_{p1})(1 + \delta_{p2}) \leq 1 + \Delta p \tag{88.36}$$

$$(1 - \delta_{p1})(1 - \delta_{p2}) \geq 1 - \Delta p \tag{88.37}$$

$$(1 + \delta_{p1})\delta_{a2} \leq \Delta a \tag{88.38}$$

$$(1 + \delta_{p2})\delta_{a1} \leq \Delta a \tag{88.39}$$

$$\delta_{a1}\delta_{a2} \leq \Delta a \tag{88.40}$$

Constraints (88.36) and (88.37) can be expressed respectively in the alternative form

$$\delta_{p1} + \delta_{p2} + \delta_{p1}\delta_{p2} \leq \Delta p \tag{88.41}$$

and

$$\delta_{p1} + \delta_{p2} - \delta_{p1}\delta_{p2} \leq \Delta p \tag{88.42}$$

Hence if (88.41) is satisfied, (88.42) is also satisfied. Similarly, constraints (88.38)–(88.40) will be satisfied if

$$\max\{(1 + \delta_{p1})\delta_{a2}, (1 + \delta_{p2})\delta_{a1}\} \leq \Delta a \tag{88.43}$$

since $(1 + \delta_{p1}) \gg \delta_{a1}$ and $(1 + \delta_{p2}) \gg \delta_{a2}$. Now if we assume that $\delta_{p1} = \delta_{p2} = \delta_p$ and $\delta_{a1} = \delta_{a2} = \delta_a$, then we can assign

$$\delta_p = (1 + \Delta p)^{1/2} - 1 \tag{88.44}$$

and

$$\delta_a = \frac{\Delta a}{(1 + \Delta p)^{1/2}} \tag{88.45}$$

so as to satisfy constraints (88.36)–(88.40). And since $\Delta p \ll 1$, we have

$$\delta_p \approx \frac{\Delta p}{2} \tag{88.46}$$

$$\delta_a \approx \Delta a \tag{88.47}$$

Consequently, if the maximum allowable passband and stopband errors Δp and Δa are specified, the maximum passband ripple A_p and the minimum stopband attenuation A_a in dB for the two 1-D filters can be determined as

$$A_p = 20 \log \frac{1}{1 - \delta_p} = 20 \log \frac{2}{2 - \Delta p} \tag{88.48}$$

and

$$A_a = 20 \log \frac{1}{\delta_p} = 20 \log \frac{1}{\Delta a} \tag{88.49}$$

Finally, if the passband and stopband edges ω_{pi} and ω_{ai} are also specified, the minimum order and the transfer function of each of the two 1-D filters can readily be obtained using any of the approaches of the previous subsections. Similar treatments for bandpass, bandstop, and high-pass filters can be carried out.

Example 3 [12]. Design a zero-phase filter whose amplitude response is specified in Fig. 88.12(a), with $A_p = 3$ dB, $A_a = 20$ dB.

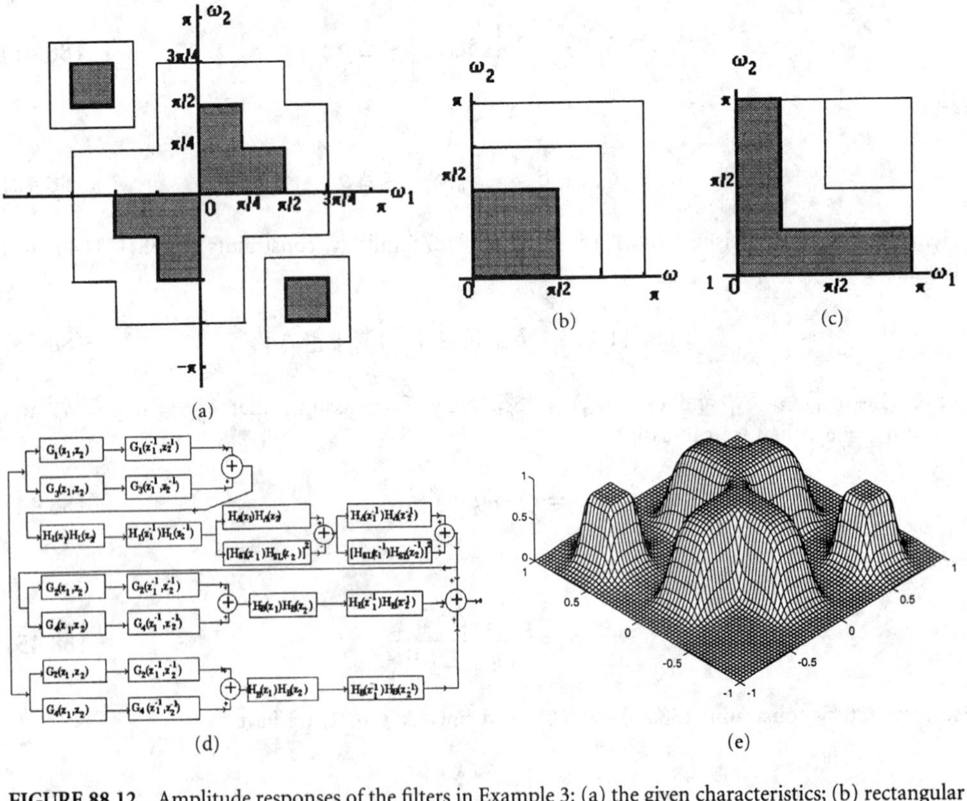

FIGURE 88.12 Amplitude responses of the filters in Example 3: (a) the given characteristics; (b) rectangular pass subfilter; (c) rectangular step subfilter; (d) final configuration of the 2-D filter; (e) 3-D plot of the amplitude response of the resulting 2-D filter.

Solution.

1. Decomposition: Because there are no contiguous passbands between the characteristics of the first and third quadrants and that of the second and fourth quadrants, in the first step of design, the required 2-D filter can be decomposed into two subfilters $H_{13}(z_1, z_2)$ and $H_{24}(z_1, z_2)$, which represent the characteristics of the first and third quadrants and that of the second and fourth quadrants, respectively. By connecting $H_{13}(z_1, z_2)$ and $H_{24}(z_1, z_2)$ in parallel, the required characteristics of the 2-D filter specified in Fig. 88.12(a) can then be realized.

$$H(z_1, z_2) = H_{13}(z_1, z_2) + H_{24}(z_1, z_2)$$

   i. Decomposition of $H_{13}(z_1, z_2)$: To accomplish the design of $H_{13}(z_1, z_2)$, further decompositions should be made. The characteristics of the first and third quadrants can be realized by cascading three subfilters, i.e., a two quadrant filter $G_{13}(z_1, z_2)$ [as shown in Fig. 88.7(a)], a low-pass filter $H_L(z_1, z_2)$ [as shown in Fig. 88.12(b)], and a rectangular stop filter $H_s(z_1, z_2)$ [as shown in Fig. 88.12(c)],

$$H_{13}(z_1, z_2) = G_{13}(z_1, z_2)H_L(z_1, z_2)H_S(z_1, z_2)$$

   furthermore, the rectangular stop filter $H_S(z_1, z_2)$ can be designed by allpass filter $H_A(z_1, z_2)$ and low-pass filter $H_{SL}(z_1, z_2)$, using (88.32)

$$H_S(z_1, z_2) = H_A(z_1, z_2) - [H_{SL}(z_1, z_2)]^2$$

   ii. Decomposition of $H_{24}(z_1, z_2)$: Similarly, $H_{24}(z_1, z_2)$ can be decomposed into the cascade of a two quadrant filter $G_{24}(z_1, z_2)$ [as is shown in Fig. 88.7(b)], and a bandpass filter $H_B(z_1, z_2)$ which can be realized by using two 1-D bandpass filters for both directions.

$$H_{24}(z_1, z_2) = G_{24}(z_1, z_2)H_B(z_1, z_2)$$

   The final configuration of the desired filter $H(z_1, z_2)$ is illustrated in Fig. 88.12(d), with the final transfer function being of the form

$$H(z_1, z_2) = H_{13}(z_1, z_2) + H_{24}(z_1, z_2) - H_{13}(z_1, z_2)H_{24}(z_1, z_2)$$

   where the purpose of the term $H_{13}(z_1, z_2)H_{24}(z_1, z_2)$ is to remove the overlap that may be created by adding $H_{13}(z_1, z_2)$ and $H_{24}(z_1, z_2)$.
2. Design of all the subfilters: At this point, the problem is to derive the two quadrant subfilters $G_{13}(z_1, z_2)$ and $G_{24}(z_1, z_2)$, the low-pass subfilters $H_L(z_1, z_2)$ and $H_{SL}(z_1, z_2)$, the allpass subfilter $H_A(z_1, z_2)$, and the bandstop filter $H_B(z_1, z_2)$.

Note that the symmetry of the given characteristics, the identical 1-D sections can be used to develop all the above 2-D subfilters, and the given specifications can easily be combined into the designs of all the 1-D sections.

3. By connecting all the 2-D subfilters in cascade or parallel as specified in Fig. 88.12(d), the required 2-D filter is obtained. The 3-D plot of the amplitude response of the final resulting 2-D filter is depicted in Fig. 88.12(e).

88.4 Design of Circularly Symmetric Filters

Design of LP Filters

As mentioned in the first subsection in Section 88.2, rotated filters can be used to design circularly symmetric filters. Costa and Venetsanopoulos [14] and Goodman [15] proposed two methods of this class, based on transforming an analog transfer function or a discrete one by rotated filter transformation, respectively. The two methods lead to filters that are, theoretically, unstable but by using an alternative transformation suggested by Mendonca *et al.* [16], this problem can be eliminated.

Design Based on 1-D Analog Transfer Function

Costa and Venetsanopoulos [14] proposed a method to design circularly symmetric filters. In their method, a set of 2-D analog transfer functions is first obtained by applying the rotated filter transformation in (88.9a) for several different values of the rotation angle β to a 1-D analog low-pass transfer function. A set of 2-D discrete low-pass functions are then deduced through the application of the bilinear transformation. The design is completed by cascading the set of 2-D digital filters obtained. The steps involved are as follows.

STEP 1. Obtain a stable 1-D analog low-pass transfer function

$$H_{A1}(s) = \frac{N(s)}{D(s)} = K_0 \frac{\prod_{i=1}^{M}(s - z_{ai})}{\prod_{i=1}^{N}(s - p_{ai})} \tag{88.50}$$

where z_{ai} and p_{ai} for $i = 1, 2, \cdots$, are the zeros and poles of $H_{A1}(s)$, respectively, and K_0 is a multiplier constant.

STEP 2. Let β_k for $k = 1, 2, \cdots, K$ be a set of rotation angles defined by

$$\beta_k = \begin{cases} \left(\dfrac{2k-1}{2K} + 1\right)\pi & \text{for even } K \\ \left(\dfrac{k-1}{K} + 1\right)\pi & \text{for odd } K \end{cases} \tag{88.51}$$

STEP 3. Apply the transformation of (88.9a) to obtain a 2-D analog transfer function as

$$H_{A2k}(s_1, s_2) = H_{A1}(s)|_{s=-s_1 \sin\beta_k + s_2 \cos\beta_k} \tag{88.52}$$

for each rotation angle β_k identified in Step 2.

STEP 4. Apply the double bilinear transformation to $H_{A2k}(s_1, s_2)$ to obtain

$$H_{D2k}(z_1, z_2) = H_{A2k}(s_1, s_2)|_{s_i = 2(z_i - 1)/T_i(z_i + 1) \quad i=1,2} \tag{88.53}$$

Assuming that $T_1 = T_2 = T$, (88.50) and (88.53) yield

$$H_{D2k}(z_1, z_2) = K_1 \prod_{i=1}^{M_0} H_{2i}(z_1, z_2) \tag{88.54}$$

where

$$H_{2i}(z_1, z_2) = \frac{a_{11i} + a_{21i}z_1 + a_{12i}z_2 + a_{22i}z_1z_2}{b_{11i} + b_{21i}z_1 + b_{12i}z_2 + b_{22i}z_1z_2} \tag{88.55}$$

$$K_1 = K_0\left(\frac{T}{2}\right)^{N-M} \tag{88.56}$$

$$\begin{aligned}
a_{11i} &= -\cos\beta_k + \sin\beta_k - \frac{Tz_{ai}}{2} \\
a_{21i} &= -\cos\beta_k - \sin\beta_k - \frac{Tz_{ai}}{2} \quad \text{for } 1 \le i \le M \\
a_{12i} &= \cos\beta_k + \sin\beta_k - \frac{Tz_{ai}}{2} \\
a_{22i} &= \cos\beta_k - \sin\beta_k - \frac{Tz_{ai}}{2} \\
a_{11i} &= a_{12i} = a_{21i} = a_{22i} = 1 \quad \text{for } M \le i \le M_0
\end{aligned} \tag{88.57a}$$

$$\begin{aligned}
b_{11i} &= -\cos\beta_k + \sin\beta_k - \frac{Tp_{ai}}{2} \\
b_{21i} &= -\cos\beta_k - \sin\beta_k - \frac{Tp_{ai}}{2} \quad \text{for } 1 \le i \le N \\
b_{12i} &= \cos\beta_k + \sin\beta_k - \frac{Tp_{ai}}{2} \\
b_{22i} &= \cos\beta_k - \sin\beta_k - \frac{Tp_{ai}}{2} \\
b_{11i} &= b_{12i} = b_{21i} = b_{22i} = 1 \quad \text{for } N \le i \le M_0
\end{aligned} \tag{88.57b}$$

and

$$M_0 = \max(M, N)$$

STEP 5. Cascade the filters obtained in Step 4 to yield an overall transfer function

$$H(z_1, z_2) = \prod_{k=1}^{K} H_{D2k}(z_1, z_2)$$

It is easy to find that, at point $(z_1, z_2) = (-1, -1)$, both the numerator and denominator polynomials of $H_{2i}(z_1, z_2)$ assume the value of zero. And thus each $H_{2i}(z_1, z_2)$ has nonessential singularity of the second kind on the unit bicircle

$$U^2 = \{(z_1, z_2)\colon\ |z_1| = 1, |z_2| = 1\}$$

The nonessential singularity of each $H_{2i}(z_1, z_2)$ can be eliminated and, furthermore, each subfilter can be stabilized by letting

$$b'_{12i} = b_{12i} + \varepsilon b_{11i} \tag{88.58a}$$

$$b'_{22i} = b_{22i} + \varepsilon b_{21i} \tag{88.58b}$$

where ε is a small positive constant. With this modification, the denominator polynomial of each $H_{2i}(z_1, z_2)$ is no longer zero and, furthermore, the stability of the subfilter can be guaranteed if

$$\text{Re}(p_{ai}) < 0 \tag{88.59}$$

and

$$270^\circ < \beta_k < 360^\circ \tag{88.60}$$

As can be seen in (88.51), half of the rotation angles are in the range $180^\circ < \beta_k < 270^\circ$ and according to the preceding stable conditions they yield unstable subfilters. However, the problem can easily be overcome by using rotation angles in the range given by (88.60) and then rotating the transfer function of the subfilter by -90° using linear transformations described in the first subsection in Section 88.2. For example, an effective rotation angle $\beta_k = 225^\circ$ is achieved by rotating the input data by 90°, filtering using a subfilter rotated by 315°, and then rotating the output data by -90°, as shown in Fig. 88.13.

In addition, a 2-D zero-phase filter can be designed by cascading subfilters for rotation angles $\pi + \beta_k$ for $k = 1, 2, \cdots, N$. The resulting transfer function is given by

$$H(z_1, z_2) = \prod_{k=1}^{K} H_{D2k}(z_1, z_2) H_{D2k}(z_1^{-1}, z_2^{-1})$$

where the noncausal sections can be realized as illustrated in Fig. 88.3.

Design Based on 1-D Discrete Transfer Function

The method proposed by Goodman [15] is based on the 1-D discrete transfer function transformation. In the method, a 1-D discrete transfer function is first obtained by applying the bilinear transformation to a 1-D analog transfer function. Then, through the application of an allpass transformation that rotates the contours of the amplitude of the 1-D discrete transfer function, a corresponding 2-D transfer function is obtained. The steps involved are as follows.

STEP 1. Obtain a stable 1-D analog low-pass transfer function $H_{A1}(s)$ of the form given by (88.50).

STEP 2. Apply the bilinear transformation to $H_{A1}(s)$ to obtain

$$H_{D1}(z) = H_{A1}(s)\big|_{s=2(z-1)/T(z+1)} \tag{88.61}$$

STEP 3. Let β_k for $k = 1, 2, \cdots, K$ be a set of rotation angles given by (88.51).

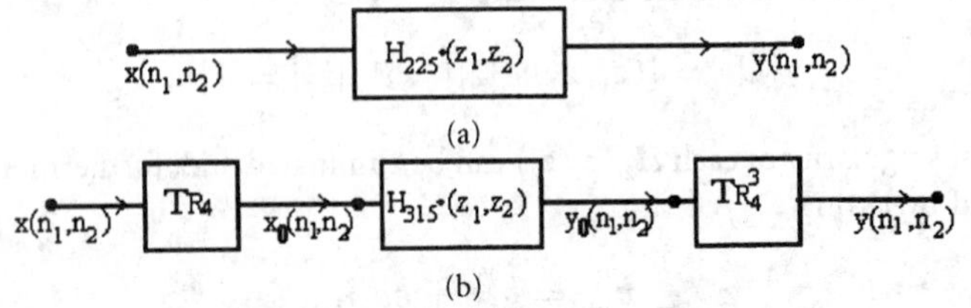

FIGURE 88.13 Realization of subfilter for rotation angle in the range $180^\circ < \beta_k < 270^\circ$.

STEP 4. Apply the allpass transformation defined by

$$z = f_k(z_1, z_2) = \frac{1 + c_k z_1 + d_k z_2 + e_k z_1 z_2}{e_k + d_k z_1 + c_k z_2 + z_1 z_2} \tag{88.62}$$

where

$$c_k = \frac{1 + \sin\beta_k + \cos\beta_k}{1 - \sin\beta_k + \cos\beta_k} \tag{88.63a}$$

$$d_k = \frac{1 - \sin\beta_k - \cos\beta_k}{1 - \sin\beta_k + \cos\beta_k} \tag{88.63b}$$

$$e_k = \frac{1 + \sin\beta_k - \cos\beta_k}{1 - \sin\beta_k + \cos\beta_k} \tag{88.63c}$$

to obtain the 2-D discrete transfer function

$$H_{D2k}(z_1, z_2) = H_{D1}(z)\big|_{z=f_k(z_1,z_2)} \tag{88.64}$$

for $k = 1, 2, \cdots, K$. The procedure yields the 2-D transfer function of (88.64), as can be easily demonstrated, and by cascading the rotated subfilters $H_{D2k}(z_1, z_2)$ the design can be completed.

The method of Goodman is equivalent to that of Costa and Venetsanopoulos and consequently, the resulting filter is subject to the same stability problem due to the nonessential singularity of the second kind at point $(z_1, z_2) = (-1, -1)$. To achieve a stable design, Goodman suggested that the transfer function $H_{D2k}(z_1, z_2)$ for $k = 1, 2, \cdots, K$ be obtained directly by minimizing an appropriate objective function subject to the constraints

$$\begin{aligned} c_k + d_k - e_k &\leq 1 - \varepsilon \\ c_k - d_k + e_k &\leq 1 - \varepsilon \\ -c_k + d_k + e_k &\leq 1 - \varepsilon \\ -c_k - d_k - e_k &\leq 1 - \varepsilon \end{aligned}$$

through an optimization procedure. If ε is a small positive constant, the preceding constraints constitute necessary and sufficient conditions for stability and, therefore, such an approach will yield a stable filter.

Elimination of Nonessential Singularities

To eliminate the nonessential singularities in the preceding two methods, Mendonca *et al.* [16] suggested a new transformation of the form

$$s = g_5(s_1, s_2) = \frac{\cos\beta_k s_1 + \sin\beta_k s_2}{1 + c s_1 s_2} \tag{88.65}$$

by combining the transformations in (88.9a) and (88.11) to replace the transformation (88.9a). If we ensure that

$$\cos\beta_k > 0, \sin\beta_k > 0, \quad \text{and } c > 0$$

then the application of this transformation followed by the application of the double bilinear transformation yields stable 2-D digital filters that are free of nonessential singularities of the second kind. If, in addition

$$c = \frac{1}{\omega_{\max}^2}$$

then local-type preservation can be achieved on the set Ω_2 given by

$$\Omega_2 = \{(\omega_1, \omega_2)\colon\ \omega_1 \geq 0, \omega_2 \geq 0, \omega_1\omega_2 \leq \omega_{\max}\}$$

and if $\omega_{\max} \to \infty$, then a global-type preservation can be approached as closely as desired.

By using the transformation of (88.65) instead of that in (88.9a) in the method of Costa and Venetsanopoulos, the transfer function of (88.54) becomes

$$\begin{aligned} H_{D2k}(z_1, z_2) &= K_1 P_{D2}(z_1, z_2) \\ &\quad \times \prod_{i=1}^{M_0} \frac{a_{11i} + a_{21i}z_1 + a_{12i}z_2 + a_{22i}z_1z_2}{b_{11i} + b_{21i}z_1 + b_{12i}z_2 + b_{22i}z_1z_2} \end{aligned} \tag{88.66}$$

where

$$K_1 = K_0\left(\frac{T}{2}\right)^{N-M} \tag{88.67a}$$

$$\begin{aligned} P_{D2}(z_1, z_2) = \bigg[1 + \frac{4c}{T^2} + \left(1 - \frac{4c}{T^2}\right)z_1 \\ + \left(1 - \frac{4c}{T^2}\right)z_2 + \left(1 + \frac{4c}{T^2}\right)z_1z_2 \bigg]^{N-M} \end{aligned} \tag{88.67b}$$

and

$$\left.\begin{aligned} a_{11i} &= -\cos\beta_k - \sin\beta_k - \left(\frac{T}{2} + \frac{2c}{T}\right)z_{ai} \\ a_{21i} &= \cos\beta_k - \sin\beta_k - \left(\frac{T}{2} - \frac{2c}{T}\right)z_{ai} \\ a_{12i} &= -\cos\beta_k + \sin\beta_k - \left(\frac{T}{2} - \frac{2c}{T}\right)z_{ai} \\ a_{22i} &= \cos\beta_k + \sin\beta_k - \left(\frac{T}{2} + \frac{2c}{T}\right)z_{ai} \end{aligned}\right\} \quad \text{for } 1 \leq i \leq M$$

$$a_{11i} = a_{21i} = a_{12i} = a_{22i} = 1 \qquad \text{for } M \leq i \leq M_0$$

$$
\begin{aligned}
b_{11i} &= -\cos\beta_k - \sin\beta_k - \left(\frac{T}{2} + \frac{2c}{T}\right)p_{ai} \\
b_{21i} &= \cos\beta_k - \sin\beta_k - \left(\frac{T}{2} - \frac{2c}{T}\right)p_{ai} \qquad \text{for } 1 \le i \le N \\
b_{12i} &= -\cos\beta_k + \sin\beta_k - \left(\frac{T}{2} - \frac{2c}{T}\right)p_{ai} \\
b_{22i} &= \cos\beta_k + \sin\beta_k - \left(\frac{T}{2} - \frac{2c}{T}\right)p_{ai} \\
b_{11i} &= b_{21i} = b_{12i} = b_{22i} = 1 \qquad \text{for } N \le i \le M_0 \\
M_0 &= \max(M, N)
\end{aligned}
$$

An equivalent design can be obtained by applying the allpass transformation of (88.62) in Goodman's method with

$$
\begin{aligned}
c_k &= \frac{1 + \cos\beta_k - \sin\beta_k - 4c/T^2}{1 - \cos\beta_k - \sin\beta_k + 4c/T^2} \\
d_k &= \frac{1 - \cos\beta_k + \sin\beta_k - 4c/T^2}{1 - \cos\beta_k - \sin\beta_k + 4c/T^2} \\
e_k &= \frac{1 + \cos\beta_k + \sin\beta_k + 4c/T^2}{1 - \cos\beta_k - \sin\beta_k + 4c/T^2}
\end{aligned}
$$

Realization of HP, BP, and BS Filters

Consider two zero-phase rotated subfilters that were obtained from a 1-D analog high-pass transfer function using rotating angles $-\beta_1$ and β_1, where $0° < \beta_1 < 90°$. The idealized contour plots of the two subfilters are shown in Fig. 88.14(a) and (b). If these two subfilters are cascaded, the amplitude response of the combination is obtained by multiplying the amplitude responses of the subfilters at corresponding points. The idealized contour plot of the composite filter is thus obtained as illustrated in Fig. 88.14(c). As can be seen, the contour plot does not represent the amplitude response of a 2-D circularly symmetric high-pass filter, and, therefore, the design of high-pass filters cannot readily be achieved by simply cascading rotated subfilters as in the case of low-pass filters. However, the design of these filters can be accomplished, through the use of a combination of cascade and parallel subfilters [16].

If the above rotated subfilters are connected in parallel, we obtain a composite filter whose contour plot is shown in Fig. 88.14(d). By subtracting the output of the cascade filter from the output of the parallel filter, we achieve an overall filter whose contour plot is depicted in Fig. 88.14(e). Evidently, this plot resembles the idealized contour plot of a 2-D circularly symmetric high-pass filter, and, in effect, following this method a filter configuration is available for the design of high-pass filters.

The transfer function of the 2-D high-pass filter is then given by

$$
\hat{H}_1 = \overline{H}_1 = H^{\beta_1} + H^{-\beta_1} - H^{\beta_1}H^{-\beta_1} \tag{88.68}
$$

where

$$
H^{\beta_1} = H_1(z_1, z_2)H_1(z_1^{-1}, z_2^{-1})
$$

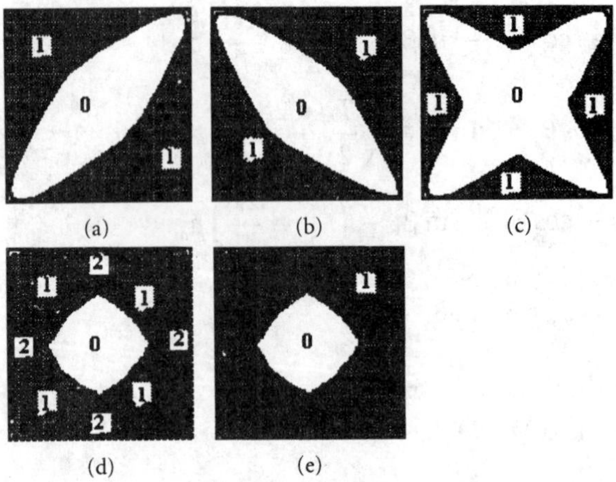

FIGURE 88.14 Derivation of 2-D high-pass filter. (a) Contour plot of subfilter rotated by $-\beta_1$. (b) Contour plot of subfilter rotated by β_1. (c) Contour plot of subfilters in (a) and (b) connected in cascade. (d) Contour plot of subfilters in (a) and (b) connected in parallel. (e) Contour plot obtained by subtracting the amplitude response in (c) from that of (d).

and

$$H^{-\beta_1} = H_1(z_1, z_2^{-1})H_1(z_1^{-1}, z_2)$$

represent zero-phase subfilters rotated by angle β_1 and $-\beta_1$, respectively.

The above approach can be extended to two or more rotation angles in order to improve the degree of circularity. For N rotation angles, $\hat{H}_N$ is given by the recursive relation

$$\hat{H}_N = \hat{H}_{N-1} + \overline{H}_N - \hat{H}_{N-1}\overline{H}_N \tag{88.69}$$

where

$$\overline{H}_N = H^{\beta_N} + H^{-\beta_N} - H^{\beta_N}H^{-\beta_N}$$

and $\hat{H}_{N-1}$ can be obtained from $\overline{H}_{N-1}$ and $\hat{H}_{N-2}$. The configuration obtained is illustrated in Fig. 88.15, where the realization of $\hat{H}_{N-1}$ is of the same as that of $\hat{H}_N$.

As can be seen in Fig. 88.15, the complexity of the high-pass configuration tends to increase rapidly with the number of rotations, and consequently, the number of rotations should be kept to a minimum. It should also be mentioned that the coefficients of the rotated filters must be properly adjusted, by using (88.65), to ensure that the zero-phase is approximated. However, the use of this transformation leads to another problem: the 2-D digital transfer function obtained has spurious zeros at the Nyquist points. These zeros are due to the fact that the transformation

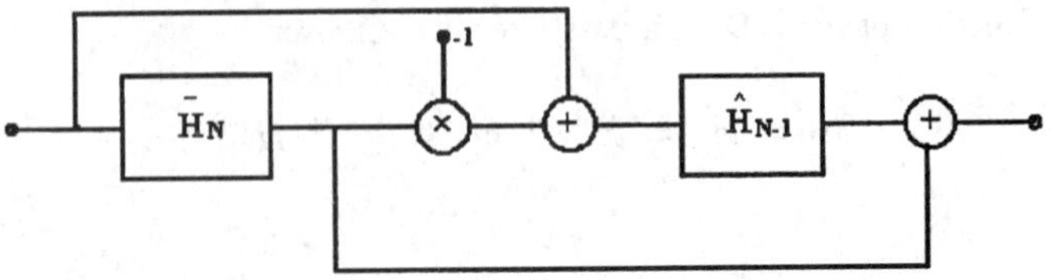

FIGURE 88.15 2-D high-pass filter configuration.

in (88.65) does not have type preservation in the neighborhoods of the Nyquist points but their presence does not appear to be of serious concern.

With the availability of circularly symmetric low-pass and high-pass filters, bandpass and bandstop filters with circularly symmetric amplitude responses can be readily obtained. A bandpass filter can be obtained by connecting a low-pass filter and a high-pass filter with overlapping passbands in cascade, whereas a bandstop filter can be realized by connecting a low-pass filter and a high-pass filter with overlapping passbands in parallel.

Design of Filters Satisfying Prescribed Specifications

A similar approach to that described in the third subsection in Section 88.3 can be used for the design of circularly symmetric filters satisfying prescribed specifications. Assume that the maximum/minimum passband and the maximum stopband gain of the 2-D filter are $(1 \pm \Delta p)$ and Δa, respectively, if K rotated filter sections are cascaded where half of the rotations are in the range of 180° to 270° and the other half are in the range 270° to 360°, then, we can assign the passband ripple δ_p and the stopband loss δ_a to be [2]

$$\delta_p = \frac{\Delta p}{K} \tag{88.70}$$

and

$$\delta_a = \Delta a^{2/K} \tag{88.71}$$

The lower (or upper) bound of the passband gain would be achieved if all the rotated sections were to have minimum (or maximum) passband gains at the same frequency point. Although it is possible for all the rotated sections to have minimum (or maximum) gains at the origin of the (ω_1, ω_2) plane, the gains are unlikely to be maximum (or minimum) together at some other frequency point and, in effect, the preceding estimate for δ_p is low. A more realistic value for δ_p is

$$\delta_p = \frac{2\Delta p}{K} \tag{88.72}$$

If Δp and Δa are prescribed, then the passband ripple and minimum stopband attenuation of the analog filter can be obtained from (88.72) and (88.71) as

$$A_p = 20 \log\left[\frac{K}{K - 2\Delta p}\right] \tag{88.73}$$

and

$$A_a = \frac{40}{K} \log\left[\frac{1}{\Delta a}\right] \tag{88.74}$$

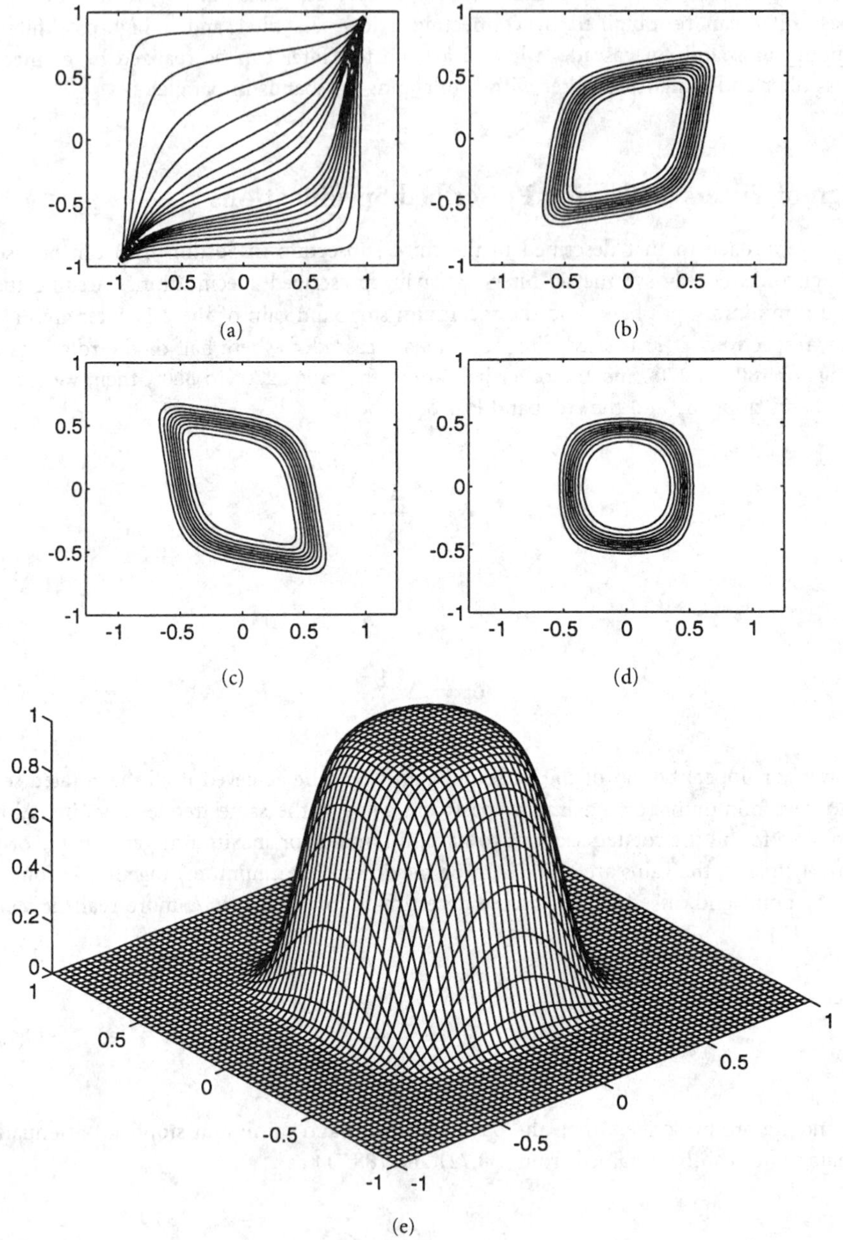

FIGURE 88.16 The amplitude response of a circularly symmetric filter in Example 4. (a) Subfilter for rotation angle of 243°. (b) Subfilters for rotation angles of 189°, 207°, 225°, 243°, and 261° in cascade. (c) Subfilters for rotation angles of 279°, 297°, 315°, 333°, and 351° in cascade. (d) All 10 subfilters in cascade. (e) 3-D plot of the amplitude response of the resulting 2-D low-pass filter.

If the passband Ω_p and stopband Ω_a are also prescribed, the minimum order and the transfer function of the analog filter can be determined using the method in preceding sections.

Example 4. Using the method of Costa and Venetsanopoulos, design a circularly symmetric low-pass filter satisfying the following specifications:

$$\omega_{s1} = \omega_{s2} = 2\pi \text{ rad/s}$$

$$\omega_p = 0.4\pi \text{ rad/s}, \qquad \omega_a = 0.6\pi \text{ rad/s}$$

$$\delta_p = \delta_a = 0.1$$

Solution. The filter satisfying prescribed specifications can be designed through the following steps:

1. Select a prototype of approximation and suitably select the number of rotations K.
2. Calculate rotation angles by (88.51).
3. Determine A_p and A_a from δ_p and δ_a, respectively, using (88.73) and (88.74). Calculate the prewarped Ω_p and Ω_a, from ω_p and ω_a, respectively.
4. Use above calculated specifications to obtain the prewarped 1-D analog transfer function.
5. Apply the transformations of (88.52), (88.53), (88.57), and (88.58) to obtain K rotated subfilters.
6. Cascade all the rotated subfilters.

The 3-D plot of the amplitude response of the resulting filter is shown in Fig. 88.16, where $K = 10$.

8.5 Design of 2-D IIR Filters by Optimization

In the preceding subsections, several methods for the solution of approximation problems in 2-D IIR filters have been described. These methods lead to a complete description of the transfer function in closed form, either in terms of its zeros and poles or its coefficients. They are, as a consequence, very efficient and lead to very precise designs. Their main advantage is that they are applicable only for the design of filters with piecewise-constant amplitude responses. In the following subsections, the optimization methods for the design of 2-D IIR filters are considered. In these methods, a discrete transfer function is assumed and an error function is formulated on the basis of some desired amplitude and/or phase response. These methods are iterative and, as a result, they usually involve a large amount of computation. However, unlike the closed-form methods, they are suitable for the design of filters having arbitrary amplitude or phase responses.

Design by Least *p*th Optimization

The least pth optimization method has been used quite extensively in the past in a variety of applications. In this approach, an objective function in the form of a sum of elemental error functions, each raised to the pth power, is first formulated and is then minimized using any one of the available unconstrained optimization methods [17].

Problem Formulation

Consider the transfer function

$$H(z_1, z_2) = H_0 \prod_{k=1}^{K} \frac{N_k(z_1, z_2)}{D_k(z_1, z_2)}$$

$$= H_0 \prod_{k=1}^{K} \frac{\sum_{l=0}^{L_{1k}} \sum_{m=0}^{M_{1k}} a_{lm}^{(k)} z_1^{-1} z_2^{-m}}{\sum_{l=0}^{L_{2k}} \sum_{m=0}^{M_{2k}} b_{lm}^{(k)} z_1^{-1} z_2^{-m}} \tag{88.75}$$

where $N_k(z_1, z_2)$ and $D_k(z_1, z_2)$ are polynomials of order equal to or less than 2 and H_0 is a constant, and let

$$\mathbf{x} = [\mathbf{a}^T \quad \mathbf{b}^T \quad H_0]^T \tag{88.76}$$

where

$$\mathbf{a} = [a_{10}^{(1)} \quad a_{20}^{(1)} \quad \cdots \quad a_{L_{11}M_{11}}^{(1)} \quad a_{10}^{(2)} \quad a_{20}^{(2)} \quad \cdots \quad a_{L_{12}M_{12}}^{(2)} \quad \cdots \quad a_{L_{1K}M_{1K}}^{(K)}]^T$$

and

$$\mathbf{b} = [b_{10}^{(1)} \quad b_{20}^{(1)} \quad \cdots \quad b_{L_{11}M_{11}}^{(1)} b_{10}^{(2)} a_{20}^{(2)} \cdots b_{L_{12}M_{12}}^{(2)} \cdots b_{L_{2K}M_{2K}}^{(K)}]^T$$

are row vectors whose elements are the coefficients of $N_k(z_1, z_2)$ and $D_k(z_1, z_2)$, respectively. An objective function can be defined in terms of the L_p norm of $\mathbf{E}(\mathbf{x})$ as

$$J(\mathbf{x}) = L_p = \|\mathbf{E}(\mathbf{x})\|_p = \left[\sum_{i=1}^{K} |E_{mn}(\mathbf{x})|^p \right]^{1/p} \tag{88.77}$$

where p is an even positive integer,

$$E_{mn}(\mathbf{x}) = M(m, n) - M_I(m, n) \tag{88.78}$$

and

$$M(m, n) = |H(e^{j\omega_{1m}T_1}, e^{j\omega_{2n}T_2})| \qquad m = 1, \cdots, M, n = 1, \cdots, N$$

are samples of the amplitude response of the filter at a set of frequency pairs $(\omega_{1m}, \omega_{2n})$, $(m = 1, \cdots, M, n = 1, \cdots, N)$ with

$$\omega_{1m} = \frac{\omega_{s1}(m-1)}{2(M-1)} \qquad \omega_{2n} = \frac{\omega_{s2}(n-1)}{2(N-1)}$$

$M_I(m, n)$ represents the desired amplitude response at frequencies $(\omega_{1m}, \omega_{2n})$.

Several special cases of the L_p norm are of particular interest. The L_1 norm, namely

$$L_1 = \sum_{i=1}^{K} |E_{mn}(\mathbf{x})|$$

is the sum of the magnitudes of the elements of $\mathbf{E}(\mathbf{x})$; the L_2 norm given by

$$L_2 = \left[\sum_{i=1}^{K} |E_{mn}(\mathbf{x})|^2\right]^{1/2}$$

is the well-known Euclidean norm; and L_2^2 is the sum of the squares of the elements of $E(\mathbf{x})$. In the case where $p \to \infty$ and

$$E_M(\mathbf{x}) = \max_{m,n}\{E_{mn}(\mathbf{x})\} \neq 0$$

we can write

$$\begin{aligned} L_\infty &= \lim_{p\to\infty}\left\{\sum_{k=1}^{K}|E_{mn}(\mathbf{x})|^p\right\}^{1/p} \\ &= E_M(\mathbf{x})\lim_{p\to\infty}\left\{\sum_{k=1}^{K}\left|\frac{E_{mn}(\mathbf{x})}{E_M(\mathbf{x})}\right|^p\right\}^{1/p} \\ &= E_M(\mathbf{x}) \end{aligned} \tag{88.79}$$

The design task at hand amounts to finding a parameter vector $\mathbf{x}$ that minimizes the least pth objective function $J(\mathbf{x})$ defined in (88.77). If $J(\mathbf{x})$ is defined in terms of L_2^2, a least-squares solution is obtained; if the L_∞ norm is used, a so-called minimax solution is obtained, since in this case the largest element in $\mathbf{E}(\mathbf{x})$ is minimized.

Quasi-Newton Algorithms

The design problem described above can be solved by using any one of the standard unconstrained optimization algorithms. A class of such algorithms that have been found to be very versatile, efficient, and robust is the class of quasi-Newton algorithms [17]–[19]. These are based on the principle that the minimum point $\mathbf{x}^*$ of a quadratic convex function $J(\mathbf{x})$ of N variables can be obtained by applying the correction

$$\boldsymbol{\delta} = -\mathbf{H}^{-1}\mathbf{g}$$

to an arbitrary point $\mathbf{x}$, that is

$$\mathbf{x}^* = \mathbf{x} + \boldsymbol{\delta}$$

where vector

$$\mathbf{g} = \nabla J(\mathbf{x}) = \left[\frac{\partial J}{\partial x_1}, \frac{\partial J}{\partial x_2}, \ldots, \frac{\partial J}{\partial x_N}\right]^T$$

and $N \times N$ matrix

$$\mathbf{H} = \begin{vmatrix} \dfrac{\partial^2 J(\mathbf{x})}{\partial x_1^2} & \dfrac{\partial^2 J(\mathbf{x})}{\partial x_1 \partial x_2} & \cdots & \dfrac{\partial^2 J(\mathbf{x})}{\partial x_1 \partial x_N} \\ \dfrac{\partial^2 J(\mathbf{x})}{\partial x_2 \partial x_1} & \dfrac{\partial^2 J(\mathbf{x})}{\partial x_2^2} & \cdots & \dfrac{\partial^2 J(\mathbf{x})}{\partial x_2 \partial x_N} \\ \cdots & \cdots & \cdots & \cdots \\ \dfrac{\partial^2 J(\mathbf{x})}{\partial x_N \partial x_1} & \dfrac{\partial^2 J(\mathbf{x})}{\partial x_N \partial x_2} & \cdots & \dfrac{\partial^2 J(\mathbf{x})}{\partial x_N^2} \end{vmatrix}$$

are the gradient vector and Hessian matrix of $J(\mathbf{x})$ at point $\mathbf{x}$, respectively.

The basic quasi-Newton algorithm as applied to the 2-D IIR filter design problem is as follows [20].

Algorithm 1: Basic Quasi-Newton Algorithm

STEP 1. Input $\mathbf{x}_0$ and ε. Set $\mathbf{S}_0 = \mathbf{I}_N$, where $\mathbf{I}_N$ is the $N \times N$ unity matrix and N is the dimension of $\mathbf{x}$, and set $k = 0$. Compute $\mathbf{g}_0 = \nabla J(\mathbf{x}_0)$.

STEP 2. Set $\mathbf{d}_k = -\mathbf{S}_k \mathbf{g}_k$ and find α_k, the value of α that minimizes $\mathbf{J}(\mathbf{x}_k + \alpha \mathbf{d}_k)$, using a line search.

STEP 3. Set $\boldsymbol{\delta}_k = \alpha_k \mathbf{d}_k$ and $\mathbf{x}_{k+1} = \mathbf{x}_k + \boldsymbol{\delta}_k$.

STEP 4. If $\|\boldsymbol{\delta}_k\|_2 < \varepsilon$, then output $\mathbf{x}^* = \mathbf{x}_{k+1}$, $\mathbf{J}(\mathbf{x}^*) = J(\mathbf{x}_{k+1})$ and stop, else go to step 5.

STEP 5. Compute $\mathbf{g}_{k+1} = J(\mathbf{x}_{k+1})$ and set $\boldsymbol{\gamma}_k = \mathbf{g}_{k+1} - \mathbf{g}_k$.

STEP 6. Compute $\mathbf{S}_{k+1} = \mathbf{S}_k + \mathbf{C}_k$, where $\mathbf{C}_k$ is a suitable matrix correction.

STEP 7. Check $\mathbf{S}_{k+1}$ for positive definiteness and if it is found to be nonpositive definite force it to become positive definite.

STEP 8. Set $k = k + 1$ and go to step 2.

The correction matrix $\mathbf{C}_k$ required in step 6 can be computed by using either the Davidon–Fletcher–Powell (DFP) formula

$$\mathbf{C}_k = \frac{\boldsymbol{\delta}_k \boldsymbol{\delta}_k^T}{\boldsymbol{\gamma}_k^T \boldsymbol{\gamma}_k} \frac{\mathbf{S}_k \boldsymbol{\gamma}_k \boldsymbol{\gamma}_k^T \mathbf{S}}{\boldsymbol{\gamma}_k^T \mathbf{S}_k \boldsymbol{\gamma}_k} \tag{88.80}$$

or the Broyden–Fletcher–Goldfarb–Shanno (BFGS) formula

$$\mathbf{C}_k = \left(1 + \frac{\boldsymbol{\gamma}_k^T \mathbf{S}_k \boldsymbol{\gamma}_k}{\boldsymbol{\delta}_k^T \boldsymbol{\delta}_k}\right) \frac{\boldsymbol{\delta}_k \boldsymbol{\delta}_k^T}{\boldsymbol{\gamma}_k^T \boldsymbol{\delta}_k} - \frac{\boldsymbol{\delta}_k \boldsymbol{\gamma}_k^T \mathbf{S}_k + \mathbf{S}_k \boldsymbol{\gamma}_k \boldsymbol{\delta}_k^T}{\boldsymbol{\gamma}_k^T \boldsymbol{\delta}_k} \tag{88.81}$$

Algorithm 1 eliminates the need to calculate the second derivatives of the objective function, in addition, the matrix inversion is unnecessary. However, matrices $\mathbf{S}_1, \mathbf{S}_2, \cdots, \mathbf{S}_k$ need to be checked for positive definiteness and may need to be manipulated. This can be easily done in practice by diagonalizing $\mathbf{S}_{k+1}$ and then replacing any nonpositive diagonal elements with corresponding positive ones. However, this would increase the computational burden quite significantly. The amount of computation required to complete a design is usually very large, due with the large numbers of variables in 2-D digital filters and the large number of sample points needed to construct the objective function. Generally, the computational load can often be reduced by starting with an approximate design based on some closed-form solution. For example, the design of circularly or elliptical symmetric filters may start with filters that have square or rectangular passbands and stopbands.

Example 5 [20]. Design a circularly symmetric low-pass filter of order (2, 2) with $\omega_{p1} = \omega_{p2} = 0.08\pi$ rad/s and $\omega_{a1} = \omega_{a2} = 0.12\pi$ rad/s, assuming that $\omega_{s1} = \omega_{s2} = 2\pi$ rad/s.

Solution.

1. Construct the ideal discrete amplitude response of the filter

$$M_I(m,n) = \begin{cases} 1 & \text{for } (\omega_{1m}^2 + \omega_{2n}^2) \le 0.08\pi \\ 0.5 & \text{for } 0.08\pi \le (\omega_{1m}^2 + \omega_{2n}^2) \le 0.12\pi \\ 0 & \text{otherwise} \end{cases}$$

where

$$\{\omega_{1m}\} = \{\omega_{2n}\} = 0, 0.02\pi, 0.04\pi, \cdots, 0.2\pi, 0.4\pi, 0.6\pi, 0.8\pi, \pi$$

2. To reduce the amount of computation, a 1-D low-pass filter with passband edge $\omega_p = 0.08\pi$ and stopband edge $\omega_a = 0.1\pi$ is first obtained with the 1-D transfer function being

$$H_1(z) = 0.11024\frac{1 - 1.64382z^{-1} + z^{-2}}{1 - 1.79353z^{-1} + 0.84098z^{-2}}$$

and then a 2-D transfer function with a square passband is obtained as

$$H(z_1, z_2) = H_1(z_1)H_1(z_2)$$

3. Construct the objective function of (88.77), using algorithm 1 to minimize the objective function $J(\mathbf{x})$. After 20 more iterations the algorithm converges to

$$H(z_1,z_2) = 0.00895 \times \frac{[1 \;\; z_1^{-1} \;\; z_1^{-2}]\begin{bmatrix} 1.0 & -1.62151 & 0.99994 \\ -1.62151 & 2.63704 & -1.62129 \\ 0.99994 & -1.62129 & 1.00203 \end{bmatrix}\begin{bmatrix} 1 \\ z_2^{-1} \\ z_2^{-2} \end{bmatrix}}{[1 \;\; z_1^{-1} \;\; z_1^{-2}]\begin{bmatrix} 1.0 & -1.78813 & 0.82930 \\ -1.78813 & 3.20640 & -1.49271 \\ 0.82930 & -1.49271 & 0.69823 \end{bmatrix}\begin{bmatrix} 1 \\ z_2^{-1} \\ z_2^{-2} \end{bmatrix}}$$

The amplitude response of the final optimal filter is depicted in Fig. 88.17.

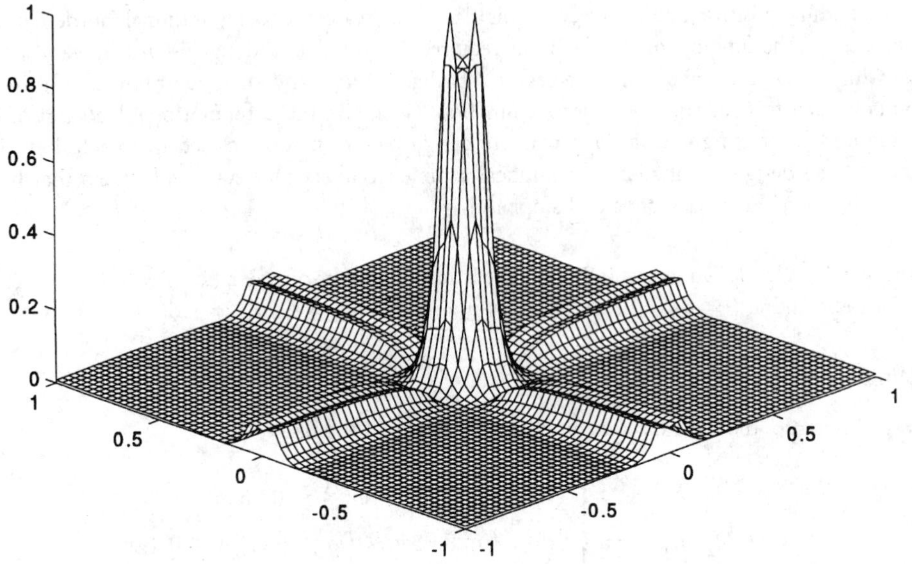

FIGURE 88.17 Amplitude response of the 2-D optimal filter in Example 5.

Minimax Algorithms

Least pth Minimax Algorithm. When an objective function is formulated in terms of the L_p norm of the error function and then minimizing $J(\mathbf{x})$ for increasing values of p, such an objective function can be obtained as

$$J(\mathbf{x}) = E_M(\mathbf{x})\left\{\sum_{i=1}^{m}\left|\frac{E(\mathbf{x}, \omega_{1i}, \omega_{2i})}{E_M(\mathbf{x})}\right|^p\right\}^{1/p} \tag{88.82}$$

where

$$E_M(\mathbf{x}) = \max_{1\le i\le m}\{E_i(\mathbf{x})\} = \max_{1\le i\le m}\{E(\mathbf{x}, \omega_{1i}, \omega_{2i})\} \tag{88.83}$$

A minimax algorithm based on $J(\mathbf{x})$ is as follows [21].

Algorithm 2: Least-*p*th Minimax Algorithm

STEP 1. Input $\mathbf{x}_0$ and ε. Set $k = 1$, $p = 2$, $\mu = 2$, $\mathbf{E}_0 = 10^{99}$.

STEP 2. Initialize frequencies ω_{1i}, ω_{2i} for $i = 1, 2, \cdots, m$.

STEP 3. Using point $\mathbf{x}_{k-1}$ as initial point, minimize $J(\mathbf{x})$ with respect to $\mathbf{x}$ to obtain $\mathbf{x}_k$. Set $\mathbf{E}_k = E(\mathbf{x}_k)$.

STEP 4. If $|\mathbf{E}_{k-1} - \mathbf{E}_k| < \varepsilon$, then output $\mathbf{x}^* = \mathbf{x}_k$ and $\mathbf{E}_k$ and stop. Else, set $p = \mu p$, $k = k + 1$ and go to step 3.

The minimization in step 3 can be carried out using Algorithm 1 or any other unconstrained optimization algorithms.

Charalambous Minimax Algorithm. The preceding algorithm gives excellent results except that it requires a considerable amount of computation. An alternative and much more efficient algorithm is the minimax algorithm proposed by Charalambous and Antoniou [22], [23]. This algorithm is based on principles developed by Charalambous [24] and involves the minimization of the objective function $J(\mathbf{x}, \zeta, \lambda)$ defined by

$$J(\mathbf{x}, \zeta, \lambda) = \sum_{i \in I_1} \frac{1}{2}\lambda_i [J_i(\mathbf{x}, \zeta)]^2 + \sum_{i \in I_2} \frac{1}{2}[J_i(\mathbf{x}, \zeta)]^2 \tag{88.84}$$

where ζ and λ_i for $i = 1, 2, \cdots, m$ are constants,

$$J_i(\mathbf{x}, \zeta) = E_i(\mathbf{x}) - \zeta$$

$$I_1 = \{i: J_i(\mathbf{x}, \zeta) > 0 \quad \text{and} \quad \lambda_i > 0\}$$

and

$$I_2 = \{i: J_i(\mathbf{x}, \zeta) > 0 \quad \text{and} \quad \lambda_i = 0\}$$

The factor 1/2 in (88.84) are included for the purpose of simplifying the gradient that is given by

$$\nabla J(\mathbf{x}, \zeta, \lambda) = \sum_{i \in I_1} \lambda_i J_i(\mathbf{x}, \zeta) \nabla J_i(\mathbf{x}, \zeta) + \sum_{i \in I_2} J_i(\mathbf{x}, \zeta) \nabla J_i(\mathbf{x}, \zeta) \tag{88.85}$$

It can be shown that, if

1. the second-order sufficient conditions for a minimum hold at x^*,
2. $\lambda_i = \lambda_i^*$, $i = 1, 2, \ldots, m$, where λ_i^* are the minimax multipliers corresponding to a minimum optimum solution $\mathbf{x}^*$, and
3. $E(\mathbf{x}^*) - \zeta$ is sufficiently small then $\mathbf{x}^*$ is a strong local minimum point of $J(\mathbf{x}, \zeta, \lambda)$.

Condition 1 are usually satisfied in practice. Therefore, a local minimum point $\mathbf{x}^*$ can be found by forcing λ_i to approach λ_i^* $(i = 1, 2, \cdots, m)$ and making $E(\mathbf{x}^*) - \zeta$ sufficiently small. These two constraints can be simultaneously satisfied by applying the following algorithm.

Algorithm 3: Charalambous Minimax Algorithm

STEP 1. Set $\zeta = 0$ and $\lambda_i = 1$ for $i = 1, 2, \cdots, m$. Initialize $\mathbf{x}$.

STEP 2. Minimize function $J(\mathbf{x}, \zeta, \lambda)$ to obtain $\mathbf{x}$.

STEP 3. Set

$$S = \sum_{i \in I_1} \lambda_i J_i(\mathbf{x}, \zeta) + \sum_{i \in I_2} J_i(\mathbf{x}, \zeta)$$

and update λ_i and ζ as

$$\lambda_i = \begin{cases} \lambda_i J_i(\mathbf{x}, \zeta)/S & \text{if } J_i(\mathbf{x}, \zeta) \geq 0, \lambda_i \geq 0 \\ J_i(\mathbf{x}, \zeta)/S & \text{if } J_i(\mathbf{x}, \zeta) \geq 0, \lambda_i = 0 \\ 0 & \text{if } J_i(\mathbf{x}, \zeta) < 0 \end{cases}$$

$$\zeta = \sum_{i=1}^{m} \lambda_i E_i(\mathbf{x})$$

STEP 4. Stop if

$$\frac{E_M(\mathbf{x}) - \zeta}{E_M(\mathbf{x})} \leq \varepsilon$$

otherwise go to step 2.

The parameter ε is a prescribed termination tolerance. When the algorithm converges, conditions 2 and 3 are satisfied and $\mathbf{x} = \mathbf{x}^*$. The unconstrained optimization in step 2 can be accomplished by applying a quasi-Newton algorithm.

Example 6 [23]. Design a 2-D circularly symmetric filter with the same specifications as in Example 5, using algorithm 3.

Solution.

1. Construct the ideal discrete amplitude response of the filter. Since the passband and stopband contours are circles, the sample points can be placed on arcs of a set of circles centered at the origin. Five circles with radii

$$r_1 = 0.3\omega_p, \quad r_2 = 0.6\omega_p, \quad r_3 = 0.8\omega_p, \quad r_4 = 0.9\omega_p, \quad \text{and} \quad r_5 = \omega_p$$

are placed in the passband and five circles with radii

$$r_6 = \omega_a, \quad r_7 = \omega_a + 0.1(\pi - \omega_a), \quad r_8 = \omega_a + 0.2(\pi - \omega_a),$$
$$r_9 = \omega_a + 0.55(\pi - \omega_a) \quad \text{and} \quad r_{10} = \pi$$

are placed in the stopband.
For circularly symmetric filters, the amplitude is uniquely specified by the amplitude response in the sector $[0°, 45°]$. Therefore, six equally spaced points on each circle described above between 0° and 45° are chosen. These points plus the origin $(\omega_1, \omega_2) = (0, 0)$ form a set of 61 sample points.
2. Select the 2-D transfer function. Because a circularly symmetric filter has a transfer function with separable denominator [24], we can select the transfer function to be of the form

$$H(z_1, z_2) = H_0(z_1, z_2)^{-1}$$
$$\times \prod_{k=1}^{K} \frac{z_1 z_2 + z_1^{-1} z_2^{-1} + a_k(z_1 + z_1^{-1} + z_2 + z_2^{-1}) + z_1^{-1} z_2 + z_1 z_2^{-1} + b_k}{(1 + c_k z_1^{-1} + d_k z_1^{-2})(1 + c_k z_2^{-1} + d_k z_2^{-2})} \tag{88.86}$$

with parameter H_0 fixed as $H_0 = (0.06582)^2$, $K = 1$, $\varepsilon = 0.01$.
3. Starting from

$$a_1^{(0)} = -1.514, \quad b_1^{(0)} = (a_1^{(0)})^2, \quad c_1^{(0)} = -1.784, \quad d_1^{(0)} = 0.8166$$

and algorithm 3 yields the solution

$$a_1^* = 1.96493, \quad b_1^* = -10.9934, \quad c_1^* = -1.61564, \quad d_1^* = 0.66781$$
$$E_M(\mathbf{x}) = 0.37995$$

The 3-D plot of the amplitude response of the resulting filter is illustrated in Fig. 88.18.

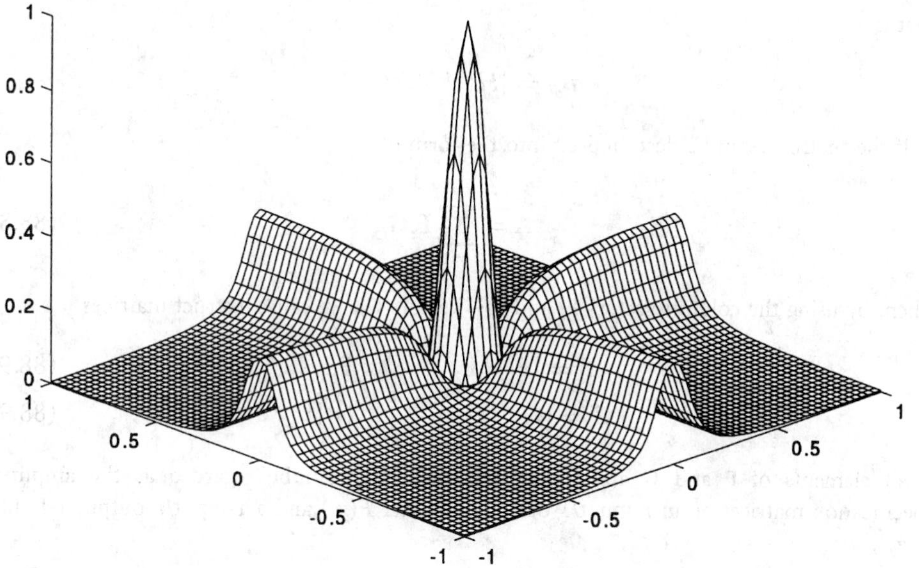

FIGURE 88.18 3-D plot of the amplitude response of the filter in Example 6.

Design by Singular-Value Decomposition

As will be seen, an important merit of the design methods of 2-D IIR filters based on singular-value decomposition (SVD) is that the required 2-D filter is decomposed into a set of 1-D digital subfilters, which are much easier to design by optimization than the original 2-D filters.

Problem Formulation

In a quadrantally symmetric filter, $H(z_1, z_2)$ has a separable denominator [24]. Therefore, it can be expressed as

$$H(z_1, z_2) = \sum_{i=1}^{K} f_i(z_1) g_i(z_2) \tag{88.87}$$

in effect, a quadrantally symmetric filter can always be realized using a set of K parallel sections where the ith section is a separable subfilter characterized by the transfer function $f_i(z_1)g_i(z_2)$.

Consider the desired amplitude response sample of 2-D filter $H(z_1, z_2)$, we form a 2-D amplitude specification matrix $\mathbf{A}$ as

$$\mathbf{A} = \begin{vmatrix} a_{11} & a_{12} & \cdots & a_{1L} \\ a_{21} & a_{22} & \cdots & a_{2L} \\ \cdots & \cdots & \cdots & \cdots \\ a_{M1} & a_{M1} & \cdots & a_{ML} \end{vmatrix} \tag{88.88}$$

where $\{a_{ml}\}$ is a desired amplitude response sampled at frequencies $(\omega_{1l}, \omega_{2m}) = (\pi\mu_l/T_1, \pi\nu_m/T_2)$, with

$$\mu_l = \frac{l-1}{L-1}, \qquad \nu_m = \frac{m-1}{M-1} \quad \text{for } 1 \le l \le L, \quad 1 \le m \le M$$

that is,

$$a_{ml} = |H(e^{j\pi\mu_l}, e^{j\pi\nu_m})|$$

If the matrix **A** can be decomposed into the form of

$$\mathbf{A} = \sum_{i=1}^{r} \mathbf{F}_i \mathbf{G}_i \tag{88.89}$$

Then, by using the column vectors $\mathbf{F}_i$ and row vectors $\mathbf{G}_i$, we can construct matrices

$$\mathbf{F} = [\mathbf{F}_1 \quad \mathbf{F}_2 \quad \cdots \quad \mathbf{F}_r] \tag{88.90}$$

$$\mathbf{G} = [\mathbf{G}_1 \quad \mathbf{G}_2 \quad \cdots \quad \mathbf{G}_r]^T \tag{88.91}$$

if all elements of **F** and **G** are nonnegative then they can be regarded as the amplitude specification matrices of an r-input/1-output 1-D filter $F(z_1)$ and a 1-input/r-output 1-D filter $G(z_2)$,

$$F(z_1) = [f_1(z_1), f_2(z_1), \cdots, f_r(z_1)] \tag{88.92}$$

$$G(z_2) = [g_1(z_2), g_2(z_2), \ldots, g_r(z_2)]^T \tag{88.93}$$

Therefore, the 2-D filter of (88.87) can be approximated by

$$H(z_1, z_2) = F(z_1)G(z_2) \tag{88.94}$$

In this subsection, two design procedures are described that can be applied to the design of 2-D IIR filters whose amplitude responses are quadrantally symmetric.

Method of Antoniou and Lu

Antoniou and Lu proposed a method of 2-D IIR filter based on the SVD of the amplitude response matrix **A** [26]. The SVD of matrix **A** gives [27]

$$\mathbf{A} = \sum_{i=1}^{r} \sigma_i \mathbf{u}_i \mathbf{v}_i^T = \sum_{i=1}^{r} \boldsymbol{\phi}_i \boldsymbol{\gamma}_i^T \tag{88.95}$$

where σ_i are the singular values of **A** such that $\sigma_1 \geq \sigma_2 \geq \cdots \geq \sigma_r \geq 0$, r is the rank of **A**, $\mathbf{u}_i$ and $\mathbf{v}_i$ are the ith eigenvector of $\mathbf{A}\mathbf{A}^T$ and $\mathbf{A}^T\mathbf{A}$, respectively, $\boldsymbol{\phi}_i = \sigma_i^{1/2}\mathbf{u}_i$, $\boldsymbol{\gamma}_i = \sigma_i^{1/2}\mathbf{v}_i$, and $\{\boldsymbol{\phi}_i: 1 \leq i \leq r\}$ and $\{\boldsymbol{\gamma}_i: 1 \leq i \leq r\}$ are sets of orthogonal L-dimensional and M-dimensional vectors, respectively.

An important property of the SVD can be stated as

$$\left\| \mathbf{A} - \sum_{i=1}^{K} \boldsymbol{\phi}_i \boldsymbol{\gamma}_i^T \right\| = \min_{\bar{\phi}, \bar{\gamma}} \left\| \mathbf{A} - \sum_{i=1}^{K} \bar{\boldsymbol{\phi}}_i \bar{\boldsymbol{\gamma}}_i^T \right\| \text{for } 1 \leq K \leq r \tag{88.96}$$

where $\bar{\boldsymbol{\phi}}_i \in R^L$, $\bar{\boldsymbol{\gamma}}_i \in R^M$.

To design a 2-D IIR filter by SVD, two steps are involved, namely

STEP 1. Design of the main section

STEP 2. Design of the error correction sections as will be detailed below.

Design of the Main Section. Note that (88.95) can be written as

$$\mathbf{A} = \boldsymbol{\phi}_1 \boldsymbol{\gamma}_1^T + \varepsilon_1 \tag{88.97}$$

where $\varepsilon_1 = \sum_{i=2}^{r} \boldsymbol{\phi}_i \boldsymbol{\gamma}_i^T$. And since all the elements of $\mathbf{A}$ are nonnegative, it follows that all elements of $\boldsymbol{\phi}_1$ and $\boldsymbol{\gamma}_1$ are nonnegative.

On comparing (88.97) with (88.95) and assuming that $K = 1$ and that $\boldsymbol{\phi}_1$, $\boldsymbol{\gamma}_1$ are sampled versions of the desired amplitude responses for the 1-D filters characterized by $f_1(z_1)$ and $g_1(z_2)$, respectively, a 2-D filter can be designed through the following procedures:

1. Design 1-D filters F_1 and G_1 characterized by $f_1(z_1)$ and $g_1(z_2)$.
2. Connect filters F_1 and G_1 in cascade, i.e.

$$H_1(z_1, z_2) = f_1(z_1)g_1(z_2)$$

Step 1 above can be carried out by using an optimization algorithm such as the quasi-Newton algorithm or the minimax algorithm.

Since $f_1(z_1)g_1(z_2)$ corresponds to the largest singular value σ_1, the subfilter characterized by $f_1(z_1)g_1(z_2)$ is said to be the main section of the 2-D filter.

Design of the Error Correction Sections The approximation error of $H_1(z_1, z_2)$ can be reduced by realizing more of the terms in (88.95) by means of parallel filter sections. From (88.97), we can write

$$\mathbf{A} = \boldsymbol{\phi}_1 \boldsymbol{\gamma}_1^T + \boldsymbol{\phi}_2 \boldsymbol{\gamma}_2^T + \varepsilon_{21} \tag{88.98}$$

Since ϕ_2 and γ_2 may have some negative components, a careful treatment in (88.98) is necessary.

Let ϕ_2^- and γ_2^- be the absolute values of the most negative components of ϕ_2 and γ_2, respectively. If

$$\mathbf{e}_\phi = [1 \quad 1 \quad \cdots \quad 1]^T \in R^L \quad \text{and} \quad \mathbf{e}_\gamma = [1 \quad 1 \quad \cdots \quad 1]^T \in R^M$$

then all components of

$$\boldsymbol{\phi}_{2p} = \boldsymbol{\phi}_2 + \phi_2^- \mathbf{e}_\phi \quad \text{and} \quad \boldsymbol{\gamma}_{2p} = \boldsymbol{\gamma}_2 + \gamma_2^- \mathbf{e}_\gamma$$

are nonnegative. If it is possible to design 1-D linear-phase or zero-phase filters characterized by $f_1(z_1)$, $g_1(z_2)$, $f_{2p}(z_1)$, and $g_{2p}(z_2)$, such that

$$f_1(e^{j\pi\mu_l}) = |f_1(e^{j\pi\mu_l})|e^{j\alpha_1\mu_l}$$
$$g_1(e^{j\pi\nu_m}) = |g_1(e^{j\pi\nu_m})|e^{j\alpha_2\nu_m}$$

and

$$f_{2p}(e^{j\pi\mu_l}) = |f_{2p}(e^{j\pi\mu_l})|e^{j\alpha_1\mu_l}$$
$$g_{2p}(e^{j\pi\nu_m}) = |g_{2p}(e^{j\pi\nu_m})|e^{j\alpha_2\nu_m}$$

for $1 \le l \le L$, $1 \le m \le M$, where

$$|f_1(e^{j\pi\mu_l})| \approx \phi_{1l}$$
$$|g_1(e^{j\pi\nu_m})| \approx \gamma_{1m}$$
$$|f_{2p}(e^{j\pi\mu_l})| \approx \phi_{2lp}$$
$$|g_{2p}(e^{j\pi\nu_m})| \approx \gamma_{2mp}$$

In above ϕ_{1l}, ϕ_{2lp}, γ_{1m} and γ_{2mp} represent the lth component of $\boldsymbol{\phi}_1$, $\boldsymbol{\phi}_{2p}$ and mth component of $\boldsymbol{\gamma}_1$ and $\boldsymbol{\gamma}_{2p}$, respectively. α_1 and α_2 are constants that are equal to zero if zero-phase filters are to be designed. Let

$$\alpha_1 = -\pi n_1, \quad \alpha_2 = -\pi n_2 \quad \text{with integers } n_1, n_2 \ge 0 \tag{88.99}$$

and define

$$f_2(z_1) = f_{2p}(z_1) - \phi_2^- z_1^{-n_1} \tag{88.100}$$
$$g_2(z_2) = g_{2p}(z_2) - \gamma_2^- z_2^{-n_2} \tag{88.101}$$

It follows that

$$f_2(e^{j\pi\mu_l}) = [f_{2p}(e^{j\pi\mu_l}) - \phi_2^-]e^{-j\pi n_1\mu_l} \approx \phi_{2l}e^{-j\pi\mu_l n_1}$$
$$g_2(e^{j\pi\nu_m}) = [g_{2p}(e^{j\pi\nu_m}) - \gamma_2^-]e^{-j\pi n_2\nu_m} \approx \gamma_{2m}e^{-j\pi\gamma_m n_2}$$

Furthermore, if we form

$$H_2(z_1, z_2) = f_1(z_1)g_1(z_2) + f_2(z_1)g_2(z_2) \tag{88.102}$$

then

$$\begin{aligned} |H_2(e^{j\pi\mu_l}, e^{j\pi\nu_m})| &= |f_1(e^{j\pi\mu_l})g_1(e^{j\pi\nu_m}) + f_2(e^{j\pi\mu_l})g_2(e^{j\pi\nu_m})| \\ &\approx |\phi_{1l}\gamma_{1m} + \phi_{2l}\gamma_{2m}| \end{aligned} \tag{88.103}$$

Follow this procedure, $K-1$ correction sections characterized by $f_2(z_1)g_2(z_2), \cdots, g_K(z_1)g_K(z_2)$ can be obtained, and $H_K(z_1, z_2)$ can be formed as

$$H_k(z_1, z_2) = \sum_{i=1}^{K} f_i(z_1)g_i(z_2) \tag{88.104}$$

and from (88.96) we have

$$\begin{aligned} \left\| \mathbf{A} - |H_K(e^{j\pi\mu_l}, e^{j\pi\nu_m})| \right\| &\approx \left\| \mathbf{A} - \left| \sum_{i=1}^{K} \boldsymbol{\phi}_i \boldsymbol{\gamma}_i^T \right| \right\| \\ &\le \|\varepsilon_K\| = \min_{\bar{\phi}_i, \bar{\gamma}_i} \left\| \mathbf{A} - \left| \sum_{i=1}^{K} \bar{\boldsymbol{\phi}}_i \bar{\boldsymbol{\gamma}}_i^T \right| \right\| \end{aligned} \tag{88.105}$$

In effect, a 2-D filter consisting of K sections is obtained whose amplitude response is a minimal mean-square-error approximation to the desired amplitude response.

The method leads to an asymptotically stable 2-D filter, provided that all 1-D subfilters employed are stable. The general configuration of the 2-D filter obtained is illustrated in Fig. 88.19, where the various 1-D subfilters may be either linear-phase or zero-phase filters.

If linear-phase subfilters are to be used, the equalities in (18.99) must be satisfied. This implies that the subfilters must have constant group delays. If zero-phase subfilters are employed, where $f_i(z_1)$ and $f_i(z_1^{-1})$, and $g_i(z_2)$ and $g_i(z_2^{-1})$ contribute equally to the amplitude response of the 2-D filter. The design can be accomplished by assuming that the desired amplitude responses for subfilters F_i, G_i are $\phi_i^{1/2}$, $\gamma_i^{1/2}$, for $i = 1, 2, \cdots, K$, respectively.

Error Compensation Procedure. When the main section and correction sections are designed by an optimization procedure as described above, approximation errors inevitably occur that will accumulate and manifest themselves as the overall error. The accumulation of error can be reduced by the following compensation procedure.

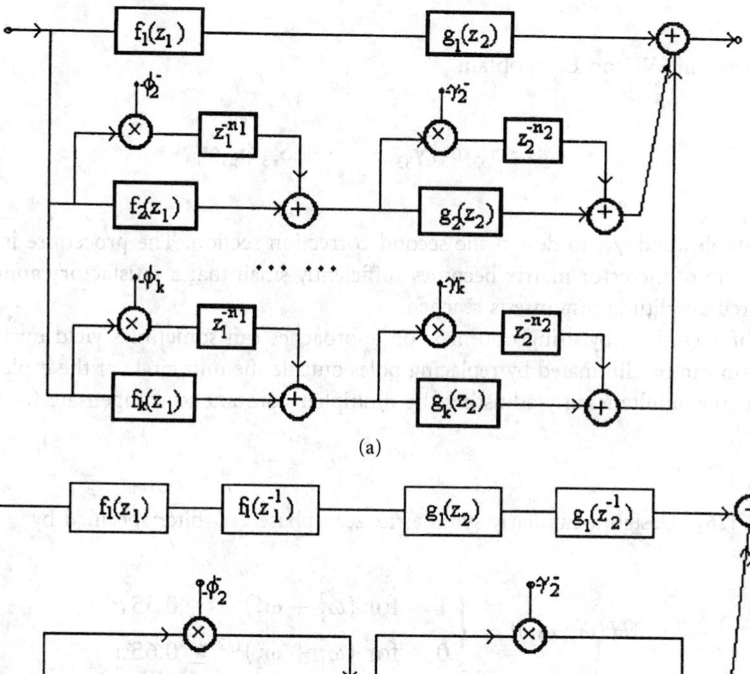

(a)

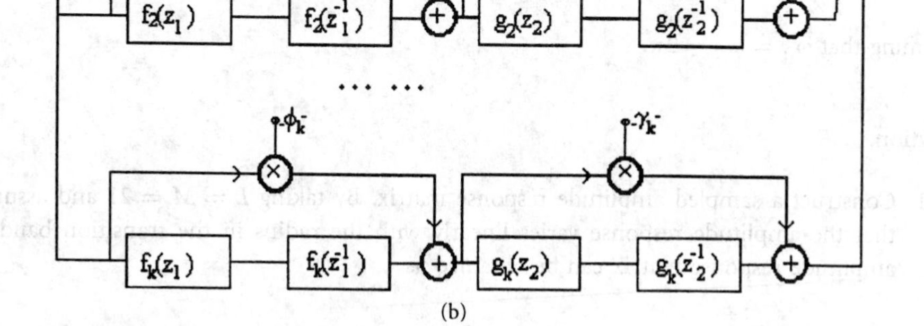

(b)

FIGURE 88.19 Configurations of 2-D IIR filter by SVD. (a) General structure of 2-D filter. (b) Structure using zero-phase IIR filters.

When filters F_1 and G_1 are designed, the approximation error matrix $\mathbf{E}_1$ can be calculated as

$$\mathbf{E}_1 = \mathbf{A} - f_1(e^{j\pi\mu_l})g_1(e^{j\pi\nu_m}) \tag{88.106}$$

and then perform SVD on $\mathbf{E}_1$ to obtain

$$\mathbf{E}_1 = S_{22}\boldsymbol{\phi}_{22}\boldsymbol{\gamma}_{22}^T + \cdots + S_{r2}\boldsymbol{\phi}_{r2}\boldsymbol{\gamma}_{r2}^T \tag{88.107}$$

Data $\boldsymbol{\phi}_{22}$ and $\boldsymbol{\gamma}_{22}$ can be used to deduce filters $f_2(z_1)$ and $g_2(z_2)$. Thus, the first correction section can be designed. Next, form the error matrix $\mathbf{E}_2$ as

$$\mathbf{E}_2 = \mathbf{E}_1 - S_{22}f_2(e^{j\pi\mu_l})g_2(e^{j\pi\nu_m}) \tag{88.108}$$

and then perform SVD on $\mathbf{E}_2$ to obtain

$$\mathbf{E}_2 = S_{33}\boldsymbol{\phi}_{33}\boldsymbol{\gamma}_{33}^T + \cdots + S_{r3}\boldsymbol{\phi}_{r3}\boldsymbol{\gamma}_{r3}^T \tag{88.109}$$

and use data $\boldsymbol{\phi}_{33}$ and $\boldsymbol{\gamma}_{33}$ to design the second correction section. The procedure is continued until the norm of the error matrix becomes sufficiently small that a satisfactory approximation to the desired amplitude response is reached.

Design of 1-D filters by using optimization approaches can sometimes yield unstable filters. This problem can be eliminated by replacing poles outside the unit circle of the z plane by their reciprocals and simultaneously adjusting the multiplier constant to compensate for the change in gain [19].

Example 7 [26]. Design a circularly symmetric, zero-phase 2-D filter specified by

$$|H(\omega_1, \omega_2)| = \begin{cases} 1 & \text{for } (\omega_1^2 + \omega_2^2)^{1/2} \leq 0.35\pi \\ 0 & \text{for } (\omega_1^2 + \omega_2^2)^{1/2} \geq 0.65\pi \end{cases}$$

assuming that $\omega_{s1} = \omega_{s2} = 2\pi$.

Solution.

1. Construct a sampled amplitude response matrix. By taking $L = M = 21$ and assuming that the amplitude response varies linearly with the radius in the transition band, the amplitude response matrix can be obtained as

$$\mathbf{A} = \begin{vmatrix} \mathbf{A}_1 & \mathbf{0} \\ \mathbf{0} & \mathbf{0} \end{vmatrix}_{21\times 21}$$

where

$$A_1 = \begin{vmatrix} 1 & 1 & 1 & 1 & 1 & 1 & 1 & 1 & 1 & 0.75 & 0.5 & 0.25 \\ 1 & 1 & 1 & 1 & 1 & 1 & 1 & 1 & 0.75 & 0.5 & 0.25 & 0 \\ 1 & 1 & 1 & 1 & 1 & 1 & 1 & 1 & 0.75 & 0.5 & 0.25 & 0 \\ 1 & 1 & 1 & 1 & 1 & 1 & 1 & 0.75 & 0.5 & 0.25 & 0 & 0 \\ 1 & 1 & 1 & 1 & 1 & 1 & 1 & 0.75 & 0.5 & 0.25 & 0 & 0 \\ 1 & 1 & 1 & 1 & 1 & 1 & 1 & 0.75 & 0.5 & 0.25 & 0 & 0 \\ 1 & 1 & 1 & 1 & 1 & 0.75 & 0.5 & 0.25 & 0 & 0 & 0 & 0 \\ 1 & 1 & 1 & 0.75 & 0.75 & 0.5 & 0.25 & 0 & 0 & 0 & 0 & 0 \\ 1 & 0.75 & 0.75 & 0.5 & 0.5 & 0.25 & 0 & 0 & 0 & 0 & 0 & 0 \\ 0.75 & 0.5 & 0.5 & 0.25 & 0.25 & 0 & 0 & 0 & 0 & 0 & 0 & 0 \\ 0.5 & 0.25 & 0.25 & 0 & 0 & 0 & 0 & 0 & 0 & 0 & 0 & 0 \\ 0.25 & 0 & 0 & 0 & 0 & 0 & 0 & 0 & 0 & 0 & 0 & 0 \end{vmatrix}$$

The ideal amplitude response of the filter is illustrated in Fig. 88.20(a).

2. Perform SVD to matrix **A** to obtain the amplitude response of the main section of the 2-D filter. It is worth noting that when a circularly symmetric 2-D filter is required,

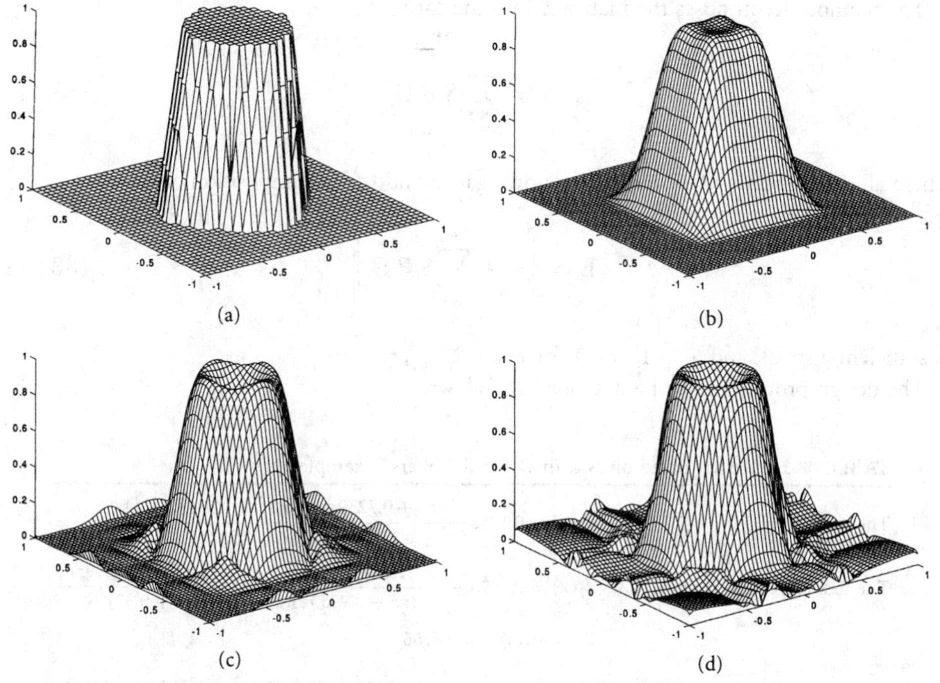

FIGURE 88.20 Amplitude responses of (a) the ideal circularly symmetric 2-D filter, (b) the main section, (c) the main section plus the first correction, (d) the main section plus the first and second correction sections.

matrix **A** is symmetric and, therefore, (88.95) becomes

$$\mathbf{A} = \sum_{i=1}^{r} S_i \boldsymbol{\phi}_i \boldsymbol{\phi}_i^T \qquad (88.110)$$

where $S_1 = 1$ and $S_i = +1$ or -1, for $2 \leq i \leq r$. This implies that each parallel section requires only one 1-D subfilter to be designed. As a consequence the design work is reduced by 50 percent.

When vector $\boldsymbol{\phi}_1$ is obtained, by selecting a fourth-order approximation and after optimization, the transfer function of the main section $f_1(z)$ is obtained.

3. Design the correction sections. Successively perform SVD to the error $\mathbf{E}_1$ and $\mathbf{E}_2$, and apply the preceding design technique, the transfer functions of the first and second correction sections can be obtained.

The transfer function of the main section and the first and second correction sections are listed in Table 88.3. And the amplitude responses of (i) the main section, (ii) the main section plus the first correction, and (iii) the main section plus the first and second correction sections are depicted in Fig. 88.20(b)–(d).

Method of Deng and Kawamata

In decomposing 2-D amplitude specifications into 1-D ones, the conventional SVD cannot avoid the problem that the 1-D amplitude specifications that result are often negative. Since negative values cannot be viewed as amplitude response, the problem of 1-D digital filter design becomes intricate. Deng and Kawamata [28] proposed a procedure that guarantees all the decomposition results to be always nonnegative and thus simplifies the design of correction sections.

The method decomposes the matrix **A** into the form

$$\mathbf{A} = \sum_{i=1}^{r} S_i \mathbf{F}_i \mathbf{G}_i \qquad (88.111)$$

where all the elements of $\mathbf{F}_i$ and $\mathbf{G}_i$ are nonnegative and the decomposition error

$$\mathbf{E} = \left\| \mathbf{A} - \sum_{i=1}^{r} S_i \mathbf{F}_i \mathbf{G}_i \right\| \qquad (88.112)$$

is sufficiently small, and $S_i = 1$ or -1 for $i = 1, 2, \cdots, r$.

The design procedure can be described as follows.

TABLE 88.3 Design Based on Fourth-Order Subfilters (Example 7)

The main section	$f_1(z) = 0.1255\dfrac{(z^2+0.7239z+1)(z^2+1.6343z+1)}{(z^2+0.1367z+1)(z^2-0.5328z+0.2278)}$
The first correction section	$f_2(z) = 0.6098\dfrac{(z^2+1.1618z+0.1661)(z^2+0.8367z+0.9958)}{(z^2+0.9953z)(z^2+0.5124z+0.32)}$
	$S_2 = -1,\ \phi_2^- = 0.6266$
The second correction section	$f_3(z) = 0.4630\dfrac{(z^2+1.5381z+0.4456)(z^2-1.397z+1.1191)}{(z^2+2.0408z-1)(z^2-0.7092z+0.6961)}$
	$S_3 = +1,\ \phi_3^- = 0.2764$

STEP 1. Let $\mathbf{A}_1^+ = \mathbf{A}$, $\mathbf{A}_1^- = 0$, and perform the SVD on $\mathbf{A}_1^+$ as

$$\mathbf{A}_1^+ = \sum_{i=1}^{r_i} \sigma_{1i}\mathbf{u}_{1i}\mathbf{v}_{1i} \approx \mathbf{F}_1^+\mathbf{G}_1^+ \tag{88.113}$$

where σ_{1i} is the ith singular value of $\mathbf{A}_1^+$ ($\sigma_{11} \geq \sigma_{12} \geq \cdots \geq \sigma_{1r_1}$) and $\mathbf{F}_1^+ = \mathbf{u}_{11}\sigma_{11}^{1/2}$, $\mathbf{G}_1^+ = \sigma_{11}^{1/2}\mathbf{v}_{11}$. Let

$$\mathbf{F}_1 = \mathbf{F}_1^+, \mathbf{G}_1 = \mathbf{G}_1^+, S_1 = 1 \tag{88.114}$$

all the elements of $\mathbf{F}_1$ and $\mathbf{G}_1$ are nonnegative.

STEP 2. Calculate the approximation error matrix $\mathbf{A}_2$ and decompose it into the sum of $\mathbf{A}_2^+$ and $\mathbf{A}_2^-$ as

$$\mathbf{A}_2 = \mathbf{A} - S_1\mathbf{F}_1\mathbf{G}_1 = \mathbf{A}_2^+ + \mathbf{A}_2^- \tag{88.115}$$

where

$$A_2^+(m,n) = \begin{cases} A_2(m,n) & \text{if } A_2(m,n) \geq 0 \\ 0 & \text{otherwise} \end{cases} \tag{88.116}$$

and

$$A_2^-(m,n) = \begin{cases} A_2(m,n) & \text{if } A_2(m,n) \leq 0 \\ 0 & \text{otherwise} \end{cases} \tag{88.117}$$

To determine S_2 and $\mathbf{F}_2$ and $\mathbf{G}_2$ for approximating $\mathbf{A}_2$ as accurately as possible, the following three steps are involved.

1. Perform the SVD on $\mathbf{A}_2^+$ and approximate it as

$$\mathbf{A}_2^+ = \sum_{i=1}^{r_2} \sigma_{2i}\mathbf{u}_{2i}\mathbf{v}_{2i} \approx \mathbf{F}_2^+\mathbf{G}_2^+ \tag{88.118}$$

where $\mathbf{F}_2^+ = \mathbf{u}_{21}\sigma_{21}^{1/2}$, $\mathbf{G}_2^+ = \sigma_{21}^{1/2}\mathbf{v}_{21}$. All the elements of $\mathbf{F}_2^+$ and $\mathbf{G}_2^+$ are nonnegative. If $\mathbf{F}_2 = \mathbf{F}_2^+$, $\mathbf{G}_2 = \mathbf{G}_2^+$, and $S_2 = 1$, the approximation error is

$$\mathbf{E}_2^+ = \left\| \mathbf{A} - \sum_{i=1}^{2} S_i\mathbf{F}_i\mathbf{G}_i \right\| \tag{88.119}$$

2. Perform the SVD on $-\mathbf{A}_2^-$ and approximate it as

$$-\mathbf{A}_2^- = \sum_{i=1}^{r_{2-}} \sigma_{2i-}\mathbf{u}_{2i-}\mathbf{v}_{2i-} \approx \mathbf{F}_2^-\mathbf{G}_2^- \tag{88.120}$$

where $\mathbf{F}_2^- = \mathbf{u}_{21-}\sigma_{21-}^{1/2}$, $\mathbf{G}_2^- = \sigma_{21-}^{1/2}\mathbf{v}_{21-}$, and r_{2-} is the rank of $-\mathbf{A}_2^-$. All the elements of $\mathbf{F}_2^-$ and $\mathbf{G}_2^-$ are nonnegative. If $\mathbf{F}_2 = \mathbf{F}_2^-$, $\mathbf{G}_2 = \mathbf{G}_2^-$, and $S_2 = -1$, the approximation error $\mathbf{E}_2^-$ is

$$\mathbf{E}_2^- = \left\| \mathbf{A} - \sum_{i=1}^{2} S_i\mathbf{F}_i\mathbf{G}_i \right\| \tag{88.121}$$

3. According to the results from steps 1 and 2, the optimal vectors $\mathbf{F}_2$ and $\mathbf{G}_2$ for approximating $\mathbf{A}$ are determined as

$$\begin{array}{ll} \mathbf{F}_2 = \mathbf{F}_2^+, \mathbf{G}_2 = \mathbf{G}_2^+, S_2 = 1 & \text{if } \mathbf{E}_2^+ \leq \mathbf{E}_2^- \\ \mathbf{F}_2 = \mathbf{F}_2^-, \mathbf{G}_2 = \mathbf{G}_2^-, S_2 = -1 & \text{if } \mathbf{E}_2^+ \geq \mathbf{E}_2^- \end{array}$$

Successively decomposing the approximation error matrices $\mathbf{A}_j(j = 3, 4, \cdots, r)$ in the same way above described, a good approximation of matrix A can be obtained as in (88.111).

STEP 3. With the matrix **A** being decomposed into nonnegative vectors, the 1-D subfilters are designed through an optimization procedure, and a 2-D filter can then be readily realized, as shown in Fig. 88.21.

It is noted that, in addition to SVD based methods as described in this subsection, design method based on other decomposition is also possible [29].

Design Based on Two-Variable Network Theory

Ramamoorthy and Bruton proposed a design method of 2-D IIR filters that always guarantees the stability of a filter and that involves the application of two-variable (2-V) strictly Hurwitz polynomials [30]. A 2-V polynomial $b(s_1, s_2)$ is said to be strictly Hurwitz if

$$b(s_1, s_2) \neq 0 \quad \text{for } \operatorname{Re}\{s_1\} \geq 0 \quad \text{and} \quad \operatorname{Re}\{s_2\} \geq 0 \qquad (88.122)$$

In their method, a family of 2-V strictly Hurwitz polynomials is obtained by applying network theory [31], [32] to the frequency-independent, 2-V lossless network illustrated in Fig. 88.22. The network has $1 + N_1 + N_2 + N_r$ ports and N_1 and N_2 are terminated in unit capacitors in complex variables s_1 and s_2, respectively, and N_r is terminated in unit resistors. Since the network is lossless and frequency independent, its admittance matrix **Y** is a real and skew-symmetric

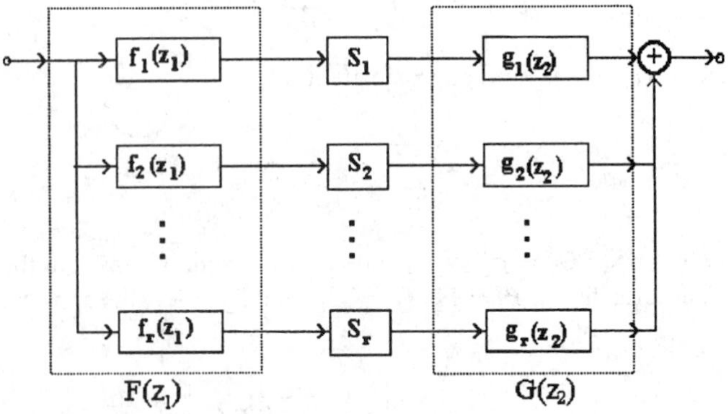

FIGURE 88.21 2-D filter realization based on iterative SVD.

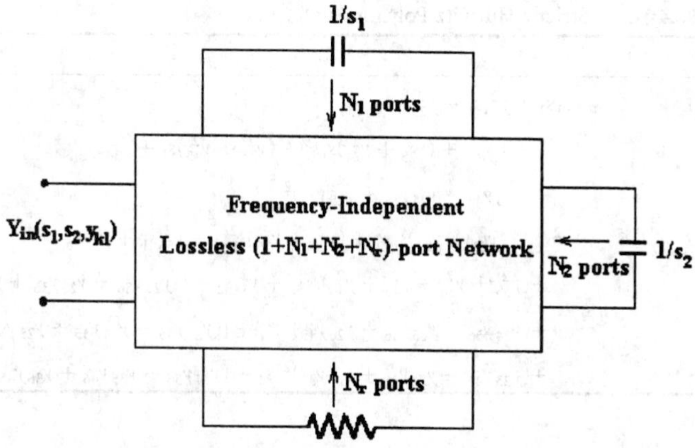

FIGURE 88.22 A $(1+N_1+N_2+N_r)$-port 2-V lossless network.

matrix given by

$$\mathbf{Y} = \begin{vmatrix} 0 & y_{12} & y_{13} & \cdots & y_{1N} \\ -y_{12} & 0 & y_{23} & \cdots & y_{2N} \\ -y_{13} & -Y_{23} & 0 & \cdots & y_{3N} \\ \cdots & \cdots & & & \\ -y_{1N} & -y_{2N} & -y_{3N} & \cdots & 0 \end{vmatrix} = \begin{vmatrix} \mathbf{Y}_{11} & \mathbf{Y}_{12} \\ -\mathbf{Y}_{12}^T & \mathbf{Y}_{22} \end{vmatrix} \qquad N = 1 + N_1 + N_2 + N_r \tag{88.123}$$

If we define

$$\mathbf{Y}_{22}(s_1, s_2, y_{kl}) = \mathbf{Y}_{22} + \text{diag}\,\{\overbrace{1 \cdots 1}^{N_r}\ \overbrace{s_1 \cdots s_1}^{N_1}\ \overbrace{s_2 \cdots s_2}^{N_2}\} \tag{88.124}$$

and

$$\Delta(s_1, s_2, y_{kl}) = \det[\mathbf{Y}_{22}(s_1, s_2, y_{kl})] \tag{88.125}$$

where $\text{diag}(1 \cdots 1 \quad s_1 \cdots s_1 \quad s_2 \cdots s_2)$ represents a diagonal matrix in which each of the first N_r elements is unity, each of the next N_1 elements is s_1, and each of the last N_2 elements is s_2. Then, from the network theory, the input admittance at port 1 is given by

$$\mathbf{Y}_{\text{in}}(s_1, s_2, y_{kl}) = \frac{\mathbf{Y}_{12}\,\text{adj}[\mathbf{Y}_{22}(s_1, s_2, y_{kl})]\mathbf{Y}_{12}^T}{\det[\mathbf{Y}_{22}(s_1, s_2, y_{kl})]} = \frac{p(s_1, s_2, y_{kl})}{\Delta(s_1, s_2, y_{kl})} \tag{88.126}$$

where $\Delta(s_1, s_2, y_{kl})$ is defined by (88.125) and is a strictly Hurwitz polynomial for any set of real values of the $(N-1)(N-2)/2$ independent parameters $\{y_{kl}: 1 < k < l \leq N\}$. Table 88.4 lists polynomial $\Delta(s_1, s_2, y_{kl})$ for $N_r = 1$ $(N_1, N_2) = (2, 1)$ and $(N_1, N_2) = (2, 2)$ [30].

TABLE 88.4 2-V Strictly Hurwitz Polynomials

N_1	N_2	N	$\Delta(s_1, s_2, y_{kl})$
1	1	4	$s_1s_2 + y_{24}^2 s_1 + y_{23}^2 s_2 + y_{34}^2$
2	1	5	$s_1^2 s_2 + y_{25}^2 s_1^2 + (y_{23}^2 + y_{24}^2) s_1 s_2 + (y_{35}^2 + y_{45}^2) s_1 + y_{34}^2 s_2$ $+ (y_{23}y_{45} - y_{24}y_{35} + y_{25}y_{34})^2$
2	2	6	$s_1^2 s_2^2 + (y_{23}^2 + y_{24}^2) s_1 s_2^2 + (y_{25}^2 + y_{26}^2) s_1^2 s_2 + y_{56}^2 s_1^2 + y_{34}^2 s_2^2$ $+ (y_{35}^2 + y_{36}^2 + y_{45}^2 + y_{46}^2) s_1 s_2 + y_{34}^2 s_2 + [(y_{23}y_{56} - y_{25}y_{36} + y_{26}y_{35})^2$ $+ (y_{24}y_{56} - y_{25}y_{46} + y_{26}y_{45})^2] s_1 + [(y_{23}y_{45} - y_{24}y_{35} + y_{25}y_{34})^2$ $+ (y_{23}y_{46} - y_{24}y_{36} + y_{26}y_{34})^2] s_2 + (y_{34}y_{56} - y_{35}y_{46} + y_{36}y_{45})^2$

Having obtained the parameterized strictly Hurwitz polynomial $\Delta(s_1, s_2, y_{kl})$, the design procedure of a 2-D IIR filter can be summarized as follows.

STEP 1. Construct a parameterized analog transfer function of the 2-D IIR filter, by using the Hurwitz polynomial $\Delta(s_1, s_2, y_{kl})$

$$H(s_1, s_2, y_{kl}, a_{ij}) = \frac{p(s_1, s_2)}{\Delta(s_1, s_2, y_{kl})} \tag{88.127}$$

where

$$p(s_1, s_2) = \sum_{i=1}^{N_1} \sum_{j=1}^{N_2} a_{ij} s_1^i s_2^j$$

is an arbitrary 2-V polynomial in s_1 and s_2 with degree in each variable not greater than the corresponding degree of the denominator.

STEP 2. Perform the double bilinear transformation to the parameterized analog transfer function obtained in step 1.

$$H(z_1, z_2, y_{kl}, a_{ij}) = \left. \frac{p(s_1, s_2)}{\Delta(s_1, s_2, y_{kl})} \right|_{s_i = 2(z_i - 1)/T_i(z_i + 1),\, i = 1, 2} \tag{88.128}$$

STEP 3. Construct an objective function according to the given design specifications and the parameterized discrete transfer function obtained in step 2.

$$J(\mathbf{x}) = \sum_{n_1} \sum_{n_2} [M(n_1, n_2) - M_I(n_1, n_2)]^p \tag{88.129}$$

where p is an even positive, $M(n_1, n_2)$ and $M_I(n_1, n_2)$ are the actual and desired amplitude responses, respectively, of the required filter at frequencies $(\omega_{1n_1}, \omega_{1n_2})$ and $\mathbf{x}$ is the vector consisting of parameters $\{y_{kl}: 1 < k < l \leq N\}$ and $\{a_{ij}, 0 \leq i \leq N_1, 0 \leq j \leq N_2\}$.

STEP 4. Apply an optimization algorithm to find the optimal vector $\mathbf{x}$ that minimizes the objective function and substitute the resulting x into (88.128) to obtain the required transfer function $H(z_1, z_2)$.

Example 8 [33]. By using the preceding approach, design a 2-D circularly symmetric low-pass filter of order (5, 5) with $\omega_p = 0.2\pi$, assuming that $\omega_{s1} = \omega_{s2} = 1.2\pi$.

Solution.

1. Construct the desired amplitude response of the desired filter

$$M_I(\omega_{1n_1}, \omega_{2n_2}) = \begin{cases} 1 & \text{for } (\omega_{1n_1}^2 + \omega_{2n_2}^2) \le 0.2\pi \\ 0 & \text{otherwise} \end{cases}$$

where

$$\omega_{1n_1} = \begin{cases} 0.01\pi n_2 & \text{for } 0 \le n_1 \le 20 \\ 0.1\pi n_1 & \text{for } 21 \le n_1 \le 24 \end{cases}$$

and

$$\omega_{2n_2} = \omega_{1(24-n_2)} \quad \text{for } 0 \le n_2 \le 24$$

2. Construct the 2-D analog transfer function and perform double bilinear transformation to obtain the discrete transfer function. The analog transfer function at hand is assumed to be an all-pole transfer function of the form

$$H(s_1, s_2, \mathbf{x}) = \frac{1}{\Delta(s_1, s_2, y_{kl})}$$

Therefore, the corresponding discrete transfer function can be written as

$$H(z_1, z_2, \mathbf{x}) = \frac{A(z_1+1)^5(z_2+1)^5}{\sum_{i=0}^{5}\sum_{j=0}^{5} b_{ij} z_1^i z_2^j} \tag{88.130}$$

where

$$\sum_{i=0}^{5}\sum_{j=0}^{5} b_{ij} z_1^i z_2^j = (z_1+1)^5(z_2+1)^5 \Delta(s_1, s_2, y_{kl})|_{s_i=(z_i-1/z_i+1),\, i=1,2}$$

contains $(N-1)(N-2)/2 = 36$ parameters.

3. Optimization: A conventional quasi-Newton algorithm has been applied to minimize the objective function in (88.129) with $p = 2$. The resulting coefficients are listed in Table 88.5. The amplitude response of the resulting filter is depicted in Fig. 88.23.

TABLE 88.5 Coefficients of Transfer Function in (88.130) $[A = 0.28627, b_{ij}: 0 \le i \le 5, 0 \le j \le 5]$

0.0652	−0.6450	3.3632	−4.8317	0.3218	−0.1645
−0.7930	7.8851	−25.871	23.838	3.4048	3.4667
4.2941	−28.734	61.551	−29.302	−13.249	−25.519
−6.3054	28.707	−33.487	−7.2275	−22.705	83.011
0.7907	1.4820	−7.4214	−33.313	136.76	−128.43
−0.4134	6.0739	−36.029	101.47	140.20	78.428

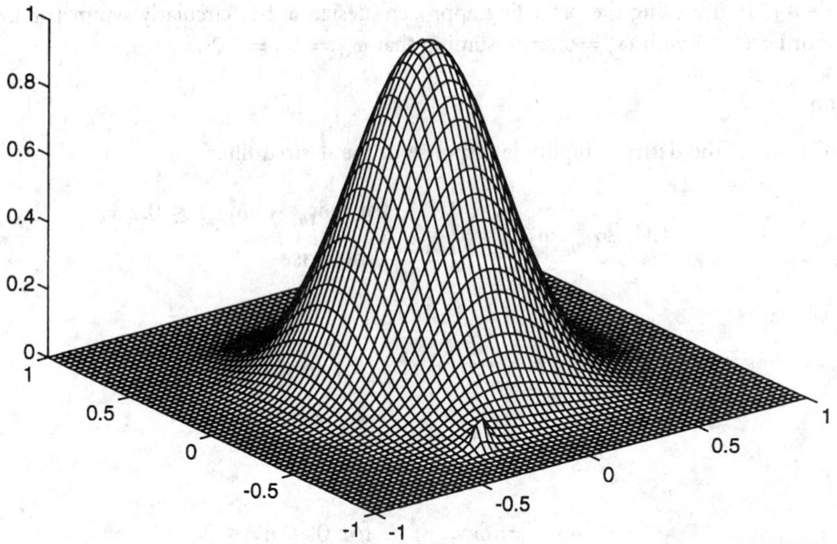

FIGURE 88.23 Amplitude response of circularly symmetric low-pass filter of Example 8.

References

[1] J. L. Shanks, S. Treitel, and J. H. Justice, "Stability and synthesis of two-dimensional recursive filters," *IEEE Trans. Audio Electroacoustic.*, vol. AU-20, pp. 115–128, June 1972.

[2] W. S. Lu and A. Antoniou, *Two-Dimensional Digital Filters*, Marcel Dekker, 1992.

[3] M. Ahmadi, A. G. Constantinides, and R. A. King, "Design technique for a class of stable two-dimensional recursive digital filters," in *Proc. 1976 IEEE Int. Conf. Acoust., Speech, Signal Processing*, 1976, pp. 145–147.

[4] A. M. Ali, A. G. Constantinides, and R. A. King, "On 2-variable reactance functions for 2-dimensional recursive filters," *Electron. Lett.*, vol. 14, pp. 12, 13, Jan. 1978.

[5] R. King *et al.*, *Digital Filtering in One and Two Dimensions: Design and Applications*, Plenum, 1989.

[6] R. A. King and A. H. Kayran, "A new transformation technique for the design of 2-dimensional stable recursive digital filters," in *Proc. IEEE Int. Symp. Circuits Syst.*, Chicago Apr. 1981, pp. 196–199, April 1981.

[7] J. M. Costa and A. N. Venetsanopoulos, "A group of linear spectral transformations for two-dimensional digital filters," *IEEE Trans. Acoust., Speech, Signal Processing*, vol. ASSP-24, pp. 424, 425, Oct. 1976.

[8] K. P. Prasad, A. Antoniou, and B. B. Bhattacharyya," On the properties of linear spectral transformations for 2-dimensional digital filters," *Circuits Syst. Signal Process.*, vol. 2, pp. 203–211, 1983.

[9] A. H. Kayran and R. A. King, "Design of recursive and nonrecursive fan filters with complex transformations," *IEEE Trans. Circuits Syst.*, vol. CAS-30, pp. 849–857, 1983.

[10] A. G. Constantinides, "Spectral transformations for digital filters," *IEE Proc.*, vol. 117, pp. 1585–1590, Aug. 1970.

[11] N. A. Pendergrass, S. K. Mitra, and E. I. Jury, "Spectral transformations for two-dimensional digital filters," *IEEE Trans. Circuits Syst.*, vol. CAS-23, pp. 26–35, Jan. 1976.

[12] K. Hirano and J. K. Aggarwal, "Design of two-dimensional recursive digital filters," *IEEE Trans. Circuits Syst.*, vol. CAS-25, pp. 1066–1076, Dec. 1978.

[13] S. K. Mitra and K. Hirano, "Digital all-pass networks," *IEEE Trans. Circuits Syst.*, vol. CAS-21, pp. 688–700, Sept. 1974.

[14] J. M. Costa and A. N. Venetsanopoulos, "Design of circularly symmetric two-dimensional recursive filters," *IEEE Trans. Acoust., Speech, Signal Processing*, vol. ASSP-22, pp. 432–443, Dec. 1974.

[15] D. M. Goodman, "A design technique for circularly symmetric low-pass filters," *IEEE Trans. Acoust., Speech, Signal Processing*, vol. ASSP-26, pp. 290–304, Aug. 1978.

[16] G. V. Mendonca, A. Antoniou, and A. N. Venetsanopoulos, "Design of two-dimensional pseudorotated digital filters satisfying prescribed specifications," *IEEE Trans. Circuits Syst.*, vol. CAS-34, pp. 1–10, Jan 1987.

[17] R. Fletcher, *Practical Methods of Optimization*, 2nd ed., New York: Wiley, 1990.

[18] S. Chakrabarti and S. K. Mitra, "Design of two-dimensional digital filters via spectral transformations," *Proc. IEEE*, vol. 65, pp. 905–914, June 1977.

[19] A. Antoniou, *Digital Filters: Analysis, Design and Applications*, 2nd ed., New York: McGraw-Hill, 1993.

[20] G. A. Maria and M. M. Fahmy, "An l_p design technique for two-dimensional digital recursive filters," *IEEE Trans. Acoust., Speech, Signal Processing*, vol. ASSP-22, pp. 15–21, Feb. 1974.

[21] C. Charalambous, "A unified review of optimization," *IEEE Trans. Microwave Theory Tech.*, vol. MTT-22, pp. 289–300, Mar. 1974.

[22] C. Charalambous and A. Antoniou, "Equalization of recursive digital filters," *IEE Proc., Pt. G*, vol. 127, pp. 219–225, Oct. 1980.

[23] C. Charalambous, "Design of 2-dimensional circularly-symmetric digital filters," *IEE Proc., Pt. G*, vol. 129, pp. 47–54, Apr. 1982.

[24] ——, "Acceleration of the least pth algorithm for minimax optimization with engineering applications," *Math Program.*, vol. 17, pp. 270–297, 1979.

[25] P. K. Rajan and M. N. S. Swamy, "Quadrantal symmetry associated with two-dimensional digital transfer functions," *IEEE Trans. Circuits Syst.*, vol. CAS-29, pp. 340–343, June 1983.

[26] A. Antoniou and W. S. Lu, "Design of two-dimensional digital filters by using the singular value decomposition," *IEEE Trans. Circuits Syst.*, vol. CAS-34, pp. 1191–1198, Oct. 1987.

[27] G. W. Stewart, *Introduction to Matrix Computations*, New York: Academic, 1973.

[28] T. B. Deng and M. Kawwamata, "Frequency-domain design of 2-D digital filters using the iterative singular value decomposition," *IEEE Trans. Circuits Syst.*, vol. CAS-38, pp. 1225–1228, 1991.

[29] T. B. Deng and T. Soma, "Successively linearized non-negative decomposition of 2-D filter magnitude design specifications," *Digital Signal Process.*, vol. 3, pp. 125–138, 1993.

[30] P. A. Ramamoorthy and L. T. Bruton, "Design of stable two-dimensional analog and digital filters with applications in image processing," *Circuit Theory Applicat.*, vol. 7, pp. 229–245, 1979.

[31] T. Koga, "Synthesis of finite passive networks with prescribed two-variable reactance matrices," *IEEE Trans. Circuit Theory*, vol. CT-13, pp. 31–52, 1966.

[32] H. G. Ansel, "On certain two-variable generalizations of circuit theory, with applications networks of transmission lines and lumped reactance," *IEEE Trans. Circuit Theory*, vol. CT-11, pp. 214–233, 1964.

[33] P. A. Ramamoorthy and L. T. Bruton, "Design of stable two-dimensional recursive filters," in *Topics in Applied Physics*, vol. 42, T. S. Huang, Ed., New York: Springer-Verlag, 1981, pp. 41–83.

Indexes

Author Index

Index of Tables

Index of Figures

Subject Index

C

D

E

F

G

H

I

J

K

L

M

N

O

P

Q

R

S

U

V